Register Now to Improve Your Grade!

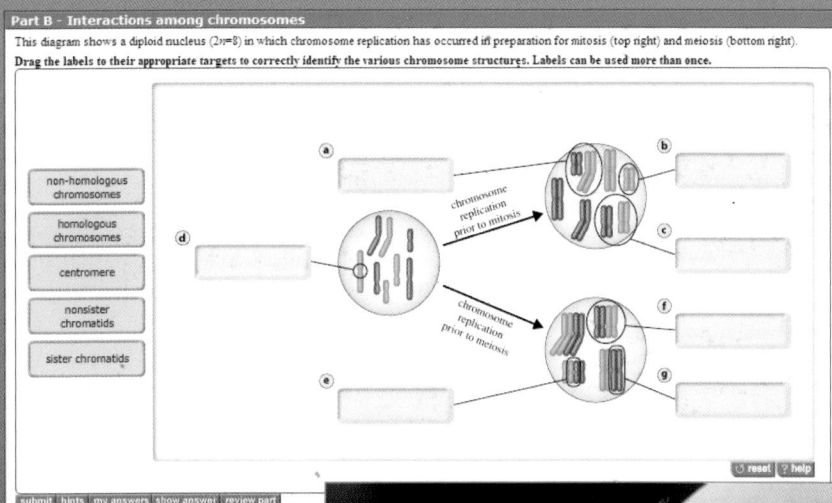
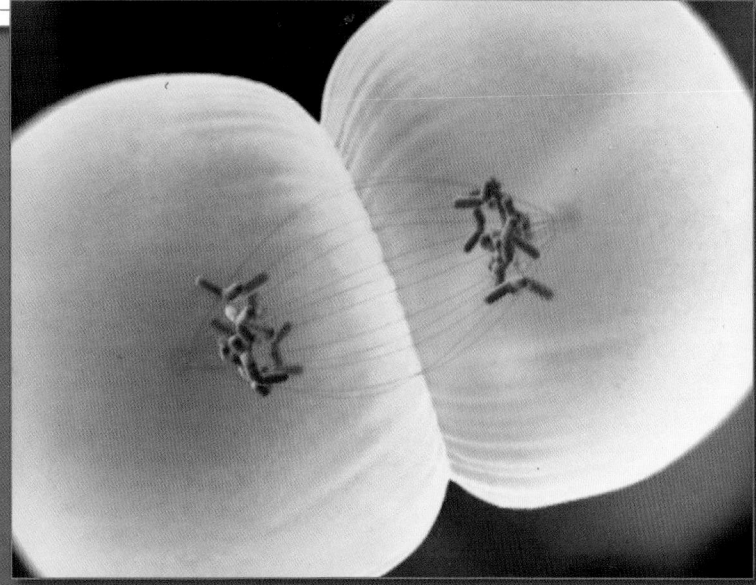

To the Student: How to Use This Book

Find important information in this book by focusing on the Gold Thread.

These black swan chicks are being introduced to the world by their mother, who trails close behind them. Likewise, this chapter introduces the key ideas that launched biological science as a discipline.

Biology and the Tree of Life

1

In essence, biological science is a search for ideas and observations that unify our understanding of the diversity of life, from bacteria living in rocks a mile underground to hedgehogs and humans. Chapter 1 is an introduction to this search.

The goals of this chapter are to introduce the nature of life and explore how biologists go about studying it. The chapter also introduces themes that will resonate throughout this book: (1) analyzing how organisms work at the molecular level, (2) understanding organisms in terms of their evolutionary history, and (3) helping you learn to think like a biologist.

Let's begin with what may be the most fundamental question of all: What is life?

1.1 What Does It Mean to Say That Something Is Alive?

An **organism** is a life-form—a living entity made up of one or more cells. Although there is no simple definition of life that is endorsed by all biologists, most agree that organisms share a suite of five fundamental characteristics.

- *Energy*   To stay alive and reproduce, organisms have to acquire and use energy. To give just two examples: plants absorb sunlight; animals ingest food.

- *Cells*   Organisms are made up of membrane-bound units called cells. A cell's membrane regulates the passage of materials between exterior and interior spaces.

- *Information*   Organisms process hereditary or genetic information, encoded in units called genes, along with information they acquire from the environment. Right now cells throughout your body are using genetic information to make the molecules that keep you alive; your eyes and brain are decoding information on this page that will help you learn some biology.

KEY CONCEPTS

- Organisms obtain and use energy, are made up of cells, process information, replicate, and as populations evolve.

- The cell theory proposes that all organisms are made of cells and that all cells come from preexisting cells.

- The theory of evolution by natural selection maintains that species change through time because individuals with certain heritable traits produce more offspring than other individuals do.

- A phylogenetic tree is a graphical representation of the evolutionary relationships between species. These relationships can be estimated by analyzing similarities and differences in traits. Species that share distinctive traits are closely related and are placed close to each other on the tree of life.

- Biologists ask questions, generate hypotheses to answer them, and design experiments that test the predictions made by competing hypotheses.

✔ When you see this checkmark, stop and test yourself. Answers are available in Appendix B.

1

Key Concepts

Start with Key Concepts on the first page of every chapter. Read these gold key points first to familiarize yourself with the chapter's big ideas.

MORE! Bulleted Lists

Take note of bulleted lists that "chunk" information and ideas. This will help you manage the information that you are learning in the course.

Gold Highlighting

Watch for important information highlighted in gold. Gold highlighting is always a signal to slow down and pay special attention.

Gold Key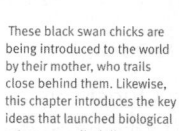

Material related to Key Concepts will be signaled with a gold key.

The best way to succeed on exams...

As you read the text and view the figures, practice with the Blue Thread.

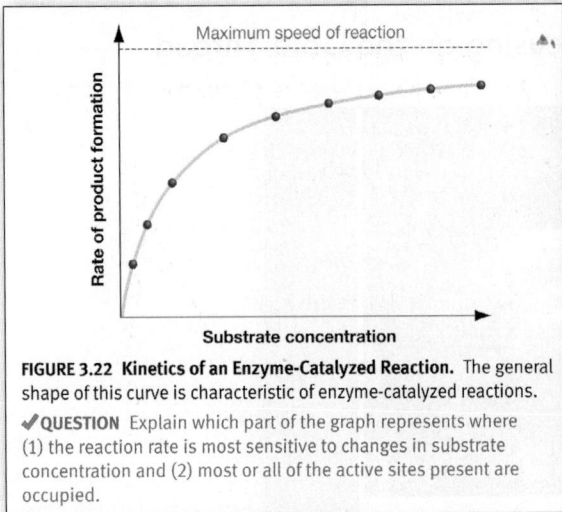

FIGURE 3.22 Kinetics of an Enzyme-Catalyzed Reaction. The general shape of this curve is characteristic of enzyme-catalyzed reactions.

✔**QUESTION** Explain which part of the graph represents where (1) the reaction rate is most sensitive to changes in substrate concentration and (2) most or all of the active sites present are occupied.

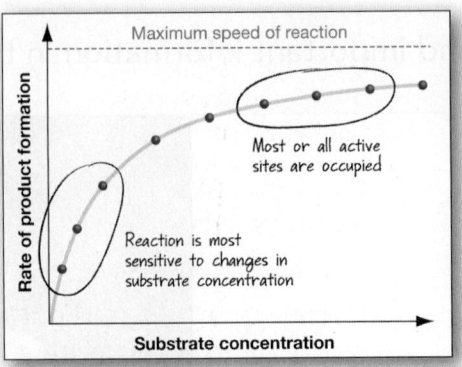

Drawing Exercises

Some Blue Thread Questions contain artwork from the textbook that you will be asked to draw on or modify.

NEW! Suggested Answers

Suggested answers for the Blue Thread Questions and Exercises are provided in Appendix B.

Blue Thread Questions

Many figures include Blue Thread Questions or Exercises to help you check your understanding of the material they present.

"You Should be Able To" Exercises

Text passages flagged with blue type and the words "you should be able to" offer exercises on concepts that professors and students have identified as most difficult. These are the topics most students struggle with on exams.

To read this graph, put your finger on the *x*-axis at time 0. Then read up the *y*-axis, and note that kernels averaged about 11 percent protein at the start of the experiment. Now read the graph to the right. Each dot is a data point, representing the average kernel protein concentration in a particular generation. (A generation in maize is one year.) The lines on this graph simply connect the dots, to make the pattern in the data easier to see. At the end of the graph, after 100 generations of selection, average kernel protein content is about 29 percent. (For more help with reading graphs, see **BioSkills 2** in Appendix A.)

This sort of change in the characteristics of a population, over time, is evolution. Humans have been practicing artificial selection for thousands of years, and biologists have now documented evolution by *natural* selection—where humans don't do the selecting—occurring in thousands of different populations, including humans.

To practice applying the principles of artificial selection, go to the online study area at *www.masteringbiology.com*.

MB **Web Activity** Artificial Selection

Evolution occurs when heritable variation leads to differential success in reproduction. ✔If you understand this concept, you should be able to describe how protein content in maize kernels changed over time, using the same *x*-axis and *y*-axis as in Figure 1.3, when researchers selected individuals with *lowest* kernel protein content to be the parents of the next generation. (This experiment was actually done, starting with the same population at the same time as selection for high protein content.)

FITNESS AND ADAPTATION Darwin also introduced some new terminology to identify what is happening during natural selection.

CHECK YOUR UNDERSTANDING

If you understand that . . .
- Natural selection occurs when heritable variation in certain traits leads to improved success in reproduction. Because individuals with these traits produce many offspring with the same traits, the traits increase in frequency and evolution occurs.
- Evolution is a change in the characteristics of a population over time.

✔ **You should be able to . . .**

On the graph you just analyzed, describe the average kernel protein content over time in a maize population where *no* selection occurred.

Answers are available in Appendix B.

1.4 The Tree of Life

Section 1.3 focused on how individual populations change through time in response to natural selection. But over the past several decades, biologists have also documented dozens of cases in which natural selection has caused populations of one species to diverge and form new species. This divergence process is called **speciation**.

Research on speciation has two important implications: All species come from preexisting species, and all species, past and present, trace their ancestry back to a single common ancestor.

The theory of evolution by natural selection predicts that biologists should be able to reconstruct a **tree of life**—a family tree

Mastering**BIOLOGY**®

Make Learning Part of the Grade®

It's possible that your professor will include these Blue Thread Questions in a graded assignment at www.masteringbiology.com.

Check Your Understanding

The blue half of the Check Your Understanding boxes asks you to do something with the information in the top half. If you can't complete these exercises, go back and re-read that section of the chapter.

...is to practice. Here's how.

NEW! Bulleted Summary of Key Concepts

The succinct Summary of Key Concepts reviews important concepts in short, manageable bullet points.

Blue Thread Exercises

End-of-chapter Blue Thread Exercises help you review the major themes of the chapter and synthesize information.

CHAPTER 6   REVIEW

For media, go to the study area at www.masteringbiology.com

Summary of Key Concepts

- **Phospholipids are amphipathic molecules—they have a hydrophilic region and a hydrophobic region. In solution, phospholipids spontaneously form bilayers that are selectively permeable—meaning that only certain substances cross them readily.**

  - The plasma membrane forms a physical barrier between the internal and external environment—often between life and nonlife.

  - The basic structure of plasma membranes is created by a phospholipid bilayer.

solutes to the region of lower water concentration and higher solute concentration.

- Osmosis is a passive process driven by an increase in entropy.

  ✔ You should be able to imagine a beaker with solutions separated by a plasma membrane, and then predict what will happen after addition of a solute to one side if the solute (1) crosses the membrane readily or (2) is incapable of crossing the membrane.

(MB) **Web Activity** Diffusion and Osmosis

Make Learning Part of the Grade®

Visit www.masteringbiology.com for practice quizzes, 3-D animations, the eText, and more.

BioFlix™

BioFlix™ 3-D animations are included in MasteringBiology's Study Area and are available as automatically graded assignments.

Analyze: Can I recognize underlying patterns and structure?	**Evaluate:** Can I make judgements on the relative value of ideas and information?	**Synthesize:** Can I put ideas and information together to create something new?

Apply: Can I use these ideas in a new situation?

Explain: Can I explain this concept in my own words?

Remember: Can I recall the key terms and ideas?

Bloom's Taxonomy

Bloom's Taxonomy categorizes six levels of learning competency. The Blue Thread Questions and Exercises in the textbook test on the higher levels of the scale—Explain, Apply, Analyze, Evaluate, and Synthesize—to help you develop critical thinking skills and prepare you for exams.

Steps to Understanding

End of Chapter questions are scaled along Bloom's Taxonomy.

✔ TEST YOUR KNOWLEDGE

Begin by testing your knowledge of new facts.

✔ TEST YOUR UNDERSTANDING

Once you're confident in your knowledge of the material, demonstrate your understanding by answering the Test Your Understanding questions.

✔ APPLYING CONCEPTS TO NEW SITUATIONS

Challenge yourself even further by applying your understanding of the concepts to new situations.

Learn to think like a scientist. Here's how.

A unique emphasis on the process of scientific discovery and experimental design teaches you how to think like a scientist as you learn fundamental biology concepts.

Make Learning Part of the Grade®

www.masteringbiology.com

NEW! Experimental Inquiry Tutorials

Experimental Inquiry Tutorials based on some of biology's most seminal experiments can be found on **www.masteringbiology.com**. Your instructor may assign these. They will give you practice analyzing the experimental design and data, and help you understand reasoning that led scientists from the data they collected to their conclusions.

Some of the topics include:

- The Process of Science
- Engelmann's Photosynthesis and Wavelengths of Light
- Morgan's Cross with White-Eyed Males
- Meselson-Stahl's Semiconservative Replication
- Steinhardt et al and Hafner et al's Polyspermy
- Grant's Changes in Finch Beak Size
- Went's Phototropism and Auxin Distribution
- Coleman's Obesity Gene
- Connell's Competition in Barnacles
- Bormann, Likens et al's Nutrient Cycling in Hubbard Brook Forest

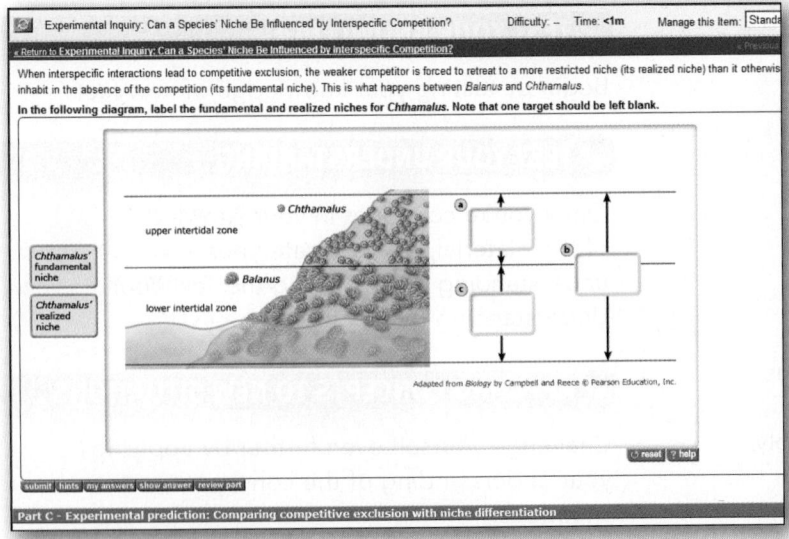

Experiment Boxes

Study Experiment Boxes to help you understand how experiments are designed and give you practice interpreting data.

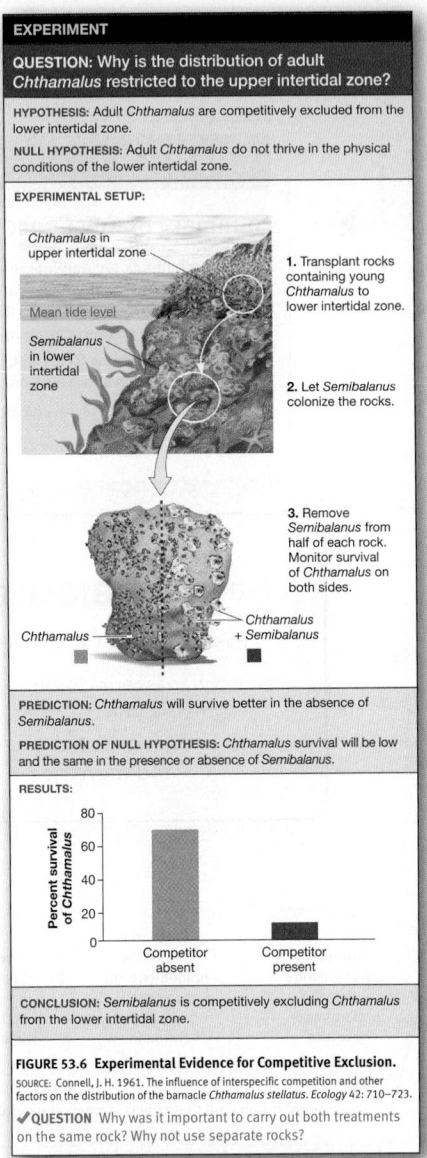

EXPERIMENT

QUESTION: Why is the distribution of adult *Chthamalus* restricted to the upper intertidal zone?

HYPOTHESIS: Adult *Chthamalus* are competitively excluded from the lower intertidal zone.

NULL HYPOTHESIS: Adult *Chthamalus* do not thrive in the physical conditions of the lower intertidal zone.

EXPERIMENTAL SETUP:

1. Transplant rocks containing young *Chthamalus* to lower intertidal zone.

2. Let *Semibalanus* colonize the rocks.

3. Remove *Semibalanus* from half of each rock. Monitor survival of *Chthamalus* on both sides.

PREDICTION: *Chthamalus* will survive better in the absence of *Semibalanus*.

PREDICTION OF NULL HYPOTHESIS: *Chthamalus* survival will be low and the same in the presence or absence of *Semibalanus*.

RESULTS:

CONCLUSION: *Semibalanus* is competitively excluding *Chthamalus* from the lower intertidal zone.

FIGURE 53.6 Experimental Evidence for Competitive Exclusion.
SOURCE: Connell, J. H. 1961. The influence of interspecific competition and other factors on the distribution of the barnacle *Chthamalus stellatus*. *Ecology* 42: 710–723.
✔**QUESTION** Why was it important to carry out both treatments on the same rock? Why not use separate rocks?

NEW! Source Citations

Each Experiment Box now cites the original research paper, encouraging you to extend your learning by exploring the primary literature.

NEW! Experiment Box Questions

Each Experiment Box now includes a question that asks students to analyze the design of the experiment.

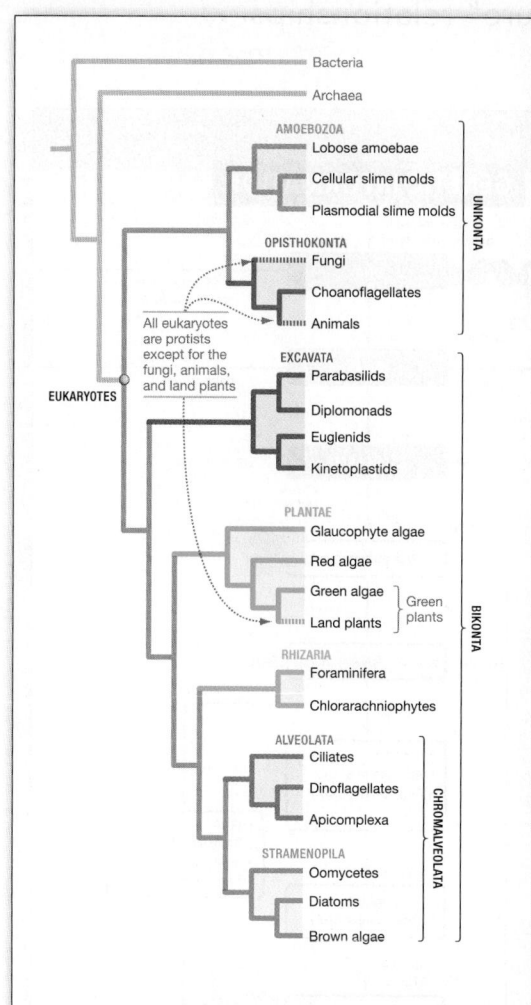

NEW! Redesigned Phylogenetic Trees

Practice "tree thinking" using these newly redesigned phylogenetic trees. Their U-shaped, top-to-bottom format is consistent with the way such trees are most commonly depicted in the scientific literature.

Expanded BioSkills Appendix

BIOSKILLS

Build skills that will be important to your success in future courses. At relevant points in the text, you'll find references to the expanded BioSkills Appendix that will help you learn and practice the following foundational skills:

- NEW! The Metric System
- Reading Graphs
- Reading a Phylogenetic Tree
- NEW! Some Common Latin and Greek Roots Used in Biology
- Using Statistical Tests and Interpreting Standard Error Bars
- Reading Chemical Structures
- Using Logarithms
- Making Concept Maps
- Separating and Visualizing Molecules
- Biological Imaging: Microscopy and X-Ray Crystallography
- NEW! Separating Cell Components by Centrifugation
- NEW! Cell Culture Methods
- Combining Probabilities
- NEW! Model Organisms

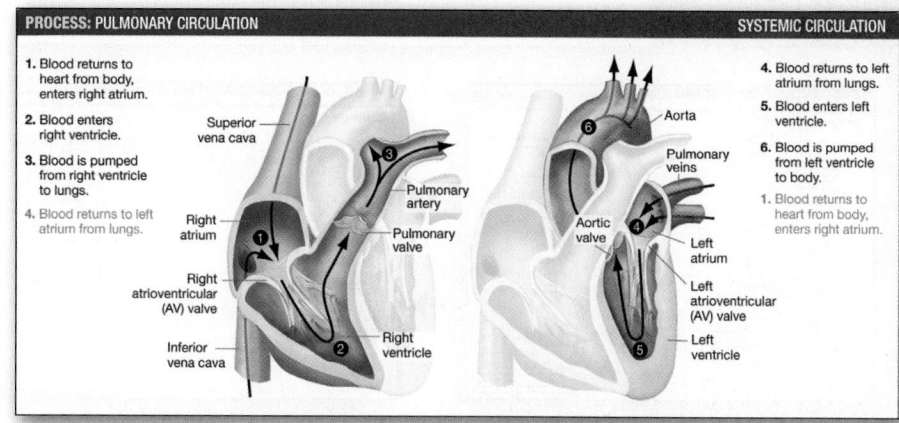

Informative Figures

Think through complex biological processes with figures that clearly define concepts.

Keep sight of the big picture.

Concept maps help you to keep sight of "big picture" relationships among biological concepts.

NEW! Big Picture Concept Maps

Four remarkable Big Picture concept maps help you synthesize information across the chapters on energy, genetics, evolution, and ecology.

Check Your Understanding

Check your understanding of these big picture relationships by answering the Blue Thread Questions.

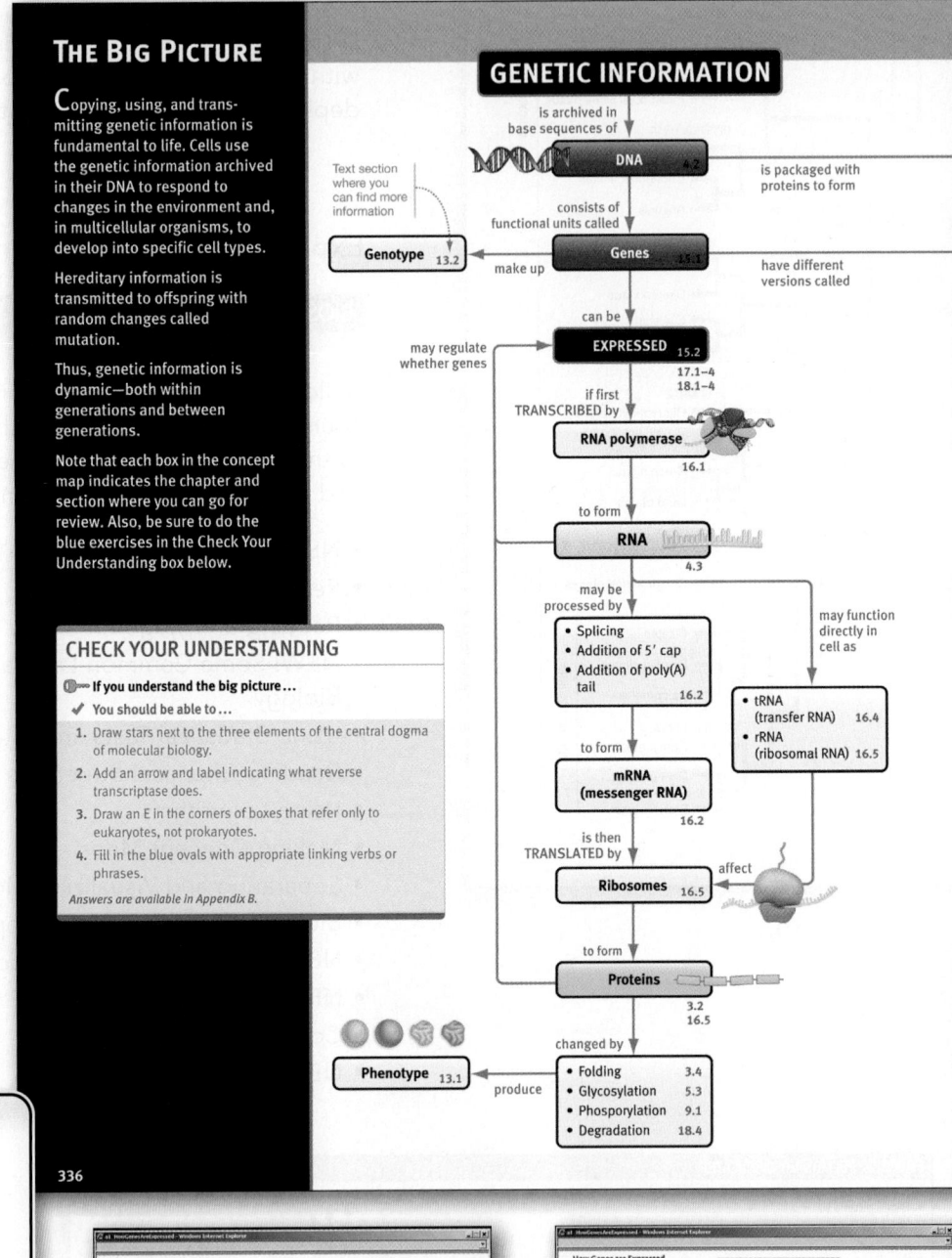

THE BIG PICTURE

Copying, using, and transmitting genetic information is fundamental to life. Cells use the genetic information archived in their DNA to respond to changes in the environment and, in multicellular organisms, to develop into specific cell types.

Hereditary information is transmitted to offspring with random changes called mutation.

Thus, genetic information is dynamic—both within generations and between generations.

Note that each box in the concept map indicates the chapter and section where you can go for review. Also, be sure to do the blue exercises in the Check Your Understanding box below.

CHECK YOUR UNDERSTANDING

If you understand the big picture...

✓ You should be able to...

1. Draw stars next to the three elements of the central dogma of molecular biology.
2. Add an arrow and label indicating what reverse transcriptase does.
3. Draw an E in the corners of boxes that refer only to eukaryotes, not prokaryotes.
4. Fill in the blue ovals with appropriate linking verbs or phrases.

Answers are available in Appendix B.

GENETIC INFORMATION
is archived in base sequences of
DNA 4.2
is packaged with proteins to form
Text section where you can find more information
consists of functional units called
Genotype 13.2   make up   **Genes** 15.1
have different versions called
can be
may regulate whether genes
EXPRESSED 15.2   17.1–4   18.1–4
if first TRANSCRIBED by
RNA polymerase 16.1
to form
RNA 4.3
may be processed by
- Splicing
- Addition of 5' cap
- Addition of poly(A) tail   16.2
may function directly in cell as
- tRNA (transfer RNA) 16.4
- rRNA (ribosomal RNA) 16.5
to form
mRNA (messenger RNA) 16.2
is then TRANSLATED by
affect
Ribosomes 16.5
to form
Proteins 3.2 16.5
changed by
Phenotype 13.1   produce
- Folding   3.4
- Glycosylation   5.3
- Phosporylation   9.1
- Degradation   18.4

336

Mastering BIOLOGY
Make Learning Part of the Grade®

Your professor may assign interactive Big Picture concept map exercises at www.masteringbiology.com.

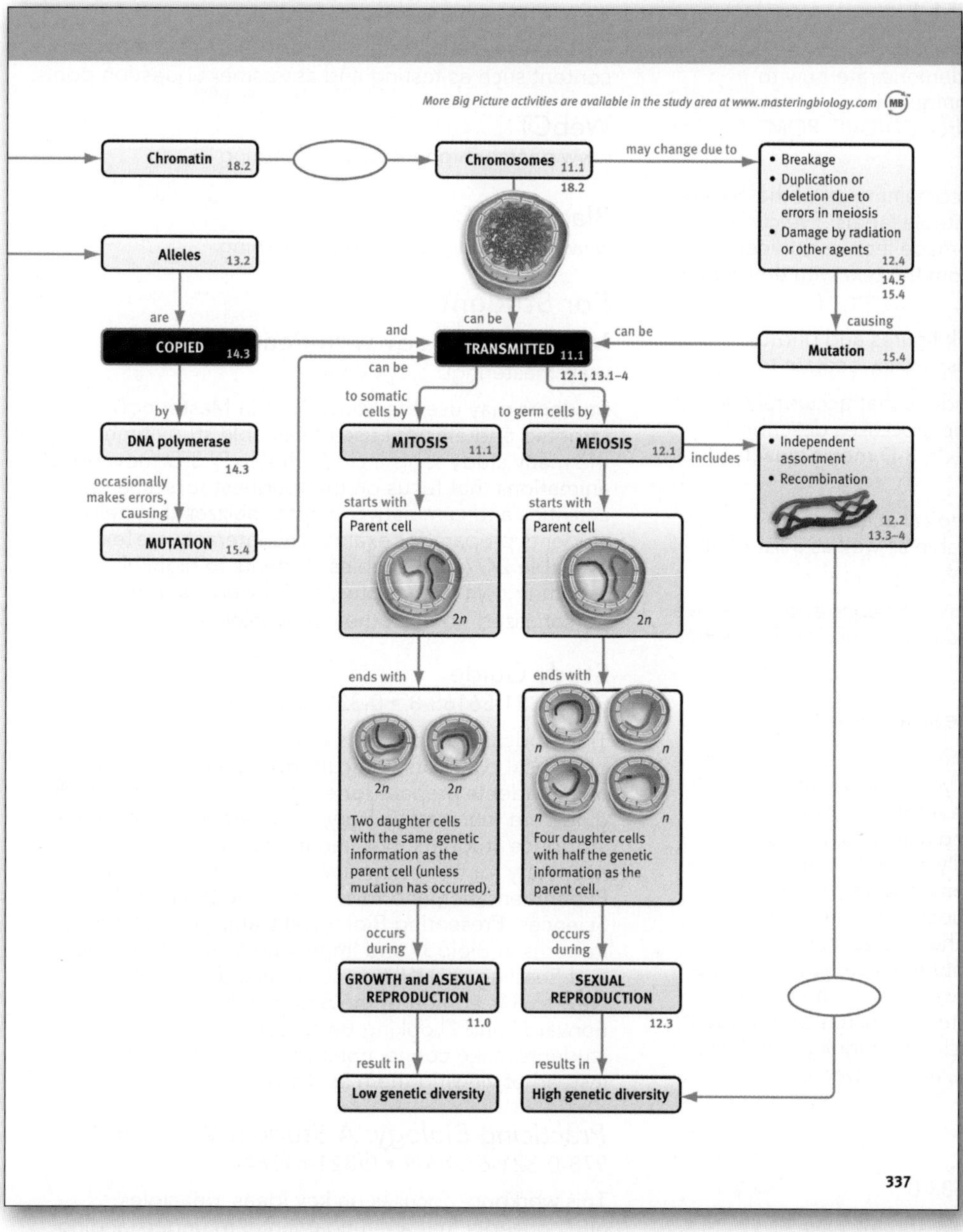

More Big Picture activities are available in the study area at www.masteringbiology.com MB

| Chromatin 18.2 | | Chromosomes 11.1 |
| 18.2 |

Chromosomes — may change due to
- Breakage
- Duplication or deletion due to errors in meiosis
- Damage by radiation or other agents 12.4
  14.5
  15.4

Alleles 13.2

are

COPIED 14.3 — and — **TRANSMITTED** 11.1 — can be — causing — **Mutation** 15.4

12.1, 13.1–4

by

DNA polymerase 14.3

occasionally makes errors, causing

MUTATION 15.4

can be to somatic cells by → **MITOSIS** 11.1

to germ cells by → **MEIOSIS** 12.1 — includes
- Independent assortment
- Recombination
  12.2
  13.3–4

starts with Parent cell 2n (mitosis)

starts with Parent cell 2n (meiosis)

ends with 2n 2n

Two daughter cells with the same genetic information as the parent cell (unless mutation has occurred).

ends with n n n n

Four daughter cells with half the genetic information as the parent cell.

occurs during **GROWTH and ASEXUAL REPRODUCTION** 11.0

occurs during **SEXUAL REPRODUCTION** 12.3

result in **Low genetic diversity**

results in **High genetic diversity**

337

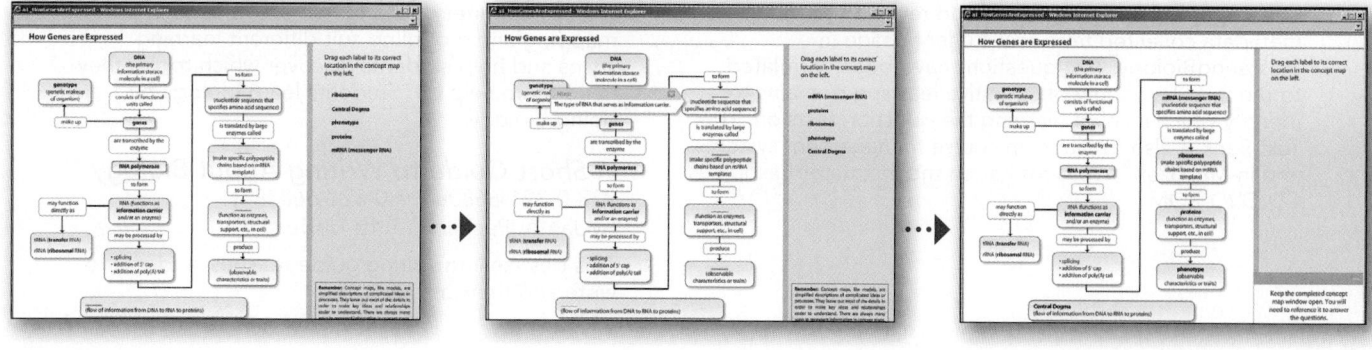

Teaching and Learning Resources

For Instructors

Instructor Resource CD/DVD-ROM
978-0-321-61351-6 • 0-321-61351-1

Everything instructors need for lectures is in one place, including video segments that demonstrate how to incorporate active-learning techniques into your own classroom. The Instructor Resource CD/DVD-ROM includes:

- PowerPoint® Lecture Tools containing all of the figures and photos, which have editable labels; five clicker questions per chapter; pre-made lecture outlines containing select images from the text with embedded animations

- JPEG images of all textbook figures and photos including printer-ready transparency acetate masters

- Over 300 animations and videos that accurately depict complex topics and dynamic processes described in the book; five new BioFlix™ 3-D movie-quality animations

- Instructor's Guides for *Biological Science* and *Practicing Biology* are available as well as a list of all primary literature citations

- The full test bank for *Biological Science* and the *Active Learning Workshop DVD* also come included with the IR-DVD

MasteringBiology® with Pearson eText
www.masteringbiology.com

Assign dynamic homework into your course with automatic grading and adaptive tutoring. Choose from a wide variety of stimulating activities, including visually stunning and scientifically accurate tutorials, ranking questions, 3-D animations, and test bank questions. The powerful gradebook compiles all your favorite teaching diagnostics—the hardest concept, class grade distribution, which students are spending the most or the least time on homework—with the click of a button. Instructors are empowered to customize the eText for themselves and students, including highlighting key text, annotating with comments, adding weblinks, and hiding chapters.

TestGen®
978-0-321-60533-7 • 0-321-60533-0

All of the exam questions in the test bank have been rigorously peer reviewed and revised using the metadata collected from real student usage in MasteringBiology. Test questions have been correlated to Bloom's Taxonomy of cognitive learning domains to identify which level of learning the question tests on. The Test Bank is also available in course management systems and in Microsoft® Word format on the Instructor Resources CD/DVD-ROM.

Course Management Options

CourseCompass™
www.pearsonhighered.com/elearning

This course management system contains preloaded content such as testing and assessment question pools.

WebCT
www.pearsonhighered.com/elearning

Blackboard
www.pearsonhighered.com/elearning

For Students

MasteringBiology with Pearson eText
www.masteringbiology.com

Students may use the Study Area in MasteringBiology for targeted and efficient use of valuable study time. Some of the many study tools include BioFlix™ 3-D movie-quality animations that focus on the toughest topics, engaging activities and cumulative chapter quizzes that help students prepare for exams. The interactive eText is available 24/7 and enables students to highlight text, add their own study notes, and review their Instructor's personalized notes at their convenience.

Study Guide
978-0-321-56168-8 • 0-321-56168-6

The Study Guide presents a breakdown of key biological concepts, difficult topics, and quizzes to help students prepare for exams. Unique to this study guide are four introductory, stand-alone chapters that introduce students to foundational ideas and skills necessary for classroom success: Introduction to Experimentation and Research in the Biological Sciences, Presenting Biological Data, Understanding Patterns in Biology and Improving Study Techniques, and Reading and Writing to Understand Biology. New to this edition of the Study Guide are "Looking Forward" and "Looking Back" sections that help students make connections across the chapters instead of viewing them as discrete entities.

Practicing Biology: A Student Workbook
978-0-321-61264-9 • 0-321-61264-7

This workbook focuses on key ideas, principles, and concepts that are fundamental to understanding biology. A variety of hands-on activities such as mapping and modeling suit different learning styles and help students discover which topics they need more help on. Students learn biology by doing biology.

A Short Guide to Writing About Biology
978-0-321-66838-7 • 0-321-66838-3
by Jan A. Pechenik, Tufts University

This best-selling writing guide teaches students to write and think as biologists.

BIOLOGICAL SCIENCE

Black Swan, *Cygnus atratus*
Male and female black swans have identical coloration—
an all-black body with white flight feathers at the tips of
the wings. Although the significance of their orange-red beak
coloration is unknown, experiments with other species have
shown that individuals with particularly bright beaks or
feathers are in exceptionally good health and are attractive
to potential mates. To explore how you might test this
hypothesis in black swans, see Chapter 25.

BIOLOGICAL SCIENCE

FOURTH EDITION

SCOTT FREEMAN

University of Washington

Benjamin Cummings

Boston   Columbus   Indianapolis   New York   San Francisco   Upper Saddle River
Amsterdam   Cape Town   Dubai   London   Madrid   Milan   Munich   Paris   Montréal   Toronto
Delhi   Mexico City   São Paulo   Sydney   Hong Kong   Seoul   Singapore   Taipei   Tokyo

VP, Editor-in-Chief, Biology: Beth Wilbur
Acquisitions Editor: Becky Ruden
Executive Director of Development, Biology:
    Deborah Gale
Editorial Project Manager: Sonia DiVittorio
Development Editors: Alice Fugate, Moira Lerner-Nelson,
    William O'Neal, Susan Teahan
Art Editor: Kelly Murphy
Assistant Editor: Brady Golden
Editorial Assistant: Leslie Allen
Senior Media Producer: Laura Tommasi
Director of Editorial Content, Mastering Biology:
    Tania Mlawer
Developmental Editor, Mastering Biology: Sarah Jensen
Director of Marketing: Christy Lawrence

Executive Marketing Manager: Lauren Harp
Director of Production, Science: Erin Gregg
Managing Editor, Biology: Michael Early
Production Supervisor: Lori Newman
Media Production Supervisor: James Bruce
Supplements Production Supervisor: Jane Brundage
Production Management and Composition:
    S4Carlisle Publishing Services
Design Manager and Interior Designer: Marilyn Perry
Cover Designer: Riezebos Holzbaur Design Group
Illustrators: Kim Quillin, Imagineering Media Services
Photo Researcher: Maureen Spuhler
Manufacturing Buyer: Michael Penne
Cover Printer: Phoenix Color Corp.
Printer and Binder: Courier, Kendallville

Cover Photo Credits: Black Swan—*Cygnus atratus*, © Eric Isselée/Fotolia (front cover); Black swan among white swans, Hokkaido, Japan, North-East Asia © Keren Su/Getty Images, Inc. (back cover)

Library of Congress Cataloging-in-Publication Data

Freeman, Scott
    Biological science / Scott Freeman.—4th ed.
        p. cm.
    Includes index.
    ISBN 978-0-32-159820-2 (student ed.)—ISBN 978-0-32-159819-6 (professional copy)—
ISBN 978-0-32-161347-9 (v. 1 : the cell, genetics, and development—ISBN 978-0-32-160530-6
(v. 2 : evolution, diversity, and ecology)—ISBN 978-0-32-157676-7 (v. 3 : how plants and animals work)
1. Biology—Textbooks. I. Title.

    QH308. 2. F73 2011
    570—dc22   2009047825

ISBN 10: 0-32-159820-2; ISBN 13: 978-0-32-159820-2 (Student edition)
ISBN 10: 0-32-159819-9; ISBN 13: 978-0-32-159819-6 (Professional copy)
ISBN 10: 0-32-161347-3; ISBN 13: 978-0-32-161347-9 (Volume 1)
ISBN 10: 0-32-160530-6; ISBN 13: 978-0-32-160530-6 (Volume 2)
ISBN 10: 0-32-157676-4; ISBN 13: 978-0-32-157676-7 (Volume 3)

2 3 4 5 6 7 8 9 10—CRK—14 13 12 11 10

Benjamin Cummings
is an imprint of

www.pearsonhighered.com

Brief Contents

1 Biology and the Tree of Life 1

| UNIT 1 | THE MOLECULES OF LIFE 15 |

2 Water and Carbon: The Chemical Basis of Life 15
3 Protein Structure and Function 38
4 Nucleic Acids and the RNA World 59
5 An Introduction to Carbohydrates 71
6 Lipids, Membranes, and the First Cells 82

| UNIT 2 | CELL STRUCTURE AND FUNCTION 102 |

7 Inside the Cell 102
8 Cell-Cell Interactions 131
9 Cellular Respiration and Fermentation 148
10 Photosynthesis 172
11 The Cell Cycle 194

| UNIT 3 | GENE STRUCTURE AND EXPRESSION 211 |

12 Meiosis 211
13 Mendel and the Gene 230
14 DNA and the Gene: Synthesis and Repair 258
15 How Genes Work 276
16 Transcription, RNA Processing, and Translation 289
17 Control of Gene Expression in Bacteria 307
18 Control of Gene Expression in Eukaryotes 319
19 Analyzing and Engineering Genes 338
20 Genomics 359

| UNIT 4 | DEVELOPMENTAL BIOLOGY 374 |

21 Principles of Development 374
22 An Introduction to Animal Development 388
23 An Introduction to Plant Development 401

| UNIT 5 | EVOLUTIONARY PROCESSES AND PATTERNS 414 |

24 Evolution by Natural Selection 414
25 Evolutionary Processes 435
26 Speciation 458
27 Phylogenies and the History of Life 474

| UNIT 6 | THE DIVERSIFICATION OF LIFE 496 |

28 Bacteria and Archaea 496
29 Protists 519
30 Green Algae and Land Plants 546
31 Fungi 579
32 An Introduction to Animals 601
33 Protostome Animals 623
34 Deuterostome Animals 646
35 Viruses 675

| UNIT 7 | HOW PLANTS WORK 695 |

36 Plant Form and Function 695
37 Water and Sugar Transport in Plants 717
38 Plant Nutrition 737
39 Plant Sensory Systems, Signals, and Responses 755
40 Plant Reproduction 783

| UNIT 8 | HOW ANIMALS WORK 803 |

41 Animal Form and Function 803
42 Water and Electrolyte Balance in Animals 822
43 Animal Nutrition 841
44 Gas Exchange and Circulation 861
45 Electrical Signals in Animals 885
46 Animal Sensory Systems and Movement 907
47 Chemical Signals in Animals 929
48 Animal Reproduction 950
49 The Immune System in Animals 973

| UNIT 9 | ECOLOGY 993 |

50 An Introduction to Ecology 993
51 Behavioral Ecology 1019
52 Population Ecology 1037
53 Community Ecology 1058
54 Ecosystems 1083
55 Biodiversity and Conservation Biology 1105

Detailed Contents

1 Biology and the Tree of Life   1

1.1   What Does It Mean to Say That Something Is Alive?   1

1.2   The Cell Theory   2
Are *All* Organisms Made of Cells?   2
Where Do Cells Come From?   2

1.3   The Theory of Evolution by Natural Selection   4
What Is Evolution?   4
What Is Natural Selection?   4

1.4   The Tree of Life   5
Using Molecules to Understand the Tree of Life   6
How Should We Name Branches on the Tree of Life?   7

1.5   Doing Biology   8
The Nature of Science   8
Why Do Giraffes Have Long Necks? An Introduction to
Hypothesis Testing   9
How Do Ants Navigate? An Introduction to Experimental
Design   10

CHAPTER REVIEW   13

UNIT 1   THE MOLECULES OF LIFE   15

2 Water and Carbon: The Chemical Basis of Life   15

2.1   Atoms, Ions, and Molecules: The Building Blocks of
Chemical Evolution   16
Basic Atomic Structure   16
How Does Covalent Bonding Hold Molecules Together?   17
Ionic Bonding, Ions, and the Electron-Sharing Continuum   18
Some Simple Molecules Formed from C, H, N, and O   19
The Geometry of Simple Molecules   20
Representing Molecules   20
Basic Concepts in Chemical Reactions   21

2.2   The Early Oceans and the Properties of Water   22
Why Is Water Such an Efficient Solvent?   22
How Does Water's Structure Correlate with Its Properties?   22
Acid–Base Reactions Involve a Transfer of Protons   25

2.3   Chemical Reactions, Chemical Evolution, and Chemical
Energy   27
How Do Chemical Reactions Happen?   27
What Is Energy?   27
Chemical Evolution: A Model System   29
How Did Chemical Energy Change during Chemical
Evolution?   33

2.4   The Importance of Carbon   33
Linking Carbon Atoms Together   34
Functional Groups   34

CHAPTER REVIEW   36

3 Protein Structure and Function   38

3.1   Early Origin-of-Life Experiments   39

3.2   Amino Acids and Polymerization   40
The Structure of Amino Acids   40
The Nature of Side Chains   40
How Do Amino Acids Link to Form Proteins?   42

3.3   Proteins Are the Most Versatile Large Molecules
in Cells   45

3.4   What Do Proteins Look Like?   45
Primary Structure   46
Secondary Structure   46
Tertiary Structure   47
Quaternary Structure   48
Folding and Function   50

3.5   Enzymes: An Introduction to Catalysis   51
Enzymes Help Reactions Clear Two Hurdles   51
How Do Enzymes Work?   53
Was the First Living Entity a Protein Catalyst?   56

CHAPTER REVIEW   57

4 Nucleic Acids and the RNA World   59

4.1   What Is a Nucleic Acid?   59
Could Chemical Evolution Result in the Production of
Nucleotides?   60
How Do Nucleotides Polymerize to Form Nucleic Acids?   61

4.2   DNA Structure and Function   62
What Is the Nature of DNA's Secondary Structure?   62
DNA Functions as an Information-Containing Molecule   65
Is DNA a Catalytic Molecule?   65

4.3   RNA Structure and Function   66
Structurally, RNA Differs from DNA   66
RNA's Structure Makes It an Extraordinarily Versatile
Molecule   67
RNA Is an Information-Containing Molecule   67
RNA Can Function as a Catalytic Molecule   68

4.4   The First Life-Form   68

CHAPTER REVIEW   69

5 An Introduction to Carbohydrates   71

5.1   Sugars as Monomers   71
How Monosaccharides Differ   72
Monosaccharides and Chemical Evolution   73

5.2   The Structure of Polysaccharides   73
Starch: A Storage Polysaccharide in Plants   74
Glycogen: A Highly Branched Storage Polysaccharide in
Animals   74
Cellulose: A Structural Polysaccharide in Plants   76

Chitin: A Structural Polysaccharide in Fungi and Animals **76**
Peptidoglycan: A Structural Polysaccharide in Bacteria **76**
Polysaccharides and Chemical Evolution **76**

5.3 What Do Carbohydrates Do? 77
The Role of Carbohydrates as Structural Molecules **77**
The Role of Carbohydrates in Cell Identity **77**
The Role of Carbohydrates in Energy Storage **78**

CHAPTER REVIEW 80

6 Lipids, Membranes, and the First Cells 82

6.1 Lipids 83
A Look at Three Types of Lipids Found in Cells **83**
The Structures of Membrane Lipids **84**

6.2 Phospholipid Bilayers 85
Artificial Membranes as an Experimental System **85**
Selective Permeability of Lipid Bilayers **86**
How Does Lipid Structure Affect Membrane Properties? **87**
How Does Temperature Affect the Fluidity and Permeability of Membranes? **88**

6.3 Why Molecules Move across Lipid Bilayers: Diffusion and Osmosis 89
Diffusion **89**
Osmosis **90**

6.4 Membrane Proteins 92
Evolution of the Fluid-Mosaic Model **92**
Systems for Studying Membrane Proteins **94**
Protein Transport I: Facilitated Diffusion via Channel Proteins **94**
Protein Transport II: Facilitated Diffusion via Carrier Proteins **96**
Protein Transport III: Active Transport by Pumps **97**
Plasma Membranes and the Intracellular Environment **98**

CHAPTER REVIEW 100

UNIT 2 CELL STRUCTURE AND FUNCTION 102

7 Inside the Cell 102

7.1 Bacterial and Archaeal Cell Structures and Their Functions 102
A Revolutionary New View **103**
Prokaryotic Cell Structures: A Parts List **103**

7.2 Eukaryotic Cell Structures and Their Functions 105
The Benefits of Organelles **107**
Eukaryotic Cell Structures: A Parts List **107**

7.3 Putting the Parts into a Whole 115
Structure and Function at the Whole-Cell Level **115**
The Dynamic Cell **116**

7.4 Cell Systems I: Nuclear Transport 116
Structure and Function of the Nuclear Envelope **116**
How Are Molecules Imported into the Nucleus? **117**

7.5 Cell Systems II: The Endomembrane System Manufactures and Ships Proteins 118
Studying the Pathway through the Endomembrane System **119**

Entering the Endomembrane System: The Signal Hypothesis **120**
Moving from the ER to the Golgi **122**
What Happens inside the Golgi Apparatus? **122**
How Do Proteins Reach Their Destinations? **122**

7.6 Cell Systems III: The Dynamic Cytoskeleton 123
Actin Filaments **123**
Intermediate Filaments **125**
Microtubules **125**
Flagella and Cilia: Moving the Entire Cell **127**

CHAPTER REVIEW 129

8 Cell-Cell Interactions 131

8.1 The Cell Surface 132
The Structure and Function of an Extracellular Layer **132**
The Cell Wall in Plants **132**
The Extracellular Matrix in Animals **133**

8.2 How Do Adjacent Cells Connect and Communicate? 134
Cell-Cell Attachments in Eukaryotes **135**
Cells Communicate via Cell-Cell Gaps **138**

8.3 How Do Distant Cells Communicate? 139
Cell-Cell Signaling in Multicellular Organisms **139**
Signal Reception **140**
Signal Processing **140**
Signal Response **144**
Signal Deactivation **144**
Cross-Talk: Synthesizing Input from Many Signals **145**
Quorum Sensing in Bacteria **145**

CHAPTER REVIEW 146

9 Cellular Respiration and Fermentation 148

9.1 The Nature of Chemical Energy and Redox Reactions 149
The Structure and Function of ATP **149**
What Is a Redox Reaction? **151**

9.2 An Overview of Cellular Respiration 153

9.3 Glycolysis: Processing Glucose to Pyruvate 155
Glycolysis Is a Sequence of 10 Reactions **155**
How Is Glycolysis Regulated? **156**

9.4 Processing Pyruvate to Acetyl CoA 156

9.5 The Citric Acid Cycle: Oxidizing Acetyl CoA to CO_2 158
How Is the Citric Acid Cycle Regulated? **158**
What Happens to the NADH and $FADH_2$? **160**

9.6 Electron Transport and Chemiosmosis: Building a Proton Gradient to Produce ATP 161
Components of the Electron Transport Chain **161**
The Chemiosmosis Hypothesis **162**
How Is the Electron Transport Chain Organized? **163**
The Discovery of ATP Synthase **164**
Organisms Use a Diversity of Electron Acceptors **165**

9.7 Fermentation 166

9.8 How Does Cellular Respiration Interact with Other Metabolic Pathways? 168
Catabolic Pathways Break Down Molecules as Fuel 168
Anabolic Pathways Synthesize Key Molecules 169

CHAPTER REVIEW 169

10 Photosynthesis 172

10.1 Photosynthesis Harnesses Sunlight to Make Carbohydrate 172
Photosynthesis: Two Linked Sets of Reactions 173
Photosynthesis Occurs in Chloroplasts 174

10.2 How Does Chlorophyll Capture Light Energy? 174
Photosynthetic Pigments Absorb Light 175
When Light Is Absorbed, Electrons Enter an Excited State 177

10.3 The Discovery of Photosystems I and II 179
How Does Photosystem II Work? 180
How Does Photosystem I Work? 182
The Z Scheme: Photosystems II and I Work Together 182

10.4 How Is Carbon Dioxide Reduced to Produce Glucose? 184
The Calvin Cycle Fixes Carbon 185
The Discovery of Rubisco 186
Carbon Dioxide Enters Leaves through Stomata 187
Mechanisms for Increasing CO_2 Concentration 187
How Is Photosynthesis Regulated? 189
What Happens to the Sugar That Is Produced by Photosynthesis? 189

CHAPTER REVIEW 190
The Big Picture: Energy for Life 192

11 The Cell Cycle 194

11.1 Mitosis and the Cell Cycle 195
What Is a Chromosome? 195
Cells Alternate between M Phase and Interphase 196
The Discovery of S Phase 196
The Discovery of the Gap Phases 196
The Cell Cycle 196

11.2 How Does Mitosis Take Place? 197
Events in Mitosis 197
Cytokinesis Results in Two Daughter Cells 200
How Do Chromosomes Move during Mitosis? 201

11.3 Control of the Cell Cycle 202
The Discovery of Cell-Cycle Regulatory Molecules 203
Cell-Cycle Checkpoints Can Arrest the Cell Cycle 204

11.4 Cancer: Out-of-Control Cell Division 206
Properties of Cancer Cells 206
Cancer Involves Loss of Cell-Cycle Control 207

CHAPTER REVIEW 209

UNIT 3 GENE STRUCTURE AND EXPRESSION 211

12 Meiosis 211

12.1 How Does Meiosis Occur? 212
Chromosomes Come in Distinct Types 212
The Concept of Ploidy 212

An Overview of Meiosis 213
The Phases of Meiosis I 216
The Phases of Meiosis II 218
A Closer Look at Prophase I 219

12.2 The Consequences of Meiosis 220
Chromosomes and Heredity 221
Independent Assortment Produces Genetic Variation 221
The Role of Crossing Over 222
How Does Fertilization Affect Genetic Variation? 222

12.3 Why Does Meiosis Exist? 223
The Paradox of Sex 223
The Purifying Selection Hypothesis 224
The Changing-Environment Hypothesis 224

12.4 Mistakes in Meiosis 225
How Do Mistakes Occur? 225
Why Do Mistakes Occur? 226

CHAPTER REVIEW 227

13 Mendel and the Gene 230

13.1 Mendel's Experimental System 230
What Questions Was Mendel Trying to Answer? 231
Garden Peas Served as the First Model Organism in Genetics 231

13.2 Mendel's Experiments with a Single Trait 232
The Monohybrid Cross 232
Particulate Inheritance 234

13.3 Mendel's Experiments with Two Traits 236
The Dihybrid Cross 236
Using a Testcross to Confirm Predictions 238

13.4 The Chromosome Theory of Inheritance 239
Meiosis Explains Mendel's Principles 240
Testing the Chromosome Theory 241

13.5 Extending Mendel's Rules 243
Linkage: What Happens When Genes Are Located on the Same Chromosome? 243
Do Heterozygotes Always Have a Dominant or Recessive Phenotype? 245
BOX 13.1 QUANTITATIVE METHODS: Linkage 245
How Many Alleles and Phenotypes Exist? 247
Does Each Gene Affect Just One Trait? 247
Are Phenotypes Determined by Genes? 247
What About Traits Like Human Height and Intelligence? 248

13.6 Applying Mendel's Rules to Humans 250
Identifying Human Alleles as Recessive or Dominant 250
Identifying Human Traits as Autosomal or Sex-Linked 251

CHAPTER REVIEW 253

14 DNA and the Gene: Synthesis and Repair 258

14.1 What Are Genes Made Of? 259
The Hershey-Chase Experiment 259
The Secondary Structure of DNA 260

14.2 Testing Early Hypotheses about DNA Synthesis: The Meselson-Stahl Experiment 261

14.3 A Comprehensive Model for DNA Synthesis 263
How Does Replication Get Started? 264
How Is the Helix Opened and Stabilized? 264
How Is the Leading Strand Synthesized? 265
How Is the Lagging Strand Synthesized? 266

14.4 Replicating the Ends of Linear Chromosomes 269

14.5 Repairing Mistakes and Damage 271
Correcting Mistakes in DNA Synthesis 271
Repairing Damaged DNA 272
Xeroderma Pigmentosum: A Case Study 272

CHAPTER REVIEW 274

15 How Genes Work 276

15.1 What Do Genes Do? 277
The One-Gene, One-Enzyme Hypothesis 277
An Experimental Test of the Hypothesis 277

15.2 The Central Dogma of Molecular Biology 279
The Genetic Code Hypothesis 279
RNA as the Intermediary between Genes and Proteins 279
Dissecting the Central Dogma 280

15.3 The Genetic Code 282
How Long Is a Word in the Genetic Code? 282
How Did Researchers Crack the Code? 283

15.4 What Is the Molecular Basis of Mutation? 285
Point Mutation 285
Chromosome-Level Mutations 286

CHAPTER REVIEW 287

16 Transcription, RNA Processing, and Translation 289

16.1 An Overview of Transcription 289
Characteristics of RNA Polymerase 290
Initiation: How Does Transcription Begin? 291
Elongation and Termination 292

16.2 RNA Processing in Eukaryotes 293
The Startling Discovery of Eukaryotic Genes in Pieces 293
RNA Splicing 294
Adding Caps and Tails to Transcripts 295

16.3 An Introduction to Translation 295
Ribosomes Are the Site of Protein Synthesis 295
Comparing Translation in Bacteria and Eukaryotes 296
How Does an mRNA Triplet Specify an Amino Acid? 297

16.4 The Structure and Function of Transfer RNA 297
What Do tRNAs Look Like? 299
How Many tRNAs Are There? 299

16.5 The Structure and Function of Ribosomes 300
Initiating Translation 301
Elongation: Extending the Polypeptide 301
Terminating Translation 302
Post-Translational Modifications 304

CHAPTER REVIEW 304

17 Control of Gene Expression in Bacteria 307

17.1 Gene Regulation and Information Flow 307
Mechanisms of Regulation—An Overview 308
Metabolizing Lactose—A Model System 309

17.2 Identifying Genes under Regulatory Control 310
Replica Plating to Find Mutant Genes 310
Different Classes of Lactose Metabolism Mutants 311
Several Genes Are Involved in Lactose Metabolism 312

17.3 Mechanisms of Negative Control: Discovery of the Repressor 312
The *lac* Operon 313
Why Has the *lac* Operon Model Been So Important? 314

17.4 Mechanisms of Positive Control: Catabolite Repression 314
The CAP Protein and Binding Site 315
How Does Glucose Influence Formation of the CAP–cAMP Complex? 316

CHAPTER REVIEW 317

18 Control of Gene Expression in Eukaryotes 319

18.1 Mechanisms of Gene Regulation in Eukaryotes— An Overview 320

18.2 Chromatin Remodeling 320
What Is Chromatin's Basic Structure? 320
Evidence That Chromatin Structure Is Altered in Active Genes 321
How Is Chromatin Altered? 322
Chromatin Modifications Can Be Inherited 323

18.3 Initiating Transcription: Regulatory Sequences and Regulatory Proteins 323
Some Regulatory Sequences Are Near the Promoter 323
Some Regulatory Sequences Are Far from the Promoter 324
The Role of Regulatory Proteins in Differential Gene Expression 326
The Initiation Complex 326

18.4 Post-Transcriptional Control 328
Alternative Splicing of mRNAs 328
mRNA Stability and RNA Interference 329

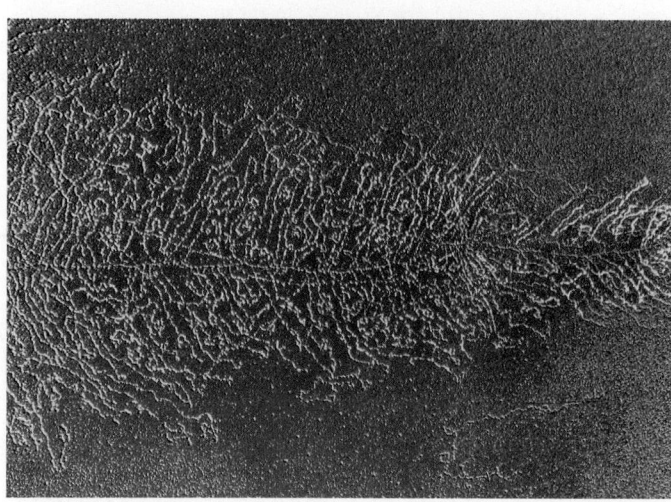

How Is Translation Controlled? **330**
Post-Translational Control **330**

18.5 How Does Gene Expression in Bacteria Compare with That in Eukaryotes? 331

18.6 Linking Cancer with Defects in Gene Regulation 332
Causes of Uncontrolled Cell Growth **332**
p53: A Case Study **332**

CHAPTER REVIEW 333

The Big Picture: Genetic Information **336**

19 Analyzing and Engineering Genes 338

19.1 Case 1—The Effort to Cure Pituitary Dwarfism: Basic Recombinant DNA Technologies 338
Why Did Early Efforts to Treat the Disease Fail? **339**
Steps in Engineering a Safe Supply of Growth Hormone **339**
Ethical Concerns over Recombinant Growth Hormone **343**

19.2 Case 2—Amplification of Fossil DNA: The Polymerase Chain Reaction 344
Requirements of PCR **344**
PCR in Action **345**

19.3 Case 3—Sanger's Breakthrough Innovation: Dideoxy DNA Sequencing 346
The Logic of Dideoxy Sequencing **346**
"Next Generation" Sequencing **347**

19.4 Case 4—The Huntington's Disease Story: Finding Genes by Mapping 348
How Was the Huntington's Disease Gene Found? **348**
What Are the Benefits of Finding a Disease Gene? **350**
Ethical Concerns over Genetic Testing **350**

19.5 Case 5—Severe Immune Disorders: The Potential of Gene Therapy 351
How Can Novel Alleles Be Introduced into Human Cells? **351**
Using Gene Therapy to Treat X-Linked Immune Deficiency **352**
Ethical Concerns over Gene Therapy **354**

19.6 Case 6—The Development of Golden Rice: Biotechnology in Agriculture 354
Rice as a Target Crop **355**
Synthesizing β-Carotene in Rice **355**
The *Agrobacterium* Transformation System **355**
Using the Ti Plasmid to Produce Golden Rice **356**

CHAPTER REVIEW 356

20 Genomics 359

20.1 Whole-Genome Sequencing 359
How Are Complete Genomes Sequenced? **360**
Which Genomes Are Being Sequenced, and Why? **361**
Which Sequences Are Genes? **362**

20.2 Bacterial and Archaeal Genomes 363
The Natural History of Prokaryotic Genomes **363**
Lateral Gene Transfer **364**
Environmental Sequencing **364**

20.3 Eukaryotic Genomes 365
Parasitic and Repeated Sequences **365**
Gene Families **367**
Insights from the Human Genome Project **368**

20.4 Functional Genomics and Proteomics 370
What Is Functional Genomics? **370**
What Is Proteomics? **371**
Applied Genomics in Action: Understanding Cancer **371**

CHAPTER REVIEW 372

UNIT 4 DEVELOPMENTAL BIOLOGY 374

21 Principles of Development 374

21.1 Shared Developmental Processes 375
Cell Proliferation **375**
Programmed Cell Death **376**
Cell Movement or Cell Growth **376**
Cell Differentiation **377**
Cell-Cell Interactions **377**

21.2 The Role of Differential Gene Expression in Development 377
Evidence That Differentiated Plant Cells Are Genetically Equivalent **377**
Evidence That Differentiated Animal Cells Are Genetically Equivalent **377**
How Does Differential Gene Expression Occur? **378**

21.3 Cell-Cell Signals Trigger Differential Gene Expression 379
Master Regulators Set Up the Major Body Axes **379**
Regulatory Genes Provide Increasingly Specific Positional Information **381**
Cell-Cell Signals and Regulatory Genes Are Evolutionarily Conserved **383**
Common Signaling Pathways Are Active in Many Contexts **383**

21.4 Changes in Developmental Pathways Underlie Evolutionary Change 384

CHAPTER REVIEW 385

22 An Introduction to Animal Development 388

22.1 Gamete Structure and Function 389
Sperm Structure and Function **389**
Egg Structure and Function **390**

22.2 Fertilization 390
How Do Gametes from the Same Species Recognize Each Other? **391**
Why Does Only One Sperm Enter the Egg? **391**

22.3 Cleavage 392
Partitioning Cytoplasmic Determinants **393**
Cleavage in Mammals **393**

22.4 Gastrulation 394
Formation of Germ Layers **394**
Definition of Body Axes **395**

22.5 Organogenesis 396
Organizing Mesoderm into Somites: Precursors of Muscle, Skeleton, and Skin **396**
Differentiation of Muscle Cells **398**

CHAPTER REVIEW 399

23 An Introduction to Plant Development 401

23.1 Gametogenesis, Pollination, and Fertilization 402
How Are Sperm and Egg Produced? 402
Pollen–Stigma Interactions 402
Double Fertilization 403

23.2 Embryogenesis 404
What Happens during Plant Embryogenesis? 404
Which Genes and Proteins Set Up Body Axes? 406

23.3 Vegetative Development 407
Meristems Provide Lifelong Growth and Development 407
Which Genes and Proteins Determine Leaf Shape? 408

23.4 Reproductive Development 409
The Floral Meristem and the Flower 409
The Genetic Control of Flower Structures 409

CHAPTER REVIEW 412

UNIT 5 EVOLUTIONARY PROCESSES AND PATTERNS 414

24 Evolution by Natural Selection 414

24.1 The Evolution of Evolutionary Thought 415
Plato and Typological Thinking 415
Aristotle and the Great Chain of Being 415
Lamarck and the Idea of Evolution as Change
through Time 415
Darwin and Wallace and Evolution by Natural Selection 415

**24.2 The Pattern of Evolution: Have Species Changed
through Time?** 416
Evidence for Change through Time 416
Evidence of Descent from a Common Ancestor 418
Evolution's "Internal Consistency"—the Importance of
Independent Datasets 422

**24.3 The Process of Evolution: How Does Natural
Selection Work?** 422
Darwin's Four Postulates 423
The Biological Definitions of Fitness and Adaptation 424

**24.4 Evolution in Action: Recent Research on Natural
Selection** 424
Case Study 1: How Did *Mycobacterium tuberculosis* Become
Resistant to Antibiotics? 424
Case Study 2: Why Are Beak Size, Beak Shape, and Body Size
Changing in Galápagos Finches? 426

**24.5 Common Misconceptions about Natural Selection
and Adaptation** 429
Selection Acts on Individuals, but Evolutionary Change Occurs
in Populations 429
Evolution Is Not Goal Directed 430
Organisms Do Not Act for the Good of the Species 430
Limitations of Natural Selection 431

CHAPTER REVIEW 432

25 Evolutionary Processes 435

**25.1 Analyzing Change in Allele Frequencies: The Hardy-
Weinberg Principle** 436
The Gene Pool Concept 436
Deriving the Hardy-Weinberg Principle 436
The Hardy-Weinberg Model Makes Important
Assumptions 437
How Does the Hardy-Weinberg Principle Serve as a Null
Hypothesis? 438

25.2 Types of Natural Selection 440
Directional Selection 440
Stabilizing Selection 441
Disruptive Selection 442
Balancing Selection 442

25.3 Genetic Drift 443
Simulation Studies of Genetic Drift 443
Experimental Studies of Genetic Drift 445
What Causes Genetic Drift in Natural Populations? 445

25.4 Gene Flow 447
Gene Flow in Natural Populations 447
How Does Gene Flow Affect Fitness? 448

25.5 Mutation 448
Mutation as an Evolutionary Mechanism 448
Experimental Studies of Mutation 449

25.6 Nonrandom Mating 450
Inbreeding 450
Sexual Selection 452

CHAPTER REVIEW 456

26 Speciation 458

26.1 How Are Species Defined and Identified? 458
The Biological Species Concept 459
The Morphospecies Concept 460
The Phylogenetic Species Concept 460
Species Definitions in Action: The Case of the Dusky
Seaside Sparrow 461

26.2 **Isolation and Divergence in Allopatry** 462
Dispersal and Colonization Isolate Populations 463
Vicariance Isolates Populations 464

26.3 **Isolation and Divergence in Sympatry** 464
Can Natural Selection Cause Speciation Even When Gene Flow Is Possible? 465
How Can Polyploidy Lead to Speciation? 465

26.4 **What Happens When Isolated Populations Come into Contact?** 468
Reinforcement 468
Hybrid Zones 468
New Species through Hybridization 470

CHAPTER REVIEW 472

27 Phylogenies and the History of Life 474

27.1 **Tools for Studying History: Phylogenetic Trees** 474
How Do Researchers Estimate Phylogenies? 475
How Can Biologists Distinguish Homology from Homoplasy? 475
Whale Evolution: A Case History 477

27.2 **Tools for Studying History: The Fossil Record** 479
How Do Fossils Form? 479
Limitations of the Fossil Record 480
Life's Time Line 481

27.3 **Adaptive Radiation** 484
Why Do Adaptive Radiations Occur? 484
The Cambrian Explosion 486

27.4 **Mass Extinction** 488
How Do Mass Extinctions Differ From Background Extinctions? 489
The End-Permian Extinction 489
What Killed the Dinosaurs? 490

CHAPTER REVIEW 492

The Big Picture: Evolution 494

UNIT 6 THE DIVERSIFICATION OF LIFE 496

28 Bacteria and Archaea 496

28.1 **Why Do Biologists Study Bacteria and Archaea?** 497
Biological Impact 497
Medical Importance 498
Role in Bioremediation 500
Extremophiles 501

28.2 **How Do Biologists Study Bacteria and Archaea?** 501
Using Enrichment Cultures 501
Using Direct Sequencing 502
Evaluating Molecular Phylogenies 503

28.3 **What Themes Occur in the Diversification of Bacteria and Archaea?** 504
Morphological Diversity 504
Metabolic Diversity 506
Ecological Diversity and Global Change 509

28.4 **Key Lineages of Bacteria and Archaea** 512
Bacteria 512
Archaea 512

■ Bacteria > Firmicutes 513
■ Bacteria > Spirochaetes (Spirochetes) 513
■ Bacteria > Actinobacteria 514
■ Bacteria > Chlamydiae 514
■ Bacteria > Cyanobacteria 515
■ Bacteria > Proteobacteria 515
■ Archaea > Crenarchaeota 516
■ Archaea > Euryarchaeota 516

CHAPTER REVIEW 517

29 Protists 519

29.1 **Why Do Biologists Study Protists?** 520
Impacts on Human Health and Welfare 520
Ecological Importance of Protists 522

29.2 **How Do Biologists Study Protists?** 524
Microscopy: Studying Cell Structure 524
Evaluating Molecular Phylogenies 525
Discovering New Lineages via Direct Sequencing 525

29.3 **What Themes Occur in the Diversification of Protists?** 526
What Morphological Innovations Evolved in Protists? 526
How Do Protists Obtain Food? 529
How Do Protists Move? 532
How Do Protists Reproduce? 533
Life Cycles—Haploid- versus Diploid-Dominated 533

29.4 **Key Lineages of Protists** 536
Amoebozoa 536
Excavata 536
Plantae 536
Rhizaria 536
Alveolata 537
Stramenopila (Heterokonta) 537
■ Amoebozoa > Myxogastrida (Plasmodial Slime Molds) 537
■ Excavata > Parabasalida 538
■ Excavata > Diplomonadida 538
■ Excavata > Euglenida 539
■ Plantae > Rhodophyta (Red Algae) 539
■ Rhizaria > Foraminifera 540
■ Alveolata > Ciliata 540
■ Alveolata > Dinoflagellata 541
■ Alveolata > Apicomplexa 541
■ Stramenopila > Oomycota (Water Molds) 542
■ Stramenopila > Diatoms 542
■ Stramenopila > Phaeophyta (Brown Algae) 543

CHAPTER REVIEW 543

30 Green Algae and Land Plants 546

30.1 **Why Do Biologists Study the Green Algae and Land Plants?** 546
Plants Provide Ecosystem Services 547
Plants Provide Humans with Food, Fuel, Fiber, Building Materials, and Medicines 548

30.2 **How Do Biologists Study Green Algae and Land Plants?** 549
Analyzing Morphological Traits 549
Using the Fossil Record 550
Evaluating Molecular Phylogenies 551

30.3 What Themes Occur in the Diversification of Land Plants? 553

The Transition to Land, I: How Did Plants Adapt to Dry Conditions? 553

Mapping Evolutionary Changes on the Phylogenetic Tree 555

The Transition to Land, II: How Do Plants Reproduce in Dry Conditions? 556

The Angiosperm Radiation 564

30.4 Key Lineages of Green Algae and Land Plants 566

Green Algae 566

Non-Vascular Plants ("Bryophytes") 567

Seedless Vascular Plants 567

Seed Plants 567

- Green Algae > Ulvophyceae (Ulvophytes) 568
- Green Algae > Coleochaetophyceae (Coleochaetes) 568
- Green Algae > Charophyceae (Stoneworts) 569
- Non-Vascular Plants > Hepaticophyta (Liverworts) 569
- Non-Vascular Plants > Bryophyta (Mosses) 570
- Non-Vascular Plants > Anthocerophyta (Hornworts) 571
- Seedless Vascular Plants > Lycophyta (Lycophytes, or Club Mosses) 571
- Seedless Vascular Plants > Psilotophyta (Whisk Ferns) 572
- Seedless Vascular Plants > Equisetophyta (or Sphenophyta) (Horsetails) 572
- Seedless Vascular Plants > Pteridophyta (Ferns) 573
- Seed Plants > Gymnosperms > Cycadophyta (Cycads) 574
- Seed Plants > Gymnosperms > Ginkgophyta (Ginkgos) 574
- Seed Plants > Gymnosperms > Redwood group (Redwoods, Junipers, Yews) 575
- Seed Plants > Gymnosperms > Pinophyta (Pines, Spruces, Firs) 575
- Seed Plants > Gymnosperms > Gnetophyta (Gnetophytes) 576
- Seed Plants > Anthophyta (Angiosperms) 576

CHAPTER REVIEW 577

31 Fungi 579

31.1 Why Do Biologists Study Fungi? 580

Fungi Provide Nutrients for Land Plants 580

Fungi Speed the Carbon Cycle on Land 580

Fungi Have Important Economic Impacts 581

31.2 How Do Biologists Study Fungi? 582

Analyzing Morphological Traits 582

Evaluating Molecular Phylogenies 584

Experimental Studies of Mutualism 586

31.3 What Themes Occur in the Diversification of Fungi? 586

Fungi Participate in Several Types of Mutualisms 586

What Adaptations Make Fungi Such Effective Decomposers? 589

Variation in Reproduction 590

Four Major Types of Life Cycles 591

31.4 Key Lineages of Fungi 594

- Fungi > Microsporidia 594
- Fungi > Chytrids 595
- Fungi > Zygomycetes 595
- Fungi > Glomeromycota 596
- Fungi > Basidiomycota (Club Fungi) 596
- Fungi > Ascomycota > Lichen-Formers 597
- Fungi > Ascomycota > Non-Lichen-Formers 598

CHAPTER REVIEW 599

32 An Introduction to Animals 601

32.1 Why Do Biologists Study Animals? 602

Biological Importance 602

Role in Human Health and Welfare 602

32.2 How Do Biologists Study Animals? 603

Analyzing Comparative Morphology 603

Evaluating Molecular Phylogenies 607

32.3 What Themes Occur in the Diversification of Animals? 610

Sensory Organs 610

Feeding 610

Movement 613

Reproduction 615

Life Cycles 615

32.4 Key Lineages of Animals: Non-Bilaterian Groups 617

- Porifera (Sponges) 618
- Cnidaria (Jellyfish, Corals, Anemones, Hydroids) 619
- Ctenophora (Comb Jellies) 620
- Acoelomorpha (Acoels) 620

CHAPTER REVIEW 621

33 Protostome Animals 623

33.1 An Overview of Protostome Evolution 624

What Is a Lophotrochozoan? 624

What Is an Ecdysozoan? 625

33.2 Themes in the Diversification of Protostomes 625

How Do Body Plans Vary among Phyla? 626

The Water-to-Land Transition 627

Adaptations for Feeding 628

Adaptations for Moving 628

Adaptations in Reproduction 630

33.3 Key Lineages: Lophotrochozoans 630

- Lophotrochozoans > Rotifera (Rotifers) 631
- Lophotrochozoans > Platyhelminthes (Flatworms) 631
- Lophotrochozoans > Annelida (Segmented Worms) 633
- Lophotrochozoans > Mollusca > Bivalvia (Clams, Mussels, Scallops, Oysters) 634
- Lophotrochozoans > Mollusca > Gastropoda (Snails, Slugs, Nudibranchs) 635
- Lophotrochozoans > Mollusca > Polyplacophora (Chitons) 636
- Lophotrochozoans > Mollusca > Cephalopoda (Nautilus, Cuttlefish, Squid, Octopuses) 636

33.4 Key Lineages: Ecdysozoans 637

- Ecdysozoans > Nematoda (Roundworms) 638
- Ecdysozoans > Arthropoda > Myriapods (Millipedes, Centipedes) 639
- Ecdysozoans > Arthropoda > Insecta (Insects) 639
- Ecdysozoans > Arthropoda > Chelicerata (Spiders, Ticks, Mites, Horseshoe Crabs, Daddy Longlegs, Scorpions) 642
- Ecdysozoans > Arthropoda > Crustaceans (Shrimp, Lobster, Crabs, Barnacles, Isopods, Copepods) 643

CHAPTER REVIEW 644

34 Deuterostome Animals 646

34.1 What Is an Echinoderm? 647
The Echinoderm Body Plan 647
How Do Echinoderms Feed? 648
Key Lineages 649
- Echinodermata > Asteroidea (Sea Stars) 649
- Echinodermata > Echinoidea (Sea Urchins and Sand Dollars) 650

34.2 What Is a Chordate? 650
Three "Subphyla" 651
Key Lineages: The Invertebrate Chordates 651
- Chordata > Cephalochordata (Lancelets) 652
- Chordata > Urochordata (Tunicates) 652

34.3 What Is a Vertebrate? 653
An Overview of Vertebrate Evolution 653
Key Innovations 655
Key Lineages 660
- Chordata > Vertebrata > Myxinoidea (Hagfish) and Petromyzontoidea (Lampreys) 661
- Chordata > Vertebrata > Chondrichthyes (Sharks, Rays, Skates) 662
- Chordata > Vertebrata > Actinopterygii (Ray-Finned Fishes) 662
- Chordata > Vertebrata > Actinistia (Coelacanths) and Dipnoi (Lungfish) 663
- Chordata > Vertebrata > Amphibia (Frogs, Salamanders, Caecilians) 664
- Chordata > Vertebrata > Mammalia > Monotremata (Platypuses, Echidnas) 665
- Chordata > Vertebrata > Mammalia > Marsupiala (Marsupials) 665
- Chordata > Vertebrata > Mammalia > Eutheria (Placental Mammals) 666
- Chordata > Vertebrata > Reptilia > Lepidosauria (Lizards, Snakes) 666
- Chordata > Vertebrata > Reptilia > Testudinia (Turtles) 667
- Chordata > Vertebrata > Reptilia > Crocodilia (Crocodiles, Alligators) 667
- Chordata > Vertebrata > Reptilia > Aves (Birds) 668

34.4 The Primates and Hominins 668
The Primates 668
Fossil Humans 670
The Out-of-Africa Hypothesis 672

CHAPTER REVIEW 673

35 Viruses 675

35.1 Why Do Biologists Study Viruses? 676
Recent Viral Epidemics in Humans 676
Current Viral Epidemics in Humans: HIV 677

35.2 How Do Biologists Study Viruses? 678
Analyzing Morphological Traits 679
Analyzing Variation in Growth Cycles: Replicative and Latent Growth 679
Analyzing the Phases of the Replicative Cycle 681

35.3 What Themes Occur in the Diversification of Viruses? 686
The Nature of the Viral Genetic Material 686
Where Did Viruses Come From? 686
Emerging Viruses, Emerging Diseases 688

35.4 Key Lineages of Viruses 689
- Double-Stranded DNA (dsDNA) Viruses 690
- RNA Reverse-Transcribing Viruses (Retroviruses) 691
- Double-Stranded RNA (dsRNA) Viruses 691
- Negative-Sense Single-Stranded RNA ([−]ssRNA) Viruses 692
- Positive-Sense Single-Stranded RNA ([+]ssRNA) Viruses 692

CHAPTER REVIEW 693

UNIT 7 HOW PLANTS WORK 695

36 Plant Form and Function 695

36.1 Plant Form: Themes with Many Variations 696
The Importance of Surface Area/Volume Relationships 696
The Root System 697
The Shoot System 699
The Leaf 701

36.2 Primary Growth Extends the Plant Body 704
How Do Apical Meristems Produce the Primary Plant Body? 704
How Is the Primary Root System Organized? 705
How Is the Primary Shoot System Organized? 706

36.3 Cells and Tissues of the Primary Plant Body 706
The Dermal Tissue System 707
The Ground Tissue System 708
The Vascular Tissue System 710

36.4 Secondary Growth Widens Shoots and Roots 712
What Is a Cambium? 712
What Does Vascular Cambium Produce? 713
What Does Cork Cambium Produce? 713
The Structure of a Tree Trunk 714

CHAPTER REVIEW 715

37 Water and Sugar Transport in Plants 717

37.1 Water Potential and Water Movement 717
What Is Water Potential? 718
What Factors Affect Water Potential? 718
Calculating Water Potential 719
Water Potentials in Soils, Plants, and the Atmosphere 720

37.2 How Does Water Move from Roots to Shoots? 721
Movement of Water and Solutes into the Root 722
Water Movement via Root Pressure 723
Water Movement via Capillary Action 723
The Cohesion-Tension Theory 724

37.3 Water Absorption and Water Loss 727
Limiting Water Loss 727
Obtaining Carbon Dioxide under Water Stress 728

37.4 Translocation 728
Tracing Connections between Sources and Sinks 728
The Anatomy of Phloem 729
The Pressure-Flow Hypothesis 730
Phloem Loading 731
Phloem Unloading 733

CHAPTER REVIEW 735

38 Plant Nutrition 737

38.1 Nutritional Requirements of Plants 738
Which Nutrients Are Essential? 738
What Happens When Key Nutrients Are in Short Supply? 740

38.2 Soil: A Dynamic Mixture of Living and Nonliving Components 741
The Importance of Soil Conservation 742
What Factors Affect Nutrient Availability? 742

38.3 Nutrient Uptake 744
Mechanisms of Nutrient Uptake 744
Mechanisms of Ion Exclusion 746

38.4 Nitrogen Fixation 748
The Role of Symbiotic Bacteria 749
How Do Nitrogen-Fixing Bacteria Colonize Plant Roots? 749

38.5 Nutritional Adaptations of Plants 750
Epiphytic Plants 750
Parasitic Plants 751
Carnivorous Plants 751

CHAPTER REVIEW 752

39 Plant Sensory Systems, Signals, and Responses 755

39.1 Information Processing in Plants 756
How Do Cells Receive and Transduce an External Signal? 756
How Are Cell-Cell Signals Transmitted? 756
How Do Cells Respond to Cell-Cell Signals? 757

39.2 Blue Light: The Phototropic Response 758
Phototropins as Blue-Light Receptors 758
Auxin as the Phototropic Hormone 759

39.3 Red and Far-Red Light: Germination and Stem Elongation 762
The Red/Far-Red "Switch" 763
Phytochromes as Red/Far-Red Receptors 763
How Were Phytochromes Isolated? 763

39.4 Gravity: The Gravitropic Response 764
The Statolith Hypothesis 764
Auxin as the Gravitropic Signal 765

39.5 How Do Plants Respond to Wind and Touch? 766
Changes in Growth Patterns 766
Movement Responses 766

39.6 Youth, Maturity, and Aging: The Growth Responses 767
Auxin and Apical Dominance 767
Cytokinins and Cell Division 768
Gibberellins and ABA: Growth and Dormancy 769
Brassinosteroids and Body Size 773
Ethylene and Senescence 773
An Overview of Plant Growth Regulators 774

39.7 Pathogens and Herbivores: The Defense Responses 776
How Do Plants Sense and Respond to Pathogens? 776
How Do Plants Sense and Respond to Herbivore Attack? 778

CHAPTER REVIEW 781

40 Plant Reproduction 783

40.1 An Introduction to Plant Reproduction 784
Sexual Reproduction 784
The Land Plant Life Cycle 784
Asexual Reproduction 786

40.2 Reproductive Structures 786
When Does Flowering Occur? 787
The General Structure of the Flower 788
How Are Female Gametophytes Produced? 790
How Are Male Gametophytes Produced? 790

40.3 Pollination and Fertilization 792
Pollination 792
Fertilization 794

40.4 The Seed 795
Embryogenesis 796
The Role of Drying in Seed Maturation 797
Fruit Development and Seed Dispersal 797
Seed Dormancy 798
Seed Germination 799

CHAPTER REVIEW 800

UNIT 8 HOW ANIMALS WORK 803

41 Animal Form and Function 803

41.1 Form, Function, and Adaptation 804
The Role of Fitness Trade-Offs 804
Adaptation and Acclimatization 804

41.2 Tissues, Organs, and Systems: How Does Structure Correlate with Function? 806
Structure-Function Relationships at the Molecular and Cellular Levels 806
Tissues Are Groups of Similar Cells That Function as a Unit 806
Organs and Organ Systems 810

41.3 How Does Body Size Affect Animal Physiology? 811
Surface Area/Volume Relationships: Theory 811
Surface Area/Volume Relationships: Data 812
Adaptations That Increase Surface Area 814

41.4 Homeostasis 814
Homeostasis: General Principles 814
The Role of Regulation and Feedback 815

41.5 How Do Animals Regulate Body Temperature? 816
Mechanisms of Heat Exchange 816
Variation in Thermoregulation 816
Endothermy and Ectothermy: A Closer Look 817
Temperature Homeostasis in Endotherms 817
Countercurrent Heat Exchangers 818

CHAPTER REVIEW 820

42 Water and Electrolyte Balance in Animals 822

42.1 Osmoregulation and Osmotic Stress 823
What Is Osmotic Stress? 823
Osmotic Stress in Seawater 824
Osmotic Stress in Freshwater 824
Osmotic Stress on Land 825
How Do Cells Move Electrolytes and Water? 825

42.2 Water and Electrolyte Balance in Aquatic Environments 826
How Do Sharks Excrete Salt? 826
How Do Freshwater Fish Osmoregulate? 827

42.3 Water and Electrolyte Balance in Terrestrial Insects 828
How Do Insects Minimize Water Loss from the Body Surface? 828
Types of Nitrogenous Wastes: Impact on Water Balance 829
Maintaining Homeostasis: The Excretory System 830

42.4 Water and Electrolyte Balance in Terrestrial Vertebrates 832
The Structure of the Kidney 832
The Function of the Kidney: An Overview 832
Filtration: The Renal Corpuscle 832
Reabsorption: The Proximal Tubule 834
Creating an Osmotic Gradient: The Loop of Henle 835
Regulating Water and Electrolyte Balance: The Distal Tubule and Collecting Duct 837

CHAPTER REVIEW 839

43 Animal Nutrition 841

43.1 Nutritional Requirements 842
Meeting Basic Needs in Humans 842
Studying Nutrient Requirements 842

43.2 Capturing Food: The Structure and Function of Mouthparts 843
Mouthparts as Adaptations 843
A Case Study: The Cichlid Jaw 844

43.3 How Are Nutrients Digested and Absorbed? 845
An Introduction to the Digestive Tract 845
An Overview of Digestive Processes 846
The Mouth and Esophagus 847
The Stomach 848
The Small Intestine 851
The Cecum and Appendix 854
The Large Intestine 854

43.4 Nutritional Homeostasis—Glucose as a Case Study 856
The Discovery of Insulin 856
Insulin's Role in Homeostasis 856
Diabetes Can Take Several Forms 856
The Type 2 Diabetes Mellitus Epidemic 857

CHAPTER REVIEW 858

44 Gas Exchange and Circulation 861

44.1 The Respiratory and Circulatory Systems 861

44.2 Air and Water as Respiratory Media 862
How Do Oxygen and Carbon Dioxide Behave in Air? 862
How Do Oxygen and Carbon Dioxide Behave in Water? 863

44.3 Organs of Gas Exchange 864
Physical Parameters: The Law of Diffusion 864
How Do Fish Gills Work? 865
How Do Insect Tracheae Work? 866
How Do Vertebrate Lungs Work? 867
Homeostatic Control of Ventilation 870

44.4 How Are Oxygen and Carbon Dioxide Transported in Blood? 870
Structure and Function of Hemoglobin 871
CO_2 Transport and the Buffering of Blood pH 873

44.5 The Circulatory System 874
What Is an Open Circulatory System? 875
What Is a Closed Circulatory System? 875
How Does the Heart Work? 877
Patterns in Blood Pressure and Blood Flow 882

CHAPTER REVIEW 883

45 Electrical Signals in Animals 885

45.1 Principles of Electrical Signaling 885
Types of Neurons in the Nervous System 886
The Anatomy of a Neuron 886
An Introduction to Membrane Potentials 887
BOX 45.1 QUANTITATIVE METHODS: Using the Nernst Equation to Calculate Equilibrium Potentials 888
How Is the Resting Potential Maintained? 888
Using Microelectrodes to Measure Membrane Potentials 890
What Is an Action Potential? 890

45.2 Dissecting the Action Potential 891
Distinct Ion Currents Are Responsible for Depolarization and Repolarization 891
How Do Voltage-Gated Channels Work? 891
How Is the Action Potential Propagated? 893

45.3 The Synapse 895
Synapse Structure and Neurotransmitter Release 895
What Do Neurotransmitters Do? 896
Postsynaptic Potentials 897

45.4 The Vertebrate Nervous System 899
What Does the Peripheral Nervous System Do? 899
Functional Anatomy of the CNS 900
How Does Memory Work? 902

CHAPTER REVIEW 904

46 Animal Sensory Systems and Movement 907

46.1 How Do Sensory Organs Convey Information to the Brain? 908
Sensory Transduction 908
Transmitting Information to the Brain 909

46.2 Hearing 909
How Do Sensory Cells Respond to Sound Waves and Other Forms of Pressure? 909
The Mammalian Ear 910
Sensory Worlds: What Do Other Animals Hear? 912

46.3 Vision 913
The Insect Eye 913
The Vertebrate Eye 914
Sensory Worlds: Do Other Animals See Color? 917

46.4 Taste and Smell 918
Taste: Detecting Molecules in the Mouth 918
Olfaction: Detecting Molecules in the Air 919

46.5 Movement 920
Skeletons 920
Muscle Types 921
How Do Muscles Contract? 922

CHAPTER REVIEW 926

47 Chemical Signals in Animals 929

47.1 Cell-to-Cell Signaling: An Overview 929
Major Categories of Chemical Signals 930
Hormone Signaling Pathways 931
What Makes Up the Endocrine System? 932
Chemical Characteristics of Hormones 933
How Do Researchers Identify a Hormone? 934

47.2 What Do Hormones Do? 935
How Do Hormones Direct Developmental Processes? 935
How Do Hormones Coordinate Responses to Environmental Change? 937
How Are Hormones Involved in Homeostasis? 938

47.3 How Is the Production of Hormones Regulated? 940
The Hypothalamus and Pituitary Gland 940
Control of Epinephrine by Sympathetic Nerves 943

47.4 How Do Hormones Act on Target Cells? 943
Steroid Hormones Bind to Intracellular Receptors 943
Hormones That Bind to Cell-Surface Receptors 945
Why Do Different Target Cells Respond in Different Ways? 947

CHAPTER REVIEW 948

48 Animal Reproduction 950

48.1 Asexual and Sexual Reproduction 950
How Does Asexual Reproduction Occur? 951
Switching Reproductive Modes: A Case History 951
Mechanisms of Sexual Reproduction: Gametogenesis 952

48.2 Fertilization and Egg Development 954
External Fertilization 954
Internal Fertilization 954
Unusual Aspects of Mating 955
Why Do Some Females Lay Eggs while Others Give Birth? 956

48.3 Reproductive Structures and Their Functions 957
The Male Reproductive System 957
The Female Reproductive System 959

48.4 The Role of Sex Hormones in Mammalian Reproduction 960
Which Hormones Control Puberty in Mammals? 961
Which Hormones Control the Menstrual Cycle in Mammals? 962

48.5 Pregnancy and Birth in Mammals 967
Gestation and Early Development in Marsupials 967
Major Events during Human Pregnancy 967
How Does the Mother Nourish the Fetus? 968
Birth 970

CHAPTER REVIEW 971

49 The Immune System in Animals 973

49.1 Innate Immunity 974
Barriers to Entry 974
The Innate Immune Response 975

49.2 The Adaptive Immune Response: Recognition 977
An Introduction to Lymphocytes 978
The Discovery of B Cells and T Cells 979
The Clonal-Selection Theory 979
How Does the Immune System Distinguish Self from Nonself? 983

49.3 The Adaptive Immune Response: Activation 984
T-Cell Activation 984
B-Cell Activation and Antibody Secretion 986

49.4 The Adaptive Immune Response: Culmination 987
How Are Bacteria and Other Foreign Cells Killed? 987
How Are Viruses Destroyed? 987
Why Does the Immune System Reject Foreign Tissues and Organs? 988
Responding to Future Infections: Immunological Memory 989

49.5 What Happens When the Immune System *Doesn't* Work Correctly? 990
Immunodeficiency Diseases 990
Allergies 990

CHAPTER REVIEW 991

50 An Introduction to Ecology 993

50.1 Areas of Ecological Study 993
Organismal Ecology 994
Population Ecology 994
Community Ecology 994
Ecosystem Ecology 995
How Do Ecology and Conservation Efforts Interact? 995

50.2 Types of Aquatic Ecosystems 995
Nutrient Availability 995
Water Flow 996
Water Depth 996
  ■ Freshwater Environments > Lakes and Ponds 997
  ■ Freshwater Environments > Wetlands 998
  ■ Freshwater Environments > Streams 999
  ■ Freshwater/Marine Environments > Estuaries 1000
  ■ Marine Environments > The Ocean 1000

50.3 Types of Terrestrial Ecosystems 1001
  ■ Terrestrial Biomes > Tropical Wet Forest 1003
  ■ Terrestrial Biomes > Subtropical Deserts 1004
  ■ Terrestrial Biomes > Temperate Grasslands 1005
  ■ Terrestrial Biomes > Temperate Forests 1006
  ■ Terrestrial Biomes > Boreal Forests 1007
  ■ Terrestrial Biomes > Arctic Tundra 1008

50.4 The Role of Climate and the Consequences of Climate Change 1008
Global Patterns in Climate 1009
How Will Global Climate Change Affect Ecosystems? 1011

50.5 Biogeography: Why Are Organisms Found Where They Are? 1013
Abiotic Factors 1013
The Role of History 1013
Biotic Factors 1015
Biotic and Abiotic Factors Interact 1015

CHAPTER REVIEW 1017

51 Behavioral Ecology 1019

51.1 An Introduction to Behavioral Biology 1019
Proximate and Ultimate Causation 1020
Conditional Strategies and Decision Making 1020
Five Questions in Behavioral Ecology 1021

51.2 What Should I Eat? 1021
Foraging Alleles in *Drosophila melanogaster* 1021
Optimal Foraging in White-Fronted Bee-Eaters 1022

51.3 Who Should I Mate With? 1022
Sexual Activity in *Anolis* Lizards 1023
How Do Female Barn Swallows Choose Mates? 1024

51.4 Where Should I Live? 1026
How Do Animals Find Their Way on Migration? 1026
Why Do Animals Move with a Change of Seasons? 1027

51.5 How Should I Communicate? 1027
Honeybee Language 1028
Modes of Communication 1029
When Is Communication Honest or Deceitful? 1029

51.6 When Should I Cooperate? 1031
Kin Selection 1031
BOX 51.1 QUANTITATIVE METHODS: Calculating the Coefficient of Relatedness 1032
Reciprocal Altruism 1033
An Extreme Case: Abuse of Non-Kin in Humans 1034

CHAPTER REVIEW 1035

52 Population Ecology 1037

52.1 Demography 1037
Life Tables 1038
The Role of Life History 1039
BOX 52.1 QUANTITATIVE METHODS: Using Life Tables to Calculate Population Growth Rates 1040

52.2 Population Growth 1041
Quantifying the Growth Rate 1041
Exponential Growth 1042
Logistic Growth 1042
BOX 52.2 QUANTITATIVE METHODS: Developing and Applying Population Growth Equations 1043
What Limits Growth Rates and Population Sizes? 1044

52.3 Population Dynamics 1046
How Do Metapopulations Change through Time? 1046
Why Do Some Populations Cycle? 1047
BOX 52.3 QUANTITATIVE METHODS: Mark-Recapture Studies 1048
How Does Age Structure Affect Population Growth? 1050
Analyzing Change in the Growth Rate of Human Populations 1052

52.4 How Can Population Ecology Help Endangered Species? 1053
Using Life-Table Data 1054
Preserving Metapopulations 1055

CHAPTER REVIEW 1056

53 Community Ecology 1058

53.1 Species Interactions 1058
Three Themes 1059
Competition 1059
Consumption 1063
Mutualism 1068

53.2 Community Structure 1070
How Predictable Are Communities? 1070
How Do Keystone Species Structure Communities? 1072

53.3 Community Dynamics 1073
Disturbance and Change in Ecological Communities 1073
Succession: The Development of Communities after Disturbance 1074

53.4 Species Richness in Ecological Communities 1077
Predicting Species Richness: The Theory of Island Biogeography 1077
Global Patterns in Species Richness 1078
BOX 53.1 QUANTITATIVE METHODS: Measuring Species Diversity 1079

CHAPTER REVIEW 1080

54 Ecosystems 1083

54.1 How Does Energy Flow through Ecosystems? 1083
Why Is NPP So Important? 1084
Solar Power: Transforming Incoming Energy to Biomass 1084
Trophic Structure 1085
Energy Transfer between Trophic Levels 1086
Trophic Cascades and Top-Down Control 1087
Biomagnification 1088
Global Patterns in Productivity 1089
What Limits Productivity? 1090

54.2 How Do Nutrients Cycle through Ecosystems? 1092
Nutrient Cycling within Ecosystems 1092
Global Biogeochemical Cycles 1094

54.3 Global Warming 1098
Understanding the Problem 1098
Positive and Negative Feedback 1099
Impact on Organisms 1099
Productivity Changes 1101

CHAPTER REVIEW 1103

55 Biodiversity and Conservation Biology 1105

55.1 What Is Biodiversity? 1106
Biodiversity Can Be Measured and Analyzed at Several Levels 1106
How Many Species Are Living Today? 1107
BOX 55.1 QUANTITATIVE METHODS: Extrapolation Techniques 1108

55.2 Where Is Biodiversity Highest? 1109
Hotspots of Biodiversity and Endemism 1109
Conservation Hotspots 1110

55.3 Threats to Biodiversity 1110
Changes in the Nature of the Problem 1110
How Can Biologists Predict Future Extinction Rates? 1114

BOX 55.2 QUANTITATIVE METHODS: Population Viability Analysis 1115

55.4 Why Is Biodiversity Important? 1117
Economic Benefits of Biodiversity 1117
Biological Benefits of Biodiversity 1117
An Ethical Dimension? 1120

55.5 Preserving Biodiversity 1120
Designing Effective Protected Areas 1120
Beyond Protected Areas: A Comprehensive Approach 1121

CHAPTER REVIEW 1123

The Big Picture: Ecology 1126

APPENDIX A: **BioSkills** B:1
1 The Metric System B:1
2 Reading Graphs B:2
3 Reading a Phylogenetic Tree B:4
4 Some Common Latin and Greek Roots Used in Biology B:6
5 Using Statistical Tests and Interpreting Standard Error Bars B:7
6 Reading Chemical Structures B:8
7 Using Logarithms B:9
8 Making Concept Maps B:10
9 Separating and Visualizing Molecules B:11
10 Biological Imaging: Microscopy and X-Ray Crystallography B:13
11 Separating Cell Components by Centrifugation B:16
12 Cell and Tissue Culture Methods B:17
13 Combining Probabilities B:19
14 Model Organisms B:19

APPENDIX B: **Answers** A:1

Glossary G:1

Credits C:1

Index I:1

About the Author

SCOTT FREEMAN received his Ph.D. in Zoology from the University of Washington and was subsequently awarded an Alfred P. Sloan Postdoctoral Fellowship in Molecular Evolution at Princeton University. His current research focuses on the scholarship of teaching and learning—specifically (**1**) how active learning and peer teaching techniques increase student learning and improve performance in introductory biology, and (**2**) how the levels of exam questions vary among introductory biology courses, standardized postgraduate entrance exams, and professional school courses. He has also done research in evolutionary biology on topics ranging from nest parasitism to the molecular systematics of the blackbird family. Scott teaches introductory biology for majors at the University of Washington and is coauthor, with Jon Herron, of the standard-setting undergraduate text *Evolutionary Analysis*.

Unit Advisors

Twelve cherished colleagues guided the revision process by synthesizing reviews, providing citations for recent high-impact publications, and drawing on their extensive teaching experience and subject-matter expertise to advise Scott on hundreds of questions, ranging from what to include to which analogies might communicate best to students. It is hard to overstate just how critical these people were to this revision. Through their own teaching and research and their work on this book, they are having a profound effect on how biology is taught.

Jason Flores, University of North Carolina, Charlotte (Unit 1)

Lisa Elfring, University of Arizona, Tucson (Unit 1)

Suzanne Simon-Westendorf, Ohio University (Unit 2)

Gregory Podgorski, Utah State University (Unit 3)

Kathleen Marrs, Indiana University–Purdue University, Indianapolis (Units 4, 6, and 7)

Jon Monroe, James Madison University (Units 4, 6, and 7)

Warren Burggren, University of North Texas (Units 4 and 8)

Joan Sharp, Simon Fraser University (Units 5 and 6)

Michael Black, California Polytechnic State University, San Luis Obispo (Units 6 and 8)

Kathleen Hunt, University of Portland (Units 6 and 9)

Emily Taylor, California Polytechnic State University, San Luis Obispo (Unit 8)

Fred Wasserman, Boston University (Unit 9)

Illustrator

KIM QUILLIN combines expertise in biology and information design to create lucid visual representations of biological principles. She received her B.A. in Biology at Oberlin College and her Ph.D. in Integrative Biology from the University of California, Berkeley (as a National Science Foundation Graduate Fellow), and taught undergraduate biology at both schools. Students and instructors alike have praised Kim's illustration programs for *Biological Science*, as well as *Biology: A Guide to the Natural World* by David Krogh and *Biology: Science for Life* by Colleen Belk and Virginia Borden, for their success in the visual communication of biology. Kim is a lecturer in the Department of Biological Sciences at Salisbury University.

Preface to Instructors

This book is a response to calls—from the National Academy of Sciences, the Howard Hughes Medical Institute and American Association of Medical Colleges, and the National Science Foundation—for changes in the way introductory biology is taught. Reports like *Biology 2010*, *Scientific Foundations for Future Physicians*, and *Vision and Change* are asking that introductory students not only learn the language of biology and understand fundamental concepts, but begin to apply those concepts in new situations, analyze experimental design, synthesize results, and evaluate hypotheses and data.

I wrote this book for instructors who embrace this challenge—who want to help their students learn how to think like a biologist. The essence of higher education is to promote higher-order thinking. Our job is to help students understand biological science at all six levels of Bloom's taxonomy of learning.

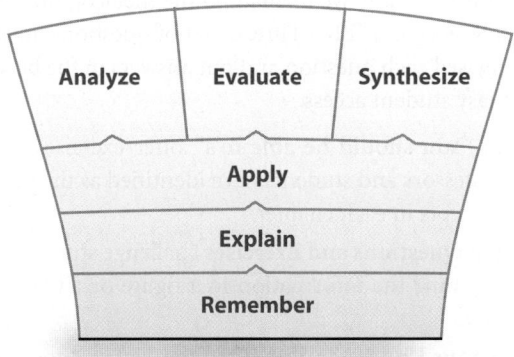

Bloom's Taxonomy. An annotated version of this graphic can be found in "To the Student: How to Use This Book" at the front of this book.

The Evolution of a Textbook

Evolution can be extremely fast in populations with short generation times and high mutation rates. Biology textbooks are no exception. Generation times have to be short because the pace of research in biology and student learning is so fast. This book, in particular, evolves quickly because it incorporates so many new ideas with each edition. Some of these "alleles" are novel mutations, but most arrive via lateral transfer—from advisors, reviewers, friends, students, and the literature.

In comparing the first edition of *Biological Science* with the fourth, the thing that jumps out is what educational researchers call "scaffolding"—tools that help professors and students teach and learn at the level demanded by the NAS, HHMI, AMC, and NSF. It's essential to have high expectations of our students, but we also need to provide the help and practice they need to meet those expectations.

What's New in This Edition

This revision was about making the book a better teaching and learning tool. To help students manage the mass of information and ideas that is contemporary biology, I broke long paragraphs into shorter paragraphs, made liberal use of numbered lists and bulleted lists to "chunk" information and ideas, and broke out dozens of new sections and subsections. I could almost hear my mother's voice: "Take small bites, and chew them well." I also made the book over 100 pages shorter by following Thoreau's dictum to "Simplify, simplify, simplify"—focusing on the most critical concepts that an introductory student needs to master.

In addition, the book team and I came up with a long list of new or expanded features.

- **The Big Picture** These new, two-page spreads are meant to help students see the forest for the trees. They are concept maps that focus on particularly critical areas—Energy, Genetic Information, Evolution, and Ecology. Each synthesizes content and concepts from an array of chapters and includes exercises for students to complete. You'll recognize these pages readily—their edges are colored black (for example, see The Big Picture: Energy on pages 192–193). In addition, the book's MasteringBiology® website has ten new concept map activities based on Big Picture content that will allow you to explore the concepts and their connections with your students during lecture.

- **BioSkills** *Biology 2010*, *Scientific Foundations*, and *Vision and Change* all place a premium on skills—the ability to read a graph, interpret an equation, understand the bands on a gel. The third edition of *Biological Science* introduced a series of appendixes focused on key skills for introductory biology students. Instructors and students found them extraordinarily helpful. New in this edition are BioSkills on using the metric system, common Latin and Greek roots, techniques for isolating and visualizing cell components, cell and tissue culture methods, and model organisms. BioSkills are located in Appendix A.

- **Answer Key** New to the Fourth Edition are suggested answers to all questions and exercises in the textbook. Students asked us to make this important change between editions to make the book a more complete study tool. The answer key will allow them to self-check their understanding while reading, and when reviewing for exams. Answers are in Appendix B.

- **Experiment Boxes** This text's hallmark has always been its emphasis on experimental evidence—on teaching how we know what we know. In the second edition, key experiments were converted to a boxed format so students could easily navigate through the logic of the question, hypothesis, and test. In this edition, I added a new question to every experiment box to encourage students to analyze some aspect of the experiment's design.

- **Art Program** Recent research shows that students are more likely to interpret phylogenetic trees correctly if the trees are designed with U-shaped branches instead of Y-shaped branches. We responded by redesigning every phylogenetic tree in the text. To make other subject areas more accessible to visual learners, we enlarged figures, replaced hundreds of photos with clearer images, and strove to streamline labels and graphics across the board. (More on improvements to the art program below.)

- **MasteringBiology Quizzes** MasteringBiology gives students round-the-clock access to quizzes. We developed 550 new assignable questions based on the book's "Blue Thread" questions (more on the "Blue Thread" and its evolution below). We also developed 1100 new quiz questions, along with a cumulative practice test to simulate what a real exam might be like. To help students keep up with their reading, we created 55 new reading quizzes—one for each chapter—that you can assign through Mastering Biology.

- **MasteringBiology Experimental Inquiry Tutorials** The call to teach students about the process of science has never been louder. In response, a team lead by Tom Owens of Cornell University developed 10 new interactive tutorials on classic scientific experiments—ranging from Meselson–Stahl on DNA replication to the Grants' work on Galápagos finches and Connell's work on competition. Students who use these interactive tutorials should be better prepared to think critically about experimental design and evaluate the wider implications of the data—preparing them to do the work of real scientists in the future.

- **MasteringBiology BioFlix Animations and Tutorials** BioFlix™ are movie-quality, 3-D animations available on MasteringBiology. They focus on the most difficult core topics and are accompanied by in-depth, online tutorials that provide hints and feedback to help guide student learning. Thirteen BioFlix were available with the third edition of *Biological Science*. Five new BioFlix 3-D animations and tutorials have been developed for this edition—on mechanisms of evolution, homeostasis, gas exchange, population ecology, and the carbon cycle.

Changes to Gold Thread Scaffolding

The third edition introduced a dramatically expanded set of tools designed to help with a chronic problem for novice learners: picking out important information. Novices highlight every line in the text and try to memorize everything mentioned in lecture; experts instinctively home in on the key unifying ideas.

For students to make the novice-to-expert transition, we have to help them with features like:

1. **Key concepts** that are declared at the start of each chapter, highlighted with a key icon within the chapter, and reviewed at the end of the chapter.
2. **In-text highlighting**, in gold, that directs their attention to particularly important ideas.
3. **Check Your Understanding boxes**, at the end of key sections, with a bulleted list of key points.
4. **Summary tables** that pull information together in a compact format that is easy to review and synthesize.

The book team and I scrubbed every aspect of the gold thread—reviewing, revising, and re-revising. The third edition was a coming-out party for the gold thread; in the Fourth Edition we wanted it to dance.

Changes to Blue Thread Scaffolding

Each edition of this text has added tools to help students with metacognition—understanding what they do and don't understand. Novices like to receive information passively, and easily persuade themselves that they know what's going on. Experts are skeptical—they want to solve some problems before they're convinced that they know and understand an idea.

In the third edition, we formalized the metacognitive tools in *Biological Science* as a "Blue Thread" set of questions; in this edition, we revised each question and put answers in the back of the book for easy student access.

1. **In-text "You should be able to's"** offer exercises on topics that professors and students have identified as the most difficult concepts in each chapter.
2. **Caption Questions and Exercises** challenge students to critically examine the information in a figure or table—not just absorb it.
3. **Check Your Understanding boxes** present two to three tasks that students should be able to complete in order to demonstrate a mastery of summarized key ideas.
4. **Chapter Summaries** include "You should be able to" problems or exercises related to each of the key concepts declared in the gold thread.
5. **End-of-Chapter Questions** are organized around Bloom's taxonomy of learning, so students can test their understanding at the knowledge, comprehension, and application levels.

The fundamental idea is that if students really understand a piece of information or a concept, they should be able to do something with it. How do you get to Carnegie Hall? Practice.

As students mature as biologists-in-training and start taking upper-division courses, most or all of this scaffolding can disappear. By the time our students are juniors and seniors, they should have enough expertise to construct a high-level understanding on their own. But if a well-designed scaffold isn't there to get them started in their first and second years, when they are novices, most will flounder. We have to help them learn how to become good students.

Supporting Visual Learners

Figures can help students, especially visual learners, at all levels of Bloom's taxonomy—not only to understand and remember the material, but also to exercise higher levels of critical thinking. The overall goal of the Fourth Edition art revision was to hone the figures for accessibility to help novice learners recognize and engage with important visual information. In addition to re-designing the previously mentioned phylogenetic trees, Kim Quillin led the effort to enhance virtually every other aspect of the visual-teaching program.

- **Art and Photos** Kim enlarged art and photographs in figures throughout the book to increase clarity by making details physically easier to see. She also reduced the amount of detail in labels and graphics to simplify, simplify, simplify.

- **Color Use** Kim continues to use color strategically to draw attention to important parts of the figures. In this revision, she boosted color contrast in many figures to make the art more vibrant and the details easier to see.

- **Molecular Icons** Kim redesigned many molecular icons to simplify their shapes. The overall contours are based on molecular coordinates, when available, to accurately represent size and geometry, but she smoothed the textures for a simpler appearance—one that is more memorable and pleasing.

- **Molecular Models** New molecular models have been introduced to help students visualize structure-function relationships. In Chapter 5, for example, redesigned 2-D line drawings of sugars are now paired with 3-D ball-and-stick models.

- **"Pointers"** The Fourth Edition figures still use pointer annotations as a "whisper in the ear" to guide students in interpreting figures, but Kim has replaced the hand with an arrow to be more precise.

Serving a Community of Teachers

I love students, but I love teachers even more. There is nothing I like better than to sit in a workshop with a bunch of biology instructors, steal some good ideas, and come home to try them out—in some cases in a framework where I can collect data and test the hypothesis that the new approaches are improving student learning.

Research on biology education is gathering momentum, trying to catch up on the trail blazed by physics education researchers, bringing the same level of rigor to our classrooms that we bring to our lab benches and field sites. I try to bring the spirit and practice of evidence-based teaching into this textbook, and welcome your comments, suggestions, and questions.

Thank you for considering this text, and for your work on behalf of your students. We have the best jobs in the world.

SCOTT FREEMAN
University of Washington

Content Highlights of the Fourth Edition

As discussed in the preface, a major focus of this revision is to enhance the pedagogical utility of *Biological Science*. Another major goal is to ensure that the content reflects the current state of science and is accurate. In addition, every chapter has been rigorously evaluated for discussions that, in the previous edition, may have been too complex or overly detailed. As a result of this scrutiny, certain sections in every chapter have been simplified, content has been pruned judiciously, and the approach to certain topics has been re-envisioned to enhance student comprehension. In this section, some of the key content improvements to the textbook are highlighted.

Unit 1 The Molecules of Life

Chapter 1 A new experiment on ant navigation and discussions of tree-based naming systems and artificial selection in maize has been added. Coverage is expanded on the definition of life and the nature of science and religion.

Chapter 2 The descriptions of bond angles and the geometry of simple molecules are simplified. Added is a discussion on the hot-start hypothesis as well as a new Key Concept on the nature of chemical energy.

Chapter 3 This chapter has been streamlined by eliminating discussion of optical isomers/chirality and reducing coverage of enzyme kinetics and reaction rates.

Chapter 4 The discussion of RNA is expanded to include recently discovered roles for RNAs in cells. Also added is a new summary table (**Table 4.1**) comparing DNA and RNA structure.

Chapter 5 A stronger emphasis on the link between electronegativity of atoms and potential energy in C−C, C−H, and C−O bonds is developed. New ball-and-stick models are added to clarify the differences in location and orientation of functional groups.

Chapter 6 Coverage of secondary active transport has been expanded. Also included in this chapter is current research on the "first cell" and a discussion of non-random distribution of membrane proteins and phospholipids.

Unit 2 Cell Structure and Function

Chapter 7 New research on bacterial cell structure has been included, and a more explicit connection between lysosomes and the endomembrane system is emphasized. Centrifugation is moved to **BioSkills 11** in Appendix A.

Chapter 8 New sections on quorum sensing in bacteria and cross-talk among signal-transduction pathways have been added.

Chapter 9 The discussions of mitochondrial structure, ATP yield from glucose oxidation, and the role of GDP in the citric acid cycle have been updated. The introductory section on cellular respiration has been simplified.

Chapter 10 A new section on regulation (inhibition) has been added. The sections on C_4 and CAM photosynthesis now emphasize the role of these pathways in increasing CO_2 concentrations versus water conservation.

Chapter 11 Micrographs have been added to the phases of mitosis figure (**Figure 11.5**). The discussion on the role of activated MPF has been updated to include the triggering M phase of the cell cycle. Animal-cell culture methods are moved to **BioSkills 12** in Appendix A.

Unit 3 Gene Structure and Expression

Chapter 12 The discussions of recombination rates and aneuploidy rates in humans are updated. New micrographs have been added to the phases of meiosis figure (**Figure 12.7**).

Chapter 13 The linkage discussion and notation in fly crosses have been simplified. Sex-linkage is moved to the Mendelian section (Section 13.4 The Chromosome Theory of Inheritance), and mapping is now covered in Box 13.1 Quantitative Methods: Linkage. A new summary table (**Table 13.3**) presenting basic vocabulary used in Mendelian genetics has been added.

Chapter 14 A new space-filling model of DNA has been added to **Figure 14.4**.

Chapter 15 Discussions on mutation in the melanocortin receptor (link to mouse-coat-color camouflage) and karyotypes of cancerous cells have been added.

Chapter 16 The sections on transcription in bacteria and eukaryotes are now combined. The structure of the translation initiation complex in bacteria has been updated to reflect current science; snRNAs have been added to the discussion of RNA splicing.

Chapter 17 The chapter was streamlined with the removal of discussions of DNA fingerprinting and the structure of the operator and DNA-binding proteins. Treatment of catabolite repression/positive control has been trimmed.

Chapter 18 Included in this chapter is a new summary table (**Table 18.1**) comparing control of gene expression in bacteria and eukaryotes. Also added are discussions on ubiquitination and protein degradation, the importance of epigenetic inheritance (chromosome structure), and the histone code hypothesis.

Chapter 19 Southern/Northern/Western blots have moved to **BioSkills 9** in Appendix A. The discussions on golden rice, the impact of GM crops, and SNP association studies for human diseases have been updated with the most recent research. Notes on "next generation" sequencing technologies have been included.

Chapter 20 Human health applications now emphasize the use of genomics and microarrays to study cancer. Several datasets are updated, including sequencing database totals. New notes on miRNA genes, metagenomics, and the definition of the gene have been added.

Unit 4 Developmental Biology

Chapter 21 The discussions of *bicoid* and regulatory gene cascades are simplified. New material on auxin as a master regulator in early development and the importance of apoptosis has been added.

Chapter 22 The discussion about sea urchin fertilization and variation has been streamlined.

Chapter 23 A new section introducing basic concepts in angiosperm gametogenesis is added.

Unit 5 Evolutionary Processes and Patterns

Chapter 24 A section on the internal consistency of diverse data as evidence for evolution, including a new phylogeny and timeline of whale evolution, has been added. **Figure 24.6**, depicting the evolution of the Galápagos mockingbird, and long-term data on ground finches (**Figure 24.17**) are updated to reflect the most current science. There is a new graph on the evolution of drug resistance in pathogenic bacteria (**Figure 24.14**).

Chapter 25 The genetic drift example has changed from breeding in a small population on Pitcairn Island to coin flips simulating mating in a single couple (using data from the author's classroom). The prairie lupine gene flow example is replaced by recent work on an island population of *Parus major*. Notes on balancing selection and interactions among evolutionary forces have been included.

Chapter 26 The speciation-by-vicariance example has been changed from ratites to snapping shrimp, and the sympatric speciation example featuring soapberry bugs has been changed to apple/hawthorn flies.

Chapter 27 The sections on adaptive radiation and mass extinction have been completely reorganized. A new hypothesis

for the cause of the Cambrian explosion is included, and detail on the "new genes, new bodies" hypothesis has been removed. Presentation of "Life's Timeline" has been significantly overhauled (see **Figures 27.8, 27.9,** and **27.10**).

Unit 6 The Diversification of Life

The model organisms have been moved to **BioSkills 14** in Appendix A. Phylogenetic trees have been redrawn to reflect a horizontal orientation with U-shaped branches for easier comprehension.

Chapter 28 New information on mechanisms of pathogenicity is added. Extensive updates include new notes on archaeon-eukaryote polymerases, the discovery of extensive biomass in the marine subfloor, an archaeon associated with a human disease, discovery of N-fixation and nitrification in archaea, and bacteriorhodopsin's role in phototrophy.

Chapter 29 A stronger emphasis on endosymbiosis as a theme in protist diversification has been threaded throughout this chapter.

Chapter 30 New content on green algae as a grade and on convergence in vascular tissue in mosses-vascular plants and gnetophytes-angiosperms has been added.

Chapter 31 The dynamic nature of mycelia, the importance of glomalin in soil, the role of mating types, and the discovery of "multigenomic" asexual glomales all have new supporting material.

Chapter 32 The treatment of embryonic tissues, developmental patterns, the coelom, and body symmetry have been updated to reflect the latest scientific thinking. A shift in emphasis to the origin of the neuron and cephalization has been implemented.

Chapter 33 New commentary on the independent transitions to land as well as a clarified discussion on the nature of the ecdysozoan-lophotrochozoan split are included. The discussion of annelids is updated to reflect recent results.

Chapter 34 The coverage of the echinoderm endoskeleton has been expanded and a phylogeny of early tetrapods has been added to the fin-to-limb transition figure (**Figure 34.16**). New data have been incorporated in the evolution-of-fishes timeline (**Figure 34.11**). The treatment of hagfish-lamprey, evolution-of-the-jaw (**Figure 34.14**) and *H. sapiens* migration (**Figure 34.40**) also include the most recent data available. The emphasis on the adaptive significance of the amniotic egg has changed from watertightness to increased size and support. Emphasis in the discussion of viviparity has changed to the adaptive advantage of embryo portability and temperature control. The recent analysis of *Ardipithecus ramidus* as the first hominin, with data on estimated body mass and braincase volume, has been included.

Chapter 35 The material on HIV phylogeny has been moved to the section on emerging viruses.

Unit 7 How Plants Work

Chapter 36 Surface area-to-volume ratios have been added as a theme in root and shoot systems. New information on contractile roots in *Ficus* and bulbs is incorporated into this chapter.

Chapter 37 New content on aquaporins and the transmembrane route to root xylem has been added, and coverage of why air has such low water pressure potential has been expanded.

Chapter 38 The description of nitrogen fixation has been clarified.

Chapter 39 **Figure 39.8** on the acid-growth hypothesis has been redesigned, and the discussion of polar auxin transport is simplified. New commentary on the role of brassinosteroids in growth regulation and on "talking trees" is included. The coverage of the receptors for GA, auxin, ABA, and brassinosteroids, and MeSA's role in the SAR has been updated with the most current research. Plant-tissue culture methods have been moved to **BioSkills 12** in Appendix A.

Chapter 40 Comments on day-length sensing and on pollination syndromes are new to this chapter.

Unit 8 How Animals Work

Chapter 41 New details on tissue types (especially connective tissue) have been incorporated. The discussion of thermoregulation has been completely reorganized for a more logical flow.

Chapter 42 The sections on the shark rectal gland and the mammalian loop of Henle have been revised to improve focus.

Chapter 43 A description of incomplete digestive systems is now included, and coverage of comparative aspects of digestive tract structure and function has been expanded.

Chapter 44 Information on the types of circulatory systems and types of blood vessels has been consolidated. Details on surface tension and lung elasticity have been removed while new content on countercurrent exchange in fish gills has been added.

Chapter 45 The chapter and section introductions have been rewritten to introduce a comparative context and to make the neuron-to-systems chapter organization more transparent. New content on interspecific variation in nervous systems has been added.

Chapter 46 The chapter has been shortened and its focus sharpened by the removal of nonessential information.

Chapter 47 New material on EPO abuse in athletes has been included.

Chapter 48 The section on sperm competition includes new data from experiments on seed beetles.

Chapter 49 The discussion of the V regions of BCRs and antibodies and recombination in BCR/TCR genes has been simplified. New content on autoimmune disorders and diseases associated with immunosuppression, allergies, and immunodeficiency diseases has been added. The discussion of vaccination has been expanded.

Unit 9 Ecology

Chapter 50 New information on the importance of nutrient availability in aquatic ecosystems, with details on lake turnover and ocean upwelling, is included. A new section on the Wallace line has also been added.

Chapter 51 The content in this chapter has been completely reorganized to increase cohesiveness. It is presented as a series of questions in behavioral ecology, with each question addressed at the proximate and ultimate levels with separate case studies. Material on modes of learning, innate behavior, bat-moth interactions, sex change in wrasses, and acoustic and visual signaling in red-winged blackbirds has been trimmed or dropped. New content on child abuse in humans is added.

Chapter 52 Discussion of the hare–lynx-cycle field experiment has been reorganized for clarity, with new supporting "Results" data added to accompanying **Figure 52.12**.

Chapter 53 New content has been added on species richness and resistance of communities to invasion, the use of predators or parasites as biocontrol agents, and character displacement in finches. The discussion of succession in Glacier Bay is reorganized and simplified. The discussion of alternative hypotheses to explain the latitudinal gradient in species richness has been expanded and clarified.

Chapter 54 The chapter was rewritten and reorganized to sharpen its focus on human impacts. Sections on trophic cascades and biomagnification have been added, as have recent data on human appropriation of NPP, sources of nutrient gain and loss, and the impact of ocean acidification on coral growth.

Chapter 55 New content on the impact of global climate change and a new section on ways to preserve biodiversity are now included. Two new boxes on quantitative methods have been added: one on estimating species numbers and species losses and the other on population viability analysis.

Acknowledgments

Reviewers

The peer review system is the key to quality and clarity in science publishing. In addition to providing a filter, the investment that respected individuals make in vetting the material—catching errors or inconsistencies and making suggestions to improve the presentation—gives authors, editors, and readers confidence that what they are publishing and reading meets rigorous professional standards.

Peer review plays the same role in textbook publishing. The time and care that this book's reviewers have invested is a tribute to their professional integrity, their scholarship, and their concern for the quality of teaching. Virtually every paragraph in this edition has been revised and improved based on insights from the following individuals.

Ann Aguanno, *Marymount Manhattan College*
Adrienne Alaie, *City University of New York, Hunter College*
John Alcock, *Arizona State University*
Sylvester Allred, *Northern Arizona University*
Suzanne Alonzo, *Yale University*
Dan Ardia, *Franklin and Marshall College*
Peter Armbruster, *Georgetown University*
David Asch, *University of Pennsylvania*
Andrea Aspbury, *Texas State University, San Marcos*
Nicanor Austriaco, *Providence College*
Mitchell Balish, *Miami University*
Elizabeth Balko, *State University of New York, Oswego*
Ralston Bartholomew, *Warren County Community College*
Christine Barton, *Centre College*
Robert Bauman, *Amarillo College*
Wayne Becker, *University of Wisconsin, Madison*
Peter Berget, *Carnegie Mellon University*
Ethan Bier, *University of California, San Diego*
Michael Black, *California Polytechnic State University, San Luis Obispo*
Anthony Bledsoe, *University of Pittsburgh*
James Bottesch, *Brevard Community College*
Scott Bowling, *Auburn University*
Robert Boyd, *Auburn University*
Ronald Breaker, *Yale University*
Diane Bridge, *Elizabethtown College*
Andrew Brower, *Middle Tennessee State University*
Rebecca Brown, *College of Marin*
Mark Browning, *Purdue University*
Arthur Buikema, *Virginia Polytechnic Institute and State University*
Warren Burggren, *University of North Texas*
Jennifer Carbrey, *Duke University*
Dale Casamatta, *University of North Florida*
Gregory Chandler, *University of North Carolina, Wilmington*
Deborah Chapman, *University of Pittsburgh*
Curt Coffman, *Vincennes University*
Patricia Colberg, *University of Wyoming*

Kathleen Cornely, *Providence College*
Elizabeth Cowles, *Eastern Connecticut State University*
Jason Curtis, *Purdue University, North Central*
Karen Curto, *University of Pittsburgh*
Farahad Dastoor, *University of Maine*
Robin Lee Davies, *Sweet Briar College*
Charles Delwiche, *University of Maryland, College Park*
Jean DeSaix, *University of North Carolina, Chapel Hill*
Hudson DeYoe, *University of Texas, Pan American*
Sunethra Dharmasari, *Texas State University, San Marcos*
Lisa Elfring, *University of Arizona, Tucson*
Eric Engstrom, *College of William and Mary*
Jean Everett, *College of Charleston*
Brent Ewers, *University of Wyoming*
Andrew Fabich, *Tennessee Temple University*
Gary Firestone, *University of California, Berkeley*
Ryan Fisher, *Salem State College*
David Fitch, *New York University*
Jason Flores, *University of North Carolina, Charlotte*
Andrew Forbes, *University of Notre Dame*
Edward Freeman, *St. John Fisher College*
Caitlin Gabor, *Texas State University, San Marcos*
George Gilchrist, *College of William and Mary*
Lynda Goff, *University of California, Santa Cruz*
Elliot Goldstein, *Arizona State University*
Andrew Goliszek, *North Carolina Agricultural & Technical State University*
Margaret Goodman, *Wittenberg University*
Joyce Gordon, *University of British Columbia*
Steve Gorsich, *Central Michigan University*
Susan Michele Green, *Texas State University, San Marcos*
Paul Greenwood, *Colby College*
Stanley Guffey, *University of Tennessee, Knoxville*
Wendy Hanna-Rose, *Pennsylvania State University*
Kiki Harbitz, *Gustavus Adolphus College*
Jeffrey Hardin, *University of Wisconsin, Madison*
Jana Henson, *Georgetown College*
Helen Hess, *College of the Atlantic*
Tracey Hickox, *University of Illinois, Urbana-Champaign*
Sara Hoot, *University of Wisconsin, Milwaukee*
Laurie Host, *Harford Community College*
Kelly Howe, *University of New Mexico, Valencia*
Kathleen Hunt, *University of Portland*
Christine Janis, *Brown University*
Eric Jellen, *Brigham Young University*
Warren Johnson, *University of Wisconsin, Green Bay*
Cindy Johnson-Groh, *Gustavus Adolphus College*
Greg M. Kelly, *University of Western Ontario*
Paul King, *Massasoit Community College*
Joel Kingsolver, *University of North Carolina, Chapel Hill*
Roger Koeppe, *Oklahoma State University*
David Kooyman, *Brigham Young University*

Jocelyn Krebs, *University of Alaska, Anchorage*
John Krenetsky, *Metro State College of Denver*
Patrick Krug, *California State University, Los Angeles*
Holly Kupfer, *Central Piedmont Community College*
Mary Rose Lamb, *University of Puget Sound*
John Lammert, *Gustavus Adolphus College*
Hans Landel, *Edmonds Community College*
Dominic Lannutti, *El Paso Community College*
Janet Lanza, *University of Arkansas, Little Rock*
Georgia Lind, *Kingsborough Community College*
Debra Linton, *Central Michigan University*
Curtis Loer, *University of California, San Diego*
Frank Logiudice, *University of Central Florida*
Barbara Lom, *Davidson College*
James Maller, *University of Colorado, Denver*
James Manser, *Harvey Mudd College (retired)*
Kathleen Marrs, *Indiana University–Purdue University, Indianapolis*
Jennifer Martin, *University of Colorado, Boulder*
Kathy Martin-Troy, *Central Connecticut State University*
John H. McDonald, *University of Delaware*
Robert McLean, *Texas State University, San Marcos*
Michael Meighan, *University of California, Berkeley*
Mariana Melo, *North Essex Community College*
Philip Meneely, *Haverford College*
Dennis Minchella, *Purdue University*
Alan Molumby, *University of Illinois, Chicago*
Vertigo Moody, *Santa Fe Community College*
Elizabeth Morgan, *Lonestar College Kingswood*
Nancy Morvillo, *Florida Southern College*
Mike Muller, *University of Illinois, Chicago*
Dennis Nyberg, *University of Illinois, Chicago*
Amanda N. Orenstein, *Centenary College of New Jersey*
Stephanie Scher Pandolfi, *Michigan State University*
Lisa Parks, *North Carolina State University*
Robert Paul, *Kennesaw State University*
Andrew Pease, *Stevenson University*
Andrew Pekosz, *Johns Hopkins University*
Nancy Pelaez, *Purdue University*
Roger Persell, *City University of New York, Hunter College*
John Peters, *College of Charleston*
Debra Pires, *University of California, Los Angeles*
Gregory Podgorski, *Utah State University*
Robert Podolsky, *College of Charleston*
Therese Poole, *Georgia State University*
Harvey Pough, *Rochester Institute of Technology*
Vanessa Quinn, *Purdue University North Central*
Stephanie Randell, *McLennan Community College*
Clifford Ross, *University of North Florida*
Michael Rutledge, *Middle Tennessee State University*
James Ryan, *Hobart and William Smith Colleges*
Margaret Saha, *College of William and Mary*
Mark Sandheinrich, *University of Wisconsin, La Crosse*
Terry Saropoulos, *Vanier College*
Thomas Sasek, *University of Louisiana, Monroe*
Jon Scales, *Midwestern State University*
Gregory Schmaltz, *Great Basin College*
Oswald Schmitz, *Yale University*
Joan Sharp, *Simon Fraser University*
Michele Shuster, *New Mexico State University*

Suzanne Simon-Westendorf, *Ohio State University*
David Skelly, *Yale University*
Meredith Somerville-Norris, *University of North Carolina, Charlotte*
Sally Sommers-Smith, *Boston University*
Eric Stavney, *Pierce College*
Scott Steinmaus, *California Polytechnic State University, San Luis Obispo*
Janet Steven, *Sweet Briar College*
Kirk Stowe, *University of South Carolina*
Christine Strand, *California Polytechnic State University, San Luis Obispo*
Cynthia Surmacz, *Bloomsburg University*
Jackie Swanik, *North Carolina Central University*
Jerilyn Swann, *Maryville College*
Brad Swanson, *Central Michigan University*
Emily Taylor, *California Polytechnic State University, San Luis Obispo*
Eric Thobaben, *Carroll University*
Ken Thomas, *North Essex Community College*
Briana Timmerman, *University of South Carolina*
Alexandru M. F. Tomescu, *Humboldt State University*
James Traniello, *Boston University*
William Velhagen, *New York University*
Sara Via, *University of Maryland, College Park*
Susan Waaland, *University of Washington*
Frederick Wasserman, *Boston University*
Elizabeth Weiss-Kuziel, *University of Texas, Austin*
Jason Wiles, *Syracuse University*
Kelly P. Williams, *Virginia Polytechnic Institute and State University*
Elizabeth Willott, *University of Arizona, Tucson*
Clifford Wilson, *Kennedy King College*
Charles Wimpee, *University of Wisconsin, Milwaukee*
Leslie Wooten-Blanks, *College of Charleston*
Todd Yetter, *University of the Cumberlands*

Correspondents

One of the most enjoyable interactions I have as a textbook author is correspondence or conversations with researchers and teachers who take the time and trouble to contact me to discuss an issue with the book, or who respond to my queries about a particular data set or study. I'm always amazed and heartened by the generosity of these individuals. They care, deeply.

Julie Aires, *Florida Community College, Jacksonville*
Göran Arnqvist, *Uppsala University*
Terry Bidleman, *Environment Canada*
Brian Buchwitz, *University of Washington*
Helaine Burstein, *Ohio University*
Curt Coffman, *Vincennes University*
Mark Cooper, *University of Washington*
Scott Creel, *Montana State University*
Laura DiCaprio, *Ohio University*
Mary Durant, *Lone Star College*
Victoria Finnerty, *Emory University*
Kathleen Foltz, *University of California, Santa Barbara*
Kathy Gillen, *Kenyon College*
Peter Grant, *Princeton University*
Rosemary Grant, *Princeton University*
Takato Imaizumi, *University of Washington*
Kathleen Janech, *College of Charleston*
Paul King, *Massosoit Community College*

John Lammert, *Gustavus Adolphus College*
Curtis Loer, *University of San Diego*
Scott Meissner, *Cornell University*
Philip Meneely, *Haverford College*
Diarmaid Ó Foighil, *University of Michigan*
John Roth, *University of California, Davis*
Brian Spohn, *Florida Community College, Jacksonville*
Fayla Schwartz, *Everett Community College*
Elizabeth Willott, *University of Arizona, Tucson*
Dan Wulff, *University at Albany, State University of New York*
Glenn Yasuda, *Seattle University*

Supplements Contributors

Instructors depend on an impressive array of support materials—in print and online—to design and deliver their courses. The student experience would be much weaker without the study guide, test bank, activities, animations, quizzes, and tutorials written by the following individuals.

Brian Bagatto, *University of Akron*
Michael Black, *California Polytechnic State University, San Luis Obispo*
Jay L. Brewster, *Pepperdine University*
Warren Burggren, *University of North Texas*
Patricia Colberg, *University of Wyoming*
Clarissa Dirks, *Evergreen State College*
Lisa Elfring, *University of Arizona, Tucson*
Brent Ewers, *University of Wyoming*
Miriam Ferzli, *North Carolina State University*
Cheryl Frederick, *University of Washington*
Cindee Giffen, *University of Wisconsin, Madison*
Kathy M. Gillen, *Kenyon College*
Mary Catherine Hager
Christopher Harendza, *Montgomery County Community College*
Laurel Hester, *University of South Carolina*
Jean Heitz, *University of Wisconsin, Madison*
Tracey Hickox, *University of Illinois, Urbana-Champaign*
Kathleen Hunt, *University of Portland*
Barbara Lom, *Davidson College*
Jennifer Nauen, *University of Delaware*
Chris Pagliarulo, *University of Arizona, Tucson*
Stephanie Scher Pandolfi, *Michigan State University*
Debra Pires, *University of California, Los Angeles*
Gregory Podgorski, *Utah State University*
Carol Pollock, *University of British Columbia*
Jessica Poulin, *University at Buffalo, the State University of New York*
Vanessa Quinn, *Purdue University North Central*
Eric Ribbens, *Western Illinois University*
Joan Sharp, *Simon Fraser University*
Suzanne Simon-Westendorf, *Ohio University*
Emily Taylor, *California Polytechnic State University, San Luis Obispo*
Fred Wasserman, *Boston University*
Cindy White, *University of Northern Colorado*

Book Team

People who watch film credits won't be surprised to learn how many talented people are involved in making a textbook happen. Ruth Steyn provided incisive comments on the earliest drafts of the revised manuscript, which were then implemented by Development Editors Alice Fugate, Moira Lerner-Nelson, Bill O'Neal, and Susan Teahan; the final version of the text was copyedited by Michael Rossa and proofread by Pete Shanks. This editorial team was directed by Executive Director of Development Deborah Gale.

As in the first three editions, the book's figures were designed by Dr. Kim Quillin. Kim's artistic sensibilities, scientific training, and teaching talent are what make the figures in this book sing. If imitation really is the sincerest form of flattery, then Kim should be blushing—we've started seeing her style copied in textbook figures and scientific illustrations by other illustrator/designers. Kim's designs were rendered by Imagineering Media Services; the final art manuscript was vetted by Art Editor Kelly Murphy and rendered art expertly proofread by Frank Purcell. Maureen Spuhler did a thorough review of the photography program and researched the hundreds of images new to the Fourth Edition.

The book's clean, elegant design is the brainchild of Marilyn Perry; the text and art were set in Marilyn's design by S4Carlisle Publishing Services. The book's production was supervised by Lori Newman and Mike Early.

The extensive supplements program was managed by Assistant Editor Brady Golden. All of the individuals I've mentioned—and more—were supported by Editorial Assistant Leslie Allen.

Creating MasteringBiology® tutorials and activities requires a talented team of people playing many different roles. Media content development was overseen by Tania Mlawer and Sarah Jensen who benefited from the program expertise of Caroline Power and Karen Sheh. Julia Henderson and Laura Tommasi worked together as Media Producers with the leadership of Senior Media Producer Deb Greco. Lauren Fogel (VP, Director, Media Development), Stacy Treco (VP, Director, Media Product Strategy), and Laura ensured that the complete media program that accompanies the Fourth Edition, including MasteringBiology, will meet the needs of the students and professors who use our offerings.

Pearson's talented Sales Reps, who listen to professors, advise the editorial staff, and get the book in students' hands, are supported by tireless Executive Marketing Manager Lauren Harp and Director of Marketing Christy Lawrence. The marketing materials that support the outreach effort were produced by Lillian Carr and her colleagues in Pearson's MarCom group.

The vision and resources required to run this entire enterprise are the responsibility of Vice President and Editor-in-Chief Beth Wilbur, Senior Vice President and Editorial Director Frank Ruggirello, President of Pearson Science Paul Corey, and President of Pearson Science & Math Linda Davis.

Finally, the driving forces behind this edition were Project Manager Sonia DiVittorio and Acquisitions Editor Becky Ruden. Sonia's razor-sharp mind, depth of heart, and dedication to excellence shine in every page. Becky's energy and belief in this book surmounted obstacles that would have felled any other editor. I thank the two of them from the bottom of my heart.

Media Guide

Students who purchase a new copy of the text receive free access to MasteringBiology® *(www.masteringbiology.com),* which contains valuable videos, animations, and practice quizzes to help students learn and prepare for exams.

THE BIG PICTURE New to the Fourth Edition, The Big Pictures are interactive concept maps based on four overarching topics in biology that help students synthesize information across broad concepts and not get lost in the details.

Energy (Chapters 9 and 10)
- How Photosynthesis Yields Sugar
- How Cellular Respiration Yields ATP
- How Photosynthesis Relates to Cellular Respiration

Genetic Information (Chapters 12–18)
- How Genes Are Expressed
- How Genetic Information Is Copied and Transmitted
- How Genetic Information Changes

Evolution (Chapters 24–27)
- How Species Evolve
- How Species Form the Tree of Life

Ecology (Chapters 50–55)
- How Organisms Interact in Their Environment
- How Energy and Nutrients Flow through Ecosystems

BIOFLIX™ BioFlix are 3-D movie-quality animations with carefully constructed student tutorials, labeled slide shows, study sheets, and quizzes which bring biology to life.

WEB ACTIVITIES Web Activities help students learn biological concepts via simple, cartoon-style animations and contain pre-quizzes and post-quizzes to test student's understanding of biology's dynamic processes and concepts.

DISCOVERY VIDEOS Brief videos from the Discovery Channel on 29 different biology topics are available for student viewing along with a corresponding video quiz.

VIDEOS Additional molecular and microscopy videos provide vivid images of processes of the cell.

BIOSKILLS BioSkills (in Appendix A) provide background on key skills and techniques for introductory biology students. New to the Fourth Edition are online questions that give students practice building their skill set.

GRAPHIT! Graphing tutorials show students how to plot, interpret, and critically evaluate real data.

Chapter 1
- An Introduction to Graphing

Chapter 50
- Animal Food Production Efficiency and Food Policy
- Atmospheric CO_2 and Temperature Changes

Chapter 52
- Age Pyramids and Population Growth

Chapter 53
- Species Area Effect and Island Biogeography

Chapter 55
- Forestation Change
- Global Fisheries and Overfishing
- Municipal Solid Waste Trends in the U.S.
- Global Freshwater Resources
- Prospects for Renewable Energy
- Global Soil Degradation

WORD STUDY TOOLS New to the Fourth Edition are Latin and Greek root word flash cards to help students practice the language of biology. In addition, an audio glossary provides correct pronunciation to help students learn key terms introduced in the book.

CUMULATIVE TEST Every chapter offers 20 Practice Test questions that students can pool from different chapters into a Cumulative Test to simulate a practice exam.

RSS FEEDS Real Simple Syndication directly links breaking news from four important sources: National Public Radio, *Scientific American, Science Daily News,* and *BioScience.* Current articles reinforce the dynamic nature of science in our daily lives.

eTEXT The eText of *Biological Science,* Fourth Edition, is available online 24/7 for students' convenience. New annotation, highlighting, and bookmarking tools allow students to personalize the material for efficient review.

	BIOFLIX	WEB ACTIVITIES	DISCOVERY VIDEOS	VIDEOS	BIOSKILLS
1 Biology and the Tree of Life		Artificial Selection; Introduction to Experimental Design	Cells; Charles Darwin; Early Life; Bacteria		The Metric System; Reading Graphs; Reading a Phylogenetic Tree; Some Common Latin and Greek Roots Used in Biology

Unit 1 The Molecules of Life

	BIOFLIX	WEB ACTIVITIES	DISCOVERY VIDEOS	VIDEOS	BIOSKILLS
2 Water and Carbon: The Chemical Basis of Life		The Properties of Water			Reading Chemical Structures; Using Logarithms; Making Concept Maps; Reading Graphs
3 Protein Structure and Function		Condensation and Hydrolysis Reactions; Activation Energy and Enzymes		An Idealized Alpha Helix (A); An Idealized Alpha Helix (B); An Idealized Beta-Pleated Sheet (A); An Idealized Beta-Pleated Sheet (B)	
4 Nucleic Acids and the RNA World		Structure of RNA and DNA		Stick Model of DNA; Surface Model of DNA	Separating and Visualizing Molecules; Biological Imaging: Microscopy and X-Ray Crystallography
5 An Introduction to Carbohydrates		Carbohydrate Structure and Function			
6 Lipids, Membranes, and the First Cells	Membrane Transport	Diffusion and Osmosis; Membrane Transport Proteins		Space-filling Model of Cholesterol; Stick Model of Cholesterol; Space-filling Model of Phosphatidylcholine; Stick Model of a Phosphatidylcholine	Biological Imaging: Microscopy and X-Ray Crystallography; Separating and Visualizing Molecules

Unit 2 Cell Structure and Function

	BIOFLIX	WEB ACTIVITIES	DISCOVERY VIDEOS	VIDEOS	BIOSKILLS
7 Inside the Cell	Tour of an Animal Cell; Tour of a Plant Cell	Transport into the Nucleus; A Pulse-Chase Experiment	Bacteria; Cells	Confocal vs. Standard Fluorescence Microscopy; Cytoplasmic Streaming; Crawling Amoeba	Separating Cell Components by Centrifugation; Biological Imaging: Microscopy and X-Ray Crystallography; Separating and Visualizing Molecules
8 Cell-Cell Interactions			Cells	Connexon Structure	Separating and Visualizing Molecules
9 Cellular Respiration and Fermentation	Cellular Respiration	Redox Reactions; Glucose Metabolism		Space-filling Model of ATP (adenosine triphosphate); Stick Model of ATP (adenosine triphosphate)	
10 Photosynthesis	Photosynthesis	Chemiosmosis; Photosynthesis; Strategies for Carbon Fixation	Space Plants	Space-filling Model of Chlorophyll	
11 The Cell Cycle	Mitosis	The Phases of Mitosis; Four Phases of the Cell Cycle	Fighting Cancer	Mitosis	Separating and Visualizing Molecules; Cell and Tissue Culture Methods

	BIOFLIX	WEB ACTIVITIES	DISCOVERY VIDEOS	VIDEOS	BIOSKILLS
Unit 3 Gene Structure and Expression					
12 Meiosis	Meiosis	Meiosis; Mistakes in Meiosis			Combining Probabilities; Using Statistical Tests and Interpreting Standard Error Bars
13 Mendel and the Gene		Mendel's Experiments; The Principle of Independent Assortment			Model Organisms; Combining Probabilities; Reading Graphs
14 DNA and the Gene: Synthesis and Repair	DNA Replication	DNA Synthesis			Separating Cell Components by Centrifugation; Cell and Tissue Culture Methods; Using Logarithms; Reading Graphs
15 How Genes Work		The One-Gene One-Enzyme Hypothesis; The Triplet Nature of the Genetic Code			
16 Transcription, RNA Processing, and Translation	Protein Synthesis	RNA Synthesis; Synthesizing Proteins		A Stick-and-Ribbon Rendering of a tRNA	
17 Control of Gene Expression in Bacteria		The *lac* Operon		Cartoon Model of the *lac* Repressor from *E. coli*	
18 Control of Gene Expression in Eukaryotes		Transcription Initiation in Eukaryotes		Cartoon Model of the DNA-Binding Portion of TATA-Binding Protein Interacting with DNA; Cartoon Model of the GAL4 Transcription Factor from the yeast *S. cerevisiae*	Biological Imaging: Microscopy and X-Ray Crystallography; Separating and Visualizing Molecules
19 Analyzing and Engineering Genes		Producing Human Growth Hormone; The Polymerase Chain Reaction	Transgenics	Cartoon Model of the BamH1a Endonuclease	Separating and Visualizing Molecules
20 Genomics		Human Genome Sequencing Strategies	DNA Forensics		Model Organisms; Using Logarithms
Unit 4 Developmental Biology					
21 Principles of Development		Early Pattern Formation in *Drosophila*	Cloning	A Cartoon and Stick Model of the Homeodomain of the *Engrailed* Protein from *Drosophila* Interacting with DNA	Model Organisms; Cell and Tissue Culture Methods
22 An Introduction to Animal Development		Early Stages of Animal Development			
23 An Introduction to Plant Development					Model Organisms
Unit 5 Evolutionary Processes and Patterns					
24 Evolution by Natural Selection		Natural Selection for Antibiotic Resistance	Charles Darwin; Antibiotics		Reading a Phylogenetic Tree; Model Organisms; Reading Graphs

	BIOFLIX	WEB ACTIVITIES	DISCOVERY VIDEOS	VIDEOS	BIOSKILLS
25 Evolutionary Processes	Mechanisms of Evolution	The Hardy-Weinberg Principle; Three Modes of Natural Selection			Combining Probabilities; Using Statistical Tests and Interpreting Standard Error Bars; Reading Graphs
26 Speciation		Allopatric Speciation; Speciation by Changes in Ploidy			Reading a Phylogenetic Tree
27 Phylogenies and the History of Life		Adaptive Radiation	Mass Extinctions		Reading a Phylogenetic Tree

Unit 6 The Diversification of Life

	BIOFLIX	WEB ACTIVITIES	DISCOVERY VIDEOS	VIDEOS	BIOSKILLS
28 Bacteria and Archaea		The Tree of Life	Bacteria; Antibiotics; Molds; Tasty Bacteria; Early Life		Reading a Phylogenetic Tree; Model Organisms
29 Protists		Alternation of Generations in a Protist		A Crawling Amoeba	Biological Imaging: Microscopy and X-Ray Crystallography; Model Organisms
30 Green Algae and Land Plants		Plant Evolution and the PhylogeneticTree	Colored Cotton; Space Plants; Plant Pollination		
31 Fungi		Life Cycle of a Mushroom	Leafcutter Ants; Molds		
32 An Introduction to Animals		The Architecture of Animals	Leafcutter Ants; Invertebrates		
33 Protostome Animals		Protostome Diversity			Model Organisms
34 Deuterostome Animals		Deuterostome Diversity	Invertebrates		
35 Viruses		The HIV Replicative Cycle	Vaccines; Emerging Diseases		Biological Imaging: Microscopy and X-Ray Crystallography; Separating and Visualizing Molecules

Unit 7 How Plants Work

	BIOFLIX	WEB ACTIVITIES	DISCOVERY VIDEOS	VIDEOS	BIOSKILLS
36 Plant Form and Function		Plant Growth			
37 Water and Sugar Transport in Plants	Water Transport in Plants	Solute Transport in Plants		Plasmolysis of Plant Cells	
38 Plant Nutrition		Soil Formation and Nutrient Uptake			
39 Plant Sensory Systems, Signals, and Responses		Sensing Light; Plant Hormones; Plant Defenses			Cell and Tissue Culture Methods
40 Plant Reproduction		Reproduction in Flowering Plants; Fruit Structure and Development	Colored Cotton; Plant Pollination		

Unit 8 How Animals Work

	BIOFLIX	WEB ACTIVITIES	DISCOVERY VIDEOS	VIDEOS	BIOSKILLS
41 Animal Form and Function		Surface Area/Volume Relationships; Homeostasis	Blood; Human Body		Using Logarithms
42 Water and Electrolyte Balance in Animals		The Mammalian Kidney			

	BIOFLIX	WEB ACTIVITIES	DISCOVERY VIDEOS	VIDEOS	BIOSKILLS
43 Animal Nutrition	Homeostasis: Regulating Blood Sugar	The Digestion and Absorption of Food; Understanding Diabetes Mellitus	Nutrition		Biological Imaging: Microscopy and X-Ray Crystallography; Separating and Visualizing Molecules
44 Gas Exchange and Circulation	Gas Exchange	Gas Exchange in the Lungs and Tissues; The Human Heart	Blood		
45 Electrical Signals in Animals	How Neurons Work; How Synapses Work	Membrane Potentials; Action Potentials	Novelty Gene; Teen Brains	The Acetylcholine Receptor	Using Logarithms
46 Animal Sensory Systems and Movement	Muscle Contraction	The Vertebrate Eye; Structure and Contraction of Muscle Fibers	Muscles & Bones	The Acetylcholine Receptor	
47 Chemical Signals in Animals		Endocrine System Anatomy; Hormone Actions on Target Cells	Endocrine System	Cartoon Model of the DNA Binding Motif of a Zinc Finger Transcription Factor Binding to DNA	Separating Cell Components by Centrifugation
48 Animal Reproduction		Human Gametogenesis; Human Reproduction			Using Logarithms; Reading a Phylogenetic Tree
49 The Immune System in Animals		The Inflammatory Response; The Adaptive Immune Response	Emerging Diseases; Vaccines	Chemotaxis of a Neutrophil	
Unit 9 Ecology					
50 An Introduction to Ecology		Tropical Atmospheric Circulation	Rain Forests; Trees; Introduced Species		
51 Behavioral Ecology		Homing Behavior in Digger Wasps	Novelty Gene		
52 Population Ecology	Population Ecology	Modeling Population Growth; Human Population Growth and Regulation			
53 Community Ecology		Life Cycle of a Malaria Parasite; Succession	Leafcutter Ants		
54 Ecosystems	The Carbon Cycle	The Global Carbon Cycle			
55 Biodiversity and Conservation Biology		Habitat Fragmentation	Rain Forests		Using Logarithms

These black swan chicks are being introduced to the world by their mother, who trails close behind them. Likewise, this chapter introduces the key ideas that launched biological science as a discipline.

Biology and the Tree of Life

1

In essence, biological science is a search for ideas and observations that unify our understanding of the diversity of life, from bacteria living in rocks a mile underground to hedgehogs and humans. Chapter 1 is an introduction to this search.

The goals of this chapter are to introduce the nature of life and explore how biologists go about studying it. The chapter also introduces themes that will resonate throughout this book: (**1**) analyzing how organisms work at the molecular level, (**2**) understanding organisms in terms of their evolutionary history, and (**3**) helping you learn to think like a biologist.

Let's begin with what may be the most fundamental question of all: What is life?

1.1 What Does It Mean to Say That Something Is Alive?

An **organism** is a life-form—a living entity made up of one or more cells. ⬤━ Although there is no simple definition of life that is endorsed by all biologists, most agree that organisms share a suite of five fundamental characteristics.

- *Energy* To stay alive and reproduce, organisms have to acquire and use energy. To give just two examples: plants absorb sunlight; animals ingest food.

- *Cells* Organisms are made up of membrane-bound units called cells. A cell's membrane regulates the passage of materials between exterior and interior spaces.

- *Information* Organisms process hereditary or genetic information, encoded in units called genes, along with information they acquire from the environment. Right now cells throughout your body are using genetic information to make the molecules that keep you alive; your eyes and brain are decoding information on this page that will help you learn some biology.

KEY CONCEPTS

⬤━ Organisms obtain and use energy, are made up of cells, process information, replicate, and as populations evolve.

⬤━ The cell theory proposes that all organisms are made of cells and that all cells come from preexisting cells.

⬤━ The theory of evolution by natural selection maintains that species change through time because individuals with certain heritable traits produce more offspring than other individuals do.

⬤━ A phylogenetic tree is a graphical representation of the evolutionary relationships between species. These relationships can be estimated by analyzing similarities and differences in traits. Species that share distinctive traits are closely related and are placed close to each other on the tree of life.

⬤━ Biologists ask questions, generate hypotheses to answer them, and design experiments that test the predictions made by competing hypotheses.

- *Replication* One of the great biologists of the twentieth century, François Jacob, said that the "dream of a bacterium is to become two bacteria." Almost everything an organism does contributes to one goal: replicating itself.

- *Evolution* Organisms are the product of evolution, and their populations continue to evolve.

You can think of this text as one long exploration of these five traits. Here's to life!

1.2 The Cell Theory

Two of the greatest unifying ideas in all of science laid the groundwork for modern biology: the cell theory and the theory of evolution by natural selection. Formally, scientists define a **theory** as an explanation for a very general class of phenomena or observations. The cell theory and theory of evolution address fundamental questions: What are organisms made of? Where do they come from?

When these concepts emerged in the mid-1800s, they revolutionized the way biologists think about the world. They established two of the five attributes of life: Organisms are cellular, and their populations change over time.

Neither insight came easily, however. The cell theory, for example, emerged after some 200 years of work. In 1665 Robert Hooke used a crude microscope to examine the structure of cork (a bark tissue) from an oak tree. The instrument magnified objects to just 30× (30 times) their normal size, but it allowed Hooke to see something extraordinary. In the cork he observed small, pore-like compartments that were invisible to the naked eye. These structures came to be called cells.

Soon after Hooke published his results, Anton van Leeuwenhoek succeeded in developing much more powerful microscopes, some capable of magnifications up to 300×. With these instruments, Leeuwenhoek inspected samples of pond water and made the first observations of human blood cells, and of sperm cells, shown in **Figure 1.1**.

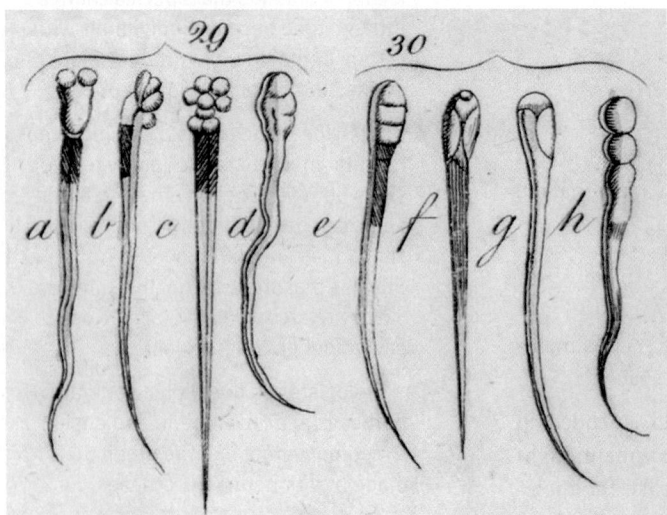

FIGURE 1.1 First Images of Cells. This drawing, by Anton van Leeuwenhoek, shows sperm cells.

In the 1670s a researcher who was studying the leaves and stems of plants with a microscope concluded that these large, complex structures are composed of many individual cells. By the early 1800s enough data had accumulated for a biologist to claim that *all* organisms consist of cells.

Are *All* Organisms Made of Cells?

The smallest organisms known today are bacteria that are barely 80 nanometers wide, or 80 *billionths* of a meter. (See **BioSkills 1** in Appendix A to review the metric system and its prefixes.[1]) It would take 12,500 of these organisms lined up end to end to span a millimeter. This is the distance between the smallest hash marks on a metric ruler.

In contrast, sequoia trees can be over 100 meters tall. This is the equivalent of a 20-story building. Bacteria and sequoias are composed of the same fundamental building block, however—the cell. Bacteria consist of a single cell; sequoias are made up of many cells.

Today, a **cell** is defined as a highly organized compartment that is bounded by a thin, flexible structure called a plasma membrane and that contains concentrated chemicals in an aqueous (watery) solution. The chemical reactions that sustain life take place inside cells. Most cells are also capable of reproducing by dividing—in effect, by making a copy of themselves.

The realization that all organisms are made of cells was fundamentally important, but it formed only the first part of the cell theory. In addition to understanding what organisms are made of, scientists wanted to understand how cells come to be.

Where Do Cells Come From?

Most scientific theories have two components: The first describes a pattern in the natural world; the second identifies a mechanism or process that is responsible for creating that pattern. Hooke and his fellow scientists articulated the pattern component of the cell theory. In 1858 Rudolph Virchow added the process component by stating that all cells arise from preexisting cells.

🔑 The complete **cell theory** can be stated as follows: All organisms are made of cells, and all cells come from preexisting cells.

TWO HYPOTHESES The cell theory was a direct challenge to the prevailing explanation of where cells come from, called spontaneous generation. At the time, most biologists believed that organisms arise spontaneously under certain conditions. For example, the bacteria and fungi that spoil foods such as milk and wine were thought to appear in these nutrient-rich media of their own accord—springing to life from nonliving materials. Spontaneous generation was a **hypothesis**: a proposed explanation.

The all-cells-from-cells hypothesis, in contrast, maintained that cells do not spring to life spontaneously but are produced only when preexisting cells grow and divide.

Biologists usually use the word theory to refer to proposed explanations for broad patterns in nature and prefer hypothesis to refer to explanations for more tightly focused questions.

---

[1]BioSkills are located in the first appendix at the back of the book. They focus on general skills that you'll use throughout this course. More than a few students have found them to be a life-saver. Please use them!

AN EXPERIMENT TO SETTLE THE QUESTION Soon after the all-cells-from-cells hypothesis appeared in print, Louis Pasteur set out to test its predictions experimentally. A **prediction** is something that can be measured and that must be correct if a hypothesis is valid.

Pasteur wanted to determine whether microorganisms could arise spontaneously in a nutrient broth or whether they appear only when a broth is exposed to a source of preexisting cells. To address the question, he created two treatment groups: a broth that was not exposed to a source of preexisting cells and a broth that was.

The spontaneous generation hypothesis predicted that cells would appear in both treatment groups. The all-cells-from-cells hypothesis predicted that cells would appear only in the treatment exposed to a source of preexisting cells.

Figure 1.2 shows Pasteur's experimental setup. Note that the two treatments are identical in every respect but one. Both used glass flasks filled with the same amount of the same nutrient broth. Both were boiled for the same amount of time to kill any existing organisms such as bacteria or fungi. But because the flask pictured in Figure 1.2a had a straight neck, it was exposed to preexisting cells after sterilization by the heat treatment. These preexisting cells are the bacteria and fungi that cling to dust particles in the air. They could drop into the nutrient broth because the neck of the flask was straight.

In contrast, the flask drawn in Figure 1.2b had a long swan neck. Pasteur knew that water would condense in the crook of the swan neck after the boiling treatment and that this pool of water

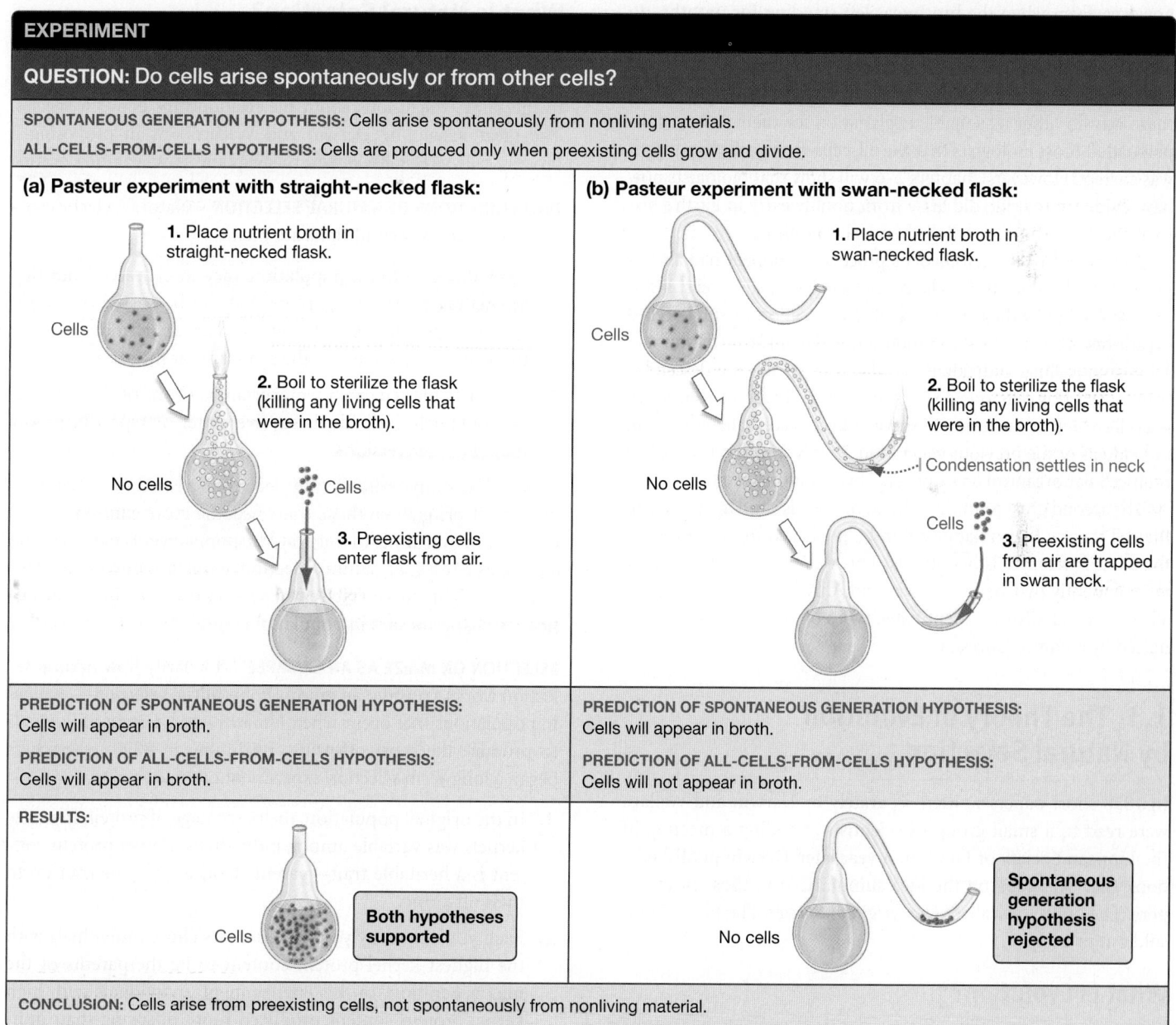

EXPERIMENT

QUESTION: Do cells arise spontaneously or from other cells?

SPONTANEOUS GENERATION HYPOTHESIS: Cells arise spontaneously from nonliving materials.

ALL-CELLS-FROM-CELLS HYPOTHESIS: Cells are produced only when preexisting cells grow and divide.

(a) Pasteur experiment with straight-necked flask:

Cells

1. Place nutrient broth in straight-necked flask.

No cells

2. Boil to sterilize the flask (killing any living cells that were in the broth).

Cells

3. Preexisting cells enter flask from air.

(b) Pasteur experiment with swan-necked flask:

Cells

1. Place nutrient broth in swan-necked flask.

No cells

2. Boil to sterilize the flask (killing any living cells that were in the broth).

Condensation settles in neck

Cells

3. Preexisting cells from air are trapped in swan neck.

PREDICTION OF SPONTANEOUS GENERATION HYPOTHESIS: Cells will appear in broth.

PREDICTION OF ALL-CELLS-FROM-CELLS HYPOTHESIS: Cells will appear in broth.

PREDICTION OF SPONTANEOUS GENERATION HYPOTHESIS: Cells will appear in broth.

PREDICTION OF ALL-CELLS-FROM-CELLS HYPOTHESIS: Cells will not appear in broth.

RESULTS:

Cells

Both hypotheses supported

No cells

Spontaneous generation hypothesis rejected

CONCLUSION: Cells arise from preexisting cells, not spontaneously from nonliving material.

FIGURE 1.2 The Spontaneous Generation and All-Cells-from-Cells Hypotheses Were Tested Experimentally.

✔**QUESTION** What problem would arise if Pasteur had (1) put different types of broth in the two treatments, (2) heated them for different lengths of time, or (3) used a ceramic flask for one treatment and a glass flask for the other?

would trap any bacteria or fungi that entered on dust particles. Thus, the contents of the swan-necked flask was isolated from any source of preexisting cells even though still open to the air.

Pasteur's experimental setup was effective because there was only one difference between the two treatments and because that difference was the factor being tested—in this case, a broth's exposure to preexisting cells.

ONE HYPOTHESIS SUPPORTED   And Pasteur's results? As Figure 1.2 shows, the treatment exposed to preexisting cells quickly filled with bacteria and fungi. This observation was important because it showed that the heat sterilization step had not altered the nutrient broth's capacity to support growth.

The treatment in the swan-necked flask remained sterile, however. Even when the broth was left standing for months, no organisms appeared in it. This result was inconsistent with the hypothesis of spontaneous generation.

Because Pasteur's data were so conclusive—meaning that there was no other reasonable explanation for them—the results persuaded most biologists that the all-cells-from-cells hypothesis was correct. However, Chapters 2–6 will show that biologists now have evidence that life did arise from nonlife early in Earth's history, through a process called chemical evolution.

The success of the cell theory's process component had an important implication: If all cells come from preexisting cells, it follows that all individuals in an isolated population of single-celled organisms are related by common ancestry. Similarly, in you and other multicellular individuals, all the cells present are descended from preexisting cells, tracing back to a fertilized egg. A fertilized egg is a cell created by the fusion of sperm and egg—cells that formed in individuals of the previous generation. In this way, all the cells in a multicellular organism are connected by common ancestry.

The second great founding idea in biology is similar, in spirit, to the cell theory. It also happened to be published the same year as the all-cells-from-cells hypothesis. This was the realization, made independently by Charles Darwin and Alfred Russel Wallace, that all species—all distinct, identifiable types of organisms—are connected by common ancestry.

1.3 The Theory of Evolution by Natural Selection

In 1858 short papers written separately by Darwin and Wallace were read to a small group of scientists attending a meeting of the Linnean Society of London. A year later, Darwin published a book that expanded on the idea summarized in those brief papers. The book was called *The Origin of Species*. The first edition sold out in a day.

What Is Evolution?

Like the cell theory, the theory of evolution by natural selection has a pattern and a process component. Darwin and Wallace's theory made two important claims concerning patterns that exist in the natural world.

- Species are related by common ancestry. This contrasted with the prevailing view in science at the time, which was that species represent independent entities created separately by a divine being.

- In contrast to the accepted view that species remain unchanged through time, Darwin and Wallace proposed that the characteristics of species can be modified from generation to generation. Darwin called this process "descent with modification."

Evolution is a change in the characteristics of a population over time. It means that species are not independent and unchanging entities, but are related to one another and can change through time.

What Is Natural Selection?

This pattern component of the theory of evolution was actually not original to Darwin and Wallace. Several scientists had already come to the same conclusions about the relationships between species. The great insight by Darwin and Wallace was in proposing a process, called **natural selection**, that explains *how* evolution occurs.

TWO CONDITIONS OF NATURAL SELECTION   Natural selection occurs whenever two conditions are met.

1. Individuals within a population vary in characteristics that are **heritable**—meaning, traits that can be passed on to offspring. A **population** is defined as a group of individuals of the same species living in the same area at the same time.

2. In a particular environment, certain versions of these heritable traits help individuals survive better or reproduce more than do other versions.

🔑   If certain heritable traits lead to increased success in producing offspring, then those traits become more common in the population over time. In this way, the population's characteristics change as a result of natural selection acting on individuals. This is a key insight: Natural selection acts on individuals, but evolutionary change occurs in populations.

SELECTION ON MAIZE AS AN EXAMPLE   To clarify how natural selection works, consider an example of **artificial selection**—changes in populations that occur when *humans* select certain individuals to produce the most offspring. Beginning in 1896, researchers began a long-term selection experiment on maize (corn).

1. In the original population, the percentage of protein in maize kernels was variable among individuals. Kernel protein content is a heritable trait—parents tend to pass the trait on to their offspring.

2. Each year for many years, researchers chose individuals with the highest kernel protein content to be the parents of the next generation. In this environment, individuals with high kernel protein content produced more offspring than individuals with low kernel protein content.

Figure 1.3 shows the results. Note that this graph plots generation number on the *x*-axis, starting from the first generation

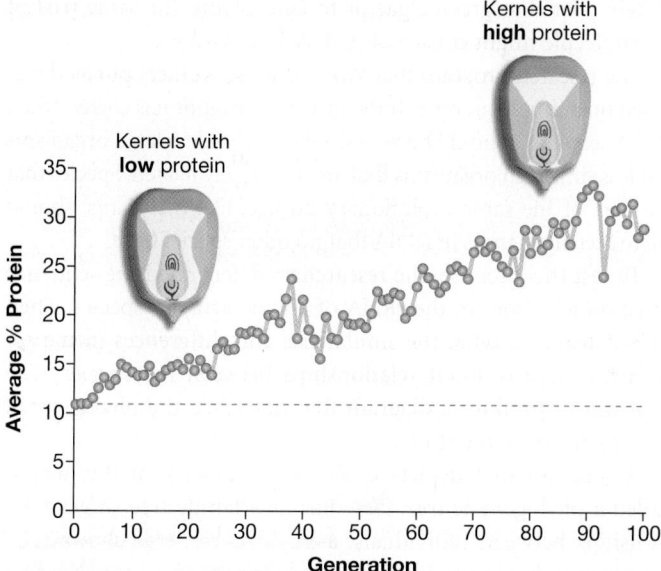

FIGURE 1.3 Response to Selection for High Kernel Protein Content in Maize. The average protein content goes down in a few years because of poor growing conditions or chance changes in how the many genes responsible for this trait interact.

("0") and continuing for 100 generations. The average percentage of protein in a kernel among individuals in this population is plotted on the *y*-axis.

To read this graph, put your finger on the *x*-axis at time 0. Then read up the *y*-axis, and note that kernels averaged about 11 percent protein at the start of the experiment. Now read the graph to the right. Each dot is a data point, representing the average kernel protein concentration in a particular generation. (A generation in maize is one year.) The lines on this graph simply connect the dots, to make the pattern in the data easier to see. At the end of the graph, after 100 generations of selection, average kernel protein content is about 29 percent. (For more help with reading graphs, see **BioSkills 2** in Appendix A.)

This sort of change in the characteristics of a population, over time, is evolution. Humans have been practicing artificial selection for thousands of years, and biologists have now documented evolution by *natural* selection—where humans don't do the selecting—occurring in thousands of different populations, including humans.

To practice applying the principles of artificial selection, go to the online study area at *www.masteringbiology.com.*

(MB) Web Activity Artificial Selection

Evolution occurs when heritable variation leads to differential success in reproduction. ✔If you understand this concept, you should be able to describe how protein content in maize kernels changed over time, using the same *x*-axis and *y*-axis as in Figure 1.3, when researchers selected individuals with *lowest* kernel protein content to be the parents of the next generation. (This experiment was actually done, starting with the same population at the same time as selection for high protein content.)

FITNESS AND ADAPTATION   Darwin also introduced some new terminology to identify what is happening during natural selection.

- In everyday English, fitness means health and well-being. But in biology, **fitness** means the ability of an individual to produce offspring. Individuals with high fitness produce many surviving offspring.

- In everyday English, adaptation means that an individual is adjusting and changing to function in new circumstances. But in biology, an **adaptation** is a trait that increases the fitness of an individual in a particular environment.

Once again, consider kernel protein content in maize: In the environment of the experiment graphed in Figure 1.3, individuals with high kernel protein content produced more offspring and had higher fitness than individuals with lower kernel protein content. In this population and this environment, high kernel protein content was an adaptation that allowed certain individuals to thrive.

Note that during this process, the amount of protein in the kernels of any individual maize plant did not change within its lifetime—the change occurred in the characteristics of the population over time.

Together, the cell theory and the theory of evolution provided the young science of biology with two central, unifying ideas:

1. The cell is the fundamental structural unit in all organisms.

2. All species are related by common ancestry and have changed over time in response to natural selection.

CHECK YOUR UNDERSTANDING

🔑 If you understand that . . .

- Natural selection occurs when heritable variation in certain traits leads to improved success in reproduction. Because individuals with these traits produce many offspring with the same traits, the traits increase in frequency and evolution occurs.

- Evolution is a change in the characteristics of a population over time.

✔ You should be able to . . .

On the graph you just analyzed, describe the average kernel protein content over time in a maize population where *no* selection occurred.

Answers are available in Appendix B.

1.4 The Tree of Life

Section 1.3 focused on how individual populations change through time in response to natural selection. But over the past several decades, biologists have also documented dozens of cases in which natural selection has caused populations of one species to diverge and form new species. This divergence process is called **speciation**.

Research on speciation has two important implications: All species come from preexisting species, and all species, past and present, trace their ancestry back to a single common ancestor.

The theory of evolution by natural selection predicts that biologists should be able to reconstruct a **tree of life**—a family tree of organisms. If life on Earth arose just once, then such a diagram would describe the genealogical relationships between species with a single, ancestral species at its base.

Has this task been accomplished? If the tree of life exists, what does it look like?

Using Molecules to Understand the Tree of Life

One of the great breakthroughs in research on the tree of life occurred when Carl Woese (pronounced *woes*) and colleagues began analyzing the chemical components of organisms, as a way to understand their evolutionary relationships. Their goal was to understand the **phylogeny** of all organisms—their actual genealogical relationships. Translated literally, phylogeny means "tribe-source."

To understand which organisms are closely versus distantly related, Woese and co-workers needed to study a molecule that is found in all organisms. The molecule they selected is called small subunit ribosomal RNA (rRNA). It is an essential part of the machinery that all cells use to grow and reproduce.

Although rRNA is a large and complex molecule, its underlying structure is simple. The rRNA molecule is made up of sequences of four smaller chemical components called ribonucleotides. These ribonucleotides are symbolized by the letters A, U, C, and G. In rRNA, ribonucleotides are connected to one another linearly, like boxcars of a freight train (**Figure 1.4**).

ANALYZING rRNA Why might rRNA be useful for understanding the relationships between organisms? The answer is that the ribonucleotide sequence in rRNA is a trait that can change during the course of evolution. Although rRNA performs the same function in all organisms, the sequence of ribonucleotide building blocks in this molecule is not identical among species.

In land plants, for example, the molecule might start with the sequence A-U-A-U-C-G-A-G. In green algae, which are closely related to land plants, the same section of the molecule might contain A-U-A-U-G-G-A-G. But in brown algae, which are not closely related to green algae or to land plants, the same part of the molecule might consist of A-A-A-U-G-G-A-C.

The research program that Woese and co-workers pursued was based on a simple premise: If the theory of evolution is correct, then rRNA sequences should be very similar in closely related organisms but less similar in organisms that are less closely related. Species that are part of the same evolutionary lineage, like the plants, should share certain changes in rRNA that no other species have.

To test this premise, the researchers determined the sequence of ribonucleotides in the rRNA of a wide array of species. Then they considered what the similarities and differences in the sequences implied about relationships between the species. The goal was to produce a diagram that described the phylogeny of the organisms in the study.

A diagram that depicts evolutionary history in this way is called a phylogenetic tree. ⚷ Just as a family tree shows relationships between individuals, a phylogenetic tree shows relationships between species. On a phylogenetic tree, branches that share a recent common ancestor represent species that are closely related; branches that don't share recent common ancestors represent species that are more distantly related.

THE TREE OF LIFE ESTIMATED FROM AN ARRAY OF GENES To construct a phylogenetic tree, researchers use a computer to find the arrangement of branches that is most consistent with the similarities and differences observed in the data.

Although the initial work was based only on the sequences of ribonucleotides observed in rRNA, biologists now use data sets that include sequences from a wide array of genes. **Figure 1.5** shows a recent tree produced by comparing these sequences. Because this tree includes such a diverse array of species, it is often called the universal tree, or the tree of life. (For help in learning how to read a phylogenetic tree, see **BioSkills 3** in Appendix A.)

The tree of life implied by rRNA and other genetic data established that there are three fundamental groups or lineages of organisms: (**1**) the Bacteria, (**2**) the Archaea, and (**3**) the Eukarya. In all **eukaryotes**, cells have a prominent component called the nucleus (**Figure 1.6a**). Translated literally, the word eukaryotes means "true kernel." Because the vast majority of bacterial and archaeal cells lack a nucleus, they are referred to as **prokaryotes** (literally, "before kernel"; see **Figure 1.6b**). The vast majority of bacteria and archaea are unicellular ("one-celled"); many eukaryotes are multicellular ("many-celled").

When these results were first published, biologists were astonished. For example:

- Prior to Woese's work and follow-up studies, biologists thought that the most fundamental division among organisms was between prokaryotes and eukaryotes. The Archaea were virtually unknown—much less recognized as a major and highly distinctive branch on the tree of life.

- Fungi were thought to be closely related to plants. Instead, they are actually much more closely related to animals.

- Traditional approaches for classifying organisms—including the system of five kingdoms divided into various classes, or-

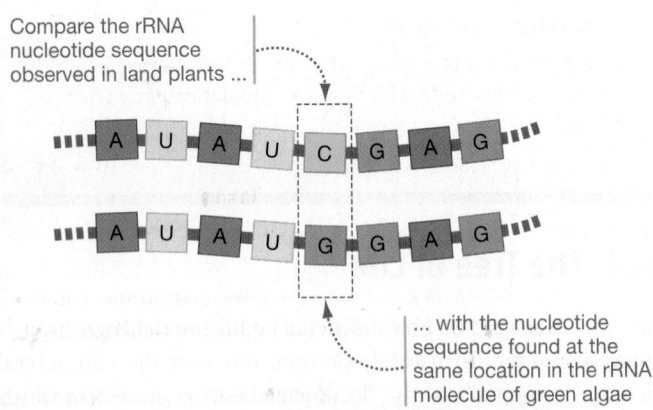

Compare the rRNA nucleotide sequence observed in land plants ...

A U A U C G A G

A U A U G G A G

... with the nucleotide sequence found at the same location in the rRNA molecule of green algae

FIGURE 1.4 RNA Molecules Are Made Up of Smaller Molecules. The complete small subunit rRNA molecule contains about 2000 ribonucleotides; just 8 are shown in this comparison.

✔ **QUESTION** Suppose that in the same portion of rRNA, molds and other fungi have the sequence A-U-A-U-G-G-A-C. According to these data, are fungi more closely related to green algae or to land plants? Explain your logic.

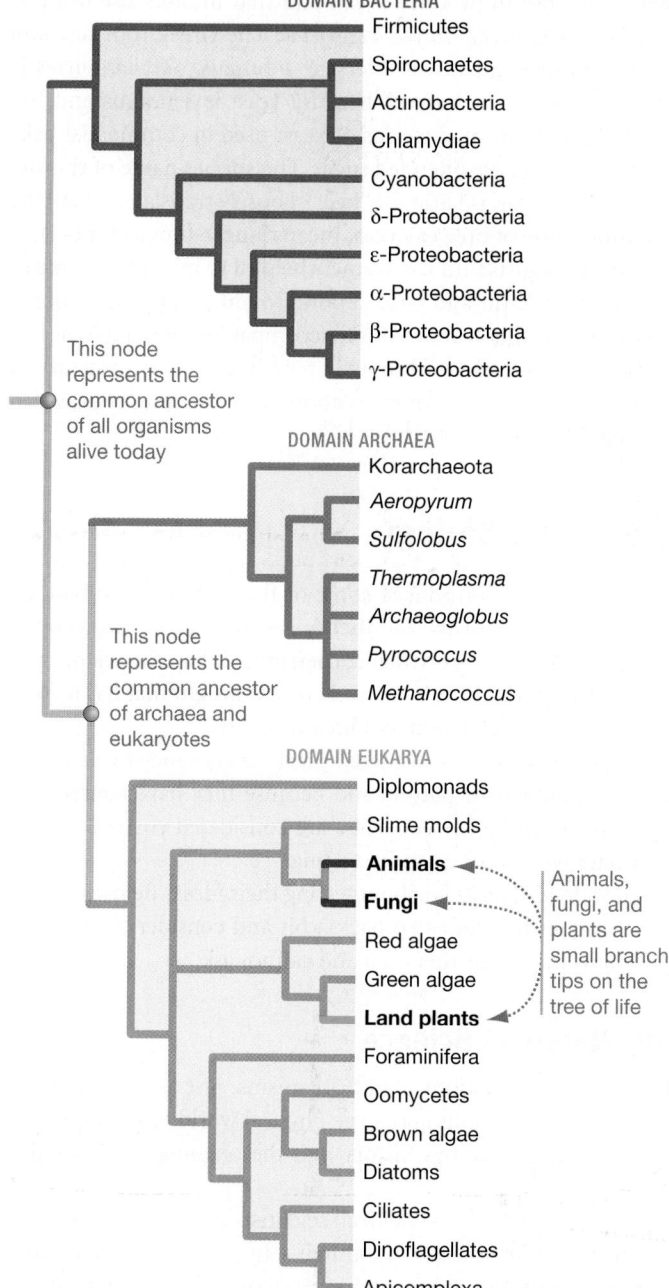

DOMAIN BACTERIA
- Firmicutes
- Spirochaetes
- Actinobacteria
- Chlamydiae
- Cyanobacteria
- δ-Proteobacteria
- ε-Proteobacteria
- α-Proteobacteria
- β-Proteobacteria
- γ-Proteobacteria

This node represents the common ancestor of all organisms alive today

DOMAIN ARCHAEA
- Korarchaeota
- *Aeropyrum*
- *Sulfolobus*
- *Thermoplasma*
- *Archaeoglobus*
- *Pyrococcus*
- *Methanococcus*

This node represents the common ancestor of archaea and eukaryotes

DOMAIN EUKARYA
- Diplomonads
- Slime molds
- **Animals**
- **Fungi**
- Red algae
- Green algae
- **Land plants**
- Foraminifera
- Oomycetes
- Brown algae
- Diatoms
- Ciliates
- Dinoflagellates
- Apicomplexa

Animals, fungi, and plants are small branch tips on the tree of life

FIGURE 1.5 The Tree of Life. "Universal tree" estimated from a large amount of gene sequence data. The three domains of life revealed by the analysis are labeled. Common names are given for most lineages in the domains Bacteria and Eukarya. Genus names are given for members of the domain Archaea, because most of these organisms have no common names.

ders, and families that you may have learned in high school—are inaccurate in many cases, because they do not reflect the actual evolutionary history of the organisms involved.

THE TREE OF LIFE IS A WORK IN PROGRESS Just as researching your family tree can help you understand who you are and where you came from, so the tree of life helps biologists understand the relationships between organisms and the history of species. The discovery of the Archaea and the accurate placement of lineages

(a) Eukaryotic cells have a membrane-bound nucleus.

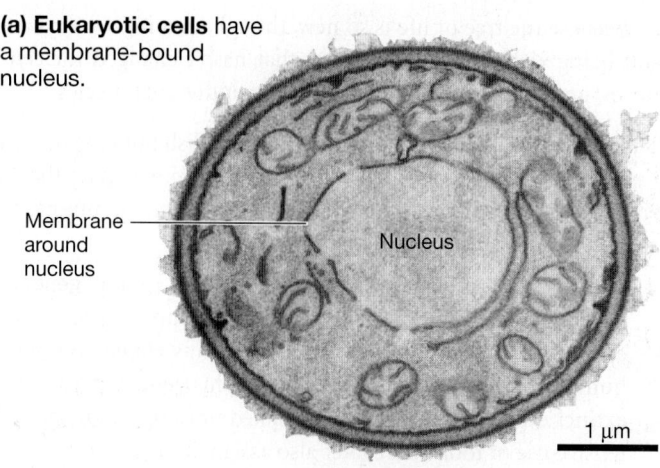

Membrane around nucleus

Nucleus

1 μm

(b) Prokaryotic cells do *not* have a membrane-bound nucleus.

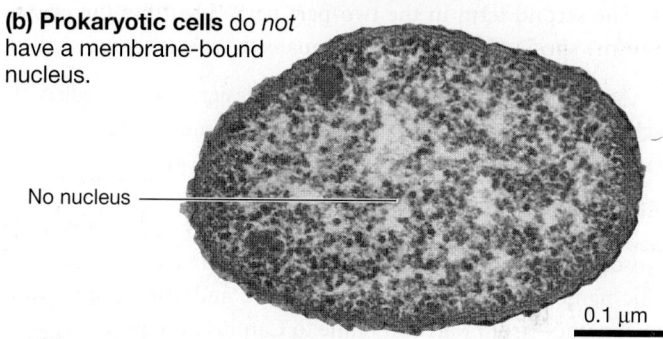

No nucleus

0.1 μm

FIGURE 1.6 Eukaryotes and Prokaryotes.

✓**QUESTION** How many times larger is the eukaryotic cell in this figure than the prokaryotic cell? (Hint: study the scale bars.)

such as the fungi qualify as exciting breakthroughs in our understanding of evolutionary history and life's diversity.

Work on the tree of life continues at a furious pace, however, and the location of certain branches on the tree is hotly debated. As databases expand and as techniques for analyzing data improve, the shape of the tree of life presented in Figure 1.5 will undoubtedly change. Our understanding of the tree of life, like our understanding of every other topic in biological science, is dynamic.

How Should We Name Branches on the Tree of Life?

In science, the effort to name and classify organisms is called **taxonomy**. Any named group is called a **taxon** (plural: **taxa**). Currently, biologists are working to create a taxonomy, or naming system, that accurately reflects the phylogeny of organisms.

For example, Woese proposed a new taxonomic category called the **domain**. The three domains of life are the Bacteria, Archaea, and Eukarya.

Biologists often use the term **phylum** (plural: **phyla**) to refer to major lineages within each domain. Although the designation is somewhat arbitrary, each phylum is considered a major branch on the tree of life. For example, within the lineage called animals, biologists name about 35 phyla—each of which is distinguished by distinctive aspects of its body structure as well as by distinctive gene sequences. The mollusks (clams, squid, octopuses) constitute a phylum, as do chordates (the vertebrates and their close relatives).

Because the tree of life is so new, though, naming systems are still being worked out. One thing that hasn't changed for centuries, however, is the naming system for individual species.

SCIENTIFIC (SPECIES) NAMES   In 1735, a Swedish botanist named Carolus Linnaeus established a system for naming species that is still in use today. Linnaeus created a two-part name unique to each type of organism.

- The first part indicates the organism's **genus** (plural: **genera**). A genus is made up of a closely related group of species. For example, Linnaeus put humans in the genus *Homo*. Although humans are the only living species in this genus, at least five extinct organisms, all of which walked upright and made extensive use of tools, were later also assigned to *Homo*.

- The second term in the two-part name identifies the organism's species. Linnaeus gave humans the species name *sapiens*.

An organism's genus and species designation is called its **scientific name** or Latin name. Scientific names are always italicized. Genus names are always capitalized, but species names are not—for instance, *Homo sapiens*.

Scientific names are based on Latin or Greek word roots or on words "Latinized" from other languages. Linnaeus gave a scientific name to every species then known, and also Latinized his own name—from Karl von Linné to Carolus Linnaeus.

Linnaeus maintained that different types of organisms should not be given the same genus and species names. Other species may be assigned to the genus *Homo*, and members of other genera may be named *sapiens*, but only humans are named *Homo sapiens*. Each scientific name is unique.

SCIENTIFIC NAMES ARE OFTEN DESCRIPTIVE   Scientific names and terms are often based on Latin or Greek word roots that are descriptive. For example, *Homo sapiens* is derived from the Latin *homo* for "man" and *sapiens* for "wise" or "knowing." The yeast

that bakers use to produce bread and that brewers use to brew beer is called *Saccharomyces cerevisiae*. The Greek root *saccharo* means "sugar," and *myces* refers to a fungus. *Saccharomyces* is aptly named "sugar fungus" because yeast is a fungus and because the domesticated strains of yeast used in commercial baking and brewing are often fed sugar. The species name of this organism, *cerevisiae*, is Latin for "beer." Loosely translated, then, the scientific name of brewer's yeast means "sugar-fungus for beer."

Most biologists find it extremely helpful to memorize some of the common Latin and Greek roots. To aid you in this process, new terms in this text are often accompanied by a reference to their Latin or Greek word roots in parentheses, and a glossary of common root words with translations and examples is provided in **BioSkills 4** in Appendix A.

1.5 Doing Biology

This chapter has introduced some of the great ideas in biology. The development of the cell theory and the theory of evolution by natural selection provided cornerstones when the science was young; the tree of life is a relatively recent insight that has revolutionized our understanding of life's diversity.

These theories are considered great because they explain fundamental aspects of nature, and because they have consistently been shown to be correct. They are considered correct because they have withstood extensive testing.

How do biologists go about testing their ideas? Before answering this question, let's step back a bit and consider the types of questions that researchers can and cannot ask.

The Nature of Science

Biologists ask questions about organisms, just as physicists and chemists ask questions about the physical world or geologists ask questions about Earth's history and the ongoing processes that shape landforms.

No matter what their field, all scientists ask questions that can be answered by measuring things—by collecting data. Conversely, scientists cannot address questions that can't be answered by measuring things.

This distinction is important. It is at the root of continuing controversies about teaching evolution in publicly funded schools. In the United States and in Turkey, in particular, some Christian and Islamic leaders have been particularly successful in pushing their claim that evolution and religious faith are in conflict. Even though the theory of evolution is considered one of the most successful and best-substantiated ideas in the history of science, they object to teaching it.

The vast majority of biologists and religious leaders reject this claim; they see no conflict between evolution and religious faith. Their view is that science and religion are compatible because they address different types of questions.

- Science is about formulating hypotheses and finding evidence that supports or conflicts with those hypotheses.

CHECK YOUR UNDERSTANDING

If you understand that . . .

- A phylogenetic tree shows the evolutionary relationships between species.
- To infer where species belong on a phylogenetic tree, biologists examine the characteristics of the species involved. Closely related species should have similar characteristics, while less closely related species should be less similar.

✔ You should be able to . . .

Examine the following sequences and draw a phylogenetic tree showing the relationships between species A, B, and C that these data imply:

Species A: A A C T A G C G C G A T
Species B: A A C T A G C G C C A T
Species C: T T C T A G C G G T A T

Answers are available in Appendix B.

- Religious faith addresses questions that cannot be answered by data. The questions addressed by the world's great religions focus on why we exist and how we should live.

Both types of questions are seen as legitimate and important.

So how do biologists go about answering questions? Let's consider two issues currently being addressed by researchers.

Why Do Giraffes Have Long Necks? An Introduction to Hypothesis Testing

If you were asked why giraffes have long necks, you might say that long necks enable giraffes to reach food that is unavailable to other mammals. This hypothesis is expressed in African folktales and has traditionally been accepted by many biologists. The food competition hypothesis is so plausible, in fact, that for decades no one thought to test it.

In the mid-1990s, however, Robert Simmons and Lue Scheepers assembled data suggesting that the food competition hypothesis is only part of the story. Their analysis supports an alternative hypothesis—that long necks allow giraffes to use their heads as effective weapons for battering their opponents.

How did biologists test the food competition hypothesis? What data support their alternative explanation? Before attempting to answer these questions, it's important to recognize that hypothesis testing is a two-step process:

1. State the hypothesis as precisely as possible and list the predictions it makes.

2. Design an observational or experimental study that is capable of testing those predictions.

If the predictions are accurate, the hypothesis is supported. If the predictions are not met, then researchers do further tests, modify the original hypothesis, or search for alternative explanations.

THE FOOD COMPETITION HYPOTHESIS: PREDICTIONS AND TESTS

The food competition hypothesis claims that giraffes compete for food with other species of mammals. When food is scarce, as it is during the dry season, giraffes with longer necks can reach food that is unavailable to other species and to giraffes with shorter necks. As a result, the longest-necked individuals in a giraffe population survive better and produce more young than do shorter-necked individuals, and average neck length of the population increases with each generation.

To use the terms introduced earlier, long necks are adaptations that increase the fitness of individual giraffes during competition for food. This type of natural selection has gone on so long that the population has become extremely long necked.

The food competition hypothesis makes several explicit predictions. For example, the food competition hypothesis predicts that

1. neck length is variable among giraffes;

2. neck length in giraffes is heritable; and

3. giraffes feed high in trees, especially during the dry season, when food is scarce and the threat of starvation is high.

The first prediction is correct. Studies in zoos and natural populations confirm that neck length is variable among individuals.

The researchers were unable to test the second prediction, however, because they studied giraffes in a natural population and were unable to do breeding experiments. As a result, they simply had to accept this prediction as an assumption. In general, though, biologists prefer to test every assumption behind a hypothesis.

What about the prediction regarding feeding high in trees? According to Simmons and Scheepers, this is where the food competition hypothesis breaks down.

Consider, for example, data collected by a different research team on the amount of time that giraffes spend feeding in vegetation of different heights. **Figure 1.7a** plots the height of vegetation versus the percentage of bites taken by a giraffe, for males and for females from the same population in Kenya. The dashed line on each graph indicates the average height of a male or female in this population.

Note that the average height of a giraffe in this population is much greater than the height where most feeding takes place. In

(a) Most feeding is done at about shoulder height.

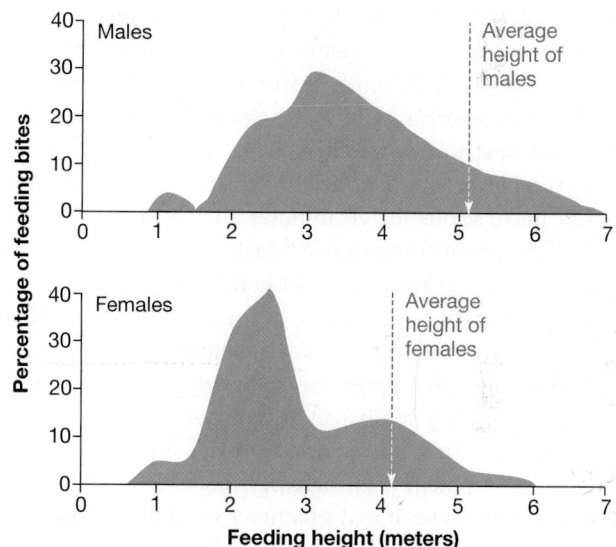

(b) Typical feeding posture in giraffes

FIGURE 1.7 Giraffes Do Not Usually Extend Their Necks to Feed.

this population, both male and female giraffes spend most of their feeding time eating vegetation that averages just 60 percent of their full height. Studies on other populations of giraffes, during both the wet and dry seasons, are consistent with these data. Giraffes usually feed with their necks bent (**Figure 1.7b**).

These data cast doubt on the food competition hypothesis, because one of its predictions does not appear to hold. Biologists have not abandoned this hypothesis completely, though, because feeding high in trees may be particularly valuable during extreme droughts, when a giraffe's ability to reach leaves far above the ground could mean the difference between life and death. Still, Simmons and Scheepers have offered an alternative explanation for why giraffes have long necks. The new hypothesis is based on the mating system of giraffes.

THE SEXUAL COMPETITION HYPOTHESIS: PREDICTIONS AND TESTS
Giraffes have an unusual mating system. Breeding occurs year round rather than seasonally. To determine when females are coming into estrus or "heat" and are thus receptive to mating, the males nuzzle the rumps of females. In response, the females urinate into the males' mouths. The males then tip their heads back and pull their lips to and fro, as if tasting the liquid. Biologists who have witnessed this behavior have proposed that the males taste the females' urine to detect whether estrus has begun.

Once a female giraffe enters estrus, males fight among themselves for the opportunity to mate. Combat is spectacular. The bulls stand next to one another, swing their necks, and strike thunderous blows with their heads. Researchers have seen males knocked unconscious for 20 minutes after being hit and have cataloged numerous instances in which the loser died. Giraffes are the only animals known to fight in this way.

These observations inspired a new explanation for why giraffes have long necks. The sexual competition hypothesis is based on the idea that longer-necked giraffes are able to strike harder blows during combat than can shorter-necked giraffes. In engineering terms, longer necks provide a longer "moment arm." A long moment arm increases the force of the impact. (Think about the type of sledgehammer you'd use to bash down a concrete wall—one with a short handle or one with a long handle?)

The idea here is that longer-necked males should win more fights and, as a result, father more offspring than shorter-necked males do. If neck length in giraffes is inherited, then the average neck length in the population should increase over time. Under the sexual competition hypothesis, long necks are adaptations that increase the fitness of males during competition for females.

Although several studies have shown that long-necked males are more successful in fighting and that the winners of fights gain access to estrous females, the question of why giraffes have long necks is not closed. With the data collected to date, most biologists would probably concede that the food competition hypothesis needs further testing and refinement and that the sexual selection hypothesis appears promising. It could also be true that both hypotheses are correct. For our purposes, the important take-home message is that all hypotheses must be tested rigorously.

In many cases in biological science, testing hypotheses rigorously involves experimentation. Experimenting on giraffes is difficult. But in the case study considered next, biologists were able to test an interesting hypothesis experimentally.

How Do Ants Navigate? An Introduction to Experimental Design

Experiments are a powerful scientific tool because they allow researchers to test the effect of a single, well-defined factor on a particular phenomenon. Because experiments testing the effect of neck length on food and sexual competition in giraffes haven't been done yet, let's consider a different question: When ants leave their nest to search for food, how do they find their way back?

The Saharan desert ant lives in colonies and makes a living by scavenging the dead carcasses of insects. Individuals leave the burrow and wander about searching for food at midday, when temperatures at the surface can reach 60°C (140°F) and predators are hiding from the heat.

Foraging trips can take the ants hundreds of meters—an impressive distance when you consider that these animals are only about a centimeter long. But when an ant returns, it doesn't follow the same long, wandering route it took on its way away from the nest. Instead, individuals return in a straight line. How do they do this?

THE PEDOMETER HYPOTHESIS Early work on navigation in desert ants showed that they use the Sun's position as a compass—meaning that they always know the approximate direction of the nest relative to the Sun. But how do they know how far to go?

After experiments had shown that the ants do not use landmarks to navigate, Matthias Wittlinger and co-workers set out to test a novel idea. The biologists proposed that Saharan desert ants know how far they are from the nest by integrating information from leg movements.

According to this pedometer hypothesis, the ants always know how far they are from the nest because they track the number of steps they have taken and their stride length. The idea is that they can make a beeline back to the burrow because they integrate information on the angles they have traveled *and* the distance they have gone—based on step number and stride length.

TESTING THE HYPOTHESIS To test their idea, Wittlinger's group allowed ants to walk from a nest to a feeder through a channel—a distance of 10 m. Then they caught ants at the feeder and created three test groups, each with 25 individuals:

- *Stumps* By cutting the lower legs of some individuals off, they created ants with shorter-than-normal legs.

- *Normal* Some individuals were left alone, meaning that they had normal leg length.

- *Stilts* By gluing pig bristles onto each leg, the biologists created ants with longer-than-normal legs.

Next they put the ants in a different channel and recorded how far they traveled before starting the characteristic set of 180° turns that ants make when they are looking for the nest hole. To see the data they collected, look at the graph on the left side of the "Results" section in **Figure 1.8**.

- *Stumps*   The ants with stumps stopped short, by about 5 m, before starting to look for the nest opening.

- *Normal*   The normal ants walked the correct distance—about 10 m.

- *Stilts*   The ants with stilts walked about 5 m too far before starting to look for the nest opening.

To check the validity of this result, they put the test ants back in the nest and recaptured them one to several days later, when they had walked to the feeder on their stumps, normal legs, or stilts. Now when the ants were put into the other channel to "walk back," they all traveled the correct distance—10 m—before starting to look for the nest (see the graph on the right side of the "Results" section in Figure 1.8).

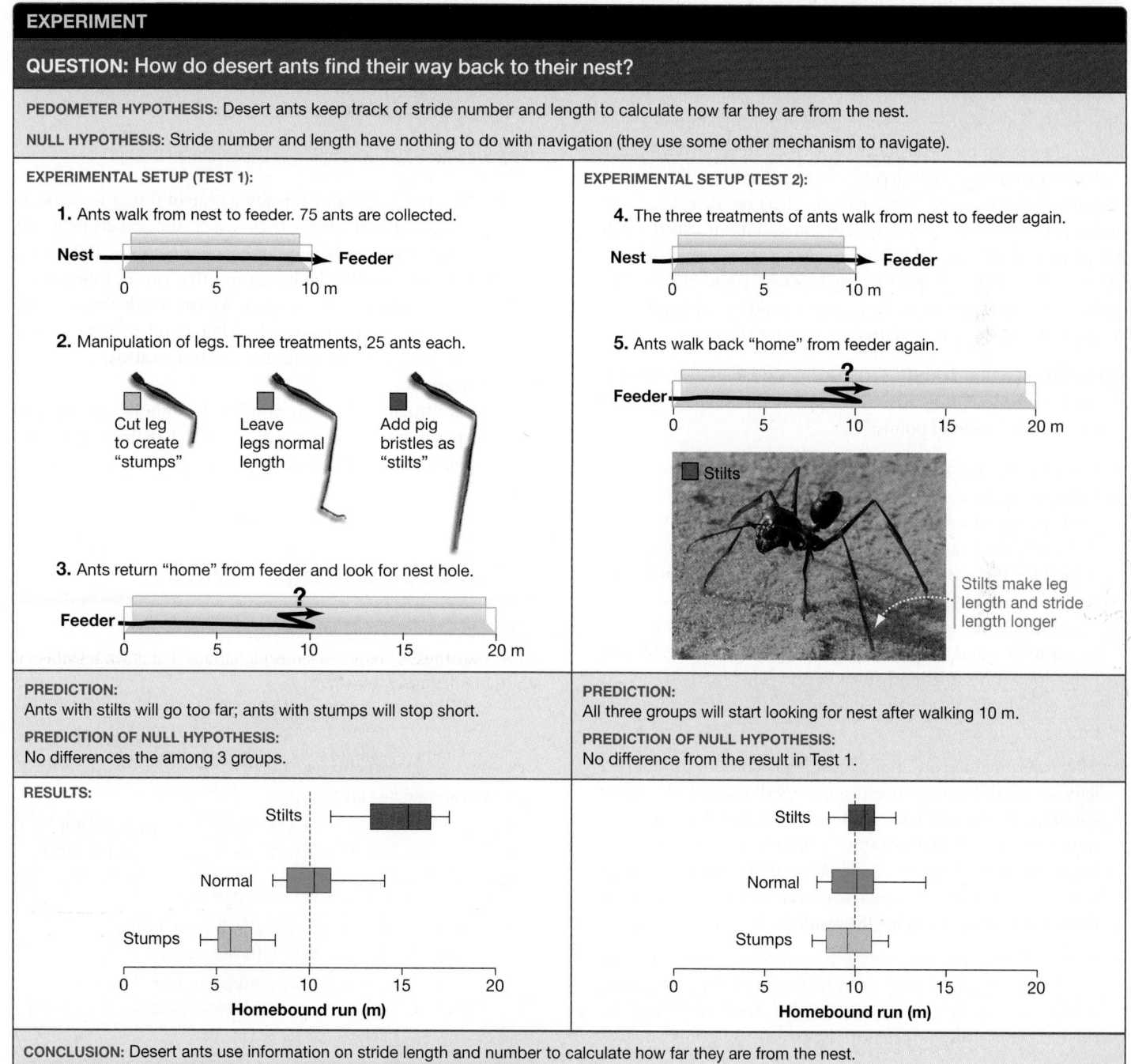

EXPERIMENT

QUESTION: How do desert ants find their way back to their nest?

PEDOMETER HYPOTHESIS: Desert ants keep track of stride number and length to calculate how far they are from the nest.

NULL HYPOTHESIS: Stride number and length have nothing to do with navigation (they use some other mechanism to navigate).

EXPERIMENTAL SETUP (TEST 1):

1. Ants walk from nest to feeder. 75 ants are collected.

2. Manipulation of legs. Three treatments, 25 ants each.

Cut leg to create "stumps"

Leave legs normal length

Add pig bristles as "stilts"

3. Ants return "home" from feeder and look for nest hole.

EXPERIMENTAL SETUP (TEST 2):

4. The three treatments of ants walk from nest to feeder again.

5. Ants walk back "home" from feeder again.

Stilts

Stilts make leg length and stride length longer

PREDICTION: Ants with stilts will go too far; ants with stumps will stop short.

PREDICTION OF NULL HYPOTHESIS: No differences the among 3 groups.

PREDICTION: All three groups will start looking for nest after walking 10 m.

PREDICTION OF NULL HYPOTHESIS: No difference from the result in Test 1.

RESULTS:

Homebound run (m)

Homebound run (m)

CONCLUSION: Desert ants use information on stride length and number to calculate how far they are from the nest.

FIGURE 1.8 An Experimental Test: Do Desert Ants Use a "Pedometer"?

SOURCE: Wittlinger, M., R. Wehner, and H. Wolf. 2006. The ant odometer: Stepping on stilts and stumps. *Science* 312: 1965–1967.

✔**QUESTION** How would you interpret the experiment if the researchers had used just one ant in each group instead of 25?

The graphs in the "Results" display "box-and-whisker" plots. Each box indicates the range of distances where 50 percent of the ants stopped to search for the nest; the whiskers indicate the range where 95 percent of the ants stopped to search. The vertical line inside each box indicates the median—meaning that half the ants stopped above this distance and half below. For more details on how biologists report averages and indicate the variability and uncertainty in data, see **BioSkills 5** in Appendix A.

INTERPRETING THE RESULTS The pedometer hypothesis predicts that an ant's ability to walk home depends on the number and length of steps taken on its outbound trip. Recall that a prediction specifies what we should observe if a hypothesis is correct. Good scientific hypotheses make testable predictions—predictions that can be supported or rejected by collecting and analyzing data. In this case, the researchers tested the prediction by altering stride length and recording the distance traveled on the return trip.

If the pedometer hypothesis is wrong, however, then stride length and step number should have no effect on the ability of an ant to get back to its nest. This latter possibility is called a **null hypothesis**. A null hypothesis specifies what we should observe when the hypothesis being tested isn't correct. Under the null hypothesis in this experiment, all the ants should have walked 10 m in the first test before they started looking for their nest.

IMPORTANT CHARACTERISTICS OF GOOD EXPERIMENTAL DESIGN In relation to designing effective experiments, this study illustrates several important points:

- It is critical to include **control** groups. A control checks for factors, other than the one being tested, that might influence the experiment's outcome. In this case, there were two controls. Including a normal, unmanipulated individual controlled for the possibility that switching the individuals to a new channel altered their behavior. In addition, the researchers had to control for the possibility that the manipulation itself—and not the change in leg length—affected the behavior of the stilts and stumps ants. This is why they did the second test, where the outbound and return runs were done with the same legs.

- The experimental conditions must be as constant or equivalent as possible. The investigators used ants of the same species, from the same nest, at the same time of day, under the same humidity and temperature conditions, at the same feeders, in the same channels. Controlling all the variables except one—leg length in this case—is crucial because it eliminates alternative explanations for the results.

- Repeating the test is essential. It is almost universally true that larger sample sizes in experiments are better. By testing many individuals, the amount of distortion or "noise" in the data caused by unusual individuals or circumstances is reduced.

✔You should be able to explain: (1) What issue would arise if in the first test, the normal individual had not walked 10 m on the return trip before looking for the nest; and (2) What you would conclude if the stilts and stumps ants had not navigated normally during the second test.

From the outcomes of these experiments, the researchers concluded that desert ants use stride length and number to measure how far they are from the nest. They interpreted their results as strong support for the pedometer hypothesis.

Biologists practice evidence-based decision-making. They ask questions about how organisms work, pose hypotheses to answer those questions, and use experimental or observational evidence to decide which hypotheses are correct.

To review the principles of experimental design, go to the study area at *www.masteringbiology.com*.

Web Activity Introduction to Experimental Design

The data on giraffes and ants are a taste of things to come. In this text you will encounter hypotheses and experiments on questions ranging from how water gets to the top of 100-meter-tall sequoia trees to why the bacterium that causes tuberculosis has become resistant to antibiotics. As you work through this book, you'll get lots of practice thinking about hypotheses and predictions, analyzing the nature of control treatments, and interpreting graphs.

A commitment to tough-minded hypothesis testing and sound experimental design is a hallmark of biological science. Understanding their value is an important first step in becoming a biologist.

CHECK YOUR UNDERSTANDING

If you understand that . . .

- Hypotheses are proposed explanations that make testable predictions.
- Predictions are observable outcomes of particular conditions.
- Well-designed experiments alter just one condition—a condition relevant to the hypothesis being tested.

✔ **You should be able to . . .**

Design an experiment to test the hypothesis that desert ants feed during the hottest part of the day because it allows them to avoid being eaten by lizards. Then answer the following questions about your experimental design:

1. How does the presence of a control group in your experiment allow you to test the null hypothesis?

2. How are experimental conditions controlled or standardized in a way that precludes alternative explanations of the data?

Answers are available in Appendix B.

Summary of Key Concepts

Organisms obtain and use energy, are made up of cells, process information, replicate, and as populations evolve.

- There is no single, well-accepted definition of life. Instead, biologists point to five characteristics that organisms share.

  ✔ You should be able to explain why the cells in a dead organism are different from the cells in a live organism.

The cell theory proposes that all organisms are made of cells and that all cells come from preexisting cells.

- The cell theory identified the fundamental structural unit common to all life.

  ✔ You should be able to describe the evidence that supported the pattern and the process components of the cell theory.

The theory of evolution by natural selection maintains that species change through time because individuals with certain heritable traits produce more offspring than other individuals do.

- The theory of evolution states that all organisms are related by common ancestry.

- Natural selection is a well-tested explanation for why species change through time and why they are so well adapted to their habitats.

  ✔ You should be able to explain why the average protein content of seeds in a natural population of a grass species would increase over time, if seeds with higher protein content survive better and grow into individuals that produce many seeds with high protein content when they mature.

   **Web Activity** Artificial Selection

A phylogenetic tree is a graphical representation of the evolutionary relationships between species. These relationships can be estimated by analyzing similarities and differences in traits. Species that share distinctive traits are closely related and are placed close to each other on the tree of life.

- The cell theory and the theory of evolution predict that all organisms are part of a genealogy of species, and that all species trace their ancestry back to a single common ancestor.

- To reconstruct this phylogeny, biologists have analyzed the sequence of components in rRNA and other molecules found in all cells.

- A tree of life, based on similarities and differences in these molecules has three major lineages: the Bacteria, Archaea, and Eukarya.

  ✔ You should be able to explain how biologists can determine whether newly discovered species are members of the Bacteria, Archaea, or Eukarya by analyzing their rRNA or other molecules.

Biologists ask questions, generate hypotheses to answer them, and design experiments that test the predictions made by competing hypotheses.

- Biology is a hypothesis-driven, experimental science.

  ✔ You should be able to explain (1) the relationship between a hypothesis and a prediction and (2) why experiments are convincing ways to test predictions.

   **Web Activity** Introduction to Experimental Design

Questions

1. Anton van Leeuwenhoek made an important contribution to the development of the cell theory. How?
   a. He articulated the pattern component of the theory—that all organisms are made of cells.
   b. He articulated the process component of the theory—that all cells come from preexisting cells.
   c. He invented the first microscope and saw the first cell.
   d. He invented more powerful microscopes and was the first to describe the diversity of cells.

2. What does it mean to say that experimental conditions are controlled?
   a. The test groups consist of the same individuals.
   b. The null hypothesis is correct.
   c. There is no difference in outcome between the control and experimental treatment.
   d. Physical conditions are identical for all groups tested.

3. What does the term *evolution* mean?
   a. The strongest individuals produce the most offspring.
   b. The characteristics of an individual change through the course of its life, in response to natural selection.
   c. The characteristics of populations change through time.
   d. The characteristics of species become more complex over time.

4. What does it mean to say that a characteristic of an organism is heritable?
   a. The characteristic evolves.
   b. The characteristic can be passed on to offspring.
   c. The characteristic is advantageous to the organism.
   d. The characteristic does not vary in the population.

5. In biology, to what does the term *fitness* refer?
   a. The degree of training and muscle mass an individual has, relative to others in the same population
   b. An individual's slimness, relative to others in the same population
   c. The longevity of a particular individual
   d. An individual's ability to survive and reproduce

6. Could *both* the food competition hypothesis and the sexual selection hypothesis explain why giraffes have long necks? Why or why not?
   a. No. In science, only one hypothesis can be correct.
   b. No. Observations have shown that the food competition hypothesis cannot be correct.
   c. Yes. Long necks could be advantageous for more than one reason.
   d. Yes. All giraffes have been shown to feed at the highest possible height and fight for mates.

1. What would researchers have to demonstrate to convince you that they had discovered life on another planet?

2. It was once thought that the deepest split between life-forms was between two groups: prokaryotes and eukaryotes. Draw and label a phylogenetic tree that represents this hypothesis. Then draw and label a phylogenetic tree that shows the actual relationships between the three domains of organisms.

3. Why was it important for Linnaeus to establish the rule that only one type of organism can have a particular genus and species name?

4. What does it mean to say that a species is adapted to a particular habitat?

5. Explain how selection occurs during natural selection. What is selected, and why?

6. The following two statements explain the logic behind the use of molecular sequence data to estimate evolutionary relationships:

   "If the theory of evolution is true, then rRNA sequences should be very similar in closely related organisms but less similar in organisms that are less closely related."

   "On a phylogenetic tree, branches that share a recent common ancestor represent species that are closely related; branches that don't share recent common ancestors represent species that are more distantly related."

   Is the logic of these statements sound? Why or why not?

1. A scientific theory is a set of propositions that defines and explains some aspect of the world. This definition contrasts sharply with the everyday usage of the word theory, which often carries meanings such as "speculation" or "guess." Explain the difference between the two definitions, using the cell theory and the theory of evolution by natural selection as examples.

2. Turn back to the tree of life shown in Figure 1.5. Note that Bacteria and Archaea are prokaryotes, while Eukarya are eukaryotes. On the simplified tree below, draw an arrow that points to the branch where the structure called the nucleus originated. Explain your reasoning.

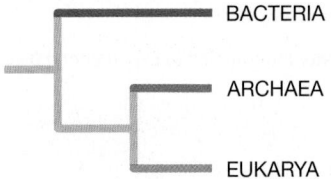

BACTERIA

ARCHAEA

EUKARYA

3. The proponents of the cell theory could not "prove" that it was correct in the sense of providing incontrovertible evidence that all organisms are made up of cells. They could state only that all organisms examined to date were made of cells. Why was it reasonable for them to conclude that the theory was valid?

4. Some humans have heritable traits that make them resistant to infection by HIV. In areas of the world where HIV infection rates are high, are human populations evolving? Explain your logic.

The mounds in the foreground, called stromatolites, are teeming with bacterial and archaeal cells. The fossil record includes 3.42-billion-year-old stromatolites that are virtually identical with the ones pictured here.

Water and Carbon: The Chemical Basis of Life 2

Chapter 1 introduced a classic experiment on spontaneous generation, which tested the idea that life arises from nonliving materials. This work helped build a consensus that spontaneous generation does not occur. But for life to exist, spontaneous generation must have occurred at least once, early in Earth's history.

How did life begin? This simple query has been called "the mother of all questions."

This chapter examines a theory, called **chemical evolution**, that is the leading scientific explanation for the origin of life. Like the theories introduced in Chapter 1, the theory of chemical evolution has a pattern component that makes a claim about the natural world and a process component that explains that pattern.

- *The pattern component* Early in Earth's history, simple chemical compounds combined to form more complex carbon-containing substances that accumulated in the oceans.

- *The process component* Complex carbon-containing compounds formed because energy in sunlight and extremely hot water was converted to chemical energy in the form of new chemical bonds.

The theory maintains that continued inputs of energy led to the formation of increasingly complex carbon-containing substances, culminating in a compound that could make a copy of itself, or self-replicate.

At this point, there was a switch from chemical evolution to biological evolution. As the original molecule multiplied, the process of evolution by natural selection took over. Eventually a descendant of the original self-replicating molecule became surrounded by a membrane. When this occurred, the five attributes of life discussed in Chapter 1 were fulfilled. Life had begun.

At first glance, the theory of chemical evolution may seem implausible. But is it? What evidence do biologists have that chemical evolution occurred? Let's start with the fundamentals: the atoms and molecules that would have combined to get chemical evolution started.

KEY CONCEPTS

- Molecules form when atoms bond to each other. Chemical bonds are based on electron sharing. The degree of electron sharing varies from nonpolar covalent bonds, to polar covalent bonds, to ionic bonds.

- Water is essential for life. Water is highly polar and readily forms hydrogen bonds. Hydrogen bonding makes water an extremely efficient solvent.

- Energy is the capacity to do work or supply heat, and can be (1) a stored potential or (2) an active motion. Chemical energy is a form of potential energy, stored in chemical bonds.

- Chemical reactions tend to be spontaneous if they lower potential energy and increase entropy (disorder). An input of energy is required for nonspontaneous reactions to occur.

- Most of the important compounds in organisms contain carbon. Key carbon-containing molecules formed early in Earth's history.

✔ When you see this checkmark, stop and test yourself. Answers are available in Appendix B.

15

2.1 Atoms, Ions, and Molecules: The Building Blocks of Chemical Evolution

Just four types of atoms—hydrogen, carbon, nitrogen, and oxygen—make up 96 percent of all matter found in organisms today. Many of the molecules found in your cells contain thousands, or even millions, of these atoms bonded together. But early in Earth's history, these elements existed only in simple substances such as water and carbon dioxide, which contain just three atoms apiece.

Two questions are fundamental to understanding how elements could have evolved into the more complex substances found in living cells:

1. What is the physical structure of the hydrogen, carbon, nitrogen, and oxygen atoms found in living cells?

2. What is the structure of the simple molecules—water, carbon dioxide, and others—that acted as the building blocks of chemical evolution?

The focus on structure follows from one of the most central themes in biology: Structure affects function. To understand how a molecule affects your body or the role it played in chemical evolution, you have to understand how it is put together.

Basic Atomic Structure

Figure 2.1a shows a simple way of depicting the structure of an atom, using hydrogen and carbon as examples. Extremely small particles called electrons orbit an atomic nucleus made up of larger particles called protons and neutrons. Protons have a positive electric charge, neutrons are electrically neutral, and electrons have a negative electric charge. Opposite charges attract; like charges repel. When the number of protons and the number of electrons in an atom (or molecule) are the same, the charges balance and the atom is electrically neutral. **Figure 2.1b** provides a sense of scale at the atomic level.

Figure 2.2 shows a segment of the periodic table of the elements. Notice that each atom of an **element** contains a characteristic number of protons, called its **atomic number**. The atomic number is given as a subscript of each element symbol in the table. The sum of the protons and neutrons in an atom is called its **mass number**, given as a superscript of each symbol in Figure 2.2.

The number of protons in an element does not vary—if the atomic number of an atom changes, then one element is transformed to another element. The number of neutrons present in an element can vary, however. Forms of an element with different numbers of neutrons are known as **isotopes** (literally, "equal-places"). For example, all atoms of the element carbon have 6 protons. But naturally occurring isotopes of carbon can have either 6 or 7 neutrons, giving them a total of 12 or 13 protons and neutrons, respectively. Isotopes have different masses.

Although the masses of protons, neutrons, and electrons can be measured in grams, the numbers involved are so small that bi-

(a) Diagrams of atoms

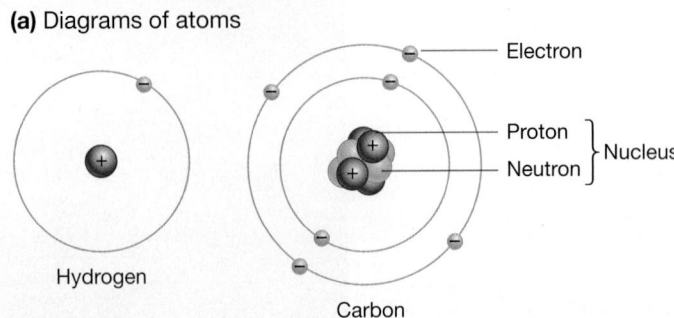

(b) Most of an atom's volume is empty space.

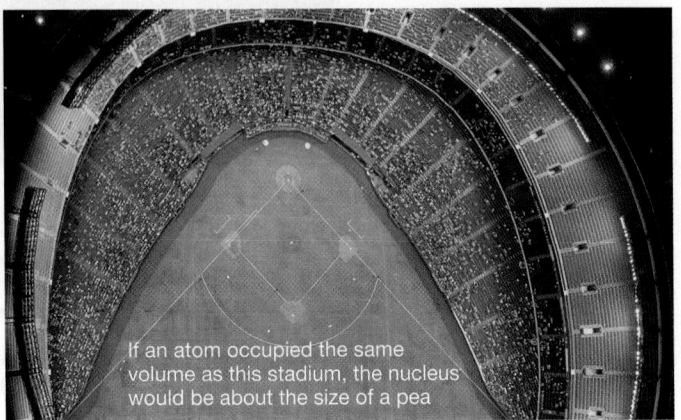

If an atom occupied the same volume as this stadium, the nucleus would be about the size of a pea

FIGURE 2.1 Parts of an Atom. The atomic nucleus, made up of protons and neutrons, is surrounded by orbiting electrons. In reality, electrons do not orbit the nucleus in circles; their actual orbits are complex.

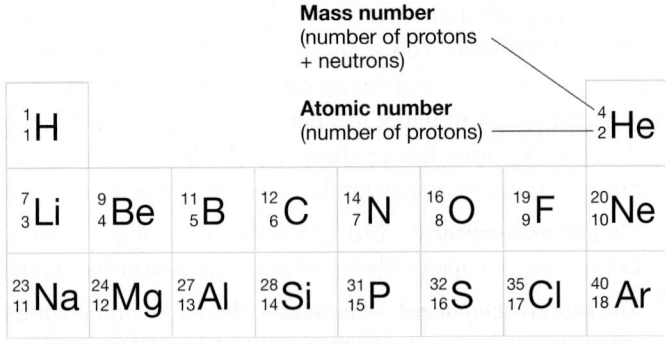

FIGURE 2.2 A Portion of the Periodic Table. Each element has a unique atomic number and is represented by a unique one- or two-letter symbol. The mass numbers given here are the most common for each element.

ologists prefer to use a special unit called the **dalton**. The masses of protons and neutrons are virtually identical and are routinely rounded to 1 dalton. A carbon atom that contains 6 protons and 6 neutrons has a mass of 12 daltons and a mass number of 12, while a carbon atom with 6 protons and 7 neutrons would have a mass number of 13. These isotopes would be written as ^{12}C and ^{13}C, respectively. The mass of an electron is so small that it is normally ignored.

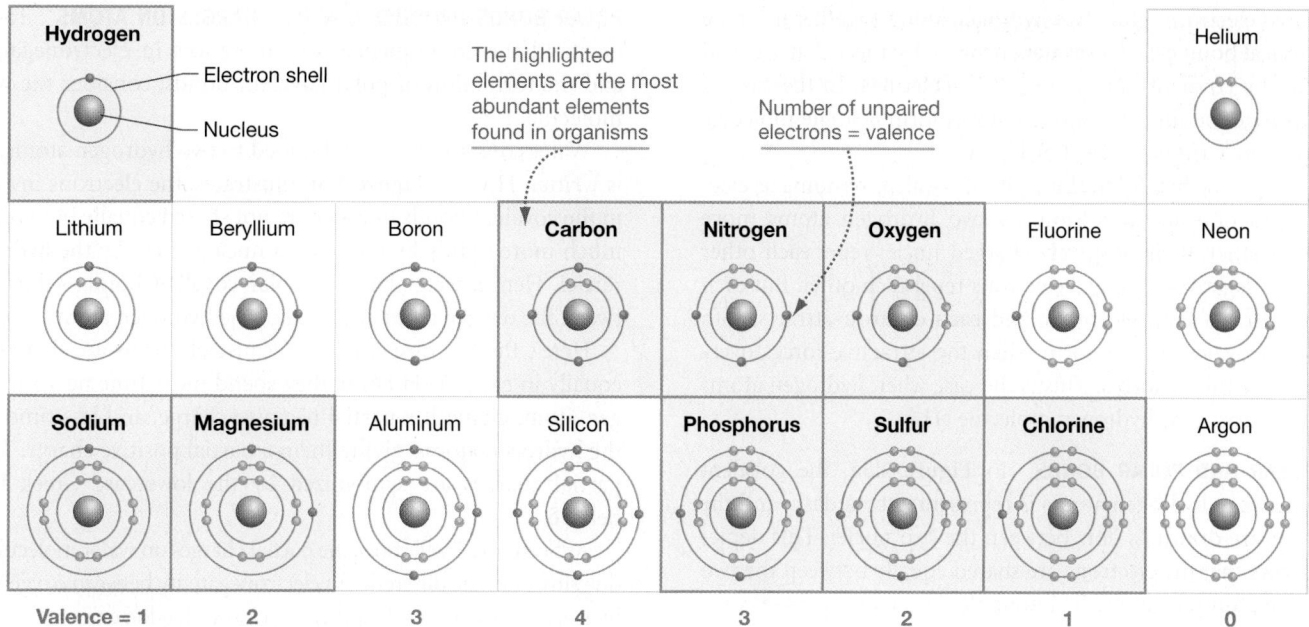

FIGURE 2.3 The Structure of Atoms Found in Organisms.

To understand how the atoms involved in chemical evolution behave, focus on how electrons are arranged around the nucleus:

- Electrons move around atomic nuclei in specific regions called **orbitals**.

- Each orbital can hold up to two electrons.

- Orbitals are grouped into levels called **electron shells**.

- Electron shells are numbered 1, 2, 3, and so on, to indicate their relative distance from the nucleus, with smaller numbers closer to the nucleus.

- Each electron shell contains a specific number of orbitals. An electron shell comprising a single orbital can hold up to two electrons; a shell with four orbitals can contain up to eight electrons.

- The electrons of an atom fill the innermost shells first, before filling outer shells.

To understand how the structures of atoms differ, take a moment to study **Figure 2.3**. This chart highlights the atoms that are most abundant in living cells. The gray ball in the center of each box represents a nucleus, and the orange circle or circles represent the electron shells around that nucleus. The small orange balls on the circles indicate how electrons are distributed in the shells of each element.

Now focus on the outermost shell of each atom. This is the element's valence shell. The electrons found in this shell are referred to as **valence electrons**. Two observations are important:

1. In each of the highlighted elements, the outermost electron shell is not full—not all the orbitals in the valence shell have two electrons. The highlighted elements have at least one unpaired valence electron.

2. The number of unpaired electrons in the valence shell varies among elements. Carbon has four unpaired electrons in its outermost shell; hydrogen has one. The number of unpaired electrons found in an atom is called its **valence**. Thus, carbon's valence is four, and hydrogen's valence is one.

These observations are significant because an atom is most stable when its valence shell is filled. One way that shells can be filled is through the formation of **chemical bonds**—strong attractions that bind atoms together.

How Does Covalent Bonding Hold Molecules Together?

To understand how atoms can become more stable by making chemical bonds, consider hydrogen. The hydrogen atom has just one electron, which resides in a shell that can hold two electrons.

Because it has an unpaired valence electron, the hydrogen atom is not very stable. But when two atoms of hydrogen come into contact, the two electrons become shared by the two nuclei (**Figure 2.4**). Both atoms now have a completely filled shell. Together, the hydrogen atoms are more stable than the two individual hydrogen atoms.

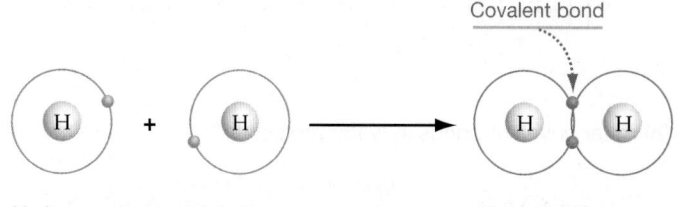

FIGURE 2.4 Covalent Bonds Result from Electron Sharing. When two hydrogen atoms come into contact, their electrons are attracted to the positive charge in each nucleus. As a result, their orbitals overlap, the electrons are shared by each nucleus, and a covalent bond forms.

Shared electrons "glue" two hydrogen atoms together in a type of chemical bond called a **covalent bond**. Substances that are held together by covalent bonds are called **molecules**. In the case of two hydrogen atoms, the bonded atoms form a single molecule of hydrogen, written as H—H or H_2.

It can also be helpful to think about covalent bonding as electrical attraction and repulsion. As two hydrogen atoms move closer together, their positively charged nuclei repel each other and their negatively charged electrons repel each other. But each proton attracts both electrons, and each electron attracts both protons. Covalent bonds form when the attractive forces overcome the repulsive forces. This is the case when hydrogen atoms interact to form the hydrogen molecule (H_2).

NONPOLAR AND POLAR BONDS In **Figure 2.5a**, the covalent bond between hydrogen atoms is represented by a dash, and the electrons are drawn as dots between the two nuclei. This depiction shows that the electrons are shared equally between the two hydrogen atoms, resulting in a nonpolar covalent bond—a covalent bond that is symmetrical.

It's important to note, though, that the electrons participating in a covalent bond are not always shared equally between the atoms involved. This happens because some atoms hold the electrons in covalent bonds much more tightly than do other atoms. Chemists call this property an atom's **electronegativity**.

Oxygen is among the most electronegative of all elements: It attracts covalently bonded electrons more strongly than does any other atom commonly found in organisms. Nitrogen's electronegativity is somewhat lower than oxygen's. Carbon and hydrogen, in turn, have relatively low and approximately equal electronegativities. Thus, the electronegativities of the four most abundant elements in organisms are related as follows: $O > N > C \cong H$.

Because carbon and hydrogen have approximately equal electronegativity, the electrons in a C—H bond are shared equally. The result is a **nonpolar covalent bond**. In contrast, asymmetric sharing of electrons results in a **polar covalent bond**. The electrons in a polar covalent bond spend most of their time close to the nucleus of the more electronegative atom. Why is this important?

(a) Nonpolar covalent bond in hydrogen molecule

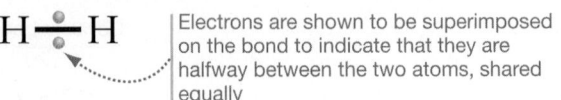

Electrons are shown to be superimposed on the bond to indicate that they are halfway between the two atoms, shared equally

(b) Polar covalent bonds in water molecule

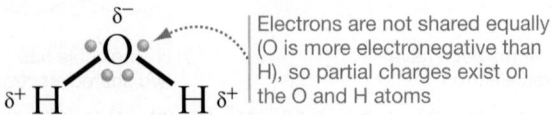

Electrons are not shared equally (O is more electronegative than H), so partial charges exist on the O and H atoms

FIGURE 2.5 Electron Sharing and Bond Polarity. Delta (δ) symbols in polar covalent bonds refer to partial positive and negative charges that arise owing to unequal electron sharing.

POLAR BONDS PRODUCE PARTIAL CHARGES ON ATOMS To understand the consequences of differences in electronegativity and the formation of polar covalent bonds, consider the water molecule.

Water consists of oxygen bonded to two hydrogen atoms, and is written H_2O. As **Figure 2.5b** illustrates, the electrons involved in the covalent bonds in water are not shared equally but are held much more tightly by the oxygen nucleus than by the hydrogen nuclei. Hence, water has two polar covalent bonds—those between the oxygen atom and each of the hydrogen atoms.

Here's the key observation: Because electrons are shared unequally in each O—H bond, they spend more time near the oxygen atom, giving it a partial negative charge, and less time near the hydrogen atoms, giving them a partial positive charge. These partial charges are symbolized by the lowercase Greek letter delta, δ.

As Section 2.2 will show, the partial charges on water molecules—due simply to the difference in electronegativity between oxygen and hydrogen—are one of the primary reasons that life exists.

Ionic Bonding, Ions, and the Electron-Sharing Continuum

Ionic bonds are similar in principle to covalent bonds, but instead of sharing electrons between two atoms, the electrons in ionic bonds are completely transferred from one atom to the other. The electron transfer occurs because it gives the resulting atoms a full outermost shell.

Sodium atoms (Na), for example, tend to lose an electron, leaving them with a full second shell. This is a much more stable arrangement, energetically, than having a lone electron in their third shell (**Figure 2.6a**). The atom that results has a net electric charge of $+1$, because it has one more proton than it has electrons.

An atom or molecule that carries a charge is called an **ion**. The sodium ion is written Na^+ and, like other positively charged ions, is called a **cation**.

Chlorine atoms (Cl), in contrast, tend to gain an electron, filling their outermost shell (**Figure 2.6b**). The ion has a net charge of -1, because it has one more electron than protons. This negatively charged ion, or **anion**, is written Cl^- and is called chlor*ide*.

When sodium and chlorine combine to form table salt (sodium chloride, NaCl), the atoms pack into a crystal structure consisting of sodium cations and chloride anions (**Figure 2.6c**). The electrical attraction between the ions is so strong that salt crystals are difficult to break apart.

This discussion of covalent and ionic bonding supports an important general observation: The degree to which electrons are shared in chemical bonds forms a continuum, from equal sharing in nonpolar covalent bonds, to unequal sharing in polar covalent bonds, to the transfer of electrons in ionic bonds.

As the left-hand side of **Figure 2.7** shows, covalent bonds between atoms with exactly the same electronegativity—for exam-

(a) A sodium ion being formed

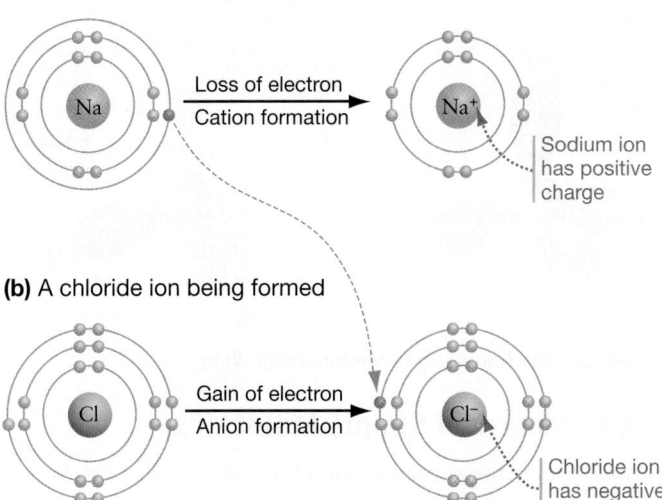

Loss of electron
Cation formation

Sodium ion has positive charge

(b) A chloride ion being formed

Gain of electron
Anion formation

Chloride ion has negative charge

(c) Table salt (NaCl) is a crystal composed of two ions.

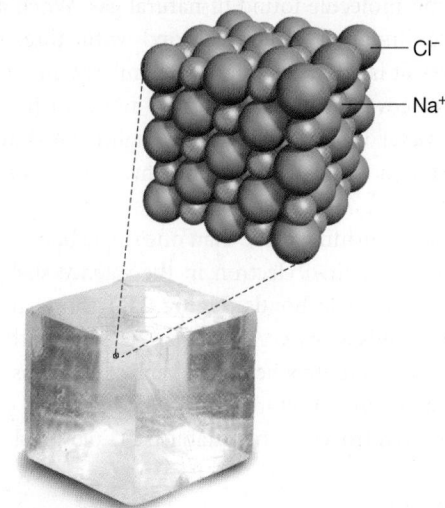

Cl⁻

Na⁺

FIGURE 2.6 Ion Formation and Ionic Bonding. The sodium ion (Na^+) and the chloride ion (Cl^-) are stable because they have full valence shells. In table salt (NaCl), sodium and chloride ions pack into a crystal structure held together by electrical attraction between their positive and negative charges.

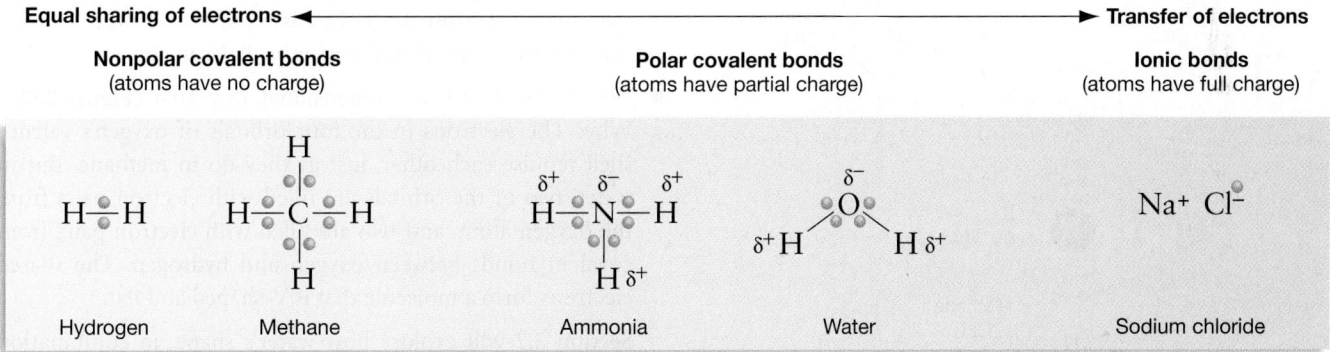

Equal sharing of electrons ◄──────────────────────────► Transfer of electrons

| Nonpolar covalent bonds (atoms have no charge) | | Polar covalent bonds (atoms have partial charge) | Ionic bonds (atoms have full charge) |

Hydrogen | Methane | Ammonia | Water | Sodium chloride

FIGURE 2.7 The Electron-Sharing Continuum. The degree of electron sharing in chemical bonds can be thought of as a continuum, from equal sharing in nonpolar covalent bonds to no sharing in ionic bonds.

✔**QUESTION** Why do most polar covalent bonds involve nitrogen or oxygen?

ple, between the atoms of hydrogen in H_2—represent one end of the continuum. The electrons in these nonpolar bonds are shared equally.

In the middle of the continuum are bonds where one atom is much more electronegative than the other. In these asymmetric bonds, substantial partial charges exist on each of the atoms. These types of polar covalent bonds occur when a highly electronegative atom such as oxygen or nitrogen is bound to an atom with a lower affinity for electrons, such as carbon or hydrogen. Ammonia (NH_3) and water (H_2O) are examples of molecules with polar covalent bonds.

At the right-hand side of the continuum are molecules made up of atoms with extreme differences in their electronegativities. In this case, electrons are transferred rather than shared, the atoms have full charges, and the bonding is ionic. Common table salt, NaCl, is a familiar example.

Most chemical bonds that occur in biological molecules are on the left-hand side and the middle of the continuum; in the molecules found in organisms, ionic bonding is rare.

Some Simple Molecules Formed from C, H, N, and O

Look back at Figure 2.3 and count the number of unpaired electrons in the valence shells of carbon, nitrogen, oxygen, and hydrogen atoms. Each unpaired electron in a valence shell can make up half of a covalent bond. It should make sense to you that a carbon atom can form a total of four covalent bonds; nitrogen can form three; oxygen can form two; and hydrogen, one.

When each of the four unpaired electrons of a carbon atom covalently bonds with a hydrogen atom, the molecule that results is written CH_4 and is called methane (**Figure 2.8a**). Methane is the most common molecule found in natural gas. When a nitrogen atom's three unpaired electrons bond with three hydrogen atoms, the result is NH_3, or ammonia. Similarly, an atom of oxygen can form covalent bonds with two atoms of hydrogen, resulting in a water molecule (H_2O). As Figure 2.4 showed, a hydrogen atom can bond with another hydrogen atom to form hydrogen gas (H_2).

In addition to forming more than one single bond, atoms with more than one unpaired electron in the valence shell can form double bonds or triple bonds. **Figure 2.8b** shows how carbon forms double bonds with oxygen atoms to produce carbon dioxide (CO_2). Triple bonds result when three pairs of electrons are shared. **Figure 2.8c** shows the structure of molecular nitrogen (N_2), which forms when two nitrogen atoms establish a triple bond.

(a) Single bonds

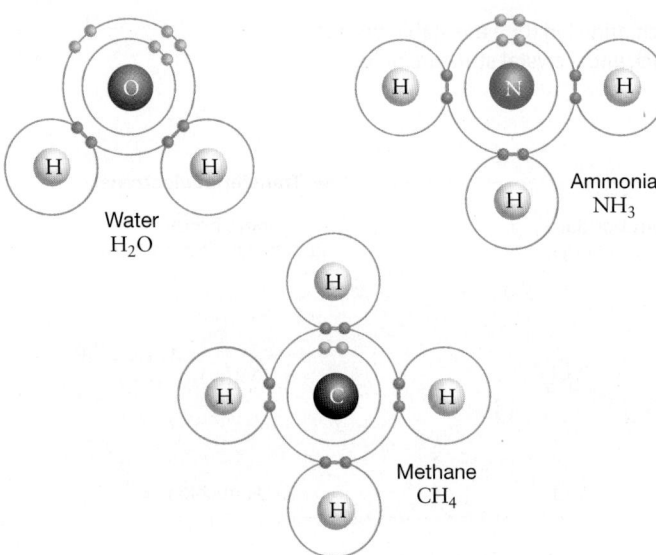

Water
H_2O

Ammonia
NH_3

Methane
CH_4

(b) Double bonds

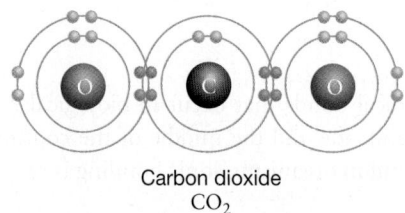

Carbon dioxide
CO_2

(c) Triple bonds

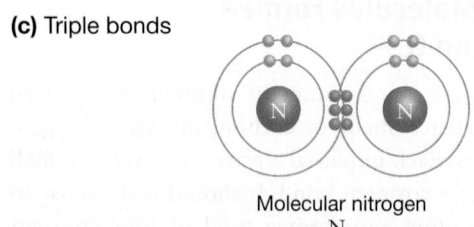

Molecular nitrogen
N_2

FIGURE 2.8 Unpaired Electrons in the Valence Shell Participate in Covalent Bonds. Covalent bonding is based on sharing of electrons in the outermost shell. Covalent bonds can be **(a)** single, **(b)** double, or **(c)** triple.

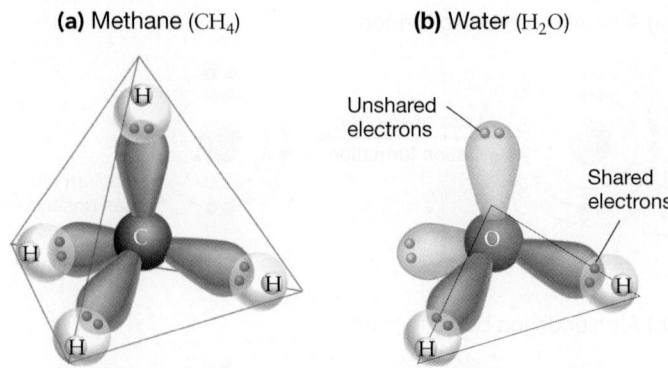

(a) Methane (CH_4)   **(b)** Water (H_2O)

Unshared electrons

Shared electrons

FIGURE 2.9 The Geometry of Methane and Water.

The Geometry of Simple Molecules

In many cases, the overall shape of a molecule dictates how it behaves. In chemistry and in biology, function is based on structure.

The shapes of the simple molecules you've just learned about are governed by the geometry of their bonds.

- Methane (CH_4) is tetrahedral—a structure with four triangular surfaces (**Figure 2.9a**). The tetrahedron forms because the electrons in the four C–H bonds repulse each other equally. The electron pairs are as far apart as they can get.

- Water is bent and two-dimensional, or planar (**Figure 2.9b**). Why? The electrons in the four orbitals of oxygen's valence shell repulse each other, just as they do in methane. But in water, two of the orbitals are filled with electron pairs from the oxygen atom, and two are filled with electron pairs from covalent bonds between oxygen and hydrogen. The shared electrons form a molecule that is V-shaped and flat.

Section 2.2 will explore how water's shape, in combination with the partial charges on the oxygen and hydrogen atoms, make it the most important molecule on Earth.

Representing Molecules

Molecules can be represented in a variety of increasingly complex ways—only some of which reflect their actual shape. Each method has advantages and disadvantages.

- **Molecular formulas** are compact but don't contain a great deal of information—they indicate only the numbers and types of atoms in a molecule (**Figure 2.10a**).

- **Structural formulas** indicate which atoms in a molecule are bonded together. Single, double, and triple bonds are represented by single, double, and triple dashes, respectively. Structural formulas also indicate geometry in two dimensions (**Figure 2.10b**). This is useful for planar molecules such as water and O_2.

- **Ball-and-stick models** take up more space than structural formulas, but provide information on the three-dimensional shape of molecules and indicate the relative sizes of the atoms involved (**Figure 2.10c**).

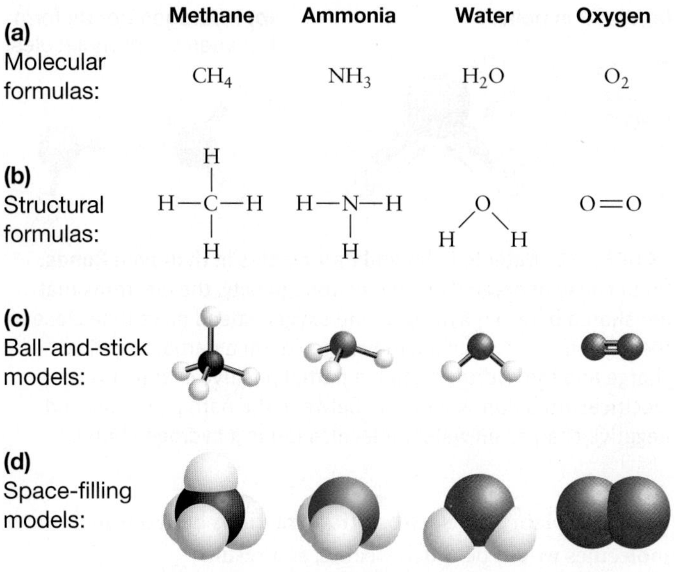

(a)
Molecular formulas:

Methane	Ammonia	Water	Oxygen
CH_4	NH_3	H_2O	O_2

(b)
Structural formulas:

(c)
Ball-and-stick models:

(d)
Space-filling models:

FIGURE 2.10 Molecules Can Be Represented Several Ways. Each method of representing a molecule has particular advantages.

- **Space-filling models** are more difficult to read than ball-and-stick models but more accurately depict the spatial relationships between atoms (**Figure 2.10d**).

In both ball-and-stick and space-filling models, biologists use certain colors to represent certain atoms. A black ball, for example, always symbolizes carbon. For more information on interpreting chemical structures, see **BioSkills 6** in Appendix A.

Basic Concepts in Chemical Reactions

The molecules you've just learned about—CH_4, NH_3, H_2O, CO_2, and N_2—are found in volcanic gases on Earth and in the atmospheres of nearby planets. Based on these observations, researchers claim that they were important components of Earth's ancient atmosphere and ocean. If so, then they provided the building blocks for chemical evolution.

The question now is: How did these simple building blocks combine to form more complex products, early in Earth's history? When a **chemical reaction** occurs, one substance is combined with others or broken down into another substance. Atoms are rearranged in molecules; in most cases, chemical bonds are broken and new bonds form.

Chemicals react in simple whole-number combinations. For example, one molecule of acetic acid (a prominent component of vinegar) reacts with one molecule of ethanol (the active ingredient in alcoholic beverages) to form one molecule of ethyl acetate.

Now suppose that you wanted to set up this reaction in an experiment. How would you know how much acetic acid and ethanol to add? The problem is that there is no simple way of counting the numbers of molecules present in a sample. Researchers solve this problem using the mole concept.

A **mole** refers to the number 6.022×10^{23}—just as the unit called the dozen refers to the number 12 or the unit million refers to the number 1×10^6. The mole is a useful unit because the mass of one mole of any molecule is the same as its molecular weight expressed in grams. **Molecular weight** is the sum of the mass numbers of all the atoms in a molecule.

For example, look back at Figure 2.2 and note that hydrogen has a mass number of 1 and oxygen has a mass number of 16. To get the molecular weight of H_2O, you sum the mass numbers of two atoms of hydrogen and one atom of oxygen, giving $1 + 1 + 16$, or a total of 18. Because mass number can also be measured in grams, it follows that if you weighed a sample of 18 grams of water, it would contain 6.022×10^{23} water molecules, or 1 mole of water molecules.

Now let's get back to your experiment. If you wanted to react one mole of acetic acid and one mole of ethanol, you'd use a balance to weigh out a number of grams of acetic acid and ethanol equal to their respective molecular weights and make up a **solution**—a homogenous (uniform) mixture of one or more substances dissolved in a liquid.

When substances are dissolved in liquid, their concentration is expressed in terms of molarity (symbolized by "M"). **Molarity** is the number of moles of the substance present per liter of solution. A 1-molar solution of acetic acid in water, for example, means that 1 mole of acetic acid is contained in 1 liter of solution.

The mole and molarity are concepts that allow you to connect what is going on in the invisible world at the atomic level to the visible world of the lab bench—where you can perform experiments with compounds such as acetic acid and ethanol.

Researchers postulate that most of the critical reactions in chemical evolution occurred in an aqueous, or water-based, solution. To understand what happened and why, let's delve into the properties of water, then start analyzing the reactions that triggered chemical evolution.

CHECK YOUR UNDERSTANDING

If you understand that . . .

- Covalent bonds are based on electron sharing. Electron sharing allows atoms to fill all the orbitals in their valence shell, making them more stable.
- Covalent bonds can be polar or nonpolar, depending on whether the electronegativities of the two atoms involved are the same or different.

✔ You should be able to . . .

Draw the structural formulas of methane (CH_4) and ammonia (NH_3) and add dots to indicate the relative locations of the covalently bonded electrons, based on the relative electronegativities of C, H, and N.

Answers are available in Appendix B.

2.2 The Early Oceans and the Properties of Water

Life is based on water. In a typical living cell, water comprises over 75 percent of the volume (**Figure 2.11**). You can survive for weeks without eating, but you aren't likely to live more than 3–4 days without drinking.

Water is vital for a simple reason: It is an excellent **solvent**—an agent for getting substances into solution. The chemical reactions occurring inside your body right now, and the reactions that caused chemical evolution some 3.5 billion years ago, depend on direct, physical interaction between the reactants. Substances are most likely to come into contact and react as **solutes**—meaning, when they are dissolved. The formation of Earth's first ocean, about 3.8 billion years ago, was a turning point in chemical evolution because it gave the process a place to happen.

Why Is Water Such an Efficient Solvent?

To understand why water is such an effective solvent, recall that:

1. Both of the O−H bonds in the molecule are polar, owing to oxygen's high electronegativity. As a result, the oxygen atom has a partial negative charge and each hydrogen atom has a partial positive charge.

2. The molecule is bent. Consequently, the partial negative charge on the oxygen atom sticks out, away from the partial positive charges on the hydrogen atoms (**Figure 2.12a**).

Figure 2.12b illustrates how water's structure affects its interactions with other water molecules. When two liquid water molecules approach each other, the partial positive charge on hydrogen attracts the partial negative charge on oxygen. This weak electrical attraction forms a **hydrogen bond** between the molecules.

✔ If you understand how water's polarity makes hydrogen bonding possible, you should be able to (1) draw a fictional version of Figure 2.12b that shows water as a linear (not bent) molecule, with partial charges on the oxygen and hydrogen atoms;

FIGURE 2.11 Water Is the Most Abundant Molecule in Organisms. Fruits shrink when they are dried because they consist primarily of water.

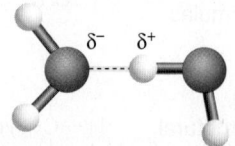

(a) Water is polar.

Electrons are pulled toward oxygen

(b) Hydrogen bonds form between water molecules.

FIGURE 2.12 Water Is Polar and Participates in Hydrogen Bonds. (a) Because of oxygen's high electronegativity, the electrons that are shared between hydrogen and oxygen spend more time close to the oxygen nucleus, giving the oxygen atom a partial negative charge and the hydrogen atom a partial positive charge. **(b)** The electrical attraction that occurs between the partial positive and negative charges on water molecules forms a hydrogen bond.

and (2) explain why electrostatic attractions between such water molecules would be much weaker, as a result.

In an aqueous solution, hydrogen bonds also form between water molecules and other polar molecules. Similar interactions occur between water and ions. Ions and polar molecules stay in solution because of their interactions with water's partial charges (**Figure 2.13a**). Substances that interact with water in this way are said to be **hydrophilic** ("water-loving").

In contrast, compounds that are uncharged and nonpolar do not interact with water through hydrogen bonding and do not dissolve in water. Because their interactions with water are minimal or non-existent, they are forced to interact with each other (**Figure 2.13b**). Substances that do not interact with water are said to be **hydrophobic** ("water-fearing"). ☞ Hydrogen bonding makes it possible for almost any charged or polar molecule to dissolve in water.

Although individual hydrogen bonds are not nearly as strong as covalent or ionic bonds, many of them occur in a solution. Hydrogen bonding is extremely important in biology owing to the sheer number of hydrogen bonds that form between water and other molecules that are polar or carry a charge.

How Does Water's Structure Correlate with Its Properties?

Water's small size, bent shape, highly polar covalent bonds, and overall polarity are unique among molecules. Because the structure of molecules routinely correlates with their function, it's not surprising that water has some remarkable properties, in addition to its extraordinary capacity to act as a solvent.

COHESION, ADHESION, AND SURFACE TENSION Binding between like molecules is called **cohesion**. Water is cohesive—meaning that it stays together—because of the hydrogen bonds that form between individual molecules.

Binding between unlike molecules, in contrast, is called **adhesion**. Adhesion is usually analyzed in regard to interactions between a liquid and a solid surface. Water adheres to surfaces that have any polar or charged components.

Cohesion and adhesion are important in explaining how water moves from the roots of plants to their leaves (see Chapter 37).

(a) Polar molecules and ions dissolve readily in water.

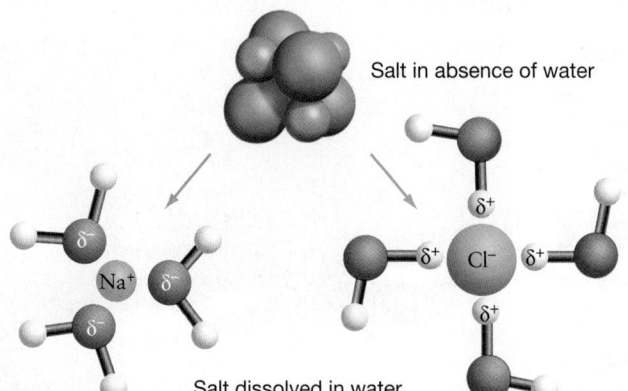

Salt in absence of water

Salt dissolved in water

(b) Nonpolar molecules do not dissolve in water.

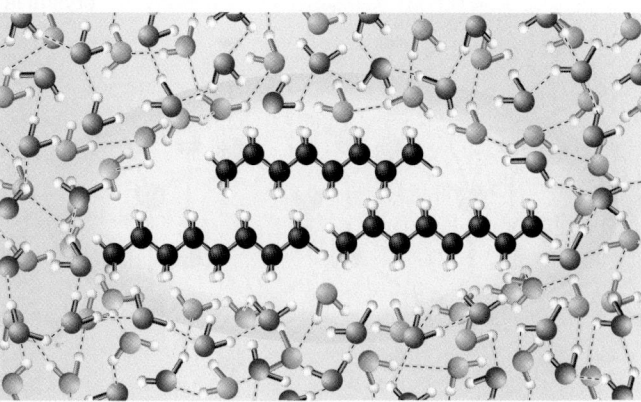

FIGURE 2.13 Water Interacts with Other Polar Molecules. (a) Water's polarity makes it a superb solvent for polar molecules and ions. **(b)** In aqueous solution, nonpolar molecules and compounds are forced to interact with each other. This occurs because water is much more stable when it interacts with itself rather than with the nonpolar molecules.

✔ **QUESTION** Explain the physical basis of the expression, "Oil and water don't mix."

But you can also see them in action in the concave surface, or meniscus, that forms in a glass of water (**Figure 2.14a**). A meniscus forms as a result of two forces:

1. Water molecules at the surface hydrogen-bond with water molecules below them, so they experience a net downward pull.

2. Water molecules at the surface also adhere to the glass, allowing them to resist the downward pull.

Cohesion is also instrumental in the phenomenon known as surface tension. Because hydrogen bonding exerts a pulling force, or tension, at the surface of any body of water, water molecules are not stable there—they are constantly being pulled away from the surface. A body of water is most stable when this source of instability is minimized—meaning that its total surface area is minimized.

This fact has an important consequence: Water resists any force that increases its surface area. More specifically, any force that depresses a water surface meets with resistance. This resistance makes a water surface act as if it had an elastic membrane (**Figure 2.14b**)—a property called **surface tension**.

All liquids have a surface tension. Water's surface tension is extraordinarily high because of the extensive hydrogen bonding that occurs between molecules. In water, the "elastic membrane" at the surface is stronger than it is in other liquids.

WATER IS DENSER AS A LIQUID THAN AS A SOLID When factory workers pour molten metal or plastic into a mold and allow it to cool to the solid state, the material shrinks. When molten lava pours out of a volcano and cools to solid rock, it shrinks. But when you fill an ice tray with water and put it in the freezer to make ice, the water expands.

Unlike most substances, water is denser as a liquid than it is as a solid. In other words, there are more molecules of water in a given volume of liquid water than there are in the same volume

(a) A meniscus forms where water meets a solid surface, as a result of two forces.

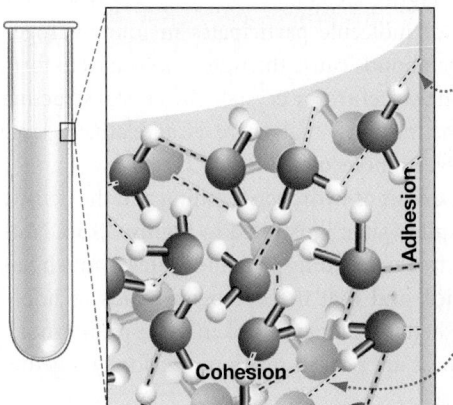

Adhesion: Water molecules that adhere to the glass resist the downward pull of cohesion

Adhesion

Cohesion: Water molecules at the surface experience a net downward pull from hydrogen bonds with water molecules below

Cohesion

(b) Water has high surface tension.

Because of surface tension, light objects do not fall through the water's surface

FIGURE 2.14 Cohesion, Adhesion, and Surface Tension. (a) Meniscus formation is based on hydrogen bonding. **(b)** Water resists forces—like the weight of an insect—that increase its surface area. The resistance is great enough that light objects do not break the surface.

(a) In ice, water molecules form a crystal lattice.

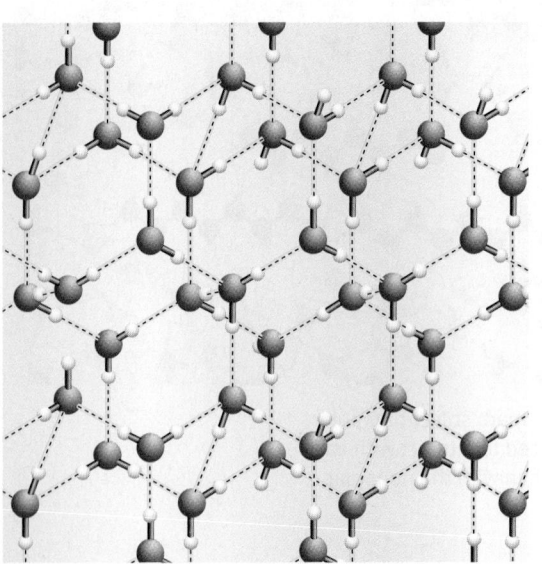

(b) In liquid water, no crystal lattice forms.

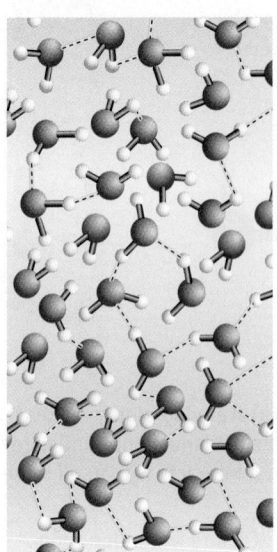

(c) Liquid water is denser than ice. As a result, ice floats.

FIGURE 2.15 Hydrogen Bonding Forms the Crystal Structure of Ice. In ice, each molecule can form four hydrogen bonds at one time. Each oxygen atom can form two; each hydrogen atom can form one.

of solid water, or ice. **Figure 2.15a** illustrates why this is so. Note that in ice, each water molecule participates in four hydrogen bonds. These hydrogen bonds cause the water molecules to form a regular and repeating structure, or crystal. The crystal structure of ice is fairly open, meaning that there is a relatively large amount of space between molecules.

Now compare the extent of hydrogen bonding and the density of ice with that of liquid water, illustrated in **Figure 2.15b**. Note that the extent of hydrogen bonding in liquid water is much less than that found in ice, and that the hydrogen bonds in liquid water are constantly being made and broken. As a result, molecules in the liquid phase are packed much more closely together than in the solid phase.

Normally, heating a substance causes it to expand because molecules begin moving faster and colliding more often and with greater force. But heating ice causes hydrogen bonds to break and the open crystal structure to collapse. In this way, hydrogen bonding explains why water is denser as a liquid than as a solid.

This property of water has an important result: Ice floats (**Figure 2.15c**). If it did not, ice would sink to the bottom of lakes, ponds, and oceans soon after it formed. The ice would stay frozen in the cold depths. If water weren't so unusual, it is almost certain that Earth's oceans would have frozen almost solid before life had a chance to start.

WATER HAS A HIGH CAPACITY FOR ABSORBING ENERGY Hydrogen bonding is also responsible for another of water's remarkable physical properties: Water has a high capacity for absorbing energy.

Specific heat, for example, is the amount of energy required to raise the temperature of 1 gram of a substance by 1°C. Water has an extremely high specific heat because when a source of energy hits it, hydrogen bonds must be broken before heat can be trans-

ferred and the water molecules begin moving faster. As **Table 2.1** indicates, it takes an extraordinarily large amount of energy to change the temperature of water and other molecules where extensive hydrogen bonding occurs.

Similarly, it takes a large amount of energy to break the hydrogen bonds in liquid water and change the molecules from the liquid phase to the gas phase. Water's **heat of vaporization**—the energy required to change 1 gram of it from a liquid to gas—is higher than that of most molecules that are liquid at room temperature. As a result, water has to absorb a great deal of energy to evaporate. Water's high heat of vaporization is the reason that sweating or dousing yourself with water is an effective way to cool off on a hot day. Water molecules absorb a great deal of energy from your body before they evaporate, so you lose heat.

TABLE 2.1 **Specific Heats of Some Liquids**

The specific heats reported in this table were measured at 25°C and are given in units of joules per gram of substance per degree Celsius. (The joule is a unit of energy.)

Liquids with extensive hydrogen bonding	Specific Heat
Ammonia (NH_3)	4.70
Water (H_2O)	4.18
Liquids with some hydrogen bonding	
Ethanol (CH_3CH_2OH)	2.44
Ethylene glycol ($HOCH_2CH_2OH$; used in antifreeze)	2.22
Liquids with little or no hydrogen bonding	
Benzene (C_6H_6)	1.80
Xylene (C_8H_{10})	1.72
Sulfuric acid (H_2SO_4)	1.40

SOURCE: J. Murray and R. C. Fay 2004. *Chemistry*, 4th ed., Prentice-Hall, Table 8.1

Water's ability to absorb energy is critical to the theory of chemical evolution. Because compounds that were important in chemical evolution dissolve readily in water, they would have formed in the ocean or rained out of the atmosphere into the ocean. Once these compounds were in aqueous solution, they were well-protected from sources of energy that could break them apart, such as intense sunlight. As a result, they would have persisted and slowly increased in concentration over time, making them more likely to react and continue the process.

Acid–Base Reactions Involve a Transfer of Protons

One other aspect of water's chemistry is important to understanding chemical evolution and how organisms work today. Water is not a completely stable molecule. In reality, water molecules continually undergo a chemical reaction with themselves. This "dissociation" reaction can be written as follows:

$$H_2O \rightleftharpoons H^+ + OH^-$$

The double arrow indicates that the reaction proceeds in both directions.

The substances on the right-hand side of the expression are the **hydrogen ion** (H^+) and the **hydroxide ion** (OH^-). A hydrogen ion is simply a proton. In reality, however, protons never exist by themselves. In water, for example, protons associate with water molecules to form the hydronium ion (H_3O^+). Thus, the dissociation of water is more accurately written as:

$$H_2O + H_2O \rightleftharpoons H_3O^+ + OH^-$$

One of the water molecules on the left-hand side of the expression has given up a proton, while the other water molecule has accepted a proton.

Substances that give up protons during chemical reactions and raise the hydrogen ion concentration of water are called **acids**; molecules or ions that acquire protons during chemical reactions and lower the hydrogen ion concentration of water are called **bases**. Most acids act only as acids, and most bases act only as bases; but water can act as both an acid and a base.

A chemical reaction that involves a transfer of protons is called an acid–base reaction. Every acid–base reaction requires a proton donor and a proton acceptor—an acid and a base, respectively.

Water is an extremely weak acid—very few water molecules dissociate to form hydronium ions and hydroxide ions. In contrast, strong acids like the hydrochloric acid in your stomach readily give up a proton when they react with water.

$$HCl + H_2O \rightleftharpoons H_3O^+ + Cl^-$$

Strong bases readily acquire a proton when they react with water. For example, sodium hydroxide (NaOH, commonly called lye) dissociates completely in water to form Na^+ and OH^-. The hydroxide ion produced by that reaction then accepts a proton from a hydronium ion in the water, forming two water molecules.

$$NaOH(aq) \longrightarrow Na^+ + OH^-$$
$$OH^- + H_3O^+ \rightleftharpoons 2\,H_2O$$

The "*aq*" in the expression indicates that NaOH is in aqueous solution.

To summarize, adding an acid to a solution increases the concentration of protons; adding a base to a solution lowers the concentration of protons. Water is both a weak acid and a weak base.

pH INDICATES THE CONCENTRATION OF PROTONS In a solution, the tendency for acid–base reactions to occur is largely a function of the number of protons present. Chemists can measure the concentration of protons in a solution directly. In a sample of pure water at 25°C, the concentration of H^+ is 1.0×10^{-7} M, or 1 ten-millionth molar.

Because the concentration of protons in water is such a small number, exponential notation is cumbersome. So chemists and biologists prefer to express the concentration of protons in a solution, and thus its acidity or alkalinity, with a logarithmic notation called **pH**.[1]

By definition, the pH of a solution is the negative of the base-10 logarithm, or log, of the hydrogen ion concentration:

$$pH = -\log[H^+]$$

(The square brackets are a standard notation for indicating "concentration of" a substance in solution.) Taking antilogs gives:

$$[H^+] = \text{antilog}(-pH) = 10^{-pH}$$

pH, then, is a convenient way to indicate the concentration of protons in a solution. If the concentration of H^+ in a sample of water is 1.0×10^{-7} M, then its pH is 7. If the solution changes to pH 9, then it has become 100 times more basic. To review logarithms, see **BioSkills 7** in Appendix A.

BUFFERS PROTECT AGAINST DAMAGING CHANGES IN pH Cells are extremely sensitive to changes in pH. Changes in hydrogen ion concentration affect the structure and function of polar or charged substances, as well as the tendency of acid–base reactions to occur.

Compounds that minimize changes in pH are called **buffers** because they buffer a solution against the damaging effects of pH change. Buffers are important in maintaining relatively constant conditions, or **homeostasis**, in cells and tissues. In cells, a wide array of naturally occurring molecules act as buffers.

Most buffers are weak acids, meaning that they are somewhat likely to give up a proton in solution. To see how they work, consider the disassociation of acetic acid in water to form acetate ions and protons.

$$CH_3COOH \rightleftharpoons CH_3COO^- + H^+$$

When acetic acid and acetate are present in about equal concentrations in a solution, they function as a buffering system. If the concentration of protons increases slightly, the protons react with acetate ions to form acetic acid and pH does not change. If the concentration of protons decreases slightly, acetic acid gives up protons and pH does not change.

---

[1]The term *pH* is derived from the French *puissance d'hydrogène*, or "power of hydrogen."

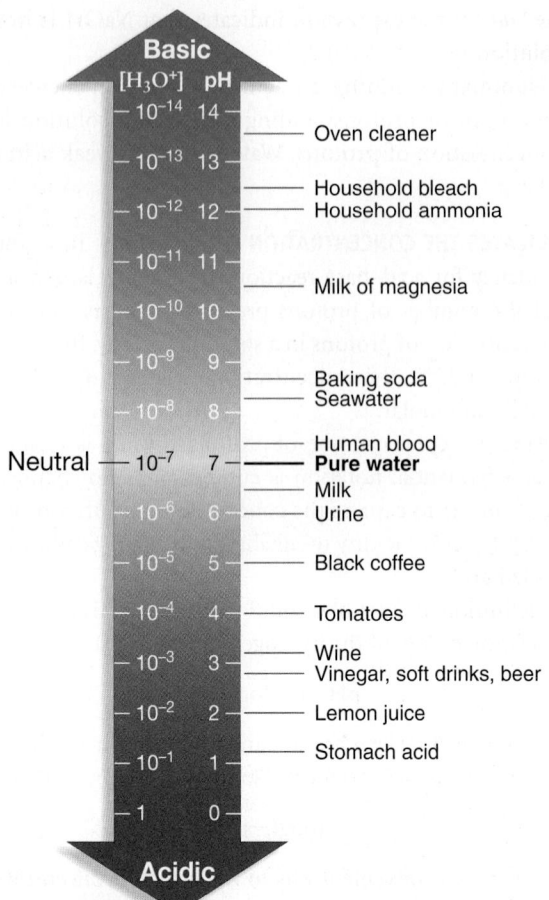

FIGURE 2.16 The pH Scale. Because the pH scale is logarithmic, a change in one unit of pH represents a change in the concentration of hydrogen ions equal to a factor of 10. Coffee has a hundred times more H^+ than pure water has.

✔ **QUESTION** How is the pH of black coffee affected if you add milk?

THE pH SCALE MEASURES THE ACIDITY OR ALKALINITY OF A SOLUTION

Figure 2.16 shows the pH scale and reports the pH of some common solutions. Note that the pH of pure water at 25°C is 7. Pure water is used as a standard, or point of reference, on the pH scale.

Solutions that contain acids have a proton concentration larger than 1×10^{-7} M and thus a pH < 7. This is because acidic molecules tend to release protons into solution. In contrast, solutions that contain bases have a proton concentration less than 1×10^{-7} M and thus a pH > 7. This is because basic molecules tend to accept protons from solution. ✔If you understand this concept, you should be able to explain the difference in hydrogen ion concentration between a slightly basic solution with pH 8 and a slightly acidic solution with pH 6.

Solutions with a pH of 7 are considered neutral—neither acidic nor basic. The solution inside living cells is about 7. Many of today's lakes also have a nearly neutral pH, which probably mimics the conditions in the early oceans—it would have taken hundreds of millions of years for rain and wind to erode rocks and send dissolved ions to the ocean, increasing its salinity and raising its pH to the current level of about 8.

As chemical evolution began, then, water provided the physical environment for key reactions to take place. In some cases, water also acted as an important reactant. Although acid–base reactions were not critical to the initial stages of chemical evolution, they became extremely important once the process was under way.

Table 2.2 summarizes some of the key properties of water, for you to use as a reference. You can also go to the study area at *www.masteringbiology.com* and review how hydrogen bonding gives water such unusual properties.

(MB) **Web Activity** The Properties of Water

Water provided a medium for chemical evolution to occur. Now let's consider what happened in solution, some 3.5 billion years ago.

SUMMARY TABLE 2.2 **Properties of Water**

Property	Cause	Biological Consequences
Solvent for charged or polar compounds		Most chemical reactions important for life take place in aqueous solution.
Denser as a liquid than a solid	As water freezes, each molecule forms a total of four hydrogen bonds, leading to the formation of the low-density crystal structure called ice.	
High specific heat	Water molecules must absorb lots of heat energy to break hydrogen bonds and experience increased movement (and thus temperature).	Oceans absorb and release heat slowly, moderating coastal climates.
High heat of vaporization		Evaporation of water from an organism cools the body.

✔**EXERCISE** You should be able to fill in the missing cells in this table. You should also be able to make a concept map relating water's structure to the properties listed here. (For an introduction to concept mapping, see **BioSkills 8** in Appendix A.) Your concept map should include the following terms or phrases: polar covalent bonds, polarity (of the water molecule), hydrogen bonding, high heat of vaporization, high specific heat, less dense as a solid, effective solvent, unequal sharing of electrons, high energy input required to break bonds, high electronegativity of oxygen.

2.3 Chemical Reactions, Chemical Evolution, and Chemical Energy

Proponents of the theory of chemical evolution contend that simple molecules present in the atmosphere and ocean of ancient Earth participated in chemical reactions that eventually produced larger, more complex molecules—such as the proteins, nucleic acids, sugars, and lipids introduced in Chapters 3–6. Currently, researchers are investigating two environments where these reactions could have occurred.

1. The atmosphere, which was probably dominated by gases ejected from volcanoes. Carbon dioxide, water vapor, and nitrogen are the dominant gases ejected from volcanoes today; a small amount of molecular hydrogen (H_2), methane (CH_4), and carbon monoxide (CO) may also be present.

2. Deep-sea vents, where extremely hot rocks from the Earth's interior contact the seafloor. In addition to the gases listed above, today's deep-sea vents are rich in metals such as iron and nickel.

When volcanic gases are put together and allowed to interact, however, very little happens. The simple molecules do not suddenly link together to create large, complex substances like those found in living cells. Instead, their bonds remain intact. How, then, did chemical evolution occur?

How Do Chemical Reactions Happen?

Chemical reactions are written in a format similar to mathematical equations, with the initial, or **reactant**, atoms or molecules shown on the left and the resulting reaction **product(s)** shown on the right. For example, the most common reaction in the mix of gases and water that emerge from volcanoes is:

$$CO_2(g) + H_2O(l) \rightleftharpoons H_2CO_3(aq)$$

This expression indicates that carbon dioxide (CO_2) reacts with water (H_2O), forming carbonic acid (H_2CO_3). The state of each reactant and product is indicated as gas (*g*), liquid (*l*), in aqueous solution (*aq*), or solid (*s*).

Note that the expression is balanced; that is, 1 carbon, 3 oxygen, and 2 hydrogen atoms are present on each side of the expression. Note also that the expression contains a double arrow, meaning that the reaction is reversible. When the forward and reverse reactions proceed at the same rate, the quantities of reactants and products remain constant, although not necessarily equal. A dynamic but stable state such as this is termed a **chemical equilibrium**.

A chemical equilibrium can be disturbed by changing the concentration of reactants or products. For example, adding CO_2 to the mixture would drive the reaction to the right, creating more H_2CO_3 until the equilibrium proportions of reactants and products are reestablished. Adding H_2CO_3 instead would drive the reaction to the left. Removing CO_2 would also drive the reaction to the left; removing H_2CO_3 would drive it to the right.

A chemical equilibrium can also be altered by changes in temperature. For example, the water molecules in this set of interacting elements, or **system**, would be present as a combination of liquid water and water vapor:

$$H_2O \rightleftharpoons H_2O(g)$$

If liquid water molecules absorb enough heat, they transform to the gaseous state. This is called an **endothermic** ("within heating") process because heat is absorbed during the process. In contrast, the transformation of water vapor to liquid water releases heat and is called **exothermic** ("outside heating"). Raising the temperature of this system drives the equilibrium to the right; cooling the system drives it to the left.

In relation to chemical evolution, though, these reactions and changes of state are not particularly interesting. Carbonic acid is not an important intermediate in the formation of more complex molecules. Interesting things do begin to happen, though, when large amounts of energy are added to mixtures of volcanic gases.

What Is Energy?

Energy can be defined as the capacity to do work or to supply heat. This capacity exists in one of two ways—as a stored potential or as an active motion.

Stored energy is called **potential energy**. An object gains or loses its ability to store energy because of its position. An electron that resides in an outer electron shell will, if the opportunity arises, fall into a lower electron shell closer to the positive charges on the protons in the nucleus. Because of its position farther from the positive charges in the nucleus, an electron in an outer electron shell has more potential energy than does an electron in an inner shell (**Figure 2.17**).

Kinetic energy is the energy of motion. Molecules have kinetic energy because they are constantly in motion.

- The kinetic energy of molecular motion is called **thermal energy**.

- The **temperature** of an object is a measure of how much thermal energy its molecules possess. If an object has a low temperature, its molecules are moving slowly. (We perceive this as "cold.") If an object has a high temperature, its molecules are moving rapidly. (We perceive this as "hot.")

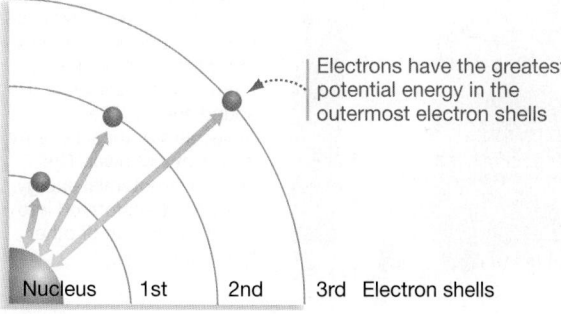

Electrons have the greatest potential energy in the outermost electron shells

Nucleus | 1st | 2nd | 3rd Electron shells

FIGURE 2.17 Potential Energy as a Function of Electron Shells. Electrons in outer shells have more potential energy than do electrons in inner shells, because the negative charges of the electrons in outer shells are farther from the positive charges of the protons in the nucleus. Each shell represents a distinct level of potential energy.

- When two objects with different temperatures come into contact, thermal energy is transferred between them. This transferred energy is called **heat**.

There are many forms of potential energy and kinetic energy, and energy can change from one form into another. To drive this point home, consider a water molecule sitting at the top of a waterfall, as in **Figure 2.18a**.

Step 1 The molecule has potential energy (E_p) because of its position.

Step 2 As the molecule passes over the waterfall, its potential energy is converted to the kinetic energy (E_k) of motion.

Step 3 When the molecule reaches the rocks below, it experiences a change in potential energy because it has changed position. The change in potential energy is transformed into an equal amount of energy in other forms: mechanical energy, which tends to break up the rocks; heat (thermal energy), which raises the temperature of the rocks and the water itself; and sound.

An electron in an outer electron shell is analogous to the water molecule at the top of a waterfall (**Figure 2.18b**). If the electron falls to a lower shell, its potential energy is converted to the kinetic energy of motion. After the electron occupies the lower electron shell, it experiences a change in potential energy. As panel 3 in Figure 2.18b shows, the change in potential energy

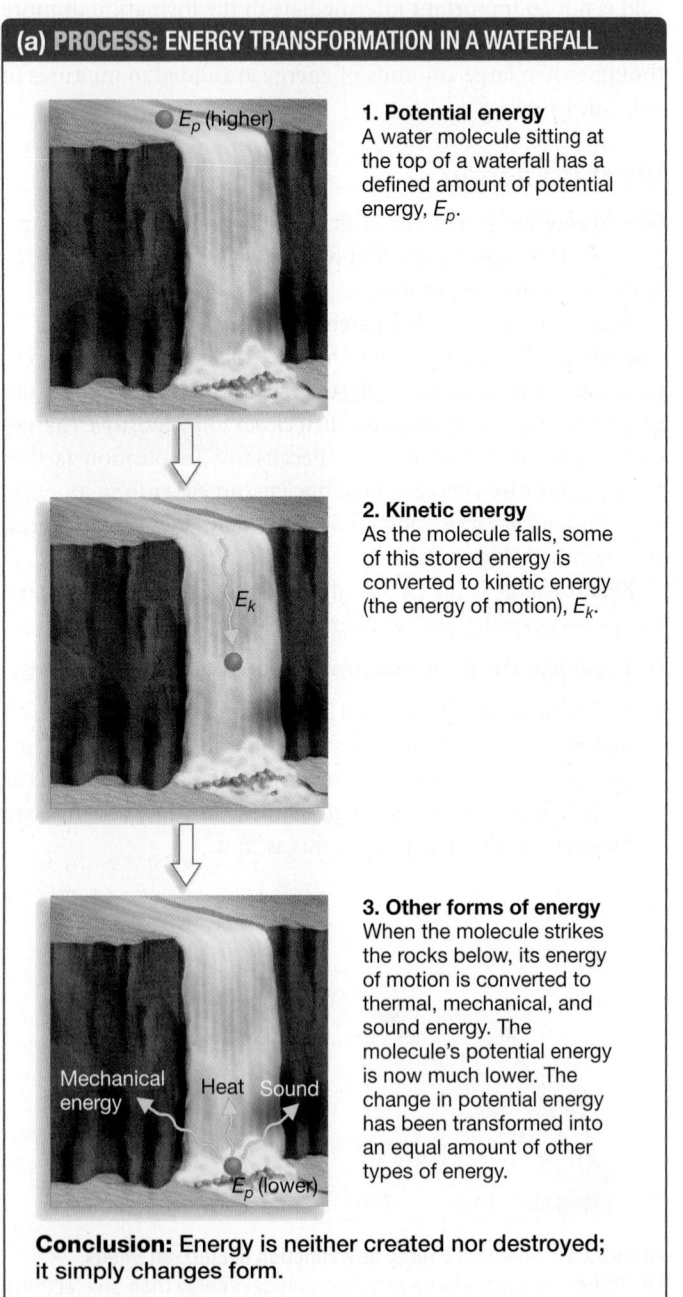

(a) PROCESS: ENERGY TRANSFORMATION IN A WATERFALL

1. Potential energy
A water molecule sitting at the top of a waterfall has a defined amount of potential energy, E_p.

2. Kinetic energy
As the molecule falls, some of this stored energy is converted to kinetic energy (the energy of motion), E_k.

3. Other forms of energy
When the molecule strikes the rocks below, its energy of motion is converted to thermal, mechanical, and sound energy. The molecule's potential energy is now much lower. The change in potential energy has been transformed into an equal amount of other types of energy.

Conclusion: Energy is neither created nor destroyed; it simply changes form.

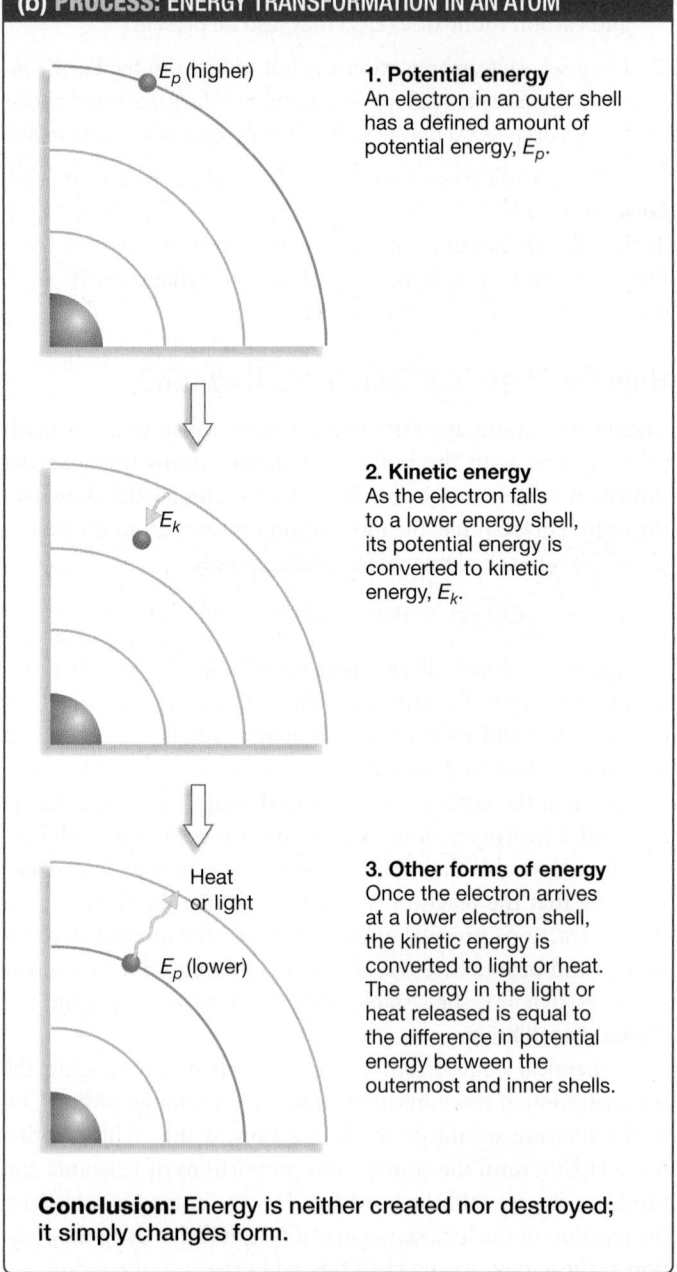

(b) PROCESS: ENERGY TRANSFORMATION IN AN ATOM

1. Potential energy
An electron in an outer shell has a defined amount of potential energy, E_p.

2. Kinetic energy
As the electron falls to a lower energy shell, its potential energy is converted to kinetic energy, E_k.

3. Other forms of energy
Once the electron arrives at a lower electron shell, the kinetic energy is converted to light or heat. The energy in the light or heat released is equal to the difference in potential energy between the outermost and inner shells.

Conclusion: Energy is neither created nor destroyed; it simply changes form.

FIGURE 2.18 Energy Transformations. During an energy transformation, the total amount of energy in the system remains constant.

is transformed into an equal amount of energy in other forms—usually thermal energy, but sometimes light.

These examples illustrate the **first law of thermodynamics**, which states that energy is conserved. Energy cannot be created or destroyed, but only transferred and transformed.

Energy transformation is the heart of chemical evolution. According to the best data available, molecules that were part of the young Earth were exposed to massive inputs of energy. Part of this energy was in the form of heat, from the gradually cooling molten mass that initially formed the planet. Part of the energy was in the form of high-energy radiation from the Sun, which today is screened out by molecules in the atmosphere. How would the application of large amounts of heat and radiation affect the course of chemical evolution?

Chemical Evolution: A Model System

To determine how energy inputs affected the simple molecules present in the early oceans and atmosphere, researchers have constructed computer models to simulate the reactions that can occur between carbon dioxide, water, ammonia, and hydrogen molecules.

The goal of one such study was quite specific: The researchers wanted to determine whether a molecule called formaldehyde (H_2CO) could be produced. Along with hydrogen cyanide (HCN), formaldehyde is a key intermediate in the creation of the larger, more complex molecules found in cells. Forming formaldehyde and hydrogen cyanide is the critical first step in chemical evolution—a trigger that could set the process in motion.

The research group began by proposing that the following reaction could take place:

$$CO_2(g) + 2 H_2(g) \longrightarrow H_2CO(g) + H_2O(g)$$

This reaction, however, doesn't occur spontaneously. Let's consider why it doesn't occur without an input of energy. We can then look at where the necessary energy input might have originated, and how conditions found on early Earth, such as temperature and the concentration of reactants, influenced the reaction.

WHAT MAKES A CHEMICAL REACTION SPONTANEOUS? When chemists say that a reaction is spontaneous, they have a precise meaning in mind: ☞ Chemical reactions are spontaneous if they proceed on their own, without any continuous external influence such as added energy. Two factors determine whether a reaction is spontaneous or nonspontaneous:

1. Reactions tend to be spontaneous if the products have lower potential energy than the reactants. If the electrons in the reaction products are held more tightly than the electrons in the reactants, then they have lower potential energy. Recall that highly electronegative atoms such as oxygen and nitrogen hold electrons much more tightly than do atoms with a lower electronegativity, such as carbon and hydrogen. For example, when natural gas burns, methane reacts with oxygen gas to produce carbon dioxide and water:

$$CH_4(g) + 2 O_2(g) \longrightarrow CO_2(g) + 2 H_2O(g)$$

The electrons involved in the C=O and O−H bonds of carbon dioxide and water are held much more tightly than they were in the C−H and O=O bonds of methane and oxygen (**Figure 2.19**). As a result, the products have much lower potential energy than the reactants. The difference in potential energy between reactants and products is given off as heat, so the reaction is exothermic. In chemical reactions, the difference in potential energy between the products and the reactants is symbolized by ΔH. (The uppercase Greek letter Δ delta, is often used in chemical and mathematical notation to represent change.) When a reaction is exothermic, ΔH is negative.

2. Reactions tend to be spontaneous when the product molecules are less ordered than the reactant molecules. TNT, or dynamite, has a highly ordered chemical structure. But when TNT explodes, gases like carbon dioxide, carbon monoxide, various nitrogen oxides, and small particulates are given off (**Figure 2.20**). These molecules are much less ordered than the reactant molecules in TNT. The amount of disorder in a group of molecules is called its **entropy**, which is symbolized by S. When the products of a chemical reaction are less ordered than the reactant molecules are, entropy increases and ΔS is positive. Reactions tend to be spontaneous if they increase entropy. The **second law of thermodynamics**, in fact, states that entropy always increases in an isolated system.

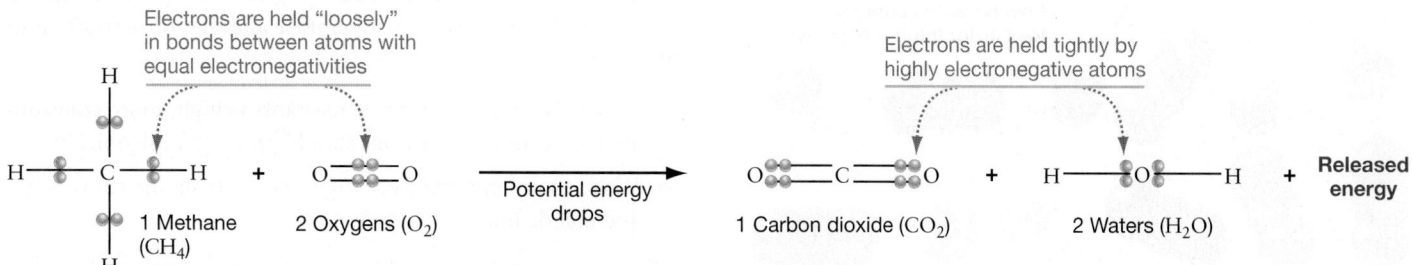

FIGURE 2.19 Potential Energy and/or Entropy May Change during Chemical Reactions. When methane burns, the products have much lower potential energy than the reactants.

✔ **EXERCISE** Label which electrons have relatively low potential energy and which electrons have relatively high potential energy.

FIGURE 2.20 Entropy May Change during Chemical Reactions. When TNT explodes, it produces carbon dioxide, water vapor, smoke, and other compounds that are much less ordered than the original system. The reaction results in an increase in entropy.

In general, physical and chemical processes proceed in the direction that results in lower potential energy and increased disorder (**Figure 2.21**). If potential energy drops, then ΔH is negative; if entropy increases, then ΔS is positive.

In the case of exploding TNT, the reaction is exothermic *and* results in higher entropy—less-ordered products. However, an

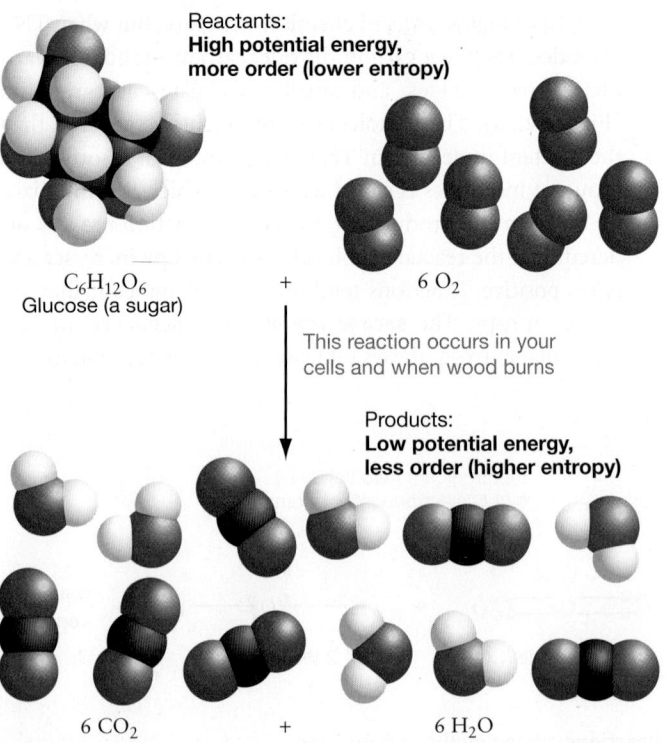

Reactants:
High potential energy, more order (lower entropy)

$C_6H_{12}O_6$
Glucose (a sugar)

$+$

$6\,O_2$

This reaction occurs in your cells and when wood burns

Products:
Low potential energy, less order (higher entropy)

$6\,CO_2$ $+$ $6\,H_2O$

FIGURE 2.21 Spontaneous Processes Result in Lower Potential Energy, Increased Disorder, or Both.

increase in entropy does not always accompany a drop in potential energy. When methane burns, for example, ΔH is negative but ΔS is essentially 0.

To determine whether a chemical reaction is spontaneous, it's necessary to assess the combined contributions of changes in heat and disorder. Chemists do this with a quantity called the **Gibbs free-energy change**, symbolized by ΔG:

$$\Delta G = \Delta H - T\Delta S$$

Here, T stands for temperature measured on the Kelvin scale (see **BioSkills 1**, in Appendix A). Water freezes at 273.15 K and boils at 373.15 K.

In words, the free-energy change in a reaction is equal to the change in potential energy minus the change in entropy multiplied by the temperature. The $T\Delta S$ term simply means that entropy becomes more important in determining free-energy change as the temperature of the molecules increases. The faster molecules are moving, the more important entropy becomes in determining the overall free-energy change.

Chemical reactions are spontaneous when ΔG is less than zero. Such reactions are said to be **exergonic**. Reactions are nonspontaneous when ΔG is greater than zero. Such reactions are termed **endergonic**. When ΔG is zero, reactions are at equilibrium. ✔ If you understand these concepts, you should be able to explain (1) why the same reaction can be nonspontaneous at low temperature but spontaneous at high temperature, and (2) why some exothermic reactions are nonspontaneous.

Free energy changes when the potential energy and/or entropy of substances changes. Spontaneous chemical reactions run in the direction that lowers the free energy of the system. Exergonic reactions are spontaneous and release energy; endergonic reactions are nonspontaneous and require an input of energy to proceed.

THE ROLES OF TEMPERATURE AND CONCENTRATION IN CHEMICAL REACTIONS Even if a chemical reaction occurs spontaneously, it may not happen quickly. The reactions that convert iron to rust or sugar molecules to carbon dioxide and water are spontaneous, but at room temperature they occur very slowly, if at all. For most reactions to proceed, one chemical bond has to break and another one has to form. For this to happen, the substances involved must collide in a specific orientation that brings the electrons involved near each other.

The number of collisions occurring between the substances in a mixture depends on the temperature and the concentrations of the reactants:

- When the concentration of reactants is high, more collisions should occur and reactions should proceed more quickly.

- When their temperature is high, reactants should move faster and collide more frequently.

Higher concentrations and higher temperatures should tend to speed up chemical reactions.

Figure 2.22 provides data from a set of experiments performed by students from Parkland College in Champaign, Illi-

QUESTION: Do chemical reaction rates increase with increased temperature and concentration?

RATE INCREASE HYPOTHESIS: Chemical reaction rates increase with increased temperature. They also increase with increased concentration of reactants.

NULL HYPOTHESIS: Chemical reaction rates are not affected by increases in temperature or concentration of reactants.

EXPERIMENTAL SETUP:

Experimental reaction: $3\,HSO_3^-\,(aq)\ +\ IO_3^-\,(aq) \rightleftharpoons 3\,HSO_4^-\,(aq)\ +\ I^-\,(aq)$

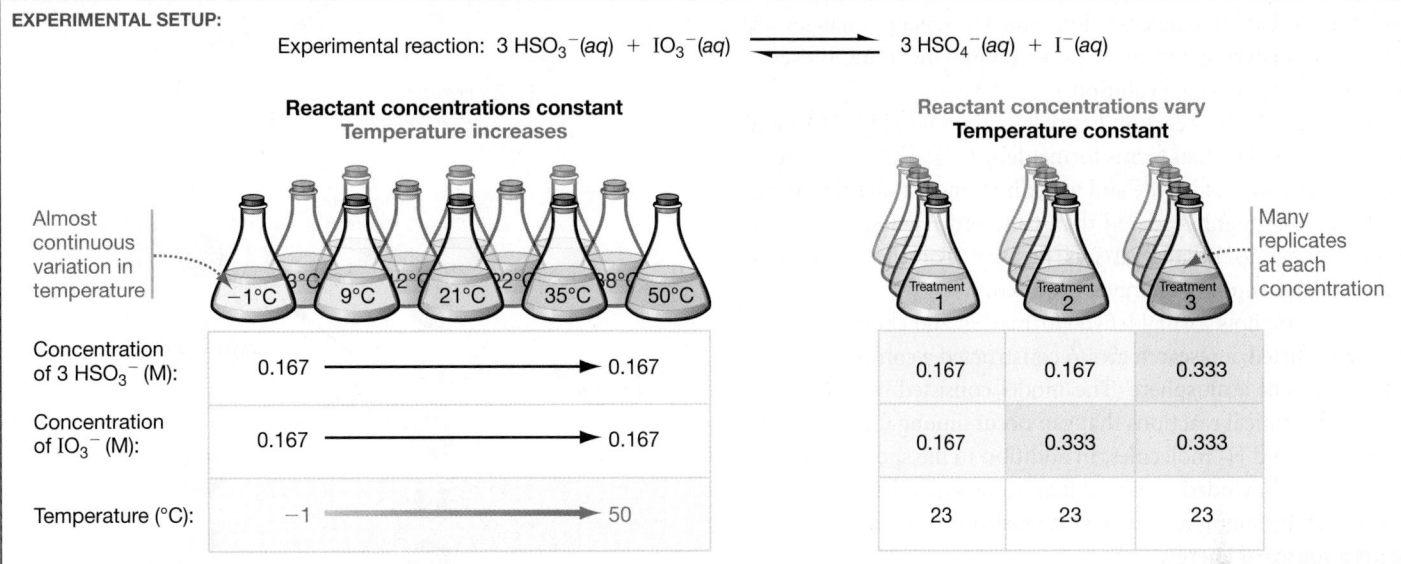

PREDICTION: Reaction rate, measured as 1/(time for reaction to go to completion), will increase with increased concentrations of reactants and increased temperature of reaction mix.

PREDICTION OF NULL HYPOTHESIS: No difference in reaction rates among treatments in each setup.

RESULTS:

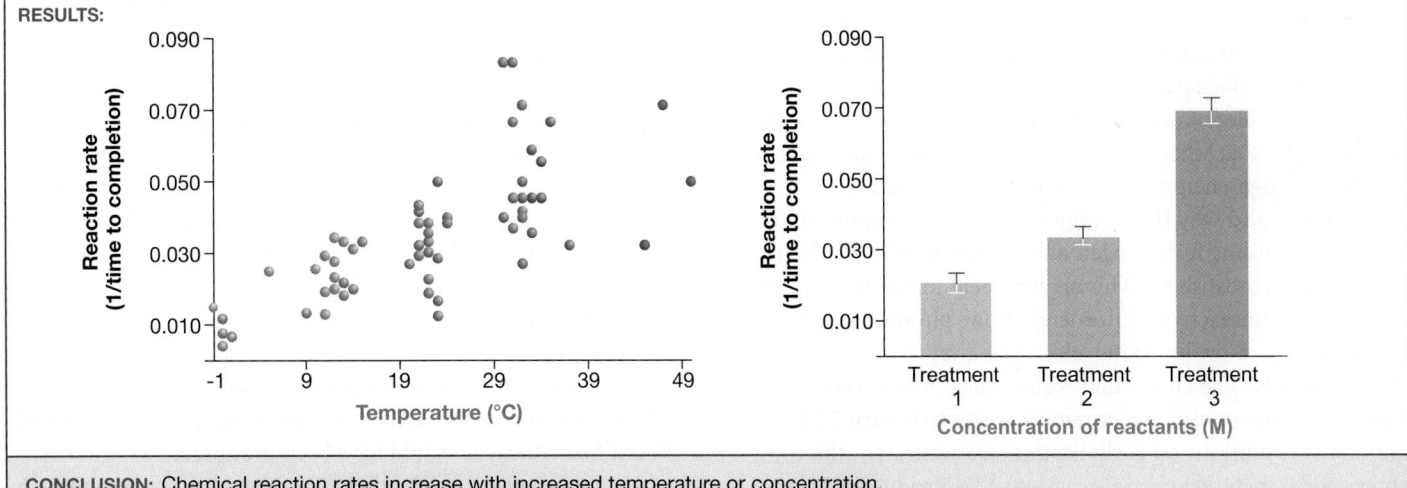

CONCLUSION: Chemical reaction rates increase with increased temperature or concentration.

FIGURE 2.22 Testing the Hypothesis That Reaction Rates Are Sensitive to Changes in Temperature and Concentration.

✔ **QUESTION** In the graph on the right side of the "Results" section, what do the heights of the orange bars

nois, to test these predictions. Pay special attention to the two graphs in the "Results" section:

- *Temperature versus reaction rate* The graph on the left is based on experiments where the concentration of the reactants was the same, but the temperature varied. Each data point represents one experiment. Note that the points rise from left to

right—meaning, in this case, that the reaction rate speeded up when the temperature of the reaction mixture was higher.

- *Concentration versus reaction rate* The graph on the right is based on experiments where the temperature was constant, but the concentration of reactants varied. Each bar represents the average reaction rate over many replicates of each treatment, or set of concentrations. The thin lines at the top of each bar indicate the standard error of the average—a measure of variability (see **BioSkills 2** in Appendix A). The take-home message of this graph is that reaction rates are higher when substrate concentrations are higher.

ENERGY INPUTS AND THE START OF CHEMICAL EVOLUTION Temperature, substrate concentration, and free energy changes are critical to analyzing any chemical reaction—including those responsible for chemical evolution.

For example, the reaction between carbon dioxide (CO_2) and hydrogen gas (H_2) that forms formaldehyde (H_2CO) and water is endergonic. Formaldehyde and water have more potential energy and are more highly ordered than CO_2 and H_2. The reaction is nonspontaneous because ΔG is positive. For the reaction to occur, a large input of energy is required.

To explore how formaldehyde formation and chemical evolution got started, a research group constructed a computer model of the ancient atmosphere. The model consisted of a list of all possible chemical reactions that can occur among CO_2, H_2O, N_2, NH_3, CH_4, and H_2 molecules. In addition to the spontaneous reactions, they included reactions that occur when these molecules are struck by sunlight. This was crucial because sunlight represents a source of energy.

The sunlight that strikes Earth is made up of packets of light energy called **photons**. The amount of light energy contained in a photon can vary widely. Today, most of the higher-energy photons in sunlight never reach Earth's lower atmosphere. Instead, they are absorbed by a molecule called ozone (O_3) in the upper atmosphere. But if Earth's early atmosphere was filled with volcanic gases released as the molten planet cooled, it is extremely unlikely that appreciable quantities of ozone existed. On the basis of this logic, researchers infer that when chemical evolution was occurring, large quantities of high-energy photons bombarded the planet.

To understand why this energy source was so important, recall that the atoms in hydrogen and carbon dioxide molecules have full outermost shells. This arrangement makes these molecules largely unreactive. But energy from photons can break molecules apart by knocking electrons away from the outer shells of atoms. The atoms that result, called **free radicals**, have unpaired electrons and are extremely reactive (**Figure 2.23**). To mimic the conditions on early Earth more accurately, the computer model included several reactions that produce highly reactive free radicals.

To understand which of the long list of possible reactions would actually occur, and to estimate how much formaldehyde could have been produced in the ancient atmosphere, the researchers needed to consider the effects of two additional factors: temperature and concentration.

THE FIRST REACTIONS IN CHEMICAL EVOLUTION To model the behavior of simple molecules in the ancient atmosphere, the researchers working on the formaldehyde synthesis reaction needed to specify temperature and the concentration of each

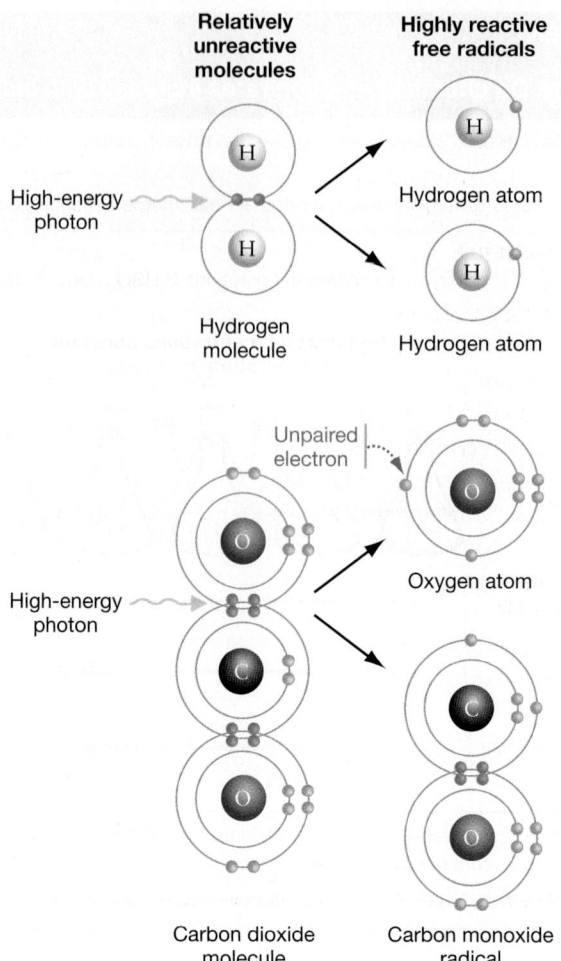

FIGURE 2.23 Free Radicals Are Extremely Reactive. When a high-energy photon strikes a hydrogen or carbon dioxide molecule, free radicals can be created. Formation of free radicals is thought to be responsible for some key reactions in chemical evolution.

molecule. In experiments like those in Figure 2.22, they measured the actual reaction rates observed at controlled temperatures and concentrations. They were then able to assign a rate to each of the reactions listed in their model.

Their result? They calculated that, under temperature and concentration conditions accepted as reasonable approximations of early Earth conditions by most scientists, appreciable quantities of formaldehyde would have been produced.

Using a similar model, other researchers have shown that significant amounts of hydrogen cyanide (HCN) could also have been produced in the ancient atmosphere. According to this research, large quantities of the critical intermediates in chemical evolution would have formed in the ancient atmosphere.

Once formaldehyde and hydrogen cyanide formed in the atmosphere, they would have rained out into the ocean. Similar reactions occur at deep-sea vents, where extremely hot water provides energy to drive endergonic reactions. As a result, organic compounds with relatively high potential energy were bubbling up from below. Once these types of compounds began to accumulate, the groundwork was in place for chemical evolution to take off.

How Did Chemical Energy Change during Chemical Evolution?

The initial products of chemical evolution are important, for a simple reason: They have more potential energy than the reactant molecules have. When formaldehyde is produced, an increase in potential energy occurs because the electrons that bond CO_2 and H_2 together are held more tightly than they are in H_2CO or H_2O. This form of potential energy—the potential energy stored in chemical bonds—is called **chemical energy**.

This observation gets right to the heart of chemical evolution: The energy in sunlight was converted to chemical energy—potential energy in chemical bonds. This energy transformation explains how chemical evolution was possible. When small, simple molecules absorb energy, chemical reactions can occur that transform the external energy into potential energy stored in chemical bonds. More specifically, the energy in sunlight was converted to chemical energy in the form of formaldehyde and hydrogen cyanide.

The complete reaction that results in the formation of formaldehyde is written as:

$$CO_2(g) + 2\,H_2(g) + \text{sunlight} \longrightarrow H_2CO(g) + H_2O(g)$$

Notice that the reaction is balanced in terms of the atoms *and* the energy involved. The sunlight on the left side of the expression balances the higher chemical energy in formaldehyde and water.

This result makes sense intuitively. Energy is the capacity to do work, and it seems logical that building larger, more complex molecules requires work to be done. Here's another way to say the same thing: The reactions involved in chemical evolution are endergonic, so inputs of energy were required. Now the question is, what happened to these first building blocks of chemical evolution?

2.4 The Importance of Carbon

Life has been called a carbon-based phenomenon, and with good reason. Except for water, almost all of the molecules found in organisms contain this atom. Molecules that contain carbon are called **organic** molecules. (Other types of molecules are referred to as *inorganic* compounds.)

Carbon has great importance in biology because it is the most versatile atom on Earth. Because of its four valence electrons, it can form many covalent bonds. With different combinations of single and double bonds, an almost limitless array of molecular shapes are possible.

You have already examined the tetrahedral structure of methane and the linear shape of carbon dioxide. When molecules contain more than one carbon atom, these atoms can be bonded to one another in long chains, as in the component of the gasoline called octane (C_8H_{18}; **Figure 2.24a**), or in a ring, as in the sugar glucose ($C_6H_{12}O_6$; **Figure 2.24b**). Carbon atoms

(a) Carbons linked in a linear molecule

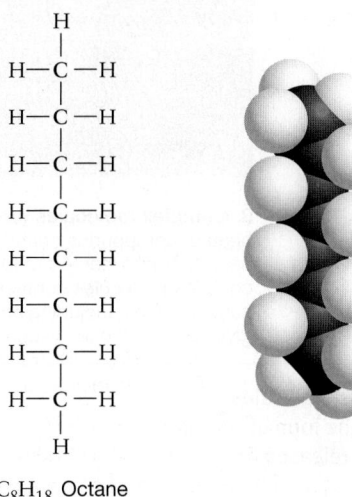

C_8H_{18} Octane

(b) Carbons linked in a ring

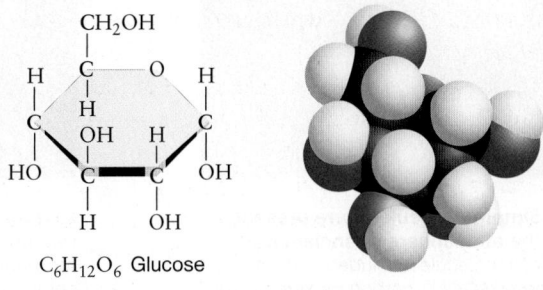

$C_6H_{12}O_6$ Glucose

FIGURE 2.24 The Shapes of Carbon-Containing Molecules. (a) Octane is a linear molecule, and one of the primary ingredients in gasoline. **(b)** Glucose is a sugar that can form the ring-like structure.

provide the structural framework for virtually all the important compounds associated with life.

Linking Carbon Atoms Together

The formation of carbon-carbon bonds was an important event in chemical evolution: It represented a crucial step toward the production of the types of molecules found in living organisms.

Once organic compounds such as formaldehyde and hydrogen cyanide had formed, continued chemical evolution could occur by the addition of heat alone. For example, when molecules of formaldehyde are heated, they react with one another to form a molecule called acetaldehyde. Acetaldehyde contains a carbon-carbon bond. With continued heating, reactions between formaldehyde and acetaldehyde molecules can produce the larger organic compounds called sugars.

Figure 2.25 summarizes these steps: the formation of building block compounds like formaldehyde and HCN, followed by reactions that resulted in small molecules with carbon-carbon bonds—including molecules that are found in organisms living today.

Functional Groups

In general, the carbon atoms in an organic molecule furnish a skeleton that gives the molecule its overall shape. But the chemical behavior of the compound—meaning the types of reactions that it participates in—is dictated by groups of H, N, or O atoms that are bonded to one of the carbon atoms in a specific way.

The critically important H-, N-, and O-containing groups found in organic compounds are called **functional groups**. The composition and properties of six prominent functional groups recognized by organic chemists are summarized in **Table 2.3**. To understand the role that organic compounds play in organisms, it is important to analyze how these functional groups behave.

- In solution, the *amino and carboxyl functional groups* tend to attract or drop a proton, respectively. Amino groups function as bases, while carboxyl groups act as acids. During chemical evolution and in organisms today, the most important types of amino- and carboxyl-containing molecules are the amino acids, which Chapter 3 analyzes in detail. Amino acids contain both an amino group and a carboxyl group. (It's common for organic compounds to contain more than one functional group.) Amino acids can be linked together by covalent bonds that form between amino- and carboxyl-containing groups. In addition, both of these functional groups participate in hydrogen-bonding.

- The *carbonyl group* is found on aldehyde and ketone molecules such as formaldehyde, acetaldehyde, and acetone. This functional group is the site of reactions that link these molecules into larger, more complex compounds, like the sugar ribose illustrated in Figure 2.25.

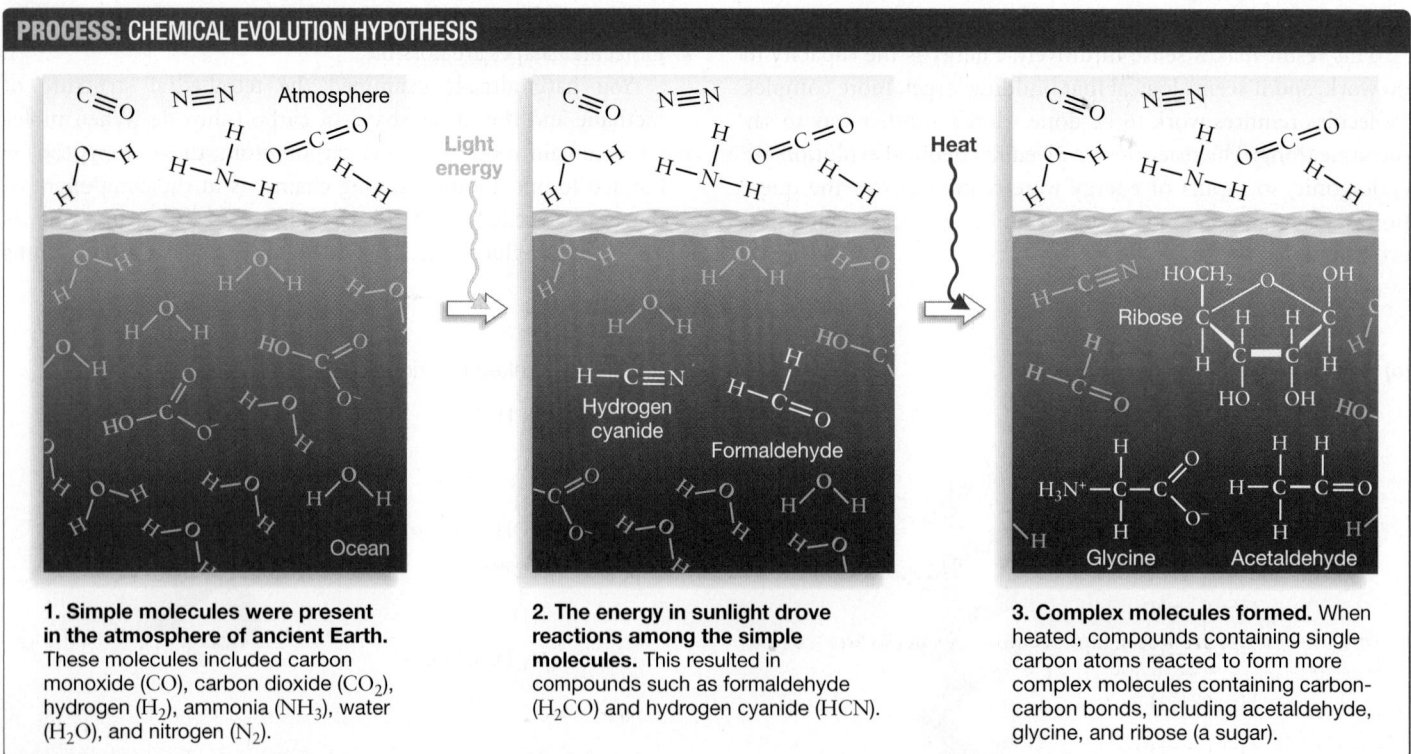

PROCESS: CHEMICAL EVOLUTION HYPOTHESIS

1. Simple molecules were present in the atmosphere of ancient Earth. These molecules included carbon monoxide (CO), carbon dioxide (CO_2), hydrogen (H_2), ammonia (NH_3), water (H_2O), and nitrogen (N_2).

2. The energy in sunlight drove reactions among the simple molecules. This resulted in compounds such as formaldehyde (H_2CO) and hydrogen cyanide (HCN).

3. Complex molecules formed. When heated, compounds containing single carbon atoms reacted to form more complex molecules containing carbon-carbon bonds, including acetaldehyde, glycine, and ribose (a sugar).

FIGURE 2.25 The Start of Chemical Evolution—An Overview. During chemical evolution, simple molecules containing C, H, O, and N reacted to form organic compounds with higher potential energy in the form of carbon-carbon bonds. The process was triggered by an energy source such as sunlight, or by the heat released from a volcanic eruption.

Functional Group	*Formula	Family of Molecules	Properties of Functional Group	Example
Amino NH_2	$H\!-\!N\!-\!R$ with H	Amines	Acts as a base—tends to attract a proton to form: $H\!-\!{}^+N\!-\!R$	Glycine (an amino acid)
Carbonyl	$R\!-\!C(\!=\!O)\!-\!H$	Aldehydes	Aldehydes, especially, react with certain compounds to produce larger molecules with form: R group from aldehyde → $R\!-\!C(OH)\!-\!H$ ← R group from another reactant	Acetaldehyde
	$R\!-\!C(\!=\!O)\!-\!R$	Ketones		Acetone
Carboxyl $COOH$	$R\!-\!C(\!=\!O)\!-\!OH$	Carboxylic acids	Acts as an acid—tends to lose a proton in solution to form: $R\!-\!C(\!=\!O)\!-\!O^-$	Acetic acid
Hydroxyl OH	$R\!-\!OH$	Alcohols	Highly polar, so makes compounds more soluble through hydrogen bonding with water; may also act as a weak acid and drop a proton	Ethanol
Phosphate PO_4	$R\!-\!O\!-\!P(\!=\!O)(O^-)\!-\!O^-$	Organic phosphates	When several phosphate groups are linked together, breaking O–P bonds between them releases large amounts of energy	3–Phosphoglyceric acid
Sulfhydryl SH	$R\!-\!SH$	Thiols	When present in proteins, can form disulfide (S–S) bonds that contribute to protein structure	Cysteine

*In these structural formulas, "R" stands for the rest of the molecule.

✔ **EXERCISE** On the basis of the electronegativities of the atoms involved, predict whether each functional group is polar or nonpolar.

- *Hydroxyl groups* are important because they act as weak acids. In many cases, the protons involved in acid–base reactions that occur in cells come from hydroxyl groups on organic compounds. Because hydroxyl groups are polar, molecules containing a number of hydroxyl groups will form hydrogen bonds and be highly soluble in water.

- *Phosphate groups* carry two negative charges. When phosphate groups are transferred from one organic compound to another, the change in charge often dramatically affects the recipient molecule.

- *Sulfhydryl groups* consist of a sulfur atom bonded to a hydrogen atom. They are important because sulfhydryl groups on different molecules can link together via disulfide (S—S) bonds.

To summarize, functional groups make things happen. The number and types of functional groups attached to a framework of carbon atoms imply a great deal about how that molecule is going to behave.

When you encounter an organic compound that is new to you, it's important to do two things: Examine its overall size and shape, and locate any functional groups. Understanding these two features will help you understand the molecule's role in chemical evolution and today's cells.

Once carbon-containing molecules with functional groups had appeared early in Earth's history, what happened next? For chemical evolution to continue, two things had to happen. First, reactions between relatively small and simple organic compounds had to produce the building blocks of the large molecules found in living cells. Second, these building blocks had to link together to form the large, complex compounds found in organisms—proteins, nucleic acids, and carbohydrates.

The next three chapters focus on how these events occurred and how proteins, nucleic acids, and carbohydrates function in organisms today. The atoms that make up your body were present in the ocean before life began, some 3.5 billion years ago.

CHAPTER 2 REVIEW

For media, go to the study area at www.masteringbiology.com

Summary of Key Concepts

🔑 **Molecules form when atoms bond to each other. Chemical bonds are based on electron sharing. The degree of electron sharing varies from nonpolar covalent bonds, to polar covalent bonds, to ionic bonds.**

- When atoms participate in chemical bonds to form molecules, the shared electrons give the atoms full valence shells and thus contribute to the atoms' stability.

- The electrons in chemical bonds may be shared equally or unequally, depending on the relative electronegativities of the two atoms involved.

- Nonpolar covalent bonds result from equal sharing; polar covalent bonds are due to unequal sharing. Ionic bonds form when an electron is completely transferred from one atom to another.

  ✔You should be able to draw the electron-sharing continuum and place molecular oxygen (O_2), carbon dioxide (CO_2), and calcium chloride ($CaCl_2$) on it.

🔑 **Water is essential for life. Water is highly polar and readily forms hydrogen bonds. Hydrogen bonding makes water an extremely efficient solvent.**

- The chemical reactions required for life take place in water.

- Water is polar—meaning that it has partial positive and negative charges—because it is bent and has two polar covalent bonds.

- Polar molecules and charged substances, including ions, interact with water and stay in solution via hydrogen bonding and electrostatic attraction.

- Water's ability to participate in hydrogen bonding also gives it an extraordinarily high capacity to absorb heat and cohere to other water molecules.

  ✔You should be able to explain how water interacts with amino, carboxyl, and hydroxyl functional groups in solution.

  🔵 **Web Activity** The Properties of Water

🔑 **Energy is the capacity to do work or supply heat, and can be (1) a stored potential or (2) an active motion. Chemical energy is a form of potential energy, stored in chemical bonds.**

- Energy comes in different forms. Although energy cannot be created or destroyed, one form of energy can be transformed into another.

- Experiments suggest that early in Earth's history, the energy in sunlight and hot water drove chemical reactions between simple molecules, resulting in the formation of more complex compounds with higher potential energy. In this way, energy in the form of sunlight or heat was transformed into chemical energy.

  ✔You should be able to explain how the heat released when methane burns represents an energy transformation.

🔑 **Chemical reactions tend to be spontaneous if they lower potential energy and increase entropy (disorder). An input of energy is required for nonspontaneous reactions to occur.**

- Spontaneous reactions do not require an input of energy to occur.

- The Gibbs free energy change, ΔG, summarizes the combined effects of changes in energy and entropy during a chemical reaction.

- Spontaneous reactions have a negative ΔG and are said to be exergonic; nonspontaneous reactions have a positive ΔG and are said to be endergonic.

  ✔You should be able to explain why cells cannot stay alive without inputs of energy.

🔑 **Most of the important compounds in organisms contain carbon. Key carbon-containing molecules formed early in Earth's history.**

- Organic molecules are critical to life because they have complex shapes provided by a framework of carbon atoms, along with complex chemical behavior due to the presence of functional groups.

- The first step in chemical evolution was the formation of the small organic compounds called formaldehyde and hydrogen cyanide from molecules such as ammonia (NH_3), methane (CH_4), molecular hydrogen (H_2), and carbon dioxide (CO_2). These reactions occur readily when a source of intense energy, such as the radiation in sunlight, is present.

  ✔You should be able to explain why molecules with carbon-carbon bonds have more potential energy and lower entropy than carbon dioxide.

Questions

1. Which of the following occurs when a covalent bond forms?
   a. The potential energy of electrons drops.
   b. Electrons in valence shells are shared between nuclei.
   c. Ions of opposite charge interact.
   d. Polar molecules interact.

2. If a reaction is exothermic, then which of the following statements is true?
   a. The products have lower potential energy than the reactants.
   b. Energy must be added for the reaction to proceed.
   c. The products have lower entropy (are more ordered) than the reactants.
   d. It occurs extremely quickly.

3. If a reaction is exergonic, then which of the following statements is true?
   a. The products have lower free energy than the reactants.
   b. Energy must be added for the reaction to proceed.
   c. The products have lower entropy (are more ordered) than the reactants.
   d. It occurs extremely quickly.

4. What is thermal energy?
   a. a form of potential energy
   b. the temperature increase that occurs when any form of energy is added to a system

   c. mechanical energy
   d. the kinetic energy of molecular motion, measured as heat

5. What determines whether a chemical reaction is spontaneous?
   a. if it increases the disorder, or entropy, of the substances involved
   b. if it decreases the potential energy of the substances involved
   c. the temperature only—reactions are spontaneous at high temperatures and nonspontaneous at low temperatures
   d. the combined effect of changes in potential energy and entropy

6. Which of the following is *not* an example of an energy transformation?
   a. A shoe drops, converting potential energy to kinetic energy.
   b. A chemical reaction converts the energy in sunlight into the chemical energy in formaldehyde.
   c. The electrical energy flowing through a lightbulb's filament is converted into light and heat.
   d. Sunlight strikes a prism and separates into distinct wavelengths.

1. Consider the reaction between carbon dioxide and water, which forms carbonic acid:

$$CO_2(g) + H_2O(l) \rightleftharpoons H_2CO_3(aq)$$

   In aqueous solution, carbonic acid immediately dissociates to form a proton and the bicarbonate ion, as follows:

$$H_2CO_3(aq) \rightleftharpoons H^+(aq) + HCO_3^-(aq)$$

   Does this second reaction raise or lower the pH of the solution? If an underwater volcano bubbled additional CO_2 into the ocean, would this sequence of reactions be driven to the left or the right? How would this affect the pH of the ocean?

2. When chemistry texts introduce the concept of electron shells, they emphasize that shells represent distinct potential energy levels. In introducing electron shells, this chapter also emphasized that they represent distinct distances from the positive charges in the nucleus. Are these two points of view in conflict? Why or why not?

3. Draw a ball-and-stick model of the water molecule, and explain why this molecule is bent. Indicate the location of the partial electric charges on it. Why do these partial charges exist?

4. Hydrogen bonds form because the opposite, partial electric charges on polar molecules attract. Covalent bonds form as a result of the electrical attraction between electrons and protons. Covalent bonds are much stronger than hydrogen bonds. Explain why, in terms of the electrical attractions involved.

5. Explain why extensive hydrogen bonding gives water an extraordinarily high specific heat.

6. Explain the relationship between the carbon framework in an organic molecule (the "R" in Table 2.3) and the functional groups on the same molecule.

1. Why isn't CO_2 bent and polar, like H_2O? Why is H_2O much more likely to participate in chemical reactions than CO_2?

2. Oxygen is extremely electronegative, meaning that its nucleus pulls in electrons shared in covalent bonds. Explain the changes in electron position that are illustrated in Figure 2.19 based on oxygen's electronegativity.

3. When nuclear fusion reactions take place, some of the mass in the atoms involved is converted to energy. The energy in sunlight is created during nuclear fusion reactions on the Sun. Explain what astronomers mean when they say that the Sun is burning down and that it will eventually burn out.

4. Why do coastal regions tend to have climates with moderate temperatures and lower annual variation in temperature than do inland areas at the same latitude?

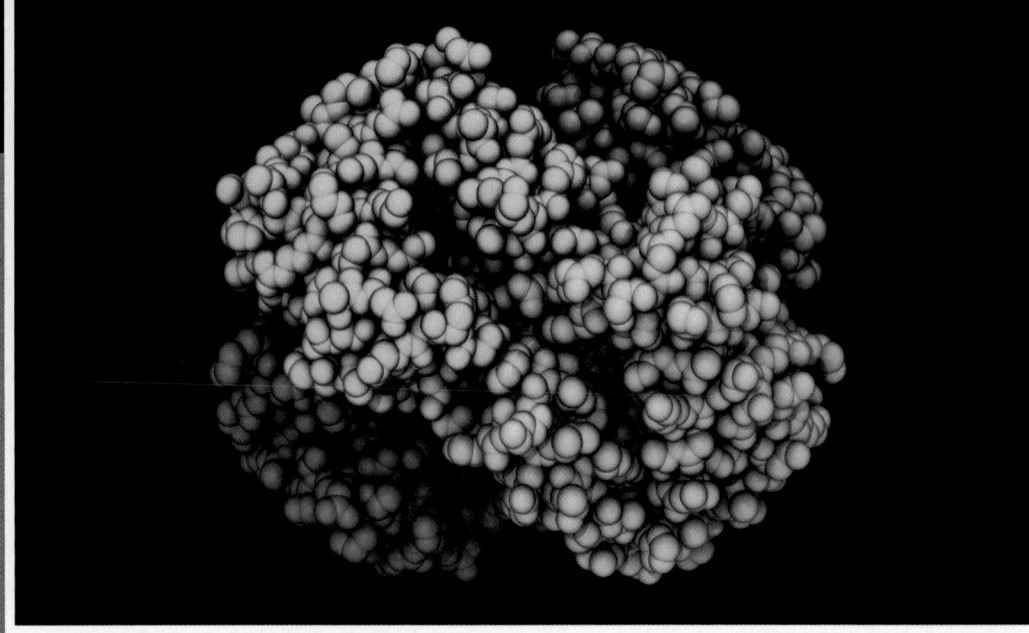

A space-filling model of hemoglobin—a protein that is carrying oxygen in your blood right now.

3 Protein Structure and Function

KEY CONCEPTS

- Most cell functions depend on proteins.

- Amino acids are the building blocks of proteins. Amino acids vary in structure and function because their side chains vary in composition.

- Proteins vary widely in structure. The structure of a protein can be analyzed at four levels that form a hierarchy—the amino acid sequence, substructures called α-helices and β-pleated sheets, interactions between amino acids that dictate a protein's overall shape, and combinations of individual proteins that make up larger, multiunit molecules.

- In cells, most proteins are enzymes that function as catalysts. Chemical reactions occur much faster when they are catalyzed by enzymes. During enzyme catalysis, the reactants bind to an enzyme's active site in a way that allows the reaction to proceed efficiently.

Chapter 2 introduced the hypothesis that chemical reactions in the atmosphere and ocean of ancient Earth led to the formation of the first complex carbon-containing compounds. This idea, called chemical evolution, was first proposed in 1923 by Alexander I. Oparin. The hypothesis was published again—independently and six years later—by J. B. S. Haldane.

Today, the Oparin-Haldane proposal can best be understood as a formal scientific theory. As Chapter 1 noted, scientific theories typically have two components: a statement about a pattern that exists in the natural world and a proposed mechanism or process that explains the pattern.

In the case of chemical evolution, the pattern is that increasingly complex carbon-containing molecules formed in the atmosphere and ocean of ancient Earth. The process responsible for this pattern was the conversion of energy, from sunlight and other sources, into chemical energy in the bonds of large, complex molecules.

Scientific theories are continuously refined as new information comes to light, and many of Oparin and Haldane's original ideas have been revised. In its current form, the theory can be broken into four steps, each requiring an input of energy:

1. Chemical evolution began with the production of small organic compounds such as formaldehyde (H_2CO) and hydrogen cyanide (HCN), from reactants such as H_2, CO_2, CH_4, and NH_3. Chapter 2 focused on this step.

2. Formaldehyde, hydrogen cyanide, and other simple organic compounds reacted to form the mid-sized molecules called amino acids, nitrogenous bases, and sugars. Amino acids are introduced in this chapter, nitrogenous bases are analyzed in Chapter 4, and sugars are discussed in Chapter 5. According to the Oparin-Haldane theory these molecules accumulated in the shallow waters of the ancient ocean, forming a complex solution called the **prebiotic soup.**

✔ When you see this checkmark, stop and test yourself. Answers are available in Appendix B.

3. Mid-sized, building-block molecules linked to form the types of large molecules found in cells today, including proteins, nucleic acids, and complex carbohydrates. Each of these large molecules is made up of distinctive chemical subunits that join together: Proteins are composed of amino acids, nucleic acids are composed of nucleotides, and complex carbohydrates are composed of sugars. Proteins, nucleic acids, and carbohydrates are featured in Chapters 3–5, respectively.

4. Life became possible when one of these large, complex molecules made a copy of itself. This self-replicating molecule began to multiply by means of chemical reactions that it controlled. At that point, life had begun. Chemical evolution gave way to biological evolution.

Which type of molecule was responsible for the origin of life? Answering this question is a recurring theme in this and the subsequent three chapters.

Researchers in several labs are pushing proteins as the candidate for the molecule that was the initial spark of life. Why?

3.1 Early Origin-of-Life Experiments

In 1953 a graduate student named Stanley Miller performed a breakthrough experiment in the study of chemical evolution. Miller wanted to answer a simple question: Can complex organic compounds be synthesized from the simple molecules present in Earth's early atmosphere and ocean? In other words, is it possible to re-create the first steps in chemical evolution by simulating ancient Earth conditions in the laboratory?

Miller's experimental setup (**Figure 3.1**) was designed to produce a microcosm of ancient Earth. The large glass flask represented the atmosphere and contained the gases methane (CH_4), ammonia (NH_3), and hydrogen (H_2), all of which have high free energy. This large flask was connected to a smaller flask by glass tubing. The small flask held a tiny ocean—200 milliliters (mL) of liquid water.

To connect the mini-atmosphere with the mini-ocean, Miller boiled the water constantly. This added water vapor to the mix of gases in the large flask. As the vapor cooled and condensed, it flowed back into the smaller flask, where it boiled again. In this way, water vapor circulated continuously through the system. This was important: If the molecules in the simulated atmosphere reacted with one another, the "rain" would carry them into the simulated ocean, forming a simulated version of the prebiotic soup.

Had Miller stopped at merely boiling the molecules, little or nothing would have happened. Even at the boiling point of water (100°C), the starting molecules used in the experiment are stable. They do not undergo spontaneous chemical reactions, even at high temperatures.

Something did start to happen in the apparatus, however, when Miller sent electrical discharges across the electrodes he'd inserted into the atmosphere. These miniature lightning bolts added a crucial element to the reaction mix—pulses of intense electrical energy. After a day of continuous boiling and sparking,

EXPERIMENT

QUESTION: Can simple molecules and kinetic energy lead to chemical evolution?

HYPOTHESIS: If kinetic energy is added to a mix of simple molecules with high free energy, reactions will occur that produce more complex molecules, perhaps including some with C-C bonds.

NULL HYPOTHESIS: Chemical evolution will not occur, even with an input of energy.

EXPERIMENTAL SETUP:

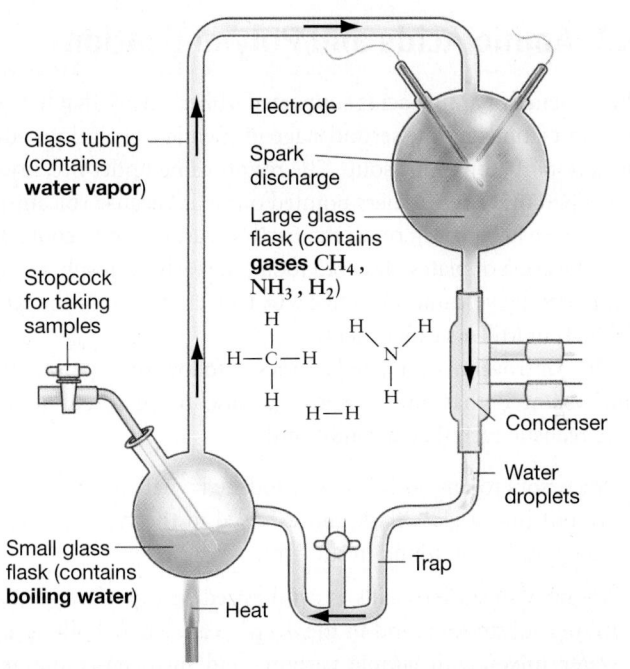

PREDICTION: Complex organic compounds will be found in the liquid water.

PREDICTION OF NULL HYPOTHESIS: Only the starting molecules will be found in the liquid water.

RESULTS

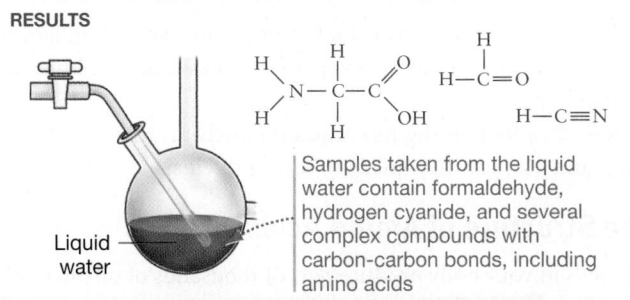

Samples taken from the liquid water contain formaldehyde, hydrogen cyanide, and several complex compounds with carbon-carbon bonds, including amino acids

CONCLUSION: Chemical evolution occurs readily if simple molecules with high free energy are exposed to a source of kinetic energy.

FIGURE 3.1 Miller's Spark-Discharge Experiment. The arrows in the "Experimental setup" diagram indicate the flow of water vapor or liquid. The condenser is a jacket with cold water flowing through it.

SOURCE: Miller, S. L. 1953. A production of amino acids under possible primitive Earth conditions. *Science* 117: 528–529.

✔**QUESTION** Which parts of the apparatus mimic the ocean, atmosphere, rain, and lightning?

the solution in the boiling flask began to turn pink. After a week, it was deep red and cloudy.

When Miller analyzed samples from the mini-ocean, he found large quantities of hydrogen cyanide (HCN) and formaldehyde (H_2CO). Chapter 2 claimed that these were key compounds in chemical evolution, because they are required for reactions that lead to the synthesis of more complex organic molecules. Indeed, some of these more complex compounds were actually present in the miniature ocean. The sparks and heating had led to the synthesis of compounds that are the building blocks of proteins: amino acids.

3.2 Amino Acids and Polymerization

The presence of amino acids supported Miller's claim that his experiment simulated the second stage in chemical evolution—the formation of a prebiotic soup. The results came under fire, however, when other researchers pointed out that because volcanism would have been common as the Earth's crust began to cool and form the crust of plates observed today, the early atmosphere was dominated by volcanic gases like CO, CO_2, and H_2, not the CH_4 and NH_3 in Miller's experiment.

The controversy stimulated a series of follow-up experiments, which showed that amino acids can also be produced under more realistic early-Earth conditions:

- Precursors to amino acids are produced when volcanic gases are put into a glass flask and exposed to the types of high-energy radiation found in sunlight.

- An array of amino acids is synthesized in experiments that mimic volcanoes found in the deep ocean, where boiling-hot water mixes with simple carbon- and nitrogen-containing molecules and metal ions.

- Several of the meteorites that have struck Earth recently contain amino acids—indicating that these molecules can be produced in outer space.

The current consensus is that amino acids were abundant in the early oceans—raining down from above and bubbling up from below.

Now let's look at the molecules themselves. What are amino acids, and how are they linked to form proteins?

The Structure of Amino Acids

The cells in your body produce tens of thousands of distinct proteins. Most of these molecules are composed of just 20 different building blocks, called **amino acids**. All 20 amino acids have a common structure.

To understand how amino acids are put together, recall that carbon atoms have a valence of four—they form four covalent bonds. Amino acids have a carbon atom that forms bonds to the four groups or atoms diagrammed in **Figure 3.2a**:

1. NH_2—the amino functional group
2. COOH—the carboxyl functional group

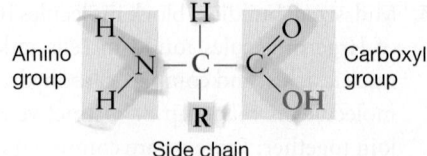

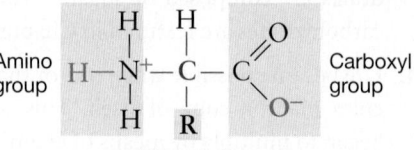

FIGURE 3.2 All Amino Acids Have the Same General Structure. The central carbon is shown in red.

3. H—a hydrogen atom
4. an "R-group"—an atom or group of atoms called a side chain

The combination of amino and carboxyl groups not only inspired the name amino acid, but is key to how these molecules behave. In water at pH 7, the concentration of protons causes the amino group to act as a base. It attracts a proton to form NH_3^+ (**Figure 3.2b**). The carboxyl group, in contrast, is acidic because its two oxygen atoms are highly electronegative. They pull electrons away from the hydrogen atom, which means that it is relatively easy for this group to lose a proton to form COO^-.

The charges on these functional groups are important for two reasons: (**1**) They help amino acids stay in solution, where they can interact with one another and with other solutes, and (**2**) they alter amino acid chemical reactivity.

The Nature of Side Chains

What about the R-group? The R-groups on amino acids vary from a single hydrogen atom to large structures containing carbon atoms linked into rings. All amino acids have a carbon atom bonded to an amino group, a hydrogen atom, and a carboxyl group, but each R-group is unique. The properties of amino acids vary because their R-groups vary.

Figure 3.3 highlights the R-groups on the 20 most common amino acids found in cells.[1] As you examine these side chains, ask yourself two questions: Is this R-group likely to participate in chemical reactions? Will it help this amino acid stay in solution?

FUNCTIONAL GROUPS AFFECT REACTIVITY Several of the side chains found in amino acids contain the carboxyl, sulfhydryl, hydroxyl, or amino functional groups introduced in Chapter 2, Table 2.3. Under the right conditions, these functional groups can participate in chemical reactions. For example, amino acids

---

[1]There are actually 22 amino acids found in proteins that occur in organisms, but two are rare.

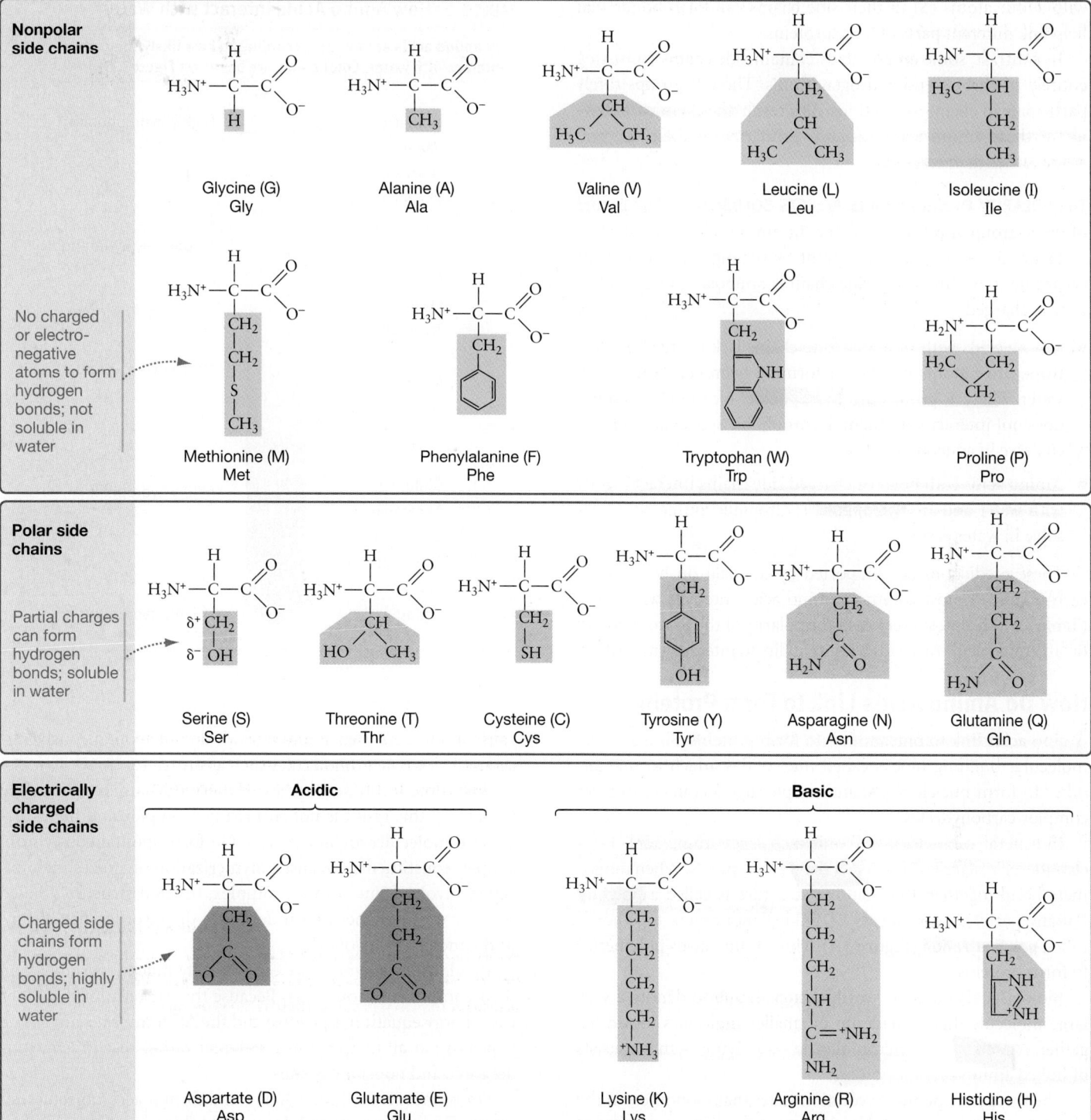

FIGURE 3.3 The 20 Major Amino Acids Found in Organisms. At the pH (about 7.0) found in cells, the 20 major amino acids found in organisms have the structural formulas shown here. The side chains are highlighted, and standard single-letter and three-letter abbreviations for each amino acid are given. For clarity, the carbon atoms in the ring structures of phenylalanine, tyrosine, tryptophan, and histidine are not shown; each bend in a ring is the site of a carbon atom. The hydrogen atoms in these structures are also not shown. A double line inside a ring indicates a double bond.

✔**EXERCISE** Explain why the green R-groups are nonpolar and why the lavender R-groups are polar, based on the relative electronegativities of O, N, C, and H. Note that sulfur (S) has an electronegativity almost equal that of carbon, and slightly higher than that of hydrogen.

with sulfur atoms (S) in their side chains can form bonds that help link different parts of large proteins.

In contrast, some amino acids contain side chains consisting entirely of carbon and hydrogen atoms. These R-groups rarely participate in chemical reactions. As a result, the chemical behavior of these amino acids depends primarily on their size and shape rather than reactivity.

THE POLARITY OF SIDE CHAINS AFFECTS SOLUBILITY   The nature of the R-group affects how soluble the amino acid is in water.

Figure 3.3 emphasizes this point by sorting the amino acids according to whether their side chain is nonpolar, polar, or electrically charged.

- Amino acids with nonpolar side chains lack charged or electronegative atoms capable of forming hydrogen bonds with water. These R-groups are **hydrophobic**, meaning that water does not interact with them. Hydrophobic side chains tend to coalesce in aqueous solution.

- Amino acids with polar or charged side chains interact readily with water and are **hydrophilic**. Hydrophilic amino acids dissolve in water easily.

These predictions are supported by the data on how readily each of the 20 most common amino acids interacts with water (**Table 3.1**). In almost every case, the polarity of the R-group found on an amino acid correlates with its ability to interact with water.

How Do Amino Acids Link to Form Proteins?

Amino acids link to one another to form proteins. Similarly, the molecular building blocks called nucleotides attach to one another to form nucleic acids, and simple sugars connect to form complex carbohydrates.

In general, a molecular subunit such as an amino acid, a nucleotide, or a sugar is called a **monomer** ("one-part"). When monomers bond together, the resulting structure is called a polymer ("many-parts"). The process of linking monomers together is called **polymerization** (**Figure 3.4**). Thus, amino acids polymerize to form proteins.

Biologists also use the word **macromolecule** to denote a very large molecule that is made up of smaller molecules joined together. A **protein** is a macromolecule—a polymer—that consists of linked amino acid monomers.

The theory of chemical evolution states that monomers in the prebiotic soup polymerized to form proteins and other types of macromolecules found in organisms. This is a difficult step, because monomers such as amino acids do not spontaneously self-assemble into macromolecules such as proteins.

According to the second law of thermodynamics reviewed in Chapter 2, this result is not surprising. Complex and highly organized molecules are not expected to form spontaneously from simpler constituents, because polymerization organizes the molecules involved into a more complex, ordered structure. Stated another way, polymerization decreases the disorder, or entropy, of the molecules involved.

In addition, polymers are energetically much less stable than their component monomers. Because the ΔH term of the Gibbs free-energy equation is positive and the ΔG term is negative, $T\Delta S$ is positive at all temperatures. Polymerization reactions are endergonic and nonspontaneous.

For monomers to link together and form macromolecules, an input of energy is required. How could this have happened during chemical evolution?

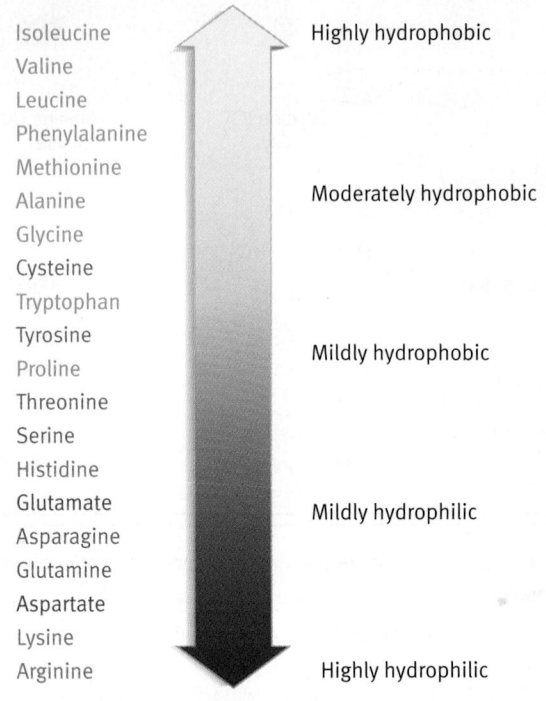

TABLE 3.1   **How Amino Acids Interact with Water**

20 amino acids are ranked according to how likely they are to interact with water. Color codes are based on Figure 3.3.

Isoleucine	Highly hydrophobic
Valine	
Leucine	
Phenylalanine	
Methionine	
Alanine	Moderately hydrophobic
Glycine	
Cysteine	
Tryptophan	
Tyrosine	
Proline	Mildly hydrophobic
Threonine	
Serine	
Histidine	
Glutamate	Mildly hydrophilic
Asparagine	
Glutamine	
Aspartate	
Lysine	
Arginine	Highly hydrophilic

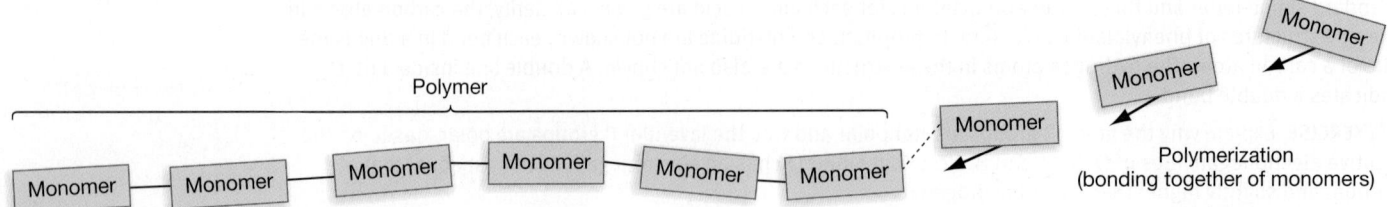

FIGURE 3.4 Monomers Are the Building Blocks of Polymers.

(a) Condensation reaction:
monomer in, water out

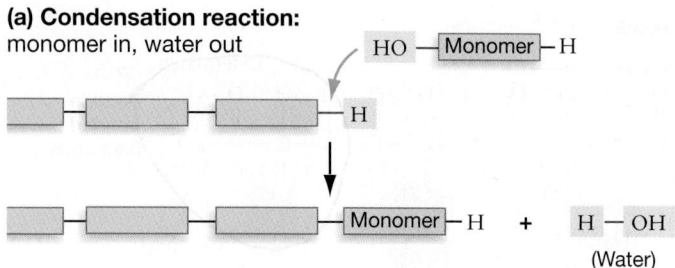

(b) Hydrolysis:
water in, monomer out

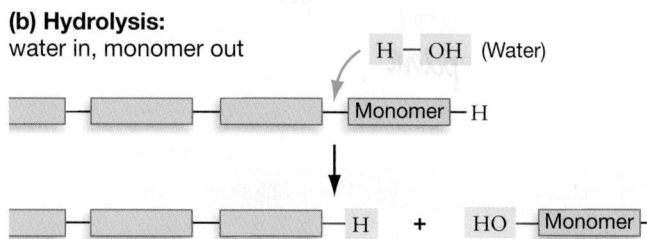

FIGURE 3.5 Polymers Can Be Extended or Broken Apart.

COULD POLYMERIZATION OCCUR IN THE ENERGY-RICH ENVIRONMENT OF EARLY EARTH? Monomers polymerize through **condensation reactions**, also known as **dehydration reactions**. These reactions are aptly named because the newly formed bond results in the loss of a water molecule (**Figure 3.5a**). The reverse reaction, called **hydrolysis**, breaks polymers apart by adding a water molecule (**Figure 3.5b**). The water molecule reacts with the bond linking the monomers, separating one monomer from the polymer chain.

In a solution such as the prebiotic soup, condensation and hydrolysis represent the forward and reverse reactions of a chemical equilibrium. Hydrolysis dominates because it is exergonic—it increases entropy and is favorable energetically.

According to recent experiments, though, there are several ways that amino acids could have polymerized early in chemical evolution.

- Researchers have been able to create stable polymers by mixing monomers with a source of chemical energy and tiny mineral particles—the size found in clay or mud. Apparently, growing macromolecules are protected from hydrolysis if they cling, or adsorb, to a mineral surface. One experiment with clay particles produced polymers that were 55 amino acids long.

- In conditions that simulate the hot, metal-rich environments of undersea volcanoes, researchers have observed not only amino acid formation but also their polymerization.

- Amino acids have also joined into polymers in experiments in cooler water if a carbon- and sulfur-containing gas—one that is commonly ejected from undersea volcanoes—is present.

The current consensus is that several mechanisms led to polymerization reactions between amino acids, early in chemical evolution. What kind of bond is responsible for linking these monomers?

THE PEPTIDE BOND As **Figure 3.6** shows, amino acids polymerize when a bond forms between the carboxyl group of one amino acid and the amino group of another. The C—N bond that results from this condensation reaction is called a **peptide bond**. Because water is lost in the condensation reaction, the carboxyl group of the amino acid is converted to a carbonyl functional group in the polymer.

To study peptide bond formation in more detail and review the general principles of polymerization reactions, go to the study area at *www.masteringbiology.com.*

(MB) **Web Activity** Condensation and Hydrolysis Reactions

Peptide bonds are unusually stable because the electrons involved are partially shared between the peptide bond and the neighboring carbonyl functional group. The degree of electron sharing is great enough that peptide bonds actually have some of the characteristics of a double bond. For example, the peptide bond is planar.

When amino acids are linked by peptide bonds into a chain, the amino acids are referred to as residues and the resulting molecule is called a **polypeptide**.

Figure 3.7a shows how the chain of peptide bonds in a polypeptide gives the molecule a structural framework, or a "backbone." There are three key points to note about the peptide-bonded backbone of a polypeptide:

1. **R-group orientation** The side chains present in each residue extend out from the backbone, making it possible for them to interact with each other and with water.

2. **Directionality** There is an amino group ($-NH_3^+$) on one end of every polypeptide chain and a carboxyl group ($-COO^-$) on the other. By convention, biologists always write amino acid sequences in the same direction. The end of the sequence that has the free amino group is placed on the left and is called the N-terminus, or amino-terminus, and the

FIGURE 3.6 Peptide Bonds Form When the Carboxyl Group of One Amino Acid Reacts with the Amino Group of a Second Amino Acid.

✔**QUESTION** Is a peptide bond a hydrogen bond, a nonpolar covalent bond, a polar covalent bond, or an ionic bond?

(a) Polypeptide chain

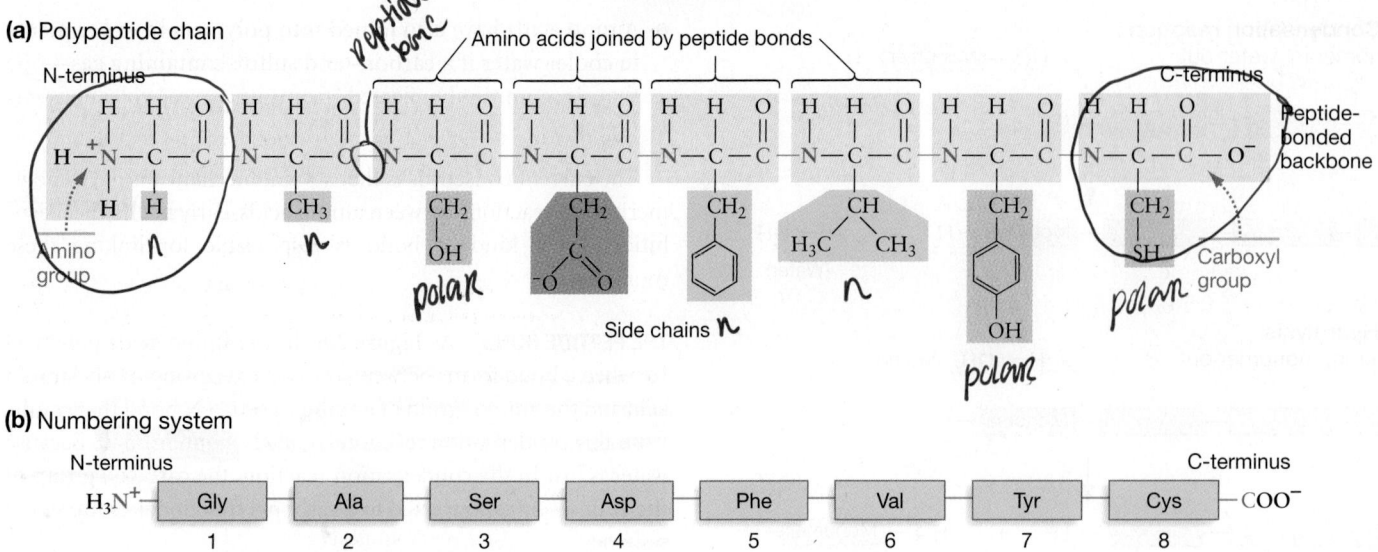

peptide bond (handwritten)

Amino acids joined by peptide bonds

N-terminus

C-terminus

Amino group

polar (handwritten, under CH₂-OH side chain)

Side chains

polar (handwritten)

polar (handwritten)

Peptide-bonded backbone

Carboxyl group

(b) Numbering system

N-terminus

H_3N^+— | Gly | Ala | Ser | Asp | Phe | Val | Tyr | Cys —COO^-

1 · 2 · 3 · 4 · 5 · 6 · 7 · 8

C-terminus

FIGURE 3.7 Amino Acids Polymerize to Form Polypeptides.

end with the free carboxyl group appears on the right-hand side of the sequence and is called the C-terminus, or carboxy-terminus. The amino acids in the chain are always numbered starting from the N-terminus (**Figure 3.7b**), because the N-terminus is the start of the chain when proteins are synthesized in cells.

3. *Flexibility* Although the peptide bond itself cannot rotate because of its double-bond nature, the single bonds on either side of the peptide bond can rotate. As a result, the structure as a whole is flexible (**Figure 3.8**).

When fewer than 50 amino acids are linked together in this way, the resulting polypeptide is called an **oligopeptide** ("few peptides") or simply a **peptide**. Polypeptides that contain 50 or more amino acids are formally called **proteins**. Proteins may consist of single polypeptides or multiple polypeptides that are bonded to one another.

Proteins are the stuff of life. Let's take a look at how they are put together, and then at what they do.

CHECK YOUR UNDERSTANDING

⊙━ If you understand that . . .

- Amino acids are small molecules with a carbon atom bonded to a carboxyl group, an amino group, a hydrogen atom, and a side chain called an R-group.
- Each amino acid has distinctive chemical properties because each has a unique R-group.
- Polypeptides are polymers made up of amino acids.
- When the carboxyl group of one amino acid reacts with the amino group of another amino acid, a strong covalent bond called a peptide bond forms. Small polypeptides are called oligopeptides, and large polypeptides are called proteins.

✔ You should be able to . . .

Draw the structural formulas of two glycine residues (glycine's R-group is an H) linked by a peptide bond, and label the amino- and carboxy-terminus.

Answers are available in Appendix B.

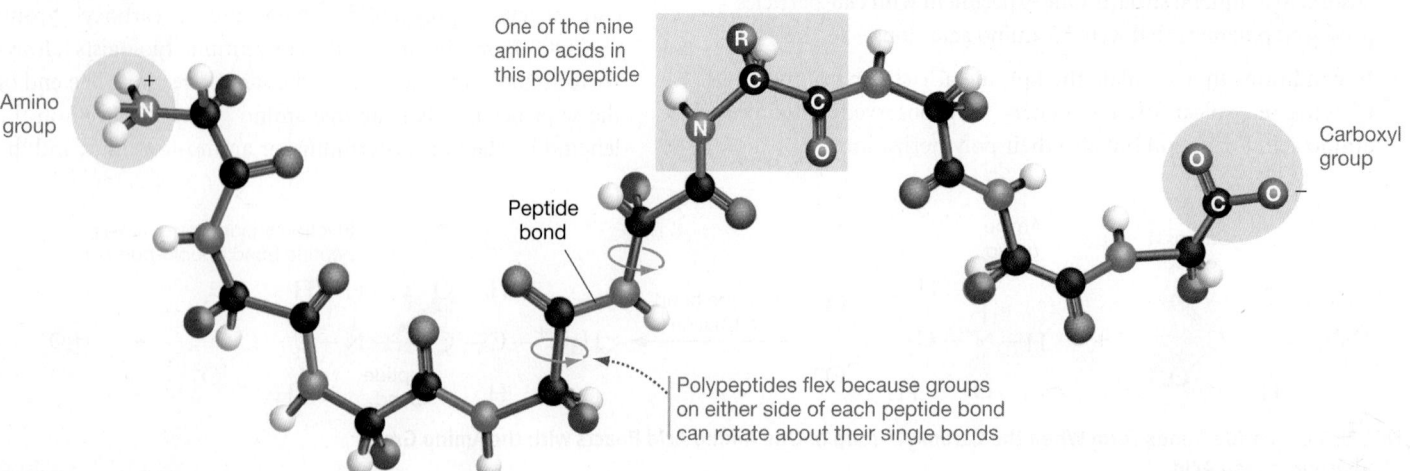

Amino group

One of the nine amino acids in this polypeptide

Peptide bond

Carboxyl group

Polypeptides flex because groups on either side of each peptide bond can rotate about their single bonds

FIGURE 3.8 Polypeptides Are Flexible.

3.3 Proteins Are the Most Versatile Large Molecules in Cells

As a group, proteins perform more types of cell functions than any other type of molecule does. It makes sense to hypothesize that life began with proteins, simply because proteins are so vital to the life of today's cells.

Consider the red blood cells that are moving through your arteries right now. Each of these cells contains about 300 million copies of the protein called hemoglobin. Hemoglobin carries oxygen from your lungs to cells throughout the body. But every red blood cell also has thousands of copies of a protein called carbonic anhydrase, which is important for moving carbon dioxide from cells back to the lungs, where it can be breathed out. Other proteins form the cell's internal skeleton or act as identification badges on its membrane.

Proteins are crucial to most tasks required for cells to exist. These tasks include:

- **Catalysis**   Many proteins are specialized to **catalyze**, or speed up, chemical reactions. A protein that functions as a catalyst is called an **enzyme**. The carbonic anhydrase molecules in red blood cells are catalysts. So is the protein called salivary amylase, found in your mouth. Salivary amylase helps begin the digestion of starch and other complex carbohydrates into simple sugars. Most chemical reactions that make life possible depend on enzymes.

- **Defense**   Proteins called antibodies and complement proteins attack and destroy viruses and bacteria that cause disease.

- **Movement**   Motor proteins and contractile proteins are responsible for moving the cell itself, or for moving large molecules and other types of cargo inside the cell. As you turn this page, for example, specialized proteins called actin and myosin will slide past one another as they work to flex or extend muscle cells in your fingers and arm.

- **Signaling**   Proteins are involved in carrying and receiving signals from cell to cell inside the body. If sugar levels in your blood are low, a small protein called glucagon will bind to receptor proteins on your liver cells, triggering enzymes inside to release sugar into your bloodstream.

- **Structure**   Structural proteins make up body components such as fingernails and hair, and define the shape of individual cells. Structural proteins keep red blood cells flexible and in their normal disc-like shape.

- **Transport**   Proteins allow particular molecules to enter or exit cells, and carry specific compounds throughout the body. Hemoglobin is a particularly well-studied transport protein, but virtually every cell is studded with membrane proteins that control the passage of specific molecules and ions.

Each of these aspects of protein function will be explored in detail later in the text. The immediate task, though, is different. Could functioning proteins be produced by chemical evolution?

3.4 What Do Proteins Look Like?

Figure 3.9 illustrates some of the diverse sizes and shapes observed in proteins. In the case of the TATA-box binding protein in Figure 3.9a and the protein called porin in Figure 3.9b, the shape of the molecule has obvious clear correlation with its function. The TATA-box binding protein has a groove where DNA molecules fit; porin has a hole that forms a pore. The groove in the TATA-box binding protein interacts with specific regions of DNA inside cells, while porin fits in cell membranes and allows certain hydrophilic molecules to pass through. But most of the proteins found in cells function as enzymes and are globular like

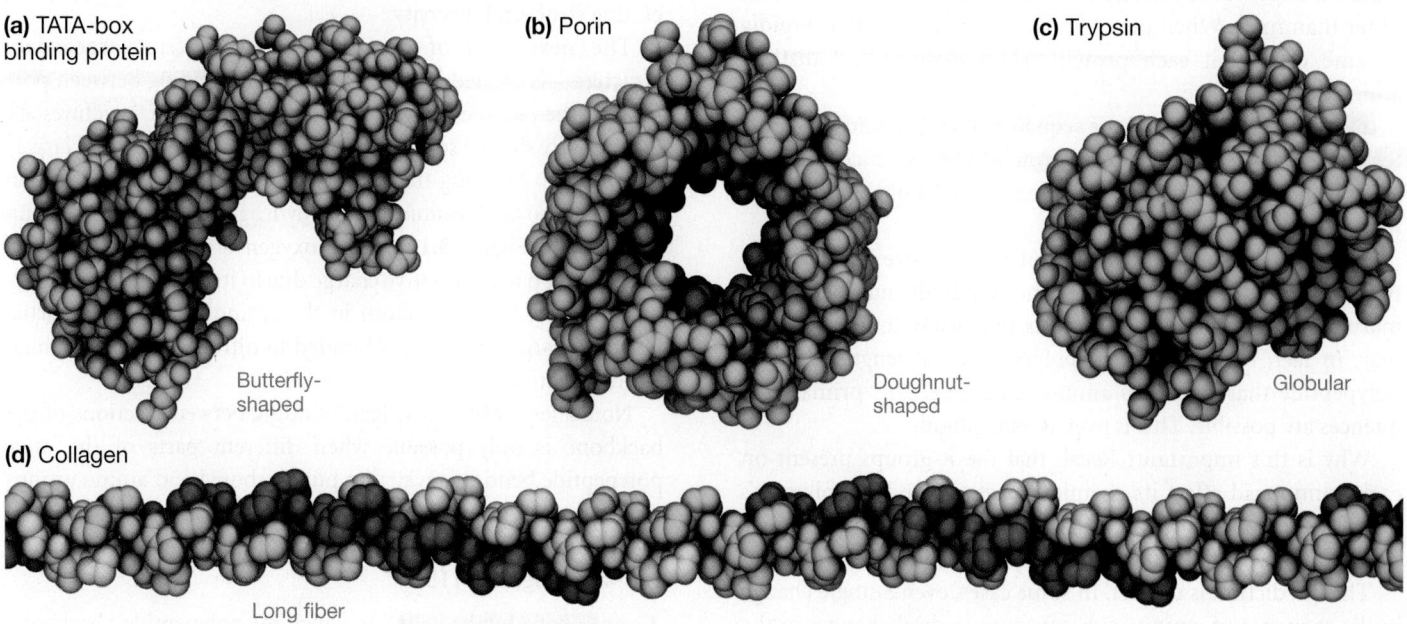

(a) TATA-box binding protein

Butterfly-shaped

(b) Porin

Doughnut-shaped

(c) Trypsin

Globular

(d) Collagen

Long fiber

FIGURE 3.9   In Overall Shape, Proteins Are the Most Diverse Class of Molecules Known.

(a) Normal amino acid sequence

	Pro	Glu	Glu	
	5	6	7	

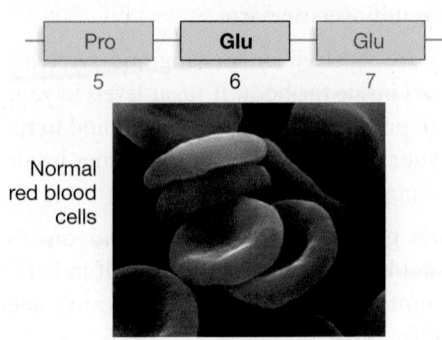

Normal
red blood
cells

(b) Single change in amino acid sequence

	Pro	Val	Glu	
	5	6	7	

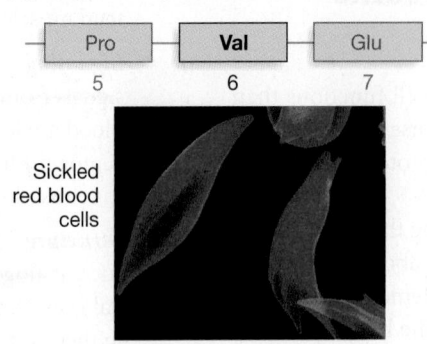

Sickled
red blood
cells

FIGURE 3.10 Changes in Primary Structure Affect Protein Function. Compare the primary structure of normal hemoglobin **(a)** with that of hemoglobin molecules of people with sickle-cell disease **(b)**. The single amino acid change causes red blood cells to change from their normal disc shape in (a) to a sickled shape in (b) when oxygen concentrations are low. Each red blood cell contains about 300 million hemoglobin molecules.

the example in Figure 3.9c. Proteins that provide structural support for cells or tissues, such as the collagen protein in Figure 3.9d, often form long fibers.

The unparalleled diversity of proteins—in size, shape, and other aspects of structure—is important because function follows from structure. 🔑 Proteins can serve diverse functions in cells because they are diverse in size and shape as well as in the chemical properties of their amino acids.

How can biologists make sense of this diversity of protein size and shape? Initially, the amount of variation seems overwhelming. Fortunately, it is not. No matter how large or complex a protein may be, its underlying structure can be broken down into just four basic levels of organization.

Primary Structure

Every protein has a unique sequence of amino acids. That simple conclusion was the culmination of 12 years of study by Frederick Sanger and co-workers during the 1940s and 1950s. Sanger's group worked out the first techniques for determining the amino acid sequence of a protein and published the completed sequence of the hormone insulin, a protein that helps regulate sugar concentrations in the blood of humans and other mammals. When other proteins were analyzed, it rapidly became clear that each protein has a definite and distinct amino acid sequence.

Biochemists call the unique sequence of amino acids in a protein the **primary structure** of that protein. The sequence of amino acids in Figure 3.7, for example, defines that polypeptide's primary structure.

With 20 types of amino acids available and size ranging from two amino acid residues to tens of thousands, the number of primary structures that are possible is practically limitless. There may, in fact, be 20^n different polypeptides of length n. For a polypeptide that is just 10 amino acids long, 20^{10} primary sequences are possible. This is over 10,000 billion.

Why is this important? Recall that the R-groups present on each amino acid affect its chemical reactivity and solubility. It's therefore reasonable to predict that the R-groups present in a polypeptide will affect that molecule's properties and function.

This prediction is correct. In some cases, even a single change in the sequence of amino acids can cause radical changes in the way the macromolecule as a whole behaves.

As an example, consider the hemoglobin protein of humans. In some individuals, hemoglobin has a valine instead of a glutamate at the amino acid numbered 6 in a strand of 146 amino acids (**Figure 3.10a**). Valine's side chain is radically different from the R-group in glutamate. The change in R-group composition produces a protein that tends to crystallize instead of staying in solution when oxygen concentrations in the blood are low. When hemoglobin crystallizes, the red blood cells that carry the protein adopt a sickled shape (**Figure 3.10b**). Sickled red blood cells get stuck in the small blood vessels called capillaries. In people whose hemoglobin contains this single amino acid change, cells downstream of blocked capillaries become starved for oxygen. A debilitating illness called sickle-cell disease results.

A protein's primary structure is fundamental to its function. Primary structure is also fundamental to the higher levels of protein structure: secondary, tertiary, and quaternary.

Secondary Structure

Even though variation in the amino acid sequence of a protein is virtually limitless, it is only the tip of the iceberg in terms of generating structural diversity.

The next level of organization in proteins—**secondary structure**—is created in part by hydrogen bonding between portions of the peptide-bonded backbone. Secondary structures are distinctively shaped sections of proteins that are stabilized largely by hydrogen bonding that occurs between the carbonyl oxygen of one amino acid residue and the hydrogen on the amino group of another (**Figure 3.11a**). The oxygen atom in the carbonyl group has a partial negative charge due to its high electronegativity, while the hydrogen atom in the amino group has a partial positive charge because it is bonded to nitrogen, which has high electronegativity.

Note a key point: Hydrogen bonding between sections of the backbone is only possible when different parts of the *same* polypeptide bend in a way that puts carbonyl and amino groups close together. In most proteins, the bending that aligns parts of the backbone and allows these bonds to form occurs in one of two ways (**Figure 3.11b**):

1. an **α-helix** (alpha helix), in which the polypeptide's backbone is coiled; or

(a) Hydrogen bonds form between peptide chains.

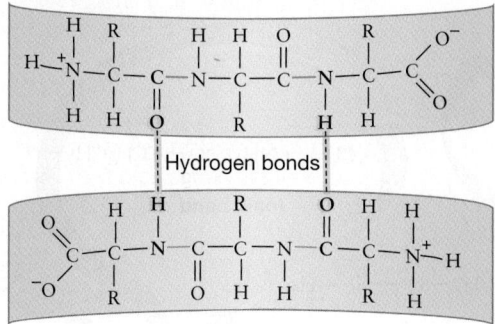

(b) Secondary structures of proteins result.

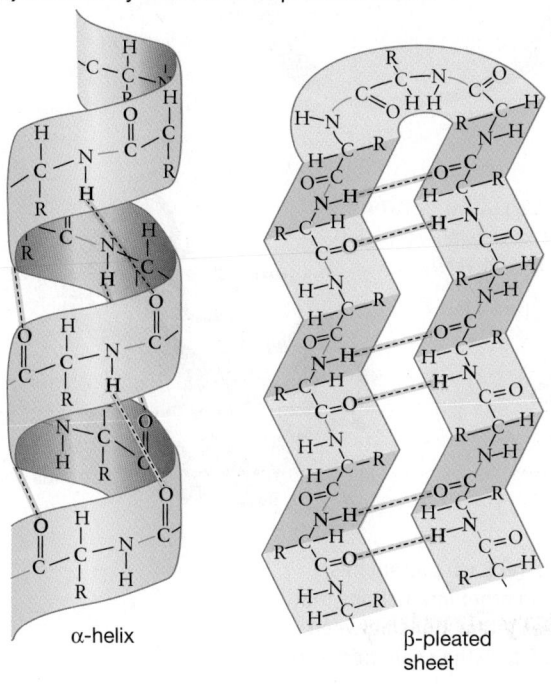

α-helix

β-pleated sheet

(c) Ribbon diagrams of secondary structure

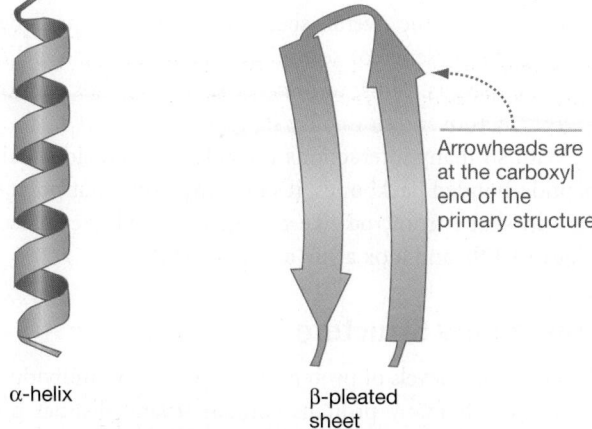

α-helix

β-pleated sheet

Arrowheads are at the carboxyl end of the primary structure

FIGURE 3.11 Secondary Structures of Proteins. A polypeptide chain can coil or fold in on itself when hydrogen bonds form between amino groups and carbonyl groups on its peptide-bonded backbone.

2. a **β-pleated sheet** (beta-pleated sheet), in which segments of a peptide chain bend 180° and then fold in the same plane.

When biologists use illustrations called ribbon diagrams to represent the shape of a protein, α-helices are shown as coils while β-pleated sheets are shown by groups of arrows in a plane (**Figure 3.11c**).

In most cases, secondary structure consists of α-helices and β-pleated sheets. Which one forms, if either, depends on the molecule's primary structure—specifically, the geometry and properties of the amino acids in the sequence. Methionine and glutamic acid, for example, are much more likely to be involved in α-helices than in β-pleated sheets. The opposite is true for valine and isoleucine. Proline, in contrast, is unlikely to be involved in either type of secondary structure.

This is a key point: Which secondary structures form depends on the protein's primary structure. A protein may have different types of secondary structure at different points along its sequence.

Although each of the hydrogen bonds in an α-helix or a β-pleated sheet is weak relative to a covalent bond, the large number of hydrogen bonds in these structures makes them highly stable. As a result, they increase the stability of the molecule as a whole and help define its shape. In terms of overall shape and stability, though, a protein's tertiary structure is even more important.

Tertiary Structure

Interactions between components of a protein's peptide-bonded backbone are responsible for α-helices and β-pleated sheets. In contrast, most of the overall shape, or **tertiary structure**, of a polypeptide results from interactions between R-groups or between R-groups and the peptide backbone. As **Figure 3.12a** shows, side chains can be involved in a wide variety of bonds and interactions. Because each contact between R-groups causes the peptide-bonded backbone to bend and fold, each contributes to the distinctive three-dimensional shape of a polypeptide.

Five types of interactions involving side chains are particularly important:

1. *Hydrogen bonding* Hydrogen bonds form between hydrogen atoms and the carbonyl group in the peptide-bonded backbone, and between hydrogen and atoms with partial negative charges in side chains.

2. *Hydrophobic interactions* In an aqueous solution, water molecules interact with hydrophilic side chains and force hydrophobic side chains to coalesce. When hydrophobic portions of proteins come together, the surrounding water molecules form more hydrogen bonds, increasing their stability. As a result, the hydrophobic side chains of proteins tend to form globular masses.

3. *van der Waals interactions* Once hydrophobic side chains are close to one another, they are stabilized by electrical attractions known as **van der Waals interactions**. These weak

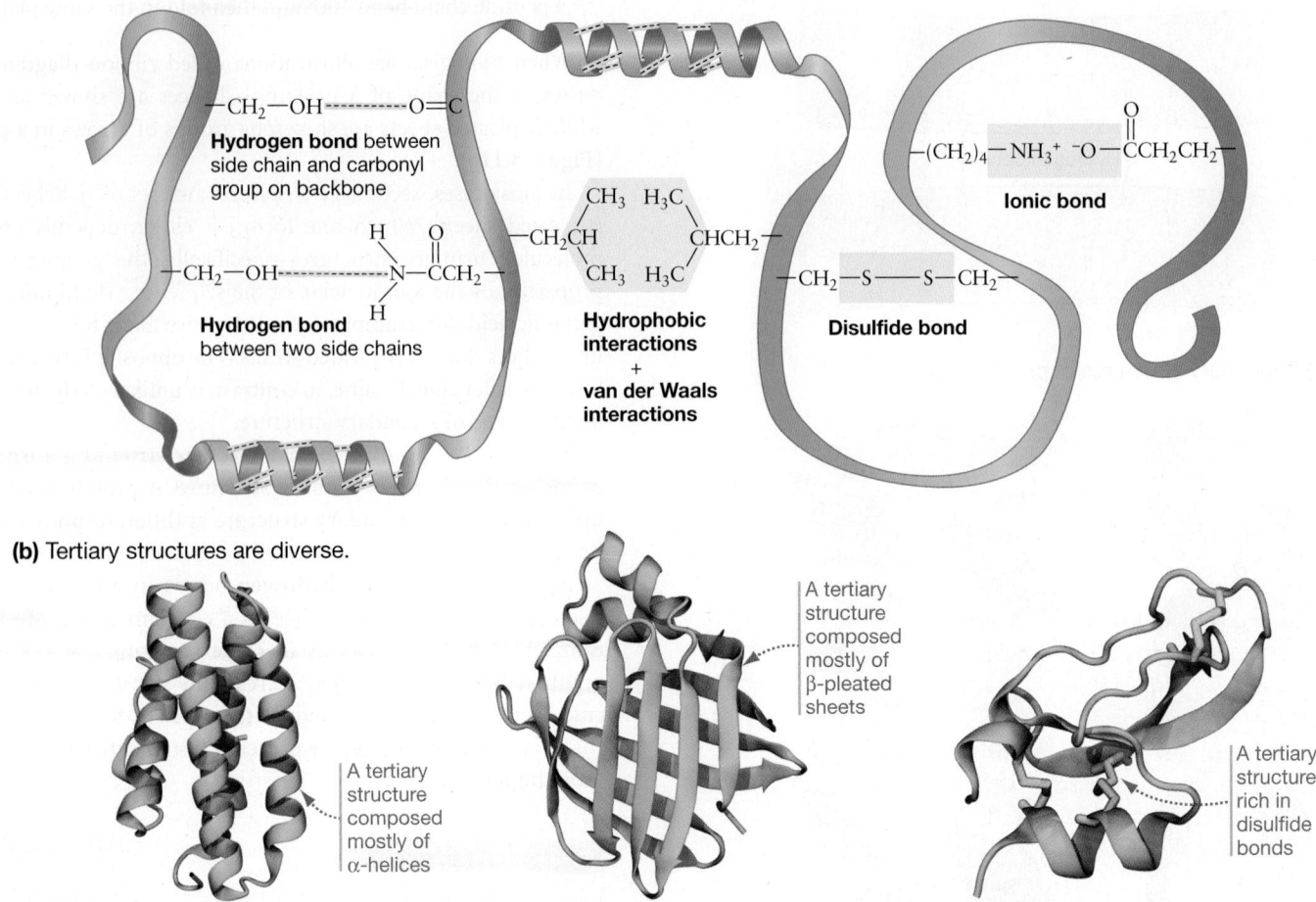

(a) Interactions that determine the tertiary structure of proteins

Hydrogen bond between side chain and carbonyl group on backbone

Hydrogen bond between two side chains

Hydrophobic interactions + **van der Waals interactions**

Ionic bond

Disulfide bond

(b) Tertiary structures are diverse.

A tertiary structure composed mostly of α-helices

A tertiary structure composed mostly of β-pleated sheets

A tertiary structure rich in disulfide bonds

FIGURE 3.12 Tertiary Structure of Proteins Results from Interactions Involving R-Groups. (a) Each protein has a unique overall shape called its tertiary structure. Tertiary structure is created by bonds and other interactions that cause proteins to fold in a precise way. **(b)** The tertiary structure of these proteins includes interactions between α-helices and β-pleated sheets. The polypeptide chains are color-coded so that you can follow the chain from the amino-end (dark blue) to the carboxy-end (red).

attractions occur because the constant motion of electrons gives molecules a tiny asymmetry in charge that changes with time. If molecules get extremely close to each other, the minute partial charge on one molecule induces an opposite partial charge in the nearby molecule and causes an attraction. Although the attraction is weak relative to covalent bonds or even hydrogen bonds, a large number of van der Waals interactions can occur in a polypeptide when many hydrophobic residues congregate. The result is a significant increase in stability.

4. ***Covalent bonding*** Covalent bonds can form between sulfur atoms when a reaction occurs between the sulfur-containing R-groups of two cysteines. These **disulfide** ("two-sulfur") **bonds** are frequently referred to as bridges because they create strong links between distinct regions of the same polypeptide.

5. ***Ionic bonding*** Ionic bonds form between groups that have full and opposing charges, such as the ionized amino and carboxyl functional groups highlighted on the right in Figure 3.12a.

In addition, the overall shape of many proteins depends in part on the presence of secondary structures like α-helices and β-pleated sheets. Thus, tertiary structure depends on both primary structure and secondary structure.

With so many interactions possible between side chains and peptide-bonded backbones, it's not surprising that polypeptides vary in shape from rod-like filaments to ball-like masses. (See **Figure 3.12b,** and look again at Figure 3.9.)

Quaternary Structure

The first three levels of protein structure involve individual polypeptides. But many proteins contain several distinct polypeptides that interact to form a single structure. The combination of polypeptide subunits gives proteins a **quaternary structure**. The individual polypeptides may be held together by bonds or other interactions among R-groups or sections of their peptide backbones.

In the simplest case, a protein with quaternary structure can consist of just two subunits that are identical. The Cro protein

(a) Cro protein, a dimer

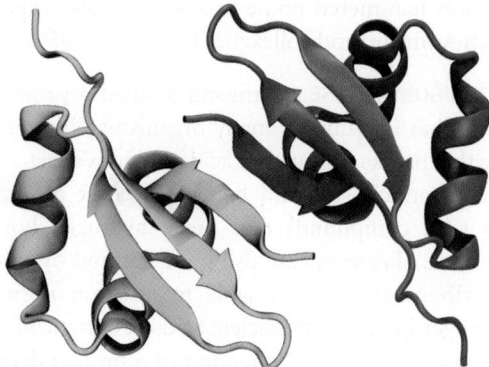

(b) Hemoglobin, a tetramer

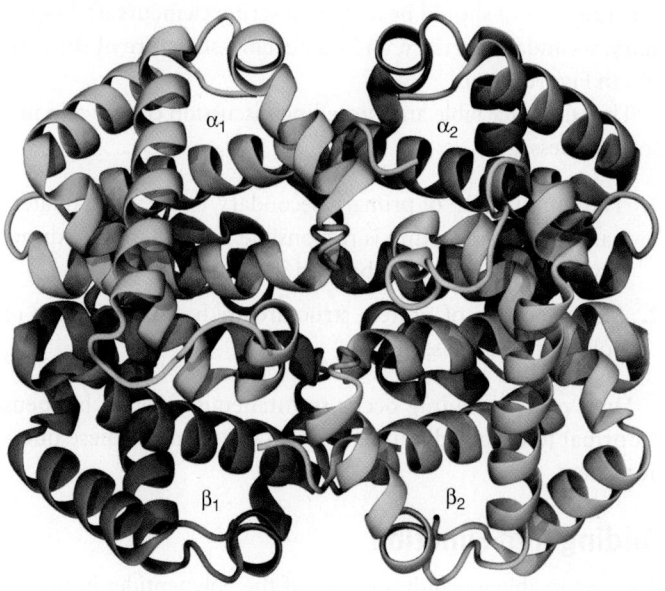

FIGURE 3.13 Quaternary Structures of Proteins Are Created by Multiple Polypeptides. These diagrams represent the primary sequence as a ribbon. **(a)** The Cro protein is a dimer—it consists of two polypeptides. The polypeptides are identical in this case but are colored yellow and green here. **(b)** Hemoglobin is a tetramer—it consists of four polypeptides. The yellow and purple polypeptides are identical; so are the blue and green polypeptides.

found in a virus called bacteriophage λ (pronounced *LAMB-da*) is an example (**Figure 3.13a**). Proteins with two polypeptide subunits are called dimers ("two-parts").

More than two polypeptides can be linked into a single protein, however, and the polypeptides involved may be distinct in primary, secondary, and tertiary structure. Hemoglobin, for example, is a tetramer ("four-parts"). It consists of two copies of two different polypeptides (**Figure 3.13b**).

Proteins that consist of a single polypeptide lack quaternary structure. Quaternary structure is common, however. Many proteins comprise several polypeptide subunits.

In addition, most cells contain at least one **multienzyme complex**: a group of enzymes, each of which catalyzes one reaction, that are physically joined to each other. The enzymes in a multi-enzyme complex are all involved in the same task. The nitrogenase complex found in certain bacteria, for example, contains the enzymes required to convert molecular nitrogen (N_2) to amine groups ($-NH_2$) used in amino acid synthesis.

Table 3.2 summarizes the four levels of protein structure, using hemoglobin as an example. The key thing to note is that protein structure is hierarchical. Quaternary structure is based on tertiary structure, which is based in part on secondary structure.

SUMMARY TABLE 3.2 **Protein Structure**

Level	Description	Stabilized by	Example: Hemoglobin	
Primary	The sequence of amino acids in a polypeptide	Peptide bonds	Gly — Ser — Asp — Cys	
Secondary	Formation of α-helices and β-pleated sheets in a polypeptide	Hydrogen bonding between groups along the peptide-bonded backbone; thus, depends on primary structure.		One α-helix
Tertiary	Overall three-dimensional shape of a polypeptide (includes contribution from secondary structures)	Bonds and other interactions between R-groups, or between R-groups and the peptide-bonded backbone; thus, depends on primary structure.		One of hemoglobin's subunits
Quaternary	Shape produced by combinations of polypeptides (thus, combinations of tertiary structures)	Bonds and other interactions between R-groups, and between peptide backbones of different polypeptides; thus, depends on primary structure.		Hemoglobin, which consists of four polypeptide subunits

All three of the higher-level structures are based on primary structure. ✔You should be able to describe elements of the primary, secondary, tertiary, and quaternary structure of the protein in Figure 3.13a.

The summary table and preceding discussion convey two important messages:

1. The combination of primary, secondary, tertiary, and quaternary levels of structure is responsible for the fantastic diversity of sizes and shapes observed in proteins.

2. Most elements of protein structure are based on folding of polypeptide chains.

Does protein folding occur spontaneously? What happens if normal folding is disrupted? Let's take a look at these questions next.

Folding and Function

If you were able to synthesize one of the polypeptides in hemoglobin from individual amino acids, and if you placed the resulting chain in water, it would spontaneously fold into the shape of the tertiary structure shown in Table 3.2.

This result seems counterintuitive. Because an unfolded protein has many more ways to move about, it has much higher entropy than the folded version. Folding *is* spontaneous in some cases, however, because the bonds, hydrophobic interactions, and van der Waals interactions that occur make the folded molecule more stable energetically than the unfolded molecule. Thus, folding may release enough free energy to be exergonic and occur spontaneously.

Folding is also crucial to the function of a completed protein. This point was hammered home in a set of classic experiments by Christian Anfinson and colleagues during the 1950s.

FOLDING IN RIBONUCLEASE Anfinson studied a protein called ribonuclease that is found in many organisms. Ribonuclease is an enzyme that breaks ribonucleic acid polymers apart. Anfinson found that ribonuclease could be unfolded, or **denatured**, by treating it with compounds that break hydrogen bonds and disulfide bonds. (Proteins can also be denatured by heat.) The denatured ribonuclease was unable to function normally—it could no longer break apart nucleic acids (**Figure 3.14**). This is not surprising, given that the function of a protein depends on its structure.

When Anfinson removed the denaturing agents, however, the molecule refolded and began to function normally again. These experiments confirmed that ribonuclease folds spontaneously and that folding is essential for normal function.

More recent work has shown that in cells, folding is often facilitated by specific proteins called **molecular chaperones**. Many molecular chaperones belong to a family of molecules called the heat-shock proteins. These compounds are produced in large quantities after cells experience high temperatures or other treatments that make proteins lose their tertiary structure. Heat-shock proteins speed the refolding of other proteins into their normal shape after denaturation has occurred.

FOLDING IN PRIONS In 1982, Stanley Prusiner published what may be the most surprising result to emerge from research on protein folding: Some improperly folded proteins act as infec-

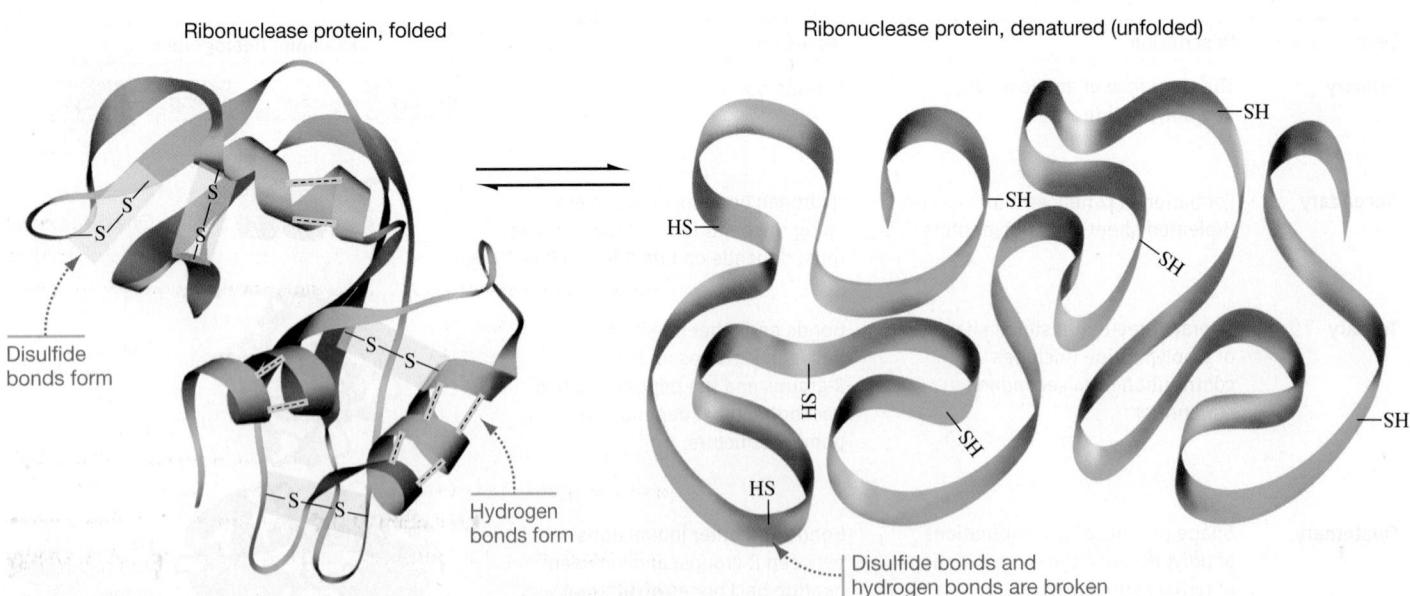

FIGURE 3.14 Proteins Fold into Their Normal, Active Shape. (a) The tertiary structure of ribonuclease is defined primarily by four disulfide bonds. The protein's primary structure is represented by the green ribbon. **(b)** When the disulfide bonds and various non-covalent bonds are broken, the protein denatures (unfolds).

tious, disease-causing agents. The proteins involved are called **prions** (pronounced *PREE-ons*), or proteinaceous infectious particles.

Follow-up work showed that prions are improperly folded forms of normal proteins that are present in healthy individuals. The infectious and normal forms do not necessarily differ in amino acid sequence, however. Instead, their *shapes* are radically different. Further, the infectious form of a protein can induce other normal protein molecules to change their shape to the altered form.

Figure 3.15 illustrates the shape differences observed in the normal and infectious forms of the first prion ever described. The molecule in Figure 3.15a is called the prion protein (PrP) and is a normal component of mammalian cells. Mutant versions of this protein, like the one in Figure 3.15b, are found in a wide variety of species and cause a family of diseases known as the spongiform encephalopathies—literally, "sponge-brain-illnesses." Hamsters, cows, goats, and humans afflicted with these diseases undergo massive degeneration of the brain. Cattle suffer from "mad cow disease"; sheep and goats acquire scrapie (so called because the animals itch so badly that they scratch off their wool or hair); humans develop kuru or Creutzfeldt-Jakob disease.

Although some spongiform encephalopathies can be inherited, in many cases the disease is transmitted when individuals eat tissues containing the infectious form of PrP. All the prion illnesses are fatal.

Prions are particularly dramatic examples of how a protein's function depends on its shape, and how the final shape of a protein depends on folding.

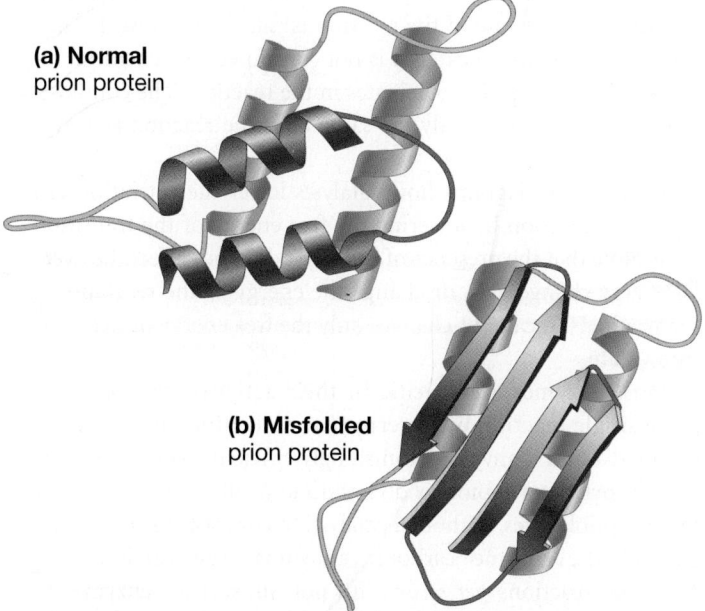

(a) Normal
prion protein

(b) Misfolded
prion protein

FIGURE 3.15 Prions Are Improperly Folded Proteins. Ribbon model of **(a)** a normal prion protein and **(b)** the misfolded form that causes mad cow disease in cattle. Secondary structure is represented by coils (α-helices) and arrows (β-pleated sheets).

🔑 **If you understand that . . .**

- Proteins have up to four levels of structure.
- Primary structure consists of the sequence of amino acids.
- Secondary structure results from bonds between atoms in the peptide-bonded backbone of the same polypeptide. These bonds produce structures such as α-helices and β-pleated sheets.
- Tertiary structure is the overall shape of a polypeptide. Most tertiary structure is a consequence of bonds or other interactions between R-groups or between R-groups and the peptide-bonded backbone.
- Quaternary structure occurs when multiple polypeptides interact to form a single protein.

✔ **You should be able to . . .**

Explain why secondary, tertiary, and quarternary structure depend on a polypeptide's primary structure.

Answers are available in Appendix B.

3.5 Enzymes: An Introduction to Catalysis

Of all the functions that proteins perform in cells, catalysis may be the most important. The reason is speed. Life, at its most basic level, consists of chemical reactions. But most of the chemical reactions in cells don't occur fast enough to support life unless a catalyst is present.

Even when life began, the first molecule that could make a copy of itself would have had to catalyze the polymerization reactions involved. Catalysis is at the heart of life's origin, and all life today. Let's dig in.

Enzymes Help Reactions Clear Two Hurdles

To appreciate how enzymes work, recall from Chapter 2 that reactions take place when reactants (**1**) collide in a precise orientation and (**2**) have enough kinetic energy to overcome repulsion between electrons that come into contact as a bond forms. Part of the reason enzymes are such effective catalysts is that they bring reactant molecules—called **substrates**—together in a precise orientation, so that the electrons involved in the reaction can interact. But enzymes can also affect the amount of kinetic energy reactants must have for a reaction to proceed.

To understand why catalysts affect the energy required for a reaction, it's critical to realize that a collision between reactants creates a combination of old and new bonds. This intermediate condition is called a **transition state**. The amount of free energy required to reach the transition state is called the **activation energy** of the reaction.

Reactions happen when reactants have enough kinetic energy to reach the transition state. The kinetic energy of molecules, in turn, is a function of their temperature. This is why

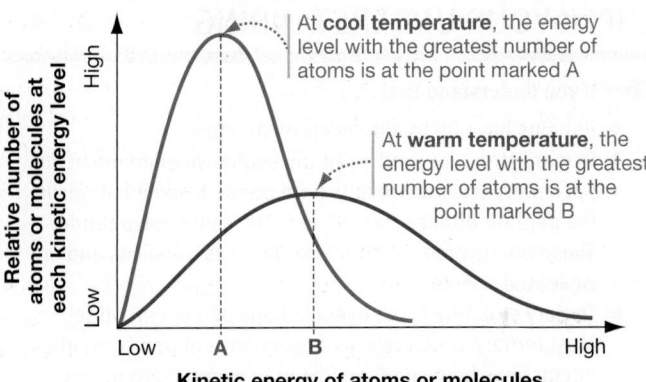

FIGURE 3.16 **Atoms and Molecules Have More Kinetic Energy at High Temperatures than at Low Temperatures.**

✔**EXERCISE** Add a vertical line two-thirds of the way along the horizontal axis. Label this line "Activation energy for reaction." At low and high temperatures, which portion of the respective curves represents molecules with enough kinetic energy for this reaction to proceed?

chemical reactions tend to proceed faster at higher temperatures, as noted in Chapter 2.

To make sure you understand these points, study **Figure 3.16**. The x-axis plots the kinetic energy of a molecule in a cell or a solution; the y-axis plots the proportion of all the molecules present that have a particular amount of kinetic energy. The blue curve shows the distribution of kinetic energy in molecules at cool temperature; the red curve shows the distribution of energy in molecules at high temperature. Even if the activation energy for a reaction involving these molecules is low, more of them will have enough kinetic energy to react if they are at high temperature versus low temperature.

Catalysts don't change the temperature of a solution, though. How do they fit in?

GRAPHING THE ENERGY OF THE TRANSITION STATE **Figure 3.17** graphs the changes in free energy that take place during the course of a chemical reaction. As you read along the x-axis from left to right, note that a dramatic rise in free energy occurs when the reactants combine to form the transition state—followed by a dramatic drop in free energy when products form. The free energy of the transition state is high because the bonds that existed in the substrate are destabilized—it is the transition point between breaking old bonds and forming new ones.

The ΔG label on the graph indicates the overall change in free energy in the reaction—that is, the energy of the products minus the energy of the reactants. In this particular case, the products have lower potential energy than the reactants, meaning that the reaction is exothermic. But because the activation energy for this reaction, symbolized by E_a, is high, the reaction would proceed slowly—even at high temperature.

This is an important point: The more unstable the transition state, the higher the activation energy and the less likely a reaction is to proceed quickly.

Reaction rates, then, depend on both the kinetic energy of the reactants and the activation energy of the particular reaction—

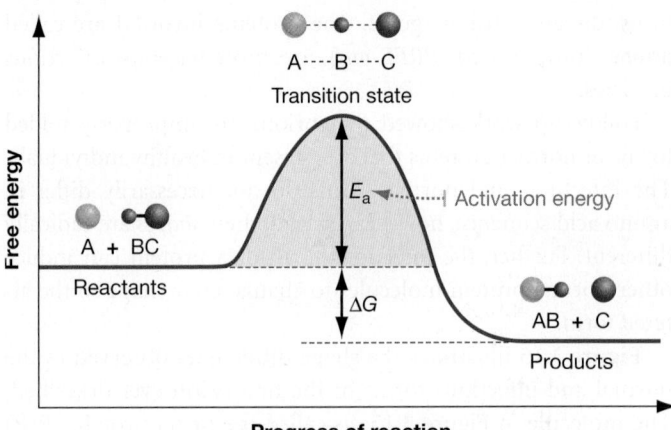

FIGURE 3.17 **Changes in Free Energy during a Chemical Reaction.** The changes in free energy that occur over the course of a hypothetical reaction between an atom A and a molecule containing atoms B and C. The overall reaction would be written as $A + BC \rightarrow AB + C$. E_a is the activation energy of the reaction, and ΔG is the overall change in free energy.

meaning the free energy of the transition state. If the kinetic energy of the participating molecules is high, then molecular collisions are likely to result in completed reactions. But if the activation energy of a particular reaction is also high, then collisions are less likely to result in completed reactions.

ENZYMES LOWER THE ACTIVATION ENERGY In many cases, the electrons in the transition-state molecule can be stabilized when they interact with another ion, atom, or molecule. When this occurs, the activation energy required for the reaction drops and the reaction rate increases.

A substance that lowers the activation energy of a reaction and increases the rate of the reaction is called a **catalyst.** It's important to note that a catalyst is not consumed in a chemical reaction, even though it participates in the reaction. The composition of a catalyst is exactly the same after the reaction as it was before.

Figure 3.18 diagrams how catalysts lower the activation energy for a reaction by lowering the free energy of the transition state. Note that the presence of a catalyst does not affect the overall energy change, ΔG, or change the energy of the reactants or the products. A catalyst changes only the free energy of the transition state.

Most enzymes are specific in their activity—they catalyze just a single reaction by lowering the activation energy that is required—and many are astonishingly efficient. Most of the important reactions in biology do not occur at all, or else proceed at imperceptible rates, without a catalyst. In contrast, a single molecule of the enzyme carbonic anhydrase can catalyze over 1,000,000 reactions *per second*. It's not unusual for enzymes to speed up reactions by a factor of a million; some enzymes make reactions go many *trillions* of times faster than they would without a catalyst.

To understand why catalysts make reactions happen so much faster than uncatalyzed reactions, go back to Figure 3.16. After

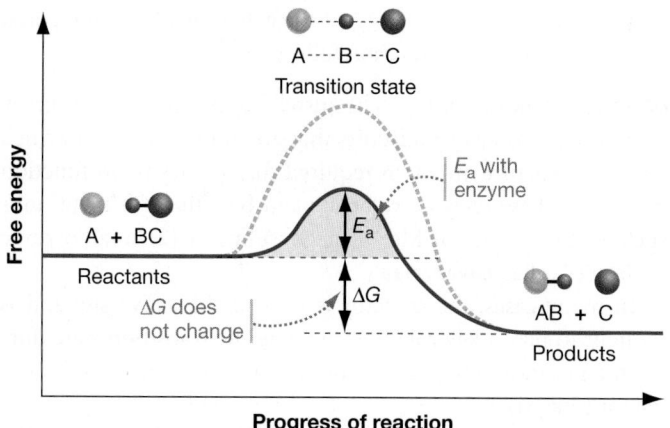

Transition state

A---B---C

E_a with enzyme

A + BC

Reactants

E_a

ΔG does not change

ΔG

AB + C

Products

Free energy

Progress of reaction

FIGURE 3.18 A Catalyst Changes the Activation Energy of a Reaction. The energy profile for the same reaction diagrammed in Figure 3.17, but with a catalyst present. Even though the energy barrier to the reaction, E_a, is much lower, ΔG does not change.

✔**QUESTION** Can a catalyst make a nonspontaneous reaction occur spontaneously? Explain why or why not.

you've done the original exercise in the caption, add another vertical line one-third of the way along the *x*-axis. Label this line "Activation energy for reaction with catalyst." Label the portion of each curve that has enough energy to undergo a reaction with a catalyst but not without a catalyst. ✔**If you understand what catalysts do, you should be able to determine whether more reactions occur with a catalyst or without, and explain why.**

🔑 To summarize, enzymes **(1)** bring reactants together in precise orientations and **(2)** stabilize transition states. Protein catalysts are important because they speed up the chemical reactions that are required for life. How do they do what they do?

How Do Enzymes Work?

The initial hypothesis for how enzymes work was proposed by Emil Fischer in 1894. According to Fischer's lock-and-key model, enzymes are rigid structures analogous to a lock. The keys are substrates that fit into the lock and then react.

Several important ideas in this model have stood the test of time. For example, Fischer was correct in proposing that enzymes bring substrates together in a precise orientation that makes reactions more likely. His model also accurately explained why most enzymes can catalyze only one specific reaction. Enzyme specificity is a product of the geometry and chemical properties of the sites where substrates bind.

As researchers began to test and extend Fischer's model, the location where substrates bind and react became known as the enzyme's **active site**. The active site is where catalysis actually occurs.

When techniques for solving the three-dimensional structure of enzymes became available, it turned out that enzymes tend to be very large relative to substrates and roughly globular, and that the active site is in a cleft or cavity within the globular shape. The enzyme hexokinase, which is at work in most cells of your body now, is a good example. (Many enzymes have names that end with *–ase*.) As the left-hand side of **Figure 3.19** shows, the active site in hexokinase is a small notch in an otherwise large, crescent-shaped enzyme.

Fischer's model had to be modified, however, as research on enzyme action progressed. Perhaps the most important realization was that enzymes are not rigid and static, but flexible and dynamic. In fact, many enzymes undergo a significant change in shape, or conformation, when reactant molecules bind to the active site. You can see this conformational change, called an **induced fit**, in the hexokinase molecule on the right of Figure 3.19. As hexokinase binds its substrate—the sugar glucose—the enzyme rocks forward over the active site.

In addition, recent research has clarified the nature of Fischer's key. When one or more substrate molecules enter the active site, they are held in place through hydrogen bonding or other electrical interactions with amino acids in the active site. Once the substrate is bound, one or more R-groups in the active site come into play. The degree of interaction between the substrate and enzyme increases and reaches a maximum when the transition state is formed. Thus, Fischer's key is actually the transition state.

Interactions with R-groups at the active site stabilize the transition state and thus lower the activation energy required for the reaction to proceed. At the atomic level, R-groups that line the

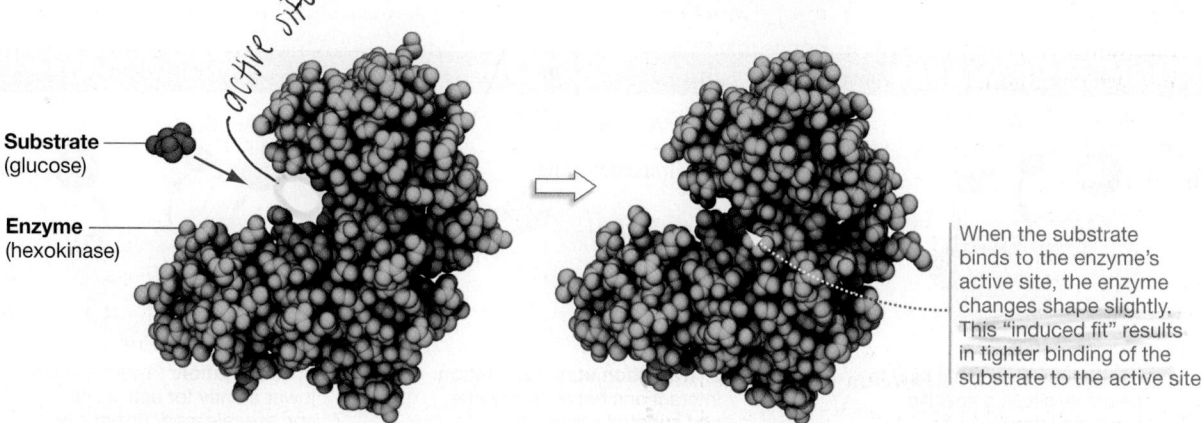

active site

Substrate (glucose)

Enzyme (hexokinase)

When the substrate binds to the enzyme's active site, the enzyme changes shape slightly. This "induced fit" results in tighter binding of the substrate to the active site

FIGURE 3.19 Reactant Molecules Bind to Specific Locations in an Enzyme. The reactant molecule, shown in red, fits into a precise location, called the active site, in the green enzyme. In this enzyme and in many others, the binding event causes the protein to change shape.

active site may form short-lived covalent bonds that assist with the transfer of atoms or groups of atoms from one reactant to another. More commonly, the presence of acidic or basic R-groups allows the reactants to lose or gain a proton more readily.

The reaction products have a much lower affinity for the active site than do either the reactants or the transition state, however. As a result, they are released from the enzyme once they form.

Figure 3.20 summarizes these principles. Enzyme catalysis can be analyzed as a three-step process:

1. **Initiation**: Instead of reactants occasionally colliding in a random fashion, enzymes orient reactants precisely as they bind at specific locations within the active site.

2. **Transition state facilitation**: The act of binding induces the formation of the transition state. In some cases the transition state is stabilized by a change in the enzyme's shape. The interaction between the substrate and R-groups in the enzyme's active site lowers the activation energy required for the reaction. Inside a catalyst's active site, then, more reactant molecules have sufficient kinetic energy to reach this lowered activation energy. Thus, the catalyzed reaction proceeds much more rapidly than the uncatalyzed reaction.

3. **Termination**: The reaction products have less affinity for the active site than the transition state does. Binding ends, the enzyme returns to its original conformation, and the products are released.

To make sure you understand how catalysts affect transition states and activation energies, go to the study area at *www.masteringbiology.com*.

(MB) Web Activity Activation Energy and Enzymes

✔If you understand the basic principles of enzyme catalysis, you should be able to complete the following sentences: (1) Enzymes speed reaction rates by _____ and lowering activation energy. (2) Activation energies drop because enzymes may destabilize bonds in the reactant, stabilize the _____, make acid–base reactions more favorable, and change the reaction mechanism through a covalent bonding interaction. (3) Enzyme specificity is a function of the active site's shape and the chemical properties of the _____ at the active site. (4) In enzymes, as in many molecules, function follows from _____.

DO ENZYMES ACT ALONE? The answer to this question, in many cases, is no. Atoms or molecules that are not part of an enzyme's primary structure are often required for an enzyme to function normally. These enzyme **cofactors** can be either (1) metal ions such as Zn^{2+} (zinc) or Mg^{2+} (magnesium) or (2) small organic molecules called **coenzymes**.

In many cases, the cofactor is part of the active site and is thought to play a key role in stabilizing the transition state during the reaction. The presence of the cofactor is therefore essential for catalysis.

To appreciate why this is important, consider that many of the vitamins in your diet are required for the production of enzyme cofactors. Vitamin deficiencies cause enzyme-cofactor deficiencies. Lack of cofactors, in turn, disrupts normal enzyme function and causes disease. For example, thiamine (vitamin B_1) is required for the production of an enzyme cofactor called thiamine pyrophosphate, which is required by three different enzymes. Lack of thiamine in the diet dramatically reduces the activity of these enzymes and causes an array of nervous system and heart disorders collectively known as beriberi.

MOST ENZYMES ARE REGULATED In addition to requiring a cofactor for activity, most enzymes associate with other molecules that regulate them. These regulatory molecules change the protein's structure in some way, and their presence either activates or inactivates the enzyme.

Most enzyme regulation occurs in one of two ways:

- Catalysis is inhibited when a molecule that is similar in size and shape to a substrate binds to the active site. This event is called **competitive inhibition**, because the molecule involved competes with the substrate for access to the enzyme's active site (**Figure 3.21a**).

- Regulatory molecules may bind at a location other than the active site. This type of regulation is called **allosteric** ("different-structure") **regulation**, because the molecule involved does not affect the active site directly. Instead, the binding event

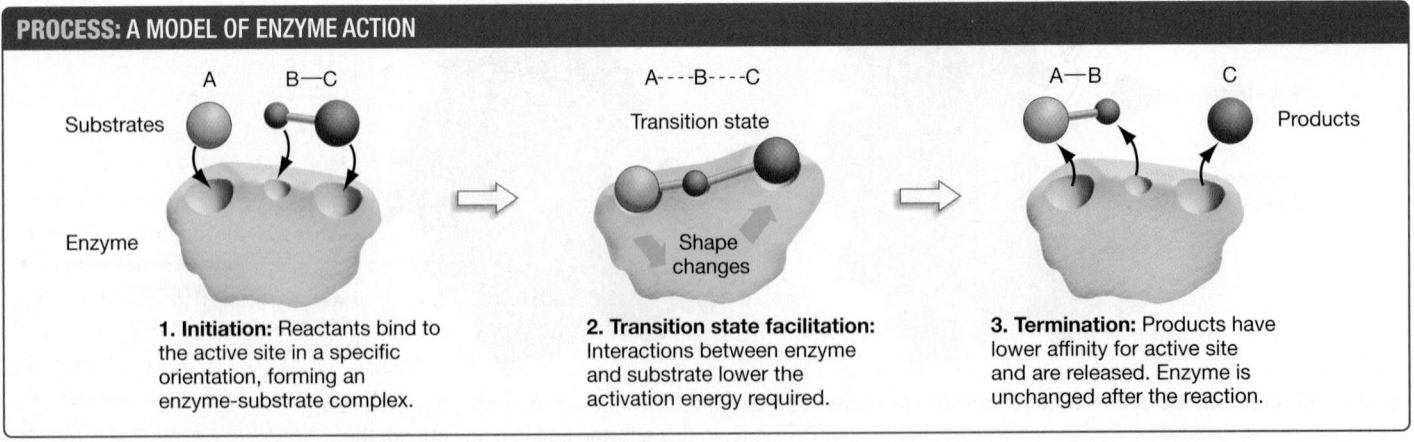

PROCESS: A MODEL OF ENZYME ACTION

Substrates / Enzyme

1. Initiation: Reactants bind to the active site in a specific orientation, forming an enzyme-substrate complex.

Transition state / Shape changes

2. Transition state facilitation: Interactions between enzyme and substrate lower the activation energy required.

Products

3. Termination: Products have lower affinity for active site and are released. Enzyme is unchanged after the reaction.

FIGURE 3.20 Enzyme Action Can Be Analyzed as a Three-Step Process.

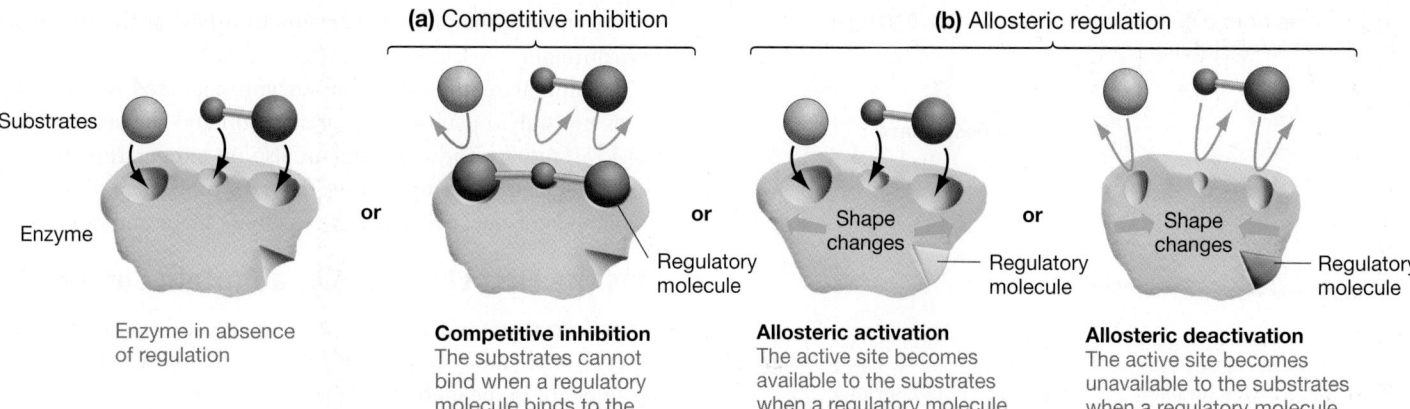

(a) Competitive inhibition

(b) Allosteric regulation

Substrates

Enzyme

or

or

Regulatory molecule

Shape changes

or

Regulatory molecule

Shape changes

Regulatory molecule

Enzyme in absence of regulation

Competitive inhibition
The substrates cannot bind when a regulatory molecule binds to the enzyme's active site.

Allosteric activation
The active site becomes available to the substrates when a regulatory molecule binds to a different site on the enzyme.

Allosteric deactivation
The active site becomes unavailable to the substrates when a regulatory molecule binds to a different site on the enzyme.

FIGURE 3.21 An Enzyme's Activity Is Precisely Regulated. Enzymes are turned on or off when specific molecules bind to them.

changes the shape of the enzyme in a way that makes the active site accessible or inaccessible (**Figure 3.21b**).

Allosteric regulation is much more common than competitive inhibition; later chapters will provide detailed examples of both processes.

WHAT LIMITS THE RATE OF CATALYSIS? For several decades after Fischer's model was published, most research on enzymes focused on rates of enzyme action, or what biologists call enzyme kinetics. Researchers observed that, when the amount of product produced per second—indicating the speed of the reaction—is plotted as a function of substrate concentration, a graph like that shown in **Figure 3.22** results.

In this graph, each data point represents an experiment where reaction rate was measured when substrates were at various concentrations. As you read the curve from left to right, note that it has three basic sections:

1. When substrate concentrations are low, the speed of an enzyme-catalyzed reaction increases in a linear fashion.

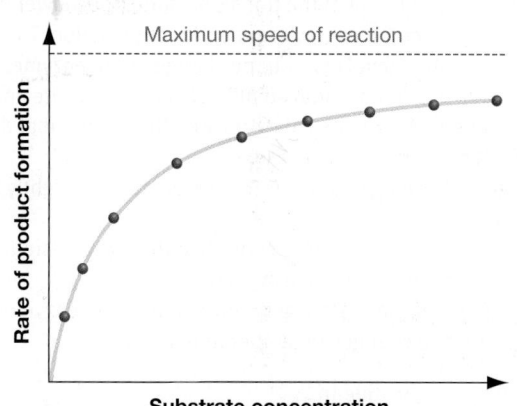

Maximum speed of reaction

Rate of product formation

Substrate concentration

FIGURE 3.22 Kinetics of an Enzyme-Catalyzed Reaction. The general shape of this curve is characteristic of enzyme-catalyzed reactions.

✔**EXERCISE** Label the parts of the graph represents where (1) the reaction rate is most sensitive to changes in substrate concentration and (2) most or all of the active sites present are occupied.

2. At intermediate substrate concentrations, the increase in speed begins to slow.

3. At high substrate concentration, the reaction rate plateaus at a maximum speed.

This pattern is in striking contrast to the situation for uncatalyzed reactions, in which reaction speed tends to show a continuing linear increase with substrate concentration. The "saturation kinetics" of enzyme-catalyzed reactions were taken as strong evidence that the enzyme-substrate complex proposed by Fischer actually exists. The idea was that, at some point, active sites cannot accept substrates any faster, no matter how large the concentration of substrates gets. Stated another way, reaction rates level off because all available enzyme molecules are being used.

HOW DO PHYSICAL CONDITIONS AFFECT ENZYME FUNCTION? Given that an enzyme's structure is critical to its function, it's not surprising that an enzyme's activity is sensitive to conditions that alter its structure. In particular, the activity of an enzyme often changes drastically as a function of temperature and pH.

Temperature affects the movement of the enzyme as well as the kinetic energy of the substrates. pH affects the makeup and charge of amino acid side chains with carboxyl or amino groups, and the active site's ability to participate in proton-transfer or electron-transfer reactions.

Do data support these assertions? **Figure 3.23a** shows how the activity of an enzyme, plotted on the *y*-axis, changes as a function of temperature, plotted on the *x*-axis. Data for the enzyme glucose-6-phosphatase, which is helping to produce usable energy in your cells right now, are shown for two species of bacteria. In this graph, each data point represents the enzyme's relative activity—meaning, the rate of the enzyme-catalyzed reaction, scaled relative to the highest rate observed—in experiments conducted under conditions that were identical except for differences in temperature.

Note that, in both bacterial species, the enzyme has a distinct optimum or peak—a temperature at which it functions best. One of the bacterial species lives inside your gut, where the temperature is about 40°C, while the other lives in hot springs, where

(a) Enzymes from different organisms may function best at different temperatures.

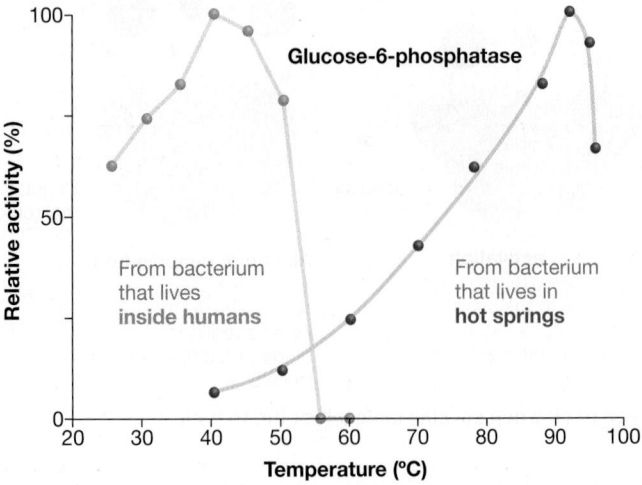

(b) Enzymes from different organisms may function best at different pHs.

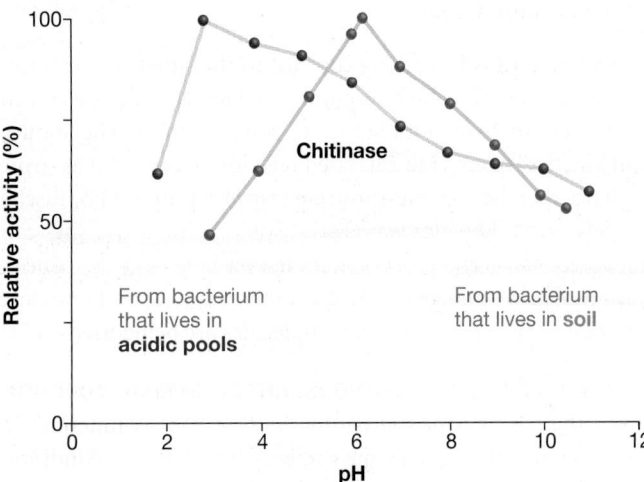

FIGURE 3.23 Enzymes Have an Optimal Temperature and pH. Enzymes are sensitive to changes in temperature **(a)** and pH **(b)**.

temperatures can be close to 100°C. The temperature optimum for the enzyme reflects these environments.

The two types of bacteria have different versions of the enzyme that differ in primary structure. Natural selection—the process introduced in Chapter 1—has favored different structures that have different functions. The enzymes are adaptations that allow each species to thrive at different temperatures.

Figure 3.23b makes the same point for pH. The enzyme in this graph, called chitinase, protects bacterial cells by digesting a molecule found in the cell walls of fungi that use bacteria as food. The data come from a species of bacterium that lives in acidic pools and a bacterial species that lives in the soil under palm trees.

Note that the organism that thrives in an acidic environment has a version of the enzyme that performs best at low pH; the organism that lives near palms has a version of the enzyme that functions best near neutral pH. Each enzyme is sensitive to changes in pH, but each species' version of the enzyme

has a structure that allows it to function best at the pH of its environment.

To summarize, the rate of an enzyme-catalyzed reaction depends not only on substrate concentration and the enzyme's intrinsic affinity for the substrate but also on temperature and pH (and other factors). Temperature affects the movement of the substrates and enzyme; pH affects the enzyme's shape and reactivity.

Was the First Living Entity a Protein Catalyst?

Several observations argue that the answer to this question is yes. Experimental studies have shown that amino acids were likely to be abundant in the prebiotic soup, and that they could have polymerized to form small proteins. In addition, proteins are the most efficient catalysts known, and a self-replicating molecule had to act as a catalyst during the assembly and polymerization of its copy.

These observations support the hypothesis that the self-replicating molecule was a polypeptide. Indeed, several laboratories that are currently working to create life have focused on synthesizing a self-replicating protein.

To date, however, attempts to simulate the origin of life with proteins have not been successful. Although it is much too early to make definitive conclusions, most origin-of-life researchers are increasingly skeptical that life began with a protein. Their reasoning is that to make a copy of something, a mold or template is required. Proteins cannot furnish this information. Nucleic acids, in contrast, can. How they do so is the subject of Chapter 4.

CHECK YOUR UNDERSTANDING

If you understand that . . .

- Most proteins are enzymes, which make specific chemical reactions occur rapidly.
- Enzymes catalyze reactions by means of a three-step mechanism:
  - **Step 1** Binding of reactants in a precise orientation inside the active site.
  - **Step 2** Facilitation of the transition state, thus lowering the activation energy required for the reaction. This step often involves a shape change in the enzyme, resulting in an "induced fit" between active site and substrate. Enzyme cofactors are often required at this step.
  - **Step 3** Release of products, which do not bind tightly to the active site.
- The activity of enzymes is controlled through allosteric regulation or competitive inhibition.
- Different enzymes work at different rates, and enzymes are sensitive to changes in temperature and pH.

✓ You should be able to . . .

1. Explain why the precise orientation of reactants in the active site is important.
2. Explain why the shape change that occurs during allosteric regulation can change an enzyme's activity.

Answers are available in Appendix B.

Summary of Key Concepts

Most cell functions depend on proteins.

- In organisms, proteins function in catalysis, defense, movement, signaling, structural support, and transport of materials.

- Proteins can have diverse functions in cells because they have such diverse structures and chemical properties.

    ✔ You should be able to give an example of how a specific protein's structure is correlated with its function.

Amino acids are the building blocks of proteins. Amino acids vary in structure and function because their side chains vary in composition.

- Amino acids have a central carbon bonded to an amino group, a hydrogen atom, a carboxyl group, and an R-group.

- The structure of the R-group affects the chemical reactivity and solubility of the amino acid.

- In proteins, amino acids are joined by a peptide bond between the carboxyl group of one amino acid and the amino group of a second amino acid.

    ✔ You should be able to explain why some R-groups make amino acids soluble versus insoluble in water.

    (MB) **Web Activity** Condensation and Hydrolysis Reactions

Proteins vary widely in structure. The structure of a protein can be analyzed at four levels that form a hierarchy—the amino acid sequence, substructures called α-helices and β-pleated sheets, interactions between amino acids that dictate a protein's overall shape, and combinations of individual proteins that make up larger, multiunit molecules.

- A protein's primary structure, or sequence of amino acids, is responsible for most of its chemical properties.

- Interactions that take place between carbonyl and amino groups in the same peptide-bonded backbone create secondary structures, which are stabilized primarily by hydrogen bonding.

- Interactions between R-groups found in the same polypeptide and between R-groups and the peptide-bonded backbone allow the protein to fold into a characteristic overall shape—its tertiary structure.

- In many cases, a complete protein consists of several different polypeptides, bonded together. The combination of polypeptides represents the protein's quaternary structure.

    ✔ You should be able to make a concept map (see **BioSkills 8** in Appendix A) that relates the four levels of protein structure to one another and to how proteins function as catalysts. Your map should include the following boxed terms: Primary structure, secondary structure, tertiary structure, quaternary structure, transition state, activation energy, amino acid sequence, active site, R-groups, helices and sheets, 3D shape.

In cells, most proteins are enzymes that function as catalysts. Chemical reactions occur much faster when catalyzed by enzymes. During enzyme catalysis, the reactants bind to an enzyme's active site in a way that allows the reaction to proceed efficiently.

- Enzymes are protein catalysts that lower activation energy by stabilizing the transition state of the reaction. This speeds reaction rates. The enzyme itself is unchanged by the reaction.

- Catalysis takes place at the enzyme's active site, which has unique chemical properties and a distinctive size and shape. As a result, most enzymes catalyze a specific reaction.

- As a group, enzymes are able to catalyze many types of reactions because the chemical and physical structures of their active sites are so diverse. This diversity is due to the variety of amino acids and the four levels of protein structure.

- Many enzymes function only with the help of cofactors.

- Virtually all enzyme activity in cells is regulated. In most cases, regulation occurs when molecules bind at the active site or at locations on the protein that induce a change in the size or shape of the active site.

- The rate at which enzymes work depends on substrate concentration, their affinity for the substrate, temperature, and pH.

    ✔ You should be able to explain why a catalyst changes only the activation energy for a reaction—not the overall free energy change.

    (MB) **Web Activity** Activation Energy and Enzymes

Questions

1. What two functional groups are present on every amino acid?
    a. a carbonyl group and a carboxyl group
    b. an amino group and a carbonyl group
    c. an amino group and a hydroxyl group
    d. an amino group and a carboxyl group

2. Twenty different amino acids are found in the proteins of cells. What distinguishes these molecules?
    a. the location of their carboxyl group
    b. the location of their amino group

    c. the composition of their side chains, or R-groups
    d. their ability to form peptide bonds

3. What determines the primary structure of a polypeptide?
    a. its sequence of amino acids
    b. hydrogen bonds that form between carboxyl and amino groups on different residues
    c. hydrogen bonds and other interactions between side chains
    d. the number, identity, and arrangement of polypeptides that make up a protein

4. In a polypeptide, what is most responsible for the secondary structure called an α-helix?

   a. the sequence of amino acids

   b. hydrogen bonds that form between carbonyl and amino groups on different residues

   c. hydrogen bonds and other interactions between side chains

   d. the number, identity, and arrangement of polypeptides that make up a protein

5. What is a transition state?

   a. the complex formed as covalent bonds are being broken and re-formed during a reaction

   b. the place where an allosteric regulatory molecule binds to an enzyme

   c. an interaction between reactants with high kinetic energy, due to high temperature

   d. the shape adopted by an enzyme that has an inhibitory molecule bound at its active site

6. By convention, biologists write the sequence of amino acids in a polypeptide in which direction?

   a. carboxy- to amino-terminus

   b. amino- to carboxy-terminus

   c. polar residues to nonpolar residues

   d. charged residues to uncharged residues

✔ TEST YOUR UNDERSTANDING

Answers are available in Appendix B

1. Explain the lock-and-key model of enzyme activity. What was incorrect about this model?

2. Isoleucine, valine, leucine, phenylalanine, and methionine are amino acids with highly hydrophobic side chains. Suppose a section of a protein contains a long series of these hydrophobic residues. How would you expect this portion of the protein to behave when the molecule is in aqueous solution?

3. Compare and contrast competitive inhibition and allosteric regulation.

4. Does it take an input of energy for polymerization reactions to proceed, or do they occur spontaneously? Why or why not?

5. A major theme in this chapter is that the structure of molecules correlates with their function. Use this theme to explain why proteins can perform so many different functions in organisms and why proteins as a group are such effective catalysts.

6. Explain why temperature and pH affect enzyme function.

✔ APPLYING CONCEPTS TO NEW SITUATIONS

Answers are available in Appendix B

1. Researchers are searching for life on other planets. Is it reasonable to expect that protein-based life-forms could exist in extraterrestrial environments that are very hot or highly acidic? Why or why not?

2. Recently, researchers were able to measure movement that occurred in a single amino acid in an enzyme as reactions were taking place in its active site. The amino acid that moved was located in the active site, and the rate of movement correlated closely with the rate at which the reaction was taking place. Discuss the significance of these findings, using the information in Figures 3.19 and 3.20.

3. Researchers can analyze the atomic structure of enzymes during catalysis. In one recent study, investigators found that the transition state included the formation of a free radical, and that a coenzyme bound to the active site donated an electron to help stabilize the free radical. How would the reaction rate and the stability of the transition state change if the coenzyme were not available?

4. Some of the most effective drugs against HIV, the virus that causes AIDS, are competitive inhibitors of an HIV enzyme. But many HIV strains have now evolved resistance to these drugs—their enzymes are not affected by the drugs. Predict which parts of the enzymes changed. Explain your reasoning.

This is part of the sheet-metal-and-wire model that James Watson and Francis Crick used to figure out the secondary structure of DNA. The large "T" stands for the nitrogen-containing base thymine.

Nucleic Acids and the RNA World

4

Chapter 3 began with experimental evidence that chemical evolution produced the monomers called amino acids and their polymers called proteins. But the chapter ended by stating that even though proteins are the workhorse molecules of today's cells, relatively few researchers favor the hypothesis that life began as a protein molecule. Instead, the vast majority of biologists contend that life began as a polymer called a nucleic acid—specifically, a molecule of ribonucleic acid (RNA). This proposal is called the RNA world hypothesis.

The RNA world hypothesis contends that chemical evolution led to the existence of an RNA molecule that could make a copy of itself. Once this self-replicating molecule existed, chance errors in the copying process would create variations that would undergo natural selection—the evolutionary process introduced in Chapter 1. Chemical evolution was over, and biological evolution was off and running.

To test this hypothesis, several labs around the world are working to synthesize a self-replicating RNA molecule in the laboratory. If they succeed, they will have created a life-form in a test tube.

This chapter is focused on the structure and function of nucleic acids. Let's begin with an analysis of nucleic acids as monomers and end with current research efforts to create a self-replicating molecule.

4.1 What Is a Nucleic Acid?

Nucleic acids are polymers, just as proteins are polymers. But instead of being made up of monomers called amino acids, **nucleic acids** are made up of monomers called **nucleotides**.

KEY CONCEPTS

○→ Nucleotides consist of a sugar, a phosphate group, and a nitrogen-containing base. Ribonucleotides polymerize to form RNA. Deoxyribonucleotides polymerize to form DNA.

○→ DNA's primary structure consists of a sequence of nitrogen-containing bases. Its secondary structure consists of two DNA strands, running in opposite directions, that are held together by complementary base pairing and twisted into a double helix. DNA's structure allows organisms to store and replicate the information needed to grow and reproduce.

○→ RNA's primary structure consists of a sequence of nitrogen-containing bases. Its secondary structure includes short regions of double helices and structures called hairpins.

○→ Because RNA molecules can carry information as well as catalyze chemical reactions, it is likely that RNA was the first self-replicating molecule and a forerunner to the first life-form.

✔ When you see this checkmark, stop and test yourself. Answers are available in Appendix B.

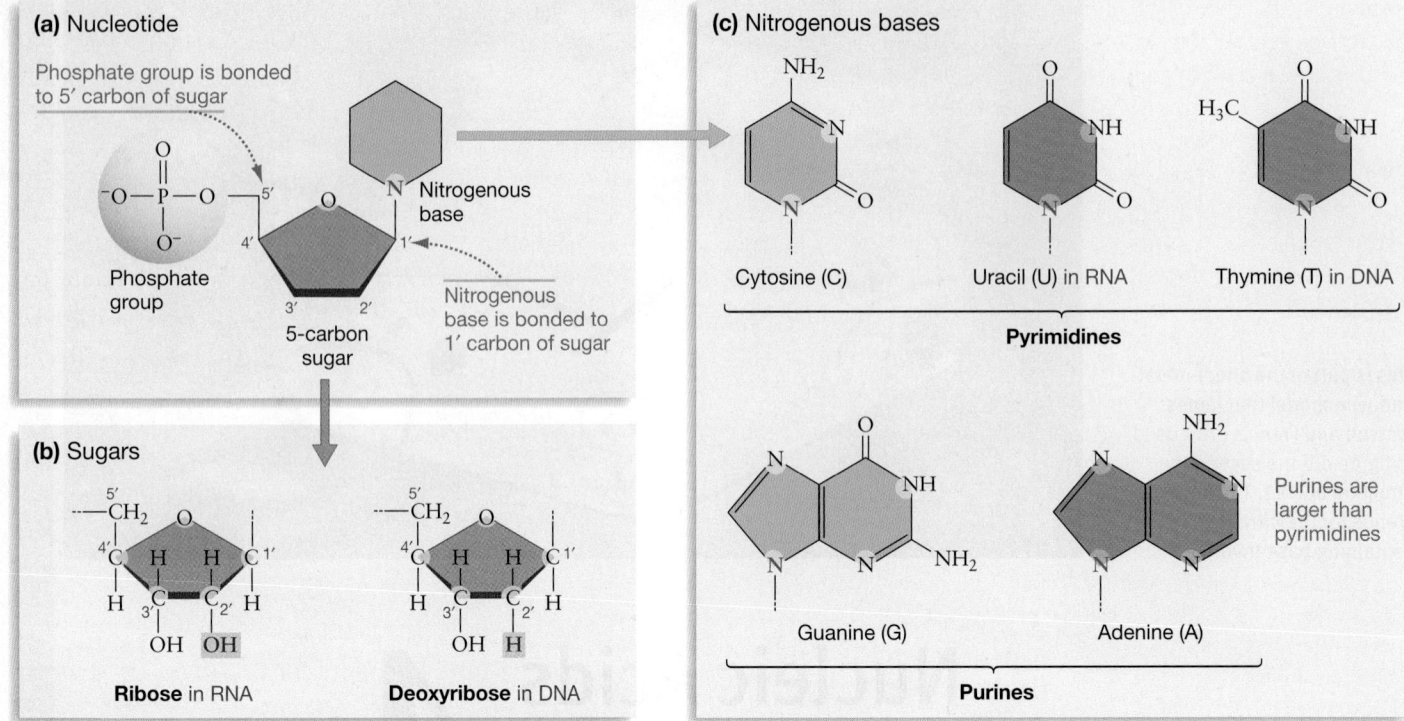

FIGURE 4.1 The General Structure of a Nucleotide. Note that in the bases, the nitrogen that bonds to the sugar is colored blue.

Figure 4.1a diagrams the three components of a nucleotide: (1) a phosphate group, (2) a sugar, and (3) a nitrogenous (nitrogen-containing) base. The phosphate is bonded to the sugar molecule, which in turn is bonded to the nitrogenous base. A **sugar** is an organic compound with a carbonyl group ($>C=O$) and several hydroxyl ($-OH$) groups.

Notice that the prime symbols ($'$) in Figure 4.1 indicate that the carbon being referred to is part of the sugar—not of the attached nitrogenous base. The phosphate group in a nucleotide is attached to the 5′ carbon.

Although a wide variety of nucleotides are found in living cells, origin-of-life researchers concentrate on two types: **ribonucleotides** and **deoxyribonucleotides**. In ribonucleotides, the sugar is ribose; in deoxyribonucleotides, it is deoxyribose. As **Figure 4.1b** shows, these two sugars differ by a single atom. Ribose has an $-OH$ group bonded to the 2′ carbon. Deoxyribose has an H instead at the same location. Note that deoxy means "lacking oxygen." Deoxyribonucleotides lack oxygen at the 2′ carbon.

Cells today have four different ribonucleotides, each of which contains a different nitrogenous base. These bases, diagrammed in **Figure 4.1c**, belong to structural groups called **purines** and **pyrimidines**. Ribonucleotides include the purines adenine (A) and guanine (G), and the pyrimidines cytosine (C) and uracil (U).

Similarly, four different deoxyribonucleotides are found in cells today and are distinguished by the structure of their nitrogenous base. Like ribonucleotides, deoxyribonucleotides include adenine, guanine, and cytosine. However, instead of uracil, a closely related pyrimidine called thymine (T) occurs in deoxyribonucleotides (Figure 4.1c).

✔You should be able to diagram a ribonucleotide and a deoxyribonucleotide, using a ball for the phosphate group, a pentagon to represent the sugar subunit, and a hexagon to represent the nitrogenous base. Label the **2′**, **3′**, and **5′** carbons on the sugar molecule, and add the atoms that are bonded to the **2′** carbon.

Could Chemical Evolution Result in the Production of Nucleotides?

Based on data presented in Chapter 3, most researchers contend that amino acids were abundant early in Earth's history. As yet, however, no one has observed the formation of a nucleotide via chemical evolution. The problem lies with mechanisms for synthesizing the sugar and nitrogenous base components of these molecules. Let's consider each issue in turn.

Laboratory simulations have shown that many sugars can be synthesized readily under conditions that mimic the prebiotic soup. Specifically, when formaldehyde (H_2CO) molecules are heated in solution, they react with one another to form almost all the sugars that have five or six carbons. (These are called pentoses and hexoses, respectively.)

In modern experiments, the various pentoses and hexoses are produced in approximately equal amounts, but ribose would have had to predominate for RNA or DNA to form in the prebiotic soup. How ribose came to be the dominant sugar during chemical evolution (i.e., what selective process was at work) is still a mystery. Origin-of-life researchers refer to this issue as the "ribose problem."

The origin of the pyrimidines is equally challenging. Simply put, origin-of-life researchers have yet to discover a plausible mechanism for the synthesis of pyrimidines (cytosine, uracil, and thymine) prior to the origin of life. Purines, in contrast, are readily synthesized by reactions among hydrogen cyanide (HCN) molecules. For example, both adenine and guanine have been found in the solutions recovered after spark-discharge experiments.

The ribose problem and the questions about the origin of pyrimidine bases are two of the most serious challenges that remain for the theory of chemical evolution. Research on these issues continues. In the meantime, let's consider the next question: Once nucleic acids formed, how did they polymerize to form RNA and DNA? This question has an answer.

How Do Nucleotides Polymerize to Form Nucleic Acids?

Nucleic acids form when nucleotides polymerize. As **Figure 4.2** shows, the polymerization reaction involves the formation of a bond between the phosphate group of one nucleotide and the hydroxyl group of the sugar component of another nucleotide. The result of this condensation reaction is called a **phosphodiester linkage**, or a phosphodiester bond.

In Figure 4.2, a phosphodiester linkage joins the 5′ carbon on the ribose of one nucleotide to the 3′ carbon on the ribose of the other. When the nucleotides involved contain the sugar ribose, the polymer that is produced is called **ribonucleic acid**, or simply **RNA**. If the nucleotides contain the sugar deoxyribose instead, then the resulting polymer is **deoxyribonucleic acid**, or **DNA**.

DNA AND RNA STRANDS ARE DIRECTIONAL   **Figure 4.3** shows how the chain of phosphodiester linkages in a nucleic acid acts as a backbone, analogous to the peptide-bonded backbone found in proteins.

The sugar-phosphate backbone of a nucleic acid is directional, as is the peptide-bonded backbone of a polypeptide. In a strand of RNA or DNA, one end has an unlinked 5′ carbon while

the other end has an unlinked 3′ carbon—meaning a carbon that is not linked to another nucleotide. By convention, the sequence of bases found in an RNA or DNA strand is always written in the 5′→3′ direction. (The system is logical because in cells, RNA and DNA are always synthesized in this direction. Bases are added at the 3′ end of the growing molecule.)

The sequence of nitrogenous bases forms the primary structure of the molecule, analogous to the sequence of amino acids in a polypeptide. When biologists write the primary structure of a stretch of DNA or RNA, they simply list the sequence of nucleotides in the 5′ to 3′ direction, using their single-letter abbreviations. For example, a six-base-long DNA sequence might be ATTAGC.

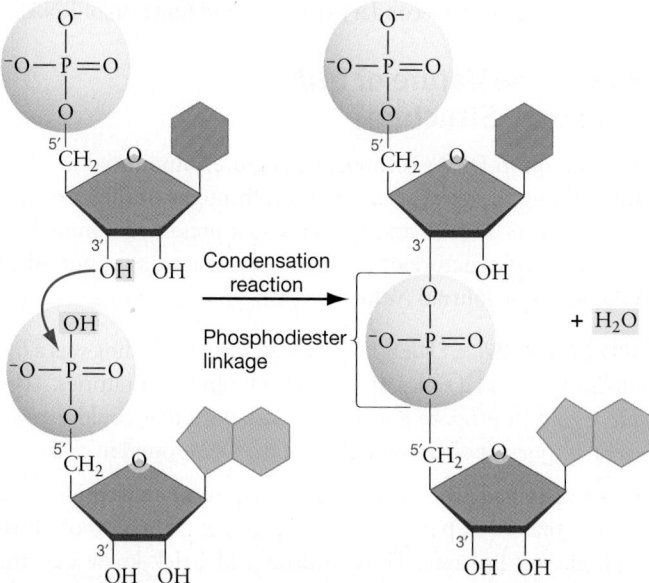

FIGURE 4.2 Nucleotides Polymerize via Phosphodiester Linkages. Ribonucleotides can polymerize via condensation reactions. The resulting phosphodiester linkage connects the 3′ carbon of one ribonucleotide and the 5′ carbon of another ribonucleotide.

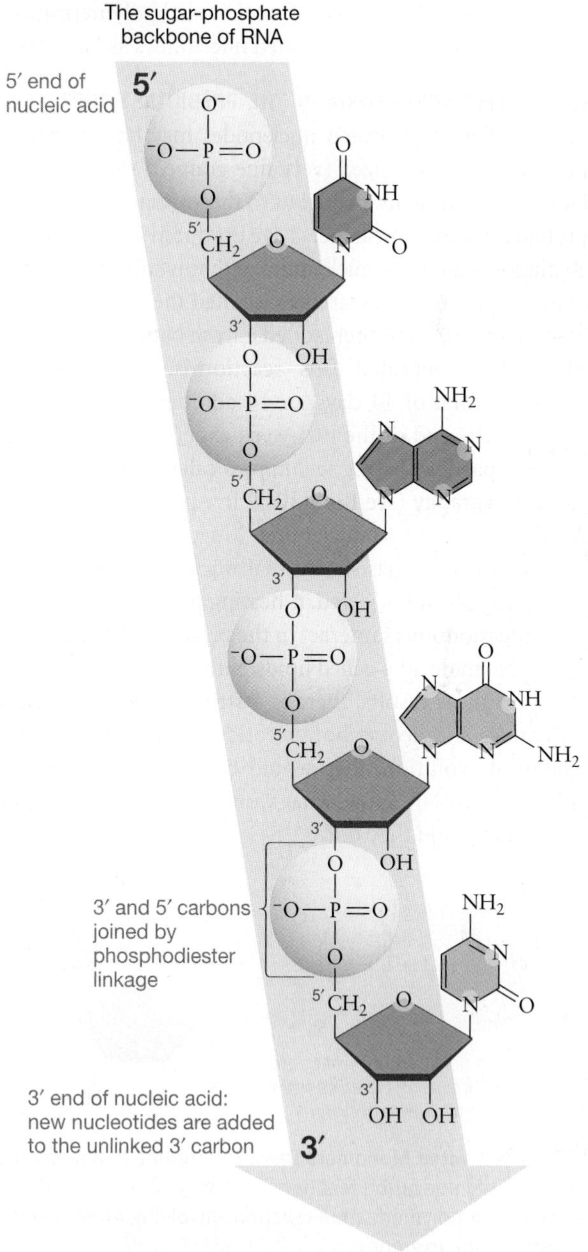

FIGURE 4.3 RNA Has a Sugar-Phosphate Backbone.

✔**EXERCISE** Identify the four bases in this RNA strand, using Figure 4.1c as a key. Then write down the base sequence, starting at the 5′ end.

POLYMERIZATION IS AN ENDERGONIC PROCESS In cells, the polymerization reactions that form nucleotides are catalyzed by enzymes. Like other polymerization reactions, the process is endergonic.

Polymerization can take place in cells because the free energy of the nucleotide monomers is first raised by reactions that add two phosphate groups to the ribonucleotides or deoxyribonucleotides, creating nucleoside triphosphates (**Figure 4.4**).[1] Molecules that have phosphate groups attached in this way are said to be **phosphorylated**.

This is a key point, and one that you will encounter again and again in this text: The addition of one or more phosphate groups raises the potential energy of substrate molecules enough to make an otherwise endergonic reaction possible. (Chapter 9 explains how this happens.) In the case of nucleic acid polymerization, researchers refer to the phosphorylated nucleotides as "activated."

COULD NUCLEIC ACIDS FORM IN THE PREBIOTIC SOUP? During chemical evolution, activated nucleotides may have polymerized on the surfaces of clay-sized (very fine grained) mineral particles. In a suite of experiments analogous to those discussed in Chapter 3 for protein synthesis, researchers have produced RNA molecules by incubating activated ribonucleotides with tiny mineral particles.

In one experiment, researchers isolated the clay particles after a day of incubation and then added a fresh batch of activated nucleotides. They repeated this reaction-isolation-reaction sequence for a total of 14 days, or 14 additions of fresh ribonucleotides. At the end of the two-week experiment, they analyzed the mineral particles, using techniques called **gel electrophoresis** and **autoradiography** (see **BioSkills 9** in Appendix A), and found RNA molecules up to 40 nucleotides long.

More recent work has shown that nucleotides can polymerize without being phosphorylated, if heating provides a source of energy and the monomers interact in the presence of nonpolar, non-water-soluble molecules called lipids, introduced in Chapter 6.

Based on these results, there is a strong consensus that if ribonucleotides and deoxyribonucleotides were able to form during chemical evolution, they would be able to polymerize and form DNA and RNA. Now, what would these nucleic acids look like, and what could they do?

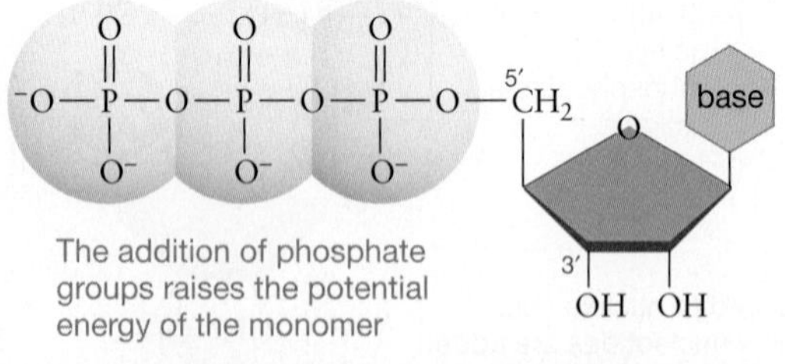

The addition of phosphate groups raises the potential energy of the monomer

FIGURE 4.4 Activated Monomers Drive Endergonic Polymerization Reactions. Polymerization reactions involving ribonucleotides are endergonic, but polymerization reactions involving ribonucleoside triphosphates are exergonic.

---

[1]A molecule consisting of a sugar and one of the bases in Figure 4.1c is called a nucleoside (a nucleotide is a sugar, a base, and one or more phosphate groups). Thus, a sugar attached to a base and three phosphate groups is called a nucleoside triphosphate.

CHECK YOUR UNDERSTANDING

If you understand that . . .

- Nucleotides are monomers that consist of a sugar, a phosphate group, and a nitrogen-containing base.
- Nucleotides polymerize to form nucleic acids through formation of phosphodiester linkages between the 3′ carbon on one nucleotide and the 5′ carbon on another.
- During polymerization, nucleotides are added only to the 3′ end of a nucleic acid strand.

✔ **You should be able to . . .**

Draw a simplified diagram of the phosphodiester linkage between two nucleotides, indicate the 5′ to 3′ polarity, and mark where the next nucleotide would be added to the growing chain.

Answers are available in Appendix B.

4.2 DNA Structure and Function

The primary structure of nucleic acids is somewhat similar to the primary structure of proteins. Proteins have a peptide-bonded backbone with a series of R-groups that extend from it. DNA and RNA molecules have a sugar-phosphate backbone, created by phosphodiester linkages, and a sequence of any of four nitrogenous bases that extend from it.

DNA and RNA also have secondary structure. But unlike the α-helices and β-pleated sheets of proteins, which are formed by hydrogen bonding between carbonyl and amino groups in the peptide-bonded backbone, the secondary structure of nucleic acids is formed by hydrogen bonding between nitrogenous bases.

Let's analyze the secondary structure and function of DNA first, and then dig into the secondary structure and function of RNA.

What Is the Nature of DNA's Secondary Structure?

The solution to DNA's secondary structure, announced in 1953, ranks among the great scientific breakthroughs of the twentieth century. James Watson and Francis Crick presented a model for the secondary structure of DNA in a one-page paper published in the scientific journal *Nature*.

EARLY DATA PROVIDE CLUES Watson and Crick's finding was a hypothesis based on a series of results from other laboratories. They were trying to propose a secondary structure that could explain several important observations about the DNA found in cells:

- Chemists had worked out the structure of nucleotides and knew that DNA polymerized through the formation of phosphodiester linkages. Thus, Watson and Crick knew that the molecule had a sugar-phosphate backbone.
- By analyzing the nitrogenous bases in DNA samples from different organisms, Erwin Chargaff had established two empirical rules: (**1**) The number of purines in a given DNA molecule

is equal to the number of pyrimidines, and (2) the number of T's and A's in DNA are equal, and the number of C's and G's in DNA are equal.

- By bombarding DNA with X-rays and analyzing how it scattered the radiation, Rosalind Franklin and Maurice Wilkins had calculated the distances between groups of atoms in the molecule (see **Bioskills 10** in Appendix A for an introduction to this technique, called **X-ray crystallography**). The scattering patterns showed that three distances were repeated many times: 0.34 nanometer (nm), 2.0 nm, and 3.4 nm. Because the measurements repeated, the researchers inferred that DNA molecules had a regular and repeat-

FIGURE 4.5 Building a Physical Model of DNA Structure. Watson (left) and Crick (right) represented the arrangement of the four deoxyribonucleotides in a double-helical arrangement, using metal plates and wires with precise lengths and geometries.

ing structure. The pattern of X-ray scattering suggested that the molecule was helical, or spiral, in nature.

Based on this work, understanding DNA's structure boiled down to understanding the nature of the helix involved. What type of helix would have a sugar-phosphate backbone and explain both Chargaff's rules and the Franklin-Wilkins measurements?

DNA STRANDS ARE ANTIPARALLEL Watson and Crick began by analyzing the size and geometry of deoxyribose, phosphate groups, and the nitrogenous bases. The bond angles and measurements suggested that the distance of 2.0 nm probably represented the width of the helix, and that 0.34 nm was likely to be the distance between bases stacked in a spiral.

How could they make sense of Chargaff's rules and the 3.4-nm distance, which appeared to be exactly 10 times the distance between a single pair of bases?

To solve this problem, Watson and Crick constructed a series of physical models such as the one pictured in **Figure 4.5**. The models allowed them to tinker with different types of helical configurations. After many false starts, something clicked:

- They arranged two strands of DNA side by side and running in opposite directions—meaning that one strand ran in the $5' \rightarrow 3'$ direction while the other strand was oriented $3' \rightarrow 5'$. Strands with this orientation are said to be **antiparallel**.

- If the antiparallel strands are twisted together to form a double helix, the coiled sugar-phosphate backbones end up on the outside of the spiral and the nitrogenous bases on the inside.

- For the bases from each backbone to fit in the interior of the 2.0-nm-wide structure, they have to form purine-pyrimidine pairs (see **Figure 4.6a**). This is key: The pairing allows hydrogen bonds to form between certain purines and pyrimidines. Adenine forms hydrogen bonds with thymine, and guanine forms hydrogen bonds with cytosine (**Figure 4.6b**).

(a) Only purine-pyrimidine pairs fit inside the double helix.

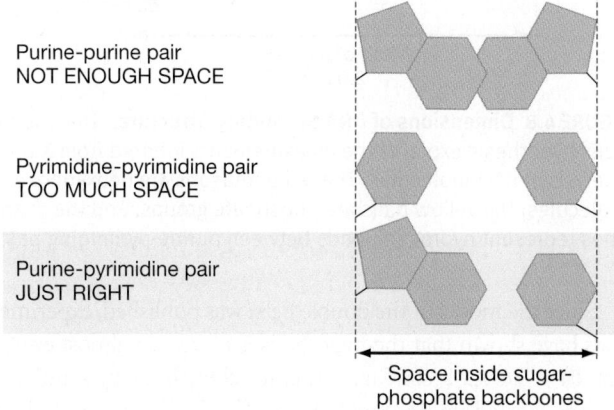

(b) Hydrogen bonds form between G-C pairs and A-T pairs.

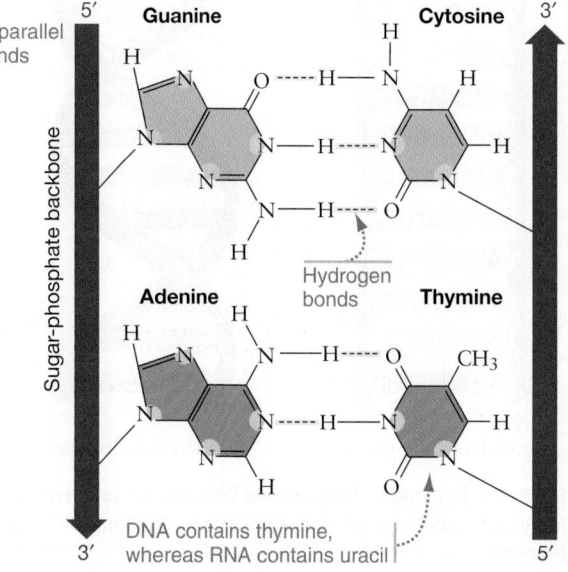

FIGURE 4.6 Complementary Base Pairing Is Based on Hydrogen Bonding.

- The A-T and G-C bases were said to be complementary. Two hydrogen bonds form when A and T pair, but three hydrogen bonds form when G and C pair. As a result, the G-C interaction is slightly stronger than the A-T bond. In contrast, A-C and G-T pairs allow no or only one hydrogen bond.

Watson and Crick had discovered **complementary base pairing**. In fact, the term **Watson-Crick pairing** is now used interchangeably with the phrase complementary base pairing.

THE DOUBLE HELIX **Figure 4.7** shows how antiparallel strands of DNA form when complementary bases line up and form hydrogen bonds. As you study the figure, notice that DNA is put together like a ladder whose ends have been twisted in opposite directions. The sugar-phosphate backbone forms the supports of the ladder; the base pairs represent the rungs of the ladder. The twisting occurs because it allows the nitrogenous bases to line up in a way that makes hydrogen bonding possible between them.

The nitrogenous bases in the middle of the DNA helix also stack tightly on top of each other. This tight packing forms a hydrophobic interior that is difficult to break apart. But the molecule as a whole is hydrophilic and water-soluble because the sugar-phosphate backbones, which face the exterior of the molecule, are negatively charged because of the negative charges on the phosphate groups. As a result, they interact with water.

It's also important to note that the outside of the helical DNA molecule forms two types of grooves. These grooves differ in size. The larger of the two is known as the major groove, and the smaller one is known as the minor groove. **Figure 4.8** highlights these grooves and illustrates how DNA's secondary structure explains the measurements observed by Franklin and Wilkins.

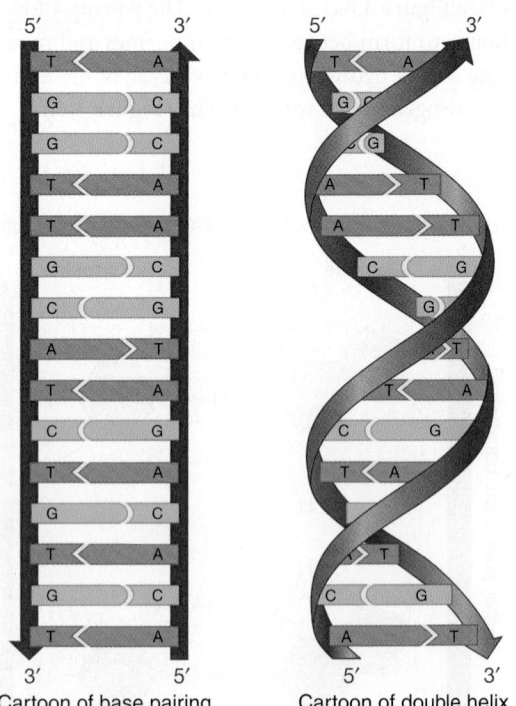

FIGURE 4.7 The Secondary Structure of DNA Is a Double Helix. Complementary base pairing is responsible for twisting DNA into a double helix.

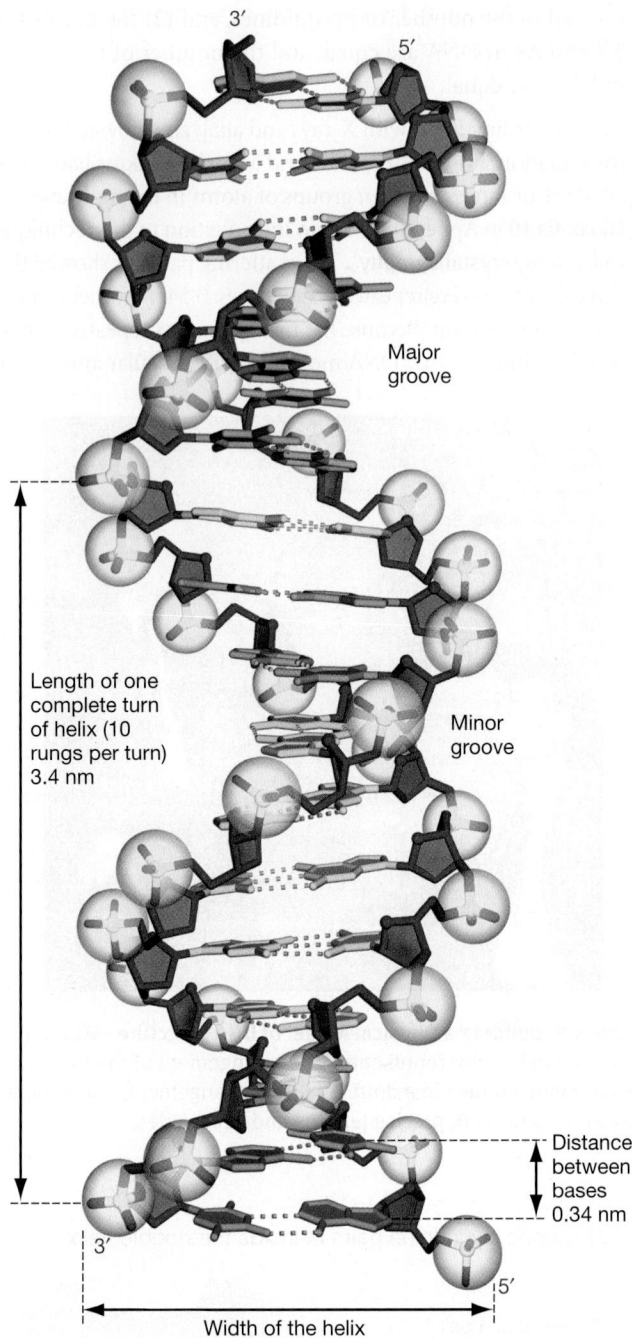

FIGURE 4.8 Dimensions of DNA Secondary Structure. The double-helix hypothesis explains the measurements inferred from X-ray analysis of DNA molecules. The red pentagons are deoxyribose molecules; the yellow balls are phosphate groups, and the dashed lines represent hydrogen bonds between purine-pyrimidine pairs.

Since the model of the double helix was published, experimental tests have shown that the hypothesis is correct in almost every detail. ▣━ DNA's secondary structure consists of two antiparallel strands twisted into a double helix. The molecule is stabilized by hydrophobic interactions in its interior and by hydrogen bonding between the complementary base pairs A-T and G-C. ✔You should be able to explain why no secondary structure forms when two DNA strands align in a parallel—instead of antiparallel—fashion.

Now the question is, How does this secondary structure affect the molecule's function?

DNA Functions as an Information-Containing Molecule

Watson and Crick's model created a sensation for a simple reason: It revealed how DNA could store and transmit biological information. In literature, information consists of letters on a page. In music, information is composed of the notes on a staff. But inside cells, information consists of a sequence of nucleotides in a nucleic acid. The four nitrogenous bases function like letters of the alphabet. A sequence of bases is like the sequence of letters in a word—it has meaning.

In all cells that have been examined to date, from tiny bacteria to gigantic redwood trees, DNA carries the information required for the organism's growth and reproduction. DNA is the basis for one of the five attributes of life introduced in Chapter 1: storing and processing genetic information. Exploring how hereditary information is encoded and translated into action is the heart of Chapters 15 through 18.

Here, however, our focus is on how life began. The theory of chemical evolution holds that life began as a self-replicating molecule—a molecule that could make a copy of itself. **Figure 4.9** shows how a copy of DNA can be made by complementary base pairing.

Step 1 Heating or enzyme-catalyzed reactions cause the double helix to separate.

Step 2 Free deoxyribonucleotides form hydrogen bonds with complementary bases on the original strand of DNA—also called a **template strand**. As they do, their sugar-phosphate groups form phosphodiester linkages to create a new strand—also called a **complementary strand**. Note that the 5′→3′ directionality of the complementary strand is opposite that of the template strand.

Step 3 Complementary base pairing allows each strand of a DNA double helix to be copied exactly, producing two daughter molecules.

Watson and Crick ended their paper on the double helix with one of the classic understatements in the scientific literature: "It has not escaped our notice that the specific pairing we have postulated immediately suggests a possible copying mechanism." Here's the key insight: DNA's primary structure serves as a mold or template for the synthesis of a complementary strand. DNA contains the information required for a copy of itself to be made.

To review DNA's primary and secondary structure and the difference between ribonucleotides and deoxyribonucleotides, go to the study area at *www.masteringbiology.com* and view the copying mechanism in action.

(MB) **Web Activity** Structure of RNA and DNA

DNA copying is the basis for a second of the five attributes of life introduced in Chapter 1: replication. But can DNA self-replicate? In today's cells and in laboratory experiments, the an-

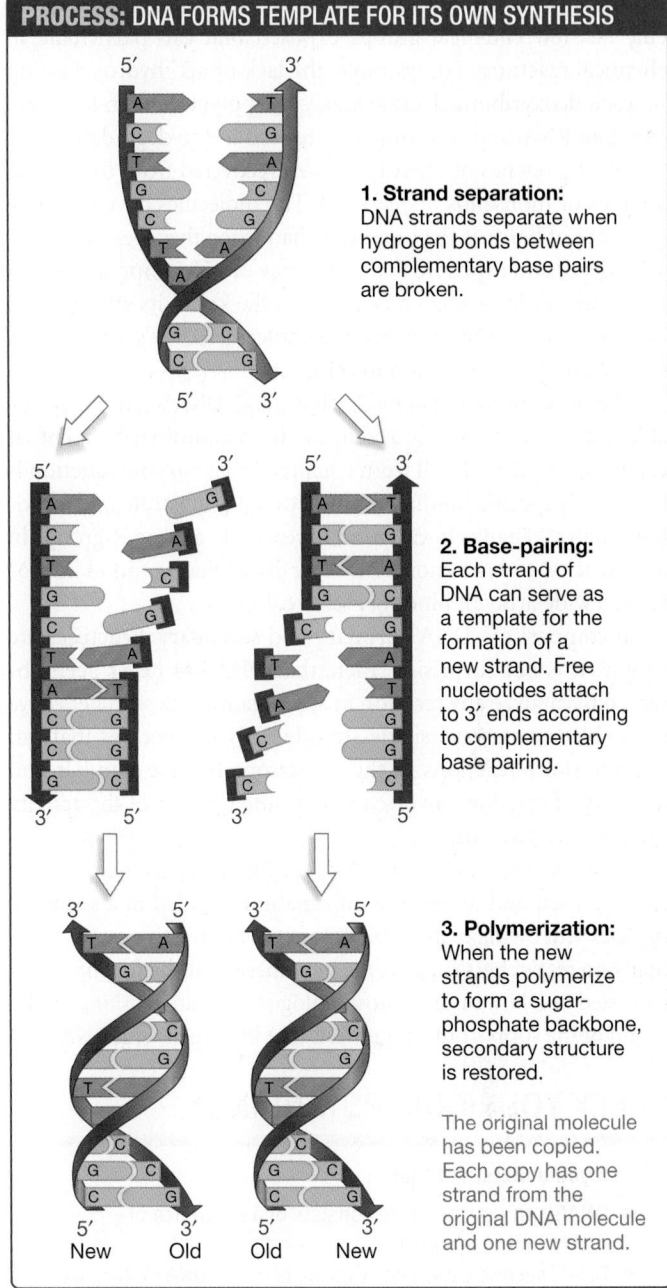

PROCESS: DNA FORMS TEMPLATE FOR ITS OWN SYNTHESIS

1. Strand separation: DNA strands separate when hydrogen bonds between complementary base pairs are broken.

2. Base-pairing: Each strand of DNA can serve as a template for the formation of a new strand. Free nucleotides attach to 3′ ends according to complementary base pairing.

3. Polymerization: When the new strands polymerize to form a sugar-phosphate backbone, secondary structure is restored.

The original molecule has been copied. Each copy has one strand from the original DNA molecule and one new strand.

FIGURE 4.9 Making a Copy of DNA. If new bases are added to each of the two strands of DNA via complementary base pairing, a copy of the DNA molecule can be produced.

✓**QUESTION** When double-stranded DNA is heated to 95°C, the bonds between complementary base pairs break and single-stranded DNA results. Considering this observation, is the reaction shown in step 1 endergonic or exergonic?

swer is no. Instead, the molecule is copied through a complicated series of energy-demanding reactions, catalyzed by a large suite of enzymes. Why can't DNA catalyze these reactions itself?

Is DNA a Catalytic Molecule?

The DNA double helix is highly structured. It is regular, symmetric, and held together by hydrogen bonding, hydrophobic interactions, and phosphodiester linkages. In addition, the molecule has few chemical groups exposed that can participate in chemical reactions. For example, the lack of a 2′ hydroxyl group on each deoxyribonucleotide makes the polymer much less reactive than RNA, and thus much more resistant to degradation.

Intact stretches of DNA have been recovered from fossils that are tens of thousands of years old. The molecules have the same sequence of bases as the organisms had when they were alive, despite death and exposure to a wide array of pH, temperature, and chemical conditions. DNA's stability is the key to its effectiveness as a reliable information-bearing molecule. DNA's structure is consistent with its function in cells.

The orderliness and stability that make DNA such a dependable information repository make it extraordinarily inept at catalysis, however. Recall from Chapter 3 that enzyme function is based on a specific binding event between a substrate and a protein catalyst. Thanks to variation in reactivity among R-groups in amino acids and the enormous diversity of shapes found in proteins, a wide array of binding events can occur.

In comparison, DNA's primary and secondary structures are simple. It is not surprising, then, that DNA has never been observed to catalyze any reaction in any organism. Researchers have been able to construct single-stranded DNA molecules that can catalyze some reactions in the laboratory, but the number and diversity of reactions involved is a minute fraction of the activity catalyzed by proteins.

In short, DNA furnishes an extraordinarily stable template for copying itself and for storing information encoded in a sequence of bases. But owing to its inability to act as an effective catalyst, virtually no researchers support the hypothesis that the first life-form consisted of DNA. Instead, most biologists who are working on the origin of life support the hypothesis that life began with RNA.

CHECK YOUR UNDERSTANDING

If you understand that . . .

- DNA's primary structure consists of a sequence of deoxyribonucleotides.
- DNA's secondary structure consists of two DNA molecules that run in opposite orientations to each other. The two strands are twisted into a double helix, and they are held together through hydrogen bonds between A-T and G-C pairs and hydrophobic interactions among atoms inside the helix.
- The sequence of deoxyribonucleotides in DNA contains information, much like the sequence of letters and punctuation on this page. Owing to complementary base pairing, each DNA strand also contains the information required to form its complementary strand.

✔ You should be able to . . .

Make a sketch of a DNA molecule and label the sugar-phosphate backbones, the hydrogen bonds between complementary bases, the location of hydrophobic interactions that stabilize the helix, and the orientation of each strand.

Answers are available in Appendix B.

4.3 RNA Structure and Function

A self-replicating molecule has to perform two key functions: it has to carry information and perform catalysis. At first glance, these two functions appear to conflict. Information storage requires regularity and stability; catalysis requires variation in chemical composition and variation and flexibility in shape.

How is it possible for a molecule to do both? The answer is structure.

Structurally, RNA Differs from DNA

Recall that most proteins have four levels of structure: the primary sequence of amino acids, secondary folds that are stabilized by hydrogen bonding between atoms in the peptide-bonded backbone, tertiary folds that are stabilized by interactions involving R-groups, and an overall or quaternary shape consisting of multiple polypeptides.

DNA has only primary and secondary structure. But RNA, like proteins, can have up to four levels of structure.

PRIMARY STRUCTURE Like DNA, RNA has a primary structure consisting of a sugar-phosphate backbone formed by phosphodiester linkages and, extending from that backbone, a sequence of four types of nitrogenous bases. But it's important to recall two significant differences between these nucleic acids:

1. The sugar in the sugar-phosphate backbone of RNA is ribose, not deoxyribose as in DNA.

2. The pyrimidine base thymine does not exist in RNA. Instead, RNA contains the closely related pyrimidine base uracil.

The first point is critical because the hydroxyl (−OH) group on the 2′-carbon of ribose is much more reactive than the hydrogen atom on the 2′-carbon of deoxyribose. This functional group can participate in reactions that tear the polymer apart. The presence of this −OH group makes RNA much less stable than DNA.

SECONDARY STRUCTURE Like DNA molecules, most RNA molecules have secondary structure that results from complementary base pairing between purine and pyrimidine bases. In RNA, adenine forms hydrogen bonds only with uracil, and guanine again forms hydrogen bonds with cytosine. (Guanine can also bond with uracil, but it does so much less effectively than with cytosine.) Thus, the complementary base pairs in RNA are A-U and G-C. Three hydrogen bonds form between guanine and cytosine, but only two form between adenine and uracil.

How do the secondary structures of RNA and DNA differ? In the vast majority of cases, the purine and pyrimidine bases in RNA undergo hydrogen bonding with complementary bases on the same strand, rather than forming hydrogen bonds with complementary bases on a different strand, as in DNA.

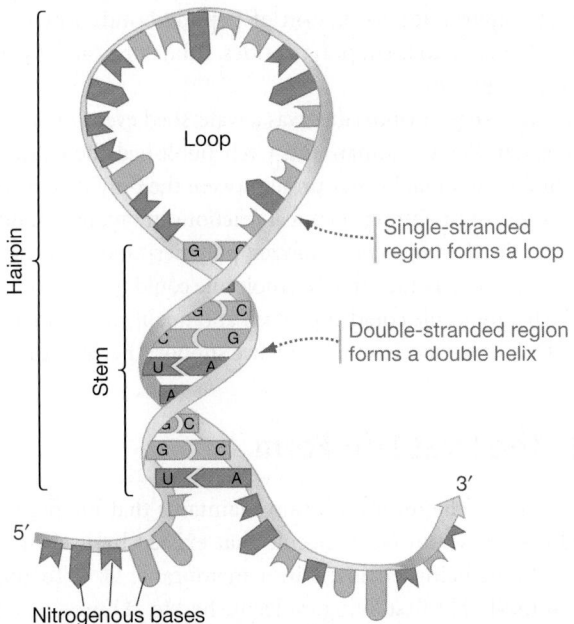

Loop

Single-stranded
region forms a loop

Double-stranded region
forms a double helix

Hairpin

Stem

G   C

G   C

G   G

U   A

A

C

G   C

U   A

3′

5′

Nitrogenous bases

FIGURE 4.10 Complementary Base Pairing and Secondary Structure in RNA: Stem-and-Loop Structures. This RNA molecule has secondary structure. The double-stranded "stem" and single-stranded "loop" form a hairpin. The bonded bases in the stem are antiparallel, meaning that they are oriented in opposite directions.

Figure 4.10 shows how the within-strand pairing works. The key is that when bases on one part of an RNA strand fold over and align with ribonucleotides on another segment of the same strand, the two sugar-phosphate strands are antiparallel. In this orientation, hydrogen bonding between complementary bases results in a stable double helix.

If the section where the fold occurs includes a large number of unbonded bases, then the stem-and-loop configuration shown in Figure 4.10 results. This type of secondary structure is called a **hairpin**. Several other types of RNA secondary structures are possible, each involving a different length and arrangement of base-paired segments.

Like the α-helices and β-pleated sheets observed in many proteins, RNA secondary structures are stabilized by hydrogen bonding and occur spontaneously. Even though hairpins and other types of secondary structure reduce the entropy of RNA molecules, they form spontaneously because the energy released in formation of H-bonds makes the overall process favorable.

Hydrogen bond formation is exothermic; overall, the formation of hairpins is exergonic.

TERTIARY AND QUATERNARY STRUCTURES RNA molecules can also have tertiary structure and quaternary structure, owing to interactions that (1) fold secondary structures into complex shapes or (2) hold different RNA strands together. As a result, RNA molecules with different base sequences can have very different overall shapes and chemical properties.

Although they do not begin to rival proteins in overall structural diversity or the presence of various functional groups, RNA molecules are much more diverse in size, shape, and reactivity than DNA molecules are. Structurally and chemically, RNA is intermediate between the complexity of proteins and the simplicity of DNA. **Table 4.1** summarizes the similarities and differences in the structures of RNA and DNA.

RNA's Structure Makes It an Extraordinarily Versatile Molecule

In terms of structure, RNA is intermediate between DNA and proteins. RNA is intermediate in terms of function as well. RNA molecules cannot archive information nearly as efficiently as DNA molecules do, but RNAs do perform key functions in information processing. Likewise, they cannot catalyze as many reactions as proteins do, but as it turns out, the reactions they do catalyze are particularly important.

In cells, RNA molecules function like a jackknife or a pocket tool with an array of attachments: They perform a wide variety of tasks reasonably well. Some of the most surprising results in the last decade of biological science, in fact, involve new insights into the diversity of roles that RNAs play in cells. These molecules process information stored in DNA, help synthesize proteins, and defend against attack by viruses, among other things.

Here let's focus on the roles that RNA could have played in the origin of life—as an information-containing entity and as a catalyst.

RNA Is an Information-Containing Molecule

Because RNA contains a sequence of bases analogous to the letters in a word, it can function as an information-containing molecule. And because hydrogen bonding occurs specifically between A-U pairs and G-C pairs in RNA, it is possible for RNA to furnish the information required to make a copy of itself.

SUMMARY TABLE 4.1 **DNA and RNA Structure**

	DNA	RNA
Primary structure	Sequence of deoxyribonucleotides; bases are A, T, G, C.	Sequence of ribonucleotides; bases are A, U, G, C.
Secondary structure	Two antiparallel strands twist into a double helix, stabilized by hydrogen bonding between complementary bases (A-T and G-C) and hydrophobic interactions.	Most common are hairpins, formed when a single strand folds back on itself to form a double-helix "stem" and a single-stranded "loop."
Tertiary structure	None*	Folds that form distinctive three-dimensional shapes.
Quaternary structure	None	Associations between several RNA molecules.

*In cells, DNA coils around proteins that bind to the double helix. In many cases the DNA-protein complex folds into highly organized, compact structures. But DNA does not form tertiary structure on its own.

The copying process occurs as shown for DNA in Figure 4.9, except that the template RNA exists as a single strand instead of a double strand. When RNA is copied, free ribonucleotides form hydrogen bonds with complementary bases on the original strand of RNA—the template strand. As they do, their sugar-phosphate groups form phosphodiester linkages to produce the complementary strand. Finally, the hydrogen bonds between the strands are broken by heating or by a catalyzed reaction. The new RNA molecule now exists independently of the original strand. If these steps were repeated with the new strand used as a template, the resulting molecule would be a copy of the original. In this way, an RNA's primary sequence serves as a mold.

Although complementary base pairing allows RNA to carry the information required for the molecule to be copied, RNA is not nearly as stable a repository for particular sequences as is DNA. In the prebiotic soup, could an RNA molecule have made a copy of itself before it would have degraded?

RNA Can Function as a Catalytic Molecule

In terms of diversity in chemical reactivity and overall shape, RNA molecules do not begin to match proteins. The primary structure of RNA molecules is much more restricted because there are only four types of nitrogenous bases in RNA versus the 20 types of amino acids found in proteins. Secondary through quaternary structure is more limited as a result, meaning that RNA cannot form the wide array of active sites observed among proteins.

But because RNA has a degree of structural and chemical complexity, it is capable of stabilizing a few transition states and catalyzing at least a limited number of chemical reactions. Sidney Altman and Thomas Cech shared the 1989 Nobel Prize in chemistry for showing that catalytic, enzyme-like RNAs, or **ribozymes**, exist in organisms.

Figure 4.11 shows the RNA molecule that Cech isolated from a single-celled organism called *Tetrahymena*. This ribozyme catalyzes both the hydrolysis and the condensation of phosphodiester linkages. Researchers have since discovered ribozymes that catalyze an array of different reactions in cells. For example, ri-

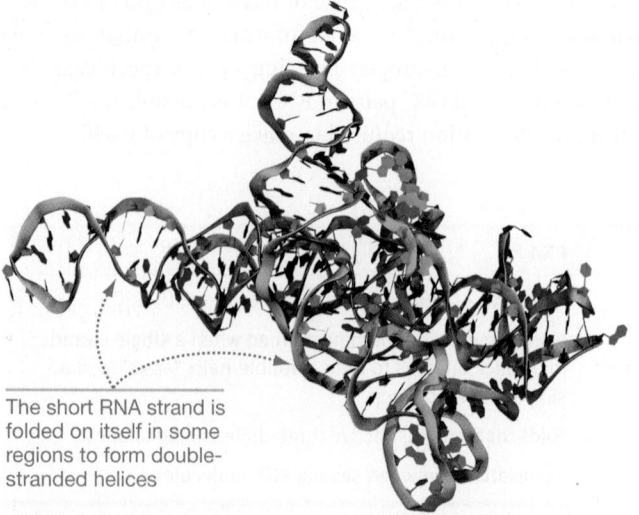

The short RNA strand is folded on itself in some regions to form double-stranded helices

FIGURE 4.11 Tertiary Structure of *Tetrahymena* Ribozyme.

bozymes catalyze the formation of peptide bonds when amino acids polymerize to form polypeptides. Ribozymes are at work in your cells right now.

The discovery of ribozymes was a watershed event in origin-of-life research. Before Altman and Cech published their discovery, most biologists thought that proteins were the only type of molecule capable of catalyzing chemical reactions in organisms. But if a ribozyme in *Tetrahymena* catalyzes polymerization reactions, it raises the possibility that an RNA molecule could make a copy of itself. Such a molecule could copy itself and qualify as the first living entity. Does any experimental evidence support this hypothesis?

4.4 The First Life-Form

The theory of chemical evolution maintains that life began as a naked self-replicator—a molecule that existed by itself in solution, without being enclosed in a membrane. To make a copy of itself, that first living molecule had to (**1**) provide a template that could be copied, and (**2**) catalyze polymerization reactions that would link monomers into a copy of that template.

Because RNA is capable of both processes, most origin-of-life researchers propose that the first life-form was made of RNA. No self-replicating RNA molecules exist today, however, so researchers test the hypothesis by trying to simulate the RNA world in the laboratory. The eventual goal is to create an RNA molecule that can catalyze its own replication.

To understand how researchers do this work, consider recent experiments by Wendy Johnston and others working in David Bartel's laboratory. This research group's goal was ambitious; they wanted to create, from scratch, a ribozyme that was capable of catalyzing the addition of ribonucleoside triphosphates to an existing strand, via complementary base pairing. Such a ribozyme would be called RNA replicase.

This research program is creating considerable excitement among biologists interested in the origin of life, because adding ribonucleotides to a growing strand is a key attribute of an RNA replicase. The logic of their experiment can be outlined as follows:

Step 1 Synthesize billions of large RNA molecules that contain randomly generated primary sequences.

Step 2 Incubate these newly synthesized, large RNAs with free ribonucleotides—including one that contains a specific molecular tag.

Step 3 Isolate large RNAs that have one of the tagged ribonucleotides attached. These molecules must be ribozymes, because they catalyzed the addition of the tagged ribonucleotide to the end of the growing RNA strand.

Step 4 Copy the ribozymes in a way that introduces a few random changes to their sequence of bases. These copied RNAs have a primary sequence that is similar to the ribozyme, but not identical.

Next Repeat steps 2–4—meaning, re-do the reaction, isolation, and copying sequence—17 times. By the end of the experiment, they had completed a total of 18 rounds of selection for ribozymes that can catalyze the formation of a phosphodiester linkage.

Here's the key to this experiment: When the ribozymes isolated in each round are copied, the molecules that result—and that furnish the starting material for the next round—are not identical in terms of their primary sequence because random changes ("mutations") were introduced during the copying step. Most of these "mutated" RNAs were worse ribozymes than the molecule that was copied, but some were better.

By isolating the best ribozymes from each round and copying them (with more "mutations" that potentially could make them even better) in the next round, the research team continually selected for ribozymes that were more and more efficient. By round 18, the group had a ribozyme that was a much better catalyst than the original molecule.

This experimental protocol was designed to mimic the process of natural selection, introduced in Chapter 1. The population of ribozymes had variable characteristics that could be copied and passed on to the next generation of ribozymes. In addition, the researchers were able to select the most efficient ribozymes to be the "parents" of the next generation.

In effect, evolution was occurring. By round 18, the researchers had a ribozyme that was reasonably proficient at adding ribonucleotides to a growing strand. As this book goes to press, the round-18 ribozyme is the closest biologists have come to creating life.

Thanks to similar efforts at other laboratories around the world, biologists have produced an increasingly impressive set of ribozymes—an array of molecules capable of catalyzing many of the key reactions responsible for replication and metabolism.

Each of these results provides support for the RNA world hypothesis. Each result also brings research teams closer to the creation of an RNA replicase. If this goal is met, human beings will have created a living entity in a test tube.

CHAPTER 4 REVIEW

For media, go to the study area at www.masteringbiology.com

Summary of Key Concepts

🔑 **Nucleotides consist of a sugar, a phosphate group, and a nitrogen-containing base. Ribonucleotides polymerize to form RNA. Deoxyribonucleotides polymerize to form DNA.**

- Ribonucleotides have a hydroxyl ($-OH$) group on their $2'$ carbon. Deoxyribonucleotides do not.

- Nucleic acids form when nucleotides polymerize. Polymerization occurs via phosphodiester linkages between the $3'$ carbon on one nucleotide and the phosphate group on another.

- Nucleic acids have a sugar-phosphate backbone with nitrogenous bases attached.

- Nucleic acids are directional: they have a $5'$ end and a $3'$ end.

- During polymerization, new nucleotides are added only to the $3'$ end.

  ✔ You should be able to compare and contrast proteins and nucleic acids in terms of (1) polymerization, (2) their directionality, and (3) the presence of a "backbone."

🔑 **DNA's primary structure consists of a sequence of nitrogen-containing bases. Its secondary structure consists of two DNA strands, running in opposite directions, that are held together by complementary base pairing and twisted into a double helix. DNA's structure allows organisms to store and replicate the information needed to grow and reproduce.**

- DNA is an extremely stable molecule that serves as a superb archive for information in the form of base sequences.

- DNA is stable because deoxyribonucleotides lack a reactive $2'$ hydroxyl group and because antiparallel DNA strands form a secondary structure called a double helix. The DNA double helix is stabilized by hydrogen bonds that form between complementary purine and pyrimidine bases and by hydrophobic interactions between bases stacked on the inside of the spiral.

- DNA's structural stability and regularity make it ineffective at catalysis.

- DNA is readily copied via complementary base pairing. Complementary base pairing occurs between A-T and G-C pairs in DNA.

  ✔ You should be able to explain why DNA molecules with a high percentage of guanine and cytosine are particularly stable.

  MB **Web Activity** Structure of RNA and DNA

🔑 **RNA's primary structure consists of a sequence of nitrogen-containing bases. Its secondary structure includes short regions of double helices and structures called hairpins.**

- RNA molecules can have secondary structure because complementary base pairing occurs between A-U and G-C pairs on the same strand. Some tertiary and quaternary structure also occurs, because hydrogen bonding allows RNA molecules to fold in precise ways, and thus interact with each other.

- RNA is versatile. The primary function of proteins is to catalyze chemical reactions, and the primary function of DNA is to carry information. But RNA is an "all-purpose" macromolecule that can do both.

  ✔ You should be able to explain why RNA molecules can have tertiary and quaternary structure, while DNA molecules do not.

🔑 **Because RNA molecules can carry information as well as catalyze chemical reactions, it is likely that RNA was the first self-replicating molecule and a forerunner to the first life-form.**

- To test the RNA world hypothesis, researchers are attempting to synthesize an RNA replicase in the laboratory. They have succeeded in isolating ribozymes that can catalyze formation of phosphodiester linkages.

  ✔ You should be able to explain why, in laboratory experiments designed to produce efficient ribozymes, it is critical to copy the RNA molecules in a way that produces random changes in the primary structure.

Questions

✓ **TEST YOUR KNOWLEDGE**   *Answers are available in Appendix B*

1. What are the four nitrogenous bases found in RNA?
   a. uracil, guanine, cytosine, thymine (U, G, C, T)
   b. adenine, guanine, cytosine, thymine (A, G, C, T)
   c. adenine, uracil, guanine, cytosine (A, U, G, C)
   d. alanine, threonine, glycine, cysteine (A, T, G, C)

2. What determines the primary structure of an RNA molecule?
   a. the sugar-phosphate backbone
   b. complementary base pairing and the formation of hairpins
   c. the sequence of deoxyribonucleotides
   d. the sequence of ribonucleotides

3. DNA attains a secondary structure when hydrogen bonds form between the nitrogenous bases called purines and pyrimidines. What are the complementary base pairs that form in DNA?
   a. A-T and G-C
   b. A-U and G-C
   c. A-G and T-C
   d. A-C and T-G

4. By convention, biologists write the sequence of bases in RNA and DNA in which direction?
   a. $3' \rightarrow 5'$
   b. $5' \rightarrow 3'$
   c. N-terminal to C-terminal
   d. C-terminal to N-terminal

5. The secondary structure of DNA is called a double helix. Why?
   a. Two strands wind around one another in a helical, or spiral, arrangement.
   b. A single strand winds around itself in a helical, or spiral, arrangement.
   c. It is shaped like a ladder.
   d. It stabilizes the molecule.

6. In RNA, when does the secondary structure called a hairpin form?
   a. when hydrophobic residues coalesce
   b. when hydrophilic residues interact with water
   c. when complementary base pairing between ribonucleotides on the same strand creates a stem-and-loop structure
   d. when complementary base pairing between two antiparallel strands forms a double helix

✓ **TEST YOUR UNDERSTANDING**   *Answers are available in Appendix B*

1. Make a concept map (see **BioSkills 8** in Appendix A) that relates DNA's primary structure to its secondary structure. Your diagram should include deoxyribonucleotides, hydrophobic interactions, purines, pyrimidines, phosphodiester linkages, DNA primary structure, DNA secondary structure, complementary base pairing, and antiparallel strands.

2. Explain why the addition of three phosphate groups to nucleosides allows otherwise-endergonic polymerization reactions to occur.

3. Explain why nucleic acids are directional. In a nucleic acid, what are the 5′ carbons bonded to? What are the 3′ carbons bonded to?

4. Why is DNA such a stable molecule compared with proteins or RNA?

5. Explain how the secondary structures called hairpins form in RNA.

6. A major theme in this chapter is that the structure of molecules correlates with their function. Explain why DNA's secondary structure limits its catalytic abilities compared with that of RNA. Why is it logical that RNA molecules can catalyze a modest but significant array of reactions? Why are proteins the most effective catalysts?

✓ **APPLYING CONCEPTS TO NEW SITUATIONS**   *Answers are available in Appendix B*

1. Viruses are particles that infect cells. In some viruses, the genetic material consists of two strands of RNA, bonded together via complementary base pairing. Would these antiparallel strands form a double helix? Explain why or why not.

2. Suppose that experiments like those reviewed in Section 4.4 succeeded in producing a molecule that could make a copy of itself. According to the criteria listed in Chapter 1, would this molecule be considered alive?

3. Before Watson and Crick published their model of the DNA double helix, Linus Pauling offered a model based on a triple helix. If the three sugar-phosphate backbones were on the outside of such a molecule, would hydrogen bonding or hydrophobic interactions be more important in keeping such a secondary structure together?

4. Origin-of-life researcher Robert Crabtree maintains that experiments simulating early Earth conditions are a valid way to test the theory of chemical evolution. Crabtree claims that if scientists working in the field agree that an experiment is a plausible reproduction of early Earth conditions, it is valid to infer that its results are probably correct—that the simulation effectively represents events that occurred some 3.5 billion years ago. Do you agree? Do you find the models and experiments presented in this chapter and previous chapters to be convincing tests of the theory? Explain your answers.

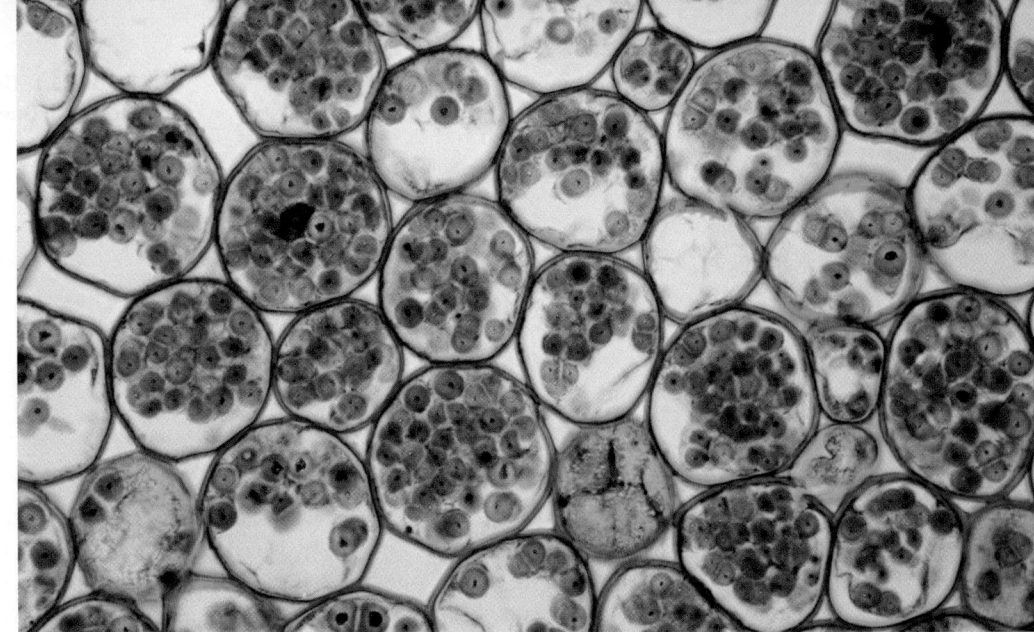

A cross section through a buttercup root. Cellulose-rich cell walls are stained green; starch-filled structures are stained purple. Cellulose is a structural carbohydrate; starch is an energy-storage carbohydrate.

An Introduction to Carbohydrates 5

This unit highlights the four types of macromolecules that are prominent in today's cells: proteins, nucleic acids, carbohydrates, and lipids. Understanding the structure and function of each of these macromolecules is a basic requirement for exploring how life began and how organisms work. Chapters 3 and 4 analyzed the way proteins and nucleic acids are put together and what they do. This chapter focuses on carbohydrates; Chapter 6 will introduce lipids.

The term **carbohydrate, or sugar**, encompasses the monomers called monosaccharides (literally, "one-sugar"), small polymers called oligosaccharides ("few-sugars"), and the large polymers called polysaccharides ("many-sugars"). The name carbohydrate is logical because the chemical formula of many of these molecules is $(CH_2O)_n$, where the n refers to the number of "carbon-hydrate" groups. The name can also be misleading, though, because carbohydrates do not consist of carbon atoms bonded to water molecules. Instead, they are molecules with a carbonyl ($>C=O$) and several hydroxyl ($-OH$) functional groups, along with several to many carbon-hydrogen (C−H) bonds.

Let's begin with monosaccharides, put them together into polysaccharides, then explore how carbohydrates figured in the origin of life and what they do in cells today. As you study this material, be sure to ask yourself the central question of biological chemistry: How does this molecule's stucture relate to its properties and function?

5.1 Sugars as Monomers

Sugars are fundamental to life. They provide chemical energy in cells and furnish some of the molecular building blocks required for the synthesis of larger, more complex compounds. Monosaccharides were important during chemical evolution, early in Earth's history. For example, the sugar called ribose is required for the formation of nucleotides. What are sugars, and how do they differ from one another?

KEY CONCEPTS

- Sugars and other carbohydrates are highly variable in structure.

- Monosaccharides are monomers that polymerize via condensation reactions to form polymers called polysaccharides. The monosaccharides in polysaccharides are joined by different types of glycosidic linkages.

- Carbohydrates have diverse functions in cells. In addition to serving as raw material for synthesizing other molecules, they provide fibrous structural materials, indicate cell identity, and store chemical energy.

✔ When you see this checkmark, stop and test yourself. Answers are available in Appendix B.

An aldose
Carbonyl group at end of carbon chain

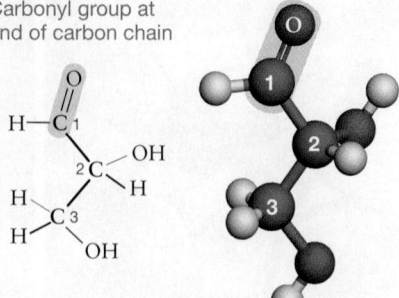

A ketose
Carbonyl group in middle of carbon chain

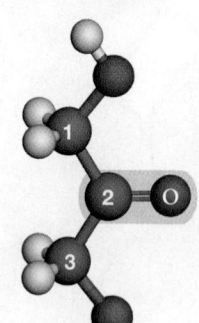

How Monosaccharides Differ

Figure 5.1 illustrates the structure of the monomer called a **monosaccharide**, or simple sugar. The carbonyl group that serves as one of monosaccharides' distinguishing features can be found either at the end of the molecule, forming an aldehyde sugar (an aldose), or within the carbon chain, forming a ketone sugar (a ketose). The presence of a carbonyl group along with multiple hydroxyl groups provides an array of functional groups in sugars. Based on this observation, it's not surprising that sugars are able to participate in a large number of chemical reactions.

The number of carbon atoms present also varies in monosaccharides. By convention, the carbons in a monosaccharide are numbered consecutively, starting with the end nearest the carbonyl group. Figure 5.1 features three-carbon sugars, or **trioses**. Ribose, which acts as a building block for nucleotides, has five carbons and is called a **pentose**; the glucose that is coursing through your bloodstream and being used by your cells right now is a six-carbon sugar, or a **hexose**.

Besides varying in the location of the carbonyl group and the total number of carbon atoms present, monosaccharides can vary in the spatial arrangement of their atoms. There is, for example, a wide array of pentoses and hexoses. Each is distinguished by the configuration of its hydroxyl functional groups. **Figure 5.2** illustrates glucose and galactose, which are six-carbon sugars. Note that the two molecules have the same chemical formula ($C_6H_{12}O_6$) but not the same structure. Both are aldose sugars with six carbons, but they differ in the spatial arrangement of the hydroxyl group at the carbon highlighted in Figure 5.2.

This is a key point: Because the structures of glucose and galactose differ, their functions differ. In cells, glucose is used as a source of chemical energy that sustains life. But for galactose to be used as a source of energy, it first has to be converted to glucose via an enzyme-catalyzed reaction. This example underscores a general theme: Even seemingly simple changes in structure—like the location of a single hydroxyl group—can have enormous consequences for function. This is because molecules interact in precise ways, based on their shape.

It's rare for sugars to exist in the form of the linear chains illustrated in Figures 5.1 and 5.2, however. In aqueous solution they tend to form ring structures. Glucose serves as the example in **Figure 5.3**. When the cyclic structure forms in glucose, the C-1 carbon (the carbon numbered 1 in the linear chain) forms a bond with the oxygen atom of the C-5 hydroxyl, and its carbonyl group becomes a hydroxyl group. This hydroxyl group can be oriented in two distinct ways: above or below the plane of the ring. The different configurations produce the molecules α-glucose and β-glucose.

To summarize, many distinct monosaccharides exist because so many aspects of their structure are variable: aldose or ketose placement of the carbonyl group, variation in carbon number, different arrangements of hydroxyl groups in space, and alternative ring forms. Each monosaccharide has a unique structure and function.

Glucose

Galactose

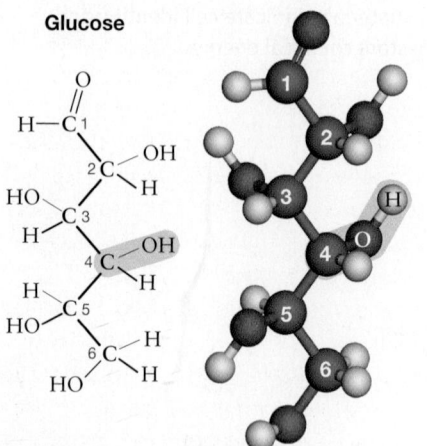

FIGURE 5.2 Sugars May Vary in the Configuration of Their Hydroxyl Groups. The two six-carbon sugars shown here vary only in the spatial orientation of their hydroxyl groups on carbon number 4.

✔ **EXERCISE** Mannose is a six-carbon sugar that is identical with glucose, except that the hydroxyl (—OH) group on carbon number 2 is switched in orientation. Circle carbon number 2 in glucose and galactose, then draw the structural formula of mannose.

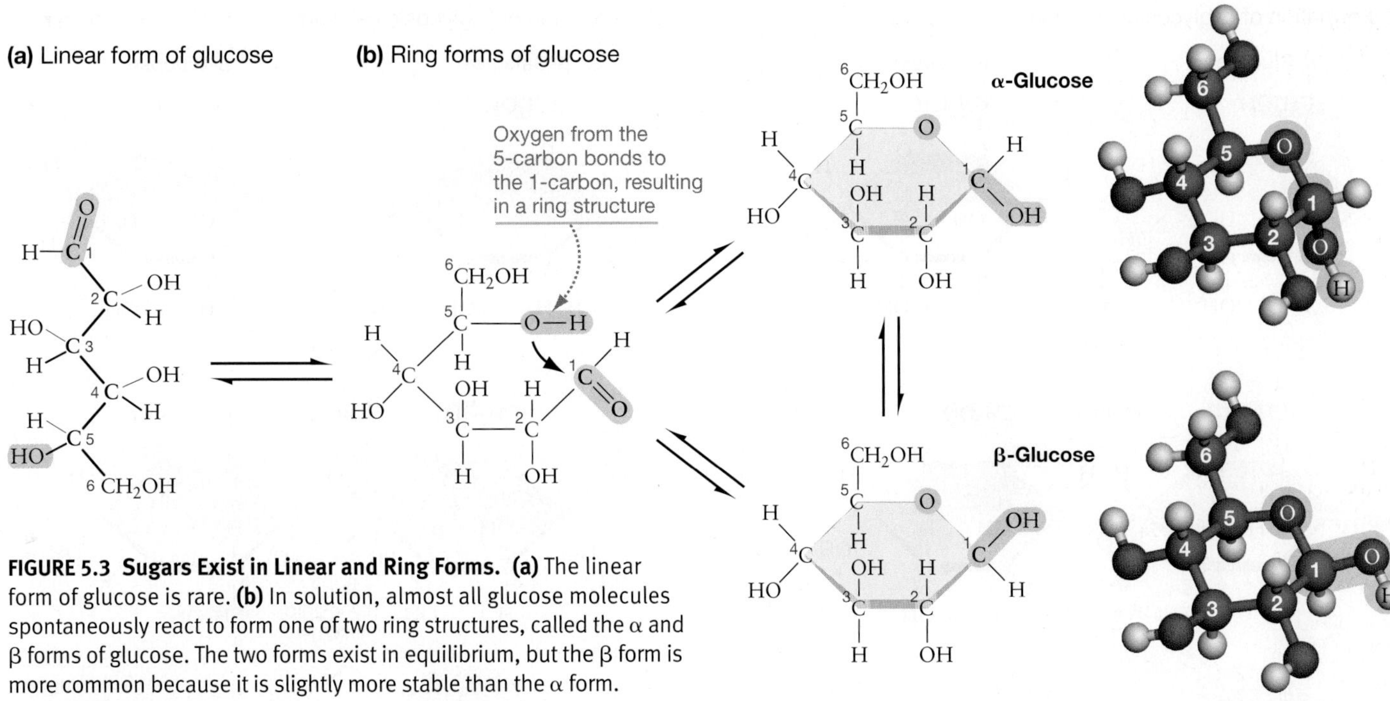

(a) Linear form of glucose   **(b)** Ring forms of glucose

Oxygen from the 5-carbon bonds to the 1-carbon, resulting in a ring structure

α-Glucose

β-Glucose

FIGURE 5.3 Sugars Exist in Linear and Ring Forms. (a) The linear form of glucose is rare. **(b)** In solution, almost all glucose molecules spontaneously react to form one of two ring structures, called the α and β forms of glucose. The two forms exist in equilibrium, but the β form is more common because it is slightly more stable than the α form.

CHECK YOUR UNDERSTANDING

If you understand that . . .

- Simple sugars differ in three respects:
  1. the location of their carbonyl group,
  2. the number of carbon atoms present, and
  3. the spatial arrangement of their atoms—particularly the relative positions of hydroxyl (−OH) groups.

✔ **You should be able to . . .**

Draw the structural formula of a three-carbon monosaccharide in linear form and then draw other sugars that differ from this one in each of the three aspects listed.

Answers are available in Appendix B.

Monosaccharides and Chemical Evolution

Laboratory simulations have shown that most monosaccharides are readily synthesized under conditions that mimic the prebiotic soup. For example, when formaldehyde (H_2CO) molecules are heated in solution, they react with one another to form almost all the pentoses and hexoses, as well as some seven-carbon sugars. In addition, researchers recently announced the discovery of the three-carbon ketose illustrated in Figure 5.1, along with a wide array of compounds closely related to sugars, on the Murchison meteorite that struck Australia in 1969.

Based on these observations, investigators suspect that sugars are synthesized on dust particles and other debris in interstellar space and could have rained down onto Earth as the planet was forming, as well as being synthesized in the hot water near undersea volcanoes. Most researchers interested in chemical evolution maintain that a wide diversity of monosaccharides existed in the prebiotic soup.

But as Chapter 4 pointed out, it remains a mystery why ribose might have predominated and made the synthesis of nucleotides possible. It also appears highly unlikely that monosaccharides were able to polymerize to form the polysaccharides found in today's cells. Let's explore why.

5.2 The Structure of Polysaccharides

Polysaccharides are polymers that form when monosaccharides are linked together. They are also known as complex carbohydrates. The simplest polysaccharides consist of two sugars and are known as **disaccharides**. The two monomers involved may be identical, as in the two α-glucose molecules that link to form maltose. Or they may be different, as in the combination of a glucose molecule and a galactose molecule that forms lactose—the most important sugar in milk.

Simple sugars polymerize when a condensation reaction occurs between two hydroxyl groups, resulting in a covalent bond called a **glycosidic linkage.** Glycosidic linkages are analogous to the peptide bonds that hold proteins together and to the phosphodiester linkages that connect the nucleotides in nucleic acids. There is an important difference, however. Peptide and phosphodiester bonds always form at the same location in their monomers. But because glycosidic linkages form between hydroxyl groups, and because every monosaccharide contains at least two hydroxyl groups, the location and geometry of glycosidic linkages can vary widely among polysaccharides.

Figure 5.4 shows two of the most common glycosidic linkages, called an α-1,4-glycosidic linkage and a β-1,4-glycosidic linkage. The numbers refer to the carbons on either side of the linkage; the α and β refer to the contrasting orientations of the linkage. As Section 5.3 will show, the differences in orientation are particularly

(a) Formation of α-glycosidic linkage

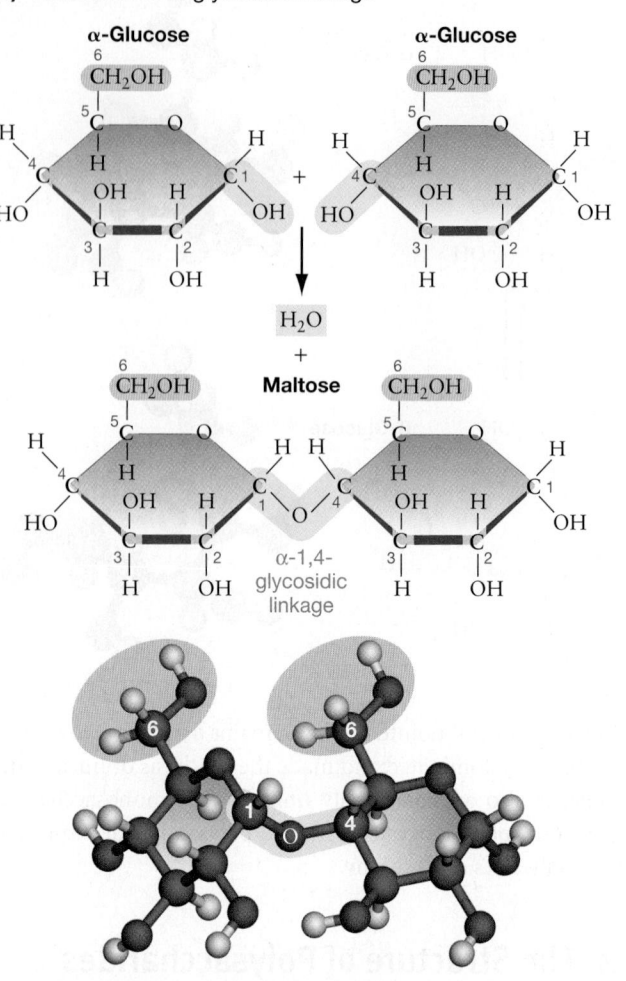

(b) Formation of β-glycosidic linkage

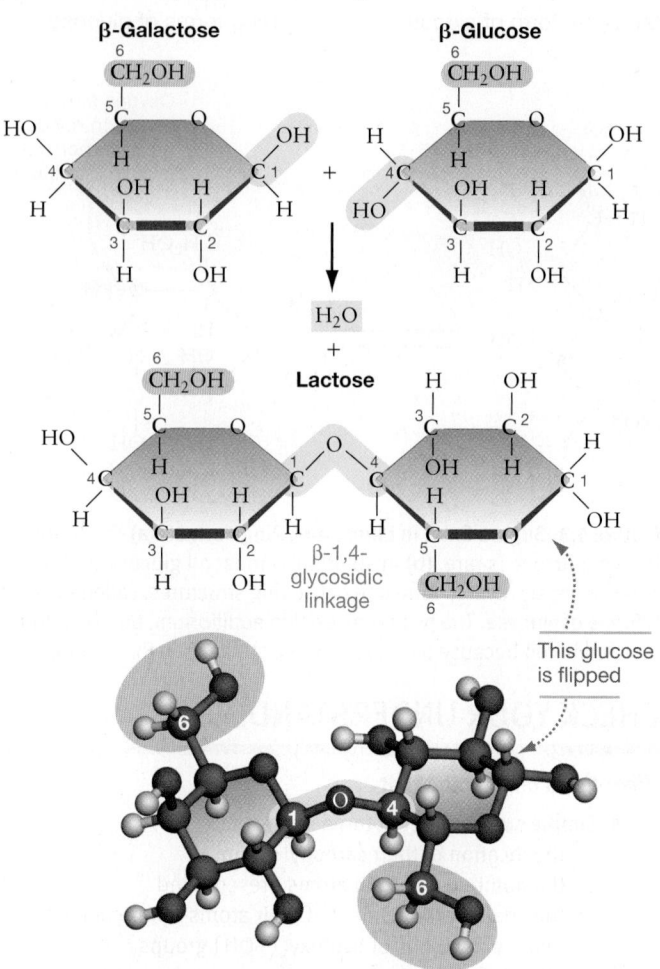

FIGURE 5.4 Monosaccharides Polymerize through Formation of Glycosidic Linkages. A glycosidic linkage occurs when hydroxyl groups on two monosaccharides undergo a condensation reaction to form a bond. Maltose and lactose are disaccharides.

important: α-linkages are easy for enzymes to break; β-linkages are difficult for enzymes to break.

To drive this point home, consider the structures of the most common polysaccharides found in organisms today: starch, glycogen, cellulose, and chitin, along with a modified polysaccharide called peptidoglycan. Each of these macromolecules can consist of a few hundred to many thousands of monomers, joined by glycosidic linkages at different locations.

Starch: A Storage Polysaccharide in Plants

In plant cells, monosaccharides are stored for later use in the form of starch. **Starch** consists entirely of α-glucose monomers that are joined by glycosidic linkages. As the top panel in **Table 5.1** shows, the angle of the linkages between carbons 1 and 4 causes the chain of glucose subunits to coil into a helix. Starch is actually a mixture of two such polysaccharides, however. One is an unbranched molecule called amylose, which contains only α-1,4-glycosidic linkages. The other is a branched molecule

called amylopectin. The branching in amylopectin occurs when glycosidic linkages form between carbon 1 of a glucose monomer on one strand and carbon 6 of a glucose monomer on another strand. In amylopectin, branches occur in about one out of every 30 monomers.

Glycogen: A Highly Branched Storage Polysaccharide in Animals

Glycogen performs the same storage role in animals that starch performs in plants. In humans, for example, glycogen is stored in the liver and in muscles. When you start exercising, enzymes begin breaking glycogen into glucose monomers, which are then processed in muscle cells to supply energy. Glycogen is a polymer of α-glucose and is nearly identical with the branched form of starch. However, instead of an α-1,6-glycosidic linkage occurring in about 1 out of every 30 monomers, a branch occurs in about 1 out of every 10 glucose subunits (see Table 5.1).

Polysaccharide	Chemical Structure	Three-dimensional Structure

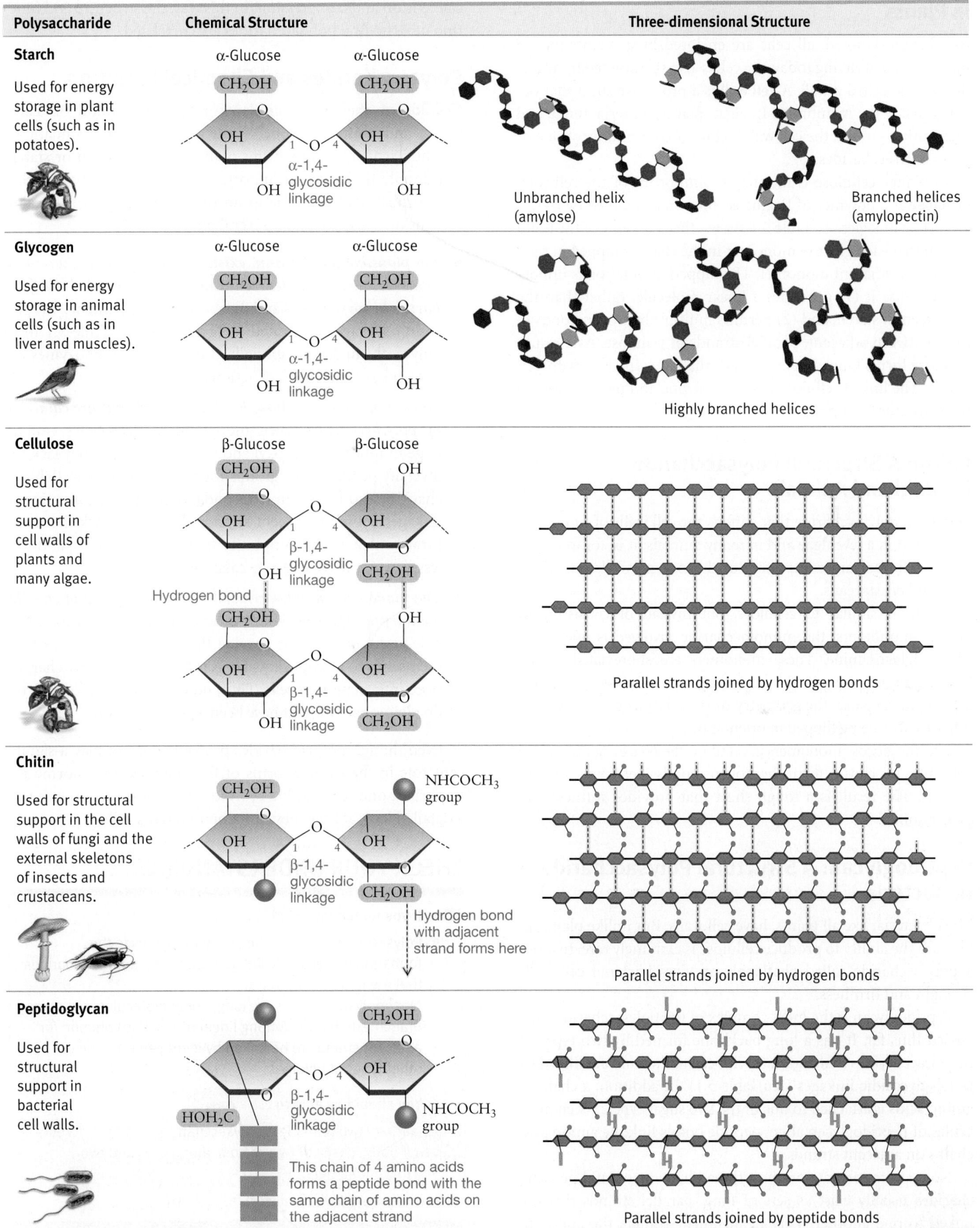

Starch

Used for energy storage in plant cells (such as in potatoes).

α-Glucose   α-Glucose

CH₂OH   CH₂OH

α-1,4-glycosidic linkage

Unbranched helix (amylose)   Branched helices (amylopectin)

Glycogen

Used for energy storage in animal cells (such as in liver and muscles).

α-Glucose   α-Glucose

CH₂OH   CH₂OH

α-1,4-glycosidic linkage

Highly branched helices

Cellulose

Used for structural support in cell walls of plants and many algae.

β-Glucose   β-Glucose

CH₂OH   OH

β-1,4-glycosidic linkage

CH₂OH

Hydrogen bond

CH₂OH   OH

β-1,4-glycosidic linkage   CH₂OH

Parallel strands joined by hydrogen bonds

Chitin

Used for structural support in the cell walls of fungi and the external skeletons of insects and crustaceans.

CH₂OH   NHCOCH₃ group

β-1,4-glycosidic linkage   CH₂OH

Hydrogen bond with adjacent strand forms here

Parallel strands joined by hydrogen bonds

Peptidoglycan

Used for structural support in bacterial cell walls.

CH₂OH

β-1,4-glycosidic linkage   OH

HOH₂C   NHCOCH₃ group

This chain of 4 amino acids forms a peptide bond with the same chain of amino acids on the adjacent strand

Parallel strands joined by peptide bonds

Cellulose: A Structural Polysaccharide in Plants

As Chapter 1 noted, all cells are enclosed by a membrane. In most organisms living today, the cell is also surrounded by a layer of material called a wall. A **cell wall** is a protective sheet that occurs outside the membrane. In algae, plants, bacteria, fungi, and many other groups, the cell wall is composed primarily of one or more polysaccharides.

In plants, cellulose is the major component of the cell wall. **Cellulose** is a polymer of β-glucose monomers, joined by β-1,4-glycosidic linkages. As Table 5.1 shows, the geometry of the bond is such that each glucose monomer in the chain is flipped in relation to the adjacent monomer. The flipped orientation is important because it (1) generates a linear molecule, rather than the helix seen in starch, and (2) permits multiple hydrogen bonds to form between adjacent, parallel strands of cellulose. As a result, cellulose forms long, parallel strands that are joined by hydrogen bonds. The linked cellulose fibers are strong and provide the cell with structural support.

Chitin: A Structural Polysaccharide in Fungi and Animals

Chitin is a polysaccharide that stiffens the cell walls of fungi. It is also found in a few algae and in many animals; it is, for example, the most important component of the external skeletons of insects and crustaceans.

Chitin is similar to cellulose, but instead of consisting of glucose monomers, the monosaccharide involved is one called *N*-acetylglucosamine. These monomers are abbreviated "NAc." NAc monomers are joined by β-1,4-glycosidic linkages (see Table 5.1). As in cellulose, the geometry of these bonds results in every other residue being flipped in orientation.

Like the glucose monomers in cellulose, the *N*-acetylglucosamine subunits in chitin form hydrogen bonds between adjacent strands. The result is a tough sheet that provides stiffness and protection.

Peptidoglycan: A Structural Polysaccharide in Bacteria

Most bacteria, like all plants, have cell walls. But unlike plants, in bacteria the ability to produce cellulose is extremely rare. Instead, a polysaccharide called **peptidoglycan** gives bacterial cell walls strength and firmness.

Peptidoglycan is the most complex of the polysaccharides discussed thus far. It has a long backbone formed by two types of monosaccharides that alternate with each other and are linked by β-1,4-glycosidic linkages (see Table 5.1). In addition, a chain of amino acids is attached to one of the two sugar types. When molecules of peptidoglycan align, peptide bonds link the amino acid chains on adjacent strands.

Note an important common thread here: Structural polysaccharides usually exist as sets of long, parallel strands that are linked to one another. This arrangement confers the ability to withstand pulling and pushing forces—what an engineer would call tension and compression. In this way, the structure and function of structural polysaccharides are correlated.

Polysaccharides and Chemical Evolution

Cellulose is the most abundant organic compound on Earth today, and chitin is probably the second most abundant by weight. Virtually all organisms manufacture glycogen or starch. But despite their current importance to organisms, polysaccharides probably played little to no role in the origin of life. This conclusion is supported by several observations:

- *No plausible mechanism exists for the polymerization of monosaccharides under conditions that prevailed early in Earth's history.* In cells and in laboratory experiments, the glycosidic linkages illustrated in Figure 5.4 and Table 5.1 form only with the aid of specialized enzymes. No ribozymes are known to catalyze these reactions.

- *To date, no reactions have been discovered that are catalyzed by polysaccharides.* Even though monosaccharides contain large numbers of hydroxyl and carbonyl groups, they lack the diversity of functional groups found in amino acids. Polysaccharides also have simple secondary structures, consisting of linkages between adjacent strands. Thus, they lack the structural and chemical complexity that makes proteins, and to a lesser extent RNA, effective catalysts.

- *The monomers in polysaccharides are not capable of complementary base pairing.* Like proteins, but unlike DNA and RNA, polysaccharides cannot provide the information required for themselves to be copied. As far as is known, no polysaccharides store information in cells. Thus, no one has proposed that the first living entity might have been a polysaccharide.

Even though polysaccharides probably did not play a significant role in the earliest forms of life, they became enormously important once cellular life evolved. In the next section let's take a detailed look at how they function in today's cells.

CHECK YOUR UNDERSTANDING

If you understand that . . .

- Polysaccharides form when enzymes catalyze the formation of glycosidic linkages between monosaccharides that are in the α or β form.
- Most polysaccharides are long, linear molecules, but some branch extensively. Among linear forms, it is common for adjacent strands to be linked by hydrogen bonding or other types of linkages.

✓ You should be able to . . .

Explain why two four-sugar polysaccharides can be different, even if both consist of two glucose monomers and two galactose monomers.

Answers are available in Appendix B.

5.3 What Do Carbohydrates Do?

Chapter 4 introduced one of the four basic functions that carbohydrates perform in organisms: serving as a substrate for synthesizing more-complex molecules. Recall that both RNA and DNA contain sugars—the five-carbon sugars ribose and deoxyribose, respectively. In nucleotides, which consist of a sugar, a phosphate group, and a nitrogenous base, the sugar itself acts as a subunit of the larger molecule.

In addition, sugars frequently furnish the raw "carbon skeletons" that are used as building blocks in the synthesis of important molecules. Amino acids are being produced by your cells right now, for example, using sugars as a starting point.

Although the details of how sugars are used in synthesizing amino acids and other complex molecules are beyond the scope of this book, you can delve into the other three major roles of carbohydrates in the rest of this chapter and in the supporting materials in the study area at *www.masteringbiology.com*.

(MB) **Web Activity** Carbohydrate Structure and Function

☞ Carbohydrates have diverse functions in cells: In addition to serving as precursors to larger molecules, they provide fibrous structural materials, indicate cell identity, and store chemical energy.

The Role of Carbohydrates as Structural Molecules

Cellulose and chitin, along with the modified polysaccharide peptidoglycan, are key structural compounds. They form fibers that give cells and organisms strength and elasticity.

To appreciate why cellulose, chitin, and peptidoglycan are effective as structural molecules, recall that they form long strands and that bonds can form between adjacent strands. In the cell walls of plants, for example, a collection of about 80 cellulose molecules are cross-linked by hydrogen bonding to create a tough fiber. These cellulose fibers, in turn, crisscross to form a tough sheet.

Besides being stiff and strong, the structural carbohydrates are durable. Almost all organisms have the enzymes required to break the α-1,4- and α-1,6-glycosidic linkages that hold starch and glycogen molecules together, but only a few organisms have enzymes capable of hydrolyzing the β-1,4-glycosidic linkages in cellulose, chitin, and peptidoglycan. The shape and orientation of β-1,4-glycosidic linkages make them difficult to break, and few enzymes have active sites with the correct geometry and reactive groups to do so. As a result, the structural polysaccharides are resistant to degradation and decay.

Ironically, the durability of cellulose is important for digestion. The cellulose that you ingest when you eat plant cells—what biologists call dietary fiber—passes through your gut without being broken down. The cellulose in dietary fiber, like the cellulose in a paper towel, absorbs water. The presence of dietary fiber adds moisture to the feces. In addition, cellulose adds bulk that helps fecal material move through the intestinal tract more quickly, preventing constipation, and other problems.

The Role of Carbohydrates in Cell Identity

Polysaccharides do not store information in cells, but they do *display* important information. More specifically, polysaccharides act as signage on the outer surface of the plasma membrane that surrounds a cell. (Chapter 6 will describe plasma membranes in detail.)

Figure 5.5 shows how this information display happens. Molecules called glycoproteins project outward from the cell surface into the surrounding environment. A **glycoprotein** is a protein that is covalently bonded to a carbohydrate—usually the relatively short chain of sugars called oligosaccharides.

Glycoproteins are key molecules in what biologists call cell-cell recognition and cell-cell signaling. Each cell in your body has glycoproteins on its surface that identify it as part of your body. Immune system cells use these glycoproteins to distinguish your body's cells from foreign cells, such as bacteria. In addition, each distinct type of cell in a multicellular organism—for example, the nerve cells and muscle cells in your body—displays a different set of glycoproteins on its surface.

In cells, glycoproteins form a "sugar coating" that acts like the magnetic stripe on the back of a credit card or the personal identification number (PIN) that you use to access a bank account—it immediately identifies the individual that bears it. The identification information displayed by glycoproteins helps cells recognize and communicate with each other.

The key point here is to recognize that the enormous number of structurally distinct monosaccharides makes it possible for an enormous number of unique oligosaccharides to exist. As a result, each cell type and each species can display a unique identity.

This is another example where a molecule's structure correlates with its function. Oligosaccharides that function in cell identity vary because their component monomers—and the linkages between them—vary.

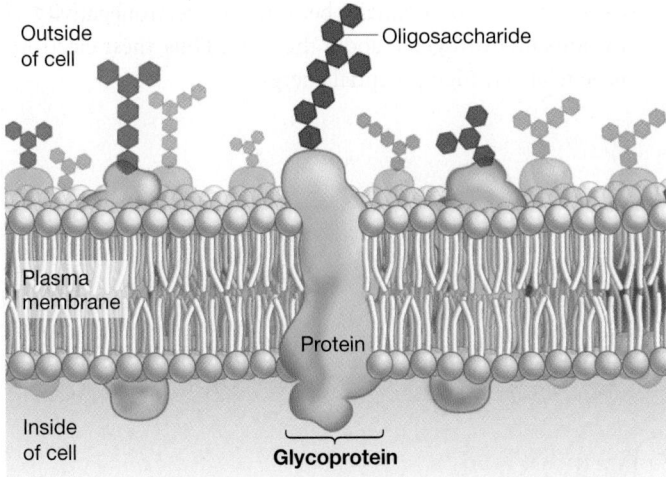

FIGURE 5.5 Carbohydrates Are an Identification Badge for Cells. Glycoproteins contain sugar groups that project outside the cell from the surface of the plasma membrane enclosing the cell. These sugar groups have distinctive structures that identify the type or species of the cell.

The Role of Carbohydrates in Energy Storage

Candy-bar wrappers promise a quick energy boost, and ads for sports drinks claim that their products provide the "carbs" needed for peak activity. If you were to ask friends or family members what carbohydrates do in your body, they would probably say something like "They give you energy." And after pointing out that carbohydrates are also used in cell identity, as a structural material, and as a source of carbon skeletons for the synthesis of other complex molecules, you'd have to agree.

Carbohydrates store and provide chemical energy in cells. What aspect of carbohydrate structure makes this function possible?

CARBOHYDRATES STORE SUNLIGHT AS CHEMICAL ENERGY Recall from earlier chapters that the essence of chemical evolution was the conversion of kinetic energy in sunlight and heat into chemical energy stored in the bonds of molecules such as formaldehyde (H_2CO) and hydrogen cyanide (HCN). Today, the kinetic energy in sunlight is converted to chemical energy stored in the bonds of carbohydrates, via the process known as **photosynthesis**.

Photosynthesis entails a complex set of reactions that can be summarized most simply as follows:

$$CO_2 + H_2O + sunlight \longrightarrow (CH_2O)_n + O_2$$

where $(CH_2O)_n$ represents a carbohydrate.

Figure 5.6 shows the structural formulas of the molecules involved in the summary reaction of photosynthesis. Note, too, that gold dots represent the relative positions of their covalently bonded electrons. The key to understanding this figure is to compare the positions of the electrons in the reactants and the products.

1. The electrons in the C=O bonds of carbon dioxide and the C−O bonds of carbohydrates are held tightly, because of oxygen's high electronegativity. Thus, they have relatively low potential energy.

2. The covalently bonded electrons in the C−H bonds of carbohydrates are shared equally because the electronegativity of carbon and hydrogen is about the same. Thus, these electrons have relatively high potential energy.

3. Electrons are also shared equally in the carbon-carbon (C−C) bonds of carbohydrates—meaning that they, too, have relatively high potential energy.

C−C and C−H bonds have much higher free energy than C−O bonds have. As a result, carbohydrates have much more free energy than carbon dioxide has.

The essence of photosynthesis, then, is that energy in sunlight is transformed into chemical energy that is stored in the C−H and C−C bonds of carbohydrates. Because carbohydrates are much more highly ordered than carbon dioxide in addition to having higher potential energy, they have much higher free energy (see Chapter 2).

Figure 5.7 summarizes and extends these points. Start by comparing the structure of carbon dioxide in Figure 5.7a with the carbohydrate in Figure 5.7b. The main difference is the presence of hydrogen atoms and C−H bonds in the carbohydrate. Now compare the carbohydrate in Figure 5.7b with the fatty acid—a subunit of a fat molecule—in Figure 5.7c. Compared with carbohydrates, fats contain many more C−C and C−H bonds and many fewer C−O bonds.

This is important. C−C and C−H bonds have high free energy because the electrons are shared equally by atoms with low electronegativities. C−O bonds, in contrast, have low free energy because the highly electronegative oxygen atom holds the electrons so tightly. Both carbohydrates and fats are used as fuel in cells, but fats store twice as much energy per gram compared with carbohydrates. Fats will be discussed in more detail in Chapter 6.

ENZYMES HYDROLYZE CARBOHYDRATES TO RELEASE GLUCOSE Starch and glycogen are efficient energy-storage molecules because they polymerize via α-glycosidic linkages instead of the β-glycosidic linkages observed in the structural polysaccharides. The α-linkages in storage polysaccharides are readily hydrolyzed to release glucose, while the β-linkages in structural polysaccharides resist enzymatic degradation.

The glucose subunits that are hydrolyzed from starch and glycogen are then processed in reactions that result in the pro-

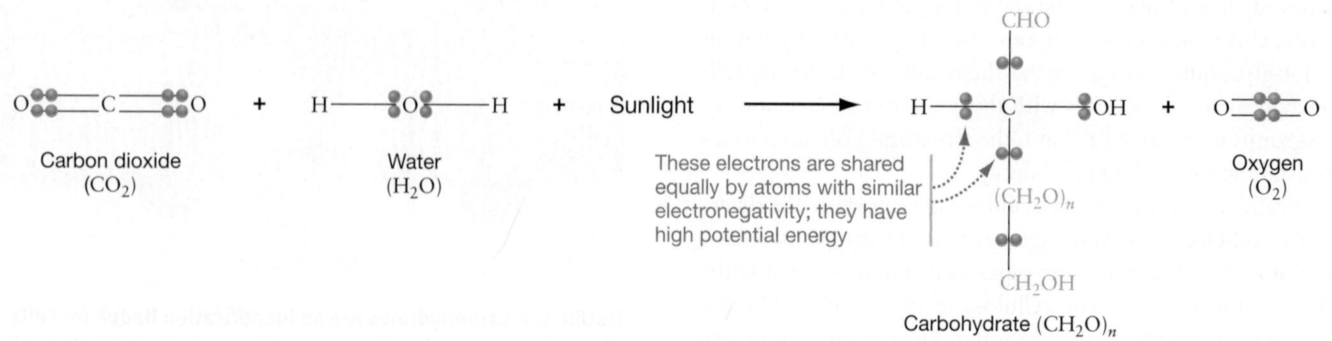

FIGURE 5.6 Carbohydrates Have High Free Energy. In these diagrams, the straight lines between atoms indicate covalent bonds. The dots show the relative positions of electrons in those bonds.

(a) Carbon dioxide

$$O = C = O$$

(b) A carbohydrate

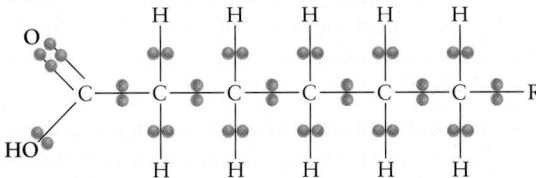

(c) A fatty acid (a component of fat molecules)

FIGURE 5.7 In Organisms, Free Energy Is Stored in C—H and C—C Bonds. (a) In carbon dioxide, the electrons involved in covalent bonds are held tightly by oxygen atoms. **(b)** In carbohydrates such as the sugar shown here, many of the covalently bonded electrons are held equally between C and H atoms. **(c)** The fatty acids found in fat molecules have more C—H bonds and fewer C—O bonds than carbohydrates do.

✔**EXERCISE** Circle the bonds in this diagram that have high free energy.

duction of chemical energy that can be used in the cell. Starch and glycogen are like a candy bar that has segments, so you can break off chunks whenever you need a boost.

The most important enzyme involved in catalyzing the hydrolysis of α-glycosidic linkages in glycogen is called **phosphorylase**. Most of your cells contain phosphorylase, so they can break down glycogen to provide glucose on demand.

The enzymes involved in breaking the α-glycosidic linkages in starch are called **amylases**. Your salivary glands and pancreas produce amylases that are secreted into your mouth and small intestine, respectively. These amylases are responsible for digesting the starch that you eat.

ENERGY STORED IN GLUCOSE IS TRANSFERRED TO ATP When a cell needs energy, exergonic reactions lead to the breakdown of the glucose and capture of the released energy through synthesis of a molecule called **adenosine triphosphate (ATP)**.

More specifically, the energy that is released when sugars are processed is used to synthesize ATP from a precursor called adenosine diphosphate (ADP) plus a free inorganic phosphate (P_i) molecule. The overall reaction can be written as follows:

$$CH_2O + O_2 + ADP + P_i \longrightarrow CO_2 + H_2O + ATP$$

To put this in words, the chemical energy stored in the C—H and C—C bonds of carbohydrate is transferred to chemical energy in the form of the third phosphate group in ATP. The free energy in ATP drives endergonic reactions like polymerization, moves your muscles, and performs other types of work in cells.

Carbohydrates are like the water that piles up behind a dam; ATP is like the electricity, generated at a dam, that lights up your home. Carbohydrates store chemical energy; ATP "spends" it.

Later chapters will analyze in detail how sugars and other carbohydrates are made in organisms, and how these carbohydrates are then broken down to provide cells with usable chemical energy in the form of ATP.

CHECK YOUR UNDERSTANDING

If you understand that . . .

- Carbohydrates provide building blocks for the synthesis of more complex compounds.
- Polysaccharides such as cellulose, chitin, and peptidoglycan form cell walls, which give cells structural strength.
- Glycoproteins project from the surface of cells. They provide a molecular PIN (personal identification number) that identifies the cell's type or species.
- Starch and glycogen store sugars for later use in reactions that produce ATP. Sugars contain large amounts of chemical energy because they contain carbon atoms that are bonded to hydrogen atoms or other carbon atoms, instead of being bonded to oxygen. The C—H and C—C bonds have high free energy because the electrons are shared equally by atoms with low electronegativity.

✔ **You should be able to . . .**

1. Identify two aspects of the structures of cellulose, chitin, and peptidoglycan that correlate with their function as structural molecules.

2. Describe how the carbohydrates you ate during breakfast today are functioning in your body right now.

Answers are available in Appendix B.

Summary of Key Concepts

🔑 Sugars and other carbohydrates are highly variable in structure.

- Carbohydrates are organic compounds that have a carbonyl group and several to many hydroxyl groups.

- Carbohydrates can occur as the simple sugars called monosaccharides or as complex polysaccharides.

- Many types of monosaccharide exist, each distinguished by one or more of the following features: (1) the location of their carbonyl group—either at the end of the molecule or within it; (2) the number of carbon atoms they contain—from three to seven, usually; and (3) the orientation of their hydroxyl groups in the linear chain or the ring form.

  ✔You should be able to explain why a relatively small difference in the location of a carbonyl or hydroxyl group can lead to dramatic changes in the properties and function of a monosaccharide.

🔑 Monosaccharides are monomers that polymerize via condensation reactions to form polymers called polysaccharides. The monosaccharides in polysaccharides are joined by different types of glycosidic linkages.

- Monosaccharides can be linked together by covalent bonds, called glycosidic linkages, that join hydroxyl groups on adjacent molecules.

- Polysaccharides are distinguished by the type of monomers involved and the location and orientation of the glycosidic linkages between the monomers.

- The most common polysaccharides in organisms today are starch, glycogen, cellulose, and chitin; peptidoglycan is an abundant polysaccharide that has short chains of amino acids attached.

  ✔You should be able to explain why different types of glycosidic linkage cause glucose polymers to form a helix (e.g., in amylase) versus a straight chain (e.g., in cellulose).

🔑 Carbohydrates have diverse functions in cells. In addition to serving as raw material for synthesizing other molecules, they provide fibrous structural materials, indicate cell identity, and store chemical energy.

- In carbohydrates, as in proteins and nucleic acids, structure correlates with function.

- Cellulose, chitin, and peptidoglycan are polysaccharides that function in support. They are made up of monosaccharides joined by β-1,4-glycosidic linkages, which are difficult for enzymes to degrade. When individual molecules of these polysaccharides align side by side, bonds form between them—resulting in strong, flexible fibers or sheets.

- The oligosaccharides on cell-surface glycoproteins can function as specific signposts or identity tags, because their constituent monosaccharides are so diverse in geometry and composition.

- Both starch and glycogen function as energy-storage molecules. They are made up of glucose molecules that are in the α ring form and that are joined by glycosidic linkages between their first and fourth carbons.

- When cells need energy, enzymes hydrolyze the α-1,4-glycosidic linkages in starch or glycogen, releasing glucose molecules. Glucose and other sugars contain a significant amount of chemical energy because they contain many C−C and C−H bonds. Sugars are processed in reactions that lead to the production ATP; chemical energy in ATP is readily usable by cells.

  ✔You should be able to describe two key differences in the structure of polysaccharides that function in energy storage versus structural support.

  MB Web Activity Carbohydrate Structure and Function

Questions

1. What is the difference between a monosaccharide, a disaccharide, and a polysaccharide?
   a. the number of carbon atoms in the molecule
   b. the type of glycosidic linkage between monomers
   c. the spatial arrangement of the various hydroxyl residues in the molecule
   d. the number of monomers in the molecule

2. What type of bond allows sugars to polymerize?
   a. glycosidic linkage
   b. phosphodiester bond
   c. peptide bond
   d. hydrogen bond

3. What holds cellulose molecules together in bundles large enough to form fibers?
   a. the cell wall
   b. peptide bonds

   c. hydrogen bonds
   d. hydrophobic interactions between different residues in the cellulose helix

4. What are the primary functions of carbohydrates in cells?
   a. energy storage, cell identity, structure, and building blocks for synthesis
   b. catalysis, structure, and transport
   c. information storage and catalysis
   d. signal reception, signal transport, and signal response

5. Why is it unlikely that carbohydrates played a large role in the origin of life?
   a. They cannot be produced by chemical evolution.
   b. They are too diverse in terms of structure and function.
   c. More types of glycosidic linkages are possible than are actually observed in organisms.
   d. They do not polymerize without the aid of enzymes.

6. What is a "quick and dirty" way to assess how much free energy an organic molecule has?
   a. Count the number of carbon atoms it contains.
   b. Compare the number of C—H and C—C bonds vs. C—O bonds it contains.
   c. Count the number of functional groups it has.
   d. Determine whether it contains a carbonyl group.

TEST YOUR UNDERSTANDING

Answers are available in Appendix B

1. Explain why the structure of carbohydrates supports their function in signaling the identity of a cell.

2. What is the difference between linking glucose molecules with α-1,4-glycosidic linkages versus β-1,4-glycosidic linkages? What are the consequences?

3. Compare and contrast the structures and functions of starch and glycogen. How are these molecules similar? How are they different?

4. Why do the bonds in a carbohydrate store a large amount of chemical energy compared with the chemical energy stored in the bonds of carbon dioxide?

5. What aspects of the structure of cellulose and chitin support their function in protecting and stiffening cells and organisms?

6. Both glycogen and cellulose consist of glucose monomers that are linked end to end. How do the structures of these polysaccharides differ? How do their functions differ?

APPLYING CONCEPTS TO NEW SITUATIONS

Answers are available in Appendix B

1. A weight-loss program for humans that emphasized minimal consumption of carbohydrates was popular in some countries in the early 2000s. What was the logic behind this diet? (Note: This diet plan caused controversy and is not endorsed by some physicians and researchers.)

2. Galactosemia is a potentially fatal disease that occurs in humans who lack the enzyme that converts galactose to glucose. To treat this disease, physicians exclude the monosaccharide galactose from the diet. Why does the disaccharide lactose also have to be excluded from the diet?

3. Amylase, an enzyme found in human saliva, catalyzes the hydrolysis of the α-1,4-glycosidic linkages in starch. If you hold a salty cracker in your mouth long enough, it will begin to taste sweet. Why?

4. Lysozyme, an enzyme found in human saliva, tears, and other secretions, catalyzes the hydrolysis of the β-1,4-glycosidic linkages in peptidoglycan. What effect does contact with this enzyme have on bacteria?

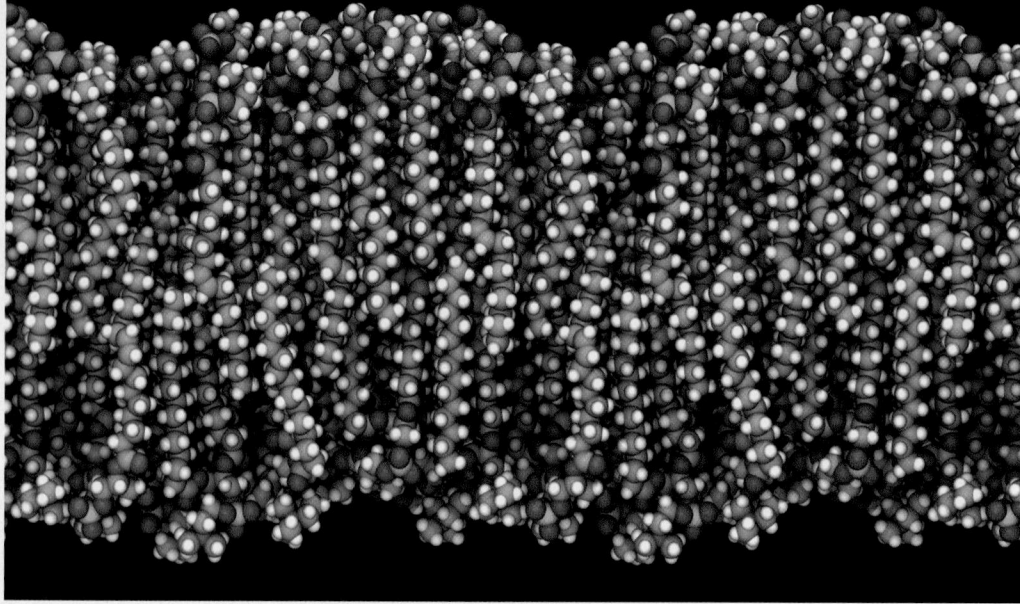

A space-filling model of a phospholipid bilayer from the membrane of a cell. In single-celled organisms, this cluster of molecules forms part of the boundary between life (inside the cell) and nonlife (outside the cell).

6 Lipids, Membranes, and the First Cells

KEY CONCEPTS

- Phospholipids are amphipathic molecules—they have a hydrophilic region and a hydrophobic region. In solution, phospholipids spontaneously form bilayers that are selectively permeable—meaning that only certain substances cross them readily.

- Ions and molecules diffuse spontaneously from regions of high concentration to regions of low concentration. Water moves across lipid bilayers from regions of high water concentration to regions of low water concentration via osmosis—a special case of diffusion.

- In cells, membrane proteins are responsible for the passage of ions, polar molecules, and large molecules that can't cross the membrane on their own because they are not soluble in lipids. Some membrane proteins form channels, some facilitate diffusion by binding to substrates, and some use energy from ATP to actively pump ions or molecules.

Currently, most biologists support the hypothesis that biological evolution began with an RNA molecule that could make a copy of itself. As the offspring of this molecule multiplied in the prebiotic soup, natural selection would have favored versions of the molecule that were particularly stable and efficient at catalysis. A second great milestone in the history of life occurred when a descendant of this replicator became enclosed within a membrane.

Why is the presence of a membrane so important? The **plasma membrane**, or **cell membrane**, separates life from nonlife. It is a layer of molecules that surrounds the cell interior and separates it from the external environment.

- The plasma membrane serves as a selective barrier: It keeps damaging compounds out of the cell and allows entry of compounds needed by the cell.

- Because the plasma membrane sequesters the appropriate chemicals in an enclosed area, reactants collide more frequently—the chemical reactions necessary for life occur much more efficiently.

The first cell functioned as an efficient and dynamic reaction vessel. It was responsible for one of the five key attributes of life introduced in Chapter 1: the presence of a barrier that defines the cell and regulates the passage of materials.

Recent research suggests that the evolution of the first self-replicating molecule and the first membrane could have occurred simultaneously. The experiments focused on nucleotides and on the "oily" or "fatty" compounds called lipids. When researchers heated nucleotides and allowed them to react in the presence of lipids, the nucleotides polymerized to form RNA-like nucleic acids—some of which become enclosed in membrane-like compartments consisting of lipids. Researchers are calling these structures "proto-cells."

Why do lipids form membranes? Which ions and molecules can pass through a membrane formed from lipids and which cannot, and why? These are some of the most fundamental questions in all of biological science. Let's delve into them further.

✔ When you see this checkmark, stop and test yourself. Answers are available in Appendix B.

6.1 Lipids

Lipid is a catch-all term for carbon-containing compounds that are found in organisms and are largely nonpolar and hydrophobic—meaning that they do not dissolve readily in water. (Recall from

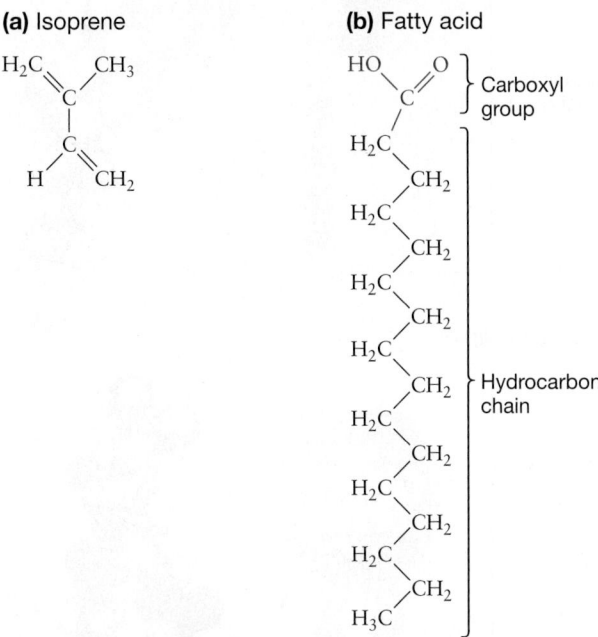

(a) Isoprene

(b) Fatty acid

Carboxyl group

Hydrocarbon chain

FIGURE 6.1 Hydrocarbon Structure. (a) Isoprene subunits like the one shown here can be linked to each other, end to end, to form long hydrocarbon chains. **(b)** Fatty acids typically contain a total of 14–20 carbon atoms, most found in their long hydrocarbon "tails."

Chapter 2 that water is a polar solvent.) Lipids do dissolve, however, in liquids consisting of nonpolar organic compounds.

To understand why lipids are insoluble in water, examine the five-carbon compound called isoprene illustrated in **Figure 6.1a**. Note that it consists of carbon atoms bonded to hydrogen atoms. Molecules that contain only carbon and hydrogen are known as **hydrocarbons**. Hydrocarbons are nonpolar because electrons are shared equally in C−H bonds—owing to the approximately equal electronegativity of carbon and hydrogen. This property makes hydrocarbons hydrophobic. Thus, the reason lipids do not dissolve in water is that they have a significant hydrocarbon component.

Figure 6.1b gives the structural formula of a **fatty acid**, which consists of a hydrocarbon chain bonded to a carboxyl (−COOH) functional group. Isoprene and fatty acids are key building blocks of the lipids found in organisms.

A Look at Three Types of Lipids Found in Cells

Unlike amino acids, nucleotides, and monosaccharides, lipids are characterized by a physical property—their solubility—instead of a shared chemical structure. The structure of lipids varies widely. For example, consider the most important types of lipids found in cells: fats, steroids, and phospholipids.

FATS **Fats** are composed of three fatty acids that are linked to a three-carbon molecule called **glycerol**. Because of this structure, fats are also called triacylglycerols or triglycerides.

As **Figure 6.2a** shows, fats form when a dehydration reaction occurs between a hydroxyl group of glycerol and the carboxyl

(a) Fats form via dehydration reactions.

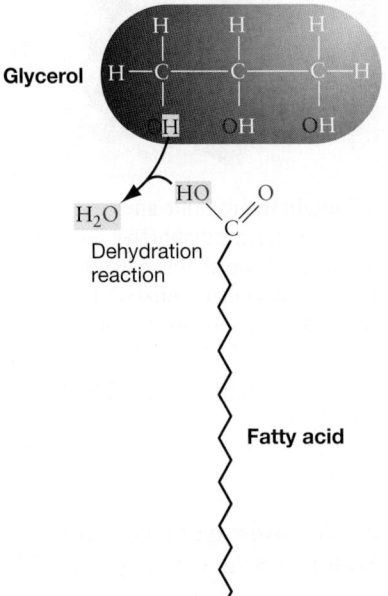

Glycerol

Dehydration reaction

Fatty acid

(b) Fats consist of glycerol linked by ester linkages to three fatty acids.

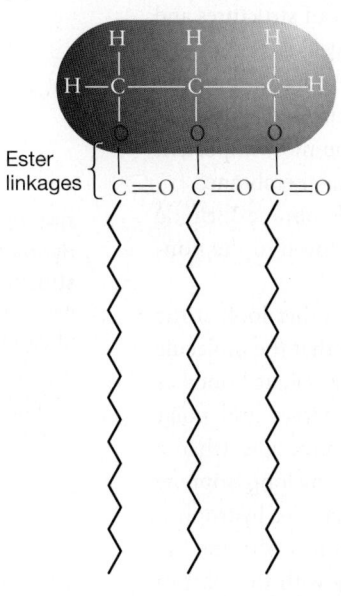

Ester linkages

FIGURE 6.2 Fats Are One Type of Lipid Found in Cells. (a) When glycerol and a fatty acid react, a water molecule leaves. The covalent bond that results from this reaction is termed an ester linkage. **(b)** The fat shown here as a structural formula and a space-filling model is tristearin, the most common type of fat in beef.

group of a fatty acid. The glycerol and fatty-acid molecules become joined by an **ester linkage**, which is analogous to the peptide bonds, phosphodiester bonds, and glycosidic linkages in proteins, nucleic acids, and carbohydrates, respectively.

Fats are not polymers, however, and fatty acids are not monomers. As **Figure 6.2b** shows, fatty acids are not linked together to form a macromolecule in the way that amino acids, nucleotides, and monosaccharides are.

✔You should be able to explain why fats store a great deal of chemical energy, and why they are hydrophobic.

STEROIDS **Steroids** are a family of lipids distinguished by the bulky, four-ring structure shown in orange in **Figure 6.3a**. The various steroids differ from one another by the functional groups or side groups attached to those rings—shown as an "R-group" in the schematic diagram in Figure 6.3a. The space-filling model in that figure is cholesterol, which has a hydrophilic hydroxyl group attached to the rings and a hydrocarbon "tail" formed of isoprene subunits. Cholesterol is an important component of plasma membranes in many organisms.

PHOSPHOLIPIDS **Phospholipids** consist of a glycerol that is linked to a phosphate group (PO_4^{3-}) and to either two chains of isoprene or two fatty acids. The phosphate group is also bonded to a small, organic molecule that is charged or polar (**Figure 6.3b**).

Phospholipids with isoprene tails are found in the domain Archaea introduced in Chapter 1; phospholipids composed of fatty acids are found in the domains Bacteria and Eukarya. In all three domains of life, phospholipids are critically important components of the plasma membrane.

The Structures of Membrane Lipids

The lipids found in organisms have a wide array of structures and functions. In addition to storing chemical energy, lipids act as pigments that capture or respond to sunlight, serve as signals between cells, form waterproof coatings on leaves and skin, and act as vitamins used in an array of cellular processes. The most important lipid function, however, is their role in the plasma membrane.

Not all lipids can form membranes. ○╼ Membrane-forming lipids have a polar, hydrophilic region—in addition to the nonpolar, hydrophobic region found in all lipids.

To better understand this structure, take another look at the phospholipid illustrated in Figure 6.3b. Notice that the molecule has a "head" region containing highly polar covalent bonds as well as positive and negative charges. The charges and polar bonds in the head region interact with water molecules when a phospholipid is placed in solution. In contrast, the long isoprene or fatty-acid tails of a phospholipid are nonpolar and hydrophobic. Water molecules cannot form hydrogen bonds with the hydrocarbon tail, and do not interact extensively with this part of the molecule.

Compounds that contain both hydrophilic and hydrophobic elements are **amphipathic** (literally, "dual-sympathy"). Phospholipids are amphipathic. As Figure 6.3a shows, cholesterol is

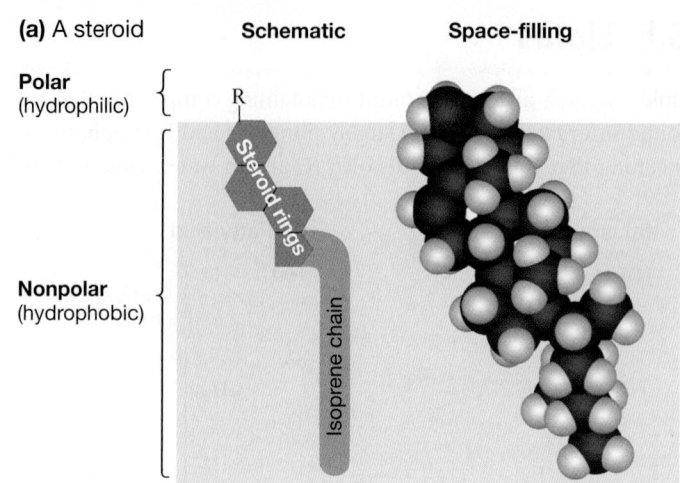

(a) A steroid

Schematic   Space-filling

Polar (hydrophilic)

R

Steroid rings

Nonpolar (hydrophobic)

Isoprene chain

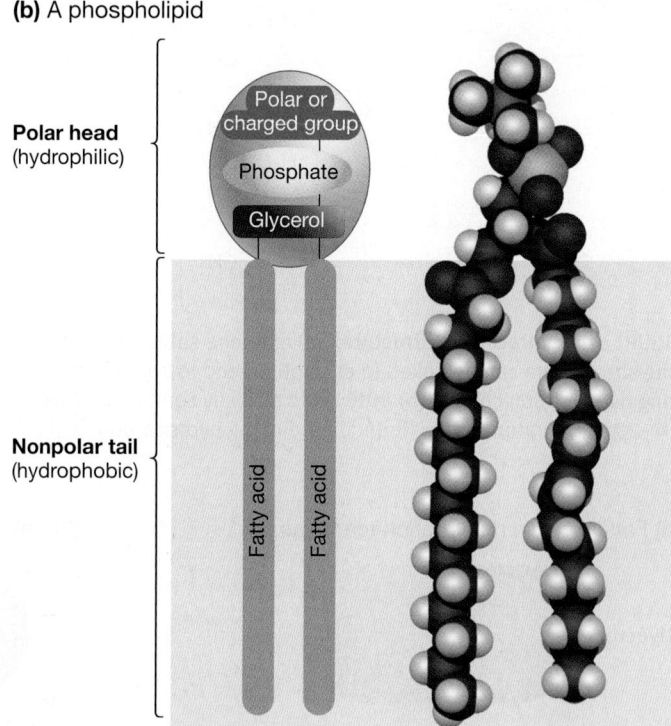

(b) A phospholipid

Polar head (hydrophilic)

Polar or charged group

Phosphate

Glycerol

Nonpolar tail (hydrophobic)

Fatty acid   Fatty acid

FIGURE 6.3 Amphipathic Lipids Contain Hydrophilic and Hydrophobic Elements. **(a)** All steroids have a distinctive four-ring structure. **(b)** All phospholipids consist of two chains of isoprene or two fatty acids that are linked to glycerol, which is linked to a phosphate group, which is linked to a small, organic molecule that is polar or charged.

✔**QUESTION** If these molecules were in solution, where would water molecules interact with them?

also amphipathic. Because it has a hydroxyl functional group attached to its rings, it has both hydrophilic and hydrophobic regions.

The amphipathic nature of phospholipids is far and away their most important feature biologically. It is responsible for their presence in plasma membranes.

CHECK YOUR UNDERSTANDING

6.2 Phospholipid Bilayers

Phospholipids do not dissolve when they are placed in water. Water molecules interact with the hydrophilic heads of the phospholipids, but not with their hydrophobic tails. The lack of interaction with water drives the hydrophobic tails together.

Instead of dissolving in water, then, phospholipids form one of two types of structures: micelles or lipid bilayers.

- Micelles (**Figure 6.4a**) are tiny droplets created when the hydrophilic heads of phospholipids face the water and the hydrophobic tails are forced together, away from the water.

- Phospholipid bilayers, or simply, **lipid bilayers**, are created when two sheets of phospholipid molecules align. As **Figure 6.4b** shows, the hydrophilic heads in each layer face a surrounding solution while the hydrophobic tails face one another inside the bilayer. In this way, the hydrophilic heads interact with water while the hydrophobic tails interact with one another. These bilayers form spontaneously.

Micelles tend to form from phospholipids with relatively short tails; bilayers tend to form from phospholipids with longer tails.

It's critical to recognize that micelles and phospholipid bilayers form spontaneously—no input of energy is required. This concept can be difficult to grasp, because entropy clearly decreases when these structures form. Micelles and lipid bilayers are much more highly organized than phospholipids floating free in the solution.

The key is to recognize that micelles and lipid bilayers are much more stable energetically than are independent phospholipids in solution. Stated another way, the structures have much lower potential energy than do the independent molecules in solution. Independent phospholipids are unstable in water because their hydrophobic tails disrupt hydrogen bonds that otherwise would form between water molecules. As a result, amphipathic molecules are much more stable in aqueous solution when their hydrophobic tails avoid water, and instead participate in the hydrophobic interactions introduced in Chapter 3.

In this case, the loss of potential energy outweighs the decrease in entropy. Overall, the free energy of the system decreases. Lipid bilayer formation is exergonic and spontaneous.

Artificial Membranes as an Experimental System

When lipid bilayers are agitated by shaking, the layers break and re-form as small, spherical structures. The resulting vesicles have water on the inside as well as the outside, because the hydrophilic heads of the lipids face outward on each side of the bilayer. Artificial membrane-bound vesicles like these are called liposomes.

To explore how plasma membranes work, researchers began creating and experimenting with liposomes (**Figure 6.5**) and

(a) Lipid micelles

Hydrophilic heads interact with water

Hydrophobic tails interact with one another

Water

(b) Lipid bilayers

Hydrophilic heads interact with water

Hydrophobic tails interact with one another

FIGURE 6.4 Phospholipids Form Micelles and Bilayers in Solution. In **(a)** a micelle or **(b)** a lipid bilayer, the hydrophilic heads of phospholipids face out, toward water; the hydrophobic tails face in, away from water. Plasma membranes consist in part of lipid bilayers.

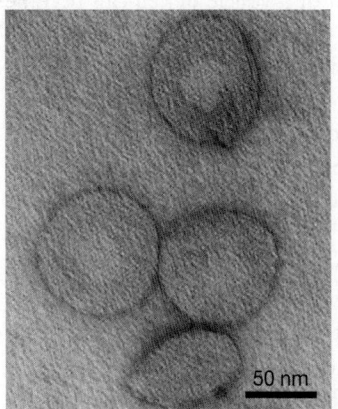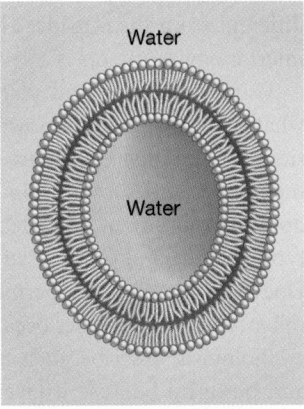

Water

Water

50 nm

FIGURE 6.5 Liposomes Are Artificial Membrane-Bound Vesicles. Electron micrograph of liposomes in cross section (left) and a cross-sectional diagram of the lipid bilayer in a liposome (right).

(a) Planar bilayers: Artificial membranes

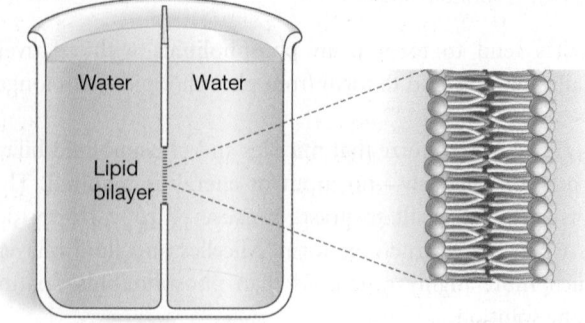

(b) Artificial-membrane experiments

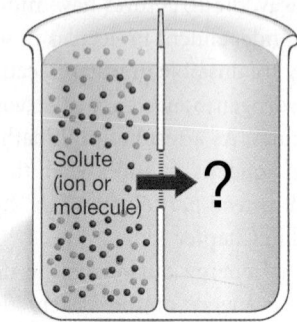

How rapidly can different solutes cross the membrane (if at all) when …

1. Different types of phospholipids are used to make the membrane?

2. Proteins or other molecules are added to the membrane?

FIGURE 6.6 Planar Bilayers Are Artificial Membranes. (a) The construction of planar bilayers across a hole in a glass wall separating two water-filled compartments (left), and a close-up sketch of the bilayer (right). **(b)** A wide variety of experiments are possible with planar bilayers; a few are suggested here.

planar bilayers—lipid bilayers constructed across a hole in a glass or plastic wall separating two aqueous solutions (**Figure 6.6a**). Some of the first questions they posed concerned the permeability of lipid bilayers. The **permeability** of a structure is its tendency to allow a given substance to pass across it.

Membrane permeability is a critical issue, because if certain molecules or ions pass through a lipid bilayer more readily than others, the internal environment of a vesicle or cell can become different from the outside. This difference between exterior and interior environments is a key characteristic of cells.

Using liposomes and planar bilayers, researchers can study what happens when a known ion or molecule is added to one side of a lipid bilayer (**Figure 6.6b**). Does the substance cross the membrane and show up on the other side? If so, how rapidly does the movement take place? What happens when a different type of phospholipid is used to make the artificial membrane? Does the membrane's permeability change when proteins or other types of molecules become part of it?

Biologists describe such an experimental system as elegant and powerful because it gives them precise control over which factor changes from one experimental treatment to the next. Control, in turn, is why experiments are such an effective way to explore scientific questions. You might recall from Chapter 1 that good experimental design allows researchers to alter one factor at

a time and determine what effect, if any, each has on the process being studied.

Selective Permeability of Lipid Bilayers

When researchers put molecules or ions on one side of a liposome or planar bilayer and measure the rate at which the molecules arrive on the other side, a clear pattern emerges: ⊶ Lipid bilayers are highly selective.

Selective permeability means that some substances cross a membrane more easily than other substances can. Small, nonpolar molecules move across bilayers quickly. In contrast, large molecules and charged substances cross the membrane slowly, if at all.

According to the data in **Figure 6.7**, small, nonpolar molecules such as oxygen (O_2) move across selectively permeable membranes more than a billion times faster than do chloride ions (Cl^-). In essence, ions cannot cross membranes at all—unless they have "help" in the form of membrane proteins introduced later in the chapter. Very small and uncharged molecules such as water (H_2O) can cross membranes relatively rapidly, even if they are polar. Small, polar molecules such as glycerol have intermediate permeability.

The leading hypothesis to explain this pattern is that charged compounds and large, polar molecules can't pass through the nonpolar, hydrophobic tails of a lipid bilayer. Because of their electrical charge, ions are more stable in solution where they interact with water than they are in the interior of membranes, which is electrically neutral. To test the hypothesis, researchers have manipulated the size and structure of the tails in liposomes or planar bilayers. ✔If you understand this hypothesis, you should be able to predict whether amino acids and nucleotides will cross a membrane readily.

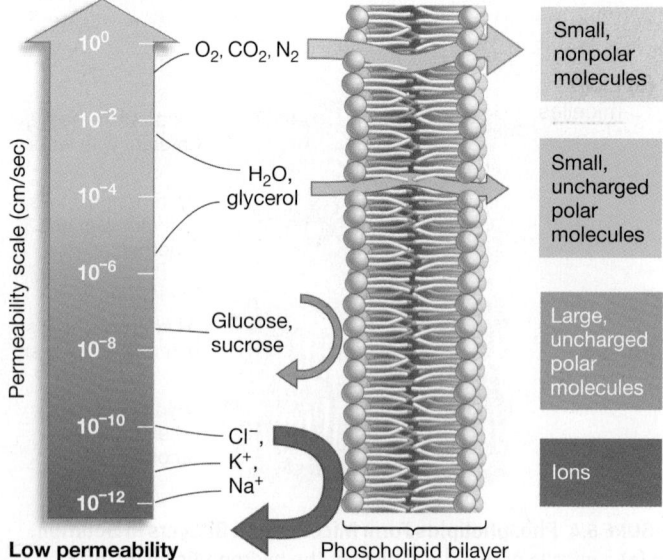

High permeability

10^0 — O_2, CO_2, N_2 — Small, nonpolar molecules

10^{-2}

10^{-4} — H_2O, glycerol — Small, uncharged polar molecules

10^{-6}

10^{-8} — Glucose, sucrose — Large, uncharged polar molecules

10^{-10} — Cl^-, K^+, Na^+ — Ions

10^{-12}

Permeability scale (cm/sec)

Low permeability — Phospholipid bilayer

FIGURE 6.7 Lipid Bilayers Show Selective Permeability. Only certain substances cross lipid bilayers readily. Size and charge affect the rate of diffusion across a membrane.

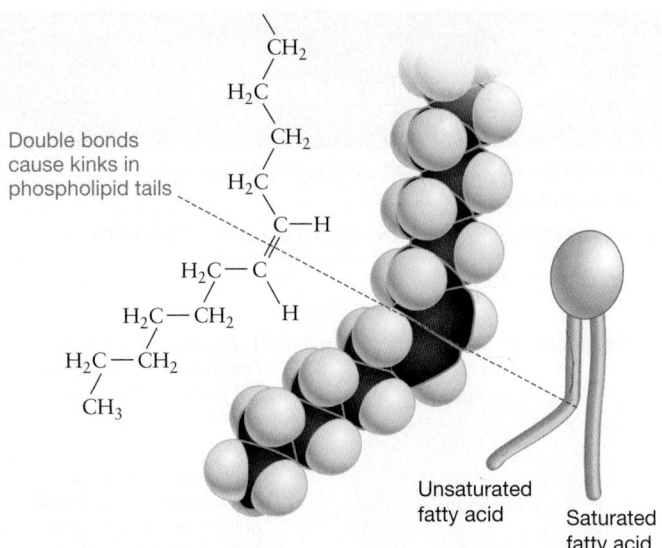

Double bonds
cause kinks in
phospholipid tails

Unsaturated
fatty acid

Saturated
fatty acid

FIGURE 6.8 Unsaturated Hydrocarbons Contain Carbon-Carbon Double Bonds. The icon on the right indicates that one of the hydrocarbon tails in a phospholipid is unsaturated and therefore kinked.

How Does Lipid Structure Affect Membrane Properties?

The correlation between the structure of a molecule and its function in cells may be the most basic theme in biological chemistry. This is certainly true of lipids. The hydrophilic-head-and-hydrophobic-tail structure of phospholipids allows them to form lipid bilayers and membranes. But the number of double bonds in a hydrophobic tail and its overall length, in addition to the number of cholesterol molecules nearby, has a profound influence on how a membrane behaves.

BOND SATURATION IS AN IMPORTANT ASPECT OF LIPID STRUCTURE

When two carbon atoms form a double bond, the attached atoms are found in a plane instead of a three-dimensional tetra-hedron. The carbon atoms involved are also locked into place. They cannot rotate freely, as they do in carbon-carbon single bonds. As a result, certain double bonds between carbon atoms produce a "kink" in an otherwise straight hydrocarbon chain (**Figure 6.8**).

Hydrocarbon chains without double bonds are said to be **saturated**. But if a hydrocarbon chain contains a double bond, the chain is said to be **unsaturated**. This choice of terms is logical. If a hydrocarbon chain does not contain a double bond, it is saturated with the maximum number of hydrogen atoms that can attach to the carbon skeleton. If it is unsaturated, then fewer than the maximum number of hydrogen atoms are attached.

Because they contain more C−H bonds, which have much more free energy than C=C bonds, saturated fats have more chemical energy than unsaturated fats do. People who are dieting are often encouraged to eat fewer saturated fats, to reduce their energy intake. Foods that contain lipids with many double bonds are said to be polyunsaturated and are advertised as healthier than foods with more-saturated fat.

BOND SATURATION AND HYDROCARBON CHAIN LENGTH CHANGE MEMBRANE FLUIDITY AND PERMEABILITY The degree of saturation in a fat—along with the length of its hydrocarbon tails—affects key aspects of a lipid's behavior in a membrane.

- When hydrophobic tails are packed into a lipid bilayer, the kinks created by double bonds produce spaces among the tightly packed tails. These spaces reduce the strength of hydrophobic interactions between the tails. These interactions are stronger among saturated hydrocarbon tails.

- Hydrophobic interactions also become stronger as saturated hydrocarbon tails increase in length.

These observations have profound impacts on membrane fluidity. Highly saturated fats, such as butter, are solid at room temperature (**Figure 6.9a**). Lipids that have extremely long hydrocarbon tails, as **waxes** do, form particularly stiff solids at room

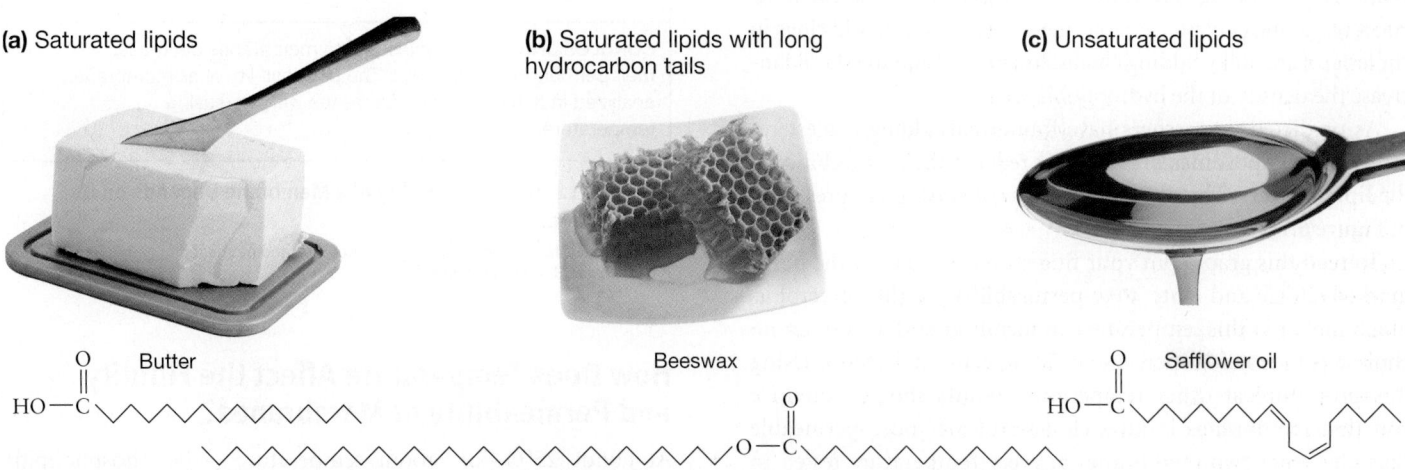

(a) Saturated lipids

Butter

(b) Saturated lipids with long hydrocarbon tails

Beeswax

(c) Unsaturated lipids

Safflower oil

FIGURE 6.9 The Fluidity of Lipids Depends on the Length and Saturation of Their Hydrocarbon Chains. (a) Butter consists primarily of saturated lipids. **(b)** Waxes are lipids with extremely long hydrocarbon chains. **(c)** Oils are dominated by "polyunsaturates"—lipids with hydrocarbon chains that contain multiple double bonds.

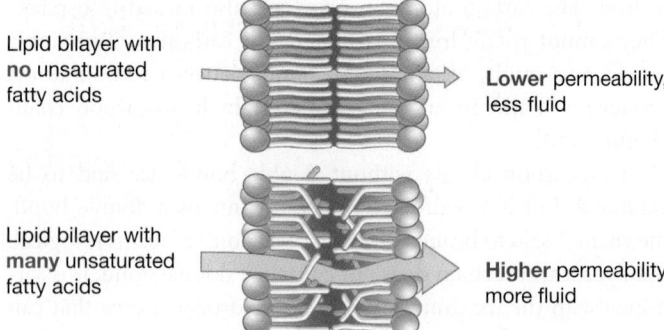

Lipid bilayer with **no** unsaturated fatty acids

Lower permeability, less fluid

Lipid bilayer with **many** unsaturated fatty acids

Higher permeability, more fluid

FIGURE 6.10 Fatty-Acid Structure Changes the Permeability of Membranes. Lipid bilayers containing many unsaturated fatty acids have more gaps and should be more permeable than bilayers with few unsaturated fatty acids.

temperature (**Figure 6.9b**). Highly unsaturated fats are liquid at room temperature (**Figure 6.9c**). Liquid triacylglycerols are called **oils**.

Membrane permeability is affected as well—fluidity and permeability are closely related. As **Figure 6.10** shows, lipid bilayers become more permeable as well as more fluid when they consist of short, unsaturated hydrocarbon tails. The membrane allows more materials to pass, because its interior is held together less tightly.

If these hypotheses are correct, then bilayers made of lipids with long, straight, saturated fatty-acid tails should be much less permeable than membranes made of lipids with short, kinked, unsaturated fatty-acid tails. Experiments on liposomes have shown exactly this pattern.

The take-home message is clear: The degree of hydrophobic interactions has a strong impact on the behavior of phospholipid bilayers.

CHOLESTEROL REDUCES MEMBRANE PERMEABILITY In addition to exploring the role of hydrocarbon chain length and degree of saturation on membrane permeability, biologists have investigated the effect of adding cholesterol molecules. Because the steroid rings in cholesterol are bulky, adding cholesterol to a membrane should increase the density of the hydrophobic section.

As predicted, researchers have found that adding cholesterol molecules to liposomes dramatically reduces the permeability of the lipid bilayers. The data behind this conclusion are presented in **Figure 6.11**.

To read this graph, put your finger on the *x*-axis at the point marked 20°C, and note that permeability to the glycerol is much higher at this temperature in membranes that contain no cholesterol versus 20 percent or 50 percent cholesterol. Using this procedure at other temperature points should convince you that membranes lacking cholesterol are more permeable than the other two membranes at every temperature tested in the experiment. Figure 6.11 also shows that for any of the three membranes, regardless of cholesterol content, permeability increases with temperature.

EXPERIMENT

QUESTION: Does adding cholesterol to a membrane affect its permeability?

HYPOTHESIS: Cholesterol reduces permeability because it fills spaces in phospholipid bilayers.

NULL HYPOTHESIS: Cholesterol has no effect on permeability.

EXPERIMENTAL SETUP:

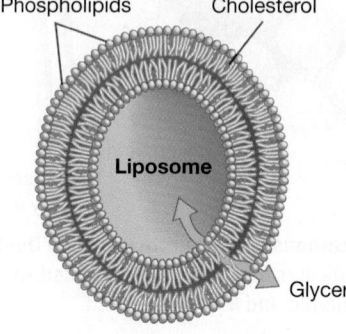

Phospholipids   Cholesterol

Liposome

Glycerol

1. Construct liposomes: Create with no cholesterol, 20% cholesterol, and 50% cholesterol.

2. Measure glycerol movement: Record how quickly glycerol moves across each type of membrane at different temperatures.

PREDICTION: Liposomes with higher cholesterol levels will have reduced permeability.

PREDICTION OF NULL HYPOTHESIS: All liposomes will have the same permeability.

RESULTS:

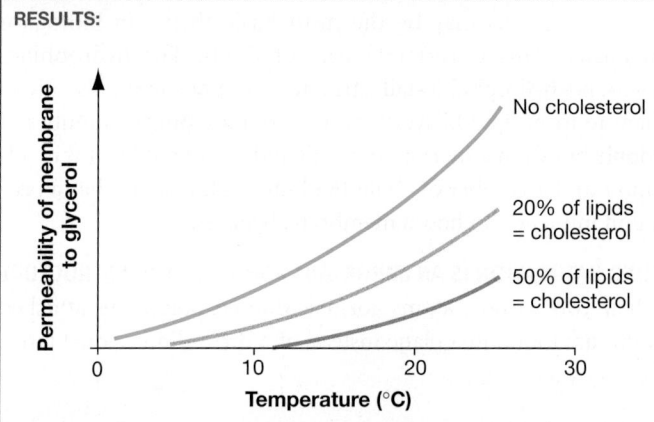

No cholesterol

20% of lipids = cholesterol

50% of lipids = cholesterol

CONCLUSION: Adding cholesterol to membranes decreases their permeability to glycerol. The permeability of all membranes analyzed in this experiment increases with increasing temperature.

FIGURE 6.11 The Permeability of a Membrane Depends on Its Composition.

SOURCE: J. de Gier et al. (1968). Lipid composition and permeability of liposomes. *Biochimica et Biophysica Acta* 150: 666–675.

How Does Temperature Affect the Fluidity and Permeability of Membranes?

At about 25°C—or "room temperature"—the phospholipids found in plasma membranes are liquid, and bilayers have the consistency of olive oil. This fluidity, as well as the membrane's permeability, decreases as temperature decreases. Why?

As temperatures drop, individual molecules in the bilayer move more slowly. As a result, the hydrophobic tails in the interior of membranes pack together more tightly. At very low temperatures, lipid bilayers begin to solidify. As the graph in Figure 6.11 indicates, low temperatures can make membranes impervious to molecules that would normally cross them readily. Put your finger on the x-axis just about the freezing point of water (0°C), and note that even membranes that lack cholesterol are almost completely impermeable to glycerol. Indeed, trace any of the three lines in Figure 6.11, and as you move to the right (increasing temperature), you also move up (increasing permeability).

The fluid nature of membranes also allows individual lipid molecules to move laterally within each layer, a little like a person moving about in a dense crowd (**Figure 6.12**). By tagging individual phospholipids and following their movement, researchers have clocked average speeds of 2 micrometers (μm)/second at room temperature. At these speeds, phospholipids could travel the length of a small bacterial cell in a second.

These experiments on lipid and ion movement demonstrate that membranes are dynamic. Phospholipid molecules whiz around each layer, while water and small, nonpolar molecules shoot in and out of the membrane. How quickly molecules move within and across membranes is a function of temperature and the structure of the hydrocarbon tails in the bilayer.

Phospholipids are in constant lateral motion, but rarely flip to the other side of the bilayer

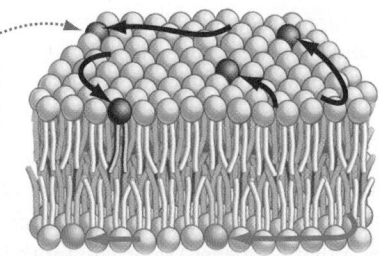

FIGURE 6.12 Phospholipids Move within Membranes. Membranes are dynamic—in part because phospholipid molecules move laterally within each layer in the structure.

CHECK YOUR UNDERSTANDING

If you understand that . . .

- In water, phospholipids form bilayers that are selectively permeable—meaning that some substances cross them much more readily than others do.
- Permeability is a function of the degree of saturation and the length of the hydrocarbon tails in membrane phospholipids, the amount of cholesterol in the membrane, and the temperature.

✓ You should be able to . . .

Fill in a chart with rows called Saturation of hydrocarbon tails, Length of hydrocarbon tails, Cholesterol, and Temperature, and columns named Factor, Effect on permeability, and Reason.

Answers are available in Appendix B.

6.3 Why Molecules Move across Lipid Bilayers: Diffusion and Osmosis

Small and uncharged molecules and hydrophobic compounds can cross membranes readily and spontaneously—without an expenditure of energy. The question now is: How is this possible? What process is responsible for the movement of molecules across phospholipid bilayers?

Diffusion

A thought experiment can help explain why molecules and ions can cross membranes spontaneously. Suppose you rack up a set of blue billiard balls on a pool table that contains many white balls, and then begin to vibrate the table.

1. Because of the vibration, the balls will move about randomly. They will also bump into one another.

2. After these collisions, some blue balls will move outward—away from their original position.

3. As movement and collisions continue, the overall or net movement of blue balls will be outward. This occurs because the random motion of the blue balls disrupts their original, nonrandom position. As the blue balls move at random, they are more likely to move away from one another than to stay together.

4. Eventually, the blue billiard balls will be distributed randomly across the table. The entropy of the blue billiard balls has increased.

Recall from Chapter 2 that entropy is a measure of the randomness or disorder in a system. The second law of thermodynamics states that in a closed system, entropy always increases.

This hypothetical example illustrates why molecules or ions located on one side of a lipid bilayer move to the other side spontaneously. The dissolved molecules and ions, or **solutes**, have thermal energy and are in constant, random motion. Movement of molecules and ions that results from their kinetic energy is known as **diffusion**. Because solutes change position randomly owing to diffusion, they tend to move from a region of high concentration to a region of low concentration. A difference in solute concentrations creates a **concentration gradient**.

Molecules and ions move randomly in all directions when a concentration gradient exists, but there is a net movement from regions of high concentration to regions of low concentration. Diffusion along a concentration gradient is a spontaneous process because it results in an increase in entropy.

Once the molecules or ions are randomly distributed throughout a solution, equilibrium is established. For example, consider two aqueous solutions separated by a lipid bilayer. **Figure 6.13** shows how molecules that can pass through the bilayer diffuse to the other side. At equilibrium, molecules continue to move back and forth across the membrane, but at equal rates—simply because each molecule or ion is equally likely to move in any direction. This means that there is no longer a net movement of molecules across the membrane.

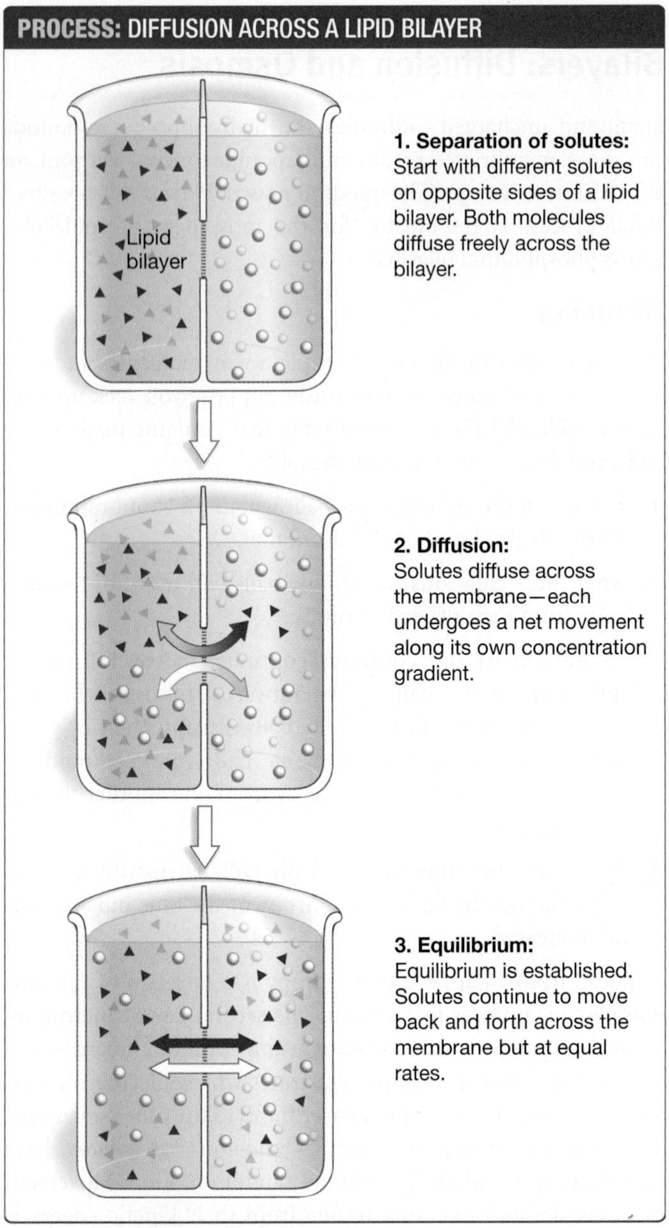

1. Separation of solutes: Start with different solutes on opposite sides of a lipid bilayer. Both molecules diffuse freely across the bilayer.

2. Diffusion: Solutes diffuse across the membrane—each undergoes a net movement along its own concentration gradient.

3. Equilibrium: Equilibrium is established. Solutes continue to move back and forth across the membrane but at equal rates.

FIGURE 6.13 Diffusion across a Selectively Permeable Membrane Establishes an Equilibrium.

Osmosis

What about water? As the data in Figure 6.7 showed, water moves across lipid bilayers relatively quickly. Like other substances that diffuse, water moves along its concentration gradient—from higher to lower concentration. The movement of water is a special case of diffusion that is given its own name: osmosis. **Osmosis** occurs only when solutions are separated by a membrane that is permeable to some molecules but not others—that is, a selectively permeable membrane.

The best way to think about water moving in response to a concentration gradient is to focus on the concentration of solutes in the solution. Let's suppose the concentration of a particular solute is higher on one side of a selectively permeable membrane

1. Unequal concentrations across membrane: Start with more solute on one side of the lipid bilayer than the other, using molecules that cannot cross the selectively permeable membrane.

2. Water movement: Water undergoes a net movement from the region of low concentration of solute (high concentration of water) to the region of high concentration of solute (low concentration of water).

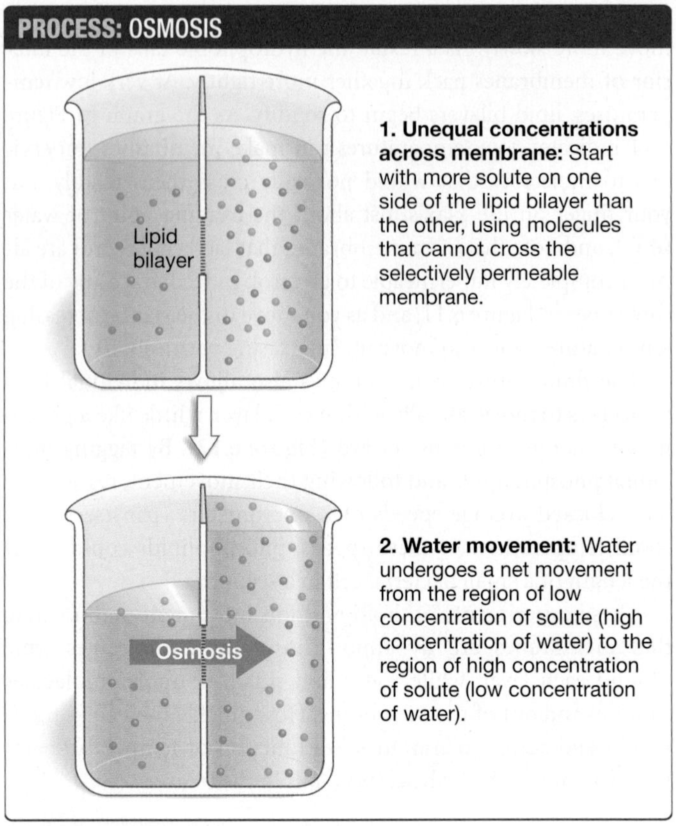

FIGURE 6.14 Osmosis Is the Diffusion of Water.

✔ **QUESTION** Suppose you doubled the number of solute molecules on the left side of the membrane (at the start). At equilibrium, would the water level on the left side be higher or lower than what is shown in the second drawing?

than it is on the other side (**Figure 6.14**, step 1). Further, suppose that this solute cannot diffuse through the membrane to establish equilibrium. What happens? Water will move from the side with a lower concentration of solute to the side with a higher concentration of solute (step 2). It dilutes the higher concentration and equalizes the concentrations on both sides.

This movement of water is spontaneous. It is driven by the increase in entropy achieved when solute concentrations are equal on both sides of the membrane.

Another way to think about osmosis is to realize that water is at higher concentration on the left side of the beaker in Figure 6.14 than it is on the right side of the beaker. As water diffuses, then, there will be net movement of water molecules from the left side to the right side: from a region of high concentration to a region of low concentration.

The movement of water by osmosis is important because it can swell or shrink a membrane-bound vesicle. Consider the liposomes illustrated in **Figure 6.15**.

- *Left* If the solution outside the membrane has a higher concentration of solutes than the interior has, and the solutes are not able to pass through the lipid bilayer, then water will move out of the vesicle into the solution outside. As a result, the

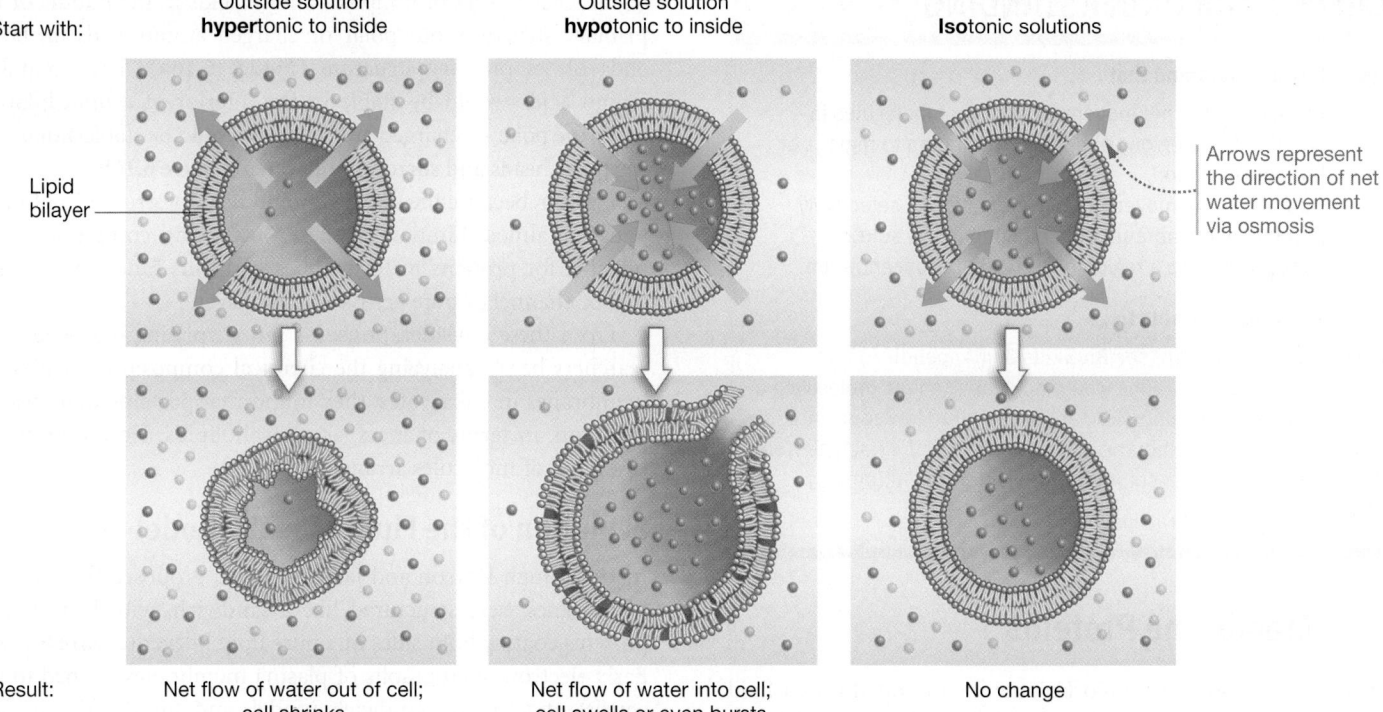

Outside solution **hyper**tonic to inside

Outside solution **hypo**tonic to inside

Isotonic solutions

Lipid bilayer

Arrows represent the direction of net water movement via osmosis

Result:

Net flow of water out of cell; cell shrinks

Net flow of water into cell; cell swells or even bursts

No change

FIGURE 6.15 Osmosis Can Shrink or Burst Membrane-Bound Vesicles.

vesicle will shrink and the membrane shrivel. The solution outside is said to be **hypertonic** ("excess-tone") relative to the inside of the vesicle. The word root hyper refers to the outside solution's containing more solutes than the solution on the other side of the membrane.

- *Middle* If the solution outside the membrane has a lower concentration of solutes than the interior has, water will move into the vesicle via osmosis. The incoming water will cause the vesicle to swell or even burst. The outside solution is termed **hypotonic** ("lower-tone") relative to the inside of the vesicle. Here the word root hypo refers to the outside solution's containing fewer solutes than the inside solution has.

- *Right* If solute concentrations are equal on either side of the membrane, the liposome will maintain its size. When the outside solution does not affect the membrane's shape, that solution is called **isotonic** ("equal-tone").

Note that the terms hypertonic, hypotonic, and isotonic are relative—they can be used only to express the relationship between a given solution and another solution. To make sure that you understand these concepts, review them in the study area at *www.masteringbiology.com*.

(MB) Web Activity Diffusion and Osmosis

✔ You should be able to (1) modify the drawing on the left side of Figure 6.15, such that the surrounding solution is hypotonic relative to the solution inside the liposome, and (2) mod-

ify the drawing in the center of Figure 6.15, such that the surrounding solution is hypertonic relative to the solution inside the liposome.

What does all this have to do with the first membranes floating in the prebiotic soup? Osmosis and diffusion tend to *reduce* differences in chemical composition between the inside and outside of membrane-bound structures. If liposome-like structures were present in the prebiotic soup, it's unlikely that their interiors offered a radically different environment from the surrounding solution. In all likelihood, the primary importance of the first lipid bilayers was simply to provide a container for self-replicating molecules.

It's important to note, though, that ribonucleotides can diffuse across lipid bilayers—meaning that monomers would be available for a ribozyme to make a copy of itself inside a vesicle. Further, experiments have shown that cell-like vesicles grow as additional lipids are added and then divide if sheared by shaking, bubbling, or wave action. On the basis of these observations, it is reasonable to hypothesize that once a self-replicating ribozyme had become surrounded by a lipid bilayer, this simple life-form and its descendants would continue to occupy cell-like structures that grew and divided.

Now let's investigate the next great event in the evolution of life: the formation of a true cell. How could lipid bilayers become a barrier capable of creating and maintaining a specialized internal environment that was conducive to life? How could an effective plasma membrane—one that admits ions and molecules needed by the replicator while excluding ions and molecules that might damage it—evolve in the first cell?

If you understand that . . .

- Diffusion is the net movement of ions or molecules in solution from regions of high concentration to regions of low concentration.
- Osmosis is the movement of water across a selectively permeable membrane, from a region of low solute concentration to a region of high solute concentration.

✓ **You should be able to . . .**

Make a concept map (see **BioSkills 8** in Appendix A) that includes the boxed terms water molecules, solute molecules, solution, osmosis, diffusion, areas of high-to-low concentration, semipermeable membrane, concentration gradients, hypertonic solutions, hypotonic solutions, and isotonic solutions.

Answers are available in Appendix B.

6.4 Membrane Proteins

What sort of molecule could become incorporated into a lipid bilayer and affect the bilayer's permeability? The title of this section gives the answer away. Proteins that are amphipathic can be inserted into lipid bilayers.

Proteins can be amphipathic because they are made up of amino acids, and because amino acids have side chains, or R-groups, that range from highly nonpolar to highly polar (see Figure 3.3 and Table 3.1). It's conceivable, then, that a protein could have a series of nonpolar amino acids in the middle of its primary structure, but polar or charged amino acids on both ends of its primary structure (**Figure 6.16a**). The nonpolar amino acids would be stable in the interior of a lipid bilayer, while the polar or charged amino acids would be stable alongside the polar heads and surrounding water (**Figure 6.16b**).

Further, because the secondary and tertiary structures of proteins are almost limitless in their variety and complexity, it is possible for proteins to form tubes and thus function as some sort of channel or pore across a lipid bilayer.

From these considerations, it's not surprising that when researchers began analyzing the chemical composition of plasma membranes in eukaryotes, they found that proteins were just as common, in terms of mass, as phospholipids. How were these two types of molecules arranged?

Evolution of the Fluid-Mosaic Model

In 1935 Hugh Davson and James Danielli proposed that plasma membranes were structured like a sandwich, with hydrophilic proteins coating both sides of a pure lipid bilayer (**Figure 6.17a**). Early electron micrographs of plasma membranes seemed to be consistent with the sandwich model, and for decades it was widely accepted.

(a) Proteins can be amphipathic.

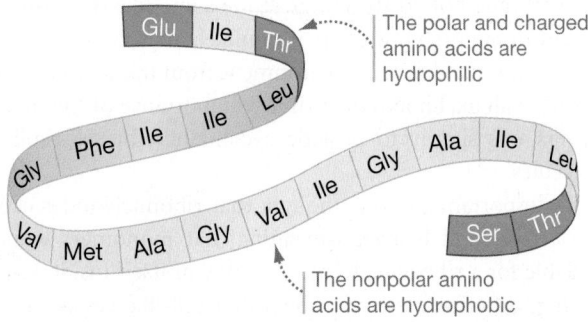

The polar and charged amino acids are hydrophilic

The nonpolar amino acids are hydrophobic

(b) Amphipathic proteins can integrate into lipid bilayers.

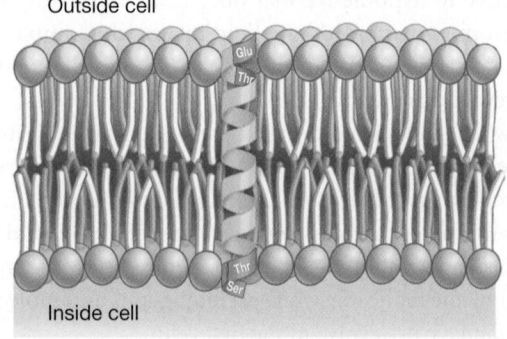

Outside cell

Inside cell

FIGURE 6.16 Amphipathic Proteins Are Stable in Lipid Bilayers.

(a) Sandwich model

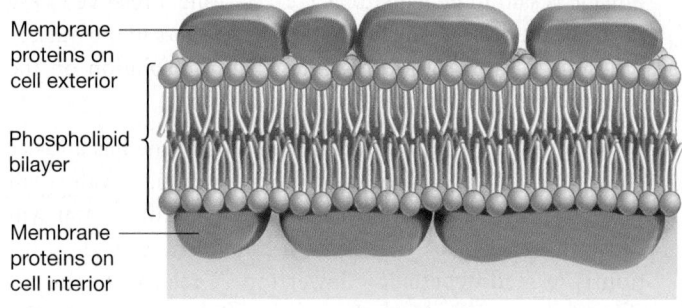

Membrane proteins on cell exterior

Phospholipid bilayer

Membrane proteins on cell interior

(b) Fluid-mosaic model

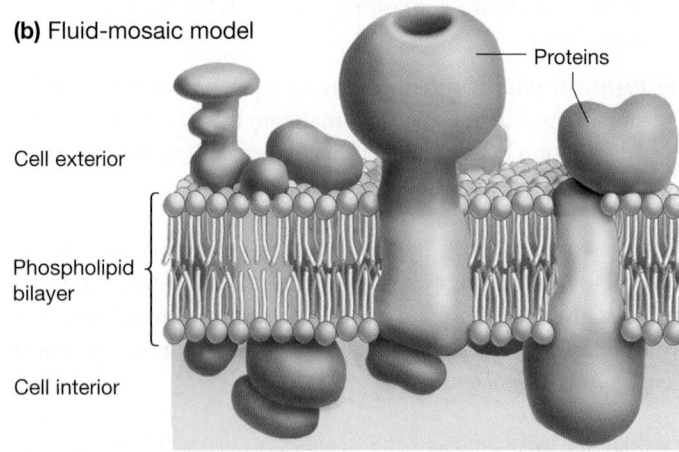

Proteins

Cell exterior

Phospholipid bilayer

Cell interior

FIGURE 6.17 Past and Current Models of Membrane Structure Differ in Where Membrane Proteins Reside. (a) The protein-lipid-lipid-protein sandwich model was the first hypothesis for the arrangement of lipids and proteins in plasma membranes. **(b)** The fluid-mosaic model was a radical departure from the sandwich hypothesis.

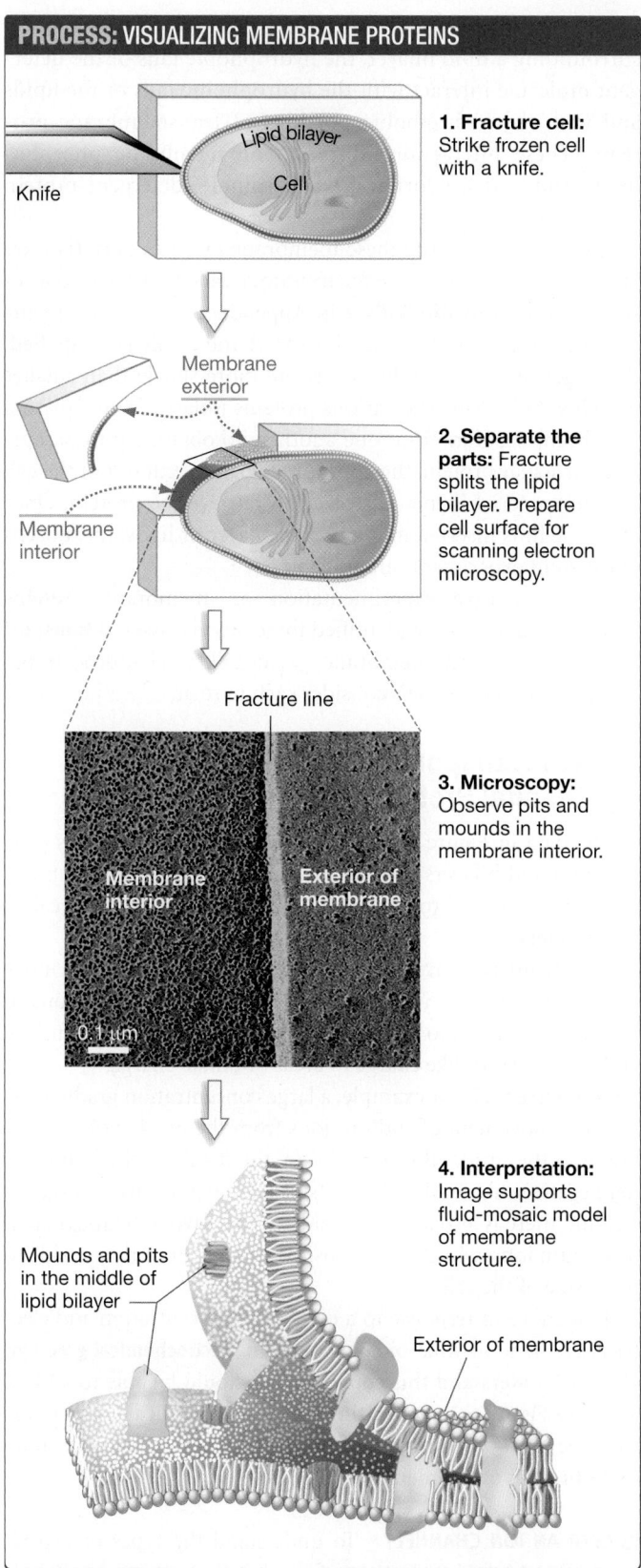

PROCESS: VISUALIZING MEMBRANE PROTEINS

Knife

Lipid bilayer

Cell

1. Fracture cell: Strike frozen cell with a knife.

Membrane exterior

Membrane interior

2. Separate the parts: Fracture splits the lipid bilayer. Prepare cell surface for scanning electron microscopy.

Fracture line

Membrane interior

Exterior of membrane

0.1 μm

3. Microscopy: Observe pits and mounds in the membrane interior.

Mounds and pits in the middle of lipid bilayer

Exterior of membrane

4. Interpretation: Image supports fluid-mosaic model of membrane structure.

FIGURE 6.18 Freeze-Fracture Preparations Allow Biologists to View Membrane Proteins. The scanning electron micrograph in Step 3 shows a freeze-fractured membrane that has been colored to match the drawing in Step 2.

The realization that membrane proteins could be amphipathic led S. Jon Singer and Garth Nicolson to suggest an alternative hypothesis, however. In 1972, they proposed that at least some proteins span the membrane instead of being found only outside the lipid bilayer. Their hypothesis was called the **fluid-mosaic model** (**Figure 6.17b**). Singer and Nicolson suggested that membranes are a mosaic of phospholipids and different types of proteins. The overall structure was proposed to be dynamic and fluid.

The controversy over the nature of the plasma membrane was resolved in the early 1970s with the development of an innovative technique for visualizing the surface of plasma membranes. The method is called freeze-fracture electron microscopy, because the steps involve freezing and fracturing the membrane before examining it with a **scanning electron microscope**, which produces images of an object's surface (see **BioSkills 10** in Appendix A).

As **Figure 6.18** shows, the freeze-fracture technique allows researchers to split plasma membranes and view the middle of the structure. The scanning electron micrographs that result show pits and mounds studding the inner surfaces of the lipid bilayer. Researchers interpret these structures as the locations of membrane proteins. As step 4 in Figure 6.18 shows, the pits and mounds are hypothesized to represent proteins that span the lipid bilayer.

These observations conflicted with the sandwich model but were consistent with the fluid-mosaic model. On the basis of these and subsequent observations, the fluid-mosaic model is now widely accepted.

Figure 6.19 summarizes the current hypothesis for where proteins and lipids are found in a plasma membrane. Note that some proteins span the membrane and have segments facing both the interior and the exterior surfaces. Proteins such as these are called **integral membrane proteins**, or **transmembrane proteins**. Other proteins, called **peripheral membrane proteins**, are found only on one side of the membrane.

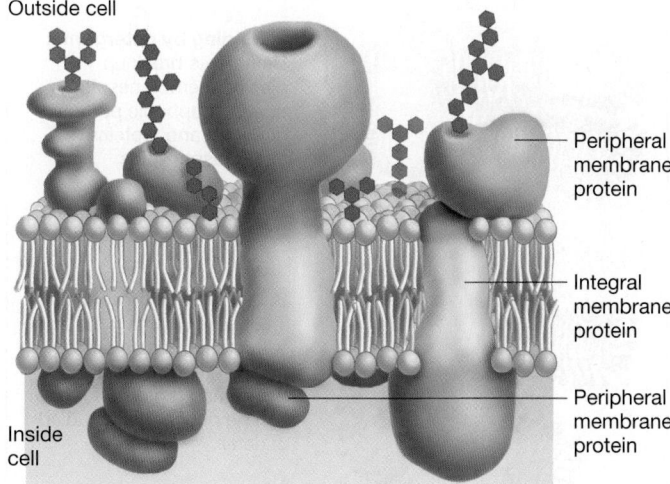

Outside cell

Peripheral membrane protein

Integral membrane protein

Peripheral membrane protein

Inside cell

FIGURE 6.19 Integral and Peripheral Membrane Proteins Individualize Bilayer Surfaces. The interior and exterior surfaces of the plasma membrane are different, because their peripheral proteins differ and the ends of the transmembrane proteins differ.

In most cells, certain peripheral proteins are found only on the interior surface of the plasma membrane and thus inside the cell, while others are found only on the exterior surface and thus face the surrounding environment. Peripheral membrane proteins are often attached to transmembrane membrane proteins.

What do all these proteins do? Later chapters will explore how certain membrane proteins act as enzymes, receive signals from other cells, or make physical connections between cells. Here, let's focus on how transmembrane proteins are involved in the transport of selected ions and molecules across the plasma membrane.

Systems for Studying Membrane Proteins

The discovery of transmembrane proteins was consistent with the hypothesis that proteins affect membrane permeability. The evidence was not considered conclusive, though, because it was also plausible to claim that transmembrane proteins were structural components that influenced membrane strength or flexibility. To test whether proteins actually do affect membrane permeability, researchers needed some way to isolate and purify membrane proteins.

Figure 6.20 outlines one method that researchers developed to separate proteins from membranes. The key to the technique is the use of detergents. A **detergent** is a small, amphi-

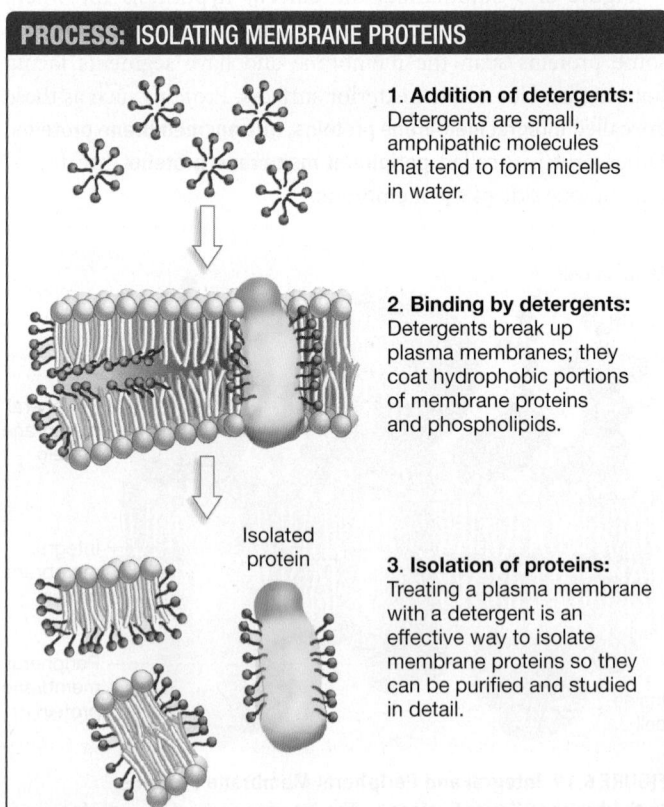

PROCESS: ISOLATING MEMBRANE PROTEINS

1. Addition of detergents: Detergents are small, amphipathic molecules that tend to form micelles in water.

2. Binding by detergents: Detergents break up plasma membranes; they coat hydrophobic portions of membrane proteins and phospholipids.

Isolated protein

3. Isolation of proteins: Treating a plasma membrane with a detergent is an effective way to isolate membrane proteins so they can be purified and studied in detail.

FIGURE 6.20 Detergents Can Be Used to Get Membrane Proteins into Solution.

pathic molecule. When detergents are added to the solution surrounding a lipid bilayer, the hydrophobic tails of the detergent molecule interact with the hydrophobic tails of the lipids and with the hydrophobic portions of transmembrane proteins. These interactions displace the membrane phospholipids and end up forming water-soluble, detergent-protein complexes.

To isolate and purify these membrane proteins once they are in solution, researchers use the technique called gel electrophoresis, introduced in **BioSkills 9** in Appendix A. When detergent-protein complexes are loaded into a gel and a voltage is applied, the larger protein complexes migrate more slowly than smaller proteins. As a result, the various proteins isolated from a plasma membrane separate from one another. To obtain a pure sample of a particular protein, the appropriate band is cut out of the gel. The gel material is then dissolved to retrieve the protein. Once this protein is inserted into a planar bilayer or liposome, dozens of different experiments are possible.

Since intensive experimentation on membrane proteins began, researchers have identified three broad classes of **transport proteins** that affect membrane permeability: channels, transporters, and pumps. Let's consider each in turn.

Protein Transport I: Facilitated Diffusion via Channel Proteins

As the data in Figure 6.7 showed, ions almost never cross pure phospholipid bilayers on their own. But in cells, ions routinely cross membranes through specialized membrane proteins called **ion channels**.

Ion channels form pores, or holes, in a membrane. Ions move through these pores in a predictable direction: from regions of high concentration to regions of low concentration via diffusion, and from areas of like charge to areas of unlike charge.

In **Figure 6.21**, for example, a large concentration gradient favors the movement of sodium ions from the outside of a membrane to the inside. But in addition, the inside of this cell has a net negative charge while the outside has a net positive charge. As a result, there is also a charge gradient that favors the movement of sodium ions, which have a positive charge, from the outside to the inside of the cell.

Ions move in response to a combined concentration and electrical gradient, or what biologists call an **electrochemical gradient.** ✔ If you understand this concept, you should be able to add an arrow to Figure 6.21, indicating the electrochemical gradient for chloride ions—assuming that chloride concentrations are equal on both sides of the membrane.

IS CFTR AN ION CHANNEL? To understand the types of experiments that biologists do to confirm that a membrane protein is an ion channel, consider work on the cystic fibrosis transmembrane conductance regulator (CFTR).

Experiments published in 1983 suggested that cystic fibrosis—the most common genetic disease in humans of Northern

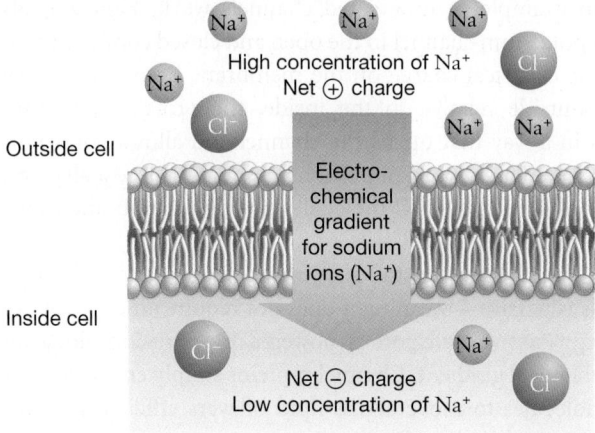

FIGURE 6.21 An Electrochemical Gradient Is a Combined Concentration and Electrical Gradient. Electrochemical gradients are established when ions build up on one side of a membrane.

✔**QUESTION** If you added a large number of sodium ions to the inside of this plasma membrane, what would happen to the electrical gradient, and to the concentration gradient for sodium?

European descent—is caused by defects in a membrane protein that allows chloride ions (Cl⁻) to move across plasma membranes. Using techniques introduced in Chapter 19, biologists were also able to (1) find the gene that is defective in people suffering from CF, and (2) use the gene to produce copies of the normal protein, which had come to be called CFTR.

Is CFTR a chloride channel? To answer this question, researchers inserted purified CFTR into planar bilayers and measured the flow of electric current across the membrane. Because ions carry a charge, ion movement across a membrane produces an electric current.

The graphs in **Figure 6.22**, which plot the amount of current flowing across the membrane over time, show their results. Note that when CFTR was absent, no electric current passed through the membrane. But when CFTR was inserted into the membrane, current began to flow. This was strong evidence that CFTR was indeed a chloride channel.

PROTEIN STRUCTURE DETERMINES CHANNEL SELECTIVITY Subsequent research has shown that cells have many different types of pore-like **channel proteins** in their membranes, including the subset called ion channels. Channel proteins are selective. Each channel protein has a structure that allows it to admit a particular type of ion or small molecule.

For example, Peter Agre and co-workers discovered channels called **aquaporins** ("water-pores") that allow water to cross the plasma membrane over 10 times faster than it does in the absence of aquaporins. Aquaporins admit water but not other small molecules or ions.

Figure 6.23a shows a cutaway view from the side of an aquaporin, indicating how it fits in a plasma membrane. Like

EXPERIMENT

QUESTION: Is CFTR a chloride channel?

HYPOTHESIS: CFTR increases the flow of chloride ions across a membrane.

NULL HYPOTHESIS: CFTR has no effect on membrane permeability.

EXPERIMENTAL SETUP:

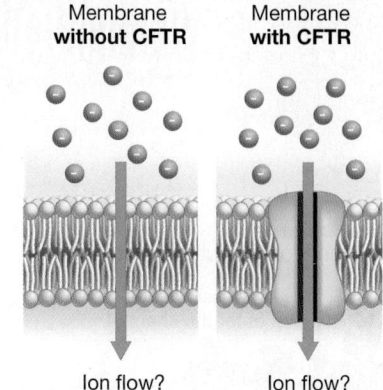

1. **Create planar bilayers** with and without CFTR.

2. **Add chloride ions** to one side of the planar bilayer to create an electrochemical gradient.

3. **Record electrical currents** to measure ion flow across the planar bilayers.

PREDICTION: Ion flow will be higher in membrane with CFTR.

PREDICTION OF NULL HYPOTHESIS: Ion flow will be the same in both membranes.

RESULTS:

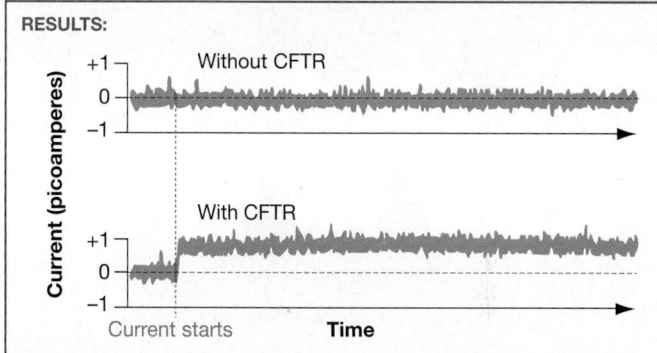

CONCLUSION: CFTR facilitates diffusion of chloride ions along an electrochemical gradient. CFTR is a chloride channel.

FIGURE 6.22 Electric Current Measurements Indicate That Ions Flow through CFTR.
SOURCE: Bear, C. A. et al. (1992). Purification and functional reconstitution of the cystic fibrosis transmembrane conductance regulator (CFTR). *Cell* 68: 809–818.

✔**QUESTION** The researchers repeated the "with CFTR" treatment 45 times, but recorded a current in only 35 of the replicates. Does this observation negate the conclusion? Explain why or why not.

other channels that have been studied in detail, aquaporins have a pore that is lined with polar regions of amino acids—in this case, functional groups that interact with water. Hydrophobic groups form the outside of channels and interact with the lipid bilayer. A channel's pore is hydrophilic; its exterior is hydrophobic.

(a) Water pores allow only water to pass through.

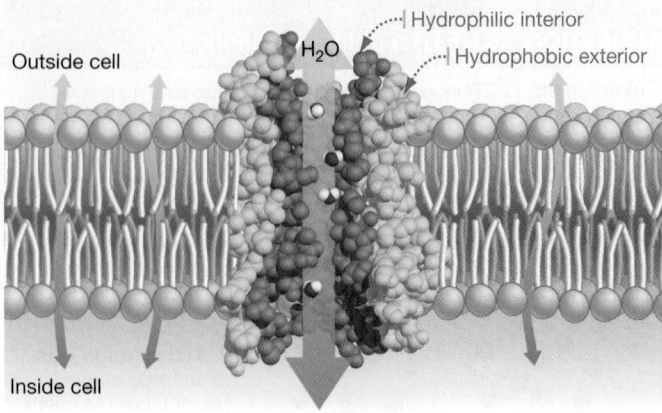

(b) Potassium channels allow only potassium ions to pass through.

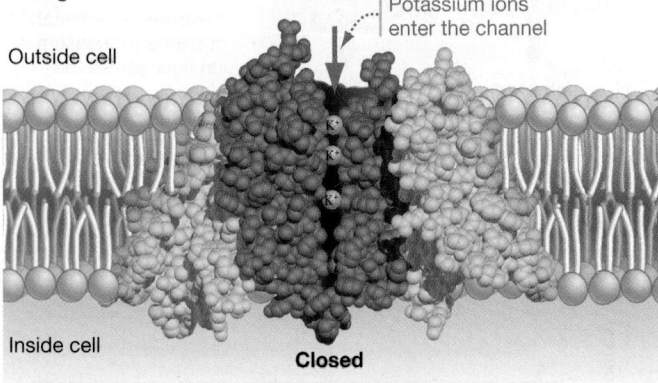

Closed

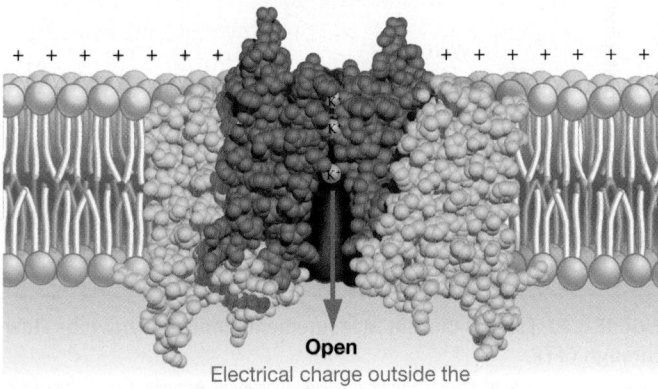

Open

Electrical charge outside the membrane triggers shape change that allows the ions to pass through

FIGURE 6.23 Membrane Channels Are Highly Selective and Highly Regulated. (a) A cutaway view looking at the inside of an aquaporin—a membrane channel that admits only water. Water moves through its pore via osmosis over 10 times faster than it can move through the lipid bilayer. **(b)** A model of a K$^+$ channel in the closed and open configurations.

MOVEMENT THROUGH MEMBRANE CHANNELS IS REGULATED Recent research has shown that the aquaporins and ion channels such as CFTR are **gated channels**—meaning that they open or close in response to the binding of a particular molecule or to a change in the electrical charge on the outside of the membrane.

As an example of how gated channels work, **Figure 6.23b** shows a potassium channel in the open and closed configuration. When the electrical charge on the membrane becomes positive on the outside relative to the inside, the protein's structure changes in a way that opens the channel and allows potassium ions to cross. The important point here is that in almost all cases, the flow of ions and small molecules through membrane channels is carefully controlled.

In all cases, however, the movement of substances through channels is passive—meaning it does not require an expenditure of energy. **Passive transport** is powered by diffusion along an electrochemical gradient. Channel proteins simply enable ions or polar molecules to move across lipid bilayers efficiently, in response to an existing gradient.

✔If you understand the nature of membrane channels, you should be able to (1) draw a channel that admits calcium ions (Ca^{2+}) when a signaling molecule binds to it, (2) label hydrophilic and hydrophobic portions of the channel, and (3) add ions to the outside and inside of a membrane containing the channel to explain why an electrochemical gradient favors entry of Ca^{2+}.

To summarize, membrane proteins such as CFTR, aquaporins, and potassium channels circumvent the lipid bilayer's impermeability to small, charged compounds. They are responsible for **facilitated diffusion**: the passive transport of substances that otherwise would not cross a membrane readily.

Protein Transport II: Facilitated Diffusion via Carrier Proteins

Even though facilitated diffusion does not require an expenditure of energy, it is facilitated—aided—by the presence of a specialized membrane protein. Facilitated diffusion can also occur through **carrier proteins**, or **transporters**, that change shape during the process. Perhaps the best-studied transporter is specialized for moving glucose into cells.

THE SEARCH FOR A GLUCOSE TRANSPORTER Next to ribose, the six-carbon sugar glucose is the most important sugar found in organisms. Virtually all cells alive today use glucose as a building block for important macromolecules and as a source of stored chemical energy. But as Figure 6.7 showed, lipid bilayers are only moderately permeable to glucose. It is reasonable to expect, then, that plasma membranes have some mechanism for increasing their permeability to this sugar.

This prediction was supported in experiments on pure preparations of plasma membranes—or "ghosts"—from human red blood cells (see **Figure 6.24**). These plasma membranes turned out to be much more permeable to glucose than are pure lipid bilayers. Why?

After isolating and analyzing many proteins from red blood cell ghosts, researchers found one protein that specifically increases membrane permeability to glucose. When this purified protein was added to liposomes, the artificial membrane transported glucose at the same rate as a membrane from a living cell.

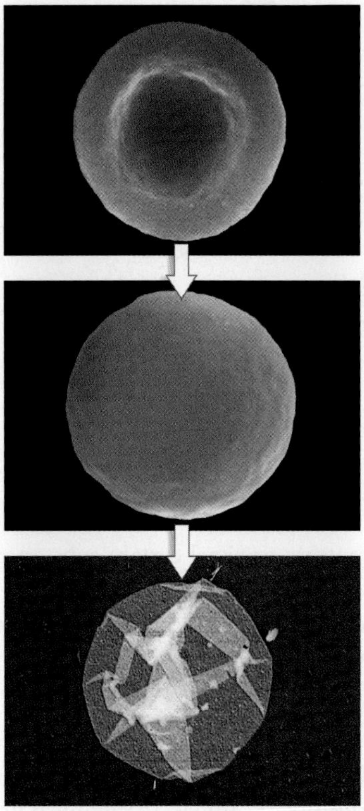

1. Start with normal red blood cells in isotonic solution.

2. Put cells in hypotonic solution. Cells swell as water enters via osmosis. Eventually the cells burst.

3. Harvest the ghosts. After the cell contents have spilled out, all that remains are cell "ghosts," which consist entirely of plasma membranes.

FIGURE 6.24 Rupture of Red Blood Cells Produces "Ghosts." Red blood cell ghosts are simple membranes that can be purified and studied in detail.

This experiment convinced biologists that the membrane protein—now called GLUT-1—was indeed responsible for transporting glucose across plasma membranes.

HOW DOES GLUT-1 WORK? **Figure 6.25** illustrates the current hypothesis for how GLUT-1 works. The idea is that glucose binds to GLUT-1 on the exterior of the membrane, and that this binding induces a conformational change in the protein which transports glucose to the interior of the cell. Recall from Chapter 3 that enzymes frequently change shape when they bind substrates and that such conformational changes are often a critical step in the catalysis of chemical reactions.

What powers the movement of molecules through transporters? The answer is diffusion. When glucose enters a cell via GLUT-1, it does so because it is following its concentration gradient. If the concentration of glucose is the same on both sides of the plasma membrane, then no net movement of glucose occurs even if the membrane contains GLUT-1. A large array of molecules moves across plasma membranes via facilitated diffusion through specific transporter proteins.

Protein Transport III: Active Transport by Pumps

Whether diffusion is facilitated by channel proteins or by transporters, it is a passive process that makes the cell interior and exterior more similar. But it is also possible for today's cells to import molecules or ions *against* their electrochemical gradient. Accomplishing this task requires energy, because the cell must counteract the entropy loss that occurs when molecules or ions are concentrated. It makes sense, then, that transport against an electrochemical gradient is called **active transport**.

In cells, the energy required to move substances against their electrochemical gradient is provided by a phosphate group (HPO_4^{2-}) from adenosine triphosphate (ATP). Recall that ATP contains three phosphate groups (Chapter 5), and that phosphate groups carry two negative charges (Chapter 2). When a phosphate group leaves ATP and binds to a protein, its negative charges are added to the protein. These charges repel other charges on the protein's amino acids. The protein's potential energy increases in response, and its shape usually changes. When a phosphate group leaves ATP, the resulting molecule is adenosine diphosphate (ADP), which has two phosphate groups.

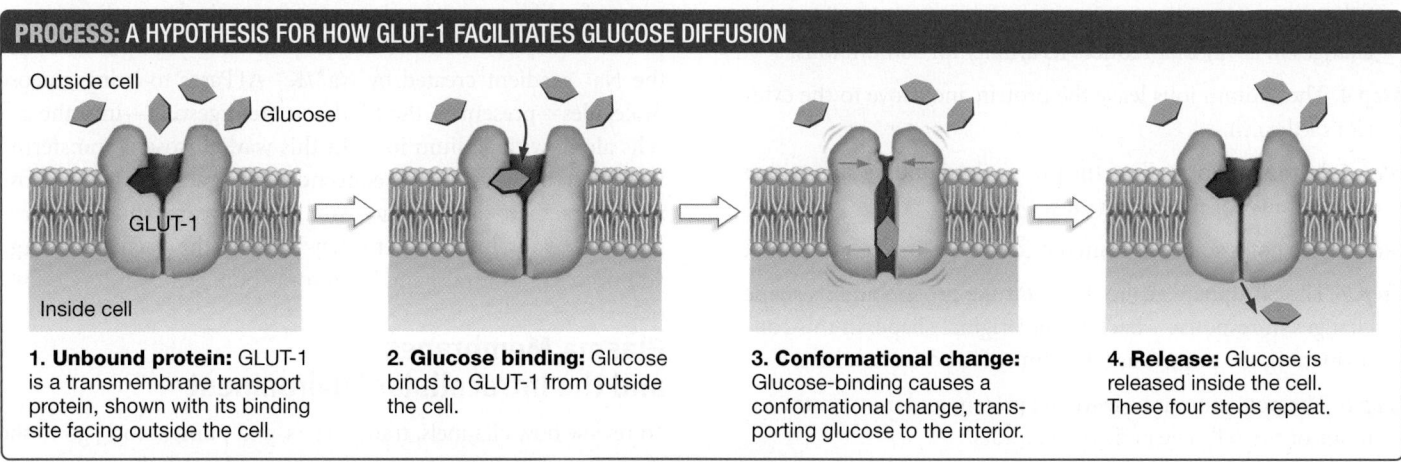

Outside cell

Glucose

GLUT-1

Inside cell

1. Unbound protein: GLUT-1 is a transmembrane transport protein, shown with its binding site facing outside the cell.

2. Glucose binding: Glucose binds to GLUT-1 from outside the cell.

3. Conformational change: Glucose-binding causes a conformational change, transporting glucose to the interior.

4. Release: Glucose is released inside the cell. These four steps repeat.

FIGURE 6.25 Membrane Transport Proteins and Enzymes May Work Similarly. This model suggests that the GLUT-1 transporter binds a substrate (in this case, a glucose molecule), undergoes a conformation change, and releases the substrate.

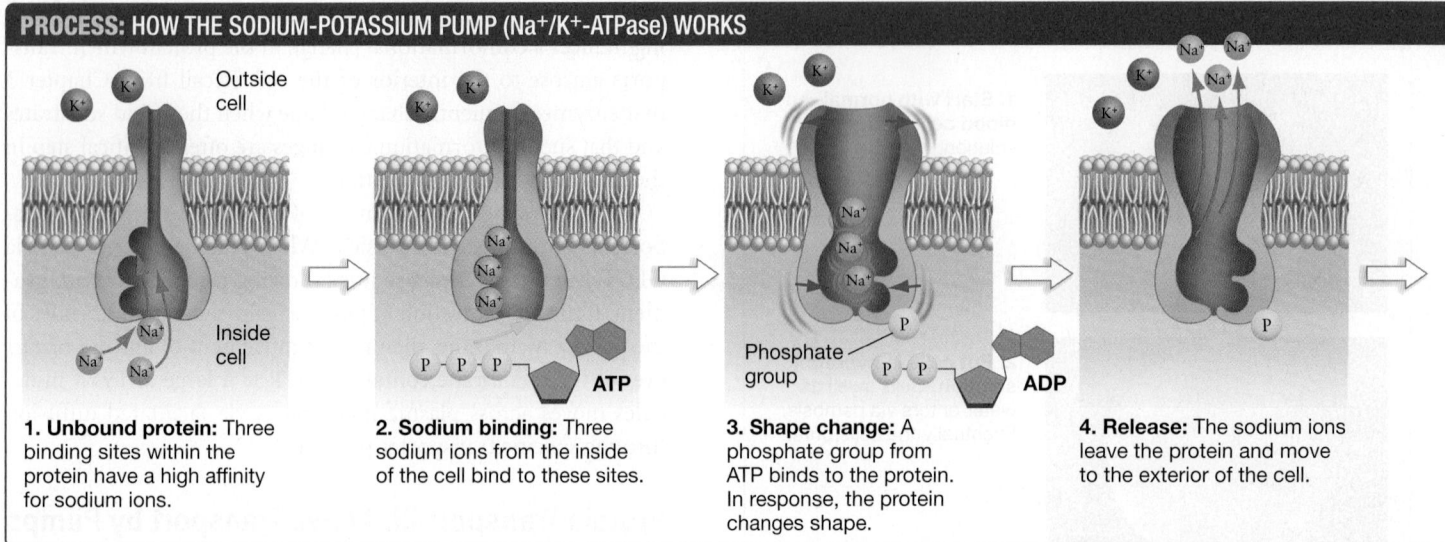

PROCESS: HOW THE SODIUM-POTASSIUM PUMP (Na⁺/K⁺-ATPase) WORKS

Outside cell

Inside cell

ATP

Phosphate group

ADP

1. Unbound protein: Three binding sites within the protein have a high affinity for sodium ions.

2. Sodium binding: Three sodium ions from the inside of the cell bind to these sites.

3. Shape change: A phosphate group from ATP binds to the protein. In response, the protein changes shape.

4. Release: The sodium ions leave the protein and move to the exterior of the cell.

FIGURE 6.26 Active Transport Depends on an Input of Chemical Energy Stored in ATP.

THE SODIUM-POTASSIUM PUMP Ions or molecules can move against an electrochemical gradient when membrane proteins called **pumps** change shape.

As an example, **Figure 6.26** highlights the first pump that was discovered and characterized: a protein called the **sodium-potassium pump**, or more formally, Na⁺/K⁺-ATPase. The Na⁺/K⁺ part of the name refers to the ions that are transported, ATP indicates that adenosine triphosphate is used, and –*ase* implies that the molecule acts like an enzyme.

As the figure indicates, sodium and potassium ions move in a multistep process:

Step 1 When Na⁺/K⁺-ATPase is in the conformation shown here, binding sites with a high affinity for sodium ions are available.

Step 2 Three sodium ions from the inside of the cell bind to these sites.

Step 3 A phosphate group from ATP then binds to the pump. When the phosphate group attaches, the pump's shape changes in a way that reduces its affinity for sodium ions.

Step 4 The sodium ions leave the protein and move to the exterior of the cell.

Step 5 In this conformation, the protein has binding sites with a high affinity for potassium ions.

Step 6 Two potassium ions from outside the cell bind to the pump.

Step 7 The phosphate group drops off the protein and its shape changes in response—back to the original shape. In this conformation, the pump has low affinity for potassium ions.

Step 8 The potassium ions leave the protein and move to the interior of the cell. The cycle then repeats.

In an analogous way, other types of pumps move protons (H⁺), calcium ions (Ca²⁺), or other ions or molecules into or out of cells, against electrical or concentration gradients.

Pumps allow cells to concentrate or expel certain substances. As a result, cells can import valuable ions or molecules that are at high concentration inside the cell but at low concentration outside the cell. They can also get rid of damaging molecules or ions, even when a concentration gradient favors diffusion of these substances into the cell.

SECONDARY ACTIVE TRANSPORT In addition to concentrating or expelling certain substances, pumps set up electrochemical gradients. For example, because the Na⁺/K⁺-ATPase exchanges three sodium ions for every two potassium ions, the outside of the membrane becomes positively charged relative to the inside. In this way, the sodium-potassium pump sets up an electrical gradient and a chemical gradient across the membrane. The electrical gradient favors a flow of anions out of the cell and cations into the cell; the chemical gradient favors a flow of sodium ions into the cell and potassium ions out of the cell.

These gradients are crucial, in part because they make it possible for cells to engage in **secondary active transport**—also known as **cotransport**. When cotransport occurs, a gradient set up by a pump provides the potential energy required to power the movement of a different molecule against its particular gradient.

For example, a transmembrane protein in your gut cells uses the Na⁺ gradient created by Na⁺/K⁺-ATPases to bring glucose molecules—present in the food you are digesting—into the gut cells along with sodium ions. In this way, glucose is transferred into your body against a concentration gradient. The glucose molecules then diffuse into your bloodstream and are transported to your brain, where they provide the chemical energy you need to stay awake and learn some biology.

Plasma Membranes and the Intracellular Environment

To review how channels, transporters, and pumps work, go to the study area at *www.masteringbiology.com* and watch them in action.

(MB) BioFlix™ Membrane Transport, **Web Activity** Membrane Transport Proteins

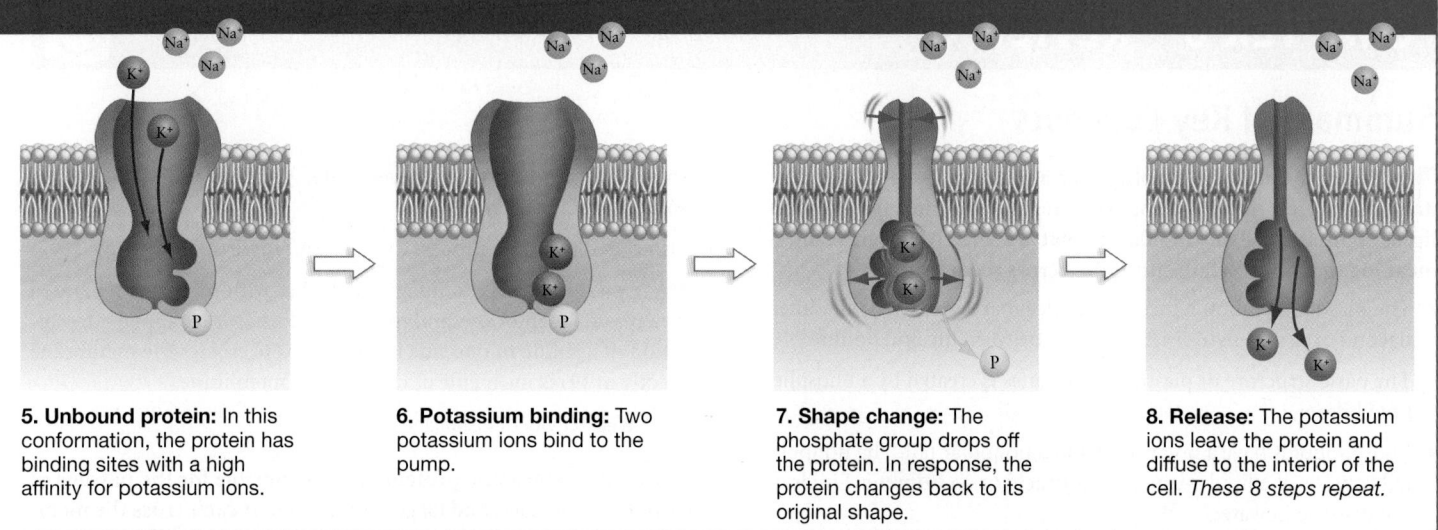

5. Unbound protein: In this conformation, the protein has binding sites with a high affinity for potassium ions.

6. Potassium binding: Two potassium ions bind to the pump.

7. Shape change: The phosphate group drops off the protein. In response, the protein changes back to its original shape.

8. Release: The potassium ions leave the protein and diffuse to the interior of the cell. *These 8 steps repeat.*

Taken together, the selective permeability of the lipid bilayer and the specificity of the proteins involved in passive transport and active transport enable cells to create an internal environment that is much different from the external one (**Figure 6.27**).

With the evolution of membrane proteins, the early cells acquired the ability to create an internal environment that was conducive to life—one that contained the substances required for manufacturing ATP and copying ribozymes. Cells with particularly efficient and selective membrane proteins would be favored by natural selection and would come to dominate the population. Cellular life had begun.

Some 3.5 billion years later, cells continue to evolve. What do today's cells look like, and how do they produce and store the chemical energy that makes life possible? Answering these and related questions is the focus of Unit 2.

CHECK YOUR UNDERSTANDING

If you understand that . . .

- Membrane proteins allow ions and molecules that ordinarily do not readily cross lipid bilayers to enter or exit cells.
- Substances may move across a plasma membrane along an electrochemical gradient, via facilitated diffusion through channel proteins or transport proteins. Or, they may move against an electrochemical gradient in response to work done by pumps.

✓ You should be able to . . .

Explain what is passive about passive transport, what is active about active transport, and what is "co" about cotransport.

Answers are available in Appendix B.

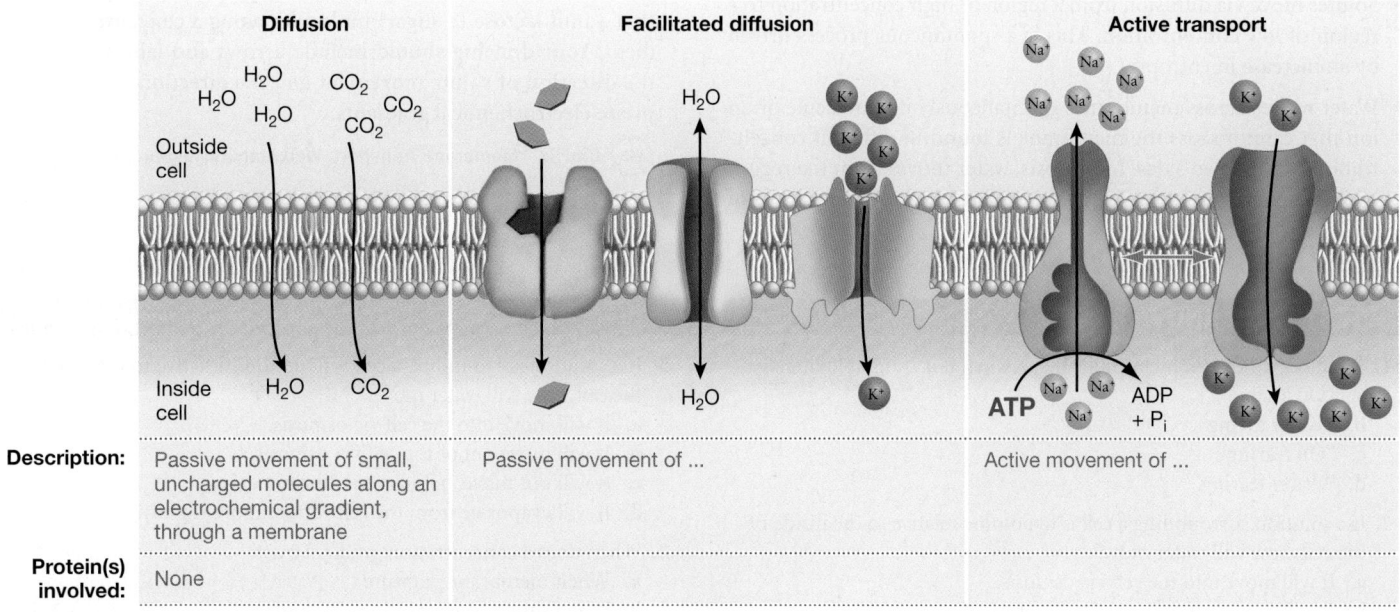

Description:	Passive movement of small, uncharged molecules along an electrochemical gradient, through a membrane	Passive movement of ...	Active movement of ...
Protein(s) involved:	None		

FIGURE 6.27 Summary of the Passive and Active Mechanisms of Membrane Transport.

✓**EXERCISE** Complete the chart.

Summary of Key Concepts

🔑 **Phospholipids are amphipathic molecules—they have a hydrophilic region and a hydrophobic region. In solution, phospholipids spontaneously form bilayers that are selectively permeable—meaning that only certain substances cross them readily.**

- The plasma membrane forms a physical barrier between the internal and external environment—often between life and nonlife.

- The basic structure of plasma membranes is created by a phospholipid bilayer.

- Phospholipids have a polar head and a nonpolar tail. The nonpolar tail consists of a lipid—usually a fatty acid or an isoprene. Lipids do not dissolve in water.

- Small, nonpolar molecules tend to move across membranes readily; ions and other charged compounds cross rarely, if at all.

- The permeability and fluidity of lipid bilayers depend on temperature and on the types of phospholipids present. Phospholipids that contain long, saturated fatty acids form a dense and highly hydrophobic membrane interior that lowers permeability, relative to phospholipids containing shorter, unsaturated fatty acids.

  ✔ You should be able to describe the structure of phospholipid bilayers that are highly permeable and fluid versus highly impermeable and lacking in fluidity.

🔑 **Ions and molecules diffuse spontaneously from regions of high concentration to regions of low concentration. Water moves across lipid bilayers from regions of high water concentration to regions of low water concentration via osmosis—a special case of diffusion.**

- Diffusion is movement of ions or molecules owing to their kinetic energy.

- Solutes move via diffusion from a region of high concentration to a region of low concentration. This is a spontaneous process driven by an increase in entropy.

- Water moves across membranes spontaneously if a molecule or an ion that cannot cross the membrane is found in different concentrations on the two sides. In osmosis, water moves from the region with a higher concentration of water and lower concentration of solutes to the region of lower water concentration and higher solute concentration.

- Osmosis is a passive process driven by an increase in entropy.

  ✔ You should be able to imagine a beaker with solutions separated by a plasma membrane, and then predict what will happen after addition of a solute to one side if the solute (1) crosses the membrane readily or (2) is incapable of crossing the membrane.

  (MB) **Web Activity** Diffusion and Osmosis

🔑 **In cells, membrane proteins are responsible for the passage of ions, polar molecules, and large molecules that can't cross the membrane on their own because they are not soluble in lipids. Some membrane proteins form channels, some facilitate diffusion by binding to substrates, and some use energy from ATP to actively pump ions or molecules.**

- The permeability of lipid bilayers can be altered significantly by membrane transport proteins.

- Channel proteins provide holes in the membrane and facilitate the diffusion of specific ions into or out of the cell.

- Transport proteins are enzyme-like proteins that facilitate the diffusion of specific molecules into or out of the cell.

- Energy-demanding pumps actively move ions or molecules against their electrochemical gradient.

- In combination, the selective permeability of phospholipid bilayers and the specificity of transport proteins make it possible to create an environment inside a cell that is radically different from the exterior.

  ✔ You should be able to draw and label the membrane of a cell that pumps hydrogen ions to the exterior, has channels that admit calcium ions along an electrochemical gradient, and has carriers that admit lactose (a sugar) molecules along a concentration gradient. Your drawing should include arrows and labels indicating the direction of solute movement and the direction of the appropriate electrochemical gradients.

  (MB) **BioFlix™** Membrane Transport, **Web Activity** Membrane Transport Proteins

Questions

✔ TEST YOUR KNOWLEDGE

Answers are available in Appendix B

1. What does the term hydrophilic mean when it is translated literally?
   a. "Oil loving"
   b. "Water loving"
   c. "Oil fearing"
   d. "Water fearing"

2. If a solution surrounding a cell is hypotonic relative to the inside of the cell, how will water move?
   a. It will move into the cell via osmosis.
   b. It will move out of the cell via osmosis.
   c. It will not move, because equilibrium exists.
   d. It will evaporate from the cell surface more rapidly.

3. If a solution surrounding a cell is hypertonic relative to the inside of the cell, how will water move?
   a. It will move into the cell via osmosis.
   b. It will move out of the cell via osmosis.
   c. It will not move, because equilibrium exists.
   d. It will evaporate from the cell surface more rapidly.

4. When does a concentration gradient exist?
   a. When membranes rupture
   b. When solute concentrations are high
   c. When solute concentrations are low
   d. When solute concentrations differ on the two sides of a membrane

5. Which of the following must be true for osmosis to occur?
   a. Water must be at room temperature or above.
   b. Solutions with the same concentration of solutes must be separated by a selectively permeable membrane.
   c. Solutions with different concentrations of solutes must be separated by a selectively permeable membrane.
   d. Water must be under pressure.

6. Why are the lipid bilayers in cells called "selectively permeable"?
   a. They are not all that permeable.
   b. Their permeability changes with their molecular composition.
   c. Their permeability is temperature-dependent.
   d. They are permeable to some substances but not others.

✓ TEST YOUR UNDERSTANDING

Answers are available in Appendix B

1. Cooking oil is composed of lipids that consist of long hydrocarbon chains. Would you expect these lipids to form membranes spontaneously? Why or why not? Describe, on a molecular level, how you would expect these lipids to interact with water.

2. Explain why phospholipids form a bilayer in solution, and why the process is spontaneous.

3. Ethanol, the active ingredient in alcoholic beverages, is a small, polar, uncharged molecule. Would you predict that this molecule crosses plasma membranes quickly or slowly? Explain your reasoning.

4. Why can osmosis occur only if solutions are separated by a selectively permeable membrane? What happens in solutions that are *not* separated by a selectively permeable membrane?

5. The text claims that the portion of membrane proteins that spans the hydrophobic tails of phospholipids is itself hydrophobic (see Figure 6.16b). Why is this logical? Look back at Figure 3.3 and Table 3.1, and make a list of amino acids you would expect to find in these regions of transmembrane proteins.

6. Examine the membrane in the accompanying figure. Label the molecules and ions that will pass through the membrane as a result of osmosis, diffusion, and facilitated diffusion. Draw arrows to indicate where each of the molecules and ions will travel.

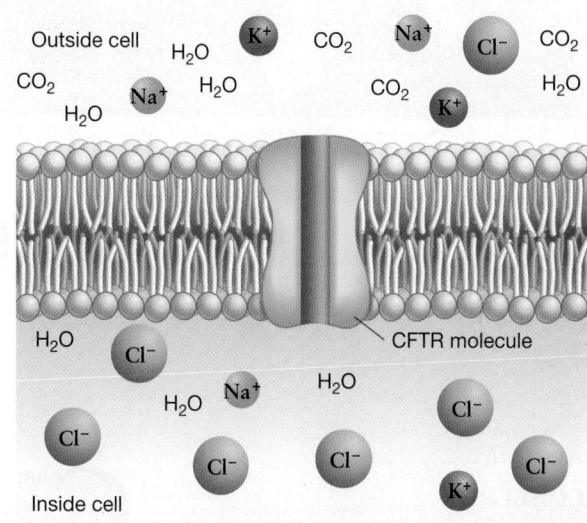

✓ APPLYING CONCEPTS TO NEW SITUATIONS

Answers are available in Appendix B

1. When phospholipids are arranged in a bilayer, it is theoretically possible for individual molecules in the bilayer to flip-flop. That is, a phospholipid could turn 180° and become part of the membrane's other surface. From what you know about the behavior of polar heads and nonpolar tails, predict whether flip-flops are frequent or rare. Then design an experiment, using a planar bilayer with one side made up of phospholipids that contain a dye molecule on their hydrophilic head, to test your prediction.

2. Unicellular organisms that live in extremely cold habitats have an unusually high proportion of unsaturated fatty acids in their plasma membranes. Some of these membranes even contain polyunsaturated fatty acids, which have more than one double bond in each hydrocarbon chain. Explain why you agree or disagree with the hypothesis that membranes with unsaturated fatty-acid tails function better at cold temperatures than do membranes with saturated fatty-

acid tails. Make a prediction about the structure of fatty acids found in organisms that live in extremely hot environments.

3. When biomedical researchers design drugs that must enter cells to be effective, they sometimes add methyl (CH_3) groups to make the drug molecules more likely to pass through plasma membranes. Conversely, when researchers design drugs that act on the exterior of plasma membranes, they sometimes add a charged group to decrease the likelihood that the drugs will pass through membranes and enter cells. Explain why these strategies are effective.

4. Advertisements frequently claim that laundry and dishwashing detergents "cut grease." The ad writers mean that the detergents surround oil droplets on clothing or dishes, making the droplets water-soluble. When this happens, the oil droplets can be washed away. Explain how this happens on a molecular level.

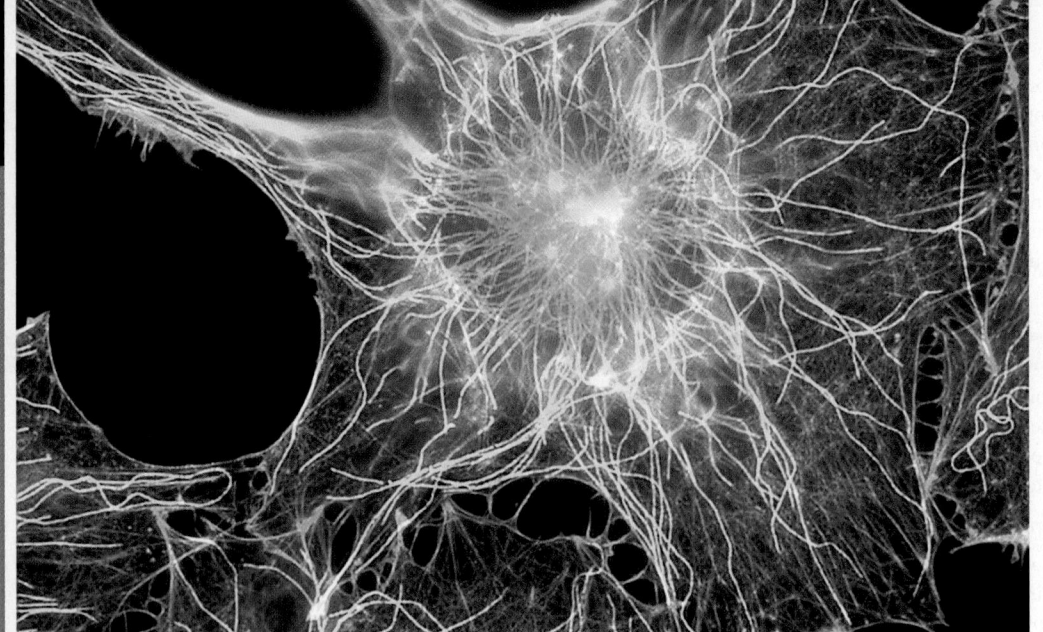

This cell has been treated with fluorescing molecules that bind to its fibrous skeleton. Microtubules (large protein fibers) are yellow; actin filaments (smaller fibers) are blue. The cell's nucleus has been stained green.

7 Inside the Cell

KEY CONCEPTS

- The structure of individual cell components is closely related to their function. The overall shape and composition of a cell is also closely related to its function.

- Inside cells, materials are transported to their destinations with the help of molecular "zip codes."

- The cytoskeleton provides a structural framework inside cells and plays a key role in cell division, movement, and transport.

- Cells are dynamic, highly integrated structures. Thousands of chemical reactions occur each second within cells; molecules constantly enter and exit across the plasma membrane; cell products are shipped along protein fibers; and elements of the cell's internal skeleton grow and shrink.

Chapter 1 introduced the cell theory, which states that all organisms consist of cells and all cells are derived from preexisting cells. Since this theory was initially developed and tested in the 1850s, an enormous body of research has confirmed that the cell is the fundamental structural and functional unit of life. Life on Earth is cellular.

Chapter 6 delved into this fundamental attribute of life by introducing the **plasma membrane**—a structure common to every cell. Thanks to the selective permeability of phospholipid bilayers and the activity of membrane transport proteins, the plasma membrane creates an internal environment that differs from conditions outside the cell.

Our task now is to explore the structures inside the plasma membrane. Let's begin by analyzing how the parts inside a cell function individually, and then explore how they work as a unit. This approach is analogous to studying individual organs in the body, and then analyzing how they work together to form the nervous system or digestive system. As you study this material, keep asking yourself some key questions: How does the structure of this part or group of parts correlate with its function? What problem does it solve?

7.1 Bacterial and Archaeal Cell Structures and Their Functions

In addition to introducing the cell theory, Chapter 1 highlighted a distinction between two fundamental types of cells observed in nature—eukaryotes and prokaryotes. Eukaryotic cells have a membrane-bound compartment called a nucleus, while prokaryotic cells do not.

According to **morphology** ("form-science"), then, species fall into two broad categories: (1) prokaryotes and (2) eukaryotes. But according to **phylogeny** ("tribe-

✔ When you see this checkmark, stop and test yourself. Answers are available in Appendix B.

source"), or evolutionary history, organisms fall into three broad domains called (1) Bacteria, (2) Archaea, and (3) Eukarya. Members of the Bacteria and Archaea are prokaryotic; members of the Eukarya—including algae, fungi, plants, and animals—are eukaryotic.

A Revolutionary New View

For almost 200 years, biologists thought that prokaryotic cells were simple in terms of their morphology, with little structural diversity among species. This was a valid conclusion at the time, given the resolution of the microscopes that were available and the number of species that had been studied.

Things have changed. The structure of archaeal cells is still poorly studied, but recent improvements in microscopy and other research tools have convinced biologists that bacterial cells are highly organized, with an array of distinctive structures found among millions of species. This conclusion represents one of the most exciting discoveries in cell biology over the past 10 years.

To keep things simple at the start, though, **Figure 7.1** offers a low-magnification, stripped-down diagram of a bacterial cell.

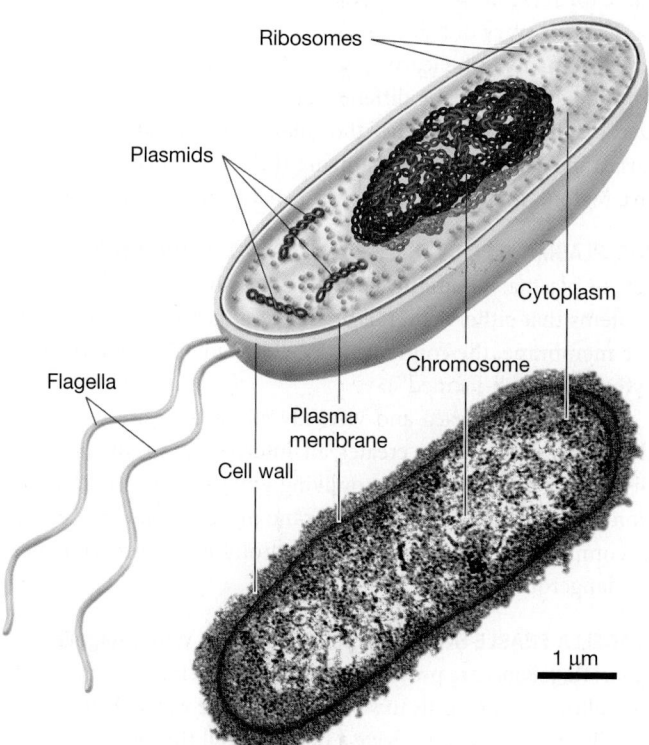

FIGURE 7.1 Overview of a Prokaryotic Cell. Prokaryotic cells are identified by a negative trait—the absence of a membrane-bound nucleus. Although there is wide variation in the size and shape of bacterial and archaeal cells, they all contain a plasma membrane, a chromosome, and protein-synthesizing ribosomes; almost all have a stiff cell wall. Some prokaryotes have flagella used in swimming and/or inner membranes where photosynthesis takes place.

✔ **EXERCISE** Label the nucleoid in the drawing and the micrograph.

Prokaryotic Cell Structures: A Parts List

The labels in Figure 7.1 highlight the components common to all or most bacteria studied to date. Let's explore these elements one by one, and also look at more specialized structures found in particular species, starting from the inside and working out.

THE CHROMOSOME IS ORGANIZED IN A NUCLEOID The most prominent structure inside a bacterial cell is the **chromosome**. Most bacterial species have a single, circular chromosome that consists of a large DNA molecule associated with a small number of proteins. The DNA molecule contains information, while the proteins provide structural support for the DNA.

Recall from Chapter 4 that the information in DNA is encoded in its sequence of nitrogenous bases, and a segment of DNA that contains the information for building an RNA molecule or a polypeptide is called a **gene**. Thus, chromosomes contain DNA, which contains genes.

In the well-studied bacterium *Escherichia coli*, the circular chromosome would be over 1 mm long if it were linear—500 times longer than the cell itself. This situation is typical in prokaryotes. To fit into the cell, the DNA double helix coils on itself with the aid of enzymes to form the highly compact, "supercoiled" structure. Supercoiled regions of DNA resemble a string that has been held at either end and then twisted until it coils back upon itself (**Figure 7.2**).

Bacterial chromosomes are found in a localized area of the cell called the **nucleoid** (pronounced *NEW-klee-oyd*). The nucleoid is usually found in the center of the cell and typically represents about 20 percent of the cell's total volume. It's important to note, though, that the genetic material in the nucleoid is not separated from the rest of the cell interior by a membrane. The organization of the nucleoid is currently the subject of intense research.

In addition to one or more main chromosomes, bacterial cells may also contain one to about a hundred small, usually circular, supercoiled DNA molecules called **plasmids**. Plasmids contain genes but are physically independent of the main, cellular

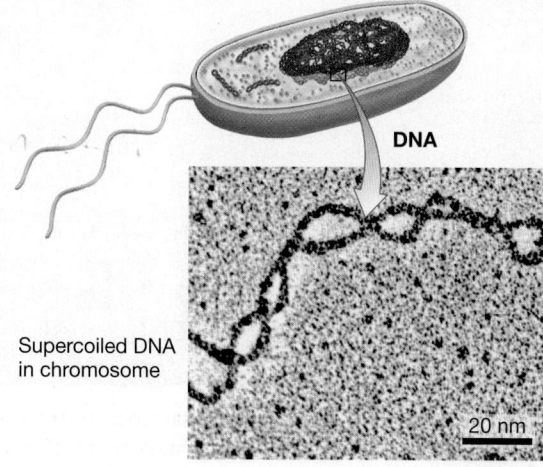

FIGURE 7.2 Bacterial DNA Is Supercoiled. The circular chromosomes of bacteria and archaea must be coiled extensively, into "supercoils," to fit in the cell.

chromosome. In most cases the genes carried by plasmids are not required under normal conditions; instead they help cells adapt to unusual circumstances, such as the sudden presence of a poison in the environment. As a result, plasmids can be considered auxiliary genetic elements.

RIBOSOMES MANUFACTURE PROTEINS   **Ribosomes** are observed in all prokaryotic cells and are found throughout the cell interior. It is not unusual for a single cell to contain 10,000 ribosomes, each of which functions as a protein-manufacturing center.

Bacterial ribosomes are complex structures consisting of RNA molecules and proteins. Biologists sometimes refer to ribosomes, along with other multicomponent complexes that perform specialized tasks, as "molecular machines." Chapter 16 analyzes the structure and function of the bacterial ribosome in detail.

INTERNAL MEMBRANE COMPLEXES IN PHOTOSYNTHETIC SPECIES In addition to the nucleoid and ribosomes found in all bacteria and archaea studied to date, it is common to observe extensive internal membranes in prokaryotes that perform photosynthesis. Photosynthesis is the suite of chemical reactions responsible for converting the energy in sunlight into chemical energy stored in sugars.

The photosynthetic membranes observed in prokaryotes contain the enzymes and pigment molecules required for these reactions to occur and develop as infoldings of the plasma membrane. In some cases, vesicles pinch off as the plasma membrane folds in. In other cases, flattened stacks of photosynthetic membrane, like those shown in **Figure 7.3**, form from the infolded sections of the plasma membrane. The extensive surface area provided by these internal membranes makes it possible for more photosynthetic reactions to occur, increasing the cell's ability to make food.

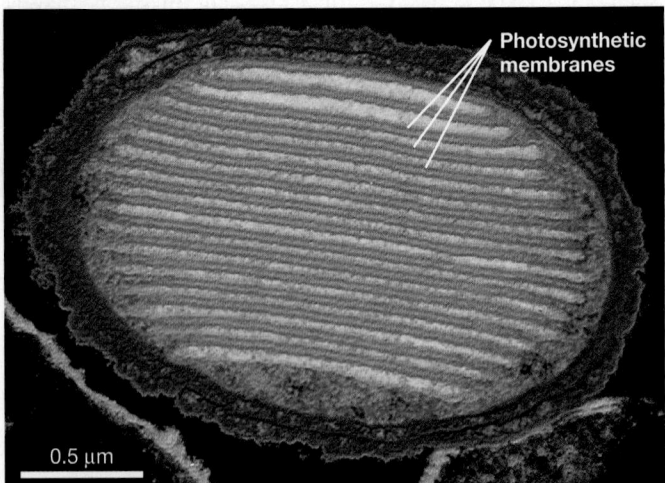

Photosynthetic membranes

0.5 μm

FIGURE 7.3 Photosynthetic Membranes in Bacteria. The green stripes in this photosynthetic bacterium are membranes that contain the pigments and enzymes required for photosynthesis. This photo is a **transmission electron micrograph** that has been colorized (see **BioSkills 10** in Appendix A).

ORGANELLES PERFORM SPECIALIZED FUNCTIONS Recent research indicates that several bacterial species have internal compartments that qualify as **organelles** ("little organs"). An organelle is a membrane-bound compartment inside the cell that contains enzymes or structures specialized for a particular function. Each type of bacterial organelle is found in certain species.

Bacterial organelles perform an array of tasks, including:

- storing calcium ions or other key molecules;
- holding crystals of the mineral magnetite, which function like a compass needle to help cells sense a magnetic field and swim in a directed way;
- organizing enzymes responsible for synthesizing complex carbon compounds from carbon dioxide; and
- sequestering enzymes that generate chemical energy from ammonium ions.

THE CYTOSKELETON STRUCTURES THE CELL INTERIOR Recent research has also shown that bacteria and archaea contain long, thin fibers that serve a structural role inside the cell. All bacterial species, for example, contain protein fibers that are essential for cell division to take place. Some species also have protein filaments that help maintain cell shape. Protein filaments such as these form the basis of the **cytoskeleton** ("cell skeleton").

The discovery of bacterial cytoskeletal elements is so new that much remains to be learned. Currently, researchers are working to understand how the different cytoskeletal elements are localized inside the cell, whether they play a role in transporting materials inside the cell or organizing the cell interior into distinctive regions, and how they move to enable the cell to divide.

THE PLASMA MEMBRANE SEPARATES LIFE FROM NONLIFE The cell, or plasma, membrane consists of a phospholipid bilayer and proteins that either span the bilayer or attach to one side. Inside the membrane, the contents of a cell are collectively termed the **cytoplasm** ("cell-formed").

Because all archaea and virtually all bacteria are unicellular, the plasma membrane creates an internal environment that is distinct from the outside, nonliving environment. In combination with the lipid bilayer, membrane proteins allow the passage of compounds required for life, and prohibit the entry of materials dangerous to life.

FLAGELLA ENABLE SOME SPECIES TO SWIM When **flagella** (singular: **flagellum**) are present in the plasma membrane, their rotation allows aquatic cells to swim through the water. At top speed, flagellar movement can drive a bacterial cell through water at 60 cell lengths per second. In contrast, the fastest land animal—the cheetah—can sprint at a mere 25 body lengths per second.

Bacterial flagella are usually few in number and are located on the surface of the cell. Over 40 different proteins are involved in building and controlling this molecular machine.

THE CELL WALL FORMS A PROTECTIVE "EXOSKELETON" Because the cytoplasm contains a high concentration of solutes, in most

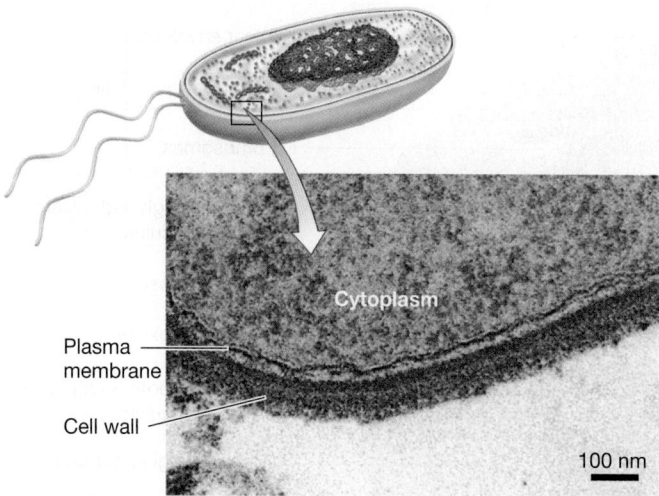

FIGURE 7.4 The Bacterial Cell Wall. In bacteria and archaea, the cell wall consists of peptidoglycan or similar polymers that are cross-linked into tough sheets. The inside of the cell wall contacts the plasma membrane, which pushes up against the wall. The outside of the cell wall makes direct contact with the outside environment, which is almost always filled with competitors and predators.

habitats it is hypertonic relative to the surrounding environment. When this is the case, water enters the cell via osmosis and makes the cell's volume expand. In virtually all bacteria and archaea, this pressure is resisted by a stiff **cell wall** (**Figure 7.4**).

Bacterial and archaeal cell walls are a tough, fibrous layer that surrounds the plasma membrane. In prokaryotes, the pressure of the plasma membrane against the cell wall is about the same as the pressure in an automobile tire.

The cell wall protects the organism and gives it shape and rigidity, much like the exoskeleton (external skeleton) of a crab or insect. In addition, many bacteria have another protective layer outside the cell wall that consists of lipids with polysaccharides attached. Lipids that contain carbohydrate groups are termed **glycolipids**.

The painting in **Figure 7.5** shows a cross section of a bacterial cell, with a close-up view of the cell wall and some of the other structures introduced in this section. One feature that prokaryotic and eukaryotic cells have in common: They are both packed with dynamic, highly integrated structures.

CHECK YOUR UNDERSTANDING

If you understand that . . .

- Each structure in a prokaryotic cell performs a function vital to the cell.

✔ You should be able to . . .

Describe the structure and function of (1) photosynthetic membranes, (2) organelles that contain magnetite, and (3) the cell wall.

Answers are available in Appendix B.

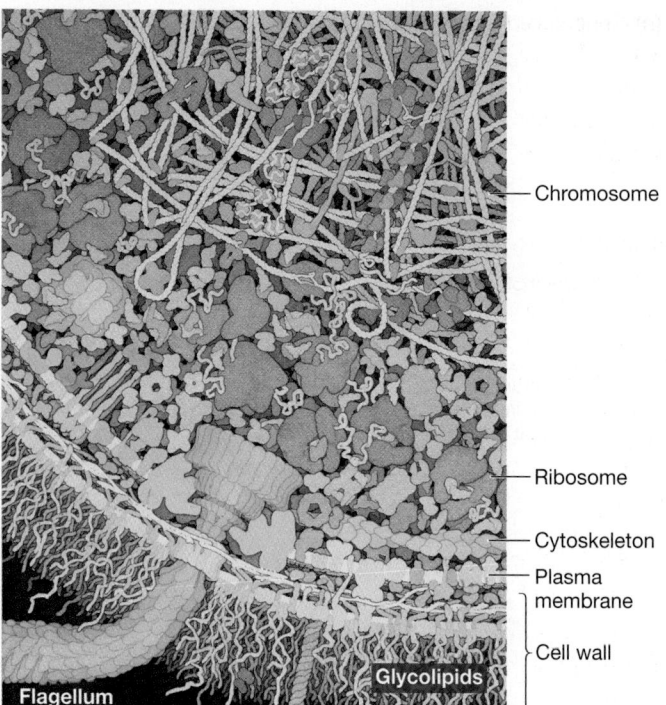

FIGURE 7.5 Close-up View of a Prokaryotic Cell. This painting is an artist's conception of a cross section through part of a bacterial cell. Note that the cell is packed with proteins, DNA, ribosomes, and other molecular machinery.

7.2 Eukaryotic Cell Structures and Their Functions

The lineage called Eukarya includes species that range from microscopic algae to 100-meter-tall redwoods. Brown algae, red algae, fungi, amoebae, and slime molds are all eukaryotic, as are green plants and animals. Although multicellularity has evolved several times among eukaryotes (see Chapter 29), many of these species are unicellular.

The first thing that strikes biologists about eukaryotic cells is how much larger they are on average than bacteria and archaea. Most prokaryotic cells measure 1 to 10 μm in diameter, while most eukaryotic cells range from about 5 to 100 μm in diameter. A micrograph of an average eukaryotic cell, at the same scale as the bacterial cell in Figure 7.3, would fill this page.

In many species of unicellular eukaryotes, large size allows them to make a living by ingesting bacteria and archaea whole. Large size has a downside, however. Ions and small molecules such as adenosine triphosphate (ATP), amino acids, and nucleotides cannot diffuse across a large volume quickly. If the ATP that supplies chemical energy in cells is used up on one side of a large cell, ATP from the other side of the cell would take a long time to diffuse to that location. Prokaryotic cells are small enough that ions and small molecules arrive where they are needed via diffusion.

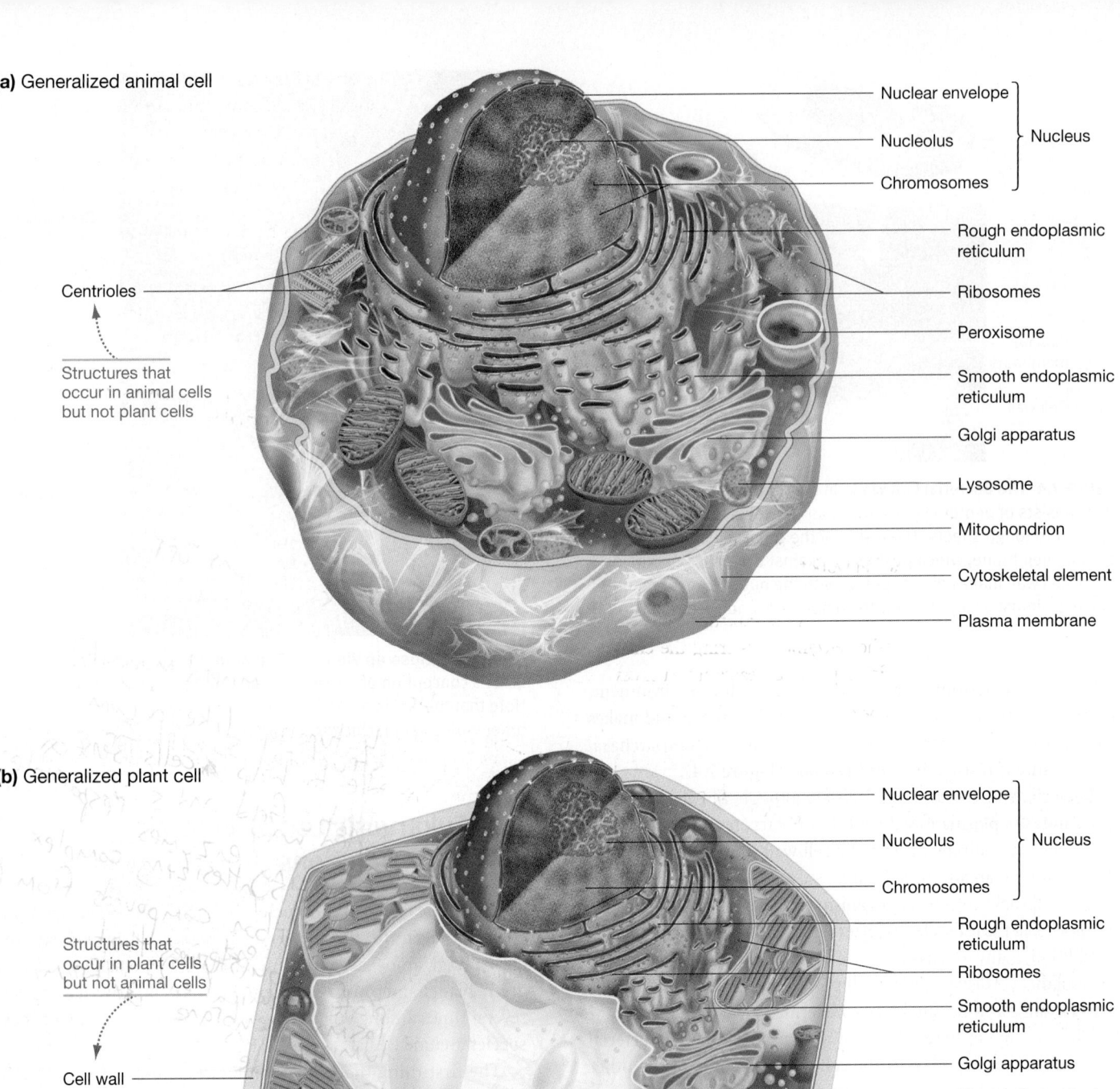

(a) Generalized animal cell

Nuclear envelope ⎤
Nucleolus ⎬ Nucleus
Chromosomes ⎦

Rough endoplasmic reticulum

Ribosomes

Peroxisome

Smooth endoplasmic reticulum

Golgi apparatus

Lysosome

Mitochondrion

Cytoskeletal element

Plasma membrane

Centrioles

Structures that occur in animal cells but not plant cells

(b) Generalized plant cell

Nuclear envelope ⎤
Nucleolus ⎬ Nucleus
Chromosomes ⎦

Rough endoplasmic reticulum

Ribosomes

Smooth endoplasmic reticulum

Golgi apparatus

Vacuole (lysosome)

Peroxisome

Mitochondrion

Plasma membrane

Cytoskeletal element

Structures that occur in plant cells but not animal cells

Cell wall

Chloroplast

On average, prokaryotes are about 10 times smaller than eukaryotic cells in diameter and about 1000 times smaller than eukaryotic cells in volume.

FIGURE 7.6 Overview of Eukaryotic Cells. Generalized, or "typical," **(a)** animal and **(b)** plant cells, color-coded for clarity. Compare with the prokaryotic cell, shown at true relative size at bottom left.

The Benefits of Organelles

How do eukaryotic cells solve the problems that size can engender? The answer lies in their numerous organelles. In effect, the huge volume inside a eukaryotic cell is compartmentalized into a large number of bacterium-sized parts. Because eukaryotic cells are subdivided, the molecules required for specific chemical reactions are often located within a given compartment or organelle.

Compartmentalization offers two key advantages:

- Incompatible chemical reactions can be separated. For example, new fatty acids can be synthesized in one organelle while excess or damaged fatty acids are degraded and recycled in a different organelle.

- Chemical reactions become more efficient. First, the substrates required for particular reactions can be localized and maintained at high concentrations within organelles. Second, if substrates are used up in a particular part of the organelle, they can be replaced by substrates that have only a short distance to diffuse. Third, groups of enzymes that work together can be clustered on internal membranes instead of floating free in the cytoplasm. When the product of one reaction is the substrate for a second reaction catalyzed by another enzyme, clustering the enzymes increases the speed and efficiency of both reaction sequences.

If bacteria and archaea can be compared to small machine shops, then eukaryotic cells resemble sprawling industrial complexes. The organelles and other structures found in eukaryotes are analogous to highly specialized buildings that act as factories, power stations, warehouses, transportation corridors, and administrative centers.

When typical prokaryotic and eukaryotic cells are compared, four key differences identified in **Table 7.1** stand out:

1. Eukaryotic chromosomes are found inside a membrane-bound compartment called the nucleus.

2. Eukaryotic cells are often much larger.

3. Eukaryotic cells contain extensive amounts of internal membrane.

4. Eukaryotic cells feature a particularly diverse and dynamic cytoskeleton.

Eukaryotic Cell Structures: A Parts List

Figure 7.6 provides a simplified view of a typical animal cell and plant cell. The artist has removed most of the cytoskeletal elements to make the organelles and other cellular parts easier to see. As you read about each cell component in the pages that follow, focus on identifying how their structure correlates with their function. Then use **Table 7.2** on page 114 as a study guide. As with bacterial cells, let's start from the inside and work out.

THE NUCLEUS The **nucleus** contains the chromosomes and functions as an information storage and processing center. Among the largest and most highly organized of all organelles (**Figure 7.7**), it is enclosed by a unique structure—a complex double membrane called the **nuclear envelope**. As Section 7.4 will detail, the nuclear envelope is studded with pore-like openings, and its inside surface is linked to fibrous proteins that form a lattice-like sheet called the **nuclear lamina**. The nuclear lamina stiffens the structure and maintains its shape.

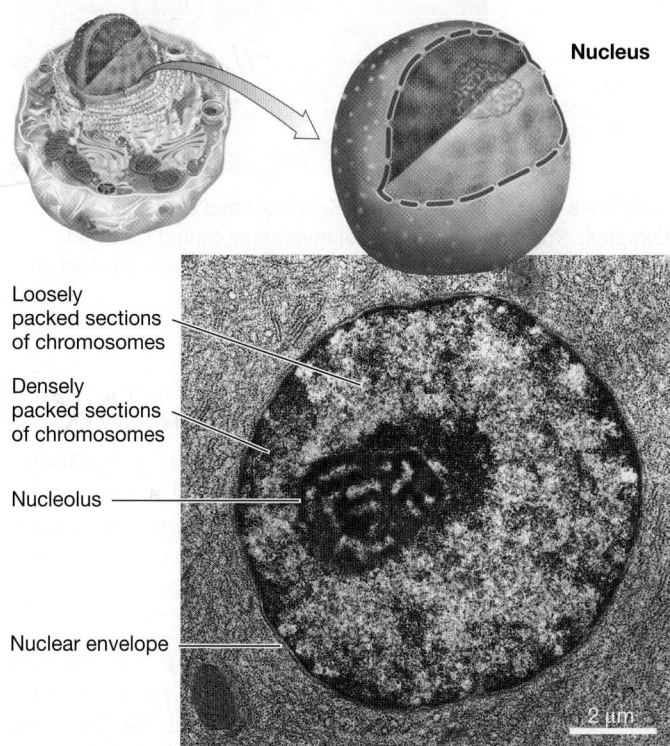

FIGURE 7.7 The Nucleus Is the Eukaryotic Cell's Information Storage and Retrieval Center. The genetic, or hereditary, information is encoded in DNA, which is a component of the chromosomes inside the nucleus.

SUMMARY TABLE 7.1 **How Do the Structures of Prokaryotic and Eukaryotic Cells Differ?**

Cell Type	Location of DNA	Internal Membranes and Organelles	Cytoskeleton	Overall Size
Bacteria and Archaea	In nucleoid (not membrane bound); plasmids also common	Extensive internal membranes only in photosynthetic species; limited types and numbers of organelles	Limited in extent, relative to eukaryotes	Usually small relative to eukaryotes
Eukaryotes	Inside nucleus (membrane bound); plasmids extremely rare	Large numbers of organelles; many types of organelles	Extensive—usually found throughout volume of cell	Most are larger than prokaryotes

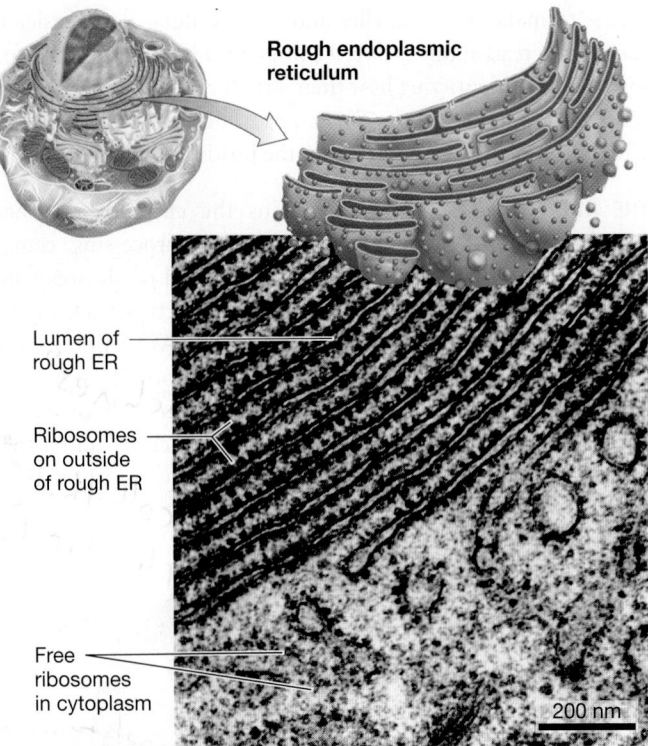

Rough endoplasmic reticulum

Lumen of rough ER

Ribosomes on outside of rough ER

Free ribosomes in cytoplasm

200 nm

FIGURE 7.8 Rough ER Is a Protein Synthesis and Processing Complex. Rough ER is a system of membrane-bound sacs and tubules with ribosomes attached. It is continuous with the nuclear envelope and with smooth ER.

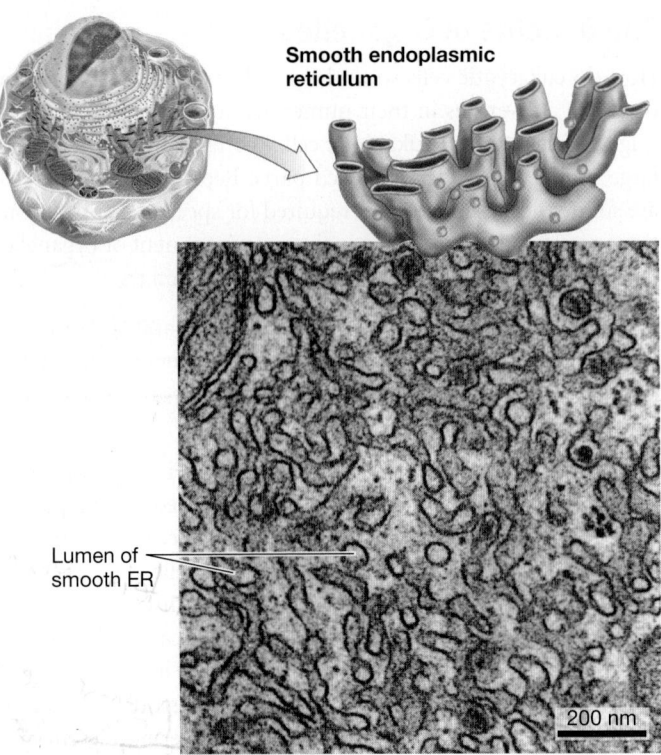

Smooth endoplasmic reticulum

Lumen of smooth ER

200 nm

FIGURE 7.9 Smooth ER Is a Lipid-Handling Center and a Storage Facility. Smooth ER is a system of membrane-bound sacs and tubules that lacks ribosomes.

Chromosomes do not float freely inside the nucleus—instead, each chromosome occupies a distinct area and is attached to the nuclear lamina in at least one location. The nucleus also contains specific sites where gene products are processed and includes a distinctive region called the **nucleolus**, where the RNA molecules found in ribosomes are manufactured and the large and small ribosomal subunits are assembled.

The nuclear envelope is continuous with an extensive series of membrane-bound sacs called the **endoplasmic reticulum** (literally, "inside-formed-network"), or ER. As Figure 7.6 shows, the ER extends from the nuclear envelope out into the cytoplasm. Although the ER is a single structure, it has regions that are distinct in structure and function. Let's consider each region in turn.

ROUGH ENDOPLASMIC RETICULUM   The **rough endoplasmic reticulum**, or **rough ER (RER)**, is named for its appearance in transmission electron micrographs (**Figure 7.8**). The knobby-looking structures in rough ER are ribosomes that attach to the membrane.

The ribosomes associated with the rough ER synthesize proteins that will be inserted into the plasma membrane, secreted to the cell exterior, or shipped to an organelle. As they are being manufactured by ribosomes, these proteins move to the interior of the sac-like component of the rough ER. The interior of the rough ER, like the interior of any sac-like structure in a cell or body, is called the **lumen**. In the lumen of the rough ER, newly manufactured proteins undergo folding and other types of processing.

The proteins produced in the rough ER have a variety of functions. Some carry messages to other cells; some act as membrane transport proteins or pumps; others are enzymes. The common theme is that rough ER products are packaged into vesicles and transported to various distant destinations—often to the surface of the cell or beyond.

SMOOTH ENDOPLASMIC RETICULUM   In electron micrographs, parts of the ER that are free of ribosomes appear smooth and even. Appropriately, these parts of the ER are called **smooth endoplasmic reticulum**, or **smooth ER** (SER; **Figure 7.9**).

The smooth ER membrane contains enzymes that catalyze reactions involving lipids. Depending on the type of cell, these enzymes may synthesize lipids needed by the organism or break down lipids that are poisonous. Smooth ER is also the manufacturing site for phospholipids used in plasma membranes. In addition to processing lipids, smooth ER functions as a reservoir for calcium ions (Ca^{2+}) that act as a signal triggering a wide array of activities inside the cell.

The structure of endoplasmic reticulum correlates closely with its function. Rough ER has ribosomes and functions primarily as a protein-manufacturing center; smooth ER lacks ribosomes and functions primarily as a lipid-processing center.

GOLGI APPARATUS   In many cases, the products of the rough ER pass through the Golgi apparatus before they reach their final

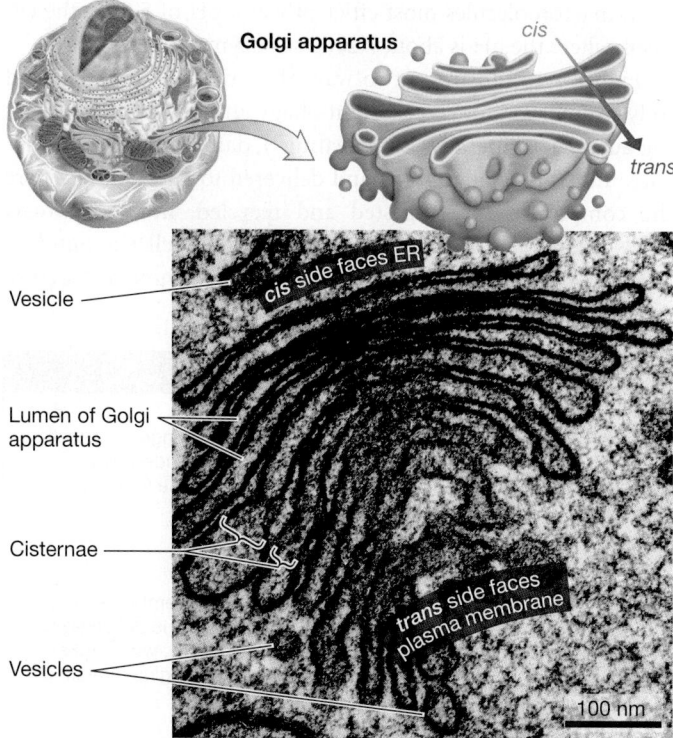

Golgi apparatus *cis*

trans

Vesicle

cis side faces ER

Lumen of Golgi
apparatus

Cisternae

trans side faces
plasma membrane

Vesicles

100 nm

FIGURE 7.10 The Golgi Apparatus Is a Site of Protein Processing, Sorting, and Shipping. The Golgi apparatus is a collection of flattened vesicles called cisternae.

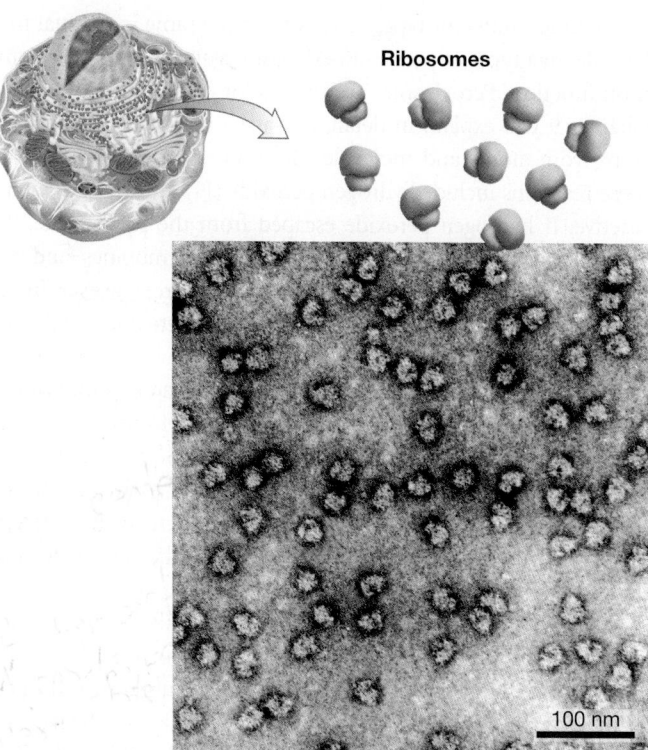

Ribosomes

100 nm

FIGURE 7.11 Ribosomes Are the Site of Protein Synthesis. Eukaryotic ribosomes are similar in structure to bacterial and archaeal ribosomes—though not identical. They are composed of large and small subunits, each of which contains both RNA molecules and proteins.

destination. The **Golgi apparatus** consists of flattened, membranous sacs called **cisternae** (singular: **cisterna**), which are stacked on top of one another (**Figure 7.10**). The organelle also has a distinct polarity, or sidedness. The *cis* ("this side") surface is closest to the rough ER and nucleus, and the *trans* ("across") surface is oriented toward the plasma membrane.

The *cis* side of a Golgi apparatus receives products from the rough ER, and the *trans* side ships them out toward the cell surface. In between, within the cisternae, the rough ER's products are processed and packaged for delivery. Micrographs often show "bubbles" on either side of the Golgi stack. These are membrane-bound vesicles that carry proteins or other products to and from the organelle. Section 7.3 analyzes the intracellular movement of molecules from the rough ER to the Golgi apparatus and beyond in more detail.

RIBOSOMES In eukaryotes, the cytoplasm consists of everything inside the plasma membrane excluding the nucleus; the fluid portion of the cytoplasm is called the **cytosol**. Although many of the cell's millions of ribosomes are attached to the RER, many are scattered throughout the cytosol (**Figure 7.11**).

Like bacterial ribosomes, eukaryotic ribosomes are complex molecular machines that manufacture proteins. They are not classified as organelles because they are not surrounded by a membrane.

PEROXISOMES Virtually all eukaryotic cells contain globular organelles called **peroxisomes** (**Figure 7.12**). These organelles have a single membrane, and originate as buds from the ER.

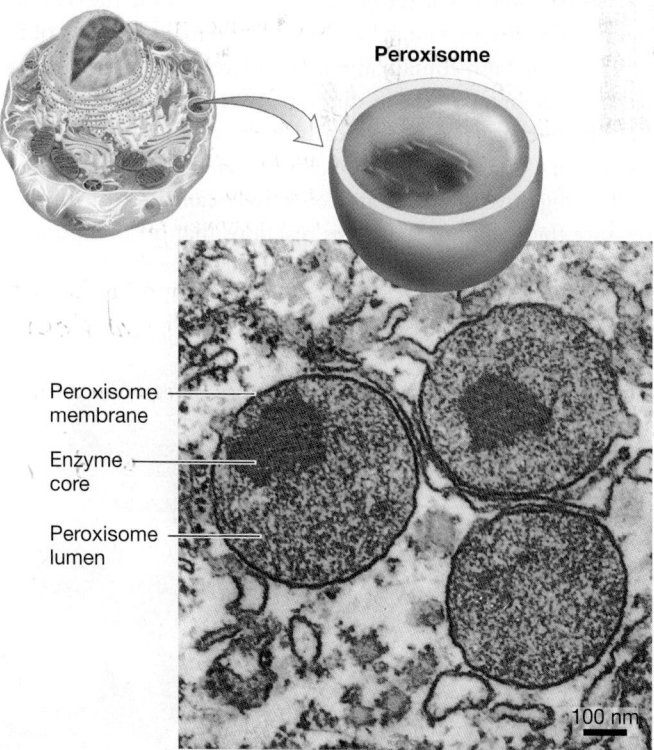

Peroxisome

Peroxisome
membrane

Enzyme
core

Peroxisome
lumen

100 nm

FIGURE 7.12 Peroxisomes Are the Site of Oxidation Reactions. Peroxisomes are globular organelles with a single membrane.

Although different types of cells from the same individual may have distinct types of peroxisomes, these organelles all share a common function: Peroxisomes are centers for oxidation reactions. As Chapter 9 will explain in detail, oxidation reactions remove electrons from atoms and molecules. In many cases the products of these reactions include hydrogen peroxide (H_2O_2), which is highly reactive. If hydrogen peroxide escaped from the peroxisome, the H_2O_2 would quickly react with organelle membranes and the plasma membrane and damage them. This is rare, however. Inside the peroxisome, the enzyme catalase quickly "detoxifies" hydrogen peroxide by converting it to water and oxygen.

Different types of peroxisomes contain different suites of oxidative enzymes. As a result, each is specialized for oxidizing particular compounds. For example, the peroxisomes in your liver cells contain enzymes that oxidize an array of toxins, including the ethanol in alcoholic beverages. The products of these oxidation reactions are usually harmless and are either excreted from the body or used in other reactions.

In plant leaves, specialized peroxisomes called **glyoxysomes** are packed with enzymes that oxidize fats to form a compound that can be used to store energy for the cell. But plant seeds have a different type of peroxisome—one that is packed with enzymes responsible for releasing energy from stored fatty acids. The young plant uses this energy as it begins to grow.

In both animals and plants, there is a clear connection between structure and function: The enzymes found inside the peroxisome make a specialized set of oxidation reactions possible.

LYSOSOMES   Animal cells contain organelles called **lysosomes** (**Figure 7.13**) which function as digestive centers. The organelle's interior, or lumen, is acidic because proton pumps in the lysosome membrane import enough hydrogen ions to maintain a pH of 5.0.

Lysosomes also contain about 40 different enzymes, each specialized for breaking up a different type of macromolecule—protein, nucleic acid, lipid, or carbohydrate—into its component monomers. The monomers are then excreted or recycled.

The digestive enzymes are collectively called acid hydrolases because they catalyze hydrolysis reactions that break monomers

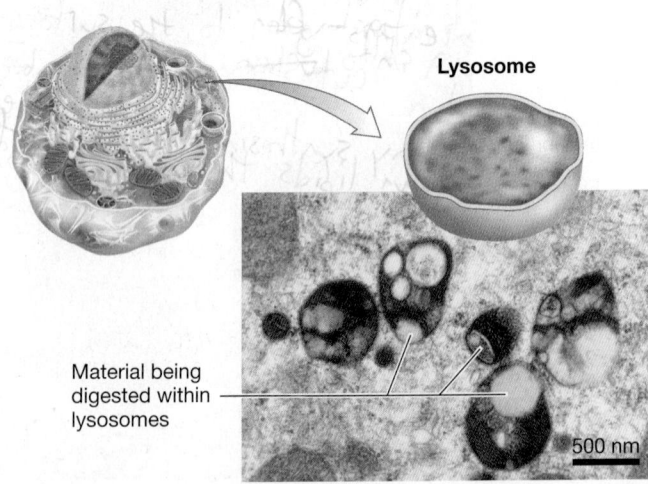

FIGURE 7.13 Lysosomes Are Recycling Centers. Lysosomes are usually oval or globular and have a single membrane.

from macromolecules most efficiently at a pH of 5.0. In the cytosol, where the pH is about 7.2, these enzymes are less active.

Figure 7.14 illustrates two ways that materials are delivered to lysosomes in animal cells: autophagy and phagocytosis. During **autophagy** (literally, "same-eating"), damaged organelles are surrounded by a membrane and delivered to a lysosome. There the components are digested and recycled. In **phagocytosis** ("eat-cell-act"), the plasma membrane of a cell surrounds a smaller cell or food particle and engulfs it, forming a structure

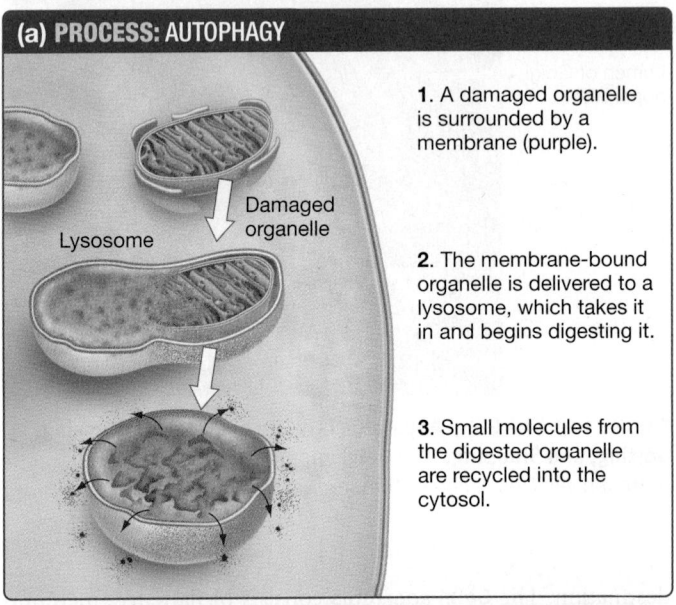

(a) **PROCESS: AUTOPHAGY**

1. A damaged organelle is surrounded by a membrane (purple).

2. The membrane-bound organelle is delivered to a lysosome, which takes it in and begins digesting it.

3. Small molecules from the digested organelle are recycled into the cytosol.

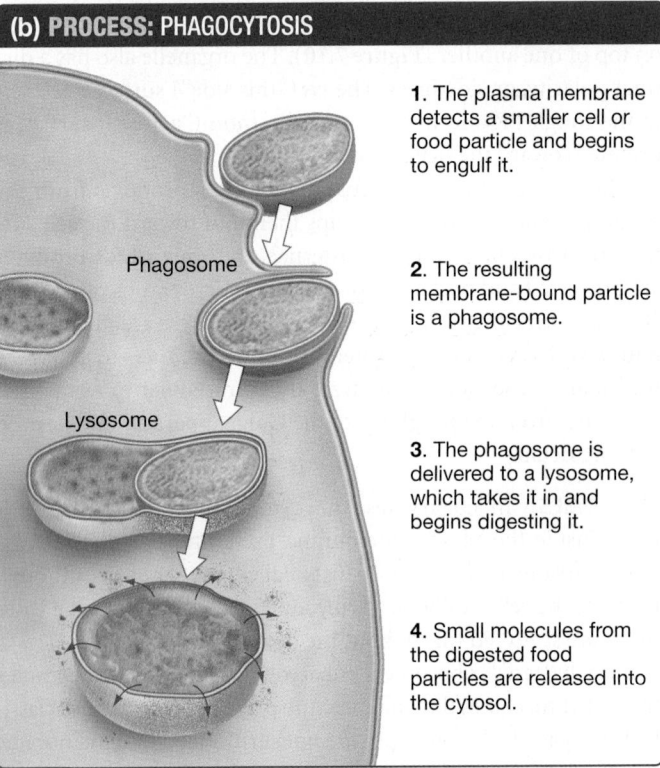

(b) **PROCESS: PHAGOCYTOSIS**

1. The plasma membrane detects a smaller cell or food particle and begins to engulf it.

2. The resulting membrane-bound particle is a phagosome.

3. The phagosome is delivered to a lysosome, which takes it in and begins digesting it.

4. Small molecules from the digested food particles are released into the cytosol.

FIGURE 7.14 Two Ways to Deliver Materials to Lysosomes. Materials can be transported to lysosomes **(a)** via autophagy or **(b)** after phagocytosis.

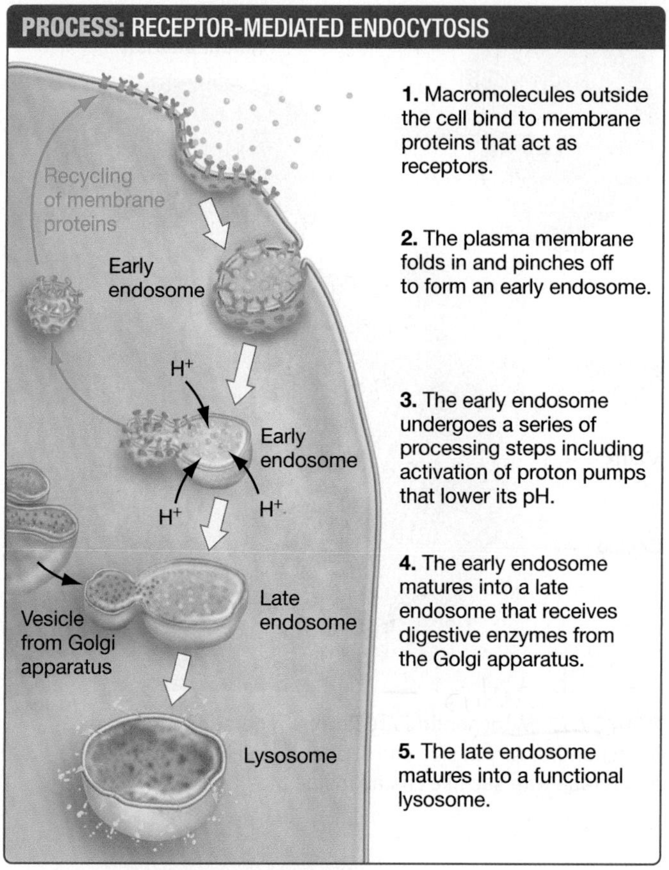

PROCESS: RECEPTOR-MEDIATED ENDOCYTOSIS

Recycling of membrane proteins

Early endosome

H⁺

Early endosome

H⁺   H⁺

Vesicle from Golgi apparatus

Late endosome

Lysosome

1. Macromolecules outside the cell bind to membrane proteins that act as receptors.

2. The plasma membrane folds in and pinches off to form an early endosome.

3. The early endosome undergoes a series of processing steps including activation of proton pumps that lower its pH.

4. The early endosome matures into a late endosome that receives digestive enzymes from the Golgi apparatus.

5. The late endosome matures into a functional lysosome.

FIGURE 7.15 Receptor-Mediated Endocytosis May Lead to Lysosome Formation. Endosomes created by receptor-mediated endocytosis may mature into lysosomes.

✔**QUESTION** Why is it significant that vesicles from the Golgi apparatus fuse with endosomes?

called a phagosome. This structure is delivered to a lysosome, where it is taken in and digested.

Figure 7.15 illustrates a third way that lysosomes process materials: **receptor-mediated endocytosis**. As its name implies, the sequence of events begins when macromolecules outside the cell bind to membrane proteins that act as receptors. More than 25 distinct receptors have now been characterized, each specialized for responding to a different macromolecule. Receptors are found in specific locations in the cell, where the membrane is underlain by a coating of specialized proteins that include a cage of clathrin molecules.

Once receptor binding occurs, the plasma membrane folds in and pinches off to form a membrane-bound vesicle called an **early endosome** ("inside-body"). The exterior of the endosome is a cage-like structure comprised of clathrins and other proteins.

Early endosomes undergo a series of processing steps that include the activation of proton pumps that gradually lower their pH. When they have matured into a **late endosome**, the structures receive digestive enzymes from the Golgi apparatus and may eventually become fully functioning lysosomes.

Regardless of whether the materials in lysosomes originate via autophagy, phagocytosis, or receptor-mediated endocytosis, the result is similar: Molecules are hydrolyzed. The amino acids, nu-

cleotides, sugars, and other molecules that result from acid hydrolysis leave the lysosome via transport proteins in the organelle's membrane. Once in the cytoplasm, they can be reused.

It is important to note, however, that not all of the materials that are surrounded by membrane and taken into a cell end up in lysosomes. **Endocytosis** ("inside-cell-act") refers to any pinching off of the plasma membrane that results in the uptake of material from outside the cell. In addition to phagocytosis and receptor-mediated endocytosis, endocytosis can occur via **pinocytosis** ("drink-cell-act"). Pinocytosis brings fluid into the cytoplasm via tiny vesicles that form from infoldings of the plasma membrane. The fluid inside these vesicles is not transported to lysosomes, but is used elsewhere in the cell. In addition, most of the macromolecules that collect in early endosomes are selectively removed and used long before the structure becomes a lysosome.

Even though lysosomes are physically separated from the Golgi apparatus and the rough and smooth endoplasmic reticulum, the organelles jointly form a key functional grouping called the **endomembrane system**. The endomembrane ("inner-membrane") system is the primary center for protein and lipid synthesis and processing in eukaryotic cells.

VACUOLES The cells of plants, fungi, and certain other groups lack lysosomes. Instead, they contain a prominent organelle called a vacuole. Compared with the lysosomes of animal cells, the **vacuoles** of plant and fungal cells are large—sometimes taking up as much as 80 percent of a plant cell's volume (**Figure 7.16**).

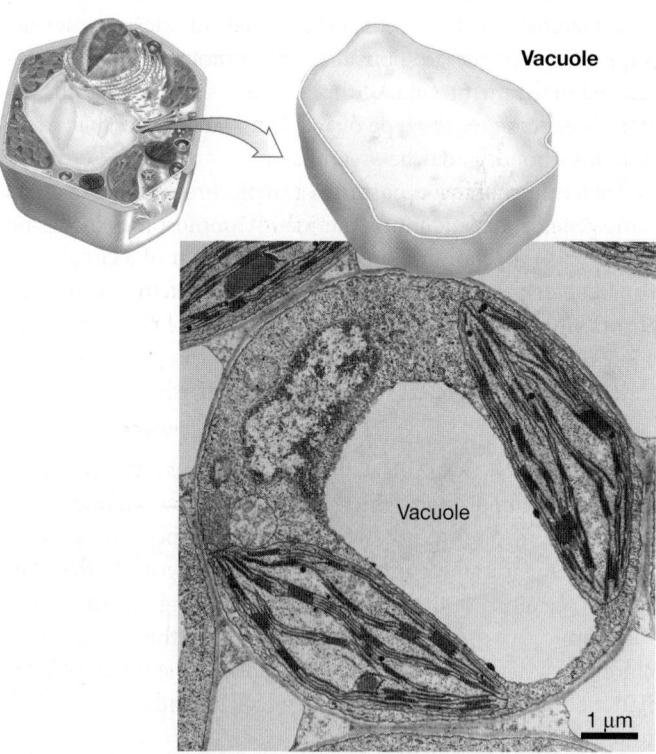

Vacuole

Vacuole

1 μm

FIGURE 7.16 Vacuoles Are Storage Centers in Plant and Fungal Cells. Vacuoles vary in size and function. Some contain digestive enzymes and serve as recycling centers; most are large storage containers.

✔**QUESTION** Why are toxins like nicotine, cocaine, and caffeine stored in vacuoles instead of the cytosol?

Although some vacuoles contain enzymes that are specialized for digestion, most of the vacuoles observed in plant and fungal cells act as storage depots. In many cases, the stored material is water, which maintains the cell's normal volume, or ions such as potassium (K^+) and chloride (Cl^-). In other cells, vacuoles have more specialized storage functions.

- Inside seeds, cells may contain a large vacuole filled with proteins. When the embryonic plant inside the seed begins to grow, enzymes begin digesting these proteins to provide amino acids for the growing individual.

- In cells that make up flower petals or fruits, vacuoles are filled with colorful pigments.

- Elsewhere, vacuoles may be packed with noxious compounds that protect leaves and stems from being eaten by predators. The type of chemical involved varies by species, ranging from bitter-tasting tannins to toxins such as nicotine, morphine, caffeine, or cocaine.

MITOCHONDRIA   The chemical energy required to build all these organelles and do other types of work comes from adenosine triphosphate (ATP), most of which is produced in the cell's **mitochondria** (singular: **mitochondrion**).

As **Figure 7.17** shows, each mitochondrion has two membranes. The outer membrane defines the organelle's surface, while the inner membrane is connected to a series of sac-like **cristae**. The solution inside the inner membrane is called the **mitochondrial matrix**. In eukaryotes, most of the enzymes and molecular machines responsible for synthesizing ATP are embedded in the membranes of the cristae or suspended in the matrix. Depending on the type of cell, from 50 to more than a million mitochondria may be present.

Each mitochondrion possesses a small chromosome that contains genes, independent of the main chromosomes in the nucleus. This mitochondrial DNA is a component of a circular and supercoiled chromosome that is similar in structure to bacterial chromosomes. Mitochondria also manufacture their own ribosomes. Like most organelles, mitochondria can grow and divide independently of nuclear division and cell division.

CHLOROPLASTS   Most algal and plant cells possess an organelle called the **chloroplast**, in which sunlight is converted to chemical energy during photosynthesis (**Figure 7.18**). The chloroplast has a double membrane around its exterior, analogous to the structure of a mitochondrion. Instead of featuring sac-like cristae that connect to the inner membrane, though, the interior of the chloroplast is dominated by hundreds of membrane-bound, flattened vesicles called **thylakoids**, which are independent of the inner membrane.

Thylakoids are stacked like pancakes into piles called **grana** (singular: **granum**). Many of the pigments, enzymes, and molecular machines responsible for converting light energy into carbohydrates are embedded in the thylakoid membranes. Certain

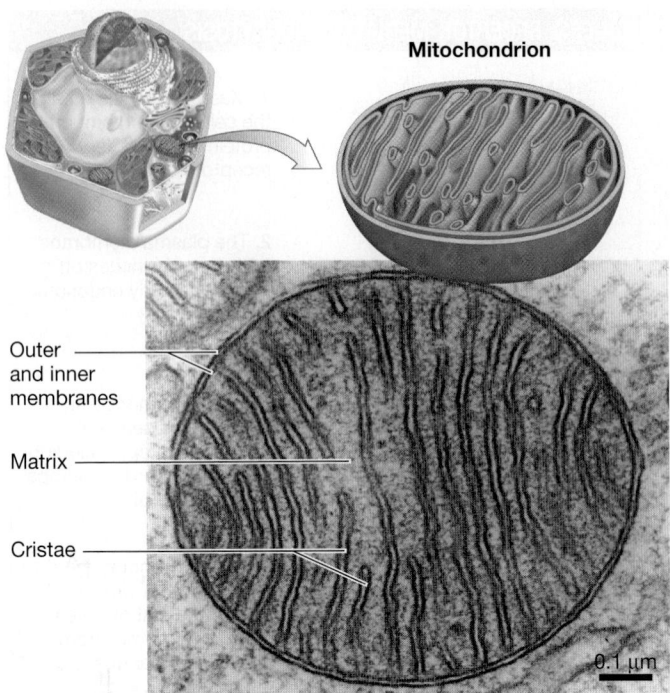

FIGURE 7.17 Mitochondria Are Power-Generating Stations. Mitochondria vary in size and shape, but all have a double membrane with sac-like cristae inside.

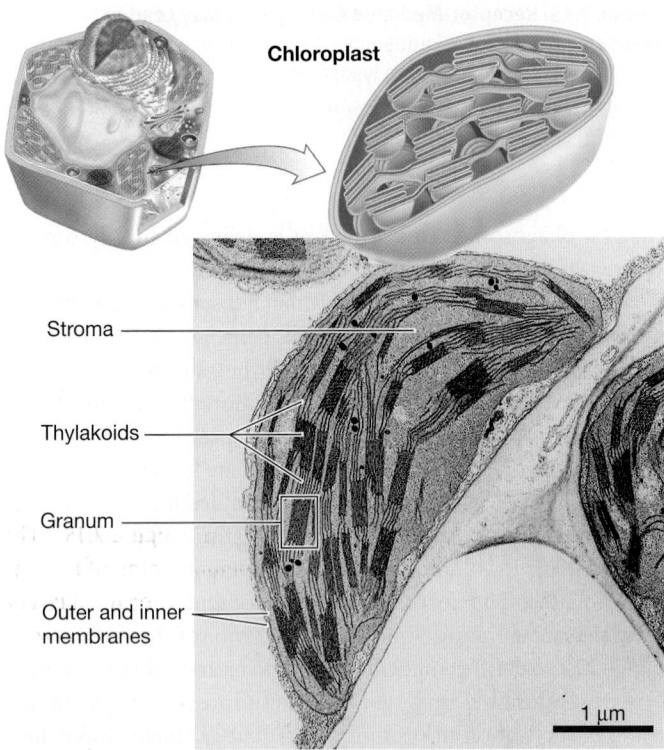

FIGURE 7.18 Chloroplasts Are Sugar-Manufacturing Centers in Plants and Algae. Many of the enzymes and other molecules required for photosynthesis are located in membranes inside the chloroplast. These membranes are folded into thylakoids and stacked into grana.

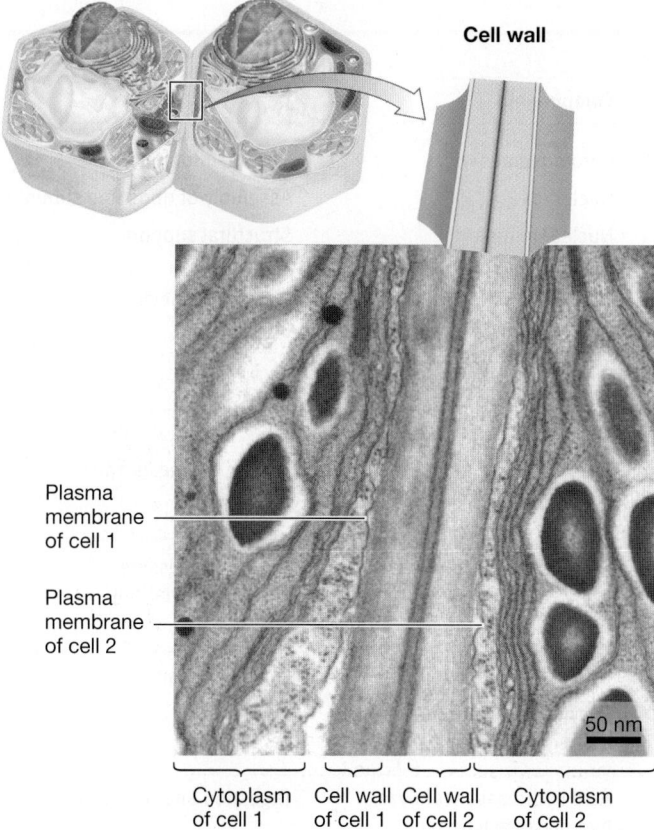

Cell wall

Plasma membrane of cell 1

Plasma membrane of cell 2

50 nm

Cytoplasm of cell 1 | Cell wall of cell 1 | Cell wall of cell 2 | Cytoplasm of cell 2

FIGURE 7.19 Cell Walls Protect Plants and Fungi. Plants have cell walls that contain cellulose; in fungi the major structural component of the cell wall is chitin.

✔ **QUESTION** Is the cell wall inside or outside the plasma membrane?

critical enzymes and substrates, however, are found outside the thylakoids in the region called the **stroma**.

The number of chloroplasts per cell varies from none to several dozen. Like mitochondria, each chloroplast contains a circular chromosome. Chloroplast DNA is independent of the main genetic material inside the nucleus. Chloroplasts also grow and divide independently of nuclear division and cell division.

THE CELL WALL In fungi, algae, and plants, cells possess an outer cell wall in addition to their plasma membrane (**Figure 7.19**). The cell wall is located outside of the plasma membrane, and furnishes a stiff, outer layer that provides structural support for the cell. The cells of animals, amoebae, and other groups lack a cell wall—their exterior surface consists of just the plasma membrane.

Although the composition of the cell wall varies among species and even between types of cells in the same individual, the general plan is similar: Rods or fibers composed of a carbohydrate run through a stiff matrix made of other polysaccharides and proteins (see Chapter 8 for details).

In addition, some plant cells produce a secondary cell wall that features a particularly tough molecule called lignin. Lignin

forms a branching, cagelike network that is almost impossible for enzymes to attack. The combination of cellulose fibers and lignin in secondary cell walls makes up most of the material we call wood.

CYTOSKELETON The final major structural feature that is common to all eukaryotic cells is the cytoskeleton, an extensive system of protein fibers. In addition to giving the cell its shape and structural stability, cytoskeletal proteins are involved in moving the cell itself and moving materials within the cell. In essence, the cytoskeleton organizes all of the organelles and other cellular structures into a cohesive whole. The painting in **Figure 7.20** is a cross section through a small part of a eukaryotic cell. Note the density of cytoskeletal elements—the long tubes colored blue in this painting. Section 7.6 will analyze the structure and functions of the cytoskeleton in detail.

Before moving on to the next section, review animations of animal and plant cells in the study area at *www.masteringbiology.com* and complete the Check Your Understanding box on page 115.

(MB) **BioFlix™** Tour of an Animal Cell, **BioFlix™** Tour of a Plant Cell

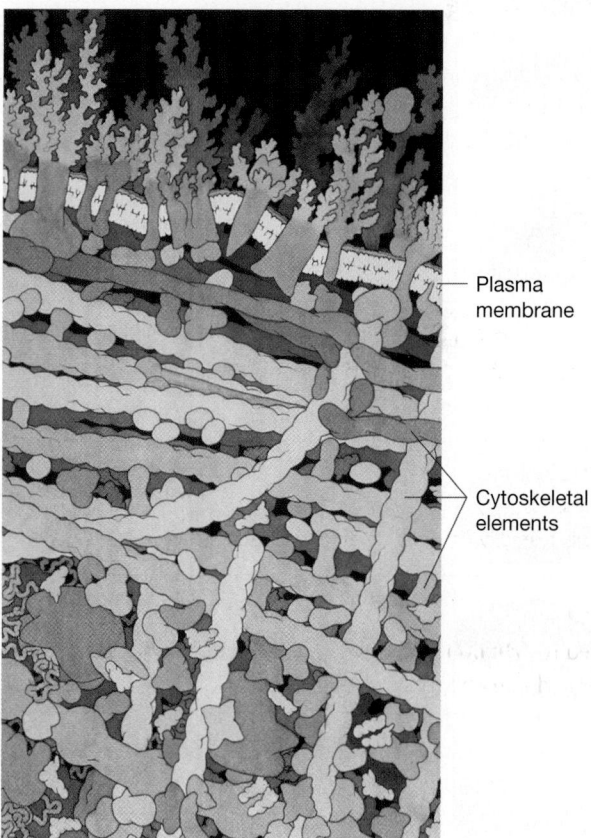

Plasma membrane

Cytoskeletal elements

FIGURE 7.20 Close-up View of a Eukaryotic Cell. Cells are highly structured entities packed with organelles, ribosomes, and cytoskeletal elements (colored blue here). Note that the phospholipid bilayer, colored yellow here, is studded with transmembrane proteins—many of which have sugar groups (in green) projecting into extracellular space.

Icons not to scale	Structure		Function
	Membrane	**Components**	
Nucleus *(handwritten: Administrative/information hub)*	Double ("envelope"); openings called nuclear pores	Chromosomes Nucleolus Nuclear lamina	Genetic information Assembly of ribosome subunits Structural support
Ribosomes *(handwritten: Protein factory)*	None	Complex of RNA and proteins	Protein synthesis
Endomembrane system			
Rough ER *(handwritten: Large molecule manufacturing + shipping, protein finishing + shipping line)*	Single; contains receptors for entry of selected proteins	Network of branching sacs Ribosomes associated	Protein synthesis and processing
Golgi apparatus *(handwritten: fat factory, waste processing centre)*	Single; contains receptors for products of rough ER	Stack of flattened cisternae	Protein processing (e.g., glycosylation)
Smooth ER *(handwritten: Fat Factory)*	Single; contains enzymes for synthesizing phospholipids	Network of branching sacs Enzymes for synthesizing lipids	Lipid synthesis
Lysosomes *(handwritten: Waste processing and recycling center)*	Single; contains proton pumps	Acid hydrolases (catalyze hydrolysis reactions)	Digestion and recycling
Peroxisomes *(handwritten: Fatty acid processing + detox center)*	Single; contains transporters for selected macromolecules	Enzymes that catalyze oxidation reactions Catalase (processes peroxide)	Oxidation of fatty acids, ethanol, or other compounds
Vacuoles *(handwritten: Warehouse)*	Single; contains transporters for selected molecules	Varies—pigments, oils, carbohydrates, water, or toxins	Varies—coloration, storage of oils, carbohydrates, water, or toxins
Mitochondria *(handwritten: Power Station)*	Double; inner contains enzymes for ATP production	Enzymes that catalyze oxidation-reduction reactions, ATP synthesis	ATP production
Chloroplasts *(handwritten: Food-manufacturing facility)*	Double; plus membrane-bound sacs in interior	Pigments Enzymes that catalyze oxidation-reduction reactions	Production of ATP and sugars via photosynthesis
Cytoskeleton *(handwritten: Support beams)*	None	Actin filaments Intermediate filaments Microtubules	Structural support; movement of materials; in some species, movement of whole cell
Plasma membrane *(handwritten: Perimeter fencing w/ secured gates)*	Single; contains transport and receptor proteins	Phospholipid bilayer with transport and receptor proteins	Selective permeability—maintains intracellular environment
Cell wall	None	Carbohydrate fibers running through carbohydrate or protein matrix	Protection, structural support

CHECK YOUR UNDERSTANDING

7.3 Putting the Parts into a Whole

Within a cell, the structure of each organelle and component correlates with its function. In the same way, the overall size, shape, and composition of a cell correlates with its function.

Cells might be analogous to machine shops or factory complexes, but clothing manufacturing centers are very different in layout and composition from airplane production facilities. How does the physical and chemical makeup of a cell correlate with its function?

Structure and Function at the Whole-Cell Level

Inside an individual plant or animal, cells are specialized for certain tasks and have a structure that correlates with those tasks. For example, the muscle cells in your upper leg are extremely long, tube-shaped structures. They are filled with protein fibers that slide past one another as the entire muscle flexes or extends. It is this sliding motion that allows your muscles to contract or extend as you run. Muscle cells are also packed with mitochondria, which produce the ATP required for the sliding motion to occur.

In contrast, nearby fat cells are rounded, globular structures that store fatty acids. They consist of little more than a plasma membrane, a nucleus, and a fat droplet. Neither cell bears a close resemblance to the generalized animal cell pictured in Figure 7.6a.

To drive home the correlation between the overall structure and function of a cell, examine the transmission electron micrographs in **Figure 7.21**.

- The animal cell in Figure 7.21a, located in the pancreas, manufactures and exports digestive enzymes. It is packed with rough ER and Golgi, which make these functions possible.

- The animal cell in Figure 7.21b is from the testis and synthesizes the steroid hormone testosterone. This cell is dominated

(a) Animal pancreatic cell: Exports digestive enzymes.

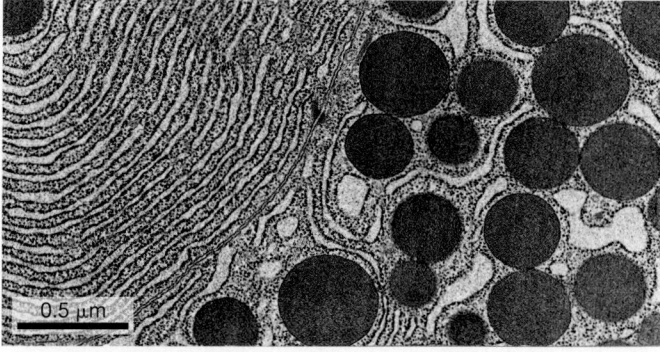

0.5 μm

(b) Animal testis cell: Exports lipid-soluble signals.

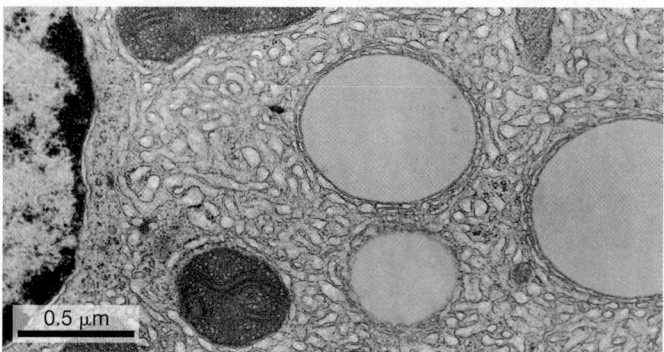

0.5 μm

(c) Plant leaf cell: Manufactures ATP and sugar.

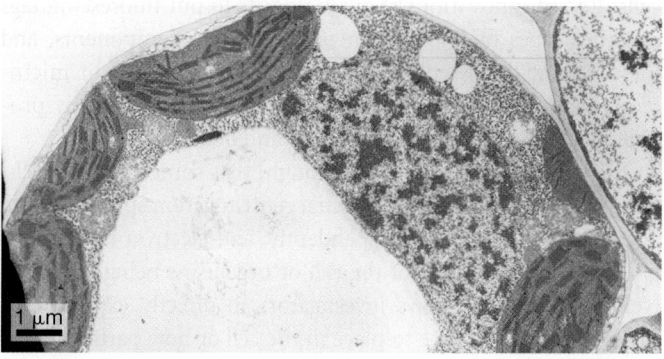

1 μm

(d) Plant root cell: Stores starch.

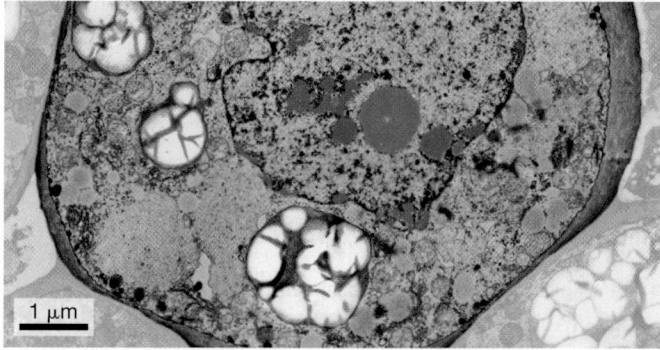

1 μm

FIGURE 7.21 Cell Structure Correlates with Function.

✔**EXERCISE** In part (a), label the rough ER and secretory vesicles. (They are dark and round.) In (b), label the smooth ER. In (c), label the chloroplasts, vacuole, and nucleus. In (d), label the starch granules.

by smooth ER, where processing of steroids and other lipids takes place.

- The plant cell in Figure 7.21c, from the leaf of a potato, has hundreds of chloroplasts and is specialized for absorbing light and manufacturing sugar.

- The plant cell in Figure 7.21d comes from a potato tuber (part of an underground stem). It has numerous vacuoles that are packed with stored starch—which shows up as white chunks in the micrograph.

In each case, the type of organelles in each cell and their size and number correlate with the cell's specialized function.

The Dynamic Cell

Biologists study the structure and function of organelles and cells with a combination of tools and approaches. For several decades, a technique called **differential centrifugation** was particularly important, because it allowed researchers to isolate particular cell components and analyze their chemical composition. As **BioSkills 11** in Appendix A explains, differential centrifugation is based on breaking cells apart to create a complex mixture and separating components in a centrifuge. The individual parts of the cell can then be purified and studied in detail, in isolation from other parts of the cell.

Historically and currently, however, the most important research in cell biology is based on imaging—simply looking at cells. Recent innovations allow biologists to put fluorescing tags or other types of markers on particular cell components, and then look at them with increasingly sophisticated light microscopes and electron microscopes. Advances in microscopy provide increasingly high magnification and better resolution.

It's important to recognize, though, that some of these techniques have limitations. Differential centrifugation splits cells into parts that are analyzed independently, and electron microscopy gives a fixed "snapshot" of the cell or organisms being observed. Neither technique allows investigators to directly explore how things move from place to place in the cell or how parts interact. The information gleaned from these techniques can make cells seem static. In reality, cells are dynamic.

The amount of chemical activity and the speed of molecular movement inside cells is nothing short of fantastic. Bacterial ribosomes add up to 20 amino acids per second to a growing polypeptide, and eukaryotic ribosomes typically add two per second. Given that there are about 15,000 ribosomes in each bacterium and possibly a million in an average eukaryotic cell, hundreds or even thousands of new protein molecules can be finished each second in every cell.

- In an average second, a typical cell in your body uses an average of 10 million ATP molecules and synthesizes just as many.

- It's not unusual for a cellular enzyme to catalyze 25,000 or more reactions per second; most cells contain hundreds or thousands of enzymes.

- A minute is more than enough time for each membrane phospholipid in your body to travel the breadth of the organelle or cell where it resides.

- The hundreds of trillions of mitochondria inside you are completely replaced about every 10 days, for as long as you live.

- The plasma membrane is fluid, and its composition is constantly changing.

Because humans are such large organisms, it's impossible for us to imagine what life is really like inside a cell. At the scale of a ribosome or an organelle or a cell, gravity is inconsequential. Instead, the dominant forces are the charge- or polarity-based electrostatic attractions between molecules and the kinetic energy of motion. At this level, events take nanoseconds, and speeds are measured in micrometers per second. This is the speed of life.

Contemporary methods for studying cells, including some of the imaging techniques featured in **BioSkills 10** in Appendix A, capture this dynamism by tracking how organelles and molecules move and interact over time. The ability to digitize video images of live cells, or take time-lapse photographs of living cells, is allowing researchers to see and study dynamic processes.

The rest of this chapter focuses on this theme of cellular dynamism and movement. Its goal is to put all the individual pieces of a cell together, and ask how they work as systems to accomplish key tasks.

To begin, let's look at how molecules move into and out of the cell's control center—the nucleus. Then we'll consider how proteins move from ribosomes into the lumen of the rough ER and then to the Golgi apparatus and beyond. The chapter closes by analyzing how cytoskeletal elements help transport cargo inside the cell or move the cell itself.

7.4 Cell Systems I: Nuclear Transport

The nucleus is the information center of eukaryotic cells—a corporate headquarters, design center, and library all rolled into one. Appropriately enough, its interior is highly organized.

The organelle's overall shape and structure are defined by the mesh-like nuclear lamina. The nuclear lamina provides an attachment point for the chromosomes, each of which occupies a well-defined region in the nucleus.

In addition, specific centers exist where the genetic information in DNA is decoded and processed. At these locations, large suites of enzymes interact to produce RNA messages from specific genes at specific times. Meanwhile, the nucleolus functions as the site of ribosome synthesis.

Structure and Function of the Nuclear Envelope

The nuclear envelope separates the nucleus from the rest of the cell. Starting in the 1950s, transmission electron micrographs of cross sections through the nuclear envelope showed that the

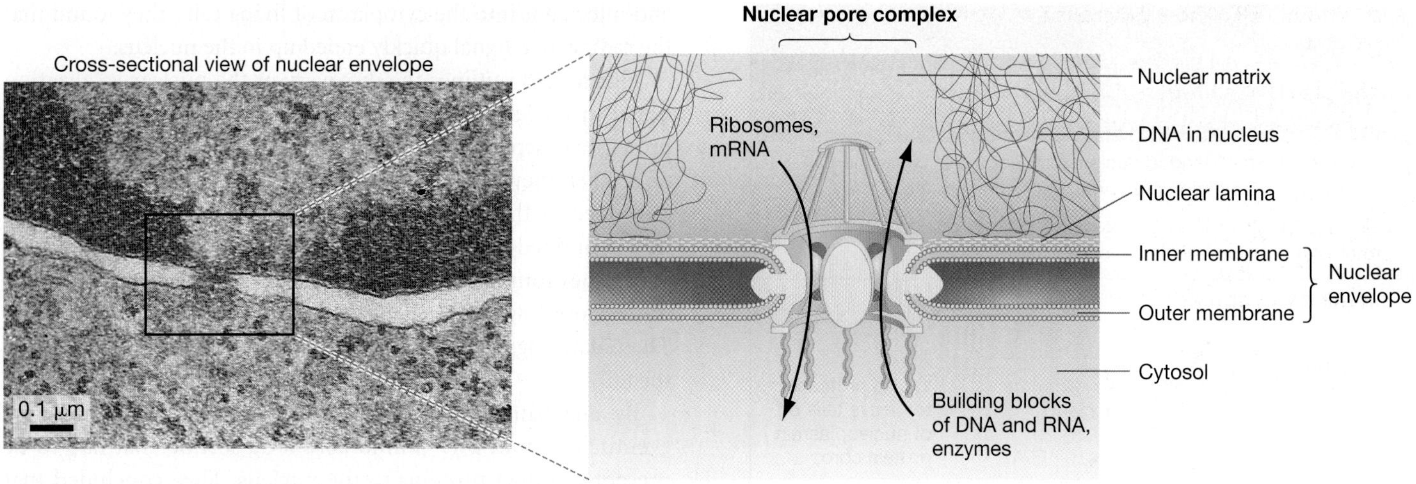

Cross-sectional view of nuclear envelope

0.1 μm

Nuclear pore complex

Nuclear matrix

DNA in nucleus

Nuclear lamina

Inner membrane
Outer membrane — Nuclear envelope

Cytosol

Ribosomes, mRNA

Building blocks of DNA and RNA, enzymes

FIGURE 7.22 Structure of the Nuclear Envelope and Nuclear Pore.

structure is supported by the fibrous nuclear lamina and bounded by two membranes, each consisting of a lipid bilayer (**Figure 7.22**).

Micrographs like the one in Figure 7.22 also show that the envelope is broken with openings called **nuclear pores**. Because these gate-like structures extend through both inner and outer nuclear membranes, they connect the inside of the nucleus with the cytosol. Follow-up research showed that each pore consists of over 50 different proteins. As the diagram on the right side of Figure 7.22 shows, these protein molecules form an elaborate structure called the **nuclear pore complex**.

Experiments in the early 1960s showed that molecules travel into and out of the nucleus through the nuclear pore complexes. The initial studies were based on injecting tiny gold particles into cells and then preparing the cells for electron microscopy. In electron micrographs, gold particles show up as black dots. One or two minutes after injection, the micrographs showed that most of the gold particles were in the cytoplasm. A few, however, were closely associated with nuclear pores. Ten minutes after injection, particles were inside the nucleus as well as in the cytoplasm.

These data supported the hypothesis that the pores function as the doors to the nucleus. Follow-up work confirmed that the nuclear pore complex is the only gate between the cytoplasm and the nucleus and only certain molecules go in and out. Passage through a nuclear pore is selective.

What substances traverse nuclear pores? DNA clearly does not—it never leaves the nucleus. But information coded in DNA is used to synthesize RNA inside the nucleus.

Several types of RNA molecules are produced, each distinguished by size and function. For example, most **ribosomal RNAs** are manufactured in the nucleolus, where they bind to proteins to form ribosomes, which are then exported to the cytoplasm. Similarly, molecules called messenger **RNAs (mRNA)** carry the information required to manufacture proteins out to the cytoplasm, where protein synthesis takes place. The outbound traffic, mostly consisting of RNAs, is intense.

Inbound traffic is also impressive. Nucleoside triphosphates that act as building blocks for DNA and RNA enter the nucleus, as do the proteins responsible for copying DNA, synthesizing RNAs, extending the nuclear lamina, assembling ribosomes, or building chromosomes.

To summarize, ribosomal subunits and various types of RNAs exit the nucleus; proteins that are needed inside enter it. In a typical cell, over 500 molecules pass through each of the 3000–4000 nuclear pores every second.

The scale of traffic through the nuclear pores is mind-boggling. How is it regulated and directed?

How Are Molecules Imported into the Nucleus?

The first experiments on how molecules move through the nuclear pore focused on proteins that are produced by viruses. **Viruses** are parasites that use the cell's machinery to make copies of themselves. When a virus infects a cell, certain of its proteins enter the nucleus.

Investigators began studying the transport of viral proteins when they noticed that if a particular amino acid in a viral protein changed, it was no longer able to pass through the nuclear pore. This simple-sounding observation led to a key hypothesis: Proteins that are synthesized by ribosomes in the cytosol but are headed for the nucleus contain a "zip code"—a molecular address tag that marks them for transport through the nuclear pore complex.

The idea was that viral proteins could enter the nucleus only if they carried the same address tag that normal cellular proteins had. Thus, the proteins with the altered amino acid were thwarted. This zip code came to be called the **nuclear localization signal (NLS)**.

A series of experiments on a protein called nucleoplasmin helped researchers understand the nature of the nuclear localization signal. Nucleoplasmin, which plays an important role in the assembly of chromosomes, has a distinctive structure: a globular protein core surrounded by a series of extended protein "tails." When researchers labeled nucleoplasmin with a radioactive atom

QUESTION: Where is the "Send to nucleus" zip code in the nucleoplasmin protein?

HYPOTHESIS: The "Send to nucleus" zip code is in either the tail region or the core region of the nucleoplasmin protein.

NULL HYPOTHESIS: The zip code is not on the nucleoplasmin protein itself, or there is no zip code.

EXPERIMENTAL SETUP:

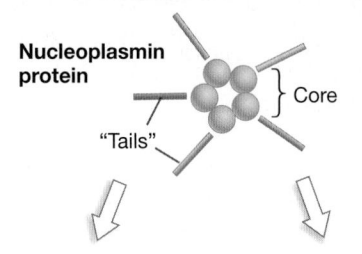

Nucleoplasmin protein

"Tails"

Core

1. Use protease to cleave tails off of nucleoplasmin protein core.

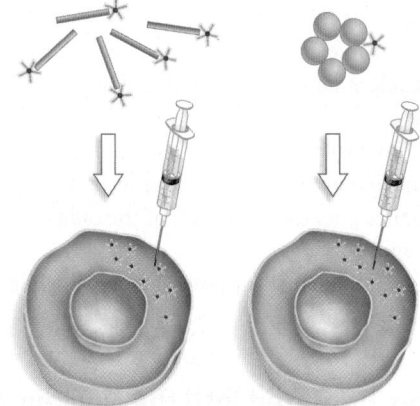

Labeled tails   Labeled cores

2. Attach radioactive labels to protein tails and cores.

3. Inject labeled tails and cores into cytoplasm of different cells.

4. Wait, then locate labeled fragments ....

PREDICTION:

PREDICTION OF NULL HYPOTHESIS:

RESULTS:

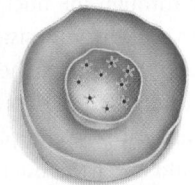
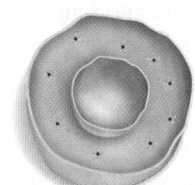

Labeled tail fragments **located in nucleus**

Labeled core fragments still **located in cytoplasm**

CONCLUSION:

FIGURE 7.23 Where Is the "Send to Nucleus" Zip Code in the Nucleoplasmin Protein?

SOURCES: Mills, A. D. et al. 1980. *Journal of Molecular Biology* 139: 561–568; Dingwall, C. et al. 1982. *Cell* 30: 449–458.

✔**EXERCISE** Without looking at the text, fill in the prediction(s) and conclusion(s) in this experiment.

and injected it into the cytoplasm of living cells, they found that the radioactive signal quickly ended up in the nucleus.

Figure 7.23 outlines the discovery of the nuclear localization signal in nucleoplasmin. First, researchers used enzymes called proteases to separate the core sections of nucleoplasmin from the tails. Then they labeled each component with radioactive atoms and injected them into the cytoplasm of different cells. When they examined the experimental cells with the electron microscope, they found that tail fragments were transported to the nucleus. Core fragments, in contrast, remained in the cytoplasm. These data suggested that the "zip code" must be somewhere in the tail.

By analyzing different stretches of the tail, the biologists eventually found a 17-amino-acid-long section that had to be present to direct proteins to the nucleus. They concluded that the nuclear localization signal consisted of 17 specific amino acids in the tail.

You can review the experimental evidence for the nuclear localization signal in the study area at *www.masteringbiology.com*.

(MB) **Web Activity** Transport into the Nucleus

Follow-up work confirmed that other proteins bound for the nucleus have similar localization signals. More recent research has shown that the movement of proteins and other large molecules into and out of the nucleus is an energy-demanding process that involves transport proteins called importins and exportins. Importins and exportins function like trucks that haul cargo into and out of the nucleus, through the nuclear pore complex.

Currently, biologists are trying to unravel how all this traffic in and out of the nucleus is regulated to avoid backups and head-on collisions. Nuclear transport is a classic research system and a key example of how organelles interact.

7.5 Cell Systems II: The Endomembrane System Manufactures and Ships Proteins

The nuclear membrane is not the only place in cells where cargo moves in a regulated and energy-demanding fashion. For example, Chapter 6 highlighted how specific ions and molecules are pumped into and out of cells or transported across the plasma membrane by specialized membrane proteins.

In addition, proteins that are synthesized by ribosomes in the cytosol for use inside mitochondria or chloroplasts contain special signal sequences, like the nuclear localization signal, that target the proteins for transport to the appropriate organelles. Ions, ATP, amino acids, and other small molecules diffuse randomly throughout the cell, but the movement of proteins and other large molecules is energy demanding and tightly regulated.

If you think about it for a moment, the need to sort proteins and ship them to specific destinations should be clear. Proteins are produced by ribosomes in the cytosol or ribosomes on the ER. Each protein that is synthesized needs to be transported to one of the many compartments inside the eukaryotic cell. Acid hydrolases need to end up in lysosomes, catalase must be shipped to peroxisomes, and ribosomal proteins require transport to the nucleolus. To get to the right location, each protein has to have an address tag and a transport and delivery system.

To get a better understanding of protein sorting and transport in eukaryotic cells, let's consider perhaps the most intricate of all manufacturing and shipping complexes: the endomembrane system. In this system, proteins that are synthesized in the rough ER move to the Golgi apparatus for processing, and from there travel to the cell surface or other destinations.

Studying the Pathway through the Endomembrane System

The idea that materials move through the endomembrane system in an orderly way was inspired by a simple observation. According to electron micrographs, cells that secrete digestive enzymes, hormones, or other products have particularly large amounts of rough ER and Golgi. This correlation led to the idea that these cells have a "secretory pathway" that starts in the rough ER and ends with products leaving the cell (**Figure 7.24**). How does this hypothesized pathway work?

THE LOGIC OF A PULSE-CHASE EXPERIMENT George Palade and colleagues did pioneering research on the secretory pathway

using a **pulse-chase experiment**. This strategy is based on two steps:

- *The "Pulse"* Expose experimental cells to a high concentration of a labeled molecule for a short time. For example, if a cell takes in a large amount of radioactively labeled amino acid for a brief time, virtually all of the proteins synthesized during that interval will be labeled.

- *The "Chase"* The pulse of labeled molecule is followed by a chase—large amounts of an unlabeled version of the same molecule, provided for a longer time. If the chase consists of unlabeled amino acid, then the proteins synthesized during the chase period will *not* be labeled.

The idea is to mark a population of molecules at a particular interval and then follow their fate over time. This approach is analogous to adding a small amount of dye to a stream and then following the movement of the dye molecules.

To understand why the chase is necessary in these experiments, imagine what would happen if you added dye to a stream continuously. Soon the entire stream would be dyed—you could no longer tell where a specific population of dye molecules were moving.

In testing the secretory pathway hypothesis, Palade's team focused on pancreatic cells that were growing in **culture**, or in

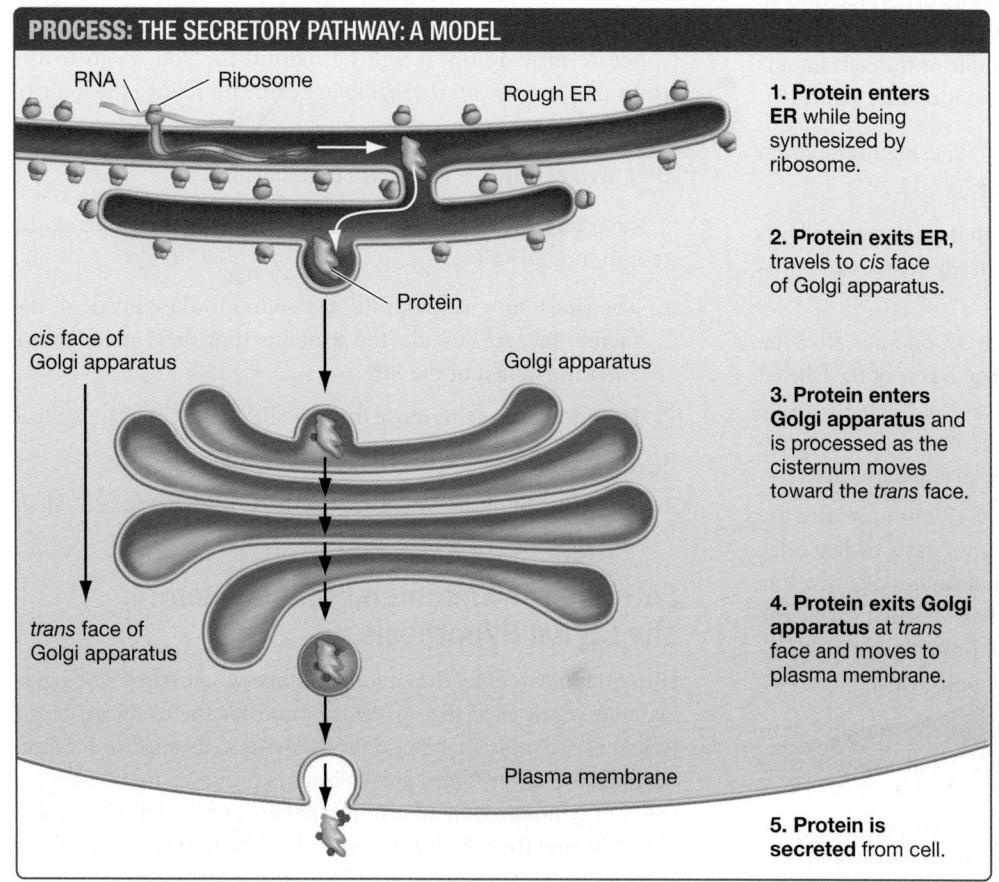

PROCESS: THE SECRETORY PATHWAY: A MODEL

RNA
Ribosome
Rough ER

Protein

cis face of Golgi apparatus

Golgi apparatus

trans face of Golgi apparatus

Plasma membrane

1. **Protein enters ER** while being synthesized by ribosome.

2. **Protein exits ER**, travels to *cis* face of Golgi apparatus.

3. **Protein enters Golgi apparatus** and is processed as the cisternum moves toward the *trans* face.

4. **Protein exits Golgi apparatus** at *trans* face and moves to plasma membrane.

5. **Protein is secreted** from cell.

FIGURE 7.24 The Secretory Pathway Hypothesis. The secretory pathway hypothesis proposes that proteins intended for secretion from the cell are synthesized and processed in a highly prescribed set of steps. Note that proteins are packaged into vesicles when they move from the RER to the Golgi and from the Golgi to the cell surface.

(a) Immediately after labeling	**(b)** 37 minutes after end of labeling	**(c)** 117 minutes after end of labeling

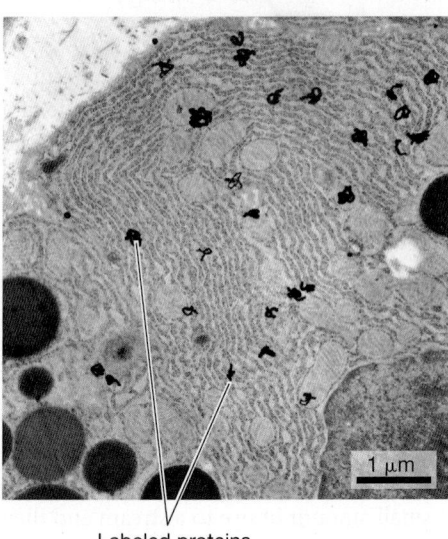 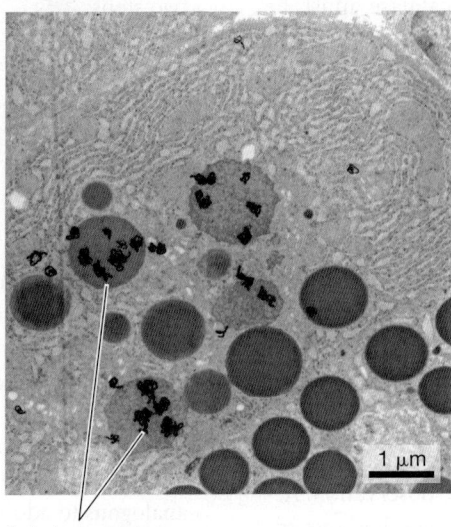 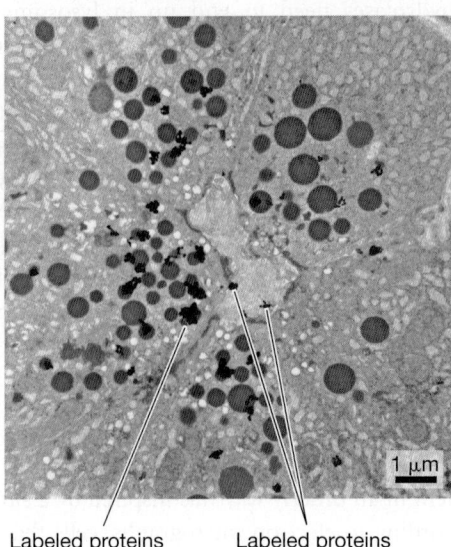

Labeled proteins in rough ER

Labeled proteins in secretory vesicles

Labeled proteins in secretory vesicles

Labeled proteins in secretory duct

FIGURE 7.25 Results of a Pulse-Chase Experiment. Proteins move **(a)** from the rough ER into **(b)** secretory vesicles before **(c)** being secreted from the cell.

vitro.[1] These cells, specialized for secreting digestive enzymes into the small intestine, are packed with rough ER and Golgi.

The basic experimental approach was to supply the cells with a 3-minute pulse of the amino acid leucine, labeled with a radioactive atom, followed by a long chase with nonradioactive leucine. Because the radioactive leucine was incorporated into all proteins being produced during the pulse, those proteins were labeled. Then the researchers prepared a sample of the cells for autoradiography and electron microscopy (see **BioSkills 9 and 10**).

RESULTS OF A PULSE-CHASE EXPERIMENT The micrographs in **Figure 7.25** show what happened over time.

- Figure 7.25a shows part of a single cell that was examined immediately after the pulse. Most of the newly synthesized proteins are inside this cell's rough ER.

- Figure 7.25b shows part of a single cell 37 minutes after the pulse ended. Now the situation has changed. Few of the labeled proteins are in the rough ER. Instead, most of them are inside structures called secretory vesicles on the *trans* side of a Golgi apparatus (some are found inside the Golgi apparatus itself).

- The micrograph in Figure 7.25c, taken 117 minutes after the pulse, is at lower magnification and shows parts of five cells. The structure in the middle is a duct that carries digestive enzymes from pancreatic cells toward their destination in the small intestine. Note that most labeled proteins are in secretory vesicles or actually outside the cell, in the duct.

Because the labeled proteins move from the rough ER to Golgi apparatus to secretory vesicles to the cell exterior over

time, the results support the hypotheses that a secretory pathway exists and that the rough ER and Golgi apparatus function as an integrated endomembrane system.

The data suggest that proteins produced in the rough ER do not drift randomly from organelle to organelle. Instead, traffic through the endomembrane system appears to be highly organized and directed.

Before moving on, it will be helpful for you to go to the study area at *www.masteringbiology.com* and review the logic of a pulse-chase experiment.

(MB) **Web Activity** A Pulse-Chase Experiment

Next, let's break the system down and examine four of the steps in more detail:

1. The ribosomes in rough ER are bound to the outside of the membrane, so how do the proteins that they manufacture enter the lumen of the ER?

2. How do the proteins move from the ER to the Golgi apparatus?

3. Once they're inside the Golgi, what happens to them?

4. And finally, how do the finished proteins reach their destinations?

Entering the Endomembrane System: The Signal Hypothesis

How do proteins enter the endomembrane system? The signal hypothesis predicted that proteins bound for the endomembrane system have a molecular zip code analogous to the nuclear localization signal. Günter Blobel and colleagues proposed that these proteins are synthesized by ribosomes that are attached to the outside of the ER, and the first few amino acids in the growing polypeptide act as a signal that brings the protein into the lumen of the ER.

---

[1]The term in vitro is Latin for "in glass." Experiments that are performed outside living organisms are done in vitro. The term in vivo, in contrast, is Latin for "in life." Experiments performed with living organisms are done in vivo.

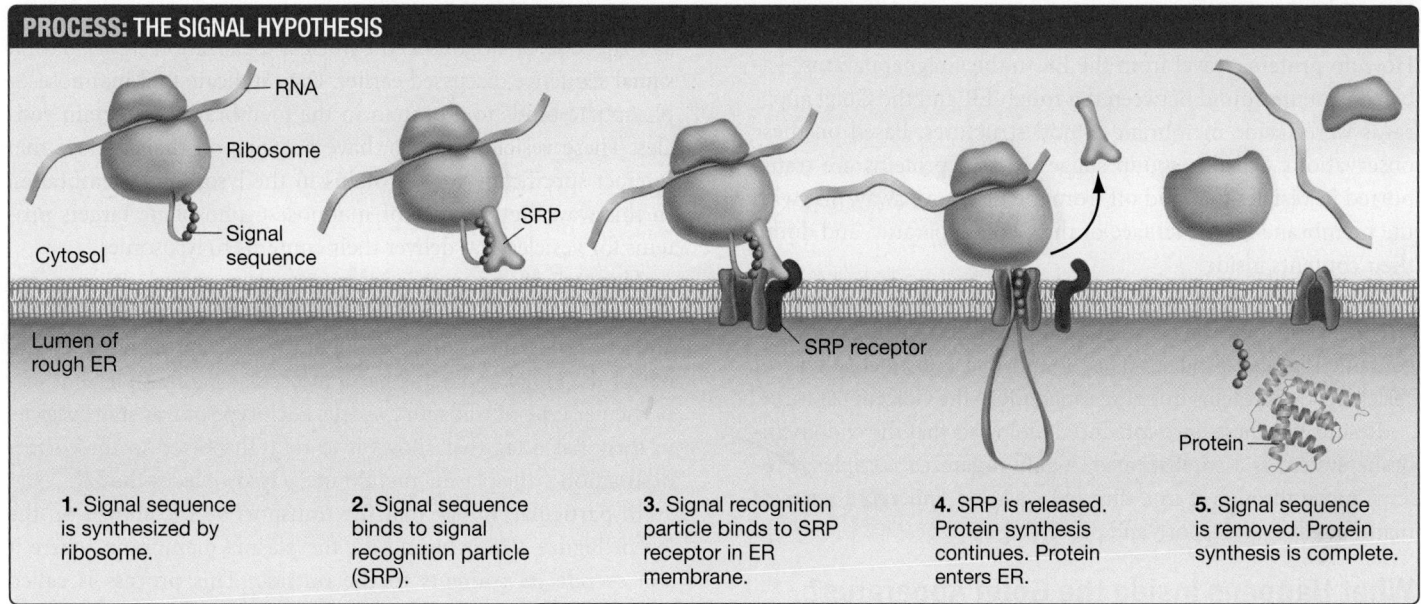

PROCESS: THE SIGNAL HYPOTHESIS

RNA
Ribosome
Signal sequence
Cytosol
SRP
Lumen of rough ER
SRP receptor
Protein

1. Signal sequence is synthesized by ribosome.

2. Signal sequence binds to signal recognition particle (SRP).

3. Signal recognition particle binds to SRP receptor in ER membrane.

4. SRP is released. Protein synthesis continues. Protein enters ER.

5. Signal sequence is removed. Protein synthesis is complete.

FIGURE 7.26 The Signal Hypothesis Explains How Proteins Destined for Secretion Enter the Endomembrane System. According to the signal hypothesis, proteins destined for secretion contain a short stretch of amino acids that interact with a signal recognition particle (SRP) in the cytoplasm. This interaction allows the protein to enter the ER.

This hypothesis received important support when researchers made a puzzling observation: When proteins that are normally synthesized in the rough ER are instead manufactured by isolated ribosomes in vitro—with *no* ER present—they are 20 amino acids longer, on average, than usual.

Blobel seized on these data. He claimed that the extra amino acids are the "send-to-ER" signal, and that the signal is removed inside the organelle. When the same protein is synthesized outside the ER, the signal is not removed.

Blobel's group went on to produce convincing data that supported the hypothesis: They identified the exact series of amino acids in the **ER signal sequence**.

More recent work has documented the mechanisms responsible for receiving the send-to-ER signal and inserting the protein into the rough ER. **Figure 7.26** illustrates the key steps involved.

Step 1 A ribosome synthesizes the ER signal sequence.

Step 2 The signal sequence binds to a **signal recognition particle (SRP)**—a complex of RNA and protein.

Step 3 The ribosome + signal sequence + SRP complex attaches to an SRP receptor in the ER membrane itself. Think of the SRP as a key that is activated by an ER signal sequence. The SRP receptor in the ER membrane is the lock.

Step 4 Once the lock (the receptor) and key (the SRP) connect, the SRP is released.

Step 5 The signal sequence is removed and protein synthesis is completed.

If the finished polypeptide will eventually be shipped to an organelle or secreted from the cell, it enters the lumen of the rough ER. If the finished polypeptide is a membrane protein, it remains in the rough ER membrane while it is being processed.

Once proteins are inside the rough ER or inserted into its membrane, they fold into their three-dimensional shape with the help of chaperone proteins (see Chapter 3). In addition, proteins that enter the lumen interact with enzymes that catalyze the addition of carbohydrate side chains. Because carbohydrates are polymers of sugar monomers, the addition of one or more carbohydrate groups is called **glycosylation** ("sugar-together"). The resulting molecule is a **glycoprotein** ("sugar-protein"; see Chapter 5).

As **Figure 7.27** shows, proteins that enter the ER often gain a specific carbohydrate that consists of 14 sugar subunits. The completed glycoproteins are ready for shipment to the Golgi apparatus.

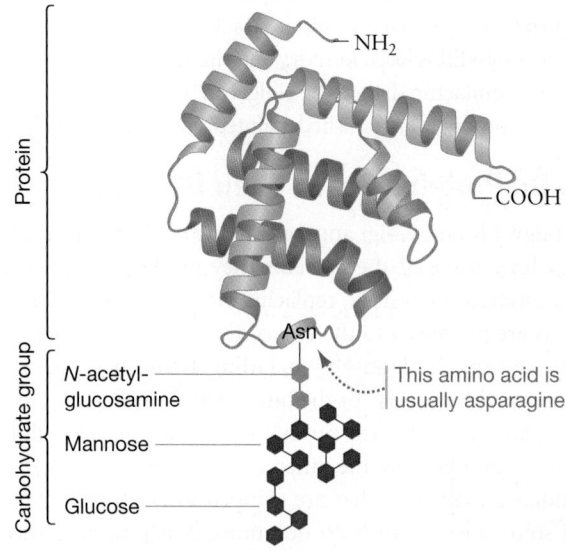

Protein

NH₂
COOH
Asn

Carbohydrate group

N-acetyl-glucosamine

This amino acid is usually asparagine

Mannose

Glucose

FIGURE 7.27 Glycosylation Adds Carbohydrate Groups to Proteins. When proteins enter the ER, most acquire the 14 sugar residues shown here. Some of these sugars may be removed or others added as proteins pass through the Golgi apparatus.

Moving from the ER to the Golgi

How do proteins travel from the ER to the Golgi apparatus? Labeled proteins found between the rough ER and the Golgi apparatus were inside membrane-bound structures. Based on these observations, Palade's group suggested that proteins are transported in vesicles that bud off from the ER, move away, fuse with the membrane on the *cis* face of the Golgi apparatus, and dump their contents inside.

This hypothesis was supported when other researchers used differential centrifugation to isolate and characterize the vesicles that contained labeled proteins. They found that a distinctive type of vesicle carries proteins from the rough ER to the Golgi apparatus.

Results like these have convinced biologists that the endomembrane system is a sophisticated, highly organized complex. Proteins move through it in a directed way and undergo a series of manufacturing, transport, and processing steps.

What Happens inside the Golgi Apparatus?

Section 7.2 indicated that the Golgi apparatus consists of a stack of flattened vesicles called cisternae, and that cargo enters one side of the organelle and exits the other. Recent research has shown that the composition of the Golgi apparatus is dynamic. New cisternae constantly form at the *cis* face, while old cisternae break apart at the *trans* face. Once a cisterna forms, it gradually moves toward the *trans* face. As it does, it changes in composition and activity.

By separating individual cisternae and analyzing their contents, researchers have found that cisternae at various stages of maturation contain different suites of enzymes. These enzymes catalyze glycosylation reactions. As a result, proteins undergo further modification as a cisterna matures.

While cisternae are still near the *cis* face, some of the proteins inside have sugar-phosphate groups added. Later, the carbohydrate group that was added in the rough ER is removed. Near the *trans* face, various types of carbohydrate chains are attached that may protect the protein or help it attach to surfaces.

If the rough ER is like a foundry and stamping plant where rough parts are manufactured, then the Golgi can be considered a finishing area where products are polished, painted, and readied for shipping.

How Do Proteins Reach Their Destinations?

The rough ER and Golgi apparatus constitute an impressive assembly line. Some of the proteins they produce stay in the endomembrane system itself, replacing worn-out molecules. But if proteins are processed to the end of the line, they will be sent to one of several destinations, including lysosomes, the plasma membrane, chloroplasts, or the outside of the cell.

How are these finished products put into the right shipping containers, and how are the different containers addressed?

Studies on enzymes that are shipped to lysosomes have provided some answers to both questions. A key finding was that lysosome-bound proteins have a phosphate group attached to a specific sugar subunit on their surface, forming the compound mannose-6-phosphate. If mannose-6-phosphate is removed from these proteins, they are not transported to a lysosome.

This is strong evidence that the phosphorylated sugar serves as a zip code, analogous to the nuclear localization signal and ER signal sequence discussed earlier. Data indicate that mannose-6-phosphate binds to a protein in the membranes of certain vesicles. These vesicles, in turn, have proteins on their surface that interact specifically with proteins in the lysosomal membranes. In this way, the presence of mannose-6-phosphate targets proteins for vesicles that deliver their contents to lysosomes.

Figure 7.28 presents a comprehensive model of how endomembrane system products are loaded into specific vesicles and shipped to their correct destinations. Each protein that comes out of the Golgi apparatus has a molecular tag that places it in a particular type of transport vesicle. Each type of transport vesicle, in turn, has a tag that allows it to be transported to the correct destination—the plasma membrane, a lysosome, or the ER.

In particular, notice that the transport vesicle shown on the left of Figure 7.28 is bound for the plasma membrane, where it will secrete its contents to the outside. This process is called **exocytosis** ("outside-cell-act"). When exocytosis occurs, the vesicle membrane and plasma membrane make contact. As the two membranes fuse, their lipid bilayers rearrange in a way that exposes the interior of the vesicle to the outside of the cell. The vesicle's contents then diffuse into the space outside the cell. This is how cells in your pancreas deliver digestive enzymes to the duct that leads to your small intestine—where food is digested.

🔑 Proteins that are synthesized in the cytoplasm also have zip codes directing them to mitochondria, chloroplasts, or other destinations. In general, the proteins produced in a cell have distinctive molecular address labels, which allow proteins to be shipped to the compartments where they function.

If vesicles function like shipping containers for products that move between organelles, do they travel along some sort of road or track? What molecule or molecules function as the delivery truck, and does ATP supply the gas? Let's delve into these questions in Section 7.6.

CHECK YOUR UNDERSTANDING

🔑 If you understand that . . .

- In cells, the transport of proteins and other large molecules is energy demanding and tightly regulated.
- Proteins must have the appropriate molecular zip code to enter or leave the nucleus, enter the lumen of the rough ER, or become incorporated into vesicles destined for lysosomes or the plasma membrane.
- In many cases, proteins and other types of cargo are shipped in vesicles that contain molecular zip codes on their surface.

You should be able to . . .

1. Predict what happens to proteins that lack an ER signal sequence.
2. Predict the outcome of an experiment where secreted proteins are placed inside vesicles with a zip code associated with shipment to lysosomes.

Answers are available in Appendix B.

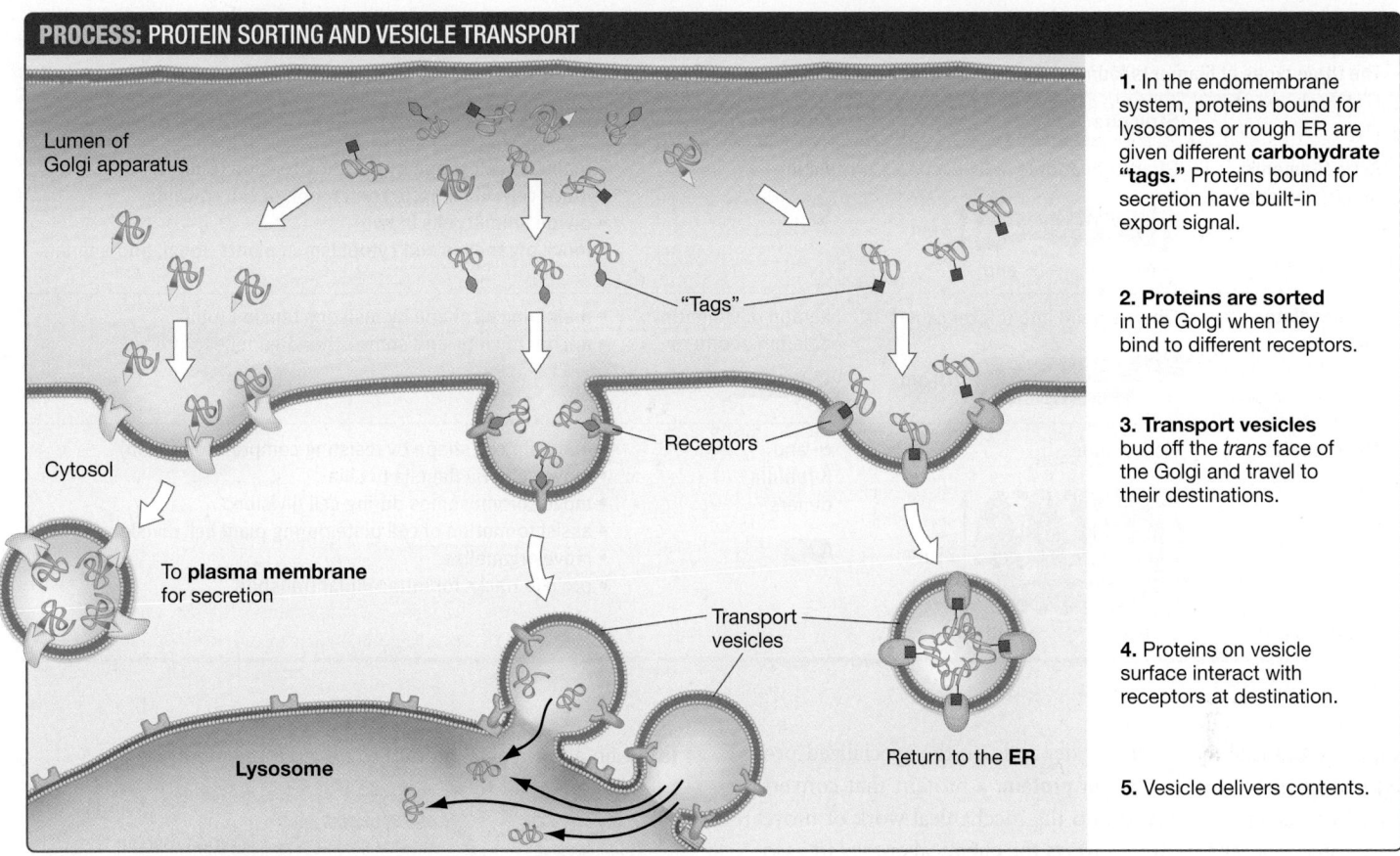

Lumen of
Golgi apparatus

Cytosol

To **plasma membrane**
for secretion

Lysosome

"Tags"

Receptors

Transport
vesicles

Return to the **ER**

1. In the endomembrane system, proteins bound for lysosomes or rough ER are given different **carbohydrate "tags."** Proteins bound for secretion have built-in export signal.

2. Proteins are sorted in the Golgi when they bind to different receptors.

3. Transport vesicles bud off the *trans* face of the Golgi and travel to their destinations.

4. Proteins on vesicle surface interact with receptors at destination.

5. Vesicle delivers contents.

FIGURE 7.28 In the Golgi Apparatus, Proteins Are Sorted into Vesicles That Are Targeted to a Destination.

7.6 Cell Systems III: The Dynamic Cytoskeleton

The endomembrane system may be the best-studied example of how individual organelles work together in a dynamic, highly integrated way. This integration depends in part on the physical relationship of organelles, which is organized by the cytoskeletal system.

The cytoskeleton is a dense and complex network of fibers that helps maintain cell shape by providing structural support. However, the cytoskeleton is not a static structure like the scaffolding used at construction sites. Its fibrous proteins move and change to alter the cell's shape, shift its contents, and move the entire structure. Like the rest of the cell, the cytoskeleton is dynamic.

As **Table 7.3** shows, there are three distinct cytoskeletal elements in eukaryotic cells: actin filaments (also known as microfilaments), intermediate filaments, and microtubules. Recent research has shown that bacterial cells have cytoskeletal elements that are extremely similar to actin filaments and microtubules.

Each of the three cytoskeletal elements found in eukaryotes has a distinct size, structure, and function. Let's look at each one in turn.

Actin Filaments

Sometimes called **microfilaments** because they are the cytoskeletal element with the smallest diameter, **actin filaments** are fibrous

structures made of the globular protein actin (Table 7.3). In animal cells, actin is often the most abundant of all proteins—typically it represents 5–10 percent of the total protein in the cell. Each of your liver cells contains about half a billion of these molecules.

ACTIN FILAMENT STRUCTURE   Actin filaments form when individual actin molecules polymerize. Because each actin subunit in the strand is asymmetrical, the structure as a whole has a distinct polarity. The two ends of an actin filament are different and are referred to as plus and minus ends. Each filament grows and shrinks as actin subunits are added to or subtracted from each end of the structure. The addition and deletion of actin subunits is called treadmilling when the rates of addition at one end and deletion at the other are about equal—the fiber stays the same length but individual subunits move as if on a treadmill.

A completed actin filament resembles two long strands that coil around each other. In general, actin filaments tend to grow at the plus end because polymerization occurs fastest there.

In animal cells, actin filaments are particularly abundant just under the plasma membrane, where they are organized into long, parallel bundles or dense, crisscrossing networks, with individual actin filaments linked to one another by other proteins. The reinforced bundles and networks of actin filaments help stiffen the cell and define its shape.

ACTIN FILAMENT FUNCTION   In addition to providing structural support, actin filaments are involved in movement. In several

The three types of filaments found in the cytoskeleton are distinguished by their size and structure, and the protein subunit of which they are made.

	Structure	Subunits	Functions
Actin filaments (microfilaments)	Strands in double helix 7 nm − end + end	Actin	• maintain cell shape by resisting tension (pull) • move cells via muscle contraction or cell crawling • divide animal cells in two • move organelles and cytoplasm in plants, fungi, and animals
Intermediate filaments	Fibers wound into thicker cables 10 nm	Keratin or vimentin or lamin or others	• maintain cell shape by resisting tension (pull) • anchor nucleus and some other organelles
Microtubules	Hollow tube 25 nm − end + end	α- and β-tubulin dimers	• maintain cell shape by resisting compression (push) • move cells via flagella or cilia • move chromosomes during cell division • assist formation of cell plate during plant cell division • move organelles • provide tracks for intracellular transport

cases, actin's role in movement depends on the specialized protein myosin. Myosin is a **motor protein**: a protein that converts the chemical energy in ATP into the mechanical work of movement, just as a car's motor converts the chemical energy in gasoline into movement.

Chapter 46 details the interaction between actin and myosin that produces movement. For now, it's enough to recognize that when ATP binds to myosin and is then hydrolyzed to ADP, the "head" region of the myosin molecule binds to actin and moves. The movement of this protein causes the actin filament to slide (**Figure 7.29a**). This type of movement is analogous to a line of people who are passing along a long log or pole. The people are myosin molecules; the log or pole is actin.

As **Figure 7.29b** shows, the ATP-powered interaction between actin and myosin is the basis for an array of cell movements:

- **Cytokinesis** ("cell-moving") is the process of cell division in animals. For these cells to divide in two, actin filaments that are arranged in a ring under the plasma membrane must slide past one another. Because they are connected to the plasma membrane, the movement of the actin fibers pinches the cell in two.

- **Cytoplasmic streaming** is the directed flow of cytosol and organelles around plant cells. The movement occurs along actin filaments and is powered by myosin. It is especially common in large cells, where the circulation of cytoplasm facilitates material transport.

In addition, the movement called **cell crawling** occurs when groups of actin filaments grow, creating bulges in the plasma membrane that extend and move the cell. Cell crawling occurs in a wide range of organisms and cell types, including amoebae, slime molds, and certain human cells.

(a) Actin and myosin interact to cause movement.

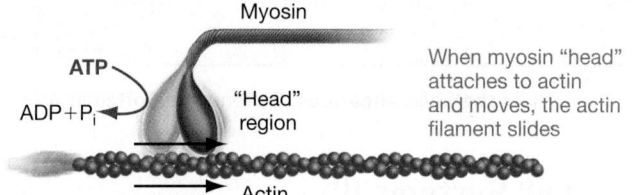

When myosin "head" attaches to actin and moves, the actin filament slides

(b) Examples of movement caused by actin-myosin interactions

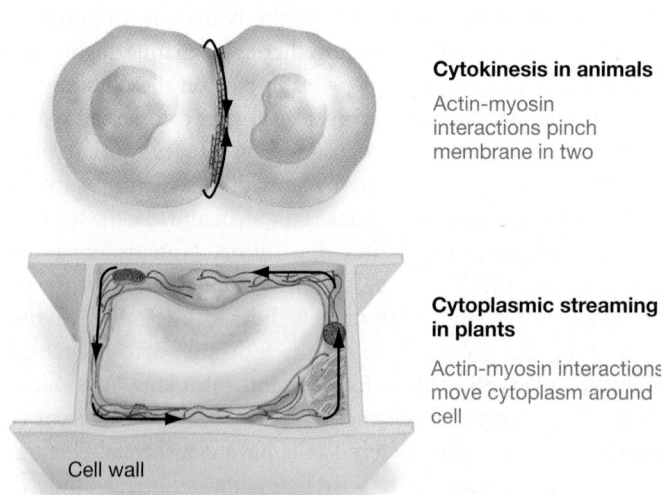

Cytokinesis in animals

Actin-myosin interactions pinch membrane in two

Cytoplasmic streaming in plants

Actin-myosin interactions move cytoplasm around cell

Cell wall

FIGURE 7.29 Many Cellular Movements Are Based on Actin-Myosin Interactions. **(a)** When the "head" region of the myosin protein interacts with ATP, myosin attaches to actin and changes shape. The movement causes the actin filament to slide. **(b)** Actin-myosin interactions can divide cells and move organelles and cytoplasm.

Intermediate Filaments

Although similar in size, many types of **intermediate filaments** exist, each consisting of a different—though structurally similar—type of protein (Table 7.3). Humans, for example, have 70 genes that code for intermediate filament proteins. This is in stark contrast to actin filaments and microtubules, which are made from the same protein subunits in all eukaryotic cells.

In addition, intermediate filaments are not polar; instead, each end of these filaments is identical. As a result, intermediate filaments do not treadmill, and they are not involved in directed movement driven by myosin or related proteins. Intermediate filaments serve a purely structural role in eukaryotic cells.

The intermediate filaments that you are most familiar with belong to a family of molecules called the keratins. The cells that make up your skin and line surfaces inside your body contain about 20 types of keratin. These intermediate filaments provide the mechanical strength required for these cells to resist pressure and abrasion. Skin cells manufacture another 10 distinct forms of keratin. Depending on the location of the skin cell and keratins involved, the secreted filaments form fingernails, toenails, or hair.

Nuclear lamins, which make up the nuclear lamina layer introduced in Section 7.4, also qualify as intermediate filaments. Nuclear lamins form a dense mesh under the nuclear envelope. Recall that in addition to giving the nucleus its shape, they anchor the chromosomes. They are also involved in the breakup and reassembly of the nuclear envelope when cells divide.

Some intermediate filaments project from the nucleus through the cytoplasm to the plasma membrane, where they are linked to intermediate filaments that run parallel to the cell surface. In this way, intermediate filaments form a flexible skeleton that helps shape the cell surface and hold the nucleus in place.

Microtubules

Microtubules are composed of two proteins called α-tubulin and β-tubulin and are the largest cytoskeletal components in terms of diameter (Table 7.3). Molecules of α-tubulin and β-tubulin bind to form **dimers** ("two-parts"), compounds formed by the joining of two monomers.

Tubulin dimers polymerize to form the large, hollow tube called a microtubule. Because each end of a tubulin dimer is different, each end of a microtubule has a distinct polarity. Like actin filaments, microtubules are dynamic and usually grow at their plus end. Microtubules grow and shrink in length as tubulin dimers are added or subtracted.

Microtubules originate from a structure called the **microtubule organizing center** and grow outward, radiating throughout the cell (see the chapter opening photograph on page 102). Although plant cells typically have hundreds of these organizing centers, most animal and fungal cells have just one.

In animals, the microtubule organizing center has a distinctive structure and is called a **centrosome**. As **Figure 7.30** shows, animal centrosomes contain two bundles of microtubules called **centrioles**.

In function, microtubules are similar to actin filaments: They provide stability and are involved in movement. Like steel girders

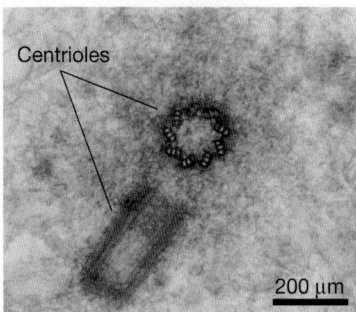 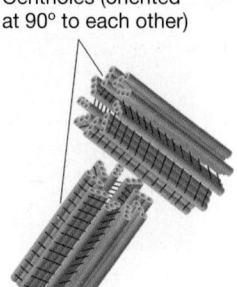

Centrosome Centrioles (oriented at 90° to each other)

Centrioles

200 μm

FIGURE 7.30 Centrosomes Are a Type of Microtubule Organizing Center. Microtubules emanate from microtubule organizing centers, which in animals are called centrosomes. The centrioles inside a centrosome are made of microtubules.

in a skyscraper, the microtubules that radiate from an organizing center stiffen the cell by resisting compression forces. Microtubules may also provide a structural framework for organelles. If microtubules are prevented from forming, the ER no longer assembles in its normal network-like configuration.

During cell division, microtubules from the organizing center move chromosomes from the original cell to each of the two resulting cells (see Chapters 11 and 12). But microtubules are involved in many other types of cellular movement as well. Let's consider their role in moving materials inside cells, then ask how the microtubules in flagella help move an entire cell.

STUDYING VESICLE TRANSPORT Materials are transported to a wide array of destinations inside cells via vesicles. To study how this movement happens, Ronald Vale and colleagues focused on the giant axon, an extremely large nerve cell in squid that runs the length of the animal's body. If the squid is disturbed, the cell signals muscles to contract so it can jet away to safety.

The researchers decided to study this particular cell for three reasons.

1. The giant axon is so large that it is relatively easy to see and manipulate.

2. Large numbers of molecules are synthesized in the cell's ER and then transported in vesicles down the length of the cell, where they are released. As a result, a large amount of cargo moves a long distance.

3. The researchers found that if they gently squeezed the cytoplasm out of the cell, vesicle transport still occurred in the cytoplasmic material. This allowed them to do experiments on vesicle transport without the plasma membrane being in the way.

In short, the squid giant axon provided a system that could be observed and manipulated efficiently in the lab. What did the biologists find out?

MICROTUBULES ACT AS "RAILROAD TRACKS" To watch vesicle transport in action, the researchers mounted a video camera to a

(a) Electron micrograph

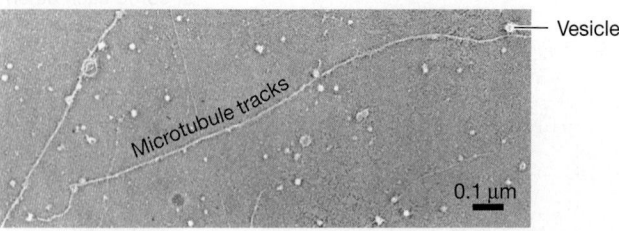

Vesicle

Microtubule tracks

0.1 μm

(b) Video image

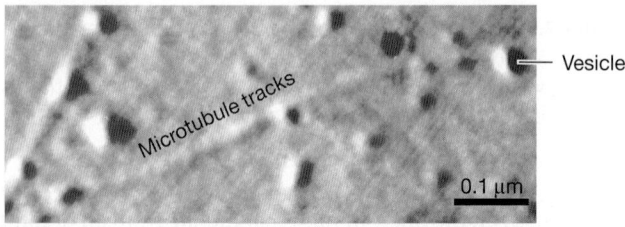

Vesicle

Microtubule tracks

0.1 μm

FIGURE 7.31 Transport Vesicles Move along Microtubule Tracks.
The images show extruded cytoplasm from a squid giant axon.
(a) An electron micrograph that allowed researchers to measure the diameter of the filaments and confirm that they are microtubules. In the upper part of this image, you can see a vesicle on a "track."
(b) A slightly fuzzy but higher-magnification videomicroscope image, in which researchers actually watched vesicles move.

microscope. As **Figure 7.31** shows, this technique allowed them to document that vesicle transport occurred along a filamentous track. A simple experiment convinced the group that this movement is an energy-dependent process: If they depleted the amount of ATP in the cytoplasm, vesicle transport stopped.

To identify the filament involved, the biologists measured the diameter of the tracks and analyzed their chemical composition. Both types of data indicated that the tracks consist of microtubules. Microtubules also appear to be required for movement of

materials elsewhere in the cell. For instance, if experimental cells are treated with a drug that disrupts microtubules, the movement of vesicles from the rough ER to the Golgi apparatus is impaired.

The general message of these experiments is that transport vesicles move through the cell along microtubules. How? Do the tracks themselves move, like a conveyer belt, or are vesicles carried along on some sort of molecular truck?

A MOTOR PROTEIN GENERATES MOTILE FORCES To study the way vesicles move along microtubules, Vale's group took the squid axon's transport system apart and put it back together.

To begin, they assembled microtubule fibers from purified α-tubulin and β-tubulin. Then they used differential centrifugation to isolate transport vesicles. But when they mixed purified microtubules and vesicles with ATP, no transport occurred. Something had been left out—but what?

To find the missing element or elements, the researchers purified one subcellular part after another, using differential centrifugation, and added it to the microtubule + vesicle + ATP system. Through trial and error, they found something that triggered movement. After further purification steps, the researchers finally succeeded in isolating a protein that generated vesicle movement. They named the molecule **kinesin**, from the Greek word *kinein* ("to move").

Like myosin, kinesin is a motor protein. Kinesin converts chemical energy in ATP into mechanical energy in the form of movement. More specifically, when ATP is added to kinesin or drops off, the protein moves.

Biologists began to understand how kinesin works when X-ray diffraction studies showed that it has three major regions: a head section with two globular pieces, a tail associated with small polypeptides, and a stalk that connects the head and tail (**Figure 7.32a**). Follow-up studies confirmed that the head region binds to the microtubule, while the tail region binds to the transport vesicle. Recent work has shown that kinesin "walks" along the microtubule when the head region binds to ATP (**Figure 7.32b**).

(a) Structure of kinesin

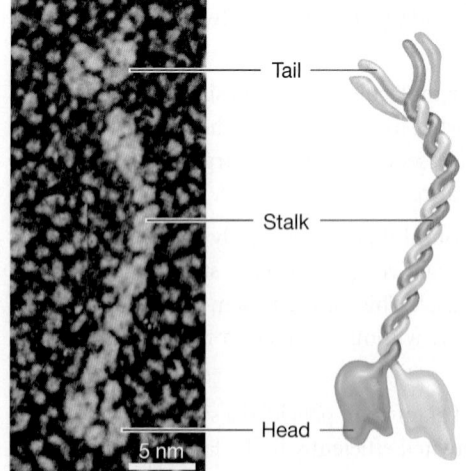

Tail

Stalk

Head

5 nm

(b) Kinesin "walks" along a microtubule track.

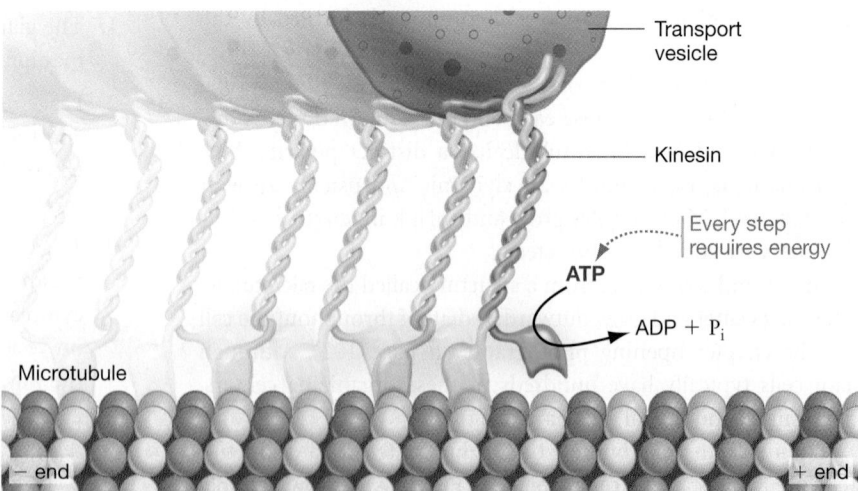

Transport vesicle

Kinesin

Every step requires energy

ATP

ADP + P$_i$

Microtubule

− end

+ end

FIGURE 7.32 A Motor Protein Moves Vesicles along Microtubules. **(a)** Kinesin has three major segments. **(b)** The current model depicting how kinesin "walks" along a microtubule track to transport vesicles. The two head segments act like feet that alternately attach and release in response to the gain or loss of a phosphate group.

A kinesin molecule is like a delivery truck that carries transport vesicles along microtubule tracks. Cells contain several different kinesin proteins, each specialized for carrying a different type of vesicle. In this way, kinesins move molecular cargo to destinations throughout the cell. How do microtubules move entire cells?

Flagella and Cilia: Moving the Entire Cell

Flagella are long, hairlike projections from the cell surface that function in movement. While many bacteria and eukaryotes have flagella, the structure is completely different in the two groups.

- Bacterial flagella are made of a protein called flagellin; eukaryotic flagella are constructed from microtubules (tubulin).

- Bacterial flagella move the cell by rotating like a ship's propeller; eukaryotic flagella move the cell by undulating—they whip back and forth.

- Eukaryotic flagella are surrounded by plasma membrane and are considered organelles; bacterial flagella are not.

Based on these observations, biologists conclude that the two structures evolved independently, even though their function is similar.

To understand how cells move, let's focus on eukaryotic flagella. Eukaryotic flagella are closely related to structures called **cilia** (singular: **cilium**), which are short, filament-like projections that are also found in some eukaryotic cells (**Figure 7.33**). Flagella are generally longer than cilia, and a cell will typically have just one or two flagella but many cilia. But when researchers examined the two structures with the electron microscope, they found that their underlying organization is identical.

HOW ARE CILIA AND FLAGELLA CONSTRUCTED? In the 1950s, anatomical studies established that most cilia and flagella have a characteristic "9 + 2" arrangement of microtubules. As **Figure 7.34a** shows, nine microtubule pairs, or doublets, surround two central microtubules. The doublets consist of one complete and one incomplete microtubule and are arranged around the periphery of the structure.

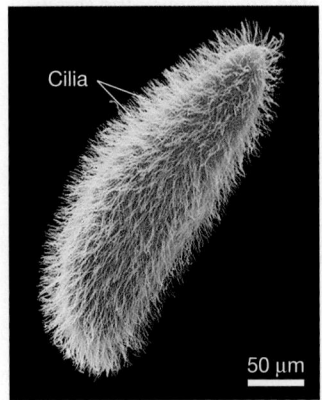
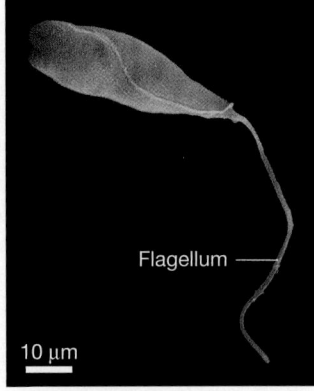

FIGURE 7.33 Cilia and Flagella Differ in Length and Number. The cells in these scanning electron micrographs have been colorized.

(a) Transmission electron micrograph of axoneme

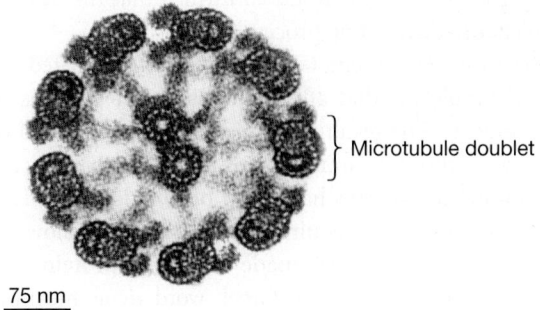

75 nm

(b) Diagram of axoneme

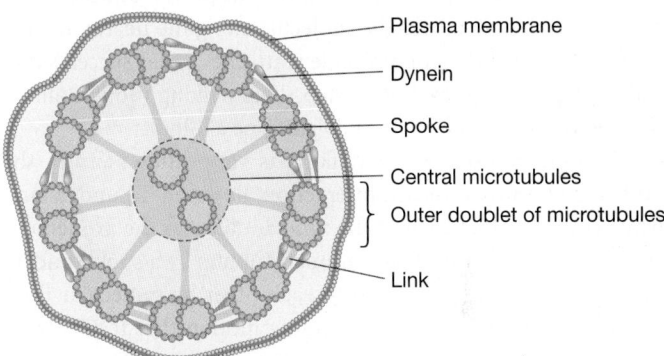

- Plasma membrane
- Dynein
- Spoke
- Central microtubules
- Outer doublet of microtubules
- Link

FIGURE 7.34 The Structure of Cilia and Flagella. (a) Transmission electron micrograph of a cross section through an axoneme. **(b)** The major structural elements in cilia and flagella. The microtubules are connected by links and spokes, and the entire structure is surrounded by the plasma membrane.

The entire 9 + 2 structure is called the **axoneme** ("axle-thread"). The axoneme attaches to the cell at a structure called the **basal body**. The basal body is identical in structure to a centriole and plays a central role in the growth of the axoneme.

As electron microscopy improved, biologists gained a more detailed view of the structure. Spoke-like structures connect each doublet to the central pair of microtubules, and molecular links connect the nine doublets to one another (**Figure 7.34b**). Each doublet has a set of arms that project toward an adjacent doublet.

Microtubules are complex. How do their components interact to generate motion?

WHAT PROVIDES THE FORCE REQUIRED FOR MOVEMENT? In the 1960s Ian Gibbons began studying the cilia of a common unicellular eukaryote called *Tetrahymena*, which lives in pond water. Gibbons found that by using a detergent to remove the plasma membrane that surrounds cilia and then subjecting the resulting solution to differential centrifugation, he could isolate axonemes.

These steps gave Gibbons a cell-free system for studying how cilia and flagella work. Cell-free systems are elegant to study because they are isolated—they are relatively easy to manipulate, and no other cell components are present that might confuse experimental results.

Gibbons found that the isolated structures would beat if he supplied them with ATP. This result confirmed that the beating of cilia is an energy-demanding process.

In another early experiment, Gibbons treated the isolated axonemes with a molecule that affects the ability of proteins to bind to one another. The resulting axonemes could not bend or use ATP. When Gibbons examined them in the electron microscope, he found that the arms had fallen off the doublets. This suggested that the arms are required for movement. Follow-up work showed that the arms are made of a large protein that Gibbons named **dynein** (from the Greek word *dyne*, meaning "force").

Like myosin and kinesin, dynein is a motor protein. Dynein changes shape when a phosphate group from ATP attaches to it. This shape change moves the molecule along the nearby microtubule. When the dynein molecule reattaches, it has succeeded in walking up the microtubule a step. This walking motion allows the microtubule doublets to slide past one another.

The outcome of dynein walking is very different from the outcome of a myosin-actin interaction, however. To understand why, remember that each of the nine doublets in the axoneme is connected to the central pair of microtubules by a spoke, and all of the doublets are connected to each other by molecular links (Figure 7.34b). As a result, the sliding motion that results from dynein walking is constrained—if one doublet slides, it transmits force to the rest of the axoneme via the links and spokes (**Figure 7.35**).

If the dynein arms on just one side of the axoneme walk and cause some of the doublets to slide while those on the other side are at rest, this constrained, localized movement results in bending. The bending of cilia or flagella results in a swimming motion. ✔If you understand this concept, you should be able to describe what would happen to the axoneme in Figure 7.35 if (1) the microtubule doublets were not connected by spokes and links, and (2) no ATP molecules were present.

Scaled for size, flagellar-powered swimming can be rapid. In terms of the number of body or cell lengths traveled per second, a sperm cell from a bull moves faster than a human world-record-holder does when swimming freestyle. At the level of the cell, life is fast paced.

🔑 Taken together, the data reviewed in this chapter can be summed up in six words: Cells are dynamic, highly integrated structures. Chemical reactions take place at mind-boggling speeds. Actin filaments and microtubules grow and shrink. The endomembrane system synthesizes, sorts, and ships an array of products in a highly regulated manner. How does all this activity inside the cell relate to what is going on outside? This is the issue we'll take up in Chapter 8.

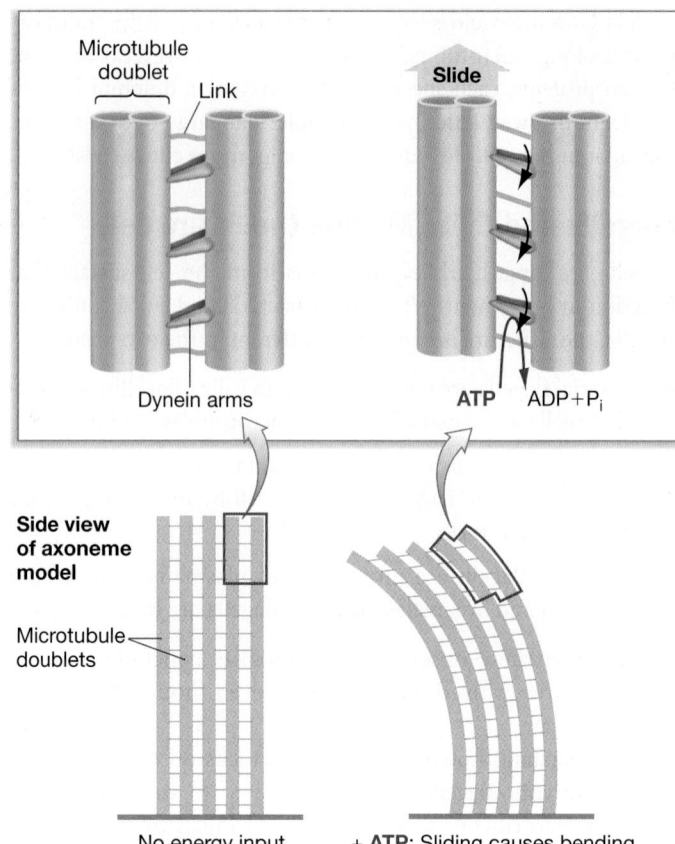

FIGURE 7.35 How Do Flagella Bend? When dynein arms walk along the microtubule doublets on one side of a flagellum, the structure bends.

CHECK YOUR UNDERSTANDING

🔑 **If you understand that . . .**

- Each component of the cytoskeleton has a unique structure and set of functions. In addition to providing structural support, actin filaments and microtubules work in conjunction with motor proteins to move the cell or materials inside the cell. Intermediate filaments provide structural support.
- Most elements of the cytoskeleton are dynamic—they grow and shrink over time.

✔ **You should be able to . . .**

Predict what will happen to endomembrane system products when experimental cells are treated with drugs that inhibit formation of microtubules.

Answers are available in Appendix B.

Summary of Key Concepts

🔑 **The structure of individual cell components is closely correlated with their function. The overall shape and composition of a cell is also closely related to its function.**

- There are two basic cellular designs: prokaryotic and eukaryotic.

- Eukaryotic cells are usually much larger and more structurally complex than prokaryotic cells.

- Most prokaryotic cells consist of a single membrane-bound compartment in which nearly all cellular functions occur.

- Eukaryotic cells contain numerous membrane-bound compartments called organelles. Organelles allow eukaryotic cells to compartmentalize functions and grow to a large size.

- Eukaryotic organelles are specialized for carrying out different functions, and their structure often correlates closely to their function.

- The defining organelle of eukaryotic cells is the nucleus, which contains the cell's chromosomes and serves as its control center.

- Rough ER is named for the ribosomes that attach to it. Ribosomes are protein-making machines, and rough ER is a site for protein synthesis and processing.

- Smooth ER lacks ribosomes because it is a center for lipid synthesis and processing.

- Mitochondria and chloroplasts have extensive internal membrane systems where the enzymes responsible for ATP generation and photosynthesis reside.

✔ You should be able to predict what would happen to cells that are exposed to (1) a drug that poisons mitochondria, (2) an enzyme that degrades the cell wall, or (3) a drug that prevents ribosomes from functioning.

(MB) **BioFlix™** Tour of an Animal Cell, **BioFlix™** Tour of a Plant Cell

🔑 **Inside cells, materials are transported to their destinations with the help of molecular "zip codes."**

- Cells have sophisticated systems for making sure that proteins and other products end up in the right place.

- Traffic across the nuclear envelope occurs through nuclear pores, which contain a multiprotein nuclear pore complex that serves as gatekeeper. Proteins can't pass through the pore and enter the nucleus unless they contain a specific molecular signal.

- Materials move through the endomembrane system inside vesicles. Before products leave the system, they are sorted with molecular "zip codes" that direct them to specific vesicles. The vesicles interact with receptor proteins at the target location so that the contents are delivered correctly.

✔ You should be able to explain, using concepts in Chapter 3, why proteins—and not RNA, DNA, carbohydrates, or lipids—are the molecules responsible for "reading" the array of molecular zip codes in cells.

(MB) **Web Activity** Transport into the Nucleus, **Web Activity** A Pulse-Chase Experiment

🔑 **The cytoskeleton provides a structural framework inside cells and plays a key role in cell division, movement, and transport.**

- The cytoskeleton is an extensive system of fibers that provides:

  (1) structural support and a framework for arranging and organizing organelles and other cell components;

  (2) paths for moving vesicles inside cells; and

  (3) machinery for moving the cell as a whole through the beating of flagella or cilia, or cell crawling.

- Movement often depends on motor proteins, which use chemical energy stored in ATP to change shape or position.

- Myosin causes actin filaments to slide, making cell division and cytoplasmic streaming possible in eukaryotes.

- Kinesin "walks" transport vesicles along microtubule tracks.

- Dynein "walks" along microtubule doublets, bending cilia and flagella so that they beat back and forth, enabling cells to swim or generate water currents.

✔ You should be able to compare and contrast the structure and function of actin filaments, intermediate filaments, and microtubules.

🔑 **Cells are dynamic, highly integrated structures. Thousands of chemical reactions occur each second within cells; molecules constantly enter and exit across the plasma membrane; cell products are shipped along protein fibers; and elements of the cell's internal skeleton grow and shrink.**

- Cells are bound by a highly selective membrane and have a tightly organized, dynamic interior.

- Cytoskeletal elements hold the nucleus, components of the endomembrane system, and other organelles in position.

- Materials are transported from one cell compartment to another along well-defined microtubule "tracks."

- Subunits are constantly being added to or removed from actin filaments and microtubules.

✔ You should be able to explain how pulse-chase experiments allowed researchers to study cells as dynamic enterprises—looking specifically at how materials move inside cells.

Questions

1. Which of the following best describes the nuclear envelope?
   a. It is continuous with the endomembrane system.
   b. It is continuous with the nucleolus.
   c. It is continuous with the plasma membrane.
   d. It contains a single membrane and nuclear pores.

2. Why is "receptor-mediated endocytosis" an appropriate term?
   a. It is the first step in autophagy.
   b. It is the first step in phagocytosis.
   c. It starts when extracellular molecules bind to receptors.
   d. It is the first step in formation of a lysosome.

3. Which of the following is *not* true of secreted proteins?
   a. They are synthesized in ribosomes.
   b. They are transported through the endomembrane system in membrane-bound transport organelles.
   c. They are transported from the Golgi apparatus to the ER.
   d. They contain a signal sequence that directs them into the ER.

4. To find the nuclear localization signal in the protein nucleoplasmin, researchers separated the molecule's core and tail segments, labeled both with a radioactive atom, and injected them into the cytoplasm.

Why did the researchers conclude that the signal is in the tail region of the protein?
   a. The protein reassembled and folded into its normal shape spontaneously.
   b. Only the tail segments appeared in the nucleus.
   c. With a confocal microscope, tail segments were clearly visible in the nucleus.
   d. The tail and head segments both appeared in the nucleus.

5. Molecular zip codes direct molecules to particular destinations in the cell. How are these signals read?
   a. They bind to receptor proteins.
   b. They enter transport vesicles.
   c. They bind to motor proteins.
   d. They are glycosylated by enzymes in the Golgi apparatus.

6. What does a motor protein do?
   a. causes actin filaments and/or microtubules to treadmill
   b. catalyzes endergonic reactions
   c. triggers receptor-mediated endocytosis
   d. changes shape in a way that moves another cell structure

1. Compare and contrast the structure of a generalized plant cell, animal cell, and prokaryotic cell. Which features are common to all cells? Which are specific to just prokaryotes, or just plants, or just animals?

2. Make a flowchart that traces the movement of a secreted protein from its site of synthesis to the outside of a eukaryotic cell. Identify all of the organelles that the protein passes through. Add notes indicating what happens to the protein at each step.

3. Describe how the motor protein kinesin can move a transport vesicle down a microtubule track. Explain why the movement requires ATP.

4. Actin filaments and microtubules may treadmill and/or undergo dramatic changes in length. How does this observation relate to the claim that the cytoskeleton is dynamic?

5. Explain why cells that are responsible for acquiring ions and other nutrients via specialized membrane proteins would be expected to have a large amount of rough ER.

6. Structurally, what is the difference between microtubules involved in structural support, vesicle movement, and flagella?

1. In addition to delivering cellular products to specific organelles, eukaryotic cells can take up material from the outside and route it to specific organelles. For example, specialized cells of the human immune system ingest bacteria and viruses and then deliver them to lysosomes for degradation. Suggest a hypothesis for how this material is tagged and directed to lysosomes. How would you test this hypothesis?

2. The enzymes found in peroxisomes are synthesized by ribosomes in the cytosol. Suggest a hypothesis for how the finished proteins find their way to the peroxisomes.

3. Propose a function for cells that contain (a) a large number of lysosomes, (b) a particularly extensive cell wall, and (c) many peroxisomes.

4. Suggest a hypothesis to explain why archaea have cell walls. Suppose that you have mutant archaeal cells that lack a cell wall. How could you use these individuals to test your idea?

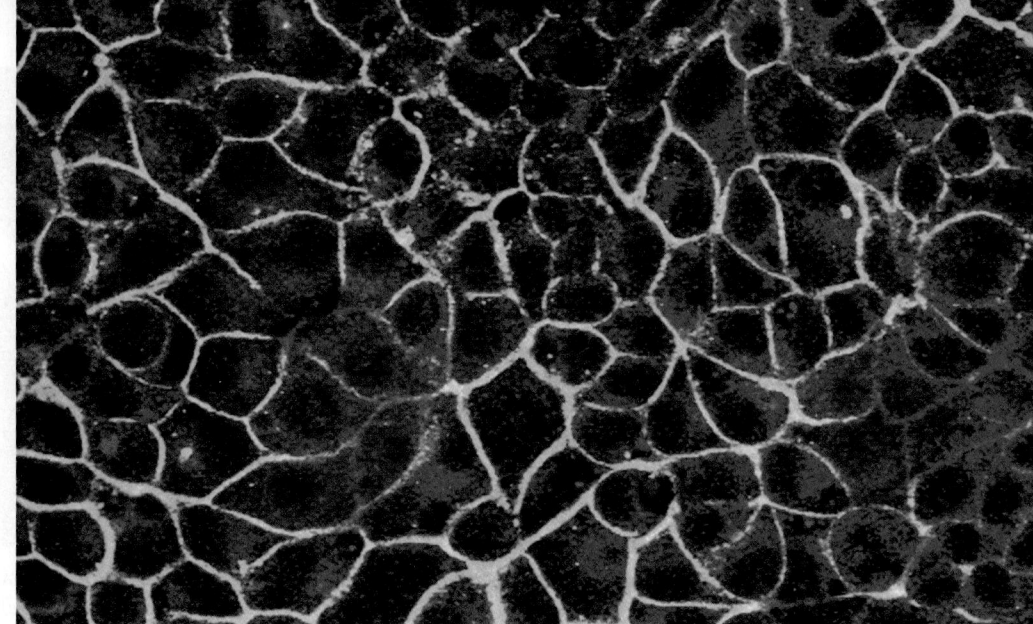

In this micrograph of skin cells, a green dye highlights proteins that help adjacent cells bind to each other. This chapter focuses on how cells in multicellular organisms stick together and communicate.

Cell-Cell Interactions 8

Chapter 6 introduced the structure and function of the plasma membrane—the defining feature of the cell. Chapter 7 surveyed the organelles, molecular machines, and cytoskeletal elements that fill the space inside that membrane and explored how cargo moves from sources to destinations within the cell. Both chapters highlighted the breathtaking speed and diversity of events that take place at the cellular level.

The cell is clearly a bustling enterprise. But it would be a mistake to think that cells are self-contained—that they are worlds in and of themselves. Instead, cells interact with other cells and the surrounding environment.

For most unicellular species, the outside environment is teeming with other organisms. Inside your gut, for example, hundreds of billions of bacterial cells are jostling for space and resources. In addition to interacting with these individuals, every unicellular organism must contend with constant shifts in the physical environment, such as heat, light, ion concentrations, and food supplies. If unicellular organisms are unable to sense these conditions and respond appropriately, they die.

In multicellular species, the environment outside the cell is made up of other cells, both neighboring and distant. The cells that make up a redwood tree, an *Amanita* mushroom, or your body are intensely social. Although biologists often study cells in isolation, an individual tree, fungus, or person is actually an interdependent community of cells. If those cells do not communicate and cooperate, the whole will break into dysfunctional parts and die.

To understand the life of a cell, then, it is critical to analyze how the cell interacts with the world outside its membrane. How do cells obtain information about the world and respond to that information? In particular, how do cells interact with other cells? To answer these questions, let's begin with the cell surface—with the molecules that separate the cell from its environment.

KEY CONCEPTS

- ⊙⇝ Extracellular material strengthens cells and helps bind them together.

- ⊙⇝ Cell-cell connections help adjacent cells adhere. Cell-cell gaps allow adjacent cells to communicate.

- ⊙⇝ Intercellular signals are responsible for creating an integrated whole from many thousands of independent parts.

- ⊙⇝ In target cells, intercellular signals are received, processed, responded to, and deactivated. If the signal is received at the cell surface, the processing step involves production of an intracellular signal.

✔ When you see this checkmark, stop and test yourself. Answers are available in Appendix B.

8.1 The Cell Surface

Chapter 6 introduced the fluid-mosaic hypothesis, the currently accepted model for the structure of the plasma membrane. Recall that the phospholipid bilayer is studded with membrane proteins that are integral, meaning that they are embedded in the bilayer, or peripheral, meaning that they are attached to one surface.

Chapter 6 also analyzed the primary function of the plasma membrane: to create an environment inside the cell that is different from conditions outside. Ions and molecules move across plasma membranes by direct diffusion through the phospholipid bilayer or via several types of membrane proteins. Transport of materials across the membrane can be energy demanding or passive, but it is always selective.

This picture—of a dynamic, complex plasma membrane that selectively admits or blocks passage of specific substances—is accurate but not complete. The plasma membrane does not exist in isolation. Cytoskeletal elements introduced in Chapter 7 attach to the interior face of the bilayer, and a complex array of extracellular structures exists outside. Let's consider the nature of the material outside the cell and then analyze how the cell interacts with other cells.

The Structure and Function of an Extracellular Layer

It is actually extremely rare for cells to be bounded simply by a plasma membrane. Most cells possess a layer or wall that forms just beyond the membrane. The extracellular material helps define the cell's shape and either attaches it to another cell or acts as a first line of defense against the outside world.

Virtually all types of extracellular structures, in turn—from the cell walls of bacteria, algae, fungi, and plants to the extracellular matrix that surrounds most animal cells—follow the same fundamental design principle. Like reinforced concrete and fiberglass, they are "fiber composites": They consist of a cross-linked network of long filaments embedded in a stiff surrounding material, or ground substance (**Figure 8.1**). The molecules that make up the filaments and the encasing material vary from group to group, but the engineering principle is the same. Why?

- The rods or filaments in a fiber composite are extremely effective at withstanding stretching and straining forces, or tension. The steel rods in reinforced concrete and the cellulose fibers in a plant cell wall are unlikely to break as a result of being pulled or pushed lengthwise.

- The stiff surrounding substance is effective at withstanding the pressing forces called compression. Concrete performs this function in highways, and a gel-forming mixture of polysaccharides achieves the same end in plant cell walls.

Thanks to the combination of tension- and compression-resisting elements, fiber composites are particularly rugged. And in many living cells, the fiber and composite elements are flexible as well as strong.

What molecules make up the rods and ground substance found on the surface of plant and animal cells? How are these extracellular layers synthesized, and what do they do?

FIGURE 8.1 Fiber Composites Resist Tension and Compression. Fiber composites such as reinforced concrete consist of a massive ground substance that fills spaces between cross-linked rods.

The Cell Wall in Plants

Virtually all plant cells are surrounded by a cell wall—a fiber composite that is the basis of major industries. The paper in this book, the threads in your cotton clothing, and the wood in your neighborhood's houses are made up primarily of plant cell walls.

Before analyzing the structure of plant cell walls in detail, it's important to note that these structures are dynamic. If they are damaged by attacking insects, they may release signaling molecules that trigger the reinforcement of walls in nearby cells. When a seed germinates (sprouts), the walls of cells that store oils or starch are actively broken down, releasing these nutrients to the growing plant. Cell walls are also degraded in a controlled way as fruits ripen, making the fruits softer and more digestible for the animals that disperse the seeds inside.

PRIMARY CELL WALLS When plant cells first form, they secrete an initial fiber composite designated a **primary cell wall**.

- The fibrous component of the primary cell wall consists of long strands of cellulose, which are cross-linked by other polysaccharide filaments and bundled into stout, cable-like structures termed microfibrils. The microfibrils are synthesized by a complex of enzymes in the plasma membrane, forming a crisscrossed network (**Figure 8.2**).

- The space between microfibrils is filled with gelatinous polysaccharides like **pectins**—the molecules that are used to thicken jams and jellies. Because the polysaccharides in pectin are hydrophilic, they attract and hold large amounts of water to keep the cell wall moist. The gelatinous components of the cell wall are synthesized in the rough endoplasmic reticulum and Golgi apparatus and secreted to the extracellular space.

The primary cell wall defines the shape of a plant cell. Under normal conditions, the nucleus and cytoplasm fill the entire volume of the cell and push the plasma membrane up against the wall. Because the concentration of solutes is higher inside the cell than outside, water tends to enter the cell via osmosis. The in-

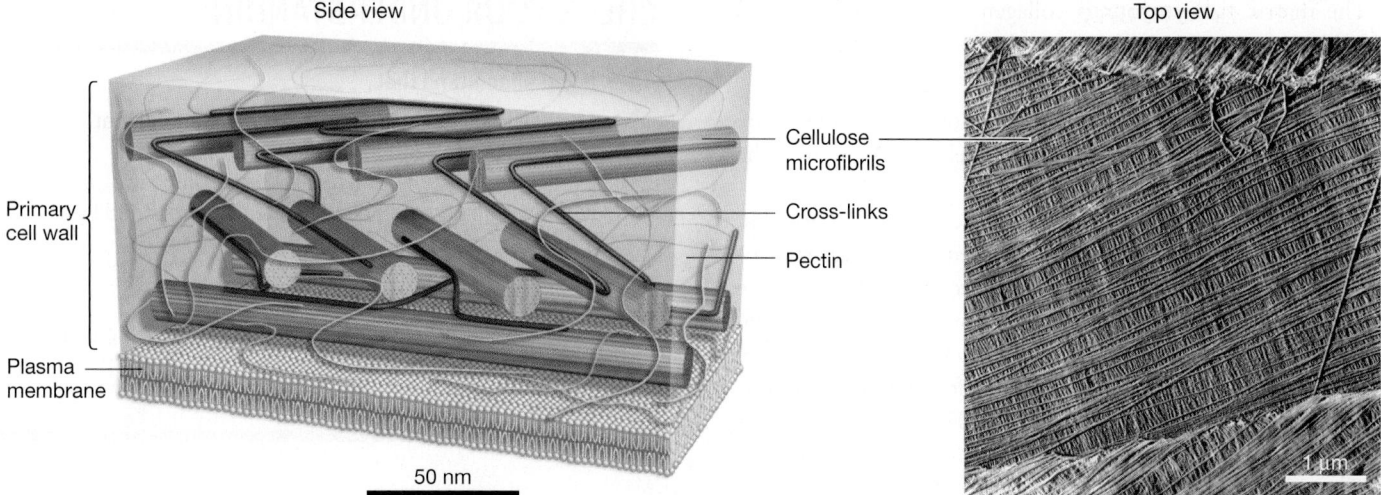

Side view | Top view

Primary cell wall
Plasma membrane

Cellulose microfibrils
Cross-links
Pectin

50 nm

1 μm

FIGURE 8.2 Primary Cell Walls of Plants Are Fiber Composites. In a plant's primary cell wall, cellulose microfibrils are cross-linked by polysaccharide chains. The spaces between the microfibrils are filled with pectin molecules, which form a gelatinous solid.

coming water inflates the plasma membrane, exerting a force against the wall that is known as **turgor pressure**.

Although plant cells experience turgor pressure throughout their lives, it is particularly important in young cells that are actively growing. Young plant cells secrete enzymes named expansins into their cell-wall matrix. Expansins catalyze reactions that allow the microfibrils in the matrix to slide past one another. Turgor pressure then forces the wall to elongate and expand. The result is cell growth.

SECONDARY CELL WALLS As plant cells mature and stop growing, they may secrete a layer of material—a **secondary cell wall**—inside the primary cell wall. The structure of the secondary cell wall varies from cell to cell in the plant and correlates with that cell's function. Cells on the surface of a leaf have cell walls that are impregnated with waxes that form a waterproof coating; while cells that support the plant's stem have secondary cell walls that contain a great deal of cellulose.

In cells that form wood, the secondary cell wall includes **lignin**, a tough substance that forms an exceptionally rigid network. Cells that have thick secondary cell walls of cellulose and lignin help plants withstand the forces of gravity and wind.

Although animal cells do not make a cell wall, they do form a fiber composite outside their plasma membrane. What is this substance, and what does it do?

The Extracellular Matrix in Animals

Most animal cells secrete a fiber composite called the **extracellular matrix (ECM).** Like the extracellular materials found in other organisms, one of the ECM's most important functions is structural support.

ECM design follows the same principles observed in the cell walls of bacteria, archaea, algae, fungi, and plants. There is a key difference, however: The animal ECM contains much more protein than a cell wall.

● The fibrous component of animal ECM is dominated by a cable-like protein termed **collagen** (**Figure 8.3a**).

(a) Collagen proteins consist of three polypeptide chains that wind around each other.

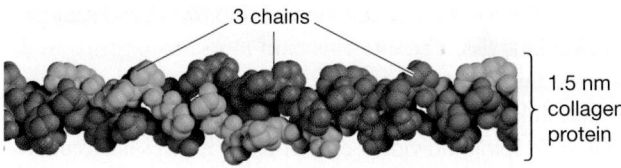

3 chains

1.5 nm collagen protein

(b) ECM containing many collagen fibrils, each made up of many collagen proteins

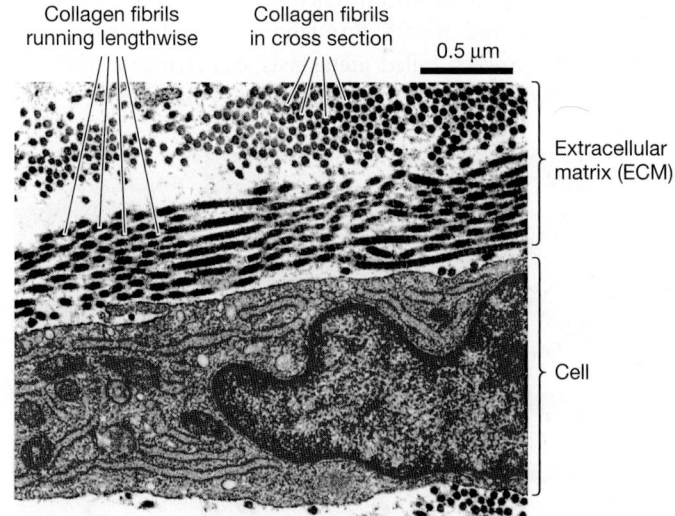

Collagen fibrils running lengthwise
Collagen fibrils in cross section
0.5 μm

Extracellular matrix (ECM)

Cell

FIGURE 8.3 The Extracellular Matrix Is a Fiber Composite.
(a) Although several types of fibrous proteins are found in the ECM of animal cells, the most abundant is collagen. Groups of collagen proteins coalesce to form collagen microfibrils, and bundles of microfibrils link to form collagen fibers. **(b)** A cross section of ECM from monkey cartilage tissue, showing a cell surrounded by abundant ECM. The spaces between the collagen fibers are filled with gelatinous polysaccharides.

- The matrix that surrounds collagen and the other fibrous components consists of gel-forming polysaccharides—most of which are attached to a protein core.

Because collagen and the other common ECM proteins are much more elastic and bendable than cellulose or lignin, the structure as a whole is relatively pliable.

Most ECM components are synthesized in the rough ER, processed in the Golgi apparatus, and secreted from the cell via exocytosis. Some of the protein-bound polysaccharides that form the composite material are synthesized by membrane proteins, however.

Even in the same organism, the amount of ECM varies among different types of cells. Bone and cartilage, for example, have relatively few cells but a large amount of ECM (**Figure 8.3b**). Skin cells, in contrast, are packed together with a minimal amount of ECM.

The composition of the ECM also varies among cell types. For example, the ECM of lung cells contains large amounts of a rubber-like protein called elastin, which allows the ECM to expand and contract during breathing. The structure of a cell's ECM correlates with that cell's function.

Although an ECM is not as stiff as a cell wall, it is strengthened by connections to transmembrane proteins. As **Figure 8.4** shows, actin filaments in the cytoskeleton are connected to transmembrane proteins called **integrins**. The integrins bind to nearby proteins in the ECM, including **fibronectins**, which in turn bind to collagen fibers. This direct linkage between the cytoskeleton and ECM is critical. In addition to keeping individual cells in place, it helps adjacent cells adhere to each other via their common connection to the ECM.

If this cytoskeleton-ECM linkage breaks down, cancer can develop. Through mechanisms that are not well understood, cells that are growing in an uncontrolled fashion, forming a tumor, can break away and migrate throughout the body, seeding new tumors. This process, called **metastasis**, can change a relatively benign and localized tumor into a life-threatening cancer.

8.2 How Do Adjacent Cells Connect and Communicate?

Although unicellular species may live in close proximity, they usually do not make direct physical connections. Physical connections between cells are the basis of **multicellularity** (**Figure 8.5**). Multicellular organisms are made up of cells that adhere to each other and have distinct structures and functions. The muscle cells in your body, for example, have distinctive shapes and are packed with proteins that are only found in muscle cells.

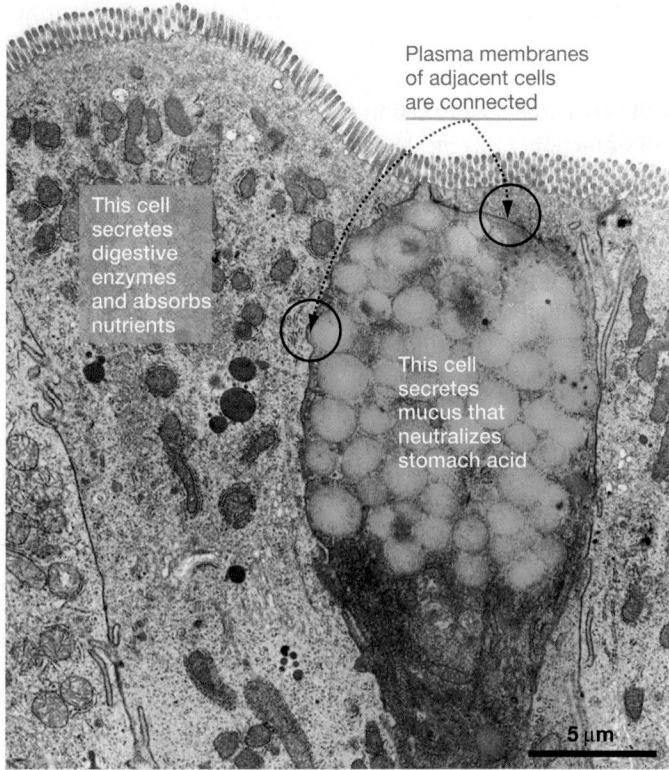

Plasma membranes of adjacent cells are connected

This cell secretes digestive enzymes and absorbs nutrients

This cell secretes mucus that neutralizes stomach acid

5 μm

FIGURE 8.5 In Multicellular Organisms, Cells Are Connected.
A cross section through cells that line the small intestine of a mammal. This transmission electron micrograph has been colorized to make the contrasting cell types more visible.

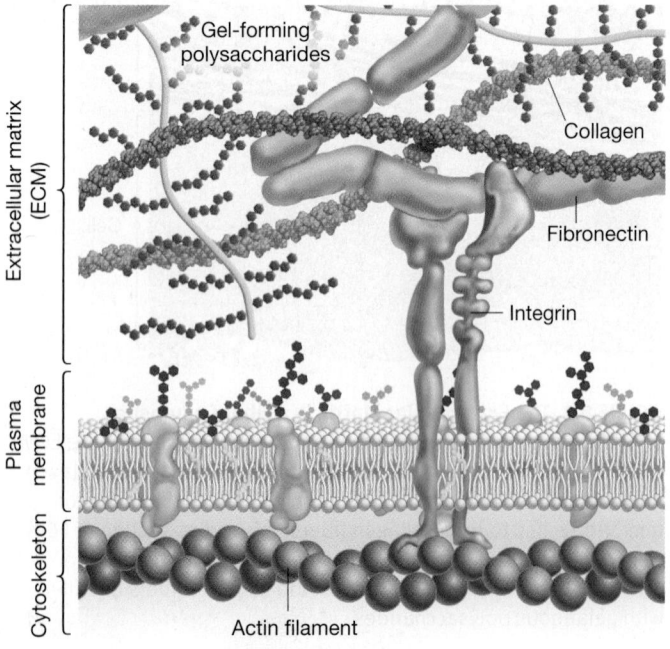

Extracellular matrix (ECM)

Gel-forming polysaccharides

Collagen

Fibronectin

Integrin

Plasma membrane

Cytoskeleton

Actin filament

FIGURE 8.4 The Extracellular Matrix Connects to the Cytoskeleton.

Most multicellular organisms also have **tissues**: groups of similar cells that perform a similar function. The individual muscle cells in your body are grouped into muscle tissue that contracts and relaxes to make movement possible.

Let's look first at the structures that attach adjacent cells to each other, then examine the openings that allow nearby cells to exchange materials and information.

Cell-Cell Attachments in Eukaryotes

The structures that hold cells together vary among multicellular organisms. To illustrate this diversity, consider the intercellular connections observed in the best-studied groups of organisms: plants and animals.

The extracellular space between adjacent plant cells comprises three layers (**Figure 8.6**). The primary cell walls of adjacent plant cells sandwich a central layer designated the middle lamella, which consists primarily of gelatinous pectins. Because this gel layer is continuous with the primary cell walls of the adjacent cells, it serves to glue them together. The two cell walls are like slices of bread; the middle lamella is like a layer of peanut butter. If enzymes degrade the middle lamella, as they do when flower petals and leaves detach and fall, the surrounding cells separate.

In many animal tissues, integrins connect the cytoskeleton of each cell to the extracellular matrix (see Section 8.1). A middle-lamella-like layer of gelatinous polysaccharides runs between adjacent animal cells, so cytoskeleton-ECM connections help hold individual animal cells together. In addition, in certain animal tissues the polysaccharide glue is reinforced by cable-like proteins that span the ECM to connect adjacent cells.

Materials and structures that bind cells together are particularly important in **epithelia** (singular: **epithelium**)—tissues that form external and internal surfaces. Epithelial cells form layers that separate organs and other structures. These epithelial layers must be sealed to prevent mixing of solutions from adjacent organs or structures.

A variety of cell-cell attachment structures exist in epithelia and other tissues. Here, however, we'll analyze just two: tight junctions and desmosomes.

TIGHT JUNCTIONS A **tight junction** is a cell-cell attachment composed of specialized proteins in the plasma membranes of adjacent animal cells (**Figure 8.7a**). As the drawing in **Figure 8.7b** indicates, these proteins line up and bind to one another. The resulting structure resembles quilting, with the proteins acting as stitches.

Because tight junctions form a watertight seal, this type of junction is commonly found in cells that form a barrier, such as the epithelial cells lining your stomach and intestines. There, they prevent the ions and molecules in your gut contents from leaking between

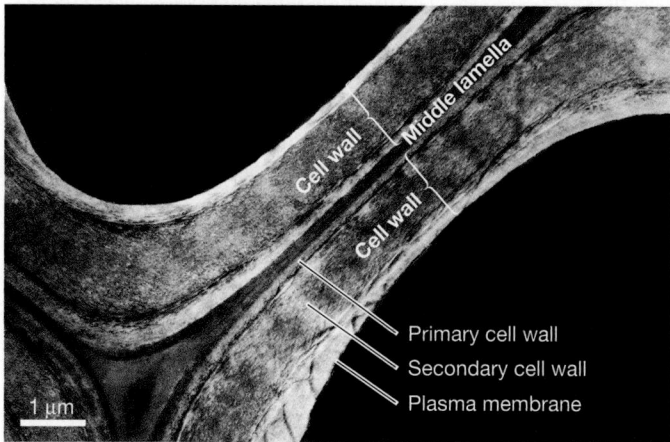

FIGURE 8.6 The Middle Lamella Connects Adjacent Plant Cells. The middle lamella contains gelatinous polysaccharides called pectins.

(a) Electron micrograph of a tight junction in longitudinal section

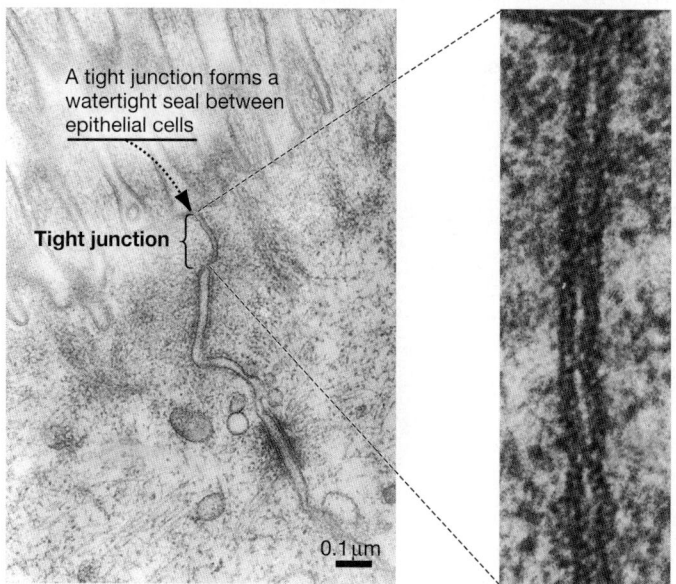

(b) Three-dimensional view of a tight junction

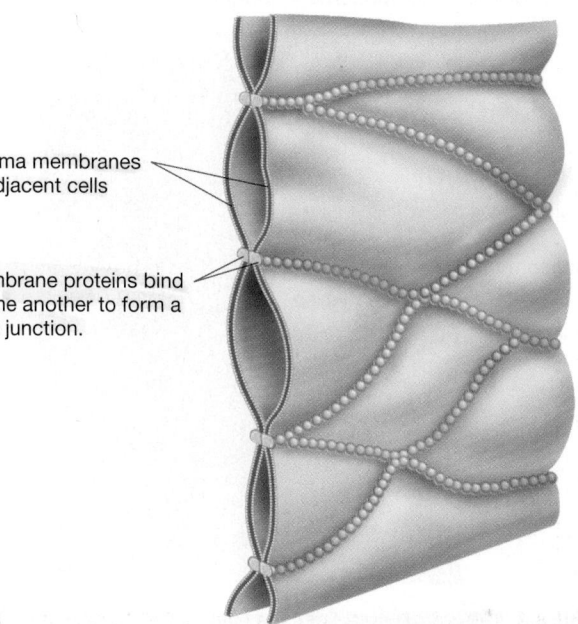

FIGURE 8.7 In Animals, Tight Junctions May Form a Seal between Adjacent Cells.

cells and diffusing into your body. Instead, only selected nutrients enter the cells. These ions and molecules are admitted via specialized transport proteins and channels in the plasma membrane.

Although tight junctions are indeed tight, they are variable. The tight junctions in the cells lining your bladder are much more snug than tight junctions in the cells lining your small intestine, because they consist of different proteins. As a result, small ions can pass through the surface of the small intestine much more easily than through the surface of the bladder—helping you absorb ions in your food and eliminate them in your waste.

Tight junctions are also dynamic. In some cases they may loosen to permit more transport between epithelial cells—for example, in the small intestine after a meal—and then "retighten" later. In this way, tight junctions can open and close in response to changes in environmental conditions.

DESMOSOMES **Figure 8.8a** illustrates **desmosomes**, cell-cell attachments particularly common in animal epithelial cells and certain muscle cells. The structure and function of a desmosome are analogous to the rivets that hold pieces of sheet metal together.

As **Figure 8.8b** indicates, desmosomes are extremely sophisticated cell-cell connections. At their heart are proteins that bind to each other and to larger proteins that anchor intermediate filaments in the cytoskeletons of the two cells. In this way, desmosomes bind together the cytoskeletons of adjacent cells.

What are these cell-attachment proteins? The answer to this question traces back to some of the first experiments conducted on cell-cell interactions.

SELECTIVE ADHESION Long before electron micrographs revealed the presence of desmosomes, biologists realized that some sort of molecule must bind animal cells to each other. This insight grew out of experiments conducted on sponges in the early 1900s.

Sponges are aquatic animals, and the sponge species used in this study consists of just two basic types of cells. When a biologist

treated adult sponges with chemicals that made the cells separate from each other, the result was a jumbled mass of individual and unconnected cells. But when normal chemical conditions were restored, the cells gradually began to move and stick to other cells.

As the experiment continued, cells of each type began to combine and aggregate because they adhered to cells of the same tissue type. This phenomenon is now called **selective adhesion**. Eventually the experimental sponge cells re-formed functional adult sponges.

In an even more dramatic experiment, the researcher dissociated the cells of adult sponges from two differently pigmented sponge species and randomly mixed them together in a culture dish. As **Figure 8.9** shows, the cells eventually sorted themselves into two distinct aggregates, each containing cells from only one species and only one cell type. How could this happen?

THE DISCOVERY OF CADHERINS What is the molecular basis of selective adhesion? The initial hypothesis, proposed in the 1970s, was that specialized membrane proteins were involved. The idea was that different types of cells produce different types of adhesion proteins in their membranes, and that the molecules interact in a way that anchors cells of the same type to each other.

This hypothesis was tested through experiments that relied on molecules called antibodies. An **antibody** is a protein that binds specifically to a section of another protein. Because they stick to certain proteins specifically, antibodies can be used to block certain portions of proteins or mark proteins so they can be seen (see **BioSkills 9** in Appendix A).

Figure 8.10 shows how researchers tested the hypothesis that cell-cell adhesion takes place via interactions between membrane proteins:

Step 1 Isolate the membrane proteins from a certain cell type. Produce pure preparations of each protein.

Step 2 Inject one of the membrane proteins into a rabbit. The rabbit's immune system cells respond by creating antibodies to the

(a) Micrograph of desmosome in longitudinal section

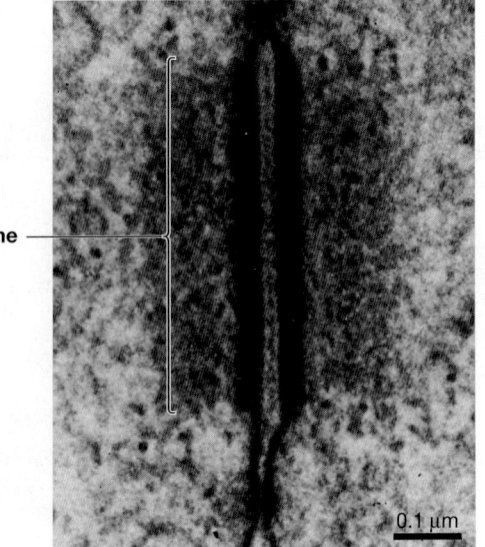

Desmosome

0.1 μm

(b) Three-dimensional view of desmosome

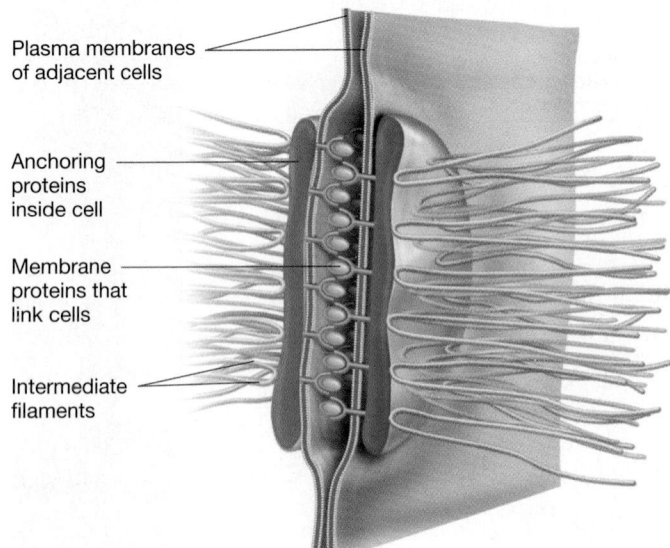

Plasma membranes of adjacent cells

Anchoring proteins inside cell

Membrane proteins that link cells

Intermediate filaments

FIGURE 8.8 Adjacent Animal Cells Are Linked by Desmosomes, Which Bind Cytoskeletons Together.

EXPERIMENT

EXPERIMENT

QUESTION: Do animal cells adhere selectively?

HYPOTHESIS: Cells of the same type and from the same species have a mechanism for selectively adhering to each other.

NULL HYPOTHESIS: Cells do not adhere or they adhere to each other randomly, not selectively.

EXPERIMENTAL SETUP:

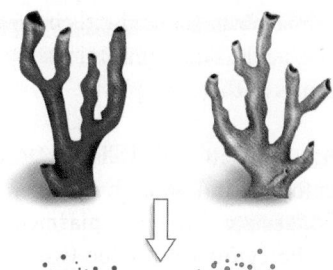

1. Start with two adult sponges of different species.

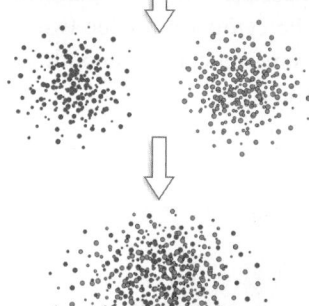

2. Use a chemical treatment to dissociate the cells.

3. Mix cells from the two species.

PREDICTION: Cells of the same cell type will adhere to one another, and cells from the same species will adhere to one another.

PREDICTION OF NULL HYPOTHESIS:

RESULTS:

Cells spontaneously reaggregate by cell type and species.

CONCLUSION: Cells of the same cell type and species have specific adhesion molecules or mechanisms.

FIGURE 8.9 Evidence for Selective Adhesion in Animal Cells.

SOURCE: Wilson, H. V. 1907. On some phenomena of coalescence and regeneration in sponges. *Journal of Experimental Zoology* 5: 245–258.

✔**EXERCISE** In the space provided, fill in the prediction made by the null hypothesis.

membrane protein. Purify those antibodies. Repeat this procedure for the other membrane proteins that were isolated. In this way, obtain a large collection of antibodies—each of which binds specifically to one (and only one) type of membrane protein.

Step 3 Add one antibody type to the mixture of dissociated cells and observe whether the cells reaggregate normally. Repeat this experiment with each of the other antibody types, one type at a time.

If treatment with a particular antibody prevents the cells from attaching to each other, the antibody is probably bound to an

EXPERIMENT

QUESTION: Do animal cells have adhesion proteins on their surfaces?

HYPOTHESIS: Selective adhesion is due to specific membrane proteins.

NULL HYPOTHESIS: Selective adhesion is not due to specific membrane proteins.

EXPERIMENTAL SETUP:

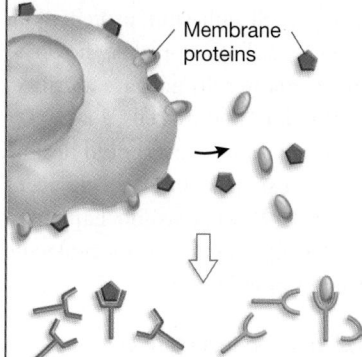

Membrane proteins

1. Isolate the membrane proteins from a certain cell type that adheres to other cells of the same type. (There are many membrane proteins; only two are shown here.)

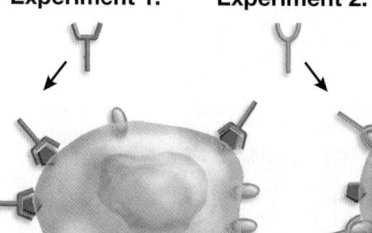

Antibodies

2. Produce antibodies that bind to specific membrane proteins. Purify the antibodies.

Experiment 1: **Experiment 2:**

3. Treat cells with an antibody, one type at a time. Wait; then observe whether cells adhere normally.

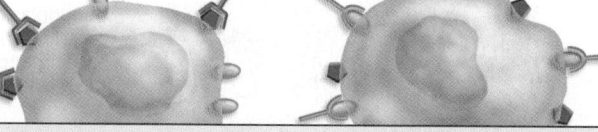

PREDICTION:

PREDICTION OF NULL HYPOTHESIS:

RESULTS:

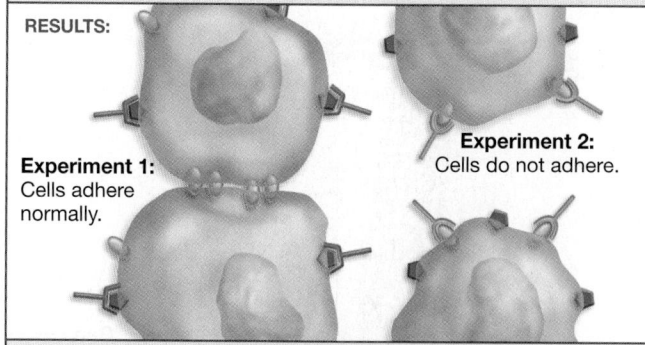

Experiment 1: Cells adhere normally.

Experiment 2: Cells do not adhere.

CONCLUSION: The protein that was blocked in experiment 2 (called a cadherin) is involved in cell-cell adhesion.

FIGURE 8.10 Evidence for Adhesion Proteins on Animal Cells.

SOURCES: Hatta, K. and M. Takeichi. 1986. Expression of N-cadherin adhesion molecules associated with early morphogenetic events in chick development. *Nature* 320: 447–449. Also Takeichi, M. 1988. The cadherins: cell-cell adhesion molecules controlling animal morphogenesis. *Development* 102: 639–655.

✔**EXERCISE** Fill in the prediction made by each hypothesis.

adhesion protein. The logic is that if the antibody "shakes hands" with the adhesion protein, the adhesion protein can't "shake hands" with other adhesion proteins and attach the cells to each other. This approach allowed biologists to identify several major classes of cell adhesion proteins, including **cadherins**—the attachment molecules in desmosomes. Different types of cells in the body have different forms of cadherin in their plasma membranes, and each cadherin can bind only to cadherins of the same type. In this way, cells of the same tissue type attach specifically to one another.

To summarize: Animal cells attach to each other in a selective manner because different types of cell adhesion proteins can bind and rivet certain cells together. Cadherins provide the physical basis for selective adhesion in many cells and are a critical component of the desmosomes that join mature cells.

✔If you understand cell-cell attachments, you should be able to predict: (1) whether molecules can pass between adjacent cells through middle lamellae, and (2) what would happen if you treated cells in a developing frog embryo with a molecule that blocked a cadherin observed in muscle tissue.

◖━ The presence of a middle lamella, continuous ECM, tight junctions, desmosomes, and cadherins bind adjacent cells to each other. But how do these cells communicate?

Cells Communicate via Cell-Cell Gaps

In both plants and animals, direct connections—holes or gaps—between cells in the same tissue help the cells work in a coordinated fashion. The presence of holes or gaps allows adjacent plant cells and adjacent animal cells to retain their own organelles, proteins, and nucleic acids, but communicate by sharing the ions and small molecules in their cytoplasm.

PLASMODESMATA In plants, gaps in cell walls create direct connections between the cytoplasm of adjacent cells. At these connections, named **plasmodesmata** (singular: **plasmodesma**), the plasma membrane and the cytoplasm of the two cells are continuous. Smooth endoplasmic reticulum (smooth ER) runs through these holes (**Figure 8.11a**).

(a) Plasmodesmata create gaps that connect plant cells.

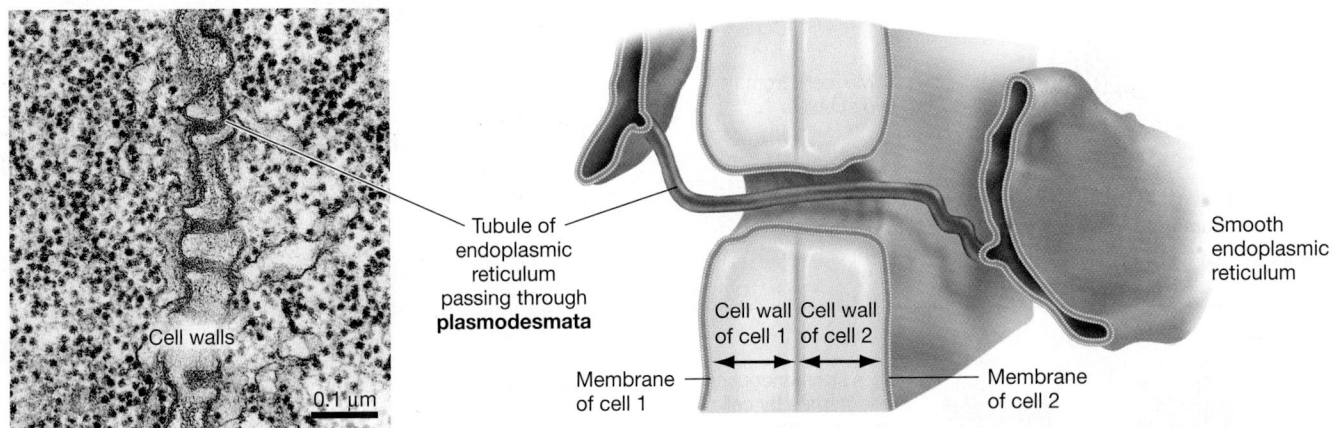

Tubule of endoplasmic reticulum passing through **plasmodesmata**

Cell walls

0.1 μm

Cell wall of cell 1 Cell wall of cell 2

Membrane of cell 1

Membrane of cell 2

Smooth endoplasmic reticulum

(b) Gap junctions create gaps that connect animal cells.

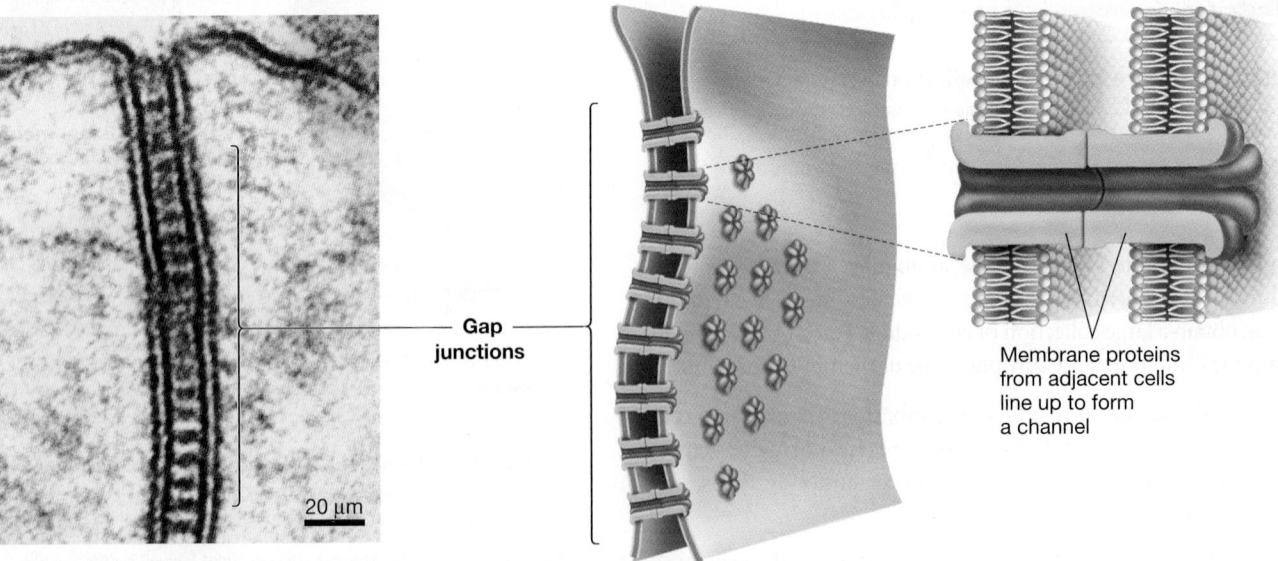

20 μm

Gap junctions

Membrane proteins from adjacent cells line up to form a channel

FIGURE 8.11 Adjacent Plant Cells and Adjacent Animal Cells Communicate Directly.

Growing evidence suggests that plasmodesmata also contain proteins that regulate the passage of specific proteins, making the connections similar in function to the nuclear pore complex introduced in Chapter 7. At least some of the proteins that are transported through plasmodesmata are involved in coordinating the activity of adjacent cells. Plasmodesmata are communication portals.

GAP JUNCTIONS   In most animal tissues, structures called **gap junctions** connect adjacent cells. The key feature of gap junctions is the specialized proteins that create channels between cells (**Figure 8.11b**). These channels allow water, ions, and small molecules such as amino acids, sugars, and nucleotides to move between adjacent cells.

The flow of small molecules through gap junctions can help adjacent cells coordinate their activities by allowing the rapid passage of regulatory ions or molecules. In the muscle cells of your heart, for example, a flow of ions through gap junctions acts as a signal that coordinates contractions. Without this cell-cell communication, a normal heartbeat would be impossible. A tissue can act as an integrated whole if its component cells are connected by gaps in their extracellular material and in their plasma membranes (**Figure 8.12**).

How do more distant cells in a multicellular organism communicate? For example, suppose that leaf cells in a maple tree are attacked by caterpillars or the muscle cells in your arm are exercising so hard that they run low on sugar. How do these cells signal tissues or organs elsewhere in the body to release materials that are needed to fend off caterpillars or exhaustion? Distant cell communication is the subject of Section 8.3.

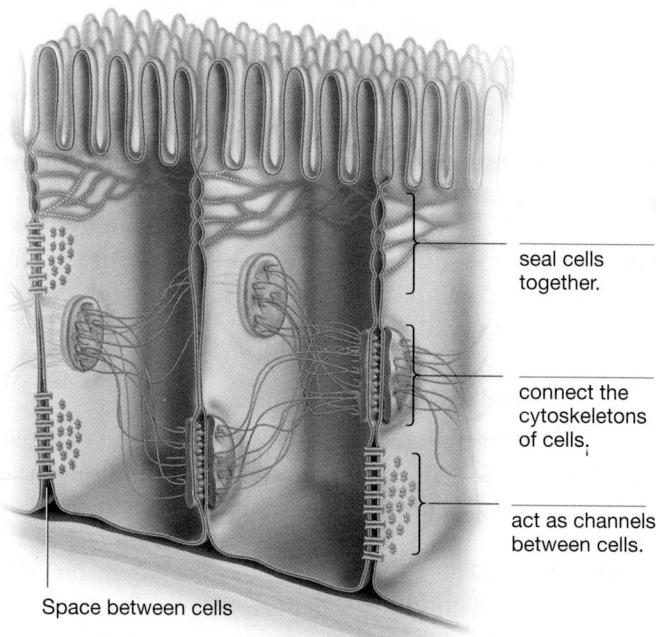

seal cells together.

connect the cytoskeletons of cells.

act as channels between cells.

Space between cells

FIGURE 8.12 An Array of Structures Is Involved in Cell-Cell Adhesion and Communication.

✔ **EXERCISE** In the blanks on the right, write the name of each type of cell-cell attachment.

8.3 How Do Distant Cells Communicate?

Cells that are not in physical contact communicate with each other. This is true for unicellular organisms, where hundreds or thousands of closely related cells may live in close proximity, as well as for multicellular organisms like humans and maple trees, which typically contain trillions of cells and dozens of tissue types.

Cell-cell communication qualifies as one of the most dynamic and important research areas in biology. Let's begin by analyzing how distant cells in humans and other multicellular eukaryotes exchange information, and close with a brief overview of how bacterial cells from the same species communicate with each other.

Cell-Cell Signaling in Multicellular Organisms

Suppose that cells in your brain sense that you are becoming dehydrated. Brain cells can't do much about the water you lose during urination, but kidney cells can. In response to dehydration, certain brain cells release a signaling molecule that travels to the kidneys. The arrival of the signal activates specialized membrane channels that prevent water from being lost in urine—an important aspect of fighting dehydration.

Thanks to cell-cell signals, the activities of cells in different parts of a multicellular body are coordinated.

Although biologists have classified many different types of signals that keep distant cells in touch, the best-studied may be **hormones**—information-carrying molecules that are secreted from a plant or animal cell, circulate in the body, and act on target cells far from the original cell that sent the signal.

Hormones are usually small molecules and are typically present in minute concentrations. Even so, they have a large impact on the activity of target cells and the condition of the body as a whole. Hormones are like a fleeting scent or whispered phrase from someone you are attracted to—a tiny signal, but one that makes your cheeks flush and your heart pound.

As **Table 8.1** indicates, hormones have a wide array of effects and chemical structures. The important point about a cell-cell signal, though, is not whether it is a gas or peptide or

TABLE 8.1 **Hormones Have Diverse Structures and Functions**

Hormone Name	Chemical Structure	Where is signal received?	Function of Signal
Auxin	Small organic compound	At plasma membrane	Signals changes in long axis of plant body
Brassinosteroids	Steroid	At plasma membrane	Stimulate plant cell elongation
Estrogens	Steroid	Inside cell	Stimulate development of female characteristics in animals
Ethylene	C_2H_4 (a gas)	At plasma membrane	Stimulates fruit ripening, regulates aging
FSH	Glycoprotein	At plasma membrane	Stimulates egg maturation, sperm production in animals
Insulin	Protein, 51 amino acids	At plasma membrane	Stimulates glucose uptake in animal bloodstream
Prostaglandins	Modified fatty acid	At plasma membrane	Perform a variety of functions in animal cells
Systemin	Peptide, 18 amino acids	At plasma membrane	Stimulates plant defenses against herbivores
Thyroxine (T4)	Modified amino acid	Inside cell	Regulates metabolism in animals

glycoprotein, but whether it is lipid soluble. The distinction is crucial because the signal has to be recognized to have an effect on a target cell. Where does this step occur—inside the cell or outside?

- Most lipid-soluble signals diffuse across the plasma membrane and enter the cytoplasm of their target cells.

- Large or hydrophilic cell-cell signals are lipid-insoluble and do not cross the plasma membrane. To affect a target cell, they have to be recognized at the cell surface.

How do cells receive and respond to signals from distant cells? The basic steps are common to all cell-cell signaling systems. Let's consider each in turn.

Signal Reception

Hormones and other types of cell-cell signals deliver their message by binding to receptor molecules. Even though the molecule that carries the message: "We're getting dehydrated—conserve water" is broadcast throughout the body, only certain kidney cells respond because only they have the appropriate receptor. The presence of an appropriate receptor dictates which cells will respond to a particular hormone. Bone and muscle cells don't respond to the "conserve water" message, because they don't have a receptor for it.

Cells in a wide array of tissues may respond to the same signal, though, if they have the appropriate receptor. If you are startled by a loud noise, cells in your adrenal glands secrete a hormone that carries the message: "Get ready to fight or run." In response, your heart rate increases, your breathing rate increases, and cells in your liver release sugars that your muscles can use to power rapid movement. This is the basis of an "adrenalin rush." Heart, lung, and liver cells respond to adrenalin because they have an adrenalin receptor. ⊙━ Identical receptors in diverse cells and tissues allow long-distance signals to coordinate the activities of cells throughout a multicellular organism.

No matter where signal receptors are located, it's critical to note two important points about these proteins:

- Receptors are dynamic. The number of receptors in a particular cell may decline if hormonal stimulation occurs at high

levels over a long period of time. The ability of a receptor molecule to bind tightly to a signal may also decline in response to intensive stimulation. As a result, the sensitivity of a cell to a particular hormone may change over time.

- Receptors can be blocked. The drugs called beta-blockers, for example, bind to certain receptors for the hormone adrenalin (also called epinephrine). When adrenalin binds to receptors in heart cells, it stimulates the cells to contract. So if a physician wants to reduce a patient's heart cell contraction as a way to lower pressure, she is likely to prescribe a beta-blocker.

In many cases, the receptors that respond to lipid-soluble signals are located inside the cell, because the signals readily diffuse through the plasma membrane. The majority of signal receptors are located in the plasma membrane, however, where they can bind to signals that cannot or do not cross the membrane.

The most important general characteristic of signal receptors, though, is that their physical conformation—meaning, overall shape—changes when a hormone binds to them. A **signal receptor** is a protein that changes its shape and activity after binding to a signaling molecule.

This is a critical event in cell-cell signaling. The change in receptor structure means that the signal has been received. It's like throwing an "on" switch.

What happens next?

Signal Processing

Once a cell receives a signal, something has to happen to initiate the cell's response. This signal-processing step happens in one of two ways, depending on whether the signal is received inside the cell or at the membrane surface.

When lipid-soluble signals enter a cell, the information they carry is processed directly—without any intermediate steps. For example, steroid hormones such as testosterone and estrogen diffuse through the plasma membrane and enter the cytoplasm, where they bind to a receptor protein. The hormone-receptor complex is then transported to the nucleus, where it triggers changes in the genes being expressed in the cell (**Figure 8.13**).

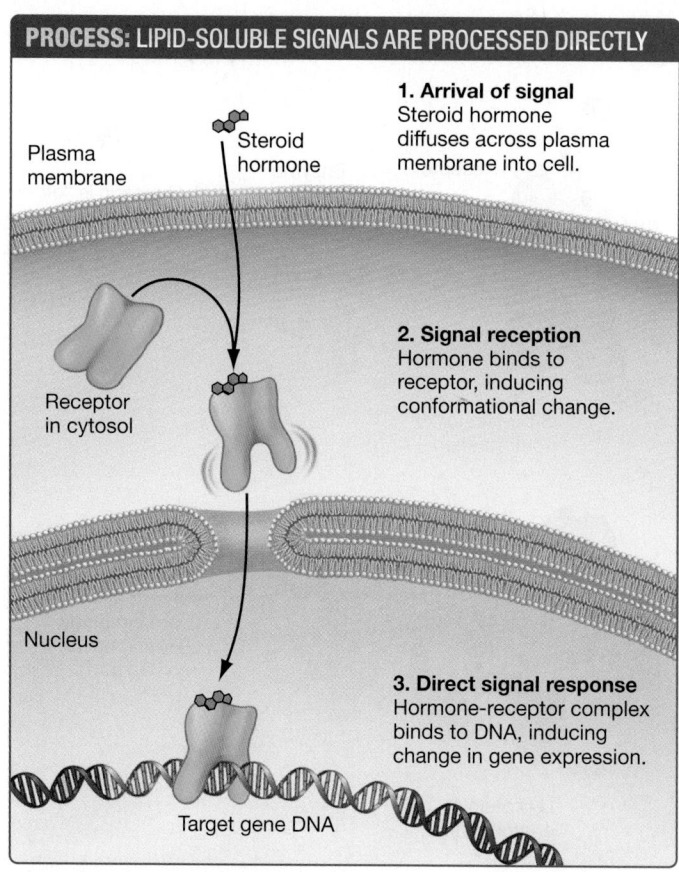

PROCESS: LIPID-SOLUBLE SIGNALS ARE PROCESSED DIRECTLY

1. Arrival of signal Steroid hormone diffuses across plasma membrane into cell.

Plasma membrane

Steroid hormone

Receptor in cytosol

2. Signal reception Hormone binds to receptor, inducing conformational change.

Nucleus

3. Direct signal response Hormone-receptor complex binds to DNA, inducing change in gene expression.

Target gene DNA

FIGURE 8.13 Some Cell-Cell Signals Enter the Cell and Bind to Receptors in the Cytoplasm. Because they are lipids, steroid hormones can diffuse across the plasma membrane and bind to signal receptors inside the cell. The hormone-receptor complex is transported to the nucleus and binds to genes, changing their activity.

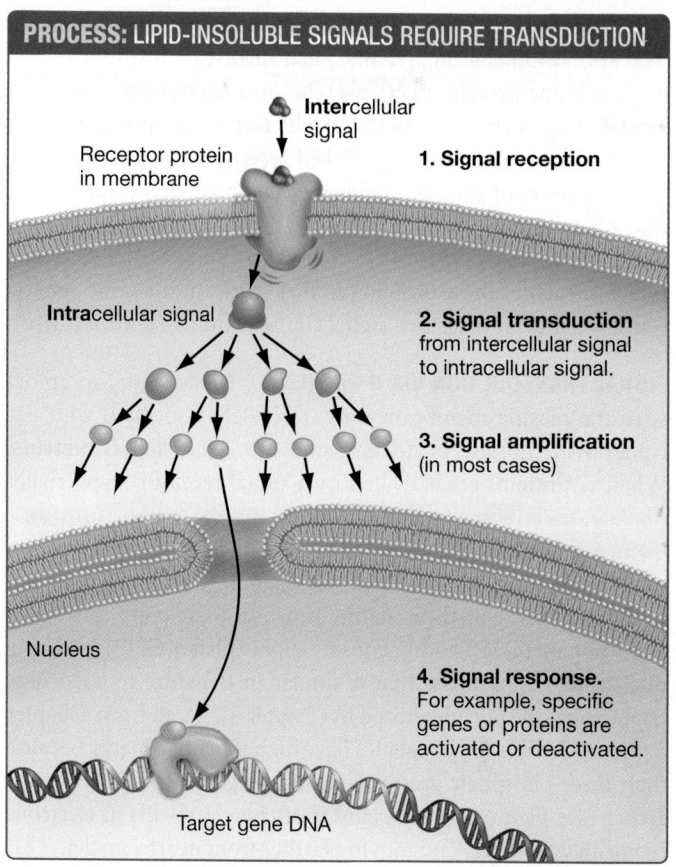

PROCESS: LIPID-INSOLUBLE SIGNALS REQUIRE TRANSDUCTION

Intercellular signal

Receptor protein in membrane

1. Signal reception

Intracellular signal

2. Signal transduction from intercellular signal to intracellular signal.

3. Signal amplification (in most cases)

Nucleus

4. Signal response. For example, specific genes or proteins are activated or deactivated.

Target gene DNA

FIGURE 8.14 Signal Transduction Converts an Intercellular Signal to an Intracellular Signal. Signal transduction is a multistep process.

Other types of lipid-soluble hormones bind to receptors that initiate a response by activating certain pumps in the plasma membrane. In each case, the hormone-receptor complex directly initiates the change by binding to the genes or to the pumps.

Hormones that *cannot* diffuse across the plasma membrane and enter the cytoplasm can't change the activity of genes or pumps directly. Instead, the signal that arrives at the surface of the cell has to be changed to an intracellular signal—the signal-processing step is indirect.

🔑 When a signal binds at the cell surface it triggers **signal transduction**—the conversion of the signal from one form to another. A long and often complex series of events ensues, collectively called a signal transduction pathway.

Signal transduction is a common occurrence in everyday life. For example, the e-mail messages you receive are transmitted from one computer to another over cables or wireless transmissions. These electronic signals can be transmitted efficiently over long distances, but would be meaningless to you. Software in your computer has to transduce, or convert, the signals into a form that you can understand and respond to, such as words on the screen.

Signal transduction pathways work the same way (**Figure 8.14**). In a cell, signal transduction converts an extracellular signal to an intracellular signal. As in an e-mail transmission, a signal that is easy to transmit is converted to a signal that is easily understood and that triggers a response.

SIGNAL AMPLIFICATION Recall that hormones are present in minuscule concentrations but trigger a large response from cells. Signal amplification is one reason this is possible. When a hormone arrives at the cell surface, the message it transmits may be amplified as it changes form. The amplifier in your portable music player performs an analogous function: Once it is amplified, a tiny sound signal can get a whole roomful of people dancing.

In cells, signal transduction begins at the plasma membrane; amplification occurs inside. Amplification may occur in a variety of ways. In general, the mechanism of amplification correlates with the mechanism of signal transduction. But the general observation is that the arrival of a single signaling molecule may result in a secondary signal that involves many ions or molecules.

For example, one major type of signal transduction system consists of membrane channels that open to allow a flow of ions into the cell, changing the electrical properties of the membrane. Chapter 45 will analyze this type of signal transduction in detail.

Here let's focus on the other major types of signal transduction and amplification systems. Each involves a distinctive class of membrane protein: (1) G proteins and (2) enzyme-linked receptors. G proteins initiate the production of an intracellular or "second" messenger. Enzyme-linked receptors trigger the activation of a series of proteins inside the cell, through the addition of phosphate groups.

There are many types of G proteins and enzyme-linked receptors, but each of these protein families works in the same general way. Let's look at these two signal transduction systems in turn.

SIGNAL TRANSDUCTION VIA G PROTEINS   Many signal receptors span the plasma membrane and are closely associated with peripheral membrane proteins inside the cell called **G proteins**. When G proteins are activated by a signal receptor, they trigger the key step in signal transduction: the production of a messenger inside the cell. They link the receipt of an extracellular signal to the production of an intracellular signal.

G proteins got their name because they bind guanosine triphosphate (GTP) and guanosine diphosphate (GDP). GTP is a nucleoside triphosphate that is similar in structure to adenosine triphosphate (ATP; introduced in Chapter 5). Recall from Chapter 4 that nucleoside triphosphates have high potential energy because their three phosphate groups have four negative charges close together (see Figure 4.4). The cluster of charges results in electrons repulsing each other and moving farther from nearby nuclei.

When GTP binds to a protein, the addition of the negative charges alters the protein's shape. Changes in shape produce changes in activity. G proteins are turned on or activated when they bind GTP; they are turned off or inactivated when a phosphate group drops away to form GDP.

To understand how G proteins fit into an overall signal transduction pathway, follow the events in **Figure 8.15**.

Step 1   A hormone arrives and binds to a receptor in the plasma membrane. Notice that the receptor is a transmembrane protein, and that it is coupled to a peripheral G protein on the membrane's inner surface. In response to hormone binding, the receptor changes shape.

Step 2   The shape change is a switch that activates its G protein. Specifically, the G protein releases the GDP molecule that kept it in an inactive state and binds GTP instead. When GTP is attached, the G protein changes shape radically: It splits into two parts.

Step 3   One part of the "split" G protein activates a nearby enzyme that is embedded in the plasma membrane. The enzyme catalyzes the production of a **second messenger**—a nonprotein signaling molecule that elicits a response to the first messenger (the signal that arrived at the cell surface).

Second messengers are effective because they are small and diffuse rapidly to spread the signal throughout the cell. In addition, they can be produced quickly in large quantities. This characteristic is important. Because the arrival of a single hormone molecule can stimulate the production of many second messen-

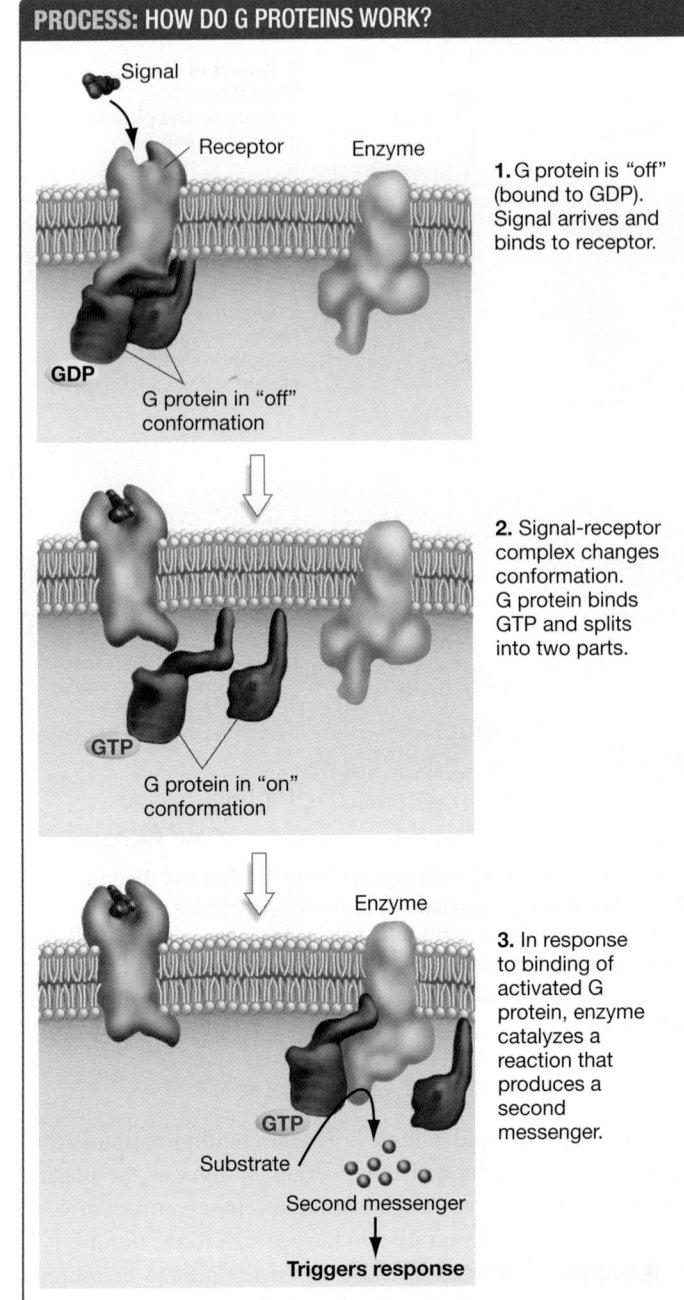

PROCESS: HOW DO G PROTEINS WORK?

1. G protein is "off" (bound to GDP). Signal arrives and binds to receptor.

2. Signal-receptor complex changes conformation. G protein binds GTP and splits into two parts.

3. In response to binding of activated G protein, enzyme catalyzes a reaction that produces a second messenger.

FIGURE 8.15   G Proteins Trigger the Production of a Second Messenger.

ger molecules, the signal transduction event amplifies the original signal.

Several types of small molecules act as second messengers in cells. **Table 8.2** lists some of the best-studied and provides an example of how cells respond to each second messenger. Note that several second messengers activate **protein kinases**—enzymes that activate or inactivate other proteins by adding a phosphate group to them.

It's also important to note that:

1. Second messengers aren't restricted to a single role or single cell type—the same second messenger can initiate dramatically different events in different cell types; and

TABLE 8.2 Examples of Second Messengers

Name	Type of Response
Cyclic guanosine monophosphate (cGMP)	Opens ion channels; activates certain protein kinases
Diacylglycerol (DAG)	Activates certain protein kinases
Inositol triphosphate (IP$_3$)	Opens calcium channels—mobilizes stored calcium ions
Cyclic adenosine monophosphate (cAMP)	Activates certain protein kinases
Calcium ions (Ca^{2+})	Binds to a receptor called calmodulin; Ca^{2+}/calmodulin complex then activates proteins

2. It is common for more than one second messenger to be involved in triggering a cell's response to the same extracellular signal.

To make sure that you understand how G proteins and second messengers work, imagine the following movie scene: A spy arrives at a castle gate. The guard receives a note from the spy, but he cannot read the coded message. Instead, the guard turns to the queen. She reads the note and summons the commander of the guard, who sends soldiers throughout the castle to warn everyone of approaching danger. ✔You should be able to identify which characters in the scene correspond to the second messenger, G protein, hormone, receptor, and enzyme activated by the G protein.

It's difficult to overstate the importance of signal transduction by G proteins. Biomedical researchers estimate that half of human drugs target signal receptors that are associated with G proteins.

SIGNAL TRANSDUCTION VIA ENZYME-LINKED RECEPTORS Instead of activating a nearby G protein, enzyme-linked receptors transduce the signal from a hormone by directly catalyzing a reaction inside the cell. **Figure 8.16** focuses on the best-studied group of enzyme-linked receptors: the **receptor tyrosine kinases (RTKs)**.

Step 1 A hormone binds to an RTK.

Step 2 The protein forms a dimer. In this conformation, the receptor has a binding site for a phosphate group from ATP inside the cell. Once it is phosphorylated, an RTK becomes an active enzyme.

Step 3 Proteins inside the cell form a bridge between the activated RTK and a peripheral membrane protein called **Ras**, which functions like a G protein. The formation of the RTK bridge activates Ras. Specifically, Ras exchanges its GDP for a GTP.

Step 4 When Ras is activated, it triggers the phosphorylation and activation of another protein.

Step 5 The phosphorylated protein then catalyzes the phosphorylation of other proteins, which phosphorylate yet another population of proteins.

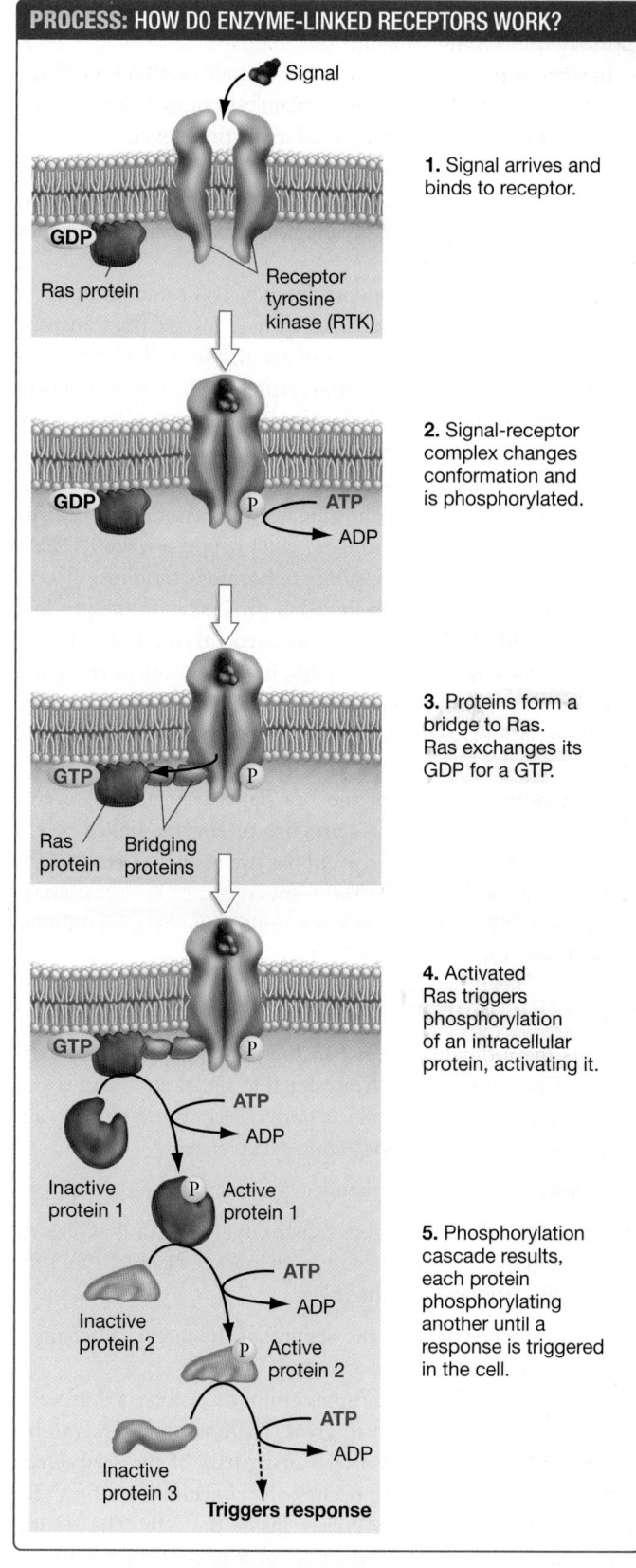

PROCESS: HOW DO ENZYME-LINKED RECEPTORS WORK?

1. Signal arrives and binds to receptor.

2. Signal-receptor complex changes conformation and is phosphorylated.

3. Proteins form a bridge to Ras. Ras exchanges its GDP for a GTP.

4. Activated Ras triggers phosphorylation of an intracellular protein, activating it.

5. Phosphorylation cascade results, each protein phosphorylating another until a response is triggered in the cell.

FIGURE 8.16 Enzyme-Linked Receptors Trigger a Phosphorylation Cascade.

This sequence of events is termed a **phosphorylation cascade**, and culminates in a response by the cell.

In some cases, each enzyme in the cascade catalyzes the phosphorylation of numerous "downstream" enzymes. When this occurs, the original signal is amplified many times over.

In most cases, though, the proteins that take part in a phosphorylation cascade are held in close physical proximity by scaffolding proteins. This structure increases the speed and efficiency of the reaction sequence.

To make sure that you understand how RTKs and phosphorylation cascades work, imagine that you have two red dominos, one black domino, and a large supply of green, blue, yellow, pink, and orange dominos. The red dominos represent the two subunits of an RTK dimer, and the black domino represents Ras. Each of the other colors represents a type of protein in a phosphorylation cascade. ✔You should be able to explain (1) how you would set up the dominos to simulate a phosphorylation cascade, and (2) why tipping the red dominos simulates what happens when RTK becomes phosphorylated in response to hormone binding.

Although this discussion linked G proteins with the production of second messengers and associated enzyme-linked receptors with phosphorylation cascades, it's important to recognize that some G proteins trigger phosphorylation cascades and some phosphorylation cascades result in the production of second messengers.

To summarize: Many of the key signal transduction events observed in cells occur via G proteins or enzyme-linked receptors. The signal transduction event has two results: **(1)** It converts an easily transmitted extracellular message into an intracellular message, and **(2)** in many cases it amplifies the original message many times over.

Signal Response

What is the ultimate response to the messages carried by hormones? The answer varies from signal to signal and from cell to cell, but fall into two general categories. Second messengers or a cascade of protein phosphorylation events may:

1. change which genes are being expressed in the target cell; or

2. activate or deactivate a particular target protein that already exists in the cell—an enzyme, a membrane channel, or a protein that activates certain genes.

Whatever the mechanism, the activity of the target cell changes dramatically after the signal arrives.

In wheat seeds, for example, embryos secrete a hormone called GA_1 when they start to grow. The hormone binds to receptors in cells near the starch-storing part of the seed. Once hormone-receptor binding occurs, the concentration of a second messenger called cGMP rises inside the cells. The second messenger triggers the production of a protein that activates the gene for a starch-digesting enzyme. Large quantities of the starch-digesting enzyme α-amylase enter the cell's endomembrane system.

Vesicles packed with α-amylase can be secreted into the starch storage area because the G protein activated by GA_1 also triggers a rise in intracellular Ca^{2+}. High Ca^{2+} concentrations allow the vesicles to fuse with the plasma membrane and release their contents. The enzymes begin breaking down the starch and releasing sucrose. In this way, the germinating embryo has signaled for the release of the nutrients it needs to grow.

We've analyzed the first three steps of long-distance cell-cell communication: signal reception, signal processing, and response. Now the question is, how is the signal turned off? Consider the flush of testosterone and estrogen that you experienced during puberty, and the morphological changes these hormones induced. Abnormalities would result if these changes continued indefinitely. What limits the response to a cell-cell signal?

Signal Deactivation

Cells have built-in systems for turning off intracellular signals. For example, activated G proteins convert bound GTP to GDP. When this reaction occurs, the G protein's conformation changes. Activation of its associated enzyme stops, and production of the second messenger ceases.

The presence of second messengers in the cytosol is also short lived. For example, pumps in the membrane of the smooth ER return calcium ions to storage, and enzymes called phosphodiesterases convert active cAMP (see Table 8.2) and cGMP to inactive AMP and GMP. When second messengers are cleared from the cytosol, the response stops.

For G proteins to stay activated and continue influencing the behavior of the cell, the extracellular signal has to continue. Otherwise, the signal transduction system quickly shuts down.

Phosphorylation cascades wind down in a similar way. Enzymes called phosphatases are always present in cells, where they catalyze reactions that remove phosphate groups from proteins. If hormone stimulation of a receptor tyrosine kinase ends, phosphatases are able to dephosphorylate enough components of the phosphorylation cascade that the response begins to slow. Eventually it stops.

Although an array of specific mechanisms are involved, here is the general observation: Signal transduction systems trigger a rapid response and can be shut down quickly. As a result, they are exquisitely sensitive to small changes in the concentration of hormones or the number and activity of signal receptors.

It is critical, though, to appreciate what happens when a signal transduction system does not shut down properly. For example, recall that Ras activates a phosphorylation cascade when it binds GTP, but is deactivated once it has broken down GTP to GDP. With surprising frequency, human cells produce defective Ras proteins that no longer convert GTP to GDP once they are activated. As a result, GTP stays bound and the defective Ras proteins stay in the "on" position. They continue stimulating a phosphorylation cascade even when no appropriate signals are present. Cells with defective Ras are likely to keep dividing, which may lead to the development of cancer. An estimated 25–30 percent of all human cancers involve defective Ras proteins. Chapters 11 and 18 explore the family of diseases called cancer in detail.

Cross-Talk: Synthesizing Input from Many Signals

Although the preceding discussion focused on how cells respond to individual signals, it's crucial to realize that every cell has an array of signal receptors on its plasma membrane and in its cytoplasm, and every cell receives an almost constant stream of different signals. You get text messages, e-mails, phone calls, and snail mail about changes in your environment; cells get an array of chemical signals about changes in their environment.

The signal transduction pathways that are triggered by these signals and receptors intersect and connect. In reality, they are not strictly linear like the pathways illustrated in Figures 8.13 through 8.16. Signal transduction pathways form a network. This complexity is important: It allows cells to respond to an array of extracellular signals in an integrated way.

The diverse signals that a cell receives are integrated by what biologists call **cross-talk**—meaning, interactions between signaling pathways (**Figure 8.17**). The key things to note are that:

1. Elements or products from one pathway may inhibit steps in a different pathway—reducing the cell's response, even though the appropriate signal is present.

2. A response from one pathway may stimulate a greater response by a protein in a different pathway, increasing the cell's response to the other signal.

3. The presence of multiple steps in each signaling pathway provides a series of points where cross-talk can regulate the flow of information. These interactions are important, because they allow the cell to respond appropriately to many signals at the same time.

Quorum Sensing in Bacteria

In eukaryotes, cell-cell signaling has been one of the hottest research areas in biological science over the past two decades. But in prokaryotes, research on cell-cell signaling is just beginning to gain momentum.

In bacteria, cell-cell communication is called **quorum sensing**. The name was inspired by the observation that bacterial cells of the same species may undergo dramatic changes in activity when their numbers reach a threshold, or quorum.

Quorum sensing is based on species-specific signaling molecules that are secreted by cells and diffuse through the environment. The response to these signals varies dramatically among species and ranges from bioluminescence—or light emission—to secretion of proteins that allow the bacteria to attack multicellular hosts and cause disease, to the production and secretion of **biofilms**—hard, polysaccharide-rich substances that encase the cells and attach them to a surface (**Figure 8.18**).

Biofilm formation has been particularly well-studied because of its medical importance.

- Biofilms that form in the human mouth are called dental plaque and contribute to tooth decay.

- When *Pseudomonas aeruginosa* cells settle on a human lung cell, they secrete a signaling molecule that recruits other *P. aeruginosa* cells to the site and triggers secretion of a biofilm. The biofilm attaches the cells to the lung surface and protects them from attack by immune system cells or antibiotics.

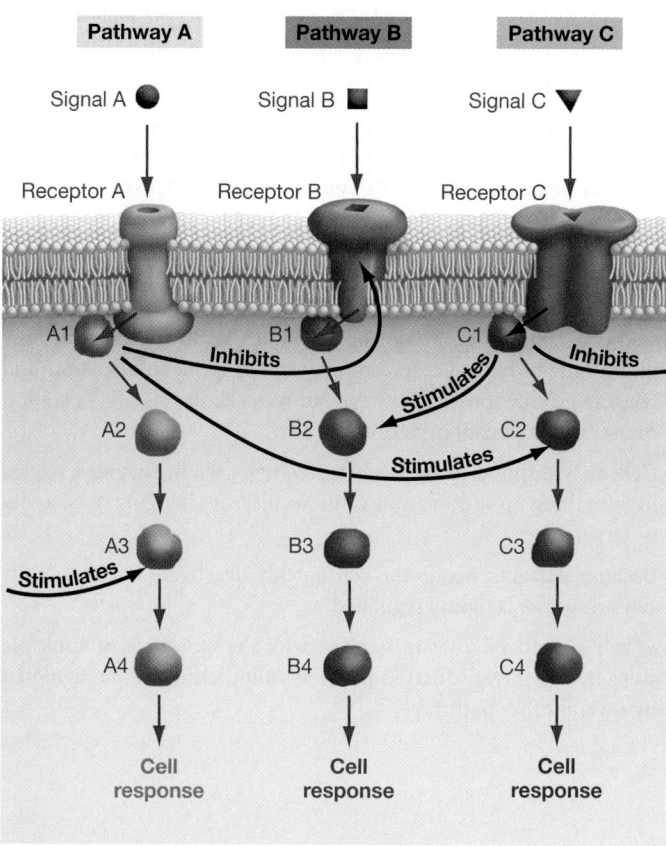

FIGURE 8.17 Signaling Pathways Interact via "Cross-Talk."

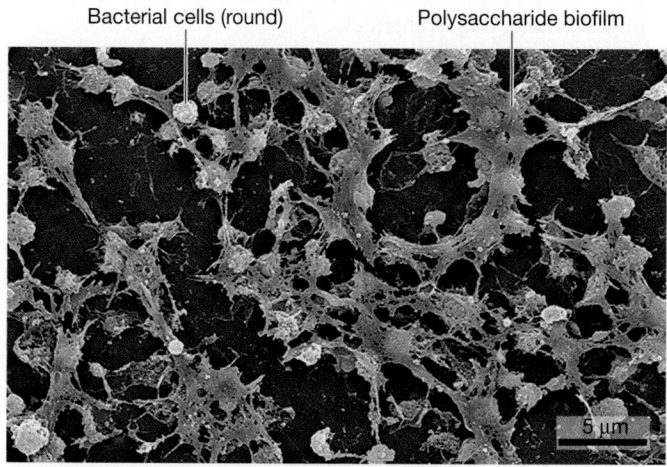

FIGURE 8.18 Groups of Unicellular Organisms May Secrete Biofilms. This micrograph shows *Staphylococcus aureus* cells (the round structures) growing inside a catheter—a tube inserted into a body so that fluids can be withdrawn or injected. The cells have secreted a sticky biofilm that helps them cling to each other and to the walls of the catheter.

In effect, quorum sensing allows closely related bacterial cells to communicate and coordinate activity. When it occurs, these single-celled organisms take on some of the characteristics of multicellular organisms.

This brief introduction to quorum sensing and cross-talk brings us to the frontier of research in cell-cell signaling. It has taken decades of painstaking research to work out each step in individual signaling pathways; now biologists are excited about probing how the major pathways interact, and how similar types of pathways operate in eukaryotes.

Although the number of molecules involved and the complexity of their interactions can seem overwhelming, the punchline is simple: In organisms ranging from bacteria to blue whales, cell-cell signaling helps organisms receive information about their environment and respond to changing conditions in an appropriate way.

CHECK YOUR UNDERSTANDING

If you understand that . . .

- Intercellular signals coordinate the activities of cells throughout the body of a multicellular organism in response to changes in internal or external conditions.
- If intercellular signals do not enter the cell, they bind to a receptor on the plasma membrane. In response, the intercellular signal is transduced to an intracellular signal that the cell responds to.

✔ **You should be able to . . .**

1. Explain why only certain cells respond to particular signals.
2. Explain how some signals are amplified.

Answers are available in Appendix B.

CHAPTER 8 REVIEW

For media, go to the study area at www.masteringbiology.com

Summary of Key Concepts

Extracellular material strengthens cells and helps bind them together.

- The vast majority of cells secrete an extracellular layer.

- In bacteria, archaea, algae, and plants, the extracellular material is stiff and is called a cell wall. In animals, the secreted layer is flexible and is called the extracellular matrix (ECM).

- Extracellular layers are fiber composites. They consist of cross-linked filaments that provide tensile strength and a ground substance that fills space and resists compression.

- In plants the extracellular filaments are cellulose microfibrils; in animals the most abundant filaments are made of the protein collagen. In both plants and animals, the ground substance is composed of gel-forming polysaccharides.

    ✔ You should be able to predict what happens to animal cells when (1) they are treated with an enzyme that cuts integrin molecules, or (2) collagen fibers degrade.

Cell-cell connections help adjacent cells adhere. Cell-cell gaps allow adjacent cells to communicate.

- In multicellular organisms, molecules in the extracellular layer and plasma membrane mediate interactions between adjacent cells.

- Adjacent cells may be physically bound to one another, by glue-like middle lamella in plants or tight junctions and desmosomes in animals.

- The cytoplasm of adjacent cells is in direct communication through openings called plasmodesmata in plants and gap junctions in animals.

    ✔ You should be able to explain the consequences of loosening the tight junctions between animal epithelial cells.

Intercellular signals are responsible for creating an integrated whole from many thousands of independent parts.

- Distant cells in multicellular organisms communicate through signaling molecules that bind to receptors found in or on specific target cells. As a result, cells and tissues throughout the body can alter their activity in response to changing conditions, and do so in a coordinated way.

- Quorum sensing allows closely related bacterial cells to coordinate changes in their activity when population density is high.

    ✔ You should be able to explain why the hormone adrenalin can stimulate cells in both the heart and liver, yet trigger different responses (increasing heart rate versus releasing glucose).

In target cells, intercellular signals are received, processed, responded to, and deactivated. If the signal is received at the cell surface, the processing step involves production of an intracellular signal.

- Cell-cell signals that are not lipid-soluble bind to receptors in the plasma membrane. The receptor then changes conformation and triggers production of a new type of intracellular signal—a second messenger or phosphorylation cascade.

- Cells may respond to signals by activating certain enzymes, releasing or taking up specific ions or molecules, or changing the activity of target genes.

- Because enzymes inside the cell quickly deactivate the signal, the cell's response is tightly regulated.

    ✔ You should be able to explain why the existence of multiple steps in signal transduction pathways allows them to be regulated by several other pathways.

Questions

1. Which of the following statements represents a fundamental difference between the fibers found in the extracellular layers of plants and those of animals?
   a. Plant fibers are thicker; they are also stronger because they have more cross-linkages.
   b. Animal fibers consist of proteins; plant fibers consist of polysaccharides instead.
   c. Plant extracellular fibers never move; animal fibers can slide past one another.
   d. Cellulose microfibrils run parallel to each other; collagen filaments crisscross.

2. In animals, where are most components of the extracellular material synthesized?
   a. smooth ER
   b. the rough ER and Golgi apparatus
   c. in the extracellular layer itself
   d. in the plasma membrane

3. Treating dissociated cells with certain antibodies makes the cells unable to reaggregate. Why?
   a. The antibodies bind to cell adhesion proteins called cadherins.
   b. The antibodies bind to the fiber component of the extracellular matrix.
   c. The antibodies bind to receptors on the cell surface.
   d. The antibodies act as enzymes that break down desmosomes.

4. What does it mean to say that a signal is transduced?
   a. The signal enters the cell directly and binds to a receptor inside.
   b. The physical form of the signal changes between the outside of the cell and the inside.
   c. The signal is amplified, such that even a single molecule evokes a large response.
   d. The signal triggers a sequence of phosphorylation events inside the cell.

5. Why are tight junctions found in only certain types of tissues, while desmosomes are found in a wide array of cells?
   a. Tight junctions are required only in cells where communication between adjacent cells is particularly important.
   b. Tight junctions are not as strong as desmosomes.
   c. Tight junctions have different structures but the same functions.
   d. Tight junctions are found only in epithelial cells that must be watertight.

6. What physical event represents the receipt of an intercellular signal?
   a. the passage of ions through a desmosome
   b. the activation of the first protein in a phosphorylation cascade
   c. the binding of a hormone to a signal receptor, which changes conformation in response
   d. the activation of a G protein associated with a signal receptor

1. Why is it difficult to damage a fiber composite?

2. How is it possible for a phosphorylation cascade to amplify an intercellular signal?

3. Compare and contrast the structure and function of tight junctions and gap junctions. Compare and contrast the structure and function of middle lamellae and plasmodesmata.

4. Animal cells adhere to each other selectively. Summarize experimental evidence that supports this statement. Explain the molecular basis of selective adhesion.

5. Make a flowchart summarizing the reception, processing, response, and deactivation steps for a signal that binds to an intracellular receptor.

6. What is the significance of the observation that many signal transduction pathways intersect or overlap, creating a network?

1. Suppose that an animal and a plant each lacked the ability to secrete an extracellular matrix. What would these organisms look like?

2. Suppose that a cell-cell signal binds to a membrane-receptor, and the cell responds without signal transduction occurring. Compared to signal transduction pathways, how would an event like this affect (a) the types of responses that are possible, (b) amplification, and (c) regulation?

3. In most species of fungi, chitin is a major polysaccharide found in cell walls. Review the structure of chitin as described in Chapter 5, and then make a sketch predicting the structure of the fungal cell wall.

4. Suppose you created an antibody that bound to the receptor illustrated in Figure 8.16. How would the signal transduction pathway be affected? How would the signal transduction pathway be affected by a drug that bound permanently to the receptor?

When table sugar is heated, it undergoes the uncontrolled oxidation reaction known as burning. Burning gives off heat. In cells, the simple sugar called glucose is oxidized through a long series of carefully regulated reactions. Instead of being given off as heat, some of the energy produced by these reactions is used to synthesize ATP.

9 Cellular Respiration and Fermentation

KEY CONCEPTS

🔑 In cells, the endergonic reactions required for life occur in conjunction with an exergonic reaction involving ATP.

🔑 Cellular respiration produces ATP from molecules with high potential energy—often glucose.

🔑 Cellular respiration has four components: (1) glycolysis, (2) pyruvate processing, (3) the citric acid cycle, and (4) electron transport and chemiosmosis.

🔑 Cellular respiration and fermentation are carefully regulated.

🔑 Fermentation pathways allow glycolysis to continue when the lack of an electron acceptor shuts down electron transport chains.

Cells are dynamic. Vesicles move cargo from the Golgi apparatus to the plasma membrane and other destinations, enzymes synthesize a complex array of macromolecules, and millions of proteins pump ions and molecules across the plasma membrane. These cell activities change constantly in response to signals from other cells or the environment.

What fuels all this action? The answer is **adenosine triphosphate (ATP)**. ATP has high potential energy and allows cells to do work. Because staying alive takes work, there is no life without ATP.

This chapter investigates how cells make adenosine triphosphate, starting from sugars and other compounds that have high potential energy. As cells process sugar, the energy that is released is used to transfer a phosphate group to **adenosine diphosphate (ADP)**, generating ATP.

The chapter begins by reviewing fundamental concepts about energy and introducing how it is used in cells. Sections 9.2 through 9.6 delve into the reactions involved in processing **glucose**, the most common fuel used by organisms, and producing ATP. Section 9.7 introduces fermentation, an alternative route for ATP production that occurs in many cells. Section 9.8 examines how cells divert certain carbon-containing compounds away from ATP production and into the synthesis of DNA, RNA, amino acids, and other molecules.

This chapter is about a key attribute of life: the ability to acquire and use energy. It is also your introduction to **metabolism**—all the chemical reactions that occur in living cells.

✔ When you see this checkmark, stop and test yourself. Answers are available in Appendix B.

9.1 The Nature of Chemical Energy and Redox Reactions

Recall from Chapter 2 that chemical energy is a form of potential energy. Potential energy is energy that is associated with position or configuration. In cells, chemical energy is stored in the position of electrons.

The amount of potential energy in an electron is based on its position relative to other electrons and the protons in the nuclei of nearby atoms. If an electron is close to negative charges on other electrons and far from the positive charges in nuclei, it has high potential energy. In general, the potential energy of a molecule is a function of the way its electrons are configured or positioned.

The Structure and Function of ATP

ATP makes things happen in cells because it has a great deal of potential energy. As **Figure 9.1a** shows, four negative charges are confined to a small area in the three phosphate groups in ATP. In part because these negative charges repel each other, the potential energy of the electrons in the phosphate groups is extraordinarily high.

When ATP reacts with water during a hydrolysis reaction, the bond between ATP's outermost phosphate group and its neighbor is broken, resulting in the formation of ADP and inorganic phosphate, P_i, which has the formula $H_2PO_4^-$ (**Figure 9.1b**). This reaction is highly exergonic. Recall that **exergonic** reactions release energy, while **endergonic** reactions require an input of energy. Under standard conditions of temperature and pressure in the laboratory, a total of 7.3 **kilocalories** of energy per mole of ATP (or 7.3 kcal/mol), is released during the reaction. A kcal of energy raises 1 g of water 1°C.

WHY DOES ATP HYDROLYSIS RELEASE ENERGY? You might remember from Chapter 2 that changes in free energy dictate whether a reaction is exergonic or endergonic. Changes in free energy, in turn, depend on the relationship between reactants and products in terms of potential energy and entropy (disorder).

The hydrolysis of ATP is exergonic because the entropy of the product molecules is much higher than that of the reactants, and because there is a large drop in potential energy when ADP and P_i are formed from ATP. The change in potential energy occurs in part because the electrons from ATP's phosphate groups are now spread between two molecules instead of being clustered on one molecule, meaning that there is less electrical repulsion. In addition, the negative charges on ADP and P_i are stabilized much more efficiently by interactions with the partial positive charges on surrounding water molecules than are the charges on ATP.

WHAT HAPPENS WHEN PROTEINS ARE PHOSPHORYLATED BY ATP? If the reaction diagrammed in Figure 9.1b occurred in a test tube, the energy released would be lost as heat. But cells don't lose that 7.3 kcal/mole as heat. Instead, they use it to make things happen. Specifically, things start to happen when ATP is hydrolyzed and the phosphate group that is released is transferred to a protein.

The addition of a phosphate group to a substrate is called **phosphorylation**. Phosphorylation of proteins is exergonic because the electrons in ADP and the phosphate group have much less potential energy than they did in ATP.

When phosphorylation adds a negative charge to a protein, the electrons in the protein change configuration. The molecule's overall shape, or conformation, usually changes as well.

(a) ATP consists of three phosphate groups, ribose, and adenine.

FIGURE 9.1 Adenosine Triphosphate (ATP) Has High Potential Energy. (a) ATP's high potential energy results, in part, from the four negative charges clustered in its three phosphate groups. The negative charges repel each other, raising the potential energy of the electrons. **(b)** When ATP is hydrolyzed to ADP and inorganic phosphate, a large free-energy change occurs.

(b) Energy is released when ATP is hydrolyzed.

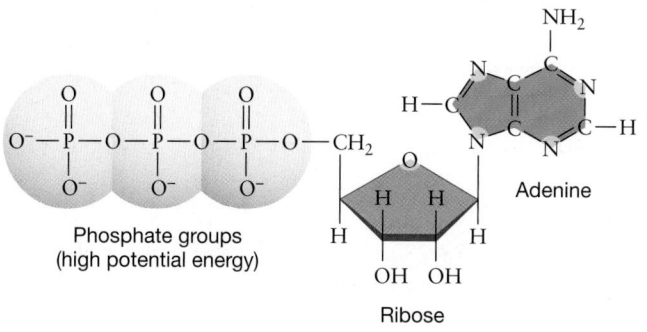

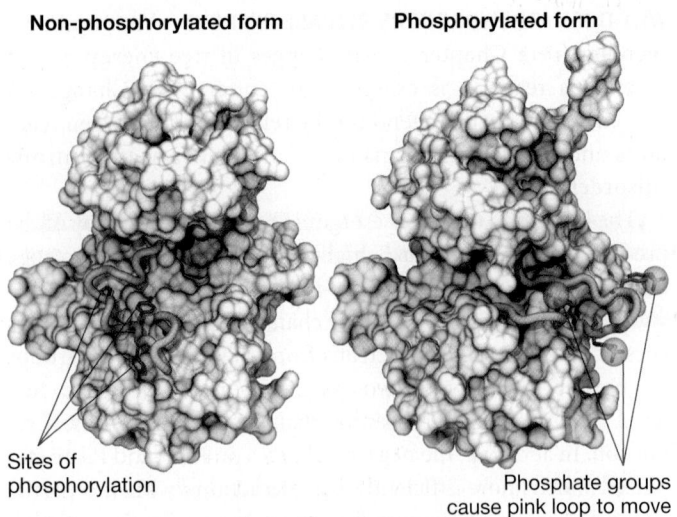

Non-phosphorylated form   **Phosphorylated form**

Sites of
phosphorylation

Phosphate groups
cause pink loop to move

FIGURE 9.2 Phosphorylation Changes the Shape and Activity of Proteins. When proteins are phosphorylated or an ATP molecule binds to them, they often change shape in a way that alters their activity. The figure shows the inactivated and activated forms of an enzyme called insulin receptor tyrosine kinase. Note that three phosphate groups (yellow) activate this molecule.

Part of the protein moves (**Figure 9.2**). Protein movement—either in response to phosphorylation or to binding of an entire ATP molecule—is what transports materials inside cells, powers flagella or cilia, and pumps ions across membranes. It also drives the endergonic reactions required for life.

HOW DOES ATP DRIVE ENDERGONIC REACTIONS? In the time it takes to read this sentence, millions of endergonic reactions have occurred in your cells. This chemical activity is possible because entire ATP molecules or phosphate groups from ATP are being added to reactant molecules or enzymes.

To see how this process works, consider an endergonic reaction between a compound A and compound B that results in a product AB needed by your cells. This reaction can happen only when ATP reacts with the substrate to produce a phosphorylated intermediate molecule and ADP. If the reactant that is phosphorylated is compound B, it is referred to as an activated substrate. Activated substrates contain a phosphate group and have high free energy. This is the critical point: Activated substrates have high enough potential energy that the reaction between compound A and the activated form of compound B is exergonic. The two compounds then go on to react and form the product molecule AB.

In some cases, the enzyme that catalyzes the reaction is phosphorylated instead of a reactant. 🔑 When either a substrate or an enzyme is phosphorylated, the exergonic phosphorylation reaction is coupled to an endergonic reaction. In many cases, phosphorylation of substrates or enzymes makes the reactions that occur in cells exergonic.

Figure 9.3 graphs how **energetic coupling** between an exergonic and endergonic reaction works. Note, on the far right, that the reaction between A and B to produce the product AB is endergonic—ΔG is positive. But when the exergonic reaction occurs that moves a phosphate group from ATP to B, the free energy of the reactants A and B is now high enough to make the reaction that forms AB exergonic. This is due to "coupling" between phosphorylation reactions and endergonic reactions.

✔If you understand the principles of energetic coupling, consider the following endergonic reaction:

$$\text{ribulose} + \text{CO}_2 \longrightarrow \text{glycerate}$$

Suppose that the enzyme that catalyzes this reaction is next to a ribulose molecule and a CO_2 molecule. You should be able to: (1) Explain why the ribulose does not react with the CO_2, even with the enzyme present. (2) Suppose that 2 ATPs add 2 phosphate groups to the ribulose. Explain why the resulting molecule, called ribulose bisphosphate, is called an activated substrate. (3) The enzyme catalyzes a reaction between the activated substrate and CO_2,

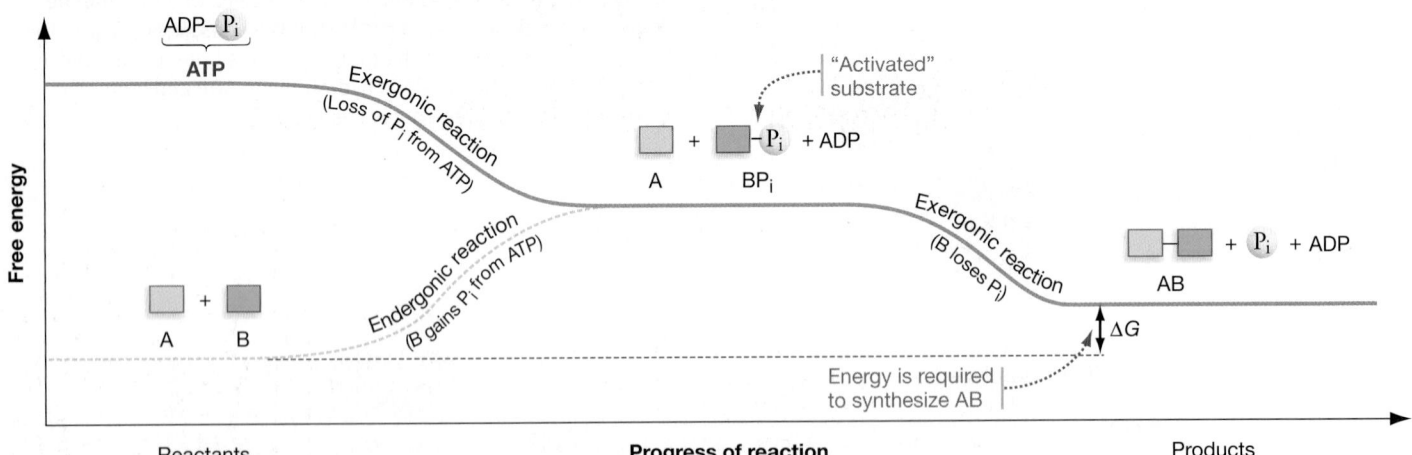

FIGURE 9.3 Exergonic Phosphorylation Reactions Are Coupled to Endergonic Reactions. In cells, many reactions only occur if one reactant (or an enzyme) undergoes phosphorylation—the addition of a phosphate group from ATP. The phosphorylated reactant molecule (or enzyme) has high enough free energy that the subsequent reaction is exergonic.

forming two molecules of glycerate (each of which has a phosphate group attached). Explain why this sequence of events represents the coupling of an exergonic reaction and endergonic reaction.

It is hard to overstate the importance of energetic coupling: Without it, life is impossible. If you ran out of ATP, enzymes and reactants could no longer be phosphorylated and you would die within minutes.

Now the question is, where do cells get ATP in the first place? A great deal of energy is required to synthesize ATP from ADP by adding P_i. Where does this energy come from? The answer is redox reactions.

What Is a Redox Reaction?

Reduction-oxidation reactions, or **redox reactions,** are a class of chemical reactions that involve the loss or gain of one or more electrons. Redox reactions are central in biology because they drive the formation of ATP.

In a redox reaction, the atom that loses one or more electrons is oxidized, and the atom that gains one or more electrons is reduced. To keep these terms straight, chemists use the mnemonic "LEO the lion goes GER"—Loss of Electrons is **Oxidation**; Gain of Electrons is **Reduction.** (An alternative is OIL RIG—Oxidation Is Loss; Reduction Is Gain.)

Oxidation events are always paired with a reduction; if one atom loses an electron, another has to gain it. Stated another way, a reactant that acts as an **electron donor** is always associated with a reactant that acts as an **electron acceptor.**

The gain or loss of an electron can be relative, however. During a redox reaction, an electron can be transferred completely from one atom to another, or an electron can simply shift its position in a covalent bond.

AN EXAMPLE OF REDOX IN ACTION   To see how redox reactions work, consider the overall reaction for photosynthesis (**Figure 9.4**). Plants take in carbon dioxide (CO_2) and water (H_2O); and with the aid of sunlight, they synthesize carbohydrate—in this example, the sugar glucose ($C_6H_{12}O_6$)—and release molecular oxygen (O_2) and water. The orange dots in the illustration represent the positions of the electrons involved in covalent bonds.

Now compare the position of the electrons in the first reactant, carbon dioxide, with their position in the first product, glucose. Notice that many of the electrons have moved closer to the carbon nucleus in glucose. This means that carbon has been reduced: it has "gained" electrons. The change occurred because the carbon and oxygen atoms in CO_2 do not share electrons equally, while the carbon and hydrogen atoms in glucose do. In CO_2, the high electronegativity of the oxygen atoms pulled electrons away from the carbon atom.

Now compare the position of the electrons in the reactant water molecules with their position in the O_2 molecules that are produced. In O_2, the electrons have moved farther from the oxygen nuclei than they were in the water molecules, meaning that the oxygen atoms have been oxidized. Oxygen has "lost" electrons. Thus, in photosynthesis, carbon atoms are reduced while oxygen atoms are oxidized.

These shifts in electron position change the amount of chemical energy in the reactants and products. When photosynthesis occurs, electrons are held much more loosely in the product molecules than in the reactant molecules. This means their potential energy has increased. The entropy of the products is also much lower than the reactants. As a result, the reaction is endergonic. It can take place only with an input of energy from sunlight.

ANOTHER APPROACH TO UNDERSTANDING REDOX   During the redox reactions that occur in cells, electrons (e^-) are often transferred from an atom in one molecule to an atom in a different molecule. When this occurs, the electron is usually accompanied by H^+.

Molecules that gain a hydrogen (H) atom in this way tend to have high potential energy, because the electrons in C–H bonds are relatively far from the positive charges in a nucleus. This observation should sound familiar, from the introduction to carbohydrates in Chapter 5. Molecules that have a large number of C–H bonds, such as carbohydrates and fats, store a great deal of potential energy.

Conversely, molecules that are oxidized in cells often lose a proton along with an electron. Instead of having many C–H bonds, oxidized molecules in cells tend to have many C–O bonds. Oxidized molecules also have lower potential energy. To understand why, remember that oxygen atoms have extremely

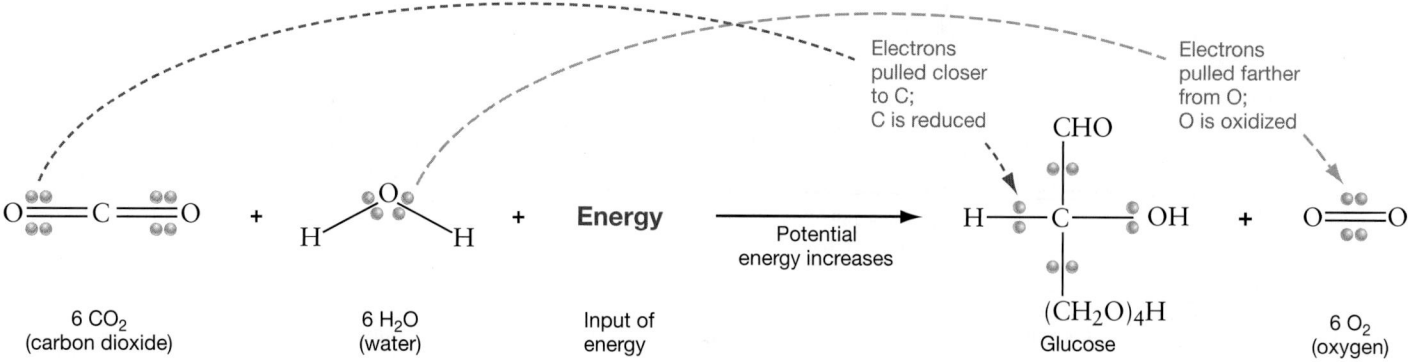

FIGURE 9.4 Redox Reactions Involve the Gain or Loss of One or More Electrons. This diagram shows how the position of electrons changes in the overall reaction of photosynthesis. During photosynthesis, carbon atoms in CO_2 are reduced to form glucose and other sugars. The process is endergonic and requires an input of energy. Glucose has much higher potential energy than carbon dioxide does.

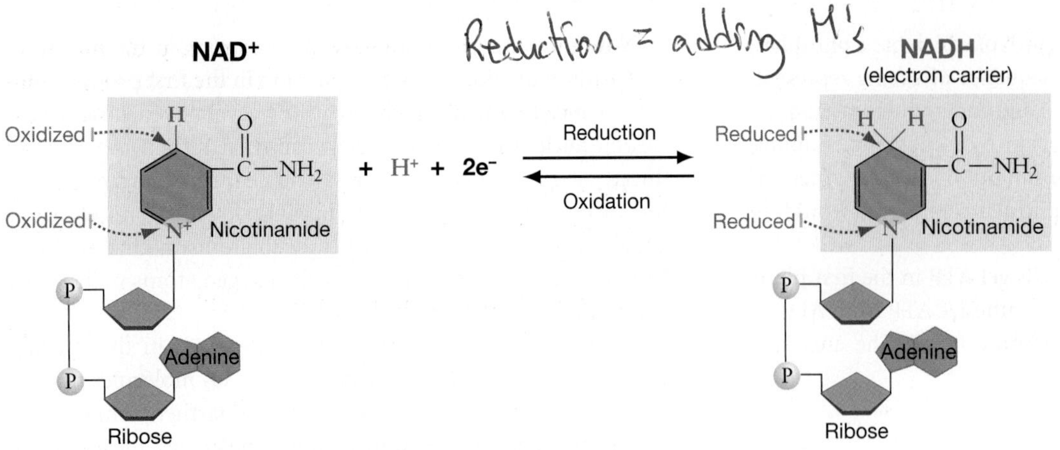

Reduction = adding H's

FIGURE 9.5 NADH is the reduced form of NAD+. NADH is an important electron carrier.

high electronegativity. Because oxygen atoms hold electrons so tightly, the electrons involved in bonds with oxygen atoms have low potential energy.

In many redox reactions in biology, understanding where oxidation and reduction have occurred becomes a matter of following hydrogen atoms—reduction means "adding H's" and oxidation means "removing H's." A good example is **nicotinamide adenine dinucleotide (NAD+)** which is reduced to form **NADH** (**Figure 9.5**). NADH readily donates electrons to other molecules. As a result, it is called an **electron carrier** and is said to have "reducing power." As you will soon see, NADH is an important electron carrier during cellular respiration.

WHAT HAPPENS WHEN GLUCOSE IS OXIDIZED? To test your understanding of redox reactions, consider what happens when glucose undergoes the uncontrolled oxidation reaction called burning:

$$C_6H_{12}O_6 + 6O_2 \longrightarrow 6CO_2 + 6H_2O + Energy$$

glucose   oxygen   carbon dioxide   water

The photograph at the start of the chapter shows this reaction occurring, and **Figure 9.6** provides an incomplete electron-sharing diagram.

✓If you understand the fundamental principles of reduction-oxidation, you should be able to complete Figure 9.6 (by filling in electron positions) and answer the following questions: (1) Are the carbon atoms in glucose oxidized or reduced?

(2) Are the oxygen atoms in the oxygen molecule (O_2) oxidized or reduced? (3) Glucose is the molecule that acts as an electron donor in this reaction. Which molecule acts as the electron acceptor? (4) Which has higher potential energy: the reactants or the products? Based on your answer, add "Energy" to the appropriate side of Figure 9.6 with a label below indicating "Input of energy" or "Release of energy."

Before going on to the next section, you may want to visit the study area at *www.masteringbiology.com* and review the principles of redox reactions.

(MB) **Web Activity** Redox Reactions

When glucose burns, the change in potential energy is converted to kinetic energy in the form of heat. More specifically, a total of 686 kcal of heat is released when one mole of this sugar is oxidized.

🔑 Glucose does not burn in cells, however. Instead, the glucose in cells is oxidized through a long series of carefully controlled redox reactions. These reactions are occurring, millions of times per minute, in your cells right now. Instead of being given off as heat, much of the energy that is released is being used to make the ATP you need to read, think, move, and stay alive. In cells, the change in free energy that occurs during the oxidation of glucose is used to synthesize ATP from ADP and P_i.

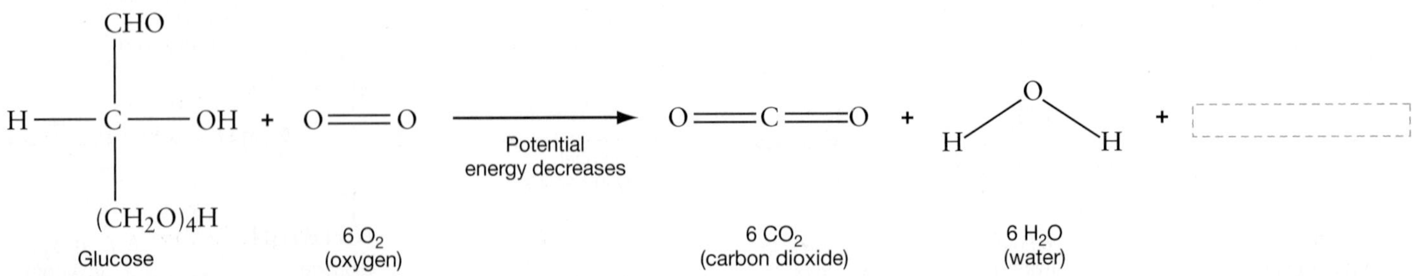

FIGURE 9.6 Tracking Electron Transfer during the Oxidation of Glucose.

✓**EXERCISE** Fill in the electron positions in each bond shown; then use these data to explain why the reaction is exergonic. (Check your work using the electron positions diagrammed in Figure 9.4).

CHECK YOUR UNDERSTANDING

9.2 An Overview of Cellular Respiration

In general, a cell contains only enough ATP to last from 30 seconds to a few minutes. Because it has such high potential energy, ATP is unstable and is not stored. Like many other cellular processes, the production and use of ATP is fast. Most cells are making ATP all the time.

Most of the glucose that is used to make ATP is produced by plants and other photosynthetic species. These organisms use the energy in sunlight to reduce carbon dioxide (CO_2) to glucose and other carbohydrates. While they are alive, photosynthetic species use the glucose that they produce to make ATP for themselves. When photosynthetic species decompose or are eaten, they provide glucose to animals, fungi, and many bacteria and archaea.

All organisms use glucose as a building block in the synthesis of fats, complex carbohydrates such as starch and glycogen, and other energy-storage compounds. The chemical energy stored in fats and in storage carbohydrates acts like a savings account. ATP, in contrast, is like cash. To make ATP and get cash, fats and storage carbohydrates have to be converted back to glucose. The glucose is then used to produce ATP through one of two general processes: cellular respiration or fermentation (**Figure 9.7**).

Because cellular respiration is much more efficient than fermentation, it is the primary source of ATP in most organisms. You can think of cellular respiration as a four-step process for producing ATP from a starting material with high potential energy—usually glucose. Each of the four steps consists of a series of chemical reactions, and each step has a distinctive starting molecule and a characteristic set of products.

1. *Glycolysis*   During **glycolysis**, one molecule of glucose is broken into two molecules of the three-carbon compound pyruvate. Two ATP molecules are produced from ADP, and one molecule of NAD^+ is reduced to form NADH.

2. *Pyruvate processing*   Pyruvate is processed to form the compound acetyl-CoA. During this step, another molecule of NADH is produced.

3. *Citric acid cycle*   Acetyl-CoA is oxidized to two molecules of CO_2. During this sequence of reactions, more ATP and NADH are produced and **flavine adenine dinucleotide (FAD)** is reduced to form another electron carrier, **FADH$_2$**.

4. *Electron transport and chemiosmosis*   Electrons from NADH and FADH$_2$ move through a series of proteins called an electron transport chain (ETC). The potential energy released during these redox reactions is used to create a proton gradient across a membrane; the ensuing flow of protons back across the membrane is used to make ATP.

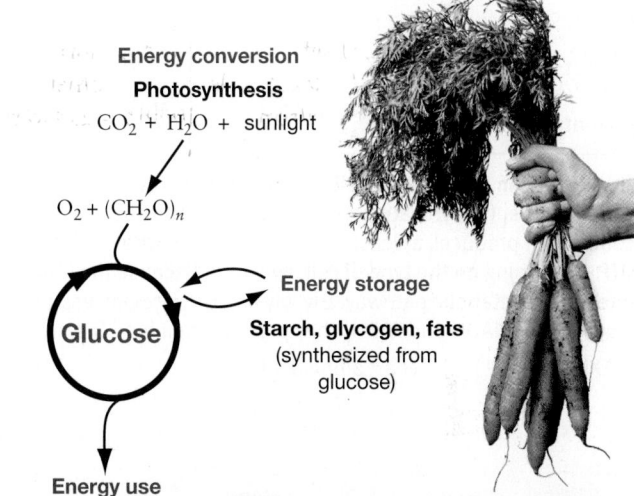

Energy conversion
Photosynthesis
$CO_2 + H_2O +$ sunlight

$O_2 + (CH_2O)_n$

Glucose

Energy storage
Starch, glycogen, fats
(synthesized from glucose)

Energy use

Cellular Respiration	**Fermentation**
Glucose + O_2 + ADP + P_i	Glucose + ADP + P_i
↓	↓
CO_2 + H_2O + **ATP**	Small organic molecules + **ATP**

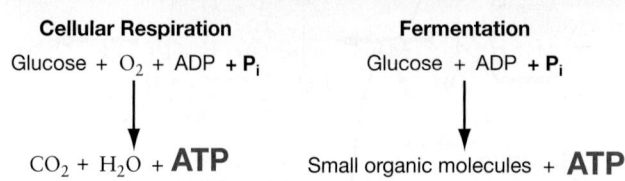

FIGURE 9.7 Glucose Is the Hub of Energy Processing in Cells. Glucose is the end-product of photosynthesis. Both plants and animals store glucose and oxidize it to provide chemical energy in the form of ATP.

Figure 9.8 summarizes the four steps in cellular respiration. Formally, **cellular respiration** is defined as any suite of reactions that produces ATP in an electron transport chain.

When you've filled in the chart at the bottom of the figure, you'll be ready to analyze each of the four steps in detail. As you delve into these details, keep asking yourself the same key questions: What goes in and what comes out? What happens to the potential energy that is released? Where does this step occur, and how is it regulated? Then take a look in the mirror. All of these processes are occurring right now, in virtually all of your cells.

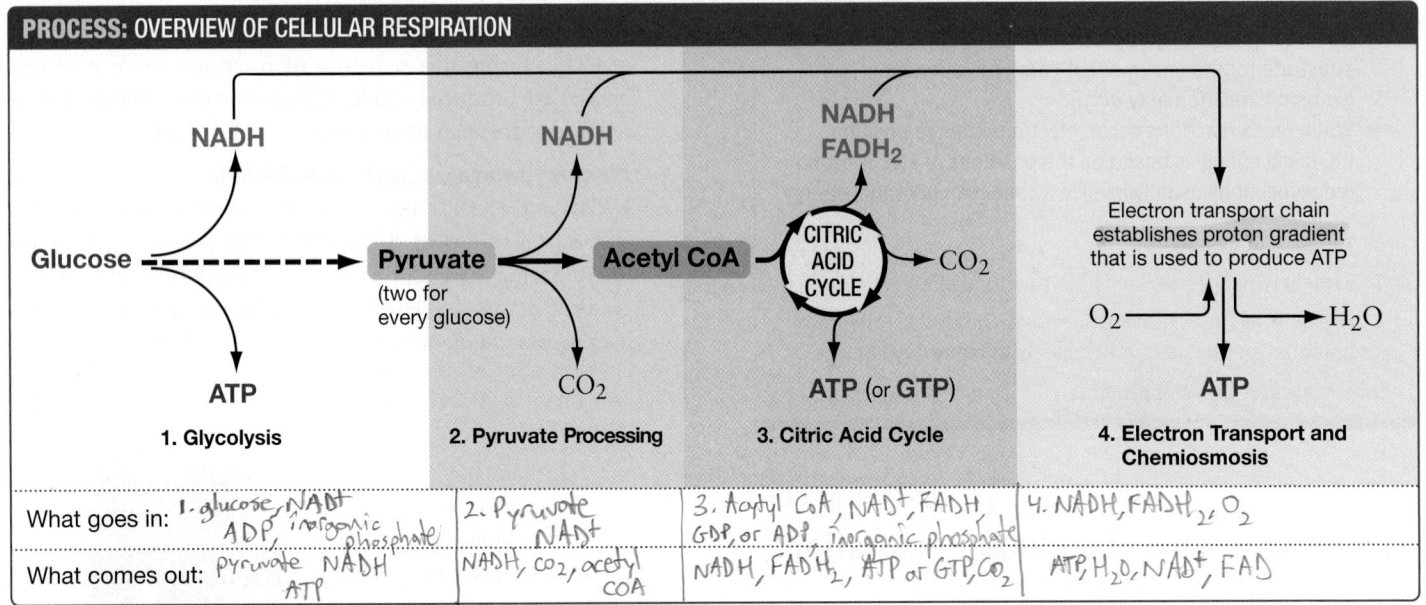

PROCESS: OVERVIEW OF CELLULAR RESPIRATION

	1. Glycolysis	2. Pyruvate Processing	3. Citric Acid Cycle	4. Electron Transport and Chemiosmosis
What goes in:	1. glucose, NAD+ ADP, inorganic phosphate	2. Pyruvate NAD+	3. Acetyl CoA, NAD+, FADH GDP, or ADP, inorganic phosphate	4. NADH, FADH$_2$, O$_2$
What comes out:	Pyruvate NADH ATP	NADH, CO$_2$, acetyl COA	NADH, FADH$_2$, ATP or GTP, CO$_2$	ATP, H$_2$O, NAD+, FAD

FIGURE 9.8 An Overview of Cellular Respiration. Cells produce ATP from glucose via a series of processes: (1) glycolysis, (2) pyruvate processing, (3) the citric acid cycle, and (4) electron transport and chemiosmosis. Each component produces at least some ATP or NADH (the citric acid cycle produces ATP or a related compound called GTP, depending on the type of cell involved). Because the four components are connected, glucose oxidation is an integrated metabolic pathway. Glycolysis, pyruvate processing, and the citric acid cycle complete the oxidation of glucose. The NADH and FADH$_2$ they produce then feed the electron transport chain.

✔**EXERCISE** Fill in the chart along the bottom.

FIGURE 9.9 Glycolysis Pathway. Glucose is oxidized to pyruvate through this sequence of 10 reactions. Each reaction is catalyzed by a different enzyme. The products are two net ATP (four ATP are produced, but two are invested), two molecules of NADH, and two molecules of pyruvate.

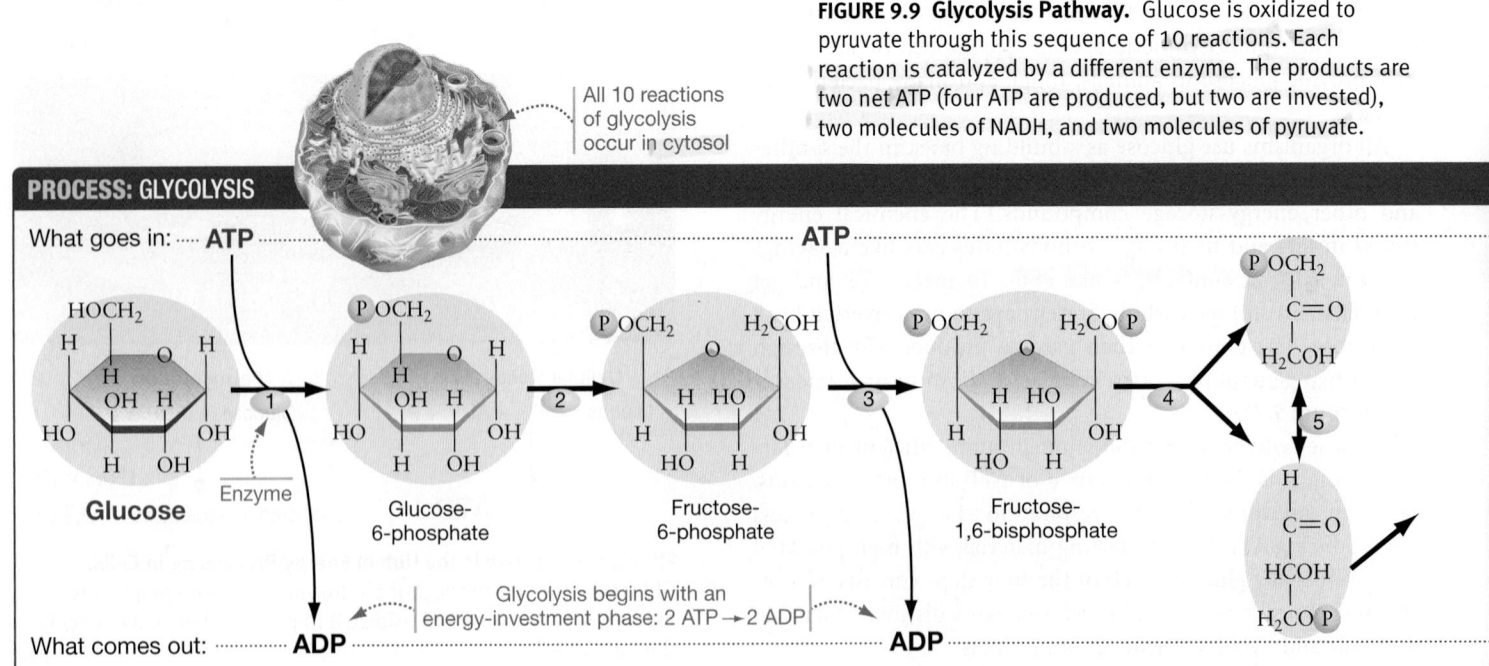

9.3 Glycolysis: Processing Glucose to Pyruvate

Because the enzymes responsible for glycolysis have been observed in nearly every bacterium, archaean, and eukaryote, it is logical to infer that the ancestor of all organisms living today made ATP by glycolysis. It's ironic, then, that the process was discovered by accident.

In the 1890s Hans and Edward Buchner were working out techniques for manufacturing extracts of baker's yeast for commercial and medicinal use. (Yeast extracts are still added to some foods as a flavor enhancer or nutritional supplement.) In one set of experiments the Buchners added sucrose, or table sugar, to their extracts. Sucrose is a disaccharide consisting of glucose and another six-carbon sugar. At the time, sucrose was commonly used as a preservative—a substance used to preserve food from decay.

Instead of preserving the yeast extracts, though, the sucrose was quickly broken down and fermented, with alcohol appearing as a by-product. This was a key finding: It showed that fermentation and other types of cellular metabolism could be studied in vitro—outside the organism. Until then, researchers thought that metabolism could take place only in intact organisms.

When researchers studied how the sugar was being processed, they found that the reactions could go on much longer than normal if inorganic phosphate was added to the mixture. This result implied that some of the compounds involved were being phosphorylated. Soon after, a molecule called fructose bisphosphate was isolated. (The prefix *bis*– means that two phosphate groups are attached to the fructose molecule at distinct locations.) Subsequent work showed that all but two of the compounds involved in glycolysis—the starting and ending molecules, glucose and pyruvate—are phosphorylated.

In 1905 researchers found that the processing of sugar by yeast extracts stopped if they boiled the reaction mix. Because enzymes were known to be inactivated by heat, this discovery suggested that enzymes were involved in at least some of the processing steps. Years later, investigators realized that each step in glycolysis is catalyzed by a different enzyme. Eventually, each of the reactions and enzymes involved was gradually worked out.

Glycolysis Is a Sequence of 10 Reactions

All 10 reactions of glycolysis occur in the cytosol (**Figure 9.9**). Note three key points about this reaction sequence:

1. Glycolysis starts by *using* ATP, not producing it. In the initial step, glucose is phosphorylated to form glucose-6-phosphate. After an enzyme rearranges this molecule to fructose-6-phosphate in the second reaction, the third reaction adds a second phosphate group, forming the fructose-1,6-bisphosphate observed by early researchers. Thus, two ATP molecules are used up before any ATP is produced.

2. Once this energy-investment phase of glycolysis is complete, the subsequent reactions represent an energy payoff phase. The sixth reaction in the sequence results in the reduction of two molecules of NAD$^+$; the seventh produces two molecules of ATP. This is where the energy "debt"—of two molecules of ATP invested early in glycolysis—is paid off. The final reaction in the sequence produces another two ATPs. For each molecule of glucose processed, the net yield is two molecules of NADH, two of ATP, and two of pyruvate.

3. In reactions 7 and 10 of Figure 9.9, an enzyme catalyzes the transfer of a phosphate group from a phosphorylated substrate to ADP, forming ATP. Enzyme-catalyzed reactions that result

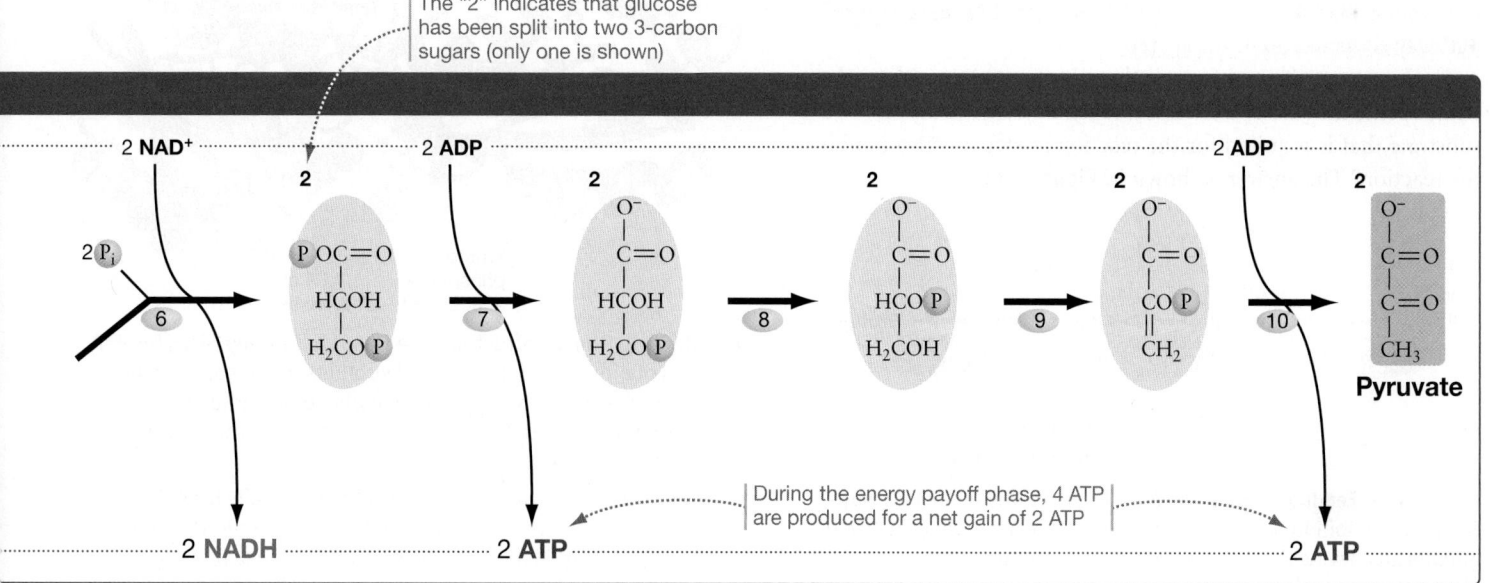

The "2" indicates that glucose has been split into two 3-carbon sugars (only one is shown)

During the energy payoff phase, 4 ATP are produced for a net gain of 2 ATP

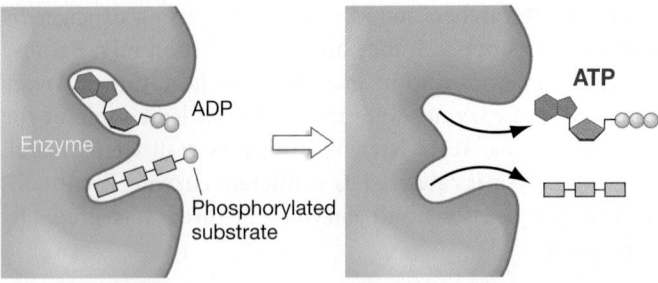

FIGURE 9.10 Substrate-Level Phosphorylation Involves an Enzyme and a Phosphorylated Substrate. Substrate-level phosphorylation occurs when an enzyme catalyzes the transfer of a phosphate group from a phosphorylated substrate to ADP, forming ATP.

in ATP production are termed **substrate-level phosphorylation** (**Figure 9.10**). The key thing to note here is that the energy to produce the ATP comes from the phosphorylated substrate—not from a proton gradient, as it does when ATP is produced by an electron transport chain.

The discovery and elucidation of the glycolytic pathway ranks as one of the great achievements in the history of biochemistry. The reactions outlined in Figure 9.9 are among the most ancient and fundamental of all life processes.

How Is Glycolysis Regulated?

An important advance occurred when biologists observed that high levels of ATP inhibit a key glycolytic enzyme called phosphofructokinase. **Phosphofructokinase** catalyzes reaction 3 in Figure 9.9—the synthesis of fructose-1,6-bisphosphate from fructose-6-phosphate. This is a crucial step in the sequence.

After reactions 1 and 2 occur, an array of enzymes can convert the products to molecules used in other metabolic pathways. Before step 3, then, the sequence can be interrupted and the intermediates used elsewhere in the cell. But once fructose-1,6-bisphosphate is synthesized, there is no point in stopping the process. Based on these observations, it makes sense that the pathway is turned on or off at step 3.

In the vast majority of cases, though, the addition of a substrate *speeds* the rate of a chemical reaction. Why would ATP—a substrate that is required for the reaction in step 3—also inhibit the reaction? The answer is shown in **Figure 9.11**.

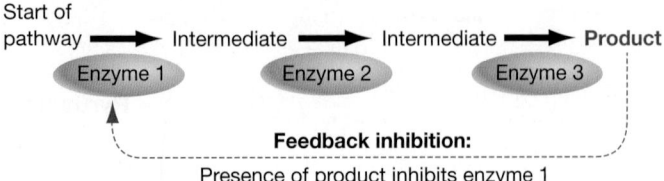

FIGURE 9.11 Feedback Inhibition May Regulate Metabolic Pathways. Feedback inhibition occurs when the product of a metabolic pathway inhibits an enzyme that is active early in the pathway.

When an enzyme in a pathway is inhibited by the product of the reaction sequence, **feedback inhibition** occurs. The product molecule "feeds back" to stop the reaction sequence when the product is abundant.

Feedback inhibition is efficient. Cells that are able to stop glycolytic reactions when ATP is abundant can conserve their stores of glucose for times when ATP is scarce. As a result, natural selection should favor individuals who have phosphofructokinase molecules that are inhibited by high concentrations of ATP.

How do high levels of the substrate inhibit the enzyme? As **Figure 9.12** shows, phosphofructokinase has two distinct binding sites for ATP. ATP can bind at the enzyme's active site or at a location that changes the enzyme's activity—a **regulatory site**.

At the active site, ATP is converted to ADP and the phosphate group is transferred to fructose-6-phosphate. This reaction results in the synthesis of fructose-1,6-bisphosphate.

🔑 If ATP concentrations are high, however, the molecule also binds at the regulatory site on phosphofructokinase. When ATP binds at this second location, the enzyme's conformation changes in a way that dramatically lowers the reaction rate at the active site. In this case, ATP acts as an allosteric regulator (see Chapter 3).

Now the question is, what happens to the pyruvate?

9.4 Processing Pyruvate to Acetyl CoA

In eukaryotes, the pyruvate produced by glycolysis is transported from the cytosol to mitochondria. As Chapter 7 noted, mitochondria are organelles found in virtually all eukaryotes.

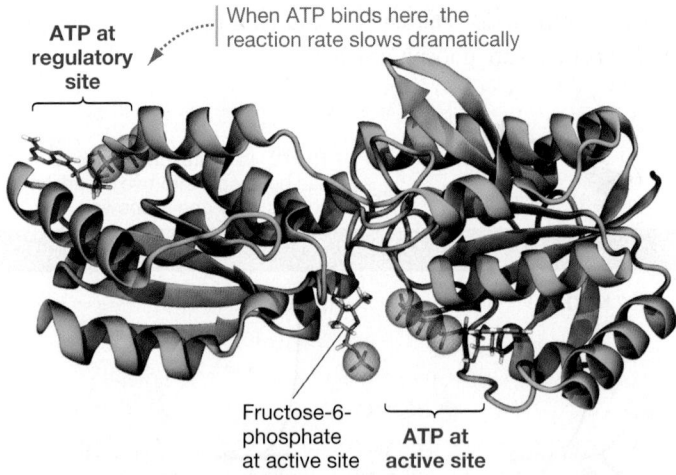

FIGURE 9.12 Phosphofructokinase Has Two Binding Sites for ATP. A model of one of the four identical subunits of phosphofructokinase. Notice the active site, where a phosphate group will be transferred from ATP to fructose-6-phosphate, and the regulatory site, where ATP binds.

✔ **QUESTION** The active site has much higher affinity for ATP than the regulatory site does. What would be the consequences if the regulatory site had higher affinity for ATP than the active site did?

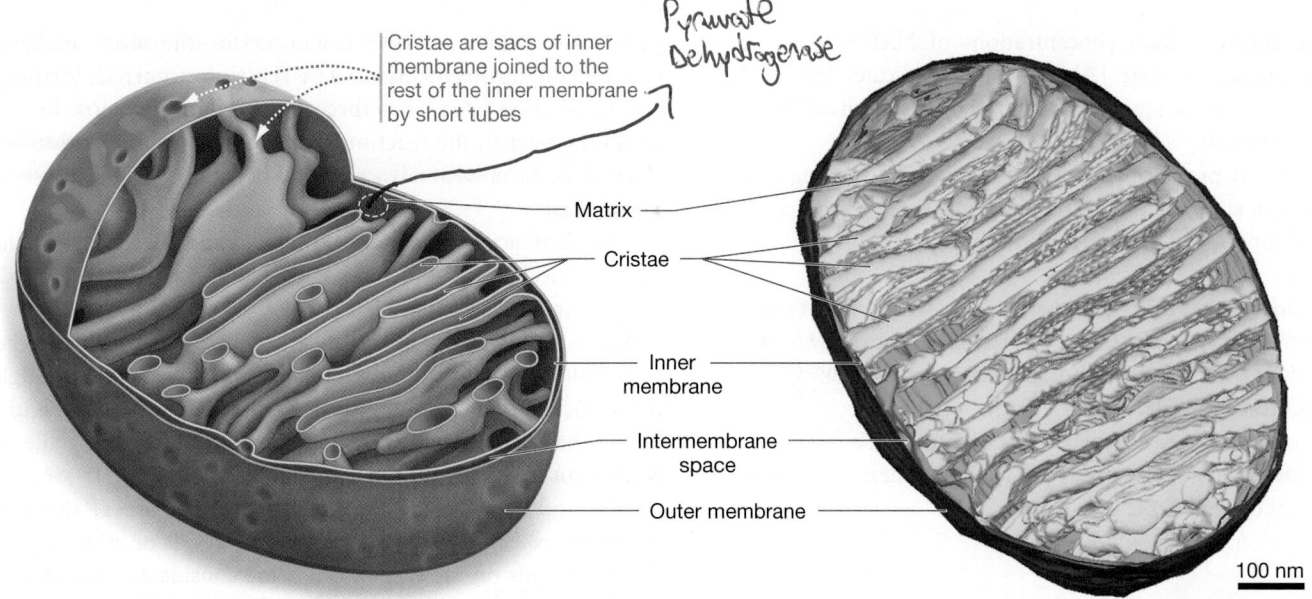

Cristae are sacs of inner membrane joined to the rest of the inner membrane by short tubes

Pyruvate Dehydrogenase

Matrix

Cristae

Inner membrane

Intermembrane space

Outer membrane

100 nm

FIGURE 9.13 The Structure of the Mitochondrion. These images are based on recent research using cryo-electron tomography (the micrograph, on the right, has been colorized). Notice that the mitochondria have outer and inner membranes, and the inner membrane is connected by short tubes to sac-like cristae. Pyruvate processing occurs within the mitochondrial matrix.

Figure 9.13 shows a diagram and an image of this organelle generated with an imaging technique called cryo-electron tomography.[1] Notice that a mitochondrion has two membranes, called the inner membrane and outer membrane. The interior of the organelle is filled with layers of sac-like structures called **cristae**. Short tubes connect the cristae to the main part of the inner membrane. The region inside the inner membrane but outside the cristae is the **mitochondrial matrix**.

Pyruvate moves across the mitochondrion's outer membrane through small pores. Entry into the matrix occurs through a membrane protein called the pyruvate carrier, located in the inner membrane. Transport into the matrix is an active process. It requires ATP and represents an energy-consuming step in glucose oxidation.

Inside the mitochondrion, pyruvate reacts with a compound called **coenzyme A (CoA)**. In this and many other reactions, CoA acts as a coenzyme by accepting and then transferring an acetyl group ($-COCH_3$) to a substrate (the "A" stands for acetylation). Pyruvate reacts with CoA, through a series of steps, to produce **acetyl CoA**.

The reaction sequence occurs inside an enormous and intricate enzyme complex called **pyruvate dehydrogenase**. In eukaryotes, pyruvate dehydrogenase is located in the mitochondrial matrix. In bacteria and archaea, pyruvate dehydrogenase is located in the cytosol.

As pyruvate is being processed, NAD^+ is reduced to NADH and one of the carbons in the pyruvate is oxidized to CO_2. The remaining two-carbon acetyl unit is transferred to CoA (**Figure 9.14**). Acetyl CoA is the final product of the pyruvate processing step in glucose oxidation. Pyruvate, NAD^+, and CoA go in; CO_2, NADH, and acetyl CoA come out.

When supplies of ATP are abundant, however, the process shuts down. Pyruvate processing stops when the pyruvate dehydrogenase complex becomes phosphorylated and changes shape. The rate of phosphorylation increases when other products—specifically acetyl CoA and NADH—are at high concentration.

These regulatory changes are more examples of feedback inhibition. Reaction products feed back to stop or slow down the pathway.

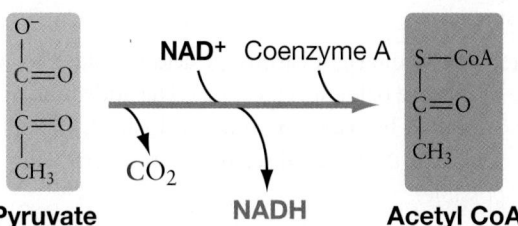

O$^-$
|
C=O
|
C=O
|
CH$_3$

Pyruvate

NAD^+ Coenzyme A

CO_2

NADH

S—CoA
|
C=O
|
CH$_3$

Acetyl CoA

FIGURE 9.14 Pyruvate Is Oxidized to Acetyl CoA. The reaction shown here is catalyzed by pyruvate dehydrogenase.

✔**EXERCISE** Above the reaction arrow, list three molecules whose presence speeds up the reaction. Label them "Positive control." Below the reaction arrow, list three molecules whose presence slows down the reaction. Label them "Negative control by feedback inhibition."

---

[1]Compared to images from transmission electron microscopy, cryo-electron tomography has provided a much more accurate picture of mitochondrial morphology. For a recent review, see C. Mannella, *Biochemica et Biophysica Acta* 1762 (2006):140–147.

On the contrary, high concentrations of NAD^+, CoA, or adenosine monophosphate (AMP)—which indicates low ATP supplies—*speed up* the reactions catalyzed by the pyruvate dehydrogenase complex.

🔑 Pyruvate processing is under both positive and negative control. Large supplies of products inhibit the enzyme complex; large supplies of reactants and low supplies of products stimulate it.

To summarize, pyruvate processing starts with pyruvate and ends with acetyl CoA, releasing CO_2. The reactions occur in the mitochondrial matrix, and the potential energy that is released is used to produce NADH. When energy supplies are high, the pyruvate dehydrogenase complex slows down; when energy supplies are low, the complex speeds up. Now the question is, what happens to the acetyl CoA?

9.5 The Citric Acid Cycle: Oxidizing Acetyl CoA to CO_2

While researchers were working out the sequence of reactions in glycolysis, biologists in other laboratories were focusing on a different set of redox reactions that take place in cells. These reactions involve small organic acids such as citrate, malate, and succinate. Because they have the form R-COOH, these molecules are called **carboxylic acids**.

In some cases, the redox reactions that produce carboxylic acids also produce carbon dioxide. Recall from Section 9.1 that carbon dioxide is a highly oxidized form of carbon and the endpoint of glucose metabolism. Thus, it was logical for researchers to propose that the oxidation of small carboxylic acids could be an important component of glucose oxidation.

Early researchers made three key observations about these reactions:

1. A total of eight small carboxylic acids are oxidized rapidly enough to imply that they are involved in glucose metabolism—the most rapid set of oxidation reactions predicted to occur in cells.

2. The eight carboxylic acids can react in sequence, from least to most oxidized.

3. When one of the eight carboxylic acids is added to cells, the rate of glucose oxidation increases. The added molecules do not appear to be used up, however. Instead, virtually all of the carboxylic acids added seem to be recovered later. How is this possible?

Hans Krebs solved the mystery when he realized that the reaction sequence might occur in a cycle instead of a linear pathway. Krebs had another crucial insight when he suggested that the reaction sequence was directly tied to the processing of pyruvate produced by glycolysis.

To test these hypotheses, Krebs and a colleague set out to determine whether pyruvate—the endpoint of the glycolytic pathway—could react with oxaloacetate—the most oxidized of the eight carboxylic acids. The reaction occurred, forming citrate. Because citrate has three carboxyl groups, most biologists now refer to the reaction sequence as the **citric acid cycle**, because it starts with citrate, which becomes citric acid when protonated.

The citric acid cycle is also known as the tricarboxylic acid (TCA) cycle, because it involves acids with three carboxyl groups, and the Krebs cycle, after its discoverer.

When radioactive isotopes of carbon became available in the early 1940s, researchers showed that carbon atoms cycle through the sequence of reactions just as Krebs had proposed (**Figure 9.15**). The energy released by the oxidation of one molecule of acetyl CoA is used to produce three molecules of NADH, one of $FADH_2$, and one of **guanosine triphosphate (GTP)** or ATP through substrate-level phosphorylation. Whether GTP or ATP is produced depends on the type of cell being considered.[2] For example, GTP appears to be produced in the liver cells of mammals while ATP is produced in muscle cells.

In bacteria and archaea, the enzymes responsible for the citric acid cycle are located in the cytosol. In eukaryotes, most of the enzymes responsible for the citric acid cycle are located in the mitochondrial matrix. Because glycolysis produces two molecules of pyruvate, the cycle turns twice for each molecule of glucose processed in cellular respiration.

How Is the Citric Acid Cycle Regulated?

By now, it shouldn't surprise you to learn that the citric acid cycle is carefully regulated. Reaction rates are high when ATP is scarce; reaction rates are low when ATP is abundant.

Figure 9.16 highlights the major control points. Notice that the enzyme that converts acetyl CoA to citrate is shut down when ATP binds to it. This is another example of feedback inhibition: Reaction products feed back to stop or slow down the pathway.

As Figure 9.16 indicates, feedback inhibition also occurs at two points later in the cycle. At the first of these two points, NADH binds to the enzyme's active site. This is an example of competitive inhibition (see Chapter 3). At the second point, ATP binds to an allosteric regulatory site.

🔑 The citric acid cycle can be turned off at multiple points, via several different mechanisms of feedback inhibition.

To summarize, the TCA cycle starts with acetyl and ends with CO_2. The reactions occur in the mitochondrial matrix, and the potential energy that is released is used to produce NADH, $FADH_2$, and ATP or GTP. When energy supplies are high, the cycle slows down. But a major question remains.

---

[2]Traditionally it was thought that the citric acid cycle produced GTP, which was later converted to ATP in the same cell. Recent work suggests that ATP is produced directly in some cell types, while GTP is produced in other cells. See C. O. Lambeth, *Biochemistry and Molecular Biology Education* 34 (2006): 21–29.

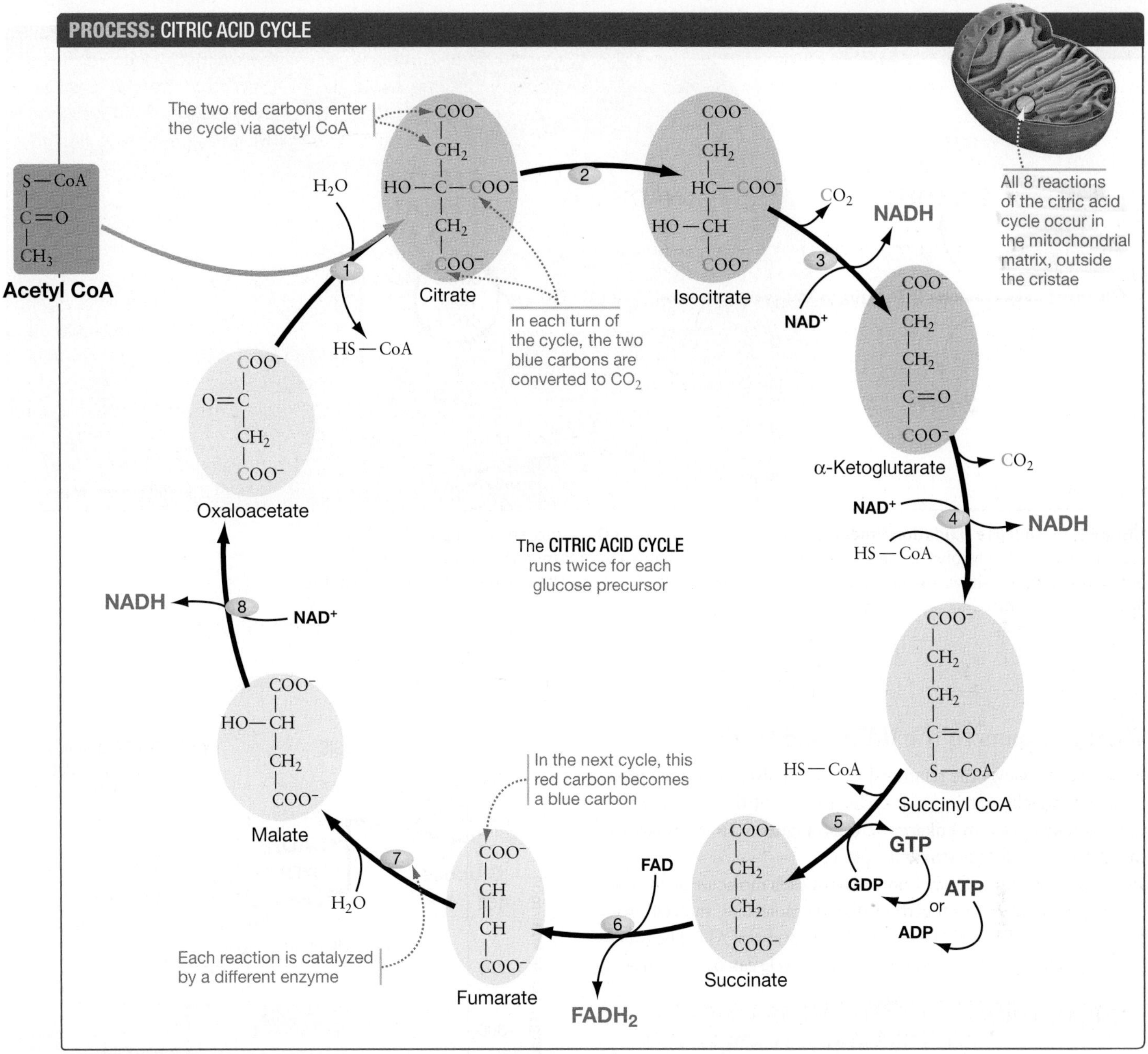

The two red carbons enter the cycle via acetyl CoA

H_2O

Acetyl CoA

1

HS — CoA

Citrate

In each turn of the cycle, the two blue carbons are converted to CO_2

2

Isocitrate

CO_2

NADH

3

NAD^+

α-Ketoglutarate

CO_2

NAD^+

4

NADH

HS — CoA

All 8 reactions of the citric acid cycle occur in the mitochondrial matrix, outside the cristae

Oxaloacetate

NADH

8

NAD^+

The **CITRIC ACID CYCLE** runs twice for each glucose precursor

Succinyl CoA

HS — CoA

5

GTP

GDP

ATP

or

ADP

Malate

7

H_2O

Fumarate

In the next cycle, this red carbon becomes a blue carbon

6

FAD

FADH$_2$

Succinate

Each reaction is catalyzed by a different enzyme

FIGURE 9.15 The Citric Acid Cycle Completes the Oxidation of Glucose. Acetyl CoA goes into the citric acid cycle, and carbon dioxide, NADH, FADH$_2$, and GTP come out. GTP or ATP are produced by substrate-level phosphorylation. If you follow individual carbon atoms around the cycle several times, you'll come to an important conclusion: each of the carbons in this cycle is eventually a "blue carbon" that is released as CO_2. This occurs because the "red carbons" in fumarate—which is symmetric—can flip position when malate forms.

FIGURE 9.16 The Citric Acid Cycle Is Regulated by Feedback Inhibition. The citric acid cycle slows down when ATP and NADH are plentiful. ATP acts as an allosteric regulator, while NADH acts as a competitive inhibitor.

✔**QUESTION** How do allosteric regulation and competitive inhibition differ?

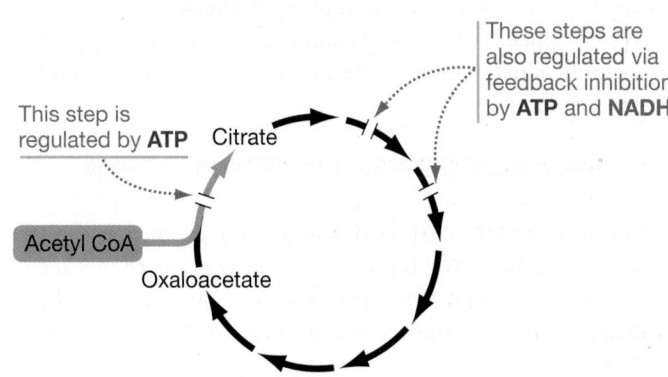

This step is regulated by **ATP**

These steps are also regulated via feedback inhibition by **ATP** and **NADH**

Citrate

Acetyl CoA

Oxaloacetate

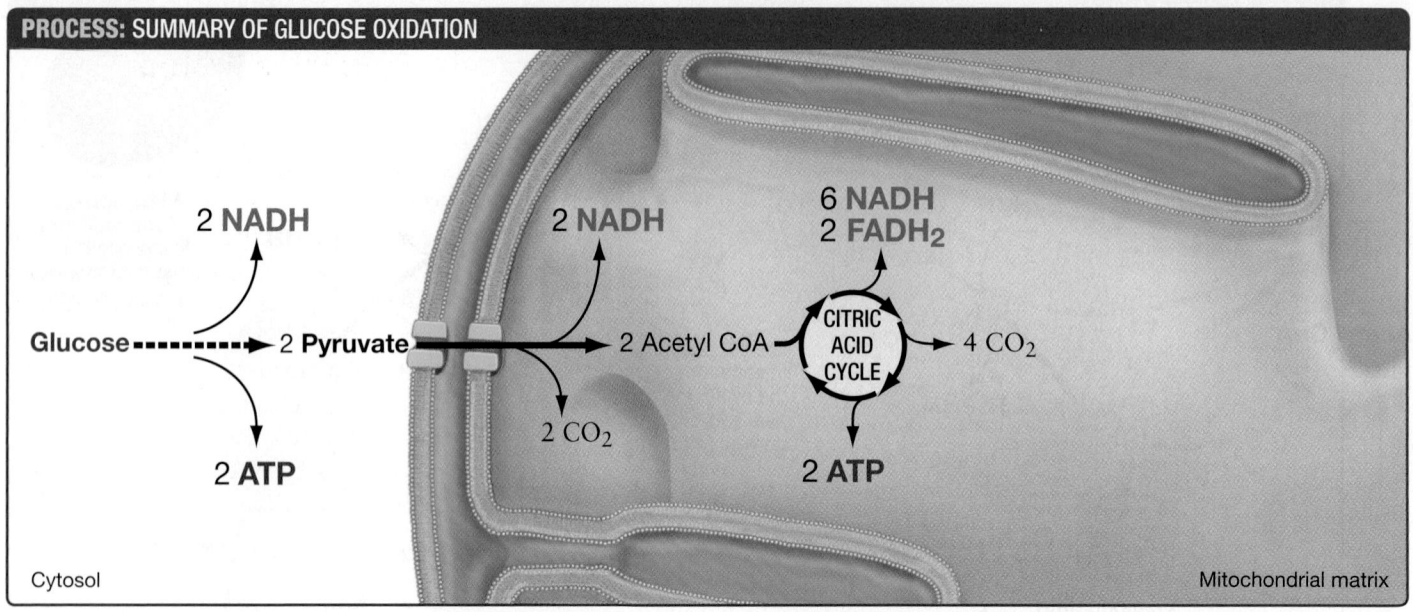

FIGURE 9.17 Glucose Oxidation Produces ATP, NADH, FADH$_2$, and CO$_2$. Glucose is completely oxidized to carbon dioxide via glycolysis, the subsequent oxidation of pyruvate, and then the citric acid cycle. In eukaryotes, glycolysis occurs in the cytosol; pyruvate oxidation and the citric acid cycle take place in the mitochondrial matrix.

What Happens to the NADH and FADH$_2$?

Figure 9.17 reviews the relationships of glycolysis, pyruvate processing, and the citric acid cycle and identifies where each process takes place in eukaryotic cells; **Figure 9.18** summarizes the free energy changes that take place.

As you study these figures, note that for each molecule of glucose that is fully oxidized to 6 carbon dioxide molecules, the cell produces 10 molecules of NADH, 2 of FADH$_2$, and 4 of ATP. The overall reaction for glycolysis and the citric acid cycle can be written as

$$C_6H_{12}O_6 + 10\ NAD^+ + 2\ FAD + 4\ ADP + 4\ P_i \longrightarrow$$
$$6\ CO_2 + 10\ NADH + 2\ FADH_2 + 4\ ATP$$

The ATP molecules are produced by substrate-level phosphorylation and can be used to drive endergonic reactions, power movement, or run membrane pumps. The carbon dioxide molecules are a gas that is disposed of as waste—you exhale it; plants release it or use it as a reactant in photosynthesis.

What happens to the NADH and FADH$_2$ produced by glycolysis and the citric acid cycle? Recall that the overall reaction for glucose oxidation is

$$C_6H_{12}O_6 + 6\ O_2 \longrightarrow 6\ CO_2 + 6\ H_2O + Energy$$

Glycolysis and the citric acid cycle account for the glucose, the CO$_2$, and—because ATP is produced—some of the chemical energy that results from the overall reaction. But the O$_2$ and the H$_2$O that appear in the overall reaction for the oxidation of

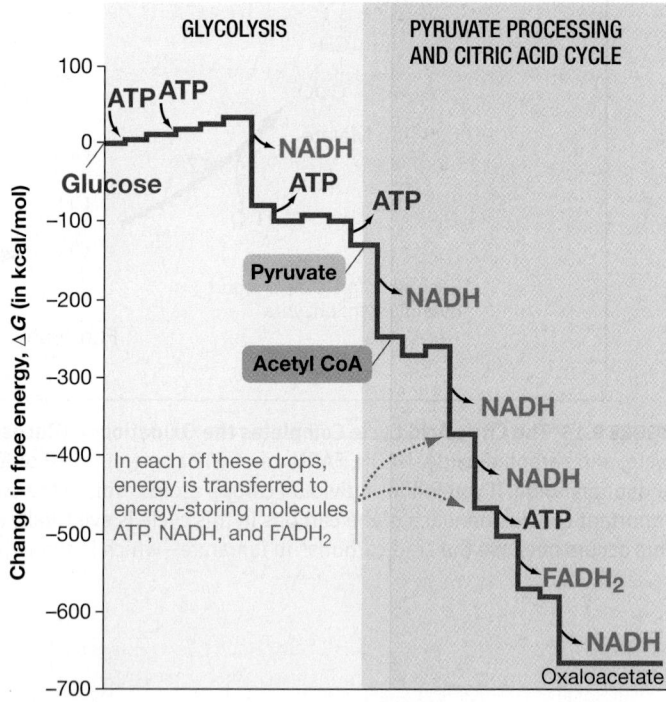

FIGURE 9.18 Free Energy Changes as Glucose Is Oxidized. If you read the vertical axis of this graph carefully, it should convince you that about 125 kcal/mol are released during glycolysis, about 125 kcal/mol during pyruvate processing, and about 430 kcal/mol during the citric acid cycle.

✔**QUESTION** Which is associated with larger changes in free energy, production of ATP or production of NADH and FADH$_2$?

glucose are still unaccounted for. As it turns out, so is much of the chemical energy. The reaction that has yet to occur is

$$NADH + FADH_2 + O_2 + ADP + P_i \longrightarrow$$
$$NAD^+ + FAD + H_2O + ATP$$

In this reaction, oxygen is reduced to form water. The electrons that drive the redox reaction come from NADH and $FADH_2$. These molecules are oxidized to $NAD^+ + FAD$.

In effect, glycolysis, pyruvate processing, and the citric acid cycle transfer electrons from glucose to NAD^+ and FAD, creating NADH and $FADH_2$. These molecules then carry the electrons to oxygen, which serves as the final electron acceptor in eukaryotic cells. When oxygen accepts electrons, water is produced. All the components of the overall reaction for glucose oxidation are accounted for.

How does this final part of the process occur? Specifically, how is ATP generated as electrons are transferred from NADH or $FADH_2$ to O_2? In the 1960s—decades after the details of glycolysis and the citric acid cycle had been worked out—a startling answer to these questions emerged.

CHECK YOUR UNDERSTANDING

If you understand that . . .

- During glycolysis, glucose is oxidized to pyruvate, in the cytosol.
- During pyruvate processing, pyruvate is oxidized to acetyl CoA, in the mitochondrial matrix.
- In the citric acid cycle, citric acid is oxidized to carbon dioxide (CO_2), in the mitochondrial matrix.
- Glycolysis, pyruvate processing, and the citric acid cycle are all regulated processes. The cell only produces ATP when ATP is needed.

✓ You should be able to . . .

Model the following components of cellular respiration by pretending that a large piece of paper is a cell. Draw a large mitochondrion inside it. Cut out small squares of paper and label them as glucose, glycolytic reactions, citric acid cycle reactions, pyruvate dehydrogenase complex, pyruvate, acetyl CoA, CO_2, ADP $\longrightarrow$ ATP, NAD^+ $\longrightarrow$ NADH, FAD $\longrightarrow$ $FADH_2$.

1. Put each of the squares in the appropriate location in the cell.
2. Draw arrows to connect the appropriate molecules and reactions.
3. Using dimes or paper circles for electrons, show how a total of 12 electrons from glucose are transferred to NADH or $FADH_2$ (two electrons should go to each NADH or $FADH_2$ formed) as glucose is oxidized to CO_2.
4. Label points where regulation occurs.

Once your model is working, you'll be ready to consider what happens to the NADH and $FADH_2$ you've produced.

Answers are available in Appendix B.

9.6 Electron Transport and Chemiosmosis: Building a Proton Gradient to Produce ATP

The answer to one fundamental question about the oxidation of NADH and $FADH_2$ turned out to be relatively straightforward. By isolating different components of mitochondria, researchers determined that NADH is oxidized in the inner membrane of the mitochondria and the membranes of cristae. In prokaryotes, the oxidation of NADH occurs in the plasma membrane.

Biologists who analyzed the components of these membranes made a key discovery when they isolated molecules that switch between a reduced and an oxidized state during respiration. The molecules were hypothesized to be the key to processing NADH and $FADH_2$. What are these molecules, and how do they work?

Components of the Electron Transport Chain

Collectively, the molecules responsible for the oxidation of NADH and $FADH_2$ are designated the **electron transport chain (ETC)**. As electrons are passed from one protein to another in the chain, the energy released by the redox reactions is used to pump protons across the inner membrane of mitochondria.

After this proton gradient is established, a stream of protons through the enzyme **ATP synthase** makes part of the protein spin, driving the production of ATP from ADP and P_i. Because this mode of ATP production links the phosphorylation of ADP with the oxidation of NADH and $FADH_2$, it is called **oxidative phosphorylation.**

Recall that when substrate-level phosphorylation occurs, a phosphate group is transferred from a phosphorylated substrate to ADP, forming ATP. This is not what happens when ATP synthase spins and synthesizes ATP from ADP and P_i.

Once the electrons at the bottom of the ETC are accepted by oxygen to form water, the oxidation of glucose is complete.

Several points are fundamental to understanding how the ETC works:

- Most of the molecules are proteins that contain distinctive chemical groups where the redox events take place. The active groups include ring-containing structures called flavins or iron-sulfur complexes or iron-containing heme groups. Each of these groups is readily reduced or oxidized.

- The inner membrane of the mitochondrion also contains a molecule called **ubiquinone**, which is not a protein. Ubiquinone got its name because it is nearly ubiquitous in organisms and belongs to a family of compounds called quinones. Also called **coenzyme Q** or simply **Q**, ubiquinone consists of a carbon-containing ring attached to a long tail made up of isoprene subunits. The structure of Q determines the molecule's function. The long, isoprene-rich tail is hydrophobic. As a result, Q is lipid soluble and can move throughout the mitochondrial membrane efficiently. In contrast, all but one of the proteins in the ETC are embedded in the membrane.

- The molecules involved in processing NADH and FADH$_2$ differ in electronegativity, or their tendency to hold electrons.

Because Q and the ETC proteins can cycle between a reduced state and an oxidized state, and because they differ in electronegativity, investigators realized that it should be possible to arrange these molecules into a logical sequence. The idea was that electrons would pass from a molecule with lower electronegativity to one with higher electronegativity, via a redox reaction.

As electrons moved through the chain, they would be held more and more tightly. A small amount of energy would be released in each reaction, and the potential energy in each successive bond would lessen.

Researchers worked out the sequence of compounds in the chain by experimenting with poisons that inhibit particular proteins in the inner membrane. For example, when an electron transport chain is treated with the drug antimycin A, cytochrome b and Q are reduced; but all of the other elements in the chain remain oxidized. This pattern only makes sense if electrons flow from NADH and FADH$_2$ to cytochrome b and Q before being passed on to other components.

Experiments with other poisons showed that NADH donates an electron to a flavin-containing protein at the top of the chain, while FADH$_2$ donates electrons to an iron-and sulfur-containing protein that then passes them directly to Q. After passing through each of the remaining components in the chain, the electrons are finally accepted by oxygen.

Figure 9.19 shows how electrons step down in potential energy from the electron carriers NADH and FADH$_2$ to O$_2$. The *x*-axis plots the sequence of redox reactions in the ETC; the *y*-axis plots the free energy changes that occur. Under standard conditions of temperature and pressure in the laboratory, the total potential energy difference from NADH to oxygen is a whopping 53 kilocalories/mole.

Once the nature of the electron transport chain became clear, biologists understood the fate of the electrons carried by NADH and FADH$_2$ and how oxygen acts as the final electron acceptor. All of the electrons that were originally present in glucose were now accounted for. This is satisfying, except for one crucial question: How is ATP produced?

The Chemiosmosis Hypothesis

Throughout the 1950s most biologists working on cellular respiration assumed that electron transport chains include enzymes that catalyze substrate-level phosphorylation. Despite intense efforts, however, no one was able to find a component of the ETC that phosphorylated ADP to produce ATP.

In 1961 Peter Mitchell broke with prevailing ideas by proposing that the connection between electron transport and ATP production is indirect. Mitchell's novel hypothesis? The real job of the electron transport chain is to pump protons from the matrix of the mitochondrion through the inner membrane and out to the intermembrane space or the interior of cristae.

According to Mitchell, the pumping activity of the electron transport chain would lead to a buildup of protons in these

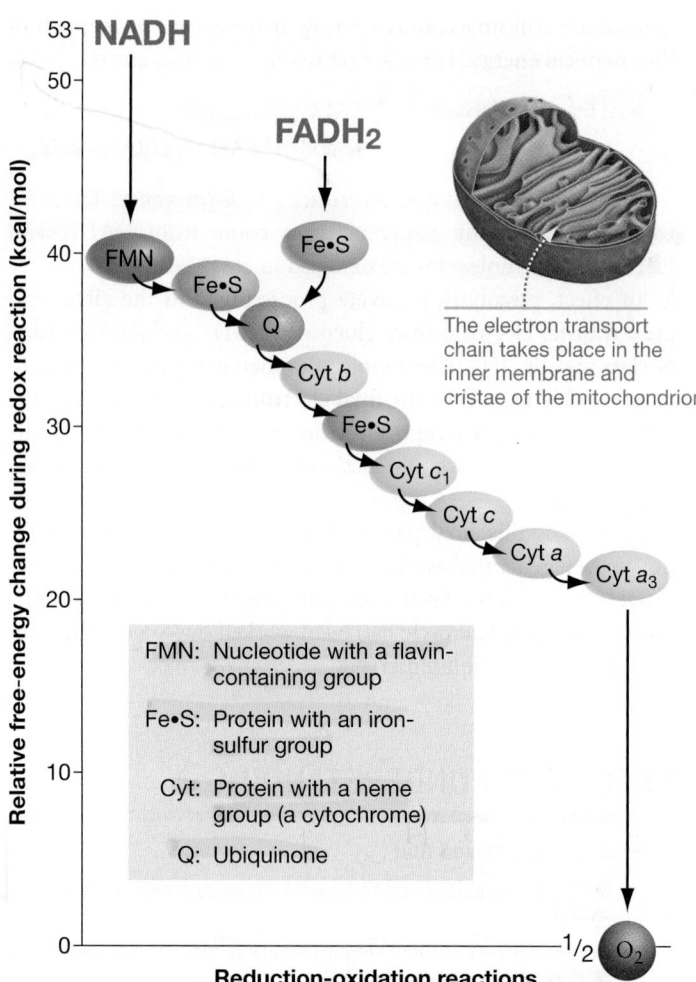

FIGURE 9.19 A Series of Reduction-Oxidation Reactions Occurs in an Electron Transport Chain. Electrons step down in potential energy from the electron carriers NADH and FADH$_2$ through an electron transport chain to a final electron acceptor. When oxygen is the final electron acceptor, water is formed. The overall free-energy change of 53 kcal/mol (from NADH to oxygen) is broken into small steps.

areas. In this way, the intermembrane space and the inside of cristae would become positively charged relative to the matrix and would have a much higher concentration of protons. The result would be a strong electrochemical gradient favoring the movement of protons back into the matrix. He hypothesized that an enzyme in the inner membrane uses this **proton-motive force** to synthesize ATP.

Mitchell called the production of ATP via a proton gradient **chemiosmosis**. Although proponents of a direct link between electron transport and substrate-level phosphorylation objected vigorously to Mitchell's idea, several key experiments supported it.

Figure 9.20 illustrates how the existence of a key element in Mitchell's hypothesis was confirmed: A mitochondrial enzyme can use a proton gradient to synthesize ATP. The researchers made vesicles from artificial membranes that contained an ATP-synthesizing enzyme found in mitochondria. Along with this en-

QUESTION: How are the electron transport chain and ATP production linked?

CHEMIOSMOTIC HYPOTHESIS: The linkage is indirect. The ETC creates a proton-motive force that drives ATP synthesis by a mitochondrial protein.

ALTERNATIVE HYPOTHESIS: The linkage is direct. The ETC is associated with enzymes that perform substrate-level phosphorylation.

EXPERIMENTAL SETUP:

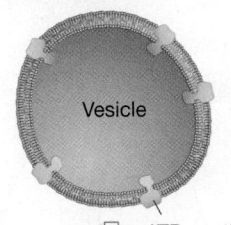

1. Produce vesicles from artificial membranes; add ATP-synthesizing enzyme found in mitochondria.

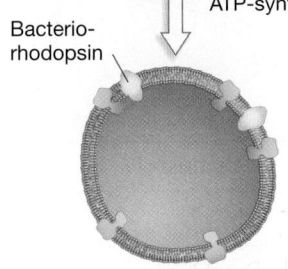

2. Add bacteriorhodopsin, a protein that acts as a light-activated proton pump.

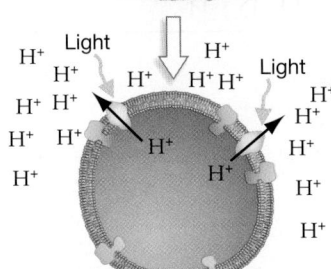

3. Illuminate vesicle so that bacteriorhodopsin pumps protons out of vesicle, creating a proton gradient.

PREDICTION OF CHEMIOSMOTIC HYPOTHESIS: ATP will be produced within the vesicle.

PREDICTION OF ALTERNATIVE HYPOTHESIS: No ATP will be produced.

RESULTS:

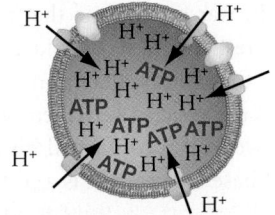

ATP is produced within the vesicle, in the absence of the electron transport chain.

CONCLUSION: The linkage between electron transport and ATP synthesis is indirect; the movement of protons drives the synthesis of ATP.

FIGURE 9.20 Evidence for the Chemiosmotic Hypothesis.

SOURCE: Racker, E. and W. Stoeckenius. 1974. Reconstitution of purple membrane vesicles catalyzing light-driven proton uptake and adenosine triphosphate formation. *Journal of Biological Chemistry.* 249: 662–663.

✔**QUESTION** Do you regard this as a convincing test of the chemiosmosis hypothesis? Why or why not?

---

zyme, they inserted bacteriorhodopsin, a well-studied membrane protein that acts as a light-activated proton pump.

When light strikes bacteriorhodopsin, it absorbs some of the light energy and changes conformation in a way that pumps protons from the interior of a membrane to the exterior. As a result, the experimental vesicles established a strong electrochemical gradient favoring proton movement to the interior. When the vesicles were illuminated to initiate proton pumping, ATP began to be produced from ADP inside the vesicles.

Mitchell's prediction was correct: In this situation, ATP production depended solely on the existence of a proton-motive force. It could occur in the *absence* of an electron transport chain. This result, and many others, have provided strong support for the hypothesis of chemiosmosis. Most ATP is produced by a flow of protons.

Chemiosmosis is like a hydroelectric dam, where the movement of water makes turbines spin and generate electricity. The electron transport chain is analogous to a series of gigantic pumps that force water up and behind the dam. The inner mitochondrial membrane functions as the dam, and ATP synthase is like the turbines inside the dam. In a mitochondrion, protons are pumped instead of water. When protons move through ATP synthase, the protein spins and generates ATP.

✔If you understand chemiosmosis, you should be able to explain why ATP production during cellular respiration is characterized as indirect. More specifically, you should be able to explain the relationship between glucose oxidation, the proton gradient, and ATP synthase.

Electron transport chains and ATP synthase occur in organisms throughout the tree of life. They are humming away in your cells now. Let's look in more detail at how they function.

How Is the Electron Transport Chain Organized?

The components of the electron transport chain are organized into four large complexes of proteins and cofactors (**Figure 9.21** on page 164). Two of the complexes pump protons. Q and the protein **cytochrome *c*** act as shuttles that transfer electrons between complexes. Q also carries a proton across the membrane along with an electron.

Research confirms that in complexes I and IV, protons actually pass directly through a sequence of electron carriers. The exact route taken by the protons is still being worked out. It is also not clear how the redox reactions taking place inside each complex—as electrons step down in potential energy—make proton movement possible.

The best-understood interaction between electron transport and proton pumping takes place in complex III. Research has shown that when Q accepts electrons from complex I or complex II, it also gains protons. The reduced form of Q then diffuses to the outer side of the inner membrane, where its electrons are used to reduce a component of complex III near the intermembrane space. The protons held by Q are released to the intermembrane space.

In this way, Q shuttles electrons and protons from one side of the membrane to the other. The electrons proceed down the

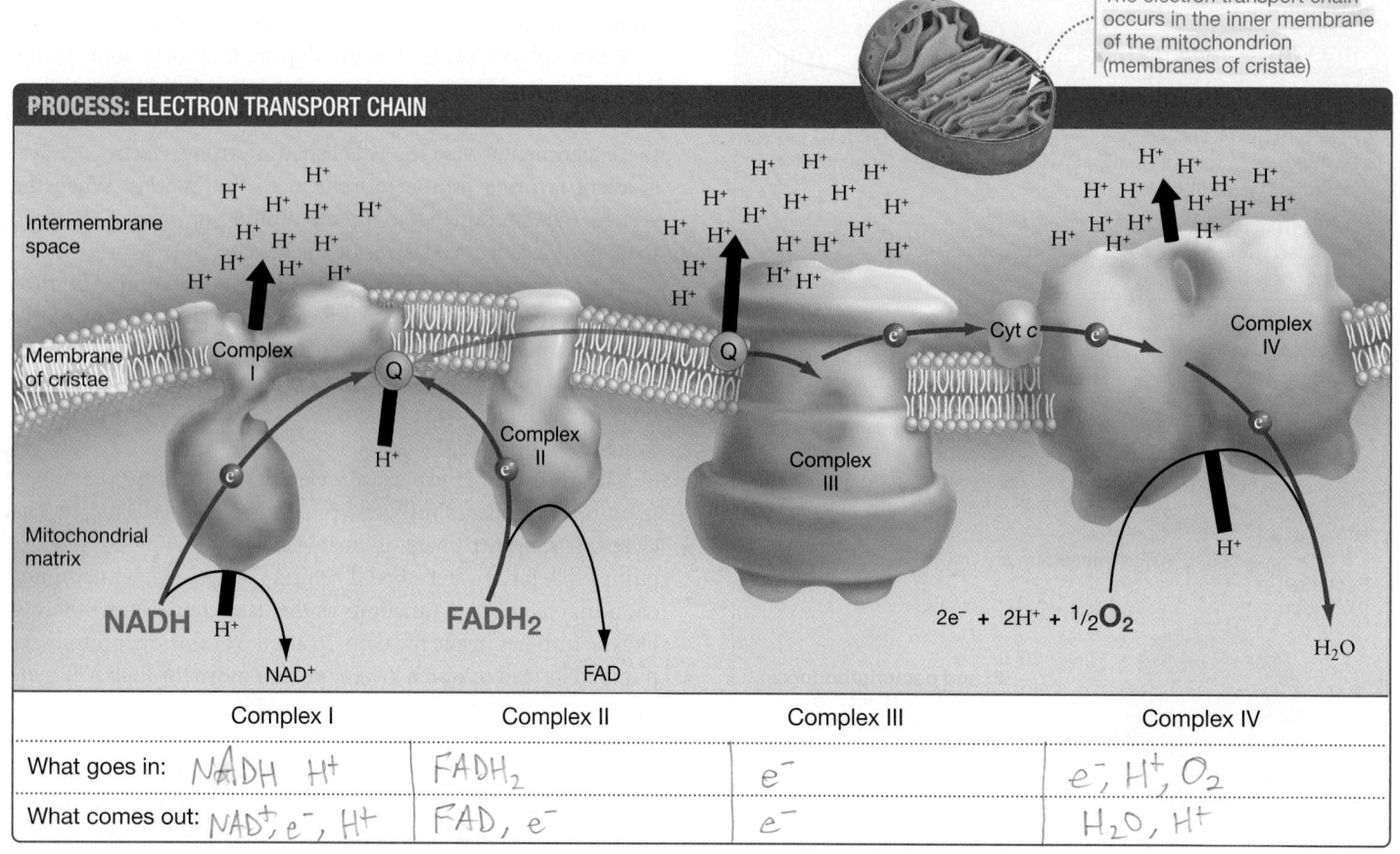

PROCESS: ELECTRON TRANSPORT CHAIN

The electron transport chain occurs in the inner membrane of the mitochondrion (membranes of cristae)

Intermembrane space

Membrane of cristae

Complex I

Complex II

Complex III

Complex IV

Mitochondrial matrix

NADH H^+ $\rightarrow$ NAD^+

FADH$_2$ $\rightarrow$ FAD

$2e^- + 2H^+ + \frac{1}{2}O_2$ $\rightarrow$ H_2O

Cyt c

	Complex I	Complex II	Complex III	Complex IV
What goes in:	NADH H^+	FADH$_2$	e^-	e^-, H^+, O_2
What comes out:	NAD$^+$, e^-, H^+	FAD, e^-	e^-	H_2O, H^+

FIGURE 9.21 How Does the Electron Transport Chain Work? The individual components of the electron transport chain diagrammed in Figure 9.19 are grouped into large multiprotein complexes. Electrons are carried from one complex to another by Q and by cytochrome *c*; Q also shuttles protons across the membrane. The orange arrow indicates Q moving back and forth. Complexes I and IV use the potential energy released by the redox reactions to pump protons from the mitochondrial matrix to the intermembrane space.

✔**EXERCISE** Add an arrow across the membrane and label it "Proton gradient." In the boxes at the bottom, list "What goes in" and "What comes out" for each complex.

transport chain, and the protons released to the intermembrane space contribute to the proton-motive force.

Now, how does this proton gradient make the production of ATP possible?

The Discovery of ATP Synthase

In 1960 Efraim Racker made several key observations about how ATP is synthesized in mitochondria. When he used mitochondrial membranes to make vesicles, Racker noticed that some formed with their membrane inside out. Electron microscopy revealed that the inside-out membranes had numerous large proteins studded along their surfaces. Each protein appeared to have a base in the membrane with a lollipop-shaped stalk and a projecting knob (**Figure 9.22a**). If the solution was vibrated or treated with a compound called urea, the stalks and knobs fell off.

Racker seized on this technique to isolate the stalks and knobs and do experiments with them. For example, he found that isolated stalks and knobs could hydrolyze ATP, forming ADP and inorganic phosphate. The vesicles that contained just the base

component, without the stalks and knobs, could not process ATP. The base components were, however, capable of transporting protons across the membrane.

Based on these observations, Racker proposed that the stalk-and-knob component of the protein was an enzyme that both hydrolyzes and synthesizes ATP. To test this idea, he added the stalk-and-knob components back to vesicles that had been stripped of them and confirmed that the vesicles were then capable of synthesizing ATP. Follow-up work also confirmed his hypothesis that the membrane-bound base component is a proton channel.

As **Figure 9.22b** shows, the structure of this protein complex is now well understood. The ATPase "knob" component is called the F_1 unit; the membrane-bound, proton-transporting base component is the F_o unit. The F_1 and F_o units are connected by a rotor, which spins the F_1 unit, and a stator, which interacts with the spinning F_1 unit. The entire complex is known as **ATP synthase**.

A flow of protons through the F_o unit causes the rotor connecting the two subunits to spin. By attaching long actin filaments to

(a) Vesicle formed from "inside-out" mitochondrial membrane

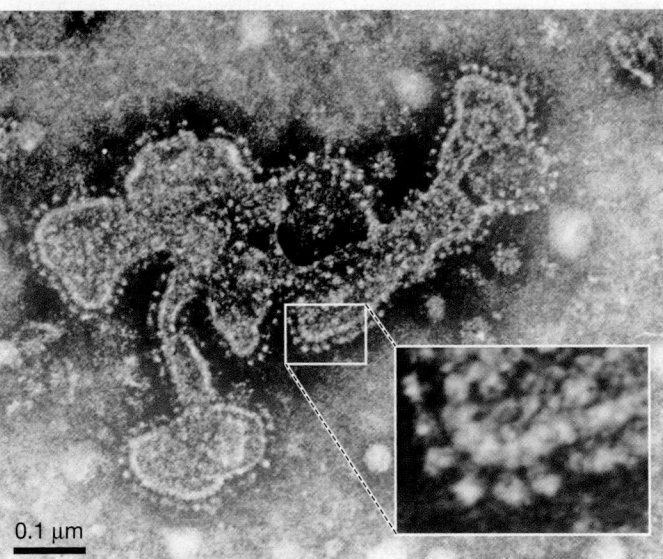

0.1 μm

FIGURE 9.22 The Structure of ATP Synthase. (a) When patches of mitochondrial membrane turn inside out and form vesicles, proteins that have a lollipop-shaped stalk-and-knob structure face outward. Normally, the stalk and knob face inward, toward the mitochondrial matrix. **(b)** ATP synthase has two major components, designated F_0 and F_1, connected by a rotor that spins.

(b) The F_0 unit is the base; the F_1 unit is the knob.

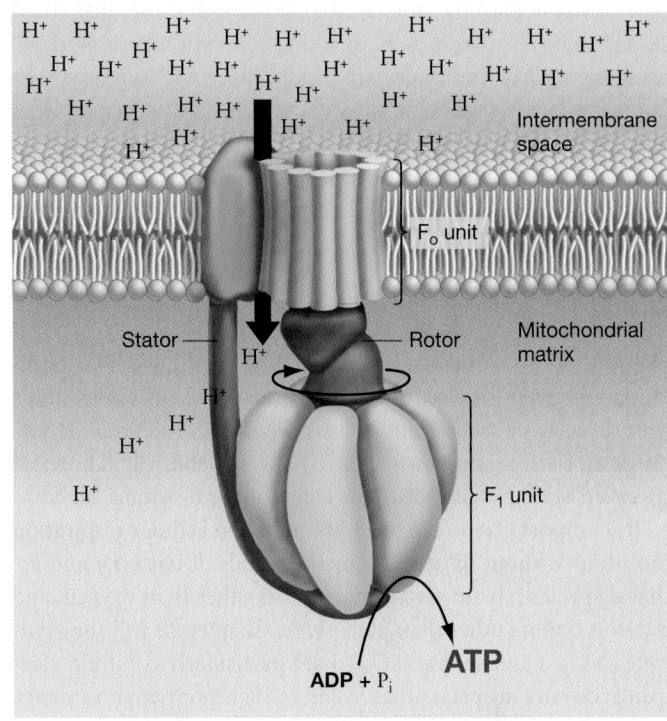

the rotor and examining them with a videomicroscope, researchers have been able to see the rotation, which can reach speeds of 350 revolutions per second. As the F_1 unit rotates along with the rotor, its subunits are thought to change conformation in a way that catalyzes the phosphorylation of ADP to ATP. Understanding how this reaction occurs is currently the focus of intense research. ATP synthase makes most of the ATP that keeps you alive.

Organisms Use a Diversity of Electron Acceptors

Figure 9.23 summarizes glucose oxidation and cellular respiration by tracing the fate of the carbon atoms and electrons in glucose. Notice that electrons from glucose are transferred to NADH and $FADH_2$, passed through the electron transport chain, and accepted by oxygen. Proton pumping during electron transport creates the proton-motive force that drives ATP synthesis.

PROCESS: SUMMARY OF CELLULAR RESPIRATION

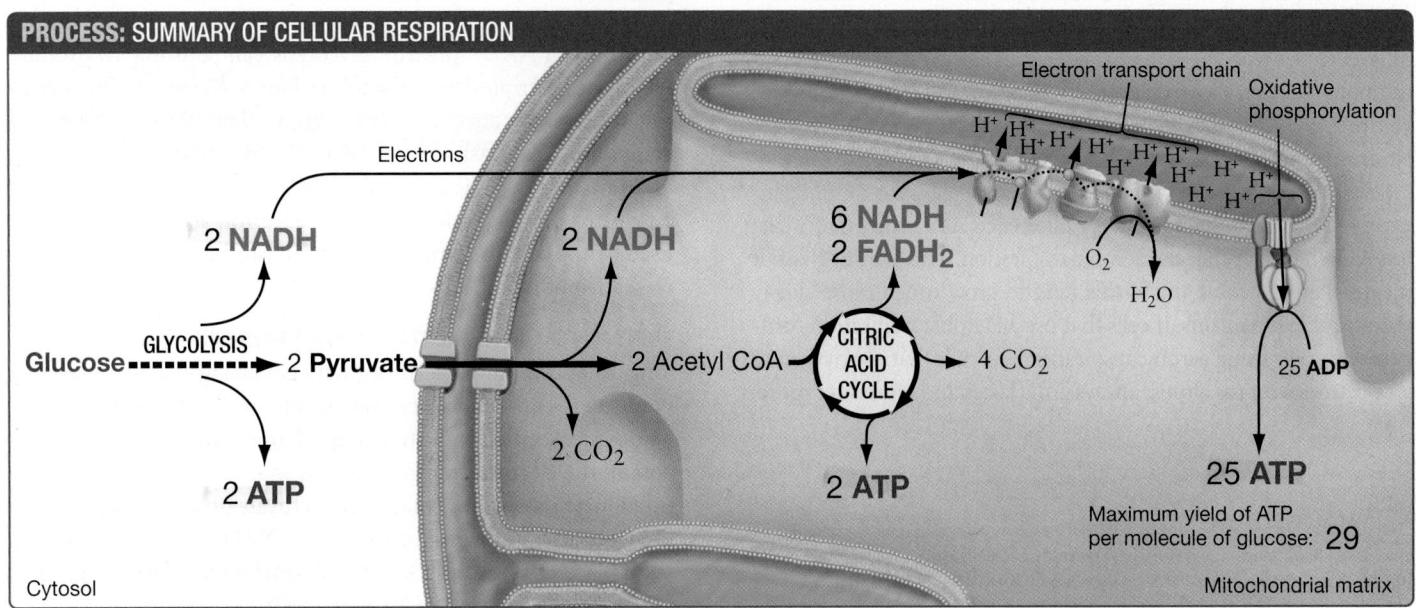

FIGURE 9.23 ATP Yield during Cellular Respiration.

The diagram also indicates the approximate yield of ATP from each component of the process. Recent research shows that about 29 ATP molecules are produced from each molecule of glucose.[3] Of these, 25 ATP molecules are produced by ATP synthase. The fundamental message here? The vast majority of the "payoff" from the oxidation of glucose occurs via oxidative phosphorylation.

If you haven't already done so, this would be a good time to review the elements of cellular respiration in the study area at *www.masteringbiology.com*.

(MB) **BioFlix™** Cellular Respiration

AEROBIC VERSUS ANAEROBIC RESPIRATION   During cellular respiration, oxygen is the electron acceptor used by all eukaryotes and a wide diversity of bacteria and archaea. Species that depend on oxygen as an electron acceptor for the ETC use **aerobic** respiration and are called aerobic organisms. (The Latin root *aero*– means "air.")

It is important to recognize, though, that cellular respiration can occur without oxygen. Many thousands of bacterial and archaeal species rely on electron acceptors other than oxygen, and electron donors other than glucose. As Chapter 28 will show, nitrate (NO_3^-) and sulfate (SO_4^{2-}) are particularly common electron acceptors in species that live in oxygen-poor environments. In addition, many bacteria and archaea use H_2, H_2S, CH_4, or other inorganic compounds as electron donors—not glucose.

Cells that depend on electron acceptors other than oxygen are said to use **anaerobic** ("no air") respiration. Even though the starting and ending points of cellular respiration are different, anaerobic cells still use electron transport chains to create a proton-motive force that drives the synthesis of ATP. In bacteria and archaea, the ETC and ATP synthase are located in the plasma membrane.

AEROBIC RESPIRATION IS EFFICIENT   Even though an array of compounds can serve as the final electron acceptor in cellular respiration, oxygen is the most efficient. Because oxygen holds electrons so tightly, the potential energy of electrons in a bond between an oxygen atom and a non-oxygen atom is low. As a result, there is a large difference between the potential energy of electrons in NADH and the potential energy of electrons bonded to an oxygen atom (see Figure 9.19). The large differential in potential energy means that the electron transport chain can generate a large proton-motive force.

Cells that do not use oxygen as an electron acceptor cannot generate such a large potential energy difference. As a result, they make less ATP than cells that use aerobic respiration. This is important: It means that anaerobic organisms tend to grow much more slowly than aerobic organisms. If cells that use anaerobic respiration compete with cells using aerobic respiration, the cells that use oxygen as an electron acceptor almost always grow faster and reproduce more.

What happens when oxygen or other electron acceptors get used up? When this happens, the electrons carried by NADH have no place to go and the electron transport chain stops. As glycolysis, pyruvate processing, and the citric acid cycle continue, all of the NAD^+ in the cell quickly becomes NADH.

This situation is life threatening. When there is no longer any NAD^+ to supply the reactions of glycolysis, no ATP can be produced. If NAD^+ cannot be regenerated somehow, the cell will die. How do cells cope?

CHECK YOUR UNDERSTANDING

⊙━ If you understand that . . .

- As electrons from NADH and $FADH_2$ move through the electron transport chain, protons are pumped into the intermembrane space of mitochondria.
- The electrochemical gradient across the inner mitochondrial membrane drives protons through ATP synthase, resulting in the production of ATP from ADP.

✓ You should be able to . . .

Add paper cutouts labeled ETC and ATP synthase to the model you made in Section 9.3, and then explain the steps in electron transport and chemiosmosis, using dimes to represent electrons and pennies to represent protons.

Answers are available in Appendix B.

9.7 Fermentation

⊙━ Fermentation is a metabolic pathway that regenerates NAD^+ from stockpiles of NADH. It allows glycolysis to continue producing ATP in the absence of the electron acceptor required by the ETC. Fermentation occurs when pyruvate or a molecule derived from pyruvate accepts electrons from NADH (**Figure 9.24**).

When NADH gets rid of electrons in this way, NAD^+ is produced. With NAD^+ present, glycolysis can continue to produce ATP via substrate-level phosphorylation. Fermentation allows the cell to stay alive and grow, even when electron transport chains are shut down for lack of an electron acceptor.

In many cases, the molecule that is formed by the addition of an electron to pyruvate (or another electron acceptor) cannot be used by the cell. In some cases, this by-product is toxic and is excreted from the cell as waste.

MANY DIFFERENT FERMENTATION PATHWAYS EXIST   If you sprint a long distance, your muscles begin metabolizing glucose so fast that your lungs and circulatory system cannot supply oxygen rapidly enough to keep electron transport chains active. When oxygen is absent, the electron transport chains shut down and NADH cannot donate its electrons there. The pyruvate produced by glycolysis then begins to accept electrons from NADH, and fermentation takes place. This process, **lactic acid fermentation**, forms a product molecule called lactate and regenerates NAD^+ (**Figure 9.25a**).

---

[3]Traditionally, biologists thought that 36 ATP would be synthesized for every mole of glucose oxidized. More recent work has shown that actual yield is only about 29 ATP [see M. Brand, *Biochemistry and Molecular Biology Education* 31 (2003): 2–4]. Also, it's important to note that yield varies with conditions in the cell.

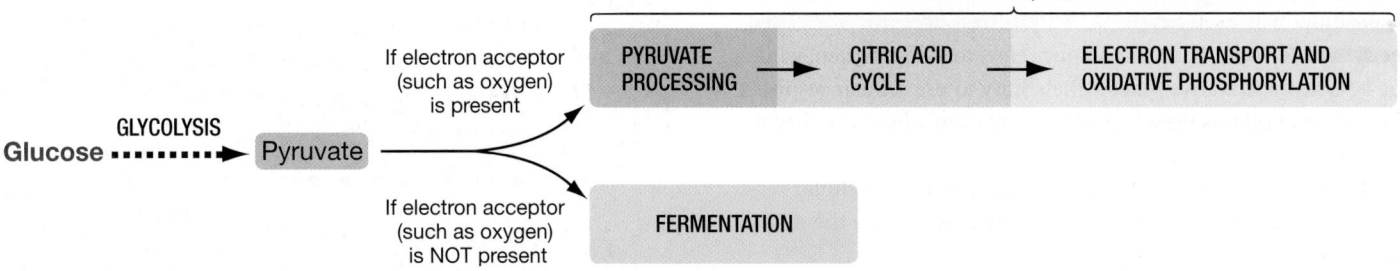

FIGURE 9.24 Cellular Respiration and Fermentation Are Alternative Pathways for Producing Energy. When oxygen or another electron acceptor used by the ETC is present in a cell, the pyruvate produced by glycolysis enters the citric acid cycle and the electron transport system is active. But if no electron acceptor is available to keep the ETC running, the pyruvate undergoes reactions known as fermentation.

Figure 9.25b illustrates a different fermentation pathway, **alcohol fermentation**, which occurs in the fungus *Saccharomyces cerevisiae*—baker's and brewer's yeast. When these fungal cells are placed in an environment such as bread dough or a bottle of grape juice and begin growing there, they quickly use up all the

(a) Lactic acid fermentation occurs in humans.

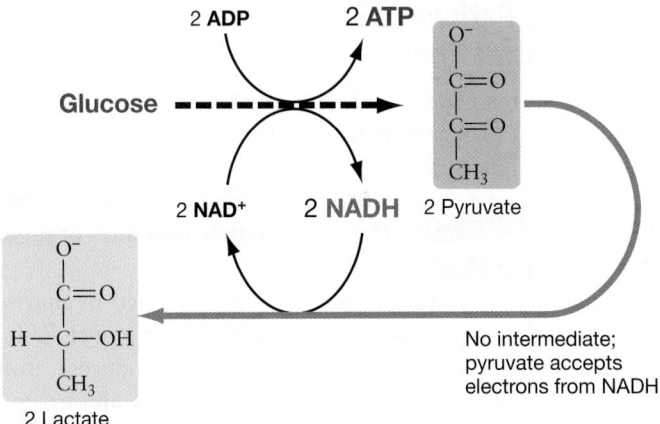

(b) Alcohol fermentation occurs in yeast.

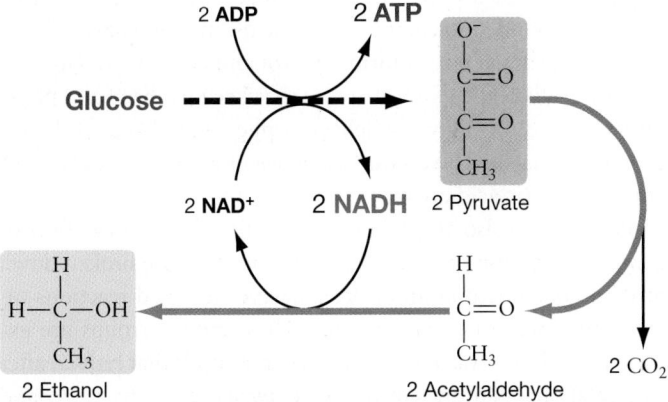

FIGURE 9.25 Fermentation Regenerates NAD⁺ So That Glycolysis Can Continue. These are just two of the many types of fermentation that occur among the bacteria, archaea, and eukaryotes.

available oxygen. They continue to use glycolysis to metabolize sugar, however, by enzymatically converting pyruvate to the two-carbon compound acetaldehyde. This reaction gives off carbon dioxide, which causes bread to rise and produces the bubbles in champagne and beer.

Acetaldehyde then accepts electrons from NADH, forming the NAD^+ required to keep glycolysis going. The addition of electrons to acetaldehyde forms ethanol as a waste product. The yeast cells excrete ethanol as waste. In essence, the active ingredient in alcoholic beverages is yeast urine.

Cells that employ other types of fermentation are used commercially in the production of soy sauce, tofu, yogurt, cheese, vinegar, and other products.

Bacteria and archaea that exist exclusively through fermentation are present in phenomenal numbers in the oxygen-free environment of your small intestine and in the rumen (first stomach) of cows. The rumen is a specialized digestive organ that contains over 10^{10} (10 billion) bacterial and archaeal cells per *milliliter* of fluid. The fermentations that occur in these cells produce an array of fatty acids. Cattle don't actually live off grass directly—they eat it to feed these bacteria and archaea, and use their fermentation by-products as a source of energy.

FERMENTATION AS AN ALTERNATIVE TO CELLULAR RESPIRATION
Even though fermentation is a widespread type of metabolism, it is extremely inefficient compared with aerobic cellular respiration. Fermentation produces just 2 molecules of ATP for each molecule of glucose metabolized, while cellular respiration produces about 29—almost 15 times more energy per glucose molecule than fermentation. The reason for this disparity is that oxygen has much higher electronegativity than electron acceptors such as pyruvate and acetaldehyde. As a result, the potential energy drop between the start and end of fermentation is a tiny fraction of the potential energy change that occurs during cellular respiration.

Based on these observations, it is not surprising that organisms capable of both processes almost never use fermentation when an appropriate electron acceptor is available for cellular respiration. In organisms that usually use oxygen as an electron acceptor, fermentation is an alternative mode of energy production when oxygen supplies temporarily run out.

Organisms that can switch between fermentation and cellular respiration that uses oxygen as an electron acceptor are called **facultative aerobes**. The term aerobe refers to using oxygen, while the adjective facultative reflects the ability to use cellular respiration when oxygen is present and fermentation when it is absent. You are a facultative aerobe.

To review the steps in glucose oxidation and how cellular respiration interacts with fermentation pathways, go to the study area at *www.masteringbiology.com*:

 Web Activity Glucose Metabolism

CHECK YOUR UNDERSTANDING

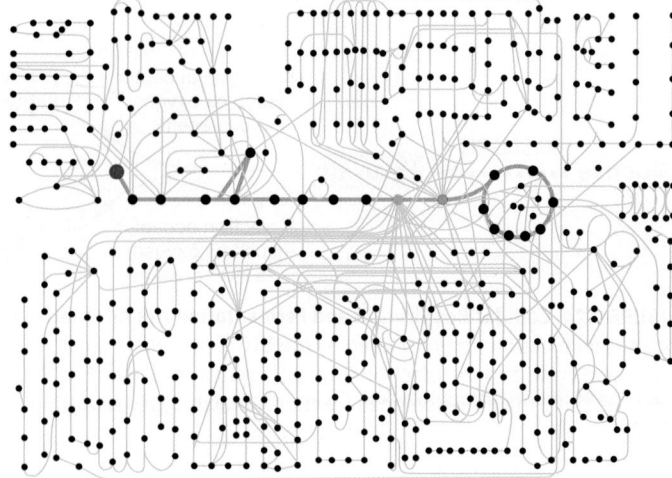

FIGURE 9.26 Metabolic Pathways Interact. A representation of a few of the thousands of chemical reactions that occur in cells. The dots represent molecules, and the lines represent enzyme-catalyzed reactions.

✔ **EXERCISE** Label the large red dot, and circle the 10 reactions of glycolysis. Draw a box around the citric acid cycle.

If you understand that . . .

- Fermentation occurs in the absence of an ETC. It consists of reactions that oxidize NADH and regenerate the NAD^+ required for glycolysis.

✔ **You should be able to . . .**

Explain why organisms that have an ETC as well as fermentation pathways never ferment pyruvate if an electron acceptor is available.

Answers are available in Appendix B.

9.8 How Does Cellular Respiration Interact with Other Metabolic Pathways?

The enzymes, products, and intermediates involved in cellular respiration and fermentation do not exist in isolation. Instead, they are part of a huge and dynamic inventory of chemicals inside the cell.

Metabolism comprises thousands of different chemical reactions, and the amounts and identities of molecules inside cells are constantly in flux. Fermentation pathways, electron transport, and other aspects of carbohydrate metabolism may be crucial to the life of a cell, but they also have to be seen as parts of a whole (**Figure 9.26**).

This complexity can be boiled down to a simple essence, however. Cells have just two fundamental requirements: energy and carbon. They need a source of high-energy electrons for generating chemical energy in the form of ATP, and a source of carbon-containing molecules that can be used to synthesize DNA, RNA, proteins, fatty acids, and other molecules.

Reactions that break down molecules and produce ATP are called **catabolic pathways**; reactions that synthesize larger molecules from smaller components are called **anabolic pathways**.

How do glycolysis and the citric acid cycle interact with other catabolic pathways and with anabolic pathways? Let's consider how eukaryotes use molecules other than carbohydrates as fuel, and then examine how molecules involved in glycolysis and the citric acid cycle are sometimes used as building blocks to synthesize cell components.

Catabolic Pathways Break Down Molecules as Fuel

Most organisms ingest, synthesize, or absorb a wide variety of carbohydrates. These molecules range from sucrose, maltose, and other simple sugars to large polymers such as glycogen and starch. As Chapter 5 noted, glycogen is the major form of stored carbohydrate in animals, while starch is the major form of stored carbohydrate in plants.

Recall that both glycogen and starch are polymers of glucose, but differ in the way their long chains of glucose branch. Using enzyme-catalyzed reactions, cells can produce glucose from glycogen, starch, and most simple sugars. Glucose and fructose can then be processed by the enzymes of the glycolytic pathway.

Carbohydrates are not the only important source of carbon compounds used in catabolic pathways, however. As Chapter 6 pointed out, fats are highly reduced macromolecules consisting of glycerol bonded to chains of fatty acids. In cells, enzymes routinely break down fats to form glycerol and acetyl CoA. Glycerol enters the glycolytic pathway once it has been oxidized and phosphorylated to form glyceraldehyde-3-phosphate—one of the intermediates in the 10-reaction sequence of glycolysis. Acetyl CoA enters the citric acid cycle.

Proteins can also be catabolized, meaning that they can be broken down and used to produce ATP. Once they are broken down to their constituent amino acids, enzyme-catalyzed reactions remove the amino ($-NH_2$) groups. These amino groups are excreted in urine as waste. The carbon compounds that remain after this catabolic step are converted to pyruvate, acetyl CoA, and other intermediates in glycolysis and the citric acid cycle.

Figure 9.27 summarizes the catabolic pathways of carbohydrates, fats, and proteins and shows how their breakdown products feed an array of steps in glucose oxidation and cellular respi-

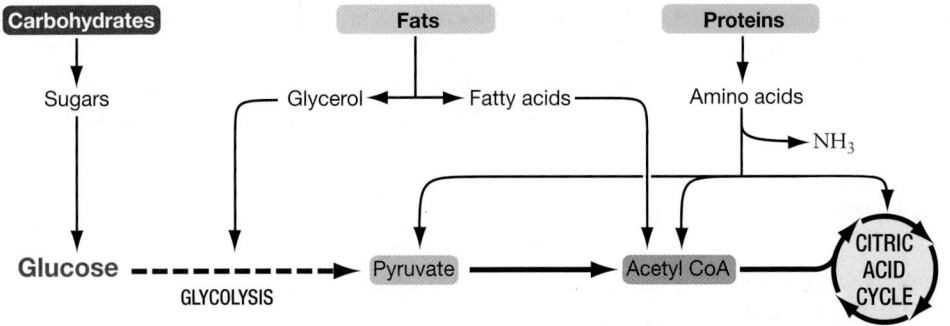

FIGURE 9.27 Catabolic Pathways: Proteins, Carbohydrates, and Fats Can All Furnish Substrates for Cellular Respiration. A variety of carbohydrates can be converted to glucose and processed by glycolysis. If carbohydrates are scarce, cells can obtain high-energy compounds from fats or even proteins for ATP production. These are catabolic reactions.

ration. When all three types of molecules are available in the cell to generate ATP, carbohydrates are used up first, then fats, and finally proteins.

Anabolic Pathways Synthesize Key Molecules

Where do cells get the precursor molecules required to synthesize amino acids, RNA, DNA, phospholipids, and other cell components? Not surprisingly, the answer often involves intermediates in carbohydrate metabolism. For example,

- In humans, about half of the 21 required amino acids can be synthesized from molecules siphoned from the citric acid cycle.
- Acetyl CoA is the starting point for anabolic pathways that result in the synthesis of fatty acids.

- The molecule that is produced by the first reaction in glycolysis can be oxidized to start the synthesis of ribose-5-phosphate—a key intermediate in the production of ribonucleotides and deoxyribonucleotides. These nucleotides, in turn, are required for manufacturing RNA and DNA.

- If ATP is abundant, pyruvate and lactate (from fermentation) can be used as a substrate in the synthesis of glucose. Excess glucose is converted to glycogen and stored.

Figure 9.28 summarizes how intermediates in carbohydrate metabolism are drawn off to synthesize macromolecules. The take-home message is that the same molecule can serve many different functions in the cell. As a result, catabolic and anabolic pathways are closely intertwined.

FIGURE 9.28 Anabolic Pathways: Intermediates in Carbohydrate Metabolism Can Be Drawn Off to Synthesize Cell Components. Several of the intermediates in carbohydrate metabolism act as precursor molecules in anabolic reactions leading to the synthesis of glycogen or starch, RNA, DNA, fatty acids, and amino acids.

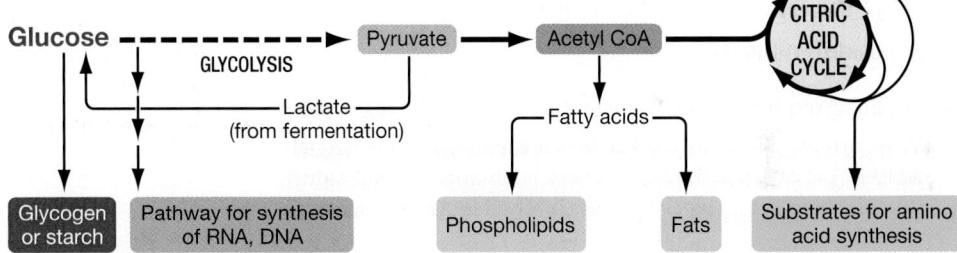

CHAPTER 9 REVIEW *For media, go to the study area at www.masteringbiology.com*

Summary of Key Concepts

🔑 **In cells, the endergonic reactions required for life occur in conjunction with an exergonic reaction involving ATP.**

- ATP is the currency that cells use to pump ions, drive endergonic reactions, move cargo, and perform other types of work.

- Almost all of the reactions that occur inside cells are endergonic.

- When ATP or a phosphate group from ATP is added to a substrate or enzyme that participates in an endergonic reaction, the potential energy of the substrate or enzyme is raised enough to make the reaction exergonic and thus spontaneous.

  ✔ You should be able to explain why the addition of phosphate groups raises the potential energy of proteins, and why phosphorylation often causes proteins to change shape.

🔑 **Cellular respiration produces ATP from molecules with high potential energy—often glucose.**

- Cells produce ATP from sugars or other compounds with high free energy by using one of two general pathways: (**1**) cellular respiration or (**2**) fermentation.

- Cellular respiration is based on redox reactions that transfer electrons from a compound with high free energy, such as glucose, to a molecule with lower free energy, such as oxygen, through an electron transport chain.

- Fermentation involves the transfer of electrons from one organic compound to another without participation by an electron transport chain.

✔ You should be able to explain why cellular respiration produces so much more ATP per mole of glucose than fermentation, in terms of the total free-energy changes involved.

(MB) **Web Activity** Redox Reactions

🗝 **Cellular respiration has four components: (1) glycolysis, (2) pyruvate processing, (3) the citric acid cycle, and (4) electron transport and chemiosmosis.**

- In eukaryotes, glycolysis takes place in the cytosol. The citric acid cycle occurs in the mitochondrial matrix, and electron transport and oxidative phosphorylation proceed in the inner membranes of mitochondria.

- Glycolysis is a 10-step reaction sequence in which glucose is broken down into two molecules of pyruvate. ATP and NADH are produced.

- During pyruvate processing, a series of reactions converts pyruvate to acetyl CoA. NADH is produced and CO_2 is released.

- The citric acid cycle is an 8-step reaction cycle that begins with acetyl CoA. $FADH_2$, NADH, and GTP or ATP are produced; CO_2 is released. At the end of the citric acid cycle, glucose is completely oxidized to CO_2.

- NADH and $FADH_2$ donate electrons to an electron transport chain, which gradually steps the electrons down in potential energy until they are transferred to a final electron acceptor (often O_2). The energy that is released pumps protons across the inner mitochondrial membrane, creating an electrochemical gradient that ATP synthase uses to produce ATP.

  ✔ You should be able to describe what would happen to NADH levels in a cell in the first few seconds after a drug has poisoned the enzyme that converts acetyl CoA to citrate.

(MB) **BioFlix™** Cellular Respiration

🗝 **Cellular respiration and fermentation are carefully regulated.**

- ATP is produced only as needed. When supplies of ATP, NADH, and $FADH_2$ are high in the cell, feedback inhibition occurs: Product molecules bind to and block enzymes involved in ATP production.

- The glycolytic pathway slows when ATP binds to phosphofructokinase.

- The pyruvate dehydrogenase complex is inhibited when it is phosphorylated by ATP. It speeds up in the presence of substrates like NAD and ADP.

- The enzyme that converts acetyl CoA to citrate slows when ATP binds to it, and certain enzymes in the citric acid cycle are inhibited when NADH or ATP bind to them.

- In eukaryotes and many bacteria, fermentation only occurs if cellular respiration stops.

  ✔ You should be able to draw a graph predicting how the rate of ATP production changes as a function of ATP concentration. (Write "ATP concentration" on the x-axis and "ATP production" on the y-axis.)

🗝 **Fermentation pathways allow glycolysis to continue when the lack of an electron acceptor shuts down electron transport chains.**

- Fermentation is an alternative method for processing glucose and making ATP—one that cells use when cellular respiration is not possible.

- If no electron acceptor is available, all NAD^+ is converted to NADH and glycolysis stops.

- Fermentation pathways regenerate NAD^+, so glycolysis can continue to make ATP and keep the cell alive. This happens when an organic molecule such as pyruvate accepts electrons from NADH.

- Depending on the molecule that acts as an electron acceptor, fermentation pathways produce lactate, ethanol, or other reduced organic compounds as a by-product.

  ✔ You should be able to explain why organisms that produce ATP only via fermentation grow much more slowly than organisms that produce ATP via cellular respiration.

(MB) **Web Activity** Glucose Metabolism

Questions

1. When does feedback inhibition occur?
   a. when lack of an appropriate electron acceptor makes an electron transport chain stop
   b. when an enzyme that is active early in a metabolic pathway is inhibited by a product of the pathway
   c. when ATP synthase reverses and begins pumping protons out of the mitochondrial matrix
   d. when cellular respiration is inhibited and fermentation begins

2. Where does the citric acid cycle occur in eukaryotes?
   a. in the cytosol
   b. in the matrix of mitochondria
   c. in the inner membrane of mitochondria
   d. in the intermembrane space of mitochondria

3. What does the chemiosmotic hypothesis claim?
   a. Substrate-level phosphorylation occurs in the electron transport chain.
   b. Substrate-level phosphorylation occurs in glycolysis and the citric acid cycle.
   c. The electron transport chain is located in the inner membrane of mitochondria.
   d. Electron transport chains generate ATP indirectly, by the creation of a proton-motive force.

4. What is the function of the reactions in a fermentation pathway?
   a. to generate NADH from NAD^+, so electrons can be donated to the electron transport chain
   b. to synthesize pyruvate from lactate
   c. to generate NAD^+ from NADH, so glycolysis can continue
   d. to synthesize electron acceptors, so that cellular respiration can continue

5. When do cells switch from cellular respiration to fermentation?
   a. when electron acceptors required by the ETC are not available
   b. when the proton-motive force runs down
   c. when NADH and FADH$_2$ supplies are low
   d. when pyruvate is not available

6. Why are NADH and FADH$_2$ said to have "reducing power"?
   a. They are the reduced forms of NAD$^+$ and FAD.
   b. They donate electrons to components of the ETC, reducing those components.
   c. They travel between the cytosol and the mitochondrion.
   d. They have the power to reduce carbon dioxide to glucose.

✔ TEST YOUR UNDERSTANDING

Answers are available in Appendix B

1. Explain why NADH and FADH$_2$ are called electron carriers. Where do these molecules get electrons, and where do they deliver them? In eukaryotes, what molecule do these electrons reduce?

2. Compare and contrast substrate-level phosphorylation and oxidative phosphorylation.

3. Why does aerobic respiration produce much more ATP than anaerobic respiration?

4. Make a flowchart indicating the relationships among the four steps of cellular respiration. Which steps are responsible for glucose oxidation? Which produce the most ATP?

5. Explain the relationship between electron transport and oxidative phosphorylation. What does ATP synthase look like, and how does it work?

6. Describe the relationship among carbohydrate metabolism, the catabolism of proteins and fats, and anabolic pathways.

✔ APPLYING CONCEPTS TO NEW SITUATIONS

Answers are available in Appendix B

1. Cyanide (C≡N$^-$) blocks complex IV of the electron transport chain. Suggest a hypothesis for what happens to the ETC when complex IV stops working. Your hypothesis should explain why cyanide poisoning in humans is fatal.

2. The presence of many sac-like cristae results in a large amount of membrane inside mitochondria. Suppose that some mitochondria had few cristae. How would their output of ATP compare with that of mitochondria with many cristae? Explain your answer.

3. When yeast cells are placed into low-oxygen environments, the mitochondria in the cells become reduced in size and number. Suggest an explanation for this observation.

4. Most agricultural societies have come up with ways to ferment the sugars in barley, wheat, rice, corn, or grapes to produce alcoholic beverages. Historians argue that this was an effective way for farmers to preserve the chemical energy in grains and fruits in a form that would not be eaten by rats or spoiled by bacteria or fungi. Why does a great deal of chemical energy remain in the products of fermentation pathways?

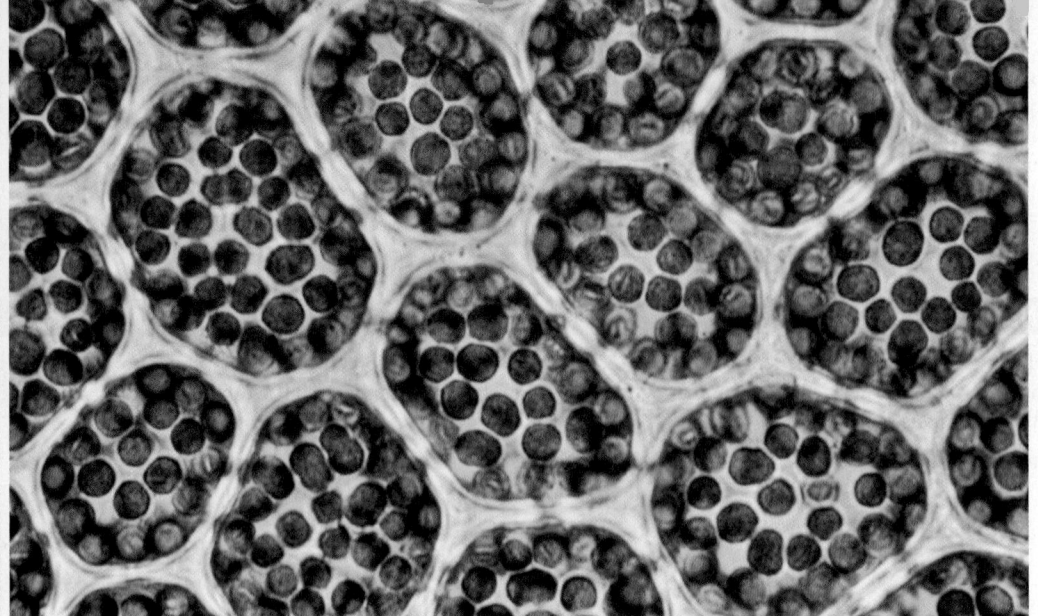

A close-up of moss cells filled with chloroplasts, where photosynthesis converts the energy in sunlight to chemical energy in the bonds of sugar. The sugar produced by photosynthetic organisms fuels cellular respiration and growth. Photosynthetic organisms, in turn, are consumed by other organisms, including you. Directly or indirectly, most organisms on Earth get their energy from photosynthesis.

10 Photosynthesis

KEY CONCEPTS

- Photosynthesis is the conversion of light energy to chemical energy stored in the bonds of carbohydrates. It consists of two linked sets of reactions.

- In the light-capturing reactions, excited electrons are used to produce the electron carrier NADPH or are donated to an electron transport chain, which results in the production of ATP via chemiosmosis.

- The reactions of the Calvin cycle start with the enzyme rubisco, which catalyzes the addition of CO_2 to a five-carbon compound. Subsequent reactions use the ATP and NADPH synthesized in the light reactions and yield a molecule required for carbohydrate production.

- In plants, CO_2 enters photosynthetic tissue through openings called stomata. The CAM and C_4 pathways increase CO_2 concentrations inside the leaves of some species and make photosynthesis more efficient.

Some three billion years ago, a novel combination of light-absorbing molecules and enzymes gave a bacterial cell the capacity to convert light energy into chemical energy in the C–C and C–H bonds of sugar. The origin of **photosynthesis**—the use of sunlight to manufacture carbohydrate—ranks as one of the great events in the history of life.

The vast majority of organisms alive today rely on photosynthesis, either directly or indirectly, to stay alive. Maples, mosses, and other photosynthetic organisms are termed **autotrophs** (literally, "self-feeders") because they make all of their own food from ions and simple molecules. Humans, houseflies, and other non-photosynthetic organisms are called **heterotrophs** ("different-feeders") because they have to obtain the sugars and many of the other macromolecules they need from other organisms.

Because there could be no heterotrophs without autotrophs, photosynthesis is fundamental to almost all life. Glycolysis may qualify as the most ancient set of energy-related chemical reactions from an evolutionary viewpoint; but ecologically—meaning, in terms of how organisms interact with each other—photosynthesis is easily the most important.

How does it happen? Let's begin with an overview, then delve into a step-by-step analysis of some of the most remarkable chemistry on Earth.

10.1 Photosynthesis Harnesses Sunlight to Make Carbohydrate

Research on photosynthesis began early in the history of biological science. Starting in the 1770s, a series of experiments showed that photosynthesis takes place only in the green parts of plants; sunlight, carbon dioxide (CO_2), and water (H_2O) are required; and oxygen (O_2) is produced as a by-product.

✔ When you see this checkmark, stop and test yourself. Answers are available in Appendix B.

By the 1840s enough was known about the process for biologists to propose that photosynthesis allows plants to convert electromagnetic energy in the form of sunlight into chemical energy in the C–C and C–H bonds of carbohydrates. When glucose is the carbohydrate that is eventually produced, the overall reaction—the sum of many independent reactions—can be written as

$$6 CO_2 + 12 H_2O + \text{light energy} \longrightarrow C_6H_{12}O_6 + 6 O_2 + 6 H_2O$$

Now read the reaction again, and note the contrast with cellular respiration. Photosynthesis is an endergonic suite of reactions that reduces carbon dioxide to glucose or other sugars. Cellular respiration is an exergonic suite of reactions that oxidizes glucose to carbon dioxide and results in the production of ATP.

Early investigators assumed that CO_2 and H_2O react directly to form the CH_2O found in carbohydrates, and that the oxygen gas (O_2) released during photosynthesis originated in the oxygen atoms of CO_2.

These researchers were wrong, though. CO_2 and H_2O participate in entirely different reactions, and the oxygen atoms in O_2 come from water.

Photosynthesis: Two Linked Sets of Reactions

During the 1930s two independent lines of research on photosynthesis converged, leading to a major advance.

The first research program, led by Cornelius van Niel, focused on photosynthesis in organisms called purple sulfur bacteria. Van Niel and his group found that these cells can grow in the laboratory on a food source that lacks sugars. Based on this observation, he concluded they must be autotrophs that manufacture their own carbohydrates. But to grow, the cells had to be exposed to sunlight and hydrogen sulfide (H_2S).

Van Niel also showed that these cells did not produce oxygen as a by-product of photosynthesis. Instead, elemental sulfur (S) accumulated in their medium. In these organisms, the overall reaction for photosynthesis was

$$CO_2 + 2 H_2S + \text{light energy} \longrightarrow (CH_2O)_n + H_2O + 2 S$$

Van Niel's work was crucial for two reasons:

1. It showed that CO_2 and H_2O do *not* combine directly during photosynthesis. Instead of acting as a reactant in photosynthesis in these species, H_2O is a product of the process.

2. It supported the hypothesis that the oxygen atoms in CO_2 are *not* released as oxygen gas (O_2). The purple sulfur bacteria produced no oxygen, even though carbon dioxide participated in the reaction—just as it did in plants.

Based on these findings, biologists hypothesized that the oxygen atoms released during plant photosynthesis must come from H_2O. This proposal was supported by experiments with isolated chloroplasts, which produced oxygen in the presence of sunlight even if no CO_2 was present.

The hypothesis was confirmed when heavy isotopes of oxygen—^{18}O compared with the normal isotope, ^{16}O—became available to researchers. Biologists exposed algae or plants to H_2O that con-

tained ^{18}O, collected the oxygen gas that was given off as a by-product of photosynthesis, and confirmed that the released oxygen gas contained the heavy isotope.

As predicted, the reaction that produced this oxygen occurred only in the presence of sunlight. The light-capturing reactions of photosynthesis result in the production of oxygen from water.

A second major line of research supported these discoveries. Between 1945 and 1955, a team led by Melvin Calvin began introducing radioactively labeled carbon dioxide ($^{14}CO_2$) to algae and identifying the molecules that subsequently became labeled with the radioisotope. These experiments allowed researchers to identify the sequence of reactions involved in reducing CO_2 to sugars.

Because Calvin played an important role in this research, the reactions that reduce carbon dioxide and produce sugar came to be known as the **Calvin cycle**. Later research showed that the Calvin cycle can function only if the light-capturing reactions are occurring.

To summarize: Early research showed that photosynthesis consists of two linked sets of reactions. One set is triggered by light; the other set—the Calvin cycle—requires the products of the light-capturing reactions. The light-capturing reactions produce oxygen from water; the Calvin cycle produces sugar from carbon dioxide.

The two reactions are linked by electrons that are released when water is split to form oxygen gas. During the light-capturing reactions, these electrons are promoted to a high energy state by light and then transferred to a phosphorylated version of NAD+, called **NADP+ (nicotinamide adenine dinucleotide phosphate)**. This reaction forms **NADPH**, which functions as an electron carrier. ATP is also produced in the light-capturing reactions (see **Figure 10.1**).

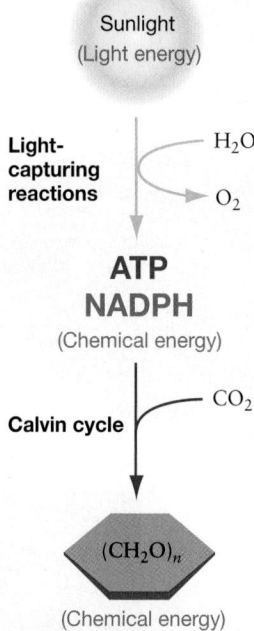

FIGURE 10.1 Photosynthesis Has Two Linked Components.
In the light-capturing reactions of photosynthesis, light energy is transformed to chemical energy in the form of ATP and NADPH. During the Calvin cycle, the ATP and NADPH produced in the light-capturing reactions are used to reduce carbon dioxide to carbohydrate.

During the Calvin cycle, the electrons in NADPH and the potential energy in ATP are used to reduce CO_2 to carbohydrate. The resulting sugars are used in cellular respiration to produce ATP for the cell. Plants oxidize sugars in their mitochondria and consume O_2 in the process, just as animals and other eukaryotes do.

Where does all this activity take place?

Photosynthesis Occurs in Chloroplasts

Once experiments had established that photosynthesis takes place only in the green portions of plants, biologists focused on

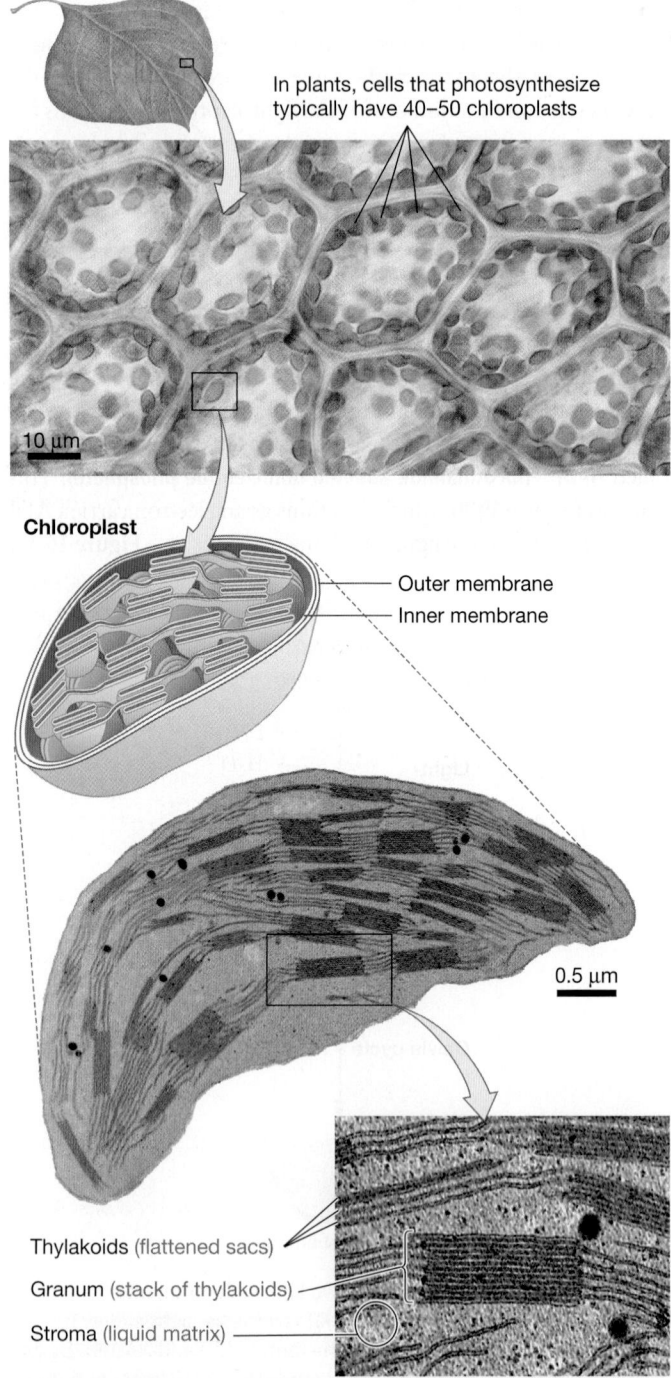

In plants, cells that photosynthesize typically have 40–50 chloroplasts

10 μm

Chloroplast

Outer membrane
Inner membrane

0.5 μm

Thylakoids (flattened sacs)
Granum (stack of thylakoids)
Stroma (liquid matrix)

FIGURE 10.2 Photosynthesis Takes Place in Chloroplasts.

the bright green organelles called **chloroplasts** ("green-formed"). One leaf cell typically contains 40 to 50 chloroplasts, and a square millimeter of leaf averages about 500,000 (**Figure 10.2**).

When membranes derived from chloroplasts were found to release oxygen after exposure to sunlight, the hypothesis that chloroplasts are the site of photosynthesis became widely accepted.

As Figure 10.2 shows, a chloroplast is enclosed by an outer membrane and an inner membrane. The interior is dominated by vesicle-like structures called **thylakoids**, which often occur in interconnected stacks called **grana** (singular: **granum**). The space inside a thylakoid is its **lumen**. (Recall that *lumen* is a general term for the interior of any sac-like structure. Your stomach and intestines have a lumen.) The fluid-filled space between the thylakoids and the inner membrane is the **stroma**.

When researchers analyzed the chemical composition of thylakoid membranes, they found huge quantities of pigments. **Pigments** are molecules that absorb only certain wavelengths of light—other wavelengths are either transmitted or reflected. Pigments have colors because we see the wavelengths that they do *not* absorb.

The most abundant pigment in the thylakoid membranes turned out to be chlorophyll ("green-leaf"), which reflects or transmits green light. As a result, it is responsible for the green color of plants, some algae, and many photosynthetic bacteria.

Before plunging into the details of how photosynthesis occurs, take a moment to consider just how astonishing the process is. Chemists have synthesized an amazing diversity of compounds from relatively simple starting materials, but their achievements pale in comparison to a cell that can synthesize sugar from just carbon dioxide, water, and sunlight. If photosynthesis is not *the* most sophisticated chemistry on Earth, it is certainly a contender.

10.2 How Does Chlorophyll Capture Light Energy?

The light-capturing reactions of photosynthesis begin with the simple act of sunlight striking chlorophyll. To understand the consequences of this event, it's helpful to review the nature of light.

Light is a type of electromagnetic radiation, a form of energy. Photosynthesis converts electromagnetic energy in the form of sunlight into chemical energy in the C–C and C–H bonds of sugar.

Physicists describe light's behavior as both wavelike and particle-like. Like water waves or airwaves, electromagnetic radiation is characterized by its **wavelength**—the distance between two successive wave crests (or wave troughs). The wavelength determines the type of electromagnetic radiation.

Figure 10.3 illustrates the range of wavelengths of electromagnetic radiation—the **electromagnetic spectrum**. **Visible light**, the electromagnetic radiation that humans can see, ranges in wavelength from about 400 to about 710 nanometers (nm, or 10^{-9} m). Shorter wavelengths of electromagnetic radiation contain more energy than longer wavelengths do. Thus, blue light and ultraviolet light contain much more energy than red light and infrared light do.

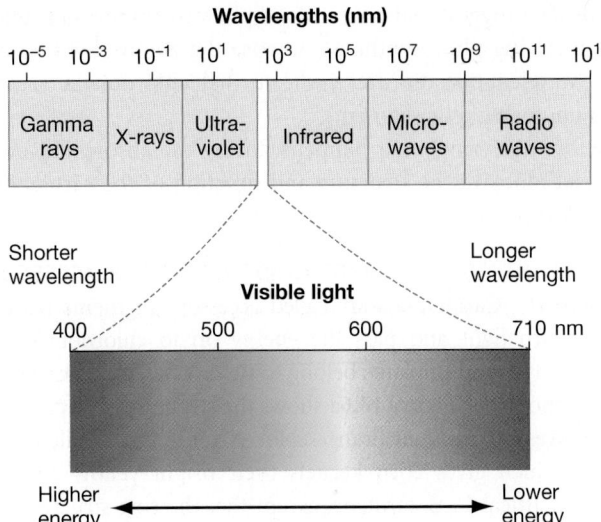

Wavelengths (nm)

Gamma rays | X-rays | Ultra-violet | Infrared | Micro-waves | Radio waves

Shorter wavelength

Visible light

Longer wavelength

400   500   600   710 nm

Higher energy ←→ Lower energy

FIGURE 10.3 The Electromagnetic Spectrum. Electromagnetic energy radiates through space in the form of waves. Humans can see radiation at wavelengths between about 400 nm to 710 nm. The shorter the wavelength of electromagnetic radiation, the higher its energy.

To emphasize the particle-like nature of light, physicists point out that it exists in discrete packets called **photons**. Each photon and each wavelength of light have a characteristic amount of energy. Pigment molecules absorb this energy. How?

Photosynthetic Pigments Absorb Light

When a photon strikes an object, the photon may be absorbed, transmitted, or reflected. A pigment molecule absorbs particular wavelengths of light. Sunlight includes white light, which consists of all wavelengths in the visible portion of the electromagnetic spectrum at once.

If a pigment absorbs all of the visible wavelengths, no visible wavelength of light is reflected back to your eye, and the pigment appears black. If a pigment absorbs many or most of the wavelengths in the blue and green parts of the spectrum but transmits or reflects red wavelengths, it appears red.

What wavelengths do various plant pigments absorb? In one approach to answering this question, researchers grind up leaves and add a liquid that acts as a solvent for lipids. The solvent extracts pigment molecules from the leaf mixture. A technique called thin layer chromatography separates the pigments in the extract (**Figure 10.4a**).

To begin, spots of raw leaf extract are placed near the bottom of a stiff support that is coated with a thin layer of silica gel, cellulose, or similar porous material. The coated support is then placed in a solvent solution. As the solvent wicks upward through the coating, it carries the pigment molecules in the mixture with it. Because the pigment molecules vary in size, solubility, or both, they are carried at different rates.

Figure 10.4b shows a chromatograph from a grass-leaf extract. Notice that this leaf contains an array of pigments. To find out which wavelengths are absorbed by each of these molecules, researchers cut out a single region (color band) of the filter paper, extract the pigment, and use an instrument to record the wavelengths absorbed.

Using thin-layer chromatography, biologists have produced data like those shown in **Figure 10.5a**. This graph is an **absorption spectrum**—a graph showing the amount of light absorbed versus wavelength. Peaks indicate wavelengths where absorbance is high; troughs indicate wavelengths where absorbance is low.

Research based on these techniques has confirmed that there are two major classes of pigment in plant leaves: chlorophylls and carotenoids.

- The **chlorophylls**, designated chlorophyll *a* and chlorophyll *b*, absorb strongly in the blue and red regions of the visible spectrum. The presence of chlorophylls makes plants look green because they reflect and transmit green light, which they do not absorb.

- **Carotenoids** absorb in the blue and green parts of the visible spectrum. Thus, carotenoids appear yellow, orange, or red.

Which of these wavelengths drive photosynthesis?

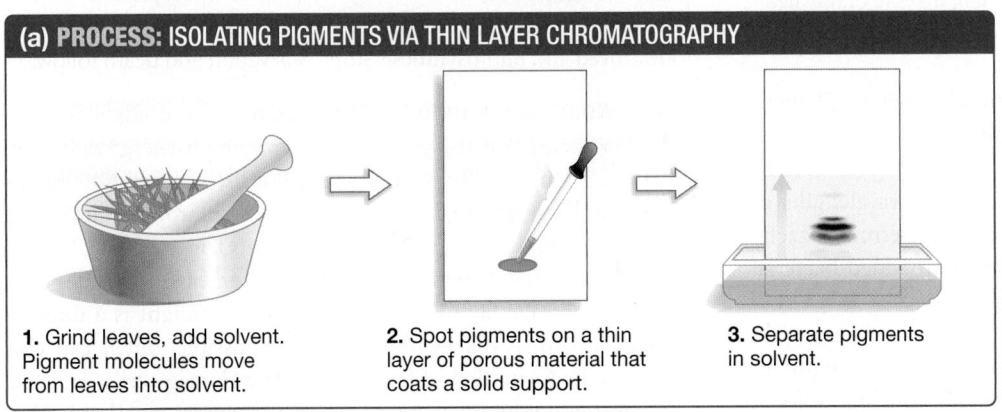

(a) PROCESS: ISOLATING PIGMENTS VIA THIN LAYER CHROMATOGRAPHY

1. Grind leaves, add solvent. Pigment molecules move from leaves into solvent.

2. Spot pigments on a thin layer of porous material that coats a solid support.

3. Separate pigments in solvent.

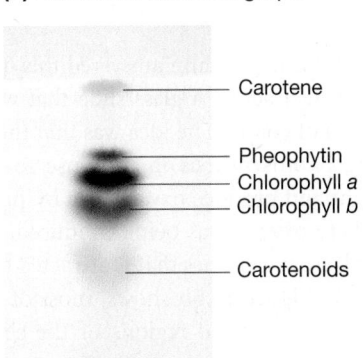

(b) A finished chromatograph

— Carotene

— Pheophytin
— Chlorophyll *a*
— Chlorophyll *b*

— Carotenoids

FIGURE 10.4 Chromatography Is a Technique for Separating Molecules. This example shows the isolation of pigments in grass leaves. Different species of photosynthetic organisms may contain different types and quantities of pigments.

(a) Different pigments absorb different wavelengths of light.

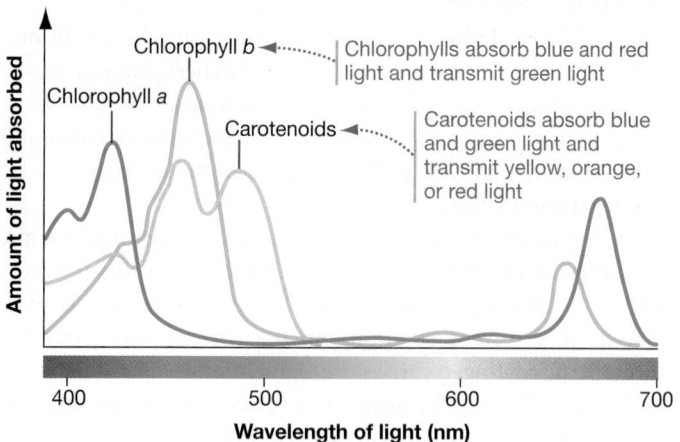

(b) Pigments that absorb blue and red photons are the most effective at triggering photosynthesis.

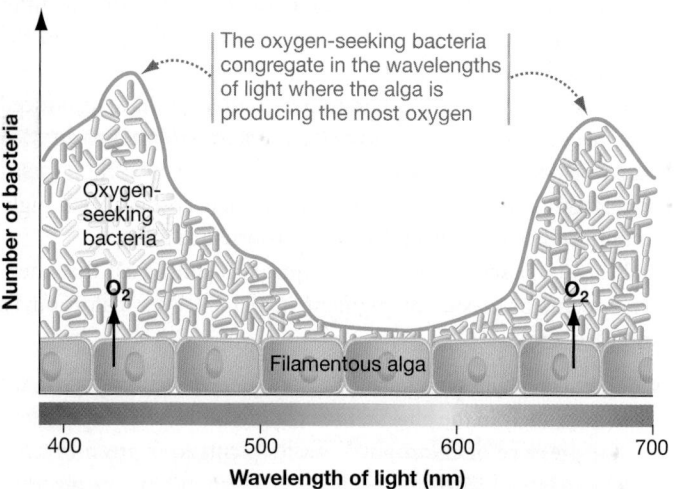

FIGURE 10.5 There Is a Strong Correlation between the Absorption Spectrum of Pigments and the Action Spectrum for Photosynthesis.

✔**EXERCISE** Some photosynthetic organisms have versions of chlorophyll pigments that absorb most strongly at 390 nm and 880 nm. Draw what the wavelength graph in part (b) would look like if these organisms were used in the experiment instead of the alga.

T. W. Engelmann answered this question by laying a filamentous alga across a glass slide that was illuminated with a spectrum of colors. The idea was that the alga would begin performing photosynthesis in response to the various wavelengths of light and produce oxygen as a by-product. To determine exactly where oxygen was being produced, Engelmann added bacterial cells from a species that is attracted to oxygen.

As **Figure 10.5b** shows, most of the bacteria congregated in the blue and red regions of the slide. Because wavelengths in these parts of the spectrum were associated with high oxygen concentrations, Engelmann concluded that they defined the **action spectrum** for photosynthesis—the wavelengths that drive the light-capturing reactions.

The data suggest that blue and red photons are the most effective at driving photosynthesis. Because the chlorophylls absorb these wavelengths, the data indicate that chlorophylls are the main photosynthetic pigments.

Before analyzing what happens during the absorption event itself, let's look at the structure and function of the carotenoids and chlorophylls.

WHAT IS THE ROLE OF CAROTENOIDS AND OTHER ACCESSORY PIGMENTS? Carotenoids are called accessory pigments because they absorb light and pass the energy on to chlorophyll. The carotenoids found in plants belong to two classes, called carotenes and xanthophylls. **Figure 10.6a** shows the structure of β-carotene, which gives carrots their orange color. A xanthophyll called zeaxanthin, which gives corn kernels their bright yellow color, is nearly identical to β-carotene, except that the ring structures on either end of the molecule contain a hydroxyl (–OH) group.

Both xanthophylls and carotenes are found in chloroplasts. In autumn, when the leaves of deciduous trees die, their chlorophyll degrades first. The wavelengths scattered by the carotenoids that remain turn forests into spectacular displays of yellow, orange, and red.

Carotenoids absorb wavelengths of light that are not absorbed by chlorophyll. As a result, they extend the range of wavelengths that can drive photosynthesis.

Researchers discovered an even more important function for carotenoids, though, by analyzing what happens to leaves when these pigments are destroyed. Many herbicides, for example, work by inhibiting enzymes that are involved in carotenoid synthesis. Plants lacking carotenoids rapidly lose their chlorophyll, turn white, and die. Based on these results, researchers have concluded that carotenoids also serve a protective function.

To understand why carotenoids are protective, recall from Chapter 2 that photons—especially the high-energy, short-wavelength photons in the ultraviolet part of the electromagnetic spectrum—contain enough energy to knock electrons out of atoms and create free radicals. Free radicals, in turn, trigger reactions that degrade molecules.

Carotenoids "quench" free radicals by accepting or stabilizing unpaired electrons. As a result, they protect chlorophyll molecules from harm. When carotenoids are absent, chlorophyll molecules are destroyed and photosynthesis stops. Starvation and death follow.

FLAVONOIDS ARE A NATURAL SUNSCREEN Flavonoids are accessory pigments that also protect plants from high-energy radiation. They are found in the vacuoles of leaf cells, and function by absorbing ultraviolet light. In their absence, chlorophyll molecules are subject to damage from the free radicals triggered by UV radiation. Flavonoids function as a sunscreen for leaves and stems.

The message here is that the energy in sunlight is a double-edged sword. It makes photosynthesis possible, but it can also lead to the formation of free radicals that damage cells. The role of carotenoids and flavonoids as protective pigments is crucial.

THE STRUCTURE OF CHLOROPHYLL As **Figure 10.6b** shows, chlorophyll a and chlorophyll b are similar in structure. Both

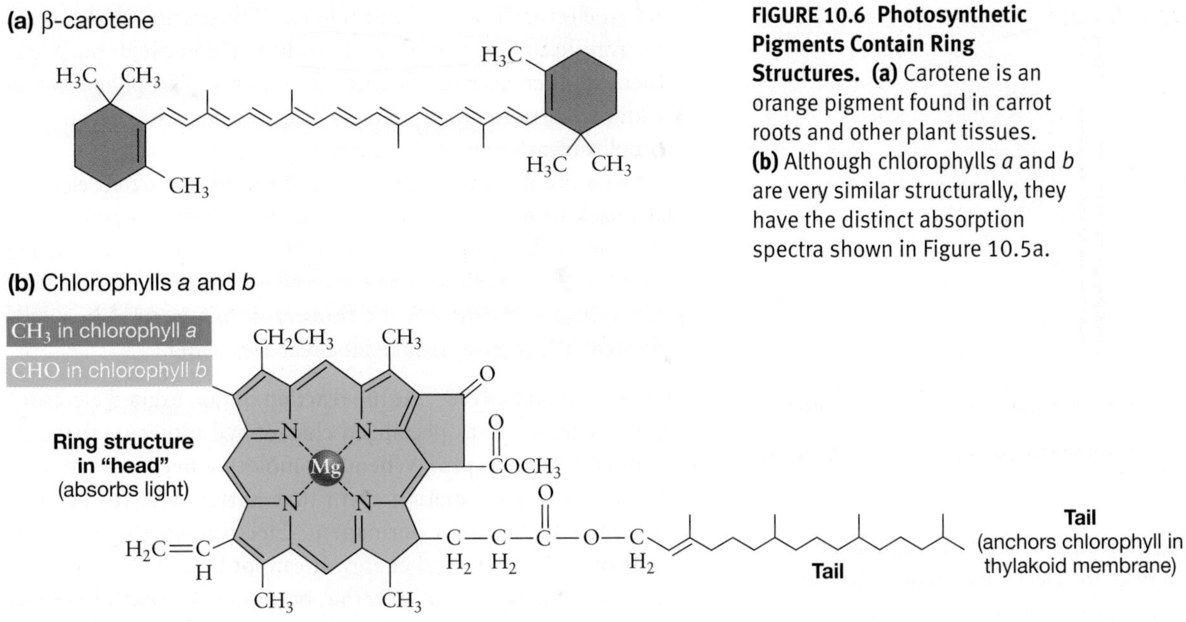

(a) β-carotene

H$_3$C CH$_3$

CH$_3$

H$_3$C

H$_3$C CH$_3$

FIGURE 10.6 Photosynthetic Pigments Contain Ring Structures. (a) Carotene is an orange pigment found in carrot roots and other plant tissues. **(b)** Although chlorophylls *a* and *b* are very similar structurally, they have the distinct absorption spectra shown in Figure 10.5a.

(b) Chlorophylls *a* and *b*

CH$_3$ in chlorophyll *a*
CHO in chlorophyll *b*

CH$_2$CH$_3$ CH$_3$

O

Ring structure in "head" (absorbs light)

N   N

Mg

—COCH$_3$

N   N

O

O

H$_2$C=C
H

C—C—C—O—C
H$_2$ H$_2$ H$_2$

Tail

Tail (anchors chlorophyll in thylakoid membrane)

CH$_3$ CH$_3$

have two fundamental parts: a long tail made up of isoprene subunits (introduced in Chapter 6) and a "head" consisting of a large ring structure with a magnesium atom in the middle. The tail keeps the molecule embedded in the thylakoid membrane; the head is where light is absorbed.

Just what is "absorption"? What happens when a photon of a particular wavelength—say, red light with a wavelength of 680 nm—strikes a chlorophyll molecule?

When Light Is Absorbed, Electrons Enter an Excited State

When a photon strikes a chlorophyll molecule, the photon's energy can be transferred to an electron in the chlorophyll molecule's head region. In response, the electron is "excited," or raised to a higher energy state.

The excited electron states that are possible in a particular pigment are discrete—meaning, incremental rather than continuous—and can be represented as lines on an energy scale. These discrete energy levels are a property of the electron configurations in a particular pigment.

Figure 10.7 shows the ground state, or unexcited state, as 0 and the higher energy states as 1 and 2. If the difference between the possible energy states is the same as the energy in the photon, the photon can be absorbed and an electron is excited to that energy state.

In chlorophyll, for example, the energy difference between the ground state and state 1 is equal to the energy in a red photon, while the energy difference between state 0 and state 2 is equal to the energy in a blue photon. Thus, chlorophyll can readily absorb red photons and blue photons.

Chlorophyll does not absorb green light well, because there is no discrete step—no difference in possible energy states for its electrons—that corresponds to the amount of energy in a green photon.

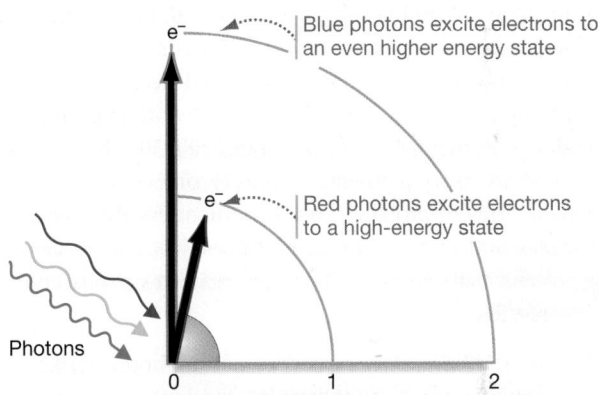

e$^-$ Blue photons excite electrons to an even higher energy state

e$^-$ Red photons excite electrons to a high-energy state

Photons

0   1   2

Energy state of electrons in chlorophyll

FIGURE 10.7 Electrons Are Promoted to High-Energy States When Photons Strike Chlorophyll. The discrete energy states labeled 1 and 2 are a property of chlorophyll's structure.

✔**QUESTION** Suppose a pigment had a discrete energy state that corresponded to the energy in green light. Where would you draw this energy state on this diagram?

Wavelengths in the ultraviolet part of the spectrum have so much energy that they may actually eject electrons from a pigment molecule and create a free radical. In contrast, wavelengths in the infrared regions have so little energy that in most cases they merely increase the movement of atoms in the pigment, generating heat—meaning molecular movement—rather than exciting electrons.

But if a pigment absorbs a photon with the right amount of energy, energy in the form of electromagnetic radiation is transferred to that electron. The electron now has high potential energy. What happens next?

If the excited electron simply falls back to its ground state, some of the absorbed energy is released as heat and the rest is released as electromagnetic radiation (light). This event is called

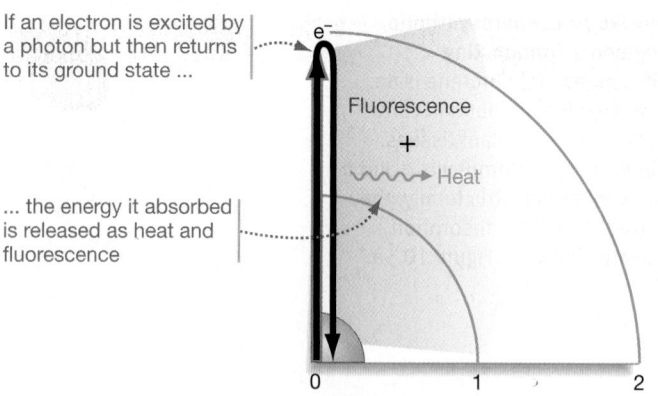

If an electron is excited by a photon but then returns to its ground state ...

... the energy it absorbed is released as heat and fluorescence

Fluorescence
+
Heat

Energy state of electrons in chlorophyll

FIGURE 10.8 Fluorescence Occurs When Excited Electrons Fall Back to the Ground State.

fluorescence (**Figure 10.8**). Because some of the original photon's energy is transformed to heat, the electromagnetic radiation that is given off during fluorescence has lower energy and a longer wavelength than the original photon did.

In a chloroplast, though, only about 2 percent of the red and blue photons that chlorophyll absorbs produce fluorescence. The other 98 percent of electrons drive photosynthesis.

To understand what happens to these excited electrons, it's important to recognize that chlorophyll molecules work in groups—not individually. In the thylakoid membrane, 200–300 chlorophyll molecules and accessory pigments such as carotenoids group together with an array of proteins, forming structures called the antenna complex and the reaction center. These complexes, along with the proteins that capture and process excited electrons, comprise a **photosystem**.

THE ANTENNA COMPLEX   When a red or blue photon strikes a pigment molecule in the **antenna complex**, the energy is absorbed and an electron is excited in response. This energy—but not the electron itself—is passed to a nearby chlorophyll molecule, where another electron is excited in response. This phenomenon is known as resonance. Chlorophyll molecules in the antenna complex transfer resonance energy.

Once the energy is transferred, the original excited electron falls back to its ground state. In this way, energy is transferred from one chlorophyll molecule to the next inside the antenna complex. Like a radio antenna, it receives a specific wavelength of electromagnetic radiation and transfers it to a receiver. In a photosystem, the receiver is called the reaction center.

THE REACTION CENTER   At the **reaction center**, excited electrons are transferred to a specialized chlorophyll molecule that acts as an electron acceptor. When this molecule becomes reduced, the energy transformation event that started with the absorption of light becomes permanent: Electromagnetic energy is transformed to chemical energy. It cannot be reemitted as fluorescence. The redox reaction that occurs in the reaction center results in the production of chemical energy from sunlight.

Note that in the absence of light, the electron acceptor does not accept electrons. It remains in an oxidized state, because the redox reactions that transfer an electron to the electron acceptor are endergonic. But when light excites electrons in chlorophyll to a high-energy state, the reactions become exergonic. In this way, the energy in light transforms an endergonic reaction to an exergonic one.

Figure 10.9 summarizes how chlorophyll interacts with light energy by illustrating the three possible fates of electrons that are excited by photons in photosynthetic pigments. The electrons can:

1. drop back down to a low energy level and cause fluorescence, or

2. excite an electron in a nearby pigment and induce resonance, or

3. be transferred to an electron acceptor in a redox reaction.

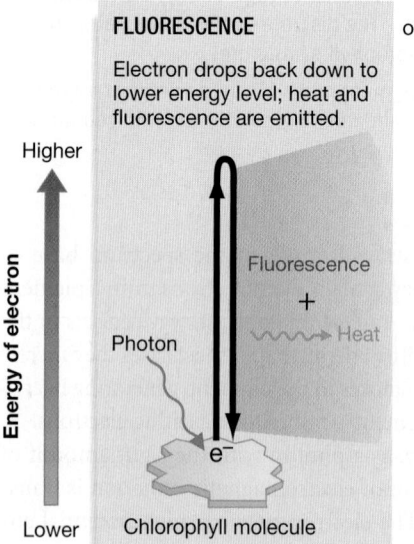

FLUORESCENCE   or

Electron drops back down to lower energy level; heat and fluorescence are emitted.

Higher

Energy of electron

Fluorescence
+
Heat

Photon

Lower   Chlorophyll molecule

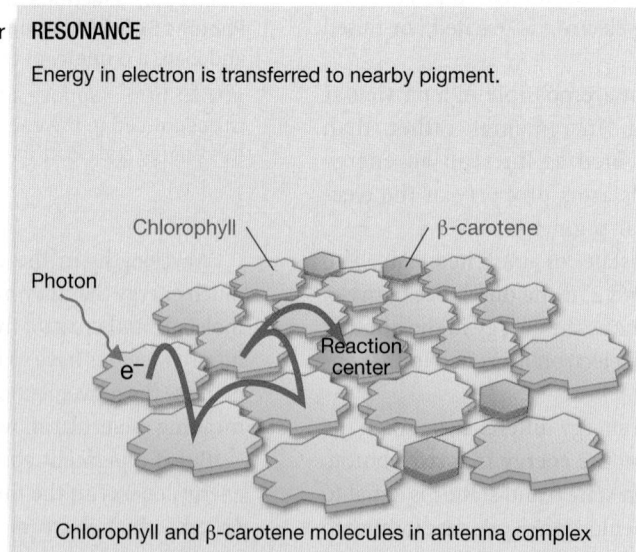

RESONANCE   or

Energy in electron is transferred to nearby pigment.

Chlorophyll   β-carotene

Photon

Reaction center

Chlorophyll and β-carotene molecules in antenna complex

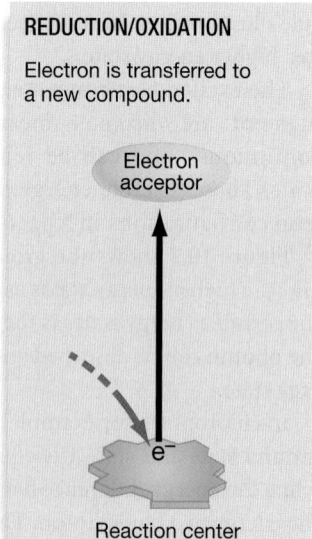

REDUCTION/OXIDATION

Electron is transferred to a new compound.

Electron acceptor

Reaction center

FIGURE 10.9 Three Fates for Excited Electrons in Photosynthetic Pigments. When sunlight promotes electrons in pigments to a high-energy state, three things can happen: They can fluoresce, pass energy to a nearby pigment via resonance, or transfer the electron to an electron acceptor.

Fluorescence is typical of isolated pigments, resonance occurs in antenna complex pigments, and redox occurs in reaction center pigments.

Now the question is, what happens to the high-energy electrons that are transferred to the reaction center? Specifically, how are they used to manufacture sugar?

CHECK YOUR UNDERSTANDING

If you understand that . . .

- Pigments absorb specific wavelengths of light.
- When a chlorophyll molecule in the antenna complex of a chloroplast membrane absorbs red or blue light, one of its electrons is promoted to a high-energy state.
- In the antenna complex, high-energy electrons transmit their energy among chlorophyll molecules. When energy is transferred to a chlorophyll molecule in the reaction center, the electron that is excited in response reduces an electron acceptor. In this way, light energy is transformed to chemical energy.

✔ You should be able to . . .

Consider what happens when you strike a tuning fork and then touch the vibrating tuning fork against another tuning fork. Explain which event is analogous to absorbing a photon, and which event is analogous to transferring resonance energy.

Answers are available in Appendix B.

10.3 The Discovery of Photosystems I and II

During the 1950s the fate of the high-energy electrons in photosystems was the central issue facing biologists interested in photosynthesis. A key breakthrough came from a simple experiment on how green algae responded to various wavelengths of light. The algal cells being studied responded to wavelengths of 700 nm and 680 nm, which are in the far-red and red portions of the visible spectrum, respectively.

Robert Emerson found that if the cells were illuminated with either far-red light or red light, the photosynthetic response was moderate (**Figure 10.10**). But if cells were exposed to a combination of far-red and red light, the rate of photosynthesis increased dramatically—much more than the sum of the rates produced by each wavelength independently. This phenomenon was called the enhancement effect. Why it occurred was a complete mystery at the time.

The puzzle was solved by Robin Hill and Faye Bendall, who proposed that green algae and plants have two distinct types of reaction centers. One reaction center, called **photosystem II**, interacts with a different reaction center, referred to as **photosystem I** (because it was discovered first). According to the two-photosystem hypothesis, the enhancement effect occurs because photosynthesis is much more efficient when both photosystems operate together.

EXPERIMENT

QUESTION: Red and far-red light each stimulate a moderate rate of photosynthesis. How does a combination of both wavelengths affect the rate of photosynthesis?

HYPOTHESIS: When red and far-red light are combined, the rate of photosynthesis will double.

NULL HYPOTHESIS: When red and far-red light are combined, the rate of photosynthesis will be the same as for each wavelength alone.

EXPERIMENTAL SETUP:

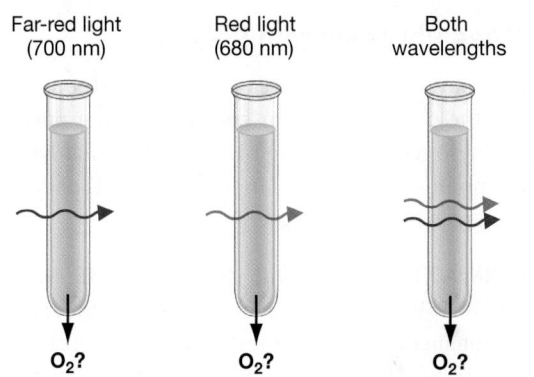

1. Expose algal cells to far-red light and then red light. Record oxygen produced as a measure of rate of photosynthesis.

2. Expose same cells to a combination of both lights.

PREDICTION: When the two wavelengths are combined, the rate of photosynthesis will double.

PREDICTION OF NULL HYPOTHESIS: When the two wavelengths are combined, the rate of photosynthesis will be the same as for each wavelength alone.

RESULTS:

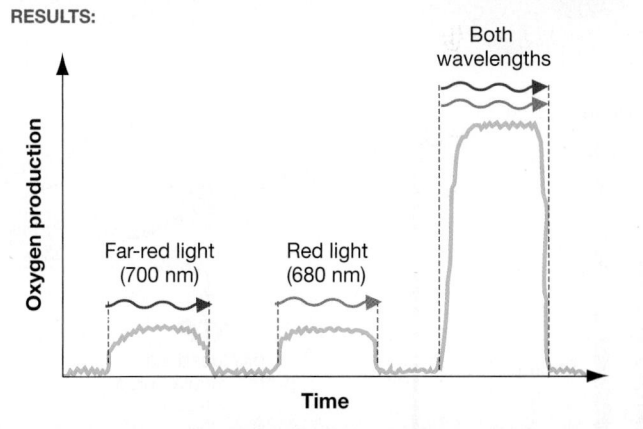

CONCLUSION: Neither hypothesis is correct. The combination of both wavelengths more than doubles the rate of photosynthesis. A new hypothesis is required to explain this enhancement effect.

FIGURE 10.10 Discovery of the "Enhancement Effect" of Red and Far-Red Light.

SOURCE: Emerson, R. and W. Arnold. 1932. The photochemical reaction in photosynthesis. *Journal of General Physiology* 16: 191–205.

✔ QUESTION Was it important for the researchers to keep the density of algal cells fairly constant in each treatment? Explain why or why not.

Subsequent work has shown that the two-photosystem hypothesis is correct. In green algae and land plants, thylakoid membranes contain photosystems that differ in structure and function but complement each other.

To figure out how the two photosystems work, investigators focused on species of photosynthetic bacteria that have photosystems similar to either photosystem I or II, but not both. Once each type of photosystem was understood in isolation, researchers explored how they work in combination. Let's do the same—we'll analyze photosystem II, then photosystem I, and then how the two interact.

How Does Photosystem II Work?

To study photosystem II, researchers focused on organisms known as the purple nonsulfur bacteria and the purple sulfur bacteria. These cells have a single photosystem that has many of the same components observed in photosystem II of cyanobacteria ("blue-green bacteria"), algae, and plants.

PHEOPHYTIN ACCEPTS ELECTRONS AND TRANSFERS THEM TO AN ELECTRON TRANSPORT CHAIN In photosystem II, the action begins when the antenna complex transmits energy to the reaction center and the molecule pheophytin comes into play (**Figure 10.11**).

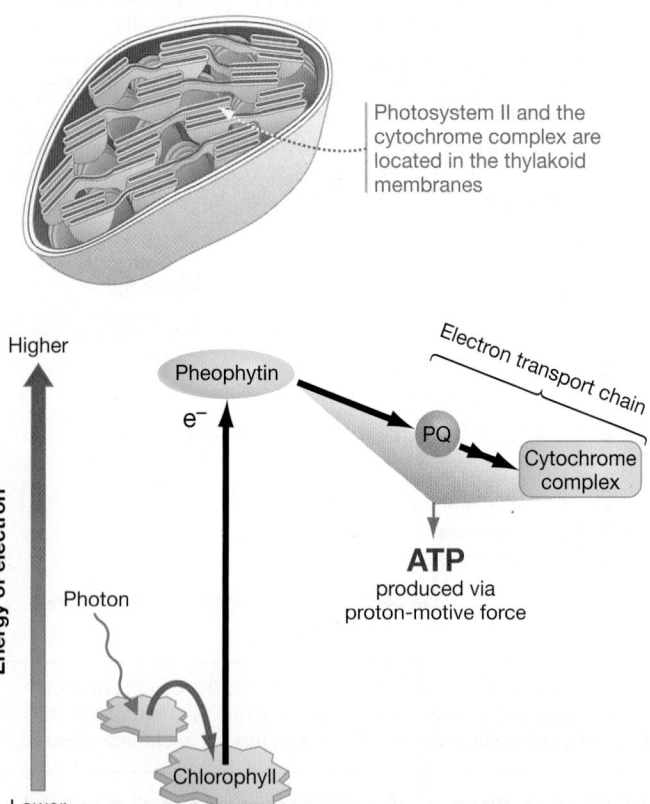

Photosystem II and the cytochrome complex are located in the thylakoid membranes

FIGURE 10.11 Photosystem II Feeds High-Energy Electrons to an Electron Transport Chain. When an excited electron leaves the chlorophyll molecule in the reaction center of photosystem II, the electron is accepted by pheophytin, transferred to plastoquinone (PQ), and then stepped down in energy along an electron transport chain.

Structurally, **pheophytin** is identical to chlorophyll except that pheophytin lacks a magnesium atom in its head region. Functionally, the two molecules are extremely different.

Instead of acting as a pigment that promotes an electron when it absorbs a photon, pheophytin acts as an electron acceptor. When an electron in the reaction center chlorophyll is excited energetically, the electron binds to pheophytin. The reduction of pheophytin (and the accompanying oxidation of the reaction center chlorophyll pigment) completes the energy transformation step that started with the absorption of light in the antenna complex.

Electrons that reach pheophytin are passed to an electron transport chain (ETC) in the thylakoid membrane. In both structure and function, this group of molecules is similar to the ETC in the inner membrane of mitochondria (Chapter 9).

- Structurally, the ETC associated with photosystem II and the ETC in the mitochondrion both contain quinones and cytochromes.

- Functionally, the redox reactions that occur in both ETCs result in protons being pumped from one side of an internal membrane to the other. The proton gradient that builds up drives ATP production via ATP synthase. Photosystem II triggers chemiosmosis and ATP synthesis in the chloroplast.

THE ELECTRON TRANSPORT CHAIN SETS UP A PROTON GRADIENT THAT DRIVES ATP SYNTHASE **Figure 10.12** explains how photosystem II works in more detail. Start by focusing on the molecule called **plastoquinone**—a quinone symbolized PQ. Note that PQ connects the excited electron in pheophytin with components of the ETC called the cytochrome complex.

Now recall from Chapter 9 that quinones are small hydrophobic molecules. Because plastoquinone is lipid soluble and not anchored to a protein, it is free to move from one side of the thylakoid membrane to the other.

When it receives electrons from pheophytin, plastoquinone carries them to the other side of the membrane and delivers them to more electronegative molecules in the chain, including a cytochrome.

In this way, plastoquinone shuttles electrons from pheophytin to the cytochrome complex. The electrons are then passed through a series of iron- and copper-containing proteins within the cytochrome complex. The potential energy released by these reactions allows protons to be added to other plastoquinone molecules, which carry them to the lumen side of the thylakoid membrane.

The protons transported by plastoquinone result in a large concentration of protons in the thylakoid lumen. When photosystem II is active, the pH of the thylakoid interior reaches 5 while the pH of the stroma hovers around 8. Because the pH scale is logarithmic, the difference of 3 units means that the concentration of H^+ is $10 \times 10 \times 10 = 1000$ times higher in the lumen than in the stroma. In addition, the stroma becomes negatively charged relative to the thylakoid lumen.

The net effect of electron transport, then, is to set up a large proton gradient that drives H^+ out of the thylakoid lumen and into the stroma. Based on your reading of Chapter 9, it should

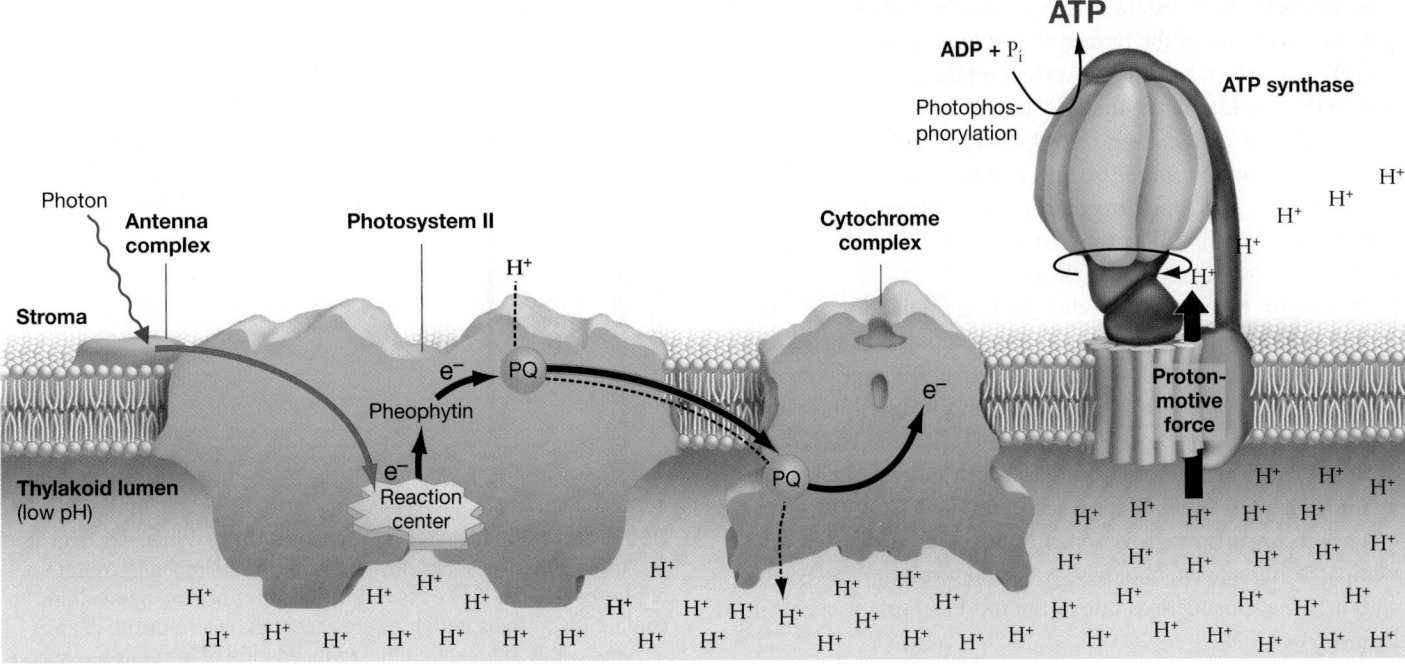

FIGURE 10.12 The ETC in PS II Pumps Protons. Plastoquinone (PQ) carries electrons from photosystem II along with protons from the stroma. The electrons are passed to the cytochrome complex, and the protons are released in the thylakoid lumen, creating a proton-motive force that drives ATP synthesis.

come as no surprise that this proton-motive force drives the production of ATP.

Specifically, proton flow is an exergonic process that drives the endergonic synthesis of ATP from ADP and P$_i$. The stream of protons through ATP synthase causes conformational changes that drive the phosphorylation of ADP. This process is called **photophosphorylation**.

Photophosphorylation is similar to the oxidative phosphorylation that occurs in plant and animal mitochondria. Both depend on chemiosmosis—a process that you can review in the study area at *www.masteringbiology.com*.

(MB) Web Activity Chemiosmosis

✔If you understand photosystem II, you should be able to compare and contrast photosystem II to the electron transfer chain of mitochondria. You also should be able to explain where high-energy electrons come from in each system and what the eventual product is.

The photosystem II story is not yet complete, however, because we haven't accounted for the electrons that flow through the system. The electron that was transferred from chlorophyll to pheophytin, in the photosystem II reaction center, needs to be replaced. In addition, the electron transport chain of photosystem II needs to donate its electrons to some final electron acceptor.

Where do the electrons required by photosystem II come from, and where do they go?

PHOTOSYSTEM II OBTAINS ELECTRONS BY OXIDIZING WATER
Think back to the overall reaction for photosynthesis:

$$6\,CO_2 + 12\,H_2O + \text{light energy} \longrightarrow C_6H_{12}O_6 + 6\,H_2O + 6\,O_2$$

In the presence of sunlight, carbon dioxide and water are used to produce carbohydrate, water, and oxygen gas.

Now recall that experiments with radioisotopes of oxygen showed that the oxygen atoms in O_2 come from water, not from carbon dioxide. As it turns out, the electrons that enter photosystem II also come from water. The oxygen-generating reaction can be written as

$$2\,H_2O \longrightarrow 4\,H^+ + 4\,e^- + O_2$$

Because electrons are removed from water, the molecule becomes oxidized. This reaction is referred to as "splitting" water. It supplies a steady stream of electrons for photosystem II and is catalyzed by enzymes that are physically integrated into the photosystem II complex.

When excited electrons leave photosystem II and enter the ETC, the photosystem becomes so electronegative that enzymes can remove electrons from water, leaving protons and oxygen. The oxygen-generating reaction is highly endergonic. It is possible only because the energy in sunlight drives it by removing electrons from photosystem II.

OXYGENIC PHOTOSYNTHESIS AND THE OXYGEN ATMOSPHERE
Among all life-forms, photosystem II is the only protein complex that can catalyze the splitting of water molecules. Organisms such as cyanobacteria, algae, and plants that have photosystem II perform **oxygenic** ("oxygen-producing") photosynthesis, because they generate oxygen as a by-product of the process. The purple sulfur and purple nonsulfur bacteria, which cannot oxidize water, perform **anoxygenic** ("no oxygen-producing") photosynthesis.

It is difficult to overstate the importance of oxygenic photosynthesis. It produced the oxygen that is keeping you alive right now. O_2 was, in fact, almost nonexistent on Earth before enzymes evolved that could catalyze the oxidation of water.

According to the geologic record, oxygen levels in the atmosphere and oceans began to rise only about 2 billion years ago, as organisms that performed oxygenic photosynthesis increased in abundance. This transition was a disaster for anaerobic organisms, because oxygen is toxic to them.

As oxygen became more abundant, certain bacterial cells evolved the ability to use it as an electron acceptor during cellular respiration. O_2 is so electronegative that it creates a huge potential energy drop for the electron transport chains involved in cellular respiration. As a result, organisms that use O_2 as an electron acceptor in cellular respiration can produce much more ATP than can organisms that use other electron acceptors (see Chapter 9).

Aerobic organisms grow so efficiently that they have long dominated our planet. Biologists rank the evolution of Earth's oxygen-rich atmosphere as one of the most important events in the history of life.

Determining exactly how photosystem II splits water and generates oxygen may be the greatest challenge currently facing researchers interested in photosynthesis. This issue has important practical applications: If human chemists could replicate the reaction, it might be possible to produce huge volumes of O_2 and hydrogen gas (H_2) from water. The resulting H_2 could provide a clean, inexpensive fuel for vehicles.

✔If you understand photosystem II, you should be able to make an energy flowchart that includes the antenna complex, ATP synthase, pheophytin, light, the proton gradient, an ETC, and a reaction center, then add notes explaining where the enzyme complex that splits water fits in.

How Does Photosystem I Work?

Recall that researchers dissected photosystem II by studying similar, but simpler, photosystems in purple nonsulfur and purple sulfur bacteria. To understand the structure and function of photosystem I, they turned to heliobacteria ("sun-bacteria").

Like purple nonsulfur and purple sulfur bacteria, heliobacteria use the energy in sunlight to promote electrons to a high-energy state. But instead of being passed to an electron transport chain that pumps protons across a membrane, the high-energy electrons in heliobacteria are used to reduce NAD^+. When NAD^+ gains two electrons and a proton, NADH is produced.

In the cyanobacteria and algae and land plants, a similar set of light-capturing reactions reduces a phosphorylated version of NAD^+, symbolized $NADP^+$, yielding NADPH. Both NADH and NADPH function as electron carriers.

Figure 10.13 explains how photosystem I works in chloroplasts—put your finger on the "2 photons" arrows and trace the steps that follow.

- Pigments in the antenna complex absorb photons and pass the energy to the reaction center.

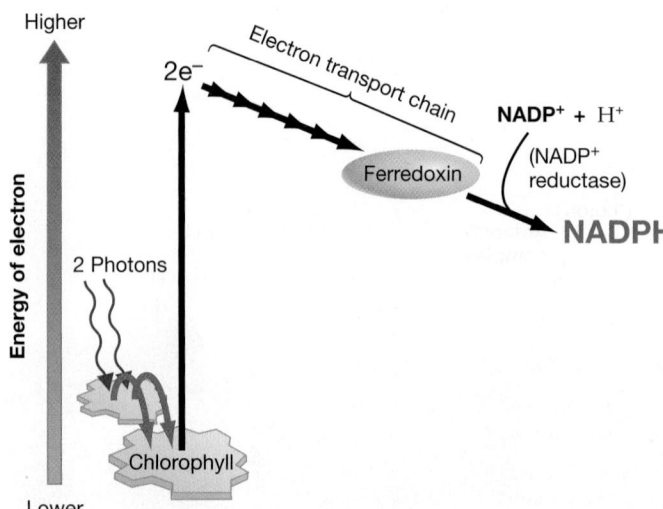

FIGURE 10.13 Photosystem I Produces NADPH. When excited electrons leave the chlorophyll molecule in the reaction center of photosystem I, they pass through a series of iron- and sulfur-containing proteins until they are accepted by ferredoxin. In an enzyme-catalyzed reaction, the reduced form of ferredoxin reacts with $NADP^+$ to produce NADPH.

- Electrons are excited in specialized chlorophyll molecules inside the reaction center.

- The high-energy electrons are passed through an electron transport chain—first to a series of iron- and sulfur-containing proteins inside the photosystem, then to a molecule called **ferredoxin**, then to the enzyme ferredoxin/$NADP^+$ oxidoreductase—also called $NADP^+$ reductase.

- $NADP^+$ reductase transfers two electrons and a proton to $NADP^+$. This reaction forms NADPH.

The photosystem itself and $NADP^+$ reductase are anchored in the thylakoid membrane; ferredoxin is closely associated with the bilayer.

🔑 To summarize: Photosystem I produces NADPH, which is similar in function to the NADH and $FADH_2$ produced by the citric acid cycle. NADPH is an electron carrier that can donate electrons to other compounds and thus reduce them. Photosystem II, in contrast, produces a proton gradient that drives the synthesis of ATP.

In combination, then, photosystems II and I produce chemical energy stored in ATP as well as reducing power in the form of NADPH. Although several groups of bacteria have just one of the two photosystems, the cyanobacteria, algae, and plants have both. In these organisms, how do the two photosystems interact?

The Z Scheme: Photosystems II and I Work Together

Figure 10.14 illustrates the **Z scheme** model for how photosystems II and I interact. The name was inspired by the proposed path of

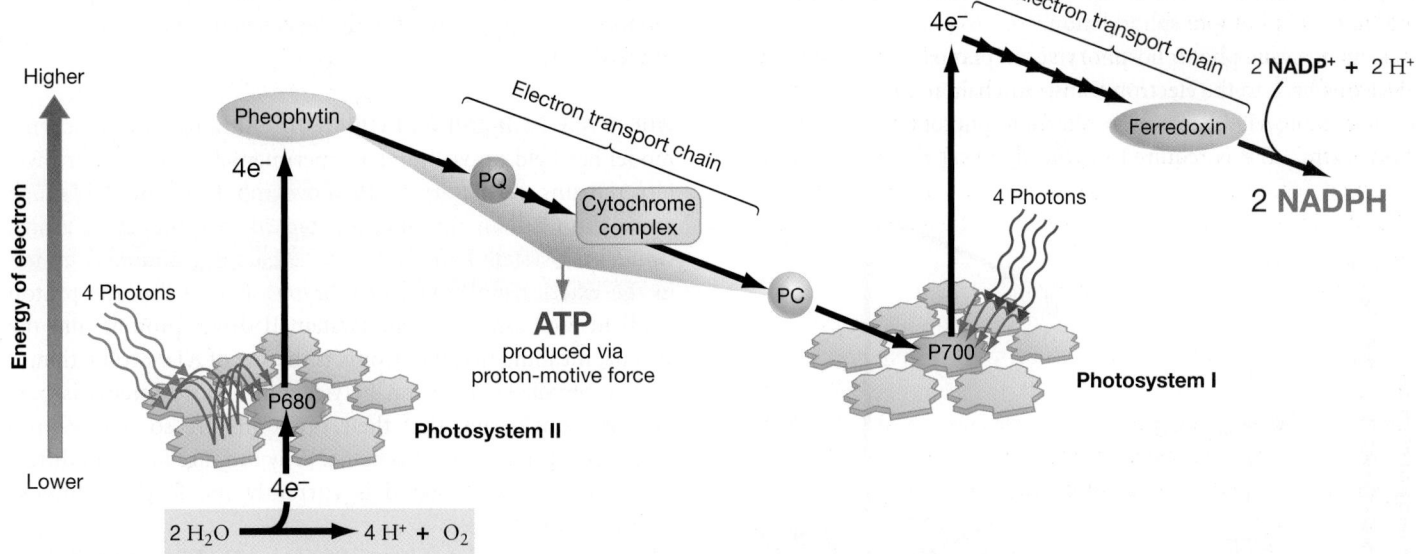

FIGURE 10.14 The Z Scheme Model Links Photosystems I and II. The Z scheme proposes that electrons from photosystem II enter photosystem I, where they are promoted to an energy state high enough to make the reduction of $NADP^+$ possible.

electrons through the two photosystems, as plotted on a vertical axis representing the changes occurring in their potential energy.

To drive home how photosynthesis works, trace the route of electrons through Figure 10.14 with your finger. Start on the lower left. The process starts when photons excite electrons in the chlorophyll molecules of photosystem II's antenna complex. When the energy in the excited electrons is transferred to the reaction center, a special pair of chlorophyll molecules named P680 passes excited electrons to pheophytin.

When pheophytin is reduced, it transfers high-energy electrons to an electron transport chain. There the electrons are gradually stepped down in potential energy through redox reactions among a series of quinones and cytochromes. Using the energy released by the redox reactions, plastoquinone (PQ) carries protons across the thylakoid membrane. ATP synthase uses the resulting proton-motive force to phosphorylate ADP, creating ATP.

When electrons reach the end of photosystem II's electron transport chain, they are passed to a small diffusible protein called **plastocyanin** (symbolized PC in Figure 10.14). Plastocyanin picks up an electron from the cytochrome complex, diffuses through the lumen of the thylakoid, and donates the electron to photosystem I.

Stop tracing for a moment, and consider the following:

- Plastocyanin is key—it forms a physical link between photosystem II and photosystem I.

- A single plastocyanin molecule can shuttle over 1000 electrons per second between photosystems.

- The flow of electrons between photosystems, by means of plastocyanin, is important because it replaces electrons that

are carried away from a chlorophyll molecule called P700 in the photosystem I reaction center.

Now keep going. The electrons that flow from photosystem II to P700, via plastocyanin, are eventually transferred to the protein ferredoxin, which passes electrons to an enzyme that catalyzes the reduction of $NADP^+$ to NADPH.

Finally, direct your attention back to the lower left portion of the figure. Note that the electrons that initially left photosystem II are replaced by electrons that are stripped away from water, producing oxygen gas as a by-product.

✔You should be able to explain (1) the role of plastocyanin in linking the two photosystems, (2) where the electrons that flow through the system have their highest potential energy, and (3) why the Z scheme is sometimes referred to as **noncyclic electron flow.**

UNDERSTANDING THE ENHANCEMENT EFFECT The Z scheme helps explain the enhancement effect documented in Figure 10.10. When algal cells are illuminated with wavelengths at 680 nm, in the red portion of the spectrum, only photosystem II can run at a maximum rate. The overall rate of electron flow through the Z scheme is moderate because photosystem I's efficiency is reduced.

Similarly, when cells receive only wavelengths at 700 nm, in the far red, only photosystem I is capable of peak efficiency; photosystem II is working at a below-maximum rate, so the overall rate of electron flow is reduced.

But when both wavelengths are available at the same time, both photosystem II and photosystem I are activated and work at a maximum rate, leading to enhanced efficiency.

CYCLIC PHOTOPHOSPHORYLATION PRODUCES ADDITIONAL ATP
Recent evidence indicates that an alternative electron path, called **cyclic photophosphorylation**, also occurs in green algae and plants

(**Figure 10.15**). In these species, ATP is produced in the Z-scheme and in cyclic photophosphorylation.

During cyclic photophosphorylation, photosystem I transfers electrons back to the electron transport chain in photosystem II, to augment ATP generation through photophosphorylation. This "extra" ATP is required for the chemical reactions that re-

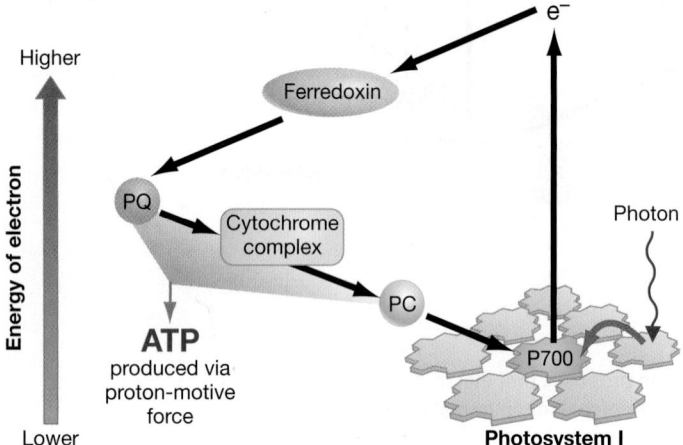

FIGURE 10.15 Cyclic Photophosphorylation Produces ATP. Cyclic electron transport is an alternative to the Z scheme. Instead of being donated to NADP⁺, electrons cycle through the system, resulting in the production of additional ATP via photophosphorylation.

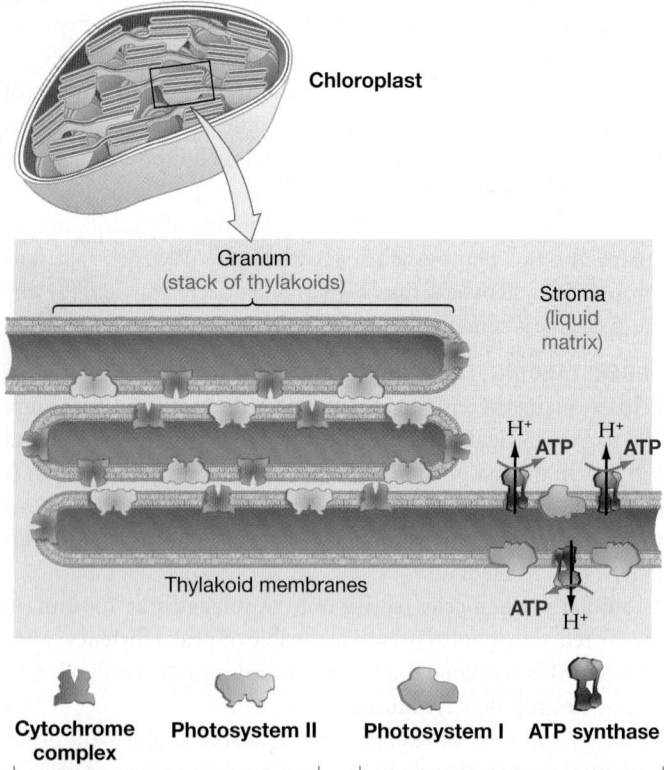

FIGURE 10.16 Photosystems I and II Occur in Separate Regions of Thylakoid Membranes within Grana. It is not yet clear how the arrangement of these components in the thylakoid relates to the paths of electrons in the Z scheme and during cyclic photophosphorylation.

duce carbon dioxide (CO_2) and produce sugars. Cyclic photophosphorylation coexists with the Z scheme and produces additional ATP.

WHERE ARE PS II AND PS I LOCATED? Although the Z-scheme model has held up well under experimental tests, a major mystery remains. As **Figure 10.16** shows, photosystem II is much more abundant in the interior, stacked membranes of grana, while photosystem I and ATP synthase are much more common in the exterior, unstacked membranes. Notice that the proton gradient established by photosystem II drives protons into the stroma, meaning that the stroma is the site of ATP production.

The physical separation between the photosystems is perplexing, considering that their functions are so tightly integrated according to the Z scheme. Why they are found in different parts of the thylakoid is currently the focus of intense research and debate.

The spatial relationships of PS II and PS I may be puzzling, but the fate of the ATP and NADPH produced by photosystems I and II is well documented. Chloroplasts use ATP and NADPH to reduce carbon dioxide to sugar. Your life, and the life of most organisms, depends on this process. How does it happen?

CHECK YOUR UNDERSTANDING

If you understand that . . .

- Photosystem II contributes high-energy electrons to an electron transport chain that pumps protons, creating a proton-motive force that drives ATP synthase.
- Photosystem I makes NADPH.

You should be able to . . .

1. Make a model of the Z scheme using paper cutouts representing the following elements: the antenna complexes of photosystems II and I, the ETCs in photosystems I and II, pheophytin, plastoquinone, plastocyanin, cytochrome complex, ferredoxin, the reaction that splits water, and the reduction of NADP⁺ to form NADPH.

2. Using dimes to represent electrons, explain how they flow through the photosystems.

Answers are available in Appendix B.

10.4 How Is Carbon Dioxide Reduced to Produce Glucose?

The reactions analyzed in Section 10.3 are triggered by light. This is logical, because their entire function is focused on transforming electromagnetic energy in the form of sunlight into chemical energy in the phosphate bonds of ATP and the electrons of NADPH. The reactions that produce sugar from carbon dioxide, in contrast, are not triggered directly by light. Instead, they depend on the ATP and NADPH produced by the light-capturing reactions of photosynthesis.

The Calvin Cycle Fixes Carbon

Carbon fixation is the addition of carbon dioxide to an organic compound. The word "fix" is appropriate because the process converts or fixes CO_2 to a biologically useful form. Once carbon atoms are fixed, they can be used to build the molecules found in cells.

Carbon fixation is a redox reaction—the carbon atom in CO_2 is reduced. Research on how this happens in chloroplasts gained momentum just after World War II, when radioactive isotopes of carbon became available for research purposes.

Melvin Calvin's group made great strides early in this effort, using a pulse-chase strategy. Recall from Chapter 7 that pulse-chase experiments introduce a pulse of labeled compound followed by a chase of unlabeled compound, and then follow the fate of the labeled compound over time.

In this case, the researchers fed green algae a pulse of $^{14}CO_2$ followed by a large amount of unlabeled CO_2 (**Figure 10.17**). After waiting a specified amount of time, they homogenized the cells by immersing them in hot alcohol, separated individual molecules in the extract via chromatography, and laid X-ray film over the chromatography surface.

If radioactively labeled molecules were present in the chromatograph, the energy they emitted would expose the film and create a dark spot. The labeled compounds could then be isolated and identified.

By varying the amount of time between starting the pulse of labeled $^{14}CO_2$ and analyzing the cells, Calvin and co-workers pieced together the sequence in which various intermediates formed. For example, when the team analyzed cells almost immediately after starting the $^{14}CO_2$ pulse, they found that the three-carbon compound 3-phosphoglycerate predominated. This result suggested that 3-phosphoglycerate was the initial product of carbon reduction. Stated another way, it appeared that carbon dioxide reacted with some unknown molecule to produce 3-phosphoglycerate.

This was an intriguing result, because 3-phosphoglycerate is also one of the 10 intermediates in glycolysis. The ATP and NADPH reactions manufacture carbohydrate; glycolysis breaks it down. Because the two processes are related in this way, it was logical that at least some intermediates in glycolysis and CO_2 reduction are the same.

RUBP IS THE INITIAL REACTANT WITH CO₂ Which compound reacts with CO_2 to produce 3-phosphoglycerate? This was the key, initial step. Calvin's group searched in vain for a two-carbon compound that might serve as the initial carbon dioxide acceptor and yield 3-phosphoglycerate.

Then, while Calvin was running errands one day, it occurred to him that the molecule reacting with carbon dioxide might contain five carbons, not two. Adding CO_2 to a five-carbon molecule would produce a six-carbon compound, which could then split in half to form 2 three-carbon molecules.

Experiments to test this hypothesis confirmed that the five-carbon compound **ribulose bisphosphate (RuBP)** is the initial reactant.

EXPERIMENT

QUESTION: What intermediates are produced as carbon dioxide is reduced to sugar?

HYPOTHESIS: No specific hypothesis.

EXPERIMENTAL SETUP:

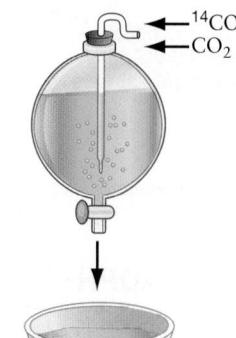

$\leftarrow$ $^{14}CO_2$
$\leftarrow$ CO_2

1. Feed algae a pulse of $^{14}CO_2$ followed by CO_2.

2. Wait 5–60 seconds; then homogenize cells by immersing in hot alcohol.

3. Separate molecules via chromatography.

4. Lay X-ray film on chromatograph to locate radioactive label.

PREDICTION: No specific prediction.

RESULTS:

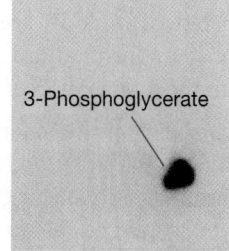

3-Phosphoglycerate

Compounds produced after 5 seconds

Compounds produced after 60 seconds

CONCLUSION: 3-Phosphoglycerate is the first intermediate product. Other intermediates appear later.

FIGURE 10.17 Experiments Revealed the Reaction Pathway Leading to Reduction of CO₂.

SOURCE: Benson, A. A., J. A. Bassham, et al. 1950. The path of carbon in photosynthesis. V. Paper chromatography and radioautography of the products. *Journal of the American Chemistry Society* 72: 1710–1718.

✓**QUESTION** Why wasn't this experiment based on a specific hypothesis and set of predictions?

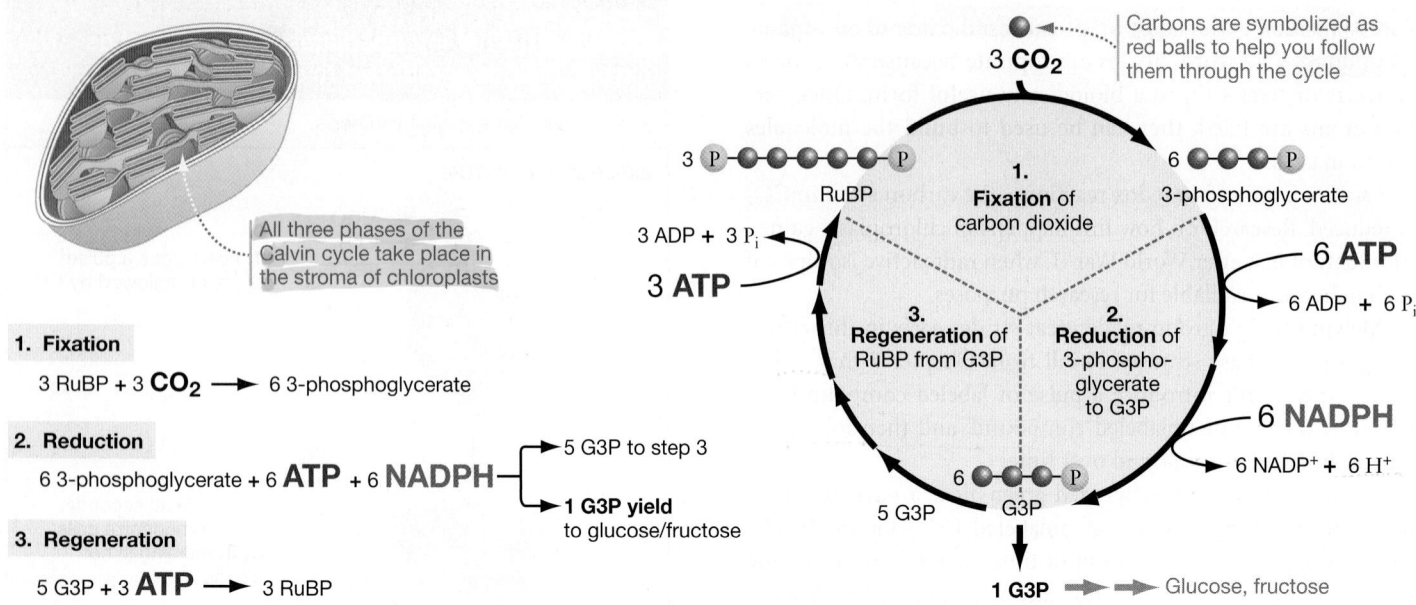

(a) The Calvin cycle has three phases.

All three phases of the Calvin cycle take place in the stroma of chloroplasts

1. Fixation

3 RuBP + 3 CO_2 ⟶ 6 3-phosphoglycerate

2. Reduction

6 3-phosphoglycerate + 6 **ATP** + 6 **NADPH** — ⟶ 5 G3P to step 3

⟶ **1 G3P yield** to glucose/fructose

3. Regeneration

5 G3P + 3 **ATP** ⟶ 3 RuBP

(b) The reaction occurs in a cycle.

Carbons are symbolized as red balls to help you follow them through the cycle

3 CO_2

3 P RuBP

6 3-phosphoglycerate

3 ADP + 3 P_i

3 **ATP**

1. Fixation of carbon dioxide

6 **ATP**

6 ADP + 6 P_i

3. Regeneration of RuBP from G3P

2. Reduction of 3-phospho-glycerate to G3P

6 **NADPH**

6 $NADP^+$ + 6 H^+

6 G3P

5 G3P G3P

1 G3P ⟹ Glucose, fructose

FIGURE 10.18 Carbon Dioxide Is Reduced in the Calvin Cycle. (a) Of the 6 G3Ps generated during the reduction phase, one is used in the synthesis of glucose—the other five are used to regenerate RuBP. Three molecules of CO_2 are required to make this happen. **(b)** The 3 RuBPs that are regenerated participate in fixation reactions, forming a cycle.

THE CALVIN CYCLE IS A 3-STEP PROCESS The complete Calvin cycle, as it came to be called, has three phases (**Figure 10.18**):

1. *Fixation phase* The Calvin cycle begins when CO_2 reacts with RuBP. This phase fixes carbon and produces two molecules of 3-phosphoglycerate.

2. *Reduction phase* 3-Phosphoglycerate is phosphorylated by ATP and then reduced by electrons from NADPH. The product is the phosphorylated sugar **glyceraldehyde-3-phosphate** (**G3P**). Some of the G3P that is synthesized is drawn off to manufacture glucose and fructose, which are linked to form the disaccharide sucrose.

3. *Regeneration phase* The rest of the G3P keeps the cycle going by serving as the substrate for the third phase in the cycle: reactions that result in the regeneration of RuBP.

All three phases take place in the stroma of chloroplasts. One turn of the Calvin cycle fixes one molecule of CO_2; three turns of the cycle are required to produce one molecule of G3P. ✔If you understand the Calvin cycle, you should be able to explain how CO_2, G3P, RuBP, and 3-phosphoglycerate are related.

🔑 The discovery of the Calvin cycle clarified how the ATP and NADPH produced by light-capturing reactions allow cells to reduce CO_2 to carbohydrate $(CH_2O)_n$. Because sugars store a great deal of potential energy, producing them takes a great deal of chemical energy. During photosynthesis, the energy required to reduce CO_2 to sugar is provided by ATP and NADPH synthesized in the light-capturing reactions.

The initial phase of the Calvin cycle—the reaction between RuBP and CO_2—is one of only two reactions that are unique to the Calvin cycle. Most reactions involved in reducing CO_2 also occur during glycolysis or other metabolic pathways.

The reaction between CO_2 and RuBP starts the transformation of carbon dioxide gas from the atmosphere into sugars. Plants use sugars to fuel cellular respiration and build leaves and other structures. Millions of non-photosynthetic organisms—from fungi to mammals—also depend on this reaction to provide the sugars they need for cellular respiration.

To review the Calvin cycle and the light-capturing reactions, go to the study area at *www.masteringbiology.com*.

(MB) **BioFlix™** Photosynthesis, **Web Activity** Photosynthesis

Ecologically, the addition of CO_2 to RuBP may be the most important chemical reaction on Earth. The enzyme that catalyzes it is fundamental to life. How does this protein work?

✗ The Discovery of Rubisco

To find the enzyme that fixes CO_2, Arthur Weissbach and colleagues ground up spinach leaves, purified a large series of proteins from the resulting cell extracts, and tested each protein to see if it could catalyze the incorporation of $^{14}CO_2$ into RuBP to form 3-phosphoglycerate. Eventually they isolated the catalyst, an enzyme which is abundant in leaf tissue. The researchers' data suggested that it constituted at least 10 percent of the total protein in spinach leaves.

The CO_2-fixing enzyme, ribulose-1,5-bisphosphate carboxylase/oxygenase (commonly referred to as **rubisco**), is found in all photosynthetic organisms that use the Calvin cycle to fix carbon, and is thought to be the most abundant enzyme on Earth. The rubisco molecule is cube-shaped and has eight active sites where CO_2 is fixed.

Despite its large number of active sites, rubisco is a slow enzyme. Each active site catalyzes just three reactions per second; other enzymes typically catalyze thousands of reactions per second. Plants synthesize huge amounts of rubisco, possibly as an adaptation compensating for its lack of speed.

Besides being slow, rubisco is extremely inefficient because it catalyzes the addition of O_2 to RuBP as well as the addition of CO_2 to RuBP. This is a key point: Oxygen and carbon dioxide compete at the enzyme's active sites, which slows the rate of CO_2 reduction.

Why would an active site of rubisco accept both O_2 and CO_2? Given rubisco's importance in producing food for photosynthetic species, this trait would appear to be **maladaptive**—it reduces the fitness of individuals.

The reaction of O_2 with RuBP actually does more than just compete with the reaction of CO_2 at the same active site. One of the molecules that results from the addition of oxygen to RuBP is processed in reactions that consume ATP and O_2 and release CO_2. Part of this pathway occurs in chloroplasts, and part in peroxisomes and mitochondria. The reaction sequence resembles respiration, because it consumes oxygen and produces carbon dioxide. As a result, it is called **photorespiration** (**Figure 10.19**).

Because photorespiration consumes energy and releases fixed CO_2, it "undoes" photosynthesis. When photorespiration occurs, the overall rate of photosynthesis declines. Explaining why rubisco is so slow and why photorespiration occurs remains an unsolved problem in biological science.

Carbon Dioxide Enters Leaves through Stomata

Atmospheric carbon dioxide is a key reactant in photosynthesizing cells. It would seem straightforward, then, for CO_2 to diffuse directly into plants along a concentration gradient. But the situation is not this simple, because plants are covered with a waxy coating called a cuticle. This lipid layer prevents water from evaporating out of tissues, but it also prevents CO_2 from entering them.

How does CO_2 get into photosynthesizing tissues? The surface of a leaf is dotted with openings bordered by two distinctively shaped cells (**Figure 10.20a**). The paired cells are called **guard cells**, the opening is called a pore, and the entire structure is a **stoma** (plural: **stomata**).

If CO_2 concentrations inside the leaf are low as photosynthesis gets under way, chemical signals activate proton pumps in the membranes of guard cells. These pumps establish a charge gradient across the membrane. In response, potassium ions (K^+) move

into the guard cells. When water follows along the newly created osmotic gradient, the guard cells swell and create a pore.

An open stoma allows CO_2 from the atmosphere to diffuse into air-filled spaces inside the leaf (**Figure 10.20b**). Eventually the CO_2 diffuses along a concentration gradient into the chloroplasts of photosynthesizing cells. A strong concentration gradient favoring entry of CO_2 is maintained by the Calvin cycle, which constantly uses up the CO_2 in chloroplasts.

Mechanisms for Increasing CO_2 Concentration

The oxygenation reaction that triggers photorespiration is favored when oxygen concentrations are high and CO_2 concentrations are low. But the atmosphere is 21 percent oxygen and only 0.03 percent carbon dioxide.

In addition, stomata are normally open during the day, when photosynthesis is occurring, and closed at night. But if the daytime is extremely hot and dry, leaf cells may lose a great deal of water to evaporation through their stomata. When this occurs, they must either close the openings and halt photosynthesis or risk death from dehydration. When conditions are hot and dry, then, CO_2 delivery stops—meaning that photosynthesis and growth slow. To make a bad situation worse, oxygen levels build

(a) Leaf surfaces contain stomata.

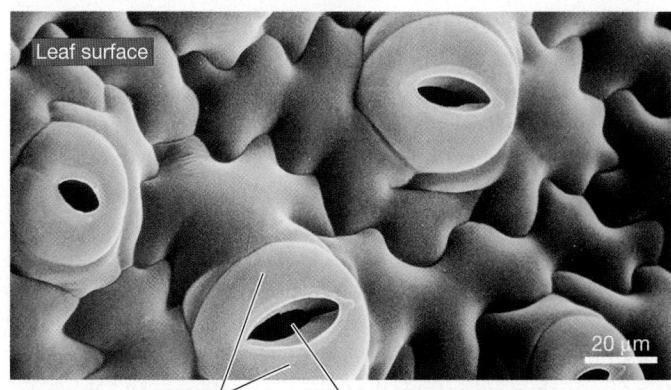

Guard cells + Pore = **Stoma**

(b) Carbon dioxide diffuses into leaves through stomata.

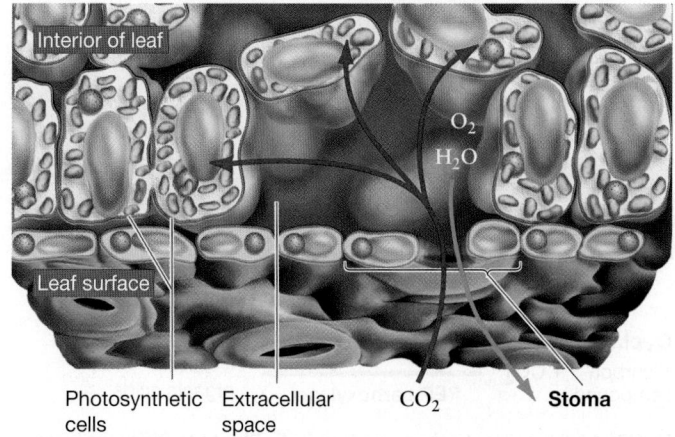

FIGURE 10.20 Leaf Cells Obtain Carbon Dioxide through Stomata.

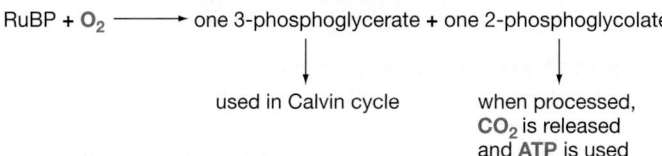

Reaction with carbon dioxide during photosynthesis:

RuBP + **CO$_2$** ⟶ two 3-phosphoglycerate

⟶ used in Calvin cycle

Reaction with oxygen during photorespiration:

RuBP + **O$_2$** ⟶ one 3-phosphoglycerate + one 2-phosphoglycolate

⟶ used in Calvin cycle

⟶ when processed, **CO$_2$** is released and **ATP** is used

FIGURE 10.19 Photorespiration Competes with Photosynthesis.
Rubisco catalyzes competing reactions with very different outcomes.

up in leaves as cellular respiration continues, meaning that photorespiration is favored.

How do plants that live in hot, dry environments keep CO_2 supplies high enough to limit the damaging effects of photorespiration? And how can photosynthesizing cells raise CO_2 concentrations to make photosynthesis more efficient? An answer emerged in a surprising experimental result.

C$_4$ PHOTOSYNTHESIS After the Calvin cycle had been worked out in algae, researchers in a variety of labs used the same pulse-chase approach to investigate how carbon fixation occurs in other species. Hugo Kortschak and colleagues and Y. S. Karpilov and associates exposed leaves of sugarcane and maize (corn) to radioactive carbon dioxide ($^{14}CO_2$) and sunlight, and then isolated and identified the product molecules.

Both research teams expected to find the first of the radioactive carbon atoms in 3-phosphoglycerate—the normal product of carbon fixation by rubisco. Instead, they found that in their species, the radioactive carbon atom ended up in four-carbon compounds such as malate and aspartate.

Instead of creating a three-carbon sugar, it appeared that these species' CO_2 fixation produced four-carbon sugars. The two pathways became known as **C$_3$ photosynthesis** and **C$_4$ photosynthesis**, respectively (**Figure 10.21**).

Researchers who followed up on the initial reports found that, in some plant species, carbon dioxide can be added to RuBP by rubisco *or* to three-carbon compounds by an enzyme called **PEP carboxylase**. They also showed that the two enzymes are found in distinct cell types within the same leaf. PEP carboxylase is common in **mesophyll cells** near the surface of leaves, while rubisco is found in **bundle-sheath cells** that surround the vascular tissue in the interior of the leaf (**Figure 10.22a**). **Vascular tissue** conducts water and nutrients in plants.

This situation contrasts with C$_3$ plants. In C$_3$ species, mesophyll cells contain chloroplasts and rubisco.

Based on the observations about C$_4$ plants, Hal Hatch and Roger Slack proposed a three-step model to explain how CO_2 that is fixed to a four-carbon sugar feeds the Calvin cycle (**Figure 10.22b**):

Step 1 PEP carboxylase fixes CO_2 in mesophyll cells.

Step 2 The four-carbon organic acids that result travel to bundle-sheath cells through plasmodesmata (Chapter 8).

Step 3 The four-carbon organic acids release a CO_2 molecule that rubisco uses as a substrate to form 3-phosphoglycerate. This step initiates the Calvin cycle.

In effect, then, the C$_4$ pathway acts as a CO_2 concentrator. The reactions that take place in mesophyll cells require energy in the form of ATP, but they increase CO_2 concentrations in cells where rubisco is active. Because it increases the ratio of carbon dioxide to oxygen in photosynthesizing cells, less O_2 binds to rubisco's active sites. As a result, the C$_4$ pathway limits the damaging effects of photorespiration.

The C$_4$ pathway is an adaptation that keeps CO_2 concentrations in leaves high, making photosynthesis more efficent. C$_4$ plants include sugarcane, maize (corn), crabgrass, and several thousand other species in 19 distinct lineages of flowering plants. This suggests that the C$_4$ pathway has evolved independently several times. It is not the only mechanism that plants use to continue growth under hot, dry conditions, however.

CAM PLANTS Researchers studying a group of flowering plants called the Crassulaceae discovered a second mechanism for limiting the effects of photorespiration. This photosynthetic pathway, **crassulacean acid metabolism**, or **CAM**, is a CO_2 concentrator that acts as an additional, preparatory step to the Calvin cycle.

Like the C$_4$ pathway, CAM increases the concentration of CO_2 inside photosynthesizing cells and involves an organic acid with

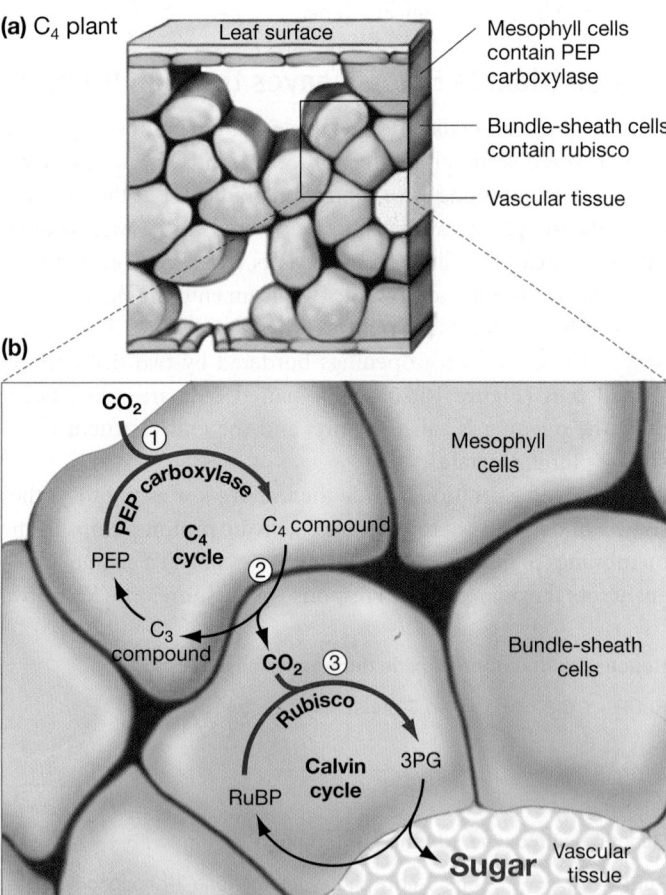

(a) C$_4$ plant

Leaf surface

Mesophyll cells contain PEP carboxylase

Bundle-sheath cells contain rubisco

Vascular tissue

(b)

CO$_2$

PEP carboxylase

C$_4$ compound

C$_4$ cycle

PEP

C$_3$ compound

CO$_2$

Rubisco

RuBP

Calvin cycle

3PG

Sugar

Mesophyll cells

Bundle-sheath cells

Vascular tissue

FIGURE 10.22 In C$_4$ plants, Carbon Fixation and the Calvin Cycle Occur in Different Cell Types. **(a)** The carbon-fixing enzyme PEP carboxylase is located in mesophyll cells, while rubisco is in bundle-sheath cells. **(b)** CO_2 is fixed to the three-carbon compound PEP by PEP carboxylase, forming a four-carbon organic acid. A CO_2 molecule from the four-carbon sugar then feeds the Calvin cycle.

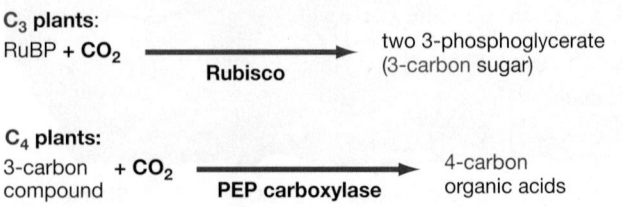

C$_3$ plants:

RuBP + CO$_2$ $\xrightarrow{\text{Rubisco}}$ two 3-phosphoglycerate (3-carbon sugar)

C$_4$ plants:

3-carbon compound + CO$_2$ $\xrightarrow{\text{PEP carboxylase}}$ 4-carbon organic acids

FIGURE 10.21 Initial Carbon Fixation in C$_4$ Plants Is Different from That in C$_3$ Plants.

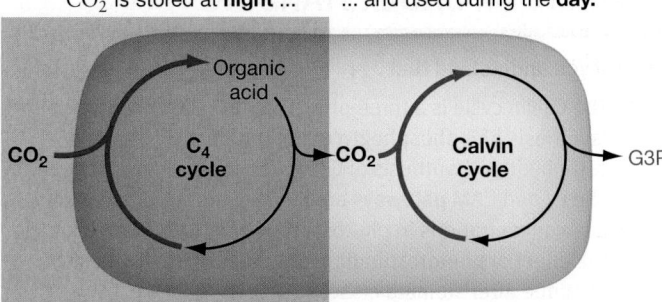

CO$_2$ is stored at **night** ...   ... and used during the **day.**

FIGURE 10.23 In CAM Plants, Carbon Fixation Occurs at Night and the Calvin Cycle Occurs during the Day.

four carbons. But unlike the C$_4$ pathway, CAM occurs at a different time than the Calvin cycle does—not in a different place.

CAM occurs in cacti and other species that routinely keep their stomata closed on hot, dry days. At night, when conditions are cooler and more moist, CAM plants open their stomata and take in huge quantities of CO$_2$. The CO$_2$ is temporarily fixed to organic acids and stored in the central vacuoles of photosynthesizing cells. During the day, the molecules are processed in reactions that release the CO$_2$ and feed the Calvin cycle (**Figure 10.23**).

C$_4$ photosynthesis and CAM function as CO$_2$ pumps. They minimize photorespiration when stomata are closed and CO$_2$ cannot diffuse in directly from the atmosphere. Both are found in species that live in hot, dry environments.

But while C$_4$ plants stockpile CO$_2$ in cells where rubisco is not active, CAM plants store CO$_2$ when rubisco is inactive. In C$_4$ plants, the reactions catalyzed by PEP carboxylase and rubisco are separated in space; in CAM plants, the reactions are separated in time. You can go to the study area at *www.masteringbiology.com* to review these two processes.

MB **Web Activity** Strategies for Carbon Fixation

How Is Photosynthesis Regulated?

Like nuclear transport, cell-cell signaling responses, and cellular respiration, photosynthesis is regulated. Although the mechanisms responsible for turning photosynthesis on or off are still under investigation, several patterns have emerged.

- The presence of light triggers the production of proteins required for photosynthesis.

- When sugar supplies are high, the production of proteins required for photosynthesis is inhibited, but the production of proteins required to process and store sugars is stimulated.

- Rubisco is activated by regulatory molecules that are produced when light is available, but inhibited in conditions of low CO$_2$ availability—when photorespiration is favored.

- Inorganic phosphate is required by the light-capturing reactions, where it is incorporated into NADPH, and is released when the sugars produced by photosynthesis are processed. Low P$_i$ levels stimulate the production of Calvin cycle proteins. When the Calvin cycle is more active and more sugar is produced and processed, inorganic phosphate is no longer tied up in NADPH and can be reused in the light-capturing reactions.

The central message here is that the rate of photosynthesis is finely tuned, to reflect changes in environmental conditions and use resources efficiently.

What Happens to the Sugar That Is Produced by Photosynthesis?

The G3P molecules that exit the Calvin cycle enter one of several reaction pathways. The most important of these reaction sequences produces the monosaccharides glucose and fructose, which combine to form the disaccharide ("two-sugar") **sucrose** (**Figure 10.24**). The process starts with G3P from the Calvin cycle, involves a series of other phosphorylated three-carbon sugars, includes the synthesis of the familiar six-carbon sugar glucose, and ends with the production of sucrose.

An alternative pathway produces glucose molecules that polymerize to form **starch**. Starch production occurs inside the chloroplast; sucrose synthesis takes place in the cytosol.

When photosynthesis is taking place slowly, almost all the glucose that is produced is used to make sucrose. Sucrose is water soluble and readily transported to other parts of the plant.

If sucrose is delivered to rapidly growing parts of the plant, it is broken down to fuel cellular respiration and growth. But if it is transported to storage cells in roots, it is converted to starch and stored for later use.

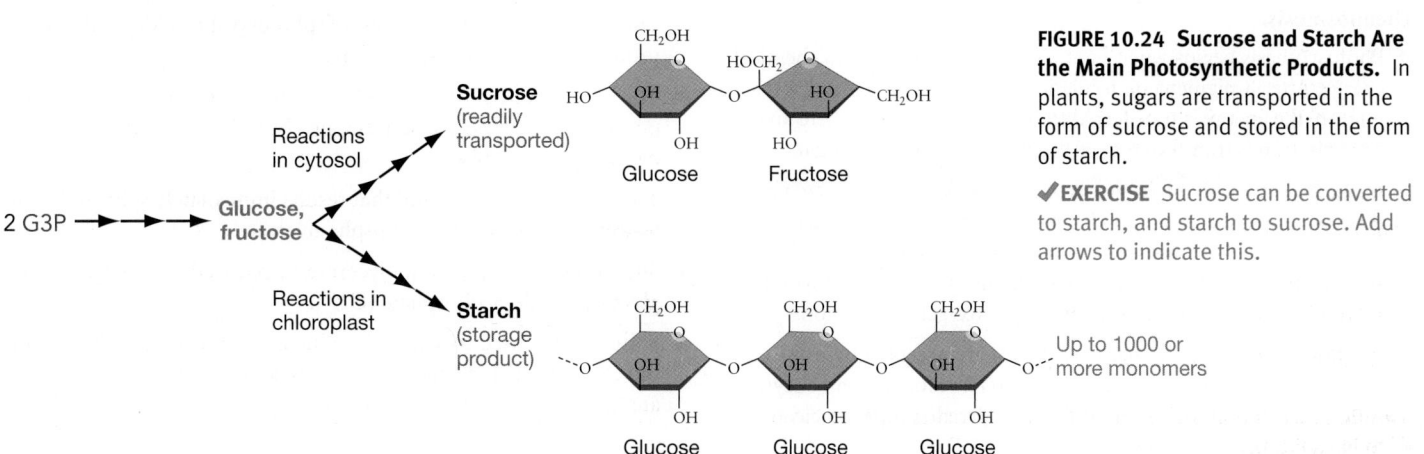

FIGURE 10.24 Sucrose and Starch Are the Main Photosynthetic Products. In plants, sugars are transported in the form of sucrose and stored in the form of starch.

✔**EXERCISE** Sucrose can be converted to starch, and starch to sucrose. Add arrows to indicate this.

When photosynthesis is proceeding rapidly and sucrose is abundant, glucose is used to synthesize starch in the chloroplasts of photosynthetic cells. Recall from Chapter 5 that starch is a polymer of glucose.

In photosynthesizing cells, starch acts as a temporary sugar-storage product. Starch is not water soluble, so it cannot be transported from photosynthetic cells to other areas of the plant. At night, the starch that is temporarily stored in leaf cells is broken down and used to manufacture sucrose molecules. The sucrose is then used by the photosynthetic cell in respiration or transported to other parts of the plant. In this way, chloroplasts provide sugars for cells throughout the plant by day and by night.

If a mouse eats the starch that is stored in a chloroplast or root cell, however, the chemical energy in the C–C and C–H bonds of the starch fuels the mouse's growth and reproduction. If an owl eats the mouse, the chemical energy in the mouse's tissues fuels the predator's growth and reproduction.

In this way, virtually all cell activity can be traced back to the chemical energy that was originally captured by photosynthesis. Photosynthesis is the staff of life.

CHECK YOUR UNDERSTANDING

If you understand that . . .

- The Calvin cycle is a three-phase process: CO$_2$ fixation (synthesis of 3-phosphoglycerate), production of carbohydrate (synthesis of G3P), and regeneration of RUBP.
- The C$_4$ and CAM pathways are mechanisms for increasing CO$_2$ concentrations in photosynthesizing cells. They limit the effect of photorespiration and allow photosynthesis to continue after stomata close.
- In photosynthesizing cells, sucrose is stored as starch. In all plant cells, sucrose is used to drive cellular respiration.

✓ **You should be able to . . .**

1. Describe how CO$_2$ is delivered to rubisco (a) via organic acids in mesophyll cells, (b) via organic acids synthesized at night and stored in vacuoles, and (c) directly.

2. Explain how the solubility of starch and sucrose affect their use by plants.

Answers are available in Appendix B.

CHAPTER 10 REVIEW

For media, go to the study area at www.masteringbiology.com

Summary of Key Concepts

Photosynthesis is the conversion of light energy to chemical energy stored in the bonds of carbohydrates. It consists of two linked sets of reactions.

- The light-capturing reactions occur in internal membranes of the chloroplast that are organized into structures called thylakoids in stacks known as grana.

- The Calvin cycle takes place in a fluid portion of the chloroplast called the stroma.

  ✓ You should be able to explain why biologists have stopped using the once-common phrase "light-independent reactions" to describe the Calvin cycle.

In the light-capturing reactions, excited electrons are used to produce the electron carrier NADPH or are donated to an electron transport chain, which results in the production of ATP via chemiosmosis.

- Photosynthesis begins when light energy is transformed to chemical energy. After a pigment molecule in an antenna complex absorbs a photon, the energy is transferred to the reaction center, where an excited electron is transferred to an electron acceptor, reducing it.

- In photosystem II, high-energy electrons are accepted by pheophytin and passed along an electron transport chain, releasing energy that pumps protons across the thylakoid membrane. The resulting proton gradient drives the synthesis of ATP by ATP synthase. PS II takes electrons from water, releasing oxygen and protons.

- In photosystem I, high-energy electrons are accepted by iron- and sulfur-containing proteins and passed to ferredoxin. In an enzyme-catalyzed reaction, the reduced form of ferredoxin passes electrons to NADP$^+$, forming NADPH.

- The Z scheme connects photosystems II and I. Plastocyanin carries electrons from the end of PS II's ETC to PS I. They are promoted to a high-energy state in PS I's reaction center, and subsequently used to reduce NADP$^+$. Electrons from PS I may occasionally be passed back to PS II's ETC instead of being used to reduce NADP$^+$. A cyclic flow of electrons between the two photosystems boosts ATP supplies.

  ✓ You should be able to explain why the rate of photosynthesis can be estimated by measuring the rate of oxygen production in chloroplasts.

  Web Activity Chemiosmosis

The reactions of the Calvin cycle start with the enzyme rubisco, which catalyzes the addition of CO$_2$ to a five-carbon compound. Subsequent reactions use the ATP and NADPH synthesized in the light reactions and yield a molecule required for carbohydrate production.

- The CO$_2$-reduction reactions of photosynthesis depend on the products of the light-capturing reactions.

- The Calvin cycle starts when CO$_2$ is attached to a five-carbon compound called ribulose bisphosphate (RuBP) in a reaction catalyzed by the enzyme rubisco.

- The six-carbon compound that results immediately splits in half to form two molecules of 3-phosphoglycerate.

- Subsequently, 3-phosphoglycerate is reduced to a sugar called glyceraldehyde-3-phosphate (G3P).

- Some G3P is used to synthesize glucose and fructose, which combine to form sucrose; the rest participates in reactions that regenerate RuBP so the cycle can continue.

✔You should be able to explain why it is accurate to call the Calvin cycle a cycle.

(MB) **BioFlix™** Photosynthesis, **Web Activity** Photosynthesis

🔑 **In plants, CO_2 enters photosynthetic tissue through openings called stomata. The CAM and C_4 pathways increase CO_2 concentrations inside the leaves of some species and make photosynthesis more efficient.**

- Rubisco catalyzes the addition of oxygen as well as carbon dioxide to RuBP. The reaction with oxygen leads to a loss of fixed CO_2 and ATP and is called photorespiration.

- CAM plants and C_4 plants fix CO_2 to organic acids, before it is transferred to rubisco. As a result, they can increase CO_2 levels in their tissues, reducing the effect of photorespiration and allowing photosynthesis to continue when stomata close.

✔You should be able to predict what would happen in a chloroplast containing an unusual form of rubisco—either a form that did not bind oxygen or a form that worked 10 times as fast as most forms of the enzyme.

(MB) **Web Activity** Strategies for Carbon Fixation

Questions

✔**TEST YOUR KNOWLEDGE**   *Answers are available in Appendix B*

1. What is resonance?
   a. when an electron is excited by a photon
   b. transformation of electromagnetic energy to chemical energy
   c. transfer of electrons among pigment molecules
   d. transfer of energy among pigment molecules

2. Why is chlorophyll green?
   a. It absorbs all wavelengths in the visible spectrum, transmitting ultraviolet and infrared light.
   b. It absorbs wavelengths only in the red and far-red portions of the spectrum (680 nm, 700 nm).
   c. It absorbs wavelengths in the blue and red parts of the visible spectrum.
   d. It absorbs wavelengths only in the blue part of the visible spectrum and transmits all other wavelengths.

3. What does it mean to say that CO_2 becomes fixed?
   a. It becomes bonded to an organic compound.
   b. It is released during cellular respiration.
   c. It acts as an electron acceptor.
   d. It acts as an electron donor.

4. What do the light-capturing reactions of photosynthesis produce?
   a. G3P   c. ATP and NADPH
   b. RuBP   d. sucrose or starch

5. Why do the absorption spectrum for chlorophyll and the action spectrum for photosynthesis coincide?
   a. Photosystems I and II are activated by different wavelengths of light.
   b. Wavelengths of light that are absorbed by chlorophyll trigger the light-capturing reactions.
   c. Energy from wavelengths absorbed by carotenoids is passed on to chlorophyll.
   d. The rate of photosynthesis depends on the amount of light received.

6. What happens when an excited electron is passed to an electron acceptor in a photosystem?
   a. It drops back down to its ground state, resulting in the phenomenon known as fluorescence.
   b. The chemical energy in the excited electron is released as heat.
   c. The electron acceptor is oxidized.
   d. Energy in sunlight is transformed to chemical energy.

✔**TEST YOUR UNDERSTANDING**   *Answers are available in Appendix B*

1. Explain how the energy transformation step of photosynthesis occurs. How is light energy converted to chemical energy in the form of ATP and NADPH?

2. Explain how the carbon reduction step of photosynthesis occurs. How is carbon dioxide fixed? Why are both ATP and NADPH required to produce sugar?

3. Compare and contrast photosystem II and photosystem I. What molecule connects the two photosystems?

4. When does photorespiration occur? What are its consequences for the plant?

5. Make a sketch showing how C_4 photosynthesis and CAM separate CO_2 acquisition from the Calvin cycle in space and time, respectively.

6. Why do plants need both chloroplasts and mitochondria?

✔**APPLYING CONCEPTS TO NEW SITUATIONS**   *Answers are available in Appendix B*

1. Compare and contrast mitochondria and chloroplasts. In what ways are their structures similar and different? What molecules or systems function in both types of organelles? Which enzymes or processes are unique to each organelle?

2. Some biologists claim that photorespiration is an evolutionary "holdover," because rubisco evolved over a billion years ago when O_2 levels were extremely low and CO_2 concentrations relatively high. Do you agree with this hypothesis? Why or why not?

3. In addition to their protective function, carotenoids absorb certain wavelengths of light and pass the energy on to the reaction centers of

photosystem I and II. Based on their function, predict exactly where carotenoids are located in the chloroplast. Explain your rationale. How would you test your hypothesis?

4. Consider plants that occupy the top, middle, or ground layer of a forest, and algae that live near the surface of the ocean or in deeper water. Would you expect the same photosynthetic pigments to be found in species that live in these different habitats? Why or why not? How would you test your hypothesis?

This chapter is part of The Big Picture. See how on pages 192–193.   **191**

It takes energy to stay alive. Use this concept map to study how the information on energy and energetics presented in this book fits together.

As you read the map, remember that chemical energy is potential energy. Potential energy is based on position of matter in space, and chemical energy is all about the position of electrons in covalent bonds. When TNT explodes, all that's happening is that electrons are moving from high-energy positions to lower-energy positions.

In essence, organisms transform energy from the Sun into chemical energy in the C–C and C–H bonds of glucose, and then into chemical energy in the P–P bonds of ATP.

The potential energy in ATP allows cells to do work: pump ions, synthesize molecules, move cargo, and send and receive signals.

Note that each box in the concept map indicates the chapter and section where you can go for review. Also, be sure to do the blue exercises in the Check Your Understanding box below.

ENERGY FOR LIFE

begins as

Electromagnetic energy in SUNLIGHT 10.2

Text section where you can find more information

drives

PHOTOSYNTHESIS (in chloroplasts) 10.1

begins with

Antenna complex
- Light excites electrons in pigment molecules 10.2

H_2O enters

donates energy from excited electrons to

donates energy from excited electrons to

Photosystem II
- "Splits" water to yield electrons
- Electron transport chain pumps H^+ 10.3

donates high-energy electrons to

Photosystem I
- Electron transport chain ends with ferrodoxin 10.3

Chemiosmosis
- H^+ gradient drives ATP synthase

releases

yields

O_2

ATP 9.1

NADPH

used in

CO_2 fixed by rubisco to start

Calvin cycle
- Series of enzyme-catalyzed reactions 10.4

yields substrate for synthesis of

stored as

Glycogen, starch 5.2

broken down to yield

GLUCOSE 5.1

CHECK YOUR UNDERSTANDING

○━ **If you understand the big picture . . .**

✓ **You should be able to . . .**

1. Explain how H_2O and O_2 are cycled between photosynthesis and cellular respiration.

2. Explain how CO_2 is cycled between photosynthesis and cellular respiration.

3. Describe what might happen to life on Earth if rubisco were suddenly unable to fix CO_2.

4. Fill in the blue ovals with appropriate linking verbs or phrases.

Answers are available in Appendix B.

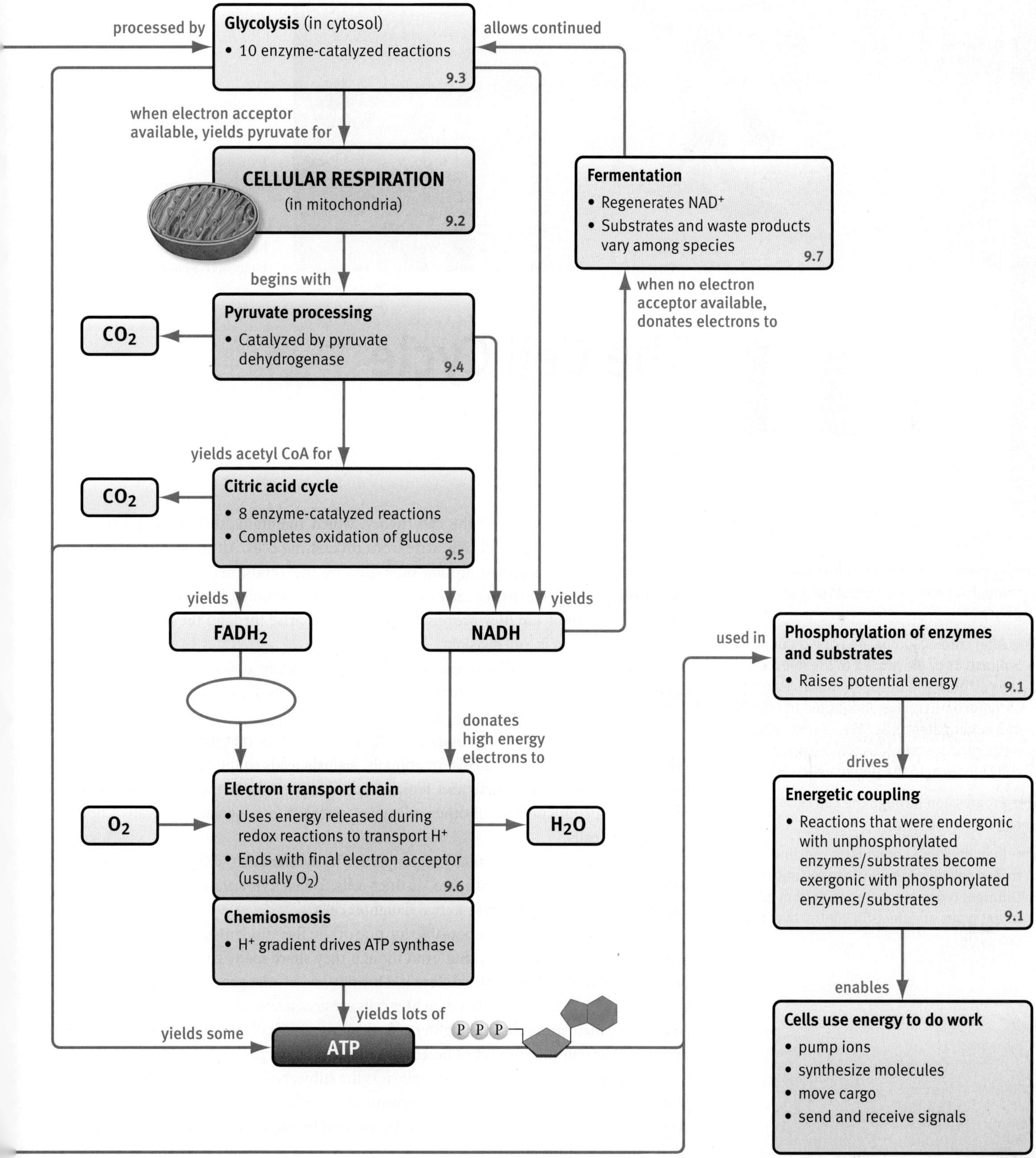

Glycolysis (in cytosol)
- 10 enzyme-catalyzed reactions
  9.3

processed by →

allows continued →

when electron acceptor available, yields pyruvate for

CELLULAR RESPIRATION
(in mitochondria)
  9.2

begins with

Pyruvate processing
- Catalyzed by pyruvate dehydrogenase
  9.4

CO_2 ←

yields acetyl CoA for

Citric acid cycle
- 8 enzyme-catalyzed reactions
- Completes oxidation of glucose
  9.5

CO_2 ←

Fermentation
- Regenerates NAD^+
- Substrates and waste products vary among species
  9.7

when no electron acceptor available, donates electrons to

yields → **FADH₂**

yields → **NADH**

Phosphorylation of enzymes and substrates
- Raises potential energy
  9.1

used in →

donates high energy electrons to

drives

Electron transport chain
- Uses energy released during redox reactions to transport H^+
- Ends with final electron acceptor (usually O_2)
  9.6

O_2 → ... → H_2O

Energetic coupling
- Reactions that were endergonic with unphosphorylated enzymes/substrates become exergonic with phosphorylated enzymes/substrates
  9.1

Chemiosmosis
- H^+ gradient drives ATP synthase

enables

yields lots of → **ATP**

yields some →

Ⓟ Ⓟ Ⓟ

Cells use energy to do work
- pump ions
- synthesize molecules
- move cargo
- send and receive signals

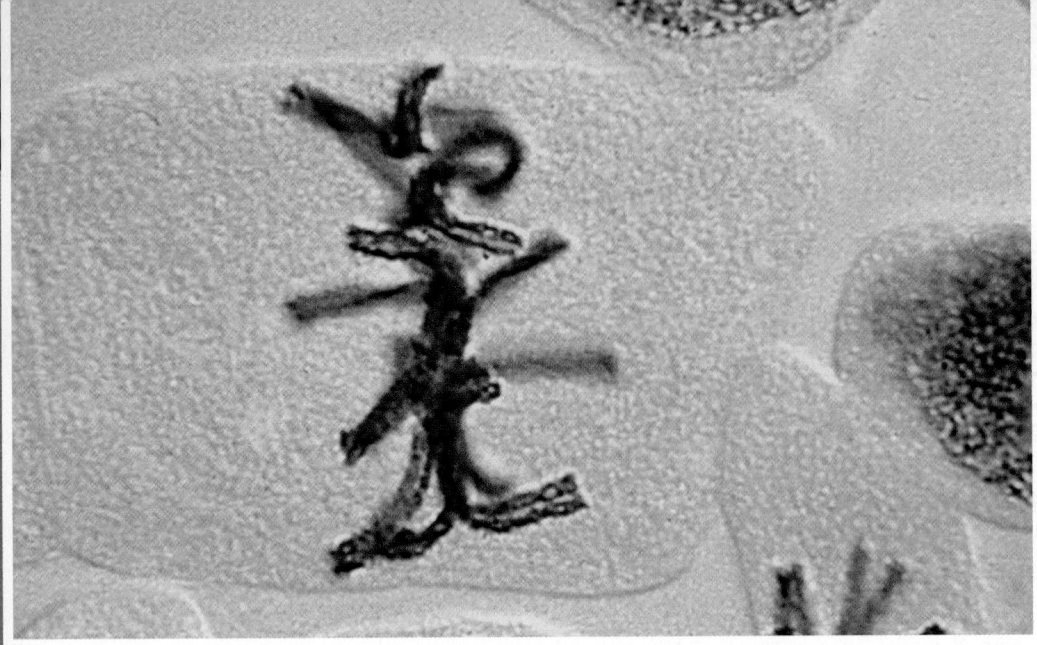

This cell, from a hyacinth, is undergoing a type of cell division called mitosis. Understanding how mitosis occurs is a major focus of this chapter.

11 The Cell Cycle

KEY CONCEPTS

- In eukaryotes, most dividing cells go through a cycle that consists of four phases.

- After chromosomes are copied during S phase, they are moved to the middle of the cell during M phase (mitosis). One chromosome copy is distributed to each of two daughter cells. Mitosis and cytokinesis produce two cells that are genetically identical to the parent cell.

- Progression through the cell cycle is carefully controlled.

- In multicellular organisms, uncontrolled cell division may lead to cancer. Different types of cancer result from different types of defects in control over the cell cycle.

Chapter 1 introduced the cell theory, which maintains that all organisms are made of cells and all cells arise from preexisting cells. Although the cell theory was widely accepted among biologists by the 1860s, most believed that new cells arose within preexisting cells by a process that resembled the growth of mineral crystals. But Rudolf Virchow proposed that new cells arise through the division of preexisting cells—that is, **cell division**.

In the late 1800s, microscopic observations of newly developing organisms, or **embryos**, confirmed Virchow's hypothesis. Multicellular eukaryotes start life as single-celled embryos and grow through a series of cell divisions.

Early studies revealed two fundamentally different ways that nuclei divide prior to cell division: meiosis and mitosis. In animals, **meiosis** leads to the production of sperm and eggs, which are the male and female reproductive cells termed **gametes**. **Mitosis** leads to the production of all other cell types, referred to as **somatic** (literally, "body-belonging") **cells**.

Mitosis and meiosis are usually accompanied by **cytokinesis** ("cell movement")—the division of the cytoplasm into two distinct cells. When cytokinesis is complete, a so-called parent cell has given rise to two daughter cells.

Mitosis and meiosis are responsible for one of the five fundamental attributes of life: reproduction (see Chapter 1). But even though they share many characteristics, mitosis and meiosis are fundamentally different. During mitosis, the genetic material is copied and then divided equally, so that daughter cells are genetically identical to the parent cell. In contrast, meiosis results in daughter cells that are genetically different from each other and that have half the amount of hereditary material as the parent cell.

This chapter focuses on mitosis; meiosis is the subject of Chapter 12. Let's begin with a look at how mitosis relates to other events in a cell's life cycle, continue with an in-depth analysis of mitosis and the cell cycle, and end by examining why uncontrolled cell division can lead to cancer.

✔ When you see this checkmark, stop and test yourself. Answers are available in Appendix B.

11.1 Mitosis and the Cell Cycle

Mitosis and cytokinesis are responsible for three key events in multicellular eukaryotes:

1. *Growth*   The trillions of genetically identical cells that make up your body are the product of mitotic divisions that started in a single fertilized egg.

2. *Wound repair*   When you suffer a scrape, mitosis and cytokinesis generate the cells that repair your skin.

3. *Reproduction*   When yeast cells grow in bread dough or in a vat of beer, they are reproducing by mitosis and cytokinesis. In yeasts and other species, mitosis followed by cytokinesis is the basis of asexual reproduction. **Asexual reproduction** produces offspring that are genetically identical to the parent.

These processes are so basic to life that mitosis has been studied for well over a century. Like much work in biology, the research began with a simple observation.

What Is a Chromosome?

As studies of cell division in eukaryotes began, biologists found that certain chemical dyes made threadlike structures visible within nuclei. In 1879 Walther Flemming documented how the threadlike structures changed as cells in salamander embryos divided. The threads first appeared in pairs just before cell division (**Figure 11.1a**), then split to produce single, unpaired threads in the daughter cells. Flemming introduced the term mitosis, from the Greek *mitos* ("thread"), to describe this division process.

Others studied the roundworm *Ascaris*, and noted that the total number of threads in a cell was the same before and after mitosis. All of the cells in a roundworm had the same number of threads.

In 1888 Wilhelm Waldeyer coined the term **chromosome** ("colored-body") to refer to these threadlike structures (**Figure 11.1b**). A chromosome consists of a single, long DNA (deoxyribonucleic acid) double helix that is wrapped around proteins in a highly organized manner. DNA encodes the cell's hereditary information, or genetic material. A gene is a length of DNA that codes for a particular protein or ribonucleic acid (RNA) found in the cell.

Prior to mitosis, each chromosome is replicated. As mitosis starts, the chromosomes condense into compact structures that can be moved around the cell efficiently. Then one of the chromosome copies is distributed to each of two daughter cells.

Figure 11.2 illustrates unreplicated chromosomes, replicated chromosomes before they have condensed prior to mitosis, and replicated chromosomes that have condensed at the start of mitosis. Each of the DNA copies in a replicated chromosome is called a **chromatid**. The two chromatids are joined together along their entire

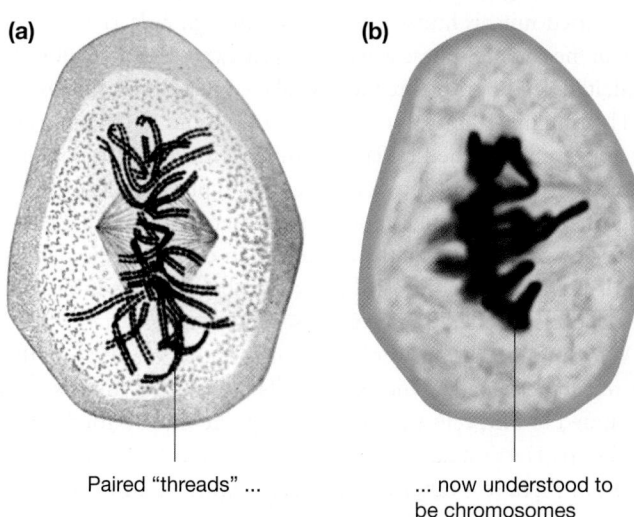

(a) **(b)**

Paired "threads" ...

... now understood to be chromosomes

FIGURE 11.1 Chromosomes Move during Mitosis. (a) Walther Flemming's 1879 drawing of mitosis in a salamander embryo. The black threads are chromosomes. **(b)** Chromosomes can be stained with dyes and observed using the light microscope.

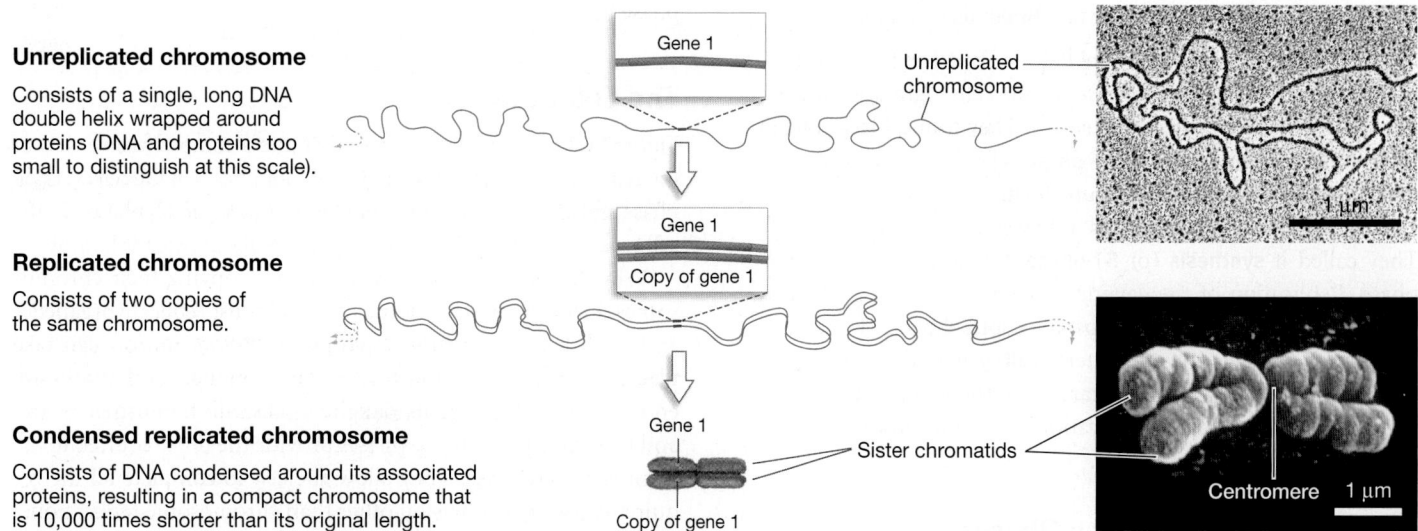

Unreplicated chromosome

Consists of a single, long DNA double helix wrapped around proteins (DNA and proteins too small to distinguish at this scale).

Gene 1

Unreplicated chromosome

Replicated chromosome

Consists of two copies of the same chromosome.

Gene 1

Copy of gene 1

Condensed replicated chromosome

Consists of DNA condensed around its associated proteins, resulting in a compact chromosome that is 10,000 times shorter than its original length.

Gene 1

Copy of gene 1

Sister chromatids

Centromere   1 μm

FIGURE 11.2 Chromosome Morphology Changes before and during Mitosis. Chromosome replication occurs prior to mitosis. The two identical copies stay attached to each other. Early in mitosis, replicated chromosomes condense. The two copies stay attached at a region called the centromere.

length as well as at a specialized region of the chromosome called the **centromere**. Chromatids from the same chromosome are referred to as **sister chromatids**. Even though a replicated chromosome consists of two chromatids, it is still considered a single chromosome.

Cells Alternate between M Phase and Interphase

Plant and animal cells do not undergo mitosis continuously. Instead, growing cells cycle between a dividing phase called the **mitotic** (or **M**) **phase** and a nondividing phase called **interphase** ("between-phase").

Interphase is an active time: the cell is either growing and preparing to divide or fulfilling its specialized function in a multicellular individual.

With a light microscope, chromosomes can be stained and observed as discrete sructures only during M phase, when they condense into compact structures. Cells actually spend most of their time in interphase, however. No dramatic changes are observed in the nucleus during interphase, when chromosomes uncoil into the extremely long, thin structures shown in Figure 11.2a.

The Discovery of S Phase

Are chromosomes copied during M phase or interphase? To answer this question, researchers exposed dividing cells to radioactive phosphorus or radioactive thymidine. Phosphorus is a component of all deoxyribonucleotides, including thymidine, and deoxyribonucleotides are components of DNA (see Chapter 4). Both phosphorus and thymidine are incorporated into DNA as it is being synthesized.

The idea was to:

1. label DNA as chromosomes were being copied,

2. wash away any radioactive isotope that hadn't been incorporated, and

3. visualize the labeled, newly synthesized DNA by exposing the treated cells to X-ray film. Emissions from radioactive phosphorus or thymidine create a black dot in the film. This is the technique called autoradiography (see **BioSkills 9** in Appendix A).

Alma Howard and Stephen Pelc looked for black dots—indicating active DNA synthesis—inside cells right after the exposure to a radioactive isotope ended. They found black dots in interphase cells, but never in M-phase cells. This was strong evidence that DNA replication occurs during interphase.

The biologists had identified a new stage in the life of a cell. They called it **synthesis** (or **S**) **phase**. S phase is part of interphase. Replication of the genetic material is separated, in time, from the partitioning of chromosome copies during M phase.

Howard and Pelc coined the term **cell cycle** to describe the orderly sequence of events that starts with the formation of a eukaryotic cell, through the duplication of its chromosomes, to the time it undergoes division itself.

The Discovery of the Gap Phases

Howard and Pelc and researchers in other labs followed up on these early results by asking how long the S phase lasted. In most cases the experiments were done on cells that were growing in culture. Cultured cells are powerful experimental tools because they can be manipulated much more easily than cells in an intact organism (see **BioSkills 12** in Appendix A).

In one experiment, researchers exposed cultured cells to radioactive thymidine. A short time later, they stopped the labeling by flooding the solution surrounding the cultured cells with nonradioactive thymidine. This pulse-chase approach—introduced in Chapter 7—labeled cells that were in any portion of S phase.

Once the pulse ended, the biologists analyzed the labeled cells 2 hours after the end of labeling, then 4 hours after labeling, 6 hours after labeling, and so on. For each batch of cells, they recorded how many were undergoing mitosis.

One striking result emerged immediately: None of the labeled cells started mitosis immediately—even though at least some had to be at the end of S-phase when they were exposed to the pulse. Instead, it took 4 hours before the first labeled cells began mitosis.

The roughly 4-hour time lag between the end of the pulse and the appearance of the first labeled mitotic nuclei corresponds to a time lag that occurs between the end of S phase and the beginning of M phase. Stated another way, there is a gap in the cell cycle. The gap represents the period when chromosome replication is complete but mitosis has not yet begun. This interval in the cell cycle is called G_2 **phase**, for second gap.

Why second? The answer is based on another observation that the researchers made. Once some cells had started mitosis, they continued to find cells that were undergoing mitosis for another 6–8 hours. Here's a key point: All of these cells had to be somewhere in S phase—ranging from just beginning DNA synthesis to almost ending—when radioactive thymidine was available. Thus, S phase must last 6 to 8 hours.

When the times for the S, G_2, and M phases are totaled and compared with the 24 hours it takes these cells to complete one cell cycle, though, there is a discrepancy of 7 to 9 hours. This discrepancy represents the G_1 **phase**, or first gap. G_1 occurs after M phase but before S phase.

The Cell Cycle

Figure 11.3 pulls these results together into a comprehensive view of the cell cycle. 🔑 There are a total of four phases in the cell cycle: M phase and an interphase consisting of the G_1, S, and G_2 phases. In the cell type diagrammed here, the G_1 phase is about twice as long as G_2.

Why do the gap phases exist? Besides copying their chromosomes during S phase, dividing cells also must replicate organelles and manufacture additional cytoplasm. Before mitosis can take place, the parent cell must grow large enough and synthesize enough organelles that its daughter cells will be normal in size and function. The two gap phases provide the time required to accomplish these tasks. They allow the cell to complete all the requirements for cell division other than chromosome replication.

Now let's turn to M phase and the process of mitosis. Once the genetic material has been copied, how do cells divide it between daughter cells?

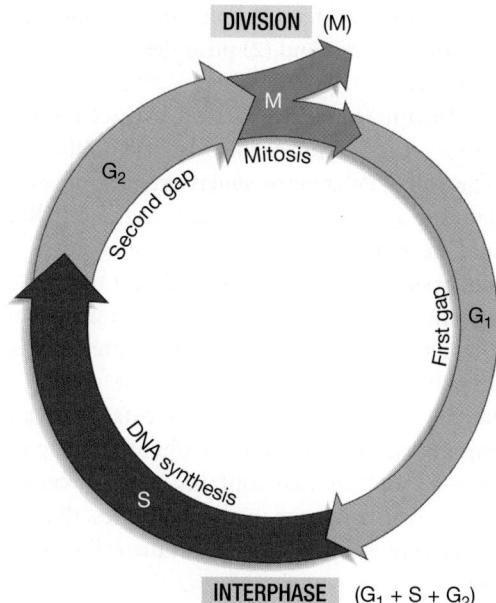

DIVISION (M)

M

Mitosis

G_2

Second gap

First gap

G_1

DNA synthesis

S

INTERPHASE ($G_1 + S + G_2$)

FIGURE 11.3 The Cell Cycle Has Four Phases. A representative cell cycle. The time required for the G_1 and G_2 phases varies dramatically among cells and organisms.

11.2 How Does Mitosis Take Place?

Mitosis results in the division of replicated chromosomes and the formation of two daughter nuclei with identical chromosomes and genes. Mitosis is usually accompanied by cytokinesis—cytoplasmic division and the formation of two daughter cells.

Figure 11.4 provides an overview of how chromosomes change before, during, and after mitosis and cytokinesis, beginning with a hypothetical plant cell or animal cell in G_1 phase. The first drawing

shows a total of four chromosomes in the cell, but the number of chromosomes present varies widely among species—chimpanzees and potato plants have a total of 48 chromosomes in each cell; a maize (corn) plant has 20, dogs have 78, and fruit flies have 8.

Eukaryotic chromosomes consist of DNA associated with globular proteins called **histones**. In eukaryotes this DNA-protein material is called **chromatin**. During interphase, most chromatin is in a "relaxed" or uncondensed state, forming long, threadlike strands (see Figure 11.2, top).

The second drawing in Figure 11.4 shows chromosomes that have been copied prior to mitosis. Each chromosome now consists of two sister chromatids. Each chromatid contains one long DNA double helix, and sister chromatids represent exact copies of the same genetic information.

At the start of M phase, then, each chromosome consists of two sister chromatids that are attached to one another at the centromere. ✔You should be able to explain the relationship between chromosomes and (1) genes, (2) chromatin, and (3) sister chromatids.

Events in Mitosis

As the third drawing in Figure 11.4 indicates, mitosis begins when chromatin condenses to form a much more compact structure. Replicated, condensed chromosomes correspond to the paired threads observed in salamander cells by early biologists (see Figure 11.1a).

🔑 The final drawing in Figure 11.4 shows that during mitosis, the two sister chromatids separate to form independent chromosomes. One copy of each chromosome goes to each of the two daughter cells. As a result, each daughter cell receives a copy of the genetic information that is contained in each chromosome.

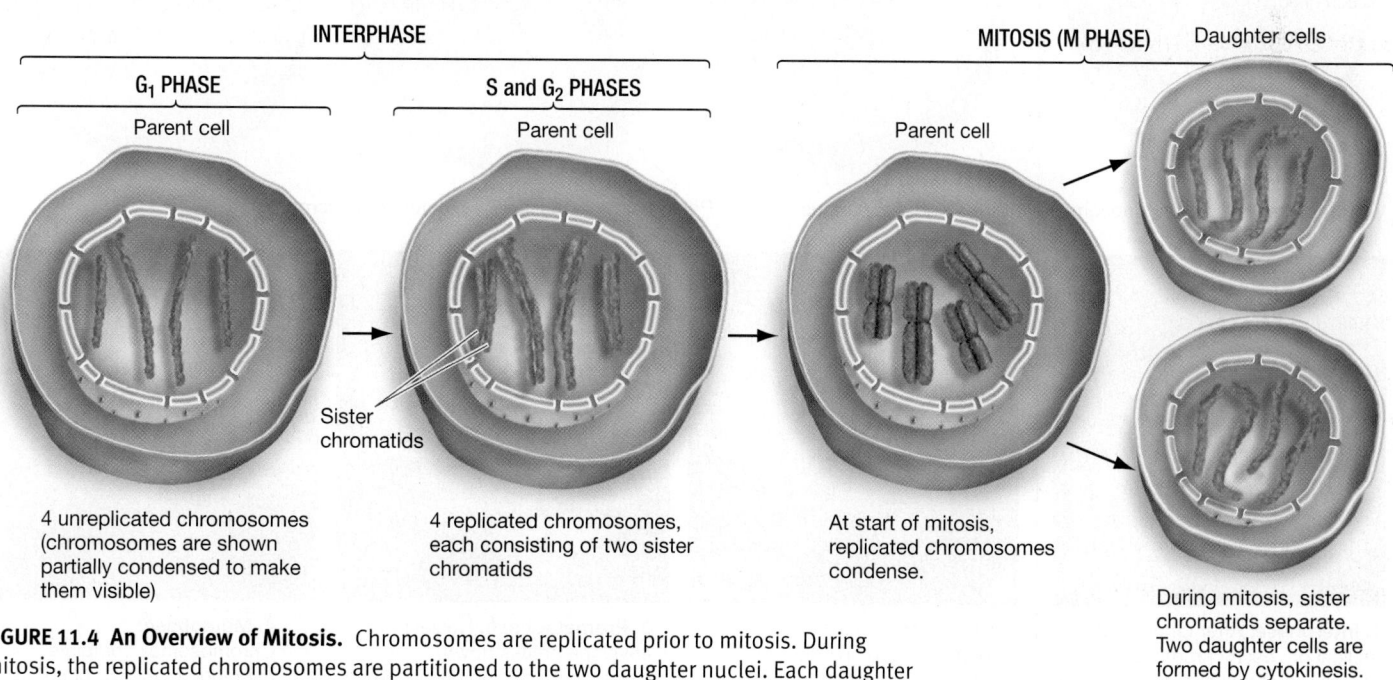

INTERPHASE

G_1 PHASE
Parent cell

S and G_2 PHASES
Parent cell

Sister chromatids

MITOSIS (M PHASE)   Daughter cells

Parent cell

4 unreplicated chromosomes (chromosomes are shown partially condensed to make them visible)

4 replicated chromosomes, each consisting of two sister chromatids

At start of mitosis, replicated chromosomes condense.

During mitosis, sister chromatids separate. Two daughter cells are formed by cytokinesis.

FIGURE 11.4 An Overview of Mitosis. Chromosomes are replicated prior to mitosis. During mitosis, the replicated chromosomes are partitioned to the two daughter nuclei. Each daughter cell contains the same complement of chromosomes as the parent cell.

✔**QUESTION** In the daughter cells of mitosis, are chromosomes replicated or unreplicated?

In mitosis, every daughter cell ends up with exactly the same complement of chromosomes as the parent cell had prior to replication. Thus, every daughter cell receives the same genetic information.

Although mitosis is a continuous process, biologists identify five subphases within M phase on the basis of distinctive events that occur. Some students use the mnemonic device IPPMAT to remember that interphase is followed by the mitotic subphases of prophase, prometaphase, metaphase, anaphase, and telophase.

Recall that chromosomes have already replicated during interphase (**Figure 11.5**, step 1), before mitosis begins. Now let's look at each subphase in turn.

PROPHASE Mitosis begins with the events of **prophase** ("before-phase," Figure 11.5, step 2), when chromosomes condense into compact structures. Chromosomes first become visible in the light microscope during prophase.

In the cytoplasm, prophase is marked by the formation of the spindle apparatus. The **spindle apparatus** is a structure that pro-

duces mechanical forces that (**1**) pull chromosomes to the poles of the cell during mitosis, and (**2**) push the poles of the cell away from each other.

The spindle apparatus consists of distinct populations of microtubules—components of the cytoskeleton that were introduced in Chapter 7. **Polar microtubules** extend from each spindle and overlap one another in the middle of the cell. **Kinetochore microtubules**, in contrast, attach to the chromosomes.

In all eukaryotes, the fibers that make up the spindle apparatus originate from a microtubule organizing center. Although the nature of this organizing center varies among plants, animals, fungi, and other eukaryotic groups, the spindle has the same function. Figure 11.5 illustrates an animal cell undergoing mitosis, so the microtubule organizing center is a **centrosome**—a structure that contains a pair of **centrioles** (see Chapter 7). During prophase in all eukaryotes, the spindles either begin moving to opposite sides of the cell or form on opposite sides.

PROMETAPHASE Once chromosomes have condensed, the nucleolus disappears and the nuclear envelope disintegrates. Kinetochore microtubules from each spindle apparatus attach to one

FIGURE 11.5 Mitosis and Cytokinesis. In the bottom micrographs, chromosomes are stained blue, microtubules are yellow/green, and actin filaments are red.

PROCESS: MITOSIS

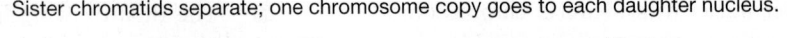
Sister chromatids separate; one chromosome copy goes to each daughter nucleus.

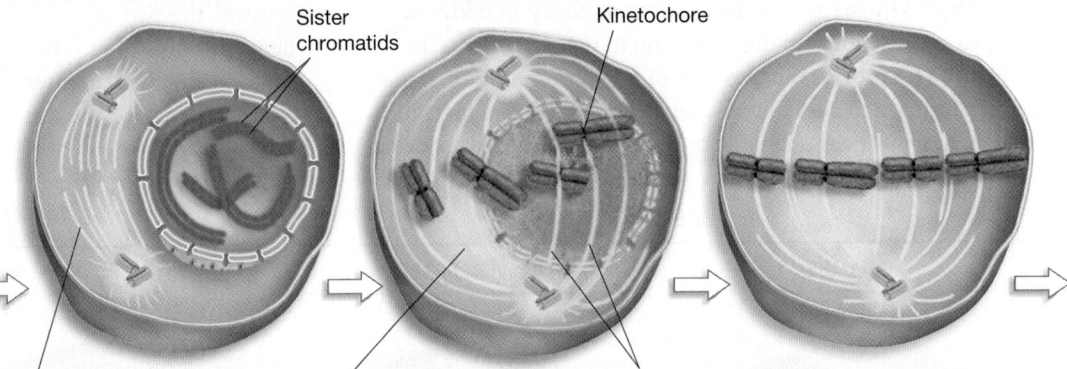

Centrioles
Centrosomes   Chromosomes   Early spindle apparatus   Polar microtubules   Kinetochore microtubules

Sister chromatids   Kinetochore

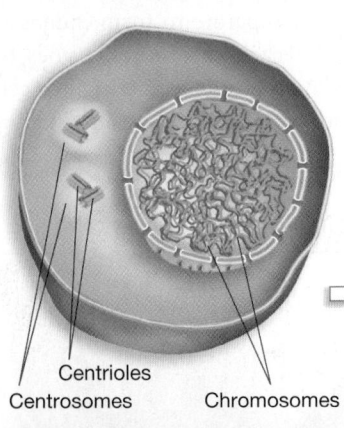
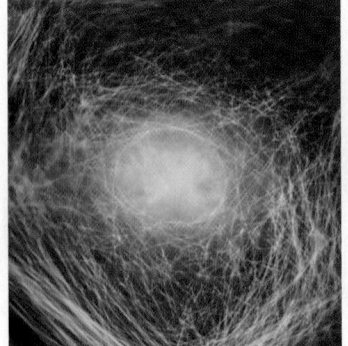
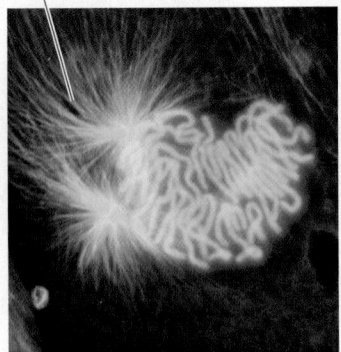
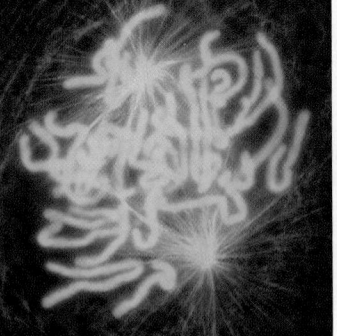

1. Interphase: After chromosome replication, each chromosome is composed of two sister chromatids. Centrosomes have replicated.

2. Prophase: Chromosomes condense, and spindle apparatus begins to form.

3. Prometaphase: Nuclear envelope breaks down. Kinetochore microtubules contact chromosomes at kinetochore.

4. Metaphase: Chromosomes complete migration to middle of cell.

of the two sister chromatids of each chromosome. These events occur during **prometaphase** ("before middle-phase"); see Figure 11.5, step 3.

The attachment between the kinetochore microtubules and each chromatid is made at a structure called the **kinetochore**. Kinetochores are located at the centromere region of the chromosome. The centromere is the region where sister chromatids attach most persistently to each other during mitosis. Each chromosome has two kinetochores where microtubules attach—one on each side.

During prometaphase in animals, the centrosomes continue their movement to opposite poles of the cell. In all groups, the microtubules that are attached to the kinetochores begin moving the chromosomes to the middle of the cell.

METAPHASE During **metaphase** ("middle-phase"), animal centrosomes complete their migration to the opposite poles of the cell (Figure 11.5, step 4). In all eukaryotes, the kinetochore microtubules finish moving the chromosomes to the middle of the cell. The polar fibers that extend from each spindle overlap in the middle of the cell, forming a pole-to-pole connection.

When metaphase ends, the chromosomes are lined up along an imaginary plane called the **metaphase plate**. At this point, the formation of the spindle apparatus is complete. Each chromosome is held by kinetochore microtubules reaching to opposite poles and exerting the same amount of tension, or pull. A tug of war is occurring, with motor proteins on kinetochore microtubules pulling each chromosome in opposite directions.

ANAPHASE At the start of **anaphase** ("against-phase"), the centromeres that are holding sister chromatids together split (Figure 11.5, step 5). Because they are under tension, sister chromatids are pulled apart to create independent chromosomes as soon as they are no longer attached.

As kinetochore microtubules shorten, motor proteins pull the chromosomes to opposite poles of the cell. The two poles of the cell are also pushed away from each other by motor proteins that push the overlapping polar fibers apart.

During anaphase, then, replicated chromosomes split into two identical sets of unreplicated chromosomes. The separation of sister chromatids to opposite poles is a critical step in mitosis, because it ensures that each daughter cell receives the same complement of chromosomes.

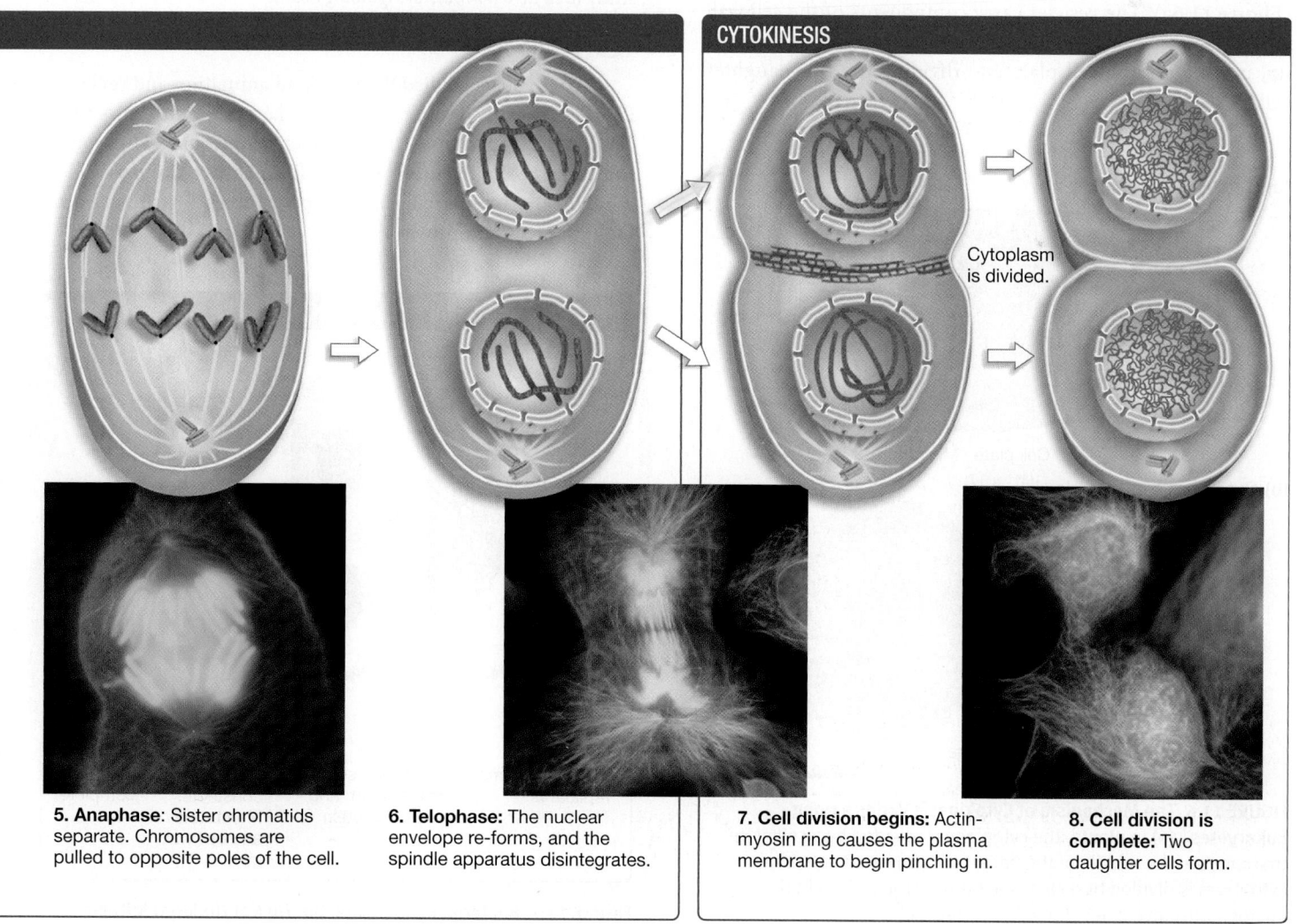

CYTOKINESIS

Cytoplasm is divided.

5. Anaphase: Sister chromatids separate. Chromosomes are pulled to opposite poles of the cell.

6. Telophase: The nuclear envelope re-forms, and the spindle apparatus disintegrates.

7. Cell division begins: Actin-myosin ring causes the plasma membrane to begin pinching in.

8. Cell division is complete: Two daughter cells form.

When anaphase is complete, each pole of the cell has an equivalent and complete collection of chromosomes that are identical to those present in the parent cell prior to chromosome replication.

TELOPHASE During **telophase** ("end-phase"), a nuclear envelope begins to form around each set of chromosomes (Figure 11.5, step 6). The spindle apparatus disintegrates, and the chromosomes begin to de-condense. Once two independent nuclei have formed, mitosis is complete.

Cytokinesis Results in Two Daughter Cells

Prior to the onset of M phase, mitochondria, lysosomes, chloroplasts, and other organelles have replicated, and the rest of the cell contents have grown. During cytokinesis (Figure 11.5, steps 7 and 8), the cytoplasm divides to form two daughter cells, each with its own nucleus and complete set of organelles. In most types of cells, mitosis is followed by cytokinesis.

In plants, cytokinesis begins when a series of microtubules and other proteins define and organize the region where the new plasma membranes and cell walls will form. Vesicles from the Golgi apparatus are transported to the middle of the dividing cell, where they form a structure called the **cell plate** (**Figure 11.6a**). The vesicles carry components of the cell wall and plasma membrane. These components gradually build up, completing the cell plate and dividing the two daughter cells.

In animals, fungi, and slime molds, cytokinesis begins with the formation of a **cleavage furrow** (**Figure 11.6b**). The furrow appears because a ring of actin filaments forms just inside the plasma membrane, in a plane that bisects the cell. A motor protein called myosin binds to these actin filaments. When myosin binds to ATP or ADP, part of the protein moves in a way that causes actin filaments to slide (see Chapter 46).

As myosin moves the ring of actin filaments on the inside of the plasma membrane, the ring shrinks in size and tightens. Because the ring is attached to the plasma membrane, the shrinking ring pulls the membrane with it. As a result, the plasma membrane pinches inward. The actin and myosin filaments continue to slide past each other, tightening the ring further, until the original membrane pinches in two and cell division is complete.

As **Figure 11.7** shows, the mechanism of cell division in bacteria is similar to cytokinesis in animals. After the bacterial chromosome has been copied, the copies move apart and cytoskeletal elements called FtsZ fibers form a ring in the middle of the cell. The ring constricts, dividing the cell in two and producing two identical daughter cells.

Table 11.1 on page 202 summarizes the key structures involved in mitosis, and you can review the process in action in the study area at *www.masteringbiology.com.*

(MB) **BioFlix™** Mitosis, **Web Activity** The Phases of Mitosis

✔ After you've studied the table and animation and reviewed Figure 11.5, you should be able to make a table with rows titled (1) spindle apparatus, (2) nuclear envelope, and (3) chromosomes, and columns titled with the five phases of mitosis. Fill in the table by summarizing what happens to each structure during each phase of mitosis.

(a) Cytokinesis in plants

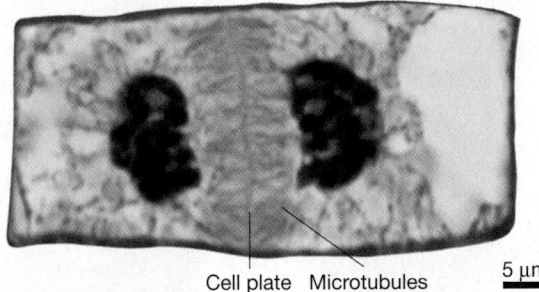

Cell plate   Microtubules   5 μm

(b) Cytokinesis in animals

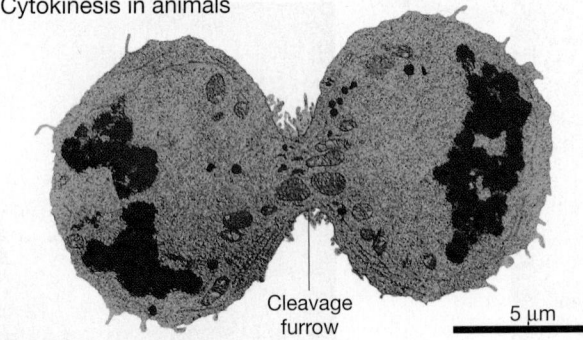

Cleavage furrow   5 μm

FIGURE 11.6 The Mechanism of Cytokinesis Varies among Eukaryotes. (a) In plants, the cytoplasm is divided by a cell plate that forms in the middle of the parent cell. **(b)** In animals, the cytoplasm is divided by a cleavage furrow. (The cells in both micrographs have been stained or colorized.)

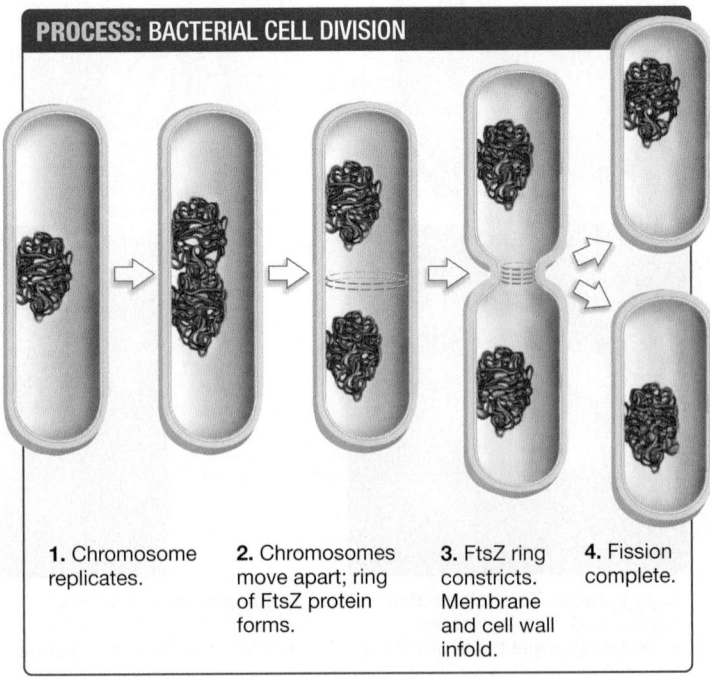

PROCESS: BACTERIAL CELL DIVISION

1. Chromosome replicates.

2. Chromosomes move apart; ring of FtsZ protein forms.

3. FtsZ ring constricts. Membrane and cell wall infold.

4. Fission complete.

FIGURE 11.7 Bacterial Cells Divide, but Do Not Undergo Mitosis.

How Do Chromosomes Move during Mitosis?

The exact and equal partitioning of genetic material to the two daughter cells is the most fundamental aspect of mitosis. How does this process occur?

To understand how sister chromatids separate and move to daughter cells, biologists have focused on understanding how the spindle apparatus functions. Do spindle microtubules act as railroad tracks, the way they do in vesicle transport? Is some sort of motor protein involved? And what is the nature of the kinetochore, where the chromosome and microtubules are joined?

MITOTIC SPINDLE FORCES   The spindle apparatus is composed of microtubules. Recall from Chapter 7 that:

- microtubules are composed of α-tubulin and β-tubulin dimers,
- the number of tubulin dimers a microtubule contains determines its length, and
- microtubules are asymmetric—meaning they have a plus end and a minus end.

During mitosis, kinetochore microtubules grow from the microtubule organizing center until their plus ends attach to a kinetochore.

These observations suggest two possible mechanisms for the movement of chromosomes during anaphase. Does the kinetochore microtubule shorten due to a loss of tubulin dimers from one end? Or do intact microtubules slide past each other, like actin filaments in a contractile ring?

To test these hypotheses, biologists introduced fluorescently labeled tubulin subunits into prophase or metaphase cells. This treatment made the entire spindle apparatus visible (**Figure 11.8**, step 1). Once anaphase began, the researchers marked a region of the spindle with a bar-shaped beam of laser light. The laser bleached the fluorescence in the exposed region, darkening it—although it was still functional (Figure 11.8, step 2).

As anaphase progressed, two things happened: (**1**) The darkened region remained stationary, and (**2**) the kinetochore microtubules got shorter between the darkened region and the kinetochore.

This result suggested that the kinetochore microtubules remain stationary during anaphase, but shorten because tubulin subunits are lost from their plus ends. As they shorten at the kinetochore, the chromosomes are pulled along. How?

A KINETOCHORE MOTOR   **Figure 11.9** on page 202 shows the current model for kinetochore structure and function during chromosome movement. Research is continuing, however, and it is likely that this model will be modified as additional data become available.

The kinetochore is thought to have a base that attaches to the centromere region of the chromosome and a "crown" of fibrous proteins projecting outward. Dyneins and other motor proteins are thought to be attached to the kinetochore's fibrous crown, where they can "walk" down microtubules—from the plus ends near the kinetochore toward the minus ends at the spindle.

EXPERIMENT

QUESTION: How do microtubules shorten to pull sister chromatids apart at anaphase?

HYPOTHESIS: Microtubules shorten at one end.

ALTERNATE HYPOTHESIS: Microtubules slide past each other like actin filaments.

EXPERIMENTAL SETUP:

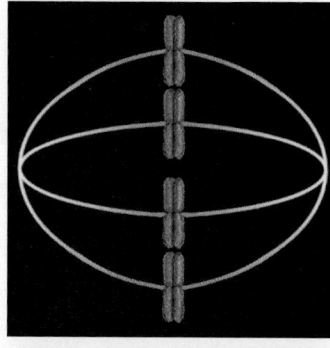

1. Use fluorescent labels to make the metaphase chromosomes fluoresce blue and the microtubules fluoresce yellow.

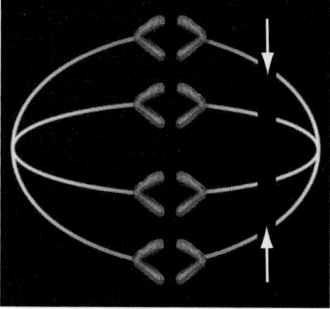

2. At the start of anaphase, darken a section of microtubules to mark them without changing their function.

PREDICTION: The darkened section will not move as chromosomes begin to move.

PREDICTION OF NULL HYPOTHESIS: The darkened section will disappear when chromosomes begin to move.

RESULTS:

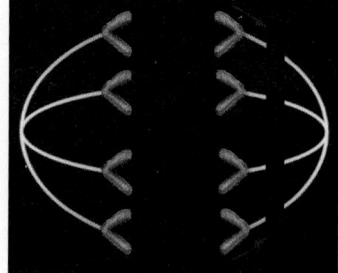

The darkened section remained visible, but the distance between chromosomes and darkened section lessened.

CONCLUSION: Microtubules shorten at one end—at the kinetochore.

FIGURE 11.8 During Anaphase, Microtubules Shorten at the Kinetochore.

SOURCE: Gorbsky, G. J., et al. 1987. Chromosomes move poleward during anaphase along stationary microtubules that coordinately disassemble from their kinetochore ends. *Journal of Cellular Biology* 104: 9–18.

✔**QUESTION** What would the outcome of the experiment be if (1) microtubules shorten at the end opposite the chromosome, or (2) microtubules slide past each other?

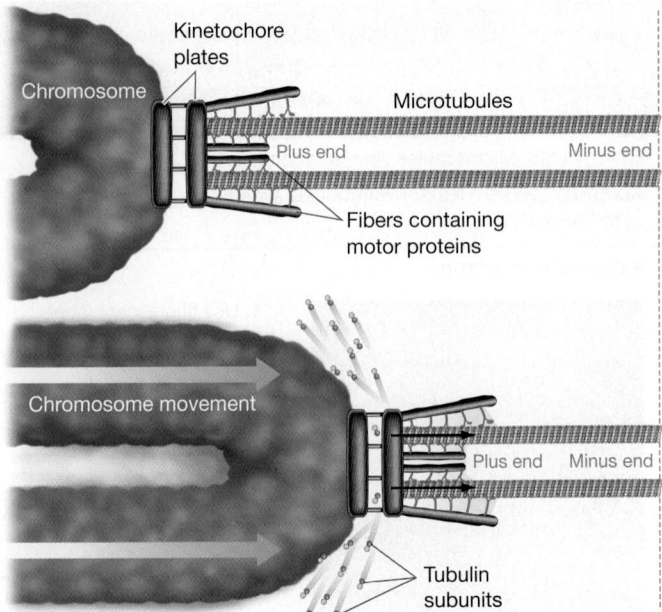

FIGURE 11.9 How Do Microtubules Move Chromosomes during Mitosis? Kinetochore microtubules shorten during anaphase, due to loss of tubulin dimers at the kinetochore. As they shorten, motor proteins walk the chromosomes down the remaining length of the kinetochore microtubules.

If these ideas are correct, the process would be similar to the way that kinesin walks down microtubules during vesicle transport or dynein walks down the microtubule doublets in flagella (see Chapter 7).

SUMMARY TABLE 11.1 **Structures Involved in Mitosis**

Structure	Definition
Chromosome	A structure composed of a DNA molecule and associated proteins
Chromatin	The material that makes up eukaryotic chromosomes; consists of a DNA molecule complexed with histone proteins
Chromatid	One strand of a replicated chromosome, with its associated proteins
Sister chromatids	The two strands of a replicated chromosome. When chromosomes are replicated, they consist of two sister chromatids. The genetic material in sister chromatids is identical. When sister chromatids separate during mitosis, they become independent chromosomes.
Centromere	The structure that joins sister chromatids
Kinetochores	The structures on sister chromatids where kinetochore microtubules attach
Microtubule organizing center	Any structure that organizes microtubules
Centrosome	The microtubule organizing center in animals
Centrioles	Cylindrical structures that comprise microtubules, located inside animal centrosomes

Biologists hypothesize that as anaphase gets under way, proteins in the kinetochore catalyze the loss of tubulin subunits at the plus end of the kinetochore microtubule, while kinetochore motor proteins walk toward the minus end. As the microtubule shortens and the motor proteins continue their walk, the chromosome is pulled to one end of the spindle apparatus.

Having explored how mitosis occurs, let's focus on how it is controlled. When does a cell divide, and when does it stop dividing?

CHECK YOUR UNDERSTANDING

If you understand that . . .

- When chromosomes replicate, each chromosome consists of two identical sister chromatids.
- After chromosomes replicate, mitosis distributes one copy of each chromosome to each daughter cell.
- Mitosis and cytokinesis produce cells with the same genetic material as that of the parent cell.

✓ **You should be able to . . .**

1. Draw an unreplicated chromosome and a replicated chromosome, and label the sister chromatids and the centromere on the replicated chromosome.
2. Explain what IPPMAT stands for, and state when sister chromatids condense, move to the middle of the cell, and break apart to become independent chromosomes.

Answers are available in Appendix B.

11.3 Control of the Cell Cycle

Although the events of mitosis are virtually identical in all eukaryotes, other aspects of the cell cycle vary. In humans, for example, intestinal cells routinely divide more than twice a day to replace tissue that is lost during digestion; mature human nerve and muscle cells do not divide at all.

Most of these differences are due to variation in the length of the G_1 phase. In rapidly dividing cells, G_1 is essentially eliminated. Most nondividing cells, in contrast, are permanently stuck in G_1. Researchers refer to this arrested stage as the G_0 state, or simply "G zero." Cells that are in G_0 have effectively exited the cell cycle and are sometimes referred to as post-mitotic. Nerve cells, muscle cells, and many other cell types enter G_0 once they have matured.

A cell's division rate can also vary in response to changing conditions. For example, human liver cells normally divide about once per year. But if part of the liver is damaged or lost, the remaining cells divide every one or two days until repair is accomplished. Cells of unicellular organisms such as yeasts, bacteria, or archaea divide rapidly only if the environment is rich in nutrients; otherwise, they enter a quiescent (inactive) state.

To explain these differences, biologists hypothesized that the cell cycle must be regulated in some way. Cell cycle control is now the most prominent issue in research on cell division—partly because defects in control can lead to uncontrolled, cancerous growth.

The Discovery of Cell-Cycle Regulatory Molecules

The first solid evidence for cell-cycle control molecules came to light in 1970, when researchers found that certain chemicals, viruses, or an electric shock could fuse the membranes of two mammalian cells that were growing in culture, forming a single cell with two nuclei.

How did cell fusion experiments relate to cell-cycle regulation? When investigators fused cells that were in different stages of the cell cycle, certain nuclei changed phases. For example, when a cell in M phase was fused with one in interphase, the nucleus of the interphase cell initiated M phase (**Figure 11.10a**). The biologists hypothesized that the cytoplasm of M-phase cells contains a regulatory molecule that induces interphase cells to enter M phase.

This hypothesis was supported by experiments on the South African claw-toed frog, *Xenopus laevis*. As an egg of these frogs matures, it changes from a cell called an **oocyte**, which is arrested in a phase similar to G_2, to a mature egg that has entered M phase. The large size of these eggs—more than 1 mm in diameter— makes it relatively easy to purify their cytoplasm and use instruments with extremely fine needles to inject the eggs with cytoplasm from eggs in different stages of development.

When biologists purified cytoplasm from M-phase frog eggs and injected it into the cytoplasm of frog oocytes arrested in the G_2 phase, the immature oocytes entered M phase (**Figure 11.10b**). But when cytoplasm from interphase cells was injected into G_2 oocytes, the cells remained in the G_2 phase. The researchers

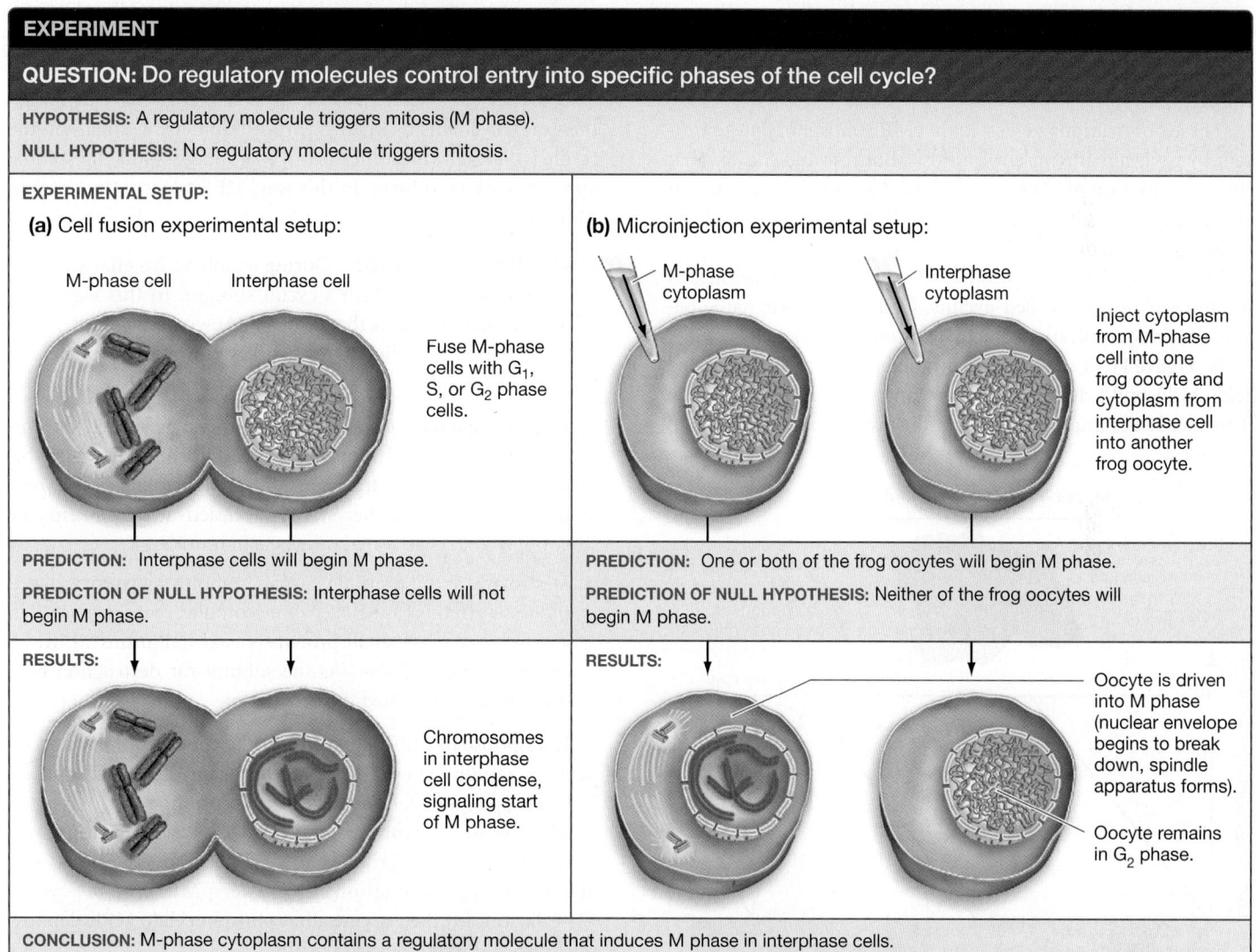

EXPERIMENT

QUESTION: Do regulatory molecules control entry into specific phases of the cell cycle?

HYPOTHESIS: A regulatory molecule triggers mitosis (M phase).

NULL HYPOTHESIS: No regulatory molecule triggers mitosis.

EXPERIMENTAL SETUP:

(a) Cell fusion experimental setup:

M-phase cell   Interphase cell

Fuse M-phase cells with G_1, S, or G_2 phase cells.

(b) Microinjection experimental setup:

M-phase cytoplasm   Interphase cytoplasm

Inject cytoplasm from M-phase cell into one frog oocyte and cytoplasm from interphase cell into another frog oocyte.

PREDICTION: Interphase cells will begin M phase.

PREDICTION OF NULL HYPOTHESIS: Interphase cells will not begin M phase.

PREDICTION: One or both of the frog oocytes will begin M phase.

PREDICTION OF NULL HYPOTHESIS: Neither of the frog oocytes will begin M phase.

RESULTS:

Chromosomes in interphase cell condense, signaling start of M phase.

RESULTS:

Oocyte is driven into M phase (nuclear envelope begins to break down, spindle apparatus forms).

Oocyte remains in G_2 phase.

CONCLUSION: M-phase cytoplasm contains a regulatory molecule that induces M phase in interphase cells.

FIGURE 11.10 Experimental Evidence for Cell-Cycle Control Molecules. (a) When M-phase cells are fused with cells in G_1, S, or G_2 phase, the interphase chromosomes condense and begin M phase. **(b)** Microinjection experiments support the hypothesis that a regulatory molecule induces M phase.

SOURCES: Rao, P. N. and R. T. Johnson. 1970. Mammalian cell fusion studies on the regulation of DNA synthesis and mitosis. *Nature* 225: 159–164. Also Masui, Y. and C. L. Markert. 1971. Cytoplasmic control of nuclear behavior during meiotic maturation of frog oocytes. *Journal of Experimental Zoology* 177: 129–145.

✔**QUESTION** In the cell fusion experiment, how would you test the hypothesis that the fusion event itself—not something in the cytoplasm—triggered the start of M phase?

concluded that the cytoplasm of M-phase cells—but not the cytoplasm of interphase cells—contains a factor that drives immature oocytes into M phase to complete their maturation.

This factor was eventually purified and is now called **mitosis-promoting factor**, or **MPF**. Subsequent experiments showed that MPF induces mitosis in all eukaryotes. For example, injecting M-phase cytoplasm from mammalian cells into immature frog eggs results in egg maturation, and human MPF can trigger mitosis in yeast cells.

MPF appears to be a general signal that says "Start mitosis." How does it work?

MPF CONTAINS A PROTEIN KINASE AND A CYCLIN MPF is made up of two distinct polypeptide subunits. One subunit is a **protein kinase**—an enzyme that catalyzes the transfer of a phosphate group from ATP to a target protein. Recall from Chapter 9 that phosphorylation may activate or inactivate proteins. As a result, protein kinases frequently act as regulatory elements in the cell.

These observations suggested that MPF phosphorylates a protein that triggers the onset of mitosis. But research showed that the concentration of MPF protein kinase is more or less constant throughout the cell cycle. How can MPF trigger mitosis if the protein kinase subunit is always present?

The answer lies in the second MPF subunit, which belongs to a family of proteins called **cyclins**. Cyclins got their name because their concentrations fluctuate throughout the cell cycle.

As **Figure 11.11** shows, concentrations of the cyclin associated with MPF build during interphase and peak during M phase. This increase is important because the protein kinase subunit in

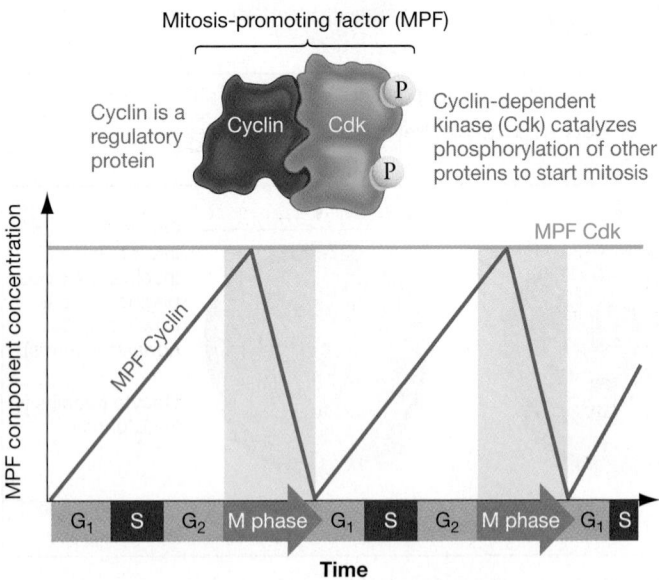

FIGURE 11.11 Cyclin Concentration Regulates MPF Concentration. Cyclin concentrations cycle in dividing cells, reaching a peak in M phase.

✔**QUESTION** Proteins that degrade cyclin are activated by events that MPF initiates. Why is this important?

MPF can be active only when it is bound to the cyclin subunit. As a result, the protein kinase subunit of MPF is called a **cyclin-dependent kinase**, or **Cdk**.

To summarize, MPF is a dimer consisting of a cyclin and a cyclin-dependent kinase. The cyclin subunit functions as a regulatory protein; the kinase subunit catalyzes the phosphorylation of other proteins to start mitosis.

HOW IS MPF ACTIVATED? According to Figure 11.11, the number of complete MPF dimers builds up steadily during interphase. Why doesn't this increasing concentration of MPF trigger the onset of M phase?

The answer is that MPF's Cdk subunit becomes phosphorylated at two sites after it binds to cyclin. When Cdk is phosphorylated at both sites, MPF is inactive. Late in G_2 phase, however, enzymes cause one of the phosphate groups on the Cdk subunit to drop off. This dephosphorylation reaction changes MPF's shape in a way that activates it.

Once MPF is activated, it triggers a chain of events. Although the exact mechanisms involved are still under investigation, the result is that chromosomes begin to condense and the spindle apparatus starts to form. In this way, MPF triggers the onset of M phase.

HOW IS MPF DEACTIVATED? During anaphase, an enzyme complex begins degrading MPF's cyclin subunit. In this way, MPF triggers a chain of events that leads to its own destruction.

MPF deactivation illustrates two key concepts about regulatory systems in cells:

- **Negative feedback** occurs when a process is slowed or shut down by one of its products. Thermostats shut down furnaces when temperatures are high; phosphofructokinase is inhibited by ATP (see Chapter 9); MPF is deactivated by an enzyme complex that is activated by events in mitosis.

- Destroying specific proteins is a common way to control cell processes. In this case, the enzyme complex that is activated in anaphase attaches small proteins called ubiquitins to MPF's cyclin subunit. This marks the subunit for destruction by a protein complex called the proteasome.

In response to MPF activity, then, the concentration of cyclin declines rapidly. Slowly, it builds up again during interphase. This sets up an oscillation in cyclin concentration.

✔If you understand this aspect of cell-cycle regulation, you should be able to describe what MPF does. You should also be able to explain the relationship between MPF and (1) cyclin, (2) Cdk, and (3) the enzymes that phosphorylate MPF, dephosphorylate MPF, and degrade cyclin.

Cell-Cycle Checkpoints Can Arrest the Cell Cycle

The dramatic oscillation in cyclin concentration and activation drives the ordered events of the cell cycle. These events are occurring in your body right now. Over a 24-hour period, you swallow millions of cheek cells and lose millions of cells from your intestinal lining as waste. To replace them, cells in your cheek and intestinal tissue are making and degrading cyclin and pushing themselves through the cell cycle.

MPF is only one of many protein complexes involved in regulating the cell cycle, however. A different cyclin and protein kinase triggers the passage from G_1 phase into S phase, and several regulatory proteins maintain the G_0 state of quiescent cells. An array of regulatory molecules holds cells in particular stages or stimulates passage to the next phase.

To make sense of these observations, Leland Hartwell and Ted Weinert introduced the concept of a **cell-cycle checkpoint**. A cell-cycle checkpoint is a critical point in the cell cycle that is regulated.

Hartwell and Weinert identified checkpoints by analyzing yeast cells with defects in the cell cycle. The defective cells kept dividing under culture conditions when normal cells stopped growing, because they lacked a specific checkpoint. In the body, cells that keep dividing in this way form a mass of cells called a **tumor**.

There are three distinct checkpoints during the four phases of the cell cycle (**Figure 11.12**). In effect, interactions among regulatory molecules at each checkpoint allow a cell to "decide" whether to proceed with division. If these regulatory molecules are defective, the checkpoint may fail and cells may start dividing in an uncontrolled fashion.

G_1 CHECKPOINT The first cell-cycle checkpoint occurs late in G_1. For most cells, this checkpoint is the most important in estab-

lishing whether the cell will continue through the cycle and divide, or exit the cycle and enter G_0. What determines whether a cell passes the G_1 checkpoint?

- *Size* Because a cell must reach a certain size before its daughter cells will be large enough to function normally, biologists hypothesize that some mechanism exists to arrest the cell cycle if the cell is too small.

- *Availability of nutrients* Unicellular organisms arrest at the G_1 checkpoint if nutrient conditions are poor.

- *Social signals* Cells in multicellular organisms pass (or do not pass) through the G_1 checkpoint in response to signaling molecules from other cells, which are termed social signals.

- *Damage to DNA* If DNA is physically damaged, the protein **p53** activates genes that either stop the cell cycle until the damage can be repaired or cause the cell's programmed, controlled destruction—a phenomenon known as **apoptosis**. In this way, p53 acts as a brake on the cell cycle.

If "brake" molecules such as p53 are defective, damaged DNA remains unrepaired. Damage in genes that regulate cell growth can lead to uncontrolled cell division. Consequently, regulatory proteins like p53 are called **tumor suppressors**.

G_2 CHECKPOINT The second checkpoint occurs after S phase, at the boundary between the G_2 and M phases. Because MPF is the key signal triggering the onset of M phase, investigators were not surprised to find that it is involved in the G_2 checkpoint.

Data suggest that if DNA is damaged or if chromosomes are not replicated correctly, the dephosphorylation and activation of MPF are blocked. When MPF is not activated, cells remain in G_2 phase. Cells at this checkpoint may also respond to signals from other cells and to internal signals relating to their size.

METAPHASE CHECKPOINT The final checkpoint occurs during mitosis. If not all chromosomes attach properly to the spindle apparatus, M phase arrests at metaphase. Specifically, anaphase is delayed until all kinetochores attach properly to the spindle apparatus. If the metaphase checkpoint did not exist, some chromosomes might not separate correctly, and daughter cells would receive too many chromosomes or not enough chromosomes.

To summarize, the three cell-cycle checkpoints have the same purpose: They prevent the division of cells that are damaged or that have other problems. The G_1 checkpoint also prevents the growth of mature cells that are in the G_0 state and should not grow any more.

You can review the cell cycle and the checkpoint concept in the study area at *www.masteringbiology.com*.

(MB) **Web Activity** Four Phases of the Cell Cycle

Understanding cell cycle regulation is fundamental. If one of the checkpoints fails, the affected cells may begin dividing in an uncontrolled fashion. For the organism as a whole, the consequences of uncontrolled cell division may be dire: cancer.

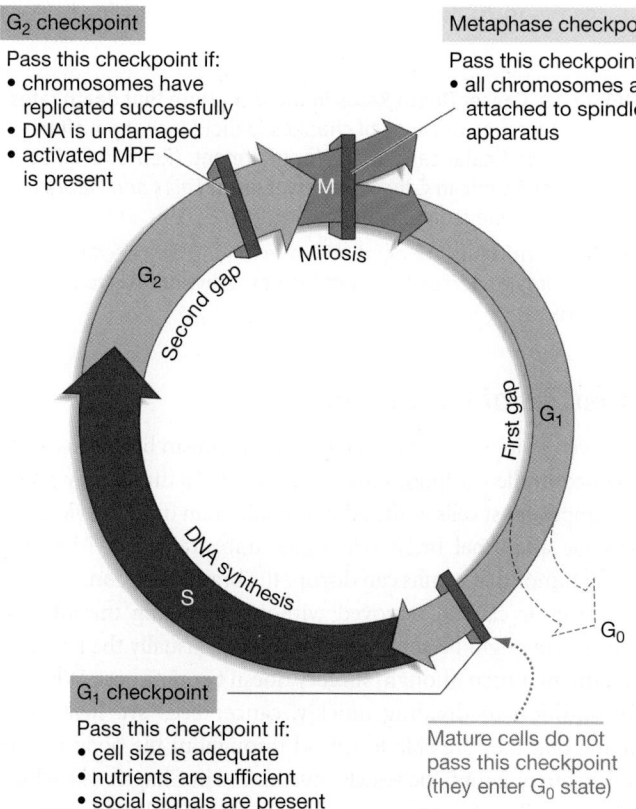

G_2 checkpoint

Pass this checkpoint if:
- chromosomes have replicated successfully
- DNA is undamaged
- activated MPF is present

Metaphase checkpoint

Pass this checkpoint if:
- all chromosomes are attached to spindle apparatus

G_1 checkpoint

Pass this checkpoint if:
- cell size is adequate
- nutrients are sufficient
- social signals are present
- DNA is undamaged

Mature cells do not pass this checkpoint (they enter G_0 state)

FIGURE 11.12 The Three Cell-Cycle Checkpoints.

11.4 Cancer: Out-of-Control Cell Division

Fifty percent of American men and 33 percent of American women will develop cancer during their lifetime. In the United States, one in four of all deaths are from cancer. It is the second leading cause of death, exceeded only by heart disease.

Cancer is a general term for disease caused by cells that divide in an uncontrolled fashion, invade nearby tissues, and spread to other sites in the body. Cancerous cells cause disease because they use nutrients and space needed by normal cells and disrupt the function of normal tissues.

Humans suffer from at least 200 types of cancer. Stated another way, cancer is not a single illness but a complex family of diseases that affect an array of organs, including the breast, colon, brain, lung, and skin. In addition, several types of cancer can affect the same organ. Skin cancers, for example, come in multiple forms.

Some cancers are relatively easy to treat; others are often fatal. **Figure 11.13** illustrates how mortality rates due to different types of cancer have changed through time in the United States.

Although cancers vary in time of onset, growth rate, seriousness, and cause, they have a unifying feature: Cancers arise from cells in which cell-cycle checkpoints have failed. Cancerous cells have two types of defects:

1. defects that make the proteins required for cell growth active when they shouldn't be, and

2. defects that prevent tumor suppressor genes from shutting down the cell cycle.

For example, Chapter 8 introduced the protein Ras as a key component in signal transduction systems—including phosphorylation cascades that trigger cell growth. Many cancerous cells have defective forms of Ras that do not become inactivated. Instead, the defective Ras constantly sends signals that trigger mitosis and cell division.

Likewise, a large percentage of cancerous cells have defective forms of the tumor suppressor p53. Instead of being arrested or destroyed, cells with damaged DNA are allowed to continue growing.

Let's review the general characteristics of cancer and then explore why regulatory mechanisms become defective.

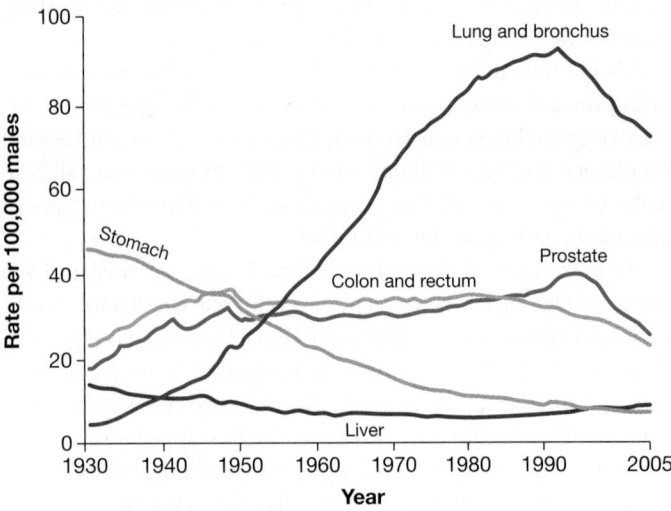

(a) Cancer death rates in **males**

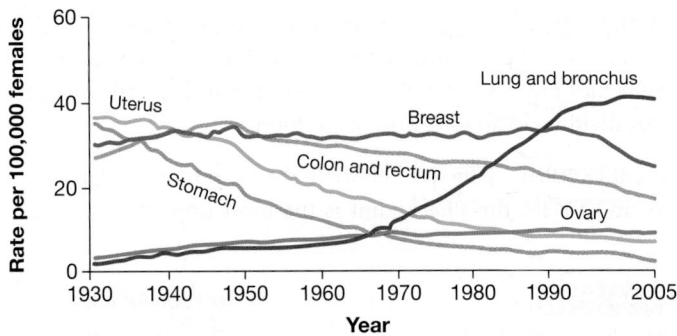

(b) Cancer death rates in **females**

FIGURE 11.13 Cancer Death Rates in the U.S. Note that death rates could vary over time because of changes in incidence (how often people get a particular cancer) and/or treatment. (Reproduced by permission of American Cancer Society. *Cancer Facts and Figures 2009.* Atlanta: American Cancer Society, Inc.)

QUESTION How has the death rate due to lung cancer changed over time in males versus females? Suggest a hypothesis to explain this pattern.

Properties of Cancer Cells

When even a single cell in a multicellular organism begins to divide in an uncontrolled fashion, a mass of cells called a tumor may result. For example, most cells in the adult human brain do not divide. But if a single abnormal brain cell begins unrestrained division, the growing tumor that results can disrupt the brain's function.

If a tumor can be removed without damaging the affected organ, a cure might be achieved, so surgery is usually the first step in treatment. Often, though, surgery doesn't cure cancer. Why?

In addition to dividing quickly, cancer cells are invasive—meaning that they are able to spread throughout the body via the bloodstream or lymphatic vessels (introduced in Chapter 49), which collect excess fluid from tissues and return it to the bloodstream.

Invasiveness is a defining feature of a **malignant tumor**—one that is cancerous. Masses of noninvasive cells are noncancerous and form **benign tumors**. Some benign tumors grow slowly and are largely

(a) Benign tumor

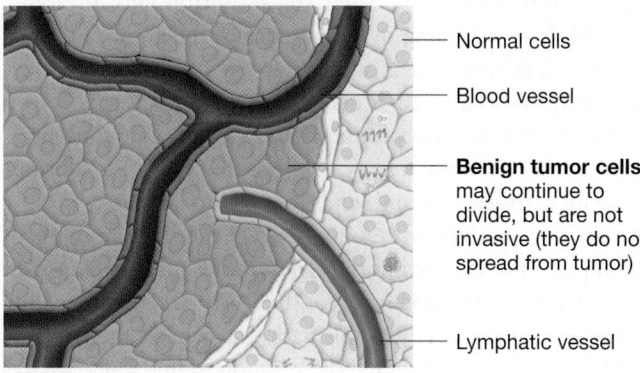

— Normal cells

— Blood vessel

— **Benign tumor cells**
may continue to
divide, but are not
invasive (they do not
spread from tumor)

— Lymphatic vessel

(b) Malignant tumor

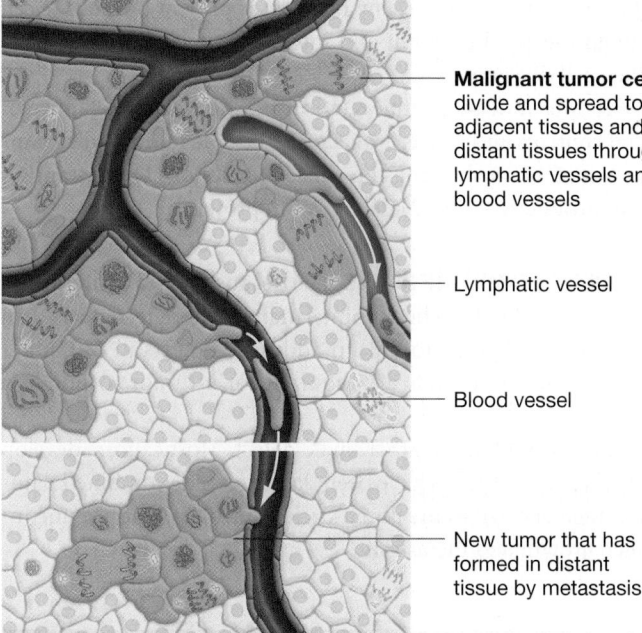

— **Malignant tumor cells**
divide and spread to
adjacent tissues and to
distant tissues through
lymphatic vessels and
blood vessels

— Lymphatic vessel

— Blood vessel

— New tumor that has
formed in distant
tissue by metastasis

FIGURE 11.14 Cancers Spread to New Locations in the Body.
(a) Benign tumors grow in a single location. **(b)** Malignant tumors are metastatic—meaning that their cells can spread to distant parts of the body and initiate new tumors. Malignant tumors cause cancer.

harmless. Others grow quickly and can cause problems if they are located in the brain or other sensitive parts of the body.

Cells become malignant and cancerous if they gain the ability to detach from the original tumor and invade other tissues. By spreading from the primary tumor site, cancer cells can establish secondary tumors elsewhere in the body (**Figure 11.14**). This process is called **metastasis**.

If metastasis has occurred by the time the original tumor is detected, secondary tumors have already formed and surgical removal of the primary tumor will not lead to a cure. This is why early detection is the key to treating cancer most effectively.

Cancer Involves Loss of Cell-Cycle Control

What causes cancer at the molecular level? Recall that when many cells mature, they enter the G_0 phase—meaning their cell cycle is arrested at the G_1 checkpoint. In contrast, cells that do pass through the G_1 checkpoint are irreversibly committed to replicating their DNA and entering G_2.

Based on this observation, biologists hypothesize that many or even most types of cancer involve defects in the G_1 checkpoint. To understand the molecular nature of the disease, then, researchers have focused on understanding the normal mechanisms that operate at that checkpoint. Cancer research and research on the normal cell cycle have become two sides of the same coin.

SOCIAL CONTROL In unicellular organisms, passage through the G_1 checkpoint is thought to depend primarily on cell size and the availability of nutrients. If nutrients are plentiful, cells pass through the checkpoint and divide rapidly.

In multicellular organisms, however, cells divide in response to signals from other cells. Biologists refer to this as *social control* over cell division. The general idea is that individual cells should be allowed to divide only when their growth is in the best interests of the organism as a whole.

Social control of the cell cycle is based on **growth factors**—polypeptides or small proteins that stimulate cell division. Many growth factors were discovered by researchers who were trying to grow cells in culture. When isolated mammalian cells were placed in a culture flask and provided with adequate nutrients, they arrested in G_1 phase. The cells began to grow again only when biologists added **serum**—the liquid portion of blood that remains after blood cells and cell fragments have been removed.

Some component of serum allowed cells to pass through the G_1 checkpoint. What was it?

One of these serum components is a protein called platelet-derived growth factor (PDGF). As its name implies, PDGF is released by blood components called **platelets**, which promote blood clotting at wound sites. PDGF binds to receptor tyrosine kinases on the surface of target cells. When they receive the signal, cells divide. This promotes wound healing.

Researchers subsequently found that PDGF is produced by an array of cell types, in addition to platelets. Investigators have identified a diverse array of other growth factors as well.

For different types of cells to grow in culture, different combinations of growth factors must be supplied. Based on this result, biologists infer that different types of cells in an intact multicellular organism are controlled by different combinations of growth factors.

Cancer cells are an exception. They can often be cultured successfully without externally supplied growth factors. This observation suggests that the normal social controls on the G_1 checkpoint have broken down in cancer cells.

HOW DOES THE G_1 CHECKPOINT WORK? To understand the G_1 checkpoint, think back to how cells progress from G_2 to M phase. The key trigger is activation of MPF. Recall that MPF is a dimer of a cyclin and a Cdk, and that Cdk stands for cyclin-dependent kinase.

In G_0 cells, the arrival of growth factors stimulates the production of a key regulatory protein called E2F. E2F is analogous to MPF. When E2F is activated, it triggers the expression of genes required for S phase.

When E2F is first produced, however, it binds to a tumor suppressor protein called Rb. **Rb protein** is one of the key molecules that enforces the G_1 checkpoint. It is called Rb because it was discovered in children with retinoblastoma, a cancer that produces malignant tumors in the light-sensing tissue, or retina, of the eye.

When E2F is bound to Rb, it is in the "off" position—it can't activate the genes required for S phase. As long as Rb stays bound to E2F, the cell remains in G_0.

But as **Figure 11.15** shows, the situation changes dramatically if growth factors continue to arrive.

Step 1 Enough growth factors arrive to override the inhibitory effects of Rb.

Step 2 In addition to stimulating production of E2F, the growth factors stimulate production of cyclins that are specific to the G_1 checkpoint.

Step 3 E2F continues to bind to Rb and remain inactive. But the cyclins—which are different from those observed in MPF during G_2—begin forming G_1 cyclin-Cdk dimers. Initially, the Cdk component is phosphorylated and inactive.

Step 4 When dephosphorylation activates G_1 cyclin-Cdk complexes, they catalyze the phosphorylation of Rb.

Step 5 The phosphorylated Rb changes shape and no longer binds to E2F.

Step 6 The unbound E2F is free to activate its target genes. Production of S-phase proteins gets S phase under way.

In this way, high concentrations of growth factors function as a social signal that says, "It's OK to override Rb. Go ahead and pass the G_1 checkpoint and divide."

WHY DO SOCIAL CONTROLS AND CELL-CYCLE CHECKPOINTS FAIL? Cells can become cancerous when social controls fail—meaning, when cells begin dividing in the absence of the go-ahead signal from growth factors. One of two things can go wrong: the G_1 cyclin is overproduced, or Rb is defective.

When cyclins are overproduced and stay at high concentrations, the Cdk that binds to cyclin phosphorylates Rb continuously. This activates E2F and sends the cell into S phase.

Cyclin overproduction results from (1) excessive amounts of growth factors or (2) cyclin production in the absence of growth signals. Cyclins are produced continuously when a signaling pathway is defective. Because this pathway includes the Ras protein highlighted in Chapter 8, it is common to find defective Ras proteins in cancerous cells.

What happens if Rb is defective? When Rb is missing or does not bind normally to E2F, any E2F that is present pushes the cell through the G_1 checkpoint and into S phase, leading to uncontrolled cell division.

✔ If you understand the relationship between the failure of social controls and cancer, you should be able to make diagrams like Figure 11.15 showing the consequences of (1) constant G_1 cyclin production and (2) defective Rb proteins.

CANCER IS A FAMILY OF DISEASES Because many proteins are essential to the G_1 checkpoint, many different defects can cause the checkpoint to fail. In addition, defects in cell adhesion or other properties are required for a tumor to undergo metastasis.

Cancer is seldom due to a single defect. Most cancers develop only after several genes have been damaged. The combined damage is then enough to break cell-cycle control and induce uncontrolled growth and metastasis.

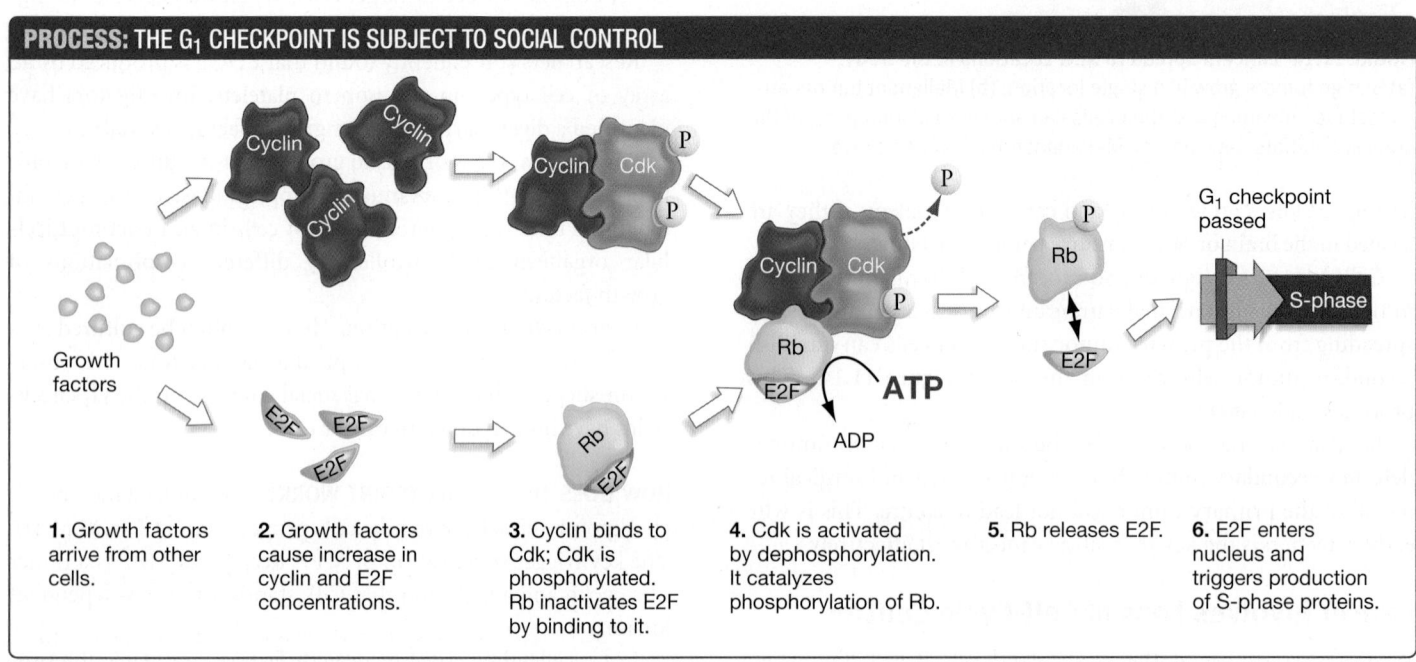

PROCESS: THE G_1 CHECKPOINT IS SUBJECT TO SOCIAL CONTROL

1. Growth factors arrive from other cells.

2. Growth factors cause increase in cyclin and E2F concentrations.

3. Cyclin binds to Cdk; Cdk is phosphorylated. Rb inactivates E2F by binding to it.

4. Cdk is activated by dephosphorylation. It catalyzes phosphorylation of Rb.

5. Rb releases E2F.

6. E2F enters nucleus and triggers production of S-phase proteins.

FIGURE 11.15 Growth Factors Move Cells through the G_1 Checkpoint.

In addition, each type of cancer is due to a unique combination of errors. Hundreds if not thousands of different defects are responsible for the many different types of cancer.

Because cancer is actually a family of diseases with a complex and highly variable molecular basis, there will be no "magic bullet," or single therapy that cures all forms of the illness. Still, recent progress in understanding the cell cycle and the molecular basis of cancer has been dramatic, and cancer prevention and early detection programs are increasingly effective. The prognosis for many cancer patients is remarkably better now than it was even a few years ago. Thanks to research, almost all of us know someone who is a cancer survivor.

CHAPTER 11 REVIEW

For media, go to the study area at www.masteringbiology.com

Summary of Key Concepts

⚷ **In eukaryotes, most dividing cells go through a cycle that consists of four phases.**

- Dividing cells alternate between the dividing phase, called M phase, and a nondividing phase known as interphase.

- Interphase consists of S phase, when chromosomes are replicated, and gap phases called G_1 and G_2, when cell growth and replication of nonnuclear cell components occur.

- The order of cell-cycle phases is $G_1 \rightarrow S \rightarrow G_2 \rightarrow M \rightarrow G_1$ and so forth. Mature cells arrest at G_1 and enter a resting phase called G_0.

   ✔You should be able to explain why the gap phases exist.

⚷ **After chromosomes are copied during S phase, they are moved to the middle of the cell during M phase (mitosis). One chromosome copy is distributed to each of two daughter cells. Mitosis and cytokinesis produce two cells that are genetically identical to the parent cell.**

- Mitosis and cytokinesis are responsible for growth in multicellular eukaryotes and reproduction in most unicellular eukaryotes and some multicellular eukaryotes.

- Mitosis can be described as a sequence of five phases:

   (1) *Prophase* Chromosomes condense. The spindle apparatus begins to form.

   (2) *Prometaphase* The nuclear envelope disintegrates. Kinetochore microtubules make contact with chromosomes.

   (3) *Metaphase* Kinetochore microtubules move chromosomes to the middle of the cell.

   (4) *Anaphase* Kinetochore microtubules pull sister chromatids apart.

   (5) *Telophase* Chromosomes are pulled to opposite poles of the cell. A nuclear envelope forms around each set.

- In most cells, mitosis is followed by cytokinesis—division of all cell contents.

✔You should be able to explain why the nuclear envelope has to disintegrate prior to kinetochore microtubules contacting chromosomes.

(MB) BioFlix™ Mitosis, **Web Activity** The Phases of Mitosis

⚷ **Progression through the cell cycle is carefully controlled.**

- Progression through the cell cycle is regulated at three checkpoints.

   (1) The G_1 checkpoint depends on sufficient nutrients, cell size, lack of DNA damage, and/or social signals.

   (2) The G_2 checkpoint delays progress until chromosome replication is complete and any damaged DNA that is present is repaired.

   (3) The M phase checkpoint delays anaphase until all chromosomes are correctly attached to the spindle apparatus.

- Cyclins and cyclin-dependent kinases (Cdks) help regulate the cell cycle. Cyclin concentrations oscillate during the cell cycle, regulating the activity of Cdks. Active Cdks phosphorylate proteins required for the next cell cycle phase.

   ✔You should be able to predict what happens if an influx of growth factors increases the production of cyclins.

(MB) Web Activity Four Phases of the Cell Cycle

⚷ **In multicellular organisms, uncontrolled cell division may lead to cancer. Different types of cancer result from different types of defects in control over the cell cycle.**

- Cancer is characterized by (1) loss of control at the G_1 checkpoint, resulting in cells that divide in an uncontrolled fashion, and (2) metastasis, or the ability of tumor cells to spread throughout the body.

- The G_1 checkpoint depends in part on Rb, which prevents progression to S phase, and G_1 cyclin-Cdk complexes that trigger progression to S phase. Defects in Rb and G_1 cyclin are common in human cancer cells.

   ✔You should be able to explain why there is unlikely to be a single cure for cancer.

Questions

✔TEST YOUR KNOWLEDGE
Answers are available in Appendix B

1. Which statement about the daughter cells of mitosis is correct?
   a. They differ genetically from one another and from the parent cell.
   b. They are genetically identical to one another and to the parent cell.
   c. They are genetically identical to one another but different from the parent cell.
   d. Only one of the two daughter cells is genetically identical to the parent cell.

2. Progression through the cell cycle is regulated by oscillations in the concentration of which type of molecule?
   a. p53, Rb, and other tumor suppressors
   b. receptor tyrosine kinases
   c. cyclin-dependent kinases
   d. cyclins

3. After replication, what comprises a single chromosome?
   a. the relaxed, uncondensed state
   b. a single strand of chromatin
   c. a single strand of DNA, complexed with proteins
   d. two sister chromatids

4. What major events occur during anaphase of mitosis?
   a. Chromosomes replicate, so each chromosome consists of two identical sister chromatids.
   b. Chromosomes condense and the nuclear envelope disappears.
   c. The chromosomes end up at opposite ends of the cell and two nuclear envelopes form.
   d. Sister chromatids separate, forming independent chromosomes.

5. What evidence suggests that during anaphase, kinetochore microtubules shorten at the kinetochore and not at the base of the spindle apparatus?
   a. Motor proteins are located at the kinetochore.
   b. Motor proteins are located at the kinetochore *and* at the base of the spindle apparatus.
   c. When fluorescing microtubules are darkened in the middle, the darkened segment stays stationary as the fibers shorten near the kinetochore.
   d. When fluorescing microtubules are darkened in the middle, the darkened segment moves toward the base of the spindle apparatus as the fibers shorten near the microtubule organizing center.

6. What happens if the sister chromatids of one chromosome fail to separate at anaphase?
   a. The kinetochore microtubules do not function properly.
   b. The polar microtubules do not function properly.
   c. One daughter cell receives too few chromosomes; the other receives a replicated chromosome.
   d. One daughter cell receives all the chromosomes; the other receives none.

✔ TEST YOUR UNDERSTANDING

Answers are available in Appendix B

1. Sketch the phases of mitosis in a cell with four chromosomes, listing the major events of each phase. Identify at least two events that must be completed successfully for daughter cells to share an identical complement of chromosomes.

2. Make a concept map illustrating normal events at the G_1 checkpoint. Your diagram should include p53, DNA damage, Rb, E2F, social signals, G_1 Cdk, G_1 cyclin, S-phase proteins, phosphorylated (inactivated) cyclin-Cdk, dephosphorylated (activated) cyclin-Cdk, phosphorylated (inactivated) Rb.

3. Explain how cell fusion and microinjection experiments supported the hypothesis that specific molecules are involved in the transition from interphase to M phase.

4. Why are most protein kinases considered regulatory proteins?

5. Why are cyclins called cyclins? Explain their relationship to cyclin-dependent kinases.

6. Early detection is the key to surviving most cancers. Why?

✔ APPLYING CONCEPTS TO NEW SITUATIONS

Answers are available in Appendix B

1. In multicellular organisms, nondividing cells stay in G_1 phase. For the cell, why is it better to be held in G_1 rather than S, G_2, or M phase?

2. When fruit fly embryos first begin to develop, mitosis occurs without cytokinesis. What is the result?

3. According to data from experiments with radioactive thymidine, the first labeled mitotic cells appear about 4 hours after the labeling period ends. From these data, researchers concluded that G_2 lasted about 4 hours. Why?

4. Cancer is primarily a disease of older people. Further, a group of individuals may share a genetic predisposition to developing certain types of cancer, yet vary a great deal in time of onset—or not get the disease at all. Discuss these observations in light of the claim that several defects usually have to occur for cancer to develop.

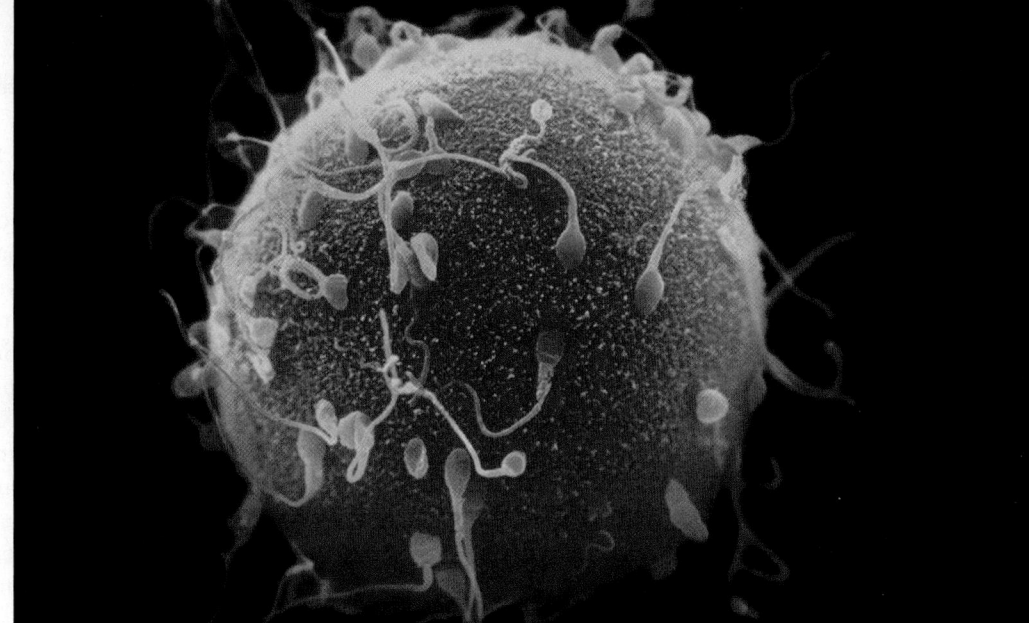

Scanning electron micrograph (with color added) showing human sperm attempting to enter a human egg. This chapter introduces the type of nuclear division called meiosis, which in animals occurs prior to the formation of sperm and eggs.

Meiosis 12

W hy sex?

Simple questions—such as why sexual reproduction exists—are sometimes the best. This chapter asks what sexual reproduction is and why some organisms employ it. The focus here is on how organisms reproduce, or replicate—one of the five fundamental attributes of life introduced in Chapter 1.

For centuries people have known that during sexual reproduction, a male reproductive cell—a **sperm**—and a female reproductive cell—an **egg**—unite to form a new individual. The process of uniting sperm and egg is called **fertilization**.

The first biologists to observe fertilization studied the large, translucent eggs of sea urchins. Due to the semitransparency of the sea urchin egg cell, researchers were able to see the nuclei of a sperm and an egg fuse.

When these observations were published in 1876, they raised an important question, because biologists had already established that the number of chromosomes is constant from cell to cell within a multicellular organism. The question is, How can the chromosomes from a sperm cell and an egg cell combine, but form an offspring that has the same chromosome number as its mother and its father?

A hint at the answer came in 1883, with the observation that cells in the body of roundworms of the genus *Ascaris* have four chromosomes, while their sperm and egg nuclei have only two chromosomes apiece.

Four years later, August Weismann formally proposed a hypothesis to explain the riddle: During the formation of **gametes**—reproductive cells such as sperm and eggs—there must be a distinctive type of cell division that leads to a reduction in chromosome number. Specifically, if the sperm and egg contribute an equal number of chromosomes to the fertilized egg, Weismann reasoned, they must each contain half of the usual number of chromosomes. Then, when sperm and egg combine, the resulting cell has the same chromosome number as its mother's cells and its father's cells have.

KEY CONCEPTS

- Meiosis is a type of nuclear division resulting in cells that have half as many chromosomes as the parent cell. In animals it leads to the formation of eggs and sperm.

- Each cell produced by meiosis receives a different combination of chromosomes. Because genes are located on chromosomes, each cell produced by meiosis receives a different complement of genes. The resulting offspring are genetically distinct from each other and from their parents.

- The leading hypothesis to explain meiosis is that genetically varied offspring are more likely to thrive in environments where parasites and disease are common.

- If mistakes occur during meiosis, the resulting egg and sperm cells may contain the wrong number of chromosomes. It is rare for offspring with an incorrect number of chromosomes to develop normally.

✔ When you see this checkmark, stop and test yourself. Answers are available in Appendix B.

In the decades that followed, biologists confirmed this hypothesis by observing gamete formation in a wide variety of plant and animal species. Eventually this form of cell division came to be called meiosis (literally, "lessening-act").

Meiosis is nuclear division that leads to a halving of chromosome number. It precedes the formation of eggs and sperm in animals. To a biologist, asking "Why sex?" is equivalent to asking "Why meiosis?" Let's delve in by looking at how meiosis happens.

12.1 How Does Meiosis Occur?

To understand meiosis, it is critical to grasp some key ideas about chromosomes. For example, when cell biologists began to study the cell divisions that lead to gamete formation, they made an important observation: Each organism has a characteristic number of chromosomes.

Consider the drawings in **Figure 12.1**, based on a paper published by Walter Sutton in 1902. They show the chromosomes of the lubber grasshopper during the cell divisions leading up to the formation of a sperm. In total, there are 23 chromosomes in the cell. Your cells have 46 chromosomes; a wheat plant has 42.

Chromosomes Come in Distinct Types

Sutton realized, however, that there are just 12 distinct types of chromosomes present in a lubber grasshopper cell. The types are distinguished by size and shape. In the cells that he studied, there were two chromosomes of each type.

Sutton designated 11 of the chromosome types by the letters *a* through *k* and the twelfth by the letter X. Some years later, Nettie

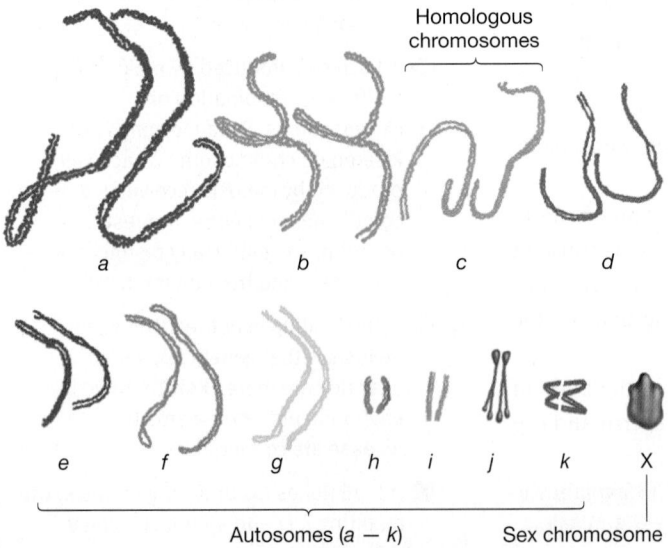

FIGURE 12.1 Cells Contain Different Types of Chromosomes, and Diploid Chromosomes Come in Pairs. Letters designate each of the 12 distinct types of chromosomes found in lubber grasshopper cells. There are two of each type of chromosome (except the X, as this is a male; a female grasshopper would have two Xs).

Stevens established that the X chromosome is associated with the sex of the individual and called it a **sex chromosome**. Non-sex chromosomes, such as *a–k* in Sutton's grasshopper cell, are known as **autosomes**.

It turns out that lubber grasshoppers have just one type of sex chromosome. In this species, females have two sex chromosomes and are designated XX; males have just one sex chromosome and are designated XO, where the O refers to the "missing" chromosome. Two types of sex chromosomes, known as X and Y, exist in humans and other mammals. Human females have two X chromosomes, while males have one X and one Y chromosome.

Sutton also introduced an important term for discussing chromosomes, whether sex chromosomes or autosomes. He referred to the two chromosomes of each type as **homologous** ("same proportion") **chromosomes**, or simply **homologs**. The two chromosomes labeled *c*, for example, have the same size and shape and are homologous.

Later work showed that homologous chromosomes are similar not only in size and shape but also in content. Homologous chromosomes carry the same genes. A **gene** is a section of DNA that influences some hereditary trait in an individual. A trait is a characteristic. For example, each copy of chromosome *c* found in lubber grasshoppers might carry genes that influence eye formation, body size, singing behavior, and jumping ability.

The versions of a gene found on homologous chromosomes may differ, however. Biologists use the term **allele** to denote different versions of the same gene. For example, the allele for eye shape on one homolog of chromosome *c* in a lubber grasshopper might contribute roundness to the eyes, whereas the allele of the eye-shape gene on the other homolog may have the effect of producing narrower eyes (**Figure 12.2**); the particular body-size alleles present will have an influence on whether the body is larger or smaller, and so on.

Homologous chromosomes carry the same genes, but each homolog may contain different alleles.

The Concept of Ploidy

At this point in his study, Sutton had succeeded in determining the lubber grasshopper's **karyotype**—meaning the number and types of chromosomes present. As karyotyping studies became more common, cell biologists realized that, like lubber grasshop-

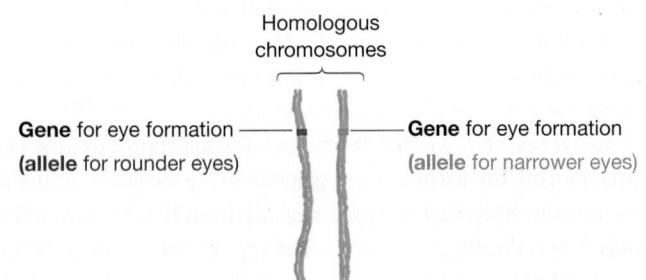

FIGURE 12.2 Homologous Chromosomes May Contain Different Alleles of the Same Gene. Only one of many possible genes is shown.

pers, the vast majority of plants and animals have more than one of each type of chromosome.

Lubber grasshoppers, humans, oak trees, and other organisms that have two versions of each type of chromosome are called **diploid** ("double-form"). Diploid organisms have two alleles of each gene—one on each of the homologous pairs of chromosomes.

Organisms whose cells contain just one of each type of chromosome—for example, bacteria, archaea, and many algae—are called **haploid** ("single-form"). Haploid organisms do not contain homologous chromosomes. They have just one allele of each gene.

Biologists use a compact notation to indicate the number of chromosomes and chromosome sets in a particular organism or type of cell:

- By convention, the letter n stands for the number of distinct types of chromosomes in a given cell and is called the **haploid number**. If sex chromosomes are present, they are counted as a single type in the haploid number. In humans, n is 23.

- To indicate the number of complete chromosome sets observed, a number is placed before the n. Thus, a cell can be n, or $2n$, or $3n$, and so on. Human autosomal cells are $2n$.

The combination of the number of sets and n is termed the cell's **ploidy**. Diploid cells or species are designated $2n$, because two chromosomes of each type are present—one from each parent. A **maternal chromosome** comes from the mother; a **paternal chromosome** comes from the father.

Humans are diploid; $2n$ is 46. Haploid cells or species are labeled simply n, because they have just one set of chromosomes—no homologs are observed. In haploid cells, the number 1 in front of n is implied and is not written out.

To summarize, the haploid number n indicates the number of distinct types of chromosomes present. In contrast, a cell's ploidy (n, $2n$, $3n$, etc.) indicates the number of each type of chromosome present. Stating a cell's ploidy is the same as stating the number of haploid chromosome sets present. ✔You should be able to state the haploid number, ploidy, and total number of chromosomes present in a female lubber grasshopper.

Later work revealed that it is common for species in some lineages—particularly certain land plants, such as ferns—to contain more than two of each type of chromosome. Instead of having two homologous chromosomes per cell, as many organisms do, **polyploid** ("many-form") species may have three or more of each type of chromosome in each cell.

Depending on the number of homologs present, polyploid species are called triploid ($3n$), tetraploid ($4n$), hexaploid ($6n$), octoploid ($8n$), and so on. Why some species are haploid versus diploid or tetraploid is currently the subject of debate and research.

Sutton and the other early cell biologists did more than just describe the karyotypes observed in their study organisms (**Table 12.1**). Through careful examination, they were able to track how chromosome numbers change during meiosis. These studies confirmed Weismann's hypothesis that a special type of cell division occurs during gamete formation.

SUMMARY TABLE 12.1 The Number of Chromosomes Found in Some Familiar Organisms

Organism	Haploid Chromosome Number (n)*	Diploid Chromosome Number ($2n$)
Humans	23	46
Domestic dog	36	72
Fruit fly	4	8
Chimpanzee	24	48
Bulldog ant	1	2
Garden pea	7	14
Corn (maize)	10	20

*Number of different types of chromosomes.

An Overview of Meiosis

Cells replicate each of their chromosomes before undergoing meiosis. At the start of meiosis, chromosomes are in the same state they are in prior to mitosis.

When chromosome replication is complete, each chromosome consists of two identical **sister chromatids**. Sister chromatids contain the same genetic information. They are physically joined at a portion of the chromosome called the **centromere** as well as along their entire length (**Figure 12.3**).

To understand meiosis, it is critical to understand the relationship between chromosomes and sister chromatids. An unreplicated chromosome consists of a single DNA molecule with its associated proteins, while a replicated chromosome consists of two sister chromatids. An unreplicated chromosome is a single thread; a replicated chromosome has paired threads.

The trick is to recognize that unreplicated and replicated chromosomes are both considered *single* chromosomes—even though the replicated chromosome comprises *two* sister chromatids.

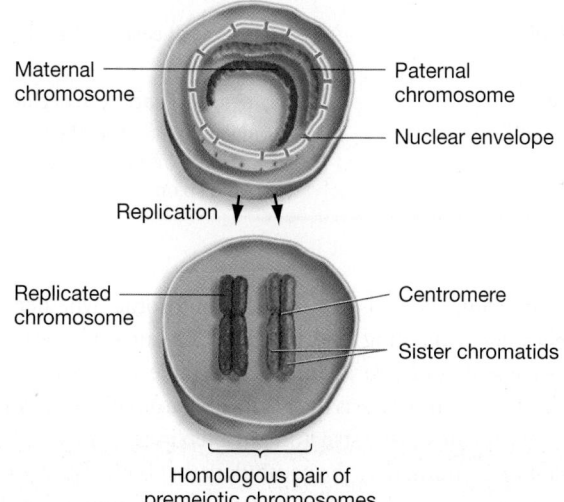

FIGURE 12.3 Each Chromosome Replicates Prior to Undergoing Meiosis.

Vocabulary for Describing the Chromosomal Makeup of a Cell

Term	Definition	Example or Comment
Chromosome	Structure made up of DNA and proteins; carries the cell's hereditary information (genes)	Eukaryotes have linear, threadlike chromosomes; most bacteria and archaea have just one, circular, chromosome
Sex chromosome	Chromosome associated with an individual's sex	X and Y chromosomes of humans (males are XY, females XX); Z and W chromosomes of birds and butterflies (males are ZZ, females ZW)
Autosome	A non-sex chromosome	Chromosomes 1–22 in humans
Unreplicated chromosome	A chromosome that consists of a single copy (in eukaryotes, a single "thread")	
Replicated chromosome	A chromosome that has been copied; consists of two linear structures joined at the centromere	Centromere
Sister chromatids	The chromosome copies in a replicated chromosome	Sister chromatids
Homologous chromosomes (homologs)	In a diploid organism, chromosomes that are similar in size, shape, and gene content	You have a chromosome 22 from your mother (red) and a chromosome 22 from your father (blue) — Homologous chromosomes
Non-sister chromatids	Chromatids belonging to homologous chromosomes	Non-sister chromatids
Tetrad	Homologous replicated chromosomes that are joined together	Tetrad
Haploid number	The number of different types of chromosomes in a cell; symbolized n	Humans have 23 different types of chromosomes ($n = 23$)
Diploid number	The number of chromosomes present in a diploid cell (see below); symbolized $2n$	In humans all cells except gametes are diploid and contain 46 chromosomes ($2n = 46$)
Ploidy	The number of each type of chromosome present	Equivalent to the number of haploid chromosome sets present
Haploid	Having one of each type of chromosome (n)	Bacteria and archaea are haploid, as are many algae; plant and animal gametes are haploid
Diploid	Having two of each type of chromosome ($2n$)	Most familiar plants and animals are diploid
Polyploid	Having more than two of each type of chromosome; cells may be triploid ($3n$), tetraploid ($4n$), hexaploid ($6n$), and so on	Seedless bananas are triploid; many ferns are tetraploid; bread wheat is hexaploid

A chromosome is still just one chromosome whether it is unreplicated—consisting of a single DNA molecule—or replicated—consisting of two identical DNA molecules. Note that an unreplicated chromosome is never called a chromatid; you can only refer to chromatids as the structures in a replicated chromosome.

Table 12.2 summarizes the vocabulary that biologists use to describe the number and types of chromosomes found in a cell and illustrates the relationship between chromosomes and chromatids. ✔If you understand this relationship, you should be able to draw the same chromosome in the replicated and unreplicated state,

explain why both structures represent a single chromosome, and then label the sister chromatids in the replicated chromosome.

MEIOSIS IS TWO CELL DIVISIONS Meiosis consists of two cell divisions, called **meiosis I** and **meiosis II**. As **Figure 12.4** shows, the two divisions occur consecutively but differ sharply.

During meiosis I, the homologs in each chromosome pair separate from each other. One homolog goes to one daughter cell; the other homolog goes to the other daughter cell. The homolog that came originally from the individual's mother is colored red in

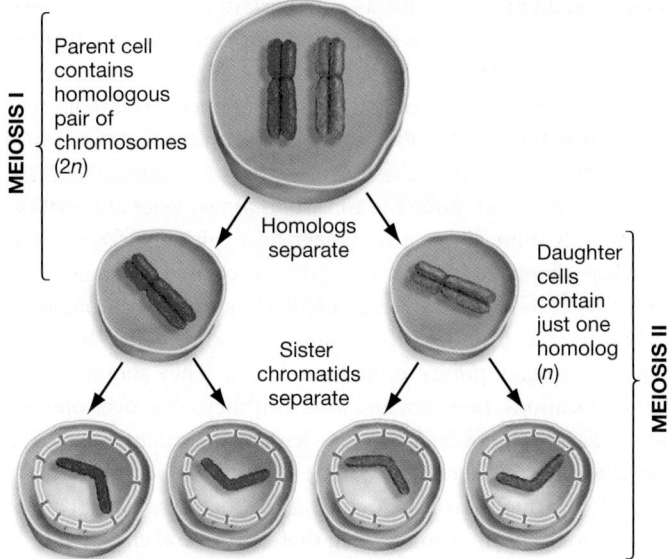

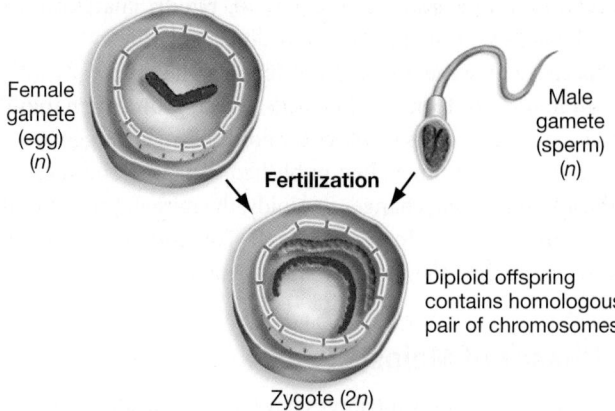

FIGURE 12.5 Fertilization Restores a Full Complement of Chromosomes.

Four daughter cells contain one chromosome each (*n*). In animals, these cells become gametes.

FIGURE 12.4 The Major Events in Meiosis. Meiosis reduces chromosome number by half. In diploid organisms, the products of meiosis are haploid. Maternal chromosomes are red, and paternal chromosomes are blue. Note that in this cell, 2*n* = 2.

Figure 12.4; the homolog that came from the father is colored blue. It is a matter of chance which daughter cell receives which homolog.

The end result of meiosis I is that each of the two daughter cells has one of each type of chromosome instead of two, and thus half as many chromosomes as the parent cell had. During meiosis I, the diploid (2*n*) parent cell produces two haploid (*n*) daughter cells. Each chromosome still consists of two sister chromatids, however—meaning that chromosomes are still replicated.

During meiosis II, sister chromatids from each chromosome separate. One sister chromatid goes to one daughter cell; the other sister chromatid goes to the other daughter cell. The cell that starts meiosis II has one of each type of chromosome, but each chromosome has been replicated (meaning it still consists of two sister chromatids). The cells produced by meiosis II also have one of each type of chromosome, but now the chromosomes are unreplicated.

To reiterate, sister chromatids separate during meiosis II, just as they do during mitosis. Meiosis II is actually equivalent to mitosis occurring in a haploid cell.

As in mitosis, chromosome movements during meiosis I and II are coordinated by kinetochore microtubules that attach at the centromere of each chromosome. Movement is driven by motor proteins located at the kinetochore—the point of attachment. (Recall from Chapter 11 that the centromere is a region on the chromosome; the kinetochore is a structure in that region.)

MEIOSIS IS A REDUCTION DIVISION Sutton and a host of other early cell biologists worked out this sequence of events through careful observation of cells with the light microscope. Based on these studies, they came to a key realization: ⊙━ The outcome of meiosis is a reduction in chromosome number. For this reason, meiosis is known as a reduction division.

In most plants and animals, the original cell is diploid and the four daughter cells are haploid. These four haploid daughter cells, each containing one of each homologous chromosome, eventually go on to form egg cells or sperm cells via a process called **gametogenesis** ("gamete-origin"), which is described in Chapter 48.

When two gametes fuse during fertilization, a full complement of chromosomes is restored (**Figure 12.5**). The cell that results from fertilization is diploid and is called a **zygote**. In this way, each diploid individual receives both a haploid chromosome set from its mother and a haploid set from its father.

Figure 12.6 puts these events into the context of an animal's **life cycle**—the sequence of events that occurs over the life span of an individual, from fertilization to the production of offspring. As you study the figure, note how ploidy changes as the result of meiosis and fertilization. In the case of the dog illustrated here, meiosis in a diploid adult results in the formation of haploid gametes, which combine to form a diploid zygote.

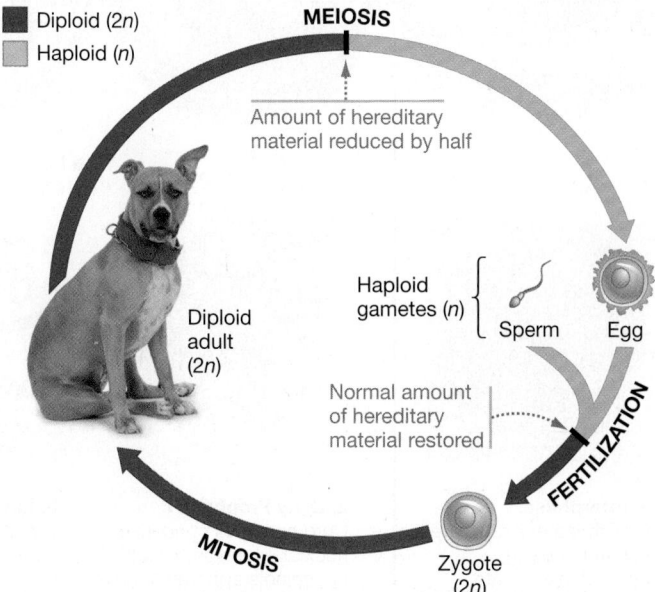

FIGURE 12.6 Ploidy Changes during the Life Cycle of a Dog. The dog is diploid throughout most of its life cycle.

✔If you understand these types of ploidy changes, you should be able to draw the "haploid-dominant" life cycle of an algal species where the only diploid phase is the zygote. Zygotes undergo meiosis to form haploid cells, which grow into haploid adults by mitosis. Adults produce haploid sperm and eggs.

Once Sutton and others had published their work on meiosis and the accompanying changes in ploidy, the mystery of fertilization was finally solved. To appreciate the consequences of meiosis fully, let's analyze the events in more detail.

The Phases of Meiosis I

Meiosis begins after chromosomes have been replicated during S phase (see Chapter 11). Prior to the start of meiosis, chromosomes are extremely long structures, just as they are during interphase of the normal cell cycle. The major steps that occur once meiosis begins are shown in **Figure 12.7**.

During early prophase I, chromosomes condense, the **spindle apparatus** forms, and the nuclear envelope begins to disappear.

Then a crucial event occurs, still during early prophase of meiosis I: Homologous chromosome pairs come together. This process is called **synapsis** and is illustrated in step 2 of Figure 12.7. Synapsis is possible because regions of homologous chromosomes that are similar at the molecular level attract one another, by means of mechanisms that are currently the subject of intense research.

The structure that results from synapsis is called a **tetrad** (*tetra* means four in Greek). A tetrad consists of two homologous chromosomes, with each homolog consisting of two sister chromatids. Chromatids from different homologs in a pair are referred to as **non-sister chromatids**. In the figure, the red-colored chromatids are non-sister chromatids with respect to the blue-colored chromatids.

During late prophase I, the non-sister chromatids begin to separate at many points along their length. They stay joined at certain locations, however, and look as if they cross over one another. Each crossover forms an X-shaped structure called a **chiasma** (plural: **chiasmata**). (In the Greek alphabet, the letter X is "chi.") Normally, at least one chiasma forms in every pair of homologous chromosomes; often there are several chiasmata.

As step 3 of Figure 12.7 shows, the chromatids that meet to form a chiasma are homologous but not sisters. Consistent with this observation, Thomas Hunt Morgan proposed that a physical exchange of paternal and maternal chromosomes occurs at chi-

FIGURE 12.7 The Phases of Meiosis. The micrographs of each phase are from a species of salamander.

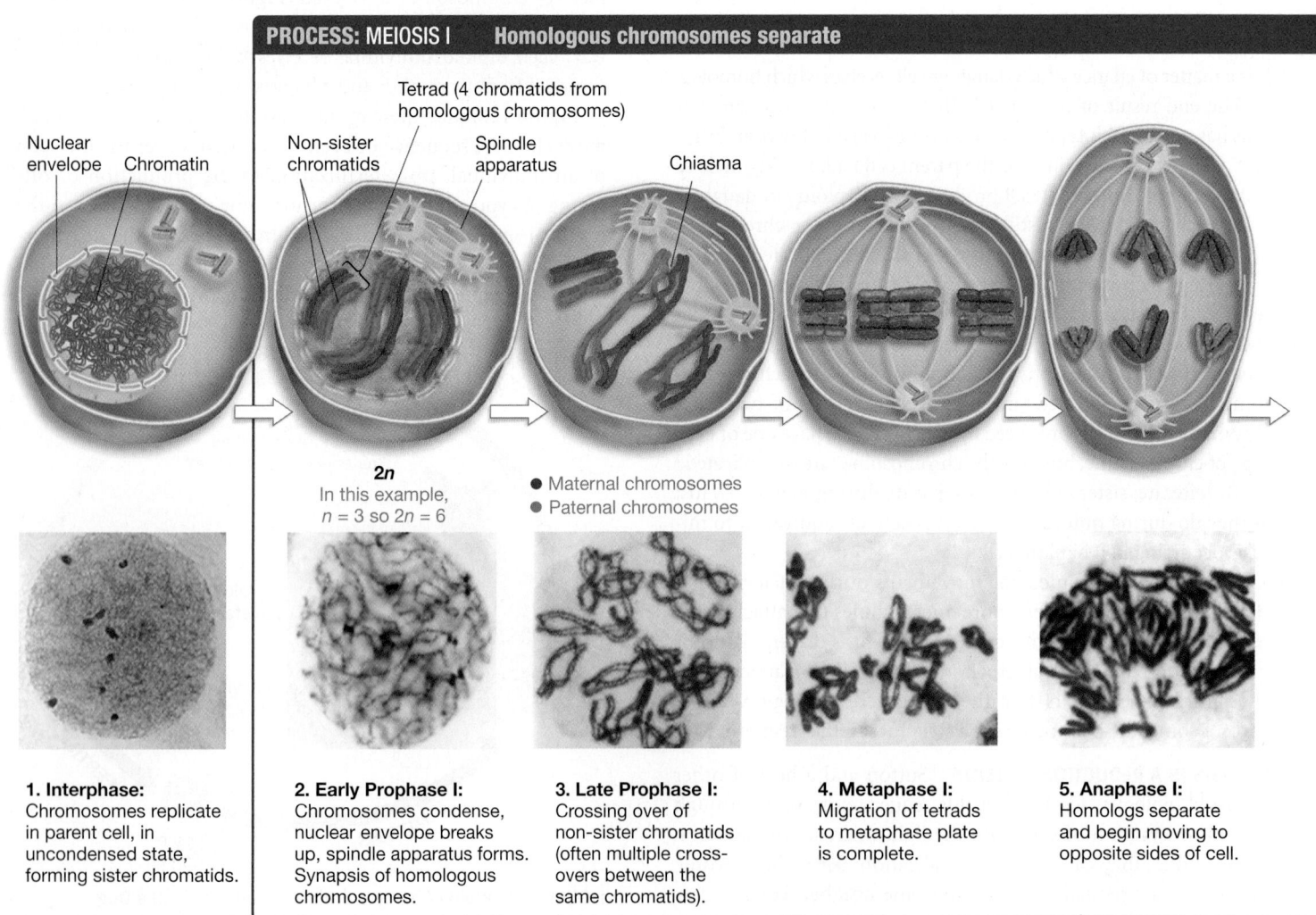

PROCESS: MEIOSIS I   Homologous chromosomes separate

Nuclear envelope   Chromatin

Tetrad (4 chromatids from homologous chromosomes)

Non-sister chromatids   Spindle apparatus

Chiasma

2n
In this example, n = 3 so 2n = 6

● Maternal chromosomes
● Paternal chromosomes

1. Interphase:
Chromosomes replicate in parent cell, in uncondensed state, forming sister chromatids.

2. Early Prophase I:
Chromosomes condense, nuclear envelope breaks up, spindle apparatus forms. Synapsis of homologous chromosomes.

3. Late Prophase I:
Crossing over of non-sister chromatids (often multiple cross-overs between the same chromatids).

4. Metaphase I:
Migration of tetrads to metaphase plate is complete.

5. Anaphase I:
Homologs separate and begin moving to opposite sides of cell.

asmata. According to this hypothesis, paternal and maternal chromatids break and rejoin at each chiasma, producing chromatids that have both paternal and maternal segments. Morgan called this process of chromosome exchange **crossing over**.

In step 4 of Figure 12.7, the result of crossing over is illustrated by chromosomes with a combination of red and blue segments. When crossing over occurs, the chromosomes that result have a mixture of maternal and paternal alleles.

The next major stage in meiosis I is metaphase I. This is when kinetochore microtubules move the pairs of homologous chromosomes (tetrads) to a region called the **metaphase plate**, in the middle of the cell (step 4). Two points are key here: Each tetrad moves to the metaphase plate independently of the other tetrads, and the alignment of maternal and paternal homologs from each chromosome is random.

During anaphase I, the homologous chromosomes in each tetrad separate and begin moving to opposite sides of the cell (step 5). Meiosis I concludes with telophase I, when the homologs finish moving to opposite sides of the cell (step 6). When meiosis I is complete, **cytokinesis** (division of cytoplasm) occurs and two haploid daughter cells form.

The end result of meiosis I is that one chromosome of each homologous pair is distributed to a different daughter cell. A reduction division has occurred: The daughter cells of meiosis I are haploid. The sister chromatids remain attached in each chromosome, however, meaning that the haploid daughter cells produced by meiosis I still contain replicated chromosomes.

The chromosomes in each cell are a random assortment of maternal and paternal chromosomes as a result of crossing over and the random distribution of maternal and paternal homologs during metaphase.

The preceding discussion shows that although meiosis I is a continuous process, biologists summarize the events by identifying distinct phases:

- *Early Prophase I* Replicated chromosomes condense, the spindle apparatus forms, and the nuclear envelope disappears. Synapsis of homologs forms pairs of homologous chromosomes, or tetrads. Kinetochore microtubules attach to the kinetochores at the centromeres of chromosomes.

- *Late Prophase I* Crossing over results in a mixing of chromosome segments from maternal and paternal chromosomes.

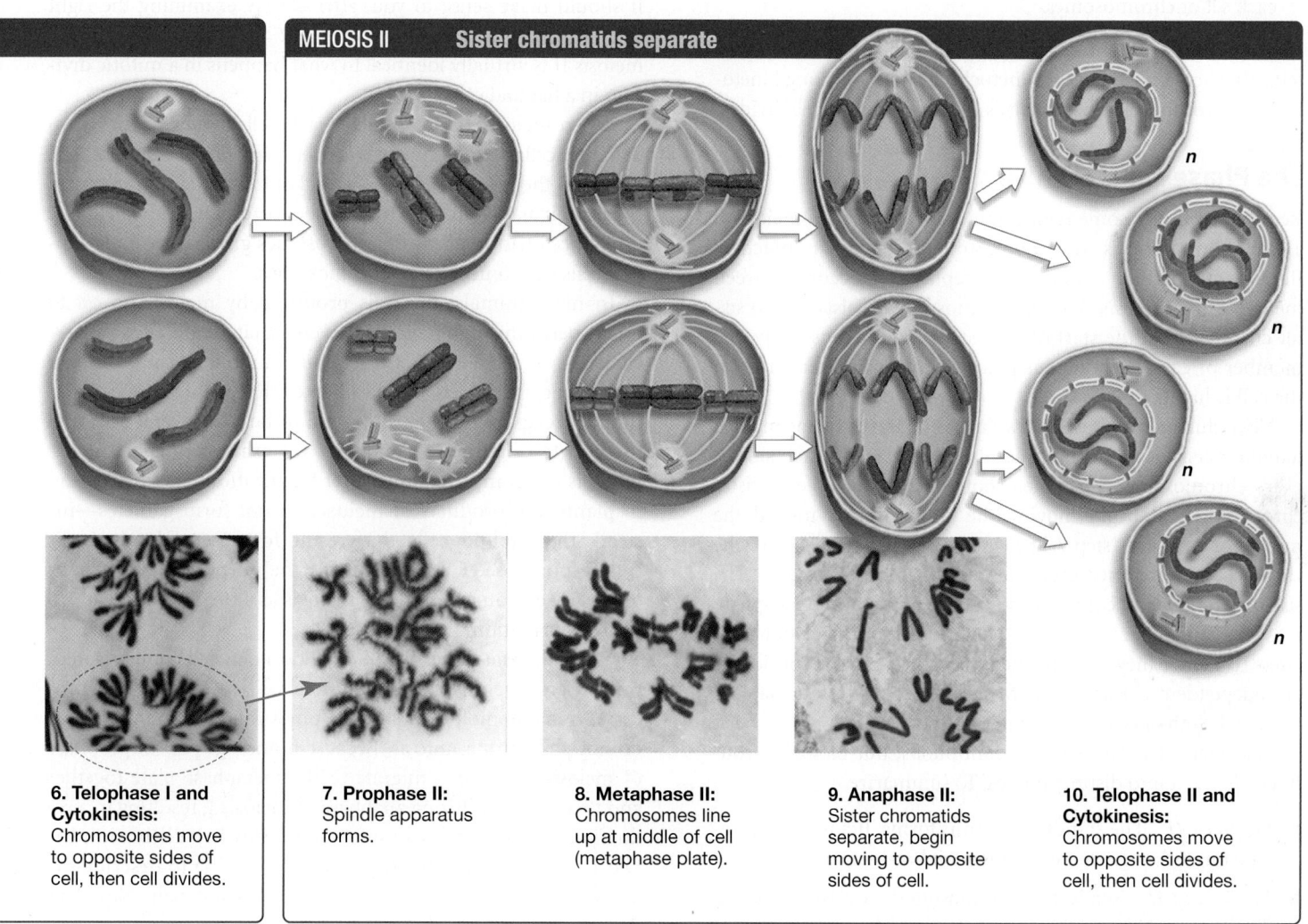

MEIOSIS II   Sister chromatids separate

n

n

n

n

6. Telophase I and Cytokinesis: Chromosomes move to opposite sides of cell, then cell divides.

7. Prophase II: Spindle apparatus forms.

8. Metaphase II: Chromosomes line up at middle of cell (metaphase plate).

9. Anaphase II: Sister chromatids separate, begin moving to opposite sides of cell.

10. Telophase II and Cytokinesis: Chromosomes move to opposite sides of cell, then cell divides.

Feature	Mitosis	Meiosis
Number of cell divisions	One	Two
Number of chromosomes in daughter cells, compared with parent cell	Same	Half
Synapsis of homologs	No	Yes
Number of crossing-over events	None	One or more per pair of homologous chromosomes
Makeup of chromosomes in daughter cells	Identical	Different—only one of each chromosome type present, paternal and maternal segments mixed within chromosomes
Role in organism life cycle	Asexual reproduction in eukaryotes; cell division for growth of multicellular organisms	Precedes production of gametes in sexually reproducing animals

- *Metaphase I* Pairs of homologous chromosomes (tetrads) move to the metaphase plate and line up.

- *Anaphase I* Homologs separate and begin moving to opposite sides of the cell.

- *Telophase I* Homologs finish moving to opposite sides of the cell. In some species, a nuclear envelope re-forms around each set of chromosomes.

Throughout, chromosome movement takes place via motor proteins that are attached to the kinetochore and walk along kinetochore microtubules. When meiosis I is complete, the cell divides.

The Phases of Meiosis II

Recall that chromosome replication occurred prior to meiosis I. Throughout meiosis I, sister chromatids remain attached. Because no further chromosome replication occurs between meiosis I and meiosis II, each chromosome consists of two sister chromatids at the start of meiosis II. And because only one member of each homologous pair of chromosomes is present, the cell is haploid.

Next, during prophase II, a spindle apparatus forms in both daughter cells. Kinetochore microtubules attach to each side of every chromosome—one kinetochore microtubule to each sister chromatid—and begin moving the chromosomes toward the middle of each cell (step 7 of Figure 12.7). In metaphase II, the chromosomes are lined up at the metaphase plate (step 8). The sister chromatids of each chromosome separate during anaphase II (step 9) and move to different daughter cells during telophase II (step 10). Once they are separated, each chromatid is considered an independent chromosome. Meiosis II results in four haploid cells, each with one chromosome of each type.

Like meiosis I, meiosis II is continuous, but biologists routinely divide it into distinct phases. To summarize,

- *Prophase II* The spindle apparatus forms. If a nuclear envelope formed at the end of meiosis I, it breaks apart.

- *Metaphase II* Replicated chromosomes, consisting of two sister chromatids, are lined up at the metaphase plate.

- *Anaphase II* Sister chromatids separate. The unreplicated chromosomes that result begin moving to opposite sides of the cell.

- *Telophase II* Chromosomes finish moving to opposite sides of the cell. A nuclear envelope forms around each haploid set of chromosomes.

It should make sense to you, after closely examining the right side of Figure 12.7, that the movement of chromosomes during meiosis II is virtually identical to what happens in a mitotic division in a haploid cell.

When meiosis II is complete, each cell divides to form two daughter cells. Because meiosis II occurs in both daughter cells of meiosis I, the process results in a total of four daughter cells from each original, parent cell. To describe meiosis in a nutshell, one diploid cell with replicated chromosomes gives rise to four haploid cells with unreplicated chromosomes.

In male animals, the cells produced by meiosis go on to form sperm through a series of events that are detailed in later chapters. In females of at least some animal species, though, meiosis begins in cells called oocytes that will eventually mature into eggs, but it then stops and restarts at several points as the oocytes mature. In mammals, for example, meiosis in females is not completed until after fertilization takes place. And in plants, the products of meiosis do not form gametes—instead, the haploid cells that result form reproductive cells called spores (see Chapter 40). For the purposes of this chapter, however, you can think of meiosis as a process that leads to gamete formation.

Table 12.3 and **Figure 12.8** provide a detailed comparison of mitosis and meiosis. The key difference between the two processes is that homologous chromosomes pair early in meiosis but do not pair during mitosis. Because homologs pair in prophase of meiosis I, they can migrate to the metaphase plate together and then separate during anaphase of meiosis I, resulting in a reduction division. ✓If you understand this key distinction between meiosis and mitosis, you should be able to describe the consequences for meiosis if homologs do not pair. Now let's delve into the details of this critical event.

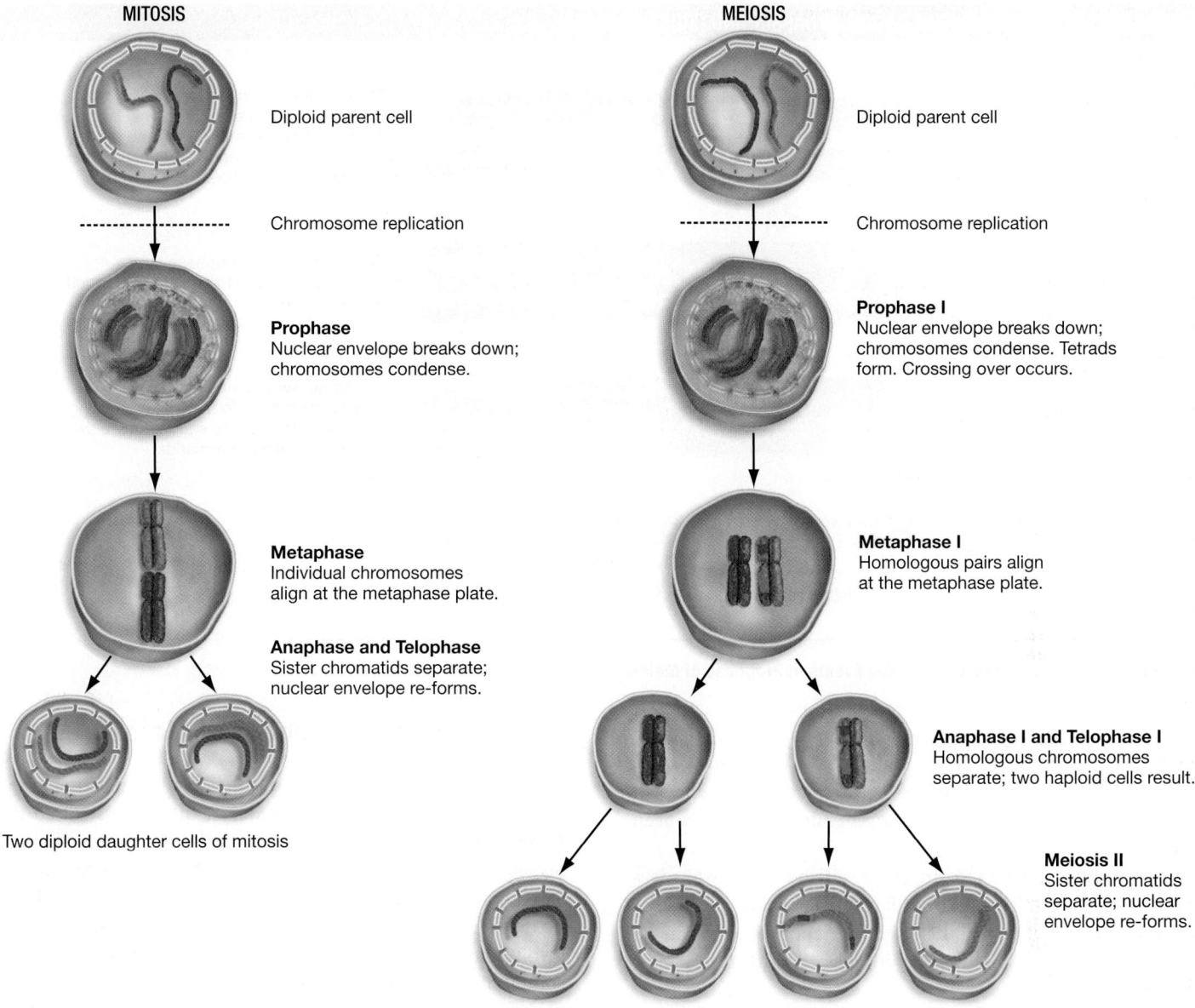

MITOSIS

Diploid parent cell

Chromosome replication

Prophase
Nuclear envelope breaks down; chromosomes condense.

Metaphase
Individual chromosomes align at the metaphase plate.

Anaphase and Telophase
Sister chromatids separate; nuclear envelope re-forms.

Two diploid daughter cells of mitosis

MEIOSIS

Diploid parent cell

Chromosome replication

Prophase I
Nuclear envelope breaks down; chromosomes condense. Tetrads form. Crossing over occurs.

Metaphase I
Homologous pairs align at the metaphase plate.

Anaphase I and Telophase I
Homologous chromosomes separate; two haploid cells result.

Meiosis II
Sister chromatids separate; nuclear envelope re-forms.

Four haploid daughter cells of meiosis

FIGURE 12.8 A Comparison of Mitosis and Meiosis. Mitosis produces two daughter cells with chromosomal complements identical to the parent cell. Meiosis produces four haploid cells with chromosomal complements unlike each other and unlike the diploid parent cell.

A Closer Look at Prophase I

Figure 12.9 on page 220 provides more detail on how several important events in prophase of meiosis I occur.

Step 1 After chromosome replication is complete, sister chromatids stay tightly joined along their entire length.

Step 2 When homologs synapse, two pairs of non-sister chromatids are brought close together and are held there by a network of proteins called the **synaptonemal complex.**

Step 3 Crossing over occurs when a complex of proteins cuts the chromosomes and then reattaches the pieces so that segments are swapped between adjacent homologs.

The key to understanding crossing over is to recognize that at each point where it occurs—as in the circles labeled "protein complex" in step 3 of Figure 12.9—the non-sister chromatids from each homolog get physically broken at the same point and *attached to each other.* As a result, segments of maternal and paternal chromosomes are swapped, as shown at the bottom of Figure 12.9.

Crossing over can occur at many locations along the length of the paired homologs, and it routinely occurs at least once between each pair of non-sister chromatids. In humans, for example, each chromosome in an egg underwent an average of 1.7 crossovers during meiosis; each chromosome in a sperm had an average of 1.1 crossovers. No one knows why crossing

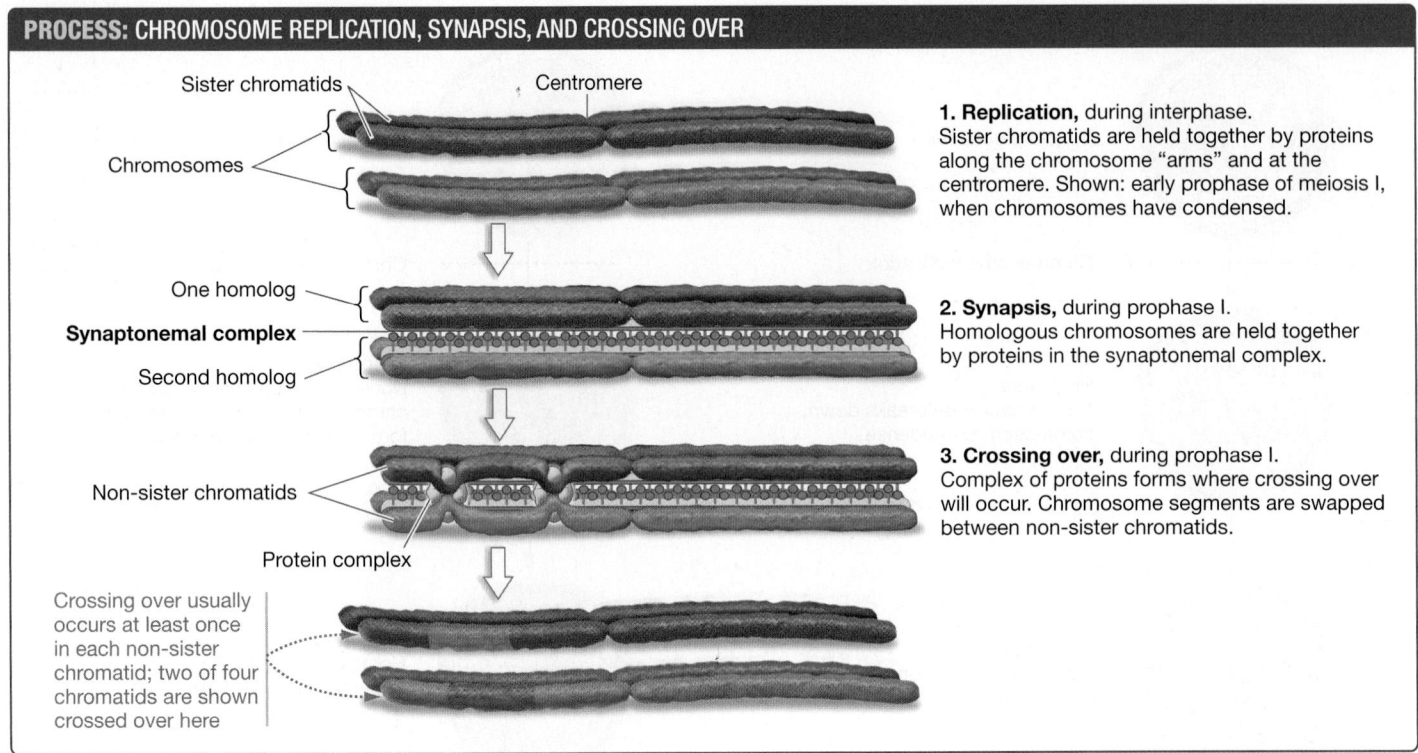

1. Replication, during interphase.
Sister chromatids are held together by proteins along the chromosome "arms" and at the centromere. Shown: early prophase of meiosis I, when chromosomes have condensed.

2. Synapsis, during prophase I.
Homologous chromosomes are held together by proteins in the synaptonemal complex.

3. Crossing over, during prophase I.
Complex of proteins forms where crossing over will occur. Chromosome segments are swapped between non-sister chromatids.

Crossing over usually occurs at least once in each non-sister chromatid; two of four chromatids are shown crossed over here

FIGURE 12.9 A Closer Look at Three Key Events in Prophase of Meiosis I.

over is so much more common in human females than in males. Recent work has shown that the rate of crossing over is also variable among individuals of the same sex. A gene that may be responsible for this variation in humans has even been identified.

Why is the rate of crossing over variable? And why does meiosis exist at all? These are fundamental questions in biological science. But before reading on to learn more about them, go to the study area at *www.masteringbiology.com* to review how meiosis occurs.

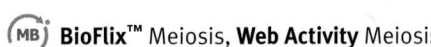 **BioFlix™** Meiosis, **Web Activity** Meiosis

CHECK YOUR UNDERSTANDING

⌾—▸ If you understand that . . .

- Meiosis is called a reduction division because the total number of chromosomes present is cut in half.
- During meiosis, a single diploid parent cell with replicated chromosomes gives rise to four haploid daughter cells.

✔ You should be able to . . .

1. Demonstrate the phases of meiosis I illustrated in Figure 12.7 by using pipe cleaners or pieces of cooked spaghetti.

2. Identify the event that makes meiosis a reduction division, unlike mitosis, and explain why it is responsible for reduction division.

Answers are available in Appendix B.

12.2 The Consequences of Meiosis

The cell biologists who worked out the details of meiosis in the late 1800s and early 1900s realized that the process solved the riddle of fertilization. Weismann's hypothesis—that a reduction division precedes gamete formation in animals—was confirmed.

By now, having come to appreciate that meiosis is an intricate, tightly regulated process, you shouldn't be surprised to learn that it involves dozens, if not hundreds, of different proteins. Given this complexity, it is logical to hypothesize that meiosis accomplishes something extremely important: Thanks to the independent shuffling of maternal and paternal chromosomes and crossing over during meiosis I, the chromosomes in gametes are different from the chromosomes in parental cells. Subsequently, fertilization brings haploid sets of chromosomes from a mother and father together to form a diploid offspring. The chromosome complement of this offspring is unlike that of either parent. It is a random combination of genetic material from each parent.

This change in chromosomal complement is crucial. The critical factor here is that changes in chromosome configuration occur only during sexual reproduction—*not* during asexual reproduction.

- **Asexual reproduction** refers to any mechanism of producing offspring that does not involve the fusion of gametes. Asexual reproduction in eukaryotes is usually based on mitosis. As Chapter 11 indicated, the chromosomes in the daughter cells of mitosis are identical to the chromosomes in the parental cell.

- **Sexual reproduction** refers to the production of offspring through the fusion of gametes. Sexual reproduction results in offspring that have chromosome complements unlike their siblings' and their parents'.

Why is this difference important?

Chromosomes and Heredity

The changes in chromosomes produced by meiosis and fertilization are significant because chromosomes contain the cell's hereditary material. Stated another way, chromosomes contain the instructions for specifying what a particular trait might be in an individual. These inherited traits range from eye color and height in humans to the number or shape of the bristles on a fruit fly's leg to the color or shape of the seeds found in pea plants.

In the early 1900s, biologists began using the term gene to refer to the inherited instructions for a particular trait. Recall from Section 12.1 that the term allele refers to a particular version of a gene and that homologous chromosomes may carry different alleles.

Chromosomes are the repositories of genes, and identical copies of chromosomes are distributed to daughter cells during mitosis. Thus, cells that are produced by mitosis are genetically identical to the parent cell, and offspring produced during asexual reproduction are genetically identical to one another as well as to their parent. The offspring of asexual reproduction are **clones**—or exact copies—of their parent.

In contrast, the offspring produced by sexual reproduction are genetically different from one another and unlike either their mother or their father.

Let's begin by analyzing two aspects of meiosis that create variation among chromosomes: (**1**) separation and distribution of homologous chromosomes, and (**2**) crossing over, and then ask how these processes interact with fertilization to produce genetically variable offspring.

Independent Assortment Produces Genetic Variation

Each somatic cell in your body contains 23 homologous pairs of chromosomes and 46 chromosomes in total. Half of these chromosomes came from your mother, and half came from your father. Each chromosome is composed of genes, and genes influence particular traits. For example, one gene that affects your eye color might be located on one chromosome, while one of the genes that affects your hair color might be located on a different chromosome (**Figure 12.10a**).

Suppose that the chromosomes you inherited from your mother contain alleles that tend to produce brown eyes and black hair, but the chromosomes you inherited from your father include alleles that tend to specify green eyes and red hair. (This is a simplification for the purpose of explanation. In reality, several genes with various alleles interact in complex ways to produce human eye color and hair color.)

Will any particular gamete you produce contain the genetic instructions inherited from your mother or the instructions inherited from your father?

To answer this question, study the diagram of meiosis in **Figure 12.10b**. It shows that when pairs of homologous chromo-

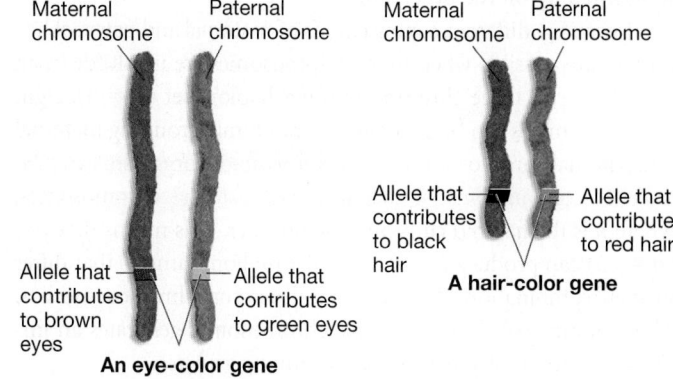

(a) Example: individual who is heterozygous at two genes

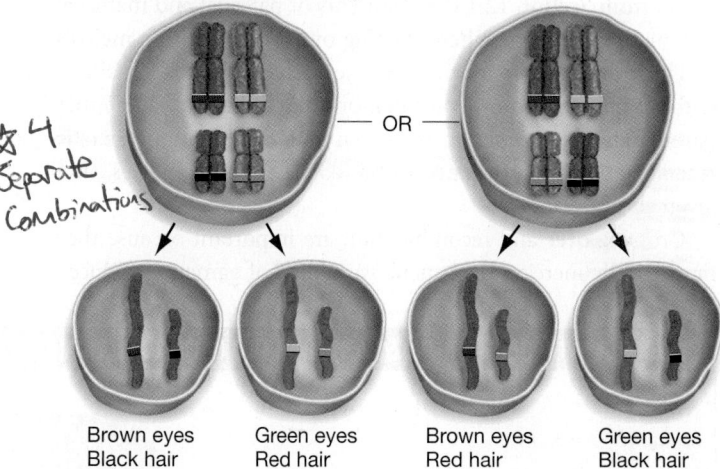

(b) During meiosis I, tetrads can line up two different ways before the homologs separate.

Brown eyes Black hair   Green eyes Red hair   Brown eyes Red hair   Green eyes Black hair

FIGURE 12.10 Independent Assortment of Homologous Chromosomes Results in Varied Combinations of Genes. (a) In this hypothetical example, genes that influence eye color and hair color in humans are on different chromosomes. **(b)** The cells along the bottom are products of meiosis. Notice that each cell has a different combination of genes, due to the separation of homologous chromosomes during meiosis I.

somes line up during meiosis I and the homologs separate, a variety of combinations of maternal and paternal chromosomes can result. Each daughter cell gets a random assortment of maternal and paternal chromosomes.

As Chapter 13 will explain in detail, this phenomenon is known as the principle of independent assortment. In the example given here, meiosis results in gametes with alleles for brown eyes and black hair, like your mother, and green eyes and red hair, like your father. But two additional combinations also occur: brown eyes and red hair, or green eyes and black hair. Four different combinations of paternal and maternal chromosomes are possible when two chromosomes are distributed to daughter cells during meiosis I.

✔If you understand how independent assortment produces genetic variation in the daughter cells of meiosis, you should be able to explain how genetic variation would be affected if maternal chromosomes always lined up together on one side of the metaphase plate during meiosis I and paternal chromosomes always lined up on the other side.

How many different combinations of maternal and paternal homologs are possible when more chromosomes are involved? In an organism with three chromosomes per haploid set ($n = 3$), eight types of gametes can be generated by randomly grouping maternal and paternal chromosomes. In general, a diploid organism can produce 2^n combinations of maternal and paternal chromosomes, where n is the haploid chromosome number. This means that you ($n = 23$) can produce 2^{23}, or about 8.4 million, gametes that differ in their combination of maternal and paternal chromosome sets. The random assortment of whole chromosomes generates an impressive amount of genetic variation among gametes.

The Role of Crossing Over

Recall from Section 12.1 that segments of paternal and maternal chromatids exchange when crossing over occurs during meiosis I. Thus, crossing over produces new combinations of alleles within a chromosome—combinations that did not exist in either parent. This phenomenon is known as recombination. **Genetic recombination** is any change in the combination of alleles on a given chromosome.

Crossing over and recombination are important because they dramatically increase the genetic variability of gametes produced by meiosis. The random assortment of homologous chromosomes during meiosis varies the combination of chromosomes present in gametes; crossing over varies the combinations of alleles within each chromosome. The number of genetically different gametes that you can produce is much more than the 8.4 million—it is virtually limitless.

How Does Fertilization Affect Genetic Variation?

Crossing over and the random mixing of maternal and paternal chromosomes ensure that each gamete is genetically unique. Even if two gametes produced by the same individual fuse to form a diploid offspring—in which case, **self-fertilization**, or "selfing," is taking place—the offspring are very likely to be genetically different from the parent (**Figure 12.11**). Selfing is common in some plant species. It also occurs in the many animal species in which single individuals contain both male and female sex organs.

Self-fertilization is rare or nonexistent in many sexually reproducing species, however. Instead, gametes from different individuals combine to form offspring. This alternative is called **outcrossing**. Outcrossing increases the genetic diversity of offspring because it combines chromosomes from different individuals, which are likely to contain different alleles.

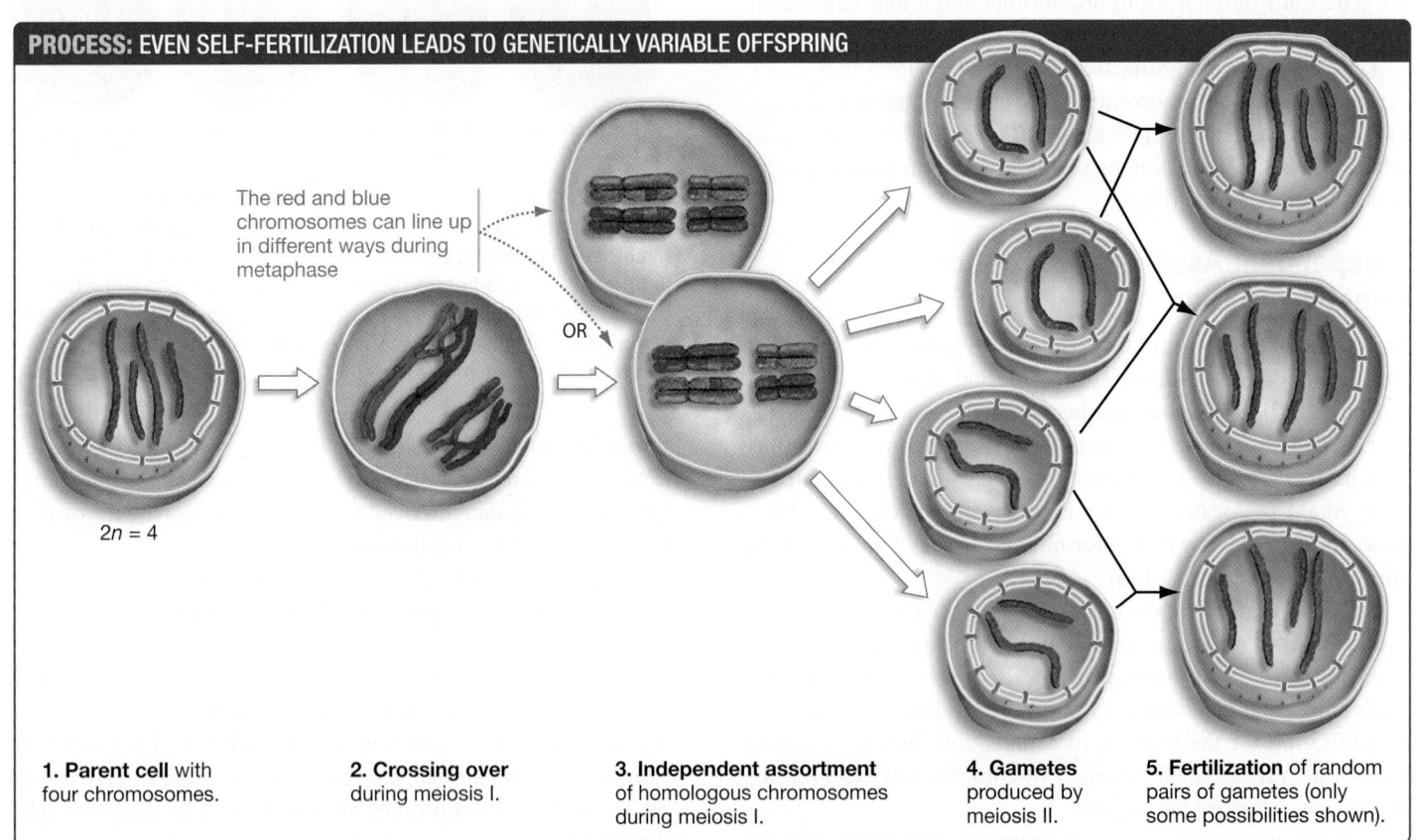

PROCESS: EVEN SELF-FERTILIZATION LEADS TO GENETICALLY VARIABLE OFFSPRING

The red and blue chromosomes can line up in different ways during metaphase

OR

$2n = 4$

1. **Parent cell** with four chromosomes.

2. **Crossing over** during meiosis I.

3. **Independent assortment** of homologous chromosomes during meiosis I.

4. **Gametes** produced by meiosis II.

5. **Fertilization** of random pairs of gametes (only some possibilities shown).

FIGURE 12.11 Crossing Over, Independent Assortment, and the Random Pairing of Gametes during Fertilization Increase Genetic Variation, Even in Offspring Produced by Self-Fertilization.

✔**EXERCISE** In step 5, only a few of the many types of offspring that could be produced are shown. Sketch two additional types that are different from those shown.

How many genetically distinct offspring can be produced when outcrossing occurs? Let's answer this question using humans as an example. Recall that a single human can produce about 8.4 million different gametes, even in the absence of crossing over. When a person mates with a member of the opposite sex, the number of possible genetic combinations that can result is equal to the product of the numbers of different gametes produced by each parent. (To understand the logic here, see **BioSkills 13** in Appendix A.) In humans this means that two parents can potentially produce 8.4 million × 8.4 million = 70.6×10^{12} genetically distinct offspring. This number is far greater than the total number of people who have ever lived—and the calculation does not even take into account variation generated by crossing over.

CHECK YOUR UNDERSTANDING

If you understand that . . .

- The daughter cells produced by meiosis are genetically different from the parent cell because (1) maternal and paternal homologs align randomly at metaphase of meiosis I and (2) crossing over leads to recombination within chromosomes.

✓ You should be able to . . .

1. Draw a diploid parent cell with $n = 3$ (three types of chromosomes), and then sketch six of the many genetically distinct types of daughter cells that may result when this parent cell undergoes meiosis.
2. Compare and contrast the degree of genetic variation that results from asexual reproduction, selfing, and outcrossing.

Answers are available in Appendix B.

The consequences of meiosis and sexual reproduction are simple: tremendous genetic diversity among offspring. Why is genetic diversity important?

12.3 Why Does Meiosis Exist?

Meiosis and sexual reproduction occur in only a small fraction of the lineages on the tree of life. Bacteria and archaea normally undergo only asexual reproduction; most algae, all fungi, and some animals and land plants reproduce asexually as well as sexually. Asexual reproduction is even observed in some vertebrates. Several species of guppy in the genus *Poeciliopsis*, for example, reproduce exclusively by mitosis.

Sexual reproduction is the major mode of reproduction in the insects, however, which number over 43 million species, as well as in species-rich groups like the mollusks (clams, snails, squid) and vertebrates. Yet although sexual reproduction plays an important role in the life of these organisms, scientists had no clear idea, until recently, of why it occurs. On the basis of theory, biologists had good reason to think that sexual reproduction should *not* exist.

The Paradox of Sex

In 1978 John Maynard Smith pointed out that the existence of sexual reproduction presents a paradox. Maynard Smith developed a mathematical model showing that because asexually reproducing individuals do not have to produce male offspring, their progeny can produce twice as many grand-offspring as can individuals that reproduce sexually. **Figure 12.12** diagrams this result by showing the number of females (♀) and males (♂) produced over several generations by asexual versus sexual reproduction.

In this example, each individual produces four offspring over the course of his or her lifetime. In the asexual population, each

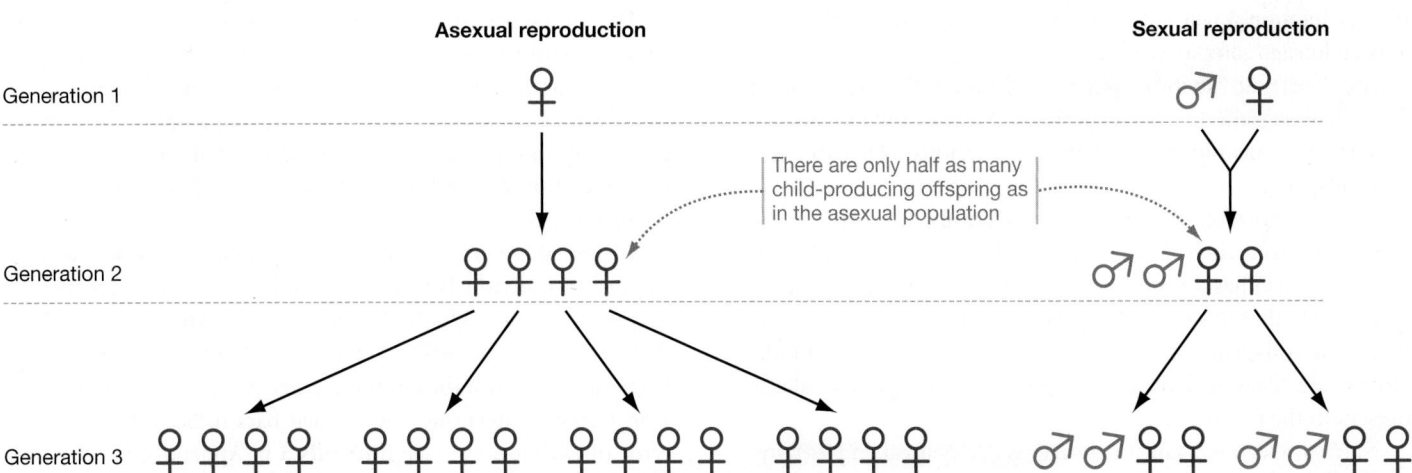

FIGURE 12.12 Asexual Reproduction Confers a Large Numerical Advantage. Each female symbol (♀) and male symbol (♂) represents an individual. This hypothetical example assumes that (1) every individual produces four offspring over the course of a lifetime, (2) sexually reproducing individuals produce half males and half females, and (3) all offspring survive to breed.

✓ QUESTION How many asexually produced offspring would be present in generation 4? How many sexually produced offspring?

individual is a female that produces four offspring. But in the sexual population, it takes two individuals—one male and one female—to produce four offspring. Thus, two out of every four children that each female produces sexually—the males—cannot have children of their own. As a result, generation 2 of the sexual population has just half as many child-producing offspring as generation 2 in the asexual population. Maynard Smith referred to this as the "two-fold cost of males." Asexual reproduction is much more efficient than sexual reproduction because no males are produced.

Based on this analysis, what will happen when asexual and sexual individuals exist in the same population and compete with one another? If all other things are equal, individuals that reproduce asexually should increase in frequency in the population while individuals that reproduce sexually should decline in frequency. In fact, Maynard Smith's model predicts that sexual reproduction is so inefficient that it should be completely eliminated.

To resolve this paradox, biologists began examining the assumption "If all other things are equal." Stated another way, biologists began looking for ways that meiosis and outcrossing could lead to the production of offspring that reproduce more than asexually produced individuals do. After decades of debate and analysis, two solid answers are beginning to emerge.

The Purifying Selection Hypothesis

The first clue to unraveling the paradox of sex is a simple observation: If a gene is damaged or changed in a way that causes it to function poorly, it will be inherited by *all* of that individual's offspring when asexual reproduction occurs. Suppose the damaged gene arose in generation 1 of Figure 12.12. If the damaged gene is important enough, it might cause the four asexual females present in generation 2 to produce fewer than four offspring apiece—perhaps because the members of generation 2 die young. If so, then generation 3 will not have twice as many individuals in the asexual lineage compared to the sexual lineage.

An allele that functions poorly and lowers the fitness of an individual is said to be deleterious. Asexual individuals are doomed to transmitting all of their deleterious alleles to all of their offspring.

Suppose, however, that the same deleterious allele arose in the sexually reproducing female in generation 1 of Figure 12.12. If the female also has a normal copy of the gene, and if she mates with a male that has normal copies of the gene, then on average half of her offspring will lack the deleterious allele. Sexual individuals are likely to have offspring that lack deleterious alleles present in the parent.

Natural selection against deleterious alleles is called purifying selection. Over time, purifying selection should steadily reduce the numerical advantage of asexual reproduction.

To test this hypothesis, researchers recently compared the same genes in two closely related species of *Daphnia*, a common planktonic inhabitant of ponds and lakes. One was a species that reproduces asexually and the other reproduces sexually. As pre-

dicted, they found that individuals in the asexual species contained many more deleterious alleles than individuals in the sexual species. Results like these have convinced biologists that purifying selection is an important factor limiting the success of asexual reproduction.

The Changing-Environment Hypothesis

The second hypothesis to explain sexual reproduction also focuses on the benefits of producing genetically diverse offspring. Here's the key idea: Offspring that are genetic clones of their parents are unlikely to thrive if the environment changes.

What type of environmental change would favor genetically diverse offspring? The leading hypothesis points to changes in parasites—bacteria, viruses, fungi, and other entities that cause disease. In your own lifetime, for example, several new disease-causing agents have emerged that afflict humans. These include the SARS virus, new strains of HIV, the parasite that causes malaria, and the tuberculosis bacterium. Hundreds of genes help defend you against these types of invaders. In many cases, certain alleles help you fight off particular strains of disease-causing bacteria, eukaryotes, or viruses.

What happens if all of the offspring produced by an individual are genetically identical? If a new strain of disease-causing agent evolves, then all of the asexually produced offspring are likely to be susceptible to that new strain. But if the offspring are genetically varied, then it is likely that at least some offspring will have combinations of alleles that enable them to fight off the new disease and produce offspring of their own.

To test this idea, a research group is studying a species of snail that is native to New Zealand. This snail lives in ponds and other freshwater habitats and is susceptible to infection by over a dozen species of parasitic trematode worms. Snails that become infected cannot reproduce—the worms eat their reproductive organs. The parasites are rare in some habitats and common in others.

The biologists are interested in this snail species because some individuals reproduce only sexually while others reproduce only asexually. If the changing-environment hypothesis for the advantage of sex is correct, then the frequency of sexually reproducing individuals should be high in habitats where parasites are common, and low in habitats where parasites are rare (**Figure 12.13**).

To test these predictions, the researchers collected a large number of individuals from different habitats and calculated the frequency of individuals that reproduce sexually versus those that reproduce asexually. On the graph in Figure 12.13, the height of each bar indicates the average percentage of individuals with parasite infections; the vertical bars indicate the standard error of each average (see **BioSkills 5** in Appendix A). The data show that habitats where parasite infection rates are high have a relatively large number of sexually reproducing individuals compared with habitats that have low parasite incidence.

This result and various other studies support the changing-environment hypothesis. Although the paradox of sex remains an active area of research, more biologists are becoming con-

QUESTION: Why does sexual reproduction occur?

HYPOTHESIS: In habitats where parasitism is common, sexually produced offspring have higher fitness than do asexually produced individuals.

NULL HYPOTHESIS: There is no relationship between mode of reproduction and the presence of parasites.

EXPERIMENTAL SETUP:

1. **Collect snails** from a wide array of habitats.

Habitat 5 Males **Habitat 5 Females**

2. **Document percentage of males in each population,** as an index of frequency of sexual reproduction. More males means that more sexual reproduction is occurring.

Habitat 5 Males **Habitat 9 Males**

3. **Compare percentage of males in both populations:** In one population, males are common; in the other, males are almost nonexistent. Infer that sexual reproduction is either common or almost nonexistent.

Parasitism rate in this population? Parasitism rate in this population?

4. **Assess infection rate:** Document percentage of individuals infected with parasites in sexually versus asexually reproducing populations.

PREDICTION: In populations where sexual reproduction is common, parasitism rates are high. In populations with only asexual reproduction, infection rates are low.

PREDICTION OF NULL HYPOTHESIS: No difference in parasitism rate between populations that reproduce sexually versus asexually.

RESULTS:

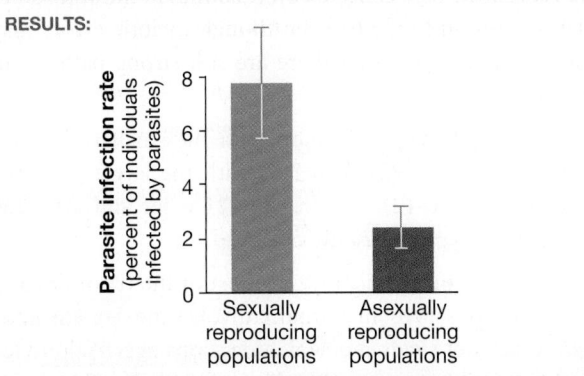

CONCLUSION: Sexual reproduction is common in habitats where parasitism is common. Asexual reproduction is common in habitats where parasitism is rare.

vinced that sexual reproduction is helpful for two reasons: (**1**) Offspring are not doomed to inherit harmful alleles, and (**2**) at least some offspring may be able to fight off new strains of disease-causing agents.

12.4 Mistakes in Meiosis

When homologous chromosomes separate during meiosis I, a complete set of chromosomes is transmitted to each daughter cell. But what happens if there is a mistake, and the chromosomes are not properly distributed? What are the consequences for offspring if gametes contain an abnormal set of chromosomes?

In 1866 Langdon Down described a distinctive suite of co-occurring conditions observed in some humans. The syndrome was characterized by mental retardation, a high risk for heart problems and leukemia, and a degenerative brain disorder similar to Alzheimer's disease. **Down syndrome**, as the disorder came to be called, is observed in about 0.15 percent of live births (1 infant in every 666).

For over 80 years the cause of the syndrome was unknown. Then, in the late 1950s, a study of the chromosome sets of nine Down syndrome children suggested that the condition is associated with the presence of an extra copy of chromosome 21. This situation is called a **trisomy** ("three-bodies")—in this case, trisomy-21—because each cell has three copies of the chromosome. The explanation proposed for the trisomy was that a mistake had occurred during meiosis in one of the parents.

How Do Mistakes Occur?

For a gamete to get one complete set of chromosomes, two steps in meiosis must be perfectly executed.

1. The chromosomes in each homologous pair must separate from each other during the first meiotic division, so that only one homolog ends up in each daughter cell.

2. Sister chromatids must separate from each other and move to opposite poles of the dividing cell during meiosis II.

If both homologs or both sister chromatids move to the same pole of the parent cell, the products of meiosis will be abnormal. This sort of meiotic error is referred to as **nondisjunction**, because the homologs fail to separate or disjoin.

FIGURE 12.13 Is Sexual Reproduction Favored When Parasite Infection Rates Are High?

SOURCE: Lively, C. and J. Jokela. 2002. Temporal and spatial distributions of parasites and sex in a freshwater snail. *Evolutionary Ecology Research* 4: 219–226.

✔**QUESTION** Why was it important to collect snails from a wide array of habitats?

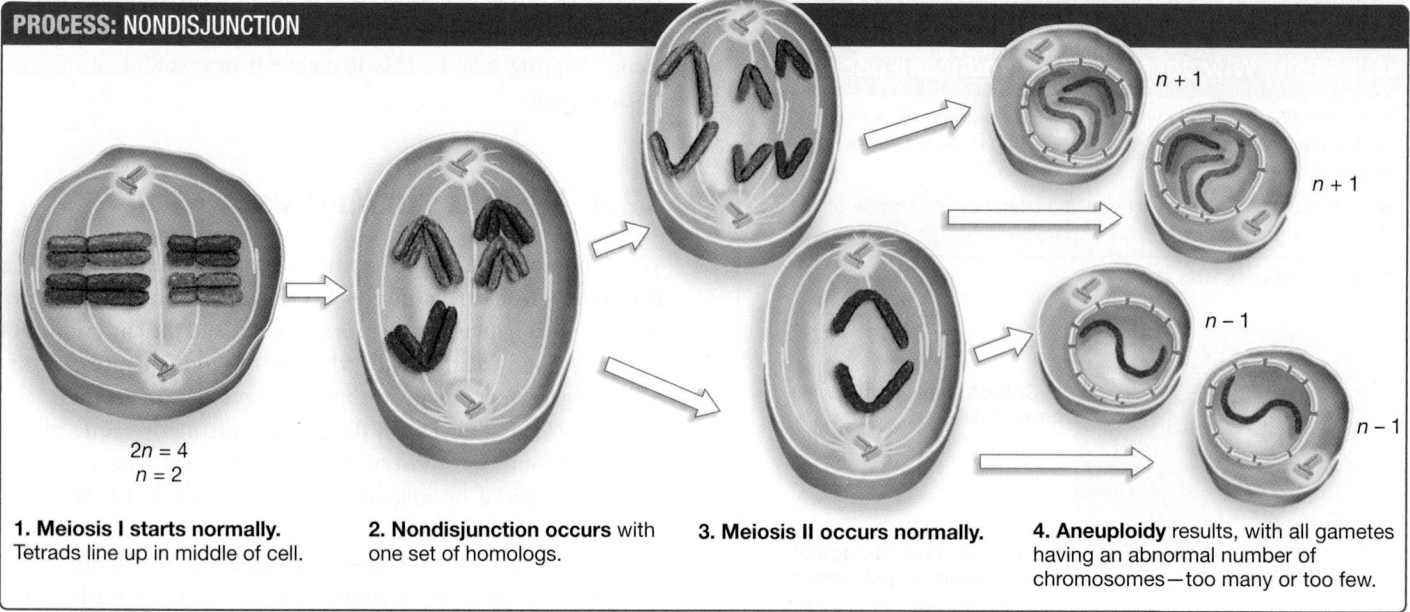

PROCESS: NONDISJUNCTION

$2n = 4$
$n = 2$

$n + 1$

$n + 1$

$n - 1$

$n - 1$

1. Meiosis I starts normally. Tetrads line up in middle of cell.

2. Nondisjunction occurs with one set of homologs.

3. Meiosis II occurs normally.

4. Aneuploidy results, with all gametes having an abnormal number of chromosomes—too many or too few.

FIGURE 12.14 Nondisjunction Leads to Gametes with Abnormal Chromosome Numbers. If homologous chromosomes fail to separate during meiosis I, the gametes that result will have an extra chromosome or will lack a chromosome.

Figure 12.14 shows what happens when homologs do not separate correctly. Notice that two daughter cells have two copies of the same chromosome—the smaller one in Figure 12.14—while the other two lack that chromosome entirely. Gametes that contain an extra chromosome are symbolized as $n + 1$; gametes that lack one chromosome are symbolized as $n - 1$.

If an $n + 1$ gamete is fertilized by a normal n gamete, the resulting zygote will be $2n + 1$. This situation is a trisomy. If the $n - 1$ gamete is fertilized by a normal n gamete, the resulting zygote will be $2n - 1$. This situation is called **monosomy**. Cells that have too many or too few chromosomes are said to be **aneuploid** ("without-form").

To review nondisjunction and its consequences, go to the study area at *www.masteringbiology.com.*

(MB) Web Activity Mistakes in Meiosis

Meiotic mistakes occur at a relatively high frequency. In humans, for example, researchers estimate that 25 percent of all conceptions produce a fertilized egg that is aneuploid. Most of these errors result from the failure of a homologous pair to separate in anaphase of meiosis I; it is relatively rare for sister chromatids to stay together during anaphase of meiosis II.

The consequences of meiotic mistakes are almost always severe when defective gametes participate in fertilization. In one study of human pregnancies that ended in early embryonic or fetal death, 38 percent of the 119 cases involved atypical chromosome complements that resulted from mistakes in meiosis. Trisomy accounted for 36 percent of the abnormal karyotypes found. It was also common to find the incorrect number of complete chromosome sets called triploidy ($3n$). Less common

were abnormally sized or shaped chromosomes and monosomy ($2n - 1$). Mistakes in meiosis are the leading cause of spontaneous abortion (miscarriage) in humans.

Why Do Mistakes Occur?

The leading hypothesis to explain the incidence of trisomy and other meiotic mistakes is that they are accidents—random errors that occur during meiosis. Consistent with this proposal, there does not seem to be any genetic or inherited predisposition to trisomy or other types of meiotic dysfunction. Most cases of Down syndrome, for example, occur in families with no history of the condition. Recent research indicates that the molecular mechanism involves misattachment between microtubules and kinetochores early in meiosis I.

Table 12.4 shows data collected on trisomies in the autosomes of human fetuses and infants. Even though meiotic errors may be random, studies show that there are still strong patterns in their occurrence:

- Trisomy is much more common with the smaller chromosomes (numbers 13–22) than it is with the larger chromosomes (numbers 1–12), and trisomy-21 is far and away the most common type of trisomy observed.

- With the exception of trisomy-21, most of the trisomies and monosomies observed in humans involve the sex chromosomes. **Klinefelter syndrome**, which develops in XXY individuals, occurs in about 1 in 1000 live male births. Trisomy X (karyotype XXX) occurs in about 1 in 1000 live births and results in females who may or may not have symptoms such as

SUMMARY TABLE 12.4 Trisomy in Human Births: Effects of Chromosome Number and Paternal versus Maternal Origin

Trisomy (chromosome number)	Number of Cases Examined	Due to Error in Sperm	Due to Error in Egg	Maternal Errors (%)
2–12	16	3	13	81
13	7	2	5	71
14	8	2	6	75
15	11	3	8	73
16	62	0	62	100
18	73	3	70	96
21	436	29	407	93
22	11	0	11	100

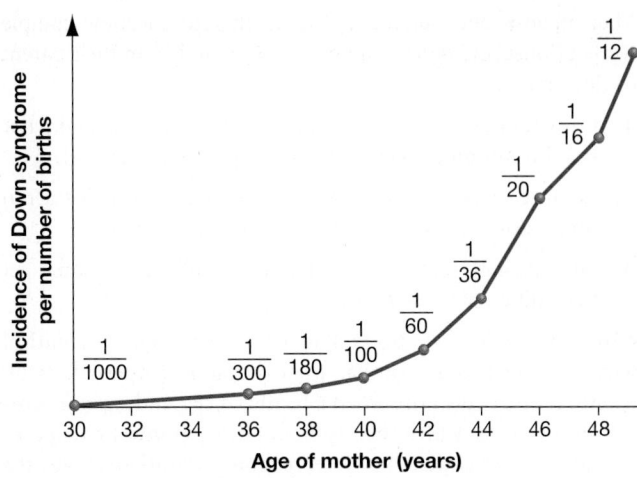

FIGURE 12.15 The Frequency of Down Syndrome Increases as a Function of a Mother's Age.

impaired mental function and sterility. **Turner syndrome** develops in XO individuals, where the "O" stands for lack of a second X, and occurs in about 1 in 5000 live births. Individuals with this syndrome are female but are sterile.

- Maternal age is an important factor in the occurrence of trisomy. In fact, maternal errors account for most incidents of trisomy. For example, over 90 percent of cases of Down syndrome are due to chromosomal defects in eggs. As **Figure 12.15** shows, the incidence of Down syndrome increases dramatically in mothers over 35 years old.

Why do these patterns occur? The frequency of nondisjunction is about equal among chromosomes, but aneuploidy tends to be lethal to embryos if it involves chromosomes that contain a large number of genes. Trisomy-21 is common because it affects the phenotype less, and is less deadly, than other trisomies. It is still a mystery, though, why such a strong correlation exists between maternal age and frequency of trisomy-21.

Studies on the mechanism of aneuploidy continue. In the meantime, one overall message is clear: Successful meiosis is critical to the health and welfare of offspring.

CHAPTER 12   REVIEW

For media, go to the study area at www.masteringbiology.com

Summary of Key Concepts

🔑 **Meiosis is a type of nuclear division resulting in cells that have half as many chromosomes as the parent cell. In animals it leads to the formation of eggs and sperm.**

- Chromosomes exist in sets. In diploid organisms, individuals have two versions of each type of chromosome. The two versions are called homologs. One homolog is inherited from the mother, and one from the father. Haploid organisms have just one of each type of chromosome.

- Each chromosome is replicated before meiosis begins. At the start of meiosis I, each chromosome consists of a pair of sister chromatids.

- Homologous pairs of chromosomes synapse early in meiosis I, forming a tetrad—a group of two homologous chromosomes. Non-sister chromatids undergo crossing over.

- When crossing over is complete, the pair of homologous chromosomes is moved to the metaphase plate.

- At the end of meiosis I, the homologous chromosomes are separated and distributed to two daughter cells. The daughter cells are haploid, because each receives one of each type of chromosome.

- During meiosis II, sister chromatids separate and are distributed to two daughter cells.

- From a diploid cell with replicated chromosomes, meiosis produces four haploid daughter cells with unreplicated chromosomes.

   ✔ You should be able to explain why meiosis does not occur in bacteria—most of which have a singular, circular chromosome.

   MB **BioFlix™** Meiosis, **Web Activity** Meiosis

🔑 **Each cell produced by meiosis receives a different combination of chromosomes. Because genes are located on chromosomes, each cell produced by meiosis receives a different complement of genes. The resulting offspring are genetically distinct from each other and from their parents.**

- When meiosis and outcrossing occur, the chromosome complements of offspring differ from one another and from their parents for three reasons:
  (1) Gametes receive a random assortment of maternal and paternal chromosomes when homologs separate in meiosis I.
  (2) Because of crossing over, each chromosome contains a random assortment of paternal and maternal alleles.
  (3) Outcrossing results in a combination of chromosome sets from different individuals.

  ✔You should be able to explain why monozygotic (identical) twins are much more similar than dizygotic (fraternal) twins. (Monozygotic twins arise when a fertilized egg undergoes mitosis to form two cells, which then split and develop into two separate individuals. Dizygotic twins develop from two different eggs that were fertilized by two different sperm.)

🔑 **The leading hypothesis to explain meiosis is that genetically varied offspring are more likely to thrive in environments where parasites and disease are common.**

- Asexual reproduction is much more efficient than sexual reproduction because males do not have to be produced and no time or energy has to be spent in courtship.

- Sexual reproduction is favored in many groups because (1) parents can produce offspring that lack deleterious alleles and (2) genetically diverse offspring are likely to include some that are better able to resist parasites.

  ✔You should be able to predict whether, in species that alternate between asexual and sexual reproduction, sexual reproduction occurs during seasons when environmental conditions are stable or seasons when conditions change rapidly.

🔑 **If mistakes occur during meiosis, the resulting egg and sperm cells may contain the wrong number of chromosomes. It is rare for offspring with an incorrect number of chromosomes to develop normally.**

- Mistakes during meiosis lead to gametes and offspring with an unbalanced set of chromosomes. Children with Down syndrome, for example, have an extra copy of chromosome 21.

- The leading hypothesis to explain meiotic mistakes is that they are random accidents resulting in a failure of homologous chromosomes or sister chromatids to separate properly during meiosis.

  ✔You should be able to explain what happens when none of the homologous chromosomes present separate at anaphase of meiosis I, but their sister chromatids separate normally at meiosis II.

(MB) **Web Activity** Mistakes in Meiosis

Questions

✔**TEST YOUR KNOWLEDGE**   *Answers are available in Appendix B*

1. In the roundworm *Ascaris*, eggs and sperm have two chromosomes, but all other cells have four. Observations such as this inspired which important hypothesis?
   a. Before gamete formation, a special type of cell division leads to a quartering of chromosome number.
   b. Before gamete formation, a special type of cell division leads to a halving of chromosome number.
   c. After gamete formation, half of the chromosomes are destroyed.
   d. After gamete formation, either the maternal or the paternal set of chromosomes disintegrates.

2. What are homologous chromosomes?
   a. chromosomes that are similar in their size, shape, and gene content
   b. similar chromosomes that are found in different individuals of the same species
   c. the two "threads" in a replicated chromosome (they are identical copies)
   d. the products of crossing over, which contain a combination of segments from maternal chromosomes and segments from paternal chromosomes

3. What is a tetrad?
   a. the X that forms when chromatids from homologous chromosomes cross over
   b. a group of four chromatids produced when homologs synapse

   c. the four points where homologous chromosomes touch as they synapse
   d. the group of four genetically identical daughter cells produced by mitosis

4. What is genetic recombination?
   a. the synapsing of homologs during prophase of meiosis I
   b. the new combination of maternal and paternal chromosome segments that results when homologs cross over
   c. the new combinations of chromosome segments that result when outcrossing occurs
   d. the combination of a haploid phase *and* a diploid phase in a life cycle

5. What is meant by a paternal chromosome?
   a. the largest chromosome in a set
   b. a chromosome that does not separate correctly during meiosis I
   c. the member of a homologous pair that was inherited from the mother
   d. the member of a homologous pair that was inherited from the father

6. Meiosis II is similar to which process?
   a. mitosis in haploid cells
   b. nondisjunction
   c. outcrossing
   d. meiosis I

1. Explain the relationship between homologous chromosomes and the relationship between sister chromatids.

2. Lay four pens and four pencils on a tabletop, and imagine that they represent replicated chromosomes in a diploid cell with $n = 2$. Explain the phases of meiosis II by moving the pens and pencils around. (If you don't have enough pens and pencils, use strips of paper or fabric.)

3. Meiosis is called a reduction division, but all of the reduction occurs during meiosis I—no reduction occurs during meiosis II. Explain why meiosis I is a reduction division but meiosis II is not.

4. Triploid ($3n$) watermelons are produced by crossing a tetraploid ($4n$) strain with a diploid ($2n$) plant. Briefly explain why this mating produces a triploid individual. Why can mitosis proceed normally in triploid cells, but meiosis cannot?

5. Some plant breeders are concerned about the susceptibility of asexually cultivated plants, such as seedless bananas, to new strains of disease-causing bacteria, viruses, or fungi. Briefly explain their concern by discussing the differences in the genetic "outcomes" of asexual and sexual reproduction.

6. Explain why nondisjunction leads to trisomy and other types of abnormal chromosome complements. In what sense are these chromosome sets "unbalanced"?

1. The gibbon has 44 chromosomes per diploid set, and the siamang has 50 chromosomes per diploid set. In the 1970s a chance mating between a male gibbon and a female siamang produced an offspring. Predict how many chromosomes were observed in the somatic cells of the offspring. Do you predict that this individual would be able to form viable gametes? Why or why not?

2. Meiosis results in a reassortment of maternal and paternal chromosomes. If $n = 3$ for a given organism, there are eight different combinations of paternal and maternal chromosomes. If no crossing over occurs, what is the probability that a gamete will receive *only* paternal chromosomes?

3. Some researchers predict that spontaneous abortion should be rare in older females, because they are less likely than young females to be able to have offspring in the future. How does this claim relate to Figure 12.15, which graphs mother's age versus the incidence of Down syndrome?

4. The data on snail populations that were used to test the changing-environment hypothesis have been criticized because they are observational and not experimental in nature. As a result, they do not control for factors other than parasites that might affect the frequency of sexually reproducing individuals.

   a. Design an experimental study that would provide stronger evidence that the frequency of parasite infection causes differences in the frequency of sexually versus asexually reproduced individuals in this species of snail.

   b. In defense of the existing data, comment on the value of observing patterns like this in nature, versus under controlled conditions in the laboratory.

This chapter is part of **The Big Picture**. See how on pages 336–337.

Experiments on garden peas and sweet peas (shown here) helped launch the science of genetics.

13 Mendel and the Gene

KEY CONCEPTS

○━ In many species, individuals have two alleles of each gene. The principle of segregation states that prior to the formation of eggs and sperm, the alleles of each gene separate so that each egg or sperm cell receives only one of them.

○━ The principle of independent assortment states that alleles of different genes are transmitted to egg cells and sperm cells independently of each other.

○━ Genes are located on chromosomes. The principle of segregation is explained by the separation of homologous chromosomes in anaphase of meiosis I. The principle of independent assortment applies to genes found on different chromosomes and is explained by chromosomes lining up randomly in metaphase of meiosis I.

○━ There are important exceptions and extensions to the basic patterns of inheritance that Mendel discovered.

The science of biology is built on a series of great ideas. Two of these—the cell theory and the theory of evolution—were introduced in Chapter 1. The cell theory describes the basic structure of organisms; the theory of evolution by natural selection clarifies why species change through time. Life is cellular; populations evolve. These are two of the five fundamental attributes of life.

This chapter introduces a third great idea in biology: the chromosome theory of inheritance. The chromosome theory explained how genetic information is transmitted from one generation to the next. It shed light on a third fundamental attribute of life: Organisms process information.

An Austrian monk named Gregor Mendel laid the groundwork for the theory in 1865, when he announced that he had worked out the rules of inheritance through a series of experiments on garden peas. Another key insight emerged during the final decades of the nineteenth century, when biologists described the details of meiosis.

The chromosome theory of inheritance, formulated in 1902 by Walter Sutton and Theodor Boveri, linked these two results. This theory contends that meiosis, introduced in Chapter 12, causes the patterns of inheritance that Mendel observed. It also asserts that the hereditary factors called genes are located on chromosomes.

The chromosome theory launched the study of **genetics**, the branch of biology that focuses on the inheritance of traits. Let's start at the beginning: What are the rules of inheritance that Mendel discovered?

13.1 Mendel's Experimental System

When biological science began to emerge as an important scientific discipline, questions about **heredity**—meaning inheritance, or the transmission of traits from parents to offspring—were primarily the concern of animal breeders and horticulturists. A **trait**

✔ When you see this checkmark, stop and test yourself. Answers are available in Appendix B.

is any characteristic of an individual, ranging from overall height to the primary structure of a particular membrane protein.

In the city where Gregor Mendel lived, there was particular interest in how selective breeding could result in hardier and more productive varieties of sheep, fruit trees, and vines; and an Agricultural Society had been formed there to promote research into making selective breeding more efficient. Mendel was an active member of this society; the monastery he belonged to was also devoted to scientific teaching and research.

What Questions Was Mendel Trying to Answer?

Mendel set out to address the most fundamental of all issues concerning heredity: What are the basic patterns in the transmission of traits from parents to offspring?

At the time, two hypotheses had been formulated to answer this question:

- *Blending inheritance* claimed that the traits observed in a mother and father blend together to form the traits observed in their offspring. As a result, an offspring's traits are intermediate between the mother's and father's traits.

- *Inheritance of acquired characters* claimed that traits present in parents are modified, through use, and passed on to their offspring in the modified form.

Each of these hypotheses made predictions. Blending inheritance contended that when black sheep and white sheep mate, their hereditary determinants blend to form a new hereditary determinant for gray wool. Therefore, their offspring should be gray. Inheritance of acquired characters predicts that if giraffes extend their necks by straining to reach leaves high in the tops of trees, they subsequently produce longer-necked offspring.

These hypotheses were being promoted by the greatest scientists of Mendel's time. Are they correct?

Garden Peas Served as the First Model Organism in Genetics

After investigating and discarding several candidate species to study, Mendel chose the garden pea *Pisum sativum*. His reasons were practical: Peas are inexpensive and easy to grow from seed, have a relatively short generation time, and produce reasonably large numbers of seeds. These features made it possible for him to continue experiments over several generations and collect data from large numbers of individuals.

Peas served as a **model organism**: a species that is used for research because it is practical and because conclusions drawn from studying it turn out to apply to many other species as well. **BioSkills 14** in Appendix A introduces some of the important model organisms used in biological science today.

Two additional features of the pea made it possible for Mendel to design his experiments: He could control which parents were involved in a mating, and he could arrange matings between individuals that differed in easily recognizable traits, such as flower color or seed shape.

HOW DID MENDEL CONTROL MATINGS? **Figure 13.1a** shows the male and female reproductive organs of a garden pea flower. Sperm cells are produced in **pollen grains**, which are small sacs that mature in the male reproductive structure of the plant. Eggs are produced in the female reproductive structure.

Under normal conditions, garden peas **self-fertilize**: a flower's pollen falls on the female reproductive organ of that same flower. As **Figure 13.1b** shows, however, Mendel could circumvent this arrangement by removing the male reproductive organs from a flower before any pollen formed. Later he could transfer pollen from another pea plant to that flower's female reproductive organ with a brush. This type of mating is referred to as a **cross-fertilization** or simply a cross. Using this technique, Mendel could control the matings of his model organism.

WHAT TRAITS DID MENDEL STUDY? Mendel conducted his experiments on varieties of peas that differed in seven traits: seed shape, seed color, pod shape, pod color, flower color, flower and pod position, and stem length. Biologists refer to the observable traits of an individual, such as the shape of a pea seed or the eye color of a human, as its **phenotype** (literally, "show-type"). In the first pea populations that Mendel studied, two distinct phenotypes existed for each of the seven traits.

Mendel began his work by obtaining individuals from what breeders called pure lines or true-breeding lines. A **pure line** consists of individuals that produce offspring identical to themselves when they are self-pollinated or crossed to another member of the pure-line population. For example, earlier breeders had developed pure lines for wrinkled seeds and round seeds. During two years of trial experiments, Mendel

(a) Self-pollination

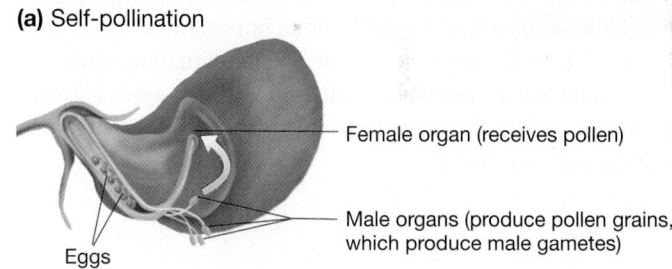

Female organ (receives pollen)

Male organs (produce pollen grains, which produce male gametes)

Eggs

(b) Cross-pollination

Collect pollen from one individual and transfer it ...

... to the female organs of an individual whose male organs have been removed.

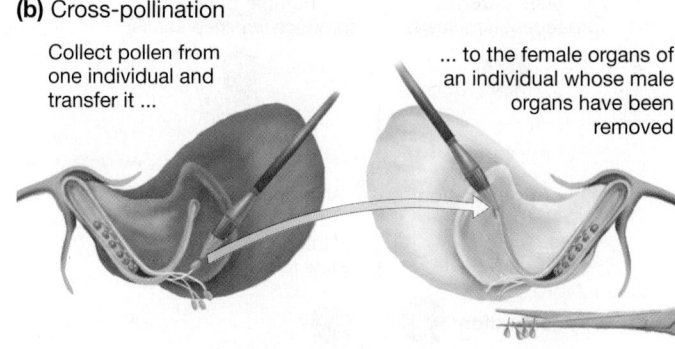

FIGURE 13.1 Peas Can Be Self-Pollinated or Cross-Pollinated.
(a) Under normal conditions, garden peas pollinate themselves.
(b) Mendel developed a method of controlling the matings of his model organism.

confirmed that individuals that germinated from his wrinkled seeds produced only wrinkled-seeded offspring when they were mated to themselves or to another pure-line individual that germinated from a wrinkled seed; and that the same was true for round seeds.

Why is this important? Remember that Mendel wanted to find out how traits are transmitted from parents to offspring. Once he had confirmed that he was working with pure lines, he could predict how matings within each line would turn out; in other words, he knew what the offspring from these matings would look like. He could then compare these results with the outcomes of crosses between individuals from different pure lines.

Suppose that Mendel arranged matings between a pure-line individual with round seeds and a pure-line individual with wrinkled seeds. He knew that one parent carried a hereditary determinant for round seeds, while the other carried a hereditary determinant for wrinkled seeds. But the offspring that resulted from this mating would have both hereditary determinants. They would be **hybrids**—offspring from matings between true-breeding parents that differ in one or more traits.

Would these hybrid offspring have wrinkled seeds, round seeds, or a blended combination of wrinkled and round? What would be the seed shape in subsequent generations when hybrid individuals self-pollinated or were crossed with members of the pure lines?

13.2 Mendel's Experiments with a Single Trait

Mendel's first set of experiments consisted of crossing pure lines that differed in just one trait. This is an important research strategy in biological science: start with a simple situation. Once you understand what's going on, you can consider more complex questions, such as, What happens in crosses between individuals that differ in two traits?

Mendel began his single-trait crosses by crossing individuals from round-seeded and wrinkled-seeded pure lines. The adults used in an initial experimental cross are the **parental generation**. Their progeny (that is, offspring) are the **F$_1$ generation**. F$_1$ stands for "first filial"; the Latin roots *fili* and *filia* mean son and daughter. Subsequent generations are called the F$_2$ generation, F$_3$ generation, and so on.

The Monohybrid Cross

In his first set of crosses, Mendel took pollen from round-seeded plants and placed it on the female reproductive organs of plants from the wrinkled-seeded line. As **Figure 13.2a** shows, all of the seeds produced by progeny from this cross were round.

This was a remarkable result, for two reasons:

1. The traits did not blend together to form an intermediate phenotype. Instead, the round-seeded form appeared intact. This result was in stark contrast to the predictions of the blending-inheritance hypothesis shown in **Figure 13.2b**.

2. The genetic determinant for wrinkled seeds seemed to have disappeared. Where did it go?

DOMINANT AND RECESSIVE TRAITS To figure out what was going on, Mendel did something that turned out to be brilliant: He planted the F$_1$ seeds and allowed the individuals to self-pollinate when they matured.

Remember that he knew that each of these individuals had inherited a genetic determinant for round seeds and a genetic determinant for wrinkled seeds. A mating like this—between parents that each carry two different genetic determinants for the same trait—is called a **monohybrid cross**.

When he collected the seeds that were produced by many plants in the resulting F$_2$ generation, he observed that 5474 were round and 1850 were wrinkled (see Figure 13.2a). This observation was striking, even astonishing. The wrinkled seed shape had reappeared in the F$_2$ generation after disappearing completely in

(a) Results of Mendel's single-trait (monohybrid) cross

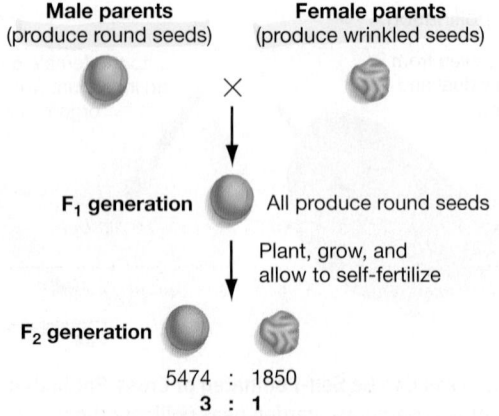

Male parents
(produce round seeds)

Female parents
(produce wrinkled seeds)

×

F$_1$ generation — All produce round seeds

Plant, grow, and allow to self-fertilize

F$_2$ generation

5474 : 1850
3 : 1

(b) Prediction of blending-inheritance hypothesis

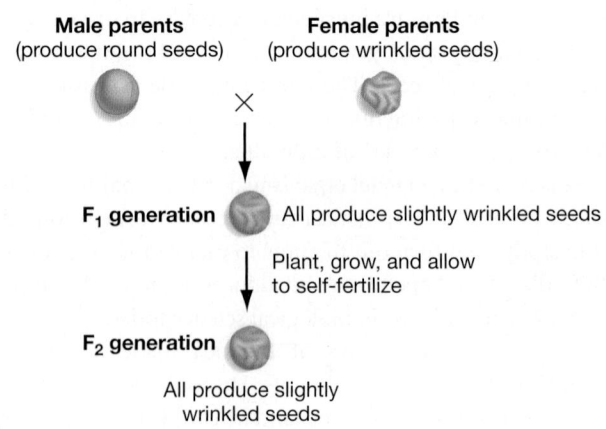

Male parents
(produce round seeds)

Female parents
(produce wrinkled seeds)

×

F$_1$ generation — All produce slightly wrinkled seeds

Plant, grow, and allow to self-fertilize

F$_2$ generation

All produce slightly wrinkled seeds

FIGURE 13.2 A Monohybrid Cross. The results of Mendel's crosses involving a single trait contrasted strongly with the predictions of the blending inheritance hypothesis.

the F_1 generation. No one had observed the phenomenon before, simply because it had been customary for biologists to stop their breeding experiments with F_1 offspring.

Mendel invented some important terms to describe this result.

- He designated wrinkled shape as a **recessive** trait relative to the round-seed trait. This was an appropriate term because none of the F_1 individuals had wrinkled seeds—meaning the wrinkled-seeds phenotype appeared to recede or temporarily become latent or hidden.

- He referred to round seeds as **dominant** to the wrinkled-seed trait. This term was apt because the round-seed phenotype appeared to dominate over the wrinkled-seed determinant when both were present.

It's important to note, though, that in genetics the term dominant has nothing to do with its everyday English usage as powerful or superior. Subsequent research has shown that individuals with the dominant phenotype do not necessarily have higher fitness than do individuals with the recessive phenotype. Nor are genetic determinants associated with a dominant phenotype necessarily more common than recessive ones. For example, a rare, dominant genetic determinant in humans causes a fatal illness—a type of brain degeneration called Huntington's disease. In genetics, the terms dominant and recessive identify *only* which phenotype is observed in individuals carrying two different genetic determinants for a given trait.

Mendel also noticed that the round and wrinkled seeds of the F_2 generation were present in a ratio of 2.96:1, or essentially 3:1. The 3:1 ratio means that for every four individuals, on average three had the dominant phenotype and one had the recessive phenotype. The results can also be stated in terms of frequencies or proportions instead of as a ratio: In this case, about 3/4 of the F_2 seeds were round and 1/4 were wrinkled.

A RECIPROCAL CROSS Mendel wanted to test the hypothesis that it mattered which parent and gamete type had a particular genetic determinant—that gender influenced the inheritance of seed shape. To do this, he performed a second set of crosses between two pure-breeding lines—this time with pollen taken from an individual from a pure line of wrinkled-seeded peas (see **Figure 13.3**).

EXPERIMENT

QUESTION: Is the inheritance of seed shape in peas affected by whether the genetic determinant is in a male or female gamete?

HYPOTHESIS: The type of gamete *does* affect the inheritance of seed shape.

NULL HYPOTHESIS: The type of gamete does *not* affect the inheritance of seed shape.

EXPERIMENTAL SETUP:

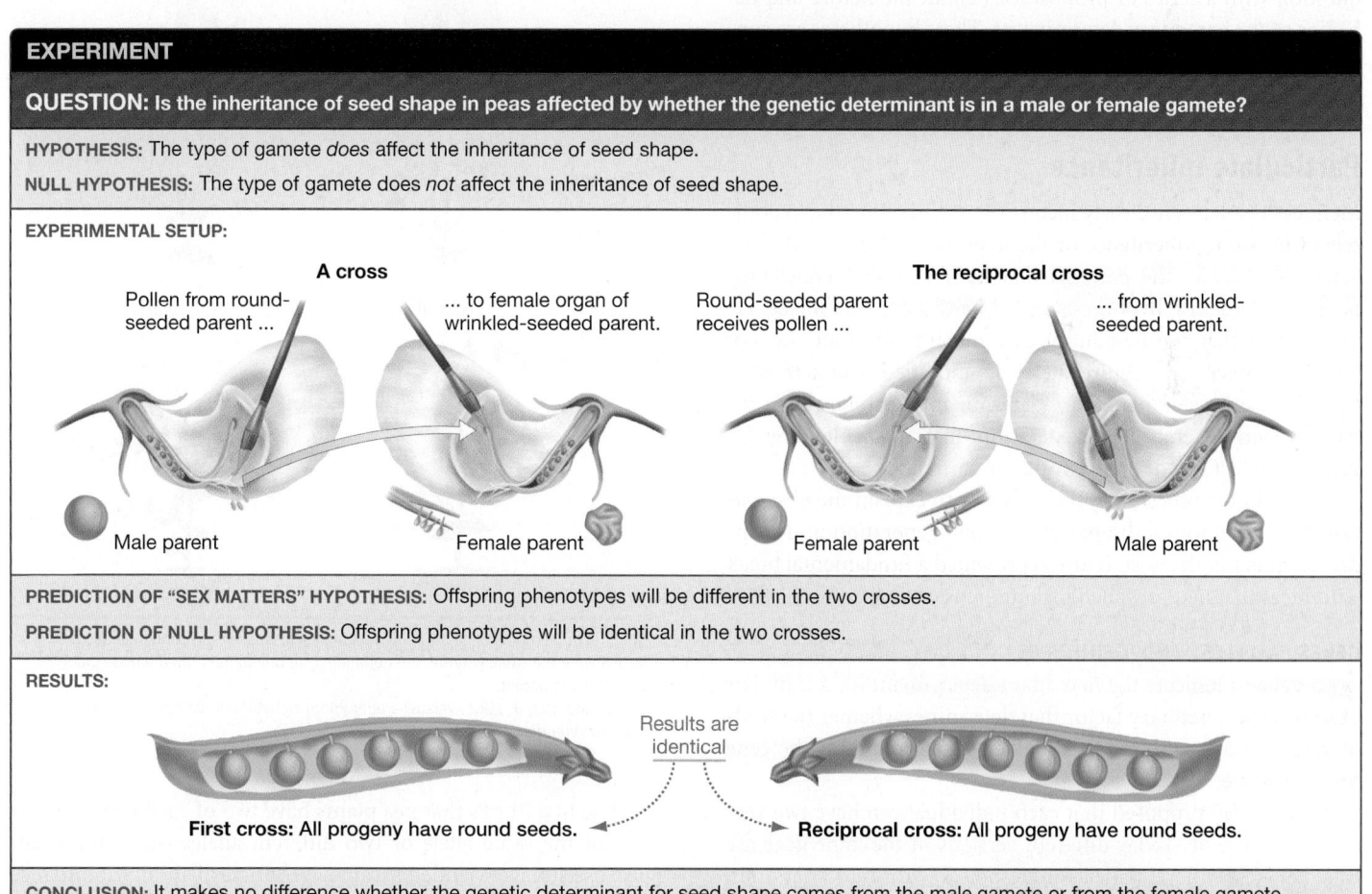

A cross

Pollen from round-seeded parent ...

... to female organ of wrinkled-seeded parent.

Male parent

Female parent

The reciprocal cross

Round-seeded parent receives pollen ...

... from wrinkled-seeded parent.

Female parent

Male parent

PREDICTION OF "SEX MATTERS" HYPOTHESIS: Offspring phenotypes will be different in the two crosses.

PREDICTION OF NULL HYPOTHESIS: Offspring phenotypes will be identical in the two crosses.

RESULTS:

Results are identical

First cross: All progeny have round seeds.

Reciprocal cross: All progeny have round seeds.

CONCLUSION: It makes no difference whether the genetic determinant for seed shape comes from the male gamete or from the female gamete.

FIGURE 13.3 A Reciprocal Cross.

SOURCE: Mendel, G. 1866. Versuche über Pflanzen-hybriden. *Verhandlungen des naturforschenden Vereines in Brünn*. 4: 3–47. English translation available from ESP: Electronic Scholarly Publishing (www.esp.org).

✔**QUESTION** Some people think that experiments are failures if the hypothesis being tested is not supported. What does it mean to say that an experiment failed? Was this experiment a failure?

These experiments completed a **reciprocal cross**—a set of matings where the mother's phenotype in the initial cross is the father's phenotype in a subsequent cross, and the father's phenotype in the initial cross is the mother's phenotype in a subsequent cross.

In this case the results of the reciprocal crosses were identical: All of the F₁ progeny in the subsequent cross had round seeds, just as in the initial cross. The reciprocal cross established that it does not matter whether the genetic determinants for seed shape are located in the male or female parent.

THE RESULTS ARE GENERAL   Before he tried to interpret this pattern, it was important for Mendel to establish that the results were not restricted to inheritance of seed shape. So he repeated the experiments with each of the six other traits listed earlier.

In each case, he obtained similar results (see **Table 13.1**): The products of reciprocal crosses were the same—one form of the trait was always dominant regardless of the parent it came from; the F₁ progeny showed only the dominant trait and did not exhibit an intermediate phenotype; and in the F₂ generation, the ratio of individuals with dominant and recessive phenotypes was about 3 to 1.

How could these patterns be explained? Mendel answered this question with a series of propositions about the nature and behavior of the hereditary determinants. These hypotheses are considered some of the most brilliant insights in the history of biological science.

Particulate Inheritance

Mendel's results were clearly inconsistent with either the hypothesis of blending inheritance or the hypothesis of acquired characters. To explain the patterns that he observed, Mendel proposed a competing hypothesis called **particulate inheritance**. He maintained that the hereditary determinants for traits do not blend together or acquire new or modified characteristics through use. In fact, hereditary determinants maintain their integrity from generation to generation. Instead of blending together, they act like discrete entities or particles.

Mendel's hypothesis was the only way to explain the observation that phenotypes disappeared in one generation and reappeared intact in the next. It also represented a fundamental break with ideas that had prevailed for hundreds of years.

GENES, ALLELES, AND GENOTYPES   Today geneticists use the word **gene** to indicate the hereditary determinant for a trait. For example, the hereditary factor that determines whether the seeds of garden peas are round or wrinkled is referred to as the gene for seed shape.

Mendel also proposed that each individual can have two versions of any gene. Today different versions of the same gene are called **alleles**. Different alleles are responsible for the variation in the traits that Mendel studied. In the case of the gene for seed shape, one allele of this gene is responsible for the round form of the seed while another allele is responsible for the wrinkled form.

The alleles that are found in a particular individual are called the **genotype**. An individual's genotype has a profound effect on the phenotype—the physical traits.

TABLE 13.1 F₂ Results from Mendel's Monohybrid Reciprocal Cross Experiments*

Trait	Dominant Phenotype	Recessive Phenotype	Ratio
Seed shape	5474 round	1850 wrinkled	2.96 : 1
Seed color	6022 yellow	2001 green	3.01 : 1
Pod shape	882 inflated	299 constricted	2.95 : 1
Pod color	428 green	152 yellow	2.82 : 1
Flower color	705 purple	224 white	3.15 : 1
Flower and pod position	651 axial	207 terminal	3.14 : 1
Stem length	787 tall	277 dwarf	2.96 : 1

*Mendel pooled the results from the reciprocal crosses for each trait because the results were the same whether the dominant trait originated from the male parent or the female parent.
SOURCE: Mendel, G. 1866. Versuche über Pflanzen-hybriden. *Verhandlungen des naturforschenden Vereines in Brünn.* 4: 3–47.

The hypothesis that pea plants have two of each gene—either two of the same allele or two different alleles—was important because it gave Mendel a framework for explaining dominance and recessiveness. He proposed that some alleles are dominant and others are recessive. Recall that dominance and recessiveness determine which phenotype actually appears in an individual when two different alleles are present. In garden peas, the allele for round seeds is dominant; the allele for wrinkled seeds is recessive. Therefore, as long as one allele for round seeds is present, seeds are round. When both alleles present are for wrinkled seeds (thus no allele for round seeds is present), seeds are wrinkled.

These hypotheses explain why the phenotype for wrinkled seeds disappeared in the F₁ generation and reappeared in the F₂ generation. But why did round- and wrinkled-seeded plants exist in a 3:1 ratio in the F₂ generation?

THE PRINCIPLE OF SEGREGATION 🗝️ To explain the 3:1 ratio of phenotypes in F₂ individuals, Mendel reasoned that the two members of each gene pair must segregate—that is, separate—into different gamete cells during the formation of eggs and sperm in the parents. As a result, each gamete contains one allele of each gene. This idea is called the **principle of segregation**.

To show how this principle works, Mendel used a letter to indicate the gene for a particular trait. For example, he used upper-case *R* to symbolize a dominant allele for seed shape and lower-case *r* to symbolize a recessive allele for seed shape. (Notice that the symbols for genes are always italicized.)

Using this notation, Mendel could describe the genotype of the individuals in the round-seed pure line as *RR* (having two of the dominant allele). The genotype of the wrinkled-seed pure line is *rr* (two of the recessive allele). Because *RR* and *rr* individuals have two copies of the same allele, they are said to be **homozygous** for the seed-shape gene (*homo* is the Greek root for "same," while *zygo* means "yoked together"). Pure-line individuals always produce offspring with the same phenotype because they are homozygous—no other allele is present.

Figure 13.4a diagrams what happened to these alleles when Mendel crossed the *RR* and *rr* pure lines. According to Mendel's analysis, *RR* parents produce eggs and sperm that all carry the *R* allele, while *rr* parents produce gametes with the *r* allele only. When two gametes—one from each parent—are fused together, they create offspring with the *Rr* genotype. Such individuals, with two different alleles for the same gene, are said to be **heterozygous** (*hetero* is the Greek root for "different"). Because the *R* allele is dominant, all of these F₁ offspring produce round seeds.

Why do the two phenotypes appear in a 3:1 ratio in the F₂ generation? Mendel proposed that during gamete formation in the F₁ (heterozygous) individuals, the paired *Rr* alleles separate into different gamete cells. As a result, about half of the gametes carry the *R* allele and half carry the *r* allele (**Figure 13.4b**). During self-fertilization, a given sperm has an equal chance of fertilizing either an *R*-bearing egg or an *r*-bearing egg. The outcome of this situation is explained in a diagram called a Punnett square.

PREDICTING OFFSPRING GENOTYPES AND PHENOTYPES WITH A PUNNETT SQUARE Years after Mendel published his work, R. C. Punnett invented a straightforward technique for predicting the genotypes and phenotypes of the offspring of different crosses. To produce a **Punnett square**, follow these steps:

1. Write each of the unique gamete genotypes produced by one parent in a horizontal row along the top of the diagram.

2. Write each of the unique gamete genotypes produced by the other parent in a vertical column down the left side of the diagram.

(a) A cross between two **homozygotes**

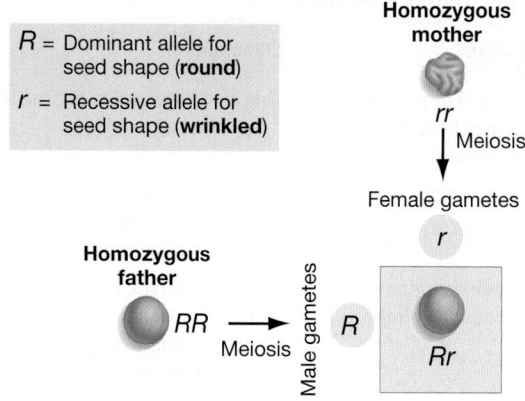

Offspring genotypes: All *Rr* (heterozygous)
Offspring phenotypes: All round seeds

(b) A cross between two **heterozygotes**

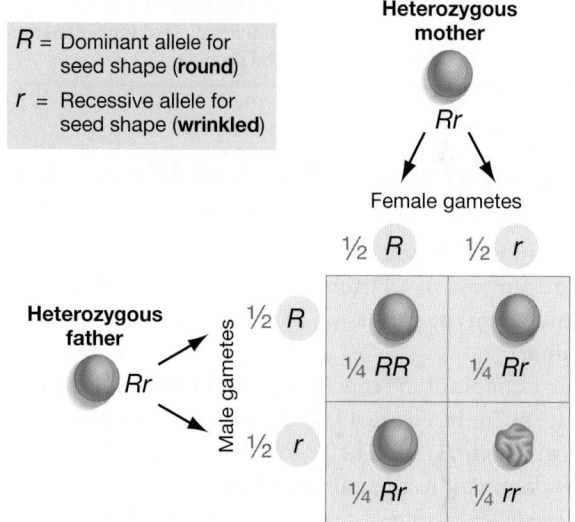

Offspring genotypes: ¼ *RR* : ½ *Rr* : ¼ *rr*
Offspring phenotypes: ¾ round : ¼ wrinkled

FIGURE 13.4 Mendel Analyzed the F₁ and F₂ Offspring of a Cross between Pure Lines. Notice that when you construct a Punnett square, you only need to list each unique type of gamete once at the head of a row or column. Therefore, for example, although the *RR* alleles segregate in the male parent of part (a), you only have to list one *R* gamete to represent the male's contribution, not two.

✔ **QUESTION** In constructing a Punnett square, does it matter whether the male or female gametes go on the left or across the top? Why or why not?

3. Draw empty boxes under the horizontal row of gametes and to the right of the vertical column of gametes.

4. Fill in each box with the parental gamete genotypes written at the top of the corresponding column and at the left of the corresponding row. This step produces the offspring genotypes that result from fusion of the parental gamete genotypes.

5. Predict the proportions or ratios of each offspring genotype and phenotype by tallying the offspring genotypes and

Mendel's Claims	Comments
1. Peas have two of each gene and thus may have two different alleles of the gene.	This also turns out to be true for many other organisms.
2. Alleles do not blend together.	The hereditary determinants maintain their integrity from generation to generation.
3. Each gamete contains one of each gene (one allele).	This is due to the principle of segregation—the members of each gene pair segregate during the formation of gametes.
4. Males and females contribute equally to the genotype of their offspring.	When gametes fuse, offspring acquire a total of two of each gene—one from each parent.
5. Some alleles are dominant to other alleles.	When a dominant allele and a recessive allele for the same gene are found in the same individual, that individual has the dominant phenotype.

resulting phenotypes produced in all the boxes. **BioSkills 13** in Appendix A explains why this tallying process works.

✔ If you understand these concepts, you should be able to state the purpose of a Punnett square and predict the phenotype and genotype ratios for a cross between *Yy* and *yy* pea individuals.

As an example of the concluding step in analyzing a cross, the Punnett square in Figure 13.4b predicts that 1/4 of the F$_2$ offspring will be *RR*, 1/2 will be *Rr*, and 1/4 will be *rr*. Because the *R* allele is dominant to the *r* allele, 3/4 of the offspring should be round seeded (the sum of the *RR* and the *Rr* offspring) and 1/4 should be wrinkled seeded (the *rr* offspring). These results are *exactly* what Mendel found in his experiments with peas. In the simplest and most elegant fashion possible, Mendel's interpretation explains the 3:1 ratio of round to wrinkled seeds observed in the F$_2$ offspring and the mysterious reappearance of the wrinkled seeds.

The term **genetic model** refers to a set of hypotheses that explains how a particular trait is inherited. **Table 13.2** summarizes Mendel's model for explaining the basic patterns in the transmission of traits from parents to offspring; the hypotheses it lists are sometimes referred to as Mendel's rules. They represent a radical break from the hypotheses of blending inheritance and inheritance of acquired characters that previously dominated scientific thinking about heredity.

13.3 Mendel's Experiments with Two Traits

Working with one trait at a time allowed Mendel to establish that blending inheritance does not occur. It also allowed him to infer that each pea plant had two of each gene and to recognize the principle of segregation.

Mendel's next step extended these results. The important question now was whether the principle of segregation holds true if individuals differ with respect to two traits, instead of just one. Do different genes segregate together, or independently?

The Dihybrid Cross

Mendel crossed a pure-line parent that produced round, yellow seeds with a pure-line parent that produced wrinkled, green seeds. According to his model, the F$_1$ offspring of this cross should be heterozygous for both genes. A mating between two such individuals—both heterozygous for two traits—is called a **dihybrid cross**.

Mendel's earlier experiments had established that the allele for yellow seeds was dominant to the allele for green seeds; these alleles were designated *Y* for yellow and *y* for green. As **Figure 13.5** indicates, two distinct possibilities existed for how the alleles of these two different genes—the gene for seed shape and the gene for seed color—would be transmitted to offspring.

- The first possibility was that the allele for seed shape and the allele for seed color present in each parent would separate from one another and be transmitted independently. This hypothesis is called independent assortment, because the two alleles would separate and be sorted into gametes independently of each other (Figure 13.5a).

- The second possibility was that the allele for seed shape and the allele for seed color would be transmitted to gametes together. This hypothesis can be called dependent assortment, because the transmission of one allele would depend on the transmission of another (Figure 13.5b).

As Figure 13.5 shows, the F$_1$ offspring of Mendel's mating are expected to have the dominant round and yellow phenotypes whether the different genes are transmitted together or independently. When Mendel did the cross and observed the F$_1$ individuals, this is exactly what he found. All of the F$_1$ offspring had round, yellow seeds.

The two hypotheses make radically different predictions, however, about what will be observed when the F$_1$ individuals are allowed to self-fertilize and produce an F$_2$ generation. If the different genes assort independently and combine randomly to form gametes, then each heterozygous parent should produce four different gamete genotypes, as illustrated in Figure 13.5a. A 4-row-by-4-column Punnett square results, and it predicts that there should be 9 different offspring genotypes and 4 phenotypes. Further, the yellow-round, green-round, yellow-wrinkled, and green-wrinkled phenotypes should be present in frequencies of 9/16, 3/16, 3/16, and 1/16, respectively. This is a ratio of 9:3:3:1.

On the other hand, if the alleles from each parent stay together, then a 2-row-by-2-column Punnett square would result and predict only three possible offspring genotypes and just two phenotypes, as Figure 13.5b shows. The hypothesis of dependent assortment predicts that F$_2$ offspring should be yellow-round or green-wrinkled, present in a ratio of 3:1.

Note that the Punnett squares make explicit predictions about the outcome of an experiment, based on a specific hypothesis

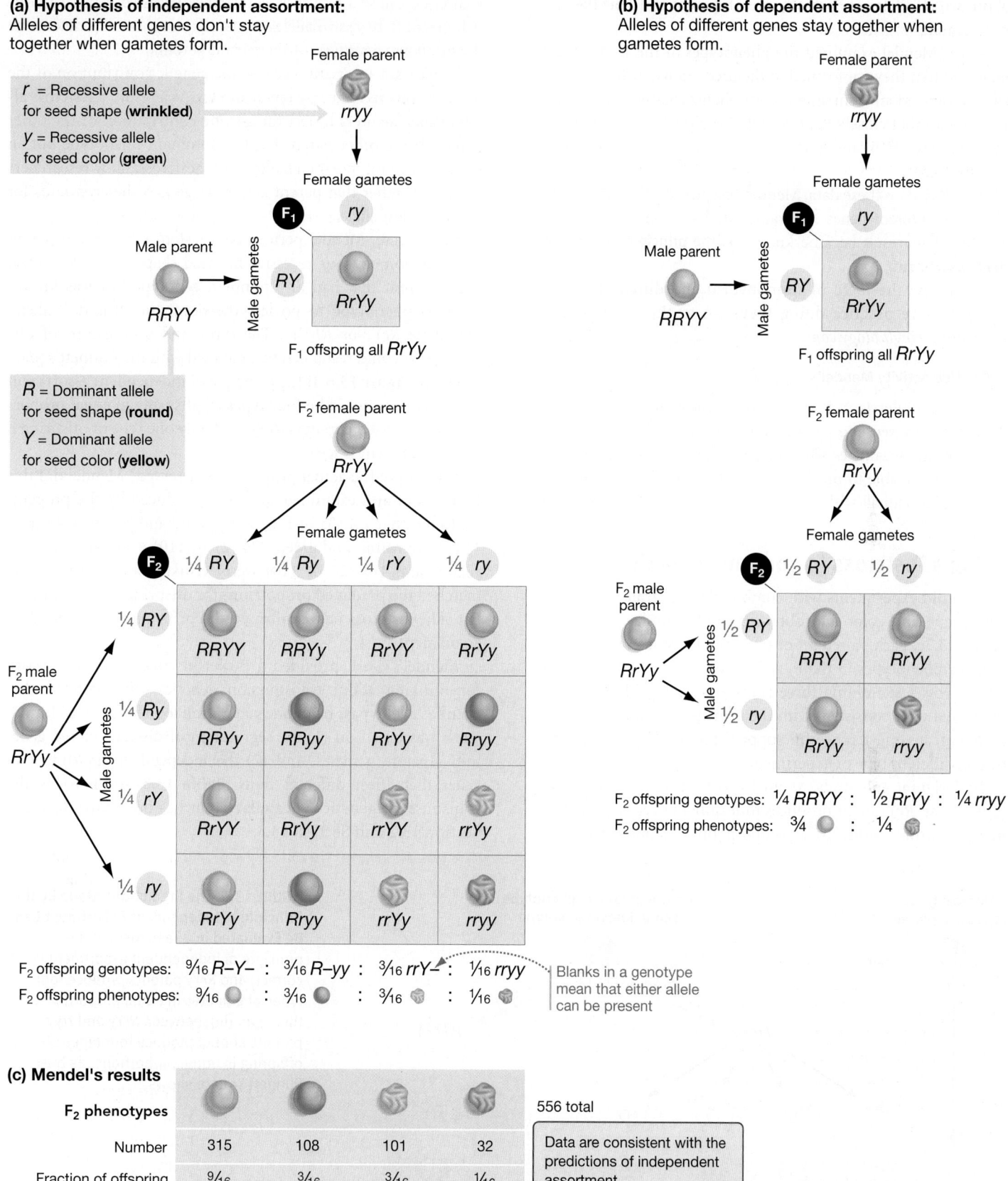

(a) Hypothesis of independent assortment: Alleles of different genes don't stay together when gametes form.

r = Recessive allele for seed shape (**wrinkled**)

y = Recessive allele for seed color (**green**)

R = Dominant allele for seed shape (**round**)

Y = Dominant allele for seed color (**yellow**)

Female parent
rryy
Female gametes
ry
F₁
Male parent
RRYY
RY
RrYy
F₁ offspring all RrYy

F₂ female parent
RrYy
Female gametes
F₂ ¼ RY   ¼ Ry   ¼ rY   ¼ ry

F₂ male parent
RrYy

	¼ RY	¼ Ry	¼ rY	¼ ry
¼ RY	RRYY	RRYy	RrYY	RrYy
¼ Ry	RRYy	RRyy	RrYy	Rryy
¼ rY	RrYY	RrYy	rrYY	rrYy
¼ ry	RrYy	Rryy	rrYy	rryy

F₂ offspring genotypes: 9/16 R–Y– : 3/16 R–yy : 3/16 rrY– : 1/16 rryy
F₂ offspring phenotypes: 9/16 : 3/16 : 3/16 : 1/16

Blanks in a genotype mean that either allele can be present

(b) Hypothesis of dependent assortment: Alleles of different genes stay together when gametes form.

Female parent
rryy
Female gametes
ry
F₁
Male parent
RRYY
RY
RrYy
F₁ offspring all RrYy

F₂ female parent
RrYy
Female gametes
F₂ ½ RY   ½ ry

F₂ male parent
RrYy

	½ RY	½ ry
½ RY	RRYY	RrYy
½ ry	RrYy	rryy

F₂ offspring genotypes: ¼ RRYY : ½ RrYy : ¼ rryy
F₂ offspring phenotypes: ¾ : ¼

(c) Mendel's results

F₂ phenotypes					556 total
Number	315	108	101	32	
Fraction of offspring	9/16	3/16	3/16	1/16	

Data are consistent with the predictions of independent assortment.

FIGURE 13.5 Mendel Analyzed the F₁ and F₂ Offspring of a Cross between Pure Lines for Two Traits. Each of two hypotheses predicted a different pattern for the outcome when alleles of different genes are transmitted to offspring: The alleles could be sorted into gametes independently of each other, or alleles from the same parent could be transmitted together, generation after generation.

about which alleles are present in each parent and how they are transmitted.

When Mendel examined the phenotypes of the F₂ offspring, he found that they conformed to the predictions of the hypothesis of independent assortment. Four phenotypes were present in frequencies that closely approximated the predicted frequencies of 9/16, 3/16, 3/16, and 1/16 and the predicted ratio of 9:3:3:1 (Figure 13.5c).

🔑 Based on these data, Mendel accepted the hypothesis that alleles of different genes are transmitted independently of one another. This result became known as the **principle of independent assortment.**

To review the logic of monohybrid and dihybrid crosses and get more practice doing them, go to the study area at *www.masteringbiology.com.*

(MB) **Web Activity** Mendel's Experiments

✔ If you understand the principle of independent assortment, it should make sense to you that an individual with the genotype *AaBb* produces gametes with the genotypes *AB, Ab, aB,* and *ab.* You should be able to predict the genotypes of the gametes produced by individuals with the genotypes *AABb, PpRr,* and *AaPpRr.*

Using a Testcross to Confirm Predictions

Mendel did experiments with combinations of traits other than seed shape and color and obtained results similar to those in Figure 13.5c. Each paired set of traits produced a 9:3:3:1 ratio of progeny phenotypes in the F₂ generation. He even did a limited set of crosses examining three traits at a time. Although all of these data were consistent with the principle of independent assortment, his most powerful support for the hypothesis came from a different type of experiment.

In designing this study, Mendel's goal was to test the prediction that an *RrYy* plant produces four different types of gametes in equal proportions. To accomplish this, Mendel invented a technique called a testcross. A **testcross** uses a parent that contributes only recessive alleles to its offspring, to help determine the unknown genotype of the second parent.

Testcrosses are useful because the genetic contribution of the homozygous recessive parent is known. As a result, a testcross allows experimenters to test the genetic contribution of the other parent. If the other parent has the dominant phenotype but an unknown genotype, the results of the testcross allow researchers to infer whether that parent is homozygous or heterozygous for the dominant allele.

In this case, Mendel performed a testcross between a parent that was homozygous recessive for seed shape and color (*rryy*) and a parent that had an unknown genotype but was known from its phenotype to possess the dominant *R* and *Y* alleles (could be *RrYy* or *RRYY*). The types and proportions of offspring that could result can be predicted with the Punnett square shown in **Figure 13.6.** If the principle of independent assortment is valid, there should be four types of offspring in equal proportions if the tested parent is *RrYy*, and only one type of offspring if the tested parent is *RRYY*.

What were the actual proportions observed? Mendel did this experiment and examined the seeds produced by the progeny. He found that 1/4 (31 of 110) were round and yellow, 1/4 (26 of 110) were round and green, 1/4 (27 of 110) were wrinkled and yellow, and 1/4 (26 of 110) were wrinkled and green, which matched the predicted proportions for offspring of an *RrYy* parent. The testcross had confirmed the principle of independent assortment.

Mendel's work provided a powerful conceptual framework for thinking about transmission genetics—the patterns that occur as alleles pass from one generation to the next. This framework was based on (1) the segregation of discrete, paired genes into separate gametes, and (2) the independent assortment of genes that affect different traits. **Table 13.3** summarizes the transmission genetics vocabulary that has been introduced in Section 13.2 and Section 13.3.

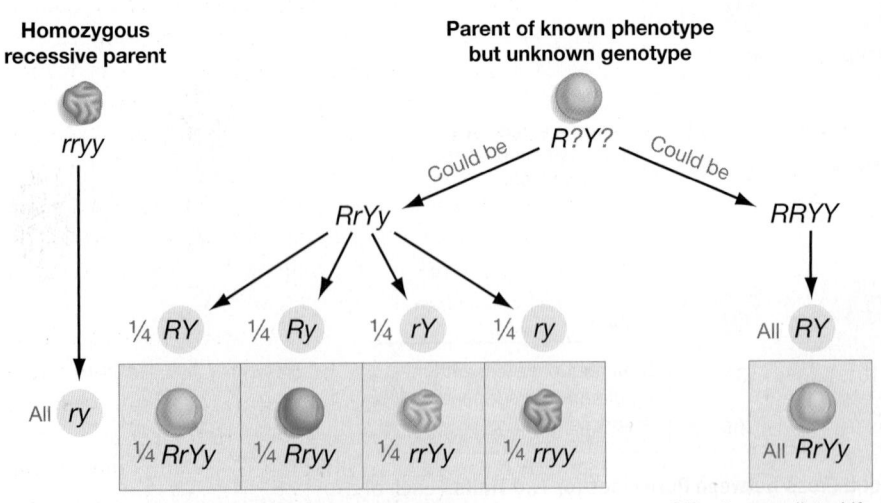

Homozygous recessive parent

rryy

Parent of known phenotype but unknown genotype

R?Y?

Could be — *RrYy*   Could be — *RRYY*

¼ *RY*   ¼ *Ry*   ¼ *rY*   ¼ *ry*   All *RY*

All *ry*

¼ *RrYy*   ¼ *Rryy*   ¼ *rrYy*   ¼ *rryy*   All *RrYy*

Offspring predicted if unknown parent is **heterozygous** at both genes

Offspring predicted if unknown parent is **homozygous dominant** at both genes

FIGURE 13.6 The Predictions Made by the Principle of Independent Assortment Can Be Evaluated in a Testcross. If the principle of independent assortment is correct, and *RrYy* parents produce four types of gametes in equal proportions, then a mating between *RrYy* and *rryy* parents should produce four types of offspring in equal proportions, as this Punnett square shows.

Term	Definition	Example or Comment
Gene	A hereditary factor that influences a particular trait.	This definition will change in later chapters (become more precise).
Allele	A version of a gene.	Diploid organisms have two alleles of each gene.
Genotype	A listing of the alleles in an individual.	In diploids, the genotype has two alleles of each gene; in haploids, one allele of each gene.
Phenotype	An individual's observable traits.	Influenced, but not dictated, by the genotype.
Homozygous	Having two of the same allele.	Refers to a particular gene.
Heterozygous	Having two different alleles.	Refers to a particular gene.
Dominant allele	An allele that produces the associated phenotype in heterozygotes.	Dominance does not imply high frequency or high fitness.
Recessive allele	An allele that produces the associated phenotype only in homozygotes.	Alleles appear to "recede" or disappear in heterozygotes.
Pure line	Individuals or populations that when crossed with individuals of the pure line, always produce offspring with that phenotype.	Pure-line individuals are homozygous for the gene in question.
Hybrid	Offspring from crosses between homozygous parents with different genotypes.	Offspring are heterozygous.
Reciprocal cross	A cross in which the phenotypes associated with the male and female in a prior cross are reversed.	If reciprocal crosses give identical results, the sex of the parent does not influence transmission of the trait.
Testcross	Experimental cross between a homozygous recessive individual and an individual with the dominant phenotype but an unknown genotype.	Usually used to determine whether a parent with a dominant phenotype is homozygous or heterozygous.

CHECK YOUR UNDERSTANDING

⊙━ If you understand that . . .

- Mendel discovered that individuals have two alleles of each gene, and that each gamete receives one of the two alleles present in a parent. This is the principle of segregation.
- Mendel found that alleles of different genes are transmitted to gametes independently of each other. This is the principle of independent assortment.
- The alleles that Mendel analyzed were either dominant or recessive, meaning heterozygous individuals had the dominant phenotype.

✔ You should be able to . . .

Use the genetic problems at the end of this chapter to practice the following skills:

1. Starting with parents of known genotypes, create and analyze Punnett squares to predict the genotypes and phenotypes that will occur in their F_1 and F_2 offspring, and then calculate the expected frequency of each genotype and phenotype. (Do **Genetics Problems 2, 6, 8, 10**.)

2. Given the outcome of a cross, infer the genotypes and phenotypes of the parents. (Do **Genetics Problems 3, 4, 13**.)

Answers are available in Appendix B.

The experiments you've just reviewed were brilliant in design, execution, and interpretation. Unfortunately, they were ignored for 34 years.

13.4 The Chromosome Theory of Inheritance

Historians of science debate why Mendel's work was overlooked for so long. It is probably true that his use of ratios and proportions were difficult for biologists of that time to understand and absorb. It may also be true that the theory of blending inheritance was so well entrenched that there was a tendency to dismiss his results as peculiar or unbelievable.

Whatever the reason, Mendel's work was not appreciated until 1900, when three groups of biologists, working with a wide variety of plants and animals, independently "discovered" Mendel's work and reached the same main conclusions.

The rediscovery of Mendel's work, 16 years after his death, ignited the young field of genetics. Mendel's experiments established the basic patterns of how traits are passed from parents to offspring. But what process is responsible for these patterns? Two biologists, working independently, came up with the answer. Walter Sutton and Theodor Boveri each realized that meiosis could be responsible for Mendel's rules. When this hypothesis was published in 1902, research in genetics exploded.

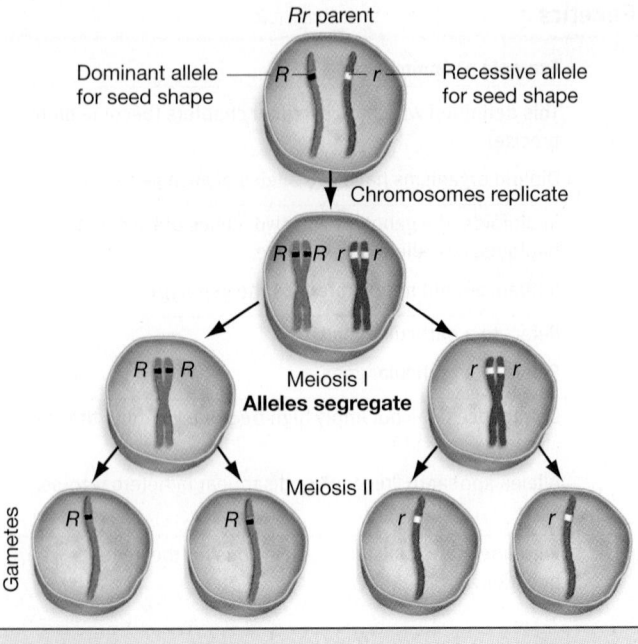

PRINCIPLE OF SEGREGATION: Each gamete carries only one allele for seed shape, because the alleles have segregated during meiosis.

FIGURE 13.7 Meiosis Is Responsible for the Principle of Segregation. The two members of a parent's gene pair segregate into different gametes, as Mendel hypothesized, because homologous chromosomes separate during meiosis I.

Meiosis Explains Mendel's Principles

Recall from Chapter 12 that meiosis precedes gamete formation. The details of the process were worked out late in the nineteenth century. What Sutton and Boveri grasped is that meiosis not only reduces chromosome number by half but also explains the principle of segregation and the principle of independent assortment.

The cell at the top of **Figure 13.7** illustrates Sutton and Boveri's central insight: Chromosomes are composed of Mendel's hereditary determinants, or genes. In this example, the gene for seed shape is shown at a particular position along a certain chromosome. This location is known as a **locus** ("place"; plural, **loci**). A genetic locus is the physical location of a gene.

The paternal and maternal chromosomes shown in Figure 13.7 happen to possess different alleles of the gene for seed shape: One allele specifies round seeds (*R*), while the other specifies wrinkled seeds (*r*).

The subsequent steps in Figure 13.7 show how these alleles segregate into different daughter cells during meiosis I, when homologous chromosomes separate. 🔑 The physical separation of alleles during anaphase of meiosis I is responsible for Mendel's principle of segregation.

Figure 13.8 follows the fates of the members of two different gene pairs—in this case, for seed shape and seed color—as meiosis proceeds. If the alleles for different genes are located on different nonhomologous chromosomes, they assort independently of

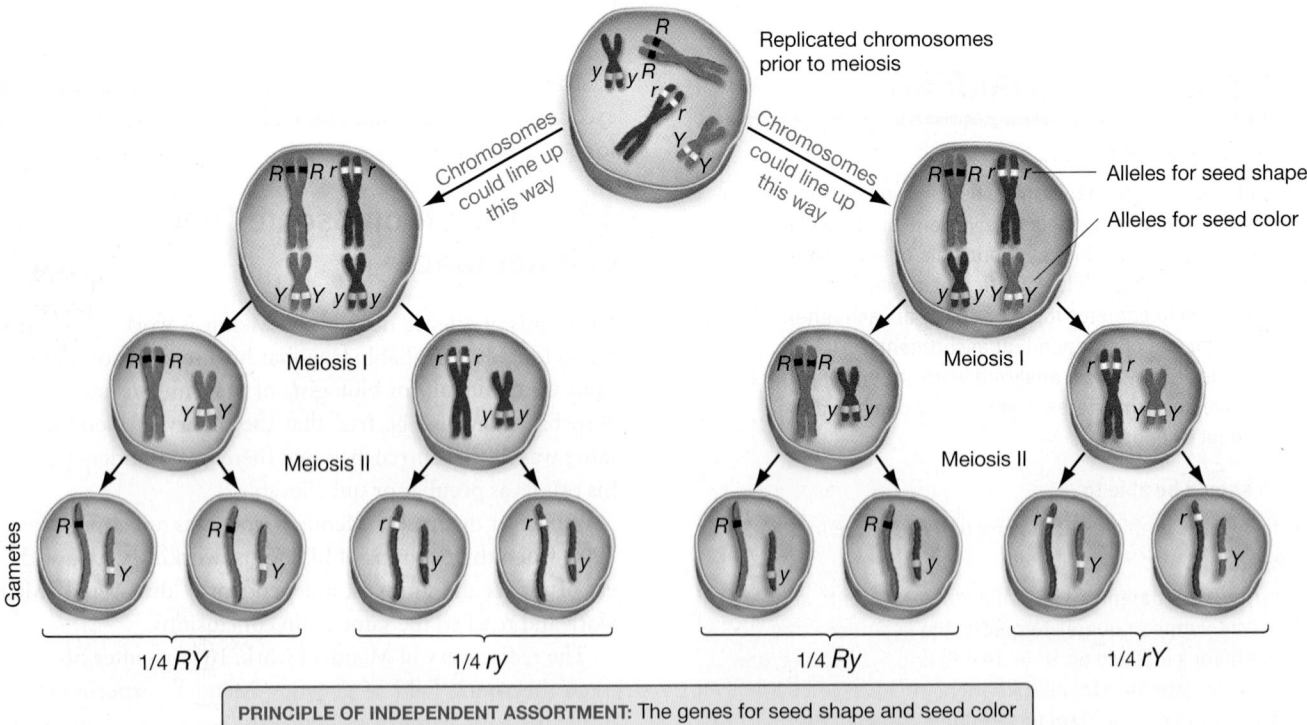

PRINCIPLE OF INDEPENDENT ASSORTMENT: The genes for seed shape and seed color assort independently, because they are located on different chromosomes.

FIGURE 13.8 Meiosis Is Responsible for the Principle of Independent Assortment. The genes for different traits assort independently, as Mendel hypothesized, because nonhomologous chromosomes assort independently during meiosis I. Maternal and paternal chromosomes are shown in different colors for clarity.

one another at meiosis I. Four types of gametes, produced in equal proportions, result. This is the physical basis of Mendel's principle of independent assortment.

Sutton and Boveri formalized these observations in the **chromosome theory of inheritance**. Like other theories in biology, the chromosome theory describes a predictable pattern—a set of observations about the natural world—and a process that explains the pattern. The chromosome theory states that Mendel's rules can be explained by the independent alignment and separation of homologous chromosomes at meiosis I.

To review why meiosis explains Mendel's findings, go to the study area at *www.masteringbiology.com*.

(MB) Web Activity The Principle of Independent Assortment

When Sutton and Boveri published their ideas, however, the hypothesis that chromosomes consist of genes was untested. What experiments confirmed that chromosomes contain genes?

Testing the Chromosome Theory

During the first decade of the twentieth century, an unassuming insect rose to prominence as a model organism for testing the chromosome theory of inheritance. This organism—the fruit fly *Drosophila melanogaster*—has been at the center of genetic studies ever since (see **BioSkills 14** in Appendix A).

Drosophila melanogaster has all the attributes of a useful model organism for experimental studies in genetics: small size, ease of rearing in the lab, a short generation time (about 10 days), and abundant offspring (up to a few hundred per mating). The elaborate external anatomy of this insect also makes it possible to identify interesting phenotypic variation among individuals (**Figure 13.9a**).

Drosophila was adopted as a model organism by Thomas Hunt Morgan and his students. But because *Drosophila* is not a domesticated species like the garden pea, Morgan was not familiar with common phenotypic variants such as Mendel's round and wrinkled seeds. Consequently, an early goal of Morgan's research was simply to find and characterize individuals with different phenotypes. In these studies, the most common phenotype for each trait was referred to as **wild type**.

THE WHITE-EYE MUTANT At one point, Morgan discovered a male fly that had white eyes rather than the wild-type red eyes (**Figure 13.9b**). This individual had a discrete and easy-to-recognize phenotype different from the normal phenotype.

Morgan inferred that the white-eyed phenotype resulted from a **mutation**—a change in a gene (in this case, a gene that affects eye color). Individuals with white eyes (or other traits attributable to mutation) are referred to as **mutants**.

To explore how the white-eye trait is inherited in fruit flies, Morgan mated a red-eyed female fly with the mutant white-eyed male fly. All of the F_1 progeny had red eyes. But when Morgan did the reciprocal cross, by mating white-eyed females to red-eyed males, he got a different result: All F_1 females had red eyes, but all F_1 males had white eyes.

(a) The fruit fly *Drosophila melanogaster*

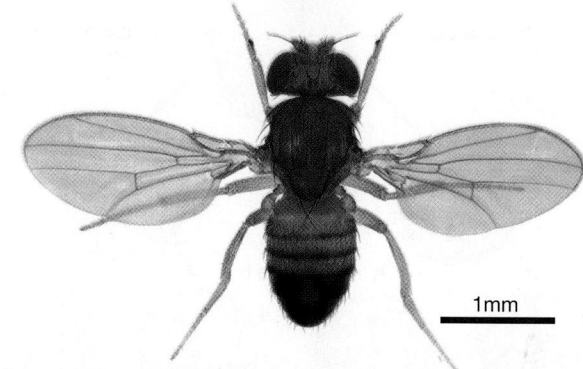

1mm

(b) Eye color is a variable trait.

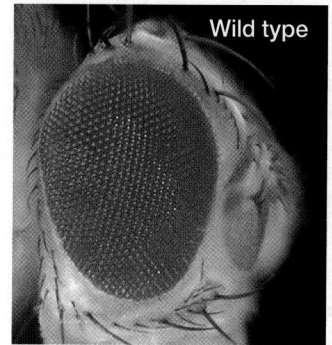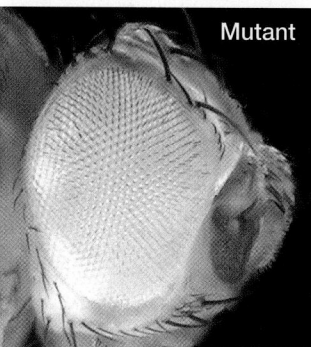

Wild type   Mutant

FIGURE 13.9 The Fruit Fly *Drosophila melanogaster* Is an Important Model Organism in Genetics.

Recall that Mendel's reciprocal crosses had always given results that were similar to each other. But Morgan's reciprocal crosses did not. The experiment suggested a definite relationship between the sex of the progeny and the inheritance of eye color. What was going on?

THE DISCOVERY OF SEX CHROMOSOMES Nettie Stevens began studying the karyotypes of insects about the time that Morgan began his work with *Drosophila*. In the beetle *Tenebrio molitor*, she noticed a striking difference in the chromosome complements of males and females. In females of this species, diploid cells contain 20 large chromosomes. But diploid cells in males contain 19 large chromosomes and 1 small one.

Stevens called the small chromosome the Y chromosome. This Y chromosome paired with one of the large chromosomes at meiosis I, which had already been named the X chromosome. The X and Y were different in size and shape, but they acted like homologs during meiosis. Later work showed that even though X and Y chromosomes contain different genes, they have regions that are similar enough to lead to proper pairing during prophase of meiosis I. The X and Y are now called sex chromosomes. Female beetles are XX; males are XY.

SEX LINKAGE AND THE CHROMOSOME THEORY *Drosophila* females, like *Tenebrio* females, have two X chromosomes; male fruit flies carry an X and a Y, just as *Tenebrio* males do. As a result,

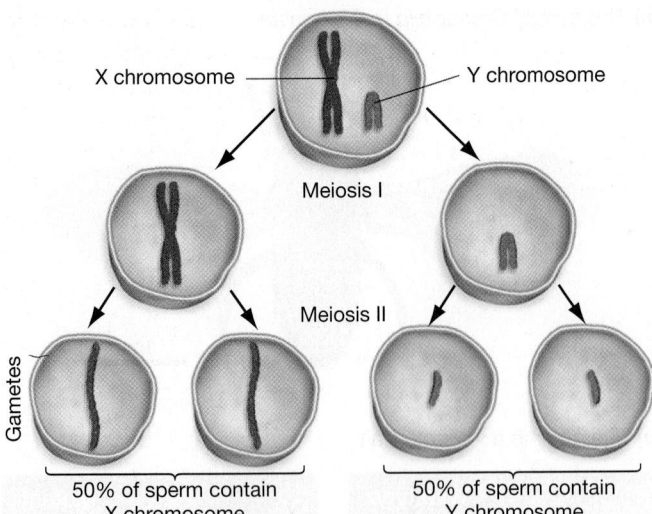

FIGURE 13.10 Sex Chromosomes Pair during Meiosis I, Then Segregate to Form X-Bearing and Y-Bearing Gametes. Sex chromosomes synapse at meiosis I in male fruit flies, even though the X and Y chromosomes differ in size and shape. No crossing over occurs between the X and Y in fruit flies. Thus, half the sperm cells that result from meiosis bear an X chromosome; half have a Y chromosome.

half the gametes produced by a male fruit fly should carry an X chromosome; the other half, a Y chromosome (**Figure 13.10**).

Morgan realized that the transmission pattern of the X chromosome in males and females explained the results of his reciprocal crosses. Specifically, he proposed that the gene for white eye color in fruit flies is located on the X chromosome and that the Y chromosome does not carry an allele of this gene.

This situation is described as **X-linked inheritance**, or simply **X-linkage**. Correspondingly, a gene residing on the Y chromosome is said to have **Y-linked inheritance**, or **Y-linkage**. The general term for such inheritance (of genes on either sex chromosome) is **sex-linked inheritance**, or **sex-linkage**.

According to the hypothesis of X-linkage, a female fruit fly has two copies of the gene that specifies eye color because she has two X chromosomes. One of these chromosomes came from her female parent, and the other from her male parent. A male, in contrast, has only one copy of the eye-color gene because he has only one X chromosome, inherited from his mother.

The Punnett squares in **Figure 13.11** show that Morgan's experimental results can be explained if the gene for eye color is located on the X chromosome, and if the allele for red color is dominant to the allele for white color. In this figure, the allele for red eyes is denoted X^W while the allele for white eyes is denoted X^w. The Y chromosome present in males is simply designated by Y. Using this notation,[1] the genotypes used in the ex-

---

[1]Scientific papers on fruit fly genetics use a different notation. The wild-type allele is designated with a superscript +, and no X is used for X-linked traits. The red-eye allele, for example, is denoted w^+; the white-eye allele w. The notation used here is simplified for use in introductory courses and conforms to conventions used in human genetics.

periment are written as $X^W X^W$ for red-eyed females; $X^w Y$ for white-eyed males; $X^w X^w$ for white-eyed females; and $X^W Y$ for red-eyed males. If you study the offspring genotypes, you should see that the results predicted by the hypothesis of X-linkage match the observed results.

When reciprocal crosses give different results, such as those illustrated in Figure 13.11, it is likely that the gene in question is located on a sex chromosome. Recall from Chapter 12 that non-sex chromosomes are called autosomes. Genes on non-sex chromosomes are said to show **autosomal inheritance**.

Morgan's discovery of X-linked inheritance carried an even more fundamental message. In *Drosophila*, the gene for white eye color is clearly correlated with inheritance of the X chromosome. This correlation was important evidence in support of the hypothesis that chromosomes contain genes. The discovery of X-linked inheritance convinced most biologists that the chromosome theory of inheritance was correct.

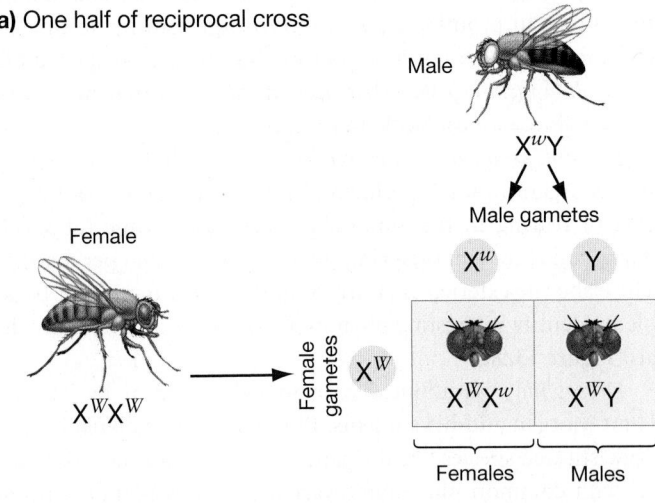

(a) One half of reciprocal cross

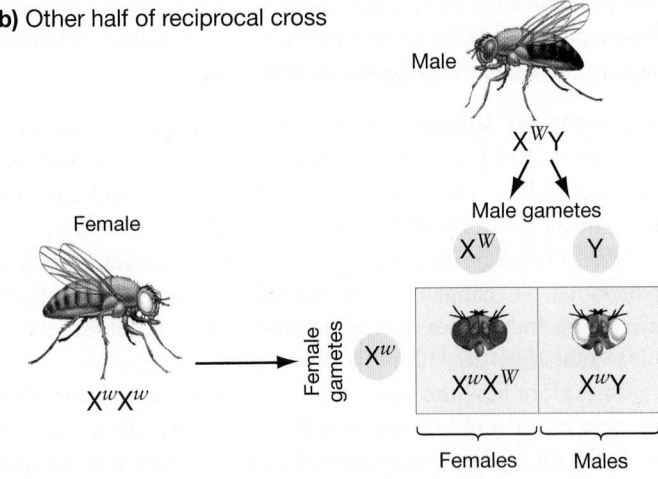

(b) Other half of reciprocal cross

FIGURE 13.11 Reciprocal Crosses Confirm That Eye Color in *Drosophila* Is an X-Linked Trait. When Morgan crossed red-eyed females with white-eyed males and then crossed white-eyed females with red-eyed males, he observed strikingly different results.

CHECK YOUR UNDERSTANDING

13.5 Extending Mendel's Rules

Biologists point out that Mendel analyzed the simplest possible genetic system. The traits that he was studying were not sex-linked. Moreover, they were influenced by just two alleles of a single gene, and each allele was completely dominant or recessive.

With these genes, Mendel was able to discover the most fundamental rules of inheritance. Investigating simple model systems is an extremely important research strategy in biological science. Researchers almost always choose to analyze the simplest situation possible before going on to explore more complicated systems. Mendel probably would have failed, as so many others had done before him, had he been trying to analyze more complex patterns of inheritance.

Once Mendel's work was rediscovered, researchers began to analyze traits and alleles whose inheritance was more complicated. If experimental crosses produced F_2 progeny that did not conform to the expected 3:1 or 9:3:3:1 ratios, researchers had a strong hint that something interesting was going on. The discovery of sex-linkage is a prominent example. How can other traits that don't appear to follow Mendel's rules contribute to a more complete understanding of heredity?

Linkage: What Happens When Genes Are Located on the Same Chromosome?

Once the chromosome theory had been tested and supported, biologists began to reevaluate Mendel's principle of independent assortment. The key issue was that genes should not undergo independent assortment if they are located on the same chromosome.

The physical association among genes on the same chromosome is called **linkage**. Notice that the terms linkage and sex-linkage are different in meaning. If two or more genes are linked, it means that they are located on the same chromosome. If a single gene is sex-linked, it means that it is located on a sex chromosome.

The first examples of linked genes were those on the X chromosome of fruit flies. After Morgan established that the white-eye gene was located on *Drosophila*'s X chromosome, he and colleagues established that one of the several genes that affects body color is also located on the X. Red eyes and gray body are the wild-type phenotypes in this species; white eyes and a yellow body occur as rare mutant phenotypes. The alleles for red eyes (X^W) and gray body (X^Y) are dominant to the alleles for white eyes (X^w) and yellow body (X^y). (Be sure not to confuse the notation for the Y chromosome in males, Y, with the yellow body allele, X^y.)

LINKED GENES AS AN EXCEPTION TO INDEPENDENT ASSORTMENT Because linked genes are physically part of the same chromosome, it is logical to predict that they should always be transmitted together during gamete formation. Stated another way, linked genes should violate the principle of independent assortment.

Recall from Section 13.4 that independent assortment is observed when genes are on different chromosomes, because the alleles of unlinked genes segregate to gametes independently of one another during meiosis I. But when genes are on the same chromosome, their alleles are carried to gametes together.

Figure 13.12 shows that a female fruit fly with one X chromosome carrying the white eye and gray body alleles, written X^{wY}, and with a second X chromosome carrying the red eye and yellow body alleles, written X^{Wy}, would be expected to generate just two classes of gametes in equal numbers during meiosis, instead of the four classes that are predicted under the principle of independent assortment. Is this what actually occurs?

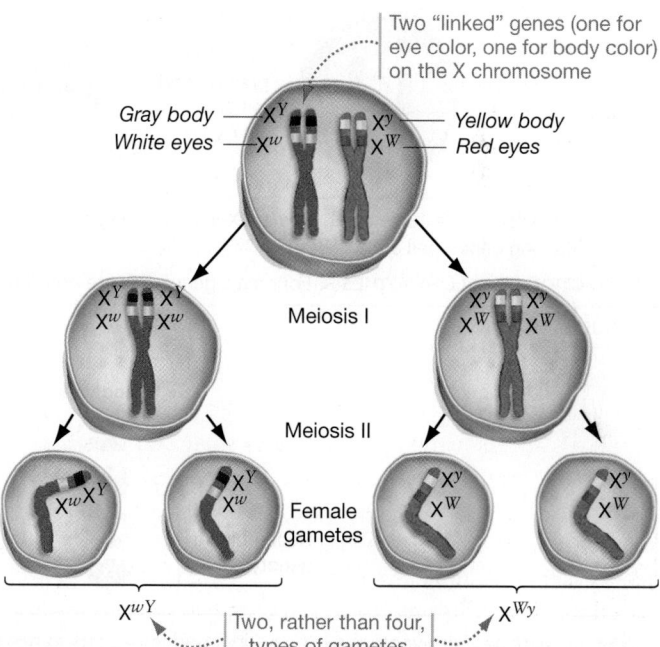

FIGURE 13.12 Linked Genes Are Inherited Together. If the *w* and *y* genes were found on different chromosomes, then this female would generate four different types of gametes instead of just two types as shown here.

✓ **EXERCISE** List the four genotypes that would be generated if the white and yellow genes were not linked.

THE ROLE OF CROSSING OVER To determine whether linked traits behave as predicted, Morgan performed crosses like the one described in the "Experimental setup" section of **Figure 13.13**. In this case, $X^{wY}X^{Wy}$ females mated with $X^{wY}Y$ males.

The Results table in Figure 13.13 summarizes the phenotypes and genotypes observed in the male offspring of this experimental cross.

- Most of these males carried an X chromosome with one of the two combinations of alleles found in their mothers: X^{wY} or X^{Wy}. In these individuals, the *white* and *yellow* alleles did not segregate independently.

- A small percentage of males had novel phenotypes and genotypes: X^{wy} and X^{WY}. Morgan referred to these individuals as **recombinant** because the combination of alleles on their X chromosome was different from the combinations of alleles present in the parental generation.

To explain this result, Morgan proposed that gametes with new, recombinant genotypes were generated when crossing over occurred during prophase of meiosis I in the females.

Recall from Chapter 12 that crossing over involves a physical exchange of segments from homologous chromosomes. Crossing over occurs at least once in every synapsed pair of homologous chromosomes, and usually multiple times. (Male fruit flies are an exception to this rule. For unknown reasons, no crossing over occurs in male fruit flies.)

As **Figure 13.14** shows, a crossing-over event occurred somewhere between the *white* and *yellow* genes in the $X^{wY}X^{Wy}$ females. The recombinant chromosomes that resulted would have the genotypes X^{wy} and X^{WY} (see the middle of the bottom row of the figure). If these chromosomes ended up in a male offspring, they would produce individuals with yellow bodies and white eyes, along with individuals with gray bodies and red eyes. This is exactly what Morgan observed.

As **Box 13.1** on page 246 explains, data on the percentage of recombinant offspring that occur in crosses like the one diagrammed in Figure 13.14 can be used to estimate the location of genes, relative to one another, on the same chromosome. Data on the frequency of crossing over can be used to create a **genetic map**—a diagram showing the relative positions of genes along a particular chromosome.

EXPERIMENT

QUESTION: Will genes undergo independent assortment if they are on the same chromosome?

LINKAGE HYPOTHESIS: Linked genes will violate the principle of independent assortment.

NULL HYPOTHESIS: Linked genes will adhere to the principle of independent assortment.

EXPERIMENTAL SETUP:

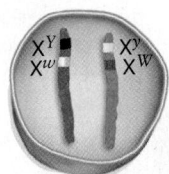
Red-eyed gray-bodied female
$X^{wY}X^{Wy}$

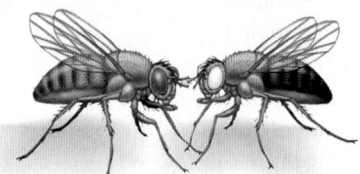

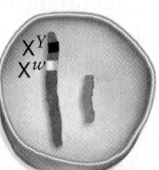
White-eyed gray-bodied male
$X^{wY}Y^{Wy}$

PREDICTION: Because these two genes are X-linked, male offspring will have only one copy of each gene, from their mother; the two possible male offspring genotypes are $X^{wY}Y$ and $X^{Wy}Y$

PREDICTION OF NULL HYPOTHESIS: Four male genotypes are possible ($X^{wY}Y : X^{Wy}Y : X^{wy}Y : X^{WY}Y$) and will occur with equal frequency.

RESULTS:

Male offspring

	Phenotype	Genotype	Number	
		$X^{wY}Y$	4292	
		$X^{Wy}Y$	4605	Four male genotypes were observed (rather than two), but not the equal frequencies predicted by independent assortment
Recombinant genotypes		$X^{wy}Y$	86	
		$X^{WY}Y$	44	

CONCLUSION: Neither hypothesis is fully supported. Independent assortment does not apply to linked genes—linked genes segregate together except when crossing over and genetic recombination have occurred.

FIGURE 13.13 Linked Genes Are Inherited Together Unless Recombination Occurs.

SOURCE: Morgan, T. H. 1911. An attempt to analyze the constitution of the chromosomes on the basis of sex-limited inheritance in *Drosophila*. *Journal of Experimental Zoology* 11: 365–414.

✔**QUESTION** Why didn't they observe equal numbers of white-eyed, yellow-bodied males and red-eyed, gray-bodied males?

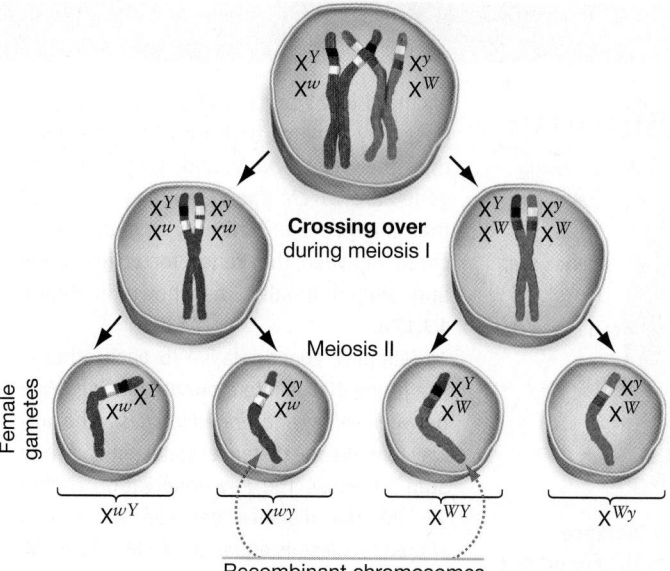

FIGURE 13.14 Genetic Recombination Results from Crossing Over. To explain the results in Figure 13.13, Morgan proposed that crossing over occurred between the *w* gene and the *y* gene in a small percentage of F₁ females during meiosis I. The recombinant chromosomes that resulted would produce the recombinant phenotypes observed in F₂ males.

The take-home message of Morgan's experiments is simple: Linked genes are inherited together unless crossing over occurs between them. When crossing over takes place, genetic recombination occurs. Linkage is an important exception to Mendel's rules.

Do Heterozygotes Always Have a Dominant or Recessive Phenotype?

The terms dominant and recessive describe which phenotype is observed when two different alleles of a gene occur in the same individual. In all seven traits that Mendel studied, only the phenotype associated with one allele—the "dominant" one—appeared in heterozygous individuals.

Not all combinations of alleles produce a completely dominant or recessive phenotype, however. In many cases, the actual phenotypes observed in heterozygous individuals conflict with the genotype and phenotype ratios that Mendel observed.

INCOMPLETE DOMINANCE Consider the flowers called four-o'-clocks, pictured in **Figure 13.15a**. In this species, biologists have developed a pure line that has purple flowers and a pure line that has white flowers. When individuals from these strains are mated, all of their offspring are lavender (**Figure 13.15b**). In Mendel's peas, crosses between purple- and white-flowered parents produced all purple-flowered offspring. Why the difference?

Biologists answered this question by examining the phenotypes of F₂ four-o'-clocks. These are the progeny of self-fertilization in lavender-flowered F₁ individuals. Of the F₂ plants, 1/4 have purple flowers, 1/2 have lavender flowers, and 1/4 have white flowers. This 1:2:1 ratio of phenotypes is unlike any we have seen to date, but it exactly matches the 1:2:1 ratio of genotypes

(a) Flower color is variable in four-o'clocks.

(b) Incomplete dominance in flower color

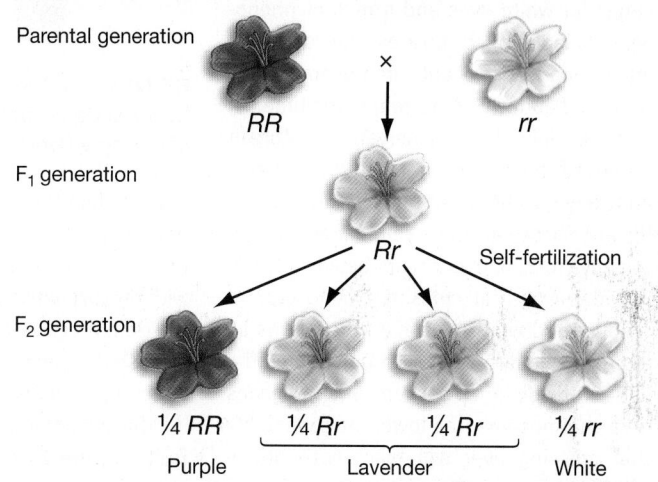

FIGURE 13.15 When Incomplete Dominance Occurs, Heterozygotes Have Intermediate Phenotypes. This cross is explained by hypothesizing that a single gene influences flower color, with alleles *R* and *r*, exhibiting incomplete dominance.

that is produced when flower color is controlled by one gene with two alleles.

To convince yourself that this explanation is sound, study the genetic model shown in Figure 13.15. According to the diagram, the inheritance of flower color genotypes in four-o'clocks and peas is identical, but the four-o'clock alleles show incomplete dominance rather than complete dominance.

When **incomplete dominance** occurs, heterozygotes have an intermediate phenotype. In the case of four-o'clocks, neither purple nor white alleles dominate. Instead, the F₁ progeny—all heterozygous—show a phenotype intermediate between the two parental strains.

Incomplete dominance illustrates an important general point: Dominance is not necessarily an all-or-none phenomenon.

CODOMINANCE Many alleles show a relationship called **codominance**. When codominance occurs, heterozygotes have the phenotype associated with each individual allele.

As an example, consider the ABO blood group in humans. The gene in question alters the polysaccharide attached to a glycoprotein (see Chapter 5) found in the plasma membranes of red blood cells. Different alleles of the *I* gene lead to the production

BOX 13.1   QUANTITATIVE METHODS: Linkage

In experiments like the one diagrammed in Figure 13.13, researchers calculate the recombination frequency as the number of offspring with recombinant phenotypes divided by the total number of offspring. With crosses involving the X-linked traits of white eyes and yellow bodies, about 1.4 percent of offspring have recombinant phenotypes and genotypes.

But in crosses with different pairs of X-linked traits, the fraction of recombinants varied. In crosses of fruit flies with X-linked genes for white eyes and a mutant phenotype called singed bristles, for example, males with recombinant chromosomes were produced about 19.6 percent of the time.

To explain these observations, Morgan proposed that genes are located in a linear array along a chromosome, and the physical distance between genes determines how frequently crossing over occurs between them. His idea was that crossing over occurs at random and can take place at locations all along the length of the chromosome. The shorter the distance between any two genes on a chromosome, the lower the probability that crossing over will take place somewhere in between (**Figure 13.16**).

A. H. Sturtevant, an undergraduate who was studying with Morgan, realized that he could figure out where genes on the same chromosome are in relation to each other based on the frequency of recombinants between various pairs. He set out to create a genetic map.

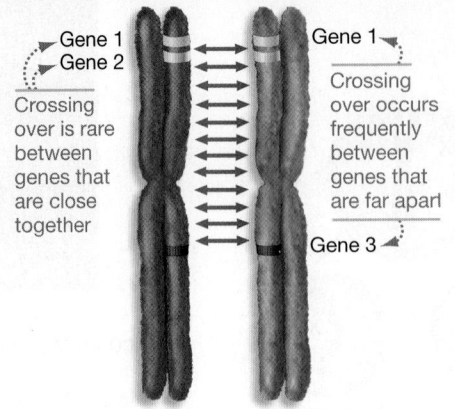

FIGURE 13.16 The Physical Distance between Genes Determines the Frequency of Crossing Over.

To define the unit of distance on his genetic map, Sturtevant used the percentage of offspring that have recombinant phenotypes with respect to two genes. He called this unit the centiMorgan (cM). One map unit, or 1 cM, represents the physical distance that produces 1 percent recombinant offspring.

The eye-color and bristle-shape genes of fruit flies are 19.6 cM apart on the X chromosome, because recombination between these genes results in 19.6 percent recombinant offspring, on average. The genes for yellow body and white eye color, in contrast, are just 1.4 cM apart.

Where is the yellow-body gene relative to the singed-bristles gene? Recombinants

occurred in 21 percent of the gametes produced by females that are $X^{ys}X^{ys}$, meaning that the yellow-body and singed-bristles genes are 21 cM apart. Sturtevant inferred that the gene for white eyes must be located between the genes for yellow body and singed bristles, as shown in **Figure 13.17a.**

Mapping genes relative to each other is like fitting pieces into a puzzle: placing white between yellow and singed bristles is the only way to make the distances between each pair sum correctly. The key observation is that 21 cM—the distance between yellow and singed bristles—is equal to 1.4 cM + 19.6 cM, or the sum of the distances between yellow and white and white and singed bristles.

Mapping is more complex when genes are farther apart. For example, crossing over occurs 50 percent of the time between genes that are 50 cM apart, making recombination indistinguishable from independent assortment. Double crossovers also become more common. To cope, researchers map genes that are relatively close, then map them to other nearby genes to gradually move down the chromosome.

Figure 13.17b provides a partial genetic map of the X chromosome in Drosophila melanogaster, along with the data on which the map positions are based. Using this logic and similar data, Sturtevant assembled the first genetic map.

(a) Mapping genetic distance

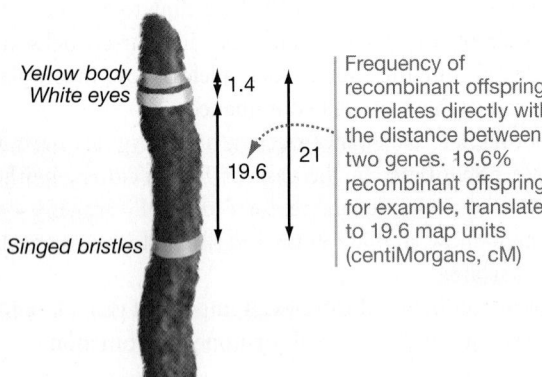

Frequency of recombinant offspring correlates directly with the distance between two genes. 19.6% recombinant offspring, for example, translates to 19.6 map units (centiMorgans, cM)

(b) Constructing a genetic map

% Frequency of crossing over between some genes on the X chromosome of fruit flies		
	Miniature Wings	Ruby Eyes
Yellow body	36.1	7.5
White eyes	34.7	6.1
Singed bristles	15.1	13.5
Miniature wings	—	28.6

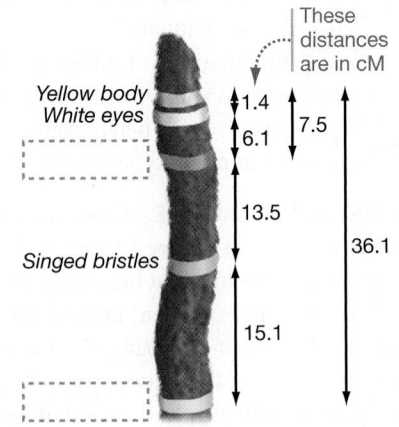

FIGURE 13.17 The Locations of Genes Can Be Mapped by Analyzing the Frequency of Recombination.
(a) The yellow body gene is known to be on the end of the fruit fly X chromosome. To explain the recombination frequencies observed in experimental crosses, the yellow body, white eyes, and singed bristles genes must be in the locations shown here. **(b)** A partial genetic map of the X chromosome in fruit flies.

✔**EXERCISE** In part (b), label the orange and blue genes. (Which is ruby and which is miniature wings?)

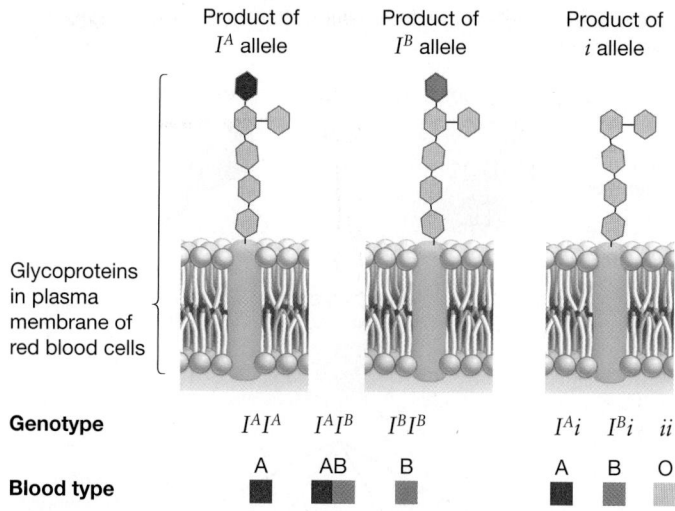

Glycoproteins in plasma membrane of red blood cells

Genotype	$I^A I^A$	$I^A I^B$	$I^B I^B$	$I^A i$	$I^B i$	ii
Blood type	A	AB	B	A	B	O

FIGURE 13.18 Phenotypes Produced by Alleles Responsible for ABO Blood Types. Alleles I^A and I^B produce a codominant phenotype when paired with each other in heterozygotes, but both produce a dominant phenotype when paired with allele i.

of membrane glycoproteins with different polysaccharides (**Figure 13.18**). $I^A I^A$ individuals have blood type A—meaning that their red blood cells only have the A-type glycoprotein in their membranes. Similarly, $I^B I^B$ individuals have blood type B. But $I^A I^B$ individuals have blood type AB. The phenotypes associated with both alleles are present because cells have both glycoproteins on their surfaces. The alleles are codominant.

There is also an additional allele of this gene called i. $I^A i$ individuals have blood type A; $I^B i$ individuals have blood type B; ii individuals have blood type O—meaning that they lack both A and B glycoproteins on the surfaces of the red blood cells. This is striking. A and B produce a codominant phenotype when paired with each other in heterozygotes, but both produce a dominant phenotype when paired with i. Dominance relationships vary among alleles.

How Many Alleles and Phenotypes Exist?

Mendel worked with a total of seven traits and just 14 alleles—two for each trait. In most populations, however, it's not unusual to find dozens of alleles of a single gene. The existence of more than two alleles of the same gene is known as **multiple allelism**.

The ABO blood group in humans is multiallelic, because most populations have the I^A, I^B, and i alleles. As a more dramatic example, consider the gene for the β-globin protein in humans. This protein makes up part of hemoglobin, which carries oxygen from the lungs to tissues. Biologists have now identified over 500 different alleles of the β-globin gene. Many of these alleles are associated with distinctive phenotypes. Some alleles produce polypeptides with normal oxygen-carrying capacity, while others lead to reduced oxygen-carrying capacity and various types of anemia. Still other β-globin alleles are associated with adaptation to living at high altitudes, decreased stability at high temperatures, or resistance to the parasites that infect red blood cells and cause malaria.

When more than two distinct phenotypes are present in a population due to multiple allelism, the trait is **polymorphic** ("many-formed"). Oxygen-carrying capacity in humans is a highly polymorphic trait, because many alleles exist for the β-globin gene.

If you were studying the inheritance of β-globin type in humans by tracking the results of a large number of matings, multiple allelism and polymorphism would make it extremely unlikely to observe the 3:1 ratios that Mendel did. Instead, you'd observe many more than two phenotypes, in unpredictable proportions.

Does Each Gene Affect Just One Trait?

As far as is known, the alleles that Mendel analyzed affect just a single trait. The gene for seed color in garden peas, for example, does not appear to affect other aspects of the individual's phenotype. In contrast, many cases have been documented in which a single allele affects a wide variety of traits.

A gene that influences many traits, rather than just one trait, is said to be **pleiotropic** ("more-turning"). The gene responsible for **Marfan syndrome** in humans, called *FBN1*, is a good example. Although current research suggests that just a single gene is involved, individuals with Marfan syndrome exhibit a wide array of phenotypic effects: increased height, disproportionately long limbs and fingers, an abnormally shaped chest, and potentially severe heart problems. A large percentage of these individuals also suffer from problems with their backbone. The gene associated with Marfan syndrome is pleiotropic.

Pleiotropy is common. In many cases a change in a single allele affects more than one trait.

Are Phenotypes Determined by Genes?

After analyzing the results of Mendel's experiments, it would be tempting to conclude that *R* alleles dictate that seeds are round and *T* alleles dictate that individual plants are tall—that there is a strict correspondence between alleles and phenotypes.

It's important to recognize, though, that when Mendel analyzed height in his experiments, he ensured that each plant received a similar amount of sunlight and grew in similar soil. This was important because even individuals with alleles for tallness will be stunted if they are deprived of nutrients, sunlight, or water—so much so that they look similar to individuals with alleles for dwarfing. For Mendel to analyze the hereditary determinants of height, he had to control the environmental determinants of height. Let's consider how two aspects of the environment affect phenotypes: (1) the individual's physical surroundings and (2) the alleles present at other gene loci.

THE PHYSICAL ENVIRONMENT EFFECTS PHENOTYPES The phenotypes produced by most genes and alleles are strongly affected by the individual's physical environment. Consequently, an individual's phenotype is often as much a product of the physical environment as it is a product of the genotype. To capture this point, biologists refer to the combined effect of genes and environment as gene-by-environment interaction.

Gene-by-environment interactions have a profound effect on how physicians treat people with the genetic disease **phenylketonuria** (**PKU**). These individuals lack an enzyme that helps convert the amino acid phenylalanine to the amino acid tyrosine. As a result, phenylalanine and a related molecule, phenylpyruvic acid, accumulate in their bodies. The molecules interfere with the development of the nervous system and produce profound mental retardation. But if PKU individuals are identified at birth and placed on a low-phenylalanine diet, then they develop normally. In many countries, newborns are routinely tested for the defect.

PKU is a genetic disease, but it is neither inevitable nor invariant. Through a simple change in their environment (their diet), individuals with a PKU genotype can have a normal phenotype.

INTERACTIONS BETWEEN GENES HAVE A PROFOUND EFFECT ON PHENOTYPES In Mendel's pea plants, a single locus influenced seed shape. Furthermore, Mendel's data showed that the seed-shape phenotype does not appear to be affected by the action of genes for seed color, seed-pod color, seed-pod shape, or other traits. The pea seeds he analyzed were round or wrinkled regardless of the types of alleles present at other loci.

In many cases, however, genes are not as independent as the gene for seed shape in peas. Consider a classic experiment published in 1905 on comb shape in chickens. The researchers, William Bateson and R. C. Punnett, crossed parents from pure-breeding lines with so-called rose and pea combs and found that the F₁ offspring had a different phenotype, called walnut combs. When these individuals bred, their offspring had walnut, rose, pea, and a fourth phenotype called single combs in a 9:3:3:1 ratio (**Figure 13.19a**). The genetic model in **Figure 13.19b** explains the data. If comb morphology results from interactions between two genes (symbolized R and P), if a dominant and a recessive allele exist for each gene, and if the four comb phenotypes are associated with the genotypes indicated at the bottom of the figure, then a cross between $RRpp$ and $rrPP$ parents would give the results that Bateson and Punnett observed.

When these types of gene-by-gene interactions occur, the phenotype produced by an allele depends on the action of alleles of other genes. If a chicken has an R allele, its phenotype depends on the allele present at the P gene.

What About Traits Like Human Height and Intelligence?

Mendel worked with **discrete traits**—characteristics that are qualitatively different from each other. In garden peas, seed color is either yellow or green—no intermediate phenotypes exist. But many traits in peas and other organisms don't fall into discrete categories. In humans, for example, height, weight, and skin color fall anywhere on a continuous scale of measurement. People are not 160 cm tall or 180 cm tall, with no other heights possible.

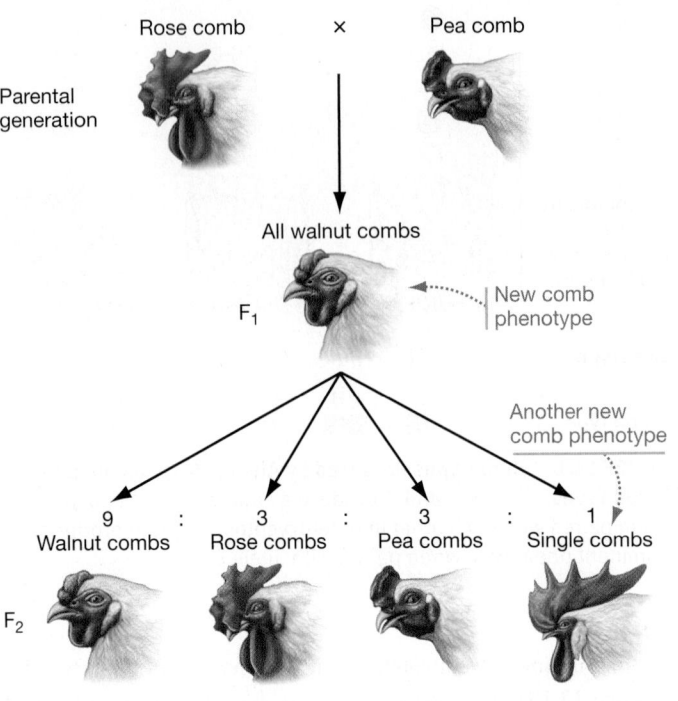

(a) Crosses between chickens with different comb phenotypes give odd results.

Rose comb × Pea comb

Parental generation

All walnut combs

F₁ — New comb phenotype

Another new comb phenotype

F₂

9 : 3 : 3 : 1
Walnut combs Rose combs Pea combs Single combs

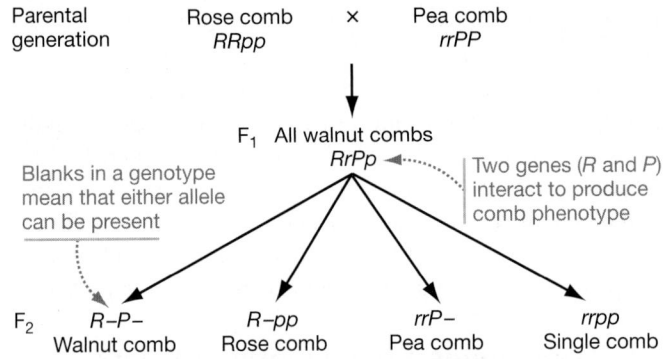

(b) A genetic model based on gene-by-gene interactions can explain the results.

Parental generation

Rose comb × Pea comb
$RRpp$ $rrPP$

F₁ All walnut combs
$RrPp$ — Two genes (R and P) interact to produce comb phenotype

Blanks in a genotype mean that either allele can be present

F₂
$R–P–$ $R–pp$ $rrP–$ $rrpp$
Walnut comb Rose comb Pea comb Single comb

FIGURE 13.19 Genes at Different Loci Can Interact to Influence a Trait. (a) This cross is unusual because new phenotypes show up in the F₁ and the F₂ generation. **(b)** To explain the results, researchers hypothesized that comb shape depends on two genes that interact. The phenotype associated with any one allele depends on the genetic environment—specifically, the alleles at a second gene.

Height and many other characteristics exhibit quantitative variation—meaning that individuals differ by degree—and are called **quantitative traits**. Like discrete traits, quantitative traits are greatly influenced by the physical environment. The effects of nutrition on human height, intelligence, and disease resistance, for example, have been well documented.

Quantitative traits share a common characteristic: When the frequencies of different values observed in a population are plot-

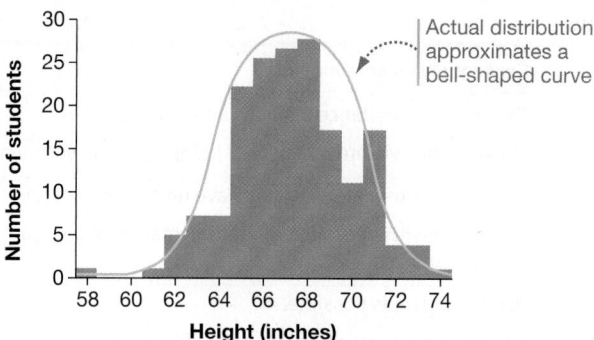

FIGURE 13.20 Quantitative Traits Have a Normal Distribution.
A histogram plotting the heights of male undergraduates at
Connecticut Agricultural College in 1914.

ted on a histogram, or frequency distribution (see **BioSkills 2** in
Appendix A), they usually form a bell-shaped curve (**Figure 13.20**).
This distribution is observed so frequently that it is often called
a normal distribution. In a normal distribution, high and low
values occur at low frequency; intermediate values occur at high
frequency.

In 1909 Herman Nilsson-Ehle had an important insight: If
many genes each contribute a small amount to the value of a
quantitative trait, then a continuous, bell-shaped (normal) distribution results for the population as a whole. Nilsson-Ehle established this finding using strains of wheat that differed in kernel color. **Figure 13.21a** includes a histogram showing the
distribution of F$_2$ phenotypes from a cross he performed between pure lines of white wheat and dark-red wheat. Notice that
the frequency of colors in F$_2$ progeny forms a bell-shaped curve.

To explain these results, Nilsson-Ehle proposed the set of hypotheses illustrated in **Figure 13.21b**:

- The parental strains differ with respect to three genes that
  control kernel color: *AABBCC* produces dark-red kernels, and
  aabbcc produces white kernels.

- The three genes assort independently. When the *AaBbCc* F$_1$
  individuals self-fertilize, white F$_2$ individuals would occur at a
  frequency of 1/4 (*aa*) $\times$ 1/4 (*bb*) $\times$ 1/4 (*cc*) = 1/64 *aabbcc*. (For
  help understanding the logic of this calculation, see **BioSkills
  13** in Appendix A.)

- The *a, b,* and *c* alleles do not contribute to pigment production, but the *A, B,* and *C* alleles contribute to pigment production in an equal and additive way. As a result, the degree of red
  pigmentation is determined by the number of *A, B,* or *C* alleles present. Each uppercase (dominant) allele that is present
  makes a wheat kernel slightly darker red.

Later work showed that Nilsson-Ehle's model hypotheses
were correct in virtually every detail. Quantitative traits are produced by the independent actions of many genes, although it is
now clear that some genes have much greater effects on the trait
in question than other genes do. As a result, the transmission of
quantitative traits is said to result from polygenic ("many-
genes") inheritance. In **polygenic inheritance**, each gene adds a
small amount to the value of the phenotype.

In the decades immediately after the rediscovery of Mendel's
work, analyses of phenomena such as sex-linkage, linkage, multiple allelism, incomplete dominance, gene-by-environment interactions, gene-by-gene interactions, and polygenic inheritance
provided a fairly comprehensive answer to the question of why

(a) Wheat kernel color is a quantitative trait.

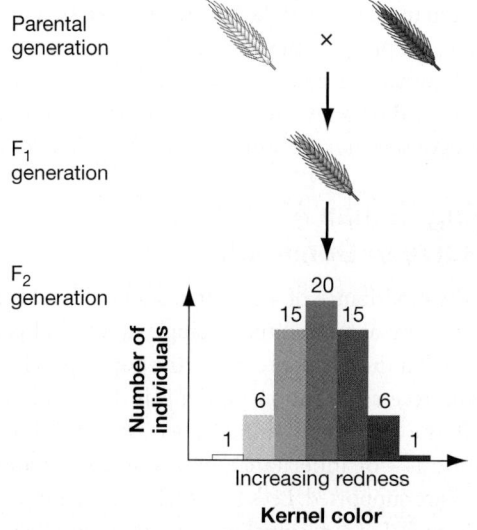

(b) Hypothesis to explain inheritance of kernel color

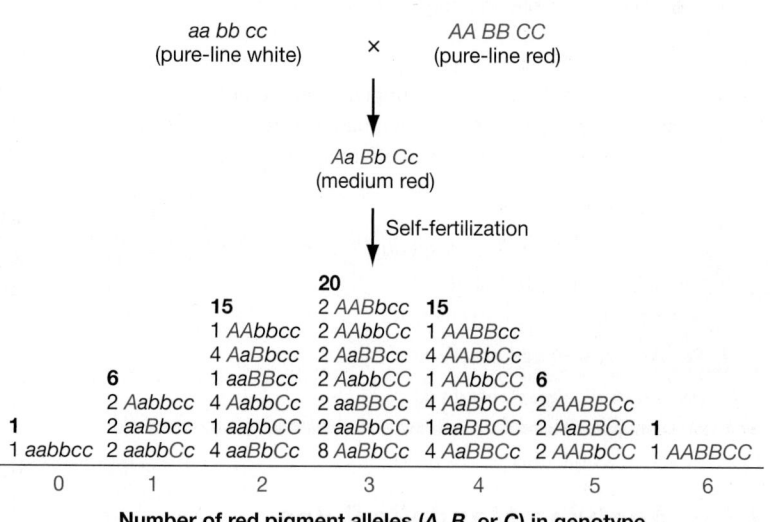

FIGURE 13.21 Quantitative Traits Result from the Action of Many Genes. **(a)** When wheat plants with white
kernels were crossed with wheat plants with red kernels, the F$_2$ offspring showed a range of kernel colors. The
frequency of these phenotypes approximates a normal distribution. **(b)** This model attempts to explain the
results of part (a).

Type of Inheritance	Definition	Consequences or Comments
Sex-linkage	Genes located on sex chromosomes.	Patterns of inheritance in males and females differ.
Linkage	Two genes found on same chromosome.	Linked genes violate principle of independent assortment.
Incomplete dominance	Heterozygotes have intermediate phenotype.	Polymorphism—heterozygotes have unique phenotype.
Codominance	Heterozygotes have phenotype of both alleles.	Polymorphism is possible—heterozygotes have unique phenotype.
Multiple allelism	In a population, more than two alleles present at a locus.	Polymorphism is possible.
Polymorphism	In a population, more than two phenotypes associated with a single gene are present.	Can result from actions of multiple alleles, incomplete dominance, and/or codominance.
Pleiotropy	A single allele affects many traits.	This is common.
Gene-by-gene interaction	In discrete traits, the phenotype associated with an allele depends on which alleles are present at another gene.	One allele can be associated with different phenotypes.
Gene-by-environment interaction	Phenotype influenced by environment experienced by individual.	Same genotypes can be associated with different phenotypes.
Polygenic inheritance of quantitative traits	Many genes are involved in specifying traits that exhibit continuous variation.	Unlike alleles that determine discrete traits, each allele adds a small amount to phenotype.

offspring resemble their parents. **Table 13.4** summarizes some of the key exceptions and extensions to Mendel's rules and gives you a chance to compare and contrast their effects on patterns of inheritance.

CHECK YOUR UNDERSTANDING

If you understand that . . .

- Genes on the same chromosome violate the principle of independent assortment. They are not transmitted to gametes independently of each other unless crossing over occurs between them.
- Sex linkage, linkage, incomplete dominance, codominance, multiple allelism, pleiotropy, environmental effects, gene interactions, and polygenic inheritance are aspects of inheritance that Mendel did not study. When they occur, monohybrid and dihybrid crosses do not result in classical Mendelian ratios of offspring phenotypes.

You should be able to . . .

Explain why the following crosses don't produce a 3:1 phenotype ratio in F_2 offspring:

1. Rose-comb × pea-comb chickens
2. Red-kernel × white-kernel wheat plants

Answers are available in Appendix B.

13.6 Applying Mendel's Rules to Humans

When researchers set out to study how a particular gene is transmitted in wheat or fruit flies or garden peas, they begin by making a series of controlled experimental crosses. For obvious rea-

sons, this research strategy is not possible with humans. But suppose that you are concerned about an illness that runs in your family, and you go to a genetic counselor to find out how likely your children are to have the disease. To advise you, the counselor needs to know how the trait is transmitted, including whether the gene involved is autosomal or sex-linked and what type of dominance is associated with the disease allele.

To understand the transmission of human traits, investigators have to analyze human genotypes and phenotypes that already exist. They do so by constructing a **pedigree**, or family tree, of affected individuals.

A pedigree records the genetic relationships among the individuals in a family along with each person's sex and phenotype with respect to the trait in question. If the trait is governed by a single gene, then analyzing the pedigree may reveal whether a given phenotype is due to a dominant or recessive allele and whether the gene responsible is located on a sex chromosome or on an autosome. Let's look at a series of specific case histories to see how this work is done.

Identifying Human Alleles as Recessive or Dominant

To analyze the inheritance of a trait that shows discrete variation, biologists begin by assuming that a single autosomal gene is responsible and that the alleles present in the population have a simple dominant–recessive relationship. This is the simplest possible situation. If the pattern of inheritance fits this model, then the assumptions—of inheritance by a single gene and simple dominance—are supported. Let's first analyze the pattern of inheritance that is typical of autosomal recessive traits and then examine patterns that emerge in pedigrees for autosomal dominant traits.

PATTERNS OF INHERITANCE: AUTOSOMAL RECESSIVE TRAITS In analyzing the inheritance of traits, it's helpful to distinguish conditions that *must* be met when a particular pattern of inheritance

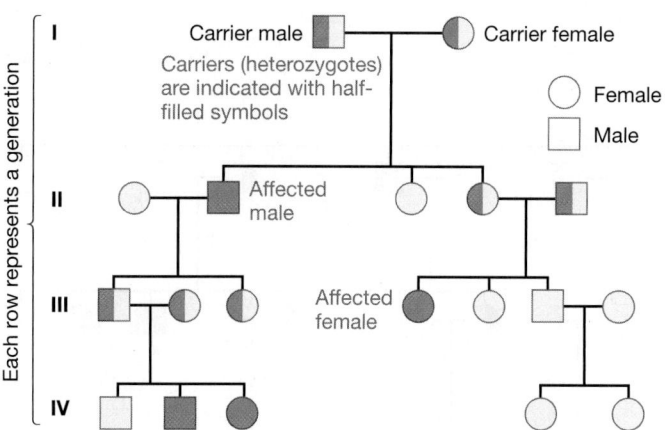

FIGURE 13.22 Pedigree of a Family with an Autosomal Recessive Disease. Diseases that are inherited as autosomal recessive traits, like sickle-cell anemia, appear in both males and females. For an individual to be affected, both parents must carry the allele responsible.

✔**QUESTION** In this pedigree, why can unaffected parents have affected offspring? If the recessive allele is rare and only one parent is affected, why are children usually unaffected?

occurs versus conditions that are *likely* to be met. For example, if a phenotype is due to an autosomal recessive allele, then

- Individuals with the trait *must* be homozygous.

- If the parents of an affected individual do not have the trait, then the parents *are likely* to be heterozygous for the trait.

Heterozygous individuals who carry a recessive allele for an inherited disease are referred to as **carriers** of the disease. These individuals carry the allele and transmit it even though they do not exhibit signs of the disease. When two carriers mate, they should produce offspring with the recessive phenotype about 25 percent of the time.

Figure 13.22 is the pedigree from a family in which an autosomal recessive trait, such as sickle-cell disease, occurs. The key feature to notice in this pedigree is that some boys and girls exhibit the trait even though their parents do not. This is the pattern you would expect to observe when the parents of an individual with the trait are heterozygous. It is also logical to observe that when an affected (homozygous) individual has children, those children do not

necessarily have the trait. This pattern is predicted if affected people marry individuals who are homozygous for the wild-type allele, and is likely to occur if the recessive allele is rare in the population.

In general, a recessive phenotype should show up in offspring only when both parents have that recessive allele and pass it on to their offspring. By definition, a recessive allele produces a given phenotype only when the individual is homozygous for that allele.

PATTERNS OF INHERITANCE: AUTOSOMAL DOMINANT TRAITS By definition, when a trait is autosomal dominant, individuals who are homozygous or heterozygous for it must have the dominant phenotype. Even if one parent is heterozygous and the other is homozygous recessive, on average half of their children should show the dominant phenotype. And unless a new mutation has occurred in a gamete, any child with the trait must have a parent with this trait. The latter observation is in strong contrast to the pattern seen in autosomal recessive traits.

Figure 13.23 shows the consequences of autosomal dominant inheritance in the pedigree of a family affected by a degenerative brain disorder called **Huntington's disease**. The pedigree has two features that indicate this disease is passed to the next generation through an autosomal dominant allele. First, if a child shows the trait, then one of its parents shows the trait as well. Second, if families have a large number of children, the trait usually shows up in every generation—due to the high probability of heterozygous parents having affected children.

Identifying Human Traits as Autosomal or Sex-Linked

When it is not possible to arrange reciprocal crosses, can data in a pedigree indicate whether a trait is autosomal or sex-linked? The answer is based on a simple premise: If a trait appears about equally often in males and females, then it is likely to be autosomal.

The data in the Huntington's pedigree (Figure 13.23) indicates that the disease appears in both males and females at about equal rates. This is strong evidence that the trait is autosomal. But if males are much more likely to have the trait in question than females are, then the allele responsible is likely to be recessive and found on the X chromosome. Because so few genes occur on the Y chromosome, Y-linked inheritance is rare.

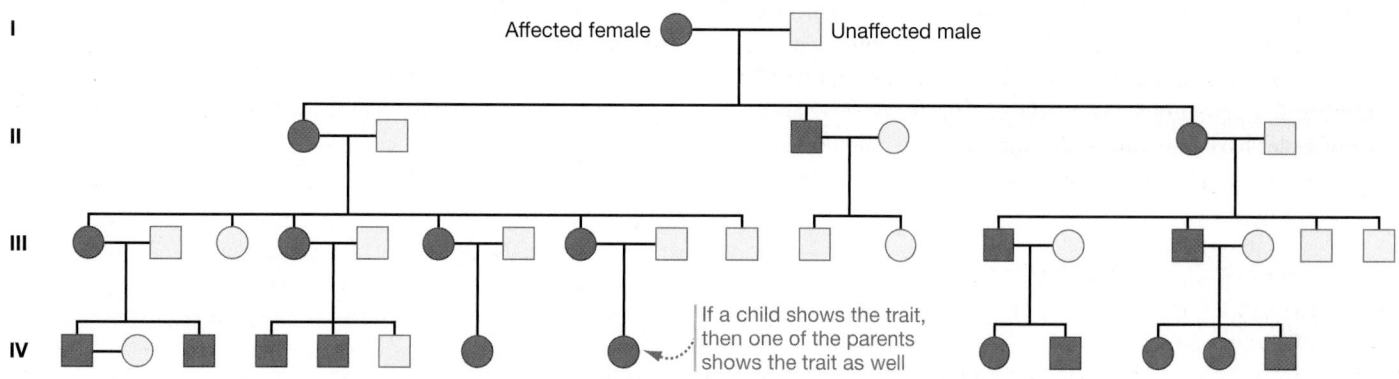

FIGURE 13.23 A Pedigree of a Family with an Autosomal Dominant Disease. Disorders like Huntington's disease, which are autosomal dominant, may appear in both males and females and tend to occur in every generation.

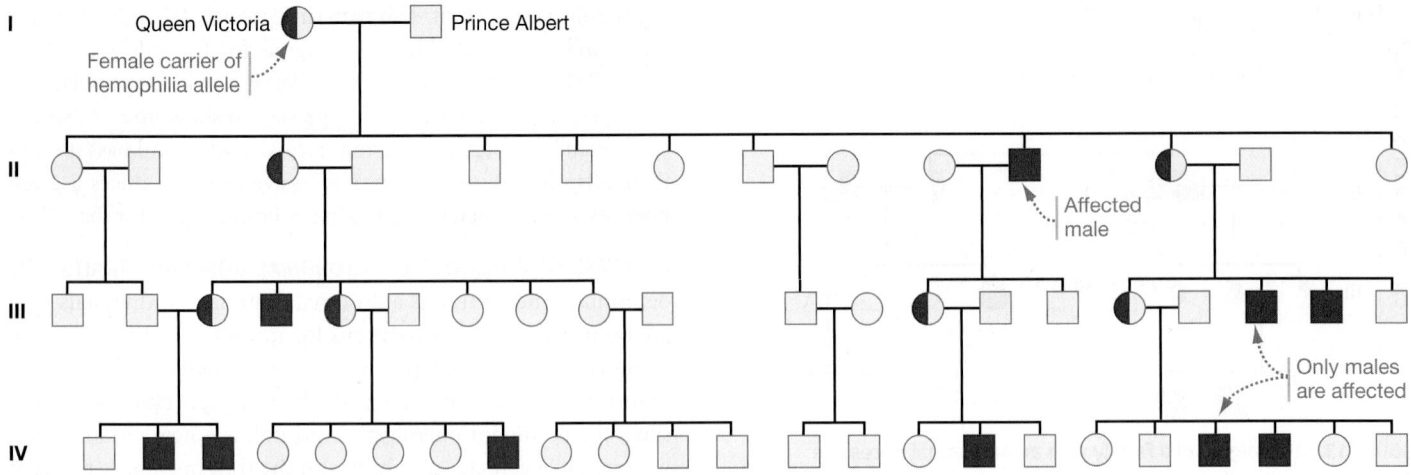

FIGURE 13.24 A Pedigree of an X-Linked Recessive Disease.

✔**QUESTION** What pattern of inheritance would you observe in a pedigree if the allele for hemophilia were X-linked dominant?

To understand why a sex bias in phenotypes implicates sex-linked inheritance, recall from Section 13.4 that sex-linked genes are located on one of the sex chromosomes. Because human males have one X chromosome and one Y chromosome, they have just one copy of each X-linked gene. But because human females have two X chromosomes, they have two copies of each X-linked gene. These simple observations are critical. In humans—just as in fruit flies and in every other species that has sex chromosomes—the pattern of inheritance in sex-linked traits is different in males and females, because the complement of sex chromosomes differs in the two sexes.

What does the pedigree of an X-linked trait look like? Let's consider the pedigree of a classic X-linked trait—the occurrence of hemophilia in the descendants of Queen Victoria, the 19th-century British monarch, and her husband Prince Albert. **Hemophilia** is caused by a defect in an important blood-clotting factor. Hemophiliacs are at a high risk of bleeding to death, because even minor injuries result in prolonged bleeding. The tip-off in the pedigree of Queen Victoria's descendants is that only males developed hemophilia (**Figure 13.24**).

Also note the affected male in generation II (the second square from the right in row II). His two sons were unaffected, but the trait reappeared in a grandson. Stated another way, the occurrence of hemophilia skipped a generation. This pattern is logical because hemophilia is due to an X-linked recessive allele. Because males have only one X chromosome, the phenotype as-

sociated with an X-linked recessive allele appears in every male that carries it. Further, the appearance of an X-linked recessive trait skips a generation in a pedigree. This pattern occurs because an affected male passes his only X chromosome on to his daughters. But because his daughters almost always received a wild-type allele from their mother, the daughters don't show the trait. They will pass the defective allele on to about half of their sons, however.

In contrast, X-linked traits that are dominant appear in every individual who has the defective allele. A good indicator of an X-linked dominant trait is a pedigree in which an affected male has all affected daughters but no affected sons. This pattern is logical because every female offspring of an affected father gets an X chromosome from him and will herself be affected, while his sons get their only X chromosome from their unaffected mother. Besides the inherited form of a bone disease called rickets, however, very few diseases are known to be due to X-linked dominant alleles.

By analyzing pedigrees in this way, biomedical researchers have been able to discover how most of the common genetic diseases in humans are inherited. As a result of such analysis, by the 1940s, the burning question in genetics was no longer the nature of inheritance but the nature of the gene itself. What are genes made of, and how are they copied so that parents pass their alleles on to their offspring? These are the questions we turn to in Chapter 14.

Summary of Key Concepts

☞ **In many species, individuals have two alleles of each gene. The principle of segregation states that prior to the formation of eggs and sperm, the alleles of each gene separate so that each egg or sperm cell receives only one of them.**

- Gregor Mendel discovered that inheritance is particulate—genes do not blend together.

- The traits that Mendel studied are specified by paired hereditary determinants that separate from each other during gamete formation.

  ✔ You should be able to predict the gamete genotypes generated by a parent with the genotype *Bb*.

☞ **The principle of independent assortment states that alleles of different genes are transmitted to egg cells and sperm cells independently of each other.**

- According to the principle of independent assortment, the segregation of alleles from one gene does not affect the segregation of other gene pairs.

  ✔ You should be able to predict the gamete genotypes generated by a parent with the genotype *BbRr*.

  (MB) **Web Activity** Mendel's Experiments, **Web Activity** The Principle of Independent Assortment

☞ **Genes are located on chromosomes. The principle of segregation is explained by the separation of homologous chromosomes in anaphase of meiosis I. The principle of independent assortment applies to genes found on different chromosomes and is explained by chromosomes lining up randomly in metaphase of meiosis I.**

- The chromosome theory of inheritance claimed that the movements of chromosomes during meiosis provide a physical basis for the principle of segregation and the principle of independent assortment.

- The chromosome theory was supported by the discovery of sex linkage. Experimental crosses with X-linked traits supported the theory's contention that genes are found on chromosomes.

✔ You should be able to use the movements of chromosomes during meiosis to explain why your answers to the above two exercises are correct.

☞ **There are important exceptions and extensions to the basic patterns of inheritance that Mendel discovered.**

- All genes follow the principle of independent assortment, but genes on the same chromosome do not follow the principle of independent assortment unless crossing over occurs.

- X-linked traits give different results in reciprocal crosses. Crosses with X-linked traits don't produce the 3:1 and 9:3:3:1 phenotypic ratios that Mendel observed.

- Not all heterozygotes show a dominant or recessive phenotype. Incomplete dominance occurs, and codominant phenotypes are common.

- Many genes are pleiotropic, meaning that they influence more than one trait. (There is not a strict one-to-one correspondence between genes and traits.)

- The phenotype associated with an allele is influenced by the environment that the individual experiences, and by the actions of alleles of other genes.

- Many traits are influenced by the action of many genes. In a population, these polygenic traits show quantitative instead of discrete variation. The frequency of each phenotype follows a normal distribution.

- Analyzing pedigrees has allowed researchers to deduce the basic modes of inheritance for most of the common genetic diseases in humans.

  ✔ You should be able to explain why crossing over is said to break up linkage between alleles.

Genetics Questions

Answers are available in Appendix B

1. In studies of how traits are inherited, what makes certain species candidates for model organisms?
   a. They are the first organisms to be used in a particular type of experiment, so they are a historical "model" of what researchers expect to find.
   b. They are easy to study because a great deal is already known about them.
   c. They are the best or most fit of their type.
   d. They are easy to maintain, have a short life cycle, produce many offspring, and yield data that are relevant to many other organisms.

2. Why is the allele for wrinkled seed shape in garden peas considered recessive?

   a. It "recedes" in the F_2 generation when homozygous parents are crossed.
   b. The trait associated with the allele is not expressed in heterozygotes.
   c. Individuals with the allele have lower fitness than that of individuals with the dominant allele.
   d. The allele is less common than the dominant allele. (The wrinkled allele is a rare mutant.)

3. The alleles found in haploid organisms cannot be dominant or recessive. Why?
   a. Dominance and recessiveness describe interactions between two alleles of the same gene in the same individual.
   b. Because only one allele is present, alleles in haploid organisms are always dominant.

c. Alleles in haploid individuals are transmitted like mitochondrial DNA or chloroplast DNA.

d. Most haploid individuals are bacteria, and bacterial genetics is completely different from eukaryotic genetics.

4. Why can you infer that individuals that are "pure line" are homozygous for the gene in question?
   a. They are highly inbred.
   b. Only two alleles are present at each gene in the populations to which these individuals belong.
   c. In a pure line, phenotypes are not affected by environmental conditions or gene interactions.
   d. No other phenotype arises in a pure-line population because no other alleles are present.

5. The genes for the traits that Mendel worked with are either located on different chromosomes or so far apart on the same chromosome that crossing over almost always occurs between them. How did this circumstance help Mendel recognize the principle of independent assortment?
   a. Otherwise, his dihybrid crosses would not have produced a 9:3:3:1 ratio of F_2 phenotypes.
   b. The occurrence of individuals with unexpected phenotypes led him to the discovery of recombination.
   c. It led him to the realization that the behavior of chromosomes during meiosis explained his results.
   d. It meant that the alleles involved were either dominant or recessive, which gave 3:1 ratios in the F_1 generation.

6. What is meant by the claim that Mendel worked with the simplest possible genetic system?
   a. Discrete traits, two alleles, simple dominance and recessiveness, no sex chromosomes, and unlinked genes are the simplest situation known.
   b. The ability to self-fertilize or cross-pollinate made it simple for Mendel to set up controlled crosses.
   c. Mendel was aware of meiosis and the chromosome theory of inheritance, so it was easy to reach the conclusions he did.
   d. Mendel's experimental designs and his rules of inheritance are actually neither complex nor sophisticated.

7. Mendel's rules do not correctly predict patterns of inheritance for tightly linked genes or the inheritance of alleles that show incomplete dominance. Does this mean that his hypotheses are incorrect?
   a. Yes, because they are relevant to only a small number of organisms and traits.
   b. Yes, because not all data support his hypotheses.
   c. No, because he was not aware of meiosis or the chromosome theory of inheritance.
   d. No, it just means that his hypotheses are limited to certain conditions.

8. The artificial sweetener NutraSweet consists of a phenylalanine molecule linked to aspartic acid. The labels of diet sodas that contain NutraSweet include a warning to people with PKU. Why?

a. NutraSweet stimulates the same taste receptors that natural sugars do.
b. People with PKU have to avoid phenylalanine in their diet.
c. In people with PKU, phenylalanine reacts with aspartic acid to form a toxic compound.
d. People with PKU cannot lead normal lives, even if their environment is carefully controlled.

9. When Sutton and Boveri published the chromosome theory of inheritance, research on meiosis had not yet established that paternal and maternal homologs assort independently of each other. Then, in 1913, Elinor Carothers published a paper about a grasshopper species with an unusual karyotype: One chromosome had no homolog (meaning no pairing partner at meiosis I); another chromosome had homologs that could be distinguished under the light microscope. If chromosomes assort independently, how often should Carothers have observed each of the four products of meiosis shown in the following figure?

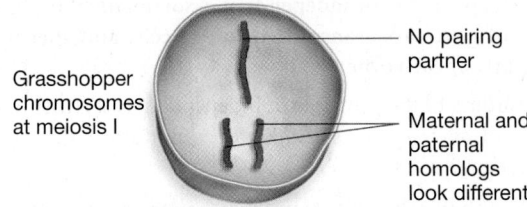

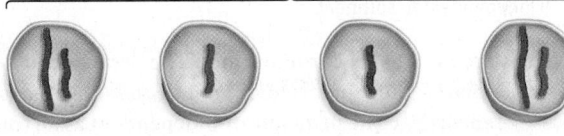

Four types of gametes possible
(each meiotic division can produce only two of the four)

a. Only the gametes with one of each type of chromosome would occur.
b. The four types of gametes should be observed to occur at equal frequencies.
c. The chromosome with no pairing partner would disintegrate, so only gametes with one copy of the other chromosome would be observed.
d. Gametes with one of each type of chromosome would occur twice as often as gametes with just one chromosome.

10. Which of the following is the strongest evidence that a trait might be influenced by polygenic inheritance?
    a. F_1 offspring of parents with different phenotypes have an intermediate phenotype.
    b. F_1 offspring of parents with different phenotypes have the dominant phenotype.
    c. The trait shows qualitative (discrete) variation.
    d. The trait shows quantitative variation.

Solving Genetics Problems

The best way to test and extend your knowledge of transmission genetics is to work problems. Most genetics problems are set up as follows: You are given some information about the genotypes or phenotypes of one or both parents, along with data on the phenotypes of F_1 or F_2 offspring. Your task is to generate a set of hypotheses—a genetic model—to explain the results. Your hypotheses should address each of the following questions:

• Is the trait under study discrete or quantitative?

- Is the phenotype a product of one gene or many genes?
- For each gene involved, how many alleles are present—one, two, or many?
- Do the alleles involved show complete dominance, incomplete dominance, or codominance?
- Are the genes involved sex-linked or autosomal?
- If more than one gene is involved, are they linked or unlinked? If they are linked, does crossing over occur frequently?

It's also helpful to ask yourself whether gene interactions or pleiotropy might be occurring and whether it is safe to assume that the experimental design carefully controlled for effects of variation in other genes and in the environment.

In working the problem, be sure to start with the simplest possible explanation. For example, if you are dealing with a discrete trait, you might hypothesize that the cross involves a single autosomal gene with two alleles that show complete dominance. Your next step is to infer what the parental genotypes would be (according to your hypothesis), if they are not already given, and then do a Punnett square to predict what the offspring phenotypes and their frequencies should be based on your hypothesis. Next, check whether these predictions match the observed results given in the problem. If the answer is yes, you have a valid solution. But if the answer is no, you need to go back and change one of your hypotheses, redo the Punnett square, and check to see if the predictions and observations match. Keep repeating these steps until you have a model that fits the data.

Example Problems

Example 1 *Plectritis congesta* plants produce fruits that either have or do not have prominent structures called wings. The alleles involved are W^+ = winged fruit; W^- = wingless fruit. Researchers collected an array of individuals from the field and performed a series of crosses. The results are given in the following table. Complete the table by writing down the genotype of the parent or parents involved in each cross.

Parental Phenotype(s)	Number of Offspring with Winged Fruits	Number of Offspring with Wingless Fruits	Parental Genotype(s)
Wingless (self-fertilized)	0	80	
Winged (self-fertilized)	90	30	
Winged × wingless	46	0	
Winged × winged	44	0	

Solution Here you're given offspring phenotypes and you're asked to infer parental genotypes. To do this you have to propose hypothetical parental genotypes to test, make a Punnett square to predict the offspring genotypes, and then see if the predicted offspring phenotypes match the data. In this case, coming up with a hypothesis for the parental genotypes is relatively straightforward, because the problem states that the trait is due to one gene and two alleles. No information on sex is given, so assume the gene is autosomal (the simplest case). Now look at the second entry in the chart. It shows

a 3:1 ratio of offspring from a winged individual that self-fertilizes. This result is consistent with the hypothesis that W^+ is dominant and W^- recessive and that this parent's genotype is W^+W^-. Now let's look at the first cross in the chart. If W^+ is dominant, then a wingless parent must be W^-W^-. When you do the Punnett square to predict offspring genotypes from selfing, you find that all the offspring will produce wingless fruits, consistent with the data. In the third cross, all the offspring make winged fruits even though one of the parents produces wingless fruits and thus is W^-W^-. This would happen only if the winged parent is W^+W^+. (If this reasoning isn't immediately clear to you, work the Punnett square.) In the fourth cross, you could get offspring that all make winged fruits if the parents were W^+W^+ and W^+W^+, or if the parents were W^+W^+ and W^+W^-. Either answer is correct. Again, you can write out the Punnett squares to see that this statement is correct.

Example 2 Two black female mice are crossed with a brown male. In several litters, female I produced 9 blacks and 7 browns; female II produced 57 blacks. What deductions can you make concerning the inheritance of black and brown coat color in mice? What are the genotypes of the parents in this case?

Solution Here you are given parental and offspring phenotypes and are asked to infer the parental genotypes. As a starting point, assume that the coat colors are due to the simplest genetic system possible: one autosomal gene with two alleles, where one allele is dominant and the other recessive. Because female II produces only black offspring, it's logical to suppose that black is dominant to brown. Let's use *B* for black and *b* for brown. Then the male parent is *bb*. To produce offspring with a 1:1 ratio of black:brown coats, female I must be *Bb*. But to produce all black offspring, female II must be *BB*. This model explains the data, so you can accept it as correct.

Genetics Problems

Answers are available in Appendix B

1. Tay-Sachs disease causes nerve cells to malfunction and results in death by age 4. Two healthy parents know from blood tests that each parent carries a recessive allele responsible for Tay-Sachs. If their first three children have the disease, what is the probability that their fourth child will not? Assuming they have not yet had a child, what is the probability that, if they have four children, all four will have the disease? If their first three children are male, what is the probability that their fourth child will be male?

2. Suppose that in garden peas the genes for seed color and seed-pod shape are linked, and that Mendel crossed *YYII* parents (which produce yellow seeds in inflated pods) with *yyii* parents (which produce green seeds in constricted pods).

- Draw the F_1 Punnett square and predict the expected F_1 phenotype(s).
- List the genotype(s) of gametes produced by F_1 individuals if no crossing over occurs.
- Draw the F_2 Punnett square if no crossing over occurs. Based on this Punnett square, predict the expected phenotype(s) in the F_2 generation and the expected frequency of each phenotype.
- If crossing over occurs during gamete formation in F_1 individuals, give the genotype of the recombinant gamete(s) that result.
- Add the recombinant gametes to the F_2 Punnett square requested above. Will any additional phenotypes be observed at low frequency in the F_2 generation? If so, what are they?

3. The smooth feathers on the back of the neck in pigeons can be reversed by a mutation to produce a "crested" appearance in which feathers form a distinctive spike at the back of the head. A pigeon breeder examined offspring produced by a single pair of non-crested birds and recorded the following: 14 non-crested and 7 crested. She then made a series of crosses using offspring from the first cross. When she crossed two of the crested birds, all 22 of the offspring were crested. When she crossed a non-crested bird with a crested bird, 7 offspring were non-crested and 6 were crested.

   - For these three crosses, provide genotypes for parents and offspring that are consistent with these results.
   - Which allele is dominant?

4. A plant with orange-spotted flowers was grown in the greenhouse from a seed collected in the wild. The plant was self-pollinated and gave rise to the following progeny: 88 orange with spots, 34 yellow with spots, 32 orange with no spots, and 8 yellow with no spots. What can you conclude about the dominance relationships of the alleles responsible for the spotted and unspotted phenotypes? For the orange and yellow phenotypes? What can you conclude about the genotype of the original plant that had orange, spotted flowers?

5. As a genetic counselor, you routinely advise couples about the possibility of genetic disease in their offspring based on their family histories. This morning you met with an engaged couple, both of whom are phenotypically normal. The man, however, has a brother who died of Duchenne-type muscular dystrophy, an X-linked condition that results in death before the age of 20. The allele responsible for this disease is recessive. His prospective bride, whose family has no history of the disease, is worried that the couple's sons or daughters might be afflicted.

   - How would you advise this couple?
   - The sister of this man is planning to marry his fiancée's brother. How would you advise this second couple?

6. Suppose you are heterozygous for two genes that are located on different chromosomes. You carry alleles *A* and *a* for one gene and alleles *B* and *b* for the other. Draw a diagram illustrating what happens to these genes and alleles when meiosis occurs in your reproductive tissues. Label the stages of meiosis, the homologous chromosomes, sister chromatids, nonhomologous chromosomes, genes, and alleles. Be sure to list all of the genetically different gametes that could form and indicate how frequently each type should be observed. On the diagram, identify the events responsible for the principle of segregation and the principle of independent assortment.

7. Review the text's description of ABO blood types. Suppose a woman with blood type O married a man with blood type AB. What phenotypes and genotypes would you expect to observe in their offspring, and in what proportions? Answer the same question for a heterozygous mother with blood type A and a heterozygous father with blood type B.

8. An alien friend named Tukan has two sets of eyes, one set forward-looking and one set backward-looking, and smooth skin. His mate, Valco, lacks eyes but has skin covered with tiny hooks that attract all sorts of debris. Tukan and Valco have thrived on earth and have had four children, all with no eyes and smooth skin. Typical of their ways, the children interbred and produced 32 children of their own.

   - Under the models of inheritance proposed by Mendel, identify which alleles are dominant and which are recessive.
   - Provide gene symbols that would reflect the dominant–recessive allelic relationships.

   - Of the 32 children, how many would you expect to have two sets of eyes and smooth skin?

9. Phenylketonuria (PKU) is a genetic disease caused by homozygosity for a recessive mutation in the enzyme that converts the amino acid phenylalanine to tyrosine. In the absence of this enzyme, phenylalanine and some of its derivatives accumulate in the body and cause mental retardation. If individuals are identified soon enough after birth, they can be treated by a low-phenylalanine diet for the early years of their lives. As adults, though, homozygous recessive individuals are allowed to adopt a diet with normal amounts of phenylalanine. Not long after such treatments were initiated, a troubling phenomenon was observed. A high number of children born to treated mothers were mentally retarded even though the children were heterozygous for the PKU gene. Children born of treated PKU males suffered no ill effects.

   - Can you offer an explanation as to why genetically heterozygous children of treated PKU mothers might be prone to mental retardation?
   - Propose a solution to reduce the likelihood of mental retardation in children of treated PKU mothers.

10. The blending-inheritance hypothesis proposed that the genetic material from parents is unavoidably and irreversibly mixed in the offspring. As a result, offspring and later descendants should always appear intermediate in phenotype to their forebears. Mendel, in contrast, proposed that genes are discrete and that their integrity is maintained in the offspring and in subsequent generations. Suppose the year is 1890. You are a horse breeder and have just read Mendel's paper. You don't believe his results, however, because you often work with cremello (very light-colored) and chestnut (reddish-brown) horses. You know that if you cross a cremello individual from a pure-breeding line with a chestnut individual from a pure-breeding line, the offspring will be palomino—meaning they have an intermediate (golden-yellow) body color. What additional crosses would you do to test whether Mendel's model is valid in the case of genes for horse color? List the crosses and the offspring genotypes and phenotypes you'd expect to obtain. Explain why these experimental crosses would provide a test of Mendel's model.

11. Two mothers give birth to sons at the same time in a busy hospital. The son of couple 1 is afflicted with hemophilia A, which is a recessive X-linked disease. Neither parent has the disease. Couple 2 has a normal son even though the father has hemophilia A. The two couples sue the hospital in court, claiming that a careless staff member swapped their babies at birth. You appear in court as an expert witness. What do you tell the jury? Make a diagram that you can submit to the jury.

12. You have crossed two *Drosophila melanogaster* individuals that have long wings and red eyes—the wild-type phenotype. In the progeny, the mutant phenotypes called curved wings and lozenge eyes appear as follows:

Females	Males
600 long wings, red eyes	300 long wings, red eyes
200 curved wings, red eyes	100 curved wings, red eyes
	300 long wings, lozenge eyes
	100 curved wings, lozenge eyes

   - According to these data, is the curved-wing allele autosomal recessive, autosomal dominant, sex-linked recessive, or sex-linked dominant?

- Is the lozenge-eye allele autosomal recessive, autosomal dominant, sex-linked recessive, or sex-linked dominant?
- What is the genotype of the female parent?
- What is the genotype of the male parent?

13. In parakeets, two autosomal genes that are located on different chromosomes control the production of feather pigment. Gene *B* codes for an enzyme that is required for the synthesis of a blue pigment, and gene *Y* codes for an enzyme required for the synthesis of a yellow pigment. Recessive, loss-of-function mutations (resulting in no production of the affected pigment) are known for both genes. Suppose that a bird breeder has two green parakeets and mates them. The offspring are green, blue, yellow, and albino (unpigmented).

- Based on this observation, what are the genotypes of the green parents? What is big genotype of each type of offspring? What fraction of the total progeny should exhibit each type of color?
- Suppose that the parents were the progeny of a cross between two true-breeding strains. What two types of crosses between true-breeding strains could have produced the green parents? Indicate the genotypes and phenotypes for each cross.

14. The pedigree shown below is for the human trait called osteopetrosis, which is characterized by bone fragility and dental abscesses. Is the gene that affects bone and tooth structure autosomal or sex-linked? Is the allele for osteopetrosis dominant or recessive?

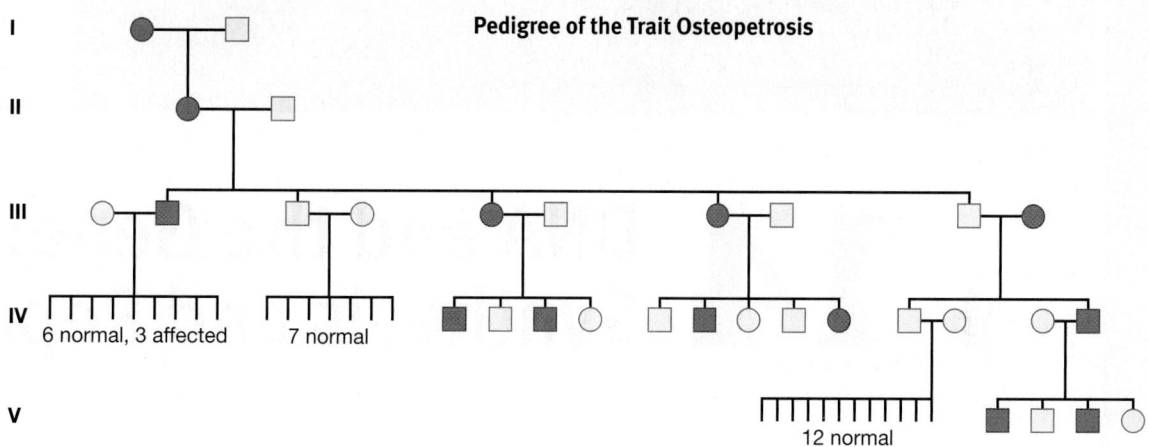

Pedigree of the Trait Osteopetrosis

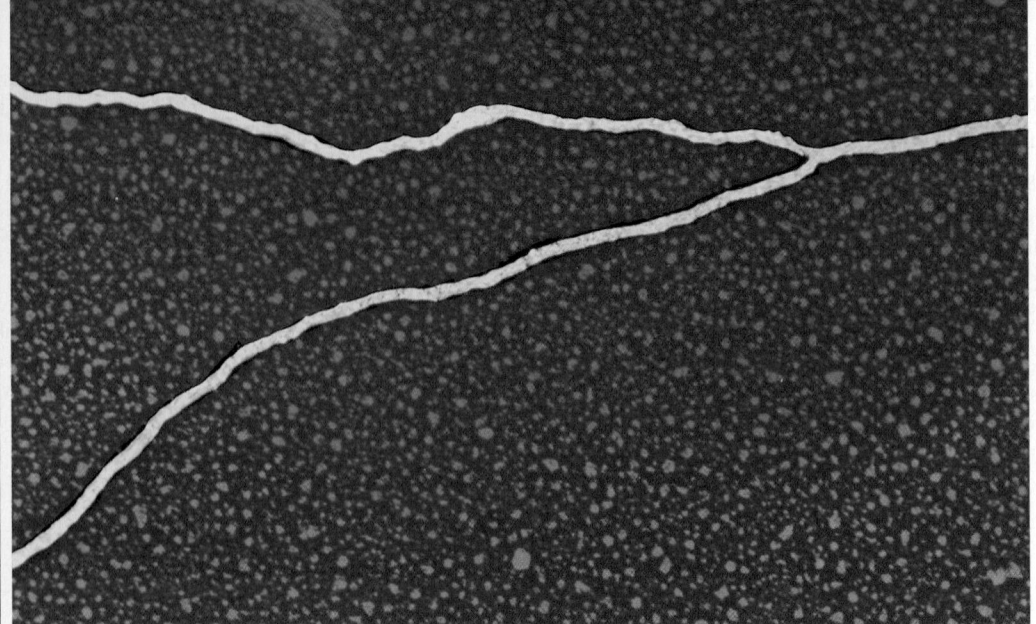

Electron micrograph (with color added) showing DNA in the process of replication. The original DNA double helix (far right) is being replicated into two DNA double helices (on the left). The point at which the two helices diverge, called the replication fork, is where DNA synthesis is taking place.

14 DNA and the Gene: Synthesis and Repair

KEY CONCEPTS

- Genes are made of DNA.

- When DNA is being copied, each strand of the double helix serves as the template for the synthesis of a complementary strand.

- DNA synthesis occurs in the $5' \rightarrow 3'$ direction only and requires a large suite of specialized enzymes. The leading strand is synthesized continuously, but the lagging strand is synthesized as a series of fragments that are then linked together.

- Specialized enzymes repair damages to DNA and fix mistakes in DNA synthesis. If these enzymes are defective, the mutation rate increases.

W hat are genes made of, and how are they copied so that they are faithfully passed on to offspring? These questions dominated biology during the middle of the twentieth century.

Since Mendel's time, the predominant research strategy in genetics had been to conduct a series of experimental crosses, create a genetic model to explain the types and proportions of phenotypes that resulted, and then test the model's predictions through reciprocal crosses, testcrosses, or other techniques. This strategy led to virtually all of the discoveries analyzed in Chapter 13, including Mendel's rules, sex linkage, linkage, and quantitative inheritance.

The chemical composition and molecular structure of Mendel's hereditary factors—which came to be called genes—remained a mystery, however. And even though biologists knew that genes and chromosomes were replicated during the cell cycle (see Chapter 11), with copies distributed to daughter cells during mitosis and meiosis, no one had the slightest clue about how the copying occurred.

The goal of this chapter is to explore how researchers solved these mysteries. The results provided a link between two of the five attributes of life introduced in Chapter 1: Processing genetic information and replication. How are genes copied, so they can be passed on to succeeding generations?

Let's begin with studies that identified the nature of the genetic material, explore how genes are copied during the synthesis phase of the cell cycle, and conclude by analyzing how damaged or incorrectly copied genes are repaired. Once the molecular nature of the gene was known, the nature of biological science changed forever.

✔ When you see this checkmark, stop and test yourself. Answers are available in Appendix B.

14.1 What Are Genes Made Of?

The chromosome theory of inheritance, introduced in Chapter 13, proposed that chromosomes are composed of genes. It had been known since the late 1800s that chromosomes are a complex of DNA and proteins. The question of what genes are made of, then, came down to a simple choice: DNA or protein?

Initially, most biologists backed the hypothesis that genes are made of proteins. The arguments in favor of this hypothesis were compelling. Hundreds, if not thousands, of complex and highly regulated chemical reactions occur in even the simplest living cells. The amount of information required to specify and coordinate these reactions is mind-boggling. With their almost limitless variation in structure and function, proteins are complex enough to contain this much information.

DNA, in contrast, was known to be composed of just four types of deoxyribonucleotides—the monomers introduced in Chapter 4. It was thought to be a simple molecule with some sort of repetitive and uninteresting structure. No one could imagine how such a simple compound could hold so much complex information.

DNA or protein? The experiment that settled the question is considered a classic in biological science.

The Hershey–Chase Experiment

Alfred Hershey and Martha Chase took up the question of whether genes were made of protein or DNA by studying how a virus called T2 infects the bacterium *Escherichia coli*. Hershey and Chase knew that T2 infections begin when the virus attaches to the cell wall of *E. coli* and injects its genes into the cell's interior (**Figure 14.1a**). These genes then direct the production of a new generation of virus particles inside the infected cell, which acts as a host for the parasitic virus. (For more information on viruses, see Chapter 35.)

During the infection, the exterior portion, or capsid, of the original, parent virus is left behind. The capsid remains attached to the exterior of the host cell as a "ghost" (**Figure 14.1b**). Hershey and Chase also knew that T2 is made up almost exclusively of protein and DNA. But was it protein or DNA that entered the host cell and directed the production of new viruses?

Hershey and Chase's strategy for determining the composition of the viral substance that enters the cell and acts as the hereditary material was based on two facts: (**1**) Proteins contain sulfur but not phosphorus, and (**2**) DNA contains phosphorus but not sulfur.

As **Figure 14.2** on page 260 shows, the researchers began their work by growing viruses in the presence of either a radioactive isotope of sulfur (^{35}S) or a radioactive isotope of phosphorus (^{32}P). Because these isotopes were incorporated into newly synthesized proteins and DNA, this step produced a population of viruses with radioactive proteins and a population with radioactive DNA.

Hershey and Chase allowed each set of radioactive viruses to infect *E. coli* cells. If genes consist of DNA, then radioactive protein should be found only in the ghost capsids outside the infected

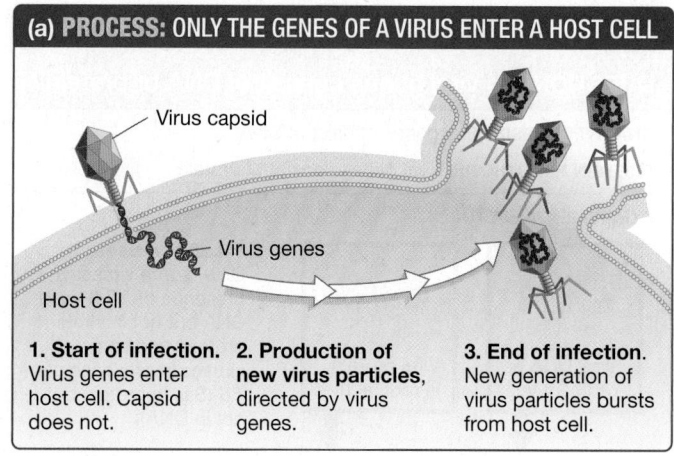

(a) PROCESS: ONLY THE GENES OF A VIRUS ENTER A HOST CELL

Virus capsid

Virus genes

Host cell

1. Start of infection. Virus genes enter host cell. Capsid does not.

2. Production of new virus particles, directed by virus genes.

3. End of infection. New generation of virus particles bursts from host cell.

(b) The virus's capsid stays outside the cell.

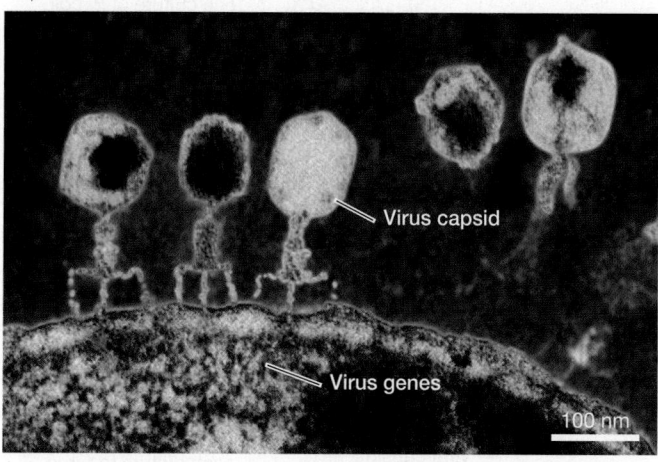

Virus capsid

Virus genes

100 nm

FIGURE 14.1 Viruses Inject Genes into Bacterial Cells and Leave a Capsid Behind. Color has been added to the transmission electron micrograph in (b) to make key structures more visible.

host cells, while radioactive DNA should be located inside the cells. But if genes consist of proteins, then only radioactive protein—and no radioactive DNA—should be found inside the cells.

To test these predictions, Hershey and Chase sheared the ghosts off the cells by agitating each of the cultures in kitchen blenders. When the researchers spun the samples in a centrifuge, the ghosts stayed in the solution while the cells formed a pellet at the bottom of the centrifuge tube (see **BioSkills 11** in Appendix A to review how centrifugation works).

As predicted by the DNA hypothesis, the biologists found that virtually all of the radioactive protein was in the ghosts, while virtually all of the radioactive DNA was inside the host cells. Because the injected component of the virus directs the production of a new generation of virus particles, it is this component that represents the virus's genes.

After these results were published, proponents of the protein hypothesis accepted that DNA, not protein, must be the hereditary material. An astonishing claim—that DNA contained all the information for life's complexity—was correct.

QUESTION: Do viral genes consist of DNA or protein?

DNA HYPOTHESIS: Viral genes consist of DNA.

PROTEIN HYPOTHESIS: Viral genes consist of protein.

EXPERIMENTAL SETUP:

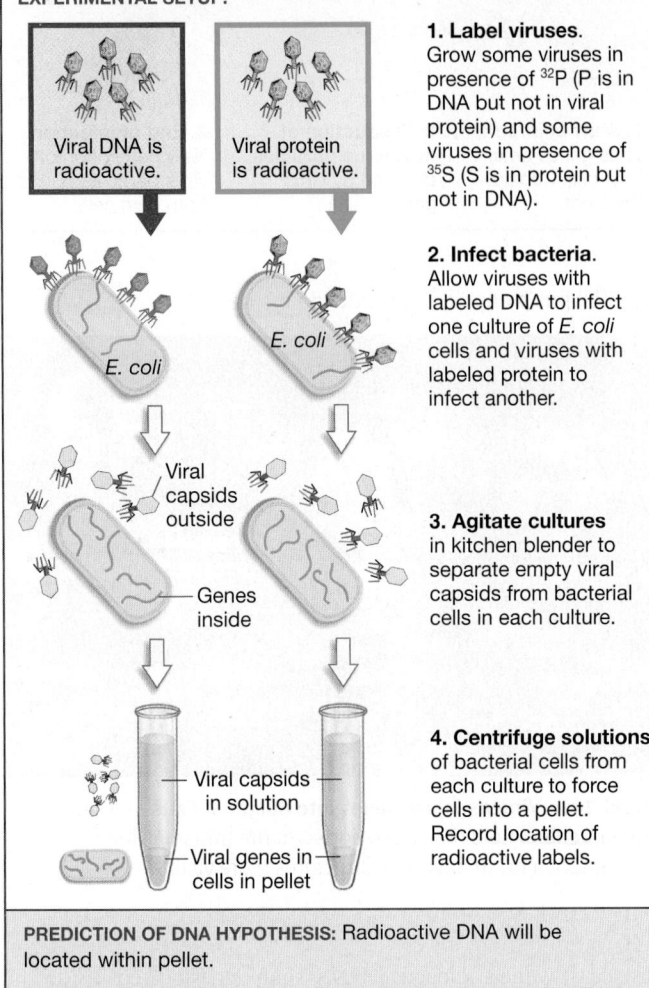

1. **Label viruses.** Grow some viruses in presence of ^{32}P (P is in DNA but not in viral protein) and some viruses in presence of ^{35}S (S is in protein but not in DNA).

2. **Infect bacteria.** Allow viruses with labeled DNA to infect one culture of *E. coli* cells and viruses with labeled protein to infect another.

3. **Agitate cultures** in kitchen blender to separate empty viral capsids from bacterial cells in each culture.

4. **Centrifuge solutions** of bacterial cells from each culture to force cells into a pellet. Record location of radioactive labels.

PREDICTION OF DNA HYPOTHESIS: Radioactive DNA will be located within pellet.

PREDICTION OF PROTEIN HYPOTHESIS: Radioactive protein will be located within pellet.

RESULTS:

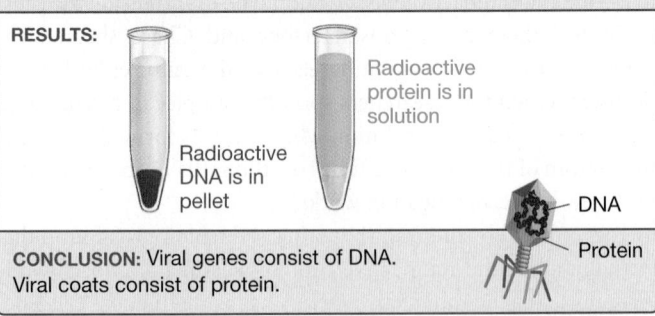

CONCLUSION: Viral genes consist of DNA. Viral coats consist of protein.

FIGURE 14.2 Experimental Evidence That DNA Is the Hereditary Material.

SOURCE: Hershey, A. D. and M. Chase. 1952. Independent functions of viral protein and nucleic acid in growth of bacteriophage. *Journal of General Physiology* 36: 39–56.

✓**QUESTION** What evidence would these investigators have to produce to convince you that the viral capsids were shaken off the bacterial cells by the agitation step?

The Secondary Structure of DNA

Chapter 4 introduced Watson and Crick's model for the secondary structure of DNA, which was proposed in 1953. Recall that DNA is a long, linear polymer made up of monomers called deoxyribonucleotides, which consist of a deoxyribose molecule (deoxyribose is a sugar), a phosphate group, and a nitrogenous base (**Figure 14.3a**). Deoxyribonucleotides link together into a polymer when a phosphodiester bond forms between a hydroxyl group on the 3′ carbon of one deoxyribose and the phosphate group attached to the 5′ carbon of another deoxyribose.

As **Figure 14.3b** shows, the primary structure of a DNA molecule has two major components: (**1**) a "backbone" made up of the sugar and phosphate groups of deoxyribonucleotides and (**2**) a series of nitrogen-containing bases that project from the backbone. A strand of DNA has a directionality, or polarity: One end has an exposed hydroxyl group on the 3′ carbon of a deoxyribose, while

(a) Structure of a deoxyribonucleotide

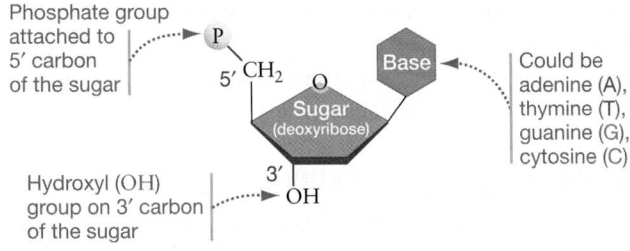

(b) Primary structure of DNA

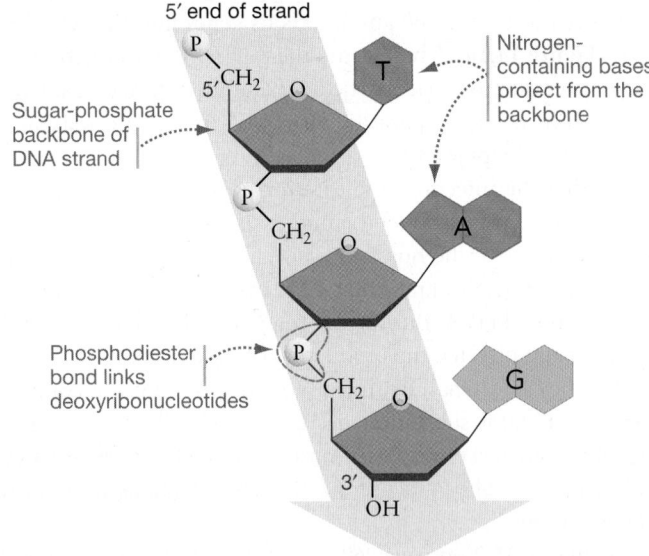

FIGURE 14.3 DNA's Primary Structure. (a) Deoxyribonucleotides are monomers that polymerize to form DNA. **(b)** DNA's primary structure is made up of a sequence of deoxyribonucleotides. Notice that the structure has a sugar-phosphate "backbone" with nitrogen-containing bases attached.

✓**EXERCISE** Write the base sequence of the DNA in part (b), in the 5′ → 3′ direction.

the other has an exposed phosphate group on a 5′ carbon. Thus, the molecule has a 3′ end and a 5′ end.

As they explored different models for the secondary structure of DNA, Watson and Crick hit on the idea of lining up two of these long strands in opposite directions, or in what is called antiparallel fashion (**Figure 14.4a**). They realized that antiparallel strands will twist around each other into a spiral or helix because certain of the nitrogen-containing bases fit together snugly in pairs inside the spiral and form hydrogen bonds (**Figure 14.4b** and **c**). The double-stranded molecule that results is called a **double helix**.

More specifically, the secondary structure is stabilized by hydrogen bonds that form between the bases adenine (A) and thymine (T) and between the bases guanine (G) and cytosine (C), along with hydrophobic interactions that the bases experience inside the helix. The pairing rules that govern this hydrogen bonding of bases are called **complementary base pairing**.

Watson and Crick realized that the A-T and G-C pairing rules suggested a way for DNA to be copied when chromosomes are replicated during S phase of the cell cycle, prior to mitosis and meiosis. They suggested that the existing strands of DNA served as a template (pattern) for the production of new strands, with bases being added to the new strands according to complementary base pairing. For example, if the template strand contained a T, then an A would be added to the new strand to pair

with that T. Similarly, a G on the template strand would dictate the addition of a C on the new strand.

14.2 Testing Early Hypotheses about DNA Synthesis: The Meselson–Stahl Experiment

Complementary base pairing provided a mechanism for DNA to be copied. But many questions remained about how the copying was done. For example, biologists had three alternative hypotheses about how the old and new strands might interact during replication:

1. If the old strands of DNA separated, each could then be used as a template for the synthesis of a new, daughter strand. This hypothesis is called semiconservative replication, because each new daughter DNA molecule would consist of one old strand and one new strand.

2. If the bases temporarily turned outward so that complementary strands no longer faced each other, they could serve as a template for the synthesis of an entirely new double helix all at once. This hypothesis, called conservative replication, results in an intact parental molecule and a

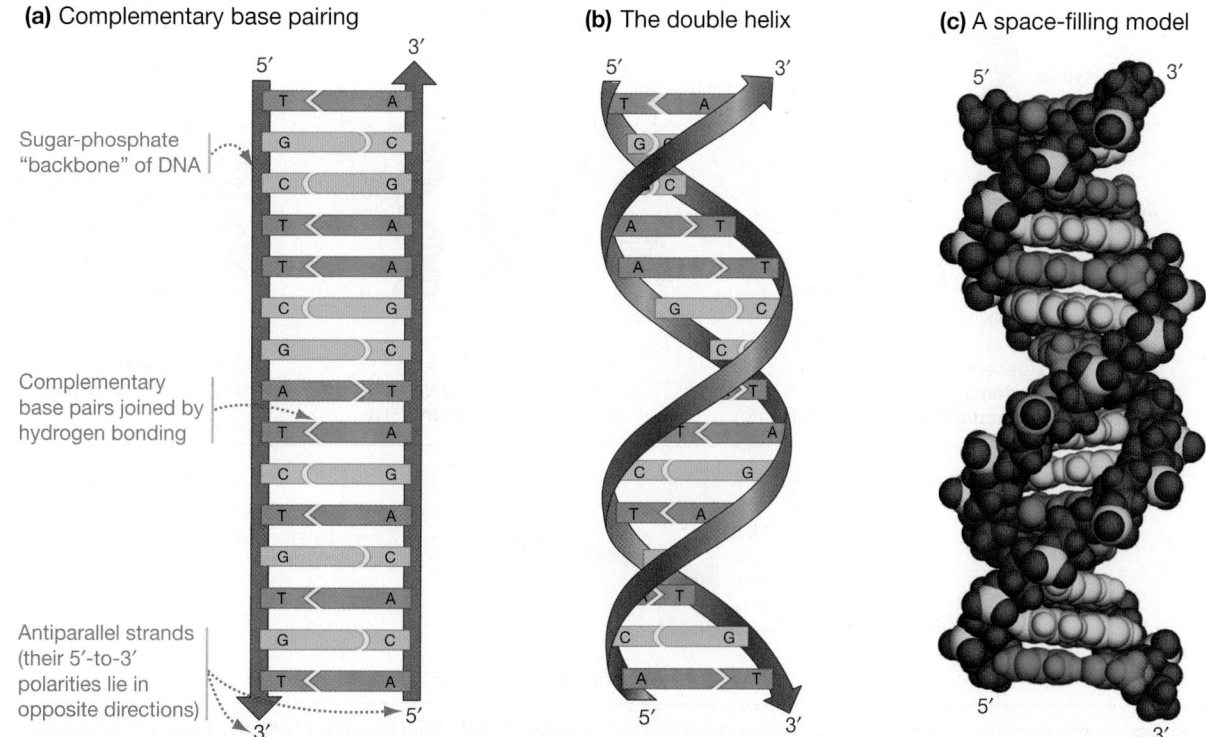

(a) Complementary base pairing   **(b)** The double helix   **(c)** A space-filling model

Sugar-phosphate "backbone" of DNA

Complementary base pairs joined by hydrogen bonding

Antiparallel strands (their 5′-to-3′ polarities lie in opposite directions)

FIGURE 14.4 DNA's Secondary Structure: The Double Helix. (a) DNA normally consists of two strands, each with a sugar-phosphate backbone. Nitrogen-containing bases project from each strand and form hydrogen bonds. Only A-T and G-C pairs fit together in a way that allows hydrogen bonding to occur between the strands. **(b)** Bonding between complementary bases twists the molecule into a double helix. **(c)** This space-filling model illustrates the tight packing of atoms in the interior of the helix, where hydrogen bonding and hydrophobic interactions help stabilize the structure.

EXPERIMENT

QUESTION: Is replication semiconservative, conservative, or dispersive?

HYPOTHESIS 1: Replication is semiconservative.	HYPOTHESIS 2: Replication is conservative.	HYPOTHESIS 3: Replication is dispersive.

EXPERIMENTAL SETUP:

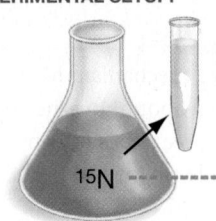

Generation 0
DNA sample

1. Grow *E. coli* cells in medium with ^{15}N as sole source of nitrogen for many generations. Collect sample and purify DNA.

Cell transfer

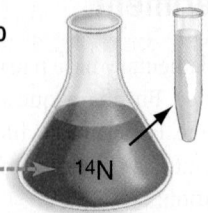

Generation 1
DNA sample

2. Transfer cells to medium containing ^{14}N. After cells divide once, collect sample and purify DNA.

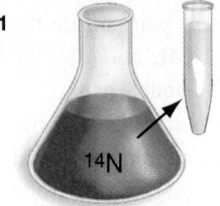

Generation 2
DNA sample

3. After cells have divided a second time in ^{14}N medium, collect sample and purify DNA.

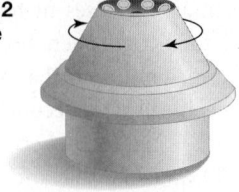

4. Centrifuge the three samples separately. Compare the locations of the DNA bands in each sample.

PREDICTIONS:

Semiconservative replication	Conservative replication	Dispersive replication

Generation 0

Generation 1

Generation 2

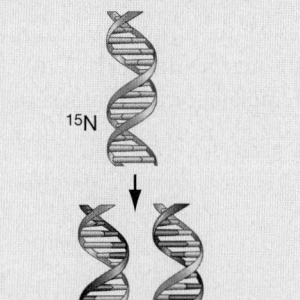

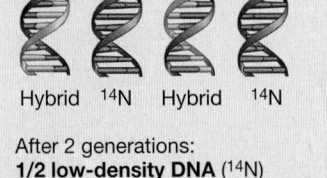

After 2 generations:
1/2 low-density DNA (^{14}N)
1/2 intermediate-density DNA (hybrid)

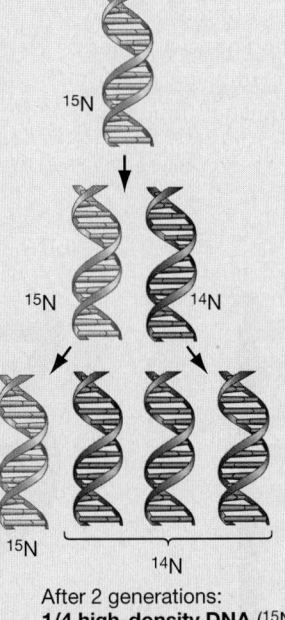

After 2 generations:
1/4 high-density DNA (^{15}N)
3/4 low-density DNA (^{14}N)

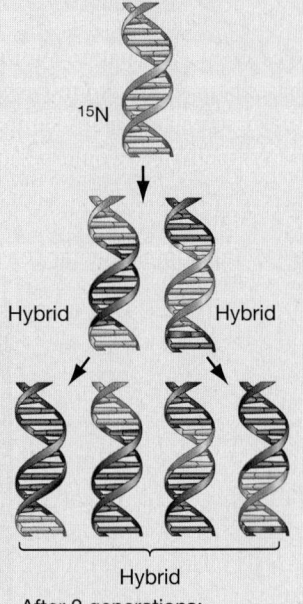

After 2 generations:
All intermediate-density DNA (hybrid)

RESULTS:

Top of centrifuge tube (lower density)

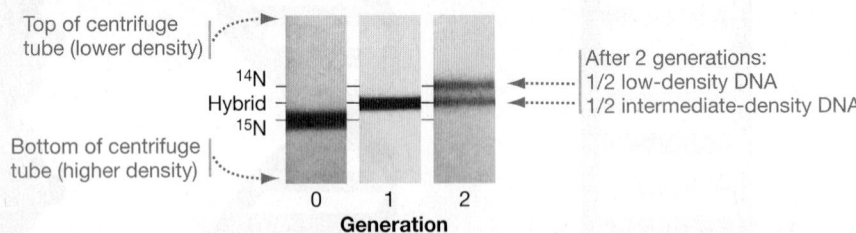

^{14}N
Hybrid
^{15}N

Bottom of centrifuge tube (higher density)

After 2 generations:
1/2 low-density DNA
1/2 intermediate-density DNA

0   1   2
Generation

CONCLUSION: Data from generation 1 conflict with conservative-replication hypothesis. Data from generation 2 conflict with dispersive-replication hypothesis. Replication is semiconservative.

FIGURE 14.5 The Meselson–Stahl Experiment.

SOURCE: Meselson, M. and F. W. Stahl. 1958. The replication of DNA in *Escherichia coli*. *Proceedings of the National Academy of Sciences USA* 44: 671–682.

✔**EXERCISE** Meselson and Stahl actually let their experiment run for a 4th generation with cultures growing in the presence of ^{14}N. Explain what data from 3rd- and 4th-generation DNA should look like—that is, where the DNA band(s) should be.

daughter DNA molecule consisting entirely of newly synthesized strands.

3. If the parent helix was cut in short sections before being unwound, copied, and put back together, then new and old segments would alternate—stretches of old DNA would be interspersed with new DNA down the length of each daughter molecule. This possibility is called dispersive replication.

Matthew Meselson and Franklin Stahl realized that if they could tag or mark parental and daughter strands of DNA in a way that would make them distinguishable from each other, they could determine whether replication was conservative, semiconservative, or dispersive.

Before they could do any tagging, however, they needed to choose an organism to study. They decided to work with the same inhabitant of the human gastrointestinal tract, the bacterium *Escherichia coli*, that Hershey and Chase did. Because *E. coli* is small and grows quickly and readily in the laboratory, it had become a favored model organism in studies of biochemistry and molecular genetics.

Like all organisms, bacterial cells copy their entire complement of DNA, or their **genome**, before every cell division. To distinguish parental strands of DNA from daughter strands when *E. coli* replicated, Meselson and Stahl grew the cells for successive generations in the presence of different isotopes of nitrogen: first ^{15}N and later ^{14}N. Because ^{15}N contains an extra neutron, it is heavier than the normal isotope, ^{14}N.

This difference in mass, which creates a difference in density between ^{14}N-containing and ^{15}N-containing DNA, was the key to the experiment summarized in **Figure 14.5**. The logic ran as follows:

● If different nitrogen isotopes were available in the growth medium when different generations of DNA were produced, then the parental and daughter strands will have different densities.

● The technique called density gradient centrifugation separates molecules based on their density (**BioSkills 11** in Appendix A). Low-density molecules cluster in bands high in the centrifuge tube; higher-density molecules cluster in bands lower in the centrifuge tube.

● When intact, double-stranded DNA molecules are subjected to density gradient centrifugation, DNA strands that contain ^{14}N should form a band higher in the centrifuge tube; DNA strands that contain ^{15}N should form a band lower in the centrifuge tube.

In short, DNA strands containing ^{14}N and strands containing ^{15}N should form separate bands. How could this tagging system be used to test whether replication is semiconservative, conservative, or dispersive?

Meselson and Stahl began by growing *E. coli* cells with nutrients that contained only ^{15}N. They purified DNA from a sample of these cells and transferred the rest of the culture to a growth medium containing only the ^{14}N isotope. After enough time had elapsed for these cells to divide once—meaning that the DNA had been copied once—they removed a sample and isolated the DNA. After the remainder of the culture had divided again, they removed another sample and isolated its DNA.

As Figure 14.5 shows, the conservative, semiconservative, and dispersive models make distinct predictions about the makeup of the DNA molecules after replication occurs in the first and second generation.

● If replication is conservative, then the daughter cells should have double-stranded DNA with either ^{14}N or ^{15}N, but not both. As a result, two distinct DNA bands should form in the centrifuge tube—one high-density band and one low-density band.

● If replication is semiconservative or dispersive, then all of the experimental DNA should contain an equal mix of ^{14}N or ^{15}N after one generation, causing all of the DNA to form one band, intermediate in density, in the centrifuge tube.

● After two generations, the situation changes. If replication is semiconservative, then half of the daughter cells should contain only ^{14}N—meaning a second, lower-density band should appear in the centrifuge tube. But the dispersive model predicts that after two generations there will still be just one band, intermediate in density.

The photograph at the bottom of Figure 14.5 shows the experiment's results. After one generation, the density of the DNA molecules was intermediate. These data suggested that the hypothesis of conservative replication was wrong. After two generations, a lower-density band appeared in addition to the intermediate-density band. This result offered strong support for the hypothesis that DNA replication is not dispersive but semiconservative. Each newly made DNA molecule comprises one old strand and one new strand.

14.3 A Comprehensive Model for DNA Synthesis

The DNA inside a cell is like an ancient text that has been painstakingly copied and handed down, generation after generation. But while the most ancient of all human texts contain messages that are thousands of years old, the DNA in living cells has been copied and passed down for billions of years. And instead of being copied by monks or clerks, DNA is replicated by molecular scribes. What molecules are responsible for copying DNA, and how do they work?

Meselson and Stahl showed that each strand of DNA is copied in its entirety each time replication occurs, but how does DNA synthesis proceed? Does it require an input of energy in the form of ATP, or it is spontaneous? Is it catalyzed by an enzyme, or does it occur quickly on its own?

The initial breakthrough in research on the reactions that make up DNA replication came with the discovery of an enzyme called **DNA polymerase**, so named because it polymerizes deoxyribonucleotides to DNA. This protein was found to catalyze DNA synthesis. Follow-up work showed that organisms contain several types of DNA polymerases. DNA polymerase III, for example, is the enzyme that is primarily responsible for copying *E. coli*'s chromosome prior to cell division.

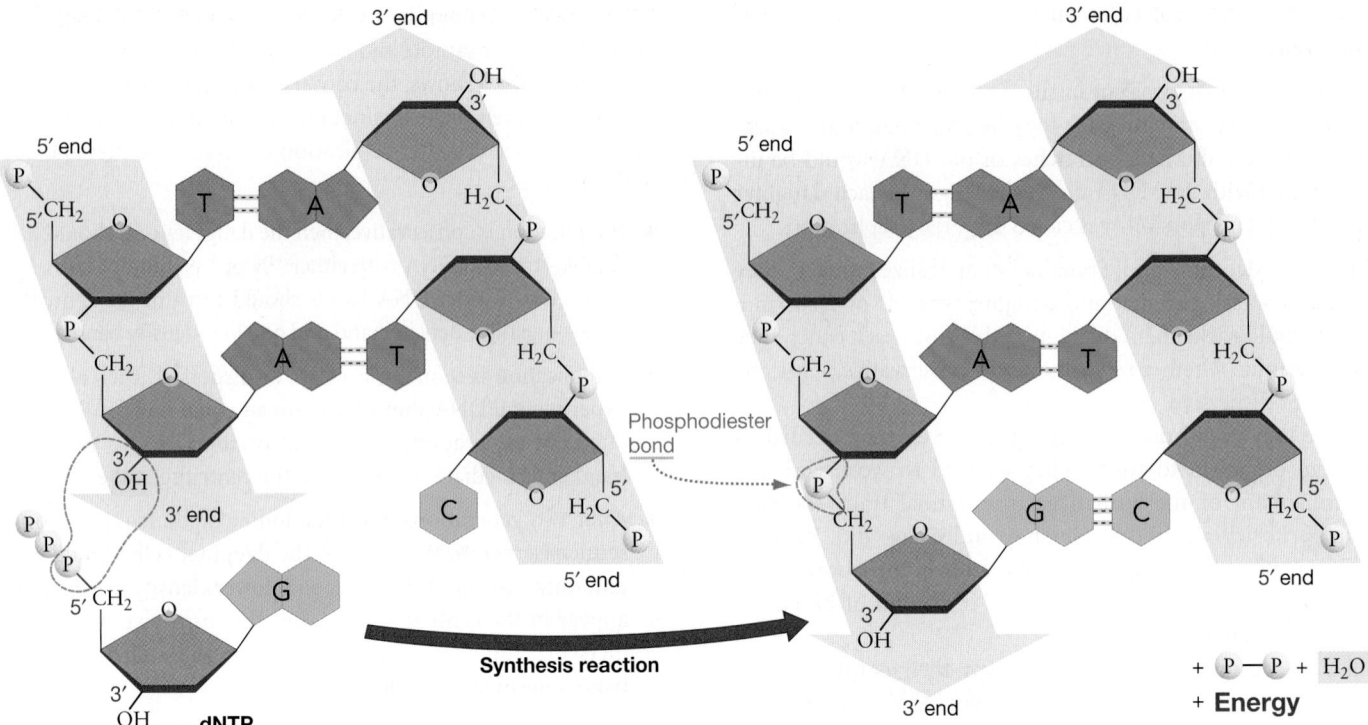

FIGURE 14.6 The DNA Synthesis Reaction. A condensation reaction results in formation of a phosphodiester bond between the 3′ carbon on the end of a DNA strand and the 5′ carbon on an incoming dNTP (deoxyribonucleoside triphosphate) monomer.

Figure 14.6 illustrates a critical characteristic of DNA polymerases: They can work in only one direction. 🔑 DNA polymerases can add deoxyribonucleotides to only the 3′ end of a growing DNA chain. As a result, DNA synthesis always proceeds in the 5′ → 3′ direction. ✔If you understand this concept, you should be able to draw two lines representing a DNA molecule, assign the 3′ to 5′ polarity of each strand, and then label the direction in which DNA synthesis will proceed for each strand.

Figure 14.6 makes another important point about DNA synthesis. You might recall from earlier chapters that polymerization reactions are endergonic. But in cells, the reaction is exergonic because the monomers that act as substrates in the reaction are **deoxyribonucleoside triphosphates (dNTPs)**. (The *N* in dNTP stands for any of the four bases found in DNA: adenine, thymine, guanine, or cytosine). Because they have three phosphate groups close together, dNTPs have high potential energy—high enough to make the formation of phosphodiester bonds in a growing DNA strand exergonic (see Chapter 9).

How Does Replication Get Started?

Another major insight into the mechanism of DNA synthesis emerged when electron micrographs caught chromosome replication in action. As **Figure 14.7a** shows, a "bubble" forms in a chromosome when DNA is actively being synthesized. Initially, the replication bubble forms at a specific sequence of bases called the **origin of replication** (**Figure 14.7b**). Bacterial chromosomes have a single location where the replication process begins, and thus a single bubble forms. Eukaryotes have multiple sites along

each chromosome where DNA synthesis begins, and thus multiple replication bubbles (**Figure 14.7c**).

Replication bubbles grow as DNA replication proceeds, because synthesis is bidirectional—that is, it occurs in both directions at the same time.

A specific suite of proteins is responsible for recognizing sites where replication begins and opening the double helix at those points. These proteins are activated by the proteins that initiate S phase in the cell cycle (see Chapter 11).

Once a replication bubble opens, a different suite of enzymes takes over and initiates replication. The action takes place in the corners of each replication bubble—at a structure called the replication fork (shown in Figure 14.7c). A **replication fork** is a Y-shaped region where the parent-DNA double helix is split into two single strands, which are then copied.

How Is the Helix Opened and Stabilized?

A battery of enzymes and specialized proteins converge on the point where the double helix opens. An enzyme called a **helicase** breaks the hydrogen bonds between deoxyribonucleotides. This reaction causes the two strands of DNA to separate. Proteins called **single-strand DNA-binding proteins** (**SSBPs**) attach to the separated strands and prevent them from snapping back into a double helix. In combination, then, the helicase and single-strand DNA-binding proteins open up the double helix and make both strands available for copying (**Figure 14.8, step 1**).

The "unzipping" process that occurs at the replication fork creates tension farther down the helix. To understand why, imag-

(a) A chromosome being replicated

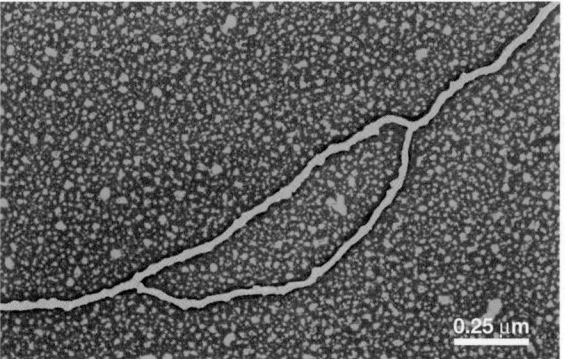

(b) Bacterial chromosomes have a single origin of replication.

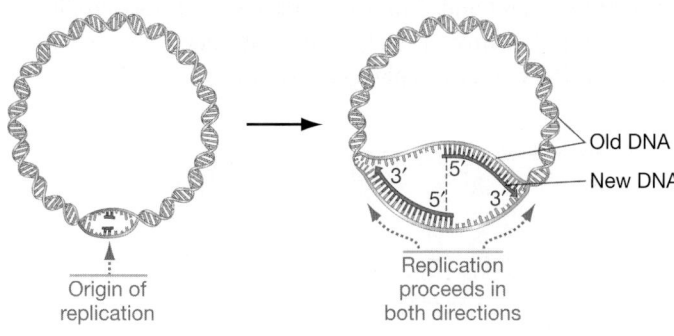

Old DNA

New DNA

Origin of replication

Replication proceeds in both directions

(c) Eukaryotic chromosomes have multiple origins of replication.

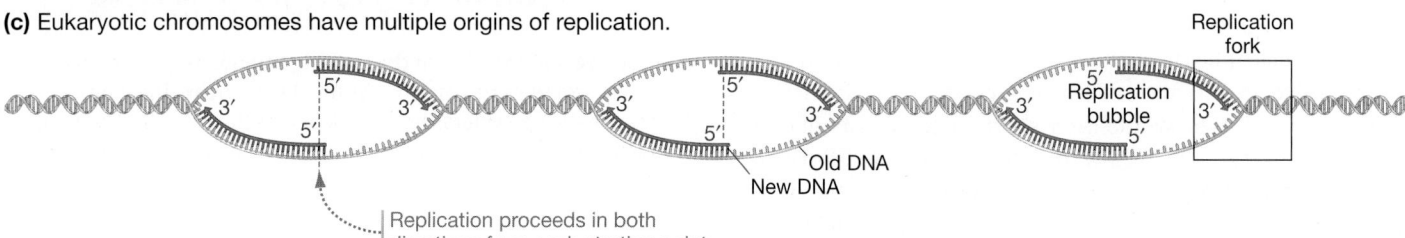

Replication fork

Replication bubble

Old DNA
New DNA

Replication proceeds in both directions from each starting point

FIGURE 14.7 DNA Synthesis Proceeds in Two Directions from a Point of Origin. Color has been added to the micrograph in part (a).

ine what would happen if you started to pull apart the twisted strands of a rope. The untwisting movements at one end would force the intact section to rotate in response. If the intact end of the rope were fixed in place, though, it would eventually begin to coil on itself and kink in response to the twisting forces. This does not happen in DNA, because the twisting stress induced by helicase is relieved by the proteins called topoisomerases. A **topoisomerase** is an enzyme that cuts DNA, allows it to unwind, and rejoins it ahead of the advancing replication fork.

Now, what happens once the DNA helix is open and has stabilized?

How Is the Leading Strand Synthesized?

The keys to understanding what happens at the start of DNA synthesis are to recall that DNA polymerase works only in the 5′ → 3′ direction and to recognize that to start synthesis, both a 3′ end and a single-stranded template are required. The single-stranded template dictates which deoxyribonucleotide should be

PROCESS: SYNTHESIS OF LEADING STRAND

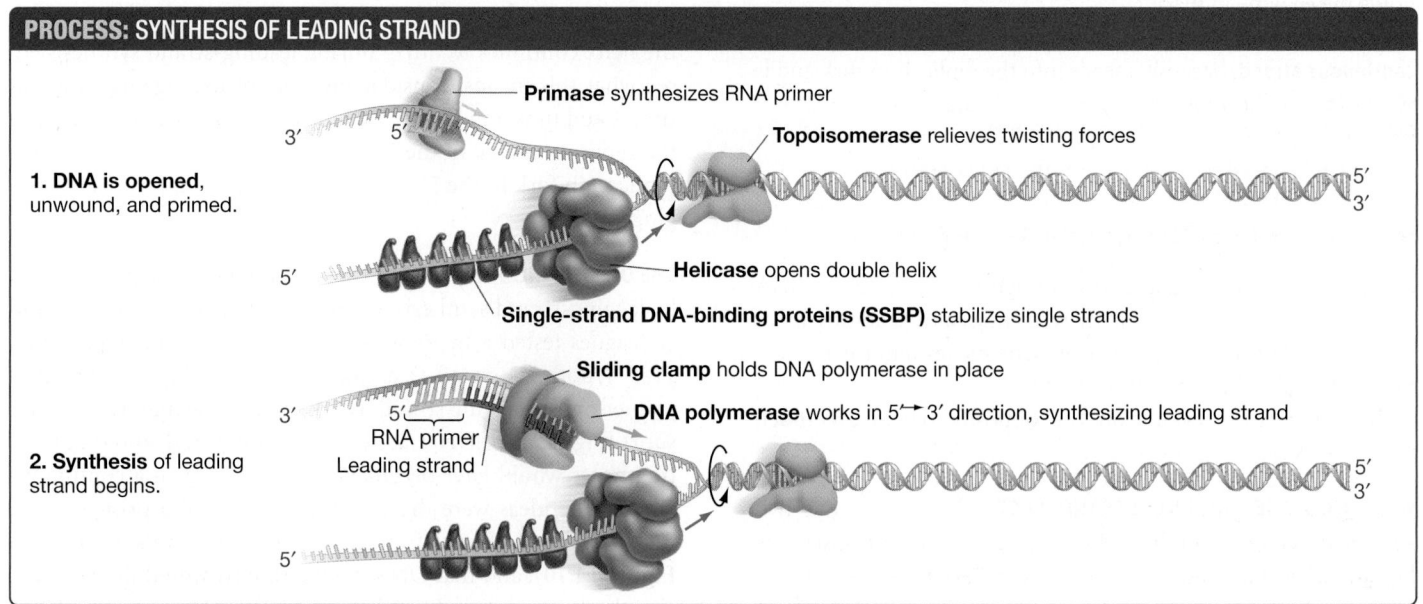

1. DNA is opened, unwound, and primed.

Primase synthesizes RNA primer

Topoisomerase relieves twisting forces

Helicase opens double helix

Single-strand DNA-binding proteins (SSBP) stabilize single strands

2. Synthesis of leading strand begins.

Sliding clamp holds DNA polymerase in place

DNA polymerase works in 5′→ 3′ direction, synthesizing leading strand

RNA primer
Leading strand

FIGURE 14.8 Leading-Strand Synthesis Is Continuous.

added next. DNA polymerase requires a **primer**—which consists of a few nucleotides bonded to the template strand—because it provides a free 3′ hydroxyl (—OH) group that can combine with an incoming dNTP to form a phosphodiester bond.

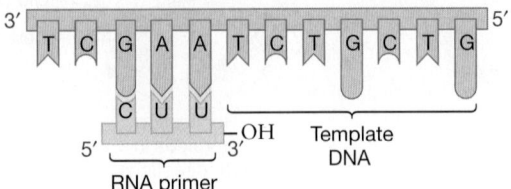

Once a primer is added to a single-stranded template, DNA polymerase begins working in the 5′ → 3′ direction and adds deoxyribonucleotides to complete the complementary strand.

Before DNA synthesis can get under way, then, an enzyme called **primase** has to synthesize a short stretch of ribonucleic acid (RNA) that acts as a primer for DNA polymerase. Primase is a type of **RNA polymerase**—an enzyme that catalyzes the polymerization of ribonucleotides into RNA (see Chapter 4 to review RNA's structure). Unlike DNA polymerases, primase and other RNA polymerases do not require a primer. These enzymes can attach ribonucleotides directly by complementary base pairing to single-stranded DNA. In this way, primase creates a primer for DNA synthesis.

Once the primer is in place, DNA polymerase begins adding deoxyribonucleotides to the 3′ end of it, producing a new strand with a sequence that is complementary to the template strand. As Figure 14.8, step 2, shows, DNA polymerase has a shape that grips the DNA strand during synthesis, a little like your hand clasping a rope. Catalysis takes place in the groove inside the enzyme, at an active site between the enzyme's "thumb" and "fingers." As DNA polymerase moves along the DNA molecule, a doughnut-shaped structure behind it, called the sliding clamp, holds the enzyme in place.

The enzyme's product is called the **leading strand**, or **continuous strand**, because it leads into the replication fork and is synthesized continuously. ✔If you understand leading-strand synthesis, you should be able to list the enzymes involved and predict the consequences if any of them are defective.

How Is the Lagging Strand Synthesized?

Synthesis of the leading strand is straightforward after an RNA primer is in place—DNA polymerase chugs along, adding bases to the 3′ end of that strand. The enzyme moves into the replication fork, which "unzips" ahead of it.

By comparison, events on the opposite strand are much more involved.

WHY DOES THE LAGGING STRAND LAG? Recall that the two strands in the DNA double helix are antiparallel—meaning they lie parallel to one another but oriented in opposite directions. DNA polymerase works in only one direction, however, so if the DNA polymerase that is synthesizing the leading strand works

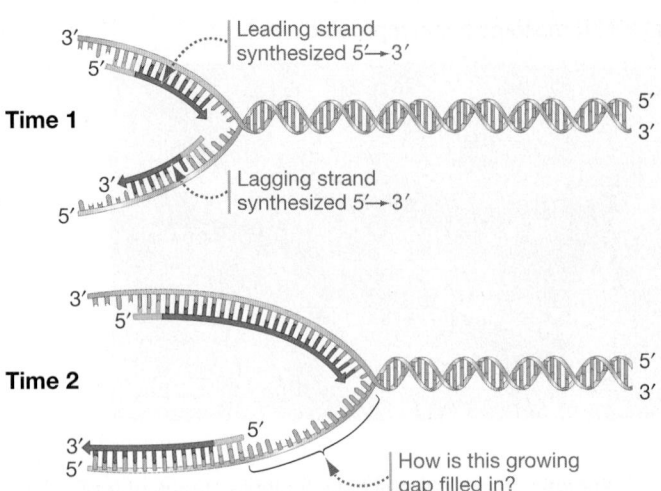

FIGURE 14.9 Gaps Form in the Lagging Strand. Gaps form in the lagging strand as helicase opens the double helix. The gaps occur because DNA polymerase can work only in the 5′ → 3′ direction, which in the lagging strand is the direction leading away from the replication fork.

into the replication fork, then a DNA polymerase must work *away* from the replication fork to synthesize the other strand in the 5′ → 3′ direction.

The strand that is synthesized in the opposite direction of the replication fork is called the **lagging strand**, because it lags behind the synthesis occurring in the fork.

The synthesis of the lagging strand starts when primase synthesizes a short stretch of RNA that acts as a primer. DNA polymerase III then adds bases to the 3′ end of the primer. The key observation is that the enzyme moves away from the replication fork. But behind it, helicase continues to open the replication fork and expose new single-stranded DNA on the lagging strand.

These events create a paradox. New single-stranded template DNA is constantly appearing behind DNA polymerase—in the direction *opposite* the direction of lagging-strand synthesis—as the helix continues to unzip during leading-strand synthesis. To see what this means, consider the state of the lagging strand at time 1 and time 2 in **Figure 14.9**. As DNA polymerase works into the replication fork in the 5′ → 3′ direction and away from the replication fork in the 5′ → 3′ direction, gaps appear in the lagging strand.

THE DISCONTINUOUS REPLICATION HYPOTHESIS The puzzle posed by lagging-strand synthesis was resolved when Reiji Okazaki and colleagues tested a hypothesis called discontinuous replication. This hypothesis stated that once primase synthesizes an RNA primer on the lagging strand, DNA polymerase might synthesize short fragments of DNA along the lagging strand, and that these fragments would later be linked together to form a continuous whole. The ideas were that primase would add a primer to the newly exposed single-stranded DNA at intervals (step 1 in **Figure 14.10**), and that DNA polymerase III would then synthesize the lagging strand until it reached the fragment produced earlier.

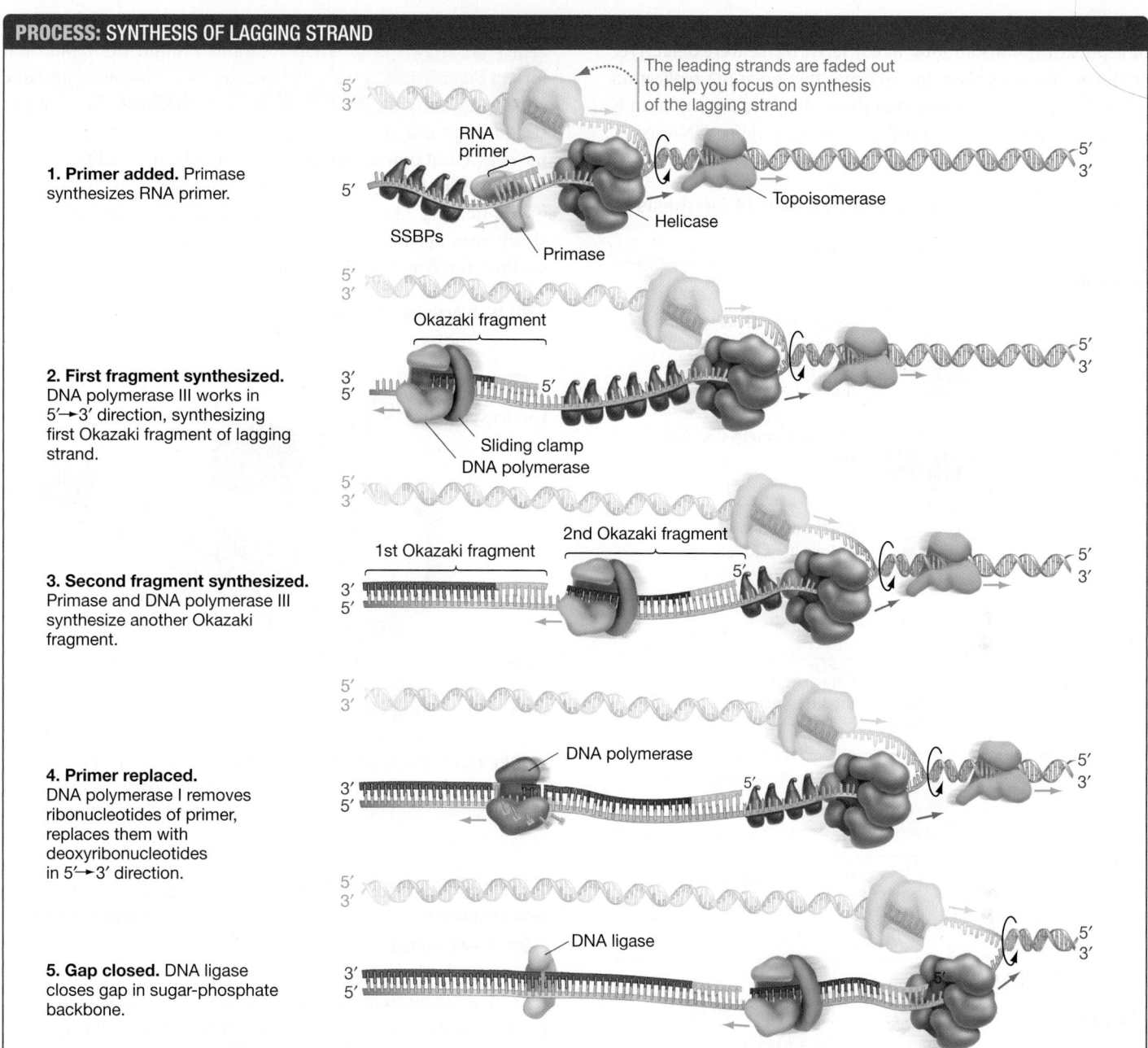

1. Primer added. Primase synthesizes RNA primer.

2. First fragment synthesized. DNA polymerase III works in 5'→3' direction, synthesizing first Okazaki fragment of lagging strand.

3. Second fragment synthesized. Primase and DNA polymerase III synthesize another Okazaki fragment.

4. Primer replaced. DNA polymerase I removes ribonucleotides of primer, replaces them with deoxyribonucleotides in 5'→3' direction.

5. Gap closed. DNA ligase closes gap in sugar-phosphate backbone.

Labels in figure: The leading strands are faded out to help you focus on synthesis of the lagging strand · RNA primer · Topoisomerase · Helicase · SSBPs · Primase · Okazaki fragment · Sliding clamp · DNA polymerase · 1st Okazaki fragment · 2nd Okazaki fragment · DNA polymerase · DNA ligase

FIGURE 14.10 The Completion of DNA Replication.

To explore this hypothesis, Okazaki's group set out to test a key prediction: Could they document the existence of short DNA fragments produced during replication? Their critical experiment was based on the pulse-chase strategy introduced in Chapter 7. Specifically, they added a short "pulse" of a radioactive deoxyribonucleotide to *E. coli* cells, followed by a large "chase" of nonradioactive deoxyribonucleotide. According to their discontinuous replication model, some of this radioactive deoxyribonucleotide should end up in short, single-stranded fragments of DNA.

THE DISCOVERY OF OKAZAKI FRAGMENTS As predicted, the researchers succeeded in finding these fragments when they purified DNA from the experimental cells and separated the molecules by centrifugation. A small number of labeled pieces of

DNA, about 1000 base pairs long, were present. These short sections came to be known as **Okazaki fragments** (see steps 2 and 3 of Figure 14.10). Subsequent work showed that Okazaki fragments in eukaryotes are smaller—just 100 to 200 base pairs long.

How are Okazaki fragments connected into a continuous whole? As step 4 of Figure 14.10 shows, a DNA polymerase removes the RNA primer at the start of each fragment and fills in the appropriate deoxyribonucleotides. Lastly, an enzyme called **DNA ligase** catalyzes the formation of a phosphodiester bond between the adjacent fragments (Figure 14.10, step 5).

Because Okazaki fragments are synthesized independently and joined together later, the lagging strand is also called the **discontinuous strand**. ✔If you understand lagging-strand synthesis, you should be able to draw what the two molecules resulting from DNA synthesis—starting from a single origin of replication—would look like if DNA ligase were defective.

In combination, then, the enzymes that open the replication fork and manage the synthesis of the leading and lagging strands succeed in producing a faithful copy of the original DNA mole-cule prior to mitosis or meiosis (**Table 14.1**). Although the enzymes are drawn at different locations around the replication fork in Figures 14.8 and 14.10, in reality most are joined into one large multi-enzyme machine called the **replisome**. The lagging strand loops out and around the complex so the replisome can move as a unit down the replication fork (**Figure 14.11**).

Before moving on to consider what happens at the ends of replicating chromosomes, you may want to go to the study area at *www.masteringbiology.com* to review the steps in leading strand and lagging strand synthesis.

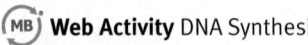
Web Activity DNA Synthesis

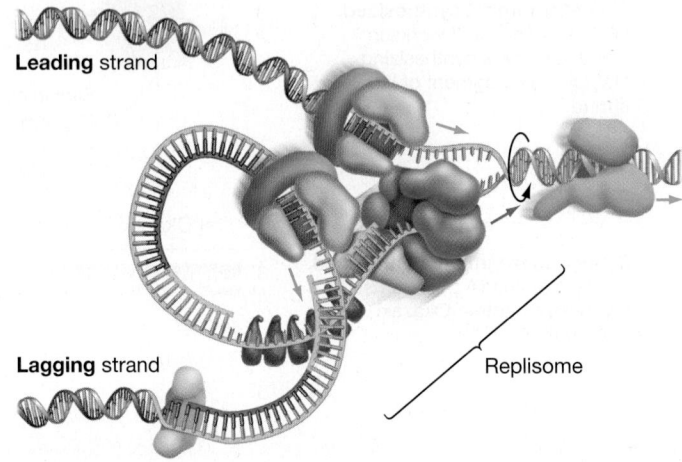

Leading strand

Lagging strand

Replisome

FIGURE 14.11 The Replisome. The enzymes required for DNA synthesis are organized into a multi-molecular machine. Note how the lagging strand loops out as the leading strand is being synthesized.

SUMMARY TABLE 14.1 **Proteins Required for DNA Synthesis**

Name	Structure	Function
Opening the helix		
Helicase		Catalyzes the breaking of hydrogen bonds between base pairs to open the double helix
Single-strand DNA-binding proteins		Stabilizes single-stranded DNA
Topoisomerase		Breaks and rejoins the DNA double helix to relieve twisting forces caused by the opening of the helix
Leading strand synthesis		
Primase		Catalyzes the synthesis of the RNA primer
DNA polymerase III		Extends the leading strand
Sliding clamp		Holds DNA polymerase in place during strand extension
Lagging strand synthesis		
Primase		Catalyzes the synthesis of the RNA primer on an Okazaki fragment
DNA polymerase III		Extends an Okazaki fragment
Sliding clamp		Holds DNA polymerase in place during strand extension
DNA polymerase I		Removes the RNA primer and replaces it with DNA
DNA ligase		Catalyzes the joining of Okazaki fragments into a continuous strand

CHECK YOUR UNDERSTANDING

If you understand that . . .

- DNA synthesis begins at specific origins of replication on the chromosome and then proceeds in both directions.
- Synthesis at the replication fork occurs in three steps:

  Step 1 Helicase opens the double helix, SSBPs stabilize the exposed single strands, and topoisomerase prevents kinks downstream of the fork;

  Step 2 DNA polymerase synthesizes the leading strand after primase has added an RNA primer; and

  Step 3 a series of enzymes synthesizes the lagging strand.
- Lagging-strand synthesis cannot be continuous, because it moves away from the replication fork. In bacteria, enzymes called primase, DNA polymerase III, DNA polymerase I, and ligase work in sequence to synthesize Okazaki fragments and link them into a continuous whole.

✓ **You should be able to . . .**

1. Explain the function of primase.
2. Explain why there is no helicase or topoisomerase required during lagging-strand synthesis.

Answers are available in Appendix B.

14.4 Replicating the Ends of Linear Chromosomes

The circular DNA molecules in bacteria and archaea can be synthesized by the events and enzymes introduced in Section 14.3, and so can the leading and lagging strands of the linear DNA molecules found in eukaryotes. But replication at the ends of linear chromosomes is another story altogether.

The region at the end of a linear chromosome is called a **telomere** (literally, "end-part"). **Figure 14.12** illustrates the problem that arises during the replication of telomeres.

- When the replication fork reaches the end of a linear chromosome, DNA polymerase synthesizes the leading strand all the way to the end of the parent DNA template (step 1 and step 2, top strand). As a result, leading-strand synthesis results in a normal copy of the DNA molecule.

- On the lagging strand, primase adds an RNA primer close to the tip of the chromosome (see step 2, bottom strand).

- DNA polymerase synthesizes the final Okazaki fragment on the lagging strand (step 3). DNA polymerase I removes the primer.

- DNA polymerase is unable to add DNA near the tip of the chromosome because there is not enough room for primase to add a new RNA primer (step 4). As a result, the single-stranded DNA that is left must stay single stranded.

The single-stranded DNA that remains at the end of the lagging strand is eventually degraded, which results in the shortening of the chromosome. If this process were to continue unabated, every chromosome would shorten by 50 to 100 deoxyribonucleotides on average each time DNA replication occurred. Over time, linear chromosomes would be expected to disappear completely.

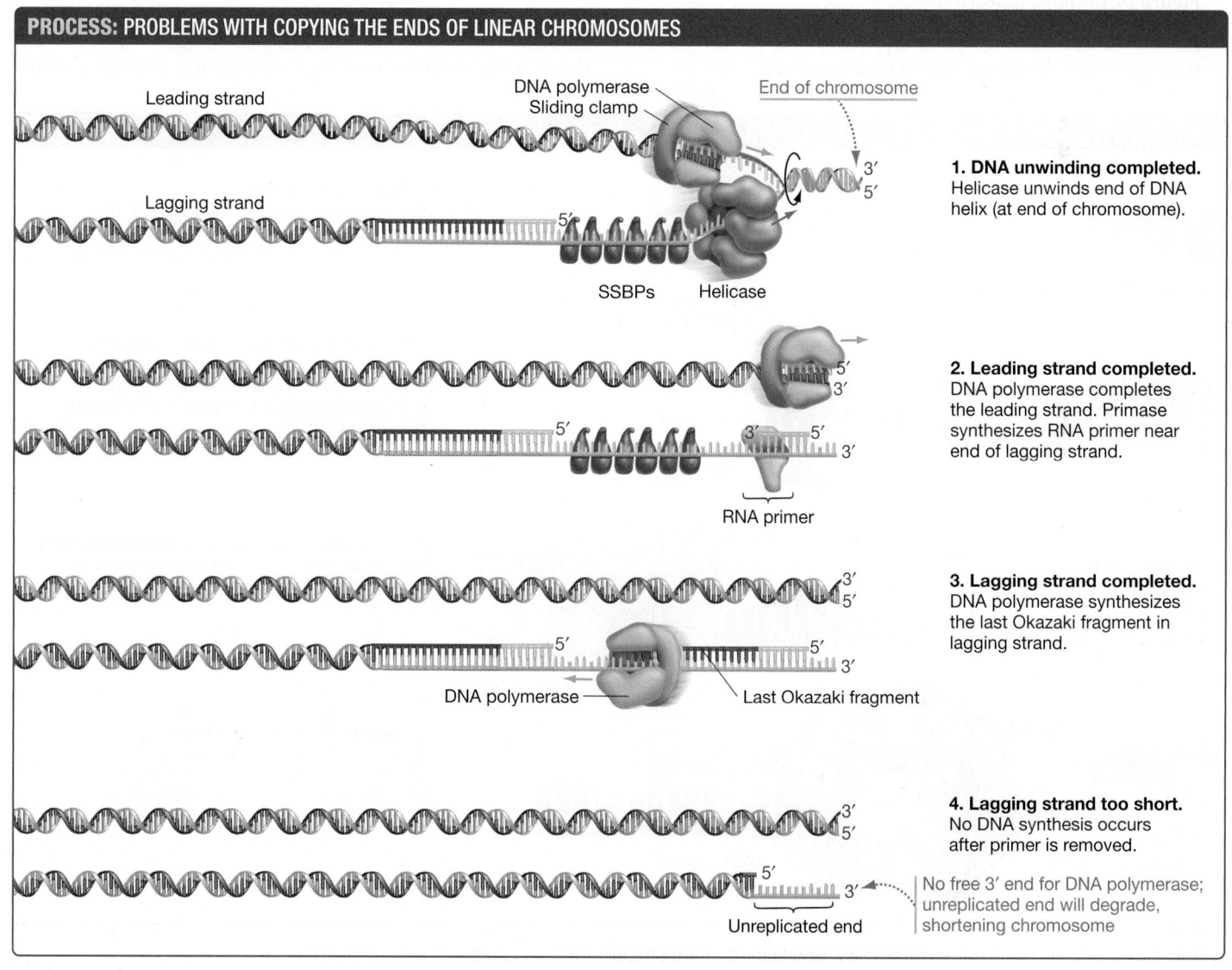

PROCESS: PROBLEMS WITH COPYING THE ENDS OF LINEAR CHROMOSOMES

Leading strand
Lagging strand
DNA polymerase
Sliding clamp
End of chromosome
3′
5′
SSBPs   Helicase

1. DNA unwinding completed. Helicase unwinds end of DNA helix (at end of chromosome).

5′
3′
5′
3′   5′
3′
RNA primer

2. Leading strand completed. DNA polymerase completes the leading strand. Primase synthesizes RNA primer near end of lagging strand.

3′
5′
5′
5′
3′
DNA polymerase   Last Okazaki fragment

3. Lagging strand completed. DNA polymerase synthesizes the last Okazaki fragment in lagging strand.

3′
5′
5′
3′
Unreplicated end

4. Lagging strand too short. No DNA synthesis occurs after primer is removed.

No free 3′ end for DNA polymerase; unreplicated end will degrade, shortening chromosome

FIGURE 14.12 Telomeres Shorten during Normal DNA Replication. An RNA primer is added to the lagging strand near the end of the chromosome. Once the primer is removed, it cannot be replaced with DNA. As a result, the chromosome shortens.

How do eukaryotes maintain the integrity of their linear chromosomes? An answer emerged after two important discoveries were made:

1. Telomeres do not contain genes that code for products needed in the cell.

2. An interesting enzyme called telomerase is involved in replicating telomeres.

Instead of genes, telomeres consist of short stretches of bases that are repeated over and over. In humans, for example, the base sequence TTAGGG is repeated thousands of times. Human telomeres consist of a total of about 10,000 deoxyribonucleotides with the sequence TTAGGGTTAGGGTTAGGG....

Telomerase is remarkable because it catalyzes the synthesis of DNA from an RNA template. In fact, the enzyme carries an RNA molecule with it that acts as a built-in template, allowing telomerase to add DNA onto the end of a chromosome and prevent it from getting shorter.

Figure 14.13 shows how telomerase works.

Step 1 The unreplicated segment of the telomere at the 3′ end of the lagging strand forms a single-strand "overhang."

Step 2 Telomerase binds to the overhanging section of single-stranded parent DNA. Once the enzyme has bound, it begins catalyzing the addition of deoxyribonucleotides, in the 5′ → 3′ direction, that are complementary to its built-in RNA template.

Step 3 Telomerase moves in the 5′ → 3′ direction and continues to catalyze the addition of deoxyribonucleotides.

Step 4 Once the single-stranded "overhang" on the lagging strand is lengthened in this way, the normal machinery of DNA synthesis—primase, DNA polymerase, and ligase—resumes synthesis of the lagging strand in the 5′ → 3′ direction. The result is that the lagging strand becomes slightly longer than it was originally.

If you go to the study area at *www.masteringbiology.com*, you'll be about to review all aspects of DNA synthesis, in action:

(MB) **BioFlix**™ DNA Replication

It is important to recognize, though, that telomerase is not active in most types of cells. In humans, for example, active telomerase is found primarily in the cells of reproductive organs—specifically, in the cells that eventually undergo meiosis and produce gametes.

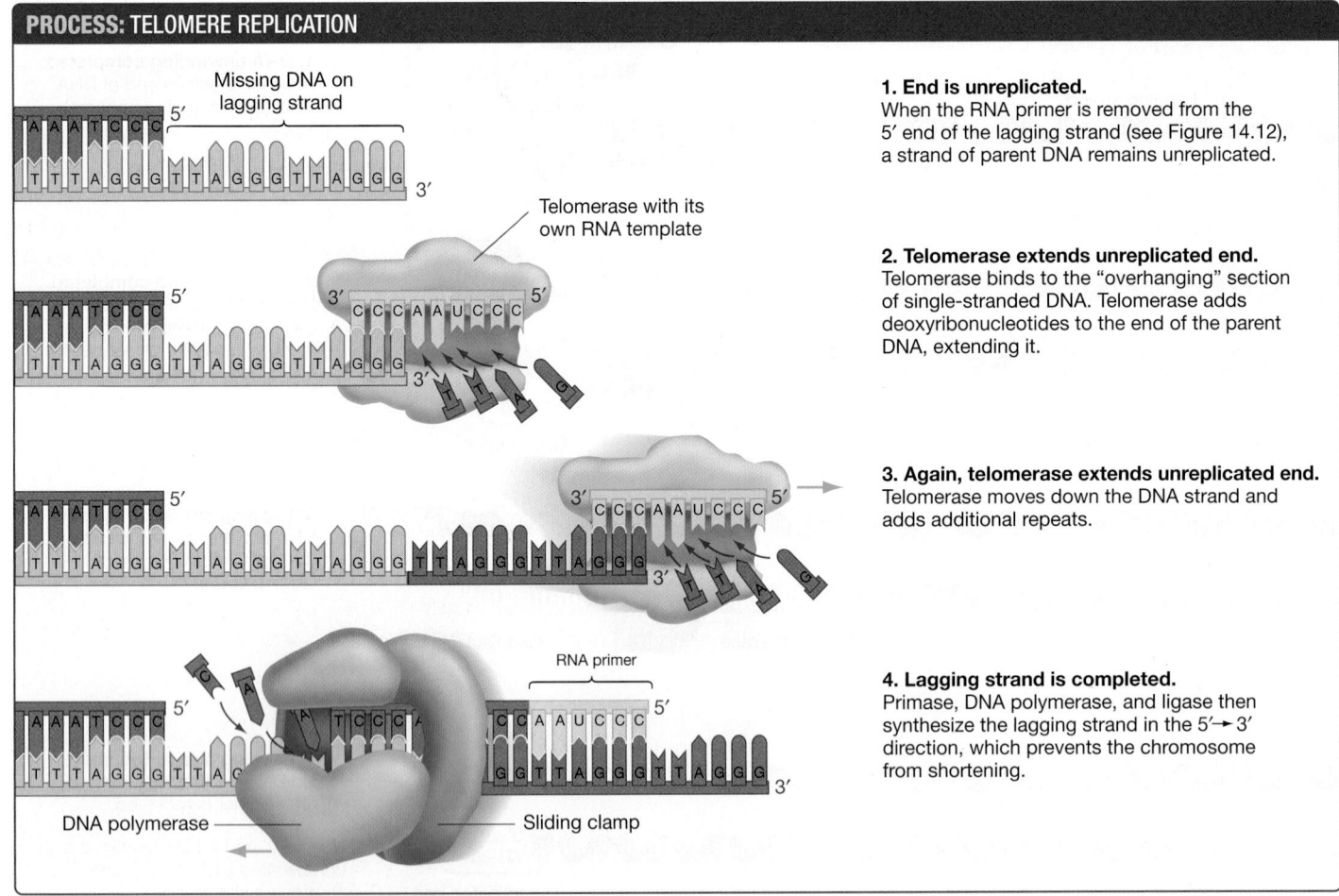

PROCESS: TELOMERE REPLICATION

Missing DNA on lagging strand

Telomerase with its own RNA template

1. End is unreplicated.
When the RNA primer is removed from the 5′ end of the lagging strand (see Figure 14.12), a strand of parent DNA remains unreplicated.

2. Telomerase extends unreplicated end.
Telomerase binds to the "overhanging" section of single-stranded DNA. Telomerase adds deoxyribonucleotides to the end of the parent DNA, extending it.

3. Again, telomerase extends unreplicated end.
Telomerase moves down the DNA strand and adds additional repeats.

RNA primer

4. Lagging strand is completed.
Primase, DNA polymerase, and ligase then synthesize the lagging strand in the 5′→ 3′ direction, which prevents the chromosome from shortening.

DNA polymerase — Sliding clamp

FIGURE 14.13 Telomerase Prevents Shortening of Telomeres during Replication. By extending the number of repeated sequences in the 5′ → 3′ direction, telomerase provides room for enzymes to add an RNA primer to the lagging strand. DNA polymerase can then fill in the missing section of the lagging strand.

✓**QUESTION** Would this telomerase work as well if its RNA template had a different sequence?

Somatic cells, meaning cells that are not involved in gamete formation, normally lack telomerase. As predicted, the chromosomes of somatic cells gradually shorten with each mitotic division, getting progressively smaller as an individual grows and ages.

These observations inspired a pair of important hypotheses. The first was that telomere shortening causes cells to stop dividing and enter the nondividing state called G_0 (see Chapter 11). The second was that if telomerase were mistakenly activated in a somatic cell, telomeres would fail to shorten. This would allow the cell to keep dividing and might possibly contribute to uncontrolled growth and cancer.

To test the first hypothesis, biologists added functioning telomerase to human cells growing in vitro. The treated cells continued dividing long past the age when otherwise identical cells stop growing. These results have convinced most biologists that telomere shortening has a role in limiting the amount of time cells remain in an actively growing state.

A link between continued telomerase activity and cancer formation has been harder to nail down, however. One suggestive observation is that many cancerous cells in humans and other organisms have functioning telomerase or some other mechanism for maintaining telomere length, while their noncancerous cells do not.

Could drugs that knock out telomerase be an effective way to fight cancer? So far, the data on this question are unclear. Research continues.

CHECK YOUR UNDERSTANDING

⊙━ If you understand that . . .

- Chromosomes shorten during replication because the end of the lagging strand lacks a primer and cannot be synthesized.
- Shortening is prevented in certain cells—particularly those that produce sperm and egg—because telomerase adds short, repeated DNA sequences to the template strand. Primase can then add an RNA primer to the lagging strand, and DNA polymerase can fill in the missing sections.

✓ You should be able to . . .

1. Explain why telomerase is not found in bacterial cells.
2. Explain why telomerase has to have a built-in template.

Answers are available in Appendix B.

14.5 Repairing Mistakes and Damage

DNA polymerases work fast. In yeast, for example, each replication fork is estimated to move at a rate of about 50 bases per second. But the replication process is also astonishingly accurate. In organisms ranging from *E. coli* to animals, the error rate during DNA replication averages less than one mistake per *billion* deoxyribonucleotides.

This level of accuracy is critical. Humans, for example, develop from a fertilized egg that has DNA containing over 6 billion deoxyribonucleotides. The DNA inside the fertilized egg is replicated over and over to create the trillions of cells that even-

tually make up the adult body. If more than one or two mutations occurred during each cell division cycle as a human grew, genes would be riddled with errors by the time the individual reached maturity. Genes that contain errors are often defective.

Based on these observations, it is no exaggeration to claim that the accurate replication of DNA is a matter of life and death. Natural selection favors individuals with enzymes that copy DNA as quickly and exactly as possible.

These observations raise a key question. How can the enzymes of DNA replication be as precise as they are?

Correcting Mistakes in DNA Synthesis

As DNA polymerase marches along a parent DNA template, hydrogen bonding occurs between incoming deoxyribonucleotides and the deoxyribonucleotides on the template strand. DNA polymerases are selective about the bases they add to a growing strand because the correct base pairings (A-T and G-C) are energetically the most favorable of all possibilities for the pairing of nitrogen-containing bases. As a result, the enzyme inserts an incorrect deoxyribonucleotide only about once every 100,000 bases added (**Figure 14.14a**).

An error rate of one in 100,000 seems low, but it is much higher than the one-in-a-billion rate claimed at the start of this section. What happens when DNA polymerase makes a mistake?

DNA POLYMERASE CAN PROOFREAD Biologists were able to study why DNA synthesis is so accurate when they found cells where DNA synthesis was *in*accurate.

Specifically, researchers found mutants in *E. coli* with error rates that were 100 times greater than normal. Recall from Chapter 13 that a mutant is an individual with a novel trait caused by mutation, and that mutation is a change in the gene responsible for that trait. Many mutations change the individual's phenotype. The change may result in a trait such as white eyes in fruit flies or an elevated

(a) DNA polymerase III adds a mismatched base...

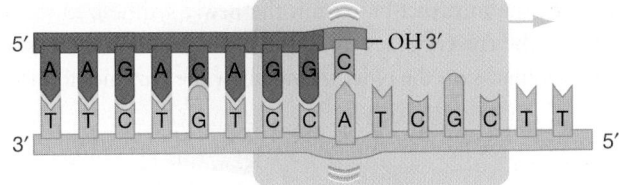

(b) ...but notices the mistake and corrects it.

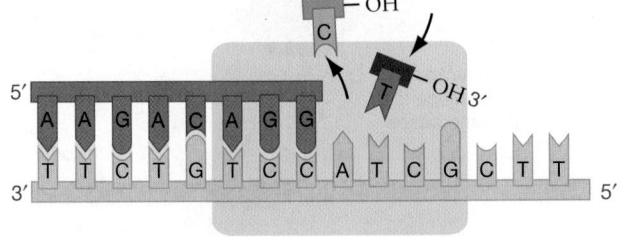

FIGURE 14.14 DNA Polymerase Can Proofread. If a mismatch such as the pairing of A with C occurs **(a)**, DNA polymerase can act as a $5' \rightarrow 3'$ exonuclease, meaning that it can remove bases in that direction **(b)**. The enzyme then adds the correct base.

mutation rate in *E. coli*. At the molecular level, a mutant phenotype usually results from a change in an enzyme or other type of protein.

In the case of *E. coli* cells with high mutation rates, biologists found that the mutation responsible for the high mutation rates was a defect in a portion of the polymerase III enzyme called the **ε** (epsilon) subunit. Further analyses showed that this subunit of the enzyme acts as an exonuclease—meaning an enzyme that removes deoxyribonucleotides from the ends of DNA strands (**Figure 14.14b**). If a new deoxyribonucleotide is not correctly hydrogen bonded to a base on the complementary strand, DNA polymerase backs up and the **ε** subunit removes it.

These findings led to the conclusion that DNA polymerase III can **proofread**. If the wrong base is added during DNA synthesis, the enzyme pauses, removes the mismatched base that was just added, and then proceeds with synthesis.

Eukaryotic DNA polymerases have the same type of proofreading ability. Typically, proofreading reduces a DNA polymerase's error rate to about 1×10^{-7} (one mistake per 10 million bases). Is this accurate enough? The answer is no.

MISMATCH REPAIR 🔑 If—in spite of its proofreading ability—DNA polymerase leaves a mismatched pair behind in the newly synthesized strand, a battery of enzymes springs into action to correct the problem. **Mismatch repair** occurs when mismatched bases are corrected after DNA synthesis is complete.

The proteins responsible for mismatch repair were discovered in the same way that the proofreading capability of DNA polymerase III was uncovered—by analyzing *E. coli* mutants. In this case, the mutants had normal DNA polymerase III but abnormally high mutation rates.

The first mutant gene that caused a deficiency in mismatch repair was identified in the late 1960s and was called *mutS*. (The *mut* is short for "mutator.") By the late 1980s, researchers had identified 10 proteins involved in the identification and repair of base-pair mismatches in *E. coli*.

These proteins recognize the mismatched pair, remove a section containing the incorrect base from the newly synthesized strand, and fill in the correct bases using the older strand as a template. (Chemical marks on the older strand allow the enzymes to distinguish the original strand from the newly synthesized strand.) The mismatch-repair enzymes are like a copy editor who corrects typos that the author—DNA polymerase, in this case—did not catch.

Repairing Damaged DNA

Even after DNA is synthesized and proofread and mismatches repaired, the job of ensuring accuracy doesn't end. Genes are under constant assault. Nucleotides are damaged by chemicals like the hydroxyl (OH) radicals produced during aerobic metabolism, the aflatoxin B1 found in moldy peanuts and corn, and the benzo[α]pyrene in cigarette smoke.

Radiation is another danger. Ultraviolet (UV) light, for example, can cause a covalent bond to form between adjacent pyrimidine bases. The thymine-thymine pair illustrated in **Figure 14.15** is an example. This defect, called a thymine dimer, creates kinks in the secondary structure of DNA. The kinks stall the movement of the replication fork during DNA replication and impair the enzymes responsible for using the information in genes. If the damage is not repaired, the cell may die.

To fix problems caused by chemical attack, radiation, or other events, cells have a wide array of damage-repair systems. As an example, consider the system called **nucleotide excision repair**. As step 1 in **Figure 14.16** shows, the symmetry and regularity of DNA's secondary structure makes it possible for repair proteins to recognize thymine dimers and other types of damaged bases that produce an irregularity in the molecule. Once a damaged region is recognized, enzymes remove a segment of single-stranded DNA containing the defective sequence (step 2). The presence of a DNA strand complementary to the damaged strand provides a template for synthesis of a corrected strand (steps 3 and 4).

In this way, DNA's secondary structure makes accurate repair possible, supporting the molecule's function in information storage and processing. But what happens when repair systems are defective?

Xeroderma Pigmentosum: A Case Study

Xeroderma pigmentosum (XP) is a rare autosomal recessive disease in humans. Individuals with this condition are extremely sensi-

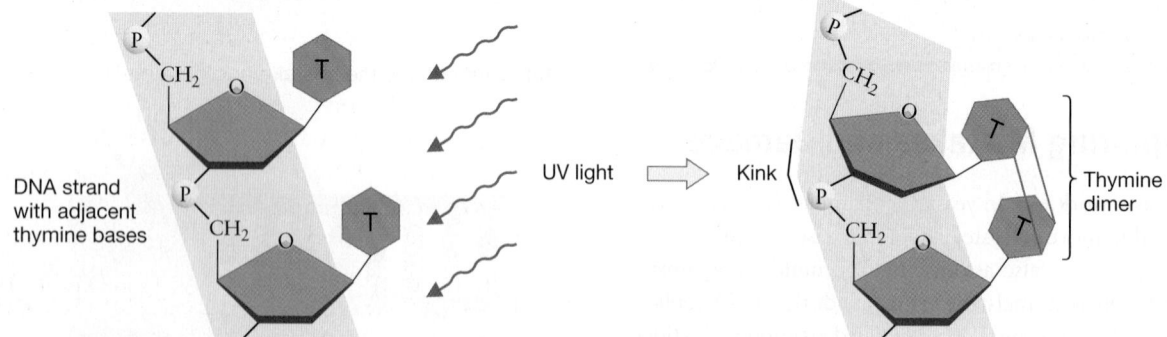

FIGURE 14.15 UV Light Damages DNA. When UV light strikes a section of DNA that has adjacent thymines, the energy can break bonds within each base and allow bonds to form *between* them. The thymine dimer that is produced causes a kink in the DNA.

✔**QUESTION** Why are infrared wavelengths much less likely than UV to damage DNA? (Hint: see Figure 10.3.)

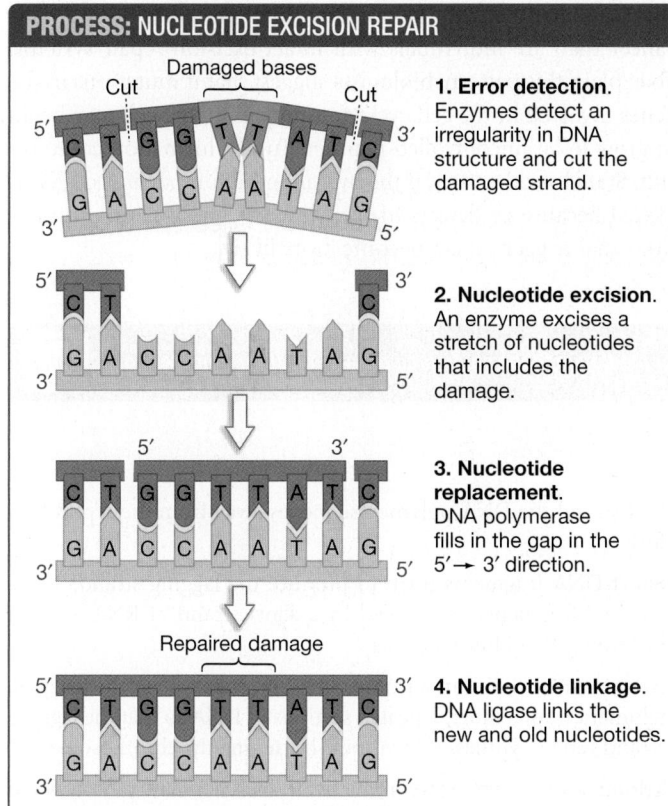

PROCESS: NUCLEOTIDE EXCISION REPAIR

1. Error detection. Enzymes detect an irregularity in DNA structure and cut the damaged strand.

2. Nucleotide excision. An enzyme excises a stretch of nucleotides that includes the damage.

3. Nucleotide replacement. DNA polymerase fills in the gap in the $5' \rightarrow 3'$ direction.

4. Nucleotide linkage. DNA ligase links the new and old nucleotides.

FIGURE 14.16 In Excision Repair, Defective Bases Are Removed and Replaced.

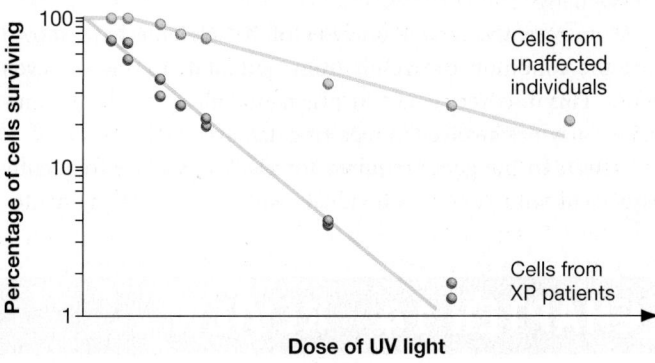

(a) Vulnerability of cells to UV light damage

Cells from unaffected individuals

Cells from XP patients

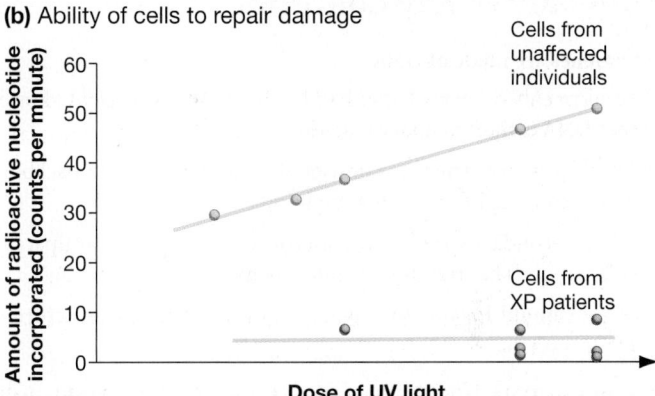

(b) Ability of cells to repair damage

Cells from unaffected individuals

Cells from XP patients

FIGURE 14.17 DNA Damage from UV Light Is Not Repaired Properly in Individuals with XP. (a) When cell cultures from unaffected individuals and from XP patients are irradiated with various doses of UV light, the percentage of cells that survive is strikingly different. **(b)** When cell cultures from unaffected individuals and from XP patients are irradiated with various doses of UV light and then fed radioactive thymidine, only unaffected individuals incorporate the labeled base into their DNA.

✔ **QUESTION** Why are light-skinned people who cultivate a skin tan increasing their risk of developing cancer? (Hint: Tanning is a response to UV light.)

tive to ultraviolet (UV) light. Their skin develops lesions after even slight exposure to sunlight. In unaffected individuals, these kinds of lesions develop only after extensive exposure to UV light, X-rays, or other forms of high-energy radiation.

In 1968 James Cleaver proposed a connection between XP and DNA excision repair systems. He knew that in *E. coli*, mutations in certain genes cause DNA nucleotide excision repair proteins to fail. Cells with these mutations have an increased sensitivity to radiation. Cleaver's hypothesis was that people with XP have similar mutations. He claimed that they are extremely sensitive to sunlight because they are unable to repair the damage that occurs when the deoxyribonucleotides in DNA absorb UV light.

Cleaver and other researchers made extensive use of cell cultures to study the hypothesized connection between DNA damage, faulty nucleotide excision repair, and XP (for an introduction to cell cultures, see **BioSkills 12** in Appendix A). For example, they collected skin cells from people with XP and from people with a normal phenotype for excision repair. When these cell populations were grown in culture, the biologists exposed them to increasing amounts of ultraviolet radiation and recorded how many died.

Figure 14.17a shows the data that resulted. Note that the intensity of the radiation is graphed on the *x*-axis, with the percentage of cells surviving graphed on the *y*-axis. Note, too, that the *y*-axis is logarithmic on this graph. (For help with logarithms, see **BioSkills 7** in Appendix A; for help with reading graphs, see **BioSkills 2**). The graph has data from both healthy cells and cells from XP pa-

tients. Cell survival declined with increasing radiation dose in both types of cells, but cells from xeroderma pigmentosum individuals died off much more rapidly than cells from unaffected individuals.

The connection to excision repair systems was confirmed when Cleaver exposed cells from unaffected individuals and cells from XP individuals to various amounts of UV light, then fed the cells a radioactive deoxyribonucleotide to label DNA synthesized during the repair period. If repair is defective in XP individuals, then their cells should incorporate virtually no radioactive deoxyribonucleotide into their DNA. Cells from unaffected individuals, in contrast, should incorporate large amounts of labeled deoxyribonucleotide into their DNA.

As **Figure 14.17b** shows, this is exactly what happens. Here the amount of radioactive dNTPs incorporated into DNA is graphed against radiation dose, for both healthy and XP cells. Increasingly large amounts of radioactive dNTPs are found in the DNA of healthy cells as UV dose increases, though no such increase occurs in XP cells. These data are consistent with the hypothesis that excision repair is virtually nonexistent in XP individuals.

More recently, genetic analyses of XP patients have shown that the condition can result from mutations in any of seven genes. This discovery is not surprising in light of the large number of enzymes involved in repairing damaged DNA.

Defects in the genes required for DNA repair are frequently associated with cancer. Individuals with xeroderma pigmento-sum, for example, are 1000 to 2000 times more likely to get skin cancer than are individuals with intact excision repair systems. To explain this pattern, biologists suggest that if mutations in the genes involved in the cell cycle go unrepaired, the cell may begin to grow in an uncontrolled manner. Tumor formation could result. Stated another way, if the overall mutation rate in a cell is elevated because of defects in DNA repair genes, then the mutations that trigger cancer become more likely.

CHAPTER 14   REVIEW

For media, go to the study area at www.masteringbiology.com

Summary of Key Concepts

🔑 **Genes are made of DNA.**

- Experiments on viruses that had labeled proteins or DNA showed that DNA is the hereditary material.

- DNA's primary structure consists of a sugar-phosphate backbone and a sequence of nitrogen-containing bases.

- DNA's secondary structure consists of two strands in an antiparallel orientation. The strands twist into a helix.

  ✔You should be able to explain why the secondary structure of DNA is stable.

🔑 **When DNA is being copied, each strand of the double helix serves as the template for the synthesis of a complementary strand.**

- By labeling DNA with ^{15}N or ^{14}N, researchers were able to validate the hypothesis that DNA replication is semiconservative, meaning that each strand is copied in its entirety.

- Each strand of a parent DNA molecule provides a template for the synthesis of a daughter strand, resulting in two complete DNA double helices.

  ✔You should be able to write a sequence of double-stranded DNA that is 10 base-pairs long, separate the strands, and, without comparing them, write in the bases that are added during DNA replication.

🔑 **DNA synthesis occurs in the $5' \rightarrow 3'$ direction only and requires a large suite of specialized enzymes. The leading strand is synthesized continuously, but the lagging strand is synthesized as a series of fragments that are then linked together.**

- DNA synthesis is an enzyme-catalyzed reaction that takes place in one direction.

- DNA synthesis requires both a template and a primer sequence, and it takes place at the replication fork, where the double helix is being opened.

- Synthesis of the leading strand in the $5' \rightarrow 3'$ direction is continuous, but synthesis of the lagging strand is more complex because on that strand the DNA polymerase moves away from the replication fork.

- Short DNA fragments form to produce the lagging strand. These Okazaki fragments are primed by a short strand of RNA and are linked together after synthesis.

- At the ends of linear chromosomes in eukaryotes, the enzyme telomerase adds short, repeated sections of DNA so that the lagging strand can be synthesized without shortening the chromosome.

- Telomerase is active in reproductive cells that eventually undergo meiosis. As a result, gametes contain chromosomes of normal length.

  ✔You should be able to compare and contrast the functions of the three polymerases introduced in this chapter: DNA polymerase, primase, and telomerase.

  (MB) Web Activity DNA Synthesis, **BioFlix™** DNA Replication

🔑 **Specialized enzymes repair damages to DNA and fix mistakes in DNA synthesis. If these enzymes are defective, the mutation rate increases.**

- DNA replication is remarkably accurate because DNA polymerase proofreads and because mismatch-repair enzymes remove incorrect bases once synthesis is complete and replace them with the correct sequence.

- DNA repair occurs after bases have been damaged by chemicals or radiation.

- Nucleotide excision repair systems cut out damaged portions of genes and replace them with correct sequences.

- Several types of human cancers are associated with defects in the genes responsible for DNA repair.

  ✔You should be able to explain the logical connections between failure of repair systems, increases in mutation rate, and high likelihood of cancer developing.

Questions

1. What does it mean to say that strands in a double helix are antiparallel?
   a. Their primary sequences consist of a sequence of *complementary* bases.
   b. They each have a sugar-phosphate backbone.
   c. They each have a $5' \rightarrow 3'$ directionality.
   d. They have opposite directionality, or polarity.

2. Which of the following is *not* a property of DNA polymerase?
   a. It catalyzes the addition of dNTPs only in the $5' \rightarrow 3'$ direction.
   b. It requires a primer to work.
   c. It is associated with a sliding clamp only on the leading strand.
   d. It can proofread because it has an exonuclease activity.

3. What is the function of topoisomerase?
   a. holding DNA polymerase steady as it moves down the leading or lagging strand
   b. opening the DNA helix at the replication fork
   c. stabilizing single strands of DNA, once the replication fork is open
   d. preventing kinks in DNA as the replication fork opens and unwinds

4. What is the function of primase?
   a. synthesis of the short section of double-stranded DNA required by DNA polymerase
   b. synthesis of a short section of RNA, complementary to single-stranded DNA

c. closing the gap at the 3′ end of DNA after excision repair
d. removing primers and synthesizing a short section of DNA to replace them

5. Where and how are Okazaki fragments synthesized?
   a. on the leading strand, in a $5' \rightarrow 3'$ direction
   b. on the leading strand, in a $3' \rightarrow 5'$ direction
   c. on the lagging strand, in a $5' \rightarrow 3'$ direction
   d. on the lagging strand, in a $3' \rightarrow 5'$ direction

6. What does telomerase do?
   a. It adds a protein primer to the ends of linear chromosomes.
   b. It adds double-stranded DNA to the "blunt end" of a linear chromosome.
   c. It adds double-stranded DNA to the lagging strand at the end of a linear chromosome.
   d. It adds single-stranded DNA to the lagging strand at the end of a linear chromosome.

✓ TEST YOUR UNDERSTANDING

Answers are available in Appendix B

1. Researchers design experiments so that only one thing is different between the treatments that are being compared. In the Hershey-Chase experiment, what was this single difference?

2. What does it mean to say that DNA replication is "bidirectional" from the point of origin?

3. Why is the synthesis of the lagging strand of DNA discontinuous? How is it possible for the synthesis of the leading strand to be continuous?

4. Explain how telomerase prevents linear chromosomes from shortening during replication.

5. List the three events that increase the accuracy of DNA replication, in chronological order. Indicate which events involve DNA polymerase III and which involve specialized repair enzymes.

6. Explain how the structure of DNA makes it relatively easy for proteins to recognize base-pair mismatches or damaged bases. How does DNA's secondary structure make it possible for damaged sections or incorrect bases to be removed and repaired?

✓ APPLYING CONCEPTS TO NEW SITUATIONS

Answers are available in Appendix B

1. If DNA polymerase III did not require a primer, which steps in DNA synthesis would differ from what is observed? Would any special enzymes be required to replicate telomeres? Explain your answers.

2. In the late 1950s Herbert Taylor grew bean root-tip cells in a solution of radioactive thymidine and allowed them to undergo one round of DNA replication. He then transferred the cells to a solution without the radioactive deoxyribonucleotide, allowed them to replicate again, and examined their chromosomes for the presence of radioactivity. His results are shown in the following figure, where red indicates a radioactive chromatid.

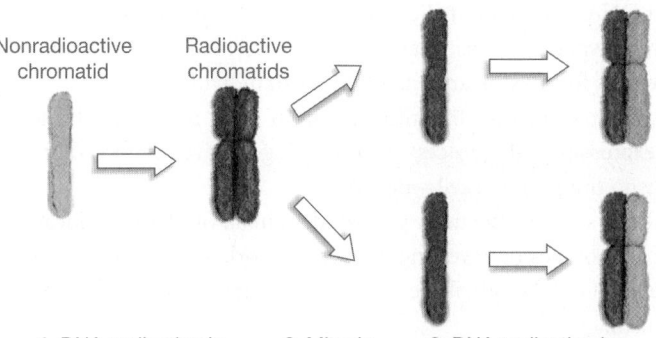

1. DNA replication in radioactive solution   **2.** Mitosis   **3.** DNA replication in nonradioactive solution

   a. Draw diagrams explaining the pattern of radioactivity observed in the sister chromatids after the first and second rounds of replication.
   b. What would the results of Taylor's experiment be if eukaryotes used a conservative mode of DNA replication?

3. The graph that follows shows the survival of four different *E. coli* strains after exposure to increasing doses of ultraviolet light. The wild-type strain is normal, but the other strains have a mutation in either a gene called *uvrA*, a gene called *recA*, or both.

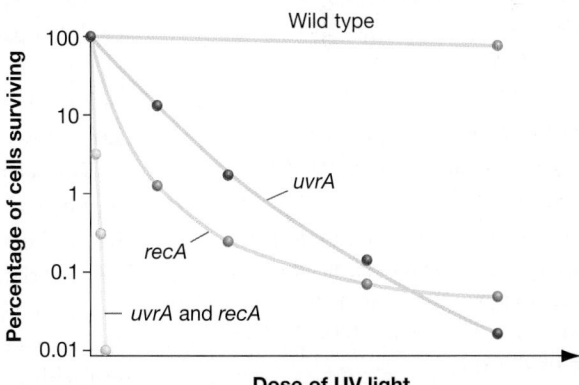

   a. Which strains are most sensitive to UV light? Which strains are least sensitive?
   b. What are the relative contributions of these genes to the repair of UV damage?

4. One widely used test to identify whether certain chemicals, such as pesticides or herbicides, might be carcinogenic (cancer causing) consists of exposing bacterial cells to the chemical and recording whether the exposure leads to an increased mutation rate. In effect, this test equates cancer-causing chemicals with mutation-causing chemicals. Why is this an informative test?

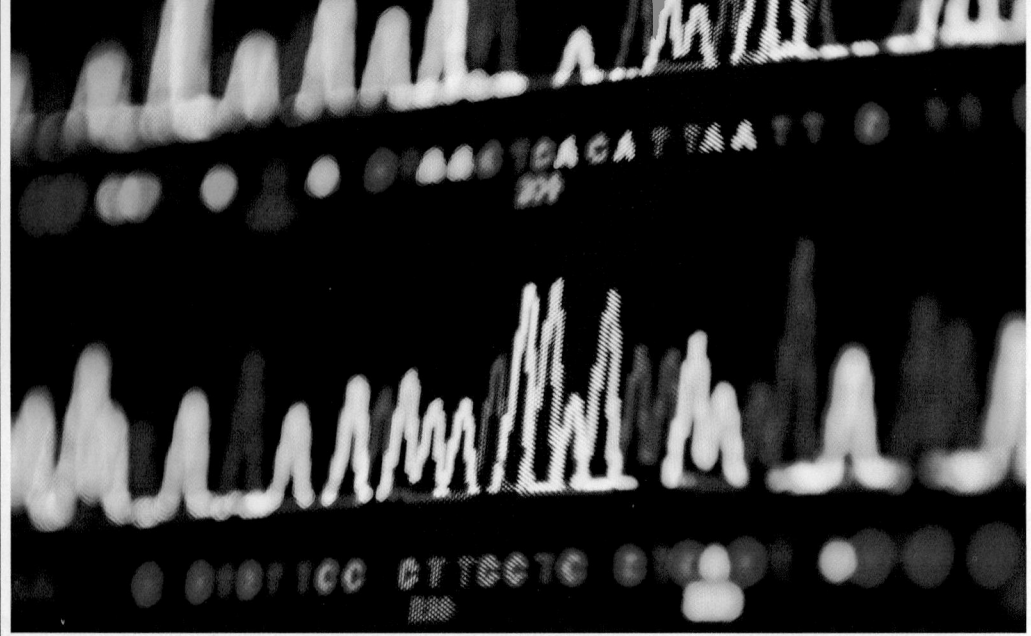

This computer screen is showing output from a DNA sequencer. Each peak corresponds to a different base in DNA—an A, T, G, or C. This chapter explores how DNA sequences in organisms are related to their phenotypes.

15 How Genes Work

KEY CONCEPTS

- ⟅⟆ Most genes code for proteins.

- ⟅⟆ DNA is transcribed to messenger RNA by RNA polymerase, and then messenger RNA is translated to proteins by ribosomes. In this way, genetic information is converted from DNA to RNA to proteins.

- ⟅⟆ Each amino acid in a protein is specified by a group of three bases in messenger RNA.

- ⟅⟆ Mutations are random changes in DNA, ranging in extent from single bases to large chromosome regions, that may or may not produce changes in the phenotype.

DNA has been called the blueprint of life. If an organism's DNA is like a set of blueprints, then its cells are like construction sites, and the enzymes inside a cell are like construction workers. But how does the DNA inside each cell assemble this team of skilled laborers and specify the construction materials needed to build and maintain the cell, and perhaps remodel it when conditions change?

Mendel provided insights that made the study of these questions possible. He discovered that particular alleles are associated with certain phenotypes and that alleles do not change when transmitted from parent to offspring. The resulting chromosome theory of inheritance established that genes are found in chromosomes, whose movement during meiosis explains Mendel's results.

The science of molecular biology began with the discovery that DNA is the hereditary material, and that DNA is a double-helical structure containing sequences of four bases. From these early advances, it was clear that genes are made of DNA and that genes carry the instructions for making and maintaining an individual.

But biologists still didn't know how the information in DNA is translated into action. How does **gene expression**—the process of converting archived information into molecules that actually do things in the cell—occur?

This chapter introduces some of the most pivotal ideas in all of biology—ideas that connect genotypes to phenotypes by revealing how genes work at the molecular level. They also speak to the heart of a key attribute of life: processing genetic information to produce a living organism.

Understanding how genes work triggered a major transition in biological science. Instead of thinking about genes solely in relation to their effects on eye color in fruit flies or on seed shape in garden peas, biologists could begin analyzing the molecular composition of genes and their products. The molecular revolution in biology took flight.

✔ When you see this checkmark, stop and test yourself. Answers are available in Appendix B.

15.1 What Do Genes Do?

Although biologists of the early twentieth century made tremendous progress in understanding how genes are inherited, an explicit hypothesis explaining what genes do did not appear until 1941. That year George Beadle and Edward Tatum published a series of breakthrough experiments on a bread mold called *Neurospora crassa*.

Beadle and Tatum's research was inspired by an idea that was brilliant in its simplicity. As Beadle said: "One ought to be able to discover what genes do by making them defective." The idea was to knock out a gene by damaging it and then infer what the gene does by observing the phenotype of the mutant individual.

Today, alleles that do not function at all are called **knock-out, null**, or **loss-of-function alleles**. Creating knock-out mutant alleles and analyzing their effects is still one of the most common research strategies in studies of gene function. But Beadle and Tatum were the pioneers.

The One-Gene, One-Enzyme Hypothesis

To start their work, Beadle and Tatum exposed a large number of *N. crassa* individuals to radiation. As Chapter 14 indicated, high-energy radiation damages the double-helical structure of DNA—often in a way that makes the affected gene nonfunctional.

Their next step was to examine the mutant individuals. Eventually they succeeded in finding mutant *N. crassa* individuals that could not make specific compounds. For example, one of the mutants could not make pyridoxine, also called vitamin B6, even though normal individuals can. Further, Beadle and Tatum showed that the inability to synthesize pyridoxine was due to a defect in a single gene, and that the inability to synthesize other molecules was due to defects in other genes.

These results inspired their **one-gene, one-enzyme hypothesis.** Beadle and Tatum proposed that the mutant *N. crassa* individual could not make pyridoxine because it lacked an enzyme required to synthesize the compound and that the lack of the enzyme was due to a genetic defect. Based on analyses of knock-out mutants, the one-gene, one-enzyme hypothesis claimed that each gene contains the information needed to make an enzyme.

An Experimental Test of the Hypothesis

Three years later, Adrian Srb and Norman Horowitz published a rigorous test of the one-gene, one-enzyme hypothesis. These biologists focused on the ability of *N. crassa* individuals to synthesize the amino acid arginine. In the lab, normal cells of this bread mold grow well on a laboratory culture medium that lacks arginine. This is possible because *N. crassa* cells are able to synthesize their own arginine.

Previous work had shown that organisms synthesize arginine in a series of steps called a **metabolic pathway**. As **Figure 15.1** shows, compounds called ornithine and citrulline are intermediate products in the metabolic pathway leading to arginine. Specific enzymes are required to synthesize ornithine, convert ornithine to citrulline, and change citrulline to arginine. Srb and Horowitz hypothesized that specific genes in *N. crassa* cells are responsible for producing each of the three enzymes involved.

To test this idea, Srb and Horowitz used radiation to create a large number of mutant individuals. High-energy radiation is equally likely to damage DNA in any part of the organism's genome, however, and most organisms have thousands or tens of thousands of genes. Of the many mutants the biologists created, how could they find the handful that specifically knocked out a step in the pathway for arginine synthesis?

To find the mutants they were looking for, the researchers performed what is now known as a genetic screen. A **genetic screen** is any technique for picking certain types of mutants out of many thousands of randomly generated mutants.

Srb and Horowitz began their screen by raising colonies of irradiated cells on a medium that included arginine. Then they transferred a sample of each colony to a medium that *lacked* arginine. If an individual could grow in the presence of arginine but failed to grow without arginine, they concluded that it couldn't make its own arginine.

The biologists followed up by confirming that the offspring of these cells also had this defect. Based on these data, they were confident that they had isolated individuals with mutations in one or more of the genes for the enzymes shown in Figure 15.1.

To test the one-gene, one-enzyme hypothesis, the biologists grew the mutants on normal media that lacked arginine and were supplemented in each case either with nothing, ornithine, citrulline, or arginine.

As **Figure 15.2** on page 278 shows, the results from these growth experiments were dramatic. Some of the mutant cells were able to grow on some of these media but not on others. More specifically, the mutants fell into three distinct classes, which the researchers called *arg1, arg2*, and *arg3*.

As the Interpretation section of the figure shows, the data make sense if each type of mutant lacked a different, specific step in a metabolic pathway because of a defect in a particular gene. In short, Srb and Horowitz had documented a correlation

Metabolic pathway for arginine synthesis: Precursor — Enzyme 1 → Ornithine — Enzyme 2 → Citrulline — Enzyme 3 → Arginine

FIGURE 15.1 Different Enzymes Catalyze Each Step in the Metabolic Pathway for Arginine.

✔**QUESTION** If a cell lacked enzyme 2 but received ornithine in its diet, could it still grow? Could it still grow if it received citrulline instead?

QUESTION: What do genes do?

HYPOTHESIS: Each gene contains the information required to make a different enzyme.

NULL HYPOTHESIS: Genes have nothing to do with making enzymes.

EXPERIMENTAL STRATEGY: Knock out specific genes. Test to see if the enzymes required for different steps in the pathway for synthesizing arginine are missing.

EXPERIMENTAL SETUP: Isolate mutant *N. crassa* that cannot synthesize arginine. Grow each type of mutant on normal medium that is:

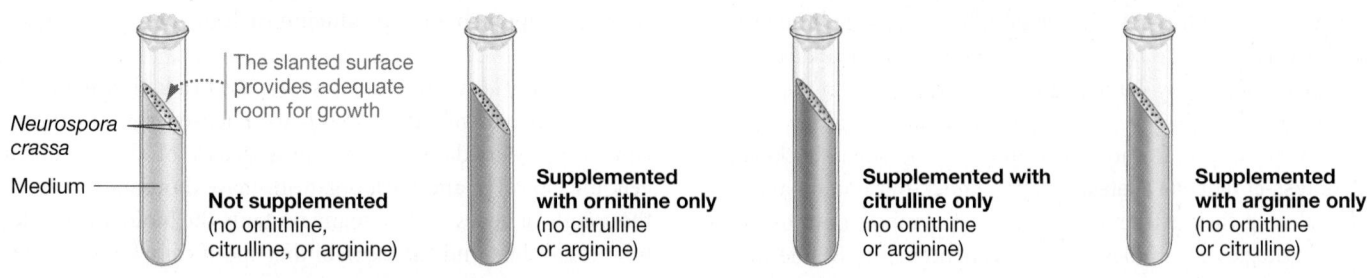

Neurospora crassa

The slanted surface provides adequate room for growth

Medium

Not supplemented
(no ornithine, citrulline, or arginine)

Supplemented with ornithine only
(no citrulline or arginine)

Supplemented with citrulline only
(no ornithine or arginine)

Supplemented with arginine only
(no ornithine or citrulline)

PREDICTION: There will be three distinct types of mutants, corresponding to defects in enzyme 1, enzyme 2, and enzyme 3 in the pathway for synthesizing arginine. Each type of mutant will be able to grow on different combinations of the four types of media.

PREDICTION OF NULL HYPOTHESIS: There will not be distinct types of mutants.

RESULTS: There are three distinct types of mutants, called *arg 1*, *arg 2*, and *arg 3*.

Supplement type

		None	Ornithine only	Citrulline only	Arginine only
	arg 1	no growth	GROWTH	GROWTH	GROWTH
Mutant type	*arg 2*	no growth	no growth	GROWTH	GROWTH
	arg 3	no growth	no growth	no growth	GROWTH

INTERPRETATION:

Precursor → Ornithine → Citrulline → Arginine

arg 1 cells lack enzyme 1

arg 2 cells lack enzyme 2

arg 3 cells lack enzyme 3

CONCLUSION: The one-gene, one-enzyme hypothesis is supported.

FIGURE 15.2 Experimental Support for the One-Gene, One-Enzyme Hypothesis. The association between specific genetic defects in *N. crassa* and specific deficits in the metabolic pathway for arginine synthesis provided evidence that the one-gene, one-enzyme hypothesis was correct.

SOURCE: Srb, A. M. and N. H. Horowitz. 1944. The ornithine cycle in *Neurospora* and its genetic control. *Journal of Biological Chemistry* 154: 129–139.

✔**QUESTION** Experimental designs must be repeatable, so that other investigators can try the experiment themselves to check the results. Name three things that these researchers would need to describe, so that others could repeat this experiment.

between a specific genetic defect and a defect at a specific point in a metabolic pathway.

This experiment convinced most investigators that the one-gene, one-enzyme hypothesis was correct. To review its design, results, and interpretation, go to the study area at *www.masteringbiology.com.*

 Web Activity The One-Gene, One-Enzyme Hypothesis

Follow-up work showed that genes dictate the structures of all the proteins produced by an organism—not just the structures of enzymes. ☞ Biologists finally understood what most genes do: They contain the instructions for making proteins.

In many cases, though, proteins are made up of different polypeptides, each of which is a product of a different gene. Consequently, for greater accuracy, the one-gene, one-enzyme hypothesis is best called the one-gene, one-polypeptide hypothesis.

15.2 The Central Dogma of Molecular Biology

How does a gene specify the production of a protein? As soon as Beadle and Tatum's hypothesis had been supported in *N. crassa* and a variety of other organisms, this question became a central one.

Part of the answer lay in the molecular structure of the gene. Biochemists knew that the primary components of DNA were four nitrogen-containing bases: the pyrimidines thymine (abbreviated T) and cytosine (C), and the purines adenine (A) and guanine (G). They also knew that these bases were connected in a linear sequence by a sugar-phosphate backbone. Watson and Crick's model for the secondary structure of the DNA molecule, introduced in Chapter 4 and reviewed in Chapter 14, revealed that two strands of DNA are wound into a double helix, held together by hydrogen bonds between the complementary base pairs A-T and G-C.

Given DNA's structure, it appeared extremely unlikely that DNA directly catalyzed the reactions that produce proteins. Its shape was too regular to suggest that it could bind a wide variety of substrate molecules and lower the activation energy for chemical reactions. So how did information translate into action?

The Genetic Code Hypothesis

Crick proposed that the sequence of bases in DNA might act as a code. His idea was that DNA was *only* an information-storage molecule. The instructions it contained would have to be read and then translated into proteins.

Crick offered Morse code as an analogy. Morse code is a message-transmission system using dots and dashes to represent the letters of the alphabet and in that way convey all the complex information of human language. Crick was proposing that different combinations of bases could specify the 20 amino acids, just as different combinations of dots and dashes specify the 26 letters of the alphabet. A particular stretch of DNA, then, could contain the information needed to produce the amino acid sequence of a particular enzyme.

In code form, the tremendous quantity of information required to build and operate a cell could be stored compactly. This information could also be copied through complementary base pairing and transmitted efficiently from one generation to the next.

It soon became apparent, however, that the information encoded in the base sequence of DNA is not translated into the amino acid sequence of proteins directly. Instead, the link between DNA as information repository and proteins as cellular machines is indirect.

RNA as the Intermediary between Genes and Proteins

The first clue that the biological information in DNA must go through an intermediary in order to produce proteins came from data on the structure of cells. In eukaryotic cells, DNA is enclosed within a membrane-bound organelle called the nucleus (see Chapter 7). But the cells' ribosomes, where protein synthesis takes place, are outside the nucleus, in the cytoplasm.

To make sense of this observation, François Jacob and Jacques Monod suggested that RNA molecules act as a link between genes and the protein-manufacturing centers. Jacob and Monod's hypothesis is illustrated in **Figure 15.3**. They predicted that short-lived molecules of RNA, which they called **messenger RNA**, or **mRNA**, carry information from DNA to the site of protein synthesis. Messenger RNA is one of several distinct types of RNA in cells.

Follow-up research confirmed that the messenger RNA hypothesis is correct. One particularly important piece of evidence was the discovery of an enzyme that catalyzes the synthesis of RNA. This protein is called **RNA polymerase**, because it polymerizes ribonucleotides into strands of RNA.

RNA polymerase synthesizes RNA molecules according to the information provided by the sequence of bases in a particular stretch of DNA. Unlike DNA polymerase, RNA polymerase does not require a primer to begin connecting ribonucleotides together to produce a strand of RNA.

To test the mRNA hypothesis more directly, researchers created a reaction mix containing three critical elements: (**1**) the enzyme RNA polymerase; (**2**) ribonucleotides containing the bases adenine (A), uracil (U), guanine (G), and cytosine (C); and (**3**) copies of a strand of synthetic DNA that contained deoxyribonucleotides in which the only base was thymine (T).

Why uracil? Recall from Chapter 4 that RNA contains the base uracil instead of thymine. If an RNA molecule binds to a DNA or RNA molecule, uracil forms complementary base pairs with adenine.

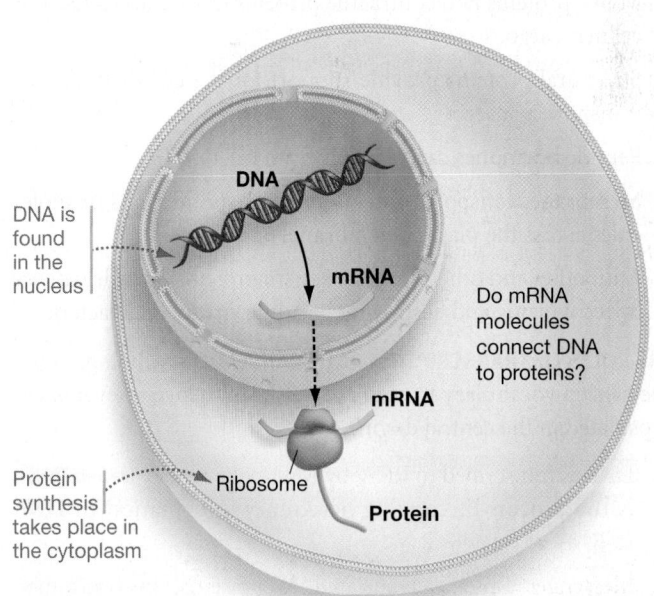

FIGURE 15.3 The Messenger RNA Hypothesis. In the cells of plants, animals, fungi, and other eukaryotes, most DNA is found only in the nucleus, but proteins are manufactured outside the nucleus, at ribosomes. Biologists proposed that the information coded in DNA is carried from inside the nucleus out to the ribosomes by messenger RNA (mRNA).

The DNA strand in the experiment had the sequence TTTTTT..., however. After allowing the polymerization reaction to proceed, the biologists isolated RNA molecules that contained only the base adenine.

This result provided strong support for the hypothesis that RNA polymerase synthesizes RNA according to the rules of complementary base pairing introduced in Chapter 4, because thymine pairs with adenine. Similar experiments showed that synthetic DNAs containing no bases other than cytosine result in the production of RNA molecules containing no bases other than guanine.

Dissecting the Central Dogma

Once the mRNA hypothesis was accepted, Francis Crick articulated what became known as the central dogma of molecular biology. The **central dogma** summarizes the flow of information in cells. It simply states that DNA codes for RNA, which codes for proteins:

$$\text{DNA} \longrightarrow \text{RNA} \longrightarrow \text{proteins}$$

Crick's simple statement encapsulates much of the research reviewed in this chapter and the preceding one. DNA is the hereditary material. Genes consist of specific stretches of DNA that code for products used in the cell. ⟐⟶ The sequence of bases in DNA specifies the sequence of bases in an RNA molecule, which specifies the sequence of amino acids in a protein. In this way, genes ultimately code for proteins.

Many proteins function as enzymes that catalyze chemical reactions in the cell. Other proteins perform the types of roles introduced in earlier chapters:

- Motor proteins and contractile proteins move the cell itself or cellular cargo.
- Structural proteins provide support for the cell or tracks for transporting cargo.
- Peptide hormones carry signals from cell to cell.
- Membrane transport proteins conduct specific ions or molecules across the plasma membrane.
- Antibodies and other immune system proteins provide defense by recognizing and destroying invading viruses and bacteria.

THE ROLES OF TRANSCRIPTION AND TRANSLATION   Biologists use specialized vocabulary to summarize the sequence of events encapsulated in the central dogma.

1. DNA is transcribed to RNA, by RNA polymerase. **Transcription** is the process of copying hereditary information in DNA to RNA.

2. Messenger RNA is translated to proteins, in ribosomes. **Translation** is the process of using the information in nucleic acids to synthesize proteins.

The term transcription is appropriate. In everyday English, transcription simply means making a copy of information. The scientific use is similar because it conveys the idea that DNA acts as a permanent record—an information archive or blueprint. This permanent record is copied, during transcription, to produce the short-lived form called mRNA.

Translation is also an appropriate term. In everyday English, translation refers to transferring information from one language to another. In biology, translation is the transfer of information from one type of molecule to another—from the "language" of nucleic acids to the "language" of proteins. Translation is also referred to simply as protein synthesis.

The following diagram summarizes the relationship between transcription and translation, as well as the relationships among DNA, RNA, and proteins:

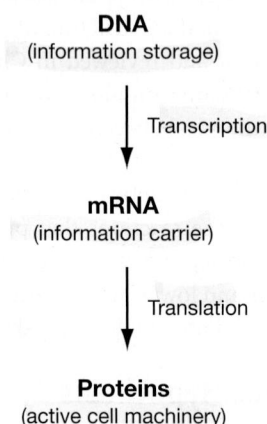

Gene expression occurs via transcription and translation.

LINKING GENOTYPES AND PHENOTYPES   According to the central dogma, an organism's genotype is determined by the sequence of bases in its DNA, while its phenotype is a product of the proteins it produces.

To appreciate this point, consider that the enzymes and other proteins encoded by genes are what make the "stuff" of the cell and dictate which chemical reactions occur inside. In populations of the Oldfield mouse native to southeastern North America, for example, individuals have a gene for a protein called the melanocortin receptor. This receptor influences how much dark pigment is deposited in fur. An important aspect of a mouse's phenotype—its coat color—is determined in part by the DNA sequence at the gene for this receptor-producing enzyme (**Figure 15.4a**).

Later work revealed that a gene's alleles differ in their DNA sequence. As a result, the proteins produced by different alleles of the gene may differ in their amino acid sequence. If the primary structures of proteins vary, their functions are likely to vary as well.

To drive this point home, look at the DNA sequence in the portion of the melanocortin receptor gene shown in **Figure 15.4b**, and compare it with the sequence in Figure 15.4a. The sequences differ—meaning that they are different alleles. Now look at the protein products of each allele, and note that one of the amino acids in the protein's primary structure differs—one allele specifies an arginine residue; the other specifies a cysteine residue.

(a) Genetic information flows from DNA to RNA to proteins.

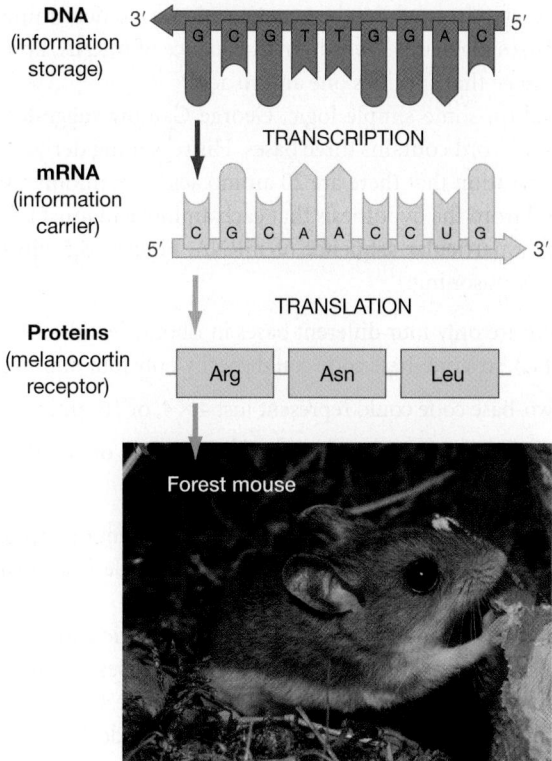

DNA (information storage) 3′ G C G T T G G A C 5′

TRANSCRIPTION

mRNA (information carrier) 5′ C G C A A C C U G 3′

TRANSLATION

Proteins (melanocortin receptor) Arg Asn Leu

Forest mouse

Mice with this DNA sequence have **dark** coats.

(b) Differences in genotype may cause differences in phenotype.

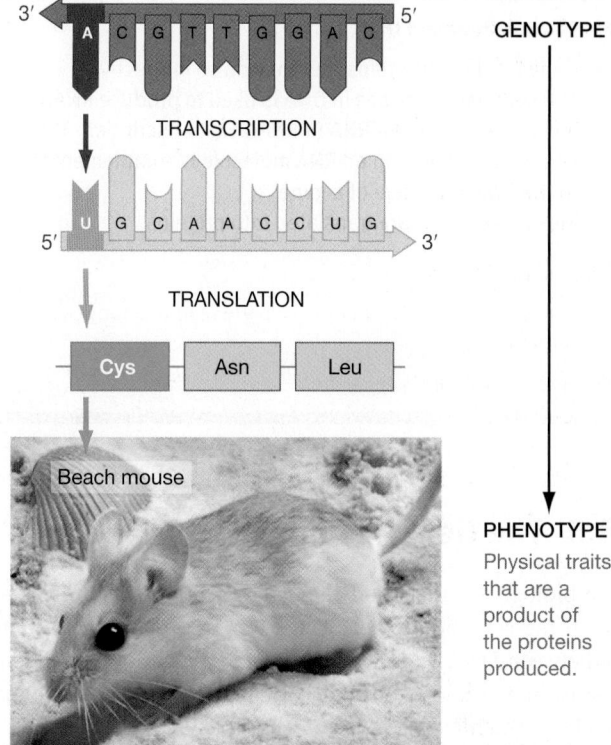

3′ A C G T T G G A C 5′ GENOTYPE

TRANSCRIPTION

5′ U G C A A C C U G 3′

TRANSLATION

Cys Asn Leu

Beach mouse

PHENOTYPE Physical traits that are a product of the proteins produced.

Mice with this DNA sequence have **light** coats.

FIGURE 15.4 The Flow of Information in the Cell. The central dogma revealed the connection between genotype and phenotype. The DNA sequences given in parts **(a)** and **(b)** are from different alleles (genotypes) that influence coat color (phenotypes) in Oldfield mice. Forest-dwelling mice are dark, which camouflages them in their forested habitats. Beach-dwelling mice are light, which camouflages them in their sandy habitat.

At the protein level, the phenotypes associated with these alleles differ. The consequences for the entire mouse are striking: Melanocortin receptors that have arginine in this location tend to deposit a large amount of pigment, but receptors that have cysteine in this location tend to deposit small amounts of pigment. Whether a mouse is dark or light depends, in large part, on a single base change in its DNA sequence. In this case, a tiny difference in genotype produces a large change in phenotype. The central dogma links genotypes to phenotypes.

EXCEPTIONS TO THE CENTRAL DOGMA The central dogma provided an important conceptual framework for the burgeoning field called molecular genetics and inspired a series of fundamental questions about how genes and cells work. But important modifications to the central dogma have occurred in the decades since Frances Crick first proposed it:

- Many genes code for RNA molecules that do not function as mRNAs—they are not translated into proteins.

- In some cases, information flows from RNA back to DNA.

The discovery of a wide array of distinct RNA types ranks among the most profound advances in the past decade of biolog-ical science. Messenger RNA is just one of seven major types currently recognized. Several types of RNA in addition to mRNA are involved in protein synthesis, while other RNAs help regulate which genes are transcribed and which proteins are active in a cell (see Chapter 18). For the genes coding for these types of RNA, information flow would be diagrammed as simply DNA → RNA.

In the early 1970s, the discovery of "reverse" information flow created the kind of excitement now being generated by the discovery of so many different kinds of RNA. Some viruses, for example, have genes comprised of RNA. When RNA viruses infect a cell, a specialized viral polymerase called **reverse transcriptase** synthesizes a DNA version of the RNA genes. In these viruses, information flows from RNA to DNA.

The human immunodeficiency virus (HIV), which causes AIDS, is an RNA virus. Several of the most-commonly prescribed drugs for AIDS patients fight the infection by poisoning the virus's reverse transcriptase. The drugs prevent viruses from replicating efficiently by disrupting reverse information flow.

The punchline? Crick's hypothesis is a central concept in biology, but cells, viruses, and researchers aren't dogmatic about it.

15.3 The Genetic Code

Once biologists understood the general pattern of information flow in the cell, the next challenge was to understand the final link between DNA and proteins. Exactly how does the sequence of bases in a strand of mRNA code for the sequence of amino acids in a protein?

If this question could be answered, biologists would have cracked the **genetic code**—the rules that specify the relationship between a sequence of nucleotides in DNA or RNA and the sequence of amino acids in a protein. Researchers from all over the world took up the challenge. A race was on.

How Long Is a Word in the Genetic Code?

The first step in cracking the genetic code was to determine how many bases make up a "word." In a sequence of mRNA, how long is a message that specifies one amino acid?

Based on some simple logic, George Gamow suggested that each code word contains three bases. His reasoning derived from the observation that there are 20 amino acids commonly used in cells and from the hypothesis that each amino acid must be specified by a particular sequence of mRNA. **Figure 15.5** illustrates Gamow's reasoning:

- There are only four different bases in ribonucleotides (A, U, G, and C), so a one-base code could specify only four amino acids.
- A two-base code could represent just 4 × 4, or 16, amino acids.
- A three-base code could specify 4 × 4 × 4, or 64, different amino acids.

A three-base code provides more than enough messages to code for all 20 amino acids. A three-base code is known as a **triplet code.**

Gamow's hypothesis suggested that the genetic code could be redundant. That is, more than one triplet of bases might specify the same amino acid. As a result, different three-base sequences in an mRNA—say, AAA and AAG—might code for the same amino acid—say, lysine.

The group of three bases that specifies a particular amino acid is called a **codon**. According to the triplet code hypothesis, many of the 64 codons that are possible might actually specify the same amino acids.

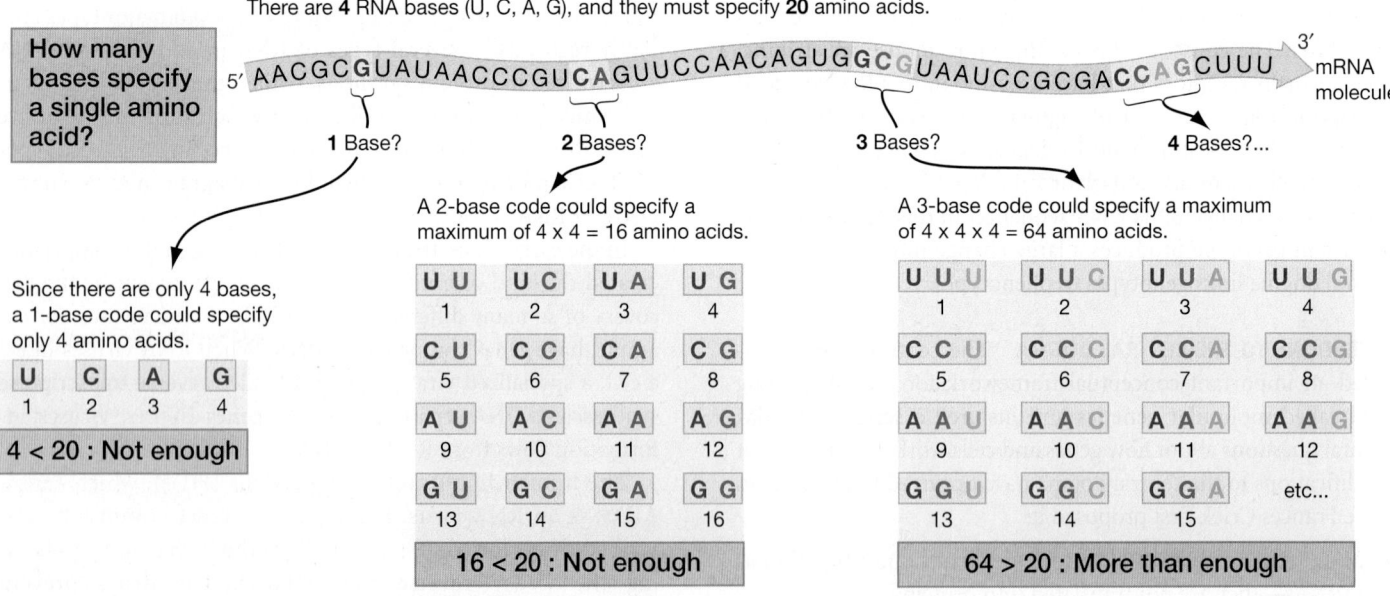

FIGURE 15.5 In the Genetic Code, How Many Bases Form a "Word"?

✔**QUESTION** How many amino acids could be specified by a four-base code? Why did biologists conclude that a four-base code was extremely unlikely? Your answer should include an explanation for why a four-base code is unlikely to evolve.

Work by Francis Crick and Sydney Brenner confirmed that codons are three bases long. Their experiments used chemicals that caused an occasional addition or deletion of a base in DNA. As predicted for a triplet code, a one-base addition or deletion in the base sequence led to a loss of function in the gene being studied. This is because a single addition or deletion mutation throws the sequence of codons, or the **reading frame**, out of register.

To understand how a reading frame works, consider the sentence

> "The fat cat ate the rat."

The reading frame of this sentence is a three-letter word and a space. If the fourth letter in this sentence—the *f* in *fat*—were deleted, the reading frame would transform the sentence into

> "The atc ata tet her at."

This is gibberish.

When the reading frame in a DNA sequence is thrown out of register by the addition or deletion of a base, the composition of each codon changes just like the letters in each word of the example sentence above. The protein produced from the altered DNA sequence has a completely different sequence of amino acids. In terms of its normal function, this protein is gibberish.

Crick and Brenner were also able to produce DNA sequences that had deletions or additions of two base pairs or three base pairs. The only time functional proteins were produced was when three bases were removed. In the sentence

> "The fat cat ate the rat."

the combination of removing one letter from each of the first three words might result in

> "Tha tca ate the rat."

Just as the altered sentence still conveys some meaning, genes with three deletion mutations were able to produce a functional protein.

The researchers interpreted these results as strong evidence in favor of the triplet code hypothesis. Most other biologists agreed. For a much more in-depth look at the experiments that confirmed the triplet code, go to the study area at *www.masteringbiology.com*.

 **Web Activity** The Triplet Nature of the Genetic Code

The confirmation of the triplet code launched a long, laborious effort to determine which amino acid is specified by each of the 64 codons. Ultimately, it was successful.

How Did Researchers Crack the Code?

The initial advance in deciphering the genetic code came in 1961, when Marshall Nirenberg and Heinrich Matthaei created a method for synthesizing RNAs of known sequence. They began by creating a long polymer of uracil-containing ribonucleotides. These synthetic RNAs were added to an in vitro system for syn-

thesizing proteins. The researchers analyzed the resulting amino acid chain and determined that it was polyphenylalanine—a polymer consisting of the amino acid phenylalanine.

This result was strong evidence that the RNA triplet UUU codes for the amino acid phenylalanine. By complementary base pairing, it was clear that the corresponding DNA sequence would be AAA. This initial work was followed by experiments using RNAs consisting of only A or C. RNAs consisting of only AAAAA... produced polypeptides consisting of only lysine; poly-C RNAs (RNAs consisting of only CCCCC...) produced polypeptides composed entirely of proline.

Nirenberg and Philip Leder later devised a system for synthesizing specific codons. With these they performed a series of experiments in which they added each codon to a cell extract containing the 20 different amino acids, ribosomes, and other molecules required for protein synthesis. As Chapter 7 noted, ribosomes are the multimolecular machines at which proteins are synthesized. Then the researchers determined which amino acid became bound to the ribosomes when a particular codon was present. For example, when the codon CAC was in the reaction mix, the amino acid histidine would bind to the ribosomes. This result confirmed that CAC codes for histidine.

These ribosome-binding experiments allowed Nirenberg and Leder to determine which of the 64 codons coded for each of the 20 amino acids.

In addition, researchers discovered that certain codons are punctuation marks signaling "start of message" or "end of message." These codons indicate that protein synthesis should start at a given codon or that the protein chain is complete.

- There is one **start codon** (AUG), which signals that protein synthesis should begin at that point on the mRNA molecule. The AUG codon codes for the amino acid methionine.

- There are three **stop codons**, also called termination codons (UAA, UAG, and UGA). The stop codons signal that the protein is complete, and end the translation process.

The complete genetic code is given in **Figure 15.6** on page 284. Deciphering it was a tremendous achievement. It represents more than five years of work by several teams of researchers.

ANALYZING THE CODE Once biologists had cracked the genetic code, they realized that it has a series of important properties.

- *It is redundant.* All amino acids except methionine and tryptophan are coded by more than one codon.

- *It is unambiguous.* A single codon never codes for more than one amino acid.

- *It is nearly universal.* With a few minor exceptions, all codons specify the same amino acids in all organisms.

- *It is conservative.* When several codons specify the same amino acid, the first two bases in those codons are almost always identical.

The last point is subtle, but important. Here's the key: If a mutation in DNA or an error in transcription or translation affects

SECOND BASE

FIRST BASE	U	C	A	G	THIRD BASE
U	UUU UUC ⎤ Phenylalanine (Phe) ⎦ UUA UUG ⎤ Leucine (Leu) ⎦	UCU UCC UCA UCG ⎤ Serine (Ser) ⎦	UAU UAC ⎤ Tyrosine (Tyr) ⎦ UAA — **Stop codon** UAG — **Stop codon**	UGU UGC ⎤ Cysteine (Cys) ⎦ UGA — **Stop codon** UGG — Tryptophan (Trp)	U C A G
C	CUU CUC CUA CUG ⎤ Leucine (Leu) ⎦	CCU CCC CCA CCG ⎤ Proline (Pro) ⎦	CAU CAC ⎤ Histidine (His) ⎦ CAA CAG ⎤ Glutamine (Gln) ⎦	CGU CGC CGA CGG ⎤ Arginine (Arg) ⎦	U C A G
A	AUU AUC AUA ⎤ Isoleucine (Ile) ⎦ AUG — Methionine (Met) **Start codon**	ACU ACC ACA ACG ⎤ Threonine (Thr) ⎦	AAU AAC ⎤ Asparagine (Asn) ⎦ AAA AAG ⎤ Lysine (Lys) ⎦	AGU AGC ⎤ Serine (Ser) ⎦ AGA AGG ⎤ Arginine (Arg) ⎦	U C A G
G	GUU GUC GUA GUG ⎤ Valine (Val) ⎦	GCU GCC GCA GCG ⎤ Alanine (Ala) ⎦	GAU GAC ⎤ Aspartic acid (Asp) ⎦ GAA GAG ⎤ Glutamic acid (Glu) ⎦	GGU GGC GGA GGG ⎤ Glycine (Gly) ⎦	U C A G

FIGURE 15.6 The Genetic Code. To read an mRNA codon, locate its first base in the red band on the left side; from there, move rightward to the box under the codon's second base in the blue band along the top; and lastly, move up and down within that box to the level of the codon's third base in the green band on the right side. The 64 codons, along with the amino acid or other signal that they specify, are displayed in the boxes. By convention, codons are always written in the 5′ → 3′ direction.

the third position in a codon, it is unlikely to change the amino acid in the final protein. This feature makes individuals less vulnerable to small, random changes or errors in their DNA sequences. Compared with randomly generated codes, the existing genetic code efficiently minimizes the phenotypic effects of small changes in DNA and errors during translation. Stated another way, the genetic code does not represent a random assemblage of bases, like letters drawn from a hat. It has been honed by natural selection, and is remarkably efficient.

USING THE CODE Using the genetic code, biologists can work forwards or backwards in the central dogma to:

1. Predict the codons and amino acid sequence encoded by a particular DNA sequence (see **Figure 15.7a**).

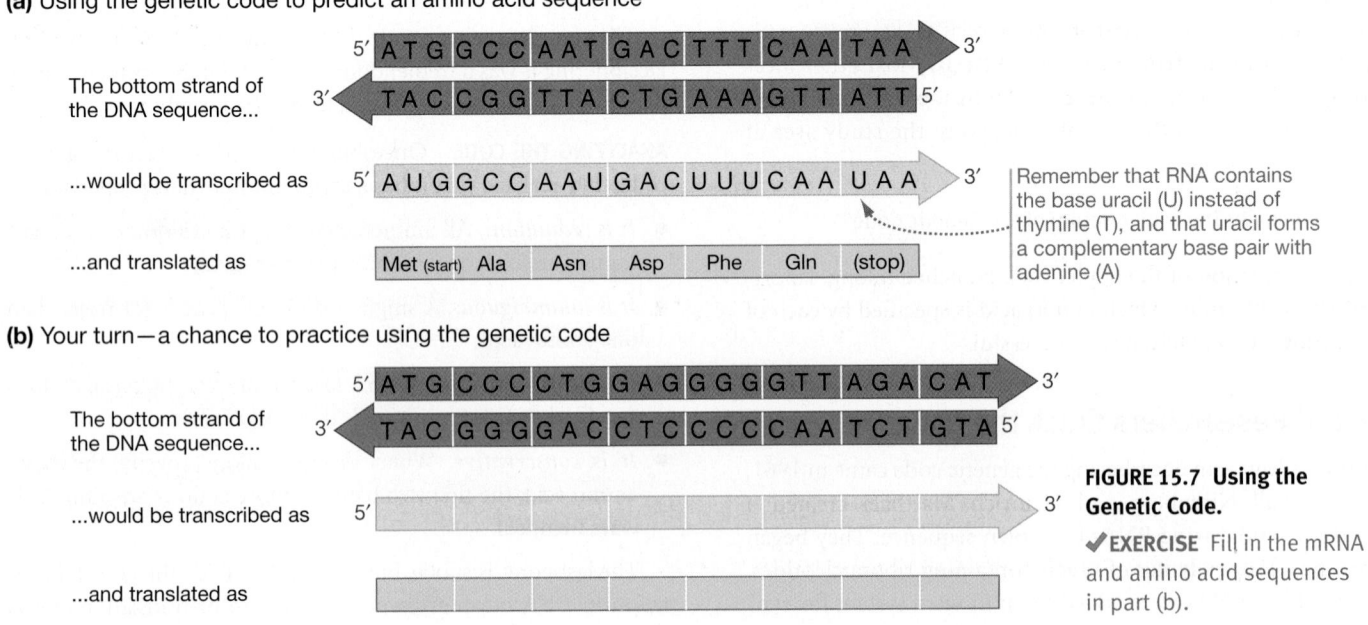

(a) Using the genetic code to predict an amino acid sequence

5′ ATG GCC AAT GAC TTT CAA TAA 3′

The bottom strand of the DNA sequence...
3′ TAC CGG TTA CTG AAA GTT ATT 5′

...would be transcribed as
5′ AUG GCC AAU GAC UUU CAA UAA 3′

Remember that RNA contains the base uracil (U) instead of thymine (T), and that uracil forms a complementary base pair with adenine (A)

...and translated as
Met (start) | Ala | Asn | Asp | Phe | Gln | (stop)

(b) Your turn—a chance to practice using the genetic code

5′ ATG CCC CTG GAG GGG GTT AGA CAT 3′

The bottom strand of the DNA sequence...
3′ TAC GGG GAC CTC CCC CAA TCT GTA 5′

...would be transcribed as
5′ _____ 3′

...and translated as
_____

FIGURE 15.7 Using the Genetic Code.

✔ **EXERCISE** Fill in the mRNA and amino acid sequences in part (b).

2. Approximate the mRNA and DNA sequence that would code for a particular sequence of amino acids.

Why can you only *approximate* an mRNA or DNA sequence from a given amino acid sequence? The answer lies in the code's redundancy. If a polypeptide contains phenylalanine, you don't know if the codon responsible is UUU or UUC.

✔️If you understand how to read the genetic code, you should be able to do the following tasks: (1) Identify the codons in Figure 15.4 and decide whether they are translated correctly. (2) Complete the Exercise for **Figure 15.7b**. (3) Write an mRNA that codes for the amino acid sequence Ala-Asn-Asp-Phe-Gln and yet is different from the one given in Figure 15.7a. Indicate the mRNA's $5' \rightarrow 3'$ polarity. Then write the double-stranded DNA that corresponds to this mRNA. Indicate the $5' \rightarrow 3'$ polarity of both DNA strands.

Once the central dogma and genetic code were understood, biologists were finally able to explore and eventually understand the molecular basis of mutation. How do novel traits—such as dwarfing in garden peas and white eye color in fuit flies—come to be?

CHECK YOUR UNDERSTANDING

◯━ If you understand that . . .

- The sequence of bases in mRNA constitutes a code. Particular combinations of three bases lead to the addition of specific amino acids to the protein encoded by the gene.
- The genetic code is redundant. It consists of 64 combinations of bases, but only 20 amino acids and a stop "punctuation mark" need to be specified.

✔ You should be able to . . .

Consider the consequences of a mutation in a DNA sequence ATA to one of the following sequences: GTA, TTA, or GCA.

1. In each case, specify the resulting change in the mRNA codon.
2. In each case, describe the effect on the resulting protein.

Answers are available in Appendix B.

15.4 What Is the Molecular Basis of Mutation?

This chapter has explored how the information archived in DNA is put into action in the form of working RNAs and proteins. Now the question is, what happens if the information in DNA changes? What are the consequences for the cell?

◯━ A **mutation** is any permanent change in an organism's DNA. It is a modification in a cell's information archive—a change in its genotype. Mutations create new alleles.

Because changes in the genotype can change the mRNA codons that are transcribed and the amino acids that are translated, mutations may lead to changes that affect the primary structure of proteins and thus the organism's phenotype.

Point Mutation

Figure 15.8 shows how a common type of mutation occurs. If DNA polymerase mistakenly inserts the wrong base as it synthesizes a new strand of DNA, and if proofreading by DNA polymerase and the mismatch repair system fail to correct the mismatched base before another round of DNA replication occurs, a change in the sequence of bases in DNA results. A single base change such as this is called a **point mutation**.

What happens when point mutations are transcribed and translated? To answer this question, look back at Figure 15.4 and recall that a change in a single base in DNA is associated with a difference in coat color in certain populations of mice. The DNA sequence in Figure 15.4a is found in dark-colored mice that live in forest habitats; the sequence in Figure 15.4b is found in light-colored mice that live in beach habitats.

Because beach-dwelling populations are younger than the nearby forest-dwelling populations, researchers hypothesize that the following sequence of events occurred:

1. Forest mice colonized beach habitats.
2. Either before or after the colonization event, a point mutation occurred that altered the melanocortin receptor gene and resulted in some offspring having light coats.

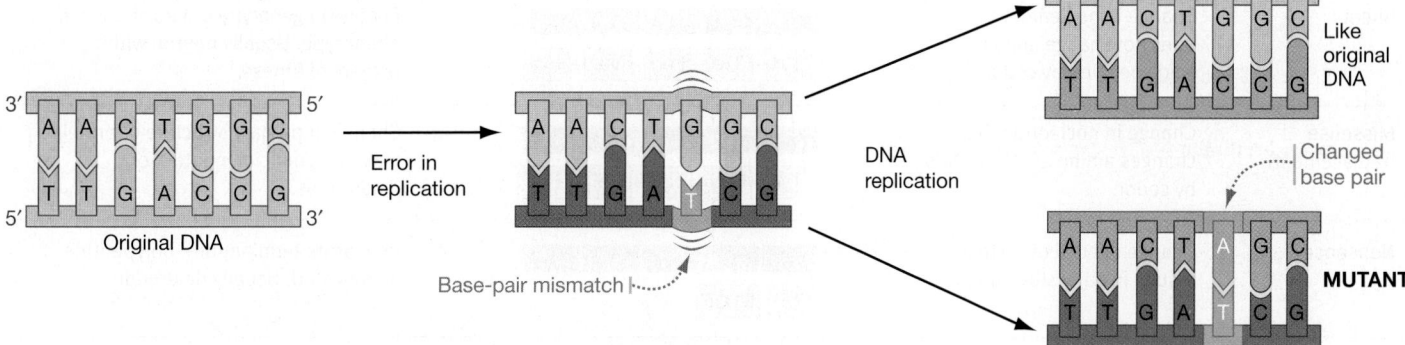

FIGURE 15.8 Unrepaired Mistakes in DNA Synthesis Lead to Point Mutations.

✔QUESTION Why is it logical that the type of mutation illustrated here is termed a point mutation?

3. Light-colored mice are camouflaged in beach habitats; in sandy environments, they suffer lower predation than dark-colored mice.

4. Over time, the allele created by the point mutation increased in frequency in beach-dwelling populations.

Point mutations that cause these types of changes in the amino acid sequence of proteins are called **missense mutations** or **replacement mutations**. But note that if the same G to A change had occurred in the third position of the same DNA codon, instead of the first position, there would have been no change in the mRNA or protein produced. The mRNA codons CGC and CGU both code for arginine. A point mutation that does not change the amino acid sequence of the gene product is called a **silent mutation**.

In terms of the impact on organisms, biologists divide mutations into three categories:

1. Beneficial Some mutations increase the fitness of the organism—meaning, its ability to survive and reproduce—in certain environments. The G-to-A mutation is beneficial in beach habitats because it camouflages mice.

2. Neutral If a mutation has no affect on fitness, it is termed neutral. Silent mutations are usually neutral.

3. Deleterious Because organisms tend to be well-adapted to their current habitat, and because mutations are random changes in the genotype, most mutations lower fitness. The G-to-A mutation would be deleterious in the forest habitat.

Recent studies indicate that the vast majority of point mutations are neutral or slightly deleterious. **Table 15.1** summarizes the types of point mutations that have been documented and reviews their consequences for the amino acid sequences of proteins and for fitness.

Chromosome-Level Mutations

Besides documenting various types of point mutations, biologists study mutations consisting of larger-scale changes in the composition of chromosomes. Chapter 12 introduced **polyploidy**, which is a change in the number of each type of chromosome present, and **aneuploidy**, the addition or deletion of a chromosome. Polyploidy, aneuploidy, and other changes in chromosome number result from chance mistakes in the partitioning of chromosomes during meiosis or mitosis.

But in addition to changes in overall chromosome number, the composition of individual chromosomes can change in important ways. For example, chromosome segments can become detached when accidental breaks in chromosomes occur. The segments may become flipped and rejoin—a phenomenon known as a chromosome **inversion**—or become attached to a different chromosome, an event called chromosome **translocation**.

Like point mutations, chromosome-level mutations can be beneficial, neutral, or deleterious. But because they represent massive changes in the genotype, chromosome alterations are usually deleterious. It is common, for example, to find that the chromosomes of cancerous cells exhibit aneuploidy, inversions, translocations, or combinations of such mutations. **Figure 15.9** drives this home by comparing the **karyotype**—the complete set of chromosomes in a cell—of a normal versus cancerous human cell.

To summarize, point mutations and chromosome-level mutations are random changes in DNA that produce new genes, new alleles, and new traits. At the level of individuals, mutations can cause disease or death or lead to increases in fitness. At the level of populations, mutations furnish the heritable variation that Mendel and Morgan analyzed and that makes evolution possible. These phenomena are explored more fully in Unit 5.

SUMMARY TABLE 15.1 **Known Types of Point Mutations**

Name	Definition	Example	Consequence
	Original DNA sequence of non-template (coding) strand	TAT TGG CTA GTA CAT — Tyr Trp Leu Val His — Original polypeptide	
Silent	Change in nucleotide that does not change amino acid specified by codon	TAC TGG CTA GTA CAT — Tyr Trp Leu Val His	Change in genotype but no change in phenotype. Usually neutral with respect to fitness.
Missense (replacement)	Change in nucleotide that changes amino acid specified by codon	TAT TGT CTA GTA CAT — Tyr Cys Leu Val His	Change in primary structure of protein may be beneficial, neutral, or deleterious.
Nonsense	Change in nucleotide that results in early stop codon	TAT TGA CTA GTA CAT — Tyr STOP	Premature termination—polypeptide is truncated. Usually deleterious.
Frameshift	Addition or deletion of a nucleotide	TAT TCG GCT AGT ACA T — Tyr Ser Ala Ser Thr	Reading frame is shifted—massive missense. Usually deleterious.

(a) Normal human karyotype

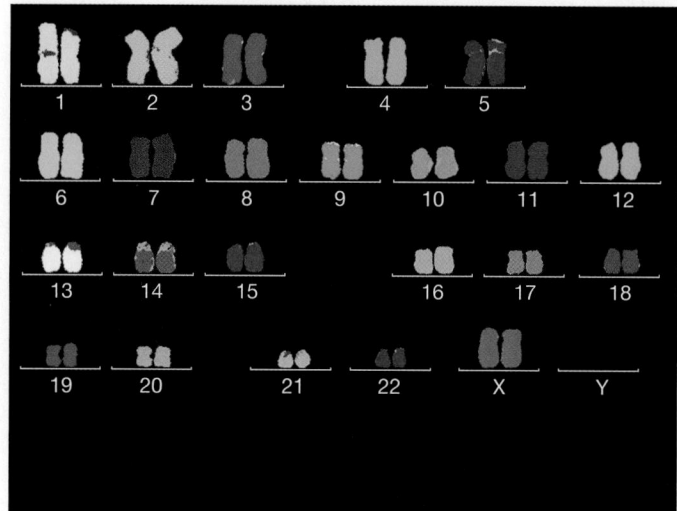

(b) Karyotype from a human cancerous cell

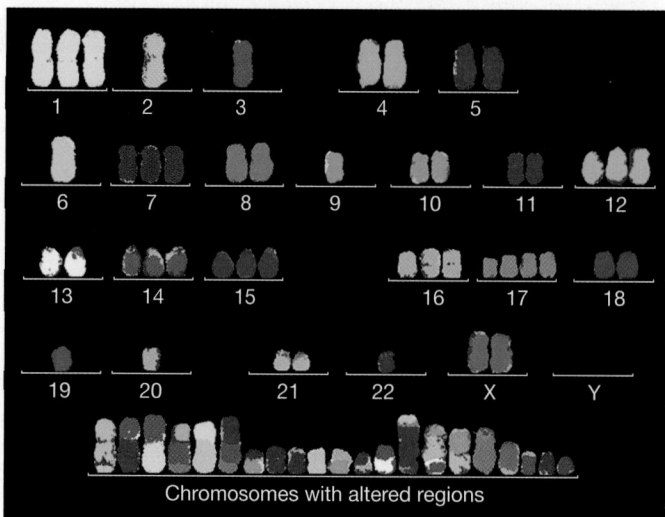

Chromosomes with altered regions

FIGURE 15.9 Chromosome-Level Mutations. Karyotypes are micrographs of all the chromosomes present in a cell. A normal human karyotype **(a)** includes pairs of 23 standard chromosomes, whereas a karyotype from a cancerous cell **(b)** exhibits many irregularities. Some whole chromosomes are missing, extra copies of others are present, and many individual chromosomes contain translocated or inverted fragments.

CHAPTER 15   REVIEW

For media, go to the study area at www.masteringbiology.com

Summary of Key Concepts

Most genes code for proteins.

- Because proteins serve so many functions in the cell, the one-gene, one-enzyme hypothesis was restated as one-gene, one-protein, or more accurately, one-gene, one-polypeptide.

  ✔You should be able to explain why the one-gene, one-polypeptide hypothesis is now considered too narrow, and use the form "one gene, . . ." to state a modern, more-accurate equivalent.

  (MB) **Web Activity** The One-Gene, One-Enzyme Hypothesis

DNA is transcribed to messenger RNA by RNA polymerase, and then messenger RNA is translated to proteins by ribosomes. In this way, genetic information is converted from DNA to RNA to proteins.

- DNA does not code for proteins directly. Instead, mRNA molecules are transcribed from DNA and then translated into proteins.

- The flow of information, from DNA to RNA to proteins, is called the central dogma of molecular biology.

- Many RNAs do not act as messengers for protein synthesis. They perform other functions in the cell.

- In some viruses, information flows from RNA to DNA.

  ✔You should be able to explain why transcription and translation are appropriate terms, in terms of how information is processed in a cell.

Each amino acid in a protein is specified by a group of three bases in messenger RNA.

- By synthesizing RNAs of known base composition and then observing the results of translation, researchers were able to unravel the genetic code.

- The genetic code is read in triplets.

- The code is redundant—meaning that most of the 20 amino acids are specified by more than one codon.

- Certain codons signal when translation starts and stops.

  ✔You should be able to explain why some changes in DNA sequences do not change the corresponding protein—meaning, why certain changes in an organism's genotype do not change its phenotype.

  (MB) **Web Activity** The Triplet Nature of the Genetic Code

Mutations are random changes in DNA, ranging in extent from single bases to large chromosome regions, that may or may not produce changes in the phenotype.

- Depending on the location and type of alteration in DNA and its impact on the resulting RNA or protein product, a mutation can be beneficial, deleterious, or neutral with respect to fitness.

- Mutations may produce novel proteins and RNAs. They are the source of the heritable variation that makes evolution possible.

  ✔You should be able to explain why a novel mutation was important during the evolution of beach-dwelling populations of Oldfield mice, and explain what would happen if the same mutation occurred in a population that lives in forest habitats.

Questions

1. What does the one-gene, one-enzyme hypothesis state?
   a. Genes are composed of stretches of DNA.
   b. Genes are made of protein.
   c. Genes code for ribozymes.
   d. A single gene codes for a single protein.

2. Which of the following is an important exception to the central dogma of molecular biology?
   a. Many genes code for RNAs that function directly in the cell.
   b. DNA is the repository of genetic information in all organisms (though not all viruses).
   c. Messenger RNA is a short-lived "information carrier."
   d. Proteins are responsible for most aspects of the phenotype.

3. DNA's primary structure is made up of just four different bases, and its secondary structure is regular and highly stable. How can a molecule with these characteristics hold all of the information required to build and maintain a cell?
   a. The information is first transcribed, then translated.
   b. The messenger RNA produced from DNA has much more complex secondary structures, and thus holds much more information.
   c. A protein produced (indirectly) from DNA has much more complex primary and secondary structures, and thus holds much more information.
   d. The information in DNA is in code form.

4. Why did researchers suspect that DNA does not code for proteins directly?
   a. In eukaryotes, DNA is found inside the nucleus, but proteins are produced outside the nucleus.
   b. In prokaryotes, DNA and proteins are never found together.
   c. When DNA was damaged by ultraviolet radiation or other sources of energy, the proteins in the cell did not change.
   d. There are several distinct types of RNA, of which only one functions as messenger RNA.

5. Which of the following describes an important experimental strategy in deciphering the genetic code?
   a. comparing the amino acid sequences of proteins with the base sequence of their genes
   b. analyzing the sequence of RNAs produced from known DNA sequences
   c. analyzing mutants that changed the code
   d. examining the polypeptides produced when RNAs of known sequence were translated

6. What is a stop codon?
   a. The place where transcription ends.
   b. The place where translation ends.
   c. The end of a chromosome.
   d. Any codon that does not code for an amino acid.

1. Explain why Morse code is an appropriate analogy for the genetic code.

2. Draw a hypothetical metabolic pathway composed of five substrates, five enzymes, and a product called Biological Sciazine. Number the substrates 1–5, and label the enzymes A–E, in order. (For instance, enzyme A catalyzes the reaction between substrates 1 and 2.)
   • Suppose a mutation made the gene for enzyme C nonfunctional. What molecule would accumulate in the affected cells?
   • Suppose some individuals can survive if given substrate 5 in the diet. But they die if given substrates 1, 2, 3, and 4. Which enzyme in the pathway is affected by this mutation?

3. Why did experiments with *Neurospora crassa* mutants support the one-gene, one-enzyme hypothesis?

4. Explain how a single-base deletion disrupts the reading frame of a gene. Include an example.

5. When researchers discovered that a combination of three deletion mutations or three addition mutations would restore the function of a gene, most biologists were convinced that the genetic code was read in triplets. Explain the logic behind this conclusion.

6. Explain why all point mutations change the genotype, but why only some point mutations change the phenotype.

1. Recall that DNA and RNA are synthesized only in the 5′ → 3′ direction and that DNA and RNA sequences are always written in the 5′ → 3′ direction. Consider the following DNA sequence:

   5′ TTGAAATGCCCGTTTGGAGATCGGGTTACAGCTAGTCAAAG 3′
   3′ AACTTTACGGGCAAACCTCTAGCCCAATGTCGATCAGTTTC 5′

   • Identify bases in the bottom strand that encode start and stop codons.
   • Write the mRNA sequence that would be transcribed between start and stop codons if the bottom strand served as the template.
   • Write the amino acid sequence that would be translated from the mRNA sequence you just wrote.

2. What problems would arise if the genetic code contained only 22 codons—one for each amino acid, a start signal, and a stop signal?

   (Hint: When DNA is copied prior to mitosis or meiosis, random errors occur that change its primary base sequence.)

3. Scientists say that a phenomenon is a "black box" if they can describe it and study its effects but don't yet know the underlying mechanism that causes it. In what sense was genetics—meaning the transmission of heritable traits—a black box before the central dogma of molecular biology was understood?

4. One of the possibilities that researchers interested in the genetic code had to consider was that the code was overlapping, meaning that a single base could be part of more than one codon. Make a diagram showing how an overlapping code would work, assuming that each codon is three bases long. As an example, use the sequence 5′ AUGUUACGGAAUUGA 3′.

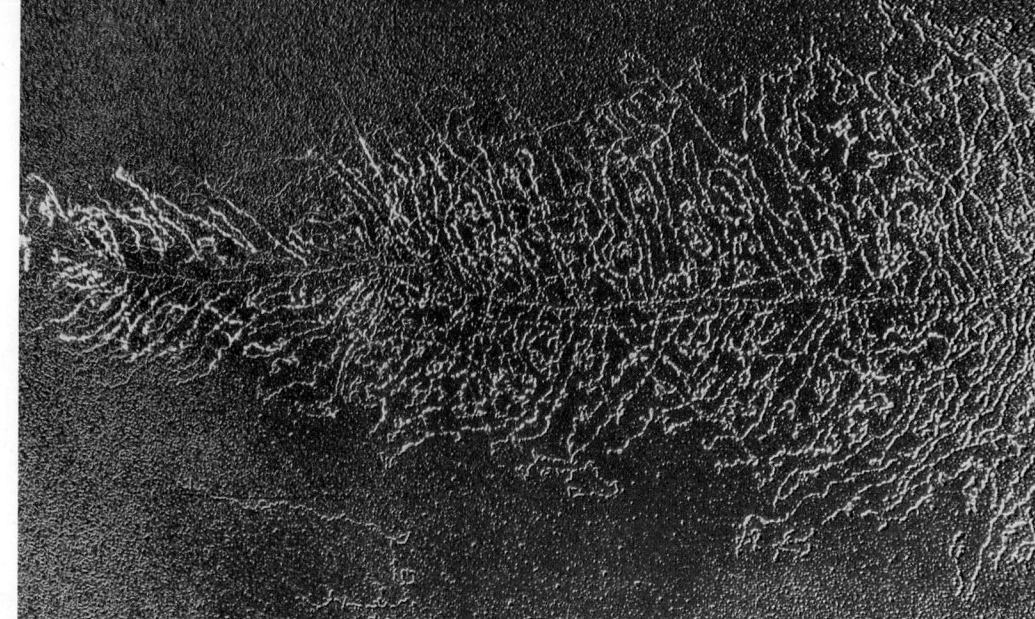

In this segment of a chromosome inside a frog egg, extensive transcription is taking place. The horizontal strand in the middle of this micrograph is DNA; the strands that have been colored yellow and red, and that are coming off on either side, are RNA molecules.

Transcription, RNA Processing, and Translation

16

roteins are the stuff of life. They give shape to our cells, control the chemical reactions that go on inside them, and regulate how materials move into, out of, and through them.

No one knows exactly how many different proteins can be made in your body's cells, but 100,000 is a reasonable estimate. Some of these proteins may not be produced at all in some types of cells; others may be present in quantities ranging from millions of copies to fewer than a dozen.

A cell builds the proteins it needs from instructions encoded in its DNA. Chapter 15 introduced the central dogma of molecular biology—the flow of information from DNA to mRNA to protein. Once this pattern of information flow had been established, biologists puzzled over how cells actually accomplish the two major steps in the process: transcription and translation. Specifically, how does RNA polymerase know where to start transcribing a gene, and where to end? And once an RNA message is produced, how is the linear sequence of ribonucleotides translated into the linear sequence of amino acids in a protein?

As Chapter 1 pointed out, processing genetic information is one of five key attributes of life. This chapter delves into the molecular mechanisms of gene expression—the blood and guts of the central dogma. It starts with the monomers that build an RNA and ends with a finished protein.

16.1 An Overview of Transcription

The first step in converting genetic information into proteins is to synthesize a messenger RNA version of the instructions archived in DNA. Enzymes called RNA polymerases, introduced in Chapter 15, are responsible for synthesizing mRNA.

KEY CONCEPTS

- After RNA polymerase binds DNA with the help of other proteins, it catalyzes the production of an RNA molecule whose base sequence is complementary to the base sequence of the DNA template strand.

- Eukaryotic genes contain regions called exons and regions called introns; during RNA processing, the regions coded by introns are removed, and the ends of the RNA receive a cap and tail.

- Ribosomes translate mRNAs into proteins with the help of intermediary molecules called transfer RNAs (tRNAs).

- Each transfer RNA carries an amino acid corresponding to the tRNA's three-base-long anticodon.

- In the ribosome, the tRNA anticodon binds to a three-base-long mRNA codon, causing the amino acid carried by the transfer RNA to be added to the growing protein.

✔ When you see this checkmark, stop and test yourself. Answers are available in Appendix B.

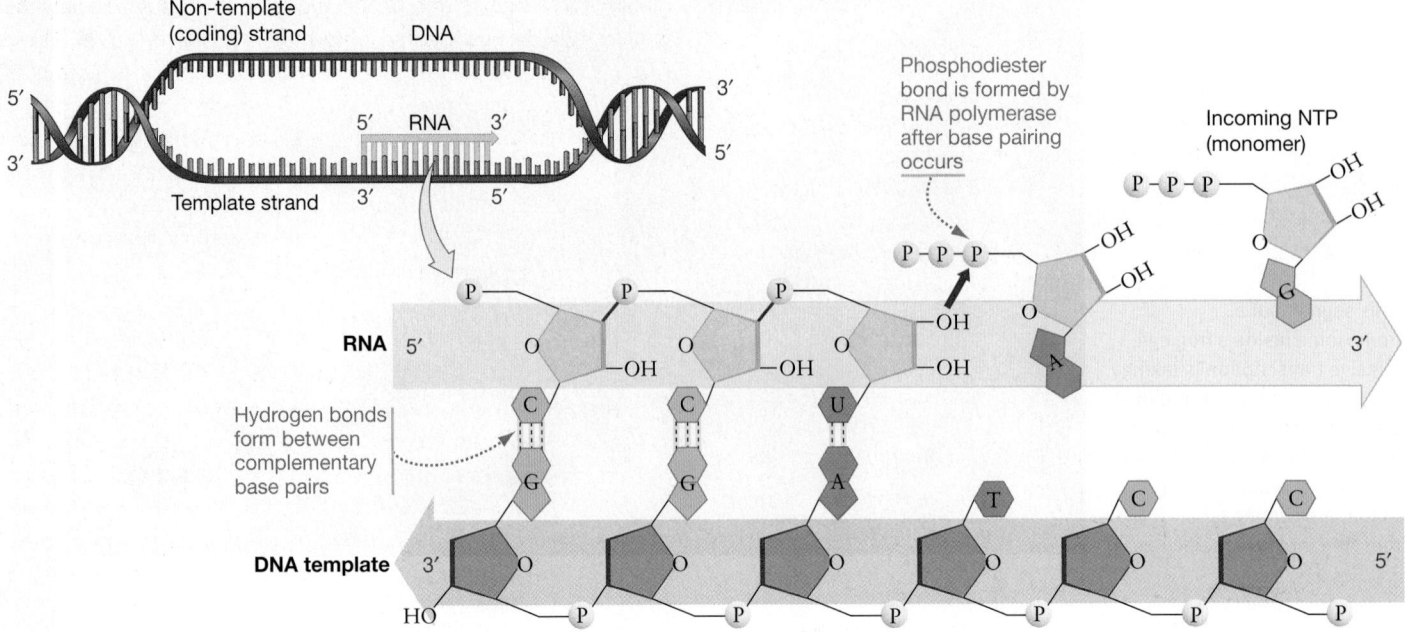

FIGURE 16.1 Transcription Is the Synthesis of RNA from a DNA Template. The reaction catalyzed by RNA polymerase (not shown) results in the formation of a phosphodiester bond between ribonucleotides. RNA polymerase produces an RNA strand whose sequence is complementary to the bases in the DNA template.

✔**QUESTION** In which direction is RNA synthesized, 5′→3′ or 3′→5′? In which direction is the DNA template "read"?

Figure 16.1 shows how the polymerization reaction occurs. Note the incoming monomer—a ribonucleotide triphosphate, or NTP—at the far right of the diagram. NTPs are like the dNTPs introduced in Chapter 14's discussion of DNA synthesis, except that they have a hydroxyl (−OH) group on the 2′ carbon. Once an NTP that matches a base on the DNA template is in place, RNA polymerase catalyzes the formation of a phosphodiester bond between the 3′ end of the growing mRNA chain and the new ribonucleotide. As this 5′→3′ matching-and-catalysis process continues, an RNA that is complementary to the gene is synthesized. This is transcription.

Notice that only one of the two DNA strands is used as a template and transcribed, or "read," by RNA polymerase.

- The strand that is read by the enzyme is called the **template strand**.
- The other strand is called the **non-template strand** or **coding strand**. Coding strand is a particularly appropriate name, because its sequence matches the sequence of the RNA that is transcribed from the template strand and codes for a polypeptide.

The coding strand and the RNA don't match exactly, however, because RNA has uracil (U) rather than the thymine (T) found in the coding strand. For the same reason, an adenine (A) in the DNA template strand specifies a U in the complementary RNA strand.

Once the basic role of RNA polymerase was understood, biologists turned to questions about how the enzyme works.

Characteristics of RNA Polymerase

🔑 Like the DNA polymerases introduced in Chapter 14, an RNA polymerase performs a template-directed synthesis in the 5′ to 3′ direction. But unlike DNA polymerases, RNA polymerases do not require a primer to begin transcription.

Bacteria have a single RNA polymerase. Eukaryotes, in contrast, have three distinct types. As **Table 16.1** shows, RNA polymerase I, II, and III—often referred to as pol I, pol II, and pol III—each transcribe only certain types of RNA in eukaryotes. RNA pol II is the only polymerase that transcribes the genes that code for proteins and produces mRNA.

TABLE 16.1 **Eukaryotic RNA Polymerases**

Name of Enzyme	Type of Gene Transcribed
RNA polymerase I (RNA pol I)	Genes that code for most of the large RNA molecules (rRNAs) found in ribosomes
RNA polymerase II (RNA pol II)	Protein-coding genes (produce mRNAs); also, genes that code for RNAs that function in ribosome assembly, and in processing and regulation of mRNAs
RNA polymerase III (RNA pol III)	Genes that code for transfer RNAs (tRNAs), for one of the small rRNAs found in ribosomes, and for noncoding RNAs (ncRNAs); also, genes that code for RNAs that function in ribosome assembly, and in processing and regulation of mRNAs

Initiation: How Does Transcription Begin?

How does RNA polymerase know where to start transcription on the DNA template? The answer to this question defined what biologists now call the **initiation** phase of transcription.

Soon after the discovery of bacterial RNA polymerase, researchers realized that the enzyme cannot initiate transcription on its own. Instead, a detachable protein subunit called sigma must bind to the polymerase before transcription can begin.

Bacterial RNA polymerase and sigma form what biologists call a **holoenzyme** (literally, "whole enzyme"; **Figure 16.2a**). A holoenzyme consists of a **core enzyme**, which contains the active site for catalysis, and other required proteins.

If bacterial RNA polymerase is the core enzyme of this holoenzyme, what does sigma do? When researchers mixed the polymerase, sigma, and DNA together, they found that the holoenzyme bound tightly to specific sections of DNA. These binding sites were named **promoters**, because they are sections of DNA where transcription begins. The discovery of promoters suggested that sigma's function is regulatory in nature. Sigma appeared to be responsible for guiding RNA polymerase to specific locations where transcription should begin.

What is the nature of these specific locations? What do promoters look like, and what do they do?

BACTERIAL AND EUKARYOTIC PROMOTERS David Pribnow offered an initial answer to these questions in the mid-1970s. When Pribnow analyzed the base sequence of promoters from various bacteria and from viruses that infect bacteria, he found that the promoters were 40–50 base pairs long and had a particular section in common: a series of bases identical or similar to TATAAT. This six-base-pair sequence is now known as the −10 box, because it is centered about 10 bases from the point where bacterial RNA polymerase starts transcription (**Figure 16.2b**).

DNA that is located in the direction RNA polymerase moves during transcription is said to be **downstream** from the point of reference; DNA located in the opposite direction is said to be **upstream**. Thus, the −10 box is centered about 10 bases upstream from the transcription start site. The place where transcription begins is called the +1 site.

Soon after the discovery of the −10 box, researchers recognized that the sequence TTGACA occurred in these same promoters and was centered about 35 bases upstream from the +1 site. This second key sequence is called the −35 box. Although all promoters have a −10 box and a −35 box, the sequences outside these boxes (but within the promoter) vary.

Eukaryotic genes have promoters that signal where transcription should begin, just as bacteria do. Promoters in eukaryotic DNA are much more diverse and complex than bacterial promoters, however. Many of the eukaryotic promoters include a unique sequence called the **TATA box**, centered about 30 base pairs upstream of the transcription start site.

THE ROLE OF SIGMA SUBUNITS AND BASAL TRANSCRIPTION FACTORS In bacteria, transcription begins when sigma, as part of the holoenzyme complex, binds to the −35 and −10 boxes. Sigma, and not RNA polymerase, makes the initial contact with DNA that starts transcription. This key observation supports the hypothesis that sigma is a regulatory protein. Sigma tells RNA polymerase where and when to start synthesizing RNA.

Recent work has shown that most bacteria have several types of sigma proteins, each with a distinct structure and function. *Escherichia coli* has seven different sigma proteins, for example, while *Streptomyces coelicolor* has more than 60. Each of these proteins binds to promoters with slightly different DNA base sequences outside of the −10 and −35 boxes. Thus, each type of sigma protein allows RNA polymerase to bind to a different type of promoter and therefore a different kind of gene.

(a) RNA polymerase and sigma form a holoenzyme.

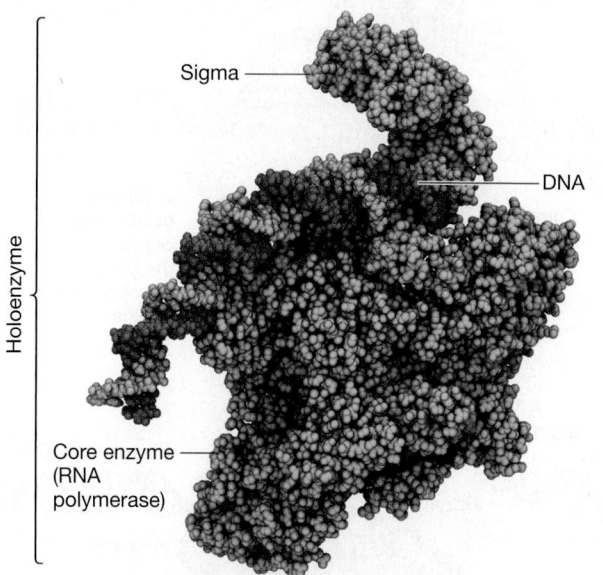

(b) Sigma recognizes and binds to the promoter.

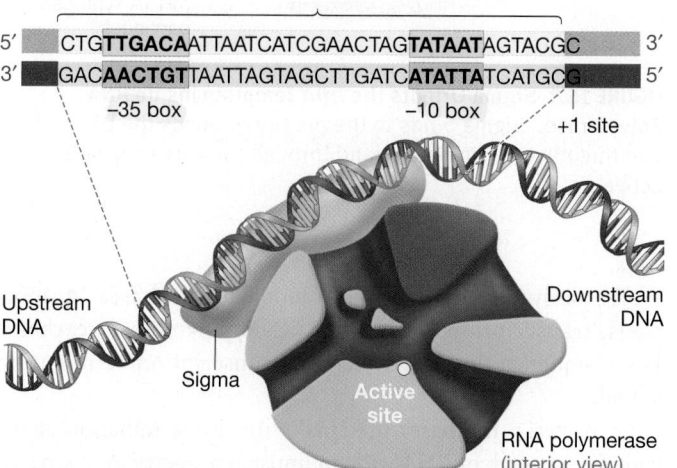

FIGURE 16.2 Sigma Is a Regulatory Subunit of RNA Polymerase.
(a) A space-filling model of RNA polymerase. **(b)** A cartoon of RNA polymerase, showing that sigma binds to the −35 box and −10 box of the promoter.

In some cases, certain sigma proteins bind to promoters for genes that have similar functions. For example, one type of sigma initiates the transcription of genes that help the cell cope with high temperatures.

The type of sigma protein in the RNA polymerase holoenzyme determines which type of genes will be transcribed. Controlling which sigma proteins are active is one of the ways that bacterial cells control which genes are expressed.

Transcription initiation in eukaryotes is similar to the process in bacteria in that RNA polymerase does not bind directly to promoter sequences by itself. Instead, proteins that came to be called **basal transcription factors** initiate eukaryotic transcription by binding to the appropriate promoter region in DNA. In eukaryotes, the function of basal transcription factors is analogous to the function of the sigma proteins in bacteria, except that many proteins are involved, instead of one, and basal transcription factors are not part of a holoenzyme. The basal transcription factors assemble at the promoter first, and RNA polymerase follows.

EVENTS INSIDE THE HOLOENZYME   Once sigma binds to a promoter for a bacterial gene, the DNA helix opens, creating two separated strands of DNA, as shown in **Figure 16.3**, steps 1 and 2. As step 2 shows, the template strand is threaded through a channel that leads to the active site inside RNA polymerase. Ribonucleoside triphosphates—the monomers that will be strung together into RNA—enter a channel at the bottom of the enzyme and diffuse to the active site.

When an incoming NTP pairs with a complementary base on the template strand of DNA, RNA polymerization begins. The reaction catalyzed by RNA polymerase is exergonic and spontaneous because NTPs have so much potential energy, owing to their three phosphate groups. As step 3 of Figure 16.3 shows, sigma is released once RNA synthesis is under way. The initiation phase of transcription is complete.

Elongation and Termination

Once RNA polymerase begins moving along the DNA template in the $3' \rightarrow 5'$ direction, synthesizing RNA in the $5' \rightarrow 3'$ direction, the **elongation** phase of transcription is under way. In the interior of the enzyme, a group of amino acids called the rudder helps steer the template and non-template strands through channels inside the enzyme (see Figure 16.3, step 3). Meanwhile, the enzyme's active site catalyzes the addition of nucleotides to the $3'$ end of the growing RNA molecule at the rate of about 50 nucleotides per second. A group of projecting amino acids called the enzyme's zipper then helps separate the newly synthesized RNA from the DNA template.

Note that during the elongation phase of transcription, all of the prominent channels and grooves in the enzyme are filled (Figure 16.3, step 3). Double-stranded DNA goes into and out of one groove; ribonucleoside triphosphates enter another; and the growing RNA strand exits to the rear. In this way, the enzyme's structure correlates closely with its function.

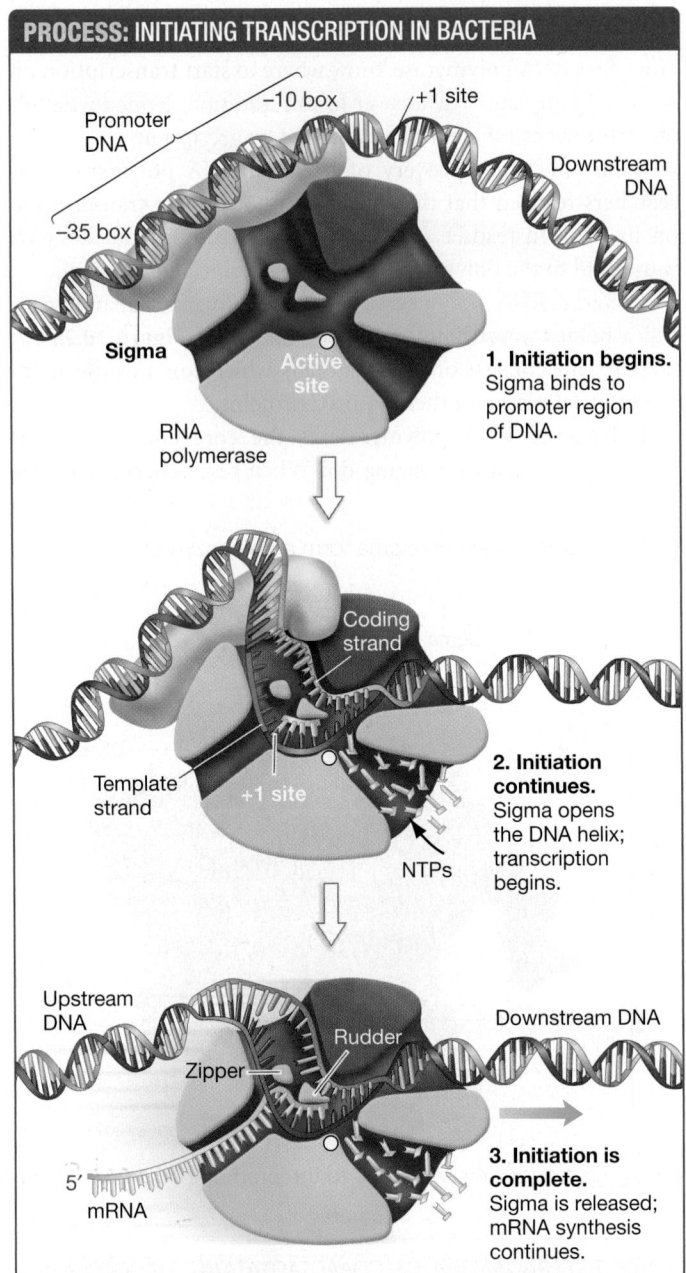

PROCESS: INITIATING TRANSCRIPTION IN BACTERIA

1. Initiation begins. Sigma binds to promoter region of DNA.

2. Initiation continues. Sigma opens the DNA helix; transcription begins.

3. Initiation is complete. Sigma is released; mRNA synthesis continues.

FIGURE 16.3 Sigma Orients the DNA Template inside RNA Polymerase. Sigma binds to the promoter, opens the DNA helix, and threads the template strand through the core enzyme's active site.

Transcription ends with a **termination** phase. In most cases, transcription stops when RNA polymerase reaches a DNA sequence that functions as a transcription-termination signal.

In bacteria, the bases that make up the termination signal code for a stretch of RNA with an unusual property: As soon as it is synthesized, the RNA sequence folds back on itself and forms a short double helix that is held together by complementary base pairing. The secondary structure that results is called a hairpin (**Figure 16.4**). The formation of the hairpin structure is thought

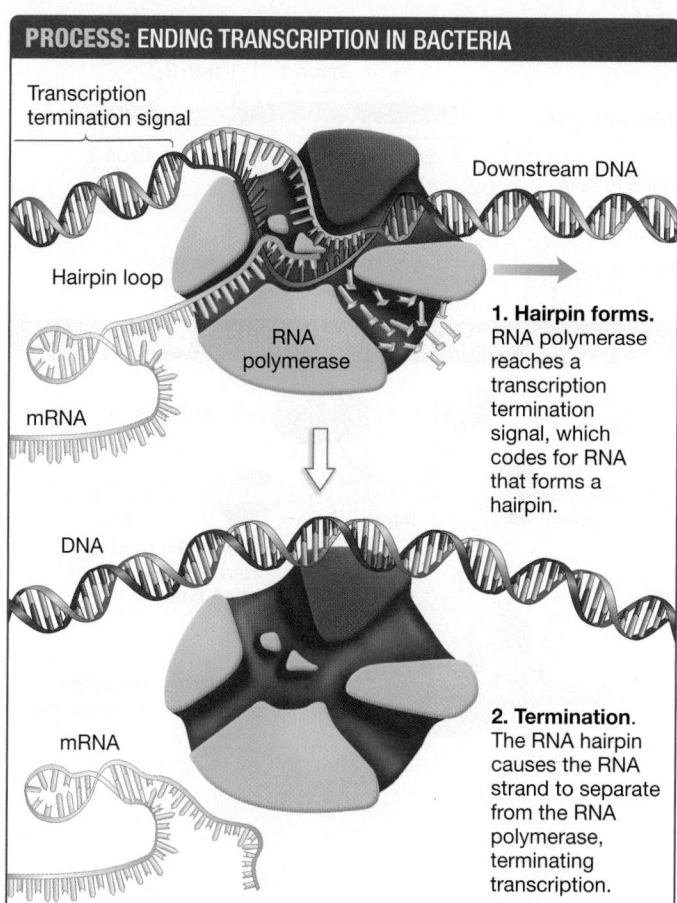

Transcription
termination signal

Downstream DNA

Hairpin loop

RNA
polymerase

mRNA

1. Hairpin forms.
RNA polymerase
reaches a
transcription
termination
signal, which
codes for RNA
that forms a
hairpin.

DNA

mRNA

2. Termination.
The RNA hairpin
causes the RNA
strand to separate
from the RNA
polymerase,
terminating
transcription.

FIGURE 16.4 Transcription Terminates When a Hairpin Forms.

CHECK YOUR UNDERSTANDING

If you understand that . . .

- Transcription initiation depends on interactions between proteins associated with RNA polymerase and a promoter sequence in DNA.
- In bacteria, sigma binds to RNA polymerase and contacts the promoter. In eukaryotes, basal transcription factors bind to the promoter and recruit RNA polymerase.
- During transcription elongation, ribonucleotide triphosphates are the substrate for a polymerization reaction catalyzed by RNA polymerase. The enzyme adds ribonucleotides that are complementary to the template strand in DNA.
- Transcription ends when a termination signal at the end of the gene leads to the disassociation of RNA polymerase and the DNA template.

✓ You should be able to . . .

1. Explain why different sigma proteins are able to bind to different promoters.

2. Explain why ribonucleotide triphosphates, rather than ribonucleotides (which have one phosphate group), are the monomers required for RNA synthesis.

Answers are available in Appendix B.

to disrupt the interaction between RNA polymerase and the RNA transcript, resulting in the physical separation of the enzyme and its product.

To review key concepts about transcription and see the initiation, elongation, and termination phases in action, go to the study area at *www.masteringbiology.com*.

(MB) Web Activity RNA Synthesis

16.2 RNA Processing in Eukaryotes

The molecular machinery required for transcription is much more complex in eukaryotes than in bacteria. But the contrast between single sigma proteins and multiple basal transcription factors, or between bacterial and eukaryotic promoters, is mild compared to what may be the most striking contrast in gene expression between bacteria and eukaryotes: In bacteria, the information in DNA is converted to mRNA directly. In eukaryotes, it isn't.

When transcription terminates in bacteria, the result is a mature mRNA that is ready to be translated into a protein. But when a eukaryotic gene is transcribed, the product is an immature **primary transcript**, or pre-mRNA. Before primary transcripts can be translated, they have to be processed in a complex series of steps.

Why? And how? And what consequences does this RNA processing requirement have for gene expression in eukaryotes?

The Startling Discovery of Eukaryotic Genes in Pieces

Eukaryotic genes do not consist of one continuous DNA sequence that codes for a product, as do bacterial genes. Instead, the regions in a eukaryotic gene that code for proteins are intermittently interrupted by stretches of hundreds or many thousands of intervening bases.

Although these intervening bases are part of the gene, they do not code for a product. To make a functional mRNA, then, eukaryotic cells must dispose of certain sequences inside the primary transcript and then combine the separated coding sections into an integrated whole.

What sort of data would provoke such a startling claim? The first evidence came from work that Phillip Sharp and colleagues carried out in the late 1970s to determine how DNA templates are transcribed.

They began one of their experiments by heating DNA molecules sufficiently to break the hydrogen bonds between complementary bases. This treatment separated the two strands. The single-stranded DNA was then incubated with the mRNA encoded by the sequence. The team's intention was to promote base pairing between the mRNA and the single-stranded DNA.

The researchers expected that the mRNA would form base pairs with the DNA sequence that acted as the template for its synthesis—that the mRNA and DNA would match up exactly. But when the team examined the DNA–RNA hybrid molecules with the electron microscope, they observed the structure shown

(a) Micrograph of DNA-RNA hybrid   **(b)** Interpretation of micrograph

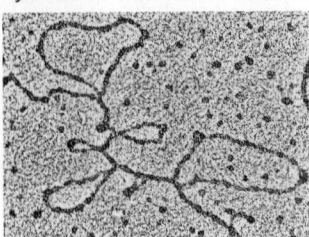
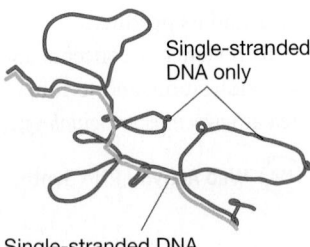

Single-stranded DNA only

Single-stranded DNA base paired with mRNA

FIGURE 16.5 The Discovery of Noncoding Regions of DNA. The loops in the micrograph and drawing represent regions of DNA that do not have an equivalent sequence in the mRNA. These intervening regions are "extra" DNA compared with the sequences in the mRNA.

✔**QUESTION** If the noncoding regions of the gene did not exist, what would the micrograph in part (a) look like?

in **Figure 16.5a**. Instead of matching up exactly, parts of the DNA formed loops.

What was going on? As **Figure 16.5b** shows, Sharp's group interpreted these loops as stretches of nucleotides that are present in the DNA template strand but are *not* in the corresponding mRNA.

Sharp's group and a team headed by Richard Roberts went on to propose that there is not a one-to-one correspondence between the nucleotide sequence of a eukaryotic gene and its mRNA. As an analogy, it could be said that eukaryotic genes do not carry messages such as "Biology is my favorite course of all time." Instead, eukaryotic genes carry messages that read something like "BIOL τηεπροτεινχοδινγρεγιονσοφγενεσOGY IS MY FAVORαρειντ ερρθπτεϑβυνονψοϑινγϑITE COURSE OF ανϑηαωετοβεσ πλιχεϑτογετηερ ALL TIME." Here the sections of noncoding sequence are represented with Greek letters. They must be removed from the mRNA before it can carry an intelligible message to the translation machinery.

When it became clear that the genes-in-pieces hypothesis was correct, Walter Gilbert suggested that regions of eukaryotic genes that are part of the final mRNA be referred to as **exons** (because they are *ex*pressed) and the sections of primary transcript not in mRNA be referred to as **introns** (because they are *int*ervening). Exons code for segments of functional proteins or RNAs; introns do not. Introns are sections of genes that are not represented in the final mRNA product. As a result, eukaryotic genes are much larger than their corresponding mature RNA transcripts.

RNA Splicing

The transcription of eukaryotic genes by RNA polymerase generates a primary RNA transcript that contains both the exon and intron regions (**Figure 16.6a**). As transcription proceeds, the introns are removed from the growing RNA strand by a process known as **splicing**. In this phase of information processing, pieces of the primary transcript are removed and the remaining segments are joined together. Splicing occurs while transcription is still under way and results in an RNA that contains an uninterrupted genetic message.

(a) Introns must be removed from eukaryotic RNA transcripts.

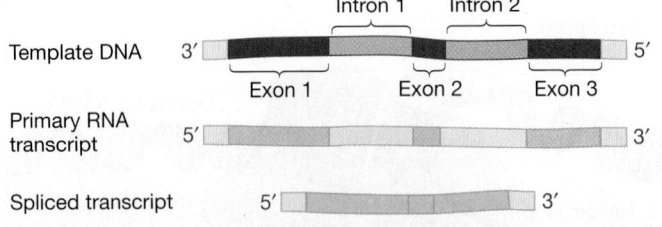

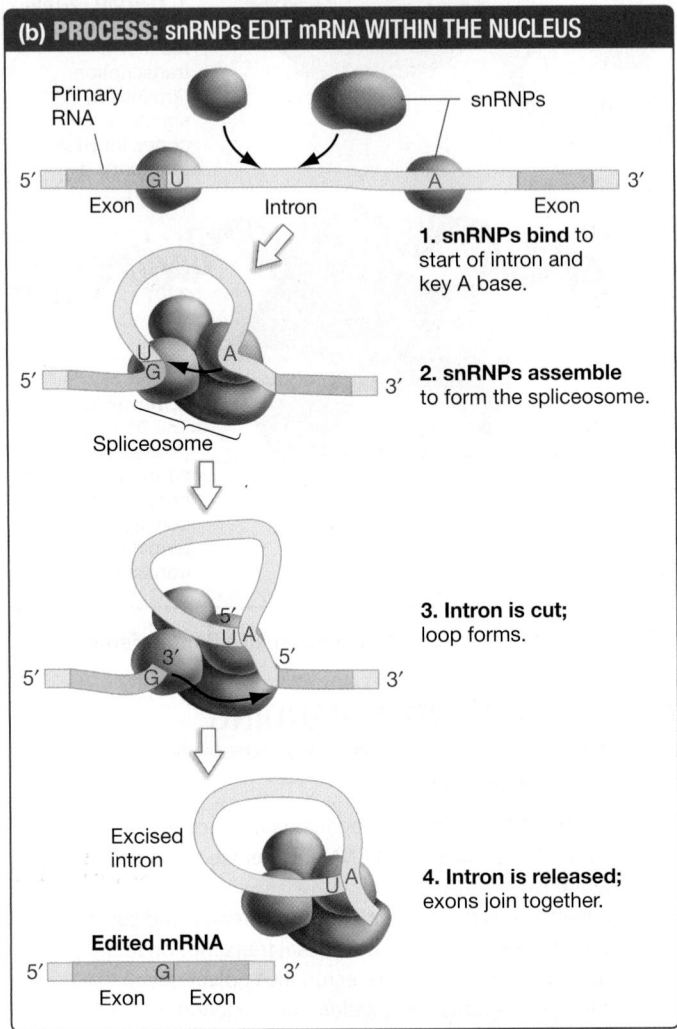

FIGURE 16.6 Introns Are Spliced Out of the Original mRNA.

Figure 16.6b provides more detail about how introns are removed from genes. In most cases, splicing is catalyzed by complexes of proteins and specialized RNAs, called small nuclear RNAs (snRNAs). These protein-plus-RNA complexes are known as **small nuclear ribonucleoproteins**, or **snRNPs** (pronounced "snurps"). The process can be broken into four steps:

1. The process begins when snRNPs bind to the 5′ exon–intron boundary, which is marked by the bases GU and a key adenine ribonucleotide near the end of the intron.

2. Once the initial snRNPs are in place, other snRNPs arrive to form a multipart complex called a **spliceosome**. The

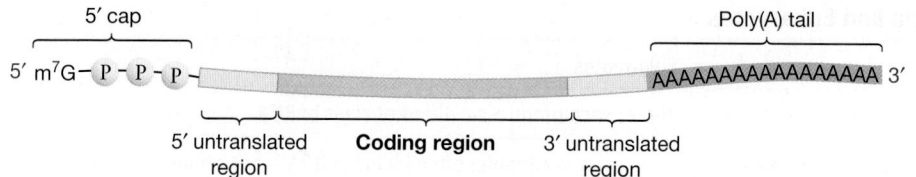

5' cap

Poly(A) tail

5' m⁷G– P – P – P – ‎ AAAAAAAAAAAAAAAAAA 3'

5' untranslated region   **Coding region**   3' untranslated region

FIGURE 16.7 In Eukaryotes, Mature mRNAs Have a Cap and a Tail. Eukaryotic mRNAs have a cap consisting of a molecule called 7-methyl-guanylate (symbolized as m⁷G) bonded to three phosphate groups; the tail is made up of a long series of adenine residues.

spliceosomes found in human cells contain about 145 different proteins and RNAs, making them the most complex molecular machines known.

3. The intron forms a loop with the key adenine at its connecting point.

4. The loop is cut out, and a phosphodiester bond links the exons on either side, producing a contiguous coding sequence.

Splicing is now complete. In most cases, the excised intron is degraded to ribonucleotide monophosphates.

Current data suggest that both the cutting and rejoining reactions that occur during splicing are catalyzed by the snRNA molecules in the spliceosome—meaning that the reactions are catalyzed by a ribozyme. Section 16.5 will demonstrate that ribozymes also play a key role in translation. As the RNA world hypothesis introduced in Chapter 4 predicts, proteins are not the only important catalysts in cells.

Adding Caps and Tails to Transcripts

Intron splicing of primary RNA transcripts in eukaryotes is followed by other important processing steps.

- As soon as the 5' end of a eukaryotic RNA emerges from RNA polymerase, enzymes add a structure called the 5' **cap** (**Figure 16.7**). The cap consists of the molecule 7-methyl-guanylate and three phosphate groups.

- An enzyme cleaves the 3' end of most RNAs once transcription is complete, and another enzyme adds a long row of 100–250 adenine nucleotides, not encoded on the DNA template strand, known as the **poly(A) tail**.

With the addition of the cap and tail and completion of splicing, processing of the primary RNA transcript is complete. The product is a mature mRNA.

Figure 16.7 also shows that in the mature RNA molecule, the coding sequence for the polypeptide is flanked by sequences that are not destined to be translated. These 5' and 3' untranslated regions (or UTRs) help stabilize the mature RNA and regulate its translation. The mRNAs in bacteria also possess 5' and 3' UTRs.

Not long after the caps and tails on eukaryotic mRNAs were discovered, evidence began to accumulate that they protect mRNAs from degradation by ribonucleases and enhance the efficiency of translation. For example:

- Experimental mRNAs that have a cap and a tail last longer when they are introduced into cells than do experimental mRNAs that lack a cap, a tail, or both a cap and a tail.

- Experimental mRNAs with caps and tails produce more proteins than do experimental mRNAs without caps and tails.

Follow-up work has shown that the 5' cap and the poly(A) tail serve as recognition signals for the translation machinery. It has also confirmed that they extend the life span of an mRNA by protecting the message from degradation by ribonucleases in the cytosol.

RNA processing is the general term for any of the modifications, such as splicing or poly(A) tail addition, needed to convert a primary transcript into a mature RNA. It is summarized in **Table 16.2** on page 296 along with other important differences in how mRNAs are produced in eukaryotes as compared to bacteria.

CHECK YOUR UNDERSTANDING

If you understand that . . .

- Eukaryotic genes consist of exons, which are parts of the primary transcript that remain in mRNA, and introns, which are regions of the primary transcript that are removed in forming mRNA.

- Introns are spliced out of primary RNA transcripts by multi-molecular machines called spliceosomes.

- Enzymes add a 5' cap and a poly(A) tail to spliced transcripts, producing a mature mRNA that is ready to be translated.

✓ You should be able to . . .

1. Explain why "ribonucleoprotein" is an appropriate name for the subunits of the spliceosome.

2. Explain the function of the 5' cap and the poly(A) tail.

Answers are available in Appendix B.

16.3 An Introduction to Translation

To synthesize a protein, the sequence of bases in a messenger RNA molecule is translated into a sequence of amino acids in a polypeptide. The genetic code presented in Chapter 15 specifies the correspondence between each triplet codon in mRNA and the amino acid it codes for. But how are the amino acids assembled into a polypeptide according to the information in messenger RNA?

Studies of translation in cell-free systems proved extremely effective at answering this question. Once in vitro translation systems had been developed from human cells, *E. coli*, and a variety of other organisms, biologists could see that the sequence of events is similar in bacteria, archaea, and eukaryotes.

Ribosomes Are the Site of Protein Synthesis

The first question that biologists answered about translation concerned where it occurs. The answer grew from a simple observation: There is a strong positive correlation between the presence

TABLE 16.2 Comparing Transcription in Bacteria and Eukaryotes

Point of Comparison	Bacteria	Eukaryotes
RNA polymerase(s)	One	Three; each produces a different class of RNA
Promoter structure	Typically contains a −35 box and a −10 box	Complex and variable; often includes a TATA box about −30 from the transcription start site
Protein(s) involved in contacting promoter	Sigma; different versions of sigma bind to different promoters	Many basal transcription factors
RNA-processing	None	Extensive; several processing steps occur in the nucleus before RNA is exported to the cytosol for translation: 1. Enzyme-catalyzed addition of 5′ cap 2. Splicing (intron removal) by spliceosome 3. Enzyme-catalyzed addition of 3′ poly(A) tail

of small structures known as **ribosomes** in a given type of cell and the rate at which that cell synthesizes proteins. For example:

- Immature human red blood cells divide rapidly, synthesize millions of copies of the protein hemoglobin, and contain large numbers of ribosomes.

- The same cells at maturity have low rates of protein synthesis and very few ribosomes.

Based on this correlation, investigators proposed that ribosomes are the site of protein synthesis in the cell.

To test this hypothesis, Roy Britten and collaborators did a pulse-chase experiment similar in design to experiments intro-

duced in Chapter 7. Recall that a pulse-chase experiment labels a population of molecules as they are being produced. The location of the tagged molecules is then followed over time.

In this case, the tagging was done by supplying a pulse of radioactive sulfur atoms that would be incorporated into the amino acids methionine and cysteine, followed by a chase of unlabeled sulfur atoms. If the ribosome hypothesis were correct, the radioactive signal should be associated with ribosomes for a short period of time—while the amino acids were being polymerized into proteins. Later, when translation was complete, all of the radioactivity should be found in proteins, not in the ribosomes.

This is exactly what the researchers found. Based on these data, biologists concluded that proteins are synthesized at ribosomes and then released.

Comparing Translation in Bacteria and Eukaryotes

About a decade after the ribosome hypothesis was confirmed, electron micrographs showed bacterial ribosomes in action (**Figure 16.8a**). The images showed that in bacteria, ribosomes attach to mRNAs and begin synthesizing proteins even before transcription is complete. In fact, multiple ribosomes attach to each

(a) Bacterial ribosomes during translation

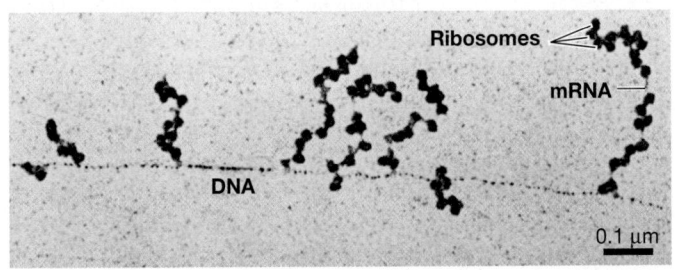

(b) In bacteria, transcription and translation are tightly coupled.

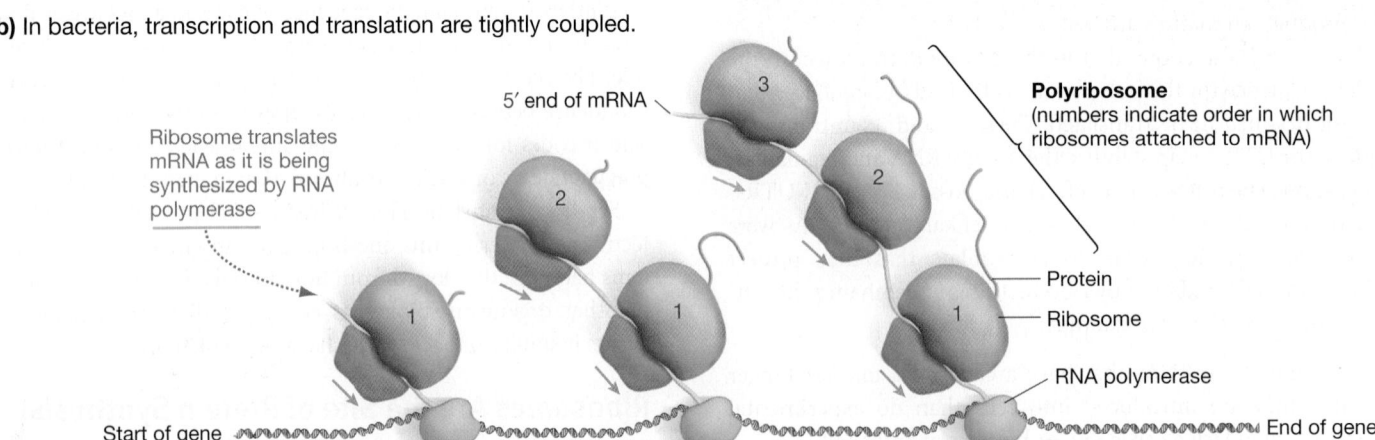

FIGURE 16.8 Transcription and Translation Occur Simultaneously in Bacteria. In bacteria, ribosomes attach to mRNA transcripts and begin translation while RNA polymerase is still transcribing the DNA template strand.

mRNA, forming a **polyribosome** (**Figure 16.8b**). In this way, many copies of a protein can be produced from a single mRNA.

Transcription and translation can occur concurrently in bacteria because there is no nuclear envelope to separate the two processes. Thus, transcription and translation are physically connected.

The situation is different in eukaryotes. In these organisms, primary transcripts are processed in the nucleus to produce a mature mRNA, which is then exported to the cytosol (**Figure 16.9a**). Once mRNAs are outside the nucleus, ribosomes attach to them and begin translation. As in bacteria, polyribosomes form (**Figure 16.9b**). But in eukaryotes, transcription and translation are separated in time and space.

(a) mRNAs are exported to the cytosol.

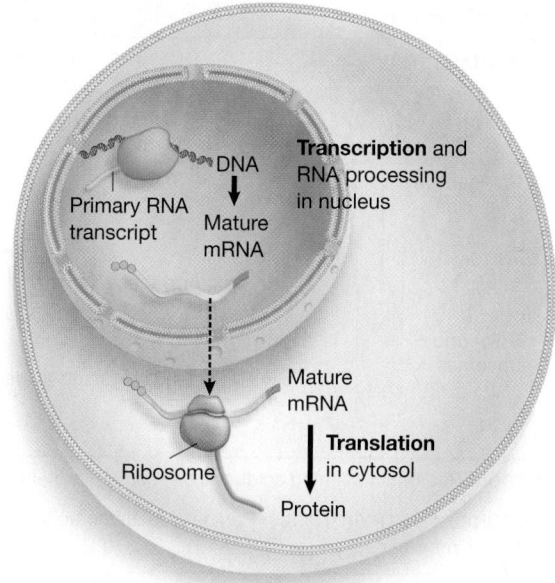

(b) Polyribosomes form in the cytosol.

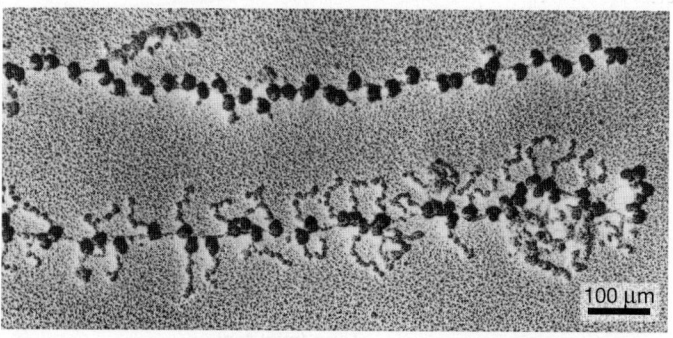

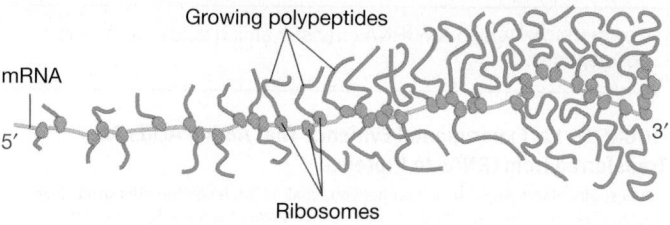

FIGURE 16.9 Transcription and Translation Are Separated in Space and Time in Eukaryotes.

How Does an mRNA Triplet Specify an Amino Acid?

When an mRNA interacts with a ribosome, hereditary instructions encoded in nucleic acids are translated into a different chemical language—the amino acid sequences found in proteins. The discovery of the genetic code revealed that triplet codons in mRNA specify particular amino acids in a protein. How does this conversion happen?

One early hypothesis was that mRNA codons and amino acids interact directly. This hypothesis proposed that the bases in a particular codon were complementary in shape or charge to the side group of a particular amino acid (**Figure 16.10a**). But Francis Crick pointed out that the idea didn't make chemical sense. For example, how could the nucleic acid bases interact with a hydrophobic amino acid side group, which does not form hydrogen bonds?

Crick proposed an alternative hypothesis. As **Figure 16.10b** shows, he suggested that some sort of adapter molecule holds amino acids in place while interacting directly and specifically with a codon in mRNA by hydrogen bonding. In essence, Crick predicted the existence of a chemical go-between that produced a physical connection between the two types of molecules. As it turns out, Crick was right.

16.4 The Structure and Function of Transfer RNA

Crick's adapter molecule was discovered by accident. Biologists were trying to work out an in vitro protein synthesis system and discovered that ribosomes, mRNA, amino acids, ATP, and a molecule called guanosine triphosphate, or GTP, had to be present for translation to occur. (GTP is similar to ATP but contains guanosine instead of adenosine.)

(a) Hypothesis 1: Amino acids interact directly with mRNA codons.

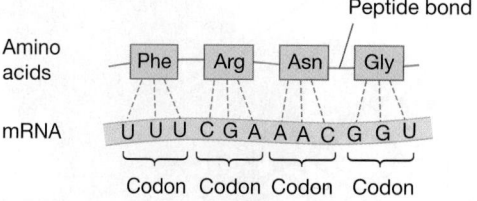

(b) Hypothesis 2: Adapter molecules hold amino acids and interact with mRNA codons.

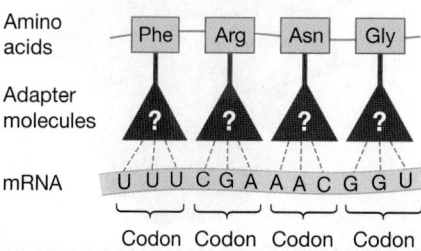

FIGURE 16.10 How Do mRNA Codons Interact with Amino Acids?

These results were logical: ribosomes provide the catalytic machinery, mRNAs contribute the message to be translated, amino acids are the building blocks of proteins, and ATP and GTP supply potential energy to drive the endergonic polymerization reactions responsible for forming proteins.

But in addition, a cellular fraction that contained a previously unknown type of RNA turned out to be indispensable. If this type of RNA is missing, protein synthesis does not occur. What is this mysterious RNA, and why is it essential to translation?

The novel class of RNAs eventually became known as **transfer RNA (tRNA)**. The role of tRNA in translation was a mystery until some researchers happened to add a radioactive amino acid—leucine—to an in vitro protein synthesis system. The treatment was actually done as a control for an unrelated experiment. To the researchers' amazement, some of the radioactive leucine attached to tRNA molecules.

Follow-up experiments revealed several key facts about this process:

- An input of energy, in the form of ATP, is required to attach an amino acid to a tRNA.

- Enzymes called **aminoacyl tRNA synthetases** catalyze the addition of amino acids to tRNAs—what biologists call "charging" a tRNA.

- For each of the 20 major amino acids, there is a different aminoacyl tRNA synthetase and one or more tRNAs.

The combination of a tRNA molecule covalently linked to an amino acid is called an **aminoacyl tRNA**. **Figure 16.11** shows an aminoacyl tRNA synthetase still bound to a tRNA that has just been charged with an amino acid. Note how tightly the two structures fit together—making it possible for the enzyme and substrate to interact in an extremely precise way.

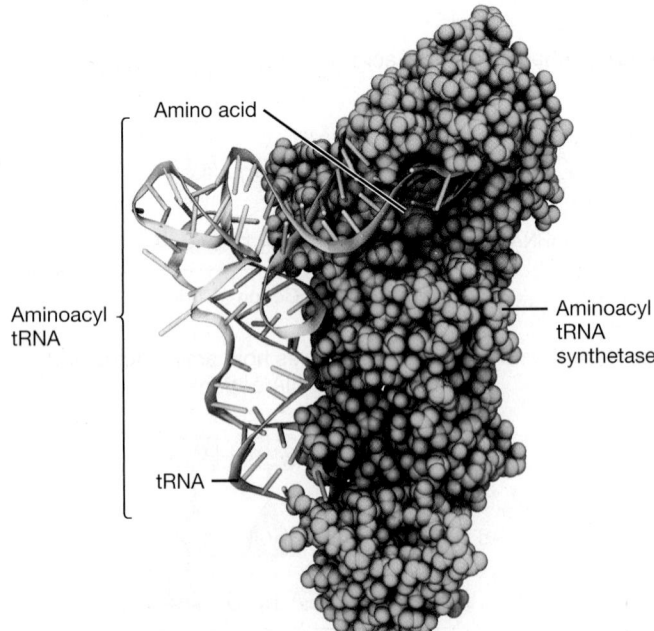

FIGURE 16.11 Aminoacyl tRNA Synthetases Load Amino Acids onto the Appropriate tRNA.

What happens to the amino acids bound to tRNAs? To answer this question, biologists tracked the fate of radioactive leucine molecules that were attached to tRNAs. They found that the amino acids are transferred from aminoacyl tRNAs to proteins.

The data supporting this conclusion are shown in **Figure 16.12**. The graph in the "Results" section of this figure shows that ra-

EXPERIMENT

QUESTION: What happens to the amino acids attached to tRNAs?

HYPOTHESIS: Aminoacyl tRNAs transfer amino acids to growing polypeptides.

NULL HYPOTHESIS: Aminoacyl tRNAs do not transfer amino acids to growing polypeptides.

EXPERIMENTAL SETUP:

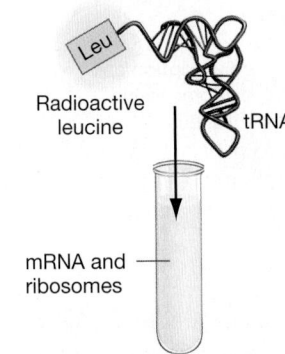

1. Attach radioactive leucine molecules to tRNAs.

2. Add these aminoacyl tRNAs to in vitro translation system. Follow fate of the radioactive amino acids.

PREDICTION: Radioactive amino acids will be found in proteins.

PREDICTION OF NULL HYPOTHESIS: Radioactive amino acids will not be found in proteins.

RESULTS:

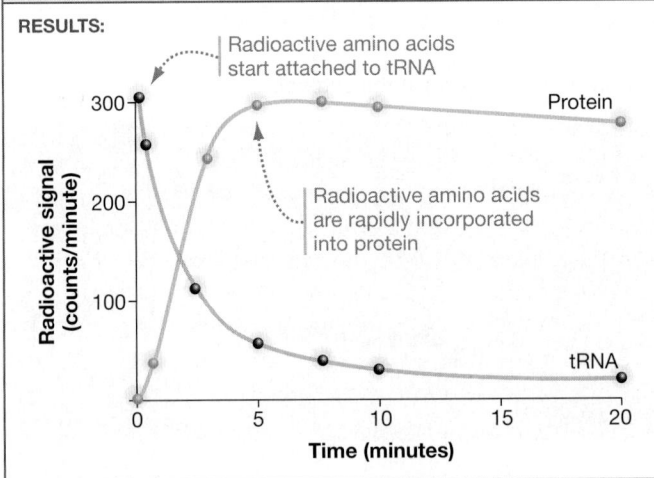

CONCLUSION: Aminoacyl tRNAs transfer amino acids to growing polypeptides.

FIGURE 16.12 Experimental Evidence That Amino Acids Are Transferred from tRNAs to Proteins.

SOURCE: Hoagland, M. B., M. L. Stephenson, et al. 1958. A soluble ribonucleic acid intermediate in protein synthesis. *Journal of Biological Chemistry* 231: 241–257.

✔ **QUESTION** What would the graphed results look like if the null hypothesis were correct?

dioactive amino acids are lost from tRNAs and incorporated into polypeptides synthesized in ribosomes. To understand this conclusion, do the following:

1. Put your finger on the point on the *x*-axis that indicates that one minute has passed since the start of the experiment.

2. Read up until you hit the green line and the gray line. The green line represents data from proteins; the gray line represents data from tRNAs.

3. Check the *y*-axis—which indicates the amount of radioactive leucine present—at each point.

4. It should be clear that early in the experiment, almost all of the radioactive leucine is attached to tRNA, not protein.

Next, do the same four steps at the point on the *x*-axis labeled 10 minutes (since the start of the experiment). Your conclusion now should be that late in the experiment, almost all of the radioactive leucine is attached to proteins, not tRNA.

🔑 These results inspired the use of "transfer" in tRNA's name, because amino acids are transferred from the RNA to the growing end of a new polypeptide. The experiment also confirmed that aminoacyl tRNAs act as the interpreter in the translation process: tRNAs are Crick's adapter molecules.

What Do tRNAs Look Like?

Transfer RNAs serve as chemical go-betweens that allow amino acids to interact with an mRNA template. But precisely how does the connection occur?

This question was answered by research on tRNA's molecular structure. The initial studies established the sequence of nucleotides in various tRNAs, or what is termed their primary structure. Transfer RNA sequences are relatively short, ranging from 75 to 85 nucleotides in length.

When biologists studied the primary sequence closely, they noticed that certain parts of the molecules can form secondary structures. Specifically, some sequences of bases in the tRNA molecule can form hydrogen bonds with complementary base sequences elsewhere in the same molecule. As a result, portions of the molecule should form the stem-and-loop structures introduced in Chapter 4. The stems are short stretches of double-stranded RNA; the loops are single-stranded.

Two aspects of this secondary structure proved especially interesting. A CCA sequence at the 3′ end of each tRNA molecule offered a binding site for amino acids, while a triplet on the loop at the far end of the structure could serve as an anticodon. An **anticodon** is a set of three ribonucleotides that forms base pairs with the mRNA codon.

Later, X-ray crystallography studies revealed that tRNAs also have tertiary structure. Recall from Chapter 3 that the tertiary structure of a molecule is the three-dimensional arrangement of its atoms and is usually a product of folding. As **Figure 16.13** shows, tRNAs fold into an L-shaped molecule. The anticodon is at one end of the structure; the CCA sequence and attached amino acid is at the other end.

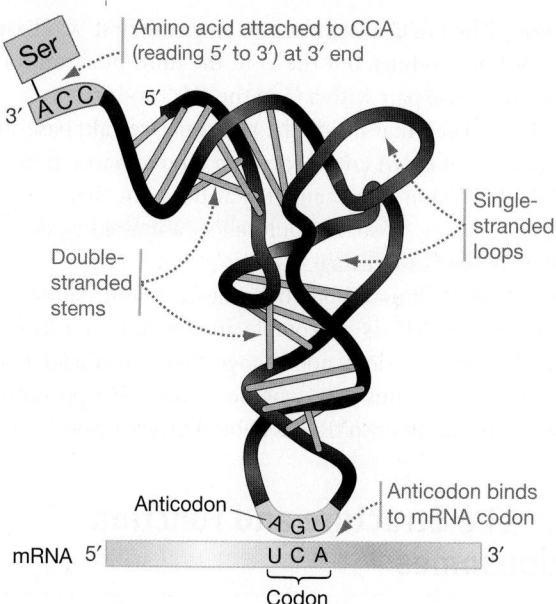

FIGURE 16.13 The Structure of Transfer RNA. The anticodon of an aminoacyl tRNA forms complementary base pairs with an mRNA codon.

🔑 All of the tRNAs in a cell have the same structure, shaped like an upside-down L. They vary at the anticodon and attached amino acid. The tertiary structure of tRNAs is important because it maintains a precise physical distance between the anticodon and amino acid. As it turns out, this separation is key to the positioning of the amino acid and the anticodon in the ribosome.

✓If you understand the structure of tRNAs, you should be able to (1) describe where on the upside-down, L-shaped structure the amino acid attaches, and (2) explain the relationship between the anticodon of a tRNA and a codon in an mRNA.

How Many tRNAs Are There?

After characterizing all of the different types of tRNAs available in cells, biologists encountered a paradox. According to the genetic code introduced in Chapter 15, the 20 most common amino acids found in proteins are specified by 61 different mRNA codons. Instead of containing 61 different tRNAs with 61 different anticodons, though, most cells contain only about 40. How can all 61 mRNA codons in the genome be translated with only two-thirds that number of tRNAs?

To resolve this paradox, Francis Crick proposed what is known as the **wobble hypothesis**. Recall from Chapter 15 that:

1. Many amino acids are specified by more than one codon.

2. Codons for the same amino acid tend to have the same nucleotides at the first and second positions but a different nucleotide at the third position.

For example, both of the codons CAA and CAG code for the amino acid glutamine. (Codons are always written in the 5′→3′ direction.) Surprisingly, experimental data have shown that a tRNA with an anticodon of GUU can base pair with both CAA and CAG in mRNA. (Anticodons are written in the 3′→5′

direction.) The GUU anticodon matches the first two bases (C and A) in both codons, but the U in the third position forms a nonstandard base pair with a G in the CAG codon.

Crick proposed that inside the ribosome, certain bases in the third position of tRNA anticodons can bind to bases in the third position of a codon in a manner that does not match Watson-Crick base pairing. If so, it would allow a limited flexibility, or "wobble," in the base pairing.

According to the wobble hypothesis, a nonstandard base pair—such as G-U—is acceptable in the third position of a codon as long as it does not change the amino acid that the codon specifies. In this way, wobble in the third position of a codon allows just 40 or so tRNAs to bind to all 61 mRNA codons.

16.5 The Structure and Function of Ribosomes

Recall that protein synthesis occurs when the sequence of bases in an RNA message is translated into a sequence of amino acids in a polypeptide. The conversion of each mRNA codon begins when the anticodon of an aminoacyl tRNA binds to the codon. The conversion is complete when a peptide bond forms between the tRNA's amino acid and the growing polypeptide chain.

Both of these events take place inside a ribosome. Biologists have known since the 1930s that ribosomes contain a considerable amount of protein along with a great deal of **ribosomal RNA (rRNA)**. Later work showed that ribosomes can be separated into two major substructures, called the large subunit and small subunit. Each ribosome subunit consists of a complex of RNA molecules and proteins. The small subunit holds the mRNA in place during translation; the large subunit is where peptide-bond formation takes place.

Figure 16.14 shows how all of the molecules required for translation fit together. Note that during protein synthesis, three distinct tRNAs are lined up inside the ribosome. All three are bound to their corresponding mRNA codon at the base of the structure.

- The tRNA that is on the right in the figure, and colored red, carries an amino acid. This tRNA's position in the ribosome is called the A site—"A" for acceptor or aminoacyl.
- The tRNA that is in the middle (green) holds the growing polypeptide chain and occupies the P site, for peptidyl, inside the ribosome. (Think of "P" for peptide-bond formation.)
- The left-hand (blue) tRNA no longer has an amino acid attached and is about to leave the ribosome. It occupies the ribosome's E site—"E" for exit.

Because all tRNAs have similar secondary and tertiary structure, they all fit equally well in the A, P, and E sites.

The ribosome is a molecular machine that synthesizes proteins in a three-step sequence:

1. An aminoacyl tRNA diffuses into the A site; its anticodon binds to a codon in mRNA.

2. A peptide bond forms between the amino acid held by the aminoacyl tRNA in the A site and the growing polypeptide, which was held by a tRNA in the P site.

3. The ribosome moves ahead, and all three tRNAs move one position down the line. The tRNA in the E site exits; the tRNA in the P site moves to the E site; and the tRNA in the A site switches to the P site.

The protein that is being synthesized grows by one amino acid each time this three-step sequence repeats. The process occurs up to 20 times per second in bacterial ribosomes and about

(a) Diagram of ribosome during translation (interior view)

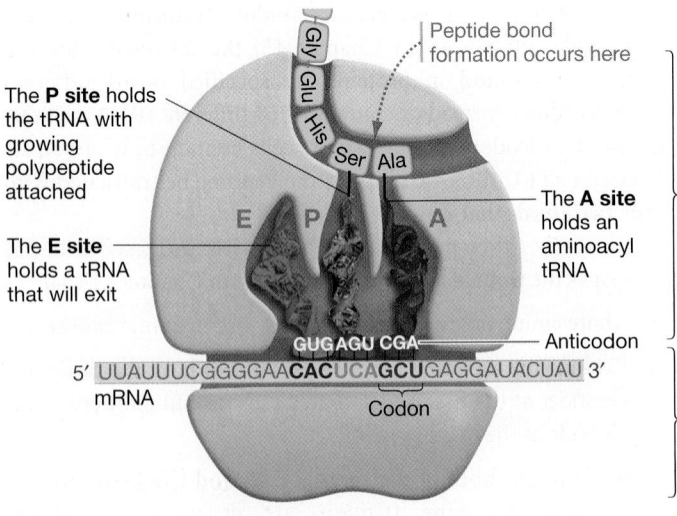

(b) Model of ribosome during translation (exterior view)

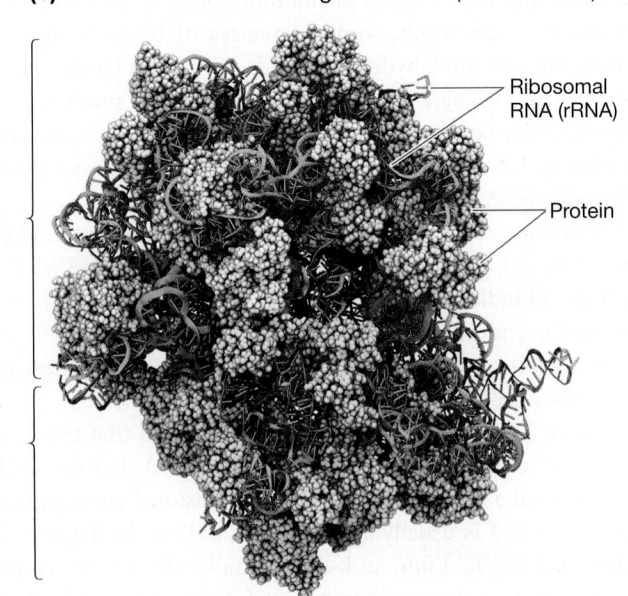

FIGURE 16.14 The Structure of the Ribosome. Ribosomes have three distinct sites in their interior where tRNAs are found.

2 times per second in eukaryotic ribosomes. Protein synthesis starts at the amino end (N-terminus) of a polypeptide and proceeds to the carboxy end (C-terminus; see Chapter 3).

This introduction to how tRNAs, mRNAs, and ribosomes interact during protein synthesis leaves several key questions unanswered, however. How do mRNAs and ribosomes get together to start the process? Once protein synthesis is under way, how is peptide-bond formation catalyzed inside the ribosome? And how does protein synthesis conclude when the ribosome reaches the end of the message? Let's consider each question in turn.

Initiating Translation

To translate an mRNA properly, a ribosome must begin at a specific point in the message, translate the mRNA up to the message's termination codon, and then stop. Using the same terminology that they apply to transcription, biologists call these three phases of protein synthesis initiation, elongation, and termination, respectively.

One key to understanding translation initiation is to recall, from Chapter 15, that a start codon (usually AUG) is found near the 5′ end of all mRNAs and that it codes for the amino acid methionine. The presence of this start codon is an aspect of initiation that is common to both bacteria and eukaryotes.

Figure 16.15 shows how translation gets under way in bacteria. The process begins when a section of rRNA in a small ribosomal subunit binds to a complementary sequence on an mRNA. The mRNA region is called the **ribosome binding site**, or **Shine-Dalgarno sequence**, after the biologists who discovered it. The site is about six nucleotides upstream from the AUG start codon. It consists of all or part of the sequence 5′-AGGAGGU-3′. The

complementary sequence in the rRNA of the small subunit reads 3′-UCCUCCA-5′.

This initial interaction between the small subunit and the message is mediated by proteins called **initiation factors** (Figure 16.15, step 1). In eukaryotes, initiation factors bind to the 5′ cap on mRNAs and guide it to the ribosome.

Once the Shine-Dalgarno sequence has attached to the small ribosomal subunit, an aminoacyl tRNA bearing a modified form of methionine called *N*-formylmethionine (abbreviated *f*-met) binds to the AUG start codon (Figure 16.15, step 2). In eukaryotes, this initial amino acid is normal methionine.

Initiation is complete when the large subunit joins the complex (Figure 16.15, step 3). When the ribosome is completely assembled, the tRNA bearing *f*-met occupies the P site.

To summarize, translation initiation is a three-step process in bacteria: (**1**) The mRNA binds to a small ribosomal subunit, (**2**) the initiator aminoacyl tRNA bearing *f*-met binds to the start codon, and (**3**) the large ribosomal subunit binds, completing the complex.

Elongation: Extending the Polypeptide

At the start of elongation, the E and A sites in the ribosome are empty of tRNAs. As a result, an mRNA codon is exposed at the base of the A site. As step 1 in **Figure 16.16** on page 302 illustrates, elongation proceeds when an aminoacyl tRNA binds to the codon in the A site by complementary base pairing between anticodon and codon.

When both the P site and A site are occupied by tRNAs, the amino acids on the tRNAs are in the ribosome's active site. This is where peptide bond formation—the essence of protein synthesis—occurs.

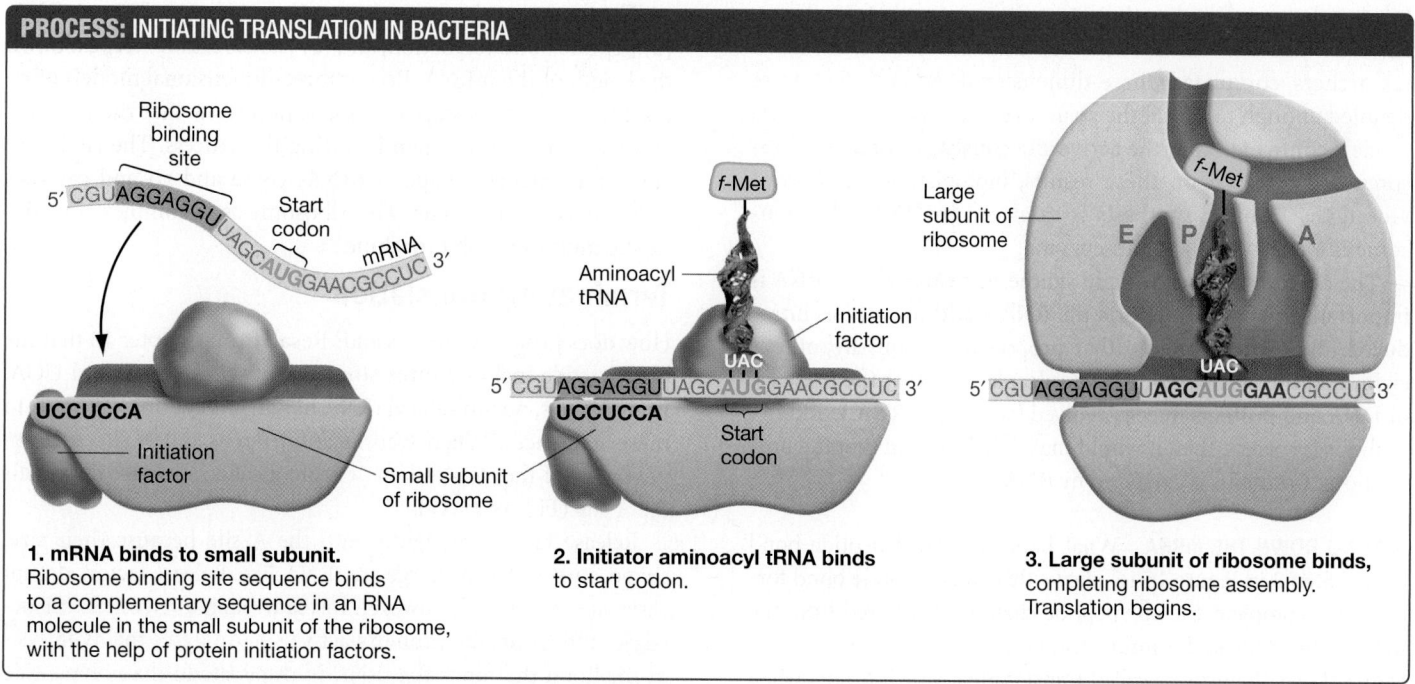

PROCESS: INITIATING TRANSLATION IN BACTERIA

1. mRNA binds to small subunit.
Ribosome binding site sequence binds to a complementary sequence in an RNA molecule in the small subunit of the ribosome, with the help of protein initiation factors.

2. Initiator aminoacyl tRNA binds to start codon.

3. Large subunit of ribosome binds, completing ribosome assembly. Translation begins.

FIGURE 16.15 Initiation Proteins Manage Assembly of Ribosomes and mRNAs.

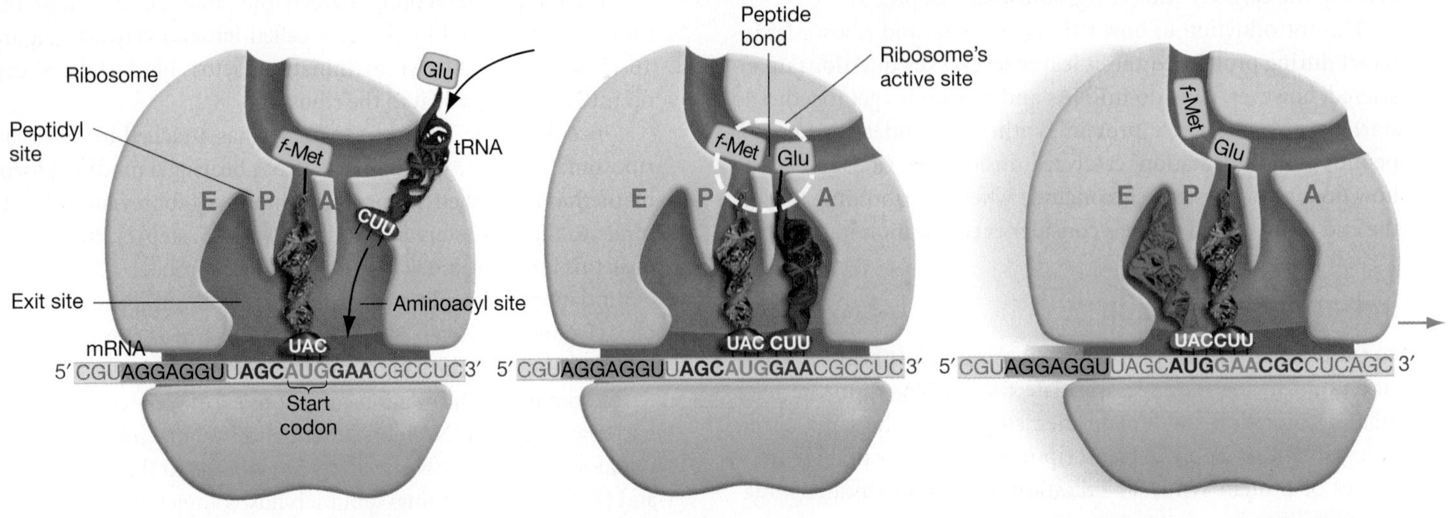

1. Incoming aminoacyl tRNA
New tRNA moves into A site, where its anticodon base pairs with the mRNA codon.

2. Peptide bond formation
The amino acid attached to the tRNA in the P site is transferred to the tRNA in the A site.

3. Translocation
mRNA is ratcheted through the ribosome by elongation factors (not shown). The tRNA attached to the polypeptide chain moves into the P site. The A site is empty.

FIGURE 16.16 During Elongation, Peptide Bond Formation Occurs inside the Ribosome.

Peptide bond formation is considered one of the most important reactions that takes place in cells, because manufacturing proteins is among the most fundamental of all cell processes. The question is, what makes it happen?

IS THE RIBOSOME AN ENZYME OR A RIBOZYME? Because ribosomes contain both proteins and RNA, researchers had argued for decades over whether the active site consisted of protein or RNA. The debate was not resolved until the year 2000, when researchers completed three-dimensional models that were detailed enough to reveal the structure of the active site. These models confirmed that the active site consists entirely of ribosomal RNA. Based on these results, biologists are now convinced that protein synthesis is catalyzed by RNA. The ribosome is a ribozyme—not an enzyme.

The observation that protein synthesis is catalyzed by RNA is important because it supports the RNA world hypothesis introduced in Chapter 4. Recall that proponents of this hypothesis claim that life began with RNA molecules and that the presence of DNA and proteins in cells evolved later. If the RNA world hypothesis is correct, then it would make sense to find that the production of proteins is catalyzed by RNA.

MOVING DOWN THE mRNA What happens after a peptide bond forms? Step 2 in Figure 16.16 shows that when peptide bond formation is complete, the polypeptide chain is transferred from the tRNA in the P site to the amino acid held by the tRNA in the A site. Step 3 shows the process called **translocation**, which occurs when proteins called **elongation factors** move the mRNA so that it ratch-

ets through the ribosome in the $5' \rightarrow 3'$ direction. Translocation is an energy-demanding event that requires GTP.

Translocation does several things: It moves the empty tRNA into the E site; it moves the tRNA containing the growing polypeptide into the P site; and it opens the A site and exposes a new mRNA codon. If the E site is occupied when translocation occurs, the tRNA there is ejected into the cytosol.

The three steps in elongation—(1) arrival of aminoacyl tRNA, (2) peptide bond formation, and (3) translocation—repeat down the length of the mRNA. Recent three-dimensional models of ribosomes in various stages of translation show that the machine as a whole is highly dynamic during the process. The ribosome constantly changes shape as tRNAs come and go and catalysis and translocation occur. The ribosome is a complex and dynamic multimolecular machine.

Terminating Translation

How does protein synthesis end? Recall from Chapter 15 that the genetic code includes three stop codons: UAA, UAG, and UGA. In most cells, no aminoacyl tRNA has an anticodon that binds to these sequences. When translocation opens the A site and exposes one of the stop codons, a protein called a **release factor** fills the A site (**Figure 16.17**).

Release factors fit tightly into the A site because their size, shape, and electrical charge are tRNA-like. Release factors do not carry an amino acid, however. Instead, when a release factor occupies the A site, the protein's active site catalyzes the hydrolysis of the bond that links the tRNA in the P site to the polypeptide chain. This reaction frees the polypeptide.

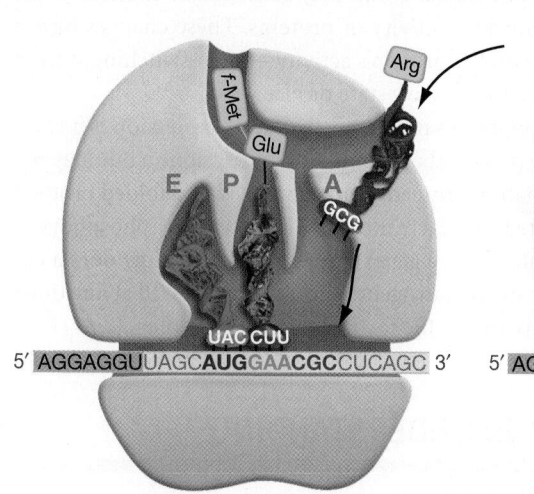

4. Incoming aminoacyl tRNA
New tRNA moves into A site, where its anticodon base pairs with the mRNA codon.

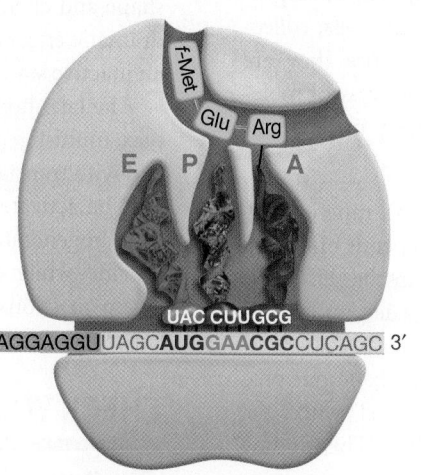

5. Peptide bond formation
The polypeptide chain attached to the tRNA in the P site is transferred to the aminoacyl tRNA in the A site.

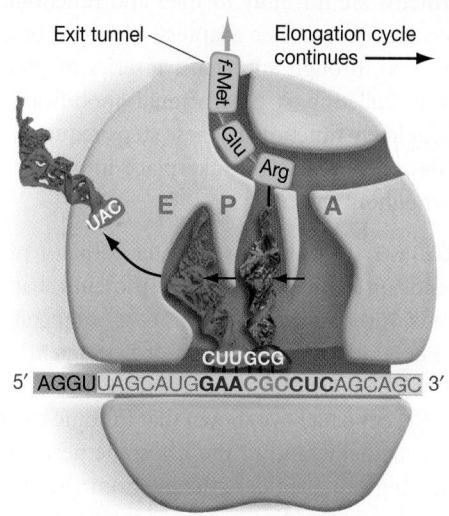

6. Translocation
mRNA is ratcheted through the ribosome again. The tRNA attached to polypeptide chain moves into P site. Empty tRNA from P site moves to E site, where tRNA is ejected. The A site is empty again.

The newly synthesized polypeptide is released from the ribosome, the ribosome separates from the mRNA, and the two ribosomal subunits dissociate. The subunits are ready to attach to the start codon of another message and start translation anew.

To review the initiation, elongation, and termination phases of translation, go to the study area at *www.masteringbiology.com*.

MB **Web Activity** Synthesizing Proteins

PROCESS: TERMINATING TRANSLATION IN BACTERIA

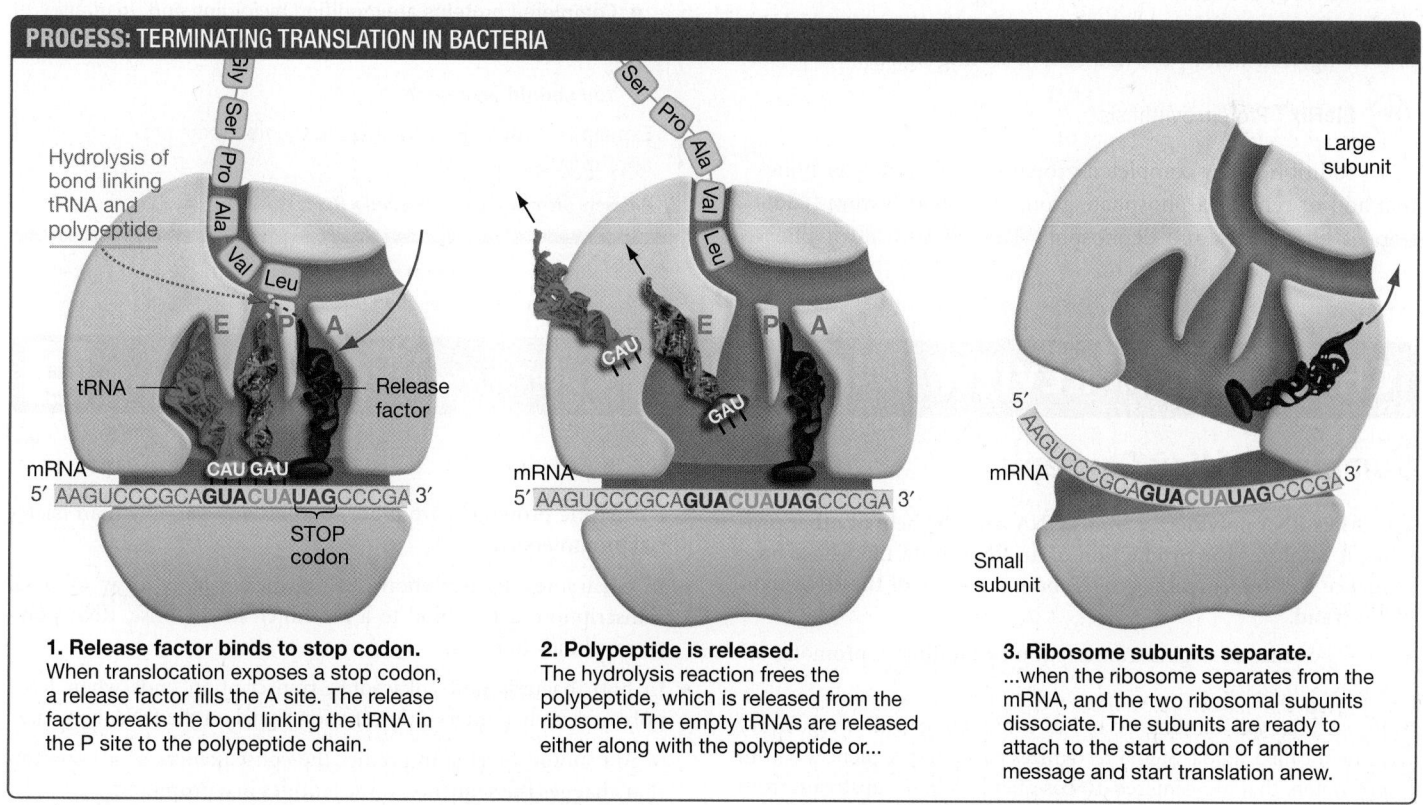

1. Release factor binds to stop codon.
When translocation exposes a stop codon, a release factor fills the A site. The release factor breaks the bond linking the tRNA in the P site to the polypeptide chain.

2. Polypeptide is released.
The hydrolysis reaction frees the polypeptide, which is released from the ribosome. The empty tRNAs are released either along with the polypeptide or...

3. Ribosome subunits separate.
...when the ribosome separates from the mRNA, and the two ribosomal subunits dissociate. The subunits are ready to attach to the start codon of another message and start translation anew.

FIGURE 16.17 Release Factors Manage Translation Termination.

Post-Translational Modifications

Proteins are not fully formed and functional when termination occurs. From earlier chapters, it should be clear that most proteins go through an extensive series of processing steps, collectively called post-translational modification, before they are completely functional. These steps require a wide array of molecules and events and take place in a wide variety of locations throughout the cell.

FOLDING   Recall from Chapter 3 that a protein's function depends on its shape, and that a protein's shape depends on how it folds. Although folding occurs spontaneously, in the sense that no energy input is required, it is frequently speeded up by proteins called **molecular chaperones**.

Recent data have shown that in some bacteria, chaperone proteins actually bind to the ribosome near the "tunnel" where the growing polypeptide emerges from the ribosome. This finding suggests that folding occurs as the polypeptide is emerging from the ribosome.

CHEMICAL MODIFICATIONS   Chapter 7 pointed out that many eukaryotic proteins are extensively modified after they are synthesized. For example, in the organelles called the rough endoplasmic reticulum and the Golgi apparatus, small chemical groups may be added to proteins—often, sugar or lipid groups that are critical for normal functioning. In some cases, the proteins receive a carbohydrate-based sorting signal that serves as an address label and ensures that the molecule will be carried to the correct location in the cell. (Other proteins have a sorting signal built into their primary structure.)

In the study area at *www.masteringbiology.com*, you'll find animations and tutorials to help you review transcription, RNA processing, translation, and addition of sorting signals.

(MB) **BioFlix™** Protein Synthesis

In addition, many completed proteins are altered by enzymes that add or remove a phosphate group. Phosphorylation (addition of phosphate) and dephosphorylation (removal of phos-

phate) of proteins were introduced in Chapters 9 and 11. Recall that because a phosphate group has two negative charges, adding or removing a phosphate group may cause major changes in the shape and chemical reactivity of proteins. These changes have a dramatic effect on the protein's activity—often switching it from an inactive state to an active state or vice versa.

The take-home message here is that gene expression is a complex, multistep process that begins with transcription but may not end with translation. Instead, even completed and folded proteins may be activated or deactivated by events such as phosphorylation. In general, how are genes turned on or off? How does a cell "decide" which of its many genes should be expressed at any time? These questions are the focus of the next two chapters.

CHECK YOUR UNDERSTANDING

If you understand that . . .

- Translation begins when (1) the ribosome binding site on an mRNA binds to an rRNA sequence in the small ribosomal subunit, (2) the initiator aminoacyl tRNA binds to the start codon in the mRNA, and (3) the large subunit of the ribosome attaches to the small subunit to complete the ribosome.
- Translation elongation occurs when (1) an appropriate aminoacyl tRNA enters the A site, (2) a peptide bond forms between the amino acid held by that tRNA in the A site and the polypeptide held by the tRNA in the P site, and (3) the ribosome moves down the mRNA one codon.
- Translation ends when the ribosome reaches a stop codon.
- Completed proteins are modified by folding and, in many cases, addition of sugar, lipid, or phosphate groups.

✓ You should be able to . . .

Explain why the E, P, and A sites in the ribosome are appropriately named.

Answers are available in Appendix B.

CHAPTER 16   REVIEW

For media, go to the study area at www.masteringbiology.com

Summary of Key Concepts

🔑 **After RNA polymerase binds DNA with the help of other proteins, it catalyzes the production of an RNA molecule whose base sequence is complementary to the base sequence of the DNA template strand.**

- RNA polymerase begins transcription by binding to promoter sequences in DNA.
- In bacteria, this binding occurs in conjunction with a regulatory protein called sigma. Sigma recognizes particular sequences within promoters that are centered 10 bases and 35 bases upstream from the start of the actual genetic message.

- Eukaryotic promoters are more complex and variable than bacterial promoters are.
- In eukaryotes, transcription begins when a large array of basal transcription factors bind to a promoter. In response, RNA polymerase binds to the site.
- In both bacteria and eukaryotes, transcription ends when RNA polymerase encounters a termination signal on the DNA template.

✓You should be able to predict the consequences of a mutation that changes the sequence of nucleotides in a promoter.

(MB) **Web Activity** RNA Synthesis

- **Eukaryotic genes contain regions called exons and regions called introns; during RNA processing, the regions coded by introns are removed, and the ends of the RNA receive a cap and tail.**

- Stretches of noncoding RNA called introns are spliced out by complex molecular machines called spliceosomes.

- A "cap" is added to the 5′ end of a primary transcript, and a poly(A) tail is added to the 3′ end.

- The cap and tail serve as recognition signals for the translation machinery and protect the message from degradation by ribonucleases.

  ✓You should be able to explain why RNA processing does not occur in bacteria.

- **Ribosomes translate mRNAs into proteins with the help of intermediary molecules called transfer RNAs.**

- Experiments with radioactively labeled amino acids confirmed that ribosomes are the site of protein synthesis.

- Experiments with radioactively labeled amino acids confirmed that transfer RNAs (tRNAs) serve as the chemical bridge between the RNA message and the polypeptide product.

  ✓You should be able to explain why the name transfer RNA is appropriate.

- **Each transfer RNA carries an amino acid corresponding to the tRNA's three-base-long anticodon.**

- tRNAs have an L-shaped tertiary structure. One leg of the L contains the anticodon, which forms complementary base pairs with the mRNA codon. The other leg holds the amino acid appropriate for that codon.

- Imprecise pairing—or "wobbling"—is allowed in the third position of a codon–anticodon pairing, so only about 40 different tRNAs are required to translate the 61 codons that code for amino acids.

✓You should be able to explain the relationships between an aminoacyl tRNA, a tRNA, an amino acid, and aminoacyl tRNA synthetase.

- **In the ribosome, the tRNA anticodon binds to a three-base-long mRNA codon, causing the amino acid carried by the transfer RNA to be added to the growing protein.**

- Protein synthesis occurs in three steps: (1) an incoming aminoacyl tRNA occupies the A site; (2) the growing polypeptide chain is transferred from a peptidyl tRNA in the ribosome's P site to the amino acid bound to the tRNA in the A site, and a peptide bond is formed; and (3) the ribosome is translocated to the next codon on the mRNA, accompanied by ejection of the empty tRNA from the E site.

- Peptide bond formation is catalyzed by a ribozyme (RNA), not an enzyme (protein).

- Proteins fold into their three-dimensional conformation (tertiary structure), sometimes with the aid of chaperone proteins.

- Proteins may be targeted to specific locations in the cell by post-translational modification that adds signal sequences. Some proteins remain inactive until modified by phosphorylation.

  ✓You should be able to create a concept map (see **BioSkills 8** in Appendix A) that describes the relationships among the following concepts and structures: translation, initiation, elongation, termination, protein folding, initial amino acid, chemical modification, mRNA, charged tRNAs in A site, growing polypeptide in P site, empty tRNA in E site, start codon, ribosome binding site, initiation factors, ribosome subunits.

(MB) **Web Activity** Synthesizing Proteins, **BioFlix™** Protein Synthesis

Questions

1. How did the A site of the ribosome get its name?
   a. It is where amino acids are affixed to tRNAs, producing aminoacyl tRNAs.
   b. It is where the amino group on the growing polypeptide chain is available for peptide bond formation.
   c. It is the site occupied by incoming aminoacyl tRNAs.
   d. It is surrounded by α-helices of ribosomal proteins.

2. How did the P site of the ribosome get its name?
   a. It is where the promoter resides.
   b. It is made up of protein.
   c. It is where peptidyl tRNAs reside.
   d. It is where a growing polypeptide chain is phosphorylated.

3. What is a molecular chaperone?
   a. a protein that recognizes the promoter and guides the binding of RNA polymerase
   b. a protein that activates or deactivates another protein by adding or removing a phosphate group
   c. a protein that is a component of the large ribosomal subunit and that assists with peptide bond formation
   d. a protein that helps newly translated proteins fold into their proper three-dimensional configuration

4. The three types of RNA polymerase found in eukaryotic cells transcribe different types of genes. What does RNA polymerase II produce?
   a. rRNAs
   b. tRNAs
   c. mRNAs
   d. spliceosomes

5. What is an anticodon?
   a. the part of an mRNA that signals translation termination
   b. the part of an mRNA that signals the start of translation
   c. the part of a tRNA that binds to a codon in mRNA
   d. the part of a tRNA that accepts an amino acid, through a reaction catalyzed by tRNA synthetase

6. What do researchers observe when mRNAs that lack a cap and tail are added to a cell—compared to identical mRNAs that have a cap and tail?
   a. Elongation factors cannot bind to the experimental mRNAs.
   b. Basal transcription factors cannot bind to the promoter.
   c. Introns cannot be spliced out (the spliceosome is inhibited).
   d. Translation rate decreases; mRNA life span decreases.

1. Explain the relationship among eukaryotic promoter sequences, basal transcription factors, and RNA polymerase. Explain the relationship among bacterial promoter sequences, sigma, and RNA polymerase.

2. According to the wobble rules, the correct amino acid can be added to a growing polypeptide chain even if the third base in the mRNA codon is not complementary to the third base in the tRNA anticodon. How do the wobble rules relate to the redundancy of the genetic code?

3. Why does splicing occur in eukaryotic mRNAs? Where does it occur, and how are snRNPs involved?

4. Describe the sequence of events that occurs during translation as a protein elongates by one amino acid and the ribosome moves down the mRNA. Your answer should specify what is happening in the ribosome's A site, P site, and E site.

5. What evidence supports the hypothesis that peptide bond formation is catalyzed by a ribozyme?

6. In an aminoacyl tRNA, why is the observed distance between the amino acid and the anticodon important?

1. The 5′ cap and poly(A) tail on eukaryotic mRNAs protect the message from degradation by ribonucleases. But why do ribonucleases exist? What function would an enzyme that destroys messages serve? Answer this question using the example of an mRNA for a hormone that causes human heart rate to increase.

2. The nucleotide shown below is called cordycepin.

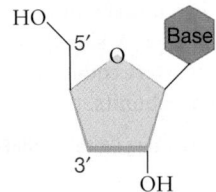

If cordycepin triphosphate, which has three phosphate groups bonded to the 5′ hydroxyl group in the figure, is added to a cell-free transcription reaction, the nucleotide is added onto the growing

RNA chain. This observation confirms that synthesis occurs in the 5′→3′ direction. Explain why cordycepin triphosphate cannot be added to the 5′ end of an RNA.

3. Certain portions of the rRNAs in the large subunit of the ribosome are very similar in all organisms. To make sense of this finding, Carl Woese suggests that the conserved sequences have an important functional role. His logic is that these conserved sequences are so important to cell function that any changes in the sequences cause death. Which specific portions of the ribosome would you expect to be identical or nearly identical in all organisms, and which would you expect to be more variable? Explain your logic.

4. Recent structural models show that a poison called α-amanitin inhibits transcription by binding to a site inside RNA pol II but not to the active site itself. Based on the model of bacterial RNA polymerase in Figure 16.2, predict where α-amanitin binds and why it inhibits transcription.

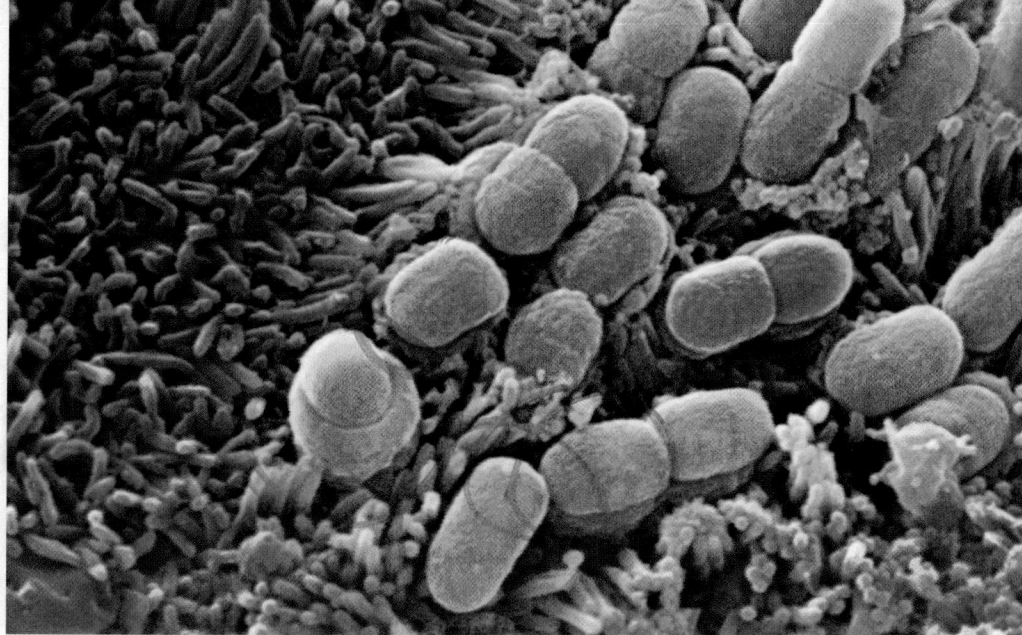

The structures that have been colored blue in this scanning electron micrograph are projections from human intestinal cells; the structures colored yellow are the bacterium *Escherichia coli*. In the intestine, the nutrients available to bacteria constantly change. This chapter explores how changes in gene expression help bacteria respond to environmental changes.

Control of Gene Expression in Bacteria

17

Imagine waiting eagerly to hear the opening lines of a wonderfully melodic symphony played by a renowned orchestra. The crowd applauds as the celebrated conductor comes onstage, then hushes as he takes the podium. He cocks the baton; the musicians raise their instruments. As the baton comes down, every instrument begins blaring a different tune at full volume. A tuba plays "Dixie"; a violinist renders "In-A-Gadda-Da-Vida"; and a cellist begins Mexico's national anthem. A snare drum lays down beats for OutKast's "Hey Ya," while the bass drum simulates the cannons in the "1812 Overture." Instead of music, there is pandemonium. The conductor staggers off stage, clutching his heart.

A cacophony like this would result if a bacterial cell "played" all its genes at full volume all the time. The *Escherichia coli* cells living in your gut right now have over 4300 genes. If all of those genes were expressed at the fastest possible rate at all times, the *E. coli* cells would stagger off the stage, too. But this does not happen. Cells are extremely selective about which genes are expressed, in what amounts, and when.

This chapter explores how bacterial cells control the activity, or expression, of their genes. Gene expression is the process of converting information that is archived in DNA into molecules that actually do things in the cell. It occurs when a protein or other gene product is synthesized and active.

Previous chapters detailed how genetic information is processed in cells; this chapter focuses on *when* genetic information is used. Let's begin by reviewing some of the environmental challenges that bacterial cells face, and then explore how these organisms meet them.

17.1 Gene Regulation and Information Flow

The bacteria that live in and on your body vastly outnumber your own cells. Consider just one of the species present: the gut-dwelling *Escherichia coli*. These cells can use a wide array of carbohydrates to supply the carbon and energy they need. But as your diet changes from day to day, the availability of different sugars in your intestines varies. Each type of nutrient

KEY CONCEPTS

- In bacteria, gene expression can be controlled at three levels: transcription, translation, or post-translation (protein activation).

- Changes in gene expression allow bacterial cells to respond to environmental changes.

- Transcriptional control can be negative or positive. Negative control occurs when a regulatory protein prevents transcription. Positive control occurs when a regulatory protein increases the transcription rate.

requires a different membrane transport protein to bring the molecule into the cell and a different suite of enzymes to process it. Precise control of gene expression gives *E. coli* the ability to use the available sugars efficiently.

To understand why precise control over gene expression is so important, you have to realize that bacterial cells from an array of species can be packed 2 cm thick along your intestinal walls. All of these organisms are competing for space and nutrients. In an environment like this, a cell has to use resources efficiently if it's going to be able to survive and reproduce. An individual that synthesizes proteins it doesn't need has fewer resources to devote to making the proteins it does need. Such cells are losers—they compete less successfully for the resources that are required to produce offspring.

Realizing this, biologists predicted that most gene expression is triggered by specific signals from the environment, such as the presence of specific sugars. Did you drink milk at your last meal, or eat French fries and a candy bar? Each type of food contains different sugars. Each sugar should induce a different response from the *E. coli* cells in your intestine. Just as a conductor needs to regulate the orchestra's musicians, cells need to regulate which proteins they produce.

Mechanisms of Regulation—An Overview

The flow of information from DNA to RNA to activation of the final gene product occurs in three steps, represented by arrows in the following diagram:

DNA $\xrightarrow{\text{transcription}}$ mRNA $\xrightarrow{\text{translation}}$ protein $\xrightarrow{\text{post-translational modifications}}$ activated protein

Gene expression can be controlled at any of these arrows. The arrow from DNA to RNA represents transcription—the making of messenger RNA (mRNA). The arrow from RNA to protein represents translation, in which ribosomes read the information in mRNA and use that information to synthesize a protein. The arrow from protein to activated protein represents post-translational modifications—including folding, addition of carbohydrate or lipid groups, or perhaps phosphorylation.

How can a bacterial cell avoid producing proteins that are not needed at a particular time, and thus use resources efficiently? A look at the flow of information from DNA to protein suggests three possible mechanisms:

1. The cell could avoid making the mRNAs for particular enzymes. If there is no mRNA, then ribosomes cannot make the gene product. **Transcriptional control** occurs when regulatory proteins affect RNA polymerase's ability to bind to a promoter and initiate transcription:

   DNA $\xrightarrow{\times}$ mRNA $\longrightarrow$ protein $\longrightarrow$ activated protein

2. If the mRNA for an enzyme has been transcribed, the cell might have a way to prevent the mRNA from being translated into protein. **Translational control** occurs when regulatory molecules alter the length of time an mRNA survives before it is degraded by ribonucleases, or affect translation initiation, or affect elongation factors and other proteins during the translation process:

   DNA $\longrightarrow$ mRNA $\xrightarrow{\times}$ protein $\longrightarrow$ activated protein

3. Chapter 16 pointed out that some proteins are manufactured in an inactive form and have to be activated by chemical modification, such as the addition of a phosphate group. This type of regulation is **post-translational control**:

   DNA $\longrightarrow$ mRNA $\longrightarrow$ protein $\xrightarrow{\times}$ activated protein

   Which of these three forms of control occur in bacteria? The short answer to this question is "all of the above." As **Figure 17.1** shows, many factors affect how much active protein is produced from a particular gene.

- Transcriptional control is particularly important due to its efficiency—it saves the most energy for the cell, because it stops the process at the earliest possible point.

- Translational control is advantageous because it allows a cell to make rapid changes in the relative amounts of different proteins.

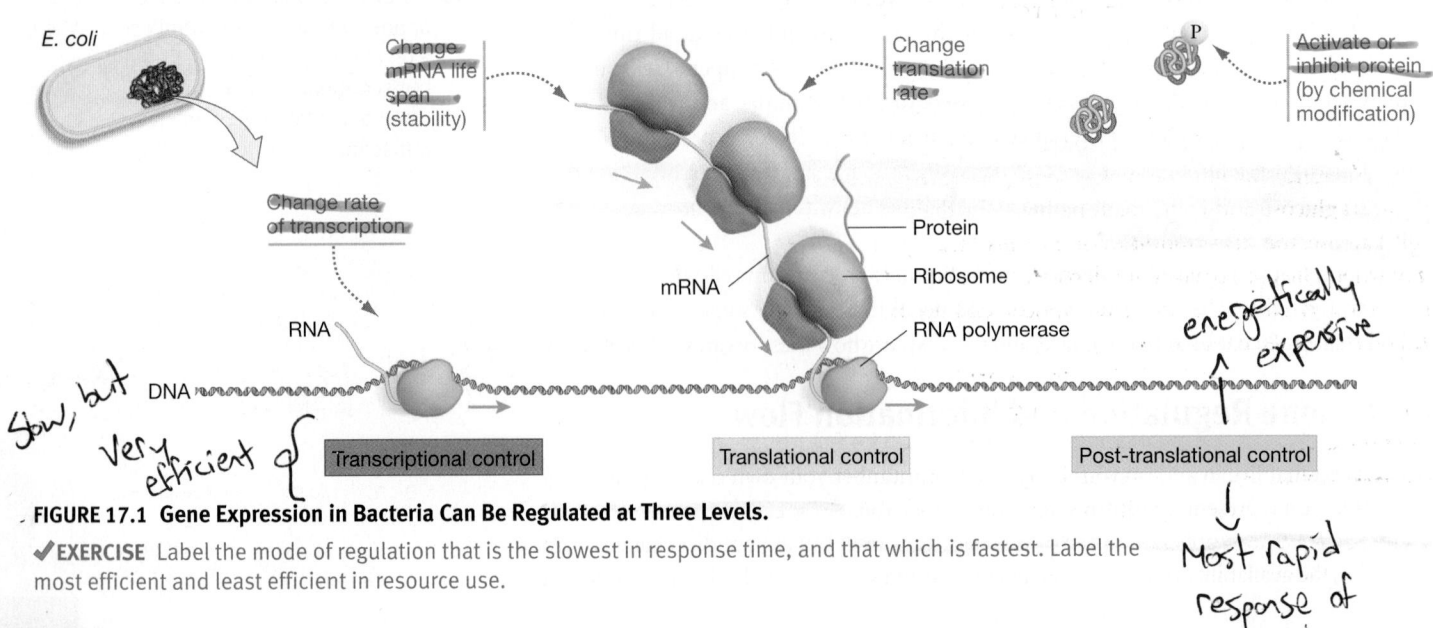

E. coli

Change mRNA life span (stability)

Change rate of transcription

Change translation rate

Activate or inhibit protein (by chemical modification)

Protein

Ribosome

mRNA

RNA polymerase

RNA

energetically expensive

Slow, but very efficient

DNA

Transcriptional control   Translational control   Post-translational control

Most rapid response of all mechanisms

FIGURE 17.1 Gene Expression in Bacteria Can Be Regulated at Three Levels.

✔**EXERCISE** Label the mode of regulation that is the slowest in response time, and that which is fastest. Label the most efficient and least efficient in resource use.

- Post-translational control is significant as well, because it provides the most rapid response of all three mechanisms.

Among these mechanisms of gene regulation, there is a clear trade-off between the speed of response and the conservation of ATP, amino acids, and other resources. Transcriptional control is slow but efficient in resource use; post-translational control is fast but energetically expensive.

Although this chapter focuses almost exclusively on mechanisms of transcriptional control, it is important to keep in mind that bacteria possess translational and post-translational controls as well, and that some genes—such as those that code for the enzymes required for glycolysis—are transcribed all the time, or **constitutively**. Finally, it is critical to realize that gene expression is not an all-or-none proposition. Genes are not just "on" or "off"—instead, the level of expression is highly variable.

🔑 Variation in gene expression allows cells to respond to changes in their environment.

Metabolizing Lactose—A Model System

As Chapters 13 through 16 have shown, many of the great advances in genetics have been achieved through the analysis of model systems. Mendel studied garden peas and discovered the fundamental patterns of gene transmission; Morgan studied fruit flies and confirmed the chromosome theory of inheritance; an array of researchers used viruses and *E. coli* to work out the mechanisms of DNA synthesis, transcription, and translation. In studies of gene regulation, the key model system has been the metabolism of the sugar lactose in *E. coli*.

Jacques Monod, François Jacob, and many colleagues introduced lactose metabolism in *E. coli* as a model system during the 1950s and 1960s. Although they worked with a single species of bacteria, their results had a profound effect on thinking about gene regulation in all organisms. Some details turned out to be specific to the *E. coli* genes responsible for lactose metabolism, but many of the results are universal.

Escherichia coli can use a wide variety of sugars for ATP production, via cellular respiration or fermentation. These sugars also serve as raw material in the synthesis of amino acids, vitamins, and other complex compounds. Glucose is *E. coli*'s preferred carbon source, however—meaning that it is the source of energy and carbon atoms that the organism uses most efficiently.

A preference for glucose makes sense, because glycolysis begins with glucose and is the main pathway for the production of ATP. Lactose, the sugar found in milk, is also used by *E. coli*, but only when glucose supplies are depleted. Recall from Chapter 5 that lactose is a disaccharide made up of one molecule of glucose and one molecule of galactose.

To use lactose, *E. coli* must first transport the sugar into the cell. Once lactose is inside the cell, the enzyme β-galactosidase catalyzes a reaction that breaks the sugar down into glucose and galactose. The glucose released by this reaction goes directly into the glycolytic pathway; other enzymes convert the galactose to a substance that can also be processed in the glycolytic pathway.

In the early 1950s, biologists discovered that *E. coli* produces high levels of β-galactosidase only when lactose is present in the environment. Based on this observation, researchers proposed that lactose itself regulates the gene for β-galactosidase—meaning that lactose acts as an inducer. An **inducer** is a substrate in a reaction, and it stimulates the expression of a specific gene or genes.

In the late 1950s Jacques Monod investigated how the presence of glucose affects the regulation of the β-galactosidase gene. Would *E. coli* produce high levels of β-galactosidase when both glucose and lactose were present in the surrounding environment? As the experiment summarized in **Figure 17.2** shows, the

EXPERIMENT

QUESTION: *E. coli* produces β-galactosidase when lactose is present. Does *E. coli* produce β-galactosidase when both glucose and lactose are present?

HYPOTHESIS: *E. coli* does not produce β-galactosidase when glucose is present, even if lactose is present. (Glucose is the preferred food source.)

NULL HYPOTHESIS: *E. coli* produces β-galactosidase whenever lactose is present, regardless of the presence or absence of glucose.

EXPERIMENTAL SETUP:

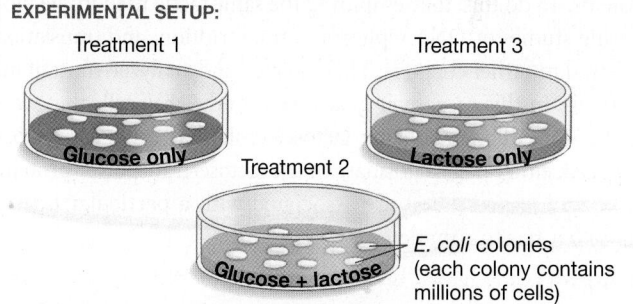

PREDICTION: β-Galactosidase will be produced only in treatment 3.

PREDICTION OF NULL HYPOTHESIS: β-Galactosidase will be produced in treatments 2 and 3.

RESULTS:

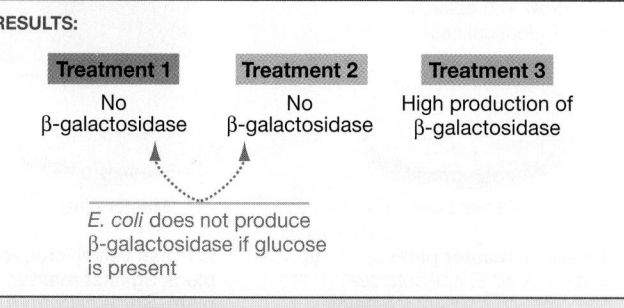

CONCLUSION: Glucose prevents expression of the gene for β-galactosidase. The presence of lactose without glucose stimulates expression of that gene.

FIGURE 17.2 Glucose Affects the Regulation of the β-Galactosidase Gene.

SOURCE: Pardee, A. B., F. Jacob, and J. Monod. 1959. The genetic control and cytoplasmic expression of "inducibility" in the synthesis of β-glactosidase by *E. coli*. *Journal of Molecular Biology* 1: 165–178.

✓**QUESTION** How would you control growing conditions in the three treatments, so that the results of this experiment are valid?

answer was no. The enzyme β-galactosidase is produced only when lactose is present and glucose is not present.

Monod teamed up with François Jacob to investigate exactly how lactose and glucose regulate the genes responsible for lactose metabolism—the gene for the membrane protein that imports lactose and the gene for β-galactosidase. Discoveries about how these genes are regulated shed light on how genes in all organisms are controlled. Research on this system is still going strong, over 50 years later.

✔You should be able to make a chart summarizing the molecules involved in regulating lactose use in *E. coli*. There should be 10 rows and 2 columns. Title the first column "Name" and the second column "Function." The rows are *lacZ*, *lacY*, operator, promoter, CAP site, repressor, CAP, cAMP, lactose, and glucose. As you read this chapter, fill in the "Function" column.

17.2 Identifying Genes under Regulatory Control

To understand how *E. coli* controls production of β-galactosidase and the membrane transport protein that brings lactose into the cell, Monod and Jacob first had to find the genes that code for these proteins. To do this, they employed the same tactic used in the pioneering studies of DNA replication, transcription, and translation reviewed in earlier chapters: They isolated and analyzed mutant individuals. In this case, their goal was to find *E. coli* cells that were not capable of metabolizing lactose. Cells that can't use lactose must lack either β-galactosidase or the lactose-transporter protein.

To find mutants that are associated with a particular trait, a researcher has to complete two steps:

1. Generate a large number of individuals with mutations at random locations in their genomes. Monod and colleagues accomplished this step by exposing *E. coli* populations to X-rays, UV light, or **mutagens**—chemicals that damage DNA and increase mutation rates.

2. Screen the mutants to find individuals with defects in the process or biochemical pathway in question—in this case, defects in lactose metabolism. Recall from Chapter 15 that a genetic screen is any technique for selecting individuals with certain types of mutations out of a large population.

The researchers were looking for cells that cannot grow in an environment that contains only lactose as an energy source. Normal cells grow well in this environment. How could the researchers select cells on the basis of *lack* of growth?

Replica Plating to Find Mutant Genes

Replica plating and growth on indicator plates were key techniques in the search for mutants with defects in lactose metabolism. **Figure 17.3** shows how **replica plating** works.

Step 1 When mutants with defects in lactose metabolism are desired, mutagenized bacteria are spread on a "master plate" filled with gelatinous agar containing glucose. It is important that the mutant cells you want to identify are capable of growing on the master plate. The bacteria are then allowed to grow, so that each cell produces a colony—a large number of identical cells descended from a single cell.

Step 2 A block covered with a piece of sterilized velvet is pressed onto the master plate. Because of the contact, cells from each colony on the master plate are transferred to the velvet.

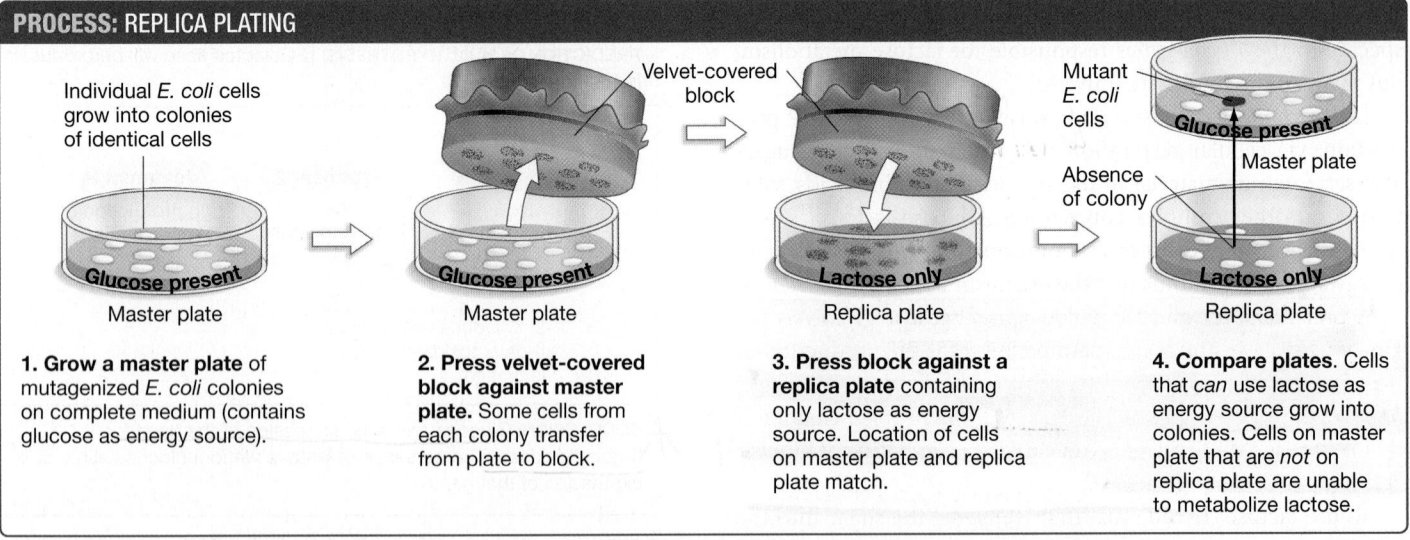

PROCESS: REPLICA PLATING

Individual *E. coli* cells grow into colonies of identical cells

Velvet-covered block

Mutant *E. coli* cells

Absence of colony

Master plate

Glucose present — Master plate

Glucose present — Master plate

Lactose only — Replica plate

Lactose only — Replica plate

1. Grow a master plate of mutagenized *E. coli* colonies on complete medium (contains glucose as energy source).

2. Press velvet-covered block against master plate. Some cells from each colony transfer from plate to block.

3. Press block against a replica plate containing only lactose as energy source. Location of cells on master plate and replica plate match.

4. Compare plates. Cells that *can* use lactose as energy source grow into colonies. Cells on master plate that are *not* on replica plate are unable to metabolize lactose.

FIGURE 17.3 Replica Plating Is a Technique for Identifying Mutant Cells. Here, replica plating is used to isolate mutant *E. coli* cells with a deficiency in lactose metabolism. There are two requirements for replica plating: (1) The location of colonies transferred from the master plate to the replica plate must match exactly on both plates, and (2) the media in the two plates must differ by just one component.

✔**QUESTION** How would you alter this protocol to isolate mutant cells with a deficiency in the enzymes required to synthesize tryptophan?

Step 3 The velvet is pressed onto a plate containing a **medium**—a liquid or solid that supports growth—that differs from the master plate by a single component. In this case, the second medium has only lactose—no glucose—as the source of carbon and energy. Cells from the velvet stick to the plate's surface, producing an exact copy of the locations of the colonies on the master plate. This copy is called the replica plate.

Step 4 After these cells grow, compare the colonies that thrive on the replica plate's medium with those on the master plate. In this example, colonies that grow on the master plate but are missing on the replica plate represent mutants deficient in lactose metabolism. By picking these particular colonies from the master plate, researchers build a collection of lactose mutants.

Monod also used an alternative strategy based on **indicator plates**, where mutants with metabolic deficiencies are observed directly. In this case, he added a compound that is acted on by β-galactosidase. The compound acts as an indicator molecule for the presence of functioning β-galactosidase because one of the molecules produced by the reaction is yellow. Colonies with normal β-galactosidase turn yellow; colonies that have a defect in the β-galactosidase enzyme or its production stay white.

Different Classes of Lactose Metabolism Mutants

The initial mutant screen yielded the three types of mutants summarized in **Table 17.1**. In one class, the mutant cells were unable to cleave the indicator molecule— even if lactose was present inside the cells to induce production of the β-galactosidase protein. The investigators concluded that these mutants must lack a functioning version of the β-galactosidase protein—meaning the gene that encodes β-galactosidase is defective. This gene was designated *lacZ*, and the mutant allele *lacZ⁻*.

In the second class of mutants, the cells failed to accumulate lactose inside the cell. In normal cells the concentration of lactose is about 100 times that of lactose in the surrounding environment, but in the mutant cells lactose concentrations were much lower. To explain this result, Jacob and Monod hypothesized that the mutant cells had defective copies of the membrane protein responsible for transporting lactose into the cell. This protein was identified and named galactoside permease; the gene that encodes it was designated *lacY*. **Figure 17.4** summarizes the functions of β-galactosidase and galactoside permease.

The third and most surprising class of mutants did not show normal regulation of the expression of β-galactosidase and galactoside permease. Instead, these mutants made the proteins all the time—even if no lactose was present.

Cells that are abnormal because they produce a product at all times are called **constitutive mutants**. The gene that mutated to produce constitutive β-galactosidase expression was named *lacI*. The letter "I" signified that these mutants did not need an inducer—lactose—to express β-galactosidase or galactoside permease.

To understand the reasoning here, recall that in normal cells, the expression of these genes is induced by the presence of lactose. But in cells with a mutant form of *lacI* (*lacI⁻* mutants), gene expression occurs with or without lactose. This means that *lacI⁻* mutants have a defect in gene regulation. In these mutants, the gene remains "on" when it should be turned off.

To pull these observations together, the researchers hypothesized that the normal product of the *lacI* gene prevents the transcription of *lacZ* and *lacY* when lactose is absent. Because lactose triggers production of β-galactosidase, it was reasonable to expect that the *lacI* gene or gene product interacts with lactose in some way. (Later work showed that the inducer is actually a derivative of lactose called *allolactose*. For the sake of historical accuracy and simplicity, however, this discussion refers to lactose itself as the inducer.)

TABLE 17.1 Three Types of Lactose Metabolism Mutants in *E. coli*

Observed Phenotype	Interpretation	Inferred Genotype
1. Cells cannot cleave indicator molecule even if lactose is present as an inducer.	No β-galactosidase; gene for β-galactosidase is defective. Call this gene *lacZ*.	*lacZ⁻*
2. Cells cannot accumulate lactose.	No membrane protein (galactoside permease) to import lactose; gene for galactoside permease is defective. Call this gene *lacY*.	*lacY⁻*
3. Cells cleave indicator molecule even if lactose is absent as an inducer.	Constitutive expression of *lacZ* and *lacY*; gene for regulatory protein that shuts down *lacZ* and *lacY* is defective—it does not need to be induced by lactose. Call this gene *lacI*.	*lacI⁻*

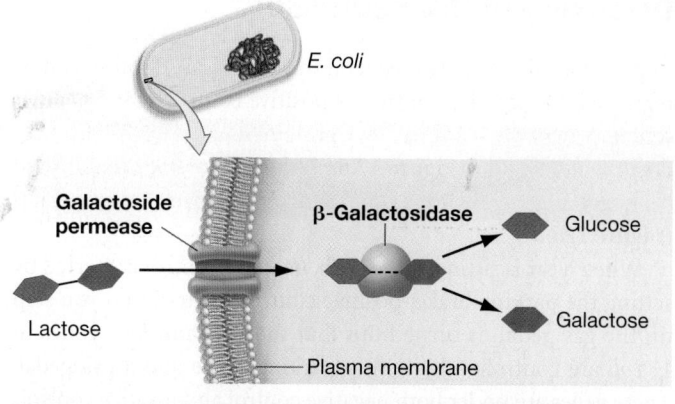

FIGURE 17.4 Two Proteins *E. coli* Needs for Using Lactose. For *E. coli* cells to use lactose, the membrane protein galactoside permease must be present to bring the sugar into the cell. Then the enzyme β-galactosidase must be present to break lactose into its glucose and galactose subunits.

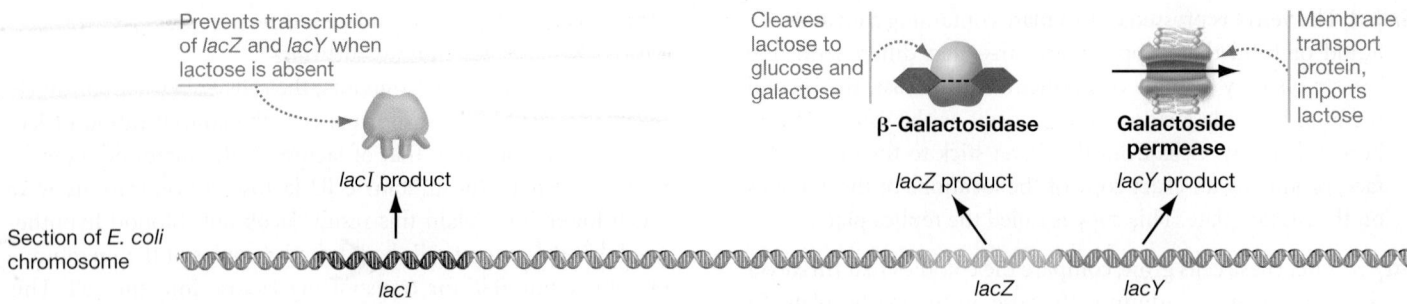

Prevents transcription of *lacI* and *lacY* when lactose is absent

lacI product

Section of *E. coli* chromosome

lacI

Cleaves lactose to glucose and galactose

β-Galactosidase

lacZ product

Membrane transport protein, imports lactose

Galactoside permease

lacY product

lacZ   *lacY*

FIGURE 17.5 The *lac* Genes Are in Close Physical Proximity.

Several Genes Are Involved in Lactose Metabolism

Jacob and Monod had succeeded in identifying three genes involved in lactose metabolism: *lacZ*, *lacY*, and *lacI*. They had concluded that *lacZ* and *lacY* code for proteins required for the metabolism and import of lactose, while *lacI* is responsible for some sort of regulatory function. When lactose is absent, the *lacI* gene or gene product shuts down the expression of *lacZ* and *lacY*. But when lactose is present, the opposite occurs—transcription of *lacZ* and *lacY* is induced.

✔If you understand the genes involved in lactose metabolism, you should be able to describe the specific function of *lacZ* and *lacY*. You should also be able to describe the effect of the *lacI* gene product when lactose is present versus absent and explain why these effects are logical.

Jacob and Monod followed up on these experiments by mapping the physical location of the three genes on *E. coli*'s circular chromosome (**Figure 17.5**). Their data showed that the genes are close together. This was a crucial finding, because it suggested that both *lacZ* and *lacY* might be controlled by *lacI*. Could one regulatory gene govern more than one protein-encoding gene? If so, how does *lacI* actually work? And why do lactose and glucose have opposite effects on it?

17.3 Mechanisms of Negative Control: Discovery of the Repressor

In principle, there are two general ways that transcription can be regulated: by negative control or positive control. **Negative control** occurs when a regulatory protein binds to DNA and shuts down transcription (**Figure 17.6a**); **positive control** occurs when a regulatory protein binds to DNA and triggers transcription (**Figure 17.6b**).

When a car is sitting at the curb, negative control is exerted by setting the parking brake; positive control occurs when you step on the gas pedal. It turned out that the *lacZ* and *lacY* genes in *E. coli* are controlled by both a parking brake and a gas pedal. These genes are under both negative control and positive control.

The hypothesis that the *lacZ* and *lacY* genes might be under negative control originated with Leo Szilard in the late 1950s. Szilard suggested to Monod that the *lacI* gene codes for a product that represses transcription of the *lacZ* and *lacY* genes.

Stated another way, the *lacI* gene produces an inhibitor that exerts negative control over the *lacZ* and *lacY* genes. This transcription inhibitor was called a **repressor**. It was thought to bind directly to DNA near or on the promoter for the *lacZ* and *lacY* genes (**Figure 17.7a**).

To explain how the presence of lactose triggers transcription in normal cells, Szilard and Monod proposed that lactose interacts with the repressor in a way that makes the repressor release from its binding site (**Figure 17.7b**). The idea was that lactose induces transcription by removing negative control. The repressor is the parking brake; lactose releases the parking brake.

(a) Negative control: Regulatory protein *shuts down* transcription.

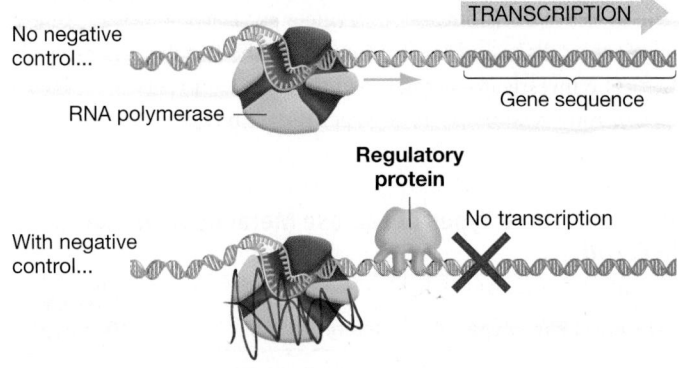

No negative control...

RNA polymerase

TRANSCRIPTION

Gene sequence

Regulatory protein

With negative control...

No transcription

(b) Positive control: Regulatory protein *triggers* transcription.

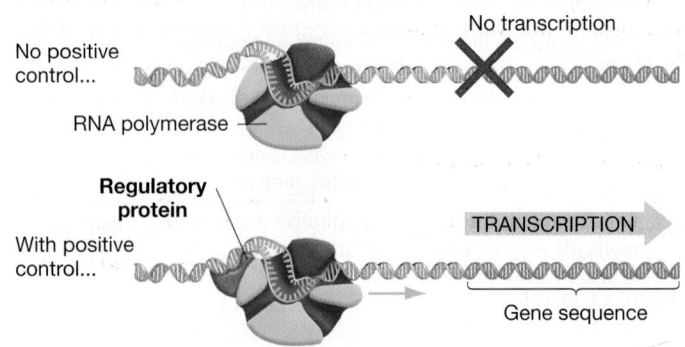

No positive control...

RNA polymerase

No transcription

Regulatory protein

With positive control...

TRANSCRIPTION

Gene sequence

FIGURE 17.6 Genes Are Regulated by Negative Control, Positive Control, or Both. Section 17.4 explores positive control and its interaction with negative control.

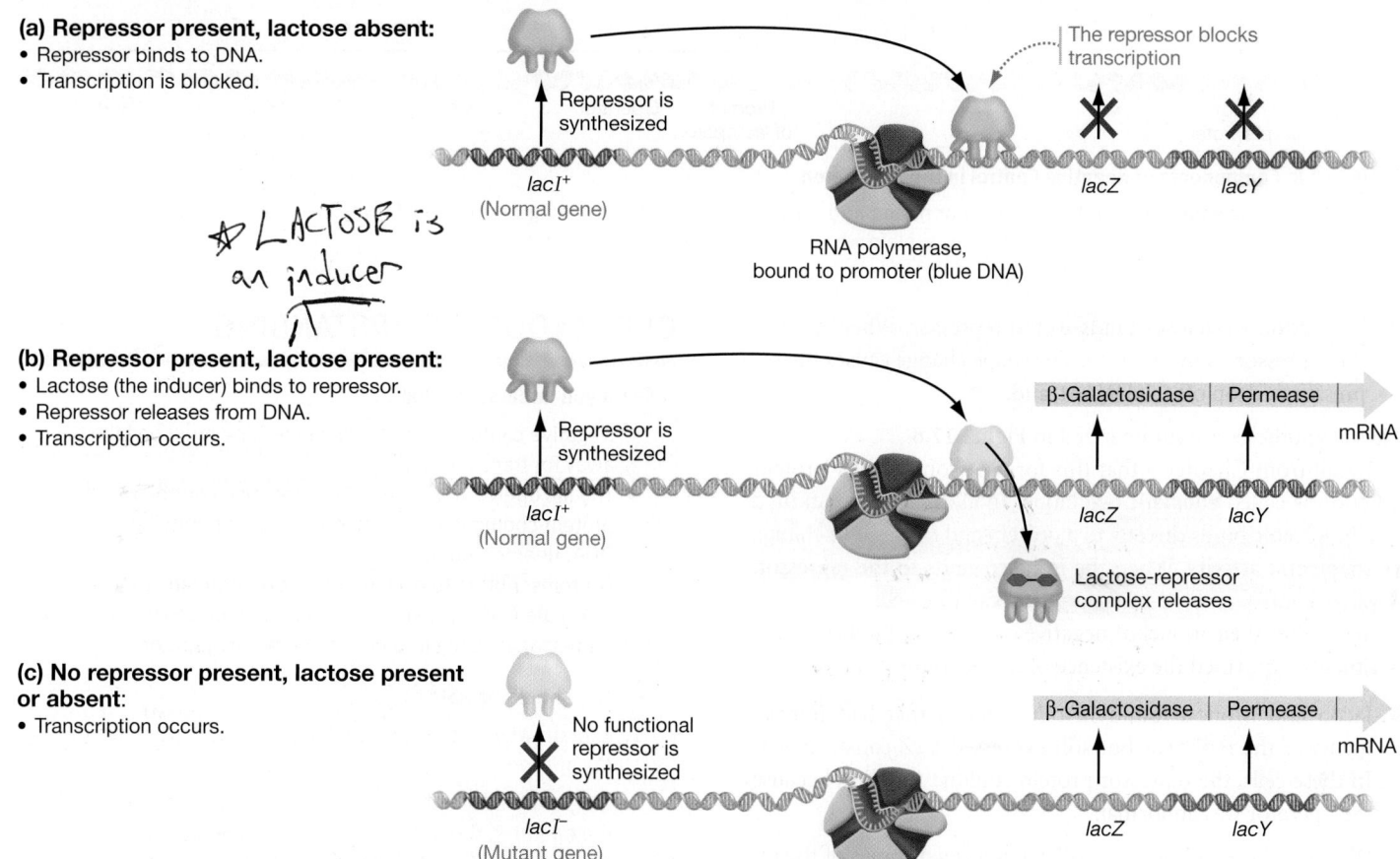

(a) Repressor present, lactose absent:
- Repressor binds to DNA.
- Transcription is blocked.

Repressor is synthesized

lacI⁺
(Normal gene)

RNA polymerase, bound to promoter (blue DNA)

The repressor blocks transcription

lacZ   *lacY*

☆ LACTOSE is an inducer

(b) Repressor present, lactose present:
- Lactose (the inducer) binds to repressor.
- Repressor releases from DNA.
- Transcription occurs.

Repressor is synthesized

lacI⁺
(Normal gene)

β-Galactosidase | Permease

mRNA

lacZ   *lacY*

Lactose-repressor complex releases

(c) No repressor present, lactose present or absent:
- Transcription occurs.

No functional repressor is synthesized

lacI⁻
(Mutant gene)

β-Galactosidase | Permease

mRNA

lacZ   *lacY*

FIGURE 17.7 The Hypothesis of Negative Control. The negative-control hypothesis maintains that **(a)** transcription of genes involved in lactose use is normally blocked by a repressor molecule that binds to DNA on or near the promoter for *lacZ* and *lacY*, **(b)** lactose induces transcription of *lacZ* and *lacY* by interacting with the repressor, and **(c)** when a functional repressor is absent, transcription proceeds.

What about the constitutive mutants? **Figure 17.7c** shows that constitutive transcription is observed in *lacI⁻* mutants because a functional repressor is absent—the parking brake is broken.

To test these hypotheses, Jacob and Monod and co-workers created *E. coli* cells that had functioning copies of the genes for β-galactosidase and galactoside permease but lacked a functional gene for the repressor. As predicted, these cells made β-galactosidase all the time. But when these cells received a functioning copy of the repressor gene, β-galactosidase production declined and then stopped.

This result supported the hypothesis that the repressor codes for a protein that shuts down transcription—that it is indeed the "parking brake" on transcription. However, if an inducer such as lactose was then added to the experimental cells, β-galactosidase activity resumed. This result supported the hypothesis that lactose removes the repressor—it releases the parking brake.

What's the take-home message? The *lacI* gene codes for a repressor protein that exerts negative control on *lacZ* and *lacY*. Lactose acts as an inducer by removing the repressor and ending negative control.

The *lac* Operon

Jacob and Monod summarized the results of their experiments with a comprehensive model of negative control that was pub-

lished in 1961. One of their key conclusions was that the genes for β-galactosidase and galactoside permease are controlled together. To encapsulate this idea, they coined the term **operon** for a set of coordinately regulated bacterial genes that are transcribed together into one mRNA. Logically enough, the group of genes involved in lactose metabolism was termed the *lac* operon.

Later, a gene called *lacA* was found to be tightly linked to *lacY* and *lacZ* and part of the same operon. The *lacA* gene codes for the enzyme transacetylase. The enzyme's function is protective in nature. It catalyzes reactions that allow certain types of sugars to be exported from the cell when they are too abundant.

Three hypotheses are central to the Jacob–Monod model of *lac* operon regulation:

1. The *lacZ*, *lacY*, and *lacA* genes are adjacent and are transcribed into one mRNA initiated from the single promoter of the *lac* operon. As a result, the expression of the three genes is coordinated.

2. The repressor is a protein encoded by *lacI* that binds to DNA and prevents transcription of *lacZ*, *lacY*, and *lacA*. Jacob and Monod proposed that lacI is expressed constitutively, and that the repressor binds to a section of DNA in the *lac* operon called the **operator**.

lac operon

DNA

lacI   lacI   Promoter   Operator   lacZ   lacY   lacA
lacI   of lac operon
promoter

FIGURE 17.8 Components of Negative Control in the lac Operon.

✔**EXERCISE** Using small, colored bits of candy or paper, add the repressor protein to the figure. Then add RNA polymerase, and then add lactose. At each step, explain what happens after the molecule is added.

3. The inducer (lactose) binds to the repressor. When it does, the repressor changes shape. The shape change causes the repressor to drop off the DNA strand.

These hypotheses are summarized in **Figure 17.8**.

Recall from Chapter 3 that this form of control over protein function is called **allosteric regulation**. In allosteric regulation, a small molecule binds directly to a protein and causes it to change its shape and activity. When the inducer binds to the repressor, negative control ends and transcription can proceed.

Years after their model of negative control was published, experiments confirmed the existence of the operator:

- Jacob and Monod found *E. coli* mutants that had normal forms of the repressor but still expressed *lacZ* constitutively. In these cells, the repressor protein couldn't function because the operator was abnormal.

- Walter Gilbert and Benno Müller-Hill tagged copies of the repressor protein with a radioactive atom and showed that it physically binds to the DNA sequences of the operator.

Why Has the *lac* Operon Model Been So Important?

The *lac* operon has been an immensely important model system in genetics, for two reasons. First, follow-up work showed that numerous bacterial genes and operons are under negative control by repressor proteins—meaning that Jacob and Monod's findings were general. Second, the *lac* operon model introduced a fundamentally important idea: Gene expression is regulated by physical contact between regulatory proteins and specific regulatory sites in DNA. Publication of the *lac* operon model was a watershed event in the history of biological science.

Besides confirming the existence of negative control, regulatory proteins, and regulatory sites in DNA, work on the *lac* operon offered an important example of post-translational control over gene expression. To understand why, you have to realize that the repressor protein is always present—because it is transcribed and translated constitutively at low levels. When a rapid change in *lac* operon activity is required, it does not occur by means of changes in the transcription or translation of new repressor proteins. Instead, the activity of *existing* repressor proteins is altered.

This turns out to be a common type of control. In virtually all cases, the activity of key regulatory proteins is controlled by post-translational modifications.

CHECK YOUR UNDERSTANDING

🗝 **If you understand that . . .**

- Negative control occurs when something must be taken away for transcription to occur.
- The *lac* operon repressor exerts negative control over three protein-coding genes by binding to the operator site in DNA, near the promoter.
- For transcription to occur in the *lac* operon, an inducer molecule (a derivative of lactose) must bind to the repressor, causing it to release from the operator.

✔ **You should be able to . . .**

1. Explain why lactose should induce transcription of the *lac* operon.

2. Diagram the *lac* operon, showing the relative positions of the operator, the promoter, and the three protein-coding genes, and indicate what is happening at the operon in the absence of lactose and in the presence of lactose.

Answers are available in Appendix B.

17.4 Mechanisms of Positive Control: Catabolite Repression

The model of negative control over the *lac* operon, summarized in Figure 17.7, is elegant and successful in explaining experimental results. But it is not complete. After studying the model, you may think of an important question that it fails to answer: Where does glucose fit in?

Transcription of the *lac* operon is drastically reduced when glucose is present in the environment—even when lactose is available to induce β-galactosidase expression. This makes sense, given that glucose is *E. coli*'s preferred carbon source. When glucose is already present, the cell doesn't need to cleave lactose as its way of acquiring glucose.

The hydrolysis of lactose into its glucose and galactose subunits, by the enzyme β-galactosidase, is an example of catabolism. In many cases, operons that encode catabolic enzymes (like β-galactosidase) are inhibited when the end product of the reaction, the catabolite, is abundant (**Figure 17.9a**). Biologists use the term **catabolite repression** to describe a situation like this. Catabolite repression is a form of feedback inhibition (see Chapter 9), sometimes called end-point or end-product inhibition.

Figure 17.9b shows how catabolite repression affects the lactose metabolism. In catabolite repression of the *lac* operon, glu-

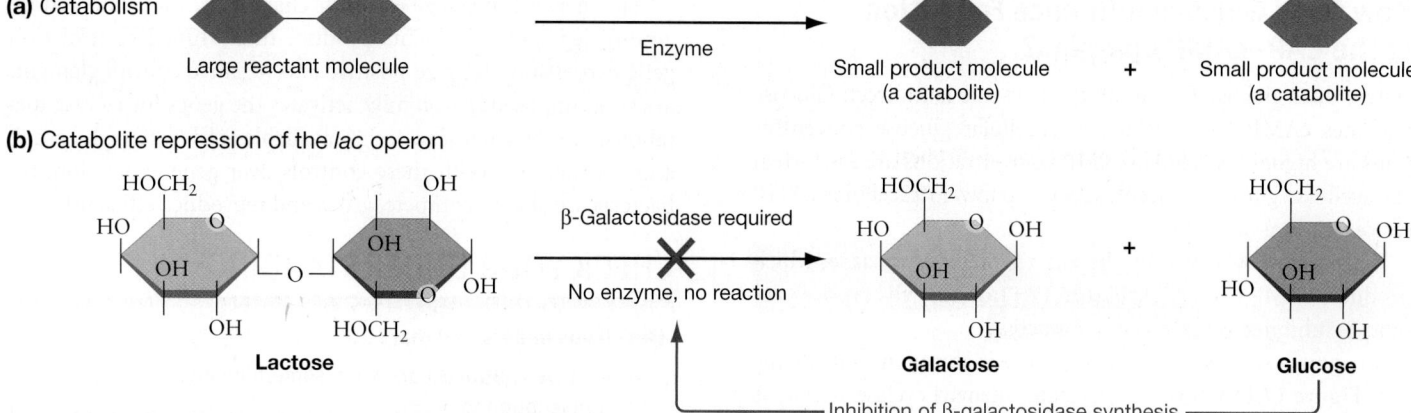

(a) Catabolism

Large reactant molecule → Enzyme → Small product molecule (a catabolite) + Small product molecule (a catabolite)

(b) Catabolite repression of the *lac* operon

Lactose → β-Galactosidase required / No enzyme, no reaction → Galactose + Glucose

Inhibition of β-galactosidase synthesis

FIGURE 17.9 Catabolite Repression Is a Mechanism of Gene Regulation. (a) A generalized example of catabolism. **(b)** Catabolite repression occurs when one of the small product molecules of catabolism represses the production of the enzyme responsible for the reaction. In the case of lactose metabolism, the production of β-galactosidase is suppressed when glucose is present.

cose is the catabolite. When glucose is abundant in the cell, transcription of the *lac* operon is decreased.

The CAP Protein and Binding Site

How does glucose prevent expression of the *lac* operon? An answer to this question began to emerge when researchers discovered a second major control element in the *lac* operon—in this case, an example of positive control.

Positive control of the *lac* operon depends on a regulatory protein called **catabolite activator protein (CAP)** and a DNA sequence known as the **CAP binding site**, which is located just upstream of the *lac* promoter. The CAP protein binds to the CAP binding site and triggers transcription of the *lac* operon.

To understand how CAP works, it's important to realize that not all promoters are created equal. Strong promoters allow effi-

cient initiation of transcription by RNA polymerase; weak ones support much less efficient initiation of transcription.

The *lac* promoter is weak. But when the CAP regulatory protein is bound to the CAP site just upstream of the *lac* promoter, the protein interacts with RNA polymerase in a way that allows transcription to begin much more frequently.

Because CAP binding greatly strengthens the *lac* promoter, CAP exerts positive control of the *lac* operon. When CAP is active, transcription increases. After the inducer removes the parking brake, CAP can push the gas pedal to the floor.

Researchers also discovered that CAP, like the repressor protein, is allosterically regulated. CAP can't bind to the CAP binding site unless the regulatory molecule **cyclic AMP (cAMP)** binds to it (**Figure 17.10**). Cyclic AMP is the green light that tells CAP to floor it.

(a) When cAMP is present:
- cAMP–CAP complex forms and binds to DNA at the CAP site.
- RNA polymerase binds the promoter efficiently.
- Transcription occurs frequently.

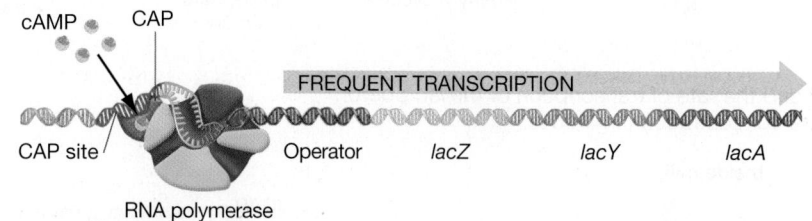

cAMP   CAP

FREQUENT TRANSCRIPTION

CAP site   Operator   *lacZ*   *lacY*   *lacA*

RNA polymerase bound **TIGHTLY** to promoter

(b) When cAMP is absent:
- CAP does not bind to DNA.
- RNA polymerase binds the promoter inefficiently.
- Transcription occurs rarely.

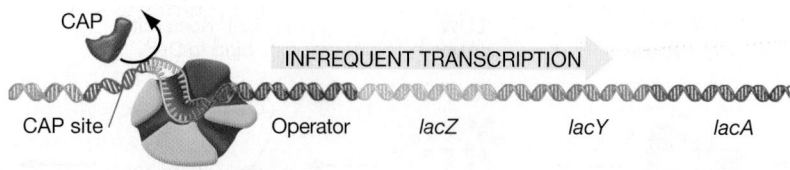

CAP

INFREQUENT TRANSCRIPTION

CAP site   Operator   *lacZ*   *lacY*   *lacA*

RNA polymerase bound **LOOSELY** to promoter

FIGURE 17.10 Positive Control of the *lac* Operon. (a) When glucose levels in an *E. coli* cell are low, cyclic AMP (cAMP) is produced. cAMP then interacts with CAP to increase transcription of the *lac* operon. **(b)** When glucose is abundant, cAMP is rare in the cell, and positive control does not occur.

✔**EXERCISE** Above each part, write "Glucose high" or "Glucose low" as appropriate.

How Does Glucose Influence Formation of the CAP–cAMP Complex?

Where does glucose fit into all this? Its role is indirect: Glucose regulates cAMP levels. When extracellular glucose concentrations are high, intracellular cAMP concentrations are low; when extracellular glucose concentrations are low, intracellular cAMP concentrations are high.

This seesaw is driven by the enzyme **adenylyl cyclase**, which produces cAMP from ATP (**Figure 17.11a**). Adenylyl cyclase's activity is inhibited by extracellular glucose.

Imagine a situation in which glucose is abundant outside the cell (**Figure 17.11b, top**). In this state, adenylyl cyclase activity is low. Therefore, cAMP levels inside the cell are low. The CAP–cAMP complex is unable to form, so CAP does not have the conformation that allows it to bind to the CAP site and stimulate *lac* operon transcription.

Conversely, when the extracellular concentration of glucose is low, the intracellular concentration of cAMP increases (**Figure 17.11b, bottom**). The CAP–cAMP complex forms, binds to the CAP site, and allows RNA polymerase to initiate transcription efficiently.

Figure 17.12 summarizes how positive control and negative control combine to regulate the *lac* operon. For even more review, go to the study area at *www.masteringbiology.com*.

(MB) Web Activity The *lac* Operon

The general message of this chapter is that interactions among regulatory elements produce finely tuned control over gene expression. Because positive and negative control elements are superimposed, *E. coli* fully activates the genes for lactose metabolism only when lactose is available and when glucose is scarce or absent. With these controls over gene expression, the bacteria are able to compete, grow, and reproduce efficiently.

CHECK YOUR UNDERSTANDING

⚷ If you understand that . . .

- Positive control occurs when something must be added for transcription to occur.
- CAP exerts positive control over the *lac* operon by binding to the CAP site and increasing the transcription rate.
- CAP can bind only when it is complexed with cAMP—a signaling molecule whose presence indicates that glucose levels are low.

✓ You should be able to . . .

1. Diagram what happens at the operon when cAMP levels are low and when cAMP levels are high.

2. Explain how positive control and negative control work together to control the *lac* operon in the presence or absence of glucose and lactose.

Answers are available in Appendix B.

(a) The enzyme adenylyl cyclase catalyzes production of cAMP from ATP.

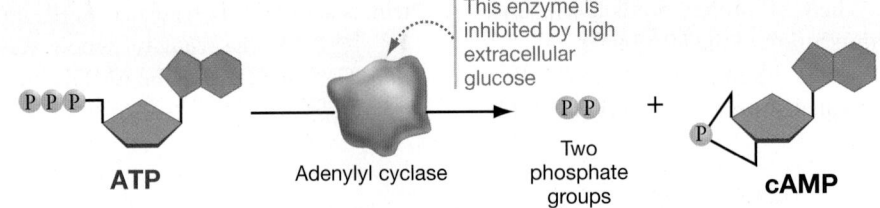

(b) The amount of cAMP and the rate of transcription of the *lac* operon are inversely related to the concentration of glucose.

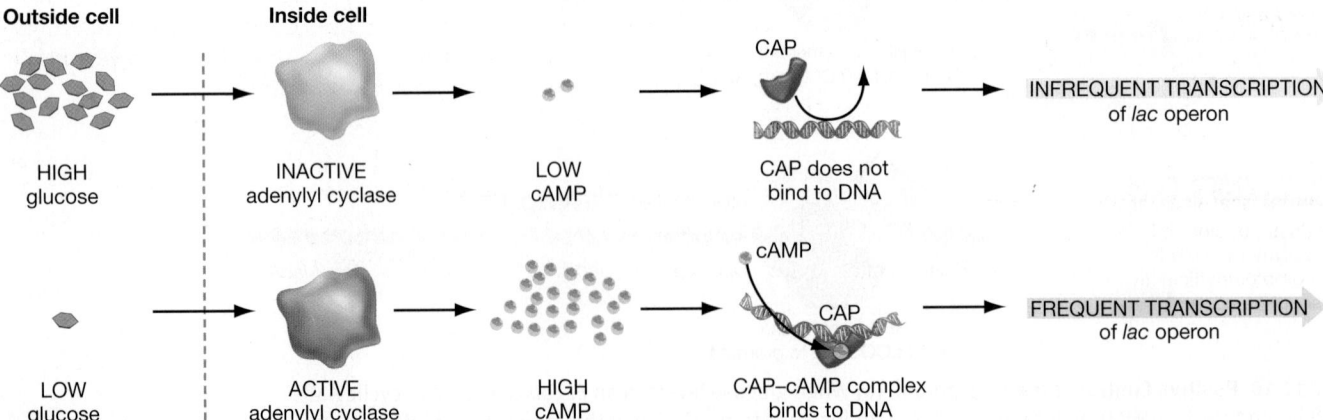

FIGURE 17.11 Cyclic AMP (cAMP) Is Synthesized When Glucose Levels Are Low.

It all has to do with glucose levels

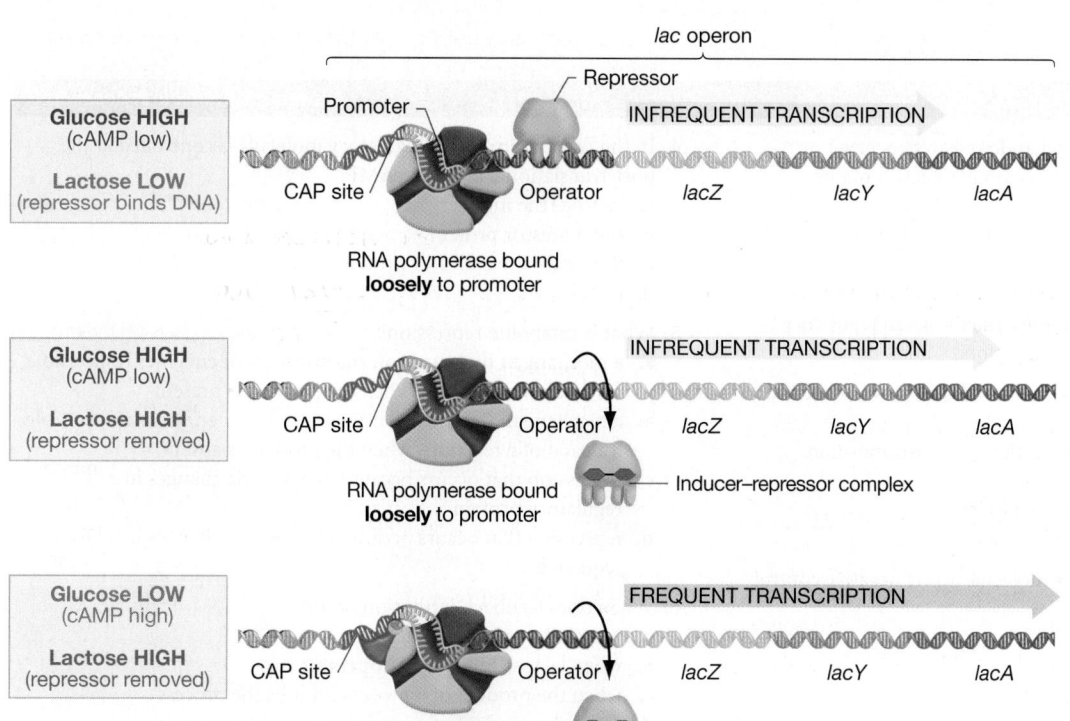

Glucose HIGH
(cAMP low)

Lactose LOW
(repressor binds DNA)

lac operon

Promoter — — Repressor   INFREQUENT TRANSCRIPTION

CAP site   Operator   lacZ   lacY   lacA

RNA polymerase bound
loosely to promoter

Glucose HIGH
(cAMP low)

Lactose HIGH
(repressor removed)

INFREQUENT TRANSCRIPTION

CAP site   Operator   lacZ   lacY   lacA

RNA polymerase bound
loosely to promoter   Inducer–repressor complex

Glucose LOW
(cAMP high)

Lactose HIGH
(repressor removed)

FREQUENT TRANSCRIPTION

CAP site   Operator   lacZ   lacY   lacA

RNA polymerase bound
tightly to promoter

FIGURE 17.12 An Overview of *lac* Operon Regulation.

✔**EXERCISE** Create a flowchart called "Release of negative control" that contains the following terms: Operator, Lactose, Transcription of *lac* operon, RNA polymerase at promoter, and Repressor. Put the terms in order and connect them with labels indicating how the elements relate to each other. Do a similar flowchart called "Elements of positive control" with the terms cAMP, RNA polymerase at promoter, CAP–cAMP, Adenylyl cyclase, CAP, CAP site, Glucose, and Transcription of *lac* operon.

CHAPTER 17   REVIEW

For media, go to the study area at www.masteringbiology.com

Summary of Key Concepts

🔑 **In bacteria, gene expression can be controlled at three levels: transcription, translation, or post-translation (protein activation).**

- Among the levels of gene regulation, there is a trade-off between the speed of response to changed conditions and the efficient use of resources.

  ✔You should be able to describe one element of the *lac* operon that is under transcriptional control and one element that is under post-translational control.

🔑 **Changes in gene expression allow bacterial cells to respond to environmental changes.**

- Transcription may be constitutive or inducible.

- Constitutive expression occurs in genes whose products are required at all times, such as genes that encode glycolytic enzymes.

- Expression of most genes is induced by environmental signals—meaning that gene products are produced or activated only when they are needed.

  ✔You should be able to predict how gene expression changes in your gut-dwelling bacteria after you eat a dessert versus drink a glass of milk.

🔑 **Transcriptional control can be negative or positive. Negative control occurs when a regulatory protein prevents transcription. Positive control occurs when a regulatory protein increases the transcription rate.**

- The *lac* operon is under both negative and positive control.

- Negative control occurs because a repressor protein binds to a DNA operator sequence near the promoter of the protein-encoding genes.

- When lactose is present, it binds to the repressor and causes it to fall off the operator, allowing transcription to occur.

- Positive control occurs because low glucose levels induce production of cAMP. After cAMP binds to CAP, the complex binds to the CAP site and triggers rapid transcription.

  ✔You should be able to extend this chapter's use of a car analogy for negative and positive control by identifying (1) how the operator, repressor, and inducer relate to a parking brake; and (2) how CAP, cAMP, and the CAP binding site relate to a gas pedal.

  (MB) **Web Activity** The *lac* Operon

Questions

1. After completing a genetic screen, what do researchers have?
   a. a master plate and a replica plate—each of which contains identical copies of bacterial colonies
   b. individuals with mutations that disable genes required for a particular process or metabolic pathway
   c. a large collection of mutagenized cells
   d. mutants with defects in genes for enzymes—*not* in regulatory sites in DNA (e.g., promoters or operators)

2. Why are the genes involved in lactose metabolism considered to be an operon?
   a. They occupy adjacent locations on the *E. coli* chromosome.
   b. They have a similar function.
   c. They are all required for normal cell function.
   d. They are under the control of the same promoter.

3. In the *lac* operon, which regulatory molecule exerts negative control by inhibiting transcription?
   a. lactose (the inducer)
   b. the repressor protein
   c. the catabolite activator protein (CAP)
   d. cAMP

4. In the *lac* operon, which regulatory molecule is controlled at the post-translational level, by cAMP?
   a. lactose (the inducer)
   b. the repressor protein
   c. the catabolite activator protein (CAP)
   d. cAMP

5. What is catabolite repression?
   a. a mechanism that turns off the synthesis of enzymes responsible for catabolic reactions when the product is present
   b. a mechanism that turns off the synthesis of enzymes responsible for catabolic reactions when the product is absent
   c. repression that occurs because of allosteric changes in a regulatory protein
   d. repression that occurs because of allosteric changes in a DNA sequence

6. When does feedback inhibition occur?
   a. when allosteric regulation occurs
   b. when lactose binds to the operator
   c. when the product of a process inhibits the process
   d. when the product of a process triggers the process

1. *E. coli* expresses genes for glycolytic enzymes constitutively. Why?

2. Explain the difference between positive and negative control over transcription.

3. Why is it advantageous for the *lac* operon in *E. coli* to be under both positive control and negative control? What would happen if only negative control occurred? What would happen if only positive control occurred?

4. In *E. coli*, rising levels of cAMP can be considered a starvation signal. Explain.

5. CAP is also known as the cAMP-receptor protein. Why?

6. The galactose released when β-galactosidase cleaves lactose enters the glycolytic pathway in *E. coli* after a series of enzyme-catalyzed, ATP-requiring reactions has converted the galactose to glucose-6-phosphate. Why, then, is glucose, not lactose, the preferred sugar in *E. coli*?

1. You are interested in using bacteria to metabolize wastes at an old chemical plant and convert them into harmless compounds. You find bacteria that are able to tolerate high levels of the toxic compounds toluene and benzene, and you suspect that it is due to the ability of the bacteria to break these compounds into less-toxic products. If that is true, these toluene- and benzene-resistant strains will be valuable for cleaning up toxic sites. How could you find out whether these bacteria are metabolizing toluene as a source of carbon compounds?

2. Assuming that the bacteria you examined in Level 3, Problem 1 do have an enzymatic pathway to break down toluene, would you predict that the genes involved are constitutively expressed, under positive control, or under negative control? Why?

3. The *lacI* gene mutants produce β-galactosidase constitutively because no repressors are present to bind to the operator. Other repressor mutants have been isolated that are called *LacI^S* mutants. These repressor proteins continue to bind to the operator, even in the presence of the inducer. How would this mutation affect the function of the *lac* operon? Specifically, how well would *LacI^S* mutants do in an environment that has lactose as its sole sugar?

4. X-gal is a colorless, lactose-like molecule that can be split into two fragments by β-galactosidase. One of these product molecules is blue. The following photograph shows *E. coli* colonies growing in a medium that contains X-gal.

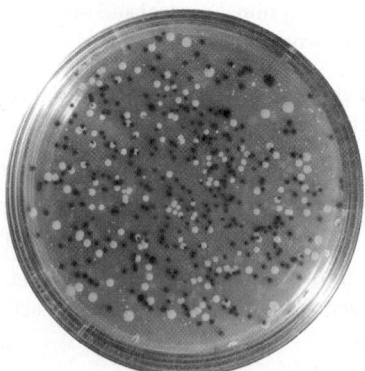

Find three colonies whose cells have functioning copies of β-galactosidase. Find three colonies whose cells have mutations in the *lacZ* locus or in one of the genes involved in regulation of *lacZ*. Suppose you could analyze the sequence of the β-galactosidase gene from each of the mutant colonies. How would these data help you distinguish which cells are structural mutants and which are regulatory mutants?

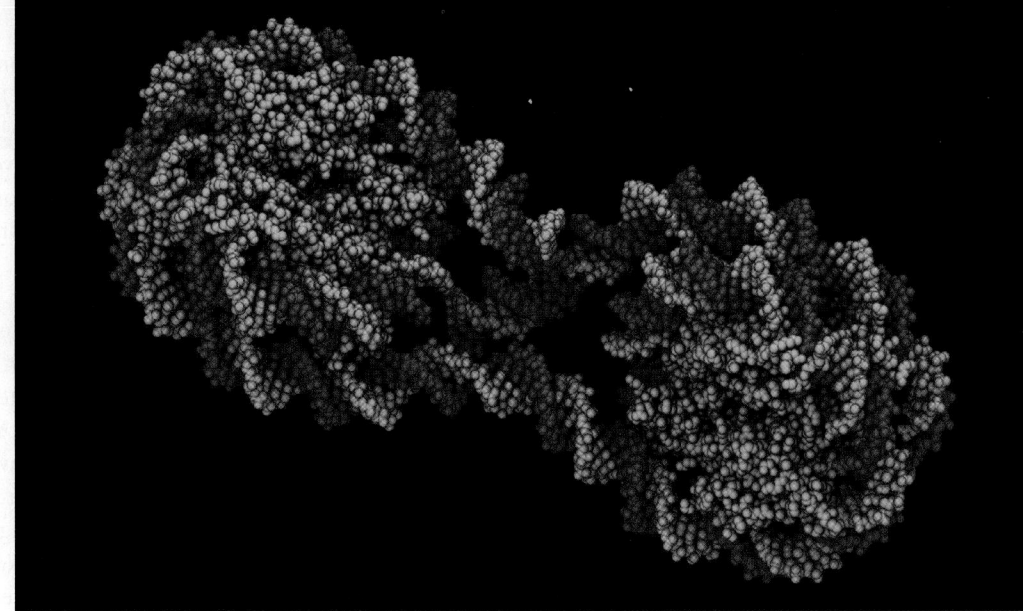

A model of eukaryotic DNA in the condensed state. The DNA (shown in red and pink) is wrapped around proteins (in green). The DNA has to be uncoiled before transcription can take place.

Control of Gene Expression in Eukaryotes

18

Bacteria regulate gene expression to respond to changes in their environment. As Chapter 17 indicated, *Escherichia coli* thrive best if the genes that are required to import and cleave lactose are expressed only when the cells are relying on lactose as a source of energy—when glucose is absent and lactose is present.

Unicellular eukaryotes face similar challenges. Consider the yeast *Saccharomyces cerevisiae*, which is used extensively in the production of beer, wine, and bread. In nature this species lives on the skins of grapes and other fruits, where temperature and humidity can change dramatically. In addition, the sugars that the cells use as food vary in type and concentration as the fruit ripens, falls, and rots. For yeast cells to grow and reproduce efficiently, gene expression has to be modified in response to these changes.

The cells that make up multicellular eukaryotes face additional challenges. Consider your body, which contains trillions of cells, each with a specialized structure and function. You have heart muscle cells, lung cells, nerve cells, skin cells, and so on. Even though these cells look different, they contain the same genes. Your bone cells and blood cells aren't different because of a difference in their genes but because they *express* different genes. Your bone cells have blood cell genes—they just don't transcribe them.

Why? The answer is that your cells respond to their environment, just as bacteria and unicellular eukaryotes do. But there's a key difference. The cells in a multicellular eukaryote express different genes in response to changes in the internal environment—specifically, to signals from other cells. As a human being or an oak tree develops, cells that are located in different parts of the organism are exposed to different cell-cell signals. As a result, they express different genes. **Differential gene expression** is responsible for creating different cell types, arranging them into tissues, and coordinating their activity to form the multicellular society we call an individual.

How does all of this regulation and differentiation happen? Unit 4 introduces the signals that trigger the formation of muscle, bone, leaf, and flower cells. In contrast, this

KEY CONCEPTS

- Changes in gene expression allow eukaryotic cells to respond to changes in the environment and cause distinct cell types to develop.

- Eukaryotic DNA is packaged with proteins into structures that must be opened before transcription can occur.

- In eukaryotes, transcription is triggered by regulatory proteins that bind to the promoter and to sequences close to and far from the promoter.

- Once transcription is complete, gene expression is controlled by (1) alternative splicing, which allows a single gene to code for several different products; (2) molecules that regulate the life span of mRNAs; and (3) activation or inactivation of protein products.

- Cancer can develop when mutations disable genes that regulate cell-cycle control genes.

✔ When you see this checkmark, stop and test yourself. Answers are available in Appendix B.

chapter focuses on what happens after a eukaryotic cell receives such a signal. Let's start with an overview of how gene expression can be controlled, and close with a look at how defects in the process can help trigger cancer.

18.1 Mechanisms of Gene Regulation in Eukaryotes—An Overview

Like bacteria, eukaryotes can control gene expression at the levels of transcription, translation, and post-translation (in the last, by protein activation or inactivation). But as **Figure 18.1** shows, three additional levels of control occur in eukaryotes as genetic information flows from DNA to proteins.

The first additional level of control involves the DNA–protein complex at the top of the figure. In eukaryotes, DNA is wrapped around proteins to create a DNA–protein complex called **chromatin**. Eukaryotic genes have promoters, just as bacterial genes do, but before transcription can begin in eukaryotes, the stretch of DNA containing the promoter must be released from tight interactions with proteins, so that RNA polymerase can make contact with the promoter. To capture this idea, biologists say that **chromatin remodeling** must occur prior to transcription.

The second level of regulation that is unique to eukaryotes is **RNA processing**. These are the steps required to produce a mature, processed mRNA from a primary RNA transcript. Recall from Chapter 16 that introns have to be spliced out of primary transcripts. In many cases, carefully orchestrated alternative splicing patterns occur—meaning that different combinations of exons are included in the mRNA. If a cell switches from one splicing pattern to another, a different gene product results.

Third, mRNA life span is regulated in eukaryotes. mRNAs that are active in the cell for a long time tend to be translated more than mRNAs that have a short life span.

Each of the six potential control points shown in Figure 18.1 is employed at certain times in a typical eukaryotic cell. This chapter explores all six—chromatin remodeling, transcription, RNA processing, mRNA stability, translation, and post-translational modification of proteins.

To appreciate the breadth and complexity of gene regulation in eukaryotes, let's follow the series of events that occur as an embryonic cell responds to a developmental signal. Suppose a molecule arrives that specifies the production of a muscle-specific protein. What happens next?

18.2 Chromatin Remodeling

For the arrival of a signaling molecule from outside the cell to result in the transcription of a particular gene, the chromatin around the target gene must be drastically remodeled. To appreciate why, consider that a typical cell in your body contains about 6 billion base pairs of DNA. Lined up end to end, these nucleotide pairs would form a double helix about 2 m (6.5 feet) long. But the nucleus that holds this DNA is only about 5 μm in diameter—less than the thickness of this page.

In eukaryotes, DNA is packed inside the nucleus so tightly that RNA polymerase can't access it. Part of this packing is done by supercoiling—meaning the DNA double helix is twisted on itself many times, just as in bacterial chromosomes (see Chapter 7). But supercoiling is just part of the packing system found in eukaryotes.

What Is Chromatin's Basic Structure?

The first data on the chemical composition of eukaryotic DNA were published in the early 1900s, when researchers established that eukaryotic DNA is intimately associated with proteins. Later work documented that the most abundant DNA-associated proteins belong to a group called the **histones**. Chromatin consists of DNA complexed with histones and other proteins.

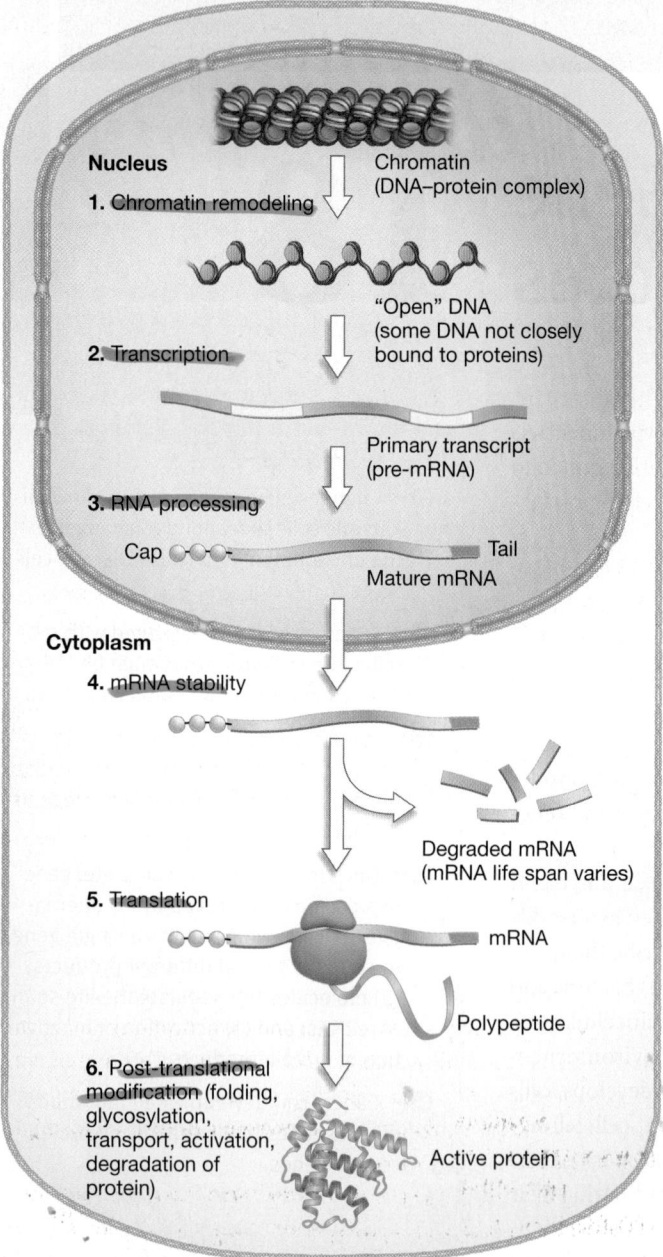

Nucleus

1. Chromatin remodeling

Chromatin (DNA–protein complex)

"Open" DNA (some DNA not closely bound to proteins)

2. Transcription

Primary transcript (pre-mRNA)

3. RNA processing

Cap — Tail

Mature mRNA

Cytoplasm

4. mRNA stability

Degraded mRNA (mRNA life span varies)

5. Translation

mRNA

Polypeptide

6. Post-translational modification (folding, glycosylation, transport, activation, degradation of protein)

Active protein

FIGURE 18.1 In Eukaryotes, Gene Expression Can Be Controlled at Many Different Levels.

In the 1970s electron micrographs like the one in **Figure 18.2a** revealed that chromatin has a regular structure. In some preparations for electron microscopy, chromatin actually looked like beads on a string. The "beads" came to be called **nucleosomes**.

More details emerged in 1984 when researchers determined the three-dimensional structure of eukaryotic DNA by using X-ray crystallography (see **BioSkills 10** in Appendix A). The X-ray crystallographic data indicated that each nucleosome consists of DNA wrapped almost twice around a core of eight histone proteins. As **Figure 18.2b** indicates, a histone called H1 "seals" DNA to each set of eight nucleosomal histones. Between each pair of nucleosomes there is a "linker" stretch of DNA.

The intimate association between DNA and histones occurs in part because DNA is negatively charged and histones are positively charged. DNA has a negative charge because of its phosphate groups; histones are positively charged because they contain many lysine or arginine residues or both (see Chapter 3).

More recent work has shown that there is another layer of complexity in eukaryotic DNA. H1 histones interact with each other and with histones in other nucleosomes to produce a tightly packed structure like that shown in Figure 18.2b. Based on its width, this structure is called the 30-nanometer fiber. (Recall that a nanometer is one-billionth of a meter and is abbreviated nm.)

Finally, the 30-nm fibers are attached, at intervals along their length, to proteins that form a scaffold or framework inside the nucleus. In this way, the entire chromosome is organized and held in place. When chromosomes condense prior to mitosis or meiosis, the scaffold proteins and 30-nm fibers are folded and packed into still larger, more-compact structures.

A eukaryotic chromosome, then, is made up of chromatin that has several layers of organization: The DNA is wrapped around histones to form nucleosomes; nucleosomes are packed into 30-nm fibers, 30-nm fibers are attached to scaffold proteins, and the entire assembly can be folded into the highly condensed structure observed during cell division.

Although recent studies have confirmed that bacterial DNA interacts with proteins that may be organized in nucleosome-like structures, nothing like the 30-nm fibers or higher-order arrangements have been observed in bacterial chromosomes.

The elaborate structure of eukaryotic chromatin does more than just package DNA so that it fits into the nucleus. Chromatin structure also has profound implications for the control of gene expression. To appreciate this point, consider the 30-nm fiber illustrated in Figure 18.2b. If this tightly packed stretch of DNA contains a promoter, how can RNA polymerase bind to it and initiate transcription?

Evidence That Chromatin Structure Is Altered in Active Genes

Once the nucleosome-based structure of chromatin was established, biologists hypothesized that the close physical interaction between DNA and histones must be altered for RNA polymerase to make contact with DNA. More specifically, biologists hypothesized that a gene could not be transcribed until the chromatin near its promoter was remodeled.

The central idea is that chromatin must be relaxed, or decondensed, for RNA polymerase to bind to the promoter. If so, then chromatin remodeling would represent the first step in the control of eukaryotic gene expression. Two types of studies have provided strong support for this hypothesis.

(a) Nucleosomes in chromatin

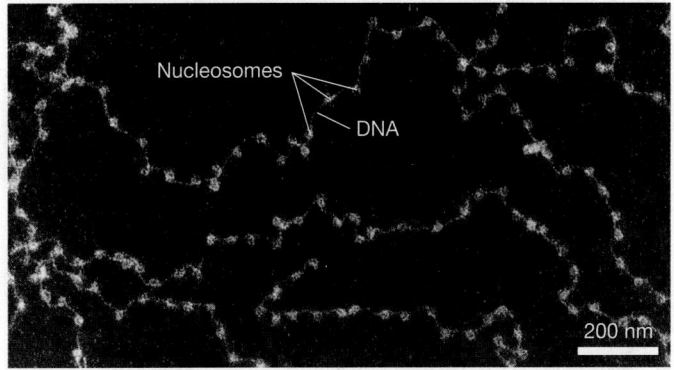

Nucleosomes

DNA

200 nm

(b) Nucleosome structure

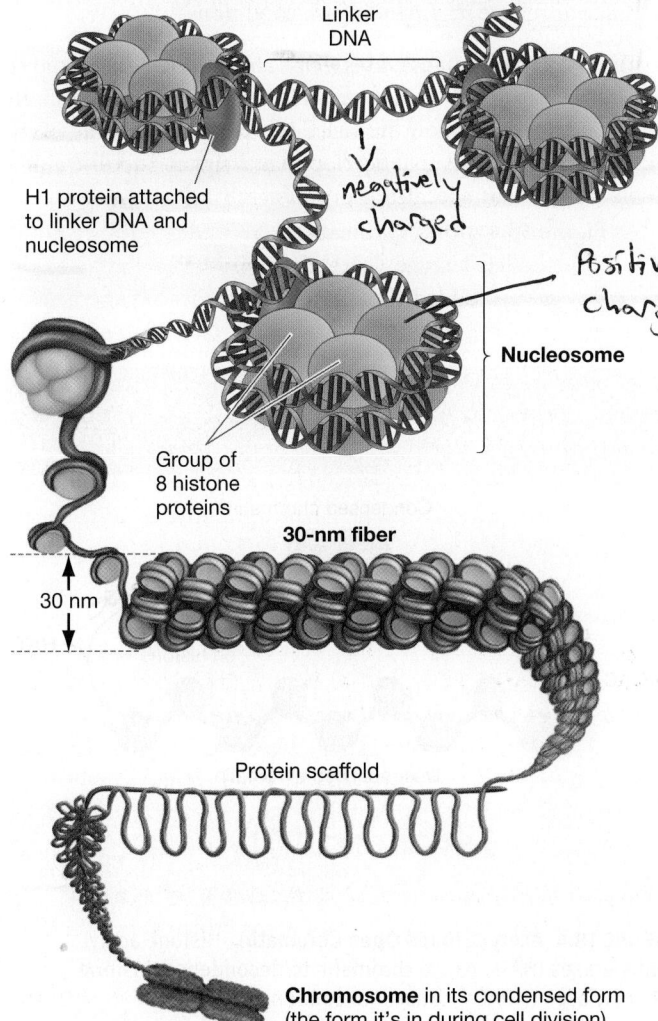

Linker DNA

H1 protein attached to linker DNA and nucleosome

negatively charged

Positively charged

Nucleosome

Group of 8 histone proteins

30-nm fiber

30 nm

Protein scaffold

Chromosome in its condensed form (the form it's in during cell division)

FIGURE 18.2 Chromatin Has Several Levels of Organization.

"CLOSED" DNA IS PROTECTED FROM DNASE DNase is an enzyme that cuts DNA at random locations. The enzyme cannot cut DNA efficiently if the molecule is tightly complexed with histones, however. As **Figure 18.3** shows, the enzyme works effectively only if DNA is in the "open" configuration.

Harold Weintraub and Mark Groudine used this observation to test the hypothesis that the DNA of actively transcribed genes is in an open configuration. They compared chromatin structure in two genes in blood cells of chickens: the β-globin and ovalbumin genes. β-Globin is a protein that is part of the hemoglobin found in red blood cells; ovalbumin is a major protein of egg white. In blood cells, the β-globin gene is transcribed at high levels, but the ovalbumin gene is not transcribed at all.

After treating blood cells with DNase and then analyzing the state of the β-globin and ovalbumin genes, the researchers found that DNase cut up the β-globin gene much more readily than the ovalbumin gene. They interpreted this finding as evidence that chromatin in blood cells was in an open configuration at the β-globin gene but closed at the ovalbumin gene. Analogous studies using DNase on different genes and in different cell types yielded similar results.

HISTONE MUTANTS The second type of evidence in support of the chromatin-remodeling hypothesis comes from studies of mutant brewer's yeast cells that do not produce the usual complement of histones. In these mutant cells, many yeast genes that are normally never transcribed are instead transcribed at high levels at all times.

To interpret this finding, biologists hypothesized that the lack of histone proteins prevented the assembly of normal chromatin. If the absence of normal histone–DNA interactions promotes transcription, then the presence of normal histone–DNA interactions must prevent it.

Taken together, the data suggest that in their normal, or default, state, eukaryotic genes are turned off. This is a new mechanism of negative control—different from the repressor proteins introduced in Chapter 17. When DNA is wrapped into a 30-nm fiber, the parking brake is on. If so, then gene expression depends on chromatin being opened up in the promoter region.

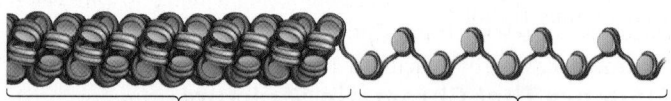

Condensed chromatin   Open chromatin

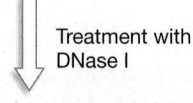

Treatment with DNase I

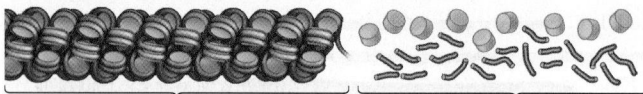

Condensed chromatin   Degraded DNA

FIGURE 18.3 DNase Assay for Chromatin Structure. DNase is an enzyme that cuts DNA at random locations. It cannot cut intact condensed chromatin.

How Is Chromatin Altered?

Research on chromatin remodeling has been proceeding at a furious pace, and has recently succeeded in identifying some of the key players. Multi-protein machines called chromatin-remodeling complexes, which reshape chromatin through a series of ATP-dependent reactions, are particularly important. In essence, the remodeling complexes make contact with DNA on the surface of the nucleosome, then twist it in a way that allows the DNA to loop out and away from the protein core, where it can be transcribed.

Other major players in chromatin-remodeling work by adding small molecules to histone proteins. To date, eight distinct types of chemical modifications have been discovered. Two of the best-studied involve adding acetyl (CH_3COOH) or methyl (CH_3) groups. These processes are known as **acetylation** and **methylation**, respectively.

- Acetylation of histones is usually associated with positive control—meaning, activation of genes through binding of a regulatory protein.

- Methylation can be correlated with either activation or inactivation, depending on which histones are altered and where the methyl groups occur. Methylation can set the parking brake or release it, depending on conditions.

The **histone acetyl transferases (HATs)** are some of the best-studied enzymes that modify chromatin. HATs acetylate the positively charged lysine residues in histones. When a HAT adds an acetyl group to selected histones, the marked proteins can act as binding sites for chromatin remodeling complexes that open the DNA.

As **Figure 18.4** shows, chromatin is "recondensed" by a group of enzymes called **histone deacetylases (HDACs)**, which remove the acetyl groups added by HATs. Histone deacetylase activity

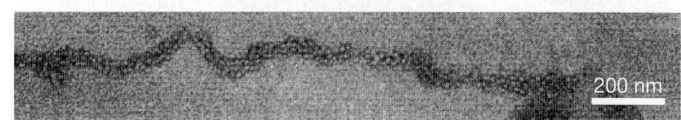

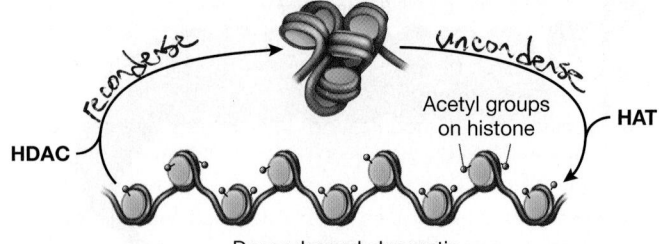

Condensed chromatin

recondense   uncondense

Acetyl groups on histone

HDAC   HAT

Decondensed chromatin

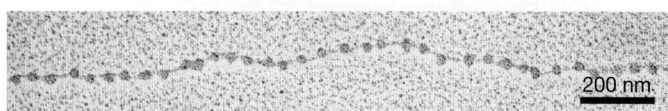

FIGURE 18.4 Acetyl Groups Open Chromatin. Histone acetyl transferases (HATs) cause chromatin to decondense; histone deacetylases (HDACs) cause it to condense.

✔ **QUESTION** Are HATs and HDACs elements in positive control or negative control? Explain your reasoning.

reverses the effects of acetylation, returning chromatin to its default condensed state. If HATs are an on switch for transcription, HDACs are the off switch.

The take-home message from work on chromatin remodeling is simple: The state of the histone proteins that are complexed with DNA is a critical determinant of whether transcription can occur.

Chromatin Modifications Can Be Inherited

Within a single individual, the pattern of chromatin modifications varies from one cell type to another. For example, suppose you analyzed the same gene in a cell that was destined to give rise to muscle and in a cell that is a precursor to part of the brain. This and other genes would likely have a completely different pattern of acetylation and methylation in the two cell types.

To capture this point, biologists propose that a **histone code** exists. The histone code hypothesis contends that precise patterns of chemical modifications of histones contain information, analogous to the way the genetic code stores information. But instead of dictating the amino acid sequence in a gene product, the histone code influences whether or not a particular gene is expressed.

The histone code hypothesis is attracting interest among biologists because of the now well-established evidence that some or most of the chemical modifications that distinguish a muscle- from a brain-associated cell are passed on to daughter cells during mitosis. This is a key observation, because it means that daughter cells inherit patterns of gene expression from their parent cells.

Histone modifications are an example of **epigenetic inheritance**, the collective term for patterns of inheritance that are not due to differences in DNA sequences. If a cell received a "become a muscle cell" signal early in development, it would not only modify its

chromatin in distinctive ways but pass those modifications on to its descendants. Muscle cells are different from brain cells in part because they inherited differently modified histones—not different types of genes.

Now the question is, what happens once a section of DNA is opened up by chromatin remodeling and exposed to RNA polymerase?

18.3 Initiating Transcription: Regulatory Sequences and Regulatory Proteins

Chapter 16 introduced the **promoter**—the site in DNA where RNA polymerase binds to initiate transcription. In position and function, eukaryotic promoters are similar to bacterial promoters.

Most eukaryotic promoters are located close to the point where RNA polymerase begins transcription, and all contain highly conserved sequences analogous to the -35 box and -10 box in bacterial promoters. To date, three such conserved sequences have been observed; each eukaryotic promoter appears to have two of the three. One of the most common of these conserved sequences is known as the **TATA box.**

Once a promoter has been exposed by chromatin remodeling, the first step in transcription is an interaction with the **TATA-binding protein (TBP).** But binding by TBP does not guarantee that a gene will be expressed. In eukaryotes, a wide array of other DNA sequences and proteins are involved.

Some Regulatory Sequences Are Near the Promoter

Regulatory sequences are sections of DNA that, like the prokaryotic CAP-binding site and operators introduced in Chapter 17, are involved in controlling the activity of genes. The first regulatory sequences in eukaryotic DNA were discovered in the late 1970s, when Yasuji Oshima and co-workers set out to understand how yeast cells control the metabolism of the sugar galactose.

When galactose is absent, *S. cerevisiae* cells produce only tiny quantities of the enzymes required to metabolize it. But when galactose is present, transcription of the genes encoding these enzymes increases by a factor of 1000.

The team's first major result was the discovery of mutant cells that failed to produce any of the five enzymes required for galactose metabolism, even if galactose was present. To interpret this observation, they hypothesized that

1. The five genes are regulated together, even though they are not on the same chromosome;

2. Normal cells have a CAP-like regulatory protein that exerts positive control over the five genes;

3. The mutant cells have a loss-of-function mutation that completely disables the regulatory protein.

Other researchers were able to isolate the regulatory protein and confirm that it binds to a short stretch of DNA located just

CHECK YOUR UNDERSTANDING

If you understand that . . .

- Eukaryotic DNA is wrapped tightly around histones, forming nucleosomes, which are then coiled into dense structures called 30-nm fibers.
- Before transcription can begin, the DNA–protein complex must be relaxed so that RNA polymerase can contact the promoter.
- Specific chemical modifications to histones play a key role in determining whether or not chromatin is opened and a gene is expressed.
- In many if not most cases, the histone modifications in a cell are passed on to its daughter cells.

✔ **You should be able to . . .**

1. Compare and contrast the structure of DNA in bacteria versus that in eukaryotes.

2. Explain why certain patterns of acetylation or methylation could influence whether a cell became a muscle cell or a brain cell.

Answers are available in Appendix B.

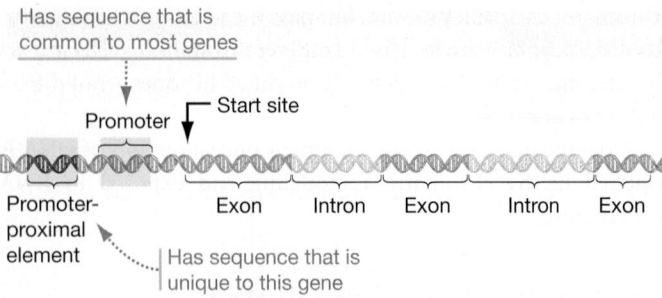

FIGURE 18.5 Promoter-Proximal Elements Regulate the Expression of Some Eukaryotic Genes. Note that exons and introns are not drawn to scale throughout this chapter. They are typically very large compared with promoters and promoter-proximal (regulatory) elements.

upstream from the promoter for the five genes required for galactose use. The location and structure of this regulatory sequence are comparable to those of the CAP-binding site in the *lac* operon of *E. coli*.

Similar regulatory sequences have now been found in a wide array of eukaryotic genes and species. Because such sequences are located close to the promoter and bind regulatory proteins, they are termed **promoter-proximal elements** (Figure 18.5).

Unlike the promoter itself, promoter-proximal elements have sequences that are unique to specific genes. In this way, they furnish a mechanism for eukaryotic cells to express certain genes but not others.

The discovery of promoter-proximal elements and a mechanism of positive control suggested a satisfying parallel between gene regulation in bacteria and in eukaryotes. This picture changed, however, when researchers discovered a new class of eukaryotic DNA regulatory sequences—sequences unlike anything in bacteria.

Some Regulatory Sequences Are Far from the Promoter

Susumu Tonegawa and colleagues made a discovery that may rank as the most startling in the history of research on gene expression. Tonegawa's group was exploring how human immune system cells regulate the genes involved in the production of antibodies. **Antibodies** are proteins that bind to specific sites on other molecules (see **BioSkills 9** in Appendix A). Produced by your immune system, antibodies bind to viruses and bacteria and mark them for destruction by other cells.

The antibody gene in question is broken into many introns and exons. Introns are DNA sequences spliced out of the primary mRNA transcript; exons are regions of eukaryotic genes that are included in the mature RNA once splicing is complete. The biologists used techniques that will be introduced in Chapter 19 to place copies of an intron in new locations and found that, when they placed the intron close to a gene, the gene's transcription rate increased. Based on this observation, Tonegawa and co-workers hypothesized that the intron contained some sort of regulatory sequence.

To test this hypothesis, the biologists performed what is now considered a classic experiment (**Figure 18.6**). The protocol was simple in concept:

Step 1 Start with a human antibody-producing gene that includes an intron flanked by two exons.

Step 2 Use enzymes to remove several different specific pieces of the intron from different samples of the gene.

Step 3 Use enzymes to ligate (link) the remaining sections of the gene back together. Each of the modified genes produced in this way is missing a different section of the intron.

Step 4 Insert the various versions of the gene into mouse cells, which do not normally contain the antibody-producing gene. Some mouse cells receive normal copies of the gene; others receive copies of the gene missing different portions of the intron; still others receive no gene.

To test their hypothesis, they analyzed the mRNAs produced by each of the experimental cells and compared them with mRNA from a normal human cell and from mouse cells that received the normal gene. If a regulatory sequence is located inside the intron, then some of the genes with modified introns should lack that sequence and fail to transcribe the antibody gene. As the "Results" section of Figure 18.6 shows, some of the modified copies of the gene were not transcribed at all. Based on these results, Tonegawa and co-workers proposed that the intron contains a regulatory sequence that is required for transcription to occur.

This result was remarkable for two reasons: (1) The regulatory sequence was thousands of bases away from the promoter, and (2) it was downstream of the promoter instead of upstream. Regulatory elements that are far from the promoter are termed **enhancers**.

Follow-up work has shown that enhancers occur in all eukaryotes and that they have several key characteristics:

- Enhancers can be more than 100,000 bases away from the promoter. They can be located in introns or in untranscribed sequences on either the 5′ or 3′ side of the gene (see **Figure 18.7**). Researchers have yet to find enhancers located in exons.

- Like promoter-proximal elements, many types of enhancers exist.

- Enhancers can work even if their normal 5′ → 3′ orientation is flipped.

- Most genes have more than one enhancer.

- Enhancers can work even if they are moved to a new location in the vicinity of the gene, on the same chromosome.

Enhancers are regulatory sequences unique to eukaryotes. When regulatory proteins bind to enhancers, transcription begins. Thus, enhancers are a gas pedal—an element in positive control.

In addition, eukaryotic genomes contain regulatory sequences that are similar in structure to enhancers but opposite in function. These sequences are **silencers**. When regulatory proteins

EXPERIMENT

QUESTION: Can a regulatory sequence in DNA be located far from the promoter?

HYPOTHESIS: A regulatory sequence exists far from the promoter—in fact, in the intron—of an antibody-producing gene.

NULL HYPOTHESIS: Regulatory sequences are located close to the promoter—not in introns or in other distant sequences.

EXPERIMENTAL SETUP:

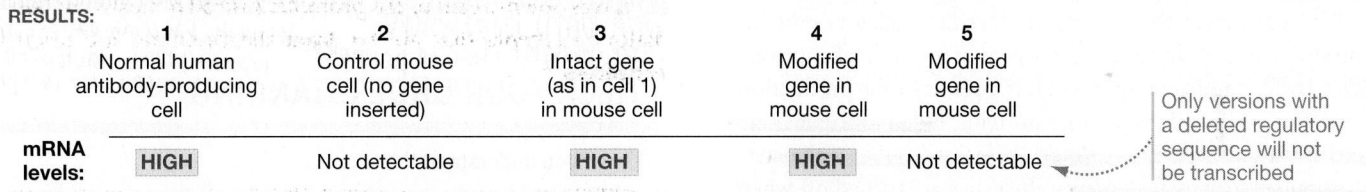

1. Isolate gene. Isolate a certain antibody-producing gene in humans that has a large intron.

2. Remove sections of intron. Use enzymes to cut out specific sections of the intron. (Several sections were cut out—only one example is shown.)

3. Reconnect fragments. Use an enzyme to put DNA fragments back together.

4. Look for transcription of gene. Introduce intact or modified genes into mouse cells. The various versions of the modified gene will be missing different sections of the intron.

PREDICTION: Transcription of antibody mRNA will be reduced when a certain part of the intron is deleted.

PREDICTION OF NULL HYPOTHESIS: Transcription of antibody mRNA will be unaffected by intron deletions.

RESULTS:

	1	2	3	4	5	
	Normal human antibody-producing cell	Control mouse cell (no gene inserted)	Intact gene (as in cell 1) in mouse cell	Modified gene in mouse cell	Modified gene in mouse cell	Only versions with a deleted regulatory sequence will not be transcribed
mRNA levels:	HIGH	Not detectable	HIGH	HIGH	Not detectable	

CONCLUSION: The deleted part of the intron must contain a regulatory sequence that is required for transcription. Thus, regulatory sequences in DNA may be located far from the promoter—in this case, in an intron.

FIGURE 18.6 Evidence That Enhancers Are Required for Transcription.

SOURCE: Hozumi, N., and S. Tonegawa. 1976. Evidence for somatic rearrangement of immunoglobulin genes coding for variable and constant regions. *Proceedings of the National Academy of Sciences, USA* 73: 3628–3632.

✔**QUESTION** Why did the researchers assess mRNA production in human cells that did *not* receive a modified or intact gene?

See answer, it's critical...

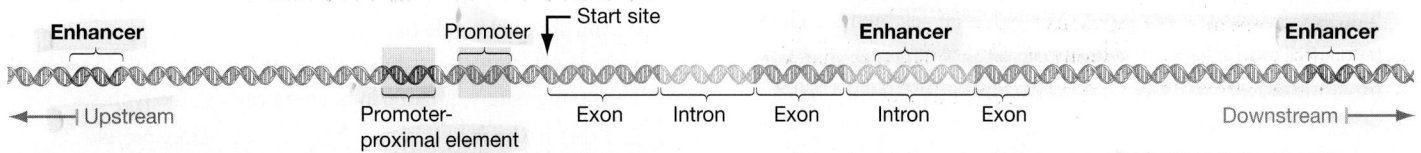

FIGURE 18.7 Enhancers Are Far from the Genes They Regulate. All of the colored sections of the DNA strand shown here (including the enhancers) are considered part of the same gene.

✔**EXERCISE** Compare and contrast the structure of this typical eukaryotic gene and the structure of a bacterial operon.

bind to silencers, transcription is shut down. Silencers are a brake—an element in negative control.

The discovery of enhancers and silencers expanded the catalog of regulatory sites known in organisms and inspired researchers to reconsider the nature of the gene. Biologists began defining the **gene** as the DNA that codes for a functional polypeptide or RNA molecule *and* the regulatory sequences required for expression.

The Role of Regulatory Proteins in Differential Gene Expression

Experiments that followed up on Tonegawa's work supported the hypothesis that enhancers are binding sites for proteins that regulate transcription. By analyzing mutant yeast, fruit flies, and roundworms that have defects in the expression of particular genes, biologists have identified a large number of regulatory proteins that bind to enhancers and silencers.

These results support one of the most general statements researchers are able to make about gene regulation in eukaryotes: In multicellular species, different types of cells express different genes because they have different histone modifications and contain different regulatory proteins. The regulatory proteins, in turn, are produced in response to signals that arrive from other cells early in embryonic development.

If a "become a muscle cell" signal arrives, for example, it triggers the production of regulatory proteins that are specific to muscle cells. Because the regulatory proteins bind to specific enhancers and silencers and promoter-proximal elements, they trigger the production of muscle-specific proteins. But if no "become a muscle cell" signal arrives, then no muscle-specific regulatory proteins are produced and no muscle-specific gene expression takes place.

Differential gene expression is a result of the production or activation of specific regulatory proteins. Eukaryotic genes are turned on when specific regulatory proteins bind to enhancers and promoter-proximal elements; the genes are turned off when regulatory proteins bind to silencers or when chromatin remains condensed. Distinctive regulatory proteins are what make a muscle cell a muscle cell and a bone cell a bone cell.

The Initiation Complex

Many questions remain about how transcription is initiated in eukaryotes. What is clear is that two broad classes of regulatory proteins interact with regulatory sequences at the start of transcription:

- **Regulatory transcription factors** are proteins that bind to enhancers, silencers, or promoter-proximal elements. These transcription factors are responsible for the expression of particular genes in particular cell types and at particular stages of development.

- **Basal transcription factors** interact with the promoter and are not restricted to particular cell types. Basal transcription factors must be present for transcription to occur, but they do not provide much in the way of regulation.

TBP, for example, is a basal transcription factor that is common to all genes. Other basal transcription factors are specific to promoters recognized by RNA polymerase I, II, or III. ✔If you understand this concept, you should be able to compare and contrast the regulatory and basal transcription factors found in muscle cells versus nerve cells.

In addition, proteins that make up the **mediator complex** have a role in starting transcription. The mediator complex does not bind to DNA. Instead, it creates a physical link between regulatory transcription factors and basal transcription factors.

Figure 18.8 summarizes a current model for how transcription is initiated in eukaryotes.

Step 1 Regulatory transcription factors bind to DNA and recruit chromatin-remodeling complexes and histone acetyl transferases (HATs).

Step 2 Once the chromatin-remodeling complexes and HATs are in place, they open a broad swath of chromatin that includes the promoter region.

Step 3 Other regulatory transcription factors bind to the newly exposed enhancers or promoter-proximal elements; basal transcription factors bind to the promoter. When mediator complexes connect the two, DNA has to loop.

Step 4 RNA polymerase II is recruited to the site, forming a multi-protein machine called the **basal transcription complex**. Transcription can begin.

✔If you understand this model, you should be able to explain why DNA forms loops near the promoter in order for transcription to begin.

[handwritten: regulatory region (silencers/enhancers) DNA forms loops when are brought close to promoter through binding of regulatory transcription factors to mediator complex]

CHECK YOUR UNDERSTANDING

If you understand that . . .

- Eukaryotic genes have regulatory sequences called promoter-proximal elements close to their promoters.
- Eukaryotic genes also have regulatory sequences called enhancers or silencers far from their promoters.
- Transcription initiation is a multistep process that begins when regulatory transcription factors bind to DNA and recruit proteins that open chromatin.
- Interactions between regulatory transcription factors and basal transcription factors result in the formation of the basal transcription complex and the arrival of RNA polymerase at the gene's start site.

✔ **You should be able to . . .**

1. Compare and contrast the nature of regulatory sequences and regulatory proteins in bacteria versus eukaryotes.

2. Explain why the presence of certain regulatory proteins could influence whether a cell became a muscle cell or a brain cell.

Answers are available in Appendix B.

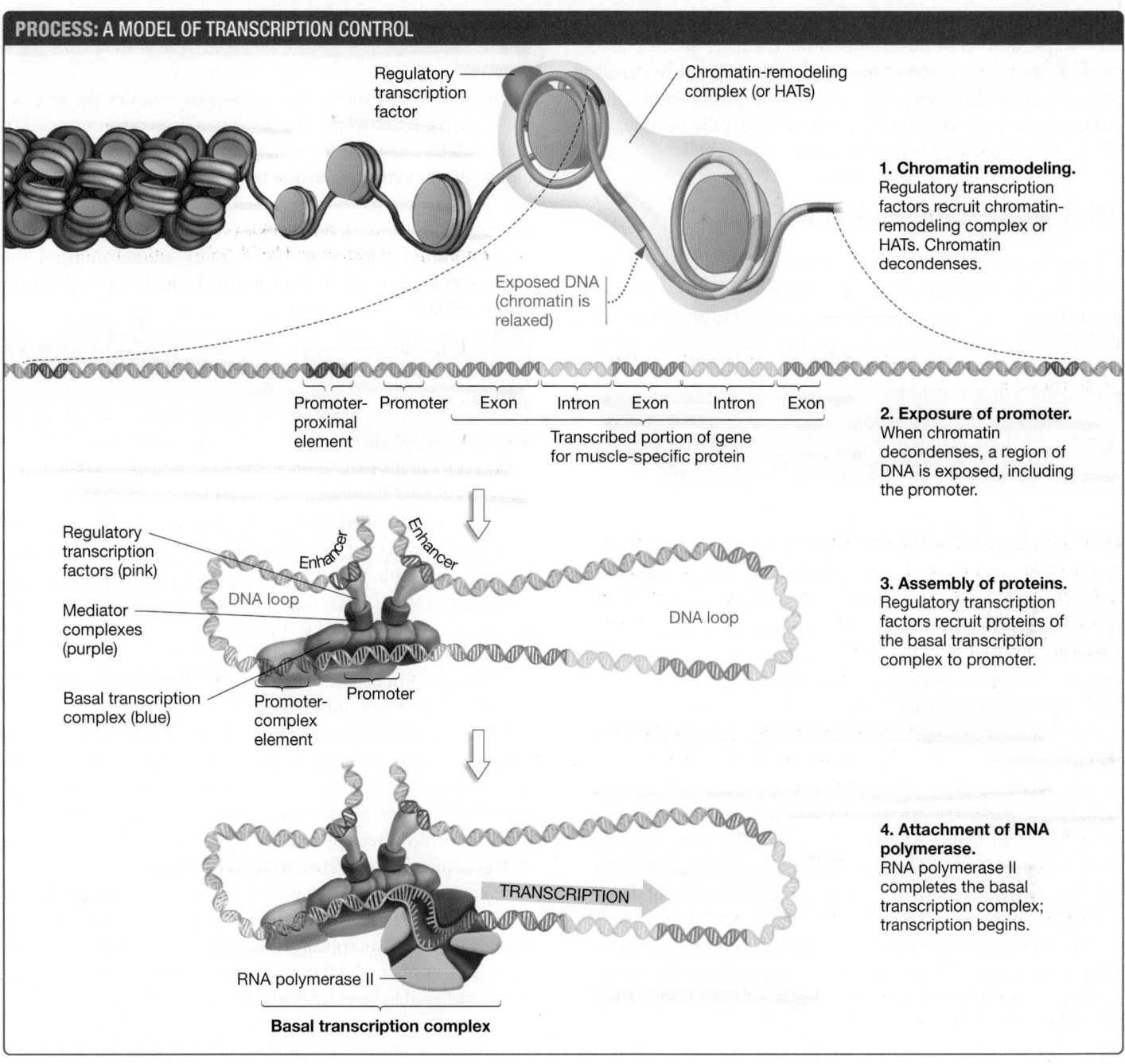

FIGURE 18.8 The Elements of Transcription Control in Eukaryotes. According to the currently accepted model, transcription is initiated through a series of steps that remodel chromatin and assemble the basal transcription complex that recruits RNA polymerase.

The assembly of the basal transcription complex depends on interactions among regulatory transcription factors that are bound to enhancers, silencers, and promoter-proximal elements. The result is a large, multimolecular machine that is positioned at the start site and able to start transcription. The completed basal transcription complex can contain as many as 60 proteins, including RNA polymerase.

Compared with what happens in bacteria, where just 3 to 5 proteins may interact at the promoter to initiate transcription, the process in eukaryotes is remarkably complicated. To review the many players involved and their interactions, go to the study area at *www.masteringbiology.com*.

(MB) Web Activity Transcription Initiation in Eukaryotes

Currently, biologists are working to understand exactly how regulatory and basal transcription factors interact to control formation of the basal transcription complex. This research is important, for transcription initiation lies at the heart of gene expression. Precise regulation of transcription is critical not only to the development of embryos but also to the daily life of eukaryotes. Right now, cells throughout your body are starting and stopping the transcription of specific genes in response to signals from nearby and distant cells. As the environment inside and outside your body continually changes, your cells continually change which genes are being transcribed.

18.4 Post-Transcriptional Control

Chromatin remodeling and transcription are just the start of the story of gene regulation. Once an mRNA is made, a series of events has to occur if the final product is going to affect the cell. Each of these events offers an opportunity to regulate gene expression, and each is used in some cells at least some of the time. The control mechanisms include (1) splicing mRNAs in various ways, (2) modifying the life span of mRNAs or altering the rate at which translation is initiated, and (3) activating or inactivating proteins after translation has occurred. Let's consider each in turn.

Alternative Splicing of mRNAs

Introns are spliced out in the nucleus as the primary RNA is transcribed. Recall from Chapter 16 that the RNA that results from splicing consists of sequences encoded by exons, it is protected by a cap on the 5′ end and a long poly(A) tail on the 3′ end, and the splicing is accomplished by the molecular machines called **spliceosomes**. What Chapter 16 did not mention, however, is that splicing provides an opportunity for the regulation of gene expression.

During splicing, changes in gene expression are possible because selected exons may be removed from the primary transcript along with the introns. As a result, the same primary RNA transcript can yield more than one kind of mature, processed mRNA, consisting of different combinations of transcribed exons.

This is important. If these mature mRNAs contain differences in their ribonucleotide sequence, then the polypeptides translated from them will likewise differ. Splicing the same primary RNA transcript in different ways to produce different mature mRNAs and thus different proteins is referred to as **alternative splicing**.

To see how alternative splicing works, consider the muscle-cell protein tropomyosin. The tropomyosin gene is expressed in both skeletal muscle cells and smooth muscle cells, which make up two distinct kinds of muscle tissue. Skeletal muscle is responsible for moving your bones; smooth muscle lines many parts of your gut and certain blood vessels.

As **Figure 18.9a** shows, the primary transcript from the tropomyosin gene contains 14 exons. In each type of muscle cell, a different subset of the 14 exons are spliced together to produce two different mRNAs (**Figure 18.9b**). As a result of alternative splicing, the tropomyosin proteins found in these two cell types are distinct. One of the many reasons skeletal muscle and smooth muscle are different is that they contain different types of tropomyosin.

Alternative splicing is controlled by proteins that bind to RNAs in the nucleus and interact with spliceosomes. When cells that are destined to become skeletal muscle or smooth muscle are developing, they receive signals leading to the production of specific proteins that are active in the regulation of splicing. Instead of transcribing different versions of the tropomyosin gene, the cells splice the same primary RNA transcript in different ways.

Before the importance of alternative splicing was widely appreciated, a gene was considered to be a nucleotide sequence that encodes one specific protein or RNA, along with its regulatory sequences. Based on this view, estimates for the number of genes in the human genome were typically in the range of 60,000 to 100,000. But once the complete human genome sequence became available, researchers realized that we may have as few as 20,000 sequences for primary mRNA transcripts.

Even though our genomes contain a relatively low number of such sequences, recent data indicate that over 90 percent of them undergo alternative splicing and produce multiple products. Thus, the number of different proteins that your cells can produce is believed to be at least 50,000.

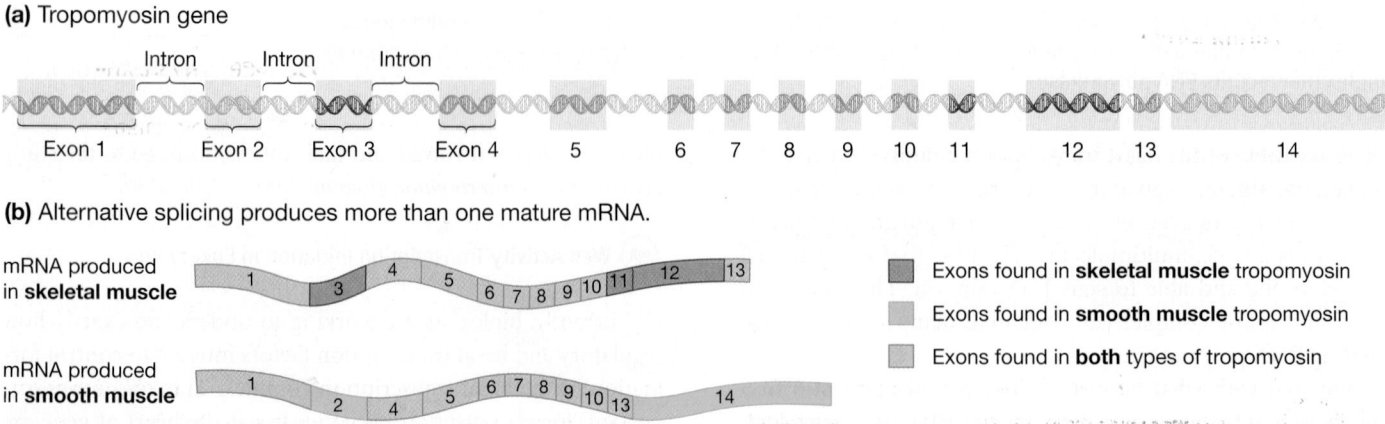

(a) Tropomyosin gene

| Intron | Intron | Intron |

Exon 1   Exon 2   Exon 3   Exon 4   5   6   7   8   9   10   11   12   13   14

(b) Alternative splicing produces more than one mature mRNA.

mRNA produced in **skeletal muscle**   1   3   4   5   6   7   8   9   10   11   12   13

mRNA produced in **smooth muscle**   1   2   4   5   6   7   8   9   10   13   14

Exons found in **skeletal muscle** tropomyosin
Exons found in **smooth muscle** tropomyosin
Exons found in **both** types of tropomyosin

FIGURE 18.9 Alternative Splicing Produces More than One Mature mRNA from the Same Gene.

Thanks to results like these, the definition of the gene is changing once again: Genes now have to be thought of as the coding and regulatory sequences that direct the production of one or more related polypeptides or RNAs.

The current record holder for the number of distinct mRNA sequences derived from one gene is the *Dscam* gene in the fruit fly *Drosophila melanogaster*. The products of this gene help to guide growing nerve cells within the embryo. Because the primary transcript is spliced into about 38,000 distinct forms of mRNA, the *Dscam* gene can produce thousands of different products.

Alternative splicing ranks as a major mechanism in the control of gene expression in multicellular eukaryotes. ✔If you understand alternative splicing, you should be able to explain why it does not occur in bacteria and describe where it occurs in Figure 18.1.

mRNA Stability and RNA Interference

Once splicing is complete and processed mRNAs are exported to the cytoplasm, new regulatory mechanisms come into play. For example, it has long been known that the life span of an mRNA in the cell can vary. The mRNA for casein—the major protein in milk—is produced in the mammary gland tissue of female mammals. Normally, many of these mRNAs persist in the cell for just an hour, and little casein protein is produced. But when a female mouse is lactating, regulatory molecules help the mRNAs persist almost 30 times longer—leading to a huge increase in the production of casein. In this instance, mRNA stability is associated with changes in the length of the poly(A) tail.

In many cases, the life span of an mRNA is controlled by tiny, single-stranded RNA molecules that bind to complementary sequences in the mRNA. Once part of an mRNA becomes double stranded in this way, specific proteins degrade the mRNA or prevent it from being translated into a polypeptide. This phenomenon is known as **RNA interference**. How does it work?

Figure 18.10 walks through the sequence of events.

Step 1 RNA interference begins when RNA polymerase transcribes DNA sequences that code for an unusual product—a small RNA molecule that doubles back on itself to form a hairpin. Hairpin formation occurs because pairs of sequences within the RNA transcript are complementary.

Step 2 Some of the RNA is trimmed by enzymes in the nucleus; then the double-stranded segment that remains is exported to the cytoplasm.

Step 3 In the cytoplasm, the double-stranded RNA sequence is cut by another enzyme into molecules that are typically about 22 nucleotides long.

Step 4 One of the strands from this short RNA is taken up by a group of proteins called the RNA-induced silencing complex, or RISC. The RNA strand held by the RISC is a **microRNA (miRNA)**, or a small interfering RNA (siRNA)

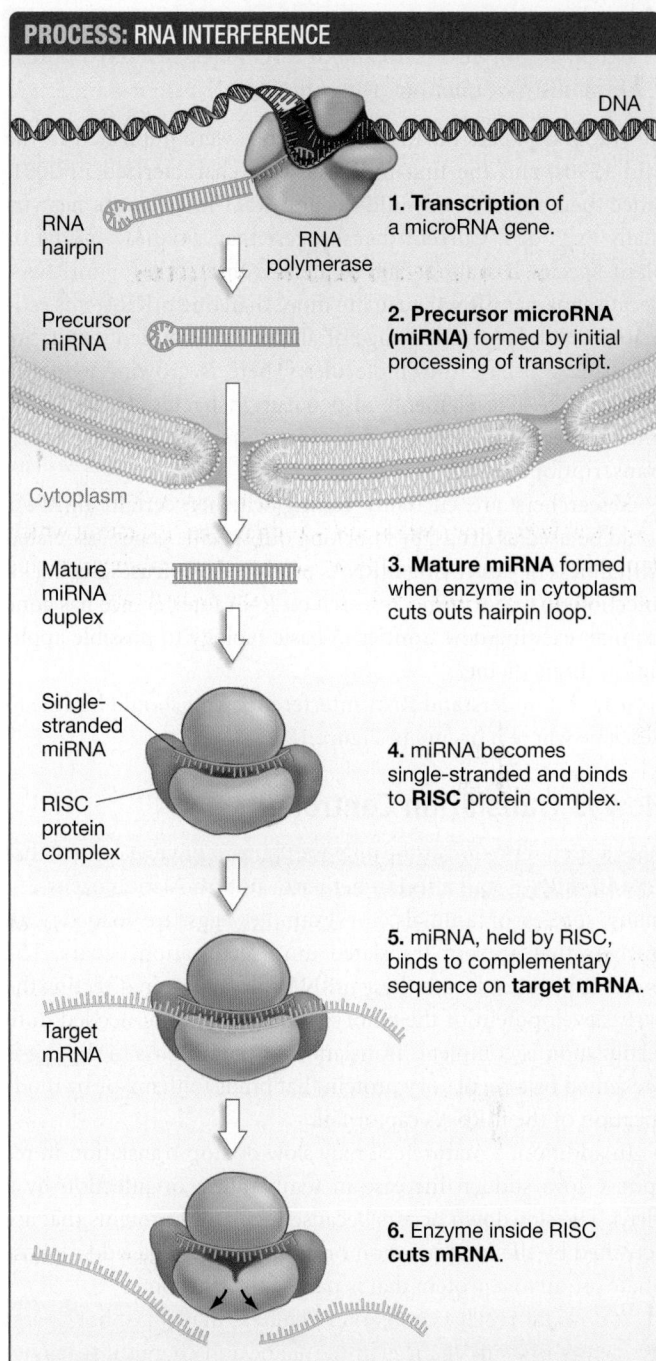

PROCESS: RNA INTERFERENCE

DNA

RNA hairpin

RNA polymerase

1. Transcription of a microRNA gene.

Precursor miRNA

2. Precursor microRNA (miRNA) formed by initial processing of transcript.

Cytoplasm

Mature miRNA duplex

3. Mature miRNA formed when enzyme in cytoplasm cuts out hairpin loop.

Single-stranded miRNA

RISC protein complex

4. miRNA becomes single-stranded and binds to **RISC** protein complex.

5. miRNA, held by RISC, binds to complementary sequence on **target mRNA**.

Target mRNA

6. Enzyme inside RISC **cuts mRNA**.

FIGURE 18.10 MicroRNAs Target Certain mRNAs for Destruction. In essence, miRNAs bind to mRNAs by complementary base pairing and target them for destruction by the RISC protein complex. The steps required to process and activate an miRNA are important because they allow the production of miRNAs to be carefully controlled.

Step 5 Once it is part of a RISC, the miRNA binds to its complementary sequences in a target mRNA.

Step 6 If the match between an miRNA and an mRNA is perfect, an enzyme in the RISC destroys the mRNA by cutting it in two. In effect, tight binding by an miRNA is a "kiss of death" for the mRNA. If the match isn't perfect, however, the mRNA is not destroyed. Instead, its translation is inhibited. Either way, miRNAs "interfere" with mRNAs.

The first papers on RNA interference were published in the mid-1990s, and the first miRNAs were characterized in 2001. Since then, research on miRNAs and RNA interference has virtually exploded. Current data suggest that a typical animal or plant species has about 500 sequences that code for miRNAs. Because many miRNAs regulate more than one mRNA, it is estimated that a large percentage of all animal and plant genes are regulated by these tiny molecules. There is growing evidence that miRNA-like elements also occur in bacterial cells. RNA interference is increasingly recognized as a key aspect of post-transcriptional control.

Researchers are currently testing whether certain miRNAs could be used as drugs, by knocking out specific genes associated with illness or destroying mRNAs produced by viruses during an infection. In a short time, research on RNA interference has gone from an exciting new frontier in basic biology to possible applications in medicine.

✔ If you understand RNA interference, you should be able to describe where it occurs in Figure 18.1.

How Is Translation Controlled?

RNA interference is not the only mechanism of gene control that acts on mRNAs and affects whether or not translation occurs. In many species of animals, for example, eggs are loaded with mRNAs that are not translated until fertilization occurs. The proteins produced from these mRNAs play a part in directing the early development of the embryo, but they are not needed until fertilization is complete. Translation of the mRNAs in the egg is prevented by a regulatory protein that binds to them, or by modification of the mRNA's cap or tail.

In addition, a mature cell may slow or stop translation in response to a sudden increase in temperature or infection by a virus. The slowdown occurs because regulatory proteins that are activated by the viral invasion or temperature spike add a phosphate group to a protein that is part of the ribosome.

You might recall from earlier chapters that phosphorylation frequently leads to changes in the shape and chemical reactivity of proteins. In the case of the phosphorylated ribosomal protein, the shape change slows or prevents translation.

For the cell, this dramatic change in gene expression can mean the difference between life and death. High temperatures disrupt protein folding, so shutting down translation prevents the production of improperly folded polypeptides. If the insult is an invading virus, the cell stops the infection because it avoids manufacturing viral proteins.

Mechanisms like these are a reminder that gene expression can be regulated at multiple points: at the level of chromatin structure, transcription initiation, RNA processing, mRNA availability, and translation rate. But that's not all. Let's look now at the last level possible: Altering protein activity, after translation is complete.

Post-Translational Control

Chapter 17 explained that in bacteria, mechanisms of post-translational regulation are important because they allow cells to respond to new conditions rapidly. The same is true for eukaryotes. Instead of waiting for transcription, RNA processing, and translation to occur, the cell can respond to altered conditions by quickly activating or inactivating existing proteins.

There is a trade-off, however: speed is gained at the expense of efficiency. Transcription, RNA processing, and translation use up energy and materials; it is wasteful to produce proteins that won't be used.

You have already encountered several important mechanisms of post-translational control over gene expression.

- Proteins are folded into their final, active conformation by chaperone proteins (see Chapter 16). Folding is required for proteins to function normally, and folding is regulated by the presence of chaperones.

- Enzymes may modify proteins by adding carbohydrate groups (see Chapter 7) or cleaving off certain amino acids.

- Phosphorylation is an extremely common mechanism for activating or deactivating proteins. You might recall that Chapter 11 featured the activation of cyclin-Cdk complexes by phosphorylation and the subsequent entry into M phase of the cell cycle.

There is yet another key mechanism of post-translational control, however: the targeted destruction of proteins. Chapter 11 introduced this phenomenon by describing the short life span of cyclin proteins. Once a cell is well into M phase, an enzyme begins adding a small polypeptide called ubiquitin to the cyclin proteins. Ubiquitin got its name because it is ubiquitous in cells. A multi-molecular machine called the **proteasome** recognizes proteins that have a ubiquitin tag, and cuts them into short segments. In this way, ubiquitinization is a mechanism for controlling the life span of proteins in a cell.

CHECK YOUR UNDERSTANDING

⊶ If you understand that . . .

- Alternative splicing allows a single gene to code for many products.
- RNA interference is one of several mechanisms for controlling an mRNA's life span and translation rate.
- Ubiquitin-tagging and destruction by proteasomes is one of several mechanisms for controlling a protein's life span or activity.

✔ You should be able to . . .

1. Explain why the discovery of alternative splicing forced biologists to change their definition of the gene.

2. Explain why "RNA interference" is aptly named.

Answers are available in Appendix B.

18.5 How Does Gene Expression in Bacteria Compare with That in Eukaryotes?

Biologists have been studying the control of gene expression for over 50 years. Almost as soon as they knew that information in DNA is transcribed into RNA and then translated into proteins, researchers began asking questions about how that flow of information is regulated. **Table 18.1** summarizes what biologists have learned over the past half century about how bacterial and eukaryotic gene expression is controlled—organized by the six steps in gene expression introduced in Figure 18.1.

How does the regulation of gene expression differ in bacteria and eukaryotes? Biologists point to four fundamental contrasts, two of which involve levels of control that exist in eukaryotes but not in bacteria:

1. *Packaging* The chromatin structure of eukaryotic DNA must be opened in order for TBP, the basal transcription complex, and RNA polymerase to gain access to genes and initiate transcription. A key insight here is that, because eukaryotic DNA is packaged so tightly, the default state of transcription in eukaryotes is the "off" state. In contrast, the default state of transcription in bacteria, which lack histone proteins and have freely accessible promoters, is "on." Chromatin structure provides a mechanism of negative control that does not exist in bacteria.

2. *Alternative splicing* Prior to translation, primary transcripts in eukaryotes must be spliced—an occurrence that is extremely rare in bacteria. The one-to-one correspondence between the number of genes and the number of gene products observed in bacteria is not seen in eukaryotes. Instead, each eukaryotic gene may code for one to thousands of distinct products.

3. *Complexity* Transcriptional control is much more complex in eukaryotes than in bacteria. The function of sigma proteins in bacteria is analogous to the role of the basal transcription complex in eukaryotes. Likewise, the function of CAP, the repressor, and other regulatory proteins is analogous to the role of regulatory transcription factors in eukaryotes. But the sheer number of eukaryotic proteins involved in regulating transcription dwarfs that in bacteria, as does the complexity of their interactions.

4. *Coordinated expression* In bacteria, genes that take part in the same cellular response are organized into operons controlled by a single promoter. Because their mRNAs are translated together, several proteins are produced in a coordinated fashion. In contrast, operons are rare in eukaryotes. Eukaryotic genes that are physically scattered can be expressed at the same time because a single set of regulatory transcription factors can trigger the transcription of several genes. For example, muscle-specific genes found on several different chromosomes can be transcribed in response to the same muscle-specific regulatory transcription factor. Recent data also suggest that in some cases, genes on different chromosomes may be physically associated inside the nucleus, and share regulatory elements. These are the means by which eukaryotes coordinate the expression of functionally related genes.

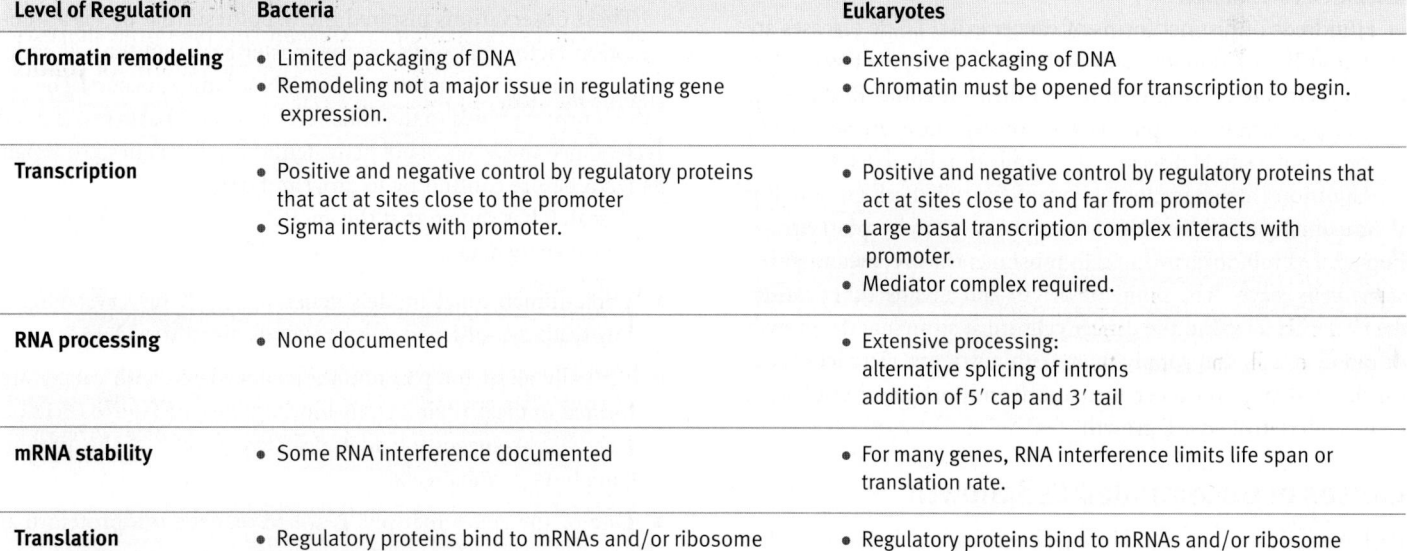

SUMMARY TABLE 18.1 **Regulating Gene Expression in Bacteria and Eukaryotes**

Level of Regulation	Bacteria	Eukaryotes
Chromatin remodeling	• Limited packaging of DNA • Remodeling not a major issue in regulating gene expression.	• Extensive packaging of DNA • Chromatin must be opened for transcription to begin.
Transcription	• Positive and negative control by regulatory proteins that act at sites close to the promoter • Sigma interacts with promoter.	• Positive and negative control by regulatory proteins that act at sites close to and far from promoter • Large basal transcription complex interacts with promoter. • Mediator complex required.
RNA processing	• None documented	• Extensive processing: alternative splicing of introns addition of 5' cap and 3' tail
mRNA stability	• Some RNA interference documented	• For many genes, RNA interference limits life span or translation rate.
Translation	• Regulatory proteins bind to mRNAs and/or ribosome and affect translation rate.	• Regulatory proteins bind to mRNAs and/or ribosome and affect translation rate.
Post-translational modification	• Folding by chaperone proteins • Chemical modification (e.g., phosphorylation) may change activity.	• Folding by chaperone proteins • Chemical modification (glycosylation, phosphorylation) • Ubiquination targets proteins for destruction by proteasome.

To date, biologists do not have a good explanation for why gene expression is so much more complex in unicellular eukaryotes than it is in bacteria. All unicellular organisms have to respond to environmental changes in an appropriate way. So why do unicellular algae, yeasts, and other eukaryotes have ways to regulate gene expression that bacteria lack? After decades of research, the answer is still not clear.

It is easier to generate a hypothesis to explain why gene expression is complex in multicellular eukaryotes. In these organisms, cells have to differentiate as an individual develops. Changes in gene expression are responsible for the differentiation of muscle cells, bone cells, leaf cells, flower cells, and so on in response to signals from other cells. The need for each cell type to have a unique pattern of gene expression may explain why control of gene expression is so much more complex in multicellular eukaryotes than in bacteria.

The effort to understand how developmental signals produce cell-specific gene expression in multicellular organisms represents one of two great frontiers in gene-expression research. The other major frontier is the quest to understand how certain defects in gene regulation result in uncontrolled cell growth and the suite of diseases called cancer.

18.6 Linking Cancer with Defects in Gene Regulation

Normal regulation of gene expression results in the orderly development of an embryo and, in juveniles and adults, appropriate responses to environmental change. Abnormal regulation of gene expression, in contrast, can lead to developmental abnormalities and diseases such as cancer.

Hundreds of distinct forms of cancer exist. These diseases are enormously varied in terms of their initial cause, the tissues they affect, their rate of progression, and their outcome. Because the underlying defects, symptoms, and consequences are so diverse, cancer is not a single disease but a family of related diseases.

All cancers are characterized by uncontrolled cell growth. But for most cancers to become dangerous, two other events are required: The rapidly growing cells must metastasize, meaning that some cells leave their point of origin and invade other tissues (see Chapter 11), and the cancer cells must stimulate the growth of blood vessels that supply them with nutrients. Here let's focus on the first step in cancer formation, and the question of what causes uncontrolled cell growth.

Causes of Uncontrolled Cell Growth

Each type of cancer is caused by a different set of genetic defects that lead to uncontrolled cell growth. ⬤➥ Many cancers are associated with mutations in regulatory transcription factors. These mutations lead to cancer when they affect one of two classes of genes: (1) genes that stop or slow the cell cycle, and (2) genes that trigger cell growth and division by initiating specific phases in the cell cycle.

Genes that stop or slow the cell cycle are **tumor suppressor** genes. The products of these genes prevent the cell cycle from progressing unless specific signals indicate that conditions are right for moving forward with mitosis and cell division. If a mutation disrupts normal function of a tumor suppressor gene, then a key brake on the cell cycle is eliminated.

Genes that encourage cell growth by triggering specific phases in the cell cycle are called **proto-oncogenes** (literally, "first-cancer-genes"). In normal cells, proto-oncogenes are required to initiate each phase in the cell cycle. They are active only when conditions are appropriate for growth, however. In cancerous cells, defects in the regulation of proto-oncogenes cause these genes to stimulate growth at all times. In such cases, a mutation has converted the proto-oncogene into an **oncogene**—an allele that promotes cancer development.

p53: A Case Study

To gain a deeper understanding of how defects in gene expression can lead to cancer, consider research on the gene that is most often defective in human cancers. The gene is called **p53** because the protein it codes for has a molecular weight of approximately 53 kilodaltons. Sequencing studies have revealed that mutant, nonfunctional forms of **p53** are found in over half of all human cancers. The **p53** gene codes for a regulatory transcription factor.

Researchers began to understand what **p53** does when they exposed normal, noncancerous human cells to UV radiation and noticed that levels of p53 protein increased markedly. Recall from Chapter 14 that UV radiation damages DNA. Follow-up studies confirmed that there is a close correlation between DNA damage and the amount of p53 in a cell. In addition, analyses of the protein's primary structure suggested that it might contain a region that binds to DNA.

These observations inspired the hypothesis that p53 is a transcription factor that serves as the master brake on the cell cycle. In this model, p53 is activated after DNA damage occurs. The activated protein binds to the enhancers of genes that arrest the cell cycle. Once these genes are activated, the cell has time to repair its DNA before continuing to grow and divide.

Research has shown that this model of p53 function is correct in almost every detail.

- Three-dimensional models generated by X-ray crystallography studies confirmed that p53 binds directly to DNA.

- Virtually all of the p53 mutations associated with cancer are located in the protein's DNA-binding site (see **Figure 18.11** on page 333)—suggesting that defective forms of the protein can't bind to enhancers.

- One of the genes that p53 regulates codes for a protein that prevents cell-cycle regulatory proteins from triggering M (mitosis) phase.

Experiments have also shown that when a cell's DNA is extensively damaged and cannot be repaired, p53 activates the transcription of genes that cause the cell to take its own life by

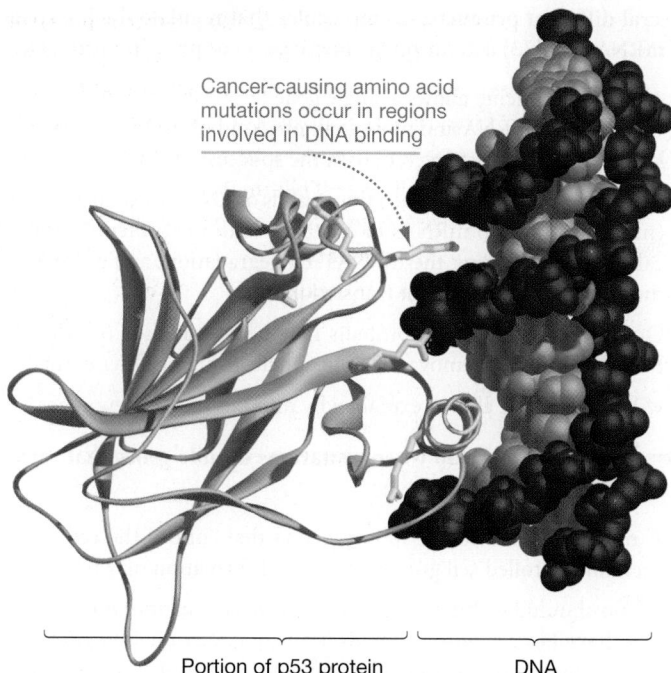

Cancer-causing amino acid mutations occur in regions involved in DNA binding

Portion of p53 protein   DNA

FIGURE 18.11 p53 Is a Transcription Factor that Serves as a Master Brake on the Cell Cycle When DNA Is Damaged. In cancer patients, the amino acids highlighted in yellow often differ from those in the normal protein, preventing p53 from binding to DNA.

apoptosis (see Chapter 11). If mutations in the *p53* gene make the protein product inactive, then damaged cells are not shut down or killed but instead continue to move through the cell cycle. They are likely to contain many mutations, however, because of the damage they have sustained to their DNA. If these

mutations create oncogenes, the cells have taken a key step on the road to cancer. The p53 protein is like a quality control officer. If it is missing, things can go downhill.

The role of *p53* in preventing cancer is so fundamental that biologists call this gene "the guardian of the genome." Currently, research is forging ahead on two fronts: Biologists are striving to (**1**) identify more of the genes that are regulated by the p53 protein, and (**2**) find molecules that could act as anticancer drugs by mimicking p53's shape and activity.

CHECK YOUR UNDERSTANDING

If you understand that . . .

- Cancer is associated with mutations that lead to loss of control over the cell cycle.
- Uncontrolled cell growth may result when a mutation in a regulatory gene creates a protein that activates the cell cycle constitutively.
- Uncontrolled cell growth may result when a mutation prevents a tumor suppressor gene product from shutting down the cell cycle in damaged cells.

✔ **You should be able to . . .**

1. Explain why cancer has a common pattern (uncontrolled cell growth), but not a common cause. Your answer should refer to the six levels of gene regulation outlined in Figure 18.1.

2. Explain why loss-of-function mutations in *p53* are observed in so many cancers.

Answers are available in Appendix B.

CHAPTER 18   REVIEW

For media, go to the study area at www.masteringbiology.com

Summary of Key Concepts

Changes in gene expression allow eukaryotic cells to respond to changes in the environment and cause distinct cell types to develop.

- In a multicellular eukaryote, cells are different not because they have different genes but because they express different genes.

- Gene expression is regulated at six distinct levels: Chromatin has to be remodeled, the transcription of specific genes may be initiated or repressed, mRNAs may be spliced in different ways to produce a different product, the life span of specific mRNAs may be extended or shortened, translation rate may be increased or decreased, and the life span or activity of particular proteins may be altered.

✔ You should be able to describe how the presence of the nuclear envelope, and the physical separation of transcription and translation, influence the levels of gene regulation observed in eukaryotes versus bacteria and archaea.

Eukaryotic DNA is packaged with proteins into structures that must be opened before transcription can occur.

- Eukaryotic DNA is wrapped around histone proteins to form bead-like nucleosomes that are then coiled into 30-nm fibers and higher-order chromatin structures.

- Transcription cannot be initiated until the interaction between DNA and histones in chromatin is relaxed.

- The state of chromatin depends on the acetylation or methylation of histones and the action of molecular machines called chromatin-remodeling complexes.

✔ You should be able to explain why chromatin remodeling has to be the first step in gene activation.

In eukaryotes, transcription is triggered by regulatory proteins that bind to the promoter and to sequences close to and far from the promoter.

- Regulatory transcription factors are proteins that bind to regulatory sequences called (1) enhancers and silencers, which are often located at a distance from the gene in question, or (2) promoter-proximal sequences, near the start of the coding sequence.
- The first regulatory transcription factors that bind to DNA recruit proteins that loosen the histones' grip on the gene, making the promoter accessible to basal transcription factors.
- Interactions between regulatory and basal transcription factors lead to the formation of the basal transcription complex.
- Once the basal transcription complex is assembled, RNA polymerase is recruited to the site and transcription begins.

   ✔ You should be able to draw a model of a eukaryotic gene undergoing transcription. Label enhancers, promoter-proximal elements, the promoter, regulatory transcription factors, basal transcription factors, and RNA polymerase.

   (MB) **Web Activity** Transcription Initiation in Eukaryotes

Once transcription is complete, gene expression is controlled by (1) alternative splicing, which allows a single gene to code for several different products; (2) molecules that regulate the life span of mRNAs; and (3) activation or inactivation of protein products.

- Alternative splicing allows a single gene to produce more than one version of an mRNA and more than one kind of protein. It is regulated by proteins that interact with the spliceosome.
- RNA interference occurs when tiny strands of RNA, called microRNAs (miRNAs), bind to mRNAs in company with the protein complex called RISC, marking the mRNAs for degradation, and also when the miRNAs merely inhibit translation.
- Once translation occurs, proteins may be activated or inactivated by the addition or removal of a phosphate group or other events.

   ✔ You should be able to explain why humans have so few genes.

Cancer can develop when mutations disable genes that regulate cell-cycle control genes.

- If mutations alter transcription factors that control the cell cycle, then uncontrolled cell growth and tumor formation may result.

   ✔ You should be able to explain why cancer is common in people who have been exposed to high levels of radiation, and why it is more common in older people than younger people.

Questions

1. What is chromatin?
   a. the protein core of the nucleosome, which consists of histones
   b. the 30-nm fiber
   c. the DNA-protein complex found in eukaryotes
   d. the histone *and* non-histone proteins in eukaryotic nuclei

2. What is a tumor suppressor?
   a. a gene associated with tumor formation when its product does not function
   b. a gene associated with tumor formation when its product functions normally
   c. a gene that accelerates the cell cycle and leads to uncontrolled cell growth
   d. a gene that codes for a transcription factor involved in tumor formation

3. Which of the following statements about enhancers is correct?
   a. They contain a unique base sequence called a TATA box.
   b. They are located only in 5'-flanking regions.
   c. They are located only in introns.
   d. They are found in a variety of locations and are functional in any orientation.

4. In eukaryotes, why are certain genes expressed only in certain types of cells?
   a. Different cell types contain different genes.
   b. Different cell types have the same genes but different promoters.
   c. Different cell types have the same genes but different enhancers.
   d. Different cell types have different regulatory transcription factors.

5. What is alternative splicing?
   a. the phosphorylation events that lead to different types of post-translational regulation
   b. mRNA processing events that lead to different combinations of exons being spliced together
   c. folding events that lead to proteins with alternative conformations
   d. action by regulatory proteins that leads to changes in the life span of an mRNA

6. What types of proteins bind to promoter-proximal elements?
   a. the basal transcription complex
   b. the basal transcription complex plus RNA polymerase
   c. basal transcription factors
   d. regulatory transcription factors

1. Compare and contrast (a) enhancers and the CAP site; (b) promoter-proximal elements and the *lac* operon operator; and (c) basal transcription factors and sigma.

2. Explain how alternative splicing could play a role in changing eukaryotic gene expression in response to changes in the environment.

3. Compare and contrast (a) enhancers and silencers; (b) promoter-proximal elements and enhancers; and (c) transcription factors and the mediator complex.

4. Explain the relationship between complementary base pairing and RNA interference.

5. Explain the concept of the histone code. Your answer should compare and contrast the structure of chromatin in muscle cells versus nerve cells.

6. Explain why mutations in *p53* can lead to loss of control over the cell cycle and to the development of cancer.

1. Histone proteins have been extremely highly conserved during evolution. The histones found in fruit flies and humans, for example, are nearly identical in amino acid sequence. Offer an explanation for this observation. (Hint: What are the consequences of a mutation in a histone?)

2. Cancers are most common in tissues where cell division is common, such as blood cells and cells in the lining of the lungs or gut. Why is this observation logical?

3. Levels of p53 protein in the cytoplasm increase after DNA damage. Design an experiment to determine whether this increase is due to increased transcription of the *p53* gene or to activation of preexisting p53 proteins by a post-translational mechanism such as phosphorylation.

4. Suggest a way that miRNAs that are complementary to viral RNAs could be useful as drugs.

THE BIG PICTURE

Copying, using, and transmitting genetic information is fundamental to life. Cells use the genetic information archived in their DNA to respond to changes in the environment and, in multicellular organisms, to develop into specific cell types.

Hereditary information is transmitted to offspring with random changes called mutation.

Thus, genetic information is dynamic—both within generations and between generations.

Note that each box in the concept map indicates the chapter and section where you can go for review. Also, be sure to do the blue exercises in the Check Your Understanding box below.

GENETIC INFORMATION

is archived in base sequences of

DNA 4.2

is packaged with proteins to form

Text section where you can find more information

consists of functional units called

Genotype 13.2 ← make up — **Genes** 15.1

have different versions called

can be

EXPRESSED 15.2
17.1–4
18.1–4

may regulate whether genes

if first TRANSCRIBED by

RNA polymerase
16.1

to form

RNA 4.3

may be processed by

may function directly in cell as

- Splicing
- Addition of 5′ cap
- Addition of poly(A) tail 16.2

- tRNA (transfer RNA) 16.4
- rRNA (ribosomal RNA) 16.5

to form

mRNA (messenger RNA) 16.2

is then TRANSLATED by

affect

Ribosomes 16.5

to form

Proteins 3.2
16.5

changed by

Phenotype 13.1 ← produce

- Folding 3.4
- Glycosylation 5.3
- Phosporylation 9.1
- Degradation 18.4

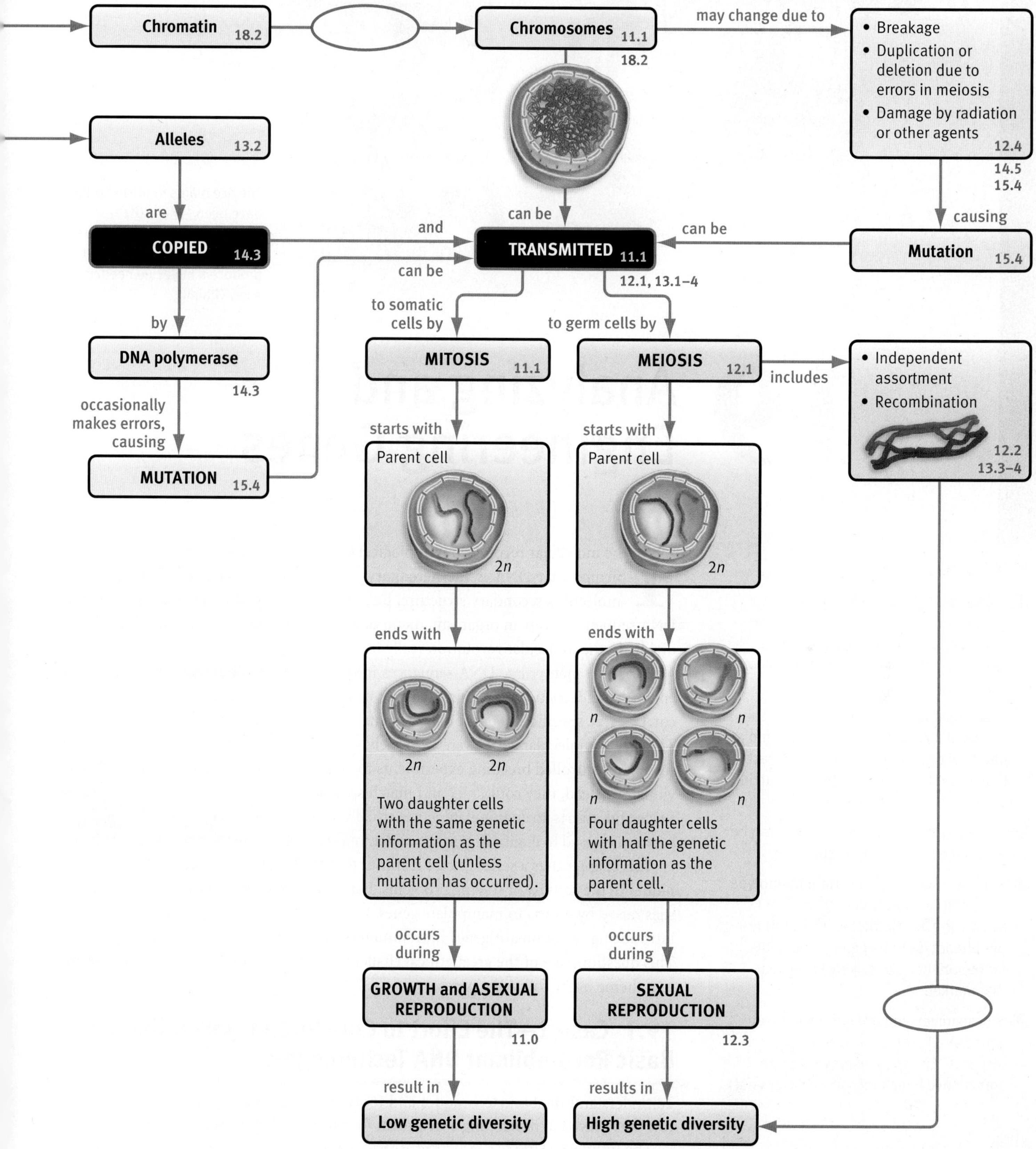

Chromatin	18.2		Chromosomes	11.1

18.2

may change due to

- Breakage
- Duplication or deletion due to errors in meiosis
- Damage by radiation or other agents 12.4
  14.5
  15.4

Alleles	13.2

are

COPIED 14.3

and

TRANSMITTED 11.1

12.1, 13.1–4

can be

can be

Mutation 15.4

causing

by

DNA polymerase

14.3

occasionally makes errors, causing

MUTATION 15.4

to somatic cells by

MITOSIS 11.1

to germ cells by

MEIOSIS 12.1

includes

- Independent assortment
- Recombination

12.2
13.3–4

starts with

Parent cell

2n

starts with

Parent cell

2n

ends with

2n   2n

Two daughter cells with the same genetic information as the parent cell (unless mutation has occurred).

ends with

n   n

n   n

Four daughter cells with half the genetic information as the parent cell.

occurs during

GROWTH and ASEXUAL REPRODUCTION 11.0

occurs during

SEXUAL REPRODUCTION 12.3

result in

Low genetic diversity

results in

High genetic diversity

The rice plants in these bottles have been genetically engineered—using techniques introduced in this chapter—to produce a molecule needed for a key vitamin.

19 Analyzing and Engineering Genes

KEY CONCEPTS

- Enzymes that cut DNA at specific locations and other enzymes that piece DNA segments back together allow biologists to move genes from one place to another.

- Biologists can obtain many identical copies of a gene by (1) inserting it into a bacterial cell that copies the gene each time the cell divides or (2) by conducting a polymerase chain reaction.

- The sequence of bases in a gene can be determined by dideoxy sequencing.

- If individuals with a certain phenotype also tend to share a genetic marker (a known site in DNA that is unrelated to the phenotype), the gene responsible for the phenotype is likely to be near that marker.

- Researchers are attempting to insert genes into humans to cure genetic diseases. Efforts to insert genes into plants have been much more successful.

The molecular revolution in biological science got its start when researchers confirmed that DNA is the hereditary material and succeeded in describing the molecule's secondary structure. But when biologists discovered how to remove DNA sequences from an organism, manipulate them, and insert them into different individuals, the molecular revolution really took off.

Efforts to manipulate DNA sequences in organisms are often referred to as genetic engineering. Genetic engineering became possible with the discovery of enzymes that cut DNA at specific sites and of other enzymes that paste DNA sequences together. These new molecular tools were extremely powerful. Biologists no longer had to rely solely on controlled breeding experiments to change the genetic characteristics of individuals. Instead, they could mix and match specific DNA sequences in the lab. Because successful efforts to manipulate genes usually result in novel combinations of DNA, techniques used to engineer genes are often referred to as **recombinant DNA technology**.

This chapter uses a series of case histories to introduce basic molecular biology techniques in the context of solving problems. It also considers the ethical and economic issues raised by efforts to manipulate genes. What are the potential perils and benefits of introducing recombinant genes into human beings, food plants, and other organisms? This question, one of the great ethical challenges of the twenty-first century, is a recurrent theme in the following pages.

19.1 Case 1—The Effort to Cure Pituitary Dwarfism: Basic Recombinant DNA Technologies

To understand the basic techniques and tools of genetic engineering, let's consider the role they played in developing a treatment for pituitary dwarfism in humans.

The pituitary gland is a structure at the base of the mammalian brain that produces several important biomolecules, including a protein that stimulates growth. This

✔ When you see this checkmark, stop and test yourself. Answers are available in Appendix B.

protein, which was found to be just 191 amino acids long, was named human growth hormone (HGH). In humans, the gene that codes for it is called *GH1*.

The discovery of growth hormone led researchers immediately to suspect that at least some forms of inherited dwarfism might be caused by a defect in the *GH1* gene. This hypothesis was confirmed by studies showing that people with certain types of dwarfism produce little growth hormone or none at all. These people have defective copies of *GH1* and exhibit pituitary dwarfism, type I (**Figure 19.1a**).

By studying the pedigrees of families in which dwarfism was common, several teams of researchers established that pituitary dwarfism, type I, is an autosomal recessive trait (see Chapter 13). In other words, affected individuals have two copies of the defective allele. Individuals who are affected by pituitary dwarfism have normal body proportions but grow more slowly than average people, reach puberty from two to ten years later than average, and are short in stature as adults—typically no more than 120 cm (4 feet) tall (**Figure 19.1b**).

Why Did Early Efforts to Treat the Disease Fail?

Once the molecular basis of pituitary dwarfism was understood, physicians began treating the disease with injections of naturally produced growth hormone. This approach was inspired by the spectacular success that had been achieved in treating type I diabetes mellitus. Diabetes mellitus is caused by a deficiency of the peptide hormone insulin, and clinicians had been able to alleviate the disease's symptoms by injecting patients with insulin from pigs.

Early trials showed that people with pituitary dwarfism could be treated successfully with growth hormone therapy, but only if the protein came from humans. Growth hormones isolated from pigs, cows, or other animals were ineffective. Until the 1980s, however, the only source of human growth hormone was pituitary glands dissected from human cadavers. As a result, the drug was extremely scarce and expensive.

Meeting demand turned out to be the least of the problems with growth hormone therapy, however. To understand why, recall from Chapter 3 that infectious proteins called prions can cause degenerative brain disorders in mammals. When some of the children treated with human growth hormone developed a prion disease in their teens and twenties, physicians realized that the supply of growth hormone was contaminated with a prion protein from the brains of the cadavers supplying the hormone. In 1984, the use of growth hormone isolated from cadavers was banned.

Steps in Engineering a Safe Supply of Growth Hormone

To replace natural sources of growth hormone, researchers turned to genetic engineering. Their plan was to insert fully functional copies of human *GH1* into the bacterium *Escherichia coli*, which they hoped would then produce huge quantities of recombinant progeny. If the plan worked, the recombinant cells would produce uncontaminated growth hormone in sufficient quantities to meet demand at an affordable price.

The plan required investigators to find *GH1*, obtain many copies of the gene, and insert them into *E. coli* cells. Their ability to do these things hinged on three of the most basic tools in molecular biology. Let's consider each in turn.

USING REVERSE TRANSCRIPTASE TO PRODUCE cDNAs Chapter 15 mentioned that an enzyme called reverse transcriptase is responsible for a major exception to the central dogma of molecular biology: It allows information to flow from RNA to DNA. More specifically, reverse transcriptase catalyzes the synthesis of DNA from an RNA template.

DNA that is produced from RNA is called **complementary DNA**, or **cDNA**. Although reverse transcriptase initially produces a

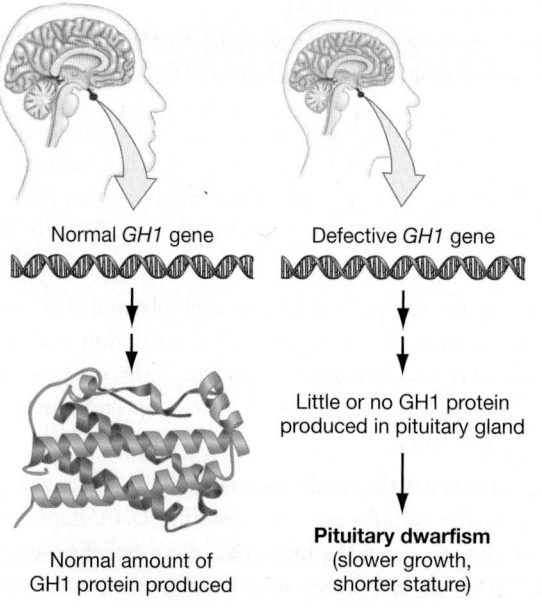

(a) *GH1* codes for a pituitary growth hormone.

Normal *GH1* gene

Defective *GH1* gene

Normal amount of GH1 protein produced

Little or no GH1 protein produced in pituitary gland

Pituitary dwarfism (slower growth, shorter stature)

(b) Normal versus GH1-deficient

FIGURE 19.1 Pituitary dwarfism is a genetic disease. (a) If mutations in the human *GH1* sequence are severe enough to knock out the gene, pituitary dwarfism may result. **(b)** William Harrison and Charles Stratton, in a photo taken about 1860. Harrison and Stratton were both celebrated comedians and performers. Stratton, whose stage name was Tom Thumb, enjoyed audiences in the White House with Abraham Lincoln and Buckingham Palace with Queen Victoria. Stratton had pituitary dwarfism; Harrison had normal height.

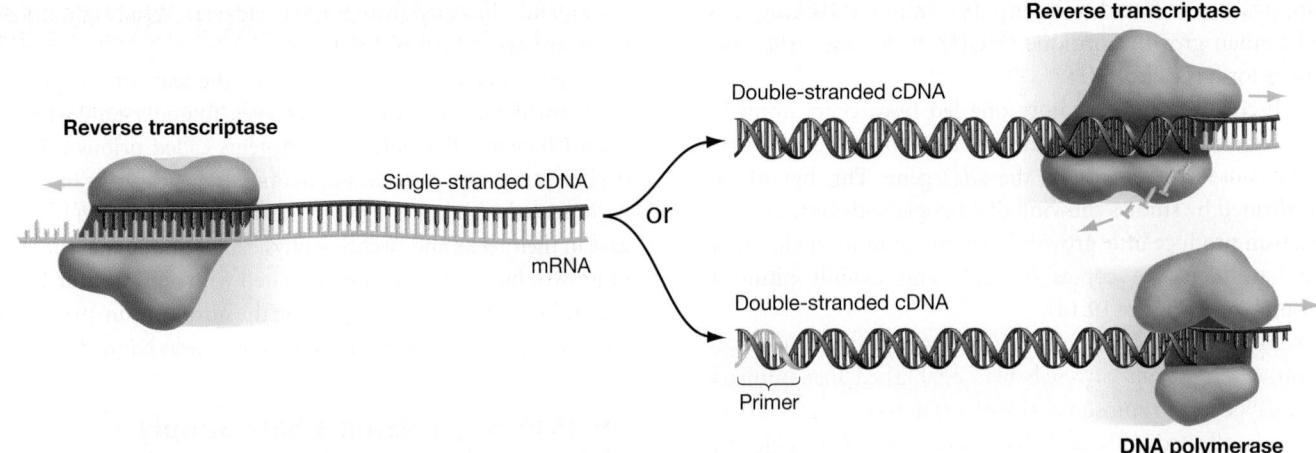

FIGURE 19.2 Reverse Transcriptase Catalyzes the Synthesis of DNA from RNA. The single-stranded DNA produced by reverse transcriptase is complementary to the RNA template. The cDNA can be made double stranded by reverse transcriptase or DNA polymerase. DNA polymerase requires a primer.

single-stranded cDNA, it is also capable of synthesizing the complementary strand to yield a double-stranded DNA. In many cases, however, researchers add a primer to single-stranded cDNAs and use DNA polymerase to synthesize the second strand (**Figure 19.2**).

Reverse transcriptase played a key role in the search for the growth hormone gene. Knowing that *GH1* is actively transcribed in cells from the pituitary gland, researchers isolated mRNAs from pituitary gland cells and used reverse transcriptase to reverse-transcribe those mRNAs to cDNAs. The reaction products could be expected to contain double-stranded cDNAs corresponding to each gene that is actively expressed in pituitary cells.

The next move? Isolating each of the cDNAs and making many identical copies of them.

USING PLASMIDS IN CLONING

Efforts to produce many copies of a gene are referred to as **DNA cloning**. If a researcher says that she has cloned a gene, it means that she has isolated it and then produced many identical copies.

In many cases, researchers can clone a gene by inserting it into a small, circular DNA molecule called a **plasmid**. You might recall from Chapter 7 that plasmids are common in bacterial cells. They are physically separate from the bacterial chromosome, and are not required by the cell for normal growth and reproduction. Most replicate independently of the chromosome. Some plasmids carry genes for antibiotic resistance or other traits that increase the cell's ability to grow in a particular environment.

Researchers realized that if they could splice a loose piece of DNA into a plasmid and then insert the modified plasmid into a bacterial cell, the engineered plasmid would be replicated and passed on to daughter cells as the bacterium grew and divided. If this recombinant bacterium were then placed in a nutrient broth and allowed to grow and reproduce overnight, billions of copies of the original cell, each containing identical modified plasmid DNA, would result. When a plasmid is used in this way—to make copies of a foreign DNA sequence—it is called a **cloning vector**, or simply a **vector**.

Biologists harvest the recombinant genes by breaking the bacteria open, isolating all of the DNA, and then separating the plasmids from the main chromosomes. But how do they insert a gene into a plasmid in the first place?

USING RESTRICTION ENDONUCLEASES AND DNA LIGASE TO CUT AND PASTE DNA

To cut a gene out for later insertion into a cloning vector, researchers use enzymes called restriction endonucleases. A **restriction endonuclease** is a bacterial enzyme that cuts DNA molecules at specific base sequences. In bacterial cells, these enzymes cut up DNA from invading viruses and prevent the cell from becoming fatally infected.

Most of the 400 known restriction endonucleases cut DNA only at sites that form palindromes. In English, a word or sentence is a palindrome if it reads the same way backward as it does forward. "Madam, I'm Adam" is an example. In biology, a stretch of double-stranded DNA forms a palindrome if the $5' \rightarrow 3'$ sequence of one strand is identical to the $5' \rightarrow 3'$ sequence on the antiparallel, complementary strand.

To insert the pituitary gland cDNAs into plasmids, researchers performed the sequence of steps outlined in **Figure 19.3**.

Step 1 The left side of the figure shows a plasmid containing a palindromic sequence that is cut by a specific restriction endonuclease. As the right side of the figure shows, the researchers attached the same palindromic sequence to the ends of each cDNA in their sample.

Step 2 They cut the recognition sites in each plasmid (left) and at the ends of each cDNA (right) with a restriction endonuclease called EcoRI. (The name stands for *Escherichia coli* restriction I, because it was the first restriction endonuclease discovered in *E. coli*.)

Step 3 Like most restriction endonucleases, EcoRI makes a staggered cut in the palindrome. The resulting DNA fragments are described as having **sticky ends**, because the single-stranded bases on one fragment are complementary to the

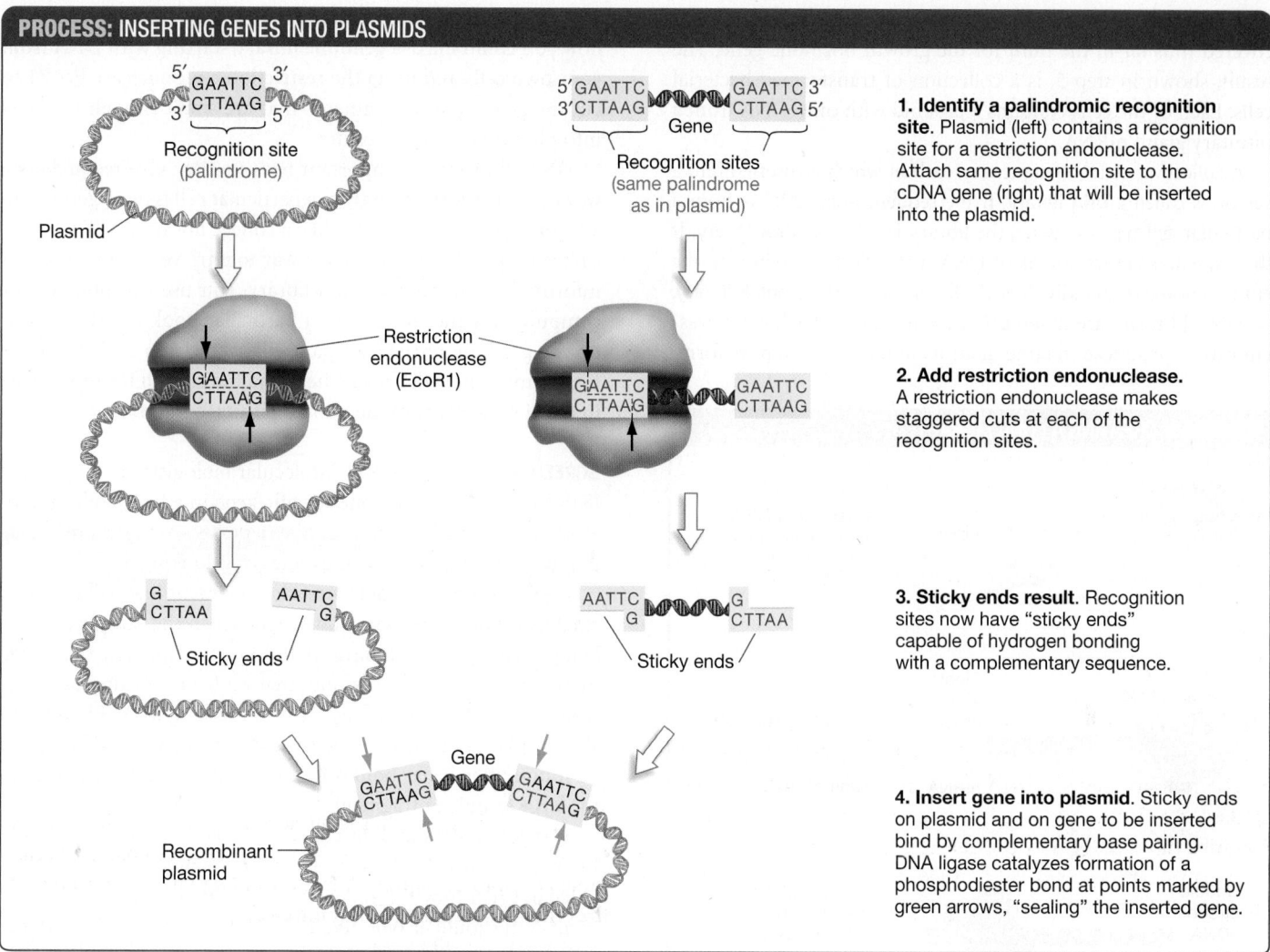

1. Identify a palindromic recognition site. Plasmid (left) contains a recognition site for a restriction endonuclease. Attach same recognition site to the cDNA gene (right) that will be inserted into the plasmid.

2. Add restriction endonuclease. A restriction endonuclease makes staggered cuts at each of the recognition sites.

3. Sticky ends result. Recognition sites now have "sticky ends" capable of hydrogen bonding with a complementary sequence.

4. Insert gene into plasmid. Sticky ends on plasmid and on gene to be inserted bind by complementary base pairing. DNA ligase catalyzes formation of a phosphodiester bond at points marked by green arrows, "sealing" the inserted gene.

FIGURE 19.3 Genes Can Be Inserted into Plasmids in Preparation for Cloning. Once a gene has been inserted into a plasmid, the recombinant plasmid can be introduced into bacterial cells that grow and divide to produce many identical copies of the gene.

single-stranded bases on the other fragment. As a result, the two ends can pair up and hydrogen bond to one another: the complementary sequences in the sticky ends of the plasmid (in black) will bind to the sticky ends in the cDNA (in red) by complementary base pairing.

Step 4 Finally, researchers used **DNA ligase**—introduced in Chapter 14 as the enzyme that connects Okazaki fragments during DNA replication—to seal the recombinant pieces of DNA together at the arrows marked in green.

The importance of the creation of sticky ends in DNA cannot be overstated. ☞ If restriction sites in different DNA sequences are cut with the same restriction endonuclease, the presence of the same sticky ends in both samples of DNA allows the resulting fragments to be spliced together by complementary base pairing. This is the essence of recombinant DNA technology—the ability to create novel combinations of DNA sequences by cutting specific sequences and pasting them into new locations.

After performing this procedure, the researchers who were hunting for the growth hormone gene had a set of recombinant plasmids. Each contained a cDNA made from one of the many human pituitary gland mRNAs.

TRANSFORMATION: INTRODUCING RECOMBINANT PLASMIDS INTO BACTERIAL CELLS ☞ If a recombinant plasmid can be inserted into a bacterial or yeast cell, the foreign DNA will be copied and transmitted to new cells as the host cell grows and divides. In this way, researchers can obtain millions or billions of copies of specific genes. How is the insertion brought about?

Cells that take up DNA from the environment and incorporate it into their genomes are said to undergo **transformation**. To transform bacterial cells with a plasmid, researchers increase the permeability of the cell's plasma membranes using a specific chemical treatment or an electrical shock.

Typically, just a single plasmid enters the cell during this treatment. The cells are then spread out on plates at a low enough density to ensure that each cell is physically isolated. The individual cells grow into colonies containing millions of identical cells.

PRODUCING A cDNA LIBRARY **Figure 19.4** summarizes the steps covered thus far in the hunt for the growth hormone gene. The result, shown in step 5, is a collection of transformed bacterial cells. Each of the cells contains a plasmid with one cDNA from a pituitary gland mRNA.

A collection of DNA sequences, each of which is inserted into a vector, is called a **DNA library**. If the sequences are cDNAs from a particular cell type or tissue, the library is called a **cDNA library**. If the sequences are fragments of DNA that collectively represent the entire genome of an individual, the library is called a **genomic library**.

DNA libraries are made up of cloned genes. Each gene present can be produced in large quantity and isolated in pure form.

✔ If you understand this concept, you should be able to describe how you could make a genomic library starting with DNA from your own cells and using the restriction endonuclease EcoR1 to cut the genome into fragments that are small enough to insert into plasmids or other vectors.

DNA libraries are important because they give researchers a way to store information from a particular cell type or genome in a form that is accessible. But like a college library, a DNA library isn't very useful unless there is a way to retrieve specific pieces of information. At your school's library, you use call numbers or computer searches to retrieve a particular book or article. How do you go about retrieving a particular gene from a DNA library? For example, how did researchers find the growth hormone gene in the cDNA library of the human pituitary gland?

SCREENING A DNA LIBRARY Molecular biologists are often faced with the task of finding one specific gene in a large collection of DNA fragments. To do this requires a **probe**—a marked molecule that binds to the molecule the biologist is looking for.

A DNA probe is a single-stranded fragment that will bind to a single-stranded complementary sequence in the sample of DNA being analyzed. By binding to the target sequence, the probe marks the fragment containing that sequence, distinguishing it from all the other DNA fragments in the sample. As **Figure 19.5** shows, a DNA probe must be labeled in some way so that it can be found after it has bound to the complementary sequence in the large sample of fragments.

✔ If you understand the concept of a DNA probe, you should be able to explain why the probe must be single stranded and labeled in order to work, and why it binds to just one specific fragment. You should also be able to indicate where a probe with the

---

PROCESS: CREATING A cDNA LIBRARY

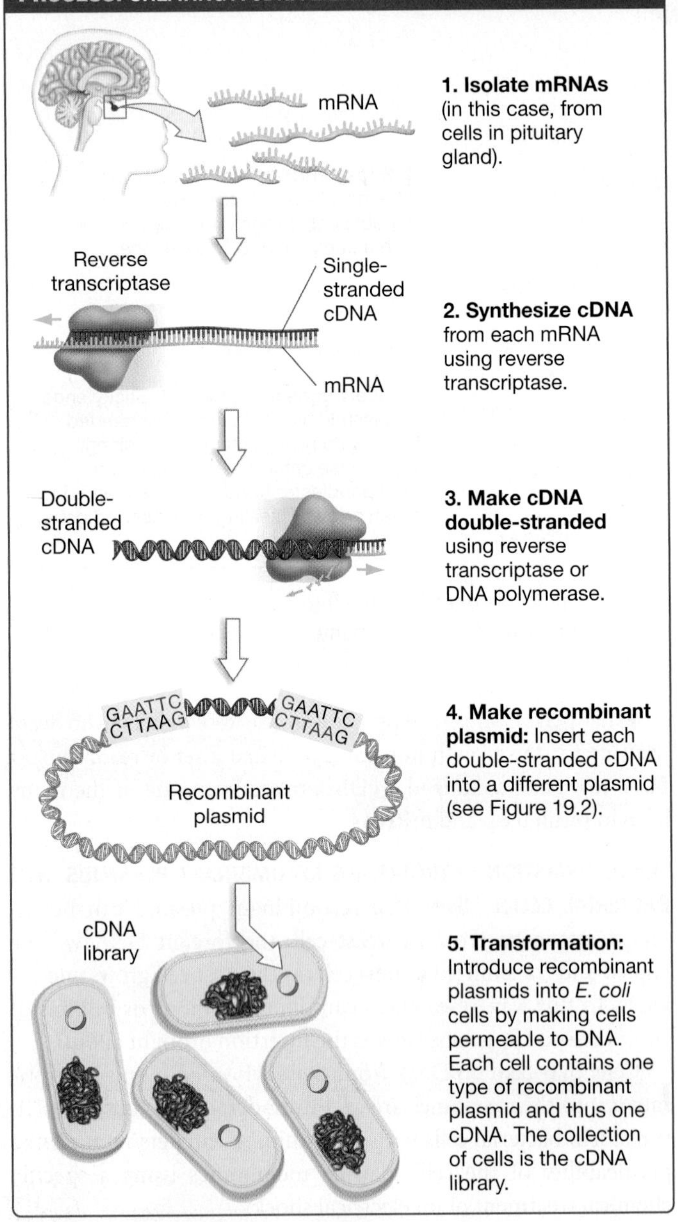

mRNA

1. Isolate mRNAs (in this case, from cells in pituitary gland).

Reverse transcriptase

Single-stranded cDNA

mRNA

2. Synthesize cDNA from each mRNA using reverse transcriptase.

Double-stranded cDNA

3. Make cDNA double-stranded using reverse transcriptase or DNA polymerase.

GAATTC
CTTAAG
GAATTC
CTTAAG

Recombinant plasmid

4. Make recombinant plasmid: Insert each double-stranded cDNA into a different plasmid (see Figure 19.2).

cDNA library

5. Transformation: Introduce recombinant plasmids into *E. coli* cells by making cells permeable to DNA. Each cell contains one type of recombinant plasmid and thus one cDNA. The collection of cells is the cDNA library.

FIGURE 19.4 Complementary DNA Libraries Represent a Collection of the mRNAs in a Cell.

✔ **QUESTION** Would each type of cDNA in the library be represented just once? Why or why not?

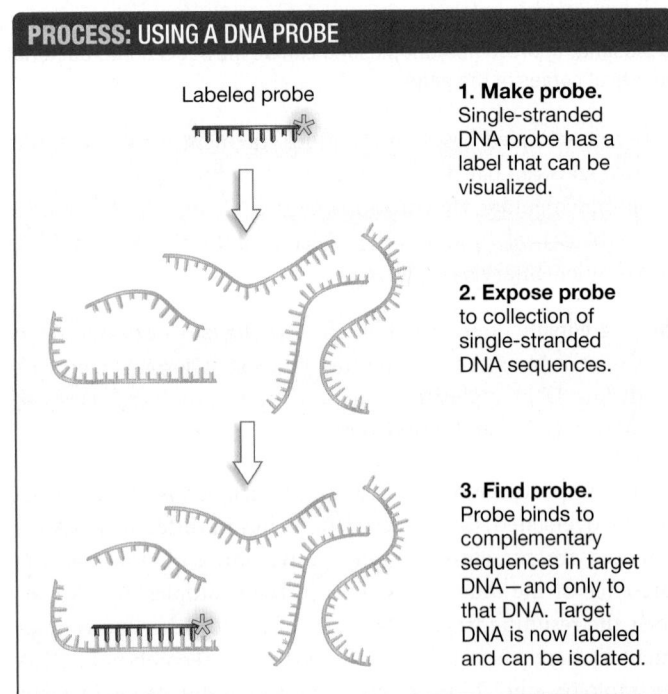

PROCESS: USING A DNA PROBE

Labeled probe

1. Make probe. Single-stranded DNA probe has a label that can be visualized.

2. Expose probe to collection of single-stranded DNA sequences.

3. Find probe. Probe binds to complementary sequences in target DNA—and only to that DNA. Target DNA is now labeled and can be isolated.

FIGURE 19.5 DNA Probes Bind to Specific Target Sequences among Many Different Sequences.

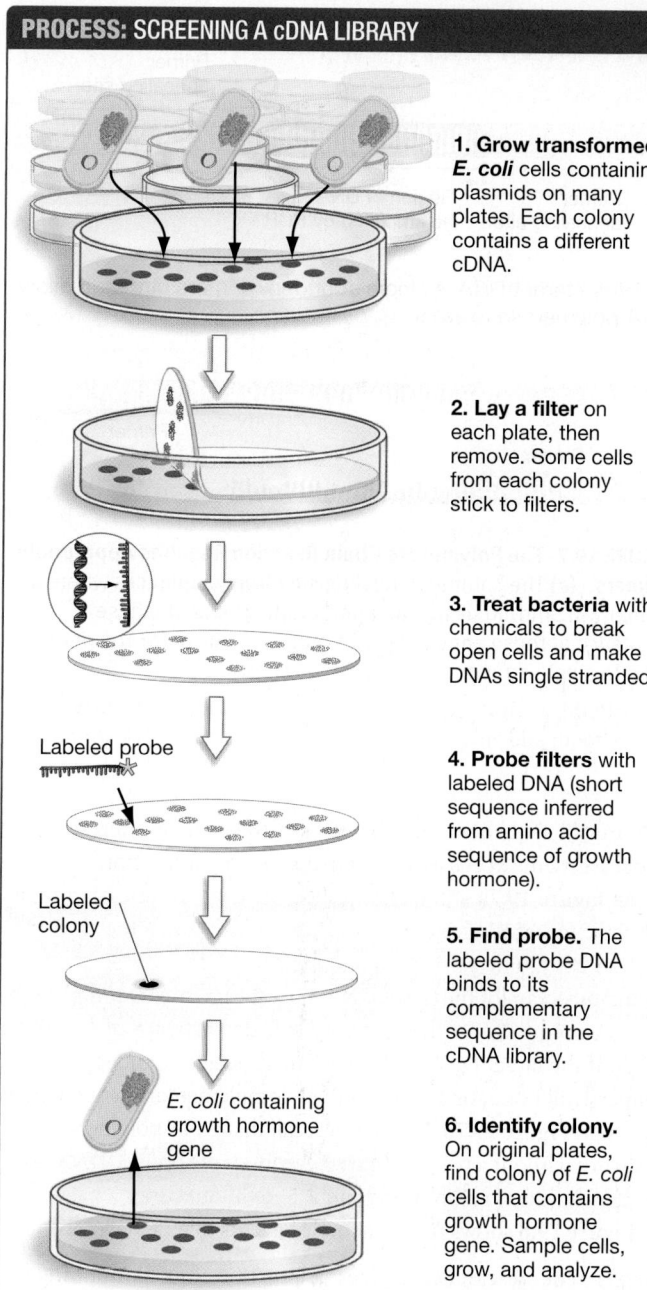

PROCESS: SCREENING A cDNA LIBRARY

1. Grow transformed *E. coli* cells containing plasmids on many plates. Each colony contains a different cDNA.

2. Lay a filter on each plate, then remove. Some cells from each colony stick to filters.

3. Treat bacteria with chemicals to break open cells and make DNAs single stranded.

Labeled probe

4. Probe filters with labeled DNA (short sequence inferred from amino acid sequence of growth hormone).

Labeled colony

5. Find probe. The labeled probe DNA binds to its complementary sequence in the cDNA library.

E. coli containing growth hormone gene

6. Identify colony. On original plates, find colony of *E. coli* cells that contains growth hormone gene. Sample cells, grow, and analyze.

FIGURE 19.6 Finding Specific Genes by Probing a cDNA Library.

sequence AATCG (recall that sequences are always written 5′ to 3′) will bind to a target DNA with the sequence TTTTACCCA TTTACGATTGGCCT (again written 5′ to 3′).

To find an appropriate probe for the human growth hormone gene, researchers began by using the genetic code to predict the approximate DNA sequence of *GH1*. This was possible because the sequence of amino acids in the polypeptide was known. Thus, the researchers could roughly infer the mRNA codon and DNA sequence that coded for each amino acid. You made similar inferences in some of the exercises in Chapter 15. But recall from Chapter 15 that the genetic code is redundant, with more than one codon for most amino acids. As a result, the sequence inferred for the growth hormone gene was actually a set of related sequences.

The next step was to synthesize many copies of a short, single-stranded stretch of DNA that was complementary to the inferred sequence. Because these molecules would bind to single-stranded fragments from the actual gene by complementary base pairing, they could act as a probe. In this case, the label the researchers attached to the probe was a radioactive atom.

Figure 19.6 shows how researchers used this probe to find the plasmid in the cDNA library that contained *GH1*. (For more information on how to use probes, see **BioSkills 9** in Appendix A.) As predicted, the labeled probe bound to its complementary sequence in the cDNA library—identifying the recombinant cell that contained human growth hormone.

MASS-PRODUCING GROWTH HORMONE To accomplish their goal of producing large quantities of the human growth hormone, the investigators used recombinant DNA techniques to transfer the growth hormone cDNA to a new plasmid. The plasmid in question contained a promoter sequence recognized by *E. coli*'s RNA polymerase holoenzyme (see Chapter 16). The recombinant plasmids were then introduced into *E. coli* cells.

The transformed *E. coli* cells that resulted contained a gene for human growth hormone attached to an *E. coli* promoter. These cells began to transcribe and translate the human growth hormone gene. Human growth hormone accumulated in the cells and was subsequently isolated and purified.

Today, bacterial cells containing the human growth hormone gene are grown in huge quantities. These cells have proved to be a safe and reliable source of the human growth hormone protein. The effort to cure pituitary dwarfism using recombinant DNA technology was a spectacular success—a triumph of applied biology, or **biotechnology**. To review this work, go to the study area at *www.masteringbiology.com*.

MB **Web Activity** Producing Human Growth Hormone

Ethical Concerns over Recombinant Growth Hormone

As supplies of growth hormone increased, physicians used it in treating not only people with pituitary dwarfism but also children of short stature who had no actual growth hormone deficiency. Even though the treatment requires several injections per week until adult stature is reached, growth hormone therapy was popular because it often increased the height of these children by a few centimeters.

In essence, growth hormone was being used as a cosmetic—a way to improve appearance in cultures where height is deemed attractive. But if short people are discriminated against in a culture, is a medical treatment a better solution than education and changes in attitudes? And what if parents wanted a tall child to be even taller, to enhance her potential success as, say, a basketball player?

Currently, the U.S. Food and Drug Administration has approved the use of human growth hormone for only the shortest 1.2 percent of children. These individuals are projected to reach adult heights of less than 160 cm (5′3″) in males and 150 cm (4′11″) in women.

Growth hormone has also become a popular performance-enhancing drug for athletes, because it improves the maintenance of bone density and muscle mass. Part of its popularity stems from the fact that it is virtually undetectable in the drug tests currently administered by governing bodies.

Should athletes be able to enhance their physical skills by taking hormones or other types of drugs? Is the drug safe at the dosages athletes are using? These questions are being debated by physicians, researchers, agencies that govern sports, and legislative bodies.

In the meantime, it is clear that while solving one important problem, recombinant DNA technology created others. One of this chapter's recurring themes is that genetic engineering has costs that must be carefully weighed against its benefits.

CHECK YOUR UNDERSTANDING

If you understand that . . .

- The essence of recombinant DNA technology is to cut DNA into fragments with a restriction endonuclease, paste specific sequences together by complementary base pairing of sticky ends and the action of DNA ligase, and insert the resulting recombinant genes into a bacterial (or yeast) cell so that the genes are expressed.
- A DNA library consists of cloned sequences that have been inserted into plasmids or other vectors. A probe can be used to find specific sequences in the library.

You should be able to . . .

1. Explain why restriction endonucleases create DNA fragments with sticky ends.
2. Explain why "probe" is an appropriate term for a labeled sequence that is used to find a particular gene in a DNA library.

Answers are available in Appendix B.

19.2 Case 2—Amplification of Fossil DNA: The Polymerase Chain Reaction

Inserting a gene into a bacterial plasmid is one method for cloning DNA. The polymerase chain reaction is another.

The **polymerase chain reaction (PCR)** is an in vitro DNA synthesis reaction in which a specific section of DNA is replicated over and over, by DNA polymerase, to amplify the number of copies of that sequence. It's a technique for generating many identical copies of a particular section of DNA.

Requirements of PCR

Although PCR is much faster and technologically easier than cloning genes into a DNA library, there is a catch: PCR is possible only when a researcher already has some information about DNA sequences near the gene in question. Sequence information is required because to do a polymerase chain reaction, you have to start by synthesizing short lengths of single-stranded DNA

(a) PCR primers must bind to sequences on either side of the target sequence, on opposite strands.

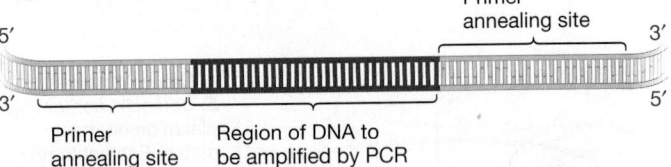

(b) When target DNA is single stranded, primers bind and allow DNA polymerase to work.

FIGURE 19.7 The Polymerase Chain Reaction Requires Appropriate Primers. (a) The "primer-annealing sites" are sequences where a primer will bind. To design an appropriate primer, the base sequence at these annealing sites must be known. **(b)** The primers bind to single-stranded target DNA, as shown.

✔ **EXERCISE** Indicate where DNA polymerase would begin to work on each strand; add an arrow indicating the direction of DNA synthesis.

that match sequences on either side of the gene of interest. These short segments act as primers for the synthesis reaction.

As **Figure 19.7a** shows, the primer sequences must be complementary to bases on either side of the target gene—the DNA you wish to copy. One primer is complementary to a sequence on one strand upstream of the target DNA; the other primer is complementary to a sequence on the other strand, downstream of the target DNA. If the target DNA molecule is made single stranded, then the primers will bond, or anneal, to their complementary sequences, as shown in **Figure 19.7b**. You might recall that DNA polymerase cannot work without a primer. Once the primers are bound, DNA polymerase can extend each strand in the 5′ to 3′ direction.

Figure 19.8 shows the sequence of events in a PCR experiment.

Step 1 The researcher creates a reaction mix containing an abundant supply of the four deoxyribonucleoside triphosphates (dNTPs; see Chapter 14), a DNA sample that includes the gene of interest, many copies of the two primers, and an enzyme called *Taq* polymerase (see below).

Step 2 The reaction mix is heated to 95°C. At this temperature, the double-stranded template DNA denatures. This means that the two DNA strands separate, forming single-stranded templates.

Step 3 The mixture is allowed to cool to 50–60°C. In this temperature range, the primers bond, or anneal, to complementary portions of the single-stranded template DNA. This step is called primer annealing.

Step 4 The reaction mix is heated to 72°C. At this temperature *Taq* polymerase synthesizes the complementary DNA strand from the dNTPs, starting at the primer. This step is called *extension*.

Step 5 Repeat steps 2 through 4.

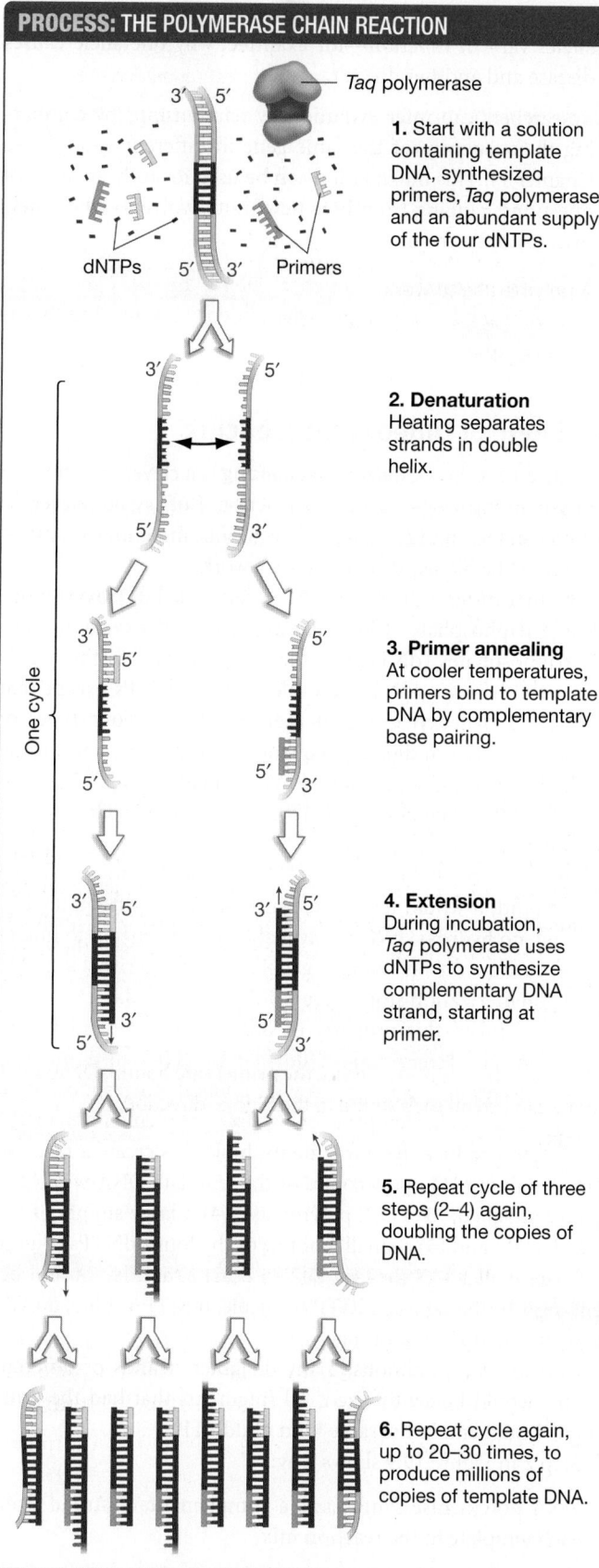

PROCESS: THE POLYMERASE CHAIN REACTION

Taq polymerase

1. Start with a solution containing template DNA, synthesized primers, *Taq* polymerase, and an abundant supply of the four dNTPs.

dNTPs   Primers

One cycle

2. Denaturation
Heating separates strands in double helix.

3. Primer annealing
At cooler temperatures, primers bind to template DNA by complementary base pairing.

4. Extension
During incubation, *Taq* polymerase uses dNTPs to synthesize complementary DNA strand, starting at primer.

5. Repeat cycle of three steps (2–4) again, doubling the copies of DNA.

6. Repeat cycle again, up to 20–30 times, to produce millions of copies of template DNA.

FIGURE 19.8 The Polymerase Chain Reaction Is a Method for Producing Many Copies of a Specific Sequence. Each PCR cycle (denaturation, primer annealing, and extension) results in a doubling of the number of target and primer sequences.

Step 6 Continue repeating steps 2 through 4 until the necessary number of copies is obtained.

Today, the temperature changes required in each step are controlled by automated PCR machines.

Taq polymerase is a DNA polymerase found in the bacterium *Thermus aquaticus*, which was originally discovered in a hot spring inside Yellowstone National Park, Wyoming. Researchers use *Taq* polymerase because the PCR reaction mix has to be heated, and *Taq* polymerase is heat stable. Most DNA polymerases are destroyed at high temperature, but *Taq* polymerase continues to function normally even when heated to 95°C.

The denaturation, primer annealing, and extension steps constitute a single PCR cycle. If one copy of the template sequence existed in the original sample, then two copies are present at the end of the first cycle (see step 4 in Figure 19.8). These two copies then act as templates for the second cycle—another round of denaturation, primer annealing, and extension—after which four copies of the target gene are present (see step 5).

Each time the cycle repeats, the amount of template sequence in the reaction mixture doubles (step 6). Doubling occurs because each newly synthesized segment of DNA serves as a template in the subsequent cycle, along with the previously synthesized segments. Starting with a single copy, successive cycles result in the production of 2, 4, 8, 16, 32, 64, 128, 256 copies, and so on. A total of n cycles can generate 2^n copies; so in just 20 cycles, one sequence can be amplified to over a million copies. By performing up to 30 cycles, researchers obtain enormous numbers of copies of the template sequence.

Before reading further, be sure to review how PCR works in the study area at *www.masteringbiology.com*.

(MB) **Web Activity** The Polymerase Chain Reaction

PCR in Action

To understand why PCR is so valuable, consider a study by biologist Svante Pääbo and colleagues, who wanted to analyze DNA recovered from the 30,000-year-old bones of a fossilized human of the species *Homo neanderthalensis*. Their goal was to determine the sequence of bases in the ancient DNA and compare it with DNA from modern humans (*Homo sapiens*).

If modern humans have sequences that are identical or almost identical to the sequences found in Neanderthals, it would suggest that some of us inherited DNA directly from a Neanderthal ancestor. That could happen only if *H. sapiens* and *H. neanderthalensis* interbred while they coexisted in Europe.

The Neanderthal bone was so old, however, that most of the DNA in it had degraded into tiny fragments. The biologists could recover only a minute amount of DNA that was still in moderate-sized pieces. Fortunately, the Neanderthal DNA sample included a few fragments of the gene region that Pääbo's team wanted to study. The researchers were able to design primers that bracketed this gene region, based on the sequence of highly conserved sections of the same gene from *H. sapiens*.

Using PCR, the researchers produced millions of copies of the Neanderthal DNA fragment. After analyzing these sequences, the team found that they differ from the same gene segment found in modern humans. Subsequent work with DNA from 14 other Neanderthal fossils, from locations throughout Europe, gave the same result. These data support the hypothesis that Neanderthals never interbred with modern humans—even though the two species lived in the same areas of Europe at the same time.

PCR has been used to study other fossil DNAs, as well. The current record holder for oldest DNA to be amplified by PCR came from 17-million-year-old magnolia trees. PCR is useful any time a researcher needs a large number of copies of a particular gene. For example,

- Forensic biologists, who use biological analyses to help solve crimes, clone DNA from tiny drops of blood or hair gathered at crime scenes. The copied DNA can then be analyzed to identify victims or implicate perpetrators.

- Genetic counselors, who advise pregnant couples on how likely their offspring are to suffer from inherited diseases, can use PCR to find out if an embryo being carried by a client has alleles associated with deadly illness.

Because the complete genomes of a wide array of organisms have now been sequenced, researchers can find appropriate primer sequences to use in cloning almost any target gene by PCR. The polymerase chain reaction is now one of the most basic and widely used techniques in molecular biology.

CHECK YOUR UNDERSTANDING

If you understand that . . .

- PCR is a technique (summarized in Figure 19.8) for amplifying a specific region of DNA into millions of copies, which can then be sequenced or used for other types of analyses.

You should be able to . . .

1. Explain the purpose of the denaturation, annealing, and extension steps in a PCR cycle, and why "chain reaction" is an appropriate part of PCR's name.
2. Write down the sequence of a DNA strand 50 base pairs long, then design 20-base-pair-long primers that would allow you to amplify the segment by PCR.

Answers are available in Appendix B.

19.3 Case 3—Sanger's Breakthrough Innovation: Dideoxy DNA Sequencing

Once researchers have cloned a gene from a DNA library or by PCR, determining the gene's base sequence is usually one of the first things they want to do. Understanding a gene's sequence is valuable for a variety of reasons. For example,

- Once a gene's sequence is known, the amino acid sequence of its product can be inferred from the genetic code. Knowing a protein's primary structure often provides clues to its function.

- Comparing sequences is fundamental to understanding why alleles vary in function—for example, why one allele causes disease and another doesn't.

- Researchers can infer evolutionary relationships by comparing the sequences of the same gene in different species (see Chapter 1). This information can be used to study an array of questions, ranging from how new traits evolve to where new diseases come from.

How do researchers sequence DNA? In 1977 Frederick Sanger published a technique called dideoxy sequencing that is still in use today.

The Logic of Dideoxy Sequencing

As **Figure 19.9** shows, **dideoxy sequencing** is a clever variation on the basic in vitro DNA synthesis reaction. But saying "clever" is an understatement. Sanger had to link three important insights to make his sequencing strategy work.

The first insight was to use monomers called dideoxyribonucleoside triphosphates (ddNTPs) along with deoxyribonucleoside triphosphates (dNTPs) in the reaction mix. As the top of Figure 19.9 shows, ddNTPs are identical to dNTPs except that they lack a hydroxyl group at their 3′ carbon. Four types of ddNTPs are used in dideoxy sequencing, each named according to whether it contains adenine (ddATP), thymine (ddTTP), cytosine (ddCTP), or guanine (ddGTP). The use of ddNTPs inspired the name "dideoxy" sequencing.

Sanger realized that if a ddNTP were added to a growing DNA strand, it would terminate synthesis. Why? After a ddNTP is added, no hydroxyl group is available on a 3′ carbon to link to the 5′ carbon on an incoming dNTP monomer. As a result, DNA polymerization stops once a ddNTP is added.

Sanger linked this property of ddNTPs to a second fundamental insight: Every time a ddNTP is added to a growing strand, the result is a fragment with a length corresponding to the position in the template of a base complementary to the ddNTP. To produce these fragments, biologists create a reaction mix containing (1) many copies of the template DNA with (2) a primer attached, (3) DNA polymerase, (4) a large supply of the four dNTPs, and (5) a small amount of the four ddNTPs (Figure 19.9, step 1). Each of the four ddNTPs carries a different fluorescent tag. (In the figure, ddGTP is purple, ddCTP is blue, ddATP is green, and ddTTP is orange.)

Under these conditions, many daughter strands of different lengths would be synthesized. All fragments that had the same length would end in the same kind of ddNTP.

Step 2 in Figure 19.9 shows why:

- DNA polymerase synthesizes a complementary strand from each template in the reaction mix.

- The synthesis of each one of these complementary strands starts at the same point—the primer.

- Because there are many dNTPs and relatively few ddNTPs in the reaction mix, dNTPs are usually incorporated opposite

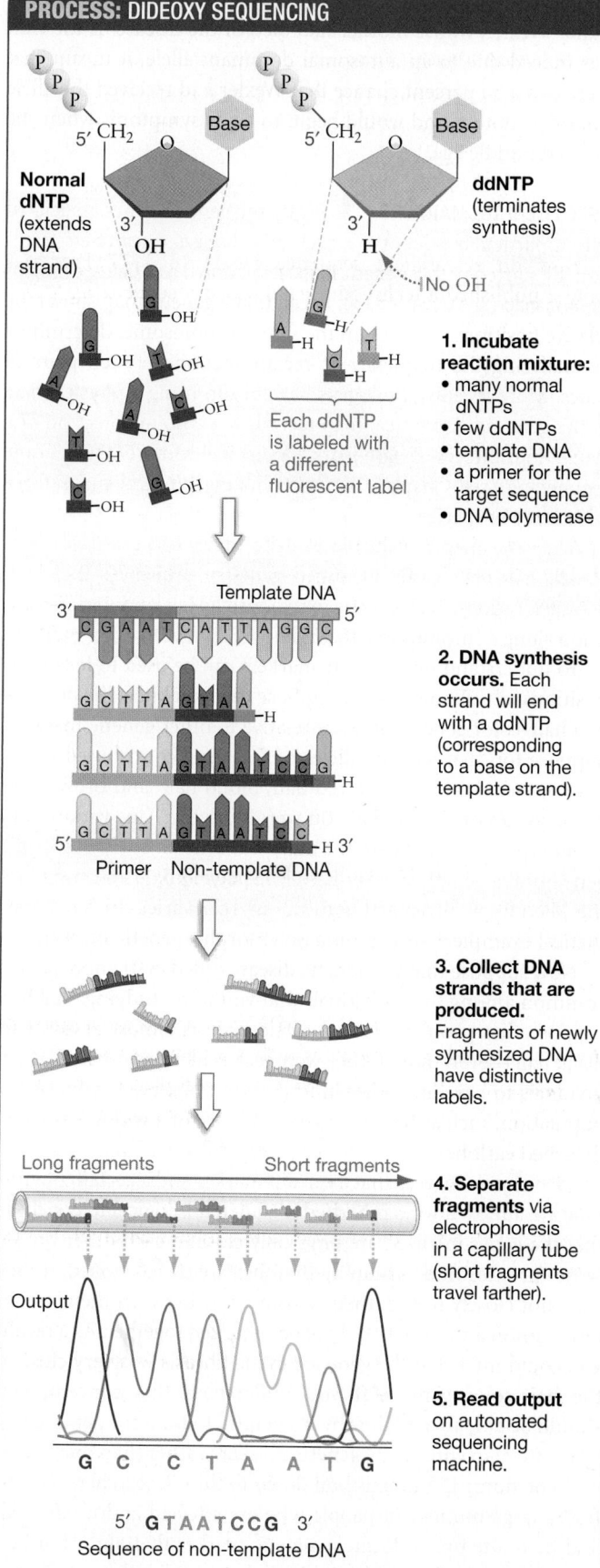

PROCESS: DIDEOXY SEQUENCING

Normal dNTP (extends DNA strand)

ddNTP (terminates synthesis)

No OH

Each ddNTP is labeled with a different fluorescent label

1. Incubate reaction mixture:
- many normal dNTPs
- few ddNTPs
- template DNA
- a primer for the target sequence
- DNA polymerase

Template DNA

3′ CGAATCATTAGGC 5′

GCTTAGTAA—H
GCTTAGTAATCCG—H
GCTTAGTAATCC—H 3′
Primer   Non-template DNA

2. DNA synthesis occurs. Each strand will end with a ddNTP (corresponding to a base on the template strand).

3. Collect DNA strands that are produced. Fragments of newly synthesized DNA have distinctive labels.

Long fragments   Short fragments

4. Separate fragments via electrophoresis in a capillary tube (short fragments travel farther).

Output

G C C T A A T G

5′ GTAATCCG 3′
Sequence of non-template DNA

5. Read output on automated sequencing machine.

FIGURE 19.9 Dideoxy Sequencing Is a Method for Determining the Base Sequence of DNA.

each complementary base on the template strand as DNA works its way along the template strand.

- Occasionally, one of the few ddNTPs present is incorporated into the growing strand, opposite the corresponding base in the template. Which complementary base in the template strand gets paired with a ddNTP is random.

- The addition of the ddNTP at this location stops further elongation.

- "Stops" of this kind happen for each base in the template strand. As a result, the overall reaction produces a collection of newly synthesized strands whose various lengths correspond to the location of each base in the template strand (see step 3 in Figure 19.9). Each fragment will fluoresce the color of its ddNTP.

Sanger's third insight? When the fragments produced by the synthesis reactions are lined up by size, the dideoxy monomers on the successive fragments reveal the sequence of bases in the template DNA. To line up fragments in order of size, biologists separate them using gel electrophoresis (step 4 in Figure 19.9). As step 5 shows, a machine can read the pattern of fluorescence, indicating the sequence of bases in the newly synthesized strand.

Dideoxy sequencing ranks among the greatest of all technological advances in the history of biological science. Its impact is comparable to the development of the light microscope, the electron microscope, microelectrodes for recording membrane potentials, and recombinant gene technology.

"Next Generation" Sequencing

Sequencing technology is advancing at a blindingly fast pace. Dideoxy sequencing remains the method of choice the first time a genome is sequenced from a particular species. But new approaches now make it possible to sequence other individuals of that species—to compare with the original—much faster and more cheaply than the dideoxy method.

Most of the newer methods are based on detecting the pyrophosphate molecule (two phosphate groups bonded together; see the "P–P" in Figure 14.6) that is released after a DNA polymerase adds a dNTP to a growing DNA strand. The pyrophosphates that are released during the synthesis reaction drive a set of other reactions that result in luminescence, which can be detected. Sequencing strategies like this don't involve any ddNTPs, and are sometimes called pyrosequencing or sequencing-by-synthesis.

Because pyrosequencing can be done with tiny quantities of template DNA and reactants, the process can be miniaturized. Hundreds of thousands of template DNA fragments can be attached to the same glass slide and sequenced at the same time—resulting in dramatic savings of time and money.

Sequencing the human genome for the first time took four years and cost $300 million U.S. dollars. Now researchers are contemplating the possibility of sequencing your genome in 4 minutes for less than $10,000.

19.4 Case 4—The Huntington's Disease Story: Finding Genes by Mapping

Mendel had no idea what a "hereditary determinant" actually was. But now we know. Biology's molecular revolution has allowed researchers to find and characterize individual genes—to explore the connection between genotype and phenotype as explicitly and directly as possible. The question is, How do researchers find the genes associated with certain traits in the first place? How do you find the gene responsible for seed shape in peas, or white eyes in flies, or DNA polymerase III in *E. coli*?

Conceptually, the answer is simple: You begin with a map of known sites in the genome, then look for an association between one of those known sites and the phenotype you're interested in. The gene that affects the phenotype is probably close to the known site.

In practice, the process is not so simple. As an example of how this type of gene hunt is done, let's consider one of the first successful searches ever conducted for a human gene—the gene associated with Huntington's disease.

How Was the Huntington's Disease Gene Found?

Huntington's disease is a rare but devastating illness. Typically, affected individuals first show symptoms between the ages of 35 and 45. At onset, an individual appears to be clumsier than normal and tends to develop small tics and abnormal movements. As the disease progresses, uncontrollable movements become more pronounced. Eventually the affected individual twists and writhes involuntarily. Personality and intelligence are also compromised—to the extent that the early stage of this disease is sometimes misdiagnosed as the brain disorder schizophrenia. The illness may continue to progress for 10 to 20 years and is eventually fatal.

Because Huntington's disease appeared to run in families, physicians suspected that it was a genetic disease. More specifically, an analysis of pedigrees from families affected by Huntington's disease suggested that the trait was due to a single, autosomal dominant allele (see Chapter 13). To test this hypothesis, researchers set out to locate and identify the gene or genes involved. It took over 10 years of intensive effort to reach this goal.

The search for the Huntington's disease gene was led by Nancy Wexler, whose mother had died of the disease. If the trait was indeed due to an autosomal dominant allele, it meant that there was a 50 percent chance that Wexler had received the allele from her mother and would begin to show symptoms when she reached middle age.

USING GENETIC MARKERS To locate the gene or genes associated with a particular phenotype, such as a disease, researchers traditionally started with a **genetic map**, also known as a **linkage map** or **meiotic map** (see Chapter 13). Recall that a genetic map shows the relative positions of genes on the same chromosome, determined by analyzing the frequency of recombination between pairs of genes. More recently, biologists have begun using a **physical map** of the genome. A physical map records the absolute position of a gene—in numbers of base pairs—along a chromosome. Genome sequencing (see Chapter 20) is producing physical maps for a wide array of species.

A genetic map is valuable in gene hunts because it contains **genetic markers**—easily identified genes or sequences that have known locations. Each genetic marker provides a landmark—a position along a chromosome that is known relative to other markers.

To understand how genetic markers can be used to locate the positions of unknown genes, suppose that you knew the position of a hair-color gene in humans relative to other genetic markers. Suppose too that various alleles of this gene contributed to the development of black hair, red hair, blond hair, and brown hair in the group of people that you were studying. This variation in phenotype associated with the marker is crucial. To be useful in a gene hunt, a genetic marker has to be **polymorphic**, meaning that the phenotype associated with the marker varies. In our hypothetical example, hair color is a polymorphic genetic marker.

Now suppose that the genetic disease called cystic fibrosis was common among the individuals that you were studying. Further, suppose that people who had cystic fibrosis almost always had black hair—even though they were just as likely as unaffected individuals to have any other inherited trait observed in the study population, such as the presence or absence of a widow's peak or detached earlobes.

If you observe that a certain marker and a certain phenotype are almost always inherited together, it is logical to conclude that the genes involved are physically close to each other on the same chromosome—meaning that they are closely linked. If they were not closely linked, then crossing over between them would be common and they would *not* be inherited together. As a result, you could infer that the gene for cystic fibrosis was very close to the hair-color gene. ✔If you understand this concept, you should be able to explain why it's helpful to hunt for genes using a genetic map with many genetic markers rather than only a few.

Gene hunts in humans boil down to this: Researchers have to find a large number of people who are affected and unaffected, and then attempt to locate a genetic marker that almost always occurs in the affected individuals but not in the unaffected people. If such a marker is found, the disease gene is almost guaranteed to be nearby.

The types of genetic markers used in gene mapping have changed over time. Today, researchers often have a large catalog of genetic markers available, including the particularly abundant markers known as **single nucleotide polymorphisms** (**SNPs**, pronounced "snips"). A SNP is a site in DNA where some individuals in the population have a different base. For example, some might have an A at a certain site while others have a C. That site would qualify as a SNP.

In the late 1970s and early 1980s, when biologists were searching for the Huntington's gene, SNPs were unknown. The best genetic markers available were restriction sites—short stretches of DNA where restriction endonucleases cut the double helix. These sequences are also known as restriction endonuclease recognition sites.

The restriction sites that Wexler's team used were polymorphic: Some individuals had sites where cuts occurred, but in other individuals, the DNA sequence at the same site was different, and no cuts occurred. Thus, each restriction site was either present in an individual or not present—just as an individual might have an A instead of a C at a certain SNP. Wexler's team was looking for restriction sites that were almost always present in diseased individuals but absent in healthy individuals, or vice versa (**Figure 19.10**).

AN ASSOCIATION STUDY Once a genetic map containing many genetic markers has been assembled, biologists need help from individuals affected by an inherited disease to find the gene in question. Recall that the fundamental goal is to find a genetic marker that is almost always inherited along with the disease-causing allele. Biologists call this an association study. Gene hunts based on association studies are more likely to be successful if large groups are involved. Large sample sizes minimize the probability that researchers will observe an association between one or more markers and the disease just by chance, rather than because they are closely linked.

Huntington's disease is rare, so obtaining a large sample size is problematic. But Wexler's Huntington's disease team was fortunate to find a large, extended family affected with the disease living along the shores of Lake Maracaibo, Venezuela.

From historical records, the researchers deduced that the Huntington's disease allele was introduced to this family by a European sailor or trader who visited the area in the early 1800s. When family members agreed to participate in the study, there were over 3000 of his descendants living in the area. One hun-

dred of these people had been diagnosed with Huntington's disease. To help in the search for the gene, family members agreed to donate skin or blood samples for DNA analysis.

When Wexler's team looked for associations between the presence or absence of the disease phenotype and the genetic markers observed in each family member, markers called A, B, C, and D turned out to be especially important. These four combinations of restriction sites represent alleles. The key finding was that the C allele was almost always found in diseased individuals. Apparently, the English sailor who introduced the Huntington's disease allele also had the C allele—this particular combination of restriction sites—in his DNA. The C allele and the Huntington's disease gene are so close together that recombination between them—which would put an A or B or D allele next to the Huntington's disease allele in his descendants—has been extremely rare.

From the genetic map of humans that was available at the time, Wexler's team knew that the C restriction sites were on chromosome 4. Eventually the team succeeded in narrowing down the location of the C sites, and thus the Huntington's disease gene, to a region about 500,000 base pairs long. Because the haploid human genome contains over 3 billion base pairs, this was a huge step in focusing the search for the gene.

PINPOINTING THE DEFECT Once the general location of the Huntington's disease gene was known, biologists looked in that region for exons that encode a functional mRNA. Then they sequenced exons from diseased and normal individuals, compared the data, and pinpointed specific bases that differed between the two groups of individuals. In this way, dideoxy sequencing played a key role in the gene hunt.

When this analysis was complete, the research team found that individuals with Huntington's disease have an unusual number of CAG codons near the 5' end of a particular gene. CAG codes for glutamine. Healthy individuals have 11–25 copies of the CAG codon at that location, while affected individuals have 42 or more copies.

When the Huntington's disease research team confirmed that the increase in the CAG codon was always observed in affected individuals, the team concluded that the long search for the Huntington's disease gene was over. They named the newly discovered gene *IT15* and its protein product huntingtin. In both affected and normal individuals, the huntingtin protein is involved in the early development of nerve cells. It is only later in life that mutant forms of the protein cause disease.

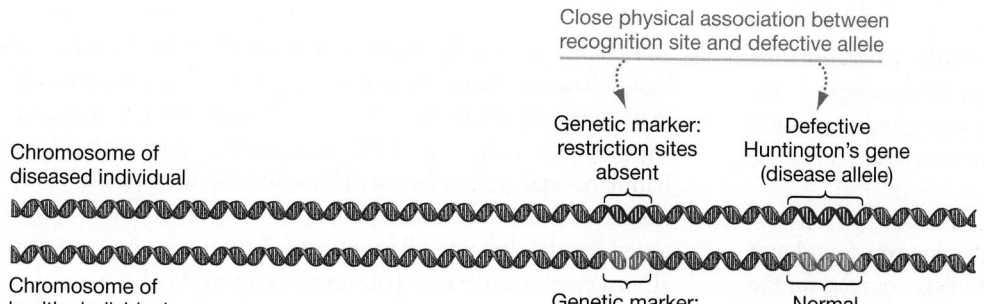

Close physical association between recognition site and defective allele

Genetic marker: restriction sites absent

Defective Huntington's gene (disease allele)

Chromosome of diseased individual

Chromosome of healthy individual

Genetic marker: restriction sites present

Normal Huntington's gene

FIGURE 19.10 Genetic Markers and Disease Alleles Are Inherited Together if They Are Closely Linked. Because of genetic recombination, genetic markers that are far from the gene of interest are equally likely to be found in both affected and unaffected individuals—there will be no association between the marker and either the normal or the disease-causing allele.

What Are the Benefits of Finding a Disease Gene?

How have efforts to find disease genes improved human health and welfare? Has the effort to locate the Huntington's disease gene helped researchers and physicians understand and treat the illness? Biomedical researchers point to three major benefits of successful disease-gene hunts.

IMPROVED UNDERSTANDING OF THE PHENOTYPE Once a disease gene is found and its sequence is known, researchers can usually figure out why its product causes disease. In the case of *IT15*, autopsies of Huntington's patients had shown that their brains decrease in size, and that the brain tissue contains clumps, or aggregates, of the huntingtin protein.

Huntingtin aggregates are a direct consequence of changes in the number of CAG repeats in the *IT15* gene. Long stretches of polyglutamine (a polymer of the amino acid glutamine) are known to result in the formation of protein aggregates. The leading hypothesis to explain Huntington's disease proposes that a gradual buildup of polyglutamine aggregates triggers neurons to undergo apoptosis, or programmed cell death.

These results explained why Huntington's disease is pleiotropic (see Chapter 13). Patients suffer from abnormal movements *and* personality changes because neurons from throughout the brain are being killed. The results also help explain why the disease takes so long to appear, and why it is progressive: The defective huntingtin proteins take time to build up to deadly levels, but then continue to increase over time. Finally, understanding the molecular mechanism responsible for the illness explained why the disease allele is dominant. One copy of the defective gene is enough to produce fatal concentrations of aggregates.

THERAPY Once *IT15* was found, biologists began a search for new therapies for Huntington's disease by introducing the defective allele into mice, using the types of genetic engineering techniques discussed in Section 19.5. These mice are called **transgenic** (literally, "across-genes"), because they have alleles that have been modified by genetic engineering.

Transgenic mice that produce defective versions of the huntingtin protein develop a version of Huntington's disease, exhibiting tremors and abnormal movements, higher-than-normal levels of aggression toward litter and cage mates, and a loss of neurons in the brain. Laboratory animals with disease symptoms that parallel those of a human disease are said to provide an **animal model** of the disease.

Animal models are valuable in disease research, because they can be used to test potential treatments before investigators try them on human patients. For example, research groups are using transgenic mice to test drugs that appear to prevent or reduce the aggregation of the huntingtin protein.

GENETIC TESTING When the Huntington's gene was found and sequenced, biologists used the knowledge to develop a test for the presence of the defective allele. The test consists of obtaining a DNA sample from an individual and using the polymerase chain reaction to amplify the chromosome region that contains the CAG repeats responsible for the disease. If the number of CAG repeats is 35 or less, the individual is not considered at risk. Forty or more repeats results in a positive diagnosis for Huntington's.

Thanks to genetic maps based on SNPs, gene hunts are increasingly successful. Biologists have recently documented alleles associated with a predisposition to developing type I diabetes, type II diabetes, breast cancer, obesity, coronary heart disease, bipolar disorder, Crohn's disease, and rheumatoid arthritis. Testing for these alleles is now available.

Three types of genetic testing are currently done for genetic diseases:

1. *Carrier testing* can determine whether an individual is a carrier for a genetic disease. Before starting their own family, people from families affected by a genetic disease frequently want to know whether they carry the allele responsible. This is especially true for diseases, such as cystic fibrosis (CF), that are due to recessive alleles. If only one of the prospective parents has the allele, then none of the children they have together should develop CF. But if both prospective parents carry the allele, then each child they produce has a 25 percent chance of having CF.

2. *Prenatal testing* analyzes DNA from fetal cells obtained early in gestation. Alleles for genetic diseases can then be amplified by PCR and sequenced. Suppose that two parents, both carrying the CF allele, decide to have children but do not want to have a child that will suffer from the disease. They can use the results of prenatal testing as a basis for deciding either to continue or terminate a pregnancy.

3. *Adult testing* is particularly important for diseases that have a strong environmental component. For example, five to ten percent of the women who develop breast cancer have inherited a faulty gene. If testing reveals that a woman has a susceptibility allele, she can be checked for the onset of cancer more frequently than normal. She may also be able to make dietary and other changes that make her less likely to actually develop the illness.

Ethical Concerns over Genetic Testing

Tables 19.1 and **19.2** on pages 352–353 summarize the tools and techniques discussed in this chapter. Although these advances have already cured serious diseases and promise additional benefits for human health and welfare, they also raise difficult ethical issues.

Genetic testing, for example, can create serious moral and legal dilemmas, as well as harrowing personal choices. Consider that some people maintain that it is morally wrong to terminate any pregnancy, even if the fetus is certain to be born with a debilitating or fatal genetic disease. Think about Nancy Wexler's position soon after the discovery of *IT15*: Would you choose to be tested for the defective allele and risk finding out that you were almost certain to develop Huntington's disease?

There are other, equally serious, questions. Should physicians agree to test people for genetic diseases that have no cure? Should it be legal for health insurance companies to test clients? If so, can companies refuse to insure people at risk for diseases that require expensive treatments? What about life insurance companies? Or employers?

These questions are being debated by political and religious leaders, health-care workers, philosophers, and the public at large. In many cases, we've yet to find answers.

CHECK YOUR UNDERSTANDING

If you understand that . . .

- Genes for particular traits can be located if they are closely linked to a known genetic marker and are thus inherited along with the marker.

✔ You should be able to . . .

Describe how you would design a study with the goal of identifying alleles associated with the development of alcoholism in humans.

Answers are available in Appendix B.

19.5 Case 5—Severe Immune Disorders: The Potential of Gene Therapy

For physicians who treat inherited diseases such as Huntington's, sickle-cell anemia, and cystic fibrosis, the ultimate goal is to replace or augment defective copies of the gene with normal alleles. This approach to treatment is called **gene therapy**.

For gene therapy to succeed, two crucial requirements must be met. First, the allele associated with the healthy phenotype must be sequenced and its regulatory sites must be understood. Second, a method must be available for introducing this allele into affected individuals. The DNA has to be introduced in a way that ensures expression of the gene in the correct tissues, in the correct amount, and at the correct time. If the defective allele is dominant, then the introduction step may be even more complicated: In at least some cases, the introduced allele must physically replace or block the expression of the undesirable dominant allele.

How Can Novel Alleles Be Introduced into Human Cells?

Section 19.1 reviewed how recombinant DNA sequences are packaged into plasmids and taken up by *E. coli* cells. Humans and other mammals lack plasmids, however, and their cells do not take up foreign DNA in response to chemical or electric treatments as efficiently as bacterial cells do. How can foreign genes be introduced efficiently into human cells?

To date, researchers have focused on packaging foreign DNA into viruses for transport into human cells. As Chapter 35 will detail, viral infection begins when a virus particle enters or attaches to a host cell and inserts its genome into that host cell. In some cases the viral DNA becomes integrated into a host-cell chromosome. This tendency enables viruses that infect human cells to be used as vectors to carry engineered alleles into the chromosomes of target cells. Potentially, the alleles delivered by the virus could be expressed and produce a product capable of curing a genetic disease.

Currently, the vectors of choice in gene therapy are retroviruses. When a **retrovirus** infects a human cell, reverse transcriptase catalyzes the production of a DNA copy of the virus's RNA genome. Other viral enzymes catalyze the insertion of the viral DNA into a host-cell chromosome. If human genes can be packaged into a retrovirus, then the virus is capable of inserting the human alleles into a chromosome in a target cell (**Figure 19.11**).

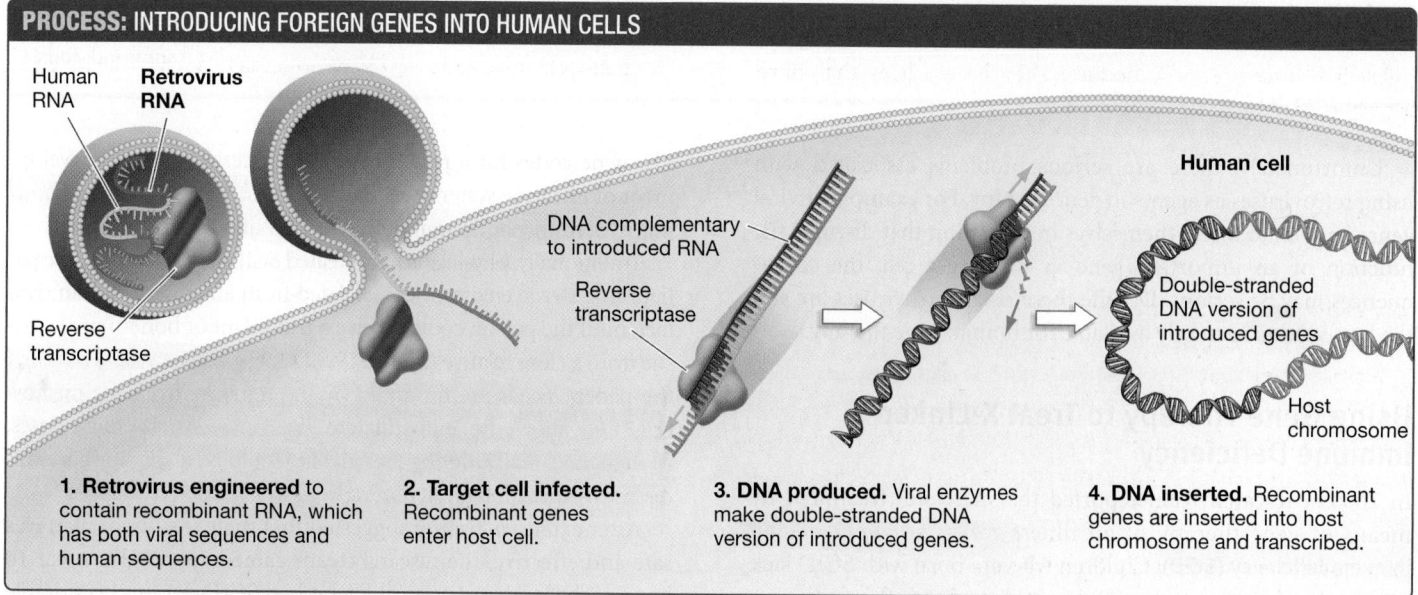

PROCESS: INTRODUCING FOREIGN GENES INTO HUMAN CELLS

Human RNA   Retrovirus RNA

Reverse transcriptase

DNA complementary to introduced RNA

Reverse transcriptase

Human cell

Double-stranded DNA version of introduced genes

Host chromosome

1. Retrovirus engineered to contain recombinant RNA, which has both viral sequences and human sequences.

2. Target cell infected. Recombinant genes enter host cell.

3. DNA produced. Viral enzymes make double-stranded DNA version of introduced genes.

4. DNA inserted. Recombinant genes are inserted into host chromosome and transcribed.

FIGURE 19.11 Retroviruses Insert Their Genes into Host-Cell Chromosomes.

✔QUESTION What happens if the recombinant DNA is inserted in the middle of a gene that is critical to normal cell function?

Tool	Description	How Used	Illustration
Reverse transcriptase	Enzyme that catalyzes synthesis of a complementary DNA (cDNA) from an RNA template.	Many applications, including making cDNAs used in constructing a genetic library.	
Restriction endonucleases	Enzymes that cut DNA at a specific sequence—often a palindromic sequence that is six base pairs long.	Allows researchers to cut DNA at specific locations. Cuts in palindromic sites create "sticky ends."	
DNA ligase	An enzyme that catalyzes the formation of a phosphodiester bond between nucleotides on the same DNA strand.	Ligates (joins) sequences that were cut with a restriction endonuclease. Gives researchers the ability to splice fragments of DNA together.	
Plasmids	Small, extrachromosomal loops of DNA found in many bacteria and in some yeast.	After a target gene is inserted into a plasmid, the recombinant plasmid serves as a vector for transferring the gene into a bacterial or yeast cell, so the gene can be cloned.	
***Taq* polymerase**	DNA polymerase from the bacterium *Thermus aquaticus*. Catalyzes synthesis of DNA from a primed DNA template; remains stable at 95°C.	Responsible for the "primer extension" step in the polymerase chain reaction. Heat stability allows enzyme to be active even after denaturation step of PCR cycle at 95°C.	
Single nucleotide polymorphisms (SNPs)	Sites in DNA where the identity of the base varies among individuals in a population.	One of many polymorphic types of DNA sequences that are useful in creating the genetic maps required for gene hunts.	

Unfortunately, there are serious problems associated with using retroviruses as agents in gene therapy. For example, if viral genes happen to insert themselves in a position that disrupts the function of an important gene in the target cell, the consequences may be serious. Despite these risks, retroviruses are still the best vectors currently available for human gene therapy.

Using Gene Therapy to Treat X-Linked Immune Deficiency

In 2000, a research team reported the successful treatment by means of gene therapy of an illness called **severe combined immunodeficiency (SCID)**. Children who are born with SCID lack a normal immune system and are unable to fight off infections.

The type of SCID the team treated is designated SCID-X1, because it is caused by mutations in a gene on the X chromosome.

The gene codes for a receptor protein necessary for the development of immune system cells called T cells. T cells develop in bone marrow, from undifferentiated cells that divide continuously.

Traditionally, physicians have treated SCID-X1 by keeping the patient in a sterile environment, isolated from any direct human contact, until the person could receive a transplant of bone-marrow tissue from a close relative (**Figure 19.12**). In most cases, the T cells that the patient needs are produced by the transplanted bone-marrow cells and allow the individual to live normally. In some cases, though, no suitable donor is available. Could gene therapy cure this disease by furnishing functioning copies of the defective gene?

After extensive testing suggested that their treatment plan was safe and effective, the research team gained approval to treat 10 boys, each less than 1 year old, who had SCID-X1 but no suitable bone-marrow donor. The researchers removed bone marrow from each patient, collected the stem cells that produce mature

Technique	Description	How Used	Illustration
Recombinant DNA technology ("genetic engineering")	Taking a copy of a gene from one individual and placing it in the genome of a different individual (often of a different species).	Many applications, including DNA cloning, gene therapy (see Section 19.5), and biotechnology (see Sections 19.1 and 19.6).	Inserted gene
Genetic libraries	A collection of all DNA sequences present in a particular source. The library consists of individual DNA fragments that are isolated and inserted into a plasmid or other vector, so they can be cloned.	cDNA libraries allow researchers to catalog the genes being expressed in a particular cell type. Genomic libraries allow researchers to archive all the DNA sequences present in a genome. Libraries can be screened to find a particular target gene.	Stored cDNA
DNA probing/screening a genetic library	Use of a labeled, known DNA fragment to hybridize (by complementary base pairing) with a collection of unlabeled, unknown fragments.	Allows a researcher to find a particular DNA sequence in a large collection of sequences.	Labeled probe
Polymerase chain reaction (PCR)	A DNA synthesis reaction that uses known primer sequences on either side of a target gene. Reaction is based on many cycles of primer annealing, primer extension, and DNA denaturation.	Produces many identical copies of a target sequence. A shortcut method for DNA cloning.	
Dideoxy sequencing	In vitro DNA synthesis reaction that includes dideoxyribonucleotide triphosphates (ddNTPs) as monomers.	Determining the base sequence of a gene or other section of DNA.	G C C T A A T G
Genetic mapping	Creation of a map showing the relative positions of genes or specific DNA sequences on chromosomes. Done by analyzing the frequency of recombination between sequences (see Chapter 13).	Many applications, including use of mapped genetic markers to find unknown genes associated with diseases or other distinctive phenotypes.	Yellow body White eyes 1.4 Ruby eyes 6.1

FIGURE 19.12 A "Bubble Child."
Children with SCID cannot fight off bacterial or viral infections. As a result, such children must live in a sterile environment.

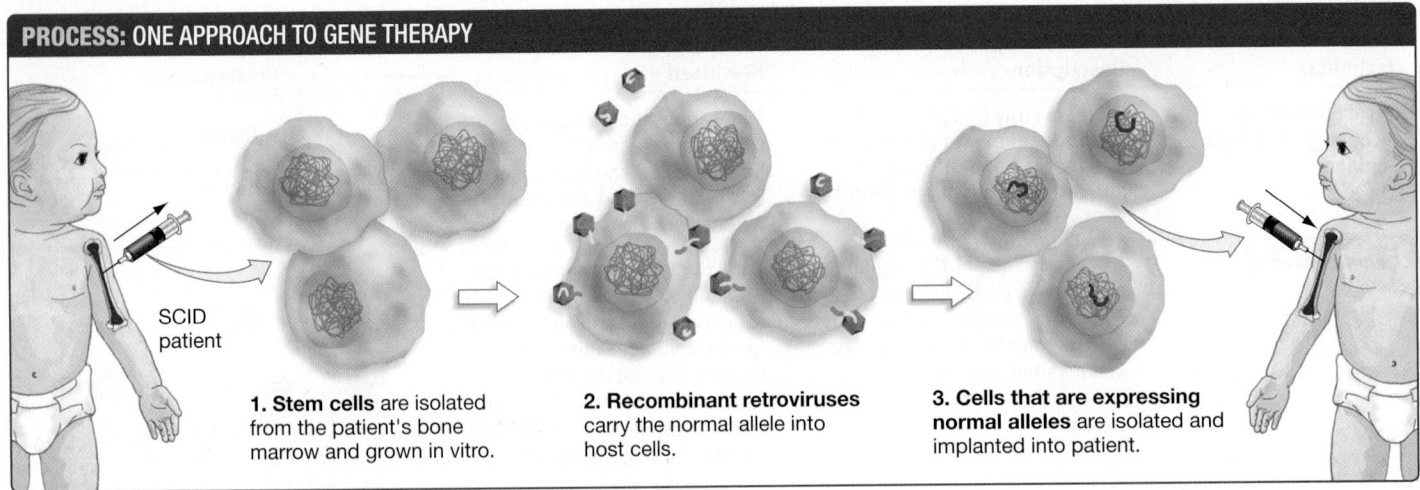

SCID patient

1. Stem cells are isolated from the patient's bone marrow and grown in vitro.

2. Recombinant retroviruses carry the normal allele into host cells.

3. Cells that are expressing normal alleles are isolated and implanted into patient.

FIGURE 19.13 Curing a Genetic Disorder Caused by a Loss-of-Function Allele. For gene therapy to work, copies of a normal allele have to be introduced into a patient's cells and be expressed.

T cells, and infected those cells with an engineered retrovirus that carried the normal receptor gene. Cells that began to produce normal receptor protein were then isolated and transferred back into the patients (**Figure 19.13**).

Within four months after reinsertion of the transformed marrow cells, nine of the boys had normal levels of functioning T cells. These patients were removed from germ-free isolation rooms and began residing at home, where they grew and developed normally.

Subsequently, however, four of the boys developed a cancer characterized by unchecked growth of T cells. Follow-up analyses of their bone-marrow cells showed that a viral-borne receptor gene had been inserted either near a gene for a regulatory transcription factor that triggers the growth of T cells, or near a gene for a cyclin that drives the cell cycle (see Chapter 11). The viral sequences apparently acted as an enhancer and led to constitutive expression of the transcription factor or cyclin.

As this book goes to press, three of the four boys have responded to cancer chemotherapy and are healthy. The fourth did not respond to treatment and died of cancer.

The tenth boy to receive gene therapy never succeeded in producing T cells at all. For unknown reasons, his recombinant stem cells failed to function normally when they were transplanted back into his bone marrow. Fortunately, physicians were later able to find a bone marrow donor whose cells were a close enough match to his to make a successful transplant possible.

Ethical Concerns over Gene Therapy

Throughout the history of medicine, efforts to test new drugs, vaccines, and surgical protocols have always carried a risk for the patients involved. Gene therapy experiments are no different. The researchers who run gene therapy trials must explain the risks clearly and make every effort to minimize them.

The initial report on the development of cancer in the boys who received gene therapy for SCID-X1 concluded with the following statement: "We have proposed . . . a halt to our trial until further evaluation of the causes of this adverse event and a careful reassessment of the risks and benefits of continuing our study of gene therapy."

When recombinant DNA technology first became possible, many researchers thought they would live to see most or all of the serious inherited diseases in humans cured by gene therapy. Several decades later, that optimism is tempered. In humans, gene therapy is still highly experimental and extremely expensive. Researchers are beginning to doubt whether gene therapy will ever be safe, effective, and affordable enough to be considered an important medical advance.

19.6 Case 6—The Development of Golden Rice: Biotechnology in Agriculture

Progress in human gene therapy has been slow, but progress in transforming crop plants with recombinant genes has been breathtakingly rapid. Worldwide, about 114 million hectares (282 million acres) of transgenic crops were grown in 2007. You have almost certainly eaten food from a genetically modified plant at some time, if not today.

Recent efforts to develop transgenic plants have focused on three general objectives:

1. *Reducing losses from herbivore damage* For instance, researchers have transferred a gene from the bacterium *Bacillis thuringiensis* into corn; the presence of the "Bt toxin" encoded by this gene protects the plant from corn borers and other caterpillar pests.

2. *Reducing competition with weeds* An example is the genetic engineering of soybeans for resistance to an herbicide—a molecule that kills plants—called glyphosate. Soybean fields with the engineered strain can be sprayed with glyphosate to kill weeds without harming the soybeans.

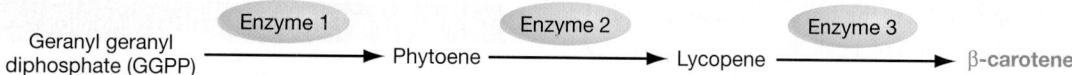

Geranyl geranyl
diphosphate (GGPP) —[Enzyme 1]→ Phytoene —[Enzyme 2]→ Lycopene —[Enzyme 3]→ β-carotene

FIGURE 19.14 Synthetic Pathway for β-Carotene. GGPP is a molecule found in rice seeds. Three enzymes are required to produce β-carotene from GGPP.

3. ***Improving the quality of the product consumed by people***
This objective is exemplified by the crop plants, including soybeans and canola, that have been engineered to produce a higher percentage of unsaturated fatty acids relative to saturated fatty acids (see Chapter 6). Saturated fatty acids can contribute to heart disease, so crops with less of them are healthier to eat.

How is this work done?

Rice as a Target Crop

Almost half the world's population depends on rice as its staple food. Unfortunately, rice is a poor source of certain vitamins and essential nutrients—including vitamin A. Vitamin A deficiency causes blindness in 250,000 Southeast Asian children each year. It also increases susceptibility to diarrhea, respiratory infections, and childhood diseases such as measles.

Humans and other mammals synthesize vitamin A from a precursor molecule known as β-carotene (beta-carotene). β-carotene belongs to a family of plant pigments called the carotenoids (see Chapter 10). Carotenoids are orange, yellow, and red and are especially abundant in carrots.

Rice plants synthesize β-carotene in their chloroplasts but not in the part of the seed that is eaten by humans. Could genetic engineering produce a strain of rice that synthesized β-carotene in the carbohydrate-rich seed tissue called endosperm, which humans do eat?

Synthesizing β-Carotene in Rice

To explore the possibility of genetically engineering rice, a research team searched for compounds in rice endosperm that could serve as precursors for the synthesis of β-carotene. They found that maturing rice endosperm contains a molecule called geranyl geranyl diphosphate (GGPP), which is an intermediate in the synthetic pathway that leads to the production of carotenoids.

As **Figure 19.14** shows, three enzymes are required to produce β-carotene from GGPP. If genes that encode these enzymes could be introduced into rice plants along with regulatory sequences that would trigger their synthesis in endosperm, the researchers could produce a transgenic strain of rice that would contain β-carotene.

Fortunately, genes that encode two of the required enzymes had already been isolated from daffodils, and the gene for the third enzyme had been purified from a bacterium. Because the sequences had been inserted into plasmids and grown in bacteria, many copies were available for manipulation. To each of the coding sequences in the plasmids, biologists added the promoter region from an endosperm-specific protein. This segment would promote transcription of the recombinant sequences in endosperm cells.

Next, the three sets of sequences had to be inserted into rice plants. How are foreign genes introduced into plants?

The *Agrobacterium* Transformation System

Agrobacterium tumefaciens is a bacterium that infects plant tissues and triggers formation of tumorlike growths called galls. When researchers looked into how these infections occur, they found that a plasmid carried by the *Agrobacterium* cells, called a **Ti (tumor-inducing) plasmid**, plays a key role (**Figure 19.15**).

Ti plasmids contain several functionally distinct sets of genes. One set encodes products that allow the bacterium to bind to the cell walls of a host. Another set, referred to as the virulence genes, encodes the proteins required to transfer part of the Ti DNA, called T-DNA (transferred DNA), into the interior of the plant cell. The T-DNA then travels to the plant cell's nucleus and integrates into the plant's chromosomal DNA (Figure 19.15, step 1).

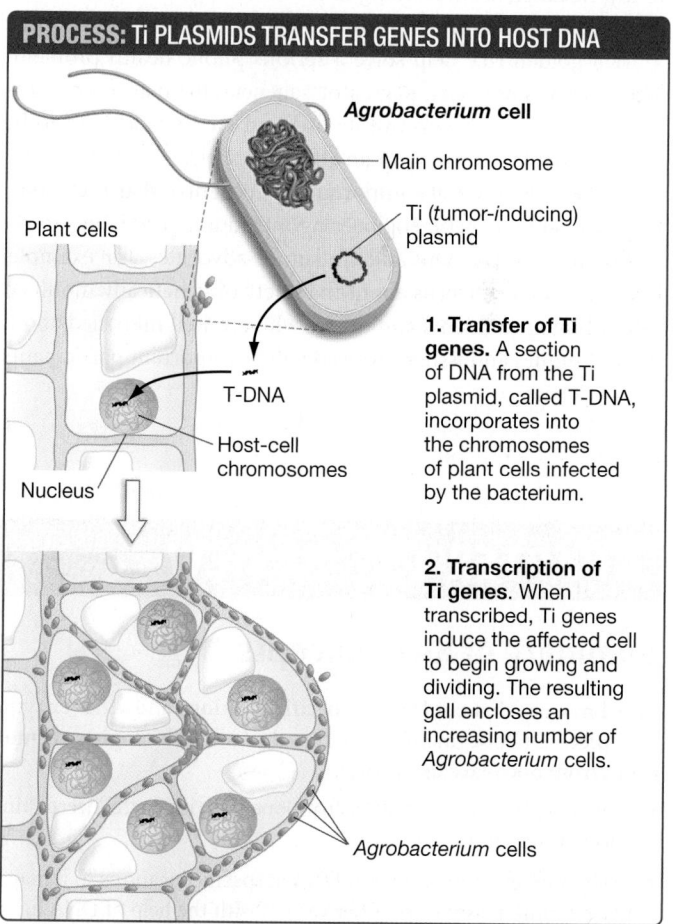

PROCESS: Ti PLASMIDS TRANSFER GENES INTO HOST DNA

Agrobacterium cell

Main chromosome

Ti (*tumor-inducing*) plasmid

Plant cells

T-DNA

Host-cell chromosomes

Nucleus

1. Transfer of Ti genes. A section of DNA from the Ti plasmid, called T-DNA, incorporates into the chromosomes of plant cells infected by the bacterium.

2. Transcription of Ti genes. When transcribed, Ti genes induce the affected cell to begin growing and dividing. The resulting gall encloses an increasing number of *Agrobacterium* cells.

Agrobacterium cells

FIGURE 19.15 Agrobacterium Infections Introduce Genes into a Host-Cell Chromosome. Ti plasmids of *Agrobacterium* cells induce gall formation—a tumorlike growth.

When transcribed, T-DNA induces the infected cell to grow and divide. The result is the formation of a gall that houses a growing population of *Agrobacterium* cells (Figure 19.15, step 2).

Researchers soon realized that the Ti plasmid offers an efficient way to introduce recombinant genes into plant cells. Follow-up experiments confirmed that recombinant genes could be added to the T-DNA that integrates into the host chromosome, that the gall-inducing genes could be removed from the T-DNA, and that the resulting sequence is efficiently transferred and expressed in its new host plant.

Using the Ti Plasmid to Produce Golden Rice

To generate a strain of rice that produces all three enzymes needed to synthesize β-carotene in endosperm, the researchers exposed embryos to *Agrobacterium* cells containing genetically modified Ti plasmids (**Figure 19.16**). The transformed plants were then grown in the greenhouse. When the transgenic individuals had matured and produced seeds, the researchers found that some rice grains contained so much β-carotene that they appeared yellow. The biologists called the engineered plants "golden rice."

Follow-up experiments used gene sequences from maize, rather than daffodil, and resulted in individuals that produce 23 times as much β-carotene in their seeds as the original transformants. Currently, researchers are working to get the new gene sequences into the rice strains that are most commonly planted in Southeast Asia.

Will golden rice help solve a serious public health problem? The answer is not clear. Regulatory agencies in an array of countries would need to approve its use, and seed would have to be made available to farmers at an affordable price.

In the meantime, it's important to recognize that each solution offered by genetic engineering introduces new issues to resolve. Some researchers and consumer advocates, for example, have expressed concerns about the safety of genetically modified foods. Biology students and others who are well informed about the techniques and issues involved will be important participants in this debate.

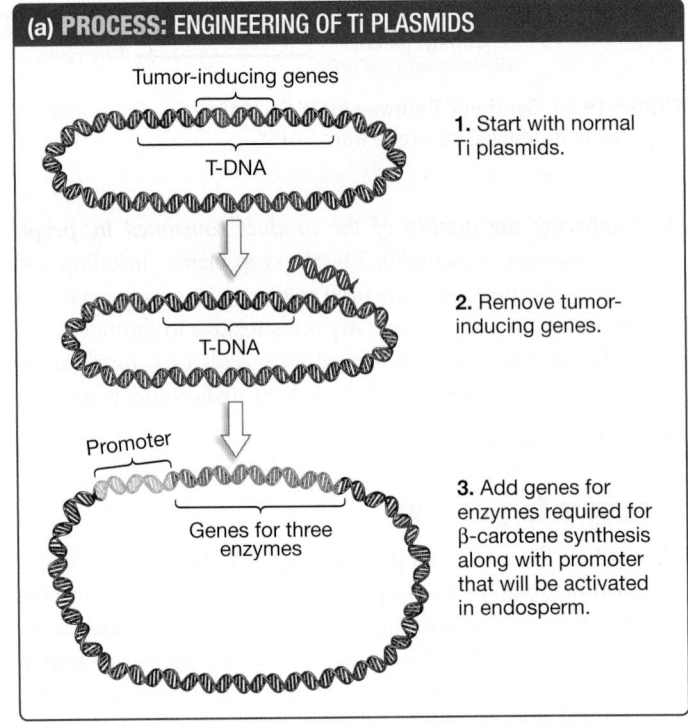

(a) PROCESS: ENGINEERING OF Ti PLASMIDS

Tumor-inducing genes

T-DNA

1. Start with normal Ti plasmids.

2. Remove tumor-inducing genes.

T-DNA

Promoter

Genes for three enzymes

3. Add genes for enzymes required for β-carotene synthesis along with promoter that will be activated in endosperm.

(b) Golden rice (right) is engineered to synthesize β-carotene.

FIGURE 19.16 Constructing a Ti plasmid that will produce "Golden Rice." Golden rice is a transgenic strain capable of synthesizing β-carotene in the endosperm of its seeds.

CHAPTER 19 REVIEW

For media, go to the study area at www.masteringbiology.com

Summary of Key Concepts

🔑 **Enzymes that cut DNA at specific locations and other enzymes that piece DNA segments back together allow biologists to move genes from one place to another.**

- In genetic engineering, alleles from one individual are inserted into another individual.

- Restriction endonucleases cut DNA at specific locations so it may be inserted into plasmids or other vectors with the help of DNA ligase.

- In many cases, engineered alleles are modified by the addition of certain types of promoters or other regulatory sequences.

✔ You should be able to explain why the discovery of DNA ligase was crucial to genetic engineering. What problem would arise for genetic engineers if the enzyme did not exist?

🔑 **Biologists can obtain many identical copies of a gene by (1) inserting it into a bacterial cell that copies the gene each time the cell divides or (2) by conducting a polymerase chain reaction.**

- To clone the gene for the human growth hormone, researchers isolated the gene, introduced it into a plasmid that was taken up by *E. coli* cells, and cultured the *E. coli* cells.

- The polymerase chain reaction (PCR) depends on having primers that bracket a target stretch of DNA. These allow *Taq* polymerase, a heat-stable form of DNA polymerase, to amplify a single target DNA sequence to millions of identical copies.

  ✔ You should be able to list the advantages and disadvantages of plasmids versus PCR for cloning genes.

  (MB) **Web Activity** Producing Human Growth Hormone, **Web Activity** The Polymerase Chain Reaction

The sequence of bases in a gene can be determined by dideoxy sequencing.

- Dideoxy sequencing is based on an in vitro synthesis reaction in which dideoxyribonucleotides stop DNA replication at each base in the sequence.

- When the DNA fragments generated by a dideoxy sequencing reaction are separated via gel electrophoresis, the sequence of nucleotides in the gene can be determined.

  ✔ You should be able to explain why the newly synthesized DNA fragments—when they are lined up by size—correspond to the sequence of bases in the template DNA.

If individuals with a certain phenotype also tend to share a genetic marker (a known site in DNA that is unrelated to the phenotype), the gene responsible for the phenotype is likely to be near that marker.

- To find the gene associated with Huntington's disease, investigators analyzed a large number of polymorphic genetic markers in affected and unaffected individuals. Their goal was to find a marker that was inherited along with the allele responsible for the disease.

- Once an association study pinpointed the general area where the gene was located, biologists could sequence DNA from the region to determine exactly where the gene of interest was located.

  ✔ You should be able to explain why genetic markers that lack polymorphism are not useful in gene hunts.

Researchers are attempting to insert genes into humans to cure genetic diseases. Efforts to insert genes into plants have been much more successful.

- Once genes are located and characterized, they can be introduced into other individuals or species in an effort to change their traits.

- Genetic transformation can occur in several ways, depending on the species involved.

- In humans, recombinant DNA must be introduced by viruses. Because this is technically difficult and possibly dangerous, progress in human gene therapy has been slow.

- Certain bacteria that infect plants have plasmids that integrate their genes into the host-plant genome. By adding recombinant alleles to these plasmids, researchers have been able to introduce alleles that improve crops.

  ✔ You should be able to explain how genes are inserted into plants.

Questions

✔ TEST YOUR KNOWLEDGE   *Answers are available in Appendix B*

1. What do restriction endonucleases do?
   a. They cleave bacterial cell walls and allow viruses to enter the cells.
   b. They join pieces of DNA by catalyzing the formation of phosphodiester bonds between them.
   c. They cut stretches of DNA at specific sites known as recognition sequences.
   d. They act as genetic markers in the chromosome maps used in gene hunts.

2. What is a plasmid?
   a. an organelle found in many bacteria and certain eukaryotes
   b. a circular DNA molecule that in some cases replicates independently of the main chromosome(s)
   c. a type of virus that has a DNA genome and that infects certain types of human cells, including lung and respiratory tract tissue
   d. a type of virus that has an RNA genome, codes for reverse transcriptase, and inserts a cDNA copy of its genome into host cells

3. When present in a DNA synthesis reaction mixture, a ddNTP molecule is added to the growing chain of DNA. No further nucleotides can be added afterward. Why?
   a. There are not enough dNTPs available.
   b. A ddNTP can be inserted at various locations in the sequence, so fragments of different length form—each ending with a ddNTP.
   c. The 5′ carbon on the ddNTP lacks a hydroxyl group, so no phosphodiester bond can form.
   d. The 3′ carbon on the ddNTP lacks a hydroxyl group, so no phosphodiester bond can form.

4. Once the gene that causes Huntington's disease was found, researchers introduced the defective allele into mice to create an animal model of the disease. Why was this model valuable?
   a. It allowed them to test potential drug therapies without endangering human patients.
   b. It allowed them to study how the gene is regulated.
   c. It allowed them to make large quantities of the huntingtin protein.
   d. It allowed them to study how the gene was transmitted from parents to offspring.

5. To begin the hunt for the human growth hormone gene, researchers created a cDNA library from cells in the pituitary gland. What did this library contain?
   a. only the sequence encoding growth hormone
   b. DNA versions of all the mRNAs in the pituitary-gland cells
   c. all of the coding sequences in the human genome, but no introns
   d. all of the coding sequences in the human genome, including introns

6. What does it mean to say that a genetic marker and a disease gene are closely linked?
   a. The marker lies within the coding region for the disease gene.
   b. The sequence of the marker and the sequence of the disease gene are extremely similar.
   c. The marker and the disease gene are on different chromosomes.
   d. The marker and the disease gene are in close physical proximity and tend to be inherited together.

1. Explain how restriction endonucleases and DNA ligase are used to insert foreign genes into plasmids and create recombinant DNA. Make a drawing that shows why sticky ends are sticky, and that identifies the exact location where DNA ligase catalyzes a key reaction.

2. Explain the function of a vector in genetic engineering. List one attribute of a "perfect" vector.

3. What is a cDNA library? Would you expect the cDNA library from a human muscle cell to be different from the cDNA library from a human nerve cell in the same individual? Explain why or why not.

4. What are genetic markers, and how are they used to create a genetic map?

5. Researchers added the promoter sequence from an endosperm-specific gene to the Ti plasmids used in creating golden rice. Why was this step important? Comment on the roles of promoter and enhancer sequences in genetic engineering in eukaryotes.

6. Compare and contrast PCR with the DNA synthesis that occurs in cells (see Chapter 14).

1. Suppose you had a large amount of sequence data, similar to the data that Nancy Wexler's team had in the region of the Huntington's disease gene, and that you knew that genes of the species being studied typically contain about 1500 bases. How would you use the genetic code (see Chapter 15) and information on the structure of promoters (see Chapters 17 and 18) to find the precise location of one or more genes in your sequence?

2. Is it possible for human gene therapy to alter germ-line cells so that individuals could pass altered alleles on to their offspring? Explain why or why not.

3. Describe similarities between how researchers screen a DNA library and how they perform a genetic screen—for example, for mutant *E. coli* cells that cannot metabolize lactose (see Chapter 17).

4. A friend of yours is doing a series of PCR reactions and comes to you for advice. She purchased two sets of primers, hoping that one set would amplify the template sequence shown here. (The dashed lines in the template sequence stand for a long sequence of bases.) Neither of the primer pairs produced any product DNA, however.

	Primer a		Primer b
Primer Pair 1:	5' CAAGTCC 3'	&	5' GCTGGAC 3'
Primer Pair 2:	5' GGACTTG 3'	&	5' GTCCAGC 3'
Template:	5' ATTCGGACTTG---GTCCAGCTAGAGG 3'		
	3' TAAGCCTGAAC---CAGGTCGATCTCC 5'		

a. Explain why each primer pair didn't work. Indicate whether both primers are at fault or just one primer is the problem.

b. Your friend doesn't want to buy new primers. She asks you whether she can salvage this experiment. What do you tell your friend to do?

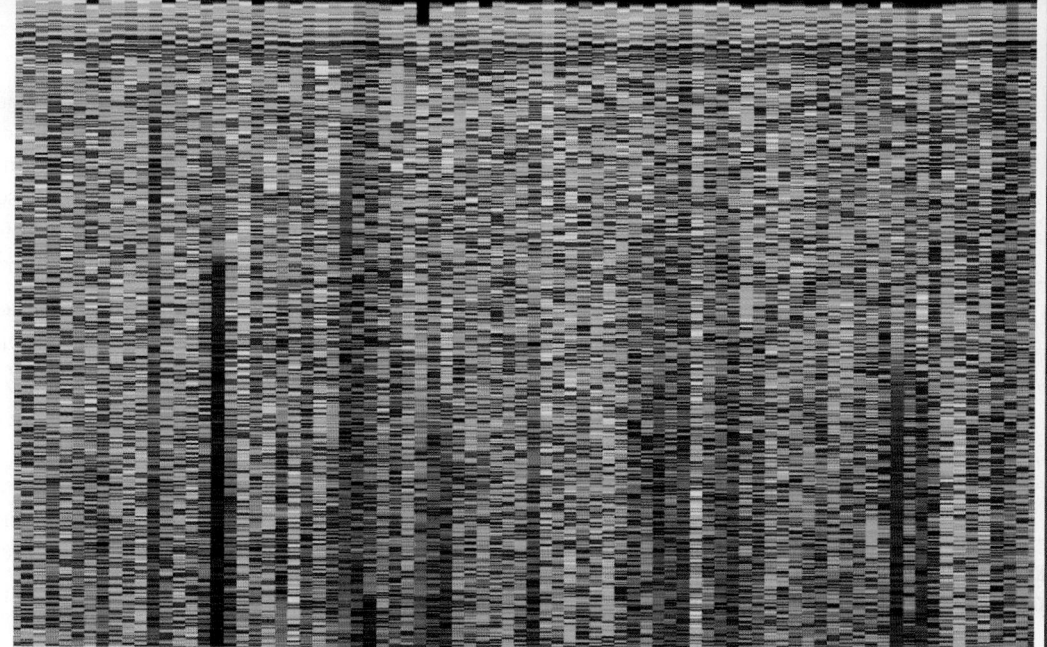

Output from an automated genome-sequencing machine, representing about 48,000 bases from the human genome. Each vertical stripe represents the sequence of a stretch of DNA.

Genomics 20

The first data sets describing the complete DNA sequence, or **genome**, of humans were published in February 2001. These papers were immediately hailed as a landmark in the history of science. In just 50 years, biologists had gone from not understanding the molecular nature of the gene to knowing the molecular makeup of every gene present in our species.

Years later, the multinational effort called the **Human Genome Project** is still producing a wide array of data on the locations and functions of genes and other types of DNA sequences found in humans. It's important to recognize, though, that research on *Homo sapiens* is part of a much larger, ongoing effort to sequence genomes from an array of other eukaryotes, hundreds of bacteria, and dozens of archaea. The effort to sequence, interpret, and compare whole genomes is referred to as **genomics**. The pace of progress in this field is nothing short of explosive.

Progress in genomics has triggered the development of a related field called functional genomics. Genomics supplies a list of the genes present in an organism; **functional genomics** answers questions about the functioning of that genome, such as when genes are expressed and how gene products interact.

As an introductory biology student, you are part of the first generation trained in the genome era. Genomics is revolutionizing biological science and will almost certainly be an important part of your personal and professional life. Let's delve in.

20.1 Whole-Genome Sequencing

Genomics has moved to the cutting edge of research in biology, largely because technological advances—including the development of the dideoxy sequencing and pyrosequencing techniques introduced in Chapter 19—have made it possible to obtain large quantities of high-quality sequence data in a reasonable amount of time, with a reasonable amount of money.

KEY CONCEPTS

○→ Once a genome has been completely sequenced, researchers use a variety of techniques to identify which sequences code for products and which act as regulatory sites.

○→ Bacterial and archaeal genomes are relatively small. Among species, there is a positive correlation between total gene number and metabolic capabilities. Gene transfer between species is also common.

○→ Eukaryotic genomes are large and complex. They include many sequences that have little to no effect on the fitness of the organism, and many transcribed sequences whose function is not known.

○→ Data and techniques derived from genome sequencing projects are being used to analyze cancer cells.

✔ When you see this checkmark, stop and test yourself. Answers are available in Appendix B.

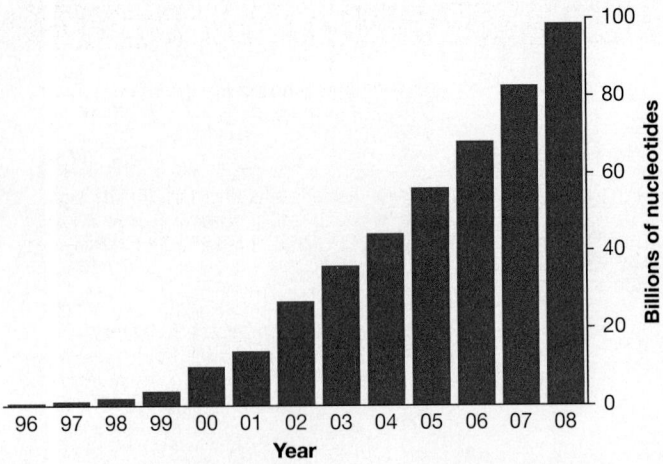

FIGURE 20.1 The Total Number of Bases Sequenced Is Growing Rapidly. Data from the DNA Data Bank of Japan.

As data become less expensive and faster to acquire, the pace of genome sequencing accelerates. The result is that an almost mind-boggling number of sequences are now being generated. As this book goes to press, the primary international repository for DNA sequence data contains over 106 *billion* nucleotides. By way of comparison, a haploid human genome contains about 3 billion bases.

Figure 20.1 gives a visual sense of the growth in sequence data by plotting time versus the number of nucleotides, in billions. There are three large international online repositories for sequence data. The numbers plotted here were compiled by one of them: the DNA Data Bank of Japan. The height of each bar on the graph represents the size of the database in that year. If you compare the increase from one year to the next to the total in the previous year, and do that for several years, you will see that the database is growing at an average rate of about 30 percent per year.

How Are Complete Genomes Sequenced?

Genomes range in size from about a half million base pairs to several billion. But even under the best conditions, a single sequencing reaction can analyze only about 1000 base pairs. How do investigators break a genome into sequencing-sized pieces and then figure out how the thousands or millions of pieces go back together?

SHOTGUN SEQUENCING When researchers set out to sequence the genome of a certain species for the first time, they usually rely on an approach known as **shotgun sequencing**. In shotgun sequencing, a genome is broken up into a set of overlapping fragments that are small enough to be sequenced. The regions of overlap are then used as guides for putting the sequenced fragments back into the correct order (**Figure 20.2**).

Step 1 High-frequency sound waves, or sonication, are used to randomly break a genome into pieces about 160 kilobases (kb) long (1 kb = 1000 bases).

Step 2 Each 160-kb piece is inserted into a plasmid called a **bacterial artificial chromosome (BAC)**. BACs are able to replicate large segments of DNA. Using techniques introduced in Chapter 19, each BAC is then inserted into a different

Escherichia coli cell, creating a **BAC library**. A BAC library is a genomic library: a set of all the DNA sequences in a particular genome, split into small segments and inserted into cloning vectors (see Chapter 19). By separating the cells in a BAC library and allowing each cell to grow into a large colony, researchers can isolate large numbers of each 160-kb fragment.

Step 3 After many copies of each 160-kb fragment have been produced, the DNA is again broken into fragments—but this time, the segments are about 1 kb long.

Step 4 These small fragments are then inserted into plasmids and placed inside bacterial cells. (Note that by this point the genome has been broken down twice, into increasingly manageable pieces: 160-kb fragments in BACs and 1-kb segments in plasmids.) The plasmids are copied many times as the bacterial cells grow into a large population. Large numbers of each 1-kb fragment are then available for sequencing reactions.

Step 5 Next, the 1-kb fragments from each 160-kb BAC clone are sequenced, and computer programs analyze regions where the ends of the various 1-kb segments overlap. Overlaps occur because there were many copies made of each 160-kb segment, and these copies were fragmented randomly by sonication.

Step 6 The computer mixes and matches segments from a single BAC clone until an alignment consistent with all available data is obtained and the BACs have been reconstructed.

Step 7 The ends of the reconstructed BACs are analyzed in a similar way. The goal is to arrange each 160-kb segment in its correct position along the chromosome, based on regions of overlap.

In essence, the shotgun strategy consists of breaking a genome into tiny fragments, sequencing the fragments, and then putting the sequence data back into the correct order. To review the logic of the shotgun sequencing approach, go to the study area at *www.masteringbiology.com*.

(MB) Web Activity Human Genome Sequencing Strategies

✔If you understand shotgun sequencing, you should be able to explain why it is essential for fragments to have regions of overlap.

THE IMPACT OF NEXT-GENERATION SEQUENCING STRATEGIES Chapter 19 introduced pyrosequencing approaches as faster and cheaper alternatives to traditional dideoxy sequencing. The speed and efficiency are due to:

- Miniaturization that makes it possible to sequence many DNA fragments at the same time.

- Avoiding slow and expensive steps like cloning fragments into BACs and plasmids to produce many copies of each fragment.

The key difference is that pyrosequencing reactions take place on a single DNA fragment—rather than many copies of the same fragment as required by dideoxy sequencing. With pyrosequencing, a genome is sheared into many tiny fragments which are then separated and sequenced directly.

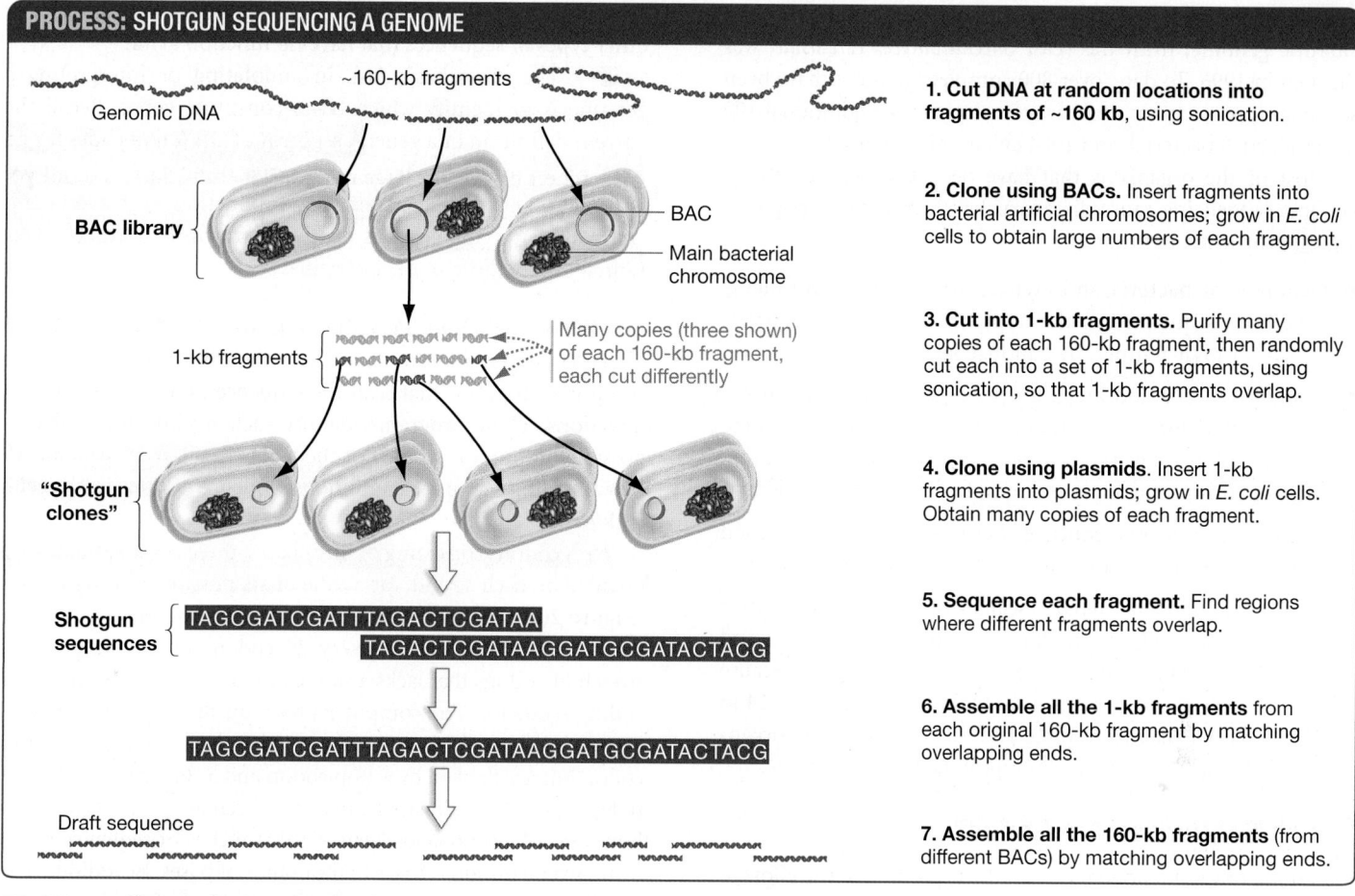

PROCESS: SHOTGUN SEQUENCING A GENOME

~160-kb fragments

Genomic DNA

BAC library

BAC

Main bacterial chromosome

1-kb fragments

Many copies (three shown) of each 160-kb fragment, each cut differently

"Shotgun clones"

Shotgun sequences

TAGCGATCGATTTAGACTCGATAA

TAGACTCGATAAGGATGCGATACTACG

TAGCGATCGATTTAGACTCGATAAGGATGCGATACTACG

Draft sequence

1. **Cut DNA at random locations into fragments of ~160 kb**, using sonication.

2. **Clone using BACs.** Insert fragments into bacterial artificial chromosomes; grow in *E. coli* cells to obtain large numbers of each fragment.

3. **Cut into 1-kb fragments.** Purify many copies of each 160-kb fragment, then randomly cut each into a set of 1-kb fragments, using sonication, so that 1-kb fragments overlap.

4. **Clone using plasmids.** Insert 1-kb fragments into plasmids; grow in *E. coli* cells. Obtain many copies of each fragment.

5. **Sequence each fragment.** Find regions where different fragments overlap.

6. **Assemble all the 1-kb fragments** from each original 160-kb fragment by matching overlapping ends.

7. **Assemble all the 160-kb fragments** (from different BACs) by matching overlapping ends.

FIGURE 20.2 Shotgun Sequencing Breaks Large Genomes into Many Short Segments.

✔**QUESTION** A shotgun blast produces many small, scattered pieces of shot. Why is "shotgun" an appropriate way to describe this sequencing strategy?

The downside of pyrosequencing is that it only works with fragments of about 100 base pairs in length. These fragments are too small to be pieced back together to reconstruct a complete genome accurately. But if a complete genome is already available for the organism, pyrosequencing offers a relatively quick and inexpensive way to sequence the entire genome from a particular individual—with all the tiny fragments being arranged in the correct order by being compared to the "master genome."

BIOINFORMATICS How do researchers piece together the millions of fragments produced by a shotgun sequencing or pyrosequencing study? And once a complete genome is assembled, how are the raw sequence data and any annotations—meaning information about the position, potential product, and function of a particular sequence—made available to the international community of researchers?

The answer is **bioinformatics**—a field that fuses computer science and biology in an effort to manage, analyze, and interpret biological information. For example, researchers in bioinformatics have created searchable databases that hold annotated sequence information, so that investigators can evaluate the similarities between newly discovered genes and genes that had been studied previously in the same or other species.

Because the amount of data is so large, the computational challenges in genomics are formidable. Thus far, sophisticated algorithms and continually improving computer hardware have allowed researchers to keep pace with the rate of data acquisition. The vast quantity of data generated by genome sequencing centers has made bioinformatics a key to continued progress.

Which Genomes Are Being Sequenced, and Why?

The first genome to be sequenced from an organism—not a virus—came from a bacterium that lives in the human upper respiratory tract. *Haemophilus influenzae* has one circular chromosome and a total of 1,830,138 base pairs of DNA. Its genome was small enough to sequence completely in a reasonable amount of time and within a reasonable budget, given the technology available in the early 1990s. *H. influenzae* was an important research subject because it causes earaches and respiratory tract infections in children. One strain is also capable of infecting the membranes surrounding the brain and spinal cord, causing meningitis.

Publication of the *H. influenzae* genome in 1995 was quickly followed by publication of complete genomes sequenced from an assortment of bacteria and archaea. Sequencing of the first eukaryotic genome, from the yeast *Saccharomyces cerevisiae*, was finished in 1996. To date, over 800 complete genomes have been sequenced; projects are underway to sequence an additional 100 archaeal, 2079 bacterial, and 1004 eukaryotic genomes.

Most of the organisms that have been selected for whole-genome sequencing cause disease or have other interesting biological properties. For example,

- Genomes of bacteria and archaea from hot environments have been sequenced in the hopes of discovering enzymes useful for high-temperature industrial applications.

- Certain bacterial and archaeal species have been chosen because they perform interesting chemical reactions, such as the synthesis of methane (natural gas; CH_4). Researchers hope that these organisms might act as a source of commercial products.

- The rice genome was sequenced because rice is the main food source for most humans.

- The fruit fly *Drosophila melanogaster*, the roundworm *Caenorhabditis elegans*, the house mouse *Mus musculus*, and the mustard plant *Arabidopsis thaliana* were analyzed because they serve as model organisms in biology (see **BioSkills 14** in Appendix A). Data from these and other well-studied organisms has helped researchers interpret the human genome.

Which Sequences Are Genes?

Obtaining raw sequence data is just the beginning of the effort to understand a genome. As researchers point out, raw sequence data are analogous to the parts list for a house. The list would read something like "windowwabeborogovestaircasedoorjubjub . . . ," however, because it has no punctuation and contains portions that appear to have no meaning.

Where do the genes for "window," "staircase," and "door" start and end? Are the segments that read "wabeborogove" and "jub-

jub" important in gene regulation, or are they simply spacers or other types of sequences that have no function at all?

🔑 The most basic task in annotating or interpreting a genome is to identify which bases constitute genes. Recall the current definition of a gene: A segment of DNA that codes for an RNA or a protein product—or a series of alternatively spliced products—and that regulates their production. In bacteria and archaea, identifying genes is relatively straightforward. The task is much more difficult in eukaryotes.

IDENTIFYING GENES IN BACTERIAL AND ARCHAEAL GENOMES To interpret bacterial and archaeal genomes, biologists begin with computer programs that scan the sequence of a genome in both directions. These programs identify each reading frame that is possible on the two strands of the DNA. Recall from Chapter 15 that a reading frame is a continuous sequence of non-overlapping codons.

With codons consisting of three bases, three reading frames are possible on each strand, for a total of six possible reading frames (**Figure 20.3**). Because randomly generated sequences contain a stop codon about one in every 20 codons on average, a long stretch of codons that lacks a stop codon is a good indication of a coding sequence. The computer programs draw attention to any "gene-sized" stretches of sequence that lack an internal stop codon but are flanked by a stop codon and a start codon. Because polypeptides range in size from a few dozen amino acids to many hundreds of amino acids, gene-sized stretches of sequence range from several hundred bases to thousands of bases. In addition, the computer programs look for sequences typical of promoters, operators, or other regulatory sites. DNA segments that are identified in this way are called **open reading frames**, or **ORFs**.

Once an ORF is found, a computer program compares its sequence with the sequences of known genes from well-studied species. If the ORF is unlike any gene that has so far been described in any species, further research is required before it can

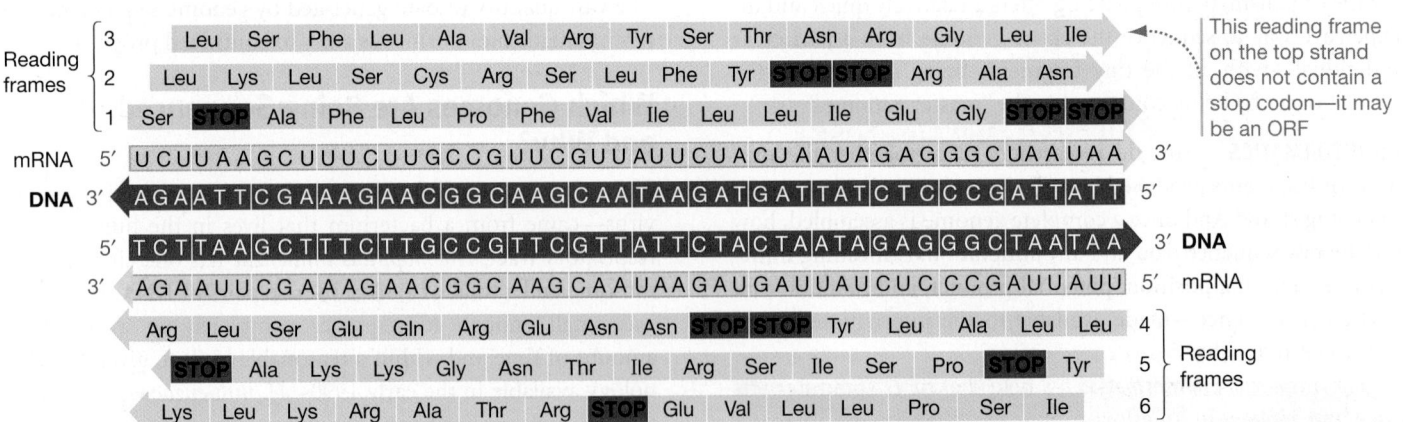

FIGURE 20.3 Open Reading Frames May Be the Locations of Genes. Computer programs scan the three possible reading frames on each strand of DNA and use the genetic code to translate each codon. A long stretch of codons that lacks a stop codon may be an open reading frame (ORF)—a possible gene.

✔**QUESTION** To predict the mRNA codons that would be produced by a particular reading frame, a computer analyzes the DNA in the 3′ to 5′ direction. Why?

actually be considered a gene. A "hit," in contrast, means that the ORF shares a significant amount of sequence with a known gene from another species.

Similarities between genes in different species are usually due to **homology**. If genes are homologous, it means they are similar because they are related by descent from a common ancestor. Homologous genes have similar base sequences and frequently the same or a similar function. For example, consider the genes introduced in Chapter 14 that code for enzymes involved in repairing mismatches in DNA. The mismatch-repair genes in *E. coli*, yeast, and humans are similar in structure, DNA sequence, and function. To explain this similarity, biologists hypothesize that the common ancestor of all cells living today had mismatch-repair genes—thus, the descendants of this ancestral species also have versions of these genes.

Researchers can confirm that an ORF is actually a gene by finding that it is homologous to a known gene. They can also analyze the product that would be produced by an ORF and see if it conforms to a known gene. It has taken much more complex analyses, though, to find eukaryotic genes.

IDENTIFYING GENES IN EUKARYOTIC GENOMES Mining eukaryotic sequence data for genes is complicated. Because coding regions are broken up by introns, for example, it is not possible to scan for long ORFs. Instead, researchers combine an array of approaches.

Perhaps the most productive gene-finding strategy has been to isolate mRNAs from cells, use reverse transcriptase to produce a cDNA version of each mRNA, and sequence a portion of the resulting molecule to produce an **expressed sequence tag**, or **EST**. ESTs represent protein-coding genes. To locate the gene responsible, researchers use the EST to find the matching sequence in genomic DNA.

Although ESTs and other gene-finding strategies have been productive, it will probably be many years before biologists are convinced they have identified all of the coding regions in even a single eukaryotic genome. As that effort continues, though, researchers are analyzing the data and making some remarkable observations. Let's first consider what genome sequencing has revealed about the nature of bacterial and archaeal genomes and then move on to eukaryotes. Is the effort to sequence whole genomes paying off?

20.2 Bacterial and Archaeal Genomes

By the time you read this paragraph, biologists will have obtained extensive, if not complete, sequence data from close to 2300 distinct bacterial and archaeal species or strains. For example, researchers have sequenced the genome of a laboratory population of *Escherichia coli*—derived from the harmless strain that lives in your gut—as well as the genome of a form that causes severe disease in humans. As a result, researchers can now compare the genomes of closely related cells that have different ways of life.

This section focuses on a simple question: Based on data published between 1995 and 2009, what general observations have biologists been able to make about the nature of all these bacterial and archaeal genomes?

The Natural History of Prokaryotic Genomes

In a sense, biologists who are working in genomics can be compared to the naturalists of the eighteenth and nineteenth centuries. These early biologists explored the globe, collecting the plants and animals they encountered. Their goal was to describe what existed. Similarly, the first task of a genome sequencer is to catalog what is in a genome—specifically, the number, type, and organization of genes. Several interesting conclusions can be drawn from relatively straightforward observations about the data obtained thus far.

- 🔑 In bacteria, there is a general correlation between the size of a genome and the metabolic capabilities of the organism. Species that live in a wide array of habitats and use a wide array of molecules for food have large genomes; parasites have small genomes (**Figure 20.4**). (A parasite lives off a host—making use of much of the host's biochemical machinery—rather than synthesizing its own molecules.)

- The function of many bacterial genes is still unknown. *E. coli* may be the most intensively studied of all organisms, but the function of over 30 percent of its genes is unknown.

- There is tremendous genetic diversity among bacteria and archaea. About 15 percent of the genes in each species' genome appear to be unique, based on the genomes sequenced to date. That is, about one in six genes in one of these species has yet to be found in a different type of prokaryote.

- Redundancy among genes within a genome is common. *E. coli* has 86 genes whose DNA sequences are nearly identical—meaning that the proteins they produce are nearly alike in structure and presumably in function. Biologists hypothesize that slightly different forms of the same protein are produced in response to slight changes in environmental conditions.

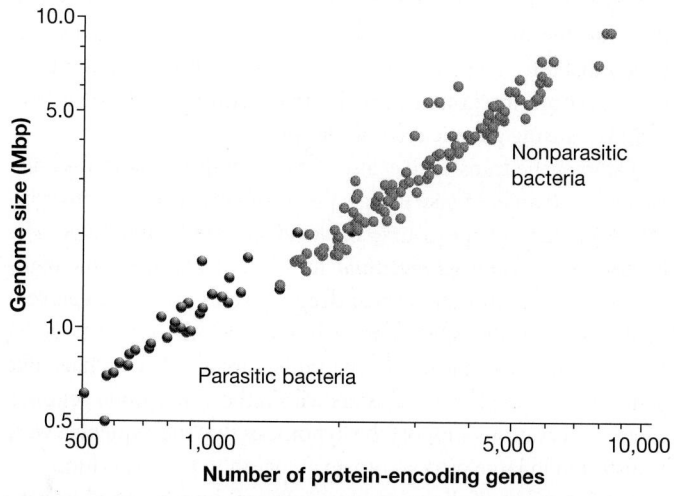

FIGURE 20.4 Parasites Have Small Genomes. When reading this graph, keep in mind that both axes are plotted on logarithmic scales (see **BioSkills 7** in Appendix A). Gene number and genome size are positively correlated—meaning that they increase together. But note that almost all of the parasitic species have genomes that are smaller than the nonparasitic species.

- Multiple chromosomes are more common than anticipated. Several species of bacteria and archaea have two different circular chromosomes instead of one. And at least some bacteria have linear chromosomes.

- Many species contain the small, extrachromosomal DNA molecules called plasmids. Recall from Chapter 19 that plasmids contain a small number of genes, though not genes that are absolutely essential for growth. Plasmids may be exchanged between cells of the same species or even of different species.

Perhaps the most surprising observation of all, however, is that in many bacterial and archaeal species, a significant proportion of the genome appears to have been acquired from other, often distantly related, species. How could this happen?

Lateral Gene Transfer

The movement of DNA from one species to another is called **lateral gene transfer**. Recent estimates suggest that over 50 percent of archaean species and 30–50% of bacterial species have at least one gene acquired by lateral gene transfer. Over the course of evolution, lateral gene transfer has been an important source of new genes and allelic diversity in bacteria and archaea.

Biologists use two general criteria to identify genes that were acquired via lateral gene transfer: (1) a gene is much more similar to genes in distantly related species than to those in closely related species and (2) the proportion of G-C base pairs to A-T base pairs in a particular gene or series of genes is markedly different from the base composition of the rest of the genome. In many cases, the proportion of G-C bases in a genome is characteristic of the particular genus or species.

How can genes move from one species to another? In at least some cases, plasmids appear to be responsible. For example, most of the genes that are responsible for conferring resistance to antibiotics are found on plasmids. Researchers have documented the transfer of plasmid-borne antibiotic-resistance genes between distantly related bacteria. In some cases, genes from plasmids become integrated into the main chromosome of a bacterium, causing genetic recombination.

Lateral gene transfer may also occur by transformation—when bacteria and archaea take up raw pieces of DNA from the environment, perhaps in the course of acquiring other molecules. The bacterium *Thermotoga maritima*, for example, occupies the high-temperature environments near deep-sea vents. Almost 25 percent of the genes in this species are extremely closely related to genes found in archaea that live in the same habitats. The archaea-like genes occur in distinctive clusters within the *T. maritima* genome. These observations support the hypothesis that the sequences were transferred in large pieces from an archaean to the bacterium.

Similar types of direct gene transfer are hypothesized to have occurred in the bacterium *Chlamydia trachomatis*. This organism is a major cause of blindness in humans from Africa and Asia; it also causes chlamydia, the most common sexually transmitted bacterial disease in the United States. The *C. trachomatis*

genome contains 35 genes that resemble eukaryotic genes in structure. Because *C. trachomatis* lives inside the cells that it parasitizes, the most logical explanation for this observation is that the bacterium occasionally takes up DNA directly from its host cell, resulting in a eukaryote-to-bacterium transfer.

Environmental Sequencing

Biologists continue to sequence the genomes of individual species and strains. But recently some research groups have taken a different approach: Cataloging all of the genes present in a community of bacteria and archaea. This type of research program is called **environmental sequencing** or **metagenomics**. The subject of these studies is genes—not organisms.

The first environmental sequencing study was conducted in the Sargasso Sea, near the islands of the Bahamas. Researchers chose the spot because it is extremely nutrient-poor and species-poor—a desert in the ocean. This is a common strategy in biological science: Start by studying a simple system, then go on to more complex situations. To inventory the complete array of bacterial genes present, the research group collected cells from different water depths and locations, isolated DNA from the samples, and sequenced the DNA.

After analyzing over 1 billion base pairs, the team concluded that at least 1800 bacterial species were present, of which 148 were previously undiscovered. They also identified more than 1.2 million alleles that had never before been characterized. These alleles included over 780 sequences that code for proteins similar to the rhodopsin found in the cells of your retina—a molecule that is absorbing the light entering your eye right now. Follow-up work suggests that most of the Sargasso Sea rhodopsin-like molecules are also absorbing light, and that bacterial cells use the energy of the light to pump protons across their plasma membranes—creating a chemiosmotic gradient that can synthesize ATP (see Chapter 9).

CHECK YOUR UNDERSTANDING

If you understand that . . .

- The size of bacterial and archaeal genomes correlates with the cell's metabolic capabilities.
- Lateral gene transfer—movement of DNA from one species to another—is extensive in prokaryotes.
- Environmental sequencing catalogs all of the genes found in a particular habitat.

✓ You should be able to . . .

1. Explain why it is logical to observe that parasitic bacteria have small genomes.

2. Explain the logic behind claiming that a gene's similarity to a gene in a distantly related species, and dissimilarity to the same gene in closely related species, is evidence for lateral gene transfer.

Answers are available in Appendix B.

A similar study of bacterial and archaeal genes in the human gut found genes for enzymes that digest the complex polysaccharides of plant cell walls. Environmental sequencing is providing new insights about the living world, from how rhodopsin-like proteins help bacteria thrive in a desert to how your lunch is being digested.

20.3 Eukaryotic Genomes

Sequencing eukaryotic genomes presents two daunting challenges. The first is size. The genomes of bacteria and archaea range from 580,070 base pairs in *Mycoplasma genitalium* to over 6.3 million base pairs in *Pseudomonas aeruginosa*, but eukaryotic genomes are even larger. The haploid genome of *Saccharomyces cerevisiae* (baker's yeast), a unicellular eukaryote, contains over 12 million base pairs. The roundworm *Caenorhabditis elegans* has a genome of 97 million base pairs; the fruit-fly genome contains 180 million base pairs; the mustard plant *Arabidopsis thaliana*'s genome has 130 million base pairs; and those of humans, rats, mice, and cattle contain roughly 3 billion base pairs each.

The second great challenge in sequencing eukaryotic genes is coping with noncoding sequences that are repeated many times. Many eukaryotic genomes are dominated by repeated DNA sequences that occur between genes or inside introns and do not code for products used by the organism. These repeated sequences greatly complicate the work of aligning and interpreting sequence data. If such sequences don't code for a product, why do they exist?

Parasitic and Repeated Sequences

In many eukaryotes, the exons and regulatory sequences associated with genes make up a relatively small percentage of the genome. Over 90 percent of a bacterial or archaeal genome consists of genes, but about 50 percent of an average eukaryotic genome consists of repeated sequences that do not code for a product used by the cell.

When noncoding and repeated sequences were discovered, they were initially considered "junk DNA" that was nonfunctional and probably unimportant and uninteresting. But subsequent work has shown that many of the repeated sequences observed in eukaryotes are actually derived from sequences known as transposable elements.

Transposable elements are segments of DNA that are capable of being inserted into new locations, or transposing, in a genome. They are similar to viruses, except that viruses leave a host cell that they have infected and find a new cell to infect. In contrast, transposable elements never leave their host cell—they simply make copies of themselves that become inserted in new locations. Transposable elements are passed from parents to offspring, generation after generation, because they are part of the genome.

A transposable element is an example of what biologists call a selfish gene: a DNA sequence that survives and reproduces but does not increase the fitness of the host genome. Transposable elements and viruses are classified as parasitic because it takes time and resources to copy them along with the rest of the genome, and because they can disrupt gene function when they insert in a new location. As a result, they decrease their host's fitness.

HOW DO TRANSPOSABLE ELEMENTS WORK? Transposable elements come in a wide variety of types and spread through genomes in a variety of ways. Different species—*E. coli*, fruit flies, yeast, and humans, for example—contain distinct types of transposable elements.

As an example of how these selfish genes work, consider a well-studied type called a **long interspersed nuclear element (LINE)** that is found in humans and other eukaryotes. Because LINEs are so similar to the retroviruses, introduced in Chapter 19, biologists hypothesize that they are derived from them evolutionarily. Your genome contains tens of thousands of LINEs, each between 1000 and 5000 bases long. **Figure 20.5** on page 366 illustrates the steps that allow an active LINE to transpose.

Most of the LINEs observed in the human genome do not actually function, however, because they don't contain a promoter or the genes for either reverse transcriptase or integrase. To make sense of this observation, researchers hypothesize that the insertion process illustrated in steps 6 and 7 of Figure 20.5 is usually disrupted in some way, leaving the inserted replica of the original line incomplete.

Virtually every prokaryotic and eukaryotic genome examined to date contains at least some transposable elements. They vary widely in type and number, however, and bacterial and archaeal genomes have relatively few transposable elements compared to most eukaryotes studied thus far. This observation has inspired the hypothesis that bacteria and archaea either have efficient means of removing parasitic sequences or can somehow thwart insertion events. To date, however, this hypothesis has yet to be tested rigorously.

Research on transposable elements and lateral gene transfer has revolutionized how biologists view the genome. Many genomes are riddled with parasitic sequences, and others have undergone radical change in response to lateral gene transfer events. In other words, genomes are much more dynamic and complex than previously thought. Their size and composition can change dramatically over time.

REPEATED SEQUENCES AND DNA FINGERPRINTING In addition to containing repeated sequences from transposable elements, eukaryotic genomes have several thousand loci called **short tandem repeats (STRs)**. These are small sequences repeated one after another contiguously along part of a chromosome.

There are two major classes of STRs:

1. Repeating units that are just 1 to 5 bases long are known as **microsatellites** or **simple sequence repeats**. The most common type of microsatellite is a repeated stretch of the dinucleotide AC, giving the sequence ACACACAC....

2. Repeating units that are 6 to 500 bases long are known as **minisatellites** or **variable number tandem repeats (VNTRs)**.

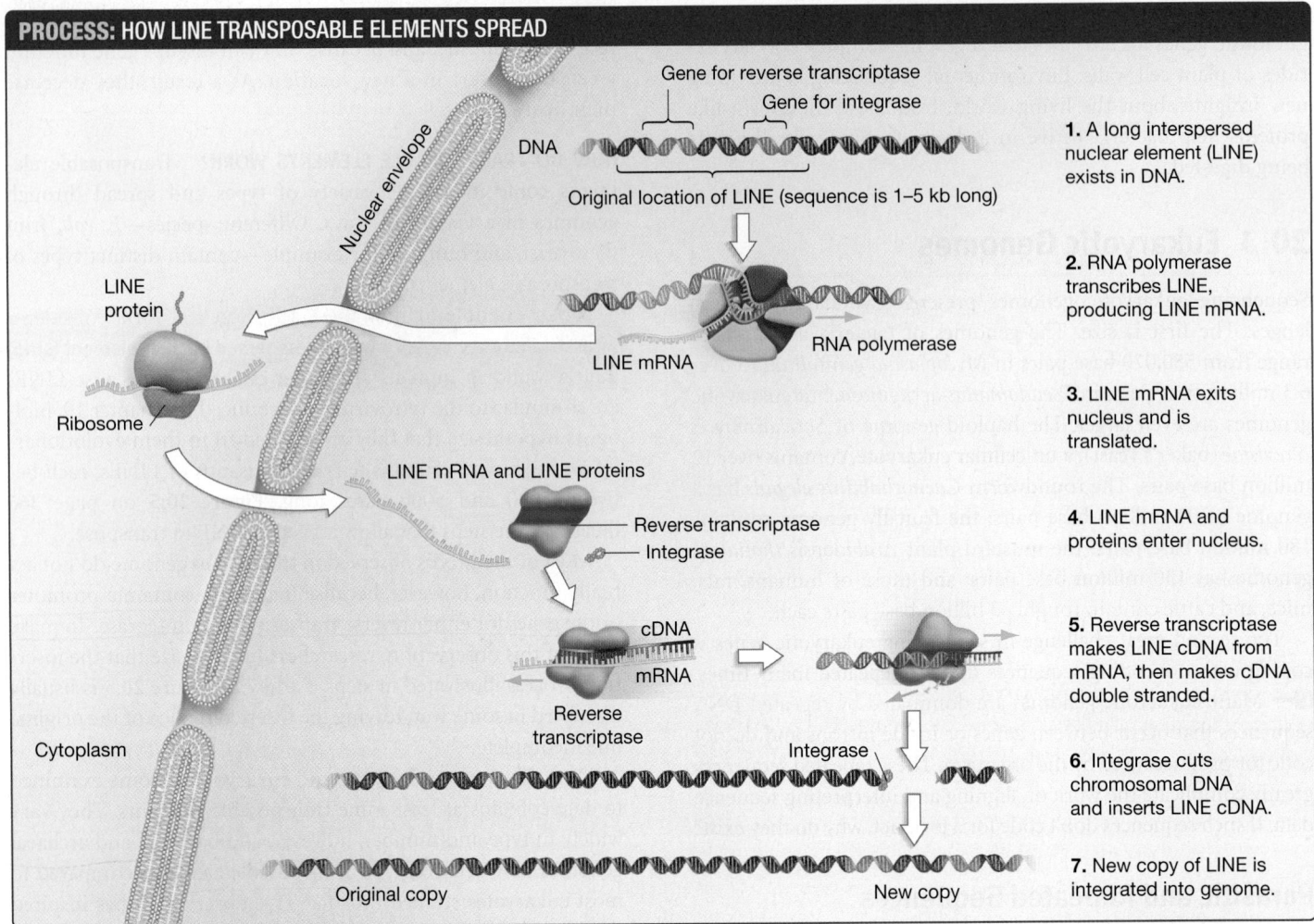

FIGURE 20.5 Transposable Elements Spread within a Genome.

Microsatellite sequences are thought to originate when DNA polymerase skips or mistakenly adds extra bases during replication; the origin of minisatellites is still unclear. Together, the two types of repeated sequences make up 3 percent of the human genome.

Soon after these sequences were first characterized, Alec Jeffreys and co-workers established that microsatellite and minisatellite loci are "hypervariable," meaning that they vary among individuals much more than any other type of sequence does.

Figure 20.6 illustrates one mechanism to explain why microsatellites and minisatellites have so many different alleles: a process called **unequal crossover**. Here's how it works: Homologous chromosomes sometimes align incorrectly during prophase of meiosis I. Instead of lining up in exactly the same location, the two chromosomes pair in a way that matches up bases in different DNA repeats. When crossover occurs, the resulting chromosomes have different numbers of repeats.

Repeated sequences are particularly prone to unequal crossover, because their homologs are so similar that they are likely to misalign. If the region in question has a unique number of repeats, it represents a unique allele. Like any other alleles, microsatellite and minisatellite alleles are transmitted from parents to offspring.

Misalignment and errors by DNA polymerase are so common in these sequences that, in most eukaryotes, the genome of virtually every individual has at least one new allele. This variation in repeat number among individuals is the basis of most DNA fingerprinting. **DNA fingerprinting** refers to any technique for identifying individuals based on the unique features of their genomes. Because microsatellite and minisatellite sequences vary so much among individuals, they are now the sequences of choice for DNA fingerprinting.

To fingerprint an individual, researchers obtain a DNA sample and perform the polymerase chain reaction (PCR), using primers that flank a region containing an STR (**Figure 20.7a**). Once the region has been cloned, it can be analyzed to determine the number of repeats present. Primers are now available for many different STR loci, so researchers can efficiently analyze the alleles present at many STRs.

These advances have important practical implications. For example, DNA fingerprinting of blood or semen found at crime scenes has been used to show that people who were accused of crimes were actually innocent. DNA fingerprinting has also been used as evidence to convict criminals or assign paternity in birds, humans, and other species that have well-characterized microsatellite or minisatellite sequences (**Figure 20.7b**).

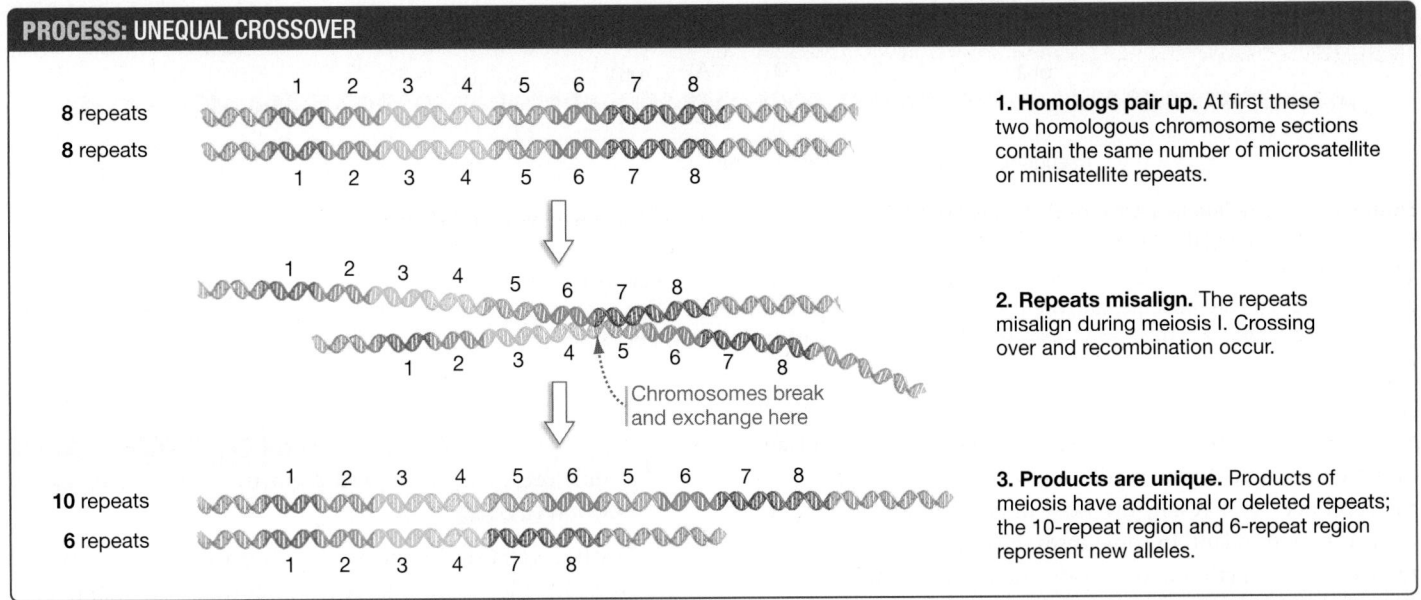

PROCESS: UNEQUAL CROSSOVER

8 repeats
8 repeats

1 2 3 4 5 6 7 8

1. Homologs pair up. At first these two homologous chromosome sections contain the same number of microsatellite or minisatellite repeats.

1 2 3 4 5 6 7 8

1 2 3 4 5 6 7 8

2. Repeats misalign. The repeats misalign during meiosis I. Crossing over and recombination occur.

Chromosomes break and exchange here

10 repeats

1 2 3 4 5 6 5 6 7 8

6 repeats

1 2 3 4 7 8

3. Products are unique. Products of meiosis have additional or deleted repeats; the 10-repeat region and 6-repeat region represent new alleles.

FIGURE 20.6 Unequal Crossover Adds or Deletes DNA Repeats.

Now that we've reviewed the characteristics of some particularly prominent types of noncoding sequences in eukaryotes, let's consider the characteristics of coding sequences in these genomes. We start with the most basic question of all: Where do eukaryotic genes come from?

Gene Families

In eukaryotes, the major source of new genes is the duplication of existing genes. Biologists infer that genes have been duplicated recently when they find groups of similar genes clustered along the same chromosome. The genes are usually similar in general structural features, such as the arrangement of exons and introns, and in their base sequence. Within a species, genes that are extremely similar to each other in structure and function are considered to be part of the same **gene family**.

The degree of sequence similarity among members of a gene family varies. In the genes that code for ribosomal RNAs (rRNAs) in vertebrates, the sequences are virtually identical—meaning that each individual has many exact copies of the same gene. In other cases, though, the proportion of bases that are identical is 50 percent or less.

HOW DO GENE FAMILIES ARISE?   Genes that make up gene families are hypothesized to have arisen from a common ancestral sequence through gene duplication. When **gene duplication** occurs, an extra copy of a gene is added to the genome.

The most common type of gene duplication results from unequal crossover during meiosis—the same process that resulted in extra microsatellite and minisatellite repeats in Figure 20.6. Gene-sized segments of chromosomes can be deleted or duplicated if homologous chromosomes misalign during prophase of meiosis I and an unequal crossover occurs. Like microsatellites or minisatellites, the duplicated segments are arranged in tandem—one after the other.

(a) Using PCR to amplify minisatellite and microsatellite loci.

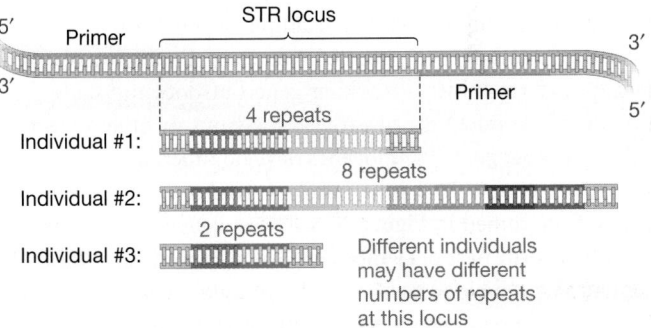

(b) Compare number of STR repeats in alleles to test paternity.

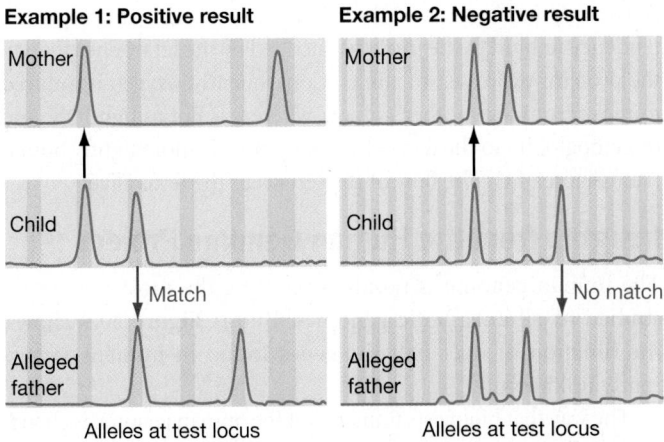

FIGURE 20.7 DNA Fingerprinting Can Be Used to Identify Parents.
(a) The lengths of minisatelite and microsatelite loci vary among individuals. **(b)** Here, the position of each peak indicates the number of repeats at a particular locus. Each individual has two alleles at each locus, and thus two peaks. The two peaks from a child should line up with one peak from each parent. Typically 6 to 16 loci are tested to determine paternity.

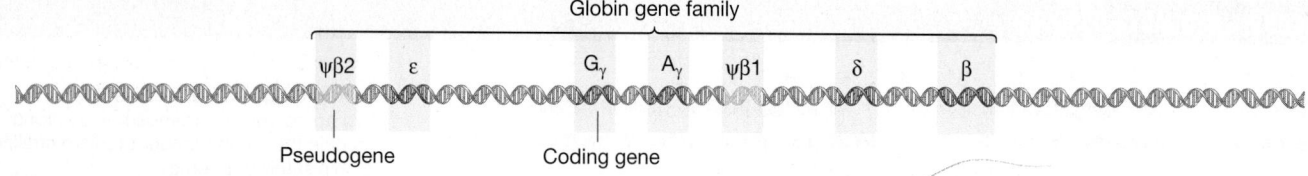

Globin gene family

Pseudogene

Coding gene

FIGURE 20.8 Gene Families Are Closely Related Genes. Most of the genes illustrated here are expressed at different times during development.

✔**EXERCISE** Suppose that during prophase of meiosis I, the β locus on one chromosome aligned with the ψβ2 locus on another chromosome, and then crossing over occurred in the noncoding sequences just to the left (as oriented in the figure) of this β-ψβ2 pairing. List the order of the globin-family genes that would result on each chromosome.

NEW GENES—NEW FUNCTIONS? Gene duplication is important because the original gene is still functional and produces a normal product. As a result, the new, duplicated stretches of sequence are redundant. In some cases the duplicated genes retain their original function and provide additional quantities of the same product. But if mutations in the duplicated sequence alter the protein product, and if the altered protein product performs a valuable new function in the cell, then an important new gene has been created.

Alternatively, mutations in the duplicated region may make its expression impossible. For example, a mutation could produce a stop codon in the middle of an exon. A member of a gene family that resembles a working gene but does not code for a functional product, due to early stop codons or other defects, is called a **pseudogene**. Pseudogenes have no function.

As an example of a gene family, consider the human globin genes diagrammed in **Figure 20.8**. These sequences code for proteins that form part of hemoglobin—the oxygen-carrying molecule in your red blood cells. Note that the globin gene family contains several pseudogenes, along with several genes that code for oxygen-transporting proteins. Each coding gene in the family serves a slightly different function. For example, some genes are active only in the fetus or the adult. Oxygen has a much greater tendency to bind to the proteins encoded by the fetal genes than to the proteins expressed in adults. Consequently, oxygen is induced to move from the mother's blood, where it is not as tightly bound to hemoglobin, to the fetus's blood, where it is more tightly bound (see Chapter 44). The flow of oxygen keeps the fetus alive.

Insights from the Human Genome Project

The human genome is rapidly becoming the most intensively studied of all eukaryotic genomes. But as **Figure 20.9** shows, the function of over half the genes found in humans is currently unknown.

The way that biologists think about the human genome is changing rapidly, however, due to two recent and dramatic discoveries:

1. Genes for miRNAs are much more common than previously thought. As Chapter 18 noted, miRNAs are small molecules involved in regulating the life span of mRNAs.

2. A much larger proportion of the genome is transcribed than previously thought. Many of these sequences produce RNAs

that never leave the nucleus. Because their role in the cell is unknown, researchers call them transcripts of unknown function (TUFs).

Clearly, a great deal remains to be learned about the human genome. In the meantime, two important questions have emerged from early studies. Let's consider each of them in turn.

WHY DO HUMANS HAVE SO FEW GENES? Of all observations about the nature of eukaryotic genomes, perhaps the most striking is that organisms with complex morphology and behavior do not appear to have particularly large numbers of genes. **Table 20.1** indicates the estimated number of genes found in selected eukaryotes. Notice that the total number of genes in *Homo sapiens*, which is considered a particularly complex organism, is about the same as in roundworms, dogs, chickens, and rats, and substantially lower than the number of genes in mice, rice, and the weedy mustard plant *Arabidopsis thaliana*.

Before the human genome was sequenced, many biologists expected that humans would have at least 100,000 genes. But the most recent estimates suggest that we have only a fifth of that number.

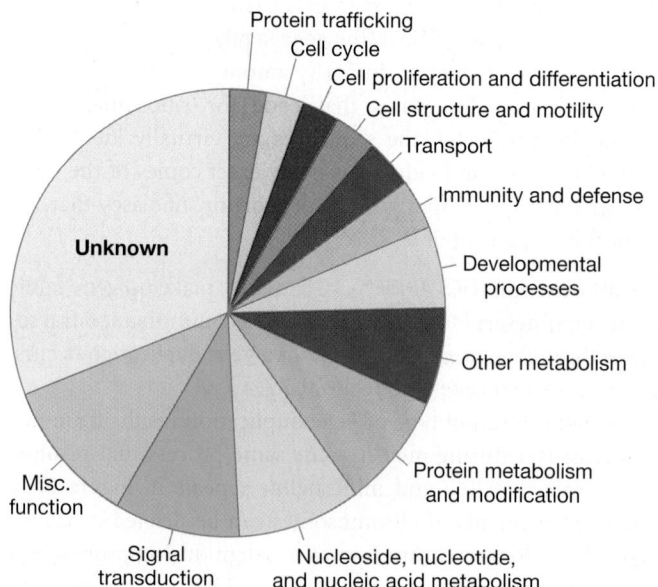

FIGURE 20.9 The Function of Many Human Protein-Coding Genes Is Unknown.

TABLE 20.1 Number of Genes in Selected Eukaryotes

Species	Description	Genome Size (millions of base pairs)	Estimated Number of Genes
Saccharomyces cerevisiae	Baker's and brewer's yeast; a unicellular fungus; an important model organism in biochemistry and genetics	12.1	6000
Plasmodium falciparum	Single-celled, parasitic eukaryote; causes malaria in humans	22.8	5268
Drosophila melanogaster	Fruit fly; an important model organism in genetics and developmental biology	120	13,600
Caenorhabditis elegans	A roundworm; an important model organism in developmental biology	97	19,000
Canis familiarus	Domestic dog	2410	19,300
Gallus gallus	Chicken	1050	21,500
Homo sapiens	Human	3000	20,500
Rattus norvegicus	Norway rat; an important model organism in physiology and behavior	2750	21,000
Mus musculus	House mouse; an important model organism in genetics and developmental biology	2500	37,000
Arabidopsis thaliana	A mustard plant; an important model organism in genetics and developmental biology	140	27,500
Oryza sativa	Rice	389	41,000

How can this be? In prokaryotes there is a correlation between genome size, gene number, a cell's metabolic capabilities, and the cell's ability to live in a variety of habitats. Similarly, it makes sense that plants have exceptionally large numbers of genes, because they synthesize so many different and complex molecules from just carbon dioxide, nitrate ions, phosphate ions, and other simple nutrients. The large numbers of genes enable plants to produce large numbers of enzymes. But why isn't there a stronger correlation between gene number and morphological and behavioral complexity in animals?

The leading hypothesis to explain this observation is based on **alternative splicing**. Recall from Chapter 18 that the exons of a particular gene can be spliced in ways that produce distinct mature mRNAs. As a result, a single eukaryotic gene can code for multiple transcripts and thus multiple proteins. The alternative-splicing hypothesis claims that multicellular eukaryotes do not need enormous numbers of distinct genes. Instead, alternative splicing creates different proteins from the same gene.

In support of the alternative-splicing hypothesis, researchers have analyzed the mRNAs produced by human genes and have estimated that each gene produces an average of more than three distinct transcripts. If this result is valid, the actual number of different proteins that can be produced is more than triple the gene number. Humans may have fewer than 20,000 genes according to current estimates, but these genes may have the ability to produce 100,000 different transcripts.

HOW CAN THE HUMAN AND CHIMP GENOMES BE SO SIMILAR?

Comparing the numbers of genes found in humans and in mice created a paradox that may be resolved by the alternative-splicing hypothesis. But comparing the base sequences of genes in humans and chimps has created another paradox.

Here is the issue: At the level of base sequences, human beings and chimpanzees are 98.8 percent identical on average. Of the homologous genes analyzed in humans and chimps, 29 percent are identical in amino acid sequence; the average difference between homologous proteins is just two amino acids. If humans and chimps are so similar genetically, why do they appear to be so different in their morphology and behavior?

The leading hypothesis to resolve this paradox cites the importance of regulatory genes and regulatory sequences. Recall from Chapter 18 that a **regulatory sequence** is a section of DNA involved in controlling the activity of other genes; it may be a promoter, a promoter-proximal element, an enhancer, or a silencer. The term **structural gene**, in contrast, refers to a sequence that codes for a tRNA, rRNA, protein, or other type of product. **Regulatory genes** code for regulatory transcription factors that alter the expression of specific genes.

To resolve the sequence-similarity paradox, biologists propose that even though many structural genes in closely related species, such as humans and chimps, are identical or nearly identical, regulatory sequences and regulatory genes in the two species might contain important differences. Suppose the structural gene for human growth hormone is identical in base sequence to that for chimp growth hormone. Even so, if differences in transcription factors, enhancers, or promoters change the pattern of expression of that gene—perhaps turning it on later and longer in humans than in chimps—then height and other characteristics will differ even though the structural gene is the same.

Based on current analyses, biologists suggest that the human genome contains about 3000 different regulatory transcription factors. Subtle mutations in these proteins and the regulatory sites that they bind to could have a significant effect on gene expression and thus on the phenotype. Subtle differences in TUFs could also play a role, if it turns out that these nucleus-restricted RNAs play a role in regulating gene expression.

These observations suggest that most of the genetic changes responsible for the rapid evolution of humans over the past 5 million years have been due to changes in regulatory genes and sequences and alternative splicing, rather than to changes in structural genes. To date, however, there are no specific examples of changes in the regulatory sequences responsible for the phenotypic differences observed between humans and chimps or other closely related species. The regulatory hypothesis still needs to be tested rigorously.

CHECK YOUR UNDERSTANDING

If you understand that . . .

- Eukaryotic genomes are riddled with parasitic sequences that do not contribute to the fitness of the organism.
- Simple repeated sequences are common in eukaryotic genomes.
- In eukaryotes, many of the coding sequences are organized into families of genes with related functions.
- The recent discovery of many miRNA genes and TUFs suggests that the regulation of eukaryotic gene expression may be much more complex than previously thought.

You should be able to . . .

1. Explain why simple tandem repeats make DNA fingerprinting possible.
2. Explain how unequal crossover leads to duplicated sequences.

Answers are available in Appendix B.

20.4 Functional Genomics and Proteomics

Eric Lander has compared the sequencing of the human genome to the establishment of the periodic table of the elements in chemistry. Once the periodic table was validated, chemists focused on understanding how the elements combine to form molecules. Similarly, biologists now want to understand how the elements of the human genome combine to produce an individual.

Remember that a genome sequence is essentially a parts list. Once that list is assembled, researchers delve deeper to understand how genes interact to produce an organism.

What Is Functional Genomics?

For decades, biologists have worked at understanding how and when individual genes are expressed. Research on the *lac* operon, reviewed in Chapter 17, is typical. But now researchers can ask how and when *all* of the genes in an organism are expressed.

Large-scale analysis of gene expression is part of functional genomics—research on how genes work together to produce a phenotype. The effort is motivated by the realization that gene products do not exist in a vacuum. Instead, groups of RNAs and proteins act together to respond to environmental challenges such as extreme heat or drought. Similarly, distinct groups of genes are transcribed at different stages as a multicellular eukaryote grows and develops.

One of the most basic tools used in functional genomics is called a microarray. A **DNA microarray** consists of a large number of single-stranded DNA segments that are permanently affixed to a glass slide. For example, the slide pictured in **Figure 20.10** contains thousands of spots, each of which contains single-stranded DNA from a unique exon found in the human genome.

A typical experiment done with a DNA microarray would follow the protocol outlined in **Figure 20.11**. For example, if the researchers' goal were to compare genes that are expressed during normal cell activity with those expressed under heat stress, the first step would be to isolate the mRNAs being produced in the cells in question under each of the two conditions: in control cells functioning at normal temperature and in cells of the same kind exposed to high temperatures.

Once they had purified mRNAs from the two populations of cells (step 1), investigators would use reverse transcriptase to make a single-stranded cDNA version of each RNA in each of the two samples (see Chapter 19). In addition to the four standard dNTPs, one of the DNA building blocks used in synthesizing the cDNA would carry a fluorescent label (step 2). The label used for the cDNA of the control cells would glow one color (let's say green), while the label chosen for the cDNA of the heat-stressed cells would glow another color (let's say red).

The labeled cDNAs of both colors would then be used to probe the microarray (step 3). As Chapter 19 noted, a probe allows an investigator to find a particular molecule in a sample containing many different molecules. In this case, the labeled cDNAs will bind to the single-stranded DNAs on the plate by complementary base pairing.

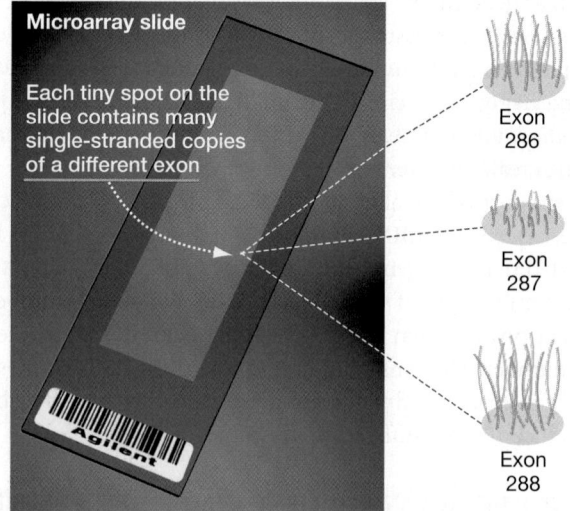

Microarray slide

Each tiny spot on the slide contains many single-stranded copies of a different exon

Exon 286

Exon 287

Exon 288

FIGURE 20.10 DNA Microarrays Represent Every Gene in a Genome. To create a DNA microarray, investigators arrange thousands of short, single-stranded DNA sequences onto spots on a glass plate. The DNAs are exons from the genome of a particular species.

Out of all the exons in the genome, then, only the exons that are being expressed by the two populations of cells will be labeled. In our example, genes that are expressed by the control cells under normal conditions will be labeled green, while those expressed by the same kind of cells during heat stress will be labeled red. If one of the exons in the microarray is expressed

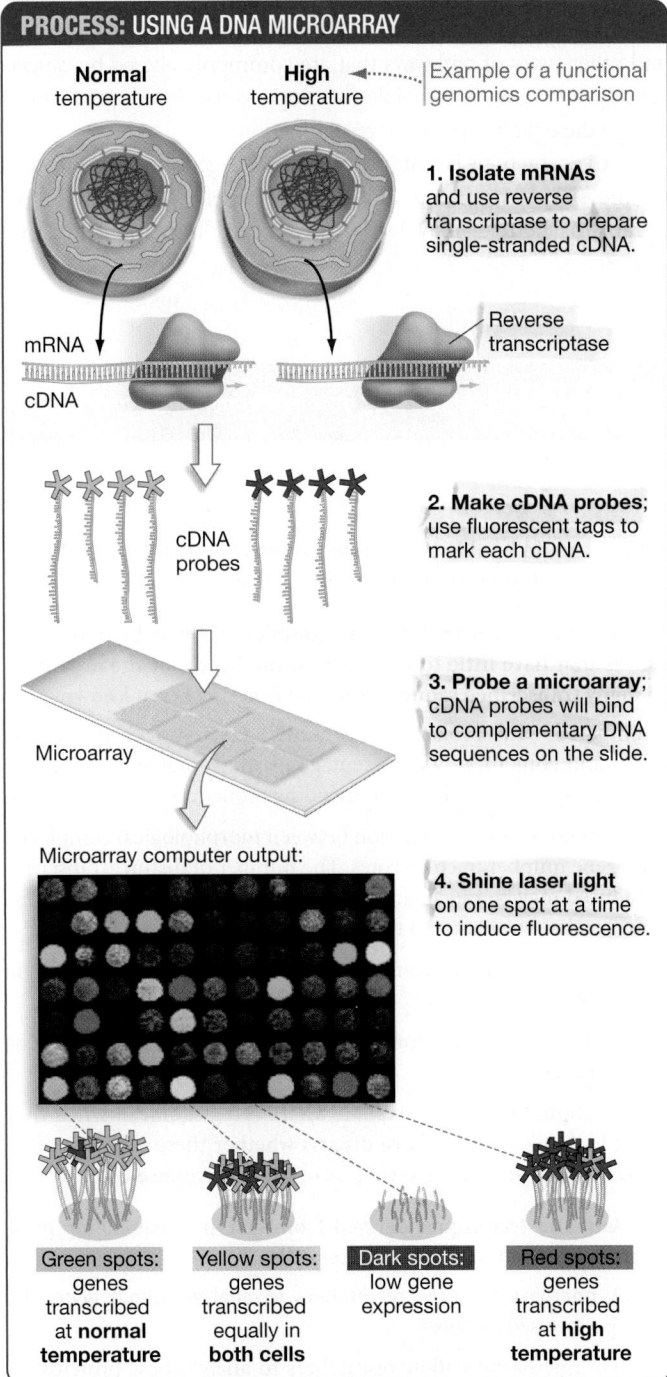

PROCESS: USING A DNA MICROARRAY

Normal temperature   High temperature ◄········· Example of a functional genomics comparison

mRNA
cDNA

Reverse transcriptase

1. Isolate mRNAs and use reverse transcriptase to prepare single-stranded cDNA.

cDNA probes

2. Make cDNA probes; use fluorescent tags to mark each cDNA.

Microarray

3. Probe a microarray; cDNA probes will bind to complementary DNA sequences on the slide.

Microarray computer output:

4. Shine laser light on one spot at a time to induce fluorescence.

Green spots: genes transcribed at **normal temperature**

Yellow spots: genes transcribed equally in **both cells**

Dark spots: low gene expression

Red spots: genes transcribed at **high temperature**

FIGURE 20.11 DNA Microarrays Are Used to Study Changes in Gene Expression. By probing a microarray with labeled cDNAs synthesized from mRNAs, researchers can identify which coding sequences are being transcribed. Here mRNAs from cells growing at normal temperature are green, while mRNAs from cells growing at high temperature are red.

under both sets of conditions, then both green- and red-labeled cDNAs will bind to that spot and make it appear yellow (step 4).

A microarray lets researchers study the expression of thousands of genes at a time. As a result, they can identify which sets of genes are expressed in concert under specific sets of conditions.

Once a microarray has been used, the bound cDNA probes can be removed. The original DNAs remain in place, so the slide may then be reused to assess gene expression in a different type of cell, or in the same cell type under different conditions. Researchers can use microarrays to establish which genes are transcribed in different organs and tissues, during cancerous growth, or in response to changes in environmental conditions—such as starvation, the presence of a toxin, or a viral infection. ✔If you understand the concept of how microarrays are used, you should be able to explain how you would use a DNA microarray to compare the genes expressed in brain cells versus liver cells of an adult human.

What Is Proteomics?

The Greek root *–ome*, meaning "all," inspired the term genome. Similarly, biologists use the term **transcriptome** in referring to the complete set of genes that are transcribed in a particular cell, and **proteome** in referring to the complete set of proteins that are produced. **Proteomics**, it follows, is the large-scale study of all the proteins in a cell or organism.

Proteomic studies begin by identifying the proteins present in a cell or organelle. Then researchers attempt to determine the locations of the proteins and how they interact. They may also try to document how the proteins that are present change through time or compare with those in other cells. Instead of studying individual proteins or how two proteins might interact, proteomics is based on studying all of the proteins present at once.

One approach to studying protein–protein interactions is similar to the use of DNA microarrays, except that large numbers of proteins, rather than DNA sequences, are affixed on a glass plate. This microarray of proteins is then treated with an assortment of proteins produced by the same organism. These proteins are labeled with a fluorescent or radioactive tag. If any labeled proteins bind to the proteins in the microarray, the two molecules may also interact in the cell. In this way, researchers hope to identify proteins that physically bind to one another—like the G proteins and associated enzymes introduced in Chapter 8, or the cyclin and Cdk molecules introduced in Chapter 11. Microarray technology is allowing biologists to study protein–protein interactions on a massive scale.

Applied Genomics in Action: Understanding Cancer

Chapter 11 and Chapter 18 introduced the family of diseases called cancer. Recall that cancer develops when an array of genes—starting with genes involved in the control of cell growth—no longer function properly, due to mutation.

Biomedical researchers have been increasingly successful at developing drugs that fight cancers effectively. Many of these drugs work by killing any rapidly dividing cell. But a variety of cells in the body have to divide rapidly in order to function normally. Cancer chemotherapy also kills these cells. This is why chemotherapy patients suffer side effects like hair loss, skin and digestive problems, and weakened immune systems.

Could safer and more-effective therapies be devised, if researchers knew exactly which genes were mutated in cancer cells? ☞ Researchers are using tools created by advances in genomics to deepen our understanding of cancer. For example,

- Investigators are using microarrays to compare which genes are expressed in cancer cells versus normal cells. A recent analysis of 40 such experiments has identified a common suite of 69 genes that are mis-expressed in the majority of cancers in the dataset.

- Using data from the human genome project, researchers have sequenced tens of thousands of genes from cancerous cells.

Comparisons with normal cells have identified common sets of genes that are mutated in cancerous cells. Some of these analyses suggest that as many as 120 distinct mutations may play a role in driving the development of different cancers.

- The complete genome sequences of cancerous and non-cancerous cells from the same person identified over 600 mutations in the cancerous cells.

Studies like these raise the possibility of identifying cell-signaling and other types of pathways that are commonly altered by cancer-causing mutations. If so, biologists hope to develop drugs that can restore these pathways to a normal function.

The general hope is that "pure research" in genomics—motivated by the simple desire to understand ourselves and our fellow organisms better—may lead to improved human health and welfare.

CHAPTER 20 REVIEW

For media, go to the study area at www.masteringbiology.com

Summary of Key Concepts

☞ **Once a genome has been completely sequenced, researchers use a variety of techniques to identify which sequences code for products and which act as regulatory sites.**

- Recent advances in dideoxy sequencing and pyrosequencing have allowed investigators to sequence DNA more rapidly and cheaply, resulting in a flood of genome data.

- Researchers annotate genome sequences by finding genes and determining their function.

- To identify genes in bacteria and archaea, researchers use computers to scan the genome for start and stop codons that are in the same reading frame and that are separated by gene-sized stretches of sequence.

- To identify genes in eukaryotes, researchers study RNAs as a way of characterizing actively transcribed sequences.

- Among species, homologous genes are identified on the basis of similarities in sequence and structure, and are inferred to have similar function.

✔ You should be able to describe how a research group that discovered a gene for coat color in mice would determine whether a homologous gene exists in the human genome.

(MB) **Web Activity** Human Genome Sequencing Strategies

☞ **Bacterial and archaeal genomes are relatively small. Among species, there is a positive correlation between total gene number and metabolic capabilities. Gene transfer between species is also common.**

- Parasitic bacteria tend to have small genomes; bacteria and archaea that live in a broad array of habitats or that use a wide variety of nutrients tend to have larger genomes.

- The function of many of the genes identified in bacteria and archaea is still unknown. Redundancy among genes is also high.

- Lateral gene transfer is common in bacteria and archaea. It is an important source of new genes in many species.

✔ You should be able to describe two mechanisms responsible for lateral gene transfer in bacteria and archaea.

Eukaryotic genomes are large and complex. They include many sequences that have little to no effect on the fitness of the organism, and many transcribed sequences whose function is not known.

- Compared with prokaryotic genomes, eukaryotic genomes are large and contain a high percentage of transposable elements, repeated sequences, and other noncoding sequences.

- There is no obvious correlation between morphological complexity and gene number in eukaryotes. The number of distinct transcripts produced may be much larger than the actual gene number in certain species, however, as a result of alternative splicing.

- Gene duplication has been an important source of new genes in eukaryotes.

- Changes in gene regulation appear to have been important during human evolution.

✔ You should be able to explain what biologists mean when they refer to "junk DNA," and to discuss whether these sequences lack function and are uninteresting, as originally proposed.

☞ **Data and techniques derived from genome sequencing projects are being used to analyze cancer cells.**

- DNA microarrays allow researchers to analyze which genes are being expressed in cells.

- Protein microarrays allow researchers to analyze how proteins interact in cells.

- DNA microarrays and DNA sequencing are being used to identify mutations that are common to many different types of cancer.

✔ You should be able to explain what you would conclude if microarray experiments indicated that a particular gene was consistently expressed in cancer cells but not in normal cells from the same tissue.

Questions

1. What is an open reading frame in bacteria?
   a. a gene whose function is already known
   b. a DNA section that is thought to code for a protein because it is similar to a complementary DNA (cDNA)
   c. a DNA section that is thought to code for a protein because it has a start codon and a stop codon flanking hundreds of base pairs
   d. any member of a gene family

2. What best describes the logic behind shotgun sequencing?
   a. Break the genome into tiny pieces. Sequence each piece. Use overlapping ends to assemble the pieces in the correct order.
   b. Start with one end of each chromosome. Sequence straight through to the other end of the chromosome.
   c. Use a variety of techniques to identify genes and ORFs. Sequence these segments—not the noncoding and repeated sequences.
   d. Break the genome into pieces. Map the location of each piece. Then sequence each piece.

3. What are minisatellites and microsatellites?
   a. small, extrachromosomal loops of DNA that are similar to plasmids
   b. parts of viruses that have become integrated into the genome of an organism
   c. incomplete or "dead" remains of transposable elements in a host cell
   d. short and simple repeated sequences in DNA

4. What is the leading hypothesis to explain the paradox that large, morphologically complex eukaryotes such as humans have relatively small numbers of genes?

   a. lateral transfer of genes from other species
   b. alternative splicing of mRNAs
   c. polyploidy, or the doubling of the genome's entire chromosome complement
   d. expansion of gene families through gene duplication

5. What evidence do biologists use to infer that a gene is part of a gene family?
   a. Its sequence is exactly identical to that of another gene.
   b. Its structure—meaning its pattern of exons and introns—is identical to that of a gene found in another species.
   c. Its composition, in terms of percentage of A-T and G-C pairs, is unique.
   d. Its sequence, structure, and composition are similar to those of another gene in the same genome.

6. What is a pseudogene?
   a. a coding sequence that originated in a lateral gene transfer
   b. a gene whose function has not yet been established
   c. a polymorphic gene—meaning that more than one allele is present in a population
   d. a gene whose sequence is similar to that of functioning genes but does not produce a functioning product

1. Explain how open reading frames are identified in the genomes of bacteria and archaea. Why is it more difficult to find open reading frames in eukaryotes?

2. Why is it logical to observe that bacterial species found in a wide array of habitats have relatively large genomes?

3. Why are LINEs and other repeated sequences referred to as "genomic parasites"?

4. Why can DNA fingerprinting help identify an individual's relatives?

5. Researchers can create microarrays of short, single-stranded DNAs that represent many or all of the exons in a genome. Explain how these microarrays are used to document changes in the transcription of genes over time or in response to environmental challenges.

6. Explain the concept of homology and how identifying homologous genes helps researchers identify the function of unknown genes. Are duplicated sequences that form gene families homologous? Explain.

1. Parasites lack genes for many of the enzymes found in their hosts. Most parasites, however, have evolved from free-living ancestors that had larger genomes. Based on these observations, W. Ford Doolittle claims that the loss of genes in parasites represents an evolutionary trend. He summarizes his hypothesis with the quip "use it or lose it." What does he mean?

2. According to eyewitness accounts, communist revolutionaries executed Nicholas II, the last czar of Russia, along with his wife and five children, the family physician, and several servants. Many decades after this event, a grave purported to hold the remains of the royal family was discovered. Biologists were asked to analyze DNA from each adult and juvenile skeleton and determine whether the bodies were indeed those of several young siblings, two parents, and several unrelated adults. If the grave was authentic, describe how similar the DNA fingerprints of each skeleton would be relative to the fingerprints of other individuals in the grave.

3. The human genome contains a gene that encodes a protein called syncytin. This gene is expressed in placental cells during pregnancy. The syncytin gene is nearly identical in DNA sequence to a gene in a virus that infects humans. In this virus, the syncytin-like gene codes for a protein found in the virus's outer envelope. State a hypothesis to explain the similarity between the two genes.

4. A recent study used microarrays to compare the patterns of expression of genes that are active in the brain, liver, and blood of chimpanzees and humans. Although the overall patterns of gene expression were similar in the liver and blood of the two species, expression patterns were strikingly different in the brain. How does this study relate to the hypothesis that most differences between humans and chimps involve changes in gene regulation?

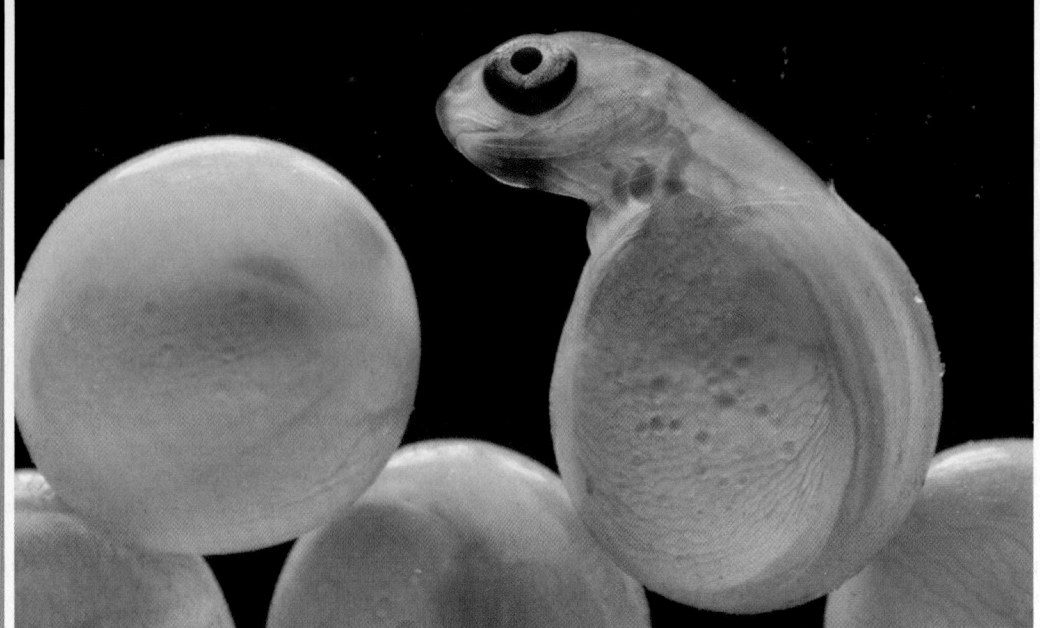

A young fish, still attached to the nutrient-filled yolk in the egg. This chapter introduces the processes responsible for transforming a fertilized egg into an individual that has specialized cells, tissues, and organs.

21 Principles of Development

KEY CONCEPTS

- During development, cells divide, die, move, or expand in a directed manner; specialize; and interact with other cells.

- Cells become specialized because they express different genes, not because they contain different genes.

- Cells interact continuously by means of cell-cell signals during development.

- When development begins, early cell-cell signals trigger a cascade of effects that cause increasing specialization as development proceeds.

- Mutations in genes responsible for development may lead to the evolution of new body sizes, shapes, and structures.

What question qualifies as the greatest current challenge in biological science? Although there are many candidates, one of the most compelling is the question addressed in Unit 4: How does a multicellular individual develop from a single cell—a fertilized egg?

It's important to pause for a moment and think about the magnitude of this problem. For example, at one time you consisted of a single cell. If you had been able to watch your own development, you would have seen that cell divide rapidly and form a ball of tiny, identical-looking cells. At that point, the fertilized egg had given rise to an **embryo**—a young, developing organism. After continued cell division, large groups of cells suddenly began moving into the interior of the embryo. Cell division continued at a rapid pace. After a week or two the embryo elongated, and a recognizable head and tail portion appeared. Tiny precursors of vertebrae became visible, along with rudimentary eyes. Eventually buds emerged and went on to form your limbs. As development continued, the embryo eventually became you.

Biologists who have watched this process in humans or other organisms never cease to marvel at it. How does all the growth and formation of distinctive body parts happen?

To understand how researchers are answering this question, you'll need to draw on what you've already learned about cell-cell interactions, the regulation of gene expression, and a host of other topics described in previous units. Part of the excitement surrounding developmental biology is that it draws on insights from biochemistry, cell biology, genetics, and evolution. It is one of the most interdisciplinary fields in all of biological science. Let's delve in.

✔ When you see this checkmark, stop and test yourself. Answers are available in Appendix B.

21.1 Shared Developmental Processes

Over one hundred years of research has culminated in one of the great insights of contemporary biology: A few fundamental principles are common to all developmental sequences observed in multicellular organisms. Their discovery has brought a unified understanding to the variation observed in how the embryos of fruit flies, oak trees, and humans grow. And it has given biologists a framework for explaining how a single cell can give rise to a complex, multicellular individual.

☞ An individual develops as cells divide, move, or expand in a directed way; begin to express certain genes rather than others; and signal to each other about where they are, what they are doing, and what type of cell they are becoming. In addition, selected cells die in a regulated manner during development (**Table 21.1**). We'll consider each of these processes, and then go on to consider more specific questions about how cells interact and specialize.

Cell Proliferation

For an embryo to grow and develop, its cells have to proliferate—they have to divide and make more cells. This statement may strike you as obvious. Less obvious, but equally important, is the following point: The location, timing, and extent of cell division have to be tightly controlled.

Chapter 11 introduced mitosis and cytokinesis, which are responsible for cell proliferation in eukaryotes. That chapter also introduced the stages of the cell cycle and how they are controlled. You might recall that cells initiate mitosis in response to a regulatory protein complex called mitosis-promoting factor (MPF), that each stage of the cell cycle has checkpoints that are carefully regulated, and that cells continue to grow or stop growing in response to what biologists call "social controls"—meaning, signals from other cells.

In both plants and animals, most cells stop growing when they mature. But both plants and animals have specific populations of undifferentiated cells that keep proliferating throughout the individual's life.

- In plants these cells are grouped into **meristems**. The meristematic tissues described in Chapter 23 are present in the same locations in embryonic and adult plants and perform the same function—giving rise to the stems, roots, leaves, flowers, and other structures that develop throughout life.

- In animals, these cells are called **stem cells**. Embryonic stem cells can give rise to almost any differentiated cell type in the body. In juveniles and adults, stem cells are found in specific locations in the body, where they proliferate to replace skin and blood and gut cells that die, repair wounds, and create a constant supply of disease-fighting cells in the immune system.

SUMMARY TABLE 21.1 **Five Essential Developmental Processes**

Cell proliferation		Cells divide by mitosis and cytokinesis. The timing, location, and amount of cell division are regulated.
Programmed cell death		The timing, location, and amount of cell death are regulated.
Cell movement or differential expansion		Cells can move past one another within a block of animal cells, causing drastic shape changes in the embryo.
		Certain cells can break away from a block of animal cells and migrate to new locations.
		Plant cells can divide along certain planes and expand in specific directions, causing dramatic changes in shape.
Cell differentiation		Undifferentiated cells specialize at specific times and places in a stepwise fashion. Cells that do not undergo differentiation are called stem cells in animals. Many plant cells are capable of de-differentiating.
Cell-cell interactions		Embryonic cells divide, die, grow, move, or differentiate in response to signals from other cells.

Programmed Cell Death

Cells grow or stop growing in response to signals from other cells. But in some cases, they also commit suicide in response to signals from other cells.

Cell death is a highly regulated aspect of plant and animal development. **Programmed cell death** is called **apoptosis** (literally, "falling away") and occurs as certain tissues and organs take shape. As the feet of a chicken embryo develop, for example, cells that are initially present between the toes must die in order for separate toes to form (**Figure 21.1a**). In plants, programmed cell death allows flower petals to fall after pollination has occurred, and causes leaves to be lost in autumn.

Some of the key initial studies on apoptosis were done on the roundworm *Caenorhabditis* (pronounced *see-nor-ab-DIE-tis*) *elegans*. This species is a popular model organism in developmental biology because its entire array of organs and tissues consists of only about a thousand cells. And because they are transparent,

(a) Chicken embryo with normal (left) and defective (right) cell-suicide genes.

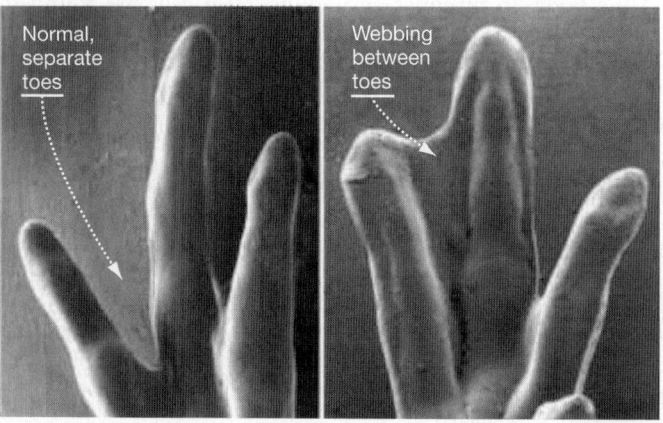

Normal, separate toes

Webbing between toes

(b) Mouse embryo with normal (left) and defective (right) cell-suicide genes.

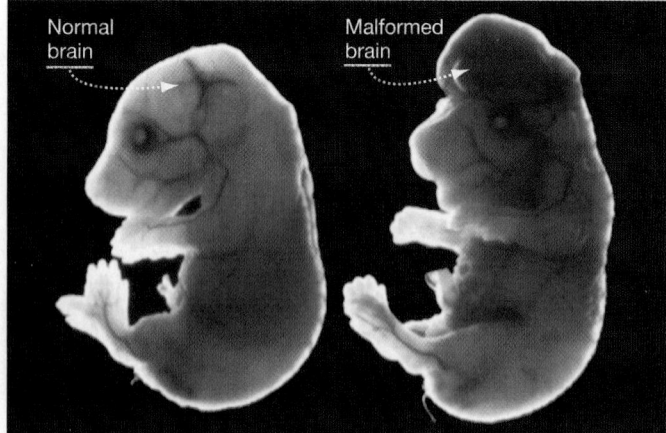

Normal brain

Malformed brain

FIGURE 21.1 Defects Occur When Programmed Cell Death Fails.

✓**QUESTION** If the defective version of the cell-suicide gene is recessive, what would the embryo on the right side of part (b) look like if it were heterozygous?

biologists can follow individual cells throughout development. (For more on *C. elegans*, see **BioSkills 14** in Appendix A.)

As a *C. elegans* individual matures, 131 of its cells undergo apoptosis. To explore how this happens, Hillary Ellis and Robert Horvitz found embryos that do *not* exhibit the normal pattern of cell deaths, and then used techniques introduced in Chapter 19 to locate the mutant genes responsible for the defect. Their initial work uncovered two genes that are essential for apoptosis. The researchers proposed that the genes are part of a genetic program for apoptosis—in short, that they are cell-suicide genes.

Follow-up work, in which researchers searched databases of known DNA sequences, confirmed that mice have similar genes in terms of their sequence and structure. When a research team used genetic engineering techniques to produce mice in which both copies of one of their cell-suicide genes were disrupted, the embryos that resulted had severe malformation of the brain (**Figure 21.1b**). The defect occurred because cells that would normally die early in development survived.

The same genes have now been found in humans and other mammals. Normal apoptosis is important in the development of human embryos, and abnormal apoptosis—either too much or too little—has been implicated in certain diseases of adults. For example, inappropriate activation of programmed cell death is involved in some neurodegenerative diseases, including ALS (Lou Gehrig's disease).

Cell Movement or Cell Growth

Besides proliferating or dying, many animal cells have to move to a new location in order for normal development to occur. In plants, the cells do not move, but changes in the orientation of cell division control the direction of subsequent cell proliferation and cell expansion.

Some of the most dramatic cell movements during animal development occur early in the process, after rapid cell divisions have produced a mass of similar-looking cells. During a sequence of events called **gastrulation**, cells in different parts of the mass rearrange themselves into three distinctive layers, which then give rise to the skin, gut, and other basic parts of the body (gastrulation is described in Chapter 22).

Later in development, certain animal cells break away from their original sites and migrate to new locations in the embryo. There they give rise to germ cells (sperm or eggs), pigment-containing cells, precursors of blood cells, or certain nerve cells. If any of these cell movements is inhibited, or if migrating cells end up in the wrong place or move at the wrong time, the embryo is likely to be deformed or die before development is complete.

Plant cells, in contrast, are encased in stiff cell walls and do not move. Instead, the directionality of cell division and cell expansion is carefully regulated. During development, changes in the direction of cell growth result in the proper formation of straight and bent stems, leaf veins, and other structures.

In plants, changes in cell size can make an individual "move." For example, Chapter 39 will detail how plants grow toward light. This response is not due to changes in the rate of cell proliferation in meristems but to changes in the stem. In response to signals from cells that receive light, cells on one side of the stem expand—causing the entire shoot to bend the other way.

Differential cell growth is a key part of plant development, just as regulated cell movement is a key part of animal development.

Cell Differentiation

As development progresses, most cells undergo **differentiation**—the process of becoming a specialized type of cell. As a result, a fertilized egg may give rise to hundreds of distinctive cell types.

In the case of your own development, some of your embryonic cells differentiated to form muscle cells that contract and relax, while others became nerve cells that conduct electrical signals throughout the body. As an oak tree develops, some cells secrete thickened walls and transport water as part of a woody stem, while others become flattened and secrete the waxy coating found on the surface of leaves.

Differentiation is a progressive, step-by-step process. Initially, cells have the capacity to differentiate into any cell type. Meristematic cells in plants and stem cells in animals remain in this state, but most cells become committed to a certain cell fate early in development and later become differentiated—meaning that they begin to look and behave like a specific cell type.

Some plant cells are also capable of "de-differentiating." They can change their structure and function, even after they have specialized. For example, de-differentiation occurs when a branch of a western redcedar tree droops down low enough to make contact with the soil. Cells in the branch de-differentiate and then re-differentiate to form root cells, resulting in the growth of a fully formed root where the branch initially rested on the ground.

These types of plant cells are said to be totipotent ("all-powerful"). Totipotent cells highlight an important difference between plant and animal development. Once differentiated, animal cells cannot de-differentiate and redifferentiate. They are what they are.

Cell-Cell Interactions

Chapter 8 introduced the topic of cell-cell interactions by examining the extracellular matrix found between cells. That chapter also explored how adjacent cells are attached and exchange materials and how cells respond to signals from other cells. During development, the most important cell-cell interactions involve sending and receiving signals.

You might recall from Section 8.3 that when a signal arrives at the surface of a cell, its message is received and processed. In most cases, cells change their activity in response to these signals. Embryonic cells grow, move, or differentiate in response to signals from other cells.

Some of the most exciting research in developmental biology is focused on how developing cells send cell-cell signals and respond to them. In many cases, the signal transduction pathways introduced in Chapter 8 trigger the production of the transcription factors introduced in Chapter 18. As a result, the arrival of cell-cell

signals changes patterns of gene expression and thus the embryonic cell's structure and behavior. In this way, the fate of a cell inside an embryo hinges on the signals it receives from other cells.

21.2 The Role of Differential Gene Expression in Development

The differentiation of a cell occurs through differential gene expression. The muscle cells in your body are different from your nerve cells because they express different genes and therefore produce different proteins. The water-transport cells in an oak tree are different from its leaf-surface cells for the same reason.

If you think about these statements, you'll realize that they have to be true. The only way that cells can have different structures and functions is if they contain different molecules. What is less obvious is whether cells express different genes because they contain different genes, or whether all the cells in a body contain the same genes but express only a specialized subset.

Evidence That Differentiated Plant Cells Are Genetically Equivalent

If cells from the stem of a cedar tree can de-differentiate to form roots, the cells involved must contain the genes required by root cells. Gardeners and farmers have known for centuries that in many plant species, complete new individuals can be produced from a small section of a root or shoot.

From observations like this, researchers strongly suspected that all plant cells contain the same genes—meaning that they are genetically equivalent. This suspicion was confirmed in the 1950s when biologists were able to grow entire tobacco plants or carrots from a single, differentiated cell taken from an adult. (For more information on how plant cells are grown in culture, see **BioSkills 12** in Appendix A.) These experiments confirmed that differentiated plant cells are genetically equivalent.

How is it possible for plant cells to de-differentiate and re-differentiate so readily? This question has been much more difficult to answer. Somehow, the processes that control gene expression—changes in chromatin structure, regulatory transcription factors, RNA processing, miRNA activity, and so on—get reprogrammed. How this happens remains a mystery.

Evidence That Differentiated Animal Cells Are Genetically Equivalent

In contrast to plants, the issue of genetic equivalence was extremely difficult to resolve for animals. Serious experimental work began in the 1950s, but the question wasn't settled until the late 1990s.

Early experiments were based on transferring nuclei from differentiated frog cells into unfertilized eggs whose nuclei had been removed. Some of these transplanted nuclei were able to direct the development of tadpoles successfully. These results provided strong evidence that all cells in the same individual are genetically equivalent.

This conclusion was confirmed in 1997, when Ian Wilmut and colleagues reported the results of nuclear transfer experiments in sheep. As **Figure 21.2** shows, the researchers removed mammary-gland cells from a 6-year-old pregnant female, grew them in culture, and fused them with eggs whose nuclei had been removed. As the upper drawings show, the eggs came from a black-faced breed of sheep, while the donor nuclei came from a white-faced breed. After developing in culture, the resulting embryos were implanted in the uteri of surrogate mothers. In one of several hundred such transfer attempts, a white-faced lamb named Dolly was born.

Genetic tests showed that Dolly was a **clone**—a copy—of the white-faced donor of the mammary-gland cell. Dolly grew into a fertile adult and, by normal mating, produced her own lamb named Bonnie. Soon after, other research groups reported similar results in mice and cows. More recently, horses, monkeys, and dogs have been cloned. The procedure remains technically difficult, however, and the vast majority of nuclear transplant experiments fail.

Taken together, research on cloning plants and animals has shown that in most cases, cellular differentiation does not involve changes in the genetic makeup of cells. Instead, it results from differential gene expression. There are some important exceptions to this rule, however. For example, small stretches of DNA are rearranged in certain immune system cells in humans and other mammals, late in development. As a result, many immune cells are genetically unique. Chapter 49 explains this phenomenon.

How Does Differential Gene Expression Occur?

Chapter 18 emphasized that eukaryotic cells control gene expression at several different levels: chromatin remodeling, chromatin modification, transcription regulation, alternative splicing of mRNAs, selective destruction of mRNAs, translation rate, and activation and deactivation of proteins after they are translated. All of these processes occur during development. Which is responsible for differentiation?

The answer is transcriptional control. To understand why, ask yourself whether a muscle cell should produce mRNAs or proteins that are specifically required by nerve cells. The answer is no. If it did, it would also have to produce microRNAs that disable nerve-cell mRNAs or regulatory proteins that keep nerve-cell proteins inactivated. It is much more logical to expect that muscle cells transcribe only genes required by muscle cells. This is exactly what researchers have found.

Transcription is the fundamental level of control in differential gene expression during development. In eukaryotes, transcription is controlled primarily by the presence of **regulatory transcription factors** that influence chromatin remodeling and bind to promoter-proximal elements, enhancers, silencers, or other regulatory sites in DNA.

This simple insight is extremely important. To understand differentiation, researchers have to understand how and why regulatory transcription factors vary among cells.

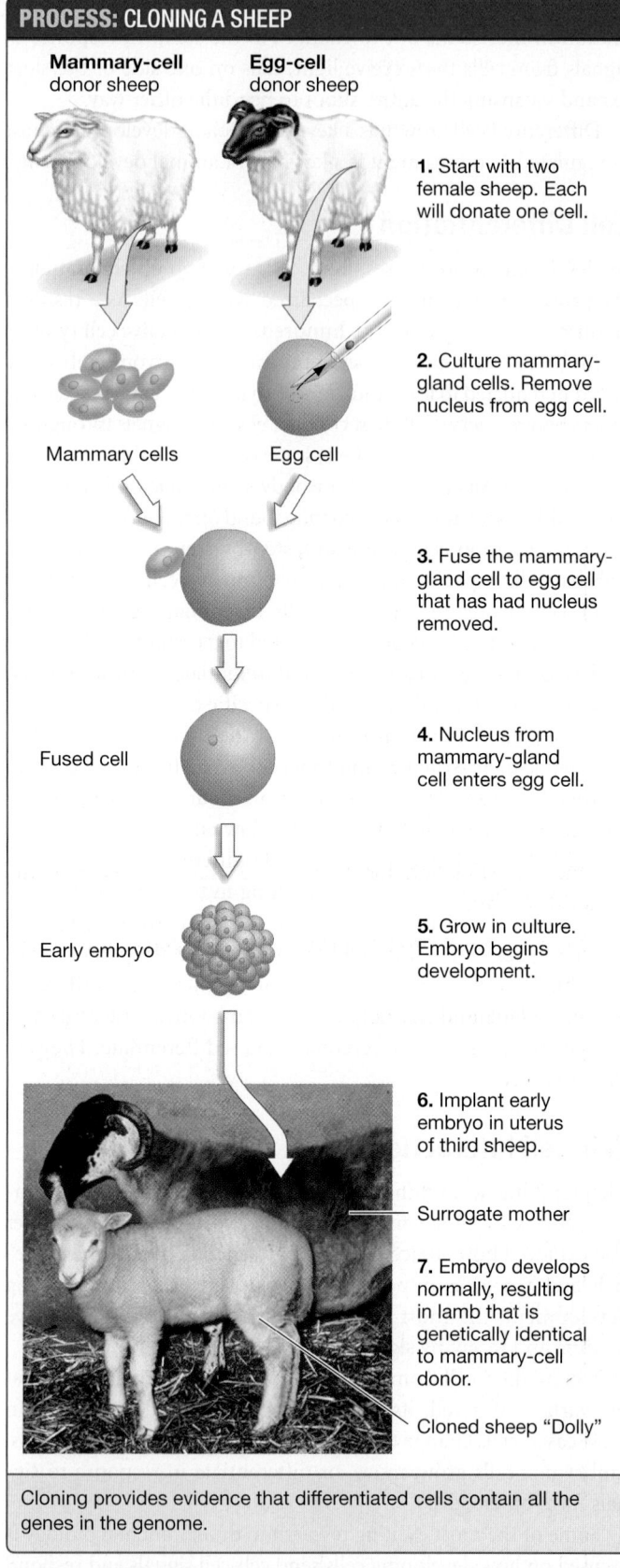

PROCESS: CLONING A SHEEP

Mammary-cell donor sheep

Egg-cell donor sheep

1. Start with two female sheep. Each will donate one cell.

Mammary cells

Egg cell

2. Culture mammary-gland cells. Remove nucleus from egg cell.

3. Fuse the mammary-gland cell to egg cell that has had nucleus removed.

Fused cell

4. Nucleus from mammary-gland cell enters egg cell.

Early embryo

5. Grow in culture. Embryo begins development.

6. Implant early embryo in uterus of third sheep.

Surrogate mother

7. Embryo develops normally, resulting in lamb that is genetically identical to mammary-cell donor.

Cloned sheep "Dolly"

Cloning provides evidence that differentiated cells contain all the genes in the genome.

FIGURE 21.2 Mammals Can Be Cloned by Transplanting Nuclei from Mature Cells. The lamb that resulted from this experiment was identical to the white-faced individual that donated the nucleus, not the black-faced egg donor or surrogate mother.

(a) The three body axes observed in humans and other animals...

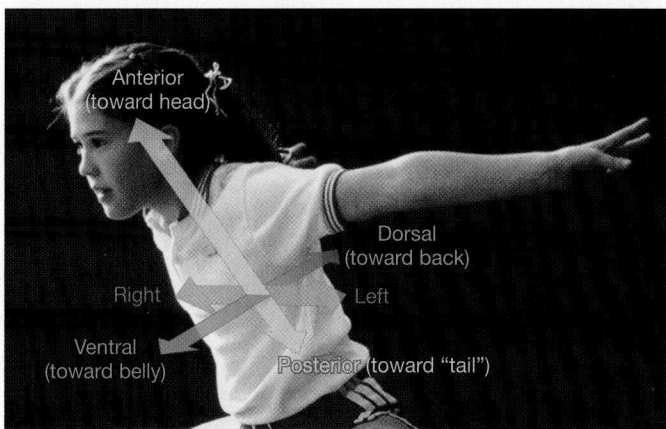

Anterior (toward head)

Dorsal (toward back)

Right

Left

Ventral (toward belly)

Posterior (toward "tail")

(b) ...are initially established in embryos.

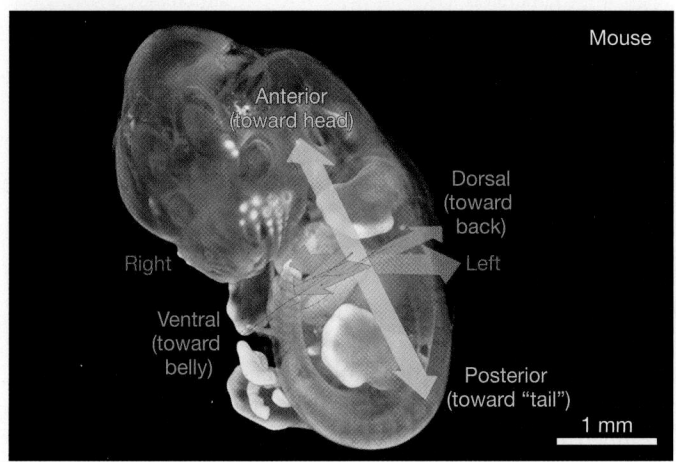

Mouse

Anterior (toward head)

Dorsal (toward back)

Right

Left

Ventral (toward belly)

Posterior (toward "tail")

1 mm

FIGURE 21.3 Most Animals Have Three Major Body Axes.

21.3 Cell-Cell Signals Trigger Differential Gene Expression

To understand development, you have to think like a cell. Suppose that you were one of the hundreds or thousands of cells in a developing animal embryo. Your fate—whether you ended up as part of an arm or a kidney, and whether you differentiated into a nerve cell or a blood-vessel cell—would depend on your location along four axes: time (meaning the organism's current stage of development) plus three spatial dimensions—the three body axes illustrated in **Figure 21.3**.

1. One axis runs **anterior**, toward the head, to **posterior**, toward the tail.

2. One axis runs **ventral**, toward the belly, to **dorsal**, toward the back.

3. One axis runs left to right.

Cell-cell signals tell cells where they are in time and space. This information activates transcription factors that turn specific genes on or off, resulting in differentiation. As development proceeds, the distinctive suites of genes that are activated at successive stages determine the fate of each cell.

Let's consider how this process happens, beginning with one of the first developmental signals ever discovered. Although you'll be analyzing what happens as a fruit-fly embryo develops, keep an important point in mind: Principles that were discovered in fruit flies are relevant to virtually all multicellular organisms studied to date—from mustard plants to humans.

Master Regulators Set Up the Major Body Axes

Biologists use the term **pattern formation** to describe the events that determine the spatial organization of an embryo. If a molecule signals that a target cell is in the embryo's head, or tail, or dorsal side, or ventral side, that molecule is involved in pattern formation.

Pattern formation is progressive. Early signals act as master regulators that set up the general anterior–posterior, dorsal–ventral, and left–right axes of an embryo. Genes activated by these master regulators send signals with more specific information about the cells' physical location. As growth continues, the process repeats: New signals arrive and activate genes that specify finer and finer control over what a cell becomes.

THE DISCOVERY OF *BICOID* The insights that led to the discovery of master regulators emerged from work on the fruit fly *Drosophila melanogaster*—a key model organism in genetics and developmental studies (see **BioSkills 14** in Appendix A).

Christiane Nüsslein-Volhard and Eric Wieschaus started this work in the 1970s. They began by exposing adult flies to treatments that cause mutations and examining their offspring for defects in development. One of the most dramatic mutations they found affected the anterior–posterior axis of the embryos. As **Figure 21.4** on page 380 shows, the mutant embryos were missing all of the structures normally found in the anterior end. Instead, the anterior end contained some structures normally found in the posterior.

The gene responsible for this phenotype was dubbed *bicoid*, meaning "two tailed." Based on its phenotype, Nüsslein-Volhard and Wieschaus suspected that the *bicoid* gene's product must provide positional information. In other words, they hypothesized

(a) A normal fruit-fly embryo

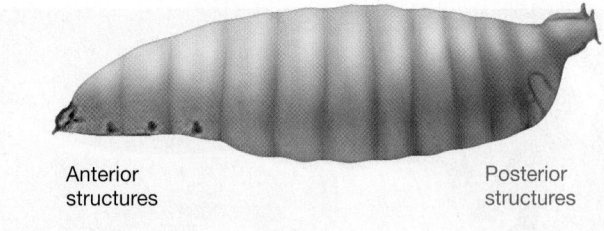

Anterior
structures

Posterior
structures

(b) A *bicoid* mutant

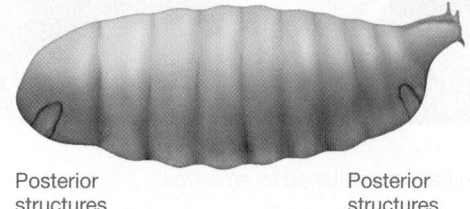

Posterior
structures

Posterior
structures

FIGURE 21.4 Pattern-Formation Mutants Have Misshapen Bodies.

that the *bicoid* gene coded for a signal that tells cells where they are located along the anterior–posterior body axis.

THE IMPORTANCE OF CONCENTRATION GRADIENTS To test their hypothesis, Nüsslein-Volhard and colleagues cloned and sequenced the gene using techniques introduced in Chapter 19. Then they used **in situ** (literally, "in place") **hybridization** to find where *bicoid* mRNAs are located in embryos. As **Figure 21.5** shows, in situ hybridization works by adding a label to single-stranded copies of DNA or RNA molecules—specifically, to molecules that are complementary in sequence to the mRNA of interest. In this case, the probes were designed to bind to *bicoid* mRNA inside the embryo. As a result, the labeled probes marked the location of the mRNAs.

When Nüsslein-Volhard's group treated eggs and early embryos with labeled copies of a probe that bound to *bicoid* mRNA, they found the mRNA to be highly localized at the anterior end (see step 5 in Figure 21.5). Follow-up work showed that when these mRNAs are translated, the protein product forms a steep concentration gradient: Bicoid protein is abundant in the anterior end but declines to progressively lower concentrations in the posterior end.

Later work showed that the Bicoid protein is a regulatory transcription factor (see Chapter 18). It binds to DNA and activates genes required for the formation of anterior structures. In effect, a high concentration of Bicoid is a signal that says, "You're in the head region." A medium concentration of Bicoid means, "You're in the middle of the body." A low concentration indicates, "You're in the posterior" (**Figure 21.6**). When Bicoid is lacking, cells throughout the embryo get the "you're in the posterior" message—leading to the mutant phenotype you saw in Figure 21.4.

To review how the *bicoid* gene works, go to the study area at *www.masteringbiology.com*.

MB **Web Activity** Early Pattern Formation in *Drosophila*

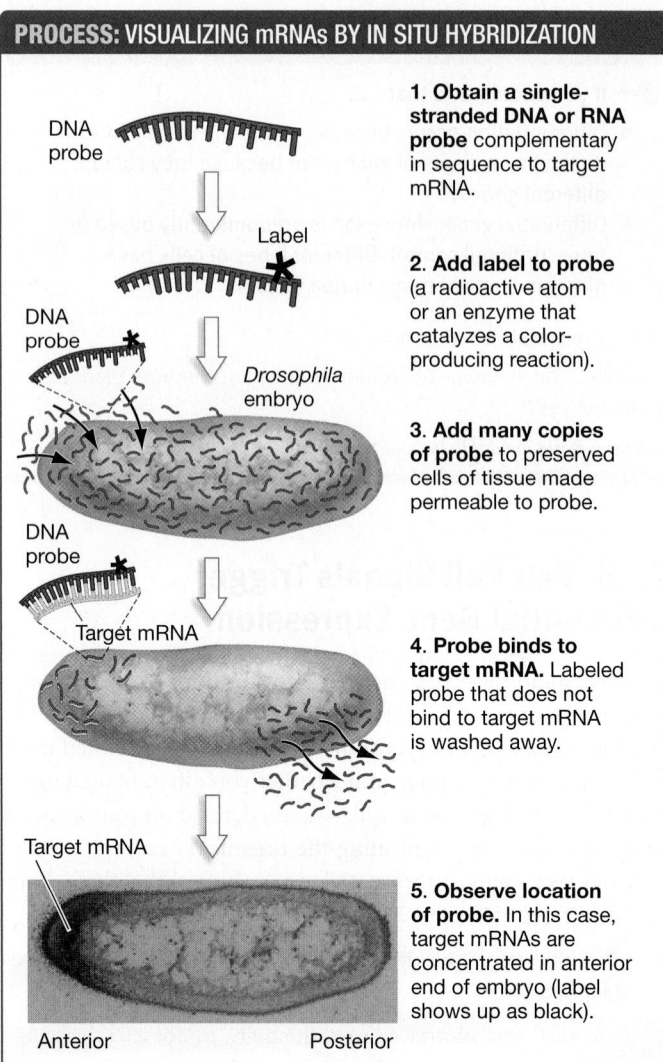

PROCESS: VISUALIZING mRNAs BY IN SITU HYBRIDIZATION

1. **Obtain a single-stranded DNA or RNA probe** complementary in sequence to target mRNA.

2. **Add label to probe** (a radioactive atom or an enzyme that catalyzes a color-producing reaction).

3. **Add many copies of probe** to preserved cells of tissue made permeable to probe.

4. **Probe binds to target mRNA.** Labeled probe that does not bind to target mRNA is washed away.

5. **Observe location of probe.** In this case, target mRNAs are concentrated in anterior end of embryo (label shows up as black).

FIGURE 21.5 In Situ Hybridization Allows Researchers to Pinpoint the Location of Specific mRNAs. The micrograph in the last step shows the location of mRNA from the *bicoid* gene in a fruit-fly embryo.

✔**QUESTION** In situ hybridization is typically used to identify cells that are expressing a particular gene. Why is it valid to claim that labeled cells are expressing the gene in question?

AUXIN'S ROLE IN PLANT DEVELOPMENT To capture Bicoid's role in the development of a fruit-fly embryo, biologists refer to it as a "master regulator." The idea is that Bicoid gets the cell differentiation process underway, by providing information on where cells are in the body, extremely early in development.

Plants also have a master regulator. But unlike Bicoid, the plant master regulator is not a transcription factor. Instead, it is a **hormone**: a signaling molecule that travels through the body and acts on distant target cells. In plant embryos, the cell-cell signal called auxin enters cells and triggers the production of transcription factors that affect differentiation.

Auxin is produced in meristematic cells at the tip, or apex, of the growing embryo—what will become the top of the stem—and is transported toward the base—what will become the root. In the process, a concentration gradient forms. A high concen-

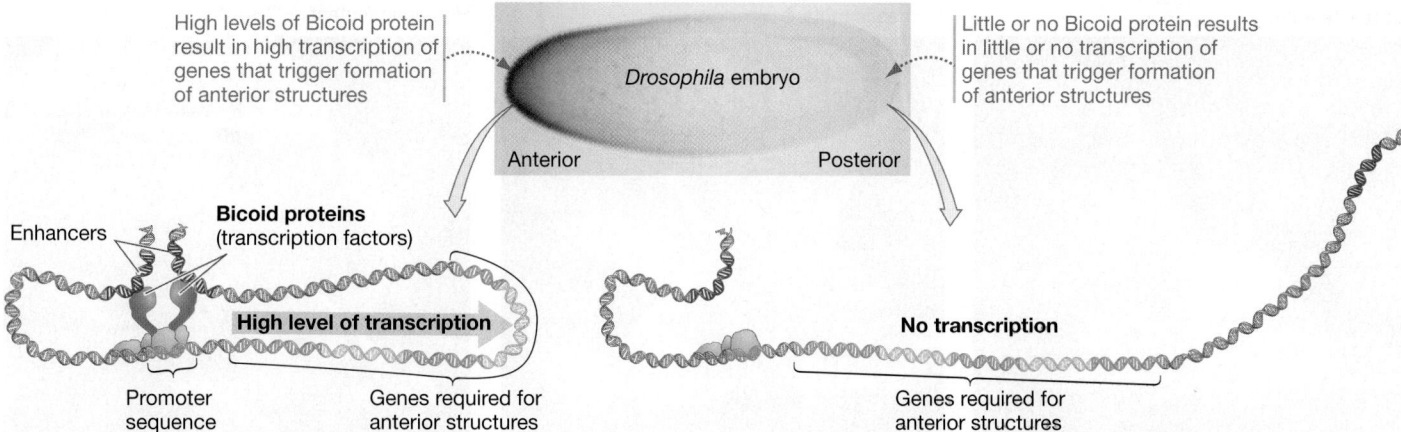

FIGURE 21.6 Differential Gene Expression Occurs through the Presence or Absence of Regulatory Transcription Factors. Bicoid protein is a regulatory transcription factor that triggers the formation of anterior structures. Because the concentration of Bicoid decreases toward the embryo's posterior, different genes are expressed in the anterior of the embryo than in the posterior.

tration of auxin is a signal that means, "You're near the top of the shoot"; when auxin accumulates at the root it signals, "You're near the base of the root."

In both plants and animals, molecules that provide spatial information during early embryonic development, via a concentration gradient, are called **morphogens** ("form-source"). Bicoid and auxin are morphogens that have a fundamental impact on early development. Because they are present in a concentration gradient, they provide cells with information about their position along the anterior–posterior or the apical–basal body axis. Other initial signals are found in concentration gradients that tell cells where they are along the other body axes.

Regulatory Genes Provide Increasingly Specific Positional Information

The initial work on *bicoid* illustrated the importance of cell-cell signals and interactions as a general theme in animal and plant development. Follow-up work was instrumental in focusing attention on a second fundamental developmental principle common to both plants and animals: Differentiation is a progressive, step-by-step process.

Along with the "two-posteriored monsters" they named *bicoid* mutants, Nüsslein-Volhard and Wieschaus found an array of embryos that had normal anterior–posterior patterning but defects in how their body segments became organized later in development. A **segment** is a region of an animal body that contains a distinct set of structures and is repeated along its length. The mutants had defective **segmentation genes**.

Researchers have now identified three classes of segmentation genes:

1. Sequences called gap genes are expressed first, in broad regions along the head-to-tail axis (**Figure 21.7a**). They define the general position of segments (what part of the body the segments are in).

Anterior   **Posterior**

(a) Gap genes

Early in development, gap genes define the *general position of head, thorax, and abdominal regions.*

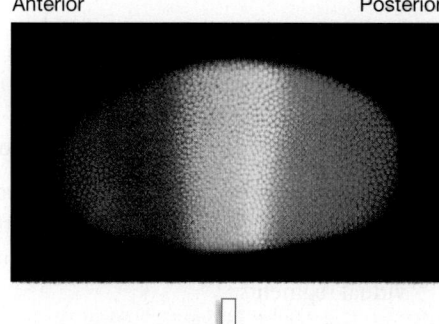

(b) Pair-rule genes

Later in development, pair-rule genes demarcate the *edges of individual segments.*

(c) Segment polarity genes

Still later, segment polarity genes delineate boundaries *within individual segments.*

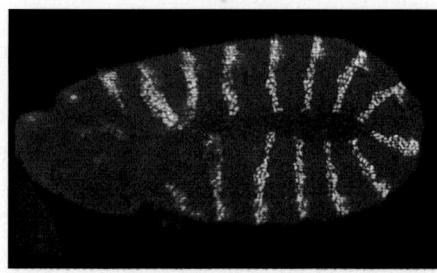

FIGURE 21.7 Sequences of Genes Demarcate Body Segments in Fruit Flies. The embryos in **(a)** and **(b)** were stained for two different gene products from segmentation genes. The embryo in **(c)** was stained for one gene product.

Normal fruit fly   Homeotic mutant   Homeotic mutant

Antennae

Haltere

Wings in place of halteres

Legs in place of antennae

FIGURE 21.8 Homeotic Mutants in *Drosophila* Have Structures in the Wrong Locations. As these colorized, scanning electron micrographs show, homeotic mutants in fruit flies include individuals with wings growing where small, stabilizing structures called halteres should be, or legs growing where antennae should be.

2. Pair-rule genes are expressed next, in alternating bands (**Figure 21.7b**). This pattern and order of expression suggest that pair-rule genes demarcate the edges of individual segments.

3. Segment polarity genes are expressed later, in more restricted bands (**Figure 21.7c**). This pattern and order of expression imply that they delineate boundaries within individual segments.

Once segmentation gene products have established the identity of each segment along the anterior–posterior axis of a fly embryo, development continues with the activation of **homeotic genes**. Segmentation gene products establish the boundaries of each segment; homeotic gene products identify each segment's structural role. More specifically, homeotic gene products trigger the development of structures that are appropriate to each type of segment, such as antennae, wings, or legs. The proteins required for these structures are produced by effector genes that are regulated by homeotic genes called the **Hox genes**.

The *Hox* genes were discovered when researchers found adult fruit flies with body parts in the wrong place. For example, a series of mutations in *Hox* genes can transform a segment in the middle part of the body to the segment just anterior. Instead of bearing the pair of small stabilizer structures called halteres, the transformed segment bears a pair of wings. The mutant has four wings instead of two (**Figure 21.8**).

This type of replacement of one structure by another is termed **homeosis** ("like-condition"). Homeosis occurs when cells get incorrect information about where they are in the body.

To put all this information into perspective, biologists recognize that the genes involved in early development form a regulatory cascade (**Figure 21.9**). Master regulators trigger the production of other regulatory signals and transcription factors, which trigger production of another set of signals and regulatory proteins, and so on.

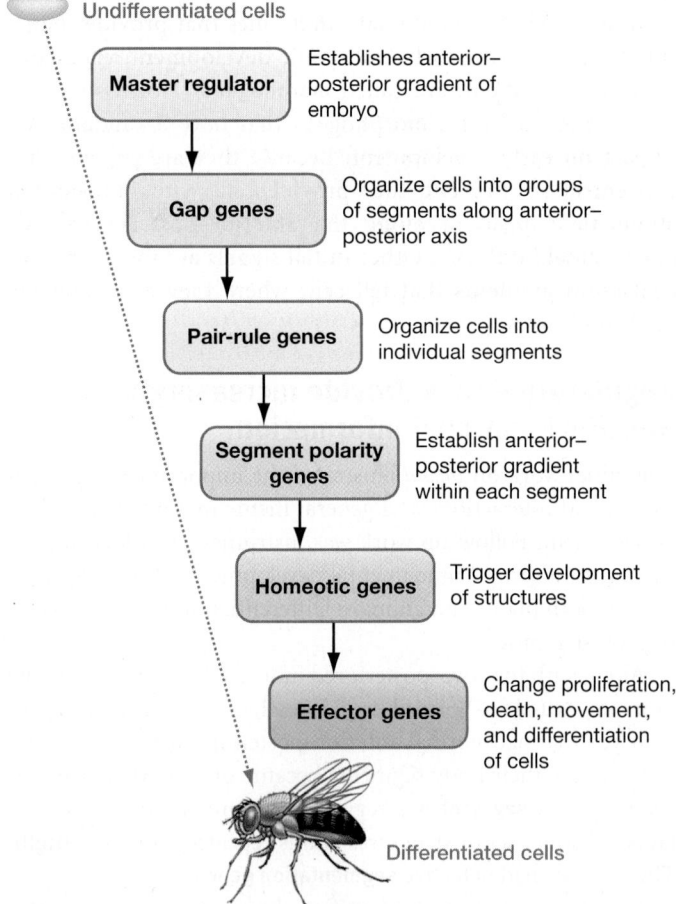

Undifferentiated cells

Master regulator — Establishes anterior–posterior gradient of embryo

Gap genes — Organize cells into groups of segments along anterior–posterior axis

Pair-rule genes — Organize cells into individual segments

Segment polarity genes — Establish anterior–posterior gradient within each segment

Homeotic genes — Trigger development of structures

Effector genes — Change proliferation, death, movement, and differentiation of cells

Differentiated cells

FIGURE 21.9 A Regulatory Gene Cascade in Fruit Flies. Thus far, researchers have described 9 gap genes, 8 pair-rule genes, 9 segment polarity genes, and 11 homeotic (*Hox*) genes in *Drosophila*. Researchers are currently trying to identify the effector genes responsible for producing key structures in this species.

Similar types of regulatory cascades are being found in the other body axes of fruit flies, and in other animal and plant species. In plants, for example, sequences called *MADS-box* genes—detailed in Chapter 23—are analogous to the homeotic genes of animals.

🔑 Regulatory genes act in a sequence, triggering gene cascades that provide progressively detailed information about where cells are located in time and space. This positional information, in turn, causes changes in cell proliferation, death, movement, differentiation, and interaction.

- Because the signals and transcription factors vary in identity and concentration along the three major body axes, cells in different locations receive unique positional information.
- Each level in a regulatory cascade provides a more specific level of information about where a cell is.
- As regulatory cascades proceed over time, a cell's fate becomes more and more finely determined.

Cell-Cell Signals and Regulatory Genes Are Evolutionarily Conserved

After homeotic genes were discovered and characterized in fruit flies, researchers began looking for similar genes in other animals. The results were striking.

Investigators have found that clusters of *Hox* genes occur in virtually every animal examined to date, including frogs, crustaceans (crabs and their relatives), birds, various types of worms, mice, and humans. Although the number of *Hox* genes varies widely among species, their chromosomal organization is similar to that of the fly homeotic genes (**Figure 21.10a**).

Recent studies of *Hox* genes have shown that they are expressed along the head-to-tail axis of the mouse embryo in the same sequence as in fruit flies (**Figure 21.10b**). In addition, experiments have shown that when mouse *Hox* genes are altered by mutation, defects in pattern formation result. Based on these data, biologists conclude that in flies, mice, humans, and most other animals, *Hox* genes play a key role in identifying the position of cells along the head-to-tail axis of the body.

This conclusion was supported in dramatic fashion when researchers in William McGinnis's lab introduced the *Hoxb6* gene from mice into fruit-fly eggs. The *Hoxb6* gene in mice is similar in structure and sequence to the *Antp* gene of flies. Because it was introduced without its normal regulatory sequences, the *Hoxb6* gene was expressed throughout the treated fly embryos. The resulting larvae had defects identical to those observed in naturally occurring fly mutants in which the *Antp* gene is mistakenly expressed throughout the embryo. This is a stunning result: A mouse allele not only affected the development of a fly but also mimicked the effect of a specific fly allele.

To interpret these observations, biologists hypothesize that the genes in *Hox* complexes of animals are homologous—meaning that they are similar because they are descended from genes in a common ancestor. If this hypothesis is correct, it

(a) Fly

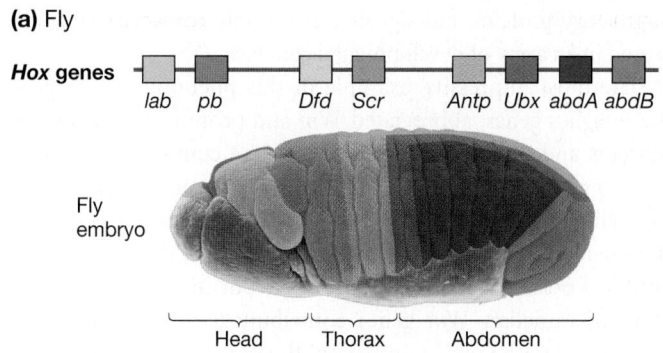

(b) Mouse

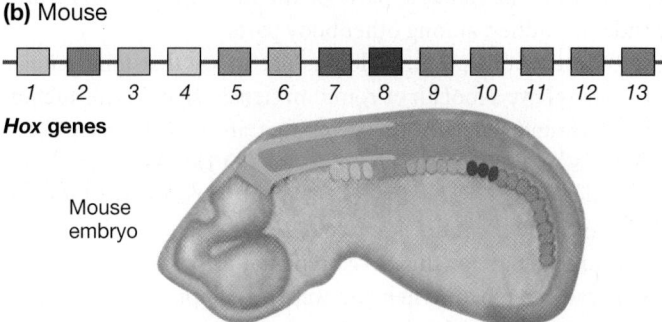

FIGURE 21.10 Organization and Expression of *Hox* Genes. The location of *Hox* genes on the chromosome correlates with their pattern of expression in the embryos of flies and mice. The genes represented by same-color boxes in parts **(a)** and **(b)** are considered homologous.

✓**QUESTION** What evidence would support the claim that these genes are homologous?

means that the first *Hox* genes arose before the origin of animals. For the past billion years, *Hox* gene products have been helping to direct the development of animals.

The take-home message from these studies is that at least some of the molecular mechanisms of pattern formation have been highly conserved during animal evolution. The discovery of these shared mechanisms is one of the most significant results to have emerged from animal development studies to date.

Although animal bodies are spectacularly diverse in size and shape, the underlying mechanisms responsible for their development are similar. Regulatory gene cascades occur in all animals, and all plants, too. Within each of these lineages, many of the elements in gene cascades are shared. What varies among species is less a matter of which genes are present and more a matter of when, where, and in what quantity similar genes are expressed.

Common Signaling Pathways Are Active in Many Contexts

Regulatory gene cascades and the evolutionary conservation of key developmental genes are general features of animal and plant development. But biologists have articulated another organizing principle as well: During development, the same regulatory transcription factors and cell-cell signals are active in a variety of contexts.

Regulatory proteins and signals are not only conserved; they are reused in an array of developmental contexts.

The most impressive example of this phenomenon may be the *wingless* genes, abbreviated *Wnt* and pronounced "wint." In humans and other mammals, the genome contains a family of 15 *Wnt* genes. Each of these genes codes for proteins responsible for cell-cell signaling during development. More specifically, these genes are part of the regulatory cascade that sets up the anterior–posterior axis in the embryo. But in addition, signals from mammalian *Wnt* genes contribute to the formation of back muscles, the midbrain region, the limbs, the gonads (testes or ovaries), hair follicles, parts of the intestine, and structures inside the kidney, among other body parts.

To capture this point, biologists like to say that multicellular organisms have a tool kit of common signals, signal-transduction pathways, and regulatory proteins that are used over and over during development. The common tool kit can direct the development of dramatically different structures because the tools are deployed at different times and in different locations.

As an analogy for this developmental principle, consider the signal called a pinch. When you were little, a pinch on the cheek from your grandmother indicated something very different from a pinch on the arm from an older sibling. It also means something different from a pinch you might receive now from someone you are romantically attracted to. In development, as in human communication, the context in which a signal is sent and received—its location, timing, and intensity—has a major effect on the signal's meaning and consequences.

Also by analogy, the elements of a regulatory gene cascade or a particular signal transduction pathway are like the tools that a carpenter uses. The same hammer and circular saw can be used to build a shack or a palace. The key is the timing and extent of hammer and saw "expression"—when and how the tools are used during the development of a structure.

CHECK YOUR UNDERSTANDING

If you understand that . . .

- Cell-cell signals trigger the production of specific sets of transcription factors that change gene expression in receiving cells.
- Different cells express different genes because they receive different sets of signals, and because they produce different sets of transcription factors in response.

✔ **You should be able to . . .**

1. Explain why the gradient of Bicoid protein delivers information about where cells are along the anterior–posterior axis of a fly embryo.

2. Describe how the cascade of transcription factors triggered by Bicoid leads to the gradual differentiation of segments along the anterior–posterior axis of a fly embryo.

Answers are available in Appendix B.

21.4 Changes in Developmental Pathways Underlie Evolutionary Change

Sections 21.1 through 21.3 have shown that for an embryo to develop, cells have to proliferate, die, move or expand, differentiate, and interact in specific ways. Differentiation is caused by differential gene expression. It results from signals that tell cells where they are in time and space, triggering a complex cascade of regulatory transcription factors.

If any of these processes is disrupted, the embryo is likely to die. But if one of these processes is modified in some slight way, the effect may be a structure with a different size or shape or activity. As a result, the embryo will develop new features, and the adult will have a novel phenotype.

Once biologists began working out the regulatory signals and cascades introduced earlier in the chapter, they realized that the genetic changes altering these developmental processes must be the foundation of evolutionary change. The increase in body size that has occurred during human evolution, for example, must have resulted from mutations that altered the signals, regulatory sequences in DNA, or transcription factors that are involved in the amount and timing of cell proliferation throughout the body.

A research field called evolutionary-developmental biology, or **evo-devo**, focuses on understanding how changes in developmentally important genes have led to the evolution of new phenotypes such as the flower, the leaf, the limbs found in tetrapods (amphibians, reptiles, mammals), and the limbs of arthropods (crabs, insects, millipedes). As an example of how this work is done, let's consider an instance of limb *loss*—in snakes.

Although some snakes do not develop any sort of forelimb or hind limb at all, boas and pythons have tiny pelvic (hip) bones and a rudimentary femur (thigh bone) and claw (**Figure 21.11**). The fossil record shows that the ancestor of all snakes had four functional legs along with feet and toes. Their closest living relatives, the lizards, also have limbs.

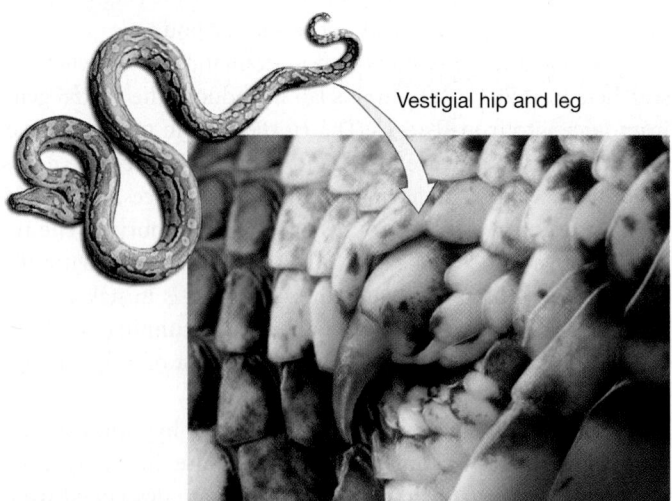

Vestigial hip and leg

FIGURE 21.11 Some Snakes Have Rudimentary Hind Limbs. This claw, of a South African python, is attached to a vestigial femur and hip.

(a) Pattern of gene expression in tetrapods with forelimbs.

(b) Pattern of gene expression in snakes (no forelimbs).

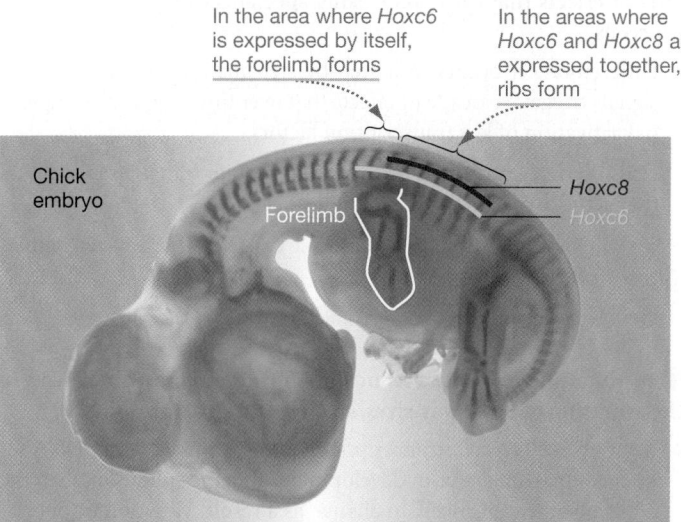

In the area where *Hoxc6* is expressed by itself, the forelimb forms

In the areas where *Hoxc6* and *Hoxc8* are expressed together, ribs form

Chick embryo

Forelimb

Hoxc8
Hoxc6

Hoxc6 and *Hoxc8* are always expressed together, so no forelimbs form

Snake embryo

Hoxc8
Hoxc6

FIGURE 21.12 Changes in Homeotic Gene Expression Led to Limb Loss in Snakes. *Hoxc6* and *Hoxc8* code for transcription factors. If both transcription factors are present in a population of cells along the anterior–posterior axis of a tetrapod, genes that lead to the formation of vertebrae and ribs are activated. But if only *Hoxc6* is expressed, then genes that lead to the formation of forelimbs are activated.

How did snakes lose their legs? What developmental changes were responsible for this dramatic evolutionary event?

Researchers were able to answer this question because the cell-cell signals and regulatory cascades responsible for limb development are well understood. In chicken embryos, for example, *Hox* genes called *Hoxc6* and *Hoxc8* are expressed together in cells where ribs form, but *Hoxc6* is expressed without *Hoxc8* in the region that gives rise to the forelimbs (**Figure 21.12a**). The situation is different in snakes. As **Figure 21.12b** shows, *Hoxc6* and *Hoxc8* are expressed together throughout the snake embryo—including in the region where forelimbs should form. These data suggest that a change in the regulation of a homeotic gene—specifically, where *Hoxc8* is expressed—led to the evolutionary loss of the forelimb. Snakes make ribs instead of forelimbs.

Hind limb loss, in contrast, is due to defects in a signaling molecule encoded by the gene *Sonic hedgehog*. This gene is homologous to one of the segment polarity genes found in fruit flies. These observations suggest that a mutation in a gene responsible for setting up the anterior–posterior axis led to the evolutionary loss of the hind limb. Recent work has shown that defects in *Sonic hedgehog* signaling were also responsible for the loss of limbs in whales.

Both in forelimb and hind limb loss, biologists can point to alterations in a specific transcription factor or cell-cell signal that explains why an important evolutionary change occurred. Changes in a regulatory cascade that is triggered early in development led to changes in the adult body of a lizard-like ancestor. Snakes have been legless ever since.

CHAPTER 21 REVIEW

For media, go to the study area at www.masteringbiology.com

Summary of Key Concepts

🔑 **During development, cells divide, die, move, or expand in a directed manner; specialize; and interact with other cells.**

- Cells have to proliferate in a regulated manner as the body develops.

- Some cells have to undergo programmed cell death as part of normal development.

- In animals, masses of cells move in a coordinated way early in development; later, certain cells move to specific locations in the body. Plant cells do not move, but normal plant development depends on precise control over the plane of cell division and directed expansion of cells.

- Cells undergo a step-by-step process that leads to differentiation, or the production of specialized cell types.

- Cell-cell signals provide a constant flow of information about where cells are in space and time.

  ✔ You should be able to predict how the development of an oak seedling would be affected by a drug that cuts the normal rate of cell proliferation by half.

⚫ **Cells become specialized because they express different genes, not because they contain different genes.**

- In animals and plants, cells differentiate due to differential gene expression, not differential loss of genes.

- Meristematic cells in plants and stem cells in animals do not differentiate, but continue proliferating throughout life.

  ✔ You should be able to explain why it's possible to propagate banana trees from cuttings—specifically, why parenchyma cells in a stem can give rise to vascular tissue in a root.

⚫ **Cells interact continuously by means of cell-cell signals during development.**

- Cells "know" where they are in the body and how far along development has progressed because they produce and receive a steady stream of signals. These signals are produced by other cells and diffuse or are transported throughout the body.

- Cells begin expressing a distinctive suite of proteins and differentiate because they receive a distinctive suite of signals. Differentiation depends on both the types and quantities of signals received.

  ✔ You should be able to explain why concentration gradients in transcription factors or cell-cell signals convey information.

⚫ **When development begins, early cell-cell signals trigger a cascade of effects that cause increasing specialization as development proceeds.**

- Differentiation occurs in a step-by-step manner because cell-cell signals trigger a cascade of effects that over time lead to the sequential activation of key transcription factors.

- Cells become specialized because they contain transcription factors that activate certain genes and not others.

  ✔ You should be able to explain the concept of a "master regulator" in early development.

  (MB) **Web Activity** Early Pattern Formation in *Drosophila*

⚫ **Mutations in genes responsible for development may lead to the evolution of new body sizes, shapes, and structures.**

- If the cell-cell signals, transcription factors, and regulatory DNA sequences that are active in development undergo mutation, then the adult phenotype is likely to change as a result. Variation in phenotypes is the basis of evolutionary change.

  ✔ You should be able to explain the evidence behind the claim that changes in cell-cell signals and transcription factors were responsible for the loss of limbs in snakes.

Questions

✔ TEST YOUR KNOWLEDGE

1. What is apoptosis?
   a. an experimental technique used to kill specific cells
   b. programmed cell death that is required for normal development
   c. a pathological condition observed only in damaged or diseased organisms
   d. a developmental mechanism unique to the roundworm *C. elegans*

2. What is the function of stem cells in adult mammals?
   a. Some of their daughter cells remain as stem cells and continue to divide throughout life.
   b. They give rise to hair, fingernails, and other structures that grow throughout life.
   c. They produce compounds that stem blood loss from wounds.
   d. They produce cells that differentiate to replace dead or damaged cells.

3. Why are in situ hybridizations such a valuable tool for studying development?
   a. They identify the location of specific mRNAs, and so provide a picture of differential gene expression.
   b. They allow researchers to understand how cell-cell signals and regulatory transcription factors interact.
   c. They provide data on homology—the presence of similar genes in different species.
   d. They can be done with RNA or DNA probes.

4. What does it mean to say that differentiation is "progressive?"
   a. Differentiation gets more efficient over time.
   b. Differentiation gets more complex over time.
   c. Cells become increasingly more specialized over time.
   d. Differentiation is triggered by master regulators.

5. What is a homeotic mutant?
   a. an individual with a structure located in the wrong place
   b. an individual with an abnormal head-to-tail axis
   c. in flies, an individual that is missing segments; in *Arabidopsis*, an individual that is missing a hypocotyl or other embryonic structure
   d. an individual with double the normal number of structures or segments

6. What is "homeosis"?
   a. replacement of one structure by another structure (a structure develops in the wrong location)
   b. the observation that the same signaling pathway is used during the development of many different structures in the same species (in different locations or at different times)
   c. the observation that the same regulatory cascades are conserved among animal or plant species
   d. similarity between species due to descent from a common ancestor

✔ TEST YOUR UNDERSTANDING

1. What does it mean to say that cell proliferation, death, movement, or expansion; differentiation; and interaction are shared developmental processes?

2. Explain the logic behind using nuclear transplant experiments to test the hypothesis that all animal cells in the body are genetically equivalent.

3. How did researchers go about looking for genes that have important roles in establishing the anterior–posterior body axis and body segments in *Drosophila*?

4. Why is it significant that many of the genes involved in development encode regulatory transcription factors?

5. Explain the connection between the existence of regulatory cascades and the observation that differentiation is a step-by-step process.

6. What evidence suggests that at least some of the molecular mechanisms responsible for pattern formation have been highly conserved over the course of animal evolution?

✓ APPLYING CONCEPTS TO NEW SITUATIONS

Answers are available in Appendix B

1. Recent research has shown that the products of two different *Drosophila* genes are required to keep *bicoid* mRNA concentrated at the anterior end of the egg. In individuals with mutant forms of these proteins, *bicoid* mRNA diffuses farther toward the posterior pole than it normally does. Predict what effect these mutations will have on segmentation of the larva.

2. In 1992 David Vaux and colleagues used genetic engineering technology to introduce an active human gene for a protein that inhibits apoptosis into embryos of the roundworm *C. elegans*. When the team examined the embryos, they found that cells that normally undergo programmed cell death survived. What is the significance of this observation?

3. Suppose that physicians implanted human stem cells into the brains of patients with Parkinson's disease, in the area of the brain where the patients had lost many nerve cells. What would have to happen for the stem cells to differentiate into functioning nerve cells?

4. Fruit flies have 6 legs; spiders have 8 legs; centipedes have many; earthworms have none. All of these species are segmented and have *Hox* genes. Based on data in this chapter, generate a hypothesis to explain how variation in leg number evolved in these species.

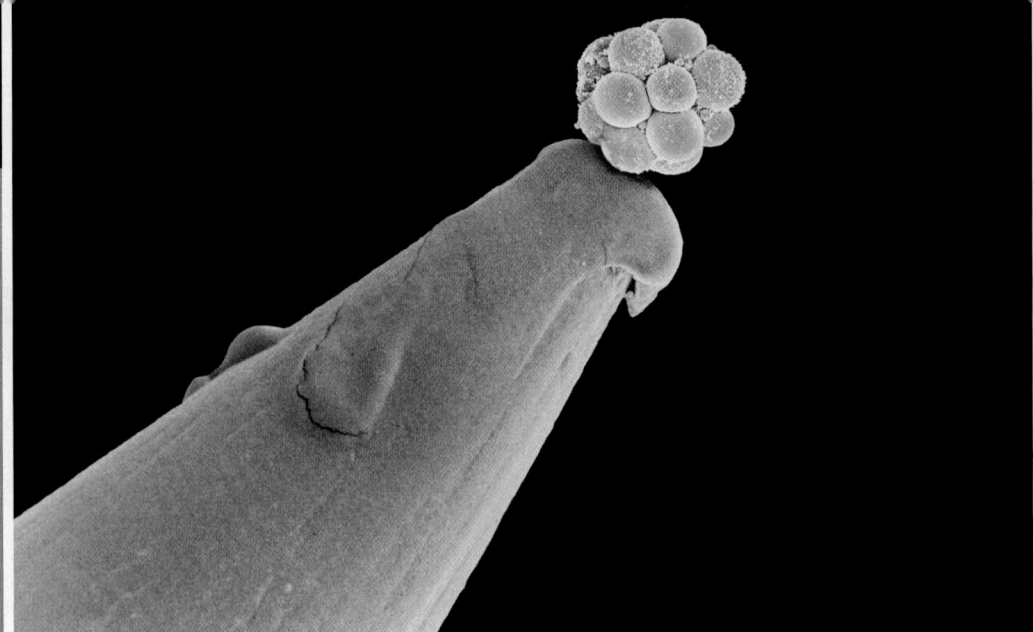

A human embryo, about 3 days old, on the tip of a pin.

22 An Introduction to Animal Development

KEY CONCEPTS

- Fertilization begins with specific interactions between proteins on the plasma membranes of sperm and egg.

- The earliest cell divisions divide the fertilized egg into a mass of cells whose individual fates depend on key regulatory molecules they contain and the signals they receive.

- Early in development, cells engage in collective, coordinated movements to form distinct tissue layers. Each layer gives rise to a different set of tissues and organs in the adult.

- As development proceeds, specialized organs and other structures form through the interacting effects of cell-cell signals, cell proliferation, cell movements, and differentiation. Differentiation is complete when cells express tissue-specific proteins.

When the physician, researcher, and writer Lewis Thomas considered how a human being develops from a fertilized egg, he could only marvel: "You start out as a single cell derived from the coupling of a sperm and an egg, this divides into two, then four, then eight, and so on, and at a certain stage there emerges a single cell which will have as all its progeny the human brain. The mere existence of that cell should be one of the great astonishments of the earth. People ought to be walking around all day, all through their waking hours, calling to each other in endless wonderment, talking of nothing except that cell. It is an unbelievable thing, and yet there it is, popping neatly into its place amid the jumbled cells of every one of the several billion human embryos around the planet, just as if it were the easiest thing in the world to do."[1]

What molecular mechanisms are responsible for the development of the brain and other tissues and organs? Chapter 21 introduced the basic developmental processes of cell proliferation, death, movement, or expansion; differentiation; and interactions, and went on to explore differentiation in depth. That chapter's central message was that cell-cell signals cause specific sets of transcription factors to be produced in various cells throughout the embryo, resulting in differential gene expression and differentiation. Brain cells "know" that they are brain cells because they have received specific types and concentrations of cell-cell signals; in response, they produce brain-specific proteins.

This chapter's goal is to use the general developmental principles presented in Chapter 21 to understand the overall sequence of events responsible for animal development. We begin with sperm and egg and end with formation of a differentiated cell—in this case, a muscle cell. At each step, remember that it is cell-cell signals and regulatory proteins that allow developing cells to divide, move, interact, and differentiate—to do the things required for them to develop correctly and contribute to the formation of a new individual.

---

[1]Lewis Thomas, *The Medusa and the Snail* (New York: Viking Press, 1979), 156. Thomas was exercising some poetic license here. The brain actually arises from a group of cells in the embryo rather than from a single cell.

✔ When you see this checkmark, stop and test yourself. Answers are available in Appendix B.

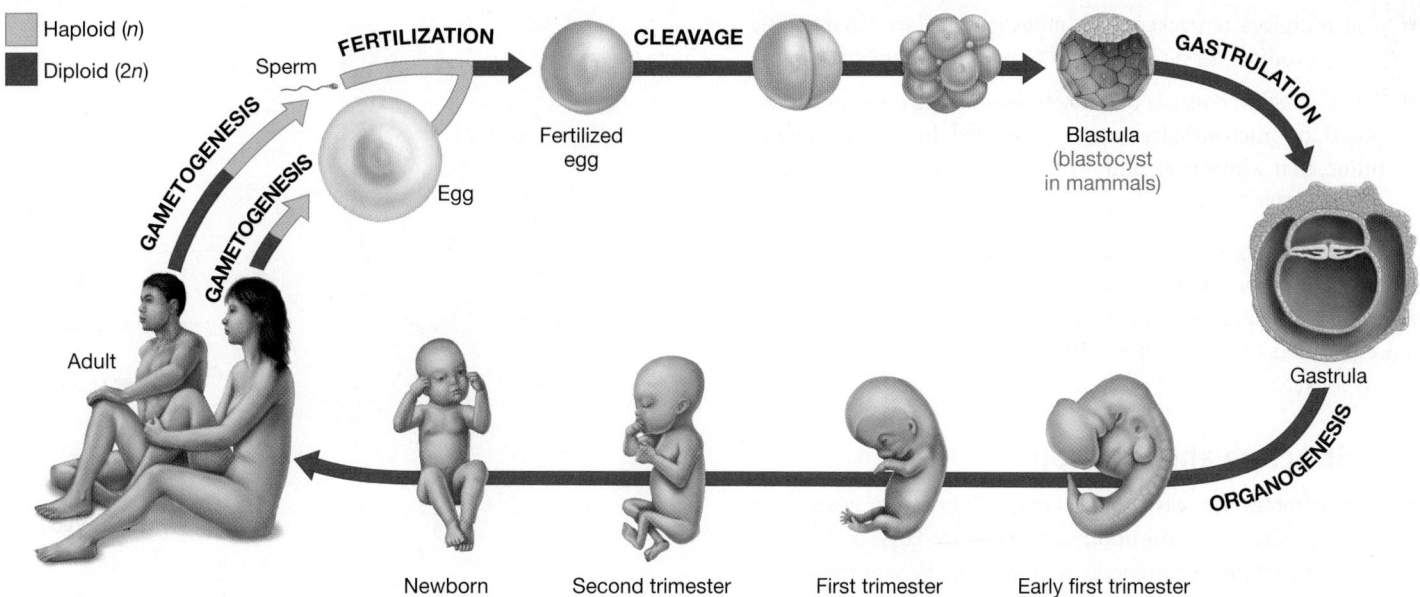

FIGURE 22.1 **Development Proceeds in Ordered Phases.** In animals, the development of a new individual starts with the formation of gametes (sperm and eggs) and continues with fertilization and early cell divisions (cleavage) that result in a blastula. Gastrulation then rearranges the blastula into a gastrula. Organogenesis follows and leads to the formation of adult tissues and structures.

22.1 Gamete Structure and Function

Recall from Chapter 12 that gametes are haploid reproductive cells. In animals and many other organisms, male gametes are called sperm and female gametes are called eggs. The sequence of mitotic and meiotic cell divisions leading up to the production of animal sperm and eggs—a process called gametogenesis—is outlined in Chapter 48.

As **Figure 22.1** shows, gametogenesis in animals is followed by **fertilization**, or the union of a sperm and an egg. Fertilization is the subject of Section 22.2; the subsequent steps in development—called cleavage, gastrulation, and organogenesis—are introduced in Sections 22.3 through 22.5.

Here let's focus on the gamete itself. Sperm and eggs form in the reproductive organs of adult organisms. The DNA and cytoplasm in these reproductive cells are the initial components of the new individual. Both sperm and egg contribute chromosomes—usually a haploid genome containing one allele of each gene—to the offspring. But the similarity ends there. Egg cells contribute much more than chromosomes, and are routinely hundreds or thousands of times larger than sperm cells.

Sperm Structure and Function

Sperm cells begin to develop after meiosis has resulted in the production of a haploid nucleus. As a mammalian sperm cell matures, it acquires the four main compartments shown in **Figure 22.2**: the head, neck, midpiece, and tail.

- The head region contains the nucleus and an enzyme-filled structure called the **acrosome**. The enzymes stored in the

acrosome allow the sperm to penetrate the barriers surrounding the egg.

- The neck encloses a **centriole** that will combine with a centriole contributed by the egg to form a centrosome (see Chapter 7). The centrosome is required for spindle formation during mitosis (see Chapter 11).

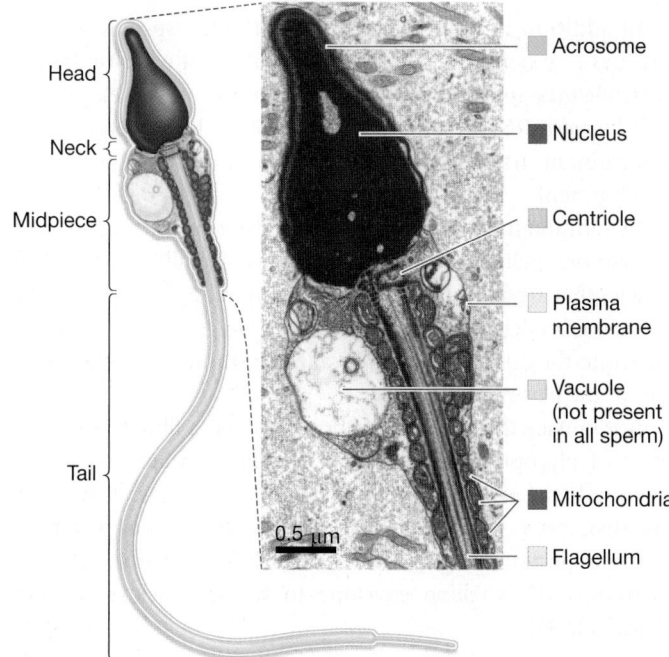

FIGURE 22.2 **Sperm Structure.** The morphology of human sperm is typical of many mammal species. The cell is specialized for motility and fusing with an egg cell.

- The midpiece is packed with **mitochondria** (see Chapter 9), which produce the ATP required to power movement.
- The tail region consists of a **flagellum**—a long structure composed of microtubules, and surrounded by plasma membrane, that whips back and forth to make swimming possible (see Chapter 7).

Sperm are race cars—stripped down, streamlined, souped-up cells that are specialized for racing other sperm to the egg. Eggs, in comparison, are like semitrailers—bulky, far less mobile storage containers that are packed with valuable merchandise and securely locked.

Egg Structure and Function

Animal gametes are easy to tell apart: Sperm are small and motile; eggs are large and nonmotile. Eggs are large mainly because they contain the nutrients required for the embryo's early development. The quantity of nutrients present in the egg varies widely among species, however.

- In mammals, embryos start to obtain nutrition through a maternal organ called the **placenta** within a week or two after fertilization. Thus, the egg only has to supply nutrients for early development and is relatively small.
- In species where females lay eggs directly into the environment, the stores in the egg are the *only* source of nutrients until organs have formed and a larva or juvenile hatches and begins to feed. In these species, the nutrients required for early development are provided by **yolk**—a fat- and protein-rich cytoplasm that is loaded into egg cells as they mature. Yolk may be present as one large mass or as many small granules.

In addition to nutrients, the eggs of many species contain key developmental regulatory molecules called **cytoplasmic determinants** that control the early events of development. The *bicoid* mRNA introduced in Chapter 21 is a cytoplasmic determinant. Its protein product acts as a master regulator in development.

In addition to yolk and cytoplasmic determinants, many eggs contain organelles called **cortical granules**—small, enzyme-filled vesicles that are activated during fertilization. As the egg matures, cortical granules are synthesized in the Golgi apparatus, transported to the cell surface, and localized just under the inner surface of the plasma membrane.

Just outside the plasma membrane of eggs, a fibrous, mat-like sheet of glycoproteins called the **vitelline envelope** forms and surrounds the egg. In the eggs of humans and other mammals, this structure is unusually thick and is called the **zona pellucida**. In some species, a large gelatinous matrix known as a jelly layer surrounds the vitelline envelope to further enclose the egg (**Figure 22.3**).

To summarize, the egg is loaded with rich stores of nutrients and guarded by one or more layers of protective material. How does a sperm cell penetrate these coatings to reach the egg's plasma membrane and fertilize the egg?

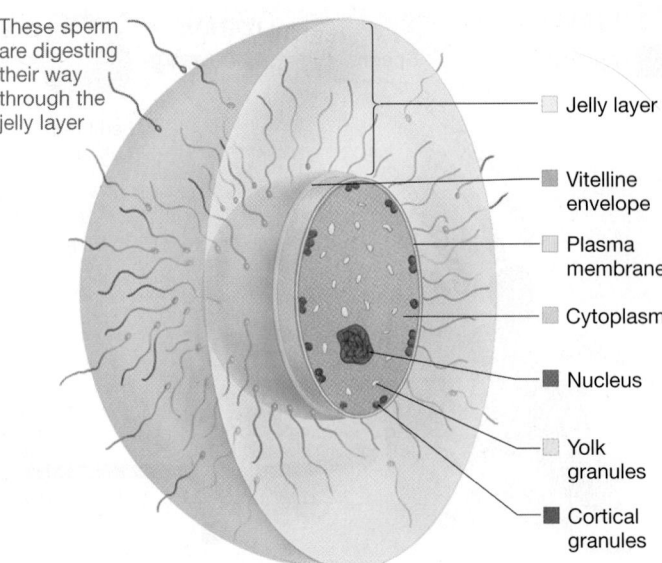

FIGURE 22.3 Sea Urchin Eggs Are Covered by a Jelly Layer. In many types of animals that lay eggs in water, the egg is surrounded with a protective jelly layer. The other features shown here are also typical of many animal eggs.

22.2 Fertilization

Fertilization may seem like a simple process: A sperm cell fuses with an egg cell to form a diploid cell known as a **zygote**—a fertilized egg. The process is actually extraordinarily complex, however.

For fertilization to take place, sperm and egg cells must be in the same place at the same time. Then they must recognize and bind to each other. Next they have to fuse—even though most of the other cells in the body do not fuse with cells they contact. In most species, fusion must also be limited to a single sperm so that the zygote does not receive extra chromosomes. Finally, the fusion of the two gametes has to trigger the onset of development. This complexity has made fertilization a fascinating research topic. The contact that takes place between sperm and egg at fertilization qualifies as the best studied of all cell-to-cell interactions.

Research on fertilization began in earnest early in the twentieth century, when biologists started to study the sperm–egg interaction in sea urchins. Like most aquatic animals, sea urchins shed their gametes into the surrounding water and fertilization occurs externally.

When the sperm and egg of sea urchins meet, the head of the sperm initially encounters the jelly layer of the egg cell. To reach the plasma membrane of the egg cell, the sperm head digests its way through the egg's jelly layer and vitelline envelope using enzymes released from the acrosome—the structure at the tip of the sperm's head. Once the sperm head contacts the egg-cell surface, the plasma membranes of the egg and sperm fuse. The sperm nucleus, mitochondria, and centriole enter the egg—though the sperm mitochondria later disintegrate—and the sperm and egg nuclei fuse to form the zygote nucleus. Fertilization is then complete.

But what prevents cross-species fertilization, and the production of dysfunctional hybrid offspring? After all, in many habitats sperm and eggs from a particular sea urchin species float in seawater along with eggs and sperm from other sea urchin species and many other organisms.

How Do Gametes from the Same Species Recognize Each Other?

In the 1970s, Victor Vacquier and co-workers succeeded in identifying a protein on the head of sea urchin sperm that binds to the surface of sea urchin eggs in a species-specific manner. They called this protein bindin. Follow-up work showed that the bindin proteins from even very closely related species are distinct. As a result, bindin should ensure that a sperm binds only to eggs from the same species.

But what does bindin bind to? If bindin acts as a key, what acts as the lock? Kathleen Foltz and William Lennarz hypothesized that egg-cell membranes contain a receptor for bindin. To test this hypothesis, they set out to isolate the bindin receptor. More specifically, they set out to isolate the part of the bindin receptor that is exposed on the outside of the egg. They predicted that this region of the protein interacts with the bindin on sperm.

Figure 22.4 illustrates Foltz and Lennarz's experimental approach. They began by treating the surface of sea urchin eggs with a **protease**—an enzyme that cleaves peptide bonds. When the investigators isolated the protein fragments that were released from the egg surface, they found one that bound to sperm and to isolated bindin molecules. Further, this binding occurred in a species-specific manner. A protein fragment from the eggs of one species bound to sperm of its own species, but did not bind to sperm of different species. Based on these observations, the biologists claimed that they had found the outward-facing portion of the egg-cell receptor protein for sperm.

🔑 During sea urchin fertilization, species-specific bindin molecules on sperm interact with species-specific receptors on the surface of the egg. This interaction is required for the plasma membranes of sperm and egg to fuse. Strong evidence exists for the same types of protein–protein interactions between the sperm and egg cells of mammals, although the exact proteins involved have yet to be identified.

✔️ If you understand the importance of protein–protein interactions for fusion of sperm and egg-cell membranes, you should be able to explain why a molecule that bound to a bindin-like protein on the human sperm head would function as an effective contraceptive.

Why Does Only One Sperm Enter the Egg?

In early studies on sea urchin fertilization, researchers noticed that only one sperm succeeded in fertilizing the egg, even when dozens or even hundreds of sperm were clustered around the vitelline envelope. This observation makes sense: Multiple fertilization, or **polyspermy**, would result in a zygote that had more

---

EXPERIMENT

QUESTION: If bindin is the protein on the surface of sperm that acts like a key, what is the "lock"?

HYPOTHESIS: The lock is a receptor protein on the surface of sea urchin eggs. Bindin binds to this receptor.

NULL HYPOTHESIS: The lock is not a receptor protein on the surface of sea urchin eggs.

EXPERIMENTAL SETUP:

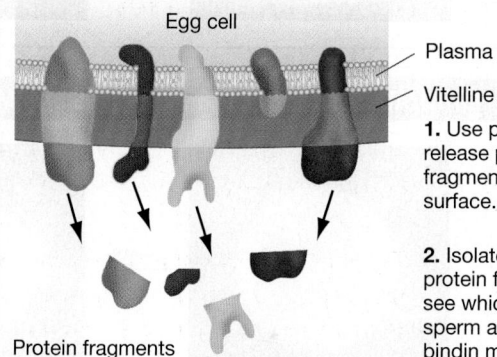

Egg cell — Plasma membrane — Vitelline envelope

1. Use protease to release protein fragments from egg surface.

2. Isolate each type of protein fragment and see which ones bind to sperm and to isolated bindin molecules.

Protein fragments

PREDICTION: One of the protein fragments isolated from egg surface will bind to sperm and to isolated bindin molecules.

PREDICTION OF NULL HYPOTHESIS: The protein fragments isolated from egg surface will not bind to sperm or to isolated bindin molecules.

RESULTS:

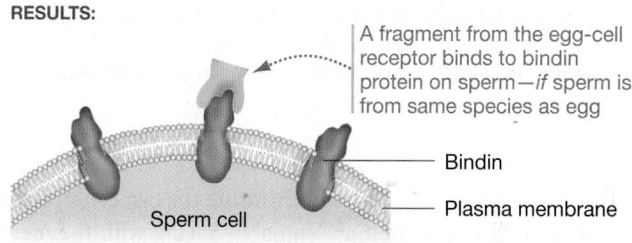

A fragment from the egg-cell receptor binds to bindin protein on sperm—*if* sperm is from same species as egg

Bindin

Sperm cell — Plasma membrane

CONCLUSION: The "lock" is a receptor protein on the surface of eggs that binds to bindin (the "key") in a species-specific manner.

FIGURE 22.4 The Egg-Cell Receptor for Sperm Was Isolated and Characterized.

SOURCE: Foltz, K. R., and W. J. Lennarz. 1990. Purification and characterization of an extracellular fragment of the sea urchin egg receptor for sperm. *Journal of Cellular Biology* 111: 2951–2959.

✔️**QUESTION** Experimental designs that depend on an exhaustive search of all possibilities are sometimes called "brute force" strategies. In what sense was this experiment a "brute force" approach?

---

than two copies of each chromosome. Sea urchin embryos with more than two copies of each chromosome die. How do animals avoid polyspermy?

In sea urchins, fertilization results in the erection of a physical barrier to sperm entry. The process begins when sperm entry causes calcium ions (Ca^{2+}) to be released from storage areas inside the egg. As **Figure 22.5a** on page 392 shows, a wave of ion release starts at the point of sperm entry and propagates throughout the egg.

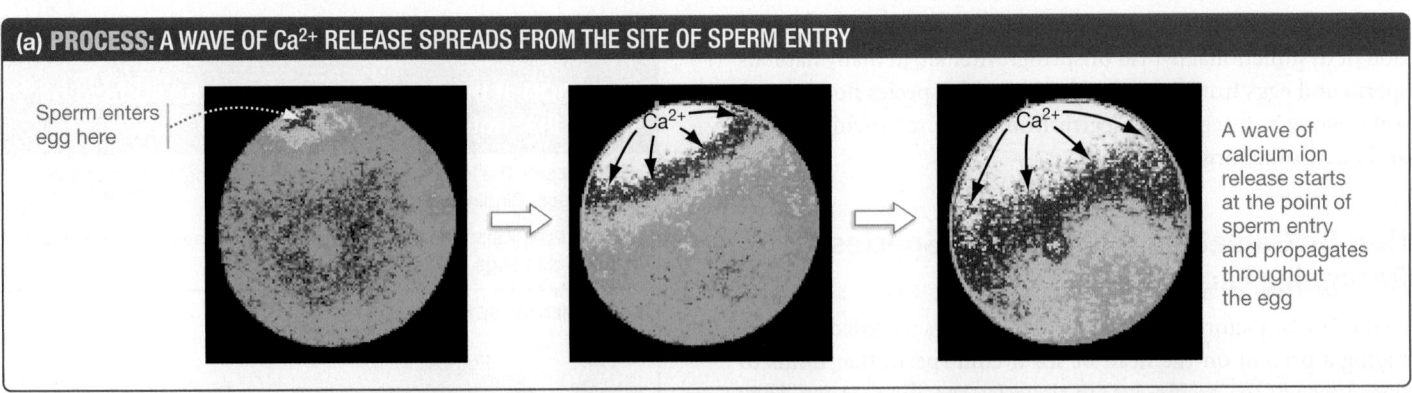

(a) PROCESS: A WAVE OF Ca²⁺ RELEASE SPREADS FROM THE SITE OF SPERM ENTRY

Sperm enters egg here

Ca²⁺

Ca²⁺

A wave of calcium ion release starts at the point of sperm entry and propagates throughout the egg

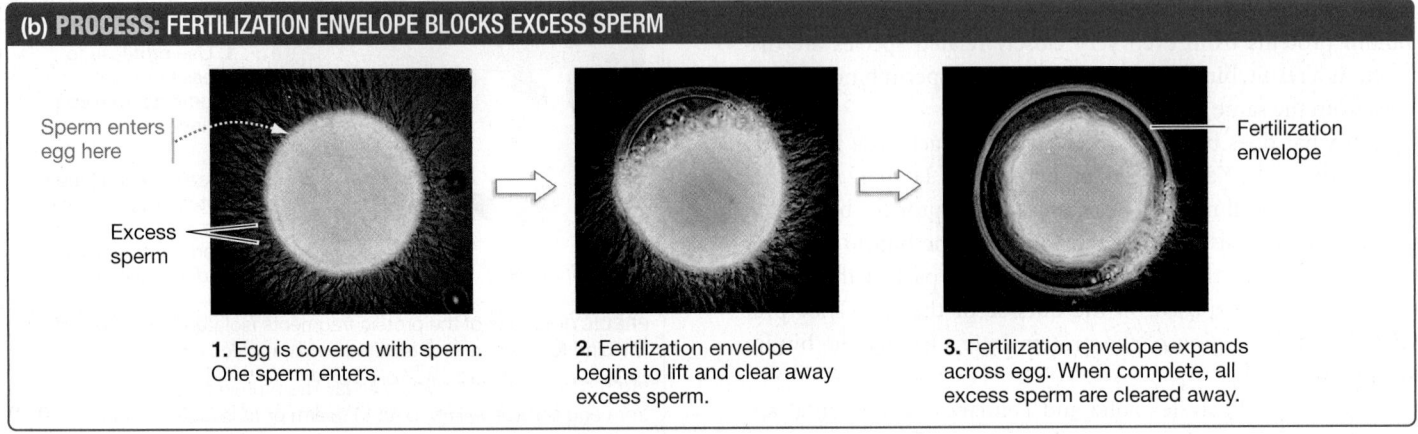

(b) PROCESS: FERTILIZATION ENVELOPE BLOCKS EXCESS SPERM

Sperm enters egg here

Excess sperm

Fertilization envelope

1. Egg is covered with sperm. One sperm enters.

2. Fertilization envelope begins to lift and clear away excess sperm.

3. Fertilization envelope expands across egg. When complete, all excess sperm are cleared away.

FIGURE 22.5 A Physical Barrier Erected after Fertilization Prevents Polyspermy. (a) During fertilization, a wave of Ca²⁺ begins at the point of sperm entry and spreads under the egg membrane in about 30 seconds. The white dots are from a reagent that fluoresces in the presence of calcium ions. **(b)** In response to increased Ca²⁺ concentrations, a fertilization envelope rises in about 40 seconds and clears away excess sperm.

The cortical granules located just inside the egg cell's plasma membrane respond to the Ca²⁺ signal by fusing with the membrane and releasing their contents to the exterior. These contents include proteases that digest the exterior-facing fragment of the egg-cell receptor for sperm, which prevents any new sperm from binding to the egg surface.

In addition, ions and other compounds released by the cortical granules accumulate between the egg cell's plasma membrane and the vitelline envelope. Because these solutes are highly concentrated, they cause water to flow into the space between the plasma membrane and vitelline envelope by osmosis (see Chapter 6). The influx of water causes the envelope matrix to lift away from the cell, forming a **fertilization envelope** (**Figure 22.5b**) that keeps additional sperm from contacting the egg's plasma membrane.

Mammal eggs do not produce a fertilization envelope, but enzymes that are released from cortical granules perform a function similar to the sea urchin enzymes that destroy the egg-cell receptor for sperm. More specifically, the mammalian enzymes modify proteins on the egg-cell surface in a way that prevents binding by additional sperm.

Once a sperm nucleus does enter a sea urchin or human egg, the two haploid nuclei fuse to form a diploid zygote. Development can begin.

CHECK YOUR UNDERSTANDING

If you understand that . . .

- Fertilization depends on specific interactions between proteins on the plasma membranes of animal sperm and egg cells.
- Enzymes that digest the egg-cell receptor for sperm help prevent polyspermy.

✓ You should be able to . . .

1. Predict the consequences of mutations that change the structure of bindin or the egg-cell receptor for sperm.
2. Describe the role of calcium signaling in processes that block polyspermy.

Answers are available in Appendix B.

22.3 Cleavage

Cleavage refers to the rapid cell divisions that take place in a zygote immediately after fertilization. Cleavage is the first step in **embryogenesis**—the process by which a single-celled zygote becomes a multicellular embryo.

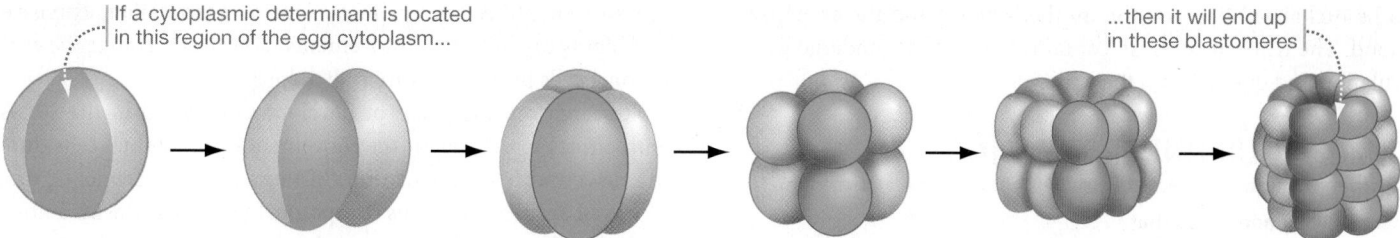

FIGURE 22.6 Cytoplasmic Determinants Are Sequestered into Certain Blastomeres during Cleavage.

In most animals, cleavage partitions the egg cytoplasm without additional cell growth taking place. The zygote simply divides into two cells, then four cells, then eight, and so on, without concurrent growth overall. The key feature of cleavage is that the cytoplasm present in the egg is divided into a larger and larger number of smaller and smaller daughter cells.

Cleavage is fast—the fastest cell divisions recorded over an individual's lifetime. In frogs, cleavage produces 37,000 cells in under two days. When cleavage occurs in fruit flies, mitosis can occur every 10 minutes, producing about 50,000 cells in half a day.

The cells that are created by cleavage divisions are called **blastomeres** (literally, "bud-part"). When cleavage is complete, the embryo consists of a mass of blastomere cells called a **blastula** ("little-sprout, bud").

Partitioning Cytoplasmic Determinants

The key to understanding cleavage is to analyze what is happening to the egg cytoplasm as it is divided up into many blastomeres. For example, think back to the Bicoid protein introduced in Chapter 21. Bicoid is a transcription factor whose distribution in the fruit-fly egg displays an anterior–posterior concentration gradient. When a fly egg undergoes cleavage, then, only nuclei in the anterior end of the egg are exposed to a high concentration of Bicoid.

As Section 22.1 indicated, a molecule that exists in eggs and helps direct early development is called a cytoplasmic determinant. It follows, then, that *bicoid* mRNA is a cytoplasmic determinant.

Cytoplasmic determinants are found in specific locations within the egg cytoplasm, so they end up in specific populations of blastomeres. As a result, certain regulatory gene cascades are triggered only in certain blastomeres (**Figure 22.6**). By dividing the egg cytoplasm to precisely distribute cytoplasmic determinants to certain cells, cleavage initiates the step-by-step process that, in combination with signals received from other cells, results in cell differentiation.

Cleavage in Mammals

In humans and other mammals, cleavage occurs in the **fallopian tube**, or **oviduct**: a structure that connects reproductive organs called the ovary and the uterus (**Figure 22.7**). The **ovary** is the organ in which the egg matures, and the **uterus** is the organ in which the embryo develops. Fertilization occurs near the ovary, and cleavage occurs as the embryo travels down the length of the fallopian tube toward the uterus.

Cleavage results in a specialized type of blastula called a **blastocyst** ("sprout-bag"), which has two major populations of cells. The exterior of the blastocyst is a thin-walled, hollow structure called the **trophoblast** ("feeding-sprout"). Inside the trophoblast there is a cluster of cells called the **inner cell mass (ICM)**.

After arriving at the uterus, the blastocyst implants into the uterine wall, and an organ called the placenta begins to form. The **placenta** is derived from a mixture of maternal cells and trophoblast cells. It has a key function: it allows nutrients and wastes

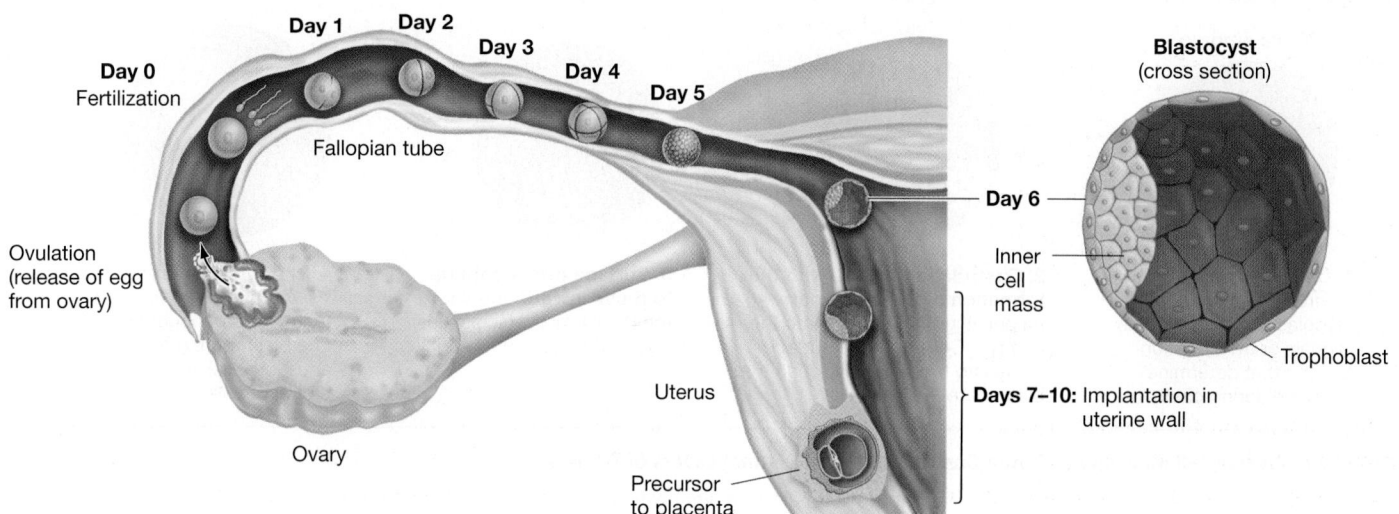

FIGURE 22.7 In Humans, Cleavage Occurs before Implantation into the Uterus.

to be exchanged between the mother's blood and the embryo's blood. The ICM, in contrast, contains the cells that undergo gastrulation and develop into the embryo.

CHECK YOUR UNDERSTANDING

22.4 Gastrulation

As cleavage nears completion, cell division slows. During the next phase of the developmental sequence, cell proliferation stops being the most important developmental process. Instead, cell movement becomes primary. During **gastrulation**, extensive and highly organized cell movements radically rearrange the embryonic cells into a structure called the gastrula.

Research on this phase of development started in the 1920s with efforts to document the movement of individual cells during newt and frog gastrulation. In these early experiments, tiny blocks of agar (a gelatinous compound) were soaked with a non-toxic dye. The dyed blocks were then pressed against the surface of blastula-stage embryos so that a small number of blastomeres became marked with dye. By allowing marked embryos to develop and then examining them at intervals during gastrulation, researchers were able to follow the movement of cells.

Formation of Germ Layers

The pattern of gastrulation varies widely among species of animals, but the general outcome is the same: Gastrulation results in the formation of embryonic tissue layers. A **tissue** is an integrated set of cells that function as a unit.

Most early animal embryos have just three primary tissue layers: (**1**) ectoderm ("outside skin"), (**2**) mesoderm ("middle skin"), and (**3**) endoderm ("inner skin"). These embryonic tissues are called **germ layers** because they give rise to the organs and tissues of the adult. Adults of most animal species have a wide array of tissues, including muscle tissue, connective tissue, and nerve tissue.

Figure 22.8 shows how the cell movements of gastrulation result in the formation of the three embryonic tissue layers in a

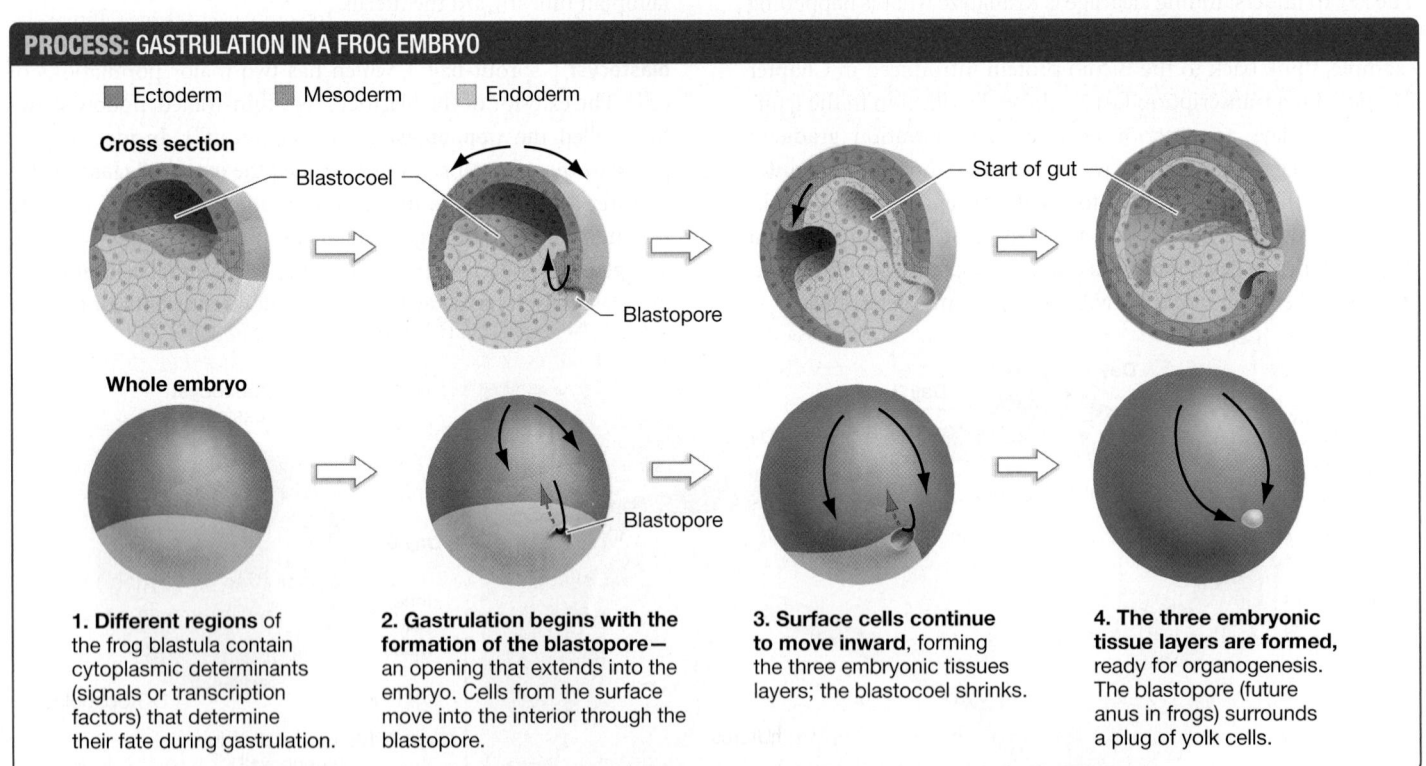

PROCESS: GASTRULATION IN A FROG EMBRYO

■ Ectoderm   ■ Mesoderm   ■ Endoderm

Cross section

Blastocoel

Blastopore

Start of gut

Whole embryo

Blastopore

1. Different regions of the frog blastula contain cytoplasmic determinants (signals or transcription factors) that determine their fate during gastrulation.

2. Gastrulation begins with the formation of the blastopore— an opening that extends into the embryo. Cells from the surface move into the interior through the blastopore.

3. Surface cells continue to move inward, forming the three embryonic tissues layers; the blastocoel shrinks.

4. The three embryonic tissue layers are formed, ready for organogenesis. The blastopore (future anus in frogs) surrounds a plug of yolk cells.

FIGURE 22.8 Precise Cell Movements during Gastrulation Form Distinct Layers of Tissues.

✔**EXERCISE** In frogs, the initial opening of the blastopore is located on what will become the dorsal side of the larva and adult. The blastopore becomes the anus of the larva and adult. On the last drawing in each row, mark where the head, tail, back, and belly will develop.

frog embryo. The drawings are based on experiments with dyed blastomeres, and they distinguish between blastomeres according to the type of embryonic tissue they will become. Cells that will become ectoderm are shown here in blue; cells destined to form mesoderm are shown in pink, and cells that will form endoderm are colored yellow. These cell fates are determined by cell-cell signals, transcription factors, and cytoplasmic determinants in the blastomeres.

Examine the steps in the figure carefully.

Step 1 The frog blastula contains a fluid-filled interior space called the **blastocoel**. The blastula of many animal species is hollow or partially hollow, as it is here.

Step 2 As gastrulation begins, cells begin moving into the blastocoel through an invagination—similar to the indentation that forms when you push a finger into a balloon. In frogs this invagination is slit-like and eventually forms the **blastopore**.

Step 3 Cells from the periphery continue to move to the interior of the embryo through the blastopore, forming a tube-like structure that will become the gut or digestive tract.

Step 4 When gastrulation is complete, the ectoderm, mesoderm, and endoderm cells are arranged in three distinct layers.

To review these steps and other features of early development, go to the study area at *www.masteringbiology.com*.

(MB) **Web Activity** Early Stages of Animal Development

Figure 22.9 shows the fate of the embryonic tissues as development proceeds: **Ectoderm** forms the outer covering of the adult body and the nervous system; **mesoderm** gives rise to muscle, most internal organs, and connective tissues such as bone and cartilage; and **endoderm** produces the lining of the digestive tract or gut, along with some of the associated organs.

Definition of Body Axes

(key) In addition to establishing the embryonic tissue layers, gastrulation has another major outcome: The major body axes become visible. In frogs, for example, the blastopore becomes the anus, and the region just above the blastopore (as drawn in Figure 22.8) becomes the dorsal, or back, side of the embryo. In this way, the anterior–posterior and dorsal–ventral axes of the body become apparent.

In frogs, the major body axes were partially determined early in development by Bicoid-like cytoplasmic determinants that were stored in the egg and partitioned into blastomeres during cleavage. These master regulators start regulatory gene cascades that begin the long, step-by-step process of differentiation introduced in Chapter 21.

Current research on gastrulation is focused on understanding how cells move in such an organized way—that is, on the mechanism responsible for the actual movement and on the mechanism by which the cells navigate. But like the embryo itself, we'll move along to the next phase in the developmental sequence—the formation of tissues, organs, and other basic structures as the body takes shape.

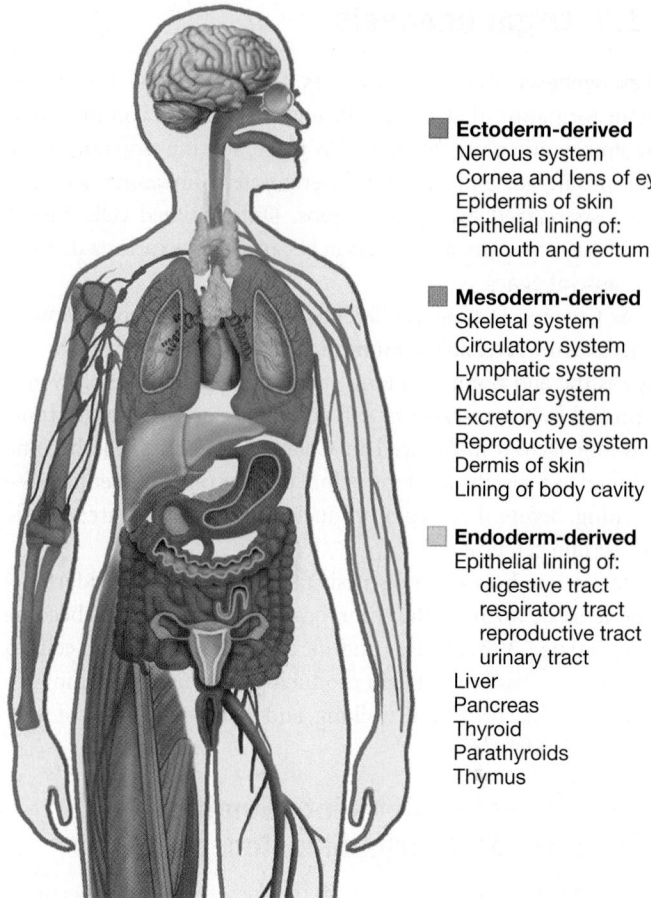

Ectoderm-derived
Nervous system
Cornea and lens of eye
Epidermis of skin
Epithelial lining of:
    mouth and rectum

Mesoderm-derived
Skeletal system
Circulatory system
Lymphatic system
Muscular system
Excretory system
Reproductive system
Dermis of skin
Lining of body cavity

Endoderm-derived
Epithelial lining of:
    digestive tract
    respiratory tract
    reproductive tract
    urinary tract
Liver
Pancreas
Thyroid
Parathyroids
Thymus

FIGURE 22.9 The Three Embryonic Tissues Give Rise to Different Adult Tissues and Organs.

✓**QUESTION** In a gastrula, mesoderm is sandwiched between ectoderm, which is on the outside of the embryo, and endoderm, which is on the inside. Relate this observation to the positions of adult tissues and organs derived from mesoderm, ectoderm, and endoderm.

CHECK YOUR UNDERSTANDING

(key) **If you understand that . . .**

- Gastrulation consists of coordinated cell movements that reorganize the embryonic cells and result in the formation of embryonic germ layers and a tube that will eventually become the gut. The embryonic tissue layers are ectoderm, mesoderm, and endoderm.
- Once gastrulation is complete, the major body axes are visible and organs and other structures can begin to form.

✓ **You should be able to . . .**

1. Name the three embryonic germ layers and describe their relative positions in the gastrula.

2. Describe the relationship between the cells that invaginate early in gastrulation, the formation of the gut, and the formation of the anterior–posterior axis of the body.

Answers are available in Appendix B.

22.5 Organogenesis

Organogenesis ("organ-origin") is the process of tissue and organ formation that begins once gastrulation is complete and the embryonic germ layers are in place. During organogenesis, cells proliferate and become differentiated—meaning they become specialized for different jobs. Differentiated cells have a distinctive structure and function because they express a distinctive suite of genes.

As Chapter 21 noted, differentiation is a progressive, stepwise process. An irreversible commitment to become a particular cell type is the end point of a long and complex sequence of events, mediated by the cascades of cell-cell signals and regulatory transcription factors introduced in Chapter 21. Thus, cells become destined to a specific fate long before they become differentiated—meaning, before they begin producing products that are specific to a certain cell type.

To explore how organogenesis takes place, let's consider how muscle tissue forms in the embryo, and how muscle cells become differentiated. What causes undifferentiated mesodermal cells to form muscle tissue and begin producing the muscle-specific proteins that make breathing, walking, and turning the pages of textbooks possible?

Organizing Mesoderm into Somites: Precursors of Muscle, Skeleton, and Skin

Figure 22.10 illustrates some of the key events that occur as organogenesis begins in frogs, humans, and other vertebrates. In this cross section of a frog embryo, the dorsal (back) side of the individual is at the top and the ventral (belly) side is at the bottom.

You should already be familiar with the ectoderm, mesoderm, and endoderm layers shown in step 1 of the figure, along with the early gut cavity. But you should also note that a new structure has appeared in the dorsal mesoderm—a rod-like element called the **notochord**. This structure is shown in cross section in the figure and runs the length of the anterior–posterior axis of the embryo, just under the dorsal surface.

The notochord is unique to the group of animals called the **chordates**, which includes humans and other vertebrates. In some species of chordates, the notochord is a long-lasting structure that functions as a simple internal skeleton—it stiffens the body and makes efficient swimming movements possible. But in species like frogs and humans, the notochord is transient. It appears only in embryos. As organogenesis proceeds, many of the cells in the notochord undergo apoptosis—programmed cell death.

NEURAL TUBE FORMATION The short-lived notochord observed in many chordate embryos functions as a key organizing element during organogenesis. As step 2 in Figure 22.10 shows, cells in or near the notochord produce signaling molecules that induce the ectoderm on the dorsal (back) side of the embryo to fold. Folding results from changes in cell shape—analogous to the differential cell expansion that occurs in plant cell development (see Chapter 21).

The folding of the ectoderm begins when signals from the notochord trigger massive changes in cytoskeletal elements inside each of the dorsal ectodermal cells. As the cytoskeleton is reorganized, the ectodermal cells extend in length and then constrict at their dorsal end and expand at their ventral end. The constriction above and expansion below makes the sheet of cells fold upward.

As folding continues, a structure called the **neural tube** forms (step 3). The neural tube is the precursor of the brain and spinal cord. Besides producing the signals that direct formation of the neural tube, the notochord furnishes physical support as the ectodermal cells fold into their final configuration.

SOMITE FORMATION As organogenesis continues, mesodermal cells near the neural tube become organized into blocks of tissue called **somites**. Somites form on both sides of the neural tube down the length of the body (**Figure 22.11**). Somite formation is

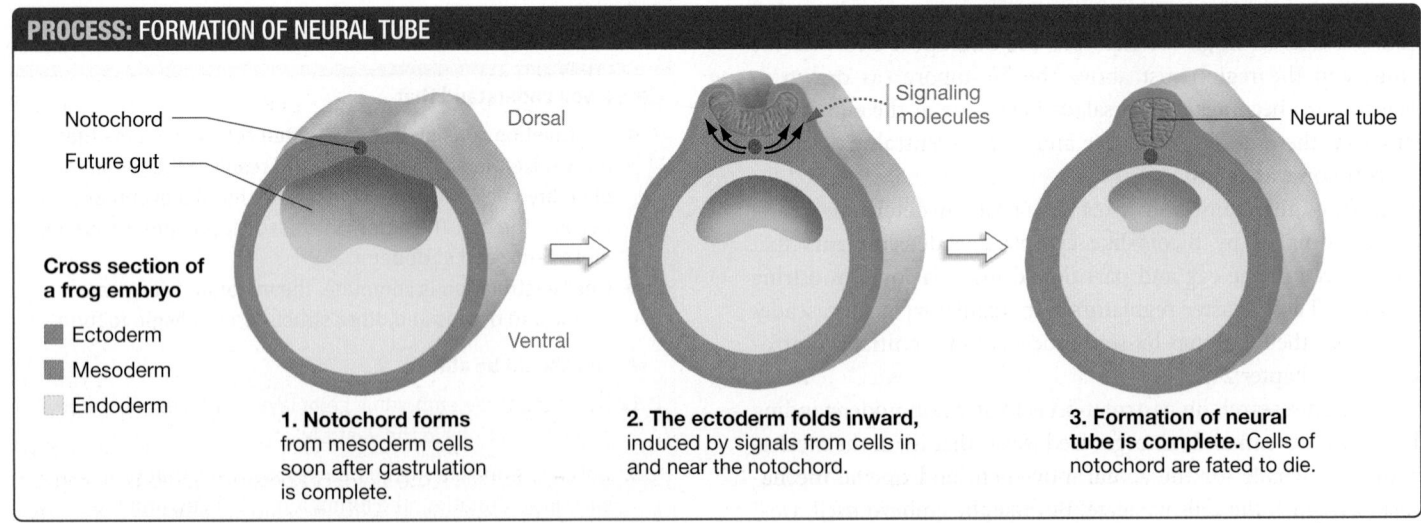

PROCESS: FORMATION OF NEURAL TUBE

Notochord

Future gut

Dorsal

Ventral

Cross section of a frog embryo

- Ectoderm
- Mesoderm
- Endoderm

Signaling molecules

Neural tube

1. Notochord forms from mesoderm cells soon after gastrulation is complete.

2. The ectoderm folds inward, induced by signals from cells in and near the notochord.

3. Formation of neural tube is complete. Cells of notochord are fated to die.

FIGURE 22.10 The Notochord and Neural Tube Form Early in Organogenesis. In vertebrates, the notochord forms from mesoderm cells soon after gastrulation is complete. Molecules produced in or near the notochord induce the formation of the neural tube and other structures along the dorsal (back) side of the embryo.

(a) Surface view of somites

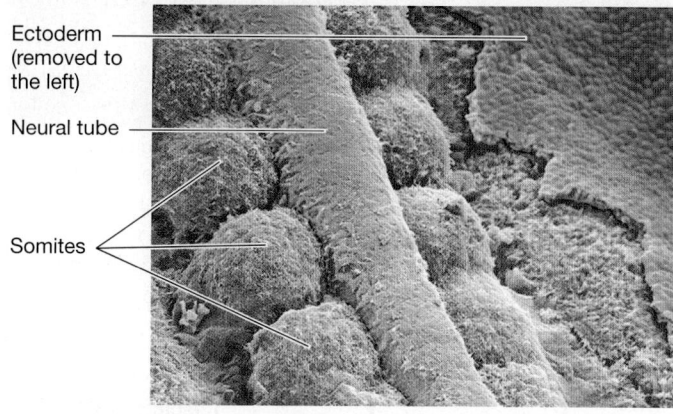

Ectoderm (removed to the left)

Neural tube

Somites

(b) Cross section of somites

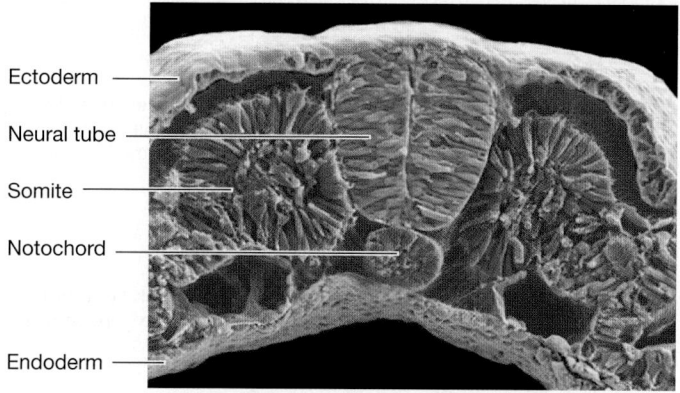

Ectoderm

Neural tube

Somite

Notochord

Endoderm

FIGURE 22.11 Somites Form on Both Sides of the Neural Tube. Somites are made of mesodermal cells (color-coded pink in these chick embryo micrographs).

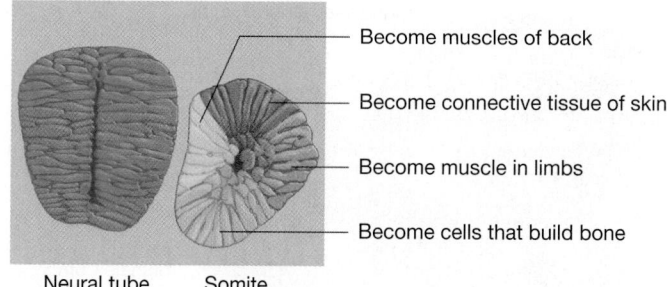

Become muscles of back

Become connective tissue of skin

Become muscle in limbs

Become cells that build bone

Neural tube   Somite

FIGURE 22.12 Within a Somite, a Cell's Position Determines Its Fate. Each somite eventually breaks up into distinct populations of cells, each of which gives rise to distinct tissue.

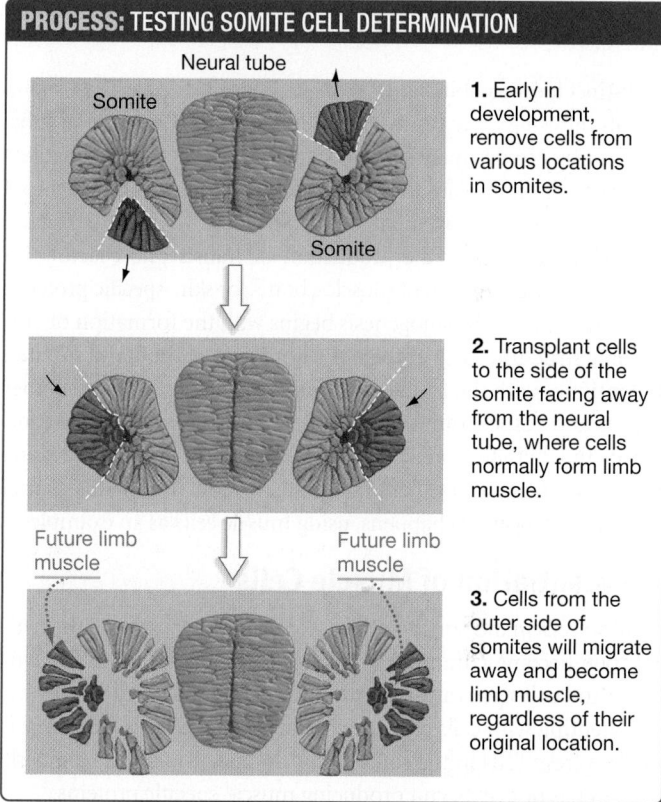

PROCESS: TESTING SOMITE CELL DETERMINATION

Neural tube

Somite

Somite

1. Early in development, remove cells from various locations in somites.

2. Transplant cells to the side of the somite facing away from the neural tube, where cells normally form limb muscle.

Future limb muscle

Future limb muscle

3. Cells from the outer side of somites will migrate away and become limb muscle, regardless of their original location.

FIGURE 22.13 Experimental Evidence for Progressive Differentiation of Mesodermal Cells in Somites. By transplanting somite cells early in somite formation to new locations within the structure, researchers showed that the cells' fates are not determined initially.

a response to changes in the cell adhesion molecules that keep mesodermal cells attached to each other (see Chapter 8).

Somites are transient structures, just as the notochord is. But by marking cells and following them over time, researchers have found that somite cells form a variety of structures.

As development proceeds, somite cells break away in distinct groups that migrate to their final location in the developing embryo, and there continue to proliferate. These cell movements are critical to organogenesis. Cells from somites build the vertebrae and ribs, the deeper layers of the skin that covers the back, and the muscles of the back, body wall, and limbs. As **Figure 22.12** shows, the destiny of a somite cell depends on its position within the somite at a particular time during organogenesis.

By transplanting cells from one location to another within the somite, researchers found that initially any of those cells can become any of the body's somite-derived elements. For example, biologists transplanted cells from various parts of the somite to the sector farthest from the neural tube. All the transplants eventually became committed to form limb muscles (**Figure 22.13**). Cells transplanted later in development, however, failed to become the cell type associated with their new position. Instead, the transplanted cells differentiated into the cell types they would

normally have formed in their original position, even though they were now in a new and inappropriate location.

CELL DETERMINATION As the somite matures, its cells become irreversibly determined—meaning that the cell type they eventually become has been decided—based on their location within the somite. Note that a cell's fate (1) becomes fixed in a step-by-step process—guided by regulatory gene cascades—and (2) is a function of its location in the embryo.

Recent studies of this process, called **determination**, have shown that distinct populations of cells in a somite differentiate in response

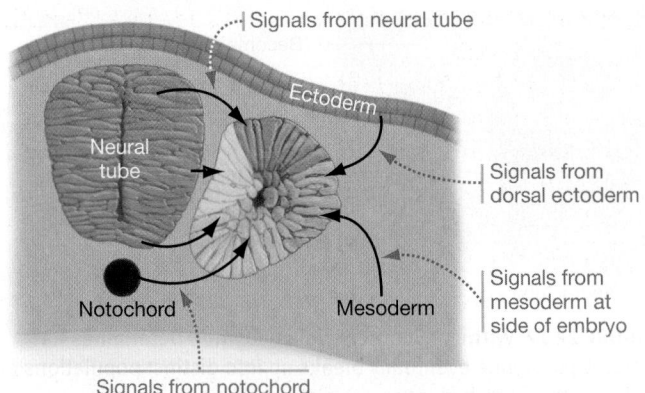

FIGURE 22.14 Somite Cells Differentiate in Response to Signals from Nearby Tissues. Somite cells become determined in response to signals from the neural tube, notochord, and nearby ectoderm and mesoderm.

to distinct local combinations of cell-cell signals. **Figure 22.14** shows signals diffusing away from cells in the notochord, the neural tube, and nearby ectoderm and mesoderm to act on specific populations of target cells in the somite. These different combinations of cell-cell signals direct the movement of somite cells when the structure breaks up and trigger the production of transcription factors required for the expression of muscle-, bone-, or skin-specific proteins.

To summarize, organogenesis begins with the formation of embryonic structures such as the notochord, neural tube, and somites. It continues with the formation of bones, muscles, skin, and other organs and tissues found in a larval or juvenile individual. At this point, the cells that make up these structures begin expressing tissue-specific proteins. Let's take a closer look at how this crucial step in organogenesis happens, using muscle cells as an example.

Differentiation of Muscle Cells

Why do cells in the portion of somite located farthest from the neural tube become committed to producing muscle? Harold Weintraub and colleagues answered this question in the late 1980s by experimenting with cells called myoblasts. A myoblast is a cell that is derived from cells in the somite. It is destined to become a muscle cell but has not yet begun producing muscle-specific proteins.

Weintraub and co-workers hypothesized that myoblasts contain at least one regulatory protein that commits them to their fate. Their idea was that myoblasts begin producing this muscle-determining protein after they receive an appropriate set of cell-cell signals from nearby tissues. In effect, they were looking for a regulatory transcription factor in myoblasts that dictates, "I will become a muscle cell."

Figure 22.15 outlines how the biologists went about searching for this hypothetical protein.

FIGURE 22.15 The Search for a Gene That Causes Muscle-Cell Differentiation.

SOURCE: Weintraub, H., et al. 1989. Activation of muscle-specific genes in pigment, nerve, fat, liver, and fibroblast cell lines by forced expression of *MyoD*. *Proceedings of the National Academy of Sciences, USA* 86: 5434–5438.

✔**QUESTION** Why did the researchers have to attach a "general purpose" promoter to the cDNAs?

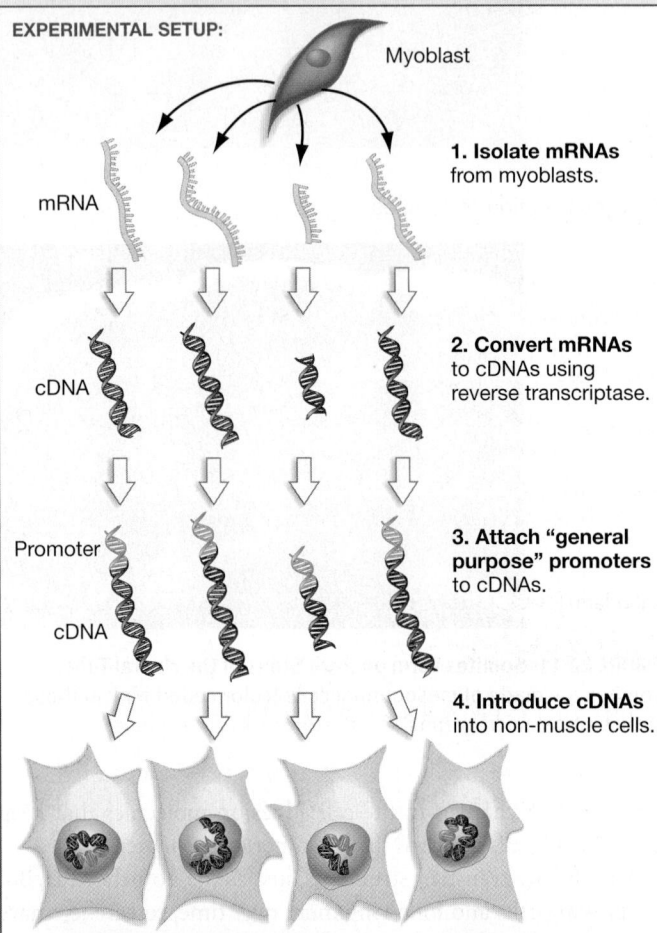

EXPERIMENT

QUESTION: Why do cells in a certain part of somites become committed to producing muscle in response to the signals they receive?

HYPOTHESIS: Cell-cell signals trigger production of a regulatory protein (or proteins) that commits myoblasts to their fate.

NULL HYPOTHESIS: Myoblasts do not contain a regulatory protein that commits them to their fate.

EXPERIMENTAL SETUP:

Myoblast

mRNA

1. Isolate mRNAs from myoblasts.

cDNA

2. Convert mRNAs to cDNAs using reverse transcriptase.

Promoter

cDNA

3. Attach "general purpose" promoters to cDNAs.

4. Introduce cDNAs into non-muscle cells.

PREDICTION: One of the myoblast-derived cDNAs will convert non-muscle cells into cells that produce muscle-specific proteins.

PREDICTION OF NULL HYPOTHESIS: None of the myoblast-derived cDNAs will convert non-muscle cells into cells that produce muscle-specific proteins.

RESULTS:

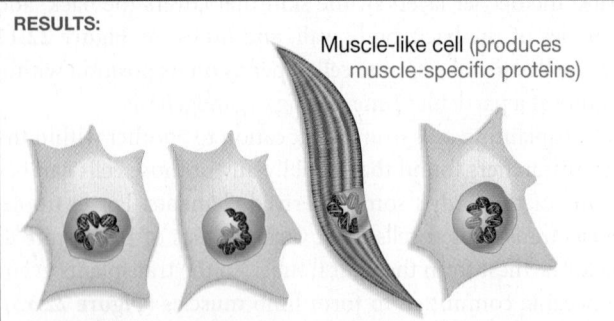

Muscle-like cell (produces muscle-specific proteins)

CONCLUSION: Certain somite cells contain a regulatory protein (later called MyoD) that commits them to differentiate into muscle.

1. They began by isolating mRNAs from myoblasts.

2. They used reverse transcriptase to convert the mRNAs to cDNAs. Because myoblasts transcribe genes required for a muscle cell to function, the cDNAs included muscle-specific genes.

3. They attached a type of promoter to the cDNAs that would ensure the genes' expression in any type of cell.

4. They introduced the recombinant genes into non-muscle cells called fibroblasts and monitored the development of the transformed cells.

As predicted, one of the myoblast-derived cDNAs converted fibroblasts into muscle-like cells. Follow-up experiments showed that the same gene product could convert pigment cells, nerve cells, fat cells, and liver cells into cells that produced muscle-specific proteins.

Weintraub's group called the protein product of this gene **MyoD**, for *myoblast* *determination*. Subsequent work showed that the *MyoD* gene encodes a regulatory transcription factor and that the MyoD protein binds to enhancer elements located upstream of muscle-specific genes (see Chapter 18).

In addition, researchers found that the MyoD protein activates further expression of the *MyoD* gene. This was a key observation because it meant that once *MyoD* is turned on, it triggers its own expression—meaning that the gene continues to be transcribed. Other researchers have found that genes closely related to *MyoD* are also required for the differentiation of muscle cells.

To put this specific example into context, think back to the sequence of events occurring in early development and to the principles of development introduced in Chapter 21. Differentiation is a step-by-step process that is complete when cells begin producing proteins that are specific to a particular cell type. By what path did the cells in your bicep become muscle cells?

- Fertilization of the egg inside your mother triggered the onset of cleavage, resulting in a blastocyst.

- Certain cells in the blastocyst began producing signals that triggered regulatory gene cascades—changes in gene expression that in turn led specific cells, during the positional changes of gastrulation, to become mesoderm in your back.

- Early in organogenesis, signals from the notochord and nearby cells induced the production of MyoD and other muscle-determining proteins in certain populations of cells from somites. In response, these target cells were committed to becoming muscle and moved into your upper arm as it formed.

- Later, the MyoD-containing cells began expressing muscle-specific proteins.

All of these steps made it possible for you to move your arms after birth. The rest, as they say, is history.

CHAPTER 22 REVIEW

For media, go to the study area at www.masteringbiology.com

Summary of Key Concepts

Fertilization begins with specific interactions between proteins on the plasma membranes of sperm and egg.

- Sperm cells only contribute a haploid genome and a centriole to the embryo.

- Eggs contribute a large amount of cytoplasm—usually containing cytoplasmic determinants and a large supply of nutrients—to the embryo in addition to a haploid genome and a centriole.

- Fertilization only occurs after a protein on the sperm head binds to a specific receptor on the egg-cell membrane.

✓You should be able to explain why bindin and the egg-cell receptor for sperm are said to interact like a lock and key.

The earliest cell divisions divide the fertilized egg into a mass of cells whose individual fates depend on key regulatory molecules they contain and the signals they receive.

- Early embryonic development begins with cleavage—a series of cell divisions that divide the egg cytoplasm into a large number of cells.

- During cleavage, an array of signals and regulatory transcription factors are apportioned to different blastomeres. As a result, different blastomeres have different fates.

- Once cleavage is complete, the embryo consists of a mass of cells.

✓You should be able to describe how a Bicoid-like master regulator might be partitioned into cells during cleavage in a human embryo.

Early in development, cells engage in collective, coordinated movements to form distinct tissue layers. Each layer gives rise to a different set of tissues and organs in the adult.

- Embryonic cells undergo dramatic movements during gastrulation.

- Gastrulation has two major consequences: (**1**) embryonic tissues are arranged into endoderm, mesoderm, and ectoderm layers; and (**2**) the back-to-belly and head-to-tail axes of the body become visible.

✓You should be able to compare and contrast a mesodermal cell in the anterior end of an embryo and a mesodermal cell in the posterior end, in terms of the cell-cell signals they receive and the regulatory transcription factors they contain.

(MB) **Web Activity** Early Stages of Animal Development

As development proceeds, specialized organs and other structures form through the interacting effects of cell-cell signals, cell proliferation, cell movements, and differentiation. Differentiation is complete when cells express tissue-specific proteins.

- Tissues, organs, and other structures form during the developmental phase called organogenesis.

- Early in vertebrate organogenesis, cells in the notochord release cell-cell signals that induce the formation of a neural tube—precursor to the brain and spinal cord.

- Blocks of mesodermal cells, called somites, form next to the neural tube.
- In response to cell-cell signals from the notochord, neural tube, and other nearby tissues, cells in the somites move to new positions, proliferate, and begin expressing tissue-specific proteins.

✔You should be able to give examples of cell interaction (via cell-cell signals), expansion, movement, proliferation, and differentiation during neural tube, somite, and muscle development in mammals.

Questions

✔TEST YOUR KNOWLEDGE

1. What is bindin, and what does it bind to?
   a. a protein on mammal embryos that allows them to attach to the uterine wall
   b. a protein on the plasma membrane of frog blastomeres that allows them to bind to each other and cohere
   c. a protein on the sea urchin sperm-cell nucleus that binds to the egg-cell nucleus
   d. a protein on the sea urchin sperm head that binds to an egg-cell receptor

2. How is calcium signaling involved in blocking polyspermy?
   a. It triggers a regulatory gene cascade.
   b. Accumulation of calcium ions causes an outward flow of water, raising the fertilization envelope.
   c. It triggers fusion of cortical granules with the plasma membrane.
   d. Calcium ions bind to egg-cell receptors for sperm, blocking binding sites.

3. What happens during cleavage?
   a. The neural tube—precursor of the spinal cord and brain—forms.
   b. Basal and apical cells—precursors of the suspensor and embryo, respectively—form.
   c. The fertilized egg divides without growth occurring, forming a mass of cells.

   d. Massive movements of cells make the primary body axes visible and organize the three embryonic tissues.

4. What happens during gastrulation?
   a. The neural tube—precursor of the spinal cord and brain—forms.
   b. Basal and apical cells—precursors of the suspensor and embryo, respectively—form.
   c. The fertilized egg divides without growth occurring, forming a ball of cells.
   d. Massive movements of cells make the primary body axes visible and organize the three embryonic tissues.

5. In animals, which adult tissues and organs are derived from ectoderm?
   a. lining of the digestive tract and associated organs
   b. blood, heart, kidney, bone, and muscle
   c. nerve cells and skin
   d. blastopore and blastocoel

6. During organogenesis in vertebrates, which of the following does *not* occur?
   a. establishment of the anterior–posterior axis
   b. differentiation of cells
   c. movement of cells into new positions
   d. extensive cell-cell signaling

✔TEST YOUR UNDERSTANDING

1. Explain why eggs are so much larger than sperm.

2. Why do elaborate mechanisms exist to prevent polyspermy?

3. Blastomeres look identical. Explain why they are not.

4. Why are ectoderm, mesoderm, and endoderm called germ layers?

5. Explain how cell transplantation experiments provided evidence that cells in a somite are determined in a step-by-step fashion.

6. What evidence supports the hypothesis that MyoD triggers differentiation of muscle cells?

✔APPLYING CONCEPTS TO NEW SITUATIONS

1. At the molecular level, explain why bindin on the sperm of the sea urchin *Strongylocentrotus purpuratus* attaches to the egg-cell receptor for sperm on eggs of *S. purpuratus*, but why bindin from other species of *Strongylocentrotus* cannot.

2. A molecule called α-amanitin inhibits translation. When researchers treat fruit-fly eggs with this compound, the early stages of cleavage proceed normally after fertilization. What can you conclude from this experiment?

3. In the marine organisms called sea squirts, eggs contain a yellow pigment. Blastomeres that contain this yellow pigment become muscle cells. State a hypothesis to explain this observation.

4. After organogenesis, frog embryos are about the same size as a fertilized egg. Why?

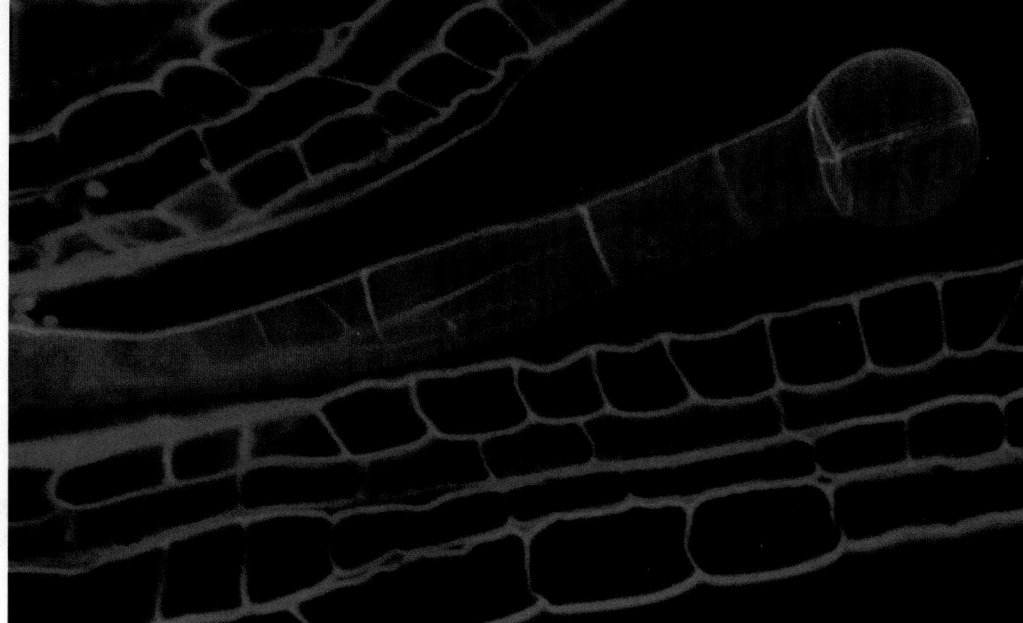

The ball-like cells on the far right of this image, colored red here, are an *Arabidopsis thaliana* embryo at the 8-cell stage. This chapter explores how plant embryos develop into adults that continue to grow throughout life and reproduce.

An Introduction to Plant Development 23

About the year 2750 B.C., perhaps more than a century before the first pyramids were constructed in ancient Egypt, a blessed event occurred on a mountainside in southwestern North America. A sperm cell and an egg cell from a bristlecone pine tree fused to form a zygote. The offspring produced by that encounter has been growing and developing ever since—for over 4700 years.

🗝 Unlike many animals, plants continue to grow and develop throughout their lives, whether that life lasts two weeks or thousands of years. In many plant species, for example, the entire aboveground shoot system dies each fall and grows back the next spring. Continuous development is a hallmark of plant biology.

In addition, many plant cells retain the ability to de-differentiate—to stop producing a certain suite of tissue-specific proteins and begin producing proteins typical of a different type of tissue. Under the right circumstances, for example, a photosynthetic cell in a plant stem can de-differentiate and give rise to cells that function as storage cells in a root. This extraordinary flexibility is another hallmark of plant development. Animal development is rigid and inflexible in comparison with that of plants.

The striking differences between plant and animal development are not surprising given their very different ways of life. Most animals move around to find food and mates, while plants stand their ground, make their own food, and use water, wind, or insects to transport their gametes.

Despite numerous differences, though, animal and plant development are both governed by the common principles described in Chapter 21. As plant cells proliferate, they divide in precise orientations and expand in specific directions. Like the cells in animal embryos, developing plant cells communicate constantly by cell-cell signals, differentiate due to specific combinations of cell-cell signals and regulatory transcription factors, and may undergo programmed cell death. This unity of underlying principles is one of the great discoveries emerging from research on plant developmental biology.

KEY CONCEPTS

🗝 In sharp contrast to animals, plants develop continuously, do not commit cells to gamete production until late in development, and produce gametes by mitosis in haploid cells.

🗝 In flowering plants, double fertilization results in the production of a zygote and a nutritive tissue that supports embryogenesis.

🗝 Embryogenesis results in the formation of the major body axes and three types of embryonic tissue.

🗝 Vegetative development is the function of meristems, where cell division continues throughout life—producing cells that go on to differentiate.

🗝 When the function of a meristem shifts from vegetative to reproductive development, key regulatory transcription factors are activated and control the position and identity of floral organs.

✔ When you see this checkmark, stop and test yourself. Answers are available in Appendix B.

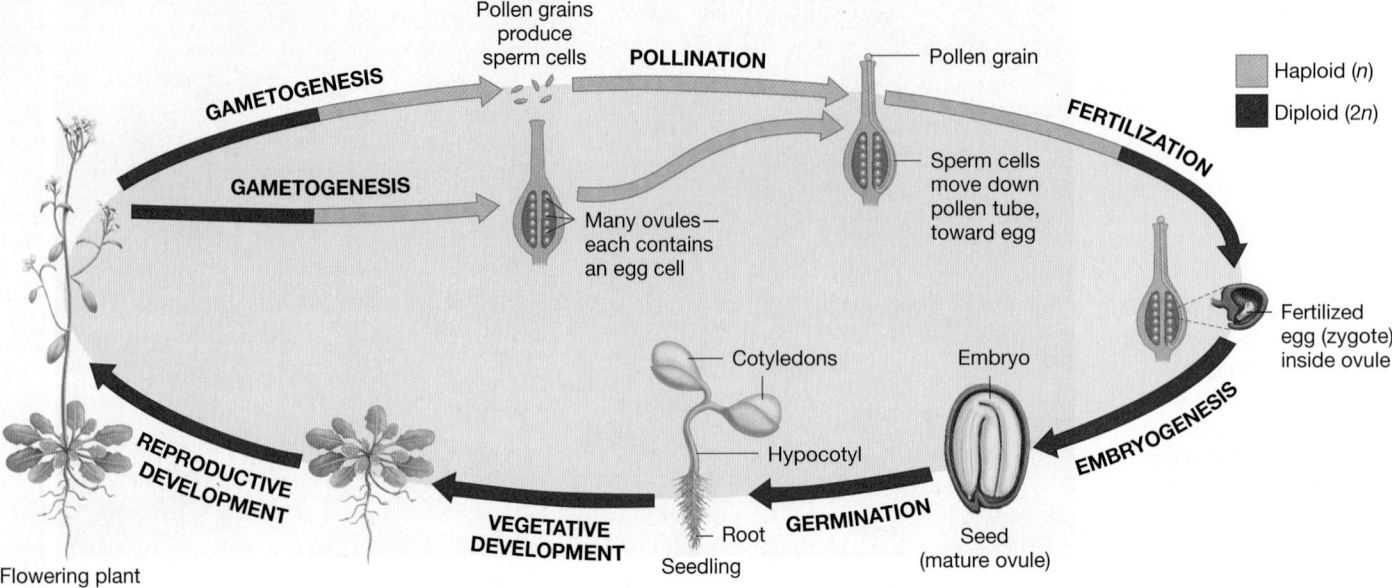

FIGURE 23.1 An Overview of Development in *Arabidopsis*. This chapter traces the events in the life of a plant, from fertilization and embryogenesis to vegetative and reproductive development.

23.1 Gametogenesis, Pollination, and Fertilization

Figure 23.1 provides an overview of development in a flowering plant—the most species-rich and abundant group of plants (see Chapter 30). In this figure and in many other places in this chapter, a small species called *Arabidopsis thaliana*—the best-studied of all land plants—will serve as a model organism (see **BioSkills 14** in Appendix A).

Note that the sequence of events, or life cycle, diagrammed in Figure 23.1 begins with gametogenesis, or gamete formation. In *Arabidopsis* and other flowering plants, an egg is fertilized inside a protective, womb-like structure called the **ovule**. Development continues inside the ovule with **embryogenesis**—literally, "embryo-origin." In many plant species, embryogenesis ends with the maturation of the ovule into a **seed**—a structure that contains the embryo and a supply of nutrients surrounded by a protective coat.

An embryo may remain in a nongrowing state inside the seed for months, years, or in some cases, centuries. When conditions are favorable, however, the seed undergoes **germination**—meaning that it resumes growth—to form a seedling. Organogenesis, which continues throughout life, then forms the three vegetative organs: roots, leaves, and stems. **Vegetative organs** are the nonreproductive portions of the plant body. Later, cells in the stem will become converted to reproductive structures, initiating gamete production and the sexual phase of the plant life cycle.

How Are Sperm and Egg Produced?

Of all the differences between plant and animal development, one of the most dramatic is seen during gametogenesis. In animals, gametes are produced by differentiation that occurs in the products of meiosis. In plants, gametes are produced by mitosis that occurs in haploid cells produced by meiosis.

Figure 23.2a introduces how the process works in the male reproductive organs of flowering plants such as *Arabidopsis*. Note that diploid cells undergo meiosis to form haploid cells, which then divide by mitosis to give rise to tiny, multicellular structures called **pollen grains**. One of the haploid cells inside the pollen grain will later divide by mitosis, giving rise to two sperm cells.

Figure 23.2b illustrates the general sequence of events leading to egg formation in *Arabidopsis* and other flowering plants. A diploid cell inside the ovule divides by meiosis, producing four daughter cells. Only one of these cells survives; the other three undergo a programmed death. The surviving cell divides by mitosis several times to produce a tiny, multicellular structure called the embryo sac. Inside the embryo sac, a haploid cell differentiates into an egg.

Because the haploid, multicellular structures called pollen grains and embryo sacs alternate with a diploid, multicellular plant as one generation gives rise to the next, this type of life cycle is called **alternation of generations**. Alternation of generations is explained more thoroughly in Chapter 29 and Chapter 30; Chapter 40 explains the steps involved in gametogenesis in plants in much more detail. Now the question is, Once a pollen grain lands on the structure that holds the egg, what happens?

Pollen–Stigma Interactions

Pollen grains are carried to a mature flower by wind, water, insects, bats, birds, or some other agent. If the pollen grains land

(a) Production of sperm in the male reproductive organs of a flowering plant.

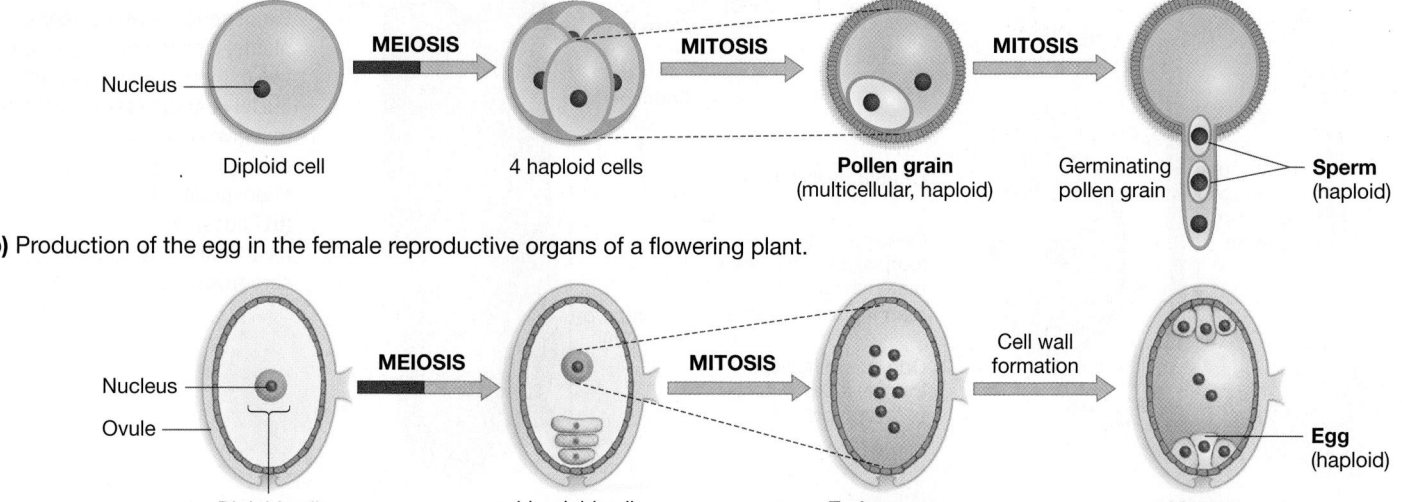

(b) Production of the egg in the female reproductive organs of a flowering plant.

FIGURE 23.2 An Overview of Gametogenesis in Flowering Plants. Unlike the situation in animals, plants produce haploid multicellular structures. A haploid cell inside the pollen grain produces sperm by mitosis; a haploid cell inside the embryo sac produces an egg by mitosis.

(a) Pollen grains interact with the stigma.

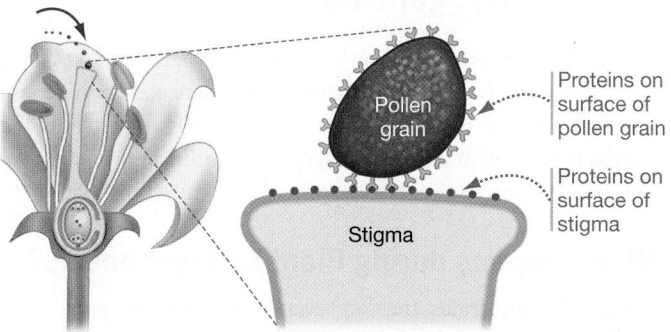

(b) Sperm move to the egg through a pollen tube.

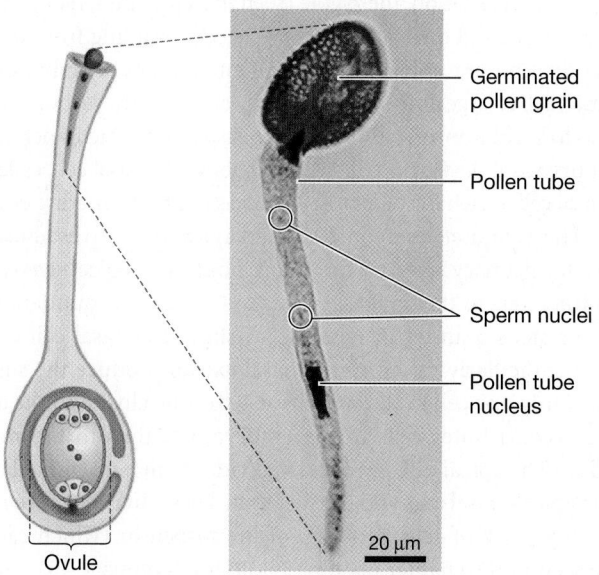

on a floral structure called the stigma, one of the best-studied cell-cell interactions in all of plant biology takes place.

For the pollination and fertilization process to continue successfully, proteins on the surface of the pollen grain have to interact with proteins on the surface of the stigma (**Figure 23.3a**). Like bindin attaching to the egg-cell receptor for sperm in sea urchins (see Chapter 22), the specificity of these interactions prevents fertilization from occurring between members of different species. In many cases, it also prevents self-fertilization—the union of a sperm and egg from the same individual (see Chapter 40 for details on these cell-cell interactions).

If the protein–protein interactions on the surface of the stigma are successful, a **pollen tube** begins to extend from the pollen grain, down the stigma, toward the egg (**Figure 23.3b**). Recent research suggests that this growth is guided by cell-cell signals released from the egg at the base of the female reproductive structure, or **carpel.**

Double Fertilization

Another dramatic contrast with animal development is seen when the pollen tube reaches the base of the carpel. The two sperm cells exit the pollen tube, pass through the wall of the ovule, and enter the embryo sac. The embryo sac contains the egg

FIGURE 23.3 The Fertilization Process Begins When Pollen and Stigma Interact.

✓**QUESTION** How does the interaction between pollen and stigma in flowering plants compare to the interaction between bindin and the egg-cell receptor for sperm in sea urchins?

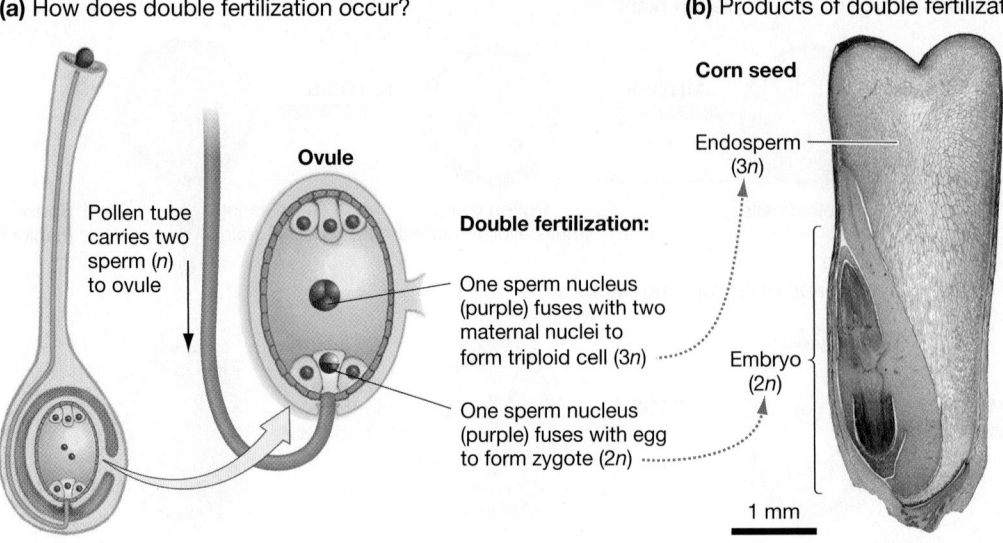

(a) How does double fertilization occur?

Pollen tube carries two sperm (*n*) to ovule

Ovule

Double fertilization:

One sperm nucleus (purple) fuses with two maternal nuclei to form triploid cell (3*n*)

One sperm nucleus (purple) fuses with egg to form zygote (2*n*)

(b) Products of double fertilization

Corn seed

Endosperm (3*n*)

Embryo (2*n*)

1 mm

FIGURE 23.4 Double Fertilization Leads to the Formation of Endosperm. **(a)** In flowering plants, a sperm (*n*) fuses with two or more haploid maternal nuclei near the egg to form a cell that gives rise to endosperm tissue. In many species, endosperm cells are 3*n*. **(b)** Endosperm is a nutritive tissue packed with proteins, carbohydrates, and fats or oils.

cell and a maternal cell that in many species contains two haploid nuclei. One sperm nucleus fuses with the egg to form the diploid zygote, while the other sperm nucleus fuses with the cell that has two maternal haploid nuclei to form a triploid (3*n*) cell. This event is known as **double fertilization**. The process is aptly named, as two cell-fusion events occur (**Figure 23.4a**).

The triploid cell divides repeatedly by mitosis to form a triploid nutritive tissue called endosperm. **Endosperm** ("inside-seed") provides the proteins, carbohydrates, and fats or oils required for embryonic development, seed germination, and early seedling growth.

In species with large seeds, the endosperm grows into a sizeable nutrient reservoir as the ovule matures (**Figure 23.4b**). When you eat wheat, rice, corn, or other grains, you are eating

primarily endosperm. Functionally, endosperm is analogous to the yolk found in most animal eggs.

Once double fertilization is complete, the stage is set for embryogenesis—the early development of a new individual.

23.2 Embryogenesis

In flowering plants such as *Arabidopsis,* embryogenesis takes place inside the ovule as the seed is maturing. In essence, embryogenesis produces a tiny, less-developed precursor to a mature plant. The process is equivalent to the cleavage, gastrulation, and organogenesis phases of early animal development introduced in Chapter 22.

What Happens during Plant Embryogenesis?

Figure 23.5 illustrates the key events in embryogenesis, using *Arabidopsis* as a model organism. This sequence of events was worked out through careful observation of different-aged embryos.

After fertilization, the zygote (seen in step 1) undergoes a highly asymmetric cell division (step 2). The cells resulting from this initial division are unlike in size, content, and fate. The bottom, or basal, cell is large and is dominated by an extensive vacuole. It gives rise to a column of cells called the suspensor, which anchors the embryo as it develops. The small cell above the basal cell, called the apical cell, is rich in cytoplasm and gives rise to the mature embryo.

The asymmetries in the basal and apical cell help establish one of the primary axes of the plant body: the **apical–basal axis**. **Apical** refers to the tip; **basal** refers to the base, or foundation.

As steps 3 and 4 in Figure 23.5 show, the basal cell divides perpendicularly to the apical–basal axis to produce the suspensor. Only one cell in the suspensor—the one closest to the apical cell—contributes cells to the embryo and thus to the mature adult. The apical cell, in contrast, divides both perpendicularly to the apical–basal axis and parallel to it. These divisions give rise to a simple ball of cells at the tip of the suspensor (which can also be seen in the chapter opening photograph on page 401). At this point, the embryo is said to be in the globular stage.

CHECK YOUR UNDERSTANDING

If you understand that . . .

- Gametes arise by mitosis in haploid cells that were produced by meiosis.
- Fertilization is preceded by specific interactions between proteins on the surfaces of plant pollen grains and stigmas.
- In flowering plants, double fertilization results in the production of a zygote and a (usually) triploid cell that grows into a nutritive tissue.

✔ **You should be able to . . .**

1. Describe when meiosis and mitosis occur during the development of a sperm cell in *Arabidopsis*.

2. Suggest a hypothesis to explain why protein–protein interactions between pollen grains and stigmas are important to the individual's ability to reproduce. (Hint: consider what might happen if the protein–protein interactions did *not* occur, and any pollen grain could germinate.)

Answers are available in Appendix B.

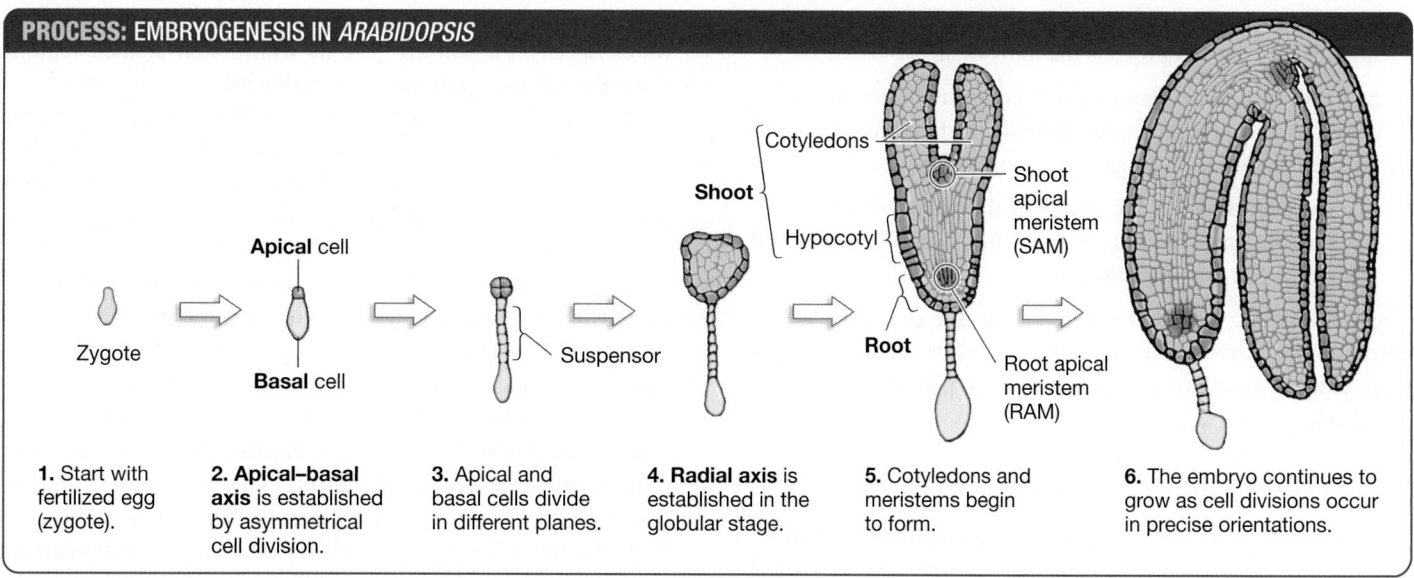

PROCESS: EMBRYOGENESIS IN *ARABIDOPSIS*

Zygote

Apical cell

Basal cell

Suspensor

Shoot

Root

Cotyledons

Shoot apical meristem (SAM)

Hypocotyl

Root apical meristem (RAM)

1. Start with fertilized egg (zygote).

2. Apical–basal axis is established by asymmetrical cell division.

3. Apical and basal cells divide in different planes.

4. Radial axis is established in the globular stage.

5. Cotyledons and meristems begin to form.

6. The embryo continues to grow as cell divisions occur in precise orientations.

FIGURE 23.5 The Stages of Embryogenesis. Embryogenesis takes place inside the developing seed.

✓**EXERCISE** Label the apical–basal and radial axes on the globular-stage embryo.

Because of the distinctive-looking cells covering its exterior, there is now a visible difference between cells in the interior of the embryo and cells on the surface. This change creates the second major body axis, the radial axis. The radial axis extends from the interior of the body out to the exterior.

The initial events in embryogenesis illustrate a general point about plant development: The fate of a plant cell can be summed up in the old quip about the three keys to success in real estate—"location, location, location." Starting with the initial division that creates the apical and basal cells, plant cells differentiate based on where they are in the body.

As the ball of cells continues to grow and develop, the embryonic leaves, or **cotyledons**, begin to take shape (step 5). The cotyledons are connected to the developing root by a stem-like structure called the **hypocotyl**. Together, the cotyledons and hypocotyl make up the **shoot**, which will become the aboveground portion of the body. The shoot system functions in photosynthesis and reproduction. The **root**, in contrast, forms the belowground portion of the body. The root system anchors the individual and functions as a water- and nutrient-gathering structure.

Once the apical–basal and radial axes are established, and as the cotyledons, hypocotyl, and root begin to take shape, groups of cells called the **shoot apical meristem (SAM)** and **root apical meristem (RAM)** form. As Chapter 21 noted, **meristem** is a tissue consisting of undifferentiated cells that divide repeatedly into daughter cells, some of which become specialized.

Meristem cells are analogous to the stem cells found in animals, except that cells derived from plant meristems differentiate into a much wider array of types. The root meristem can form all the underground portions of the plant, and the shoot meristem can form all the aerial portions, including reproductive structures. Throughout a plant's life, meristematic tissues continue to produce cells that can differentiate into adult tissues and structures.

Note that all of this growth and development takes place without the aid of the cell migration seen in gastrulation in animals. Because plant cells have stiff cell walls, they do not move. Consequently, for the cotyledons and other embryonic structures to take shape, cell divisions must occur in precise orientations (step 6 of Figure 23.5), and the resulting cells must exhibit differential growth—meaning that some cells grow larger than others.

🔑 In addition to establishing the two body axes, early development in *Arabidopsis* produces three embryonic tissues (**Figure 23.6**).

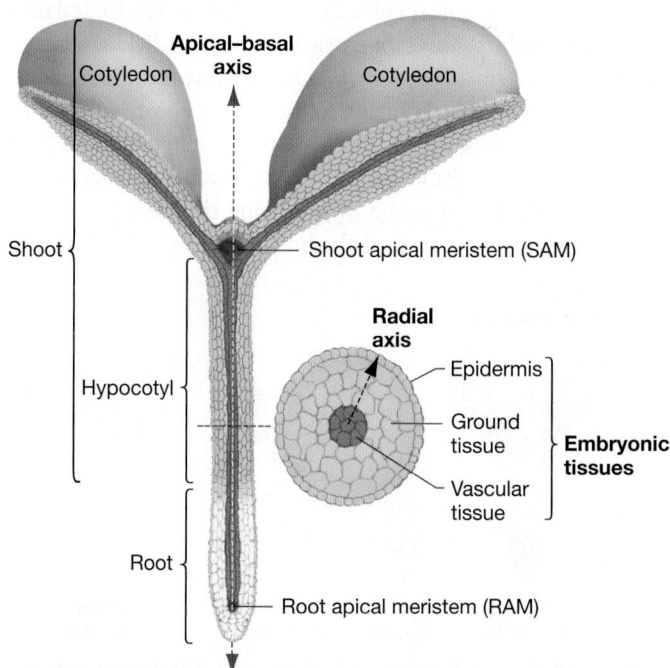

Cotyledon

Apical–basal axis

Cotyledon

Shoot

Shoot apical meristem (SAM)

Radial axis

Hypocotyl

Epidermis

Ground tissue

Vascular tissue

Embryonic tissues

Root

Root apical meristem (RAM)

FIGURE 23.6 Embryogenesis Establishes the Two Body Axes and Three Embryonic Tissues.

- The **epidermis** (literally, "over-skin") is an outer covering of specialized cells that protects the individual.

- Inside the epidermal layer of cells is **ground tissue**, a mass of cells that may later differentiate into cells that are specialized for photosynthesis, food storage, or other functions.

- The **vascular tissue** in the center of the plant will eventually differentiate into specialized cells that transport food and water between root and shoot.

These embryonic tissues are analogous to the ectoderm, mesoderm, and endoderm of developing animals (see Chapter 22). The three tissue systems are arranged in a radial pattern, echoing the radial axis.

Which Genes and Proteins Set Up Body Axes?

The genetic approach to development that was pioneered with research on *Drosophila melanogaster* has also proven to be a powerful way of studying plant embryogenesis. Although the specific genes involved are different in plants than in animals, the basic mechanism by which genes direct the earliest events in development is similar.

Research on the genetics of early development in plants was pioneered by Gerd Jürgens and colleagues in the 1990s. This research group set out to identify genes that are transcribed in the zygote or embryo of *Arabidopsis* and that are responsible for establishing the apical–basal axis of the plant body. It's no coincidence that this effort was similar to the work on anterior–posterior pattern formation in *Drosophila* introduced in Chapter 21. Jürgens had participated in the work with flies.

The biologists' initial goal was to identify individuals with developmental defects at the seedling stage. More specifically, they were looking for mutants that lacked particular regions along the apical–basal axis of the body. The team succeeded in finding several bizarre-looking mutants (**Figure 23.7**). Individuals called apical mutants lacked the first leaves, or cotyledons. Some individuals lacked the embryonic stem, or hypocotyl, and were named central mutants. Individuals dubbed basal mutants lacked both hypocotyls and roots.

To interpret these results, the researchers suggested that each type of *Arabidopsis* mutant had a defect in a different gene and that each gene played a role in specifying the position of cells along the apical–basal axis of the body. They hypothesized that these genes are analogous to the segmentation genes of fruit flies, which specify the destiny of cells within well-defined regions along the anterior–posterior axis of insects.

What are these *Arabidopsis* genes, and how do they exert their effects? To answer these questions, consider the gene responsible for the mutants lacking hypocotyls and roots. This gene has been cloned and sequenced and named *MONOPTEROS*. Because its DNA sequence indicates that its protein product has a DNA-binding domain, *MONOPTEROS* is hypothesized to encode a transcription factor that regulates the activity of target genes.

The MONOPTEROS protein, in turn, is activated in response to signals from auxin. Auxin is a cell-to-cell signal molecule that was introduced in Chapter 21. It is produced in the shoot apical meristem and transported toward the basal parts of the individual. (For more detail on auxin's prominent role as a long-distance signal, or hormone, see Chapter 39.) Recall that the concentration of auxin along the apical–basal axis of a plant forms a concentration gradient that provides positional information, not unlike the Bicoid concentration gradient found in fruit-fly embryos (see Chapter 21).

The take-home message from these results is that the auxin signal is part of a regulatory cascade that triggers the activation of MONOPTEROS and other regulatory transcription factors specific to cells in the developing hypocotyl and roots. Although the genes and proteins involved in forming the cotyledons, hypocotyl, and root of *Arabidopsis* are not yet understood in as much detail as the segmentation genes of *Drosophila*, several important similarities are clear. Both developmental pathways are based on cell-to-cell signals and regulatory cascades that result in the step-by-step specification of a cell's position and fate.

Many questions remain about embryogenesis in *Arabidopsis*. How is auxin production turned on as the shoot apical meristem first begins to form in embryos? Once production of the *MONOPTEROS* gene product begins, what target genes are affected? What genes other than *MONOPTEROS* are found in the regulatory cascade responsible for development along the apical–basal axis? These questions present a host of interesting challenges to current researchers.

CHECK YOUR UNDERSTANDING

If you understand that . . .

- Early embryonic development results in the formation of the apical–basal and radial axes of the plant body and three embryonic tissues.
- The early structures of the shoot and root systems form along the body's apical–basal axis.
- The genes responsible for setting up the body axes are currently the focus of intense research.

✔ You should be able to . . .

1. Relate the "location, location, location" quip to the differentiation of epidermal, ground, and vascular tissue.
2. Predict the effect on the *MONOPTEROS* gene of adding auxin to embryonic root cells.

Answers are available in Appendix B.

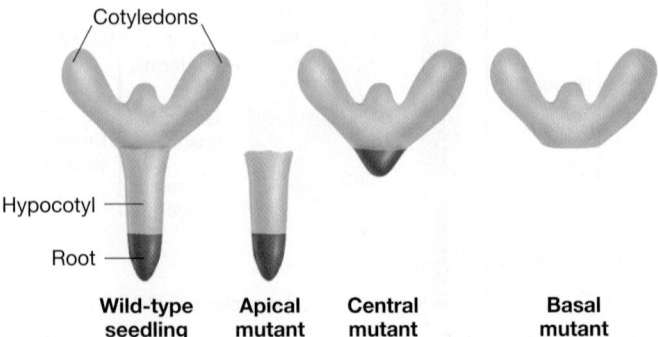

FIGURE 23.7 Mutant *Arabidopsis* Embryos with Misshapen Bodies. Researchers have identified *Arabidopsis* mutant individuals missing specific sections of the body along the apical–basal axis.

23.3 Vegetative Development

For a plant to thrive, it has to adjust to constantly changing conditions. Consider just one such condition—the availability of light.

You might recall from Chapter 10 that plants use wavelengths in the blue and red portions of the spectrum to drive photosynthesis. Now suppose that you are an oak tree with a life expectancy of 300 years. The quality and quantity of light that your leaves receive depends on where you happen to germinate. Are you growing on flat ground? With a southern exposure in full sun? With a northern exposure rarely exposed to full sun?

In addition, from the time you emerge from an acorn to the time of your death, the light you receive will depend on changes in climate and weather as well as on your size relative to the size and proximity of plants that compete with you for light and shade you. Finally, the leaves in your bottommost branches experience a different light regime from the leaves at your apex.

How do plants cope with all this variation in their living conditions? Unlike most animals, they don't move around to find a place that suits their requirements. Instead, they adjust to their immediate surroundings, in large part through the continuous growth and development of roots, stems, and leaves. If an oak tree is heavily shaded on one side, it stops growing in that direction and extends branches on the other side. If it is heavily shaded on all sides, its growth is directed upward. ☞ This constant adjustment to changing environmental conditions is possible because of the meristems that are located at the tips of shoots and roots.

Meristems Provide Lifelong Growth and Development

When embryonic development is complete, the basic body axes are established and the initial structures in the root and shoot systems have formed. For the rest of the individual's life, further development is driven by the meristems (**Figure 23.8a**). Meristematic tissue is located at each tip in the shoot and at the tips of root systems. As a result, the individual is capable of growing in any direction aboveground or belowground, depending on conditions.

Figure 23.8b provides a close-up view of a shoot apical meristem, or SAM. The cells within the meristem are small and undifferentiated. Within each meristem, the rate and direction of cell growth are dictated by cell-cell signals produced in response to environmental cues, such as the arrival of spring, the presence of abundant water, or the amount of light striking the plant. Just below the meristem, daughter cells produced by mitosis and cytokinesis in the meristem initially differentiate into epidermal, ground, or vascular tissue. Eventually these cells will differentiate into more specialized cell types.

Careful microscopy allowed biologists to tease out the sequence of events that occur as meristems grow, and intense research continues to explore how interactions between auxin and other cell-cell signals influence the fate of cells produced by meristems.

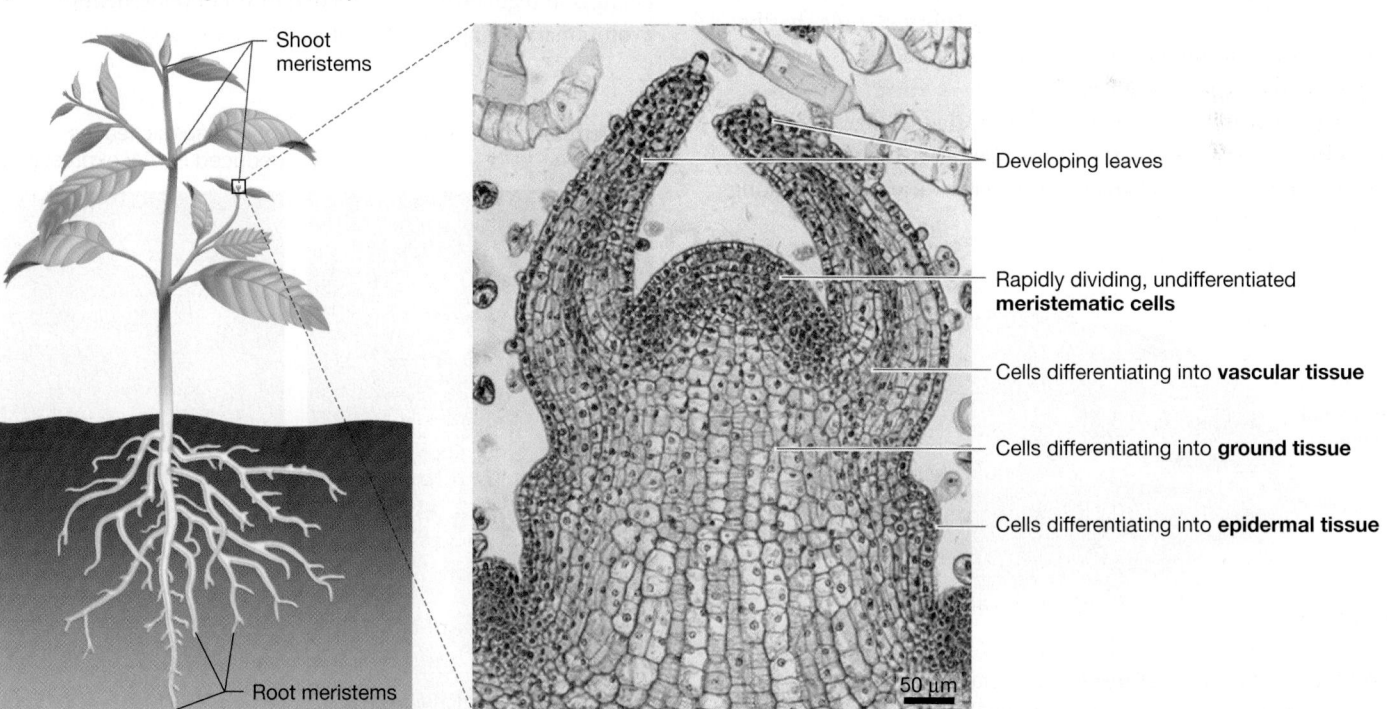

(a) Apical meristems are located at specific points throughout the body.

Shoot meristems

Root meristems

(b) Close-up cross section of a shoot apical meristem

Developing leaves

Rapidly dividing, undifferentiated **meristematic cells**

Cells differentiating into **vascular tissue**

Cells differentiating into **ground tissue**

Cells differentiating into **epidermal tissue**

50 μm

FIGURE 23.8 Meristems Are Where Development Takes Place. (a) Each tip in the root and shoot system contains a meristem. The individual can grow in any direction to which a meristem is oriented. **(b)** When meristem cells divide, the daughter cells either remain undifferentiated and continue to function as meristem cells or differentiate into new epidermal, ground, or vascular cells.

Recently researchers have also taken up the question of which genes respond to these cell-cell signals. In particular, which genes and gene products direct the formation of specific structures during vegetative development? Let's consider one especially well studied example—the genetic control of shape in developing leaves.

Which Genes and Proteins Determine Leaf Shape?

Applying tiny amounts of auxin to cells within a SAM induces leaves to grow there. This observation suggests that the initiation of a leaf depends on the concentration of auxin in parts of a SAM, although other types of cell-cell signals are undoubtedly involved as well. Once a leaf begins to grow, the next key event in its development is the formation of three axes: the proximal–distal, lateral, and adaxial–abaxial (upper–lower) axes shown in **Figure 23.9**. Proximal is toward the main body, while distal is away from the main body; the lateral axis runs from the middle of a leaf toward its margin. The amount and direction of growth along these axes determines the shape of the leaf.

Recent research has begun to identify the genes responsible for specifying these three leaf axes. For example, analyses of mutant snapdragons—a flowering plant you may have seen growing in a garden near your home—and other species has shown that a gene called *PHANTASTICA* (abbreviated *PHAN*) is critical in setting up the adaxial–abaxial axis of leaves.

The protein product of *PHAN* has a DNA-binding domain and acts as a regulatory transcription factor. *PHAN* triggers the expression of genes that cause cells to form the upper surface of leaves and suppresses transcription of genes required for forming the lower leaf surface. It is part of a regulatory cascade that begins with auxin and other cell-cell signals and ends with the growth of a normal-shaped leaf.

Recent research on tomatoes suggests that changes in *PHAN* expression may also underlie at least some of the evolutionary changes observed in leaf shape. Leaf shape varies widely among species (see Chapter 36). Simple leaves consist of a single blade, but as **Figure 23.10a** shows, tomatoes have compound leaves—each leaf blade is divided into smaller units called leaflets. Other species have palmately compound leaves, meaning that leaflets radiate from a single point.

To explore whether changes in *PHAN* might have a role in the evolution of various leaf shapes, a team of biologists used techniques introduced in Chapter 19 to create transgenic tomato plants. In these individuals, the *PHAN* gene product was blocked to a moderate or large extent.

As **Figure 23.10b** shows, leaf shape in the transgenic individuals was dramatically different. Some of the individuals had simple leaves that were cup-shaped, while others had several leaflets emerging from the same point. Although it is still uncertain why changing the specification of the upper–lower leaf axis affects overall shape, it appears clear that *PHAN* expression plays a role.

These results have inspired the hypothesis that at least some evolutionary changes in leaf size and shape are due to mutations that created new alleles of genes that regulate *PHAN* expression. Alleles that result in lowered *PHAN* expression might lead to simple leaves with a single blade; alleles that increase the extent of *PHAN* expression might result in compound leaves like those of normal tomatoes.

Mutations that alter development alter phenotypes. Variation in phenotype is the raw material of evolution (see Chapter 1 and Chapter 24).

By experimentally altering the genes that regulate development, researchers are beginning to understand the genetic changes leading to novel types of leaves. If this research continues to be productive, biologists will have another example of how changes in regulatory pathways that direct development underlie evolutionary change (see Chapter 21).

FIGURE 23.9 Leaves Have Three Axes. Overall, the plant body has just two axes: apical–basal and radial. Individually, however, every leaf has the three axes shown here. (To remember the difference between adaxial and abaxial, pretend that the *b* in *abaxial* stands for "below.")

✔ **QUESTION** How do the body axes in plants compare and contrast with the body axes in animals?

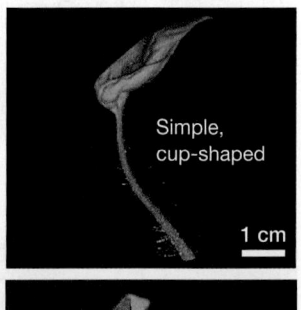

FIGURE 23.10 Changes in Genes That Establish Leaf Axes Can Change Leaf Shape.

SOURCE: Reprinted by permission from *Nature* (Vol 424, July 24, 2003, p. 439, by Kim, M., et al.). © 2004 Macmillan Publishers Ltd. Photographs supplied by Dr. Neelima Sinah.

23.4 Reproductive Development

Among the many startling contrasts observed between animal and plant development, one of the most important concerns the development of reproductive tissues and organs. In animals, the cells that give rise to sperm and egg cells are set aside, or sequestered, early in development. These **germ cells** migrate to the ovaries or testes once the reproductive organs have developed. As a result, the cells that give rise to animal gametes undergo relatively few rounds of mitosis—perhaps 20 to 50—prior to meiosis and gametogenesis.

In contrast, plants do not have germ cells that are set aside early in development. Instead, flowering and gametogenesis occur when a SAM converts from vegetative development to reproductive development.

As a result, meristematic cells that have divided hundreds of times can give rise to the reproductive organs of plants, and eventually to sperm and eggs. Because mutations occur during each cell cycle, plants generate much more genetic variation by mutation than animals do.

Although biologists have only begun to explore the consequences of this fact, research on the mechanisms responsible for the formation of reproductive structures has been intense. To introduce this huge body of work, let's consider some highlights of research on *Arabidopsis*.

The Floral Meristem and the Flower

When specialized proteins in *Arabidopsis* sense that nights are getting shorter and the temperature is favorable, they trigger the production of signals that convert SAMs from vegetative to reproductive development. A SAM converted in this way is called a **floral meristem**; instead of vegetative structures, it produces flowers, which contain the plant's reproductive organs. The genes that take part in the regulatory cascade responsible for the maturation of a floral meristem are now well characterized.

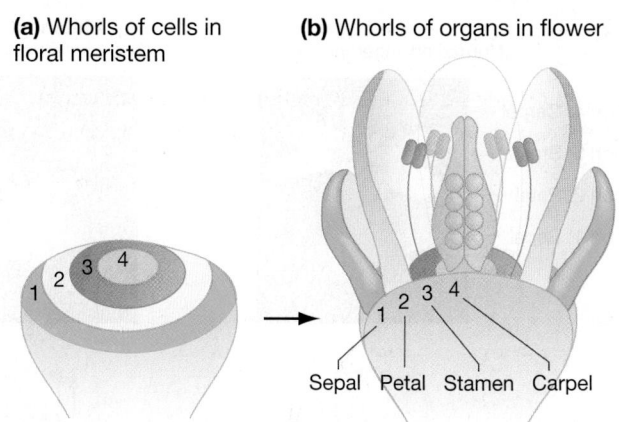

(a) Whorls of cells in floral meristem

(b) Whorls of organs in flower

Sepal Petal Stamen Carpel

FIGURE 23.11 Flowers Are Composed of Four Organs.

As an *Arabidopsis* flower develops, the floral meristem produces four kinds of organs: (1) sepals, (2) petals, (3) stamens, and (4) carpels. Each of these organs is a modified leaf (**Figure 23.11**).

- **Sepals** are located around the outside of the flower and provide protection as the organ develops. In some species, sepals are also colorful and function in attracting pollinators.

- Inside the sepals is a whorl, or circular arrangement, of **petals**, which enclose the male and female reproductive organs. If insects or other animals have to be attracted to pollinate the species in question, the petals may be colored to help advertise the reproductive structures.

- The pollen-producing organs, or **stamens**, are located in a whorl inside the petals.

- In the center of the entire structure are egg-producing reproductive organs, or **carpels**. (Ovules are located at the base of carpels.)

The question is, How does the floral meristem produce these four organs in the characteristic pattern of whorls within whorls?

The first hint of an answer came in the late 1800s, when researchers realized that several types of mutant flowering plants—including some popular garden plants—were homeotic. In the mutant individuals, one kind of floral organ was replaced by another. For example, one homeotic mutant had flowers with sepals, petals, another ring of petals, and carpels instead of having sepals, petals, stamens, and carpels. These mutants were similar to the *Drosophila* homeotic mutants described in Chapter 21, where individuals have legs or antennae growing in the wrong location—in place of the appropriate structure.

Just as an analysis of homeotic mutants in fruit flies triggered a breakthrough in understanding the genetic control of body axis formation and segmentation in animals, an analysis of homeotic floral mutants in *Arabidopsis* triggered a breakthrough in understanding the genetic control of flower structure in plants.

The Genetic Control of Flower Structures

Over 100 years after floral homeotic mutants were first described, Elliot Meyerowitz and colleagues assembled a large collection of *Arabidopsis* individuals with homeotic mutations in flower

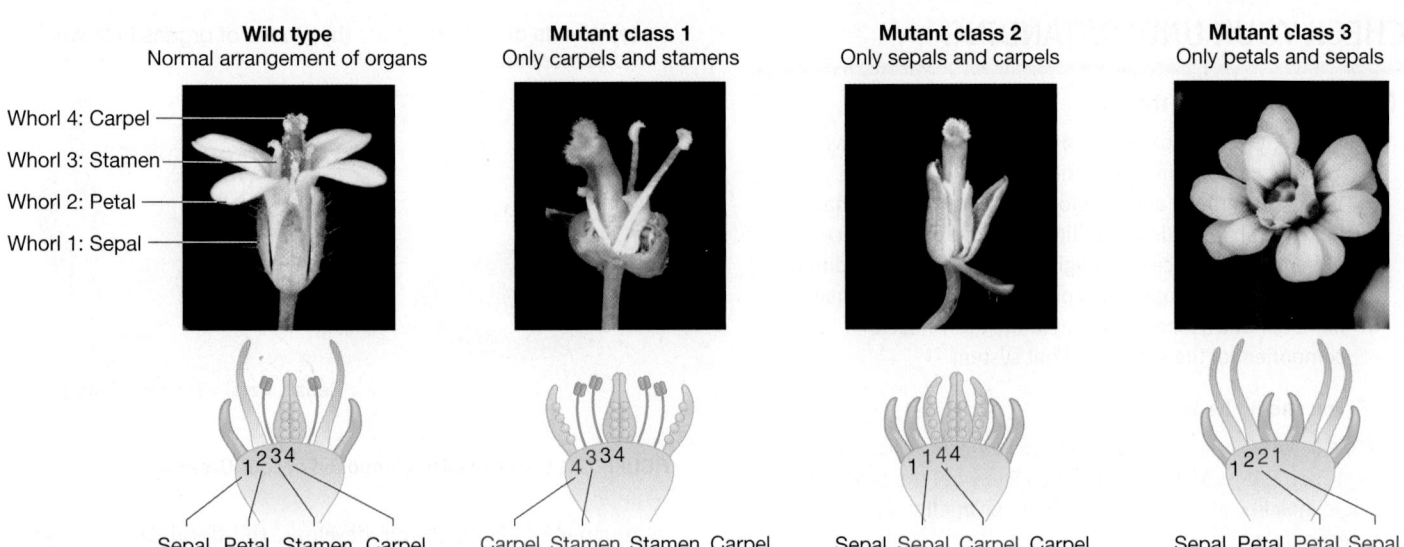

FIGURE 23.12 Homeotic Mutants in *Arabidopsis* Flowers Have Organs in the Wrong Locations.

structure. The researchers' goal was to identify and characterize the genes responsible for specifying the four floral organs.

The group found that the mutants could be sorted into the three general classes shown in **Figure 23.12**. The classes were distinguished by the type of homeotic transformation that occurred. Some mutants had only carpels and stamens; others had only sepals and carpels; still others had only petals and sepals. The key observation was that each type of mutant lacked the elements found in *two* of the four whorls.

What was going on? To begin searching for an answer, the biologists hypothesized that each of the three classes of homeotic mutants was caused by a defect in a single gene. They reasoned that if three genes are responsible for setting up the pattern of a flower, the mutants suggested a hypothesis for how the three gene products interact. Because they referred to the three hypothetical genes as *A*, *B*, and *C*, the hypothesis is called the ABC model.

THE ABC MODEL Three basic ideas underlie the ABC model (**Figure 23.13a**):

- Each of the three genes involved is expressed in two adjacent whorls.

- Because each gene is expressed in two adjacent whorls, a total of four different combinations of gene products can occur.

- Each of these four combinations of gene products triggers the development of a different floral organ.

Specifically, the Meyerowitz group proposed that (**1**) the A protein alone causes cells to form sepals, (**2**) a combination of A and B proteins sets up the formation of petals, (**3**) B and C combined specify stamens, and (**4**) the C protein alone designates cells as the precursors of carpels.

Does this model explain how the three classes of homeotic mutants occur? The answer is yes, if two additional elements are added to the model:

- The A protein inhibits production of the C protein.

- The C protein inhibits production of the A protein.

Then the patterns of gene expression correspond to the mutant phenotypes, as shown in **Figure 23.13b**.

For example, if the *A* gene is disabled by mutation, then it no longer inhibits the expression of the *C* gene and all cells produce the C protein. As a result, cells in the outermost whorl express only C protein and develop into carpels, while cells in the whorl just to the inside produce B and C proteins and develop into stamens.

✔If you understand the ABC model, you should be able to explain why biologists did not hypothesize that four genes are involved—one that specifies each of the four floral organs—and why they proposed that each gene involved is expressed in two adjacent whorls.

TESTING THE MODEL Although the ABC model is plausible and appeared to explain the data, it needed to be tested. To accomplish this, Meyerowitz and co-workers mapped the genes responsible for the mutant phenotypes and cloned the appropriate DNA sequences, using techniques introduced in Chapter 19. Once they had isolated the genes, they were able to obtain and use single-stranded DNAs to perform in situ hybridizations (see Chapter 21). The goal was to document the pattern of expression of the *A*, *B*, and *C* genes and see if that pattern corresponded to the model's predictions.

As anticipated, the mRNAs for each of the three genes showed up in the sets of whorls predicted by the model. The *A* gene is expressed in the outer two whorls, the *B* gene is expressed in the middle two whorls, and the *C* gene is expressed in the inner two whorls.

This result strongly supported the validity of the ABC model. Just as different combinations of *Hox* gene products specify the identity of fly segments, different combinations of floral identity genes specify the parts of a flower.

MADS-BOX GENES Similarities to principles of animal development did not end there. When Meyerowitz and others analyzed the

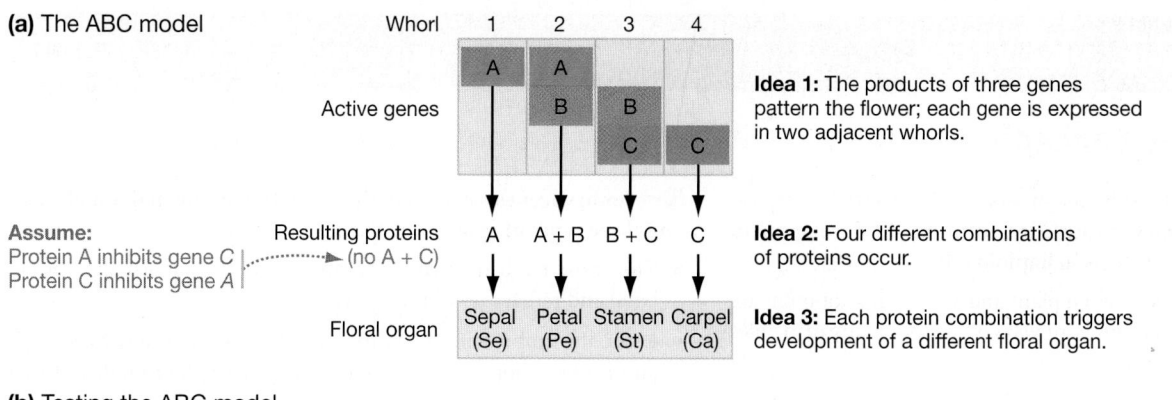

(a) The ABC model

Idea 1: The products of three genes pattern the flower; each gene is expressed in two adjacent whorls.

Assume:
Protein A inhibits gene C
Protein C inhibits gene A

Idea 2: Four different combinations of proteins occur.

Idea 3: Each protein combination triggers development of a different floral organ.

(b) Testing the ABC model

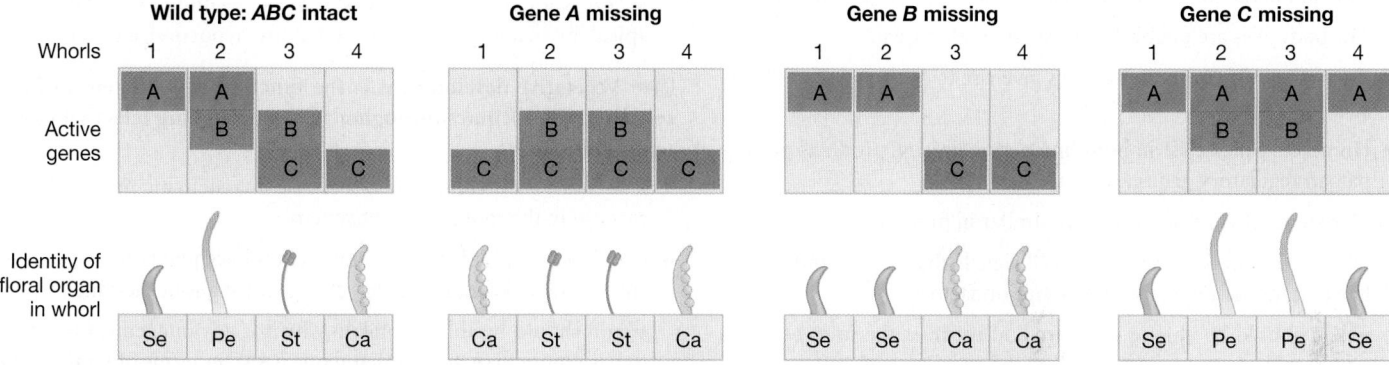

FIGURE 23.13 The ABC Model for Genetic Control of Flower Development. The ABC model is a hypothesis to explain why three types of homeotic mutants exist in *Arabidopsis* flowers.

DNA sequences of the floral organ identity genes, they discovered that all three genes contained a segment that coded for a long sequence of 58 amino acids, called the **MADS box**, that binds to DNA. Based on this observation, the researchers hypothesized that the floral genes encode regulatory transcription factors that bind to enhancers or other regulatory sequences and trigger the expression of genes required for sepal, petal, carpel, and stamen formation.

The researchers found many remarkable similarities between the *MADS-box* genes and the homeotic genes found in *Drosophila* and other animals.

- The *Hox* genes of animals have a region called the homeobox that is functionally similar to the MADS box: It encodes a DNA-binding domain in Hox proteins.

- Both *Hox* and *MADS-box* gene products are parts of regulatory cascades that lead to the specification of important structures. *Hox* genes regulate the expression of genes responsible for forming limbs or other structures in specific parts of the body; *MADS-box* genes regulate the expression of genes responsible for forming flowers.

- Both sets of genes tell cells where they are in the body and can produce homeotic mutants if they do not work properly.

Although different genes are involved, the underlying genetic principles for putting a multicellular body together are similar in plants and animals.

Biologists are currently working to identify the genes targeted by the ABC proteins, just as they are working to identify the

genes controlled by the *Hox* products in animals. Eventually, they hope to understand the complete regulatory cascade from initiation of the floral meristem to the expression of proteins that are specific to petals or other floral organs.

Ultimately, the goal is to explore how changes in this cascade could have led to the evolution of the spectacular diversity of flowers observed today. Were mutations in *MADS-box* genes, or the effector genes that they control, responsible for the flamoyant petals of the lady-slipper orchid, the broad, red sepals of the pointsettia, and the elaborate stamens of a sugar maple tree? Stay tuned.

CHECK YOUR UNDERSTANDING

If you understand that . . .

- Reproductive development in plants begins when SAMs are converted to floral meristems.
- In *Arabidopsis*, development of the four floral organs depends on the expression of regulatory transcription factors encoded by the *A*, *B*, and *C* genes.

✔ You should be able to . . .

1. Describe a mechanism whereby an A protein could inhibit expression of a *C* gene.

2. Compare and contrast the *MADS-box* genes of plants with the *Hox* genes of animals.

Answers are available in Appendix B.

Summary of Key Concepts

👉 **In sharp contrast to animals, plants develop continuously, do not commit cells to gamete production until late in development, and produce gametes by mitosis in haploid cells.**

- The striking differences between plant and animal development are not surprising given their different ways of life. But many principles are shared in the two groups.

- In both plants and animals, fertilization is mediated by protein–protein interactions between cells produced by each parent.

- The body axes are established early in development.

- Three types of embryonic tissues give rise to the array of adult tissues and organs.

- Homeotic genes exist in both lineages; their gene products participate in regulatory gene cascades.

- Meristem cells and stem cells are similar in function.

- Differentiation is based on cell-cell signals that trigger and modulate cascades of regulatory transcription factors.

  ✔You should be able to compare and contrast the roles of auxin and Bicoid.

👉 **In flowering plants, double fertilization results in the production of a zygote and a nutritive tissue that supports embryogenesis.**

- The pollination and fertilization process begins with interactions between proteins on the surface of pollen grains and proteins on the surface of the stigma.

- Pollen tubes grow toward the egg and form a conduit for cells.

- In flowering plants, one sperm nucleus fertilizes the egg to form a zygote while a second sperm nucleus fuses with a (usually) diploid cell to form a triploid cell.

- The triploid cell goes on to produce the nutritive tissue called endosperm.

  ✔You should be able to explain what's "double" about double fertilization.

👉 **Embryogenesis results in the formation of the major body axes and three types of embryonic tissue.**

- The earliest cell divisions in embryogenesis establish the apical–basal and radial axes of the individual.

- Later, the embryonic structures called the cotyledons, hypocotyl, and root develop, and the embryonic epidermal, ground, and vascular tissues form.

  ✔You should be able to sketch a seedling and label the apical–basal and radial axes, cotyledons, hypocotyl, and root.

👉 **Vegetative development is the function of meristems, where cell division continues throughout life—producing cells that go on to differentiate.**

- Plants can grow continuously because meristematic tissue exists at each tip in the root and shoot system.

- Continuous growth and development allows plants to respond to environmental conditions that change throughout their life.

  ✔You should be able to explain why it is advantageous for plants to have more than one SAM and one RAM, in terms of their ability to grow and reproduce.

👉 **When the function of a meristem shifts from vegetative to reproductive development, key regulatory transcription factors are activated and control the position and identity of floral organs.**

- Meristems that carry out vegetative development can convert to reproductive development instead, as a response to changes in day length or other environmental cues.

- Once a floral meristem is established in *Arabidopsis*, combinations of regulatory transcription factors encoded by A, B, and C genes interact to produce the flower's sepals, petals, stamens, and carpels.

  ✔You should be able to explain why individuals with defective alleles at the B gene are considered homeotic mutants.

Questions

✔ **TEST YOUR KNOWLEDGE**   *Answers are available in Appendix B*

1. What is the fate of the two cells found inside pollen grains prior to germination?
   a. One cell directs development of the pollen tube; the other gives rise to sperm cells by mitosis.
   b. One cell directs development of the pollen tube; the other gives rise to sperm cells by meiosis.
   c. One cell fertilizes the egg; the other fuses with a diploid cell to form triploid endosperm.
   d. One cell initiates germination by interacting with proteins on the surface of the stigma; the other gives rise to sperm cells.

2. Which of the following does *not* represent a contrast between plant and animal development?
   a. Under certain conditions, plant cells can "de-differentiate" readily.
   b. The fate of a cell is determined in part by its location in the embryo.
   c. Germ cells are set aside early in development.
   d. Plant cells do not move.

3. What evidence suggests that auxin concentrations help determine where leaves form near SAMs?
   a. Auxin is produced in SAMs and transported from there toward the root.

b. Auxin is present in a concentration gradient, with higher concentrations apically and lower concentrations basally.

c. Auxin concentrations are relatively constant along the radial axis of the body, and leaves form along the radial axis.

d. Addition of small quantities of auxin to a SAM can induce leaf development.

4. Which of the following does *not* occur during embryogenesis?

a. formation of the radial axis

b. production of the suspensor

c. formation of the cotyledons and hypocotyl

d. formation of the lateral and proximal–distal axes

5. When does the apical–basal axis first become apparent?

a. when the epidermal, ground, and vascular tissues form

b. when the cotyledons, hypocotyl, and root form

c. when the first cell division produces the apical cell and basal cell

d. during the globular stage, when the suspensor is complete

6. What evidence suggests that changes in the way that the *PHAN* gene is expressed could be partly responsible for evolutionary changes in leaf shape?

a. If *PHAN* gene expression is manipulated experimentally, individuals produce leaf types found in different species.

b. Experiments have shown that *PHAN* plays a role in establishing the upper–lower surface axis in leaves.

c. Sequencing studies and other data have shown that *PHAN* encodes a regulatory transcription factor.

d. All plant species surveyed to date have a gene homologous to *PHAN*.

✓ TEST YOUR UNDERSTANDING

Answers are available in Appendix B

1. When do important protein–protein interactions occur as a pollen grain interacts with a stigma? Describe the major type of cell-cell communication that occurs.

2. In plants, reproductive tissues may develop late in life, from cells that have undergone mitosis hundreds or thousands of times. How does this differ from animals, and what are the consequences?

3. Compare and contrast the stem cells of animals with the meristems of plants. How are these cells similar? How are they different?

4. In what sense are the epidermal, ground, and vascular tissues produced in the SAMs and RAMs of a 300-year-old oak tree "embryonic"?

5. When in situ hybridization experiments documented where *A*, *B*, and *C* genes were expressed in developing *Arabidopsis* flowers, it was considered strong support for the ABC model. Explain why.

6. Give an example of how each of the fundamental developmental processes—cell proliferation, death, expansion, interaction, and differentiation—plays a role in plant development.

✓ APPLYING CONCEPTS TO NEW SITUATIONS

Answers are available in Appendix B

1. When growing conditions are extremely poor, many plant species stop vegetative growth and put all of their energy into reproductive growth. Propose a hypothesis to explain why.

2. When growth occurs in an unlimited or unrestricted way, it is said to be indeterminate. But when growth is of limited duration and then stops, it is said to be determinant. Which process is observed in vegetative development and which in reproductive development? Explain your logic.

3. Leaves that grow at the top of a tree are typically smaller than leaves that grow at the bottom of the same tree. Small leaves lose less water in bright sunlight; large leaves capture more light in shade. Explain

how this size difference might develop, in terms of changes in gene expression.

4. Make a sketch showing the locations of SAM and RAM in a young oak tree. Suppose that a concrete sidewalk gets added on the left side of the tree, preventing water penetration below. A billboard goes up on the right side, cutting off light but not water penetration from that direction. But after a few years, an underground water pipe under the billboard begins to leak, providing abundant water year round. Draw the expected locations of SAM and RAM in the same tree 50 years after your initial sketch. Explain your logic.

Natural selection acts on individuals in populations such as these sea stars, but only populations evolve. One of Darwin's greatest contributions to science was the introduction of population thinking to the theory of evolution.

24 Evolution by Natural Selection

KEY CONCEPTS

🔑 Populations and species evolve, meaning that their heritable characteristics change through time. Evolution is change in allele frequencies over time.

🔑 Evolution by natural selection occurs when individuals with certain alleles produce the most surviving offspring in a population. An adaptation is a genetically based trait that increases an individual's ability to produce offspring in a particular environment.

🔑 Evolution by natural selection is not progressive, and it does not change the characteristics of the individuals that are selected—it changes only the characteristics of the population. Animals do not do things for the good of the species, and not all traits are adaptive. All adaptations are constrained by trade-offs and genetic and historical factors.

This chapter is about one of the great ideas in science. The theory of evolution by natural selection, formulated independently by Charles Darwin and Alfred Russel Wallace, explains how organisms have come to be adapted to environments ranging from arctic tundra to tropical wet forest. It revealed one of the five attributes of life introduced in Chapter 1: Populations of organisms evolve—meaning that they change through time.

As an example of a revolutionary breakthrough in our understanding of the world, the theory of evolution by natural selection ranks alongside Copernicus's theory of the Sun as the center of our solar system, Newton's laws of motion and theory of gravitation, the germ theory of disease, the theory of plate tectonics, and Einstein's general theory of relativity. These theories are the foundation stones of modern science; all are accepted on the basis of overwhelming evidence.

Evolution by natural selection is one of the best-supported and most important theories in the history of scientific research. But like most scientific breakthroughs, this one did not come easily. When Darwin published his theory in 1859 in a book called *On the Origin of Species by Means of Natural Selection*, it unleashed a firestorm of protest throughout Europe. At that time, the leading explanation for the diversity of organisms was a theory called special creation.

To understand the contrast between the theory of special creation and the theory of evolution by natural selection more thoroughly, recall from Chapter 1 that scientific theories usually have two components: a pattern and a process.

● The pattern component is a statement that summarizes a series of observations about the natural world. The pattern component is about facts—about how things *are* in nature.

● The process component is a mechanism that produces that pattern or set of observations.

✔ When you see this checkmark, stop and test yourself. Answers are available in Appendix B.

The pattern component of the theory of special creation held that: (1) All species are independent, in the sense of being unrelated to each other; (2) life on Earth is young—perhaps just 6000 years old; and (3) species are immutable, or incapable of change. The process that explained this pattern was the instantaneous and independent creation of living organisms by a supernatural being.

Darwin's ideas were radically different. What are the pattern and process components of the theory of evolution by natural selection?

24.1 The Evolution of Evolutionary Thought

People often use the word revolutionary to describe the theory of evolution by natural selection. Revolutions overturn things—they replace an existing entity with something new and often radically different. A political revolution removes the ruling class or group and replaces it with another. The industrial revolution replaced small shops for manufacturing goods by hand with huge, mechanized assembly lines.

A scientific revolution, in contrast, overturns an existing idea about how nature works and replaces it with another, radically different, idea. The idea that Darwin and Wallace overturned—that species were specially, not naturally, created—had dominated thinking about the nature of organisms for over 2000 years.

Plato and Typological Thinking

The Greek philosopher Plato claimed that every organism was an example of a perfect essence, or type, created by God, and that these types were unchanging. Plato acknowledged that the individual organisms present on Earth might deviate slightly from the perfect type, but he said this deviation was similar to seeing the perfect type in a shadow on a wall. The key to understanding life, in Plato's mind, was to ignore the shadows and focus on understanding each type of unchanging, perfect essence.

Today, philosophers and biologists refer to ideas like this as typological thinking. Typological thinking is based on the idea that species are unchanging types and that variations within species are unimportant or even misleading. Typological thinking also occurs in the Bible's book of Genesis, where God creates one of each type of organism.

Aristotle and the Great Chain of Being

Not long after Plato developed his ideas, Aristotle ordered the types of organisms known at the time into a linear scheme called the great chain of being, also called the scale of nature (**Figure 24.1**). Aristotle proposed that species were organized into a sequence based on increased size and complexity, with humans at the top. He also claimed that the characteristics of species were fixed—they did not change through time.

In the 1700s Aristotle's ideas were still popular in scientific and religious circles. The central claims were that (1) species are fixed types, and (2) some species are higher—in the sense of being more complex or "better"—than others.

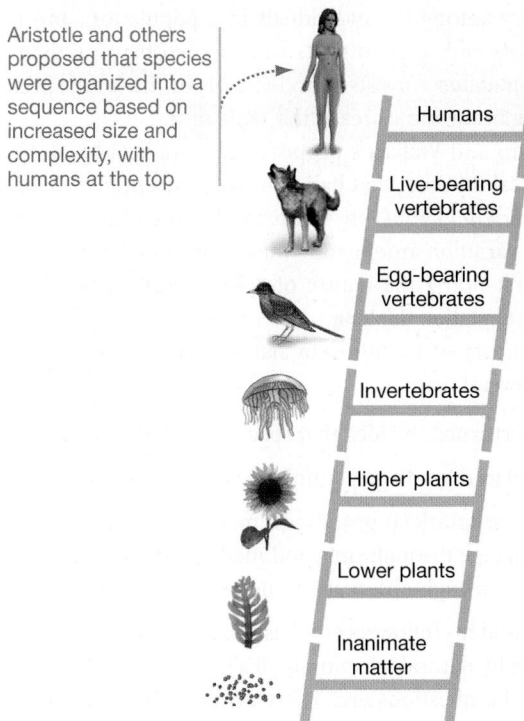

Aristotle and others proposed that species were organized into a sequence based on increased size and complexity, with humans at the top

Humans

Live-bearing vertebrates

Egg-bearing vertebrates

Invertebrates

Higher plants

Lower plants

Inanimate matter

FIGURE 24.1 The Great Chain of Being, or Scale of Nature

Lamarck and the Idea of Evolution as Change through Time

Typological thinking eventually began to break down, however. In 1809 the biologist Jean-Baptiste de Lamarck proposed a formal theory of evolution—that species are not static but change through time. The pattern component of Lamarck's theory was initially based on the great chain of being, however.

When he started his work on evolution, Lamarck claimed that simple organisms originate at the base of the chain by spontaneous generation (see Chapter 1) and then evolve by moving up the chain over time. Thus, Lamarckian evolution is progressive in the sense of always producing larger and more complex, or "better," species. To capture this point, biologists often say that Lamarck turned the ladder of life into an escalator.

Lamarck also contended that species change through time via the inheritance of acquired characters. The idea here is that an individual's phenotype changes as they develop in response to challenges posed by the environment, and they pass on these phenotypic changes to offspring. A classic Lamarckian scenario is that giraffes develop long necks as they stretch to reach leaves high in treetops, and they then produce offspring with elongated necks.

Darwin and Wallace and Evolution by Natural Selection

As his thinking matured, Lamarck eventually abandoned his linear and progressive view of life. Darwin and Wallace concurred. What is more important, they emphasized that the process responsible for change through time—evolution—occurs because traits vary among the individuals in a population, and because individuals with certain traits leave more offspring than others do. A **population** consists of individuals of the same species that are living in the same area at the same time.

Darwin and Wallace's proposal was a radical break from the typological thinking that had dominated scientific thought since Plato. Darwin claimed that instead of being unimportant or an illusion, variation among individuals in a population was the key to understanding the nature of species. Biologists refer to this view as **population thinking**.

The theory of evolution by natural selection was revolutionary for several reasons:

1. It overturned the idea that species are static and unchanging.

2. It replaced typological thinking with population thinking.

3. It was scientific. It proposed a mechanism that could account for change through time and made predictions that could be tested through observation and experimentation.

Plato and his followers emphasized the existence of fixed types; evolution by natural selection is all about change and diversity.

Now the questions are: What evidence backs the claim that species are not fixed types? What data convince biologists that the theory of evolution by natural selection is correct?

24.2 The Pattern of Evolution: Have Species Changed through Time?

In *On the Origin of Species*, Darwin repeatedly described evolution as **descent with modification**. He meant that species that lived in the past are the ancestors of the species existing today, and that species and their descendant species change through time. Descendant species are modified.

This view was a radical departure from the pattern of independently created and immutable species embodied in Plato's work and in the theory of special creation. ☞ In essence, the pattern component of the theory of evolution by natural selection makes two statements about the nature of species:

1. Species change through time.

2. Species are related by common ancestry.

Let's consider the evidence for each of these claims in turn.

Evidence for Change through Time

When Darwin began his work, biologists and geologists had just begun to assemble and interpret the fossil record. A **fossil** is any trace of an organism that lived in the past. These traces range from bones and branches to shells, tracks or impressions, and dung (**Figure 24.2**). The **fossil record** consists of all the fossils that have been found and described in the scientific literature.

Why did data in the fossil record support the hypothesis that species have changed through time? And what data from **extant species**—those living today—support the claim that they are modified forms of ancestral species?

THE VASTNESS OF GEOLOGIC TIME Initially, fossils were organized according to their relative ages. This was possible because geologists had created the **geologic time scale**: a sequence of named intervals called eons, eras, and periods that represented the major events in Earth history (see Chapter 27).

The geologic time scale, in turn, was based on a series of principles derived from observations about rock formation. **Sedimentary rocks**, for example, form from sand or mud or other materials deposited at locations such as beaches or river mouths. Sedimentary rocks, along with rocks derived from volcanic ash or lava, are known to form in layers—with younger layers deposited on top of older layers. In a similar vein, rock deposits that contained boulders were inferred to be younger than the rocks that the boulder material originated from.

As the geologic record was being established, then, researchers placed fossils in a younger-to-older sequence, based on their relative position in layers of sedimentary rock. They also realized that vast amounts of time were required to form the thick layers of sedimentary rock that they were studying, because erosion and deposition of sediments are such slow processes.

(a) 180-million-year-old ammonite shells   **(b)** 210-million-year-old bird tracks   **(c)** 13,000-year-old giant sloth dung

1 cm   5 cm   1 cm

FIGURE 24.2 A Fossil Is *Any* Trace of an Organism That Lived in the Past. In addition to **(a)** body parts such as shells or bones or branches, fossils may consist of **(b)** tracks or impressions, or even **(c)** pieces of dung.

This was an important insight. The geologic record indicated that the Earth was much, much older than the 6000 years claimed by proponents of the theory of special creation.

After the discovery of radioactivity in the late 1800s, researchers realized that radioactive decay—the steady rate at which unstable or "parent" atoms are converted into more stable "daughter" atoms—furnished a way to assign absolute ages, in years, to the relative ages in the geologic time scale.

Radiometric dating is based on three pieces of information:

1. Observed decay rates of parent to daughter atoms.

2. The ratio of parent to daughter atoms present in newly formed rocks—such as the amount of uranium atoms versus lead atoms when uranium-containing molten rock first cools. (Uranium decays to form lead.)

3. The ratio of parent to daughter atoms present in a particular rock sample.

Combining information from these two ratios with information on the decay rate allows researchers to estimate how long ago a rock formed. According to data from radiometric dating, Earth is about 4.6 billion years old, and the earliest signs of life appear in rocks that formed 3.4–3.8 billion years ago.

Data from relative and absolute dating techniques agree: Life on Earth is ancient. There has been a great deal of time for change through time to occur.

EXTINCTION CHANGES THE SPECIES PRESENT OVER TIME In the early nineteenth century, researchers began discovering fossil bones, leaves, and shells that were unlike structures from any known animal or plant. At first, many scientists insisted that living examples of these species would be found in unexplored regions of the globe. But as research continued and the number and diversity of fossil collections grew, the argument became less and less plausible.

The issue was finally settled when Baron Georges Cuvier published a detailed analysis of an **extinct** species—that is, a species that no longer exists—called the Irish "elk" in 1812. Scientists accepted the fact of extinction because this gigantic deer was judged to be too large to have escaped discovery and too distinctive to be classified as a large-bodied population of an existing species (**Figure 24.3**).

Advocates of the theory of special creation argued that the fossil species were victims of the flood at the time of Noah. Darwin, in contrast, interpreted extinct forms as evidence that species are not static, immutable entities, unchanged since the moment of special creation. His reasoning was that if species have gone extinct, then the array of species living on Earth has changed through time.

Recent analyses of the fossil record suggest that over 99 percent of all the species that have ever lived are now extinct. The data also indicate that species have gone extinct continuously throughout Earth's history—not just in one or even a few catastrophic events.

TRANSITIONAL FEATURES LINK OLDER AND YOUNGER SPECIES Long before Darwin published his theory, researchers reported striking resemblances between the fossils found in the rocks underlying

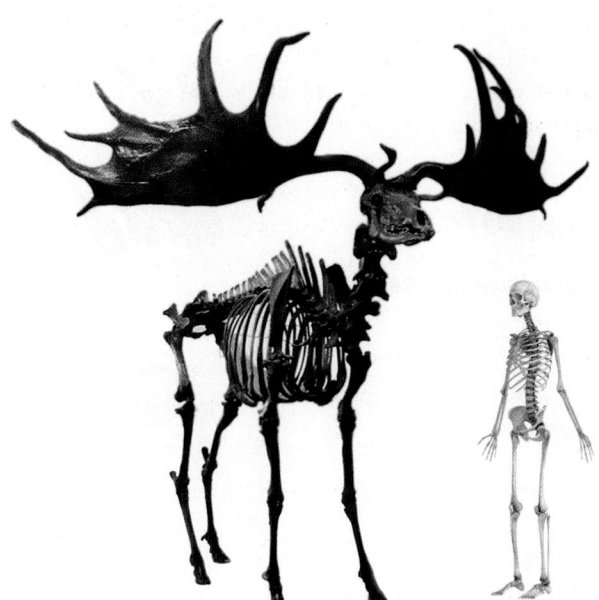

FIGURE 24.3 Evidence of Extinction. The skeleton of the Irish "elk" dwarfs a human. Scientists agreed that the deer was too large and unique to be overlooked if it were alive; it must have gone extinct.

certain regions and the living species found in the same geographic areas. The pattern was so widespread that it became known as the "law of succession." The general observation was that extinct species in the fossil record were succeeded, in the same region, by similar species.

Early in the nineteenth century, the pattern was simply reported and not interpreted. But later, Darwin pointed out that it provided strong evidence in favor of the hypothesis that species had changed through time. His idea was that the extinct forms and living forms were related—that they represented ancestors and descendants.

As the fossil record improved, researchers discovered species with characteristics that broadened the scope of the law of succession. A **transitional feature** is a trait in a fossil species that is intermediate between those of older and younger species. For example, intensive work over the past several decades has yielded fossils that document a gradual change over time from aquatic animals that had fins to terrestrial animals that had limbs (**Figure 24.4** on page 418). Over a period of about 25 million years, the fins of species similar to today's lungfish changed into limbs similar to those found in today's amphibians, reptiles, and mammals—a group called the tetrapods (literally, "four-footed").

These observations support the hypothesis that an ancestral lungfish-like species began living on land, and that their descendants became more and more like today's tetrapods in appearance and lifestyle. Lungfish and tetrapod species have clearly changed through time.

Similar sequences of transitional features document changes that led to the evolution of feathers and flight in birds, stomata and vascular tissue in plants, upright posture, flattened faces, and large brains in humans, jaws in vertebrates (animals with

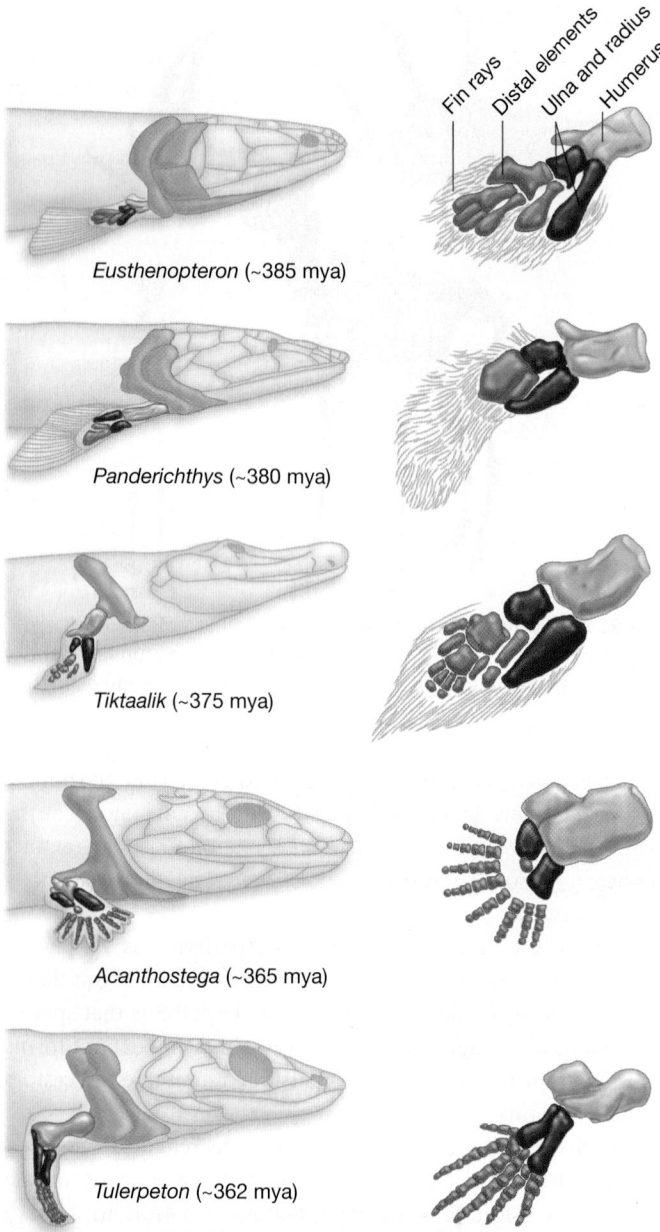

Eusthenopteron (~385 mya)

Panderichthys (~380 mya)

Tiktaalik (~375 mya)

Acanthostega (~365 mya)

Tulerpeton (~362 mya)

Fin rays · Distal elements · Ulna and radius · Humerus

FIGURE 24.4 Transitional Features during the Evolution of the Tetrapod Limb. Fossil species similar to today's lungfish and tetrapods have fin and limb bones that are transitional features. *Eusthenopteron* was aquatic; *Tulerpeton* was probably semi-aquatic (mya = million years ago).

✔**QUESTION** How would observations of transitional features be explained under the theory of special creation?

backbones), the loss of limbs in snakes, and other traits. Data like these are consistent with predictions from the theory of evolution: If the traits observed in more recent species evolved from traits in more ancient species, then intermediate forms are expected to occur in the appropriate time sequence.

The fossil record provides compelling evidence that species have evolved. What data from extant forms supports the hypothesis that the characteristics of species change through time?

VESTIGIAL TRAITS ARE EVIDENCE OF CHANGE THROUGH TIME

Darwin was the first to provide a widely accepted interpretation of vestigial traits. A **vestigial trait** is a reduced or incompletely developed structure that has no function or reduced function, but is clearly similar to functioning organs or structures in closely related species.

Biologists have documented thousands of examples of vestigial traits.

- The genomes of humans and other organisms contain hundreds of pseudogenes—the functionless DNA sequences introduced in Chapter 20.
- Bowhead whales and rubber boas have tiny hip and leg bones that do not help them swim or slither.
- Ostriches and kiwis have reduced wings and cannot fly.
- Blind cave-dwelling fish still have eye sockets.
- Even though marsupial mammals give birth to live young, an eggshell forms briefly early in their development; in some species, newborns have a non-functioning "egg tooth" similar to those seen in birds and reptiles.
- Monkeys and many other primates have long tails; but our coccyx, illustrated in **Figure 24.5a**, is too small to help us maintain balance or grab tree limbs for support.
- Many mammals, including primates, are able to erect their hair when they are cold or excited. But our sparse fur does little to keep us warm, and goose bumps are largely ineffective in signaling our emotional state (**Figure 24.5b**).

The existence of vestigial traits is inconsistent with the theory of special creation, which maintains that species were perfectly designed by a supernatural being and that the characteristics of species are static. Instead, vestigial traits are evidence that the characteristics of species have changed over time.

CURRENT EXAMPLES OF CHANGE THROUGH TIME Biologists have documented hundreds of contemporary populations that are changing in response to changes in their environment. Bacteria have evolved resistance to drugs; insects have evolved resistance to pesticides; weedy plants have evolved resistance to herbicides. Section 24.4 provides a detailed analysis of research on two examples of evolution in action.

To summarize, change through time continues and can be measured directly. Evidence from the fossil record and living species indicates that life is ancient, that species have changed through the course of Earth's history, and that species continue to change. The take-home message is that species are dynamic—not static, unchanging, and fixed types, as claimed by Plato, Aristotle, and the theory of special creation.

Evidence of Descent from a Common Ancestor

Data from the fossil record and contemporary species refute the hypothesis that species are immutable. What about the claim that species were created independently—meaning that they are unrelated to each other?

(a) The human tailbone is a vestigial trait.

(b) Goose bumps are a vestigial trait.

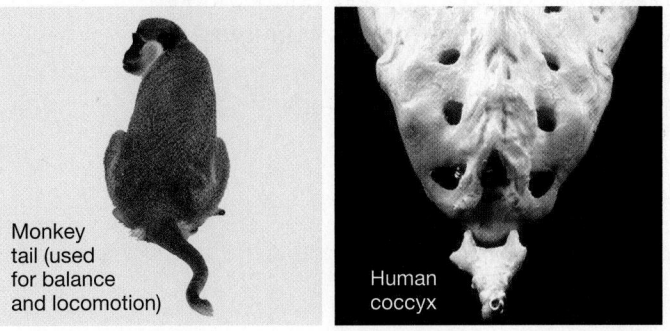

Monkey tail (used for balance and locomotion)

Human coccyx

Erect hair on chimp (insulation, emotional display)

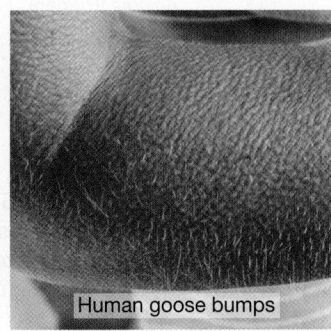

Human goose bumps

FIGURE 24.5 Vestigial Traits Are Reduced Versions of Traits in Other Species. **(a)** The tailbone and **(b)** goose bumps are human traits that have reduced function. They are similar to larger, fully functional structures in other species.

✓**QUESTION** How would observations of vestigial traits be explained if evolution occurred via inheritance of acquired characters?

SIMILAR SPECIES ARE FOUND IN THE SAME GEOGRAPHIC AREA Charles Darwin began to realize that species are related by common ancestry, just as individuals within a family are, during a five-year voyage he took aboard the English naval ship HMS *Beagle*. While fulfilling its mission to explore and map the coast of South America, the *Beagle* spent considerable time in the Galápagos Islands off the coast of present-day Ecuador. Darwin

had taken over the role of ship's naturalist and as the first scientist to study the area, gathered extensive collections of the plants and animals found in these islands. Among the birds he collected were what came to be known as the Galápagos mockingbirds, pictured in **Figure 24.6a**.

Several years after Darwin returned to England, a biologist pointed out that the mockingbirds collected on different islands

(a) Pattern: Although the Galápagos mockingbirds are extremely similar, distinct species are found on different islands.

Nesomimus parvulus

Nesomimus trifasciatus

Nesomimus melanotis

Nesomimus macdonaldi

(b) Recent data support Darwin's hypothesis that the Galápagos mockingbirds share a common ancestor.

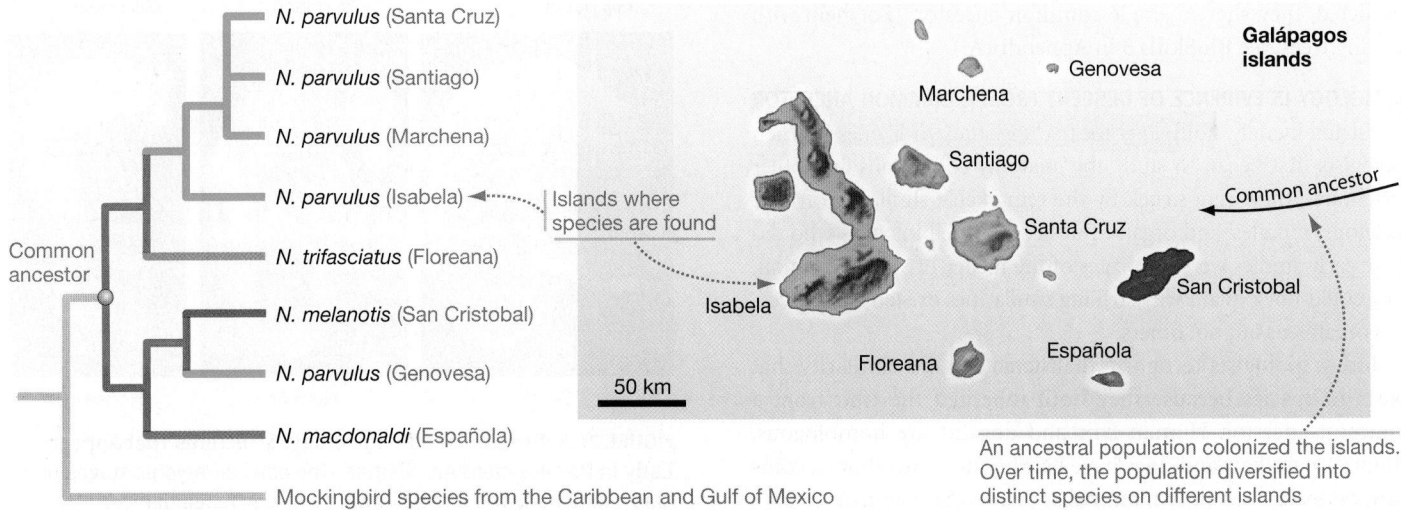

FIGURE 24.6 Close Relationships among Island Forms Argue for Shared Ancestry.

Gene:	Amino acid sequence (single-letter abbreviations):
Aniridia (Human)	LQRNRTSFTQEQIEALEKEFERTHYPDVFARERLAAKIDLPEARIQVWFSNRRAKWRREE
eyeless (Fruit fly)	LQRNRTSFTNDQIDSLEKEFERTHYPDVFARERLAGKIGLPEARIQVWFSNRRAKWRREE

Only six of the 60 amino acids in these sequences are different. The two sequences are 90% identical

FIGURE 24.7 Genetic Homology: Genes from Different Species May Be Similar in DNA Sequence or Other Attributes. Amino acid sequences from a portion of the *Aniridia* gene product found in humans are almost identical to those found in the *Drosophila eyeless* gene product. For a key to the single-letter abbreviations used for the amino acids, see Figure 3.3.

were distinct species, based on differences in coloration and beak size and shape. This struck Darwin as remarkable. Why would species that inhabit neighboring islands be so similar, yet clearly distinct? This turns out to be a widespread pattern: In island groups across the globe, it is routine to find similar but distinct species on neighboring islands.

Darwin realized that this pattern—puzzling when examined as a product of special creation—made perfect sense when interpreted in the context of evolution, or descent with modification. He proposed that the mockingbirds were similar because they had descended from the same common ancestor. Instead of being created independently, he proposed that mockingbird populations that colonized different islands had changed through time and formed new species (**Figure 24.6b**).

Recent analyses of DNA sequences in these mockingbirds support Darwin's hypothesis. Specifically, the molecular data are consistent with the prediction that the mockingbirds are part of a **phylogeny**—a family tree of populations or species.

Using techniques that will be introduced in Chapter 27, researchers have placed the mockingbirds on a phylogenetic tree—a branching diagram that describes the ancestor–descendant relationships among species. A phylogenetic tree is similar to a genealogy describing the ancestor–descendant relationships among individual humans. As Figure 24.6b shows, the Galápagos mockingbirds are each others' closest living relatives. As Darwin predicted, they share a single common ancestor. (For help with reading trees, see **BioSkills 3** in Appendix A.)

HOMOLOGY IS EVIDENCE OF DESCENT FROM A COMMON ANCESTOR
Translated literally, homology means "the study of likeness." When biologists first began to study the anatomy of humans and other vertebrates, they were struck by the remarkable similarity of their skeletons, muscles, and organs. But because the biologists who did these early studies were advocates of the theory of special creation, they could not explain why striking similarities existed among certain organisms but not others.

Today, biologists recognize that **homology** is a similarity that exists in species because they both inherited the trait from a common ancestor. Human hair and dog fur are homologous. Humans have hair and dogs have hair because they share a common ancestor—an early mammal species—that had hair.

Homology can be recognized and studied at three levels:

- **Genetic homology** occurs in DNA sequences. For example, the *eyeless* gene in fruit flies and the *Aniridia* gene in humans are so similar that their protein products are 90 percent identical in amino acid sequence (**Figure 24.7**). Both genes act in determining where eyes will develop—even though fruit flies have a compound eye with many lenses and humans have a camera eye with a single lens.

- **Developmental homology** is recognized in embryos. For example, early chick, human, and cat embryos have tails and structures called gill pouches (**Figure 24.8**). Later, gill pouches are lost in all three species and tails are lost in humans. But in fish, the embryonic gill pouches stay intact and give rise to functioning gills in adults. To explain this observation, biologists hypothesize that gill pouches and tails exist in chicks, humans, and cats because they existed in the fishlike species that was the common ancestor of today's vertebrates. Embryonic gill pouches are a vestigial trait in chicks, humans, and cats; embryonic tails are a vestigial trait in humans.

- **Structural homology** is a similarity in adult **morphology**, or form. A classic example is the common structural plan observed in the limbs of vertebrates (**Figure 24.9**). In Darwin's

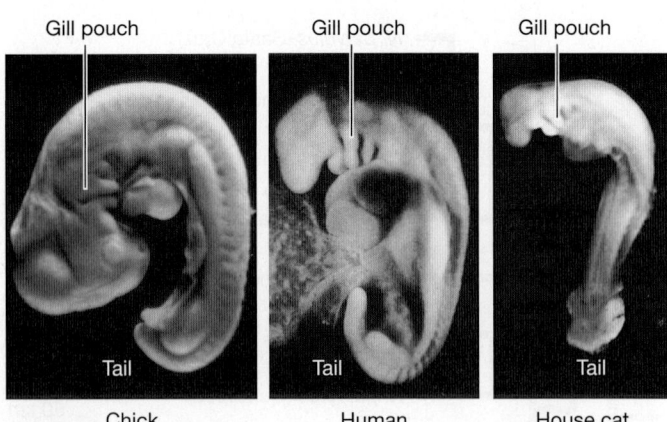

Gill pouch   Gill pouch   Gill pouch

Tail   Tail   Tail

Chick   Human   House cat

FIGURE 24.8 Developmental Homology: Structures That Appear Early in Development Are Similar. The early embryonic stages of a chick, a human, and a cat, showing a strong resemblance.

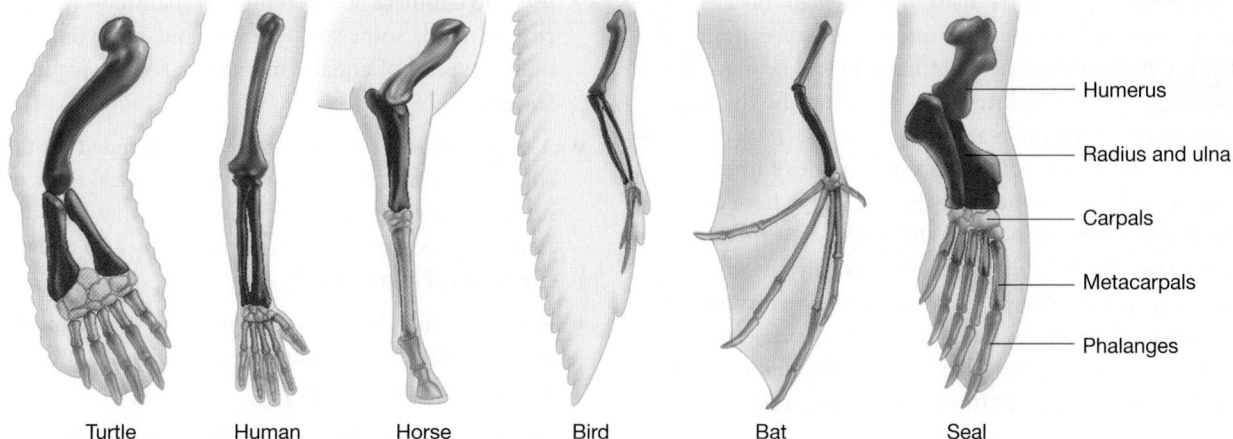

Turtle   Human   Horse   Bird   Bat   Seal

— Humerus
— Radius and ulna
— Carpals
— Metacarpals
— Phalanges

FIGURE 24.9 Structural Homology: Limbs with Different Functions Have the Same Underlying Structure. Even though their function varies, all vertebrate limbs are modifications of the same number and arrangement of bones. Darwin interpreted structural homologies like these as a product of descent with modification. (These limbs are not drawn to scale.)

own words, "What could be more curious than that the hand of a man, formed for grasping, that of a mole for digging, the leg of the horse, the paddle of the porpoise, and the wing of the bat, should all be constructed on the same pattern, and should include the same bones, in the same relative positions?" An engineer would never use the same underlying structure to design a grasping tool, a digging implement, a walking device, a propeller, and a wing. Instead, the structural homology exists because mammals evolved from the lungfish-like ancestor in Figure 24.4, which had the same general arrangement of bones in its fins.

The three levels of homology interact. Genetic homologies cause the developmental homologies observed in embryos, which then lead to the structural homologies recognized in adults. Perhaps the most fundamental of all homologies is the genetic code. With a few minor exceptions, all organisms use the same rules for transferring the information coded in DNA into proteins (see Chapter 15).

In some cases, hypotheses about homology can be tested experimentally. For example, researchers (**1**) isolated genes from mice or squid that were thought to be homologous to the fruit fly *eyeless* gene, (**2**) inserted the mouse or squid gene into fruit fly embryos, (**3**) stimulated expression of the foreign gene in locations that normally give rise to appendages, and (**4**) observed formation of eyes on legs and antennae (**Figure 24.10**). The inserted genes' function was identical to the function of *eyeless*. This result was strong evidence that the fruit fly, mouse, and squid genes are homologous, as predicted from their sequence similarity.

Homology is a key concept in contemporary biology:

- Chemicals that are cancer-causing in humans can be identified by testing their effects on mutation rates in bacteria, because the molecular machinery responsible for copying and repairing DNA is homologous in all organisms (see Chapter 14).

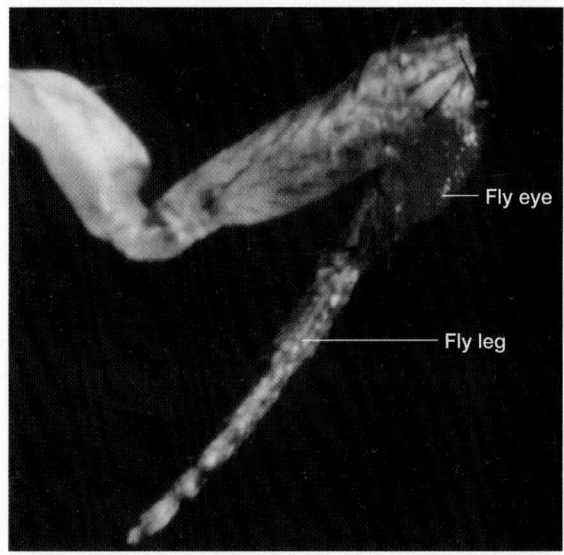

— Fly eye
— Fly leg

FIGURE 24.10 Evidence for Homology: A Mouse Gene Expressed in Fruit Flies. As an embryo, this fruit fly received a mouse gene that signals where eyes should form. A fruit fly eye formed in the location where the mouse gene was expressed.

- Drugs intended for human use can be tested on mice or rabbits if the molecules targeted by the drugs are homologous.

- Unknown sequences in the human, rice, or other genomes can be identified if they are homologous to known sequences in yeast, fruit flies, *Arabidopsis thaliana*, or other well-studied model organisms (see Chapter 20 and **BioSkills 14** in Appendix A).

The theory of evolution by natural selection predicts that homologies will occur. If species were created independently of one another, as the theory of special creation claims, these types of similarities would not occur.

CURRENT EXAMPLES Biologists have documented dozens of contemporary populations that are undergoing speciation—a process that results in one species splitting into two or more descendant species. Chapter 26 introduces what may be the best-studied example of speciation in action.

In most cases, the identity of the ancestral species and the descendant species is known—meaning that biologists have established a direct link between ancestral and descendant species. In addition, the reason for the splitting event is usually known.

The contemporary examples of new species being formed are powerful evidence that species living today are the descendants of species that lived in the past. They support the claim that all organisms are related by descent from a common ancestor.

Evolution's "Internal Consistency"— the Importance of Independent Datasets

Biologists draw upon data from several sources to challenge the hypothesis that species are immutable and were created independently. The data support the idea that species have descended, with modification, from a common ancestor. **Table 24.1** summarizes this evidence.

Perhaps the most powerful evidence for any scientific theory, including evolution by natural selection, is what scientists call "internal consistency." This is the observation that data from independent sources agree in supporting predictions made by a theory.

As an example, consider the evolution of whales and dolphins—a group called the cetaceans.

- The fossil record contains a series of species that are clearly identified as cetaceans, on the basis of unusual ear bones found only in this group. Some of the species have the long legs and compact bodies typical of mammals that live primarily on land; some are limbless and have the streamlined bodies typical of aquatic mammals; some have intermediate features.

- A phylogeny of the fossil cetaceans, estimated on the basis of similarities and differences in morphological traits other than limbs and overall body shape, indicates that a gradual transition occurred between terrestrial forms and aquatic, whale-like forms (**Figure 24.11**).

- Relative dating, based on the positions of sedimentary rocks where the fossils were found, agrees with the order of species indicated in the phylogeny.

- Absolute dating, based on analyses of radioactive atoms in rocks in or near the layers where the fossils were found, also agrees with the order of species indicated in the phylogeny.

- A phylogeny of living whales and dolphins, estimated from similarities and differences in DNA sequences, indicates that hippos—which spend much of their time in shallow water—are the closest living relative of cetaceans. This observation supports the hypothesis that whales and hippos shared a common ancestor that was semi-aquatic.

- Some whales have vestigial hip and limb bones as adults, and some dolphin embryos have vestigial hindlimb buds—outgrowths where legs form in other mammals.

The general message here is that many independent lines of evidence converge on the same conclusion: Whales gradually evolved from a terrestrial ancestor, over the span of about 12 million years.

As you evaluate the evidence supporting the pattern component of the theory of evolution, though, it's important to recognize that no single observation or experiment instantly "proved" the fact of evolution and swept aside belief in special creation. Rather, data from many different sources are much more consistent with evolution than with special creation. Descent with modification is a more successful and powerful scientific theory because it explains observations—such as vestigial traits and the close relationships among species on neighboring islands—that special creation does not.

What about the process component of the theory of evolution by natural selection? If the limbs of bats and humans were not created independently and recently, how did they come to be?

24.3 The Process of Evolution: How Does Natural Selection Work?

Darwin's greatest contribution did not lie in recognizing the fact of evolution. Lamarck and other researchers had already proposed evolution as a pattern in nature long before Darwin began his work. Instead, Darwin's crucial insight lay in recognizing a process, called **natural selection**, that could explain the pattern of descent with modification.

TABLE 24.1 **Evidence for Evolution**

Prediction 1: Species Are Not Static, but Change through Time

- Most species have gone extinct.
- Fossil (extinct) species frequently resemble living species found in the same area.
- Transitional features document change in traits through time.
- Vestigial traits are common.
- The characteristics of populations can be observed changing today.

Prediction 2: Species Are Related, Not Independent

- Closely related species often live in the same geographic area.
- Homologous traits are common and are recognized at three levels:
  1. genetic (gene structure and the genetic code)
  2. developmental (embryonic structures and processes)
  3. structural (morphological traits in adults)
- The formation of new species, from preexisting species, can be observed today.

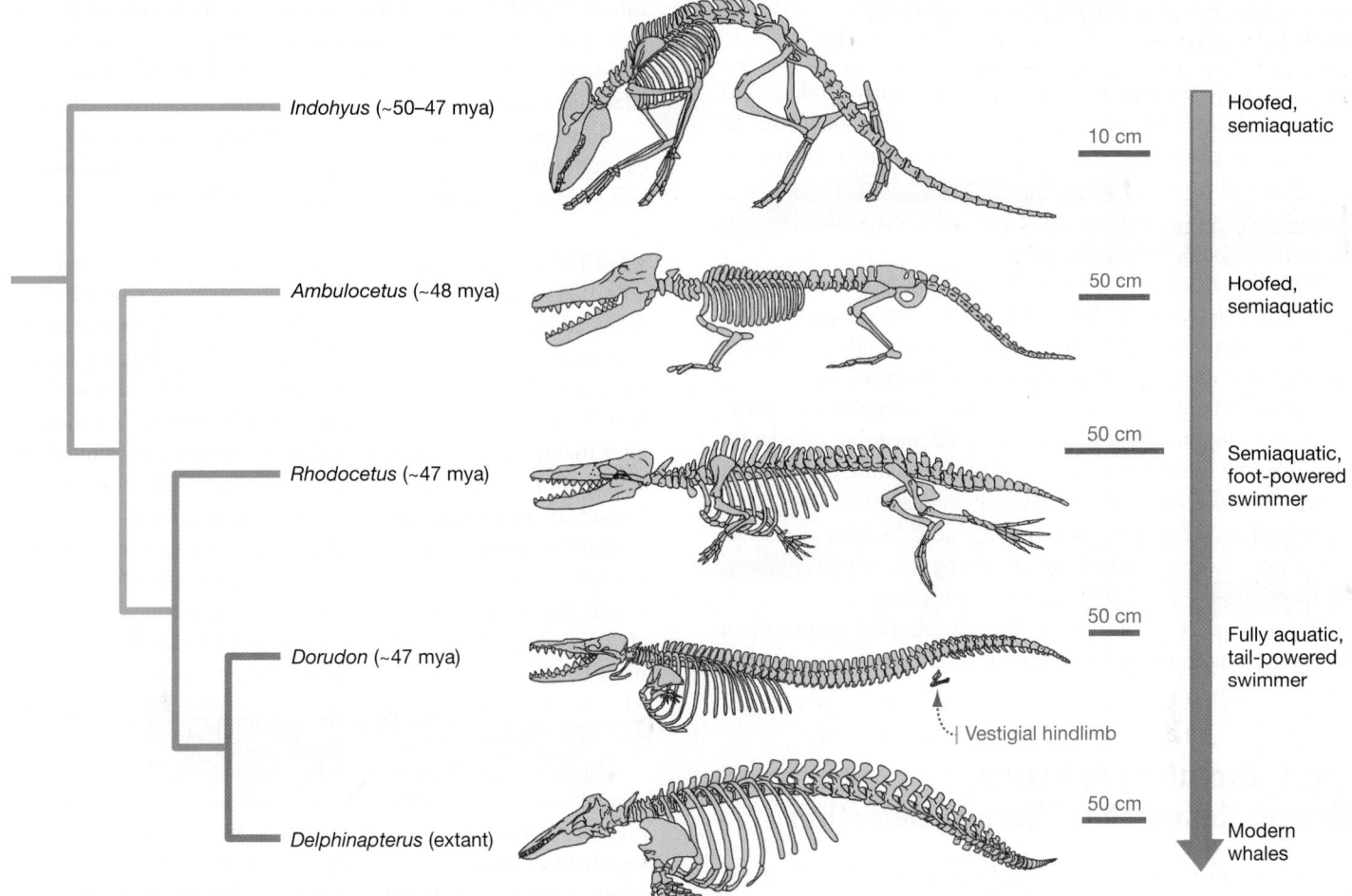

FIGURE 24.11 Data on Evolution from Independent Sources Are Consistent. This phylogeny of fossil cetaceans is consistent with data from relative dating, absolute dating, and phylogenies estimated from molecular traits in living species—all agree that whales evolved from terrestrial ancestors that were related to today's hippos.

Darwin's Four Postulates

In his original formulation, Darwin broke the process of evolution by natural selection into four simple postulates—meaning, steps in a logical sequence:

1. The individual organisms that make up a population vary in the traits they possess, such as their size and shape.

2. Some of the trait differences are heritable, meaning that they are passed on to offspring genetically. For example, tall parents may tend to have tall offspring.

3. In each generation, many more offspring are produced than can possibly survive. Thus, only some individuals in the population survive long enough to produce offspring; and among the individuals that produce offspring, some will produce more than others.

4. The subset of individuals that survive best and produce the most offspring is not a random sample of the population. Instead, individuals with certain heritable traits are more likely to survive and reproduce. Natural selection occurs when individuals with certain characteristics produce more offspring than do individuals without those characteristics. The individuals are selected naturally—meaning, by the environment.

Because the selected traits are passed on to offspring, the frequency of the selected traits increases from one generation to the next. Evolution is simply the outcome of these four steps. Each of the steps, and evolution, is observable. Evolution is defined as a change in allele frequencies in a population over time.

In studying these steps, you should realize that variation among individuals is essential if evolution is to occur. You should also recognize that Darwin had to introduce population thinking into biology because it is populations that change over time when evolution occurs. To come up with these postulates and understand their consequences, Darwin had to think in a revolutionary way.

Today, biologists usually condense Darwin's four postulates into a two-part statement that communicates the essence of evolution by natural selection more forcefully: Evolution by natural selection occurs when (**1**) heritable variation leads to (**2**) differential reproductive success.

The Biological Definitions of Fitness and Adaptation

To explain the process of natural selection, Darwin referred to successful individuals as "more fit" than other individuals. In doing so, he gave the word fitness a definition different from its everyday English usage. **Biological fitness** is the ability of an individual to produce surviving offspring, relative to that ability in other individuals in the population.

Note that fitness is a measurable quantity. When researchers study a population in the lab or in the field, they can estimate the relative fitness of individuals by counting the number of offspring each individual produces and comparing the data.

The concept of fitness, in turn, provides a compact way of formally defining adaptation. The biological meaning of adaptation, like the biological meaning of fitness, is different from its normal English usage. In biology, an adaptation is a heritable trait that increases the fitness of an individual in a particular environment relative to individuals lacking the trait. Adaptations increase fitness—the ability to produce offspring.

To summarize, evolution by natural selection occurs when heritable variation in traits leads to differential reproductive success among individuals.

24.4 Evolution in Action: Recent Research on Natural Selection

The theory of evolution by natural selection is testable. If the theory is correct, biologists should be able to test the validity of each of Darwin's postulates—documenting heritable variation and differential reproductive success in a wide array of natural populations.

This section summarizes two examples in which evolution by natural selection is being observed in nature. Literally hundreds of other case studies are available, involving a wide variety of traits and organisms. To begin, let's explore the evolution of drug resistance: one of the great challenges facing today's biomedical researchers and physicians.

Case Study 1: How Did *Mycobacterium tuberculosis* Become Resistant to Antibiotics?

Mycobacterium tuberculosis, the bacterium that causes **tuberculosis**, or TB, has long been a scourge of humankind. In Europe and the U.S., TB was once as great a public health issue as cancer is now. It receded in importance during the early 1900s, though, for two reasons:

1. Advances in nutrition made people better able to fight off most *M. tuberculosis* infections quickly.

2. The development of antibiotics such as rifampin allowed physicians to stop even advanced infections.

In the late 1980s, however, rates of *M. tuberculosis* infection surged in many countries, and in 1993 the World Health Organization (WHO) declared TB a global health emergency. Physicians were particularly alarmed because the strains of *M. tuberculosis* responsible for the increase were largely or completely resistant to rifampin and other antibiotics that were once extremely effective.

How and why did the evolution of drug resistance occur? The case of a single patient—a young man who lived in Baltimore—will illustrate what is happening all over the world.

A PATIENT HISTORY The story begins when the individual was admitted to the hospital with fever and coughing. Chest X-rays, followed by bacterial cultures of fluid ejected from the lungs, showed that he had an active TB infection. He was given several antibiotics for 6 weeks, followed by twice-weekly doses of rifampin and isoniazid for an additional 33 weeks. Ten months after therapy started, bacterial cultures from his chest fluid indicated no *M. tuberculosis* cells. His chest X-rays were also normal. The antibiotics seemed to have cleared the infection.

Just two months after the TB tests proved normal, however, the young man was readmitted to the hospital with a fever, severe cough, and labored breathing. Despite being treated with a variety of antibiotics, including rifampin, he died of respiratory failure 10 days later. Samples of material from his lungs showed that *M. tuberculosis* was again growing actively there. But this time the bacterial cells were completely resistant to rifampin.

Drug-resistant bacteria had killed this patient. Where did they come from? Is it possible that a strain that was resistant to antibiotic treatment evolved *within* him? To answer this question, a research team analyzed DNA from the drug-resistant strain and compared it with stored DNA from *M. tuberculosis* cells that had been isolated a year earlier from the same patient. After examining extensive stretches from each genome, the biologists were able to find only one difference: a point mutation in a gene called *rpoB*.

A MUTATION IN A BACTERIAL GENE CONFERS RESISTANCE The *rpoB* gene codes for a component of the enzyme RNA polymerase. Recall from Chapter 16 that RNA polymerase transcribes DNA to mRNA, and that a point mutation is a single base change in DNA (see Chapter 15). In this case, the mutation in the bacterial gene changed a cytosine to a thymine, altering the normal codon TCG to a mutant one, TTG (**Figure 24.12**). As a result, the bacterial RNA polymerase produced by the drug-resistant strain had leucine instead of serine at the 153rd amino acid in the polypeptide chain.

This result is meaningful. Rifampin, the drug that was being used to treat the patient, works by binding to the RNA polymerase of *M. tuberculosis*. When the drug enters an *M. tuberculosis* cell and binds to RNA polymerase, it interferes with transcription. If sufficient quantities of rifampin are present for long enough and if the drug binds tightly, bacterial cells don't make proteins efficiently and they produce few offspring. But apparently the substitution of a leucine for a serine prevents rifampin from binding efficiently. Consequently, cells with the C → T mutation continue to produce offspring efficiently even in the presence of the drug.

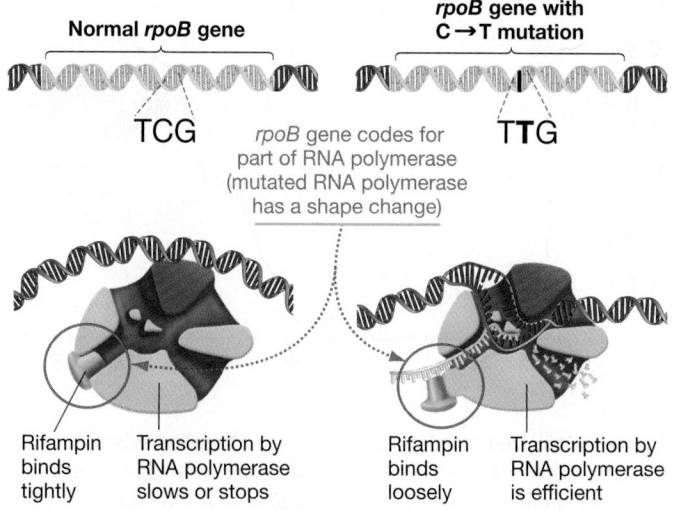

FIGURE 24.12 Mutation in *rpoB* gene Confers Drug Resistance. The drug rifampin cannot bind tightly to RNA polymerase in *M. tuberculosis* cells with the C → T mutation in the *rpoB* gene. As a result, the drug-resistant bacteria continue to reproduce efficiently, even in the presence of the drug.

Normal *rpoB* gene

TCG

rpoB gene codes for part of RNA polymerase (mutated RNA polymerase has a shape change)

rpoB gene with C → T mutation

TTG

Rifampin binds tightly — Transcription by RNA polymerase slows or stops

Rifampin binds loosely — Transcription by RNA polymerase is efficient

These results suggest that a chain of events led to this patient's death (**Figure 24.13**).

1. By chance, one or a few of the cells present in the patient, before drug therapy, started, happened to have an *rpoB* gene with the C → T mutation. Under normal conditions, mutant forms of RNA polymerase do not work as well as the more common form, so cells with the C → T mutation would not produce many offspring and would stay at low frequency—even while the overall population grew to the point of inducing symptoms that sent the young man to the hospital.

2. Therapy with rifampin began. In response, cells in the population with normal RNA polymerase began to grow much more slowly or to die outright. As a result, the overall bacterial population declined in size so drastically that the patient appeared to be cured—his symptoms began to disappear.

3. Cells with the C → T mutation had an advantage in the new environment. They began to grow more rapidly than the normal cells and continued to increase in number after therapy ended. Eventually the *M. tuberculosis* population regained its former abundance, and the patient's symptoms reappeared.

4. Drug-resistant cells now dominated the population, so the second round of rifampin therapy was futile.

Note that in most individuals, the immune system is able to eliminate the few bacteria that remain at step 2. This individual had AIDS, however, so his immune system was damaged.

✔ If you understand these concepts, you should be able to explain: (1) Why the relapse in step 3 occurred, and (2) whether a family member or health care worker who got TB from this patient at step 3 or step 4 would respond to drug therapy.

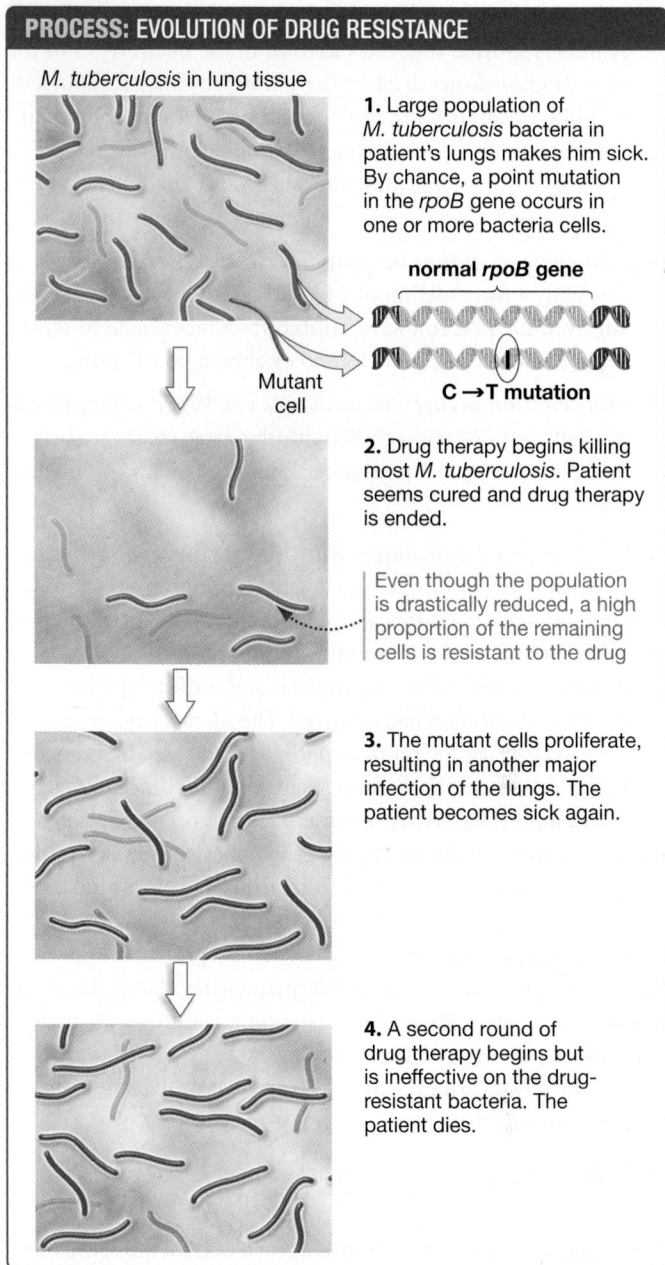

PROCESS: EVOLUTION OF DRUG RESISTANCE

M. tuberculosis in lung tissue

1. Large population of *M. tuberculosis* bacteria in patient's lungs makes him sick. By chance, a point mutation in the *rpoB* gene occurs in one or more bacteria cells.

Mutant cell

normal *rpoB* gene

C → T mutation

2. Drug therapy begins killing most *M. tuberculosis*. Patient seems cured and drug therapy is ended.

Even though the population is drastically reduced, a high proportion of the remaining cells is resistant to the drug

3. The mutant cells proliferate, resulting in another major infection of the lungs. The patient becomes sick again.

4. A second round of drug therapy begins but is ineffective on the drug-resistant bacteria. The patient dies.

FIGURE 24.13 Alleles That Confer Drug Resistance Increase in Frequency When Drugs Are Used.

TESTING DARWIN'S POSTULATES Does the sequence of events illustrated in Figure 24.13 mean that evolution by natural selection occurred? One way of answering this question is to review Darwin's four postulates and test whether each was verified:

1. ***Did variation exist in the population?*** The answer is yes. Due to mutation, both resistant and nonresistant strains of TB were present prior to administration of the drug. Most *M. tuberculosis* populations, in fact, exhibit variation for the trait; studies on cultured *M. tuberculosis* show that a mutation conferring resistance to rifampin is present in one out of every 10^7 to 10^8 cells.

2. **Was this variation heritable?** The answer is yes. The researchers showed that the variation in the phenotypes of the two strains—from drug susceptibility to drug resistance—was due to variation in their genotypes. Because the mutant *rpoB* gene is passed on to daughter cells when a *Mycobacterium* replicates, the allele and the phenotype it produces—drug resistance—are passed on to offspring.

3. **Was there variation in reproductive success?** The answer is yes. Only a tiny fraction of *M. tuberculosis* cells in the patient survived the first round of antibiotics long enough to reproduce. Most cells died and left no or almost no offspring.

4. **Did selection occur?** The answer is yes. When rifampin was present, certain cells—those with the drug-resistant allele—had higher reproductive success than cells with the normal allele.

M. tuberculosis individuals with the mutant *rpoB* gene had higher fitness in an environment where rifampin was present. The mutant allele produces a protein that is an adaptation when the cell's environment contains the antibiotic.

This study verified all four postulates and confirmed that evolution by natural selection had occurred. The *M. tuberculosis* population evolved because the mutant *rpoB* allele increased in frequency.

It is critical to note, however, that the individual cells themselves did not evolve. When natural selection occurred, the individual cells did not change through time; they simply survived or died, or produced more or fewer offspring. This is a fundamentally important point: Natural selection acts on individuals, because individuals experience differential reproductive success. But only populations evolve. Allele frequencies change in populations, not in individuals. Understanding evolution by natural selection requires population thinking.

To review how drug resistance evolves, go to the study area at *www.masteringbiology.com*.

 Web Activity Natural Selection for Antibiotic Resistance

A WIDESPREAD PROBLEM The events reviewed for a single patient have occurred many times in other patients. Recent surveys indicate that drug-resistant strains now account for about 10 percent of the *M. tuberculosis*–causing infections throughout the world.

Unfortunately, the emergence of drug resistance in TB is far from unusual. Resistance to a wide variety of insecticides, fungicides, antibiotics, antiviral drugs, and herbicides has evolved in hundreds of insects, fungi, bacteria, viruses, and plants. In every case, evolution has occurred because individuals with the heritable ability to resist some chemical compound were present in the original population. As the susceptible individuals die from the pesticide, herbicide, or drug, the resistance alleles increase in frequency.

To drive home the prevalence of evolution in response to drugs and other human-induced changes in the environment, consider the data in **Figure 24.14**. The graph shows changes through time in the percentage of infections, in intensive care

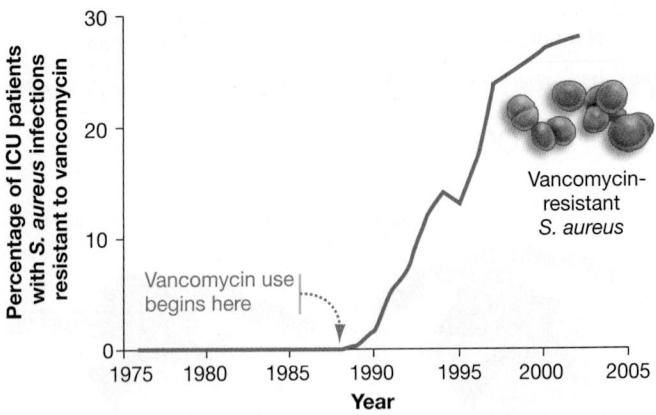

FIGURE 24.14 Trends in Infections Due to Antibiotic-resistant Bacteria. These data are from large hospitals in the USA. The line indicates changes in the percentage of *S. aureus* infections—acquired in the hospitals—that are resistant to the antibiotic vancomycin.

units in the United States, caused by strains of the bacterium *Staphylococcus aureus* that are resistant to the antibiotic vancomycin. Most of these *S. aureus* cells are also resistant to methicillin and other antibiotics as well—a phenomenon known as multi-drug resistance. In some cases, physicians have no effective antibiotics available to treat these infections.

Case Study 2: Why Are Beak Size, Beak Shape, and Body Size Changing in Galápagos Finches?

Can biologists study evolution in response to natural environmental change—when humans are not involved? The answer is yes. As an example, consider research led by Peter and Rosemary Grant. These biologists have been investigating changes in beak size, beak shape, and body size that have occurred in finches native to the Galápagos Islands.

The medium ground finch makes its living by eating seeds. Finches crack seeds with their beaks. For well over three decades, the population of medium ground finches on Isle Daphne Major of the Galápagos has been studied intensively by the Grants' team (**Figure 24.15a**). Because Daphne Major is small—about the size of 80 football fields (see **Figure 24.15b**)—the researchers have been able to catch, weigh, and measure all individuals and mark each one with a unique combination of colored leg bands.

Early studies of the finch population established that beak size and shape and body size vary among individuals, and that beak morphology and body size are heritable. Stated another way, parents with particularly deep beaks tend to have offspring with deep beaks. Large parents also tend to have large offspring. Beak size and shape and body size are traits with heritable variation.

SELECTION DURING DROUGHT CONDITIONS Not long after the team began to study the population, a dramatic selection event occurred. In the annual wet season of 1977, Daphne Major re-

(a) Medium ground finches (*Geospiza fortis*)

Male

Female

(b) Isle Daphne Major of the Galápagos

FIGURE 24.15 Studying Evolution-in-Action on the Galápagos. The Galápagos Islands are the tops of undersea volcanoes.

ceived just 24 mm of rain instead of the 130 mm that normally falls. During the drought, few plants were able to produce seeds, and 84 percent (about 660 individuals) of the medium ground finch population disappeared.

Two observations support the hypothesis that most or all of these individuals died of starvation:

● The researchers found a total of 38 dead birds, and all were emaciated.

● None of the missing individuals were spotted on nearby islands, and none reappeared once the drought had ended and food supplies returned to normal.

The research team realized that the die-off was a **natural experiment**. Instead of comparing groups created by direct manipulation under controlled conditions, natural experiments allow researchers to compare treatment groups created by an unplanned change in conditions. In this case, the Grants' team could test whether natural selection occurred by comparing the population before and after the drought.

Were the survivors different from nonsurvivors? The histograms in **Figure 24.16** show the distribution of beak sizes in the population before and after the drought. (For more on how histograms are constructed, see **BioSkills 2** in Appendix A.) On average, survivors tended to have much deeper beaks than did the birds that died.

EXPERIMENT

QUESTION: Did natural selection on ground finches occur when the environment changed?

HYPOTHESIS: Beak characteristics changed in response to a drought.

NULL HYPOTHESIS: No changes in beak characteristics occurred in response to a drought.

EXPERIMENTAL SETUP:

Weigh and measure all birds in the population before and after the drought.

PREDICTION:

PREDICTION OF NULL HYPOTHESIS:

RESULTS:

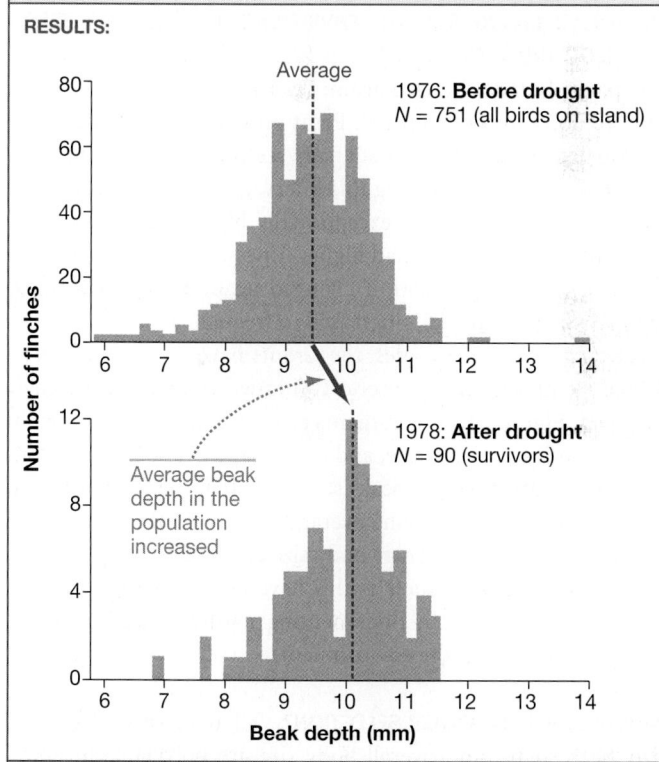

CONCLUSION: Natural selection occurred. The characteristics of the population have changed.

FIGURE 24.16 A Natural Experiment: Changes in a Medium Ground Finch Population in Response to a Change in the Environment (a Drought). The results show the distribution of beak depth in the population of medium ground finches on Daphne Major before and after the drought of 1977. *N* is the sample size.

SOURCE: Boag, P. T. and P. R. Grant. 1981. Intense natural selection in a population of Darwin's finches (Geospizinae) in the Galápagos. *Science* 214: 82-85.

✔**EXERCISE** Fill in the predictions made by the two hypotheses.

This was an important finding, because the type of seeds available to the finches had changed dramatically as the drought continued. At the drought's peak, most seed sources were absent and the tough fruits of a plant called *Tribulus cistoides* served as the finches' primary food source. These fruits are so difficult to crack that they are ignored in years when food supplies are normal. The group hypothesized that individuals with particularly large and deep beaks were more likely to crack these fruits efficiently enough to survive.

At this point, the Grants had shown that natural selection led to an increase in average beak depth in the population. When breeding resumed in 1978, the offspring that were produced had beaks that were half a millimeter deeper, on average, than those in the population that existed before the drought. This result confirmed that evolution had occurred.

In only one generation, natural selection led to a measurable change in the characteristics of the population. Alleles that led to the development of deep beaks had increased in frequency in the population. Large, deep beaks were an adaptation for cracking large fruits and seeds.

CONTINUED CHANGES IN THE ENVIRONMENT, CONTINUED SELECTION, CONTINUED EVOLUTION   In 1983, the environment on the Galápagos Islands changed again. Over a seven-month period, a total of 1359 mm of rain fell. Plant growth was luxuriant, and finches fed primarily on small, soft seeds that were being produced in abundance. During this interval, small individuals with small, pointed beaks had exceptionally high reproductive success—meaning that they had higher fitness. As a result, the characteristics of the population changed again. Alleles associated with small, pointed beaks increased in frequency.

Over subsequent decades, the Grants have documented continued evolution in response to continued changes in the environment. **Figure 24.17** documents changes that have occurred in average body size, beak size, and beak shape over 35 years. From 1972 to 2006, average beak size declined. Beak shape also changed dramatically—on average, finch beaks got much pointier. In addition, average body size got smaller.

Long-term studies such as this have been powerful, because they have succeeded in documenting natural selection in response to changes in the environment.

WHICH GENES ARE UNDER SELECTION?   Characteristics like beak size, beak shape, and overall body size are polygenic, meaning that many genes—each one exerting a relatively small effect—influence the trait (see Chapter 13). Because many genes are involved, it can be difficult for researchers to know exactly which alleles are changing in frequency when polygenic traits evolve.

To explore which medium ground finch genes might be under selection, researchers in Clifford Tabin's lab began studying beak development in an array of Galápagos finch species. More specifically, they looked for variation in the pattern of expression of cell-cell signals that had already been identified as important in the development of chicken beaks. The hope was that homologous genes might affect beak development in finches.

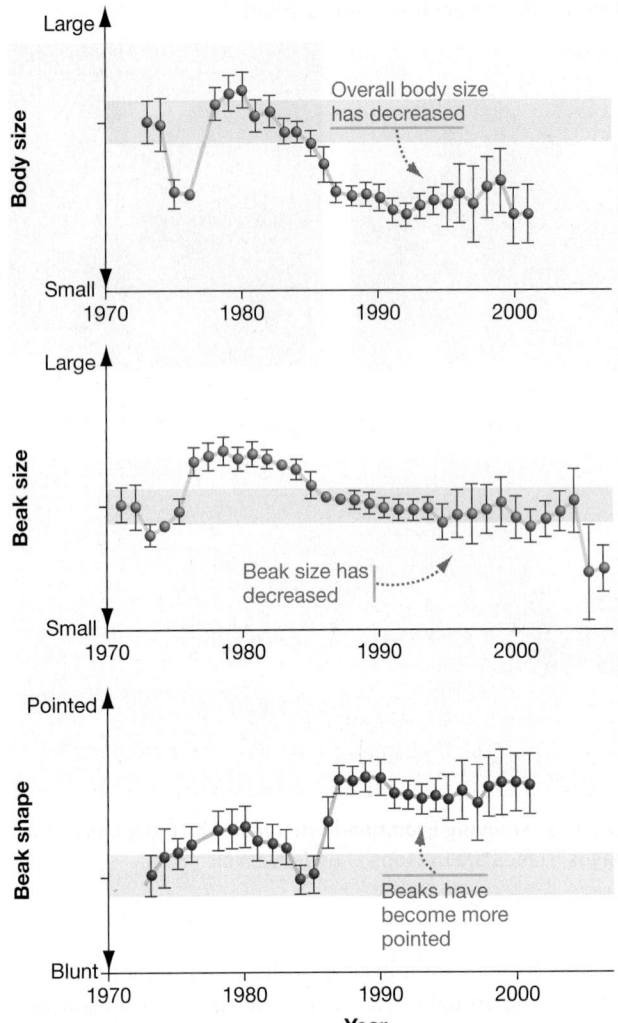

FIGURE 24.17 Body Size, Beak Size, and Beak Shape in Finches Changed over a 35-Year Interval.

The researchers struck pay dirt when they did in situ hybridizations—a technique featured in Chapter 21—showing where a cell-cell signal gene called *Bmp4* is expressed.

- There is a strong correlation between the amount of *Bmp4* expression when beaks are developing in young Galápagos finches and the width and depth of adult beaks (**Figure 24.18**).

- When the researchers experimentally increased *Bmp4* expression in young chickens, they found that beaks got wider and deeper than normal.

Similar experiments suggest that variation in alleles for a molecule called calmodulin, which is involved in calcium signaling during development, affects beak length.

Based on these data, biologists suspect that alleles associated with *Bmp4* and calmodulin expression may be under selection in the population of medium ground finches that the Grants are studying. If so, the research community will have made a direct connection between natural selection on phenotypes and evolutionary change in genotypes.

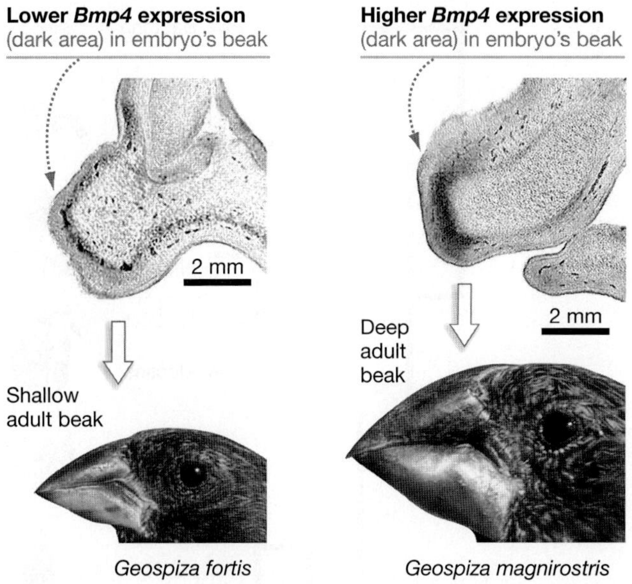

Lower *Bmp4* expression (dark area) in embryo's beak

Higher *Bmp4* expression (dark area) in embryo's beak

2 mm

2 mm

Shallow adult beak

Deep adult beak

Geospiza fortis

Geospiza magnirostris

FIGURE 24.18 Changes in *Bmp4* Expression Change Beak Depth and Width. These micrographs are in situ hybridizations (see Chapter 21) showing the location and extent of *Bmp4* expression in young *Geospiza fortis* and *G. magnirostris*. In these and four other species that were investigated, the amount of Bmp4 protein produced correlates with the depth and width of the adult beak.

CHECK YOUR UNDERSTANDING

If you understand that . . .

- If individuals with certain alleles produce the most offspring in a population, then those alleles increase in frequency over time. Evolution—a change in allele frequencies—results from this process (natural selection on heritable variation).

✓ You should be able to . . .

1. List Darwin's four postulates, and indicate which are related to heritable variation and which are related to differential reproductive success.

2. Explain how data on beak size and shape and body size of Galápagos finch populations provide examples of heritable variation and differential reproductive success.

Answers are available in Appendix B.

24.5 Common Misconceptions about Natural Selection and Adaptation

Evolution by natural selection is a simple process—just the logical outcome of some straightforward postulates. Ironically, it can be extremely difficult to understand.

Research has shown that evolution by natural selection is often misunderstood. To help clarify how the process works, let's consider some of the more common misconceptions about natural selection in light of data on drug resistance in the TB bacterium and changes in finch populations.

Selection Acts on Individuals, but Evolutionary Change Occurs in Populations

Perhaps the most important point to clarify about natural selection is that during the process, individuals do not change—only the population does. During the drought, the beaks of individual finches did not become deeper. Rather, the average beak depth in the population increased over time, because deep-beaked individuals produced more offspring than shallow-beaked individuals did. Natural selection acted on individuals, but the evolutionary change occurred in the characteristics of the population.

In the same way, individual bacterial cells did not change when rifampin was introduced to their environment. Each *M. tuberculosis* cell had the same polymerase alleles all its life. But because the mutant allele increased in frequency over time, the characteristics of the bacterial population changed.

This point should make sense, given that evolution is defined as changes in allele frequencies. An individual's allele frequencies cannot change over time—it has the alleles it was born with all its life.

A CONTRAST WITH "LAMARCKIAN" INHERITANCE There is a sharp contrast between evolution by natural selection and evolution by the inheritance of acquired characters—the hypothesis promoted by Jean-Baptiste de Lamarck. If you recall, Lamarck proposed that (1) individuals change in response to challenges posed by the environment, and (2) the changed traits are then passed on to offspring. The key claim is that the important evolutionary changes occur in individuals.

In contrast, Darwin realized that individuals do not change when they are selected. Instead, they simply produce more offspring than other individuals do. When this happens, alleles found in the selected individuals become more frequent in the population.

Darwin was correct: there is no mechanism that makes it possible for natural selection to change the nature of an allele inside an individual. An individual's heritable characteristics don't change when natural selection occurs. Natural selection just sorts existing variants—it doesn't change them.

ACCLIMATION IS *NOT* ADAPTATION The issue of change in individuals is tricky because individuals often *do* change in response to changes in the environment. For example, wood frogs native to northern North America are exposed to extremely cold temperatures as they overwinter. When ice begins to form in their skin, their bodies begin producing a sort of natural antifreeze—molecules that protect their tissues from being damaged by the ice crystals. These individuals are changing in response to a change in temperature.[1] You may have observed changes in your own body as you got accustomed to living at high elevation or in a particularly hot or cold environment.

---

[1]In some species of frogs, so much extracellular fluid freezes during cold snaps that individuals appear to be frozen solid. Their hearts also stop beating. When temperatures warm in the spring, their hearts start beating again, their tissues thaw, and they resume normal activities.

Biologists use the term **acclimation** to describe changes in an individual's phenotype that occur in response to changes in environmental conditions. The key is to realize that phenotypic changes due to acclimation are not passed on to offspring, because no alleles have changed in composition. As a result, acclimation does not cause evolution. ✓ If you understand this concept, you should be able to (1) explain the difference between the biological definition of adaptation and its use in everyday English, and (2) explain the difference between acclimation and adaptation.

Evolution Is Not Goal Directed

It is tempting to think that evolution by natural selection is goal directed. For example, you might hear a fellow student say that *M. tuberculosis* cells "wanted" or "needed" the mutant, drug-resistant allele so that they could survive and continue to reproduce in an environment that included rifampin. This does not happen. The mutation that created the mutant allele occurred randomly, due to an error during DNA synthesis, and it just happened to be advantageous when the environment changed.

Stated another way, the mutation that conferred resistance did not occur because of the presence of the drug. It just happened. Every mutation is equally likely to occur in every environment. There is no mechanism that makes it possible for the environment to direct which mistakes DNA polymerase makes when it copies genes. Adaptations do not occur because organisms want or need them.

EVOLUTION IS NOT PROGRESSIVE It is often tempting to think that evolution by natural selection is progressive—meaning organisms have gotten "better" over time. (In this context, *better* usually means bigger, stronger, or more complex.) It is true that the groups appearing later in the fossil record are often more morphologically complex or "advanced" than closely related groups that appeared earlier. Flowering plants are considered more complex than mosses, and most biologists would agree that the morphology of mammals is more complex than that of the first vertebrates in the fossil record. But there is nothing predetermined or absolute about this tendency.

In fact, complex traits are routinely lost or simplified over time as a result of evolution by natural selection. You've already analyzed evidence on limb loss in snakes (Chapter 21) and whales (this chapter). Chapter 20 presented data indicating that parasitic bacteria often lose large portions of their genomes.

Populations that become parasitic are particularly prone to loss of complex traits. Tapeworms, for example, lack a mouth and digestive system. As parasites that live in the intestines of humans and other mammals, they simply absorb nutrients directly from their environment, across their plasma membranes. But tapeworms evolved from species with a sophisticated digestive tract. Tapeworms lost their digestive tract as a result of evolution by natural selection, but they did not become bigger or stronger or more complex.

THERE IS NO SUCH THING AS A HIGHER OR LOWER ORGANISM The nonprogressive nature of evolution by natural selection contrasts

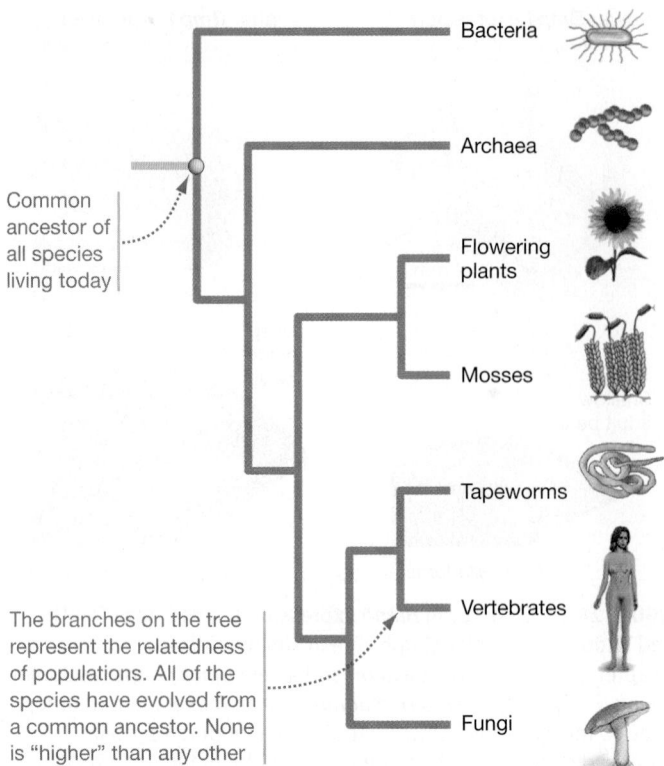

The branches on the tree represent the relatedness of populations. All of the species have evolved from a common ancestor. None is "higher" than any other

FIGURE 24.19 Evolution Produces a Tree of Life, Not a Progressive Ladder of Life. Under evolution by natural selection, species are related by common ancestry and all have evolved through time. (Not all branches of the tree of life are shown.)

sharply with Lamarck's conception of the evolutionary process, in which organisms progress over time to higher and higher levels on a chain of being.

Under Aristotle's and Lamarck's hypothesis, it is sensible to refer to "higher" and "lower" organisms. But under evolution by natural selection, there is no such thing as a higher or lower organism (**Figure 24.19**). Mosses may be a more ancient group than flowering plants, but neither group is higher or lower than the other. Mosses simply have a different suite of adaptations than do flowering plants, so they thrive in different types of environments. A human is no higher than its tapeworm parasite; each is well adapted to its environment.

All populations have evolved by natural selection based on their ability to gather resources and produce offspring. All organisms are adapted to their environment, and are related by common ancestry.

Natural selection is not goal directed or progressive. It simply favors individuals that happen to be better adapted to the environment existing at the time.

Organisms Do Not Act for the Good of the Species

Consider the widely circulated story that rodents called lemmings sacrifice themselves for the good of their species. The story claims that when lemming populations are high, overgrazing is so exten-

FIGURE 24.20 Self-Sacrificing Behavior Cannot Evolve if "Selfish" Alleles Exist. Suppose that most individuals in a lemming population had an allele that led to self-sacrificing behavior, or even suicide, for "the good of the species." Now suppose that this individual had alleles that prevented self-sacrifice. Which alleles would increase in frequency in the population over time?

sive that the entire species is threatened with starvation and extinction. In response, some individuals throw themselves into the sea and drown. This lowers the overall population size and allows the vegetation to recover enough to save the species. Even though individuals suffer, the good-of-the-species hypothesis maintains that the behavior evolved because the group benefits.

The lemming suicide story is false. Although lemmings do disperse from areas of high population density in order to find habitats with higher food availability, they do not throw themselves into the sea.

To understand why this type of self-sacrificing behavior does not occur, suppose that certain alleles predispose lemmings to sacrifice themselves for others. But consider what happens if alleles exist that prevent this type of behavior—what biologists call a "selfish" allele (**Figure 24.20**). Individuals with self-sacrificing alleles die and do not produce offspring. But individuals with selfish, cheater alleles survive and produce offspring. As a result, selfish alleles increase in frequency while self-sacrificing alleles decrease in frequency. Thus, it is not possible for individuals to sacrifice themselves for the good of the species.

No instance of purely self-sacrificing behavior—where the individual received no fitness benefit in return—has ever been recorded in nature. Chapter 52 provides additional details on this point.

Limitations of Natural Selection

Although organisms are often exquisitely adapted to their environment, adaptation is far from perfect. A long list of circumstances limits the effectiveness of natural selection; only a few of the most important are discussed here.

NON-ADAPTIVE TRAITS Vestigial traits such as the human coccyx (tailbone), goose bumps, and appendix do not increase the fitness of individuals with those traits. The structures are not adaptive. They exist simply because they were present in the ancestral population.

Vestigial traits are not the only types of structures with no function. Some adult traits exist as holdovers from structures that appear early in development. For example, human males have rudimentary mammary glands. The structures are not adaptive. They exist only because nipples form in the human embryo before sex hormones begin directing the development of male organs instead of female organs.

Perhaps the best example of nonadaptive traits involves evolutionary changes in DNA sequences. Recall from Chapter 16 that mutation may change a base in the third position of a codon without changing the amino acid sequence of the protein encoded by that gene. Changes such as these are said to be silent. They occur because of the redundancy of the genetic code (see Chapter 16). Silent changes in DNA sequences are extremely common. But because they don't change the phenotype, they can't be acted on by natural selection.

The general point here is that not all traits are adaptive. Evolution by natural selection does not lead to "perfection." Besides carrying an array of traits that have no function, the adaptations that organisms have are constrained in a variety of important ways.

GENETIC CONSTRAINTS The Grants' team analyzed data on the characteristics of finches that survived the 1977 drought, and the team made an interesting observation: Although individuals with deep beaks survived better than individuals with shallow beaks, birds with particularly narrow beaks survived better than individuals with wider beaks.

This observation made sense because finches crack *Tribulus* fruits by twisting them. Narrow beaks concentrate the twisting force more efficiently than wider beaks, so they are especially useful for cracking the fruits. But narrower beaks did not evolve in the population.

To explain why, the biologists noted that parents with deep beaks tend to have offspring with beaks that are both deep and wide. This is a common pattern. Many alleles that affect body size have an effect on all aspects of size—not just one structure or dimension. As a result, selection for increased beak depth overrode selection for narrow beaks, even though a deep and narrow beak would have been more advantageous.

The general point here is that selection was not able to optimize all aspects of a trait. In the case of the finches, wider beaks were not the best possible beak shape for individuals living in an arid habitat. Wider beaks evolved anyway, due to a type of constraint called a **genetic correlation**. Genetic correlations occur because of pleiotropy (see Chapter 13)—in which a single allele affects multiple traits. In this case, selection on alleles for one trait (increased beak depth) caused a correlated, though suboptimal, increase in another trait (beak width).

Genetic correlations are not the only genetic constraint on adaptation. Lack of genetic variation is also important. Consider that salamanders have the ability to regrow severed limbs. Some eels and sharks can sense electric fields. Birds can sense magnetic fields and see ultraviolet light. Even though it is possible that these traits would confer increased reproductive success in humans, they do not exist—because the requisite genes are lacking.

FITNESS TRADE-OFFS In everyday English, the term trade-off refers to a compromise between competing goals. It is difficult to design a car that is both large and fuel efficient, a bicycle that is both rugged and light, or a plane that is both fast and maneuverable.

In nature, selection occurs in the context of fitness trade-offs. A **fitness trade-off** is a compromise between traits, in terms of how those traits perform in the environment. During the drought in the Galápagos, for example, medium ground finches with large bodies had an advantage because they won fights over the few remaining sources of seeds. But individuals with large bodies also require large amounts of food to maintain their mass; they also tend to be slower and less nimble than smaller individuals. When food is short, large individuals are more prone to starvation. Even if large size is advantageous in an environment, there is always counteracting selection that prevents individuals from getting even bigger.

Biologists have documented trade-offs between the size of eggs or seeds that an individual makes and the number of offspring it can produce, between rapid growth and long life span, and between bright coloration and tendency to attract predators. The message of this research is simple: Because selection acts on many traits at once, every adaptation is a compromise.

HISTORICAL CONSTRAINTS In addition to being constrained by genetic correlations, lack of genetic variation, and fitness trade-offs, adaptations are constrained by history. The reason is simple: All traits have evolved from previously existing traits.

Natural selection acts on structures that originally had a very different function. For example, the tiny hammer, anvil, and stirrup bones found in your middle ear evolved from bones that were part of the jaw and braincase in the ancestors of mammals. These bones now function in the transmission and amplification of sound from your outer ear to your inner ear. Biologists rou-tinely interpret these bones as adaptations that improve your ability to hear airborne sounds. But are the bones a "perfect" so-lution to the problem of transmitting sound from the outside of the ear to the inside? The answer is no. They are the best solution possible, given an important historical constraint. Other verte-brates have different structures involved in transmitting sound to the ear. In at least some cases, those structures may be more effi-cient than our hammer, anvil, and stirrup.

To summarize, not all traits are adaptive, and even adaptive traits are constrained by genetic and historical factors. In addi-tion, natural selection is not the only process that causes evolu-tionary change. Chapter 25 introduces three other processes that change allele frequencies over time. Compared with natural se-lection, these processes have very different consequences.

CHECK YOUR UNDERSTANDING

If you understand that. . .

- Selection by drugs on the TB bacterium and changes in seed availability to finches in the Galápagos are well-studied examples of evolution by natural selection.
- Evolution by natural selection is simple in concept but widely misunderstood.

✔ You should be able to. . .

1. Explain why individuals do not change when natural selection occurs.
2. Explain why trade-offs and genetic and historical constraints prevent adaptations from being "perfect."

Answers are available in Appendix B.

CHAPTER 24 REVIEW

For media, go to the study area at www.masteringbiology.com

Summary of Key Concepts

Populations and species evolve, meaning that their heritable characteristics change through time. Evolution is change in allele frequencies over time.

- Data on (1) the resemblance of modern to fossil forms; (2) transitional features in fossils; (3) the fact of extinction; (4) the presence of vestigial traits; and (5) change in contemporary populations are inconsistent with the claim that species have remained unchanged through time.

- Data on (1) the geographic proximity of closely related species such as the Galápagos mockingbirds; (2) the existence of structural, de-velopmental, and genetic homologies; and (3) contemporary for-mation of new species are inconsistent with the theory that species were formed instantaneously, recently, and independently by a di-vine being.

- Evidence for evolution is internally consistent—meaning that data from several independent sources is mutually reinforcing.

✔You should be able to predict how changes in *Mycobacterium* populations would be explained under the theory of special cre-ation and under evolution by inheritance of acquired characters.

Evolution by natural selection occurs when individuals with certain alleles produce the most surviving offspring in a population. An adaptation is a genetically based trait that in-creases an individual's ability to produce offspring in a particu-lar environment.

- Evolution by natural selection occurs whenever genetically based differences among individuals lead to differences in their ability to reproduce—when heritable variation leads to differential repro-ductive success.

- Alleles or traits that increase the reproductive success of an individ-ual are said to increase the individual's fitness.

- A trait that leads to higher fitness, relative to individuals without the trait, is an adaptation.

- If a particular allele increases fitness and leads to adaptation, the allele will increase in frequency in the population.

- Evolution is an outcome of natural selection. Evolution by natural selection has been confirmed by a wide variety of studies and has long been considered to be the central organizing principle of biology.

  ✔ You should be able to explain the difference between the biological and everyday English definitions of fitness.

  (MB) **Web Activity** Natural Selection for Antibiotic Resistance

🔑 **Evolution by natural selection is not progressive, and it does not change the characteristics of the individuals that are selected—it changes only the characteristics of the population. Animals do not do things for the good of the species, and not all traits are adaptive. All adaptations are constrained by trade-offs and genetic and historical factors.**

- Individuals that are naturally selected are not changed by the process—they simply produce more offspring than other individuals do.

- Traits that increase in frequency under natural selection do so because they improve fitness, not because they are necessarily larger or more complex.

- Because natural selection acts only on existing traits, it does not lead to perfection.

  ✔ You should be able to discuss how an adaptation—such as the large brains of *Homo sapiens* or the ability of falcons to fly very fast—is constrained.

Questions

1. How can biological fitness be estimated?
   a. Document how long different individuals in a population survive.
   b. Count the number of offspring produced by different individuals in a population.
   c. Determine which individuals are strongest.
   d. Determine which phenotype is the most common one in a given population.

2. Why are some traits considered vestigial?
   a. They improve the fitness of an individual who bears them, compared with the fitness of individuals without those traits.
   b. They change in response to environmental influences.
   c. They existed long ago.
   d. They are reduced in size, complexity, and function compared with traits in related species.

3. What is an adaptation?
   a. a trait that improves the fitness of its bearer, compared with individuals without the trait
   b. a trait that changes in response to environmental influences within the individual's lifetime
   c. an ancestral trait—one that was modified to form the trait observed today
   d. the ability to produce offspring

4. Why does the presence of extinct forms and transitional features in the fossil record support the pattern component of the theory of evolution by natural selection?
   a. It supports the hypothesis that individuals change over time.
   b. It supports the hypothesis that weaker species are eliminated by natural selection.
   c. It supports the hypothesis that species evolve to become more complex and better adapted over time.
   d. It supports the hypothesis that species have changed through time.

5. Why are homologous traits similar?
   a. They are derived from a common ancestor.
   b. They are derived from different ancestors.
   c. They result from convergent evolution.
   d. Their appearance, structure, or development is similar.

6. According to data presented in this chapter, which of the following statements is correct?
   a. When individuals change in response to challenges from the environment, their altered traits are passed on to offspring.
   b. Species are created independently of each other and do not change over time.
   c. Populations—not individuals—change when natural selection occurs.
   d. The Earth is young, and most of today's landforms were created during the floods at the time of Noah.

1. Compare and contrast the theory of evolution by natural selection and the theory of special creation and evolution by inheritance of acquired characters. What testable predictions does each make?

2. Some biologists encapsulate evolution by natural selection with the phrase "mutation proposes, selection disposes." Mutation is a process that creates heritable variation. Explain what the phrase means.

3. Review the section on the evolution of drug resistance in *Mycobacterium tuberculosis.*

- What evidence do researchers have that a drug-resistant strain evolved in the patient analyzed in their study, instead of having been transmitted from another infected individual?

- If the antibiotic rifampin were banned, would the mutant *rpoB* gene have lower or higher fitness in the new environment? Would strains carrying the mutation continue to increase in frequency in *M. tuberculosis* populations?

4. Compare and contrast typological thinking with population thinking. Why was Darwin's emphasis on the importance of variation among individuals so crucial to his theory, and why was it a revolutionary idea in Western science?

5. The evidence supporting the pattern component of the theory of evolution can be criticized on the grounds that it is indirect. For example, no one has directly observed the formation of a vestigial trait over time. Is indirect evidence for a scientific theory legitimate? Why or why not?

6. Why isn't evolution by natural selection progressive? Why don't the biggest and strongest individuals in a population always produce the most offspring?

✓ APPLYING CONCEPTS TO NEW SITUATIONS

Answers are available in Appendix B

1. The geneticist James Crow wrote that successful scientific theories have the following characteristics: (1) They explain otherwise puzzling observations; (2) they provide connections between otherwise disparate observations; (3) they make predictions that can be tested; and (4) they are heuristic, meaning that they open up new avenues of theory and experimentation. Crow added two other elements that he considered important on a personal, emotional level: (5) They should be elegant, in the sense of being simple and powerful; and (6) they should have an element of surprise. How well does the theory of evolution by natural selection fulfill these six criteria?

2. The average height of humans has increased steadily for the past 100 years in industrialized nations. This trait has clearly changed over time. Most physicians and human geneticists hypothesize that the change is due to better nutrition and a reduced incidence of disease. Has human height evolved?

3. Genome sequencing projects may dramatically affect how biologists analyze evolutionary changes in quantitative traits. For example, suppose that the genomes of many living humans are sequenced and that genomes could be sequenced from many people who lived 100 years ago. (That might be possible with preserved tissue.) If 20 genes have been shown to influence height, how could you use the sequence data from these genes to test the hypothesis that human height has evolved in response to natural selection?

4. In some human populations, individuals tan in response to exposure to sunlight. Tanning is an acclimation response to a short-term change in the environment. It is adaptive because it prevents sunburn; it may also prevent a vitamin called folate from being destroyed by sunlight. The ability to tan varies among individuals in these populations, however. Is the ability to tan an acclimation or an adaptation? Explain your logic.

A male raggiana bird of paradise, left, displays for a female, right. His long, colorful feathers and dramatic behavior result from sexual selection—a process introduced in this chapter.

Evolutionary Processes 25

C hapter 24 defined evolution as a change in allele frequencies. One of the key concepts from that chapter was that even though natural selection acts on individuals, evolutionary change occurs in **populations**. A population is a group of individuals from the same species that live in the same area and regularly interbreed.

Natural selection is not the only process that causes evolution, however. There are actually four mechanisms that shift allele frequencies in populations:

1. *Natural selection* increases the frequency of certain alleles—the ones that contribute to reproductive success in a particular environment.

2. *Genetic drift* causes allele frequencies to change randomly. In some cases, drift may cause alleles that decrease fitness to increase in frequency.

3. *Gene flow* occurs when individuals leave one population, join another, and breed. Allele frequencies may change when gene flow occurs, because arriving individuals introduce alleles to their new population and departing individuals remove alleles from their old population.

4. *Mutation* modifies allele frequencies by continually introducing new alleles. The alleles created by mutation may be beneficial or detrimental or have no effect on fitness.

This chapter has two fundamental messages: Natural selection is not the only agent responsible for evolution, and each of the four evolutionary processes has different consequences. Natural selection is the only mechanism that acting alone can result in adaptation. Mutation, gene flow, and genetic drift do not favor certain alleles over others. Mutation and drift introduce a nonadaptive component into evolution.

Let's take a closer look at the four evolutionary processes by examining a null hypothesis—what happens to allele frequencies when the evolutionary mechanisms are *not* operating.

KEY CONCEPTS

- The Hardy-Weinberg principle acts as a null hypothesis when researchers want to test whether evolution or nonrandom mating is occurring at a particular gene.

- Each of the four evolutionary mechanisms has different consequences. Only natural selection produces adaptation. Genetic drift causes random fluctuations in allele frequencies. Gene flow equalizes allele frequencies between populations. Mutation introduces new alleles.

- Inbreeding changes genotype frequencies but does not change allele frequencies.

- Sexual selection leads to the evolution of traits that help individuals attract mates. It is usually stronger on males than on females.

✔ When you see this checkmark, stop and test yourself. Answers are available in Appendix B.

25.1 Analyzing Change in Allele Frequencies: The Hardy-Weinberg Principle

To study how the four evolutionary processes affect populations, biologists take a three-pronged approach.

1. They create mathematical models that predict the fate of alleles over time under various conditions.

2. They collect data to test predictions made by the models.

3. They apply the results to solve problems in human genetics, conservation of endangered species, or other fields.

This research strategy began in 1908, when G. H. Hardy and Wilhelm Weinberg each published a major result independently. At the time, it was commonly believed that changes in allele frequency occur simply as a result of sexual reproduction—meiosis followed by the random fusion of gametes (eggs and sperm) to form offspring. Some biologists claimed that dominant alleles inevitably increase in frequency when gametes combine at random. Others predicted that two alleles of the same gene inevitably reach a frequency of 0.5.

To test these hypotheses, Hardy and Weinberg analyzed what happens to the frequencies of alleles when many individuals in a population mate and produce offspring. Instead of thinking about the consequences of a mating between two parents with a specific pair of genotypes, as we did with Punnett squares in Chapter 13, Hardy and Weinberg wanted to know what happened in an entire population, when *all* of the individuals—and thus all possible genotypes—bred. Like Darwin, Hardy and Weinberg were engaged in population thinking.

The Gene Pool Concept

To analyze the consequences of matings among all of the individuals in a population, Hardy and Weinberg invented a novel approach: They imagined that all of the gametes produced in each generation go into a single group called the **gene pool** and then combine at random to form offspring. Something very much like this happens in species like clams and sea stars and sea urchins, which release their gametes into the water, where they mix randomly with gametes from other individuals in the population and combine to form zygotes.

To determine which genotypes would be present in the next generation and in what frequency, Hardy and Weinberg simply had to calculate what happened when two gametes were plucked at random out of the gene pool, many times, and each of these gamete pairs was then combined to form offspring. These calculations would predict the genotypes of the offspring that would be produced, as well as the frequency of each genotype.

Deriving the Hardy-Weinberg Principle

Hardy and Weinberg began by analyzing the simplest situation possible—that just two alleles of a particular gene exist in a population. Let's call these alleles A_1 and A_2. We'll use p to symbolize the frequency of A_1 alleles in the gene pool and q to symbolize

the frequency of A_2 alleles in the same gene pool. Because there are only two alleles, the two frequencies must add up to 1; that is, $p + q = 1$. Now follow the steps in **Figure 25.1**:

Step 1 Although p and q can have any value between 0 and 1, let's suppose that the initial frequency of A_1 is 0.7 and that of A_2 is 0.3.

Step 2 In this gene pool, 70 percent of the gametes carry A_1 and 30 percent carry A_2.

Step 3 Each time a gamete is involved in forming an offspring, there is a 70 percent chance that it carries A_1 and a 30 percent chance that it carries A_2. In general, there is a p chance that it carries A_1 and a q chance that it carries A_2.

Step 4 Because only two alleles are present, three genotypes are possible in the offspring generation: A_1A_1, A_1A_2, and A_2A_2. What will the frequency of these three genotypes be? According to the logic of Hardy's and Weinberg's result:

- The frequency of the A_1A_1 genotype is p^2.
- The frequency of the A_1A_2 genotype is $2pq$.
- The frequency of the A_2A_2 genotype is q^2.

The genotype frequencies in the offspring generation must add up to 1, which means that $p^2 + 2pq + q^2 = 1$. In our numerical example, $0.49 + 0.42 + 0.09 = 1$.

The last two steps in the figure show how the frequencies of alleles A_1 and A_2 are calculated from these genotype frequencies:

Step 5 The easiest way to calculate the allele frequencies in the offspring is to imagine that they form gametes that go into a gene pool. All of the gametes from A_1A_1 individuals carry A_1, so 49 percent (q^2) of the gametes in the gene pool will carry A_1. But half of the gametes from A_1A_2 will also carry A_1, so an additional $\frac{1}{2}(0.42) = 0.21$ (this is $\frac{1}{2} \times 2pq = pq$) gametes in the gene pool will carry A_1, for a total of $0.49 + 0.21 = 0.70$ or $q^2 + pq = q$ $(q + p) = q$. Use the same logic to figure out the frequency of A_2.

Step 6 In our example, the frequency of allele A_1 in the offspring generation is still 0.7 and the frequency of allele A_2 is still 0.3. Thus, the frequency of allele A_1 is still p and the frequency of allele A_2 is still q.

No allele frequency change occurred. Even if A_1 is dominant to A_2, it does not increase in frequency. And there is no trend toward both alleles reaching a frequency of 0.5. This result is called the **Hardy-Weinberg principle**.

Figure 25.2 illustrates the same result a little differently. The figure uses a Punnett square to predict the outcome of random mating—meaning, random combinations of all gametes in a population. (Recall that in Chapter 13 you used Punnett squares to predict the outcome of a mating between two individuals.) The outcome of this analysis is the same as in Figure 25.1.

The Hardy-Weinberg principle makes two fundamental claims:

1. If the frequencies of alleles A_1 and A_2 in a population are given by p and q, then the frequencies of genotypes A_1A_1, A_1A_2, and A_2A_2 will be given by p^2, $2pq$, and q^2 for generation after generation.

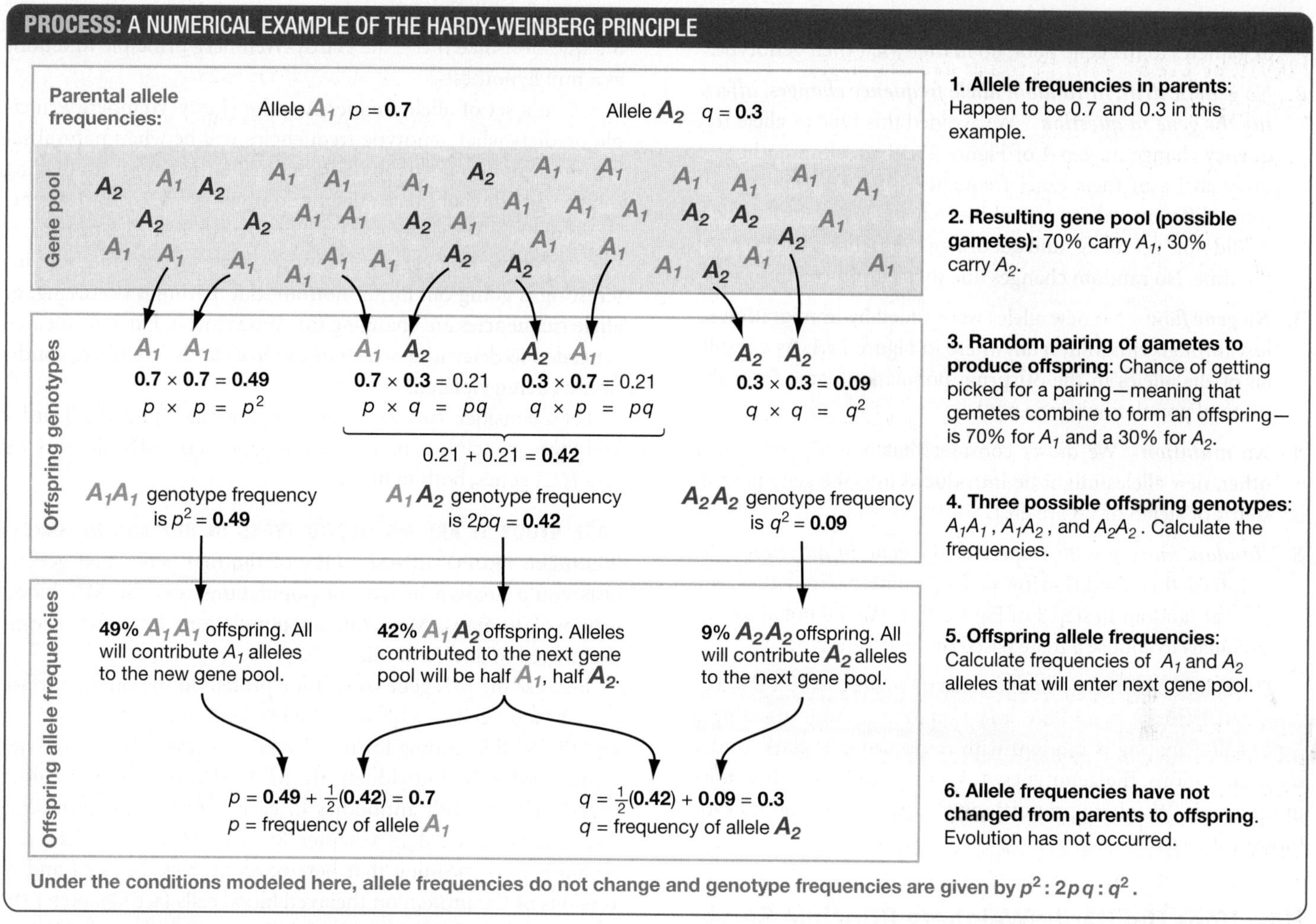

Parental allele frequencies: Allele A_1 $p = 0.7$ Allele A_2 $q = 0.3$

1. Allele frequencies in parents: Happen to be 0.7 and 0.3 in this example.

Gene pool

A_2 A_1 A_2 A_1 A_1 A_1 A_2 A_1 A_1 A_1 A_1 A_1 A_1

A_2 A_2 A_1 A_1 A_2 A_1 A_1 A_1 A_1 A_2 A_2 A_1

A_1 A_1 A_1 A_1 A_1 A_2 A_2 A_1 A_1 A_1 A_2 A_1

A_1

2. Resulting gene pool (possible gametes): 70% carry A_1, 30% carry A_2.

Offspring genotypes

A_1 A_1 A_1 A_2 A_2 A_1 A_2 A_2

$0.7 \times 0.7 = 0.49$ $0.7 \times 0.3 = 0.21$ $0.3 \times 0.7 = 0.21$ $0.3 \times 0.3 = 0.09$

$p \times p = p^2$ $p \times q = pq$ $q \times p = pq$ $q \times q = q^2$

$0.21 + 0.21 = 0.42$

A_1A_1 genotype frequency is $p^2 = 0.49$ A_1A_2 genotype frequency is $2pq = 0.42$ A_2A_2 genotype frequency is $q^2 = 0.09$

3. Random pairing of gametes to produce offspring: Chance of getting picked for a pairing—meaning that gemetes combine to form an offspring—is 70% for A_1 and a 30% for A_2.

4. Three possible offspring genotypes: A_1A_1, A_1A_2, and A_2A_2. Calculate the frequencies.

Offspring allele frequencies

49% A_1A_1 offspring. All will contribute A_1 alleles to the new gene pool.   42% A_1A_2 offspring. Alleles contributed to the next gene pool will be half A_1, half A_2.   9% A_2A_2 offspring. All will contribute A_2 alleles to the next gene pool.

5. Offspring allele frequencies: Calculate frequencies of A_1 and A_2 alleles that will enter next gene pool.

$p = 0.49 + \frac{1}{2}(0.42) = 0.7$ $q = \frac{1}{2}(0.42) + 0.09 = 0.3$

$p = $ frequency of allele A_1 $q = $ frequency of allele A_2

6. Allele frequencies have not changed from parents to offspring. Evolution has not occurred.

Under the conditions modeled here, allele frequencies do not change and genotype frequencies are given by $p^2 : 2pq : q^2$.

FIGURE 25.1 Deriving the Hardy-Weinberg Principle. To understand the logic behind calculating the frequency of A_1A_2 genotypes in step 4, see **BioSkills 13** in Appendix A.

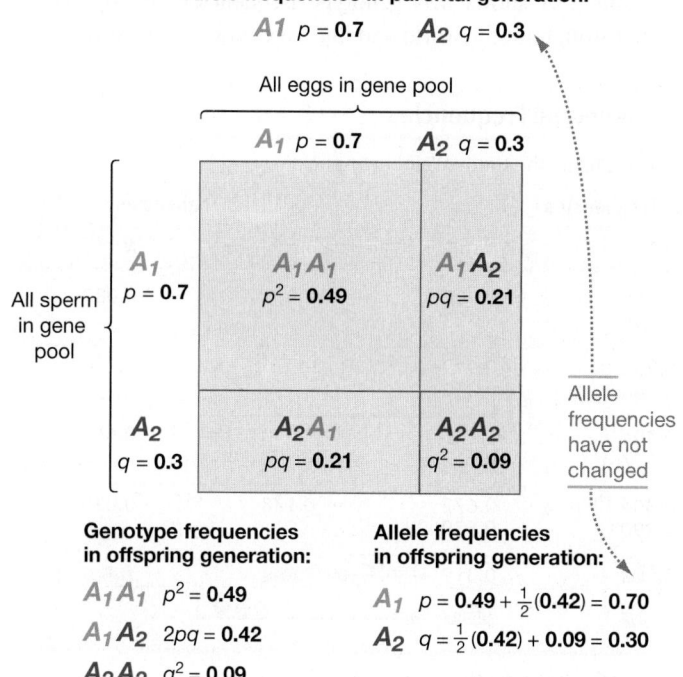

Allele frequencies in parental generation:

$A1$ $p = 0.7$ A_2 $q = 0.3$

All eggs in gene pool

A_1 $p = 0.7$ A_2 $q = 0.3$

All sperm in gene pool

	A_1 $p = 0.7$	A_2 $q = 0.3$
A_1 $p = 0.7$	A_1A_1 $p^2 = 0.49$	A_1A_2 $pq = 0.21$
A_2 $q = 0.3$	A_2A_1 $pq = 0.21$	A_2A_2 $q^2 = 0.09$

Allele frequencies have not changed

Genotype frequencies in offspring generation:

A_1A_1 $p^2 = 0.49$
A_1A_2 $2pq = 0.42$
A_2A_2 $q^2 = 0.09$

Allele frequencies in offspring generation:

A_1 $p = 0.49 + \frac{1}{2}(0.42) = 0.70$
A_2 $q = \frac{1}{2}(0.42) + 0.09 = 0.30$

2. When alleles are transmitted via meiosis and random combination of gametes, their frequencies do not change over time. For evolution to occur, some other factor or factors must come into play.

What are these other factors?

The Hardy-Weinberg Model Makes Important Assumptions

The Hardy-Weinberg model is based on important assumptions about how populations and alleles behave. Specifically, for a population to conform to the Hardy-Weinberg principle, none of the four mechanisms of evolution can be acting on the population. In addition, the model assumes that mating is random with respect to the gene in question. Thus, here are the five assumptions that must be met:

1. *No natural selection at the gene in question* In step 2 of Figure 25.1, the model assumed that all members of the

FIGURE 25.2 A Punnett Square Illustrates the Hardy-Weinberg Principle.

parental generation survived and contributed equal numbers of gametes to the gene pool, no matter what their genotype.

2. **No genetic drift, or random allele frequency changes, affecting the gene in question** We avoided this type of allele frequency change in step 4 of Figure 25.1 by assuming that we drew alleles in their exact frequencies p and q, and not at some different values caused by chance. For example, allele A_1 did not "get lucky" and get drawn more than 70 percent of the time. No random changes due to luck occurred.

3. **No gene flow** No new alleles were added by immigration or lost through emigration anywhere in Figure 25.1. As a result, all of the alleles in the offspring population came from the original population's gene pool.

4. **No mutation** We didn't consider that new A_1s or A_2s or other, new alleles might be introduced into the gene pool in step 2 or step 5 of Figure 25.1.

5. **Random mating with respect to the gene in question** We enforced this condition by picking gametes from the gene pool at random in step 3 of Figure 25.1. We did not allow individuals to choose a mate based on their genotype.

The Hardy-Weinberg principle tells us what to expect if selection, genetic drift, gene flow, and mutation are not affecting a gene, *and* if mating is random with respect to that gene. Under these conditions, the genotypes A_1A_1, A_1A_2, and A_2A_2 should be in the Hardy-Weinberg proportions p^2, $2pq$, and q^2, and no evolution will occur.

How Does the Hardy-Weinberg Principle Serve as a Null Hypothesis?

Recall from Chapter 1 that a null hypothesis predicts there are no differences among the treatment groups in an experiment. Biologists often want to test whether natural selection is acting on a particular gene, nonrandom mating is occurring, or one of the other evolutionary mechanisms is at work. In addressing questions like these, the Hardy-Weinberg principle functions as a null hypothesis.

Given a set of allele frequencies, the Hardy-Weinberg Principle predicts what genotype frequencies will be when natural selection, mutation, genetic drift, and gene flow are not affecting the gene; and when mating is random with respect to that gene. If biologists observe genotype frequencies that do not conform to the Hardy-Weinberg prediction, it means that something interesting is going on: Either nonrandom mating is occurring, or allele frequencies are changing for some reason. Further research is needed to determine which of the five Hardy-Weinberg conditions is being violated.

Let's consider two examples to illustrate how the Hardy-Weinberg principle is used as a null hypothesis: MN blood types and *HLA* genes, both in humans.

CASE STUDY 1: ARE MN BLOOD TYPES IN HUMANS IN HARDY-WEINBERG PROPORTIONS? One of the first genes that geneticists could analyze in natural populations was the MN blood group of humans. Most human populations have two alleles, designated *M* and *N*, at this gene.

Because the *MN* gene codes for a protein found on the surface of red blood cells—with the M allele coding for the M version and the N allele coding for the N version—researchers could determine whether individuals are *MM*, *MN*, or *NN* by treating blood samples with antibodies to each protein (this technique was first introduced in Chapter 8). The *M* and *N* alleles are codominant—meaning that heterozygotes have both M and N versions of the protein on their red blood cells (see Chapter 13).

To estimate the frequency of each genotype in a population, geneticists obtain data from a large number of individuals and then divide the number of individuals with each genotype by the total number of individuals in the sample.

Table 25.1 shows *MN* genotype frequencies for populations from throughout the world and illustrates how observed genotype

TABLE 25.1 **The MN Blood Group of Humans: Observed and Expected Genotype Frequencies**

The expected genotype frequencies are calculated from the observed allele frequencies, using the Hardy-Weinberg principle.

Population and Location	Data Type	Genotype Frequencies			Allele Frequencies	
		MM	**MN**	**NN**	**M**	**N**
Inuit (Greenland)	Observed	0.835	0.156	0.009	0.913	0.087
	Expected	0.834	0.159	0.008		
Native Americans (U.S.)	Observed	0.600	0.351	0.049	0.776	0.224
	Expected	0.602	0.348	0.050		
Caucasians (U.S.)	Observed	0.292	0.494	0.213	0.540	0.460
	Expected	0.290	0.497	0.212		
Aborigines (Australia)	Observed	0.025	0.304	0.672	0.178	0.825
	Expected	0.031	0.290	0.679		
Ainu (Japan)	Observed — Step ❶ →	(0.179	0.502	0.319) Step ❷ →	( )	( )
	Expected	(		) ← Step ❸		

✔**EXERCISE** Fill in the values for observed allele frequencies and expected genotype frequencies for the Ainu people of Japan.

frequencies are compared with the genotype frequencies expected if the Hardy-Weinberg principle holds. The analysis is based on the following steps:

Step 1 Estimate genotype frequencies by observation—in this case, by testing many blood samples for the *M* and *N* alleles. These frequencies are given in the rows labeled "observed" in Table 25.1.

Step 2 Calculate observed allele frequencies from the observed genotype frequencies. In this case, the frequency of the *M* allele is the frequency of *MM* homozygotes plus half the frequency of *MN* heterozygotes; the frequency of the *N* allele is the frequency of *NN* homozygotes plus half the frequency of *MN* heterozygotes. (You can review the logic behind this calculation in steps 5 and 6 of Figure 25.1.)

Step 3 Use the observed allele frequencies to calculate the genotypes expected according to the Hardy-Weinberg principle. Under the null hypothesis of no evolution and random mating, the expected genotype frequencies are $p^2 : 2pq : q^2$.

Step 4 Compare the observed and expected values. Researchers use statistical tests to determine whether the differences between the observed and expected genotype frequencies are small enough to be due to chance or large enough to reject the null hypothesis of no evolution and random mating.

Although using statistical testing is beyond the scope of this text (see **BioSkills 5** in Appendix A for a brief introduction to the topic), you should be able to inspect the numbers and comment on them. In these populations, for example, the observed and expected *MN* genotype frequencies are almost identical. (A statistical test shows that the small differences observed are probably due to chance.) For every population surveyed, genotypes at the *MN* gene are in Hardy-Weinberg proportions. As a result, biologists conclude that the assumptions of the Hardy-Weinberg model are valid for this locus.

The results imply that when these data were collected, the *M* and *N* alleles in these populations were not being affected by the four evolutionary mechanisms and that mating was random with respect to this gene—meaning that humans were not choosing mates on the basis of their *MN* genotype.

Before moving on, however, it is important to note that a study such as this does not mean that the *MN* gene has never been under selection or subject to nonrandom mating or genetic drift. Even if selection has been very strong for many generations, one generation of no evolutionary forces and of random mating will result in genotype frequencies that conform to Hardy-Weinberg expectations.

The Hardy-Weinberg principle is used to test the hypothesis that currently no evolution is occurring at a particular gene and that in the previous generation, mating was random with respect to the gene in question.

CASE STUDY 2: ARE *HLA* GENES IN HUMANS IN HARDY-WEINBERG EQUILIBRIUM?

A research team collected data on the genotypes of 122 individuals from the Havasupai tribe native to Arizona. These biologists were studying two genes that are important in the functioning of the human immune system. More specifically, the genes that they analyzed code for proteins that help immune system cells recognize and destroy invading bacteria and viruses.

Previous work had shown that different alleles exist at both the *HLA-A* and *HLA-B* genes, and that the alleles at each gene code for proteins that recognize proteins from slightly different disease-causing organisms. Like the *M* and *N* alleles, *HLA* alleles are codominant.

As a result, the research group hypothesized that individuals who are heterozygous at one or both of these genes may have a strong fitness advantage. The logic is that heterozygous people have a wider variety of HLA proteins, so their immune systems can recognize and destroy more types of bacteria and viruses. They should be healthier and have more offspring than homozygous people do.

To test this hypothesis, the researchers used their data on observed genotype frequencies to determine the frequency of each allele present. When they used these allele frequencies to calculate the expected number of each genotype according to the Hardy-Weinberg principle, they found the observed and expected values reported in **Table 25.2**.

When you inspect these data, notice that there are many more heterozygotes and many fewer homozygotes than expected under Hardy-Weinberg conditions. Statistical tests show it is extremely unlikely that the difference between the observed and expected numbers could occur purely by chance.

These results supported the team's prediction and indicated that one of the assumptions behind the Hardy-Weinberg principle was being violated. But which assumption? The researchers argued that mutation, migration, and drift are negligible in this case and offered two competing explanations for their data:

1. **Mating may not be random with respect to the HLA genotype.** Specifically, people may subconsciously prefer mates with *HLA* genotypes unlike their own and thus produce an excess of heterozygous offspring. This hypothesis is plausible. For example, experiments have shown that college students can distinguish each others' genotypes at genes related to *HLA* on the basis of body odor. Individuals in this study were more attracted to the smell of genotypes unlike their own. If this is true among the Havasupai, then nonrandom mating

TABLE 25.2 *HLA* Genes of Humans: Observed and Expected Genotypes

The expected numbers of homozygous and heterozygous genotypes are calculated from observed allele frequencies, according to the Hardy-Weinberg principle.

Gene	Data Type	Genotype counts ($n = 122$)	
		Homozygotes	Heterozygotes
HLA-A	Observed	38	84
	Expected	48	74
HLA-B	Observed	21	101
	Expected	30	92

SOURCE: T. Markow et al., 1993. *HLA* polymorphism in the Havasupai: Evidence for balancing selection. *American Journal of Human Genetics* 53: 943–952, Table 3.

would lead to an excess of heterozygotes compared with the proportion expected under Hardy-Weinberg.

2. **Heterozygous individuals may have higher fitness.** This hypothesis is supported by data collected by a different research team, who studied the Hutterite people living in South Dakota. In the Hutterite population, married women who have the same *HLA*-related alleles as their husbands have more trouble getting pregnant and experience higher rates of spontaneous abortion than do women with *HLA*-related alleles different from those of their husbands. The data suggest that homozygous fetuses have lower fitness than do fetuses heterozygous at these genes. If this were true among the Havasupai, selection would lead to an excess of heterozygotes relative to Hardy-Weinberg expectations.

Which explanation is correct? It is possible that both are. But the fact is, no one knows. Using the Hardy-Weinberg principle as a null hypothesis allowed biologists to detect an interesting pattern in a natural population. Research continues on the question of why the pattern exists.

To review how Hardy and Weinberg derived their principle and how it is used as a null hypothesis, go to the study area at *www.masteringbiology.com*.

 Web Activity The Hardy-Weinberg Principle

CHECK YOUR UNDERSTANDING

⊙━ If you understand that . . .

- The Hardy-Weinberg principle functions as a null hypothesis when researchers test whether nonrandom mating or evolution is occurring at a particular gene.

✔ You should be able to . . .

Analyze whether a gene suspected of causing hypertension in humans is in Hardy-Weinberg proportions, and comment on why or why not. In one study, the observed genotype frequencies were A_1A_1 0.574; A_1A_2 0.339; A_2A_2 0.087. (Note: The sample size in this study was so large that a difference of 3 percent or more in any of the observed versus expected frequencies indicated a statistically significant difference—meaning, a difference that was not due to chance.)

Answers are available in Appendix B.

25.2 Types of Natural Selection

Natural selection occurs when individuals with certain phenotypes produce more offspring than individuals with other phenotypes do. If certain alleles are associated with the favored phenotypes, they increase in frequency while other alleles decrease in frequency. The result is evolution. To use the language introduced in Chapter 24, evolution by natural selection occurs when heritable variation leads to differential success in survival and reproduction.

Although you should have a solid understanding of why evolution by natural selection occurs, it is important to recognize that natural selection occurs in a wide variety of patterns. Each of these patterns has different causes and consequences.

When biologists analyze the consequences of different patterns of selection, they often focus on **genetic variation**—the number and relative frequency of alleles that are present in a particular population. The reason is simple: Lack of genetic variation in a population is usually a bad thing.

To understand why this is so, recall from Chapter 24 that selection can occur only if heritable variation exists in a population. If genetic variation is low and the environment changes—perhaps due to the emergence of a new disease-causing virus, a rapid change in climate, or a reduction in the availability of a particular food source—it is unlikely that any alleles will be present that have high fitness under the new conditions. As a result, the average fitness of the population will decline. If the environmental change is severe enough, the population may even be faced with extinction.

Let's examine some of the different types of natural selection with this question in mind: How do they affect the level of genetic variation in the population?

Directional Selection

According to the data introduced in Chapter 24, natural selection has increased the frequency of drug-resistant strains of the tuberculosis bacterium and caused changes in beak shape and body size in medium ground finches. This type of natural selection is called **directional selection**, because the average phenotype of the populations changed in one direction.

DIRECTIONAL SELECTION TENDS TO REDUCE GENETIC VARIATION The top graph in **Figure 25.3a** plots the value of a trait on the *x*-axis and the number of individuals with a particular value of that trait is plotted on the *y*-axis. (In a histogram like this, the *y*-axis could also plot the frequency of individuals with a particular trait value—see **BioSkills 2** in Appendix A.) Note that the trait in question has a bell-shaped, normal distribution in this hypothetical population. Recall from Chapter 13 that when many different genes influence a trait, the distribution of phenotypes in the population tends to form a bell-shaped curve like this.

The middle and bottom graphs in the figure show what happens when directional selection acts on this trait. Note that in cases like this, directional selection is acting on many different genes at once. In contrast, selection on drug resistance in the TB bacterium was acting on a single gene.

Directional selection tends to reduce the genetic diversity of populations. If directional selection continues over time, the favored alleles will eventually approach a frequency of 1.0 while disadvantageous alleles will approach a frequency of 0.0. Alleles that reach a frequency of 1.0 are said to be fixed; those that reach a frequency of 0.0 are said to be lost. When disadvantageous alleles decline in frequency, **purifying selection** is said to occur.

(a) Directional selection changes the average value of a trait.

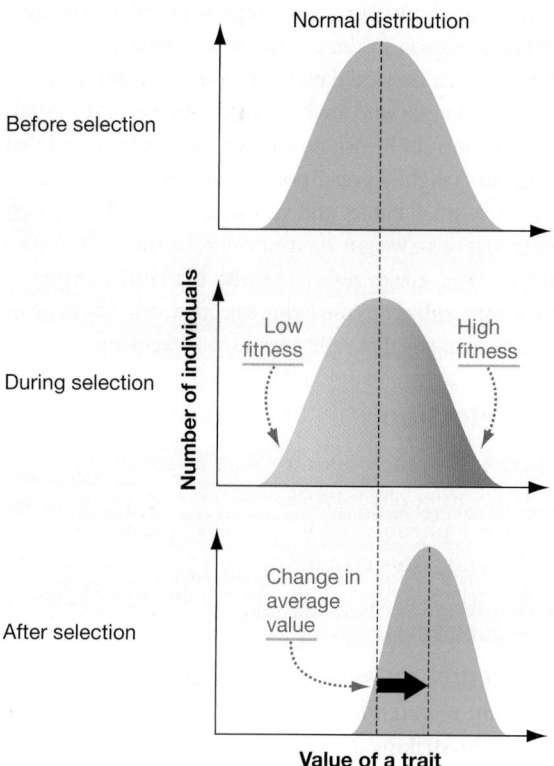

(b) For example, directional selection caused average body size to increase in a cliff swallow population.

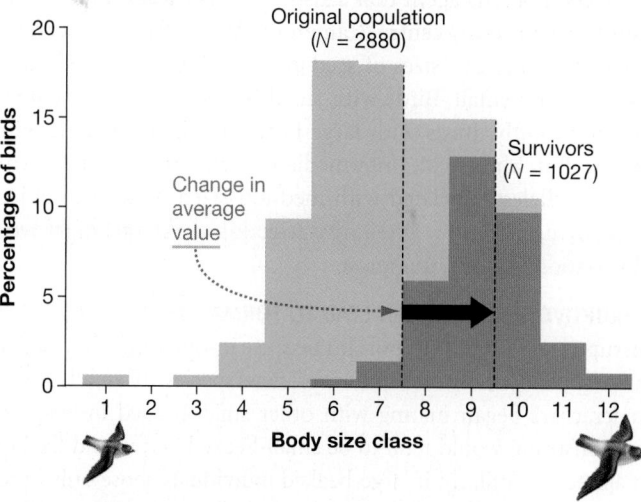

FIGURE 25.3 Directional Selection. (a) When directional selection acts on traits that have a normal distribution, individuals at one end of the distribution experience poor reproductive success. **(b)** The light green histogram shows the distribution of overall body size in cliff swallows prior to an extended cold snap that killed many individuals. (The various body-size classes were calculated from measurements of wing length, tail length, leg length, and beak size.) The dark green histogram shows the size distribution of individuals from the same population that survived the cold spell. Here *N* indicates the sample size.

Fixation and loss may not always occur under directional selection, however. To appreciate why, consider recent data on the body size of swallows.

DIRECTIONAL SELECTION ON BODY SIZE IN CLIFF SWALLOWS In 1996 a population of cliff swallows native to the Great Plains of North America endured a six-day period of exceptionally cold, rainy weather. Cliff swallows feed by catching mosquitoes and other insects in flight. Insects disappeared during this cold snap, however, and the biologists recovered the bodies of 1853 swallows that died of starvation.

As soon as the weather improved, the researchers caught and measured the body size of 1027 survivors from the same population. As the histograms in **Figure 25.3b** show, survivors were much larger on average than the birds that died.

Directional selection, favoring large body size, had occurred. To explain this observation, the investigators suggest that larger birds survived because they had larger fat stores and did not get as cold as the smaller birds. As a result, the larger birds were less likely to die of exposure to cold and more likely to avoid starvation until the weather warmed up and insects were again available.

If the differences in body size among individuals were due in part to differences in their genotypes—meaning that heritable variation in body size existed—then the population evolved. If so, and if this type of directional selection continued, then alleles that contribute to small body size would quickly decline in frequency in the cliff swallow population.

It is not clear that this will be the case, however, because directional selection is rarely constant throughout a species' range and through time. By examining weather records, the researchers established that cold spells as severe as the one that occurred in 1996 are rare.

COUNTERVAILING SELECTION AND FITNESS TRADE-OFFS Research on other swallow species suggests that smaller birds are more maneuverable in flight and thus more efficient when they feed. If so, then selection for feeding efficiency could counteract selection by cold weather. When this is the case, individuals with intermediate body size should be favored.

In swallows and many other species, it is common to find that one cause of directional selection on a trait is counterbalanced by a different factor that causes selection in the opposite direction. This concept, known as a fitness trade-off, was introduced in Chapter 24. In such cases, the optimal phenotype is intermediate. The same pattern can result from a different pattern of natural selection, called stabilizing selection.

Stabilizing Selection

When cliff swallows were exposed to cold weather, selection greatly reduced one extreme in the range of phenotypes and resulted in a directional change in the average characteristics of the population. But selection can also reduce both extremes in a

(a) Stabilizing selection reduces the amount of variation in a trait.

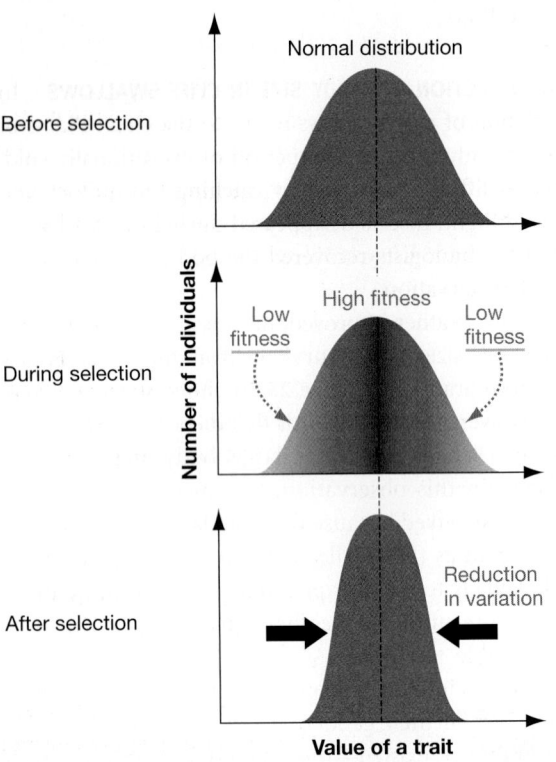

Normal distribution

Before selection

High fitness

Low fitness   Low fitness

During selection

Number of individuals

After selection

Reduction in variation

Value of a trait

(b) For example, very small and very large babies are the most likely to die, leaving a narrower distribution of birth weights.

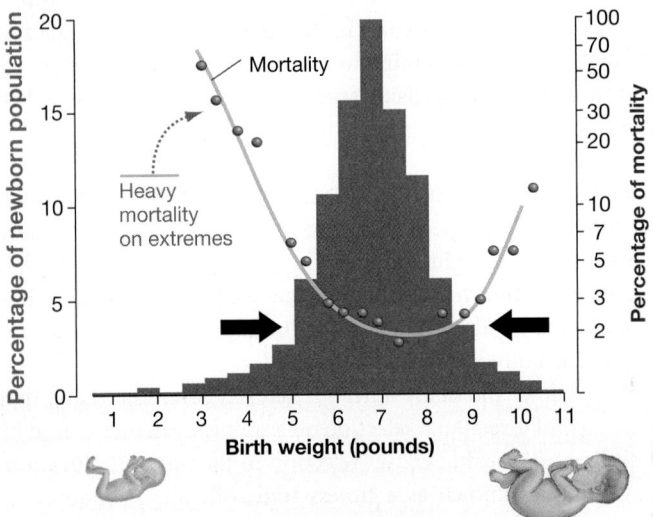

Mortality

Heavy mortality on extremes

Percentage of newborn population

Percentage of mortality

Birth weight (pounds)

FIGURE 25.4 Stabilizing Selection. (a) When stabilizing selection acts on normally distributed traits, individuals with extreme phenotypes experience poor reproductive success. (b) Histogram showing the percentages of newborns with various birth weights on the left-hand axis. The purple dots indicate the percentage of newborns in each weight class that died, plotted on the logarithmic scale shown on the right.

population, as illustrated in **Figure 25.4a**. This pattern of selection is called **stabilizing selection**. It has two important consequences: There is no change in the average value of a trait over time, and genetic variation in the population is reduced.

Figure 25.4b shows a classical data set in humans illustrating stabilizing selection. Biologists who analyzed birth weights and mortality in 13,730 babies born in British hospitals in the 1950s found that babies of average size (slightly over 7 pounds) survived best. Mortality was high for very small babies and very large babies. This is persuasive evidence that birth weight was under strong stabilizing selection in this population. Alleles associated with high birth weight or low birth weight were subject to purifying selection, and alleles associated with intermediate birth weight increased in frequency.

Disruptive Selection

Disruptive selection has the opposite effect of stabilizing selection. Instead of favoring phenotypes near the average value and eliminating extreme phenotypes, it eliminates phenotypes near the average value and favors extreme phenotypes (**Figure 25.5a**). When disruptive selection occurs, the overall amount of genetic variation in the population is maintained.

DISRUPTIVE SELECTION ON BEAK SIZE IN BLACK-BELLIED SEED-CRACKERS Recent research has shown that disruptive selection is responsible for the striking distribution of bills of black-bellied seedcrackers (**Figure 25.5b**). The data plotted in the graph show that individuals with either very short or very long beaks survive best and that birds with intermediate phenotypes are at a disadvantage.

In this case, the agent that causes natural selection is food. At a study site in south-central Cameroon, West Africa, a researcher found that only two sizes of seed are available to the seedcrackers: large and small. Birds with small beaks crack and eat small seeds efficiently. Birds with large beaks handle large seeds efficiently. But birds with intermediate beaks have trouble with both, so alleles associated with medium-sized beaks are subject to purifying selection. Disruptive selection maintains high overall variation in this population.

DISRUPTIVE SELECTION CAN LEAD TO FORMATION OF NEW SPECIES Disruptive selection is important because it sometimes plays a part in speciation, or the formation of new species. If small-beaked seedcrackers began mating with other small-beaked individuals, their offspring would tend to be small-beaked and would feed on small seeds. Similarly, if large-beaked individuals chose only other large-beaked individuals as mates, they would tend to produce large-beaked offspring that would feed on large seeds.

In this way, selection would result in two distinct populations. Under some conditions, the populations may eventually form two new species. The process of species formation, based on disruptive selection and other mechanisms, is explored in detail in Chapter 26.

Balancing Selection

Directional selection, stabilizing selection, and disruptive selection describe how natural selection can act on polygenic traits in

(a) Disruptive selection increases the amount of variation in a trait.

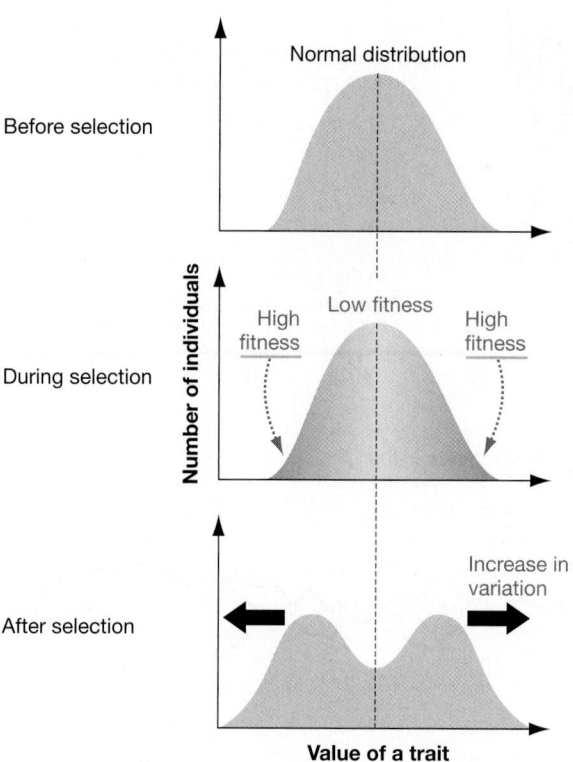

Normal distribution

Before selection

Number of individuals

During selection

High fitness — Low fitness — High fitness

After selection

Increase in variation

Value of a trait

(b) For example, only juvenile black-bellied seedcrackers that had very long or very short beaks survived long enough to breed.

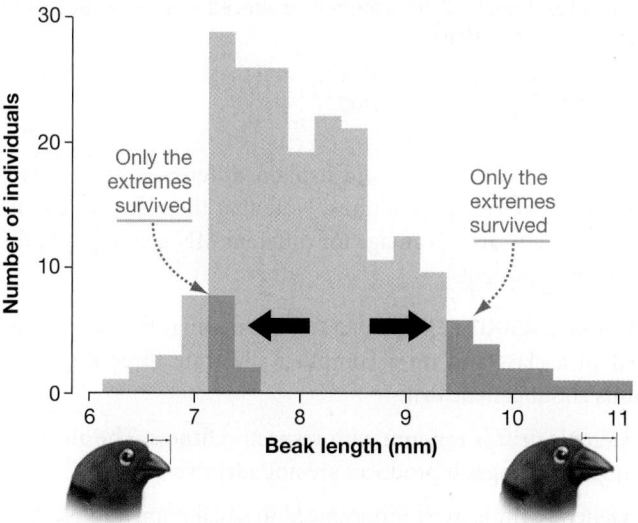

Only the extremes survived

Only the extremes survived

Beak length (mm)

FIGURE 25.5 Disruptive Selection. (a) When disruptive selection occurs on traits with a normal distribution, individuals with extreme phenotypes experience high reproductive success. **(b)** Histogram showing the distribution of beak length in a population of black-bellied seedcrackers. The light orange bars represent all juveniles; the dark orange bars, juveniles that survived to adulthood.

a single generation or episode. To review how they work, go to the study area at *www.masteringbiology.com*.

Web Activity Three Modes of Natural Selection

It's important to recognize, though, that they are not the only patterns of selection. For example, the second explanation for the data in Table 25.2 is a pattern of natural selection called **heterozygote advantage**. When selection operates in this way, heterozygous individuals have higher fitness than homozygous individuals do. The consequence of this pattern is that genetic variation is maintained in populations.

Heterozygote advantage is one mechanism responsible for a more general phenomenon known as **balancing selection**. When balancing selection occurs, no single allele has a distinct advantage and increases in frequency. Instead, there is a balance among several alleles in terms of their fitness and frequency. Balancing selection also occurs when:

1. The environment varies over time or in different geographic areas occupied by a population—meaning that certain alleles are favored by natural selection at different times or in different places. As a result, overall genetic variation in the population is maintained or increased.

2. Certain alleles are favored when they are rare, but not when they are common—a pattern known as **frequency-dependent selection**. For example, rare alleles responsible for coloration in guppies are favored because predators learn to recognize common color patterns. Alleles for common colors get eliminated; alleles for rare colors increase in frequency. As a result, overall genetic variation in the population is maintained or increased.

No matter how natural selection occurs, though, its most fundamental attribute is the same: It increases fitness and leads to adaptation.

25.3 Genetic Drift

Natural selection is not random. It is directed by the environment and results in adaptation. Genetic drift, in contrast, is undirected and random.

Genetic drift is defined as any change in allele frequencies in a population that is due to chance. The process is aptly named, because it causes allele frequencies to drift up and down randomly over time. When drift occurs, allele frequencies change due to blind luck—what is formally known as **sampling error**. Drift occurs in every population, in every generation.

Simulation Studies of Genetic Drift

To understand why genetic drift occurs, imagine a couple marooned on a deserted island. Suppose that at gene *A*, the wife's genotype is A_TA_H and the husband is also A_TA_H. In this population, the two alleles are each at a frequency of 0.5.

Now suppose that the couple produce five children over their lifetime. Half of the eggs produced by the wife carry allele A_T and half carry allele A_H. Likewise, half of the sperm produced by the husband carry allele A_T and half carry allele A_H. To simulate which sperm and which egg happen to combine to produce each of the five offspring, you can flip a coin for each sperm and each egg, with tails standing for allele A_T and heads standing for allele A_H.

The following coin flips were done by a pair of students in a recent biology class:

	Sperm	Egg	Genotype
First offspring	A_H	A_H	$A_H A_H$
Second offspring	A_T	A_T	$A_T A_T$
Third offspring	A_T	A_H	$A_H A_T$
Fourth offspring	A_H	A_H	$A_H A_H$
Fifth offspring	A_T	A_H	$A_H A_T$

When the parents die, there are a total of 10 alleles in the population. But note that the allele frequencies have changed. In this generation, six of the 10 alleles (60 percent) are A_H; four of the 10 alleles (40 percent) are A_T. Evolution—a change in allele frequencies in a population—occurred due to genetic drift.

Instead of each allele being sampled in exactly its original frequency when offspring formed, as the Hardy-Weinberg principle assumes, a chance sampling error occurred. Allele A_H got lucky; allele A_T was unlucky.

COMPUTER SIMULATIONS **Figure 25.6** shows what happens when a computer simulates the same process of random combinations in gametes over time. The program that generated the graphs combines the alleles in a gene pool at random to create an offspring generation, calculates the allele frequencies in the offspring generation, and uses those allele frequencies to create a new gene pool.

In this example, the process was continued for 100 generations. The x-axis on each graph plots time in generations; the y-axis plots the frequency of one of the two alleles present at the A gene in a hypothetical population.

The top graph shows eight replicates of this process with a population size of 4; the bottom graph shows eight replicates with a population of 400. Notice (**1**) the striking differences between the effects of drift in the small versus large population and (**2**) the consequences for genetic variation when alleles drift to fixation or loss.

Given enough time, drift can be an important factor even in large populations. To drive this point home, consider two types of alleles that were introduced in earlier chapters and that have no effect on fitness. Recall from Chapter 15 that alleles containing silent mutations, usually in the third position of a codon, do not change the gene product. As a result, most of these alleles have little or no effect on the phenotype. Yet these alleles routinely drift to high frequency or even fixation over time. Similarly, recall from Chapter 20 that pseudogenes do not code for a product. Although their presence does not affect an individual's phenotype, dozens of pseudogenes in the human genome have reached fixation, due to drift.

✔ If you understand genetic drift, you should be able to examine the MN blood group genotype frequencies in Table 25.1

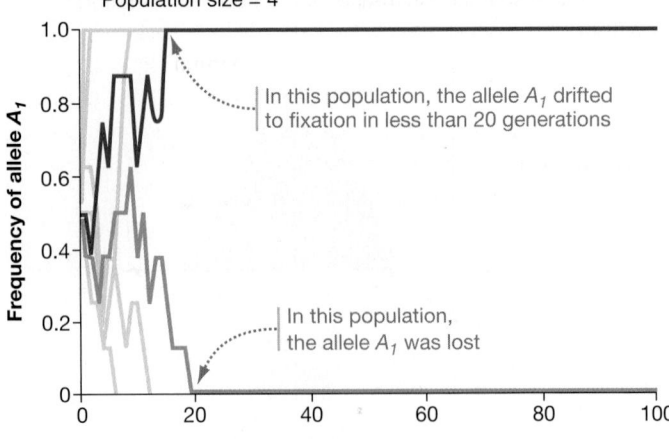

In this population, the allele A_1 drifted to fixation in less than 20 generations

In this population, the allele A_1 was lost

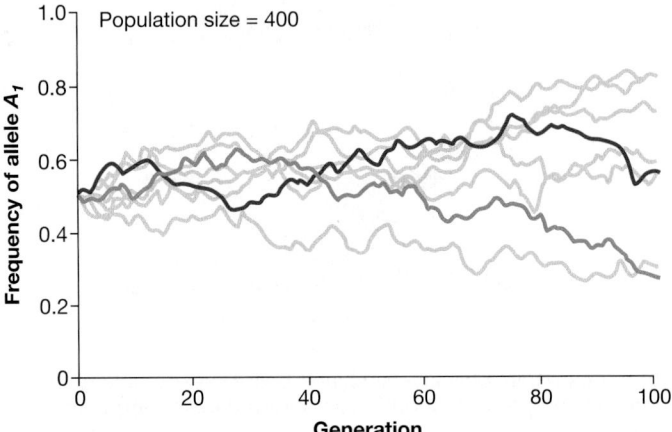

FIGURE 25.6 Genetic Drift Is More Pronounced in Small Populations than Large Populations.

✔**EXERCISE** Draw graphs predicting what these graphs would look like for a population size of 4000.

and describe how drift could explain differences in genotype frequencies among populations. Note that there are no data indicating a selective advantage for different MN genotypes in different environments.

KEY POINTS ABOUT GENETIC DRIFT The matings that were simulated in a class and on a computer illustrate three important points about genetic drift:

- Genetic drift is random with respect to fitness. The allele frequency changes it produces are not adaptive.

- Genetic drift is most pronounced in small populations. In the computer simulation, allele frequencies changed much less in the large population than the small population. And if the couple on the deserted island had produced 50 children instead of five, it is almost certain that allele frequencies in the next generation would have been much closer to 0.5.

- Over time, genetic drift can lead to the random loss or fixation of alleles. In the computer simulation with a population of 4, it took at most 20 generations for one allele to be fixed or lost. When random loss or fixation occurs, genetic variation in the population declines.

The importance of drift in small populations is a particular concern for conservation biologists, because many populations are being drastically reduced in size by habitat destruction and other human activities. Small populations that occupy nature reserves or zoos are particularly susceptible to genetic drift. If drift leads to a loss of genetic diversity, it could darken the already bleak outlook for some endangered species.

Experimental Studies of Genetic Drift

Research on genetic drift began with theoretical work in the 1930s and 1940s, which used mathematical models to predict the effect of genetic drift on allele frequencies and genetic variation. In the mid-1950s, Warwick Kerr and Sewall Wright did an experiment to show how drift works in practice.

Kerr and Wright started with a large laboratory population of fruit flies that contained a **genetic marker**—a specific allele that causes a distinctive phenotype. In this case, the marker was the morphology of bristles. Fruit flies have bristles on their bodies that can be either straight or bent (**Figure 25.7a**). This difference in bristle phenotype depends on a single gene. Kerr and Wright's lab population contained just two alleles—normal (straight) and "forked" (bent).

(a) Bristle shape is a useful genetic marker in fruit flies.

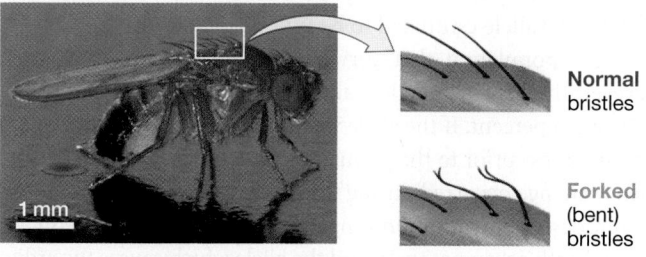

(b) Genetic drift reduced allelic diversity in most populations.

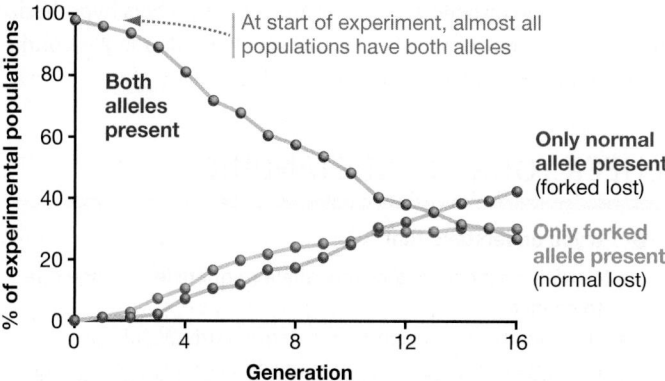

FIGURE 25.7 An Experiment on the Effects of Genetic Drift in Small Populations. At the start of an experiment on fruit flies, the frequencies of the wild-type (normal) bristle alleles and forked-bristle alleles were each 50 percent in all of 96 separate populations. Population size was kept at 8 individuals for 16 generations. By the end of the experiment, 73 percent of the 96 populations had lost either the normal or the forked allele; 26 percent of the populations still had both alleles. The only evolutionary force acting in these populations was genetic drift.

- The researchers set up 96 cages in their lab.
- They placed four adult females and four adult males of the fruit fly *Drosophila melanogaster* in each.

They chose individual flies to begin these experimental populations so that the frequency of the normal and forked alleles in each of the 96 starting populations was 0.5. The two alleles do not affect the fitness of flies in the lab environment, so Kerr and Wright could be confident that if changes in the frequency of normal and forked phenotypes occurred, they would not be due to natural selection.

- After these first-generation adults bred, Kerr and Wright reared their offspring.
- In the offspring (F_1) generation, they randomly chose four males and four females—meaning that they simply grabbed individuals without regard to whether their bristles were normal or forked—from each of the 96 offspring populations and allowed them to breed and produce the next generation.
- They repeated this procedure until all 96 populations had undergone a total of 16 generations.

During the entire course of the experiment, no migration from one population to another occurred. Previous studies had shown that mutations from normal to forked bristles (and forked to normal) are rare. Thus, the only evolutionary process operating during the experiment was genetic drift. It was as if random accidents claimed the lives of all but eight individuals in each generation, so that only eight bred.

Their result? After 16 generations, the 96 populations fell into three groups. Forked bristles were found on all of the individuals in 29 of the experimental populations. Due to drift, the forked allele had been fixed in these 29 populations and the normal allele had been lost (**Figure 25.7b**). In 41 other populations, however, the opposite was true: All individuals had normal bristles. In these populations, the forked allele had been lost due to chance. Both alleles were still present in 26 of the populations.

The message of the study is startling: In 73 percent of the experimental populations (70 out of the 96), genetic drift had reduced allelic diversity at this gene to zero.

As predicted, genetic drift decreased genetic variation within populations and increased genetic differences between populations. Is drift important in natural populations as well?

What Causes Genetic Drift in Natural Populations?

The sampling process that occurs during fertilization occurs in every population in every generation in every species that reproduces sexually. Similarly, accidents that remove individuals at random occur in every population in every generation.

It is important to realize, though, that because drift is caused by sampling error, it can occur by *any* process or event that involves sampling—not just the sampling of gametes that occurs during fertilization or the loss of unlucky individuals due to accidents. Let's consider two such examples, called founder effects and bottlenecks.

FOUNDER EFFECTS ON THE GREEN IGUANAS OF ANGUILLA When a group of individuals immigrates to a new geographic area and establishes a new population, a founder event is said to occur. If the group is small enough, the allele frequencies in the new population are almost guaranteed to be different from those in the source population—meaning, the population in the place from which the group emigrated—due to sampling error. A change in allele frequencies that occurs when a new population is established is called a **founder effect** (**Figure 25.8a**).

In 1995, fishermen on the island of Anguilla in the Caribbean witnessed a founder event involving green iguanas. A few weeks after two major hurricanes swept through the region, a large raft composed of downed logs tangled with other debris floated onto a beach on Anguilla. The fishermen noticed green iguanas on the raft and several on shore. Because green iguanas had not previously been found on Anguilla, the fishermen notified biologists. The researchers were able to document that at least 15 individuals had arrived; two years later they were able to confirm that at least some of the individuals were breeding. A new population had formed.

During this founder event, it is extremely unlikely that allele frequencies in the new Anguilla population of green iguanas exactly matched those of the source population, thought to be on the islands of Guadeloupe.

Colonization events like these have been the major source of populations that occupy islands all over the world, as well as island-like habitats such as mountain meadows, caves, and ponds.

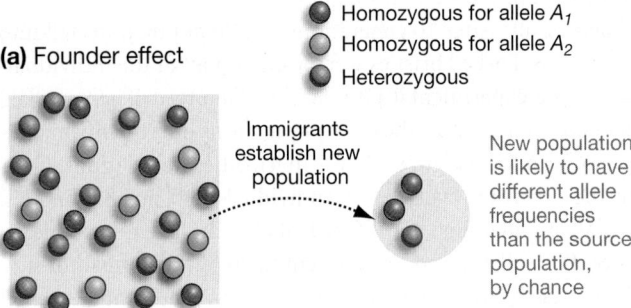

(a) Founder effect

- ● Homozygous for allele A_1
- ○ Homozygous for allele A_2
- ◐ Heterozygous

Immigrants establish new population

New population is likely to have different allele frequencies than the source population, by chance

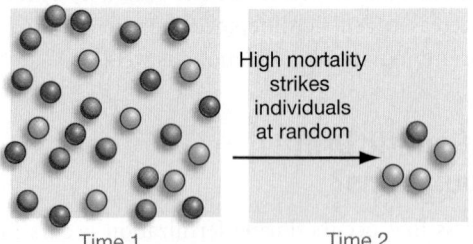

(b) Genetic bottleneck

High mortality strikes individuals at random

Bottlenecked population is likely to have different allele frequencies than original population, by chance

Time 1   Time 2

FIGURE 25.8 Two Causes of Genetic Drift in Natural Populations. The smaller the new population, the higher the likelihood that genetic drift will result not only in differences in allele frequencies, but also loss of alleles.

✔ **EXERCISE** The original population in (a) consists of 9 A_1A_1, 11 A_1A_2, and 7 A_2A_2 individuals (this hypothetical population is very small for simplicity). Compare the frequencies of the A_1 allele in the original and new populations.

Each time a founder event occurs, a founder effect is likely to accompany it, changing allele frequencies through genetic drift.

GENETIC BOTTLENECK ON PINGELAP ATOLL If a large population experiences a sudden reduction in size, a population bottleneck is said to occur. The term comes from the metaphor of a few individuals passing through the neck of a bottle, by chance. Disease outbreaks, natural catastrophes such as floods or fires or storms, or other events can cause population bottlenecks.

Genetic bottlenecks follow population bottlenecks, just as founder effects follow founder events. A **genetic bottleneck** is a sudden reduction in the number of alleles in a population. Drift occurs during genetic bottlenecks and causes a change in allele frequencies. **Figure 25.8b** provides a hypothetical example; for an example from a natural population consider the humans who occupy Pingelap Atoll in the South Pacific.

On this island, only about 20 people out of a population of several thousand managed to survive the effects of a typhoon and a subsequent famine that occurred around 1775. The survivors apparently included at least one individual who carried a loss-of-function allele at a gene called *CNGB 3*, which codes for a protein involved in color vision.

The *CNGB 3* allele is recessive, and when it is homozygous it causes a serious vision deficit called achromatopsia. The condition is extremely rare in most populations, with the frequency of the *CNGB 3* allele estimated to be under 1.0 percent.

In the population that survived the Pingelap Atoll disaster, however, the loss-of-function allele was at a frequency of about 1/40, or 2.5 percent. If the allele was at the typical frequency of 1.0 percent or less prior to the population bottleneck, then a large frequency change occurred during the bottleneck, due to drift.

In today's population on Pingelap Atoll, over 1 in 20 people is afflicted with achromatopsia, and the allele which causes the affliction is at a frequency of well over 20 percent. Because it is extremely unlikely that the loss-of-function allele is favored by directional selection or heterozygote advantage, researchers hypothesize that the frequency of the allele in this small population has continued to increase over the past 230 years due to drift.

CHECK YOUR UNDERSTANDING

If you understand that . . .

- Genetic drift occurs any time allele frequencies change due to chance.
- Drift violates the assumptions of the Hardy-Weinberg principle and occurs during many different types of events, including random fusion of gametes at fertilization, founder events, and population bottlenecks.

✔ **You should be able to . . .**

1. Explain why drift leads to a random loss or fixation of alleles.
2. Explain why drift is particularly important as an evolutionary force in small populations.

Answers are available in Appendix B.

25.4 Gene Flow

Gene flow is the movement of alleles from one population to another. It occurs when individuals leave one population, join another, and breed.

As an evolutionary mechanism, gene flow usually has one outcome: It equalizes allele frequencies between the source population and the recipient population. When alleles move from one population to another, the populations tend to become more alike. To capture this point, biologists say that gene flow homogenizes allele frequencies among populations (**Figure 25.9**).

Gene Flow in Natural Populations

Theoretical work quickly established how gene flow should affect populations. Studying gene flow in natural populations has been more challenging, however. To see how researchers are documenting gene flow's effects, consider a population of birds called great tits, native to Western Europe.

On the island of Vlieland off the coast of the Netherlands, great tits breed in two sets of woodlands. One lies to the west; one to the east. In both areas, mated pairs nest in boxes that have been maintained by biologists since 1955. For many decades, all of the great tits hatched on the island have been marked with leg bands.

Recently, biologists analyzed data on nesting that took place between 1975 and 1995 and found that individuals hatched in the eastern woodlands laid smaller clutches—groups of eggs that are incubated together—than individuals hatched in the western woodlands. Because the difference occurred whether the birds actually bred in the eastern or western habitats, the researchers inferred that it was due to a genetic difference between the eastern and western subpopulations.

Further, the data indicated that females hatched in the eastern woodlands survive much better than females hatched in the western habitats. The eastern birds appear to be much better adapted to the island environment.

Why aren't individuals from the western population as well-adapted to local conditions? The answer is gene flow. In the western woodlands, 43 percent of individuals breeding for the first time are unbanded—meaning that they just arrived from the mainland (**Figure 25.10a**). But in the eastern woodlands, only 13 percent of the individuals breeding for the first time are immigrants. Gene flow is over three times as high in the west as it is in the east.

To see the impact of gene flow on this population, study the data graphed in **Figure 25.10b**. The x-axis plots the number of grandparents that a female has that were hatched on the island. This is a measure of how local an individual's ancestry is. The y-axis plots the average chance that a female survives from one year to the next. This is a measure of fitness. Note that if a female has 0–2 grandparents from the island, they have a 30–35% chance of surviving from year to year. But if they have 3 or 4 grandparents who were local, their chance of surviving jumps to about 45 percent. Females with more island-born grandparents have higher fitness.

(a) Gene flow from mainland to Vlieland

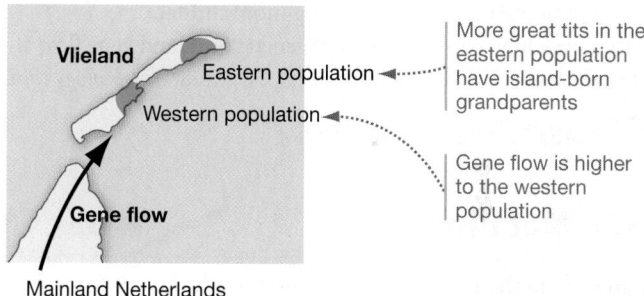

Mainland Netherlands

(b) In this case, survival rate is highest where gene flow is lowest.

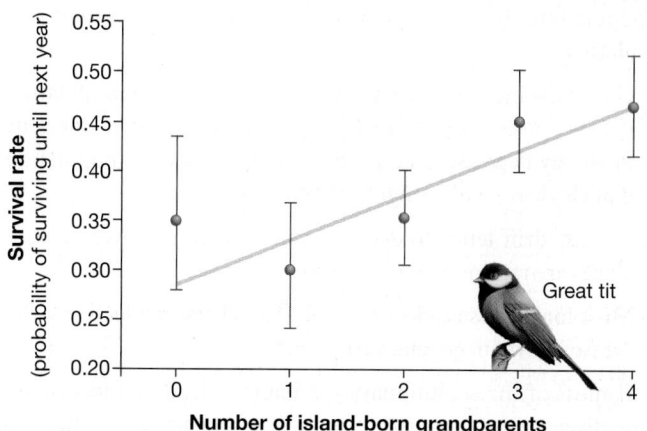

FIGURE 25.10 Gene Flow Reduces Fitness in a Population of Great Tits. In this study, the fewer immigrant grandparents an island bird has, the higher its survival rate.

FIGURE 25.9 Gene Flow Makes Allele Frequencies More Similar between Populations. Gene flow can occur in one or both directions. If gene flow continues and no selection occurs, allele frequencies will eventually be identical.

How Does Gene Flow Affect Fitness?

In the case of great tits on Vlieland, immigrants bring in alleles that have relatively low fitness in the island environment. Natural selection and gene flow are working in opposition: Selection favors small clutch sizes and other traits that affect survival in the island environment; gene flow constantly introduces alleles that have high fitness in mainland habitats but low fitness on the island. The homogenizing effect of gene flow is bad for island birds.

It is not true, though, that gene flow always reduces fitness. If a population has lost alleles due to genetic drift, then the arrival of new alleles via gene flow should increase genetic diversity. If increased genetic diversity results in better resistance to infections by bacteria or viruses or other parasites, gene flow would increase the average fitness of individuals. Chapter 55 provides an example of how gene flow aided an endangered population.

Gene flow is random with respect to fitness—the arrival or departure of alleles can increase or decrease average fitness, depending on the situation. But in every case, a movement of alleles between populations tends to reduce genetic differences between them.

This latter generalization is particularly important in our own species right now. Large numbers of people from Africa, the Middle East, Mexico, Central America, and Asia are immigrating to the countries of the European Union and the United States. Because individuals from different cultural and ethnic groups are intermarrying frequently and having offspring, allele frequencies in human populations are becoming more similar.

To review how natural selection, genetic drift, and gene flow affect populations, go to the study area at *www.masteringbiology.com*.

 BioFlix™ Mechanisms of Evolution

25.5 Mutation

To appreciate the role of mutation as an evolutionary force, let's return to one of the central questions that biologists ask about an evolutionary mechanism: How does it affect genetic variation in a population? To appreciate why this is an important question, recall that:

- Gene flow increases genetic diversity in a recipient population if new alleles arrive with immigrating individuals. But gene flow may decrease genetic variation in the source population if alleles leave with emigrating individuals.

- Genetic drift tends to decrease genetic diversity over time, as alleles are randomly lost or fixed.

- Most forms of selection favor certain alleles and lead to a decrease in overall genetic variation.

If most of the evolutionary mechanisms lead to a loss of genetic diversity over time, what restores it? In particular, where do entirely new alleles come from? The answer to both of these questions is **mutation**.

As Chapter 15 noted, mutations occur when DNA polymerase makes an error as it copies a DNA molecule, resulting in a change in the sequence of deoxyribonucleotides. If a mutation occurs in a stretch of DNA that codes for a protein, the changed codon may result in a polypeptide with a novel amino acid sequence. Mutations also occur at the level of chromosomes—when the number of chromosomes changes due to errors in meiosis or when chromosome segments detach and change position, due to damage.

Because errors and chromosome damage are inevitable, mutation constantly introduces new alleles into populations in every generation. Mutation is an evolutionary mechanism that increases genetic diversity in populations.

Despite the fact that it consistently leads to an increase in genetic diversity in a population, mutation is random with respect to the affected allele's impact on the fitness of the individual. Changes in the makeup of chromosomes or in specific DNA sequences do not occur in ways that tend to increase fitness or decrease fitness. Mutation just happens.

But because most organisms are well adapted to their current habitat, random changes in genes usually result in products that do not work as well as the alleles that currently exist. Stated another way, most mutations in sequences that code for a functional protein or RNA result in **deleterious** alleles—alleles that lower fitness. Deleterious alleles tend to be eliminated by purifying selection.

On rare occasions, however, mutation in these types of sequences produces a **beneficial** allele—an allele that allows individuals to produce more offspring. Beneficial alleles should increase in frequency in the population due to natural selection.

Because mutation produces new alleles, it can in principle change the frequencies of alleles through time. But does mutation occur often enough to make it an important factor in changing allele frequencies? The short answer is no.

Mutation as an Evolutionary Mechanism

To understand why mutation is not a significant mechanism of evolutionary change by itself, consider that the highest mutation rates that have been recorded at individual genes in humans are on the order of 1 mutation in every 10,000 gametes produced by an individual. This rate means that for every 10,000 alleles produced, on average one will have a mutation at the gene in question.

When two gametes combine to form an offspring, then, at most about 1 in every 5000 offspring will carry a mutation at a particular gene. Now suppose that 195,000 humans live in a population; that 5000 offspring are born one year; and that at the end of that year, the population numbers 200,000. Humans are diploid, so in a population this size, there is a total of 400,000 copies of each gene. Only one of them is a new allele created by mutation, however. Over the course of a year, the allele frequency change introduced by mutation is 1/400,000, or 0.0000025 (2.5×10^{-6}). At this rate, it would take 4000 years for mutation to produce a change in allele frequency of 1 percent.

These calculations support the conclusion that mutation does little to change allele frequencies on its own. Although mutation can be a significant evolutionary force in bacteria and archaea, which have extremely short generation times, mutation in eukaryotes rarely causes a change from the genotype frequencies expected under the Hardy-Weinberg principle. As an evolutionary mechanism, mutation is slow compared with selection, genetic drift, and gene flow.

But mutation still plays a role in evolution. Humans have over 20,000 genes, so each individual carries at least 40,000 alleles. Although the rate of mutation per allele may be very low, the total number of alleles is high. Multiplying the estimated number of genes in a human by the average mutation rate per gene suggests that an average person contains about 1.1 new alleles created by mutation.

Mutation introduces new alleles into every individual in every population in every generation. And in species that undergo sexual reproduction, meiosis and genetic recombination create variation in terms of the allele combinations present in each individual. Even if selection and drift are eliminating genetic diversity, mutation renews it.

Experimental Studies of Mutation

To get a better appreciation for how mutation affects evolution, consider an experiment conducted by Richard Lenski and colleagues. The experiment was designed to evaluate the role that mutation plays in many genes over many generations.

EXPERIMENTAL EVOLUTION Lenski's group focused on *Escherichia coli*, a bacterium that is a common resident of the human intestine. To begin, they set up a large series of populations, each founded with a single cell (see **Figure 25.11**).

To track the evolution of the experimental populations over time, the researchers transferred a small number of cells from each of the populations into a new batch of the same growth medium, under the same light and temperature conditions, every day for over four years. In this way, each population grew continuously.

Over the course of the experiment, the researchers estimated that each population underwent a total of 10,000 generations. This is the equivalent of over 200,000 years of human evolution. In addition, the biologists saved a sample of cells from each population at regular intervals and stored them in a freezer. Because frozen *E. coli* cells resume growth when they are thawed, the frozen cells archived individuals that existed over the 10,000 generation time interval.

The strain of *E. coli* used in the experiment is completely asexual and reproduces by cell division. Thus, mutation was the only source of genetic variation in these populations. Although no

FIGURE 25.11 Testing Changes in Fitness over Time.

✔ **QUESTION** The authors of this study claim that the frozen cells from each generation form a fossil record, except that the fossils can be brought back to life. Is their analogy valid?

SOURCE: Elena, S. F., V. S. Cooper, and R. E. Lenski. 1996. Punctuated evolution caused by selection of rare beneficial mutations. *Science* 272: 1802–1804.

EXPERIMENT

QUESTION: How does average fitness in a population change over time?

HYPOTHESIS: Average fitness increases over time.

NULL HYPOTHESIS: Average fitness does not increase over time.

EXPERIMENTAL SETUP:

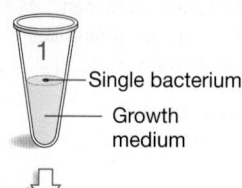

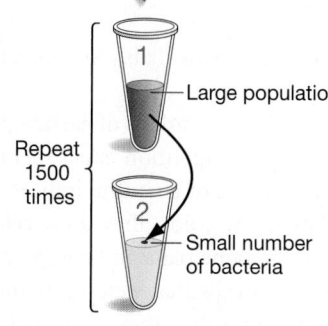

1. Place 10 mL of identical growth medium into many replicate tubes with one bacterium in each.

2. Incubate overnight. Average population in each tube is now 5×10^8 cells.

3. Remove 0.1 mL from each tube and move to 10 mL of fresh medium. Freeze remaining cells for later analysis.

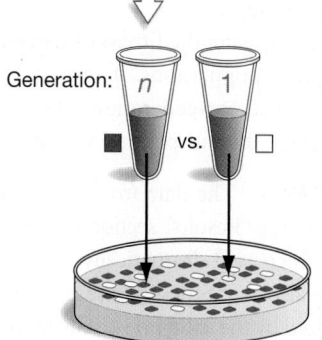

4. Competition experiment. Put an equal number of cells from generation 1 and a later generation (*n*) in fresh growth medium.

5. Incubate overnight and count the cells (color indicator distinguishes generations). Which are more numerous?

PREDICTION: Descendant populations have higher average fitness. (There will be more individuals from descendant than ancestral populations on the plates.)

PREDICTION OF NULL HYPOTHESIS: There will be no difference in fitness between descendant and ancestral populations.

RESULTS:

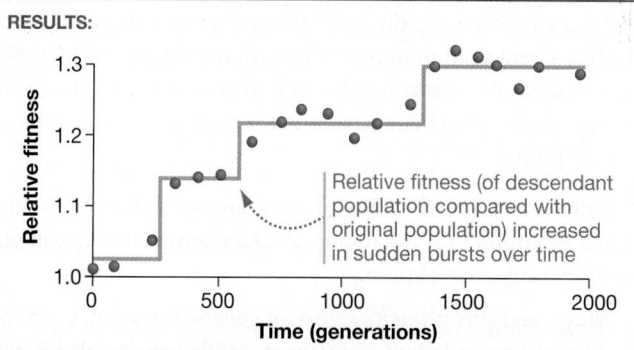

Relative fitness (of descendant population compared with original population) increased in sudden bursts over time

CONCLUSION: Descendant populations have higher fitness than do ancestral populations.

	Definition and Notes	Effect on Genetic Variation	Effect on Average Fitness
Natural selection	Certain alleles are favored	Can lead to maintenance, increase, or reduction of genetic variation	Can produce adaptation
Genetic drift	Random changes in allele frequencies; most important in small populations	Tends to reduce genetic variation via loss or fixation of alleles	Random with respect to fitness; usually reduces average fitness
Gene flow	Movement of alleles between populations; reduces differences between populations	May increase genetic variation by introducing new alleles; may decrease it by removing alleles	Random with respect to fitness; may increase or decrease average fitness by introducing high- or low-fitness alleles
Mutation	Production of new alleles	Increases genetic variation by introducing new alleles	Random with respect to fitness; most mutations in coding sequences lower fitness

gene flow occurred, both selection and genetic drift were operating in each population.

Were cells from the older and newer generations of each population different? Lenski's group used competition experiments to address this question. Taking advantage of a neutral indicator that can distinguish two populations by color, they grew cells from two different generations on the same plate and compared their growth rates. The populations of cells that were more numerous had grown the fastest, meaning that they were better adapted to the experimental environment.

In this way the researchers could measure the fitness of descendant populations relative to ancestral populations. If relative fitness was greater than 1, it meant that recent-generation cells outnumbered older-generation cells when the competition was over.

FITNESS INCREASED IN FITS AND STARTS The data from the competition experiments are graphed in the "Results" section of Figure 25.11. Notice that relative fitness increased dramatically—almost 30 percent—over time. But notice also that fitness increased in fits and starts. This pattern is emphasized by the solid line on the graph, which represents a mathematical function fitted to the data points.

What caused this stair-step pattern? Lenski's group hypothesizes that genetic drift was relatively unimportant in this experiment because population sizes were so large. Instead, they propose that each jump was caused by a novel mutation that conferred a fitness benefit. Their interpretation is that cells that happened to have the beneficial mutation grew rapidly and came to dominate the population.

After a beneficial mutation occurred, the fitness of the population stabilized—sometimes for hundreds of generations—until another random but beneficial mutation occurred and produced another jump in fitness.

TAKE-HOME MESSAGES The data on mutation rates and experimental evolution in *E. coli* reinforce fundamental messages about mutation's role in evolution.

1. Mutation is the ultimate source of genetic variation. Crossing over and independent assortment shuffle existing alleles into new combinations, but only mutation creates new alleles.

2. If mutation did not occur, evolution would eventually stop. Recall that natural selection and genetic drift tend to eliminate

alleles. Without mutation, eventually there would be no variation for selection and drift to act on. Without mutation, there would be no fitness increases observed in the *E. coli* experiment.

3. Mutation alone is usually inconsequential in changing allele frequencies at a particular gene. When considered across the genome and when combined with natural selection, however, it becomes an important evolutionary mechanism—it is the basis of fitness increases like those documented in the *E. coli* populations.

Table 25.3 summarizes these points and similar conclusions about the four evolutionary mechanisms. Each of the four evolutionary forces has different consequences for allele frequencies.

If one or more of these processes affects a gene, then genotypes will not be in Hardy-Weinberg proportions. But we've yet to consider the effects of another assumption in the Hardy-Weinberg model—that mating takes place at random with respect to the gene in question. What happens when the random mating assumption is violated?

25.6 Nonrandom Mating

In the Hardy-Weinberg model, gametes were picked from the gene pool at random and paired to create offspring genotypes. In nature, however, matings between individuals may not be random with respect to the gene in question. Even in species like clams that simply broadcast their gametes into the surrounding water, gametes from individuals who live close to each other are more likely to combine than gametes from individuals that live farther apart. In insects, vertebrates, and many other animals, females don't mate at random but actively choose certain males.

Two mechanisms that violate the Hardy-Weinberg assumption of random mating have been studied intensively. One is the phenomenon known as inbreeding; the other is the process called sexual selection.

Inbreeding

Mating between relatives is called **inbreeding**. By definition, relatives share a recent common ancestor. Individuals that inbreed are likely to share alleles they inherited from their common ancestor.

HOW DOES INBREEDING AFFECT ALLELE FREQUENCIES AND GENOTYPE FREQUENCIES? To understand how inbreeding affects populations, let's follow the fate of alleles and genotypes when inbreeding occurs. Focus again on a single locus with two alleles, A_1 and A_2, and suppose that these alleles initially have equal frequencies of 0.5.

Now imagine that the gametes produced by individuals in the population don't go into a gene pool. Instead, individuals self-fertilize. Many flowering plants, for example, contain both male and female organs and routinely self-pollinate. Self-fertilization, or selfing, is the most extreme form of inbreeding.

As **Figure 25.12a** shows, homozygous parents that self-fertilize produce all homozygous offspring. Heterozygous parents, in contrast, produce homozygous and heterozygous offspring in a 1:2:1 ratio.

Figure 25.12b shows the outcome for the population as a whole. In this figure, the width of the boxes represents the frequency of the three genotypes, which start out at the Hardy-Weinberg ratio of $p^2 : 2pq : q^2$. Notice that the homozygous proportion of the population increases each generation, while the heterozygous proportion is halved. At the end of the four generations illustrated, heterozygotes are rare. The same outcomes occur, more slowly, with less-extreme forms of inbreeding.

This simple exercise demonstrates two fundamental points about inbreeding:

1. Inbreeding increases homozygosity. In essence, inbreeding takes alleles from heterozygotes and puts them into homozygotes.

2. Inbreeding does not cause evolution, because allele frequencies do not change in the population as a whole.

🔑 Inbreeding and other forms of nonrandom mating change genotype frequencies—not allele frequencies. ✔If you understand these concepts, you should be able to predict how observed genotype frequencies should differ from those expected under the Hardy-Weinberg principle, when inbreeding is occurring.

WHY DOES INBREEDING DEPRESSION OCCUR? Inbreeding has attracted a great deal of attention because of a phenomenon called inbreeding depression. **Inbreeding depression** is a decline in average fitness that takes place when homozygosity increases and heterozygosity decreases in a population. Inbreeding depression results from two causes:

1. ***Many recessive alleles represent loss-of-function mutations.*** Because these alleles are usually rare, there are normally very few homozygous recessive individuals in a population. Instead, most loss-of-function alleles exist in heterozygous individuals. The alleles have little or no effect when they occur in heterozygotes, because one normal allele usually produces enough functional protein to support a normal phenotype. But inbreeding increases the frequency of homozygous recessive individuals. Loss-of-function mutations are usually deleterious or even lethal when they are homozygous.

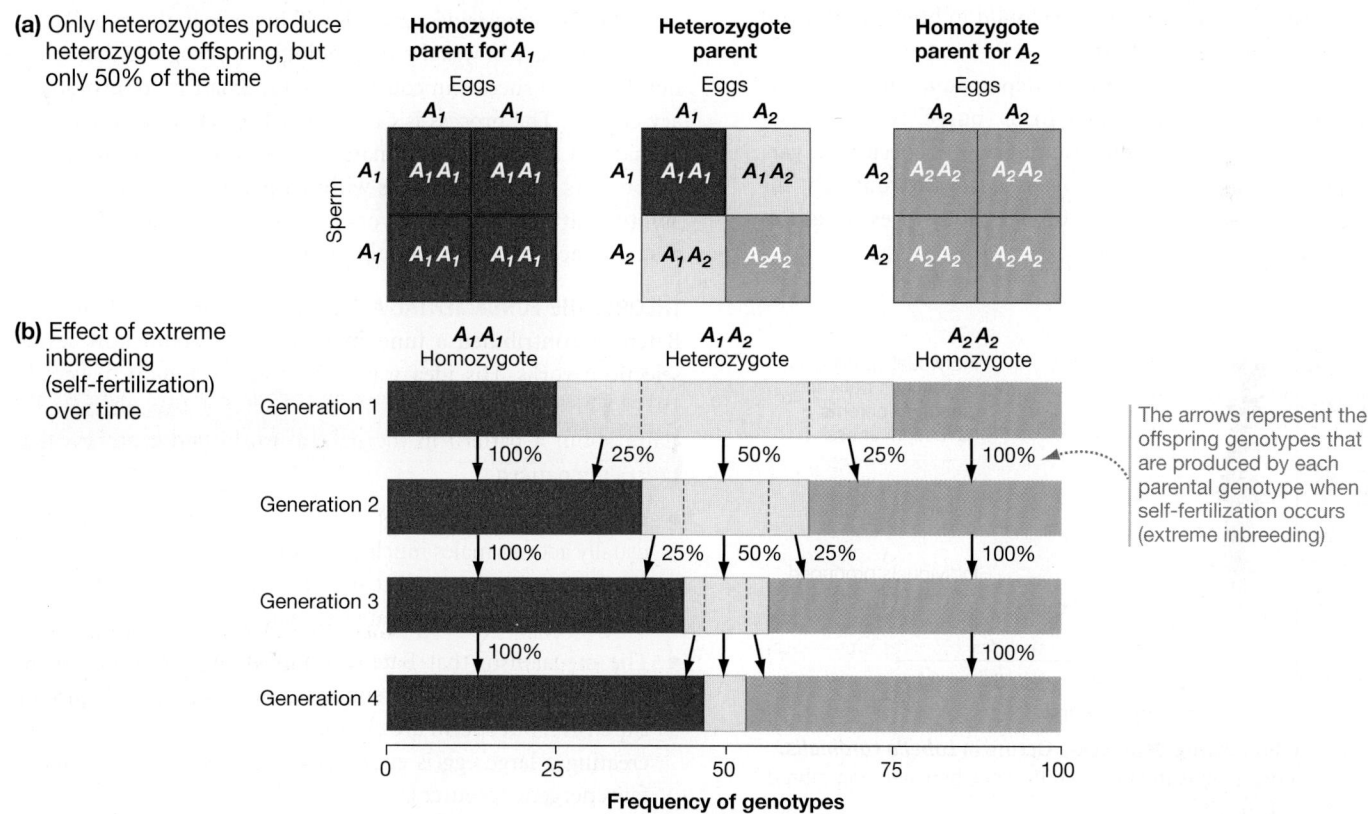

FIGURE 25.12 Inbreeding Increases Homozygosity and Decreases Heterozygosity. **(a)** Heterozygous parents produce homozygous and heterozygous offspring in a 1:2:1 ratio. **(b)** The width of the boxes corresponds to the frequency of each genotype.

2. *Many genes—especially those involved in fighting disease—are under intense selection for heterozygote advantage.* If an individual is homozygous at these genes, then fitness declines.

The upshot here is that the offspring of inbred matings are expected to have lower fitness than the progeny of outcrossed matings. This prediction has been verified in a wide variety of species. **Figure 25.13** shows data from controlled matings between *Lobelia cardinalis* individuals, with the fitness of offspring plotted against their age. Note that the two sets of data points compare the fitnesses of offspring from inbred and non-inbred matings. **Table 25.4** provides data on inbreeding depression in an array of human populations. Because inbreeding has such deleterious consequences in humans, it is not surprising that many contemporary human societies have laws forbidding marriages between individuals who are related as first cousins or closer.

The trickiest point to grasp about inbreeding is that even though it does not cause evolution directly—because it does not change allele frequencies—it can speed the rate of evolutionary change. More specifically, it increases the rate at which purifying selection eliminates recessive deleterious alleles from a population.

A moment's thought should convince you why this is so. Deleterious recessives are usually very rare in populations, because they lower fitness. Rare alleles are usually found in heterozygotes because it is more likely for individuals to have one copy of a rare allele than two. When no inbreeding is occurring, the recessive deleterious alleles found in heterozygotes cannot be eliminated by natural selection. But when inbreeding occurs, more recessive deleterious alleles are found in homozygotes and are quickly eliminated by selection.

✔If you understand inbreeding depression, you should be able to explain (1) why inbreeding helps "purge" recessive deleterious alleles, and (2) why it does not occur in species like garden peas, where self-fertilization has occurred routinely for many generations and most or all deleterious recessive alleles have been eliminated.

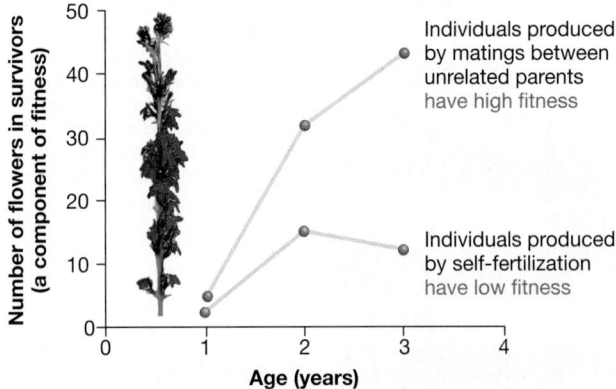

FIGURE 25.13 Inbreeding Depression Occurs in *Lobelia cardinalis*. Inbreeding depression is the fitness difference between non-inbred and inbred individuals.

✔**EXERCISE** Does inbreeding depression increase with age in this species or remain constant throughout life?

TABLE 25.4 **Inbreeding Reduces Fitness in Humans**

The percentages reported here give the mortality rate of children produced by first-cousin marriages versus marriages between nonrelatives. In every study, children of first-cousin marriages have a higher mortality rate.

| | | Deaths (%) | |
Age of Children	Time Period	Children of First Cousins	Children of Nonrelatives
Under 20 years (U.S.)	18th–19th century	17.0	12.0
Under 10 years (U.S.)	1920–1956	8.1	2.4
At or before birth (France)	1919–1950	9.3	3.9
Children (France)	1919–1950	14.0	10.0
Under 1 year (Japan)	1948–1954	5.8	3.5
1–8 years (Japan)	1948–1954	4.6	1.5

SOURCE: C. Stern. 1973. *Principles of Human Genetics*, 3rd ed., W. H. Freeman and Company, Table 5.8.

Sexual Selection

If female peacocks choose males with the longest and most iridescent tails as mates, then nonrandom mating is occurring. Due to this type of nonrandom mating, the frequency of alleles that contribute to long, iridescent tails will increase in the population.

Charles Darwin was the first biologist to recognize that selection based on success in courtship is a mechanism of evolutionary change. The process is called **sexual selection**, and it can be considered a special case of natural selection. Sexual selection occurs when individuals within a population differ in their ability to attract mates. It favors individuals with heritable traits that enhance their ability to obtain mates.

THEORY: THE FUNDAMENTAL ASYMMETRY OF SEX In 1948 A. J. Bateman contributed a fundamental insight about how sexual selection works. His idea was elaborated by Robert Trivers in 1972. The Bateman-Trivers theory contains two elements: a claim about a pattern in the natural world and a process that causes the pattern.

● The pattern component of their theory is that sexual selection usually acts on males much more strongly than on females. As a result, traits that attract members of the opposite sex are much more highly elaborated in males.

● The mechanism that Bateman and Trivers proposed to explain this pattern can be summarized with a quip: "Eggs are expensive, but sperm are cheap." That is, the energetic cost of creating a large egg is enormous, whereas a sperm contains few energetic resources.

In most species, females invest much more in their offspring than do males. This phenomenon is called the fundamental asymmetry of sex. It is characteristic of almost all sexual species and has two important consequences:

1. Because eggs are large and energetically expensive, females produce relatively few young over the course of a lifetime. A female's fitness is limited not by the ability to find a mate but primarily by her ability to gain the resources needed to produce more eggs and healthier young.

2. Sperm are so simple to produce that a male can father an almost limitless number of offspring. Thus, a male's fitness is limited not by the ability to acquire the resources needed to produce sperm but by the number of females he can mate with.

PREDICTIONS OF THE BATEMAN-TRIVERS THEORY The theory of sexual selection makes strong predictions.

- If females invest a great deal in each offspring, then they should protect that investment by being choosy about their mates. Conversely, if males invest little in each offspring, then they should be willing to mate with almost any female.

- If there are an equal number of males and females in the population, and if males are trying to mate with any female possible, then males will compete with each other for mates.

Stated another way, there should be two broad types of sexual selection: female choice and male-male competition. In addition:

- If male fitness is limited by access to mates, then any allele that increases a male's attractiveness to females or success in male-male competition should increase rapidly in the population. Thus, sexual selection should act more strongly on males than on females.

Stated another way, traits that evolve due to sexual selection—meaning traits that are useful only in courtship or in competition for mates—should be found primarily in males.

Do data from experimental or observational studies agree with these predictions? Let's consider each of them in turn.

FEMALE CHOICE FOR "GOOD ALLELES" If females are choosy about which males they mate with, what criteria do females use to make their choice? Recent experiments have shown that in several bird species, females prefer to mate with males that are well fed and in good health. These experiments were motivated by three key observations:

1. In many bird species, the existence of colorful feathers or a colorful beak is due to the presence of the red and yellow pigments called carotenoids.

2. Carotenoids protect tissues and stimulate the immune system to fight disease more effectively.

3. Animals cannot synthesize their own carotenoids, but plants can. To obtain carotenoids, animals have to eat carotenoid-rich plant tissues.

These observations suggest that the healthiest and best-nourished birds in a population have the most colorful beaks

and feathers. Sick birds have dull coloration because they are using all of their carotenoids to stimulate their immune system. Poorly fed birds have dull coloration because they have few carotenoids available. By choosing a colorful male as the father of her offspring, a female is likely to have offspring with alleles that will help the offspring fight disease effectively and feed efficiently.

To test the hypothesis that females prefer to mate with colorful males, a team of researchers experimented with zebra finches (**Figure 25.14a**). They fed one group of male zebra finches a diet that was heavily supplemented with carotenoids, and they fed a second group of male zebra finches (the control group) a diet that was similar in every way except for the additional carotenoids.

To control for other differences between the groups, the researchers used individuals that had been raised and maintained in environments that were as similar as possible. In addition, they identified pairs of brothers, and randomly assigned one brother to the control group and one brother to the treatment group. Thus, there should have been no pronounced genetic differences between individuals in the treatments.

As predicted, the males eating the carotenoid-supplemented diet developed more colorful beaks than did the males fed the carotenoid-poor diet (**Figure 25.14b**). When given a choice of mating with either of the two brothers, 9 out of the 10 females tested preferred the more-colorful male. These results are strong evidence that females of this species are choosy about their mates and that they prefer to mate with healthy, well-fed males.

Enough experiments have been done on other bird species to support a general conclusion: Colorful beaks and feathers, along with songs and dances and other types of courtship displays, carry the message "I'm healthy and well fed because I have good alleles. Mate with me."

FEMALE CHOICE FOR PATERNAL CARE Choosing "good alleles" is not the entire story in sexual selection via female choice. In many species, females prefer to mate with males that care for young or that provide the resources required to produce eggs.

(a) Male zebra finch

(b) Effect of carotenoids

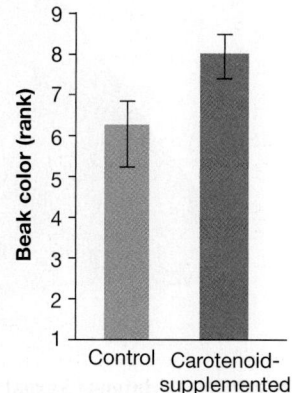

FIGURE 25.14 If Male Zebra Finches Are Fed Carotenoids, Their Beaks Get More Colorful. In this experiment, beak color was ranked on a scale where higher numbers indicate more intense orange color.

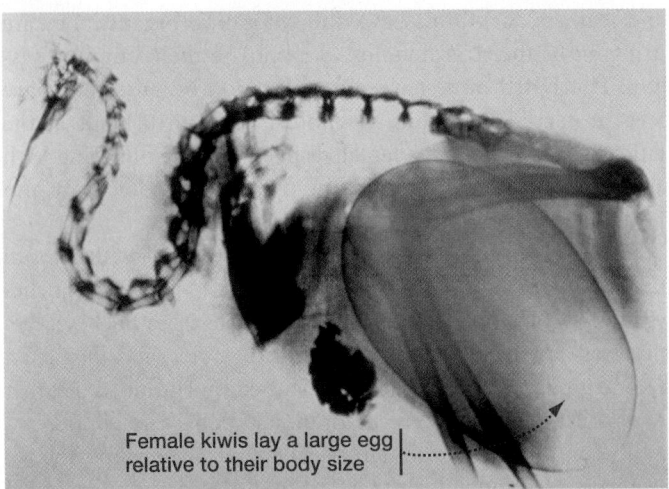

Female kiwis lay a large egg relative to their body size

FIGURE 25.15 In Many Species, Females Make a Large Investment in Each Offspring. X-ray of a female kiwi, ready to lay an egg.

Brown kiwi females make an enormous initial investment in their offspring—their eggs routinely represent over 15 percent of the mother's total body weight (**Figure 25.15**)—but choose to mate with males that take over all of the incubation and other care of the offspring. It is common to find that female fish prefer to mate with males that protect a nest site and care for the eggs until

they hatch. In humans and many species of birds, males provide food, protection, and other resources required for rearing young.

To summarize, females may choose mates on the basis of (**1**) physical characteristics that signal male genetic quality, (**2**) resources or parental care provided by males, or (**3**) both. In some species, however, females do not have the luxury of choosing a male. Instead, competition among males is the primary cause of sexual selection.

MALE-MALE COMPETITION As an example of research on how males compete for mates, consider data from a long-term study of a northern elephant seal population breeding on Año Nuevo Island, off the coast of California.

Elephant seals feed on marine fish and spend most of the year in the water. But when females are ready to mate and give birth, they haul themselves out of the water onto land. Females prefer to give birth on islands, where newborn pups are protected from terrestrial and marine predators. Because elephant seals have flippers that are ill suited for walking, females can haul themselves out of the water only on the few beaches that have gentle slopes. As a result, large numbers of females congregate in tiny areas to breed.

Male elephant seals establish territories on breeding beaches by fighting (**Figure 25.16a**). A **territory** is an area that is actively defended and that provides exclusive use by the owner. Males

(a) Males compete for the opportunity to mate with females.

(b) Variation in reproductive success is high in males.

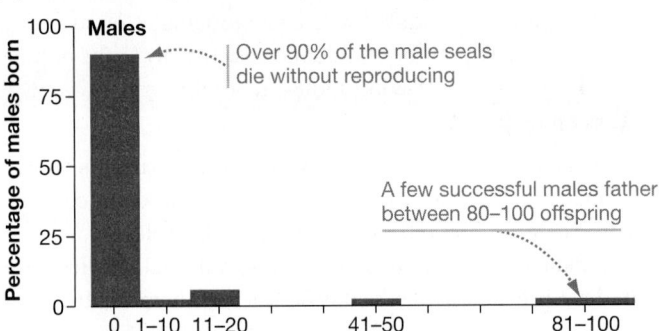

(c) Variation in reproductive success is relatively low in females.

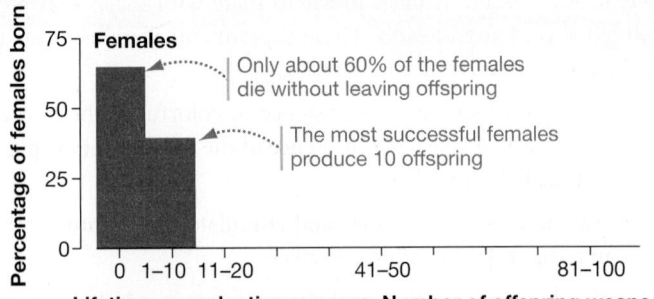

FIGURE 25.16 Intense Sexual Selection on Male Elephant Seals. The histograms show that variation in lifetime reproductive success is even higher **(b)** in male northern elephant seals than it is **(c)** in females.

✓**QUESTION** Consider an allele that increases reproductive success in elephant seal males versus an allele that increases reproductive success in females. Which allele will increase in frequency faster, and why?

that win battles with other males monopolize matings with the females residing in their territories. Females don't choose among males—they simply mate with the winning male. Males that lose battles are relegated to territories with few females or are excluded from the beach. Fights are essentially slugging contests and are usually won by the larger male. The males stand face to face, bite each other, and land blows with their heads.

Based on these observations, it is not surprising that male northern elephant seals frequently weigh 2700 kg (5940 lbs) and are over four times more massive, on average, than females. The logic here runs as follows:

- Males that own beaches with large congregations of females father large numbers of offspring. Males that lose fights father few or no offspring.

- The alleles of territory-owning males rapidly increase in frequency in the population.

- If the ability to win fights and produce offspring is determined primarily by body size, then alleles for large body size have a significant fitness advantage, leading to the evolution of large male size. The fitness advantage is due to sexual selection.

Figure 25.16b provides evidence for intense sexual selection in males. Biologists have marked many of the individuals in the

seal population on Año Nuevo to track the lifetime reproductive success of a large number of individuals. The x-axis indicates fitness, plotted as 0 offspring produced over a lifetime, 1–10, 11–20, and so on. The y-axis indicates the percentage of males in the population that achieved each category of offspring production. As the data show, in this population a few males father a large number of offspring, while most males father few or none.

Among females, variation in reproductive success is also high; but it is much lower than in males (**Figure 25.16c**). In this species, most sexual selection is driven by male-male competition rather than female choice.

WHAT ARE THE CONSEQUENCES OF SEXUAL SELECTION? In elephant seals and most other animals studied, most females that survive to adulthood get a mate. In contrast, many males do not. Because sexual selection tends to be much more intense in males than females, males tend to have many more traits that function only in courtship or male-male competition. Stated another way, sexually selected traits often differ sharply between the sexes.

Sexual dimorphism (literally, "two-forms") refers to any trait that differs between males and females. **Figure 25.17** illustrates sexually dimorphic traits. They range from weapons that males use to fight over females, such as antlers and horns, to the elaborate ornamentation and behavior used in courtship displays. Humans are sexually dimorphic in size, distribution of body hair, and many other traits.

Like inbreeding and other forms of nonrandom mating, sexual selection violates the assumptions of the Hardy-Weinberg principle. Unlike inbreeding, however, it causes certain alleles to increase or decrease in frequency and results in evolution.

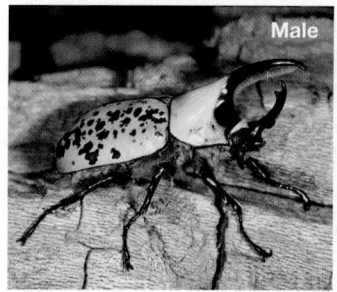 

Beetle. During breeding season, males of the beetle *Dynastes granti* use their elongated horns to fight over females.

Lion. Male lions are larger than female lions and have an elaborate ruff of fur called a mane.

FIGURE 25.17 Sexually Selected Traits Are Used to Compete for Mates. Males often have exaggerated traits that they use in fighting or courtship. In many species, females lack these traits.

CHECK YOUR UNDERSTANDING

If you understand that . . .

- Inbreeding and sexual selection are forms of nonrandom mating.
- Inbreeding is mating between relatives. It is not an evolutionary process, because it changes genotype frequencies—not allele frequencies. Inbreeding increases homozygosity and may lead to inbreeding depression.
- Sexual selection is based on differential success in obtaining mates. It causes evolution by increasing the frequency of alleles associated with successful courtship.

✔ You should be able to . . .

1. Define the fundamental asymmetry of sex.
2. Explain why males are usually the sex with exaggerated traits used in courtship.

Answers are available in Appendix B.

Summary of Key Concepts

○━ **The Hardy-Weinberg principle acts as a null hypothesis when researchers want to test whether evolution or nonrandom mating is occurring at a particular gene.**

- The Hardy-Weinberg principle can serve as a null hypothesis in evolutionary studies because it predicts what genotype and allele frequencies are expected to be if mating is random with respect to the gene in question and none of the four evolutionary processes is operating on that gene.

  ✔You should be able to predict how genotype frequencies differ from Hardy-Weinberg proportions under directional selection, when alleles are codominant.

  MB **Web Activity** The Hardy-Weinberg Principle

○━ **Each of the four evolutionary mechanisms has different consequences. Only natural selection produces adaptation. Genetic drift causes random fluctuations in allele frequencies. Gene flow equalizes allele frequencies between populations. Mutation introduces new alleles.**

- Directional selection favors phenotypes at one end of a distribution; stabilizing selection eliminates phenotypes with extreme characteristics. Both decrease allelic diversity in populations.

- Genetic drift causes random changes in allele frequency, is particularly important in small populations, and tends to reduce overall genetic diversity.

- Gene flow homogenizes allele frequencies among populations and can serve as an important source of new alleles.

- Mutation is too infrequent to be a major cause of allele frequency change, but is the only mechanism that creates new alleles.

  ✔You should be able to consider how the evolutionary forces affect the management of endangered species. What effect will drift

have? Predict how selection will affect allele frequencies in populations that are being maintained in captivity.

MB **Web Activity** Three Modes of Natural Selection, **BioFlix™** Mechanisms of Evolution

○━ **Inbreeding changes genotype frequencies but does not change allele frequencies.**

- Inbreeding, or mating among relatives, is a form of nonrandom mating.

- Inbreeding does not change allele frequencies, so it is not an evolutionary mechanism.

- Inbreeding changes genotype frequencies. It leads to an increase in homozygosity and a decrease in heterozygosity.

- These patterns can accelerate natural selection and can cause inbreeding depression.

  ✔You should be able to predict how extensive inbreeding during the 1700s and 1800s affected the royal families of Europe.

○━ **Sexual selection leads to the evolution of traits that help individuals attract mates. It is usually stronger on males than on females.**

- Sexual selection is a form of nonrandom mating.

- Sexual selection occurs when certain traits help males succeed in contests over mates or when certain traits are attractive to females.

- Sexual selection is responsible for the evolution of phenotypic differences between males and females.

  ✔You should be able to predict whether males or females are the most brightly colored in birds called red-necked phalaropes. In this species, males make a much larger investment in offspring than females do because they build the nest and incubate the eggs.

Questions

✔**TEST YOUR KNOWLEDGE**   *Answers are available in Appendix B*

1. Why isn't inbreeding considered an evolutionary mechanism?
   a. It does not change genotype frequencies.
   b. It does not change allele frequencies.
   c. It does not occur often enough to be important in evolution.
   d. It does not violate the assumptions of the Hardy-Weinberg principle.

2. Why is genetic drift aptly named?
   a. It causes allele frequencies to drift up or down randomly.
   b. It is the ultimate source of genetic variability.
   c. It is an especially important mechanism in small populations.
   d. It occurs when populations drift into new habitats.

3. How do sexual selection and inbreeding differ?
   a. Unlike inbreeding, sexual selection changes allele frequencies and affects only genes involved in attracting mates.

   b. Unlike sexual selection, inbreeding changes allele frequencies and involves any mating between relatives—not just self-fertilization.
   c. Unlike inbreeding, sexual selection results from the random fusion of gametes during fertilization. It is particularly important in small populations, where few mates are available.
   d. Inbreeding occurs only in small populations, while sexual selection can occur in any size population.

4. What does it mean when an allele reaches "fixation"?
   a. It is eliminated from the population.
   b. It has a frequency of 1.0.
   c. It is dominant to all other alleles.
   d. It is adaptively advantageous.

5. In what sense is the Hardy-Weinberg principle a null hypothesis, similar to the control treatment in an experiment?
   a. It defines what genotype frequencies should be if nonrandom mating is occurring.
   b. Expected genotype frequencies can be calculated from observed allele frequencies and then compared with the observed genotype frequencies.
   c. It defines what genotype frequencies should be if natural selection, genetic drift, gene flow, or mutation is occurring *and* if mating is random.
   d. It defines what genotype frequencies should be if evolutionary mechanisms are *not* occurring.

6. Mutation is the ultimate source of genetic variability. Why is this statement correct?
   a. DNA polymerase (the enzyme that copies DNA) is remarkably accurate.
   b. "Mutation proposes and selection disposes."
   c. Mutation is the only source of new alleles.
   d. Mutation occurs in response to natural selection. It generates the alleles that are required for a population to adapt to a particular habitat.

✔ TEST YOUR UNDERSTANDING
Answers are available in Appendix B

1. Consider selection, drift, gene flow, mutation, and inbreeding. Which of the five processes decreases genetic variation? Which increases it?

2. Directional selection can lead to the fixation of favored alleles. When this occurs, genetic variation is zero and evolution stops. Explain why this rarely occurs.

3. Why does sexual selection often lead to sexual dimorphism?

4. In what sense does the Hardy-Weinberg principle function as a null hypothesis?

5. Explain why small populations become inbred.

6. How can allele frequencies change under stabilizing selection, even if the average phenotype in the population does not?

✔ APPLYING CONCEPTS TO NEW SITUATIONS
Answers are available in Appendix B

1. In humans, albinism is caused by loss-of-function mutations in genes involved in the synthesis of melanin, the dark pigment in skin. Only people homozygous for a loss-of-function allele have the relevant phenotype. In Americans of northern European ancestry, albino individuals are present at a frequency of about 1 in 10,000 (or 0.0001). Knowing this genotype frequency and assuming that genotypes are in Hardy-Weinberg proportions, we can calculate the frequency of the loss-of-function alleles. If we let p_2 stand for this frequency, we know that $p_2^2 = 0.0001$; therefore $p_2 = \sqrt{0.0001} = 0.01$. By subtraction, the frequency of normal alleles is 0.99. Under the Hardy-Weinberg principle, what is the frequency of "carriers"—or people who are heterozygous for this condition? Your answer estimates the percentage of Caucasians in the United States who carry an allele for albinism.

2. Conservation managers frequently use gene flow, in the form of transporting individuals or releasing captive-bred young, to counteract the effects of drift on small, endangered populations. Explain how gene flow can also mitigate the effects of inbreeding.

3. Suppose you were studying several species of human. In one, males never lifted a finger to help females raise children. In another, males did just as much parental care as females except for actually carrying the baby during pregnancy. How does the fundamental asymmetry of sex compare in the two species? How would you expect sexual dimorphism to compare between the two species?

4. You are a conservation biologist charged with creating a recovery plan for an endangered species of turtle. The turtle's habitat has been fragmented into small, isolated but protected areas by suburbanization and highway construction. Some evidence indicates that certain turtle populations are adapted to marshes that are normal, whereas others are adapted to acidic wetlands or salty habitats. Further, some turtle populations number less than 25 breeding adults, making genetic drift and inbreeding a major concern. In creating a recovery plan, the tools at your disposal are captive breeding, the capture and transfer of adults to create gene flow, or the creation of habitat corridors between wetlands to make migration possible. How would you use gene flow to help this species?

New sunflower species have formed recently in southwestern North America. This chapter explains how.

26 Speciation

Although Darwin called his masterwork *On the Origin of Species by Means of Natural Selection*, he actually had little to say about how new species arise. Instead, his data and analyses focused on the process of natural selection and the changes that occur within populations over time. He spent much less time considering changes that occur *between* populations.

Since Darwin, however, biologists have realized that populations of the same species may diverge from each other when they are isolated in terms of gene flow. Recall from Chapter 25 that gene flow makes allele frequencies more similar among populations. 🔑 If gene flow ends, allele frequencies in isolated populations are free to diverge—meaning that the populations begin to evolve independently of each other. For example, when a new mutation creates an allele that changes the phenotype of individuals in one population, there is no longer any way for that allele to appear in the other population. If mutation, selection, and genetic drift cause isolated populations to diverge sufficiently, distinct types, or species, form—that is, the process of speciation takes place.

Speciation is a splitting event that creates two or more distinct species from a single ancestral group (**Figure 26.1**). When speciation is complete, a new branch has been added to the tree of life.

In essence, then, speciation results from genetic isolation and genetic divergence. Genetic isolation results from lack of gene flow, and divergence occurs because selection, genetic drift, and mutation proceed independently in the isolated populations. How does genetic isolation come about? And how do selection, drift, and mutation cause divergence?

26.1 How Are Species Defined and Identified?

Like the Galápagos finches in Figure 26.1, species are distinct from one another in appearance, behavior, habitat use, or other traits. These characteristics differ among species because their genetic characteristics differ. Genetic distinctions occur because

✔ When you see this checkmark, stop and test yourself. Answers are available in Appendix B.

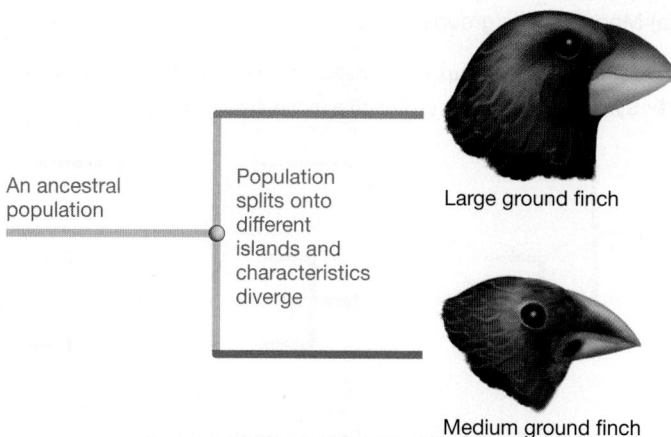

An ancestral population

Population splits onto different islands and characteristics diverge

Large ground finch

Medium ground finch

FIGURE 26.1 Speciation Creates Evolutionarily Independent Populations. The large ground finch and medium ground finch are derived from the same ancestral population. This ancestral population split into two populations isolated by lack of gene flow. Because the populations began evolving independently, over time they acquired the distinctive characteristics observed today.

mutation, selection, and drift act on each species independently of what is happening in other populations.

Formally, then, a **species** is defined as an evolutionarily independent population or group of populations. Even though this definition sounds straightforward, it can be exceedingly difficult to put into practice.

How can evolutionarily independent populations be identified in the field and in the fossil record? There is no single, universal answer. Even though biologists agree on the definition of a species, they frequently have to use different sets of criteria to identify them. Three criteria for identifying species are in common use: (1) the biological species concept, (2) the morphospecies concept, and (3) the phylogenetic species concept.

The Biological Species Concept

According to the **biological species concept**, the critical criterion for identifying species is reproductive isolation. This is a logical yardstick because no gene flow occurs between populations that are reproductively isolated from each other. Specifically, if two different populations do not interbreed in nature, or if they fail to produce viable and fertile offspring when matings take place, then they are considered distinct species. Groups that naturally or potentially interbreed, and that are reproductively isolated from other groups, belong to the same species. Biologists can be confident that reproductively isolated populations are evolutionarily independent.

Reproductive isolation can result from a wide variety of events and processes. To organize the various mechanisms that stop gene flow between populations, biologists distinguish

- **prezygotic** (literally, "before-zygote") **isolation**, which prevents individuals of different species from mating; and

- **postzygotic** ("after-zygote") **isolation**, in which the offspring of matings between members of different species do not survive or reproduce.

Table 26.1 outlines some of the more important mechanisms of prezygotic and postzygotic isolation.

TABLE 26.1 **Mechanisms of Reproductive Isolation**

	Process	Example
Prezygotic Isolation		
Temporal	Populations are isolated because they breed at different times.	Bishop pines and Monterey pines release their pollen at different times of the year.
Habitat	Populations are isolated because they breed in different habitats.	Parasites that begin to exploit new host species are isolated from their original population.
Behavioral	Populations do not interbreed because their courtship displays differ.	To attract male fireflies, female fireflies give a species-specific sequence of flashes.
Gametic barrier	Matings fail because eggs and sperm are incompatible.	In sea urchins, a protein called bindin allows sperm to penetrate eggs. Differences in the amino acid sequence of bindin cause matings to fail between closely related populations.
Mechanical	Matings fail because male and female reproductive structures are incompatible.	In alpine skypilots (a flowering plant), the length of the floral tube varies. Bees can pollinate in populations with short tubes, but only hummingbirds can pollinate in populations with long tubes.
Postzygotic Isolation		
Hybrid viability	Hybrid offspring do not develop normally and die as embryos.	When ring-necked doves mate with rock doves, less than 6 percent of eggs hatch.
Hybrid sterility	Hybrid offspring mature but are sterile as adults.	Eastern meadowlarks and western meadowlarks are almost identical morphologically, but hybrid offspring are largely infertile.

Although the biological species concept has a strong theoretical foundation, it has disadvantages. The criterion of reproductive isolation cannot be evaluated in fossils or in species that reproduce asexually. In addition, it is difficult to apply when closely related populations do not happen to overlap with each other geographically. In this case, biologists are left to guess whether interbreeding and gene flow would occur if the populations happened to come into contact.

The Morphospecies Concept

How do biologists identify species when the criterion of reproductive isolation cannot be applied? Under the **morphospecies** ("form-species") **concept**, researchers identify evolutionarily independent lineages by differences in size, shape, or other morphological features. The logic behind the morphospecies concept is that distinguishing features are most likely to arise if populations are independent and isolated from gene flow.

The morphospecies concept is compelling simply because it is so widely applicable. It is a useful criterion when biologists have no data on the extent of gene flow, and it is equally applicable to sexual, asexual, or fossil species. Its disadvantages are that: (1) it cannot identify **cryptic species**, which differ in traits other than morphology, such as the meadowlarks in Table 26.1, and (2) the morphological features used to distinguish species are subjective. In the worst case, different researchers working on the same populations disagree on the characters that distinguish species. Disagreements like these often end in a stalemate, because no independent criteria exist for resolving the conflict.

The Phylogenetic Species Concept

The **phylogenetic species concept** is a recent addition to the tools available for identifying evolutionarily independent lineages and is based on reconstructing the evolutionary history of populations. Proponents of this approach argue that it is widely applicable and precise.

The reasoning behind the phylogenetic species concept begins with Darwin's claim that all species are related by common ancestry. In modern terms, Darwin was suggesting that all species form a monophyletic ("one-tribe") group—the tree of life.

A **monophyletic group**, also called a **clade** or **lineage**, consists of an ancestral population, all of its descendants, and *only* those descendants. On any given evolutionary tree (whether the tree of life, or any smaller part of the tree of life), there are many monophyletic groups (**Figure 26.2a**).

Monophyletic groups, in turn, are identified by traits called synapomorphies ("unique-forms"). A **synapomorphy** is a trait found in certain groups of organisms that exists in no others. It is a homologous trait—one inherited from a common ancestor—that is unique to certain populations or lineages. Fur and lactation, for example, are synapomorphies that identify mammals as a monophyletic group. The placenta is a synapomorphy that distinguishes the placental mammals as a monophyletic group within mammals, distinct from the monophyletic group called marsupials.

(a) Monophyletic groups

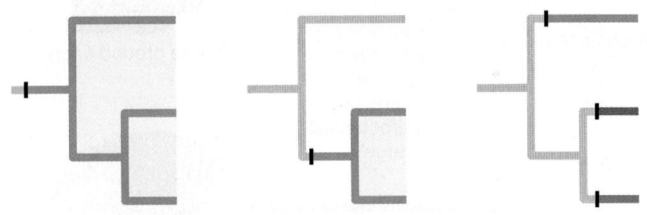

— **Monophyletic group:** an ancestral population and all descendants
⊢ **Synapomorphy:** trait unique to a monophyletic group

(b) Phylogenetic species: smallest monophyletic groups

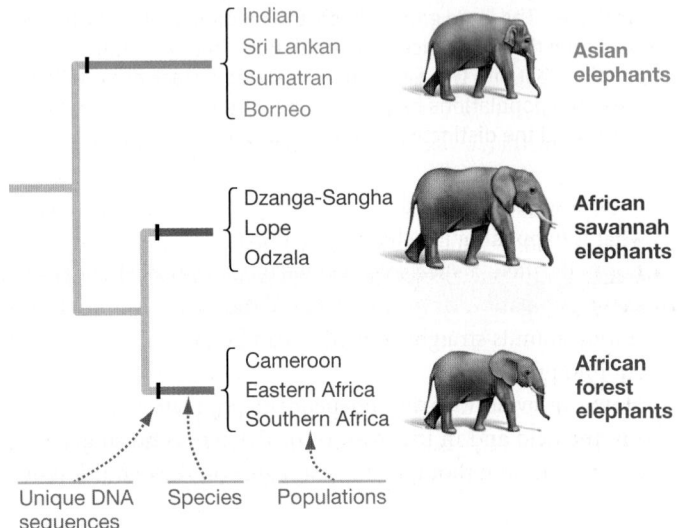

Indian
Sri Lankan
Sumatran
Borneo
— Asian elephants

Dzanga-Sangha
Lope
Odzala
— African savannah elephants

Cameroon
Eastern Africa
Southern Africa
— African forest elephants

Unique DNA sequences | Species | Populations

FIGURE 26.2 The Phylogenetic Species Concept Is Based on Monophyletic Groups. Although there are several isolated populations of elephants on two continents, the phylogenetic species concept identifies no more or less than three species of elephants. The populations within species are too similar to form separate branches on the tree. This tree was created by comparing DNA sequences. Thus, the synapomorphies in this case are DNA sequences unique to each species.

Like other homologous traits, synapomorphies can be identified at the genetic, developmental, or structural level. In many cases, researchers use DNA sequence data to identify synapomorphies and estimate phylogenetic trees. Chapter 27 explores the data and logic that biologists use to reconstruct phylogenies, and **BioSkills 3** in Appendix A provides more help on how to read phylogenetic trees.

Under the phylogenetic species concept, species are defined as the smallest monophyletic groups on the tree of life. Phylogenetic species are made up of populations that share one or more unique synapomorphies.

The tree in **Figure 26.2b**, for example, is based on DNA sequence data from an array of elephant populations; but many of the populations do not have unique synapomorphies—they are not distinctive enough to represent separate branches. The phylogenetic species on the tree are labeled Asian elephants, African forest elephants, and African savanna elephants. These are the smallest monophyletic groups on the tree.

The names at some of the tips on this tree (Cameroon, Eastern Africa, Southern Africa, etc.) represent populations within species. These populations may be separated geographically, but their characteristics are so similar that they do not form independent branches on the tree. They are simply part of the same monophyletic group containing other populations.

The phylogenetic species concept has two distinct advantages: (1) It can be applied to any population (fossil, asexual, or sexual), and (2) it is logical because different species have different synapomorphies only if they are isolated from gene flow and have evolved independently. The approach has a distinct disadvantage, however: Carefully estimated phylogenies are available only for a tiny (though growing) subset of populations on the tree of life.

Critics of this approach also point out that it would probably lead to recognition of many more species than either the morphospecies or biological species concept. Proponents counter that, far from being a disadvantage, the recognition of increased numbers of species might better reflect the extent of life's diversity.

🔑 In actual practice, researchers use all three species concepts to identify evolutionarily independent populations in nature. The concepts are summarized in **Table 26.2**. Conflicts have occurred, however, when different species concepts are applied to the real world. To appreciate this point, consider the case of the dusky seaside sparrow.

Species Definitions in Action: The Case of the Dusky Seaside Sparrow

Seaside sparrows live in salt marshes along the Atlantic and Gulf Coasts of the United States. The scientific name of this species is *Ammodramus maritimus*. (Recall from Chapter 1 that scientific names consist of a genus name followed by a species name.)

Using the morphospecies concept, researchers had traditionally named a variety of seaside sparrow "subspecies."

Subspecies are populations that live in discrete geographic areas and have distinguishing features, such as coloration or calls, but are not considered distinct enough to be called separate morphospecies.

Because salt marshes are often destroyed for agriculture or oceanfront housing, by the late 1960s biologists began to be concerned about the future of some seaside sparrow populations. A subspecies called the dusky seaside sparrow (*A. m. nigrescens*) was in particular trouble; by 1980 only six individuals from this population remained. All were males.

At this point government and private conservation agencies sprang into action under the auspices of the Endangered Species Act, a law whose goal is to prevent the extinction of species. The law uses the biological species concept to identify species and calls for the rescue of endangered species through active management.

Because current populations of seaside sparrows are physically isolated from one another, and because young seaside sparrows tend to breed near where they hatched, researchers believed that little to no gene flow occurred among populations. Under the biological species concept and morphospecies concept, there may be as many as six species of seaside sparrow (**Figure 26.3a** on page 462). The dusky seaside sparrow subspecies became a priority for conservation efforts because it was reproductively isolated.

To launch the rescue program, the remaining male dusky seaside sparrows were taken into captivity and bred with females from a nearby subspecies: *A. m. peninsulae*. Officials planned to use these hybrid offspring as breeding stock for a reintroduction program. The goal was to preserve as much genetic diversity as possible by reestablishing a healthy population of dusky-like birds. The plan was thrown into turmoil, however, when a different group of biologists estimated the phylogeny of the seaside sparrows by comparing gene sequences.

The tree in **Figure 26.3b** shows that seaside sparrows represent just two distinct monophyletic groups: one native to the Atlantic Coast and the other native to the Gulf Coast. Under the phylogenetic species concept, only two species of seaside sparrow exist. Far from being an important, reproductively isolated population, the phylogeny showed that the dusky sparrow is part of

SUMMARY TABLE 26.2 **Species Concepts**

	Criterion for Identifying Populations as Species	Advantages	Disadvantages
Biological	Reproductive isolation between populations (they don't breed and don't produce viable offspring)	Reproductive isolation = evolutionary independence	Not applicable to asexual or fossil species; difficult to assess if populations do not overlap geographically.
Morphospecies	Morphologically distinct populations	Widely applicable	Subjective (researchers often disagree about how much or what kinds of morphological distinction indicate speciation); misses cryptic species.
Phylogenetic	Smallest monophyletic group on phylogenetic tree	Widely applicable; based on testable criteria	Relatively few well-estimated phylogenies are currently available.

(a) Each subspecies of seaside sparrow has a restricted range.

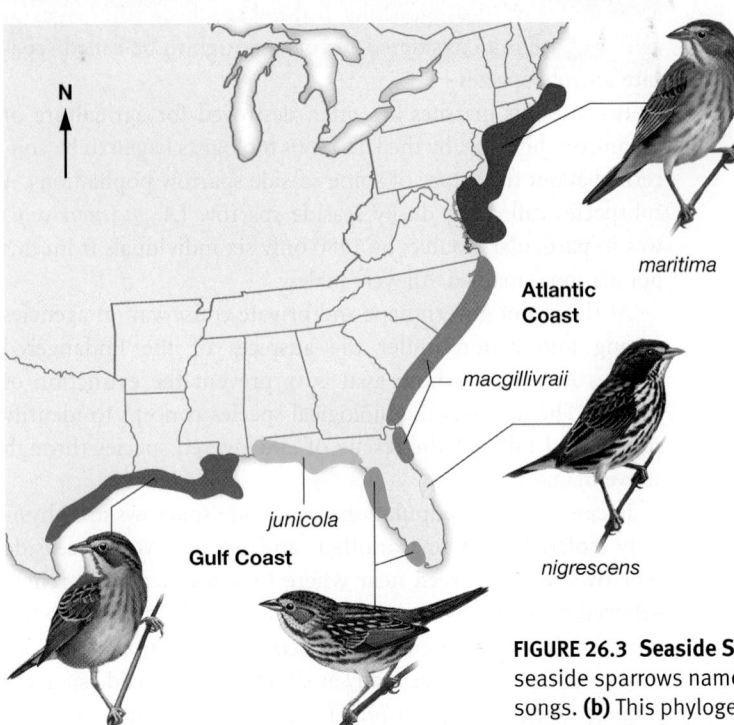

(b) The six subspecies form two monophyletic groups when DNA sequences are compared.

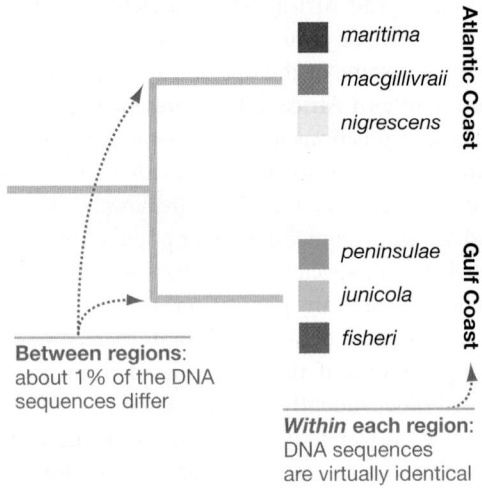

Between regions: about 1% of the DNA sequences differ

Within each region: DNA sequences are virtually identical

FIGURE 26.3 Seaside Sparrows Form Two Monophyletic Groups. (a) The subspecies of seaside sparrows named on this map are distinguished by their distinctive coloration and songs. **(b)** This phylogenetic tree was constructed by comparing DNA sequences. The tree shows that seaside sparrows represent two distinct monophyletic groups, one native to the Atlantic Coast and the other native to the Gulf Coast.

✔**QUESTION** If you were a conservation biologist and could save only two subspecies of seaside sparrows from extinction, would you choose two subspecies from the Atlantic Coast, two from the Gulf Coast, or one from the Atlantic and one from the Gulf? Explain why.

the same monophyletic group that includes the other Atlantic Coast sparrows.

Further, officials had unwittingly crossed the dusky males (*A. m. nigrescens* from the Atlantic Coast) with females from the Gulf Coast lineage. Because the goal of the conservation effort was to preserve existing genetic diversity, this was the wrong population to use.

The researchers who did the phylogenetic analysis maintained that the biological and morphospecies concepts had misled a well-intentioned conservation program. Under the phylogenetic species concept, they claimed that officials should have allowed the dusky sparrow to go extinct and then concentrated their efforts on simply preserving one or more populations from each coast. In this way, the two monophyletic groups of sparrows—and the most genetic diversity—would be preserved.

Under the morphospecies concept, however, officials did the right thing by preserving distinct types. They argue that dusky seaside sparrows had distinctive, heritable traits like coloration and songs that are now lost forever.

When conservation funding is scarce, life-and-death decisions like these are crucial. Now our task is to consider an even more fundamental question: How do isolation and divergence produce the event called speciation?

CHECK YOUR UNDERSTANDING

🔑 If you understand that . . .

- Species are evolutionarily independent because no gene flow occurs between them and other species.
- Biologists use an array of criteria to identify evolutionarily independent groups.

✔ **You should be able to . . .**

1. Explain why the criteria invoked by the biological, morphological, and phylogenetic species concepts allow biologists to identify evolutionarily independent groups.

2. Describe the disadvantages of the biological, morphological, and phylogenetic species concepts.

Answers are available in Appendix B.

26.2 Isolation and Divergence in Allopatry

Speciation begins when gene flow between populations is reduced or eliminated. 🔑 Genetic isolation happens routinely when populations become physically separated. Physical isolation, in turn, occurs in one of two ways: dispersal or vicariance.

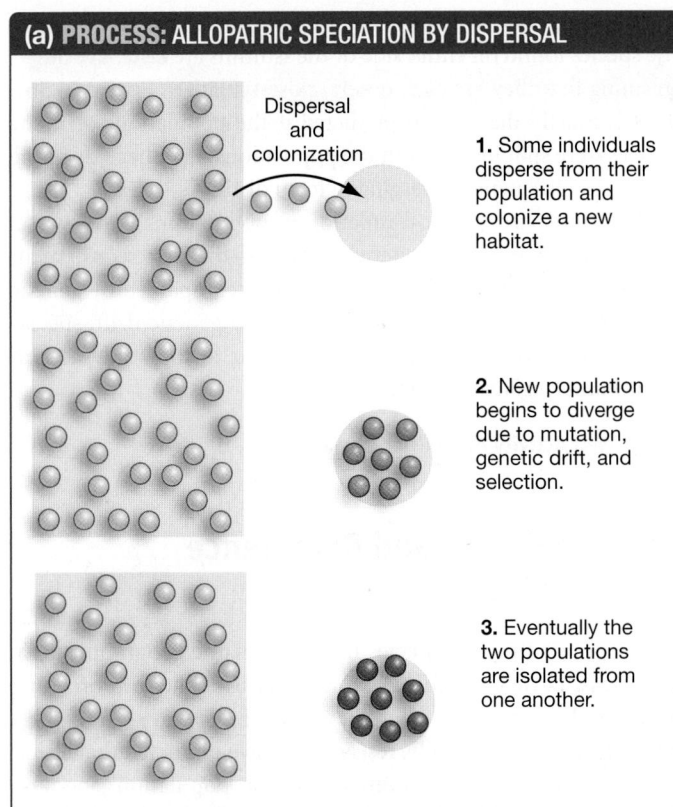

(a) PROCESS: ALLOPATRIC SPECIATION BY DISPERSAL

1. Some individuals disperse from their population and colonize a new habitat.

Dispersal and colonization

2. New population begins to diverge due to mutation, genetic drift, and selection.

3. Eventually the two populations are isolated from one another.

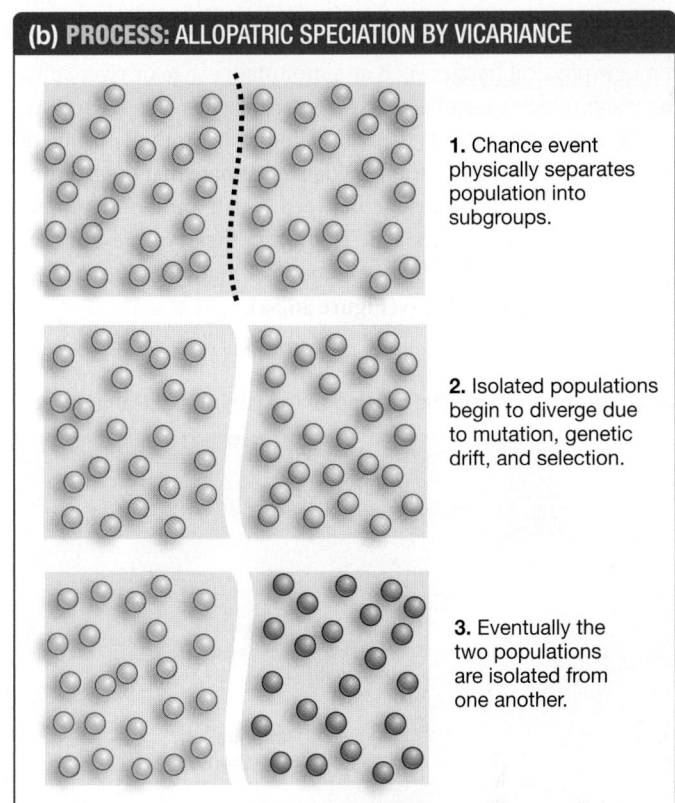

(b) PROCESS: ALLOPATRIC SPECIATION BY VICARIANCE

1. Chance event physically separates population into subgroups.

2. Isolated populations begin to diverge due to mutation, genetic drift, and selection.

3. Eventually the two populations are isolated from one another.

FIGURE 26.4 Allopatric Speciation Begins via Dispersal or Vicariance. (a) When dispersal occurs, colonists establish a new population in a novel location. **(b)** In vicariance, a widespread population becomes fragmented into isolated subgroups.

As **Figure 26.4a** illustrates, a population can disperse to a new habitat, colonize it, and found a new population. Alternatively, a new physical barrier can split a widespread population into two or more subgroups that are physically isolated from each other (**Figure 26.4b**). A physical splitting of habitat is called **vicariance**.

Speciation that begins with physical isolation via either dispersal or vicariance is known as **allopatric** ("different-homeland") **speciation**. Populations that live in different areas are said to be in **allopatry**.

The case studies that follow address two questions: How do colonization and range-splitting events occur? Answering this question takes us into the field of **biogeography**—the study of how species and populations are distributed geographically. Once populations are physically isolated, how do genetic drift and selection produce divergence?

Dispersal and Colonization Isolate Populations

Peter Grant and Rosemary Grant witnessed a colonization event while working in the Galápagos Islands off the coast of South America. Recall from Chapter 24 that the Grants have been studying medium ground finches on the island of Daphne Major since 1971. In 1982 five members of a new species, called the large ground finch, arrived and began nesting. These colonists

had apparently dispersed from a population that lived on a nearby island in the Galápagos. Because finches normally stay on the same island year-round, the colonists represented a new population, allopatric with their source population.

Could this colonization event lead to speciation? To evaluate this question, Grant and Grant caught, weighed, and measured most of the parents and offspring produced on Daphne Major over the succeeding 12 years. When they compared these data with measurements of large ground finches in other populations, they discovered that the average beak size in the new population was much larger.

Two evolutionary processes could be responsible for the change in beak size:

1. Genetic drift produced a colonizing population that happened to have particularly large beaks relative to the source population.

2. Natural selection in the new environment could favor alleles associated with large beaks.

The general message here is that the characteristics of a colonizing population are likely to be different from the characteristics of the source population due to chance—that is, genetic drift. Subsequent natural selection may extend the rapid divergence that begins with genetic drift.

Colonization, followed by genetic drift and natural selection, is thought to be responsible for speciation in Galápagos finches and many other island groups.

Vicariance Isolates Populations

If a new physical barrier such as a mountain range or river splits the geographic range of a species, vicariance has taken place. An example of vicariance in speciation comes from the Isthmus of Panama in Central America. Geologists estimate that the isthmus closed, forming a land bridge between North and South America, about 3 million years ago. This was a vicariance event that split Caribbean and Pacific populations of many marine species, including snapping shrimp (**Figure 26.5a**).

(a) Vicariance event: The closing of the Isthmus of Panama

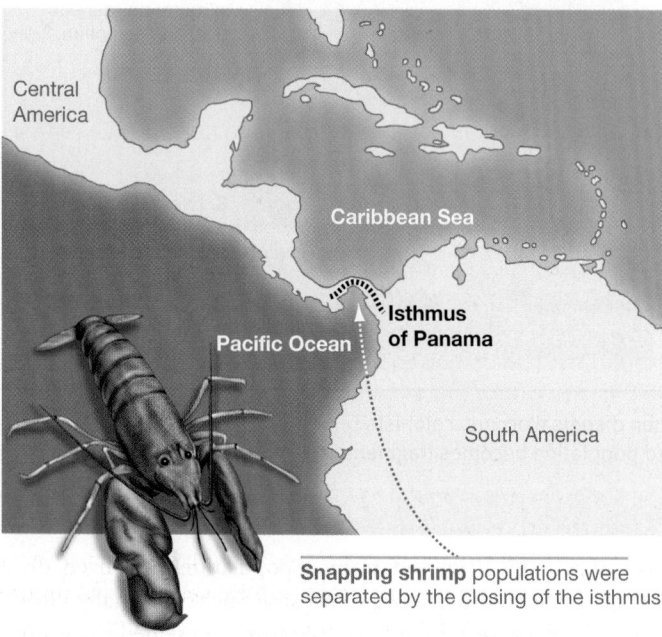

Snapping shrimp populations were separated by the closing of the isthmus

(b) Result: Pairs of sister species straddling the isthmus

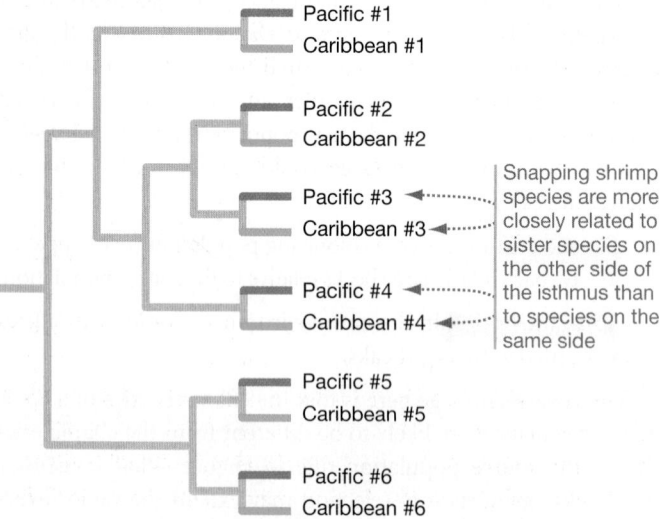

Snapping shrimp species are more closely related to sister species on the other side of the isthmus than to species on the same side

FIGURE 26.5 Evidence for Speciation by Vicariance in Snapping Shrimp. The most logical interpretation of this phylogenetic tree is that 6 species of snapping shrimp split into 12 species when the Isthmus of Panama formed and separated the Caribbean Sea from the Pacific Ocean.

A phylogenetic tree of the snapping shrimp shows that many of the species found on either side of the Isthmus are **sister species**—meaning that they are each others' closest relative (**Figure 26.5b**). This is exactly the pattern predicted if the populations of many species were split in two, with the populations on either side of the Isthmus subsequently diverging to form distinct species.

To summarize, physical isolation of populations via dispersal or vicariance produces genetic isolation—the first requirement of speciation. When genetic isolation is accompanied by genetic divergence due to mutation, selection, and genetic drift, speciation results. To review these concepts, go to the study area at *www.masteringbiology.com*:

 Web Activity Allopatric Speciation

26.3 Isolation and Divergence in Sympatry

When populations or species live in the same geographic area, or at least close enough to one another to make interbreeding possible, biologists say that they live in **sympatry** ("together-homeland"). Traditionally, researchers have predicted that speciation could not occur among sympatric populations, because gene flow is possible (**Figure 26.6**). The prediction was that gene

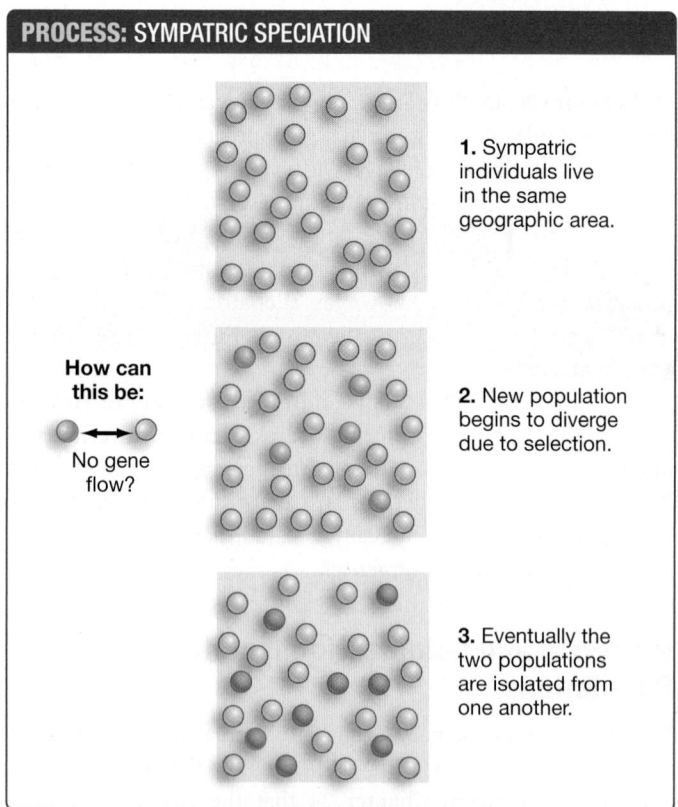

FIGURE 26.6 Sympatric Speciation Has Long Perplexed Researchers. Recent studies have documented several mechanisms that reduce gene flow and result in sympatric speciation.

flow would easily overwhelm any differences among populations created by genetic drift and natural selection. As Chapter 25 showed, gene flow can homogenize gene frequencies even between populations that are allopatric, such as an island population close to a population on a continent.

In general, gene flow overwhelms the diversifying force of natural selection and prevents speciation. Is this always the case?

Can Natural Selection Cause Speciation Even When Gene Flow Is Possible?

Recently, several well-documented examples have upset the traditional view that **sympatric speciation**—speciation that occurs even though gene flow is possible—is rare or nonexistent. These studies are fueling a growing awareness that under certain circumstances, natural selection that causes populations to diverge can overcome gene flow and cause speciation.

The key realization is that even though sympatric populations are not physically isolated, they may be isolated by preferences for different habitats. As an example, let's consider research on speciation in apple maggot flies.

After apple maggot flies court and mate on apple fruits, females lay an egg inside. After the egg hatches, the larva eats and grows as the fruit ripens and drops to the ground. The larva leaves, burrows into the ground, and pupates—meaning that it secretes a protective case and undergoes metamorphosis (see Chapter 32). Individuals emerge as adults the following spring, starting the cycle anew.

Apple trees were introduced to North America from Europe less than 300 years ago, however. Where did apple maggot flies come from?

Phylogenetic trees, estimated from synapomorphies in DNA sequence data, indicate that apple maggot flies are extremely closely related to hawthorn flies, which are native to North America. Hawthorn flies lay their eggs in hawthorn fruits.

Hawthorn trees and apple trees often grow almost side by side. But by following marked individuals in the field, biologists have determined that only about 6 percent of the total matings observed are between apple flies and hawthorn flies. The data in **Figure 26.7** show why. The bars on the graphs indicate the percentage of apple flies (top) or hawthorn flies (bottom) that land on a surface containing scents from apple, hawthorn, both apple and hawthorn, or neither apple nor hawthorn, in laboratory tests. Note that:

- Apple flies respond most strongly to apple scents; hawthorn flies respond most strongly to hawthorn scents.

- In both types of flies, there is no difference in the response to a mix of both scents and no scent at all.

- Apple flies avoid hawthorn scent, and hawthorn flies avoid apple scent—both types respond less to the "wrong" scent than to no scent at all.

Other experiments have (1) established that a fly's ability to discriminate scents has a genetic basis—meaning that apple flies and hawthorn flies have different alleles associated with attraction to fruit; (2) identified the specific odor receptor cells responsible

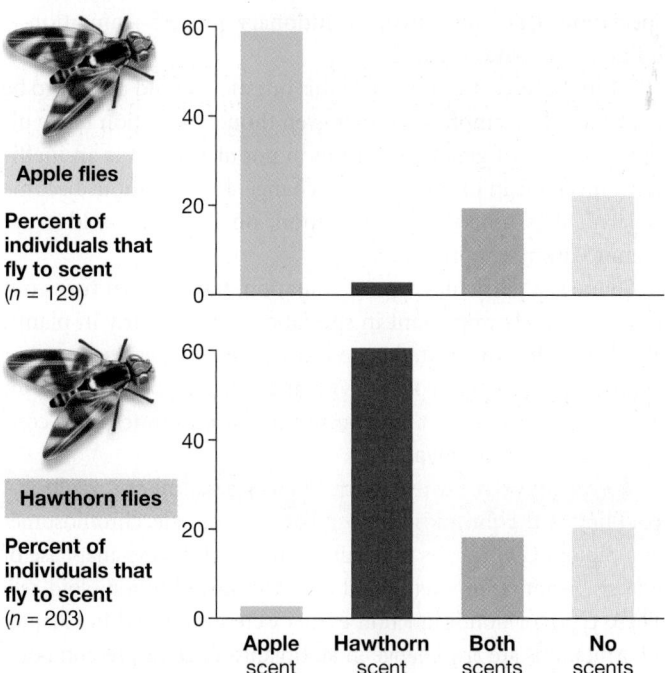

FIGURE 26.7 Disruptive Selection on Fruit Preference in Flies. Each fly was tested with four types of scent, one at a time, in a lab setting.

✓**QUESTION** To promote divergence, why is it important for apple maggot flies to mate on their food plants?

for the difference in scent response, and (3) shown that hybrid individuals cannot discriminate between apple and hawthorn fruit.

The upshot is that apple flies mate on apples and hawthorn flies mate on hawthorn fruits. Disruptive selection—a process introduced in Chapter 25—is occurring. The two populations are becoming genetically isolated.

Although they are not yet separate species on the basis of the biological, morphological, or phylogenetic species concepts, apple flies and hawthorn flies are diverging. They are currently in the process of becoming distinct species.

✓If you understand the speciation process occurring in these flies, you should be able to predict why natural selection would favor divergence based on two observations: (1) the sugars and other molecules in apple fruit and hawthorn fruit are different, and (2) apple fruits drop much earlier than hawthorn fruits—meaning that apple larvae develop in warmer temperatures and endure a longer winter before emerging.

Although the apple maggot fly's story might seem localized and specific, the events may be common. Biologists currently estimate that over 3 million insect species exist. Most of these species are associated with specific host plants. Based on the data from apple maggot flies, it is reasonable to hypothesize that switching host plants has been a major trigger for speciation throughout the course of insect evolution.

How Can Polyploidy Lead to Speciation?

Based on the theory and data reviewed thus far, it is clear that gene flow, genetic drift, and natural selection play important roles in speciation. Can the fourth evolutionary process—mutation—influence speciation as well?

At first glance, the answer to this question would appear to be no. Chapter 25 emphasized that even though mutation is the ultimate source of genetic variation in populations, it is an inefficient mechanism of evolutionary change. If populations become isolated, it is unlikely that mutation, on its own, could cause them to diverge appreciably.

There is a particular type of mutation, though, that turns out to be extremely important in speciation—particularly in plants. The key is that the mutation reduces gene flow between mutant and normal, or wild-type, individuals. It does so because mutant individuals have more than two sets of chromosomes. This condition is known as **polyploidy**.

Polyploidy occurs when an error in meiosis or mitosis results in a doubling of the chromosome number. For example, chromosomes in a diploid ($2n$) species may fail to pull apart during anaphase of mitosis, resulting in a tetraploid cell ($4n$) instead of a diploid cell. These types of nondisjunction events were introduced in Chapter 12. Mutations are any change in an organism's DNA present; polyploidization is a massive mutation that affects entire chromosomes.

To understand why polyploid individuals are genetically isolated from wild-type individuals, consider what happens when that tetraploid cell undergoes meiosis to form gametes and mates with a diploid individual (**Figure 26.8**).

- By meiosis, diploid individuals produce haploid gametes and tetraploid individuals produce diploid gametes. These gametes unite to form a triploid ($3n$) zygote.

- Even if this offspring develops normally and reaches sexual maturity, its three homologous chromosomes cannot synapse and separate correctly during meiosis. Thus, they are distributed to daughter cells unevenly. Virtually all of the gametes produced by the triploid individual end up with an uneven number of chromosomes.

- Because almost all of its gametes contain a dysfunctional set of chromosomes, the triploid individual is virtually sterile.

🔑 Tetraploid and diploid individuals rarely produce fertile offspring when they mate. As a result, tetraploid and diploid populations are reproductively isolated.

How do the polyploid individuals involved in speciation form? There are two general mechanisms:

1. **Autopolyploid** ("same-many-form") individuals are produced when a mutation results in a doubling of chromosome number and the chromosomes all come from the same species.

2. **Allopolyploid** ("different-many-form") individuals are created when parents that belong to different species mate and produce an offspring where chromosome number doubles. Allopolyploid individuals have chromosome sets from different species.

Let's consider specific examples to illustrate how speciation by polyploidy occurs.

AUTOPOLYPLOIDY Although autopolyploidy is thought to be much less common than allopolyploidy, biologists recently doc-

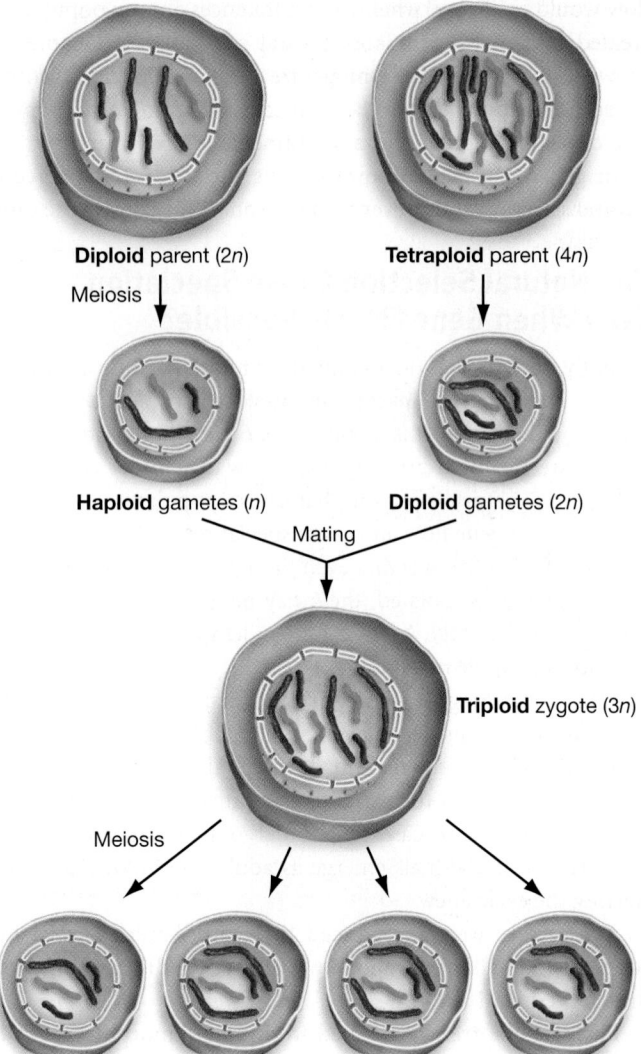

Diploid parent ($2n$) **Tetraploid** parent ($4n$)

Meiosis

Haploid gametes (n) **Diploid** gametes ($2n$)

Mating

Triploid zygote ($3n$)

Meiosis

Gametes have uneven chromosome numbers

FIGURE 26.8 Polyploidy Can Lead to Reproductive Isolation. The mating diagrammed here illustrates why tetraploid individuals are reproductively isolated from diploid individuals.

umented autopolyploidy in the maidenhair fern. This plant inhabits woodlands across North America. During the normal life cycle of a fern, individuals alternate between a haploid (n) stage and a diploid ($2n$) stage.

Biologists initially set out to do a routine survey of allelic diversity in a population of these ferns. They happened to be examining individuals in the haploid stage and found several individuals that had *two* versions of each gene instead of just one. These individuals were diploid even though they had the "haploid" growth form. The biologists followed these individuals through their life cycle and confirmed that when the ferns mated, they produced offspring that were tetraploid ($4n$). The researchers had stumbled upon polyploid mutants within a normal population.

To follow up on the observation, researchers located the parent of the mutant individuals. The parent turned out to have a defect in meiosis that caused non-disjunction of chromosomes. Instead of producing normal, haploid cells as a result of meiosis, the mutant individual produced diploid cells. These diploid cells eventually led to the production of diploid gametes.

Because maidenhair ferns can self-fertilize, the diploid gametes could combine to form tetraploid offspring. The tetraploid offspring could then self-fertilize or mate with their tetraploid parent or each other. If the process continued, a polyploid population of maidenhair ferns would be established.

Polyploid individuals are genetically isolated from the original diploid population and thus evolutionarily independent, because tetraploid individuals can breed with other tetraploids but not with diploids. If genetic drift and selection then caused the two populations to diverge, speciation would be under way.

✔If you understand how autopolyploidy works, you should be able to create a scenario explaining how the process gave rise to a tetraploid grape with extra-large fruit, from a diploid population with smaller fruit. (You've probably seen both types of fruit in the supermarket.)

This autopolyploidy study documented the critical first step in speciation—the establishment of genetic isolation. In this population of maidenhair ferns, as in apple maggot flies, speciation is under way right before our eyes.

ALLOPOLYPLOIDY New tetraploid species may be created when two diploid species hybridize. **Figure 26.9a** shows how. The top three drawings show a diploid offspring forming from a mating between two different species. Because the offspring has chromosomes that do not pair normally during meiosis, it is sterile. But if a mutation occurs that doubles the chromosome number in this individual prior to or during meiosis, then each chromosome gains a homolog. When these homologs synapse, meiosis can proceed, and diploid gametes are produced. When diploid gametes fuse during self-fertilization, a tetraploid individual results.

Exactly this chain of events occurred after three European species of weedy plants in the genus *Tragopogon* were introduced to western North America in the early 1900s. In 1950 a biologist described the first of two new tetraploid species that have been discovered. Based on an analysis of their chromosomes, both were clearly the descendants of matings between the introduced diploids. Follow-up work has shown that at least one of the new tetraploid species is expanding its geographic range (**Figure 26.9b**).

✔If you understand how alloploidy works, you should be able to create a scenario explaining how a cross between a tetraploid population called Emmer wheat and a wild, diploid wheat gave rise to the hexaploid bread wheat grown throughout the world today.

WHY IS SPECIATION BY POLYPLOIDY SO COMMON IN PLANTS? The claim that speciation by polyploidization has been particularly important in plants is backed by the observation that many diploid species have close relatives that are polyploid. Three

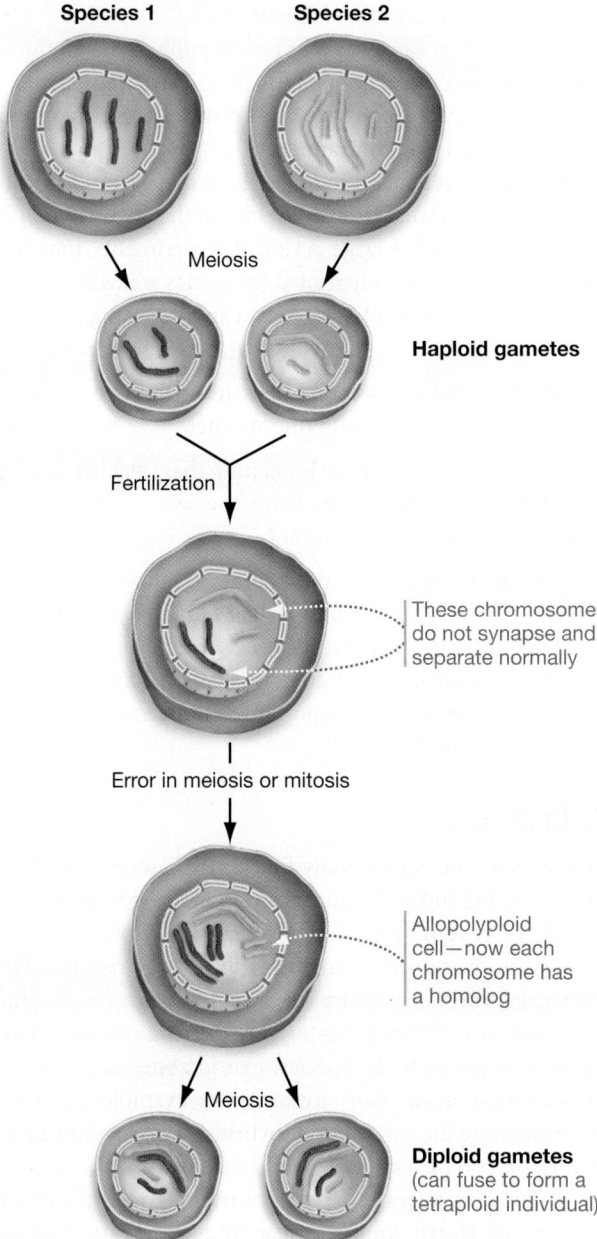

(a) If chromosome doubling occurs, allopolyploid offspring can be fertile and form new species.

Species 1　　　Species 2

Meiosis

Haploid gametes

Fertilization

These chromosomes do not synapse and separate normally

Error in meiosis or mitosis

Allopolyploid cell—now each chromosome has a homolog

Meiosis

Diploid gametes (can fuse to form a tetraploid individual)

FIGURE 26.9 Allopolyploids Can Form New Species.

(b) An allopolyploid species that formed recently.

One of the **diploid species** introduced to North America (*Tragopogon dubius*)

New tetraploid species resulting from allopolyploidy (*Tragopogon mirus*)

properties of plants have been noteworthy in making this mode of speciation possible:

1. In plants, somatic cells that have undergone many rounds of mitosis can undergo meiosis and produce gametes. If sister chromatids separate during anaphase of one of these mitotic divisions but do not migrate to opposite poles, the result can be a tetraploid daughter cell that later undergoes meiosis to form diploid gametes.

2. The ability of some plant species to self-fertilize makes it possible for diploid gametes to fuse and create genetically isolated tetraploid populations.

3. Hybridization between plant species is common, creating opportunities for speciation via formation of allopolyploids.

To summarize, speciation by polyploidization is driven by chromosome-level mutations and occurs in sympatry (**Table 26.3** on page 471). Compared to the gradual process of speciation by geographic isolation or by disruptive selection in sympatry, speciation by polyploidy is virtually instantaneous. It is fast, sympatric, and common. Before moving on, be sure to go to the study area at *www.masteringbiology.com* and review how polyploidization can trigger speciation.

 Web Activity Speciation by Changes in Ploidy

CHECK YOUR UNDERSTANDING

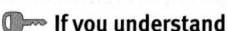

🔑 **If you understand that . . .**

- Speciation occurs when populations become isolated genetically and then diverge due to selection, genetic drift, or mutation.

✔ **You should be able to . . .**

Give an example of an event that can lead to the genetic isolation of populations, and explain why selection and drift would cause the populations to diverge.

Answers are available in Appendix B.

26.4 What Happens When Isolated Populations Come into Contact?

Suppose two populations that have been isolated come into contact again. If divergence has taken place and if divergence has affected when, where, or how individuals in the populations mate, then it is unlikely that interbreeding will take place. In cases such as this, prezygotic isolation exists. When it does, mating between the populations is rare, gene flow is minimal, and the populations continue to diverge.

But what if prezygotic isolation does not exist, and the populations begin interbreeding? The simplest outcome is that the populations fuse over time, as gene flow erases any distinctions between them. Several other possibilities exist, however. Let's explore three of them: reinforcement, hybrid zones, and speciation by hybridization.

Reinforcement

If two populations have diverged extensively and are distinct genetically, it is reasonable to expect that their hybrid offspring will have lower fitness than their parents. The logic here is that if populations are well-adapted to different habitats, then hybrid offspring will not be well-adapted to either habitat. If the two populations have diverged enough genetically, hybrid offspring also may fail to develop normally or may be infertile.

When postzygotic isolation occurs, there should be strong selection against interbreeding because hybrid offspring represent a wasted effort on the part of parents. Individuals that do not interbreed, due to a different courtship ritual or pollination system or other form of prezygotic isolation, should be favored because they produce more viable offspring.

Natural selection for traits that isolate populations in this way is called **reinforcement**. The name is descriptive because the selected traits reinforce differences that evolved while the populations were isolated from one another.

Some of the best data on reinforcement come from laboratory studies of closely related fruit fly species in the genus *Drosophila*. Researchers analyzed a large series of experiments that tested whether members of closely related fly species are willing to mate with one another. The biologists found an interesting pattern:

- If closely related species are sympatric—meaning that they live in the same area—individuals from the two species are seldom willing to mate with one another.

- If the species are allopatric—meaning that they live in different areas—then individuals from the two species are often willing to mate with one another.

The pattern is logical because natural selection can act to reduce mating between species only if their ranges overlap. Thus, it is reasonable to find that sympatric species exhibit prezygotic isolation but that allopatric species do not. There is a long-standing debate, however, over just how important reinforcement is in groups other than the genus *Drosophila*.

Hybrid Zones

Hybrid offspring are not always dysfunctional. In some cases they are capable of mating and producing offspring and have features that are intermediate between those of the two parental populations. When this is the case, hybrid zones can form. A **hybrid zone** is a geographic area where interbreeding occurs and hybrid offspring are common.

Depending on the fitness of hybrid offspring and the extent of breeding between parental species, hybrid zones can be narrow or wide, and long or short lived. As an example of how researchers analyze the dynamics of hybrid zones, let's consider recent work on two bird species.

Townsend's warblers and hermit warblers live in the coniferous forests of North America's Pacific Northwest. In western

(a) Hybrids have intermediate characteristics.

Townsend's warbler ▣

Townsend's-hermit hybrid ▨

Hermit warbler ▤

(b) Hybrids inherit species-specific mtDNA sequences from their mothers.

Pacific Ocean

N ↑

Individuals that look like Townsend's warblers but have hermit mtDNA

● All individuals have Townsend's mtDNA

◑ Some individuals have Townsend's mtDNA, others have hermit mtDNA

○ All individuals have hermit mtDNA

Present hybrid zone where two ranges meet

FIGURE 26.10 Analyzing a Hybrid Zone. (a) When Townsend's warblers and hermit warblers hybridize, the offspring have intermediate characteristics. **(b)** Map showing the current range of Townsend's and hermit warblers. The small pie charts show the percentage of individuals with Townsend's warbler mtDNA (in black) and hermit warbler mtDNA (in white).

Washington State, where their ranges overlap, the two species hybridize extensively. As **Figure 26.10a** shows, hybrid offspring have characteristics that are intermediate relative to the two parental species.

To explore the dynamics of this hybrid zone, a team of biologists examined gene sequences in the mitochondrial DNA (mtDNA) of a large number of Townsend's, hermit, and hybrid warblers collected from forests throughout the region. The team found that each of the parental species has certain species-specific mtDNA sequences. This result allowed the researchers to infer how hybridization was occurring.

To grasp the reasoning here, it is critical to realize that mtDNA is maternally inherited in most animals and plants. If a hybrid individual has Townsend's mtDNA, its mother had to be a Townsend's warbler while its father had to be a hermit warbler. In this way, identifying mtDNA types allowed the research team

to infer whether Townsend's females were mating with hermit males, or vice versa, or both.

Their data presented a clear pattern: Most hybrids form when Townsend's males mate with hermit warbler females. One of the investigators followed up on this result with experiments showing that Townsend's males are extremely aggressive in establishing territories and that they readily attack hermit warbler males. The data suggest that Townsend's males invade hermit territories, drive off the hermit males, and mate with hermit females.

The team also found something completely unexpected. When they analyzed the distribution of mtDNA types along the Pacific Coast and in the northern Rocky Mountains, they found that many Townsend's warblers actually had hermit mtDNA. **Figure 26.10b** shows that in some regions—such as the larger islands off the coast of British Columbia—*all* of the warblers had hermit mtDNA, even though they looked like full-blooded Townsend's warblers.

How could this be? To explain the result, the team hypothesized that hermit warblers were once found as far north as Alaska and that Townsend's warblers have gradually taken over their range. Their logic is that repeated mating with Townsend's warblers over time made the hybrid offspring look more and more like Townsend's, even while maternally inherited mtDNA kept the genetic record of the original hybridization event intact.

If this hypothesis is correct, then the hybrid zone should continue moving south. If it does so, hermit warblers may eventually become extinct. In many cases, however, hybridization does not lead to extinction but rather leads to the opposite—the creation of new species.

New Species through Hybridization

A team of researchers recently examined the relationships of three sunflower species native to the American West: *Helianthus annuus*, *H. petiolaris*, and *H. anomalus*. The first two of these species are known to hybridize in regions where their ranges overlap. The third species, *H. anomalus*, resembles these hybrids and is pictured at the start of this chapter.

DNA sequencing studies have shown that some gene regions in *H. anomalus* are remarkably similar to those found in *H. annuus*, while other gene sequences are almost identical to those found in *H. petiolaris*. Based on these data, biologists hypothesize that *H. anomalus* originated in hybridization between *H. annuus* and *H. petiolaris*. All three species have the same number of chromosomes, so neither allopolyploidy nor autopolyploidy was involved. Instead, the chromosomes of *H. annuus* and *H. petiolaris* must be similar enough that they can synapse and undergo normal meiosis in hybrid offspring.

If this interpretation is correct, then hybridization must be added to the list of ways that new species can form. The specific hypothesis here is that *H. annuus* and *H. petiolaris* were isolated and diverged as separate species, and later began interbreeding. The hybrid offspring created a third, new species that had unique combinations of alleles from each parental species and therefore different characteristics.

This hypothesis is supported by the observation that *H. anomalus* grows in much drier habitats than either of the parental species—suggesting that a unique combination of alleles allowed *H. anomalus* to thrive in certain habitats.

Biologists set out to test the hybridization hypothesis by trying to re-create the speciation event experimentally (**Figure 26.11**).

Step 1 They mated individuals from the two parental species and raised the offspring in a greenhouse.

Step 2 When these hybrid individuals were mature, the researchers either mated the plants to other hybrid individuals or "backcrossed" them to individuals from one of the parental species.

Step 3 This breeding program continued for four more generations before the experiment ended. Ultimately, the experimental lines were backcrossed twice, and they were mated to other hybrid offspring three times.

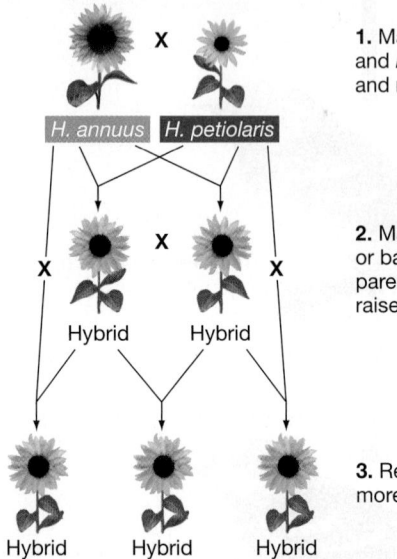
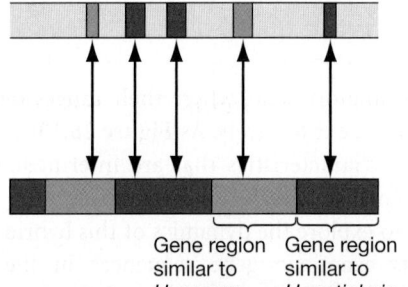
FIGURE 26.11 Experimental Evidence That New Species Can Originate in Hybridization Events.

SOURCE: Rieseberg, L. H., B. Sinervo, et al. 1996. Role of gene interactions in hybrid speciation: Evidence from ancient and experimental hybrids. *Science* 272: 741–745.

✔ **QUESTION** Why it is valid to use experiments with living organisms—like this one—to infer what happened during historical events?

SUMMARY TABLE 26.3 **Mechanisms of Sympatric Speciation**

SUMMARY TABLE 26.3 **Mechanisms of Sympatric Speciation**

	Process	Example
Disruptive selection	Genetic divergence is caused by natural selection for different habitats or resources. Must be accompanied by some mechanism of genetic isolation.	Apple maggot flies that are adapted to breed on apple versus hawthorn fruits mate on those fruits, so little or no interbreeding occurs.
Polyploidization	Genetic isolation is created by formation of polyploid individuals (mutants with more than two sets of chromosomes) that can breed only with each other.	Particularly common in plants because frequent hybridization occurs between species and many mitotic divisions occur prior to meiosis.
• *Autopolyploidy*	Autopolyploids gain duplicate chromosome sets from the same species due to chromosome doubling during mitosis or meiosis.	A maidenhair fern individual became tetraploid due to an error in meiosis.
• *Allopolyploidy*	Allopolyploids gain duplicate chromosome sets from different species due to a hybridization event, followed by chromosome doubling.	*Tragopogon* species introduced to North America hybridized and formed offspring that became tetraploid and formed new species.

The goal of these crosses was to simulate matings that might have occurred naturally.

The experimental hybrids looked like the natural hybrid species, but did they resemble them genetically? To answer this question, the research team constructed genetic maps of each population, using a large series of genetic markers similar to the types of markers introduced in Chapter 19 and Chapter 20. Because each parental population had a large number of unique markers in their genomes, the research team hoped to identify which genes found in the experimental hybrids came from which parental species.

Some of their data are diagrammed in the Results section of Figure 26.11. The bottom bar in the illustration represents a region called S in the genome of the naturally occurring species *Helianthus anomalus*. As the legend indicates, this region con-tains three sections of sequences that are also found in *H. petiolaris* (indicated with the color red) and two that are also found in *H. annuus* (indicated with the color orange). The top bar shows the composition of this same region in the genome of the experimental hybrid lines.

The key observation here is that the genetic composition of the synthesized hybrids matches that of the naturally occurring hybrid species. In effect, the researchers had succeeded in re-creating a speciation event. Their results provide strong support for the hybridization hypothesis for the origin of *H. anomalus*.

Secondary contact of two populations can produce a dynamic range of possible outcomes: fusion of the populations, reinforcement of divergence, founding of stable hybrid zones, extinction of one population, or the creation of new species. **Table 26.4** summarizes the outcomes of secondary contact.

SUMMARY TABLE 26.4 **Possible Outcomes of Secondary Contact between Populations**

	Process	Example
Fusion of the populations	The two populations freely interbreed.	Occurs whenever populations of the same species come into contact.
Reinforcement of divergence	If hybrid offspring have low fitness, natural selection favors the evolution of traits that prevent interbreeding between the populations.	Appears to be common in fruit fly species that occupy the same geographic areas.
Hybrid zone formation	There is a well-defined geographic area where hybridization occurs. This area may move over time or be stable.	Many stable hybrid zones have been described; the hybrid zone between hermit and Townsend's warblers appears to have moved over time.
Extinction of one population	If one population or species is a better competitor for shared resources, then the poorer competitor may be driven to extinction.	Townsend's warblers may be driving hermit warblers to extinction.
Creation of new species	If the combination of genes in hybrid offspring allows them to occupy distinct habitats or use novel resources, they may form a new species.	Hybridization between sunflowers gave rise to a new species with unique characteristics.

Summary of Key Concepts

Speciation occurs when populations of the same species become genetically isolated by lack of gene flow and then diverge from each other due to selection, genetic drift, or mutation.

- Speciation is a splitting event in which one lineage gives rise to two or more independent descendant lineages.

- Speciation begins when populations become genetically isolated—meaning that gene flow does not occur.

- Once populations are isolated, natural selection, genetic drift, and mutation act on them independently. As a result, the populations diverge over time.

- Eventually, the populations become so different that they can be recognized as distinct species—evolutionarily independent populations.

  ✔ You should be able to design an experiment that would, given enough time, result in the production of two species from a single ancestral population.

Populations can be recognized as distinct species if they are reproductively isolated from each other, if they have distinct morphological characteristics, or if they form independent branches on a phylogenetic tree.

- Researchers use several criteria to test whether populations represent distinct species.

- The biological species concept focuses on the degree of hybridization between species to determine whether gene flow is occurring.

- The morphospecies concept infers that speciation has occurred if populations have distinctive morphological traits.

- The phylogenetic species concept identifies species as the smallest monophyletic groups on the tree of life.

  ✔ You should be able to explain whether human populations would be considered separate species under the biological, morphological, and phylogenetic species concepts.

Populations can become genetically isolated from each other if they occupy different geographic areas, if they use different habitats

or resources within the same area, or if one population is polyploid and cannot breed with the other.

- Speciation often begins when small groups of individuals colonize a new habitat or when a large, continuous population becomes fragmented into isolated habitats.

- When populations are sympatric, speciation can occur when disruptive selection favors individuals that breed in different habitats.

- Mutations that produce polyploidy can trigger rapid speciation in sympatry because they lead to reproductive isolation between diploid and tetraploid populations.

  ✔ You should be able to evaluate whether your experiment on speciation (see above) represents a case of dispersal, vicariance, different habitat use, or polyploidy.

(MB) **Web Activity** Allopatric Speciation, **Web Activity** Speciation by Changes in Ploidy

When populations that have diverged come back into contact, they may fuse, continue to diverge, stay partially differentiated, or have offspring that form a new species.

- If gene flow occurs, populations that have diverged may fuse into a single species.

- If prezygotic isolation exists, populations that come back into contact will probably continue to diverge.

- Secondary contact can lead to reinforcement—the evolution of mechanisms that prevent hybridization.

- Gene flow between different species can lead to the formation of hybrid zones that move over time or are stable.

- In some cases, hybridization between species can create new species with unique combinations of traits.

  ✔ You should be able to predict how a hybrid zone will change over time when hybrid offspring have higher fitness than the parental populations.

Questions

1. What distinguishes a morphospecies?
   a. It has distinctive characteristics, such as size, shape, or coloration.
   b. It represents a distinct twig in a phylogeny of populations.
   c. It is reproductively isolated from other species.
   d. It is a fossil from a distinct time in Earth history.

2. When does vicariance occur?
   a. Small populations coalesce into one large population.
   b. A population is fragmented into isolated subpopulations.
   c. Individuals colonize a novel habitat.
   d. Individuals disperse and found a new population.

3. Why is "reinforcement" an appropriate name for the concept that natural selection should favor divergence and genetic isolation if populations experience postzygotic isolation?
   a. Selection should reinforce high fitness for hybrid offspring.
   b. Selection should reinforce the fact that they are "good species" under the morphological species concept.
   c. Selection acts because hybrid offspring do not develop at all or are sterile when mature.
   d. It reinforces selection for divergence that began when the species were geographically isolated.

4. The biological species concept can be applied only to which of the following groups?
   a. bird species living today
   b. dinosaurs
   c. bacteria
   d. archaea

5. Why are genetic isolation and genetic divergence occurring in apple maggot flies, even though populations occupy the same geographic area?
   a. Different populations feed and mate on different types of fruit.
   b. One population is tetraploid; others are diploid.
   c. The introduction of a nonnative host plant caused vicariance.
   d. Responses to scents have changed due to disruptive selection.

6. When the ranges of different species meet, a stable "hybrid zone" occupied by hybrid individuals may form. How is this possible?
   a. Hybrid individuals may have intermediate characteristics that are advantageous in a given region.
   b. Hybrid individuals are always allopolyploid and are thus unable to mate with either of the original species.
   c. Hybrid individuals may have reduced fitness and thus be strongly selected against.
   d. One species has a selective advantage, so as hybridization continues, the other species will go extinct.

TEST YOUR UNDERSTANDING

Answers are available in Appendix B

1. Which studies in this chapter represent direct observation of speciation? Which are indirect studies of historical events?

2. In the case of the seaside sparrow, how did the species identified by the biological species concept, the morphospecies concept, and the phylogenetic species concept conflict?

3. Explain why genetic drift occurs during colonization events. Explain why natural selection occurs after colonization events.

4. Explain how isolation and divergence are occurring in apple maggot flies. Of the four evolutionary processes (mutation, gene flow, drift, and selection), which two are most important in causing this event?

5. Unlike animal gametes, plant reproductive cells do not differentiate until late in life. Explain why plants are much more likely to produce diploid gametes and polyploid offspring than are animals.

6. Individuals from closely related fruit fly species are less likely to mate if the related species are sympatric versus allopatric. Why is this considered strong evidence for reinforcement?

APPLYING CONCEPTS TO NEW SITUATIONS

Answers are available in Appendix B

1. A large amount of gene flow is now occurring among human populations due to intermarriage among people from different ethnic groups and regions of the world. Is this phenomenon increasing or decreasing racial differences in our species? Explain.

2. Humans have introduced thousands of species to new locations around the globe, forming geographically isolated populations. Few, if any, of these colonization events have resulted in speciation. Use these data to evaluate the hypothesis that founder events trigger speciation.

3. A friend says that apple maggot flies prefer apple fruit scents because they need to, in order to survive. Another agrees and adds that the flies acquire the ability to distinguish the apple scents by spending time on the fruit, and that's why their offspring prefer apples. What's wrong with these statements?

4. All over the world, natural habitats are being fragmented into tiny islands as suburbs, ranches, and farms expand. Explain why this fragmentation process could lead to extinction. Then explain how it could lead to speciation.

A fossilized trilobite. The last trilobites disappeared during a mass extinction event analyzed in Section 27.4.

27 Phylogenies and the History of Life

KEY CONCEPTS

- Phylogenetic trees document the evolutionary relationships among organisms and are estimated from data.

- The fossil record provides physical evidence of organisms that lived in the past.

- Adaptive radiations are a major pattern in the history of life. They are instances of rapid diversification associated with new ecological opportunities and new morphological innovations.

- Mass extinctions have occurred repeatedly throughout the history of life. They are environmental catastrophes that rapidly eliminate most of the species alive.

This chapter is about time and change. More specifically, it's about vast amounts of time and profound change in organisms. Both of these topics can be difficult for humans to grasp. Our lifetimes are measured in decades, and our knowledge of history is usually measured in centuries or millennia. But this chapter analyzes events that occurred over millions and even billions of years. A million years is completely beyond our experience and almost beyond our imagination.

It takes practice to get comfortable analyzing the profound changes that occur in organisms over deep time. To help you get started, the chapter begins by introducing the two major analytical tools that biologists use to reconstruct the history of life: phylogenetic trees and the fossil record. The remaining two sections explore episodes called adaptive radiations, which produce large numbers of highly diverse new species, and mass extinctions, which wipe out large numbers of species.

27.1 Tools for Studying History: Phylogenetic Trees

The evolutionary history of a group of organisms is called its **phylogeny**. Phylogenies are usually summarized and depicted in the form of a phylogenetic tree. A **phylogenetic tree** shows the ancestor-descendant relationships among populations or species, and clarifies who is related to whom. In a phylogenetic tree:

- a **branch** represents a population through time;

- the point where two branches diverge, called a **node** (or fork), represents the point in time when an ancestral species split into two or more descendant species; and

- a **tip** (or terminal node), the endpoint of a branch, represents a group (a species or larger taxon) that is living today or ended in extinction.

✔ When you see this checkmark, stop and test yourself. Answers are available in Appendix B.

Evolutionary trees have been introduced at various points in earlier chapters—often with just enough information to help you understand a new concept. **BioSkills 3** in Appendix A introduces the parts of a phylogenetic tree and how to read one. (It would be a *very* good idea to review **BioSkills 3**, right now.) Here let's focus on how biologists go about building them.

How Do Researchers Estimate Phylogenies?

Phylogenetic trees are an extremely effective way of summarizing data on the evolutionary history of a group of organisms. But like any other pattern or measurement in nature, from the average height of a person in a particular human population to the speed of a passing airplane, the genealogical relationships among species cannot be known with absolute certainty. Instead, the relationships depicted in an evolutionary tree are estimated from data.

To infer the historical relationships among species, researchers analyze the species' morphological or genetic characteristics, or both. For example, to reconstruct relationships among fossil species of humans, scientists analyze aspects of tooth, jaw, and skull structure. To reconstruct relationships among contemporary human populations, investigators usually compare the sequences of bases in a particular gene.

The fundamental idea in phylogeny inference is that closely related species should share many of their characteristics, while distantly related species should share fewer characteristics. But there are two general strategies for using data to estimate trees: the phenetic approach and the cladistic approach.

The **phenetic approach** to estimating trees is based on computing a statistic that summarizes the overall similarity among populations, based on the data. For example, researchers might use gene sequences to compute an overall "genetic distance" between two populations. A genetic distance summarizes the average percentage of bases in a DNA sequence that differ between two populations. A computer program then builds a tree that clusters the most similar populations and places more-divergent populations on more-distant branches.

The **cladistic approach** to inferring trees is based on the realization that relationships among species can be reconstructed by identifying shared derived characters in the species being studied—the synapomorphies introduced in Chapter 26. Recall that a synapomorphy is a trait that certain groups of organisms have that exists in no others. Synapomorphies allow biologists to recognize monophyletic groups—also called clades or lineages. Synapomorphies are characteristics that are shared because they are derived from traits that existed in their common ancestor.

Figure 27.1 illustrates the logic behind a cladistic analysis. When the ancestral population at the left of the figure splits into two descendant lineages at node A, each descendant group begins evolving independently and acquires unique traits. These traits are derived from their common ancestor via mutation, selection, and genetic drift.

An **ancestral trait** is a characteristic that existed in an ancestor; a **derived trait** is one that is a modified form of the ancestral trait,

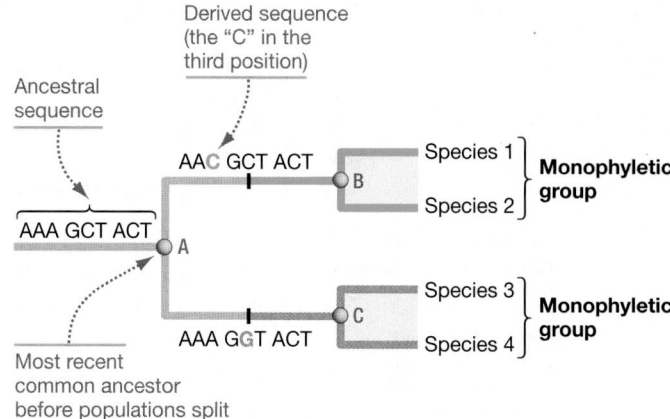

FIGURE 27.1 Synapomorphies Identify Monophyletic Groups. In this example, synapomorphies are changes in a DNA sequence.

found in a descendant. It's important to recognize that ancestral and derived traits are relative. If you are comparing mammals with the fossil forms called mammal-like reptiles, then fur and lactation are derived traits. But if you are comparing whales and humans, then fur and lactation are ancestral traits.

As an example, the traits used in a cladistic analysis might be a particular region of DNA that changes as follows:

AAA GCT ACT	ancestral population
AAC GCT ACT	a descendant population
AAA GGT ACT	another descendant population

When the two lineages themselves split at nodes B and C, the species that result share the derived characteristics. In this way, the ancestor at node B and species 1 and 2 can be recognized as a monophyletic group. Similarly, the ancestor at node C and species 3 and 4 can be recognized as a different monophyletic group. When many such traits have been measured, a computer program can be used to identify which traits are unique to each monophyletic group and then place the groups in a tree in the correct relationship to each other.

How Can Biologists Distinguish Homology from Homoplasy?

Although the logic behind phenetic and cladistic analyses is elegant, problems arise. The issue is that traits can be similar in two species not because those traits were present in a common ancestor, but because similar traits evolved independently in two distantly related groups.

In the example given in Figure 27.1, it is possible that species 2 is not at all closely related to species 1. Its ancestors may have had the sequence TAT GGT AGT, which happened to change to AAC GCT ACT due to mutation, selection, and drift that took place independently of the changes that took place in the ancestors of species 1.

WHAT IS HOMOPLASY? **Homology** (literally, "same-source") occurs when traits are similar due to shared ancestry; **homoplasy**

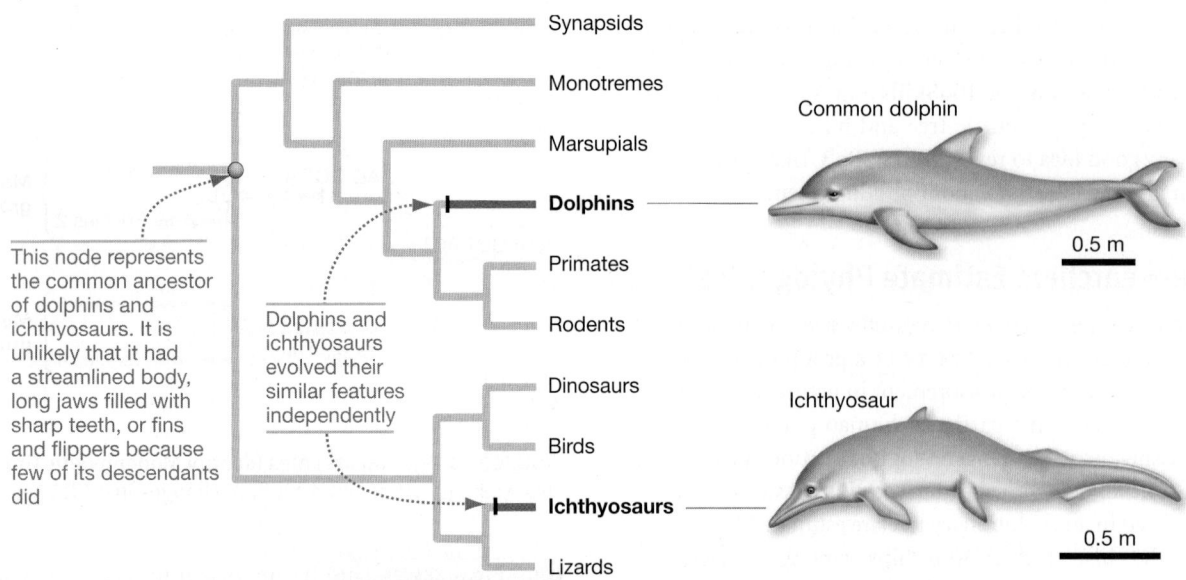

This node represents the common ancestor of dolphins and ichthyosaurs. It is unlikely that it had a streamlined body, long jaws filled with sharp teeth, or fins and flippers because few of its descendants did

Dolphins and ichthyosaurs evolved their similar features independently

Synapsids
Monotremes
Marsupials
Dolphins — Common dolphin
Primates
Rodents
Dinosaurs
Birds — Ichthyosaur
Ichthyosaurs
Lizards

0.5 m

0.5 m

FIGURE 27.2 Homoplasy: Traits Are Similar but Not Inherited from a Common Ancestor. Dolphins and ichthyosaurs look similar but are not closely related—dolphins are mammals; ichthyosaurs are reptiles.

("same-form") occurs when traits are similar for reasons other than common ancestry. For example, the aquatic reptiles called ichthyosaurs were strikingly similar to modern dolphins (**Figure 27.2**). Both are large marine animals with streamlined bodies and large dorsal fins. Both chase down fish and capture them between elongated jaws filled with dagger-like teeth. But no one would argue that ichthyosaurs and dolphins are similar because the traits they share existed in a common ancestor. As the phylogeny in Figure 27.2 shows, analyses of other traits show that ichthyosaurs are reptiles whereas dolphins are mammals.

Based on these data, it is logical to argue that the similarities between ichthyosaurs and dolphins result from a common cause of homoplasy called convergent evolution. **Convergent evolution** occurs when natural selection favors similar solutions to the problems posed by a similar way of making a living. But convergent traits, or what biologists once called analogous traits, do not occur in the common ancestor of the similar species. Streamlined bodies and elongated jaws filled with sharp teeth are adaptations that help *any* species—whether it is a reptile or a mammal—chase down fish in open water.

EVIDENCE FOR HOMOLOGY In many cases, homology and homoplasy are much more difficult to distinguish than in the ichthyosaur and dolphin example. How do biologists recognize homology in such cases? As an example, consider the *Hox* genes of insects and vertebrates introduced in Chapter 21. Even though insects and vertebrates last shared a common ancestor some 600–700 million years ago, biologists argue that their *Hox* genes are derived from the same ancestral sequences. There are several lines of evidence to support this hypothesis:

- Groups of *Hox* genes are organized on chromosomes in a similar way. **Figure 27.3** shows that *Hox* genes in both insects and vertebrates are found in gene complexes, with similar genes

found adjacent to one another on the chromosome. Recall from Chapter 20 that genes with these characteristics are called gene families. The organization of the gene families is nearly identical among insect and vertebrate species.

- All of the *Hox* genes share a 180-base-pair sequence called the homeobox. The portion of a protein encoded by the homeobox (introduced in Chapter 23) is almost identical in insects and vertebrates and has a similar function. It binds to DNA and regulates the expression of other genes.

- The products of the *Hox* genes have similar functions (see Chapter 21): identifying the locations of cells in embryos. They are also expressed in similar patterns in time and space.

In addition, many other animals, on lineages that branched off between insects and mammals, have similar genes. This is a crucial observation: If similar traits found in distantly related lineages are indeed similar due to common ancestry, then similar traits should be found in many intervening lineages on the tree of life—because all of the species in question inherited the trait from the same common ancestor.

Now suppose that a researcher set out to infer a phylogenetic tree from a set of morphological traits or DNA sequences. Without already having a tree in hand, it can be difficult or impossible to tell which traits are homologous and qualify as synapomorphies. Just as any data set contains unavoidable errors, or "noise," the data sets used to infer phylogenies inevitably include homoplasy.

USING PARSIMONY TO MINIMIZE THE IMPACT OF "NOISE" IN THE DATA To reduce the chance that homoplasy will lead to erroneous conclusions about which species are most closely related, biologists who are using cladistic approaches invoke the logical principle of **parsimony**. Under parsimony, the most likely explanation or pattern is the one that implies the least amount of change.

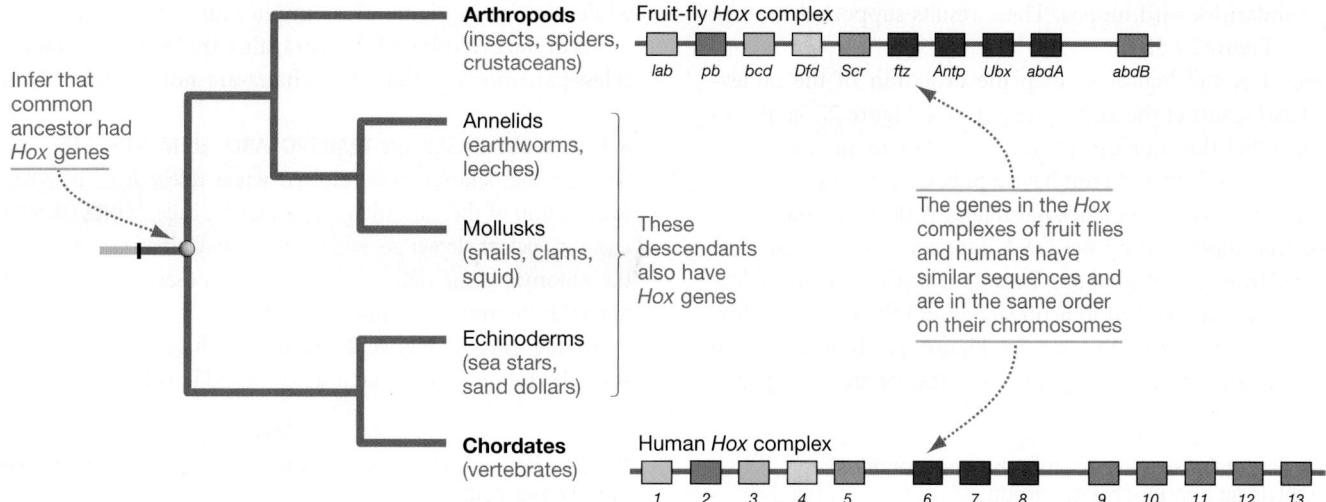

FIGURE 27.3 Homology: Similarities Are Inherited from a Common Ancestor. All of the animal groups illustrated on this phylogeny have *Hox* complexes that are similar to those illustrated for fruit flies and humans.

To implement a parsimony analysis, a computer compares the branching patterns that are theoretically possible and counts the number of changes in DNA sequences required to produce each pattern. For example, the tree in **Figure 27.4a** requires two changes in base sequence; the tree in **Figure 27.4b** requires four.

Convergent evolution and other causes of homoplasy should be rare compared with similarity due to shared descent. Thus,

the tree that implies the fewest overall evolutionary changes should be the one that most accurately reflects what really happened during evolution.

If the branching pattern in Figure 27.4a is the most parsimonious of all possible trees, then biologists conclude that it is the most likely representation of actual phylogeny based on the data in hand.

Whale Evolution: A Case History

As an example of how a cladistic approach works, consider the evolutionary relationships of the lineage of mammals called the Artiodactyla and the whales. Chapter 24 introduced the early evolution of whales—specifically how data from the fossil record support the hypothesis that whales evolved from terrestrial mammals. That chapter claimed that hippos, which are semi-aquatic, are the closest living relative of today's whales. What data back that claim?

A PHYLOGENY BASED ON MORPHOLOGICAL TRAITS Hippos, along with cows, deer, pigs, and camels, are artiodactyls. Members of this group have hooves and an even number of toes. They also share another feature: the unusual pulley shape of an ankle bone called the astragalus. Along with having feet with hooves and an even number of toes, the shape of the astragalus is a synapomorphy that identifies the artiodactyls as a monophyletic group.

These data support the tree shown in **Figure 27.5a** on page 478. Note that whales do not have an astragalus and are shown as an **outgroup** on this tree—that is, a species or group that is closely related to the monophyletic group but not part of it. It is logical to map the gain of the pulley-shaped astragalus in the ancestral population marked by a black bar in Figure 27.5a, because all of the descendants of that ancestor have the trait, but members of the outgroup do not.

A PHYLOGENY BASED ON DNA SEQUENCE DATA When researchers began comparing DNA sequences of artiodactyls and other species of mammals, however, the data showed that whales share

(a) Two changes

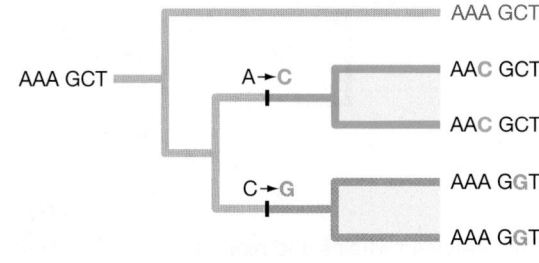

(b) Four changes

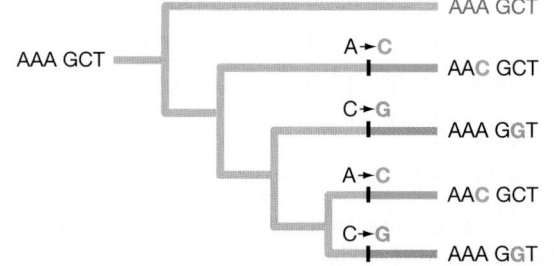

FIGURE 27.4 Parsimony Is One Method for Choosing among the Many Possible Trees. If you were using DNA sequence data to infer the phylogeny of five species, many different trees are possible. (Only two are shown here.) Under parsimony, the best tree is the one that requires the fewest changes to explain the sequences observed in the five species.

many similarities with hippos. These results supported the tree shown in **Figure 27.5b**.

Here, it is still logical to map the evolution of the pulley-shaped astragalus at the same ancestor as in Figure 27.5a. But to recognize that the trait had to have been lost in an ancestor of today's whales—which do not have a pulley-shaped astragalus—the tree contains another black bar mapping the trait loss.

The tree supported by the DNA data conflicts with the tree supported by morphological data because it implies that the pulley-shaped astragalus evolved in artiodactyls and then was lost during whale evolution. The tree in Figure 27.5b implies two changes in the astragalus (a gain *and* a loss of the astragalus),

while the tree in Figure 27.5a implies just one (a gain only). In terms of the evolution of the astragalus, the "whale + hippo" tree is less parsimonious than the "whales-are-not-artiodactyls" tree.

A PHYLOGENY BASED ON TRANSPOSABLE ELEMENTS The conflict between the data sets was resolved when researchers analyzed the distribution of the parasitic gene sequences called **SINEs (short interspersed nuclear elements)**, which occasionally insert themselves into the genomes of mammals. SINEs are transposable elements, similar to the LINEs introduced in Chapter 20.

As the data in **Figure 27.5c** show, whales and hippos share several types of SINES that are not found in other groups.

(a) The astragalus is a synapomorphy that identifies artiodactyls as a monophyletic group.

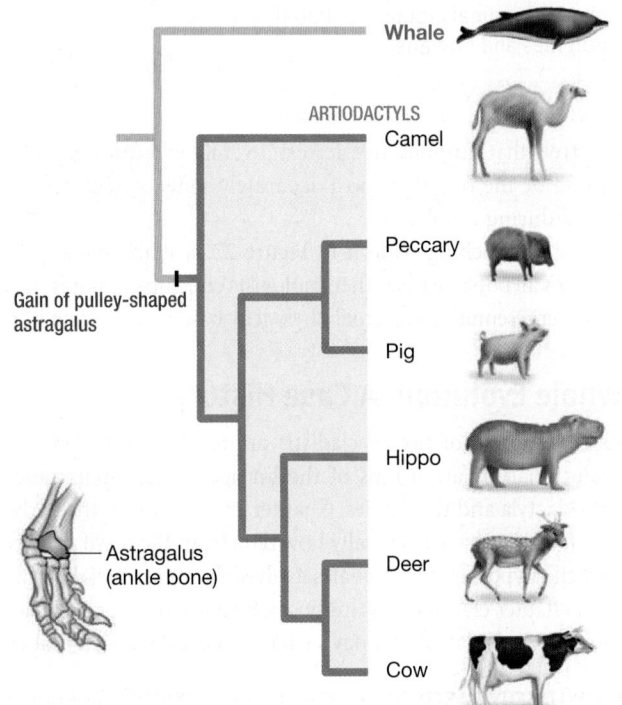

(b) If whales are related to hippos, then two changes occurred in the astragalus.

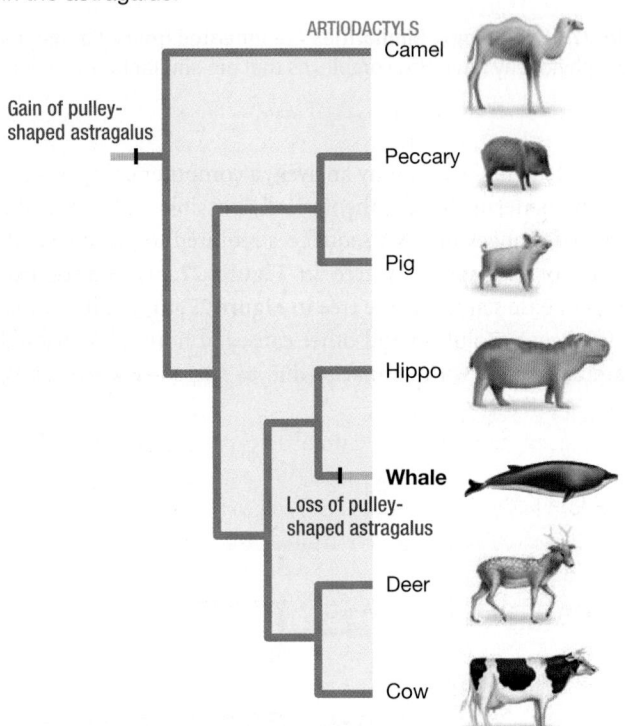

(c) Data on the presence and absence of SINE genes support the close relationship between whales and hippos.

Locus	1	2	3	4	5	6	7	8	9	10	11	12	13	14	15	16	17	18	19	20
Cow	0	0	0	0	0	0	0	1	1	1	1	1	1	1	1	1	1	1	0	0
Deer	0	0	0	0	0	0	0	1	?	1	1	1	1	1	1	?	1	1	0	0
Whale	1	1	1	1	1	1	1	0	?	1	0	1	1	0	0	0	?	1	0	0
Hippo	0	?	0	1	1	1	1	0	1	1	0	1	1	0	0	0	?	1	0	0
Pig	0	0	0	?	0	0	0	0	?	0	0	0	?	?	0	0	0	1	1	1
Peccary	?	?	?	?	?	?	?	?	?	?	?	?	?	?	?	?	?	?	1	1
Camel	0	0	0	0	0	0	0	0	0	0	0	0	0	0	0	0	0	0	0	0

1 = gene present
0 = gene absent
? = still undetermined

Whales and hippos share four *unique* SINE genes (4, 5, 6, and 7)

FIGURE 27.5 Evidence That Whales and Hippos Form a Monophyletic Group. Based on a parsimony analysis of the pulley-shaped astragalus, biologists favored **(a)** a tree excluding whales from the Artiodactyla over **(b)** the hypothesis that whales are artiodactyls and are closely related to hippos. **(c)** Data on the presence and absence of SINE genes support the tree in part (b).

✔**EXERCISE** The presence of SINE genes 4, 5, 6, 7 identifies hippos and whales as part of a monophyletic group. The presence of SINE genes 8, 11, 14, 15, 17 identifies deer and cows as part of a monophyletic group. SINE genes 10, 12, 13 identify hippos, whales, deer, and cows as part of a monophyletic group. Map the origin of these SINE genes on the tree in part (b). Where did SINE genes 19 and 20 first insert themselves into the genomes of artiodactyls?

- Whales and hippos share the SINEs numbered 4, 5, 6, and 7.
- Other SINE genes are present in some artiodactyls but not in others.
- Camels have no SINE genes at all.

To explain these data, biologists hypothesize that no SINEs were present in the population that is ancestral to all of the species in the study. Then, after the branching event that led to the split between the camels and all the other artiodactyls, different SINEs became inserted into the genomes of descendant populations.

If this hypothesis is correct, then the presence of a particular SINE represents a derived character. Because whales and hippos share four of these derived characters, it is logical to conclude that these animals are closely related.

✔ If you understand this concept, you should be able to explain why SINES numbered 4–7 are synapomorphies that identify whales and hippos as a monophyletic group, and why the similarity in these SINES is unlikely to represent homoplasy.

Based on these data, most biologists accepted the phylogeny shown in Figure 27.5b as the most accurate estimate of evolutionary history. According to this phylogeny, whales are artiodactyls and share a relatively recent common ancestor with hippos. This observation inspired the hypothesis that both whales and dolphins are descended from a population of artiodactyls that spent most of their time feeding in shallow water, much as hippos do today.

In 2001, the discovery of the fossil artiodactyls featured in Figure 24.11 supported this hypothesis in spectacular fashion. These fossil species were clearly related to whales—they have an unusual ear bone found only in whales—and yet had a pulley-shaped astragalus.

The combination of DNA sequence data and data from the fossil record has clarified how a particularly interesting group of mammals evolved. What else do fossils have to say?

CHECK YOUR UNDERSTANDING

If you understand that . . .
- Phylogenies can be estimated by finding synapomorphies that identify monophyletic groups.

✔ You should be able to . . .

Explain whether the following traits represent homoplasy or homology: hair in humans and whales; extensive hair loss in humans and whales; limbs in humans and whales; social behavior in certain whales (e.g., dolphins) and humans.

Answers are available in Appendix B.

27.2 Tools for Studying History: The Fossil Record

Phylogenetic analyses are powerful ways to infer the order in which events occurred during evolution and to understand how particular groups of species are related. ☞ But only the fossil record provides direct evidence about what organisms that lived in the past looked like, where they lived, and when they existed.

A **fossil** is a piece of physical evidence from an organism that lived in the past. The **fossil record** is the total collection of fossils that have been found throughout the world. The fossil record is housed in thousands of private and public collections.

Let's review how fossils form, analyze the strengths and weaknesses of the fossil record, and then summarize major events that have taken place in life's approximately 3.5-billion-year history. (Although most evidence suggests that life originated 3.4–3.6 billion years ago, research continues.)

How Do Fossils Form?

Most of the processes that form fossils begin when part or all of an organism is buried in ash, sand, mud, or some other type of sediment. **Figure 27.6** illustrates the leaves of a tree falling onto a patch of mud, where they are buried by soil and debris before they decay. Pollen and seeds settle into the muck at the bottom of

PROCESS: HOW FOSSILIZATION OCCURS

1. A tree lives in a swampy habitat. The tree drops leaves, pollen, and seeds into the mud, where decomposition is slow.

Seeds   Pollen   Leaves

2. The tree falls. The trunk and branches break up as they rot.

3. Flooding brings in sand and mud, burying the remains of the tree.

4. Over millions of years, the mountains erode and the swamp is filled with sediment. The habitat dries.

Sand and gravel
Buried material from swamp
Bedrock

FIGURE 27.6 Fossilization Preserves Traces of Organisms That Lived in the Past. Fossilization occurs most readily when the remains of an organism are buried in sediments, where decay is slow.

the swamp, where decomposition is slow. The stagnant water is too acidic and too oxygen-poor to support large populations of bacteria and fungi, so much of this material is buried intact before it decomposes. The trunk and branches that sit above the water line rot fairly quickly, but as pieces break off they, too, sink to the bottom and are buried.

PRESERVATION AFTER BURIAL Once burial occurs, several things can happen.

- If decomposition does not occur, the organic remains can be preserved intact—like the fossil pollen in **Figure 27.7a**.

- If sediments accumulate on top of the material and become cemented into rocks such as mudstone or shale, the sediments' weight can compress the organic material below into a thin, carbonaceous film. This happened to the leaf in **Figure 27.7b**.

- If the remains decompose *after* they are buried—as did the branch in **Figure 27.7c**—the hole that remains can fill with dissolved minerals and faithfully create a **cast** of the remains.

- If the remains rot extremely slowly, dissolved minerals can gradually infiltrate the interior of the cells and harden into stone, forming a permineralized fossil, such as petrified wood (**Figure 27.7d**).

After many centuries have passed, fossils can be exposed at the surface by erosion, a road cut, quarrying, or other processes. If researchers find a fossil, they can prepare it for study by painstakingly clearing away the surrounding rock.

If the species represented is new, researchers describe its morphology in a scientific publication, name the species, estimate the fossil's age based on dates assigned to nearby rock layers, and add the specimen to a collection so that it is available for study by other researchers. It is now part of the fossil record.

FOSSILIZATION IS A RARE EVENT The scenario just presented is based on conditions that are ideal for fossilization: The tree fell into an environment where decomposition was slow and burial was rapid. In most habitats the opposite situation occurs—decomposition is rapid and burial is slow. In reality, then, fossilization is an extremely rare event.

To appreciate this point, consider that there are 10 specimens of the first bird to appear in the fossil record, *Archaeopteryx*. All were found at the same site in Germany where limestone is quarried for printmaking (the bird's specific name is *lithographica*). If you accept an estimate that crow-sized birds native to wetland habitats in northern Europe would have a population size of around 10,000 and a life span of 10 years, and if you accept the current estimate that the species existed for about 2 million years, then you can calculate that about 2 billion *Archaeopteryx* lived. But as far as researchers currently know, only 1 out of every 200,000,000 individuals fossilized. For this species, the odds of becoming a fossil were almost 40 times worse than your odds are of winning the grand prize in a state lottery.

Limitations of the Fossil Record

Before looking at how the fossil record is used to answer questions about the history of life, it is essential to review the nature of this archive and recognize several features.

HABITAT BIAS Because burial in sediments is so crucial to fossilization, there is a strong habitat bias in the database. Organisms that live in areas where sediments are actively being deposited—including beaches, mudflats, and swamps—are much more likely to form fossils than are organisms that live in other habitats.

Within these habitats, burrowing organisms such as clams are already underground—pre-buried—at death and are therefore much more likely to fossilize. Organisms that live aboveground in dry forests, grasslands, and deserts are much less likely to fossilize.

TAXONOMIC AND TISSUE BIAS Slow decay is almost always essential to fossilization, so organisms with hard parts such as bones or shells are most likely to leave fossil evidence. This requirement introduces a strong taxonomic bias into the record. Clams, snails, and other organisms with hard parts have a much higher tendency to be preserved than do worms.

A similar bias exists for tissues within organisms. For instance, pollen grains are encased in a tough outer coat that resists decay, so they fossilize much more readily than do flowers. Shark teeth are abundant in the fossil record; but shark skeletal elements, which are made of cartilage, are almost nonexistent.

TEMPORAL BIAS Recent fossils are much more common than ancient fossils. This causes a temporal bias in the fossil record.

(a) Intact fossil (pollen) **(b)** Compression fossil (leaf) **(c)** Cast fossil (bark) **(d)** Permineralized fossil (trunk)

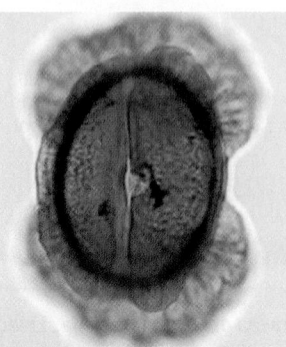

FIGURE 27.7 Fossils Are Formed in Several Ways. Different preservation processes give rise to different types of fossils.

To understand why, consider that when two of Earth's tectonic plates converge, the edge of one plate usually sinks beneath the other plate. The rocks composing the edge of the descending plate are either melted or radically altered by the increased heat and pressure they encounter as they move downward into Earth's interior. These alterations obliterate any fossils in the rock.

In addition, fossil-bearing rocks on land are constantly being broken apart and destroyed by wind and water erosion. The older a fossil is, the more likely it is to be demolished.

ABUNDANCE BIAS The fossil record has an abundance bias; it is weighted toward common species. Organisms that are abundant, widespread, and present on Earth for long periods of time leave evidence much more often than do species that are rare, local, or ephemeral.

To summarize, the fossil database represents a highly nonrandom sample of the past. **Paleontologists**—scientists who study fossils—recognize that they are limited to asking questions about tiny and scattered segments on the tree of life.

And yet, as this chapter shows, the record is a scientific treasure trove. Analyzing fossils is the only way scientists have of examining the physical appearance of extinct forms and inferring how they lived. The fossil record is like an ancient library, filled with volumes that give us glimpses of what life was like millions of years before humans appeared.

Life's Time Line

The best data available indicate that the Earth started to form about 4.6 billion years ago, and that life began around 3.5 billion years ago. To organize the tremendous sweep of time between then and now, researchers divide Earth history into segments called eons, eras, and periods.

Originally, geologists used distinctive rock formations or fossilized organisms to identify the boundaries between named time intervals. Later, researchers were able to use radiometric dating to assign absolute dates—expressed as years before the present—to events in the fossil record. Radiometric dating is based on the well-studied decay rates of certain radioactive isotopes (see Chapter 24). By dating rocks near fossils, researchers can also assign an absolute age to many of the species in the fossil record.

To summarize the history of life, researchers create time lines that record key "evolutionary firsts"—the appearance of important new lineages or innovations. It's important to recognize, though, that the times assigned to these first appearances underestimate when lineages appeared and events occurred. The reason is simple: It is possible for a particular species or lineage to exist for millions of years before leaving fossil evidence. The fossil record and efforts to date fossils are constantly improving, so time lines are always a work in progress.

PRECAMBRIAN **Figure 27.8** is a time line for the interval between the formation of Earth about 4.6 billion years ago and the

FIGURE 27.8 Major Events of the Precambrian. Life, photosynthesis, and the oxygen atmosphere all originated in the Precambrian.

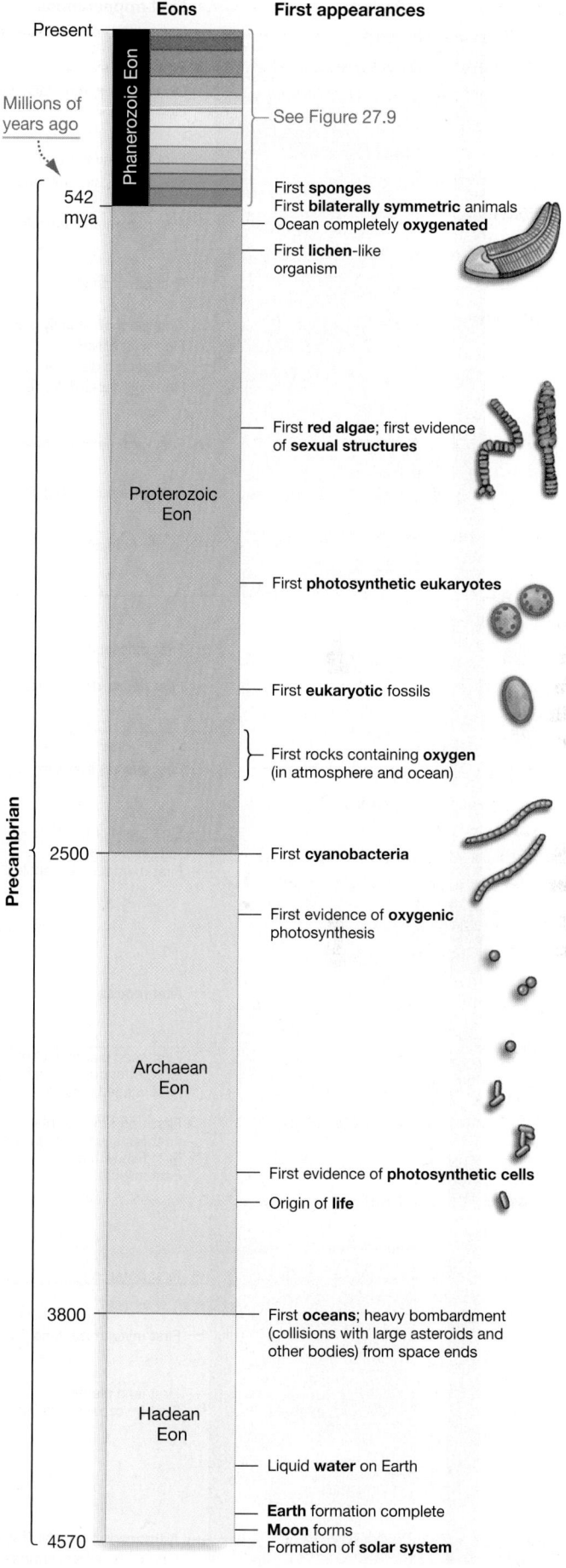

Eons **First appearances**

Present

Millions of years ago

Phanerozoic Eon

542 mya
— See Figure 27.9

First **sponges**
First **bilaterally symmetric** animals
Ocean completely **oxygenated**

First **lichen**-like organism

Proterozoic Eon

First **red algae**; first evidence of **sexual structures**

First **photosynthetic eukaryotes**

First **eukaryotic** fossils

First rocks containing **oxygen** (in atmosphere and ocean)

2500

First **cyanobacteria**

First evidence of **oxygenic** photosynthesis

Archaean Eon

First evidence of **photosynthetic cells**

Origin of **life**

3800

First **oceans**; heavy bombardment (collisions with large asteroids and other bodies) from space ends

Hadean Eon

Liquid **water** on Earth

Earth formation complete
Moon forms
Formation of **solar system**

4570

Precambrian

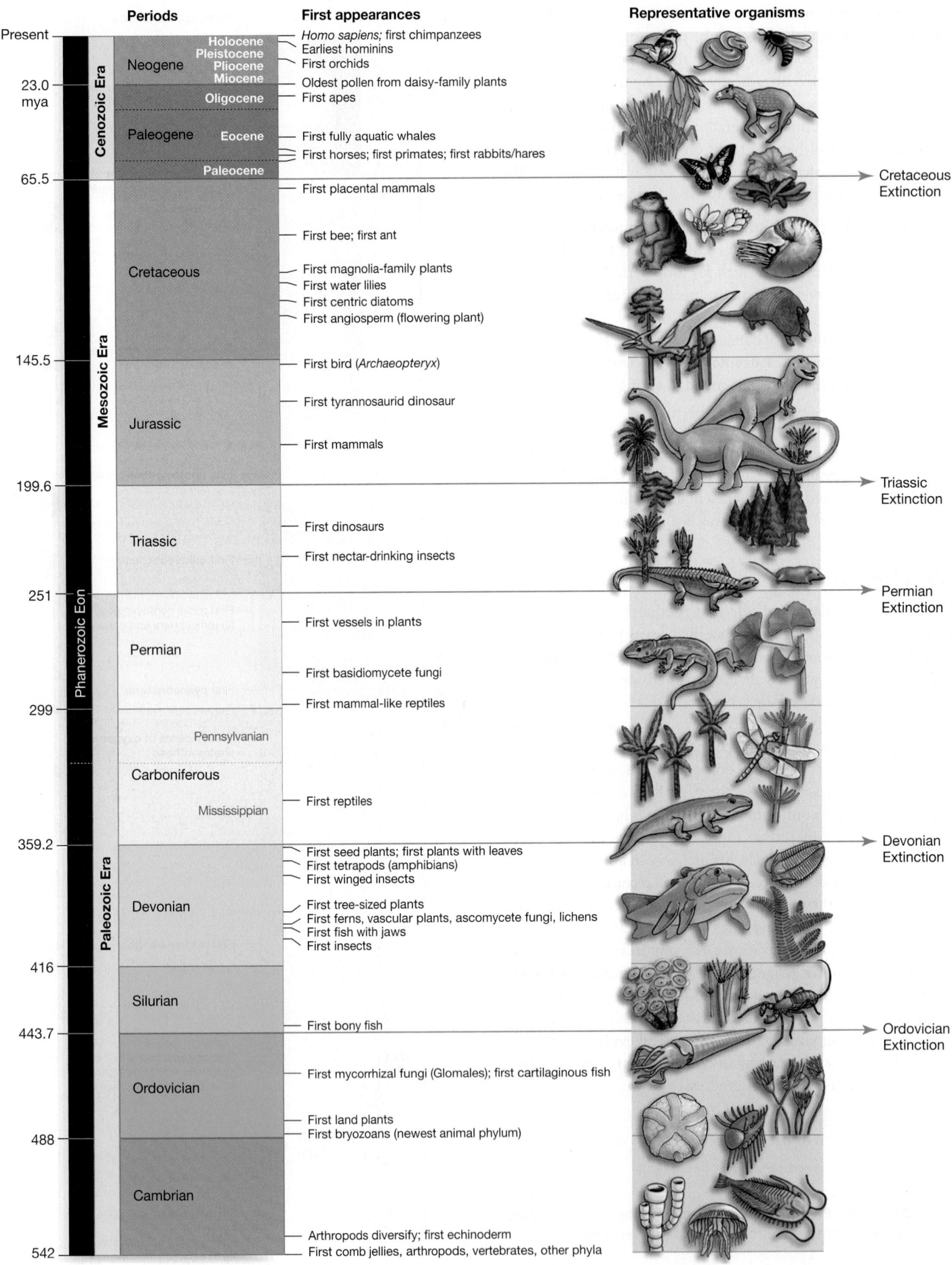

Periods

First appearances

Representative organisms

Present

Holocene
Pleistocene
Pliocene
Miocene

Neogene

Homo sapiens; first chimpanzees
Earliest hominins
First orchids

23.0 mya

Oldest pollen from daisy-family plants

Oligocene

First apes

Paleogene

Eocene

First fully aquatic whales

First horses; first primates; first rabbits/hares

Paleocene

65.5

First placental mammals

Cretaceous

First bee; first ant

First magnolia-family plants
First water lilies
First centric diatoms
First angiosperm (flowering plant)

145.5

First bird (*Archaeopteryx*)

Jurassic

First tyrannosaurid dinosaur

First mammals

199.6

Triassic

First dinosaurs

First nectar-drinking insects

251

Permian

First vessels in plants

First basidiomycete fungi

299

First mammal-like reptiles

Pennsylvanian

Carboniferous

Mississippian

First reptiles

359.2

First seed plants; first plants with leaves
First tetrapods (amphibians)
First winged insects

Devonian

First tree-sized plants
First ferns, vascular plants, ascomycete fungi, lichens
First fish with jaws
First insects

416

Silurian

443.7

First bony fish

Ordovician

First mycorrhizal fungi (Glomales); first cartilaginous fish

First land plants
First bryozoans (newest animal phylum)

488

Cambrian

542

Arthropods diversify; first echinoderm
First comb jellies, arthropods, vertebrates, other phyla

Cretaceous Extinction

Triassic Extinction

Permian Extinction

Devonian Extinction

Ordovician Extinction

Cenozoic Era

Mesozoic Era

Paleozoic Era

Phanerozoic Eon

FIGURE 27.9 Major Events of the Phanerozoic Eon (at left). The Phanerozoic began with the initial diversification of animals, continued with the evolution and early diversification of land plants and fungi, and includes the subsequent movement of animals to land. A total of five mass extinctions occurred during the Eon.

appearance of most animal groups about 542 million years ago (abbreviated mya). The entire interval is called the **Precambrian**; it is divided into the Hadean, Archaean, and Proterozoic eons.

The important things to note about the Precambrian are that:

- life was exclusively unicellular for most of Earth's history, and
- oxygen was virtually absent from the oceans and atmosphere for almost 2 billion years after the origin of life.

PHANEROZOIC EON   The interval between 542 mya and the present is called the Phanerozoic eon and is divided into three eras (**Figure 27.9**). Each of these eras is further divided into intervals called periods.

1. The **Paleozoic** ("ancient life") **era** begins with the appearance of most major animal lineages and ends with the obliteration of almost all multicellular life-forms at the end of the Permian period. The Paleozoic saw the origin and initial diversification of the animals, land plants, and fungi, as well as the appearance of land animals.

2. The **Mesozoic** ("middle life") **era** begins with the end-Permian extinction events and ends with the extinction of the dinosaurs and other groups at the boundary between the Cretaceous period and Paleogene period. In terrestrial environments of the Mesozoic, gymnosperms were the most important plants and dinosaurs were the most important vertebrates.

3. The **Cenozoic** ("recent life") **era** is divided into the Paleogene period and the Neogene period. On land, angiosperms were the most important plants and mammals were the most important vertebrates. Events that occur today are considered to be part of the Cenozoic era.

CHANGES IN THE OCEANS AND CONTINENTS   The changes in environments and life-forms that are recorded in Figures 27.8 and 27.9 took place in the context of radical changes in climate and in the extent and location of Earth's oceans and continents. Earth's crust is broken into enormous plates that are in constant motion, driven by heat rising from the planet's core.

Figure 27.10 summarizes the most recent data on how the positions of the continents and oceans have changed through time, starting at the beginning of the Phanerozoic Eon. The figure also includes notes on the prevailing climate. Over the past 542 million

FIGURE 27.10 Continental Positions during the Phanerozoic (at right). Dramatic changes in the extent and position of the continents took place during the Phanerozoic. These changes affected the total amount of land area, the relative amounts of land in the tropics versus northern latitudes, and the nature of ocean currents.

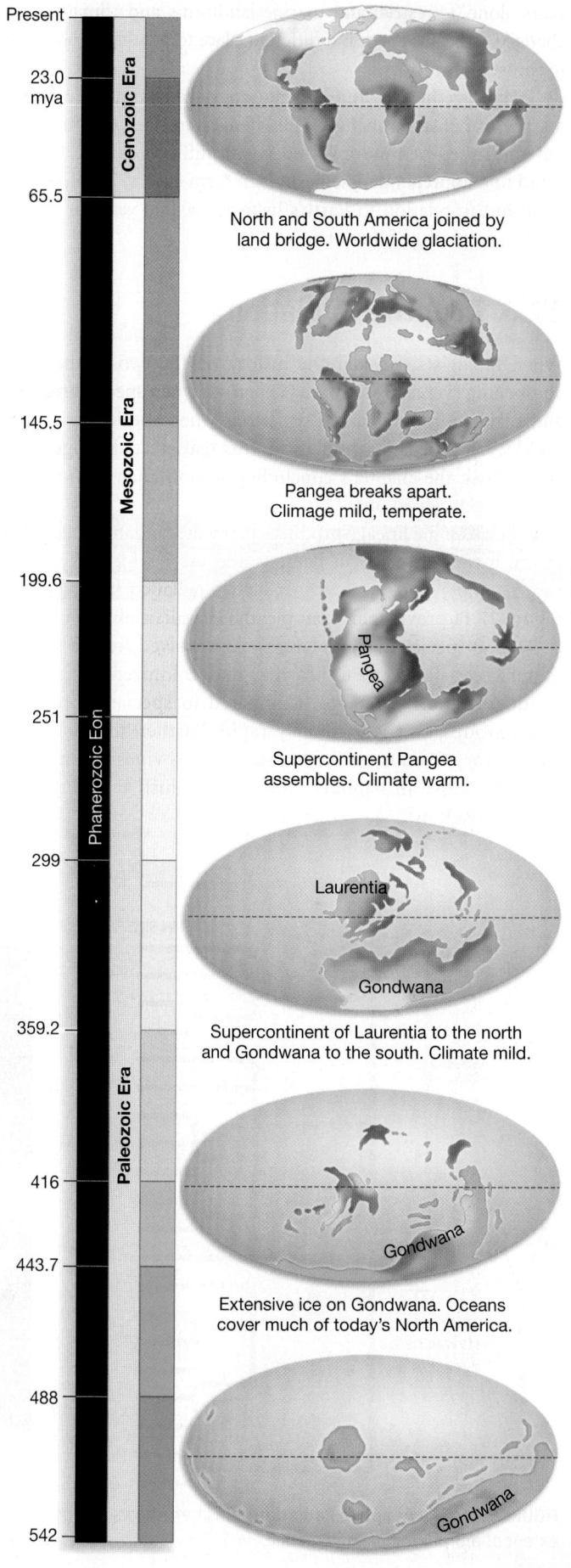

North and South America joined by land bridge. Worldwide glaciation.

Pangea breaks apart. Climage mild, temperate.

Supercontinent Pangea assembles. Climate warm.

Supercontinent of Laurentia to the north and Gondwana to the south. Climate mild.

Extensive ice on Gondwana. Oceans cover much of today's North America.

years alone, terrestrial and marine landforms and climates have changed radically. Earth is a different place today than it has been for most of its history.

Taken together, the fundamental message in time-line data is a story of constant change. The changes are well documented, but the sweep of time involved is still difficult for the human mind to comprehend. A semester can seem long to a college student; but in relation to Earth's history, 100,000 years or even a million years is the blink of an eye.

27.3 Adaptive Radiation

When biologists consider the history of life, two of the most compelling events to study are (1) periods when species originate and diversify rapidly, and (2) periods when species go extinct rapidly. Let's focus on dramatic events that create biological diversity first; the chapter's concluding section analyzes how that diversity gets wiped out.

When a single lineage produces many descendant species that live in a wide diversity of habitats and use a wide array of resources, biologists say that an **adaptive radiation** has occurred. **Figure 27.11** provides an example: the Hawaiian silverswords.

The 30 species in this plant lineage evolved from a species of tarweed, native to California, that colonized the islands about 5 million years ago. Compared to speciation rates in other groups, this is extremely rapid. Further, today's silverswords vary from mosslike mat-formers to vines to shrubs to trees. They live in habitats ranging from lush rain forests to austere lava flows.

The Hawaiian silverswords fulfill the three hallmarks of an adaptive radiation: (1) they are a monophyletic group, (2) they speciated rapidly, and (3) they diversified ecologically—meaning, in terms of the resources they use and the habitats they occupy. Biologists use the term **niche** (pronounced *nitch*) to describe the range of resources that a species can use and the range of conditions that it can tolerate. Silverswords occupy a wide array of niches.

Why Do Adaptive Radiations Occur?

Adaptive radiations are a major pattern in the history of life. But why do some lineages diversify rapidly, while others do not?

Two general mechanisms can trigger adaptive radiations: new resources, and new ways to exploit resources. Let's consider each in turn.

ECOLOGICAL OPPORTUNITY Ecological opportunity—meaning the availability of new or novel types of resources—has driven a wide array of adaptive radiations. For example, biologists explain the diversification of silverswords by hypothesizing that few other flowering plant species were present on the Hawaiian Islands 5 million years ago. With few competitors, the descendants of the colonizing tarweed were able to grow in a wide range of habitats. Over time, some became specialized for growth on dry or wet sites; others evolved the different growth forms illustrated in Figure 27.11.

The same type of ecological opportunity was hypothesized to explain the adaptive radiation of *Anolis* lizards on islands in the

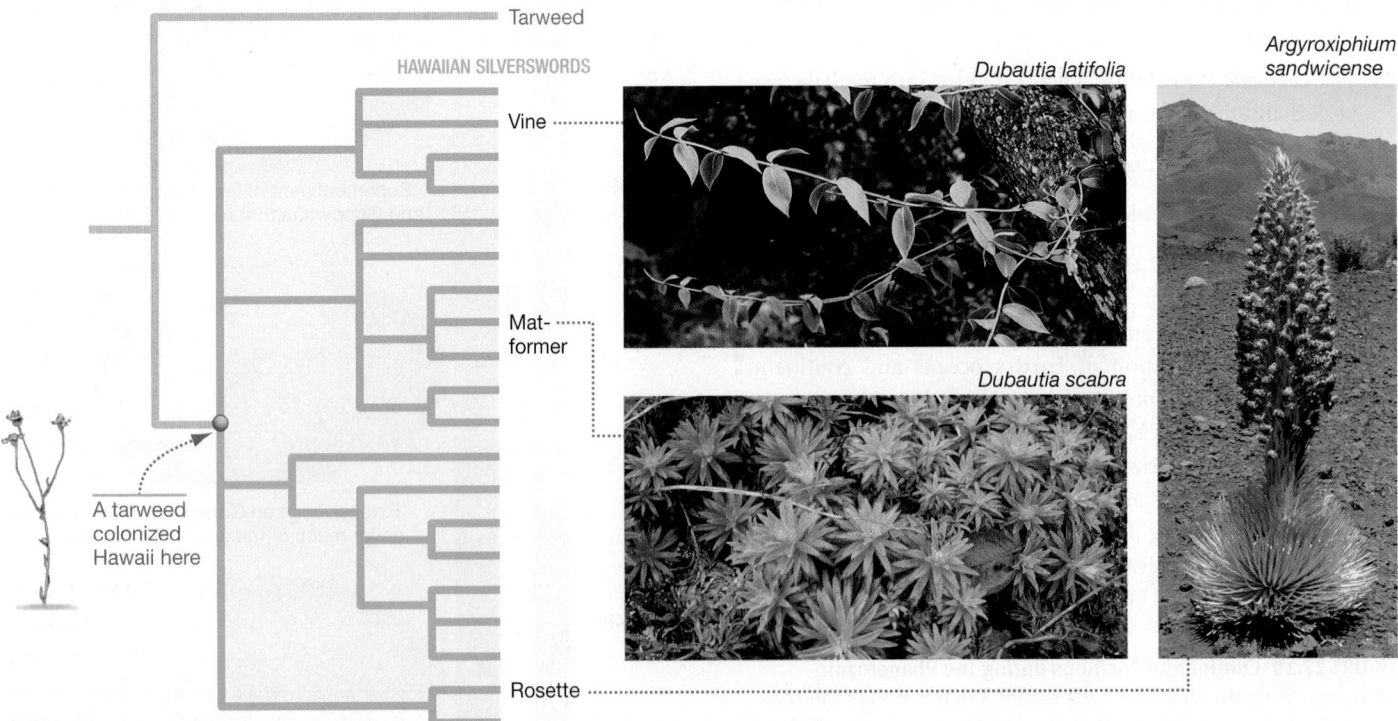

FIGURE 27.11 An Adaptive Radiation. This tree shows a subset of the Hawaiian silverswords, illustrating the extent of morphological divergence.

(a) Short-legged lizard species spend most of their time on the twigs of trees and bushes.

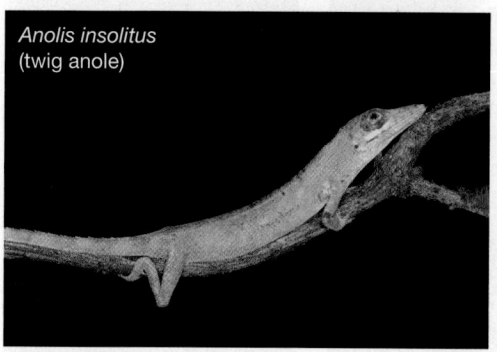

Anolis insolitus
(twig anole)

(b) Long-legged lizard species live on tree trunks and the ground.

Anolis cybotes
(trunk/ground anole)

FIGURE 27.12 Adaptive Radiations of *Anolis* Lizards. (a, b) Species of *Anolis* lizards vary in leg length and tail length. **(c)** Evolutionary relationships among lizard species on the islands of Hispaniola and Jamaica. The initial colonist species was different on these islands; but in terms of how they look and where they live, a similar suite of four species evolved.

(c) The same adaptive radiation of *Anolis* has occurred on different islands, starting from different types of colonists.

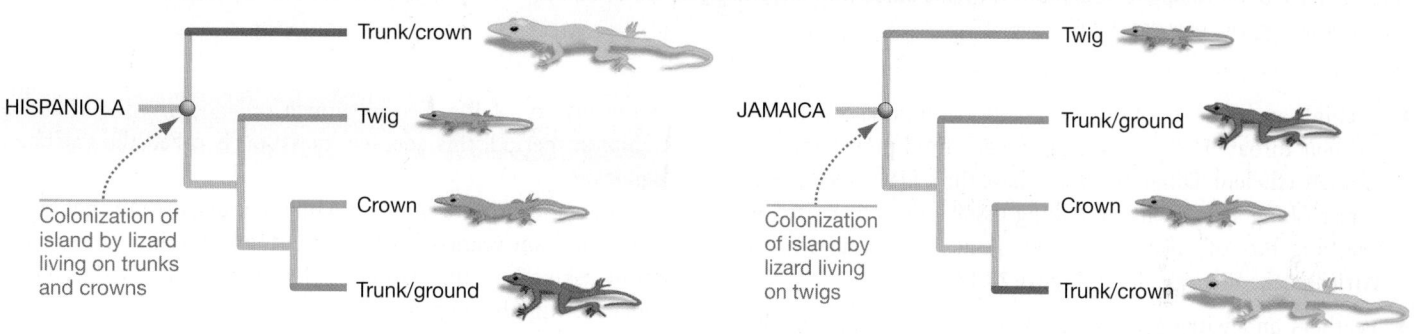

Caribbean. The lineage includes 150 species. They thrive in a wide array of habitats and have diverse body sizes and shapes. And in most cases, a lizard's size and shape are correlated with the habitat it occupies. For example:

- Species that live on tree twigs have short legs and tails that allow them to move efficiently on narrow surfaces (**Figure 27.12a**).

- Species that spend most of their time clinging to broad tree trunks or running along the ground have long legs and tails, making them fast and agile on broad surfaces (**Figure 27.12b**).

Most islands in the Caribbean have a distinct suite of lizard species. And in most cases, each island has a species that lives only in the twigs, the ground, or other distinctive habitats. The classical explanation for this pattern was that a mini-radiation occurred on each island: An original colonizing group encountered no competitors, and diversified in a way that led to efficient use of the available resources by a group of descendant species.

To test this hypothesis rigorously, a group of biologists estimated the phylogeny of *Anolis* from DNA sequence data. The results supported a key claim of the ecological opportunity hypothesis: The lizards on each island were monophyletic.

The critical data, though, are shown in **Figure 27.12c**. These phylogenetic trees, for species found on two different islands, are typical. The key observation is that the original colonist on each island was specialized for a different niche. The initial species on Hispaniola lived on the trunks and crowns of trees, while the original colonist on Jamaica occupied twigs.

From different evolutionary starting points, then, an adaptive radiation filled the same niches on both islands. This is exactly what the ecological opportunity hypothesis predicts. In *Anolis* lizards in the Caribbean, a series of mini-radiations were triggered when colonists arrived on an island that had food, space, and other resources, but lacked competitors.

✔If you understand this concept, you should be able to generate a hypothesis to explain why tarweeds in California and *Anolis* lizards on the mainland are not particularly species-rich or ecologically diverse.

MORPHOLOGICAL INNOVATION The evolution of a key morphological trait—one that allowed descendants to live in new areas, exploit new sources of food, or move in new ways—triggered many of the important diversification events in the history of life. For example:

- The evolution of wings, three pairs of legs, complex mouthparts, and a strong external skeleton helped insects move and feed efficiently. Today insects are the most diverse lineage on Earth, with perhaps well over 3 million species in existence (**Figure 27.13a** on page 486).

- Flowers are a unique reproductive structure that helped trigger the diversification of angiosperms (flowering plants). Because flowers are particularly efficient at attracting pollinators, the evolution of the flower made angiosperms more efficient in reproduction. Today angiosperms are far and away the most species-rich lineage of land plants. Over 250,000 species are known (**Figure 27.13b** on page 486).

(a) Insect body plan

(b) Angiosperm flowers

(c) Cichlids "throat jaws"

(d) Dinosaur feathers

FIGURE 27.13 Some Adaptive Radiations Are Associated with Morphological Innovations.
[Part (c) ©Don P. Northup, www.africancichlidphotos.com]

- Cichlids are a lineage of fish that evolved a unique set of jaws in their throat. These second jaws make food processing extremely efficient. Different species have throat jaws specialized for crushing snail shells, shredding tissue from other fish, or mashing bits of algae. Over 300 species of cichlid live in Africa's Lake Victoria alone (**Figure 27.13c**).

- Feathers and wings gave some dinosaurs the ability to fly. Feathers originally evolved for display or insulation; later they were used in gliding and in powered flight. The dinosaur in **Figure 27.13d**, *Archaeopteryx*, was covered with complex feathers and could probably fly, at least short distances. Today the lineage called birds contains about 10,000 species, with representatives that live in virtually every habitat on the planet.

In each case, the evolution of new morphological features is hypothesized to have supported rapid speciation and ecological divergence.

To review fundamental concepts about adaptive radiation, go to the study area at *www.masteringbiology.com*.

 Web Activity Adaptive Radiation

Then you'll be ready to analyze what may be the most spectacular adaptive radiation in the history of life.

The Cambrian Explosion

Almost all life-forms were unicellular for almost 3 *billion* years after the origin of life. The exceptions were several lineages of small multicellular algae, which show up in the fossil record about 1 billion years ago. Then, about 565 million years ago, the first animals—sponges, jellyfish, and perhaps simple worms—appear in the fossil record. A mere 50 million years later, virtually every major group of animals had appeared.

In a relatively short time, creatures with shells, exoskeletons, internal skeletons, legs, heads, tails, eyes, antennae, jaw-like mandibles, segmented bodies, muscles, and brains had evolved. It was arguably the most spectacular period of evolutionary change

in the history of life. Because much of it occurred during the Cambrian period, this adaptive radiation is called the **Cambrian explosion**.

All organisms were small for 3 billion years, then life got big in just 50 million years—1/60th of the total time life had existed. Before asking how this happened, let's explore the fossilized evidence for what happened.

THE DOUSHANTUO, EDIACARAN, AND CAMBRIAN FOSSILS The Cambrian explosion is documented by three major fossil assemblages that record the state of animal life at 570 mya, at 565 to 542 mya, and at 525 to 515 mya. As **Figure 27.14a** indicates, the species collected from each of these intervals are referred to respectively as the Doushantuo fossils (from the Doushantuo formation in China), Ediacaran fossils (from Ediacara Hills, Australia), and Burgess Shale fossils (from British Columbia, Canada). Fossils from the Ediacaran interval and Burgess Shale interval have now been found at localities throughout the world.

Each fossil assemblage records a distinctive **fauna**—or collection of animal species (**Figure 27.14b**):

- *Doushantuo microfossils* include tiny sponges, less than 1 mm across, and clusters of cells interpreted as animal embryos because they resemble the stages of cleavage observed in today's animals (see Chapter 22). These creatures—the first animals on Earth—probably made their living by filtering organic debris from the water.

- *Ediacaran faunas* include sponges, jellyfish, and comb jellies as well as fossilized burrows, tracks, and other traces from unidentified animals. As the scale bars in Figure 27.14 show, these organisms were small; none have shells, limbs, heads, mouths, or feeding appendages. It is likely that Ediacaran animals simply filtered or absorbed organic material as they burrowed in sediments, sat immobile on the seafloor, or floated in the water.

- *Burgess Shale faunas* are among the most sensational additions ever made to the fossil record. Sponges, jellyfish, and comb jel-

(a) A time line of early animal evolution

(b) Early animals from three major fossil assemblages

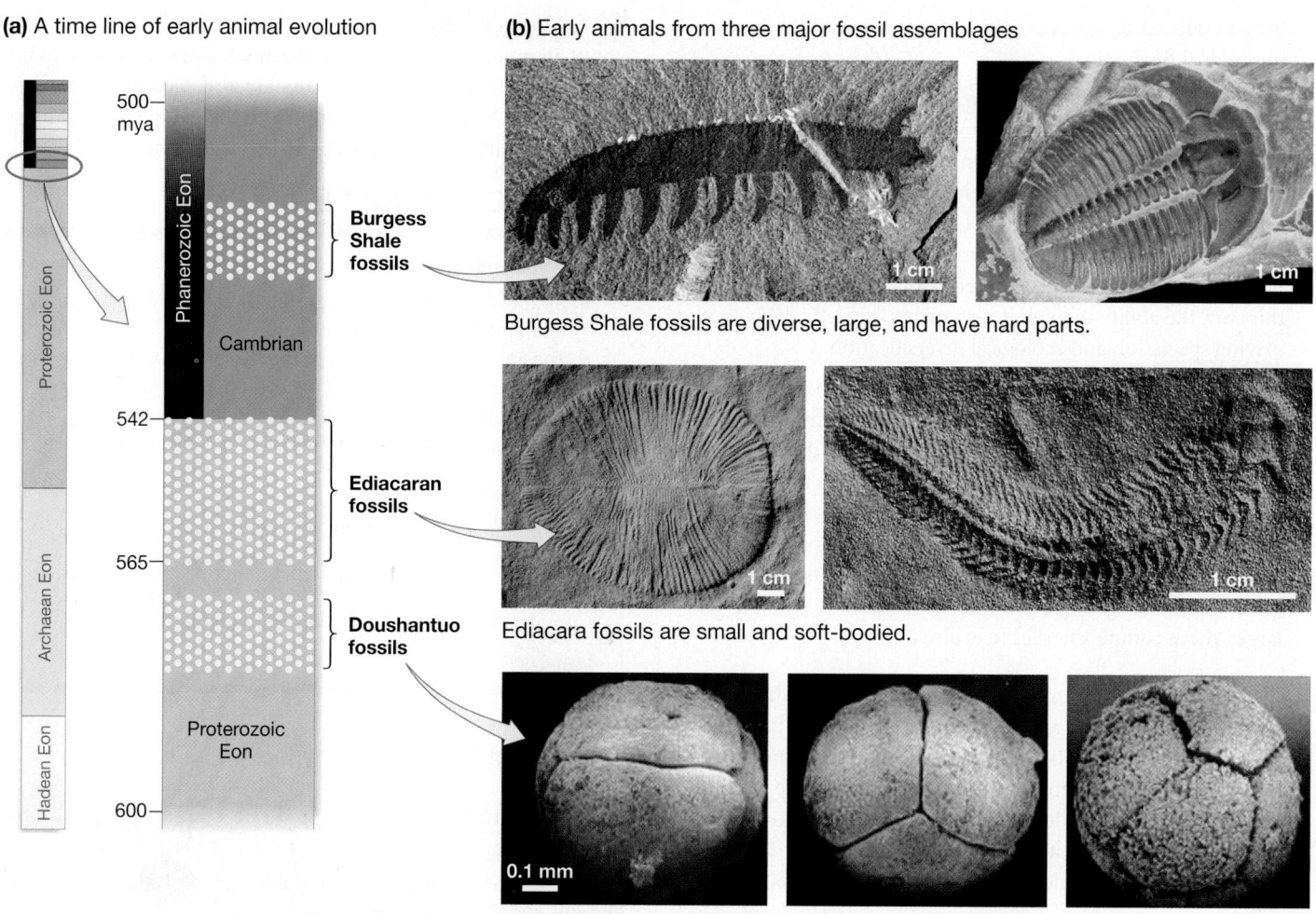

Burgess Shale fossils are diverse, large, and have hard parts.

Ediacara fossils are small and soft-bodied.

Doushantuo fossils are microscopic (shown: early embryos).

FIGURE 27.14 Fossils Document the Cambrian Explosion. The origin of animals and their diversification during the Cambrian explosion is documented by three major fossil assemblages.

lies are abundant in these rocks; but so are arthropods and mollusks. Today, the arthropods include the spiders, insects, and crustaceans (crabs, shrimp, and lobsters); mollusks include the clams, mussels, squid, and octopi. Echinoderms (sea stars and sea urchins), several types of worm and wormlike creatures, and even vertebrates—in short, virtually every major animal lineage—are found in these fossil faunas. A tremendous increase in the size and morphological complexity of animals occurred, accompanied by diversification in how they made a living. Species in this fauna had eyes, mouths, limbs, and shells. They swam, burrowed, walked, ran, slithered, clung, or floated; there were predators, scavengers, filter feeders, and grazers. The diversification created and filled many of the ecological niches found in marine habitats today.

The Cambrian explosion still echoes. Animals that fill today's teeming tide pools, beaches, and mudflats trace their ancestry to species preserved in the Burgess Shale.

WHAT TRIGGERED THE CAMBRIAN EXPLOSION? The Doushantuo, Ediacaran, and Burgess Shale faunas document what happened during the Cambrian explosion. Now the question is: *How* did all

this speciation, morphological change, and ecological diversification come about?

To answer this question, biologists point to an array of datasets and hypotheses:

- *Higher oxygen levels* By analyzing the composition of rocks formed during the Proterozoic, geologists have established that oxygen levels gradually rose in the atmosphere and ocean. Increased oxygen levels make aerobic respiration (see Chapter 9) more efficient; increased aerobic respiration is required to support larger bodies and more active movements. Some biologists suggest that oxygen levels reached a critical threshold at the start of the Cambrian explosion, making the evolution of big, mobile animals possible.

- *The evolution of predation* Prior to the Cambrian explosion, animals made their living by eating organic material that settled on the seafloor or filtering cells and debris from the water. But Cambrian fossils include shelled animals with holes in the shells—evidence that a predator drilled through and ate the animal inside. When predators evolved, they exerted selection pressure for shells, hard exoskeletons, rapid movement, and other adaptations for prey to defend themselves. This selection would drive morphological divergence.

- *New niches beget more new niches*   The vast majority of Doushantuo and Ediacaran animals lived on the ocean floor—in what biologists call benthic habitats. Once animals could move off this substrate, they could exploit algae and other resources that were available above the ocean floor. The presence of animals at an array of depths created selection pressure for the evolution of species that could eat them. In this way, the ability to exploit new niches created new niches, driving speciation and ecological diversification.

- *New genes, new bodies*   Figure 27.15 shows the *Hox* genes found in some major animal groups. As Chapter 21 pointed out, *Hox* genes play a key role in organizing the development of the animal body by signaling where cells are in the embryo. The key observation is that the earliest animals in the fossil record had few or no *Hox* genes; most groups that appear later have more. The idea here is that mutations increased the number of *Hox* genes in animals and made it possible for larger, more complex bodies to evolve.

It's important to recognize that most or all of these hypotheses could be correct. They are not mutually exclusive. If increased oxygen levels made larger bodies and more rapid movement possible, then animals could move into new habitats off the seafloor and become large enough to eat smaller animals. Selection would favor individuals with mutations in *Hox* genes and other DNA sequences that make the development of a large, complex body possible.

About 100 million years after the Cambrian explosion, a similar adaptive radiation occurred after the first plants—the descendants of green algae—adapted to life on land. In the span of about 56 million years, an array of growth forms and most major lineages of plants appear in the fossil record. Chapter 30 features more on this "Devonian explosion" and the morphological innovations that allowed green plants to thrive on land.

CHECK YOUR UNDERSTANDING

If you understand that . . .

- Adaptive radiations are triggered by ecological opportunity and morphological innovation.
- The Cambrian explosion saw the rise of virtually every major animal lineage, and the evolution of a wide array of morphological innovations and food-getting strategies, in the relatively short time frame of 50 million years.

✓ You should be able to . . .

1. Compare and contrast the animals documented in the Doushantuo, Ediacara, and Burgess Shale faunas.
2. Explain the role of ecological opportunity and morphological innovation in the Cambrian explosion.

Answers are available in Appendix B.

27.4 Mass Extinction

A **mass extinction** refers to the rapid extinction of a large number of lineages scattered throughout the tree of life. More specifically, a mass extinction occurs when at least 60 percent of the species present are wiped out within 1 million years.

Mass extinction events are evolutionary hurricanes. They buffet the tree of life, snapping twigs and breaking branches. They

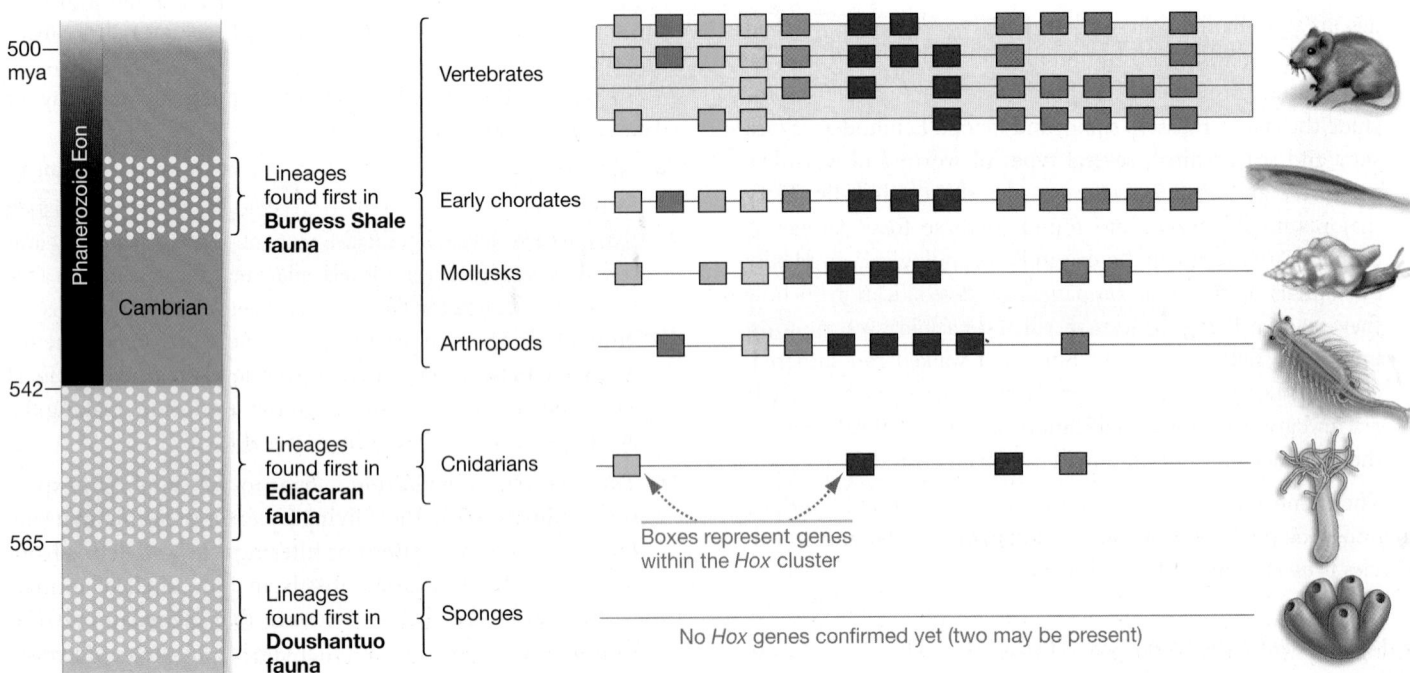

FIGURE 27.15 *Hox* **Genes May Have Been Important in Animal Diversification.**

are catastrophic episodes that wipe out huge numbers of species and lineages in a short time, giving the tree of life a drastic pruning. They are the polar opposite of adaptive radiation.

Before analyzing two of the best-studied mass extinctions in the fossil record, let's step back and ask how they differ from normal extinction events.

How Do Mass Extinctions Differ From Background Extinctions?

Mass extinction events are distinguished from background extinctions. **Background extinction** refers to the lower, average rate of extinction observed when a mass extinction is not occurring.

Although there is no hard-and-fast rule for distinguishing between the two extinction rates, paleontologists traditionally recognize and study five mass extinction events. **Figure 27.16**, for example, plots the percentage of plant and animal lineages called families that died out during each stage in the geologic time scale

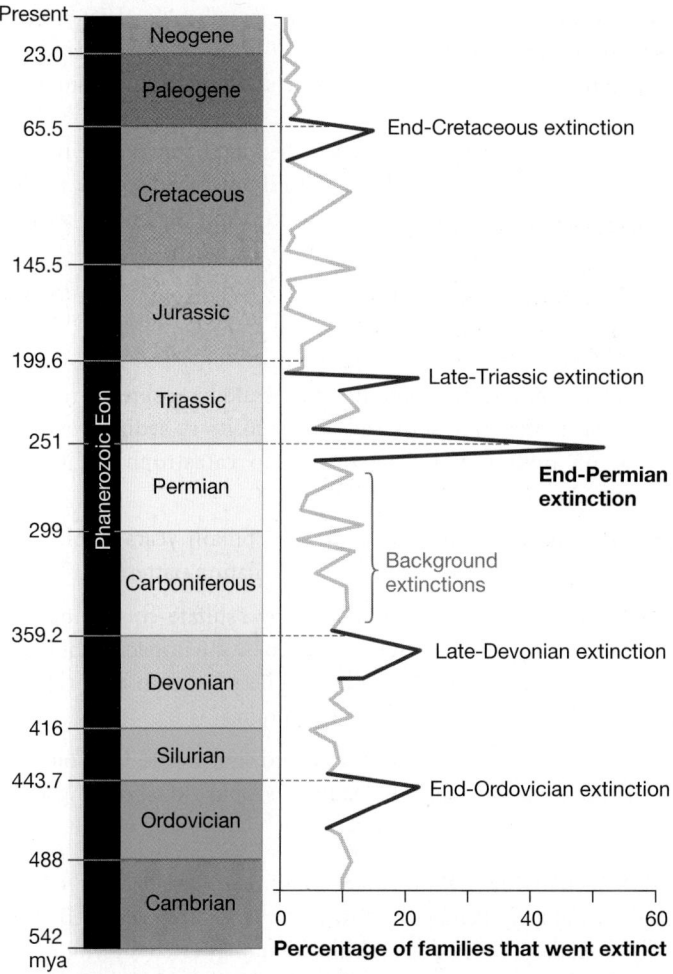

FIGURE 27.16 The Big Five Mass Extinction Events. This graph shows the percentage of lineages called families that went extinct over each interval in the fossil record since the Cambrian explosion. Over 50 percent of families and 90 percent of species went extinct during the end-Permian extinction.

✔**QUESTION** Which extinction event ended the era of the dinosaurs 65 million years ago?

since the Cambrian explosion. Five spikes in the graph—denoting a large number of extinctions within a short time—are drawn in red. These are referred to as "The Big Five."

Biologists are interested in distinguishing between background and mass extinctions because these events have contrasting causes and effects.

- *Background extinctions* are thought to occur when normal environmental change, emerging diseases, or competition with other species reduces certain populations to zero.

- *Mass extinctions* result from extraordinary, sudden, and temporary changes in the environment. During a mass extinction, species do not die out because individuals are poorly adapted to normal or gradually changing environmental conditions. Rather, species die out from exposure to exceptionally harsh, short-term conditions—such as huge volcanic eruptions or catastrophic sea-level changes.

🔑 In a general sense, background extinctions are thought to result primarily from natural selection. Mass extinctions, in contrast, function like genetic drift. The extinctions they cause are largely random with respect to the fitness of individuals under normal conditions.

To drive these points home, consider a mass extinction event that nearly uprooted the tree of life entirely. The end-Permian extinction, which occurred about 251 million years ago, came close to ending multicellular life on Earth.

The End-Permian Extinction

The end-Permian has been called the Mother of Mass Extinctions. To appreciate the scale of what happened, imagine that you took a walk along a seashore and identified 100 different species of algae and animals living on the beach and tide pools and the shallow water offshore. Now imagine that you snapped your fingers and 90 of those species disappeared forever. Only 10 species are left. An area that was teeming with diverse forms of life would look barren.

This is what happened, all over the world, during the end-Permian extinction. The event was a catastrophe of almost unimaginable proportions. On a personal level, it would be like nine of your ten best friends dying.

Although biologists have long appreciated the scale of the end-Permian extinction, research on its causes is ongoing. Consider the following:

- Flood basalts are outpourings of molten rock that flow across the Earth's surface. The largest flood basalts on Earth, called the Siberian traps, occurred during the end-Permian. They added enormous quantities of heat, CO_2, and sulfur dioxide to the atmosphere. The CO_2 led to intense global warming, and sulfur dioxide reacted with water to form sulfuric acid, which is toxic to most organisms.

- Rocks that formed during the interval indicate that the oceans became completely or largely anoxic—meaning that they lacked oxygen. These conditions are fatal to organisms that rely on aerobic respiration.

- There is convincing evidence that sea level dropped dramatically during the extinction event, reducing the amount of habitat available for marine organisms.

- Terrestrial animals may have been restricted to small patches of low-elevation habitats, due to low oxygen concentrations and high CO_2 levels in the atmosphere.

In short, both marine and terrestrial environments deteriorated dramatically for organisms that depend on oxygen to live. A prominent researcher has captured this point by naming the suite of killing mechanisms the "world went to hell hypothesis."

What biologists don't understand is *why* the environment changed so radically, and so quickly. For example, no one has yet been able to establish a convincing connection between the eruption of the Siberian traps and the other, more global, environmental changes that occurred.

The cause of the end-Permian extinction may be the most important unsolved question in research on the history of life. The cause of the dinosaur's demise, in contrast, is settled.

What Killed the Dinosaurs?

The end-Cretaceous extinction of 65 million years ago is as satisfying a murder mystery as you could hope for, but the butler didn't do it. The **impact hypothesis** for the extinction of the dinosaurs, first put forth in the early 1970s, proposed that an asteroid struck Earth and snuffed out an estimated 60–80 percent of the multicellular species alive.

EVIDENCE FOR THE IMPACT HYPOTHESIS The impact hypothesis was intensely controversial at first. As researchers set out to test its predictions, however, support began to grow:

- Worldwide, sedimentary rocks that formed at the Cretaceous-Paleogene (K-P)[1] boundary were found to contain extremely high quantities of the element iridium. Iridium is vanishingly rare in Earth rocks, but it is an abundant component of asteroids and meteorites (**Figure 27.17a**).

- Shocked quartz and microtektites are minerals that are found only at documented meteorite impact sites (**Figure 27.17b**). Shocked quartz forms when shock waves from an asteroid impact alter the structure of sand grains. Microtektites form when minerals are melted at an impact site and then cool and resolidify. In Haiti and an array of other locations, both shocked quartz and microtektites have been discovered in abundance in rock layers dated to 65 million years ago.

- A crater the size of Sicily was found just off the northwest coast of Mexico's Yucatán peninsula (**Figure 27.17c**). Microtektites are abundant in sediments from the crater's walls, and the crater dates to the K-P boundary.

Taken together, these data provided conclusive evidence in favor of the impact hypothesis. Researchers agree that the mystery is solved.

[1]Geologists use *K* to abbreviate Cretaceous, because *C* refers to the Cambrian period.

NATURE OF THE IMPACT The shape of the Yucatán crater and the distribution of shocked quartz and microtektites dated to 65 mya indicates that the asteroid hit Earth at an angle and splashed material over much of southeastern North America. Based on currently available data, astronomers and paleontologists estimate that the asteroid was about 10 km across.

To get a sense of the event's scale, consider that Mt. Everest is about 10 km above sea level and that planes cruise at an altitude of about 10 km. Imagine Earth being hit by a rock the size of Mt. Everest, or a rock that would fill the space between you and a jet in the sky.

To understand the impact's consequences, consider the results of the Tunguska event. On June 30, 1908, a piece of a comet about 30 m across and about 2 megatons in mass exploded at least 5 km above Earth's surface near the Tunguska River in Siberia. The explosion released about 1000 times as much energy as the atomic bomb that destroyed Hiroshima in World War II. The event incinerated vegetation over hundreds of square kilometers, leveled trees over thousands of square kilometers, and significantly increased dust levels across the Northern Hemisphere. A deafening blast was heard 500 km (300 mi) away; people standing 60 km away were thrown to the ground or knocked unconscious.

These events pale in comparison to what happened 65 million years ago. The energy the K-P asteroid unleashed was 37 *million* times greater than the Tunguska event. And the energy was released at the surface—not 5 km above the Earth.

According to both computer models and geologic data, the consequences of the K-P asteroid strike were nothing short of devastating.

- A tremendous fireball of hot gas would have spread from the impact site; large soot and ash deposits in sediments dated to 65 million years ago testify to catastrophic wildfires, worldwide.

- The largest tsunami in the last 3.5 billion years would have disrupted ocean sediments and circulation patterns.

- The impact site itself is underlain by a sulfate-containing rock called anhydrite. The SO_4^{2-} released by the impact would have reacted with water in the atmosphere to form sulfuric acid (H_2SO_4), triggering extensive acid rain.

- Massive quantities of dust, ash, and soot would have blocked the Sun for long periods, leading to rapid global cooling and a crash in plant productivity.

SELECTIVITY OF THE EXTINCTIONS The asteroid impact did not kill indiscriminately. Perhaps by chance, certain lineages escaped virtually unscathed while others vanished. Among vertebrates, for example, the dinosaurs, pterosaurs (flying reptiles), and all of the large-bodied marine reptiles (mosasaurs, ichthyosaurs, and plesiosaurs) expired; mammals, crocodilians, amphibians, and turtles survived.

Why? Answering this question has sparked intense debate. For years the leading hypothesis was that the K-P extinction event was size selective. The logic here was that the extended darkness and cold would affect large organisms disproportionately, be-

(a) Iridium is present at high concentration in rocks formed 65 million years ago.

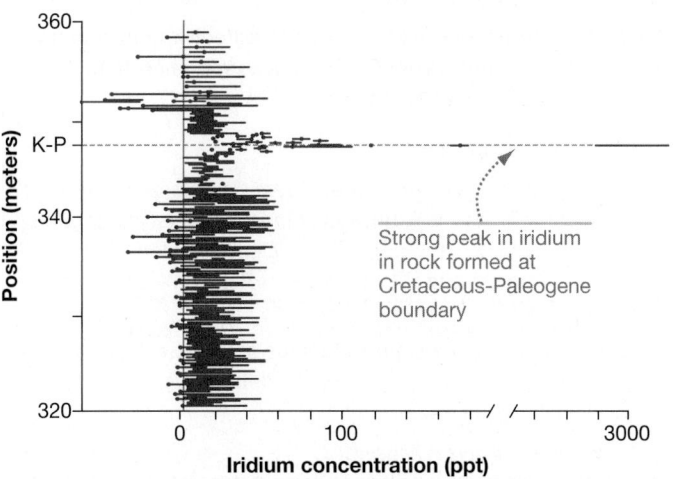

(b) Minerals that form during asteroid impacts

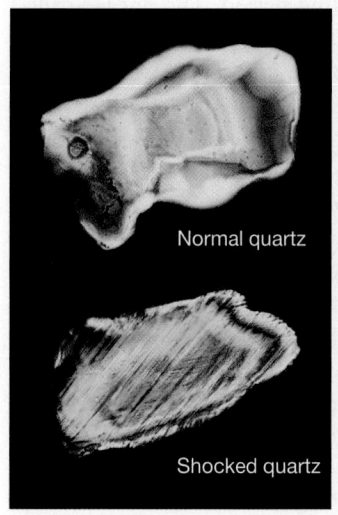

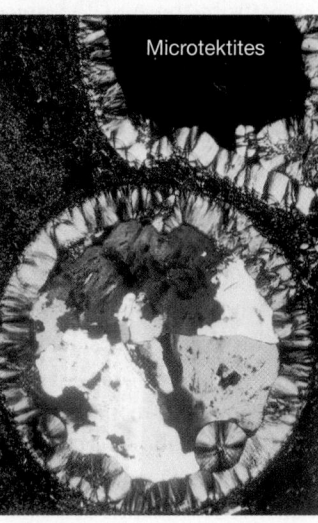

Normal quartz

Shocked quartz

Microtektites

(c) The asteroid left a crater 180 km (112 miles) wide.

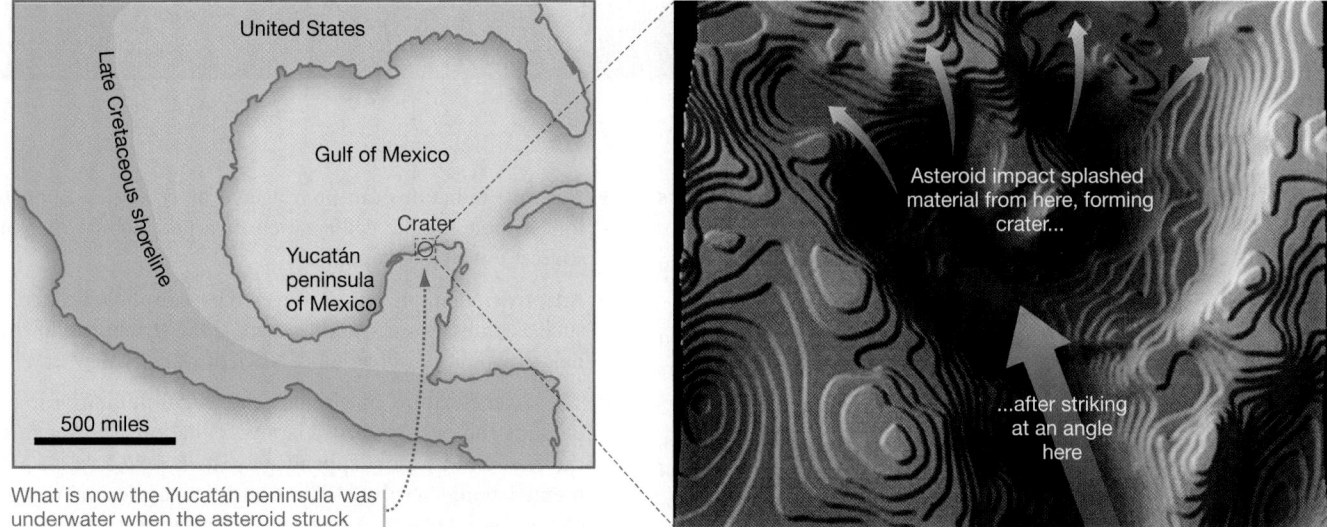

United States

Gulf of Mexico

Late Cretaceous shoreline

Crater

Yucatán peninsula of Mexico

500 miles

What is now the Yucatán peninsula was underwater when the asteroid struck

Asteroid impact splashed material from here, forming crater...

...after striking at an angle here

FIGURE 27.17 Evidence of an Asteroid Impact 65 Million Years Ago. (a) The concentration of iridium in rocks that formed on either side of the Cretaceous-Paleogene (K-P) boundary, in parts per trillion (ppt). On the *y*-axis, the position in meters refers to the location of rocks on either side of the rocks dated to the K-P boundary. Note that the *x*-axis is broken—otherwise the value of 3000 ppt at the K-P boundary would be far off the page. **(b)** Normal quartz grains are markedly different from shocked quartz grains. The striations in the shocked quartz are caused by a sudden increase in pressure. Microtektites (right) are tiny glass particles formed when minerals melt at an impact site and then recrystallize. **(c)** Geologists have identified the walls of a crater off the northwest coast of the Yucatán Peninsula, dated to 65 million years ago.

cause they require more food than do small organisms. But extensive data on the survival and extinction of marine clams and snails have shown no hint of size selectivity, and small-bodied and juvenile dinosaurs perished along with large-bodied and adult forms.

One hypothesis currently being tested is that organisms that were capable of inactivity for long periods—by hibernating or resting as long-lived seeds or spores—were able to survive the catastrophe. But this aspect of the mystery is still unsolved.

RECOVERY FROM THE EXTINCTION After the K-P extinction, fern fronds and fern spores dominate the plant fossil record from North America and Australia. These data suggest that extensive stands of ferns replaced diverse assemblages of cone-bearing and flowering plants after the impact. The fundamental message here is that terrestrial ecosystems around the world were radically simplified. In marine environments, some invertebrate groups do not exhibit normal levels of species diversity in the fossil record until 4–8 million years past the K-P boundary. Recovery was slow.

The organisms present in the Paleogene were markedly different from those of the preceding period. The lineage called Mammalia, which had consisted largely of rat-sized predators and scavengers in the heyday of the dinosaurs, exploded after the impact and took the place of the dinosaurs. Within 10–15 million years, all of the major mammalian orders observed today had appeared—from pigs to primates. Why? A major branch on the tree of life had disappeared. With competitors removed, mammals flourished.

The change in the terrestrial vertebrate fauna was not due to a competitive superiority conferred by adaptations such as fur and lactation. Rather, it was due to a chance event: a once-in-a-billion-years collision with a massive rock from outer space.

The diversification of lineages across the tree of life—including mammals—is the subject of Unit 6. A mass extinction that is currently occurring, and pruning the tree of life before our very eyes, is analyzed in Chapter 55.

CHECK YOUR UNDERSTANDING

If you understand that . . .

- Mass extinctions have occurred repeatedly throughout the history of life, and are due to cataclysmic, short-term changes in the environment.

✔ **You should be able to . . .**

1. Evaluate the strength of the evidence supporting the impact hypothesis for the extinction of the dinosaurs, and identify which evidence is strongest. (In science, evidence is considered strong if it cannot be explained by alternative hypotheses.)

2. Explain why mass extinctions wipe out species more or less randomly—much the way genetic drift affects changes in allele frequencies.

Answers are available in Appendix B.

CHAPTER 27 REVIEW

For media, go to the study area at www.masteringbiology.com

Summary of Key Concepts

🔑 **Phylogenetic trees document the evolutionary relationships among organisms and are estimated from data.**

- Phylogenetic trees document the genealogical relationships among species and identify the order in which events occurred.

- Phylogenetic trees can be estimated by grouping species based on overall similarity in traits or by analyzing shared, derived characters (synapomorphies) that identify monophyletic groups.

- To minimize the impact of homoplasy when inferring phylogenies, researchers use parsimony or other approaches to decide which of the many trees that are possible is most likely to reflect actual evolutionary history.

  ✔ You should be able to explain why upright posture and bipedalism (walking on two legs) is a synapomorphy that defines a monophyletic group called hominins, which includes humans.

🔑 **The fossil record provides physical evidence of organisms that lived in the past.**

- The fossil record is the only direct source of data about what extinct organisms looked like and where they lived.

- The fossil record is critical to understanding the history of life, but it is biased: common and recent species that burrow and that have hard parts are most likely to be present in the record.

  ✔ You should be able to explain why teeth and pollen grains are the most common types of vertebrate and plant fossils, respectively.

🔑 **Adaptive radiations are a major pattern in the history of life. They are instances of rapid diversification associated with new ecological opportunities and new morphological innovations.**

- Speciation events and morphological change occur rapidly during an adaptive radiation, as a single lineage diversifies into a wide variety of ecological roles.

- Adaptive radiations can be triggered by ecological opportunity—for example, by the colonization of a new habitat that offers resources but lacks competitors.

- Adaptive radiations can be triggered by morphological innovations such as limbs, flowers, and feathers, if the structure allows individuals to exploit resources more efficiently or in a new way.

- The diversification of animals over a 50-million-year period is perhaps the best-studied adaptive radiation. During this period the first heads, tails, appendages, shells, exoskeletons, and segmented bodies evolved. The Cambrian explosion was probably caused by increases in oxygen and expansion of *Hox* and other gene sequences, which made the evolution of large, motile bodies possible.

  ✔ You should be able to explain why the evolution of animals that could move throughout marine environments—not just in benthic habitats—created an ecological opportunity.

  MB **Web Activity** Adaptive Radiation

🔑 **Mass extinctions have occurred repeatedly throughout the history of life. They are environmental catastrophes that rapidly eliminate most of the species alive.**

- Mass extinctions have altered the course of evolutionary history at least five times.

- They prune the tree of life more or less randomly and have marked the end of several prominent lineages and the rise of new branches.

- The end-Permian extinction killed over 90 percent of existing species; its underlying cause is unknown.

- The end-Cretaceous extinction killed 60 to 80 percent of existing species and was caused by an asteroid impact.

- After the devastation of a mass extinction, the fossil record indicates that it can take 10–15 million years for ecosystems to recover their former levels of diversity.

✔ You should be able to evaluate whether environmental changes caused by humans are eliminating species in a random manner, as opposed to those species being poorly adapted to the environment.

Questions

1. Choose the best definition of a fossil.
   a. any trace of an organism that has been converted into rock
   b. a bone, tooth, shell, or other hard part of an organism that has been preserved
   c. any trace of an organism that lived in the past
   d. the process that leads to preservation of any body part from an organism that lived in the past

2. What is the difference between a branch and a node on a phylogenetic tree?
   a. A branch is a population through time; a node is where a population splits into two independent populations.
   b. A branch represents a common ancestor; a node is a species.
   c. A branch is a lineage; a node is any named taxonomic group.
   d. A branch is a population through time; a node is the common ancestor or all species or lineages present on a particular tree.

3. Which of the following best characterizes an adaptive radiation?
   a. Speciation occurs extremely rapidly, and descendant populations occupy a large geographic area.
   b. A single lineage diversifies rapidly, and descendant populations occupy many habitats and ecological roles.
   c. Natural selection is particularly intense, because disruptive selection occurs.
   d. Species recover after a mass extinction.

4. Which of the following is most accurate?
   a. Mass extinctions are due to asteroid impacts; background extinctions may have a wide variety of causes.
   b. Mass extinctions focus on particularly prominent groups, such as dinosaurs; background extinctions affect species from throughout the tree of life.
   c. Only five mass extinctions have occurred, but hundreds of background extinctions have occurred.
   d. Mass extinctions extinguish groups rapidly and randomly; background extinctions are slower and often result from natural selection.

5. **BioSkills 3** in Appendix A recommends a "one-snip test" to identify monophyletic groups—meaning that if you cut any branch on a tree, everything that "falls off" is a monophyletic group. Why is this valid?
   a. Monophyletic groups are nested on a tree—meaning that they are hierarchical.
   b. Monophyletic groups can also be called clades or lineages.
   c. Species are the smallest monophyletic groups on the tree of life.
   d. One snip gets an ancestor and all of its descendants.

6. What is homoplasy?
   a. similarity not due to homology
   b. similarity not due to inheritance from a common ancestor
   c. "noise" in the data sets used to infer phylogenies
   d. all of the above

1. The text claims that the fossil record is biased in several ways. What are these biases? If the database is biased, is it still an effective tool to use in studying the diversification of life? Explain.

2. The initial diversification of animals took place over some 50 million years, at the start of the Cambrian period. Why is the diversification called an "explosion"?

3. Why is parsimony usually a reliable way to minimize the effect of homoplasy in assessing which phylogenetic tree is most accurate? Why was parsimony misleading in the case of the astragalus during evolution of artiodactyls?

4. Give an example of an adaptive radiation that occurred after a colonization event and after a morphological innovation. In each case, provide a hypothesis to explain why the adaptive radiation occurred.

5. What is the "world went to hell" hypothesis for the end-Permian extinction. What underlying mechanism was responsible?

6. Why are monophyletic groups identified by shared, derived traits?

1. Suppose that the dying wish of a famous eccentric was that his remains be fossilized. His family has come to you for expert advice. What steps would you recommend to maximize the chances that his wish will be fulfilled?

2. Using data from the molecular phylogenies presented in this section and data on the fossil record of whales presented in Chapter 24, summarize how whales evolved from the common ancestor they share with today's hippos.

3. Some researchers contend that the end-Permian extinction event was also caused by an impact with a large extraterrestrial object. List the

evidence, ordered from least convincing to most convincing, that you would like to see before you accept this hypothesis. Explain your rankings.

4. One of the "triggers" proposed for the Cambrian explosion is a dramatic rise in oxygen concentrations that occurred in the oceans about 800 mya. Review material in Chapter 9 on how oxygen compares with other atoms or compounds as an electron acceptor during cellular respiration (look near the end of Section 9.6, where aerobic and anaerobic respiration are compared). Then explain the logic behind the "oxygen-trigger" hypothesis for the Cambrian explosion.

THE BIG PICTURE

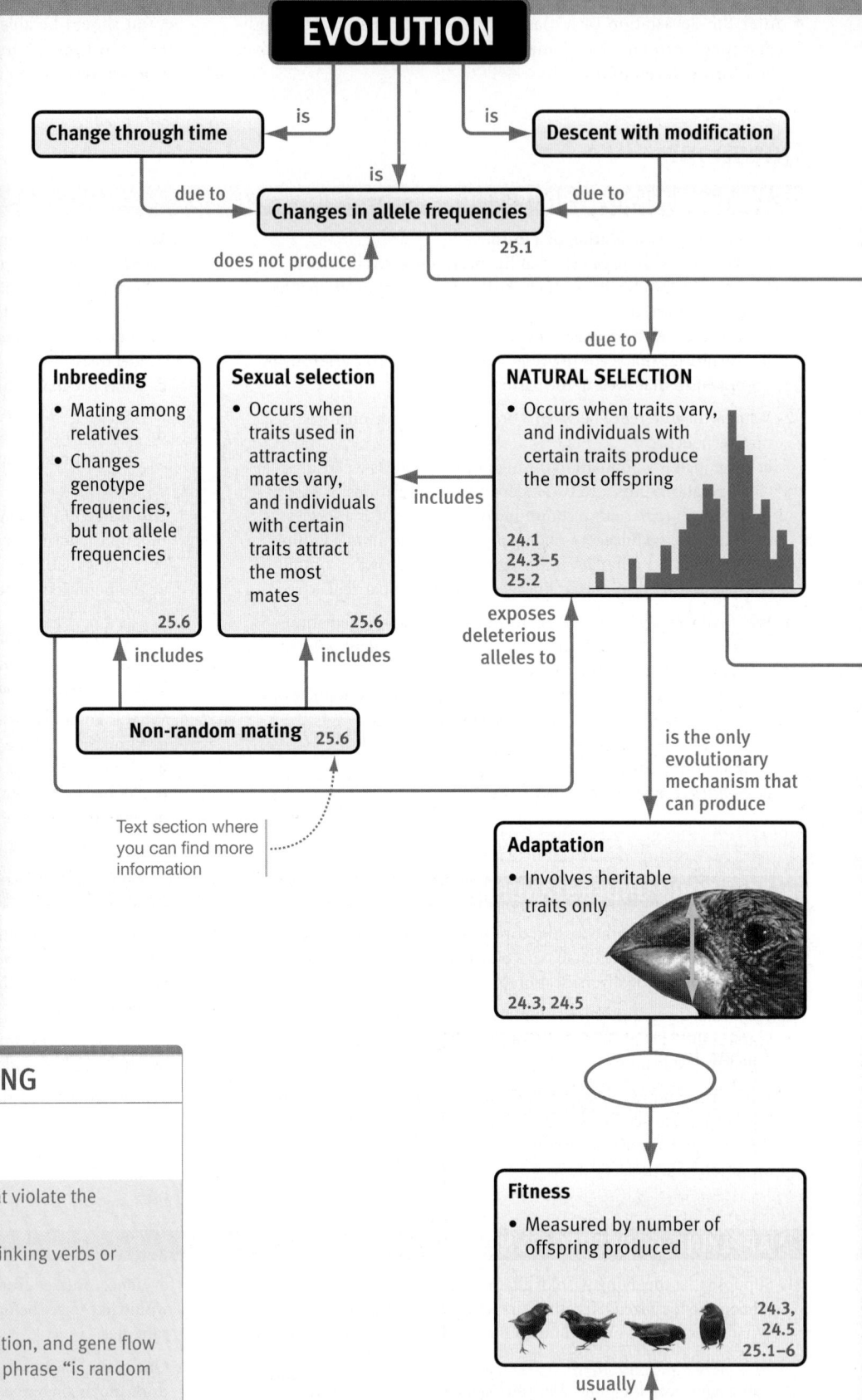

EVOLUTION

Change through time ← is

is → Descent with modification

is ↓

due to → Changes in allele frequencies   25.1 ← due to

does not produce

Inbreeding
- Mating among relatives
- Changes genotype frequencies, but not allele frequencies   25.6

Sexual selection
- Occurs when traits used in attracting mates vary, and individuals with certain traits attract the most mates   25.6

includes ←

NATURAL SELECTION
- Occurs when traits vary, and individuals with certain traits produce the most offspring
24.1
24.3–5
25.2

due to ↓

exposes deleterious alleles to

includes ↑   includes ↑

Non-random mating   25.6

Text section where you can find more information

is the only evolutionary mechanism that can produce ↓

Adaptation
- Involves heritable traits only
24.3, 24.5

Fitness
- Measured by number of offspring produced
24.3, 24.5
25.1–6

usually reduces ↑

CHECK YOUR UNDERSTANDING

If you understand the big picture . . .

✔ **You should be able to . . .**

1. Draw a circle around the processes that violate the Hardy-Weinberg principle.
2. Fill in the blue ovals with appropriate linking verbs or phrases.
3. Give an example of a key innovation.
4. Draw arrows linking genetic drift, mutation, and gene flow to the appropriate box using the linking phrase "is random with respect to."

Answers are available in Appendix B.

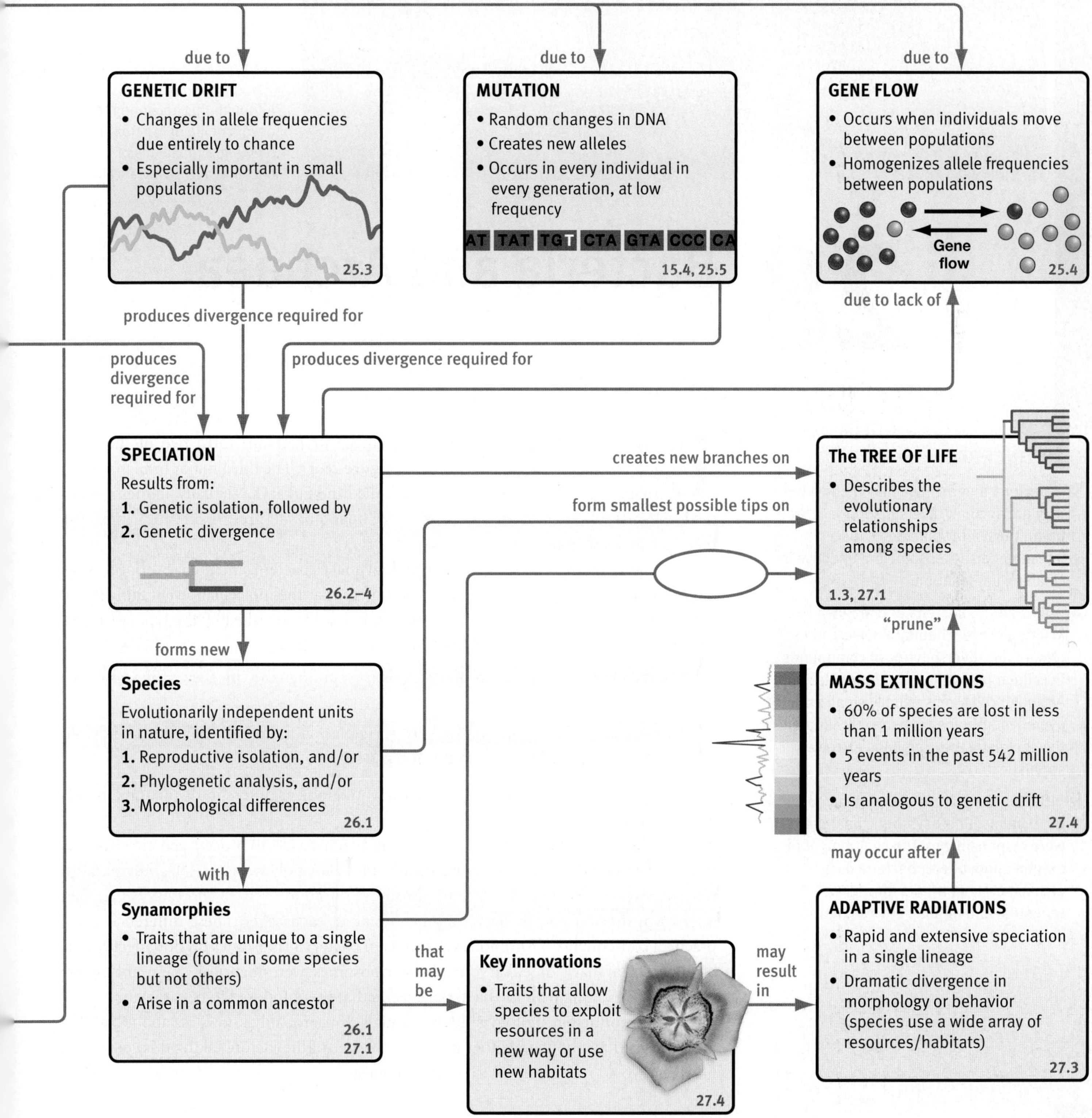

due to

due to

due to

GENETIC DRIFT

- Changes in allele frequencies due entirely to chance
- Especially important in small populations

25.3

MUTATION

- Random changes in DNA
- Creates new alleles
- Occurs in every individual in every generation, at low frequency

AT TAT TGT CTA GTA CCC CA

15.4, 25.5

GENE FLOW

- Occurs when individuals move between populations
- Homogenizes allele frequencies between populations

Gene flow

25.4

due to lack of

produces divergence required for

produces divergence required for

produces divergence required for

SPECIATION

Results from:
1. Genetic isolation, followed by
2. Genetic divergence

26.2–4

creates new branches on

form smallest possible tips on

The TREE OF LIFE

- Describes the evolutionary relationships among species

1.3, 27.1

"prune"

forms new

Species

Evolutionarily independent units in nature, identified by:
1. Reproductive isolation, and/or
2. Phylogenetic analysis, and/or
3. Morphological differences

26.1

MASS EXTINCTIONS

- 60% of species are lost in less than 1 million years
- 5 events in the past 542 million years
- Is analogous to genetic drift

27.4

may occur after

with

Synamorphies

- Traits that are unique to a single lineage (found in some species but not others)
- Arise in a common ancestor

26.1
27.1

that may be

Key innovations

- Traits that allow species to exploit resources in a new way or use new habitats

27.4

may result in

ADAPTIVE RADIATIONS

- Rapid and extensive speciation in a single lineage
- Dramatic divergence in morphology or behavior (species use a wide array of resources/habitats)

27.3

Although this hot spring looks devoid of life, it is actually teeming with billions of bacterial and archaeal cells.

28 Bacteria and Archaea

Bacteria and Archaea (usually pronounced *ar-KAY-ah*) form two of the three largest branches on the tree of life (**Figure 28.1**). The third major branch, or domain, consists of eukaryotes and is called the Eukarya. Virtually all members of the Bacteria and Archaea domains are unicellular, and all are prokaryotic—meaning that they lack a membrane-bound nucleus.

Although their relatively simple morphology makes bacteria and archaea appear similar to the untrained eye, they are strikingly different at the molecular level (**Table 28.1**). Organisms in the Bacteria and Archaea domains are distinguished by the types of molecules that make up their plasma membranes and cell walls:

- **Bacteria** have a unique compound called peptidoglycan in their cell walls (see Chapter 5).

- **Archaea** have unique phospholipids in their plasma membranes—the hydrocarbon tails of the phospholipids contain isoprene (see Chapter 6).

If you were unicellular, bacteria and archaea would look as different to you as mammals and fish do now.

In addition, the machinery that bacteria and archaea use to process genetic information is strikingly different. More specifically, the DNA polymerases, RNA polymerases, transcription-initiation proteins, and ribosomes found in Archaea and Eukarya are distinct from those found in Bacteria and similar to each other. These differences have practical consequences: Antibiotics that poison bacterial ribosomes do not affect the ribosomes of archaea or eukaryotes. If all ribosomes were identical, these antibiotics would kill you along with the bacterial species that was supposed to be targeted.

As an introductory biology student, though, you probably know less about bacteria and archaea than about any other group on the tree of life. Before taking this course, it's likely that you'd never even heard of the Archaea.

✔ When you see this checkmark, stop and test yourself. Answers are available in Appendix B.

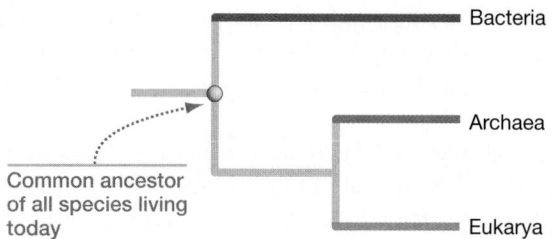

FIGURE 28.1 Bacteria, Archaea, and Eukarya Are the Three Domains of Life. Archaea are more closely related to eukaryotes than they are to bacteria.

✔**QUESTION** Was the common ancestor of all species living today prokaryotic or eukaryotic? Explain your reasoning.

This chapter's goal is to change this state of affairs, and convince you that even though bacteria and archaea are tiny, they have an enormous impact on you and the planet in general. By the time you finish reading the chapter, you should understand why a researcher summed up the bacteria and archaea by claiming, "They run this joint."

28.1 Why Do Biologists Study Bacteria and Archaea?

Biologists study bacteria and archaea for the same reasons they study any organisms. First, they are intrinsically fascinating. Discoveries such as finding bacterial cells living a kilometer underground or in 95°C hot springs keep biologists awake at night, staring at the ceiling. They can't wait to get into the lab in the morning and figure out how those cells are staying alive.

Second, there are practical benefits to understanding the species that share the planet with us. Understanding bacteria and archaea is particularly important—both in terms of understanding life on Earth and improving human health and welfare.

Biological Impact

🔑 The lineages in the domains Bacteria and Archaea are ancient, diverse, abundant, and ubiquitous. The oldest fossils of any type found to date are 3.5-billion-year-old carbon-rich deposits derived from bacteria. Because eukaryotes do not appear in the

SUMMARY TABLE 28.1 **Characteristics of Bacteria, Archaea, and Eukarya**

	Bacteria	Archaea	Eukarya
Nuclear envelope? (see Chapter 7)	No	No	Yes
Circular chromosome?	Yes (but linear in some species)	Yes	No
DNA associated with histone proteins? (see Chapter 18)	No	Yes	Yes
Organelles present? (see Chapter 7)	Some in limited number of species	None described to date	Extensive in number and density
Flagella present? (see Chapter 7)	Yes—spin like propeller	Yes; spin like bacterial flagella, but distinctive in molecular composition	Yes, but undulate back and forth and have completely different molecular composition compared with bacteria and archaea
Unicellular or multicellular?	Almost all unicellular	All unicellular	Many multicellular
Gene exchange between individuals? (see Chapter 12)	Transfer from viruses (transduction), environment (transformation), or direct contact from another cell (conjugation)	Transfer from viruses (transduction), environment (transformation), or direct contact from another cell (conjugation)	Sexual reproduction in some species—fusion of haploid genomes. Transduction and transformation can also occur.
Structure of lipids in plasma membrane (see Chapter 6)	Glycerol bonded to straight-chain fatty acids via ester linkage	Glycerol bonded to branched fatty acids (synthesized from isoprene subunits) via ether linkage	Glycerol bonded to straight-chain membrane fatty acids via ester linkage
Cell-wall material (see Chapter 5, Chapter 8)	Almost all include peptidoglycan, which contains muramic acid	Varies, but no peptidoglycan and no muramic acid	When present, usually made of cellulose or chitin
Information processing: DNA synthesis, transcription, and translation machinery (see Chapter 15, Chapter 16)	One relatively simple RNA polymerase; translation begins with formylmethionine; translation poisoned by several antibiotics that do not affect archaea or eukaryotes	DNA polymerase is eukaryote-like; one relatively complex RNA polymerase; eukaryote-like basal transcription complex; translation begins with methionine	Several relatively complex RNA polymerases; translation begins with methionine

✔**EXERCISE** Using the data in this table, add labeled marks to Figure 28.1 indicating where the following traits evolved: peptidoglycan in cell wall, archaeal-type plasma membrane, archaeal and eukaryote-type ribosomes/DNA polymerase/transcription machinery, nuclear envelope.

fossil record until 1.75 billion years ago, biologists infer that prokaryotes were the only form of life on Earth for at least 1.7 billion years.

Just how many bacteria and archaea are alive today? Although a mere 5000 species have been formally named and described to date—most by the morphological species concept introduced in Chapter 26—it is virtually certain that millions exist. Consider that over 400 species of prokaryotes are living in your small intestine right now, with another 128 in your stomach lining. About 500 species live in your mouth, of which just 300 have been described and named. Norman Pace points out that there may be tens of millions of different insect species but notes, "If we squeeze out any one of these insects and examine its contents under the microscope, we find hundreds or thousands of distinct microbial species." Most of these **microbes** (microscopic organisms) are bacteria or archaea. Virtually all are unnamed and undescribed. If you want to discover and name new species, then study bacteria or archaea.

ABUNDANCE In addition to recognizing how diverse bacteria and archaea are in terms of numbers of species, it's critical to appreciate their abundance.

- The approximately 10^{13} (10 trillion) cells in your body are vastly outnumbered by the bacterial and archaeal cells living on and in you. An estimated 10^{12} bacterial cells live on your skin, and an additional 10^{14} bacterial and archaeal cells occupy your stomach and intestines. You are a walking, talking habitat—one that is teeming with bacteria and archaea.

- A mere teaspoon of good-quality soil contains *billions* of microbial cells, most of which are bacteria and archaea.

- In sheer numbers, species in a lineage called the Group I marine archaea may be the most successful organisms on Earth. Biologists routinely find these cells at concentrations of 10,000 to 100,000 individuals per milliliter in most of the world's oceans. At these concentrations, a drop of seawater contains a population equivalent to that of a large human city. Yet this lineage was first described in the early 1990s.

- Recent research has found enormous numbers of bacterial and especially archaeal cells in rocks and sediments as much as 1600 meters underneath the world's oceans. Although recently discovered, the bacteria and archaea living under the ocean may make up 10 percent of the world's total mass of living material.

- Biologists estimate the total number of individual bacteria and archaea alive today at over 5×10^{30}. If they were lined up end to end, they would make a chain longer than the Milky Way galaxy. These cells contain 50 percent of all the carbon and 90 percent of all the nitrogen and phosphorus found in organisms.

In terms of the total volume of living material on our planet, bacteria and archaea are dominant life-forms.

HABITAT DIVERSITY Bacteria and archaea are found almost everywhere. They live in environments as unusual as oxygen-free mud, hot springs, and salt flats. In seawater they are found from the surface to depths of 10,000 m, at temperatures ranging from near 0°C in Antarctic sea ice to over 121°C near submarine volcanoes.

Although there are far more prokaryotes than eukaryotes, much more is known about eukaryotic diversity than about prokaryotic diversity. Researchers who study prokaryotic diversity are exploring one of the most wide-open frontiers in all of science. So little is known about the extent of these domains that recent collecting expeditions have turned up entirely new **phyla** (singular: **phylum**). These are names given to major lineages within each domain. To a biologist, this achievement is equivalent to the sudden discovery of a new group of eukaryotes as distinctive as flowering plants or animals with backbones.

The physical world has been explored and mapped, and many of the larger plants and animals are named. But in **microbiology**—the study of organisms that can be seen only with the aid of a microscope—this is an age of exploration and discovery.

Medical Importance

The first paper documenting that an archaeon was associated with a human disease—a dental condition called periodontitis—was published in 2004. But biologists have been studying disease-causing bacteria for over a century.

Of the hundreds or thousands of bacterial species living in and on your body, a tiny fraction can disrupt normal body functions enough to cause illness. Bacteria that cause disease are said to be **pathogenic** (literally, "disease-producing"). Pathogenic bacteria have been responsible for some of the most devastating epidemics in human history.

Table 28.2 lists some of the bacteria that cause illness in humans. The important things to note are that:

- Pathogenic forms come from several different lineages in the domain Bacteria.

- Pathogenic bacteria tend to affect tissues at the entry points to the body, such as wounds or pores in the skin, the respiratory and gastrointestinal tracts, and the urogenital canal.

KOCH'S POSTULATES Robert Koch was the first person to establish a link between a particular species of bacterium and a specific disease. When Koch began his work on the nature of disease in the late 1800s, microscopists had confirmed the existence of the particle-like organisms we now call bacteria, and Louis Pasteur had shown that bacteria and other microorganisms are responsible for spoiling milk, wine, broth, and other foods. Koch hypothesized that bacteria might also be responsible for causing infectious diseases, which spread by being passed from an infected individual to an uninfected individual.

Koch set out to test this hypothesis by identifying the organism that causes anthrax. Anthrax is a disease of cattle and other

TABLE 28.2 Some Bacteria That Cause Illness in Humans

Lineage	Species	Tissues Affected	Disease
Firmicutes	*Clostridium botulinum*	Gastrointestinal tract, nervous system	Food poisoning (botulism)
	Clostridium tetani	Wounds, nervous system	Tetanus
	Staphylococcus aureus	Skin, urogenital canal	Acne, boils, impetigo, toxic shock syndrome
	Streptococcus pneumoniae	Respiratory tract	Pneumonia
	Streptococcus pyogenes	Respiratory tract	Strep throat, scarlet fever
Spirochaetes	*Borrelia burgdorferi*	Skin and nerves	Lyme disease
	Treponema pallidum	Urogenital canal	Syphilis
Actinomycetes	*Mycobacterium leprae*	Skin and nerves	Leprosy
	Mycobacterium tuberculosis	Respiratory tract	Tuberculosis
	Propionibacterium acnes	Skin	Acne
Chlamydiales	*Chlamydia trachomatis*	Urogenital canal	Genital tract infection
ε-Proteobacteria	*Helicobacter pylori*	Stomach	Ulcer
β-Proteobacteria	*Neisseria gonorrhoeae*	Urogenital canal	Gonorrhea
γ-Proteobacteria	*Haemophilus influenzae*	Ear canal, nervous system	Ear infections, meningitis
	Pseudomonas aeruginosa	Urogenital canal, eyes, ear canal	Infections of eye, ear, urinary tract
	Salmonella enterica	Gastrointestinal tract	Food poisoning
	Vibrio parahaemolyticus	Gastrointestinal tract	Food poisoning
	Yersinia pestis	Lymph and blood	Plague

grazing mammals that can result in fatal blood poisoning. The disease also occurs infrequently in humans and mice.

To establish a causative link between a specific microbe and a specific disease, Koch proposed that four criteria had to be met:

1. The microbe must be present in individuals suffering from the disease and absent from healthy individuals. By careful microscopy, Koch was able to show that the bacterium *Bacillus anthracis* was always present in the blood of cattle suffering from anthrax, but absent from healthy individuals.

2. The organism must be isolated and grown in a pure culture away from the host organism. Koch was able to grow pure colonies of *B. anthracis* in glass dishes on a nutrient medium, using gelatin as a substrate.

3. If organisms from the pure culture are injected into a healthy experimental animal, the disease symptoms should appear. Koch demonstrated this crucial causative link in mice injected with *B. anthracis*. The symptoms of anthrax infection appeared, and then the infected mice died.

4. The organism should be isolated from the diseased experimental animal, again grown in pure culture, and demonstrated by its size, shape, and color to be the same as the original organism. Koch did this by purifying *B. anthracis* from the blood of diseased experimental mice.

These criteria, now called **Koch's postulates**, are still used to confirm a causative link between new diseases and a suspected infectious agent.

THE GERM THEORY Koch's experimental results also became the basis for the **germ theory of disease**.

- The pattern component of this theory is that certain diseases are infectious—meaning that they can be passed from person to person.

- The process responsible for this pattern is the transmission and growth of certain bacteria and viruses.

(Viruses are acellular particles that parasitize cells and are analyzed in detail in Chapter 35.)

The germ theory of disease laid the foundation for modern medicine. Initially its greatest impact was on sanitation—efforts to prevent transmission of pathogenic bacteria. During the American civil war, for example, it was common for surgeons to sharpen their scalpels on their shoe leather, after walking in horse manure. During that conflict, records indicate that more soldiers died of dysentery and typhoid fever, contracted from drinking water contaminated with human feces, than from wounds in battle.

Fortunately, improvements in sanitation and nutrition have caused dramatic reductions in mortality rates due to infectious diseases in the industrialized countries. For example, the graph

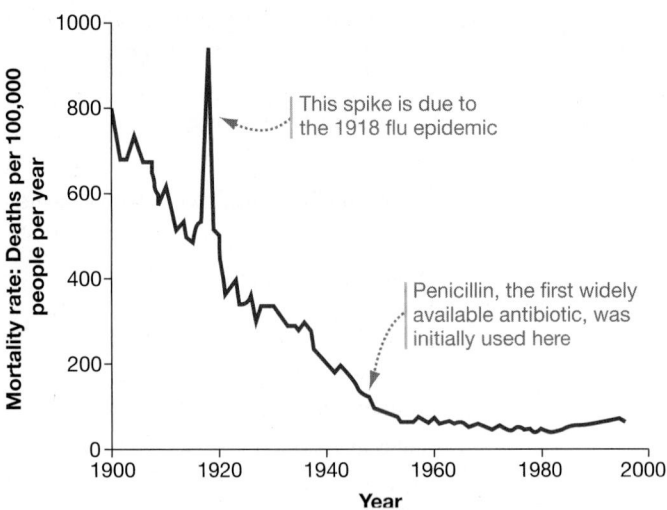

FIGURE 28.2 Deaths Due to Bacterial Infections Have Declined Dramatically in Some Countries. These data are from a country that industrialized in the late 1800s (the United States).

✔**QUESTION** Why did the death rate due to infectious disease drop to low levels, long before antibiotics were available?

In **Figure 28.2** shows the annual death rate due to infectious diseases—meaning, bacterial or viral infections—in the United States between 1900 and the late 1990s. Except for a large spike in the death rate due to a devastating flu epidemic in 1918, the death rate dropped steadily and dramatically during this period. The vast majority of this decline occurred long before antibiotics were introduced.

WHAT MAKES SOME BACTERIAL CELLS PATHOGENIC? Virulence, or the ability to cause disease, is a heritable trait that varies among individuals in a population. Most *Escherichia coli*, for example, are harmless inhabitants of the gastrointestinal tract of humans and other mammals. But some *E. coli* cells cause potentially fatal food poisoning.

What makes some cells of the same species pathogenic, while others are harmless? Biologists have answered this question for *E. coli* by sequencing the entire genome of a harmless lab strain and the pathogenic strain called O157. The genome of the virulent strain is much larger, and contains genes for proteins that allow the cells to adhere to host cells and secrete toxins that disrupt host cells. The diarrhea that results is advantageous to the bacterial cells—they are shed into the environment. If sanitation is poor, the virulent cells may infect many new hosts.

Similar types of studies are identifying the genes responsible for virulence in a wide array of pathogenic bacteria.

THE PAST, PRESENT, AND FUTURE OF ANTIBIOTICS **Antibiotics** are molecules that kill bacteria. They are produced naturally by a wide array of soil-dwelling bacteria and fungi. In these environments, antibiotics are hypothesized to be a weapon that helps cells reduce competition for nutrients and other resources.

The discovery of antibiotics by biomedical researchers in 1928, their development over subsequent decades, and wide-spread use starting in the late 1940s gave physicians effective tools to combat many bacterial infections.

Unfortunately, extensive use of antibiotics in the late twentieth century in clinics and animal feed led to the evolution of drug-resistant strains of pathogenic bacteria (see Chapter 24). One recent study found that there are now soil-dwelling bacteria in natural environments that—far from being killed by antibiotics—actually use them as food.

Coping with antibiotic resistance in pathogenic bacteria has become a great challenge of modern medicine. Some researchers even claim that we may be entering the "post-antibiotic era" in medicine.

Role in Bioremediation

Only a tiny proportion of bacteria and archaea cause disease in humans or other organisms. In the vast majority of cases, bacteria and archaea have no direct impact on humans or are beneficial. For example, researchers are using bacteria and archaea to clean up sites polluted with organic solvents—an effort called **bioremediation.**

Throughout the industrialized world some of the most serious pollutants in soils, rivers, and ponds consist of organic compounds that were originally used as solvents or fuels but leaked or were spilled into the environment. Most of these compounds are highly hydrophobic. Because they do not dissolve in water, they tend to accumulate in sediments. If the compounds are subsequently ingested by burrowing worms or clams or other organisms, they can be passed along to fish, insects, humans, birds, and other species.

At moderate to high concentrations, these pollutants are toxic to eukaryotes. Petroleum from oil spills and compounds that contain ring structures and chlorine atoms, such as the family of compounds called dioxins, are particularly notorious because of their toxicity to humans.

At sites like these, bioremediation is often based on complementary strategies:

- *Fertilizing contaminated sites to encourage the growth of existing bacteria and archaea that degrade toxic compounds.* After several recent oil spills, researchers added nitrogen to affected sites as a fertilizer, but left nearby beaches untreated as controls. Dramatic increases occurred in the growth of bacteria and archaea that use hydrocarbons in cellular respiration, probably because the cells used the added nitrogen to synthesize enzymes and other key compounds. In at least some cases, the fertilized sediments cleaned up much faster than the unfertilized sites (**Figure 28.3**).

- *"Seeding," or adding, specific species of bacteria and archaea to contaminated sites.* Seeding shows promise of alleviating pollution in some situations. For example, researchers have recently discovered bacteria that are able to render certain chlorinated, ring-containing compounds harmless. Instead of being poisoned by the pollutants, these bacteria use the chlorinated compounds as electron acceptors during cellular respiration. In at least some cases, the by-product is dechlorinated and nontoxic to humans and other eukaryotes.

FIGURE 28.3 Bacteria and Archaea Can Play a Role in Cleaning Up Pollution. On the left is a rocky coast that was polluted by an oil spill but fertilized to promote the growth of oil-eating bacteria. The portion of the beach on the right was untreated.

To follow up on these discoveries, researchers are now growing the bacteria in quantity and testing them in the field, to test the hypothesis that seeding can speed the rate of decomposition in contaminated sediments. Initial reports suggest that seeding may help clean up at least some polluted sites.

Extremophiles

Bacteria or archaea that live in high-salt, high-temperature, low-temperature, or high-pressure habitats are known as **extremophiles** ("extreme-lovers"). Studying them has been extraordinarily fruitful for understanding the tree of life, developing industrial applications, and exploring the structure and function of enzymes.

As an example of these habitats, consider hot springs at the bottom of the ocean, where water as hot as 300°C emerges and mixes with 4°C seawater. At locations like these, archaea are abundant forms of life. Researchers recently discovered an archaeon that grows so close to these hot springs that its surroundings are at 121°C—a record for life at high temperature. This organism can live and grow in water that is heated past its boiling point (100°C) and at pressures that would instantly destroy a human being. Other recently discovered bacteria and archaea can grow at:

- a pH less than 1.0;

- seawater at a depth of over 2500 m; and

- anoxic ("no-oxygen") habitats deep in the Mediterranean Sea, where the water is virtually saturated with salt.

Extremophiles have become a hot area of research. The genomes of a wide array of extremophiles have been sequenced, and expeditions regularly seek to characterize new species. Why?

- *Origin of life* Based on models of conditions that prevailed early in Earth's history, it appears likely that the first forms of life lived at high temperature and pressure in environments that lacked oxygen—conditions that we would call extreme.

Thus, understanding extremophiles may help explain how life on Earth began.

- *Extraterrestrial life?* In a similar vein, many astrobiologists ("space-biologists") use extremophiles as model organisms in the search for extraterrestrial life. The idea is that if bacteria and archaea can thrive in extreme habitats on Earth, it is possible that cells might be found in similar environments on other planets or moons of planets.

- *Commercial applications* Because enzymes that function at extreme temperatures and pressures are useful in many industrial processes, extremophiles are of commercial interest as well. Chapter 19 introduced *Taq* polymerase—a DNA polymerase that is stable up to 95°C. Recall that *Taq* polymerase is used to run the polymerase chain reaction (PCR) in research and commercial settings. This enzyme was isolated from a bacterium called *Thermus aquaticus* ("hot water"), which was discovered in hot springs in Yellowstone National Park.

Bacteria and archaea may be small, but they thrive in an amazing range of conditions.

28.2 How Do Biologists Study Bacteria and Archaea?

Our understanding of the domains Bacteria and Archaea is advancing more rapidly right now than at any time during the past 100 years—and perhaps faster than our understanding of any other lineages on the tree of life.

As an introduction to the domains Bacteria and Archaea, let's examine a few of the techniques that biologists use to answer questions about them. Some of these research strategies have been used since bacteria were first discovered; some were invented less than 10 years ago.

Using Enrichment Cultures

Which species of bacteria and archaea are present at a particular location, and what do they use as food? To answer questions like these, biologists rely heavily on their ability to culture organisms in the lab. Of the 5000 species of bacteria and archaea that have been described to date, almost all were discovered when they were isolated from natural habitats and grown under controlled conditions in the laboratory.

One classical technique for isolating new types of bacteria and archaea is called **enrichment culture**. Enrichment cultures are based on establishing a specified set of growing conditions—temperature, lighting, substrate, types of available food, and so on. Cells that thrive under the specified conditions increase in numbers enough to be isolated and studied in detail.

To appreciate how this strategy works in practice, consider research on bacteria that live deep below Earth's surface. One study began with samples of rock and fluid from drilling operations in Virginia and Colorado. The samples came from sedimentary

QUESTION: Can bacteria live a mile below Earth's surface?

HYPOTHESIS: Bacteria are capable of cellular respiration deep below Earth's surface by using H_2 as an electron donor and Fe^{3+} as an electron acceptor.

NULL HYPOTHESIS: Bacteria from this environment are not capable of using H_2 as an electron donor and Fe^{3+} as an electron acceptor.

EXPERIMENTAL SETUP:

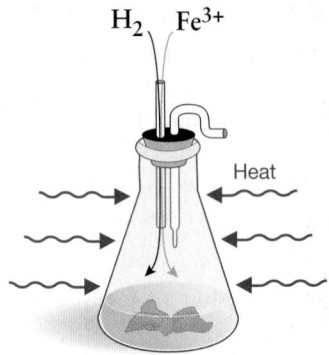

H_2 Fe^{3+}

Heat

Rock and fluid samples

1. Prepare enrichment culture abundant in H_2 and Fe^{3+}; raise temperatures above 45°C.

2. Add rock and fluid samples extracted from drilling operations at depths of about 1000 m below Earth's surface.

PREDICTION: Black, magnetic grains of magnetite (Fe_3O_4) will accumulate because Fe^{3+} is reduced by growing cells and shed as waste product. Cells will be visible.

PREDICTION OF NULL HYPOTHESIS: No magnetite will appear. No cells will grow.

RESULTS: Cells are visible, and magnetite is detectable.

1 μm

CONCLUSION: At least one bacterial species that can live deep below Earth's surface grew in this enrichment culture. Different culture conditions might result in the enrichment of different species present in the same sample.

FIGURE 28.4 Enrichment Cultures Isolate Large Populations of Cells That Grow under Specific Conditions.

✔**QUESTION** Suppose no organisms had grown in this culture. Explain why the lack of growth would be strong evidence or weak evidence on the question of whether organisms live a mile below the Earth's surface.

SOURCE: Liu, S. V., et al. 1997. Thermophilic Fe(III)-reducing bacteria from the deep subsurface: the evolutionary implications. *Science* 277: 1106–1109.

rocks at depths ranging from 860 to 2800 meters below the surface, where temperatures are between 42°C and 85°C. The questions posed in the study were simple: Is anything alive down there? If so, what do the organisms use to fuel cellular respiration?

The research team hypothesized that if organisms were living deep below the surface of the Earth, the cells might use hydrogen molecules (H_2) as an electron donor and the ferric ion (Fe^{3+}) as an electron acceptor (**Figure 28.4**). (Recall from Chapter 9 that most eukaryotes use sugars as electron donors and use oxygen as an electron acceptor during cellular respiration.) Fe^{3+} is the oxidized form of iron, and it is abundant in the rocks the biologists collected from great depths. It exists at great depths below the surface in the form of ferric oxyhydroxide. The researchers predicted that if an organism in the samples reduced the ferric ions during cellular respiration, then a black, oxidized, and magnetic mineral called magnetite (Fe_3O_4) would start appearing in the cultures as a by-product of cellular respiration.

What did their enrichment cultures produce? In some cultures, a black compound began to appear within a week. Using a variety of tests, the biologists confirmed that the black substance was indeed magnetite. As the "Results" section of Figure 28.4 shows, microscopy revealed the organisms themselves—previously undiscovered bacteria. Because they grow only when incubated at between 45°C and 75°C, these organisms are considered **thermophiles** ("heat-lovers"). The discovery was spectacular—it was one of the first studies demonstrating that Earth's crust is teeming with organisms to depths of over a mile below the surface. Enrichment culture continues to be a productive way to isolate and characterize new species of bacteria and archaea.

Using Direct Sequencing

Researchers estimate that of all the bacteria and archaea living today, less than 1 percent have been grown in culture. To augment research based on enrichment cultures, researchers are employing a technique called direct sequencing. **Direct sequencing** is a strategy for documenting the presence of bacteria and archaea that cannot be grown in culture. It is based on identifying phylogenetic species—populations that have enough distinctive characteristics to represent an independent branch on an evolutionary tree (see Chapter 26).

Direct sequencing allows biologists to identify and characterize organisms that have never been seen. The technique has revealed huge new branches on the tree of life and produced revolutionary data on the habitats where archaea are found.

Figure 28.5 outlines the steps performed in a direct sequencing study. The essence of the approach is to use the polymerase chain reaction, introduced in Chapter 19, to isolate specific genes from the sample. After sequencing these genes, biologists compare the data with sequences in existing databases. If the sequences are markedly different, the sample probably contains previously undiscovered organisms.

Direct sequencing studies have produced new and sometimes startling results. For two decades after the discovery of the Ar-

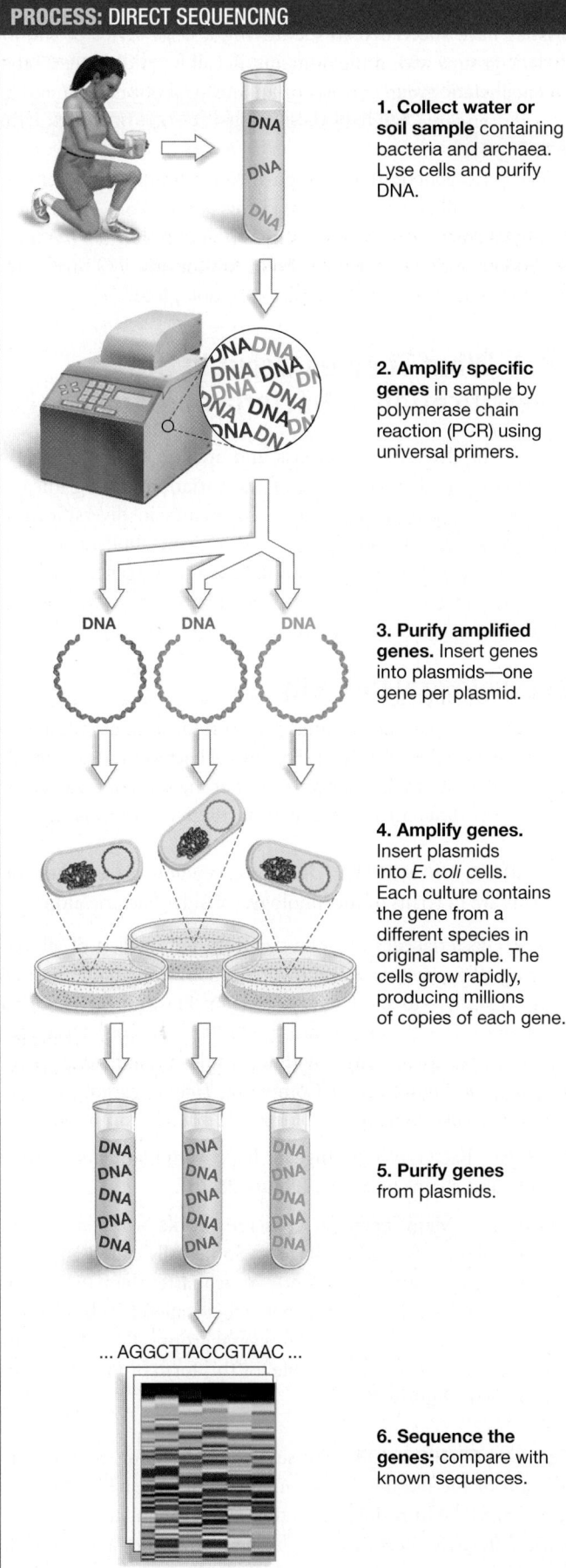

PROCESS: DIRECT SEQUENCING

1. **Collect water or soil sample** containing bacteria and archaea. Lyse cells and purify DNA.

2. **Amplify specific genes** in sample by polymerase chain reaction (PCR) using universal primers.

3. **Purify amplified genes.** Insert genes into plasmids—one gene per plasmid.

4. **Amplify genes.** Insert plasmids into *E. coli* cells. Each culture contains the gene from a different species in original sample. The cells grow rapidly, producing millions of copies of each gene.

5. **Purify genes** from plasmids.

... AGGCTTACCGTAAC ...

6. **Sequence the genes;** compare with known sequences.

chaea, for example, researchers thought that these organisms could be conveniently grouped into just four categories:

- extreme **halophiles** ("salt-lovers") that live in salt lakes, salt ponds, and salty soils;

- **sulfate reducers**—cells that produce hydrogen sulfide (H_2S) as a by-product of cellular respiration (H_2S may be familiar because it smells like rotten eggs);

- **methanogens** produce methane (natural gas; CH_4) as a by-product of cellular respiration; and

- extreme **thermophiles** grow best at temperatures above 80°C.

Direct sequencing studies threw this generalization out the window, revealing archaea in habitats as diverse as rice paddies and the Arctic Ocean, and entirely new lineages and metabolic capabilities within the Archaea.

Direct sequencing has also revealed previously unrecognized diversity in the bacteria and archaea found on and in the human body. In combination with environmental genomics—a technique based on sequencing all or most of the genes found in a particular habitat (see Chapter 20)—direct sequencing is revolutionizing our understanding of bacterial and archaeal diversity.

Evaluating Molecular Phylogenies

To put data from enrichment culture and direct sequencing studies into context, biologists depend on the accurate placement of species on phylogenetic trees. Recall from Chapter 1, Chapter 27, and **BioSkills 3** in Appendix A that phylogenetic trees illustrate the evolutionary relationships among species and lineages. They are a pictorial summary of which species are more closely or distantly related to others.

Some of the most useful phylogenetic trees for the Bacteria and the Archaea have been based on studies of the RNA molecules found in the small subunit of ribosomes, or what biologists call 16S and 18S RNA. (See Chapter 16 for more information on the structure and function of ribosomes.) In the late 1960s Carl Woese and colleagues began a massive effort to determine and compare the base sequences of 16S and 18S RNA molecules from a wide array of species. The result of their analysis was the **universal tree**, or the **tree of life**, illustrated in Figure 28.1. To review the results of their work, go to the study area at *www.masteringbiology.com.*

(MB) Web Activity The Tree of Life

Woese's tree is now considered a classic result. Prior to its publication, biologists thought that the major division among organisms was between prokaryotes and eukaryotes.

FIGURE 28.5 Direct Sequencing Allows Researchers to Identify Species That Have Never Been Seen. Direct sequencing isolates specific genes from organisms in a sample. The polymerase chain reaction generates many copies of the genes from each species, so they can be sequenced.

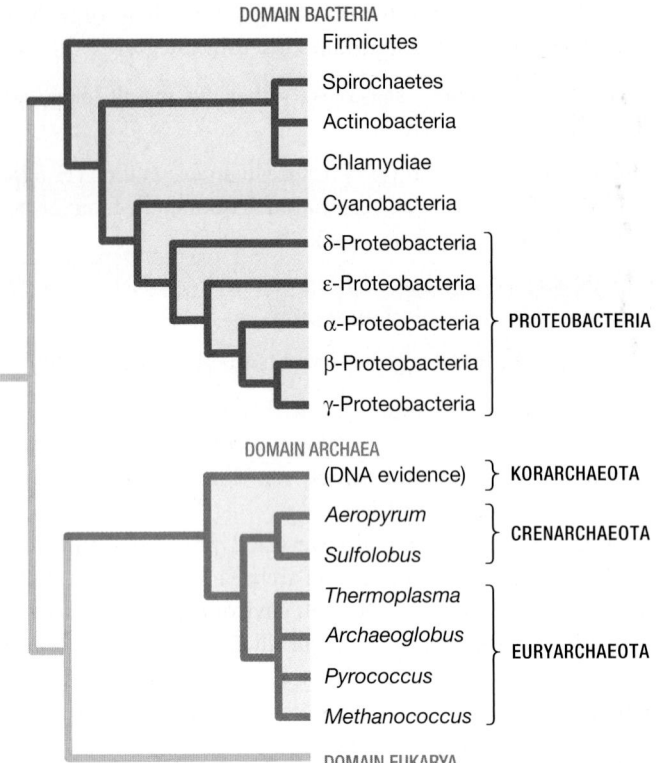

FIGURE 28.6 Phylogeny of Some Major Lineages in Bacteria and Archaea.

But based on data from the ribosomal RNA molecule, the major divisions of life-forms are actually the Bacteria, Archaea, and Eukarya. Follow-up work documented that Bacteria were the first of the three lineages to diverge from the common ancestor of all living organisms—meaning that the Archaea and Eukarya are more closely related to each other than they are to the Bacteria.

CHECK YOUR UNDERSTANDING

If you understand that . . .

- Enrichment cultures isolate cells that grow in response to specific conditions. They create an abundant sample of bacteria that thrive under particular conditions, allowing further study.
- Direct sequencing is based on isolating DNA from samples taken directly from the environment, purifying and sequencing specific genes, and then analyzing where those DNA sequences are found on the phylogenetic tree of Bacteria and Archaea.

✔ **You should be able to . . .**

1. Design an enrichment culture that would isolate species that could be used to clean up oil spills.

2. Outline a study designed to identify the bacterial and archaeal species present in a soil sample near the biology building on your campus.

Answers are available in Appendix B.

More recent analyses of morphological and molecular characteristics have succeeded in identifying a large series of monophyletic groups within the domains. Recall from Chapter 27 that a **monophyletic group** consists of an ancestral population and all of its descendants. Monophyletic groups can also be called clades or lineages.

The phylogenetic tree in **Figure 28.6** summarizes recent results but is still considered highly provisional. Work on molecular phylogenies continues at a brisk pace. Section 28.4 will explore some of these lineages in detail, but for now let's turn to the question of how all this diversification took place.

28.3 What Themes Occur in the Diversification of Bacteria and Archaea?

Initially, the diversity of bacteria and archaea can seem almost overwhelming. To make sense of the variation among lineages and species, biologists focus on two themes in diversification: morphology and metabolism. Regarding metabolism, the key question is which molecules are used as food. Bacteria and archaea are capable of living in a wide array of environments because they vary in cell structure and in how they make a living.

Morphological Diversity

Because we humans are so large, it is hard for us to appreciate the morphological diversity that exists among bacteria and archaea. To us, they all look small and similar. But at the scale of a bacterium or archaean, different species are wildly diverse in morphology.

SIZE, SHAPE, AND MOTILITY To appreciate how diverse these organisms are in terms of morphology, consider bacteria alone:

- *Size* Bacterial cells range in size from the smallest of all free-living cells—bacteria called mycoplasmas with volumes as small as 0.15 μm³—to the largest bacterium known, *Thiomargarita namibiensis*, with volumes as large as 200×10^6 μm³. Over a billion *Mycoplasma* cells could fit inside an individual *Thiomargarita* (**Figure 28.7a**). Comparing *Thiomargarita* and *E. coli* cell size is like comparing a blue whale to a newborn mouse.

- *Shape* Bacterial cells range in shape from filaments, spheres, rods, and chains to spirals (**Figure 28.7b**).

- *Motility* Many bacterial cells are motile, with swimming movements powered by flagella. Some cells can swim 10 or more body lengths per second—much faster than any human can sprint. Gliding movement, which allows cells to creep along a surface, also occurs in several groups, though the molecular mechanism responsible for this form of motility is still unknown (**Figure 28.7c**).

CELL WALL COMPOSITION For single-celled organisms, the composition of the plasma membrane and cell wall are particularly important. The introduction to this chapter highlighted the dramatic differences between the plasma membranes and cell walls of bacteria versus archaea.

(a) Size varies.

Most bacteria are about 1 μm in diameter.

Smallest (*Mycoplasma mycoides*)

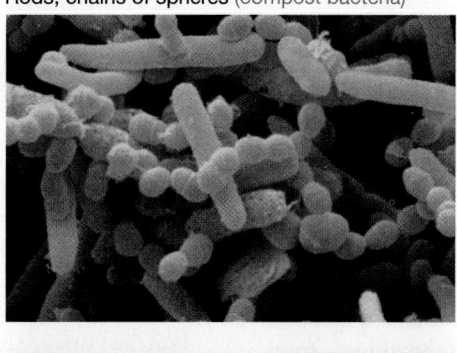

0.5 μm

Compare sizes

100 μm

Largest (*Thiomargarita namibiensis*)

(b) Shape varies...

... from rods to spheres to spirals to filaments. In some species, cells adhere to form chains.

Rods, chains of spheres (compost bacteria)

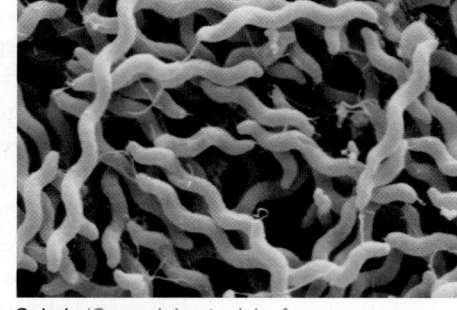

Spirals (*Campylobacter jejuni*)

(c) Mobility varies.

Some bacteria are immotile, but swimming and gliding are common.

Swimming (*Pseudomonas aeruginosa*)

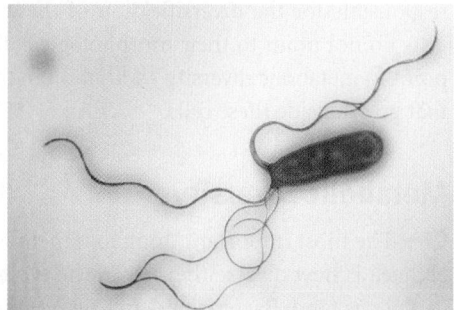

Gliding (*Oscillatoria limosa*)

FIGURE 28.7 Morphological Diversity among Bacteria Is Extensive. Some of the cells in these micrographs have been colorized to make them more visible.

Within bacteria, biologists distinguish two general types of cell wall using a dyeing system called the **Gram stain**. As **Figure 28.8a** shows, Gram-positive cells look purple but Gram-negative cells look pink.

At the molecular level, cells that are **Gram-positive** have a plasma membrane surrounded by a cell wall with extensive peptidoglycan (**Figure 28.8b**). You might recall from Chapter 5 that peptidoglycan is a complex substance composed of carbohydrate strands that are cross-linked by short chains of amino acids. Cells that are **Gram-negative**, in contrast, have a plasma membrane surrounded by a cell wall that has two components—a thin gelatinous layer containing peptidoglycan and an outer phospholipid bilayer (**Figure 28.8c**).

Analyzing cell cultures with the Gram stain can be an important preliminary step in treating bacterial infections. Because they contain so much peptidoglycan, Gram-positive cells may respond to treatment by penicillin-like drugs that disrupt peptidoglycan synthesis. Gram-negative cells, in contrast, are more likely to be affected by erythromycin or other types of drugs that poison bacterial ribosomes.

(a) Gram-positive cells stain more than Gram-negative cells.

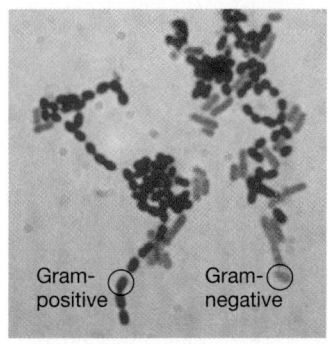

Gram-positive   Gram-negative

(b) Gram-positive cell wall

(c) Gram-negative cell wall

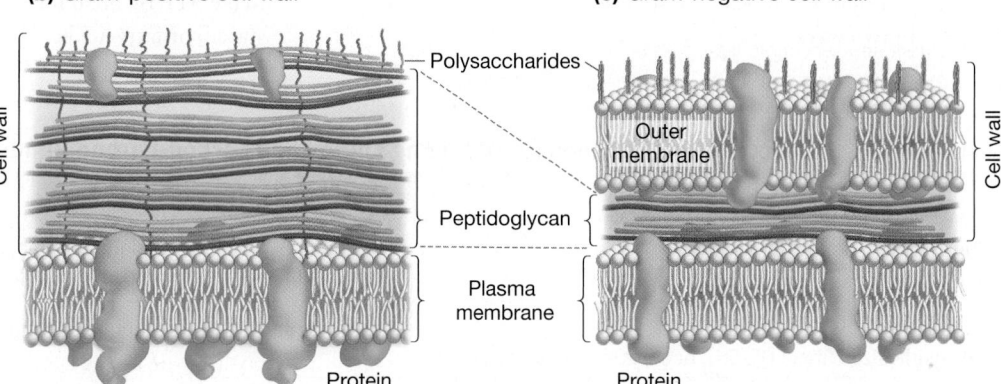

Polysaccharides

Cell wall

Outer membrane

Peptidoglycan

Plasma membrane

Protein

Cell wall

Protein

FIGURE 28.8 Gram Staining Distinguishes Two Types of Cell Walls in Bacteria. Cells with extensive peptidoglycan retain a large amount of stain and look purple; others retain little stain and look pink, as can be seen in part (a).

To summarize, members of the Bacteria and the Archaea are remarkably diverse in their overall size, shape, and motility, as well as in the composition of their cell walls and plasma membranes. But when asked to name the innovations that were most responsible for the diversification of these two domains, biologists do not point to their morphological diversity. Instead, they point to metabolic diversity—variation in the chemical reactions that go on inside these cells.

Metabolic Diversity

👈 The most important thing to remember about bacteria and archaea is how diverse they are in the types of compounds they can use as food. Bacteria and archaea are the masters of metabolism. Taken together, they can subsist on almost anything—from hydrogen molecules to crude oil. Bacteria and archaea look small and relatively simple to us in their morphology, but their biochemical capabilities are dazzling.

Just how varied are bacteria and archaea when it comes to making a living? To appreciate the answer, recall from Chapters 9 and 10 that organisms have two fundamental nutritional needs—acquiring chemical energy in the form of adenosine triphosphate (ATP) and obtaining molecules with carbon-carbon bonds that can be used as building blocks for the synthesis of fatty acids, proteins, DNA, RNA, and other large, complex compounds required by the cell.

Bacteria and archaea produce ATP in three ways:

1. **Phototrophs** ("light-feeders") use light energy to promote electrons to the top of electron transport chains. ATP is produced by photophosphorylation (see Chapter 10).

2. **Chemoorganotrophs** oxidize organic molecules with high potential energy, such as sugars. ATP may be produced by cellular respiration—with sugars serving as electron donors—or via fermentation pathways (see Chapter 9).

3. **Chemolithotrophs** ("rock-feeders") oxidize inorganic molecules with high potential energy, such as ammonia (NH_3) or methane (CH_4). ATP is produced by cellular respiration, with inorganic compounds serving as the electron donor.

Bacteria and archaea fulfill their second nutritional need—obtaining building block compounds with carbon-carbon bonds—in two ways:

1. By synthesizing their own from simple starting materials such as CO_2 and CH_4. Organisms that manufacture their own building-block compounds are termed **autotrophs** ("self-feeders").

2. By absorbing ready-to-use organic compounds from their environment. Organisms that acquire building-block compounds from other organisms are called **heterotrophs** ("other-feeders").

Because there are three distinct ways of producing ATP and two general mechanisms for obtaining carbon, there are a total of six methods for producing ATP and obtaining carbon. The names that biologists use for organisms that use these six "feeding strategies" are given in **Table 28.3**.

Of the six possible ways of producing ATP and obtaining carbon, just two are observed among eukaryotes. But bacteria and archaea do them all. In addition, certain species can switch among modes of living, depending on environmental conditions. In their metabolism, eukaryotes are simple compared with bacteria and archaea. ✔If you understand the essence of metabolic diversity in bacteria and archaea, you should be able to match the six example species described in **Table 28.4** to the appropriate category in Table 28.3.

What makes this remarkable diversity possible? Bacteria and archaea have evolved dozens of variations on the basic processes you learned about in Chapters 9 and 10. They use compounds with high potential energy to produce ATP via cellular respiration (electron transport chains) or fermentation, they use light to produce high-energy electrons, and they reduce carbon from CO_2 or other sources to produce sugars or other building-block molecules with carbon-carbon bonds.

The story of bacteria and archaea can be boiled down to two sentences: The basic chemistry required for photosynthesis, cellular respiration, and fermentation originated in these lineages. Then the evolution of variations on each of these processes allowed prokaryotes to diversify into millions of species that occupy diverse habitats. Let's take a closer look.

PRODUCING ATP THROUGH CELLULAR RESPIRATION: VARIATION IN ELECTRON DONORS AND ACCEPTORS Millions of bacterial, archaeal, and eukaryotic species—including animals and plants—are organotrophs. These organisms obtain the energy required to make ATP by breaking down organic compounds such as sugars, starch, or fatty acids.

SUMMARY TABLE 28.3 **Six General Methods for Obtaining Energy and Carbon-Carbon Bonds**

Source of C–C Bonds (for synthesis of complex organic compounds)	Source of Energy (for synthesis of ATP)		
	Phototrophs: from sunlight	Chemoorganotrophs: from organic molecules	Chemolithotrophs: from inorganic molecules
Autotrophs: self-synthesized from CO_2, CH_4, or other simple molecules	photoautotrophs	chemoorganoautotrophs	chemolitho[auto]trophs
Heterotrophs: from molecules produced by other organisms	photoheterotrophs	chemoorganoheterotrophs	chemolithotrophic heterotrophs

TABLE 28.4 **Six Examples of Metabolic Diversity**

Example	How ATP Is Produced	How Building-Block Molecules Are Synthesized
Cyanobacteria	via photosynthesis	from CO_2 via the Calvin cycle
Clostridium aceticum	fermentation of glucose	from CO_2 via reactions called the acetyl-CoA pathway
Nitrifying bacteria (e.g., *Nitrosomonas* sp.)	via cellular respiration, using ammonia (NH_3) as an electron donor	from CO_2 via the Calvin cycle
Heliobacteria	via photosynthesis	absorb carbon-containing building-block molecules from the environment
Escherichia coli	fermentation of organic compounds or cellular respiration, using organic compounds as electron donors	absorb carbon-containing building-block molecules from the environment
Beggiatoa	via cellular respiration, using hydrogen sulfide (H_2S) as an electron donor	absorb carbon-containing building-block molecules from the environment

As Chapter 9 showed, cellular enzymes can strip electrons from organic molecules that have high potential energy and then transfer these high-energy electrons to the electron carriers NADH and $FADH_2$. These compounds feed electrons to an electron transport chain (ETC), where electrons are stepped down from a high-energy state to a low-energy state (**Figure 28.9a**).

The energy that is released allows components of the ETC to generate a proton gradient across the plasma membrane (**Figure 28.9b**). The resulting flow of protons through the enzyme ATP synthase results in the production of ATP, via the process called chemiosmosis.

The essence of this process, called **cellular respiration**, is that a molecule with high potential energy serves as an original electron donor and is oxidized, while a molecule with low potential energy

serves as a final electron acceptor and becomes reduced. Much of the potential energy difference between the electron donor and electron acceptor is transformed into chemical energy in the form of ATP.

Chapter 9 focused on how eukaryotes perform cellular respiration. In eukaryotes:

- Organic compounds with high potential energy—often glucose—serve as the original electron donor. When cellular respiration is complete, glucose is completely oxidized to CO_2, which is given off as a by-product.

- Oxygen is the final electron acceptor, and water is also produced as a by-product.

Many bacteria and archaea also rely on these molecules.

(a) Model of Electron Transport Chain (ETC)

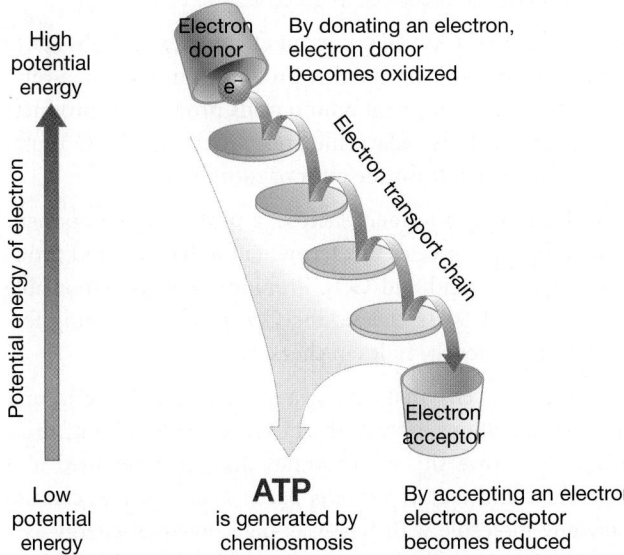

(b) ETC generates proton gradient across plasma membrane.

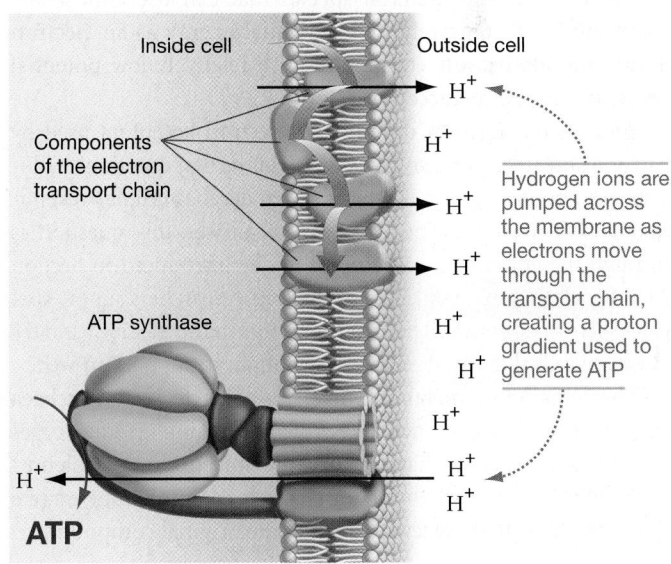

FIGURE 28.9 Cellular Respiration Is Based on Electron Transport Chains.

✔**EXERCISE** In part (a), add the chemical formula for a specific electron donor, electron acceptor, and reduced by-product for a species of bacteria or archaea. Then write in the electron donor, electron acceptor, and reduced by-product observed in humans.

TABLE 28.5 **Some Electron Donors and Acceptors Used by Bacteria and Archaea**

| Electron Donor | Electron Acceptor | By-Products | | Category* |
		From Electron Donor	From Electron Acceptor	
Sugars	O_2	CO_2	H_2O	Organotrophs
H_2 or organic compounds	SO_4^{2-}	H_2O or CO	H_2S or S^{2-}	Sulfate reducers
H_2	CO_2	H_2O	CH_4	Methanogens
CH_4	O_2	CO_2	H_2O	Methanotrophs
S^{2-} or H_2S	O_2	SO_4^{2-}	H_2O	Sulfur bacteria
Organic compounds	Fe^{3+}	CO_2	Fe^{2+}	Iron reducers
NH_3	O_2	NO_2^-	H_2O	Nitrifiers
Organic compounds	NO_3^-	CO_2	N_2O, NO, or N_2	Denitrifiers (or nitrate reducers)
NO_2^-	O_2	NO_3^-	H_2O	Nitrosifiers

*The name biologists use to identify species that use a particular metabolic strategy.

✔ **QUESTION** Explain why the terms organotrophs, sulfate reducers, and methanogens are appropriate. (The word root *–gen* means source or origin; *–troph* refers to feeding.)

It is common, however, to find bacteria and archaea that employ an electron donor other than sugars and an electron acceptor other than oxygen during cellular respiration. These species produce by-products other than carbon dioxide and water (**Table 28.5**). In some bacteria and archaea:

- Inorganic ions or molecules with high potential energy serve as electron donors. The substances used as electron donors range from hydrogen molecules (H_2) and hydrogen sulfide (H_2S) to ammonia (NH_3) and methane (CH_4).

- Compounds with relatively low potential energy—including sulfate (SO_4^{2-}), nitrate (NO_3^-), carbon dioxide (CO_2), or ferric ions (Fe^{3+})—act as electron acceptors.

It is only a slight exaggeration to claim that researchers have found bacterial and archaeal species that can use almost any compound with relatively high potential energy as an electron donor and almost any compound with relatively low potential energy as an electron acceptor.

Because the electron donors and electron acceptors used by bacteria and archaea are so diverse, one of the first questions biologists ask about a species is whether it undergoes cellular respiration and if so, how. The best way to answer this question is through the enrichment culture technique introduced in Section 28.2. Recall that in an enrichment culture, researchers supply specific electron donors and electron acceptors in the medium and try to isolate cells that can use those compounds to support growth.

The remarkable metabolic diversity of bacteria and archaea explains why they play such a key role in cleaning up some types of pollution. Species that use organic solvents or petroleum-based fuels as electron donors or electron acceptors may excrete waste products that are less toxic than the original compounds.

PRODUCING ATP VIA FERMENTATION: VARIATION IN SUBSTRATES
Chapter 9 introduced **fermentation** as a strategy for making ATP that does not involve electron transport chains. In fermentation, no outside electron acceptor is used.

Because fermentation is a much less efficient way to make ATP compared with cellular respiration, in many species it occurs as an alternative metabolic strategy when no electron acceptors are available to make cellular respiration possible. In other species, fermentation does not occur at all. But in many bacteria and archaea, fermentation is the only way that cells make ATP.

Although the presentation in Chapter 9 focused on how glucose is fermented to ethanol or lactic acid, some bacteria and archaea are capable of using other organic compounds as the starting point for fermentation. Bacteria and archaea that produce ATP via fermentation are still classified as organotrophs, but they are much more diverse in the substrates used. For example:

- The bacterium *Clostridium aceticum* can ferment ethanol, acetate, and fatty acids as well as glucose.

- Other species of *Clostridium* ferment complex carbohydrates (including cellulose or starch), proteins, purines, or amino acids. Species that ferment amino acids produce by-products with names such as cadaverine and putrescine. These molecules are responsible for the odor of rotting flesh.

- Other bacteria can ferment lactose, a prominent component of milk. In some species this fermentation has two end products: propionic acid and CO_2. Propionic acid is responsible for the taste of Swiss cheese; the CO_2 produced during fermentation creates the holes in cheese.

The diversity of enzymatic pathways observed in bacterial and archaeal fermentations extends the metabolic repertoire of these organisms. The diversity of substrates that are fermented also supports the claim that as a group, bacteria and archaea can use virtually any molecule with relatively high potential energy as a source of high-energy electrons for producing ATP.

PRODUCING ATP VIA PHOTOSYNTHESIS: VARIATION IN ELECTRON SOURCES AND PIGMENTS Instead of using molecules as a source of high-energy electrons, phototrophs pursue a radically different strategy: **photosynthesis**. Among bacteria and archaea, photosynthesis can happen in one of three ways:

- Light activates a pigment called bacteriorhodopsin, which uses the absorbed energy to transport protons across a membrane. The resulting flow of protons drives the synthesis of ATP via chemiosmosis (see Chapter 9).

- A recently discovered bacterium that lives near hydrothermal vents on the ocean floor performs photosynthesis not by absorbing light, but by absorbing geothermal radiation.

- Pigments that absorb light raise electrons to high-energy states. As these electrons are stepped down to lower energy states by electron transport chains, the energy released is used to generate ATP.

Chapter 10 introduced an important feature of this last mode of photosynthesis: The process requires a source of electrons. Recall that in cyanobacteria and plants the required electrons come from water. When these organisms "split" water molecules apart to obtain electrons, they generate oxygen as a by-product. Species that use water as a source of electrons for photosynthesis are said to complete **oxygenic** photosynthesis.

In contrast, many phototrophic bacteria use a molecule other than water as the source of electrons. In many cases, the electron donor is hydrogen sulfide (H_2S); a few species can use the ion known as ferrous iron (Fe^{2+}). Instead of producing oxygen as a by-product of photosynthesis, these cells produce elemental sulfur (S) or the ferric ion (Fe^{3+}). They are said to complete **anoxygenic** photosynthesis. They live in habitats where oxygen is rare.

Chapter 10 also introduced the photosynthetic pigments found in plants and explored the light-absorbing properties of chlorophylls *a* and *b*. Cyanobacteria have these two pigments. But researchers have isolated seven additional chlorophylls from bacterial phototrophs. Each major group of photosynthetic bacteria has one or more of these distinctive chlorophylls, and each type of chlorophyll absorbs light best at a different wavelength. As a result, a diverse array of photosynthetic bacteria can live in the same habitat without competing for light.

OBTAINING BUILDING-BLOCK COMPOUNDS: VARIATION IN PATHWAYS FOR FIXING CARBON In addition to acquiring energy, organisms must obtain building-block molecules that contain carbon-carbon bonds. Chapters 9 and 10 introduced the two mechanisms that organisms use to procure usable carbon—either making their own or getting it from other organisms. Autotrophs make their own building-block compounds; heterotrophs don't.

In many autotrophs, including cyanobacteria and plants, the enzymes of the Calvin cycle transform carbon dioxide (CO_2) to organic molecules that can be used in synthesizing cell material. The carbon atom in CO_2 is reduced during the process and is said to be "fixed." Animals and fungi, in contrast, obtain carbon from living plants or animals or by absorbing the organic compounds released as dead tissues decay.

Bacteria and archaea pursue these same two strategies. Some interesting twists occur among bacterial and archaeal autotrophs, however. Not all of them use the Calvin cycle to make building-block molecules, and not all start with CO_2 as a source of carbon atoms. For example:

- Several groups of bacteria fix CO_2 using pathways other than the Calvin cycle. Three of these distinctive pathways have been discovered to date.

- Some proteobacteria are called **methanotrophs** ("methane-eaters") because they use methane (CH_4) as their carbon source. (They also use CH_4 as an electron donor in cellular respiration.) Methanotrophs process CH_4 into more complex organic compounds via one of two enzymatic pathways, depending on the species.

- Some bacteria can use carbon monoxide (CO) or methanol (CH_3OH) as a starting material.

These observations drive home an important message from this chapter: Compared with eukaryotes, the metabolic capabilities of bacteria and archaea are remarkably sophisticated and complex.

Ecological Diversity and Global Change

The metabolic diversity observed among bacteria and archaea explains why these organisms can thrive in such a wide array of habitats.

- The array of electron donors, electron acceptors, and fermentation substrates exploited by bacteria and archaea allows the heterotrophic species to live just about anywhere.

- The evolution of three distinct types of photosynthesis—based on bacteriorhodopsin, geothermal energy, or pigments that donate high-energy electrons to ETCs—extends the types of habitats that can support phototrophs. In addition, the remarkable diversity of bacterial chlorophylls allows photosynthetic species with different absorption spectra to live together without competing for light.

The complex chemistry that these cells carry out, combined with their numerical abundance, has made them potent forces for global change throughout Earth's history. Bacteria and archaea have altered the chemical composition of the oceans, atmosphere, and terrestrial environments for billions of years. They continue to do so today.

THE OXYGEN REVOLUTION Today, oxygen represents almost 21 percent of the molecules in Earth's atmosphere. But researchers who study the composition of the atmosphere are virtually certain that no free molecular oxygen (O_2) existed for the first 2.3 billion years of Earth's existence. This conclusion is based on two observations:

1. There was no plausible source of oxygen at the time the planet formed.

2. The oldest Earth rocks indicate that the first signs of oxygen appear many years after the formation of the planet. Even then, virtually all of the oxygen present reacted immediately with iron atoms to produce iron oxides, such as hematite (Fe_2O_3) and magnetite (Fe_3O_4).

Early in Earth's history, the atmosphere was dominated by nitrogen and carbon dioxide. Where did the oxygen we breathe come from? The answer is cyanobacteria.

FIGURE 28.10 Cyanobacteria Were the First Organisms to Perform Oxygenic Photosynthesis.

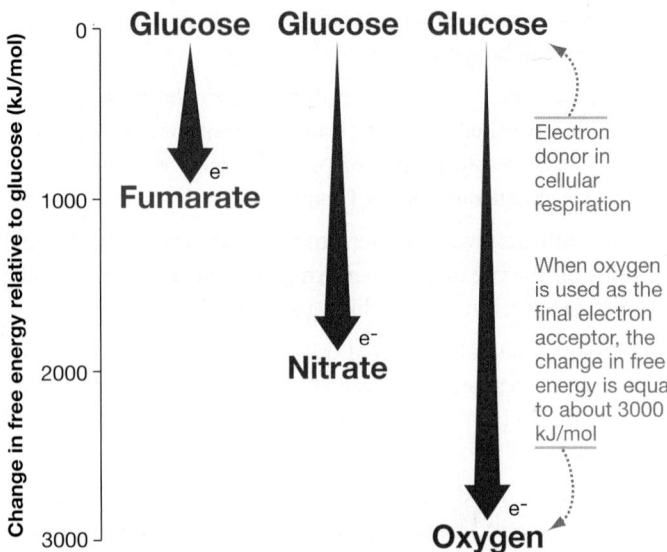

FIGURE 28.11 Cellular Respiration Can Produce More Energy When Oxygen Is the Final Electron Acceptor. Because oxygen has such high electronegativity, the potential energy of the electrons used in cellular respiration is much lower when oxygen is the final acceptor compared to other molecules or ions—resulting in the release of a larger amount of free energy.

✔**QUESTION** Which organisms grow faster—those using aerobic respiration or those using anaerobic respiration? Explain your reasoning.

Cyanobacteria are a lineage of photosynthetic bacteria. According to the fossil record, species of cyanobacteria first became numerous in the oceans about 2.7–2.55 billion years ago. Their appearance was momentous because cyanobacteria were the first organisms to perform oxygenic ("oxygen-producing") photosynthesis (**Figure 28.10**).

The fossil record and geological record indicate that oxygen concentrations in the oceans and atmosphere began to increase 2.3–2.1 billion years ago. Once oxygen was common in the oceans, cells could begin to use it as the final electron acceptor during cellular respiration. **Aerobic** respiration was now a possibility. Prior to this, organisms had to use compounds other than oxygen as a final electron acceptor—only **anaerobic** respiration was possible.

The evolution of aerobic respiration was a crucial event in the history of life. Because oxygen is extremely electronegative, it is an efficient electron acceptor. Much more energy is released as electrons move through electron transport chains with oxygen as the ultimate acceptor than is released with other substances as the electron acceptor.

To drive this point home, study the graph in **Figure 28.11**. Note that the vertical axis plots free energy changes; the graph shows the energy released when glucose is oxidized with fumarate, nitrate, or oxygen as the final electron acceptor. Once oxygen was available, then, cells could produce much more ATP for each electron donated by NADH or FADH$_2$. As a result, the rate of energy production could rise dramatically.

To summarize, data indicate that cyanobacteria were responsible for a fundamental change in Earth's atmosphere—a high concentration of oxygen. Never before, or since, have organisms done so much to alter the nature of our planet.

NITROGEN FIXATION AND THE NITROGEN CYCLE In many environments, fertilizing forests or grasslands with nitrogen results in increased growth. Researchers infer from these results that plant growth is limited by the availability of nitrogen.

Organisms must have nitrogen to synthesize proteins and nucleic acids. Although molecular nitrogen (N$_2$) is extremely abundant in the atmosphere, most organisms cannot use it. To incorporate nitrogen atoms into amino acids and nucleotides, all eukaryotes and many bacteria and archaea have to obtain N in a form such as ammonia (NH$_3$) or nitrate (NO$_3^-$).

Certain bacteria and archaea are the only species that are capable of converting molecular nitrogen to ammonia. The steps in the process, called **nitrogen fixation**, are complex and highly endergonic reduction-oxidation (redox) reactions (see Chapter 9). The enzymes required to accomplish nitrogen fixation are found only in selected bacterial and archaeal lineages.

- Some cyanobacteria that live in surface waters of the ocean or in association with water plants are capable of fixing nitrogen.

- In terrestrial environments, nitrogen-fixing bacteria live in close association with plants—often taking up residence in special root structures called nodules (**Figure 28.12**).

If bacteria and archaea could not fix nitrogen, it is virtually certain that only a tiny fraction of life on Earth would exist today. Large or multicellular organisms would probably be rare to nonexistent, because too little nitrogen would be available to make large quantities of proteins and build a large body.

Nitrogen-fixation is only the beginning of the story, however. A quick glance back at Table 28.5 should convince you that bacteria and archaea use a wide array of nitrogen-containing compounds as electron donors and electron acceptors during cellular respiration.

To understand why this is important, consider that the nitrite (NO$_2^-$) that some bacteria produce as a by-product of respiration does not build up in the environment. Instead, it is used as an electron acceptor by other species and converted to molecular nitrate (NO$_3^-$). Nitrate, in turn, is converted to molecular nitrogen (N$_2$) by yet another suite of bacterial and archaeal species. In

FIGURE 28.12 Root nodules form a protective structure for bacteria that fix nitrogen.

this way, bacteria and archaea are responsible for driving the movement of nitrogen atoms through ecosystems around the globe (**Figure 28.13**).

Similar types of interactions occur with molecules that contain phosphorus, sulfur, and carbon. In this way, bacteria and archaea play a key role in the cycling of nitrogen and other nutrients.

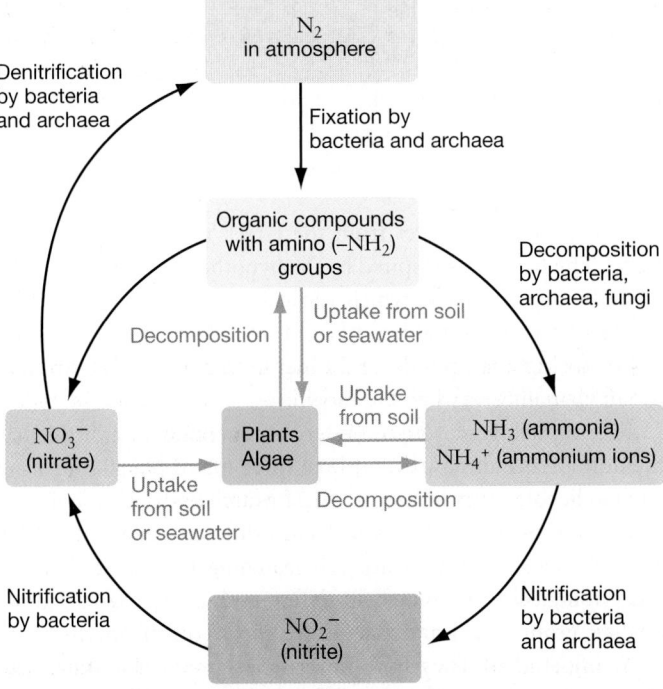

FIGURE 28.13 Bacteria and Archaea Drive the Movement of Nitrogen Atoms through Ecosystems. Nitrogen atoms cycle in different molecular forms. In addition to the conversions shown here, some bacteria can also convert nitrate to ammonium ions; others combine ammonium and nitrate ions to form nitrogen gas (N_2).

NITRATE POLLUTION Corn, rice, wheat, and many other crop plants do not live in association with nitrogen-fixing bacteria. To increase yields of these crops, farmers use fertilizers that are high in nitrogen. In parts of the world, massive additions of nitrogen in the form of ammonia are causing serious pollution problems.

Figure 28.14 shows why. When ammonia is added to a cornfield—in midwestern North America, for example—much of it never reaches the growing corn plants. Instead, a significant

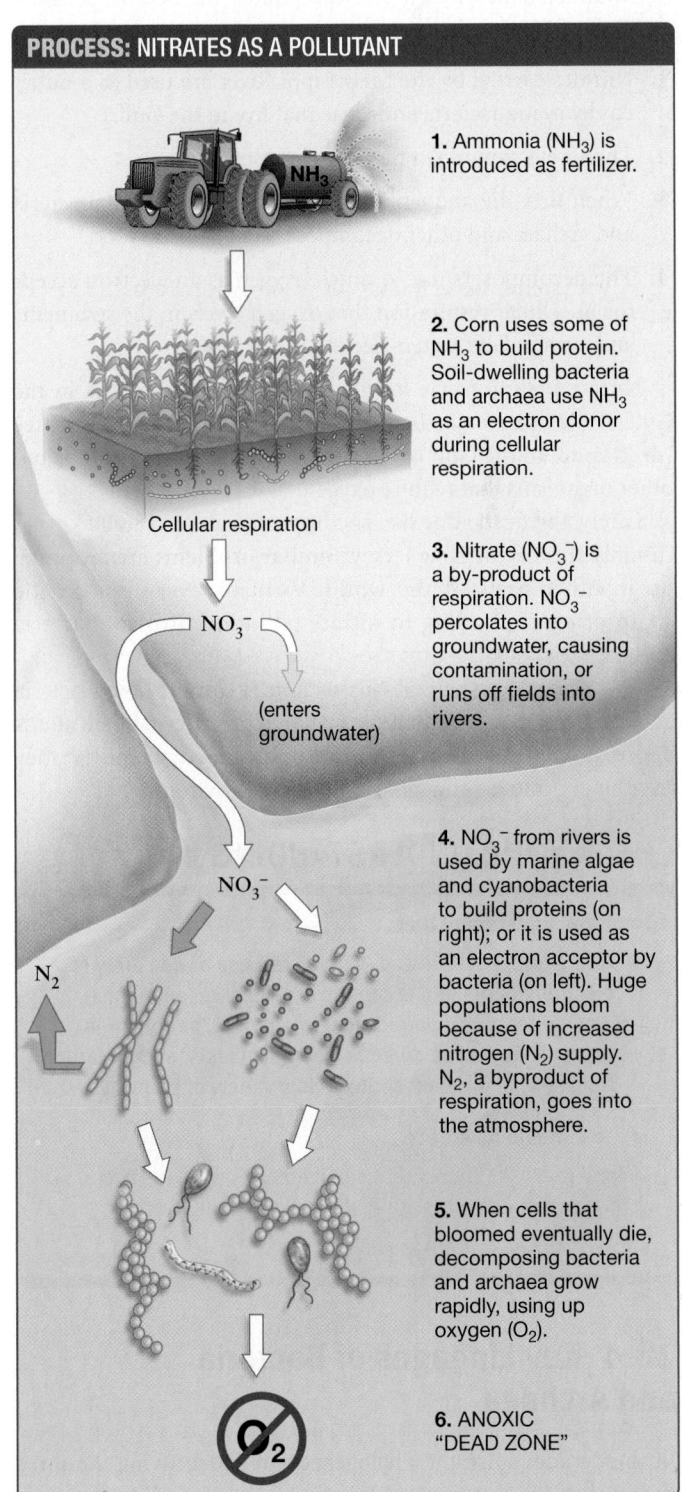

FIGURE 28.14 Nitrates Act as a Pollutant in Aquatic Ecosystems.

fraction of the ammonia molecules is used as food by bacteria in the soil. Bacteria that use ammonia as an electron donor to fuel cellular respiration release nitrite (NO_2^-) as a waste product. Other bacteria use nitrite as an electron donor and release nitrate (NO_3^-). Nitrate molecules are extremely soluble in water and tend to be washed out of soils into groundwater or streams. From there they eventually reach the ocean, where they can cause pollution.

To understand why nitrates can pollute the oceans, consider the Gulf of Mexico:

1. Nitrates carried by the Mississippi River are used as a nutrient by cyanobacteria and algae that live in the Gulf.

2. These cells explode in numbers in response.

3. When they die and sink to the bottom of the Gulf, bacteria and archaea and other decomposers use them as food.

4. The decomposers use so much oxygen as an electron acceptor in cellular respiration that oxygen levels in the sediments and even in Gulf waters decline.

Nitrate pollution has been so severe that large areas in the Gulf of Mexico are anoxic (lacking in oxygen). The oxygen-free "dead zone" in the Gulf of Mexico is devoid of fish, shrimp, and other organisms that require oxygen.

Lately, the dead zone has encompassed about 18,000 km^2— roughly the size of New Jersey. Similar problems are cropping up in other parts of the world. Virtually every link in the chain of events leading to nitrate pollution involves bacteria and archaea.

The general message of this section is simple: 🔑 Bacteria and Archaea may be small in size, but because of their abundance, ubiquity, and ability to do sophisticated chemistry, they have an enormous influence on the global environment.

CHECK YOUR UNDERSTANDING

🔑 **If you understand that . . .**

- As a group, Bacteria and Archaea can use a wide array of electron donors and acceptors in cellular respiration, a diverse set of compounds in fermentation, perform non-oxygenic as well as oxygenic photosynthesis, and fix carbon from several different sources via a variety of pathways.

✔ **You should be able to . . .**

Defend the claim that in terms of metabolism, bacteria and archaea are much more sophisticated than eukaryotes.

Answers are available in Appendix B.

28.4 Key Lineages of Bacteria and Archaea

In the decades since the phylogenetic tree identifying the three domains of life was first published, dozens of studies have con-

firmed the result. It is now well established that all organisms alive today belong to one of the three domains, and that archaea and eukaryotes are more closely related to each other than either group is to bacteria.

Although the relationships among the major lineages within Bacteria and Archaea are still uncertain in some cases, many of the lineages themselves are well studied. Let's survey the attributes of species from selected major lineages within the Bacteria and Archaea, with an emphasis on themes explored earlier in the chapter: their morphological and metabolic diversity, their impacts on humans, and their importance to other species and to the environment.

Bacteria

The name *bacteria* comes from the Greek root *bacter*, meaning "rod" or "staff." The name was inspired by the first bacteria to be seen under a microscope, which were rod shaped. But as the following descriptions indicate, bacterial cells come in a wide variety of shapes.

Biologists who study bacterial diversity currently recognize at least 16 major lineages, or phyla, within the domain. Some of these lineages were recognized by distinctive morphological characteristics and others by phylogenetic analyses of gene sequence data. The lineages reviewed here are just a sampling of bacterial diversity.

- Bacteria > Firmicutes

- Bacteria > Spirochaetes (Spirochetes)

- Bacteria > Actinobacteria

- Bacteria > Chlamydiae

- Bacteria > Cyanobacteria

- Bacteria > Proteobacteria

Archaea

The name *archaea* comes from the Greek root *archae*, for "ancient." The name was inspired by the hypothesis that this is a particularly ancient group, which turned out to be incorrect. Also incorrect was the initial hypothesis that archaeans are restricted to hot springs, salt ponds, and other extreme habitats. Archaea live in virtually every habitat known.

Recent phylogenies based on DNA sequence data indicate that the domain is composed of at least three major phyla, called the Crenarchaeota, Euryarchaeota, and Korarchaeota. Although it is clear that these three groups are highly differentiated at the DNA sequence level, biologists are still searching for shared, derived morphological traits that help define each lineage as a monophyletic group. The Korarchaeota are known only from direct sequencing studies. They have never been grown in culture and almost nothing is known about them.

- Archaea > Crenarchaeota

- Archaea > Euryarchaeota

The Firmicutes have also been called "low-GC Gram positives" because their cell walls react positively to the Gram stain and because they have a relatively low percentage of guanine and cytosine (G and C) in their DNA. In some species, G and C represent less than 25 percent of the bases present. There are over 1100 species. ✔You should be able to mark the origin of the Gram-positive cell wall and low-GC genome on Figure 28.6 (only Firmicutes have a low-GC genome; Actinobacteria are the only other Gram-positive lineage).

Morphological diversity Most are rod shaped or spherical. Some of the spherical species form chains or tetrads (groups of four cells). A few form a durable resting stage called a spore. One subgroup lacks cell walls entirely; another synthesizes a cell wall made of cellulose.

Metabolic diversity Some species can fix nitrogen; some perform non-oxygenic photosynthesis. Others make all of their ATP via various fermentation pathways; still others perform cellular respiration, using hydrogen gas (H_2) as an electron donor.

Human and ecological impacts Recent direct sequencing studies have shown that members of this lineage are extremely common in the human gut. Species in this group cause a variety of diseases, including anthrax, botulism, tetanus, walking pneumonia, boils, gangrene, and strep throat. *Bacillus thuringiensis* produces a toxin that is one of the most important insecticides currently used in farming. Species in the lactic acid bacteria group are used to ferment milk products into yogurt or cheese (**Figure 28.15**). Species in this group are important components of soil, where they speed the decomposition of dead plants, animals, and fungi.

Lactobacillus bulgaricus (rods) and *Streptococcus thermophilus*

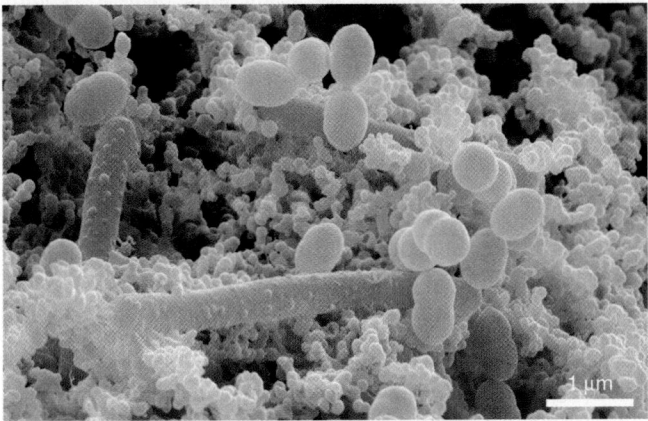

FIGURE 28.15 Firmicutes in Yogurt. (The cells in this micrograph have been colorized—see **BioSkills 10** in Appendix A.)

Bacteria > Spirochaetes (Spirochetes)

The spirochetes are one of the smaller bacterial phyla in terms of numbers of species: only 13 genera and a total of 62 species have been described to date. Recent analyses suggest that the spirochete lineage branched near the base of the bacterial phylogenetic tree.

Morphological diversity Spirochetes are distinguished by their unique corkscrew shape and flagella (**Figure 28.16**). Instead of extending into the water surrounding the cell, spirochete flagella are contained within a structure called the outer sheath, which surrounds the cell. When the flagella beat, the cell lashes back and forth and swims forward. ✔You should be able to mark the origin of the spirochete flagellum on Figure 28.6.

Metabolic diversity Most spirochetes manufacture ATP via fermentation. The substrate used in fermentation varies among species and may consist of sugars, amino acids, starch, or the pectin found in plant cell walls. A spirochete that lives only in the hindgut of termites can fix nitrogen.

Treponema pallidum

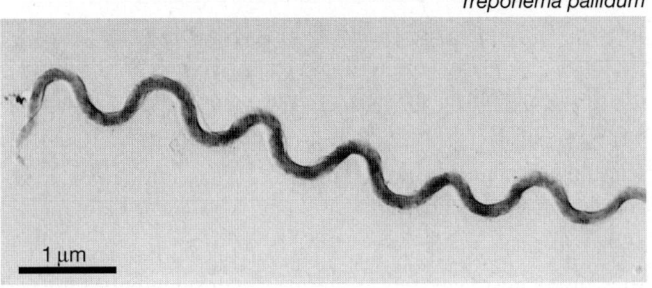

FIGURE 28.16 Spirochetes Are Corkscrew-Shaped Cells Inside an Outer Sheath.

Human and ecological impacts The sexually transmitted disease syphilis is caused by a spirochete. So is Lyme disease, which is transmitted to humans by deer ticks. Spirochetes are extremely common in freshwater and marine habitats; many live only under anaerobic conditions.

Actinobacteria are sometimes called the "high-GC Gram positives" because (1) their cell-wall material appears purple when treated with the Gram stain—meaning that they have a peptidoglycan-rich cell wall and lack an outer membrane—and (2) their DNA contains a relatively high percentage of guanine and cytosine. In some species, G and C represent over 75 percent of the bases present. Over 1100 species have been described to date (**Figure 28.17**). ✔You should be able to mark the origin of the high-GC genome in Actinobacteria on Figure 28.6.

Morphological diversity Cell shape varies from rods to filaments. Many of the soil-dwelling species are found as chains of cells that form extensive branching filaments called **mycelia**.

Metabolic diversity Many are heterotrophs that use an array of organic compounds as electron donors and oxygen as an electron acceptor. There are a handful of parasitic species. Like other parasites, they get most of their nutrition from host organisms.

Human and ecological impacts Over 500 distinct antibiotics have been isolated from species in the genus *Streptomyces*; 60 of these—including streptomycin, neomycin, tetracycline, and erythromycin—are now actively prescribed to treat diseases in humans or domestic livestock. Tuberculosis and leprosy are caused by members of the Actinobacteria. One actinobacterium species is critical to the manufacture of Swiss cheese. Species in the genus *Streptomyces* and *Arthrobacter* are abundant in soil and are vital as decomposers of dead plant and animal material. Some species in these genera live in association with plant roots and fix nitrogen; others can break down toxins such as herbicides, nicotine, and caffeine.

Streptomyces griseus

5 µm

FIGURE 28.17 A *Streptomyces* Species That Produces the Antibiotic Streptomycin.

In terms of numbers of species living today, Chlamydiae may be the smallest of all major bacterial lineages. Although the group is highly distinct phylogenetically, only 13 species are known. All are Gram-negative.

Morphological diversity Chlamydiae are spherical. They are tiny, even by bacterial standards.

Metabolic diversity All known species in this phylum live as parasites *inside* host cells and are termed **endosymbionts** ("inside-together-living"). Chlamydiae acquire almost all of their nutrition from their hosts. In **Figure 28.18**, the chlamydiae have been colored red; the animal cells that they live in are colored brown. ✔You should be able to mark the origin of the endosymbiotic lifestyle in this lineage on Figure 28.6. (The endosymbiotic lifestyle has also arisen in other bacterial lineages, independently of Chlamydiae.)

Human and ecological impacts *Chlamydia trachomatis* infections are the most common cause of blindness in humans. When the same organism is transmitted from person to person via sexual intercourse, it can cause serious urogenital tract infections. One species causes epidemics of a pneumonia-like disease in birds.

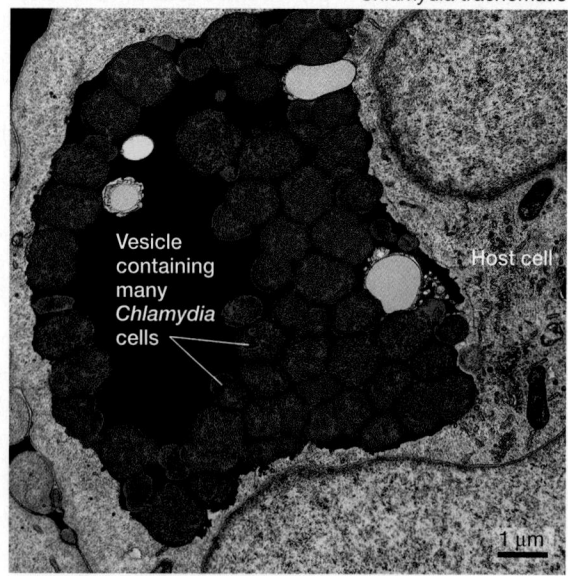

Chlamydia trachomatis

Vesicle containing many *Chlamydia* cells

Host cell

1 µm

FIGURE 28.18 Chlamydiae Live Only inside Animal Cells.

The cyanobacteria were formerly known as the "blue-green algae"—even though algae are eukaryotes. Only about 80 species of cyanobacteria have been described to date, but they are among the most abundant organisms on Earth. In terms of total mass, cyanobacteria dominate the surface waters in many marine and freshwater environments.

Morphological diversity Cyanobacteria may be found as independent cells, in chains that form filaments (**Figure 28.19**), or in the loose aggregations of individual cells called colonies. The shape of colonies varies from flat sheets to ball-like clusters of cells.

Metabolic diversity All perform oxygenic photosynthesis; many can also fix nitrogen. Because cyanobacteria can synthesize virtually every molecule they need, they can be grown in culture media that contain only CO_2, N_2, H_2O, and a few mineral nutrients. ✔You should be able to mark the origin of oxygenic photosynthesis on Figure 28.6.

Human and ecological impacts If cyanobacteria are present in high numbers, their waste products can make drinking water smell bad. Some species release molecules called microcystins that are toxic to plants and animals. Cyanobacteria were responsible for the

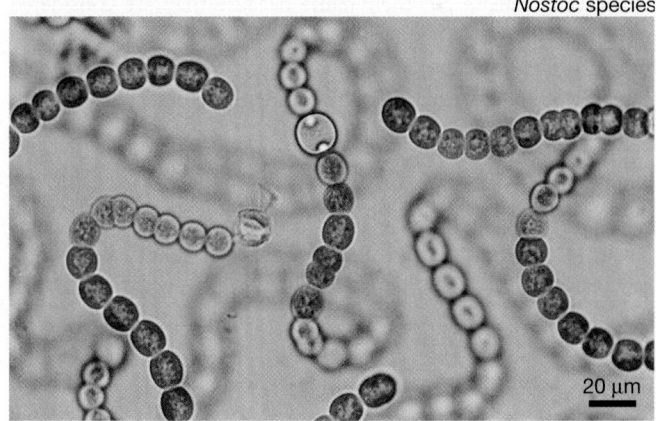

Nostoc species

FIGURE 28.19 Cyanobacteria Contain Chlorophyll and Are Green.

origin of the oxygen atmosphere on Earth. Today they still produce much of the oxygen and nitrogen and many of the organic compounds that feed other organisms in freshwater and marine environments. A few species live in association with fungi, forming lichens.

The approximately 1200 species of proteobacteria form five major subgroups, designated by the Greek letters α (alpha), β (beta), γ (gamma), δ (delta), and ϵ (epsilon). Because they are so diverse in their morphology and metabolism, the lineage is named after the Greek god Proteus, who could assume many shapes.

Morphological diversity Proteobacterial cells can be rods, spheres, or spirals. Some form stalks (**Figure 28.20a**). Some are motile. In one group, cells may move together to form colonies, which then transform into the specialized cell aggregate shown in **Figure 28.20b**. This structure is known as a **fruiting body**. Cells that are surrounded by a durable coating are produced at the tips of fruiting bodies. These spores sit until conditions improve, and then they resume growth.

Metabolic diversity Proteobacteria make a living in virtually every way known to bacteria—except that none perform oxygenic photosynthesis. Various species may perform cellular respiration by using organic compounds, nitrite, methane, hydrogen gas, sulfur, or ammonia as electron donors and oxygen, sulfate, or sulfur as an electron acceptor. Some perform non-oxygenic photosynthesis.

Human and ecological impacts *Escherichia coli* may be the best-studied of all organisms and is a key species in biotechnology (see Chapter 19 and **BioSkills 14** in Appendix A). Pathogenic proteobacteria cause Legionnaire's disease, cholera, food poisoning, dysentery, gonorrhea, Rocky Mountain spotted fever, typhus, ulcers, and diarrhea. *Wolbachia* infections are common in insects and are often transmitted from mothers to offspring via eggs. Biol-

(a) Stalked bacterium

Caulobacter crescentus

Stalk   1 µm

(b) Fruiting bodies

Chondromyces crocatus

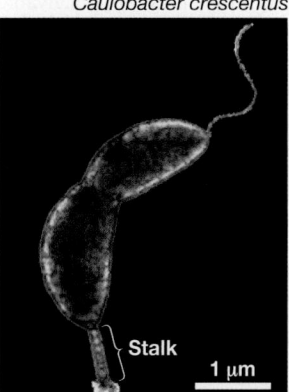

Spores filled with bacteria

50 µm

FIGURE 28.20 Some Proteobacteria Grow on Stalks or Form Fruiting Bodies. The stalked bacterium (a) has been colorized.

ogists use *Agrobacterium* cells to transfer new genes into crop plants. Certain acid-loving species of proteobacteria are used in the production of vinegars. Species in the genus *Rhizobium* (α-proteobacteria) live in association with plant roots and fix nitrogen. The bdellovibrios are a group in the δ-proteobacteria that are predators—they drill into other proteobacterial cells and digest them. Because some species use nitrogen-containing compounds as electron acceptors, proteobacteria are critical players in the cycling of nitrogen atoms through terrestrial and aquatic ecosystems.

The Crenarchaeota got their name because they are considered similar to the oldest archaeans; the word root *cren-* refers to a source or fount. Although only 37 species have been named to date, it is virtually certain that thousands are yet to be discovered.

Morphological diversity Crenarchaeota cells can be shaped like filaments, rods, discs, or spheres. One species that lives in extremely hot habitats has a tough cell wall consisting solely of glycoprotein.

Metabolic diversity Depending on the species, cellular respiration can involve organic compounds, sulfur, hydrogen gas, ammonia, or Fe^{2+} ions as electron donors and oxygen, nitrate, sulfate, sulfur, carbon dioxide, or Fe^{3+} ions as electron acceptors. Some species make ATP exclusively through fermentation pathways.

Human and ecological impacts Crenarchaeota have yet to be used in the manufacture of commercial products. In certain extremely hot, high-pressure, cold, or acidic environments, crenarchaeota may be the only life-form present (**Figure 28.21**). Acid-loving species thrive in habitats with pH 1–5; some species are found in ocean sediments at depths ranging from 2500 to 4000 m below the surface.

Sulfolobus species

0.5 μm

FIGURE 28.21 Some Crenarchaeota Live in Sulfur-Rich Hot Springs. The cells in the micrograph have been colorized to make them more visible.

The Euryarchaeota are aptly named, because the word root *eury-* means "broad." Members of this phylum live in every conceivable habitat. Some species are adapted to high-salt habitats with pH 11.5—almost as basic as household ammonia (**Figure 28.22**). Other species are adapted to acidic conditions with a pH as low as 0. Species in the genus *Methanopyrus* live near hot springs called black smokers that are 2000 m (over 1 mile) below sea level. About 170 species have been identified thus far, and more are being discovered each year.

Morphological diversity Euryarchaeota cells can be spherical, filamentous, rod shaped, disc shaped, or spiral. Rod-shaped cells may be short or long or arranged in chains. Spherical cells can be found in ball-like aggregations. Some species have several flagella. Some species lack a cell wall; others have a cell wall composed entirely of glycoproteins.

Metabolic diversity The group includes a variety of methane-producing species. These methanogens can use up to 11 different organic compounds as electron acceptors during cellular respiration; all produce CH_4 as a by-product of respiration. In other species of Euryarchaeota, cellular respiration is based on hydrogen gas or Fe^{2+} ions as electron donors and nitrate or sulfate as electron acceptors. Species that live in high-salt environments can use the molecule retinal—which is responsible for light reception in your eyes—to capture light energy and perform photosynthesis.

Human and ecological impacts Species in the genus *Ferroplasma* live in piles of waste rock near abandoned mines. As a by-product of metabolism, they produce acids that drain into streams and pollute them. Methanogens live in the soils of swamps and the guts of mammals (including yours). They are responsible for adding about 2 billion tons of methane to the atmosphere each year. A methanogen in this phylum was also recently implicated in gum disease.

Euryarchaeota cells tint these salt crystals different colors

Salt ponds where seawater is evaporating

FIGURE 28.22 Some Euryarchaeota Live in High-Salt Habitats.

Summary of Key Concepts

🔑 **Bacteria and archaea may be tiny, but they are ancient, diverse, abundant, and ubiquitous. A few bacteria cause infectious disease; some are effective at cleaning up pollution.**

- Bacteria and archaea are the most abundant organisms on Earth and are found in every habitat that has been sampled.

- Bacteria cause some of the most dangerous human diseases, including plague, syphilis, botulism, cholera, and tuberculosis.

- Enrichment cultures are used to grow large numbers of bacterial or archaeal cells that thrive under specified conditions.

- Biologists can study bacteria and archaea that cannot be cultured by extracting DNA directly from an environment, sequencing target genes, and using the data to place the organisms present on the tree of life.

  ✔ You should be able to explain the difference between a bacterium, an archaeon, and a eukaryote.

  MB **Web Activity** The Tree of Life

🔑 **As a group, bacteria and archaea live in virtually every habitat known and use remarkably diverse types of compounds in cellular respiration and fermentation. Although cell size is usually small and overall morphology is relatively simple, the chemistry they can do is extremely sophisticated.**

- Metabolic diversity and complexity are the hallmarks of the bacteria and archaea, just as morphological diversity and complexity are the hallmarks of the eukaryotes.

- Among bacteria and archaea, a wide array of inorganic or organic compounds with high potential energy can serve as electron donors in cellular respiration, and a wide variety of inorganic or organic molecules with low potential energy may serve as electron acceptors. Dozens of distinct organic compounds are fermented.

- Photosynthesis is widespread in bacteria. In cyanobacteria, water is used as a source of electrons and oxygen gas is generated as a by-product. But in other species, the electron excited by photon capture comes from a source other than water, and no oxygen is produced.

- To acquire building-block molecules containing carbon-carbon bonds, some species use the enzymes of the Calvin cycle to reduce CO_2. But three other biochemical pathways found in bacteria and archaea can also reduce simple organic compounds to sugars or carbohydrates.

  ✔ You should be able to explain why species that release H_2S as a by-product and that use H_2S as an electron donor often live side by side.

🔑 **Bacteria and archaea play a key role in ecosystems. Photosynthetic bacteria were responsible for the evolution of the oxygen atmosphere; bacteria and archaea cycle nutrients through terrestrial and aquatic environments.**

- Because of their metabolic diversity, bacteria and archaea play a large role in global change and nutrient cycling.

- Nitrogen-fixing species provide nitrogen in forms that can be used by many other species, including plants and animals.

  ✔ You should be able to explain what the composition of the atmosphere and what the nitrogen cycle would be like if bacteria and archaea did not exist.

Questions

1. How do molecules that function as electron donors and those that function as electron acceptors differ?
   a. Electron donors are almost always organic molecules; electron acceptors are always inorganic.
   b. Electron donors are almost always inorganic molecules; electron acceptors are always organic.
   c. Electron donors have relatively high potential energy; electron acceptors have relatively low potential energy.
   d. Electron donors have relatively low potential energy; electron acceptors have relatively high potential energy.

2. What do some photosynthetic bacteria use as a source of electrons instead of water?
   a. oxygen (O_2)
   b. hydrogen sulfide (H_2S)
   c. organic compounds (e.g., CH_3COO^-)
   d. nitrate (NO_3^-)

3. What is distinctive about the chlorophylls found in different photosynthetic bacteria?
   a. their membranes
   b. their role in acquiring energy
   c. their role in carbon fixation
   d. their absorption spectra

4. What are organisms called that use inorganic compounds as electron donors in cellular respiration?
   a. phototrophs
   b. heterotrophs
   c. organotrophs
   d. lithotrophs

5. What has direct sequencing allowed researchers to do for the first time?
   a. sample organisms from an environment and grow them under defined conditions in the lab
   b. sample organisms from an environment and sequence the entire genomes present
   c. study organisms that cannot be cultured (grown in the lab)
   d. identify important morphological differences among species

6. Koch's postulates outline the requirements for which of the following?
   a. showing that an organism is autotrophic
   b. showing that a bacterium's cell wall lacks an outer membrane and consists primarily of peptidoglycan
   c. showing that an organism causes a particular disease
   d. showing that an organism can use a particular electron donor and electron acceptor

✔TEST YOUR UNDERSTANDING

Answers are available in Appendix B

1. Biologists often use the term energy source as a synonym for "electron donor." Why?

2. The text claims that the tremendous ecological diversity of bacteria and archaea is possible because of their impressive metabolic diversity. Do you agree with this statement? Why or why not?

3. How is it possible for direct sequencing to identify species that no one has ever seen?

4. The text claims that the evolution of an oxygen atmosphere paved the way for increasingly efficient cellular respiration and higher growth rates in organisms. Explain.

5. Look back at Table 28.5 and note that the by-products of respiration in some organisms are used as electron donors or acceptors by other organisms. In the table, draw lines between the dual-use molecules listed in the "Electron Donor," "Electron Acceptor," and "By-Products" columns.

6. Explain the statement, "Prokaryotes are a paraphyletic group."

✔APPLYING CONCEPTS TO NEW SITUATIONS

Answers are available in Appendix B

1. The researchers who observed that magnetite was produced by bacterial cultures from the deep subsurface carried out a follow-up experiment. These biologists treated some of the cultures with a drug that poisons the enzymes involved in electron transport chains. In cultures where the drug was present, no more magnetite was produced. Does this result support or undermine their hypothesis that the bacteria in the cultures perform cellular respiration? Explain your reasoning.

2. *Streptococcus mutans* obtains energy by oxidizing sucrose. This bacterium is abundant in the mouths of Western European and North American children and is a prominent cause of cavities. The organism is virtually absent in children from East Africa, where tooth decay is rare. Propose a hypothesis to explain this observation. Outline the design of a study that would test your hypothesis.

3. Suppose that you've been hired by a firm interested in using bacteria to clean up organic solvents found in toxic waste dumps. Your new employer is particularly interested in finding cells that are capable of breaking a molecule called benzene into less toxic compounds. Where would you go to look for bacteria that can metabolize benzene as an energy or carbon source? How would you design an enrichment culture capable of isolating benzene-metabolizing species?

4. Would you predict that disease-causing bacteria, such as those listed in Table 28.2, obtain energy from light, organic molecules, or inorganic molecules? Explain your answer.

The red algae shown here live attached to rocks in shallow ocean waters. The algae and other species featured in this chapter are particularly abundant in the world's oceans.

Protists 29

This chapter introduces the third domain on the tree of life: the **Eukarya**. Eukaryotes range from single-celled organisms that are the size of bacteria to sequoia trees and blue whales. The largest and most morphologically complex organisms on the tree of life—algae, plants, fungi, and animals—are eukaryotes.

Although eukaryotes are astonishingly diverse, they share fundamental features that distinguish them from bacteria and archaea:

- The nuclear envelope is a synapomorphy that defines the Eukarya.

- Compared to bacteria and archaea, most eukaryotic cells are large, have many more organelles, and have a much more extensive system of structural proteins called the cytoskeleton.

- Multicellularity is rare in bacteria and unknown in archaea, but has evolved multiple times in eukaryotes.

- Bacteria and archaea reproduce asexually by fission; many eukaryotes reproduce asexually via mitosis.

- Many eukaryotes undergo meiosis and reproduce sexually.

One of this chapter's fundamental goals is to explore how these morphological innovations—features like the nuclear envelope and organelles—evolved. Another goal is to analyze how morphological innovations allowed eukaryotes to pursue novel ways of performing basic life tasks such as feeding, moving, and reproducing.

In introducing the Eukarya, this chapter focuses on an informal grab bag of lineages known as the protists. The term **protist** refers to all eukaryotes that are not green plants, fungi, or animals. Protist lineages are colored orange in **Figure 29.1** on page 520.

🗝️ Protists do not make up a monophyletic group. Instead, they refer to a **paraphyletic group**—they represent some, but not all, of the descendants of a single common ancestor. To use the vocabulary introduced in Chapter 27, no synapomorphies

KEY CONCEPTS

🗝️ Protists are a paraphyletic grouping that includes all eukaryotes except the land plants, fungi, and animals. Biologists study them because they are important medically, ecologically, and evolutionarily.

🗝️ Key morphological innovations occurred as protists diversified: the nuclear envelope, multicellularity, and an array of structures that function in support and protection. In addition, the mitochondrion and chloroplast arose by endosymbiosis.

🗝️ Protists vary widely in terms of how they obtain food. Many species are photosynthetic; while others obtain carbon compounds by ingesting food or parasitizing other organisms.

🗝️ Protists vary widely in terms of how they reproduce. Sexual reproduction evolved in protists, and many protist species can reproduce both sexually and asexually.

✔ When you see this checkmark, stop and test yourself. Answers are available in Appendix B.

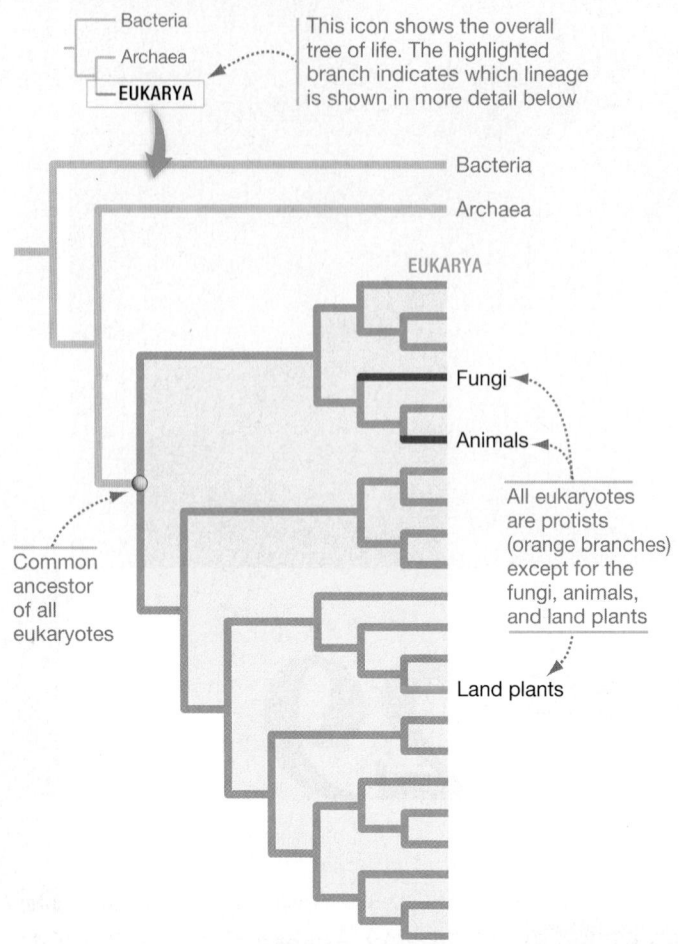

This icon shows the overall tree of life. The highlighted branch indicates which lineage is shown in more detail below

Bacteria

Archaea

EUKARYA

Fungi

Animals

All eukaryotes are protists (orange branches) except for the fungi, animals, and land plants

Common ancestor of all eukaryotes

Land plants

FIGURE 29.1 Protists Are Paraphyletic. The group called protists includes some, but not all, descendants of a single common ancestor.

define the protists. There is no trait that is found in protists but no other organisms.

By definition, then, the protists are a diverse lot. The common feature among protists is that they tend to live in environments where they are surrounded by water (**Figure 29.2**). Most plants, fungi, and animals are terrestrial, but protists are found in wet soils, aquatic habitats, or the bodies of other organisms—including, perhaps, you.

29.1 Why Do Biologists Study Protists?

Biologists study protists for three reasons, in addition to their intrinsic interest: (**1**) they are important medically, (**2**) they are important ecologically, and (**3**) they are critical to understanding the evolution of plants, fungi, and animals. The remainder of the chapter will focus on why protists are interesting in their own right and how they evolved; here let's consider their impact on the environment and human health.

Impacts on Human Health and Welfare

The most spectacular crop failure in history, the Irish potato famine, was caused by a protist. In 1845 most of the 3 million acres that had been planted to grow potatoes in Ireland became infested with *Phytophthora infestans*—a parasite that belongs to a lineage of protists called Oomycota (pronounced *oo-oh-my-COTE-ah*). Potato tubers that were infected with *P. infestans* rotted in the fields or in storage.

As a result of crop failures in Ireland for two consecutive years, an estimated 1 million people out of a population of less

Open ocean: Surface waters teem with microscopic protists, such as these diatoms.

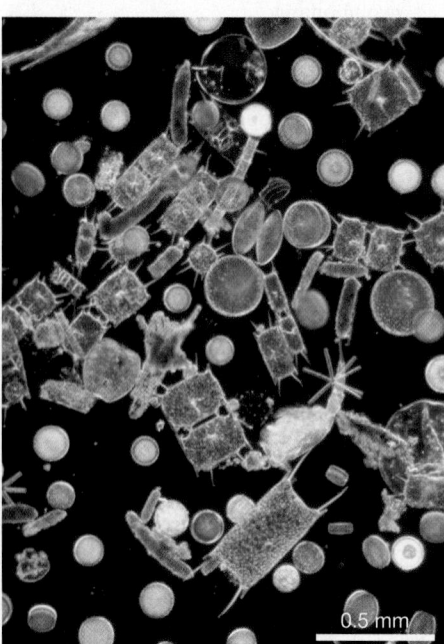

0.5 mm

Shallow coastal waters: Gigantic protists, such as these kelp, form underwater forests.

Intertidal habitats: Protists such as these red algae are particularly abundant in tidal habitats.

5 cm

FIGURE 29.2 Protists Are Particularly Abundant in Aquatic Environments.

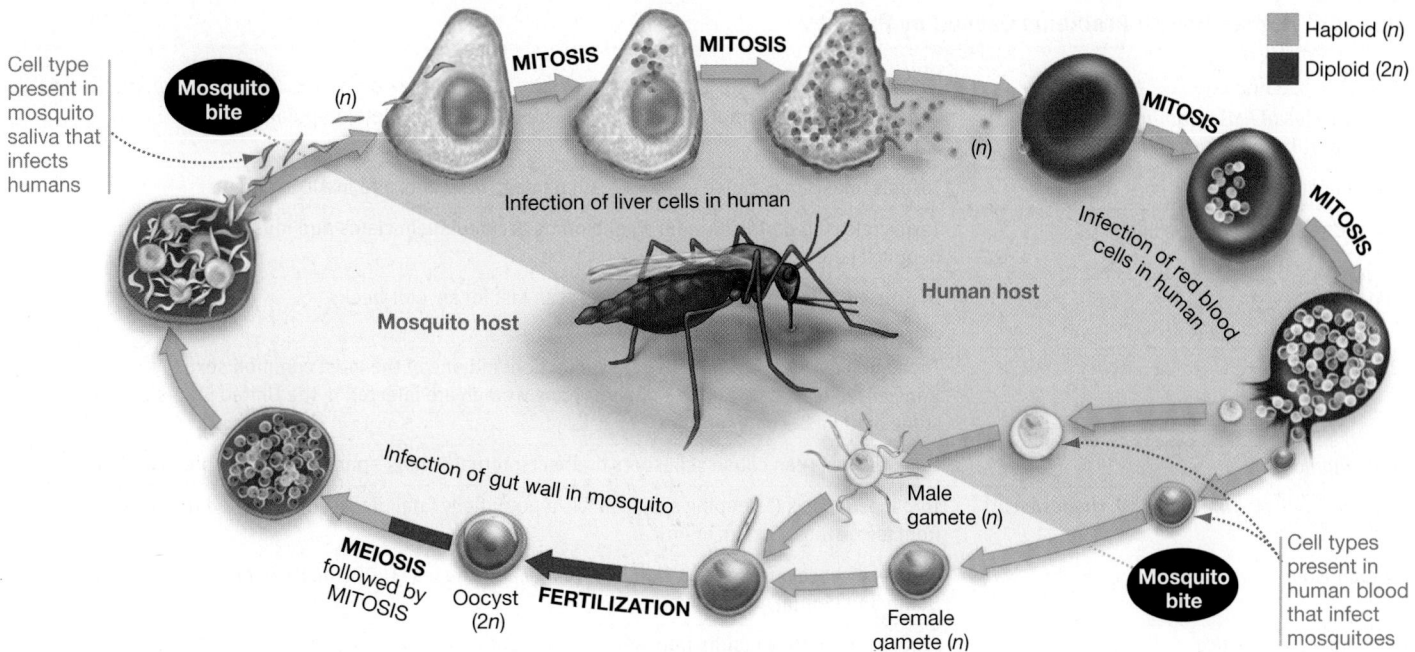

Haploid (n)
Diploid (2n)

Cell type present in mosquito saliva that infects humans

Mosquito bite
(n)
MITOSIS
MITOSIS

Infection of liver cells in human

(n)

Infection of red blood cells in human

MITOSIS

MITOSIS

Human host

Mosquito host

Infection of gut wall in mosquito

Male gamete (n)

Cell types present in human blood that infect mosquitoes

MEIOSIS followed by MITOSIS

Oocyst (2n)

FERTILIZATION

Female gamete (n)

Mosquito bite

FIGURE 29.3 *Plasmodium* Lives in Mosquitoes and in Humans, where It Causes Malaria. Over the course of its life cycle, *Plasmodium falciparum* develops into a series of distinct cell types. Each type is specialized for infecting a different host cell in mosquitoes or humans. In humans, it infects and kills liver cells and red blood cells, contributing to anemia and high fever. In mosquitoes, the protist lives in the gut and salivary glands.

than 9 million died of starvation or starvation-related illnesses. Several million others emigrated. Many people of Irish heritage living in North America, New Zealand, and Australia trace their ancestry to relatives who left Ireland to evade the famine. As devastating as the potato famine was, however, it does not begin to approach the misery caused by the protist *Plasmodium*.

MALARIA Physicians and public health officials point to three major infectious diseases that are currently afflicting large numbers of people worldwide: tuberculosis, HIV, and malaria. Tuberculosis is caused by a bacterium and was introduced in Chapter 24; HIV is caused by a virus and is analyzed in Chapter 35. Malaria is caused by a protist—specifically, by species in the eukaryotic lineage called Apicomplexa.

Malaria ranks as the world's most chronic public health problem. In India alone, over 30 million people each year suffer from debilitating fevers caused by malaria. At least 300 million people worldwide are sickened by it each year, and over 1 million die from the disease annually. The toll is equivalent to eight 747s, loaded with passengers, crashing every day. Most of the dead are children of preschool age.

Four species of the protist *Plasmodium* are capable of parasitizing humans. Infections start when *Plasmodium* cells enter a person's bloodstream during a mosquito bite. As **Figure 29.3** shows, *Plasmodium* initially infects liver cells; later, some of the *Plasmodium* cells change into a distinctive cell type that infects the host's red blood cells as well. The *Plasmodium* cells multiply inside the host cells, which are killed as parasite cells exit to infect additional liver cells or red blood cells.

If infected red blood cells are transferred to a mosquito during a bite, they differentiate to form gametes. Inside the mosquito, gametes fuse to form a diploid cell called an oocyst, which undergoes meiosis. The haploid cells that result from meiosis can infect a human when the mosquito bites again.

Although *Plasmodium* is arguably the best studied of all protists, researchers have still not been able to devise effective and sustainable measures to control it.

- Natural selection has favored mosquito strains that are resistant to the insecticides that have been sprayed in their breeding habitats, in attempts to control malaria's spread.

- *Plasmodium* has evolved resistance to most of the drugs used to control its growth in infected people.

- Efforts to develop a vaccine against *Plasmodium* have been fruitless to date, in part because the parasite evolves so quickly (flu virus and HIV pose similar problems; see Chapter 49).

Unfortunately, malaria is not the only important human disease caused by protists. **Table 29.1** on page 522 lists protists that have been the cause of human suffering and economic losses.

HARMFUL ALGAL BLOOMS When a unicellular species experiences rapid population growth and reaches high densities in an aquatic environment, it is said to "bloom." Unfortunately, a handful of the many protist species involved in blooms can be harmful.

Harmful algal blooms are usually due to photosynthetic protists called dinoflagellates. Certain dinoflagellates synthesize toxins to protect themselves from predation by small animals called copepods. Because toxin-producing dinoflagellates have high

TABLE 29.1 Human Health Problems Caused by Protists

Species	Disease
Four species of *Plasmodium*, primarily *P. falciparum* and *P. vivax*	Malaria has the potential to affect 40 percent of the world's total population.
Toxoplasma	Toxoplasmosis may cause eye and brain damage in infants and in AIDS patients.
Many species of dinoflagellates	Toxins released during harmful algal blooms accumulate in clams and mussels and poison people if eaten.
Giardia	Diarrhea due to giardiasis (beaver fever) can last for several weeks.
Trichomonas	Trichomoniasis is a reproductive tract infection and one of the most common sexually transmitted diseases. About 2 million young women are infected in the United States each year; some of them become infertile.
Leishmania	Leishmaniasis can cause skin sores or affect internal organs—particularly the spleen and liver.
Trypanosoma gambiense and *T. rhodesiense*	Trypanosomiasis ("sleeping sickness") is a potentially fatal disease transmitted through bites from tsetse flies. Occurs in Africa.
Trypanosoma cruzi	Chagas disease affects 16–18 million people and causes 50,000 deaths annually, primarily in South and Central America.
Entamoeba histolytica	Amoebic dysentery results from severe infections.
Phytophthora infestans	An outbreak of this protist wiped out potato crops in Ireland in 1845–1847, causing famine.

concentrations of accessory pigments called xanthophylls, their blooms can sometimes discolor seawater (**Figure 29.4**).

Algal blooms can be harmful to people because clams and other shellfish filter photosynthetic protists out of the water as food. During a bloom, high levels of toxins can build up in the flesh of these shellfish. Typically, the shellfish themselves are not harmed. But if a person eats contaminated shellfish, several types of poisoning can result.

Paralytic shellfish poisoning, for example, occurs when people eat shellfish that have fed heavily on protists that synthesize poisons called saxitoxins. Saxitoxins block ion channels that have to open for electrical signals to travel through nerve cells (see Chapter 45). In humans, high dosages of saxitoxins cause unpleasant symptoms such as prickling sensations in the mouth or even life-threatening symptoms such as muscle weakness and paralysis.

Ecological Importance of Protists

As a whole, the protists represent just 10 percent of the total number of named eukaryote species. Although the species diversity of protists may be relatively low, their abundance is extraordinarily high. The number of individual protists found in some habitats is astonishing.

- One milliliter of pond water can contain well over 500 single-celled protists that swim with the aid of flagella.

- Under certain conditions, dinoflagellates can reach concentrations of 60 million cells per liter of seawater.

Why is this important?

PROTISTS PLAY A KEY ROLE IN AQUATIC FOOD CHAINS Photosynthetic protists take in carbon dioxide from the atmosphere and reduce, or "fix," it to form sugars or other organic compounds with high potential energy (Chapter 10). Photosynthesis transforms some of the energy in sunlight into chemical energy that organisms can use to grow and produce offspring.

Species that produce chemical energy in this way are called **primary producers**. Diatoms, for example, are photosynthetic protists that rank among the leading primary producers in the oceans, simply because they are so abundant. Primary production in the ocean represents almost half of the total carbon dioxide that is fixed on Earth.

Diatoms and other small organisms that live near the surface of oceans or lakes are called **plankton**. The sugars and other or-

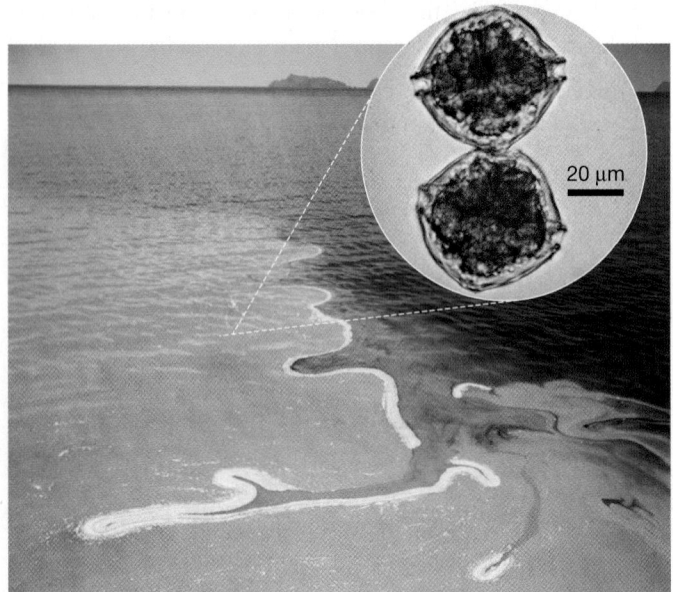

FIGURE 29.4 Harmful Algal Blooms Are Caused by Dinoflagellates.

20 μm

ganic compounds produced by photosynthetic plankton are the basis of food chains in freshwater and marine environments.

A **food chain** describes nutritional relationships among organisms, and thus how chemical energy flows within ecosystems. In this case, photosynthetic protists and other primary producers are eaten by primary consumers, many of which are protists. Primary consumers are eaten by fish, shellfish, and other secondary consumers, which in turn are eaten by tertiary consumers—whales, squid, and large fish (such as tuna).

Many of the species at the base of food chains in aquatic environments are protists. Without protists, most food chains in freshwater and marine habitats would collapse.

COULD PROTISTS HELP REDUCE GLOBAL WARMING? Carbon dioxide levels in the atmosphere are increasing rapidly because humans are burning fossil fuels and forests (see Chapter 54). Carbon dioxide traps heat that is radiating from Earth, so high CO_2 levels in the atmosphere contribute to global warming—an issue that many observers consider today's most pressing environmental problem.

The carbon atoms in carbon dioxide molecules move to organisms in the soil or the ocean and then back to the atmosphere, in what researchers call the **global carbon cycle**. To reduce global warming, researchers are trying to figure out ways to decrease carbon dioxide concentrations in the atmosphere and increase the amount of carbon stored in terrestrial and marine environments.

To understand how this might be done, consider the marine carbon cycle diagrammed in **Figure 29.5**. The cycle starts when CO_2 from the atmosphere dissolves in water and is taken up by primary producers called phytoplankton and converted to organic matter. The phytoplankton are eaten by primary consumers, die and are consumed by decomposers or scavengers, or die and sink to the bottom of the ocean. There they may enter one of two long-lived repositories:

1. *Sedimentary rocks* Several lineages of protists have shells made of calcium carbonate ($CaCO_3$). When these shells rain down from the ocean surface and settle in layers at the bottom, the deposits that result are compacted by the weight of the water and by sediments accumulating above them. Eventually the deposits turn into rock. The limestone used to build the pyramids of Egypt consists of protist shells.

2. *Petroleum* Although the process of petroleum (oil) formation is not well understood, it begins with accumulations of dead bacteria, archaea, and protists at the bottom of the ocean.

Recent experiments have shown that the carbon cycle speeds up when habitats in the middle of the ocean are fertilized with iron. Iron is a critical component of the electron transport chains responsible for photosynthesis and respiration, but it is in particularly short supply in the open ocean. After iron is added to ocean waters, it is not uncommon to see populations of protists and other primary producers increase by a factor of 10.

Some researchers hypothesize that when these blooms occur, the amount of carbon that rains down into carbon sinks in the form of shells and dead cells may increase. If so, then fertilizing

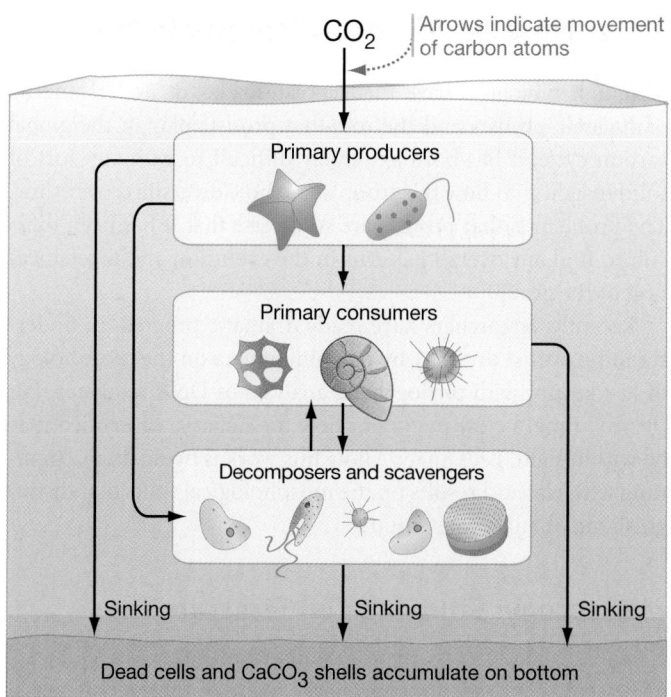

FIGURE 29.5 Protists Play a Key Role in the Marine Carbon Cycle. At the surface, carbon atoms tend to shuttle quickly among organisms. But if carbon atoms sink to the bottom of the ocean in the form of shells or dead cells, they may be locked up for long periods in carbon sinks. (The bottom of the ocean may be miles below the surface.)

the ocean to promote blooms might be an effective way to reduce CO_2 concentrations in the atmosphere.

The effectiveness of iron fertilization is hotly debated, however. Fertilizing the ocean with iron might lead to large accumulations of dead organic matter and the formation of anaerobic dead zones like those described in Chapter 28. But if further research shows that iron fertilization is safe and effective, it could be added to the list of approaches to reduce global warming.

CHECK YOUR UNDERSTANDING

If you understand that . . .

- Malaria is caused by a protist that lives in mosquitoes and in humans in different parts of its life cycle.
- Harmful algal blooms are caused by protists that produce a toxin as a defense against predation.
- Protists are key primary producers in aquatic environments. As a result, they play a key role in the global carbon cycle.

You should be able to . . .

1. Explain why public health workers are promoting the use of insecticide-treated sleeping nets as a way of reducing malaria.
2. Make a flowchart showing the chain of events that would start with massive iron fertilization and end with large deposits of carbon-containing compounds on the ocean floor.

Answers are available in Appendix B.

29.2 How Do Biologists Study Protists?

Although biologists have made great strides in understanding pathogenic protists and the role that protists play in the global carbon cycle, it has been extremely difficult to gain any sort of solid insight into how the group as a whole diversified over time. The problem is that protists are so diverse that it has been difficult to find any overall patterns in the evolution and diversification of the group.

Recently, researchers have made dramatic progress in understanding protist diversity by combining data on the morphology of key groups with phylogenetic analyses of DNA sequence. For the first time, a clear picture of how the Eukarya diversified may be within sight. Let's analyze how this work is being done, beginning with classical results on the morphological traits that distinguish major eukaryote groups.

Microscopy: Studying Cell Structure

Using light microscopy (see **BioSkills 10** in Appendix A), biologists were able to identify and name many of the protist species known today. When transmission electron microscopes became available, a major breakthrough in understanding protist diversity occurred: Detailed studies of cell structure revealed that protists could be grouped according to characteristic overall form, according to organelles with distinctive features, or both.

For example, both light and electron microscopy confirmed that the species that caused the Irish potato famine has reproductive cells with an unusual type of **flagellum**. Flagella are organelles that project from the cell and whip back and forth to produce swimming movements (Chapter 7). In reproductive cells of this species, one of the two flagella present has tiny, hollow, hairlike projections. Biologists noted that kelp and other forms of brown algae also have cells with this type of flagellum.

To make sense of these results, researchers interpreted these types of distinctive morphological features as **synapomorphies**—

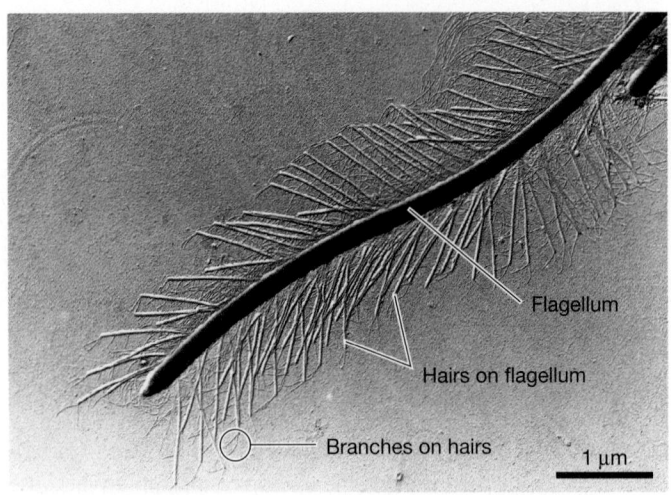

FIGURE 29.6 Species in the Lineage Called Stramenopiles Have a Distinctive Flagellum. The unusual, hollow "hairs" that decorate the flagella of stramenopiles often have three branches at the tip.

shared, derived traits that distinguish major monophyletic groups (see Chapter 27). Species that have a flagellum with hollow, hairlike projections became known as stramenopiles (literally, "straw-hairs"); the hairs typically have three branches at the tip (**Figure 29.6**).

In recognizing this group, investigators hypothesized that because an ancestor had evolved a distinctive flagellum, all or most of its descendants also had this trait. The qualifier most is important, because it is not unusual for certain subgroups to lose particular traits over the course of evolution, much as humans are gradually losing fur and tailbones.

Eventually, seven major groups of eukaryotes came to be identified on the basis of diagnostic morphological characteristics. These groups and the synapomorphies that identify them are listed in **Table 29.2**. Note that in almost every case, the synapomorphies that define eukaryotic lineages are changes in structures that protect or support the cell or that influence the

SUMMARY TABLE 29.2 **Major Lineages of Eukaryotes**

Lineage	Distinguishing Morphological Features (synapomorphies)
Amoebozoa	Cells lack cell walls. When portions of the cell extend outward to move the cell, they form large lobes.
Opisthokonta	Reproductive cells have a single flagellum at their base. The cristae inside mitochondria are flat, not tube shaped as in other eukaryotes. (This lineage includes protists as well as the fungi and the animals. Fungi and animals are discussed in detail in Chapters 31 through 34.)
Excavata	Cells have a pronounced "feeding groove" where prey or organic debris is ingested. No functioning mitochondria are present, although genes derived from mitochondria are found in the nucleus.
Plantae	Cells have chloroplasts with a double membrane.
Rhizaria	Cells lack cell walls, although some produce an elaborate shell-like covering. When portions of the cell extend outward to move the cell, they are slender in shape.
Alveolata	Cells have sac-like structures called alveoli that form a continuous layer just under the plasma membrane. Alveoli are thought to provide support.
Stramenopila	If flagella are present, cells usually have two—one of which is covered with hairlike projections.

organism's ability to move or feed. The plants, fungi, and animals analyzed in Chapters 30 through 34 represent subgroups within two of the seven major eukaryotic lineages.

Evaluating Molecular Phylogenies

When researchers began using DNA sequence data to estimate the evolutionary relationships among eukaryotes, the analyses suggested that the seven groups identified on the basis of distinctive morphological characteristics were indeed monophyletic. This was important support for the hypothesis that the distinctive morphological features were shared, derived characters that existed in a common ancestor of each lineage.

The phylogenetic tree in **Figure 29.7** is the current best estimate of the group's evolutionary history. As you read this tree, note that:

- The Amoebozoa and the Opisthokonta—which include fungi and animals—form a monophyletic group called the Unikonta.

- The other five major lineages form a monophyletic group called the Bikonta.

- The Alveolata and Stramenopila form a monophyletic group called the Chromalveolata.

Understanding where the root or base of the tree lies has been more problematic. The latest data suggest that once the first eukaryotic organism evolved, the major split was between unikonts and bikonts—meaning, between eukaryotes that have one flagellum versus two flagella. This finding is tentative and controversial, however, given the data analyzed to date. Researchers continue to work on the issue of placing the root of the Eukarya.

Discovering New Lineages via Direct Sequencing

The effort to refine the phylogeny of the Eukarya is ongoing. But of all the research frontiers in eukaryotic diversity, the most exciting may be the one based on the technique called direct sequencing.

You might recall from Chapter 28 that **direct sequencing** is based on sampling soil or water, analyzing the DNA sequence of specific genes in the sample, and using the data to place the organisms in the sample on a phylogenetic tree. Direct sequencing led to the discovery of previously unknown but major lineages of Archaea. To the amazement of biologists all over the world, the same thing happened when researchers used direct sequencing to survey eukaryotes.

The first direct sequencing studies that focused on eukaryotes were published in 2001. One study sampled organisms at depths from 250 to 3000 m below the surface in waters off Antarctica; another focused on cells at depths of 75 m in the Pacific Ocean, near the equator. Both studies detected a wide array of species that were new to science.

Investigators who examined the samples under the microscope were astonished to find that many of the newly discovered eukaryotes were tiny—from 0.2 μm to 5 μm in diameter. Subsequent work has confirmed the existence of protists that are

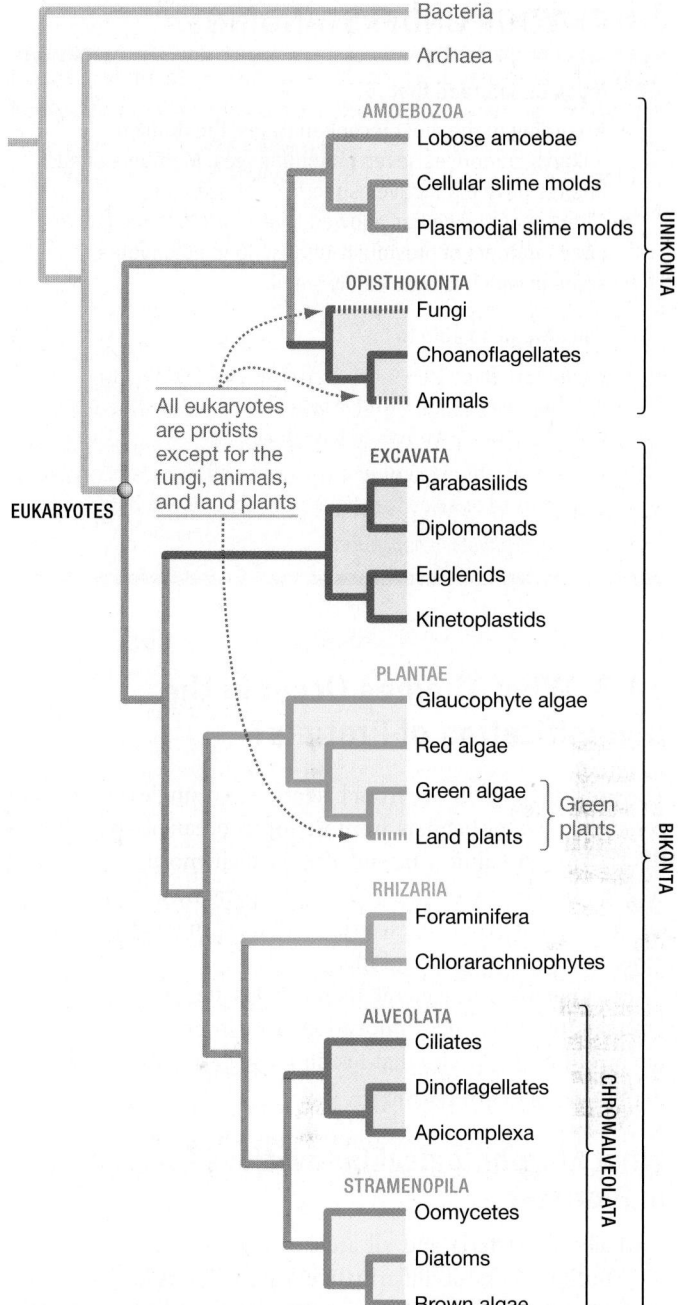

FIGURE 29.7 Phylogenetic Analyses Have Identified Seven Major Lineages of Eukaryotes. This tree shows selected subgroups from the seven major lineages discussed in this chapter. Many other lineages of protists have been identified. Note that kinetoplastids are not discussed in detail elsewhere in the chapter, but include the pathogens *Trypanosoma* and *Leishmania* described in Table 29.1.

less than 0.2 μm in diameter. These cells overlap in size with bacteria, which typically range from 0.5 μm to 2 μm in diameter.

Eukaryotic cells are much more variable in size than previously imagined. A whole new world of tiny protists has just been discovered.

CHECK YOUR UNDERSTANDING

If you understand that . . .

- According to the most recent analyses, the domain Eukarya comprises seven major lineages. Members of each lineage have distinctive aspects of cell structure.
- Direct sequencing has allowed investigators to recognize large numbers of previously undescribed eukaryotes, some of which are extremely small.

✓ **You should be able to . . .**

1. Explain why opisthokonts, alveolates, and stramenopiles got their names. (The root *opistho* refers to the back of a cell; the root *kont* refers to a flagellum.)
2. Explain why direct sequencing studies allow researchers to characterize species that have never been seen before.

Answers are available in Appendix B.

29.3 What Themes Occur in the Diversification of Protists?

The protists range in size from bacteria-sized single cells to giant kelp. They live in habitats from the open oceans to dank forest floors. They are almost bewildering in their morphological and ecological diversity. Because they are a paraphyletic group, they do not share derived characteristics that set them apart from all other lineages on the tree of life.

Fortunately, one general theme helps tie protists together. Once an important new innovation arose in protists, it triggered the evolution of species that live in a wide array of habitats and make a living in diverse ways.

What Morphological Innovations Evolved in Protists?

Virtually all bacteria and all archaea are unicellular. Given the distribution of multicellularity in eukaryotes, it is logical to conclude that the first eukaryote was also a single-celled organism.

Further, all eukaryotes alive today have (**1**) a nucleus and endomembrane system, (**2**) mitochondria or genes that are normally found in mitochondria, and (**3**) a cytoskeleton. Based on the distribution of cell walls in living eukaryotes, though, it is likely that the first eukaryotes lacked this feature.

Based on these observations, biologists hypothesize that the earliest eukaryotes were probably single-celled organisms with a nucleus and endomembrane system, mitochondria, and a cytoskeleton, but no cell wall.

It is also likely that the first eukaryotic cells swam using a novel type of flagellum. Eukaryotic flagella are completely different structures from bacterial flagella and evolved independently. The eukaryotic flagellum is made up of microtubules, and dynein is the major motor protein. An undulating motion occurs as dynein molecules walk down microtubules (see Chapter 7).

The flagella of bacteria and archaea, in contrast, are composed primarily of a protein called flagellin. Instead of undulating, these flagella rotate to produce movement.

✓ If you understand the synapomorphies that identify the eukaryotes as a monophyletic group, you should be able to map the origin of the nuclear envelope and the eukaryotic flagellum on Figure 29.7. Once you've done that, let's consider how several of these key new morphological features arose and influenced the subsequent diversification of protists, beginning with the trait that defines the Eukarya.

THE NUCLEAR ENVELOPE The leading hypothesis to explain the origin of the nuclear envelope is based on infoldings of the plasma membrane. As the drawings in **Figure 29.8** show, a stepwise process could give rise to small infoldings that were elaborated by mutation and natural selection over time, with the infolding eventually becoming detached from the plasma membrane. Note that the infoldings would have given rise to the nuclear envelope and the endoplasmic reticulum (ER) together.

Two lines of evidence support this hypothesis: Infoldings of the plasma membrane occur in some bacteria living today, and the nuclear envelope and ER of today's eukaryotes are continuous (see Chapter 7). ✓ If you understand the infolding hypothesis, you should be able to explain why these observations support it.

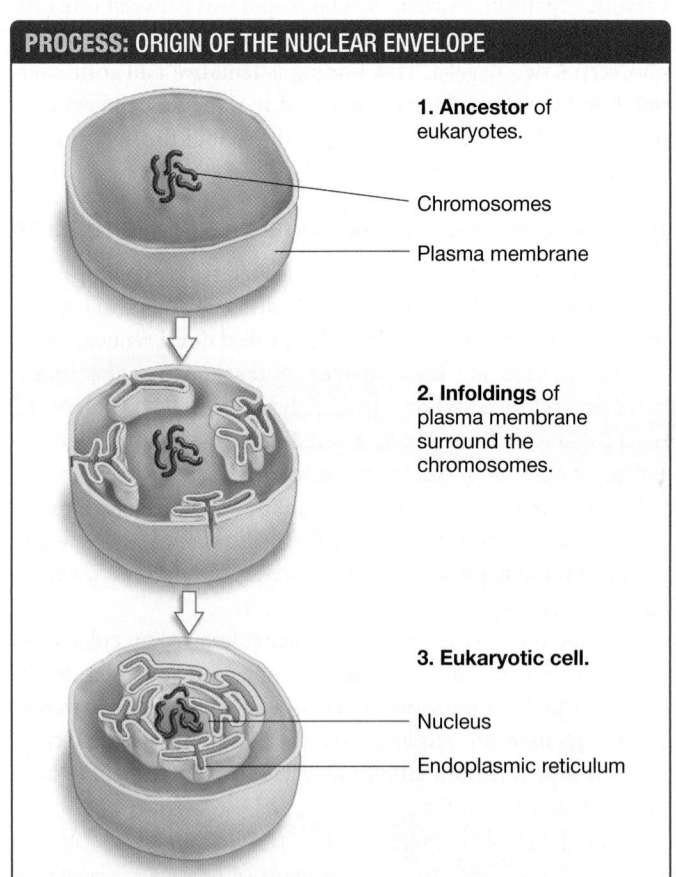

PROCESS: ORIGIN OF THE NUCLEAR ENVELOPE

1. **Ancestor** of eukaryotes.

Chromosomes

Plasma membrane

2. **Infoldings** of plasma membrane surround the chromosomes.

3. **Eukaryotic cell.**

Nucleus

Endoplasmic reticulum

FIGURE 29.8 A Hypothesis for the Origin of the Nuclear Envelope. Infoldings of the plasma membrane, analogous to those shown here, have been observed in bacteria living today.

According to current thinking, the evolution of the nuclear envelope was advantageous because it separated transcription and translation. Recall from Chapter 16 that RNA transcripts are processed inside the nucleus but translated outside the nucleus. In bacteria and archaea, transcription and translation occur together.

Once a simple nuclear envelope was in place, alternative splicing and other forms of RNA processing could occur—giving the early eukaryotes a novel way to control gene expression (see Chapter 18). The take-home message here is that an important morphological innovation gave the early eukaryotes a new way to manage and process genetic information.

Once a nucleus had evolved, it underwent diversification. In some cases, unique types of nuclei are associated with the founding of important lineages of protists.

- Ciliates have a diploid micronucleus that is involved only in reproduction and a polyploid macronucleus where transcription occurs.

- Diplomonads have two nuclei that look identical; it is not known how they interact.

- In foraminifera, red algae, and plasmodial slime molds, certain cells may contain many nuclei.

- Dinoflagellates have chromosomes that lack histones and attach to the nuclear envelope.

In each case, the distinctive structure of the nucleus is a synapomorphy that allows us to recognize these lineages as distinct monophyletic groups.

ENDOSYMBIOSIS AND THE ORIGIN OF THE MITOCHONDRION Mitochondria are organelles that generate ATP using pyruvate as an electron donor and oxygen as the ultimate electron acceptor (see Chapter 9).

In 1981 Lynn Margulis expanded on a radical hypothesis—first proposed in the nineteenth century—to explain the origin of mitochondria. The **endosymbiosis theory** proposes that mitochondria originated when a bacterial cell took up residence inside a eukaryote about 2 billion years ago.

The theory's name was inspired by the Greek word roots *endo*, *sym*, and *bio* (literally, "inside-together-living"). **Symbiosis** is said to occur when individuals of two different species live in physical contact; **endosymbiosis** occurs when an organism of one species lives inside an organism of another species.

In its current form, the endosymbiosis theory proposes that mitochondria evolved through a series of steps, beginning with a eukaryotic cell that was capable only of anaerobic fermentation—meaning it could not use oxygen as an electron acceptor in cellular respiration (**Figure 29.9**):

1. The eukaryotic cell used its cytoskeletal elements to surround and engulf smaller prey.

2. Instead of fusing with a lysosome and being digested (see Chapter 7), an engulfed bacterium began to live inside a eukaryotic cell.

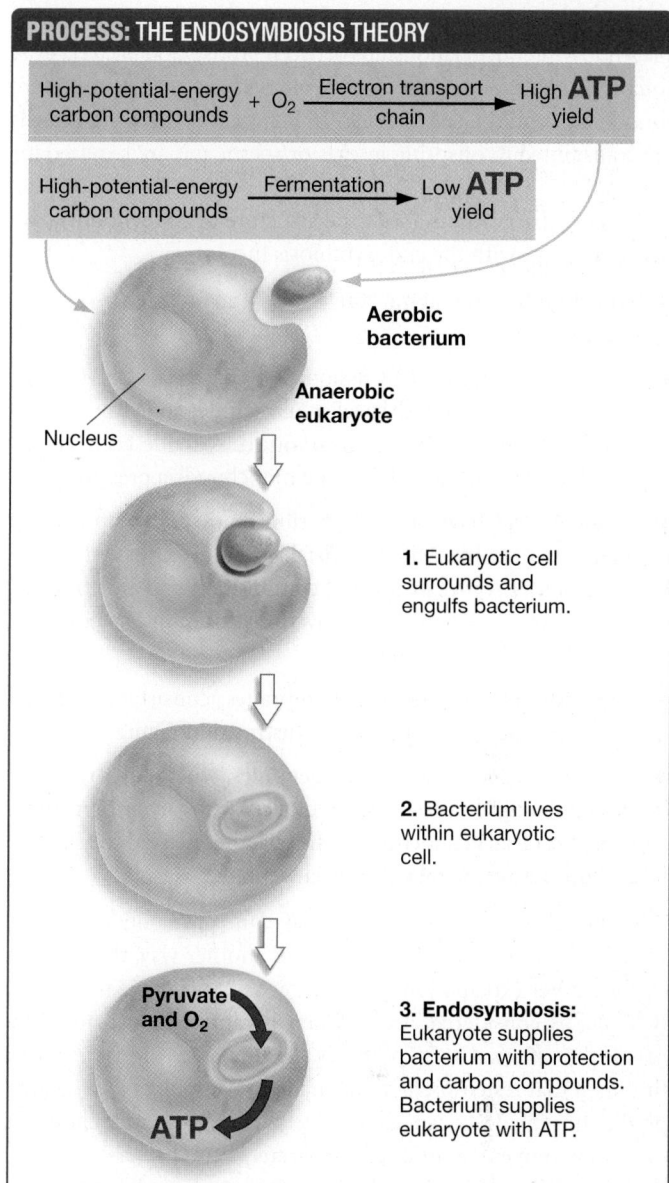

PROCESS: THE ENDOSYMBIOSIS THEORY

High-potential-energy carbon compounds + O_2 → (Electron transport chain) → High **ATP** yield

High-potential-energy carbon compounds → (Fermentation) → Low **ATP** yield

Aerobic bacterium

Anaerobic eukaryote

Nucleus

1. Eukaryotic cell surrounds and engulfs bacterium.

2. Bacterium lives within eukaryotic cell.

Pyruvate and O_2

ATP

3. Endosymbiosis: Eukaryote supplies bacterium with protection and carbon compounds. Bacterium supplies eukaryote with ATP.

FIGURE 29.9 Proposed Initial Steps in the Evolution of the Mitochondrion.

✓**QUESTION** According to this hypothesis, how many membranes should surround a mitochondrion? Explain your logic.

3. The engulfed cell survived by absorbing carbon molecules with high potential energy from the eukaryotic cell and oxidizing them, using oxygen as a final electron acceptor.

The relationship between the host and the engulfed cell was presumed to be stable because a mutual advantage existed between them: The host supplied the bacterium with protection and carbon compounds from its other prey, while the bacterium produced much more ATP than the host cell could synthesize on its own. Cells that can use oxygen during cellular respiration are able to produce much more ATP than are cells that cannot (see Chapter 28).

When Margulis first began promoting the theory, it met with a storm of criticism—largely because it seemed slightly preposterous.

But gradually biologists began to examine it rigorously. For example, endosymbiotic relationships between protists and bacteria exist today. Among the α-proteobacteria alone, three major groups are found *only* inside eukaryotic cells. In many cases, the bacterial cells are transmitted to offspring in eggs or sperm and are required for survival.

Several observations about the structure of mitochondria are also consistent with the endosymbiosis theory:

- Mitochondria are about the size of an average α-proteobacterium.

- Mitochondria replicate by fission, as do bacterial cells. The duplication of mitochondria takes place independently of division by the host cell. When eukaryotic cells divide, each daughter cell receives some of the many mitochondria present.

- Mitochondria have their own ribosomes and manufacture their own proteins. Mitochondrial ribosomes closely resemble bacterial ribosomes in size and composition and are poisoned by antibiotics such as streptomycin that inhibit bacterial, but not eukaryotic, ribosomes.

- Mitochondria have double membranes, consistent with the engulfing mechanism of origin illustrated in Figure 29.9.

- Mitochondria have their own genomes, which are organized as circular molecules—much like a bacterial chromosome. Mitochondrial genes code for the enzymes needed to replicate and transcribe the mitochondrial genome.

Although these data are impressive, they are only consistent with the endosymbiosis theory. Stated another way, they do not exclude other explanations. This is a general principle in science: Evidence is considered strong when it cannot be explained by reasonable alternative hypotheses. In this case, the key was to find data that tested predictions made by Margulis's idea against predictions made by an alternative theory: that mitochondria evolved within eukaryotic cells, separately from bacteria.

A breakthrough occurred when researchers realized that according to the "within-eukaryotes" theory, the genes found in mitochondria are derived from nuclear genes in ancestral eukaryotes. Margulis's theory, in contrast, proposed that the genes found in mitochondria were bacterial in origin.

These predictions were tested by studies on the phylogenetic relationships of mitochondrial genes. Specifically, researchers compared gene sequences isolated from the nuclear DNA of eukaryotes, mitochondrial DNA from eukaryotes, and DNA from several species of bacteria. Exactly as the endosymbiosis theory predicted, mitochondrial gene sequences are much more closely related to the sequences from the α-proteobacteria than to sequences from the nuclear DNA of eukaryotes.

The result diagrammed in **Figure 29.10** was considered overwhelming evidence that the mitochondrial genome came from an α-proteobacterium rather than from a eukaryote. The endosymbiosis theory was the only reasonable explanation for the data. The results were a stunning vindication of a theory that had once been intensely controversial. Mitochondria evolved via endosymbiosis.

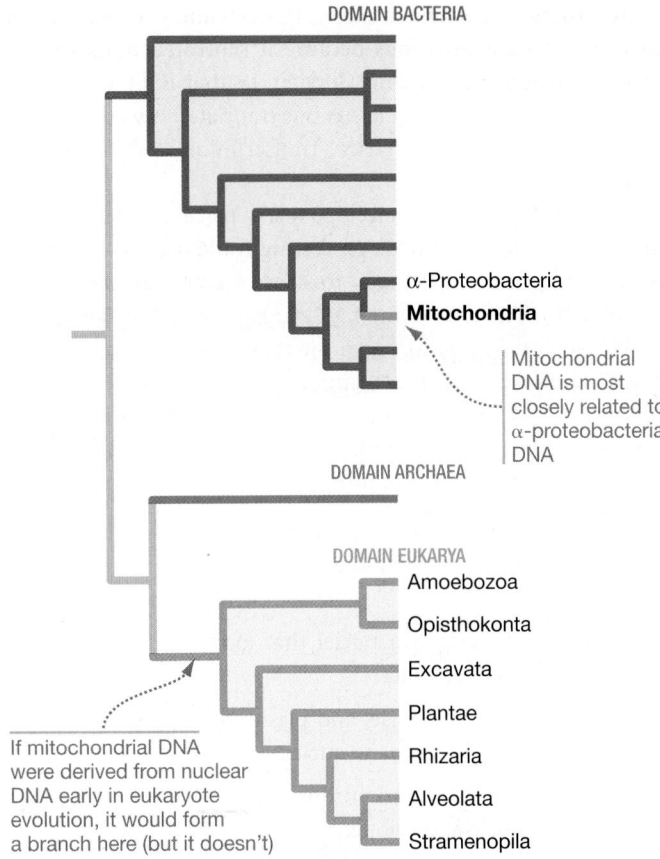

FIGURE 29.10 Phylogenetic Data Support the Endosymbiosis Theory.

✔If you understand the endosymbiosis theory and the evidence for it, you should be able to describe how the chloroplast—the organelle where photosynthesis takes place in eukaryotes—could arise via endosymbiosis.

Once the mitochondrion was present in the ancestor of today's protists, it underwent diversification.

- In Diplomonads and Parabasalids, the organelle was lost entirely or is now a vestigial trait.

- In Euglenida and related species, the sac-like cristae inside the mitochondrion—where the electron transport chain and ATP synthase are located—are disc-shaped.

- In the Opisthokonta, the cristae are flattened.

Presumably, variation in the presence or structure of mitochondria reflects variation in how ATP is produced by protists. In these cases, the mitochondrion's absence or structure qualifies as a synapomorphy that identifies a monophyletic group. The mitochondrion was a morphological innovation that subsequently diversified.

STRUCTURES FOR SUPPORT AND PROTECTION   As far as is known, the basic structure of the cytoskeleton does not vary much among protists. In contrast, other structures that provide support and protection for the cell vary a great deal.

Many protists have cell walls outside their plasma membrane; others have hard external structures called a **shell**; others have rigid structures inside the plasma membrane. In many cases, these novel structures represent synapomorphies that identify monophyletic groups among protists. For example:

- Diatoms are surrounded by a glass-like, silicon-dioxide cell wall (**Figure 29.11a**). The cell wall is made up of two pieces that fit together in a box-and-lid arrangement, like the petri plates you may have seen in lab.

- Dinoflagelletes have a cell wall made up of cellulose plates (**Figure 29.11b**).

- Within Foraminifera, some lineages secrete an intricate, chambered calcium carbonate shell (**Figure 29.11c**).

- Members of other Foraminifera lineages, and some amoebae, cover themselves with tiny pebbles.

- The parabasalids have a unique internal support rod, consisting of cross-linked microtubules that run the length of the cell.

- The euglenids have a collection of protein strips located just under the plasma membrane. The strips are supported by microtubules and stiffen the cell.

- The alveolates have distinctive sac-like structures called alveoli, located just under the plasma membrane, that help stiffen the cell.

In many cases, the diversification of protists has been associated with the evolution of innovative structures for support and protection.

MULTICELLULARITY   Multicellular individuals contain more than one cell. At least some of the cells are attached to each other, and some are specialized for different functions. In the simplest multicellular species, certain cells are specialized for producing or obtaining food while other cells are specialized for reproduction. The key point about **multicellularity** is that not all cells express the same genes.

A few species of bacteria are capable of aggregating and forming structures called fruiting bodies (see Chapter 28). Because cells in the fruiting bodies of these bacteria differentiate into specialized stalk cells and spore-forming cells, they are considered multicellular. But the vast majority of multicellular species are members of the Eukarya. Multicellularity arose independently in a wide array of eukaryotic lineages: the green plants, fungi, animals, brown algae, slime molds, and red algae.

To summarize, an array of novel morphological traits played a key role as protists diversified: the nucleus and endomembrane system, the mitochondrion, structures for protection and support, and multicellularity. Evolutionary innovations allowed protists to build and manage the eukaryotic cell in new ways.

Once a new type of eukaryotic cell or multicellular individual existed, subsequent diversification was often triggered by novel ways of finding food, moving, or reproducing. Let's consider each of these life processes in turn.

How Do Protists Obtain Food?

According to Chapter 28, bacteria and archaea can use a wide array of molecules as electron donors and electron acceptors during cellular respiration. They get these molecules by absorbing them directly from the environment. Other bacteria don't absorb their nutrition—instead, they make their own food via photosynthesis. Many groups of protists are similar to bacteria in the way they find food: They perform photosynthesis or absorb their food directly from the environment.

But one of the most important stories in the diversification of protists was the evolution of a novel method for finding food. Many protists ingest their food—they eat bacteria, archaea, or even other protists whole.

When ingestive feeding occurs, an individual takes in packets of food much larger than individual molecules. Thus, protists feed by (1) ingesting packets of food, (2) absorbing organic molecules directly from the environment, or (3) performing photosynthesis.

Some protists ingest food as well as performing photosynthesis—meaning that they use a combination of feeding strategies. In

(a) Diatom

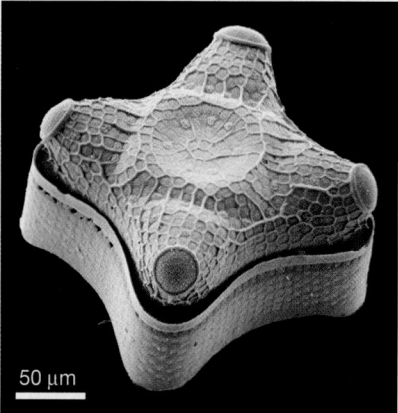

50 µm

Glassy cell wall made of silicon dioxide

(b) Dinoflagellate

10 µm

Tough plates in cell wall made of cellulose

(c) Foraminiferan

50 µm

Chalky, chambered shell made of calcium carbonate

FIGURE 29.11 Hard Outer Coverings in Protists Vary in Composition.

addition, it's important to recognize that all three lifestyles—ingestive, absorptive, and photosynthetic—can occur within a single clade.

To drive this last point home, consider the monophyletic group called the alveolates. There are three major subgroups of this lineage, called dinoflagellates, apicomplexa, and ciliates. About half of the dinoflagellates are photosynthetic, while many others are parasitic. Apicomplexa are parasitic. Ciliates include many species that ingest prey, but some ciliates live in the guts of cattle or the gills of fish and absorb nutrients from their hosts. Other ciliate species make a living by holding algae or other types of photosynthetic symbionts inside their cells.

The punchline? Within each of the seven major lineages of eukaryotes, different methods for feeding helped trigger diversification.

INGESTIVE FEEDING Ingestive lifestyles are based on eating live or dead organisms or on scavenging loose bits of organic debris. Protists such as the cellular slime mold *Dictyostelium discoideum* are large enough to engulf bacteria and archaea; many protists are large enough to surround and ingest other protists or microscopic animals.

Feeding by engulfing is possible in protists that lack a cell wall. A flexible membrane and dynamic cytoskeleton give these species the ability to surround and "swallow" prey using long, fingerlike projections called **pseudopodia** ("false-feet"). The engulfing process is illustrated in **Figure 29.12a**.

Losing a cell wall and gaining the ability to move their plasma membrane around prey items or food particles was an important innovation during the evolution of protists:

- Bacteria and archaea are abundant in wet soils and aquatic habitats, where protists also live. Instead of competing with prokaryotes for sunlight or food molecules, protists could eat them.

- It was a prerequisite for the endosymbiosis events that led to the evolution of the mitochondrion and chloroplast.

Although many ingestive feeders actively hunt down prey and engulf them, others do not. Instead of taking themselves to food, these species attach themselves to a surface. Protists that feed in this way have cilia that surround the mouth and beat in a coordinated way. The motion creates water currents that sweep food particles into the cell (**Figure 29.12b**).

ABSORPTIVE FEEDING Absorptive feeding occurs when nutrients are taken up directly from the environment, across the plasma membrane. Absorptive feeding is common among protists.

Some protists that live by absorptive feeding are **decomposers**, meaning that they feed on dead organic matter, or **detritus**. But many of the protists that absorb their nutrition directly from the environment live inside other organisms. If they damage their host, the absorptive species is called a **parasite**.

PHOTOSYNTHESIS—ENDOSYMBIOSIS AND THE ORIGIN OF THE CHLOROPLAST You might recall from Chapter 10 and Chapter 28 that photosystems I and II evolved in bacteria, and that both photosystems occur in cyanobacteria. None of the basic machinery required for photosynthesis evolved in eukaryotes. Instead, they "stole" it—via endosymbiosis.

The endosymbiosis theory contends that the eukaryotic chloroplast originated when a protist engulfed a cyanobacterium. Once inside the protist, the photosynthetic bacterium provided its eukaryotic host with oxygen and glucose in exchange for protection and access to light. If the endosymbiosis theory is correct, today's chloroplasts trace their ancestry to cyanobacteria.

All of the photosynthetic eukaryotes have both a chloroplast and a mitochondrion. The evidence for an endosymbiotic origin for the chloroplast is even more persuasive than for mitochondria.

- Chloroplasts have the same list of bacteria-like characteristics presented earlier for mitochondria.

- There are many examples of endosymbiotic cyanobacteria living inside protists or animals today.

- The DNA sequences inside chloroplasts are extremely similar to cyanobacterial genes.

(a) Pseudopodia engulf food

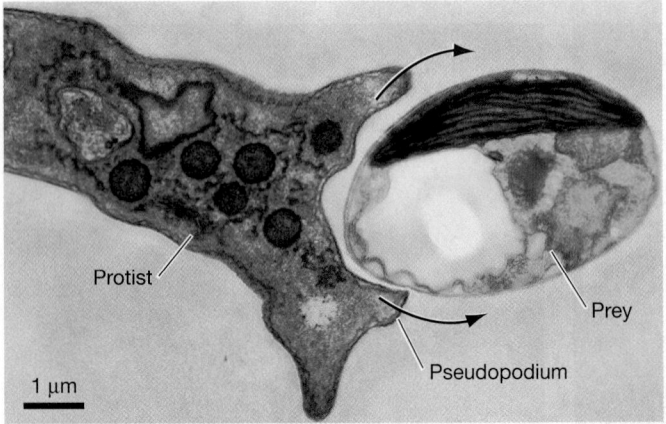

(b) Ciliary currents sweep food into gullet

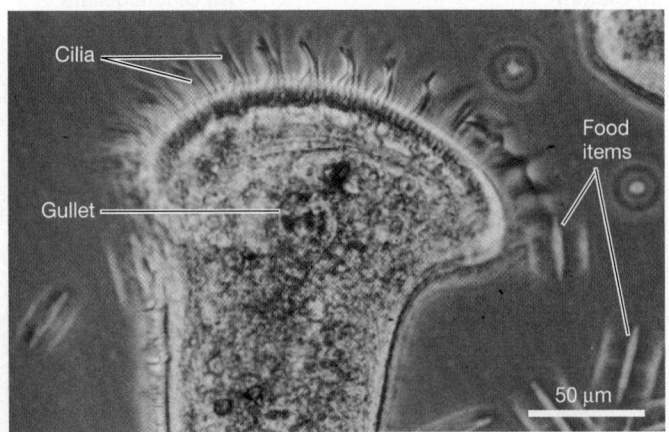

FIGURE 29.12 Ingestive Feeding. Methods of prey capture vary among ingestive protists. **(a)** Some predators engulf prey with pseudopodia; **(b)** other predators sweep them into their gullets with water currents set up by the beating of cilia. Note that the cells in part (a) have been colorized.

- The photosynthetic organelle of one group of protists, called the glaucophyte algae, has an outer layer containing the same constituent (peptidoglycan) found in the cell walls of cyanobacteria.

- Like chloroplasts, the cyanobacterium *Prochloron* contains chlorophylls *a* and *b* and has a system of internal membranes where the photosynthetic pigments and enzymes are located.

✔If you understand the evidence for the endosymbiotic origin of the chloroplast, you should be able to add a label indicating the location of chloroplast genes on the phylogenetic tree in Figure 29.7.

PHOTOSYNTHESIS—PRIMARY VERSUS SECONDARY ENDOSYMBIOSIS

Which eukaryote originally obtained a photosynthetic organelle? Because all of the species in the Plantae have chloroplasts with two membranes, biologists infer that the original, or primary, endosymbiosis occurred in their common ancestor—a species that eventually gave rise to the glaucophyte algae, red algae, and green plants (green algae and land plants).

But chloroplasts also occur in three of the other major lineages of protists—the Excavata, Chromalveolates, and Rhizaria. In these species, the chloroplast is surrounded by more than two membranes—usually four.

To explain this observation, researchers hypothesize that the ancestors of these groups acquired their chloroplasts by ingesting photosynthetic protists that already had chloroplasts. This process, called secondary endosymbiosis, occurs when an organism engulfs a photosynthetic eukaryotic cell and retains its chloroplasts as intracellular symbionts. Secondary endosymbiosis is illustrated in **Figure 29.13**.

Figure 29.14 shows where primary and secondary endosymbiosis occurred on the phylogenetic tree of eukaryotes. Once

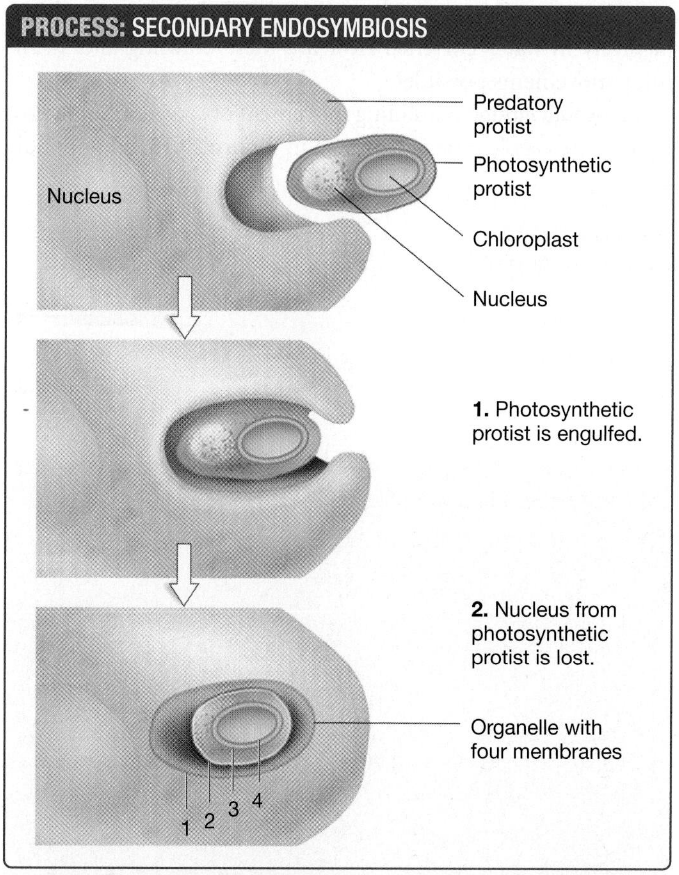

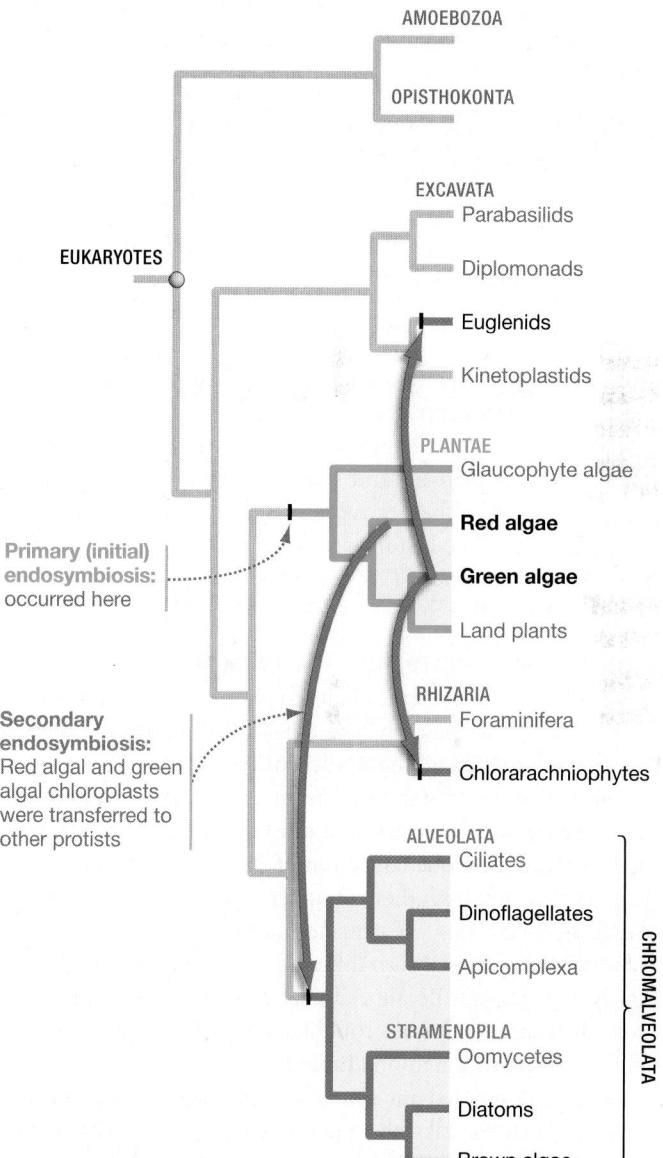

FIGURE 29.14 Photosynthesis Arose in Protists by Primary Endosymbiosis, Then Spread among Lineages via Secondary Endosymbiosis. Biochemical similarities link the chloroplasts found in chromalveolates with red algae and the chloroplasts found in euglenids and chlorarachniophytes with green algae. Note that among the chromalveolates, only the dinoflagellates, diatoms, and brown algae have photosynthetic species. In ciliates, apicomplexa, and oomycetes, the chloroplast has been lost or changed function.

FIGURE 29.13 Secondary Endosymbiosis Leads to Organelles with Four Membranes. The chloroplasts found in some protists have four membranes and are hypothesized to be derived by secondary endosymbiosis. In species where chloroplasts have three membranes, biologists hypothesize that secondary endosymbiosis was followed by the loss of one membrane.

TABLE 29.3 **Pigments in Photosynthetic Protists**

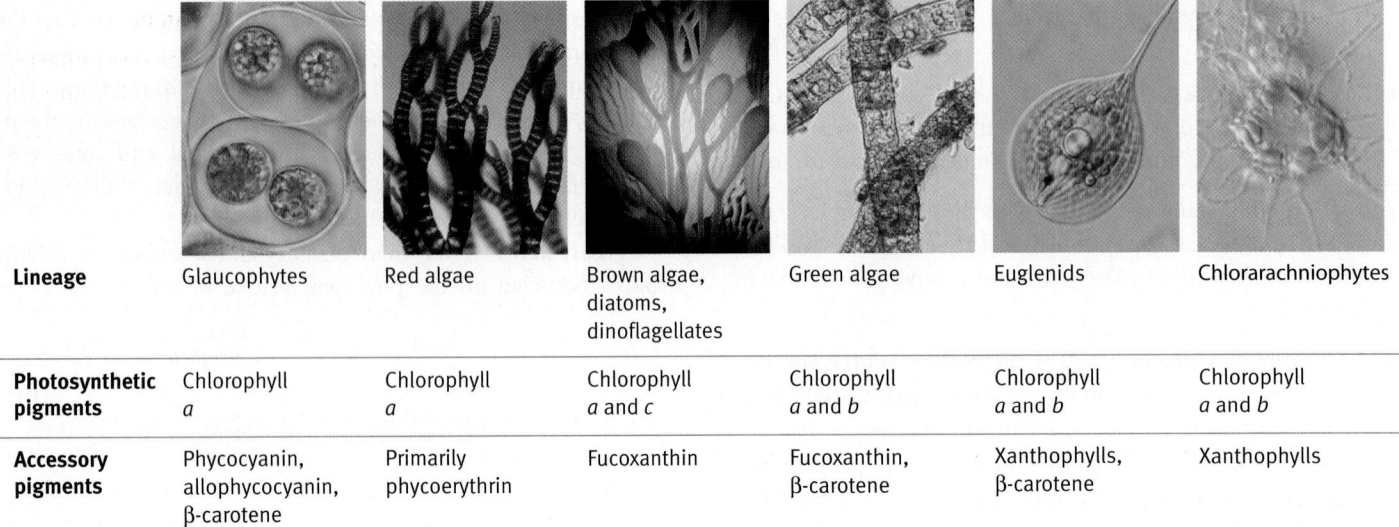

Lineage	Glaucophytes	Red algae	Brown algae, diatoms, dinoflagellates	Green algae	Euglenids	Chlorarachniophytes
Photosynthetic pigments	Chlorophyll *a*	Chlorophyll *a*	Chlorophyll *a* and *c*	Chlorophyll *a* and *b*	Chlorophyll *a* and *b*	Chlorophyll *a* and *b*
Accessory pigments	Phycocyanin, allophycocyanin, β-carotene	Primarily phycoerythrin	Fucoxanthin	Fucoxanthin, β-carotene	Xanthophylls, β-carotene	Xanthophylls

✔**QUESTION** Are the pigments listed in this table consistent with the primary endosymbiosis and secondary endosymbiosis hypotheses described in Figures 29.9 and 29.13? Explain your answer.

protists obtained the chloroplast, it was "swapped around" to new lineages via secondary endosymbiosis.

✔If you understand endosymbiosis, you should be able to explain why the primary and secondary endosymbiosis events introduced in this chapter represent the most massive lateral gene transfers in the history of life, in terms of the number of genes moved at once. (To review lateral gene transfer, see Chapter 20.)

PHOTOSYNTHESIS—DIVERSIFICATION IN PIGMENTS Many lineages of protists acquired the ability to photosynthesize independently of each other. In each case, the acquisition of a chloroplast triggered a radiation of photosynthetic species.

The key to understanding the diversification of photosynthetic eukaryotes is to realize that most of the photosynthetic lineages contain a unique collection of photosynthetic pigments. A particular photosynthetic pigment absorbs specific wavelengths of light. Thus, the presence of different pigments means that different species absorb different wavelengths in the electromagnetic spectrum. Because diverse combinations of pigments evolved, eukaryotic species could harvest unique wavelengths of light and avoid competition (**Table 29.3**).

Chapter 28 pointed out a similar phenomenon among photosynthetic bacteria. Like the nuclear envelope, mitochondrion, multicellularity, ingestive feeding, and other traits that originated with protists, the acquisition of the chloroplast was an innovation that triggered subsequent diversification.

How Do Protists Move?

Many protists actively move to find food. Predators such as the slime mold *Dictyostelium discoideum* crawl over a substrate in search of prey. Most of the unicellular, photosynthetic species are

capable of swimming to sunny locations, though others drift passively in water currents. How are these crawling and swimming movements possible?

Amoeboid motion is a sliding movement observed in some protists. In the classic mode illustrated in **Figure 29.15**, pseudopodia

Amoeboid motion via **pseudopodia**

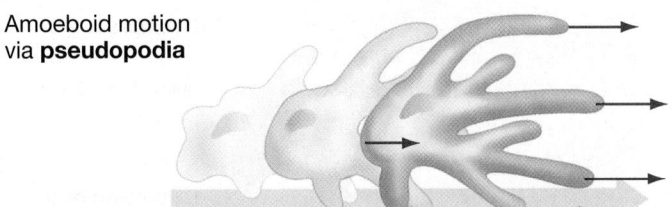

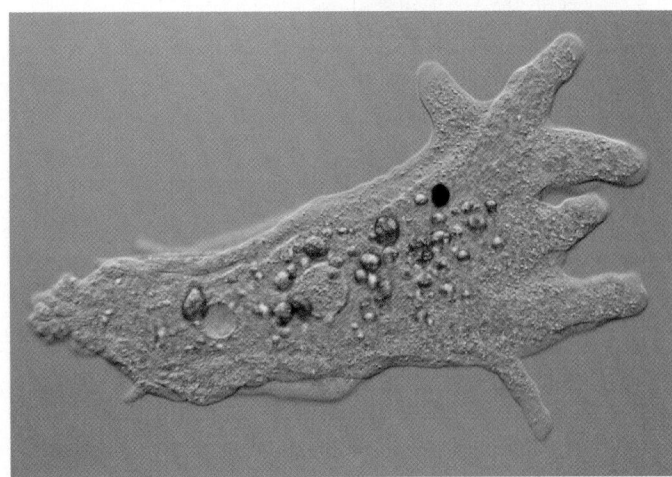

FIGURE 29.15 Amoeboid Motion Is Possible in Species That Lack Cell Walls. In amoeboid motion, long pseudopodia stream out from the cell. The rest of the cytoplasm, organelles, and external membrane follow.

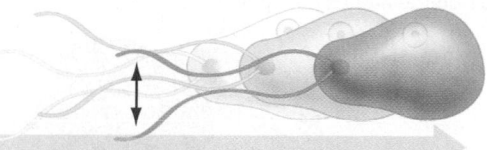

(a) Swimming via **flagella**

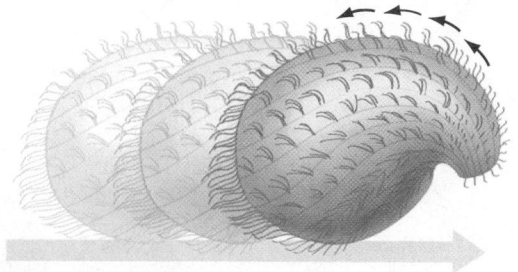

(b) Swimming via **cilia**

FIGURE 29.16 Many Protists Swim Using Flagella or Cilia.
(a) Flagella are long and few in number. **(b)** Cilia are short and numerous. In many cases they are used in swimming.

stream forward over a substrate, with the rest of the cytoplasm, organelles, and plasma membrane following. The motion requires ATP and involves interactions between proteins called actin and myosin inside the cytoplasm. The mechanism is related to muscle movement in animals, which is detailed in Chapter 46. But at the level of the whole cell, the precise sequence of events during amoeboid movement is still uncertain. The issue is attracting attention, because key immune system cells in humans use amoeboid motion as they hunt down and destroy disease-causing agents.

The other major mode of locomotion in protists involves flagella or cilia (**Figure 29.16a, and b**). Recall from Chapter 7 that flagella and cilia both consist of nine sets of doublet (paired) microtubules arranged around two central, single microtubules. Flagella are long and are usually found alone or in pairs; cilia are short and usually occur in large numbers on any one cell. Flagella and cilia can also be distinguished by the types of structures associated with the basal bodies where they originate.

Even closely related protists can use radically different forms of locomotion. For example, the ciliates swim by beating their cilia, the dinoflagellates swim by whipping their flagella, and mature cells in apicomplexa move by amoeboid motion (though their gametes swim via flagella). It is also common to find protists that do not exhibit active movement but instead float passively in water currents.

Movement is yet another example of the extensive diversification that occurred within each of the seven major monophyletic groups of eukaryotes.

How Do Protists Reproduce?

Sexual reproduction evolved in protists. As Chapter 12 pointed out, sexual reproduction can best be understood in contrast to asexual reproduction. The key issues are the type of nuclear division involved and the consequences for genetic diversity in the offspring produced.

- Asexual reproduction is based on mitosis in eukaryotic organisms and on fission in bacteria and archaea. It results in offspring that are genetically identical to the parent.

- Sexual reproduction is based on meiosis and fusion of gametes. It results in offspring that are genetically different from their parents and from each other.

Most protists undergo asexual reproduction routinely. But sexual reproduction occurs only intermittently in many protists—often at one particular time of year, or when individuals are crowded or food is scarce.

☞ The evolution of sexual reproduction ranks among the most significant evolutionary innovations observed in eukaryotes.

SEXUAL VERSUS ASEXUAL REPRODUCTION The leading hypothesis to explain why meiosis evolved states that genetically variable offspring may be able to thrive if the environment changes. For example, offspring with genotypes different from those of their parents may be better able to withstand attacks by parasites that successfully attacked their parents (see Chapter 12). This is a key point because many types of parasites, including bacteria and viruses, have short generation times and evolve quickly.

Because the genotypes and phenotypes of parasites are constantly changing, natural selection is constantly favoring host individuals with new genotypes. The idea is that new offspring genotypes generated by meiosis may contain combinations of alleles that allow hosts to withstand attack by new strains of parasites. In short, many biologists view sexual reproduction as an adaptation to fight disease.

✔If you understand the changing-environment hypothesis for the evolution of sex, you should be able to explain whether it is consistent with the observation that many protists undergo meiosis when food is scarce or the population density is high. In addition, if a damaging mutation occurs in a parent, it will be passed along to offspring that are produced asexually. But if sexual reproduction occurs, some offspring may be free of the mutation.

Life Cycles—Haploid- versus Diploid-Dominated

A life cycle describes the sequence of events that occur as individuals grow, mature, and reproduce. The evolution of meiosis introduced a new event in the life cycle of protist species; what's more, it created a distinction between haploid and diploid phases in the life of an individual.

Recall from Chapter 12 that diploid individuals have two of each type of chromosome inside each cell, while haploid individuals have just one of each type of chromosome inside each cell. When meiosis occurs in diploid cells, it results in the production of haploid cells.

The life cycle of most bacteria and archaea is extremely simple: A cell divides, feeds, grows, and divides again. Bacteria and archaea are always haploid.

In contrast, virtually every aspect of a life cycle is variable among protists—whether meiosis occurs, whether asexual reproduction occurs, and whether the haploid or the diploid phase of the life cycle is the longer and more prominent phase.

(a) A life cycle dominated by haploid cells (species shown here is the dinoflagellate *Gymnodinium fuscum*)

Haploid (*n*)
Diploid (2*n*)

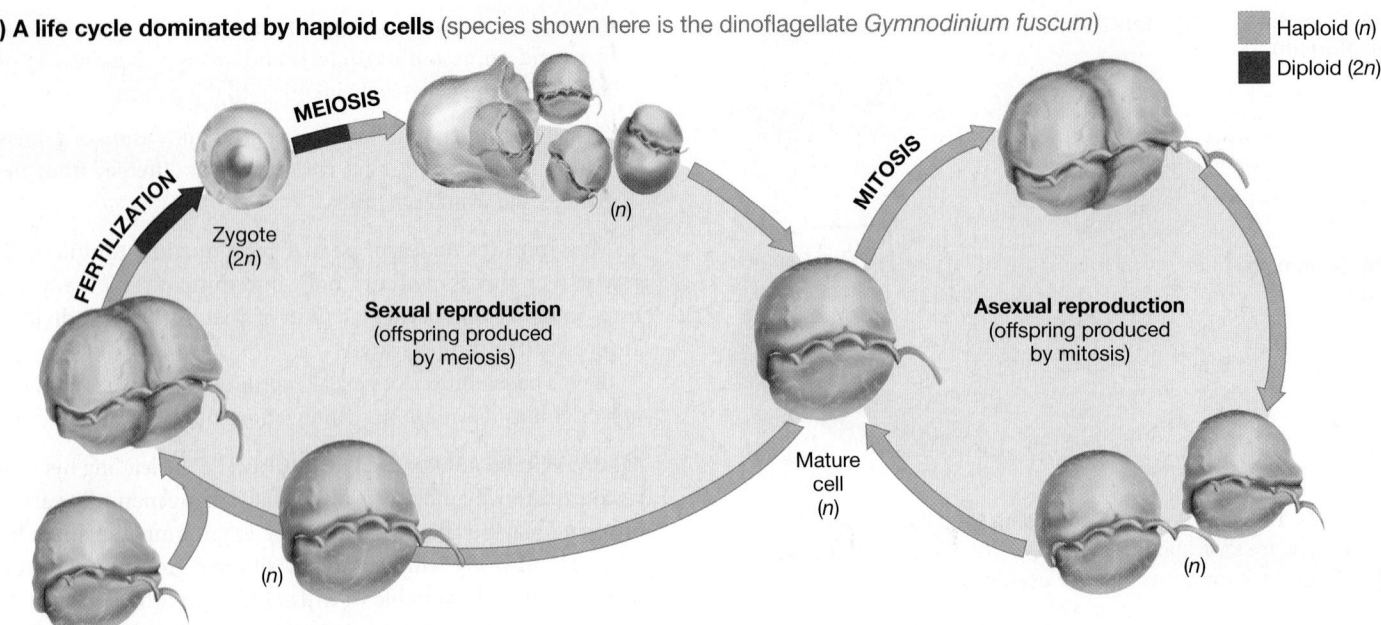

(b) A life cycle dominated by diploid cells (species shown here is the diatom *Cyclotella meneghiniana*)

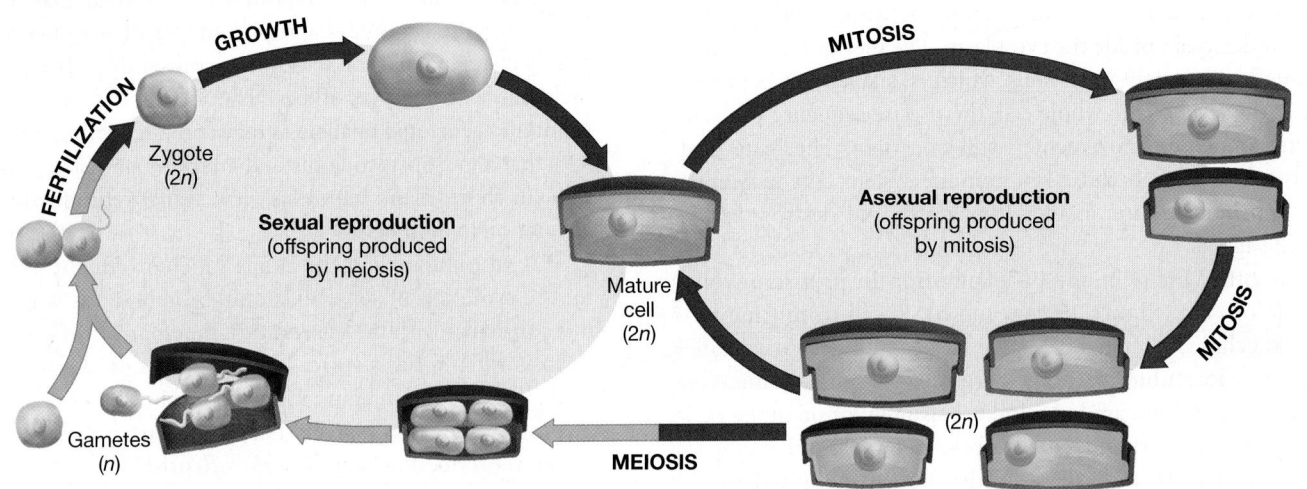

FIGURE 29.17 Life Cycles Vary Widely among Unicellular Protists. Many unicellular protists can reproduce by both sexual reproduction and asexual reproduction. The cell may be **(a)** haploid for most of its life or **(b)** diploid for most of its life.

Figure 29.17 illustrates some of this variation. Figure 29.17a depicts the haploid-dominated life cycle observed in many unicellular protists. The specific example given here is the dinoflagellate *Gymnodinium fuscum*.

To analyze a life cycle, start with **fertilization**—the fusion of two gametes to form a diploid zygote. Then trace what happens to the zygote. In this case, the diploid zygote undergoes meiosis. The haploid products of meiosis then grow into mature cells that eventually undergo asexual reproduction or produce gametes by mitosis.

Contrast the dinoflagellate cycle with the diploid-dominated life cycle of Figure 29.17b. The specific organism shown here is the diatom *Cyclotella meneghiniana*. Note that after fertilization, the diploid zygote develops into a sexually mature, diploid adult cell. Meiosis occurs in the adult and results in the formation of haploid

gametes, which then fuse to form a diploid zygote. The important contrasts are **(1)** that meiosis occurs in the adult cell rather than in the zygote and **(2)** that gametes are the only haploid cells in the life cycle.

LIFE CYCLES—ALTERNATION OF GENERATIONS In contrast to the relatively simple life cycles of Figure 29.17, many multicellular protists have one phase in their life cycle that is based on a multicellular haploid form and another phase in their life cycle that is based on a multicellular diploid form. This alternation of multicellular haploid and diploid forms is known as **alternation of generations**.

● The multicellular haploid form is called a **gametophyte**, because specialized cells in this individual produce gametes by mitosis.

(a) Alternation of generations in which multicellular haploid and diploid forms look identical (*Ectocarpus siliculosus*)

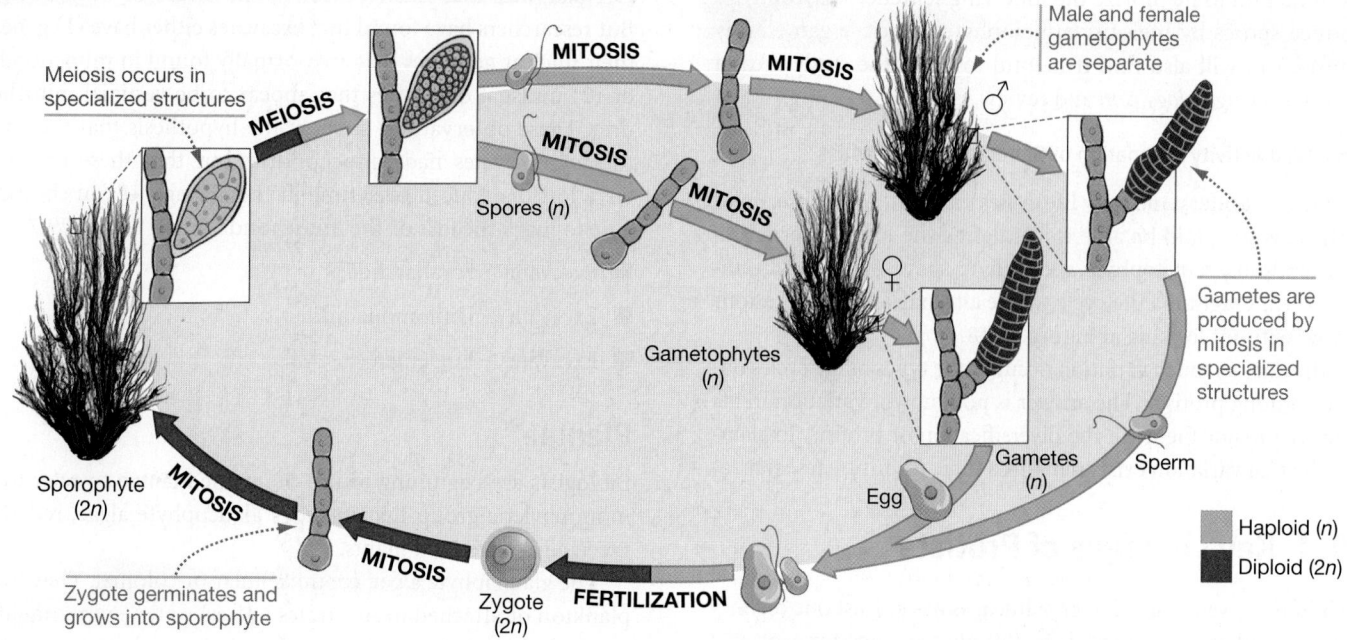

(b) Alternation of generations in which multicellular haploid and diploid forms look different (*Laminaria solidungula*)

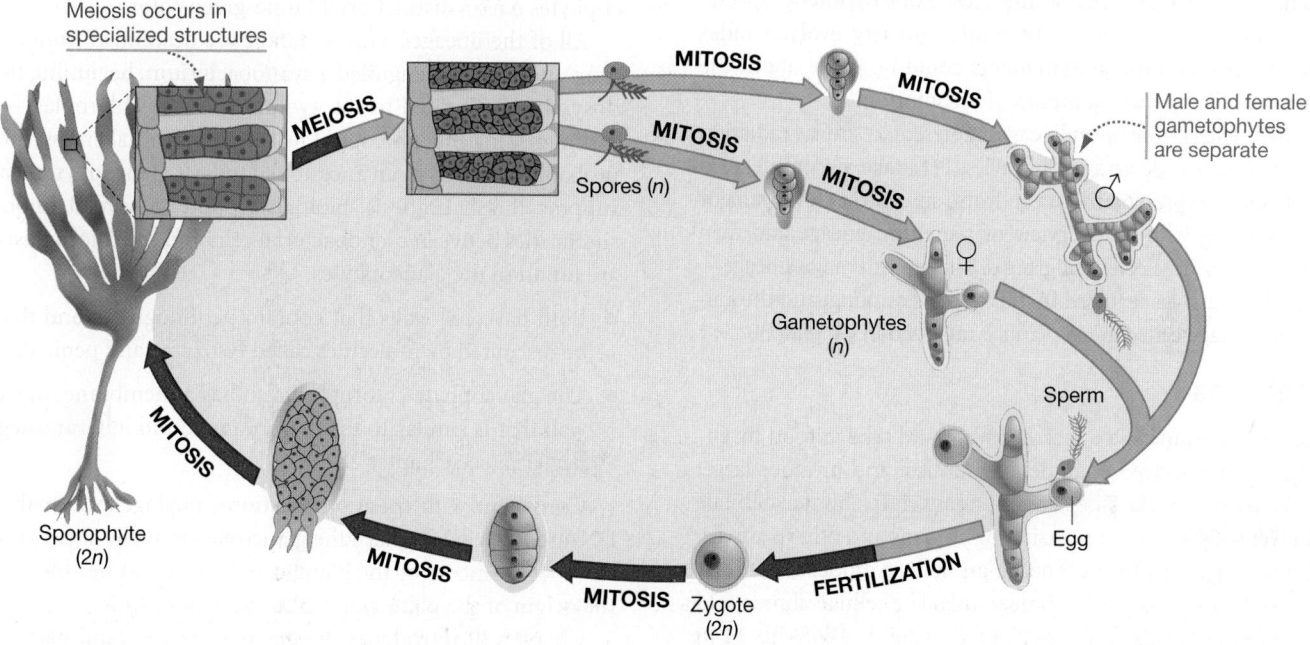

FIGURE 29.18 Alternation of Generations Occurs in Many Multicellular Protists.

- The multicellular diploid form is called a **sporophyte**, because it has specialized cells that undergo meiosis to produce haploid cells called spores.

- A spore is a single cell that develops into an adult organism but is not a product of fusion by gametes.

When alternation of generations occurs, a spore divides by mitosis to form a haploid, multicellular gametophyte. The haploid gametes produced by the gametophyte then fuse to form a diploid zygote, which grows into the diploid, multicellular sporophyte.

Note that gametophytes and sporophytes may be identical in appearance, as in the brown alga called *Ectocarpus siliculosus* (**Figure 29.18a**). In many cases, however, the gametophyte and sporophyte look different, as in the brown alga called *Laminaria solidungula* (**Figure 29.18b**). Among the protists, alternation of generations evolved independently in brown algae, red algae, and other groups.

To understand alternation of generations, some students find it helpful to memorize the following sentence: Sporophytes produce spores by meiosis; gametophytes produce gametes by mitosis. You will also find it helpful to go to the study area at *www.masteringbiology.com* and review.

MB **Web Activity** Alternation of Generations in a Protist

✔If you understand how life cycles vary among multicellular protists, you should be able to (**1**) define the terms alternation of generations, gametophyte, sporophyte, spore, zygote, and gamete and (**2**) diagram a life cycle where alternation of generations occurs, without looking at Figure 29.18.

Why does so much variation occur in the types of life cycles observed among protists? The answer is not known. Variation in life cycles is a major theme in the diversification of protists. Explaining why that variation exists remains a topic for future research.

29.4 Key Lineages of Protists

Each of the seven major Eukarya lineages has at least one distinctive morphological characteristic. But once an ancestor evolved a distinctive cell structure, its descendants diversified into a wide array of lifestyles. For example, parasitic species evolved independently in all seven major lineages. Photosynthetic species exist in most of the seven, and multicellularity evolved independently in four. Similar statements could be made about the evolution of life cycles and modes of locomotion.

In effect, each of the seven lineages represents a similar radiation of species into a wide array of lifestyles. Let's take a more detailed look at some representative taxa from six of the seven major groups, starting with an overview of the entire lineage and then looking in detail at selected subgroups. The seventh major lineage—the opisthokonts—is featured in the chapters on fungi and the animals. The lineage called green plants is analyzed in Chapter 30.

Amoebozoa

Species in the Amoebozoa lack cell walls and take in food by engulfing it. They move via amoeboid motion and produce large, lobe-like pseudopodia. They are abundant in freshwater habitats and in wet soils; some are parasites of humans and other animals.

Major subgroups in the lineage are lobose amoebae, cellular slime molds, and plasmodial slime molds. The cellular slime mold *Dictyostelium discoideum* is described in detail in **BioSkills 14** in Appendix A. ✔You should be able to mark the origin of this lineage's amoeboid form on Figure 29.7 and explain whether it evolved independently of the amoeboid form in Rhizaria.

■ Amoebozoa > Myxogastrida (Plasmodial Slime Molds)

Excavata

The unicellular species that form the Excavata are named for the morphological feature that distinguishes them—an "excavated" feeding groove found on one side of the cell. Because all excavates

lack mitochondria, they were once thought to trace their ancestry to eukaryotes that existed prior to the origin of mitochondria. But researchers have found that excavates either have (**1**) genes in their nuclear genomes that are normally found in mitochondria, or (**2**) unusual organelles that appear to be vestigial mitochondria. These observations support the hypothesis that the ancestors of excavates had mitochondria, but that these organelles were lost or reduced over time in this lineage. ✔You should be able to mark the loss of the mitochondrion on Figure 29.7.

■ Excavata > Parabasalida

■ Excavata > Diplomonadida

■ Excavata > Euglenida

Plantae

Biologists are beginning to use the name **Plantae** to refer to the monophyletic group that includes glaucophyte algae, red algae, green algae, and land plants.

The glaucophyte algae are unicellular or colonial. They live in plankton or attached to substrates in freshwater environments—particularly in bogs or swamps. Some glaucophyte species have flagella or produce flagellated spores, but sexual reproduction has never been observed in the group. The chloroplasts of glaucophytes have a distinct bright blue-green color.

All of the lineages within Plantae are descended from a common ancestor that engulfed a cyanobacterium, beginning the endosymbiosis that led to the evolution of the chloroplast—their distinguishing morphological feature. This initial endosymbiosis probably occurred in an ancestor of today's glaucophyte algae. To support this hypothesis, biologists point to several important similarities between cyanobacterial cells and the chloroplasts that are found in the glaucophytes.

● Both have cell walls that contain peptidoglycan and that can be disrupted by molecules called lysozyme and penicillin.

● The glaucophyte chloroplast also has a membrane outside its wall that is similar to the membrane found in Gram-negative bacteria (see Chapter 28).

Consistent with these observations, phylogenetic analyses of DNA sequence data place the glaucophytes as the sister group to all other members of the Plantae. ✔You should be able to mark the origin of the plant chloroplast on Figure 29.7.

Chapter 30 introduces the green algae and land plants; here we consider just the red algae in detail.

■ Plantae > Rhodophyta (Red Algae)

Rhizaria

The rhizarians are single-celled amoebae that lack cell walls, though some species produce elaborate shell-like coverings. They move by amoeboid motion and produce long, slender pseudopodia. Over 11 major subgroups have been identified and named, including the planktonic organisms called actinopods, which synthesize glassy, silicon-rich skeletons, and the chlorarachniophytes, which obtained a chloroplast via secondary endosymbiosis and are photosynthetic. The best-studied and most abundant group is the Foraminifera. ✔You should be able to mark the origin of the rhizarian amoeboid form on Figure 29.7.

- ■ Rhizaria > Foraminifera

Alveolata

Alveolates are distinguished by small sacs, called alveoli, that are located just under their plasma membranes. Although all members of this lineage are unicellular, the groups highlighted below—the Ciliata, the Dinoflagellata, and the Apicomplexa—are remarkably diverse in morphology and lifestyle. ✔You should be able to mark the origin of alveoli on Figure 29.7.

- ■ Alveolata > Ciliata
- ■ Alveolata > Dinoflagellata
- ■ Alveolata > Apicomplexa

Stramenopila (Heterokonta)

Stramenopiles are sometimes called heterokonts, which translates as "different hairs." At some stage of their life cycle, all stramenopiles have flagella that are covered with distinctive hollow "hairs." The structure of these flagella is unique to the stramenopiles. ✔You should be able to mark the origin of the "hairy" flagellum on Figure 29.7. The lineage includes a large number of unicellular forms, although the brown algae are multicellular and include the world's tallest marine organisms, the kelp.

- ■ Stramenopila > Oomycota (Water Molds)
- ■ Stramenopila > Diatoms
- ■ Stramenopila > Phaeophyta (Brown Algae)

Amoebozoa > Myxogastrida (Plasmodial Slime Molds)

The plasmodial slime molds got their name because individuals form a large, weblike structure (**Figure 29.19**) that consists of a single cell containing many diploid nuclei. Like oomycetes, the plasmodial slime molds were once considered fungi on the basis of their general morphological similarity (see Chapter 31).

Morphology The huge "supercell" form, with many nuclei in a single cell, occurs in few protists other than plasmodial slime molds. ✔You should be able to mark the origin of the supercell on Figure 29.7.

Feeding and locomotion Myxogastrida cells feed on decaying vegetation and move by amoeboid motion.

Reproduction When food becomes scarce, part of the amoeba forms a stalk topped by a ball-like structure in which nuclei undergo meiosis and form spores. The spores are then dispersed to new habitats by the wind or small animals. After spores germinate to form amoebae, two amoebae fuse to form a diploid cell that begins to feed and eventually grows into a supercell.

Human and ecological impacts Like cellular slime molds, plasmodial slime molds are important decomposers in forests. They help break down leaves, branches, and other dead plant material, releasing nutrients that they and other organisms can use.

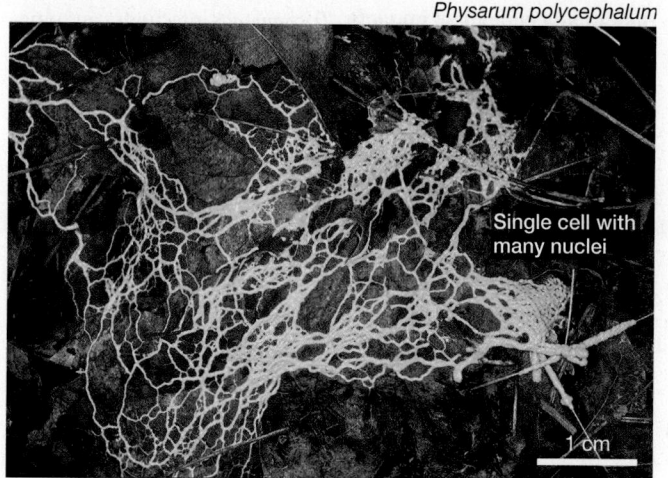

Physarum polycephalum

Single cell with many nuclei

1 cm

FIGURE 29.19 Plasmodial Slime Molds Are Important Decomposers in Forests.

No free-living parabasalids are known; all of the species described to date live inside animals. Of the 300 species of parabasalid, several live only in the guts of termites. Termites eat wood but cannot digest it themselves. Instead, the parabasalids produce enzymes that digest the cellulose in wood and release compounds that can be used by the termite host. The relationship between termites and parabasalids is considered mutualistic, because both parties benefit from the symbiosis.

Morphology Parabasalid cells lack a cell wall and mitochondria, and they have a single nucleus. A distinctive rod of cross-linked microtubules runs the length of the cell. The rod is attached to the basal bodies where a cluster of flagella arise (**Figure 29.20**). Although the number of flagella present varies widely, four or five is typical. ✔ You should be able to mark the origin of the parabasalid structure for supporting flagella on Figure 29.7.

Feeding and locomotion Parabasalids feed by engulfing bacteria, archaea, and organic matter. They swim using their flagella.

Reproduction All parabasalids reproduce asexually. Sexual reproduction also has been observed in a few species.

Human and ecological impacts *Trichomonas* infections can sometimes cause reproductive tract problems in humans, although members of this genus may also live in the gut or mouth of humans without causing harm. *Histomonas meleagridis* is an agricultural pest that causes disease outbreaks on chicken and turkey farms.

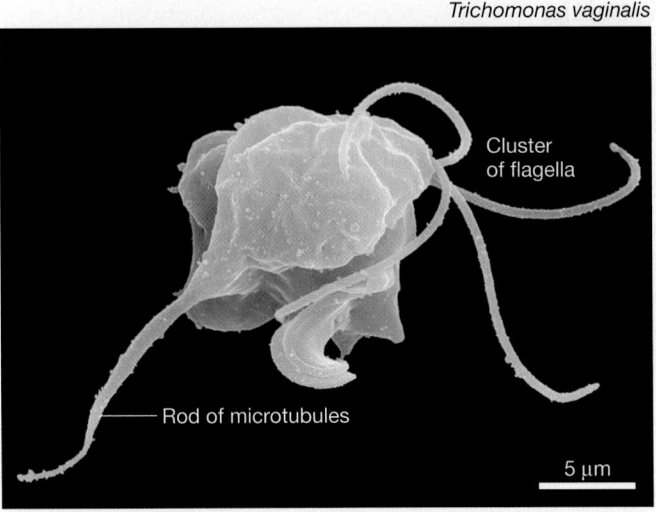

Trichomonas vaginalis

Cluster of flagella

Rod of microtubules

5 μm

FIGURE 29.20 *Trichomonas* **Causes the Sexually Transmitted Disease Trichomoniasis.** The cell in this micrograph has been colorized.

The first species of diplomonad known was described by the early microscopist Anton van Leeuwenhoek, who found *Giardia intestinalis* while examining samples of his own feces. About 100 species have been named to date. Many live in the guts of animal species without causing harm to their host; other species live in stagnant water habitats. In both types of environment, oxygen availability tends to be very low.

Morphology Each cell has two nuclei, which resemble eyes when viewed under the microscope (**Figure 29.21**). Each nucleus is associated with four flagella, for a total of eight flagella per cell. Some species lack the organelles called peroxisomes and lysosomes in addition to lacking functional mitochondria. All diplomonads lack a cell wall. ✔ You should be able to mark the origin of the double nucleus on Figure 29.7.

Feeding and locomotion Some are parasitic, but most ingest bacteria whole. They swim using their flagella.

Reproduction Only asexual reproduction occurs; meiosis has yet to be observed in members of this lineage.

Human and ecological impacts Both *Giardia intestinalis* and *G. lamblia* are common intestinal parasites in humans and cause giardiasis, or beaver fever. *Hexamida* is an agricultural pest that causes heavy losses in turkey farms each year.

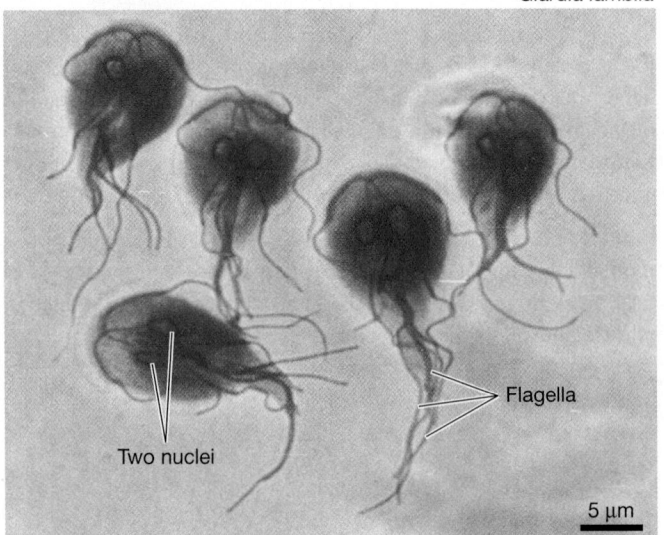

Giardia lamblia

Two nuclei

Flagella

5 μm

FIGURE 29.21 *Giardia* **Causes Intestinal Infections in Humans.**

There are about 1000 known species of euglenid. Although most live in freshwater, a few are found in marine habitats. Fossil euglenids have been found in rocks over 410 million years old.

Morphology Most euglenids lack an external wall but have a unique system of interlocking protein molecules lying under the plasma membrane, which stiffen and support the cell.

Feeding and locomotion About one-third of the species have chloroplasts and perform photosynthesis, but most ingest bacteria and other small cells or particles. ✔ You should be able to explain how the observation of ingestive feeding in euglenids relates to the hypothesis that this lineage gained chloroplasts via secondary endosymbiosis. Instead of storing starch, the photosynthetic species synthesize a unique storage carbohydrate called paramylon. Some cells have a light-sensitive "eyespot" and use flagella to swim toward light (**Figure 29.22**). Other euglenids can inch along substrates using a unique form of movement based on extension of microtubules.

Reproduction Only asexual reproduction is known to occur among euglenids.

Human and ecological impacts Euglenids are important components of freshwater plankton and food chains.

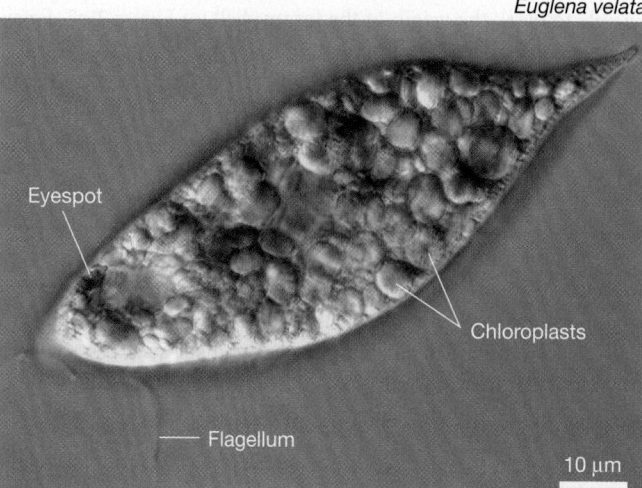

Euglena velata

FIGURE 29.22 **Euglenids Are Common in Ponds and Lakes.**

Plantae > Rhodophyta (Red Algae)

The 6000 species of red algae live primarily in marine habitats. One species lives over 200 m below the surface; another is the only eukaryote capable of living in acidic hot springs. Although their color varies, many species are red because their chloroplasts contain large amounts of the accessory pigment phycoerythrin, which absorbs strongly in the blue and green portions of the visible spectrum. Because blue light penetrates water better than other wavelengths, red algae are able to live at considerable depth in the oceans. ✔ You should be able to mark the origin of high phycoerythrin concentrations on Figure 29.7.

Morphology Red algae cells have walls that are composed of cellulose and other polymers. A few species are unicellular, but most are multicellular. Many of the multicellular species are filamentous, but others grow as thin, hard crusts on rocks or coral. Some species have erect, leaf-like structures called thalli (**Figure 29.23**). Some species have cells with many nuclei.

Feeding and locomotion The vast majority of red algae are photosynthetic, though a few parasitic species have been identified. Red algae are the only type of algae that lack flagella.

Reproduction Asexual reproduction occurs through production of spores by mitosis. Alternation of generations is common, but the types of life cycles observed in red algae are extremely variable.

Human and ecological impacts On coral reefs, some red algae become encrusted with calcium carbonate. These species contribute to reef building and help stabilize the entire reef structure. Cultivation of *Porphyra* (or nori) for sushi and other foods is a billion-dollar-per-year industry in East Asia.

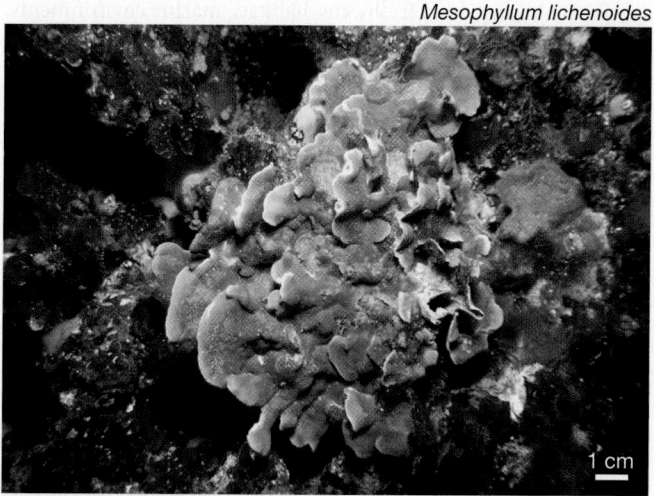

Mesophyllum lichenoides

FIGURE 29.23 **Red Algae Adopt an Array of Growth Forms.**

Foraminifera, or forams, got their name from the Latin *foramen*, meaning "hole." Forams produce shells that have holes (see **Figure 29.24**) through which pseudopodia emerge. ✔ You should be able to mark the origin of the foram shell on Figure 29.7. Fossil shells of foraminifera are abundant in marine sediments—there is a continuous record of fossilized forams that dates back 530 million years. A foram species was recently found living in sediments at a depth of 11,000 m below sea level. They are abundant in marine plankton as well as bottom habitats.

Morphology Foraminifera cells generally have multiple nuclei. The shells of forams are usually made of organic material stiffened with calcium carbonate ($CaCO_3$), and most species have several chambers. One species known from fossils was 12 cm long, but most species are much smaller. The size and shape of the shell are traits that distinguish foram species from each other.

Feeding and locomotion Like other rhizarians, forams feed by extending their pseudopodia and using them to capture and engulf bacterial and archaeal cells or bits of organic debris, which are digested in food vacuoles. Some species have symbiotic algae that perform photosynthesis and contribute sugars to their host. Forams simply float in the water.

Reproduction Asexual reproduction occurs by mitosis. When meiosis occurs, the resulting gametes are released into the open water, where pairs fuse to form a new individual.

Human and ecological impacts The shells of dead forams commonly form extensive sediment deposits when they settle out of the water, producing layers that eventually solidify into chalk, limestone, or marble. Geologists use the presence of certain foram species to date rocks—particularly during petroleum exploration.

Unidentified species

0.1 mm

FIGURE 29.24 Forams Are Shelled Amoebae.

Ciliates were named for the cilia that cover them. Some 12,000 species are known from freshwater habitats, marine environments, and wet soils.

Morphology Ciliata cells have two distinctive nuclei: a large macronucleus and a small micronucleus (both are stained dark red in **Figure 29.25**). The macronucleus is polyploid and is actively transcribed. The micronucleus is diploid and is involved only in gene exchange between individuals. A few ciliates secrete an external skeleton. ✔ You should be able to mark the origin of the macronucleus/micronucleus structure on Figure 29.7.

Feeding and locomotion Depending on the species, ciliates may be filter feeders, predators, or parasites. They use cilia to swim and have a mouth area where food is ingested.

Reproduction Ciliates divide to produce daughter cells asexually. They also undergo an unusual type of gene exchange called conjugation. During conjugation in ciliates, two cells line up side by side and physically connect. Micronuclei exchange between cells and fuse. The resulting nucleus eventually forms a new macronucleus and micronucleus.

Human and ecological impacts Ciliates are abundant in marine plankton and are important consumers. They are common in the digestive tracts of goats, sheep, cattle, and other grazers, where they feed on plant matter and help the host animal digest it.

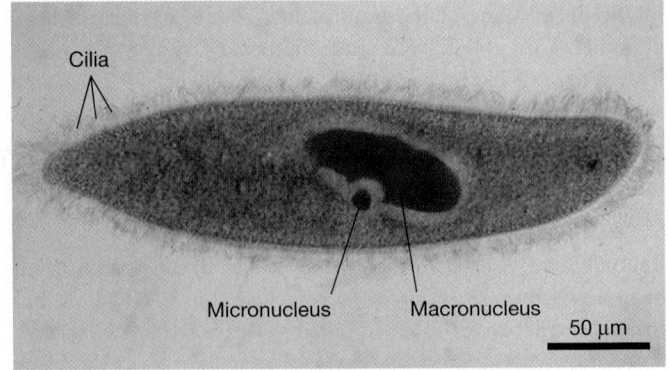

Paramecium caudatum

Cilia

Micronucleus   Macronucleus

50 μm

FIGURE 29.25 Ciliates Are Abundant in Freshwater Plankton.

Most of the 4000 known species of dinoflagellates are ocean-dwelling plankton, although they are abundant in freshwater as well. Some species are capable of **bioluminescence**, meaning they emit light via an enzyme-catalyzed reaction (**Figure 29.26**).

Morphology Most dinoflagellates are unicellular, although some live in the aggregations of individual cells called colonies. Each species has a distinct shape, in some maintained by plates of cellulose near the surface of the cell. Unlike chromosomes of other eukaryotes, chromosomes in dinoflagellates are attached to the nuclear envelope at all times and do not contain histones.

Feeding and locomotion About half of the dinoflagellates are photosynthetic. Some of the photosynthetic species live in association with corals or sea anemones, but most are planktonic. The other species are predatory or parasitic. Dinoflagellates are distinguished by the arrangement of their two flagella: One flagellum projects out from the cell while the other runs around the cell in a groove. The two flagella are perpendicular to each other. Cells swim in a spinning motion using the two flagella. ✔ You should be able to mark the origin of perpendicularly oriented flagella on Figure 29.7.

Reproduction Both asexual and sexual reproduction occur. Cells that result from sexual reproduction may form tough cysts. A cyst is a resistant structure that can remain dormant until environmental conditions improve.

Human and ecological impacts Photosynthetic dinoflagellates are important primary producers in marine ecosystems—second only to diatoms in the amount of carbon they fix per year. A few species are responsible for harmful algal blooms (see Figure 29.4).

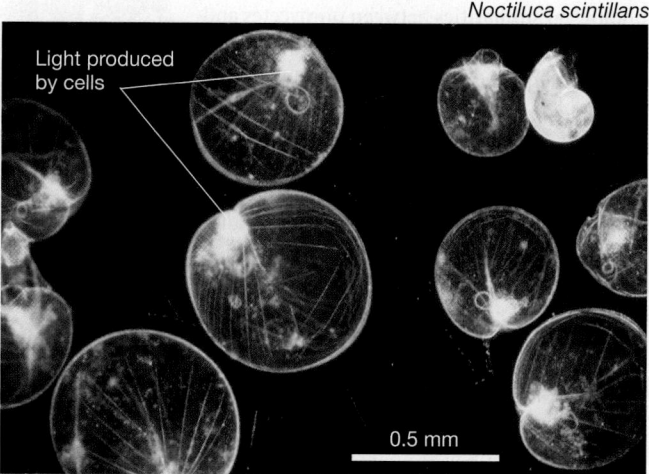

Noctiluca scintillans

Light produced by cells

0.5 mm

FIGURE 29.26 Some Dinoflagellate Species Are Bioluminescent.

All of the 5000 known species of Apicomplexa are parasitic. Species in the genus *Plasmodium* are well studied because they cause malaria in humans and other vertebrates.

Morphology Apicomplexa cells have a system of organelles at one end, called the apical complex, that is unique to the group. The apical complex allows the apicomplexan to penetrate the plasma membrane of its host. ✔ You should be able to mark the origin of the apical complex on Figure 29.7. The cells have a chloroplast-derived, nonphotosynthetic organelle with four membranes, indicating that apicomplexans descended from an ancestor that gained a chloroplast via secondary endosymbiosis.

Feeding and locomotion All absorb nutrition directly from their host. They lack cilia or flagella, but some species can move by amoeboid motion.

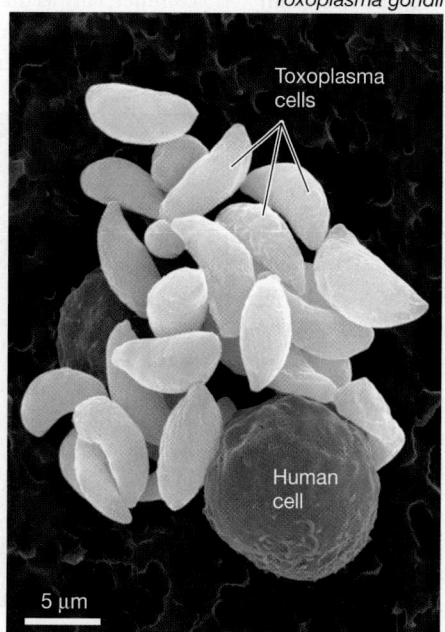

Toxoplasma gondii

Toxoplasma cells

Human cell

5 μm

FIGURE 29.27 *Toxoplasma* Causes Infections in AIDS patients.

Reproduction Apicomplexans can reproduce sexually or asexually. In some species, the life cycle involves two distinct hosts, and cells must be transmitted from one host to the next.

Human and ecological impacts Various species in the genus *Plasmodium* infect birds, reptiles, or mammals and cause malaria. *Toxoplasma* is an important pathogen in people infected with HIV (**Figure 29.27**), and chicken farmers estimate that they lose $600 million per year controlling the apicomplexan *Eimeria*. Apicomplexans that parasitize insects have been used to control pests.

Based on morphology, oomycetes were thought to be fungi and were given fungus-like names, such as downy mildew and water molds. They are readily distinguished from fungi at the DNA level, are primarily aquatic, have cellulose instead of chitin as the primary carbohydrate in the cell walls, and have gametes with stramenopile-like flagella. The morphological similarities between oomycetes and fungi result from convergent evolution, because both groups make their living absorbing nutrition from living or dead hosts. ✔ You should be able to mark the origin of the absorptive lifestyle in oomycetes on Figure 29.7.

Morphology Some oomycetes are unicellular and some form long, branching filaments called **hyphae**. Species that form hyphae often have multinucleate cells. Cells have walls containing cellulose.

Feeding and locomotion Most species feed on decaying organic material in freshwater environments; a few are parasitic. Mature individuals are **sessile**—that is, permanently fixed to a substrate.

Phytophthora infestans

Structures that produce spores

Potato leaf

50 μm

FIGURE 29.28 *Phytophthora infestans* **Infects Potatoes.** This is a colorized scanning electron micrograph.

Reproduction Most species are diploid throughout the majority of their life cycle. In aquatic species, spores that are produced via asexual or sexual reproduction have flagella and swim to find new food sources; in terrestrial species, spores swim in rainwater or are dispersed by the wind in dry conditions. Spores form in special structures, like those shown in **Figure 29.28**.

Human and ecological impacts Oomycetes are extremely important decomposers in aquatic ecosystems. Along with certain bacteria and archaea, they are responsible for breaking down dead organisms and releasing nutrients for use by other species. The parasitic species can be harmful to humans, however. The organism that caused the Irish potato famine was an oomycete. An oomycete parasite almost wiped out the French wine industry in the 1870s, and other species are responsible for epidemic diseases of trees, including the diebacks currently occurring in oaks found in Europe and the western United States, and in eucalyptus in Australia.

Diatoms are unicellular or form chains of cells. About 10,000 species have been described to date, but some researchers claim that millions of species exist and have yet to be discovered.

Morphology Diatom cells are supported by silicon-rich, glassy cell walls (**Figure 29.29**) that sometimes form a box-and-lid arrangement (see Figure 29.11a). ✔ You should be able to mark the origin of the glassy cell wall on Figure 29.7.

Feeding and locomotion Diatoms are photosynthetic. Only their sperm cells have flagella and are capable of powered movement. In many species, the adult cells float in the water. But species living on a surface can glide via microtubules that project from their cell walls and that move in response to motor proteins.

Reproduction Diatoms divide by mitosis to reproduce asexually or by meiosis to form gametes that fuse to form a new individual. Many species can produce spores that are dormant during unfavorable growing conditions.

Human and ecological impacts In abundance, diatoms dominate the plankton of cold, nutrient-rich waters. They are found in virtually all aquatic habitats and are considered the most important producer of carbon compounds in fresh and salt water. Their cell walls settle into massive accumulations that are mined and sold commercially as diatomaceous earth, which is used in filtering applications and as an ingredient that adds bulk to polishes, paint, cosmetics, and other products.

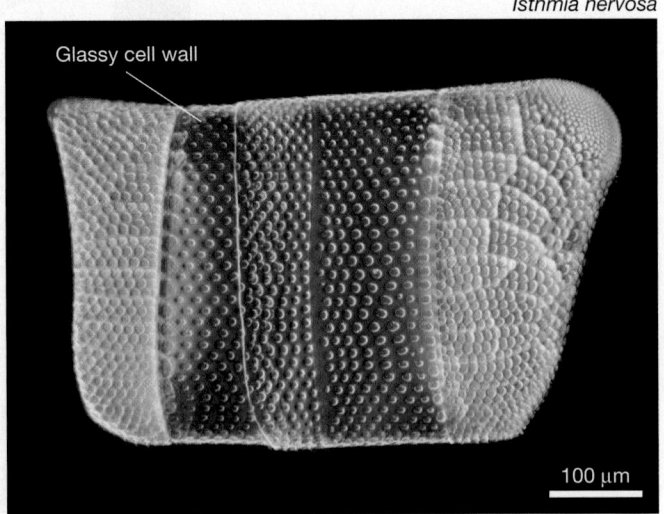

Isthmia nervosa

Glassy cell wall

100 μm

FIGURE 29.29 Diatoms Have Glass-Like Cell Walls.

The color of brown algae is due to their unique suite of photosynthetic pigments. Over 1500 species have been described, most of them living in marine habitats.

Morphology The walls of Phaeophyta cells contain cellulose in addition to other complex polymers. All species are multicellular, with the body typically consisting of leaflike blades, a stalk known as a stipe, and a rootlike holdfast, which attaches the individual to a substrate (**Figure 29.30**). ✔ You should be able to mark the origin of multicellularity and brown algal photosynthetic pigments on Figure 29.7.

Feeding and locomotion Brown algae are photosynthetic and sessile, though reproductive cells may have flagella and be motile—capable of locomotion.

Reproduction Sexual reproduction occurs via the production of swimming gametes, which fuse to form a zygote. Most species exhibit alternation of generations. Depending on the species, the gametophyte and sporophyte stages may look similar or different.

Human and ecological impacts In many coastal areas with cool water, brown algae form forests or meadows that are important habitats for a wide variety of animals. In the Sargasso Sea, off Bermuda, floating brown algae form extensive rafts that harbor an abundance of animal species. The compound algin is purified from kelp and used in the manufacture of cosmetics and paint.

Durvillaea species
— Blade
— Stipe
— Holdfast

5 cm

FIGURE 29.30 Many Brown Algae Have a Holdfast, Stalk, and Leaflike Blades.

CHAPTER 29 REVIEW

For media, go to the study area at www.masteringbiology.com

Summary of Key Concepts

⟐⟶ Protists are a paraphyletic grouping that includes all eukaryotes except the land plants, fungi, and animals. Biologists study them because they are important medically, ecologically, and evolutionarily.

- Parasitic protists cause several important diseases in humans, including malaria.

- Toxin-producing protists that grow to high densities result in a harmful algal bloom.

- Protists are often tremendously abundant in marine and freshwater plankton and other habitats. Protists provide food for many organisms in aquatic ecosystems and fix so much carbon that they have a large impact on the global carbon budget.

    ✔ You should be able to explain what is "primary" about primary production by photosynthetic protists.

⟐⟶ Key morphological innovations occurred as protists diversified: the nuclear envelope, multicellularity, and an array of structures that function in support and protection. In addition, the mitochondrion and chloroplast arose by endosymbiosis.

- Several morphological features are common to all or almost all eukaryotes, including the nucleus, endomembrane system, extensive cytoskeleton, and flagellum.

- Multicellularity evolved in several different protist groups independently.

- Structures that provide support or protection vary among lineages, and include cell walls, shells, internal support rods, and alveoli.

- Eukaryotes contain mitochondria or have genes indicating that their ancestors once contained mitochondria.

- Several types of data support the hypothesis that mitochondria originated as an endosymbiotic α-proteobacterium.

- The chloroplast's size, DNA structure, ribosomes, double membrane, and evolutionary relationships are consistent with the hypothesis that this organelle originated as an endosymbiotic cyanobacterium.

- After primary endosymbiosis occurred, chloroplasts were "passed around" to new lineages of protists via secondary endosymbiosis.

    ✔ You should be able to explain the original source of (**1**) the two membranes in a plant chloroplast, and (**2**) the four membranes in the chloroplast-derived organelle of *Plasmodium*.

- **Protists vary widely in the way they obtain food. Many species are photosynthetic; others obtain carbon compounds by ingesting food or parasitizing other organisms.**

- Protists exhibit predatory, parasitic, or photosynthetic lifestyles, which evolved in many groups independently.

- The evolution of ingestive feeding was important for two reasons: (1) It allowed eukaryotes to obtain resources in a new way—by eating bacteria, archaea, and other eukaryotes; and (2) it made endosymbiosis and the evolution of mitochondria and chloroplasts possible.

  ✔ You should be able to propose a hypothesis to explain why the wide diversity of photosynthetic pigments observed among protists limits competition.

- **Protists vary widely in terms of how they reproduce. Sexual reproduction evolved in protists, and many protist species can reproduce both sexually and asexually.**

- Protists undergo cell division based on mitosis and reproduce asexually.

- Many protists also undergo meiosis and sexual reproduction at some phase in their life cycle.

- Alternation of generations is common in multicellular species—meaning there are separate haploid and diploid forms of the same species. When alternation of generations occurs, haploid gametophytes produce gametes by mitosis; diploid sporophytes produce spores by meiosis.

  ✔ You should be able to explain how you can tell a gametophyte from a sporophyte, in terms of ploidy.

  (MB) **Web Activity** Alternation of Generations in a Protist

Questions

✔ TEST YOUR KNOWLEDGE

Answers are available in Appendix B

1. Why are protists considered paraphyletic?
   a. They include many extinct forms, including lineages that no longer have any living representatives.
   b. They include some but not all descendants of their most recent common ancestor.
   c. They represent all of the descendants of a single common ancestor.
   d. Not all protists have all of the synapomorphies that define the Eukarya, such as a nucleus.

2. What material is *not* used by protists to manufacture hard outer coverings?
   a. cellulose
   b. lignin
   c. glass-like compounds that contain silicon
   d. mineral-like compounds such as calcium carbonate ($CaCO_3$)

3. What does amoeboid motion result from?
   a. interactions among actin, myosin, and ATP
   b. coordinated beats of cilia
   c. the whiplike action of flagella
   d. action by the mitotic spindle, similar to what happens during mitosis and meiosis

4. According to the endosymbiosis theory, what type of organism is the original ancestor of the chloroplast?
   a. a photosynthetic archaean
   b. a cyanobacterium
   c. an algal-like, primitive photosynthetic eukaryote
   d. a modified mitochondrion

5. Multicellularity is defined in part by the presence of distinctive cell types. At the cellular level, what does this criterion imply?
   a. Individual cells must be extremely large.
   b. The organism must be able to reproduce sexually.
   c. Cells must be able to move.
   d. Different cell types express different genes.

6. Why are protists an important part of the global carbon cycle and marine food chains?
   a. They have high species diversity.
   b. They are numerically abundant.
   c. They have the ability to parasitize humans.
   d. They have the ability to undergo meiosis.

✔ TEST YOUR UNDERSTANDING

Answers are available in Appendix B

1. Why is an extensive cytoskeleton and the lack of a cell wall required for ingestive feeding by engulfing?

2. Explain the logic behind the claim that the nuclear envelope is a synapomorphy that defines eukaryotes as a monophyletic group.

3. What is the relationship between meiosis and the alternation of generations?

4. Consider the endosymbiosis theory for the origin of the mitochondrion. What did each partner provide the other, and what did each receive in return?

5. Why was finding a close relationship between mitochondrial DNA and bacterial DNA considered particularly strong evidence in favor of the endosymbiosis theory?

6. The text claims that the evolutionary history of protists can be understood as a series of morphological innovations that established seven distinct lineages, each of which subsequently diversified based on innovative ways of feeding, moving, and reproducing. Explain how the Alveolata support this claim.

1. Consider the following:
   - All living eukaryotes have mitochondria or have evidence in their genomes that they once had these organelles. Thus it appears that eukaryotes acquired mitochondria very early in their history.
   - The first eukaryotic cells in the fossil record correlate with the first appearance of rocks formed in an oxygen-rich ocean and atmosphere.

   How are these observations connected? (Hint: Before answering, glance at Chapter 9 and remind yourself what happens in a mitochondrion.)

2. Consider the following:
   - *Plasmodium* has an unusual organelle called an apicoplast. Recent research has shown that apicoplasts are derived from chloroplasts via secondary endosymbiosis and have a large number of genes encoded by chloroplast DNA.
   - Glyphosate is one of the most widely used herbicides. It works by poisoning an enzyme encoded by a gene in chloroplast DNA.
   - Biologists are testing the hypothesis that glyphosate could be used as an antimalarial drug in humans.

   How are these observations connected?

3. Suppose a friend says that we don't need to worry about global warming. Her claim is that increased temperatures will make planktonic algae grow faster and that carbon dioxide (CO_2) will be removed from the atmosphere faster. According to her, this carbon will be buried at the bottom of the ocean in calcium carbonate shells. As a result, the amount of carbon dioxide in the atmosphere will decrease and global warming will decline. Comment.

4. Biologists are beginning to draw a distinction between "species trees" and "gene trees." A species tree is a phylogeny that describes the actual evolutionary history of a lineage. A gene tree, in contrast, describes the evolutionary history of one particular gene, such as a gene required for the synthesis of chlorophyll *a*. In some cases, species trees and gene trees don't agree with each other. For example, the species tree for green algae indicates that their closest relatives are protists and plants. But the gene tree based on chlorophyll *a* from green algae suggests that this gene's closest relative is a bacterium, not a protist. What's going on? Why do these types of conflicts exist?

Earth has been called the Blue Planet, but it would be just as accurate to call it the Green Planet.

30 Green Algae and Land Plants

KEY CONCEPTS

- Green algae are an important source of oxygen and provide food for aquatic organisms; land plants release oxygen, hold soil and water in place, build soil, moderate extreme temperatures and winds, and provide food for terrestrial organisms.

- Land plants were the first multicellular organisms that could live with most of their tissues exposed to the air. A series of key adaptations allowed them to survive on land. In terms of total mass, plants dominate today's terrestrial environments.

- Once plants were able to grow on land, a sequence of important evolutionary changes made it possible for them to reproduce efficiently—even in extremely dry environments.

I n terms of their total mass and their importance to other organisms, the green plants dominate terrestrial and freshwater habitats. When you walk through a forest or meadow, you are surrounded by green plants. If you look at pond or lake water under a microscope, green plants are everywhere.

The green plants comprise two major types of organisms: the green algae and the land plants. Green algae are important photosynthetic organisms in aquatic habitats—particularly lakes, ponds, and other freshwater settings. Land plants are the key photosynthesizers in terrestrial environments.

Although green algae have traditionally been considered protists, it is logical to study them along with land plants for two reasons: (**1**) They are the closest living relative to land plants and form a monophyletic group with them, and (**2**) the transition from aquatic to terrestrial life occurred when land plants evolved from green algae.

Land plants were the first organisms that could thrive with their tissues completely exposed to the air instead of being partially or completely submerged. Before they evolved, it is likely that the only life on the continents consisted of bacteria, archaea, and single-celled protists that thrive in wet soils. By colonizing the continents, plants transformed the nature of life on Earth. It was, in the words of Karl Niklas, "one of the greatest adaptive events in the history of life." Land plants made the Earth green.

30.1 Why Do Biologists Study the Green Algae and Land Plants?

Biologists study the green plants because they are fascinating, and because we couldn't live without them. Along with most other animals and fungi, humans are almost completely dependent on land plants for food. People rely on plants for other necessities of life as well—oxygen, fuel, building materials, and the fibers used in making clothing,

✔ When you see this checkmark, stop and test yourself. Answers are available in Appendix B.

paper, ropes, and baskets. But we also rely on land plants for important intangible value such as the aesthetic appeal of landscaping and bouquets. To drive this point home, consider that the sale of cut flowers generates over $1 billion each year in the United States alone.

Based on these observations, it is not surprising that agriculture, forestry, and horticulture are among the most important endeavors supported by biological science. Tens of thousands of biologists are employed in research designed to increase the productivity of plants and to create new ways of using them for the benefit of people. Research programs also focus on two types of land plants that cause problems for people: weeds that decrease the productivity of crop plants and newly introduced species that invade and then degrade natural areas.

Plants Provide Ecosystem Services

An **ecosystem** consists of all the organisms in a particular area, along with physical components of the environment such as the atmosphere, precipitation, surface water, sunlight, soil, and nutrients. Green algae and land plants provide **ecosystem services** because they enhance the life-supporting attributes of the atmosphere, surface water, soil, and other physical components of an ecosystem. Green plants alter the environment in ways that benefit many other organisms:

PLANTS PRODUCE OXYGEN   Recall from Chapter 10 that plants perform oxygenic (literally, "oxygen-producing") photosynthesis. In this process, electrons that are removed from water molecules are used to reduce carbon dioxide (CO_2). In the process of stripping electrons from water, plants release oxygen molecules (O_2) as a by-product.

As Chapter 28 noted, oxygenic photosynthesis evolved in cyanobacteria and was responsible for the origin of an oxygen-rich atmosphere. The evolutionary success of plants continued this trend because plants add huge amounts of oxygen to the atmosphere.

Without the green plants, we and other aerobic organisms would be in danger of suffocating for lack of oxygen.

PLANTS BUILD AND HOLD SOIL   Leaves and roots and stems that are not eaten when they are alive fall to the ground and provide food for worms, fungi, bacteria, archaea, protists, and other decomposers in the soil. These organisms add organic matter to the soil, which improves soil structure and the ability of soils to hold nutrients and water.

The extensive network of fine roots produced by trees, grasses, and other land plants helps hold soil particles in place. And by taking up nutrients in the soil, plants prevent the nutrients from being blown or washed away.

When areas are devegetated by grazing, farming, logging, or suburbanization, large quantities of soil and nutrients are lost to erosion by wind and water (**Figure 30.1**).

PLANTS HOLD WATER AND MODERATE CLIMATE   Plant tissues take up and retain water. Intact forests, prairies, and wetlands also prevent rain from quickly running off a landscape: Plant leaves

FIGURE 30.1 Devegetation Leads to Erosion.

soften the physical impact of rainfall on soil, and plant organic matter builds the soil's water-holding capacity.

When areas are devegetated, streams are more prone to flooding and groundwater is not replenished efficiently. It is common to observe streams alternately flooding and then drying up completely when the surrounding area is deforested.

By providing shade, plants reduce temperatures beneath them and increase relative humidity. They also reduce the impact of winds that dry out landscapes.

When plants are removed from landscapes to make way for farms or suburbs, habitats become much dryer and are subject to more extreme temperature swings.

PLANTS AS PRIMARY PRODUCERS   Land plants are the dominant primary producers in terrestrial ecosystems. As Chapter 29 indicated, primary producers convert energy in sunlight into chemical energy. The sugars that land plants produce by photosynthesis support virtually all of the other organisms present in terrestrial habitats.

Plants are eaten by herbivores ("plant-eaters"), which range in size from insects to elephants. These consumers are eaten by carnivores ("meat-eaters"), ranging in size from the tiniest spiders to polar bears. Humans are an example of omnivores ("all-eaters"), organisms that eat both plants and animals. Omnivores feed at several different levels in the food chain. For example, people consume plants, herbivores such as chicken and cattle, and carnivores such as salmon and tuna.

Because of their role in primary production, land plants are the key to the carbon cycle on the continents. Plants take CO_2 from the atmosphere and reduce it to make sugars. Although green algae and land plants also produce a great deal of CO_2 as a result of cellular respiration, they fix much more CO_2 than they release.

The loss of plant-rich prairies and forests, due to suburbanization or conversion to agriculture has contributed to increased concentrations of CO_2 in the atmosphere. Higher carbon dioxide levels, in turn, are responsible for the rapid warming that is occurring worldwide (see Chapter 54).

Plants Provide Humans with Food, Fuel, Fiber, Building Materials, and Medicines

It is difficult to overstate the importance of plant research to the well-being of human societies. Plants provide our food supply as well as a significant percentage of the fuel, fibers, building materials, and medicines that we use.

FOOD Agricultural research began with the initial domestication of crop plants, which occurred independently at several locations around the world between 10,000 and 2000 years ago (**Figure 30.2a**). By actively selecting individuals with the largest and most nutritious seeds, leaves, or other plant parts year after year, our ancestors gradually changed the characteristics of certain wild species. This process is called **artificial selection**.

Figure 30.2b compares kernel size in modern maize and its wild ancestor—a grass called teosinte that is native to Mexico. Chapter 1 presented data indicating that artificial selection has been responsible for dramatic increases in the oil content of maize kernels over the past 100 years. Chapter 19 highlighted a current focus in agricultural research—the improvement of crop varieties through genetic engineering.

(a) Plants were domesticated at an array of locations.

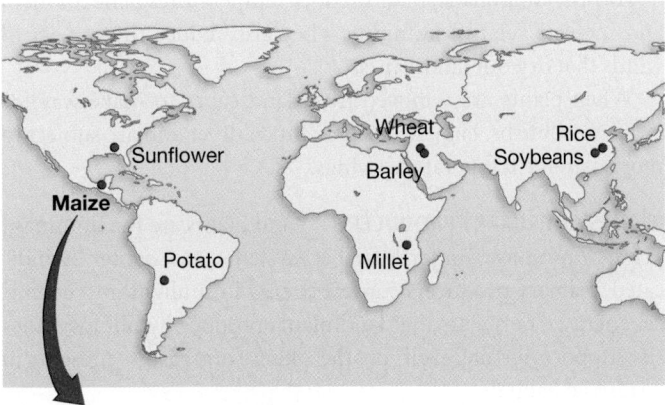

(b) Artificial selection changes the traits of domesticated species.

FIGURE 30.2 Crop Plants Are Derived from Wild Species via Artificial Selection. (a) Crop plants have originated on virtually every continent. **(b)** In some cases, artificial selection has changed domesticated species so radically that they bear little resemblance to their wild ancestors.

FUEL For perhaps 100,000 years, wood burning was the primary source of energy used by all humans. As **Figure 30.3** shows, however, wood has been replaced in the industrialized countries by other sources of energy. This graph shows the percentage of total energy needs in the United States fulfilled by wood, coal, and petroleum or natural gas, over time. Note that the first fuel to replace wood in the United States and many other industrialized countries has been coal, which forms from partially decayed plant material that is compacted over time by overlying sediments and hardened into rock.

Starting in the mid-1800s, people in England, Germany, and the United States began to mine coal deposits that originally formed during the Carboniferous period some 350–275 million years ago. The coal fueled blast furnaces that smelted vast quantities of iron ore into steel and powered the steam engines that sent trains streaking across Europe and North America. It is no exaggeration to claim that the organic compounds synthesized by plants more than 300 million years ago laid the groundwork for the Industrial Revolution.

Coal still supplies about 20 percent of the energy used in Japan and Western Europe and 80 percent of the energy used in China. Current research is focused on finding cleaner and more efficient ways of mining and burning coal.

FIBER AND BUILDING MATERIALS Although nylon and polyester derived from petroleum are increasingly important in manufacturing, cotton and other types of plant fibers are still important sources of raw material for clothing, rope, and household articles like towels and bedding.

Woody plants also provide lumber for houses and furniture. Relative to its density, wood is a stiffer and stronger building material than concrete, cast iron, aluminum alloys, or steel.

Woody plants also provide most of the fibers used in papermaking. The cellulose fibers refined from trees or bamboo and then used in paper manufacturing are stronger under tension (pulling) than nylon, silk, chitin, collagen, tendon, or bone—even though cellulose is 25 percent less dense. One line of research is focused on bioengineering reduced lignin content of

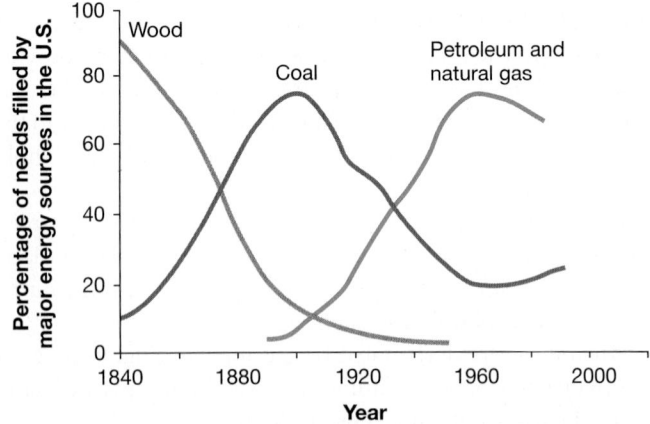

FIGURE 30.3 Changes over Time in Fuel Sources Used by an Industrialized Country. Although wood has declined in importance in the industrialized countries, it is still the primary cooking and heating fuel in many areas of the world.

TABLE 30.1 Some Drugs Derived from Land Plants

Compound	Source	Use
Atropine	Belladonna plant	Dilating pupils during eye exams
Codeine	Opium poppy	Pain relief, cough suppressant
Digitalin	Foxglove	Heart medication
Ipecac	Ipecac	Treating amoebic dysentery, poison control
Menthol	Peppermint	Cough suppressant, relief of stuffy nose
Morphine	Opium poppy	Pain relief
Papain	Papaya	Reduce inflammation, treat wounds
Quinine	Quinine tree	Malaria prevention
Quinidine	Quinine tree	Heart medication
Salicin	Aspen, willow trees	Pain relief (aspirin)
Steroids	Wild yams	Precursor compounds for manufacture of birth control pills and cortisone (to treat inflammation)
Taxol	Pacific yew	Ovarian cancer
Tubocurarine	Curare vine	Muscle relaxant used in surgery
Vinblastine, vincristine	Rosy periwinkle	Leukemia (cancer of blood)

certain woody species, so fewer pollutants need to be used in extracting lignin during the papermaking process.

PHARMACEUTICALS In both traditional and modern medicine, plants are a key source of drugs. **Table 30.1** lists some of the more familiar medicines derived from land plants; overall, it has been estimated that about 25 percent of the prescriptions written in the United States each year include at least one molecule derived from plants.

In most cases, plants synthesize these compounds in order to repel insects, deer, or other types of herbivores. For example, experiments have confirmed that morphine, cocaine, nicotine, caffeine, and other toxic compounds found in plants are effective deterrents to insect or mammalian consumers. Researchers continue to isolate and test new plant compounds for medicinal use in humans and domesticated animals.

30.2 How Do Biologists Study Green Algae and Land Plants?

Given the importance of plants to the planet in general and humans in particular, it is not surprising that knowing as much as possible about plants, including how they evolved, is a key component of biological science. To understand how green plants originated and diversified, biologists analyze (1) morphological traits, (2) the fossil record, and (3) phylogenetic trees estimated from similarities and differences in DNA sequences from homologous genes.

The three approaches are complementary and have produced a remarkably clear picture of how land plants evolved from green algae and then diversified. Let's consider each of these research strategies.

Analyzing Morphological Traits

The green algae include species that are unicellular, colonial, or multicellular and that live in marine, freshwater, or moist terrestrial habitats. The vast majority are aquatic. Although some land plants live in ponds or lakes or rivers, the vast majority live on land.

SIMILARITIES BETWEEN GREEN ALGAE AND LAND PLANTS The green algae have long been hypothesized to be closely related to land plants, because key morphological traits are similar in the two groups.

- Their chloroplasts contain the photosynthetic pigments chlorophyll *a* and *b* and the accessory pigment β-carotene.

- They have similar arrangements of the internal, membrane-bound sacs called thylakoids (see Chapter 10).

- Their cell walls, sperm, and peroxisomes are similar in structure and composition. (Recall from Chapter 7 that peroxisomes are organelles in which specialized oxidation reactions take place.)

- Their chloroplasts synthesize starch as a storage product.

Of all the green algal groups, the two most similar to land plants are the Coleochaetophyceae (coleochaetes) and Charophyceae (stoneworts; **Figure 30.4** on page 550). Because the species that make up these groups are multicellular and live in ponds and other types of freshwater environments, biologists hypothesize that land plants evolved from multicellular green algae that lived in freshwater habitats.

MAJOR MORPHOLOGICAL DIFFERENCES AMONG LAND PLANTS Based on morphology, the land plants were traditionally clustered into three broad categories:

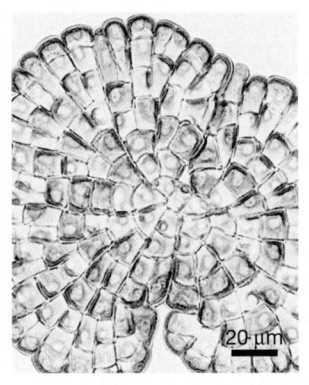

Coleochaetophyceae (coleochaetes)   Charophyceae (stoneworts)

FIGURE 30.4 Most Green Algae Are Aquatic. Examples of species from the green algal lineages most closely related to the land plants.

1. Non-vascular plants (**bryophytes**) lack **vascular tissue**—specialized groups of cells that conduct water and nutrients from one part of the plant body to another. Mosses are bryophytes (**Figure 30.5a**).

2. Seedless vascular plants have vascular tissue but do not make seeds. A **seed** consists of an embryo and a store of nutritive tissue, surrounded by a tough protective layer. Ferns are seedless vascular plants (**Figure 30.5b**).

3. Seed plants have vascular tissue. The flowering plants, or **angiosperms** (literally, "encased seeds"), are seed plants (**Figure 30.5c**).

How are non-vascular plants, seedless vascular plants, and seed plants related to each other, and to green algae?

Using the Fossil Record

The first green plants in the fossil record are green algae in 700–725 million-year-old rocks. The first land plant fossils are found in rocks that are about 475 million years old. Because green algae appear long before land plants, the data support the hypothesis that land plants are derived from green algae.

At roughly the same time that green algae appeared and began to diversify, the oceans and atmosphere were starting to become oxygen-rich—as never before in Earth's history. Based on this time correlation, it is reasonable to hypothesize that the evolution of green algae contributed to the rise of oxygen levels on Earth. The origin of the oxygen atmosphere occurred not long before the appearance of animals in the fossil record and may have played a role in their origin and early diversification.

The fossil record of the land plants themselves is massive. In an attempt to organize and synthesize the database, **Figure 30.6** breaks it into five time intervals—each encompassing a major event in the diversification of land plants.

ORIGIN OF LAND PLANTS The oldest interval begins about 475 million years ago (mya), spans some 60 million years, and documents the origin of the group.

Most of the fossils dating from this period are microscopic. They consist of the reproductive cells called spores and sheets of a waxy coating called cuticle. Several observations support the hypothesis that these fossils came from green plants that were growing on land.

1. Cuticle is a watertight barrier that coats today's land plants and helps them resist drying.

2. The fossilized spores are surrounded by a sheetlike coating. Under the electron microscope, the coating material appears almost identical in structure to a watertight material called **sporopollenin**, which encases spores and pollen from modern land plants and helps them resist drying.

3. Fossilized spores that are 475 million years old have recently been found in association with spore-producing structures, called **sporangia** (singular: **sporangium**), that are similar to sporangia observed in some non-vascular plants.

SILURIAN-DEVONIAN EXPLOSION The second major interval in the fossil record of land plants is called the "Silurian-Devonian explosion." In rocks dated 416–359 mya, biologists find fossils from most of the major plant lineages. Virtually all of the adapta-

(a) Non-vascular plants   **(b)** Seedless vascular plants   **(c)** Seed plants

Do not have vascular tissue to conduct water and provide support (for example, mosses)

Have vascular tissue but do not make seeds (for example, ferns)

Have vascular tissue and make seeds (for example, flowering plants, or angiosperms)

FIGURE 30.5 Morphological Diversity in Land Plants.

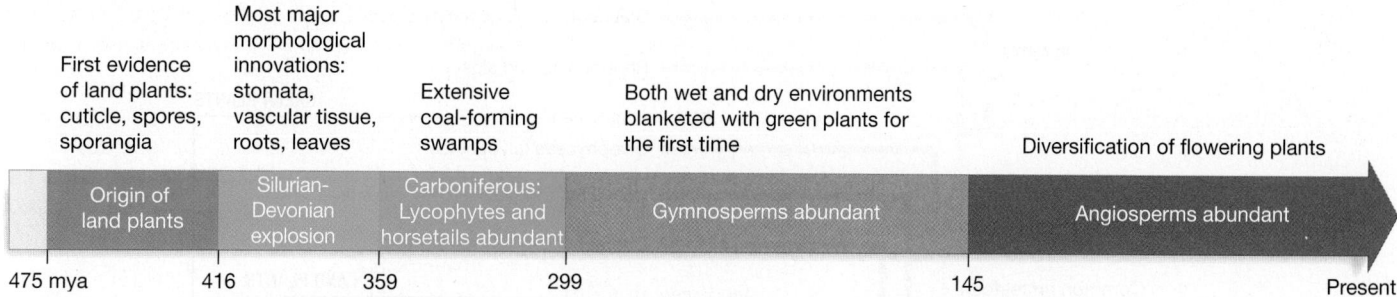

FIGURE 30.6 The Fossil Record of Land Plants Can Be Broken into Five Major Intervals. Note that the first insects are found in the fossil record at about 400 mya; the first terrestrial vertebrate animals at 365 mya, and the first mammals at 190 mya.

tions that allow plants to occupy dry, terrestrial habitats are present, including water-conducting tissue and roots.

According to the fossil record, plants colonized the land in conjunction with fungi that grew in a mutually beneficial association. The fungi grew belowground and helped provide land plants with nutrients from the soil; in return, the plants provided the fungi with sugars and other products of photosynthesis (see Chapter 31).

THE CARBONIFEROUS PERIOD   The third interval in the fossil history of plants spans the aptly named Carboniferous period. In sediments dated from about 359 to 299 mya, biologists find extensive deposits of coal. Coal is a carbon-rich rock packed with fossil spores, branches, leaves, and tree trunks.

Most of these fossils are derived from the Lycophyta (the lycophytes or club mosses), Equisetophyta (horsetails), and ferns. Although the only living lycophytes and horsetails are small, during the Carboniferous these groups were species rich and included a wide array of tree-sized forms.

Because coal formation is thought to start only in the presence of water, the Carboniferous fossils indicate the presence of extensive forested swamps.

DIVERSIFICATION OF GYMNOSPERMS   The fourth interval in land plant history is characterized by seed plants called **gymnosperms** ("naked-seeds"). Five major groups of gymnosperms are living today: Cycadophyta (cycads), Ginkgophyta (ginkgoes), Gnetophyta (gnetophytes), Pinophyta (pines, spruces, and firs), and other cone-bearing species (redwoods, junipers, yews).

Because gymnosperms grow readily in dry habitats, biologists infer that both wet and dry environments on the continents became blanketed with green plants for the first time during this interval. Gymnosperms are particularly prominent in the fossil record from 251 mya to 145 mya.

DIVERSIFICATION OF FLOWERING PLANTS   The fifth interval in the history of land plants is still under way. This is the age of flowering plants—the angiosperms. The first flowering plants in the fossil record appear about 150 mya. The plants that produced the first flowers are the ancestors of today's grasses, orchids, daisies, oaks, maples, and roses.

According to the fossil record, then, the green algae appear first, followed by the non-vascular plants, seedless vascular plants, and seed plants. Organisms that appear late in the fossil record are often much less dependent on moist habitats than are groups that appear earlier. For example, the sperm cells of mosses and ferns swim to accomplish fertilization, while gymnosperms and angiosperms produce pollen grains that are transported via wind or insects and that then produce sperm.

These observations support the hypotheses that green plants evolved from green algae and that in terms of habitat use, the evolution of green plants occurred in a wet-to-dry trend.

To test the validity of these observations, biologists analyze data sets that are independent of the fossil record. Foremost among these are DNA sequences used to infer phylogenetic trees. Does the phylogeny of land plants confirm or contradict the patterns in the fossil record?

Evaluating Molecular Phylogenies

The phylogenetic tree in **Figure 30.7** on page 552 is a recent version of results emerging from laboratories around the world. The black bars across some branches show when key innovations occurred.

Each of the following bullets states a key observation about this tree, followed by a highlighted interpretation—often, a hypothesis that is supported by the observation.

- The green plants are monophyletic.
  Interpretation: A single common ancestor gave rise to all of the green algae and land plants.

- The initial splitting events on the tree, near the root, lead to lineages of green algae.
  Interpretation: Land plants evolved from green algae.

- Green algae are paraphyletic.
  Interpretation: The green algae include some but not all of the descendants of a single common ancestor.

- Charophyceae are the closest living relative to land plants.
  Interpretation: Land plants evolved from a multicellular green alga that lived in freshwater habitats.

- Land plants are monophyletic.
  Interpretation: There was only one transition from freshwater environments to land.

- The bryophytes or non-vascular plants are the earliest-branching groups among land plants.

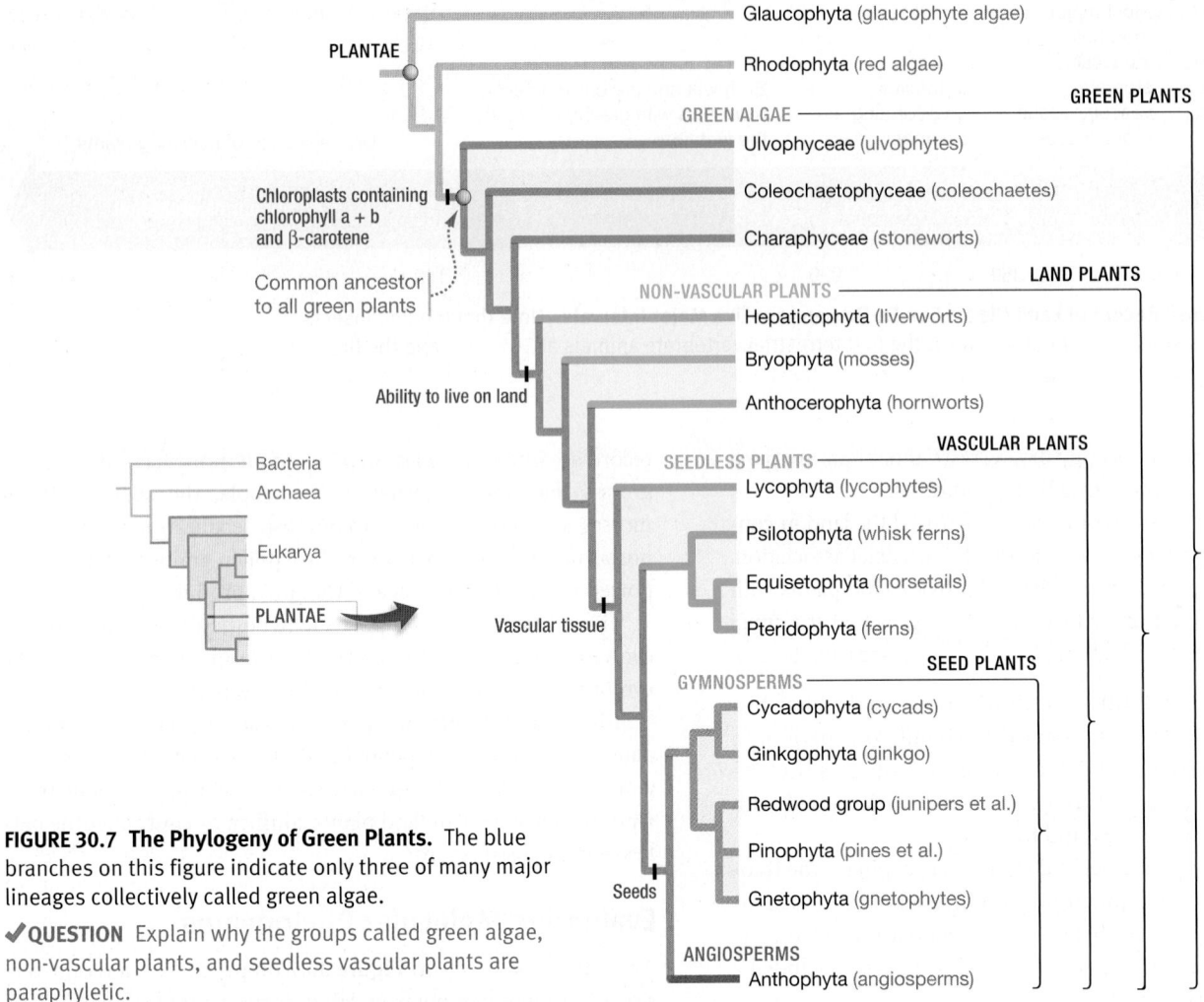

FIGURE 30.7 The Phylogeny of Green Plants. The blue branches on this figure indicate only three of many major lineages collectively called green algae.

✔**QUESTION** Explain why the groups called green algae, non-vascular plants, and seedless vascular plants are paraphyletic.

Interpretation: The non-vascular plants are the most ancient living group of plants.

- The non-vascular plants form a grade—meaning a sequence of lineages.
Interpretation: The non-vascular plants are paraphyletic.

- The seedless vascular plants form a grade, but the vascular plants as a whole are monophyletic.
Interpretation: Vascular tissue evolved once.

- The seed plants—the gymnosperms plus angiosperms—are monophyletic.
Interpretation: The seed evolved once.

- The gymnosperms are a monophyletic group, as are the angiosperms.
Interpretation: Among seed plants, there was a major divergence in how seeds develop—either "naked" (in gymnosperms) or protected inside a capsule (in angiosperms).

Although the tree in Figure 30.7 will undoubtedly change and improve as additional data accumulate, biologists are confident about its most fundamental message: The fossil record and the phylogenetic tree based on DNA sequence data agree on the order in which groups appeared. Land plant evo-lution began with non-vascular plants, proceeded to seedless vascular plants, and continued with the evolution of seed plants.

CHECK YOUR UNDERSTANDING

🔑 **If you understand that . . .**

- Biologists use the fossil record and phylogenetic analyses to study how green plants diversified.
- The data analyzed to date support the hypotheses that green plants are monophyletic and that land plants evolved from multicellular green algae that inhabited freshwater.
- The fossil record and molecular analyses agree that the non-vascular plants evolved first, followed by the seedless vascular plants and the seed plants.

✔ **You should be able to . . .**

Explain why (1) morphological data, (2) the fossil record, and (3) molecular phylogenies all support the hypothesis that land plants evolved from green algae.

Answers are available in Appendix B.

30.3 What Themes Occur in the Diversification of Land Plants?

Land plants have evolved from algae that grew on the muddy shores of ponds 475 million years ago to organisms that enrich the soil, produce much of the oxygen you breathe and most of the food you eat, and serve as symbols of health, love, and beauty. How did this happen?

Answering this question begins with recognizing the most striking trend in the phylogeny and fossil record of green plants: The most ancient groups in the lineage are dependent on wet habitats, while more recently evolved groups are tolerant of dry—or even desert—conditions. The story of land plants is the story of adaptations that allowed photosynthetic organisms to move from aquatic to terrestrial environments.

Let's first consider adaptations that allowed plants to grow in dry conditions without drying out and dying, and then analyze the evolution of traits that allowed plants to reproduce efficiently on land. This section closes with a brief look at the radiation of flowering plants, which are the most important plants in many of today's terrestrial environments.

The Transition to Land, I: How Did Plants Adapt to Dry Conditions?

For aquatic green algae, terrestrial environments are deadly. Compared with a habitat in which the entire organism is bathed in fluid, in terrestrial environments only a portion, if any, of the plant's tissues are in direct contact with water. Tissues that are exposed to air tend to dry out and die.

Once green plants made the transition to survive out of water, though, growth on land offered a bonanza of resources.

- *Light*   The water in ponds, lakes, and oceans absorbs and reflects light. As a result, the amount of light available to drive photosynthesis is drastically reduced even a meter or two below the water surface.

- *Carbon dioxide* CO_2—the most important molecule required by photosynthetic organisms—is more abundant in the atmosphere and diffuses more readily there than it does in water

Natural selection favored early land plants with adaptations that solved the drying problem. These adaptations arose in two steps: (1) preventing water loss, which kept cells from drying out and dying; and (2) moving water from tissues with direct access to water to tissues without direct access. Let's examine each in turn.

PREVENTING WATER LOSS: CUTICLE AND STOMATA   Section 30.2 pointed out that sheets of the waxy substance called cuticle are present early in the fossil record of land plants, along with encased spores. This observation is significant because the presence of cuticle in fossils is a diagnostic indicator of land plants.

Cuticle is a watertight sealant that covers the aboveground parts of plants and gives them the ability to survive in dry environments (**Figure 30.8a**). If biologists had to point to one innovation that made the transition to land possible, it would be the random mutations that led to the production of cuticle.

Covering surfaces with wax creates a problem, however, regarding the exchange of gases across those surfaces. Plants need to take in carbon dioxide (CO_2) from the atmosphere in order to perform photosynthesis. But cuticle is almost as impervious to CO_2 as it is to water.

Most modern plants solve this problem with a structure called a **stoma** ("mouth"; plural: **stomata**). A stoma consists of an opening surrounded by specialized **guard cells** (**Figure 30.8b**). The opening,

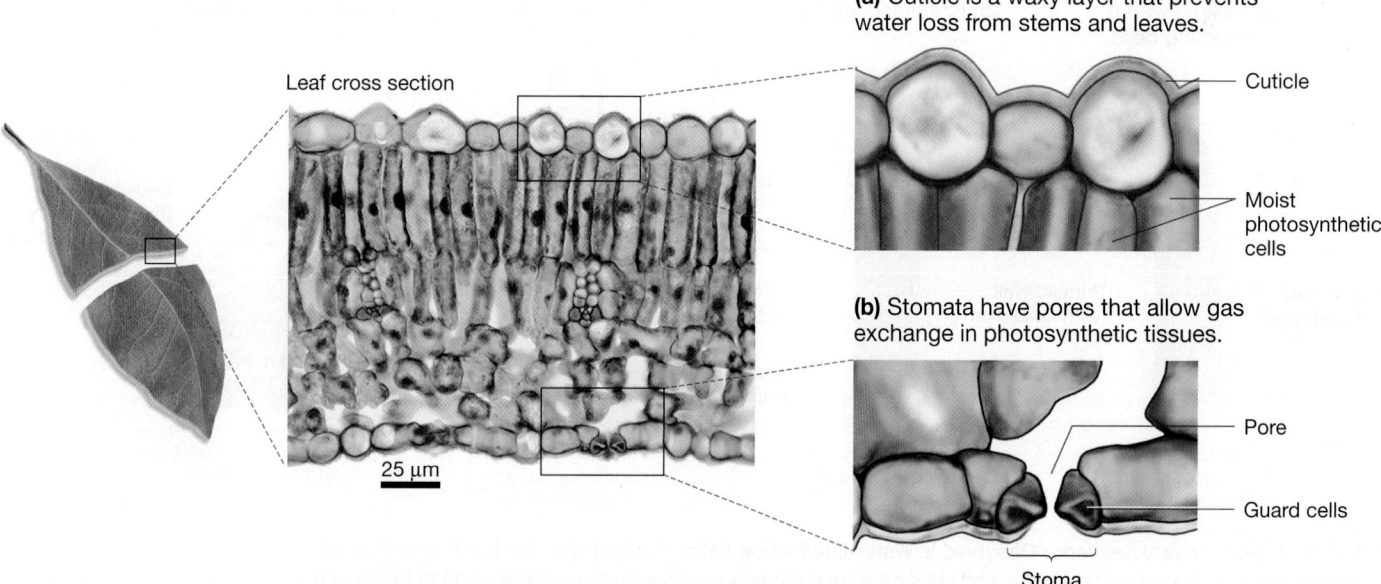

(a) Cuticle is a waxy layer that prevents water loss from stems and leaves.

Leaf cross section

25 μm

Cuticle

Moist photosynthetic cells

(b) Stomata have pores that allow gas exchange in photosynthetic tissues.

Pore

Guard cells

Stoma

FIGURE 30.8 Cuticle and Stomata Are the Most Fundamental Plant Adaptations to Life on Land. In these micrographs, leaf cells have been stained blue to make their structure more visible. **(a)** The interior of plant leaves and stems is extremely moist; cuticle prevents water from evaporating away. **(b)** Stomata create pores to allow CO_2 to diffuse into the interior of leaves and stems where cells are actively photosynthesizing, and to allow excess O_2 to diffuse out.

called a **pore**, opens or closes as the guard cells change shape. When guard cells become flaccid or "limp" due to an outflow of water, they close the stomata. Pores are closed in this way to limit water loss from the plant. But when guard cells become turgid or "taut" due to an inflow of water, they open the pore. Open stomata allow CO_2 to diffuse into the interior of leaves and stems where cells are actively photosynthesizing. They also allow excess O_2 to diffuse out. (The mechanism behind guard-cell movement is explored in Chapter 39.)

Stomata are present in all land plants except the liverworts, which have pores but no guard cells. These data suggest that the earliest land plants evolved pores that allowed gas exchange to occur at breaks in the cuticle-covered surface. Later, the evolution of guard cells gave land plants the ability to regulate gas exchange—and control water loss—by opening and closing their pores.

THE IMPORTANCE OF UPRIGHT GROWTH   Once cuticle and stomata had evolved, plants could keep from drying out and thus keep photosynthesizing while exposed to air. Cuticle and stomata allowed plants to grow on the saturated soils of lake or pond edges. The next challenge? Defying gravity.

Multicellular green algae can grow erect because they float. They float because the density of their cells is similar to water's density. But outside of water, the body of a multicellular green alga collapses. The water that fills its cells is 1000 times denser than air. Although the cell walls of green algae are strengthened by the presence of cellulose, their bodies lack the structural support to withstand the force of gravity and to keep an individual erect in air.

Based on these observations, biologists hypothesize that the first land plants were small or had a low, sprawling growth habit. Besides lacking rigidity, the early land plants would have had to obtain water through pores or through a few cells that lacked cuticle—meaning they would have had to grow in a way that kept many or most of their tissues in direct contact with moist soil. If this hypothesis is correct, then competition for space and light would have become intense soon after the first plants began growing on land.

In a terrestrial environment, individuals that can grow erect have much better access to sunlight than individuals that are incapable of growing erect. But two problems have to be overcome for a plant to grow erect: (**1**) transporting water from tissues that are in contact with wet soil to tissues that are in contact with dry air, against the force of gravity; and (**2**) becoming rigid enough to avoid falling over in response to gravity and wind. As it turns out, vascular tissue solved both problems.

THE ORIGIN OF VASCULAR TISSUE   Paul Kenrick and Peter Crane explored the origin of water-conducting cells and erect growth in plants by examining the extraordinary fossils found in a rock formation in Scotland called the Rhynie Chert. These rocks formed about 400 million years ago and contain some of the first large plant specimens in the fossil record—as opposed to the microscopic spores and cuticle found in older rocks.

The Rhynie Chert contains numerous plants that fossilized in an upright position. This indicates that many or most of the Rhynie plants grew erect. How did they stay vertical?

By examining fossils with the electron microscope, Kenrick and Crane established that species from the Rhynie Chert contained elongated cells that were organized into tissues along the length of the plant. Based on these data, the biologists hypothesized that the elongated cells were part of water-conducting tissue and that water could move from the base of the plants upward to erect portions through these specialized water-conducting cells.

- Some of the fossilized water-conducting cells had simple, cellulose-containing cell walls like the water-conducting cells found in today's mosses (**Figure 30.9a**).

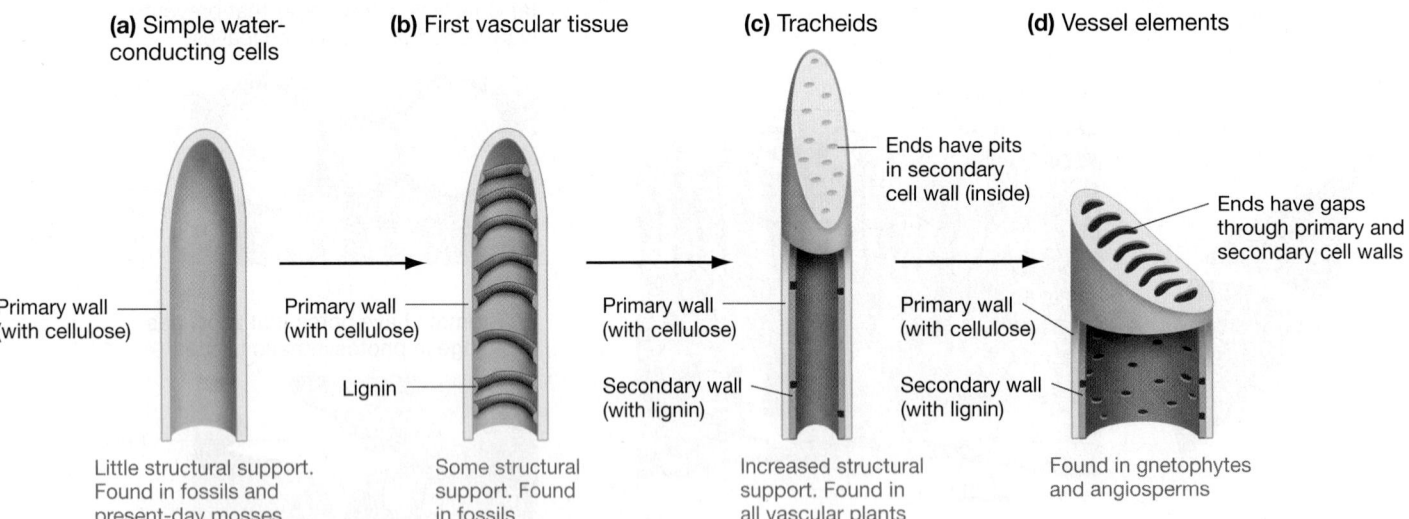

(a) Simple water-conducting cells   **(b)** First vascular tissue   **(c)** Tracheids   **(d)** Vessel elements

Primary wall (with cellulose)

Primary wall (with cellulose)

Lignin

Primary wall (with cellulose)

Secondary wall (with lignin)

Ends have pits in secondary cell wall (inside)

Primary wall (with cellulose)

Secondary wall (with lignin)

Ends have gaps through primary and secondary cell walls

Little structural support. Found in fossils and present-day mosses

Some structural support. Found in fossils

Increased structural support. Found in all vascular plants

Found in gnetophytes and angiosperms

FIGURE 30.9 Evolutionary Sequence Observed in Water-Conducting Cells. According to the fossil record and the phylogeny of green plants, water-conducting cells became stronger over time due to the evolution of lignin and secondary cell walls. Efficient water transport was maintained through pits where the secondary cell wall is missing, or gaps where both the secondary and primary cell wall are absent.

✔**QUESTION** Biologists claim that vessels are more efficient than tracheids at transporting water, in part because vessels are shorter and wider than tracheids. Why does this claim make sense?

- Some of the water-conducting cells present in the early fossils had cell walls with thickened rings containing a molecule called lignin (**Figure 30.9b**).

Lignin is a complex polymer built from six-carbon rings. It is extraordinarily strong for its weight and is particularly effective in resisting compressing forces such as gravity.

These observations inspired the following hypothesis: The evolution of lignin rings gave stem tissues the strength to remain erect in the face of wind and gravity. Today, the presence of lignin in the cell walls of water-conducting cells is considered the defining feature of vascular tissue. The evolution of vascular tissue allowed early plants to support erect stems and transport water from roots to aboveground tissues.

ELABORATION OF VASCULAR TISSUE: TRACHEIDS AND VESSELS
Once simple water-conducting tissues evolved, evolution by natural selection favored more complex types that were more efficient in providing support and transport.

In rocks that are about 380 million years old, biologists find the advanced water-conducting cells called tracheids. **Tracheids** are long, thin, tapering cells that have:

- a thickened, lignin-containing **secondary cell wall** in addition to a cellulose-based **primary cell wall**; and

- pits in the sides and ends of the cell where the secondary cell wall is absent, where water can flow efficiently from one tracheid to the next (**Figure 30.9c**).

The secondary cell wall gave tracheids the ability to provide better structural support, but water could still move through the cells easily because of the pits. Today, all vascular plants contain tracheids.

In fossils dated to 250–270 million years ago, biologists have documented the most advanced type of water-conducting cells observed in plants. **Vessel elements** are shorter and wider than tracheids, and their upper and lower ends have gaps where both the primary and secondary cell wall are missing. The width of vessels and the presence of open gaps reduces resistance and makes water movement extremely efficient (**Figure 30.9d**). In vascular tissue, vessel elements are lined up end to end to form a continuous pipelike structure.

In the stems and branches of some vascular plant species, tracheids or a combination of tracheids and vessels can form the extremely strong support material called **wood**. The anatomy of wood is explained in detail in Chapter 36.

All of the cell types shown in Figure 30.9 are dead when they mature, which means they lack cytoplasm. This feature allows water to move through the cells more efficiently (see Chapter 37). Taken together, the data summarized in the figure indicate that vascular tissue evolved in a series of gradual steps that provided increased structural support and increased efficiency in water transport.

Mapping Evolutionary Changes on the Phylogenetic Tree

Cuticle, stomata, and vascular tissue, were key adaptations that allowed early plants to colonize land. **Figure 30.10** summarizes how land plants adapted to dry conditions by mapping where major innovations occurred as the group diversified. To review the concepts involved, go to the study area at *www.masteringbiology.com*.

Web Activity Plant Evolution and the Phylogenetic Tree

As you study the tree in Figure 30.10, note that fundamentally important adaptations to dry conditions—such as cuticle, pores, stomata, vascular tissue, and tracheids—evolved just once. But convergent evolution, which was introduced in Chapter 27, also occurred. When convergence occurs, similar traits evolve independently in two distinct lineages.

- Water-conducting cells evolved independently in mosses and in the vascular plants.

- Vessels evolved independently in gnetophytes and angiosperms.

As predicted by the convergence hypothesis, there are important differences in the development and anatomy in the water-conducting cells and vessels found in the different lineages.

The evolution of cuticle, stomata, and vascular tissue made it possible for plants to avoid drying out and to grow upright, while

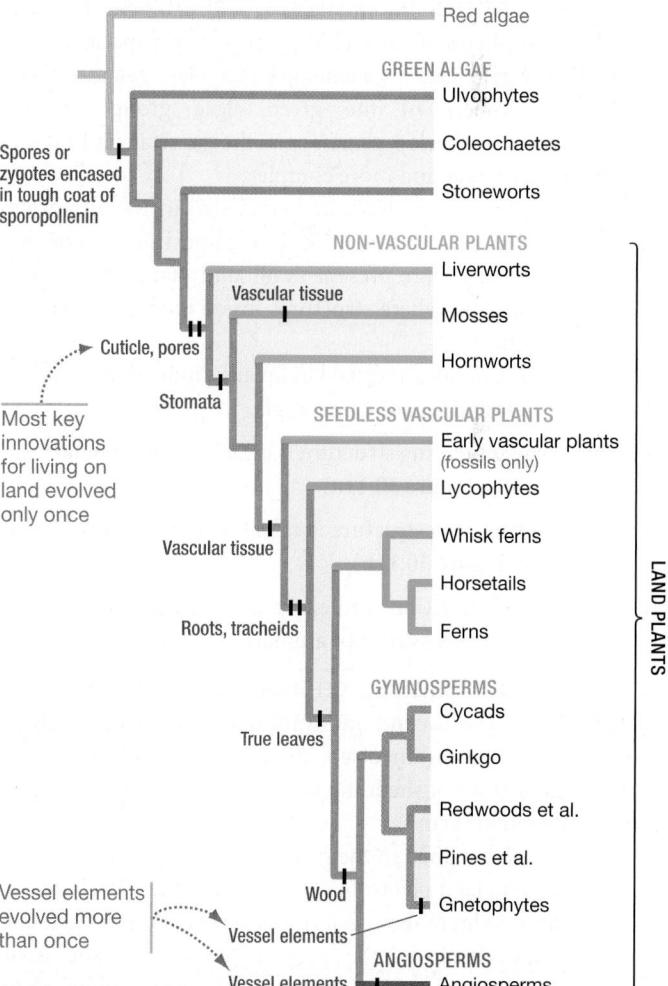

FIGURE 30.10 A Series of Evolutionary Innovations Allowed Plants to Adapt to Life on Land.

✔**QUESTION** Explain the logic that biologists used to map the location of the innovations on phylogenetic trees.

moving water from the base of the plant to its apex. Plants gained adaptations that allowed them not only to survive on land, but thrive. Now the question is, how did they reproduce?

The Transition to Land, II: How Do Plants Reproduce in Dry Conditions?

Section 30.2 introduced one of the key adaptations for reproducing on land: spores that resist drying because they are encased in a tough coat of sporopollenin. Sporopollenin-like compounds are found in the walls of some green algal zygotes; thick-walled, sporopollenin-rich spores appear early in the fossil record of land plants and occur in all land plants living today. Based on these observations, biologists infer that sporopollenin-encased spores were one of the innovations that made the initial colonization of land possible.

Two other innovations occurred early in land plant evolution and were instrumental for efficient reproduction in a dry environment: (1) Gametes were produced in complex, multicellular structures; and (2) the embryo was retained on the parent (mother) plant and was nourished by it.

PRODUCING GAMETES IN PROTECTED STRUCTURES The fossilized gametophytes of early land plants contain specialized reproductive organs called **gametangia** (singular: **gametangium**). Although members of the green algae group Charales (stoneworts) also develop gametangia, the gametangia found in land plants are larger and more complex.

The evolution of an elaborate gametangium was important because it protected gametes from drying and from mechanical damage. Gametangia are present in all land plants living today except angiosperms, where structures inside the flower perform the same functions.

In both the Charales and the land plants, individuals produce distinctive male and female gametangia.

● The sperm-producing structure is called an **antheridium** (plural: **antheridia**; **Figure 30.11a**).

● The egg-producing structure is called an **archegonium** (plural: **archegonia**; **Figure 30.11b**).

In terms of their function, antheridia and archegonia are analogous to the testes and ovaries of animals.

RETAINING AND NOURISHING OFFSPRING: LAND PLANTS AS EMBRYOPHYTES The second innovation that occurred early in land plant evolution involved the eggs that formed inside archegonia. Instead of shedding their eggs into the water or soil, land plants retain them.

Eggs are also retained in the green algal lineages that are most closely related to land plants: In Charales and other closely related groups, sperm swim to the egg, fertilization occurs, and the resulting zygote stays attached to the parent. Either before or after fertilization, the egg or zygote receives nutrients from the mother plant. But the parent plant dies each autumn as the temperature drops. The zygote remains on the dead parental tissue, settles to the bottom of the lake or pond, and overwinters. In spring, meiosis occurs, and the resulting spores develop into haploid adult plants.

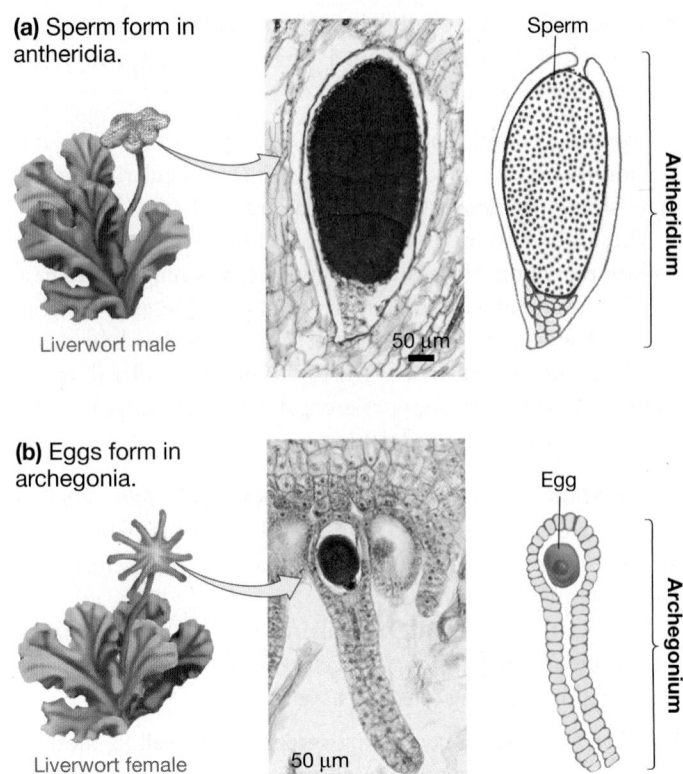

(a) Sperm form in antheridia.

Liverwort male

Sperm

50 μm

Antheridium

(b) Eggs form in archegonia.

Liverwort female

Egg

50 μm

Archegonium

FIGURE 30.11 In All Land Plant Groups but Angiosperms, Gametes Are Produced in Gametangia. Gametangia are complex, multicellular structures that protect developing gametes from drying and mechanical damage.

In land plants, the zygote is also retained on the parent plant after fertilization. But in contrast to the zygotes of most green algae, the zygotes of all land plants begin to develop on the parent plant, forming a multicellular embryo that remains attached to the parent and can be nourished by it. This is important because land plant embryos do not have to manufacture their own food early in life. Instead, they receive most or all of their nutrients from the parent plant.

The retention of the embryo was such a key event in land plant evolution that the formal name of the group is **Embryophyta**—literally, the "embryo-plants." The retention of the fertilized egg in **embryophytes** is analogous to pregnancy in mammals, where offspring are retained by the mother and nourished through the initial stages of growth. Land plant embryos even have specialized **transfer cells**, which make physical contact with parental cells and facilitate the transfer of nutrients (**Figure 30.12**) much like the placenta that develops in a pregnant mammal.

Thick-walled spores, elaborate gametangia, and the embryophyte condition weren't the only key innovations associated with reproducing on land, though. In addition, all land plants undergo the phenomenon known as **alternation of generations**, introduced in Chapter 29.

ALTERNATION OF GENERATIONS When alternation of generations occurs, individuals represent a multicellular haploid phase or a multicellular diploid phase. The multicellular haploid stage

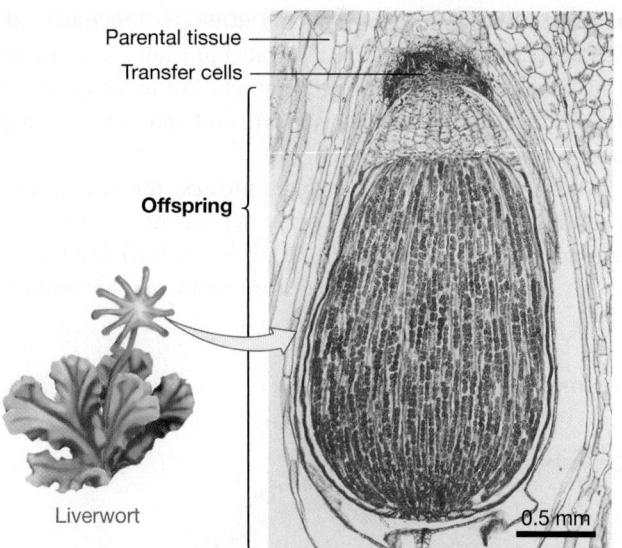

FIGURE 30.12 Land Plants Are Also Known as Embryophytes because Parents Nourish Their Young. In land plants, fertilization and early development take place on the parent plant. As embryos develop, transfer cells carry nutrients from the mother to the offspring.

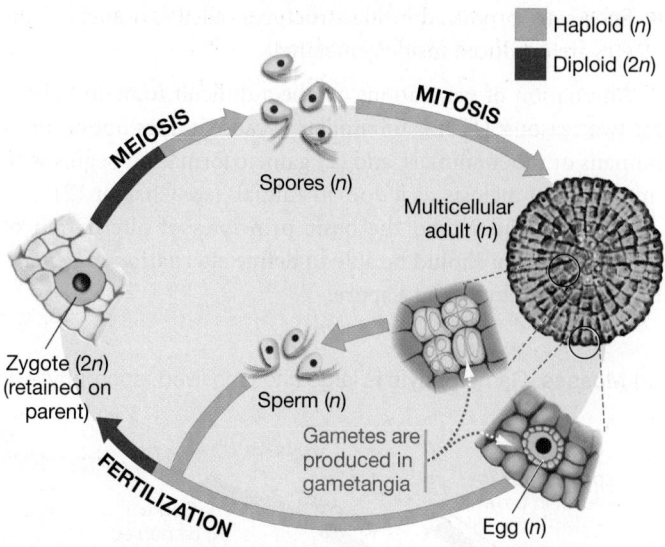

FIGURE 30.13 In Green Algae That Are Closely Related to Land Plants, Only the Zygote Is Diploid. The coleochaetes and stoneworts do not have alternation of generations. The multicellular stage is haploid.

is called the **gametophyte**; the multicellular diploid stage is called the **sporophyte**. The two phases of the life cycle are connected by distinct types of reproductive cells—gametes and spores.

Although alternation of generations is observed in a wide array of eukaryotic lineages and in some groups of green algae, it does not occur in the algal groups most closely related to land plants. In the coleochaetes and stoneworts, the multicellular form is haploid. Only the zygote is diploid. As **Figure 30.13** shows, the zygote undergoes meiosis to form haploid spores. After dispersing with the aid of flagella, the spores begin dividing by mitosis and eventually grow into an adult, haploid individual. You might recall that this haploid-dominant life cycle is common in protists and was diagrammed in Figure 29.17a.

These data suggest that alternation of generations originated in land plants independently of its evolution in other groups of eukaryotes, and that it originated early in their history—soon after they evolved from green algae. The adaptive significance of alternation of generations and its role in the successful colonization of land is still being debated, however. Keep this in mind as you study alternation of generations in land plants: You are analyzing one of the great unsolved problems in contemporary biology.

Alternation of generations always involves the same basic sequence of events, illustrated in **Figure 30.14**. To review how this type of life cycle works, put your finger on the sporophyte in the figure and trace the cycle clockwise to find the following five key events:

1. The sporophyte produces spores by meiosis. Spores are haploid.

2. Spores divide by mitosis and develop into a haploid gametophyte.

3. Gametophytes produce gametes by mitosis. Both the gametophyte and the gametes are haploid.

4. Two gametes unite during fertilization to form a diploid zygote.

5. The zygote divides by mitosis and develops into a multicellular, diploid sporophyte.

Once you've traced the cycle successfully, take a moment to compare and contrast zygotes and spores.

- Zygotes and spores are both single cells that divide by mitosis to form a multicellular individual.

- Zygotes result from the fusion of two cells, such as a sperm and an egg, but spores are not formed by the fusion of two cells.

- Zygotes produce sporophytes; spores produce gametophytes.

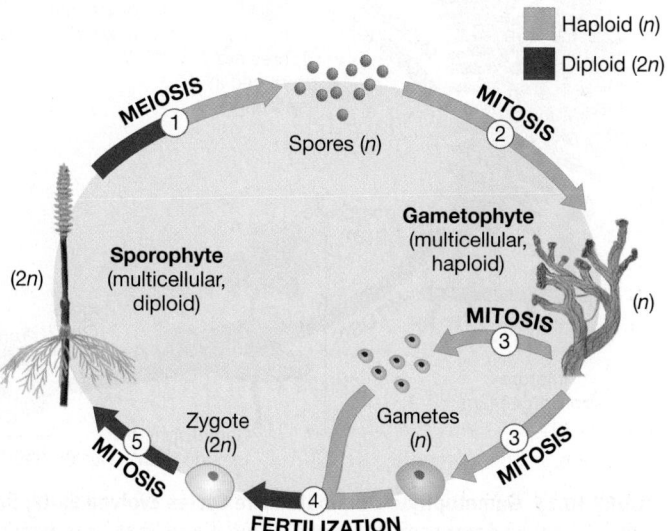

FIGURE 30.14 All Land Plants Undergo Alternation of Generations. Alternation of generations always involves the same sequence of five events.

- Spores are produced inside structures called sporangia; gametes are produced inside gametangia.

Alternation of generations can be a difficult topic to master, for two reasons: (1) It is unfamiliar because it does not occur in humans or other animals, and (2) gamete formation begins with mitosis—not meiosis, as it does in animals (see Chapter 12).

✔If you understand the basic principles of alternation of generations, you should be able to define alternation of generations, a sporophyte, and a spore.

THE GAMETOPHYTE-DOMINANT TO SPOROPHYTE-DOMINANT TREND IN LIFE CYCLES The five steps illustrated in Figure 30.14 occur in all species with alternation of generations. But in land plants, the relationship between the gametophyte and sporophyte is highly variable.

In non-vascular plants such as mosses, the sporophyte is small and short lived and is largely dependent on the gametophyte for nutrition (**Figure 30.15a**). When you see leafy-looking mosses growing on a tree trunk or on rocks, you are looking at

(a) Mosses: Gametophyte is large and long lived; sporophyte depends on gametophyte for nutrition.

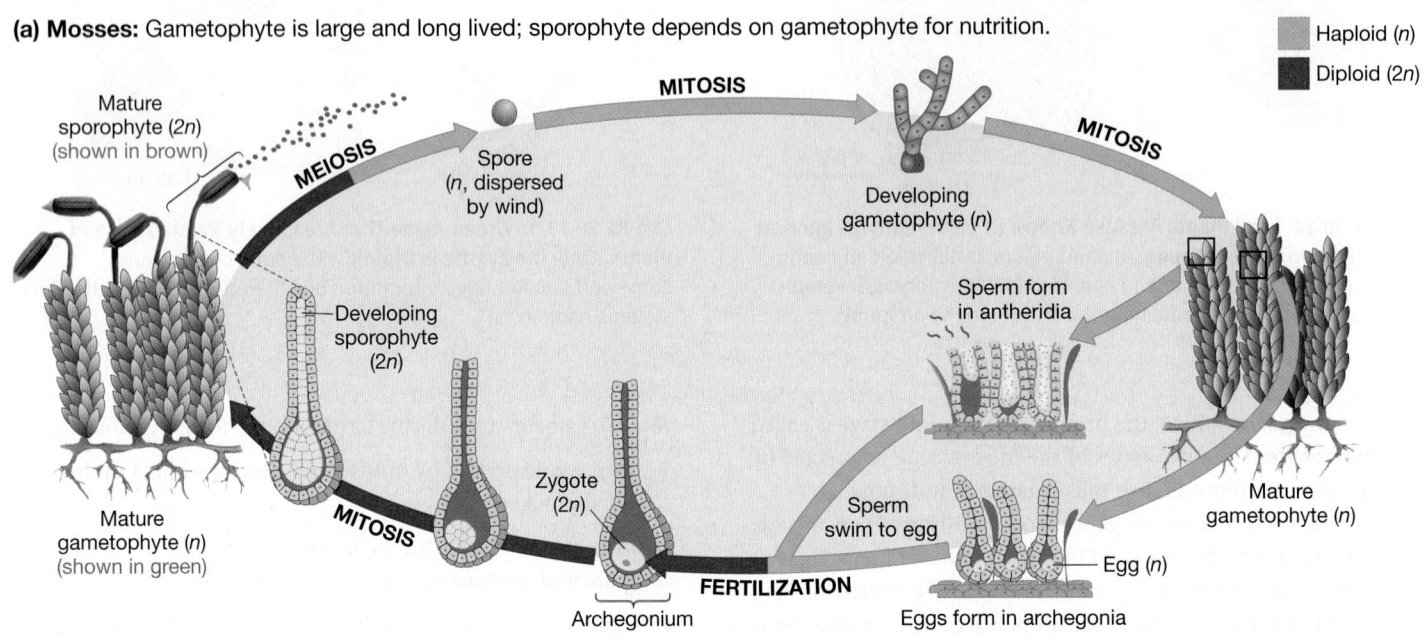

(b) Ferns: Sporophyte is large and long lived but, when young, depends on gametophyte for nutrition.

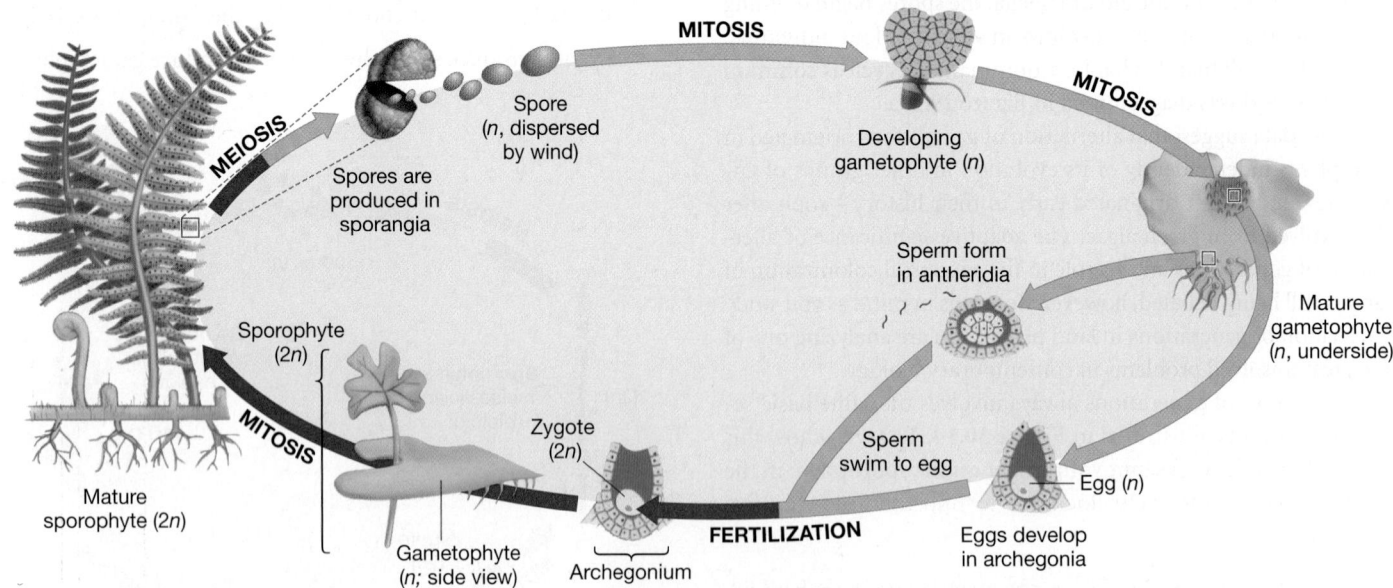

FIGURE 30.15 Gametophyte-Dominated Life Cycles Evolved Early; Sporophyte-Dominated Life Cycles Evolved Later. Like today's mosses, the earliest land plants in the fossil record have gametophytes that are much larger and longer lived than the sporophyte. In lineages that evolved later, such as ferns, the sporophyte is much larger and longer lived than the gametophyte.

✔**QUESTION** How can you tell that alternation of generations occurs in mosses and ferns?

gametophytes. Because the gametophyte is long lived and produces most of the food required by the individual, it is considered the dominant part of the life cycle.

In contrast, in ferns and other vascular plants the sporophyte is much larger and longer lived than the gametophyte (see **Figure 30.15b**). The ferns you see growing in gardens or forests are sporophytes. You'd have to hunt on your hands and knees to find their gametophytes, which are typically just a few millimeters in diameter. As you'll learn later in the chapter, the gametophytes of gymnosperms and angiosperms are even smaller—they are microscopic. Ferns and other vascular plants are said to have a sporophyte-dominant life cycle.

✔If you understand the difference between gametophyte-dominant and sporophyte-dominant life cycles, you should be able to examine the photos of hornworts (a non-vascular plant) and horsetails (a seedless vascular plant) in **Figure 30.16**, and identify which is the gametophyte and which is the sporophyte.

(a) Hornwort gametophytes and sporophytes

(b) Horsetail gametophyte and sporophyte

FIGURE 30.16 The Reduction of the Gametophyte Is One of the Strongest Trends in Land Plant Evolution. **(a)** A hornwort with spike-like sporophytes emerging from the gametophyte. **(b)** A horsetail species, both as a tiny, microscopic gametophyte and as a large, macroscopic sporophyte.

The transition from gametophyte-dominated life cycles to sporophyte-dominated life cycles is one of the most striking of all trends in land plant evolution. To explain why it occurred, biologists hypothesize that sporophyte-dominated life cycles were advantageous because diploid cells can respond to varying environmental conditions more efficiently than haploid cells can—particularly if the individual is heterozygous at many genes.

This idea has yet to be tested rigorously, however. If you came up with explanations for why alternation of generations evolved in land plants and why the gametophyte-dominant to sporophyte-dominant trend occurred, and if your ideas stood up to testing, the result would be celebrated worldwide.

HETEROSPORY In addition to sporophyte-dominated life cycles, another important innovation found in seed plants is called **heterospory**—the production of two distinct types of spore-producing structures and thus two distinct types of spores.

All of the non-vascular plants and most of the seedless vascular plants are **homosporous**—meaning that they produce a single type of spore. (Among the seedless vascular plants, some lycophytes and a few ferns are heterosporous.) Homosporous species produce spores that develop into bisexual gametophytes. Bisexual gametophytes produce both eggs and sperm (**Figure 30.17a**).

The two types of spore-producing structures found in heterosporous species are often found on the same individual (**Figure 30.17b**).

● **Microsporangia** are spore-producing structures that produce microspores. **Microspores** develop into male gametophytes, which produce the small gametes called sperm.

● **Megasporangia** are spore-producing structures that produce megaspores. **Megaspores** develop into female gametophytes, which produce the large gametes called eggs.

Thus, the gametophytes of seed plants are either male or female, but never both.

The evolution of heterospory was a key event in land plant evolution because it made possible one of the most important adaptations for life in dry environments—pollen.

POLLEN The non-vascular plants and the seedless vascular plants have male gametes that swim to the egg to perform fertilization. For a sperm cell to swim to the egg and fertilize it, there has to be a

(a) Non-vascular plants and most seedless vascular plants are **homosporous.**

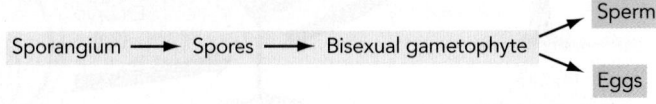

(b) Seed plants are **heterosporous.**

Microsporangia → Microspores → Male gametophyte → Sperm
Megasporangia → Megaspores → Female gametophyte → Eggs

FIGURE 30.17 Heterosporous Plants Produce Male and Female Spores That Are Morphologically Distinct.

continuous sheet of water between the male and female gameto-phyte, a raindrop has to splash sperm onto a female gametophyte, or a wet insect has to carry sperm to a female gametophyte.

In species that live in dry environments, these conditions are rare. The land plants made their final break with their aquatic origins and were able to reproduce efficiently in dry habitats when a structure evolved that could move their gametes without the aid of water.

In heterosporous seed plants, the microspore germinates to form a tiny male gametophyte that is surrounded by a tough coat of sporopollenin, resulting in a **pollen grain**.

Pollen grains can be exposed to the air for long periods of time without dying from dehydration. They are also tiny enough to be carried to female gametophytes by wind or animals. Upon landing near the egg, the male gametophyte releases the sperm cells that accomplish fertilization.

When pollen evolved, then, heterosporous plants lost their de-pendence on water to accomplish fertilization. Instead of swim-ming to the egg as a naked sperm cell, their tiny gametophytes took to the skies.

SEEDS The evolution of large gametangia protected the eggs and sperm of land plants from drying. Embryo retention allowed offspring to be nourished directly by their parent, and pollen en-abled fertilization to occur in the absence of water.

Retaining embryos has a downside, however: In ferns and horsetails, sporophytes have to live in the same place as their par-ent gametophyte. Seed plants overcome this limitation. Their embryos are portable and can disperse to new locations.

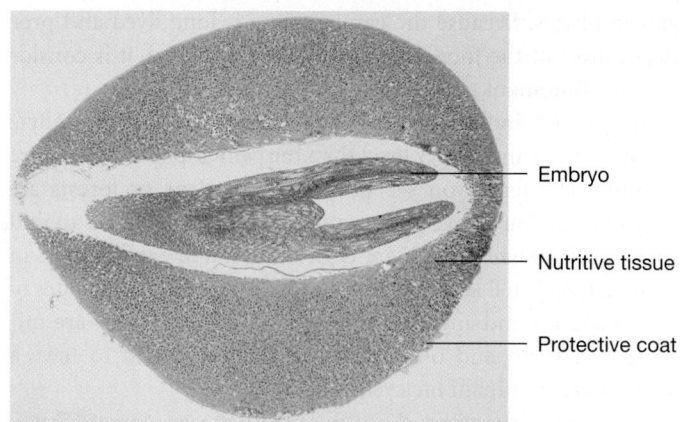

FIGURE 30.18 Seeds Contain an Embryo and a Food Supply and Can Be Dispersed. (Note that this specimen has been stained.)

A **seed** is a structure that includes an embryo and a food supply surrounded by a tough coat (**Figure 30.18**). Seeds allow embryos to be dispersed to a new habitat, away from the parent plant. Like a bird's egg, the seed provides a protective case for the embryo and a store of nutrients provided by the mother. Spores are an effective dispersal stage in non-vascular plants and seedless vascular plants, but they lack the stored nutrients found in seeds.

The evolution of heterospory, pollen, and seeds triggered a dra-matic radiation of seed plants starting about 299 million years ago. To make sure that you understand these key processes and struc-tures, study the life cycle of the pine tree pictured in **Figure 30.19**.

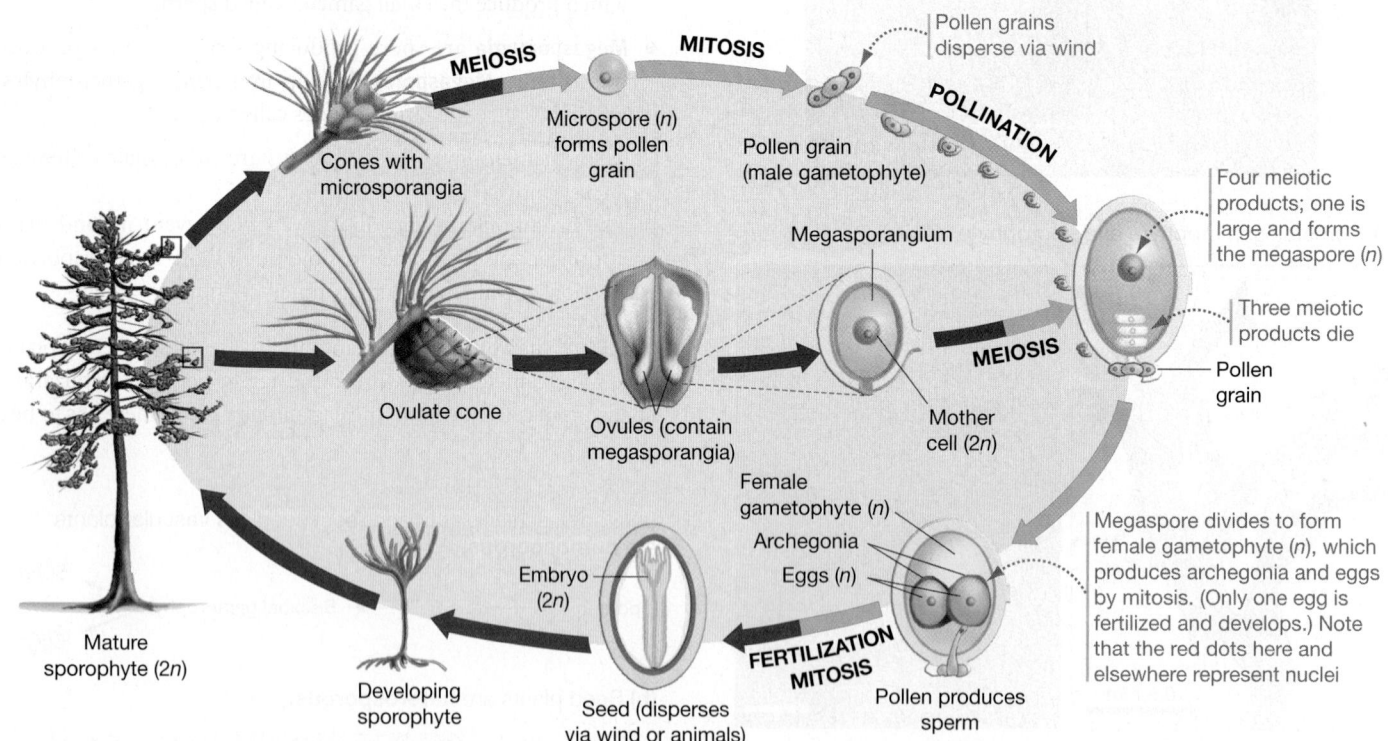

FIGURE 30.19 Heterospory in Gymnosperms: Microspores Produce Pollen Grains; Megaspores Produce Female Gametophytes.

✓**QUESTION** Compare the life cycle of the pine tree in Figure 30.19 with that of the fern pictured in Figure 30.15b. Is the gymnosperm gametophyte larger than, smaller than, or about the same size as a fern gametophyte? Compared with ferns, is the gymnosperm gametophyte more or less dependent on the sporophyte for nutrition?

- Starting with the sporophyte on the left, note that this and many other gymnosperm species have separate structures, called cones, where microsporangia and megasporangia develop. In this case, the two types of spores associated with heterospory develop in separate cones.

- The microsporangia contain a cell that divides by meiosis to form microspores, which then divide by mitosis to form pollen grains—tiny male gametophytes.

- Megasporangia are found inside protective structures called ovules. Megasporangia contain a mother cell that divides by meiosis to form a megaspore.

- The megaspore undergoes mitosis to form the female gametophyte, which contains egg cells.

- The female gametophyte stays attached to the sporophyte as pollen grains arrive and produce sperm that fertilize the eggs.

- Seeds mature as the embryo develops. Inside the seed, cells derived from the female gametophyte become packed with nutrients provided by the sporophyte.

When the seed disperses and germinates, the cycle of life begins anew.

FLOWERS Flowering plants, or angiosperms, are the most diverse land plants living today. About 250,000 species have been described, and more are discovered each year. Their success in terms of geographical distribution, number of individuals, and number of species revolves around a reproductive organ—the **flower**.

Flowers contain two key reproductive structures: the stamens and carpels illustrated on the left-hand side of **Figure 30.20**. Stamens and carpels are responsible for heterospory.

- A **stamen** includes a structure called an anther, where microsporangia develop. Meiosis occurs inside the microsporangia, forming microspores. Microspores then divide by mitosis to form pollen grains.

- A **carpel** contains a protective structure called an **ovary** where the ovules are found.

The presence of enclosed ovules inspired the name angiosperm ("encased-seed") as opposed to gymnosperm ("naked-seed").

As in gymnosperms, ovules contain the megasporangia. A cell inside the megasporangium divides by meiosis to form the megaspore, which then divides by mitosis to form the female gametophyte. When a pollen grain lands on a carpel and produces sperm, fertilization takes place, as shown on the right-hand side of Figure 30.20.

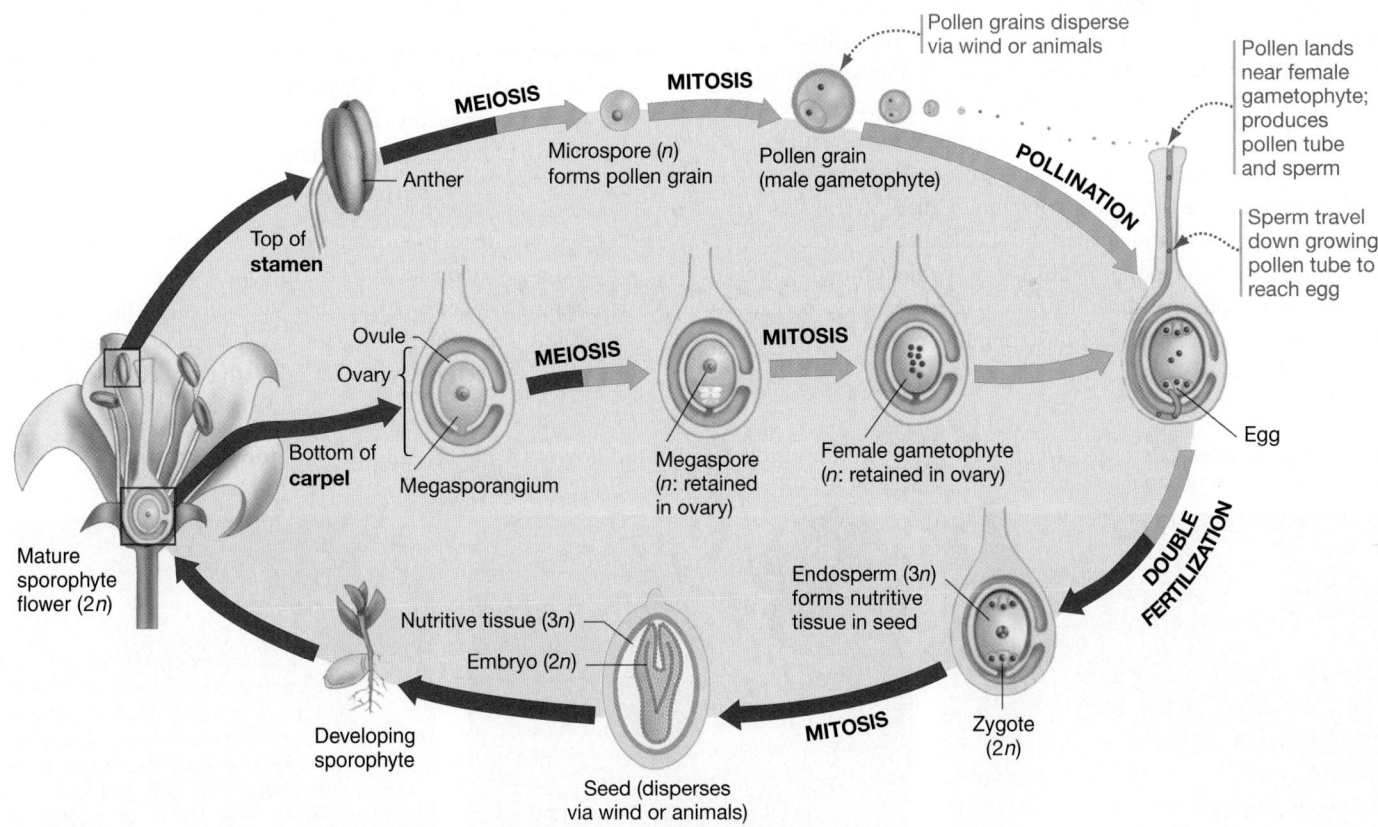

FIGURE 30.20 Heterospory in Angiosperms: Flowers Contain Microspores and Megaspores.

✔**QUESTION** Gymnosperm pollen grains typically contain from 4 to 40 cells; mature angiosperm pollen grains contain three cells. Gymnosperm female gametophytes typically contain hundreds of cells; angiosperm female gametophytes typically contain seven. In the fossil record, gymnosperms appear long before angiosperms. Do these observations conflict with the trend of reduced gametophytes during land plant evolution, or are they consistent with it? Explain your logic.

Angiosperm fertilization is unique, however, because it involves *two* sperm cells. One sperm fuses with the egg to form the zygote, while a second sperm fuses with two nuclei in the female gametophyte to form a triploid (3*n*) nutritive tissue called **endosperm**. The involvement of two sperm nuclei is called **double fertilization**.

The evolution of the flower, then, is an elaboration of heterospory. The key innovation was the evolution of the ovary, which helps protect female gametophytes from insects and other predators.

Double fertilization is another striking innovation associated with the flower, but its adaptive significance is still not well understood. Explaining the significance of double fertilization is another major challenge for biologists interested in understanding how land plants diversified.

POLLINATION BY INSECTS AND OTHER ANIMALS The story of the flower doesn't end with the ovary. Once stamens and carpels evolved, they became enclosed by modified leaves called **sepals** and **petals**. The four structures then diversified to produce a fantastic array of sizes, shapes, and colors—from red roses to blue violets. Specialized cells inside flowers also began producing a wide range of scents.

To explain these observations, biologists hypothesize that flowers are adaptations to increase the probability that an animal will perform **pollination**—the transfer of pollen from one individual's stamen to another individual's carpel. Instead of leaving pollination to an undirected agent such as wind, the hypothesis is that natural selection favored structures that reward an animal—usually an insect—for carrying pollen directly from one flower to another.

Under the directed-pollination hypothesis, natural selection has favored flower colors and shapes and scents that are successful in attracting particular types of pollinators. A pollinator is an animal that disperses pollen. Pollinators are attracted to flowers because flowers provide the animals with food in the form of protein-rich pollen or a sugar-rich fluid known as **nectar**. In this way, the relationship between flowering plants and their pollinators is mutually beneficial. The pollinator gets food; the plant gets sex (fertilization).

What evidence supports the hypothesis that flowers vary in size, structure, scent, and color in order to attract different pollinators?

The first type of evidence on this question is correlational in nature. In general, the characteristics of a flower correlate closely with the characteristics of its pollinator. A few examples will help drive this point home:

- *Scent* The carrion flower in **Figure 30.21a** produces molecules that smell like rotting flesh. The scent attracts carrion flies, which normally lay their eggs in animal carcasses. In effect, the plant tricks the flies. While looking for a place to lay their eggs on a flower, the flies become dusted with pollen. If the flies are already carrying pollen from a visit to a different carrion flower, they are likely to deposit pollen grains near the plant's female gametophyte. In this way, the carrion flies pollinate the plant.

- *Flower shape* Flowers that are pollinated by hummingbirds typically have petals that form a long, tubelike structure corresponding to the size and shape of a hummingbird's beak (**Figure 30.21b**). Nectar-producing cells are located at the base of the tube. When hummingbirds visit the flower, they insert their beaks and harvest the nectar. In the process, they transfer pollen grains attached to their throats or faces.

- *Flower color* Hummingbird-pollinated flowers tend to have red petals (Figure 30.21b), while bee-pollinated flowers tend to be purple or yellow (**Figure 30.21c**). Hummingbirds are attracted to red; bees have excellent vision in the purple and ultraviolet end of the spectrum.

The directed-pollination hypothesis also has strong experimental support. Consider, for example, recent work on a South African orchid called *Disa draconis*. The length of the long tube, or spur, located at the back of this flower varies among populations of this species. As predicted by the directed-pollination

(a) Carrion flowers: Smell like rotting flesh and attract carrion flies

(b) Hummingbird-pollinated flowers: Red, long tubes with nectar at the base

(c) Bee-pollinated flowers: Often bright purple

FIGURE 30.21 Flowers with Different Fragrances, Shapes, and Colors Attract Different Pollinators.

hypothesis, the length of the spur correlates with the length of the proboscis found in the insect pollinator. A proboscis is a specialized mouthpart found in some insects. When extended, it functions like a straw in sucking nectar or other fluids.

- Short-spurred *D. draconis* that grow in mountain habitats are pollinated by insect species that have a relatively short proboscis.

- Long-spurred members of this species that grow on low-lying sandplain habitats are pollinated by insects that have a particularly long proboscis.

To test the hypothesis that spur length affects pollination success, researchers artificially shortened the spurs of individuals in a long-spurred population (**Figure 30.22**). The biologists did this by tying off the spurs of randomly selected flowers with a piece of yarn, so that pollinating insects could not reach the end of the spur. The idea was that insects would not make contact with the flower's reproductive organs when they inserted their proboscis into the shortened tube. The biologists left nearby flowers alone but attached a piece of yarn near the spur, as a control treatment.

As the data in the "Results" box of Figure 30.22 indicate, flowers with short spurs received much less pollen and set much less seed than did flowers with normal-length spurs. These data strongly support the hypothesis that spur length is an adaptation that increases the frequency of pollination by particular insects.

Based on results like this, biologists contend that the spectacular diversity of angiosperms resulted, at least in part, from natural selection exerted by the equally spectacular diversity of insect, mammal, and bird pollinators. Mutations that change flower shape can change gene flow and start speciation (see Chapter 26).

FRUITS The evolution of the ovary was an important event in land plant diversification, but not only because it protected the female gametophytes of angiosperms. It also made the evolution of fruit possible.

A **fruit** is a structure that is derived from the **ovary** and encloses one or more seeds (**Figure 30.23a** on page 564). Tissues derived from the ovary are often nutritious and brightly colored (**Figure 30.23b**). Animals eat these types of fruits, digest

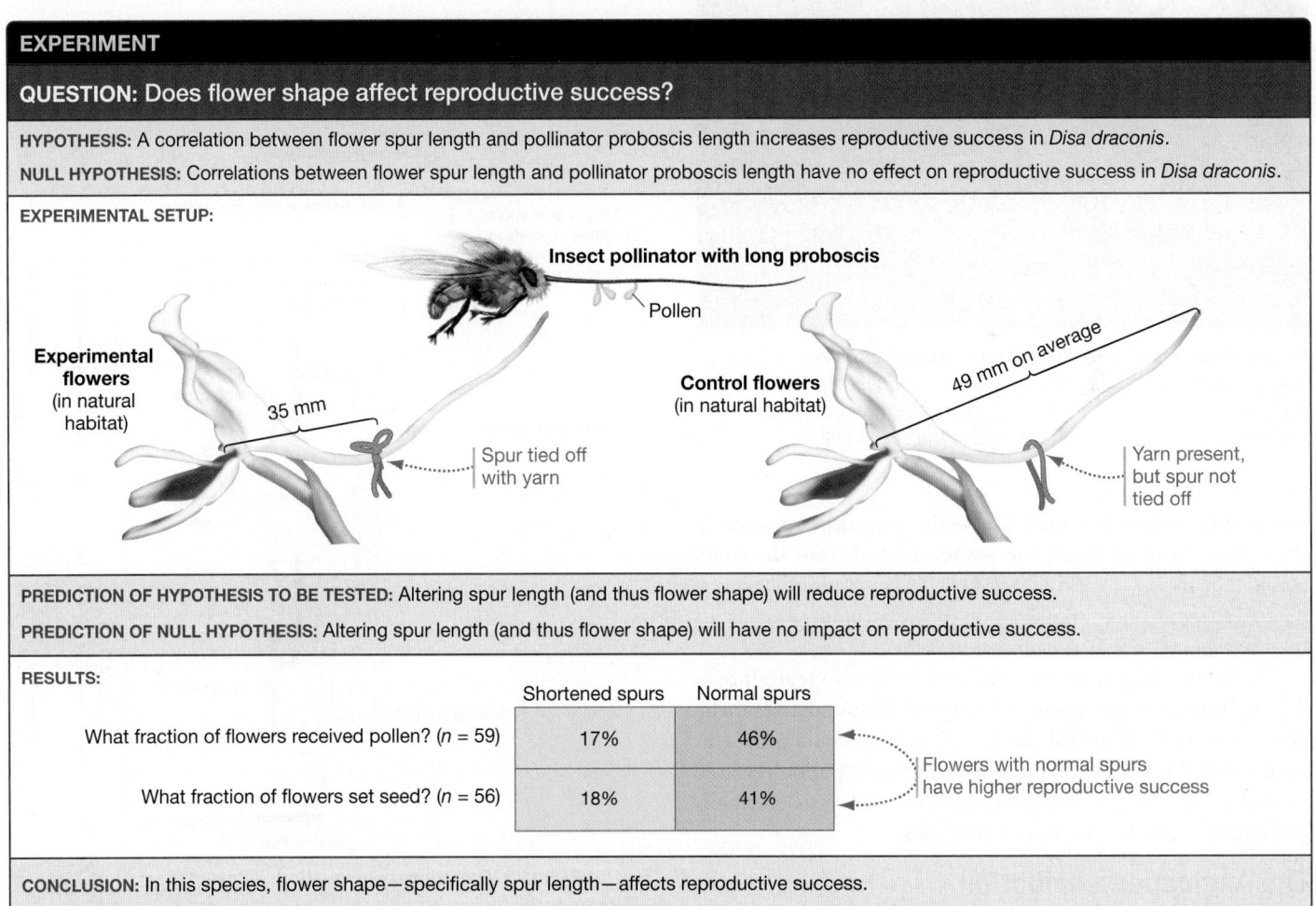

EXPERIMENT

QUESTION: Does flower shape affect reproductive success?

HYPOTHESIS: A correlation between flower spur length and pollinator proboscis length increases reproductive success in *Disa draconis*.

NULL HYPOTHESIS: Correlations between flower spur length and pollinator proboscis length have no effect on reproductive success in *Disa draconis*.

EXPERIMENTAL SETUP:

Insect pollinator with long proboscis

Pollen

Experimental flowers (in natural habitat)

35 mm

Spur tied off with yarn

Control flowers (in natural habitat)

49 mm on average

Yarn present, but spur not tied off

PREDICTION OF HYPOTHESIS TO BE TESTED: Altering spur length (and thus flower shape) will reduce reproductive success.

PREDICTION OF NULL HYPOTHESIS: Altering spur length (and thus flower shape) will have no impact on reproductive success.

RESULTS:

	Shortened spurs	Normal spurs
What fraction of flowers received pollen? (n = 59)	17%	46%
What fraction of flowers set seed? (n = 56)	18%	41%

Flowers with normal spurs have higher reproductive success

CONCLUSION: In this species, flower shape—specifically spur length—affects reproductive success.

FIGURE 30.22 The Adaptive Significance of Flower Shape: An Experimental Test.

SOURCE: Johnson, S. D. and K. E. Steiner. 1997. Long-tongued fly pollination and evolution of floral spur length in the *Disa draconis* complex (Orchidaceae). *Evolution* 51: 45–53.

✔**QUESTION** Why did the researchers bother to put yarn around the control flowers?

(a) Fruits are derived from ovaries and contain seeds.

Seed

Wall of ovary

(b) Many fruits are dispersed by animals.

FIGURE 30.23 Fruits Are Derived from Ovaries Found in Angiosperms. (a) A pea pod is one of the simplest types of fruit. **(b)** The ovary wall often becomes thick, fleshy, and nutritious enough to attract animals that disperse the seeds inside.

the nutritious tissue around the seeds, and disperse seeds in their feces. In other cases, the tissues derived from the ovary help fruits disperse via wind or water. The evolution of flowers made efficient pollination possible; the evolution of fruits made efficient seed dispersal possible.

The list of adaptations that allow land plants to reproduce in dry environments is impressive; **Figure 30.24** summarizes them. ⚿ Once land plants had vascular tissue and could grow efficiently in dry habitats, the story of their diversification revolved around traits that allowed sperm cells to reach eggs efficiently and helped seeds disperse to new locations.

The Angiosperm Radiation

For the past 125 million years, land plant diversification has really been about angiosperms. The Anthophyta or angiosperms represent one of the great adaptive radiations in the history of life. As Chapter 27 noted, an **adaptive radiation** occurs when a

single lineage produces a large number of descendant species that are adapted to a wide variety of habitats.

The diversification of angiosperms is associated with three key adaptations: (**1**) vessels, (**2**) flowers, and (**3**) fruits. In combination, these traits allow angiosperms to transport water, pollen, and seeds efficiently. Based on these observations, it is not surprising that most land plants living today are angiosperms.

On the basis of morphological traits, the 250,000 species of angiosperms identified to date have traditionally been classified into two major groups: the monocotyledons, or **monocots**, and the dicotyledons, or **dicots**. Some familiar monocots include the grasses, orchids, palms, and lilies; familiar dicots include the roses, buttercups, daisies, oaks, and maples.

The names of the two groups were inspired by differences in a structure called the cotyledon. A **cotyledon** is the first leaf that is

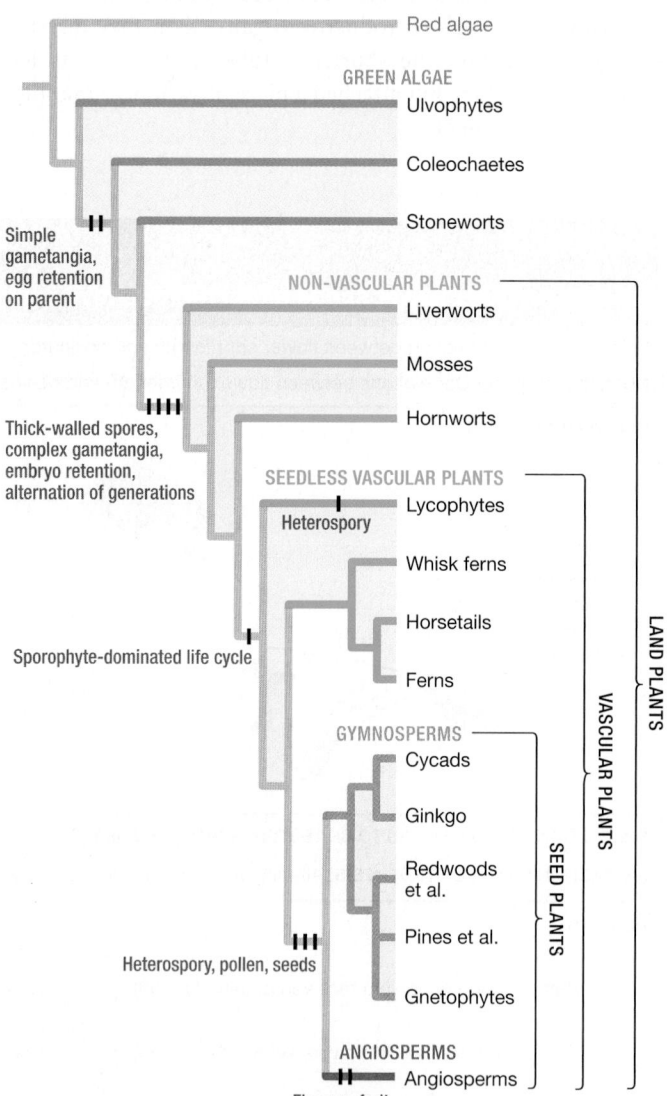

FIGURE 30.24 Evolutionary Innovations Allowed Plants to Reproduce Efficiently on Land.

✔**EXERCISE** Redwoods, pines, gnetophytes, and angiosperms are the only land plants that do not have flagellated sperm that swim to the egg (at least a short distance). Mark the loss of flagellated sperm on Figure 30.7.

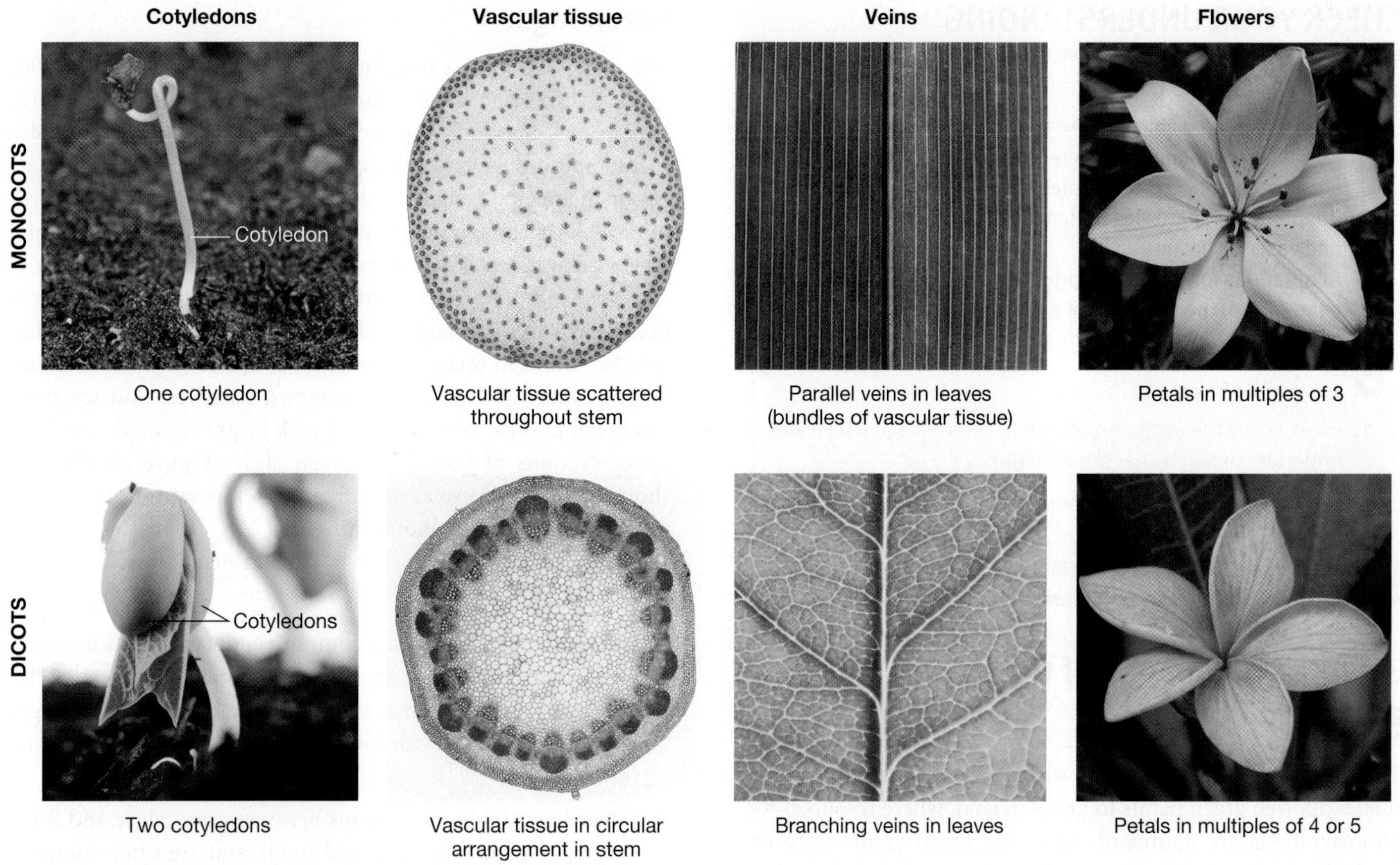

Cotyledons	Vascular tissue	Veins	Flowers

MONOCOTS

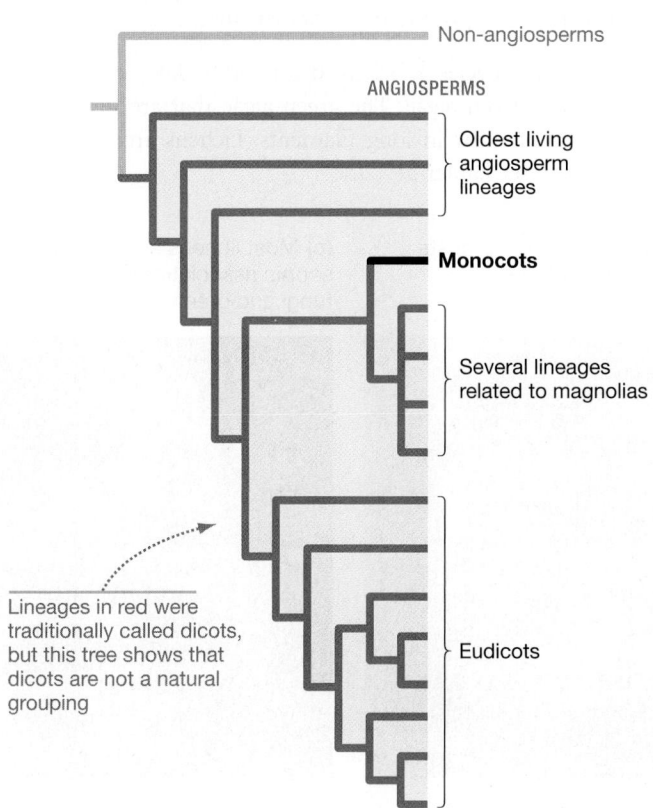

One cotyledon · Vascular tissue scattered throughout stem · Parallel veins in leaves (bundles of vascular tissue) · Petals in multiples of 3

DICOTS

Two cotyledons · Vascular tissue in circular arrangement in stem · Branching veins in leaves · Petals in multiples of 4 or 5

FIGURE 30.25 Four Morphological Differences between Monocots and Dicots. Note that the stem cross-sections, showing vascular tissue, have been stained.

FIGURE 30.26 Monocots Are Monophyletic, Dicots Are Paraphyletic.

Lineages in red were traditionally called dicots, but this tree shows that dicots are not a natural grouping

Non-angiosperms
ANGIOSPERMS
Oldest living angiosperm lineages
Monocots
Several lineages related to magnolias
Eudicots

formed in an embryonic plant. As **Figure 30.25** shows, monocots have a single cotyledon (hence the "mono") while dicots have two cotyledons (hence the "di"). The figure also highlights other major morphological differences observed in monocots and dicots, including the arrangement of vascular tissue and leaf veins and the characteristics of flowers.

It would be misleading, however, to think that all species of flowering plants fall into one of these two groups—either monocots or dicots. Recent work has shown that dicots do not form a natural group consisting of a common ancestor and all of its descendants.

To drive this point home, consider the phylogeny illustrated in **Figure 30.26**. These relationships were estimated by comparing the sequences of several genes that are shared by all angiosperms. Notice that species with dicot-like characters are scattered around the angiosperm phylogenetic tree. Based on this analysis, biologists have concluded that although monocots are monophyletic, dicots are not. Dicots are paraphyletic.

Biologists have adjusted the names assigned to angiosperm lineages to reflect this new knowledge of phylogeny. The most important of these changes was identifying the **eudicots** ("true dicots") as a lineage that includes roses, daisies, and maples. Plant systematists continue to work toward understanding relationships throughout the angiosperm phylogenetic tree; there will undoubtedly be more name changes as knowledge grows.

CHECK YOUR UNDERSTANDING

If you understand that . . .

- Land plants were able to make the transition to growing in terrestrial environments, where sunlight and carbon dioxide are abundant, based on a series of evolutionary innovations.
- Adaptations for growing on land included cuticle, stomata, and vascular tissue.
- Adaptations for effective reproduction on land included gametangia, the retention of embryos on the parent, pollen, seeds, flowers, and fruits.

✓ **You should be able to . . .**

1. Explain why the evolution of cuticle and vascular tissue was important in survival or reproduction.
2. On Figure 30.7, map where the origin of pollen, flowers, and fruits occurred.

Answers are available in Appendix B.

30.4 Key Lineages of Green Algae and Land Plants

The evolution of cuticle, pores, stomata, and water-conducting tissues allowed green plants to grow on land, where resources for photosynthesis are abundant. Once the green plants were on land, the evolution of gametangia, retained embryos, pollen, seeds, and flowers enabled them to reproduce efficiently even in dry environments. The adaptations reviewed in Section 30.3 allowed the land plants to make the most important water-to-land transition in the history of life.

To explore green plant diversity in more detail, let's take a closer look at some major groups of green algae and land plants. The initial part of this section considers broad groupings of lineages; the "index cards" that follow focus on particular monophyletic groups.

Green Algae

The **green algae** are a paraphyletic group that totals about 7000 species. Their bright green chloroplasts are similar to those found in land plants. Specifically, green algal chloroplasts have a double membrane and contain chlorophylls *a* and *b* but relatively few accessory pigments. And like land plants, green algae synthesize starch in the chloroplast as a storage product of photosynthesis. They also have a cell wall that is composed primarily of cellulose.

Green algae are important primary producers in nearshore ocean environments and in all types of freshwater habitats. They are also found in several types of more exotic environments, including snowfields at high elevations, pack ice, and ice floes. These habitats are often splashed with bright colors due to large concentrations of unicellular green algae (**Figure 30.27a**). Although these cells live at near-freezing temperatures, they make all their own food via photosynthesis.

In addition, green algae live in close association with an array of other organisms.

- Unicellular green algae are common endosymbionts in planktonic protists that live in lakes and ponds (**Figure 30.27b**). The association is considered mutually beneficial: The algae supply the protists with food; the protists provide protection to the algae.

- **Lichens** are stable associations between green algae and fungi or between cyanobacteria and fungi, and are often found in terrestrial environments that lack soil, such as tree bark or bare rock (**Figure 30.27c**). The algae or cyanobacteria in a lichen are protected from drying by the fungus; in return they provide sugars produced by photosynthesis.

Of the 17,000 species of lichens described to date, about 85 percent involve green algae. The green algae that are involved are unicellular or grow in long filaments. Lichens are explored in more detail in Chapter 31.

(a) Green algae with red carotenoid pigments are responsible for pink snow.

(b) Many unicellular protists harbor green algae.

(c) Most lichens are a microscopic association between fungi and green algae.

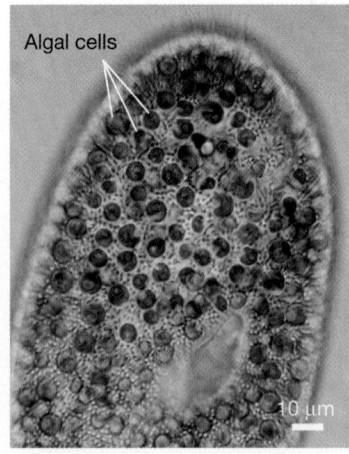

Algal cells

10 μm

1 cm

FIGURE 30.27 Some Green Algae Live in Unusual Environments.

Green algae are a large and fascinating group of organisms. In the section that closes the chapter, we'll take a closer look at just three of the many lineages.

- Green Algae > Ulvophyceae (Ulvophytes)
- Green Algae > Coleochaetophyceae (Coleochaetes)
- Green Algae > Charophyceae (Stoneworts)

Non-Vascular Plants ("Bryophytes")

The initial lineages to branch off the phylogeny of living land plants are sometimes known as the bryophytes or non-vascular plants.

All of the non-vascular plant species present today have a low, sprawling growth habit. In fact, it is unusual to find bryophytes more than 5 to 10 centimeters tall.

Individuals are anchored to soil, rocks, or tree bark by structures called **rhizoids**. Although simple water-conducting cells and tissues are found in some mosses, no bryophytes have vascular tissue with lignin-reinforced cell walls.

All bryophytes have flagellated sperm that swim to eggs through raindrops or small puddles on the plant surface. Spores are dispersed by wind.

- Non-Vascular Plants > Hepaticophyta (Liverworts)
- Non-Vascular Plants > Bryophyta (Mosses)
- Non-Vascular Plants > Anthocerophyta (Hornworts)

Seedless Vascular Plants

The seedless vascular plants are a paraphyletic group that forms a grade between the non-vascular plants and the seed plants. All species of seedless vascular plants have conducting tissues with cells that are reinforced with lignin, forming vascular tissue. Tree-sized lycophytes and horsetails are abundant in the fossil record, and tree ferns are still common inhabitants of certain habitats, such as mountain slopes in the tropics.

The sporophyte is the larger and longer-lived phase of the life cycle in all of the seedless vascular plants. The gametophyte is physically independent of the sporophyte, however. Eggs are retained on the gametophyte, and sperm swim to the egg with the aid of flagella. Thus, seedless vascular plants depend on the presence of water for reproduction—they need enough water

to form a continuous layer that "connects" gametophytes and allows sperm to swim to eggs. Sporophytes develop on the gametophyte and are nourished by the gametophyte when they are small.

- Seedless Vascular Plants > Lycophyta (Lycophytes, or Club Mosses)
- Seedless Vascular Plants > Psilotophyta (Whisk Ferns)
- Seedless Vascular Plants > Equisetophyta (or Sphenotophyta) (Horsetails)
- Seedless Vascular Plants > Pteridophyta (Ferns)

Seed Plants

The seed plants are a monophyletic group consisting of the gymnosperms—cycads, ginkgo, redwoods, pines, and gnetophytes—and the angiosperms. The group is defined by two key synapomorphies: the production of seeds and the production of pollen grains.

- Seeds are a specialized structure for dispersing embryonic sporophytes to new locations. Seeds are the mature form of a fertilized ovule, the female reproductive structure that encloses the female gametophyte and egg cell.
- Pollen grains are tiny, sperm-producing gametophytes that are easily dispersed through air as opposed to water.

Seed plants are found in virtually every type of habitat, and they adopt every growth habit known in land plants. Their forms range from mosslike mats to shrubs and vines to 100-meter-tall trees. Seed plants can be **annual** (have a single growing season) or **perennial** (live for many years), with life spans ranging from a few weeks to almost five thousand years. The structure and function of seed plants is the focus of Chapters 36 through 40.

- Seed Plants > Gymnosperms > Cycadophyta (Cycads)
- Seed Plants > Gymnosperms > Ginkgophyta (Ginkgoes)
- Seed Plants > Gymnosperms > Redwood Group (Redwoods, Junipers, Yews)
- Seed Plants > Gymnosperms > Pinophyta (Pines, Spruces, Firs)
- Seed Plants > Gymnosperms > Gnetophyta (Gnetophytes)
- Seed Plants > Anthophyta (Angiosperms)

The Ulvophyceae are a monophyletic group composed of several diverse and important subgroups, with a total of about 4000 species. Members of this lineage range from unicellular to multicellular.

Many of the large green algae in habitats along ocean coastlines are members of the Ulvophyceae. *Ulva*, the sea lettuce (**Figure 30.28**), is a representative marine species. But there are also large numbers of unicellular or small multicellular species that inhabit the plankton of freshwater lakes and streams.

Reproduction Most ulvophytes reproduce both asexually and sexually. Asexual reproduction often involves production of spores that swim with the aid of flagella. Sexual reproduction usually results in production of a resting stage—a cell that is dormant in winter. In many species the gametes are not called eggs and sperm, because they are the same size and shape. In most species gametes are shed into the water, so fertilization takes place away from the parent plants.

Life cycle Many unicellular forms are diploid only as zygotes. Alternation of generations occurs in multicellular species. When alternation of generations occurs, gametophytes and sporophytes may look identical or different.

Human and ecological impacts Ulvophyceae are important primary producers in freshwater environments and in coastal areas of the oceans.

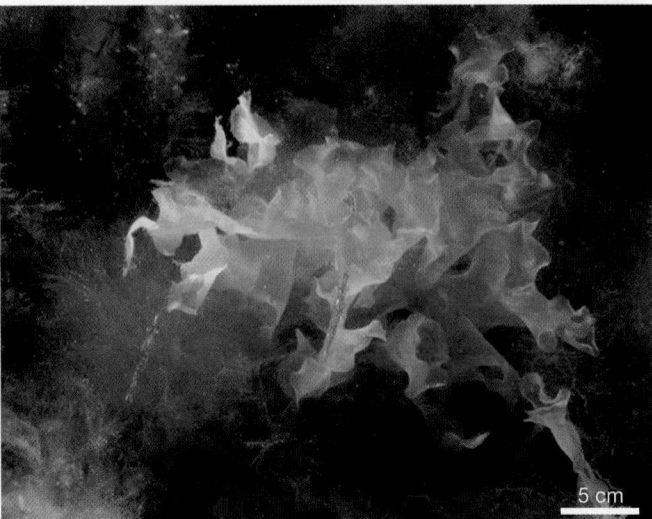

Ulva lactuca

5 cm

FIGURE 30.28 Green Algae Are Important Primary Producers in Aquatic Environments.

There are 19 species in this group. Most coleochaetes are barely visible to the unaided eye and grow as flat sheets of cells (**Figure 30.29**). They are considered multicellular because they have specialized photosynthetic and reproductive cells and because they contain **plasmodesmata**—structures introduced in Chapter 8 that connect adjacent cells.

The coleochaetes are strictly freshwater algae. They grow attached to aquatic plants such as water lilies and cattails or over submerged rocks in lakes and ponds. When they grow near beaches, they are often exposed to air when water levels drop in late summer.

Reproduction Asexual reproduction is common in coleochaetes and involves production of flagellated spores. During sexual reproduction, eggs are retained on the parent and are nourished after fertilization with the aid of transfer cells—a situation similar to that observed in land plants. In some species certain individuals are male and produce only sperm, while other individuals are female and produce only eggs.

Life cycle Alternation of generations does not occur. Multicellular individuals are haploid; the only diploid stage in the life cycle is the zygote.

Human and ecological impacts Because they are closely related to land plants, coleochaetes are studied intensively by researchers interested in how land plants made the water-to-land transition.

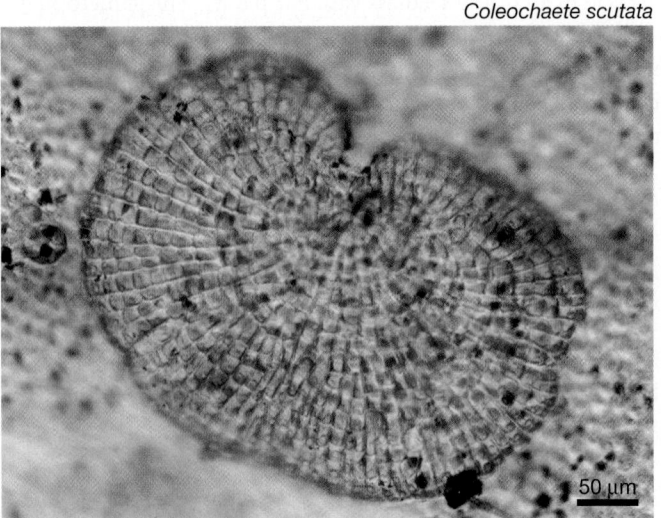

Coleochaete scutata

50 μm

FIGURE 30.29 Coleochaetes Are Thin Sheets of Cells.

There are several hundred species in this group. They are collectively known as stoneworts, because they commonly accumulate crusts of calcium carbonate ($CaCO_3$) over their surfaces. Like the coleochaetes, they have plasmodesmata and are multicellular. ✔You should be able to mark the origin of plasmodesmata on Figure 30.7. (They do not occur in ulvophytes.) Some species of stonewort can be a meter or more in length.

The stoneworts are freshwater algae. Certain species are specialized for growing in relatively deep waters, though most live in shallow water near lake beaches or pond edges.

Reproduction Sexual reproduction is common and involves production of prominent, multicellular gametangia similar to those observed in early land plants. In stoneworts, as in coleochaetes, the eggs are retained on the parent plant, which supplies eggs with nutrients prior to fertilization. ✔You should be able to mark the origin of egg retention on Figure 30.7. (Egg retention does not occur in ulvophytes.)

Life cycle Alternation of generations does not occur. Multicellular individuals are haploid; the only diploid stage in the life cycle is the zygote.

Human and ecological impacts Some species form extensive beds in lake bottoms or ponds and provide food for ducks and geese as well as food and shelter for fish (**Figure 30.30**). They are a good indicator that water is not polluted.

Chara species

5 cm

FIGURE 30.30 Stoneworts Can Form Beds on Lake Bottoms.

Liverworts got their name because some species native to Europe have liver-shaped leaves. According to the medieval *Doctrine of Signatures*, God indicated how certain plants should be used by giving them a distinctive appearance. Thus, liverwort teas were hypothesized to be beneficial for liver ailments. (They are not.) About 6500 species are known. They are commonly found growing on damp forest floors or riverbanks, often in dense mats (**Figure 30.31**), or on the trunks or branches of tropical trees.

Adaptations to land Liverworts are covered with cuticle. Some species have pores that allow gas exchange; in species that lack pores, the cuticle is thin.

Reproduction Asexual reproduction occurs when fragments of a plant are broken off and begin growing independently. Some species also produce small structures called **gemmae** asexually, during the gametophyte phase. Mature gemmae are knocked off the parent plant by rain and grow into independent gametophytes. During sexual reproduction, sperm and eggs are produced in gametangia.

Life cycle The gametophyte is the largest and longest-lived phase in the life cycle. Sporophytes are small, grow directly from the gametophyte, and depend on the gametophyte for nutrition. Spores are shed from the sporophyte and are carried away by wind or rain.

Human and ecological impacts When liverworts grow on bare rock or tree bark, their dead and decaying body parts contribute to the initial stages of soil formation. The liverwort *Marchantia* is an important model organism in plant biology.

Marchantia polymorpha

1 cm

FIGURE 30.31 Liverworts Thrive in Moist Habitats.

Over 12,000 species of mosses have been named and described to date, and more are being discovered every year—particularly in the tropics. Mosses are informally grouped with other "bryophytes" (liverworts and hornworts) but are formally classified in their own monophyletic group: the phylum Bryophyta.

Although mosses are common in moist forests, they can also be abundant in more extreme environments, such as deserts and windy, treeless habitats in the Arctic, Antarctic, or mountaintops. In these severe conditions, mosses are able to thrive because their bodies can become extremely dry without dying. When the weather makes photosynthesis difficult, individuals dry out and become dormant, or inactive. Then when rains arrive or temperatures warm, the plants rehydrate and begin photosynthesis and reproduction.

Adaptations to land One subgroup of mosses contains simple conducting tissues consisting of cells that are specialized for the transport of water or food. But because these cells do not have walls that are reinforced by lignin, they are not considered true vascular tissue. ✔You should be able to mark the origin of the simple water-conducting cells and tissues in this moss subgroup on Figure 30.7. Because they lack true vascular tissue, most mosses are not able to grow much taller than a few centimeters.

Reproduction Asexual reproduction often occurs by fragmentation, meaning that pieces of gametophytes that are broken off by wind or a passing animal can begin growing independently. In many species, sexual reproduction cannot involve self-fertilization because the sexes are separate—meaning that an individual plant produces only eggs in archegonia or only sperm in antheridia. A typical sporophyte produces up to 50 million tiny spores. Spores are usually distributed by wind.

Life cycle The moss life cycle is similar to that of liverworts and hornworts: The sporophyte is retained on the much larger and longer-lived gametophyte and gets most of its nutrition from the gametophyte.

Human and ecological impacts Species in the genus *Sphagnum* are often the most abundant plant in wet habitats of northern environments (**Figure 30.32a**). Because *Sphagnum*-rich environments account for 1 percent of Earth's total land area, equivalent to half the area of the United States, *Sphagnum* species are among the most abundant plants in the world. *Sphagnum*-rich habitats are water-logged, nitrogen-poor, and often anaerobic, however, so the decomposition of dead mosses and other plants is slow. As a result, large deposits of semi-decayed organic matter, known as **peat**, accumulate. Researchers estimate that the world's peatlands store about 400 billion metric tons of carbon. If peatlands begin to burn or decay rapidly due to global warming, the CO_2 released will exacerbate the warming trend (see Chapter 54).

Peat is harvested as a traditional heating and cooking fuel in some countries (**Figure 30.32b**). It is also widely used as a soil additive in gardening, because *Sphagnum* can absorb up to 20 times its dry weight in water. This high water-holding capacity is due to the presence of large numbers of dead cells in the leaves of these mosses, which readily fill with water via pores in their walls.

(a) *Sphagnum* moss is abundant in northern wet habitats.

5 mm

(b) Semi-decayed *Sphagnum* moss forms peat.

FIGURE 30.32 *Sphagnum* **Mosses Are among the Most Abundant Plants in the World.**

Hornworts got their name because their sporophytes have a horn-like appearance (**Figure 30.33**) and because wort is the Anglo-Saxon word for plant. About 100 species have been described to date.

Adaptations to land Hornwort sporophytes have stomata. Research is under way to determine if they can open their pores to allow gas exchange or close their pores to avoid water loss during dry intervals.

Reproduction Depending on the species, gametophytes may contain only egg-producing archegonia or only sperm-producing antheridia, or both. Stated another way, individuals of some species are either female or male, while in other species each individual has both types of reproductive organs.

Life cycle The gametophyte is the longest-lived phase in the life cycle. Although sporophytes grow directly from the gametophyte, they are green because their cells contain chloroplasts. Sporophytes manufacture some of their own food but also get nutrition from the gametophyte. Spores disperse from the parent plant via wind or rain.

Human and ecological impacts Some species harbor symbiotic cyanobacteria that fix nitrogen.

Anthoceros laevis

5 mm

FIGURE 30.33 Hornworts Have Horn-Shaped Sporophytes.

Seedless Vascular Plants > Lycophyta (Lycophytes, or Club Mosses)

Although the fossil record documents lycophytes that were 2 m wide and 40 m tall, the 1000 species of lycophytes living today are all small in stature (**Figure 30.34**). Most live on the forest floor or on the branches or trunks of tropical trees. Because of their appearance, they are often called ground pines or **club mosses**—even though they are neither pines nor mosses.

Adaptations to land Lycophytes are the most ancient land plant lineage with **roots**—a belowground system of tissues and organs that anchors the plant and is responsible for absorbing water and mineral nutrients. Roots differ from the rhizoids observed in bryophytes, because roots contain vascular tissue and thus are capable of conducting water and nutrients from belowground to the upper reaches of the plant. Unusual leaves called microphylls, which extend from the stems, are a synapomorphy found in lycophytes. ✔You should be able to mark the origin of microphylls on Figure 30.7.

Reproduction Asexual reproduction can occur by fragmentation or gemmae. During sexual reproduction, spores of some species give rise to bisexual gametophytes—meaning that each gametophyte produces both eggs and sperm. Self-fertilization is extremely rare in most of these species, however. In the genera *Selaginella* and *Isoetes*, in contrast, heterospory occurs and gametophytes are either male or female.

Life cycle The gametophytes of some species live entirely underground and get their nutrition from symbiotic fungi. In certain species, gametophytes live 6 to 15 years and give rise to a large number of sporophytes over time.

Human and ecological impacts Tree-sized lycophytes were abundant in the coal-forming forests of the Carboniferous period. In coal-fired power plants today, electricity is being generated by burning fossilized lycophyte and fern tissues.

Lycopodium species

5 cm

FIGURE 30.34 Lycophytes Living Today Are Small in Stature.

Only two genera of whisk ferns are living today, and there are perhaps six distinct species. Whisk ferns are restricted to tropical regions and have no fossil record. They are extremely simple morphologically, with aboveground parts consisting of branching stems that have tiny, scale-like outgrowths instead of leaves (**Figure 30.35**).

Because they lack roots and leaves, the whisk ferns were considered an evolutionary "throwback"—a group that retained traits found in the earliest vascular plants. Molecular phylogenies challenge this hypothesis and support an alternative hypothesis: The morphological simplicity of whisk ferns is a derived trait—meaning that complex structures have been lost in this lineage.

Adaptations to land Whisk ferns lack roots. Some species gain most of their nutrition from fungi that grow in association with the whisk ferns' extensive underground stems called **rhizomes**. Other species grow in rock crevices or are **epiphytes** ("upon-plants"), meaning that they grow on the trunks or branches of other plants—in this case, in

Psilotum nudum

1 cm

FIGURE 30.35 Psilotophytes Are Extremely Simple Morphologically.

the branches of tree ferns. ✔After reviewing where leaves and roots originated during land plant evolution (see Figure 30.10), you should be able to mark the loss of leaves and roots in whisk ferns on Figure 30.7.

Reproduction Asexual reproduction occurs via the extension of rhizomes and the production of new aboveground stems. When spores mature, they are dispersed by wind and germinate into gametophytes that contain both archegonia and antheridia.

Life cycle Sporophytes may be up to 30 cm tall, but gametophytes are less than 2 mm long and live under the soil surface. Gametophytes absorb nutrients directly from the surrounding soil and from symbiotic fungi. Fertilization takes place inside the archegonium, and the sporophyte develops directly on the gametophyte.

Human and ecological impacts Some whisk fern species are popular landscaping plants, particularly in Japan. The same species can be a serious pest in greenhouses.

Although horsetails are prominent in the fossil record of land plants, just 15 species are known today. All 15 are in the genus *Equisetum*. Translated literally, *Equisetum* means "horse-bristle." Both the scientific name and the common name, horsetail, come from the brushy appearance of the stems and branches in some species (**Figure 30.36**). Horsetails may be locally abundant in wet habitats such as stream banks or marsh edges.

Adaptations to land Horsetails have an adaptation that allows them to flourish in waterlogged, oxygen-poor soils: their stems are hollow, so oxygen readily diffuses down the stem to reach roots that cannot obtain oxygen from the surrounding soil. Horsetails are also distinguished by having whorled leaves and branches. ✔You should be able to mark the origin of hollow stems, whorled branches, and whorled leaves on Figure 30.7.

Reproduction Asexual reproduction is common in sporophytes and occurs via

Equisetum telmateia

1 cm

FIGURE 30.36 Horsetails Have a Distinctive Brushy Appearance and Are Prominent in the Fossil Record.

fragmentation or the extension of rhizomes. From these rhizomes, two types of erect, specialized stems may grow—stems that contain tiny leaves and chloroplast-rich branches and that are specialized for photosynthesis, or stems that bear clusters of sporangia and produce huge numbers of spores by meiosis.

Life cycle Gametophytes perform photosynthesis but are small and short lived. They normally produce both antheridia and archegonia, but in most cases the sperm-producing structure matures first. This pattern is thought to be an adaptation that minimizes self-fertilization and maximizes cross-fertilization.

Human and ecological impacts Horsetail stems are rich in silica granules. The glass-like deposits not only strengthen the stem but also made these plants useful for scouring pots and pans prior to the invention of other scrubbing tools—hence these plants are often called "scouring rushes."

With 12,000 species, ferns are by far the most species-rich group of seedless vascular plants. They are particularly abundant in the tropics. About a third of the tropical species are epiphytes, usually growing on the trunks or branches of trees. Species that can grow epiphytically live high above the forest floor, where competition for light is reduced, without making wood and growing tall themselves. The growth habits of ferns are highly variable among species, however, and ferns range in size from rosettes the size of your smallest fingernail to 20-meter-tall trees (**Figure 30.37**).

Adaptations to land Ferns are the only seedless vascular plants that have large, well-developed leaves—commonly called **fronds** because it is not yet clear whether they are homologous to the leaves found in seed plants. Leaves give the plant a large surface area, allowing it to capture sunlight for photosynthesis efficiently.

Reproduction In a few species, gametophytes reproduce asexually via production of gemmae. Typically, species that can reproduce via gemmae never produce gametes or sporophytes. In most species, however, sexual reproduction is the norm. Ferns are homosporous, but gametophytes develop as males or as females only, depending on light intensity, proximity to other gametophytes, and other environmental conditions.

Life cycle Although fern gametophytes contain chloroplasts and are photosynthetic, the sporophyte is typically the larger and longer-lived phase of the life cycle. In mature sporophytes, sporangia are usually found in clusters called **sori** on the undersides of

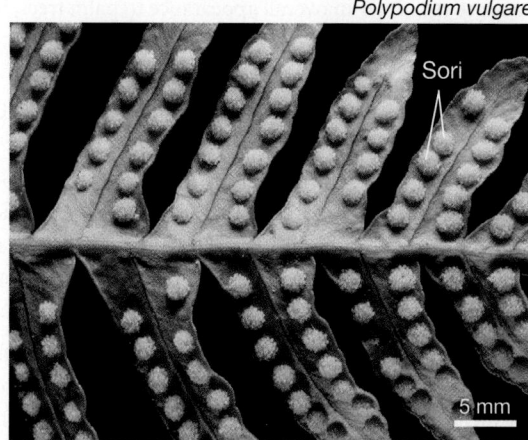

Polypodium vulgare

Sori

5 mm

FIGURE 30.38 Fern Sporangia Are Often in Clusters (Sori) on the Underside of Leaves.

leaves (**Figure 30.38**). The structure of the sporangia is a distinctive feature of ferns: It arises from a single cell and has a wall composed of a single cell layer. ✔You should be able to mark the evolution of the distinctive fern sporangium on Figure 30.7.

Human and ecological impacts In many parts of the world, people gather the young, unfolding fronds, or "fiddleheads," of ferns in spring as food. Ferns are also widely used as ornamental plants in landscaping.

Gonocormus minutus

Cyathea medullaris

1 m

FIGURE 30.37 Ferns Vary in Size and Are the Most Species-Rich Group of Seedless Vascular Plants.

The cycads are so similar in overall appearance to palm trees, which are angiosperms, that cycads are sometimes called "sago palms." Although cycads were extremely abundant when dinosaurs were present on Earth 150–65 million years ago, only about 140 species are living today. Most are found in the tropics (**Figure 30.39**).

Adaptations to land Cycads do not make wood but are supported by stiff stems. They are unique among gymnosperms in having compound leaves—meaning that each leaf is divided into many smaller leaflets. ✓You should be able to mark the origin of the distinctive cycad leaf on Figure 30.7.

Reproduction and life cycle Like other seed plants, cycads are heterosporous. Each sporophyte individual bears either microsporangia or megasporangia, but not both. Pollen is carried by insects (usually beetles or weevils) or, in some species, wind. Cycad seeds are large and often brightly colored. The colors attract birds and mammals, both of which eat and disperse the seeds.

Human and ecological impacts Cycads harbor large numbers of symbiotic cyanobacteria in specialized, aboveground root structures. The cyanobacteria are photosynthetic and fix nitrogen. The nitrogen acts as an important nutrient for nearby plants as well as the cycads themselves. Cycads are popular landscaping plants in some parts of the world.

Encephalartos transvenosus

10 cm

FIGURE 30.39 Cycads Resemble Palms but Are Not Closely Related to Them.

Although ginkgoes have an extensive fossil record, just one species is alive today. Leaves from the ginkgo, or maidenhair, tree are virtually identical in size and shape to those observed in fossil ginkgoes that are 150 million years old (**Figure 30.40**).

(a) Fossil ginkgo

Ginkgo huttoni

(b) Living ginkgo

Ginkgo biloba

2 cm

5 cm

FIGURE 30.40 The Ginkgo Tree Is a "Living Fossil."

Adaptations to land Unlike most gymnosperms, the ginkgo is **deciduous**—meaning that it loses its leaves each autumn. This adaptation allows plants to be dormant during the winter, when photosynthesis and growth are difficult.

Reproduction and life cycle Sexes are separate—individuals are either male or female. Pollen is transported by wind. Sperm have flagella, however. Once pollen grains land near the female gametophyte and mature, the sperm cells leave the pollen grain and swim to the egg cells.

Human and ecological impacts Although today's ginkgo trees are native to southeast China, they are planted widely as an ornamental. They are especially popular in urban areas, as they are tolerant of air pollution. In some countries, the inside of the seed is eaten as a delicacy.

The species in this lineage vary in growth form from sprawling juniper shrubs to the world's largest plants. Redwood trees growing along the Pacific Coast of North America can reach heights of over 115 m (375 ft) and trunk diameters of over 9.5 m (30 ft). These species were recently confirmed as a monophyletic group, independent of the Pinophyta, which also reproduce via cone-bearing structures, but the lineage still does not have a formal name.

Adaptations to land All of the species in this lineage are trees or large shrubs. Most have narrow leaves, which in many cases are arranged in overlapping scales (**Figure 30.41**). Narrow leaves have a small amount of surface area, which is not optimal for capturing sunlight and performing photosynthesis. But because the small surface area reduces water loss from leaves, many species in this lineage thrive in dry habitats or in cold environments where water is often frozen.

Thuja plicata

5 cm

FIGURE 30.41 Some Species in the Redwood Group Have Scale-Like Leaves.

Reproduction and life cycle The species in this group are wind pollinated. As in all seed plants, the female gametophyte is retained on the parent. Thus, fertilization and seed development take place in the female cone. Like other gymnosperms, seeds do not form inside an encapsulated structure. Depending on the species, the seeds are dispersed by wind or by seed-eating birds or mammals.

Human and ecological impacts Redwoods, redcedar, whitecedar, and yellowcedar have wood that is highly rot-resistant and thus prized for making furniture, decks, house siding, or other applications where wood is exposed to the weather. Yew wood is often preferred for making traditional archery bows, and the berry-like cones of juniper are used to flavor gin. The anticancer drug taxol was originally found in Pacific yew trees.

The gymnosperms include two major lineages of cone-bearing species: the pines and allies discussed here, and the group that includes redwoods, junipers, yews, and cypresses. Species in both lineages have a reproductive structure called the cone, in which microsporangia and megasporangia are produced (**Figure 30.42**).

Pinophyta include the familiar pines, spruces, firs, Douglas fir, tamaracks, and true cedars. These are among the largest and most abundant trees on the planet, as well as some of the most long-lived. One of the bristlecone pines native to southwestern North America is at least 4750 years old.

Adaptations to land Pinophyta have needle-like leaves, with a small surface area that allows them to thrive in habitats where water is scarce. Pines are common on sandy soils that have poor water-holding capacity, and spruces and firs are common in cold environments where water is often frozen. All of the living species make wood as a support structure.

Reproduction and life cycle Sexes are separate, and pollen is transferred to female cones by the wind. Female cones take two years to mature, and are usually found high in trees. Male cones are usually found lower in the tree—possibly to reduce self-pollination from falling pollen.

Human and ecological impacts In terms of biomass, pines, spruces, firs, and other species in this group dominate forests that grow at high latitudes and high elevations—as well as sandy sites in warmer regions. Their seeds are key food sources for a variety of birds, squirrels, and mice, and their wood is the basis of the building products and paper industries in many parts of the world. The paper in this book was made from species in this group.

(a) Cones that produce microsporangia and pollen
Picea abies

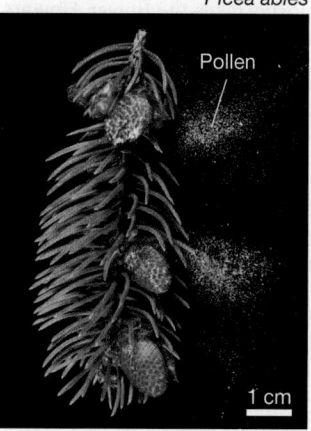
Pollen

1 cm

(b) Cones that produce macrosporangia and eggs
Picea abies

5 cm

FIGURE 30.42 Pollen-Bearing Cones Produce Microsporangia; Ovulate Cones Produce Megasporangia.

The gnetophytes comprise about 70 species in three genera: *Gnetum* comprises vines and trees from the tropics; *Ephedra* is made up of desert-dwelling shrubs, including what may be the most familiar gnetophyte—the shrub called Mormon tea, which is common in the deserts of southwestern North America; *Welwitschia* contains a single species that probably qualifies as the world's most bizarre plant (**Figure 30.43**)—*W. mirabilis*, which is native to the deserts of southwest Africa. Although it has large belowground structures, the aboveground part consists of just two strap-like leaves, which grow continuously from the base and die at the tips. The leaves also split lengthwise as they grow and age.

Adaptations to land Gnetophytes have vessel elements in addition to tracheids. All of the living species make wood as a support structure.

Reproduction and life cycle The microsporangia and megasporangia are arranged in clusters at the end of stalks, similar to the way flowers are clustered in some angiosperms. Pollen is transferred by the wind or by insects. Double fertilization occurs in two of the three genera. As in other gymnosperms, seeds do not form inside an encapsulated structure.

Human and ecological impacts The drug ephedrine was originally isolated from species of *Ephedra* that are native to northern China and Mongolia. Ephedrine is used in the treatment of hay fever, colds, and asthma.

Welwitschia mirabilis

0.5 m

FIGURE 30.43 *Welwitschia* **Is an Unusual Plant.**

The flowering plants, or angiosperms, are far and away the most species-rich lineage of land plants. Over 250,000 species have already been described. They range in size from *Lemna gibba*—a floating, aquatic species that is less than half a millimeter wide—to massive oak trees. Angiosperms thrive in desert to freshwater to rain forest environments and are found in virtually every habitat except the deep oceans. They are the most common and abundant plants in most terrestrial environments.

The defining adaptation of angiosperms is the flower. **Flowers** are reproductive structures that hold either pollen-producing microsporangia or the megasporangia that produce megaspores and eggs, or both. Nectar-producing cells are often present at the base of the flower, and the color of petals helps to attract insects, birds, or bats that carry pollen from one flower to another (**Figure 30.44a**). Some angiosperms are pollinated by wind, however. Wind-pollinated flowers lack both colorful petals and nectar-producing cells (**Figure 30.44b**).

Adaptations to land In addition to flowers, angiosperms evolved vessels, the conducting cells that make water transport particularly efficient. Most angiosperms contain both tracheids and vessels.

Reproduction and life cycle Unlike gymnosperms, angiosperms have a carpel, a structure within the flower that contains an ovary.

The ovary encloses one or more ovules, each of which encloses the female gametophyte. In most cases, male gametophytes are carried to female gametophytes by animal pollinators that are inadvertently dusted with pollen as they visit flowers to find food. Depending on the angiosperm species, self-fertilization may be common or absent. When the egg produced by the female gametophyte is fertilized, the ovule develops into a seed. When the ovary matures it forms a fruit, which contains the seed or seeds.

Human and ecological impacts It is almost impossible to overstate the importance of angiosperms to humans and other organisms. In most terrestrial habitats today, angiosperms supply the food that supports virtually every other species. For example, many insects eat flowering plants. Historically, the diversification of angiosperms correlated closely with the diversification of insects, which are by far the most species-rich lineage on the tree of life. It is not unusual for a single tropical tree to support dozens or even hundreds of insect species. Angiosperm seeds and fruits have also supplied the staple foods of virtually every human culture that has ever existed.

(Continued on next page)

(a) Animal-pollinated flower (this species produces both pollen and eggs in the same flower)

Ornithogalum dubium

0.5 cm

(b) Wind-pollinated flower (this species has separate male and female flowers)

Acer negundo

Male flower

1 cm

Acer negundo

Female flower

1 cm

FIGURE 30.44 Wind-Pollinated Flowers Lack the Colorful Petals and Nectar Found in Most Animal-Pollinated Species.

CHAPTER 30 REVIEW

For media, go to the study area at www.masteringbiology.com

Summary of Key Concepts

🗝️ **Green algae are an important source of oxygen and provide food for aquatic organisms; land plants release oxygen, hold soil and water in place, build soil, moderate extreme temperatures and winds, and provide food for terrestrial organisms.**

- Plants improve the quality of the environment for other organisms.

- Humans depend on plants for food, fiber, and fuel.

  ✔️ You should be able to predict how the current and massive loss of plant species and plant communities will affect soils around the world.

🗝️ **Land plants were the first multicellular organisms that could live with most of their tissues exposed to the air. A series of key adaptations allowed them to survive on land. In terms of total mass, plants dominate today's terrestrial environments.**

- The land plants evolved from green algae and colonized terrestrial environments in conjunction with fungi.

- The evolution of cuticle allowed plant tissues to be exposed to air without dying.

- The evolution of pores provided breaks in the cuticle and facilitated gas exchange, with CO_2 diffusing into leaves and O_2 diffusing out.

- The evolution of guard cells allowed plants to control the opening and closing of pores in a way that maximizes gas exchange and minimizes water loss.

- Vascular tissue conducts water and provides structural support that makes erect growth possible.

- True vascular tissue has cells that are dead at maturity and that have secondary cell walls reinforced with lignin.

- Erect growth is important because it reduces competition for light.

- Tracheids are water-conducting cells found in all vascular plants; in addition, angiosperms and gnetophytes have water-conducting cells called vessels.

  ✔️ You should be able to explain why algae made the transition to terrestrial life just once, in light of the number of adaptations that were required.

MB **Web Activity** Plant Evolution and the Phylogenetic Tree

🗝️ **Once plants were able to grow on land, a sequence of important evolutionary changes made it possible for them to reproduce efficiently—even in extremely dry environments.**

- All land plants are embryophytes, meaning that eggs and embryos are retained on the parent plant. Consequently, the developing embryo can be nourished by its mother.

- Seed plant embryos are dispersed from the parent plant to a new location, encased in a protective housing, and supplied with a store of nutrients.

- All land plants have alternation of generations.

- Over the course of land plant evolution, the gametophyte phase became reduced in terms of size and life span and the sporophyte phase became more prominent. In seed plants, male gametophytes are reduced to pollen grains and female gametophytes are reduced to tiny structures that produce an egg.

- The evolution of pollen was an important breakthrough in the history of life, because sperm no longer needed to swim to the egg—tiny male gametophytes could be transported through the air via wind or animals.

✔ You should be able to discuss (compare and contrast) the advantages and disadvantages of having spores or seeds serve as the dispersal stage in a plant life cycle.

Questions

✔ TEST YOUR KNOWLEDGE

1. Which of the following groups is definitely monophyletic?
   a. non-vascular plants
   b. green algae
   c. green plants
   d. seedless vascular plants

2. What is a difference between tracheids and vessels?
   a. Tracheids are dead at maturity; vessels are alive and are filled with cytoplasm.
   b. Vessels have openings (gaps) in the primary and secondary cell wall; tracheids have openings (pits) only in the secondary cell wall.
   c. Only tracheids have a thick secondary cell wall containing lignin.
   d. Only vessels have a thick secondary cell wall containing lignin.

3. Which of the following statements is *not* true?
   a. Green algae in the lineage called Charales are the closest living relatives of land plants.
   b. "Bryophytes" is a name given to the land plant lineages that do not have vascular tissue.
   c. The horsetails and the ferns form a distinct clade, or lineage. They have vascular tissue but reproduce via spores, not seeds.

   d. According to the fossil record and phylogenetic analyses, angiosperms evolved before the gymnosperms. Angiosperms are the only land plants with vessels.

4. The appearance of cuticle and stomata correlated with what event in the evolution of land plants?
   a. the first erect growth forms
   b. the first woody tissues
   c. growth on land
   d. the evolution of the first water-conducting tissues

5. What do seeds contain?
   a. male gametophyte and nutritive tissue
   b. female gametophyte and nutritive tissue
   c. embryo and nutritive tissue
   d. mature sporophyte and nutritive tissue

6. What is a pollen grain?
   a. male gametophyte
   b. female gametophyte
   c. male sporophyte
   d. sperm

✔ TEST YOUR UNDERSTANDING

1. Soils, water, and the atmosphere are major components of the abiotic (nonliving) environment. Describe how green plants affect the abiotic environment in ways that are advantageous to humans.

2. The evolution of cuticle presented land plants with a challenge that threatened their ability to live on land. Describe this challenge and explain why stomata represent a solution. Compare and contrast stomata with the pores found in liverworts. Explain why it is logical to observe that liverworts that lack pores have extremely thin cuticle.

3. Why was the evolution of lignin-reinforced cell walls significant?

4. Land plants may have reproductive structures that (1) protect gametes as they develop; (2) nourish developing embryos, (3) allow sperm to

be transported in the absence of water, (4) provide stored nutrients and a protective coat so that offspring can be dispersed away from the parent plant, and (5) provide nutritious tissue around seeds that facilitates dispersal by animals. Name each of these five structures, and state which land plant group or groups have each structure.

5. What does it mean to say that a life cycle is gametophyte-dominant versus sporophyte-dominant?

6. Explain the difference between homosporous and heterosporous plants. Where are the microsporangium and megasporangium found in a tulip? What happens to the spores that are produced by these structures?

✔ APPLYING CONCEPTS TO NEW SITUATIONS

1. What is the significance of the observation that some members of the coleochaetes and stoneworts synthesize sporopollenin and/or lignin?

2. Vessel elements transport water much more efficiently than tracheids, but are much more susceptible than tracheids to being blocked by air bubbles. Suggest a hypothesis to explain why the vascular tissue of angiosperms consists of a combination of vessel elements and tracheids.

3. Angiosperms such as grasses, oaks, and maples are wind pollinated. The ancestors of these subgroups were probably pollinated by insects, however. As an adaptive advantage, why might a species "revert" to wind pollination? (Hint: Think about the costs and benefits of being pollinated by insects versus wind.) Why is it logical

to observe that wind-pollinated species usually grow in dense stands containing many individuals of the same species? Why is it logical to observe that in wind-pollinated deciduous trees, flowers form very early in spring—before leaves form?

4. You have been hired as a field assistant for a researcher interested in the evolution of flower characteristics in orchids. Design an experiment to determine whether color, size, shape, scent, or amount of nectar is the most important factor in attracting pollinators to a particular species. Assume that you can change any flower's color with a dye and that you can remove petals or nectar stores, add particular scents, add nectar by injection, or switch parts among species by cutting and gluing.

The jack-o-lantern mushroom (*Omphalotus illudens*) is common on rotting logs and stumps in eastern North America. It is toxic to humans, but a derivative of one of the molecules responsible for its toxicity is being tested as an anticancer drug.

Fungi **31**

ungi are eukaryotes that grow as single cells or as large, branching networks of multicellular filaments. Familiar fungi include the mushrooms you've encountered in woods or lawns, the molds and mildews in your home, the organism that causes athlete's foot, and the yeasts used in baking and brewing.

Along with the land plants and animals, the **fungi** are one of three major lineages of large, multicellular eukaryotes that occupy terrestrial environments. When it comes to making a living, the species in these three groups use radically different strategies. Land plants make their own food through photosynthesis. Animals eat plants, protists, fungi, or each other. Fungi absorb their nutrition from other organisms—dead or alive.

Fungi that absorb nutrients from dead organisms are the world's most important decomposers. Although a few types of organisms are capable of digesting the cellulose in plant cell walls, fungi and a handful of bacterial species are the only organisms capable of completely digesting both the lignin and cellulose that make up wood. Without fungi, Earth's surface would be piled so high with dead tree trunks and branches that there would be almost no room for animals to move or plants to grow.

Other fungi specialize in absorbing nutrients from living organisms. When fungi absorb these nutrients without providing any benefit in return, they lower the fitness of their host organism and act as parasites. If you've ever had athlete's foot or a vaginal yeast infection, you've hosted a parasitic fungus.

The vast majority of fungi that live in association with other organisms benefit their hosts, however. In these cases, fungi are not parasites but **mutualists**. The roots of virtually every land plant in the world are colonized by an array of mutualistic fungi. In exchange for sugars that are synthesized by the host plant, the fungi provide the plant with water and key nutrients such as nitrogen and phosphorus. Without these nutrients, the host plants grow much more slowly or even starve.

KEY CONCEPTS

◖➤ Fungi are important in part because many species live in close association with land plants. They supply plants with key nutrients and decompose dead wood. They are the master recyclers of nutrients in terrestrial environments.

◖➤ All fungi make their living by absorbing nutrients from living or dead organisms. Fungi secrete enzymes so that digestion takes place outside their cells. Their morphology provides a large amount of surface area for efficient absorption.

◖➤ Many fungi have unusual life cycles. It is common for species to have a long-lived heterokaryotic stage, in which cells contain haploid nuclei from two different individuals. Although most species reproduce sexually, few species produce gametes.

✔ When you see this checkmark, stop and test yourself. Answers are available in Appendix B.

579

It is not possible to overstate the importance of these relationships between living land plants and the fungi that live in their roots. In the soils beneath every prairie, forest, and desert, an underground economy is flourishing. Plants are trading the sugar they manufacture for nitrogen or phosphorus atoms that are available from fungi. These plant-fungal associations are the world's most extensive bartering system. The soil around you is alive with an enormous network of fungi that are fertilizing the plants you see aboveground.

In short, fungi are the master traders and recyclers in terrestrial ecosystems. Some fungi release nutrients from dead plants and animals; others transfer nutrients they obtain to living plants. 🔑 Because they recycle key elements such as carbon, nitrogen, and phosphorus and because they transfer key nutrients to plants, fungi have a profound influence on productivity and biodiversity. In terms of nutrient cycling on the continents, fungi make the world go around.

31.1 Why Do Biologists Study Fungi?

Given their importance to life on land and their intricate relationships to other organisms, it's no surprise that fungi are fascinating to biologists. But there are important practical reasons for humans to study fungi as well. They nourish the plants that nourish us. They affect global warming, because they are critical to the carbon cycle on land. Unfortunately, a handful of species can cause debilitating diseases in humans and crop plants. Let's take a closer look at some of the ways that fungi affect human health and welfare.

Fungi Provide Nutrients for Land Plants

Fungi that live in close association with plant roots are said to be **mycorrhizal** (literally, "fungal-root"; see **Figure 31.1a**). When biologists first discovered how extensive these fungal-plant associations are, they asked an obvious question: Does plant growth suffer if mycorrhizal fungi are absent?

Figure 31.1b shows a result typical of many experiments. In this case, seedlings were grown in the presence and absence of the mycorrhizal fungi normally found on their roots. The photographs document that this species grows three to four times faster in the presence of its normal fungal associates than it does without them.

For farmers, foresters, and ranchers, the presence of normal mycorrhizal fungi can mean the difference between profit and loss. Fungi are critical to the productivity of forests, croplands, and rangelands.

Fungi Speed the Carbon Cycle on Land

Fungi that make their living by digesting dead plant material are called **saprophytes** ("rotten-plants"). To understand why saprophytic fungi play a key role in today's terrestrial environments, recall from Chapter 30 that cells in the vascular tissues of land plants have secondary cell walls containing both lignin and cellulose. Wood forms when stems grow in girth by adding layers of lignin-rich vascular tissue.

(a) Mycorrhizal fungi form extensive networks in soil.

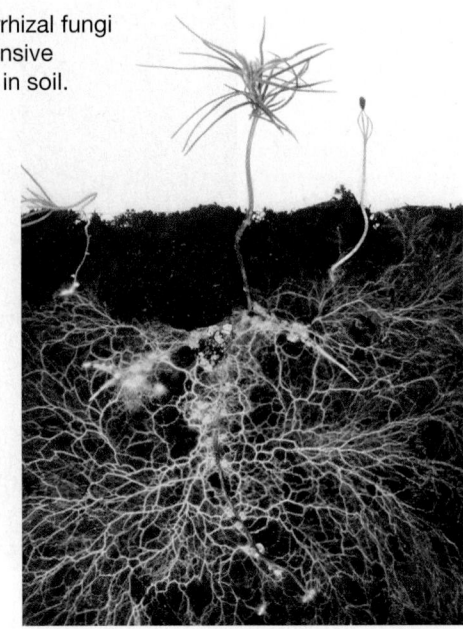

(b) Mycorrhizal fungi increase plant growth.

With mycorrhizal fungi   Without

FIGURE 31.1 Plants Grow Better in the Presence of Mycorrhizal Fungi. (a) Root system of a larch tree seedling, with the mycelium from a mycorrhizal fungus visible. **(b)** Typical experimental results when plants are grown with and without their normal mycorrhizal fungi. (Fungi are not visible in the photo.)

When trees die, fungi are the organisms that break down wood into sugars and other small organic compounds. Fungi use these molecules as food. When fungi die or are eaten, the molecules are passed along to a wide array of other organisms.

Figure 31.2 highlights the role that fungi play as carbon atoms cycle through today's terrestrial environments. Note that there are two basic components of the **carbon cycle** on land:

1. the fixation of carbon by land plants—meaning that carbon in atmospheric CO_2 is reduced to cellulose, lignin, and other complex organic compounds in the bodies of plants; and

2. the release of CO_2 from plants, animals, and fungi as the result of cellular respiration—meaning the oxidation of glucose and production of the ATP that sustains life.

FIGURE 31.2 Fungi Speed Up the Cycling of Carbon Atoms through Terrestrial Ecosystems.

(labels within figure:)
Atmospheric CO_2
Live plants
Release of CO_2 from cellular respiration by plants, animals, fungi, and other organisms
Fixation of CO_2 by plants
Plants
Fungi
Animals
Dead plants
Fungi are eaten
Fungal mycelia digest cellulose and lignin (from wood) to obtain sugars and other small organic compounds

(a) Parasitic fungi infect corn and other crop plants.

(b) Saprophytic fungi rot fruits and vegetables.

FIGURE 31.3 Fungi Cause Problems with Crop Production and Storage. (a) A wide variety of grain crops are parasitized by fungi. Corn smut is a serious disease in sweet corn, although in Mexico the smut fungus is eaten as a delicacy. **(b)** Fungi decompose fruits and vegetables as well as leaves and tree trunks.

The fundamental point is that, for most carbon atoms, fungi connect the two parts of the cycle.

If fungi had not evolved the ability to digest lignin and cellulose soon after land plants evolved the ability to make these compounds, carbon atoms would have been sequestered in wood for millennia instead of being rapidly recycled into glucose molecules and CO_2. Terrestrial environments would be radically different than they are today and probably much less productive. On land, fungi make the carbon cycle turn much more rapidly than it would without fungi.

Fungi Have Important Economic Impacts

In humans, parasitic fungi cause athlete's foot, vaginitis, diaper rash, ringworm, pneumonia, and thrush, among other miseries. But even though these maladies can be serious, in reality only about 31 species of fungi—out of the hundreds of thousands of existing species—regularly cause illness in humans. Compared with the frequency of diseases caused by bacteria, viruses, and protists, the incidence of fungal infections in humans is low.

It would be easy to argue, in fact, that fungi have done more to promote human health than degrade it. The first antibiotic that was widely used, penicillin, was isolated from a fungus, and soil-dwelling fungi continue to be the source of many of the most important antibiotics prescribed against bacterial infections.

The major destructive impact that fungi have on people is through the food supply. Fungi known as rusts, smuts, mildews, wilts, and blights cause annual crop losses computed in the bil-

lions of dollars. These fungi are particularly troublesome in wheat, corn, barley, and other grain crops (**Figure 31.3a**). Saprophytic fungi are also responsible for enormous losses due to spoilage—particularly for fruit and vegetable growers (**Figure 31.3b**).

In nature, epidemics caused by fungi have killed 4 billion chestnut trees and tens of millions of American elm trees in North America. The fungal species responsible for these epidemics were accidentally imported on species of chestnut and elm native to other regions of the world. When the fungi arrived in North America and began growing in chestnuts and elms native to North America, the results were catastrophic. The local chestnut and elm populations had virtually no genetic resistance to the pathogens and quickly succumbed. The epidemics radically altered the composition of upland and floodplain forests in the eastern United States. Before these fungal epidemics occurred, chestnuts and elms dominated these habitats.

Fungi also have important positive impacts on the human food supply:

- Mushrooms are consumed in many cultures; in the industrialized nations they are used in sauces, salads, and pizza.

- The yeast *Saccharomyces cerevisiae* was domesticated thousands of years ago; today it and other fungi are essential to the manufacture of bread, soy sauce, tofu, cheese, beer, wine, whiskey, and other products. In most cases, domesticated fungi are used in conditions where the cells grow via fermentation, creating by-products like the CO_2 that causes bread to rise and beer and champagne to fizz.

- Enzymes derived from fungi are used to improve the characteristics of foods ranging from fruit juice and candy to meat.

To summarize, biologists study fungi because they affect a wide range of species in nature, including humans. What tools are helping researchers understand the diversity of fungi?

31.2 How Do Biologists Study Fungi?

About 80,000 species of fungi have been described and named to date, and about 1000 more are discovered each year. But the fungi are so poorly studied that the known species are widely regarded as a tiny fraction of the actual total.

To predict the actual number of fungal species alive today, David Hawksworth looked at the ratio of vascular plant species to fungal species in the British Isles—the area where the two groups are the most thoroughly studied. According to Hawksworth's analysis, there is an average of six species of fungus for every species of vascular plant on these islands. If this ratio holds worldwide, then the estimated total of 275,000 vascular plant species implies that there are 1.65 million species of fungi.

Although this estimate sounds large, recent data on fungal diversity suggest that it may be an underestimate. Consider what researchers found when they analyzed fungi growing on Barro Colorado Island, Panama: Living on the healthy leaves of just two tropical tree species were a total of 418 distinct morphospecies of fungi. (Recall from Chapter 26 that morphospecies are distinguished from each other by some aspect of morphology.) Because over 310 species of trees and shrubs grow on Barro Colorado, the data suggest that tens of thousands of fungi may be native to this island alone. If further work on fungal diversity in the tropics supports these conclusions, there may turn out to be many millions of fungal species.

This viewpoint of fungal diversity was reinforced by an analysis of the fungi living in conjunction with the roots of a single species of grass native to Eurasia. In this study, researchers used the direct sequencing approach, introduced in Chapter 28, to analyze the gene that codes for the RNA molecule in the small subunit of fungal ribosomes. The data showed that 49 phylogenetic species were living in conjunction with the grass roots. (The phylogenetic species concept was introduced in Chapter 26.) Most of the species had never before been described, and several represented completely new lineages of fungi. Biologists are only beginning to realize the extent of species diversity in fungi.

Let's consider how biologists are working to make sense of all this diversity, beginning with an overview of fungal morphology.

Analyzing Morphological Traits

Compared with animals and land plants, fungi have simple bodies. Only two growth forms occur among them: (1) single-celled forms called **yeasts** (**Figure 31.4a**), and (2) multicellular, filamentous structures called **mycelia** (singular: **mycelium**; **Figure 31.4b**). Many species of fungus grow either as a yeast or as a mycelium, but some regularly adopt both growth forms.

Because most fungi form mycelia and because this body type is so fundamental to the absorptive mode of life, most studies of fungal morphology have focused on them.

(a) Single-celled fungi are called yeasts.

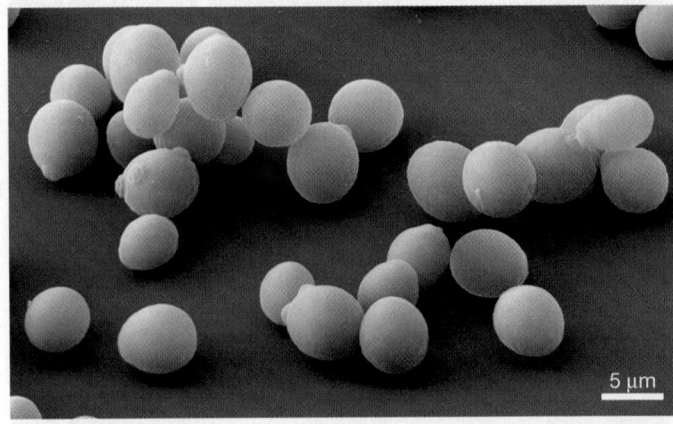

(b) Multicellular fungi have weblike bodies called mycelia.

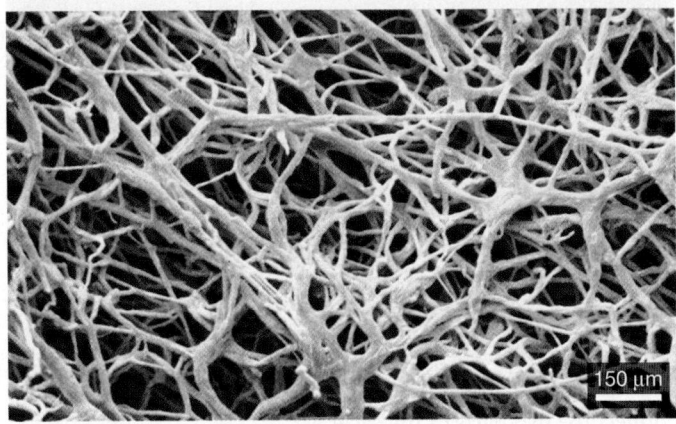

FIGURE 31.4 Fungi Have Just Two Growth Forms. Fungi grow **(a)** as single-celled yeasts and/or **(b)** as multicellular mycelia made up of long, thin, highly branched filaments. Both of these scanning electron micrographs are colorized.

THE NATURE OF THE FUNGAL MYCELIUM If food sources are plentiful, mycelia can be long lived and grow to be extremely large. Researchers discovered a mycelium growing across 1310 acres (6.5 km²) in Oregon. This is an area substantially larger than most college campuses. The biologists estimated the individual's weight at hundreds of tons and its age at thousands of years, making it one of the largest and oldest organisms known.

Although most mycelia are much smaller and shorter lived than the individual in Oregon, all mycelia are dynamic. Mycelia constantly grow in the direction of food sources and die back in areas where food is running out. The body shape of a fungus can change almost continuously throughout its life. Recent data indicate that the individual filaments that make up a mycelium live, on average, only about five days.

THE NATURE OF HYPHAE The filaments within a mycelium are called **hyphae** (singular: **hypha**). Hyphae may be haploid or **heterokaryotic** ("different-kernel"), meaning that each cell contains several haploid nuclei from different parents. Most heterokaryotic hyphae are **dikaryotic** ("two-kernel"). In this case, two haploid nuclei, one from each parent, are present.

(a) Both the reproductive structure and mycelium are composed of hyphae.

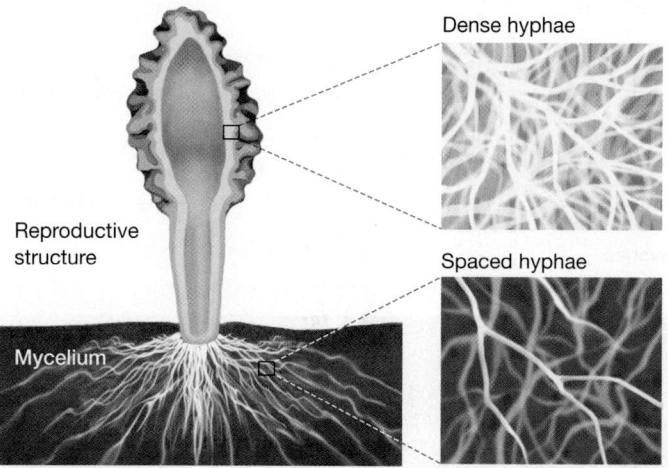

Reproductive structure

Dense hyphae

Spaced hyphae

Mycelium

(b) Hyphae are usually broken into compartments by septa.

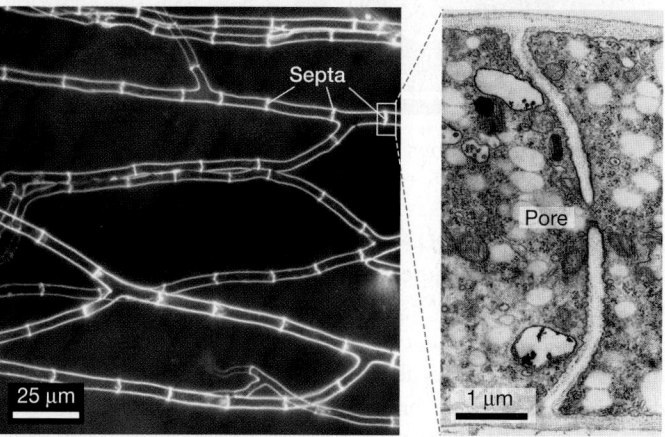

Septa

Pore

25 µm

1 µm

FIGURE 31.5 Multicellular Fungi Have Unusual Bodies. (a) The feeding structure of a fungus is a mycelium, which is made up of hyphae. In some species, hyphae come together to form multicellular structures such as mushrooms, brackets, or morels that emerge from the ground. **(b)** Hyphae are often divided into cell-like compartments by partitions called septa, which are broken by pores. As a result, the cytoplasm of different compartments is continuous.

As **Figure 31.5a** shows, hyphae are long, narrow filaments that branch frequently. In most fungi, each filament is broken into cell-like compartments by cross-walls called **septa** (singular: **septum**; **Figure 31.5b**).

Septa do not close off segments of hyphae completely. Instead, gaps called pores enable a wide variety of materials, even organelles and nuclei, to flow from one compartment to the next. Septa may have single large openings or a series of small gaps that give the septum a sieve-like appearance. Because nutrients, mitochondria, and even genes can flow though the entire mycelium—at least to a degree—the fungal mycelium is intermediate between a multicellular land plant or animal and an enormous single-celled organism.

Some fungal species are even **coenocytic** ("common-celled"; pronounced *see-no-SIT-ick*)—meaning that they lack septa en-

tirely. Coenocytic fungi have many nuclei scattered throughout the mycelium. In effect, they are a single, gigantic cell.

It's also important to appreciate just how thin hyphae are. Plant root tips are typically about 1 mm in diameter, but fungal hyphae are typically less than 10 µm in diameter. Hyphae are 1/100th the size of a root tip. Fungal mycelia can penetrate tiny fissures in soil and absorb nutrients that are inaccessible to plant roots.

MYCELIA HAVE A LARGE SURFACE AREA Perhaps the most important aspect of mycelia and hyphae, however, is their shape. Because mycelia are composed of complex, branching networks of extremely thin hyphae, fungi have the highest surface-area-to-volume ratios observed in multicellular organisms.

To drive this point home, consider that the hyphae found in any fist-sized ball of rich soil typically have a surface area equivalent to half a page of this book. This surface area is important because fungi make their living via absorption, and a large surface area makes nutrient absorption extremely efficient.

The extraordinarily high surface area in a mycelium has a downside, however. The amount of water that evaporates from an organism is a function of its surface area—meaning that these organisms are prone to drying out. As a result, fungi are most abundant in moist habitats.

When conditions dry, the fungal mycelium may die back partially or completely. The reproductive cells called **spores** that are produced by sexual or asexual reproduction are resistant to drying, however. As a result, spores can endure dry periods, then germinate and resume growth when conditions improve. Mycelial growth is dynamic, and changes with moisture availability or food supply.

REPRODUCTIVE STRUCTURES Mycelia are an adaptation that supports the absorptive lifestyle of fungi. The only thick, fleshy structures that fungi produce are reproductive organs.

Mushrooms, puffballs, and other dense, multicellular structures that arise from mycelia do not absorb food. Instead, they function in reproduction. Typically they are the only part of a fungus that is exposed to the atmosphere, where drying is a problem. The mass of filaments on the inside of mushrooms is protected from drying by the densely packed hyphae forming the surface.

Relatively few species of fungi make the reproductive organs called mushrooms, however. Instead, most fungal species that undergo sexual reproduction produce one of four types of distinctive reproductive structures—only one of which is found inside mushrooms.

1. ***Swimming gametes and spores*** In certain species that live primarily in water or wet soils, the spores that are produced during asexual reproduction have flagella, as do the gametes produced during sexual reproduction (**Figure 31.6a** on page 584). These are the only motile cells known in fungi. Species with swimming gametes are traditionally known as chytrids (pronounced *KYE-trids*).

2. ***Zygosporangia*** In some species, haploid hyphae from two individuals meet and become yoked together as shown in **Figure 31.6b**. Cells from yoked hyphae fuse to form a distinctive

(a) Swimming gametes and spores

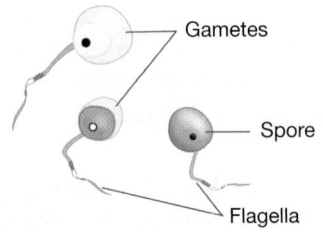

- Gametes
- Spore
- Flagella

(b) Zygosporangia: spore-producing structures formed when hyphae yolk together

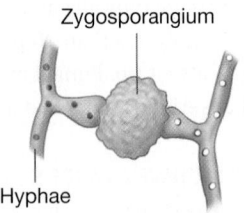

- Zygosporangium
- Hyphae

(c) Basidia: "little pedestals" where meiosis occurs and spores form

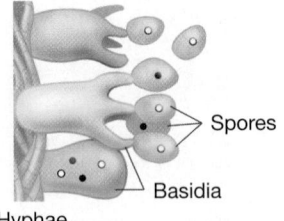

- Spores
- Basidia
- Hyphae

(d) Asci: sacs where meiosis occurs and spores form

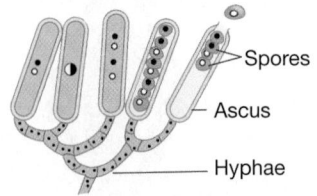

- Spores
- Ascus
- Hyphae

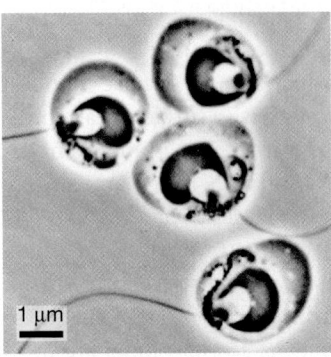

1 μm

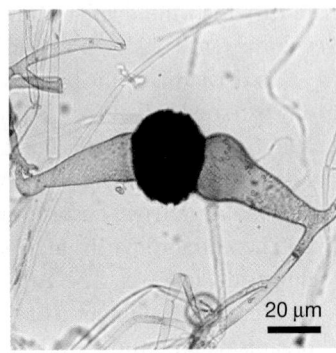

20 μm

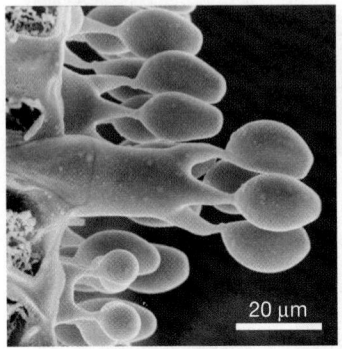

20 μm

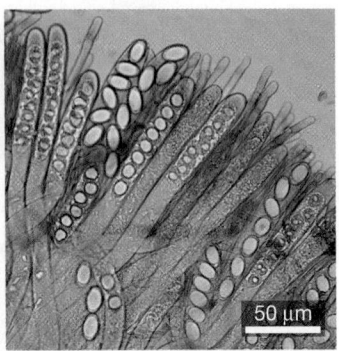

50 μm

FIGURE 31.6 Four Distinct Reproductive Structures Are Observed in Fungi. The dots in the illustrations represent nuclei.

spore-producing structure called a **zygosporangium** (plural: **zygosporangia**; the Greek root *zygos* means to be yoked together like oxen). Species with a zygosporangium are traditionally known as zygomycetes.

3. *Basidia*   Inside a mushroom, bracket, or puffball, specialized cells called **basidia** (singular: **basidium**; "little-pedestal") form at the ends of hyphae and produce spores (**Figure 31.6c**). Species with basidia are traditionally called basidiomycetes.

4. *Asci*   Inside cups, morels, and some other types of aboveground reproductive structures, specialized cells called **asci** (singular: **ascus**) form at the ends of hyphae and produce spores (**Figure 31.6d**). Species with asci are traditionally known as ascomycetes.

In sum, morphological studies allowed biologists to describe and interpret the mycelia growth habit as an adaptation that makes nutrient absorption extremely efficient. Careful analyses of morphological features also allowed researchers to identify four major types of reproductive structures.

Now the question is, which eukaryotes are most closely related to the fungi? And within the Fungi, do species that produce swimming gametes and spores, zygosporangia, basidia, and asci each form monophyletic groups—meaning that these distinctive reproductive structures evolved just once?

Evaluating Molecular Phylogenies

Researchers have sequenced and analyzed an array of genes to establish where fungi fit on the tree of life. The position of fungi as a whole is now well established; the position of lineages within fungi is still the subject of intense research.

FUNGI ARE CLOSELY RELATED TO ANIMALS   **Figure 31.7** shows that fungi are much more closely related to animals than they are to land plants.

The close evolutionary relationship between fungi and animals explains why fungal infections in humans are much more difficult to treat than bacterial infections. Fungi and humans shared a common ancestor relatively recently. As a result, their enzymes and cell components are similar in structure and function. Drugs that disrupt fungal enzymes and cells are also likely to damage humans.

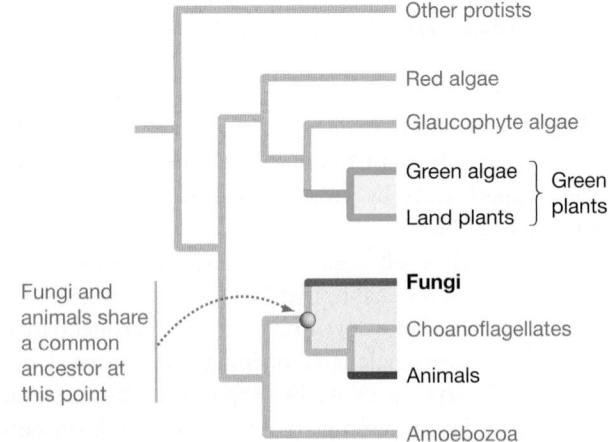

- Other protists
- Red algae
- Glaucophyte algae
- Green algae ⎫ Green
- Land plants ⎭ plants
- **Fungi**
- Choanoflagellates
- Animals
- Amoebozoa

Fungi and animals share a common ancestor at this point

FIGURE 31.7 Fungi Are More Closely Related to Animals than to Land Plants. Phylogenetic tree showing the evolutionary relationships among the green plants, animals, fungi, and some groups of protists. (Choanoflagellates are solitary or colonial protists found in freshwater; they are introduced in Chapter 32.)

In addition to the DNA sequence data, three key morphological traits link animals and fungi:

1. Most animals and fungi synthesize the tough structural material called chitin (see Chapter 5). Chitin is a prominent component of the cell walls of fungi.

2. The flagella that develop in chytrid spores and in chytrid gametes are similar to those observed in animals: As in animals, the flagella in chytrids are single, are located at the back of reproductive cells, and move in a whiplash manner.

3. Both animals and fungi store food by synthesizing the polysaccharide glycogen. (Green plants, in contrast, synthesize starch as their storage product.)

WHAT IS THE RELATIONSHIP AMONG THE MAJOR FUNGAL GROUPS? To understand the relationships among species with swimming gametes, zygosporangia, basidia, and asci, biologists have sequenced a series of genes from an array of fungal species and used the data to estimate the phylogeny of the group. The results, shown in **Figure 31.8**, support several important conclusions:

- The single-celled eukaryotes called microsporidians are actually fungi.
  Interpretation: They are not distant relatives, as initially thought. This is important. Researchers are now testing the hypothesis that **fungicides**—molecules that are lethal to fungi—can cure microsporidian infections in bee colonies, silkworm colonies, and AIDS patients.

- The chytrids and zygomycetes form a paraphyletic group that is poorly resolved—meaning that the actual order of branching events among lineages with these reproductive structures is still not known (in Figure 31.8, they are collapsed into a polytomy—see **BioSkills 3** in Appendix A).
  Interpretation: Swimming gametes and the zygosporangium evolved more than once. Or, both structures were present in a common ancestor but then were lost in certain lineages.

- An important group called the Glomeromycota is monophyletic.
  Interpretation: The adaptations that allow these species to live in association with plant roots (discussed in Section 31.3) evolved once.

- The basidiomycetes are monophyletic—they form a lineage called Basidiomycota, or club fungi.
  Interpretation: The basidium evolved once.

- The ascomycetes are monophyletic—they form a lineage called Ascomycota, or sac fungi.
  Interpretation: The ascus evolved once.

- Together, the Basidiomycota and Ascomycota form a monophyletic group.
  Interpretation: Because basidiomycetes and ascomycetes are both dikaryotic, this growth habit evolved once.

- The sister group to fungi consists of protists called choanoflagellates and the animals.
  Interpretation: Because choanoflagellates and the most an-

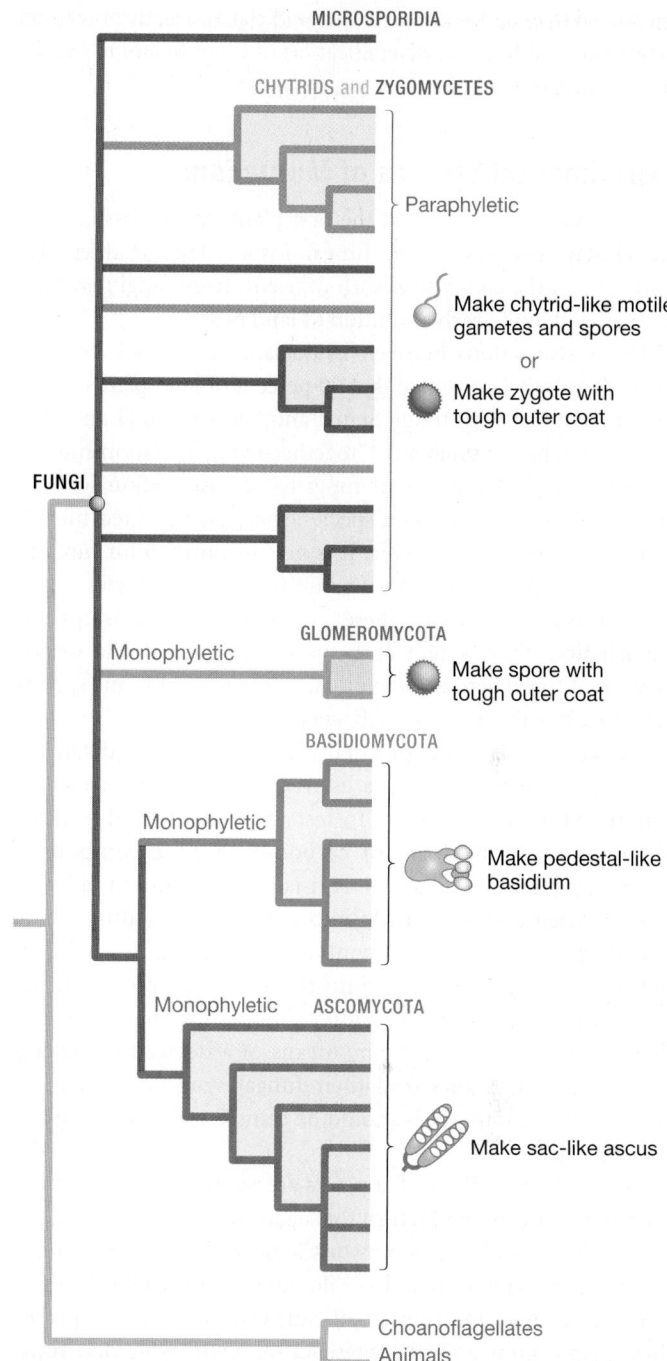

FIGURE 31.8 Phylogeny of the Fungi. A recent phylogenetic tree based on analyses of DNA sequence data. The icons represent the types of sexual reproductive structures observed in each major lineage.

cient groups of animals are aquatic, and because chytrids are aquatic, it is reasonable to conclude that the earliest fungi were aquatic, and that the switch to terrestrial life occurred early in the evolution of the Fungi.

Although progress on understanding the evolutionary history of fungi has been rapid, the phylogenetic tree in Figure 31.8 is still a work in progress. For example, it is not yet clear where microsporidians are placed relative to several lineages of chytrids and zygomycetes. (Microsporidians lack both swimming gametes and yoked hyphae.) Future work should clarify exactly how fungi diversified and how the diversification of fungi relates to the diversification of land plants.

Experimental Studies of Mutualism

Chapter 30 pointed out that the first plants in the fossil record are closely associated with fungal fossils. That chapter also claimed that the ability to absorb nutrients from fungi may have been crucial in the early evolution of land plants.

Close associations between land plants and fungi continue today. Researchers estimate that 90 percent of land plants live in physical contact with fungi. Stated another way, fungi and land plants often have a **symbiotic** ("together-living") relationship.

Although some species of fungi live in association with an array of different land plant species, some documented fungal-plant associations are specific. It is not uncommon for fungi to live in only a particular type of tissue, in one plant species.

Scientists categorize these symbiotic relationships as **mutualistic** if they benefit both species, **parasitic** if one species benefits at the expense of the other, or **commensal** if one species benefits while the other is unaffected.

To explore the nature of fungi-plant symbioses in detail, researchers have used isotopes as tracers for specific elements (**Figure 31.9**). For example, to test the hypothesis that fungi obtain food in the form of carbon-containing compounds from their plant associates, biologists have introduced radioactively labeled carbon dioxide into the air surrounding plants that do or do not contain symbiotic fungi. The labeled CO_2 molecules are incorporated into the sugars produced during photosynthesis, and the location of the radioactive atoms can then be followed over time by means of a device that detects radioactivity. If plants feed their fungal symbionts, then labeled carbon compounds should be transferred from the plant to the fungi.

To test the hypothesis that plants are receiving nutrients from their symbiotic fungi in return for sugars, researchers have added radioactive phosphorus atoms or the heavy isotope of nitrogen (^{15}N) to potted plants that do or do not contain symbiotic fungi. If fungi facilitate the transfer of nutrients from soil to plants, then plants grown in the presence of their symbiotic fungi should receive much more of the radioactive phosphorus or heavy nitrogen than do plants grown in the absence of fungi.

Experiments with isotopes used as tracers have shown that sugars and other carbon-containing compounds produced by plants via photosynthesis are transferred to their fungal symbionts. In some cases, as much as 20 percent of the sugars produced by a plant end up in their symbiotic fungi. In exchange, the symbiotic fungi facilitate the transfer of phosphorus or nitrogen—or both—from soil to the plant.

Because phosphorus and nitrogen are in extremely short supply in most environments, the nutrients supplied by symbiotic fungi are critical to the success of the plant. In this way, studies with isotopes have supported the hypothesis that most relationships between fungi and land plants are mutually beneficial.

CHECK YOUR UNDERSTANDING

If you understand that . . .
- The bodies of fungi are either single-celled yeasts or multicellular mycelia.
- During sexual reproduction, different groups of fungi produce distinct reproductive structures.

✓ **You should be able to . . .**
1. Explain why mycelia are interpreted as an adaptation to an absorptive lifestyle.
2. Identify the four types of reproductive structures observed in fungi.

Answers are available in Appendix B.

31.3 What Themes Occur in the Diversification of Fungi?

Why are there so many different species of fungi? This question is particularly puzzling given that fungi share a common attribute: They all make their living by absorbing food directly from their surroundings. In contrast to the diversity of food-getting strategies observed in bacteria, archaea, and protists, all fungi make their living in the same basic way. In this respect, fungi are like plants—virtually all of which make their own food via photosynthesis.

Chapter 30 showed that the diversification of land plants was driven not by novel ways of obtaining food, but by adaptations that allowed plants to grow and reproduce in a diverse array of terrestrial habitats. What drove the diversification of fungi? The answer is the evolution of novel methods for absorbing nutrients from a diverse array of food sources.

This section introduces a few of the ways that fungi go about absorbing nutrients from different food sources, as well as how they produce offspring. Let's explore the diversity of ways that fungi do what they do.

Fungi Participate in Several Types of Mutualisms

Not long after associations between fungi and the roots of land plants were discovered and shown to be mutualistic, researchers found that two types of plant-mycorrhizal interactions are particularly common, involving **ectomycorrhizal fungi (EMF)** and **arbuscular mycorrhizal fungi (AMF)**. The two major types of mycorrhizae have distinctive morphologies, geographic distributions, and functions.

- EMF are usually species from the Basidiomycota, though some ascomycetes participate.
- AMF include species from the Glomeromycota.

Mycorrhizae aren't the only type of symbiotic fungi found in plants, however. Researchers have also become interested in fungi that live in close association with the aboveground tissues of land plants—their leaves and stems. Fungi that live in the aboveground parts of plants are said to be **endophytic** ("inside-plants").

QUESTION: Are mycorrhizal fungi mutualistic?

HYPOTHESIS: Host plants provide mycorrhizal fungi with sugars and other photosynthetic products. Mycorrhizal fungi provide host plants with phosphorus and/or nitrogen from the soil.

NULL HYPOTHESIS: No exchange of food or nutrients occurs between plants and mycorrhizal fungi. The relationship is not mutualistic.

EXPERIMENTAL SETUP:

Labeled carbon treatment:
Labeled CO_2
added to air around plant

Mycorrhizal fungi present

Labeled carbon control:
Labeled CO_2
added to air around plant

No fungi present

Labeled P or N treatment:

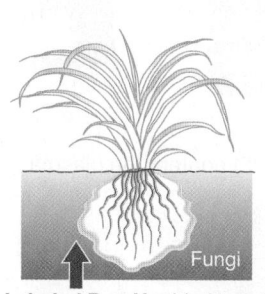

Fungi

Labeled P or N added to soil

Labeled P or N control:

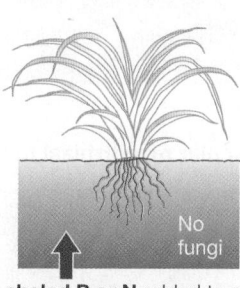

No fungi

Labeled P or N added to soil

PREDICTION FOR LABELED CARBON: A large percentage of the labeled carbon taken up by the plant will be transferred to mycorrhizal fungi. In the control, little labeled carbon will be present in the soil surrounding the roots.

PREDICTION OF NULL HYPOTHESIS, LABELED CARBON: There will be no difference in the localization of carbon in the two treatments.

PREDICTION FOR LABELED P OR N: A large percentage of the labeled P or N taken up by the fungi will be transferred to the plant. In the control, little or no labeled P or N will be taken up by the plant.

PREDICTION OF NULL HYPOTHESIS, LABELED P OR N: There will be no difference between amounts of labeled P or N found in plant in presence or absence of fungi.

RESULTS:

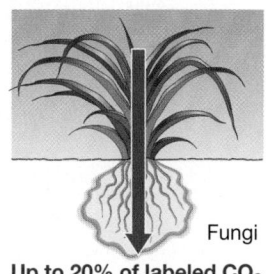

Fungi

Up to 20% of labeled CO_2
taken up by plant is transferred
to mycorrhizal fungi

No fungi

Little to no labeled carbon
is found in soil

Large amounts of labeled P or N
are found in host plant

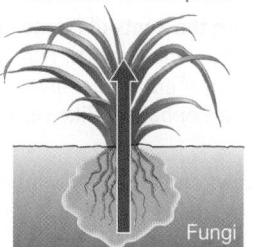

Fungi

Little labeled P or N
is found in host plant

No fungi

CONCLUSION: The relationship between plants and mycorrhizal fungi is mutualistic. Plants provide mycorrhizal fungi with carbohydrates. Mycorrhizal fungi supply host plants with nutrients.

FIGURE 31.9 Experimental Evidence That Mycorrhizal Fungi and Plants Are Mutualistic. Sugars flow from plants to mycorrhizal fungi; key nutrients flow from mycorrhizal fungi to plants.

SOURCES: Bücking, H. and W. Heyser. 2001. Microautoradiographic localization of phosphate and carbohydrates in mycorrhizal roots of *Populus tremula* x *Populus alba* and the implications for transfer processes in ectomycorrhizal associations. *Tree Physiology* 21: 101–107.

✔**QUESTION** What do these "labeled nutrient" experiments tell you that you didn't already know from experiments like Figure 31.1, where plants are grown with and without mycorrhizae?

Recent research has shown that endophytic fungi are much more common and diverse than previously suspected. Further, data indicate that at least some species of endophytes are mutualistic.

These results support the general realization that most plants are covered with fungi—from the tips of their branches to the base of their roots. Throughout their lives, many or even most plants are involved in several distinct types of mutualistic relationships with fungi.

ECTOMYCORRHIZAL FUNGI (EMF) EMF like the one shown in **Figure 31.10a** on page 588 are found on many of the tree species in the temperate regions of the world (see Chapter 50), where warm summers alternate with cold winters. In this type of association, hyphae form a dense network that covers a plant's root tips. As the cross section in Figure 31.10a shows, individual hyphae penetrate between cells in the outer layer of the root, but hyphae do not enter the root cells.

(a) Ectomycorrhizal fungi (EMF) form sheaths around roots and penetrate between root cells.

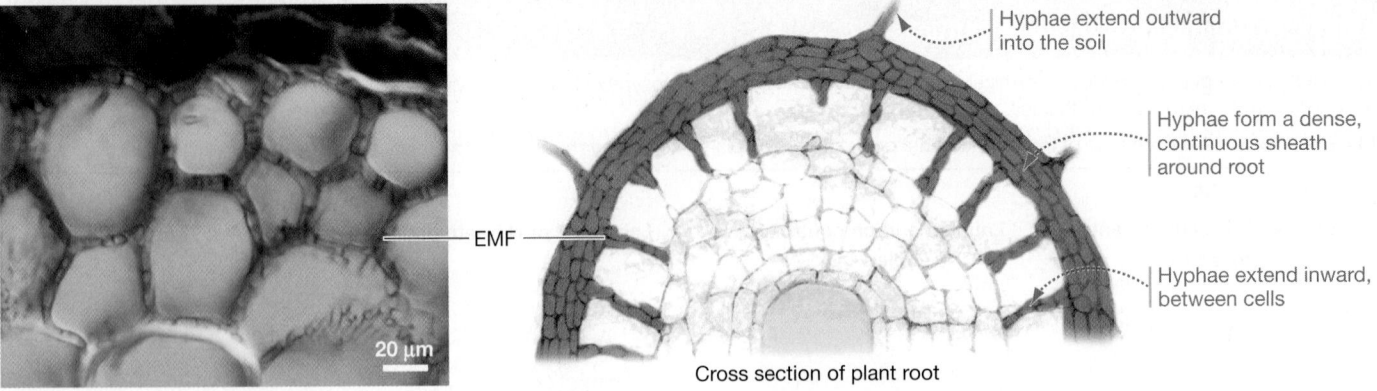

Hyphae extend outward into the soil

Hyphae form a dense, continuous sheath around root

EMF

Hyphae extend inward, between cells

20 μm

Cross section of plant root

(b) Arbuscular mycorrhizal fungi (AMF) contact the plasma membranes of root cells.

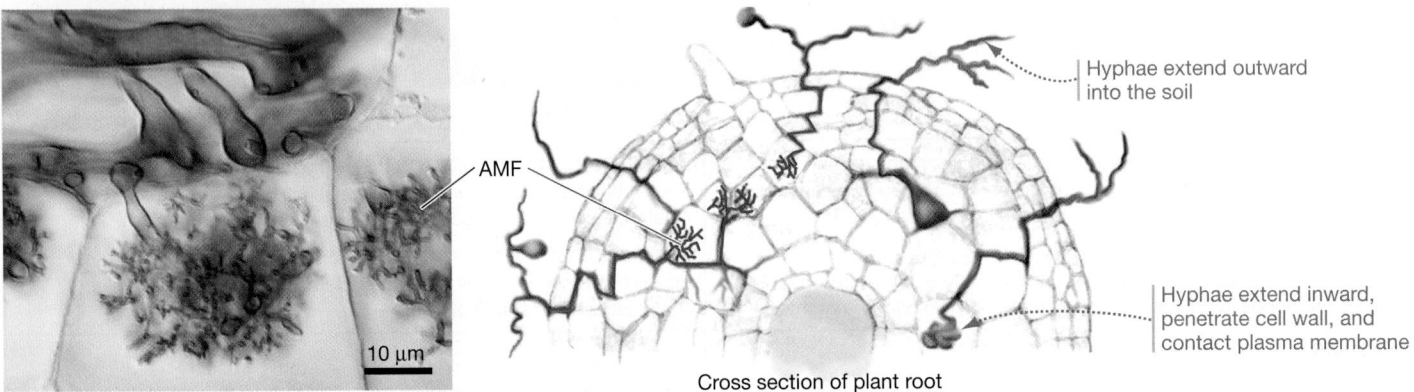

Hyphae extend outward into the soil

AMF

Hyphae extend inward, penetrate cell wall, and contact plasma membrane

10 μm

Cross section of plant root

FIGURE 31.10 Mutualistic Fungi Interact with the Roots of Plants in Two Distinct Ways. (a) Ectomycorrhizal fungi (EMF) form a dense network around the roots of plants. Their hyphae penetrate the intercellular spaces of the root but do not enter the root cells. **(b)** The hyphae of arbuscular mycorrhizal fungi (AMF) penetrate the walls of root cells, where they branch into bushy structures or balloon-like vesicles that are in close contact with the root cell's plasma membrane.

✔**EXERCISE** In the cross sections, add arrows and labels showing the direction of movement of N (nitrogen-containing compounds), P (phosphorus-containing compounds), and C (carbon-containing compounds).

The Greek root *ecto*, which refers to "outer," describes this association accurately: the fungi form an outer sheath on root tips that is often 0.1 mm thick. Hyphae also extend out from the sheath-like portion of the mycelium into the soil.

A recent genome sequencing project found that an EMF species contains genes for a large array of small proteins that are secreted from hyphae. This finding prompted the hypothesis that the secreted proteins act as signals to potential host plants. If so, establishing an association between an EMF and its host plant may involve a complex array of signals, which biologists are only beginning to understand.

How do these trees and fungi interact, once they start living together? In the habitats where EMF are abundant, nitrogen atoms tend to remain tied up in dead tissues, in amino acids and nucleic acids, instead of being available in the soil. The hyphae of EMF penetrate decaying material and release enzymes called peptidases that cleave the peptide bonds between amino acids in dead tissues. The amino acids released by this reaction are absorbed by the hyphae and transported to spaces between the root cells of trees, where they can be absorbed by the plant. EMF are also able to acquire phosphate ions that are bound to soil particles and transfer the ions to host plants. In return, the fungi receive sugars and other complex carbon compounds from the tree.

Researchers have found that when birch tree seedlings are grown with and without their normal EMF in pots filled with forest soil, only the seedlings with EMF are able to acquire significant quantities of nitrogen and phosphorus. Inspired by such data, a biologist has referred to EMF as the "dominant nutrient-gathering organs in most temperate forest ecosystems."

The hyphae of EMF are like an army of miners that discover, excavate, and deliver precious nuggets of nitrogen to trees. The productivity of the world's most important commercial forests depends on EMF.

ARBUSCULAR MYCORRHIZAL FUNGI (AMF) In contrast to the hyphae of EMF, the hyphae of arbuscular mycorrhizal fungi (AMF) grow *into* the cells of root tissue. The name *arbuscular* ("little-tree") was inspired by the bushy, highly branched hyphae, shown in **Figure 31.10b**, that form inside root cells.

AMF are also called **endomycorrhizal fungi**, because they penetrate the interior of root cell walls, or vesicular-arbuscular mycorrhizae (VAM), because the hyphae of some species form large, balloon-like vesicles inside root cells.

The key point is that the hyphae of AMF penetrate the cell wall and contact the plasma membrane of root cells directly. The highly branched hyphae inside the plant cell wall are thought to be an adaptation that increases the surface area available for exchange of molecules between the fungus and its host. However, AMF do not form a tight sheath around roots, as do EMF. Instead, they form a pipeline extending from inside plant cells in the root to the soil well beyond the root.

AMF are found in a whopping 80 percent of all land plant species. They are particularly common in grasslands and in the forests of tropical habitats. They are also widespread in temperate climates.

What do AMF do? Plant tissues decompose quickly in the grasslands and tropical forests where AMF flourish because the growing season is long and warm. As a result, nitrogen is often readily available to plants. Phosphorus is usually in short supply, though, because it tends to leach out of soils that experience high rainfall.

Based on these observations, biologists hypothesized that the most important function of AMF is to transfer phosphorus atoms from the soil to the host plant. Experiments with radioactive atoms confirmed the phosphate-transfer hypothesis by showing that AMF supply host plants with particularly large amounts of phosphorus, along with other nutrients. In return, host plants supply AMF with sugars and other forms of reduced carbon.

AMF are also extremely important in soil formation. The cell walls of their hyphae contain large quantities of a glycoprotein called **glomalin**. When the cells die, the glomalin enriches the organic matter in soil and helps bind organic compounds to sand or clay particles. Some recent estimates suggest that over 25 percent of all of the organic matter in soil consists of glomalin.

ARE ENDOPHYTES MUTUALISTS? Although endophytic fungi are relatively new to science, they are turning out to be both extremely common and highly diverse.

- Biologists in Brazil are examining tree leaves for the presence of fungi; each time they do, they are discovering several new species of endophytes.

- Recall from Section 31.2 that a study in Panama found hundreds of fungal species living in the leaves of just two tree species.

These newly discovered species are endophytes.

Recent research has shown that the endophytes found in some grasses produce compounds that benefit plants. The compounds deter or even kill herbivores. In exchange for the compounds, endophytes absorb sugars from the plant.

Based on these results, biologists have concluded that the relationship between endophytes and grasses is mutualistic. Similar types of anti-herbivore compounds have recently been documented in an endophyte that lives on morning glories.

In other types of plants, however, researchers have not been able to document benefits for the plant host. The current consensus is that at least some endophytic fungi may be commensals—meaning the fungi and the plants simply coexist with no observable effect, either deleterious or beneficial, on the host plant.

MUTUALISMS WITH OTHER SPECIES Do fungi take up residence with species other than land plants? The answer is yes.

- **Lichens** are a mutualistic partnership between a species of ascomycete and either a cyanobacterium or an alga. The nature of this relationship is explored in more detail in Section 31.4.

- Some ant species actively farm fungi inside their colonies. The ants fertilize and "weed" the fungal gardens, then harvest the fungi for food (see Chapter 32).

EMF mine nitrogen and some phosphorus for plants in temperate forests; AMF mine phosphorus for plants in grasslands and tropics as well as temperate regions. Other species of fungi engage in a wide array of symbiotic relationships.

What Adaptations Make Fungi Such Effective Decomposers?

The saprophytic fungi are master recyclers. Although bacteria and archaea are also important decomposers in terrestrial environments, fungi and a few bacterial species are the only organisms that can digest wood completely. Given enough time, fungi can turn even the hardest, most massive trees into soft soils.

How do fungi do it? You've already been introduced to two key adaptations:

- The large surface area of a mycelium makes nutrient absorption exceptionally efficient.

- Saprophytic fungi can grow toward the dead tissues that supply their food.

What other adaptations help fungi decompose plant tissues?

EXTRACELLULAR DIGESTION Large molecules such as starch, lignin, cellulose, proteins, and RNA cannot diffuse across the plasma membranes of hyphae. Only sugars, amino acids, nucleic acids, and other small molecules can enter the cytoplasm through hyphae. As a result, fungi have to digest their food before they can absorb it.

🔑 Instead of digesting food inside a stomach or food vacuole, as most animals and some protists do, respectively, fungi synthesize digestive enzymes and then secrete them outside their hyphae, into their food. Fungi perform **extracellular digestion**—digestion that takes place outside the organism. The simple compounds that result from enzymatic action are then absorbed by the hyphae.

As an example of how this process occurs, consider the enzymes responsible for digesting lignin and cellulose—the two most abundant organic molecules on Earth.

- Lignin comprises a family of extremely strong, complex polymers built from monomers that contain six-carbon rings. Recall from Chapter 30 that most lignin is found in the secondary cell walls of plant vascular tissues, where it furnishes structural support.

- Cellulose is a polymer of glucose and is found in the primary and secondary cell walls of plant cells.

Basidiomycetes can degrade lignin completely—to CO_2 and H_2O—as well as digest cellulose. Let's take a closer look at how they do it.

LIGNIN DEGRADATION   Biologists have been keenly interested in understanding how basidiomycetes digest lignin. Paper manufacturers are also interested in this process because they need safe, efficient ways to degrade lignin in order to make soft, absorbent paper products.

To find out how lignin-digesting fungi do it, biologists began analyzing the proteins that these species secrete into extracellular space. After purifying these molecules, the investigators tested each protein for the ability to degrade lignin. Using this approach, investigators from two labs independently discovered an enzyme called lignin peroxidase.

Lignin peroxidase catalyzes the removal of a single electron from an atom in the ring structures of lignin. This oxidation step creates a free radical—an atom with an unpaired electron (see Chapter 2). This is an extremely unstable electron configuration, and leads to a series of uncontrolled and unpredictable reactions. These follow-on reactions split the polymer into smaller units.

Biologists have referred to this mechanism of lignin degradation as enzymatic combustion. The phrase is apt: The uncontrolled oxidation reactions triggered by lignin peroxidase are analogous to the uncontrolled oxidation reactions that occur when gasoline burns in a car engine.

The nonspecific nature of the reaction is remarkable, because virtually all of the other reactions catalyzed by enzymes are extremely specific. The lack of specificity makes sense, however. Unlike proteins, nucleic acids, and most other polymers with a regular and predictable structure, lignin is extremely heterogeneous. Over 10 types of covalent linkages are routinely found between the monomers that make up lignin. But once lignin peroxidase has created a free radical in the ring structure, any of these linkages can be broken.

The uncontrolled nature of the reactions has an important consequence. Instead of being stepped down an orderly electron transport chain as described in Chapter 9, the electrons involved in these reactions lose their potential energy in large, unpredictable jumps. As a result, the oxidation of lignin cannot be harnessed to drive the production of ATP and fuel the growth of hyphae. This conclusion is supported by an experimental observation: Fungi cannot grow with lignin as their sole source of food.

If wood-rotting fungi don't use lignin as food, why do they produce enzymes to digest it? The answer is simple. In wood, lignin forms a dense matrix around long strands of cellulose. Degrading the lignin matrix gives hyphae access to huge supplies of energy-rich cellulose. Saprophytic fungi are like miners. But instead of seeking out rare, gem-like nitrogen or phosphorus atoms as do EMF and AMF, the saprophytes use lignin peroxidase to blast away enormous lignin molecules, exposing rich veins of cellulose that can fuel growth and reproduction.

CELLULOSE DIGESTION   Once lignin peroxidase has softened wood by stripping away its lignin matrix, the long strands of cellulose that remain can be attacked by enzymes called cellulases.

Like lignin peroxidase, cellulases are secreted into the extracellular environment by fungi. But unlike lignin peroxidase, cellulases are extremely specific in their action.

Biologists have purified seven different cellulases from the fungus *Trichoderma reesei*. Two of these enzymes catalyze a critical early step in digestion—they cleave long strands of cellulose into a disaccharide called cellobiose. The other cellulases are equally specific and also catalyze hydrolysis reactions. In combination, the suite of seven enzymes in *T. reesei* transforms long strands of cellulose into a simple monomer—glucose—that the fungus uses as a source of food.

Variation in Reproduction

Recall from Section 31.2 that fungi may produce swimming gametes and spores, yoked hyphae in which nuclei from different individuals fuse to form a zygote inside a protective structure, or specialized spore-producing cells called basidia and asci. As fungi diversified, they evolved an array of ways to reproduce as well as an array of sources for absorbing nutrition.

SPORES AS KEY REPRODUCTIVE CELLS   The spore is the most fundamental reproductive cell in fungi. Spores are the dispersal stage in the fungal life cycle and are produced during both asexual and sexual reproduction. (Recall from Chapter 12 that asexual reproduction is based on mitosis, while sexual reproduction is based on meiosis.) Fungi produce spores in such prodigious quantities that it is not unusual for them to outnumber pollen grains in air samples.

If a spore falls on a food source and is able to germinate, a mycelium begins to form. As the fungus expands, hyphae grow in the direction in which food is most abundant. But if food begins to run out, mycelia respond by making spores, which are dispersed by wind or animals.

Why would mycelia reproduce when food is low? The leading hypothesis to answer this question is that spore production allows starving mycelia to disperse offspring to new habitats where more food might be available. Thus, spore production is favored by natural selection when individuals are under nutritional stress.

MULTIPLE MATING TYPES   In many fungal lineages, hyphae come in several or many different mating types. Instead of having morphologically distinct males and females that produce sperm and eggs, hyphae of different mating types look identical.

Hyphae of the same mating type will not combine during sexual reproduction. When zygomycete hyphae grow close to each other, for example, they will not fuse unless the individuals have different alleles of one or more genes involved in mating. If chemical messengers released by two hyphae indicate that they are of different mating types, then fusion and zygosporangium formation follows.

In this way, mating types function as sexes. Instead of having just two sexes, a single fungal species may have tens of thousands. The basiomycete *Schizophyllum commune*, for example, is estimated to have 28,000 mating types.

Why so many? The leading hypothesis to explain the existence of mating types is that it helps generate genetic diversity in offspring. Genetic diversity, in turn, is known to be advantageous in fighting off infections and responding to changes in the environment (see Chapter 12).

HOW DOES FERTILIZATION OCCUR? Compared with green plants, protists, and animals, fertilization in fungi has important unique features:

- 🔑 Only members of the Chytridiomycota produce gametes, and no fungus produces gametes that are different enough in size to be called sperm and egg.

- Fertilization occurs in two distinct steps in many fungi: (**1**) fusion of cells, and (**2**) fusion of nuclei from the fused cells. These two steps can be separated by long time spans and even long distances.

In many fungi, the process of sexual reproduction begins when hyphae from two individuals fuse to form a hybrid hypha. When the cytoplasm of two individuals fuses in this way, **plasmogamy** is said to occur (**Figure 31.11**).

In some cases, the nuclei from the two or more individuals stay independent after plasmogamy and grow into a heterokary-

otic mycelium. The distinct nuclei function independently, even though gene expression must be coordinated in order for growth and development to occur. For example, the two nuclei divide as the hyphae expand, so each compartment that is divided by a septum contains one of each of the two nuclei. How the activities of the two nuclei are coordinated is a mystery.

In a heterokaryotic mycelium, one or more pairs of unlike nuclei may eventually fuse to form a diploid zygote. The fusion of nuclei is called **karyogamy**. The nuclei that are produced by karyogamy then divide by meiosis to form haploid spores.

As Chapter 29 indicated, most eukaryotes have life cycles dominated by haploid cells or diploid forms. Many fungi, in contrast, have a life cycle dominated by heterokaryotic cells.

✔️ If you understand the relationship between plasmogamy, heterokaryosis, and karyogamy, you should be able to compare and contrast these events with the life cycles of other eukaryotes. For example, which human cells undergo plasmogamy and karyogamy? Does heterokaryosis occur?

ASEXUAL REPRODUCTION As the left side of Figure 31.11 indicates, fungal species can reproduce asexually as well as sexually. There are, in fact, large numbers of ascomycetes that have never been observed to reproduce sexually.

During asexual reproduction, spore-forming structures are produced by a haploid mycelium, and spores are generated by mitosis. As a result, offspring are clones—meaning that they are genetically identical to their parent.

Four Major Types of Life Cycles

Among the sexually reproducing species of fungi, the presence of a heterokaryotic stage and the morphology of the spore-producing structure vary. Let's take a closer look at each of the four major types of life cycles that have been observed in fungi.

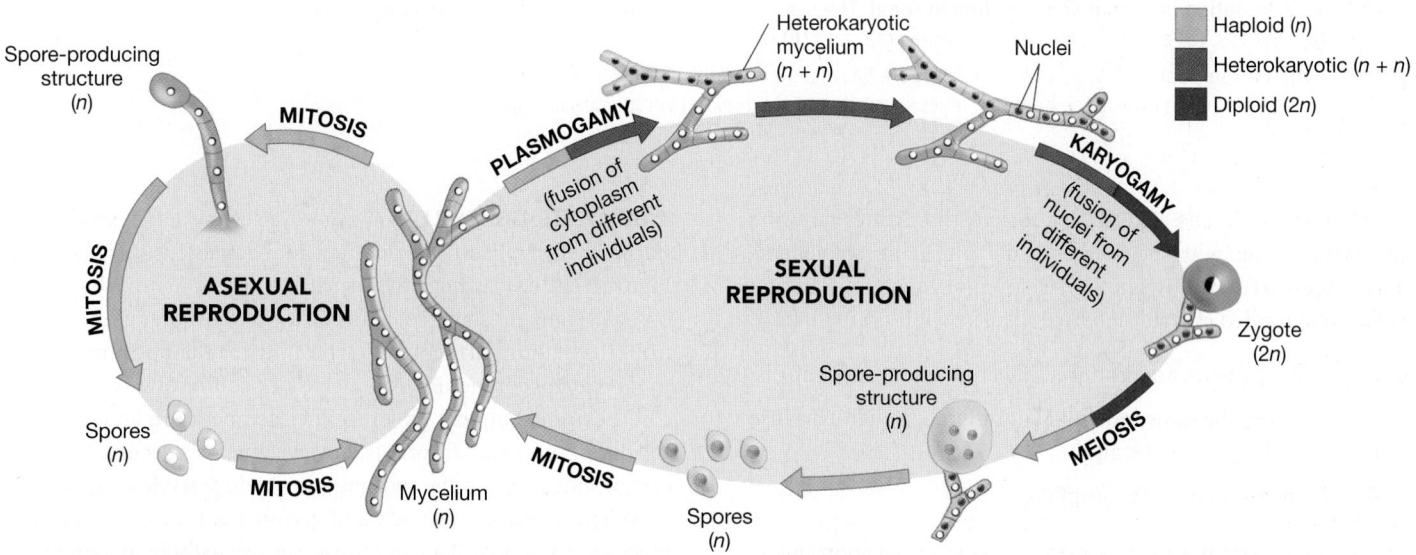

FIGURE 31.11 Fungi Have Unusual Life Cycles. A generalized fungal life cycle, showing both asexual and sexual reproduction.

✔️**QUESTION** Fungi spend most of their lives feeding. Which is the longest-lived component of this life cycle?

(a) Chytrids include the only fungi in which alternation of generations occurs.

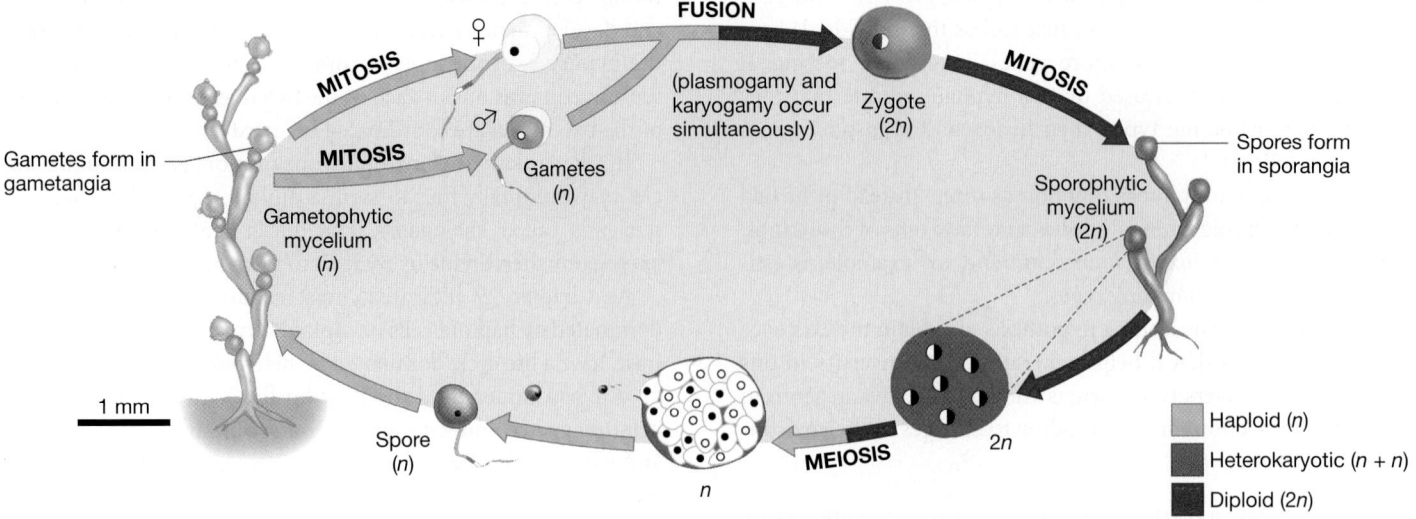

(b) Zygomycetes form yoked hyphae that produce a spore-forming structure (zygosporangium).

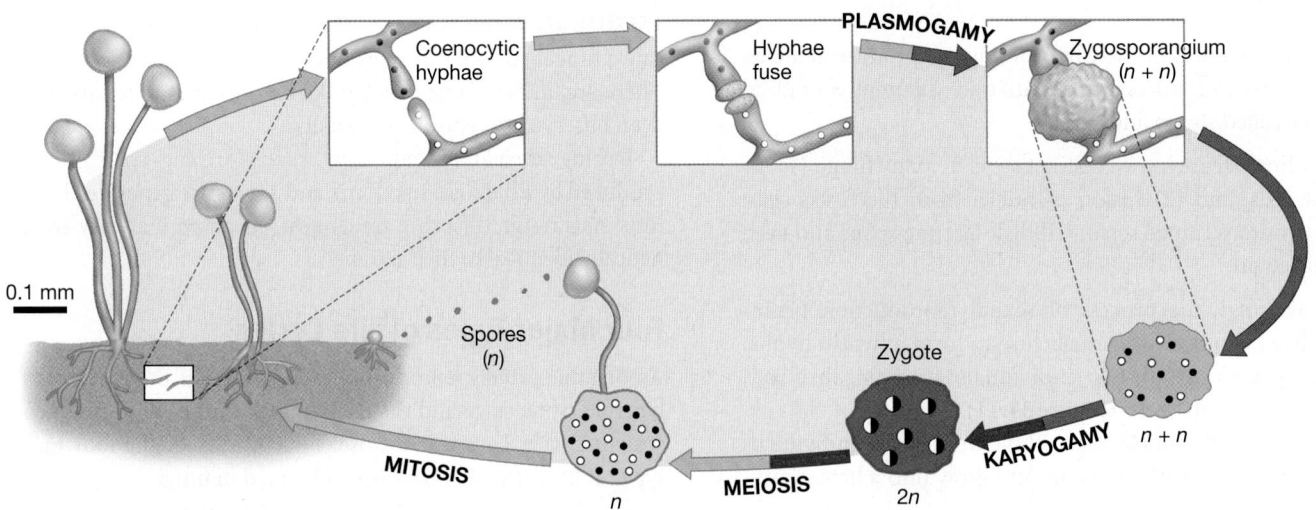

FIGURE 31.12 Variation in Sexual Reproduction in Fungi The sexual part of the life cycle in the four major groups of fungi: **(a)** chytrids, **(b)** zygomycetes, **(c)** Basidiomycota, and **(d)** Ascomycota.

✔**QUESTION** Asexual reproduction, as shown in Figure 31.11, is extremely common in zygomycetes and ascomycetes. What is the difference between mycelia produced via asexual versus sexual reproduction?

CHYTRIDIOMYCETE LIFE CYCLE The chytridiomycetes are the only type of fungi with species that exhibit alternation of generations. **Figure 31.12a** shows how alternation of generations occurs in the well-studied chytrid *Allomyces*. The key points are that:

- swimming gametes are produced in haploid adults by mitosis;
- gametes from the same individual or different individuals fuse to form a diploid zygote; and
- the zygote grows into a sporophyte.

The life cycle continues when meiosis occurs in the sporophyte, inside a structure called a **sporangium**. The haploid spores produced by meiosis disperse by swimming.

ZYGOMYCETE LIFE CYCLE In zygomycetes, sexual reproduction starts when hyphae from different individuals fuse, as shown in **Figure 31.12b**. Plasmogamy forms a zygosporangium that develops a tough, resistant coat. Inside the zygosporangium, nuclei from the mating partners fuse—meaning that karyogamy occurs.

The zygosporangium can persist if conditions become too cold or dry to support growth. When temperature and moisture conditions are again favorable, however, meiosis occurs. The meiotic products within the sporangium produce haploid spores.

When spores are released and germinate, they grow into new mycelia. These mycelia can reproduce asexually by making sporangia, with haploid spores being produced by mitosis and dispersed by wind.

(c) Basidiomycota have reproductive structures with many spore-producing basidia.

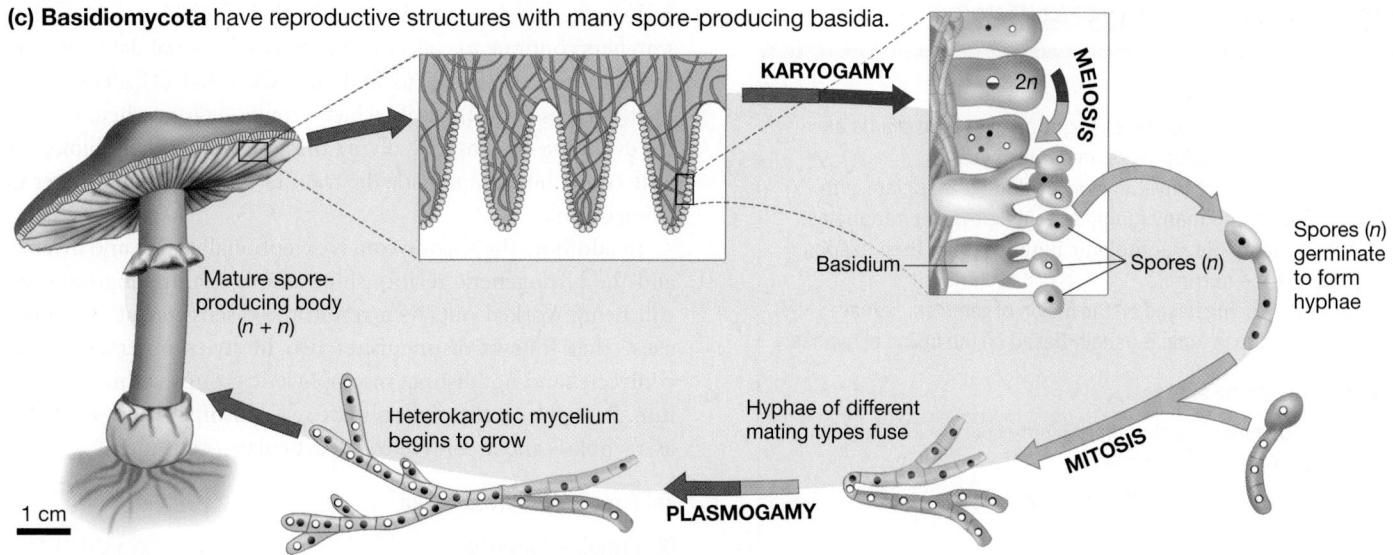

KARYOGAMY

2n

MEIOSIS

Basidium

Spores (n)

Spores (n) germinate to form hyphae

Mature spore-producing body (n + n)

1 cm

Heterokaryotic mycelium begins to grow

Hyphae of different mating types fuse

MITOSIS

PLASMOGAMY

(d) Ascomycota have reproductive structures with many spore-producing asci.

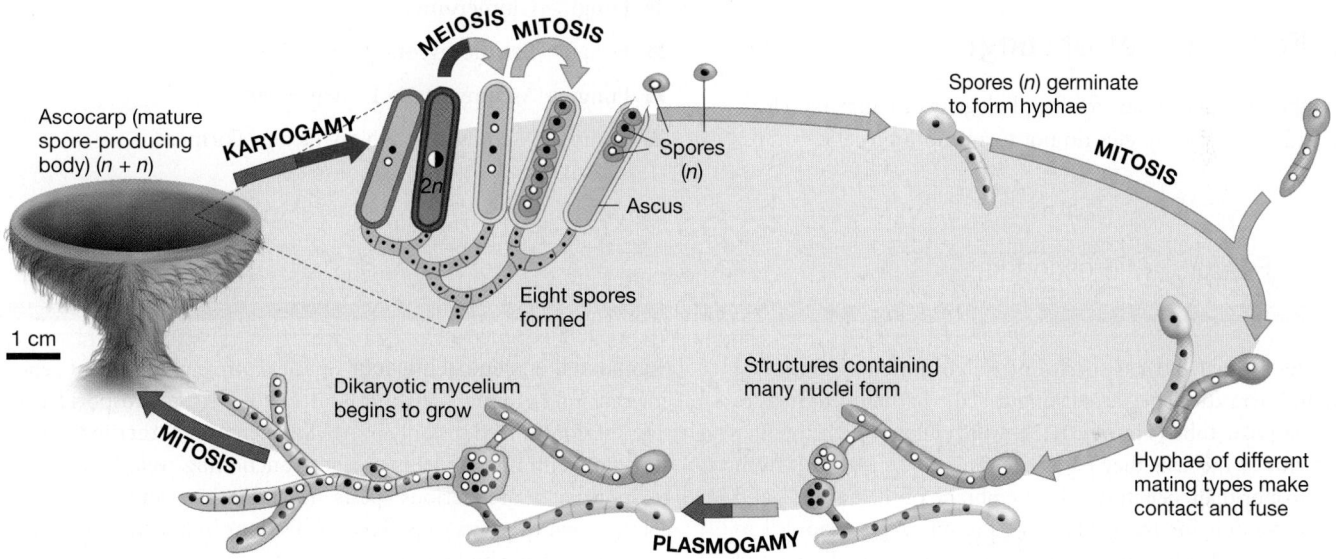

MEIOSIS MITOSIS

Ascocarp (mature spore-producing body) (n + n)

KARYOGAMY

2n

Spores (n)

Ascus

Eight spores formed

Spores (n) germinate to form hyphae

MITOSIS

1 cm

Dikaryotic mycelium begins to grow

MITOSIS

PLASMOGAMY

Structures containing many nuclei form

Hyphae of different mating types make contact and fuse

BASIDIOMYCOTA LIFE CYCLE Mushrooms, bracket fungi, and puffballs are reproductive structures produced by members of the Basidiomycota (**Figure 31.12c**). Even though their size, shape, and color vary enormously from species to species, all basidiomycete reproductive structures originate from the heterokaryotic hyphae of mated individuals.

Inside a mushroom or bracket or puffball, the pedestal-like, spore-producing cells called basidia form at the ends of heterokaryotic hyphae. Karyogamy occurs within the basidia. The diploid nucleus that results undergoes meiosis, and four haploid spores mature.

Spores are eventually ejected from the end of the basidia and are dispersed by the wind. It is not unusual for a single puffball or mushroom to produce a billion spores.

To review this sequence of life cycle events, go to the study area at *www.masteringbiology.com*.

(MB) Web Activity The Life Cycle of a Mushroom

ASCOMYCOTA LIFE CYCLE The reproductive structure in Ascomycota is produced by a dikaryotic hypha.

As **Figure 31.12d** illustrates, the process usually begins when hyphae or specialized structures from the same ascomycete species but from different mating types fuse, forming a cell containing many independent nuclei. A short dikaryotic hypha, containing one nucleus from each parent, emerges and eventually grows into a complex reproductive structure whose hyphae have the sac-like, spore-producing structures called asci at their tips.

After karyogamy occurs inside each ascus, meiosis takes place and haploid spores are produced. When the ascus matures, the spores inside are forcibly ejected. The spores are often picked up by the wind and dispersed.

CHECK YOUR UNDERSTANDING

31.4 Key Lineages of Fungi

Based on the current estimate for the phylogeny of Fungi, it is clear that chytrids and zygomycetes do not form monophyletic groups and should not be named as single phyla (see Figure 31.8). Researchers continue to collect and analyze additional data—mostly DNA sequences—trying to find characters that (1) identify distinct lineages and (2) clarify which groups evolved when, early in the evolution of fungi. Resolving this issue would help biologists understand how fungi made the transition from living in water to living on land.

In addition, the Ascomycota is exceptionally large and diverse, and the phylogenetic relationships among major subgroups are still being worked out. As a result, the discussion of "key lineages" that follows distinguishes two lifestyles observed in ascomycetes and not distinct monophyletic groups within the phylum. Research on the phylogenetic relationships within the fungi as a whole—and ascomycetes in particular—continues.

- ■ Fungi > Microsporidia
- ■ Fungi > Chytrids
- ■ Fungi > Zygomycetes
- ■ Fungi > Glomeromycota
- ■ Fungi > Basidiomycota (Club Fungi)
- ■ Fungi > Ascomycota > Lichen-Formers
- ■ Fungi > Ascomycota > Non-Lichen-Formers

Fungi > Microsporidia

All of the estimated 1200 species of microsporidians are single celled and parasitic. They are distinguished by a unique structure called the polar tube (**Figure 31.13**), which allows them to enter the interior of the cells they parasitize. The tube shoots out from the microsporidian, penetrates the membrane of the host cell, and acts as a conduit for the contents of the microsporidian cell to enter the host cell. Once inside, the microsporidian replicates and produces a generation of daughter cells, which go on to infect other host cells.

Microsporidians have a dramatically reduced genome. One species has the smallest genome known among eukaryotes—the total number of base pairs present is just half the number found in *E. coli*. Unlike other fungi, microsporidians lack functioning mitochondria. Like other fungi, the polysaccharide chitin is a prominent component of their cell wall. ✔ You should be able to mark the origin of the polar tube and the loss of mitochondria on Figure 31.8.

Absorptive lifestyle Most microsporidians parasitize insects or fish. Because they enter the interior of host cells, they are called intracellular parasites.

Life cycle Variation in life cycles is extensive. Some species appear to reproduce only asexually. Others produce several different types of sexual or asexual spores, and some must successfully infect several different host species to complete their life cycle.

Human and ecological impacts Species from eight different genera can infect humans. In most cases, however, microsporidians cause serious infections only in AIDS patients and other individuals whose immune systems are not functioning well. Some microsporidians are serious pests in honeybee and silkworm colonies, while others cause severe infections in grasshoppers and are marketed as a biological control agent.

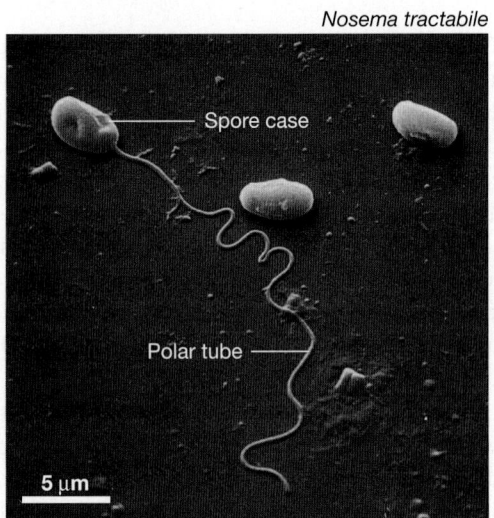

Nosema tractabile

Spore case

Polar tube

5 μm

FIGURE 31.13 Microsporidia Infect Other Cells via a Polar Tube.

Chytrids are largely aquatic and are common in freshwater environments. Some species live in wet soils, and a few have been found in desert soils that are wet only during a rainy season. Species found in dry soils have tough spores that endure harsh conditions. Spores from chytrids have been shown to germinate after a resting period of 31 years. Members of this group are the only fungi that produce motile cells. In each case, the motile cells use flagella to swim.

Absorptive lifestyle Many species of chytrids have enzymes that allow them to digest cellulose. As a result, these species are important decomposers of plant material in wet soils, ponds, and lakes. Many of the freshwater species are parasitic, and on occasion parasitic chytrids cause disease epidemics in algae or aquatic insects (including mosquitoes). Other species parasitize mosses, ferns, or flowering plants. Mutualistic chytrids are among the most important of the many organisms living in the guts of deer, cows, elk, and other mammalian herbivores, because the chytrids help these mammals digest their food (**Figure 31.14**). Chytrids produce cellulases that degrade the cell walls of grasses and other plants ingested by the animal, releasing sugars that are used by both the chytrids and their hosts.

Life cycle During asexual reproduction, most chytrid species produce spores that swim to new habitats via a flagellum. A few species reproduce sexually as well as asexually and exhibit alternation of generations. (Recall from Section 31.3 that these chytrids are the only fungal species to do so.) In species that reproduce sexually, plasmogamy and karyogamy may occur in a variety of ways: through fusion of hyphae, gamete-forming structures, or gametes.

Human and ecological impacts Chytrids are important decomposers in aquatic habitats and key mutualists in the guts of large, plant-eating mammals. Parasitic chytrids are also common. Biologists are investigating the possibility of using a chytrid that parasitizes mosquito larvae as a biological control agent; potato tubers are sometimes invaded and spoiled by a parasitic chytrid that causes black wart disease. Parasitic chytrids are largely responsible for catastrophic declines that have been occurring recently in amphibian populations all over the world.

Allomyces macrogynus

250 μm

FIGURE 31.14 A Chytrid That is a Common Inhabitant of Soils and Ponds Worldwide.

Fungi > Zygomycetes

The zygomycetes ("yoked-fungi") are primarily soil-dwellers. Their hyphae yoke together and fuse during sexual reproduction and then form a durable, thick-walled zygosporangium.

Absorptive lifestyle Many members of zygomycete lineages are saprophytes and live in plant debris. Some parasitize other fungi, however, or are important parasites of insects and spiders.

Life cycle Asexual reproduction is extremely common. Ball-like sporangia are produced at the tips of hyphae that form stalks, and mitosis results in the production of spores which are dispersed by wind. During sexual reproduction, fusion of hyphae occurs only between individuals of different mating types. Fusion of hyphae from the same individual mycelium does not occur.

Human and ecological impacts The common bread mold *Rhizopus stolonifer* is a frequent household pest—probably the zygomycete that is most familiar to you. Saprophytic and parasitic members of these lineages are responsible for rotting strawberries and other fruits and vegetables and causing large losses in the fruit and vegetable industries. The black bread mold pictured in **Figure 31.15** is a familiar example of a saprophytic zygomycete. Some

species of *Mucor* are used in the production of steroids for medical use. Species of *Rhizopus* and *Mucor* are used in the commercial production of organic acids, pigments, alcohols, and fermented foods.

(a) Black bread mold
Rhizopus nigricans

Sporangia

Mycelium   100 μm

(b) A sporangium

Spores

20 μm

FIGURE 31.15 Black bread molds are zygomycetes.

Recent phylogenetic analyses have shown that the arbuscular mycorrhizal fungi (AMF) form a distinct phylum, indicating that they are a major monophyletic group. ✔ You should be able to mark the origin of the AMF association on Figure 31.8.

Absorptive lifestyle Recall from Section 31.3 that AMF absorb phosphorus-containing ions or molecules in the soil and transfer them, along with other nutrients and water, into the roots of trees, grasses, and shrubs in grassland and tropical habitats (**Figure 31.16**). In exchange, the host plant provides the symbiotic fungi with sugars and other organic compounds.

Life cycle Most species form spores underground. Glomeromycetes are difficult to grow in the laboratory, so their life cycle is not well known. No one has yet discovered a sexual phase in these species.

Human and ecological impacts Because grasslands and tropical forests are among the most productive habitats on Earth, the AMF are enormously important to both human and natural economies.

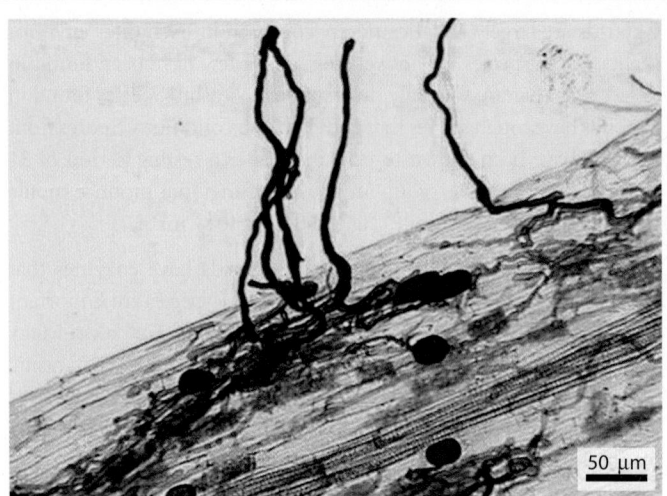

50 μm

FIGURE 31.16 AMF Penetrate the Walls of Plant Root Cells.

Fungi > Basidiomycota (Club Fungi)

Although most basidiomycetes form mycelia and produce multicellular reproductive structures, some species have the unicellular growth form. The group is named for basidia, the club-like or pedestal-like cells where meiosis and spore formation occur. ✔ You should be able to mark the origin of the basidium on Figure 31.8. About 31,000 species of basidiomycetes have already been described, and more are being discovered each year.

Absorptive lifestyle Basidiomycetes are important saprophytes. Along with a few soil-dwelling bacteria, they are the only organisms capable of synthesizing lignin peroxidase and completely digesting wood. Some basidiomycetes are ectomycorrhizal fungi (EMF) that associate with trees in temperate forests. One subgroup consists entirely of parasitic forms called rusts, including species that cause serious infections in wheat and rye fields. The plant parasites called smut fungi are also basidiomycetes. Smuts specialize in infecting grasses; a few infect other fungi. Thus, the entire array of absorptive lifestyles found in fungi—saprophytic, mutualistic, and parasitic—is found within Basidiomycota.

Life cycle Asexual reproduction through production of spores is common in the Basidiomycota, although not as prevalent as in other

Lycoperdon species

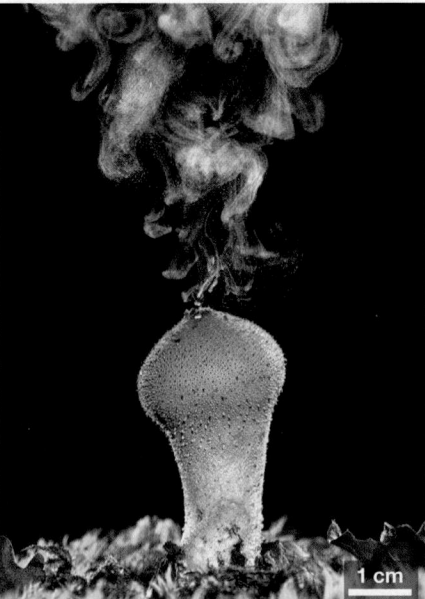

1 cm

FIGURE 31.17 Puffballs Can Produce Billions of Spores.

groups of fungi. Asexual reproduction also occurs through growth and fragmentation of mycelia in the soil or in rotting wood, resulting in genetically identical individuals that are physically independent. During sexual reproduction, all basidiomycetes—even unicellular ones—produce basidia. In the largest subgroup in this lineage, basidia form in large, aboveground reproductive structures called mushrooms, brackets, earthstars, or puffballs (**Figure 31.17**). Basidiomycota and Ascomycota are the only fungi that have heterokaryotic (dikaryotic) mycelia. ✔ You should be able to mark the origin of the dikaryotic condition on Figure 31.8.

Human and ecological impacts EMF are enormously important in forestry. The temperate forests where these fungi are particularly abundant provide most of the hardwoods and softwoods used in building construction, furniture-making, and paper-making. Throughout the world, mushrooms are cultivated or collected from the wild as a source of food. The white button, crimini, and portabella mushrooms you may have seen in grocery stores are all varieties of the same species, *Agaricus bisporus*. Some of the toxins found in poisonous mushrooms are used in biological research; others have hallucinogenic effects on people and are used and traded illegally.

About half of the ascomycetes grow in symbiotic association with cyanobacteria and/or single-celled members of the green algae, forming the structures called lichens. Over 15,000 different lichens have been described to date; in most, the fungus involved is an ascomycete (although a few basidiomycetes participate as well). To name a lichen, biologists use the genus and species name assigned to the fungus that participates in the association. Most lichen-formers are found in a single monophyletic group within Ascomycota. ✔ Based on this observation, you should be able to mark the origin of the lichen-forming habit on Figure 31.8.

Absorptive lifestyle The fungus in lichens appears to protect the photosynthetic bacterial or algal cells. The fungal hyphae form a dense protective layer that shields the photosynthetic species and reduces water loss. In return, the cyanobacterium or alga provides carbohydrates that the fungus uses as a source of carbon and energy. However, the hyphae of some lichen-forming fungi have been observed to invade algal cells and kill them. This observation suggests a partially parasitic relationship in at least some lichens. The nature of lichen-forming associations is the subject of ongoing research.

Life cycle As **Figure 31.18a** shows, many lichens reproduce asexually via the production of small "mini-lichen" structures called soredia that contain both symbionts. Soredia disperse to a new location via wind or water and then develop into a new, mature individual via the growth of the algal or bacterial and fungal symbionts. In addition, the fungal partner may form asci. Spores that are shed from asci germinate to form a small mycelium. If the growing hyphae encounter enough of the appropriate algal or bacterial cells, a new lichen can form.

Human and ecological impacts In terms of their abundance and diversity, lichens dominate the Arctic and Antarctic tundras and are extremely common in boreal forests (see Chapter 50). They are the major food of caribou, as well as the most prevalent colonizers of bare rock surfaces throughout the world (**Figure 31.18b**). Rock-dwelling lichens are significant, because they break off mineral particles from the rock surface as they grow—launching the first step in soil formation. About 10,000 tons of lichens are processed annually and used in perfume production—either as a source of fragrant molecules or as a source of molecules that keep the fragrant components of perfumes from evaporating too rapidly.

(a) Cross section of a lichen, showing three layers

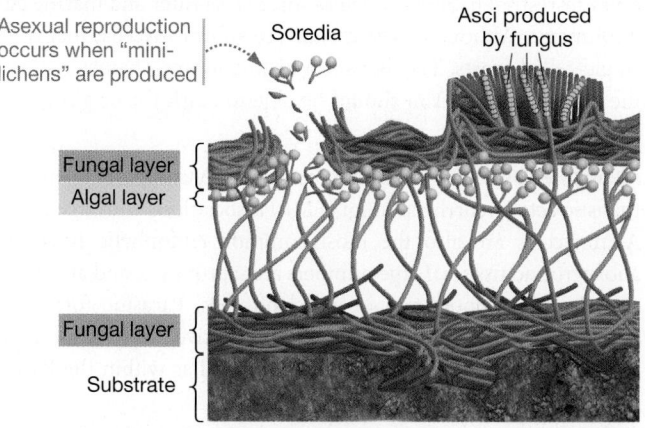

(b) Top view of lichen on a rock

FIGURE 31.18 Lichens Are Associations between a Fungus and a Cyanobacterium or Green Alga. (a) In a lichen, cyanobacteria or green algae are enmeshed in a dense network of fungal hyphae. **(b)** Lichens often colonize surfaces, such as tree bark or bare rock, where other organisms are rare.

The ascomycetes that do not form lichens are found in virtually every terrestrial habitat, as well as some freshwater and marine environments. Although most ascomycetes form mycelia, many are single-celled yeasts. The ascus is a distinguishing characteristic of the Ascomycota. ✔ You should be able to mark the origin of the ascus on Figure 31.8.

Absorptive lifestyle A few members of the Ascomycota form mutualistic ectomycorrhizal fungi (EMF) associations with tree roots. Ascomycetes are also the most common endophytic fungi on aboveground tissues. Large numbers are saprophytic and are abundant in forest floors and in grassland soils. Parasitic forms are common as well. The entire array of absorptive lifestyles found in fungi has evolved within Ascomycota as well as within the Basidiomycota.

In addition, about 65 species in one ascomycete lineage are predatory—primarily on amoebae and other unicellular protists. Some are large enough to capture the microscopic animals called roundworms, however. The predatory ascomycetes trap their prey by means of sticky substances on their cell walls or catch prey in snares consisting of looped hyphae (**Figure 31.19**). Once a prey individual is captured, hyphae invade its body, digest it, and absorb the nutrients that are released. A predatory ascomycete has been found in fossilized material that is 100 million years old. ✔ You should be able to mark the origin of predation on Figure 31.8.

Life cycle The aboveground, ascus-bearing reproductive structures of these fungi may be a cup or saucer shape called an **ascocarp** (**Figure 31.20**). For that reason, these ascomycetes are known as **cup fungi**. It is also routine for spores called conidia to be produced asexually, at the ends of specialized hyphae called conidiophores.

Human and ecological impacts Some saprophytic ascomycetes can grow on jet fuel or paint and are used to help clean up contaminated sites. *Penicillium* is an important source of antibiotics, and *Aspergillus* produces citric acid used to flavor soft drinks and candy. Truffles and morels are so highly prized that they can fetch $1000 per pound, and the multibillion-dollar brewing, baking, and wine-making industries would collapse without the yeast *Saccharomyces cerevisiae*. A few parasitic ascomycetes cause infections in humans and other animals. In land plants, parasites from this group cause diseases including Dutch elm disease and chestnut blight.

Arthrobotrys anchonia

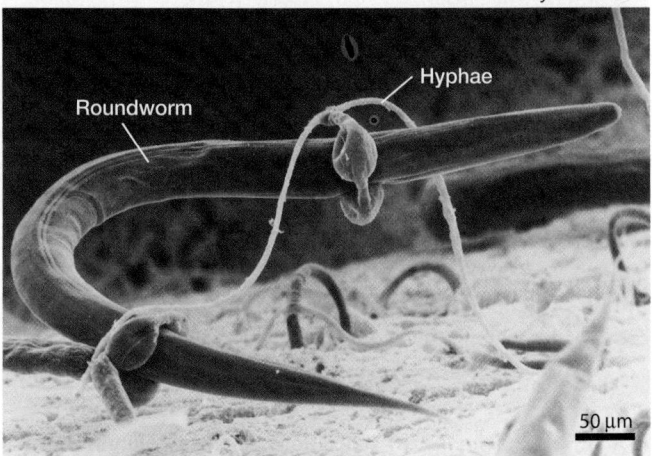

Roundworm

Hyphae

50 μm

FIGURE 31.19 Some Ascomycetes Are Predatory.

Cookeina sulcipes

1 cm

FIGURE 31.20 Ascomycetes Are Sometimes Called Cup Fungi.

Summary of Key Concepts

👉 **Fungi are important in part because many species live in close association with land plants. They supply plants with key nutrients and decompose dead wood. They are the master recyclers of nutrients in terrestrial environments.**

- Living plants are colonized by fungi.

- The roots of grasses and trees that grow in warm or tropical habitats are infiltrated by symbiotic glomeromycetes that supply the plant with phosphorus in exchange for sugars and other products of plant photosynthesis.

- The roots of trees in temperate habitats are colonized by basidiomycetes that exchange nitrogen for photosynthetic products.

- Many fungi live in leaves and stems; in some grasses and other species, these endophytic fungi secrete toxins that discourage herbivores.

- Parasitic fungi are responsible for devastating blights in crops and other plants.

- Once plants die, saprophytic fungi degrade the lignin and cellulose in wood and use nutrients from decaying plant material.

- Because they free up carbon atoms that would otherwise be locked up in wood, fungi speed up the carbon cycle in terrestrial habitats.

  ✔You should be able to predict the results of using fungicides to experimentally exclude fungi from 10 m × 10 m plots in a forest or grassland near your campus.

👉 **All fungi make their living by absorbing nutrients from living or dead organisms. Fungi secrete enzymes so that digestion takes place outside their cells. Their morphology provides a large amount of surface area for efficient absorption.**

- Several adaptations make fungi exceptionally effective at absorbing nutrients from the environment.

- The two growth habits found among Fungi—single-celled "yeasts" or mycelia composed of long, filamentous hyphae—give fungal cells extremely high surface-area-to-volume ratios.

- Extracellular digestion, in which enzymes are secreted into food sources, enables fungi to break down extremely large molecules without ingesting them.

- Lignin decomposes through a series of uncontrolled oxidation reactions triggered by the enzyme lignin peroxidase.

- Cellulose digestion occurs in a carefully regulated series of steps, each catalyzed by a specific cellulase.

  ✔You should be able to explain why a large surface area increases the efficiency of absorption.

👉 **Many fungi have unusual life cycles. It is common for species to have a long-lived heterokaryotic stage, in which cells contain haploid nuclei from two different individuals. Although most species reproduce sexually, very few species produce gametes.**

- Many species of fungi have never been observed to reproduce sexually. Most species can produce haploid spores either sexually or asexually, however.

- Sexual reproduction usually starts when hyphae from different individuals fuse—an event called plasmogamy. If the fusion of nuclei, or karyogamy, does not occur immediately, a heterokaryotic mycelium forms. Heterokaryotic cells may eventually produce spore-forming structures where karyogamy and meiosis take place.

- Four distinctive types of reproductive structures are known in fungi: (1) chytrids are aquatic fungi with motile gametes; (2) zygomycetes are soil-dwelling fungi with tough sporangia; (3) Basidiomycota have club-like, spore-forming structures; and (4) Ascomycota have sac-like, spore-forming structures.

- Recent analyses of DNA sequence data have revealed that the Basidiomycota, Ascomycota, and Glomeromycota each form monophyletic groups. Researchers are still trying to understand how the various groups of chytrids and zygomycetes are related to each other and to the microsporidians.

  ✔You should be able to explain why most fungi don't need to have gametes to accomplish sexual reproduction.

  (MB) **Web Activity** The Life Cycle of a Mushroom

Questions

✔ TEST YOUR KNOWLEDGE

Answers are available in Appendix B

1. The mycelial growth habit leads to a body with a high surface-area-to-volume ratio. Why is this important?
   a. Mycelia have a large surface area for absorption.
   b. The hyphae that make up mycelia are long, thin tubes.
   c. Most hyphae are broken up into compartments by walls called septa, although some exist as single, gigantic cells.
   d. Hyphae can infiltrate living or dead tissues.

2. What is plasmogamy?
   a. production of gametes, after fusion of hyphae from different individuals
   b. exchange of nutrients between symbiotic fungi and hosts
   c. fertilization—the fusion of cytoplasm and nuclei from different individuals
   d. fusion of the cytoplasm from different individuals, without nuclear fusion

3. The Greek root *ecto* means "outer." Why are ectomycorrhizal fungi, or EMF, aptly named?
   a. Their hyphae form tree-like branching structures inside plant cell walls.
   b. They are mutualistic.
   c. Their hyphae form dense mats that envelop roots but do not penetrate the walls of cells inside the root.
   d. They transfer nitrogen from outside their plant hosts to the interior.

4. The hyphae of AMF form bushy or balloon-like structures after making contact with the plasma membrane of a root cell. Why?
   a. They anchor the fungus inside the root, so the association is more permanent.
   b. They increase the surface area available for the transfer of nutrients.
   c. They produce toxins that protect the plant cells against herbivores.
   d. They break down cellulose and lignin in the plant cell wall.

5. What does it mean to say that a hypha is dikaryotic or heterokaryotic?
   a. Two nuclei fuse during sexual reproduction to form a zygote.
   b. Two or more independent nuclei, derived from different individuals, are present.
   c. The nucleus is diploid or polyploid—not haploid.
   d. It is extremely highly branched, which increases its surface area and thus absorptive capacity.

6. Very few organisms besides fungi have a heterokaryotic stage in their life cycle. Which of the following is another unusual aspect of the fungal life cycle?
   a. Some fungi exhibit alternation of generations—meaning that there is a multicellular diploid stage and a multicellular haploid stage.
   b. They produce eggs and sperm in approximately equal numbers, instead of many sperm and a few eggs.
   c. Spores have to fuse with each other before they develop into a new mycelium.
   d. Most varieties undergo sexual reproduction without producing eggs or sperm.

✔ TEST YOUR UNDERSTANDING

Answers are available in Appendix B

1. Explain why fungi that degrade dead plant materials are important to the global carbon cycle. Do you accept the text's statement that, without these fungi, "Terrestrial environments would be radically different than they are today and probably much less productive"? Why or why not?

2. Lignin and cellulose provide rigidity to the cell walls of plants. But in most fungi, chitin performs this role. Why is it logical that most fungi don't have lignin or cellulose in their cell walls?

3. Biologists claim that EMF and AMF species are better than plants at acquiring nutrients because they have a higher surface area and because they are more effective at acquiring phosphorus (P) and/or nitrogen (N). Compare and contrast the surface area of mutualistic fungi and plant roots. Explain why fungi are particularly efficient at acquiring P and N, compared to plants.

4. Using information from Chapters 3 through 6, list three key macromolecules found in plants that contain phosphorus. Explain why plant growth might be limited by access to P.

5. Compare and contrast the way that fungi degrade lignin with the way that they digest cellulose.

6. How is it possible for thousands of mating types to exist in a single species of fungus, instead of just two sexes like other eukaryotes?

✔ APPLYING CONCEPTS TO NEW SITUATIONS

Answers are available in Appendix B

1. The box on chytrids in Section 31.4 mentions that they may be responsible for massive die-offs currently occurring in amphibians. Review Koch's postulates in Chapter 28, then design a study showing how you would use Koch's postulates to test the hypothesis that chytrid infections are responsible for the frog deaths.

2. Some biologists contend that the ratio of plant species to fungus species worldwide is on the order of 1:6. Explain why you agree or disagree with this claim. In doing so, consider the analyses of endophytic, parasitic, lichen-forming, mycorrhizal, and saprophytic strategies presented in this chapter. Also, consider the diversity of tissues and organs available in plants.

3. Experiments indicate that cellulase genes are transcribed and translated together. If cells are selected to be extremely efficient at digesting cellulose, is this result logical? Would you predict that the gene that codes for lignin peroxidase is transcribed along with the cellulase genes? How would you test your prediction?

4. Many mushrooms are extremely colorful. Fungi do not see, so it is unlikely that colorful mushrooms are communicating with one another. One hypothesis is that the colors serve as a warning to animals that eat mushrooms, much like the bright yellow and black stripes on wasps. Design an experiment capable of testing this hypothesis.

Jellyfish such as this hydromedusa are among the most ancient of all animals—they appear in the fossil record over 560 million years ago.

An Introduction to Animals

32

The **animals** are a monophyletic group of eukaryotes that share a series of traits:

- They are multicellular, with cells that lack cell walls but have an extensive extracellular matrix (ECM; see Chapter 8). The ECM includes proteins specialized for cell-cell adhesion and communication.

- They are heterotrophs, meaning that they obtain the carbon compounds they need from other organisms. Most ingest their food, rather than absorbing it across the body surface.

Animals are the only multicellular heterotrophs on the tree of life that ingest their food. Fungi are multicellular heterotrophs, but they absorb nutrients. Some slime molds could also be classified as multicellular heterotrophs that ingest food, but they are multicellular only during the reproductive phase of their life cycle—they are unicellular when they feed (see Chapter 29). As a result, animals are the largest predators, herbivores, and detritivores on Earth.

In addition, all animals move under their own power at some point in their life cycle, and all animals other than sponges have: **(1)** specialized cells called **neurons**—nerve cells—that transmit electrical signals to other cells; and **(2)** muscle cells that can change the shape of the body by contracting. In most animals, neurons connect to each other, forming a nervous system, and some neurons connect to muscle cells—which may contract in response to electrical signals from neurons. Muscles and neurons are adaptations that allow a large, multicellular body to move efficiently.

Over 1.2 million species of animals have been described and given scientific names to date, and biologists predict that tens of millions more have yet to be discovered. To analyze the almost overwhelming number and diversity of species in this lineage, this chapter presents a broad overview of how they diversified, along with more detailed information on the characteristics of the lineages that appear first in the fossil record. Chapters 33 and 34 follow up by focusing on the most species-rich groups.

KEY CONCEPTS

- Animals are multicellular, heterotrophic eukaryotes that lack cell walls and ingest their prey.

- Fundamental changes in morphology and development occurred as animals diversified.

- Recent phylogenetic analyses have shown that there are four major groups of animals: non-bilaterian lineages, two protostome groups (Lophotrochozoa and Ecdysozoa), and the deuterostomes.

- Within major groups of animals, evolutionary diversification was based on innovative ways of sensing the environment, feeding, and moving.

- Methods of sexual reproduction vary widely among animal groups, and many species can reproduce asexually. It is common for individuals to undergo metamorphosis during their life cycle.

✔ When you see this checkmark, stop and test yourself. Answers are available in Appendix B.

32.1 Why Do Biologists Study Animals?

If you ask biologists why they study animals, the first answer they'll give is, "Because they're fascinating." It's hard to argue with this statement. Consider ants.

- Ants live in colonies that may include millions of individuals. But colony-mates cooperate so closely in feeding, defense, and rearing young that each ant seems like a cell in a multicellular organism instead of an individual.

- Some ants live in trees, and protect their host trees by attacking giraffes and other grazing animals a million times their size. The trees provide oil-rich growths that the ants eat.

- Rancher ants protect the plant-sucking insects called aphids from spiders and other predators, and eat the sugar-rich honeydew that aphids secrete from their anus (**Figure 32.1a**).

- Farmer ants eat fungi that they plant, fertilize, and harvest in underground gardens (**Figure 32.1b**).

- Parasitic ants look and smell like a host ant species but enslave them, forcing the hosts to rear the young of the parasitic species instead of their own.

Based on observations like these, most people would agree that ants—and by extension, other animals—are indeed fascinating. But beyond pure intellectual interest, there are compelling practical reasons to study animals.

Biological Importance

Animals are key consumers in virtually every ecosystem—from the deep oceans to alpine ice fields and from tropical forests to arctic tundras. It is not possible to understand or preserve these ecosystems without understanding and preserving animals.

In addition to their ecological importance, animals are an extraordinarily diverse and species-rich lineage on the tree of life. Current estimates suggest that there are between 10 million and 50 million animal species living today. These species range in size and complexity from tiny sponges, which attach to a substrate and contain just a few cell types and simple tissues, to blue whales, which migrate tens of thousands of kilometers each year in search of food and contain trillions of cells, dozens of distinct tissues, an elaborate skeleton, and highly sophisticated sensory and nervous systems. A great deal of diversification has occurred in this lineage. To understand the history of life, biologists have to understand how animals came to be so diverse.

Role in Human Health and Welfare

The vast majority of research funded by government agencies, foundations, and corporations is motivated by a simple desire: improving people's standard of living. In some cases, understanding animals has a direct economic benefit; in other situations, the gains are through improving human health.

FOOD, MATERIALS, AND TRANSPORTATION Humans depend on wild and domesticated animals for food. In addition to domesti-

(a) "Rancher ants" tend aphids and eat their sugary secretions.

(b) "Farmer ants" cultivate fungi in gardens.

FIGURE 32.1 Examples of Sophisticated Behavior in Ants.

cated mammals and birds, large industries are organized around the harvest of wild fish, mollusks (clams and mussels), and crustaceans (lobster, shrimp, crab).[1] In 2006, the global market value of fish and shellfish (mollusks and crustaceans) was $91.2 billion U.S. dollars. Most commercial fruit growers rely on bees and other animals to pollinate their crops.

Animals are also an important source of materials: fibers used to make clothing, blankets, and other articles, as well as the leather used in shoe-making and other industries.

In preindustrial societies, domesticated animals are the principal source of transportation and power. Horses, donkeys, oxen, and other domesticated animals transport people and cargo and do work such as plowing fields.

DISEASE TRANSMISSION Many animals play an important role in transmitting human diseases. Examples include the mosquito species that harbor the malaria parasite, and the rats that in earlier centuries carried the bacterium responsible for deadly epidemics of bubonic plague.

---

[1] In Chapters 32–34, common names are often given in parentheses after the formal name of a lineage. The common names indicate some of the more familiar members of the lineage in question.

When animals act as a transmission vector for a human disease, biologists work with physicians and public health officials on strategies to reduce animal–human transmission.

ANIMALS AS MODEL ORGANISMS Humans are animals, so efforts to understand human biology depend on advances in animal biology.

- Humans share a large portion of their genome with other animals. As a result, research on animals like fruit flies and zebrafish has revealed fundamental aspects of cellular and developmental biology that are shared by humans.

- Because genetic homologies between humans and other animals are so extensive, most drug testing and other types of biomedical research can be done on mice, rats, or primates instead of humans.

Tens of thousands of biologists are employed in studying or managing animals. How do researchers go about understanding the most fundamental aspects of animal diversity?

32.2 How Do Biologists Study Animals?

Biologists currently recognize about 30 **phyla**, or major lineages, of animals—the exact number is being debated and revised as additional information comes to light. **Table 32.1** lists 29 phyla that are included in most published analyses.

Each animal phylum has distinct morphological features—synapomorphies that identify it as a monophyletic group. At a more general level, however, groups of phyla are characterized by fundamental aspects of morphology and development that changed as animals diversified. What are these, and why are they important?

Analyzing Comparative Morphology

In essence, animals are moving and eating machines. To understand animal diversity, then, one of the basic tasks is to understand how these machines are put together. 🔑 The origin and early evolution of animals was based on four aspects of the fundamental architecture, or **body plan**, of animals:

1. the origin and elaboration of tissues—especially the tissues found in embryos;

2. the origin and elaboration of the nervous system and the subsequent evolution of a cephalized body—one with a distinctive head region;

3. the evolution of a fluid-filled body cavity; and

4. variation in the events of early embryonic development.

Let's consider each in turn.

THE ORIGIN AND DIVERSIFICATION OF TISSUES All animals have groups of similar cells that are organized into the tightly integrated structural and functional units called **tissues**. Although sponges lack many types of tissues found in other animals, recent research has shown that one group of sponges have

TABLE 32.1 **An Overview of Major Animal Phyla**

Group and Phylum	Common Name or Example Taxa	Estimated Number of Species
Non-bilaterian groups		
Porifera	Sponges	7,000
Placozoa	Placozoans	1
Cnidaria	Jellyfish, corals, anemones, hydroids, sea fans	10,000
Ctenophora	Comb Jellies	100
Acoelomorpha	Acoelomate worms	10
Protostomes (lacks typical protostome development)		
Chaetognatha	Arrow Worms	100
Protostomes: Lophotrochozoa		
Rotifera	Rotifers	1,800
Platyhelminthes	Flatworms	20,000
Nemertea	Ribbon worms	900
Gastrotricha	Gastrotrichs	450
Acanthocephala	Acanthocephalans	1,100
Entoprocta	Entoprocts, kamptozoans	150
Gnathostomulida	Gnathostomulids	80
Annelida	Segmented worms	16,500
Molluska	Mollusks (clams, snails, octopuses)	94,000
Phoronida	Horseshoe worms	20
Bryozoa	Bryozoa, ectoprocts, moss animals	4,500
Brachiopoda	Brachiopods; lamp shells	335
Protostomes: Ecdysozoa		
Nematoda	Roundworms	25,000
Kinorhyncha	Kinorhynchs	150
Nematomorpha	Hair worms	320
Priapula	Priapulans	16
Onychophora	Velvet worms	110
Tardigrada	Water bears	800
Arthropoda	Arthropods (spiders, insects, crustaceans)	1,100,000
Deuterostomes		
Echinodermata	Echinoderms (sea stars, sea urchins, sea cucumbers)	7,000
Xenoturbellida	Xenoturbillidans	2
Hemichordata	Acorn worms	85
Chordata	Chordates (tunicates, lancelets, sharks, bony fish, amphibians, reptiles, mammals)	50,000

epithelium—a layer of tightly joined cells that covers the surface.[2] Other animals have an array of other tissue types as well as epithelium.

In animals other than sponges, the number of tissues that exist in an embryo varies. Animals whose embryos have two types of tissue are called **diploblasts** (literally, "two-buds"); animals whose embryos have three types are called **triploblasts** ("three-buds").

You might recall from Chapter 22 that these embryonic tissues are organized in layers, called **germ layers**. In diploblasts these germ layers are called **ectoderm** and **endoderm** (**Figure 32.2**). The Greek roots *ecto* and *endo* refer to outer and inner, respectively; the root *derm* means "skin." In most cases the outer and inner "skins" of diploblast embryos are connected by a gelatinous material that may contain some cells. In triploblasts, however, there is a germ layer called **mesoderm** between the ectoderm and endoderm. (The Greek root *meso* refers to middle.)

The embryonic tissues found in animals develop into distinct adult tissues, organs, and organ systems (see Chapter 22). In triploblasts, for example:

- *Ectoderm* gives rise to skin and the nervous system.

- *Endoderm* gives rise to the lining of the digestive tract.

- *Mesoderm* gives rise to the circulatory system, muscle, and internal structures such as bone and most organs.

In general, then, ectoderm produces the covering of the animal and endoderm generates the digestive tract. Mesoderm gives rise to the tissues in between. The same pattern holds in diploblasts, except that (**1**) muscle is simpler in organization and is derived from ectoderm, and (**2**) reproductive tissues are derived from endoderm.

---

[2]For discussion and citations, see Nichols, S. and G. Wörheide. 2005. *Integrative and Comparative Biology* 45: 333–334; Elliott, G. R. D. and S. P. Leys. 2007. *Journal of Experimental Biology* 210: 3736–3748; Philippe, H., et al. 2009. *Current Biology* 19: 706–712.

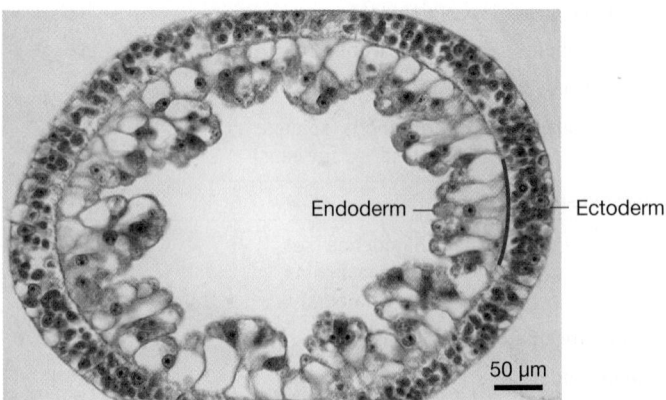

FIGURE 32.2 Diploblastic Animals Have Bodies Built from Ectoderm and Endoderm. This is a cross section through the tube-shaped portion of a hydra's body. The cells have been stained to make them more visible.

Traditionally, two groups of animals have been recognized as diploblasts: the Cnidaria (pronounced *ni-DARE-ee-uh*)–which include the jellyfish, corals, sea pens, hydra, and anemones (**Figure 32.3a**)—and the Ctenophora (pronounced *ten-AH-for-ah*), or comb jellies (**Figure 32.3b**). Recent data, however, suggest that at least some cnidarians have mesoderm and are triploblastic. All other animals, from leeches to humans, are triploblastic.

The evolution of mesoderm was important because it gave rise to the first complex muscle tissue used in movement. Sponges lack muscle and are sessile, or nonmoving, as adults; their larvae (immature forms) move via cilia. Ctenophores have muscle cells that can change the body's shape, but both larvae and adults swim using cilia.

NERVOUS SYSTEMS, BODY SYMMETRY, AND CEPHALIZATION

Sponges lack neurons, but cnidarians and ctenophores have nerve cells that are organized into a diffuse arrangement called a **nerve net** (**Figure 32.4a**). All other animals have a **central nervous system**, or CNS. In a CNS, some neurons are clustered into one or more large tracts or cords that project throughout the body; others are clustered into masses called **ganglia** (**Figure 32.4b**).

Nerve nets and central nervous systems are associated with two contrasting types of body symmetry: radial (literally, "spoke") symmetry or bilateral ("two-sides") symmetry. A body is symmetrical if it can be divided by a plane such that the resulting pieces are nearly identical.

Cnidarians and ctenophores, along with many sponges, have **radial symmetry**—meaning that they have at least two planes of symmetry. For example, almost any plane sectioned through the center of the hydra in Figure 32.4a produces two identical halves.

(a) Cnidaria include jellyfish, corals, and sea pens (shown).

(b) Ctenophora are the comb jellies.

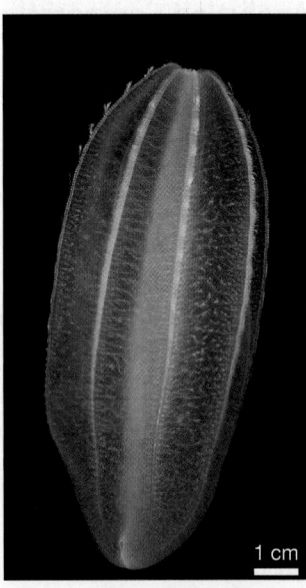

FIGURE 32.3 Cnidarians and Ctenophorans. (a) Like most members of the Cnidaria, this sea pen lives in marine environments. **(b)** Comb jellies are a major component of planktonic communities in the open ocean.

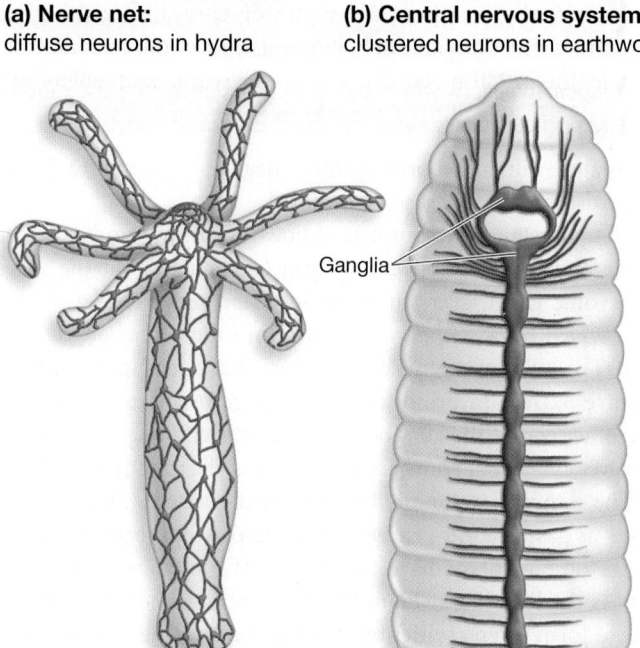

(a) Nerve net: diffuse neurons in hydra

(b) Central nervous system: clustered neurons in earthworm

Ganglia

FIGURE 32.4 Associations between Body Symmetry and the Nervous System. (a) Radially symmetric animals, like this hydra, have a nerve net. **(b)** Bilaterally symmetric animals, like this earthworm, have a central nervous system.

Many of the radially symmetric animals living today either float in water or live attached to a substrate.

Organisms with **bilateral symmetry**, in contrast, have one plane of symmetry and tend to have a long, narrow body. The earthworm in Figure 32.4b, for example, only has one plane of symmetry—running lengthwise down its middle. Most of the bilaterally symmetric animals living today move through water or air or along or through a substrate.

How are symmetry and nervous systems related? The function of neurons and nervous systems is to transmit and process information in the form of electrical signals (see Chapter 45).

• *Radially symmetric organisms* are equally likely to encounter prey and other aspects of the environment in any direction. As a result, a diffuse nerve net can receive and send signals efficiently.

• *Bilaterally symmetric organisms* tend to encounter prey and other aspects of the environment at one end. As a result, it is advantageous to have many neurons concentrated at that end, with nerve tracts that carry information from there down the length of the body.

The evolution of bilateral symmetry occurred along with **cephalization**: the evolution of a head, or anterior region, where structures for feeding, sensing the environment, and processing information are concentrated. The large mass of neurons that is located in the head, and that is responsible for processing information from throughout the body, is called the cerebral ganglion or **brain**.

All triploblastic animals have bilateral symmetry except for species in the phylum Echinodermata (pronounced *ee-KINE-oh-der-ma-ta*)—a group where radial symmetry evolved independently of the sponges, cnidaria, and ctenophores. The echinoderms include species such as sea stars, sea urchins, feather stars, and brittle stars. Although their larvae are bilaterally symmetric, adult echinoderms are radially symmetric.

To explain the pervasiveness of bilateral symmetry, biologists point out that locating and capturing food is particularly efficient when movement is directed by a distinctive head region and powered by the rest of the body. In combination with the origin of mesoderm, a bilaterally symmetric body plan enabled rapid, directed movement and hunting. Lineages with a triploblastic, bilaterally symmetric, cephalized body had the potential to diversify into an array of formidable eating and moving machines.

WHY WAS THE EVOLUTION OF A BODY CAVITY IMPORTANT? A third architectural element that distinguishes animal phyla is the presence of an enclosed, fluid-filled cavity called a **coelom** (pronounced *SEE-lum*). The coelom creates a container for the circulation of oxygen and nutrients, along with space where internal organs can move independently of each other.

Although cnidarians and ctenophores have a central canal that functions in digestion and circulation, they do not have a coelom. The handful of triploblasts that do not have a coelom are called **acoelomates** ("no-cavity-form"; see **Figure 32.5a**); those that possess a coelom are known as **coelomates** (**Figure 32.5b**).

The coelom was a critically important innovation during animal evolution because an enclosed, fluid-filled chamber can act as an efficient **hydrostatic skeleton**. Soft-bodied animals with hydrostatic skeletons can move effectively even if they do not have fins or limbs.

Using a nematode (roundworm) as an example, movement is possible because the fluid inside the coelom stretches the body

(a) Acoelomates have no enclosed body cavity.

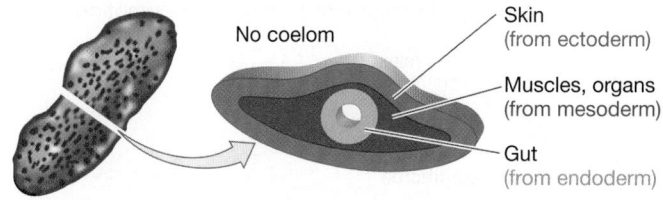

No coelom

Skin (from ectoderm)

Muscles, organs (from mesoderm)

Gut (from endoderm)

(b) Coelomates have an enclosed body cavity completely lined with mesoderm.

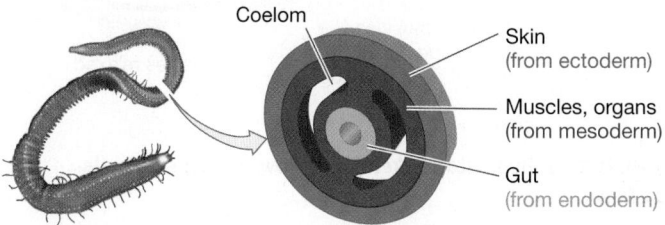

Coelom

Skin (from ectoderm)

Muscles, organs (from mesoderm)

Gut (from endoderm)

FIGURE 32.5 Animals May or May Not Have a Body Cavity.

wall—much like a water balloon—meaning that it is under tension (**Figure 32.6a**). This force exerts pressure on the fluid inside the coelom. When muscles in the body wall contract against the pressurized fluid, the force is transmitted through the fluid, changing the body's shape.

Figure 32.6b shows how coordinated muscle contractions and relaxations produce changes in the shape of a hydrostatic skeleton. The coordinated changes in body shape make writhing or swimming movements possible. ✔If you understand how a hydrostatic skeleton works, you should be able to demonstrate its function with a long, tube-shaped water balloon, using your fingers to pinch one side and then another to simulate muscle contractions.

By providing a hydrostatic skeleton, the evolution of the coelom gave bilaterally symmetric organisms the ability to move efficiently in search of food.

WHAT ARE THE PROTOSTOME AND DEUTEROSTOME PATTERNS OF DEVELOPMENT? Except for adult echinoderms, all of the coelomates—including juvenile forms of echinoderms—are bilaterally symmetric. This huge group of organisms is formally called the **Bilateria**. Based on distinctive events that occur early in the development of the embryo, the bilaterians can be split into two subgroups:

1. **protostomes**, in which the mouth develops before the anus, and blocks of mesoderm hollow out to form the coelom; and

2. **deuterostomes**, in which the anus develops before the mouth, and pockets of mesoderm pinch off to form the coelom.

Translated literally, protostome means "first-mouth" and deuterostome means "second-mouth."

To understand the contrasts in how protostome and deuterostome embryos develop, recall from Chapter 22 that the three embryonic germ layers form during a process called gastrulation, which begins when cells move into the center of the embryo. The cell movements create a pore that opens to the outside (**Figure 32.7a**). In protostomes, this pore becomes the mouth. The other end of the gut, the anus, forms later. In deuterostomes, however, this initial pore becomes the anus; the mouth forms later.

The other developmental difference between the groups arises as gastrulation proceeds and the coelom begins to form. Protostome embryos have two blocks of mesoderm beside the gut. As the left side of **Figure 32.7b** indicates, their coelom begins to form when cavities open within each of the two blocks of

(a) Hydrostatic skeleton of a nematode

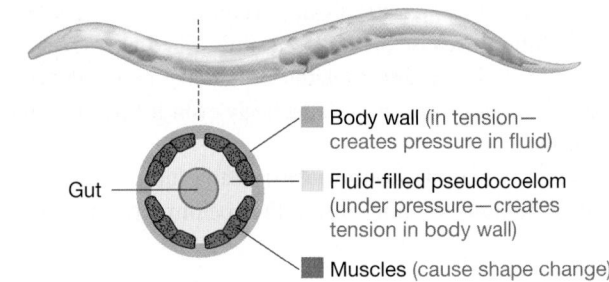

(b) Coordinated muscle contractions result in locomotion.

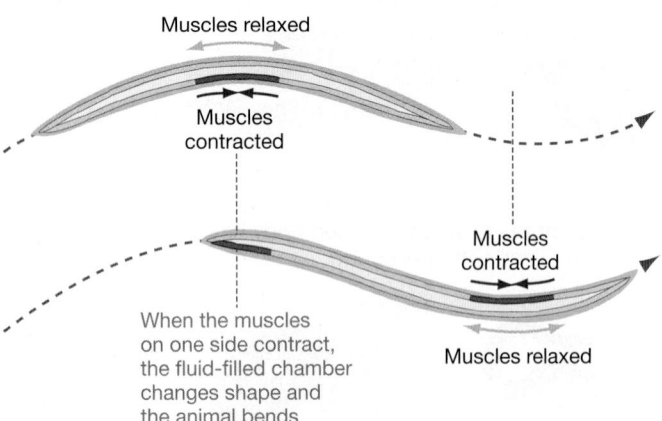

Muscles relaxed

Muscles contracted

Muscles contracted

When the muscles on one side contract, the fluid-filled chamber changes shape and the animal bends

Muscles relaxed

FIGURE 32.6 Hydrostatic Skeletons Allow Limbless Animals to Move.

✔**QUESTION** Suppose that all of the longitudinal muscles of this nematode contracted at the same time. What would happen?

(a) Gastrulation (formation of gut and embryonic germ layers)

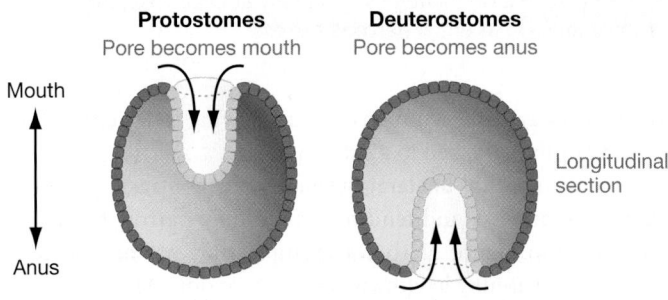

(b) Formation of coelom (body cavity lined with mesoderm)

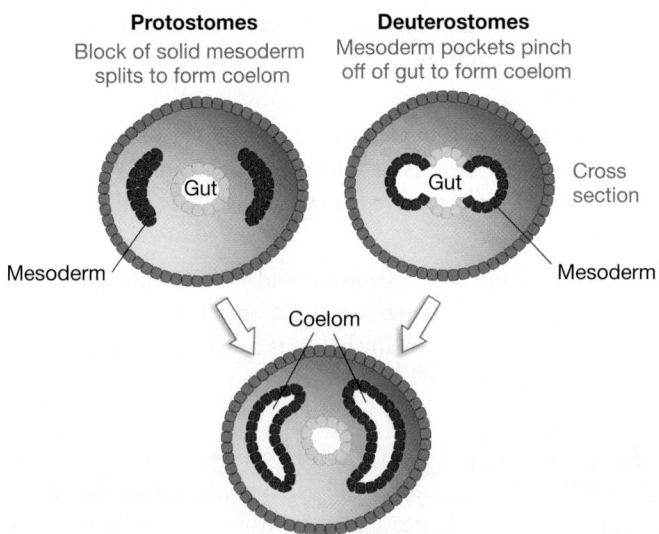

FIGURE 32.7 In Protostomes and Deuterostomes, Events in Early Development Differ. The differences between protostomes and deuterostomes show that there is more than one way to build a bilaterally symmetric, coelomate body plan.

mesoderm. In contrast, deuterostome embryos have layers of mesodermal cells located on either side of the gut. As the right side of Figure 32.7b shows, their coelom begins to form when these layers bulge out and pinch off to form fluid-filled pockets lined with mesoderm.

The vast majority of animal species, including the arthropods (insects, spiders, crustaceans), mollusks (clams, squid, octopuses), and annelids (segmented worms), are protostomes; humans and other chordates (fishes, amphibians, mammals) are deuterostomes.

In essence, the protostome and deuterostome patterns of development represent two distinct ways of achieving the same end—the construction of a bilaterally symmetric body that contains a cavity lined with mesoderm. The functional or adaptive significance of the differences—if any—is not known.

THE TUBE-WITHIN-A-TUBE DESIGN   Over 99 percent of the animal species alive today are bilaterally symmetric triploblasts that have coeloms and follow either the protostome or deuterostome pattern of development. This combination of features turned out to be a spectacularly successful way to organize a moving and eating machine. To review basic features of the animal body plan, go to the study area at *www.masteringbiology.com*.

(MB) Web Activity The Architecture of Animals

Although it sounds complex to describe a certain animal as a "bilaterally symmetric, coelomic triploblast with protostome [or deuterostome] development," the bodies of most animals are actually extremely simple in form. The basic animal body is a tube within a tube. The inner tube is the individual's gut, and the outer tube forms the body wall (**Figure 32.8a**). The mesoderm in between forms muscles and organs.

In several animal phyla, individuals have long, thin, tubelike bodies that lack limbs. Animals with this body shape are commonly called **worms**. There are many wormlike phyla, including the nemerteans and chaetognaths shown in **Figure 32.8b**.

What about more complex-looking animals, such as grasshoppers and lobsters and horses? The body plan of these animals can also be thought of as a tube within a tube, except that the tube is mounted on legs. Consider that most animals with complex-looking bodies are relatively long and thin. They have an outer body wall that is more or less tubelike and an internal gut that runs from mouth to anus. The body cavity itself is filled with muscles and organs derived from mesoderm. Wings and legs are just efficient ways to move a tube-within-a-tube body around the environment.

Once evolution by natural selection produced the basic tube-within-a-tube design, most of the diversification of animals was triggered by the evolution of novel types of structures for moving and capturing food. Do data from phylogenetic analyses of DNA sequences support this view?

Evaluating Molecular Phylogenies

Perhaps the most influential paper ever published on the phylogeny of animals appeared in 1997. Using sequences from the gene that codes for the RNA molecule in the small subunit of the ribosome,

(a) The tube-within-a-tube body plan in an earthworm

Body wall derived from ectoderm   Muscles and organs derived from mesoderm   Gut derived from endoderm

(b) Many animal phyla have wormlike bodies.

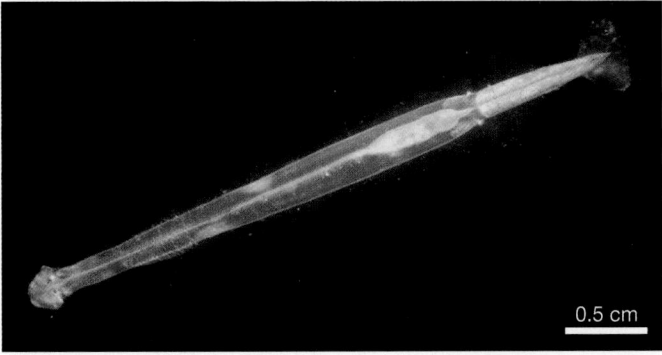

1 cm

Ribbon worm (Nemertea)

0.5 cm

Arrow worm (Chaetognatha)

FIGURE 32.8 The Tube-within-a-Tube Body Plan Is Common in Animals.

Anna Marie Aguinaldo and colleagues estimated the phylogeny of species from 14 animal phyla. The results were revolutionary.

The phylogenetic tree in **Figure 32.9** on page 608 is an updated version of the 1997 result, based on further studies of the genes for ribosomal RNA and a large series of proteins. To analyze it, start from the root, work your way toward the tips, and note several key points:

• A group of protists called the choanoflagellates are the closest living relatives of animals. Choanoflagellates are not animals because they are not multicellular, but they share several key characteristics with sponges. Both are **sessile**, meaning that

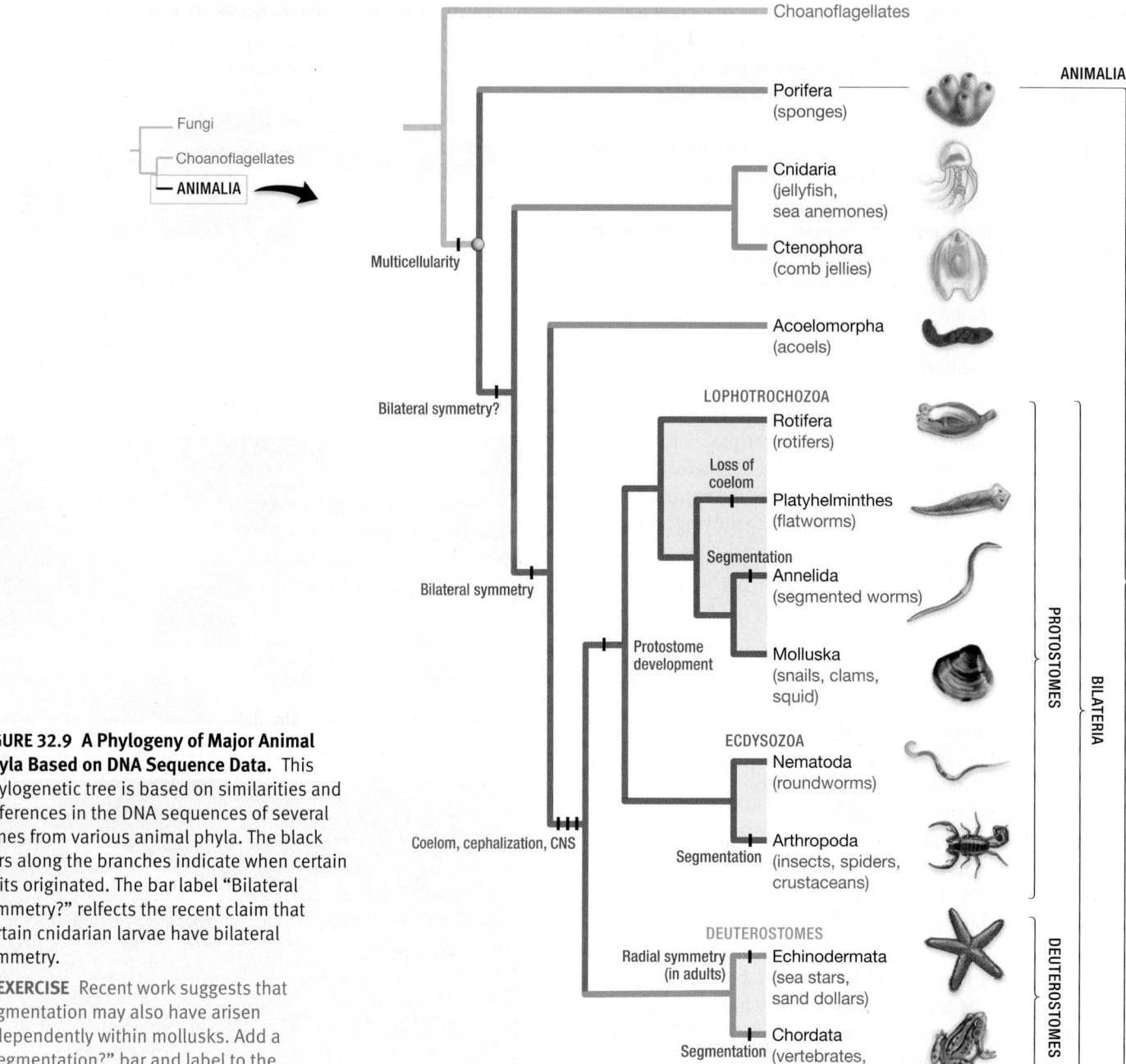

FIGURE 32.9 A Phylogeny of Major Animal Phyla Based on DNA Sequence Data. This phylogenetic tree is based on similarities and differences in the DNA sequences of several genes from various animal phyla. The black bars along the branches indicate when certain traits originated. The bar label "Bilateral symmetry?" reflects the recent claim that certain cnidarian larvae have bilateral symmetry.

✔**EXERCISE** Recent work suggests that segmentation may also have arisen independently within mollusks. Add a "Segmentation?" bar and label to the tree to indicate this possibility.

adults live permanently attached to a substrate. They also feed in a similar way, using cells with nearly identical morphology. As **Figure 32.10** shows, the beating of flagella creates water currents that bring organic debris toward the feeding cells of choanoflagellates and sponges. Sponge feeding cells are called **choanocytes**. In these feeding cells, food particles are trapped and ingested.

- Sponges are the sister group to all other animals. Stated another way, the initial split that occurred during animal diversification produced a lineage that evolved into today's sponges, and a lineage that evolved into all other animals living today.

- Ctenophora (jellyfish, hydra, anemones) and Cnidaria (comb jellies) are a monophyletic group.

- The placement of cnidarians and ctenophores as the sister group to acoelomorphs and bilaterians implies that endoderm and ectoderm were the first embryonic tissue types to evolve, and that radial symmetry evolved before bilateral symmetry.

- The closest living relative of bilaterally symmetric triploblasts, the Acoelomorpha, lack a coelom. This result indicates that the evolution of mesoderm preceded the evolution of the coelom.

(a) Choanoflagellates are sessile protists; some are colonial.

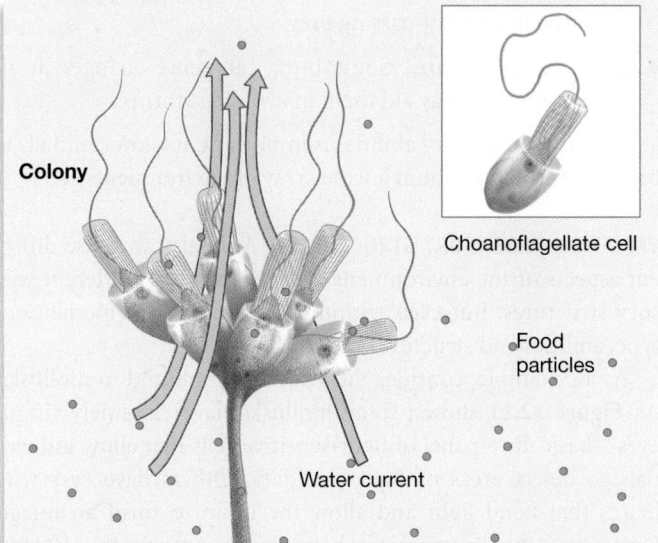

(b) Sponges are multicellular, sessile animals.

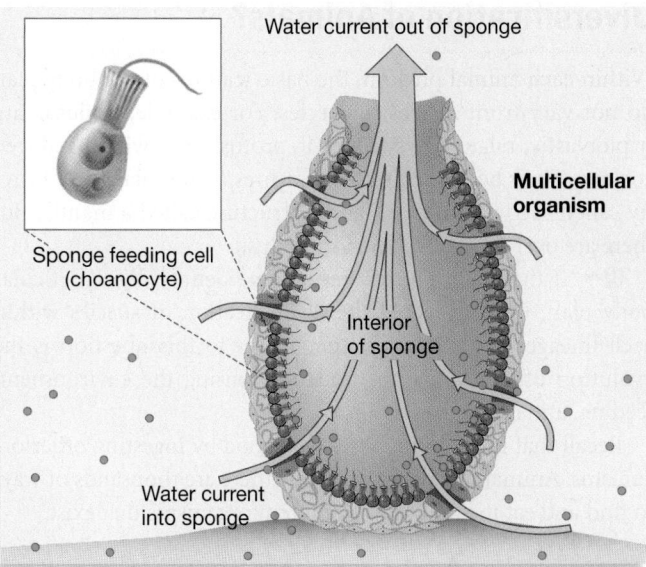

FIGURE 32.10 Choanoflagellates and Sponge Feeding Cells Are Almost Identical in Structure and Function.
(a) Choanoflagellates are suspension feeders. **(b)** A cross section of a simple sponge as it suspension feeds. The beating of flagella produces a water current that brings food into the body of the sponge, where it can be ingested by feeding cells.

- The split between the protostomes and the deuterostomes was a major event in the evolution of Bilateria.

- ⌐━ A fundamental split occurred within protostomes, forming two major subgroups with protostome development: (**1**) the Lophotrochozoa (pronounced *low-foe-tro-ko-ZOH-ah*) includes the mollusks and the annelids; and (**2**) the Ecdysozoa (pronounced *eck-die-so-ZOH-ah*) includes the arthropods and the nematodes. **Ecdysozoans** grow by shedding their external skeletons or outer coverings and expanding their bodies. **Lophotrochozoans** grow continuously when conditions are good.

- Species in the phylum Platyhelminthes (flatworms) are protostomes but lack a coelom—meaning that the coelom was lost during flatworm evolution. As Chapters 33 and 34 will detail, the coelom has also been reduced in an array of lineages—specifically, in groups that have limbs and thus do not rely on a hydrostatic skeleton for movement.

- During the evolution of bilaterians, segmentation arose several times independently. **Segmentation** is defined by the presence of repeated body structures, such as an earthworm's segments or a fish's vertebral column and ribs. Segmentation evolved independently in annelids (earthworms and other segmented worms) and arthropods (insects, spiders, and crustaceans), as well as in vertebrates. **Vertebrates** are a monophyletic lineage within the Chordata that is defined by the presence of a skull; many vertebrate species also have a backbone (see Chapter 34). Recent evidence suggests that segmentation also evolved independently in mollusks (snails, clams, squid). Segmentation is thought to be a particularly efficient way to organize a bilaterally symmetric body.

- **Invertebrates**, traditionally defined as all animals that are not vertebrates, are paraphyletic. The group includes some, but not all, of the descendants of a common ancestor.

Although biologists are increasingly confident that most or all of these conclusions are correct, the phylogeny of animals is still a work in progress. As data sets expand, it is likely that new analyses will not only confirm or challenge these results but also contribute other important insights into animal evolution. Stay tuned.

CHECK YOUR UNDERSTANDING

⌐━ **If you understand that . . .**

- The origin and early diversification of animals was marked by changes in four fundamental features: the number of embryonic tissues present; the evolution of nervous systems and a bilaterally symmetric, cephalized body; the evolution of a body cavity; and protostome versus deuterostome patterns of development.

✓ **You should be able to . . .**

1. Explain why bilateral symmetry in combination with triploblasty and a coelom is responsible for the "tube-within-a-tube" design observed in most animals living today.

2. Explain why cephalization was important.

Answers are available in Appendix B.

32.3 What Themes Occur in the Diversification of Animals?

Within each animal phylum, the basic features of the body plan do not vary from species to species. For example, **mollusks** are triploblastic, bilaterally symmetric protostomes with a reduced coelom. Their body architecture features a muscular foot, a cavity called the visceral mass, and a structure called a mantle. But there are over 100,000 species of mollusk.

🔑 If the major animal lineages are defined by a particular body plan, what triggered the diversification of species within each lineage? In most cases, the answer to this question is the evolution of innovative methods for sensing the environment, feeding, and moving.

Recall that most animals get their food by ingesting other organisms. Animals are diverse because there are thousands of ways to find and eat the millions of different organisms that exist.

Sensory Organs

The evolution of a cephalized body was a major breakthrough in the evolution of animals. Along with a mouth and brain, a concentration of sensory organs in the head region—where the animal initially encounters the environment—is a key aspect of cephalization.

Chapter 46 explores the cells and organs responsible for sensing light (vision), specific molecules (smell and taste), and sound (hearing) in detail. Here the key point is to appreciate the diversity of sensory abilities and structures found in animals.

VARIATION IN SENSORY ABILITIES Certain senses are almost universal in animals. The common senses include touch and balance—meaning that animals can sense objects and gravity—along with a sense of smell, taste, and hearing. At least some ability to sense light and time is also common.

But as animals diversified, a wide array of more specialized sensory abilities evolved, including:

- *Magnetism* Many birds, sea turtles, sea slugs, and other animals can detect magnetic fields and use Earth's magnetic field as an aid in navigation.

- *Electric fields* Some aquatic predators, such as sharks, are so sensitive to electric fields that they can detect electrical activity in the muscles of passing prey.

- *Barometric pressure* Some birds can sense changes in air pressure, which may aid them in avoiding storms.

Variation in sensory abilities is important: it allows animals to collect information about a wide array of environments.

VARIATION IN SENSORY STRUCTURES Animals can sense different aspects of the environment because they have different sensory structures. But even within the same sensory modality or type, abilities and structures vary.

As an example, consider the array of eyes found in mollusks. As **Figure 32.11** shows, some mollusks have extremely simple eyes—basically a panel of light-sensitive cells that allow individuals to detect areas of light and dark. Others have eyes with lenses that bend light and allow the brain to form an image. Some have two eyes; some have many. Among the 100,000 species of mollusk, the structure of the eye is highly variable. In most cases, this variation in structure helps each species function efficiently in the environment it occupies.

Feeding

To organize the diversity of ways that animals find food, biologists distinguish *how* individuals eat from *what* they eat. In many cases, species that pursue different feeding strategies and food sources are found within a single lineage.

HOW ANIMALS FEED: FOUR GENERAL TACTICS Two simple ideas are key to understanding how animals from the same lineage can have the same basic body architecture but feed in radically different ways: (1) Their mouthparts vary, and (2) the structure of an animal's mouthparts correlates closely with its method of feeding. Keep these concepts in mind as you review the four general tactics that animals use to obtain food.

1. **Suspension feeders**, *also known as* **filter feeders**, *capture food by filtering out or concentrating particles floating in water or drifting through the air.*

(a) Simple group of light-sensitive cells in limpets

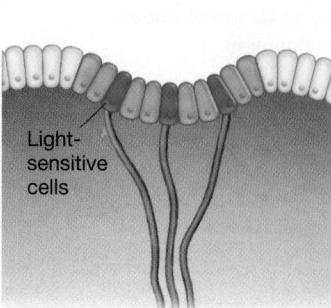

(b) Fluid-filled eye, but no lens in abalone and nautilus

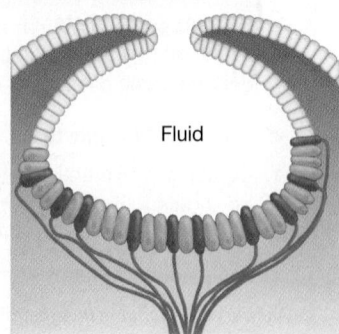

(c) Fixed lens in snails and scallops

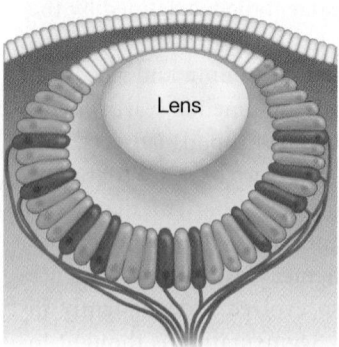

(d) Lens that changes shape in octopus and squid

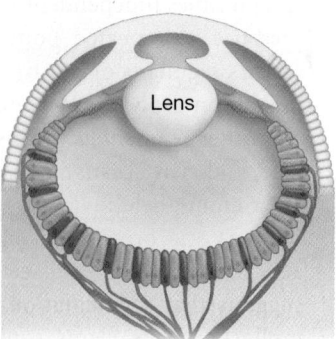

FIGURE 32.11 Variation in Eye Structure among Mollusk Eyes. Some mollusks, such as clams and mussels, lack eyes.

(a) Earthworms (Annelida) eat organic material within soil and detritus such as leaves on the surface of soil.

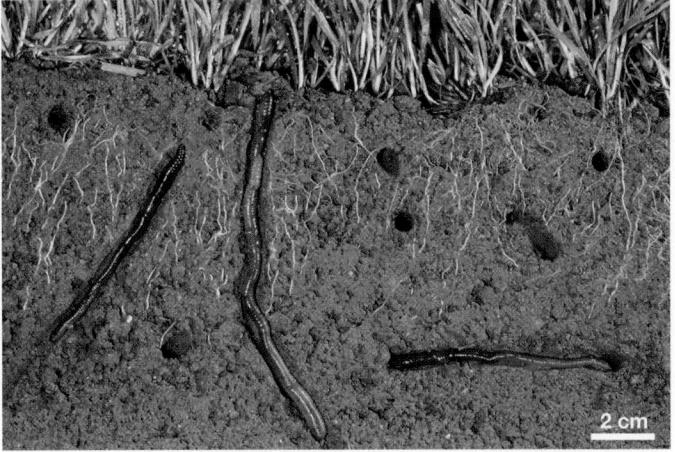

2 cm

(b) Sea cucumbers (Echinodermata) use feeding tentacles to move detritus into their mouths.

Feeding tentacles

1 cm

FIGURE 32.12 Deposit Feeders Eat Organic Material in Sediments and Other Deposits.

Suspension feeders employ a wide array of structures to trap suspended particles—usually small algae or animals or bits of detritus—and ingest them. Sponges, like the individual illustrated in Figure 32.10, are suspension feeders. So are clams and mussels, which pump water through their bodies and trap suspended food on their feathery gills—structures that also function in gas exchange. Baleen whales suspension feed by gulping water, squeezing it out between the horny baleen plates that line their mouths, and trapping shrimp-like organisms called krill inside.

Because particles float in water much more readily than in air, suspension feeding is particularly common in aquatic environments. Many suspension feeders are sessile.

2. **Deposit feeders** *ingest organic material that has been deposited within a substrate or on its surface.*

Many deposit feeders digest organic matter in the soil; their food consists of soil-dwelling bacteria, archaea, protists, and fungi, along with detritus that settles on the surface of the soil. Earthworms, for example, are annelids that swallow soil as well as leaves and other detritus on the surface of the soil (**Figure 32.12a**).

The seafloor is also rich in organic matter, which rains down from the surface and collects in food-rich deposits. These deposits are exploited by a wide array of segmented worms (annelids) as well as the echinoderms called sea cucumbers (**Figure 32.12b**).

Unlike suspension feeders, which are diverse in size and shape and use various trapping or filtering systems, deposit feeders are similar in appearance. They usually have simple mouthparts and their body shape is wormlike. Like suspension feeders, however, deposit feeders occur in a wide variety of lineages.

3. **Fluid feeders** *suck or mop up liquids like nectar, plant sap, blood, or fruit juice.*

Fluid feeders range from butterflies that feed on nectar with a straw-like proboscis (**Figure 32.13a**) to blowflies that feed on rotting fruit using a sponge-like mouthpart (**Figure 32.13b**). Fluid feeders are found in a wide array of lineages and often have mouthparts that allow them to pierce seeds, stems, skin, or other structures in order to withdraw the fluids inside.

4. **Mass feeders** *take chunks of food into their mouths.*

(a) Butterflies have an extensible, hollow proboscis.

(b) Blowflies have an extensible, sponge-like mouthpart.

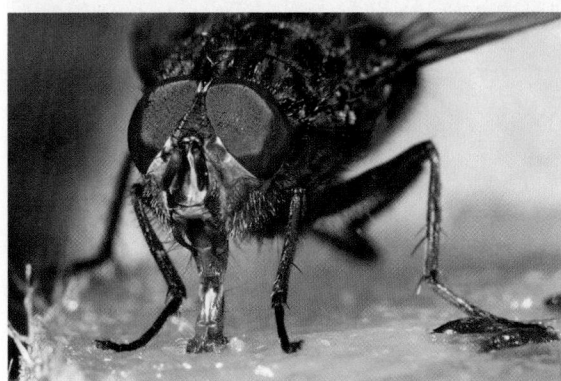

FIGURE 32.13 Fluid Feeders Drink Liquids.

(a) Horses feed on grass stems.

Molars grind stems

Front teeth nip stems

(b) Terrestrial snails feed on leaves.

Radula

50 µm

Radula scrapes off pieces of leaf

FIGURE 32.14 Food-Mass Feeders Ingest Bites or Lumps of Food.

In mass feeders, the structure of the mouthparts correlates with the type of food pieces that are harvested and ingested. Horses, for example, have sharp teeth in the front of the jaw for biting off grass stems and broad, flat molar teeth in the back of the jaw for mashing the coarse stems into a soft wad that can be swallowed (**Figure 32.14a**). In many snails, a feeding structure called a **radula** functions like a rasp or a file. The sharp plates on the radula move back and forth to scrape material away from a plant or alga so that it can be ingested (**Figure 32.14b**). Among vertebrates and snails, variation in tooth or radula structure correlates with the food source.

WHAT ANIMALS EAT: THREE GENERAL SOURCES Whether they feed by filtering, eating through deposits, or taking in fluids or masses, animals can be classified as (**1**) **herbivores** that feed on plants or algae, (**2**) **carnivores** that feed on animals, or (**3**) **detritivores** that feed on dead organic matter. Animals that eat both plants and animals are called **omnivores**.

In addition, herbivores and carnivores can be subclassified as (**1**) predators or (**2**) parasites.

1. **Predators** *are usually larger than their prey and kill them quickly.*

Predators use an array of mouthparts and hunting strategies. Many types of frogs, for example, are sit-and-wait predators. They sit still and wait for an insect or worm to move close, then capture it with a lightning-quick extension of their long, sticky tongue (**Figure 32.15a**). In contrast, wolves hunt by locating a prey organism and then running it down during an extended, long-distance chase (**Figure 32.15b**).

2. **Parasites** *are usually much smaller than their victims and often harvest nutrients without causing death.*

Endoparasites live inside their hosts and usually have simple, wormlike bodies. Tapeworms, of the phylum platyhelminthes

(a) Many frogs (Chordata) sit and wait for prey.

(b) Wolves (Chordata) chase prey.

FIGURE 32.15 Predators Kill and Eat Organisms.

(a) Tapeworms (Platyhelminthes) are endoparasites.

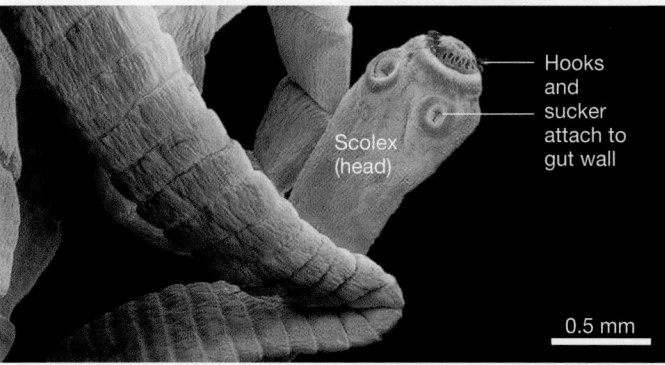

Scolex (head)

Hooks and sucker attach to gut wall

0.5 mm

(b) Lice (Arthropoda) are ectoparasites.

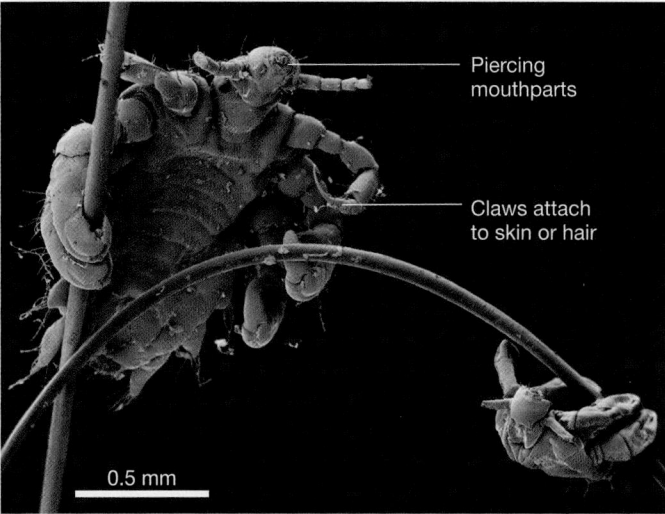

Piercing mouthparts

Claws attach to skin or hair

0.5 mm

FIGURE 32.16 Parasites Take Nutrients from Living Animals.
(a) Tapeworms are common intestinal parasites of humans and other vertebrates. They absorb nutrition directly across their body wall. **(b)** The louse *Phthirus pubis* uses its clawlike legs to attach to the pubic region of humans, pierces the host's skin with its mouthparts, and feeds by sucking body fluids.

(flatworms), are endoparasites with no digestive system. Instead of a mouth, they have hooks or other structures on their head, called a scolex, that attach to their host's intestinal wall (**Figure 32.16a**). Instead of digesting food themselves, they absorb nutrients directly from their surroundings.

Ectoparasites live outside their hosts. Ectoparasites usually have limbs or mouthparts that allow them to grasp the host and mouthparts that allow them to pierce their host's skin and suck the nutrient-rich fluids inside (**Figure 32.16b**).

Movement

Animal locomotion has an array of important functions: finding food, finding mates, escaping from predators, and dispersing to new habitats. The ways that animals move in search of food and sex are highly variable; they burrow, slither, swim, fly, crawl, walk, or run. Movement is powered by cilia, flagella, or muscles that attach to a hard skeleton or compress a hydrostatic skeleton, enabling wriggling movements.

The hydrostatic skeleton is essential to locomotion in the many animal phyla with wormlike bodies. Another major innovation occurred as animals diversified, however. The limb made highly controlled, rapid movement possible.

TYPES OF LIMBS: UNJOINTED AND JOINTED Limbs are a prominent feature of species in many phyla. They are particularly important in two major lineages: the ecdysozoans and the vertebrates.

Some members of the Ecdysozoa, such as onychophorans (velvet worms), have unjointed, lobe-like limbs; others, such as arthropods, have more complex, jointed limbs (**Figure 32.17**). Jointed limbs make fast, precise movements possible and are a prominent limb type in vertebrates and arthropods.

ARE ALL ANIMAL APPENDAGES HOMOLOGOUS? Chapter 24 introduced the concept of homology, which is defined as similarity in traits due to inheritance from a common ancestor. Traditionally, biologists have hypothesized that the major types of jointed and unjointed animal limbs were not homologous.

To appreciate the logic behind this hypothesis, it's important to recognize just how diverse animal appendages are: They

(a) Velvet worms (Onychophora) have lobe-like limbs.

1 cm

(b) Crabs (Arthropoda) have jointed limbs.

1 cm

FIGURE 32.17 Unjointed and Jointed Limbs May Be Used in Locomotion.

(a) Polychaetes (Annelida) have parapodia.

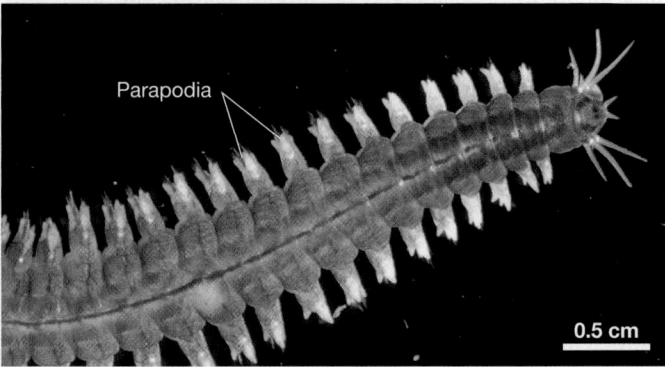

Parapodia

0.5 cm

(b) Sea urchins (Echinodermata) have tube feet.

Tube feet

1 cm

FIGURE 32.18 Some Species "Walk" on Parapodia or Tube Feet Instead of Legs.

range from the human arm to bristle-like structures called parapodia in segmented worms (**Figure 32.18a**) and the soft, extensible tube feet found in echinoderms (**Figure 32.18b**). Because the structure of animal appendages is so diverse, it was logical to maintain that at least some appendages evolved independently of each other. As a result, biologists predicted that completely different genes are responsible for each major type of appendage.

Recent results have challenged this view, however. The experiments in question involve a gene called *Distal-less*, which was originally discovered in fruit flies. (*Distal* means "away from the body.") *Distal-less*, or *Dll*, is aptly named. In fruit flies that lack this gene's normal protein product, only the most rudimentary limb buds form. The mutant limbs are "distal-less." The protein seems to deliver a simple message as a fruit-fly embryo develops: "Grow appendage out this way."

Biologists in Sean B. Carroll's lab set out to test the hypothesis that *Dll* might be involved in the initial phase of limb or appendage formation in other animals. As **Figure 32.19** shows, they used a fluorescent marker that sticks to the *Dll* gene product to locate tissues where the gene is expressed. When they introduced the fluorescent marker into embryos from annelids, arthropods, echinoderms, chordates, and other phyla, they found that it bound to *Dll* in all of them. More important, the highest concentrations of *Dll* gene products were found in cells that form appendages—even in phyla with

EXPERIMENT

QUESTION: Is the gene *Dll* involved in limb formation in species other than insects?

HYPOTHESIS: In all animals, *Dll* signals "grow appendage out here."

NULL HYPOTHESIS: *Dll* is not involved in the development of appendages in species other than insects.

EXPERIMENTAL SETUP:
Add a stain to developing embryos that will attach to *Dll* gene products (proteins), revealing their location. The stain can be fluorescent green or dark brown.

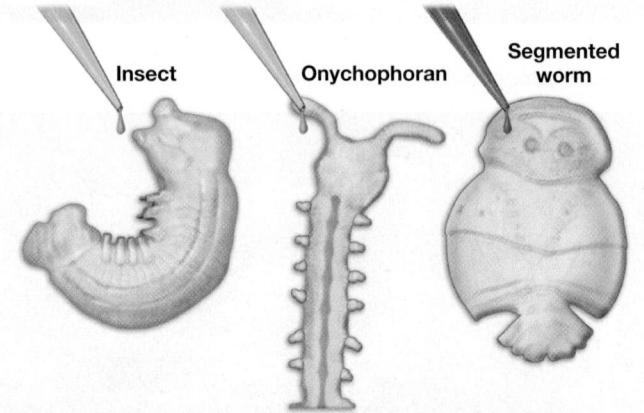

Insect   Onychophoran   Segmented worm

PREDICTION: In embryos from a wide array of species, stained *Dll* gene products will be localized to areas where appendages are forming.

PREDICTION OF NULL HYPOTHESIS: Stained *Dll* gene products will be localized to areas where appendages are forming only in insects.

RESULTS:

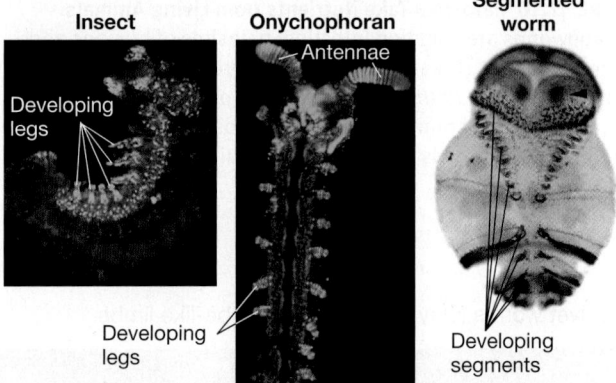

Insect   Onychophoran   Segmented worm

Antennae

Developing legs

Developing legs

Developing segments

In species representing both Ecdysozoa (e.g., insect and onychophoran) and Lophotrochozoa (e.g., segmented worm), *Dll* is localized in areas of the embryo where appendages are forming.

CONCLUSION: The gene *Dll* is involved in limb formation in diverse species. The results suggest that all animal appendages may be homologous.

FIGURE 32.19 Experimental Evidence That All Animal Appendages Are Homologous.

SOURCE: Panganiban, G., et al. 1997. The origin and evolution of animal appendages. *Proceedings of the National Academy of Sciences USA* 94: 5162–5166.

✔**QUESTION** What results would have supported the null hypothesis?

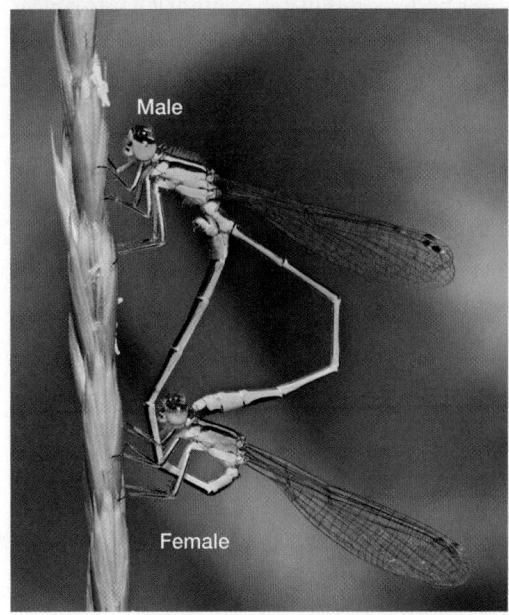

(a) Internal fertilization in damselflies

(b) External fertilization in giant clams

FIGURE 32.20 Fertilization Can Be Internal or External in Animals. **(a)** When damselflies copulate, the male holds the female just behind her head with claspers and the female places the tip of her abdomen against the male's sperm transfer organ. The two can fly in this position. **(b)** A male giant clam releasing sperm into the ocean.

wormlike bodies that have extremely simple appendages. Other experiments have shown that *Dll* is also involved in limb formation in vertebrates.

Based on these findings, biologists are concluding that the protein product of the same gene marks the initial site of appendage growth in most if not all animals. The hypothesis is that all animal appendages have some degree of genetic homology and that they are all derived from appendages that were present in a common ancestor. The idea is that a simple appendage evolved early in the history of the Bilateria and that, subsequently, evolution by natural selection produced the diversity of limbs, antennae, and wings observed today.

Reproduction

An animal may be efficient at moving and eating, but if it does not reproduce, the alleles responsible for its effective locomotion and feeding will not increase in frequency in the population. As Chapter 24 emphasized, natural selection occurs when individuals with certain alleles produce more offspring than other individuals do. Organisms live to reproduce.

Given the array of habitats and lifestyles pursued by animals, it's not surprising that they exhibit a high degree of variation in how they reproduce. Chapter 48 details how gamete formation, fertilization, and early development occur in animals, but a few examples here will help drive home just how diverse animal reproduction is.

DOES FERTILIZATION OCCUR? At least some species in most animal phyla can reproduce asexually, through mitosis, as well as sexually (via meiosis and fusion of gametes).

In the lophotrochozoan phylum Rotifera, an entire lineage called the bdelloids (pronounced *DELL-oyds*) reproduces only asexually. Even certain fish, lizard, and snail species have never been observed to undergo sexual reproduction.

WHERE DOES FERTILIZATION OCCUR? When sexual reproduction does occur, fertilization may be internal or external.

When internal fertilization takes place, males typically insert a sperm-transfer organ into the body of a female (**Figure 32.20a**). In some cases, males produce sperm in packets, which females then pick up and insert into their own bodies. But in seahorses, females insert eggs into the male's body, where they are fertilized. (The male is pregnant for a time and then gives birth to live young.)

External fertilization is extremely common in aquatic species. Females lay eggs onto a substrate or into open water. Males shed sperm, which swim, on or near the eggs (**Figure 32.20b**).

WHERE DO EMBRYOS DEVELOP? Eggs or embryos may be retained in the female's body during development, or eggs may be laid outside to develop independently of the mother.

Mammals and other species that nourish embryos inside the body and give birth to live young are said to be **viviparous** ("live-bearing"); species that deposit fertilized eggs are **oviparous** ("egg-bearing"); and some species are **ovoviviparous** ("egg-live-bearing").

In ovoviviparous species, the females retain eggs inside their body during early development; but the growing embryos are nourished by yolk inside the egg and not by nutrients transferred directly from the mother, as in viviparous species. Ovoviviparous females then give birth to well-developed young.

Most mammals and a few species of sea stars, onychophorans, sharks, fish, amphibians, and lizards are viviparous; some snails, insects, reptiles, fishes, and sharks are ovoviviparous. But the vast majority of animals are oviparous.

Life Cycles

Besides reproducing in a variety of ways, animal life cycles are diverse. ⌐ Perhaps the most spectacular innovation in animal life cycles involves the phenomenon known as **metamorphosis** ("change-form")—a change from an immature body type to an adult body type.

In discussing metamorphosis and other types of animal life cycles, biologists distinguish three types of life stages:

- **Larvae** (singular: **larva**) look radically different from adults, live in different habitats, and eat different foods. They are sexually immature—meaning that their reproductive organs are undeveloped.

- **Juveniles** look like adults and live in the same habitats and eat the same foods as adults, but are sexually immature.

- **Adults** are the reproductive stage in the life cycle.

In insects, either a larval or a juvenile stage occurs.

TWO TYPES OF INSECT METAMORPHOSIS   In insects, the presence of a larval versus juvenile stage defines two distinct types of metamorphosis.

In **hemimetabolous** ("half-change") **metamorphosis**—also called **incomplete metamorphosis**—young are juveniles called nymphs that look like miniature versions of the adult. The aphid nymphs in **Figure 32.21a**, for example, shed their external skeletons several times and grow—gradually changing from wingless, sexually immature nymphs to sexually mature adults, some of which can fly. But throughout their life, aphids live in the same habitats and feed on the same food source in the same way: They suck sap.

In **holometabolous** ("whole change") **metamorphosis**—also called **complete metamorphosis**—young are larvae. As an example, consider the life cycle of the mosquito, illustrated in **Figure 32.21b**.

- Newly hatched mosquitoes live in quiet bodies of freshwater, where they suspension feed on bacteria, algae, and detritus.

- When a larva has grown sufficiently, the individual stops feeding and moving and secretes a protective case. The indi-

vidual is now known as a **pupa** (plural: **pupae**). During pupation, the pupa's body is completely remodeled into a new, adult form.

- The adult mosquito flies and gets its nutrition as a parasite—taking blood meals from mammals and nectar from flowers.

WHAT IS THE ADAPTIVE SIGNIFICANCE OF METAMORPHOSIS?   In insects, holometabolous metamorphosis is 10 times more common than hemimetabolous metamorphosis. Why?

The leading hypothesis is based on efficiency in feeding. Because juveniles and adults from holometabolous species feed on different materials in different ways and sometimes even in different habitats, they do not compete with each other.

An alternative hypothesis is based on the advantages of specialization. In many moths and butterflies, larvae are specialized for feeding, whereas adults are specialized for mating and feed rarely, if ever. Larvae are largely sessile, whereas adults are highly mobile. If specialization leads to higher efficiency in feeding and reproduction and thus higher fitness, then complete metamorphosis would be advantageous.

These hypotheses are not mutually exclusive, however—meaning that both could be correct, depending on the species being considered. Research continues.

A COMPLEX LIFE CYCLE IN CNIDARIA   Metamorphosis is extremely common in marine animals, as well as in insects. Many marine fish, for example, have a larval stage—a period where young live near the ocean surface, far from the habitats where adults are found, and feed on microscopic prey.

In marine species that have limited or no movement as adults, larvae function as a dispersal stage. They are a little like the seeds of land plants—a life stage that allows individuals to move to new habitats, where they will not compete with their parents for space and other resources.

(a) Aphid:
Hemimetabolous metamorphosis

Nymphs look like miniature versions of adults and eat the same foods.

(b) Mosquito:
Holometabolous metamorphosis

Larvae look substantially different from adults and eat different foods.

FIGURE 32.21 During Insect Metamorphosis, Individuals May or May Not Change Form Completely.

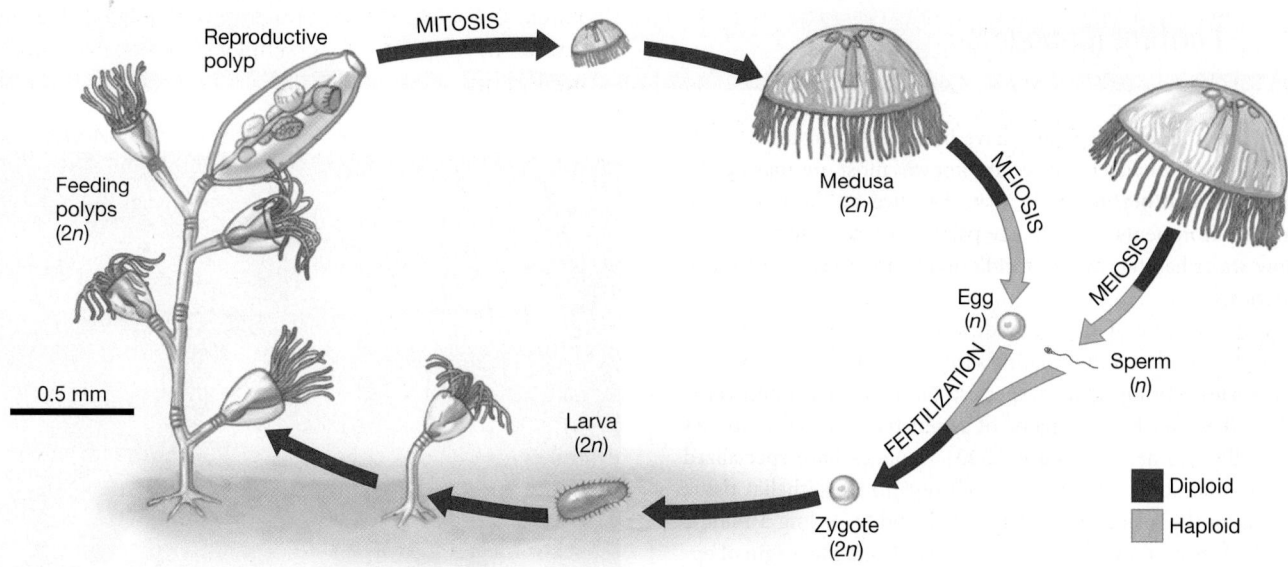

FIGURE 32.22 Cnidarian Life Cycles May Include a Polyp and Medusa Form. This is an example of a hydrozoan called *Obelia*. Colonies are often found attached to seaweed, shells, or rocks and contain hundreds of polyps.

✔ **QUESTION** Does a protist- or plant-like alternation of generations occur in this species? Explain why or why not.

An even more complex life cycle is found in some cnidarians, which have three distinct body types during their life cycle: (**1**) A largely sessile form called a **polyp** that reproduces asexually, (**2**) a free-floating stage called a **medusa** (plural: **medusae**) that reproduces sexually, and (**3**) a larval form (**Figure 32.22**).

Polyps usually live attached to a substrate, suspension feed on detritus or small organisms, and frequently form large clusters of individuals called colonies. A **colony** is a group of identical individuals that are physically attached. Medusae, in contrast, float freely in the plankton and feed on crustaceans and other animals. Because polyps and medusae live in different habitats, the two stages of the life cycle exploit different food sources.

CHECK YOUR UNDERSTANDING

🔑 If you understand that . . .

- The story of animal evolution is based on two themes: (1) the evolution of a small suite of basic body plans, and (2) a diversification of species with the same basic body architecture, based on the evolution of innovative structures and methods for sensing the environment, capturing food, and moving.
- Methods for fertilization and parental care of embryos vary widely.
- Metamorphosis is common, with larvae often serving as a feeding or dispersal stage.

✔ You should be able to . . .

1. Explain how the mouthparts of deposit feeders and mass feeders are expected to differ.
2. Explain why external fertilization is particularly common in aquatic environments, and internal fertilization is particularly common in terrestrial environments.

Answers are available in Appendix B.

32.4 Key Lineages of Animals: Non-Bilaterian Groups

The goal of this chapter is to provide a broad overview of how animals diversified into such a morphologically diverse lineage. According to recent phylogenetic analyses, the closest living relatives to the animals are the choanoflagellates, a group of protists. The fossil record indicates that the phyla Porifera (sponges), Cnidaria (jellyfish and others), and Ctenophora are the most ancient of all animal lineages. The phylogeny of animals is consistent with the fossil record—the first sponges, cnidarians, and ctenophores appear at the base of the tree.

Let's explore the origins of animals by taking a more detailed look at lineages with similar-looking relatives that appear early in the fossil record, along with the Acoelomorpha—the lineages other than the Bilateria. The protostomes and deuterostomes are considered in Chapters 33 and 34, respectively.

- ■ Porifera (Sponges)
- ■ Cnidaria (Jellyfish, Corals, Anemones, Hydroids)
- ■ Ctenophora (Comb Jellies)
- ■ Acoelomorpha (Acoels)

About 7000 species of sponges have been described to date. Although a few freshwater species are known, most are marine. All sponges are **benthic**, meaning that they live at the bottom of aquatic environments. Sponges are particularly common in rocky, shallow-water habitats of the world's oceans and in coastal areas of Antarctica.

The architecture of sponge bodies is built around a system of tubes and pores that create channels for water currents. Body symmetry varies among sponge species; most are asymmetrical—meaning that they have no plane of symmetry—but some species are radially symmetric (**Figure 32.23**). Sponges have specialized cell types and some species have well-organized epithelial tissue layers lining the inside and outside of the body, sealing a middle layer in between. ✔ You should be able to indicate the origin of epithelial tissue, which occurs in all animals, on Figure 32.9. Sponges lack the other tissue types found in other animals, however. The "looseness" of sponge tissues is thought to be an adaptation that facilitates regeneration after wounding, the ability to remodel the body if water currents change, and the ability to move into and colonize new substrates as they become available. In many species, collagen fibers are augmented by **spicules**—stiff spikes of silica or calcium carbonate ($CaCO_3$)—to provide structural support for the body. One sponge species native to the Caribbean can grow to heights of 2 m.

Sponges have commercial and medical value to humans. The dried bodies of certain sponge species are able to hold large amounts of water and thus are prized for use in bathing and washing. In addition, researchers are increasingly interested in the array of toxins that sponges produce to defend themselves against predators and bacterial parasites—possibly for use in cancer chemotherapy.

Feeding Most sponges are suspension feeders. Their cells beat in a coordinated way to produce a water current that flows through small pores in the outer body wall, into chambers inside the body, and out through a single larger opening. As water passes by feeding cells, organic debris and bacteria, archaea, and small protists are filtered out of the current and then digested. Some deep-sea sponges are predators, however—they capture small crustaceans on hooks that project from the body.

Movement Most adult sponges are sessile, though a few species are reported to move at rates of up to 4 mm per day. Most species

Agelas species

10 cm

FIGURE 32.23 Some Sponges Form Radially Symmetric Tubes.

produce larvae that swim with the aid of cilia. Recent research has confirmed that at least one species can contract its body to expel waste products.

Reproduction Asexual reproduction occurs in a variety of ways, depending on the species. Some sponge cells are totipotent, meaning that small groups of adult cells have the capacity to develop into a complete adult organism. Thus, a fragment that breaks off an adult sponge has the potential to grow into a new individual. Although individuals of most species produce both eggs and sperm, self-fertilization is rare because individuals release their male and female gametes at different times. Fertilization usually takes place in the water, but some ovoviviparous species retain their eggs and then release mature, swimming larvae after fertilization and early development have occurred.

Although a few species of Cnidaria inhabit freshwater, the vast majority of the 11,000 species are marine. They are found in all of the world's oceans, occupying habitats from the surface to the substrate, and are important predators. The phylum comprises four main lineages: Hydrozoa (hydroids), Cubozoa (box jellyfish), Scyphozoa (jellyfish), and Anthozoa (anemones, corals, and sea pens).

Many cnidarians are radially symmetric diploblasts consisting of ectoderm and endoderm layers that sandwich gelatinous material known as **mesoglea**, which contains a few scattered ectodermal cells. Recent research indicates that at least some jellyfish are triploblastic and have bilaterally symmetric larvae. Cnidarians have a gastrovascular cavity instead of a flow-through gut—meaning there is only one opening to the environment for both ingestion and elimination of wastes.

Some cnidarians have a life cycle that includes both a sessile polyp form and a free-floating medusa (see Figure 32.22). Anemones and coral, however, exist only as polyps—never as medusae. Reef-building corals secrete outer skeletons of calcium carbonate that create the physical structure of a coral reef—one of the world's most productive habitats (see Chapter 54).

Feeding The morphological innovation that triggered the diversification of the cnidarians is a specialized cell, called a **cnidocyte**, which is used in prey capture. When cnidocytes sense a fish or other type of prey, the cells forcibly eject a barbed, spear-like structure, which may contain toxins. The barbs hold the prey, and the toxins subdue it until it can be brought to the mouth and ingested. Cnidocytes are commonly located near the mouths of cnidarians or on elongated structures called tentacles. Cnidarian toxins can be deadly to humans as well as to prey organisms; in Australia, twice as many people die each year from stings by box jellyfish as from shark attacks (**Figure 32.24**). ✔You should be able to indicate the origin of cnidocytes on Figure 32.9. Besides capturing prey actively, most species of coral and many anemones host photosynthetic dinoflagellates. The relationship is mutually beneficial, because the protists supply the cnidarian host with food in exchange for protection.

Movement Both polyps and medusae have simple, muscle-like tissue derived from ectoderm or endoderm, or in some cases mesoderm. In polyps, the gut cavity acts as a hydrostatic skeleton that works in conjunction with the muscle-like cells to contract or extend the body. Many polyps can also creep along a substrate, using muscle cells at their base. In medusae, the bottom of the bell structure is ringed with muscle-like cells. When these cells contract rhythmically, the bell pulses and the medusa moves by jet propulsion—meaning a forcible flow of water in the opposite direction of movement. Cnidarian larvae swim by means of cilia.

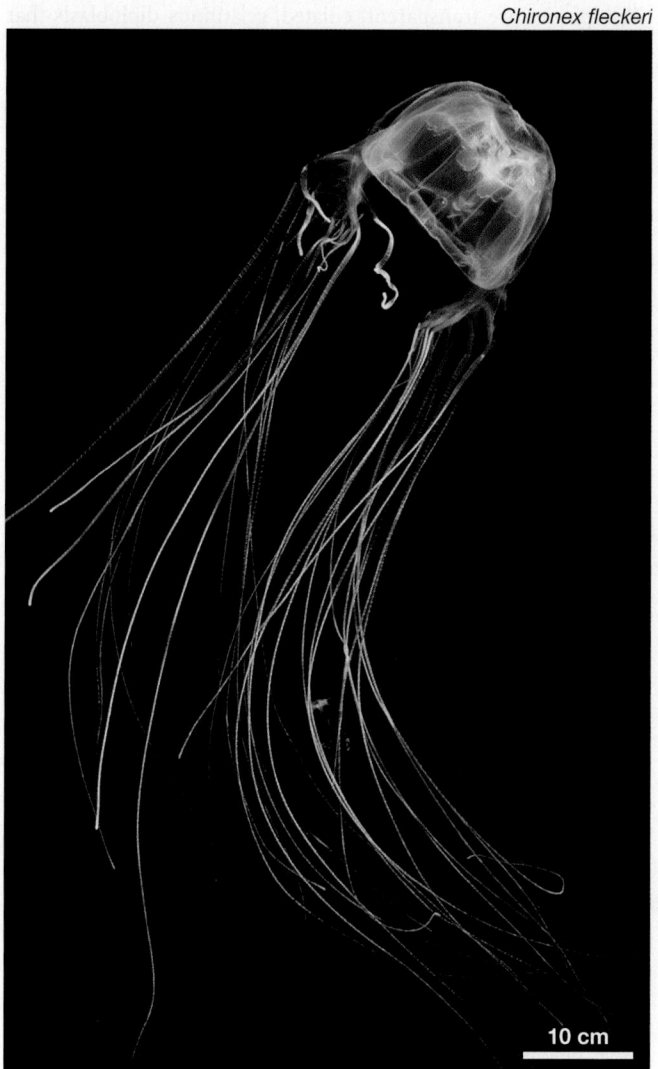

Chironex fleckeri

10 cm

FIGURE 32.24 Some Jellyish Have Long Tentacles, Packed with Cnidocysts That Sting Prey.

Reproduction Polyps may produce new individuals asexually by (**1**) budding, in which a new organism grows out from the body wall of an existing individual; (**2**) fission, in which an existing adult splits lengthwise to form two individuals; or (**3**) fragmentation, in which parts of an adult regenerate missing pieces to form a complete individual. During sexual reproduction, gametes are usually released from the mouth of a polyp or medusa and fertilization takes place in the open water. Eggs hatch into larvae that become part of the plankton before settling and developing into a polyp.

Ctenophores are transparent, ciliated, gelatinous diploblasts that live in marine habitats (**Figure 32.25**). Although a few species live on the ocean floor, most are planktonic—meaning that they live near the surface. Only about 100 species have been described to

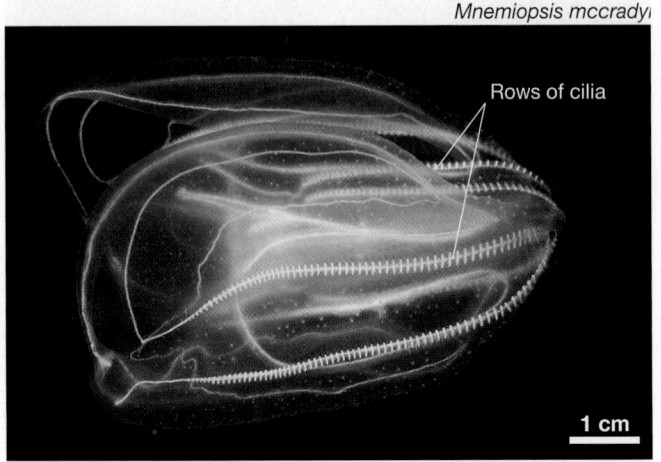

Mnemiopsis mccradyi

Rows of cilia

1 cm

FIGURE 32.25 Ctenophores Are Planktonic Predators.

date, but some are abundant enough to represent a significant fraction of the total planktonic biomass. Accidental introductions of the ctenophore *Mnemiopsis leidyi*, which preys on fish larvae, have devastated fish production in the Black Sea and Caspian Sea.

Feeding Ctenophores are predators. Feeding occurs in several ways, depending on the species. Some comb jellies have long tentacles covered with cells that release an adhesive when they contact prey. These tentacles are periodically wiped across the mouth so that captured prey can be ingested. In other species, prey can stick to mucus on the body and be swept toward the mouth by cilia. Still other species ingest large prey whole.

Movement Adults move via the beating of cilia, which occur in comblike plates. The plates form rows that run the length of the body. Ctenophores are the largest animals known to use cilia for locomotion. ✔You should be able to indicate the origin of swimming via rows of coordinated cilia on Figure 32.9.

Reproduction Most species have both male and female organs and routinely self-fertilize, though fertilization is external. Larvae are free swimming. The few species that live on the ocean floor undergo internal fertilization and brood their embryos until they hatch into larvae.

Acoelomorpha (Acoels)

As their name implies, the acoelomorphs lack a coelom. They are bilaterally symmetric worms that have distinct anterior and posterior ends and are triploblastic. Their nervous system is a nerve net, however, and they are not cephalized. They have a simple cavity where digestion occurs or no digestive system at all. In some species with a digestive cavity, the mouth is the only opening for the ingestion of food and elimination of waste products. Species that lack a digestive cavity absorb nutrients across their body wall. Most acoelomorphs are only a couple of millimeters long and live in mud or sand in marine environments (**Figure 32.26**).

Feeding Acoelomorphs feed on detritus or prey on small animals or protists that live in mud or sand.

Movement Acoelomorphs swim, glide along the surface, or burrow through substrates with the aid of cilia that cover either the entire body or the ventral surface.

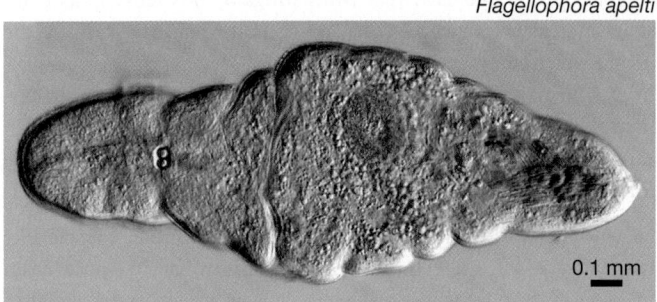

Flagellophora apelti

0.1 mm

FIGURE 32.26 Acoelomorphs Are Small Worms That Live in Mud or Sand.

Reproduction Adults can reproduce asexually by fission (splitting in two) or by direct growth (budding) of a new individual from the parent's body. Individuals produce both sperm and eggs. Fertilization is internal, and fertilized eggs are laid outside the body.

Summary of Key Concepts

○━ **Animals are multicellular, heterotrophic eukaryotes that lack cell walls and ingest their prey.**

- The animals comprise 30–35 phyla and may number 10 million or more species.

- Biologists study animals because they are key consumers and because humans depend on them for food, materials, transportation, and power.

  ✔ You should be able to predict how the total amount of plant material in a forest would change if all animals were excluded for one year.

○━ **Fundamental changes in morphology and development occurred as animals diversified.**

- Sponges lack highly organized, complex tissues other than epithelium.

- Most cnidarians and ctenophores have radial symmetry, two embryonic germ layers, and neurons organized into a nerve net.

- Members of the Acoelomorpha have bilateral symmetry and three embryonic tissues but lack a body cavity, or coelom.

- Most animal species have bilateral symmetry, three embryonic tissues, and a coelom—features that gave rise to a "tube-within-a-tube" body plan—and a central nervous system.

- Bilaterians have cephalized bodies—meaning that a distinctive head region contains the mouth, brain, and sensory organs.

- Depending on the species involved, the tube-within-a-tube design is built in one of two fundamental ways—via the protostome or deuterostome patterns of development.

  ✔ You should be able to draw a bilaterian that lacks limbs, and label the gut, outer body wall, muscle layers, head region, mouth, anus, brain, and major nerve tracts.

  (MB) **Web Activity** The Architecture of Animals

○━ **Recent phylogenetic analyses have shown that there are four major groups of animals: non-bilaterian lineages, two protostome groups (Lophotrochozoa and Ecdysozoa), and the deuterostomes.**

- Choanoflagellates—unicellular or colonial protists that filter feed—are the closest living relative to animals.

- Radial symmetry and diploblasty evolved before bilateral symmetry and mesoderm.

- Although the coelom was an important innovation in animal evolution, as it provided worm-like animals with a hydrostatic skeleton, it was lost or reduced in an array of important lineages.

- Segmented body plans arose independently as animals diversified.

  ✔ You should be able to describe how protostomes and deuterostomes differ.

○━ **Within major groups of animals, evolutionary diversification was based on innovative ways of sensing the environment, feeding, and moving.**

- Sensory abilities and sensory structures vary among species and correlate with their habitat and feeding method.

- Animals capture food in one of four ways: Suspension feeders filter organic material or small organisms from water; deposit feeders swallow sediments; fluid feeders lap or suck up liquids; food-mass feeders harvest packets of material.

- Even though the types of appendages used in animal locomotion range from simple lobe-like limbs to complex lobster legs, the genes that indicate where appendages develop may be homologous.

  ✔ You should be able to state a hypothesis for why the structure of the eye varies so much among mollusks.

○━ **Methods of sexual reproduction vary widely among animal groups, and many species can reproduce asexually. It is common for individuals to undergo metamorphosis during their life cycle.**

- Sexual reproduction may involve laying eggs or giving birth to well-developed young.

- In most cases, metamorphosis involves a dramatic change between the larval and adult body form, as well as stark contrasts between larvae and adults in the habitat used and feeding strategy.

  ✔ You should be able to distinguish between a juvenile and larval form in insects and explain why the difference is important.

Questions

✔ TEST YOUR KNOWLEDGE   *Answers are available in Appendix B*

1. What synapomorphy distinguishes animals as a monophyletic group, distinct from choanoflagellates?
   a. multicellularity
   b. movement via a hydrostatic skeleton
   c. growth by molting
   d. ingestive feeding

2. Which of the following patterns in animal evolution is correct?
   a. All triploblasts have a coelom.
   b. All triploblasts evolved from an ancestor that had a coelom.
   c. Sponges have epithelial tissues that line an enclosed, fluid-filled body cavity.
   d. Bilateral symmetry and cephalization evolved once.

3. Which of the following patterns in animal evolution is correct?
   a. Segmentation evolved once.
   b. The coelom was lost or reduced in many lineages.
   c. Sponges lack tissues and are asymmetrical.
   d. Radial symmetry evolved once.

4. Why do some researchers maintain that the limbs of all animals are homologous?
   a. Homologous genes, such as *Dll*, are involved in their development.
   b. Their structure—particularly the number and arrangement of elements inside the limb—is the same.
   c. They all function in the same way—in locomotion.
   d. Animal appendages are too complex to have evolved more than once.

5. In a "tube-within-a-tube" body plan, what is the interior tube?
   a. ectoderm
   b. mesoderm
   c. the coelom
   d. the gut

6. How are choanoflagellates and sponges similar?
   a. Both are multicellular.
   b. Both have a single tissue type: epithelium.
   c. They share distinctive flagellated cells that function in suspension feeding.
   d. Both groups are animals that appear early in the fossil record of the lineage.

TEST YOUR UNDERSTANDING

Answers are available in Appendix B

1. Explain the difference between a diploblast and a triploblast. How was the evolution of mesoderm associated with the evolution of a coelom?

2. Explain how a hydrostatic skeleton works.

3. Why is it significant that animals are heterotrophic *and* multicellular?

4. Compare and contrast the types of mouthparts you would expect to find in herbivorous insect species that (1) suck plant fluids from stems, (2) eat their way through decaying plant material on the forest floor, (3) bite leaves, and (4) suck nectar from flowers.

5. Explain how an animal mother nourishes an embryo in oviparous species versus viviparous species.

6. Why are nerve nets associated with radial symmetry, while a CNS is associated with bilateral symmetry?

APPLYING CONCEPTS TO NEW SITUATIONS

Answers are available in Appendix B

1. Suppose that a gene originally identified in nematodes (roundworms) is found to be homologous with a gene that can cause developmental abnormalities in humans. Would it be possible to study this same gene in fruit flies? Explain.

2. Ticks are arachnids (along with spiders and mites); mosquitoes are insects. Both ticks and mosquitoes are ectoparasites that make their living by extracting blood meals from mammals. Ticks undergo incomplete metamorphosis, while mosquitoes undergo complete metamorphosis. Based on these observations, would you predict ticks or mosquitoes to be the more successful group in terms of number of species, number of individuals, and geographic distribution? Explain why. How could you test your prediction?

3. Radial symmetry evolved in echinoderms (sea stars and sea urchins) from bilaterally symmetric ancestors. Predict whether the echinoderm nervous system is centralized or diffuse. Explain your logic.

4. Why are spiders unusual, in terms of the four general feeding tactics outlined in this chapter?

In numbers of individuals and species richness, protostomes are the most abundant and diverse of all animals—and the arthropods are by far the most abundant and diverse of the protostomes. The arthropods shown here come from all over the world.

Protostome Animals   33

Protostomes include some of the most familiar organisms on Earth. The phylum Arthropoda, for example, includes the insects, chelicerates (spiders and mites), crustaceans (shrimp, lobster, crabs, barnacles), and myriapods (millipedes, centipedes); the Mollusca comprises the snails, clams, chitons, and cephalopods (octopuses and squid).

Some protostome phyla are also particularly species-rich. Over 93,000 mollusks have been named thus far, and biologists estimate that there are over 10 million arthropod species living today—most of them unnamed and undescribed. Among insects alone, about 925,000 species have been formally identified to date.

Certain protostome lineages are also extremely abundant. A typical acre of pasture in England is home to almost 18 *million* individual beetles; the world population of ants is estimated to be 1 million billion individuals.

It is important to recognize, though, that not all protostome lineages have been as spectacularly successful as the Arthropoda and Mollusca. There are over 20 phyla of protostomes; some qualify as particularly obscure lineages on the tree of life. There are just 150 species in the phylum Kinorhyncha (mud dragons); the 80 species in the Gnathostomulida (pronounced *nath-oh-stoh-MEW-lida*) and the 450 species in the Gastrotricha (*GAS-troh-trika*) average less than 2 mm long.

Figure 33.1 on page 624 is a pie chart showing the relative numbers of species in various animal phyla. Note that the wedges labeled Chordata, Cnidaria, Echinodermata, Porifera, and "Other invertebrates" are small. Almost all animal species are protostomes.

As a group, protostomes live in just about every habitat that you might explore. They also include some of the most important model organisms in all of biological science: the fruit fly *Drosophila melanogaster* and the roundworm *Caenorhabditis elegans*. If you walk into a biology building on any university campus around the world, you are almost certain to find at least one lab where fruit flies or roundworms are being studied. If one of biology's most fundamental goals is to understand the diversity of life on Earth, then protostomes—particularly the arthropods and mollusks—demand our attention.

KEY CONCEPTS

- Molecular phylogenies support the hypothesis that protostomes are a monophyletic group divided into two major subgroups: the Lophotrochozoa and the Ecdysozoa.

- Although the members of many protostome phyla have limbless, wormlike bodies and live in marine sediments, the most diverse and species-rich lineages—Mollusca and Arthropoda—have bodies with distinctive, complex features. Mollusks and arthropods inhabit a wide range of environments.

- Key events triggered the diversification of protostomes, including several lineages making the water-to-land transition, a diversification in appendages and mouthparts, and the evolution of metamorphosis in both marine and terrestrial forms.

✔ When you see this checkmark, stop and test yourself. Answers are available in Appendix B.

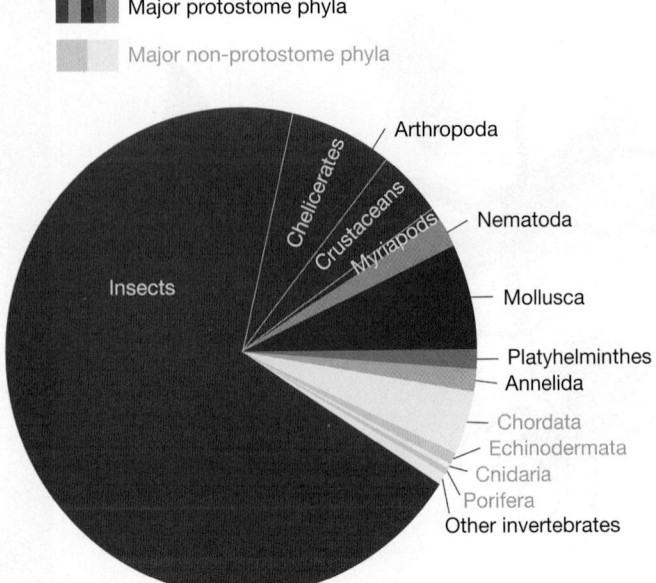

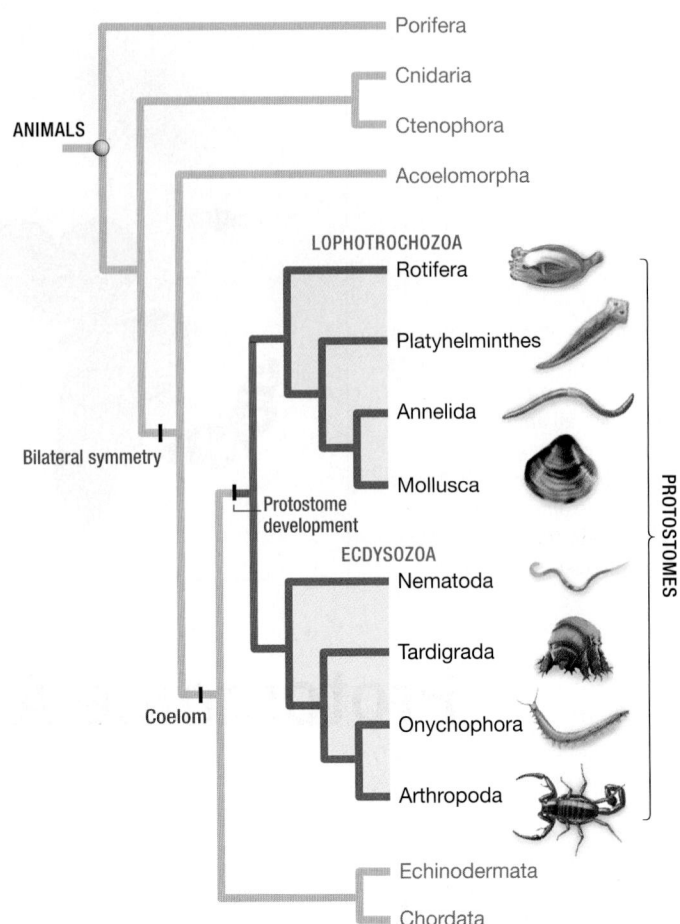

FIGURE 33.1 **The Relative Abundances of Animal Lineages.** Most animals are protostomes. About 70 percent of all known species of animals on Earth are insects, most of them beetles. (Humans and other vertebrates are deuterostomes, in the phylum Chordata.) This chapter focuses on the most species-rich lineages of protostomes.

FIGURE 33.2 **Protostomes Are a Monophyletic Group Comprising Two Major Lineages.** There are 22 phyla of protostomes, but the eight major phyla shown account for about 99.5 percent of the known species.

33.1 An Overview of Protostome Evolution

There are two major groups of bilaterally symmetric, triploblastic, coelomate animals: the protostomes and deuterostomes. You might recall from Chapter 32 that protostome and deuterostome embryos develop in dramatically different ways. The protostome and deuterostome patterns of early development represent distinctive pathways for building a bilaterally symmetric body with a coelom.

When gastrulation occurs in protostomes, the initial pore that forms in the embryo becomes the mouth. If a **coelom** (a body cavity) forms later in development, it forms from openings that arise within blocks of mesodermal tissue. ⧈⊸ Phylogenetic studies have long supported the hypothesis that protostomes are a monophyletic group, meaning the protostome developmental sequence arose just once.

More recent analyses of DNA sequence data indicated that two major subgroups exist within the protostomes (**Figure 33.2**). The two monophyletic groups of protostomes are called the Lophotrochozoa and Ecdysozoa. (The lineages are pronounced *low-foe-tro-ko-ZOH-ah* and *eck-die-so-ZOH-ah*.)

What Is a Lophotrochozoan?

The 13 phyla of lophotrochozoans include the mollusks, annelids, and flatworms (Platyhelminthes; pronounced *plah-tee-hell-MIN-theez*). The group's name was inspired by morphological traits that are found in some, but not all, of the phyla in the lineage:

1. a feeding structure called a lophophore, which is found in three phyla; and

2. a type of larva called a trochophore, which is common to many of the phyla in the lineage.

As **Figure 33.3a** shows, a **lophophore** (literally, "tuft-bearer") is a specialized structure that rings the mouth and functions in suspension feeding. Lophophores are found in bryozoans (moss animals), brachiopods (lamp shells), and phoronids (horseshoe worms).

Trochophores are a type of larvae common to marine mollusks, annelids that live in the ocean, and several other phyla in the Lophotrochozoa. As **Figure 33.3b** shows, a **trochophore** (literally, "wheel-bearer") larva has a ring of cilia around its middle. These cilia allow swimming and in some species sweep food particles into the mouth. Recent analyses are suggesting that the trochophore originated early in the evolution of Lophotrochozoans, though different larval types evolved later in some groups.

Neither lophophores nor trochophores qualify as a synapomorphy in Lophotrochozoa, however. To date, biologists have yet

(a) Lophophores function in suspension feeding in adults.

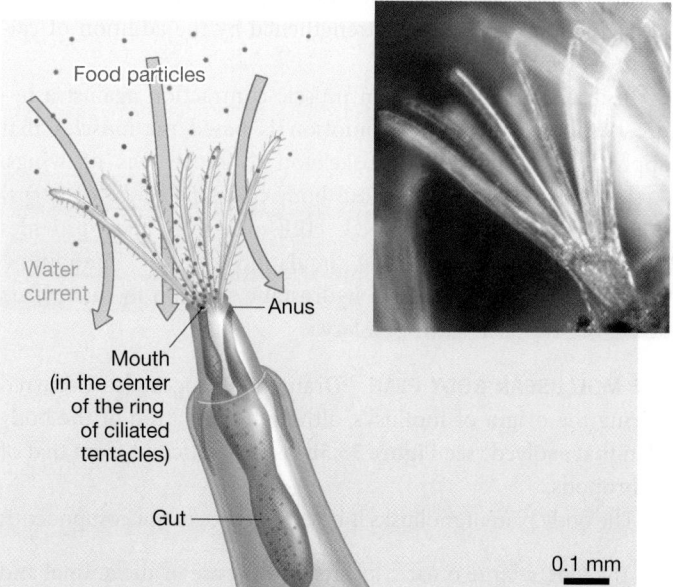

Food particles

Water current

Anus

Mouth (in the center of the ring of ciliated tentacles)

Gut

0.1 mm

(b) Trochophore larvae swim and may feed.

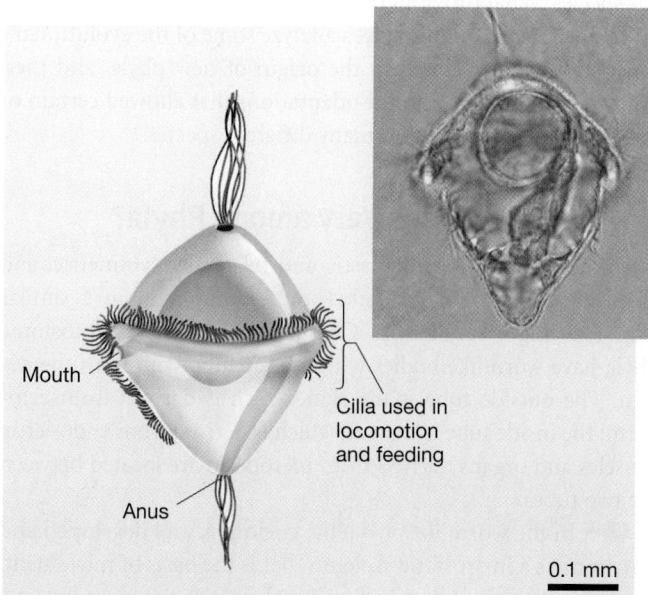

Mouth

Cilia used in locomotion and feeding

Anus

0.1 mm

FIGURE 33.3 Lophotrochozoans Have Distinctive Traits. (a) Three phyla of lophotrochozoans have the feeding structure called a lophophore. **(b)** Many phyla of lophotrochozoans have the type of larva called a trochophore. In some cases the trochophore is a stage during larval development.

✔**EXERCISE** The word root *lopho–* means tuft; *trocho–* means wheel. Label the tuft and wheel in this figure.

to discover a morphological synapomorphy that distinguishes species in this lineage from all other protostomes.

In contrast, Ecdysozoa are defined by a clear synapomorphy: their method of growth. Like other animals, lophotrochozoans grow continuously and incrementally (**Figure 33.4a**). But ecdysozoans do not.

What Is an Ecdysozoan?

The most prominent of the seven ecdysozoan phyla are the roundworms (Nematoda) and arthropods (Arthropoda). All ecdysozoans grow by **molting**—that is, by shedding an exoskeleton or external covering. The Greek root *ecdysis*, which means "to slip out or escape," is appropriate because during a molt, an individual sheds its outer layer, or **cuticle**—called an **exoskeleton** if it is hard—and slips out of it (**Figure 33.4b**). Once the old covering is gone, the body expands and a larger cuticle or exoskeleton then forms. As ecdysozoans grow and mature, they undergo a succession of molts.

Growth by molting was required in ecdysozoans once they had evolved tough outer cuticles or thick exoskeletons. A stiff body covering was advantageous because it provides an effective structure for muscle attachment and affords protection. To support this claim, biologists point to the secretive behavior of ecdysozoans while individuals are molting. When crabs and other crustaceans have shed an old exoskeleton, their new exoskeleton takes several hours to harden. During this interval, individuals hide and do not feed or move about. In addition, experiments have shown that it is much easier for predators to attack and subdue individuals that are not protected by an exoskeleton during the intermolt period.

(a) Lophotrochozoans grow incrementally.

(b) Ecdysozoans grow by molting.

Growth bands

FIGURE 33.4 Lophotrochozoans and Ecdysozoans Differ in Their Mechanism of Growth. (a) Lophotrochozoans do not molt. The growth bands on this clam show periods of slow and rapid incremental growth. **(b)** Once an ecdysozoan such as this cicada has left its old exoskeleton, hours or days pass before the new exoskeleton is hardened. During this time, the individual is highly susceptible to predation or injury.

33.2 Themes in the Diversification of Protostomes

Protostomes have diverged into more than 22 phyla that are recognized by distinctive body plans or specialized mouthparts used in feeding. Some of these phyla then split into millions of different species. What drove all this diversification?

To answer this question, let's analyze some of the evolutionary innovations that resulted in the origin of new phyla, and then follow up by delving into the adaptations that allowed certain of these phyla to diversify into many different species.

How Do Body Plans Vary among Phyla?

All protostomes are triploblastic and bilaterally symmetric, and all protostomes undergo embryonic development in a similar way. You might recall from Chapter 32 that most protostome phyla have wormlike bodies with a basic tube-within-a-tube design. The outside tube is the skin, which is derived from ectoderm; the inside tube is the gut, which is derived from endoderm. Muscles and organs derived from mesoderm are located between the two tubes.

In the wormlike phyla, the coelom is well developed and functions as a hydrostatic skeleton that is the basis of movement. But the coelom is absent in flatworms, and in the most species-rich and morphologically complex protostome phyla—the Arthropoda and the Mollusca (snails, clams, squid)—it is drastically reduced.

A fully functioning coelom has two roles: providing space for fluids to circulate among organs, and providing a hydrostatic skeleton for movement. In arthropods and mollusks, other structures fulfill these functions. Species in these phyla don't have a fully functioning coelom for a simple reason: They don't need one. Let's take a closer look.

THE ARTHROPOD BODY PLAN Arthropods have segmented bodies that are organized into prominent regions called tagmata. In the grasshopper shown in **Figure 33.5a**, these regions are called the head, thorax, and abdomen. In addition, arthropods are dis-tinguished by their jointed limbs and an exoskeleton made primarily of the polysaccharide chitin (see Chapter 5). In crustaceans, the exoskeleton is strengthened by the addition of calcium carbonate ($CaCO_3$).

Instead of being based on muscle contraction against a hydroskeleton, arthropod locomotion is based on muscles that apply force against the exoskeleton to move legs or wings. Arthropods also have a spacious body cavity called the **hemocoel** ("blood-hollow"; pronounced HEE-mah-seal) that provides space for internal organs and circulation of fluids. In addition, the hemocoel functions as a hydrostatic skeleton in caterpillars and other types of arthropod larva.

THE MOLLUSCAN BODY PLAN Dramatic changes also occurred during the origin of mollusks, although the nature of the body plan that evolved (see **Figure 33.5b**) is very different from that of arthropods.

The body plan of mollusks is based on three major components:

1. the **foot**, a large muscle located at the base of the animal and usually used in movement;

2. the **visceral mass**, the region containing most of the main internal organs; and

3. the **mantle**, an outgrowth of the body wall that covers the visceral mass, forming an enclosure called the mantle cavity.

Although the structure of the foot, visceral mass, and mantle vary widely among mollusk species, in all mollusks the visceral mass provides space for organs and the circulation of fluids. In some species the fluid enclosed by the visceral mass also functions in movement as a hydrostatic skeleton.

In many species the mantle secretes a shell made of calcium carbonate. Some mollusk species have a single shell; others have two, eight, or none at all.

(a) Arthropod body plan (external view)

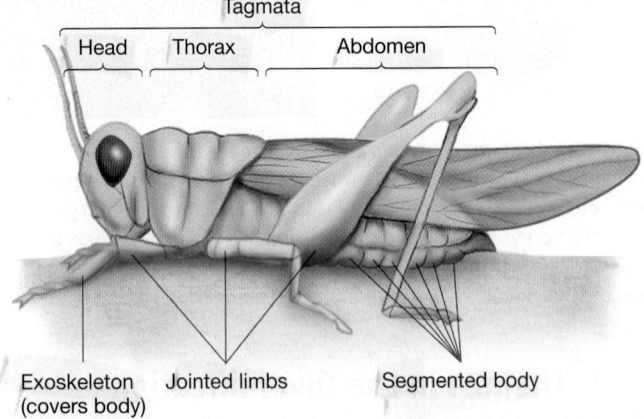

(b) A generalized mollusk body plan (internal view)

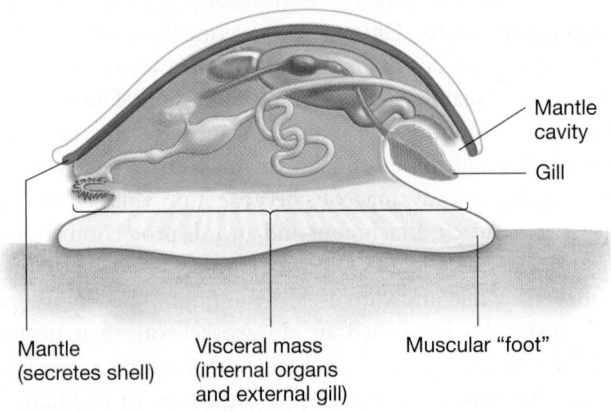

FIGURE 33.5 Arthropods and Mollusks Have Specialized Body Plans. (a) Arthropods have segmented bodies and jointed limbs, which enable these animals to move despite their hard outer covering, the exoskeleton. **(b)** The mollusk body plan is based on a foot, a visceral mass, and a mantle. Gills are located inside a cavity created by the mantle.

(a) Echiurans ("spoon worms")

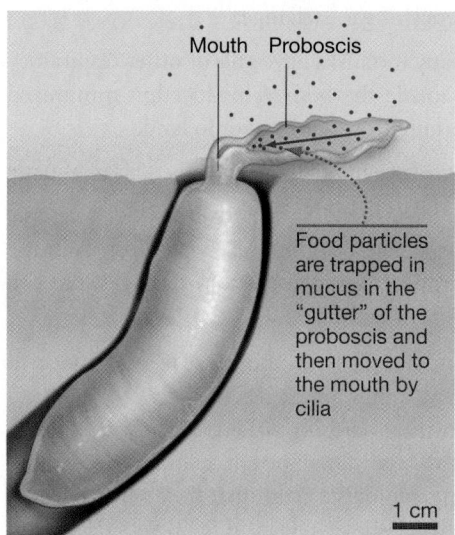

Mouth Proboscis

Food particles are trapped in mucus in the "gutter" of the proboscis and then moved to the mouth by cilia

1 cm

(b) Priapulids ("penis worms")

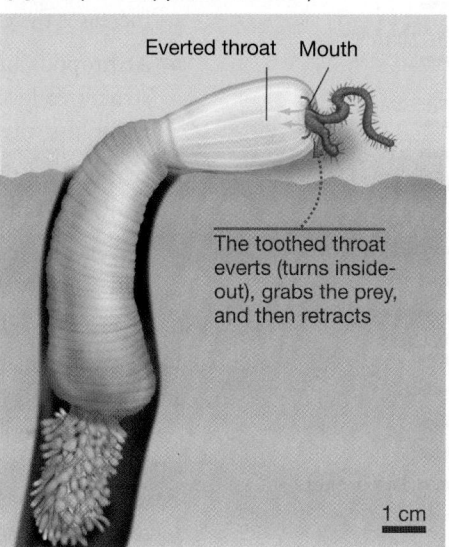

Everted throat Mouth

The toothed throat everts (turns inside-out), grabs the prey, and then retracts

1 cm

(c) Nemerteans ("ribbon worms")

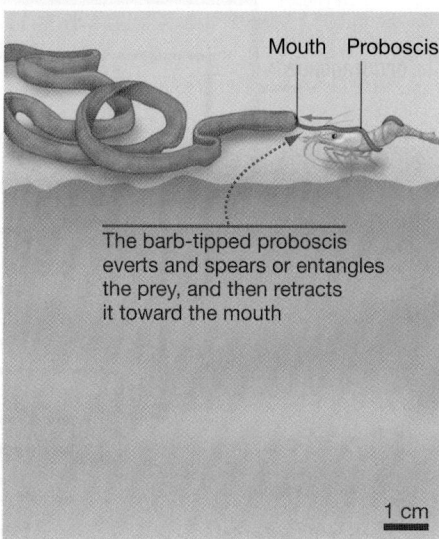

Mouth Proboscis

The barb-tipped proboscis everts and spears or entangles the prey, and then retracts it toward the mouth

1 cm

FIGURE 33.6 Some Protostome Phyla with Wormlike Bodies Have Specialized Mouthparts.

✔**QUESTION** It is common to find species from an array of wormlike phyla in the same habitat. Suggest a hypothesis to explain why this is possible.

VARIATION AMONG BODY PLANS OF THE WORMLIKE PHYLA New types of body plans distinguish most of the phyla illustrated in Figure 33.2. In contrast, the general body plan is similar in most of the protostome lineages that have wormlike bodies. In many cases the wormlike lineages are distinguished by specialized mouthparts or feeding structures. To drive this point home, consider the three phyla illustrated in **Figure 33.6**.

- *Annelids (segmented worms)* Annelids include a group called the echiurans or spoon worms, which burrow into marine mud. Echiurans deposit feed using an extended structure called a **proboscis**, which forms a gutter leading to the mouth (**Figure 33.6a**). Cells in the gutter secrete mucus, which is sticky enough to capture pieces of detritus. The combination of mucus and detritus is then swept into the mouth by cilia on cells in the gutter.

- *Priapulids (penis worms)* Priapulids also burrow into the substrate, but act as sit-and-wait predators. When a polychaete (Annelida) or other prey item approaches, the priapulid everts its toothed, cuticle-lined throat—meaning that it turns its throat inside out—grabs the prey item, and retracts the structure to take in the food (**Figure 33.6b**).

- *Nemerteans (ribbon worms)* Nemerteans are active predators that move around the ocean floor in search of food. Instead of everting their throat to capture prey, they have a proboscis that can extend or retract (**Figure 33.6c**). Nemerteans spear small animals with their proboscis or wrap the extended proboscis around prey. The food is then pulled into the mouth.

 Once a specific body plan had evolved, subsequent diversification was largely driven by adaptations that allowed protostomes to live on land or feed, move, or reproduce in novel ways. Recall that an **adaptation** is a trait that increases the fitness (reproductive success) of individuals relative to individuals without the trait.

Before going on to review the adaptations that drove diversification within certain lineages, go to the study area at *www.masteringbiology.com* to get an overview of variation among protostomes.

(MB) **Web Activity** Protostome Diversity

The Water-to-Land Transition

Protostomes are the most abundant animals in the surface waters and substrates of many marine and freshwater environments. But protostomes are common in virtually every terrestrial setting as well. Like the land plants and fungi, protostomes made the transition from aquatic to terrestrial environments.

To help put this achievement in perspective, recall from Chapter 30 that green plants made the move from freshwater to land just once. Data in Chapter 31 show that it is not yet clear whether fungi moved from aquatic habitats to land once or several times. Chapter 34 will show that only one lineage among deuterostomes moved onto land. But a water-to-land transition occurred multiple times as protostomes diversified.

EVIDENCE FOR MULTIPLE TRANSITIONS The evidence for multiple water-to-land transitions in protostomes is based on phylogenetic analyses, which support the hypothesis that the ancestors of the terrestrial lineages in each major subgroup of protostomes were aquatic.

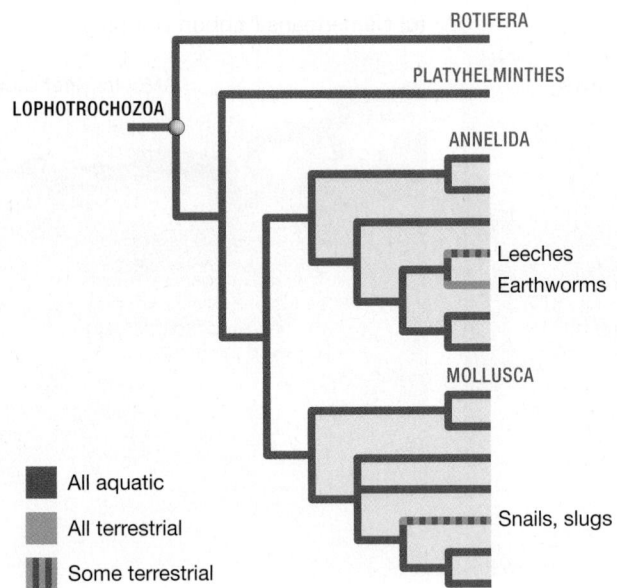

FIGURE 33.7 Evidence for a Water-to-Land Transition in the Lophotrochozoa. The appearance of isolated terrestrial groups among marine groups also occurs in the Ecdysozoa.

Figure 33.7, for example, shows a recent phylogeny of the annelids and mollusks within the Lophotrochozoa. Because almost all of the species in these two groups are aquatic, and because the outgroup species, the rotifers and platyhelminths are also aquatic, it is logical to infer that the ability to live on land evolved in the lineages leading to the snails/slugs and the earthworms. The alternative hypothesis—that terrestrial living was the ancestral state—would require that aquatic living arose fourteen times independently. Similar analyses show that terrestrial living evolved several times independently in the Ecdysozoa.

ADAPTATIONS TO TERRESTRIAL ENVIRONMENTS Why were so many different protostome groups able to move from water to land independently? The short answer is that it was easier for protostomes to accomplish this than for plants to do so. For example, land plants had to evolve roots and vascular tissue to transport water and support their stems. Similarly, the first terrestrial animals had to be able to support their body weight on land and move about. But the protostome groups that made the water-to-land transition already had hydrostatic skeletons, exoskeletons, appendages, or other adaptations for support and locomotion that happened to work on land as well as water.

To make the transition to land, new adaptations allowed protostomes to (1) exchange gases and (2) avoid drying out. As Chapter 44 will detail, land animals exchange gases with the atmosphere readily as long as a large, moist surface area is exposed to the air. The bigger challenge is to prevent the gas-exchange surface and other parts of the body from drying out. How do terrestrial protostomes solve this problem?

- Roundworms and earthworms live in humid soils or other moist environments and exchange gases across their body surface. They have a high surface-area-to-volume ratio, which increases the efficiency of gas exchange.

- Arthropods and many mollusks have gills or other respiratory structures located inside the body. This location minimized water loss when certain groups moved onto land.

- In mollusks, the mantle cavity that encloses the gills of aquatic snails evolved into the lung found in terrestrial snails.

- Insects evolved a waxy layer to minimize water loss from the body surface, with openings to respiratory passages that can be closed if the environment dries.

Unlike land plants and fungi, land animals can also move to moister habitats if the area they are in gets too dry. Water-to-land transitions are important because they open up entirely new habitats and new types of resources to exploit. Based on this reasoning, biologists claim that the ability to live in terrestrial environments was a key event in the diversification of several protostome phyla.

Adaptations for Feeding

Protostomes include suspension, deposit, liquid, and food-mass feeders. Besides exploiting detritus, they prey on or parasitize plants, algae, or other animals. Exploiting a diversity of foods is possible because protostomes have such a wide variety of mouthparts for capturing and processing food.

Figure 33.6 highlighted mouthparts that distinguish some of the wormlike phyla. But within phyla, arthropods take the prize for mouthpart diversity. The mouthparts observed in this phylum vary in structure from tubes to pincers and allow the various species to pierce, suck, grind, bite, mop, chew, engulf, cut, or mash (**Figure 33.8**). All arthropods have the same basic body plan, but their mouthparts and food sources are highly diverse.

In many species of arthropods, the jointed limbs also play a key role in getting food. The crustaceans called krill and barnacles use their legs to sweep food toward their mouths as they suspension feed. Certain insects, crustaceans, spiders, and mollusks use their appendages to capture prey or hold food as it is being chewed or bitten by the mouthparts.

In most cases, larval and adult forms of the same species exploit different food sources. Metamorphosis is extremely common in protostomes and usually results in larvae and adults that live in different habitats and have different overall morphology and mouthparts.

Adaptations for Moving

Protostomes move in virtually every way known among animals. Aquatic larvae move with the aid of cilia, and wormlike protostomes that lack limbs move with the aid of a coelom that functions as a hydrostatic skeleton. Caterpillars, grubs, maggots, and other types of insect larvae with wormlike bodies have an enclosed, fluid-filled body cavity—the hemocoel—that also functions as a hydrostatic skeleton.

(a) Leaf-cutter ants cut leaves.

(b) Sheep ticks pierce skin.

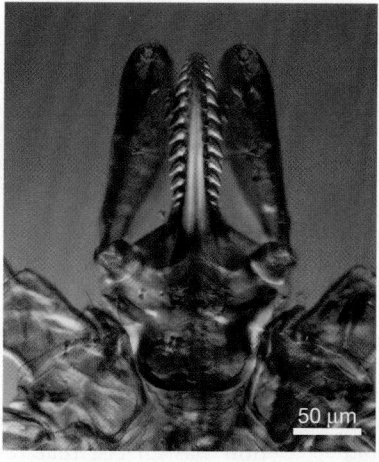

(c) Nut weevils bore into hard fruit.

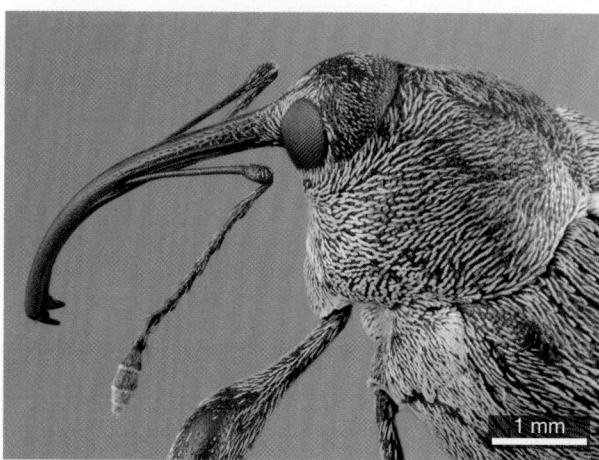

FIGURE 33.8 Arthropod Feeding Structures Are Diverse. Even though their mouthparts and food sources are diverse, all of these arthropods have segmented bodies that are organized into tagmata, with an exoskeleton and jointed appendages.

As arthropods diversified, however, a number of important evolutionary innovations allowed them to move in unique ways:

- *Jointed limbs in arthropods* Jointed limbs made rapid, precise walking, running, and jumping movements possible (**Figure 33.9a**). The jointed limb is a major reason arthropods have been so spectacularly successful in terms of species diversity, abundance of individuals, geographic range, and duration in the fossil record.

- *The insect wing* The insect wing is one of the most important adaptations in the history of life (**Figure 33.9b**). About two-thirds of the multicellular species living today are winged insects. In many cases, larvae lack wings and crawl or swim—only adults fly. According to data in the fossil record, insects were the first organisms that had wings and could fly.

Like most insects today, the earliest insects had two pairs of wings. In most four-winged insects living today, however, the four wings function as two. Beetles fly with only their hindwings; butterflies, moths, bees, and wasps have hooked structures that make their two pairs of wings move together.

Flies, in contrast, have a single pair of large wings that power flight. They also have a pair of small, winglike structures called halteres that provide stability during flight.

- *The mollusk foot* Snails and chitons are mollusks that have a muscular foot at the base of the body. Waves of muscle contractions sweep up or down the length of the large, muscular foot, allowing individuals to crawl along a surface (**Figure 33.9c**). Cells in the foot secrete a layer of mucus, which reduces friction and increases the efficiency of movement.

(a) Walking, running, and jumping

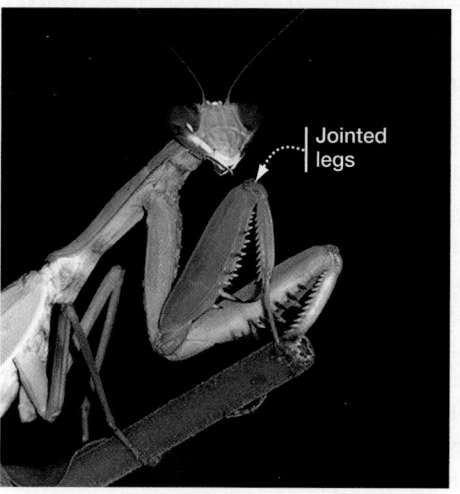

Jointed legs

(b) Flying

Wings (most insects have two pairs)

(c) Gliding and crawling

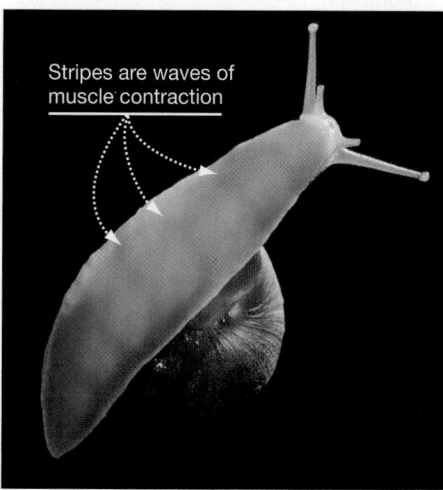

Stripes are waves of muscle contraction

FIGURE 33.9 Protostome Locomotion Is Diverse. The evolution of **(a)** jointed limbs and **(b)** wings were key innovations in arthropod movement. **(c)** A wave of muscle contractions, up or down the length of the foot, allows mollusks like this snail to creep along a substrate.

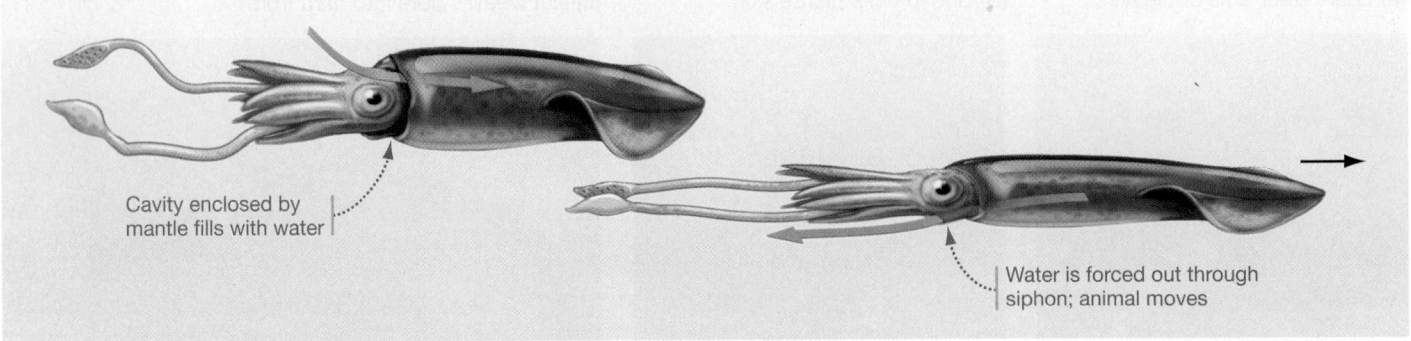

Cavity enclosed by
mantle fills with water

Water is forced out through
siphon; animal moves

FIGURE 33.10 Jet Propulsion in Mollusks. In jet propulsion, muscular contractions force the water out through a movable siphon.

- *Jet propulsion* Cephalopods (squid and octopus) are mollusks that have a mantle lined with muscle. When the cavity surrounded by the mantle fills with water and the mantle muscles contract, a stream of water is forced out of a tube called a **siphon**. The force of the water propels the squid (**Figure 33.10**). This mechanism, jet propulsion, evolved in squid long before human engineers thought of using the same principle to power aircraft.

Adaptations in Reproduction

When it comes to variation in reproduction and life cycles, protostomes do it all.

- Asexual reproduction by splitting the body lengthwise or by fragmenting the body is common in many of the wormlike phyla.

- Many crustacean and insect species reproduce asexually via **parthenogenesis** ("virgin-origin")—the production of unfertilized eggs, by mitosis, that develop into offspring.

- Sexual reproduction is often based on external fertilization in clams, bryozoans, brachiopods, and other groups.

- Sexual reproduction often begins with copulation and internal fertilization in groups that are capable of movement—such as crustaceans, snails, and insects—because males and females can meet.

- Females of ovoviviparous insects and snails bear live young. They do so by retaining fully formed eggs inside their bodies and then nourishing them via the nutrient-rich yolk that is inside the egg (see Chapter 32).

Although asexual reproduction is common, sexual reproduction is the predominant mode of producing offspring in most protostome groups.

Metamorphosis is another important aspect of reproduction in protostomes. In marine species, metamorphosis is hypothesized to be an adaptation that allows larvae to disperse to new habitats by floating or swimming in the plankton, where they can feed on food sources that are unavailable to adults. Metamorphosis is also common in insects, where it is hypothesized to be an adaptation that reduces competition for food between larvae and adults. The pupal stage that occurs in insects—when the larval body is remodeled into an adult (see Chapter 32)—does not occur in marine species, however.

In terrestrial environments, though, a more critical adaptation was an egg that resists drying. Insect eggs have a thick membrane that keeps moisture in, and the eggs of slugs and snails have a thin calcium carbonate shell that helps retain water. Desiccation-resistant eggs evolved repeatedly in populations that made the transition to life on land.

CHECK YOUR UNDERSTANDING

If you understand that . . .

- In protostomes with wormlike bodies, phyla vary in mouthpart structure and mode of feeding.
- Arthropods and mollusks have specialized body plans.
- A water-to-land transition occurred in several lineages independently.

✓ You should be able to . . .

1. Describe the major features of the arthropod body and explain the advantage of a jointed limb versus an unjointed limb.

2. Name three problems that arise during a water-to-land transition in animals.

Answers are available in Appendix B.

33.3 Key Lineages: Lophotrochozoans

The lineages within Lophotrochozoa are united by shared mechanisms of early development and growth, but they are highly diverse in terms of their resulting morphology. To drive this point home, this section details key points about the biology of four key lophotrochozoan phyla: (**1**) Rotifera, (**2**) Platyhelminthes, (**3**) Annelida, and (**4**) Mollusca.

Because the Mollusca is so species-rich and diverse, the detailed descriptions provided later in this section consider the four most important lineages of mollusks separately: (**1**) **bivalves** (clams and mussels), (**2**) **gastropods** (slugs and snails), (**3**), **chitons**, and (**4**) **cephalopods** (squid and octopuses).

Unlike the lophotrochozoans as a whole, the mollusks are identified by distinctive characteristics. In addition to their lophotrochozoan mode of growth and development, mollusks have a specialized body plan based on a muscular foot, a visceral mass, and a mantle that in most species secretes a calcium carbonate shell. The coelom is much reduced in mollusks and functions only in reproduction and excretion of wastes. ✔ You should be able to indicate the origin of the molluscan body plan on Figure 33.2.

A great deal of diversification occurred once the basic molluskan body plan had evolved. Although most live in marine environments, there are some terrestrial and freshwater forms. Most bivalves are suspension feeders; the other three groups of mollusks are herbivores or predators.

To review the diversity across and within lophotrochozoan phyla efficiently, the remainder of the section is organized as follows:

- ■ Lophotrochozoans > Rotifera (Rotifers)
- ■ Lophotrochozoans > Platyhelminthes (Flatworms)
- ■ Lophotrochozoans > Annelida (Segmented Worms)
- ■ Lophotrochozoans > Mollusca > Bivalvia (Clams, Mussels, Scallops, Oysters)
- ■ Lophotrochozoans > Mollusca > Gastropoda (Snails, Slugs, Nudibranchs)
- ■ Lophotrochozoans > Mollusca > Polyplacophora (Chitons)
- ■ Lophotrochozoans > Mollusca > Cephalopoda (Nautilus, Cuttlefish, Squid, Octopuses)

Lophotrochozoans > Rotifera (Rotifers)

The 1800 rotifer species that have been identified thus far live in damp soils as well as marine and freshwater environments. They are important components of the plankton in freshwater and in brackish waters, where rivers flow into the ocean and water is slightly salty. Rotifers have a coelom, and most are less than 1 mm long. Although rotifers do not have a lophophore or a trochophore larval stage, their mode of growth and extensive similarities in DNA sequence identify them as a member of the lophotrochozoan lineage.

Feeding Rotifers have a cluster of cilia at their anterior end called a **corona** (**Figure 33.11**). In many species, the beating of the cilia in the corona makes suspension feeding possible by creating a current that sweeps microscopic food particles into the mouth. The corona is the signature morphological feature of

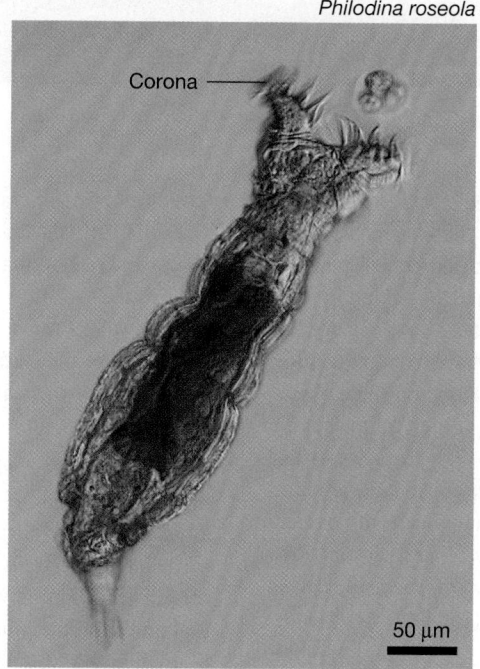

Philodina roseola

Corona

50 μm

FIGURE 33.11 Rotifers Are Tiny, Aquatic Suspension Feeders.

this group. ✔ You should be able to indicate the origin of the corona on Figure 33.2.

Movement Although a few species of rotifers are sessile, most swim via the beating of cilia in the corona.

Reproduction Females produce unfertilized eggs by mitosis; the eggs then hatch into new, asexually produced individuals. Recall that the production of offspring via unfertilized eggs is termed parthenogenesis. In a group of rotifers called the bdelloids, only females have been found and all reproduction appears to be by parthenogenesis. Both sexual reproduction and asexual reproduction are observed in most rotifer species, however. Development is direct, meaning that fertilized eggs hatch and grow into adults without going through metamorphosis.

Lophotrochozoans > Platyhelminthes (Flatworms)

The flatworms are a large and diverse phylum. More than 400,000 species have been described in four major lineages within the phylum: (1) the free-living species called Turbellaria (**Figure 33.12a**, on page 632), (2) the endoparasitic tapeworms called Cestoda (**Figure 33.12b**), (3) the endoparasitic or ectoparasitic flukes, called Trematoda (**Figure 33.12c**), and (4) ectoparasites called Monogenea.

Although a few turbellarian species are terrestrial, most live on the substrates of freshwater or marine environments. Tapeworms and other cestodes parasitize fish, mammals, or other vertebrates. Flukes parasitize vertebrates or mollusks; most monogeneans parasitize fish. In humans, a fluke is responsible for schistosomiasis—a serious public health issue in many developing nations.

(Continued on next page)

(a) Turbellarians are free living.

Pseudoceros ferrugineus

1 cm

(b) Cestodes are endoparasitic.

Taenia species

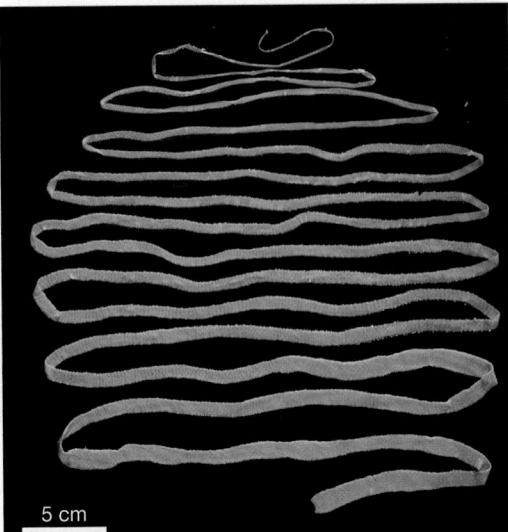

5 cm

(c) Trematodes are endoparasitic.

Dicrocoelium dendriticun

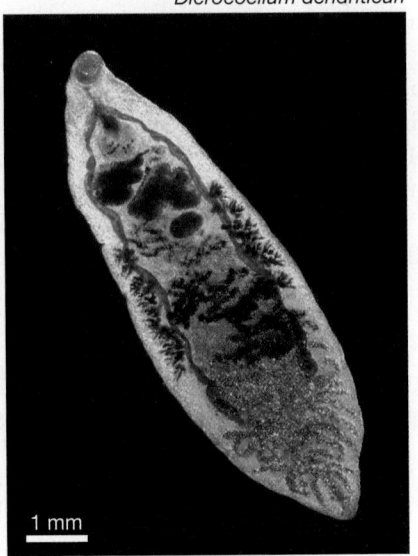

1 mm

FIGURE 33.12 Flatworms Have Simple, Flattened Bodies.

Flatworms are named for the broad, flattened shape of their bodies. (The Greek roots *platy* and *helminth* mean "flat-worm.") Species in the Platyhelminthes are unsegmented and lack a coelom. They also lack structures that are specialized for gas exchange—taking in oxygen and ridding the body of carbon dioxide. Instead, gas exchange occurs across the body wall. Further, they do not have blood vessels for circulating oxygen and nutrients to their cells. Based on these observations, biologists interpret the flattened bodies of these animals as an adaptation that gives flatworms an extremely high surface-area-to-volume ratio. Because they have so much surface area, a large amount of gas exchange can occur directly across their body wall, with oxygen diffusing into the body from the surrounding water and carbon dioxide diffusing out. Because the volume of the body is so small relative to the surface area available, nutrients and gases can diffuse efficiently to all of the cells inside the animal. This body plan has a downside, however: Because the body surface has to be moist for gas exchange to take place, flatworms are restricted to environments where they are surrounded by fluid. ✔You should be able to indicate the origin of the flattened, acoelomate body plan on Figure 33.2.

Feeding Platyhelminthes lack a lophophore and have a digestive tract that is "blind"—meaning it has only one opening for ingestion of food and elimination of wastes.

Most turbellarians are hunters that prey on protists or small animals; others scavenge dead animals.

Tapeworms are strictly parasitic and feed on nutrients provided by hosts. They do not have a mouth or a digestive tract, and obtain nutrients by diffusion across their body wall.

Flukes are parasites and feed by gulping host tissues and fluids through a mouth. They have a blind digestive tract.

Movement In general, flatworms do not move much.

Some turbellarians can swim by undulating their bodies, and most can creep along substrates with the aid of cilia on their ventral surface.

Tapeworms move little; adult cestodes have hooked attachment structures at their anterior end that permanently attach them to the interior of their host.

Flukes move little.

Reproduction Turbellarians can reproduce asexually by splitting themselves in half. If they are fragmented as a result of a predator attack, the body parts can regenerate into new individuals. Most turbellarians contain both male and female organs and reproduce sexually by aligning with another individual and engaging in mutual and simultaneous fertilization.

In flukes and tapeworms, reproduction is similar: Individuals reproduce sexually and either cross-fertilize or self-fertilize. The reproductive systems and life cycles of flukes and tapeworms are extremely complex, however. In many cases the life cycle involves two or even three distinct host species, with sexual reproduction occurring in the **definitive host** and asexual reproduction occurring in one or more **intermediate hosts**. For example, humans are the definitive host of the blood fluke *Schistosoma mansoni*. Fertilized eggs are shed in a human host's feces and enter aquatic habitats if sanitation systems are poor. The eggs develop into larvae that infect snails. Inside the snail, asexual reproduction results in the production of a different type of larva, which emerges from the snail and burrows into the skin of humans who wade in infested water. Once inside a human, the parasite lives in blood and develops into a sexually mature adult.

Most **annelids** have a segmented body and a coelom that functions as a hydrostatic skeleton. ✔ You should be able to indicate the origin of annelid segmentation on Figure 33.2. The 16,500 species that have been described thus far were traditionally divided into groups called Polychaeta (*pol-ee-KEE-ta*) and Clitellata. Recent analyses have shown the Polychaeta are paraphyletic, however. The current understanding of annelid phylogeny is consistent with the following points:

- The common ancestor of the annelids had a key synapomorphy in addition to segmentation: numerous, bristlelike extensions called **chaetae** that extend from appendages called **parapodia** (**Figure 33.13a**). (The name polychaetes means "many bristles".)

- Lineages that retained many chaetae are mostly marine. They range in size from species that are less than 1 mm long to species that grow to lengths of 3.5 m. They include a large number of burrowers in mud or sand, sedentary forms that secrete chitinous tubes, and active, mobile species.

- Chaetae were reduced or lost in a lineage called Clitellata.

- The Clitellata includes monophyletic groups called the oligochaetes (*oh-LIG-oh-keetes*) and leeches, or Oligochaeta and Hirudinea, respectively.

- Oligochaetes ("few-bristles") include the earthworms, which burrow in moist soils; an array of freshwater species; and a few marine forms. Oligochaetes lack parapodia and, as their name implies, have many fewer chaetae than do polychaetes (**Figure 33.13b**).

- The coelom of leeches is much smaller than that of other annelids and consists of a series of connected chambers. Leeches live in freshwater as well as marine and environments.

- Lineages called Sipunculida (peanut worms) and Echiura (spoon worms), which were traditionally thought to be independent phyla, are actually groups within Annelida. Segmentation and chaetae were lost in these species.

Feeding The polychaete lineages have a wide variety of methods for feeding: The burrowing forms deposit feed or suspension feed using mucous-lined nets; the sedentary forms suspension feed with the aid of a dense crown of tentacles; and the active forms graze on algae or hunt small animals, which they capture by everting their throats. Virtually all oligochaetes, in contrast, make their living by deposit feeding in soils. The tunnels that they make are critically important in aerating soil, and their feces contribute large amounts of organic matter. About half of the leeches are ectoparasites that attach themselves to fish or other hosts and suck blood and other body fluids (**Figure 33.13c**). Hosts are usually unaware of the attack, because leech saliva typically contains an anesthetic. The host's blood remains liquid as the leech feeds, because the parasite's saliva also contains an anticoagulant. Parasitic leeches are used by physicians to reduce swelling after surgery to reattach severed fingers or other body parts. The nonparasitic leech species are predators or scavengers.

Movement Polychaetes and oligochaetes crawl or burrow with the aid of their hydrostatic skeletons; the parapodia of polychaetes also act as paddles or tiny feet that aid in movement. Many polychaetes are excellent swimmers. Leeches can swim by using their hydrostatic skeletons to make undulating motions of the body.

Reproduction Asexual reproduction occurs in polychaetes and oligochaetes by transverse fission or fragmentation—meaning that body parts can regenerate a complete individual. Sexual reproduction in polychaetes may begin with internal or external fertilization, depending on the species. Most of the polychaete lineages have separate sexes and usually release their eggs directly into the water. Some species produce eggs that hatch into trochophore larvae, then go on to develop into more-complex larval forms. In oligochaetes and leeches, individuals produce both sperm and eggs and engage in mutual, internal cross-fertilization. Eggs are enclosed in a mucus-rich, cocoon-like structure; after fertilization, offspring develop directly into miniature versions of their parents.

(a) Polychaetes have many bristle-like chaetae.

Hermodice carunculata

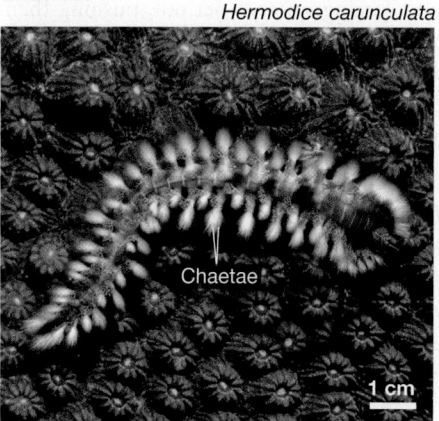

(b) Oligochaetes have few or no chaetae.

Chaetogaster species

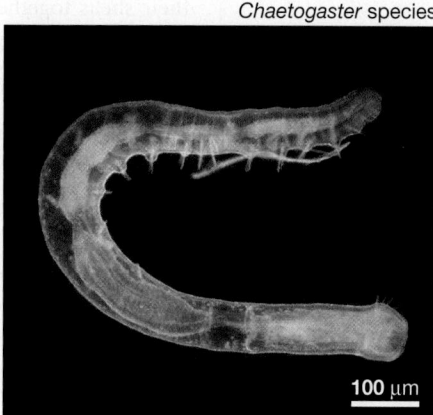

(c) Some leeches are ectoparasites.

Haemadipsa species

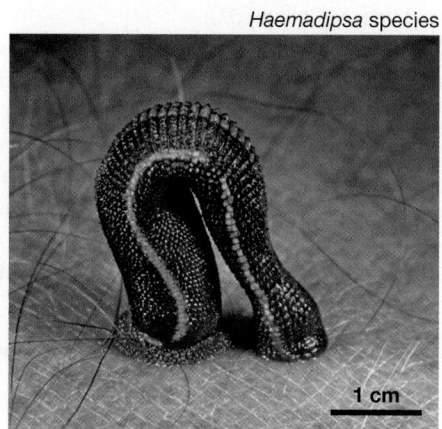

FIGURE 33.13 Annelids Are Segmented Worms.

The bivalves are so named because they have two separate shells made of calcium carbonate secreted by the mantle. The shells are hinged, and they open and close with the aid of muscles attached to them (**Figure 33.14a**). When the shell is closed, it protects the mantle, visceral mass, and foot. The bivalve shell is an adaptation that reduces predation.

Most bivalves live in the ocean, though there are many freshwater forms. Clams burrow into mud, sand, or other soft substrates and are sedentary as adults. Oysters and mussels are largely sessile as adults, but most of them live above the substrate, attached to rocks or other solid surfaces. Scallops are mobile and live on the surface of soft substrates. The smallest bivalves are freshwater clams that are less than 2 mm long; the largest bivalves are giant marine clams that may weigh more than 400 kg. All bivalves can sense gravity, touch, and certain chemicals; scallops even have eyes.

Because most bivalves live on or under the ocean floor and because they are covered by a hard shell, their bodies are often buried in sediment after death. Bivalves thus fossilize readily, and the Bivalvia lineage has the most extensive fossil record of any animal, plant, or fungal group. This large database has allowed biologists to conduct thorough studies of the evolutionary history of bivalves.

Bivalves are important commercially. Clams, mussels, scallops, and oysters are farmed or harvested from the wild in many parts of the world and used as food by humans. Pearl oysters that are cultivated or collected from the wild are the most common source of natural pearls used in jewelry.

Feeding Most bivalves are suspension feeders that take in any type of small animal or protist or detritus. Suspension feeding is based on a flow of water across gas-exchange structures called **gills**. The gills lie in the mantle cavity, between the mantle and the visceral mass, and consist of a series of thin membranes where particles are trapped. In many cases the water current flows through siphons, which are tubes formed by edges of the mantle—extending from the shell and forming a plumbing system. The siphons conduct water into the gills and then back out of the body, powered by the beating of cilia on the gills (**Figure 33.14b**). Bivalves are the only major group of mollusks that lack the feeding structure called a radula (see Chapter 32).

The efficiency of filter feeding in bivalves is illustrated by the impact that zebra mussels have had on certain lakes and streams in North America. After being introduced from Eurasia, zebra mussel populations have exploded in some aquatic ecosystems. The mussels are so efficient at removing organic matter that little food is left for native fish, whose populations decline dramatically as a result. Conversely, loss of native oysters from Chesapeake Bay has resulted in large accumulations of organic matter that have reduced water quality there.

Movement Clams burrow with the aid of their muscular foot, but are otherwise sedentary. Scallops are able to swim by clapping

(a) Scallops live on the surface of the substrate and suspension feed.

Chlamys varia

5 mm

(b) Most clams burrow in soft subtrates and suspension feed.

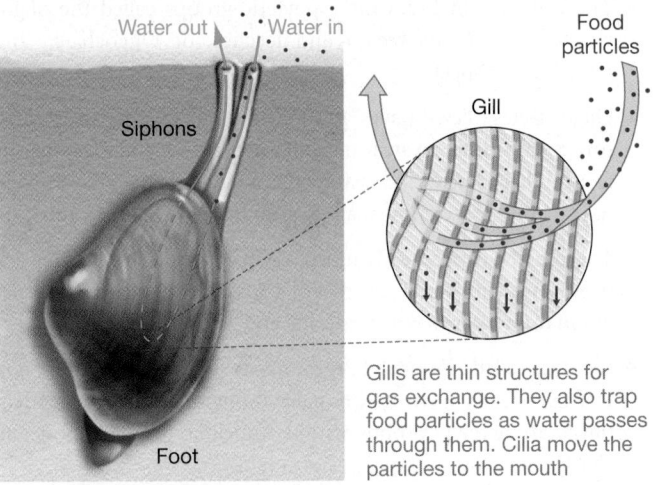

Water out   Water in

Food particles

Siphons

Gill

Foot

Gills are thin structures for gas exchange. They also trap food particles as water passes through them. Cilia move the particles to the mouth

FIGURE 33.14 Bivalves Have Two Shells.

their shells together and forcing water to jet out, pushing them along. Scallop locomotion is similar to the swimming of cnidarian medusae, which move when muscular contractions force water out of their bell (see Chapter 32). Bivalves produce a swimming trochophore larva that is responsible for dispersing individuals to new locations.

Reproduction Most bivalves reproduce sexually. Eggs and sperm are shed into the water, and fertilized eggs develop into trochophore larvae. Trochophores then metamorphose into a distinct type of larva called a **veliger**, which continues to feed and swim before settling to the substrate and metamorphosing into an adult form that secretes a shell.

The gastropods ("belly-feet") are named for the large, muscular foot on their ventral side. Most snails can retract their foot and body into a shell when they are attacked or when their tissues begin to dry out (**Figure 33.15a**). Land slugs and nudibranchs (pronounced *NEW-da-branks*)—also called sea slugs—lack shells, however. The shell was lost independently in the two groups; individuals often contain toxins or foul-tasting chemicals to protect them from being eaten. The bright colors of nudibranchs are thought to act as a warning to potential predators (**Figure 33.15b**). Gastropods are used as food in some cultures and are important in human disease because they serve as a host for flukes that also infect humans before completing their life cycle. About 70,000 species of gastropods are known.

Feeding Gastropods and other mollusks have a unique structure in their mouths called a radula. Recall from Chapter 32 that in many species the **radula** functions like a rasp to scrape away algae, plant cells, or other types of food. It is usually covered with teeth that are made of chitin and that vary in size and shape among species. Although most gastropods are herbivores or detritivores, specialized types of teeth allow some gastropods to act as predators. Species called drills, for example, use their radula to bore a hole in the shells of oysters or other mollusks and expose the visceral mass, which they then eat. Cone snails have highly modified, harpoon-like "teeth" mounted at the tip of an extensible proboscis and armed with poison. When a fish or worm passes by, the proboscis shoots out. The prey is speared by the tooth, subdued by the poison, and consumed.

Movement Waves of contractions down the length of the foot allow gastropods to move by creeping (see Figure 33.9). Sea butterflies are gastropods with a reduced or absent shell but a large, winglike foot that flaps and powers swimming movements.

Reproduction Females of some gastropod species can reproduce asexually by producing eggs parthenogenetically, but most reproduction is sexual. Sexual reproduction in some gastropods begins with internal fertilization. Most marine gastropods produce a veliger larva that may disperse up to several hundred kilometers from the parent. But in some marine species and all terrestrial forms, larvae are not free living. Instead, offspring remain in an egg case while passing through several larval stages and then hatch as miniature versions of the adults.

(a) Snails have a single shell, which they use for protection.

Maxacteon flammea

5 mm

(b) Land and sea slugs (nudibranchs) have lost their shells.

Chromodoris annae

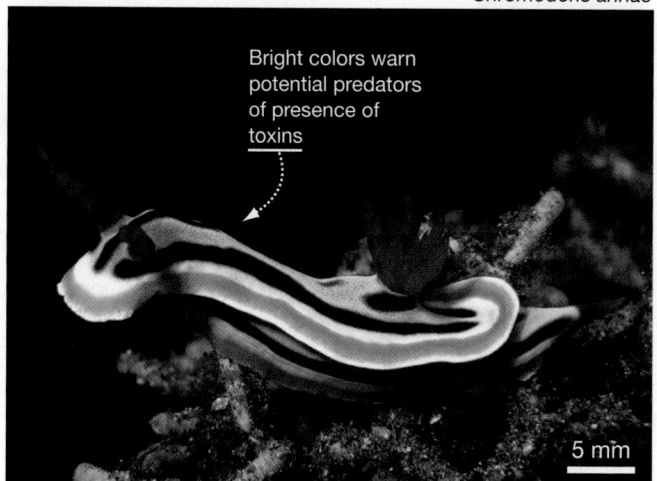

Bright colors warn potential predators of presence of toxins

5 mm

FIGURE 33.15 Gastropods Have a Single Shell or Lack Shells.

The Greek word roots that inspired the name *Polyplacophora* mean "many-plate-bearing." The name is apt because chitons (pronounced *KITE-uns*) have eight calcium carbonate plates along their dorsal side (**Figure 33.16**). The plates form a protective shell. The approximately 1000 species of chitons are marine. They are usually found on rocky surfaces in the intertidal zone, where rocks are periodically exposed to air at low tides.

Feeding Chitons have a radula and use it to scrape algae and other organic matter off rocks.

Movement Chitons move by creeping on their broad, muscular foot—as gastropods do.

Reproduction In most chiton species the sexes are separate and fertilization is external. In some species, however, sperm that are shed into the water enter the female's mantle cavity and fertilize eggs inside the body. Depending on the species involved, eggs may be enclosed in a membrane and released or retained until hatching and early development are complete. Most species have trochophore larvae.

Tonicella lineata

1 cm

FIGURE 33.16 Chitons Have Eight Shell Plates.

The cephalopods ("head-feet") have a well-developed "head" consisting of the visceral mass and a foot that is modified to form **tentacles:** long, thin, muscular extensions that aid in movement and prey capture (**Figure 33.17**). Except for the nautilus, the cephalopod species living today have either highly reduced shells or none at all. Most have large brains and image-forming eyes with sophisticated lenses.

Feeding Cephalopods are highly intelligent predators that hunt by sight and use their tentacles to capture prey—usually fish or crustaceans. They have a radula as well as a structure called a **beak**, which can exert powerful biting forces. Some cuttlefish and octopuses also inject poisons into their prey to subdue them.

Movement Cephalopods can swim by moving their fins to "fly" through the water, or by jet propulsion using the mantle cavity and siphon. They draw water into their mantle cavity and then force it out through a siphon (see Figure 33.10). Squid are built for speed and hunt small fish by chasing them down. Octopuses, in contrast, crawl along the substrate using their tentacles. They chase down crabs or other crustaceans, or they pry mussels or clams from the substrate and then use their beaks to crush the exoskeletons of their prey.

Reproduction Cephalopods have separate sexes, and some species have elaborate courtship rituals that involve color changes and inter-action of tentacles. When a male is accepted by a female, he deposits sperm that are encased in a structure called a **spermatophore**. The spermatophore is transferred to the female, and fertilization is internal. Females lay eggs. When they hatch, juveniles develop directly into adults.

Octopus bimaculatus

10 cm

FIGURE 33.17 Cephalopods Have Highly Modified Bodies.

33.4 Key Lineages: Ecdysozoans

The ecdysozoans were first recognized as monophyletic when investigators began using DNA sequence data to estimate the phylogeny of protostomes. Seven phyla are currently recognized in the lineage (see Table 32.1), including the Nematoda, Onychophora (*on-ee-KOFF-er-uh*) and the Tardigrada. Onychophora and Tardigrada are not species-rich phyla, but both are closely related to arthropods. They are similar to arthropods in having a segmented body and limbs. Unlike arthropods, their limbs are not jointed and they do not have an exoskeleton.

The onychophorans, or velvet worms, are small, caterpillar-like organisms that live in moist leaf litter and prey on small invertebrates. Onychophorans have lobe-shaped appendages and segmented bodies with a hemocoel (**Figure 33.18a**).

The tardigrades, or water bears, are microscopic animals that live in benthic (bottom) habitats of marine or freshwater environments. Large numbers of water bears can also be found in the film of water that covers moss or other land plants in moist habitats. Tardigrades have a reduced coelom but a prominent hemocoel, and they walk on their clawed, lobe-shaped legs (**Figure 33.18b**). Most feed by sucking fluids from plants or animals; others are **detritivores**. Tardigrades are notable for their ability to withstand extreme drying and pressure. In response to drying or other adverse conditions, individuals may reduce metabolism to near zero for a decade or more, then "re-animate" and resume normal activities when conditions improve.

In terms of duration in the fossil record, species diversity, and abundance of individuals, however, arthropods are the most important phylum within Ecdysozoa. They appear in the fossil record over 520 million years ago and have long been the most abundant animals observed in both marine and terrestrial environments. Well over a million living species have been described, and biologists estimate that millions or perhaps even tens of millions of arthropod species have yet to be discovered. In terms of species diversity, arthropods are easily the most successful lineage of eukaryotes.

Morphologically, **arthropods** are distinguished by segmented bodies and a complex, jointed exoskeleton. They have a highly reduced coelom but possess an extensive body cavity called a hemocoel. They also have an exoskeleton and paired, jointed appendages. The body is organized into distinct head and trunk tagmata in all arthropods; in many species there is an additional grouping of segments into two distinct trunk regions, usually called an abdomen and a thorax. ✔You should be able to indicate the origin of lobe-shaped limbs and jointed limbs on Figure 33.2.

Metamorphosis is common in arthropods. Their larvae have segmented bodies but may lack a hardened exoskeleton. As in other ecdysozoans, larval and adult forms grow by molting.

At least some segments along the arthropod body produce paired, jointed appendages. Arthropod appendages have an array of functions: sensing aspects of the environment, exchanging gases, feeding, or locomotion via swimming, walking, running, jumping, or flying. Arthropod appendages provide the ability to sense environmental stimuli and make sophisticated movements in response.

Most species also have sophisticated, image-forming compound eyes. A **compound eye** contains many lenses, each associated with a light-sensing, columnar structure. (Human and cephalopod eyes are **simple eyes**, meaning they have just one lens.) Most arthropods also have a pair of antennae on the head. **Antennae** are long, tentacle-like appendages that contain specialized receptor cells that function in the senses of touch and smell.

Although the phylogeny of arthropods is still being worked out, most data sets agree that the phylum as a whole is monophyletic; that the myriapods (millipedes and centipedes), chelicerates (spiders), insects, and crustaceans represent four major lineages within the phylum; and that crustaceans and insects are closely related.

Let's take a closer look at the Nematoda—another particularly diverse and species-rich phylum within Ecdysozoa—and the four major lineages within Arthropoda:

- Ecdysozoans > Nematoda (Roundworms)

- Ecdysozoans > Arthropoda > Myriapods (Millipedes, Centipedes)

- Ecdysozoans > Arthropoda > Insecta (Insects)

- Ecdysozoans > Arthropoda > Chelicerata (Spiders, Ticks, Mites, Horseshoe Crabs, Daddy Longlegs, Scorpions)

- Ecdysozoans > Arthropoda > Crustaceans (Shrimp, Lobster, Crabs, Barnacles, Isopods, Copepods)

(a) Onychophorans are called velvet worms.

Peripatus species

0.5 cm

(b) Tardigrades are called water bears.

Echiniscus species

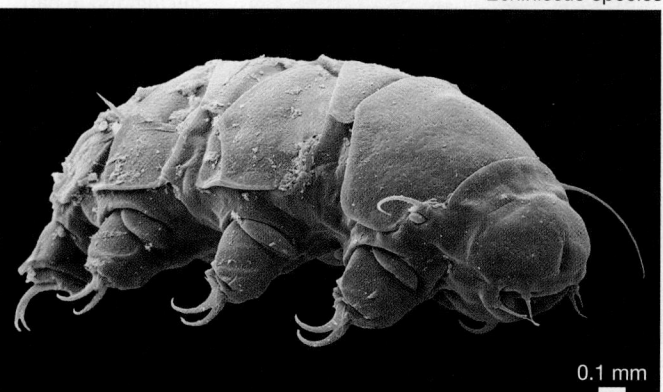

0.1 mm

FIGURE 33.18 Onychophorans and Tardigrades Have Lobe-Shaped Limbs with Claws. The Onychophora and Tardigrada are terrestrial—though tardigrades live in wet habitats. The scanning electron micrograph in part (b) has been colorized.

Species in the phylum Nematoda are commonly called **nematodes** or **roundworms** (**Figure 33.19**). Roundworms are unsegmented worms with a coelom, a tube-within-a-tube body plan, and no appendages. They have no muscles that can change the diameter of the body—their body musculature consists solely of longitudinal muscles that shorten or lengthen the body upon contracting or relaxing, respectively. ✔ You should be able to indicate the loss of circular muscles on Figure 33.2. Although some nematodes can grow to lengths of several meters, the vast majority of species are tiny—most are much less than 1 mm long. They lack specialized systems for exchanging gases and for circulating nutrients. Instead, gas exchange occurs across the body wall, and nutrients move by simple diffusion.

The nematode *Caenorhabditis elegans* is one of the most thoroughly studied model organisms in biology (see **BioSkills 14** in Appendix A). Species that parasitize humans have also been studied intensively. Pinworms, for example, infect about 40 million people in the United States alone, and *Onchocerca volvulus* causes an eye disease that infects around 20 million people in Africa and Latin America. Advanced infections by the roundworm *Wuchereria bancrofti* result in the blockage of lymphatic vessels, causing fluid accumulation and massive swelling—the condition known as elephantiasis. The disease trichinosis is caused by species in the genus *Trichinella* (**Figure 33.19a**). The parasites are passed from infected animals to humans by ingesting raw or undercooked meat. An array of symptoms results, including discomfort caused by *Trichinella* larvae encasing themselves in cysts within muscle or other types of tissue (**Figure 33.19b**).

Parasites represent a relatively small fraction of the 25,000 nematode species that have been described to date, however. The free-living nematodes are important ecologically because they are found in virtually every habitat known.

Nematodes are also fabulously abundant. Biologists have found 90,000 roundworms in a single rotting apple and have estimated that rich farm soils contain up to 9 billion roundworms per acre. Although they are not the most species-rich animal group, they are among the most abundant. The simple nematode body plan has been extraordinarily successful.

Feeding Roundworms feed on a wide variety of materials, including bacteria, fungi, plant roots, small protists or animals, or detritus. In most cases, the structure of their mouthparts is specialized in a way that increases the efficiency of feeding on a particular type of organism or material.

Movement Roundworms move with the aid of their hydrostatic skeleton. Most roundworms live in soil or inside a host, so when contractions of their longitudinal muscles cause them to wriggle, the movements are resisted by a stiff substrate. As a result, the worm pushes off the substrate, and the body moves.

Reproduction Sexes are separate in most nematode species. Sexual reproduction begins with internal fertilization and culminates in egg laying and the direct development of offspring. Individuals go through a series of four molts over the course of their lifetime.

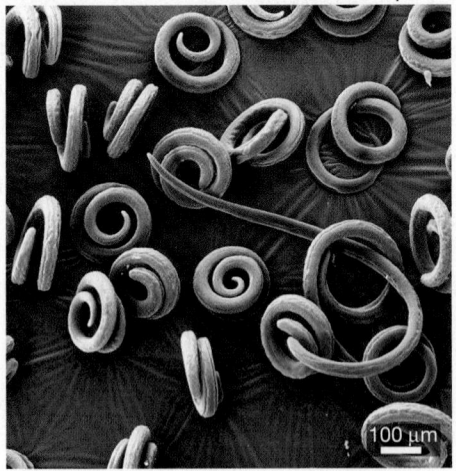

(a) A nematode causes the parasitic disease trichinosis.
Trichinella spiralis

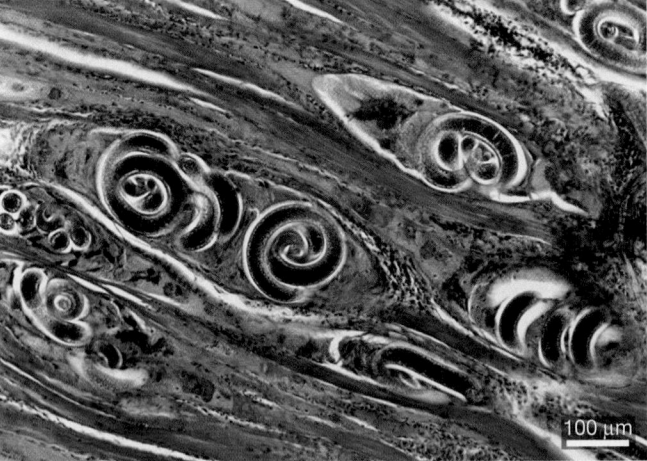

(b) *Trichinella spiralis* larvae encysted in muscle tissue

FIGURE 33.19 Most Nematodes Are Free Living, but Some Are Parasites. The scanning electron micrograph in part (a) has been colorized.

The **myriapods** have relatively simple bodies, with a head region and a long trunk featuring a series of segments, each bearing one or two pairs of legs (**Figure 33.20**). If eyes are present, they consist of a few to many simple structures clustered on the sides of the head. The 11,600 species that have been described to date inhabit terrestrial environments all over the world.

Feeding Millipedes and centipedes have mouthparts that can bite and chew. These organisms live in rotting logs and other types of dead plant material that litters the ground in forests and grasslands. Millipedes are detritivores. Centipedes, in contrast, use a pair of poison-containing fangs just behind the mouth to hunt an array of insects. Large centipedes can inject enough poison to debilitate a human.

Movement Myriapods walk or run on their many legs; a few species burrow. Some millipedes have over 190 trunk sections, each with two pairs of legs, for a total of over 750 legs. Centipedes usually have fewer than 30 segments, with one pair of legs per segment.

Reproduction Myriapod sexes are separate, and fertilization is internal. Males deposit sperm in packets that are picked up by the female or transferred to her by the male. After females lay eggs in the environment, the eggs hatch into juveniles that develop into adults via a series of molts.

Scolopendra species

FIGURE 33.20 Myriapods Have One or Two Pairs of Legs on Each Body Segment.

About 925,000 species of insects have been named thus far, but it is certain that many more exist. In terms of species diversity and numbers of individuals, insects dominate terrestrial environments. In addition, the larvae of some species are common in freshwater streams, ponds, and lakes. **Table 33.1** on pages 640–641 provides detailed notes about eight of the most prominent lineages, called orders, of insects. The insect *Drosophila melanogaster* is one of the most thoroughly studied model organisms in biology (see **BioSkills 14** in Appendix A).

Insects are distinguished by having three tagmata—the head, **thorax**, and **abdomen** (see Figure 33.5). Three pairs of walking legs are located on the lateral surface of the thorax. Most species have one or two pairs of wings, mounted on the dorsal (back) side of the thorax. Typically the head contains four mouthpart structures, a pair of antennae that are used to touch and smell, and a pair of compound eyes.

Feeding Most insect species have four mouthpart structures (labrum, paired mandibles, paired maxillae, and labium) which have diversified in structure and function in response to natural selection. As a result, insects are able to feed in every conceivable manner and on almost every type of food source available on land.

In species with holometabolous metamorphosis (see Chapter 32), larvae have wormlike bodies; most tunnel their way through food sources, though some feed by biting leaves. Adults, however, are predators or parasites on plant or animal tissues. Thus, diversification in food sources occurs even within species. Because so many insects make their living by feeding on plant tissues or fluids, the diversification of insects was closely correlated with the diversification of land plants. Insect predators usually eat other insects; insect parasites usually victimize other arthropods or mammals.

Movement Insects use their legs to walk, run, or swim, or they use their wings to fly. When insects walk or run, the sequence of movements usually results in three of their six legs maintaining contact with the ground at all times.

Reproduction Insect sexes are separate. Mating usually takes place through direct copulation, with the male inserting a sperm-transfer organ into the female. Most females lay eggs, but in a few species eggs are retained until hatching. Many species are also capable of reproducing asexually, through the production of unfertilized eggs via mitosis. In the vast majority of species, either incomplete or complete metamorphosis occurs, with complete metamorphosis being by far the most common.

TABLE 33.1 **Prominent Orders of Insects**

Order	Common Name	Number of Known Species	Description
Coleoptera ("sheath-winged")	Beetles	350,000	**Key traits:** Hardened forewings, called elytra, protect the membranous hindwings that power flight. During flight, the elytra are held to the side and act as stabilizers. **Feeding:** Adults are important predators and scavengers. Larvae are called grubs and often have chewing mouthparts. **Reproduction:** All have complete metamorphosis. **Notes:** The most species-rich lineage on the tree of life—about 25 percent of all described species are beetles. Range in length from 0.25 mm to 10 cm.
Lepidoptera ("scale-winged")	Butterflies, moths	180,000	**Key traits:** Wings are covered with tiny, often colorful scales. The forewings and hindwings hook together and move as a unit in many groups. **Feeding:** Larvae usually have chewing mouthparts and either bore through food material or bite leaves; adults often feed on nectar. **Reproduction:** All have complete metamorphosis. **Notes:** Some species migrate long distances.
Diptera ("two-winged")	Flies (including mosquitoes, gnats, midges)	120,000	**Key traits:** Reduced hindwings, called halteres, act as stabilizers during flight. **Feeding:** Adults are usually liquid feeders; larvae are called maggots and are often parasitic or feed on rotting material. **Reproduction:** All have complete metamorphosis. **Notes:** First flies in fossil record appear 240 million years ago.
Hymenoptera ("membrane-winged")	Ants, bees, wasps	115,000	**Key traits:** Membranous forewings and hindwings lock together via tiny hooks, thus acting as a single wing during flight. **Feeding:** Most ants feed on plant material; most bees feed on nectar; wasps are predatory and often have parasitic larvae. **Reproduction:** All have complete metamorphosis. Males are haploid (they hatch from unfertilized eggs) and females are diploid (they hatch from fertilized eggs). Females deposit eggs via an ovipositor, which in some species is modified into a stinger used in defense. Larvae are often fed and protected by adults. **Notes:** Most species live in colonies, and many are "eusocial"—meaning that some individuals in the colony help raise the queen's offspring but never reproduce themselves.

TABLE 33.1 **Prominent Orders of Insects** (continued)

Order	Common Name	Number of Known Species	Description
Hemiptera ("half-winged")	Bugs (including leaf hoppers, aphids, cicadas, scale insects)	85,000	**Key traits:** Some have a thickened forewing with a membranous tip; also mouthparts that are modified for piercing and sucking. **Feeding:** Most suck plant juices, but some are predatory. **Reproduction:** All have complete metamorphosis. **Notes:** Range in length from 1 mm to 11 cm.
Orthoptera ("straight-winged")	Grasshoppers, crickets	20,000	**Key traits:** Large, muscular hind legs power movement by jumping. **Feeding:** Most have chewing mouthparts and are leaf eaters. **Reproduction:** All have incomplete metamorphosis. In many species, males give distinctive songs to attract mates. **Notes:** Range in length from 5 mm to 11.5 cm.
Trichoptera ("hairy-winged")	Caddisflies	12,000	**Key traits:** Wings are covered with minute "hairs"—projections of the cuticle. **Feeding:** Adults have mouthparts adapted for sucking liquids, but have rarely been observed feeding; larvae are all aquatic and feed on detritus. Most larvae build a protective case of tiny stones and sticks. **Reproduction:** All have complete metamorphosis. **Notes:** Larvae and adults are important food sources for fish.
Odonata ("toothed")	Dragonflies, damselflies	6,500	**Key traits:** Four membranous wings and long, slender abdomens. **Feeding:** Larvae and adults are predatory, often on flies. The "odon" in their name refers to strong, toothlike structures on their mandibles. **Reproduction:** Metamorphosis is hemimetabolous: nymphs are aquatic, adults are terrestrial. **Notes:** They hunt by sight and have reduced antennae (used for touch and smell) but enormous eyes with up to 28,000 lenses.

Most of the 70,000 species of chelicerates are terrestrial, although the horseshoe crabs and sea spiders are marine. The most prominent lineage within Chelicerata is the Arachnida (spiders, scorpions, mites, and ticks).

The chelicerate body consists of anterior and posterior regions (**Figure 33.21a**). The anterior region lacks antennae for sensing touch or odor but usually contains eyes. The group is named for a pair of appendages called **chelicerae**, found near the mouth. Depending on the species, the chelicerae are used in feeding, defense, copulation, movement, or sensory reception. Chelicerates have a total of 6 pairs of appendages.

Feeding Spiders, scorpions, and daddy longlegs capture insects or other prey. Although some of these species are active hunters, most spiders are sit-and-wait predators. They create sticky webs to capture prey, which fly or walk into the web and are subsequently trapped. The spider senses the vibrations of the struggling prey, pounces on it, and administers a toxic bite. Mites and ticks are ectoparasitic and use their piercing mouthparts to feed on host animals, or in the case of dust mites, on their dead skin (**Figure 33.21b**). Horseshoe crabs eat a variety of animals as well as algae and detritus. Most scorpions feed on insects; the largest scorpion species occasionally eat snakes and lizards.

Movement Like other arthropods, chelicerates move with the aid of muscles attached to an exoskeleton. They walk or crawl on their four pairs of jointed walking legs; some species are also capable of jumping. Horseshoe crabs and some other marine forms can swim slowly. Newly hatched spiders spin long, silken threads that serve as balloons and that can carry them on the wind more than 400 kilometers from the point of hatching.

Reproduction Sexual reproduction is the rule in chelicerates, and fertilization is internal in most groups. Courtship displays are extensive in many arthropod groups and may include both visual displays and the release of chemical odorants. In spiders, males use appendages called pedipalps to transfer sperm to females. These organs fit into the female reproductive tract in a "lock-and-key" fashion. Differences in the size and shape of male genitalia are often the only way to identify closely related species of spider.

Development is direct, meaning that metamorphosis does not occur. In spiders, males may present a dead insect as a gift that the female eats as they mate; in some species, the male himself is eaten as sperm is being transferred. In scorpions, females retain fertilized eggs. After the eggs hatch, the young climb on the mother's back. They remain there until they are old enough to hunt for themselves.

(a) Spider, showing general chelicerate features

Holconia immanis

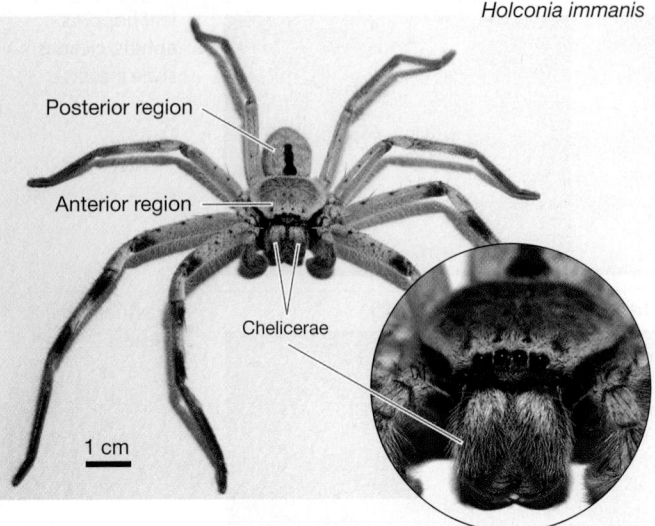

Posterior region

Anterior region

Chelicerae

1 cm

(b) Mites are ectoparasitic.

Dermatophagoides species

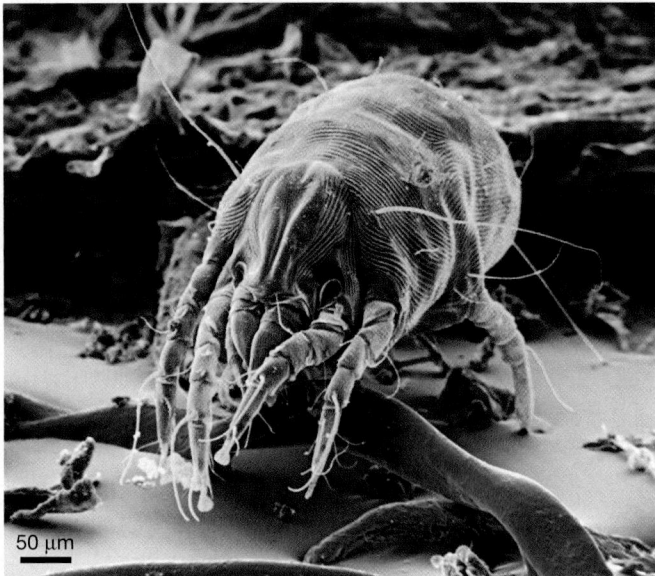

50 μm

FIGURE 33.21 Chelicerates Have Two Body Regions. The scanning electron micrograph in part (b) has been colorized.

The 67,000 species of **crustaceans** that have been identified to date live primarily in marine and freshwater environments. A few species of crab and some isopods are terrestrial, however. (Terrestrial isopods are known as sowbugs, pillbugs, or roly-polies). Crustaceans are common in surface waters, where they are important consumers. They are also important grazers and predators in shallow-water benthic environments.

The segmented body of most crustaceans is divided into two distinct regions: (**1**) the cephalothorax, which combines the head and thorax, and (**2**) the abdomen. Many crustaceans have a **carapace**—a platelike section of their exoskeleton that covers and protects the cephalothorax (**Figure 33.22a**). They are the only type of arthropod with two pairs of antennae, and they have sophisticated, compound eyes—usually mounted on stalks (**Figure 33.22b**). Crustaceans and trilobites—a major lineage that did not survive the end-Permian mass extinction, 250 million years ago (see Chapter 27)—were the first arthropods to appear in the fossil record.

Feeding   Most crustaceans have 4–6 pairs of mouthparts that are derived from jointed appendages, and as a group they use every type of feeding strategy known. Barnacles (**Figure 33.23**) and many shrimp are suspension feeders that use feathery structures located on head or body appendages to capture passing prey. Crabs and lobsters are active hunters, herbivores, and scavengers. Typically they have a pair of mouthparts called **mandibles** that bite or chew. Individuals capture and hold their food source with claws or other types of feeding appendages near their mouth and then use their mandibles to shred the food into small bits that can be ingested. As herbivores or detritivores, many species of crustaceans depend on algae for food.

Movement   Crustacean limbs are highly diverse: species have many pairs of limbs and it is common for a species to have more than one type of limb. Limb structures in crustaceans include paddle-shaped forms used in swimming, feathery structures used in capturing food that is suspended in water, and slender legs that make sophisticated walking or running movements possible. Barnacles are one of the few types of sessile crustaceans. Adult barna-

Lepas species

Legs

Shell

1 mm

FIGURE 33.23 Barnacles Stand on Their Heads and Kick Food into Their Mouths.

cles cement their heads to a rock or other hard substrate, secrete a protective shell of calcium carbonate, and use their legs to capture food particles and transfer them to the mouth.

Reproduction   Most individual crustaceans are either male or female, and sexual reproduction is the norm. Fertilization is usually internal, and eggs are usually retained by the female until they hatch. Most crustaceans pass through several distinct larval stages; many species include a larval stage called a **nauplius**, which is usually planktonic. A nauplius has a single eye and appendages that develop into the two pairs of antennae, the mouthparts, and swimming legs of the adult.

(a) Spiny lobster

Carapace

(b) Fiddler crab

Compound eyes on stalks

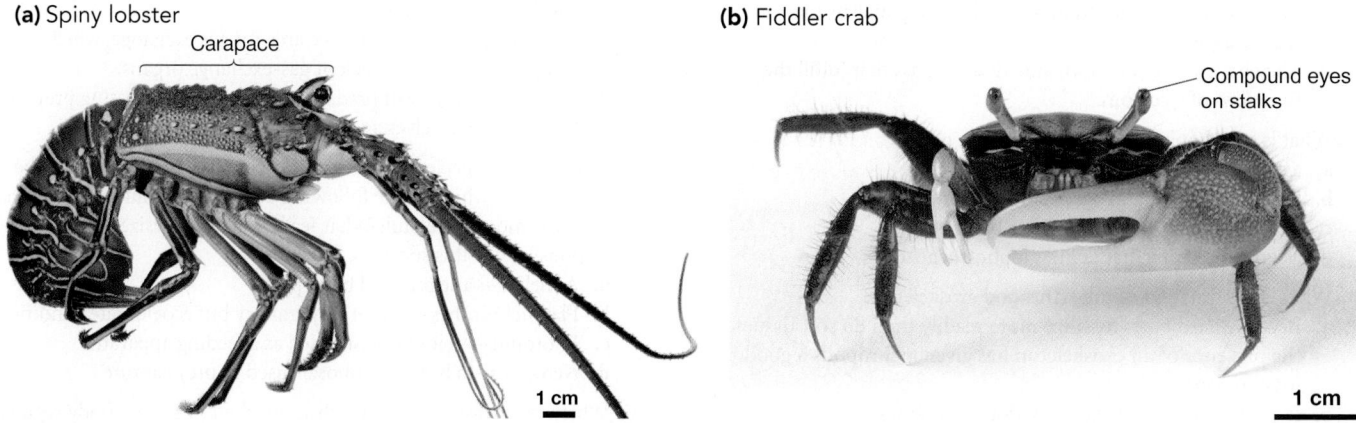

1 cm

1 cm

FIGURE 33.22 Crustaceans Are Named for the Hard, Calcified Exoskeltons of Lobsters and Crabs.

✔**EXERCISE** Circle the lobster's cephalothorax (combined head and thorax).

Summary of Key Concepts

◑━ Molecular phylogenies support the hypothesis that protostomes are a monophyletic group divided into two major subgroups: the Lophotrochozoa and the Ecdysozoa.

- The protostomes comprise about 20 phyla and were originally identified because their embryos undergo early development in the same way.

- Lophotrochozoans grow incrementally. Some phyla in the lineage also have characteristic feeding structures called lophophores; many have a distinctive type of larvae called trochophores.

- Species in the Ecdysozoa grow by molting—meaning that they shed their old external covering and grow a new, larger one.

  ✔ You should be able to explain the logic behind the claim that the protostome pattern of development is a synapomorphy that identifies a monophyletic group.

  MB **Web Activity** Protostome Diversity

◑━ Although the members of many protostome phyla have limbless, wormlike bodies and live in marine sediments, the most diverse and species-rich lineages—Mollusca and Arthropoda—have body plans with distinctive, complex features. Mollusks and arthropods inhabit a wide range of environments.

- All protostomes are bilaterally symmetric and triploblastic, but the nature of the body cavity is variable.

- The wormlike phyla of protostomes have well-developed coeloms that provide a hydrostatic skeleton used in locomotion. The coelom is absent in flatworms and reduced in Mollusca and Arthropoda.

- The mollusk's body includes a muscular foot, a visceral mass of organs, and a protective mantle.

- The arthropod body is segmented, divided into specialized tagmata such as the head, thorax, and abdomen of insects, and protected by an exoskeleton made primarily of chitin.

  ✔ You should be able to explain why flatworms lost a coelom, and why arthropods have a drastically reduced coelom.

◑━ Key events triggered the diversification of protostomes, including several lineages making the water-to-land transition, a diversification in appendages and mouthparts, and the evolution of metamorphosis in both marine and terrestrial forms.

- The transition to living on land occurred several times independently during the evolution of protostomes. It was facilitated by watertight shells and exoskeletons, the evolution of specialized structures to minimize water loss and manage gas exchange, and the ability to move to moist locations.

- Diversification in appendages and mouthparts gave protostomes the ability to move and find food in innovative ways.

- Complete metamorphosis allows offspring in sessile species to disperse to new habitats and enables larvae and adults to find food in different ways.

  ✔ You should be able to suggest two hypotheses for why complete metamorphosis is advantageous in clams.

Questions

1. Why is it logical to observe that mollusks have a drastically reduced coelom or no coelom at all?
   a. They evolved from flatworms, which also lack a coelom.
   b. They are the most advanced of all the protostomes.
   c. Their bodies are encased by a mantle, which may or may not secrete a shell.
   d. They have a muscular foot and visceral mass that fulfill the functions of a coelom.

2. What is a lophophore?
   a. a specialized filter-feeding structure
   b. the single opening in species with a blind gut
   c. a distinctive type of larva with a band of cilia
   d. a synapomorphy that defines lophotrochozoans

3. What is the function of the arthropod exoskeleton?
   a. Because hard parts fossilize more readily than do soft tissues, the presence of an exoskeleton has given arthropods a good fossil record.
   b. It has no well-established function. (Trilobites had an exoskeleton, and they went extinct.)
   c. It provides protection and functions in locomotion.
   d. It makes growth by molting possible.

4. Why is it logical that Platyhelminthes have flattened bodies?
   a. They have simple bodies and evolved early in the diversification of protostomes.
   b. They lack a coelom, so their body cannot form a rounded tube-within-a-tube design.
   c. A flat body provides a surface area for gas exchange, which compensates for their lack of gas-exchange organs.
   d. They are sit-and-wait predators that hide from passing prey by flattening themselves against the substrate.

5. In number of species, number of individuals, and duration in the fossil record, which of the following phyla of wormlike animals has been the most successful? What feature is hypothesized to be responsible for this success?
   a. Annelida—a segmented body plan
   b. Platyhelminthes—bilateral symmetry but acoelomate condition
   c. Phoronida—a lophophore used as a feeding apparatus
   d. Nemertea—a barbed proboscis used in prey capture

6. Which protostome phylum is distinguished by having body segments organized into tagmata?
   a. Mollusca   c. Annelida
   b. Arthropoda   d. Nematoda

1. Describe the traits that distinguish the Lophotrochozoa and Ecdysozoa. Describe the traits that are common to both groups.

2. Did segmentation evolve once or multiple times during the evolution of protostomes? Explain the logic that justifies your answer.

3. Why were water-to-land transitions an important aspect of protostome diversification?

4. Explain why the evolution of the exoskeleton was such an important event in the evolution of arthropods.

5. Suggest a hypothesis to explain why the evolution of the wing was such an important event in the evolution of insects.

6. All of the wormlike protostome phyla have similar body plans. But there are many different worm-like phyla. What factors are responsible for this diversification?

1. Recall that the phylum Platyhelminthes includes three major groups: the free-living turbellarians, the parasitic cestodes, and the parasitic trematodes. Because the parasitic forms are so simple morphologically, researchers suggest that they are derived from more complex, free-living forms. Draw a phylogeny of Platyhelminthes that would support the hypotheses that ancestral flatworms were morphologically complex, free-living organisms and that parasitism is a derived condition.

2. Brachiopoda is a phylum within the Lophotrochozoa. Even though they are not closely related to mollusks, brachiopods look and act like bivalve mollusks (clams or mussels). Specifically, brachiopods suspension feed, live inside two calcium carbonate shells that hinge together in some species, and attach to rocks or other hard surfaces on the ocean floor. How is it possible for brachiopods and bivalves to be so similar if they did not share a recent common ancestor?

3. The Mollusca includes a group of about 370 wormlike species called the Aplacophora. Although aplacophorans do not have a shell, their cuticle secretes calcium carbonate scales or spines. They lack a well-developed foot, and some species have a simple radula.

   - Predict where Aplacophora are found in the phylogenetic tree of Mollusca relative to bivalves, gastropods, chitons, and cephalopods. Explain your reasoning.

   - Predict where Aplacophora live and how they move. Explain your reasoning.

4. Consider the following: (a) The first arthropods to appear in the fossil record are marine crustaceans and trilobites, (b) terrestrial insects appeared relatively late in arthropod evolution, (c) although most crustaceans are aquatic, the crustacean groups called crabs and isopods each include both aquatic and terrestrial forms, and (d) analyses of DNA sequence data suggest that crustaceans and insects are closely related. Based on these observations, do you agree or disagree with the hypothesis that arthropods made the water-to-land transition more than once? Explain your reasoning.

In most habitats the "top predators"—animals that prey on other animals and aren't preyed upon themselves—are deuterostomes.

34 Deuterostome Animals

KEY CONCEPTS

○━ Echinoderms are radially symmetric as adults and have an endoskeleton and water vascular system. They are among the most important predators and herbivores in marine environments.

○━ All vertebrates have a skull and an extensive endoskeleton made of cartilage or bone; their diversification was driven in part by the evolution of the jaw and limbs. Vertebrates are the most important large-bodied predators and herbivores in marine and terrestrial environments.

○━ Humans are a tiny twig on the tree of life. Chimpanzees and humans diverged from a common ancestor that lived in Africa 6–7 million years ago. Since then, at least 14 humanlike species have existed.

The **deuterostomes** include the largest-bodied and some of the most morphologically complex of all animals. They range from the sea stars that cling to dock pilings, to the fish that dart in and out of coral reefs, to the wildebeests that migrate across the Serengeti Plains of East Africa.

The deuterostomes were initially grouped together because they all undergo early embryonic development in a similar way. When a humpback whale, sea urchin, or human is just beginning to grow, the gut starts developing from posterior to anterior—with the anus forming first and the mouth second. A coelom, if present, develops from outpocketings of mesoderm (see Chapter 32).

Today, most biologists recognize four phyla of deuterostomes: the Echinodermata, Hemichordata, Xenoturbellida, and Chordata (**Figure 34.1**).

- The echinoderms include the sea stars and sea urchins.

- The hemichordates, or "acorn worms," are probably unfamiliar to you—they burrow in marine sands or muds and make their living by deposit feeding or suspension feeding.

- A lone genus with two wormlike species, called *Xenoturbella,* was recognized as a distinct phylum in 2006.

- The chordates include the **vertebrates**, or animals with backbones. The vertebrates, in turn, comprise the hagfish, lampreys, sharks, bony fishes, amphibians, mammals, and reptiles (including birds).

Animals that are not vertebrates are collectively known as **invertebrates**.

Although deuterostomes share important features of embryonic development, their adult body plans and their feeding methods, modes of locomotion, and reproductive strategies are highly diverse. This chapter explores deuterostome diversity by introducing the echinoderms and chordates, and then taking a more in-depth look at the vertebrates and our own closest ancestors: the hominins.

✔ When you see this checkmark, stop and test yourself. Answers are available in Appendix B.

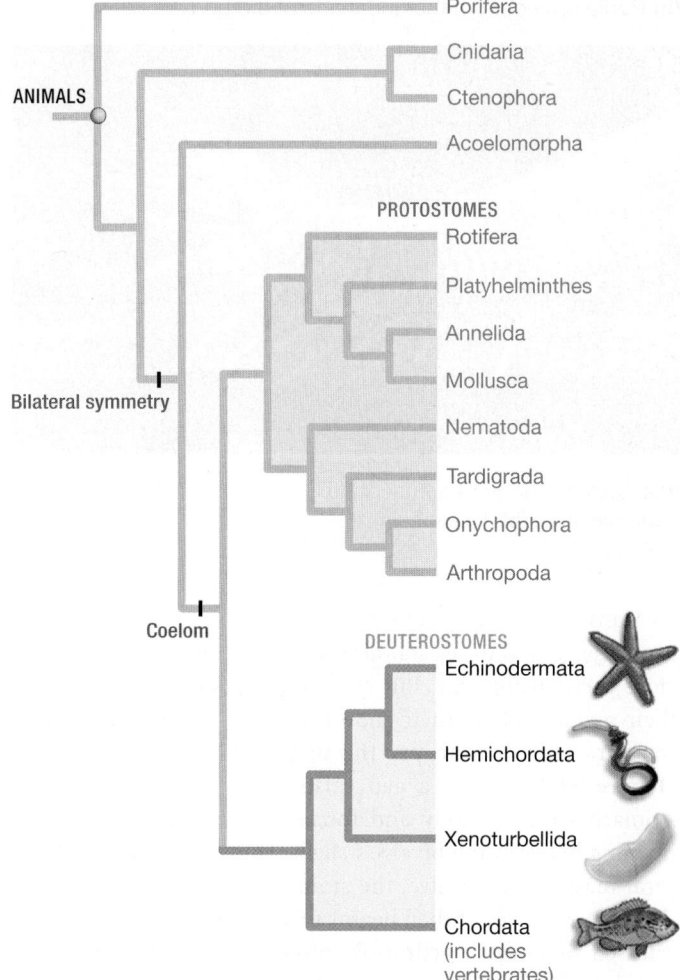

FIGURE 34.1 There Are Four Phyla of Deuterostomes. Vertebrates are in the phylum Chordata. Animals that are not vertebrates are collectively known as invertebrates. Over 95 percent of the known animal species are invertebrates.

✔**QUESTION** Are invertebrates monophyletic or paraphyletic?

34.1 What Is an Echinoderm?

The echinoderms (literally, "spiny-skins") were named for the spines or spikes observed in many species. Echinodermata is a large and diverse phylum but is exclusively marine.

In terms of numbers of species and range of habitats occupied, echinoderms are the second-most successful lineage of deuterostomes—next to the vertebrates. To date, biologists have cataloged about 7000 species of echinoderms.

Echinoderms are also abundant. In some deepwater environments, species in this phylum represent 95 percent of the total mass of organisms.

How are these animals put together?

The Echinoderm Body Plan

All deuterostomes are considered bilaterians, because they evolved from an ancestor that was bilaterally symmetric (see

(a) Echinoderm larvae are bilaterally symmetric.

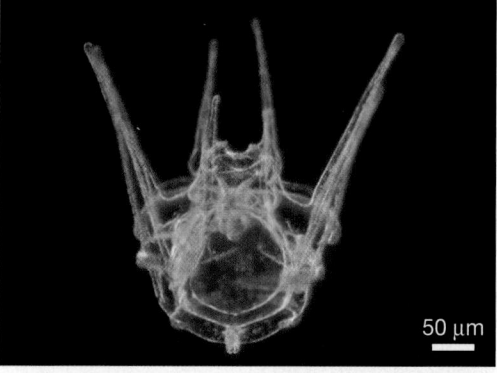

(b) Adult echinoderms are radially symmetric.

FIGURE 34.2 Body Symmetry Differs in Adult and Larval Echinoderms. Bilaterally symmetric sea urchin larvae **(a)** undergo metamorphosis and emerge as radially symmetric adults **(b)**.

Chapter 32). Echinoderm larvae are also bilaterally symmetric (**Figure 34.2a**). But a remarkable event occurred early in the evolution of **echinoderms**: the origin of five-sided radial symmetry, called pentaradial symmetry, in adults.

Recall from Chapter 32 that radially symmetric animals are not cephalized. As a result, they tend to interact with the environment in all directions at once instead of facing the environment in one direction. If adult echinoderms are capable of movement, they tend to move equally well in all directions.

A second noteworthy feature of the echinoderm body is its **endoskeleton**: a hard structure, located just inside a thin layer of epidermal tissue, that provides protection and support (**Figure 34.2b**). As an individual is developing, cells secrete plates of calcium carbonate inside the skin. Depending on the species involved, the plates may remain independent and result in a flexible structure or fuse into a rigid case. If the plates do not fuse, they are connected by an unusual tissue that can be stiff or flexible—depending on conditions.

Another remarkable event in echinoderm evolution was the origin of a unique morphological feature: a series of branching, fluid-filled tubes and chambers called the **water vascular system**. In some groups, one of the tubes is open to the exterior where it meets the body wall, so seawater can flow into and out of the system. Inside, fluids move via the beating of cilia that line the interior of the tubes and chambers.

(a) Echinoderms have a water vascular system.

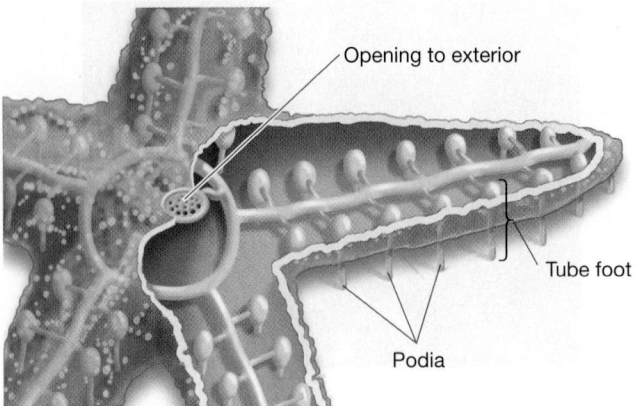

(b) Podia project from the underside of the body.

FIGURE 34.3 Echinoderms Have a Water Vascular System. (a) The water vascular system is a series of tubes and chambers that radiates throughout the body, forming a sophisticated hydrostatic skeleton. **(b)** Podia aid in movement because they extend from the body and can grab and release the substrate.

Figure 34.3a highlights a particularly important part of the water vascular system called tube feet. **Tube feet** are elongated, fluid-filled appendages. **Podia** (literally, "feet") are sections of the tube feet that project outside the body and make contact with the substrate (**Figure 34.3b**). The water vascular system forms a specialized hydrostatic skeleton that operates the podia. As podia extend and contract in a coordinated way along the base of an echinoderm, they alternately grab and release from the substrate. As a result, the individual moves.

🔑 Radial symmetry in adults, an endoskeleton of calcium carbonate, and the water vascular system are all synapomorphies—traits that identify echinoderms as a monophyletic group.

How Do Echinoderms Feed?

Depending on the lineage and species in question, echinoderms make their living by suspension feeding, deposit feeding, or harvesting algae or other animals. In most cases, an echinoderm's podia play a key role in obtaining food.

Many sea stars, for example, prey on bivalves. Clams and mussels respond to sea star attacks by contracting muscles that close their shells tight. But by clamping onto each shell with their podia, making their endoskeleton rigid, and pulling, sea stars are often able to pry the shells apart a few millimeters (**Figure 34.4a**). Once a gap exists, the sea star extrudes its stomach from its body and forces the stomach through the opening between the bivalve's shells. Upon contacting the visceral mass of the bivalve, the stomach of the sea star secretes digestive enzymes. It then begins to absorb the small molecules released by enzyme action. Eventually, only the shells of the prey remain.

Podia are also used in suspension feeding (**Figure 34.4b**). In most cases, podia are extended out into the water. When food particles contact them, the podia flick the food down to cilia, which sweep the particles toward the animal's mouth.

In deposit feeders, podia secrete mucus that is used to sop up food material on the substrate. The podia then roll the food-laden mucus into a ball and move it to the mouth.

(a) Sea star podia adhere to bivalve shells and pull them apart.

(b) Feather star podia trap particles during suspension feeding.

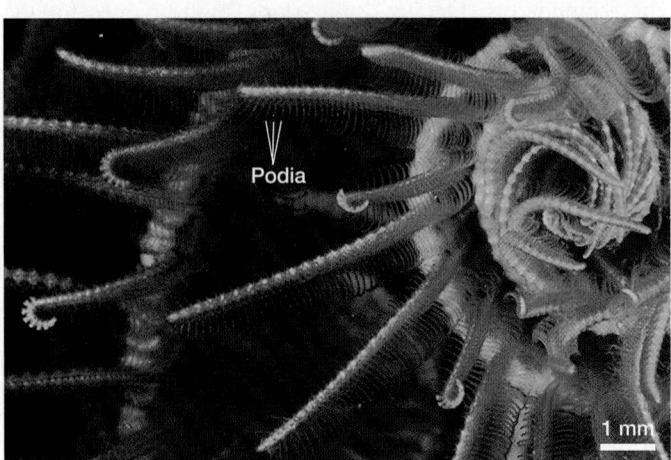

FIGURE 34.4 Echinoderms Use Their Podia in Feeding.

Key Lineages

The echinoderms living today make up five major lineages: (1) feather stars and sea lilies, (2) sea stars, (3) brittle stars and basket stars, (4) sea urchins and sand dollars, and (5) sea cucumbers (**Figure 34.5**). ✔You should be able to indicate the origin of pentaradial symmetry in adults, the water vascular system, and the echinoderm endoskeleton on Figure 34.5.

- Most feather stars and sea lilies are sessile suspension feeders.

- Brittle stars and basket stars have five or more long arms that radiate out from a small central disk. They use these arms to suspension feed, deposit feed by sopping up material with mucus, or capture small prey animals.

- Sea cucumbers are sausage-shaped animals that suspension feed or deposit feed with the aid of modified tube feet called tentacles that are arranged in a whorl around their mouths.

Sea stars and the sea urchins and sand dollars are described in detail in the boxes that follow.

▪ Echinodermata > Asteroidea (Sea Stars)

▪ Echinodermata > Echinoidea (Sea Urchins and Sand Dollars)

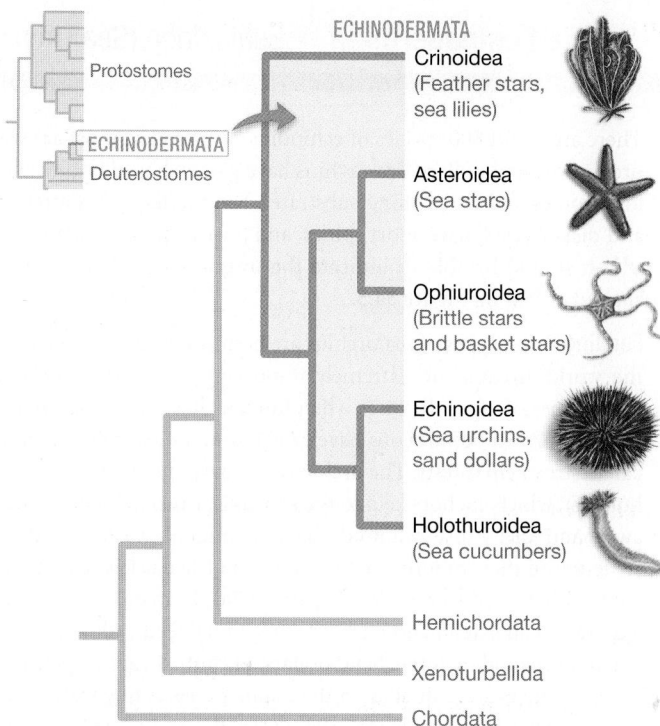

FIGURE 34.5 There Are Five Major Lineages of Echinoderms.

Echinodermata > Asteroidea (Sea Stars)

The 1700 known species of sea stars have bodies with five or more long arms—in some species up to 40—radiating from a central region that contains the mouth, stomach, and anus (**Figure 34.6**). Unlike brittle stars, though, the sea star's arms are not set off from the central region by clear, joint-like articulations. ✔You should be able to indicate the origin of arms, and the origin of arms that are continuous with the central region as in sea stars, on Figure 34.5.

When fully grown, sea stars can range in size from less than 1 cm in diameter to 1 m across. They live on hard or soft substrates along the coasts of all the world's oceans. Although the spines that are characteristic of some echinoderms are reduced to knobs on the surface of most sea stars, the crown-of-thorns star and a few other species have prominent, upright, movable spines.

Feeding Sea stars are predators or scavengers. Some species pull bivalves apart with their tube feet and feed by everting their stomach into the prey's visceral mass. In some cases, sea star predation on bivalves opens up space for other organisms—increasing biodiversity (see Chapter 53). Sponges, barnacles, and snails are also common prey. The crown-of-thorns sea star specializes in feeding on corals and is native to the Indian Ocean and western Pacific Ocean. Its population has skyrocketed recently—possibly because people are harvesting their major predator, a large snail called the triton, for its shell. Large crown-of-thorns star populations have led to the destruction of large areas of coral reef.

Movement Sea stars crawl with the aid of their tube feet.

Reproduction Sexual reproduction predominates in sea stars, and sexes are separate. At least one sea star arm is filled with reproductive organs that produce massive amounts of gametes—millions of eggs per female, in some species. Some species care for their offspring by holding fertilized eggs on their body until the eggs hatch. Larvae are long-lived and may spend weeks or months in the plankton, possibly dispersing thousands of kilometers. Most sea stars are capable of regenerating arms that are lost in predator attacks or storms. Some species can reproduce asexually by dividing the body in two, with each of the two individuals then regenerating the missing half.

Crossaster papposus

FIGURE 34.6 Sea Stars May Have Many Arms.

There are about 800 species of echinoids living today; most are sea urchins or sand dollars. Sea urchins have globe-shaped bodies and long spines and crawl along substrates. Sand dollars are flattened and disk-shaped, have short spines, and burrow in soft sediments. ✔You should be able to indicate the origin of globular or disc-shaped bodies on Figure 34.5.

Feeding Most types of sea urchins are herbivores. In some areas of the world, urchins are extremely important grazers on kelp and other types of algae. In cases when humans have removed predators and urchin populations have grown in response, their grazing can destroy kelp forests. The urchins chew away a structure called a holdfast, which anchors kelp to the substrate, causing them to float away and die. Most echinoids have a unique, jaw-like feeding structure in their mouths that is made up of five calcium carbonate teeth attached to muscles (**Figure 34.7a**). In many species, this apparatus can extend and retract as the animal feeds, allowing it to reach more food material. Sand dollars, in contrast, are suspension feeders. The beating of cilia on their spines creates tiny water currents which bring organic material toward the body. The food particles are then trapped in mucus and transported in long strings to the mouth. In this way, they harvest nutritious detritus found in sand or in other soft substrates (**Figure 34.7b**).

Movement Using their spines, sea urchins crawl and sand dollars burrow. Some species also use their podia to aid movement.

Reproduction Sexual reproduction predominates in sea urchins and sand dollars. Fertilization is external, and sexes are separate. As in sea stars, larvae are long-lived and may disperse long distances. The larvae of some species undergo asexual reproduction by dividing in two or budding off new individuals, before they undergo metamorphosis.

(a) Sea urchin
Lytechinus variegatus
Teeth at center of underside
5 mm

(b) Sand dollar
Echinarachnius parma
1 cm

FIGURE 34.7 Sea Urchins and Sand Dollars Are Closely Related.

CHECK YOUR UNDERSTANDING

🔑 If you understand that . . .

- Echinoderms have a distinctive body plan: pentaradial symmetry in adults, an endoskeleton, and a water vascular system.
- Most echinoderms use their podia to move, but they feed in a wide variety of ways—including using their podia to pry open bivalves, suspension feed, or deposit feed.

✔ You should be able to . . .

Suppose you were walking along the ocean and encountered an animal you had never seen before. Explain how you would figure out whether it was an echinoderm or not.

Answers are available in Appendix B.

34.2 What Is a Chordate?

The **chordates** are defined by the presence of four morphological features:

1. openings into the throat called **pharyngeal gill slits**;

2. a dorsal hollow **nerve cord** that runs the length of the body, comprised of projections from neurons;

3. a stiff and supportive but flexible rod called a **notochord**, which runs the length of the body; and

4. a muscular, post-anal tail—meaning a tail that contains muscle and extends past the anus.

Electrical signals carried by the dorsal hollow nerve cord coordinate muscle movement, and the notochord stiffens the muscular tail. Together, these traits create a "torpedo"—a long, streamlined animal that can swim forward rapidly. Because pharyngeal gill slits often function in suspension feeding, chordates are yet another illustration of a fundamental theme in animal evolution: diversification was driven by innovations that affect feeding and movement.

Three "Subphyla"

The phylum Chordata is made up of three major lineages: the (1) cephalochordates, (2) urochordates, and (3) vertebrates. All four defining characteristics of chordates—pharyngeal gill slits, dorsal hollow nerve cords, notochords, and muscular tails that extend past the anus—are found in these species at some stage in their life cycle.

THE CEPHALOCHORDATE BODY PLAN Cephalochordates are also called lancelets or amphioxus; they are small, mobile, torpedo-shaped animals with a "fish-like" appearance, and make their living by suspension feeding (**Figure 34.8a**). Adult cephalochordates live in ocean-bottom habitats, where they burrow in sand and suspension feed with the aid of their pharyngeal gill slits.

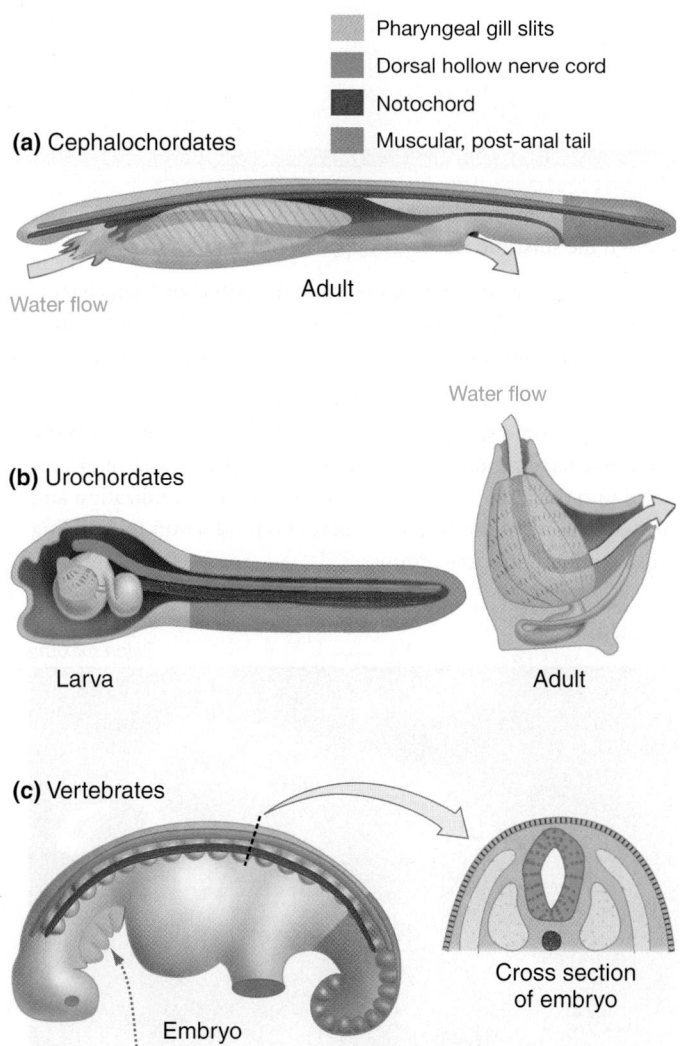

(a) Cephalochordates

- Pharyngeal gill slits
- Dorsal hollow nerve cord
- Notochord
- Muscular, post-anal tail

Water flow

Adult

Water flow

(b) Urochordates

Larva

Adult

(c) Vertebrates

Embryo

Cross section of embryo

Pharyngeal pouches become gill slits in aquatic vertebrates

FIGURE 34.8 Four Traits Distinguish the Chordates.

The cephalochordates also have a notochord that functions as an endoskeleton. Because it stiffens their bodies, muscle contractions on either side result in fishlike movement when they swim during dispersal or mating.

THE UROCHORDATE BODY PLAN | **Urochordates** are also called tunicates or sea squirts. As **Figure 34.8b** shows, pharyngeal gill slits are present in both larvae and adults and function in both feeding and gas exchange.

The dorsal hollow nerve cord, notochord, and tail are present in larvae and in the sexually mature forms of species that move as adults. Because the notochord stiffens the tail, muscular contractions on either side of a larva's tail wag it back and forth and result in swimming movements. As larvae swim or float in the upper water layers of the ocean, they drift to new habitats where food might be more abundant. As a result, larvae function as a dispersal stage in the life cycle.

THE VERTEBRATE BODY PLAN Vertebrates (**Figure 34.8c**) include the hagfish, lamprey, sharks, several lineages of fishes, amphibians, reptiles (including birds), and mammals. In vertebrates, the dorsal hollow nerve cord is the familiar spinal cord—a bundle of nerve cells that runs from the brain to the posterior of the body. Electrical signals from the spinal cord control movements and the function of organs (see Chapter 45).

Structures called pharyngeal pouches are homologous to pharyngeal gill slits and are present in all vertebrate embryos. In aquatic species, the creases between pouches open into gill slits and develop into part of the main gas-exchange organ—the **gills**. In terrestrial species, however, gill slits do not develop after the pharyngeal pouches form. In these species, the pharyngeal pouches are a vestigial trait (see Chapter 24).

A notochord appears in all vertebrate embryos. It no longer functions in body support and movement, however, as these roles are filled by the vertebral column. Instead, the notochard helps organize the body plan, early in development. You might recall from Chapter 22 that cells in the notochord secrete proteins that help induce the formation of somites—segmented blocks of tissue that form along the length of the body. Although the notochord itself disappears, cells in the somites later differentiate into the vertebrae, ribs, and skeletal muscles of the back, body wall, and limbs. In this way, the notochord is instrumental in the development of the trait that gave vertebrates their name.

Key Lineages: The Invertebrate Chordates

There are about 1600 species of tunicates, 24 species of lancelets, and over 50,000 vertebrates. Because they are so species-rich and diverse, vertebrates rate their own section in this chapter (Section 34.3). Here let's focus on the tunicates and lancelets. These lineages are sometimes referred to as the invertebrate chordates—meaning, members of the Chordata that lack vertebrae.

- Chordata > Cephalochordata (Lancelets)

- Chordata > Urochordata (Tunicates)

About two dozen species of cephalochordate have been described to date, all of them found in marine sands. Lancelets—also called amphioxus—have several characteristics that are intermediate between invertebrates and vertebrates. Chief among these is a notochord that is retained in adults, where it functions as an endoskeleton.

Feeding Adult cephalochordates feed by burrowing in sediment until only their head is sticking out into the water. They take water in through their mouth and trap food particles with the aid of mucus on their pharyngeal gill slits (**Figure 34.9**).

Movement Adults have large blocks of muscle arranged in a series along the length of the notochord. Lancelets are efficient swimmers because the flexible, rod-like notochord stiffens the body, making it wriggle when the blocks of muscle contract.

Reproduction Asexual reproduction is unknown, and individuals are either male or female. Gametes are released into the environment and fertilization is external.

Branchiostoma lanceolatum

FIGURE 34.9 Lancelets Look Like Fish but Are Not Vertebrates.

The urochordates are also called tunicates and are comprised of two major sub-lineages: the sea squirts (or ascidians) and salps. All of the species described to date live in the ocean. Sea squirts live on the ocean floor (**Figure 34.10a**); salps live in open water (**Figure 34.10b**).

The synapomorphies that define the urochordates include an exoskeleton-like coat of polysaccharide, called a tunic, that covers and supports the body; a U-shaped gut; and two body openings, called siphons, where water enters and leaves during feeding.

Feeding Adult urochordates use their pharyngeal gill slits to suspension feed. A mucous sheet on the inside of the pharynx traps particles present in the water that enters one siphon, passes through the slits, and leaves through the other siphon.

Movement Larvae swim with the aid of the notochord, which stiffens the body and functions as a simple endoskeleton. Larvae are a dispersal stage and do not feed. Adults are sessile or float in currents.

Reproduction In most species, individuals produce both sperm and eggs. In some species, both sperm and eggs are shed into the water and fertilization is external; in other species, sperm are released into the water but eggs are retained, so that fertilization and early development are internal. Asexual reproduction by budding is also common in some groups.

(a) Sea squirt

Rhopalaea crassa

(b) Salp

Salpa maxima

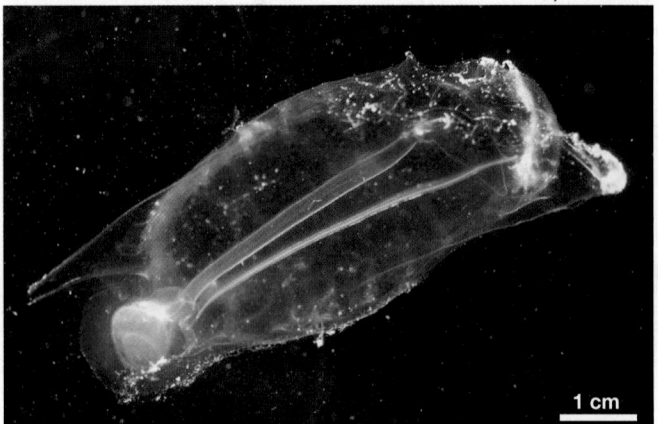

FIGURE 34.10 Sea Squirts and Salps Live in Different Habitats.

34.3 What Is a Vertebrate?

⊙━ The vertebrates are a monophyletic group distinguished by two synapomorphies:

- a column of cartilaginous or bony structures called **vertebrae**, which form along the dorsal sides of the body; and

- a **cranium**, or skull—a bony, cartilaginous, or fibrous case that encloses the brain.

The vertebral column is important because it protects the spinal cord. The cranium is important because it protects the brain along with sensory organs such as eyes. Together, the vertebrae and cranium protect the central nervous system (CNS) and key sensory structures (see Chapter 32).

The vertebrate brain develops as an outgrowth of the most anterior end of the dorsal hollow nerve cord, and is important to the vertebrate lifestyle. Vertebrates are active predators and herbivores that can make rapid, directed movements with the aid of their endoskeleton. Coordinated movements are possible in part because vertebrates have large brains divided into three distinct regions:

1. a **forebrain**, housing the sense of smell;

2. a **midbrain**, associated with vision; and

3. a **hindbrain**, responsible for balance and, in some species, hearing.

Part of the forebrain is elaborated into a large structure called the **cerebrum**. In the jawed vertebrates, or **gnathostomes** (pronounced *NATH-oh-stomes*) the hindbrain consists of enlarged regions called the **cerebellum** and **medulla oblongata**. (The structure and functions of the vertebrate brain are analyzed in detail in Chapter 45.) The evolution of a large, three-part brain, protected by a hard cranium, was a key innovation in vertebrate evolution.

An Overview of Vertebrate Evolution

Vertebrates have been the focus of intense research for well over 300 years, in part because they are large and conspicuous and in part because they include the humans. All this effort has paid off in an increasingly thorough understanding of how the vertebrates diversified. Let's consider what data from the fossil record have to say about key events in vertebrate evolution and then examine the current best estimate of the phylogenetic relationships among vertebrates.

THE VERTEBRATE FOSSIL RECORD Both echinoderm and vertebrate fossils are present in rocks that formed during the Cambrian explosion (see Chapter 27). The vertebrate fossils in these rocks show that the earliest members of this lineage lived in the ocean about 540 million years ago, had streamlined, fishlike bodies, and appear to have had a skull made of cartilage. **Cartilage** is a strong but flexible tissue—found in your earlobes and the tip of your nose—that consists of scattered cells in a gel-like matrix of polysaccharides and protein fibers.

A cartilaginous endoskeleton is a basic vertebrate feature; only in the bony fishes and their descendants, the **tetrapods** (literally, "four-footed"), does the skeleton become composed of

bone in the adult. **Bone** is a tissue consisting of cells and blood vessels encased in a matrix made primarily of a calcium- and phosphate-containing compound called hydroxyapatite, along with a small amount of protein fibers. In species with bony skeletons, growing bones are cartilaginous first and only become bony later in development.

Following the appearance of vertebrates, the fossil record documents a series of key innovations that occurred as this lineage diversified (**Figure 34.11**):

- *Bony exoskeleton* Fossil vertebrates from the early part of the Ordovician period, about 480 million years ago, are the first fossils to have bone. When bone first evolved, it did not occur in the endoskeleton. Instead, bone was deposited in scalelike plates that formed an exoskeleton—a hard structure that envelops the body. Most of your skull is made up of this type of dermal ("skin") bone—making your skull similar to the "head shields" in Ordovician fishes. Based on the fossils' overall morphology, biologists infer that these animals swam with the aid of a notochord and that they breathed and fed by gulping water and filtering it through their pharyngeal gill slits. Presumably, the bony plates helped provide protection from predators.

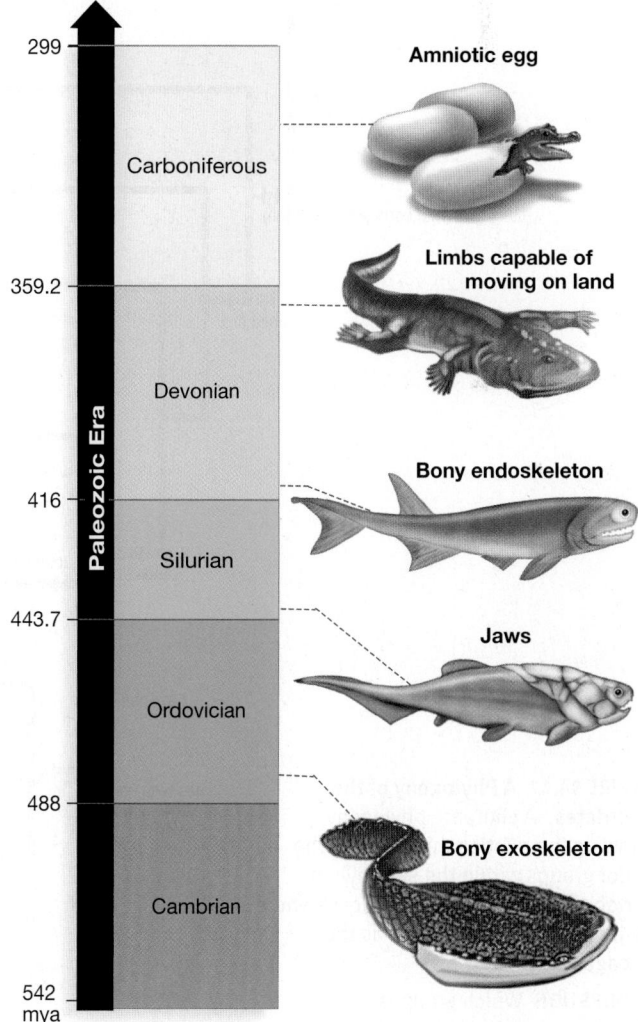

FIGURE 34.11 Timeline of the Early Vertebrate Fossil Record.

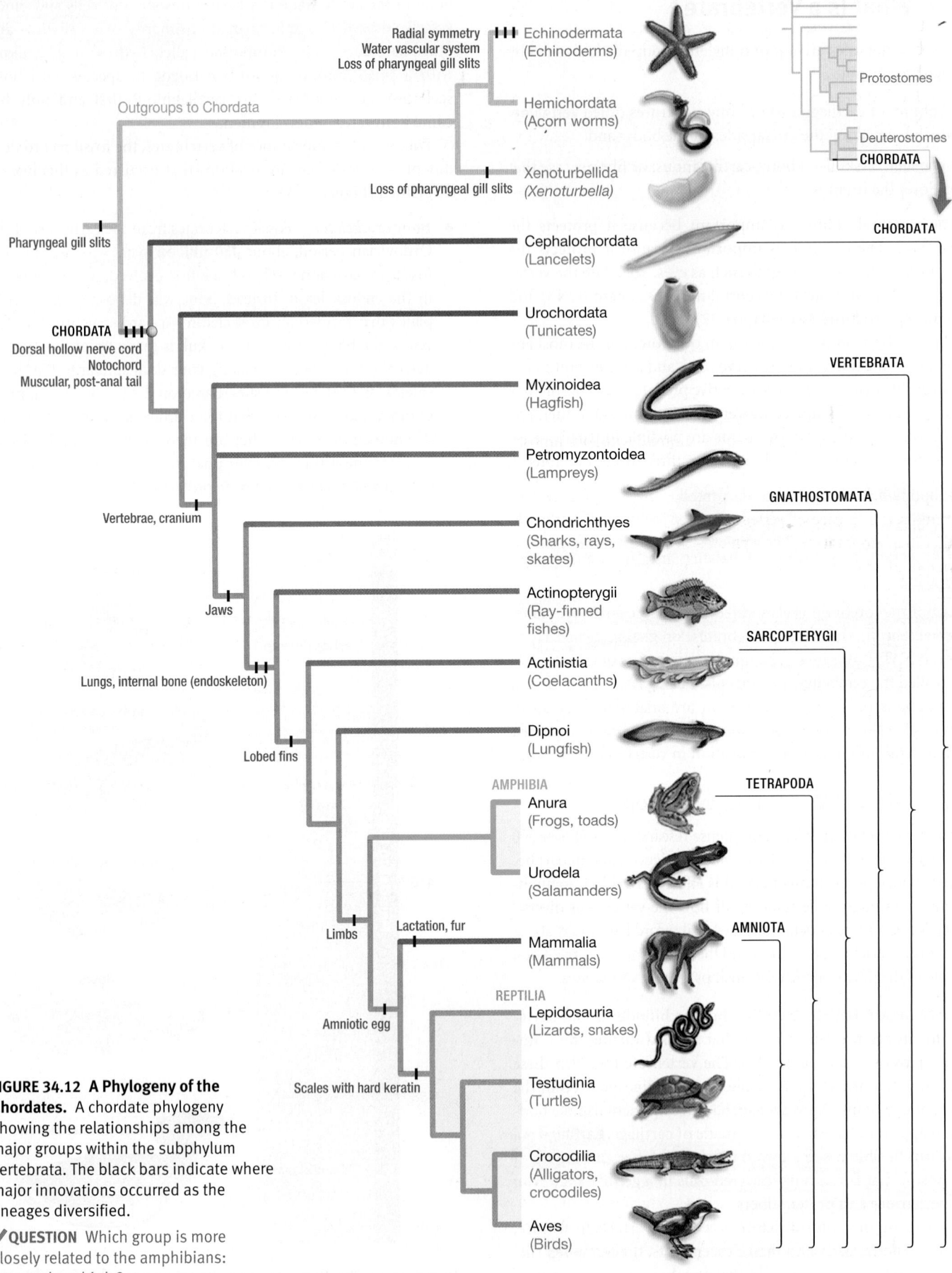

FIGURE 34.12 A Phylogeny of the Chordates. A chordate phylogeny showing the relationships among the major groups within the subphylum vertebrata. The black bars indicate where major innovations occurred as the lineages diversified.

✔**QUESTION** Which group is more closely related to the amphibians: mammals or birds?

- **Jaws** The first fish with jaws show up early in the Silurian, about 440 million years ago. The evolution of jaws was significant because it gave vertebrates the ability to bite, meaning that they were no longer limited to suspension feeding or deposit feeding. Instead, they could make a living as herbivores or predators. Soon after, jawbones with teeth appear in the fossil record. With jaws and teeth, vertebrates became armed and dangerous. The fossil record shows that a spectacular radiation of jawed fishes followed, filling marine and freshwater habitats.

- **Bony endoskeleton** In some lineages of fish found early in the Devonian period, the cartilaginous endoskeleton began to be stiffened by the deposition of bone. Unlike the dermal bone that had evolved earlier, the bony endoskeleton functioned to support movement—rapid swimming.

- **Limbs capable of moving on land** The next great event in the evolution of vertebrates was the transition to living on land. The first animals that had limbs and were capable of moving on land are dated to about 365 million years ago, late in the Devonian. These were the first of the tetrapods—animals with four limbs.

- **Amniotic egg** About 20 million years after the appearance of tetrapods in the fossil record, the first amniotes are present. The Amniota is a lineage of vertebrates that includes all tetrapods other than amphibians. The **amniotes** are named for a signature synapomorphy and adaptation: the amniotic egg. An **amniotic egg** is an egg that has membranes surrounding a food supply, a water supply, and a waste repository. These membranes provided support and extra surface area for gas exchange, allowing amniotes to produce larger, better-developed young.

To summarize, the fossil record indicates that vertebrates evolved through a series of major steps, beginning about 540 million years ago with vertebrates whose endoskeleton consisted of a notochord. The earliest vertebrates gave rise to cartilaginous fishes (sharks and rays) and several lineages of fish with bony skeletons. After the tetrapods emerged and amphibians resembling salamanders began to live on land, the evolution of the amniotic egg paved the way for the diversification of amniotes—specifically, reptiles and the animals that gave rise to mammals. Do phylogenetic trees estimated from analyses of DNA sequence data agree or conflict with these conclusions?

EVALUATING MOLECULAR PHYLOGENIES **Figure 34.12** provides a phylogenetic tree that summarizes the relationships among vertebrates, based on DNA sequence data. The labeled bars on the tree indicate where major innovations occurred. Although the phylogeny of deuterostomes continues to be a topic of intense research, researchers are increasingly confident that the relationships described in Figure 34.12 are accurate. Note that according to recent data, the tunicates are the closest living relatives of the vertebrates.

The overall conclusion from this tree is that the branching sequence inferred from morphological and molecular data correlates with the sequence of first appearances of groups in the fossil record. The sister groups to vertebrates lack the skull and vertebral column that define the vertebrates, and the sister groups to jawed vertebrates lack jaws and bony skeletons.

SEVERAL LINEAGES ARE FISHES The tree also indicates that there is no monophyletic group that includes all of the fishlike lineages. Instead, "fishy" organisms are a series of independent monophyletic groups. Taken together, they form what biologists call a **grade**—a sequence of lineages that are paraphyletic. You might recall from Chapter 30 that the non-vascular plants and seedless vascular plants also form grades and are paraphyletic.

✔ If you understand the difference between a paraphyletic and monophyletic group, you should be able to draw what Figure 34.12 would look like if there actually were a monophyletic group called "The Fish."

To understand what happened during the subsequent diversification of vertebrates, let's explore some of the major innovations involved in feeding, movement, and reproduction in more detail.

Key Innovations

Among vertebrates, the most species-rich and ecologically diverse lineages are the ray-finned fishes and the tetrapods. Ray-finned fishes occupy habitats ranging from deepwater environments, which are perpetually dark, to shallow ponds that dry up each year. Tetrapods include the large herbivores and predators in terrestrial environments all over the world. About 27,000 species of ray-finned fish and about 27,000 species of tetrapods are living today (**Figure 34.13**).

THE VERTEBRATE JAW The most ancient vertebrates in the fossil record have relatively simple mouthparts. Today's hagfish and lampreys retain this trait: They lack jaws and cannot bite algae, plants, or animals efficiently. They have to make their living as ectoparasites or as deposit feeders that scavenge dead animals. Vertebrates were not able to harvest food by biting until jaws evolved.

The leading hypothesis for the origin of the jaw proposes that natural selection acted on mutations that affected the morphology of **gill arches**, which are curved regions of tissue between the

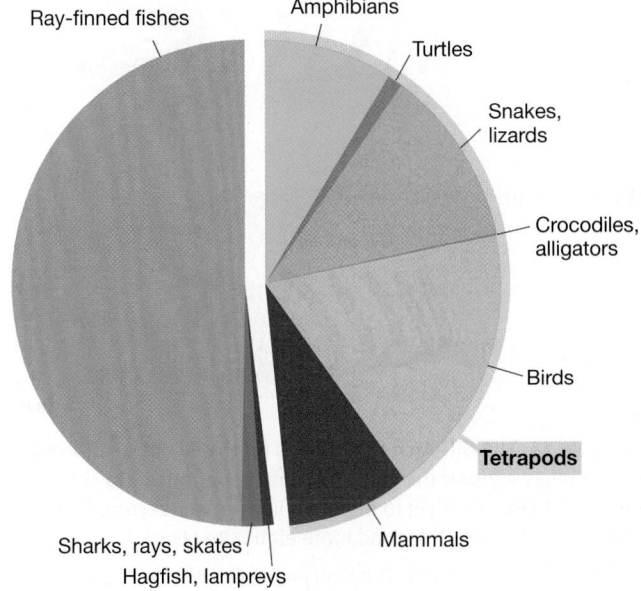

FIGURE 34.13 Relative Species Abundance among Vertebrates.

gills. The jawless vertebrates have bars of cartilage that stiffen these gill arches. The gill-arch hypothesis proposes that mutation and natural selection increased the size of the most anterior arch and modified its orientation slightly, producing the first working jaw (**Figure 34.14**). Three lines of evidence, drawn from comparative anatomy and embryology, support this hypothesis:

1. Both gill arches and jaws consist of flattened bars of bony or cartilaginous tissue that hinge and bend forward.

2. During development, the same population of cells gives rise to the muscles that move jaws and the muscles that move gill arches.

3. Unlike most other parts of the vertebrate skeleton, both jaws and gill arches are derived from specialized cells in the embryo called neural crest cells.

Taken together, these data support the hypothesis that jaws evolved from gill arches.

To explain why ray-finned fishes are so diverse in their feeding methods, biologists point to important modifications of the jaw:

- In most ray-finned fishes, the jaw is protrusible—meaning it can be extended to nip or bite out at food.

(a) Jawless vertebrate

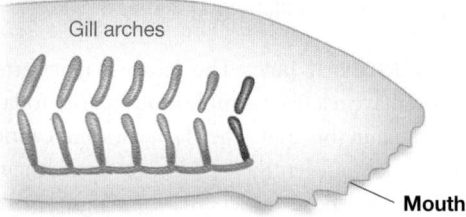

(b) Intermediate form (basal gnathostomes)

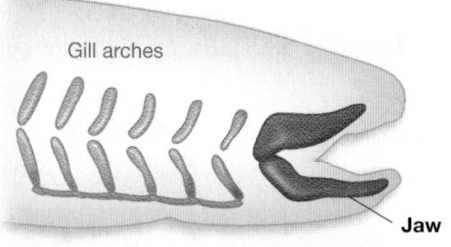

(c) Fossil shark

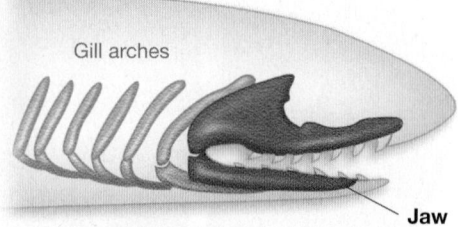

FIGURE 34.14 The Gill-Arch Hypothesis for the Evolution of the Jaw. **(a)** Gill arches support the gills in jawless vertebrates. **(b)** This intermediate form has yet to be found in the fossil record. **(c)** Fossil sharks that appeared later had more elaborate jaws.

✔**QUESTION** The transition from gill arches to jaws is complex. Would intermediate stages in the evolution of the jaw have any function?

- Several species-rich lineages of ray-finned fishes have a **pharyngeal** ("throat") **jaw**, consisting of modified gill arches, that functions as a second set of jaws. They are located in the back of the throat and make food processing particularly efficient.

(For more on the structure and function of pharyngeal jaws, see Chapter 43.)

To summarize, the radiation of ray-finned fishes was triggered in large part by the evolution of the jaw, by modifications that made it possible to protrude the jaw, and by the origin of the pharyngeal jaw. The story of tetrapods is different, however. Although jaw structure varies somewhat among tetrapod groups, the adaptation that triggered their initial diversification involved the ability to move and get to food, not to bite it and process it.

THE TETRAPOD LIMB To understand how tetrapods made the water-to-land transition, consider the morphology and behavior of their closest living relatives, the lungfish (**Figure 34.15**). Most living species of lungfish inhabit shallow, oxygen-poor water. To supplement the oxygen taken in by their gills, they have lungs and breathe air. (Note that many other fish have lungs, which function in gas exchange or maintaining buoyancy in water.) Some lungfish also have fleshy fins supported by bones and are capable of walking short distances along watery mudflats or the bottoms of ponds. In addition, some species can survive extended droughts by burrowing in mud.

Lungfish have a series of adaptations that allow them to survive on land for short periods. Could a lungfish-like organism have evolved into the first land-dwelling vertebrate with limbs?

Fossils provide strong links between the ancestors of today's lungfish and the earliest land-dwelling vertebrates. **Figure 34.16** shows a phylogeny of the species involved. Note that during the Devonian, the fossil record documents a series of species that indicate a gradual transition from a lobe-like fin to a limb that could support walking on land. The fossil record is actually more complete than shown here—the figure is just a sample of the tetrapod and tetrapod-like species known from this interval.

The figure highlights the number and arrangement of bones in the fossil fish fin and in early tetrapods. The color coding emphasizes that each fin or limb has a single bony element that is proximal (closest to the body; in blue) and then two bones that

FIGURE 34.15 Lungfish Have Limb-like Fins. Some species of lungfish can walk or crawl short distances on their fleshy fins.

are distal (farther from the body) and arranged side by side (these are shown in red), followed by a series of distal elements (in orange). Because the structures are similar, and because no other animal groups have limb bones in this arrangement, the evidence for homology is strong.

To summarize, data from the fossil record support the hypothesis that mutation and natural selection gradually transformed fins into limbs as the first tetrapods became more and more dependent on terrestrial habitats.

The hypothesis that tetrapod limbs evolved from fish fins has also been supported by molecular genetic evidence. Recent work has shown that several regulatory proteins involved in the development of cartilaginous fish fins and ray-finned fish fins and the upper parts of mammal limbs are homologous. For example, the proteins produced by several different *Hox* genes (see Chapter 21) are found at the same times and in the same locations in fins and limbs.

These data suggest that these appendages are patterned by the same genes. As a result, the data support the hypothesis that tetrapod limbs evolved from fins.

FEATHERS AND FLIGHT Once the tetrapod limb evolved, natural selection elaborated it into structures that are used for running, gliding, crawling, burrowing, or swimming. In addition, wings and the ability to fly evolved independently in three lineages of tetrapods: the extinct flying reptiles called pterosaurs (pronounced *TARE-oh-sors*), the bats, and the birds.

How did flight evolve? The best data sets on this question involve feathered flight in birds. In the early 2000s, for example, Xing Xu and colleagues announced the discovery of a spectacular series of feathered dinosaur fossils (**Figure 34.17**). The newly discovered species address key questions about the evolution of birds, feathers, and flight:

- *Did birds evolve from dinosaurs?* On the basis of skeletal characteristics, all of these recently discovered fossil species clearly belong to a lineage of dinosaurs called the dromaeosaurs. Birds are part of the monophyletic group called dinosaurs.

- *How did feathers evolve?* The early fossils have an array of feather types that support Xu's model of feathers evolving in a series of steps, beginning with simple projections from the skin and culminating with the complex structures observed in today's birds.

- *Did birds begin flying from the ground up or from the trees down?* More specifically, did flight evolve with running species that began to jump and glide or make short flights, with the aid of feathers to provide lift? Or did flight evolve from tree-dwelling species that used feathers to glide from tree to tree, much as flying squirrels do today? This question is still unresolved.

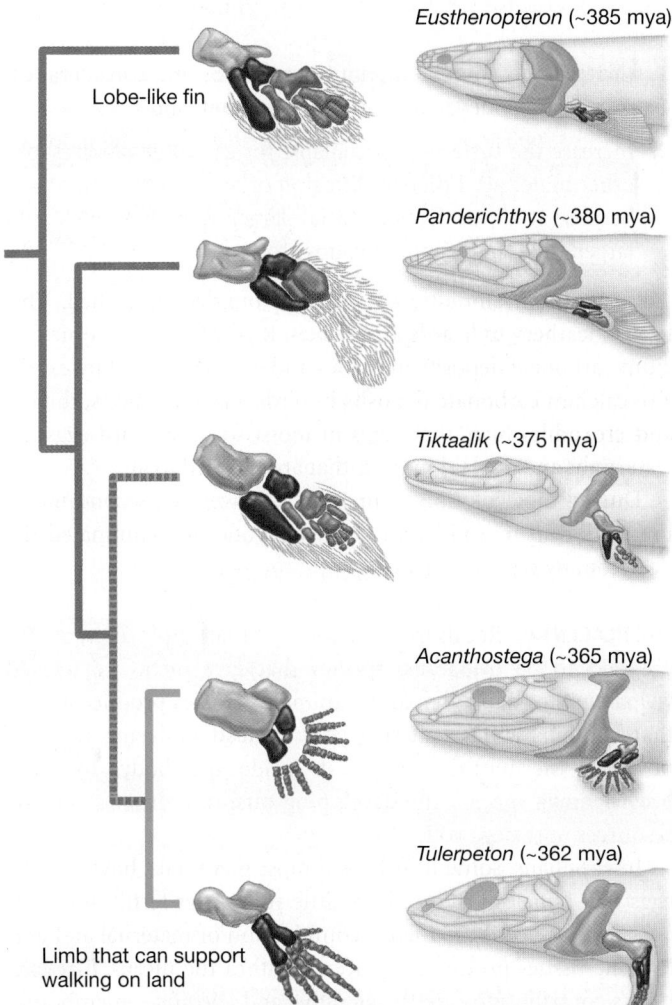

FIGURE 34.16 Evidence for a Fin-to-Limb Transition. The fossil record documents a series of species that indicate a gradual transition from a lobe-like fin to a limb that could support walking on land.

✔ **QUESTION** This phylogeny was estimated from traits other than the morphology of the limb. Why is this important, in terms of addressing how limbs evolved?

Labels in Figure 34.16:
Lobe-like fin
Eusthenopteron (~385 mya)
Panderichthys (~380 mya)
Tiktaalik (~375 mya)
Acanthostega (~365 mya)
Tulerpeton (~362 mya)
Limb that can support walking on land

FIGURE 34.17 Feathers Evolved in Dinosaurs. An artist's depiction of what *Caudipteryx,* a feathered dinosaur, might have looked like in life.

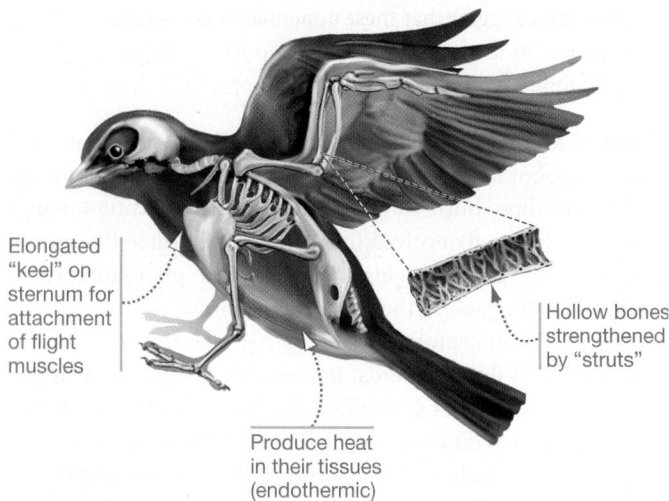

FIGURE 34.18 In Addition to Feathers, Birds Have Several Adaptations That Allow for Flight.

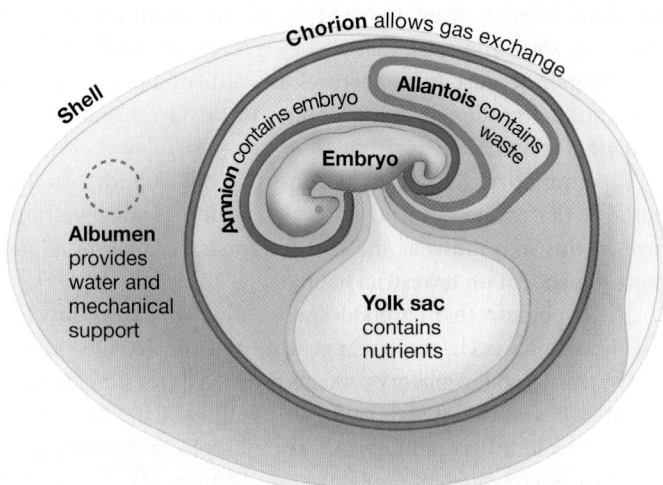

FIGURE 34.19 An Amniotic Egg. Amniotic eggs have four membrane-bound sacs.

Once dinosaurs evolved feathers and took to the air, the fossil record shows that a series of adaptations made powered, flapping flight increasingly efficient (**Figure 34.18**).

- Most dinosaurs have a flat sternum (breastbone), but the bird sternum has a projection called the keel, which provides a large surface area to which flight muscles attach.

- Birds are extraordinarily light for their size, primarily because they have a drastically reduced number of bones and because their larger bones are thin-walled and hollow—though strengthened by bony "struts."

- Birds are capable of sustained activity year-round because they are endothermic—they maintain a high body temperature by producing heat in their tissues.

From dinosaurs that jumped and glided or floated from tree to tree, birds have evolved into extraordinary flying machines.

To summarize, the evolution of the jaw gave tetrapods the potential to capture and process a wide array of foods. With limbs, they could move efficiently on land—or even fly—in search of food. What about the other major challenge of terrestrial life? How did tetrapods produce offspring that could survive out of water?

THE AMNIOTIC EGG Fish and amphibians (frogs, toads, salamanders, and limbless species called caecilians) produce eggs that have a single membrane and lay their eggs in water. In contrast, reptiles (including birds) and the egg-laying mammals produce an **amniotic egg**, which has an external membrane and three internal membranes. Species that produce amniotic eggs lay them outside of water.

As **Figure 34.19** shows, the outermost membrane of an amniotic egg encloses a supply of water in a protein-rich solution called **albumen**. Albumen cushions the developing embryo and provides nutrients. The inner membranes surround the embryo itself, the yolk provided by the mother, and waste from the embryo. The additional membranes are thought to be advantageous because they:

1. provide mechanical support—an important consideration outside of a buoyant aquatic environment; and

2. increase the surface area available for exchange of gases and other materials. Efficient diffusion of molecules is important because it allows females to lay larger eggs that hatch into larger, more independent young.

In addition, amniotic eggs are surrounded by a shell. This layer is leathery in lizards and snakes. It is stiffened by some calcium carbonate deposits in turtles and crocodiles, and by extensive calcium carbonate deposits in birds. Lizards, snakes, turtles, and crocodiles bury their eggs in moist soils, but bird eggs are watertight and are laid in nests that are exposed to air.

During the evolution of mammals, however, a second major innovation in reproduction occurred—one that eliminated the need for any type of egg laying.

THE PLACENTA Recall from Chapter 32 that egg-laying animals are said to be **oviparous**; species that give birth are termed **viviparous**. In many viviparous animals, females produce an egg that contains a nutrient-rich yolk. Instead of laying the egg, however, the mother retains it inside her body. In these **ovoviviparous** species, the developing offspring depends on the resources in the egg yolk.

In caecilians, some lizards, and most mammals, however, the eggs that females produce have little yolk. After fertilization occurs and the egg is retained, a combination of maternal and embryonic tissues produces a placenta within the uterus. The embryo's contributions are the allantois and chorion—membranes that are involved in gas exchange in an amniotic egg. In the placenta, tissues derived from the allantois and chorion are also involved in the diffusion of gases, nutrients, and wastes.

The **placenta** is an organ that is rich in blood vessels and that facilitates a flow of oxygen and nutrients from the mother to the developing offspring (**Figure 34.20**). After a development period called **gestation**, the embryo emerges from the mother's body.

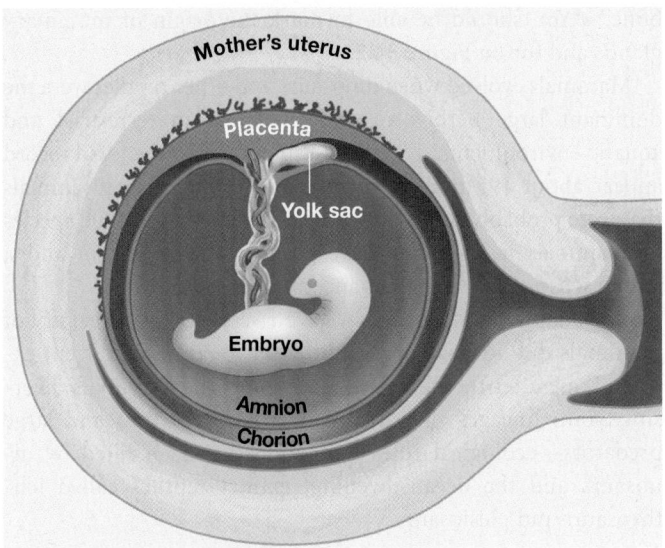

FIGURE 34.20 A Placenta Nourishes a Fetus Internally.

✔**QUESTION** Compare the relative size of the yolk sac here with that in the amniotic egg (Figure 34.19). Are they the same or different?

Why did viviparity and the placenta evolve? The leading hypotheses are that retaining the embryo inside the body has several advantages:

1. Offspring develop at a more constant, favorable temperature.

2. Offspring are protected.

3. Offspring are portable—mothers are not tied to a nest.

In effect, a placenta and viviparity are mechanisms for increasing the level of **parental care**: behavior by a parent that improves the ability of its offspring to survive. What other types of parental care evolved in vertebrates?

PARENTAL CARE   In species that provide extensive parental care—even if only in the form of exceptionally large amounts of yolk in eggs—mothers participate in a fitness trade-off (see Chapter 24 and Chapter 41). They can produce fewer offspring, but the offspring that they do produce have a higher probability of survival.

In animals, mechanisms of parental care include incubating eggs to keep them warm during early development, keeping young warm and dry, supplying young with food, and protecting them from danger. In some insect and frog species, mothers carry eggs or newly hatched young on their bodies. In fishes, parents commonly guard eggs during development and fan them with oxygen-rich water.

Mammals and birds provide particularly extensive parental care. In both groups, the mother (and often the father, in birds) continues to feed and care for individuals after birth or hatching—sometimes for many years (**Figure 34.21**). Female mammals also **lactate**—meaning that they produce milk and use it to feed their offspring after birth.

Among large animals, the evolution of extensive parental care is hypothesized to be a major reason for the evolutionary success of mammals and birds. In both lineages, mothers produce relatively small numbers of large, high-quality offspring.

(a) Mammal mothers feed and protect newborn young.

(b) Many bird species have extensive parental care.

FIGURE 34.21 Parental Care in Mammals and Birds. (a) Female mammals feed and protect embryos inside their bodies until the young are well developed. Once the offspring is born, the mother feeds it milk until it is able to eat on its own. In some species, parents continue to feed and protect young for years. **(b)** In birds, one or both parents may incubate the eggs, protect the nest, and feed the young after hatching occurs.

CHECK YOUR UNDERSTANDING

If you understand that . . .

- Vertebrates have a distinctive body plan: bilateral symmetry with vertebrae that protect a spinal cord and a cranium that protects the brain.
- An array of key innovations occurred during the evolution of vertebrates: Jaws made it possible to bite and process food, limbs allowed tetrapods to move on land, and amniotic eggs were advantageous in terrestrial environments.

✔ **You should be able to . . .**

1. Explain the adaptive significance of the jaw.

2. Explain why the additional membranes in an amniotic egg allowed eggs to be larger.

Answers are available in Appendix B.

Key Lineages

The fossil record provides an increasingly clear picture of early vertebrate evolution; intensive and continuing research has explored the origins of key vertebrate innovations such as the jaw, the tetrapod limb, and the amniotic egg. Now that you've been introduced to key vertebrate features, let's take a detailed look at specific vertebrate lineages, beginning with an overview of the mammals and reptiles.

MAMMALIA **Mammals** are a monophyletic group, today comprising three major lineages: (**1**) the monotremes, (**2**) marsupials, and (**3**) eutherians (**Figure 34.22**). Monotremes lay eggs; marsupials have a poorly developed placenta but well-developed pouch for rearing offspring; eutherians have a well-developed placenta and extended pregnancy.

Mammals are easily recognized by the presence of hair or fur, which serves to insulate the body. Like birds, mammals are endotherms that maintain high body temperatures by oxidizing large amounts of food and generating large amounts of heat. Instead of insulating themselves with feathers, though, mammals retain heat because the body surface is covered with layers of hair or fur.

Endothermy evolved independently in birds and mammals. In both groups, endothermy is thought to be an adaptation that enables individuals to maintain high levels of activity—particularly at night or during cold weather.

In addition to being endothermic and having fur, mammals have **mammary glands**—a unique structure that makes lactation possible. The evolution of mammary glands gave mammals the ability to provide their young with particularly extensive parental care. Mammals are also the only vertebrates with facial muscles and lips—traits that make suckling milk possible—and the only vertebrates that have a lower jaw formed from a single bone. ✔You should be able to mark the origin of mammary glands and fur on Figure 34.22.

Mammals evolved when dinosaurs and other reptiles were the dominant large herbivores and predators in terrestrial and aquatic environments. The earliest mammals in the fossil record appear about 195 million years ago; most were small animals that were probably active only at night. Many of the 4800 species of mammals living today have good nocturnal vision and a strong sense of smell, as their ancestors presumably did.

The adaptive radiation that gave rise to today's diversity of mammals did not take place until after the dinosaurs went extinct. Long after the dinosaurs were gone, the mammals diversified into lineages that included large herbivores and large predators—ecological roles that had once been filled by dinosaurs and the ocean-dwelling, extinct reptiles called ichthyosaurs and plesiosaurs.

REPTILIA The **reptiles** are a monophyletic group and represent one of the two major living lineages of amniotes—the other lineage consists of the extinct mammal-like reptiles and today's mammals. The major feature distinguishing the reptilian and mammalian lineages is the number and placement of openings in the side of the skull. These skull openings are important: Jaw muscles required for biting and chewing pass through them and attach to bones on the upper part of the skull.

Several features adapt reptiles for life on land. Their skin is made watertight by a layer of scales made of the protein keratin. Reptiles breathe air through well-developed lungs and lay shelled, amniotic eggs. In snakes and lizards, the egg has a leathery shell; in other reptiles, the shell includes some calcium carbonate.

In many reptiles, the sex of an individual is determined by the environment it experiences during early development. In certain species, for example, high temperatures produce mostly males while low temperatures produce mostly females.

The reptiles include the dinosaurs, pterosaurs (flying reptiles), and other extinct lineages that flourished from about 250 million years ago until the mass extinction at the end of the Cretaceous period, 65 million years ago. Today the Reptilia are represented by four major lineages: (**1**) lizards and snakes, (**2**) turtles, (**3**) crocodiles and alligators, and (**4**) birds.

Except for birds, almost all of the reptiles living today are **ectothermic** ("outside-heated")—meaning that individuals do not use internally generated heat to regulate their body temperature. It would be a mistake, however, to conclude that reptiles other than birds do not regulate their body temperature closely. Reptiles bask in sunlight, seek shade, and perform other behaviors to keep their body temperature at a preferred level.

To review the synapomorphies that biologists use to identify the major deuterostome and vertebrate groups, go to the study area at *www.masteringbiology.com*.

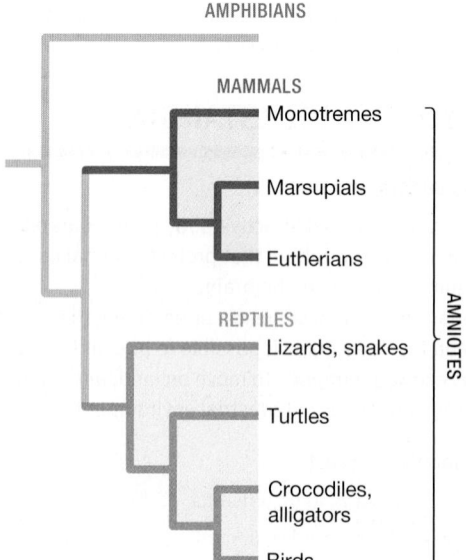

FIGURE 34.22 Mammals and Reptiles Are Monophyletic Groups.

 Web Activity Deuterostome Diversity

Now let's delve into a more detailed look at an array of fish lineages and major groups within Mammalia and Reptilia. All of the following groups are members of the Chordata and Vertebrata.

- Myxinoidea (Hagfish) and Petromyzontoidea (Lampreys)
- Chondrichthyes (Sharks, Rays, Skates)
- Actinopterygii (Ray-Finned Fishes)
- Actinistia (Coelacanths) and Dipnoi (Lungfish)
- Amphibia (Frogs, Salamanders, Caecilians)
- Mammalia > Monotremata (Platypuses, Echidnas)
- Mammalia > Marsupiala (Marsupials)
- Mammalia > Eutheria (Placental Mammals)
- Reptilia > Lepidosauria (Lizards, Snakes)
- Reptilia > Testudinia (Turtles)
- Reptilia > Crocodilia (Crocodiles, Alligators)
- Reptilia > Aves (Birds)

Chordata > Vertebrata > Myxinoidea (Hagfish) and Petromyzontoidea (Lampreys)

Although recent phylogenetic analyses indicate that hagfish and lampreys may belong to two independent lineages, some data suggest that they are a single group called the Cyclostomata ("round-mouthed"). The issue is not resolved, and will only be clarified with additional research. Because these animals are the only vertebrates that lack jaws, the 110 species in the two groups are still referred to as the jawless fishes.

The hagfish and lampreys are the only surviving members of the earliest branches at the base of the Vertebrata. Hagfish and lampreys have long, slender bodies and are aquatic. Hagfish are strictly marine; lampreys have both marine and freshwater forms. Most species are less than a meter long when fully grown. Adult hagfish lack any sort of vertebral column, but adult lampreys have small pieces of cartilage along the length of their dorsal hollow nerve cord. Both hagfish and lampreys have brains protected by a cranium, as do all vertebrates.

Feeding Hagfish are scavengers and predators (**Figure 34.23a**). They deposit feed on the carcasses of dead fish and whales, and some are thought to burrow through ooze at the bottom of the ocean, feeding on polychaete worms and other buried prey. Lampreys, in contrast, are ectoparasites. They attach to fish or other hosts by suction, and then use spines in their mouth and tongue to rasp a hole in the side of their victim (**Figure 34.23b**). Once the wound is open, they suck blood and other body fluids.

Movement Hagfish and lampreys have a well-developed notochord and swim by making undulating movements. Lampreys can also move themselves upstream, against the flow of water, by attaching their suckers to rocks and looping the rest of their body forward, like an inchworm. Both groups have tail fins and lamprey have a dorsal fin, but neither have the paired lateral appendages—meaning fins or limbs that emerge from each side—found in other vertebrates.

Reproduction Virtually nothing is known about hagfish mating or embryonic development. Some lampreys live in freshwater; others are **anadromous**—meaning they spend their adult life in the ocean, but swim up streams to breed. Fertilization is external, and adults die after breeding once. Lamprey eggs hatch into larvae that look and act like lancelets. The larvae burrow into sediments and suspension feed for several years before metamorphosing into free-swimming adults.

(a) Hagfish

Eptatretus stoutii

(b) Lampreys feeding on fish

Petromyzon marinus

FIGURE 34.23 Hagfish and Lampreys Are Jawless Vertebrates.

The 970 species in this lineage are distinguished by a specialized type of reinforced cartilaginous skeleton (*chondrus* is the Greek word for cartilage). Along with jawed vertebrates, or gnathostomes (pronounced *NATH-oh-stomes*), sharks also have jaws. They, like the other fish-like gnathostomes, also have paired fins. Paired fins were an important evolutionary innovation because they stabilize the body during rapid swimming—keeping it from pitching up or down, yawing to one side or the other, or rolling. ✔You should be able to indicate the origin of paired fins and the chondrichthyan form of cartilage on Figure 34.12.

Most sharks, rays, and skates are marine, though a few species live in freshwater. Sharks have streamlined, torpedo-shaped bodies and an asymmetrical tail—the dorsal portion is longer than the ventral portion (**Figure 34.24a**). In contrast, the dorsal–ventral plane of the body in rays and skates is strongly flattened (**Figure 34.24b**).

Feeding A few species of ray and shark suspension feed on plankton, but most species in this lineage are predators. Skates and rays lie on the ocean floor and ambush passing animals; electric rays capture their prey by stunning them with electric discharges of up to 200 volts. Most sharks, in contrast, are active hunters that chase down prey in open water and bite them. The larger species of shark feed on large fish or marine mammals. Sharks are referred to as the "top predator" in many marine ecosystems, because they are at the top of the food chain—nothing eats them. Yet the largest of all sharks, the whale shark, is a suspension feeder. Whale sharks filter plankton out of water as it passes over their gills.

Movement Rays and skates swim by flapping their greatly enlarged pectoral fins. (Pectoral fins are located on the sides of an organism; dorsal fins are located on the dorsal surface.) Sharks swim by undulating their bodies from side to side and beating their large tails.

Reproduction Sharks use internal fertilization, and fertilized eggs may be shed into the water or retained until the young are hatched and well developed. In some viviparous species, embryos are attached to the mother by specialized tissues in a placenta, where the exchange of gases, nutrients, and wastes takes place. (A placenta evolved in certain shark lineages independently of its evolution elsewhere on the tree of life.) Skates are oviparous, but rays are viviparous.

(a) Sharks are torpedo shaped.

Prionace glauca

Asymmetrical tail

Dorsal fin

Pectoral fin

(b) Skates and rays are flat.

Dasyatis americana

Pectoral fin

FIGURE 34.24 Sharks and Rays Have Cartilaginous Skeletons.

Chordata > Vertebrata > Actinopterygii (Ray-Finned Fishes)

Actinopterygii (pronounced *ack-tin-op-teh-RIJ-ee-i*) means "ray-finned." Logically enough, these fish have fins that are supported by long, bony rods arranged in a ray pattern, in addition to a skeleton made of bone. Their bodies are covered with interlocking scales that provide a stiff but flexible covering, and many have a gas-filled **swim bladder**, which evolved from the lungs found in the earliest lineages of fish. The evolution of the swim bladder was an important innovation because it allowed ray-finned fishes to avoid sinking. Tissues are heavier than water, so the bodies of aquatic organisms tend to sink. Sharks and rays, for example, have to swim to avoid sinking. But ray-finned fishes have a bladder that changes in volume, depending on the individual's position. Gas is added to the bladder when a ray-finned fish swims down; gas is removed when the fish swims up. In this way, ray-finned fishes maintain neutral buoyancy in water of various depths and thus various pressures. ✔You should be able to indicate the origin of rayed fins and the swim bladder on Figure 34.12.

The actinopterygians are the most successful vertebrate lineage based on number of species, duration in the fossil record, and extent of habitats occupied. Almost 27,000 species of ray-finned

(Continued on next page)

fishes are known, including the smallest known vertebrate—a species where adult females average less than 8 mm in length.

The most important major lineage of ray-finned fishes is the Teleostei. About 96 percent of all living fish species, including familiar groups like the tuna, trout, cod, and goldfish, are teleosts (**Figure 34.25**). The teleosts underwent an adaptive radiation about the same time that mammals did.

Feeding Teleosts can suck food toward their mouths, grasp it with their protrusible jaws, and then process it with teeth on their jaws and with pharyngeal jaws in their throat. The size and shape of the mouth, the jaw teeth, and the pharyngeal jaw teeth all correlate with the type of food consumed. For example, most predatory teleosts have long, spear-shaped jaws armed with spiky teeth, as well as bladelike teeth on their pharyngeal jaws. Besides being major predators, ray-finned fishes are the most important large herbivores in both marine and freshwater environments.

Movement Like other fish, ray-finned fishes swim by alternately contracting muscles on the left and right sides of their bodies from head to tail, resulting in rapid, side-to-side undulations. Their bodies are streamlined to reduce drag in water. Teleosts have a flexible, symmetrical tail, which reduces the need to use their pectoral (side) fins as steering and stabilizing devices during rapid swimming.

Reproduction Most ray-finned fish species rely on external fertilization and are oviparous; some species have internal fertilization with external development; still others have internal fertilization and are

Holocentrus rufus

Bony rods in fin

FIGURE 34.25 In Ray-Finned Fishes, Fins Are Supported by Long, Bony Rods.

viviparous. Although it is common for fish eggs to be released in the water and left to develop on their own, parental care occurs in some species. Parents may carry fertilized eggs on their fins, in their mouth, or in specialized pouches to guard them until the eggs hatch. In other cases the eggs are laid into a nest that is actively guarded and cared for. In freshwater teleosts, offspring develop directly; but marine species have larvae that are very different from adult forms. As they develop, marine fish larvae undergo a metamorphosis to the juvenile form, which then grows into an adult.

Chordata > **Vertebrata** > Actinistia (Coelacanths) and Dipnoi (Lungfish)

Although coelacanths (pronounced *SEEL-uh-kanths*) and lungfish represent independent lineages, they are sometimes grouped together and called **lobe-finned fishes**. Lobe-finned fishes are common and diverse in the fossil record in the Devonian period, about 400 million years ago, but only eight species are living today. They are important, however, because they represent a crucial evolutionary link between the fishes and the tetrapods. Early fish in the fossil record have fins supported by stiff structures that extend from a base of bone. In ray-finned fish, the bony elements are reduced. But in lobefins, the bony elements extend down the fin, and branch—similar to the bony elements in the limbs of tetrapods (**Figure 34.26**). ✔You should be able to indicate the origin of extensive fin bones on Figure 34.12.

Coelacanths are marine and occupy habitats 150–700 m below the surface. In contrast, lungfish live in shallow, freshwater ponds and rivers (see Figure 34.15). As their name implies, lungfish have lungs and breathe air when oxygen levels in their habitats drop. Some species burrow in mud and enter a quiescent, sleeplike state when their habitat dries up during each year's dry season.

Feeding Coelacanths prey on fish. Lungfish are **omnivorous** ("all-eating"), meaning that they eat algae and plant material as well as animals.

Movement Coelacanths and lungfish swim by waving their bodies; some lungfish can also use their fins to walk along pond bottoms.

Reproduction Sexual reproduction is the rule, with fertilization internal in coelacanths and external in lungfish. Coelacanths are ovoviviparous; lungfish lay eggs. Lungfish eggs hatch into larvae that resemble juvenile salamanders.

Latimeria menadoensis

Fleshy lobes supported by bones

FIGURE 34.26 Coelacanths Are Lobe-Finned Fishes.

Amphibians are found throughout the world and occupy ponds, lakes, or moist terrestrial environments (**Figure 34.27a**). Translated literally, their name means "both-sides-living." The name is appropriate because adults of most species of amphibian feed on land but lay their eggs in water. In many species of amphibians, gas exchange occurs exclusively or in part across their moist, mucus-covered skin. ✔ You should be able to indicate the origin of "skin-breathing" on Figure 34.12. The 5500 species of **amphibians** living today form three distinct clades: (**1**) frogs and toads, (**2**) salamanders, and (**3**) caecilians (pronounced *suh-SILL-ee-uns*).

Feeding Adult amphibians are carnivores.

- Most frogs are sit-and-wait predators that use their long, extendable tongues to capture passing prey.

- Salamanders also have an extensible tongue, which some species use in feeding.

- Terrestrial caecilians prey on earthworms and other soil-dwelling animals; aquatic forms eat vertebrates and small fish.

Movement Most amphibians have four well-developed limbs.

- In water, frogs and toads move by kicking their hind legs to swim; on land they kick their hind legs out to jump or hop.

- Salamanders walk on land; in water they undulate their bodies to swim.

- Caecilians lack limbs and eyes; terrestrial forms burrow in moist soils (**Figure 34.27b**).

Reproduction Reproduction is sexual, breeding occurs in water, and larvae undergo a dramatic metamorphosis into land-dwelling adults. For example, the fishlike tadpoles of frogs and toads develop limbs, and their gills are replaced with lungs.

- Frogs are oviparous and have external fertilization. In some species of frogs, parents may guard or even carry eggs. In many frogs, young develop in the water and suspension feed on plant or algal material.

- Salamanders have internal fertilization, and most species are oviparous. Salamander larvae are carnivorous.

- Caecilians have internal fertilization, and many species are viviparous.

(a) Frogs and other amphibians lay their eggs in water.

Bufo periglenes

Eggs

(b) Caecilians are legless amphibians.

Gymnophis multiplicata

1 cm

FIGURE 34.27 Amphibians Are the Most Ancient Tetrapods.

The **monotremes** are the most basal lineage of mammals living today and are found only in Australia. They lay eggs and have metabolic rates—meaning a rate of using oxygen and oxidizing sugars for energy—that are lower than other mammals. Three species exist: one species of platypus and two species of echidna.

Feeding Monotremes have a leathery beak or bill. The platypus feeds on insect larvae, mollusks, and other small animals in streams (**Figure 34.28a**). Echidnas feed on ants, termites, and earthworms (**Figure 34.28b**).

Movement Platypuses swim with the aid of their webbed feet and walk when on land. Echidnas walk on their four legs.

Reproduction Platypuses lay their eggs in a burrow, while echidnas keep their eggs in a pouch on their belly. Young monotremes hatch quickly, and the mother must continue keeping them warm and dry for another four months. Like other mammals, monotremes produce milk and nurse their young. They lack well-defined nipples, however, and instead secrete milk from glands in the skin.

(a) Platypus

Ornithorhynchus anatinus

(b) Echidna

Tachyglossus aculeatus

FIGURE 34.28 Platypuses and Echidnas Are Egg-Laying Mammals.

The 275 known species of **marsupials** live in the Australian region and the Americas (**Figure 34.29**) and include the familiar opossums, kangaroos, wallabies, and koala. Although females have a placenta that nourishes embryos during development, the young are born after a short embryonic period and are poorly developed. They crawl from the opening of the female's reproductive tract to the female's nipples, where they suck milk. They stay attached to their mother until they grow large enough to move independently. ✔You should be able to indicate the origin of the placenta and viviparity—traits that are also found in Eutherian mammals—on Figure 34.22.

Feeding Marsupials are herbivores, omnivores, or carnivores. In many cases, convergent evolution has resulted in marsupials that are similar to placental species in morphology and way of life. For example, a recently extinct marsupial called the Tasmanian wolf was a long-legged hunter similar to the wolves of North America and northern Eurasia. A species of marsupial native to Australia specializes in eating ants and looks and acts much like the South American anteater, which is not a marsupial.

Movement Marsupials move by crawling, gliding, walking, running, or hopping.

Reproduction Marsupial young spend more time developing while attached to their mother's nipple than they do inside her body being fed via the placenta.

Didelphis virginiana

FIGURE 34.29 Marsupials Give Birth after a Short Embryonic Period. Opossums are the only marsupials in North America.

The approximately 4300 species of **placental mammals**, or **eutherians**, are distributed worldwide. They are far and away the most species-rich and morphologically diverse group of mammals.

Biologists group placental mammals into 18 lineages called orders. The six most species-rich orders are the rodents (rats, mice, squirrels; 1814 species), bats (986 species), insectivores (hedgehogs, moles, shrews; 390 species), artiodactyls (pigs, hippos, whales, deer, sheep, cattle; 293 species), carnivores (dogs, bears, cats, weasels, seals; 274 species), and primates (lemurs, monkeys, apes, humans; 235 species).

Feeding The size and structure of the teeth correlate closely with the diet of placental mammals. Herbivores have large, flat teeth for crushing leaves and other coarse plant material; predators have sharp teeth that are efficient at biting and tearing flesh. The structure of the digestive tract also correlates with the placental mammals' diet. In some plant-eaters, for example, the stomach hosts unicellular organisms that digest cellulose and other complex polysaccharides.

Pongo abelii

FIGURE 34.30 Eutherians Are the Most Species-Rich and Diverse Group of Mammals.

Movement In placental mammals, the structure of the limb correlates closely with the type of movement performed. Eutherians fly, glide, run, walk, swim, burrow, or swing from trees (**Figure 34.30**). Hindlimbs are reduced or lost in aquatic groups such as whales and dolphins, which swim by undulating their bodies.

Reproduction Eutherians have internal fertilization and are viviparous. An extensive placenta develops from a combination of maternal and fetal tissues; and at birth, young are much better developed than in marsupials—some are able to walk or run minutes after emerging from the mother. ✔You should be able to indicate the origin of delayed birth (extended development prior to birth) on Figure 34.22. All eutherians feed their offspring milk until the young have grown large enough to process solid food. A prolonged period of parental care, extending beyond the nursing stage, is common as offspring learn how to escape predators and find food on their own.

Most lizards and snakes are small reptiles with elongated bodies and scaly skin. ✔You should be able to indicate the origin of scaly skin on Figure 34.22. Most lizards have well-developed jointed legs, but snakes are limbless (**Figure 34.31**). The hypothesis that snakes evolved from limbed ancestors is partially supported by the presence of vestigial hip and leg bones in boas and pythons. There are about 7000 species of lizards and snakes alive now.

Feeding Small lizards prey on insects. Although most of the larger lizard species are herbivores, the 3-meter-long monitor lizard from the island of Komodo is a predator and scavenger that can eat deer. Snakes are carnivores; some subdue their prey by injecting poison through modified teeth called fangs. Snakes prey primarily on small mammals, amphibians, and invertebrates, which they swallow whole—usually headfirst.

Movement Lizards crawl or run on their four limbs. Snakes and lizards that are limbless burrow through soil, crawl over the ground, or climb trees by undulating their bodies.

Reproduction Although most lizards and snakes lay eggs, many are ovoviviparous; viviparity has also evolved numerous times in

Morelia viridis

FIGURE 34.31 Snakes Are Limbless Predators.

lizards. Most species reproduce sexually, but asexual reproduction, via the production of eggs by mitosis, is known to occur in six groups of lizards and one snake lineage.

The 300 known species of turtles inhabit freshwater, marine, and terrestrial environments throughout the world. The testudines are distinguished by a shell composed of bony plates that fuse to the vertebrae and ribs (**Figure 34.32**). The shell functions in protection from predators—when threatened, turtles can withdraw their head and legs into it. ✔You should be able to indicate the origin of the turtle shell on Figure 34.22. The turtles' skulls are highly modified versions of the skulls of other reptiles. Turtles lack teeth, but their jawbone and lower skull form a bony beak. They range in size from the 2-m long, 900 kg leatherback sea turtle to the 8-cm long, 140 g speckled padloper tortoise.

Feeding Turtles are either carnivorous—feeding on whatever animals they can capture and swallow—or herbivorous. They may also scavenge dead material. Most marine turtles are carnivorous. Leatherback turtles, for example, feed primarily on jellyfish, and they are only mildly affected by the jellyfish's stinging cnidocytes (see Chapter 32). In contrast, species in the lineage of terrestrial turtles called the tortoises are plant-eaters.

Testudo hermanni

FIGURE 34.32 Turtles Have a Shell Consisting of Bony Plates.

Movement Turtles swim, walk, or burrow. Aquatic species usually have feet that are modified to function as flippers.

Reproduction All turtles are oviparous. Other than digging a nest prior to depositing eggs, parental care is lacking.

Only 24 species of crocodile and alligator are known. Most are tropical and live in freshwater or marine environments. They have eyes located on the top of their heads and nostrils located at the top of their long snouts—adaptations that allow them to sit semisubmerged in water for long periods of time, breathing air and monitoring activity around them by sight (**Figure 34.33**).

Feeding Crocodilians are predators. Like other toothed vertebrates, except mammals, their jaws are filled with conical teeth that are continually replaced as they fall out during feeding. Their usual method of killing small prey is by biting through the body wall. Large prey are usually subdued by drowning. Crocodilians eat amphibians, turtles, fish, birds, and mammals; one of their common hunting strategies is to leap out of the water to snatch unwary prey that have come to drink.

Movement Crocodiles and alligators walk on land. In water they swim with the aid of their large, muscular tails.

Reproduction Although crocodilians are oviparous, parental care is extensive. Eggs are laid in earth-covered nests that are guarded by the parents. When young inside the eggs begin to vocalize, parents dig them up and carry the newly hatched young inside their mouths to nearby water. Crocodilian young can hunt

Alligator mississippiensis

FIGURE 34.33 Alligators Are Adapted for Aquatic Life.

when newly hatched but stay near their mother for up to three years. ✔You should be able to indicate the origin of extensive parental care—which also occurs in birds (and dinosaurs)—on Figure 34.22.

The fossil record provides conclusive evidence that birds descended from a lineage of dinosaurs that had a unique trait: **feathers.** In dinosaurs, feathers are hypothesized to have functioned as insulation and in courtship or aggressive displays. In birds, feathers insulate and are used for display but also furnish the lift, power, and steering required for flight. Birds have many other adaptations that make flight possible, including large breast muscles used to flap the wings. Bird bodies are lightweight because they have a reduced number of bones and organs and because their hollow bones are filled with air sacs linked to the lungs. Instead of teeth, birds have a horny beak. They are endotherms ("within-heating"), meaning that they have a high metabolic rate and use the heat produced, along with the insulation provided by feathers, to maintain a constant body temperature. The 9100 bird species alive today occupy virtually every habitat, including the open ocean (**Figure 34.34a**). ✔ You should be able to indicate the origin of feathers, endothermy, and flight on Figure 34.22.

Feeding Plant-eating birds usually feed on nectar or seeds. Some birds are omnivores, although many are predators that capture insects, small mammals, fish, other birds, lizards, mollusks, or crustaceans. The size and shape of a bird's beak correlate closely with its diet. For example, predatory species such as falcons have sharp, hook-shaped beaks; finches and other seedeaters have short, stocky bills that can crack seeds and nuts; fish-eating species such as the great blue heron have spear-shaped beaks.

Movement Almost all species can fly, although flightlessness has evolved repeatedly in certain groups (**Figure 34.34b**). The size and shape of birds' wings correlate closely with the type of flying they do. Birds that glide or hover have long, thin wings; species that specialize in explosive takeoffs and short flights have short, stocky wings. Many seabirds are efficient swimmers, using their webbed feet to paddle or flapping their wings to "fly" underwater. Some ground-dwelling birds such as ostrich can run long distances at high speed.

Reproduction Birds are oviparous but provide extensive parental care. In most species, one or both parents build a nest and incubate the eggs. After the eggs hatch, parents feed offspring until they are large enough to fly and find food on their own.

(a) Albatross use their long, narrow wings to glide.

Diomedea salvini

(b) Flightless cormorants are native to the Galápagos Islands.

Phalacrocorax harrisi

FIGURE 34.34 Birds Are Feathered Descendants of Dinosaurs.

34.4 The Primates and Hominins

Although humans occupy a tiny twig on the tree of life, there has been a tremendous amount of research on human origins. This section introduces the lineage of mammals called the Primates, the fossil record of human ancestors, and data on the relationships among human populations living today.

The Primates

The lineage known as Primates consists of two main groups: prosimians and anthropoids.

- The **prosimians** ("before-monkeys") consist of the lemurs, found in Madagascar, and the tarsiers, pottos, and lorises of Africa and south Asia (**Figure 34.35a**). Most prosimians living today are relatively small in size, reside in trees, and are active at night.

- The Anthropoidea or **anthropoids** ("human-like") include the New World monkeys found in Central and South America, the Old World monkeys that live in Africa and tropical regions of Asia, the gibbons of the Asian tropics, and the Hominidae, or **great apes**—orangutans, gorillas, chimpanzees, and humans (**Figure 34.35b**).

(a) Prosimians live in Africa, Madagascar, and south Asia.

Lemur catta

(b) Anthropoids include New World monkeys, Old World monkeys, gibbons, and Great Apes.

Cacajao calvus

FIGURE 34.35 There Are Two Main Lineages of Primates.

The phylogenetic tree in **Figure 34.36** shows the evolutionary relationships among these groups.

WHAT MAKES A PRIMATE A PRIMATE? **Primates** tend to have hands and feet that are efficient at grasping, flattened nails instead of claws on the fingers and toes, brains that are large relative to overall body size, color vision, complex social behavior, and extensive parental care of offspring.

Along with other mammal groups that live in trees or make their living by hunting, primates have eyes located on the front of the face. Eyes that look forward provide better depth perception than do eyes on the sides of the face. The hypothesis here is that good depth perception is important in species that run or swing through trees and/or attack prey.

WHAT MAKES A GREAT APE A GREAT APE? The great apes are also called **hominids.** Compared with most types of primate, the hominids are relatively large bodied and have long arms, short legs, and no tail.

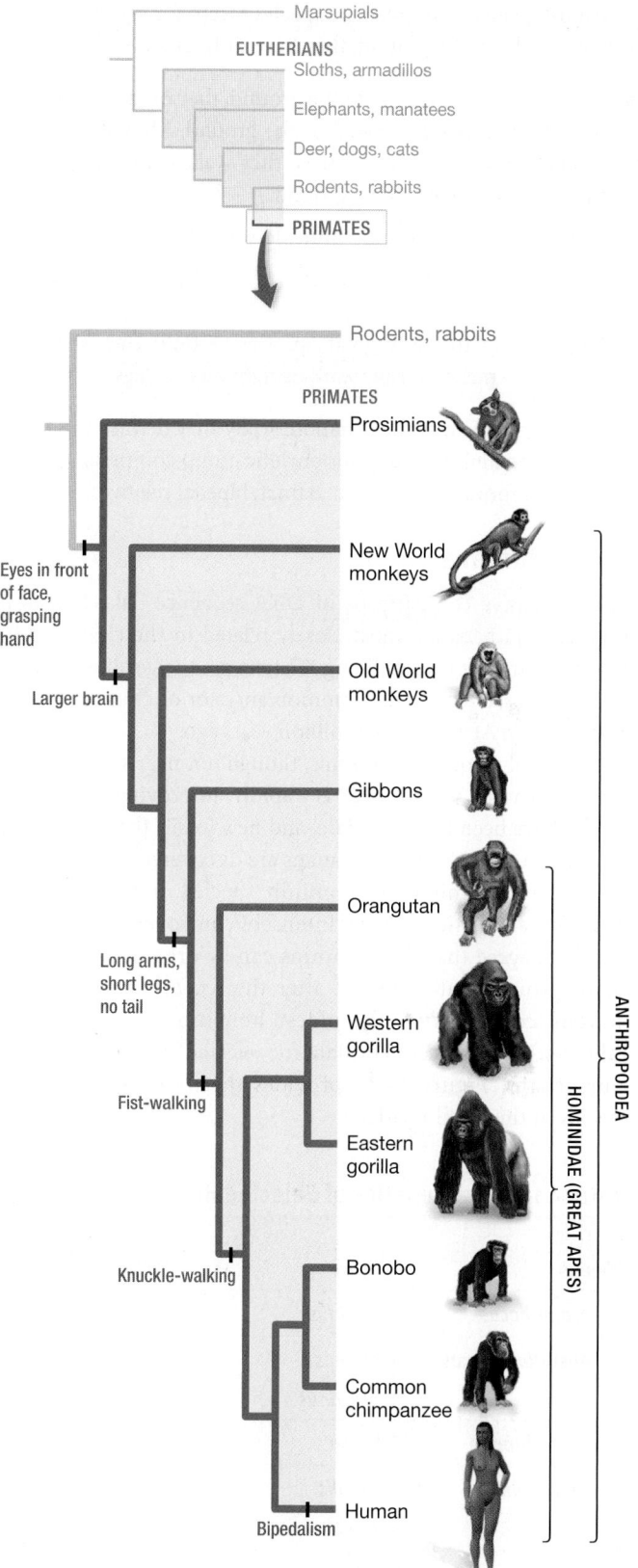

FIGURE 34.36 A Phylogeny of the Primates. Phylogenetic tree estimated from extensive DNA sequence data. According to the fossil record, humans and chimps shared a common ancestor 6 to 7 million years ago.

Although all of the great ape species except for the orangutans live primarily on the ground, they have distinct ways of walking.

- When orangutans come to the ground, they occasionally walk with their knuckles pressed to the ground. More commonly, though, they fist-walk—that is, they walk with the backs of their hands pressed to the ground.

- Gorillas and chimps only knuckle-walk. They also occasionally rise up on two legs—usually in the context of displaying aggression.

- Humans are the only great ape that is fully **bipedal** ("two-footed")—meaning they walk upright on two legs.

🔑 Bipedalism is the synapomorphy that defines the hominins. The **hominins** are a monophyletic group comprising *Homo sapiens* and more than a dozen extinct, bipedal relatives.

Fossil Humans

From extensive comparisons of DNA sequence data, it is now clear that humans are most closely related to the chimpanzees and that our next nearest living relatives are the gorillas. According to the fossil record, the common ancestor of chimps and humans lived in Africa about 7 million years ago.

The fossil record of hominins, though not nearly as complete as investigators would like, is rapidly improving. About 14 species have been found to date, and new fossils that inform the debate over the ancestry of humans are discovered every year.

Although naming the hominin species and interpreting their characteristics remain intensely controversial, most researchers agree that the hominins can be organized into four major groups that appeared after the recently characterized *Ardipithecus ramidus*—the oldest hominin known to date. **Table 34.1** summarizes key data for selected species within the four groups. **Figure 34.37** provides the time range of each species in the fossil record.

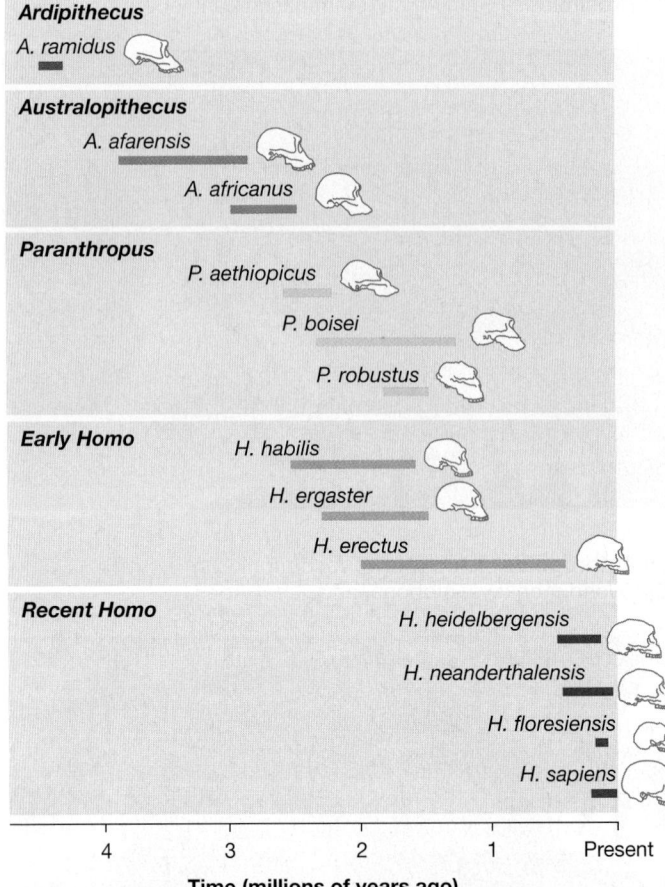

FIGURE 34.37 A Timeline of Human Evolution. The colored bars indicate the first appearance and last appearance in the fossil record of each species.

✔ **QUESTION** How many species of hominin existed 2.2 million years ago (mya), 1.8 mya, and 100,000 years ago?

TABLE 34.1 **Characteristics of Selected Hominins**

Genus	Species	Location of Fossils	Estimated Average Braincase Volume (cm³)	Estimated Average Body Size (kg)	Associated with Stone Tools?
■ **Ardipithecus**	A. ramidus	Africa	325	50	no
■ **Australopithecus**	A. afarensis	Africa	450	36	no
	A. africanus	Africa	450	36	no
■ **Paranthropus**	P. boisei	Africa	510	44	no?
■ **Early *Homo***	H. habilis	Africa	550	34	yes
	H. ergaster	Africa	850	58	yes
	H. erectus	Africa, Asia	1000	57	yes
■ **Recent *Homo***	H. heidelbergensis	Africa, Europe	1200	62	yes
	H. neanderthalensis	Middle East, Europe, Asia	1500	76	yes
	H. floresiensis	Flores (Indonesia)	380	28	yes
	H. sapiens	Middle East, Europe, Asia	1350	53	yes

GRACILE AUSTRALOPITHECINES Four species of small hominins called gracile australopithecines have been identified thus far. The adjective *gracile*, or "slender," is appropriate because these organisms were slightly built. Adult males were about 1.5 meters tall, with an estimated weight of about 36 kg.

The gracile australopithecines are placed in the genus *Australopithecus* ("southern ape"). The name was inspired by the earliest specimens, which came from South Africa.

Several lines of evidence support the hypothesis that the gracile australopithecines were bipedal. The shape of the australopithecine knee and hip are consistent with bipedal locomotion, and the hole in the back of their skulls where the spinal cord connects to the brain is oriented downward (**Figure 34.38a**), just as it is in our species, *Homo sapiens*. In chimps, gorillas, and other vertebrates that walk on four feet, this hole is oriented backward.

ROBUST AUSTRALOPITHECINES Three species are grouped in the genus *Paranthropus* ("beside-human"). Like the gracile australopithecines, these robust australopithecines were bipedal. They were much stockier than the gracile forms, however—about the same height but an estimated 8–10 kilograms (20 pounds) heavier on average. In addition, their skulls were much broader and more robust.

All three species had massive cheek teeth and jaws, very large cheekbones, and a sagittal crest—a flange of bone at the top of the skull (**Figure 34.38b**). Because muscles that work the jaw attach to the sagittal crest and cheekbones, researchers conclude that these organisms had tremendous biting power and made their living by crushing large seeds or coarse plant materials. One species is nicknamed "nutcracker man."

The name *Paranthropus* was inspired by the hypothesis that the three known species are a monophyletic group that was a side branch during human evolution—an independent lineage that went extinct.

EARLY *HOMO* Species in the genus *Homo* are called **humans**. As **Figure 34.38c** shows, species in this genus have flatter and narrower faces, smaller jaws and teeth, and larger braincases than the earlier hominins do. (The **braincase** is the portion of the skull that encloses the brain.)

The appearance of early members of the genus *Homo* in the fossil record coincides closely with the appearance of tools made of worked stone—most of which are interpreted as handheld choppers or knives. Although the fossil record does not exclude the possibility that *Paranthropus* made tools, many researchers favor the hypothesis that extensive toolmaking was a diagnostic trait of early *Homo*.

RECENT *HOMO* The recent species of *Homo* date from 1.2 million years ago to the present. As **Figure 34.38d** shows, these species have even flatter faces, smaller teeth, and larger braincases than the early *Homo* species do. The 30,000-year-old fossil in the figure, for example, is from a population of *Homo sapiens* (our species) called the **Cro-Magnons**.

The Cro-Magnons were accomplished painters and sculptors who buried their dead in carefully prepared graves. There is also evidence that another species, the **Neanderthal** people (*Homo neanderthalensis*) made art and buried their dead in a ceremonial fashion.

(a) Gracile australopithecines (*Australopithecus africanus*)

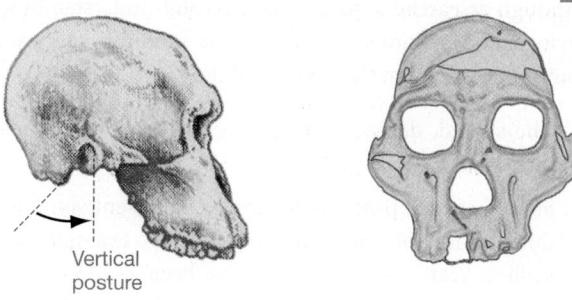

Vertical posture

(b) Robust australopithecines (*Paranthropus robustus*)

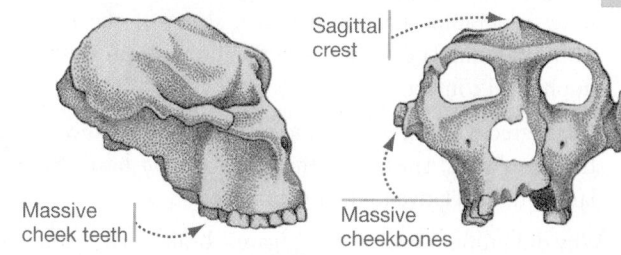

Sagittal crest

Massive cheek teeth

Massive cheekbones

(c) Early *Homo* (*Homo habilis*)

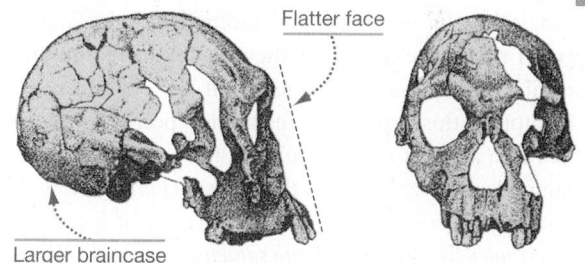

Flatter face

Larger braincase

(d) Recent *Homo* (*Homo sapiens*)

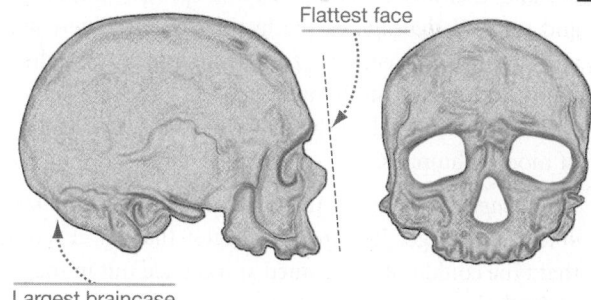

Flattest face

Largest braincase

FIGURE 34.38 African Hominins Comprise Four Major Groups.

✓**QUESTION** The skulls are arranged as they appear in the fossil record, from most ancient to most recent—(a) to (d). How did the forehead and brow ridge of hominins change through time?

Perhaps the most striking recent *Homo*, though, is *H. floresiensis*. This species has been found only on the island of Flores in Indonesia, which was also home to a species of dwarfed elephants. *H. floresiensis* consisted of individuals that had braincases smaller than those of gracile australopithecines and were about a meter tall. Fossil finds suggest that the species inhabited the island from about 100,000 to 12,000 years before present and that dwarfed elephants were a major source of food.

WHAT CAN BE DEDUCED FROM THE HOMININ FOSSIL RECORD?

Although researchers do not have a solid understanding of the phylogenetic relationships among the hominin species, several points are clear from the available data.

1. The shared, derived character that defines the hominins is bipedalism.

2. Several species from the lineage were present simultaneously during most of hominin evolution. For example, about 1.8 million years ago there may have been as many as five hominin species living in eastern and southern Africa.

3. Fossils from more than one species have been found in the same geographic location in rock strata of the same age. Thus, it is almost certain that different hominin species lived in physical contact.

4. Compared with the gracile and robust australopithecines and the great apes, species in the genus *Homo* have extremely large brains relative to their overall body size.

Why did humans evolve such gigantic brains? The leading hypothesis on this question is that early *Homo* began using symbolic spoken language along with initiating extensive tool use. The logic here is that increased toolmaking and language use triggered natural selection for the capacity to reason and communicate, which required a larger brain.

To support this hypothesis, researchers point out that, relative to the brain areas of other hominins, the brain areas responsible for language were enlarged in the earliest *Homo* species. There is even stronger fossil evidence for extensive use of speech in *Homo neanderthalensis* and early *Homo sapiens*:

- The hyoid bone is a slender bone in the voice box of modern humans that holds muscles used in speech. In Neanderthals and early *Homo sapiens*, the hyoid is vastly different in size and shape from a chimpanzee's hyoid bone. Researchers have found an intact hyoid bone associated with a 60,000-year-old Neanderthal individual that is virtually identical to the hyoid of modern humans.

- *Homo sapiens* colonized Australia by boat between 60,000 and 40,000 years ago. Researchers suggest that an expedition of that type could not be planned and carried out in the absence of symbolic speech.

To summarize, *Homo sapiens* is the sole survivor of an adaptive radiation that took place over the past 7 million years. Why all but one species went extinct is still a mystery, though competition for food and space may have played a part. Neanderthals, for example, flourished in Europe and central Asia prior to the arrival of *Homo sapiens* from Africa.

The Out-of-Africa Hypothesis

Our own species, *Homo sapiens*, is the only primate with a chin. The first *H. sapiens* fossils appear in African rocks that date to about 195,000 years ago. For some 130,000 years thereafter, the fossil record indicates that our species occupied Africa while *H. neanderthalensis* resided in Europe and the Middle East. Some evidence suggests that *H. erectus* may still have been present in Asia at that time.

In rocks dated between 60,000 and 30,000 years ago, however, *H. sapiens* fossils are found throughout Europe, Asia, Africa, and Australia. *H. erectus* had disappeared by this time, and *H. neanderthalensis* went extinct after coexisting with *H. sapiens* in Europe for thousands of years.

Phylogenetic trees that show the relationships among human populations living today agree with the pattern in the fossil record. As **Figure 34.39** shows, the first lineages to branch off lead to descendant populations that live in Africa today. Based on this observation, it is logical to infer that the ancestral population of modern humans also lived in Africa.

The tree shows that lineages subsequently branched off, leaving descendants that today live in Europe, central Asia, Polynesia, the Americas, and east Asia (**Figure 34.40**). The data suggest that modern humans originated in Africa, and that a population that left Africa split into a group that colonized Europe and Russia and a group that eventually spread throughout the rest of the world.

This scenario for the evolution of *H. sapiens* is called the **out-of-Africa hypothesis**. It contends that *H. sapiens* evolved independently of the earlier European and Asian species of *Homo*—meaning there was no interbreeding between *H. sapiens* and Neanderthals, *H. erectus*, or *H. floresiensis*. According to the out-of-Africa hypothesis, *H. sapiens* evolved its distinctive traits in Africa and then dispersed throughout the world.

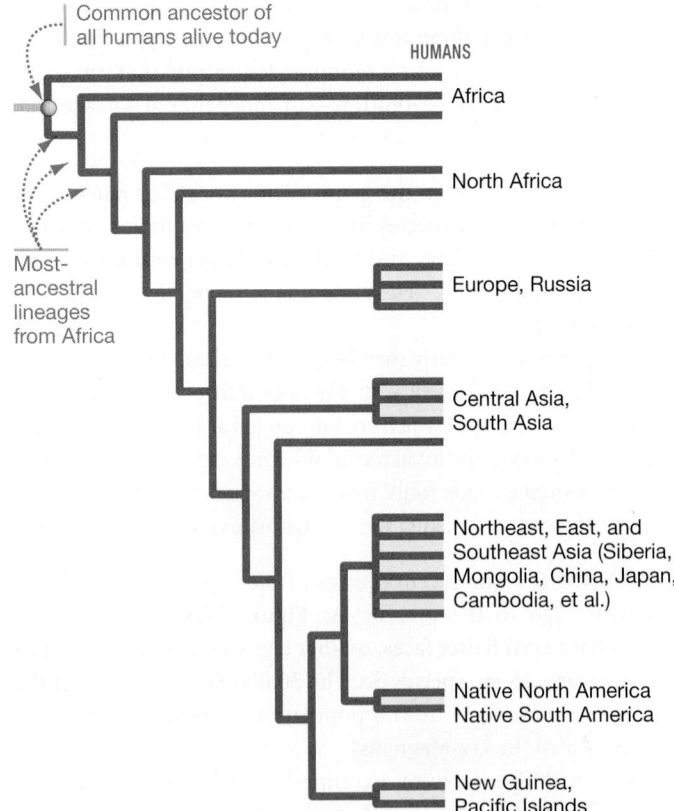

FIGURE 34.39 Phylogeny of Human Populations Living Today.
The phylogeny was estimated from DNA sequence data.

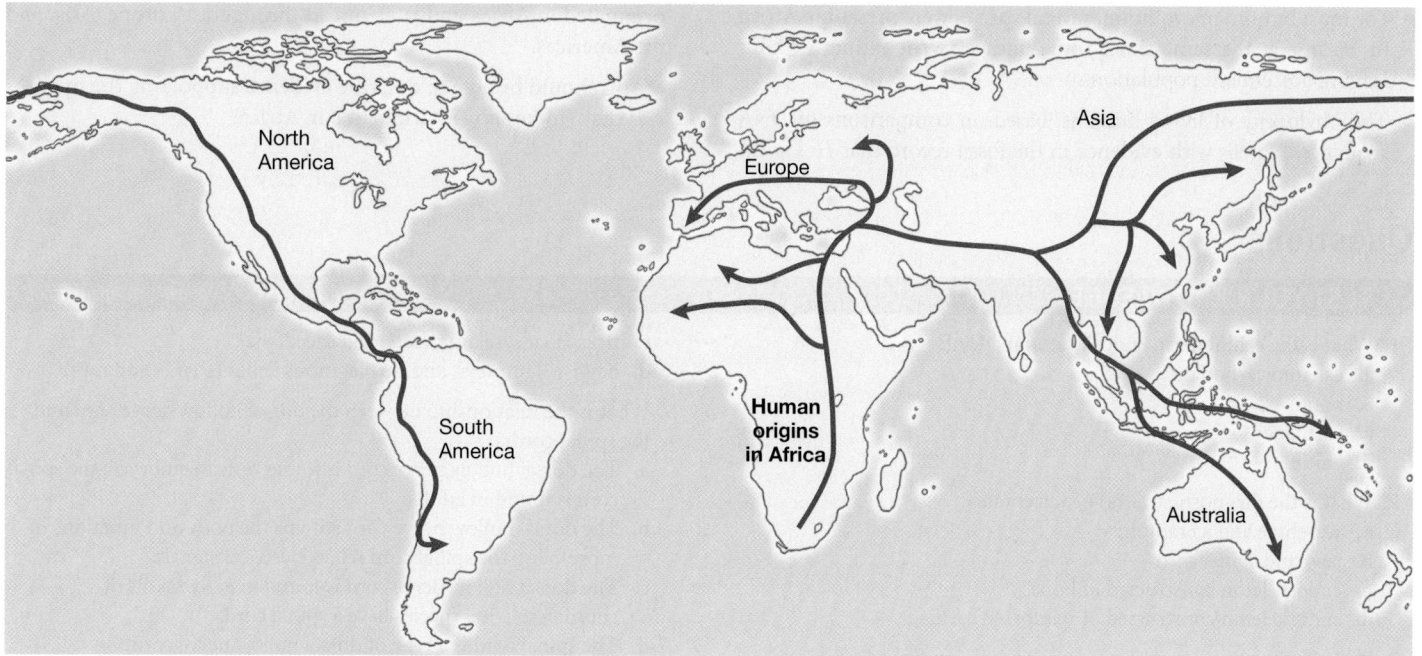

FIGURE 34.40 *Homo sapiens* **Originated in Africa and Spread throughout the World.** The phylogeny in Figure 34.39 supports the hypothesis that humans originated in Africa and spread out along the routes shown here.

CHAPTER 34 REVIEW

For media, go to the study area at www.masteringbiology.com

Summary of Key Concepts

🔑 **Echinoderms are radially symmetric as adults and have an endoskeleton and water vascular system. They are among the most important predators and herbivores in marine environments.**

- Echinoderm larvae are bilaterally symmetric but undergo a metamorphosis into radially symmetric adults.

- The echinoderm water vascular system is composed of fluid-filled tubes and chambers and extends from the body wall in projections called podia.

- Echinoderm podia can extend and retract in response to muscle contractions that move fluid inside the water vascular system. Many echinoderms move via their podia, and many species suspension feed, deposit feed, or act as predators with the aid of their podia.

  ✔ You should be able to predict what happens when (**1**) sea stars that prey on mussels and clams are excluded from experimental plots, and (**2**) sea urchins that graze on kelp are excluded from experimental plots.

🔑 **All vertebrates have a skull and an extensive endoskeleton made of cartilage or bone; their diversification was driven in part by the evolution of the jaw and limbs. Vertebrates are the most important large-bodied predators and herbivores in marine and terrestrial environments.**

- Chordates are distinguished by the presence of pharyngeal gill slits, a dorsal hollow nerve cord, a notochord, and a muscular tail that extends past the anus.

- Vertebrates are distinguished by the presence of a cranium and vertebrae. In some groups of vertebrates, the body plan features an extensive endoskeleton composed of bone.

- Ray-finned fishes and tetrapods use their jaws to bite food and process it with teeth.

- Ray-finned fishes and tetrapods move when muscles attached to their endoskeletons contract or relax.

- Tetrapods can move on land because their limbs enable walking, running, or flying.

- The evolution of the amniotic egg allowed tetrapods to lay large eggs on land.

- Parental care was an important adaptation in some groups of ray-finned fishes and tetrapods—particularly mammals.

  ✔ You should be able to explain why the tetrapod limb could have evolved in an animal that was aquatic.

  (MB) **Web Activity** Deuterostome Diversity

🔑 **Humans are a tiny twig on the tree of life. Chimpanzees and humans diverged from a common ancestor that lived in Africa 6–7 million years ago. Since then, at least 14 humanlike species have existed.**

- The fossil record of the past 3.5 million years contains at least 14 distinct species of hominins. *Homo sapiens* is the sole surviving representative of an adaptive radiation.

- For most of human evolution, several species were present in Africa or Europe at the same time. Some lineages went extinct without leaving descendant populations.

- The phylogeny of living humans, based on comparisons of DNA sequences, agrees with evidence in the fossil record that *H. sapiens*

originated in Africa and later spread throughout Europe, Asia, and the Americas.

✔You should be able to describe evidence supporting the hypothesis that *Homo sapiens* originated in Africa.

Questions

✔TEST YOUR KNOWLEDGE

Answers are available in Appendix B

1. What is the echinoderm endoskeleton made of?
   a. calcium carbonate
   b. bone
   c. cartilage
   d. chitin

2. What is the diagnostic trait(s) of vertebrates?
   a. vertebrae and a cranium
   b. jaws and spinal cord
   c. endoskeleton constructed of bone
   d. endoskeleton constructed of reinforced cartilage

3. Why are the pharyngeal jaws found in many ray-finned fishes important?
   a. They allow the main jaw to be protrusible (extendable).
   b. They make it possible for individuals to suck food toward their mouths.
   c. They give rise to teeth that are found on the main jawbones.
   d. They help process food.

4. Which of the following lineages make up the living Amniota?
   a. reptiles and mammals
   b. viviparous fishes

   c. frogs, toads, salamanders, and caecilians
   d. hagfish, lampreys, and cartilaginous fishes (sharks and rays)

5. What is the relationship between the dorsal hollow nerve cord and the spinal cord?
   a. The dorsal hollow nerve cord is found only in embryos; the spinal cord is found in adults.
   b. The dorsal hollow nerve cord stiffens the body and functions in movement; the spinal cord relays electrical signals.
   c. The dorsal hollow nerve cord is found in early fossils of chordates; living species have a spinal cord.
   d. The spinal cord is a type of dorsal hollow nerve cord.

6. Most species of hominins are known only from Africa. Which species have been found in other parts of the world as well?
   a. early *Homo*—*H. habilis* and *H. ergaster*
   b. *H. erectus*, *H. neanderthalensis*, and *H. floresiensis*
   c. gracile australopithecines
   d. robust australopithecines

✔TEST YOUR UNDERSTANDING

Answers are available in Appendix B

1. Explain how the water vascular system of echinoderms functions as a type of hydrostatic skeleton.

2. Explain why each of the four key synapomorphies that distinguish chordates is important in feeding or movement. Why is it possible to say that an animal is an "invertebrate chordate"?

3. The cells that make up jaws and gill arches are derived from the same population of embryonic cells. Why does this observation support the hypothesis that jaws evolved from gill arches in fish?

4. Describe genetic evidence that supports the hypothesis that the tetrapod limb evolved from the fins of lobe-finned fishes.

5. The text claims that "*Homo sapiens* is the sole survivor of an adaptive radiation that took place over the past 7 million years." Do you agree with this statement? Why or why not?

6. Explain how the evolution of the placenta and lactation in mammals improved the probability that their offspring would survive, compared to species without parental care.

✔APPLYING CONCEPTS TO NEW SITUATIONS

Answers are available in Appendix B

1. There is some evidence that pharyngeal gill slits occur in certain species of echinoderms that appear early in the fossil record. Explain the significance of this observation.

2. When feathers first evolved, did they function in flight? Explain your logic.

3. Xenoturbellidans have extremely simple, wormlike bodies. For example, they have a blind gut (only one opening) and no brain.

Propose a hypothesis to explain how this simple body plan evolved in this lineage.

4. Mammals and birds are endothermic. Did they inherit this trait from a common ancestor, or did endothermy evolve independently in these two lineages? Provide evidence to support your answer.

Photomicrograph created by treating seawater with a fluorescing compound that binds to nucleic acids. The smallest, most abundant dots are viruses. The larger, numerous spots are bacteria and archaea. The largest splotches are protists.

Viruses 35

I f you have ever been laid low by a high fever, cough, scratchy throat, body ache, and debilitating lack of energy, you may have wondered what hit you. What hit you was probably a **virus**: an obligate, intracellular parasite.

Why obligate? Viral replication is *completely* dependent on host cells. Why intracellular? Viruses must enter a host cell for replication to occur. Why parasite? Viruses reproduce at the expense of their host cells.

Viruses can also be defined by what they are not.

● They are not cells and are not made up of cells, so they are not considered organisms.

● They cannot manufacture their own ATP or amino acids or nucleic acids, and they cannot produce proteins on their own.

Viruses enter a **host cell**, take over its biosynthetic machinery, and use that machinery to manufacture a new generation of viruses. Outside of host cells, viruses simply exist.

When you have the flu, influenza viruses enter the cells that line your respiratory tract and use the machinery inside to make copies of themselves. Every time you cough or sneeze, you eject millions or billions of their offspring into the environment. If one of those viruses is lucky enough to be breathed in by another person, it may enter their respiratory tract cells and start a new infection.

Because they are not organisms, viruses are not given scientific (genus + species) names. Most biologists would argue that viruses are not alive, because they lack the five attributes of life introduced in Chapter 1. Yet viruses have a genome, they are superbly adapted to exploit the metabolic capabilities of their host cells, and they evolve. **Table 35.1** on page 676 summarizes some characteristics of viruses.

The diversity and abundance of viruses almost defy description. ⎝⊸ Each type of virus infects a specific unicellular species or cell type in a multicellular species, and nearly all organisms examined thus far are parasitized by at least one kind of virus. The bacterium

KEY CONCEPTS

⎝⊸ Viruses are tiny, noncellular parasites that infect virtually every type of cell known. They cannot perform metabolism on their own—meaning outside a parasitized cell—and are not considered to be alive.

⎝⊸ Although viruses are diverse morphologically, they can be classified as two general types: enveloped and nonenveloped.

⎝⊸ The viral replication cycle can be broken down into six steps: (1) attachment and entry into a host cell, (2) production of viral proteins, (3) replication of the viral genome, (4) assembly of a new generation of virions, (5) exit from the infected cell, and (6) transmission to a new host.

⎝⊸ In terms of diversity, the key feature of viruses is the nature of their genetic material.

✔ When you see this checkmark, stop and test yourself. Answers are available in Appendix B.

Characteristics	Viruses	Organisms
Hereditary material	DNA or RNA; can be single stranded or double stranded	DNA; always double stranded
Plasma membrane present?	No	Yes
Can carry out transcription independently?	No—even if a viral polymerase is present, transcription of viral genomes requires use of ATP and nucleotides provided by host cell	Yes
Can carry out translation independently?	No	Yes
Metabolic capabilities	Virtually none	Extensive—synthesis of ATP, reduced carbon compounds, vitamins, lipids, nucleic acids, etc.

Escherichia coli, which resides in the human intestine, is afflicted by several dozen types of viruses. The surface waters of the world's oceans teem with bacteria and archaea, yet viruses outnumber them in this habitat by a factor of 10 to 1. If you leaned over a boat and filled a wine bottle with seawater, it would contain about 10 billion virus particles, or **virions**—close to one and a half times the world's population of humans.

35.1 Why Do Biologists Study Viruses?

Any study of life's diversity would be incomplete unless it included a look at the acellular parasites that exploit that diversity. But viruses are also important from a practical standpoint. To health-care workers, agronomists, and foresters, these parasites are a persistent—and sometimes catastrophic—source of misery and economic loss.

The nature of viruses has been understood only since the 1940s, but they have been the focus of intense research ever since. Biologists study viruses because they cause illness and death. In the human body, virtually every system, tissue, and cell can be infected by one or more kinds of virus (**Figure 35.1**). Much research on viruses is motivated by the desire to minimize the damage they can cause.

Recent Viral Epidemics in Humans

Physicians and researchers use the term **epidemic** (literally, "upon-people") to describe a disease that rapidly affects a large number of individuals over a widening area. Viruses have caused the most devastating epidemics in recent human history. During the eighteenth and nineteenth centuries, it was not unusual for

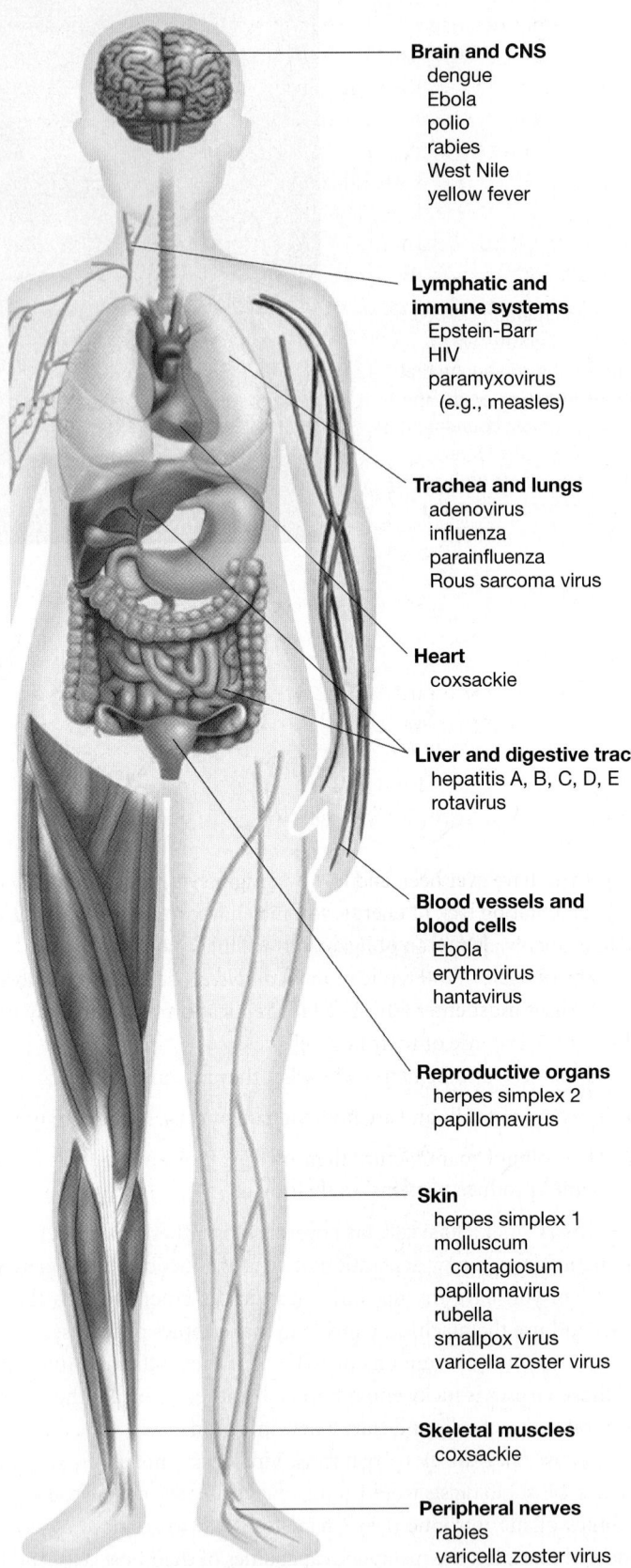

Brain and CNS
dengue
Ebola
polio
rabies
West Nile
yellow fever

Lymphatic and immune systems
Epstein-Barr
HIV
paramyxovirus
(e.g., measles)

Trachea and lungs
adenovirus
influenza
parainfluenza
Rous sarcoma virus

Heart
coxsackie

Liver and digestive tract
hepatitis A, B, C, D, E
rotavirus

Blood vessels and blood cells
Ebola
erythrovirus
hantavirus

Reproductive organs
herpes simplex 2
papillomavirus

Skin
herpes simplex 1
molluscum
contagiosum
papillomavirus
rubella
smallpox virus
varicella zoster virus

Skeletal muscles
coxsackie

Peripheral nerves
rabies
varicella zoster virus

FIGURE 35.1 Human Organs and Systems That Are Parasitized by Viruses.

✔**EXERCISE** Choose two viruses that you are familiar with. Name the location of the infection, and describe how the virus is transmitted.

Native American tribes to lose 90 percent of their members over the course of a few years to measles, smallpox, and other viral diseases spread by contact with European settlers. To appreciate the impact of these epidemics, think of 10 close friends and relatives—then remove nine.

An epidemic that is worldwide in scope is called a **pandemic**. The influenza outbreak of 1918–1919, called "Spanish flu," qualifies as the most devastating pandemic recorded to date. The strain of influenza virus that emerged in 1918 infected people worldwide and was particularly **virulent**—meaning it tended to cause severe disease. Within hours of showing symptoms, the lungs of previously healthy people often became so heavily infected that affected individuals suffocated to death. Most victims were between the ages of 20 and 40. The viral outbreak occurred just as World War I was drawing to a close and killed far more people than did the conflict itself. For example, ten times as many Americans died of influenza than were killed in combat in the war. Worldwide, the Spanish flu is thought to have killed 20–50 million people.

Current Viral Pandemics in Humans: HIV

In terms of the total number of people affected, the measles and smallpox epidemics among native peoples of the Americas and the 1918 influenza outbreak are almost certain to be surpassed by the incidence of AIDS. **Acquired immune deficiency syndrome (AIDS)** is an affliction caused by the **human immunodeficiency virus (HIV)**.

HIV is now the most intensively studied of all viruses. Since the early 1980s, governments and private corporations from around the world have spent hundreds of millions of dollars on HIV research. Given this virus's current and projected impact on human populations around the globe, the investment is justified.

HOW DOES HIV CAUSE DISEASE? Like other viruses, HIV parasitizes specific types of cells. The cells most affected by HIV are called helper T lymphocytes and macrophages (**Figure 35.2**). These cells are components of the **immune system**, which is the body's defense system against disease. Chapter 49 explains just how crucial helper T lymphocytes and macrophages are to the immune system's response to invading bacteria and viruses.

If an HIV virion succeeds in infecting a helper T lymphocyte or macrophage and reproduces inside, the cell dies as hundreds of new virions are released and infect more cells. Although the body continually replaces helper T lymphocytes and macrophages, the number produced does not keep pace with the number being destroyed by HIV. As a result, the total number of helper T lymphocytes in the bloodstream gradually declines as an HIV infection proceeds (**Figure 35.3**).

When the T-cell count drops, the immune system's responses to invading bacteria and viruses become less and less effective. Eventually, too few helper T lymphocytes are left to fight off pathogens efficiently, and bacteria and viruses begin to multiply unchecked. In almost all cases, one or more of these infections proves fatal. HIV kills people indirectly—by making them susceptible to pneumonia, an array of eukaryotic parasites, and unusual types of cancer.

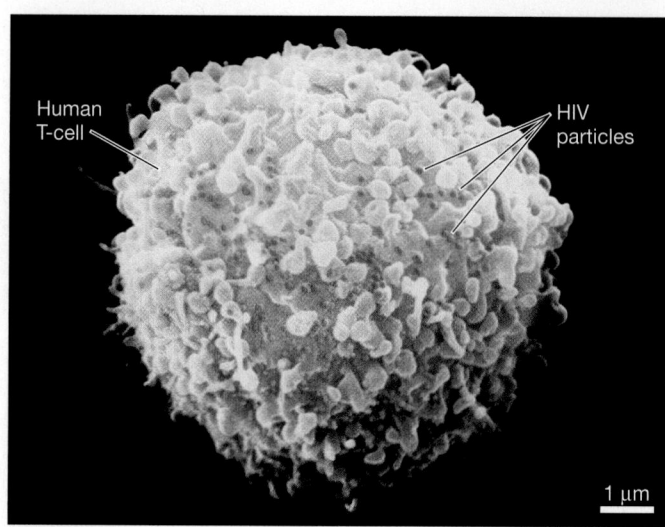

FIGURE 35.2 HIV-Infected T cell. Scanning electron micrograph showing HIV particles (false-colored green) emerging from an infected human T lymphocyte (a type of immune system cell).

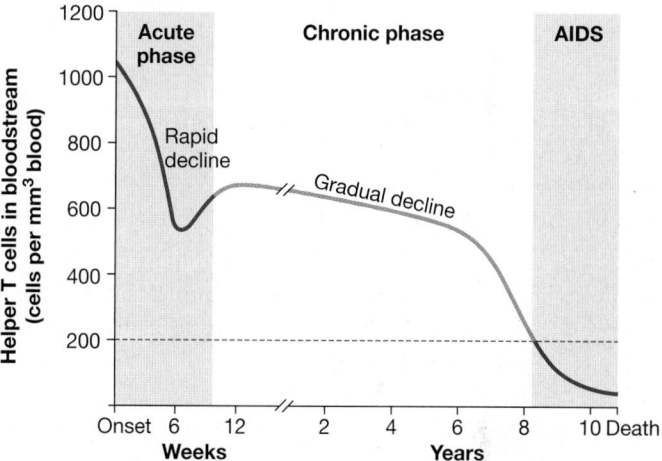

FIGURE 35.3 T-Cell Counts Decline during an HIV Infection. Graph of changes in the number of T cells that are present in the bloodstream over time, based on data from a typical patient infected with HIV. The acute phase may be associated with symptoms such as fever. Few or no disease symptoms occur in the chronic phase. AIDS typically occurs when T-cell counts dip below 200/mm³ of blood.

WHAT IS THE SCOPE OF THE AIDS PANDEMIC? Researchers with the United Nations AIDS program estimate that AIDS has already killed 28 million people worldwide. HIV infection rates have been highest in east and central Africa, where one of the greatest public health crises in history is now occurring. In Botswana, for example, blood-testing programs have confirmed that over 20 percent of individuals carry HIV. Although there may be a lag of as much as 8–12 years between the initial infection and the onset of illness, virtually all people who become infected with the virus will die of AIDS.

Currently, the UN estimates the total number of HIV-infected people worldwide at about 33 million. An additional 2.7 million

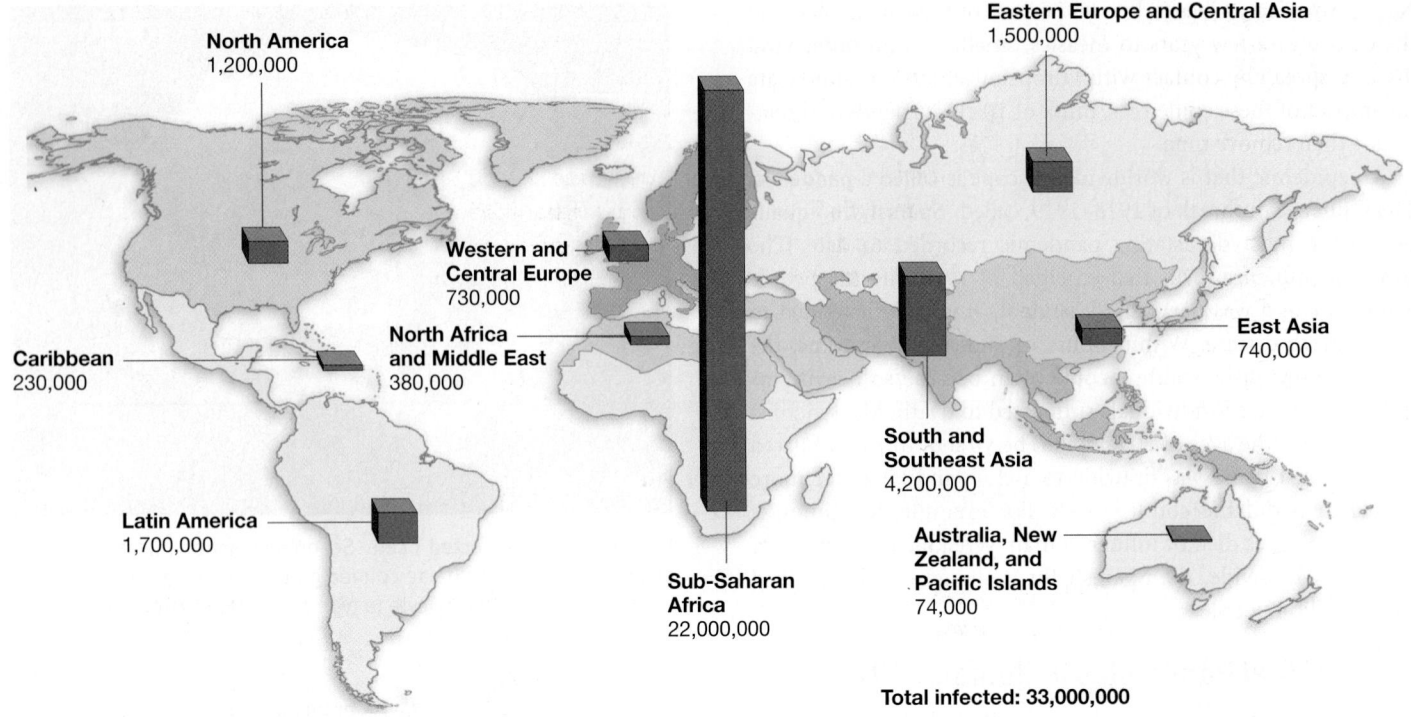

North America
1,200,000

Eastern Europe and Central Asia
1,500,000

Western and Central Europe
730,000

Caribbean
230,000

North Africa and Middle East
380,000

East Asia
740,000

Latin America
1,700,000

Sub-Saharan Africa
22,000,000

South and Southeast Asia
4,200,000

Australia, New Zealand, and Pacific Islands
74,000

Total infected: 33,000,000

FIGURE 35.4 Geographic Distribution of HIV Infections. Data, compiled by the United Nations AIDS program, showing the numbers of people living with HIV in December 2008, by geographic area.

people are infected each year, and the pandemic is growing. Researchers are particularly alarmed because the focus of the epidemic is shifting from its historical center of incidence—central and southern Africa—to south and east Asia (**Figure 35.4**). Infection rates are growing in some of the world's most populous countries—particularly India and China.

Most viral and bacterial diseases afflict the very young and the very old. But because HIV is primarily a sexually transmitted disease, young adults are most likely to contract the virus and die. People who become infected in their late teens or twenties die of AIDS in their twenties or thirties. Tens of millions of people are being lost in the prime of their lives. Physicians, politicians, educators, and aid workers all use the same word to describe the epidemic's impact: staggering.

35.2 How Do Biologists Study Viruses?

Many researchers who study viruses focus on two goals: (**1**) developing vaccines that help hosts fight off disease if they become infected and (**2**) developing antiviral drugs that prevent a virus from replicating efficiently inside the host. Both types of research begin with attempts to isolate the virus in question.

Isolating viruses takes researchers into the realm of nanobiology, in which structures are measured in billionths of a meter. (One nanometer, abbreviated nm, is 10^{-9} meter.) Viruses range from about 20 to 300 nm in diameter. They are dwarfed by eukaryotic cells and even by bacterial cells (**Figure 35.5**). Millions of viruses can fit on the period at the end of this sentence.

If virus-infected cells can be grown in culture or harvested from a host individual, researchers can usually isolate the virus by passing the cells through a filter. The filters used to study viruses have pores that are large enough for viruses to pass through but are too small to admit cells.

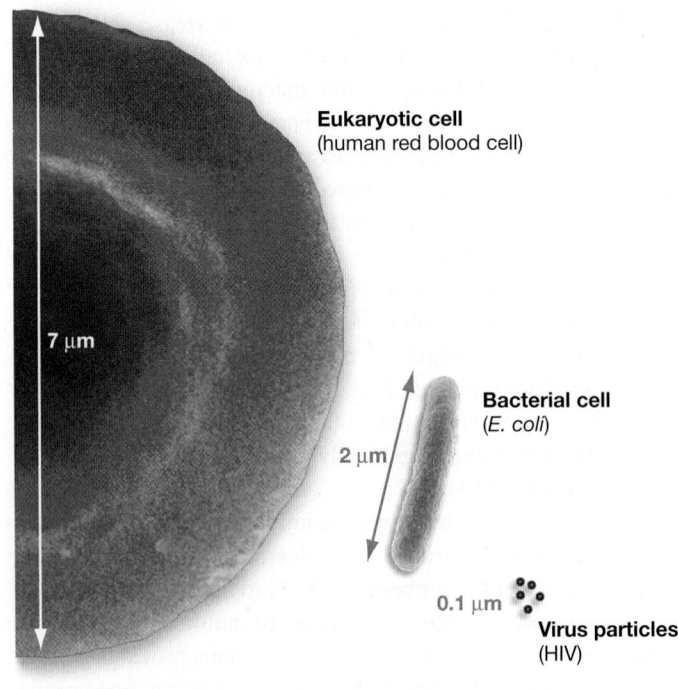

Eukaryotic cell
(human red blood cell)

7 µm

Bacterial cell
(*E. coli*)

2 µm

0.1 µm

Virus particles
(HIV)

FIGURE 35.5 Viruses Are Tiny.

(a) Tobacco mosaic virus **(b)** Adenovirus **(c)** Rotavirus **(d)** Bacteriophage T4

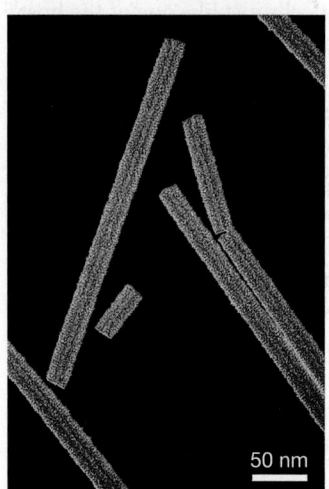
50 nm

100 nm

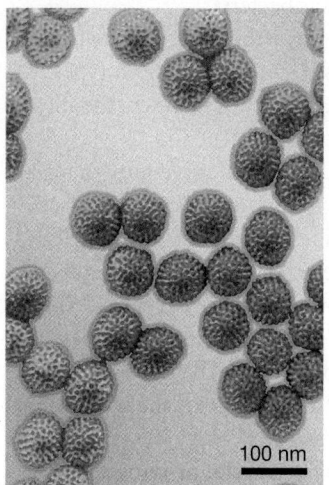
100 nm

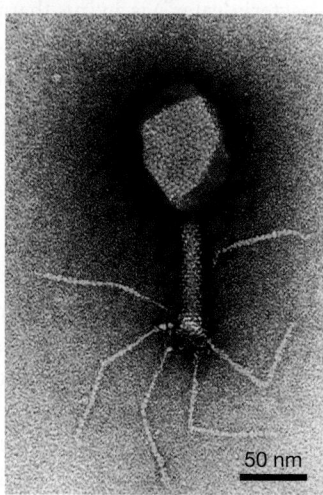
50 nm

FIGURE 35.6 Viruses Vary in Size and Shape. Virus shapes include **(a)** rods, **(b)** polyhedrons, **(c)** spheres, and **(d)** complex shapes with "heads" and "tails." The viruses in these micrographs have been colorized.

To test the hypothesis that the solution passing through the filter contains viruses that cause a specific disease, researchers expose uninfected host cells to this filtrate. If exposing host cells to the filtrate results in infection, then the hypothesis that the virus caused the disease is supported.

In this way, researchers can isolate a virus and confirm that it is the causative agent of infection. Recall from Chapter 28 that these steps are inspired by Koch's postulates, which established the criteria for linking a specific infectious agent with a specific disease.

Once biologists have isolated a virus, how do they study and characterize it? Let's begin with morphological traits, then consider how viral replication cycles vary.

Analyzing Morphological Traits

To see a virus, researchers usually rely on transmission electron microscopy (see **BioSkills 10** in Appendix A). Only the very largest viruses, such as the smallpox virus, are visible with a light microscope. Electron microscopy has revealed that viruses come in a wide variety of shapes, and many viruses can be identified by shape alone (**Figure 35.6**).

🔑 In terms of overall structure, most viruses fall into just two general categories. Most viruses are (1) enclosed by just a shell of protein called a **capsid** or (2) enclosed by both a capsid and a membrane-like **envelope**. Regarding their morphology, then, the important distinction among viruses is whether they are nonenveloped or enveloped.

Nonenveloped viruses consist of genetic material and possibly one or more enzymes inside a capsid—a protein coat. The nonenveloped virus illustrated in **Figure 35.7a** is an adenovirus. You undoubtedly have adenoviruses on your tonsils or in other parts of your upper respiratory passages right now. As the micrograph in Figure 35.6d shows, the morphology of nonenveloped viruses may be complex.

Enveloped viruses also have genetic material inside a capsid or bound to capsid proteins, but the capsid is surrounded by an enve-

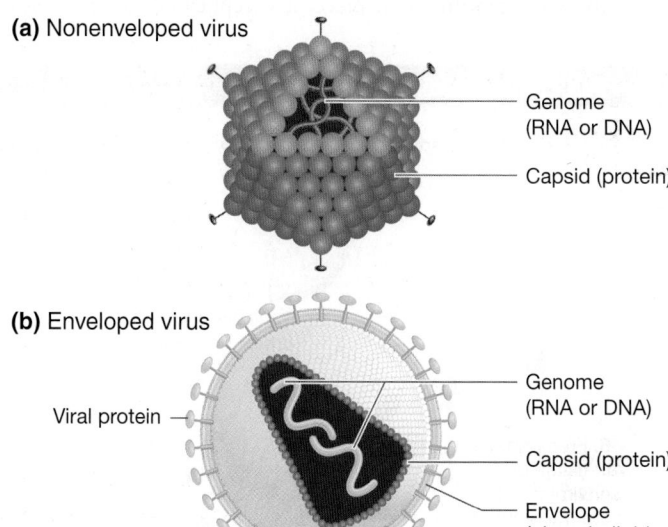

(a) Nonenveloped virus

Genome (RNA or DNA)

Capsid (protein)

(b) Enveloped virus

Viral protein

Genome (RNA or DNA)

Capsid (protein)

Envelope (phospholipid bilayer)

FIGURE 35.7 Viruses Are Nonenveloped or Enveloped. (a) A protein coat forms the exterior of nonenveloped viruses. **(b)** In enveloped viruses, the exterior is composed of a membranous sphere. Inside this envelope, the hereditary material is enclosed by a protein coat or bound to proteins.

lope. The envelope consists of viral proteins embedded in a phospholipid bilayer derived from a membrane found in a host cell—specifically, the host cell in which the virion was manufactured (**Figure 35.7b**). Later sections in the chapter will detail how most of these viruses obtain their envelope from an infected host cell.

Analyzing Variation in Growth Cycles: Replicative and Latent Growth

Although it is likely that millions of types of virus exist, they all infect their host cells in one of two general ways: via replicative growth or in a dormant form referred to as latency in animal viruses or lysogeny in bacteriophages. A **bacteriophage** (literally, "bacteria-eater") is a virus that infects bacteria.

All viruses undergo replicative growth; some can halt the replication cycle and enter a dormant state as well. Both types of viral infection begin when part or all of a virion enters the interior of a host cell.

Figure 35.8a shows the **lytic cycle**, or replicative growth, of a bacteriophage. Note that the viral genome enters the host cell and it is transcribed, using nucleoside triphosphates provided by the host. The host cell manufactures viral proteins and begins to replicate the viral genome using host or viral enzymes. When synthesis of the viral genome and viral proteins is complete, a new generation of virions assembles inside the host cell. The replicative cycle is complete when the virions exit the cell—usually killing the host cell in the process.

Figure 35.8b diagrams the **lysogenic cycle**, or lysogeny, of a bacteriophage. Note that only certain types of viruses are capable of this type of latent infection. During lysogenic growth, viral DNA becomes incorporated into the host's chromosome. Once the viral genome is in place, it is replicated by the host's

DNA polymerase each time the cell divides. Copies of the viral genome are passed on to daughter cells just like one of the host's own genes.

In the lysogenic state, no new virions are being produced and no unrelated cells are being infected. The virus is simply transmitted from one generation to the next along with the host's genes.

✓If you understand this concept, you should be able to explain the contrast in how bacteriophage genes are replicated via lytic versus lysogenic growth.

In some viruses that infect animal cells, a state called **latency** can exist. When an animal virus is latent, the genome just resides in the cell without replicating or producing new virions. Depending on the virus involved, the genome may or may not be integrated into the host chromosome.

It is not possible to treat a latent infection with drugs called **antivirals**, because the viruses are quiescent. But even lytic infections are notoriously difficult to treat because viruses use so many of the host cell's enzymes during the lytic replication cycle. Drugs that disrupt these enzymes usually harm the host much more than they harm the virus.

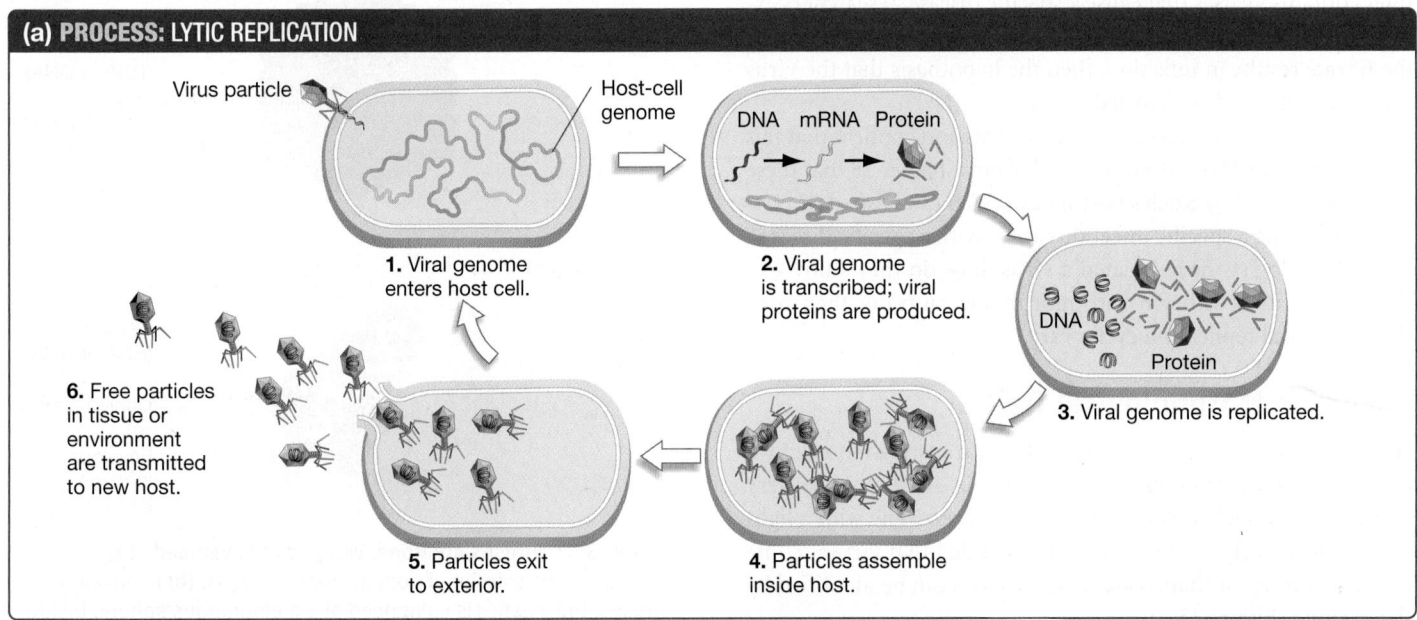

(a) PROCESS: LYTIC REPLICATION

Virus particle — Host-cell genome

1. Viral genome enters host cell.

DNA mRNA Protein

2. Viral genome is transcribed; viral proteins are produced.

DNA ... Protein

3. Viral genome is replicated.

4. Particles assemble inside host.

5. Particles exit to exterior.

6. Free particles in tissue or environment are transmitted to new host.

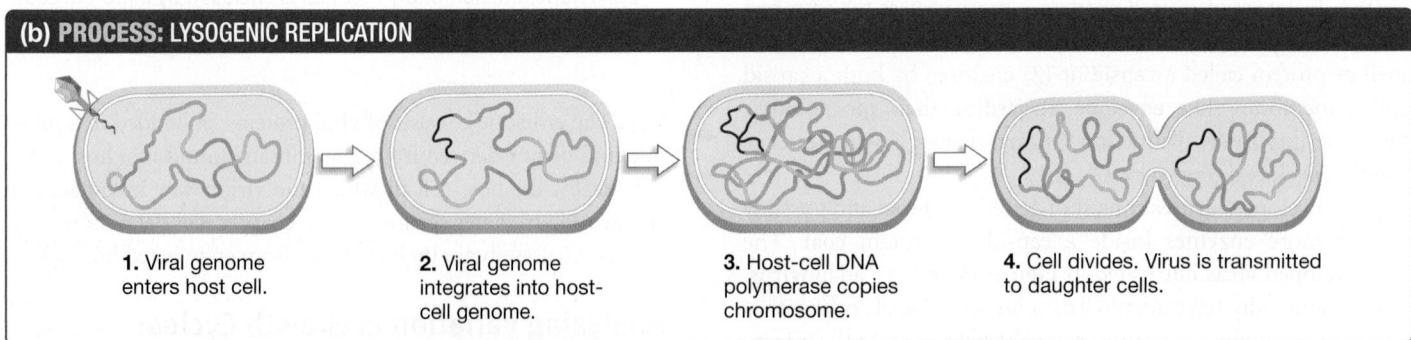

(b) PROCESS: LYSOGENIC REPLICATION

1. Viral genome enters host cell.

2. Viral genome integrates into host-cell genome.

3. Host-cell DNA polymerase copies chromosome.

4. Cell divides. Virus is transmitted to daughter cells.

FIGURE 35.8 Bacteriophages May Replicate via Lytic or Lysogenic Cycles, or Both. (a) All viruses follow the same general lytic replication cycle, which results in a new generation of virions and host cell death. **(b)** Some bacteriophages are also capable of lysogeny, meaning their genome can become integrated into the host-cell chromosome.

✓**EXERCISE** Compare and contrast a lysogenic bacteriophage to the transposable elements introduced in Chapter 20.

To understand viral diversity and how antiviral drugs are developed, let's consider each phase of the lytic cycle in more detail.

Analyzing the Phases of the Replicative Cycle

🔑 Six phases are common to replicative growth in virtually all viruses: (1) attachment onto a host cell and entry into the interior, (2) transcription of the viral genome and production and processing of viral proteins, (3) replication of the viral genome, (4) assembly of a new generation of virions, (5) exit from the infected cell, and (6) transmission to a new host. The corresponding steps are depicted in Figure 35.8a.

Each virus has a particular way of entering a host cell and completing the subsequent phases of the cycle. Let's take a closer look.

HOW DO VIRUSES ENTER A CELL? The replication cycle of a virus begins when a free virion enters a target cell. This is no simple task. All cells are protected by a plasma membrane, and many cells also have a cell wall. How do viruses breach these defenses, insert themselves into the cytoplasm inside, and begin an infection?

Most plant viruses enter host cells after a sucking or biting insect has disrupted the host cell wall with its mouthparts. In contrast, viruses that parasitize bacterial cells or that attack animal cells gain entry by binding to a specific molecule on the cell wall or plasma membrane.

After binding to a bacterial cell wall, bacteriophages use an enzyme called lysozyme to degrade the wall. (Lysozyme is also found in human tears, where it acts as an antibiotic in the eye.) The genome of the nonenveloped virus enters the host cell through the hole created by lysozyme. A portion of the capsid seals the hole, and the remainder of the capsid remains on the cell wall or membrane.

Enveloped viruses and noneveloped viruses that attack animal cells bind to one or more specific proteins in the host cell's plasma membrane. When this happens, the capsid of a nonenveloped virus is taken up or the viral envelope and the host's plasma membrane fuse, and the capsid enters the cell.

To appreciate how investigators identify the proteins that viruses use to enter host cells, consider research on HIV. In 1981— right at the start of the AIDS epidemic—biomedical researchers realized that people with AIDS had few or no T lymphocytes possessing **CD4**, a particular membrane protein. These cells, called helper T lymphocytes, are symbolized $CD4^+$. This discovery led to the hypothesis that CD4 functions as the receptor or "doorknob" that HIV uses to enter host cells. The doorknob hypothesis predicts that if CD4 is blocked, then HIV will not be able to enter host cells.

Two teams tested this hypothesis (**Figure 35.9**):

- They grew large populations of T lymphocytes in culture and created 160 separate samples.

- They added an antibody to one of the cell-surface proteins found on T lymphocytes, with each of the 160 samples receiving a different antibody. Recall from earlier chapters that **antibodies** are proteins that bind to specific parts of specific proteins (see **BioSkills 9** in Appendix A).

- They added HIV virions to each sample.

EXPERIMENT

QUESTION: Does the CD4 protein function as the "doorknob" that HIV uses to enter host cells?

HYPOTHESIS: CD4 is the membrane protein that HIV uses to enter cells.

NULL HYPOTHESIS: CD4 is not the membrane protein that HIV uses to enter cells.

EXPERIMENTAL SETUP:

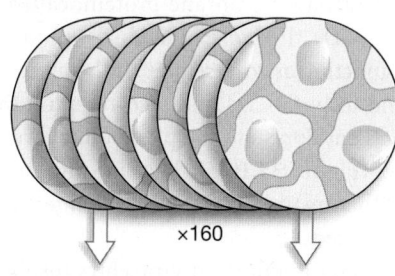

1. Take 160 identical samples of helper T cells from a large population of T cells growing in culture.

×160

Antibody to protein other than CD4 — Antibody to CD4

2. Add a different antibody to each sample of cells — each antibody will "block" a specific membrane protein.

Add HIV — Add HIV

3. Add a constant number of HIV virions to all samples. Incubate cultures under conditions optimal for virus entry.

PREDICTION: HIV will not infect cells with antibody to CD4 but will infect other cells.

PREDICTION OF NULL HYPOTHESIS: HIV will infect cells with antibody to CD4.

RESULTS:

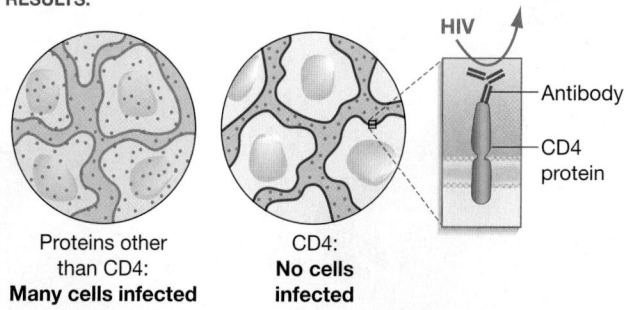

HIV

Antibody

CD4 protein

Proteins other than CD4: **Many cells infected**

CD4: **No cells infected**

CONCLUSION: HIV uses CD4 proteins as the "doorknob" to enter helper T cells. Thus, only cells with CD4 on their surface can be infected by HIV.

FIGURE 35.9 Experiments Confirmed That CD4 Is the Receptor Used by HIV to Enter Host Cells. In this experiment, the antibodies added to each culture bound to a specific protein found on the surface of T lymphocytes. Antibody binding blocked the membrane protein, so the protein could not be used by HIV to gain entry to the cells.

SOURCE: Maddon, P. J., et al. 1986. The T4 gene encodes the AIDS virus receptor and is expressed in the immune system and the brain. Cell 47: 333–348.

✔**QUESTION** Does this experiment show that CD4 is the only membrane protein required for HIV entry? Explain why or why not.

The key point here is that each of the 160 antibodies blocked a different cell-surface protein. If one of the antibodies used in the experiment happened to bind to the receptor used by HIV, that antibody would cover up the receptor. In this way, the antibody would prevent HIV from entering the cell and protect that cell from infection.

This approach led both research teams to reach exactly the same result: Only antibodies to CD4 protected the cells from viral entry.

Later work confirmed that HIV virions can enter cells only if the virions also bind to a second membrane protein, called a **co-receptor**, in addition to CD4. In most individuals, proteins called CXCR4 and CCR5 function as co-receptors. If the proteins in a virion's envelope bind to both CD4 and a co-receptor, the lipid bilayers of the virion's envelope and the plasma membrane of the T lymphocyte fuse (**Figure 35.10**). When fusion occurs, HIV has breached the cell boundary. The viral capsid enters the cytoplasm and infection proceeds.

The discovery of the proteins required for viral entry inspired a search for compounds that would block them and prevent HIV from entering cells. Drugs that act in this way are called fusion inhibitors. Molecules that block the HIV envelope protein or CCR5 are currently in use, and molecules that block CD4 are now being tested.

PRODUCING VIRAL PROTEINS Viruses lack the ribosomes and tRNAs necessary for translating their own mRNAs into proteins. For a virus to make the proteins required to produce a new generation of virions, it must exploit the host cell's biosynthetic machinery.

Production of viral proteins begins soon after a virus enters a cell and continues after the viral genome is replicated. Viral mRNAs and proteins are produced and processed in one of two ways, depending on whether the proteins end up in the outer envelope of a virion or in the capsid.

RNAs that code for a virus's envelope proteins are translated into proteins as if they were the RNAs of the cell's own membrane proteins (see Chapter 7). As **Figure 35.11a** shows, these viral mRNAs are translated by ribosomes attached to the endoplasmic reticulum (ER). Afterward the resulting proteins are transported to the Golgi apparatus, where carbohydrate groups are attached, producing glycoproteins. In some viruses, such as HIV, the finished glycoproteins are then transported to the plasma membrane, where they are ready to be assembled into new virions.

In contrast, a different route is taken by RNAs that code for proteins that make up the capsid or inner core of a virion (see **Figure 35.11b**). These RNAs are translated by ribosomes in the cytoplasm, just as non–membrane-bound cellular mRNAs are. In some viruses, long polypeptide sequences called polyproteins are later cut into functional proteins by a viral enzyme called **protease**. This enzyme cleaves viral polyproteins at specific locations—a critical step in the production of finished viral proteins. In HIV, the resulting protein fragments are assembled into new viral capsids near the host cell's plasma membrane.

The discovery that HIV produces a protease triggered a search for drugs that would block the enzyme. This search got a huge boost when researchers succeeded in visualizing the three-dimensional structure of HIV's protease, using the X-ray crystallographic techniques introduced in **BioSkills 10** in Appendix A. The enzyme has an opening in its interior adjacent to the active site (**Figure 35.12a**). Viral polyproteins fit into the opening and are cleaved at the active site.

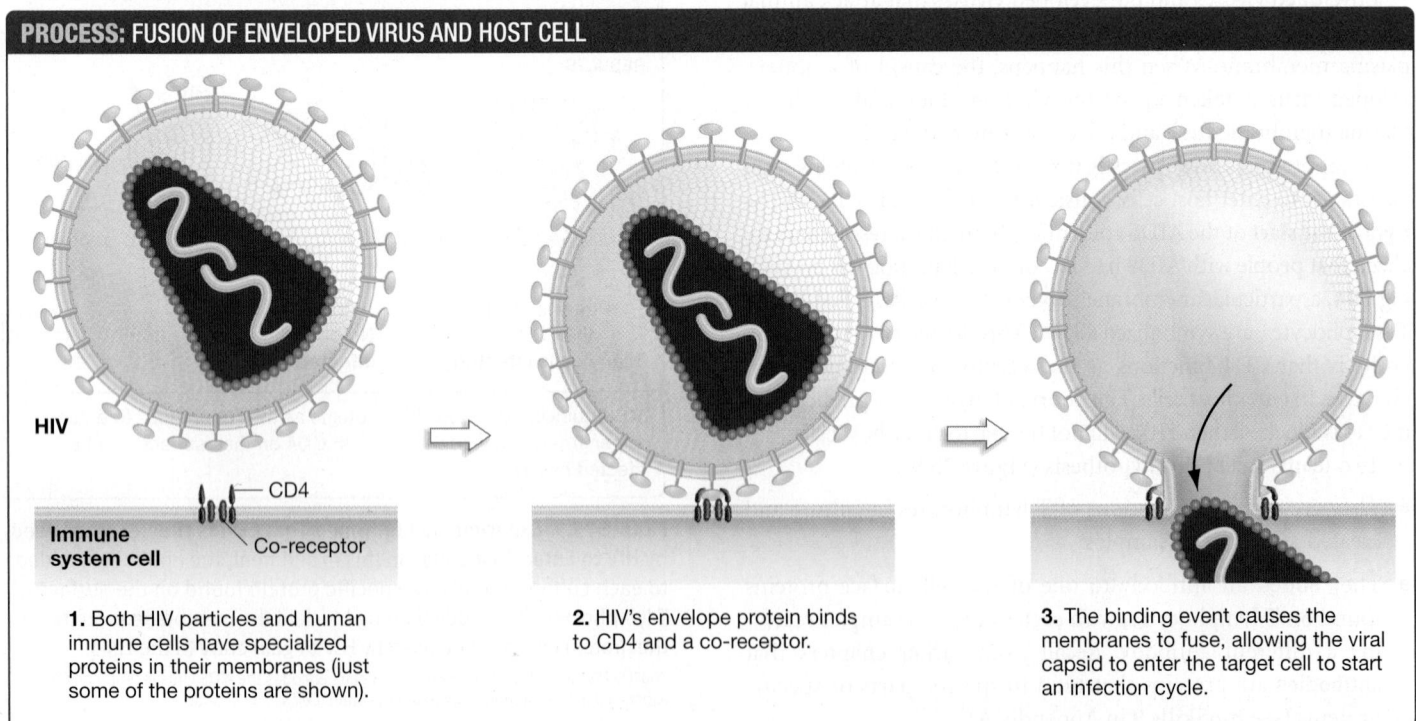

PROCESS: FUSION OF ENVELOPED VIRUS AND HOST CELL

HIV

Immune system cell — CD4 — Co-receptor

1. Both HIV particles and human immune cells have specialized proteins in their membranes (just some of the proteins are shown).

2. HIV's envelope protein binds to CD4 and a co-receptor.

3. The binding event causes the membranes to fuse, allowing the viral capsid to enter the target cell to start an infection cycle.

FIGURE 35.10 Enveloped Viruses Bind to Membrane Proteins of Target Cells.

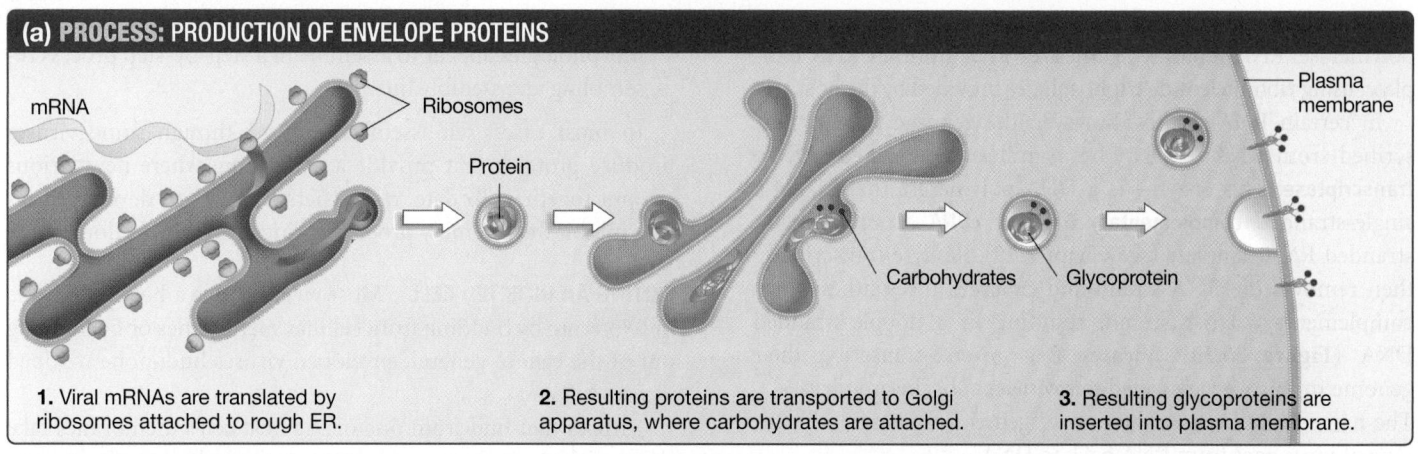

(a) PROCESS: PRODUCTION OF ENVELOPE PROTEINS

mRNA

Ribosomes

Protein

Plasma membrane

Carbohydrates   Glycoprotein

1. Viral mRNAs are translated by ribosomes attached to rough ER.

2. Resulting proteins are transported to Golgi apparatus, where carbohydrates are attached.

3. Resulting glycoproteins are inserted into plasma membrane.

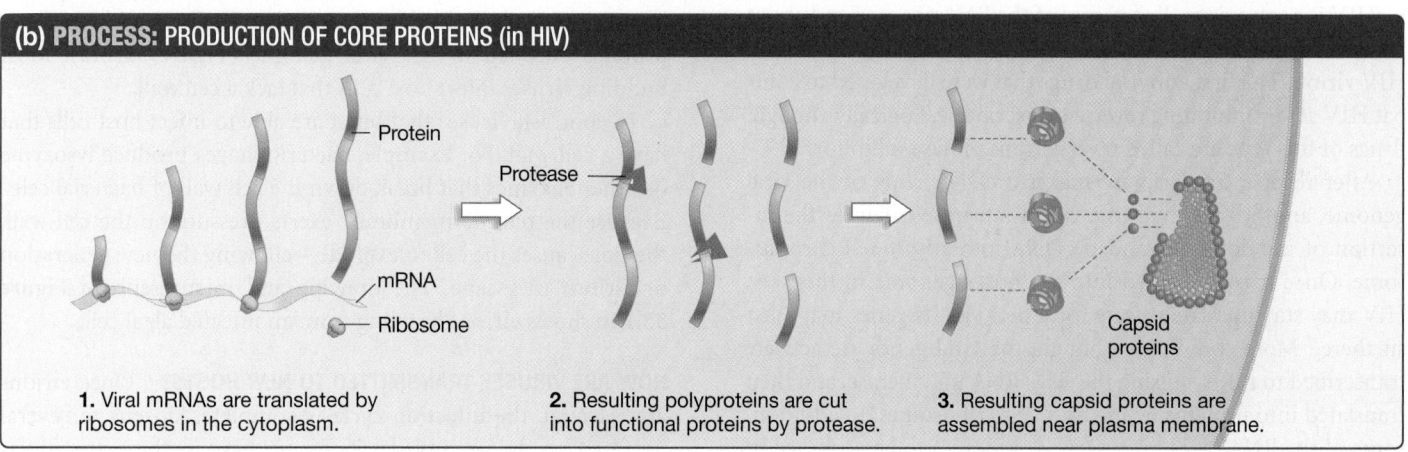

(b) PROCESS: PRODUCTION OF CORE PROTEINS (in HIV)

Protein

Protease

mRNA

Ribosome

Capsid proteins

1. Viral mRNAs are translated by ribosomes in the cytoplasm.

2. Resulting polyproteins are cut into functional proteins by protease.

3. Resulting capsid proteins are assembled near plasma membrane.

FIGURE 35.11 Production of Viral Proteins.

Based on these data, researchers immediately began searching for molecules that could fit into the opening and prevent protease from functioning by blocking the active site (**Figure 35.12b**). Several HIV protease inhibitors are being used to reduce viral replication.

HOW DO VIRUSES COPY THEIR GENOMES?   Viruses must copy their genes to make a new generation of virions and continue an infection. Many viruses contain a polymerase that copies the viral genome inside the infected host cell, using nucleoside triphosphates provided by the host cell. Some DNA viruses, for example, contain a viral DNA polymerase that makes a copy of the viral genome.

In other viruses, however, the genome consists of RNA. In most viruses that have an RNA genome, copies of the genome are

(a) HIV's protease enzyme

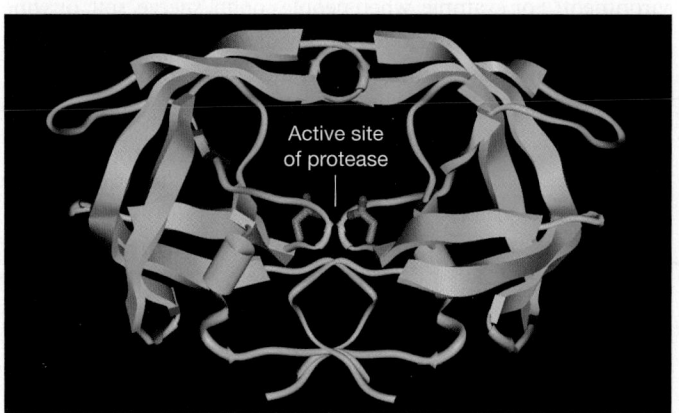

Active site of protease

(b) Could a drug block the active site?

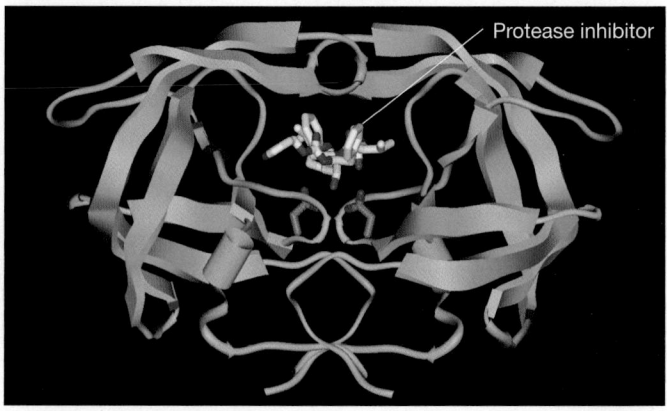

Protease inhibitor

FIGURE 35.12 The Three-Dimensional Structure of HIV's Protease. (a) Ribbon diagram depicting the three-dimensional shape of HIV's protease enzyme. **(b)** Once protease's structure was solved, researchers began synthesizing compounds that were predicted to fit into the active site and prevent the enzyme from working.

synthesized by the viral enzyme **RNA replicase**, which is an RNA polymerase. RNA replicase synthesizes RNA from an RNA template, using ribonucleotide triphosphates provided by the host cell.

In certain RNA viruses, however, the genome is first transcribed from RNA to DNA by a viral enzyme called **reverse transcriptase**. This enzyme is a DNA polymerase that makes a single-stranded **complementary DNA**, or **cDNA**, from a single-stranded RNA template (see Chapter 19). Reverse transcriptase then removes the RNA strand and catalyzes the synthesis of a complementary DNA strand, resulting in a double-stranded DNA (**Figure 35.13**). Viruses that reverse-transcribe their genome in this way are called **retroviruses** ("backward viruses"). The name is apt because the flow of genetic information in this type of virus goes from RNA back to DNA.

HIV is a retrovirus. Two copies of the RNA genome and about 50 molecules of reverse transcriptase lie inside the capsid of each HIV virion. The first antiviral drugs that were developed to combat HIV act by inhibiting reverse transcriptase. Logically enough, drugs of this type are called reverse transcriptase inhibitors.

After reverse transcriptase makes a cDNA copy of the viral genome, another viral enzyme called integrase catalyzes the insertion of the double-stranded cDNA into a host-cell chromosome. Once it is integrated into the host's genome in this way, HIV may stay in a latent state for a period—its genes may "just sit there." More commonly, though, the viral genes are actively transcribed to mRNA, using the cell's RNA polymerase, and then translated into proteins by the host cell's ribosomes. In addition, some of the RNA transcripts serve as genomes to be packaged in the next generation of HIV virions.

ASSEMBLY OF NEW VIRIONS Once the viral genome has been replicated and viral proteins are produced, they are transported to locations where a new generation of virions assembles inside the infected host cell.

During assembly, the capsid forms around the viral genome. In many cases, copies of non-capsid proteins like viral DNA or RNA polymerases or reverse transcriptases are also packaged inside the capsid. And in enveloped viruses, envelope proteins become inserted into the host cell's plasma membrane.

In many cases, the details of the assembly process are not yet well understood.

- HIV and other enveloped viruses appear to assemble while attached to the inside surface of the host's plasma membrane.

- Bacteriophages and other nonenveloped viruses with complex morphologies appear to assemble in a step-by-step process resembling an assembly line.

In most cases, self-assembly occurs—though some viruses produce proteins that provide a scaffolding where new virions are put together. To date, researchers have yet to develop drugs that inhibit the assembly process during an HIV infection.

EXITING AN INFECTED CELL Most viruses leave a host cell in one of two ways: by budding from cellular membranes or by bursting out of the cell. In general, enveloped viruses bud; nonenveloped viruses burst.

Viruses that bud from one of the host cell's membranes take some of that membrane with them. As a result, they incorporate host-cell phospholipids into their envelope, along with envelope proteins encoded by the viral genome (**Figure 35.14a**). Most budding viruses infect host cells that lack a cell wall.

In contrast, viruses that burst are able to infect host cells that have a cell wall. For example, bacteriophages produce lysozyme or other enzymes that break down the cell wall of bacterial cells. Because the plasma membrane exerts pressure on the cell wall, the hole causes the cell to explode—allowing the new generation of virions to escape. The drawing and micrograph in **Figure 35.14b** shows virions bursting from an infected algal cell.

HOW ARE VIRUSES TRANSMITTED TO NEW HOSTS? Once virions are released, the infection cycle is complete. Dozens to several hundred newly assembled virions are now in the extracellular space. What happens next?

If the host cell is part of a multicellular organism, the new generation of virions begins traveling through the body—often via the bloodstream or lymphatic system. There, they may be bound by antibodies produced by the immune system. In vertebrates, this binding marks the virions for destruction. But if a virion contacts an appropriate host cell before it encounters antibodies, then the virion will infect that cell. This starts the replication cycle anew.

What if the virus has infected a unicellular organism, or if the virus leaves a multicellular host entirely and enters the external environment? For example, when people cough, sneeze, spit, or wipe a runny nose, they help rid their body of viruses and bacteria. But they also project the pathogens into the environment, sometimes directly onto an uninfected host.

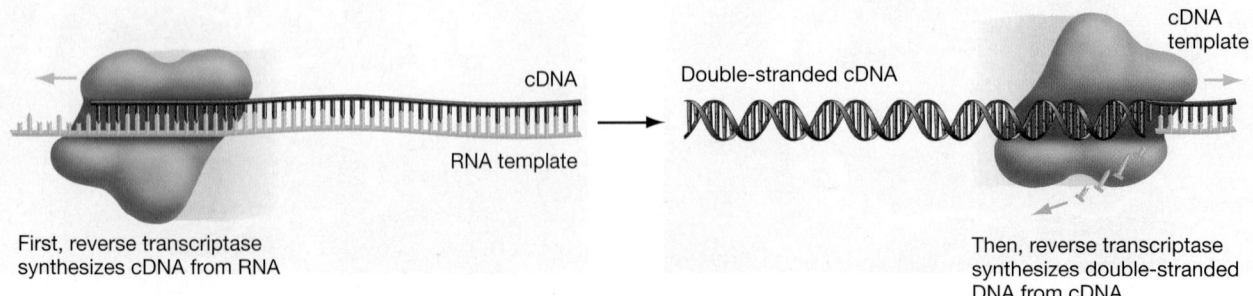

FIGURE 35.13 **Reverse Transcriptase Catalyzes Synthesis of a Double-Stranded DNA from an RNA Template.** The DNA produced by reverse transcriptase is called a cDNA because its base sequence is complementary to the RNA template.

(a) Budding of enveloped viruses

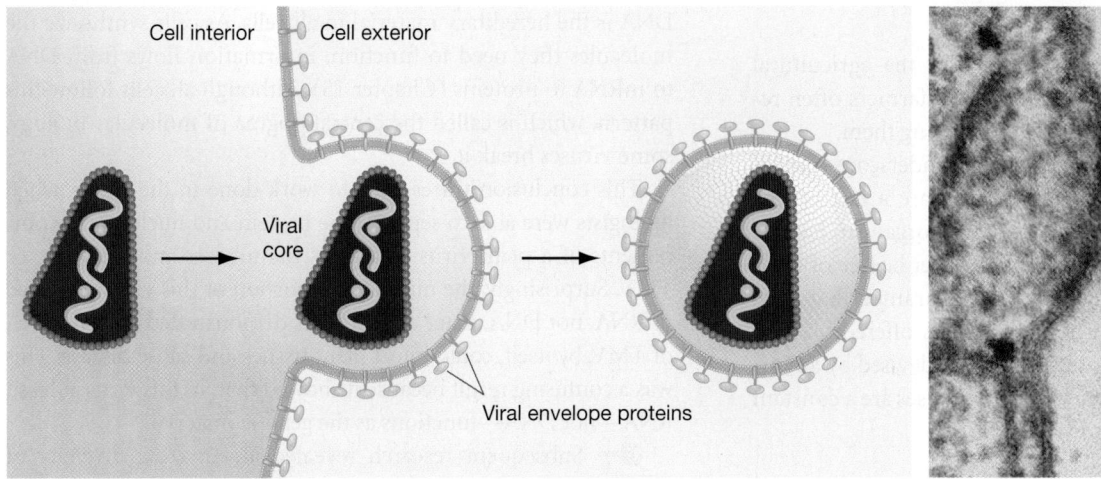
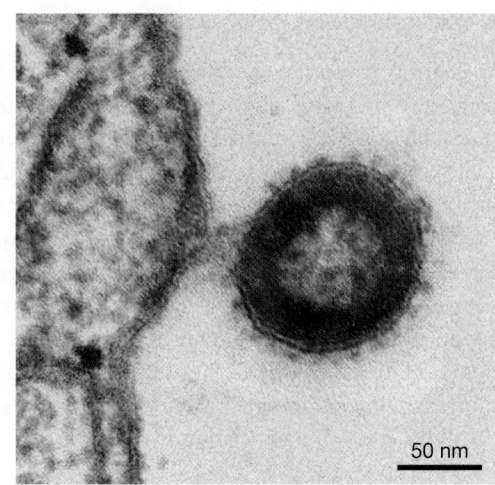

(b) Bursting of nonenveloped viruses

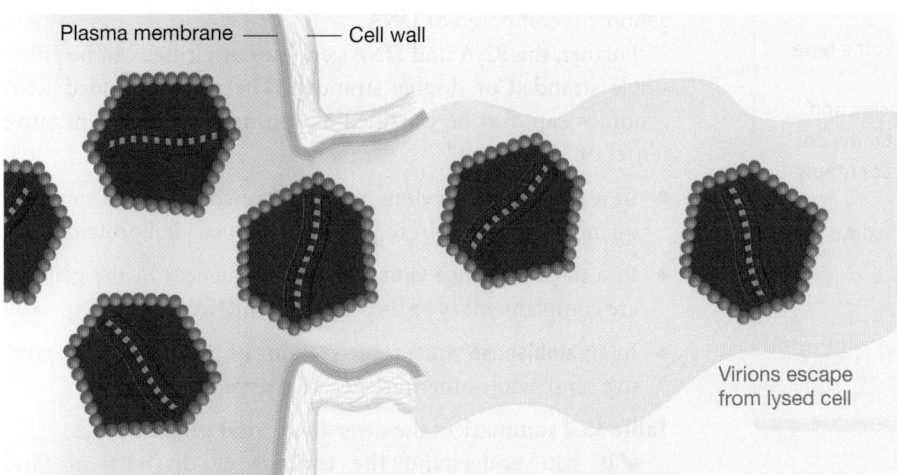
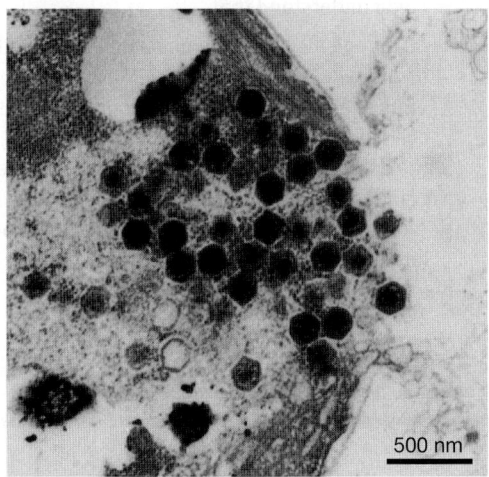

FIGURE 35.14 Viruses Leave Infected Cells by Budding or Bursting. Bursting kills the host cell; budding kills the host cell if it is extensive.

✔**QUESTION** Propose a hypothesis to explain why budding can be fatal to the host cell.

From the virus's point of view, a new host represents an unexploited habitat brimming with resources in the form of target cells. The situation is analogous to that of a multicellular animal dispersing to a new habitat and colonizing it. Viruses that successfully colonize a new host replicate and increase in number. The alleles carried by these successful colonists increase in frequency in the total population. In this way, natural selection favors alleles that allow viruses to do two things: (**1**) replicate within a host and (**2**) be transmitted to new hosts.

To review the phases of the HIV replicative cycle, including transmission, go to the study area at *www.masteringbiology.com*.

(MB) **Web Activity** The HIV Replicative Cycle

For agricultural scientists, physicians, and public health officials, reducing the likelihood of **transmission**—the spread of pathogens from one individual to another—is often an effective way to combat a viral disease. For example, HIV is trans-

mitted from person to person via body fluids such as blood, semen, or vaginal secretions. Faced with decades of disappointing results in drug and vaccine development, public health officials are aggressively promoting preventive medicine, through:

● condom use;

● aggressive treatment of venereal diseases, because the lesions caused by chlamydia, genital warts, and gonorrhea encourage the transmission of HIV-contaminated blood during sexual intercourse; and

● promoting sexual abstinence or monogamy.

In the case of viruses that are transmitted through more casual person-to-person contact, the most effective means of reducing transmission involve the simple steps that public health officials recommend during flu and cold outbreaks.

● frequent and thorough hand washings,

- covering the mouth and nose when coughing or sneezing, and
- staying home during illness.

Halting transmission is also important in the agricultural context. When virus outbreaks threaten crops, farmers often respond by removing infected plants and destroying them.

The effectiveness of preventive medicine underscores one of this chapter's fundamental messages: Viruses are a fact of life. Every organism is victimized by viruses; every organism has defenses against them. But the tree of life will never be free of these parasites. Mutation and natural selection guarantee that viral genomes will continually adapt to the defenses offered by their hosts, regardless of whether those defenses are devised by an immune system or by biomedical researchers. Viruses are a constant threat for every organism alive.

CHECK YOUR UNDERSTANDING

⦿━ If you understand that . . .

- All organisms and cell types are parasitized by some type of virus.
- Viruses are specialized for infecting certain species and cell types, because for a virus to enter a cell, proteins on the exterior of the virus have to bind to proteins or other components on the surface of the host cell.
- After infecting a cell, viruses may replicate or become latent.

✔ You should be able to . . .

Explain how HIV enters a host cell.

Answers are available in Appendix B.

35.3 What Themes Occur in the Diversification of Viruses?

If viruses can infect virtually every type of organism and cell known, how can biologists possibly identify themes that help organize viral diversity? The answer is that, in addition to being identified as enveloped or nonenveloped, viruses can be categorized by the nature of their hereditary material—in essence, the type of molecule that serves as the genome. The single most important aspect of viral diversity is the variation that exists in their genetic material.

Two other points are critical to recognize about viral diversity: (1) Biologists do not have a solid understanding of how viruses originate, but (2) it is certain that viruses will continue to diversify. Because viral polymerases have high error rates and because viruses lack error repair enzymes, mutation rates are extremely high. Viruses change constantly—giving them the potential to evolve rapidly.

Let's analyze the diversity of molecules that make up viral genes, then consider hypotheses to explain where viruses come from and recent data on how viruses are evolving before our eyes.

The Nature of the Viral Genetic Material

DNA is the hereditary material in all cells. As cells synthesize the molecules they need to function, information flows from DNA to mRNA to proteins (Chapter 15). Although all cells follow this pattern, which is called the central dogma of molecular biology, some viruses break it.

This conclusion traces back to work done in the 1950s, when biologists were able to separate the protein and nucleic acid components of a plant virus known as the tobacco mosaic virus, or TMV. Surprisingly, the nucleic acid portion of this virus consisted of RNA, not DNA. Later experiments demonstrated that the RNA of TMV, by itself, could infect plant tissues and cause disease. This was a confusing result because it showed that, in this virus at least, RNA—not DNA—functions as the genetic material.

⦿━ Subsequent research revealed an amazing diversity of viral genome types. In some groups of viruses, such as the agents that cause measles and flu, the genome consists of RNA. In others, such as the viruses that cause herpes and smallpox, the genome is composed of DNA.

Further, the RNA and DNA genomes of viruses can be either single stranded or double stranded. The single-stranded RNA genomes can also be classified as "positive sense" or "negative sense" or "ambisense."

- In a **positive-sense virus**, the genome contains the same sequences as the mRNA required to produce viral proteins.

- In a **negative-sense virus**, the base sequences in the genome are complementary to those in viral mRNAs.

- In an **ambisense virus**, some sections of the genome are positive sense while others are negative sense.

Table 35.2 summarizes the diversity of viral genome types.

✔ If you understand the concept of diversity in viral genomes, you should be able to name two types of viral genomes that are not based on double-stranded DNA. For each of these genomes, you should be able to explain the steps in information flow from gene to protein.

Finally, the number of genes found in viruses varies widely. The tymoviruses that infect plants contain as few as three genes, but the genome of smallpox can code for up to 353 proteins.

Where Did Viruses Come From?

No one knows how viruses originated. To address this question, biologists are currently considering three hypotheses to explain where viruses came from.

ORIGIN IN PLASMIDS AND TRANSPOSABLE ELEMENTS? Like viruses, the plasmids introduced in Chapter 19 and the transposable elements introduced in Chapter 20 are acellular, mobile genetic elements that replicate with the aid of a host cell. Certain viruses are actually indistinguishable from plasmids except for one feature: They encode proteins that form a capsid and allow the genes to exist outside of a cell.

Some biologists hypothesize that simple viruses, plasmids, and transposable elements represent "escaped gene sets." This hy-

Key: ss = single stranded; ds = double stranded; (+) = positive sense (genome sequence is the same as viral mRNA);
(−) = negative sense (genome sequence is complementary to viral mRNA)

Genome		Example(s)	Host	Result of Infection	Notes
(+)ssRNA	+	TMV	Tobacco plants	Tobacco mosaic disease (leaf wilting)	TMV was the first RNA virus to be discovered.
(−)ssRNA	−	Influenza	Many mammal and bird species	Influenza	The negative-sense ssRNA viruses transcribe their genomes to mRNA via RNA replicase.
dsRNA	+ / −	Phytovirus	Rice, corn, and other crop species	Dwarfing	Double-stranded RNA viruses are transmitted from plant to plant by insects. Many can also replicate in their insect hosts.
ssRNA that requires reverse transcription for replication	+	Rous sarcoma virus	Chickens	Sarcoma (cancer of connective tissue)	These are called retroviruses. Rous sarcoma virus was identified as a cancer-causing agent in 1911—decades before any virus was seen.
ssDNA—can be (+), (−), or (+) and (−)	+ or −	φX174	Bacteria	Death of host cell	The genome for φX174 is circular and was the first complete genome ever sequenced.
dsDNA that is replicated through an RNA intermediate	+ / −	Hepatitis B virus	Humans	Hepatitis	These are called "reversiviruses."
dsDNA that is replicated by DNA polymerase	+ / −	Baculovirus Smallpox Bacteriophage	Insects Humans Bacteria	Death Smallpox Death	These include the largest viruses in terms of genome size and overall size.

pothesis states that mobile genetic elements are descended from clusters of genes that physically escaped from bacterial or eukaryotic chromosomes long ago.

According to this hypothesis, the escaped gene sets took on a mobile, parasitic existence because they happened to encode the information needed to replicate themselves at the expense of the genomes that once held them. In the case of viruses, the hypothesis is that the escaped genes included the instructions for making a protein capsid and possibly envelope proteins. According to the escaped-gene hypothesis, it is likely that each of the distinct types of RNA viruses and some of the DNA viruses represent distinct "escape events."

To support the escaped-genes hypothesis, researchers would probably have to discover a brand-new virus that originated in this way, or viruses that had so recently derived from intact bacterial or eukaryotic genes that the viral DNA sequence still strongly resembled the DNA sequence of those genes.

ORIGIN IN SYMBIOTIC BACTERIA? Some researchers contend that DNA viruses with large genomes trace their ancestry back to free-living bacteria that once took up residence inside eukaryotic cells. The idea is that these organisms degenerated into viruses by gradually losing the genes required to synthesize ribosomes, ATP, nucleotides, amino acids, and other compounds.

Although this idea sounds speculative, it cannot be dismissed lightly. Chapter 28 introduced species in the genus *Chlamydia*, which are bacteria that live as parasites inside animal cells. And Chapter 29 provided evidence that the organelles called mitochondria and chloroplasts, which reside inside eukaryotic cells, originated as intracellular symbionts. Investigators contend that, instead of evolving into intracellular symbionts that aid their host cell, DNA viruses became parasites capable of replication and transmission from one host to another.

To support the degeneration hypothesis, researchers have pointed to mimivirus, which contains some genes involved in protein synthesis. It is still not clear, though, whether these genes are remnants from an ancestral cell or were acquired from a host cell that was infected relatively recently. If the degeneration hypothesis is correct, then mimivirus should eventually lose these genes, and become completely dependent on host cell machinery to make its proteins.

ORIGIN AT THE ORIGIN OF LIFE? Recently, some researchers have started to discuss a third alternative to explain the origin of viruses: that they trace their ancestry back to the first, RNA-based forms of life on Earth. If this hypothesis is correct, then the RNA genomes of some viruses are descended from genes found in early inhabitants of the RNA world (see Chapter 4).

To support this hypothesis, advocates point to the ubiquity of viruses—which suggests that they have been evolving along with organisms since life began. But currently, there is no widely accepted view of where viruses came from. Because viruses are so diverse, all three hypotheses may be valid.

Emerging Viruses, Emerging Diseases

Although it is not known how the various types of virus originated, it is certain that viruses will continue to diversify. With alarming regularity, the front pages of newspapers carry accounts of deadly viruses that are infecting humans for the first time.

- In 1993 a hantavirus that normally infects mice suddenly afflicted dozens of people in the southwestern United States. Nearly half of the people who developed hantavirus pulmonary syndrome died.

- In 1995 the Ebola virus, a variant of a monkey virus, caused a wave of infections in the Democratic Republic of Congo. By the time the outbreak subsided, over 200 cases had been reported; 80 percent were fatal.

- Numerous reports of avian flu infecting and killing humans caused worldwide alarm in 2005–2006.

HIV, hantavirus pulmonary syndrome, Ebola, and avian flu are examples of **emerging diseases**: new illnesses that suddenly affect significant numbers of individuals in a host population. In these cases, the causative agents were considered emerging viruses because they had switched from their traditional host species to a new host—humans.

USING PHYLOGENETIC TREES TO UNDERSTAND EMERGING VIRUSES

How do researchers know that an emerging virus has "jumped" to a new host? The answer is to analyze the evolutionary history of the virus in question, and estimate a phylogenetic tree that includes its close relatives.

HIV, for example, belongs to a group of viruses called the lentiviruses, which infect a wide range of mammals, including house cats, horses, goats, and primates. (*Lenti* is a Latin root that means "slow"; here it refers to the long period observed between the start of an infection by these viruses and the onset of the diseases they cause.) Consider the conclusions that can be drawn from the phylogenetic tree of HIV, shown in **Figure 35.15**.

1. ***There are immunodeficiency viruses.*** Many of HIV's closest relatives parasitize cells that are part of the immune system. Several of them cause diseases with symptoms reminiscent of AIDS. For unknown reasons, though, HIV's closest relatives don't appear to cause disease in their hosts. These viruses infect monkeys and chimpanzees and are called simian immunodeficiency viruses (SIVs).

2. ***There are two HIVs.*** There are two distinct types of human immunodeficiency viruses, called HIV-1 and HIV-2. Although both can cause AIDS, HIV-1 is far more virulent and is the better studied of the two.

    HIV-1's closest known relatives are immunodeficiency viruses isolated from chimpanzees that live in central Africa.

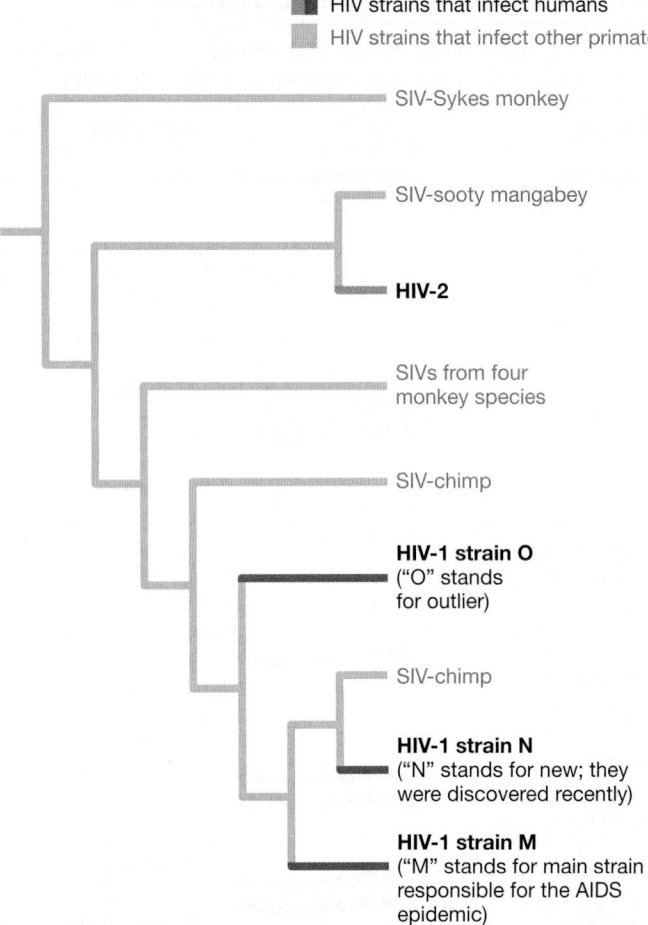

■ HIV strains that infect humans
■ HIV strains that infect other primates

- SIV-Sykes monkey
- SIV-sooty mangabey
- **HIV-2**
- SIVs from four monkey species
- SIV-chimp
- **HIV-1 strain O** ("O" stands for outlier)
- SIV-chimp
- **HIV-1 strain N** ("N" stands for new; they were discovered recently)
- **HIV-1 strain M** ("M" stands for main strain responsible for the AIDS epidemic)

FIGURE 35.15 Phylogeny of HIV Strains and Types. Phylogenetic tree showing the evolutionary relationships among some of the immunodeficiency viruses that infect primates—including chimpanzees, humans, and several species of monkeys.

✔**EXERCISE** On the appropriate branches, indicate where an SIV jumped to humans (draw and label bars across the appropriate branches).

In contrast, HIV-2's closest relatives are immunodeficiency viruses that parasitize monkeys called sooty mangabeys.

In central Africa, where HIV-1 infection rates first reached epidemic proportions, contact between chimpanzees and humans is extensive. Chimps are hunted for food and kept as pets. Similarly, sooty mangabeys are hunted and kept as pets in west Africa, where HIV-2 infection rates are highest.

To make sense of these observations, biologists suggest that HIV-1 is a descendant of viruses that infect chimps, and HIV-2 is a descendant of viruses that infect sooty mangabeys. The "jumps" between host species probably occurred when a human cut up a monkey or chimpanzee for use as food.

3. ***Multiple "jumps" have occurred.*** Several strains of HIV-1 exist. A virus **strain** consists of populations that have similar characteristics. The most important HIV-1 strains are called O for outlier (meaning, the most distant group relative to other

strains), N for new, and M for main. Because few mechanisms exist for the virus to jump from humans back to chimps or monkeys, it is likely that each of these strains represents an independent origin of HIV from a chimp SIV strain.

The last point is particularly important. The existence of distinct strains suggests that HIV-1 has jumped from chimps to humans several times. It may do so again in the future.

RESPONDING TO AN OUTBREAK Physicians become alarmed when they see a large number of patients with identical and unusual disease symptoms in the same geographic area over a short period of time. The physicians report these cases to public health officials, who take on two urgent tasks: (1) identifying the agent that is causing the new illness and (2) determining how the disease is being transmitted.

Several strategies can be used to identify the pathogen responsible for an emerging disease. In the case of the outbreak of hantavirus pulmonary syndrome, officials recognized strong similarities between symptoms in the U.S. cases and symptoms caused by a hantavirus called Hantaan virus native to northeast Asia. The Hantaan virus rarely causes disease in North America; its normal host is rodents.

To determine whether a Hantaan-like virus was responsible for the U.S. outbreak, researchers began capturing mice in the homes and workplaces of afflicted people. About a third of the captured rodents tested positive for the presence of a Hantaan-like virus. Genome sequencing studies confirmed that the virus was a previously undescribed type of hantavirus. Further, the sequences found in the mice matched those found in infected patients. Based on these results, officials were confident that a rodent-borne hantavirus was causing the wave of infections.

The next step in the research program, identifying how the agent is being transmitted, is equally critical. If a virus that normally parasitizes a different species suddenly begins infecting humans, if it can be transmitted efficiently from person to person, and if it causes serious illness, then the outbreak has the potential to become an epidemic. But if transmission takes place only between the normal host and humans, as is the case with rabies, then the number of cases will probably remain low. To date, for example, avian flu has not been transmitted efficiently from person to person—only inefficiently from birds to people.

Determining how a virus is transmitted takes old-fashioned detective work. By interviewing patients about their activities, researchers called epidemiologists decide whether each patient could have acquired the virus independently. Were the individuals infected with hantavirus in contact with mice? Was the illness showing up in health-care workers who were in contact with infected individuals, implying that it was being transmitted from human to human?

In the hantavirus outbreak, public health officials concluded that no human-to-human transmission was taking place. Health-care workers did not become ill, and because patients had not had extensive contact with one another, it was likely that each had acquired the virus independently. The outbreak also coincided with a short-term, weather-related explosion in the local mouse population.

The most likely scenario was that people had acquired the pathogen by inhaling dust or handling food that contained remnants of mouse feces or urine. In short, hantavirus did not have the potential to cause an epidemic. The best medicine was preventive: Homeowners were advised to trap mice in living areas, wear dust masks while cleaning any area where mice might have lived, and store food in covered jars.

The Ebola virus, in contrast, was clearly transmitted from person to person. Many doctors and nurses were stricken after tending to patients with the virus. Infections that originate in hospitals usually spread when carried from patient to patient on the hands of caregivers. But by carefully observing the procedures that were being followed by hospital staff, researchers concluded that the Ebola virus was being transmitted only through direct contact with body fluids (blood, urine, feces, or sputum).

The Ebola outbreak was brought under control when hospital workers raised their standards of hygiene and insisted on the immediate disposal or disinfection of all contaminated bedding, utensils, and equipment. The situation with the avian flu outbreak of 2005–2006 was similar—person-to-person transmission was extremely rare and based only on extensive, direct contact with body fluids.

Had the Ebola virus or avian flu virus been transmitted by casual contact, such as touch or inhalation—as is the common cold virus—a massive epidemic could have ensued.

CHECK YOUR UNDERSTANDING

If you understand that . . .

- Among viruses, several different types of molecules serve as the genetic material.
- Emerging viruses have switched host species. They become dangerous if they are rapidly transmitted among host individuals.

✓ You should be able to . . .

1. Explain how the negative-sense, single-stranded RNA genome of influenza viruses, which is complementary in sequence to viral mRNA, is copied during the replicative cycle.

2. State whether a mutation that allowed avian flu virus to spread via airborne virions coughed out by infected individuals would make the virus more or less dangerous to humans. Explain your logic.

Answers are available in Appendix B.

35.4 Key Lineages of Viruses

Because scientists are almost certain that viruses originated multiple times throughout the history of life, there is no such thing as the phylogeny of all viruses. Stated another way, there is no single phylogenetic tree that represents the evolutionary history of viruses as there is for the organisms discussed in previous chapters. Instead, researchers focus on comparing base sequences in the genetic material of small, closely related groups of viruses and using these data to reconstruct the phylogenies of particular lineages, exemplified by the tree in Figure 35.15.

To organize the diversity of viruses on a larger scale, researchers group them into seven general categories based on the nature of their genetic material. This approach to organizing viruses is called the Baltimore classification, because it was proposed by David Baltimore in 1971. Within these seven general categories, biologists also identify a total of about 70 virus families that are distinguished by (1) the structure of the virion (often whether it is enveloped or nonenveloped), and (2) the nature of the host species.

Although they do not have formal scientific names, viruses within families are grouped into distinct genera for convenience.

Within genera, biologists identify and name types of virus, such as HIV, the measles virus, and smallpox. Within each of these viral types, populations with distinct characteristics may be identified and named as strains. The strain is the lowest, or most specific, level of taxonomy for viruses.

To get a sense of viral diversity, let's survey some of the most important groups that can be identified by the nature of their genetic material.

■ Double-Stranded DNA (dsDNA) Viruses

■ RNA Reverse-Transcribing Viruses (Retroviruses)

■ Double-Stranded RNA (dsRNA) Viruses

■ Negative-Sense Single-Stranded RNA ([−]ssRNA) Viruses

■ Positive-Sense Single-Stranded RNA ([+]ssRNA) Viruses

Double-Stranded DNA (dsDNA) Viruses

The double-stranded DNA viruses are a large group, composed of some 21 families and 65 genera. Smallpox (**Figure 35.16**) is perhaps the most familiar of these viruses. Although smallpox had been responsible for millions of deaths throughout human history, it was eradicated by vaccination programs. Smallpox is currently extinct in the wild; the only remaining samples of the virus are stored in research labs.

Genetic material As their name implies, the genes of these viruses consist of a single molecule of double-stranded DNA. The molecule may be linear or circular.

Host species These viruses parasitize hosts from throughout the tree of life, with the notable exception of land plants. They include virus families called the T-even and λ bacteriophages, some of which infect *E. coli*. In addition, the pox viruses, herpesviruses, and adenoviruses—which include types that parasitize humans—have genomes consisting of double-stranded DNA.

Replication cycle In most double-stranded DNA viruses that infect eukaryotes, viral genes have to enter the nucleus to be replicated. In many DNA viruses, the viral genes are replicated only during S phase, when the host cell's chromosomes are being replicated. As a result, these types of viruses can sustain an infection only in cells that are actively dividing, such as the cells lining the respiratory tract or urogenital canal. Some of these viruses are capable of inducing replication in infected cells.

FIGURE 35.16 Smallpox Is a Double-Stranded DNA Virus.

The genomes of the RNA reverse-transcribing viruses are composed of single-stranded RNA. There is only one family, called the retroviruses.

Genetic material Virions have two copies of their single-stranded RNA genome.

Host species Species in this group are known to parasitize only vertebrates—specifically birds, fish, or mammals. HIV is the most familiar virus in this group. The Rous sarcoma virus, the mouse mammary tumor virus, and the murine (mouse) leukemia virus are other retroviruses that have also been studied intensively. Rous sarcoma virus was the first virus shown to be associated with the development of cancer (in chickens); the mouse viruses were the first viruses known to increase the risk of cancer in mammals (**Figure 35.17**). In most cases, viruses that are associated with cancer development carry genes that contribute to uncontrolled growth of the cells they infect.

Replication cycle Retroviruses contain the enzyme reverse transcriptase inside their capsid. When the virus's RNA genome and reverse transcriptase enter a host cell's cytoplasm, the enzyme catalyzes the synthesis of a single-stranded cDNA from the original RNA. Re-

verse transcriptase then makes this cDNA double stranded. The double-stranded DNA version of the genome enters the nucleus with a viral protein called integrase. Integrase catalyzes the integration of the viral genes into a host chromosome. The virus may remain quiescent for a period, but eventually the genes are transcribed to RNA to begin lytic growth and the production of a new generation of virions.

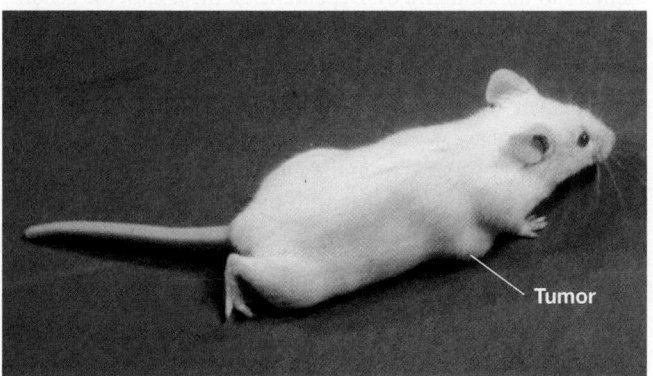

FIGURE 35.17 Some Retroviruses, Such as the Mouse Mammary Tumor Virus, Are Associated with Cancer.

Double-Stranded RNA (dsRNA) Viruses

There are 7 families of double-stranded RNA viruses and a total of 22 genera. Most of the viruses in this group are nonenveloped.

Genetic material In some families, virions have a genome consisting of 10–12 double-stranded RNA molecules; in other families, the genome is composed of just 1–3 RNA molecules.

Host species A wide variety of hosts, including fungi, land plants, insects, vertebrates, and bacteria, are victimized by viruses with double-stranded RNA genomes. Particularly prominent are viruses that cause disease in rice, corn, sugarcane, and other crops. **Figure 35.18** shows rice plants that are being attacked by a double-stranded RNA virus. Infections are also common in *Penicillium*, a filamentous fungus that produces the antibiotic penicillin. Reovirus and rotavirus infections, the leading cause of infant diarrhea in humans, are responsible for over 110 million cases and 440,000 deaths each year.

Replication cycle Once inside the cytoplasm of a host cell, the double-stranded genome of these viruses serves as a template for the synthesis of single-stranded RNAs, which are then translated into viral proteins. Some of the proteins form the capsids for a new generation of virions. Copies of the genome are created when a viral enzyme converts the single-stranded transcripts into double-stranded RNA.

FIGURE 35.18 dsRNA Viruses, Such as the Ragged Stunt Virus, Parasitize a Wide Array of Organisms—Here, Rice.

Negative-Sense Single-Stranded RNA ([−]ssRNA) Viruses

There are 7 families and 30 genera in this group. Most members of this group are enveloped, but some negative-sense single-stranded RNA viruses lack an envelope.

Genetic material The sequence of bases in a negative-sense RNA virus is opposite in polarity to the sequence in a viral mRNA. Stated another way, the single-stranded virus genome is complementary to the viral mRNA. Depending on the family, the genome may consist of a single RNA molecule or up to eight separate RNA molecules.

Host species A wide variety of plants and animals are parasitized by viruses that have negative-sense single-stranded RNA genomes. If you have ever suffered from the flu, the mumps, or the measles, then you are painfully familiar with these viruses (**Figure 35.19**). The Ebola, Hantaan, and rabies viruses also belong to this group.

Replication cycle When the genome of a negative-sense single-stranded RNA virus enters a host cell, a viral RNA polymerase uses that genome as a template to make new viral mRNA. Certain viral transcripts are then translated to form viral proteins. The viral mRNAs also serve as a template for the synthesis of new copies of the negative-sense single-stranded RNA genome.

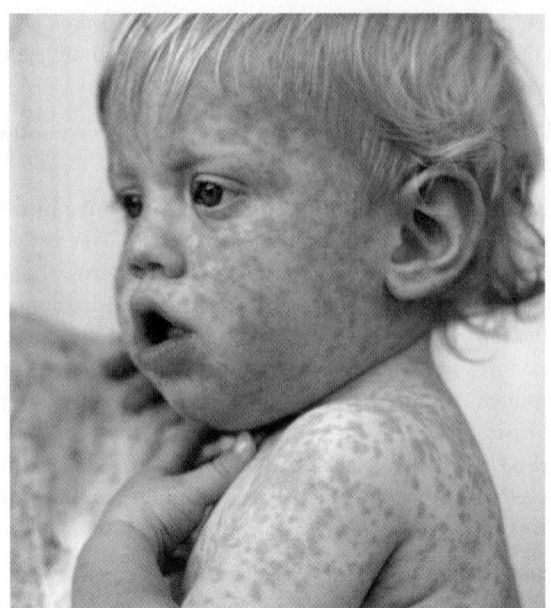

FIGURE 35.19 (−)ssRNA Viruses, Such as the Measles Virus, Cause Some Common Childhood Diseases.

Positive-Sense Single-Stranded RNA ([+]ssRNA) Viruses

This is the largest group of viruses known, with 21 families and 81 genera.

Genetic material The sequence of bases in a positive-sense RNA virus is the same as that of a viral mRNA. Stated another way, the genome does not need to be transcribed in order for proteins to be produced. Depending on the species, the genome consists of one to three RNA molecules.

Host species Most of the commercially important plant viruses belong to this group. Because they kill groups of cells in the host plant and turn patches of leaf or stem white, they are often named mottle viruses, spotted viruses, chlorotic (meaning, lacking chlorophyll) viruses, necrotic (meaning, killed cells surrounded by intact tissue) viruses, or mosaic viruses. **Figure 35.20** shows a healthy cowpea leaf and a cowpea leaf that has been attacked by a positive-sense single-stranded RNA virus. Some species in this group of viruses specialize in parasitizing bacteria, fungi, or animals, however. A variety of human maladies, including the common cold, polio, and hepatitis A, C, and E, are caused by positive-sense RNA viruses.

Replication cycle When the genome of these viruses enters a host cell, the single-stranded RNA is immediately translated into a protein, often a polyprotein, which is cleaved to generate other func-

tional distinct, individual proteins. These proteins include enzymes that make copies of the genome. The new generation of virions is assembled within the host cell.

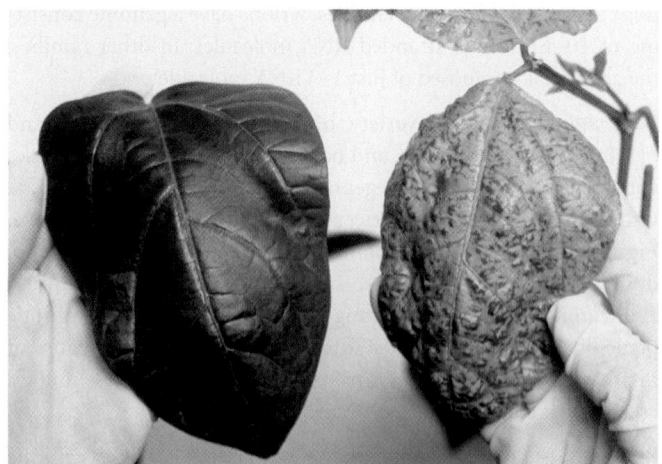

Healthy leaf   Leaf infected with virus

FIGURE 35.20 (+)ssRNA Viruses, Such as the Cowpea Mosaic Virus, Cause Important Plant Diseases.

Summary of Key Concepts

🔑 **Viruses are tiny, noncellular parasites that infect virtually every type of cell known. They cannot perform metabolism on their own—meaning outside a parasitized cell—and are not considered to be alive.**

- Viruses cause illness and death in plants, fungi, bacteria, and archaea, as well as in humans and other animals.

- Viruses are specialists—different types of viruses infect particular species and types of cells.

  ✔You should be able to explain how it is possible for viruses to evolve by natural selection if they are not alive. (Hint: review Darwin's four postulates in Chapter 24.)

🔑 **Although viruses are diverse morphologically, they can be classified as two general types: enveloped and nonenveloped.**

- Both nonenveloped and enveloped viruses have a capsid made of protein.

- A capsid usually encloses viral enzymes as well as the viral genome.

- Nonenveloped viruses consist of a naked capsid, but in enveloped viruses, the capsid is surrounded by a membranous envelope.

  ✔You should be able to explain how enveloped viruses obtain their envelope.

🔑 **The viral replication cycle can be broken down into six steps: (1) attachment and entry into a host cell, (2) production of viral proteins, (3) replication of the viral genome, (4) assembly of a new generation of virions, (5) exit from the infected cell, and (6) transmission to a new host.**

- Viral entry into a host depends on specific interactions between viral proteins and compounds on the surface of the host cell.

- The protein-production and genome-replication phases of the replication cycle depend on enzymes, chemical energy, and molecules provided by the host cell.

- The assembly phase of the cycle is not well understood.

- Nonenveloped viruses exit a cell by lysis; enveloped viruses exit a cell by budding. Host cells are usually killed in the process.

- Once they are released, virions can infect new cells in the same multicellular organism or be transmitted to a new host.

- Some viruses may enter a lysogenic or latent phase, when they do not replicate by making more virions, but may transmit genetic material to daughter cells when the host divides.

  ✔You should be able to explain how at least three of the six phases of a viral replicative cycle can be stopped, using HIV as an example.

  (MB) **Web Activity** The HIV Replicative Cycle

🔑 **In terms of diversity, the key feature of viruses is the nature of their genetic material.**

- The genomes of viruses may consist of double-stranded DNA, single-stranded DNA, double-stranded RNA, or one of several types of single-stranded RNA.

- Viral genomes are small. They do not code for ribosomes, and most viral genomes do not code for the enzymes needed to translate their own proteins or to perform other types of biosynthesis.

  ✔You should be able to explain how a virus that has a single-stranded DNA genome is able to replicate its genes.

Questions

1. How do viruses that infect animals enter an animal's cells?
   a. The viruses pass through a wound.
   b. The viruses bind to a membrane component.
   c. The viruses puncture the cell wall.
   d. The viruses lyse the cell.

2. What does reverse transcriptase do?
   a. It synthesizes proteins from mRNA.
   b. It synthesizes tRNAs from DNA.
   c. It synthesizes DNA from RNA.
   d. It synthesizes RNA from DNA.

3. What do host cells provide for viruses?
   a. nucleoside triphosphates and amino acids
   b. ribosomes
   c. ATP
   d. all of the above

4. When do most enveloped virions acquire their envelope?
   a. during entry into the host cell
   b. during budding from the host cell
   c. as they burst from the host cell
   d. as they integrate into the host cell's chromosome

5. What reaction does protease catalyze?
   a. polymerization of amino acids into peptides
   b. cutting of polyprotein chains into functional proteins
   c. folding of long peptide chains into functional proteins
   d. assembly of virions

6. What feature distinguishes the seven major categories of viruses?
   a. the method of exiting a cell
   b. the method of entering a cell
   c. the nature of the host
   d. the nature of the genome

1. The outer surface of a virus consists of either a membrane-like envelope or a protein capsid. Which type of outer surface does HIV have? Which type does adenovirus have? How does the outer surface of a virus correlate with its mode of exiting a host cell, and why?

2. Compare the morphological complexity of HIV with that of bacteriophage T4. Which virus would you predict has the larger genome? Explain the logic behind your hypothesis.

3. Compare and contrast bacteriophage growth rates during lytic versus lysogenic growth. Is it possible for viral populations to increase if viral DNA remains in the lysogenic state?

4. Explain why viral diseases are more difficult to treat than diseases caused by bacteria.

5. Why does the diversity of viral genome types support the hypothesis that viruses originated several times independently?

6. What types of data convinced researchers that HIV originated when a simian immunodeficiency virus "jumped" to humans?

1. Suppose you could isolate a virus that parasitizes the pathogen *Staphylococcus aureus*—a bacterium that causes acne, boils, and a variety of other afflictions in humans. How could you test whether this virus might serve as a safe and effective treatment?

2. If you were in charge of the government's budget devoted to stemming the AIDS epidemic, would you devote most of the resources to drug development, vaccine development, or preventive medicine? Defend your answer.

3. Bacteria fight viral infections with restriction endonucleases (bacterial enzymes; introduced in Chapter 19). Restriction endonucleases cut up, or break, viral DNA at specific sequences. The enzymes do not cut a bacterium's own DNA, because the bases in the bacterial genome are protected from the enzyme by methylation (the addition of a CH_3 group). Generate a hypothesis to explain why members of the Eukarya do not have restriction endonucleases.

4. Consider these two contrasting definitions of life:
   a. An entity is alive if it is capable of replicating itself via the directed chemical transformation of its environment.
   b. An entity is alive if it is an integrated system for the storage, maintenance, replication, and use of genetic information.

   According to these definitions, are viruses alive? Explain.

All plants are able to harvest diffuse resources and concentrate them in cells and tissues, but their forms and strategies are diverse. These baobab trees in Madagascar may live to be hundreds of years old despite drought conditions, in part by storing water in their enormous trunks.

Plant Form and Function

36

Photosynthetic plants do the most remarkable chemistry of any terrestrial organism. Using the energy in sunlight and the simplest of starting materials—carbon dioxide, water, and ions containing nitrogen, phosphorus, potassium, and other key atoms—plants synthesize thousands of different carbohydrates, proteins, nucleic acids, and lipids. They use these compounds to build bodies that may live for thousands of years.

This feat is even more impressive when you consider that the simple starting materials that plants need to grow are tiny and diffuse—carbon dioxide molecules, water molecules, ammonium ions, and other resources are usually found at low concentrations over a large area. To gather the raw materials required for their sophisticated biosynthetic machinery, a plant's roots and shoots grow outward, extending the individual into the soil and atmosphere.

In essence, a plant's body harvests diffuse resources and concentrates them in cells and tissues. The structure of its body is dynamic, because most plants exhibit **indeterminate growth**, that is, they grow throughout their lives. A 4750-year-old bristlecone pine has roots and shoots that are still growing. In response to favorable conditions, a plant sends roots and shoots in the most promising directions, seeking light and the simple compounds it requires.

The contrast between the plant and animal way of life is striking. Most animals move around and eat concentrated sources of food. But plants stay in one place, extend their roots and shoots to harvest diffuse resources, and make their own food.

This chapter focuses on three fundamental questions:

1. How is the plant body organized?

2. Why are plants so diverse in size and shape?

3. How do plants grow throughout their lives?

KEY CONCEPTS

- The vascular plant body consists of (1) a root system that anchors the individual and absorbs water and key ions, and (2) a shoot system that absorbs carbon dioxide and sunlight. Both systems are dynamic—they grow and change throughout life.

- Variation in body size and shape allows different species to harvest water, light, and other resources in unique ways.

- Primary growth occurs when cells located at the tips of each root and shoot divide and enlarge. Primary growth lengthens roots and shoots and gives rise to three primary tissue systems that are specialized for protection, food production and storage, and transport.

- In some species, secondary growth occurs when cells near the perimeter of a root or shoot divide and enlarge, widening the structure. Secondary growth adds transport tissue and provides structural support.

✔ When you see this checkmark, stop and test yourself. Answers are available in Appendix B.

Instead of surveying the entire catalog of land plants, though, the focus here is on the angiosperms. Recall from Chapter 30 that angiosperms, the flowering plants, are the most recent major group of plants to appear in the fossil record and the most abundant, species-rich, and geographically widespread on Earth today. It is also difficult to overstate their economic and medical importance to humans. Most of the food we eat and the drugs we use are derived from angiosperms.

If you look outside or down the produce aisle of a grocery store, you will see a wide diversity of plants and plant products. By the time you finish this chapter, you'll understand how these plant bodies are put together and how they grow. Exploring questions about the anatomy of flowering plants is vital to understanding the world at large as well as the other chapters in this unit.

36.1 Plant Form: Themes with Many Variations

Chapter 10 detailed how plants—along with algae, cyanobacteria, and a variety of protists—obtain the energy and carbon they need to grow and reproduce. Plants use light energy (photons) to synthesize carbohydrates using carbon dioxide from the air and water from the soil.

For photosynthesis to occur, plants need large amounts of light and carbon dioxide, along with water as an electron source (**Figure 36.1**). Plants also need large amounts of water to fill their cells and maintain them at normal volume and pressure.

To synthesize nucleic acids, enzymes, phospholipids, and the other macromolecules needed to build and run cells, plants must obtain nitrogen (N), phosphorus (P), potassium (K), magnesium (Mg), and a host of other nutrients. Most of these key elements exist in nature as ions that dissolve in water found in soil.

Figure 36.2 labels the major structures in the two basic systems that plants use to acquire the resources they need for photosynthesis. ○▬ A belowground portion called the **root system** anchors the plant and takes in water and nutrients from the soil; an aboveground portion called the **shoot system** harvests light and carbon dioxide from the atmosphere to produce sugars. Both systems grow throughout the life of the individual, allowing the plant to increase in size, outcompete other individuals, and acquire resources.

In most plants, vascular tissue connects the root and shoot systems. Water is transported from roots to shoots through vascular tissue; sugars and other nutrients are transported in both directions.

The Importance of Surface Area/Volume Relationships

Before exploring the nature of root and shoot systems in more detail, it's important to recognize a key physical relationship that connects their structure to their function.

Root and shoot systems both function in absorption—of light or key ions and molecules. Absorption takes place across a surface. But

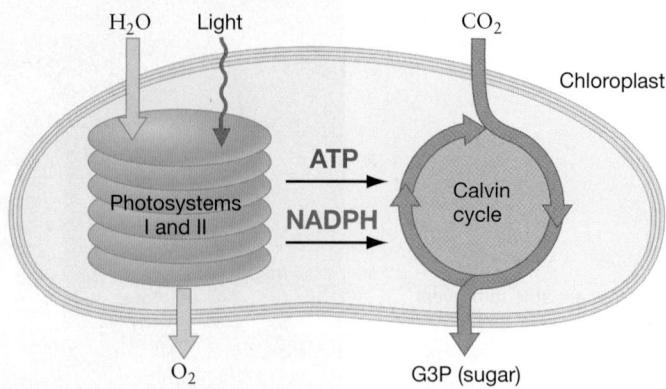

FIGURE 36.1 Plants Need Resources to Perform Photosynthesis. Plants perform some of the most remarkable biochemistry of any organism—manufacturing sugars from water, carbon dioxide, and sunlight. In the process, plants absorb diffuse resources and concentrate them in the form of complex organic compounds.

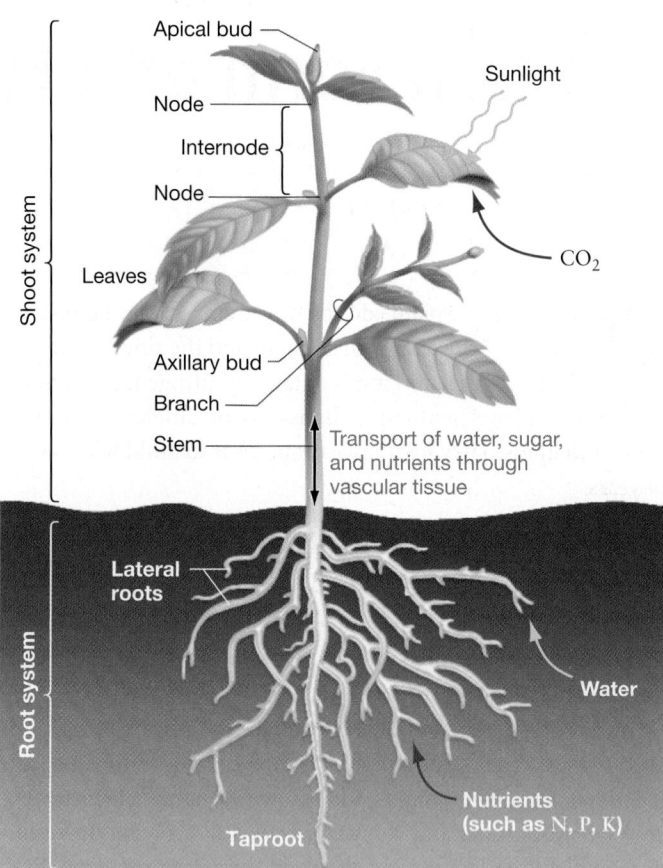

FIGURE 36.2 Root and Shoot Systems Acquire and Transport Resources. Most root systems have the same general structures, and most shoot systems have the same general structure. Shoot systems are specialized for harvesting light and CO_2. Root systems absorb water and key nutrients such as nitrogen (N), phosphorus (P), and potassium (K).

✔**QUESTION** Suppose that this plant's growth was limited by access to light and nutrients, and that there was a new deposit of nutrient-rich soil to the left and much more sunlight was suddenly available to the right. Explain what you would expect this individual to look like in one month.

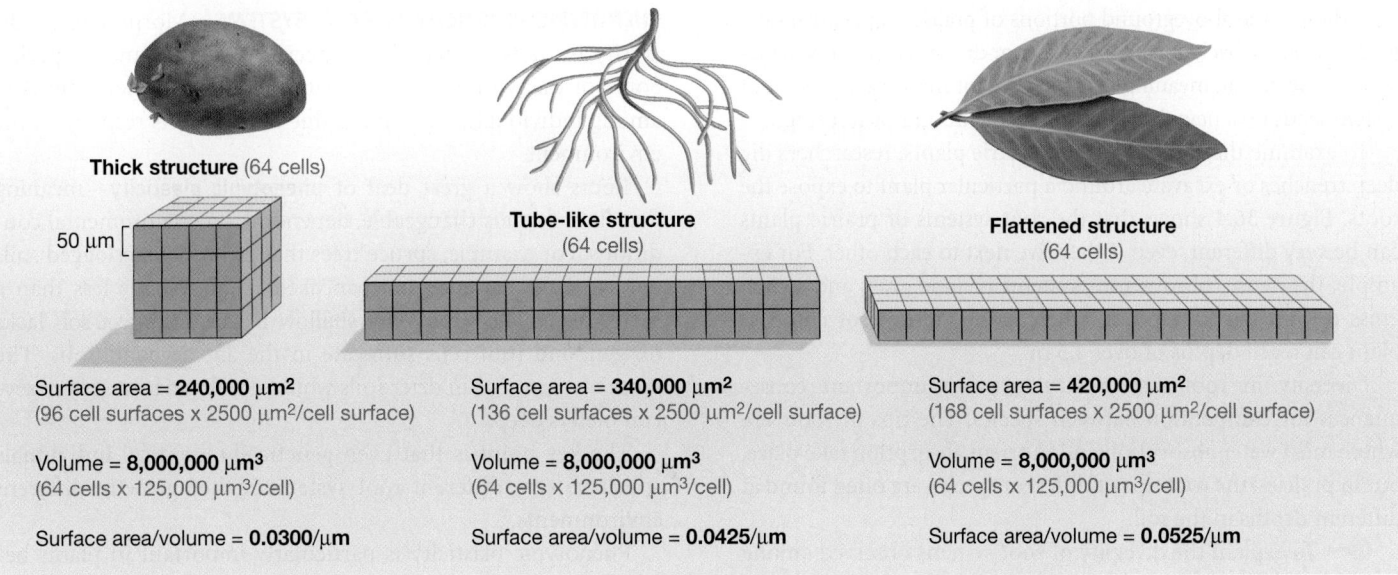

Thick structure (64 cells)

50 μm

Surface area = **240,000 μm²**
(96 cell surfaces x 2500 μm²/cell surface)

Volume = **8,000,000 μm³**
(64 cells x 125,000 μm³/cell)

Surface area/volume = **0.0300/μm**

Tube-like structure
(64 cells)

Surface area = **340,000 μm²**
(136 cell surfaces x 2500 μm²/cell surface)

Volume = **8,000,000 μm³**
(64 cells x 125,000 μm³/cell)

Surface area/volume = **0.0425/μm**

Flattened structure
(64 cells)

Surface area = **420,000 μm²**
(168 cell surfaces x 2500 μm²/cell surface)

Volume = **8,000,000 μm³**
(64 cells x 125,000 μm³/cell)

Surface area/volume = **0.0525/μm**

FIGURE 36.3 The Morphology of Roots and Shoots Gives Them a High Surface-Area-to-Volume Ratio. In this example, the "thick structure" represents a tree trunk or potato-like storage organ; the "tube-like structure" represents a root; the "flattened structure" represents a leaf. Note that each schematic structure has the same number of cells and the same total volume—but a very different amount of surface area.

the cells that use the absorbed light and molecules occupy a volume. Thus, a plant body is more efficient as an absorbance-and-synthesis machine when it has a large surface area relative to its volume.

Figure 36.3 illustrates this point. In this example, the cells in a plant are represented by cubes; the side of each cell is 50 μm long. Thus, each face of a cell has a surface area of $50 \times 50 = 2500$ μm²; each cell has a volume of $50 \times 50 \times 50 = 125,000$ μm³. Follow the calculations in the figure and note that:

- If 64 cells are arranged in a thick block, the surface area/volume relationship is 0.0300/μm.

- If 64 cells are arranged in a long tube, the surface area/volume relationship is 0.0425/μm.

- If 64 cells are arranged in a flat sheet, the surface area/volume relationship is 0.0525/μm.

This simple exercise has an important punchline: tubes and sheets have much more surface area relative to their volume than squares. It's no surprise, then, that the absorptive regions of a root system are tubelike, and the absorptive regions of a shoot system are the flattened structures called leaves. Storage tissues such as tubers and seeds have a low surface-area-to-volume ratio because they are not involved in absorption.

The Root System

Many root systems have a vertical **taproot**, as well as numerous **lateral roots** that run more or less horizontally. The root system anchors the plant in soil, absorbs water and ions from the soil, conducts water and selected ions to the shoot, and stores material produced in the shoot for later use.

Root systems can be impressive in extent. For example, a researcher grew a winter rye plant in a container full of soil for four

months, then unearthed the plant and meticulously measured its roots. The root system of this single individual contained more than 13 million identifiable structures with a combined length of over 11,000 km—almost one-third of Earth's circumference! A root system like this contains an enormous surface area for absorbing diffuse resources located underground.

Other studies have shown that (**1**) the roots of trees routinely extend wider than their aboveground canopy, and (**2**) it is not unusual for a plant's root system to represent over 80 percent of its total mass. Many plants devote a great deal of energy and resources to the growth of their root systems.

Although most root systems contain the same general structures, the root systems observed in different species are diverse, as well. This diversity can be analyzed on three levels:

1. morphological diversity among species;

2. phenotypic plasticity, or changes in the structure of an individual's root system over time; and

3. modified roots that are specialized for unusual functions.

Let's consider each level in turn.

MORPHOLOGICAL DIVERSITY IN ROOT SYSTEMS As an example of the range of morphological diversity observed in the root systems of angiosperms, consider prairie plants.

Prairies are grassland ecosystems found in areas of the world such as central North America, the Serengeti Plain of East Africa, the Pampas region of Argentina, and the steppes of central Asia. Rain is abundant enough in these areas to support a lush growth of **herbaceous plants**—meaning, seed plants that lack woody tissue. Rain is scarce enough to exclude trees and most shrubs, however. The growth of woody species is also discouraged by fires that regularly sweep through these ecosystems.

Although the aboveground portions of prairie plants burn during fires and die back during the winter or dry season, their root systems are **perennial**, meaning that they live for many years. The root system sends up a new shoot system after a fire and each spring.

To examine the root systems of prairie plants, researchers dig deep trenches or excavate around a particular plant to expose the roots. **Figure 36.4** shows that the root systems of prairie plants can be very different, even if they live next to each other. For example, the dense, fibrous root systems of june grass and switch grass do not have a taproot, whereas the taproots of compass plant can reach depths of over 4.5 m.

Diversity in root system structure has important consequences for competition between species. The tips of roots are where most water absorption and nutrient absorption take place, but in prairies, the root tips of different species are often found at different depths in the soil.

🗝 To explain the diversity of root systems observed among species that grow in the same habitat, biologists suggest that natural selection has favored structures that minimize competition for water and nutrients.

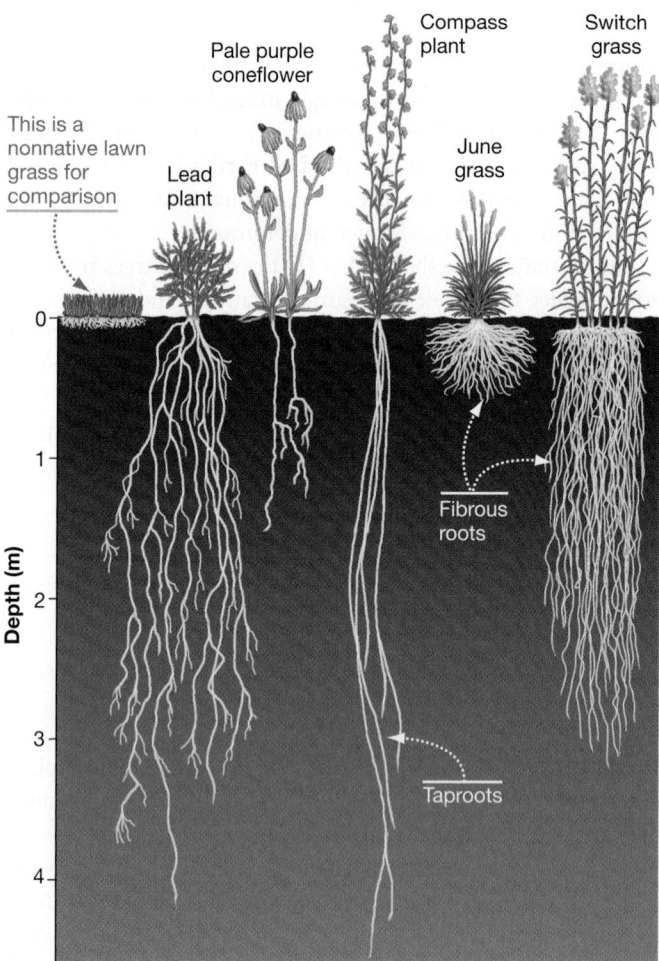

FIGURE 36.4 Plants Have Diverse Root Systems. The roots of prairie plants that live side by side can be very different.

✔**QUESTION** Why are lawns not drought tolerant?

PHENOTYPIC PLASTICITY IN ROOT SYSTEMS Morphological diversity in roots occurs within species as well as among species. Some of the within-species variation is due to genetic diversity among individuals, but some is due to how roots respond to the environment.

Roots show a great deal of **phenotypic plasticity**—meaning that their form is changeable, depending on environmental conditions. For example, spruce trees that grow in waterlogged soils tend to have flattened or "pancaked" root systems less than a meter deep. Their roots are shallow because the wet soil lacks oxygen, and root cells suffocate in the anoxic conditions. The same tree growing in drier soils would develop a root system several meters deep.

The key point is that even genetically identical individuals will have very different root systems if they grow in different environments.

Phenotypic plasticity is particularly important in plants because they grow throughout their lives. Thanks to this mechanism, plants can respond when environmental conditions change over the course of their lifetime. Root systems that grow into nutrient-rich septic fields or sewer pipes that leak human waste are a prime example. Roots actively grow into areas of soil where resources are abundant; roots stop growing or die back in areas where resources are used up or lacking.

MODIFIED ROOTS The taproots and fibrous roots illustrated earlier do not begin to exhaust the types of roots found among plants. For example, some roots are **adventitious**, meaning they develop from the shoot system instead of the root system.

- In ivy, adventitious roots that grow from nodes in the shoot system help individuals cling to brick walls or other structures.

- The prop roots of corn are adventitious roots that help brace individuals in windy weather (**Figure 36.5a**).

Even roots that are part of the root system can have specialized functions—meaning that they do things other than absorbing water and nutrients and anchoring the shoot system.

- The pneumatophores of mangroves in the genus *Avicennia* are specialized lateral roots that function in gas exchange (**Figure 36.5b**). These mangroves grow in habitats where fine silt is deposited, cutting off oxygen from their roots. Their root cells do not suffocate, however, because oxygen from the atmosphere can diffuse into the root system through the pneumatophores. These roots grow upward—not downward—in response to gravity.

- The roots of some plants are contractile, meaning that their cells can shorten much the way an animal muscle cell does. Some species of *Ficus* that grow as vines have adventitious roots that grow from the shoot system down into the ground, and then contract to form a tight supporting structure. In the greenhouse, contractile roots from *Ficus bengalensis* have been known to grow into a bucket of soil and lift it off the ground as they contract. In many cases, the

(a) Prop roots support.

(b) Pneumatophores function in gas exchange.

FIGURE 36.5 Modified Roots Have Unusual Structures or Functions. (a) The prop roots of corn plants that help stabilize the stem are adventitious—they develop from the shoot system. **(b)** The pneumatophores of mangrove trees allow gas exchange to occur between root tissues and the atmosphere.

✔**QUESTION** Why do root cells need oxygen?

roots of plants with bulbs contract, pulling the bulb deeper into the soil over time—in effect, planting themselves.

The Shoot System

As Figure 36.2 indicated, the shoot system has an array of important anatomical features.

- The shoot system consists of one or more **stems**, which are vertical aboveground structures.

- A stem consists of **nodes**, where leaves are attached, and **internodes**, or segments between nodes.

- A **leaf** is an appendage that projects from a stem laterally. Leaves usually function as photosynthetic organs.

- The nodes where leaves attach to the stem are also the site of **axillary** (or **lateral**) **buds**, which form just above the leaf.

- If conditions are appropriate, an axillary bud may grow into a **branch**—a lateral extension of the shoot system.

- The tip of each stem and branch contains an **apical bud**, where growth occurs that extends the length of the stem or branch.

- If conditions are appropriate, apical or axillary buds may develop into flowers or other reproductive structures.

In essence, the shoot system is a repeating series of nodes, internodes, leaves, and apical and axillary buds. As plants grow, the number of nodes, internodes, and leaves increases.

After an initial period of growth, however, an internode does not increase much in size over time. The shoot system of a plant grows by adding more parts rather than by increasing the size of each part.

As with root systems, diversity in shoots can be analyzed on three levels: morphological diversity among species, phenotypic plasticity within individuals, and modified shoots with specialized functions.

MORPHOLOGICAL DIVERSITY IN SHOOT SYSTEMS The shoot systems of land plants range in size from species like the tiny (<5-mm diameter) duckweed that you may have seen growing on the surface of stagnant ponds to redwood trees that reach heights of over 100 m (300 ft) and giant sequoia trunks that weigh 2.6 million kg (over 5.7 million lbs)—about the same as 10 diesel locomotives.

The shape of the shoot system also varies a great deal among species. For example, the manner in which new branches are added as the shoot system grows affects the shape of the individual and its ability to compete for light. As **Figure 36.6** shows, a plant growing with wide branching angles and short internodes has a very different shape than that of a plant with narrow branching angles and long internodes.

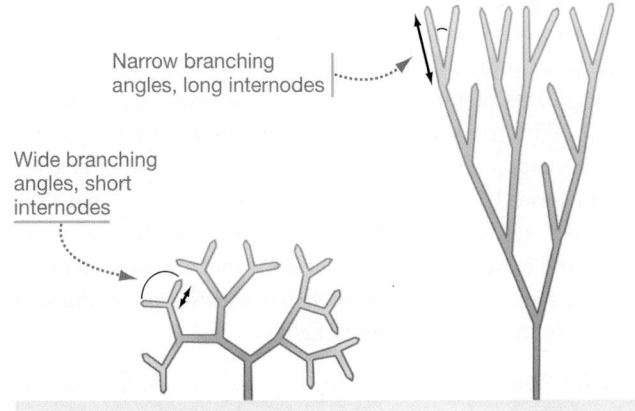

Narrow branching angles, long internodes

Wide branching angles, short internodes

FIGURE 36.6 Plant Form Can Vary as Function of Branch Angle and Internode Length.

Variation in the size and shape of the shoot system is important: it allows plants of different species to harvest light at different locations and thus minimize competition. It also allows them to thrive in a wide array of habitats.

As an example of how the shape of a shoot system varies among species in different environments, consider the silversword plants native to Hawaii. You might recall from Chapter 27 that all of the silverswords are descended from the same ancestor—a species of tarweed that arrived in Hawaii from the west coast of North America, about 5 million years ago.

Silverswords represent an adaptive radiation: a lineage that rapidly split into many species occupying a wide array of habitats. Their shoot systems are particularly diverse in size, shape, and growth habit (see Figure 27.11). Some silverswords grow low to the ground in dense mats; some form bunched rosettes of leaves; some are vines; others are woody shrubs or even small-to-medium-sized trees.

Biologists interpret this diversity of shoot systems as a suite of adaptations for harvesting light and carbon dioxide in different environments. In lush habitats, where competition for light is intense, woody individuals grow tall and are favored by natural selection. But in dry, windblown habitats, individuals with short stems or rosettes thrive because they require less water than taller individuals do, and they don't blow over. The adaptive radiation of silverswords has been based in part on diversification in shoot systems.

PHENOTYPIC PLASTICITY IN SHOOT SYSTEMS The size and shape of an individual's shoot system can vary dramatically based on variation in growing conditions: temperature, exposure to wind, and availability of water, nutrients, and light.

This conclusion was driven home in an experiment conducted by Jens Clausen and colleagues in the late 1930s. These biologists transplanted several species of herbaceous plants between sites along an elevational gradient: from sea level to alpine habitats. In each case, the transplanted individuals were propagated from cuttings—meaning that they were genetically identical to individuals growing at the other locations. As the "Results" section in **Figure 36.7** shows, the overall size and shape of the shoot system varied markedly among locations.

Because an individual's shoot system continues to grow over the course of its lifetime, it can respond to changes in environmental conditions just as the root system can. Experiments highlighted in Chapter 39, for example, established that shoot systems can bend toward light if an individual is shaded on one side. Plants also undergo differential growth, producing more branches and leaves in regions of the body that are exposed to the highest light levels. A plant's shoot system grows in directions that maximize its chances of capturing light.

MODIFIED SHOOTS Even though they are a single lineage, silverswords illustrate many of the modified shoot systems found among land plants. But still other variations occur. Not all stems grow vertically, and not all stems acquire carbon dioxide and photons.

EXPERIMENT

QUESTION: How much does a plant's growth form depend on its environment?

HYPOTHESIS: (No explicit hypothesis—the goal of this experiment was to explore the interaction between genetic makeup versus environmental influence on size and shape.)

EXPERIMENTAL SETUP:

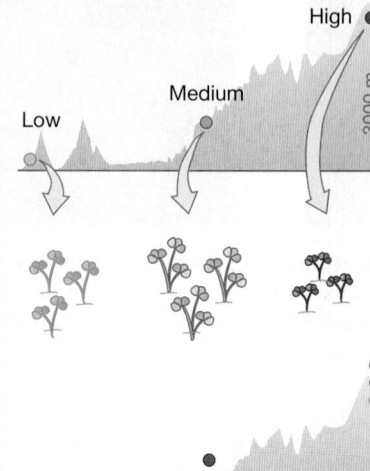

1. **Take cuttings** from individuals of *Potentilla glandulosa* growing at low, medium, and high elevation habitats in the Sierra Nevada mountains.

2. **Propagate** genetically indentical individuals.

3. **Transplant** individuals from each source population into each habitat (low, medium, and high elevation). Allow to grow and observe mature plants.

PREDICTION: (No explicit predictions.)

RESULTS:

Examples of mature plants observed:

| "High" plant grown at low elevation | "High" plant grown at medium elevation | "High" plant grown at high elevation |

CONCLUSION: Environmental conditions have a profound influence on body size and shape (genetically identical plants look different at each site). BUT, genetic makeup also has a large influence on plant morphology (plants from each source population look different, even when grown in the same habitat).

FIGURE 36.7 Experimental Evidence for Phenotypic Plasticity in Shoot Systems.

SOURCE: Clausen, J., D. D. Keck, and W. M. Hiesey. 1945. Experimental studies on the nature of species. II. Plant evolution through amphiploidy and autoploidy, with examples from the Madiinae. Washington D.C., Carnegie Institution of Washington.

✔ **QUESTION** Why was it important for the researchers to propagate the individuals from cuttings?

(a) Cactus stems (shown here in cross section) store water.

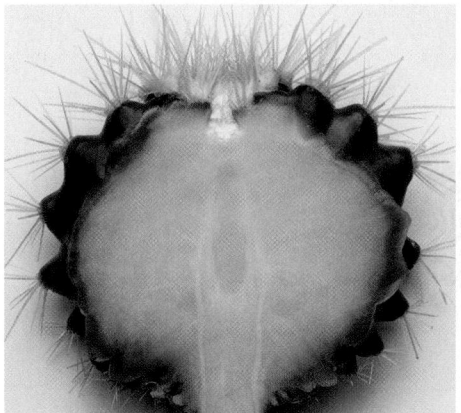

(b) Stolons produce new individuals at nodes aboveground.

Stolon

(c) Rhizomes produce new individuals at nodes belowground.

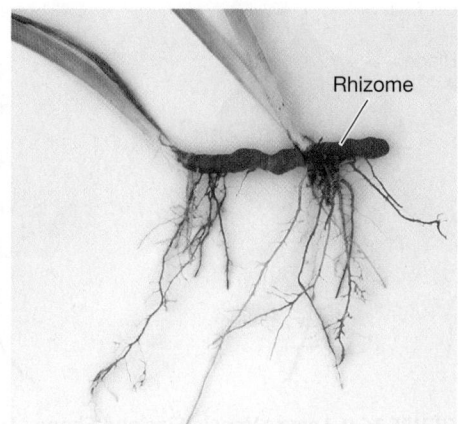

Rhizome

(d) Tubers store carbohydrates.

Tuber

(e) Thorns provide protection.

Thorn

FIGURE 36.8 Modified Stems Have Unusual Structures or Functions. Instead of providing support, the stems of some species have been modified to function as water- or starch-storage tanks, a means of asexual reproduction, or a defense weapon.

- Many desert cacti have highly modified stems. Instead of functioning primarily to support leaves, cactus stems often enlarge into water-storage organs (**Figure 36.8a**). Water accounts for up to 98 percent of the weight of a cactus stem. A cactus stem also contains the plant's photosynthetic tissue. Instead of being the main food-producing organ, its leaves are modified into protective structures called **spines**.

- **Stolons** are modified stems that grow over the soil surface, producing adventitious roots and leaves at each node (**Figure 36.8b**). Because new plants form at these nodes, stolons function in asexual reproduction (see Chapter 12).

- Like stolons, **rhizomes** are stems that grow horizontally instead of vertically. They produce new plants at nodes and thus participate in asexual reproduction. But while stolons grow aboveground, rhizomes spread belowground (**Figure 36.8c**). Rhizomes also store starch.

- **Tubers** are underground, swollen ends of rhizomes that function as carbohydrate-storage organs (**Figure 36.8d**). The eyes of a potato—a typical tuber—are nodes in the stem where new branches may arise.

- **Thorns** are modified stems that help protect the plant from attacks by large **herbivores**, or plant-eaters, such as deer, giraffe, or cattle (**Figure 36.8e**).

The Leaf

In most plant species, the vast majority of photosynthesis occurs in leaves. The total area of leaf produced by a single plant can be enormous—a single tree can have hundreds of thousands of leaves with a total leaf surface area equivalent to that of a football field. All of this area is available for absorbing photons and supporting photosynthesis.

A simple leaf (**Figure 36.9a** on page 702) is composed of just two major structures: an expanded portion called the **blade** and a stalk called the **petiole**. But leaves exhibit many variations on the central theme of a flattened structure specialized for performing photosynthesis.

MORPHOLOGICAL DIVERSITY IN LEAVES Glance outside or stroll through a garden, and you'll find many types of simple leaves with an easily recognizable blade and petiole. (Grass leaves will stump you, though, because they lack petioles entirely.) You

(a) Simple leaves have a petiole and a single blade.

Blade —

— Petiole

(b) Compound leaves have blades divided into leaflets.

(c) Doubly compound leaves are large yet rarely damaged by wind or rain.

(d) Species from very cold or hot climates have needlelike leaves.

FIGURE 36.9 Leaves Vary in Size and Shape. The structures in parts **(a)** through **(c)** represent single leaves; part **(d)** shows two leaves.

✔**QUESTION** For capturing photons, what is the advantage of having a leaf with a large surface area? In terms of wind damage and water loss, what is the disadvantage of having a leaf with a large surface area?

will also find compound leaves that have blades divided into a series of leaflets (**Figure 36.9b**). You may even encounter doubly compound leaves, which have leaflets that are again divided (**Figure 36.9c**).

Not all leaf blades are thin with a large surface area, however. For example, plants that thrive in deserts and in cold, dry habitats tend to have needle-shaped leaves (**Figure 36.9d**). The leading hypothesis to explain this pattern is based on two observations: (**1**) Water is often in short supply in these environments because it is absent in deserts or frozen and thus unavailable in cold habitats, and (**2**) leaves with large surface areas lose large amounts of water through an evaporative process called **transpiration** (discussed in Chapter 37).

Thus, needlelike leaves are interpreted as adaptations that minimize transpiration in water-scarce habitats. Small, narrow leaves are also much less susceptible to wind damage than are large, broad leaves.

The arrangement of leaves on a stem can vary as much as leaf shape. For example, leaves can be:

- paired opposite each other on the stem (**Figure 36.10a**);
- arranged in a whorl (**Figure 36.10b**);
- arranged to alternate on either side of the stem (**Figure 36.10c**);
- found in a compact basal arrangement where internodes are extremely short—leading to the rosette growth form (**Figure 36.10d**).

PHENOTYPIC PLASTICITY IN LEAVES Even though leaves do not grow continuously, they exhibit phenotypic plasticity just as root and shoot systems do.

Leaves from the same individual that grow in sun versus shade are a prominent example of phenotypic plasticity in leaf morphology. As the oak tree leaves in **Figure 36.11** show:

(a) Opposite leaves

(b) Whorled leaves

(c) Alternate leaves

(d) Rosette

FIGURE 36.10 The Arrangement of Leaves on Stems Varies.

FIGURE 36.11 Phenotypic Plasticity in Leaves. These leaves came from the same tree.

- *Sun leaves* have a relatively small surface area, which reduces water loss in areas of the body where light is abundant.
- *Shade leaves* are relatively large and broad, providing a high surface area that maximizes absorption of rare photons.

Water loss is less of a problem for shade leaves, because temperatures are cooler in shade than in bright sun.

MODIFIED LEAVES Not all leaves function primarily in photosynthesis; some perform other roles.

- Cactus spines are modified leaves that protect the stem (see Figure 36.8a).
- Onion bulbs consist of thickened leaf bases, separated by highly condensed internodes, that store nutrients (**Figure 36.12a**).
- The thick leaves of species called succulents, such as aloe vera, store water (**Figure 36.12b**).
- The tendrils that enable garden peas and other vines to climb are modified leaflets or leaves (**Figure 36.12c**).
- The bright red leaves of poinsettias attract pollinators to the tiny yellow flowers that they surround (**Figure 36.12d**).
- The tube-like leaves of the pitcher plant trap insects (**Figure 36.12e**). When insects enter, they are discouraged from flying out by the dark "hood" that covers the opening. As they feed on the plant's nectar, they appear to become dizzy. Eventually they fall into the bottom of the tube and drown in water that has accumulated. The nutrients are digested by enzymes secreted by the plant and taken up by epidermal cells.

The variability of plant root systems, shoot systems, and leaves is impressive. Diversity, plasticity, and dynamism are recurring themes in the study of plant anatomy.

(a) Onion leaves store food.

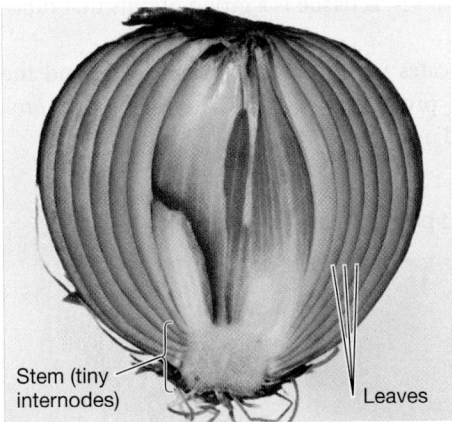

(b) Aloe vera leaves store water.

(c) Pea tendrils aid in climbing.

(d) Poinsettia leaves attract pollinators.

(e) Pitcher plant leaves trap insects.

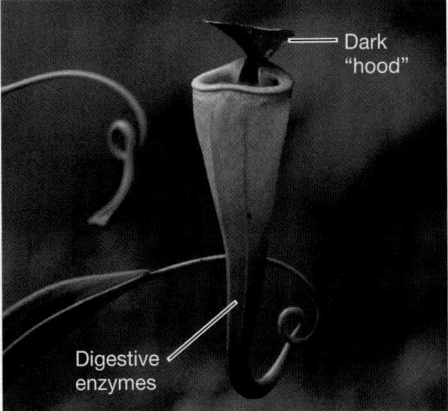

FIGURE 36.12 Modified Leaves Have Unusual Structures or Functions.
Instead of functioning primarily as the site of photosynthesis, the leaves of some species have been modified to function as water- or starch-storage tanks or in climbing, sexual reproduction, or nutrient acquisition.

If you understand that . . .

- The plant body is organized into a root system and a shoot system.
- Roots and shoots explore the environment via continuous growth and efficiently absorb diffuse resources like water, ions, carbon dioxide, and sunlight.
- Roots and shoots may also function to anchor the plant, store water, produce offspring asexually, provide protection, or store carbohydrates.
- Leaves vary among species and within individuals, and may be modified to store food or water, capture insects, or attract pollinators.

✓ **You should be able to . . .**

1. Diagram a generalized version of the angiosperm body, labeling each major part.
2. Provide two examples each of root systems, shoot systems, and leaves that differ from the generalized body shown in Figure 36.2, in structure and/or function.

Answers are available in Appendix B.

36.2 Primary Growth Extends the Plant Body

Plants grow continuously because they have **meristems**—populations of undifferentiated cells that retain the ability to undergo mitosis and produce new cells. When meristematic cells divide, some of the daughter cells remain in the meristem, allowing the meristem to persist. Other cells, though, undergo differentiation. You might recall from Chapter 21 that differentiation is a developmental process that produces a specialized cell—one that expresses only certain genes and has a distinctive structure and function.

Apical meristems are located at the tip of each root and shoot. As cells in apical meristems divide, enlarge, and differentiate, root and shoot tips extend the plant body outward, allowing it to explore new space.

This process is **primary growth**. The major consequence of primary growth is to increase the length of the root and shoot systems. Cells that are derived from apical meristems form the primary plant body.

To understand how primary growth occurs, let's look at the overall organization of the primary plant body and then delve into a detailed look at its tissues and cells.

How Do Apical Meristems Produce the Primary Plant Body?

Whether located in the root or the shoot, apical meristems give rise to three distinct populations of cells: protoderm, ground meristem, and procambium. These cells are partially differentiated but retain the character of meristematic cells because they keep dividing.

The three types of primary meristematic cells are important because they give rise to three major tissue systems that extend throughout the plant body. A **tissue** is a group of cells that functions as a unit.

Figure 36.13 indicates where the apical meristems and the primary meristems—protoderm, ground meristem, and procambium—are found in shoots and roots.

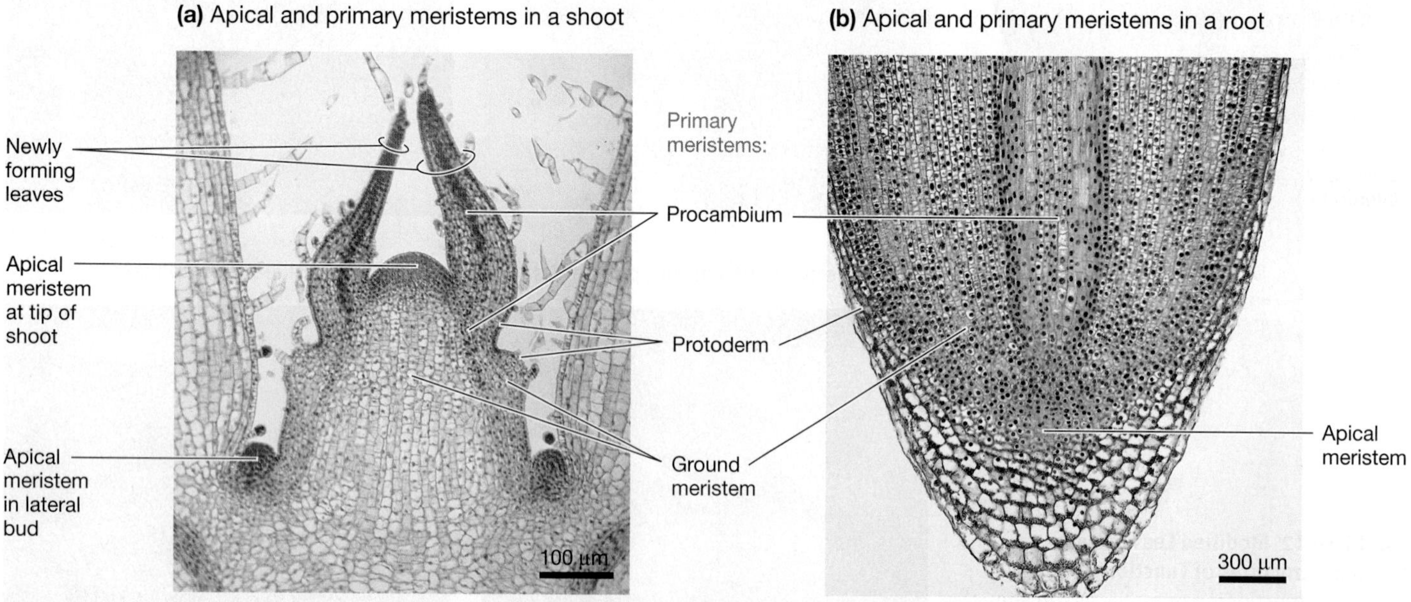

(a) Apical and primary meristems in a shoot

(b) Apical and primary meristems in a root

Newly forming leaves

Apical meristem at tip of shoot

Apical meristem in lateral bud

Primary meristems:

Procambium

Protoderm

Ground meristem

100 μm

Apical meristem

300 μm

FIGURE 36.13 The Structure of Apical Meristems in the Shoot and Root. Apical meristems consist of small, similar-looking cells that divide when water and nutrients are plentiful. Three types of cells—protoderm, ground meristem, and procambium—are derived from the apical meristem and consist of partially differentiated cells that can still divide.

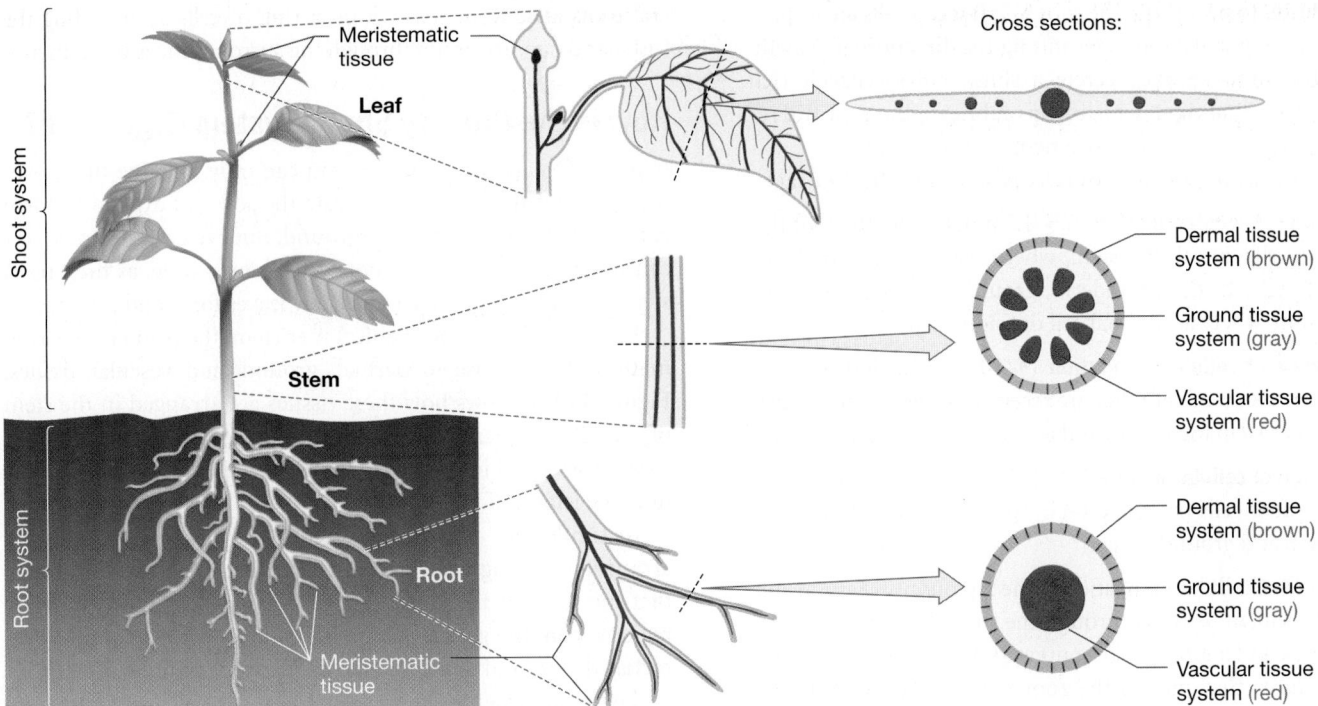

FIGURE 36.14 The Primary Plant Body Comprises the Dermal, Ground, and Vascular Tissue Systems. The dermal, ground, and vascular tissue systems arise from the protoderm, ground meristem, and procambium illustrated in Figure 36.13.

- **Protoderm** gives rise to the **dermal** (literally, "skin") **tissue system**. The dermal tissue system, or **epidermis**, is a single layer of cells that covers the plant body and protects it.

- **Ground meristem** gives rise to the **ground tissue system**, which makes up the bulk of the plant body and is responsible for photosynthesis and storage.

- **Procambium** gives rise to the **vascular tissue system**, which provides support and transports water, nutrients, and photosynthetic products between the root system and shoot system. Vascular tissue runs through ground tissue, so the cells that make up ground tissue are adjacent to cells that conduct the water and nutrients they need.

Figure 36.14 shows how the dermal, ground, and vascular tissues are distributed in the plant body. In contrast to these tissues, meristematic cells are highly localized at the tips of shoots and roots.

The key point to remember is that the dermal, ground, and vascular tissue systems are originally derived from cells in apical meristems. Thus, they represent the primary plant body.

How Is the Primary Root System Organized?

Roots have several features that allow them to grow into new regions of the soil, so they can furnish cells throughout the body with water and key nutrients.

As **Figure 36.15** shows, a group of cells called the **root cap** protects the root apical meristem. Cells produced by the meristem constantly replenish the cap, which regularly loses cells.

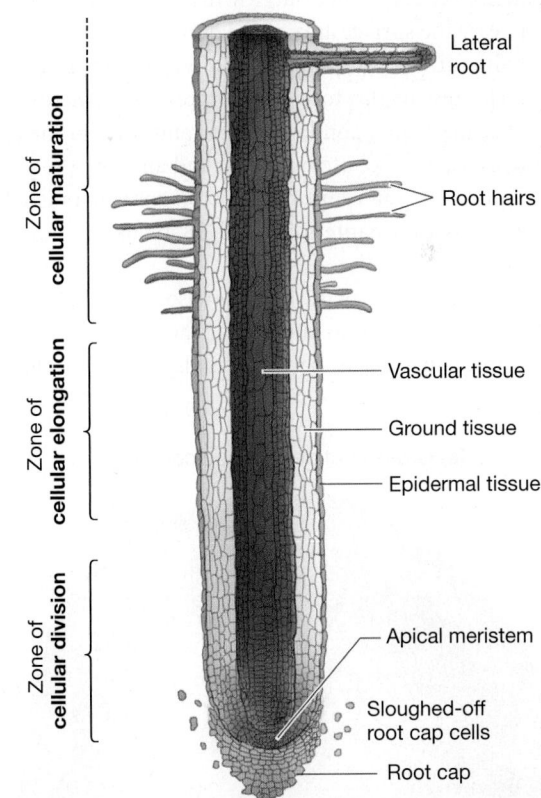

FIGURE 36.15 Roots Extend into the Soil via Growth of Apical Meristems and Cell Elongation. This is a longitudinal (lengthwise) section. The zone of cellular maturation is actually much larger than can be shown here. Most absorption of water and nutrients occurs at root hairs.

In addition to protecting the root tip, root cap cells are important in sensing gravity and determining the direction of growth. They also synthesize and secrete a slimy, polysaccharide-rich substance called **mucigel**, which helps lubricate the root tip, reducing friction and making movement more efficient.

Three distinct populations of cells exist behind the root cap:

1. The **zone of cellular division** (0.5–1.5 mm behind the root tip) contains the apical meristem, where cells are actively dividing, along with the protoderm, ground meristem, and procambium, where additional cell division occurs.

2. The **zone of cellular elongation** (4–10 mm behind the root tip) is made up of cells that are recently derived from the primary meristematic tissues and actively increasing in length.

3. The **zone of cellular maturation** (1–5 cm from the root tip) is where older cells complete their differentiation into dermal, vascular, and ground tissues.

The zone of cellular elongation is the region most responsible for the movement of roots through the soil. The cells in this region increase in length by taking up water. Their expansion provides the force that pushes the root cap and apical meristem through the soil. When conditions are good, roots can extend by as much as 4 centimeters per day.

The zone of cellular maturation is the most important root segment in terms of water and nutrient absorption. In this region, epidermal cells produce outgrowths called **root hairs**, which greatly increase the surface area of the dermal tissue.

Root hairs furnish the actual sites of water and nutrient absorption. The rest of the root system provides structural support for the root hairs, conducts water and ions to the shoot, stores the products of photosynthesis, and anchors the plant in the soil. Uptake of water and nutrients in root hairs is vital to plants; portions of Chapters 37 and 38 focus on how these processes occur.

The zone of cellular maturation is also where lateral roots begin to grow. In contrast to lateral branches in the shoot, which arise from apical meristems in axillary buds (see Figure 36.2), lateral roots arise from cells within a ring of cells surrounding the vascular tissue, and erupt through the surrounding ground tissue.

How Is the Primary Shoot System Organized?

If you visit a garden regularly, you can only imagine the movement of root tips as they penetrate the soil and expand to form complex networks deep underground. But even a casual observer can watch the growth of shoot systems over time, as the tips of stems extend and branch and as new leaves form and expand.

Just behind each shoot apical meristem, the primary meristematic cells give rise to dermal, ground, and vascular tissues. **Figure 36.16a** shows how these tissues are arranged in the stem of a sunflower plant when it matures. Note that the vascular tissues are grouped into **vascular bundles**, which form strands running the length of the stem.

In sunflowers and other eudicots, the vascular bundles are arranged in a ring near the stem's perimeter. The ground tissue that the vascular tissue runs through is divided into two major regions: **pith**, the ground tissue inside the vascular bundles, and **cortex**, the ground tissue outside the vascular bundles.

The arrangement of the vascular bundles and ground tissue is dramatically different in the stems of monocots, however. As **Figure 36.16b** shows, vascular bundles tend to be scattered throughout the ground tissue of monocot stems.

Now let's drill down a bit deeper, and look at the composition of the dermal, ground, and vascular tissue systems. Each of these tissue systems is made up of an array of distinct cell and tissue types. What are they?

36.3 Cells and Tissues of the Primary Plant Body

Recall from Chapter 7 that plant cells and animal cells share most of their key characteristics: both have chromosomes enclosed in a nuclear envelope, a plasma membrane studded with proteins

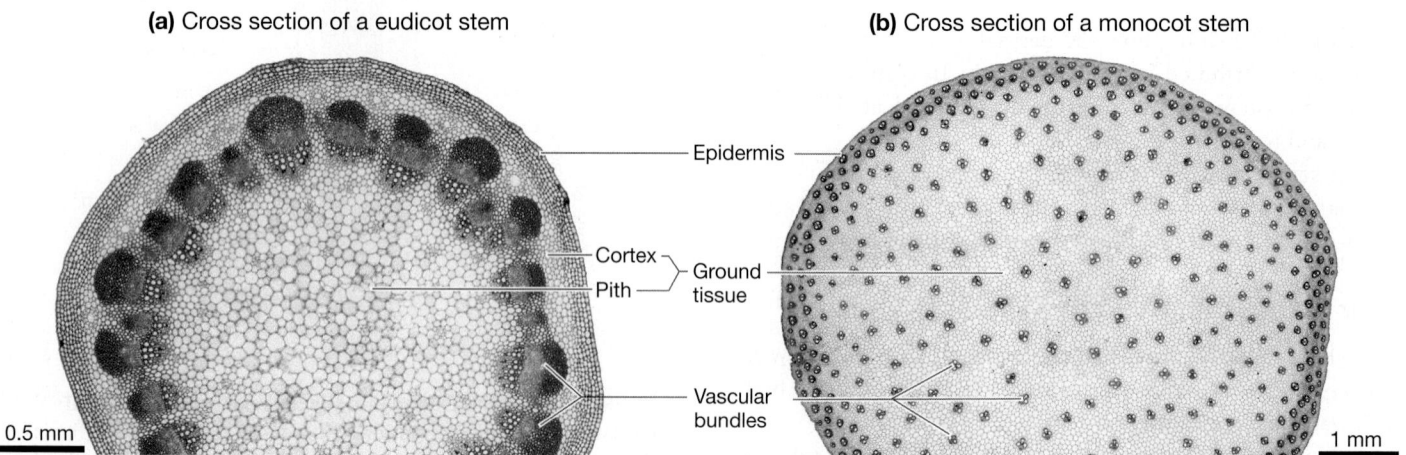

(a) Cross section of a eudicot stem   **(b)** Cross section of a monocot stem

Epidermis

Cortex
Pith — Ground tissue

Vascular bundles

0.5 mm   1 mm

FIGURE 36.16 Stems Contain a Variety of Cell and Tissue Types. As these cross sections show, vascular bundles are **(a)** arranged in a ring near the perimeter of eudicot stems but **(b)** scattered throughout the pith in monocots.

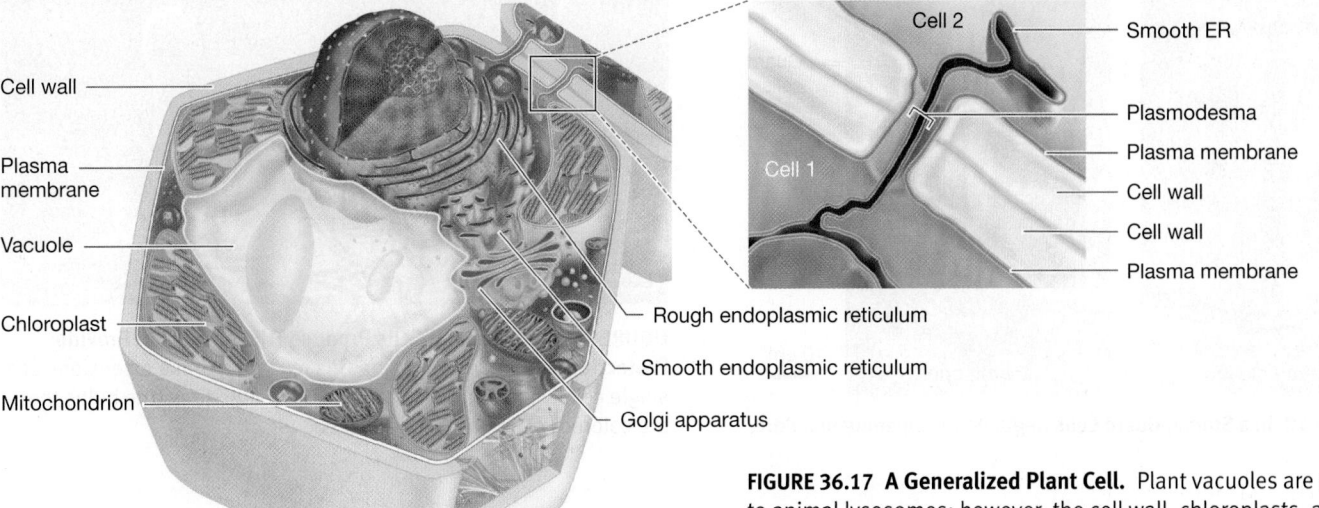

(a) Plant cells have cell walls, vacuoles, and chloroplasts.

Cell wall

Plasma membrane

Vacuole

Chloroplast

Mitochondrion

Rough endoplasmic reticulum

Smooth endoplasmic reticulum

Golgi apparatus

(b) Adjacent plant cells are connected by plasmodesmata.

Cell 2

Cell 1

Smooth ER

Plasmodesma

Plasma membrane

Cell wall

Cell wall

Plasma membrane

FIGURE 36.17 A Generalized Plant Cell. Plant vacuoles are similar to animal lysosomes; however, the cell wall, chloroplasts, and plasmodesmata are unique to plants. Unlike the extracellular matrix of animals, the plant cell wall is rigid.

that regulate the passage of materials in and out, mitochondria that produce ATP by oxidizing sugars, and an array of other organelles that synthesize or degrade key molecules.

In addition, plant cells have several features that are absent in animal cells (**Figure 36.17a**):

1. All plant cells are surrounded by a stiff, cellulose-rich **cell wall** that supports the cell and defines its shape.

2. The cytoplasm of adjacent plant cells is connected via **plasmodesmata** (singular: **plasmodesma**; see Chapter 8). Plasmodesmata consist of cytoplasm and segments of smooth endoplasmic reticulum (smooth ER) that run through tiny, membrane-lined gaps in the cell wall (**Figure 36.17b**).

3. Plant cells often contain several types of organelles that are not found in animals—specifically chloroplasts and a large, membrane-bound organelle called a vacuole, which fills most of the cell's volume.

Chloroplasts are the site of photosynthesis (see Chapter 10). Non-photosynthetic cells found in roots, seeds, flower petals, and other locations may have organelles that are related to chloroplasts but are specialized for storing pigments, starch, oils, or protein.

Vacuoles, which contain an aqueous solution called **cell sap**, store wastes and in some cases also digest wastes, as do animal lysosomes. In addition, plant vacuoles store water and nutrients. They may also hold pigments that provide color or poisons that deter herbivores.

Another important distinction between plant cells and animal cells is that plant cells do not change position once they form. Some animal cells change positions either early in the development of an individual or as mature cells.

During both plant and animal development, a cell's location in the body determines which tissue system it contributes to and what type of cell it becomes. Let's examine the cells and tissues of the three primary tissue systems in turn.

The Dermal Tissue System

Dermal tissue is the interface between the individual and the external environment. Its primary function is to protect the plant body—from water loss, disease-causing agents, and herbivores.

Most tissues are made up of several different cell types, each of which has a distinct structure and function. Let's consider the cell types found in dermal tissue.

EPIDERMAL CELLS PROTECT THE SURFACE Most of the cells in the dermal tissue system are epidermal cells, which are flattened and usually lack chloroplasts. Epidermal cells in the root are responsible for absorbing water and nutrients. In addition, epidermal cells in both the root and shoot systems play a key role in protecting the plant.

Epidermal cells in the shoot system fulfill their protective role in part by secreting the **cuticle**: a waxy layer that forms a continuous sheet on the surface (see Chapter 30). Waxes are lipids and are thus highly hydrophobic. As a result, the presence of cuticle on stems and leaves drastically reduces the amount of water that is lost by evaporation.

From a human perspective, the water-repellent properties of cuticle make it a valuable ingredient in polishes and lipsticks. The carnauba wax used in car and floor polishes, for example, is secreted by epidermal cells in the leaves of carnauba palms native to Brazil.

Besides minimizing water loss, cuticle forms a barrier to protect the plant from viruses, bacteria, and the spores or growing filaments of parasitic fungi. In this way, the plant epidermis forms the first line of defense against disease-causing agents, or **pathogens**. For pathogens to enter the plant body and initiate an infection, they must either secrete enzymes that digest the cuticle or enter via a wound where the cuticle has been torn away.

The waxes found in cuticle can also be detrimental to the plant, however, by reducing gas exchange. This can be a serious

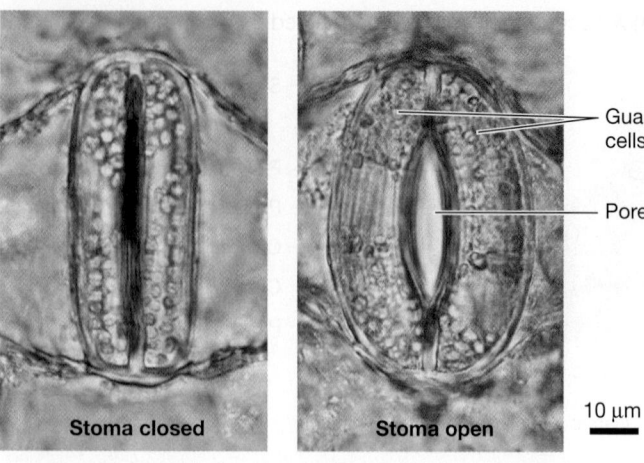

FIGURE 36.18 In a Stoma, Guard Cells Regulate the Opening of a Pore.

Guard cells

Pore

Stoma closed

Stoma open

10 μm

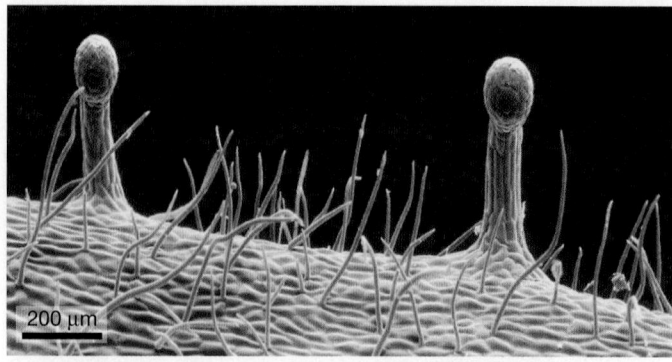

200 μm

FIGURE 36.19 Epidermal Cells Produce Trichomes That Provide Protection. Some trichomes on this leaf are hairlike extensions of a single cell, while others are multicellular structures (the bulbs that are colored orange here hold toxins).

problem because photosynthesis depends on the free flow of carbon dioxide to photosynthetic cells. The problem is solved by specialized structures in dermal tissue called stomata.

STOMATA REGULATE GAS EXCHANGE AND WATER LOSS Most land plants have structures called **stomata** (singular: **stoma**) that allow carbon dioxide to enter photosynthetically active tissues. A stoma consists of two specialized **guard cells**, which change shape to open or close an opening in the epidermis known as a **pore** (**Figure 36.18**).

When stomata are open, CO_2, O_2, water vapor, and other gases can move between the atmosphere and the interior of the plant by diffusion. Stomata open when CO_2 is needed. They close when CO_2 is not needed or when large amounts of water are lost by transpiration. Chapter 39 explores the molecular mechanisms responsible for stomatal opening and closing.

TRICHOMES PERFORM AN ARRAY OF FUNCTIONS In addition to minimizing water loss and regulating gas exchange, cells in dermal tissue may protect the individual from the damaging effects of intense sunlight and attacks by herbivores.

Trichomes are protective, hairlike appendages made up of specialized epidermal cells. They are found in shoot systems, and come in a wide variety of shapes, sizes, and abundance.

Depending on the species, trichomes may (1) keep the leaf surface cool by reflecting sunlight, (2) reduce water loss by forming a dense mat that limits transpiration, (3) provide barbs, or store toxic compounds that thwart herbivores (**Figure 36.19**), or even (4) trap and digest insects.

The Ground Tissue System

Most photosynthesis, as well as most carbohydrate storage, takes place in ground tissue. Cells in ground tissue are also responsible for most of the synthesis and storage of specialized products such as colorful pigments, the chemical signals called hormones, and toxins required for defense.

If the primary business of the dermal tissue system is protection, the ground tissue is all about producing and storing valuable molecules.

Ground tissue is made up of three distinct tissue types: parenchyma (pronounced *pa-REN-ki-ma*), collenchyma (*ko-LEN-ki-ma*), and sclerenchyma (*skle-REN-ki-ma*).

PARENCHYMA ARE "WORKHORSE" CELLS **Parenchyma cells** have relatively thin primary cell walls and are the most abundant and versatile plant cells. Groups of parenchyma cells form parenchyma tissue.

The parenchyma tissue in leaves consists of parenchyma cells filled with chloroplasts, and is the primary site of photosynthesis (**Figure 36.20a**). But in other organs, parenchyma cells store starch granules (**Figure 36.20b**). When you eat a salad, a potato, or an apple, you are ingesting primarily parenchyma cells in ground tissue.

Many parenchyma cells are **totipotent**, meaning they retain the capacity to divide and develop into a complete, mature plant. The totipotency of parenchyma cells is important in healing wounds and in reproducing asexually via stolons or rhizomes. In each case, parenchyma cells may begin to divide, grow, and differentiate to form new roots and shoots.

The totipotency of parenchyma cells also allows gardeners to clone plants by making cuttings. For example, if you cut a piece of coleus stem and place it in water, parenchyma cells will divide to produce a mass of undifferentiated cells called a **callus**. Roots develop from the callus (**Figure 36.21**), and the new individual can be planted in soil.

Bananas, seedless grapes, and several other commercially important strains cannot undergo sexual reproduction to produce seeds. Instead, they are propagated entirely by cuttings.

COLLENCHYMA CELLS FUNCTION PRIMARILY IN SHOOT SUPPORT **Collenchyma cells** have primary cell walls that are thicker in some areas than others, and their overall shape is longer and thinner than that of parenchyma cells. Groups of collenchyma cells form collenchyma tissue that supports the plant body.

Even when collenchyma cells are mature, their cell walls retain the ability to stretch and elongate. As a result, collenchyma cells can continue to lengthen as they provide structural support to the growing regions of shoots.

(a) In leaves: photosynthesis and gas exchange

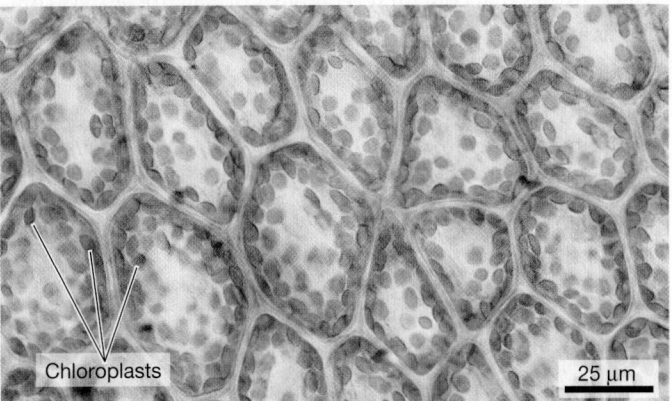

Chloroplasts

25 μm

(b) In roots: carbohydrate storage

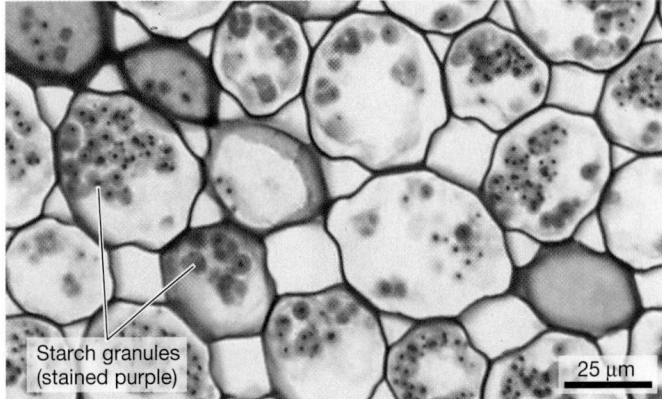

Starch granules
(stained purple)

25 μm

FIGURE 36.20 Parenchyma Cells Perform a Wide Array of Tasks.

✔**EXERCISE** Give an example of a gene that is likely to be expressed in these leaf cells, but not in the root cells.

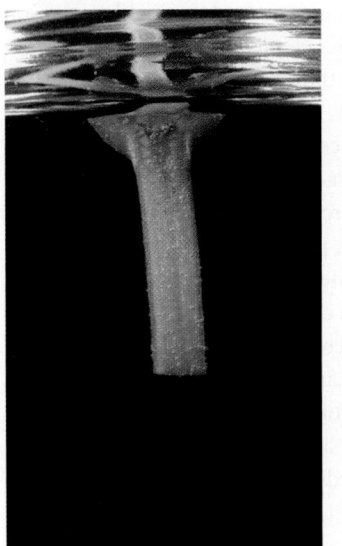

FIGURE 36.21 Parenchyma Cells in Cut Stems Can Form Adventitious Roots. Parenchyma cells in a cut coleus stem (left) divide to form a mass of undifferentiated cells called a callus, which then sprouts roots (right).

Because they can elongate, collenchyma cells are particularly abundant in growing stems and in the stalk portions of leaves. The "strings" you may have peeled from a stalk of celery or rhubarb—which is actually the petiole of a celery or rhubarb leaf—include many strands of collenchyma cells (**Figure 36.22**).

SCLERENCHYMA: TWO TYPES OF SPECIALIZED SUPPORT CELLS The cells that are classified as **sclerenchyma** produce a thick **secondary cell wall** in addition to the relatively thin primary cell wall found in all cells. Unlike a primary cell wall, the secondary cell wall contains the tough, rigid compound **lignin** in addition to cellulose (see Chapter 30).

Collenchyma cells can support actively growing parts of the plant because they have an expandable primary cell wall. In contrast, the nonexpandable secondary cell wall of sclerenchyma cells specializes them for supporting stems and other structures after active growth has ceased. Another key difference between collenchyma and sclerenchyma is that the cells in sclerenchyma tissue are usually dead at maturity—meaning they contain no cytoplasm.

Ground tissue typically contains two types of sclerenchyma cells: fibers and sclereids.

(a) Cross section of celery stalk   **(b)** Close-up of "string," in cross section   **(c)** Collenchyma cells, in cross section

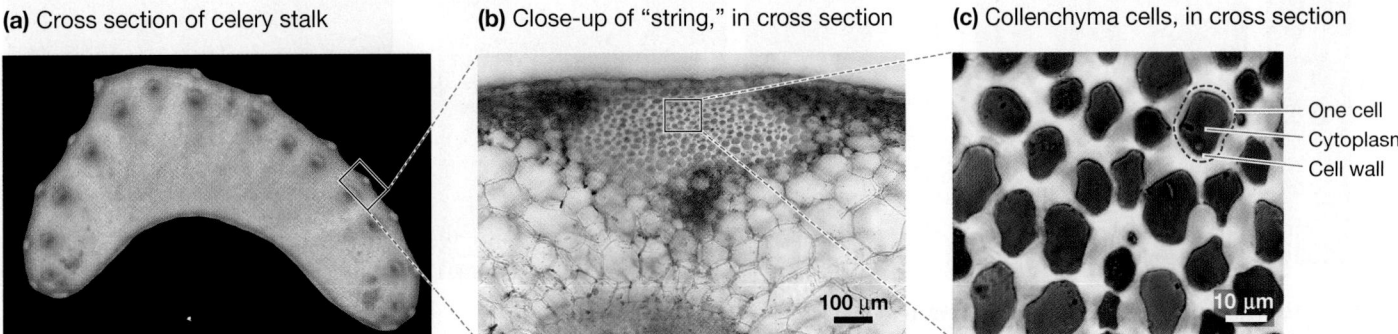

100 μm

10 μm

One cell
Cytoplasm
Cell wall

FIGURE 36.22 Collenchyma Cells Support Growing Tissues. A celery stalk is actually a petiole; the strands you can peel from it are columns of collenchyma cells.

(a) Fibers

(b) Sclereids

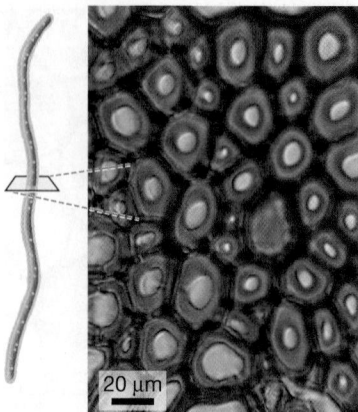
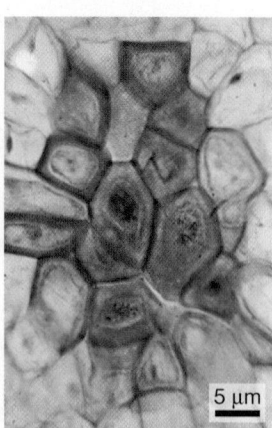

20 µm

5 µm

FIGURE 36.23 Sclerenchyma Cells Support Mature Tissues.
(a) Fibers and **(b)** sclereids have thickened secondary cell walls.
These cells provide support for tissues that are no longer growing.

- **Fibers** are extremely elongated. The fiber cells from ramie plants, for example, can be over half a meter long. Fiber cells are important in the manufacture of paper, hemp or jute ropes, or linen and other fabrics (**Figure 36.23a**).

- **Sclereids** are relatively short, have variable shapes, and often function in protection. The tough coats of seeds and the thick shells of nuts are composed of sclereids; these cells are also responsible for the gritty texture of pears (**Figure 36.23b**).

The Vascular Tissue System

The vascular tissue system functions in support and long-distance transport of water and dissolved nutrients. It moves the products made and stored in ground tissue.

Plant tissues that consist of a single cell type are called simple tissues; tissues that contain several types of cells are termed complex tissues. The vascular tissue system is made up of two complex tissues, xylem and phloem.

- **Xylem** (pronounced *ZYE-lem*) conducts water and dissolved ions in one direction: from the root system to the shoot system.

- **Phloem** (*FLO-em*) conducts sugar, amino acids, chemical signals, and other substances in two directions: from roots to shoots and from shoots to roots.

XYLEM STRUCTURE   The most important cell types in xylem tissue are tracheids and vessel elements.

- In all vascular plants, xylem contains water-conducting cells called **tracheids** (*TRAY-kee-ids*).

- In angiosperms and species in the group Gnetophyta, xylem also contains conducting cells called **vessel elements**.

Tracheids and vessel elements have thick, lignin-containing secondary cell walls that are often deposited in ringlike or spiral patterns. Both tracheids and vessel elements are dead at maturity. As a result, they are filled with the fluids that they conduct instead of with cytoplasm.

Tracheids are long, slender cells with tapered ends (**Figure 36.24a**). The sides and ends of tracheids have **pits**, which are gaps in the secondary cell wall where only the primary cell wall is present. Because the cell is dead, pits have no plasma membrane spanning the opening. When water is moving up a plant through tracheids, it moves from cell to cell both vertically and laterally through pits, because that is where resistance to flow is lowest.

Vessel elements, in contrast, are shorter and wider than tracheids (**Figure 36.24b**). In addition to having pits, vessel elements have **perforations**—openings that lack both primary and secondary cell walls. In some species, the ends of vessel elements lack any cell

(a) Tracheids are spindle shaped and have pits.

(b) Vessel elements are short and wide and have perforations as well as pits.

(c) Tracheids and vessel elements together in vascular tissue

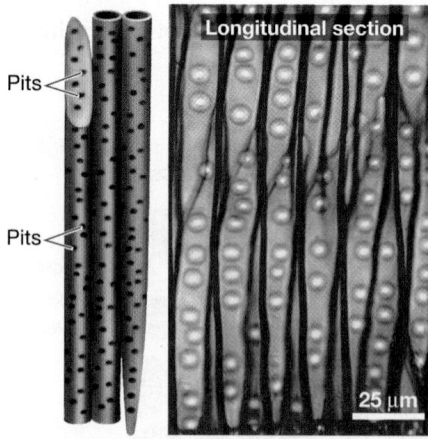
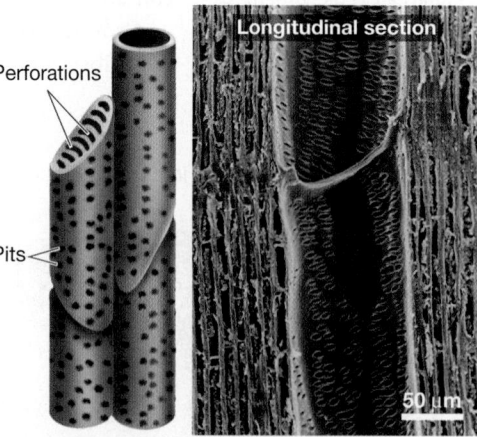
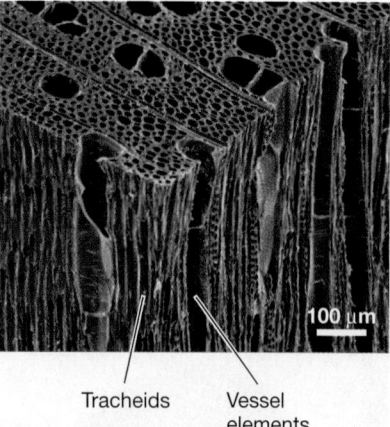

Pits

Pits

Longitudinal section

25 µm

Perforations

Pits

Longitudinal section

50 µm

100 µm

Tracheids

Vessel elements

FIGURE 36.24 Xylem May Contain Two Types of Water-Conducting Cells. **(a)** Tracheids are long and thin compared to **(b)** vessel elements, which are much shorter and wider. **(c)** Both types of water-conducting cells are found in the vascular tissue of angiosperms. The image in (a) is a light micrograph of stained tissue; the images in (b) and (c) are colorized scanning electron micrographs.

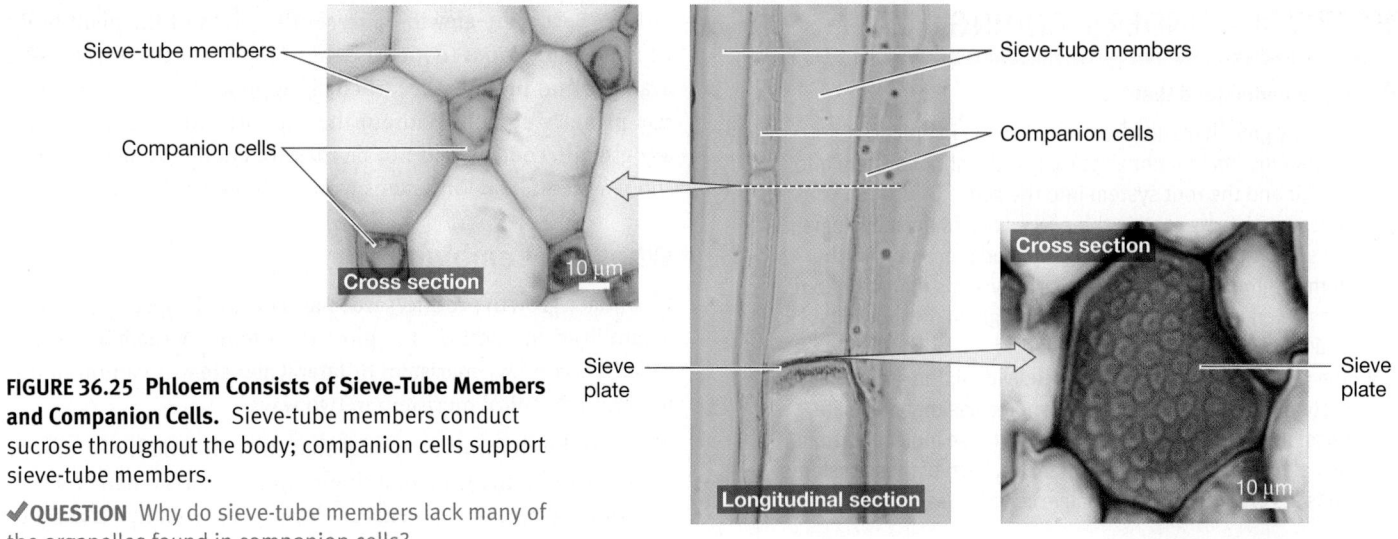

Sieve-tube members

Companion cells

Cross section

10 μm

Sieve-tube members

Companion cells

Sieve
plate

Cross section

Longitudinal section

Sieve
plate

10 μm

FIGURE 36.25 Phloem Consists of Sieve-Tube Members and Companion Cells. Sieve-tube members conduct sucrose throughout the body; companion cells support sieve-tube members.

✔**QUESTION** Why do sieve-tube members lack many of the organelles found in companion cells?

wall at all, and stacked cells form open pipes called vessels. Vessel elements conduct water much more efficiently than do tracheids because their width and perforations offer less resistance to flow.

In angiosperms, tracheids and vessel elements are found adjacent to each other (**Figure 36.24c**). Xylem also contains some parenchyma cells that transport materials laterally in the stem—not vertically.

PHLOEM STRUCTURE Phloem is made up primarily of two specialized types of parenchyma cells: sieve-tube members and companion cells. Both sieve-tube members and companion cells are alive at maturity, lack lignified secondary cell walls, and arise from division of a common precursor cell.

- **Sieve-tube members** (also called sieve-tube elements) are long, thin cells that have perforated ends called **sieve plates** (**Figure 36.25**). They are responsible for transporting sugars and other nutrients.

- **Companion cells** are not conducting cells, but instead provide materials to maintain the cytoplasm and plasma membrane of sieve-tube members.

Sieve-tube members lack nuclei and most other major organelles, but are directly connected to adjacent companion cells by means of numerous plasmodesmata. Companion cells contain all the organelles normally found in a plant cell.

As Chapter 37 will show, companion cells are also involved in loading and unloading carbohydrates and other nutrients from the solution inside sieve-tube members. Phloem also contains fibers and sclereids, which provide support.

Table 36.1 summarizes the major tissue and cell types found in the dermal, ground, and vascular tissue systems. Once you've mastered the structure and function of the primary plant body, you're ready to consider the next level of complexity: secondary growth.

SUMMARY TABLE 36.1 **Components of the Primary Plant Body**

Tissues Present	Description of Tissue	Function
Dermal Tissue System (arises from protoderm)		
Epidermal	Complex tissue consisting of epidermal cells, guard cells, and trichome cells	Shoots: Protection, gas exchange Roots: Protection, water and nutrient absorption
Ground Tissue System (arises from ground meristem)		
Parenchyma	Simple tissue consisting of parenchyma cells	Synthesis and storage of sugars and other compounds
Collenchyma	Simple tissue consisting of collenchyma cells	Support (expandable in size)
Sclerenchyma	Simple tissues consisting of sclereids or fibers	Support (fixed in size)
Vascular Tissue System (arises from procambium)		
Xylem	Complex tissue consisting of tracheids, vessels, and parenchyma cells and sclerenchyma cells (fibers)	Transport of water and ions; support
Phloem	Complex tissue consisting of parenchyma cells (sieve-tube members, companion cells) and sclerenchyma cells (fibers, sclereids)	Transport of sugars, amino acids, hormones, etc.; support

CHECK YOUR UNDERSTANDING

If you understand that . . .

- Primary growth results from cell division in apical meristems. Its function is to extend the shoot system into the air and the root system into the soil.
- Apical meristems contain three types of primary meristematic cells: protoderm, ground meristem, and procambium. The dermal, ground, and vascular tissue systems that arise from these meristematic cells extend throughout the individual and make up the primary plant body.
- The dermal system protects the individual; the ground system produces and stores the molecules that make life possible; the vascular system moves those molecules from place to place and holds the plant up. Each of these systems consists of an array of distinctive cell and tissue types.

✓ **You should be able to . . .**

1. Explain the relationship between an apical meristem and the three primary meristematic tissues.
2. Describe the structure and function of epidermal cells, parenchyma cells, tracheids, and sieve-tube elements.

Answers are available in Appendix B.

36.4 Secondary Growth Widens Shoots and Roots

Primary growth increases the length of roots and shoots; its major function is to extend the reach of the root and shoot system and thus increase a plant's ability to absorb photons and acquire carbon dioxide, water, and ions.

Secondary growth increases the width of the plant body. Its major function is to increase the amount of conducting tissue available and provide the structural support required for extensive primary growth. Without the support provided by secondary growth, roots would not be massive enough to anchor large shoot systems, and long stems would fall over or break.

What Is a Cambium?

Secondary growth produces **wood** and occurs in species that have a cambium in addition to apical meristems. A **cambium** is also called a secondary meristem or **lateral meristem**. A cambium differs from an apical meristem in two ways:

1. A cambium forms a cylinder that runs the length of a root or stem and is made up of a single layer of meristematic cells. In contrast, apical meristems are localized at root tips and shoot tips and are dome shaped.

2. In a cambium, cells divide in a way that increases the width of roots and shoots (**Figure 36.26a**). Cells in an apical meristem divide in a way that extends the root and shoot tips.

As **Figure 36.26b** shows, there are two distinct types of cambium in plants that undergo secondary growth.

- A ring of meristematic cells called the **vascular cambium** is located between the secondary xylem and phloem, inside the stem.

- A second ring of meristematic cells called the **cork cambium** is located near the perimeter of the stem.

One other observation is critical to understanding how lateral meristems work: The cork cambium produces new cells primarily to the outside. Vascular cambium, in contrast, generates new layers of cells both to the inside and outside. The new cells

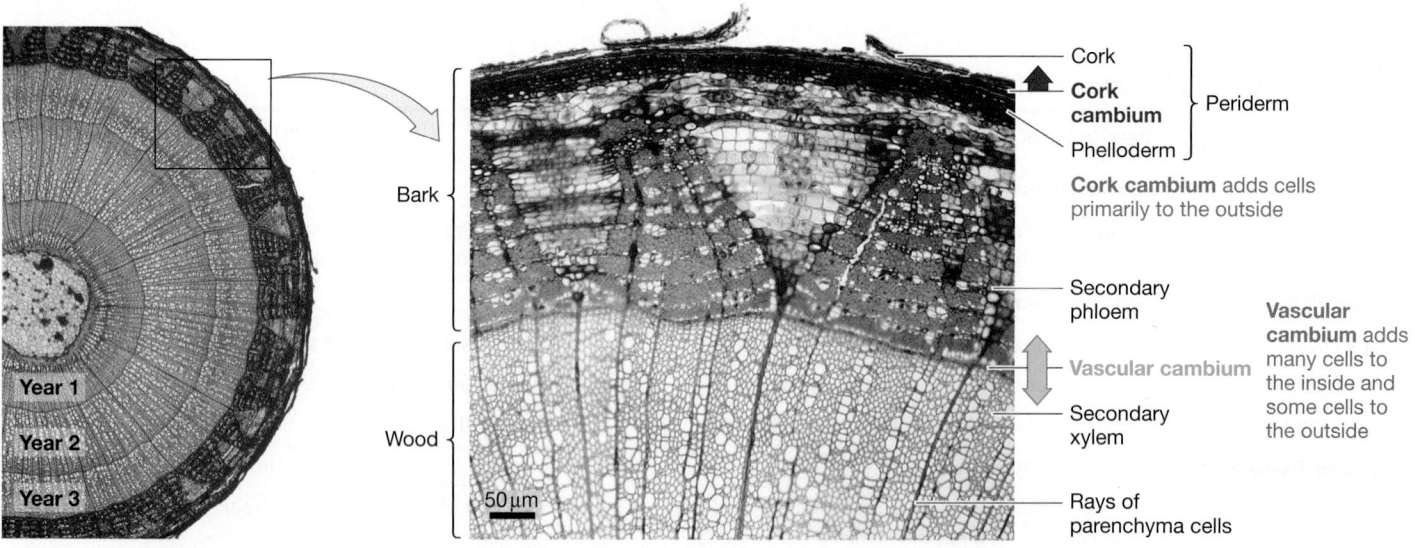

FIGURE 36.26 Vascular Cambium Is Responsible for Secondary Growth. Tree trunks contain two types of lateral meristems: cork cambium and vascular cambium. Bark consists primarily of the cork cells produced by the cork cambium. Wood consists of the secondary xylem and the parenchyma cells produced by vascular cambium.

formed to the inside push all of the other cells toward the outside, causing an increase in girth.

✔If you understand this concept, you should be able to draw a cross section of a stem, draw rings representing the vascular cambium and the cork cambium, and add arrows showing the direction of growth in each meristem.

Now let's consider how each of these meristems works.

What Does Vascular Cambium Produce?

Vascular cambium produces both phloem and xylem (see Figure 36.26b). New cells that are produced to the outside of the meristem differentiate into phloem; new cells produced to the inside differentiate into xylem.

More specifically, cells produced by vascular cambium develop into secondary phloem and secondary xylem. (In contrast, the procambium at each apical meristem produces primary phloem and primary xylem.) Secondary phloem and secondary xylem cells are not always produced simultaneously. In most cases, the vascular cambium produces many more secondary xylem cells than secondary phloem cells.

✔If you understand this concept, you should be able to add "Secondary phloem" and "Secondary xylem" labels to the arrows on your stem cross-section diagram. (Make one arrow fatter than the other to reflect the relative amount of cell division).

Primary phloem and xylem are found throughout the roots and shoots of all vascular plants, but secondary phloem and xylem are found only in some lycophytes, the gymnosperms, and certain angiosperms.

Structurally, primary and secondary phloem and primary and secondary xylem are complex tissues, made up of more than one cell type. Functionally, primary and secondary phloem are similar; primary and secondary xylem are also similar.

- Secondary phloem functions in sugar transport. In combination with cork cambium tissues, it forms bark.

- Secondary xylem functions in water transport and structural support, forming the structural material called wood.

Besides producing conducting cells such as sieve-tube members, tracheids, and vessel elements, the vascular cambium produces fibers for additional strength, along with parenchyma cells. The parenchyma cells radiate laterally across the xylem and form structures called **rays** (see Figure 36.26b), which transport water and nutrients laterally across the stem.

Secondary growth in roots is similar to secondary growth in stems. In both portions of the plant, width increases as cells produced on the inside of the vascular cambium form secondary xylem, and cells produced on the outside form secondary phloem.

It's important to realize, though, that the results of cell division in lateral meristems are highly asymmetrical. As the vascular cambium grows, all of the secondary xylem is retained and accumulates but the primary xylem eventually dies and may rot away. In addition, the outermost secondary phloem and cork layers are sloughed off as the stem increases in diameter. As a result, mature woody roots and stems are dominated by secondary xylem, or wood.

Table 36.2 summarizes the major tissue types and cell types involved in secondary growth. You can also go to the study area at *www.masteringbiology.com* and review both primary and secondary growth.

 **Web Activity** Plant Growth

What Does Cork Cambium Produce?

The cork cambium produces **cork cells** to the outside and a smaller layer of cells called the **phelloderm** ("cork-skin") to the inside (Figure 36.26b). Together, the cork cambium, cork cells, and phelloderm make up the tissue called **bark**.

✔If you understand the structure of bark, you should be able to add labels to your stem cross-section diagram that read "Cork

SUMMARY TABLE 36.2 **Components of Secondary Growth**

Region	Tissue Type	Cell Composition	Function
Bark (arises from and includes cork cambium)	Cork	Cork cells	Protection
	Cork cambium	Meristematic cells	Production of cork and phelloderm
	Phelloderm	Parenchyma cells	Synthesis and storage
Secondary phloem (arises from vascular cambium)	Phloem	Parenchyma cells (sieve-tube members, companion cells) and sclerenchyma cells (fibers, sclereids)	Transport of sugars, amino acids, hormones, etc.; support
Secondary xylem* (arises from vascular cambium)	Xylem	Tracheids, vessels, parenchyma cells (arranged in rays) and sclerenchyma cells (fibers)	Transport of water and ions; support

*Secondary xylem is also called wood.

cells" and "Phelloderm." You should also be able to make one arrow from the cork cambium fatter than the other to reflect the relative amount of cell division that occurs.

Bark is important because it protects the woody stem as it increases in girth. As a woody stem or root matures, the epidermal tissue produced by the apical meristem during primary growth is replaced by the bark, which takes over the role of preventing water loss and protecting the stem and root from pathogens and herbivores. In some species, exceptionally thick bark can even protect the shoot system from fire damage. Redwood trees, for example, which grow in fire-prone habitats, can have bark that is 20 cm (12 in.) thick.

Bark provides a particularly tough barrier in species where cork cells secrete a strong secondary cell wall containing lignin. Bark also helps prevent water loss because cork cells produce a layer of wax and other molecules inside their cell walls, making them impermeable to water and gases. Gas exchange can still occur between the atmosphere and living tissues inside the stem, though—through small, spongy segments of the bark called **lenticels**.

Cork cells die when they mature. As a stem continues to widen, the cork layer often cracks and flakes.

✔If you understand the structure of the cork cambium and the cells it produces, you should be able to add a label to your stem cross-section diagram that reads "Bark." You should also be able to label the areas of the stem that function in structural support, transport of water and nutrients, and protection.

The Structure of a Tree Trunk

Trees are perennial plants that live for many years. As a tree matures and grows in width, the innermost xylem layers stop transporting water—only the xylem from the most recent years actually transports fluid.

HEARTWOOD AND SAPWOOD Xylem that no longer transports begins accumulating protective compounds secreted by other tissues. These compounds form resins, gums, and other complex mixtures. The deposition of these molecules causes the oldest portions of secondary xylem to become darker than the younger portions.

The darker-colored, inner xylem region is called **heartwood** while the lighter-colored, outer xylem is called **sapwood** (**Figure 36.27a**). If you look closely at wood furniture or flooring, you may see a color difference between sapwood and heartwood in some boards.

ANNUAL GROWTH RINGS Another important phenomenon occurs in environments where the vascular cambium stops growing for a portion of each year. This period of **dormancy** occurs during the winter in cold climates and during the dry season in tropical habitats.

When the vascular cambium resumes growth in the spring or at the start of the rainy season, it produces large, relatively thin-walled cells. As the growing season nears its end, conditions tend to dry out or become cooler; the secondary xylem cells that are produced at this time tend to be smaller, thicker walled, and

(a) Heartwood and sapwood have different functions.

Heartwood   Sapwood Bark

Provides structural support but no longer transports water | Includes active water-conducting xylem tissue

(b) Growth rings result from variation in cell size.

Early wood   Late wood

One growth ring

(c) Patterns in growth rings can tell a tree's history.

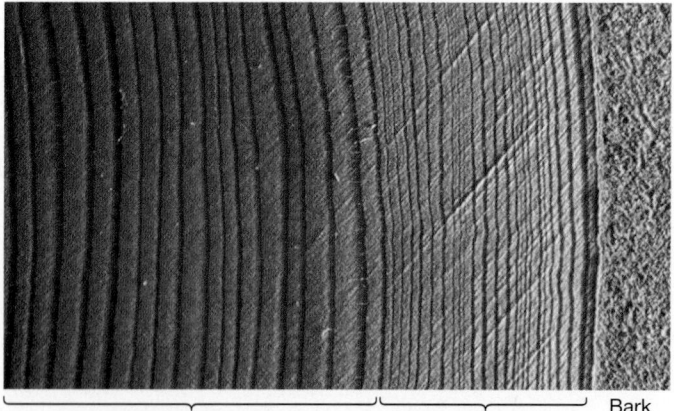

Bark

Thick growth rings before onset of acid rain | Thin growth rings after onset of acid rain

FIGURE 36.27 Anatomy of a Tree Trunk. (a) Unstained section of wood. **(b)** Section of wood, stained to show individual cells. **(c)** Unstained section through a fir tree from Germany's Black Forest.

✔**EXERCISE** In part (c), pick a thick growth ring and label the early wood and late wood.

darker. Thus when growth is seasonal, regions of large, thin-walled cells alternate with layers of small, thick-walled cells, resulting in annual growth rings (**Figure 36.27b**).

Analyzing patterns in tree growth rings is an important field of study in biology. Because trees grow faster when moisture and nutrients are plentiful, wide tree rings are reliable indicators of wet years. In contrast, narrow rings signal drought years—or in the case of the fir tree shown in **Figure 36.27c**, years when abundant acid rain, due to air pollution, reduced growth.

By studying the growth rings in fossil trees and extremely old living trees, biologists can often assemble a continuous record that dates back thousands of years. In doing so, they gain a better understanding of climate changes that occurred in the past. With continued research, researchers also hope to predict how forests might respond to the global warming that is currently under way.

CHAPTER 36 REVIEW

For media, go to the study area at www.masteringbiology.com

Summary of Key Concepts

The vascular plant body consists of (1) a root system that anchors the individual and absorbs water and key ions, and (2) a shoot system that absorbs carbon dioxide and sunlight. Both systems are dynamic—they grow and change throughout life.

- The root and shoot systems of plants are specialized for harvesting the light, water, and nutrients required for performing photosynthesis. Structures involved in absorption have a high surface-area-to-volume ratio.

- Roots extract water and nutrients such as nitrogen, phosphorus, and potassium from the soil.

- Shoots capture light and carbon dioxide from the atmosphere.

- Leaves, the major organ for performing photosynthesis, usually consist of a flattened blade that extends from a petiole.

- Roots, stems, and leaves may be modified to perform a variety of other functions, however, including nutrient storage, water storage, protection, and asexual reproduction.

- Because roots and shoots grow throughout life, a plant is able to respond appropriately to changes in environmental conditions.

  ✔ You should be able to explain why phenotypic plasticity in roots, shoots, and leaves is expected to be more important (**1**) in environments where conditions are variable versus stable, and (**2**) in long-lived versus short-lived species.

Variation in body size and shape allows different species to harvest water, light, and other resources in unique ways.

- The overall morphology of root and shoot systems varies widely among plant species.

- In prairie plants, root systems range from long, linear taproots to shallow, dense mats.

- Among the silverswords of Hawaii, shoot systems vary from low mats or rosettes to woody, highly branched tree trunks.

- Among species, variation in plant size and shape allows individuals to reduce competition for resources and thrive in a particular habitat.

  ✔ You should be able to describe a habitat where (**1**) plants would be expected to have relatively large root systems and small shoot systems, and (**2**) plants would be expected to have relatively small root systems and large shoot systems.

Primary growth occurs when cells located at the tips of each root and shoot divide and enlarge. Primary growth lengthens roots and shoots and gives rise to three primary tissue systems that are specialized for protection, food production and storage, and transport.

- Each apical meristem gives rise to three primary meristematic tissues: protoderm, ground meristem, and procambium.

- The primary meristematic tissues give rise to the dermal, ground, and vascular tissue systems, which extend throughout the plant body.

- The dermal tissue system is usually one cell layer thick and plays a role in protection, water conservation, and water absorption.

- The ground tissue system performs photosynthesis and stores carbohydrates and other compounds.

- The vascular tissue system transports materials throughout the plant. Within the vascular system, xylem tissue transports water and dissolved ions up the plant; phloem tissue transports sugars up and down.

- Each tissue system is made up of simple tissues that contain a single cell type, complex tissues that contain two or more cell types, or a combination of simple and complex tissues.

- Ground tissue contains (**1**) parenchyma cells, which function in material synthesis and storage, (**2**) collenchyma cells, which provide structural support for growing regions, and (**3**) the fiber and sclereid cells of sclerenchyma tissue, which strengthen regions of the body that have stopped growing.

  ✔ You should be able to predict the results of an experiment where a drug was used to (**1**) poison an apical meristem in the shoot and an apical meristem in the root, and (**2**) selectively poison protoderm cells in a shoot apical meristem.

🔑➤ **In some species, secondary growth occurs when cells near the perimeter of a root or shoot divide and enlarge, widening the structure. Secondary growth adds transport tissue and provides structural support.**

- In some plant species, shoots and roots are widened by lateral meristems that produce secondary xylem, secondary phloem, and bark.

- Wood consists of secondary xylem, while bark consists of all tissue outside of the vascular cambium.

  ✔ You should be able to predict the results of an experiment where a drug was used to (**1**) slow the growth of the vascular cambium but not the cork cambium, and (**2**) slow the growth of both the vascular and cork cambia on one side of a tree trunk.

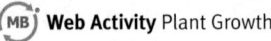 **Web Activity** Plant Growth

Questions

✔ TEST YOUR KNOWLEDGE

Answers are available in Appendix B

1. Which of the following functions is not performed by parenchyma cells?
   a. synthesizing key molecules
   b. transporting water
   c. performing photosynthesis
   d. storing nutrients

2. How do tracheids differ from vessel elements, in addition to their overall shape?
   a. Tracheids are stacked end to end to form continuous, open columns.
   b. In tracheids, water flows from cell to cell primarily through gaps in the secondary cell wall called pits.
   c. Tracheids are dead at maturity.
   d. Tracheids have secondary cell walls reinforced with lignin.

3. What is a sieve-tube member?
   a. the sugar-conducting cell found in phloem
   b. the widened, perforation-containing, water-conducting cell found only in angiosperms
   c. the nutrient- and water-absorbing cell found in root hairs
   d. the nucleated and organelle-rich support cell found in phloem

4. What is an adventitious root?
   a. one that performs a function other than anchoring the plant or absorbing water and ions
   b. one that erupts through the root cortex and spreads laterally
   c. a long, filamentous extension that increases the surface area of the root system, for efficient absorption
   d. one that arises from the shoot system

5. Which statement best characterizes primary growth?
   a. It does not occur in roots, only in shoots.
   b. It leads to the development of cork.
   c. It produces the dermal, ground, and vascular tissues.
   d. It produces rings of xylem and phloem tissue as well as rings of cork tissue.

6. Which statement best characterizes secondary growth?
   a. It results from divisions of the vascular cambium cells.
   b. It increases the length of the plant stem.
   c. It results from divisions in the apical meristem cells.
   d. It often produces phloem cells to the inside and xylem cells to the outside of the vascular cambium.

✔ TEST YOUR UNDERSTANDING

Answers are available in Appendix B

1. Describe the general function of the shoot system and the general function of the root system. Which tissues are continuous throughout these two systems? Suggest a hypothesis to explain why the shoot and root systems of different species are so variable in size and shape.

2. Explain why continuous growth enhances the phenomenon known as phenotypic plasticity.

3. To illustrate the concept that a plant structure can be modified for many different functions, give an example of a leaf and a stem that protect individuals against large herbivores.

4. What does cuticle do? What do stomata do? Predict how the thickness of cuticle and the number of stomata differ in plants from wet habitats versus dry habitats.

5. Compare and contrast the roles of parenchyma cells in the ground tissue system versus the vascular tissue system.

6. Describe how the vascular cambium produces secondary xylem and secondary phloem.

✔ APPLYING CONCEPTS TO NEW SITUATIONS

Answers are available in Appendix B

1. The shoot systems of domesticated broccoli family plants (broccoli, cauliflower, cabbage, kohlrabi, and others) changed in response to artificial selection from the wild mustard that was their ancestor. Would you predict that leafy populations such as cabbage, kale, and savoy have more or fewer sclerenchyma cells in their stems and leaves than the wild population has? Explain your logic.

2. Identify the structure you are consuming when you eat the following vegetables: asparagus, Brussels sprouts, celery, spinach, carrots, potato.

3. Why do trees that grow in tropical rain forests lack growth rings?

4. Trees can be killed by girdling—meaning the removal of bark and vascular cambium in a ring all the way around the tree. Explain why.

This chapter explores how plants move water from their roots to their leaves and how they transport sugars to all of their tissues—sometimes over great distances.

Water and Sugar Transport in Plants

37

O n a hot summer day, a large deciduous tree can lose enough water to fill three 55-gallon drums. To understand why, recall from Chapter 36 that the surfaces of leaves are dotted with stomata, which open during the day so gas exchange can occur between the atmosphere and the cells inside the leaf. This exchange is crucial. For photosynthesis and food production to continue, leaf cells must acquire carbon dioxide (CO_2).

There's a catch, however. While stomata are open, the moist interior of the leaf is exposed to the dry atmosphere. As a result, large quantities of water evaporate from the leaf.

In essence, then, water loss is an inevitable consequence of a plant's need to obtain carbon and release oxygen. Water loss is a side effect of photosynthesis.

Evaporation from leaves can actually be beneficial under some conditions, because it cools the plant, just as sweating cools your body. Heavy rates of evaporation can lower leaf temperatures by as much as 10–15°C.

If the lost water is not replaced, however, plant cells will dry out and die. In the case of a redwood tree, the leaves that lose water may be 100 m from the root hairs that absorb water.

How do plants transport water against the force of gravity—in some cases, the length of a football field? And how do plants move the sugar they produce from active photosynthetic sites to storage sites in roots? These questions are the heart and soul of this chapter. Answering them is a fundamental part of understanding how plants work.

37.1 Water Potential and Water Movement

Loss of water via evaporation from the aerial parts of a plant is called **transpiration**. Transpiration occurs whenever two conditions are met: (**1**) Stomata are open, and (**2**) the air surrounding leaves is drier than the air inside leaves.

KEY CONCEPTS

- Water moves from areas of high water potential to areas of low water potential. Water's potential energy in plants is a combination of (1) its tendency to move in response to differences in solute concentration and (2) the pressure exerted on it.

- Plants lose water to transpiration when stomata are open and photosynthesis is occurring, but they do not expend energy to replace it. Instead, water moves from soil and roots to leaves along a water-potential gradient. Evaporation of water from leaves, driven by the Sun, creates a negative pressure (tension) that pulls water up.

- In phloem, sugars are transported from "sources"—tissues that release sugars for use elsewhere—to "sinks"—tissues in which sugars are being used or stored. Movement occurs because large amounts of sucrose move into phloem cells near source tissues. Water follows by osmosis, creating a pressure gradient that favors the movement of water and sucrose to sinks.

✔ When you see this checkmark, stop and test yourself. Answers are available in Appendix B.

The first condition is usually met during the day, when photosynthesis takes place. The second condition occurs whenever atmospheric humidity is less than 100 percent.

Plants replace water that is lost from leaves with water that is absorbed by roots. One of the most astonishing observations in biology is that water moves from roots to leaves passively—that is, with no expenditure of ATP. Plants do not need a heart muscle to pump water from roots to shoots. Even in 100-m-tall redwood trees, water flows passively from the root system to the shoot system. This movement occurs because of differences in the potential energy of water.

What Is Water Potential?

Recall from Chapter 2 that potential energy is stored energy. Changes in potential energy are associated with changes in position, such as the position of a molecule or an electron.

Biologists use the term **water potential** to indicate the potential energy that water has in a particular environment compared with the potential energy of pure water at room temperature and atmospheric pressure. Under these conditions, pure water has a water potential of 0.

Water potential is symbolized by the Greek letter ψ (psi, pronounced *sigh*). Differences in water potential determine the direction that water moves. Water always flows from areas of high water potential to areas of lower water potential.

What Factors Affect Water Potential?

To understand how water moves from cell to cell in a plant, consider the cell in the beaker on the left in **Figure 37.1a**. Notice that it is sitting in a **solution**—a homogenous, liquid mixture containing several substances. In this case, the solution consists of water and dissolved substances, or **solutes**.

In the beaker on the left, the solute concentrations in the cell and in the surrounding solution are the same. Recall from Chapter 6 that such a solution is **isotonic** to the cell. When two solutions are isotonic, there is no net movement of water between them.

THE ROLE OF SOLUTE POTENTIAL What happens when the cell is transferred to the beaker on the right in Figure 37.1a? This beaker contains pure water, which has no solutes. As a result, the solution surrounding the cell is strongly **hypotonic** relative to the cell.

The concentration of the solution is important, because when solutions are separated by a selectively permeable membrane, such as the plasma membrane of a cell, water passes through the membrane from regions of low solute concentration to regions of high solute concentration. This movement of water across membranes, in response to differences in water potential, is called **osmosis** (see Chapter 6).

The tendency for water to move by osmosis, in response to differences in solute concentrations, is called the **solute potential (ψ_S) or osmotic potential**.

The solute potential of a solution is defined by its solute concentration relative to pure water. If water contains a high concentration of solutes, then it has a low solute potential compared with pure water.

(a) Solute potential is the tendency of water to move by osmosis.

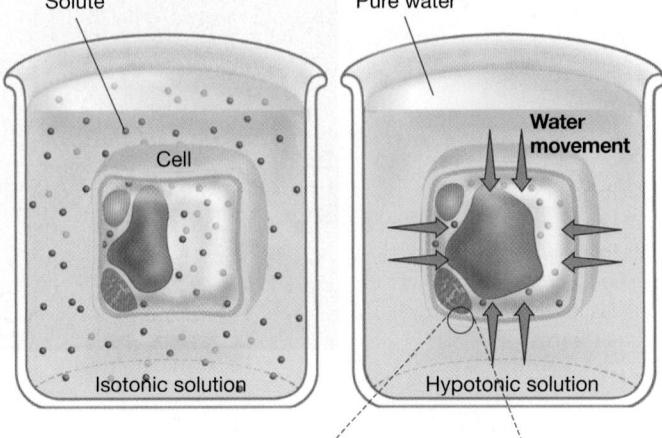

Solute potential inside cell and in surrounding solution is the same. No net movement of water.

Cell is placed in pure water. The cell's solute potential is low relative to its surroundings. Water moves into cell via osmosis.

(b) Pressure potential is the tendency of water to move in response to pressure.

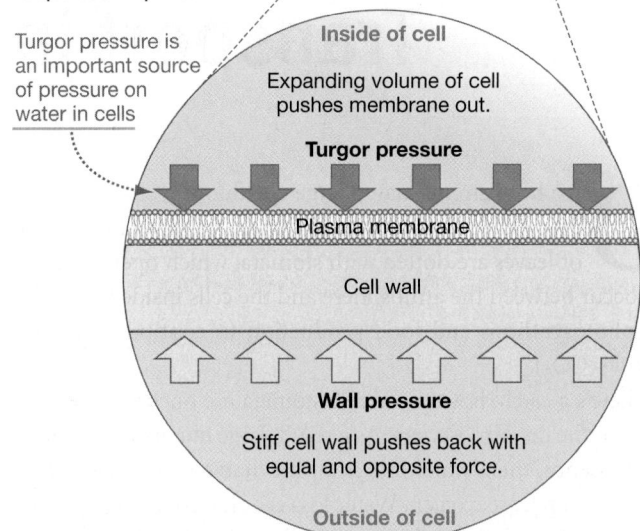

Turgor pressure is an important source of pressure on water in cells

FIGURE 37.1 Water Potential Has Two Major Components: Solute Potential and Pressure Potential.

THE ROLE OF PRESSURE POTENTIAL When an animal cell is placed in a hypotonic solution and water enters the cell via osmosis, the volume of the cell increases until the cell bursts. This does not happen to plant cells, however.

If a plant cell swells in response to incoming water, its plasma membrane pushes against the relatively rigid cell wall. The cell wall resists expansion of the cell volume by pushing back, much as the walls of a basketball push back when the ball is inflated.

The force exerted by the wall is called **wall pressure** (**Figure 37.1b**). As water moves into the cell, the pressure inside the cell, known as **turgor pressure**, increases until wall pressure is induced. Cells that are firm and that experience wall pressure are said to be **turgid**.

Turgor pressure is important, because it counteracts the movement of water due to osmosis. In the right-hand example in Figure 37.1a, the solute potential favors water moving into the cell. However, the rigid cell walls limit the amount of water that can enter the cell.

Pressure potential (ψ_P) refers to any kind of physical pressure on water. Inside a cell, the pressure potential consists of turgor pressure.

When osmosis and pressure affect a cell at the same time, how do biologists determine the direction of water movement?

Calculating Water Potential

As a form of potential energy, water potential (ψ) can be thought of as water's stored energy or its tendency to move to a new position. Thus, a water potential summarizes the stored energy that will tend to make water move—in response to the combined effects of a pressure potential and a solute potential.

When selectively permeable membranes are present, water tends to move by osmosis from areas of high solute potential to areas of low solute potential. When no membranes are present to stop it, water moves from areas of high pressure potential to areas of low pressure potential.

If we ignore the effects of gravity, water potential is defined algebraically as

$$\psi = \psi_P + \psi_S$$

🔑 In words, the potential energy of water in a particular location is the sum of the pressure potential and the solute potential that it experiences.

ASSIGNING UNITS OF PRESSURE AND SIGNS Water potential is measured in units called **megapascals** (**MPa**, 10^6 Pa). A **pascal** (**Pa**) is a unit of measurement commonly applied to pressures—force per unit area. A car tire is inflated to about 0.2 MPa, and the water pressure in home plumbing is usually 0.2 to 0.3 MPa.

Solute potentials (ψ_S) are always negative, because they are measured relative to the solute potential of pure water, which is 0 Mpa because it contains no dissolved substances. And because there are always some solutes inside a cell, the water inside always has a solute potential lower than that of pure water. Therefore pure water will tend to move *into* the cell. Increasing the concentration of solutes in a cell lowers its water potential even more—making it more negative.

In contrast to the solute potential in cells, the pressure potential (ψ_P) from turgor pressure is positive inside cells. It increases the potential energy of the water inside by exerting pressure on the water, making it more likely to move *out of* the cell. But pressure potential can also be negative.

WATER MOVEMENT IN THE ABSENCE OF PRESSURE In the U-shaped tube on the left side of **Figure 37.2a**, two solutions are separated by a selectively permeable membrane. The system is open to the atmosphere and thus is not under additional pressure, meaning $\psi_P = 0$ MPa.

Note that the left side of the tube contains pure water, which has a ψ_S of 0 MPa. The ψ_S for the solution on the right side of the mem-

(a) Solute potentials differ.

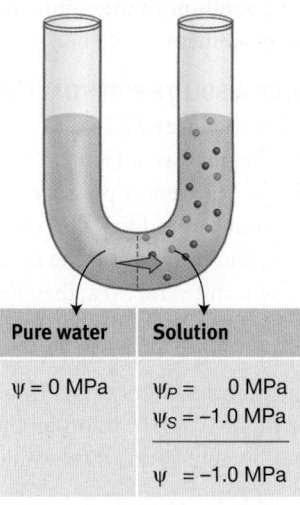
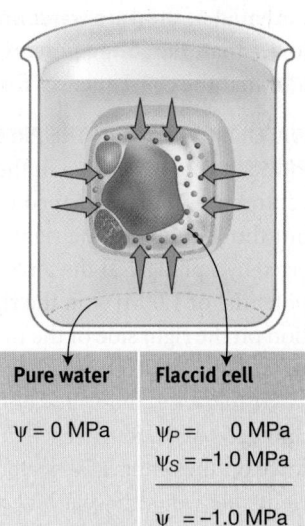

Pure water	Solution		Pure water	Flaccid cell
$\psi = 0$ MPa	$\psi_P = \quad 0$ MPa		$\psi = 0$ MPa	$\psi_P = \quad 0$ MPa
	$\psi_S = -1.0$ MPa			$\psi_S = -1.0$ MPa
	$\psi = -1.0$ MPa			$\psi = -1.0$ MPa

Water moves left to right—from area with high water potential to area with low water potential

Water moves into cell—from area with high water potential to area with low water potential

(b) Solute and pressure potentials differ.

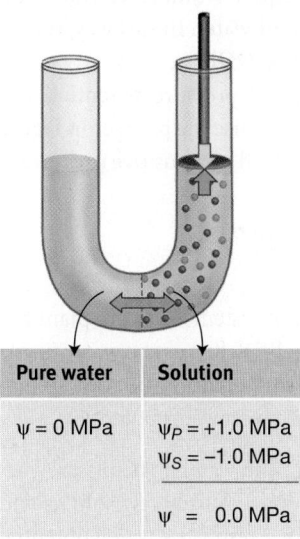

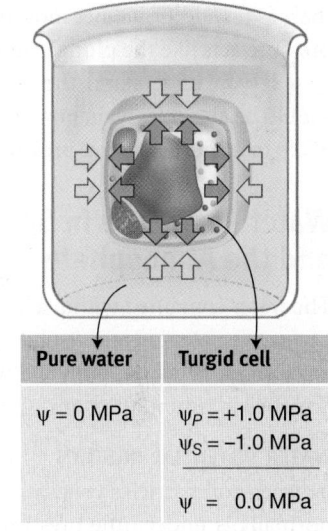

Pure water	Solution		Pure water	Turgid cell
$\psi = 0$ MPa	$\psi_P = +1.0$ MPa		$\psi = 0$ MPa	$\psi_P = +1.0$ MPa
	$\psi_S = -1.0$ MPa			$\psi_S = -1.0$ MPa
	$\psi = \quad 0.0$ MPa			$\psi = \quad 0.0$ MPa

Water potentials are equal—no net movement

Water potentials are equal—no net movement

FIGURE 37.2 Solute Potential and Pressure Potential Interact.

✔ **QUESTION** In the left side of part (a), does the solute potential on the right side of the tube increase, decrease, or stay the same when water flows from left to right by osmosis?

brane is -1.0 MPa. Because water potential is higher on the left side of the tube than on the right side, water moves from left to right.

The right side of Figure 37.2a models the same situation with a cell that is **flaccid**—meaning it has no turgor pressure and thus a pressure potential of 0. Note that the cell has been placed in a solution of pure water. Because the cell has low solute potential (-1.0 MPa) and the pure water has a higher water potential than the cell, water enters the cell via osmosis.

✔ If you understand the concept of solute potential, you should be able to explain (**1**) how you would change the solute potential of the pure water on the left of Figure 37.2a to make it lower than the solute potential in the solution to the right, and (2) what the consequences for water movement would be.

WATER MOVEMENT IN THE PRESENCE OF A SOLUTE POTENTIAL AND PRESSURE POTENTIAL

On the left side of **Figure 37.2b**, the concentrations in the U-shaped tube are the same as in Figure 37.2a, but the solution on the right-hand side experiences pressure exerted by a plunger. If the force on the plunger produces a pressure potential of 1.0 MPa on the right side, and if the ψ_S for the solution on the right side of the membrane is still -1.0 MPa, then the water potential of the right side is -1.0 MPa $+ 1.0$ MPa $= 0$.

In this case, the water potential on both sides of the membrane is equal and there will be no net movement of water. If the force on the plunger is greater than 1.0 MPa, the solution on the right side would have a *higher* water potential and water would flow from right to left.

The right side of Figure 37.2b models this situation in a cell. In this case, the incoming water creates turgor pressure. When the positive turgor pressure ($+1.0$ MPa) plus the cell's negative solute potential (-1.0 MPa) equals 0 MPa—the water potential of pure water—the system reaches equilibrium. At equilibrium, there is no additional net movement of water. In this way, turgor pressure acts like the plunger in Figure 37.2b.

✔ If you understand the concept of pressure potential, you should be able to explain how you would use the plunger in Figure 37.2b to create a negative pressure instead of a positive pressure.

Water Potentials in Soils, Plants, and the Atmosphere

The water contained within a leaf, root system, or entire plant has a pressure potential and a solute potential, just as the water inside a cell does. Likewise, both the soil surrounding the root system and the air around the shoot system have a water potential.

WATER POTENTIAL IN SOILS

In soil, the water that fills crevices between soil particles usually contains relatively few solutes and normally is under little pressure. As a result, its water potential tends to be high relative to the water potential found in a plant's roots, which is high in solutes.

There are important exceptions to this rule, however.

- *Salty soils* The soils near ocean coastlines and in deserts may have water potentials as low as -4 MPa or less due to high solute concentrations. This is much lower than the water potential typically found inside plant roots.

- *Dry soils* When soils dry, water no longer floats freely in the spaces between soil particles. All of the remaining water clings tightly to soil particles, creating a negative pressure that lowers the pressure potential of soil water.

When the water potential in soil drops, water is less likely to move from soil into plants. If soil water potential is low enough, water may even move from plants to the soil.

This is an enormously important issue for world agriculture. When soils are irrigated to boost crop yields, much of the water evaporates. The solutes in the irrigation water are left behind and tend to collect in the first few inches below the surface. Over time, then, irrigated soils tend to become salty. In some parts of the world, formerly productive soils have become so salty that they are now abandoned as cropland.

HOW DO PLANTS THAT ARE ADAPTED TO SALTY OR DRY HABITATS COPE?

Salt-adapted species respond to low water potentials in soil by lowering the solute potential of their root cells. These plants have enzymes and transport proteins that increase the concentration of sugars and other organic molecules in the cytoplasm. As a result, they can keep the water potential of their tissues even lower than the water potential of salty soils.

Species that are adapted to dry sites cope by tolerating low solute potentials. **Figure 37.3**, for example, plots changes in solute potential that were recorded in tissue from a shrub species called ninebark. In this graph, each data point represents the solute potential recorded on a particular day; the lines between the data points are drawn simply to make the trend clear.

Ninebarks thrive on dry sites in the Rocky Mountains of western North America. Notice that the solute potential of tissue is relatively high in June, at the start of the growing season. In the Rockies, June tends to be rainy and cool. July and August are progressively hotter and drier, however. As the graph indicates, the solute potential in tissue drops dramatically during this period.

This is a key observation. As the summer progresses and water potential in soil drops, ninebark shrubs are able to keep acquiring water and grow because the solute potentials of their tissues can drop to match a decline in soil water potential.

Plants that are adapted to wetter sites cannot tolerate such low solute potentials in their tissues. When conditions get hot and dry, they have to close their stomata and stop photosynthesis—meaning that growth will stop or slow down dramatically compared to dry-adapted plants.

If plants keep their stomata open and leaf cells lose water faster than the water is replaced, plasma membranes contract. If the cells do not regain turgor soon, they are at risk of dehydra-

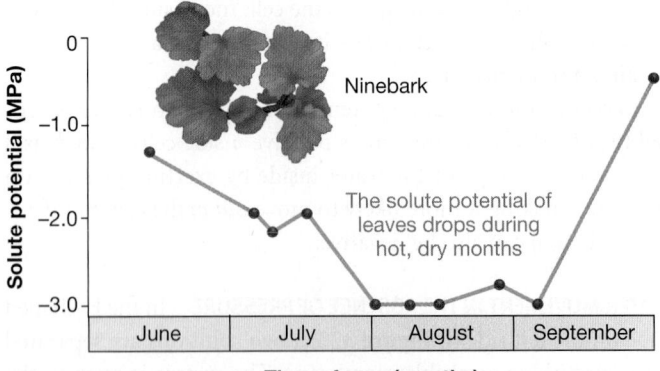

FIGURE 37.3 Plants with Low Solute Potentials Can Grow in Dry Soils. As soils dry, their water potential declines. But if a plant's solute potential also drops, it can maintain a water-potential gradient that continues to bring water into the plant.

tion and death as cells shrivel like grapes drying into raisins. When an entire tissue loses turgor, it will **wilt** (**Figure 37.4**).

Turgor pressure is required for growth to occur; without it, cells cannot expand once cell division is complete. In addition, turgor pressure provides structural support—which is why wilted plants droop. Unless corrected, extensive wilting may lead to the death of the tissue and, eventually, the plant.

FIGURE 37.4 Wilting Occurs when Water Loss Leads to Loss of Turgor Pressure. Wilting is a life-threatening condition in plants, analogous to severe dehydration in humans.

Low water potential
Atmosphere ψ: –95.2 MPa
(Changes with humidity; usually very low)

Leaf ψ: –0.8 MPa
(Depends on transpiration rate; low when stomata are open)

Root ψ: –0.6 MPa
(Medium–high)

Soil ψ: –0.3 MPa
(High if moist; low if extremely dry)

High water potential

FIGURE 37.5 A Water-Potential Gradient Exists between Soil, Plants, and Atmosphere. Water moves from regions of high water potential to regions of low water potential.

WATER POTENTIAL IN AIR In the atmosphere, water exists as a vapor with a solute potential of 0 MPa. The pressure exerted by water vapor in the atmosphere can be low or high, depending on conditions.

- When air is dry, there are few water molecules present and the pressure they exert is low.

- When air is warm, water molecules move farther apart and exert lower pressure.

Warm, dry air has an extremely low water potential. When it is rainy or foggy, however, the water potential of the atmosphere may be equal to the water potential inside a leaf. But otherwise, the water potential of the atmosphere is much lower than the water potential inside a leaf.

In most cases, water potential is high in soil and roots but low in leaves and the atmosphere. This situation sets up a **water-potential gradient** between roots and shoots.

To move *up* a plant, water moves *down* the water-potential gradient that exists between the soil, its tissues, and the atmosphere. When it does so, it replaces the water lost to transpiration (**Figure 37.5**).

CHECK YOUR UNDERSTANDING

If you understand that . . .

- Water moves along a water-potential gradient, from areas of high water potential to low water potential.
- In areas separated by a selectively permeable membrane, part of water's potential energy is made up of its solute potential—its tendency to move via osmosis.
- Water also has a pressure potential. In plant cells, for example, the cell wall can exert pressure on water and affect its pressure potential.
- Soils, plant tissues, and the atmosphere all have a water potential made up of a solute potential and a pressure potential.

✓ You should be able to . . .

Compare and contrast:

1. The pressure potential of wet soils versus dry soils.

2. The solute potential of salty soils versus more-typical soils.

Answers are available in Appendix B.

37.2 How Does Water Move from Roots to Shoots?

Suppose that you are caring for the wilted plant in Figure 37.4. If you add water to the soil, the water potential of the soil increases and water will move into the plant along a water-potential gradient. Biologists have tested three major hypotheses for how the water could be transported:

1. Root pressure—a pressure potential that develops in roots—could drive water up against the force of gravity.

2. Capillary action could draw water up the cells of xylem.

3. Cohesion-tension, a force generated in leaves, could pull water up from roots.

Note that all three hypotheses could be correct—they are not mutually exclusive.

Let's begin by considering how water and solutes move from the soil across the root and into the vascular tissue. Then we can analyze each of the three hypotheses for how water and solutes finish the trip up the xylem to the shoot system.

Movement of Water and Solutes into the Root

To understand how water enters a root, consider the cross section through a young buttercup root shown in **Figure 37.6**. Starting at the outside of the root and working inward, notice that several distinct tissues are present:

- The **epidermis** (literally, "outside skin") is a single layer of cells. In addition to protecting the root, some epidermal cells produce **root hairs**, which greatly increase the total surface area of the root (see Chapter 36).

- The **cortex** consists of ground tissue—usually parenchyma cells—and stores carbohydrates.

- The **endodermis** ("inside skin") is a cylindrical layer of cells that forms a boundary between the cortex and the vascular tissue. The function of the endodermis is to control ion uptake.

- The **pericycle** ("around-circle") is a layer of cells that can become meristematic and produce lateral roots.

- Conducting cells function in transport and are located in the center of roots in buttercups and other eudicots. Notice that, in these species, phloem is situated between each of four arms formed by xylem, which is arranged in a cross-shaped pattern.

THREE ROUTES THROUGH ROOT CORTEX TO XYLEM   When water enters a root along a water-potential gradient, it does so through root hairs. As water is absorbed, it moves through the root cortex toward the xylem through three distinct routes (**Figure 37.7a**).

1. The **transmembrane route** is based on flow through aquaporin proteins—water channels located in the plasma membranes of root cells (see Chapter 6), as well as some direct diffusion across plasma membranes.

2. The **apoplastic** ("away-from-particle") pathway is outside the plasma membrane. It consists of cell walls, which are porous, and the spaces that exist between cells. The name apoplastic is apt because movement takes place outside the plasma membrane and cell interior.

3. The **symplastic** ("with-particle") pathway consists of the continuous connection through cells that exists via plasmodesmata (see Chapter 8).

In essence, water can flow through cells via aquaporins, around cells, or through cells via plasmodesmata.

THE ROLE OF THE CASPARIAN STRIP   The situation changes when water reaches the endodermis, however. Endodermal cells are tightly packed and secrete a narrow band of wax called the **Casparian strip**. This layer is composed primarily of a compound called **suberin**, which forms a water-repellent cylinder at the endodermis.

The Casparian strip blocks the apoplastic pathway by preventing water from moving through the walls of endodermal cells and proceeding to the vascular tissue (**Figure 37.7b**). The Casparian strip does not affect water that is moving through the symplastic pathway.

The Casparian strip is important because it means that for water and solutes to reach vascular tissue, they have to move into the cytoplasm of an endodermal cell. Endodermal cells, in turn, act as filters.

Endodermal cells allow ions such as potassium (K^+) that are needed by the plant to pass through to the vascular tissue. In contrast, these cells can prevent the passage of ions such as sodium (Na^+) that are not needed or may be harmful.

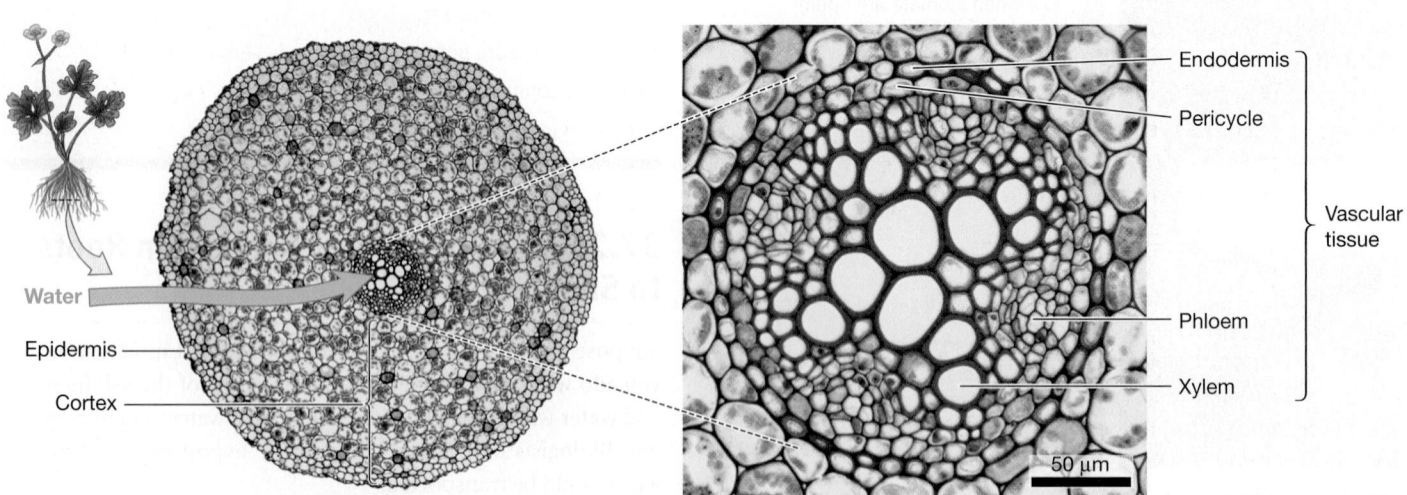

FIGURE 37.6 In Roots, Water Has to Travel through Several Tissue Layers to Reach Vascular Tissue.
Cross section through a buttercup root, showing the anatomy that is typical of roots in eudicots.

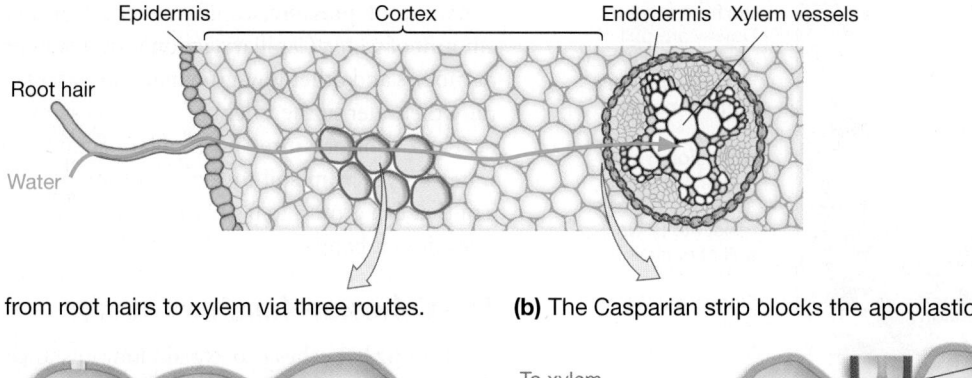

Epidermis   Cortex   Endodermis Xylem vessels

Root hair

Water

(a) Water travels from root hairs to xylem via three routes.

(b) The Casparian strip blocks the apoplastic route at the endodermis.

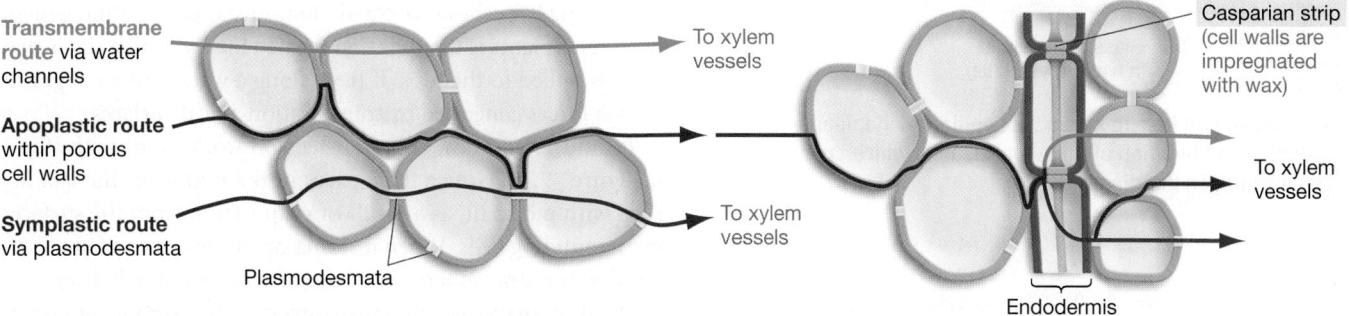

Transmembrane route via water channels

Apoplastic route within porous cell walls

Symplastic route via plasmodesmata

Plasmodesmata

To xylem vessels

To xylem vessels

Casparian strip (cell walls are impregnated with wax)

To xylem vessels

Endodermis

FIGURE 37.7 How Water Travels from Root Hair to Xylem. Water in the apoplastic pathway *must* pass into the cytoplasm of endodermal cells before entering xylem.

✔**EXERCISE** Add dots representing aquaporins in the transmembrane route.

✔If you understand this concept, you should be able to predict what would happen if a plant had a mutation that prevented synthesis of suberin and formation of the Casparian strip.

Water Movement via Root Pressure

Movement of ions and water into the root cortex is responsible for the phenomenon known as **root pressure**. Recall that root pressure is one of three hypothesized mechanisms for moving water up xylem, from root to shoot.

Stomata normally close during the night, when photosynthesis is not occurring and CO_2 is not needed. Their closure minimizes water loss and slows the movement of water into roots. But roots often continue to accumulate ions that their epidermal cells acquire from the soil as nutrients. These nutrients move into xylem. The influx of ions lowers the water potential of xylem below the water potential in the surrounding cells.

As water flows into xylem from other root cells in response, a positive pressure is generated that forces fluid up the xylem. More water moves up xylem and into leaves than is being transpired from the leaves.

In low-growing plants, enough water can move to force water droplets out of the leaves, a phenomenon known as **guttation**. If you are up early in the morning, you may observe water drops on leaf edges formed by guttation (**Figure 37.8**).

At one time, positive root pressure was a leading hypothesis to explain how water moves from roots to leaves in trees. However, research showed that over long distances, such as a tree trunk, the force of root pressure is not enough to overcome the force of gravity on the water inside xylem.

In addition, researchers demonstrated that cut stems, which have no contact with the root system, are still able to transport water to leaves. Biologists concluded that there must be some other mechanism involved in the long-distance transport of water. What is it?

Water Movement via Capillary Action

Researchers have also evaluated a hypothesis based on the phenomenon of **capillarity**, or movement of water up a narrow tube. When a thin glass tube is placed upright in a pan of water, water

FIGURE 37.8 Root Pressure Causes Guttation. When ions accumulate in the xylem of roots at night, enough water may enter xylem via osmosis to force water up and out of low-growing leaves.

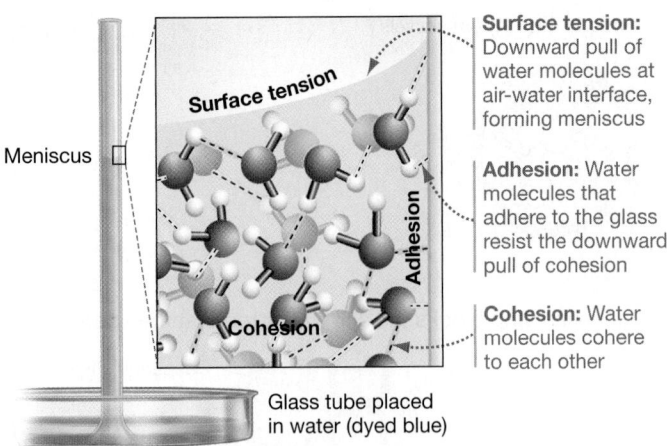

Meniscus

Surface tension: Downward pull of water molecules at air-water interface, forming meniscus

Adhesion: Water molecules that adhere to the glass resist the downward pull of cohesion

Cohesion: Water molecules cohere to each other

Glass tube placed in water (dyed blue)

FIGURE 37.9 Water Can Rise in Xylem via Capillarity. Capillarity occurs through a combination of three forces, all of which are generated by hydrogen bonding.

creeps up the tube (**Figure 37.9**). The movement occurs in response to three forces: (**1**) surface tension, (**2**) adhesion, and (**3**) cohesion. Let's consider each force in turn.

Surface tension is a downward pull that exists on water molecules at an air-water interface. In the body of a solution, all the water molecules present are surrounded by other water molecules and form hydrogen bonds in all directions. The water molecules at a surface, however, can form hydrogen bonds in one direction only—with the water molecules below them. As a result, the topmost layer of water molecules is pulled downward by the bonds. Because pulling forces create tension, the pulling force on the water molecules at the air-water interface is called surface tension.

Adhesion is a molecular attraction among unlike molecules. In this case, water interacts with a solid substrate—such as the glass walls of a capillary tube or the cell walls of tracheids or vessel elements—through hydrogen bonding. As water molecules bond to each other and adhere to the side of the tube, they are pulled upward.

Cohesion is a molecular attraction among like molecules, such as the hydrogen bonding that occurs among molecules in water. Because water molecules cohere, the tension at the surface of a thin tube is transmitted downward through the water column. Water molecules at the surface are pulled down; the water molecules below them are pulled up.

Surface tension creates a pull at the surface; adhesion creates a pull at the water-container surface; cohesion transmits both forces to the water below. The result is capillarity.

Capillarity is resisted by the force of gravity and by surface tension. The interaction between cohesion and surface tension is also responsible for the formation of a concave boundary layer called a **meniscus** (plural: **menisci**). A meniscus forms at most air-water interfaces—including those found in narrow tubes. Menisci form because cohesion pulls water molecules up along the tube-water surface, while surface tension pulls the surface down in the middle.

Like root pressure, capillarity can transport only a limited amount of water. Capillary action moves water along the surfaces of mosses and other low-growing, non-vascular plants; but it can raise the water in the xylem of a vertical stem only about 1 m.

Thus, root pressure and capillary action cannot explain how water moves from soil to the top of a redwood tree—an organism that can grow 5 to 6 stories higher than the Statue of Liberty. How does it happen?

The Cohesion-Tension Theory

The leading hypothesis to explain long-distance water movement in vascular plants is the **cohesion-tension theory**, which states that water is pulled to the tops of trees along a water-potential gradient, via forces generated by transpiration at leaf surfaces.

To understand how cohesion-tension works, start with step 1 in **Figure 37.10**. Notice that spaces in the middle of the leaf are filled with moist air, as a result of evaporation from the surfaces of surrounding cells. When a stoma opens, this humid air is exposed to the atmosphere, which in most cases is much drier.

Open stomata usually create steep concentration gradients between the leaf interior and its surroundings. The steeper the gradient, the faster water vapor diffuses out through the stomata.

Step 2 shows that as water is lost from the leaf to the atmosphere, the humidity of the gas-filled space inside the leaf drops. In response, more water evaporates from the walls of the parenchyma cells, where menisci exist at the air-water interface.

If water molecules leave menisci rapidly due to a steep water-potential gradient with the atmosphere, then fewer water molecules are available at the surface than before. The water molecules that remain at the surface are pulled downward more strongly, the meniscus deepens, and the water molecules below the surface are pulled up more strongly.

In 1894 Henry Dixon and John Joly hypothesized that the formation of steep menisci produces a force capable of pulling water up from the roots, dozens or hundreds of meters into the air. Is this really possible?

THE ROLE OF SURFACE TENSION IN WATER TRANSPORT ▢— The key concept in the cohesion-tension theory is that the negative force or pull (tension) generated at the air-water interface is transmitted through the water outside of leaf cells (step 3 in Figure 37.10), to the water in xylem (step 4), to the water in the vascular tissue of roots (step 5), and finally to the water in the soil (step 6).

The continuous transmission of pulling force from the leaf surface to the root is possible because (**1**) there is water throughout the plant, and (**2**) all of the water molecules present hydrogen bonds to one another in a continuous fashion (cohesion).

Note that the plant does not expend energy to create the pulling force. The force is generated by energy from the Sun, which drives evaporation from the leaf surface. Water transport is solar-powered.

In effect, the cohesion-tension theory of water movement states that, because of the hydrogen bonding between water molecules, water is pulled up through xylem in continuous columns.

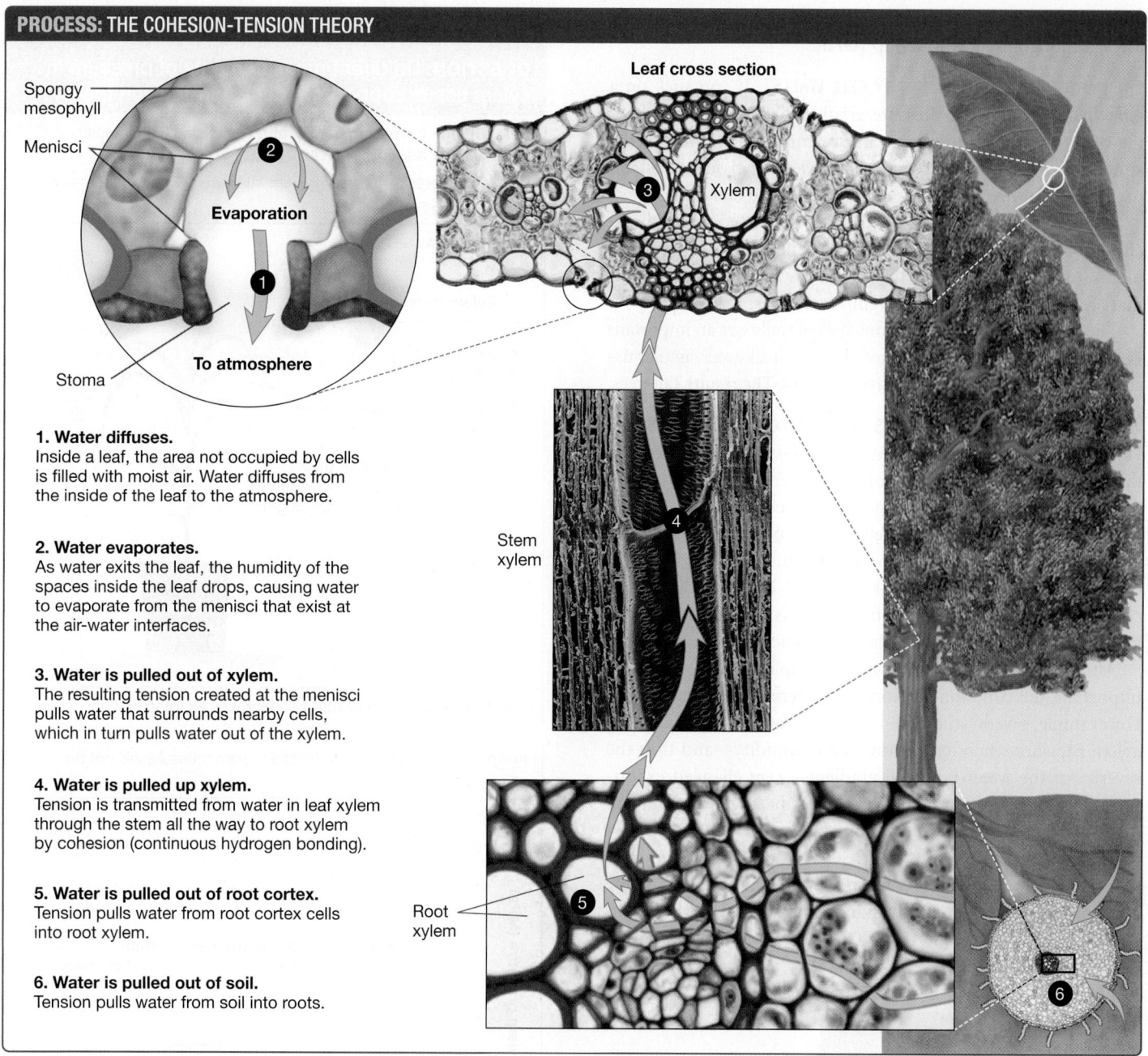

PROCESS: THE COHESION-TENSION THEORY

Spongy mesophyll

Menisci

2

Evaporation

1

Stoma

To atmosphere

Leaf cross section

3

Xylem

Stem xylem

4

Root xylem

5

6

1. Water diffuses.
Inside a leaf, the area not occupied by cells is filled with moist air. Water diffuses from the inside of the leaf to the atmosphere.

2. Water evaporates.
As water exits the leaf, the humidity of the spaces inside the leaf drops, causing water to evaporate from the menisci that exist at the air-water interfaces.

3. Water is pulled out of xylem.
The resulting tension created at the menisci pulls water that surrounds nearby cells, which in turn pulls water out of the xylem.

4. Water is pulled up xylem.
Tension is transmitted from water in leaf xylem through the stem all the way to root xylem by cohesion (continuous hydrogen bonding).

5. Water is pulled out of root cortex.
Tension pulls water from root cortex cells into root xylem.

6. Water is pulled out of soil.
Tension pulls water from soil into roots.

FIGURE 37.10 Transpiration Creates Tension That Is Transmitted from Leaves to Roots.
✔**EXERCISE** Using the data in Figure 37.5, label typical values for the water potential present in soil, roots, leaves, and the atmosphere.

CREATING A WATER-POTENTIAL GRADIENT You can also think about the cohesion-tension theory in terms of water potentials. Note that because tracheids and vessels are dead at maturity, the water in xylem does not cross plasma membranes. As a result, water does not move between cells by osmosis. In xylem, water movement is driven entirely by differences in pressure potential.

The pulling force generated at menisci lowers the pressure potential of water in leaves. Even though the tension created at each meniscus is relatively small, there are millions or billions of menisci in the leaves of the entire plant. The tension created by summing many small pulling forces is remarkable. It creates

a water-potential gradient between leaves and roots that is steep enough to overcome the force of gravity and pull water up long distances.

To appreciate just how great the forces involved are, think of the vessel elements or tracheids in xylem as groups of straws. When you use a straw, the vacuum that you create causes liquid to rise. Sucking on a straw creates a pressure difference of about -0.1 MPa, which can draw water up a maximum of about 10 m. In contrast, the negative pressure exerted by the menisci in leaves can be as high as -2.0 MPa. The tension in xylem tissue may be ten times the amount of pressure on a fully inflated car tire. (But note that the pressure on a car tire is positive, not negative.) The force is enough to draw water up 100 m.

THE IMPORTANCE OF SECONDARY CELL WALLS If you suck on a straw hard enough, the pressure gradient between the inside of the straw and the atmosphere can overcome the stiffness of the straw and cause it to collapse. How can vascular tissue withstand negative pressures as large as -2.0 MPa without collapsing?

The answer is found in a key adaptation: The secondary thickenings characteristic of the cell walls in tracheids and vessel elements. As Chapter 36 noted, the cells in vascular tissue have walls that are reinforced with tough lignin molecules.

The evolution of lignified secondary cell walls was an important event in the evolution of land plants, because it allowed vascular tissue to withstand extremely negative pressures. The result? Tall trees.

WHAT EVIDENCE DO BIOLOGISTS HAVE FOR THE COHESION-TENSION THEORY? If the cohesion-tension theory is correct, the water present in xylem should experience a strong pulling force. A simple experiment supports this prediction.

If you find a leaf that is actively transpiring and cut its petiole, the watery fluid in the xylem, or **xylem sap**, withdraws from the edge toward the inside of the leaf (**Figure 37.11**). According to the cohesion-tension theory, this observation is due to a transpirational pull at the air-water interface in parenchyma cells of leaves.

Although the observation that xylem sap is under tension was important, the cohesion-tension theory remained controversial. For example, several studies failed to document rapid changes in xylem pressure when temperature and humidity—and thus the severity of the water-potential gradient—were changed experimentally. Advocates of the theory blamed the negative results on instruments that could not document small rapid changes in pressure potential. Who was right?

Chunfang Wei, Melvin Tyree, and Ernst Steudle answered the question in the late 1990s with an instrument called a xylem pressure probe (**Figure 37.12**). A xylem pressure probe includes an oil-

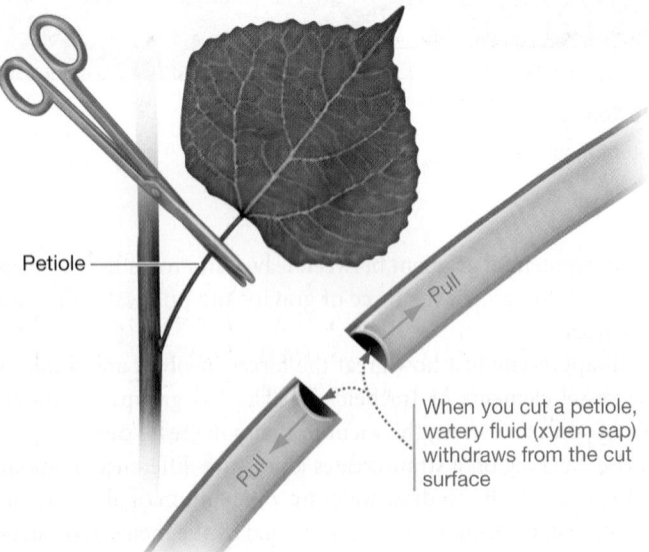

FIGURE 37.11 Xylem Sap in Cut Stems "Snaps Back."

Petiole

Pull

Pull

When you cut a petiole, watery fluid (xylem sap) withdraws from the cut surface

QUESTION: Do direct measurements of pressure in xylem tissue support the cohesion-tension theory?

HYPOTHESIS: Increasing transpiration by raising light intensity will lower xylem pressure in leaves.

NULL HYPOTHESIS: Increasing light intensity will not affect xylem pressure in leaves.

EXPERIMENTAL SETUP:

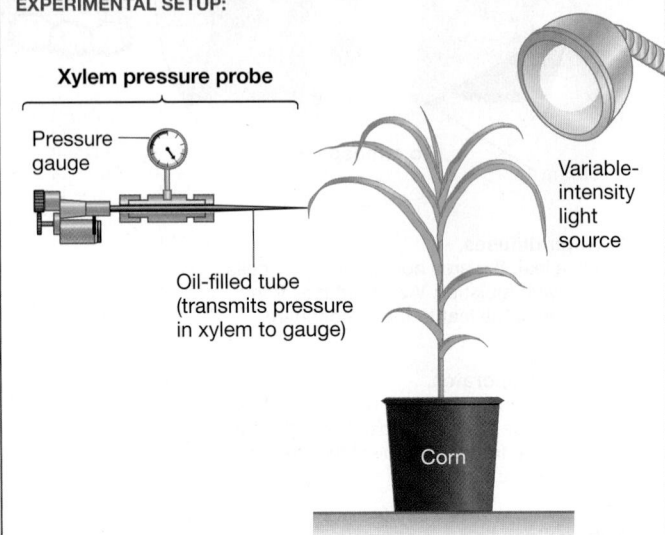

Xylem pressure probe

Pressure gauge

Oil-filled tube (transmits pressure in xylem to gauge)

Variable-intensity light source

Corn

PREDICTION: Xylem pressure will decrease as transpiration increases with higher light levels.

PREDICTION OF NULL HYPOTHESIS: Xylem pressure will not be affected by light intensity.

RESULTS:

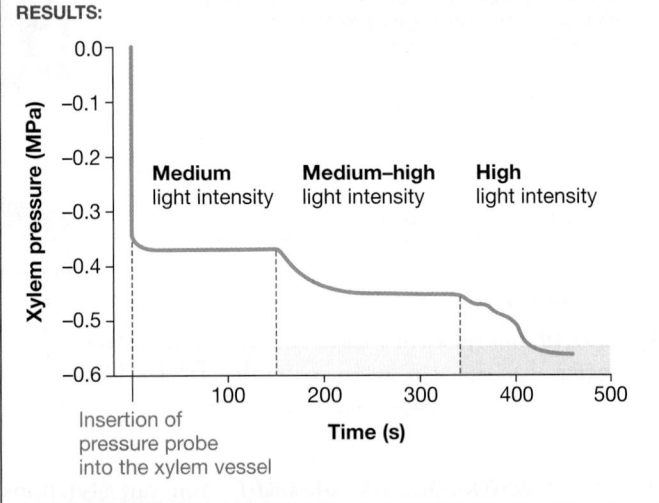

Medium light intensity

Medium–high light intensity

High light intensity

Insertion of pressure probe into the xylem vessel

Time (s)

Xylem pressure (MPa)

CONCLUSION: Xylem pressure decreases when light intensity increases. The data support the cohesion-tension theory.

FIGURE 37.12 Measuring Changes in Pressure inside Xylem.

SOURCE: Wei, C., M. T. Tyree, and E. Steudle. 1999. Direct measurement of xylem pressure in leaves of intact maize plants. A test of the cohesion-tension theory taking hydraulic architecture into consideration. *Plant Physiology* 121: 1191–1205.

✓**QUESTION** Suppose the researchers had chosen to plot changes in the water-potential gradient between the xylem in roots and leaves on the *y*-axis, during the same experiment. What would the graph look like?

filled glass tube that can be inserted directly into the xylem of a leaf. The oil transmits changes in pressure within the xylem to a gauge, allowing researchers to record changes in xylem pressure instantly and directly.

To test the cohesion-tension theory, the researchers altered xylem pressure by raising light levels to alter transpiration rates in the leaves of corn plants. Their results?

- The graph in Figure 37.12 plots how xylem pressure changed as the researchers changed light levels, over a 7-minute period. Note that as light intensity increased, the xylem pressure probe documented increased tension, or pull—negative pressure.

- In addition, higher light levels reduced the weight of the entire plant—suggesting that higher transpiration rates caused water loss.

Both observations are consistent with predictions that follow from the cohesion-tension theory.

These experiments convinced most biologists that increased transpiration leads to increased tension on xylem sap. Rising tension, in turn, lowers the water potential of leaves and exerts a pull on water in the roots and soil, where the water potential is high. On the basis of these and other results, most biologists now accept the cohesion-tension theory.

To review how water moves from soil to root cells and from root to shoots, go to the study area at *www.masteringbiology.com*.

 BioFlix™ Water Transport in Plants

CHECK YOUR UNDERSTANDING

⊙━ If you understand that . . .

- Water can move a short distance in xylem via root pressure or capillarity.
- Long-distance transport of water depends on movement along a steep water-potential gradient. This gradient is created primarily by the negative pressure potential of water in leaves, due to surface tension that develops in response to transpiration.

✔ You should be able to . . .

1. Explain how asymmetrical hydrogen bonding and transpiration at the surface of a meniscus near a stoma creates a pull on the water in leaves and xylem.

2. Predict what happens to the meniscus when each of the following occurs: a nearby stoma closes, a rain shower starts, and weather changes and dry air blows in.

Answers are available in Appendix B.

37.3 Water Absorption and Water Loss

One of the most important features of cohesion-tension is that it does not require plants to expend energy. Instead, the Sun furnishes the energy required to pull water from roots to shoots—not ATP supplied by the plant.

1. Energy from the Sun heats water molecules at the air-water interface inside leaves enough to break the hydrogen bonds between them and cause transpiration.

2. Rapid transpiration creates deep menisci in the walls of leaf cells, causing tension that lowers the water potential of leaves.

3. Hydrogen bonding between water molecules transmits this tension down to water molecules in the root and soil.

Xylem acts as a passive conduit—a set of pipes that allows water to move from a region of high water potential (the soil) to a region of low water potential (the leaves). Water flows from roots to shoots as long as the water-potential gradient—from soil to root to leaf to atmosphere—is intact.

When soils begin to dry, however, it becomes difficult for plants to replace water being lost via transpiration. If water is not replaced fast enough, the solute potentials of leaves drop and leaves and branches begin to wilt. In response, stomata may close down partially or completely to reduce transpiration rates and conserve water. However, closing stomata affect the ability of plants to carry on photosynthesis, because CO_2 acquisition slows or stops.

The balance between conserving water and maximizing photosynthesis is termed the photosynthesis-transpiration compromise. This compromise is particularly delicate for species that grow on dry sites. How do they cope?

Limiting Water Loss

Plants that thrive in dry sites have several adaptations that help them slow transpiration and limit water loss. Consider the oleander plant, which is native to the dry shrub-grassland habitats of southern Eurasia. The micrograph in **Figure 37.13** on page 728 shows a cross section through an oleander leaf, which has the following special features:

- A particularly thick cuticle covers the upper surface of oleander leaves. This waxy layer minimizes water loss from cells that are directly exposed to sunlight. In general, species that are adapted to dry soils have much thicker cuticles than do species adapted to wet soils.

- The epidermis of most plants is a single cell layer thick, but the epidermis in oleanders is several cell layers deep. The thick epidermis of species found in dry environments is thought to reduce water loss from parenchyma cells in the leaf, where most photosynthesis takes place.

- The stomata of oleanders are located on the undersides of their leaves, inside deep pits in the epidermis. Hairlike extensions of epidermal cells called trichomes (see Section 36.3), shield these pits from the atmosphere. The leading hypothesis to explain these traits is that they slow the loss of water vapor from stomata to the dry air surrounding the leaf.

Other species have other adaptations for limiting water loss. For example, recall from Chapter 36 that many species adapted to water-short habitats—either cold environments where water is often frozen or deserts where rainfall is rare—have needlelike leaves. Long, thin leaf shapes minimize the surface area exposed

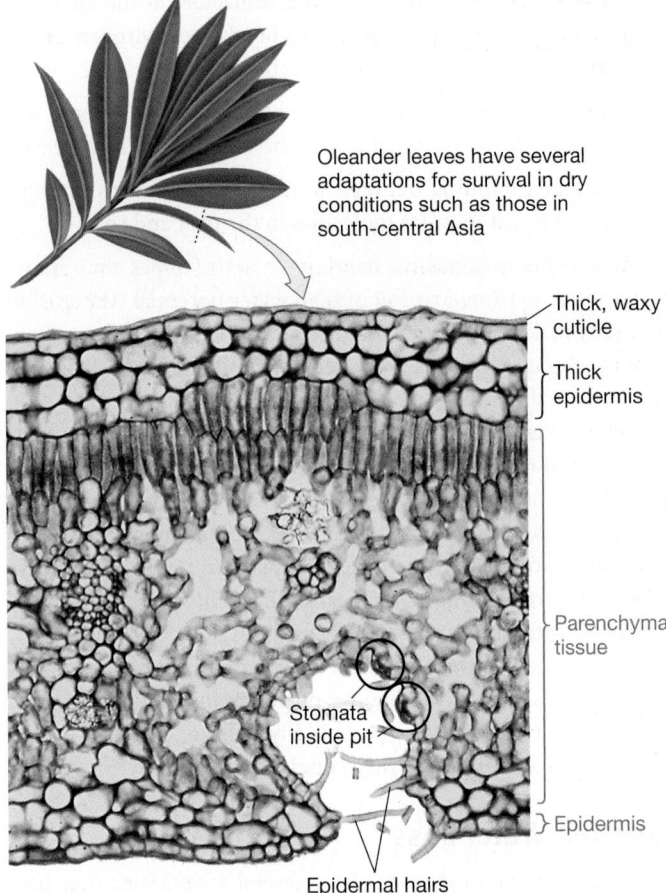

Oleander leaves have several adaptations for survival in dry conditions such as those in south-central Asia

Thick, waxy cuticle

Thick epidermis

Parenchyma tissue

Stomata inside pit

Epidermis

Epidermal hairs

FIGURE 37.13 Species That Are Adapted to Dry Habitats Have Modified Leaf Structures.

to sunlight and thus minimize transpiration. Plants adapted to wetter habitats, in contrast, tend to have broad leaves with a large surface area.

Obtaining Carbon Dioxide under Water Stress

Many of the species that thrive in deserts and other hot, dry habitats can continue photosynthesizing even when soil moisture content is low. Recall from Chapter 10 that two novel biochemical pathways, **crassulacean acid metabolism** (**CAM**) and **C_4 photosynthesis**, allow plants to increase CO_2 concentrations in their leaves and conserve water.

CAM plants open their stomata at night and store the CO_2 that diffuses into their tissues by adding the carbon dioxide molecules to organic acids. When sunlight is available during the day and photosynthesis begins, the CO_2 molecules are released from the organic acids and transferred to **rubisco**—the enzyme that initiates the Calvin cycle. In this way, CAM plants can photosynthesize and grow even with their stomata closed during the day.

C_4 plants minimize the extent to which their stomata open because they use CO_2 so efficiently. Mesophyll cells in C_4 plants take up CO_2 and add it to organic acids. The CO_2 is then transferred to specialized cells called **bundle-sheath cells** (see Chapter 10), where

rubisco is abundant. In effect, the C_4 pathway is a mechanism for concentrating carbon dioxide in cells deep inside the leaf, so stomata do not have to be wide open continuously.

Like the cuticle and stomata-containing pits of oleanders, CAM and C_4 photosynthesis are adaptations that help plants conserve water by limiting transpiration.

37.4 Translocation

Translocation is the movement of sugars throughout a plant—specifically, from sources to sinks. In vascular plants, a **source** is a tissue where sugar enters the phloem; a **sink** is a tissue where sugar exits the phloem. Sources contain a high concentration of sugar; sinks have a low concentration of sugar.

Where do sources and sinks occur in a plant? The answer often depends on the time of year.

- *During the growing season* Mature leaves and stems that are actively photosynthesizing produce sugar in excess of their own needs. These tissues act as sources. Sugar moves from leaves and stems to a variety of sinks, where sugar use is high and production is low. Apical meristems, lateral meristems, developing leaves, flowers, developing seeds and fruits, and storage cells in roots all act as sinks (**Figure 37.14**).

- *Early in the growing season* When a plant resumes growth after the winter or the dry season, sugars move from storage areas to growing areas. Storage cells in roots act as sources; developing leaves act as sinks.

Tracing Connections between Sources and Sinks

To explore the relationship between sources and sinks in more detail, consider research on sugar-beet plants that were exposed to carbon dioxide molecules containing the radioactive isotope ^{14}C. The goal was to track where carbon atoms moved after they were incorporated into sugars via photosynthesis.

The location of ^{14}C atoms inside a plant can be documented in two ways: (**1**) by measuring the number of radioactive emis-

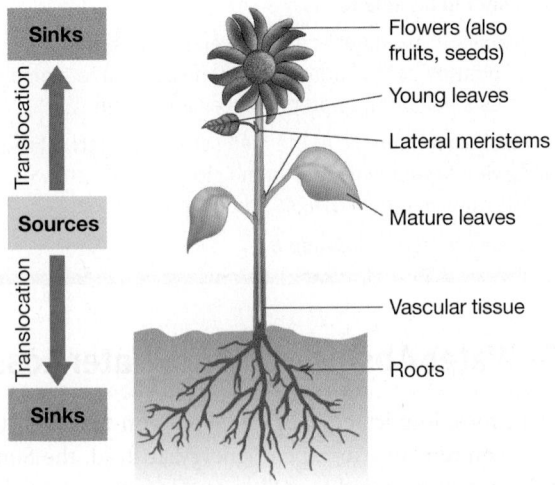

Sinks

Translocation

Sources

Translocation

Sinks

Flowers (also fruits, seeds)

Young leaves

Lateral meristems

Mature leaves

Vascular tissue

Roots

FIGURE 37.14 Sugars Move from Sources to Sinks.

sions emanating from different tissues, or **(2)** by laying plant parts on X-ray film and allowing the radioactivity to expose and blacken the film.

Researchers have enclosed individual leaves of intact plants in a bag and introduced a fixed amount of radioactive CO_2 for a fixed amount of time. One typical experiment documented that mature leaves retained just over 9 percent of the labeled carbon. In contrast, growing leaves retained 67 percent. These data are consistent with the prediction that fully expanded leaves act as sources of sugar, while actively growing leaves and roots act as sinks.

In similar experiments, researchers have exposed all the leaves on a growing plant to labeled carbon. One experiment like this found that over 16 percent of the total carbon was translocated to root tissue within 3 hours. This result is consistent with the prediction that, during the growing season, roots also act as sinks.

(a) Source leaves send sugar to the same side of the plant.

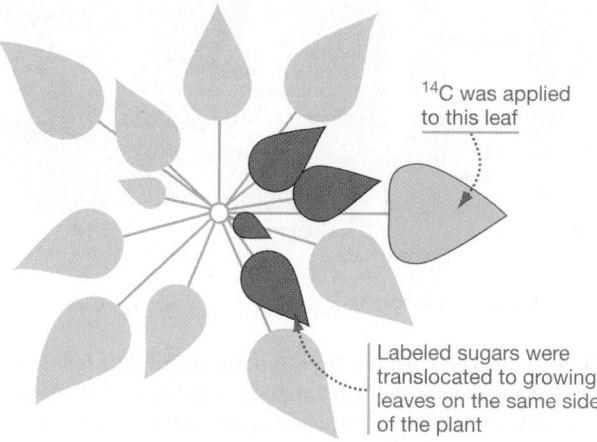

¹⁴C was applied to this leaf

Labeled sugars were translocated to growing leaves on the same side of the plant

(b) Source leaves send sugar to tissues on the same end of the plant.

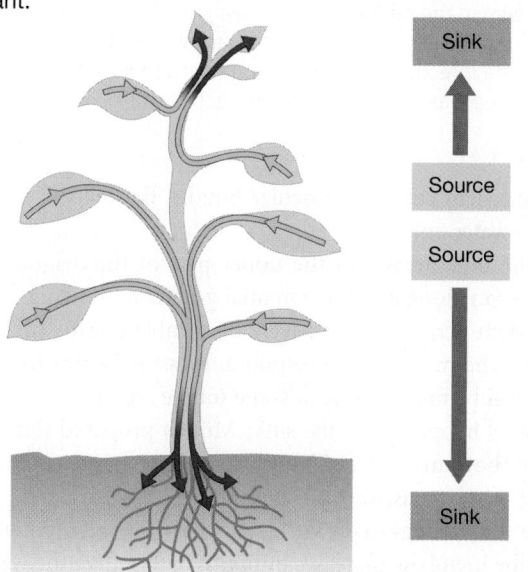

Sink

Source

Source

Sink

FIGURE 37.15 Sources Supply Sinks on the Same Side and Same End of the Body.

Similar experiments support two generalizations.

1. Sugars can be translocated rapidly—typically 50–100 cm/hr.

2. There is a strong correspondence between the physical locations of sources and sinks.

The second point is particularly interesting. For example, mature leaves that act as sources send sugar to tissues on the same side of the plant (**Figure 37.15a**). In addition, experiments with tall herbaceous plants show that leaves on the upper part of the stem send sugar to apical meristems, but leaves on the lower part of the plant send sugar to the roots (**Figure 37.15b**).

Why would leaves send sugar to tissues on a certain side or part of the body? The answer hinges on understanding the structure of phloem.

The Anatomy of Phloem

Chapter 36 introduced the two specialized parenchyma cell types that make up phloem: **sieve-tube members** and **companion cells**. Unlike the tracheids and vessel elements that make up most of the xylem, sieve-tube members and companion cells are alive at maturity.

Recall that, in most plants, sieve-tube members lack nuclei and many major organelles. They are connected to one another, end to end, by perforated **sieve plates** (**Figure 37.16**). The pores

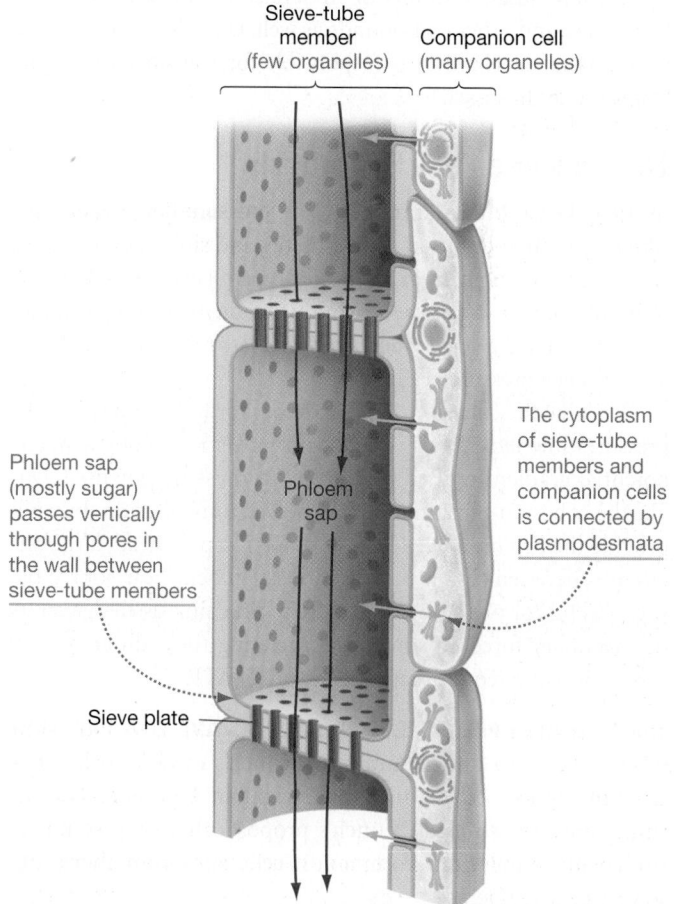

Sieve-tube member (few organelles)

Companion cell (many organelles)

Phloem sap (mostly sugar) passes vertically through pores in the wall between sieve-tube members

Phloem sap

The cytoplasm of sieve-tube members and companion cells is connected by plasmodesmata

Sieve plate

FIGURE 37.16 Sieve-Tube Members Are Connected by Pores.

create a direct connection between the cytoplasms of adjacent cells. Companion cells, in contrast, have nuclei and a rich assortment of ribosomes, mitochondria, and other organelles. Companion cells function as "support staff" for sieve-tube members.

You might also recall from Chapter 36 that secondary phloem is part of a tree's bark. In addition, it's important to recognize that the phloem in secondary vascular tissue is continuous throughout the plant—meaning that there is a direct anatomical connection to the phloem in trunks, stems, branches, and roots. The sieve-tube members in bark phloem represent a continuous system for transporting sugar throughout the plant body.

As Chapter 36 noted, however, phloem is restricted to discrete vascular bundles in tissues that do not form wood. Each vascular bundle runs the length of stems and roots, and certain bundles extend into specific branches, leaves, and lateral roots. In primary vascular tissue, phloem sap does not move from one vascular bundle to another—instead, each bundle is independent.

Based on these results, the physical relationships observed between sources and sinks in herbaceous plants are logical. For example, the phloem in the leaves on one side of an herbaceous plant connects directly with the phloem of branches, stems, and roots on the same side of the individual, through a specific set of vascular bundles.

The phloem sap that flows through vascular tissue is often dominated by the disaccharide sucrose—table sugar. Phloem sap can contain small amounts of minerals, amino acids, mRNAs, hormones, and other compounds as well. How does this solution move? What mechanism is responsible for translocating sugars from sources to sinks?

The Pressure-Flow Hypothesis

In 1926 Ernst Münch proposed the **pressure-flow hypothesis**, which states that events at source tissues and sink tissues create a steep pressure potential gradient in phloem (**Figure 37.17**). The water in phloem sap moves down this gradient, and sugar molecules are carried along by **bulk flow**—a mass movement of molecules along a pressure gradient.

🔑 Like the cohesion-tension theory for water transport, the pressure-flow hypothesis is based on movement along a water-potential gradient created by changes in pressure potential. Unlike the cohesion-tension model, however, transpiration does not provide the driving force to move phloem sap. Instead, large differences between turgor pressure in the phloem near source tissues and turgor pressure in the phloem near sink tissues generate the necessary force. In some cases, creating these differences in turgor pressure requires an expenditure of ATP.

CREATING HIGH PRESSURE NEAR SOURCES AND LOW PRESSURE NEAR SINKS To understand how Münch's model works, start with the source cell at the upper right in Figure 37.17. The small red arrows reflect Münch's proposal that sucrose moves from source cells into companion cells and from there into sieve-tube members.

Because of this phloem loading, the phloem sap near the source has a high concentration of sucrose. Compared to the water in the

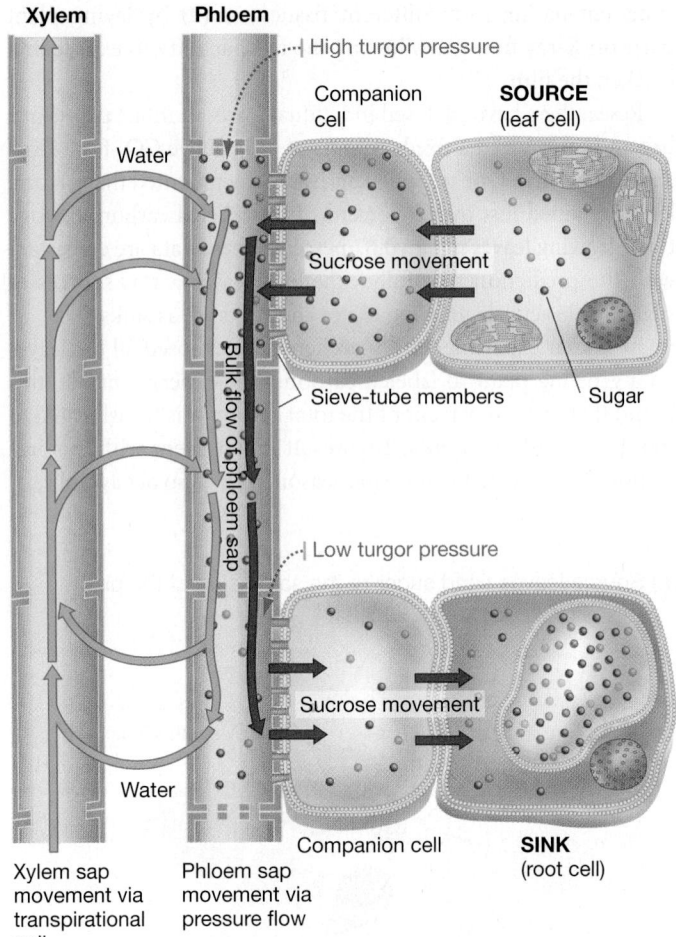

FIGURE 37.17 The Pressure-Flow Hypothesis: High Turgor Pressure Near Sources Causes Phloem Sap to Flow to Sinks. The pressure-flow hypothesis predicts that water cycles between xylem and phloem and that water movement in phloem is a response to a gradient in pressure potential.

✔**EXERCISE** This diagram shows how water moves between xylem and phloem in the middle of the growing season, when leaves are sources and roots are sinks. Add new arrows, in new colors, to indicate the direction of water and phloem sap flow in spring, when roots act as sources and leaves act as sinks.

adjacent xylem cells in a vascular bundle, the phloem sap has a very low water potential.

As the blue arrows in the upper part of the diagram show, water moves along a water-potential gradient—flowing passively from xylem across the selectively permeable plasma membrane of sieve-tube members. In response, pressure begins to build in the sieve-tube members nearest the source region.

What is happening at the sink? Münch proposed that cells in the sink (bottom right in Figure 37.17) remove sucrose from the phloem sap by passive or active transport. As a result of this phloem unloading—a loss of solutes—the water potential in sieve-tube members increases until it is higher than the water potential in adjacent xylem cells. As the blue arrows at the bottom of the figure show, water flows across the selectively permeable membranes of sieve-tube members into xylem along a water-potential gradient. In response, turgor pressure in the sieve-tube members near the sink drops.

The net result of these events is high turgor pressure in phloem near the source and low turgor pressure in phloem near the sink, created by the loading and unloading of sugars. This difference in pressure potential drives phloem sap from source to sink via bulk flow. There is a one-way flow of sucrose and a continuous loop of water movement, with water being supplied to and from the xylem.

TESTING THE PRESSURE-FLOW MODEL The pressure-flow hypothesis is logical, given the anatomy of vascular tissue and the principles that govern water movement. But has any experimental work supported the theory? Some of the best tests have relied on aphids—small insects that make their living ingesting phloem sap.

Aphids insert a syringe-like mouthpart, called a stylet, into sieve-tube members. The pressure on the fluid in these cells forces it through the stylet, into the aphid's digestive tract, and out its anus as droplets of "honeydew" (**Figure 37.18**).

If the aphids are then severed from their stylets, sap continues to flow out through the stylets. This phenomenon allows researchers to collect phloem sap efficiently for analysis. It also confirms that the aphids do not actively suck the fluid. As predicted by the pressure-flow model, phloem is indeed under pressure.

This observation supports one of the fundamental predictions of the pressure-flow hypothesis. Now the question is, How does sucrose enter and leave phloem in a way that sets up the water-potential gradient?

Phloem Loading

In contrast to the cohesion-tension model of water movement in xylem, pressure flow often requires that plants expend energy to set up a water-potential gradient in phloem. To establish a high pressure potential in sieve-tube members near source cells, large amounts of sugar have to be transported into the phloem sap—enough to raise the solute concentration of sieve-tube members. This requirement is illustrated in Figure 37.17, top right.

In some cases, loading is active—it requires an expenditure of ATP and some sort of membrane transport system. But when su-

crose concentrations in source cells are extremely high, movement into sieve-tube members can also occur via passive diffusion through plasmodesmata.

Conversely, sugar must sometimes be unloaded against its concentration gradient at sinks. When active unloading occurs, it requires an expenditure of ATP and a second membrane transport mechanism. How do phloem loading and unloading occur? What specific membrane proteins are involved?

To answer these questions, let's start by reviewing how transport proteins make it possible for sugars and other large or charged substances to cross a phosholipid bilayer.

HOW ARE SUCROSE AND OTHER SOLUTES TRANSPORTED ACROSS MEMBRANES? Passive transport (**Figure 37.19**) occurs when ions or molecules move across a plasma membrane by diffusion—that is, with their electrochemical gradient. The adjective passive is appropriate because no expenditure of energy is required for the movement to occur.

Recall from Chapter 6 that small, nonpolar molecules diffuse across phospholipid bilayers rapidly (Figure 37.19, left). But ions and many large molecules diffuse across phospholipid bilayers slowly if at all, even when their movement is favored by a strong electrochemical gradient. To diffuse rapidly, they must avoid direct contact with the phospholipid bilayer by passing through a membrane protein.

As Chapter 6 explained, two types of membrane protein—channels and carriers—permit the passive diffusion of specific ions or molecules.

- **Channels** form pores that selectively admit certain ions (Figure 37.19, middle).

- **Carriers** work like enzymes, undergoing a conformational change that transports a bound substrate across the lipid bilayer (Figure 37.19, right).

Channels and carriers are responsible for **facilitated diffusion**.

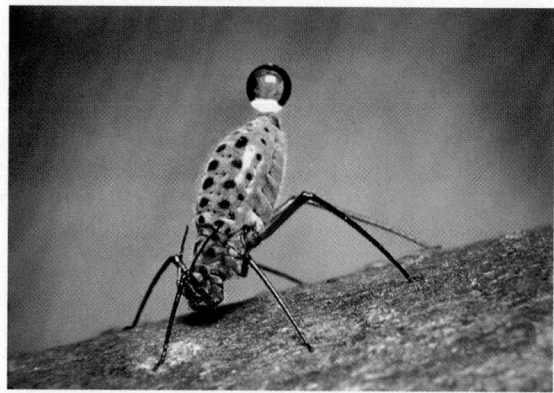

FIGURE 37.18 Aphids Feed on Phloem Sap. The tip of this aphid's mouthpart (the stylet) is in a sieve-tube member within the plant stem. The droplet emerging from the aphid's anus is honeydew, which consists of sugary phloem sap.

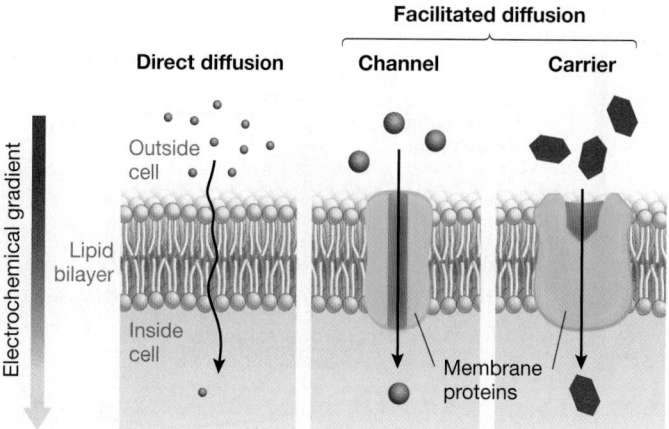

FIGURE 37.19 Passive Transport Is Based on Diffusion. In passive transport, ions or molecules diffuse across membranes—meaning they follow their electrochemical gradient. The movement can occur directly through the phospholipid bilayer or be facilitated by a channel or carrier protein.

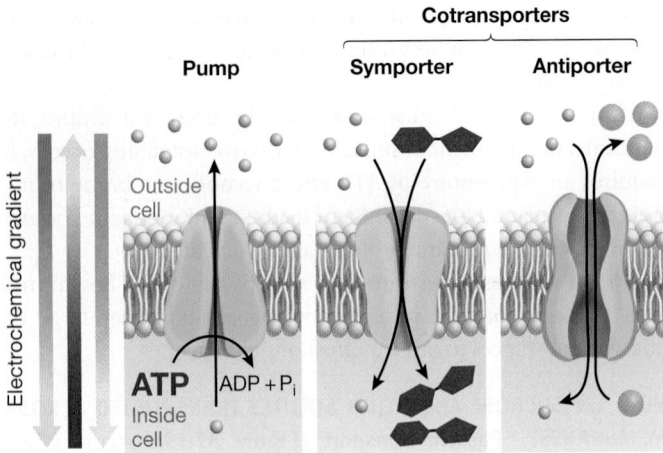

Cotransporters

Pump | Symporter | Antiporter

Outside cell

ATP | ADP + P$_i$

Inside cell

Electrochemical gradient

FIGURE 37.20 Active Transport Moves Ions or Molecules against an Electrochemical Gradient. All forms of active transport require an expenditure of ATP. Pumps use ATP directly; cotransporters use ATP indirectly. Cotransport depends on a previous expenditure of ATP by a pump.

Active transport (**Figure 37.20**) occurs when ions or molecules move across a plasma membrane against their electrochemical gradient. The adjective active is appropriate because cells must expend energy in the form of ATP to move solutes in an energetically unfavorable direction.

Active transport always involves membrane proteins. **Pumps** are proteins that change shape when they bind ATP or a phosphate group from ATP. As they move, pumps transport ions or molecules against an electrochemical gradient.

Pumps can establish an electrochemical gradient that favors movement of an ion or molecule across the plasma membrane. For example, the pump on the left side of Figure 37.19b has established an electrochemical gradient for bringing the ions or molecules symbolized by the light gray balls into the cell.

In many cases, the electrochemical gradients established by pumps are used to transport other molecules or ions by two types of membrane proteins called **cotransporters**.

- **Symporters** transport solutes *against* a concentration gradient, using the energy released when a different solute moves in the same direction *down* its electrochemical gradient. The red molecules in the middle of Figure 37.20 are moving through a symporter.

- **Antiporters** work in a similar way, except that the solute being transported against its concentration gradient moves in the direction *opposite* that of the solute moving down its concentration gradient. The orange ions in Figure 37.20, right, are moving through an antiporter.

When solutes or ions move through cotransporters, **secondary active transport** occurs (see Chapter 6).

Active transport, secondary active transport, and passive transport are all involved in moving sugars around plants. Let's look first at events at source tissues, where the active transport of sucrose into sieve-tube members results in a high pressure potential and high water potential. The chapter concludes with a look at how the same molecules are unloaded to maintain low water potentials at sinks.

HOW ARE SUGARS CONCENTRATED IN SIEVE-TUBE MEMBERS AT SOURCES? Because sucrose may be more highly concentrated in companion cells than in photosynthetic cells where it is produced, researchers hypothesized that sucrose transport from source cells into companion cells may be active. Another key observation—that strong pH differences exist between the interior and exterior of phloem cells—suggested that sucrose might enter companion cells with protons.

Figure 37.21 explains the logic behind this hypothesis. Note a key claim: that a membrane protein in companion cells hy-

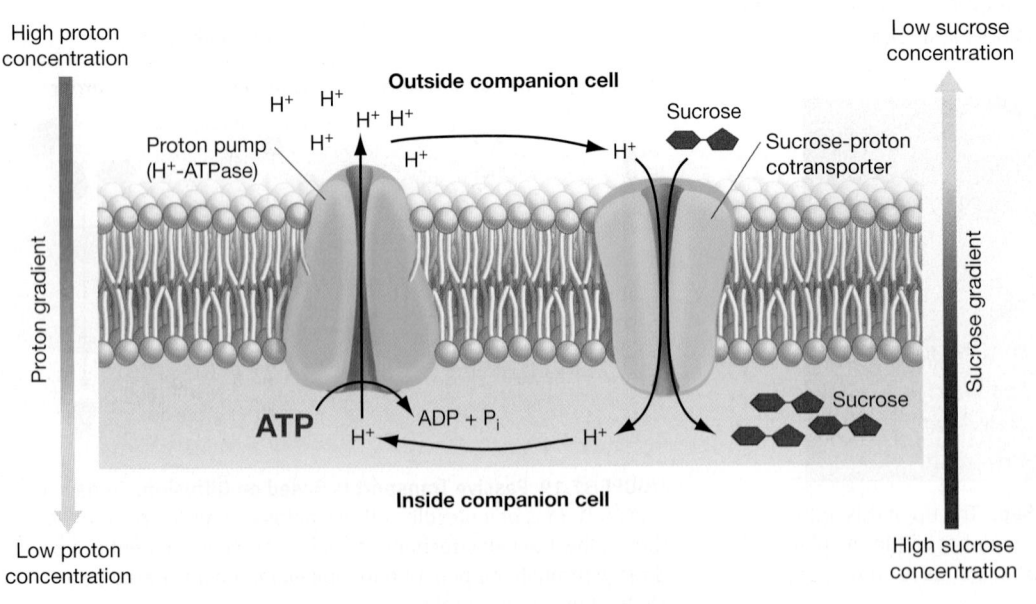

High proton concentration

Outside companion cell

Proton pump (H$^+$-ATPase)

H$^+$ H$^+$ H$^+$ H$^+$ H$^+$ H$^+$ H$^+$

Sucrose

Sucrose-proton cotransporter

Proton gradient

ATP | ADP + P$_i$

H$^+$ H$^+$

Inside companion cell

Low proton concentration

Low sucrose concentration

Sucrose gradient

Sucrose

High sucrose concentration

FIGURE 37.21 A Model for Cotransport of Protons and Sucrose. According to the model of cotransport, a proton pump hydrolyzes ATP to move hydrogen ions to the exterior of the cell. The resulting high concentration of H$^+$ establishes an electrochemical gradient that allows the transport of sucrose into the cell against its concentration gradient.

drolyzes ATP and uses the energy that is released to transport protons (H^+) across the membrane to the exterior of the cell. Proteins like these are called **proton pumps**, or more formally, **H^+-ATPases.**

Proton pumps establish a large difference in charge and in hydrogen ion concentration on the two sides of the membrane. The resulting electrochemical gradient favors the entry of protons into the cell.

The right side of the figure shows the second key claim: A membrane protein called a cotransporter acts as a conduit for protons and sucrose to enter the cell together. Protons move along their electrochemical gradient; sucrose moves against its concentration gradient.

If phloem loading depends on the activity of a proton pump, researchers should be able to find and characterize the pump proteins.

WHERE ARE H^+-ATPases LOCATED? Proton pumps are found in the plasma membranes of a wide variety of organisms, including bacteria, fungi, and animals. To analyze the proton pumps in plants, researchers from several laboratories focused on a small member of the mustard family called *Arabidopsis thaliana.*

Researchers determined the amino acid sequence of a proton pump purified from the plasma membranes of *Arabidopsis* and used these data to infer the DNA sequence of the corresponding gene. They found that the *Arabidopsis* genome actually codes for 10 different proton pump proteins. One of these genes, *AHA3*, appeared to be expressed primarily in vascular tissues.

These observations led Natalie DeWitt and Michael Sussman to hypothesize that *AHA3* encodes the proton pump responsible for phloem loading. To test this hypothesis, they produced antibodies to the AHA3 protein. You might recall from earlier chapters that an antibody is a protein that binds to a specific location on a molecule—typically, a specific protein.

To track the antibody, the investigators attached gold particles to it. When viewed with the electron microscope, a gold particle looks like a black dot. DeWitt and Sussman's goal was to treat *Arabidopsis* leaves with the AHA3 antibody, examine treated leaves under the electron microscope, and determine where the pump proteins are located.

As the bar graphs in **Figure 37.22** show, the proton pumps responsible for phloem loading were found almost exclusively in the plasma membranes of companion cells. This result supported the following model for phloem loading:

1. Proton pumps in the membranes of companion cells create a strong gradient that favors a flow of protons into companion cells.

2. A cotransporter protein in the membranes of companion cells uses the proton gradient to bring sucrose into companion cells from the surrounding cell walls and intercellular spaces around source cells.

3. Once inside companion cells, sucrose travels into sieve-tube members via plasmodesmata.

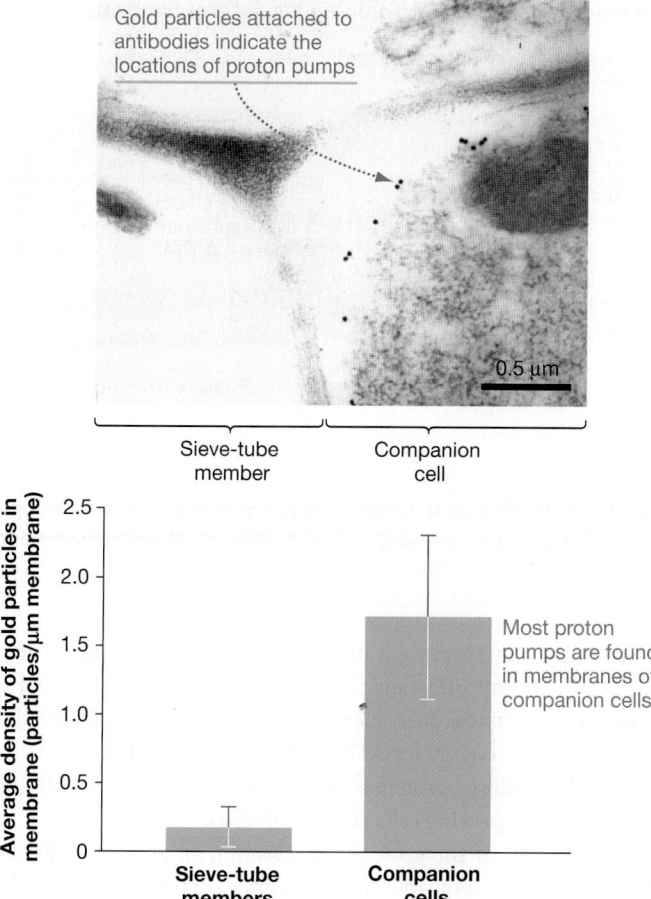

FIGURE 37.22 The Membranes of Companion Cells Contain H^+-ATPases. The micrograph shows that the proton pumps responsible for phloem loading are located in the plasma membranes of companion cells. The histogram compares the average density of proton pumps in the membranes of companion cells versus those of sieve-tube members. These pumps create a proton gradient that allows sucrose to be transported into companion cells against a concentration gradient.

Although work on the mechanism of phloem loading in *Arabidopsis* and other species continues, most researchers are convinced that proton pumps and proton-sucrose cotransporters play a key role. Now, once sucrose has been loaded into sieve-tube members near sources and follows a water-potential gradient to sinks, how is it unloaded?

Phloem Unloading

The membrane proteins that are involved in transporting sucrose molecules, and the mechanism of movement, vary among different types of sinks within the same plant. Mechanisms for phloem unloading also vary among different species.

To appreciate this diversity, consider how sucrose is unloaded in the phloem of sugar beets—a crop grown for the storage tissues in its root, which are a major source of the granulated and powdered sucrose sold in grocery stores.

(a) Phloem unloading into growing leaves of sugar beets

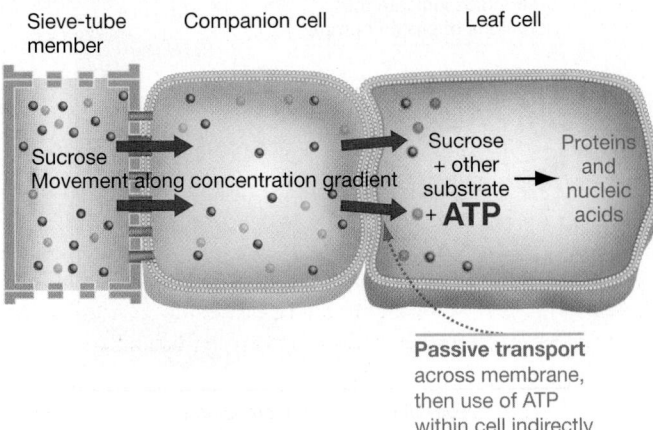

(b) Phloem unloading into roots of sugar beets

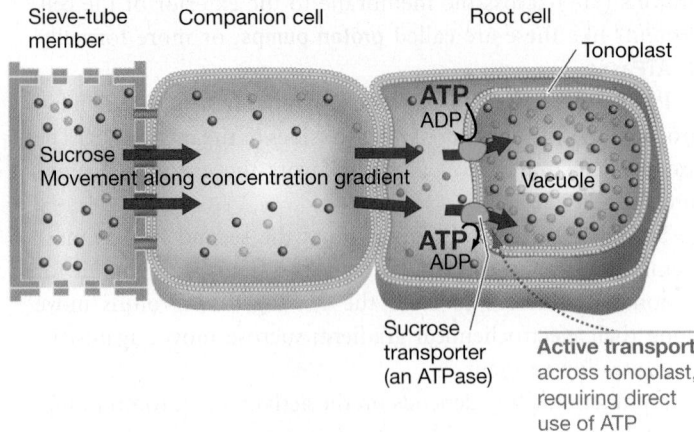

FIGURE 37.23 Phloem Unloading Occurs when Sucrose Is Taken Up by Cells. The mechanism of phloem unloading can vary from sink to sink, such as in **(a)** young leaves and **(b)** roots of the same plant.

In sugar beets, sucrose is unloaded along a concentration gradient into an important sink: young, growing leaves. The passive transport occurs because sucrose is rapidly used up inside the cells to provide energy for ATP synthesis and carbon for the synthesis of cellulose, proteins, nucleic acids, and phospholipids needed by the growing cells (**Figure 37.23a**).

In the roots of the same plant, however, an entirely different mechanism is responsible for offloading sucrose. Root cells in this species have a large vacuole that stores sucrose. The membrane surrounding this organelle is called the **tonoplast**. It contains a protein that hydrolyzes ATP and uses the energy released to transport sucrose into the vacuole, against its concentration gradient (**Figure 37.23b**). The active transport of sucrose into the vacuole allows sucrose to move passively from phloem into the storage cells, keeping the water potential of the phloem sap near the sink low—as required by the Münch pressure-flow model.

To summarize, more than seven decades of research provide convincing evidence that the pressure-flow hypothesis is fundamentally correct. To see the pressure-flow mechanism in action and review the cohesion-tension theory for water movement in xylem, go to the study area at *www.masteringbiology.com*.

(MB) Web Activity Solute Transport in Plants

The cohesion-tension theory for water movement and the pressure-flow model for phloem sap movement represent major advances in our understanding of how plants work.

CHECK YOUR UNDERSTANDING

If you understand that . . .

- Phloem sap moves from areas of high water potential to areas of low water potential.
- In phloem, high water potential is due to the high turgor pressure observed in the sieve-tube members near source cells. This pressure is created by pumps that actively load sucrose into companion cells against a concentration gradient. Water follows by osmosis, creating high turgor pressure inside sieve tubes.
- At sinks, turgor pressure is much lower than it is at sources because storage cells or growing tissues remove sucrose from phloem sap. The decreased concentration of solutes in phloem causes water to leave phloem and enter xylem.

✔ You should be able to . . .

Predict the water potential of phloem sap near leaves and near roots in a deciduous tree under the following conditions:

1. at the start of growth in spring
2. at midday when growing conditions are ideal in summer

Answers are available in Appendix B.

Summary of Key Concepts

🗝 **Water moves from areas of high water potential to areas of low water potential. Water's potential energy in plants is a combination of (1) its tendency to move in response to differences in solute concentration and (2) the pressure exerted on it.**

- Plants lose water as an inevitable side effect of exchanging gases with the atmosphere. The flow of water from soil to air via plant tissues follows a water-potential gradient.

- Water potential (ψ) is a measure of the tendency of water to move down its potential energy gradient.

- In plants, water potential has two components: **(1)** a solute potential, formed by the concentration of solutes in a cell or tissue; and **(2)** a pressure potential, provided by the cell wall and other factors. The water potential of a cell, tissue, or plant is the sum of its solute potential and pressure potential.

- When selectively permeable membranes are present, water moves by osmosis from areas of high potential to areas of low potential.

- When no membranes are present, water moves by bulk flow from areas of high pressure to areas of low pressure, independently of differences in solute potential.

   ✔ You should be able to explain why water in the xylem of a root moves up through the shoot in response only to a pressure gradient—not a solute gradient.

🗝 **Plants lose water to transpiration when stomata are open and photosynthesis is occurring, but they do not expend energy to replace it. Instead, water moves from soil and roots to leaves along a water-potential gradient. Evaporation of water from leaves, driven by the Sun, creates a negative pressure (tension) that pulls water up.**

- According to the cohesion-tension theory, water is pulled in one continuous column from the soil to roots to shoots, against the force of gravity, by the surface tension caused by transpiration from leaves.

- Surface tension occurs at menisci that form as water evaporates from the walls of leaf cells, and is transmitted downward via hydrogen bonding between water molecules. In this way, the energy in sunlight is responsible for the movement of water from roots to shoots.

- Plants that occupy dry habitats have traits that limit the amount of water they lose to transpiration. In some species, stomata are located in pits on the undersides of their leaves, where humidity is higher. The CAM and C_4 photosynthetic pathways are adaptations that limit water loss and photorespiration in dry habitats.

   ✔ You should be able to explain how the pressure potential of water in a leaf, and thus the rate of water movement up a stem, changes when the Sun goes behind a cloud.

   🅑 **BioFlix™** Water Transport in Plants

🗝 **In phloem, sugars are transported from "sources"—tissues that release sugars for use elsewhere—to "sinks"—tissues in which sugars are being used or stored. Movement occurs because large amounts of sucrose move into phloem cells near source tissues. Water follows by osmosis, creating a pressure gradient that favors the movement of water and sucrose to sinks.**

- Translocation is the movement of sucrose and other products through the plant.

- According to the Münch pressure-flow model, sugars move from sources to sinks via bulk flow along a pressure gradient that develops in phloem.

- A pressure gradient is generated by the transport of sugars into sieve-tube members in source tissues, coupled with the transport of sucrose out of sieve-tube members at sink tissues. Water moves from xylem into sieve-tube members near sources and cycles back to xylem near sinks.

   ✔ You should be able to explain why the pressure-flow model would not work if xylem and phloem were not bundled together.

   🅑 **Web Activity** Solute Transport in Plants

Questions

1. Under what conditions does the rate of transpiration increase?
   a. in species in which stomata are located in pits on the bottom of leaves
   b. when stomata close at night
   c. during rainstorms, when atmospheric pressure is low
   d. when the weather changes and air becomes drier

2. Which of the following does *not* affect the pressure potential of water?
   a. High sucrose concentrations in companion cells create turgor pressure.
   b. Menisci around spongy mesophyll cells create tension when transpiration rates are high.

   c. Few ions or other solutes are present in soil water.
   d. Water tends to adhere tightly to soil particles when soils dry.

3. The cells of a certain plant species can tolerate extremely high solute potentials. Which of the following statements is correct?
   a. The plant's transpiration rates will tend to be extremely low.
   b. The plant can compete for water effectively and live in dry soils.
   c. The plant will grow most effectively in soils that are saturated with water year round.
   d. The plant's leaves will wilt easily.

4. What forces are responsible for capillarity?
   a. adhesion of water molecules to the sides of xylem cells, cohesion of water molecules to each other, and surface tension
   b. surface tension created by transpiration and cohesion of water molecules in a continuous flow from leaf to root
   c. high solute potentials created by the entry of ions during the night, when transpiration rates are low, followed by an influx of water
   d. gravity and wall pressure (from the sides of xylem cells)

5. What is a proton pump?
   a. a membrane protein that transports sucrose against a concentration gradient
   b. a membrane protein that transports protons against an electrochemical gradient
   c. a membrane protein that transports protons *with* an electrochemical gradient and sucrose *against* a concentration gradient
   d. any membrane protein that acts as a channel—meaning it does not consume ATP

6. Why is the transport of phloem sap considered an active process?
   a. The manufacture of sucrose via photosynthesis is driven by the energy in sunlight.
   b. Transpiration is driven by the energy in sunlight.
   c. ATP is used to transport sucrose into companion cells near sources against a concentration gradient.
   d. In spring, phloem sap moves against the force of gravity.

✔ TEST YOUR UNDERSTANDING

Answers are available in Appendix B

1. Draw a plant cell in pure water. Add dots to indicate solutes inside the cell. Now add dots to indicate an increase in solute potential inside the cell. Add an arrow showing the direction of water movement in response. Add arrows showing the direction of wall pressure and turgor pressure in response to water movement. Repeat the same exercise, but this time add solutes to the solution outside the cell.

2. Compare and contrast the forces involved in transporting water in xylem via root pressure, capillarity, and transpiration. Which of these mechanisms are passive?

3. Why are "cohesion-tension" and "pressure-flow" sensible names for the hypotheses analyzed in this chapter?

4. Suppose Aphid A and Aphid B are sitting on the same plant. A inserts her stylet into the phloem at the base of a large, mature leaf, while B prefers to probe the phloem of the young growing tissue near the shoot's apical meristem. Which aphid is attacking a source, and which is preying on a sink? Which aphid is getting a higher concentration of sugar, A or B? Where would you expect aphids to be found on a plant growing in the wild?

5. How does cotransport result in phloem loading?

6. A seed is a sink when it is forming inside the parent plant. When is it a source?

✔ APPLYING CONCEPTS TO NEW SITUATIONS

Answers are available in Appendix B

1. The text claims that water loss is an inevitable side effect of gas exchange. What data or observations support or challenge this claim? Would the same statement be true in terrestrial animals?

2. Suppose that plants over 1 meter tall had to expend energy to transport water from their roots to their leaves. What would be the consequences in terms of growth rates and overall height?

3. When young trees are transplanted to a new site, it takes several weeks or months for their root systems to grow and establish a high capacity to take up water. If a heat wave occurs during this period, the trees are likely to die—but not of starvation or loss of turgor. What kills them?

4. A recent paper indicates that the aquaporins in plasma membranes of cells throughout a plant close in response to drought stress. How does the closing of these channels help plant cells maintain turgor?

In most plants, roots obtain the water and key elements required for individuals to survive and thrive.

Plant Nutrition

38

The most urgent tasks facing any organism are to acquire (1) carbon-containing molecules that will be used as cellular building blocks and (2) the chemical energy required to make ATP. Plants acquire both by producing sugar through the process of photosynthesis.

Yet plants cannot live on sugar alone. Besides making the carbohydrates they need, plants synthesize all of their own nucleic acids, amino acids, enzymes, chlorophylls, enzyme cofactors, and other molecules. Plants do some of the world's most impressive synthetic organic chemistry.

A plant's ability to perform sophisticated reactions depends on its capacity to harvest a wide variety of simple ions and elements as raw materials. In addition to carbon dioxide and water, plants have to obtain nitrogen, phosphorus, potassium, sulfur, magnesium, and other elements. Soil provides most of these nutrients, the majority existing as ions that are dissolved in soil water at low—sometimes extremely low—concentrations. Once these ions are inside root cells, the one-way flow of water up xylem carries the nutrients throughout the plant body.

The plant body is an efficient machine for harvesting these diffuse resources and concentrating them in cells and tissues. Chapter 36 introduced how the organization and growth of the plant root and shoot systems make resource acquisition possible; Chapter 37 focused on how water and nutrients are transported throughout the plant body. This chapter concentrates on how plants take up simple ions and elements from the soil, so that these nutrients can be transported to the cells that need them.

Questions about nutrition are fundamental to understanding how plants work, increasing agricultural productivity, and maintaining the productivity of forests that supply lumber and fuel—as well as mitigate global warming (see Chapter 54) by transforming atmospheric CO_2 into wood. Let's begin by analyzing the basic nutritional needs of plants—the equivalent of the minimum daily requirements in humans.

KEY CONCEPTS

- In addition to needing carbon dioxide and water, plants require an array of essential nutrients to support growth. These nutrients are available as ions dissolved in soil water and are taken up by roots.

- Nutrient absorption occurs via specialized proteins in the plasma membranes of root cells. Most plants also obtain nitrogen or phosphorus from fungi associated with their roots. Toxins that enter roots are either excluded or actively transported into cell vacuoles and stored.

- Some species of plants have specialized methods of obtaining nutrients, including associations with nitrogen-fixing bacteria, parasitism, and carnivory.

✔ When you see this checkmark, stop and test yourself. Answers are available in Appendix B.

38.1 Nutritional Requirements of Plants

What do plants need to live? In the early 1600s, Jean-Baptiste van Helmont performed a classic experiment designed to answer this question. Van Helmont wanted to know where the mass of a growing plant comes from, and he used a willow tree as a study organism.

As **Figure 38.1** shows, he began by placing 200 pounds of soil in a pot with a 5-pound willow sapling. He allowed the plant to grow for five years, adding only water. At the end of the experiment, he weighed the willow and the soil. The willow weighed 169 lb, 3 ounces; the soil weighed 199 lb, 14 ounces.

Where had the additional 164 pounds, 3 ounces, of tree come from? Because he was not aware that gases have mass, van Helmont hypothesized that the new plant material came from water. He also ignored the loss of 2 ounces in the soil, chalking it up to measurement error.

As it turned out, van Helmont's measurements were not the problem—his conclusions were. Most of the mass of the tree came from carbon dioxide in the atmosphere. The 2 ounces removed from the soil contained vital ions and elements—the nutrients that are the focus of this chapter. What are they?

About half the elements in the periodic table—more than 60—can be found in the tissues of one or more plant species. The question that biologists, farmers, and foresters ask is, Which of these elements are essential for growth and reproduction in most species, and in what quantities?

Which Nutrients Are Essential?

Biologists define an **essential nutrient** as an element or compound that fulfills two criteria:

1. It is required for normal growth and reproduction—meaning that the plant cannot complete its life cycle without this nutrient.

2. It is required for a specific structure or metabolic function.

Researchers test whether a nutrient is essential by denying a specific element to plants and documenting the effects or lack of effect. For most vascular plants, 17 elements are essential. Just three of these—carbon, hydrogen, and oxygen—typically make up about 96 percent of the dry weight of a plant. The remaining 14 elements are sometimes called mineral nutrients, because they originate in soil.

Although different classification schemes for the essential elements have been proposed, the most common is based on distinguishing nutrients that are obtained from water or carbon dioxide versus soil, and then dividing soil nutrients into macronutrients and micronutrients (**Table 38.1**).

MACRONUTRIENTS Certain elements in the soil are required during the synthesis of nucleic acids, proteins, carbohydrates, phospholipids, and other key molecules. Because they are required in relatively large quantities, these elements are called **macronutrients**.

Among the macronutrients, nitrogen (N), phosphorus (P), and potassium (K) are particularly important because they often act as **limiting nutrients**, meaning their availability limits plant growth. If N, P, and/or K are added to soil as fertilizer, plant growth usually increases. This observation explains why the leading ingredients in virtually every commercial fertilizer are N, P, and K—usually listed in that order on the container.

EXPERIMENT

QUESTION: Where does the mass of a growing plant come from?

HYPOTHESIS: The mass of a growing plant comes from soil.

NULL HYPOTHESIS: The mass of a growing plant does not come from soil.

EXPERIMENTAL SETUP:

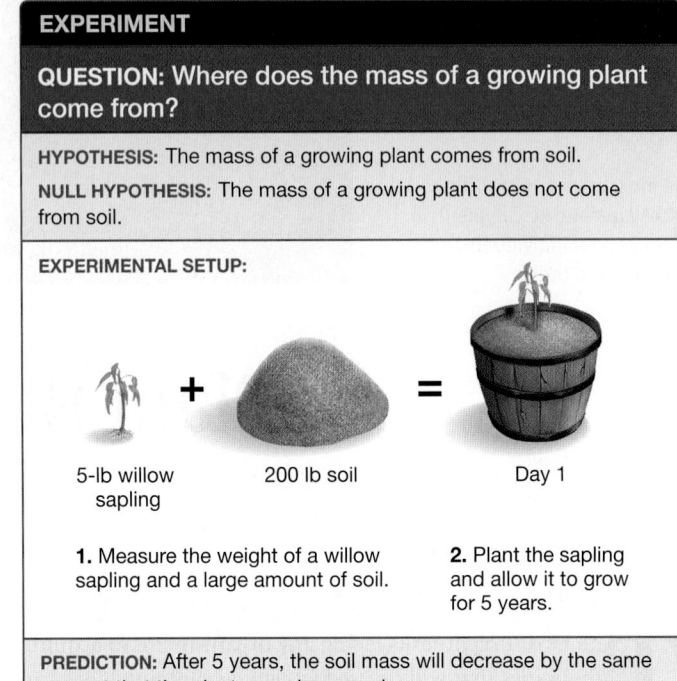

5-lb willow sapling   +   200 lb soil   =   Day 1

1. Measure the weight of a willow sapling and a large amount of soil.

2. Plant the sapling and allow it to grow for 5 years.

PREDICTION: After 5 years, the soil mass will decrease by the same amount that the plant mass increased.

PREDICTION OF NULL HYPOTHESIS: The soil mass will not decrease.

RESULTS:

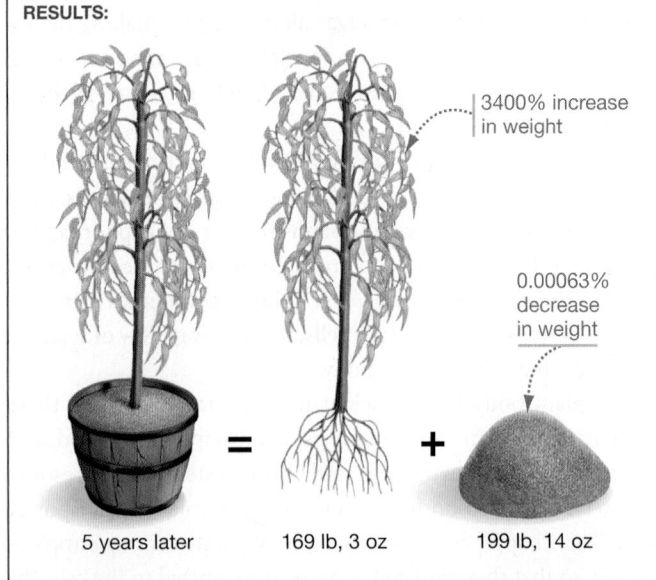

3400% increase in weight

0.00063% decrease in weight

5 years later   =   169 lb, 3 oz   +   199 lb, 14 oz

CONCLUSION: The mass of a growing plant does not come from soil.

FIGURE 38.1 An Early Experiment on the Role of Soil in Plant Nutrition.

✓**QUESTION** Many nonbiologists think that most of a plant's mass comes from soil or water. Describe an experiment that would convince someone that most of a plant's mass comes from CO_2.

TABLE 38.1 Essential Nutrients

Element	Form Available to Plants	Functions	Average % Dry Weight*	Deficiency Symptoms
Obtained from H₂O or CO₂				
Oxygen	O_2, H_2O	Electron acceptor in cellular respiration; major component of organic compounds	45	Usually affects roots—cells suffocate, leading to root rot and wilting
Carbon	CO_2	Substrate for photosynthesis; major component of organic compounds	45	Slow growth (starvation)
Hydrogen	H_2O	Major component of organic compounds; electrical balance and establishment of electrochemical gradients	6	Slow growth due to cell death (desiccation)
Obtained from Soil: Macronutrients				
Nitrogen	NO_3^- (nitrate) NH_4^+ (ammonium ion)	Component of nucleic acids, ATP, chlorophyll, proteins, hormones, and coenzymes	1.5	Failure to thrive; chlorosis (yellowing of older leaves)
Potassium	K^+	Cofactor for many enzymes; necessary for osmotic adjustment in cells; required for synthesis of organic molecules	1.0	Chlorosis at margins of leaves or in mottled pattern; weak stems; short internodes
Calcium	Ca^{2+}	Regulatory functions; role in cell wall structure; stabilizes membranes; second messenger in signal transduction	0.5	Necrosis (small spots of dead cells) in meristems; deformation of young leaves; stunted, highly branched root system
Magnesium	Mg^{2+}	Chlorophyll component; activates many enzymes	0.2	Chlorosis between leaf veins; premature leaf drop
Phosphorus	$H_2PO_4^-$ (dihydrogen phosphate ion) HPO_4^{2-} (hydrogen phosphate ion)	Component of ATP nucleic acids, phospholipids, and several coenzymes	0.2	Stunted growth in young plants; dark green leaves with necrosis
Sulfur	SO_4^{2-} (sulfate ion)	Component of protein and coenzymes	0.1	Stunted growth; chlorosis
Obtained from Soil: Micronutrients				
Chlorine	Cl^- (chloride ion)	Needed for water-splitting step of photosynthesis; functions in water balance and electrical balance	0.01	Wilting at leaf tips; general chlorosis and necrosis of leaves or development of bronze color
Iron	Fe^{3-} (ferric ion) Fe^{2-} (ferrous ion)	Necessary for chlorophyll synthesis; component of cytochromes and ferredoxin; enzyme cofactor	0.01	Chlorosis between veins of young leaves
Manganese	Mn^{2+}	Involved in photosynthetic O_2 evolution; enzyme activator; important in electron transfer	0.005	Chlorosis between leaf veins and small necrotic spots
Zinc	Zn^{2+}	Involved in synthesis of the plant hormone auxin; maintenance of ribosome structure; enzyme activation	0.002	Small internodes; stunted and distorted ("puckered") leaves
Boron	$H_2BO_3^-$ (borate ion)	Strengthens cell walls; required for pollen tube growth and normal membrane function	0.002	Black necrosis in young leaves and buds
Copper	Cu^+ (cuprous ion) Cu^{2+} (cupric ion)	Cofactor of some enzymes; present in lignin of xylem	0.0006	Light-green leaves with necrotic spots; twisted and malformed leaves
Nickel	Ni^{2+}	Cofactor for enzyme functioning in nitrogen metabolism	[no data]	Necrosis at leaf tips
Molybdenum	MoO_4^{2-} (molybdate ion)	Cofactor in nitrogen reduction; essential for nitrogen fixation	0.00001	Chlorosis between veins; necrosis of older leaves

*These percentages were obtained by drying vascular plants and documenting what proportion of the waterless mass consists of various elements.

MICRONUTRIENTS In contrast to macronutrients, **micronutrients** are required in small quantities. When plant tissues are dried and analyzed, micronutrients are typically present in 1–100 parts per trillion. Instead of acting as components of macromolecules, micronutrients usually function as cofactors for specific enzymes—substances that are required for normal enzyme function (Chapter 3).

It's important not to underestimate the importance of micronutrients, even though only tiny amounts are needed. For example, a typical plant contains just one molybdenum atom for every 60 million hydrogen atoms in its body, not including water. Yet plants die without molybdenum, because it functions as a cofactor for several enzymes involved in nitrogen processing. What happens to plants when other essential nutrients are missing?

What Happens When Key Nutrients Are in Short Supply?

In some cases, biologists can examine a plant that is growing poorly and diagnose a nutrient deficiency (**Figure 38.2**). If older leaves are in poor condition, the problem is probably due to lack of N, P, K, or magnesium. These elements are mobile—meaning they are readily transferred from older leaves to newer leaves when they are in short supply—so older leaves deteriorate first when these atoms are scarce. Immobile nutrients like iron or calcium, in contrast, stay tied up in older leaves. When they are in short supply, newer leaves are the first to show symptoms.

In large part, the ability to diagnose nutrient deficiencies is based on studies involving hydroponic growth systems. **Hydroponic growth** takes place in liquid cultures, without soil, so researchers can precisely control the availability of nutrients.

Consider an experiment on copper deficiency in tomatoes (**Figure 38.3**). Researchers grew seedlings in two types of treatments. One treatment consisted of flasks containing water and all the essential nutrients in the relative concentrations that are optimal for tomato growth. The second treatment was identical, except that the nutrient solution lacked copper.

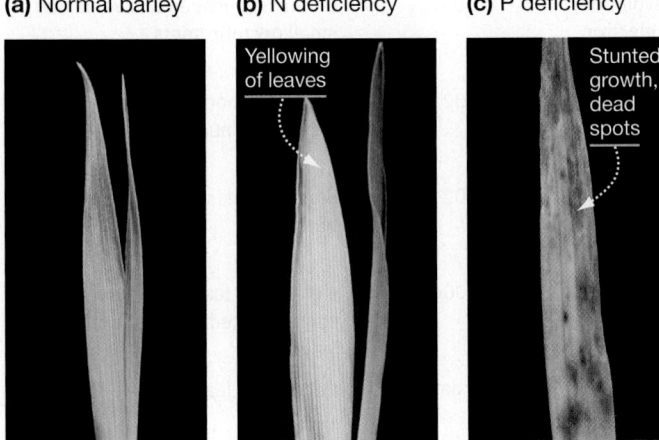

(a) Normal barley   **(b)** N deficiency   **(c)** P deficiency

Yellowing of leaves

Stunted growth, dead spots

FIGURE 38.2 Nutrient Deficiencies Can Have Distinctive Symptoms.

✔**QUESTION** When nitrogen is in short supply, which molecules or processes are affected?

EXPERIMENT

QUESTION: How does copper deficiency affect plants?

HYPOTHESIS: Plants denied copper will grow poorly.

NULL HYPOTHESIS: Plants denied copper will grow normally.

EXPERIMENTAL SETUP:

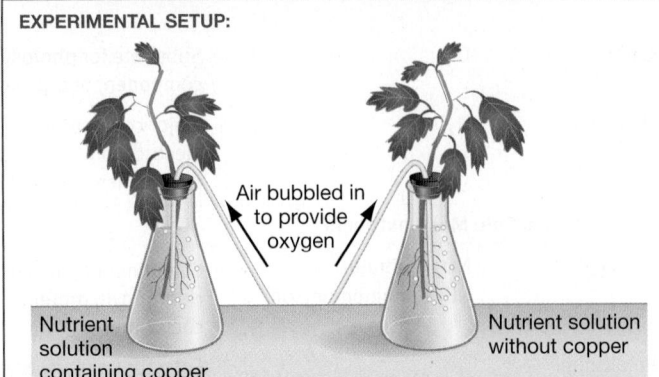

Air bubbled in to provide oxygen

Nutrient solution containing copper

Nutrient solution without copper

PREDICTION: The copper-deficient plant will grow less than the normal plant.

PREDICTION OF NULL HYPOTHESIS: Both plants will grow the same.

RESULTS:

Normal   Copper-deficient

CONCLUSION: Copper deficiency leads to poor growth. All tissues appear to be affected adversely.

FIGURE 38.3 Hydroponics Is Used to Study Nutrient Deficiencies.

SOURCE: Arnon, D. I. and P. R. Stout. 1939. The essentiality of certain elements in minute quantity for plants with special reference to copper. *Plant Physiology* 14: 371–375.

✔**QUESTION** What problem would arise if this experiment had been done in soil with and without added copper?

As Figure 38.3 shows, copper-deprived individuals have stunted shoots, unnaturally light foliage, and curled leaves. Given copper's role as a cofactor or component of several enzymes involved in redox reactions required for ATP production, it is understandable that all tissues in the plant were severely affected. And because copper is a micronutrient, it is reasonable to expect that a relatively small amount would cure the deficiency. In line with this prediction, the researchers found that the symptoms were prevented if the plants were cultured in a solution containing just 0.002 mg/L of copper. Analogous studies have been done on the other essential nutrients.

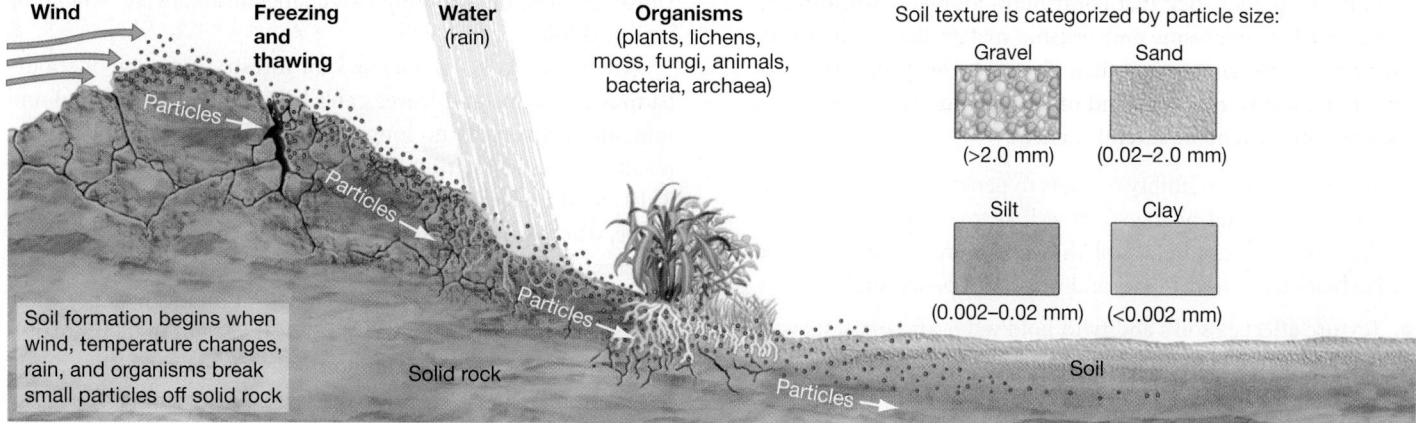

FIGURE 38.4 Soil Formation Begins with Erosion of Rock.

For farmers, foresters, and plant ecologists, understanding which nutrients are essential, and why, is basic to understanding why certain plants thrive and others fail. Now, where do these nutrients come from? The answer—soil—is simple. But soil itself is astonishingly complex.

38.2 Soil: A Dynamic Mixture of Living and Nonliving Components

The process of soil building begins with solid rock. As **Figure 38.4** shows, **weathering**—the forces applied by rain, running water, and wind—continually breaks tiny pieces off large rocks. The weathering process is accelerated if small cracks develop in the rock. If plant roots grow into the crack, they expand as they grow, widen the crack, and break off small flakes or pebbles. A

similar effect occurs in high latitudes or at high elevations when water enters the cracks, freezes in winter, expands, and breaks off pieces.

Depending on their size and composition, the particles resulting from these processes are called gravel, sand, silt, or clay. These rock fragments are the first ingredient in soil. As organisms occupy the substrate, they add dead cells and tissues and feces. This decaying organic matter is called **humus** (pronounced *HEW-muss*).

With time, soil eventually becomes a complex and dynamic mixture of inorganic particles, organic particles, and living organisms. It is commonplace to find thousands of species living in the top few centimeters of a single square meter of soil. In addition to plants, soil-dwelling organisms include a variety of fungi and animals, along with vast numbers of bacteria, archaea, and microscopic protists (**Figure 38.5**).

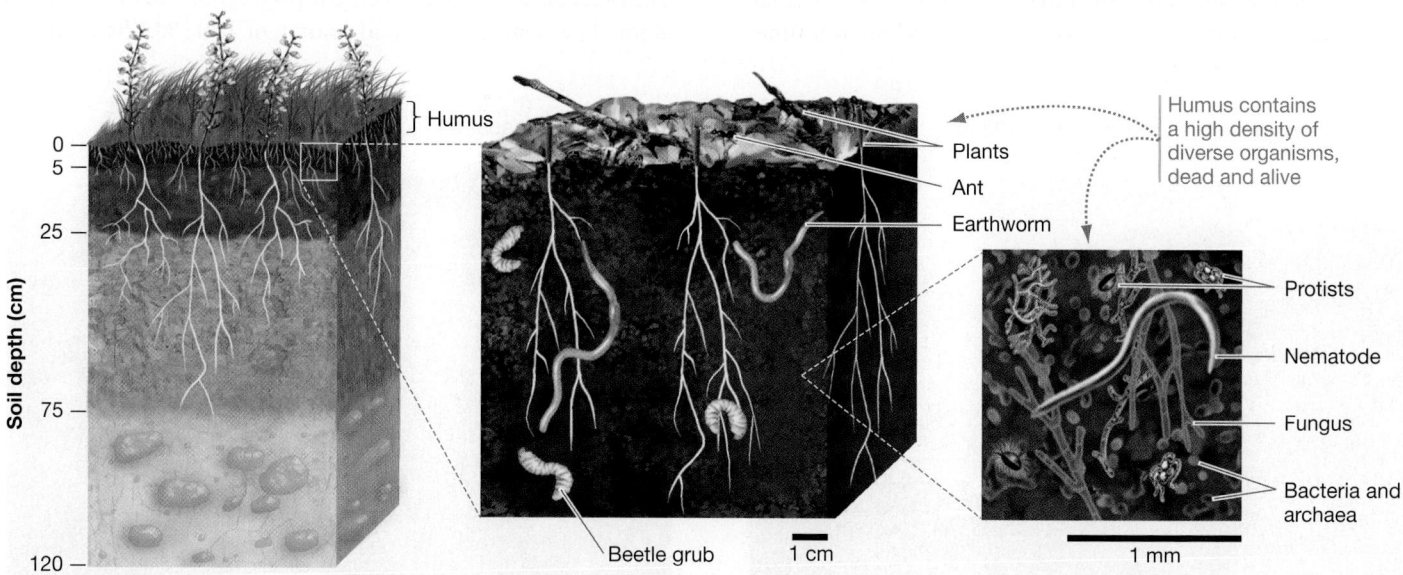

FIGURE 38.5 Mature Soils Are a Complex Mixture of Organic and Inorganic Components. In addition to mineral particles, soil contains humus—organic material derived from dead organisms—and a wide array of living organisms.

Both the parent rock that contributes inorganic soil components and the organisms and organic matter that occupy soils vary from one site to another. **Texture**—the proportions of gravel, sand, silt, and clay—and other qualities vary as well. Soil texture is important for several reasons:

- Texture affects the ability of roots to penetrate more deeply to obtain water and nutrients, as well as to anchor and support the body. For example, soil that is dominated by clay-sized particles tends to compact and resist root penetration.

- Texture affects a soil's ability to hold water and make it available to plants. Water tends to adhere to clay and silt particles but runs through sand and gravel.

- A soil's texture and water content dictate the availability of oxygen. Like other eukaryotes, plants have to take in oxygen to use as an electron acceptor during cellular respiration. The oxygen used by plant root cells is found in air pockets among soil particles. This explains why overwatering a plant is just as detrimental as underwatering it: Overwatering drowns a plant's roots.

The best soils, called loams, contain roughly equal amounts of sand, silt, and clay along with a high proportion of humus.

Loams and other topsoils that have good texture and large amounts of organic matter can take thousands of years to develop through the weathering of rocks and the continual addition of humus. Unfortunately, it can take just a few years of abuse by humans for them to blow or wash away.

The Importance of Soil Conservation

Soil erosion occurs when soil is carried away from a site by wind or water. Soil erosion occurs naturally, such as when rivers cut away at their banks and carry material downstream. In most natural environments, though, the rate of soil formation exceeds the rate of soil erosion, so soils build up over time.

Unfortunately, the situation can change dramatically when humans exploit an area.

When plant cover is removed for forestry, farming, or suburbanization, stems and leaves can't lessen the force of wind and rain, and plant roots no longer hold soil particles in place. The results can be devastating. At some locations in the United States, 8–10 cm of topsoil were blown away during the Dust Bowl of the 1930s, when drought and poor farming practices left thousands of acres of soil unprotected (**Figure 38.6a**).

U.S. soil erosion rates have declined dramatically since then. Still, researchers estimate that almost 30 percent of all croplands in the United States are eroding too fast to maintain their long-term productivity. Deforestation is also exposing forest soils, contributing to disasters such as the mudslides and flooding that occurred in the Dominican Republic and Haiti in 2004, killing close to 5000 people. **Figure 38.6b** provides an aerial view of devastating slides that occurred in the U.S. state of Washington, after recent deforestation. Worldwide, it is estimated that 36 billion tons of soil are lost to erosion every year.

If soil is managed carefully, however, it can be a renewable resource. Techniques that maintain long-term soil quality and productivity are the basis of **sustainable agriculture** and sustainable forestry. Farmers can reduce soil loss dramatically by:

- planting rows of trees as windbreaks;
- using techniques that minimize the amount of plowing and tilling needed to control weeds; and
- planting crops in strips that follow the contour of hillsides.

They can also maintain soil quality by adding organic material in the form of manures and planting cover crops that are plowed in and allowed to decompose.

What Factors Affect Nutrient Availability?

The elements required for plant growth are found in the soil not as atoms, but as **ions**. The second column of Table 38.1 lists some of

(a) Wind erosion in the United States, 1930s

(b) Mudslides caused by deforestation

FIGURE 38.6 Soil Erosion Can Have Devastating Consequences.

the ions in soil that contain essential nutrients. Notice that some of these nutrients are available as elemental ions, such as K^+ or Cl^-, while others exist as molecular ions, such as HPO_4^{2-} or NO_3^-.

ANIONS AND CATIONS BEHAVE DIFFERENTLY The ions present in soil tend to behave in one of two ways, depending on their charge (**Figure 38.7**). Anions—ions with negative charges—usually dissolve in soil water, because they interact with water molecules via hydrogen bonding. (Phosphate ions, an exception to this rule, tend to form insoluble complexes with iron, aluminum, calcium, or other positively charged cations.)

Because they exist as solutes, negatively charged anions are readily available to plants for absorption. They are also easily washed out of the soil by rain, however. The loss of nutrients via the movement of water through soil is called **leaching**.

Cations—ions with positive charges—dissolve in soil water but are not as immediately available as anions. In solution, cations interact with the negative charges found on two types of soil particles: (**1**) organic matter that is rich in negatively charged organic acids; and (**2**) the surfaces of the tiny, sheetlike particles called clay, which are rich in mineral anions (see Figure 38.7).

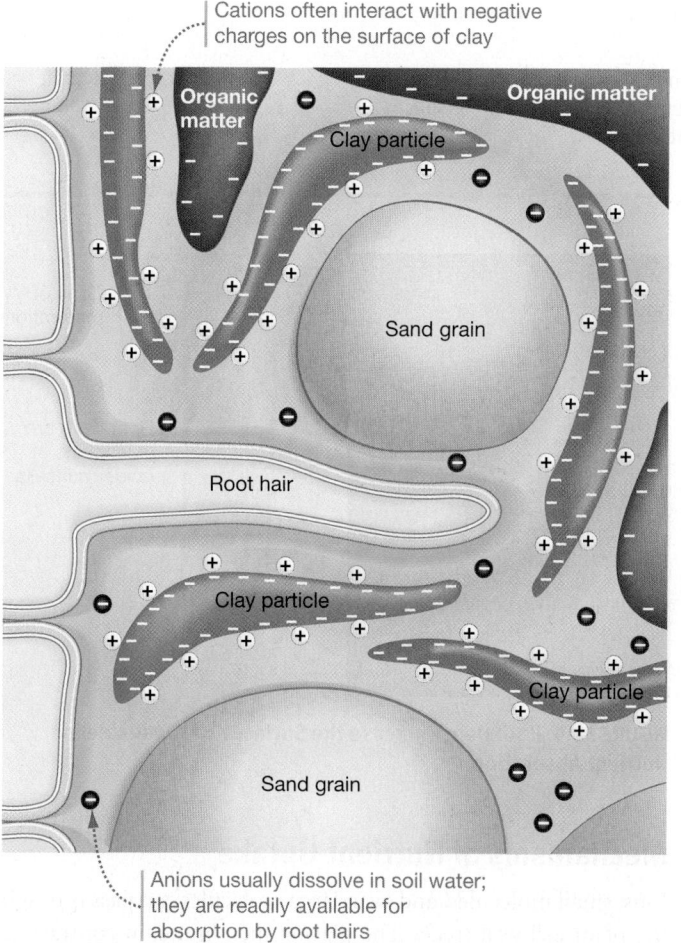

Cations often interact with negative charges on the surface of clay

Organic matter

Clay particle

Organic matter

Sand grain

Root hair

Clay particle

Clay particle

Sand grain

Anions usually dissolve in soil water; they are readily available for absorption by root hairs

FIGURE 38.7 Cations Tend to Bind to Soil Particles; Anions Stay in Solution. Cations bind to organic matter in soil as well as to clay particles. Anions, in contrast, tend to go into solution.

Organic soils that contain clay tend to retain nutrients, because few positively charged cations leach away and because these soils hold water (and thus anions) better than sandy soils do. The presence of clay makes cations more difficult for plants to extract and use, however, because the cations are tightly bound.

THE ROLE OF SOIL PH In addition to soil texture, other factors can influence the availability of essential elements. Perhaps the most important of these is soil pH.

Recall from Chapter 2 that the pH scale indicates the relative concentration of hydrogen ions in a solution, and that it ranges from 0 to 14. Soils with low pH have a relatively high concentration of hydrogen ions and are considered acidic. Soils with a high pH contain relatively few hydrogen ions and are termed basic or alkaline.

Acidic soils are found in regions such as conifer forests, where the decomposition of organic matter produces carbonic acid, phosphoric acid, or nitric acid. Alkaline soils, in contrast, are common in regions where limestone ($CaCO_3$) is abundant. When limestone reacts with water, the calcium ions that are released take the place of protons that cling to soil particles. The protons then react with CO_3^{2-} to form bicarbonate ions (HCO_3^-), lowering the hydrogen ion concentration of the soil and raising its pH.

Most plants prefer soils with a relatively neutral pH—between 6 and 7. But some plants, such as blueberries, grow best in acidic soils (pH around 3 to 5); others require an alkaline soil (pH above 8). Certain species have enzymes that are adapted to work well in low-pH or high-pH environments.

CATION EXCHANGE Soil pH affects the availability of plant nutrients in a number of ways. For example, the presence of protons in soil water can cause the release of cations that are bound to soil particles. The process responsible is called **cation exchange**. Cation exchange occurs when protons or other cations bind to negative charges on soil particles and cause bound cations, such as magnesium or calcium, to be released (**Figure 38.8a** on page 744). Nutrients that are released by cation exchange become available for uptake by nearby plant roots (**Figure 38.8b**).

Plants influence cation exchange because root cells release CO_2 as a by-product of cellular respiration (see Chapter 9). The CO_2 reacts with H_2O to form carbonic acid, which releases protons.

If soil is too acidic, however, rain may wash cations away before the roots can take up nutrients (**Figure 38.8c**). In wet environments, acidic soils tend to be nutrient-poor.

To summarize, anions stay in solution in soil water. They are readily available to plants but may wash away easily. Positive ions, in contrast, tend to bind to soil particles but can be released by cation exchange.

Table 38.2 on page 744 details how the presence of sand, clay, and organic matter affects ion availability and other soil properties, including water availability.

Nutrients are found at extremely low concentrations in soil but at high concentrations in plant cells. How are plants able to bring N, P, K, and other key elements into their bodies against a concentration gradient?

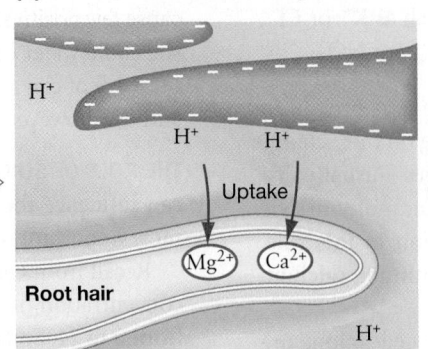

(a) Cation exchange releases nutrients...

H⁺

Clay or organic matter

Mg²⁺ → Ca²⁺

→ H⁺ → H⁺ → Released nutrients

Protons in soil water

H⁺

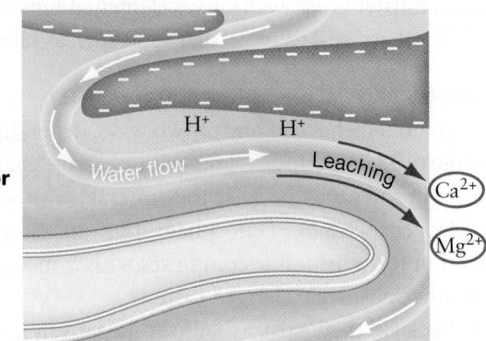

(b) ... which are absorbed by roots...

H⁺

H⁺   H⁺

Uptake

Root hair

Mg²⁺ Ca²⁺

H⁺

or

(c) ... or leached in heavy rains.

H⁺   H⁺

Water flow

Leaching

Ca²⁺

Mg²⁺

FIGURE 38.8 Cation Exchange Releases Nutrients Bound to Soil Particles. When cation exchange occurs, a proton binds to negative charges on clay or organic matter, releasing bound cations.

✔**QUESTION** Why is cation exchange an appropriate term?

TABLE 38.2 **Effects of Soil Composition on Soil Properties**

Composition	Water Availability	Nutrient Availability	Oxygen Availability	Root Penetration Ability
Sand	*Low:* water drains through	*Low:* poor capacity for cation exchange; anions leave in solution	*High:* many air-containing spaces	*High:* does not pack tight
Clay	*High:* water clings to charged surface	*High:* large capacity for cation exchange; anions remain in solution	*Low:* few air-containing spaces	*Low:* packs tight
Organic matter	*High:* water clings to charged surface	*High:* source of nutrients; large capacity for cation exchange; anions remain in solution	*High:* many air-containing spaces	*High:* does not pack tight

38.3 Nutrient Uptake

In most species of plants, the root system is the site of nutrient uptake. Chapter 36 introduced the general features of root anatomy, and Chapter 37 analyzed the role of roots in water uptake. Now we need to explore the detailed anatomy of roots and analyze events that occur in the plasma membranes of their epidermal cells.

Nutrient uptake occurs just above the growing root tip, in the region called the **zone of maturation**. Recall from Chapter 36 that epidermal cells in this part of the root have extensions called **root hairs** (**Figure 38.9**). Root hairs dramatically increase the surface area available for nutrient and water absorption. For example, the root system of a single annual rye plant can have a total surface area the size of a basketball court. Most of this area—60 percent—is found in an estimated 10¹⁰ root hairs.

Root hairs are so numerous and so efficient at absorbing nutrients from soil that, over time, they create a "zone of nutrient depletion" in soil immediately surrounding them. The creation of this mined-out region is why continued root growth is vital to a plant's health. Because roots continue to grow throughout a plant's life, the zone of maturation is continually entering new and potentially nutrient-rich areas of soil.

If an epidermal cell encounters soil where nutrients are available, what happens? How do ions enter?

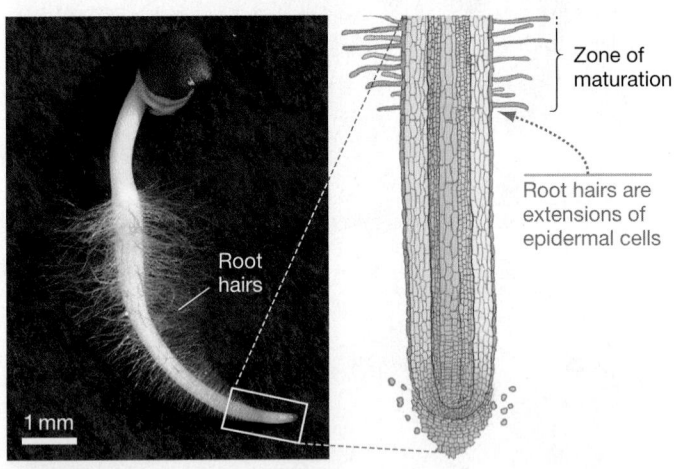

Zone of maturation

Root hairs are extensions of epidermal cells

Root hairs

1 mm

FIGURE 38.9 Root Hairs Increase the Surface Area Available for Nutrient Absorption.

Mechanisms of Nutrient Uptake

Ions, small molecules, and even large molecules can pass through the plant cell wall freely. The plasma membrane, in contrast, is highly selective. Recall from Chapter 6 that the plasma membrane is a fluid, sheetlike structure consisting of a phospholipid bilayer studded with proteins. Because the interior of the phospholipid bilayer is nonpolar and hydrophobic, it resists the passage of ions—even ions that are required by the plant. Some membrane proteins span the bilayer, however, and allow specific ions to cross the membrane.

Because root hairs have such a large surface area, they have large numbers of membrane proteins that bring nutrients into the cortex of the root. These membrane proteins cannot import the ions that the plant needs completely on their own, however.

ESTABLISHING A PROTON GRADIENT Plants harvest diffuse nutrients and concentrate them in their tissues. In an epidermal cell, ions such as potassium (K^+), hydrogen phosphate (HPO_4^{2-}), and nitrate (NO_3^-) are many times more concentrated than they are in the soil water outside the cell. For these and other ions to enter the cell, they have to cross the plasma membrane against a strong concentration gradient. How is this possible?

As **Figure 38.10a** shows, the answer hinges on **proton pumps**, or H^+-ATPases, in the plasma membranes of epidermal cells. These H^+-ATPases are similar to the pumps that make it possible for companion cells to load sucrose into phloem against a strong concentration gradient. Recall from Chapter 37 that when a phosphate group from ATP binds to the pumps, they change conformation in a way that allows them to transport protons to the exterior of the cell.

The activity of proton pumps leads to a strong excess of protons on the exterior of the plasma membrane relative to the interior. In the case of root hairs, this differential results in a strong concentration gradient favoring the movement of protons into the epidermal cell. In addition, the outside of the epidermal membrane becomes positively charged relative to the inside. Stated another way, there is a separation of charge—a **voltage**—across the membrane.

Proton pumps are found in all types of plant cells, and all plant cell membranes carry a voltage. Because voltage is a form of potential energy, the charges that are separated by the membrane create a **membrane potential**, or difference in electrical charge across a cell membrane. By convention, membrane potentials are expressed as inside-cell relative to outside-cell. Typically, the membrane potential of an active plant cell is about −200 mV (millivolts).

USING A PROTON GRADIENT TO IMPORT CATIONS In the membranes of root hairs, the electrical gradient established by proton pumps, which favors the entry of positive ions, is strong enough to overcome the concentration gradient, which opposes the entry of these cations. In essence, plant cells are batteries that are charged up to attract nutritionally necessary cations.

To drive this point home, consider potassium cations. Potassium is an essential nutrient in plants because it is required as a cofactor by over 40 enzymes. In addition, it is found at relatively high concentrations inside plant cells and plays a key role in bringing water into cells via osmosis and maintaining normal turgor pressure.

Researchers who added radioactive potassium ions to the solution outside a root cell and followed their movement found that K^+ does indeed flow into cells along an electrochemical gradient (**Figure 38.10b**). In follow-up experiments, biologists were able to isolate the membrane protein responsible for the uptake of K^+ and sequence the gene that encodes the protein.

✔If you understand this concept, you should be able to explain what happens to cation uptake if (**1**) H^+-ATPases fail, and (**2**) a molecule blocks a cation channel.

Cations like K^+ enter root hairs through membrane proteins along an electrochemical gradient, but how is it possible for anions to enter? The negatively charged interior of the cell should repel these ions.

USING A PROTON GRADIENT TO IMPORT ANIONS **Figure 38.10c** shows why anions such as NO_3^- are able to enter root hairs against their electrochemical gradient. The key is that anions enter through membrane transport proteins called cotransporters (see Chapter 37), which transport two solutes at once. In this case,

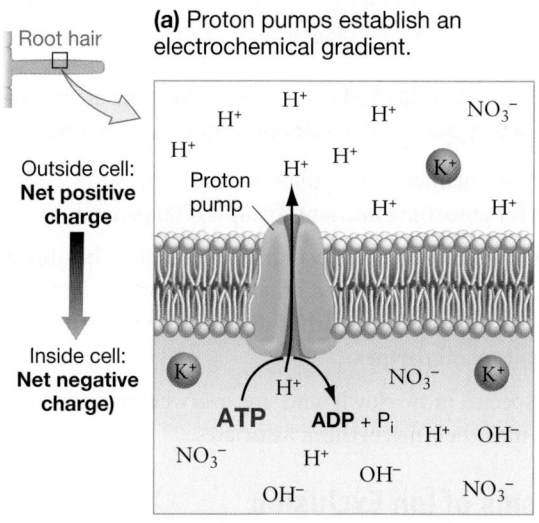
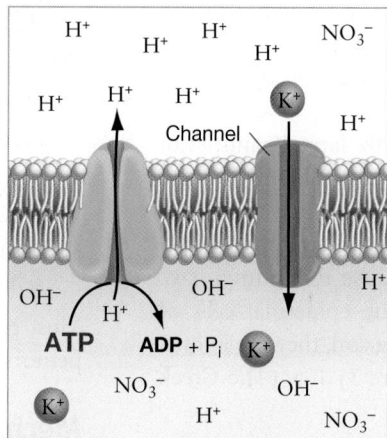
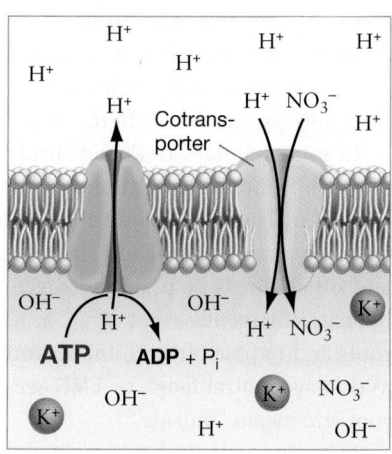

(a) Proton pumps establish an electrochemical gradient.

(b) Cations enter root hairs via channels.

(c) Anions enter root hairs via cotransporters.

Root hair

Outside cell: **Net positive charge**

Inside cell: **Net negative charge)**

FIGURE 38.10 Ions Enter Roots along Electrochemical Gradients Created by Proton Pumps.

✔**EXERCISE** In part (b), add an arrow labeled "Proton gradient." In part (c), add an arrow labeled "Electrical gradient" and indicate its positive-negative polarity.

the cotransporter is a symporter: It brings NO_3^- into the cell along with a proton, and both solutes move in the same direction.

Here's the key observation: So much energy is released when a proton enters the cell along its electrochemical gradient that nitrate, phosphate ions, or other anions can be cotransported *against* their electrochemical gradients.

Symporters are also involved in bringing an array of cations into root cells. Ions that are missed by root hairs and that move into the root through cell walls can be taken into cortical cells or endodermal cells in the same fashion.

🗝 To summarize, the electrochemical gradient set up by proton pumps makes it possible for plant roots to absorb key cations and anions via ion channels and symporters. ✔ If you understand this concept, you should be able to explain what happens to anion uptake if (**1**) H^+-ATPases fail, and (**2**) a molecule blocks an anion symporter.

To review how soils form and how nutrients enter via passive and active transport, go to the study area at *www.masteringbiology.com*.

MB Web Activity Soil Formation and Nutrient Uptake

Once ions have entered the root, they are transported across the root cortex via the symplastic, transmembrane, or apoplastic pathways introduced in Chapter 37. To enter xylem, however, the presence of the Casparian strip (see Chapter 37) means that nutrients moving through the apoplastic pathway have to pass through the plasma membranes of endodermal cells. From there they diffuse into the xylem tubes. Once they become part of xylem sap, the ions are transported passively to tissues throughout the plant.

NUTRIENT TRANSFER VIA MYCORRHIZAL FUNGI To synthesize proteins and nucleic acids, plants need to extract large quantities of nitrogen and phosphorus from the soil. The vast majority of plants take up nutrients by establishing a proton gradient across the membranes of their root hairs. But most can only absorb enough N and P to satisfy their nutritional needs with help from fungi that live in close association with their roots.

You might recall from Chapter 31 that fungi that live in association with plant roots are called **mycorrhizae** (literally, "fungus-root"). Mycorrhizae and plants are **symbiotic** ("living together"), meaning that they live in physical contact with each other. Biologists estimate that more than 80 percent of all vascular plant species associate with mycorrhizae.

In some habitats, trees and shrubs receive large quantities of nitrogen from mycorrhizal fungi. The fungal symbionts in these associations are particularly efficient at digesting amino acids in decaying plant material and absorbing ammonium ions (NH_4^+) and other forms of usable nitrogen. Because these fungi have hyphae—filaments—that wrap around the epidermal cells of roots and radiate out into the surrounding soil, they are known as **ectomycorrhizal fungi**, or **EMF**; see Figure 31.10a. (The Greek root *ecto* means "outside.")

In contrast, plants that live in grasslands and tropical forests receive much of the phosphorus and nitrogen they need from species of fungi whose hyphae can actually penetrate the walls of plant root cells. Because some of these fungal hyphae form branched

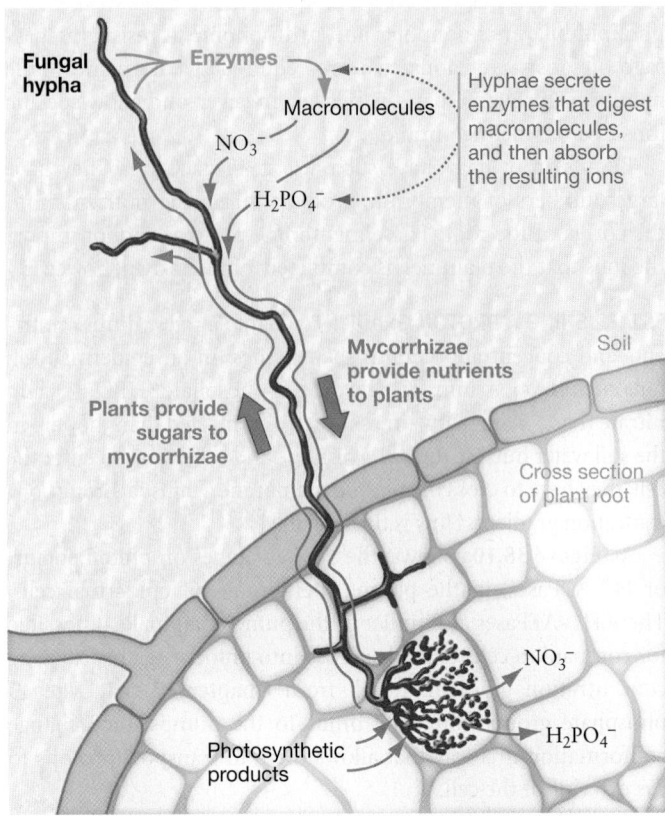

FIGURE 38.11 Most Mycorrhizae and Plants Are Mutualists. The AMF illustrated here provides nutrients to its host plant in exchange for photosynthetic products.

clusters that look like little trees when viewed under a microscope, these symbionts are called **arbuscular mycorrhizal fungi**, or **AMF**; see Figure 31.10b. (The Latin root *arbor* means "tree.")

Chapter 31 presented experimental evidence that EMF and AMF transfer nitrogen and phosphorus from soil to plant roots. In exchange, the plants that they associate with transfer sugars and other photosynthetic products to the fungi. Because there is a reciprocal exchange of nutrients, the symbiotic relationship is considered **mutualistic**, or mutually beneficial (**Figure 38.11**).

Chapter 31 also emphasized two reasons why fungi are particularly efficient at acquiring the nutrients required by plants:

1. Networks of filamentous hyphae increase the surface area available for absorbing nutrients by up to 700 percent.

2. Fungi synthesize and secrete digestive enzymes that break down dead plant material and other sources of nutrients, releasing ions that can be absorbed and used by the fungus itself or by its plant partner.

Most plant species grow slowly and are overwhelmed by competitors if denied their mycorrhizal associates.

Mechanisms of Ion Exclusion

Plants have sophisticated systems for absorbing nutrients via the proton gradients they generate and mutualistic relationships with fungi. Not all ion uptake is beneficial, however.

Certain types of natural soils—as well as soils that have been contaminated by waste products from mining or smelting operations—contain enough cadmium, zinc, nickel, lead, or other metals to poison enzymes in most plants. Sodium is also detrimental at high concentrations: Too much Na^+ inside cells can disrupt enzyme function; and too much sodium in extracellular spaces can create a solute potential that pulls enough water out of cells to result in a loss of turgor.

Sodium poisoning is a key issue in an array of environments, including (1) ocean coastlines, (2) habitats near roads that are treated with salt to melt ice and snow, and (3) irrigated farmlands. When soils are irrigated, solutes are left behind when water evaporates from the surface—leading to salt buildup over time.

How do plants exclude ions that are detrimental?

PASSIVE EXCLUSION Many ions do not enter the transmembrane or symplastic routes through roots (see Chapter 37), because the epidermal and cortical cells lack the requisite membrane channels. And ions that enter roots via the apoplastic pathway may never make it to xylem. To understand why, recall from Chapter 37 that the Casparian strip forces all solutes that travel through the root apoplast to cross the plasma membrane of endodermal cells, before moving further into the root. Ions that cannot enter endodermal cells in this way are excluded from entering the rest of the plant.

In the apoplastic pathway, ion exclusion occurs at endodermal cells because their plasma membranes contain only certain types of transporters and channels. The presence of the Casparian strip allows endodermal cells to act as a selective filter—preventing some ions from entering the symplast and reaching the xylem. Because the excluded ions are not actively pumped out of the root cortex, the Casparian strip qualifies as a mechanism for the passive exclusion of metals, sodium, or other ions (**Figure 38.12**).

A similar type of passive exclusion also occurs in root hairs. If root hairs lack the membrane protein required for a certain ion to enter the cell, the ion won't enter. For example, salt-tolerant strains of corn transport much less salt into their roots than salt-intolerant strains—possibly because of genetic variation in the number of sodium channels found in root hairs. Among species, rice is notoriously sensitive to salt buildup. Barley, in contrast, is relatively tolerant to high salt concentrations in soil.

ACTIVE EXCLUSION BY METALLOTHIONEINS Plants also have mechanisms for coping with toxins once they are inside their cells. This is important, because all nutrients are toxic at high enough concentrations.

Copper channels, for example, admit the small amounts of copper ions required for normal cell function. But plants that grow on soils near copper-mining operations experience large concentration gradients that favor an influx of this nutrient, so a surplus is likely to build up inside the plant body. How do plants neutralize copper and other types of excess nutrients before they poison key enzymes?

One mechanism for coping with toxic concentrations of metals involves small proteins called **metallothioneins**. Metallothioneins bind to metal ions and prevent them from acting as a poison. Producing metallothionein proteins requires an expenditure of energy and represents a form of active exclusion.

Genes for metallothioneins have been found in a wide variety of organisms—bacteria, fungi, and animals as well as plants. Recent research on the mustard-family species *Arabidopsis thaliana* has shown that individuals from populations with a high tolerance for copper produce many more metallothionein proteins than do individuals from populations with a low tolerance for copper.

ACTIVE EXCLUSION BY ANTIPORTERS A second mechanism for actively neutralizing specific toxins involves transport proteins located in the **tonoplast**—the membrane surrounding the vacuole. Proteins in the tonoplast membrane allow plants to actively remove toxic substances from the cytosol and store them in the vacuole.

Perhaps the best-studied example involves proteins that move sodium ions from the cytosol into vacuoles, where they cannot

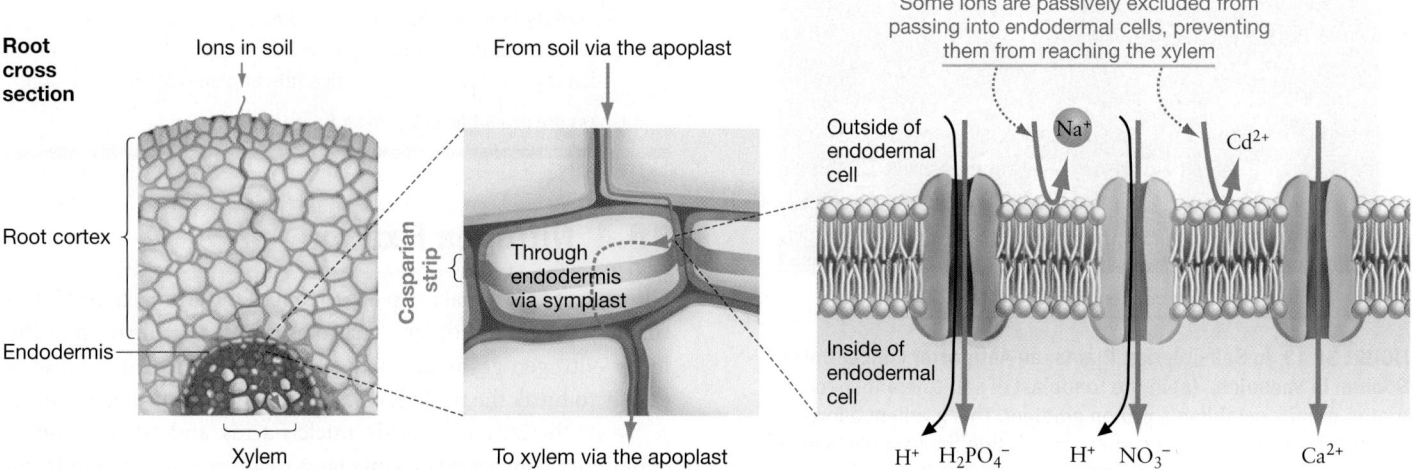

FIGURE 38.12 Passive Exclusion Occurs in Endodermal Cells. Certain ions are excluded from endodermal cells because it is difficult for them to cross the plasma membrane.

(a) In the tonoplast, antiporters send H^+ out and Na^+ in.

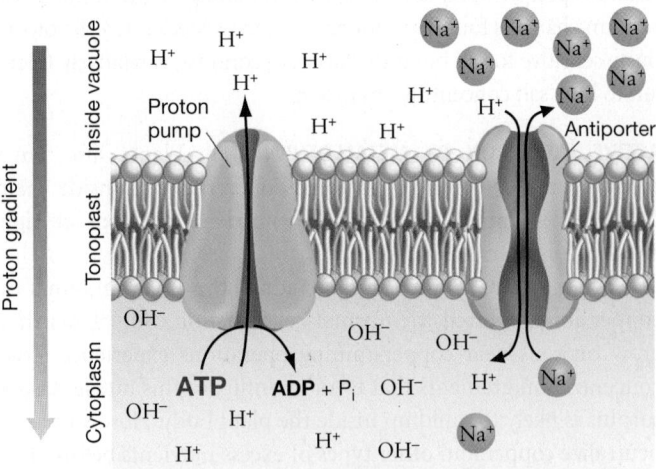

(b) Plants with H^+/Na^+ antiporters tolerate salt.

Transformed plants: antiporter gene present

Control plants: no antiporter gene

Salt concentration in water

Low ◄──────────────────► High

FIGURE 38.13 In Salt-Tolerant Plants, an Antiporter Concentrates Sodium in Vacuoles. (a) In the tonoplast of salt-tolerant species, proton pumps establish a proton gradient. This gradient allows antiporters to concentrate sodium ions inside the vacuole, where they cannot poison enzymes in the cytoplasm. **(b)** *Arabidopsis* plants that were transformed with the gene for the antiporter protein grow well even when watered with salty water.

poison enzymes (**Figure 38.13a**). H^+-ATPases in the tonoplast move protons into the vacuole, creating an electrochemical gradient that favors the movement of H^+ out of the vacuole. A transport protein that functions as an **antiporter** then uses this gradient to conduct protons *out* of the vacuole, along their electrochemical gradient, and bring sodium ions *into* the vacuole—against the sodium concentration gradient.

Recall from Chapter 37 that an antiporter is a cotransporter where the solutes move in opposite directions instead of the same direction—as they do through symporters.

This H^+/Na^+ antiporter has created a great deal of excitement among biologists, because it offers a way to genetically engineer crop plants that may grow well in salty soils created by poor irrigation practices. As an example, consider an experiment in which normal *Arabidopsis* plants and individuals transformed with copies of the gene for the H^+/Na^+ antiporter were exposed to different levels of NaCl. As **Figure 38.13b** shows, the genetically transformed individuals were able to grow efficiently even when exposed to high concentrations of salt.

CHECK YOUR UNDERSTANDING

If you understand that . . .

- Plants absorb most of the nutrients they need via membrane proteins in root hairs.
- Proteins in the plasma membranes of root-hair cells pump protons out of the cells, creating an electrochemical gradient that favors the entry of selected ions via ion channels or cotransporters.
- If no membrane proteins in the epidermal, cortex, or endodermal cells of a root admit a certain type of ion, that ion is excluded from entering the root.
- In many species, mycorrhizal fungi are important for bringing ions that contain nitrogen or phosphorus atoms into the root.

✓ You should be able to . . .

1. Explain how plants generate and use electrical power to import cations and anions.

2. Make a diagram that traces a cadmium ion (Cd^{2+}) from the soil into the plant, showing how it is excluded at different locations by passive and active mechanisms of ion exclusion.

Answers are available in Appendix B.

38.4 Nitrogen Fixation

Nitrogen gas (N_2) makes up 80 percent of the atmosphere. Unfortunately, plants and other eukaryotes cannot use nitrogen in this form. Nitrogen gas is unreactive because it takes a great deal of energy to break the triple bond between the two nitrogen atoms.

To synthesize amino acids, nucleic acids, and other nitrogen-containing compounds, plants have to absorb nitrogen in forms such as ammonium or nitrate ions. But these ions are in short supply in many soils, meaning that plant growth is often restricted by the availability of usable nitrogen. Most plants grow much faster when they receive a nitrogen-containing fertilizer.

Nitrogen-based fertilizers have drawbacks, however. Fertilizer production is extremely energy-intensive; and in many parts of the world, ammonia and other nitrogen-based fertilizers are too expensive for farmers. In more affluent regions, these fertilizers are used so extensively that they are causing serious pollution problems (see Chapter 28). For these and other reasons, there is intense interest in understanding the molecular basis of a phenomenon called biological nitrogen fixation.

The Role of Symbiotic Bacteria

Among all the organisms on the tree of life, only a few species of bacteria and archaea are able to absorb N_2 from the atmosphere and convert it to ammonia, nitrites, or nitrates. This process is called **nitrogen fixation**.

🔑 Nitrogen fixation requires a series of specialized enzymes and cofactors, including a large multi-enzyme complex called nitrogenase. The process is extremely energy demanding. An expenditure of 16 ATP molecules is required for nitrogenase to reduce one molecule of N_2 to two molecules of NH_3. The process can be summarized as follows:

$$N_2 + 8\,e^- + 8\,H^+ + 16\,ATP \longrightarrow 2\,NH_3 + H_2 + 16\,ADP + P_i$$

In many cases, bacterial cells that are capable of nitrogen fixation take up residence *inside* plant root cells. Although several different bacteria and plant hosts can be involved, the best-studied nitrogen-fixing bacteria are members of the genus *Rhizobium* that associate with plants in the pea family. Members of the genus *Rhizobium* and closely related species are often called **rhizobia**; pea family plants are often called **legumes**.

As **Figure 38.14** shows, the root cells of legumes form distinctive structures called **nodules**, where nitrogen-fixing rhizobia are found. The nodules are pink because they contain an iron-containing molecule called leghemoglobin (short for legume hemoglobin). **Leghemoglobin** is related to the hemoglobin that carries oxygen in your blood. Like hemoglobin, leghemoglobin binds oxygen.

Leghemoglobin is important because nitrogenase—the enzyme complex responsible for nitrogen fixation—is poisoned by the presence of oxygen. Inside root nodules, oxygen molecules bind to leghemoglobin instead of binding to nitrogenase. In this way, leghemoglobin keeps levels of free oxygen low enough to keep nitrogenase functioning, while maintaining a source of oxygen for electron transport in cortex cells, where intense cellular respiration and ATP synthesis are occurring.

Root cortex cells begin producing leghemoglobin when they are colonized by rhizobia. How does this colonization process start?

How Do Nitrogen-Fixing Bacteria Colonize Plant Roots?

Like mycorrhizal fungi and their host plants, legumes and rhizobia have a mutualistic relationship. The nitrogen-fixing bacteria

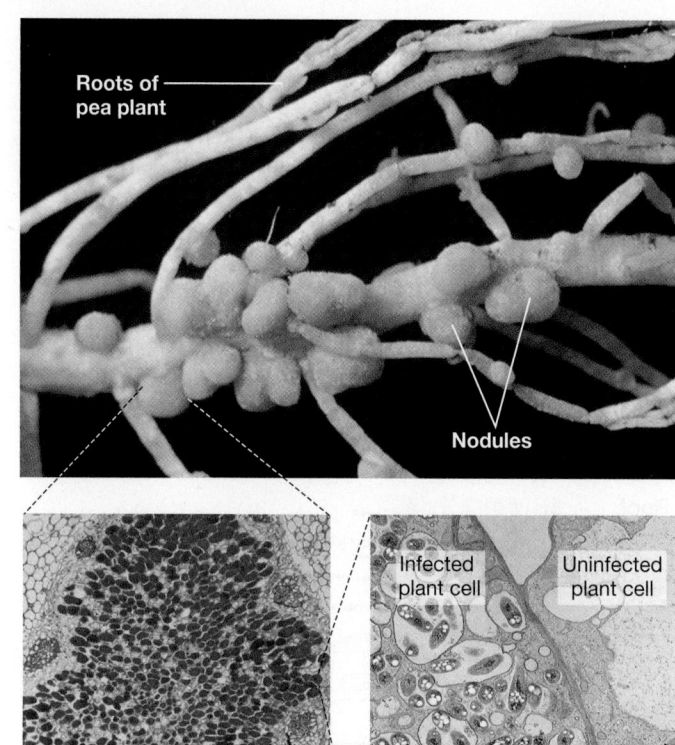

FIGURE 38.14 In Some Plants, Roots Form Nodules where Nitrogen-Fixing Bacteria Live.

provide the plant with ammonia, while the legume provides the bacteria with carbohydrates and protection. The association is costly for the host plant, however. For the bacteria to synthesize the enormous amounts of ATP required to manufacture ammonia, plants have to supply them with large quantities of sugar.

Given the cost of the plant-bacterium interaction and its importance to both species, it's not surprising that the colonization process is carefully regulated.

When a pea seed germinates, its roots do not contain a population of rhizobia. Instead, the root makes contact with bacterial cells existing in the soil, the rhizobia colonize root cells, and the root cells grow into a nodule.

The first event in colonization is a recognition step that occurs between a pea family plant and its symbiotic bacterium. First, young roots release compounds called flavonoids. When rhizobia contact the flavonoids, the bacteria respond by producing sugar-containing molecules called **Nod factors** (for *nod*ule-formation). Nod factors, in turn, bind to proteins on the membrane surface of root hairs.

The recognition step is specific to the species involved. Each legume species produces a different flavonoid that acts as a recognition signal, and each rhizobium species responds with one or more unique Nod factors. When researchers have switched recognition signals or Nod factors between species, the recognition step fails.

When Nod factors bind to the root-hair surface, they set off a chain of events that leads to dramatic morphological changes in the host legume, as **Figure 38.15** shows.

1. The bacterial cells contact the root-hair surface.

2. Rhizobia begin to multiply and move, through an invagination of the root-hair membrane called an **infection thread**.

3. The infection thread extends toward the root, invading the cortex.

4. Portions of the infection thread bud off, forming membrane-bound clusters within cortex cells, each filled with rhizobia.

5. The infected cortex cells divide rapidly, forming root nodules.

The interaction between rhizobia and legumes is one of the most complex and best-studied of all mutualisms.

CHECK YOUR UNDERSTANDING

If you understand that. . .

- Certain species of bacteria infect plant root cells and fix nitrogen, which is then made available to the plant.

✓ **You should be able to. . .**

Generate hypotheses addressing two additional questions about nitrogen fixation:

1. Why don't all plants have symbiotic bacteria that fix nitrogen?

2. Why is the interaction that establishes the infection by nitrogen-fixing bacteria so complex?

Answers are available in Appendix B.

38.5 Nutritional Adaptations of Plants

Based on the data available so far, over 95 percent of vascular plants import nutrients from soil. And over 80 percent of all plants supplement their "diet" with nutrients acquired from mycorrhizal fungi, while a small but significant fraction associate with nitrogen-fixing bacteria. Perhaps 99 percent of all living plant species make their own sugar through the process of photosynthesis.

What about the small number of plant species that don't follow these rules? Some appear to live on air, some parasitize other plants, and others catch insects and digest them.

Epiphytic Plants

Species from a diverse array of plant lineages do not absorb nutrients from soil. In fact, these species never even make contact with soil. As **Figure 38.16a** shows, they often grow on the leaves or branches of trees. For this reason they are called **epiphytes** ("upon-plants").

In northern forests, it is common for mosses and ferns to grow epiphytically on tree trunks. The so-called Spanish moss that hangs from oak trees in the southern United States is another familiar epiphyte—although this species is actually not a moss but a bromeliad, a relative of the pineapple. In the tropics and subtropics, one-third of all ferns grow as epiphytes; and there are thousands of species of epiphytic orchids, bromeliads, and lycophytes.

Epiphytes absorb most of the water and nutrients they need from rainwater, dust, and particles that collect in their tissues or in the crevices of bark. As **Figure 38.16b** shows, some epiphytes have leaves that grow in rosettes and form "tanks" that collect

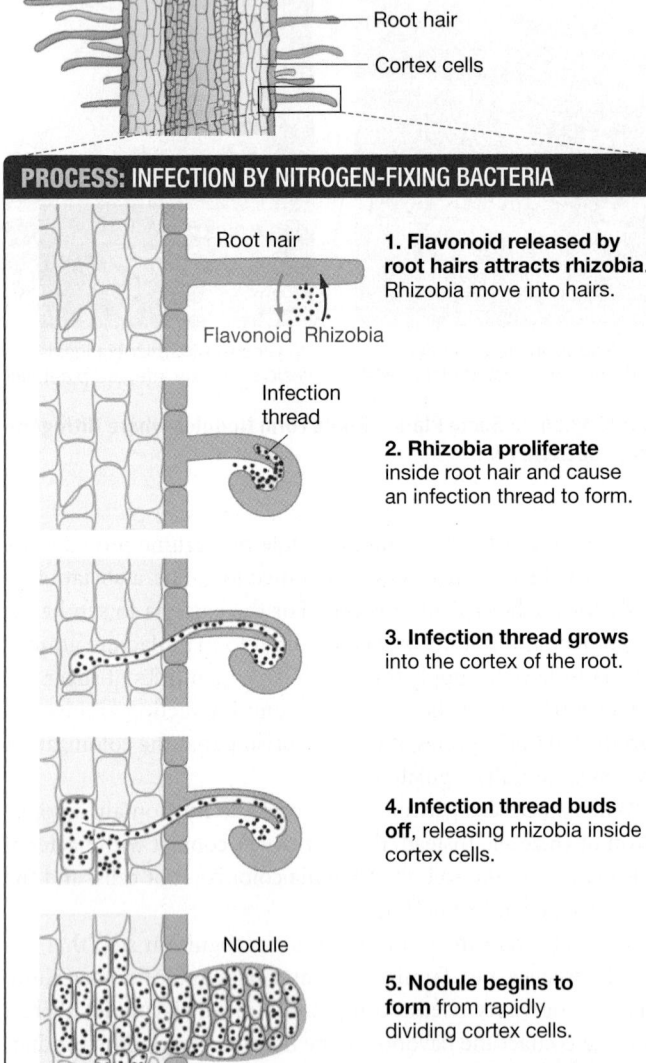

FIGURE 38.15 Infection by Nitrogen-Fixing Bacteria Is a Multistep Process. After rhizobia bind to a root hair, they enter the cytoplasm and travel down an infection thread into the root cortex, where they enter cortex cells.

In the figure:

Root

Root hair

Cortex cells

PROCESS: INFECTION BY NITROGEN-FIXING BACTERIA

Root hair

Flavonoid   Rhizobia

1. Flavonoid released by root hairs attracts rhizobia. Rhizobia move into hairs.

Infection thread

2. Rhizobia proliferate inside root hair and cause an infection thread to form.

3. Infection thread grows into the cortex of the root.

4. Infection thread buds off, releasing rhizobia inside cortex cells.

Nodule

5. Nodule begins to form from rapidly dividing cortex cells.

(a) Epiphytes grow on trees.

(b) Water-holding "tanks" formed by leaves of an epiphyte

Tanks

FIGURE 38.16 Epiphytes Are Adapted to Grow in the Absence of Soil. The environments where many epiphytes grow—tree trunks and branches—are dry and nutrient poor.

water and organic debris. In such cases, nutrients are actually absorbed through the leaves themselves.

Parasitic Plants

Parasites are organisms that live in close physical contact with individuals from another species and that lower the fitness of those individuals, usually by obtaining water or nutrients from them. Biologists estimate that about 3000 species of angiosperms are parasitic—less than 1 percent of the plant species that have been studied and named to date.

Some plant parasites are non-photosynthetic and obtain all of their nutrition by tapping into the vascular tissue of the host individual. But most parasitic plants make their own sugars through photosynthesis and tap the xylem of other species for water and essential nutrients.

For example, the mistletoe (**Figure 38.17**) is green and photosynthetic but has structures called haustoria that penetrate a host's xylem and extract water and ions that the mistletoe uses as nutrients. In the forests of western North America, mistletoe infection is a serious cause of economic loss.

Carnivorous Plants

Carnivorous plants trap insects and other animals, kill them, and absorb the prey's nutrients. Carnivorous species make their own carbohydrates via photosynthesis but use carnivory to supplement the nitrogen available in the environment. Most are found in bogs or other habitats where nitrogen is scarce or unavailable.

MODIFIED LEAVES FORM AN ARRAY OF TRAPPING MECHANISMS Carnivorous plants use modified leaves or roots to trap insects. Chapter 36 highlighted the leaves of pitcher plants, which form tubes (see Figure 36.12e). Prey are enticed to the tube by an attractive odor, then have a difficult time climbing back out. Eventually they fall into the pool of water below, where they drown. Enzymes

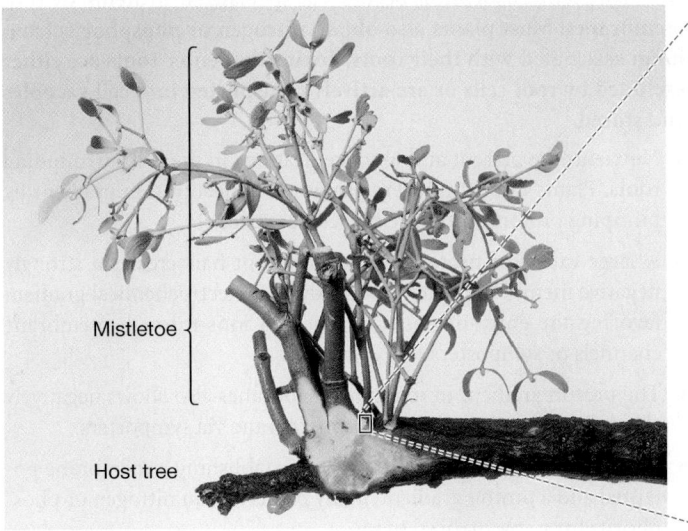

Mistletoe

Host tree

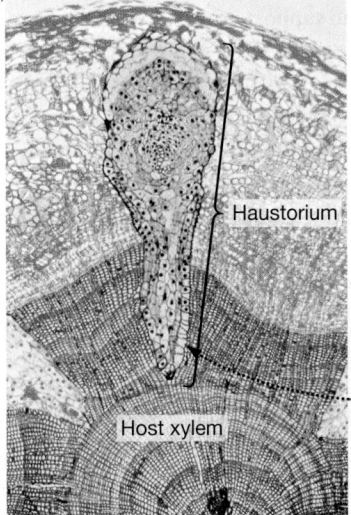

Haustorium

Host xylem

FIGURE 38.17 Some Plant Parasites Tap into the Vascular Tissue of Their Hosts.

✔**QUESTION** The bromeliads in Figure 38.16 and mistletoe are epiphytes. How do they differ in their relationship to the trees on which they grow?

Mistletoe haustoria penetrate host xylem and extract water and ions released by the plant digest the prey, and the plant absorbs the nutrients obtained from their dead bodies.

Sundews have an alternative hunting strategy: They have modified leaves that function like flypaper. The leaf surfaces develop hairs that exude a sticky substance and trap insects (**Figure 38.18**). After the insects die, glands near the sundew's trap release enzymes that slowly digest the prey. The leaf then absorbs the nutrients that are released.

In the Venus flytrap, modified leaves trap insects mechanically. Sensory hairs protrude from the epidermis of each leaf. When an insect lands on the trap and bumps two or three of these hairs, the hair cells produce an electrical signal and the leaf responds by snapping shut (see Chapter 39). If the prey continues to stimulate hairs, the leaf secretes digestive enzymes that slowly digest the prey.

COSTS AND BENEFITS OF CARNIVORY Initially, research on carnivorous plants focused on documenting that meat-eating was adaptive. Early experiments in natural environments supported this hypothesis by showing that, compared to individuals of the same species that are not fed fruit flies, carnivorous plants that were fed fruit flies grew faster and flowered more often.

Follow-up work explored the costs as well as the benefits of carnivory. For example, several teams of researchers showed that when carnivorous plants are provided with nitrogen-based fertilizers, they produce fewer of their specialized insect-trapping leaves and a higher proportion of leaves that function primarily in photosynthesis. These results suggest carnivory is a trait that shows phenotypic plasticity (see Chapter 36): Plants increase investment in prey-capture devices when nitrogen is rare, but they decrease investment in prey-capturing structures when nitrogen is readily available.

Trapped insect

FIGURE 38.18 Sundews Have Modified Leaves with Sticky Surfaces That Catch Insects.

Currently, some research groups are doing "evo-devo" research (see Chapter 21) on carnivory. They are working to identify the novel alleles that made the evolution of carnivory possible in certain groups. Meat-eating may be relatively rare in plants, but it has inspired interesting new directions for research.

CHAPTER 38 REVIEW

For media, go to the study area at www.masteringbiology.com

Summary of Key Concepts

🔑 **In addition to needing carbon dioxide and water, plants require an array of essential nutrients to support growth. These nutrients are available as ions dissolved in soil water and are taken up by roots.**

- About 96 percent of the dry weight of a typical plant consists of carbon, hydrogen, and oxygen. Plants obtain these elements by absorbing carbon dioxide from the atmosphere and water from the soil.

- The other 4 percent of the plant body consists of a complex suite of elements, acquired from soil, that are required for normal growth and reproduction. These essential elements are usually absorbed in the form of ions like nitrate and phosphate.

- Soil is a complex and dynamic mixture of inorganic particles such as clay and sand, organic matter, and organisms. Soil provides plants with oxygen, water, and nutrients as well as a physical substrate for anchoring and supporting the body.

  ✔ You should be able to design a field experiment that would determine whether nitrogen is a limiting nutrient in a habitat near your campus.

🔑 **Nutrient absorption occurs via specialized proteins in root membranes. Most plants also obtain nitrogen or phosphorus from fungi associated with their roots. Toxins that enter roots are either excluded by root cells or are actively transported into cell vacuoles and stored.**

- Nutrients are present at low concentrations in the soil surrounding roots. Plants import nutrients against a concentration gradient by pumping protons into the extracellular space.

- A large excess of protons outside the root hair creates a strongly negative membrane potential and thus an electrochemical gradient favoring the entry of positively charged ions through membrane channels or symporters.

- The proton gradient in root hair membranes also allows negatively charged ions to cross the plasma membrane via symporters.

- In addition to acquiring nutrients by establishing a membrane potential and a proton gradient, many plants obtain nitrogen or phosphorus from mycorrhizal fungi.

- The relationship between plants and the fungi is mutually beneficial. The host plant provides the fungi with carbohydrates; the fungi harvest N or P from the surrounding soil and transport it to the root system.

- Passive and active systems also exist for excluding certain ions. This is important because all nutrients are toxic in high concentrations.

- Passive exclusion occurs when ions cannot pass through the Casparian strip.

- Active exclusion often results from the production of specialized proteins that bind a particular ion, or the active transport of toxic ions into vacuoles.

  ✓You should be able to predict the consequences of mutations that allow root cells to establish a membrane voltage that is much more negative than normal, without a large increase in energy expenditure.

  (MB) **Web Activity** Soil Formation and Nutrient Uptake

🔑 **Some species of plants have specialized methods of obtaining nutrients, including associations with nitrogen-fixing bacteria, parasitism, and carnivory.**

- Certain plants, including those in the pea family, are capable of symbiosis with nitrogen-fixing bacteria. The bacteria colonize the interior of root cells after exchanging species-specific signals.

- Nitrogen-fixing bacteria receive protection and sugar from the host plant, in exchange for producing ammonia from nitrogen gas.

- Some parasitic plants produce their own carbohydrates through photosynthesis but steal water and nutrients by infecting the xylem of host plants.

- Carnivorous plants have evolved mechanisms for trapping and digesting insects and other animals. The primary benefit of carnivory is to obtain nitrogen in low-nitrogen environments.

  ✓You should be able to design an experiment, using radioactive carbon and the heavy isotope of nitrogen (^{15}N), that would test whether the rhizobia–pea plant interaction is mutualistic.

Questions

✓TEST YOUR KNOWLEDGE
Answers are available in Appendix B

1. Which of the following characteristics defines an element as essential for a particular species?
   a. It has to be added as fertilizer to achieve maximum seed production.
   b. If it is missing, a plant cannot grow or reproduce normally.
   c. If it is present in high concentration, plant growth increases.
   d. If it is absent, other nutrients may be substituted for it.

2. Why is the presence of clay particles important in soil?
   a. They provide macronutrients—particularly nitrogen, phosphorus, and potassium.
   b. They bind metal ions, which would be toxic if absorbed by plants.
   c. They allow water to percolate through the soil, making oxygen-rich air pockets available.
   d. The negative charges on clay bind to positively charged ions and prevent them from leaching.

3. Where does most nutrient uptake occur in roots?
   a. at the root cap, where root tissue first encounters soil away from the zone of nutrient depletion
   b. at the Casparian strip, where ions must enter the symplast prior to entering xylem cells
   c. in the symplastic and apoplastic pathways
   d. in root hairs, in the zone of maturation

4. Why is the activity of proton pumps in root hair membranes important?
   a. It creates a strongly positive membrane potential (voltage).
   b. It allows toxins to be concentrated in vacuoles, so the toxins do not poison enzymes in the cytoplasm.
   c. It sets up an electrochemical gradient that makes it possible to absorb cations and anions.
   d. It sets up the membrane voltage required for action potentials to occur.

5. Why is the relationship between most mycorrhizal fungi and their host plants considered mutualistic?
   a. They live in close physical association.
   b. Both species benefit from the association.
   c. The host plant cannot live without the mycorrhizae.
   d. The mycorrhizae cannot live without the host plant.

6. How do most epiphytes obtain nutrients?
   a. by trapping and digesting insects
   b. from their host plant
   c. via symbiotic bacteria called rhizobia
   d. from dust or other airborne particles

✓TEST YOUR UNDERSTANDING
Answers are available in Appendix B

1. A farmer is concerned that her corn crop is suffering from iron deficiency. Design an experiment that uses hydroponic cultures to identify (1) the symptoms of iron deficiency in corn and (2) the minimum amount of iron required to support normal growth rates.

2. Explain why carnivorous and parasitic plants are most common in nutrient-poor habitats.

3. Would you expect plant productivity to be higher in sandy soils or in soils containing both sand and clay? Why?

4. Sketch the plasma membrane of a root hair. Include proton pumps, ion channels, and cotransporters. Now add ATP to the system, then

protons in their correct relative concentration, and ions that move across the membrane via channels or cotransporters. Explain why plant cells have a negative membrane potential.

5. Why is it important for plants to exclude certain ions? Summarize the difference between active and passive exclusion mechanisms.

6. Why is it logical for biologists to (1) distinguish nutrients that are obtained from the atmosphere or water (C, H, and O) from mineral nutrients obtained from soil, and (2) distinguish macronutrients from micronutrients?

1. There is a conflict between van Helmont's data on willow tree growth and the data on essential nutrients listed in Table 38.1. According to the table, nutrients other than C, H, and O should make up about 4 percent of a willow tree's weight. Most or all of these nutrients should come from soil. But van Helmont claimed that the soil in his experiment lost just 2 ounces, while the tree gained 2627 ounces. If so, then soil contributed just 0.08 percent of the added weight instead of 4 percent. State at least one hypothesis to explain the conflict. How would you test this hypothesis?

2. Acid rain occurs when sulfur oxides and nitrous oxides released by cars or factories react with water vapor, forming sulfuric and nitric acids, respectively. In areas affected by acid rain, cations quickly leach out of the soil. Explain why.

3. Design an experiment using the drug vanadate, which poisons proton pumps, to test the hypothesis that phosphate ions enter cells via an H^+/HPO_4^{2-} cotransporter.

4. Alder trees have nitrogen-fixing bacteria associated with their roots and have a much higher proportion of nitrogen in their tissues than spruce trees that grow in the same habitat. Which species decays faster after death: alder or spruce? Explain your logic.

Plants have sophisticated information processing systems. These radish seedlings sense the presence of light and are bending toward it.

Plant Sensory Systems, Signals, and Responses

39

I magine standing in place for several hundred years, like an oak tree. Each spring you produce flowers. All summer you absorb light from the Sun, carbon dioxide from the atmosphere, and water and nutrients from the soil. Water, ions, and sugars flow up and down your vascular tissue. For six months or more your body grows upward, outward, and downward. In fall, thousands of your offspring drop to the ground as acorns. Then as winter approaches, your metabolism slows and you stop growing. Like a hibernating animal, you spend the long, hard months of winter in a state of suspended animation, before awakening the following spring.

To stay alive, an oak needs to gather information about its environment. It has to sense the season of the year, the time of day, the pull of gravity, the force of wind, and attacks by enemies. It needs to sense when its leaves are being shaded and which leaves are receiving the wavelengths of light that are required to support photosynthesis.

Plants may not have eyes or ears, but they can sense light, gravity, pressure, and wounds. They have the equivalent of a sense of smell, because they can perceive certain airborne molecules. It could even be argued that they have a sense of taste, because their roots sense the presence of nutrients in the soil.

In addition to gathering information about the conditions around them, plants have to respond in an appropriate way. They do not jump or swim or run, but their shoot systems grow toward light or end up shorter and stockier in response to wind. In response to gravity, shoots grow up and roots grow down. In response to touch, the modified leaves of a Venus flytrap shut fast enough to catch flying insects. If a plant is being attacked or if it senses that a neighboring individual is under attack, it may lace its tissues with toxic compounds or mobilize other defenses.

The message of this chapter is simple: 🔑 Plants have sophisticated systems for collecting information about their environment and responding in ways that maximize their chances of surviving, thriving, and producing offspring. The ability to gather information and respond to it is one of the five fundamental attributes of life (see Chapter 1).

KEY CONCEPTS

🔑 Plants are selective about the information they process. They perceive a wide variety of environmental stimuli that affect their ability to grow and reproduce.

🔑 When sensory cells receive a stimulus, they transduce the signal and respond by producing hormones that carry information to target cells elsewhere in the body.

🔑 Target cells respond to hormonal stimulation in ways that increase the ability of the plant to survive and reproduce.

🔑 Hormones are also responsible for regulating how plants grow throughout their lives—especially in response to changes in environmental conditions. Each type of plant growth regulator plays a general role in the life of a plant, and growth responses are usually affected by interactions among several different hormones.

✔ When you see this checkmark, stop and test yourself. Answers are available in Appendix B.

39.1 Information Processing in Plants

Every environment is full of information. But, like other organisms, plants monitor only aspects of the environment that matter to them—that affect their ability to stay alive and produce offspring.

Figure 39.1 summarizes the three-step process by which plants gather, process, and respond to the information they monitor:

1. Receptor cells receive an external signal and change it to an intracellular signal.

2. Receptor cells sends a signal to cells in other parts of the body that can respond to the information.

3. Responder cells receive this signal, transduce it to an intracellular signal, and change their activity in a way that produces an appropriate response.

Let's briefly consider how each of these steps works, then delve into the details of how plants sense and respond to light, gravity, and other information.

How Do Cells Receive and Transduce an External Signal?

When you text a friend, the signal travels from you to the receiver via airwaves. When the message arrives, the receiving cell phone changes the information in the airwaves into electrical signals and then into words that your friend can understand.

Plants work in much the same way. When light strikes a sensory cell, for example, the information that it carries has to be changed into a form that is meaningful to that cell. Signals from the environment are usually received by a protein specialized for that function. Receptor proteins change shape in response to an environmental stimulus, such as being struck by a particular wavelength of light, having pressure applied, or binding to a particular type of molecule.

When a receptor changes shape in response to a stimulus, the information changes form—from an external signal to an intracellular signal. This process is called **signal transduction** (**Figure 39.2**); the verb transduce means "to convert energy from one form to another." Once information has been transduced to an intracellular form, it travels down what biologists call a signal transduction pathway.

You might recall from Chapter 8 that there are two basic types of signal transduction pathways: phosphorylation cascades and second messengers. Both begin with a receptor protein in the plasma membrane.

- **Phosphorylation cascades** are triggered when the change in the receptor protein's shape leads to the addition of a phosphate group (PO_4^{3-}) from ATP to the receptor or an associated protein. Phosphorylation activates proteins involved in signal transduction cascades, causing them to phosphorylate and activate a different set of proteins, which in turn catalyze the phosphorylation and activation of still other proteins, and so on.

- **Second messengers** are produced when hormone binding results in the production of intracellular signals or their release from storage areas. Calcium ions (Ca^{2+}) stored in the vacuole, ER, or cell wall are one of several ions or molecules that function as the second messenger in plants.

In some cases, phosphorylation cascades and second messengers interact.

Signal transduction primes the receptor cell for action. In many cases, though, the cells that receive information from the environment are located in a part of the body that is distant from the cells that need to respond to the information. How does information from an activated receptor cell get to responder cells?

In most cases, the answer is a **hormone**—an organic compound that is produced in small amounts in one part of a plant and transported to target cells, where it causes a physiological response. ☞ Signal transduction in a receptor cell often results in the release of a hormone that carries information to responder cells.

How Are Cell-Cell Signals Transmitted?

Because plants perceive such a wide array of stimuli, they have a wide array of hormones coursing through their bodies. These signals may be transmitted from cell to cell by specialized transport proteins in cell membranes, in xylem sap or phloem sap, or by simple diffusion from the originating cell. In most cases the signals act on target tissues throughout the body.

Plant cells routinely receive information from several different hormones at the same time, so it is common for different types of hormones to interact with each other and modulate the cell's response.

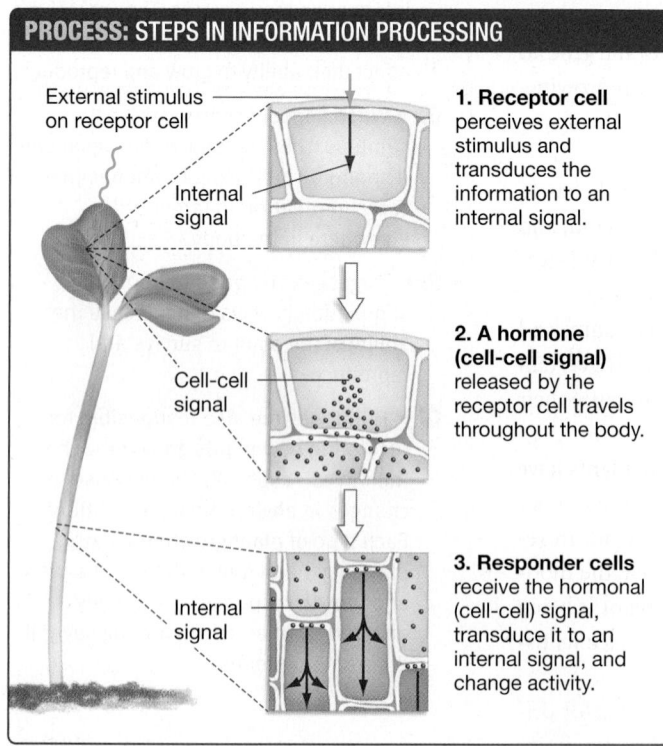

PROCESS: STEPS IN INFORMATION PROCESSING

External stimulus on receptor cell

Internal signal

Cell-cell signal

Internal signal

1. Receptor cell perceives external stimulus and transduces the information to an internal signal.

2. A hormone (cell-cell signal) released by the receptor cell travels throughout the body.

3. Responder cells receive the hormonal (cell-cell) signal, transduce it to an internal signal, and change activity.

FIGURE 39.1 In Many Cases, Hormones Link a Stimulus and a Response.

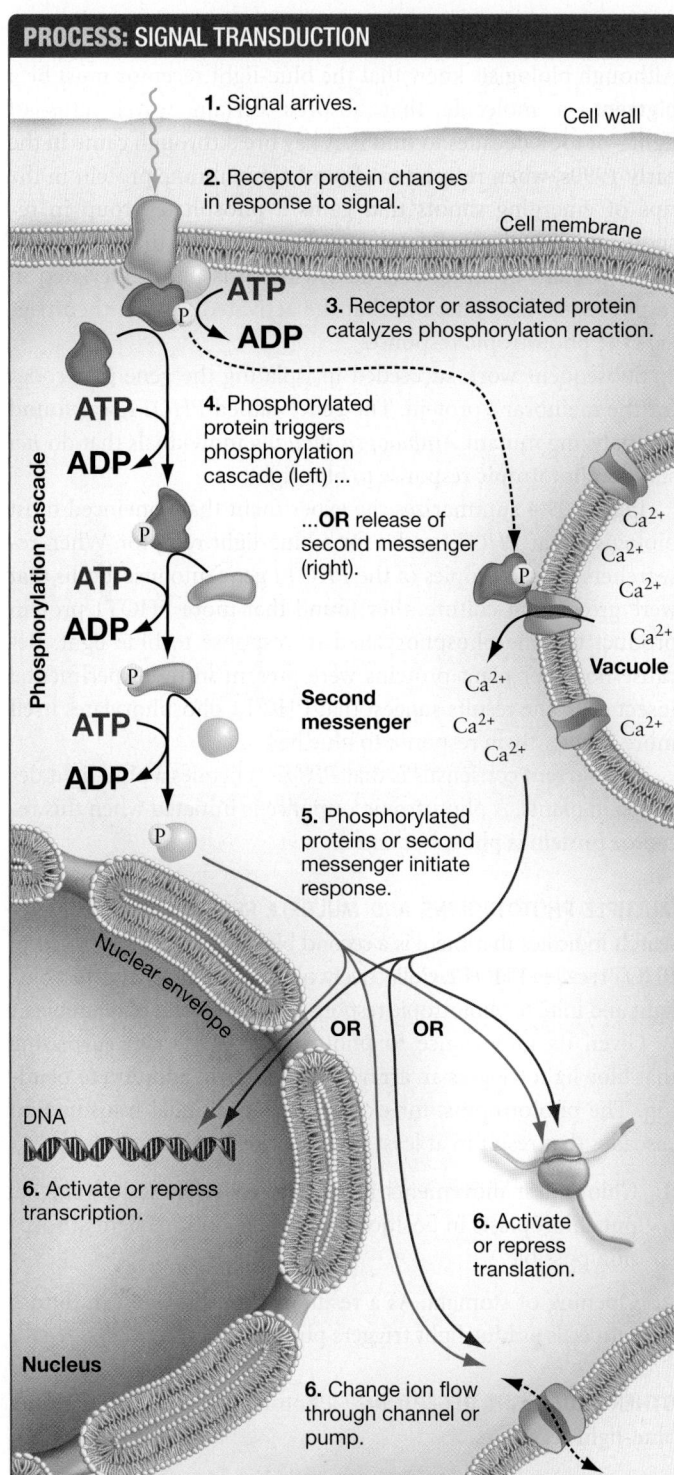

PROCESS: SIGNAL TRANSDUCTION

1. Signal arrives.

Cell wall

2. Receptor protein changes in response to signal.

Cell membrane

ATP

P

ADP

3. Receptor or associated protein catalyzes phosphorylation reaction.

ATP

ADP

4. Phosphorylated protein triggers phosphorylation cascade (left) ...

Phosphorylation cascade

P

...**OR** release of second messenger (right).

ATP

ADP

P

Ca²⁺
Ca²⁺
Ca²⁺
Ca²⁺

Vacuole

ATP

ADP

P

Second messenger

Ca²⁺
Ca²⁺
Ca²⁺
Ca²⁺

P

5. Phosphorylated proteins or second messenger initiate response.

Nuclear envelope

OR   **OR**

DNA

6. Activate or repress transcription.

6. Activate or repress translation.

Nucleus

6. Change ion flow through channel or pump.

FIGURE 39.2 Signal Transduction Changes an External Signal to an Internal Signal. After an environmental or cell-cell signal is transduced to an intracellular signal, it triggers a change in the cell's activity.

How Do Cells Respond to Cell-Cell Signals?

Cells are exposed to a constant stream of hormones. Many of these signals have little to no effect on what is happening inside. The reason is simple: Hormones can elicit a response only if a cell has an appropriate receptor.

But if a receptor exists on or in the cell, an arriving hormone binds to it. When the receptor changes shape in response, the effect is like a sharp rap on the door of your room. The signal—often received at the cell's periphery—is rapidly transduced to a series of phosphorylated proteins or the release of a second messenger inside the cell. The result may be a dramatic change in the cell's activity.

To understand why these binding events can have such significant effects on cell activity, look again at Figure 39.2. Activation of a signal transduction cascade results in the production of many phosphorylated proteins, or the release of many second messengers. These events amplify the original signal many times. In this way, low concentrations of plant hormones can have a large impact on target cells. Hormones are tiny molecules in tiny concentrations, but because their signal is amplified during signal transduction, they produce big results.

As Figure 39.2 shows, the response to hormone binding can include:

- activation of membrane transport proteins (Chapter 6), which produces a change in the membrane's electrical potential or the cell wall's pH; or

- changes in gene expression—via alterations in transcription activators or repressors (Chapter 18) or the translation machinery—that result in new suites of proteins or RNAs in the cell.

The response to the signal is important. 🔑 When cells respond to a hormone, the change in their activity helps the plant cope with the environmental change sensed by the receptor cell.

CHECK YOUR UNDERSTANDING

🔑 **If you understand that . . .**

- Plants can respond to changes in their environment because sensory cells receive information from the environment, sensory cells produce hormones or other cell-cell signals, and target cells respond to hormones.
- Information processing in sensory cells and the cells that respond to cell-cell signals involves three steps:

  Step 1 A receptor molecule changes in response to the signal.

  Step 2 A signal transduction pathway transforms the hormonal signal into an intracellular signal.

  Step 3 The intracellular signal triggers a response—transcription of target genes, activation of specific enzymes or transport proteins, or other changes in cell activity.

✔ **You should be able to . . .**

1. Explain why only certain cells in the body—not all—respond to an environmental signal, and why only certain cells respond to a hormone.

2. Explain how cell-cell signals are amplified inside a target cell.

Answers are available in Appendix B.

39.2 Blue Light: The Phototropic Response

Most of the general principles of plant communication emerged from studies of how plants respond to light. For example, consider the claim that plants are highly selective about the information they process. Light is made up of a wide array of wavelengths (see Chapter 10), but plants sense and respond to only a few.

This conclusion traces back to experiments that Charles Darwin and his son Francis did in 1881 with coleoptiles of a plant called reed canary grass. A **coleoptile** is a modified leaf that forms a sheath protecting the emerging shoots of young grasses.

The Darwins germinated seeds in the dark, placed the young, straight shoots next to a light source, and noted that the shoots grew toward the light—a response, due to differential cell elongation, called bending. You have probably seen the same response in houseplants that are near a window (see the photo at the start of the chapter). Directed movement in response to light is called **phototropism** (literally, "light-turn").

When the Darwins exposed shoots to light filtered through a solution of potassium dichromate, however, the shoots did not bend toward the light. Potassium dichromate solutions filter out wavelengths in the blue part of the visible spectrum. The photo in **Figure 39.3** shows the results of a follow-up experiment, with coleoptiles that have been exposed to blue, yellow, green, orange, and red light. Bending occurs only toward light that contains blue wavelengths.

To understand the specificity of the response is important, recall from Chapter 10 that (1) chlorophylls *a* and *b* are the primary photosynthetic pigments, and (2) these pigments absorb strongly in the blue and red parts of the spectrum. Plants exhibit a phototropic response if blue wavelengths are available, but show no response if blue wavelengths are not present. Plants move toward blue light because it is important in photosynthesis.

Let's look at what happens when blue light strikes a receptor cell, and follow through to the cells that undergo differential growth in response.

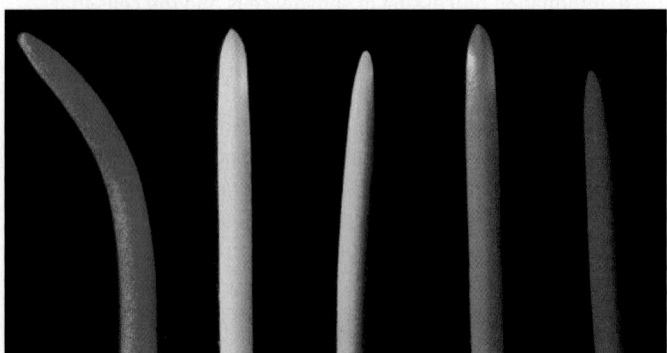

FIGURE 39.3 Experimental Evidence That Plants Sense Specific Wavelengths of Light. Each of these shoots was exposed to a different wavelength of light. In each case, the light source was on the left as the shoots are shown here; only the shoot exposed to blue light bent toward the source.

Phototropins as Blue-Light Receptors

Although biologists knew that the blue-light receptor must be a **pigment**—a molecule that absorbs certain wavelengths of light—it took decades to find it. A key breakthrough came in the early 1990s, when researchers found a membrane protein in the tips of emerging shoots that gains a phosphate group in response to blue light. Researchers hypothesized that the membrane protein becomes activated when it is phosphorylated in response to blue light, and that the activated protein then triggers the phototropic response.

Subsequent work succeeded in isolating the gene that codes for the membrane protein. The gene, named *PHOT1*, was found by analyzing mutant *Arabidopsis thaliana* individuals that do *not* show a phototropic response to blue light.

Figure 39.4 summarizes the experiment that convinced most biologists that *PHOT1* codes for a blue-light receptor. When researchers inserted copies of the *PHOT1* gene into insect cells that were growing in culture, they found that more PHOT1 protein product became phosphorylated in response to blue light. Because no other plant proteins were present in the experimental insect cells, the results suggest that PHOT1 phosphorylates itself more frequently in response to blue light.

The current consensus is that *PHOT1* encodes a blue-light detector in plants. A phototropic response is initiated when this receptor protein is phosphorylated.

MULTIPLE PHOTOTROPINS AND MULTIPLE RESPONSES Recent research indicates that there is a second blue-light receptor related to PHOT1, called PHOT2. Collectively, photoreceptors that detect blue light and initiate phototropic responses are known as **phototropins**.

Given its importance to photosynthesis, it's not surprising that blue light triggers an array of responses in addition to bending. The phototropins, for example, trigger signal transduction cascades that result in at least two other responses.

1. Chloroplast movements inside leaf cells. These movements put chloroplasts in positions that make optimal light absorption possible.

2. Opening of stomata. As a result, carbon dioxide can diffuse into cells as blue light triggers photosynthesis.

OTHER BLUE-LIGHT RECEPTORS Phototropins are not the only blue-light receptors.

- The carotenoid pigment zeaxanthin absorbs blue light and, along with phototropins, is involved in the opening of stomata in response to blue light.

- A group of photoreceptors called cryptochromes (literally, "hidden-colors") are blue-light photoreceptors involved in stem growth under shady conditions and the developmental change that leads to flower production.

Phototropism, however, ranks as the best-studied of all blue-light responses. How is the signal from PHOT proteins transmitted to the cells that cause bending?

EXPERIMENT

QUESTION: Does *PHOT1* encode a blue-light receptor?

HYPOTHESIS: The *PHOT1* gene codes for a blue-light receptor.

NULL HYPOTHESIS: The *PHOT1* gene does not code for a blue-light receptor.

EXPERIMENTAL SETUP:

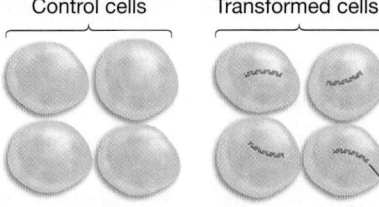

Control cells   Transformed cells

1. Transform cells. Insert *PHOT1* genes into half of insect cells in culture. These cells should produce PHOT1 protein. The other half serve as the control.

PHOT1 gene

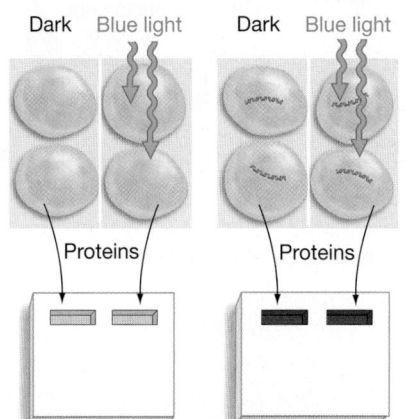

Dark   Blue light   Dark   Blue light

Proteins   Proteins

2. Expose to light. Grow cells in a medium containing radioactive phosphorus and subject cells to darkness or to blue light.

3. Isolate proteins and separate via electrophoresis. See BioSkills 9 for an introduction to this technique.

PREDICTION: Insect cells containing the *PHOT1* gene will have a radiolabeled band when treated with blue light.

PREDICTION OF NULL HYPOTHESIS: Insect cells containing the *PHOT1* gene, like control cells, will not have a radiolabeled band.

RESULTS:

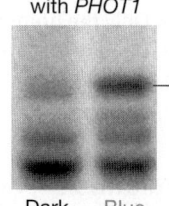

Control insect cells   Insect cells transformed with *PHOT1*

PHOT1 band (appears only if phosphorylation has occurred)

Dark   Blue light   Dark   Blue light

CONCLUSION: Much more PHOT1 is phosphorylated in response to blue light than in darkness. Because no other plant proteins were present in the experimental cells, PHOT1 must phosphorylate itself.

FIGURE 39.4 Experimental Evidence That PHOT1 Is a Blue-Light Receptor That Phosphorylates Itself.

SOURCE: Christie, J. M., P. Reymond, G. K. Powell, P. Bernasconi, A. A. Raibekas, E. Liscum, and W. R. Briggs. 1998. *Arabidopsis* NPH1: A flavoprotein with the properties of a photoreceptor for phototropism. *Science* 282: 1698–1701.

✔**QUESTION** Why did the researchers bother to analyze cells that were not transformed with the *PHOT1* gene and that were not exposed to blue light?

Auxin as the Phototropic Hormone

Long before the phototropins were identified, biologists knew that receptor cells responded to blue light by releasing a hormone. The Darwins established this result when they followed up on their initial experiments. **Figure 39.5** shows their experimental setup and results.

- If they removed the tips of coleoptiles, they found that the decapitated seedlings stopped bending toward the light.
- If they covered the tips of coleoptiles with opaque material, the seedlings did not bend toward light.

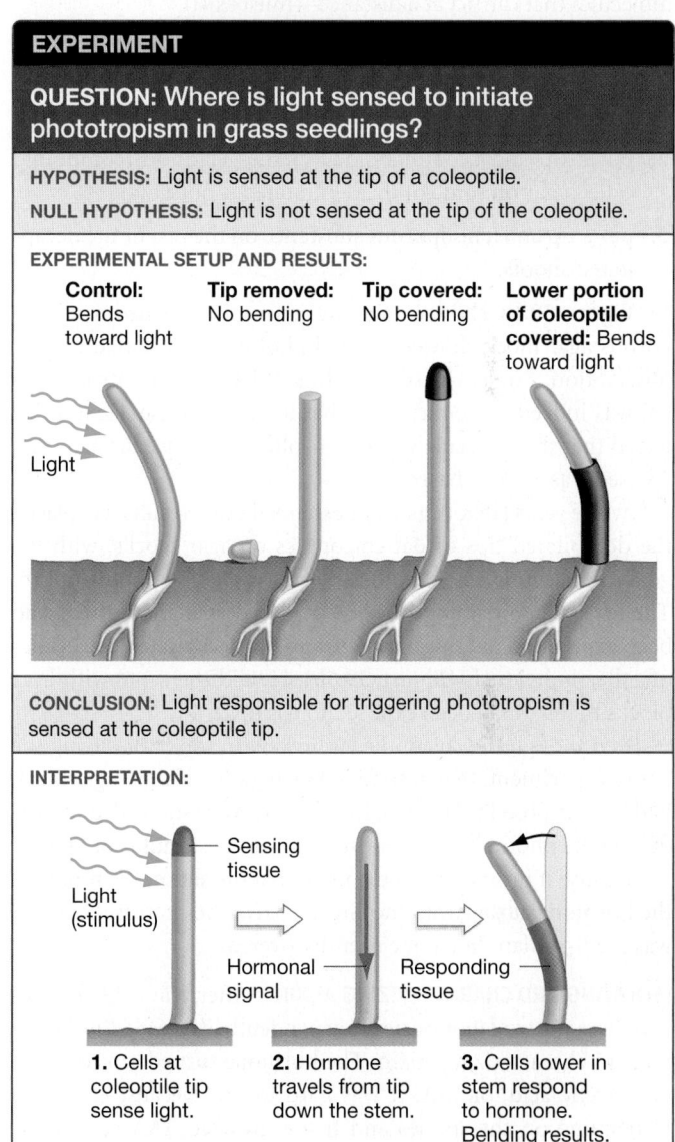

EXPERIMENT

QUESTION: Where is light sensed to initiate phototropism in grass seedlings?

HYPOTHESIS: Light is sensed at the tip of a coleoptile.

NULL HYPOTHESIS: Light is not sensed at the tip of the coleoptile.

EXPERIMENTAL SETUP AND RESULTS:

Control: Bends toward light

Tip removed: No bending

Tip covered: No bending

Lower portion of coleoptile covered: Bends toward light

Light

CONCLUSION: Light responsible for triggering phototropism is sensed at the coleoptile tip.

INTERPRETATION:

Light (stimulus)

Sensing tissue

Hormonal signal

Responding tissue

1. Cells at coleoptile tip sense light.

2. Hormone travels from tip down the stem.

3. Cells lower in stem respond to hormone. Bending results.

FIGURE 39.5 The Sensory and Response Cells Involved in Phototropism Are Not the Same.

SOURCE: Darwin, C. and F. Darwin. 1897. *The Power of Movement in Plants* (D. Appleton & Co. New York).

✔**QUESTION** A critic could argue that this experiment lacked appropriate controls for the treatments labeled "Tip removed" and "Tip covered." Suggest better controls for these treatments than the unmanipulated individual at the far left.

- If they put opaque collars below the tips, in the area where bending occurs, the seedlings bent toward light normally.

These data provided convincing evidence that the blue-light sensors were located in the tips of the coleoptiles.

How did the sensory cells in the tip communicate with the cells that actually elongate? The Darwins proposed that phototropism depends on "some matter in the upper part which is acted on by light, and which transmits its effects to the lower part." Their hypothesis was that a substance produced at the tip of the coleoptile acts as a signal and is transported to the area of bending. This was the first explicit hypothesis stating that hormones—signaling molecules that can act at a distance—must exist.

The hormone hypothesis was not tested rigorously until 1913, when Peter Boysen-Jensen:

1. cut the tips off young oat shoots;

2. put a tip and a porous block of the gelatinous compound called agar on some of the decapitated shoots; and

3. put a tip and a nonporous substance on the rest of the decapitated shoots.

As **Figure 39.6a** shows, only the coleoptiles treated with the porous agar block showed normal phototropism. Based on this observation, Boysen-Jensen concluded that the phototropic signal was indeed a chemical and that it could diffuse. He also inferred that the molecule was water soluble, because the agar that he used was a water-based gelatin.

Twelve years later, Frits Went extended these results. He placed the decapitated tips of oat coleoptiles on agar blocks, with the goal of collecting the hypothesized hormone for phototropism. Then he did something clever: He placed agar blocks that had been exposed to oat tips off-center on the decapitated coleoptiles of other individuals (**Figure 39.6b**). He did the same with agar blocks that had *not* been exposed to oat tips.

Even though the coleoptiles were kept in the dark during the entire experiment, they responded by bending if their agar block had been exposed to oat tips. In this way, Went succeeded in producing the phototropic response without the stimulus of light.

Because it promotes cell elongation in the shoot, Went named the hormone **auxin** (from the Greek *auxein*, "to increase"). Auxin was the first plant hormone ever discovered.

ISOLATING AND CHARACTERIZING AUXIN After years of effort, researchers in two laboratories independently succeeded in isolating and characterizing auxin. The hormone turned out to be indole acetic acid, or IAA. It was hard to find because it is rare. Depending on the species and tissue involved, IAA concentrations range from 3 to 500 nanograms per gram of tissue. (The prefix *nano–* refers to billionths.)

Like the other plant hormones introduced in this chapter, auxin is a small molecule with a relatively simple structure. It is present in quantities so small that its concentration is difficult to measure. Yet its impact is huge. Auxin can bend stems—producing, in some cases, tree trunks that are permanently bowed.

(a) The phototropic signal is a chemical.

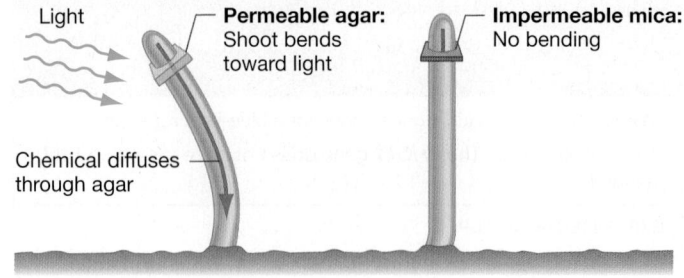

(b) The hormone can cause bending in darkness.

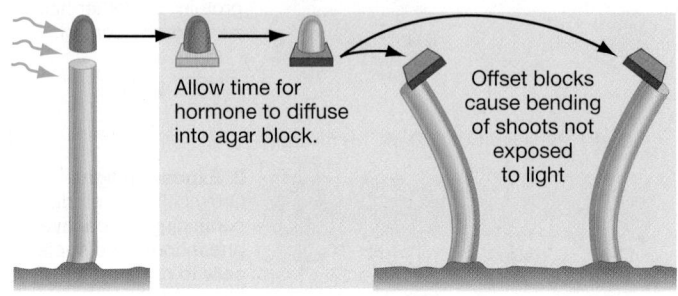

(c) The hormone causes bending by elongating cells.

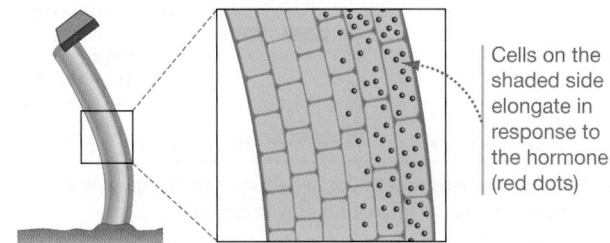

Cells on the shaded side elongate in response to the hormone (red dots)

FIGURE 39.6 Experimental Evidence Supports the Hormone Hypothesis for Phototropism. **(a)** Coleoptiles bend in response to light if substances from the tip are allowed to move downward. **(b)** If bending can take place in darkness, then light is not directly required for the response. Only the hormone is required. **(c)** During the phototropic response, bending occurs because cells on the shaded side of the shoot elongate.

THE CHOLODNY-WENT HYPOTHESIS Went's experiments were a breakthrough in research on information processing: They confirmed the hormone hypothesis and led to the discovery and characterization of IAA. But Went's experiments also inspired an important hypothesis for *how* the hormone produces the bending response. Working independently, both N. O. Cholodny and Went proposed that phototropism results from an asymmetric distribution of auxin.

The Cholodny-Went hypothesis contends that:

1. Auxin produced in the tips of coleoptiles is shunted from one side of the tip to the other in response to light.

2. Auxin is then transported straight down one side of the shoot.

3. The asymmetric distribution of auxin causes cells on the shaded side of the coleoptile to elongate more than cells on the illuminated side (**Figure 39.6c**).

Bending results. In essence, bending results from differential elongation.

To test the Cholodny-Went model, Winslow Briggs divided coleoptile tips completely or partially in half with a thin piece of mica, then exposed one side to sunlight (**Figure 39.7**). Mica is impermeable to dissolved molecules. Briggs's idea was that the movement of auxin would be stopped in the completely divided tips and agar blocks, but it would not be stopped in the partially divided tips and agar blocks.

Next, Briggs placed the resulting blocks on one side of decapitated shoots and recorded the bending response.

- If the tip was completely divided, there was no difference in bending induced by the sunlit or shaded side of the tip.

- If the tip was partially divided, the bending responses differed in the sunlit and shaded sides. The side away from light induced much more bending, indicating that auxin had been transported from one side of the tip to the other.

The Cholodny-Went asymmetric distribution model was correct. It explained how auxin leads to asymmetric cell elongation, and thus the bending response called phototropism.

THE CELL-ELONGATION RESPONSE   How do cells in the stem respond to auxin? Experiments in corn plants and *Arabidopsis* suggest that proteins called TIR1 and auxin-binding protein 1, or ABP1, are auxin receptors found in stem and leaf cells. Researchers proposed that once TIR1 or ABP1 has bound to auxin, the signal transduction cascade that follows increases the number of membrane H^+-ATPases, or proton pumps, in the plasma membrane.

Recall from Chapter 38 that **proton pumps** use the energy in ATP to drive protons out of the cell against an electrochemical gradient. Because the pH of the cell wall decreases when H^+-ATPases are active, the idea that these pumps are responsible for cell elongation became known as the **acid-growth hypothesis**.

To understand the rationale behind the acid-growth hypothesis, it's important to realize that two things have to happen for a plant cell to get larger:

1. The cell wall has to expand to create a larger volume.

2. Water has to enter the cell and generate turgor pressure on the cell wall to make an increase in volume possible.

FIGURE 39.7 Testing the Auxin Redistribution Hypothesis.

SOURCE: Baskin, T. I., M. Iino, P. G. Green, and W. R. Briggs. 1985. High-resolution measurements of growth during first positive phototropism in maize. *Plant, Cell & Environment* 8: 595–603.

✔**QUESTION** Consider the hypothesis that cutting coleoptile tips and inserting mica sheets disrupts normal cell activity. What prediction does this hypothesis make?

EXPERIMENT

QUESTION: How does an asymmetric distribution of auxin in shoot tips, which causes bending, develop?

HYPOTHESIS: Auxin moves from the sunny side to the shady side of shoot tip, resulting in an asymmetric distribution.

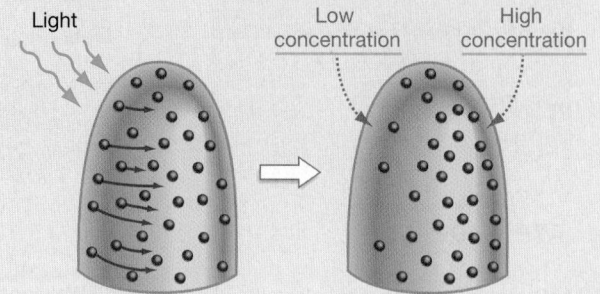

Light

Low concentration — High concentration

NULL HYPOTHESIS: Auxin does not move from the sunny side to the shady side of the shoot tip.

EXPERIMENTAL SETUP:

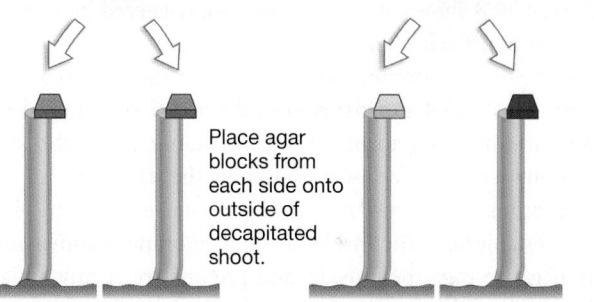

Light — Mica sheet completely divides tip

Light — Mica sheet partially divides tip

Place agar blocks from each side onto outside of decapitated shoot.

PREDICTION: In completely divided tip, both agar blocks will elicit the same degree of bending. In partially divided tip, block receiving the most light will elicit less bending.

PREDICTION OF NULL HYPOTHESIS: The same degree of bending will occur in all treatments.

RESULTS:

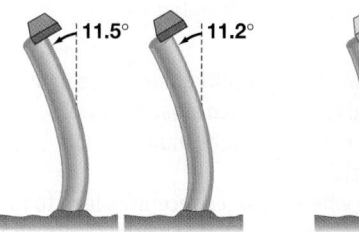 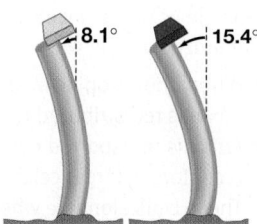

11.5°   11.2°   8.1°   15.4°

Interpretation: Mica prevented flow of auxin to shaded side, and shoots bent the same amount.

Interpretation: Auxin was redistributed to shaded side, which bent more than sunny side.

CONCLUSION: Asymmetric distribution of auxin results from lateral redistribution in tip.

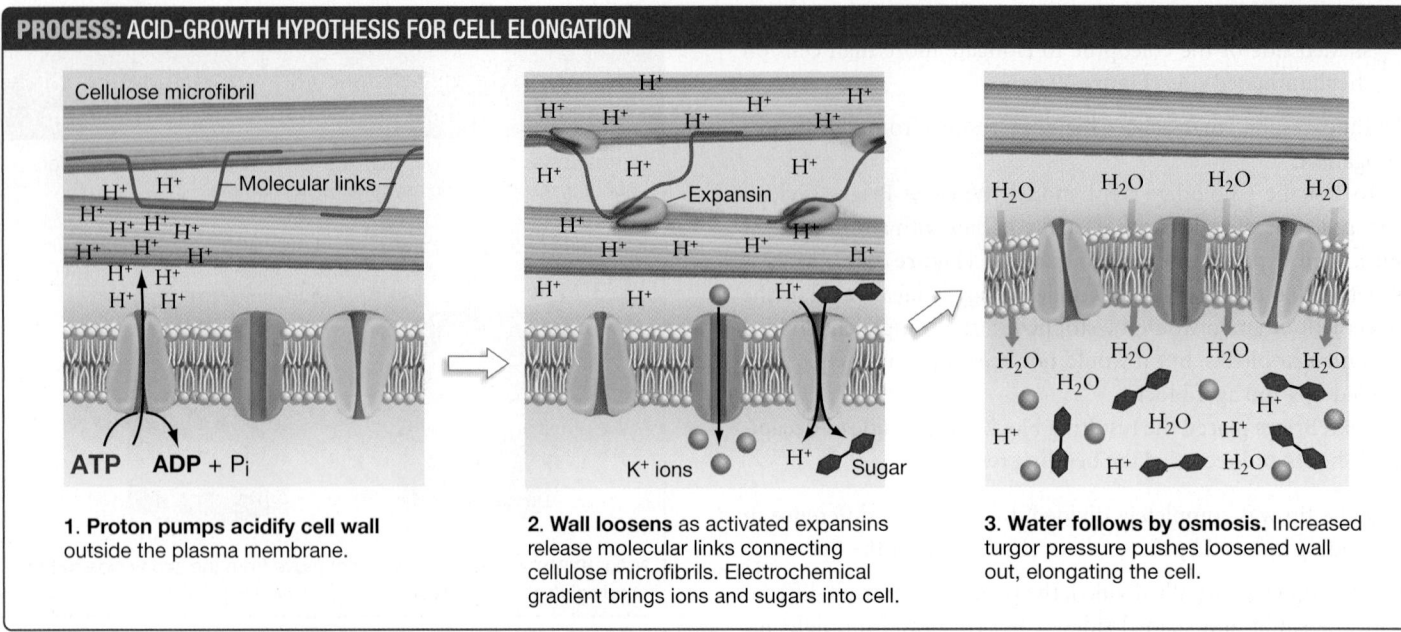

PROCESS: ACID-GROWTH HYPOTHESIS FOR CELL ELONGATION

Cellulose microfibril — Molecular links

ATP → ADP + P$_i$

1. Proton pumps acidify cell wall outside the plasma membrane.

Expansin

K$^+$ ions — H$^+$ — Sugar

2. Wall loosens as activated expansins release molecular links connecting cellulose microfibrils. Electrochemical gradient brings ions and sugars into cell.

H$_2$O

H$^+$

3. Water follows by osmosis. Increased turgor pressure pushes loosened wall out, elongating the cell.

FIGURE 39.8 The Acid-Growth Hypothesis Requires Expansion of Cell Wall and Intake of Water. When activated proton pumps lower the pH outside the cell membrane in response to auxin, a series of events leads to elongation of the cell.

As **Figure 39.8** shows, both processes are triggered by pumping protons into the cell wall.

When proton pumping lowers the pH of the wall to 4.5, cell-wall proteins called **expansins** are activated. Expansins "unzip" molecular links that form between cellulose microfibrils and other polymers in the cell wall, loosening the structure.

As protons are pumped out of the cell, an electrochemical gradient is established. The inside of the membrane becomes much more negative than the outside, and protons are at much higher concentrations outside than inside. The gradient favors the entry of potassium (K$^+$) or other positively charged ions, as well as

sugars that enter via proton cotransporters. As the concentration of solutes increases inside the cell, water follows via osmosis.

This is a key point: Cells don't move water directly. Instead, they create an osmotic gradient that favors water movement. Pumping protons *out* of a cell is a way to bring water *into* the cell.

The incoming water increases turgor pressure, which pushes out the loosened cell wall. The cell gets bigger.

The upshot? When cell walls on one side of a stem are acidified in response to a signal from auxin, bending results. Auxin's role in phototropism may qualify as the best-understood example of information processing in plants.

CHECK YOUR UNDERSTANDING

If you understand that . . .

The chain of events involved in phototropism can be summarized as follows:

- When phototropins in shoot-tip cells absorb blue light, auxin is redistributed to the shaded side of the tip.
- Auxin is transported down the shoot and binds to receptors in target cells.
- These cells elongate when activated receptors lead to the activation of proton pumps, acidification of the cell wall, and activation of expansin proteins.

✔ **You should be able to . . .**

Assuming that you have a large supply of purified auxin, state how you would manipulate a large bed of roses so their stems bend toward the east.

Answers are available in Appendix B.

39.3 Red and Far-Red Light: Germination and Stem Elongation

Plants are sensitive to wavelengths in the red and far-red portions of the visible spectrum, as well as to blue light. This sensitivity is interesting, because red wavelengths (about 660 to 700 nm) and far-red wavelengths (over 710 nm) signal very different things to a plant.

- Red light drives photosynthesis, just as blue light does.
- Far-red wavelengths are not absorbed strongly by photosynthetic pigments, so they tend to pass through leaves. As a result, far-red wavelengths are prominent in light that is filtered through tree leaves before it reaches the forest floor. Far-red light indicates shade.

To review experimental work on the differences between red and far-red light, go to the study area at *www.masteringbiology.com*.

 Web Activity Sensing Light

The Red/Far-Red "Switch"

The first hint that plants monitor red and far-red light emerged from studies on how lettuce seeds germinate. By exposing lettuce seeds to various wavelengths of light and plotting their frequency of germination, researchers discovered that germination rates peak when seeds receive red light (about 660 nm).

This observation made sense, because lettuce thrives best when it grows in bright sunlight. But the stimulatory effect of red light disappeared if seeds were later exposed to far-red light.

This observation also made sense, because far-red light indicates that the seeds are shaded. Wavelengths near 735 nm inhibit germination the most effectively.

Follow-up experiments showed that red and far-red light act like an on-off switch for lettuce seed germination (**Table 39.1**). Red light promotes lettuce germination; far red inhibits it.

The key observation, though, is that the last wavelength sensed by the seed determines whether germination occurs at a high rate. This result implies that plants sense red and far-red light *together*. How could this happen?

Phytochromes as Red/Far-Red Receptors

To interpret the red/far-red switch in seed germination, biologists hypothesized that the same pigment absorbs both wavelengths. Further, they suggested that the pigment exists in two shapes, or conformations: one shape absorbs red light, and one shape absorbs far-red light.

The idea was that switching behavior, or **photoreversibility**, occurs because light absorption makes the photoreceptor pigment change shape, like a light switch moving up or down in response to touch. Each conformation would be responsible for a different response.

TABLE 39.1 **How Do Red Light and Far-Red Light Affect the Germination of Lettuce Seeds?**

Biologists exposed moistened lettuce seeds to flashes of light containing one of two wavelengths: red or far-red (FR). After exposure to light, the seeds were held in the dark for several days.

Light Exposure	Germination (%)
None (control)	9
Red	98
Red → FR	54
Red → FR → Red	100
Red → FR → Red → FR	43
Red → FR → Red → FR → Red	99
Red → FR → Red → FR → Red → FR	54
Red → FR → Red → FR → Red → FR → Red	98

SOURCE: H. A. Borthwick et al. 1952. A reversible photoreaction controlling seed germination. *PNAS* 38: 662–666, Table 1.

✔**QUESTION** According to the data above, what is the average germination rate of lettuce seeds that were last exposed to red light? To far-red light? How do these values compare with the germination rate of seeds that are buried underground and receive no light at all?

Biologists called the hypothesized pigment **phytochrome** ("plant-color"). Phytochrome was thought to be a specialized light receptor, different from any of the pigments involved in absorbing light during photosynthesis.

Figure 39.9 illustrates the photoreversibility hypothesis. One conformation of phytochrome called P_r (phytochrome red) absorbs red light. Another conformation of the same molecule, called P_{fr} (phytochrome far-red), absorbs far-red light. According to the photoreversibility hypothesis, each conformation switches to the other when it absorbs its preferred wavelength.

How Were Phytochromes Isolated?

Young corn and bean plants lengthen their stems in response to light deprivation or exposure to excessive far-red light. If you have grown these seeds indoors, you have seen this response to far-red light firsthand. The shoots get long and spindly.

Corn and beans normally grow in open sunlight. If they are grown under incandescent light indoors, the plants react as though they are being shaded—they attempt to grow high enough to reach full sunlight. This behavior suggested that young corn and bean plants have a receptor protein for far-red light.

To follow up on this observation, researchers purified proteins from corn coleoptiles and succeeded in isolating one that was photoreversible. Specifically, when the protein was placed in solution and exposed to alternating red and far-red light, the color of the solution switched from blue to blue-green and back. The color switches supported the hypothesis that the same protein absorbed red as well as far-red light. Phytochrome was found.

Follow-up work showed that a region of the phytochrome molecule changes shape in response to red and far-red light. The

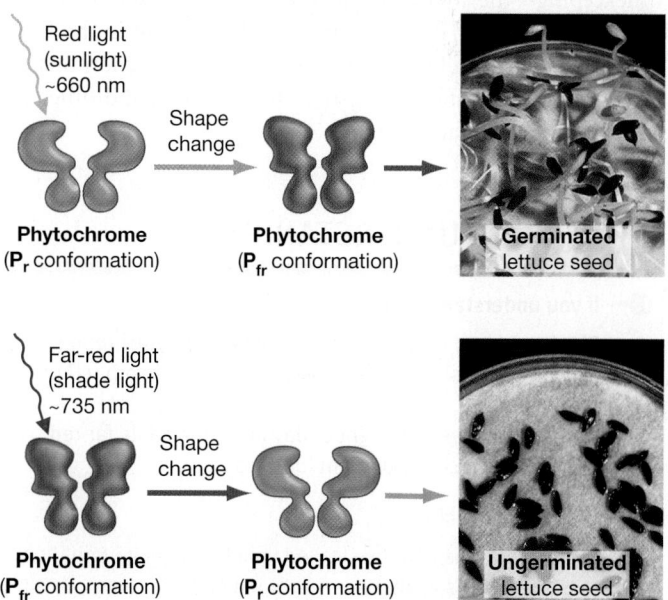

FIGURE 39.9 The Photoreversibility Hypothesis for Phytochrome Behavior. According to the photoreversibility hypothesis, phytochrome switches between the P_r conformation and the P_{fr} conformation when it absorbs red or far-red light respectively. Each form of the protein has a different effect on germination.

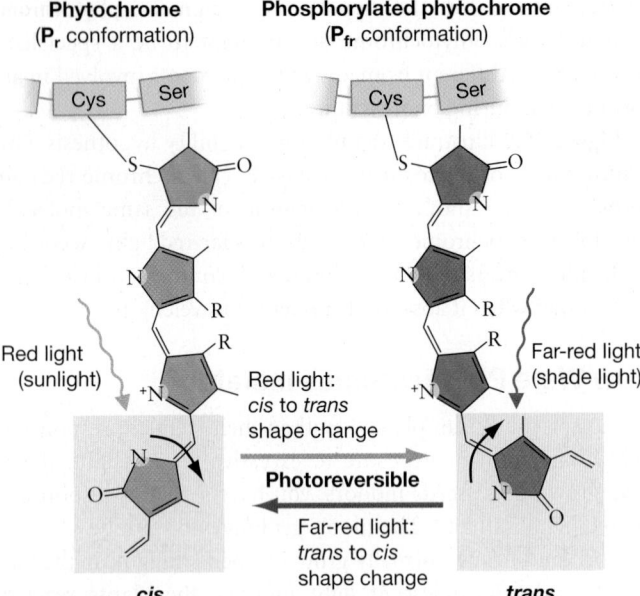

Phytochrome
(P_r conformation)

Phosphorylated phytochrome
(P_fr conformation)

Red light
(sunlight)

Red light:
cis to *trans*
shape change

Photoreversible

Far-red light:
trans to *cis*
shape change

Far-red light
(shade light)

cis

trans

FIGURE 39.10 Phytochrome Changes Shape after Absorbing Red and Far-Red Light. This diagram shows how a subunit of phytochrome changes shape in response to absorbing red or far-red light. The shape changes correspond to phytochrome's P_r and P_{fr} conformations.

✔ **QUESTION** How does this shape change relate to signal transduction?

shape changes cause phytochrome to take on or lose a phosphate group, just as the shape changes triggered by blue light result in phosphorylation of the phototropins. Conformational changes in proteins are frequently associated with a change in phosphorylation.

As Chapter 40 will show, phytochrome also plays a key role in timekeeping—the ability of plants to track changes in the length of nights and days. How is the information present in the P_r/P_{fr} switch, illustrated in **Figure 39.10**, translated into action that affects germination, stem elongation, timekeeping, and other responses to red and far-red light?

CHECK YOUR UNDERSTANDING

If you understand that . . .

- Plants respond to red and far-red light via phytochromes, which change shape and activity when they absorb red or far-red light.
- Red light acts as a sunlight or day indicator, while far-red light acts as a shade or night indicator.

✔ **You should be able to . . .**

1. Explain why it is adaptive for red light to trigger germination in lettuce seeds while far-red light inhibits it.
2. Explain why it is adaptive for plants that normally grow in sunny habitats to elongate their stems in response to far-red light.

Answers are available in Appendix B.

This question brings us to the forefront of research on phytochromes. The answer appears to be complex; one study suggests that almost one-third of all *Arabidopsis* genes are involved in the elongated growth characteristic of shade avoidance. Work on linking the phytochrome switch to the elongation and germination responses continues.

39.4 Gravity: The Gravitropic Response

The wavelengths, quantity, and direction of light that a plant receives change with the season, weather, time of day, and shading by other plants. However, gravity is constant and unidirectional. Light means food; gravity provides information about how the plant should orient itself in space.

Shoots usually respond to gravity by growing in an upward direction; roots usually respond by growing downward or laterally. How do plants sense gravity, so that it can be used as a signal to orient the body?

In 1881 Charles and Francis Darwin published one of the first experimental results on **gravitropism** ("gravity-turn")—the ability to move in response to gravity. Recall from Chapter 36 that the ends of root tips are covered by a protective collection of cells called the **root cap**. The Darwins found that roots stop responding to gravity if the root caps are removed. This observation suggested that gravity sensing occurs somewhere inside the root cap.

Recently biologists demonstrated precisely which cells are involved in gravity sensing in *Arabidopsis* roots. By killing tiny blocks of cells with laser beams, researchers showed that the cells illustrated in **Figure 39.11**, at the center of the root cap, are the most important for regulating the gravitropic response. Cells in the tips of roots respond to gravity and initiate gravitropism. How do root cap cells sense the force of gravity?

The Statolith Hypothesis

The leading explanation for how plants sense gravity—the **statolith hypothesis**—is based on two interconnected ideas:

(a) Root tips have a protective cap.

(b) Gravity-sensing cells are in the center of the cap.

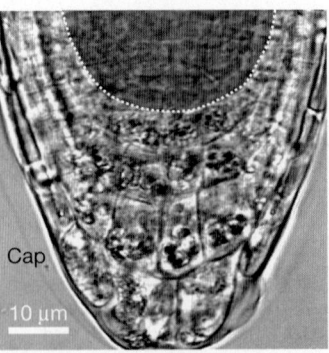

Cap

10 μm

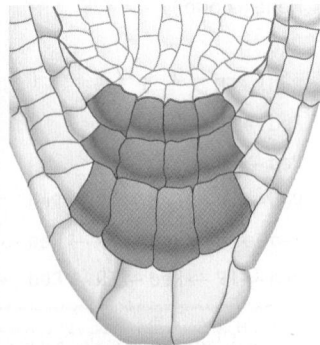

FIGURE 39.11 Gravity-Sensing Occurs in the Root Cap. The root cap is a protective structure. When the cells marked in orange are killed experimentally, the gravitational response in roots is dramatically reduced.

1. Dense, starch-storing organelles called **amyloplasts** respond to gravity by being pulled to the bottom of root cap cells (**Figure 39.12**).

2. The weight of the amyloplasts activates sensory proteins located in the plasma membrane. These sensory proteins initiate the gravitropic response.

The statolith hypothesis was inspired by animals that use dense particles to sense gravity. Lobsters, for example, take up grains of sand that become positioned in specialized gravity-sensing organs in their antennae. The grains of sand are called **statoliths** ("place-stones"). When the animal tilts or flips over, the statolith moves in response to gravity. Inside the organ, the sand grain ends up pushing against a sensory cell. When this cell is activated, it indicates that the animal is no longer upright.

According to the statolith hypothesis, the same thing happens in root cap cells. If the wind tips a plant over, for example, the amyloplasts settle onto the new "lower" cell walls. The weight activates a new set of receptors, which signal that the root no longer faces in the correct direction.

Although recent experiments strongly support the statolith hypothesis, the search for the gravity receptor itself continues. In contrast, the second and third steps of information processing in response to gravity—the production of a cell-cell signal and the response of target cells—is much better understood.

Auxin as the Gravitropic Signal

Root cap cells that sense changes in the direction of gravitational pull respond by changing the distribution of auxin in the root tip. **Figure 39.13** illustrates this chain of events.

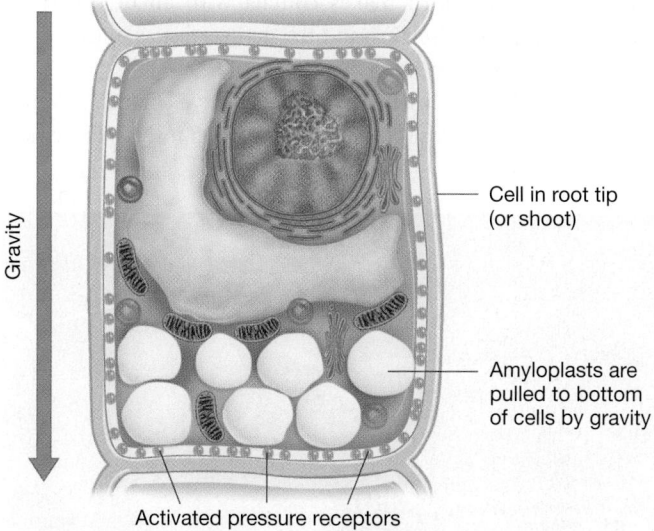

FIGURE 39.12 The Statolith Hypothesis States That Amyloplasts Stimulate Sensory Cells. Amyloplasts are filled with starch. They are dense, so they sink in response to gravity. The statolith hypothesis predicts that pressure receptors in the plasma membrane become activated as a result.

✔**EXERCISE** Assuming that the receptor is a protein, suggest a hypothesis to explain how signal transduction occurs in response to pressure from a statolith.

Step 1 Under normal conditions, auxin flows down the middle of the root and then toward the perimeter and away from the root cap.

Step 2 If the root is tipped, sensory receptors trigger changes in transport proteins that redistribute auxin.

Step 3 Auxin is redistributed: The lower portion of the root receives increased concentrations of auxin; the upper portion receives lower concentrations. The redistribution is mediated by changes in the location of auxin-binding PIN proteins.

Step 4 In response to the differences in auxin concentration, cells in the lower portion of the root grow more slowly and cells in the upper portion grow more quickly. In roots, high auxin concentrations inhibit growth. The result is bending.

Note that the way root cells respond to auxin redistribution during the gravitropic response is opposite to the way that cells in the stem respond during phototropism. In stems, high concentrations of auxin lead to *increased* cell elongation and bending. In roots, high concentrations of auxin lead to *decreased* cell division and elongation. Roots bend as cells on the other side of the zones of cellular division and elongation

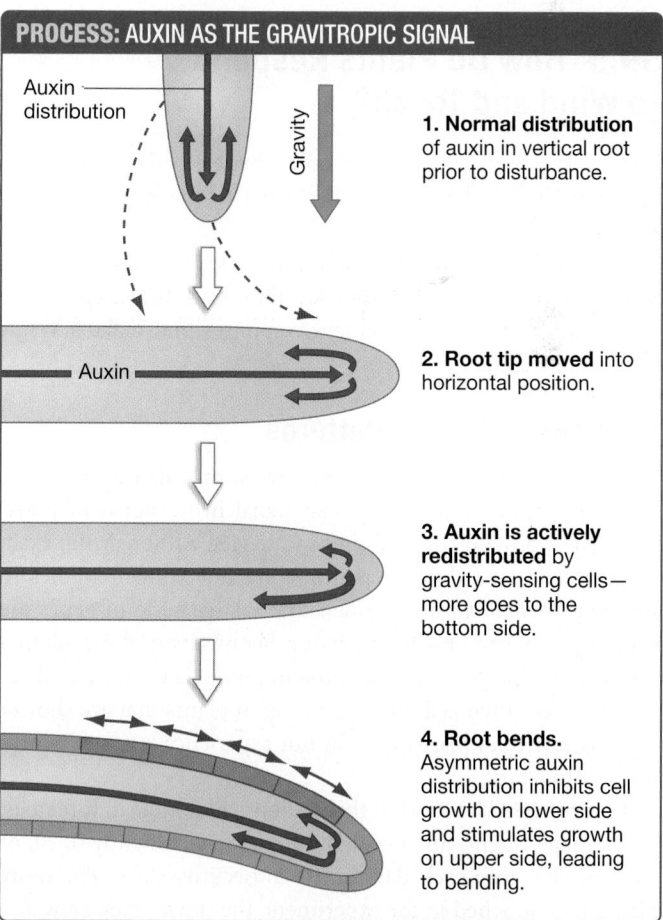

FIGURE 39.13 The Auxin Redistribution Hypothesis for Gravitropism. This sequence of events might begin when a growing root tip hits a rock and is displaced horizontally or when a plant is tipped in a windstorm and partially uprooted.

(see Chapter 36) continue to grow, in response to low auxin concentrations. The mechanism responsible for this difference is still unknown.

To review how auxin is redistributed during phototropism and gravitropism, go to the study area at *www.masteringbiology.com*.

(MB) **Web Activity** Plant Hormones

CHECK YOUR UNDERSTANDING

If you understand that . . .

- Cells in root caps sense gravity via pressure that amyloplasts exert on receptors.
- Changes in gravity sensing result in a redistribution of auxin and changes in the growth rate of root tips.

✔ **You should be able to . . .**

1. Explain the parallel that exists between auxin redistribution in shoots and roots and the responses called phototropism and gravitropism.
2. Explain why auxin could be considered a gravitropic hormone.

Answers are available in Appendix B.

39.5 How Do Plants Respond to Wind and Touch?

Light and gravity are not the only physical forces that plants sense and respond to. Plants also react to mechanical stresses such as wind and touch.

Let's analyze two of the best-studied responses to pushing or pulling forces: growth responses that lead to exceptionally stout, stiff stems, and movement responses that make a Venus flytrap snap shut.

Changes in Growth Patterns

When plants are buffeted by wind, receptor cells transduce the mechanical force into an internal signal in the form of phosphorylated proteins or a second messenger. Although the exact receptor and transduction pathway are not known, studies in *Arabidopsis thaliana* have shown that a large suite of genes are transcribed in response to touch or other mechanical stimuli that mimic the effect of wind. The protein products of some of these genes act to stiffen cell walls, resulting in plants that are shorter and stockier than plants that do not experience repeated vibrations or touching.

Figure 39.14 shows what this response looks like in tomatoes. In this experiment, tomato plants were touched lightly 0, 10, or 20 times per day, each day for 10 consecutive days. The more plants were touched in the experiment, the slower they grew and the stockier they became.

In response to wind, then, plants change their growth patterns in ways that make them more likely to withstand the force, stay upright, and live long enough to produce flowers and fruit.

Number of touchings per day

FIGURE 39.14 Plant Growth Changes in Response to Wind or Touch. The tomato plants shown here were touched lightly 0, 10, or 20 times per day each day for 10 consecutive days.

✔ **QUESTION** What is the adaptive significance of this response? Give an example of when it would occur in nature.

Movement Responses

In some cases, plants respond to touch by moving. This response, **thigmotropism** ("touch-bending"), can be fast. For example, species that grow by climbing objects or other plants may have modified leaves or stems that form long, thin structures called tendrils. When a tendril makes contact with an object, it responds by wrapping itself around the item as fast as one or more times per hour (**Figure 39.15**).

Movement is even faster in "touch-sensitive" plants. A Venus flytrap, for example, closes fast enough to catch insects. Extremely

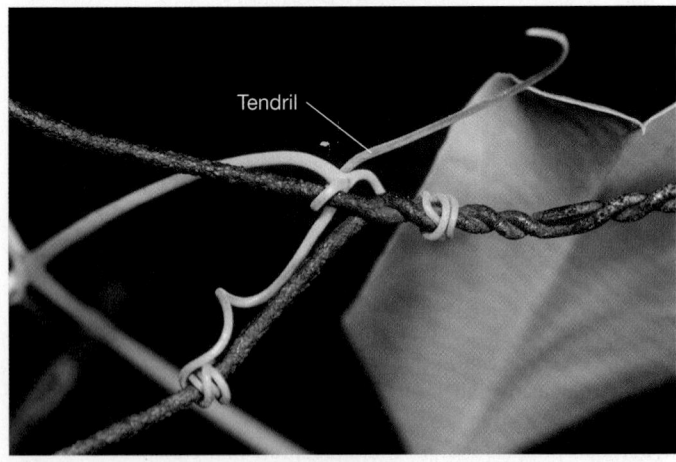

Tendril

FIGURE 39.15 Thigmotropism Is Movement in Response to Touch. Portions of pea plants called tendrils wind around support structures after contacting them. The attachment provided by the tendrils allows peas to climb.

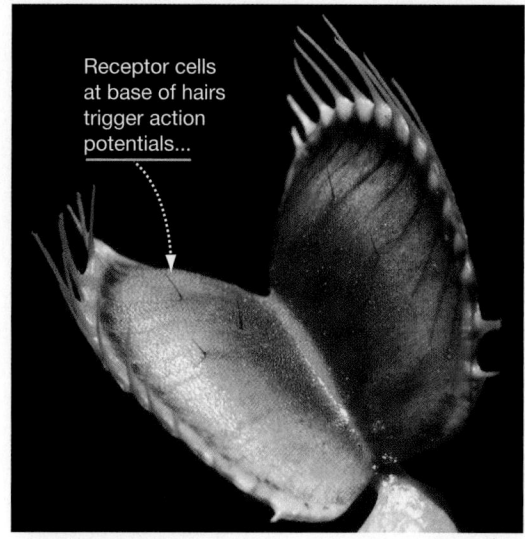

Receptor cells at base of hairs trigger action potentials...

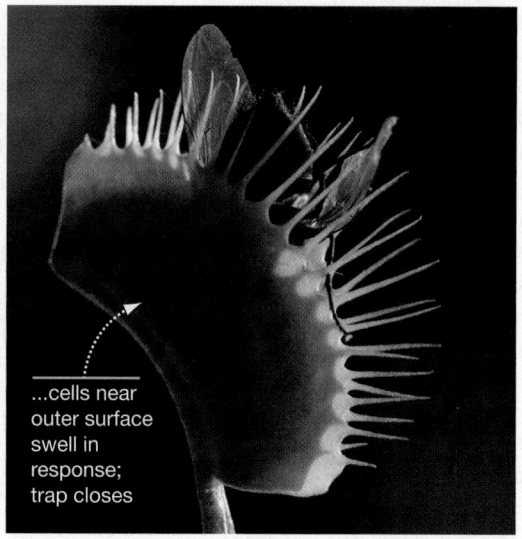

...cells near outer surface swell in response; trap closes

FIGURE 39.16 Venus Flytraps Close in Response to Action Potentials Generated by Sensory Hairs.

rapid movements like this are possible when a touch-receptor cell transduces the mechanical signal to an electrical signal.

To understand how plants use electrical signaling, recall from Chapter 38 that proton pumps in the plasma membrane of most plant cells give them a negative charge relative to the exterior environment. The separation of charges creates a membrane voltage, or **membrane potential**. If a receptor protein responds to touch by allowing ions to flow across the membrane—which changes the amount of charge on either side—then the membrane potential changes. In this way, the mechanical signal (touch) can be transduced to an electrical signal.

To travel from a sensory cell to a response cell, electrical signals are propagated in a characteristic form called an action potential. (Chapter 45 analyzes the action potentials that occur in animal nerve cells in detail.) In a Venus flytrap, action potentials race across the leaf at a rate of about 10 cm/sec.

When the action potentials reach cells on the outer surface of the trap, the cells change shape and push the trap shut (**Figure 39.16**). The response is rapid enough to resemble the way an animal's muscle contracts in response to an action potential.

Although biologists have been fascinated by the speed of the trap closing for decades, the molecular mechanism involved remains controversial. Rapid water loss from cells along the inside of the hinge is the most probable cause, but how water moves in response to the change in membrane voltage remains a mystery.

39.6 Youth, Maturity, and Aging: The Growth Responses

Plants grow throughout their lives, from the time they germinate until the time they die. As the plant body grows, it matures into an efficient machine for absorbing sunlight, water, nutrients, and other diffuse resources. But growth is not constant. It speeds up when water, light, and nutrients are abundant and slows or stops when conditions are poor, or when it is time for leaves to drop or fruits to ripen.

Controlling growth in response to changes in age or environmental conditions is one of the most important aspects of information processing in plants. Hormones play a key role in regulating growth.

To explore how plants grow in response to changing conditions, let's consider five of the best-studied hormones involved in growth responses. We'll start with auxin—the molecule responsible for phototropism and gravitropism.

Auxin and Apical Dominance

When **apical dominance** occurs, growth is restricted to the main stems, and the lateral buds in the axils of each leaf remain dormant. But if the apical bud dies, the dormancy of the lateral buds is broken and lateral branches begin to grow. **Figure 39.17** shows this phenomenon in action.

(a) Apical meristem intact

(b) Apical meristem cut off

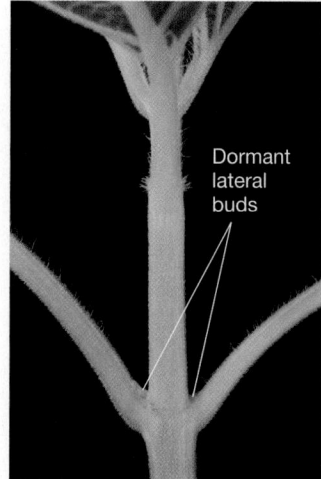

Dormant lateral buds

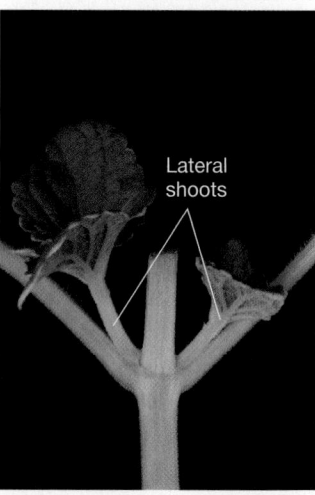

Lateral shoots

FIGURE 39.17 When Apical Dominance Occurs, Growth of Lateral Buds Is Suppressed. (a) The stem of a coleus plant is shown still intact. **(b)** The same plant, several weeks after the stem was cut. The lateral shoots will orient themselves vertically.

What caused the change? Because auxin is produced in shoot tips, researchers suspected that it might have a role in apical dominance as well as phototropism. This hypothesis was confirmed when it was shown that apical dominance could be sustained by adding auxin to a shoot's cut surface after its tip had been removed.

Auxin's role in apical dominance suggests that tip cells send a constant stream of information down to other organs and tissues. If the signal stops, it means that apical growth has been interrupted. In response, lateral branches sprout and begin to take over for the main shoot. Now the question is, How does this signal move?

POLAR TRANSPORT OF AUXIN Auxin transport is **polar**, or unidirectional. If radioactively labeled auxin is added to the top of a cut stem, the hormone is transported toward the base. But if labeled auxin is added to the base of a cut stem, it is not transported toward the apex. Auxin is the only plant hormone known to be transported in one direction only.

Studies with labeled auxin have also shown that the hormone is transported all the way down the stem through the root, via parenchyma cells in the ground tissue and vascular tissue. Auxin enters the apical end of cells via a specialized membrane protein, diffuses to the other end of the cell, and then is transported out by carrier proteins located in the basal portion of the plasma membrane. Labeled auxin moves from cell to cell at about 10 cm/hr—approximately 10 times more slowly than substances traveling in the phloem or xylem.

Because enzymes destroy some auxin molecules as they travel down the long axis of the plant, polar transport sets up a strong gradient in auxin concentration. Auxin concentrations are much higher in shoots than they are in roots.

WHAT IS AUXIN'S OVERALL ROLE? Auxin clearly plays a key role in controlling growth via apical dominance, phototropism, and gravitropism. But this chemical messenger has other important effects as well:

- Auxin produced by seeds within the fruit influences fruit development.

- Falling auxin concentrations are involved in the **abscission**, or shedding of leaves and fruits, associated with the genetically programmed aging process called senescence.

- The presence of auxin in growing roots and shoots is essential not only for the proper differentiation of xylem and phloem cells in vascular tissue but also for the development of vascular cambium.

- Auxin stimulates the development of adventitious roots in tissue cultures and cuttings.

Auxin has so many different effects on plants that it has been difficult for biologists to understand its overall role. Recently, several investigators have proposed that auxin's overall function is to signal where cells are in space. The idea is that auxin concentration identifies where a cell is located relative to the long axis of the plant body—the axis that runs from shoot to root.

If conditions relating to the long axis change—for instance, a windstorm tips the plant or a deer eats the shoot apex—changes in auxin concentration effectively signal how the individual's tissues should respond. Phototropism, gravitropism, apical dominance, and the production of adventitious roots are all ways of coping with changes in the long axis of the plant.

Cytokinins and Cell Division

Cytokinins are a group of plant hormones that promote cell division. (*Cyto* is the Greek root for cell; *kinin* refers to kinesis, meaning "movement" or "division.")

THE DISCOVERY OF CYTOKININS When biologists were first attempting to grow plant cells and embryos in culture, they found that coconut milk, which stores nutrients used by growing coconut embryos, promoted cell division. This was the first hint that certain molecules can promote cell division in plants. Later experiments showed that molecules derived from the nitrogenous base adenine also stimulate the growth of cells in culture.

Eventually, naturally occurring adenine derivatives that stimulate growth were discovered in corn and apples, and were named cytokinins. Zeatin, the cytokinin that has been found in the most species, is derived from adenine.

Cytokinins are synthesized in root tips, young fruits, seeds, growing buds, and other developing organs. But most of the zeatin and other cytokinins that are active in plants are synthesized in the apical meristems of roots and transported up into the shoot system via the xylem. Biologists still add cytokinins to plant cells growing in culture, to stimulate cell division (see **BioSkills 12** in Appendix A).

HOW DO CYTOKININS PROMOTE CELL DIVISION? After years of searching, a group of closely related proteins that act as cytokinin receptors has now been isolated and characterized. When cytokinins bind to these receptors in the plasma membranes of target cells, the receptors activate genes that regulate cell division.

Recent research on cytokinins has explored whether they affect molecules that regulate the cell cycle. Chapter 11 introduced some of these cell-cycle regulators, including the cyclins and the cyclin-dependent kinases (Cdks). Recall that activated cyclins and Cdks allow cells to progress through checkpoints in the cell cycle and continue dividing.

To assess whether cytokinins affect cell-cycle genes, researchers grew *Arabidopsis* cells in culture so the nutrients and other molecules available could be carefully controlled. The biologists starved the cells of cytokinins for a day, then added the hormones again to half of the cells. When they assessed the level of mRNA from a cyclin gene called *CycD3*, they documented significant increases in the cells that were exposed to cytokinins again compared with the level in cells that were not reexposed to cytokinins.

This is strong evidence that cytokinins regulate growth by activating genes that keep the cell cycle going. In the absence of cytokinins, cells arrest at the G_2 checkpoint in the cell cycle and stop dividing (**Figure 39.18**).

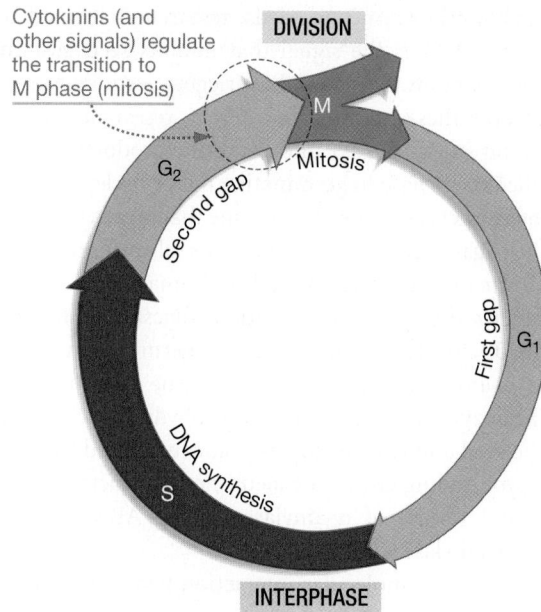

Cytokinins (and other signals) regulate the transition to M phase (mitosis)

DIVISION

M

Mitosis

G₂

Second gap

First gap

G₁

DNA synthesis

S

INTERPHASE

FIGURE 39.18 Cytokinins Affect the Cell Cycle.

Gibberellins and ABA: Growth and Dormancy

In high latitudes and at high elevations, most seeds and mature plants start growing in spring. Conditions for growth are good at that time of year, because temperatures are warming and soil moisture levels are usually at their peak. Seedlings and mature plants continue to grow throughout the summer and early fall if moisture and nutrients are still available.

During drought conditions, however, growth stops. Growth also stops in the embryos inside seeds. Embryos begin to develop as seeds mature, but cease this initial growth and remain dormant throughout the cold winter months. **Dormancy** is a temporary state of reduced metabolic activity or no metabolic activity.

Which signals initiate growth in response to changing environmental conditions, and which signals stop it? Two hormones provide the answer. **Gibberellins**, a large family of closely related compounds, stimulate growth in plants. **Abscisic acid**, commonly abbreviated **ABA**, inhibits growth. In at least some cases, the two hormones interact like start and stop signals.

THE DISCOVERY OF GIBBERELLINS Over 100 years ago, Japanese farmers noticed that some of their rice seedlings grew exceptionally quickly but fell over before they could be harvested. Biologists found that the diseased plants were infected with the fungus *Gibberella fujikuroi*.

Researchers confirmed a causal connection between *Gibberella* infection and rapid stem elongation when they treated rice seedlings with an extract from the fungus. As predicted, the treated seedlings produced abnormally long shoots.

The active component in the extract was eventually isolated and named gibberellic acid (GA), which is a gibberellin. Follow-up research showed that rice plants produce their own gibberellin but respond to applications of additional hormone by elongating their stems. In effect, the infected rice seedlings were suffering from a gibberellin overdose.

Gibberellins are found in a wide array of fungi and plants. Most plant species produce several different gibberellins that are active as hormones.

Even though gibberellins have dramatic effects on growth, they are present in vanishingly small concentrations. In growing stems and leaves, active forms of gibberellin may be present in concentrations of about 10 nanograms per gram of tissue.

DEFECTIVE GIBBERELLIN GENES CAUSE DWARFING To find the genes that are responsible for producing gibberellins, biologists analyzed mutant plants with abnormal stem length (**Figure 39.19**). Recall from Chapter 13 that Gregor Mendel analyzed the transmission of two alleles at a single gene that affected stem height in garden peas. One allele was associated with tall stems; the other was associated with dwarfed growth. The tall allele was dominant to the dwarf allele.

The gene responsible for the stem-length differences in garden peas is known as *Le* (for *le*ngth). Early work on dwarf mutants showed that they attain normal height if they are treated with the gibberellin called GA₁. This observation suggested that dwarf peas can respond to gibberellins normally—meaning that the problem is not with a hormone receptor.

Follow-up experiments treated dwarf peas with a radioactively labeled molecule used in the synthesis of GA₁. These plants did not produce radioactively labeled GA₁, even though plants

Individual with dwarfing alleles, "rescued" by application of gibberellins

Dwarfed plant

FIGURE 39.19 Dwarfed Individuals May Have Mutations That Affect Gibberellins. Dwarfed plants have much shorter stems than do normal individuals of the same age. By analyzing dwarfed individuals, biologists were able to identify genes involved in gibberellin synthesis.

✓**QUESTION** If dwarfed individuals receive the same amount of sun and thus perform as much photosynthesis as taller individuals, they often produce more flowers and seeds. Why?

with the normal allele did. Based on these results, researchers became convinced that the *Le* locus encodes an enzyme involved in GA synthesis.

Investigators recently provided support for this hypothesis by finding a gene in peas that encodes an enzyme called 3β-hydrolase. This enzyme adds a hydroxyl group (–OH) to a gibberellin called GA_{20}, producing the biologically active molecule GA_1. Researchers who compared the DNA sequences of this gene from normal and dwarf pea plants found an important difference: In a part of the enzyme near the active site, the mutant DNA sequence codes for the amino acid threonine instead of alanine. Follow-up tests showed that the mutant enzymes are unable to convert GA_{20} to GA_1.

These experiments provided strong evidence that the *Le* gene encodes the enzyme GA 3β-hydroxylase, and that a single amino acid change renders the enzyme largely ineffective and causes dwarfing. Over 100 years after Mendel did his experiments, biologists finally understood the molecular basis of the dwarfing phenotype he studied.

In stems, gibberellins appear to promote both cell elongation and rates of cell division. But it is well established that auxin also promotes cell elongation, and that cytokinins also promote cell division. Research continues on how GAs, cytokinins, and auxin interact on the molecular level to control plant growth and development.

GIBBERELLINS AND ABA INTERACT DURING SEED DORMANCY AND GERMINATION Many plants produce seeds that have to undergo a period of drying or a period of cold, wet conditions before they are able to germinate in response to warm, wet conditions. A requirement for drying ensures that mature seeds will not sprout on the parent plant; an obligatory cold period prevents seeds from germinating just before the onset of winter. Some seeds also have to receive a dose of red light, which indicates that they are in a sunny location.

In essence, then, seeds have an "off" setting that discourages germination and an "on" setting that initiates growth. The appropriate state for the on/off switch is determined by environmental cues such as temperature, moisture, and light.

By applying hormones to seeds, researchers learned that in many plants, ABA is the signal that inhibits seed germination and gibberellins are the signal that triggers germination. To understand how these messengers interact, researchers have concentrated on studying a specific event: the production of an enzyme called α-amylase in germinating oat or barley seeds.

α-Amylase acts as a digestive enzyme that breaks the bonds between the sugar subunits of starch. (Your saliva contains an amylase that acts on the starch in food. In humans and other mammals, this enzyme initiates carbohydrate digestion in the mouth.)

Figure 39.20 shows that, during the germination of a barley seedling, α-amylase is released from a tissue called the aleurone layer. The enzyme diffuses into the carbohydrate-rich storage tissue in the seed and releases sugars that can be used by the growing embryo. Adding GA to the aleurone layer increases the production and release of α-amylase; adding ABA to that layer decreases α-amylase levels.

Research on the molecular interaction between GA and ABA carries several important messages:

1. A cell's response to a hormone often occurs because specific genes are turned on or off.

2. Hormones don't act on genes directly. Instead, a receptor on the surface of a cell or in the cytosol receives the message and responds by initiating a signal transduction cascade, which activates specific gene regulatory proteins—the transcription activators and repressors introduced in Chapter 18.

3. Different hormones interact at the molecular level because they induce different gene regulatory proteins, which increase or decrease expression of key genes.

The logic runs as follows: Different hormones trigger the production of different regulatory transcription factors. Hormone concentration affects the amount of each transcription factor produced. Changes in transcription factors are responsible for changes in gene expression.

GA's role in seed germination has important commercial applications. Brewers, for example, routinely use gibberellins in the

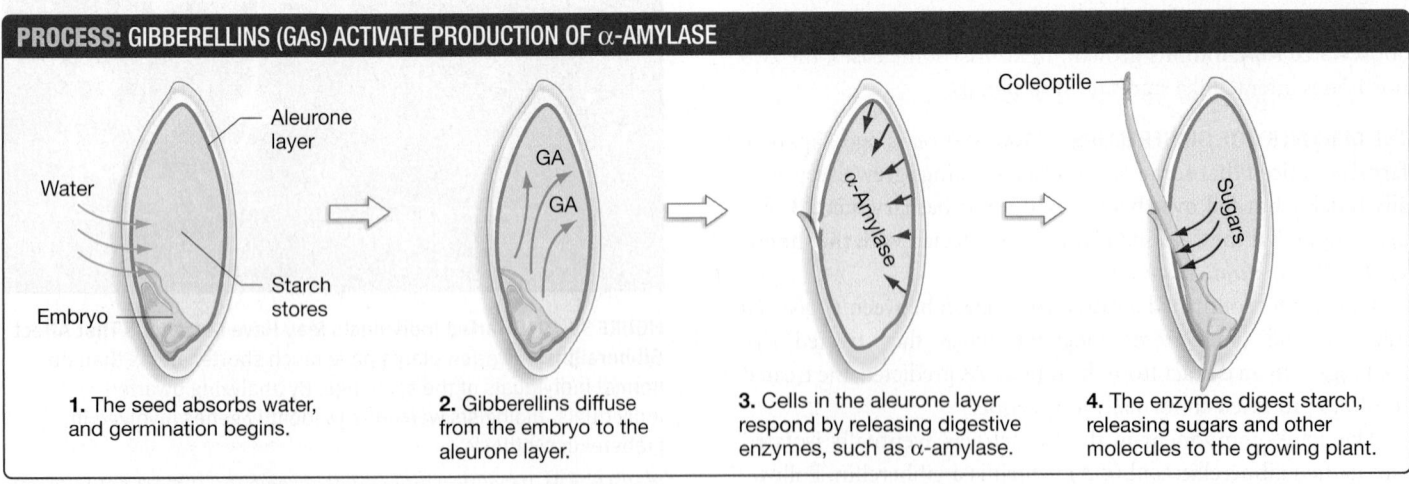

PROCESS: GIBBERELLINS (GAs) ACTIVATE PRODUCTION OF α-AMYLASE

Aleurone layer · Water · Starch stores · Embryo

GA · GA

α-Amylase

Coleoptile · Sugars

1. The seed absorbs water, and germination begins.

2. Gibberellins diffuse from the embryo to the aleurone layer.

3. Cells in the aleurone layer respond by releasing digestive enzymes, such as α-amylase.

4. The enzymes digest starch, releasing sugars and other molecules to the growing plant.

FIGURE 39.20 The Molecular Mechanism of Gibberellin Action.

malting process—the conversion of starches stored in barley seeds to sugars. The sugars that are released in response to GA treatment support fermentation by yeast and provide flavor in the finished beer.

ABA CLOSES GUARD CELLS IN STOMATA Chapter 37 introduced one of the major problems faced by land plants: replacing water that is lost to the atmosphere when **stomata** are open. Because stomata open in response to blue light, they allow gas exchange to occur while the plant is receiving the wavelengths of light used in photosynthesis.

However, if plant roots are unable to obtain enough water to replace the fluid being lost at the leaves, stomata close. Closing stomata is an adaptive response when roots cannot find adequate water, because continued transpiration would lead to wilting and potential tissue damage.

Early work on the mechanism of stomatal closing suggested that ABA is involved. For example, applying ABA to the exterior of stomata causes them to close.

To explore the hypothesis that a hormone regulates stomatal closing, researchers performed the experiments summarized in **Figure 39.21**. The fundamental idea was to grow plants whose roots had been divided. Only one side of the experimental plants was watered, while both sides of the control plants were watered. During this treatment, investigators documented that the water potential of the leaves remained the same in both control and experimental plants. Yet the stomata of experimental plants began to close. This result suggested that roots from the dry side of the pot were signaling drought stress, even though the leaves were not actually experiencing a water shortage.

Follow-up experiments have supported two important predictions: ABA concentrations in roots on the dry side of the pot are extraordinarily high relative to the watered side, and ABA concentrations in the leaves of experimental plants are much higher than in the leaves of control plants.

These results suggest that ABA from roots is transported to leaves and that it serves as an early warning of drought stress. In doing so, ABA overrides the signal from the blue-light photoreceptors introduced earlier in this chapter.

THE MOLECULAR MECHANISM OF GUARD CELL CLOSURE To understand how stomata open and close, recall from Chapter 36 that a stoma consists of two **guard cells**. When the vacuoles of guard cells are filled with water, the cells are turgid. The shape of turgid guard cells results in an open pore, which allows gas exchange between the atmosphere and the interior of the leaf. But when vacuoles lose water and guard cells become flaccid and lack turgor, cell shape changes in a way that closes the pore and stops both gas exchange and loss of water via transpiration.

Based on these observations, the question of how stomata open and close is the same as asking how guard cells become turgid or flaccid—meaning, how water flows into or out of vacuoles. Activation of PHOT by blue light leads to water entry and

---

EXPERIMENT

QUESTION: Can roots communicate with shoots?

HYPOTHESIS: Roots that are dry can signal shoots to close stomata.

NULL HYPOTHESIS: Roots cannot communicate with the shoot.

EXPERIMENTAL SETUP:

1. Divide roots of many plants into two sides.

Water

Dry | Wet   Wet | Wet

2. In experimental group, water one side.

3. In control group, water both sides.

4. In both groups, measure water potential of leaves and observe stomata.

PREDICTION: Stomata in experimental plants will close; stomata in control plants will stay open.

PREDICTION OF NULL HYPOTHESIS: Stomata in both experimental and control plants will stay open.

RESULTS: No difference between experimental and control plants in water potential of leaves.

Stomata began to close.

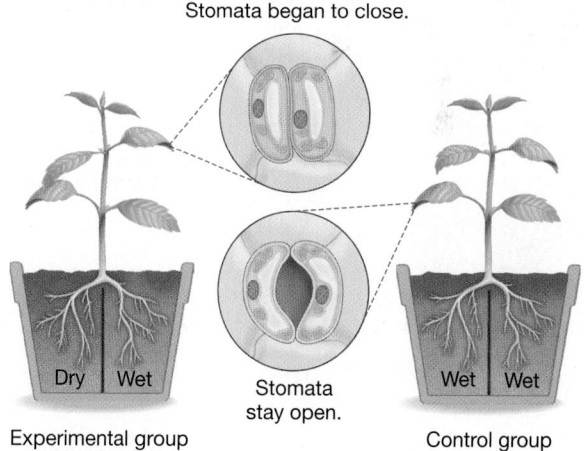

Dry | Wet   Stomata stay open.   Wet | Wet

Experimental group   Control group

CONCLUSION: Roots can communicate with shoots. Dry roots signal the shoot and cause stomata to close, even though leaves are receiving sufficient water (from roots on the wet side of the plant).

FIGURE 39.21 Experimental Evidence That Roots Produce an "It's Too Dry" Signal.

SOURCE: Blackman, P. G. and W. J. Davies. 1985. Root to shoot communication in maize plants of the effects of soil drying. *Journal of Experimental Botany* 36: 39–48.

✔**QUESTION** Why was it important to show that the water potential of leaves was the same in the two treatments?

stomatal opening; activation of ABA receptors leads to water exit and stomatal closing.

Remember that cells do not transport water directly. Instead, they change ion concentrations, creating osmotic gradients that result in water movement.

Based on this observation, it shouldn't be surprising to learn that guard cell opening and closing is based on changes in the activity of H$^+$-ATPases in the plasma membrane.

When PHOTs or cryptochromes are stimulated by blue light, large numbers of protons are pumped out of each guard cell. As **Figure 39.22a** shows, increased H$^+$-ATPase activity creates a strong electrochemical gradient that brings potassium and chloride ions into the interior of the guard cells. Water fol-lows the incoming ions via osmosis. The upshot? The cells swell and pores open.

But as **Figure 39.22b** shows, guard cells respond to ABA in a very different way. When ABA reaches the guard cells, two things happen:

1. Channels that allow chloride and other anions to leave along their electrochemical gradients are opened.

2. H$^+$-ATPases and inward-directed potassium channels are inhibited.

When the anions leave guard cells, the change in membrane potential causes outward-directed potassium channels to open. Large amounts of K$^+$ leave the cells, with water following by osmosis. The result is a loss of turgor and closing of the pore.

(a) PROCESS: STOMATA OPEN IN RESPONSE TO BLUE LIGHT

Blue light strikes photoreceptor.

H$^+$ H$^+$

H$^+$ H$^+$

1. Pumping by H$^+$-ATPases increases. Protons leave guard cells.

K$^+$ Cl$^-$ H$^+$ K$^+$ Cl$^-$ H$^+$

2. K$^+$ and Cl$^-$ enter cells along electrochemical gradients via inward-directed K$^+$ channels and H$^+$/Cl$^-$ cotransporter.

H$_2$O   H$_2$O

H$_2$O   H$_2$O

3. H$_2$O follows by osmosis.

4. Cells swell. Pore opens.

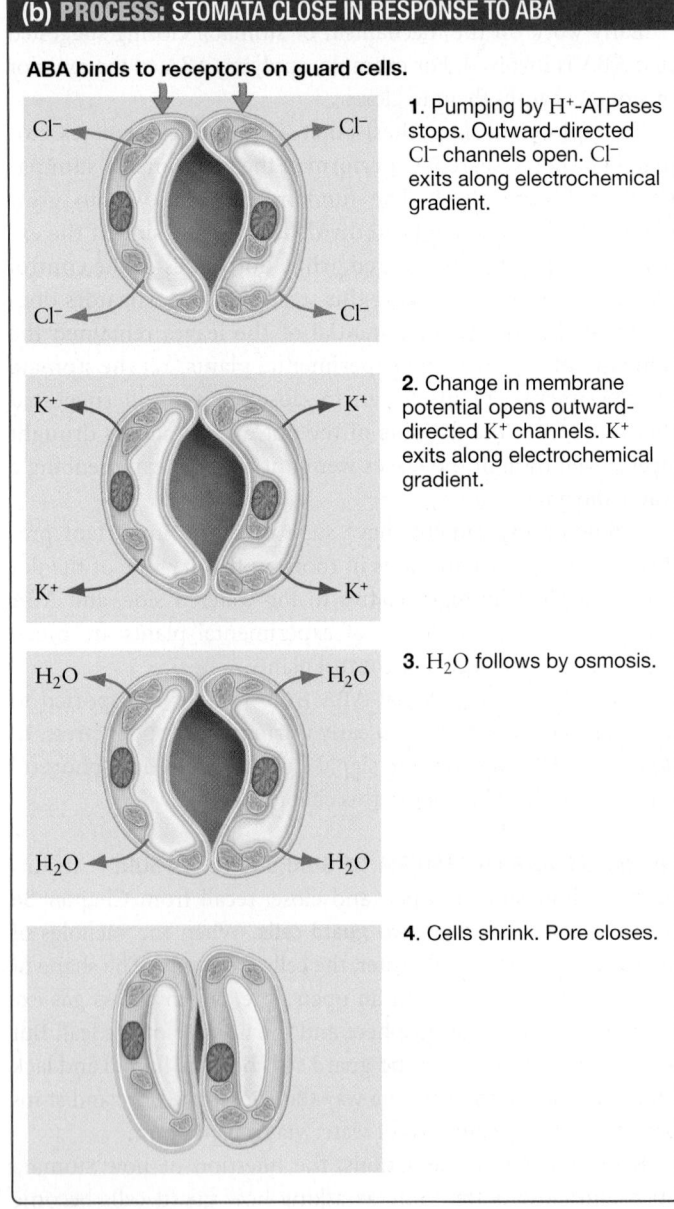

(b) PROCESS: STOMATA CLOSE IN RESPONSE TO ABA

ABA binds to receptors on guard cells.

Cl$^-$ Cl$^-$

Cl$^-$ Cl$^-$

1. Pumping by H$^+$-ATPases stops. Outward-directed Cl$^-$ channels open. Cl$^-$ exits along electrochemical gradient.

K$^+$ K$^+$

K$^+$ K$^+$

2. Change in membrane potential opens outward-directed K$^+$ channels. K$^+$ exits along electrochemical gradient.

H$_2$O   H$_2$O

H$_2$O   H$_2$O

3. H$_2$O follows by osmosis.

4. Cells shrink. Pore closes.

FIGURE 39.22 Changes in Ion Flows Are Responsible for Opening and Closing Stomata. (a) Activated blue-light receptors trigger ion flows into guard cells. The cells swell when water follows by osmosis. **(b)** Activated ABA receptors trigger ion flows out of guard cells. The cells shrink when water follows by osmosis.

Whether it acts on guard cells or seeds, ABA fulfills a general role in plants as a dormancy or "no-growth" signal. In many cases, its action depends on input from other hormones and photoreceptors. To survive and reproduce successfully, plants have to integrate information from a variety of sources.

Brassinosteroids and Body Size

When you went through puberty, you underwent a growth spurt triggered by surges in steroid hormones called testosterone and estradiol. In plants, growth spurts are triggered by surges in steroid hormones called **brassinosteroids**.

The name brassinosteroid was inspired by two observations:

1. The hormones were initially discovered in *Brassica napus*—a crop plant that is the source of the canola oil you may use in cooking.

2. They are steroids—part of a family of lipid-soluble compounds introduced in Chapter 6.

Brassinosteroids promote growth and are a key regulator of overall body size in plants. In the model organism *Arabidopsis thaliana*, for example, mutant individuals that cannot synthesize brassinosteroids are extremely dwarfed.

Recent research has highlighted a fascinating difference between the brassinosteroids and the steroid hormones found in animals. As Chapter 47 will show, steroid signals in animals act by entering cells, binding to receptors in the cytosol, and forming a hormone-receptor complex that enters the nucleus, binds to DNA, and directly changes gene expression. Because steroids are lipid-soluble, researchers were not surprised to find that testosterone and estradiol cross the plasma membrane before binding to a receptor.

Brassinosteroids, in contrast, never enter the cell. Instead, they bind to receptors on the plasma membrane and activate signal transduction events—probably phosphorylation cascades—that lead to changes in gene expression.

The genes for the synthesis of brassinosteroids appear to be homologous with the genes required for steroid hormone in animals, meaning that they are derived from a similar gene in the common ancestor of plants and animals. So why do they have such different modes of action? And how do brassinosteroids interact with auxin, cytokinins, and gibberellins to regulate growth and body size? These are questions for future research.

Ethylene and Senescence

Senescence is a regulated process of aging and eventual death of an entire organism or organs such as fruits and leaves. Like most aspects of plant growth and development, senescence is regulated by complex interactions between several hormones in response to changes in temperature, light, and other factors.

The hormone most strongly associated with senescence is **ethylene**. Like other plant hormones, ethylene is simple in structure and active at small concentrations. Unlike other plant hormones, however, ethylene is a gas at normal temperatures.

Ethylene is synthesized from the amino acid methionine and is strongly involved in three aspects of senescence in plants:

1. fruit ripening, which eventually leads to the aging and rotting of fruit;

2. flowers fading; and

3. leaf abscission—meaning their detachment and fall.

In addition, ethylene influences plant growth, and is a stress hormone induced by drought and other conditions. Ethylene regulates a surprisingly large range of physiological responses.

THE DISCOVERY OF ETHYLENE Ethylene was initially discovered in ancient China, when fruit growers noticed that burning incense in closed rooms made pears ripen faster. Westerners made a similar observation in the late 1800s, when gas street lamps came into wide use in cities and plants growing near leaky gas lines dropped their leaves prematurely.

Researchers showed that ethylene in lamp gas was the molecule responsible for the leaf loss; ethylene is also present in incense smoke. In the 1930s ethylene was found in the gases that are released by ripening apples.

Subsequently, biologists documented sharp spikes in ethylene production during fruit ripening in tomatoes, bananas, and certain other species in addition to apples. Follow-up research on these species showed that ethylene induces (1) the production of some of the enzymes required for the ripening process, and (2) an increase in cellular respiration, which furnishes ATP.

ETHYLENE AND FRUIT RIPENING During ripening, stored starch is converted to sugar, enhancing sweetness; protective toxins are removed or destroyed; cell walls are degraded, softening the fruit; chlorophyll is broken down; and pigments and aromas that signal ripeness are produced. Biologists interpret fruit ripening as an adaptation that enhances its attractiveness to birds, mammals, and other animals that disperse seeds to new locations.

Today, fruit growers manipulate ethylene levels to control fruit ripening. For example, they treat green bananas with ethylene after the bananas have been shipped to encourage ripening (**Figure 39.23**). Conversely, apples are stored in warehouses with

+ Ethylene

FIGURE 39.23 Ethylene Speeds Ripening and Other Aspects of Senescence. These bananas are identical, except that the bunch on the right was exposed to the plant growth regulator ethylene.

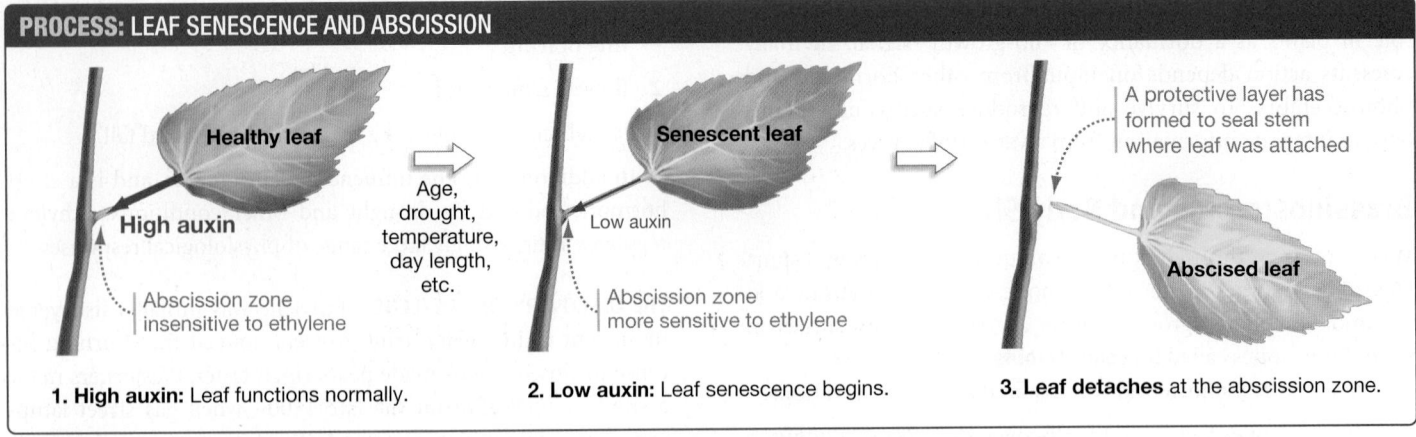

PROCESS: LEAF SENESCENCE AND ABSCISSION

Healthy leaf

High auxin

Abscission zone insensitive to ethylene

Age, drought, temperature, day length, etc.

Senescent leaf

Low auxin

Abscission zone more sensitive to ethylene

A protective layer has formed to seal stem where leaf was attached

Abscised leaf

1. High auxin: Leaf functions normally.

2. Low auxin: Leaf senescence begins.

3. Leaf detaches at the abscission zone.

FIGURE 39.24 Leaves Drop in Response to Signals from Auxin and Ethylene. Young leaves produce much larger amounts of auxin than old leaves do. The combination of low auxin and high ethylene concentrations triggers leaf senescence and abscission.

✔**QUESTION** When ethylene levels in leaves are high relative to auxin, (1) nutrients are transported from leaves to the stem and (2) chlorophyll synthesis stops (this is why leaves change color in the fall). What is the adaptive significance of these two events?

high concentrations of CO_2 and low concentrations of O_2, which inhibits ethylene production in the fruit. Apples stored under these conditions can be sold long after their original harvest date, when untreated fruits have rotted.

ETHYLENE AND LEAF ABSCISSION Ethylene's effects on leaf senescence and leaf abscission involve complex interactions with auxin and cytokinins. In addition to being sequestered in apical meristems, auxin is synthesized in healthy leaves. It is then transported from the leaf to the stem through the petiole. In response to age or to changes in ambient temperature or day length, leaves produce less auxin (**Figure 39.24**). As a result, cells in a region of the leaf petiole called the **abscission zone** become more sensitive to ethylene in the tissue.

Increased ethylene sensitivity activates enzymes that weaken the cell walls of cells near the base of the petiole. At the same time, chlorophyll in the leaf degrades, and nutrients are withdrawn and stored in parenchyma cells in the stem. Eventually the cell walls at the base of the petiole degrade enough that the leaf falls.

Applications of cytokinins, in contrast, reverse these effects and dramatically extend the life span of leaves. As a result, ethylene and cytokinins are thought to have opposite effects on at least some of the processes involved in senescence. With two hormones involved, the process can be sped up or slowed down.

An Overview of Plant Growth Regulators

Understanding how different signals interact is an exciting frontier in research on plant growth regulators. And work continues on characterizing the genes that are directly regulated by auxin, cytokinins, ABA, GAs, brassinosteroids, and ethylene. Although progress has been rapid, a great deal remains to be learned about the response step in information processing during growth responses.

Table 39.2 provides notes on the structure and function of hormones discussed in this section. As you study this table, two key observations should emerge:

1. It is common for a single hormone to affect many different target tissues. This means that there can be an array of responses to the same cell-cell signal. To interpret this pattern, biologists point out that hormones may carry a common message to a variety of tissues and organs. Auxin can define the long axis of the body; gibberellins trigger stem growth; cytokinins promote cell division; ABA slows or prevents growth; ethylene signals senescence; brassinosteroids increase overall mass.

2. In most cases, several hormones affect the same response. Stated another way, hormones do not work independently—they interact with each other. To make sense of this pattern, biologists point out that individual hormones tend to be produced by an environmental cue at a certain location, such as water availability at root tips. Many environmental cues may be changing at the same time, however. For plants to respond appropriately, they need to integrate information from various environmental cues perceived at various locations in the body.

Chapter 8 introduced the concept of **cross-talk**: interactions between the signal transduction cascades triggered by different hormones. The key insight is that signaling systems form communication networks. Cross-talk is the molecular mechanism responsible for integrating information from many sensory cells and signals.

The complex interactions among hormones involved in the growth response have a purpose: allowing individuals to survive and thrive long enough to reproduce. The same can be said for the hormones involved in protecting plants from danger.

Hormone	Function(s)	Notes	Chemical Structure
Auxin	• Helps define long axis of body (phototropism and gravitropism responses) • Involved in cell elongation and apical dominance • Promotes cell division • Induces ethylene production • Development of adventitious roots and secondary growth • Differentiation of xylem and phloem	• First plant hormone ever characterized and isolated • Produced in shoot apical meristems and young leaves • Receptor discovered 2005; receptor structure solved 2007	
Cytokinins	• Promote cell division in the presence of auxin • Promote chloroplast development and break lateral bud dormancy • Delay senescence (aging)	• New data indicate they may act on cell cycle regulators • Produced in root apical meristems, many other tissues • 3 distinct receptors have been identified; each found in a different location in cells	
Gibberellins (GAs)	• Promote stem growth via both cell elongation and division • Encourage seed germination • Involved in flowering	• Fungi that produce gibberellins infect rice plants and induce hyper-elongated stems • Analysis of these fungi led to discovery of gibberellins • Produced in apical meristems, immature seeds, anthers (pollen-producing organs) • Receptor discovered 2005	
Abscisic Acid (ABA)	• Inhibits bud growth and seed germination • Induces closure of stomata in response to water stress	• Acts as a stress hormone analogous to cortisol in humans • Produced in almost all cells	
Brassinosteroids	• Promote cell elongation in stems and leaves (mutants that lack these hormones or their receptors are dwarfed)	• First steroid hormones discovered in plants • Structurally related to steroid hormones in animals • Produced in almost all tissues • Act on receptor at cell surface	
Ethylene	• Involved in fruit ripening • Induces senescence of fruits, flowers, and leaves • Produced when plants are under stress	• A gas; first identified through unusual morphology of plants growing near gas lines for illuminating streets • Produced in all organs but highest in aging tissues and fruits	

CHECK YOUR UNDERSTANDING

39.7 Pathogens and Herbivores: The Defense Responses

Plants cannot run away from danger. Instead, they have to stand and fight.

Like humans and other animals, plants are constantly threatened by an array of disease-causing viruses, bacteria, and parasitic fungi. In addition, plant roots are susceptible to attacks by nematodes—the soil-dwelling roundworms introduced in Chapter 33.

Disease-causing agents are termed **pathogens**; the ability to cause disease is called **virulence**. If plants were not able to sense attacks by pathogens and respond to them quickly and effectively, the landscape would be littered with dead and dying vegetation.

The waxy cuticle that covers epidermal cells is an effective barrier to viruses, bacteria, fungi, and other disease-causing agents, and the structures called thorns, spines, and trichomes help protect leaves and stems from damage by herbivores (see Chapter 36).

In addition, many plants lace their tissues with **secondary metabolites**—defense molecules that are closely related to compounds in key synthetic pathways. Some secondary metabolites poison herbivores.

- The flavorful oils in peppermint, lemon, basil, and sage have insect repellent properties.
- The pitch that oozes from pines and firs contains a molecule called pinene, which is toxic to bark beetles.
- The pyrethroids produced by *Chrysanthemum* plants are a common ingredient in commercial insecticides.

- Molecules called tannins are found in a wide array of plant species; when they are ingested by animals, tannins bind to digestive enzymes and make the herbivore sick.
- Compounds like opium, caffeine, cocaine, nicotine, and tetrahydrocannabinol (THC) disrupt the nervous systems of plant-eating insects and vertebrates.

Although these defenses are effective, they are also expensive to produce in terms of the ATP and materials invested. It is not surprising that plants may produce defenses or increase their existing defenses only in direct response to attacks by pathogens or herbivores.

Responses to attacks are called **inducible defenses**, because they are induced by the presence of a threat. Let's first consider how plants sense and respond to viruses and other pathogens, then explore what they do when attacked by insects and other herbivores.

How Do Plants Sense and Respond to Pathogens?

If a virus, bacterium, or fungus is able to get inside a plant, the cells at the infection site respond by committing suicide. The rapid and localized death of one or a few infected cells is called the **hypersensitive response (HR)**.

In several respects, the hypersensitive response in plants is similar to the cell-mediated immune response in mammals, which leads to the death of infected cells (see Chapter 49). The HR is also extremely effective; when individuals mount a hypersensitive response, they rarely succumb to disease. How do cells sense the presence of pathogens, so that the HR can kill them?

THE GENE-FOR-GENE HYPOTHESIS In the early twentieth century, crop breeders established that plants have disease-resistance genes, known as **resistance (R) genes**, that are inherited according to Mendel's rules. Follow-up research showed that many of the R genes are responsible for sensing the presence of pathogens and triggering the HR.

The fungi that cause disease in wheat, flax, barley, and other crops have alleles that make individuals either virulent or avirulent (not virulent) in certain strains or varieties of these crops. The genes associated with virulence in pathogens are known as **avirulence (avr) genes**.

In 1956 H. H. Flor published data demonstrating a one-to-one correspondence between the resistance alleles found in host plants and the avirulence alleles found in pathogens. Each R allele in the host corresponded to an avr allele in the pathogen.

Flor's **gene-for-gene hypothesis** expanded on this observation by proposing that R and avr gene products interact in a specific way. What molecular mechanism could be responsible? Researchers who followed up on Flor's work suggested that the HR begins when proteins produced by host plants bind to proteins or other molecules produced by the pathogen (**Figure 39.25**). More specifically, the hypothesis was that R genes

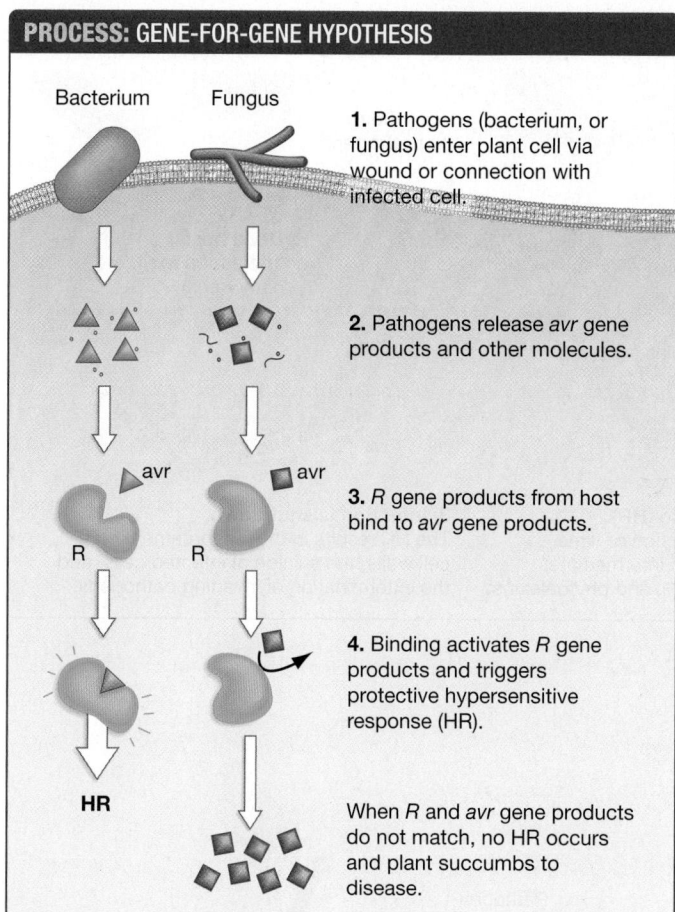

PROCESS: GENE-FOR-GENE HYPOTHESIS

Bacterium   Fungus

1. Pathogens (bacterium, or fungus) enter plant cell via wound or connection with infected cell.

2. Pathogens release *avr* gene products and other molecules.

avr   avr

R   R

3. *R* gene products from host bind to *avr* gene products.

4. Binding activates *R* gene products and triggers protective hypersensitive response (HR).

HR

When *R* and *avr* gene products do not match, no HR occurs and plant succumbs to disease.

FIGURE 39.25 The Hypersensitive Response Begins when *R* Gene Products Bind to *avr* Gene Products. The gene-for-gene hypothesis predicts that *R* gene products act as receptors for *avr* gene products and initiate a defense response.

produce receptors and *avr* genes produce **ligands**—molecules that bind to receptors.

The first breakthrough in testing the gene-for-gene hypothesis occurred when researchers cloned and sequenced a series of *R* genes from crop plants and *avr* genes from bacterial and fungal pathogens. These results confirmed that *R* genes and *avr* genes exist and that they code for products that could interact at the start of an infection.

A second major advance, published in 1996, supported the gene-for-gene hypothesis by showing that *R* products and *avr* products actually do bind to one another. *R* gene products function as "pathogen receptors."

WHY IS THE EXISTENCE OF SO MANY RESISTANCE GENES AND ALLELES SIGNIFICANT? A large number of *R* genes have been identified in *Arabidopsis*, tomato, flax, tobacco, and other plants. Two general patterns are emerging as data on these genes accumulate:

1. *R* genes that are similar in sequence and structure tend to cluster together on the same chromosome.

2. Within a population of plants, there are usually many different alleles at each *R* locus. To use a term introduced earlier in the text, *R* genes are highly polymorphic.

These observations are important because they provide hints about the history and function of these genes. For example, clusters of similar genes, or what biologists call **gene families**, are thought to originate through errors in recombination that result in **gene duplication** events (see Chapter 20). Mutations in duplicated genes can result in novel gene products.

The idea here is that duplicated *R* genes change over time due to mutation, and begin to produce new proteins that give individuals the ability to recognize and respond to novel *avr* products and thus new pathogens.

Why is it significant that many different alleles exist at each *R* locus? The hypothesis here is that different alleles allow plants to recognize different proteins from the same pathogen. Plants are diploid or polyploid, so they have at least two copies of each *R* gene. If many different alleles exist in a population, each individual is likely to have at least two different alleles of each gene. The different alleles allow the host to recognize different *avr* products. This is important because new *avr* products constantly arise in pathogen populations via mutation.

Plants with different alleles for each of many *R* genes should be able to sense a wide variety of disease-causing agents and respond by triggering the HR. ✔If you understand this concept, you should be able to explain why bananas—which are propagated asexually and are thus genetically identical—are considered particularly vulnerable to disease epidemics.

THE HYPERSENSITIVE RESPONSE Once an *R* gene product is activated by binding to a ligand from a pathogen, what happens? Recent work has established that the HR consists of two key events (**Figure 39.26** on page 778):

- Production of nitric oxide (NO) and **reactive oxygen intermediates (ROIs)** such as hydrogen peroxide (H_2O_2) and superoxide ions (O_{2-}). ROIs (1) trigger reactions that help reinforce cell walls, and (2) kill host cells at the point of infection by disrupting their enzymes. Immune system cells in your body also use a lethal combination of NO and ROIs to kill cells that have become infected.

- Production of antibacterial and antifungal compounds that are collectively known as **phytoalexins**.

As a result, the HR leads to the walling-off of the infected area, the direct killing of pathogens by phytoalexins, and the starvation of pathogens as nearby cells commit suicide. Although the signal transduction events responsible for these events are not well understood, they are currently the focus of intense research.

AN ALARM HORMONE EXTENDS THE HR Once the HR is under way in a localized area of infection, a hormone produced at the infection site travels throughout the body and triggers a slower and more widespread set of events called **systemic acquired resistance (SAR)**. Over the course of several days, SAR primes cells throughout the root or shoot system for resistance to assault by a pathogen—even cells that have not been directly exposed to the disease-causing agent.

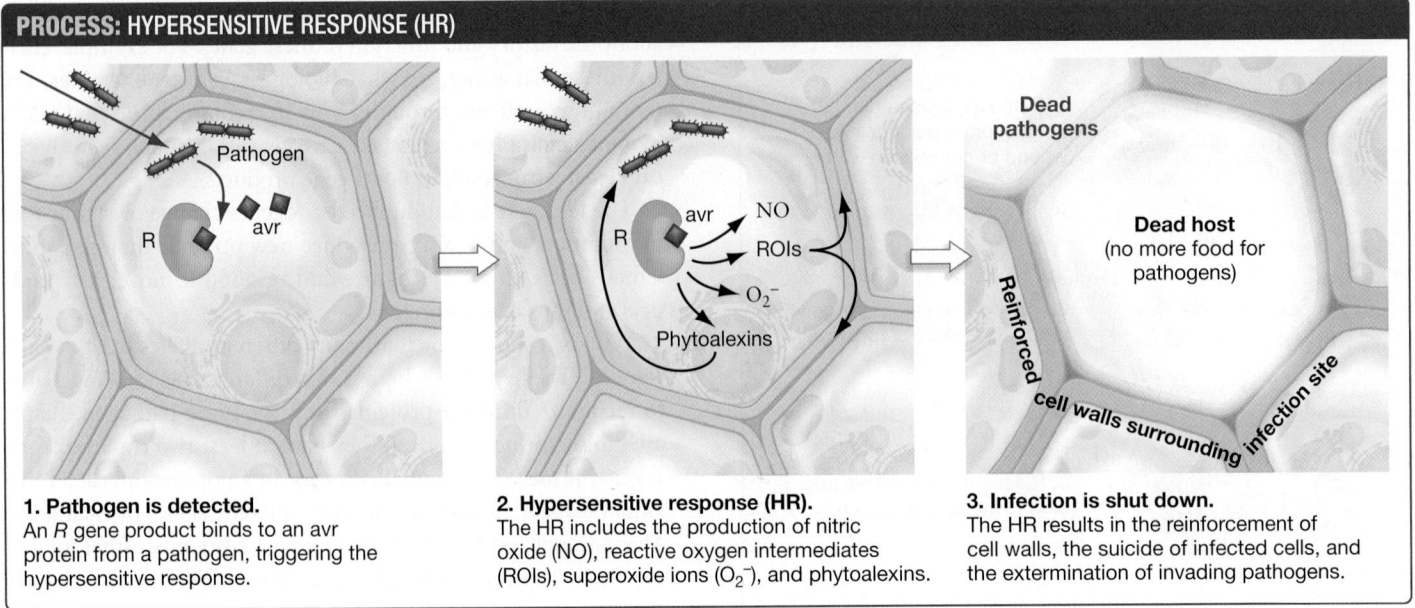

PROCESS: HYPERSENSITIVE RESPONSE (HR)

1. Pathogen is detected.
An *R* gene product binds to an avr protein from a pathogen, triggering the hypersensitive response.

2. Hypersensitive response (HR).
The HR includes the production of nitric oxide (NO), reactive oxygen intermediates (ROIs), superoxide ions (O_2^-), and phytoalexins.

3. Infection is shut down.
The HR results in the reinforcement of cell walls, the suicide of infected cells, and the extermination of invading pathogens.

FIGURE 39.26 The Hypersensitive Response Protects Plants from Pathogens.

Figure 39.27 illustrates how the HR and SAR are thought to work together. In addition to triggering the HR, interaction between the *R* and *avr* gene products releases a hormone that initiates SAR. This signal acts globally as well as locally—that is, at the point of infection—and results in the expression of a large suite of genes called the pathogenesis-related (*PR*) genes.

When biologists set out to locate the hormone responsible for SAR, they found that levels of **methyl salicylate (MeSA)**—a molecule derived from salicylic acid—increase dramatically after tissues are infected with a pathogen. Follow-up work showed that:

- Phloem sap leaving infected sites has elevated levels of MeSA.
- Treatments that reduce MeSA reduce or abolish SAR.
- Adding MeSA to the lower leaves of tobacco plants leads to SAR in the upper, untreated leaves.

MeSA is a short-lived molecule, however, and it is still not clear that MeSA is transported throughout the plant. Researchers continue to look for the defense hormone that travels from the site of infection, through phloem sap, to the rest of the body—where it sounds a Paul Revere-like warning to prepare for attack.

How Do Plants Sense and Respond to Herbivore Attack?

Over a million species of insects have been discovered and named so far. Most of them make their living by eating leaves, stems, phloem sap, seeds, roots, or pollen. Plants have effective induced defenses in response to pathogens like viruses, bacteria, and fungi. But can they ramp up defenses in response to insect attack?

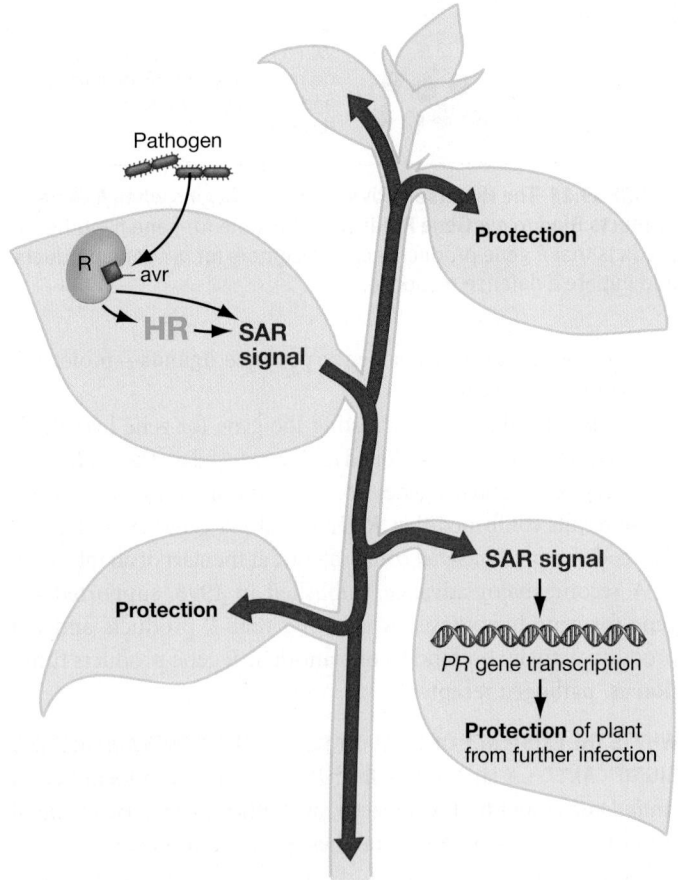

FIGURE 39.27 The Hypersensitive Response Produces a Signal That Induces Systemic Acquired Resistance. This diagram summarizes the current consensus on how the HR and SAR interact.

THE ROLE OF PROTEINASE INHIBITORS When researchers started studying why some plant tissues are more palatable and digestible than others, biochemists discovered that many seeds and some storage organs, such as potato tubers, contain proteins called **proteinase inhibitors**.

Proteinase inhibitors block the enzymes—found in the mouths and stomachs of animals—that are responsible for digesting proteins. When an insect or a mammalian herbivore ingests a large dose of a proteinase inhibitor, the herbivore gets sick. As a result, herbivores learn to detect proteinase inhibitors by taste and avoid plant tissues containing high concentrations of these molecules.

Although many plant tissues contain proteinase inhibitors in low concentrations, biologists wanted to test the hypothesis that these proteins might also be part of an induced defense by the plant. To evaluate this idea, researchers allowed herbivorous beetles to attack one leaf on each of several potato plants.

- In liquid extracted from the other leaves on the attacked plants, proteinase inhibitor concentrations averaged 336 μg per mL.

- In leaves of control plants, where no insect damage had occurred, proteinase inhibitor levels averaged just 103 μg per mL of leaf juice.

This result supported the hypothesis that a hormone produced by wounded cells travels to undamaged tissues and induces the production of proteinase inhibitors.

THE DISCOVERY OF SYSTEMIN Biologists isolated the wound-response hormone by purifying the compounds found in tomato leaves and testing them for the ability to induce proteinase inhibitor production. The hormone that is active in tomato plants turned out to be **systemin**, a polypeptide just 18 amino acids long. Systemin was the first peptide hormone ever described in plants.

Researchers who labeled copies of systemin with a radioactive carbon atom, injected the hormone into plants, and then monitored its location confirmed that systemin moves from damaged tissues to undamaged tissues.

Currently, work on the production of systemin and proteinase inhibitors focuses on determining each step in the signal transduction pathway that alerts undamaged cells to danger (**Figure 39.28**). The data indicate that:

1. Systemin is released from damaged cells.

2. Systemin travels through the body via phloem and binds to membrane receptors on target cells.

3. The activated receptor triggers a long series of chemical reactions that eventually synthesize a molecule called jasmonic acid.

4. Jasmonic acid activates the production of at least 15 new gene products, including proteinase inhibitors.

In this way, plants build potent concentrations of insecticides in tissues that are in imminent danger of attack.

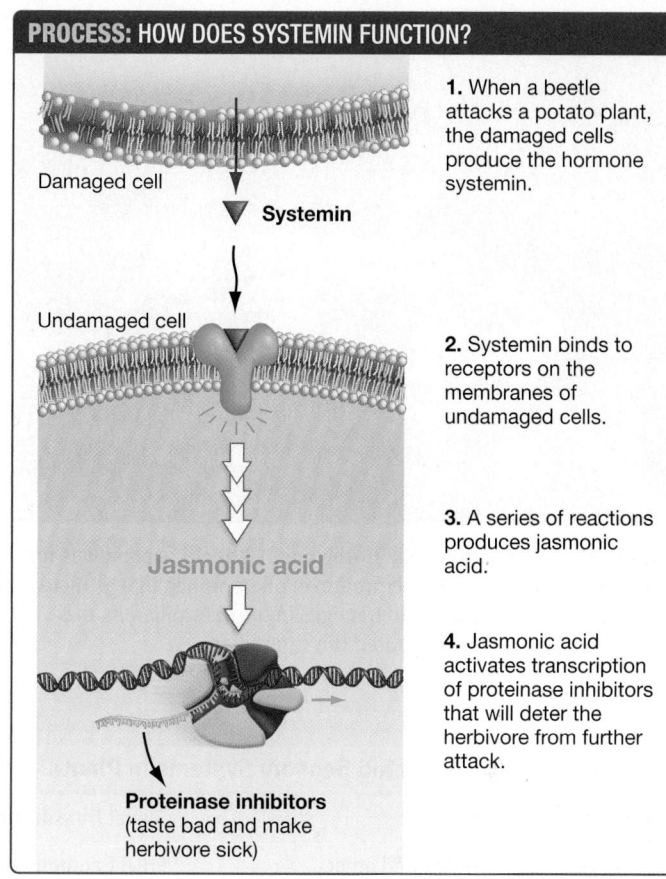

PROCESS: HOW DOES SYSTEMIN FUNCTION?

Damaged cell

Systemin

Undamaged cell

Jasmonic acid

Proteinase inhibitors
(taste bad and make herbivore sick)

1. When a beetle attacks a potato plant, the damaged cells produce the hormone systemin.

2. Systemin binds to receptors on the membranes of undamaged cells.

3. A series of reactions produces jasmonic acid.

4. Jasmonic acid activates transcription of proteinase inhibitors that will deter the herbivore from further attack.

FIGURE 39.28 Signals from Insect-Damaged Cells Prepare Other Cells for Attack. Systemin, a hormone produced by herbivore-damaged cells, initiates a protective response in undamaged cells.

✔**QUESTION** Why is "systemin" an appropriate name for this hormone?

To review SAR and the systemin-based systems for plant defense, go to the study area at *www.masteringbiology.com*.

(MB) **Web Activity** Plant Defenses

"TALKING TREES": RESPONSES FROM NEARBY PLANTS When an insect starts munching on a leaf, volatile compounds evaporate from the surface and travel through the air. In the 1980s, researchers began to suspect that individuals growing near plants under herbivore attack "eavesdrop" on these volatiles. In response, they increase their own defenses—even though they've yet to be attacked.

The "talking trees" hypothesis has received extensive support, primarily through experiments based on exposing plants to volatiles in the absence of any insects. Even if the volatiles come from an entirely different species, research has shown that plants can sense the presence of these chemicals and respond by increasing the production of proteinase inhibitors and other defense compounds.

In some cases, plants that are under attack even call for help from other organisms.

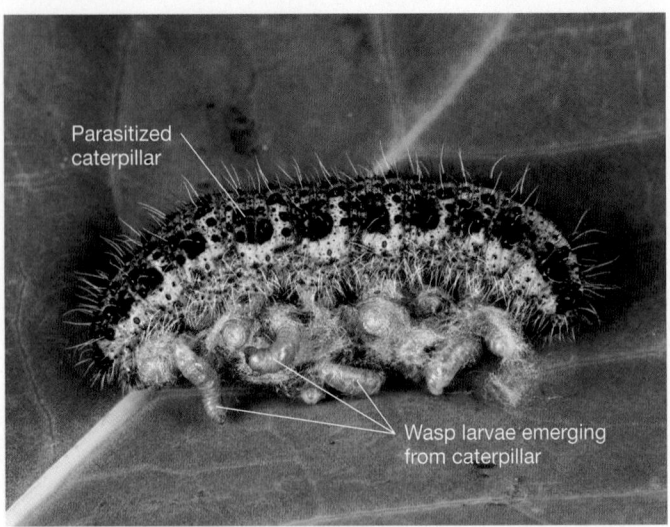

FIGURE 39.29 Parasitoids Kill Herbivores. When this plant was first attacked by this caterpillar, it produced pheromones that attracted a female wasp. The wasp laid her eggs in the caterpillar. As the wasp larvae grew, they devoured the caterpillar.

PHEROMONES RELEASED FROM PLANT WOUNDS RECRUIT HELP FROM WASPS Caterpillars and other herbivorous insects have enemies of their own—often wasps that lay their eggs in the insects' bodies. When wasp eggs hatch inside a caterpillar, the wasp larvae begin eating the host from the inside out. An organism that is free living as an adult but parasitic as a larva, such as these wasps, is called a **parasitoid** (**Figure 39.29**).

Biologists observed that parasitoids are particularly common when insect outbreaks occur in croplands. They wondered whether wounded plants release compounds that actively recruit parasitoids. More specifically, they hypothesized that plants produce **pheromones**—chemical messengers that are synthesized by an individual, released into the environment, and elicit a response from a different individual. Hormones act on cells inside an individual; pheromones act on another individual.

To explore this idea, researchers analyzed compounds that were released from corn seedlings during attacks by caterpillars. The insect-damaged leaves produced 11 molecules that were not produced by undamaged leaves. These compounds were not

SUMMARY TABLE 39.3 **Selected Sensory Systems in Plants**

Stimulus	Receptor	Signal Transduction	Response	Adaptive Significance
Blue light	PHOT1 and other phototropins in stem and leaves	PHOT1 autophosphorylates; remainder of signal transduction systems unknown	Phototropism occurs; also involved in stomatal opening	Stems grow toward light with wavelengths needed for photosynthesis
Blue light	Cryptochromes in stems and leaves	Details under investigation	Inhibits stem elongation; regulates daily rhythms and flowering response to day length	Keeps stems short if light is abundant; allows plant to time events
Red light	Phytochrome in seeds and elsewhere	Phytochrome changes to P_{fr} form and activates responses	Seed germinates	Sunlight triggers germination
Far-red light	Phytochrome in stem and elsewhere	P_r moves into nuclei and induces genetic responses	Stems lengthen	Species that require full sunlight attempt to escape shade from leaves
Gravity	Proteins located in plasma membrane?	Details unknown	Cells on opposite side of root or shoot elongate; tissue curves	Roots grow down; shoots grow up
Touch or wind	Stretch receptors; location unknown	Details unknown, but result is transcription activation in target genes	Stems grow shorter and thicker; in some cases vining (twining) growth	Individual is more resistant to damage; plants grow toward light
Touch	Receptor hair cell in Venus flytrap	Electrical changes in receptor cell's plasma membrane trigger action potentials	Target cells change shape; trap shuts	Plant can capture prey
Pathogens	R gene products	Details unknown	Hypersensitive response (HR); death of infected cells	Pathogens starve, so infection is slowed or stopped
Herbivores	Unknown; activated in response to molecule from herbivore	Details unknown	Insecticide production; signals to parasitoids	Herbivores are sickened or killed

released by leaves that had been cut with scissors or crushed with a tool; only insect damage triggered their production.

To follow up on this result, the investigators put female wasps in an arena that contained leaves damaged by insects or leaves that had suffered mechanical damage. In more than two-thirds of the tests that were performed, the wasps preferred to fly toward the insect-damaged leaves.

These results support the hypothesis that plants produce wasp attractants in response to attack by caterpillars. Biologists are increasingly convinced that plants can produce pheromones that recruit help in the form of egg-laden wasps.

From the research that has been done on plant sensory systems, it is abundantly clear that plants don't just sit there. These organisms may be stationary, but they constantly monitor and respond to a wide array of information about their environment (**Table 39.3**). Gaining a better understanding of phototropism, gravitropism, response to disease, and other aspects of plant behavior is an exciting frontier in biology.

CHECK YOUR UNDERSTANDING

If you understand that . . .

- The presence of *R* gene products allows plants to recognize attacks by parasites such as viruses, bacteria, and fungi.
- When the protein product of an *R* allele binds to a specific *avr* gene product from a pathogen, a signal transduction sequence results in the hypersensitive response.
- When plants are attacked by insect herbivores, damaged tissues release signals that trigger the production of toxins in undamaged leaves and that recruit enemies of the insects.

✓ You should be able to . . .

1. Explain the fitness advantage of having a wide array of *R* alleles in a single individual.

2. Explain why plants do not maintain high proteinase inhibitor concentrations in their tissues at all times.

Answers are available in Appendix B.

CHAPTER 39 REVIEW

For media, go to the study area at www.masteringbiology.com

Summary of Key Concepts

Plants are selective about the information they process. They perceive a wide variety of environmental stimuli that affect their ability to grow and reproduce.

- In most cases, information processing starts when a receptor protein changes shape in response to a stimulus. For example, a portion of the phytochrome protein changes shape when it absorbs red light. The same protein changes to an alternate shape when it absorbs far-red light.

- Plants monitor information about the nature and amount of light they receive, the direction of gravity, mechanical forces, and attacks by pathogens and herbivores.

  ✓ You should be able to state a hypothesis to explain why animals do not have phytochromes (a red/far-red light switch).

  MB Web Activity Sensing Light

When sensory cells receive a stimulus, they transduce the signal and respond by producing hormones that carry information to target cells elsewhere in the body.

- When signal transduction occurs, an external signal is changed into an internal signal.

- In receptor cells, signal transduction culminates in the production of hormones that are transported throughout the plant body.

  ✓ You should be able to explain the analogy between how a plant sensory cell works and the following events: A person sees a barn on fire and calls the rural fire department; the dispatcher rings a siren; in response, members of the volunteer fire department race to the station to get the pumper truck.

Target cells respond to hormonal stimulation in ways that increase the ability of the plant to survive and reproduce.

- If a hormone binds to a receptor on a target cell, signal transduction occurs and culminates in changes in gene expression, altered translation rates, or changes in the activity of specific membrane pumps, channels, or ion carriers.

- Cells near the tips of shoots sense changes in blue light and respond by altering the distribution of the hormone auxin. Cells on one side of the shoot elongate in response to auxin much more than do cells on the other side of the stem. In this way, plants bend toward sunlight.

  ✓ You should be able to predict whether the phototropic response differs in plants that require high-light conditions versus plants that thrive best in low-light conditions.

Hormones are also responsible for regulating how plants grow throughout their lives—especially in response to changes in environmental conditions. Each type of plant growth regulator plays a general role in the life of a plant, and growth responses are usually affected by interactions among several different hormones.

- Auxin establishes and maintains the long axis of the plant body, playing a key role in phototropism and gravitropism and maintaining apical dominance. All of these responses rely on the polar transport of auxin, which establishes a gradient in auxin from the plant's apex to its roots.

- Gibberellins signal that conditions for growth are good and promote the initiation or continuation of growth and development.

- Abscisic acid (ABA) signals that environmental conditions are bad by suppressing growth and enforcing dormancy.

- Regulation of dormancy and growth by ABA and by gibberellins (GAs) are examples of how hormones interact—allowing plants to integrate information from several different stimuli and respond appropriately.

✔ You should be able to suggest a hypothesis explaining why some hormones are not transported in a single direction.

(MB) **Web Activity** Plant Hormones, **Web Activity** Plant Defenses

Questions

1. Which of the following statements about phytochrome is *not* correct?
   a. It is photoreversible.
   b. Its function was understood long before the protein itself was isolated.
   c. The P_{fr} form activates the responses to red light.
   d. It is involved in guard-cell opening.

2. Why was it logical to predict that amyloplasts function as statoliths?
   a. They are dense and settle to the bottom of gravity-sensing cells.
   b. They are only present in gravity-sensing cells.
   c. They make a direct physical connection with membrane proteins called integrins, which have been shown to be the gravity receptor molecule.
   d. Their density changes in response to gravity.

3. If a plant is touched repeatedly over the course of many days or if it experiences long-term exposure to wind, what happens?
   a. Growth of the root system is accelerated, making the plant more stable.
   b. Large-scale changes in gene expression occur, resulting in shoots that are short and stout.
   c. Electrical signals cause leaves to fold up, avoiding damage.
   d. Continued mechanical stimulation indicates that the individual is threatened with destruction, so it initiates flowering in an attempt to reproduce before it dies.

4. Which of the following statements about hormones is *not* correct?
   a. They tend to be small molecules.
   b. They exert their effects only on the same cells that produce them.
   c. They can exert strong effects even when they are present in extremely low concentrations.
   d. They trigger a response by binding to receptors in target cells.

5. In order for auxin to stimulate cell elongation, what two things have to happen?
   a. Transport proteins on the apical and basal ends of the cell must be activated.
   b. Proteins in the aleurone layer must be activated and α-amylase transcription increased.
   c. Water must flow into the cell, and the cell wall must become more extensible.
   d. Water must flow out of the cell, and the cell wall must become more extensible.

6. What evidence suggests that ABA from roots can signal guard cells to close?
   a. If roots are given sufficient water, guard cells close anyway.
   b. If roots are dry, guard cells begin to close—even though leaves are not experiencing water stress.
   c. Applying ABA on guard cells directly causes them to close.
   d. If roots are dry, ABA concentrations in leaf cells drop dramatically.

1. Phytochromes can be considered "shade detectors," while phototropins such as PHOT1 can be considered "sunlight detectors." Explain why these characterizations are valid.

2. In the experiment that confirmed the self-phosphorylation hypothesis for the blue-light receptor, researchers inserted the *PHOT1* gene into insect cells. Why was it important for the gene to be expressed in an organism other than a plant?

3. What does *transduce* mean? Give an example of a signal transduction event in plant sensory systems.

4. A plant's response to a given hormone depends on the cells or tissues that receive the signal, the plant's developmental stage or age, the concentration of the hormone, and the concentration of other plant hormones that are present. Provide examples that support each of these claims.

5. How do changes in hormonal signals allow plants to respond to changes in their physical environment over the course of their lives? Answer this question using phototropism as an example.

6. Discuss the general role that ethylene serves in plants. Provide evidence that supports your claim.

1. In general, small seeds that have few food reserves must be exposed to red light before they will germinate. (Lettuce is an example.) In contrast, large seeds that have substantial food reserves typically do not depend on red light as a stimulus to trigger germination. State a hypothesis to explain these observations.

2. To explore how hormones function, researchers have begun to transform plants with particular genes. In one experiment, a gene involved in cytokinin synthesis was introduced into tobacco plants. When the recombinant individuals matured, they produced more than the usual number of lateral branches. How does this experiment inform our understanding of how cytokinins function?

3. In many species native to tropical wet forests, seeds do not undergo a period of dormancy. Instead, they germinate immediately. Make a prediction about the role of ABA in these seeds. How would you test your predictions?

4. Researchers have shown that stomata do not close if ABA is injected into guard cells. Stomata do close, however, if ABA is applied to the surface of guard cells. Based on these results, researchers claim that the ABA receptor must be on the surface of guard cells, not in the interior. Do you agree? Why or why not?

This chapter focuses on the structure and function of plant reproductive structures, such as those of this flower.

Plant Reproduction

40

It would be difficult to overemphasize the importance of the reproductive organs and processes that are analyzed in this chapter—for plants, for biologists, and for you.

- For plants, every structure in the body and every physiological process—from water transport to photosynthesis—exist for one reason: to maximize the chances that the individual will produce offspring. As Chapter 1 pointed out, reproduction is the unconscious goal of everything that an organism does.

- For biologists, plant reproduction is not only fundamental to understanding how plants work but also the basis for major industries. Agriculture, horticulture, forestry, biotechnology, and ecological restoration draw extensively on what biologists know about plant reproduction.

- For you, plant reproduction means food. Human diets are based on consuming plant reproductive structures—primarily the seeds and fruits derived from flowers. A **flower** is a reproductive structure that produces gametes, attracts gametes from other individuals, nourishes embryos, and develops seeds and fruits. **Seeds** consist of an embryo and nutrient stores surrounded by a protective coat. **Fruits** develop from the flower's seed-producing organ and contain seeds.

This chapter's analysis of plant reproduction focuses on angiosperms, for three reasons: (**1**) Angiosperms represent over 85 percent of the land plants described to date; (**2**) virtually every important domesticated plant is an angiosperm; and (**3**) Chapter 30 introduced aspects of reproduction in other land plant lineages. By the end of this chapter, you'll appreciate the practical aspects of flowers as well as their beauty.

KEY CONCEPTS

- Plants undergo alternation of generations, in which a diploid sporophyte phase alternates with a haploid gametophyte phase. Sporophytes produce spores by meiosis. Gametophytes produce gametes by mitosis.

- In angiosperms, male and female gametophytes are microscopic and are produced inside flowers. Male gametophytes (pollen grains) are portable. Female gametophytes are encased in an ovary and are retained in the flower. When pollen grains land on a flower, they deliver sperm cells that fertilize the egg produced by the female gametophyte.

- Seeds contain an embryo and a food supply surrounded by a coat. In angiosperms, the walls of the ovary develop into a fruit that encloses the seed or seeds. In many cases, fruits function in seed dispersal.

✔ When you see this checkmark, stop and test yourself. Answers are available in Appendix B.

40.1 An Introduction to Plant Reproduction

Plant reproductive structures and processes vary among species. Consider just one aspect of reproductive organs—size. Flowers vary from microscopic to the size of a small child; seeds and fruits range from dustlike particles to coconuts.

Fortunately for students of plant biology, several basic principles unify this diversity of reproductive systems. Let's begin by defining sex.

Sexual Reproduction

Most plants reproduce sexually. As Chapter 12 pointed out, **sexual reproduction** is based on meiosis and fertilization, and results in offspring that are genetically unlike each other and unlike their parents. In plants, meiosis and fertilization occur in alternate phases of a life cycle.

To review briefly, **meiosis** is a type of nuclear division that results in four daughter cells, each of which has half the number of chromosomes present in the parent cell. **Fertilization** is the fusion of haploid cells termed **gametes**. The result of fertilization is the production of a single, diploid cell called the **zygote**, which will develop into a multicellular individual. Male gametes, or **sperm**, are small cells that contribute genetic information in the form of DNA but few or no nutrients to the offspring. Female gametes, or **eggs**, also contain DNA. But the female parent contributes nutrient stores as well. The important point is that although both sperm and egg contribute DNA, the male and female parent contribute vastly different amounts of other resources to the offspring.

Meiosis and fertilization result in offspring that are genetically unlike the parents. Sexual reproduction produces genetically diverse offspring that may be much more successful at warding off attacks from viruses, bacteria, and other pathogens than their parents (see Chapter 12).

The advantages of sexual reproduction are common to all eukaryotes that undergo meiosis. When and where meiosis occurs, however, is highly variable. Let's take a closer look.

The Land Plant Life Cycle

In most animals, meiosis leads directly to the formation of gametes. In plants, the situation is much different.

🔑 Land plants are characterized by a life cycle with two distinct multicellular forms—one diploid and one haploid. An individual in the diploid phase of the life cycle is called a **sporophyte**, while an individual in the haploid phase of the life cycle is called a **gametophyte**.

WHAT IS ALTERNATION OF GENERATIONS? This type of life cycle, **alternation of generations**, was introduced in Chapters 29 and 30 and is reviewed in **Figure 40.1**. Figures in those chapters detail how alternation of generations works in various protists and land plant groups. It is a life cycle that has evolved several times independently during the history of life.

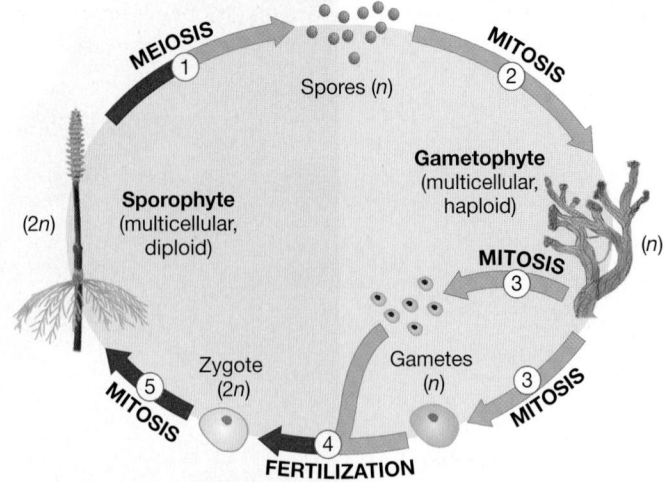

FIGURE 40.1 All Plants Undergo Alternation of Generations.

When alternation of generations occurs, meiosis does not lead directly to the formation of gametes as it does in humans and other animals. Instead it leads to the production of haploid cells called spores. A **spore** is a cell that grows directly into an adult individual. Several structures and processes are common to all land plant life cycles:

1. Meiosis occurs in sporophytes and results in the production of haploid spores. Unlike zygotes, spores are not produced by the fusion of two cells. Unlike gametes, spores produce an adult without fusing with another cell. Meiosis and spore production occur inside structures called **sporangia**.

2. Spores divide by mitosis to form multicellular, haploid gametophytes.

3. Gametophytes produce gametes by mitosis.

4. Fertilization occurs when two gametes fuse to form a diploid zygote.

5. The zygote grows by mitosis to form the sporophyte.

A good way to keep these terms straight is to remember that sporophyte means "spore-plant," while gametophyte means "gamete-plant." Sporophytes produce spores by meiosis. Gametophytes produce gametes by mitosis.

VARIATION IN LAND PLANT LIFE CYCLES Chapter 30 pointed out that land plant species vary a great deal in how large and long lived the gametophyte and sporophyte are relative to each other. The examples in **Figure 40.2** represent the extremes of the variation in life cycles observed within land plants.

- Figure 40.2a diagrams the life cycle of a liverwort. Notice that the largest stage in the liverwort life cycle is the gametophyte. The sporophyte is dependent on the gametophyte for nutrition.

- Figure 40.2b diagrams the life cycle of an angiosperm. Notice that the male and female gametophytes are microscopic, physically separate, and completely dependent on the sporophyte for nutrition.

(a) Liverworts: Gametophytes are large and long lived; the sporophyte is small and short lived.

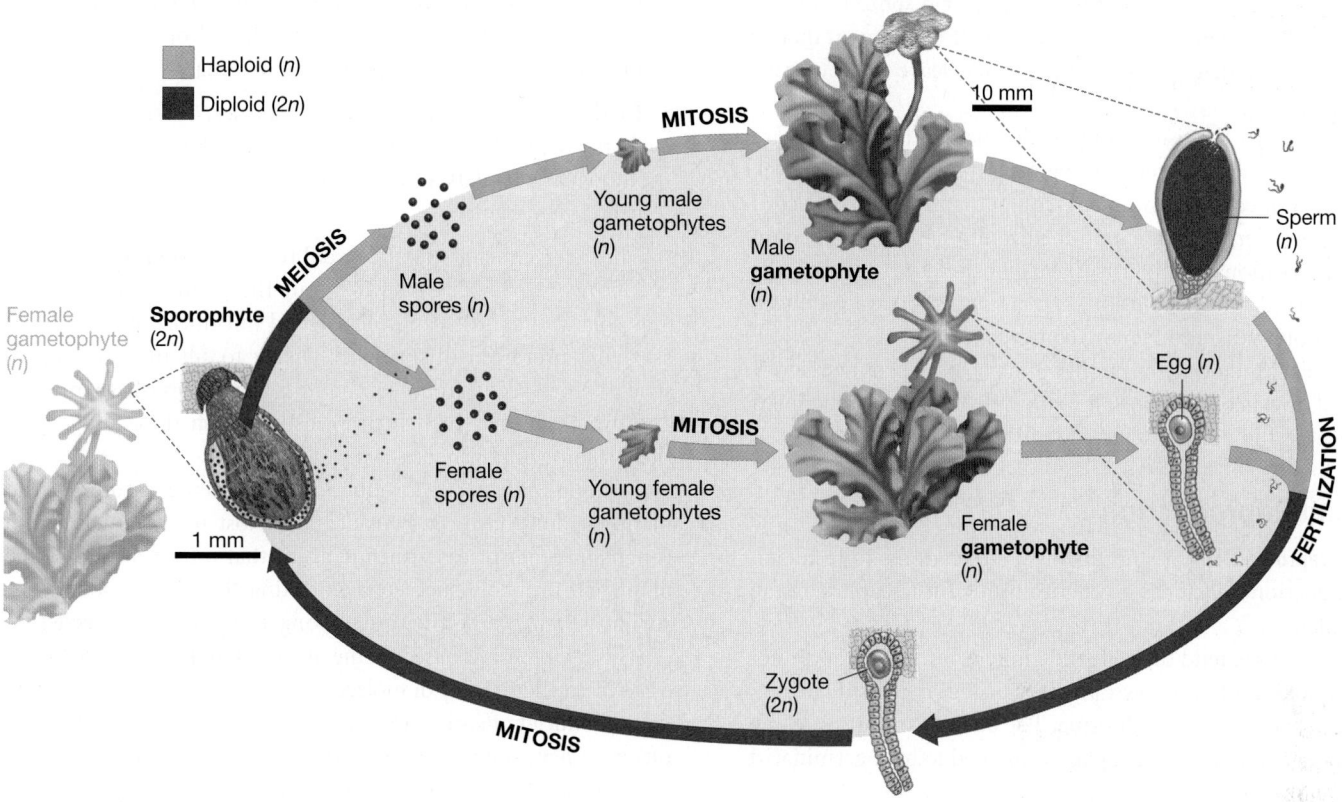

(b) Angiosperms: Sporophyte is large and long lived; gametophytes are small (microscopic) and short lived.

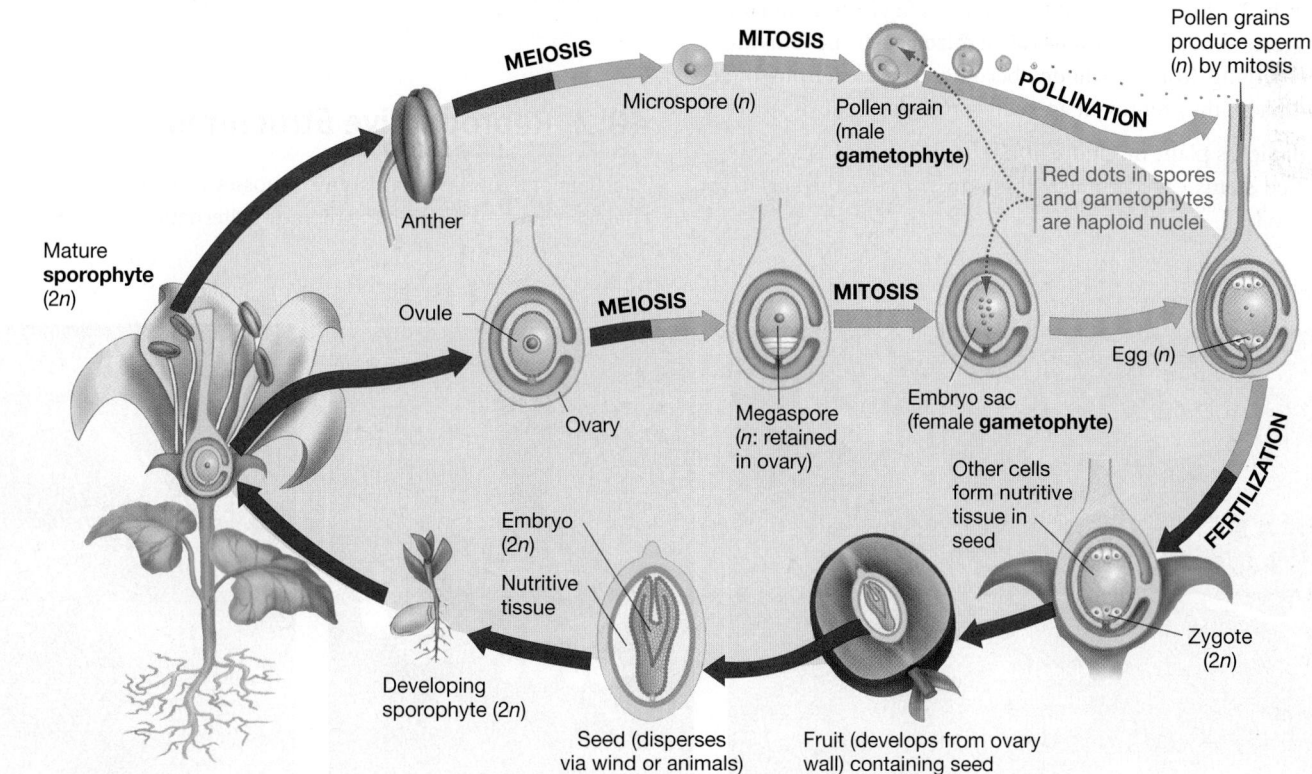

FIGURE 40.2 There Is Wide Variation in Plant Life Cycles. (a) In liverworts and other basal groups of land plants, the sporophyte depends on the gametophyte for nutrition. **(b)** In angiosperms and other more recent groups of land plants, the gametophytes depend on the sporophyte for nutrition.

✔If you understand the angiosperm life cycle, you should be able to (1) identify the male spore and female spore in Figure 40.2b, and (2) provide evidence to support the statement that in angiosperms, female gametophytes never leave their parent plant.

Liverworts are the sister group to all other land plants, and are found early in the fossil record. Angiosperms, in contrast, are the most recent lineage of land plants to appear in the fossil record and the most derived branch on the evolutionary tree. Based on these observations, biologists have concluded that over the course of land plant evolution, gametophytes became reduced while sporophytes became more conspicuous.

Why this trend occurred is still unknown. The adaptive significance of the gametophyte and sporophyte phases remains a challenge for biologists interested in plant reproduction.

Asexual Reproduction

Asexual reproduction does not involve fertilization and results in the production of **clones**—genetically identical copies of the parent plant.

Some plants extend their life indefinitely by asexual reproduction. The oldest of all known plants is a ring of creosote bushes in the Mojave Desert of California. The bushes comprise a clone that originated from a parent plant estimated to have germinated 12,000 years ago.

Although all asexual reproduction is based on mitosis, a wide array of mechanisms is involved.

- **Figure 40.3a** shows shoots and roots emerging from nodes on underground horizontal stems called **rhizomes**. If the individuals emerging from the nodes become separated from the parent plant, they represent asexually produced offspring.

- The gladiolus plant in **Figure 40.3b** has propagated itself via modified stems called **corms**, which grow under the surface of the soil.

- The kalanchoe in **Figure 40.3c** produces **plantlets**, which form from meristematic tissue located along the margins of its leaves. When the plantlets mature, they drop off the parent plant and grow into independent individuals.

- In dandelion and certain other species, mature seeds can form without fertilization occurring. This phenomenon, known as **apomixis**, results in seeds that are genetically identical to the parent.

The key characteristic of asexual reproduction is efficiency. If an herbivore or a disease wipes out the plants surrounding a grass plant, the grass can quickly send out horizontal stems. Its asexually produced offspring are likely to fill the unoccupied space before seeds from competitors can establish themselves and grow. The parent plant can also nourish these progeny as they become established.

Although asexual reproduction is extremely common in plants, it does have a downside. The most important is that a fungus or other disease-causing agent that infects an individual plant will probably succeed in infecting the plant's cloned offspring as well, even if they are no longer physically connected.

This hypothesis is based on the observation that plants fight disease with a wide variety of molecules (see Chapter 39). Because sexually produced offspring are genetically unlike their parents, these offspring have unique combinations of disease-fighting molecules and may be able to resist infections that devastate their parents.

This is an important point in agriculture and horticulture because asexually propagated apples, bananas, and other crops are more susceptible to epidemics than are sexually propagated species. How does sexual reproduction occur?

40.2 Reproductive Structures

Each major group of plants, from mosses to angiosperms, has a characteristic variation on the theme of alternation of generations,

(a) Rhizome

Node with root and shoot emerging

(b) Corm

Parent corm

New corms

(c) Plantlets

FIGURE 40.3 The Mechanisms of Asexual Reproduction Are Diverse. These are just three of many mechanisms of asexual reproduction in plants.

as well as characteristic male and female reproductive structures (see Chapter 30). Here, though, the focus is on the flower.

Let's start with a simple question: When do plants produce them?

When Does Flowering Occur?

Anatomically, a flower is a modified shoot that develops from a compressed stem and highly modified leaves. In essence, flower formation begins when an apical meristem stops making energy-harvesting stems and leaves and begins to produce the modified stems and leaves that make up flowers. Instead of making more food through photosynthesis, a sporophyte commits to investing energy in sexual reproduction—by producing gametophytes that will produce gametes.

When does this happen? Early experiments on the environmental signals that promote flowering focused on the number of hours of light and dark during a day.

RESPONDING TO CHANGES IN PHOTOPERIOD **Photoperiodism** is any response by an organism that is based on photoperiod—the relative lengths of day and night. In plants, the ability to measure photoperiod is important because it allows individuals to respond to seasonal changes in climate—for example, to flower when pollinators are available and when resources for producing seeds are abundant.

Experiments on photoperiodism in plants have shown that, with respect to flowering, plants fall into three main categories:

1. **Long-day plants** bloom in midsummer, when days are longest and nights shortest. Radishes, lettuce, spinach, corn, irises, and other long-day plants flower only when days are longer than a certain length—usually between 10 and 16 hours, depending on the species.

2. **Short-day plants** bloom in spring, late summer, or fall. Asters, chrysanthemums, poinsettias, and other short-day plants flower only if days are shorter than a certain species-specific length.

3. **Day-neutral plants** flower without regard to photoperiod. Day length has no effect on flowering in roses, snapdragons, dandelions, tomatoes, cucumbers, and many **weeds**—plants that are adapted to grow in soils that have been disturbed enough to remove or damage existing vegetation.

How do plants sense changes in day length and initiate the chain of events that leads to flowering? Early experiments suggested that phytochrome—the photoreversible pigment introduced in Chapter 39—is involved.

Figure 40.4 summarizes some of the key experiments. Researchers showed that interrupting the night period with a light flash changed the flowering response. In short-day plants, for example, flowering was inhibited if the dark period was interrupted by red light—wavelengths around 660 nm. But a subsequent flash of far-red light—wavelengths around 735 nm—erased the effect. In some way, a red/far-red switch linked changes in day length to flowering.

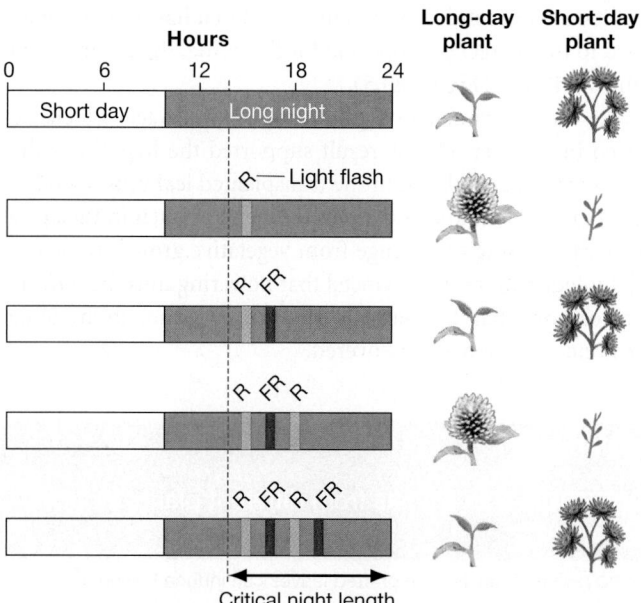

FIGURE 40.4 Flashes of Red Light and Far-Red Light Switch the Photoperiod Response On and Off. If flashes of red light (R) and far-red light (FR) are alternated during the night, the plant's flowering response correlates with the last light it experienced. A flash of red light turns a long night into a short night; a flash of far-red light restores the perception of a long night.

More recent research has shown that phytochrome's effect is closely tied to the molecular mechanisms of timekeeping in plants. Plants have a clock that is reset each morning. Clock proteins rise during the day and trigger expression of a gene called *CONSTANS* (*CO*). The CO protein is a transcription factor that affects the production of a flowering hormone. When phytochrome is activated by light, it stabilizes CO—so that CO accumulates in cells.

- In long-day plants, high levels of CO stimulate production of the flowering hormone.

- In short-day plants, high levels of CO *inhibit* production of the flowering hormone.

Many questions remain, however, and research is continuing. In the meantime, it is clear that there is a complex interaction between clock proteins, phytochrome activation, and the production of the signal that initiates flowering. What is that signal?

THE DISCOVERY OF THE FLOWERING HORMONE Since the 1930s, biologists have known that exposing even one leaf on a plant to the conditions necessary to induce flowering may result in the whole plant flowering. This result suggests that the signal to flower comes from leaves and travels to the apical meristem.

Grafting experiments, in which an organ from one individual is physically attached to a different individual, support this result. If you provide an experimental plant with the appropriate flower-triggering night length and then cut off a leaf or stem and graft it onto a second, experimental plant that has never been exposed to the correct photoperiod for flowering, the experimental plant will flower (**Figure 40.5**).

Like the experiments on phototropism with agar blocks reviewed in Chapter 39, this result supported the hypothesis that some substance produced in the transplanted leaf must travel up the recipient plant and trip a developmental switch in the apical meristem, causing the change from vegetative growth to flowering. Biologists were so convinced that flowering must be induced by a hormone that they named it **florigen**, even though the actual hormone had not been discovered.

EXPERIMENT

QUESTION: Can signals from different plants induce flowering?

HYPOTHESIS: Signals from grafted leaves can induce flowering.

NULL HYPOTHESIS: Signals from grafted leaves cannot induce flowering.

EXPERIMENTAL SETUP:

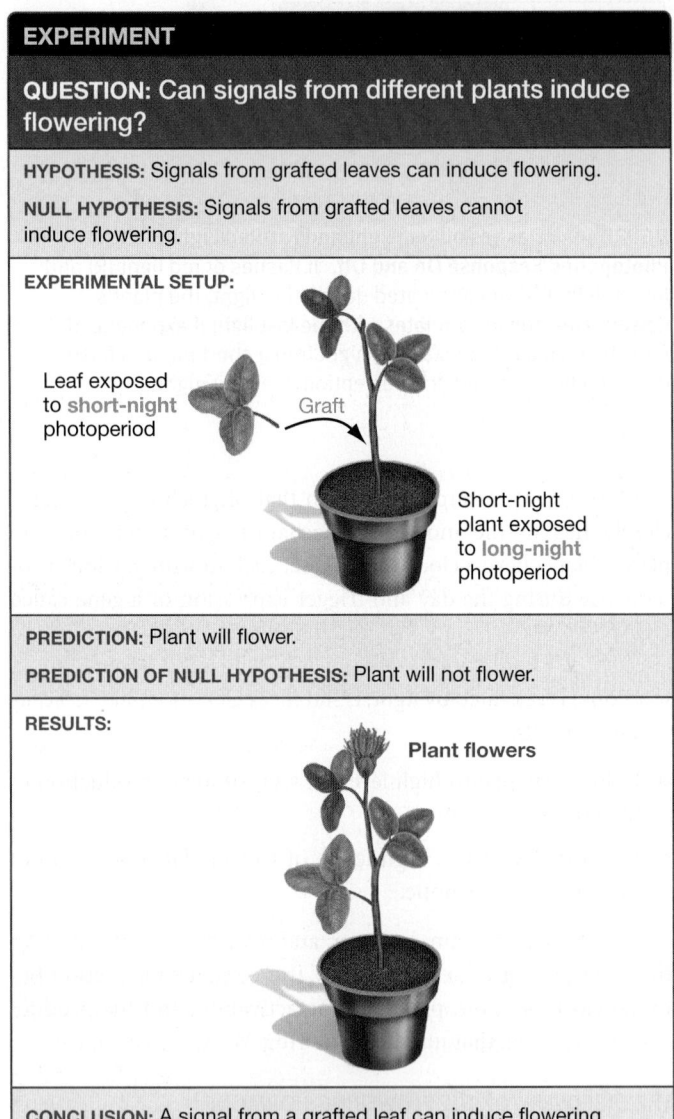

Leaf exposed to **short-night** photoperiod

Graft

Short-night plant exposed to **long-night** photoperiod

PREDICTION: Plant will flower.

PREDICTION OF NULL HYPOTHESIS: Plant will not flower.

RESULTS:

Plant flowers

CONCLUSION: A signal from a grafted leaf can induce flowering.

FIGURE 40.5 Experimental Support for the Hypothesis That a Hormone for Flowering Exists.

SOURCE: Lang, A., M. K. Chailakhyan, and I. A. Frolova. 1977. Promotion and inhibition of flower formation in a day-neutral plant in grafts with a short-day plant and a long-day plant. *Proceedings of the National Academy of Sciences, USA* 74: 2412–2416.

✔**QUESTION** For this experiment to support the claim that a flowering hormone exists, a control treatment needs to be done. What is this control treatment? If the hypothesis is correct, what is the predicted outcome of the control treatment?

Almost 80 years later, researchers finally found the florigen molecule. This work began by focusing on the *FLOWERING LOCUS T (FT)* gene in *Arabidopsis thaliana*, which is known to promote flowering when activated. When researchers exposed leaves to short nights, they found that the gene was expressed in the leaf vascular tissue. Subsequently, some research indicates that the protein product of the *FT* gene is transported from leaves to the shoot apical meristem, and that the protein's presence triggers the activation of genes required for converting the stem to a flower.

Now the question is, What does the flower itself look like?

The General Structure of the Flower

Structurally, all flowers are variations on a theme. They are made up of four basic organs that are essentially modified leaves: (**1**) sepals, (**2**) petals, (**3**) stamens, and (**4**) one or more carpels. These organs are attached to a compressed portion of stem called the receptacle (**Figure 40.6a**).

Not all four organs are present in all flowers, however; and as **Figure 40.6b** shows, the coloration, size, and shape of these four components are fabulously diverse. Let's consider each of the four parts, in turn.

SEPALS FORM AN OUTER, PROTECTIVE WHORL **Sepals** are leaflike structures that make up the outermost parts of a flower. Sepals are usually green and photosynthetic, and they are relatively thick compared with other parts of the flower.

Because they attach to the receptacle in a circle or whorled arrangement, sepals enclose the flower bud as it develops and grows—protecting young buds from damage by insects or disease-causing agents. The entire group of sepals in the flower is called the **calyx**.

PETALS FURNISH A VISUAL ADVERTISEMENT Like sepals, **petals** are arranged around the receptacle in a whorl. Often brightly colored and scented, petals function to advertise the flower to bees, flies, hummingbirds, and other pollinators.

In some cases, the color of the petals correlates with the visual abilities of particular animals. Bees, for example, respond strongly to wavelengths in the blue and purple regions of the light spectrum, as well as yellow (they don't see red well). Flowers that attract bees, in turn, often have yellow, blue, or purple petals with ultraviolet patches.

The ultraviolet sections of petals in "bee flowers" frequently highlight the center of the flower (**Figure 40.7**), where the stamens and carpels are located. Why? In these flowers, the base of the petals contains a gland called a **nectary**. The nectary produces the sugar-rich fluid **nectar**, which is harvested by many of the animals that visit flowers, along with pollen. In the process of collecting pollen or nectar, the visiting animal usually deposits pollen from a different plant on the stamens—accomplishing pollination.

The entire group of petals in a flower is called the **corolla**. In some species, the petals within the corolla vary in size, shape, and function:

(a) Basic parts of a flower

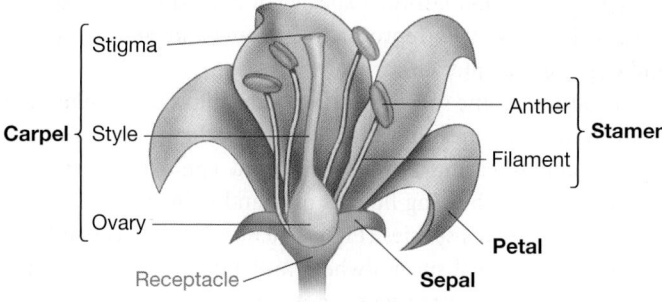

(b) Examples of flower diversity

FIGURE 40.6 The Basic Structures in Flowers Are Highly Variable.
(a) Flowers comprise sepals, petals, stamens, and carpels. **(b)** The four parts vary among species.

(a) What you and a bee see   **(b)** What a bee sees in addition

FIGURE 40.7 Insects See in the Ultraviolet Range.
(a) The inflorescence (flower cluster) of a black-eyed Susan, seen by the unaided human eye. **(b)** The same structure, photographed with a camera that records ultraviolet wavelengths that are visible to bees but invisible to humans.

- Flattened petals may provide a landing pad for flying insects.
- Elongated, tubelike petals frequently have a nectary at their base that can be reached only by animals with a long beak or tongue-like proboscis.
- Some petals protect the reproductive organs located inside the corolla.
- Specialized cells in some petals synthesize and release molecules that provide a scent attractive to certain species of pollinating insects.

In contrast, wind-pollinated angiosperms such as oaks, birches, pecans, and grasses have flowers that have small petals or no petals at all, and that lack nectaries. These species do not invest in structures that aren't required for pollination.

STAMENS PRODUCE POLLEN **Stamens** are reproductive structures that produce male gametophytes—also known as pollen grains. The male gametophytes, in turn, produce sperm.

Each stamen consists of two components:

1. a slender stalk termed the **filament**, and

2. the pollen-producing organs called **anthers** (see Figure 40.6a).

The anther is the business end of the stamen—where meiosis and pollen formation take place. The function of the filament is to hold the stamen in a place where wind, insects, hummingbirds, or other agents can make contact with the pollen grains produced in the anther.

CARPELS PRODUCE OVULES The fourth reproductive structure is the **carpel**, which produces female gametophytes. A carpel consists of three regions:

1. The **stigma** is a moist tip that receives pollen.

2. The **style** is a slender stalk.

3. The **ovary** is an enlarged structure at the base of the carpel (see Figure 40.6a).

Inside the ovary, female gametophytes are produced in structures called **ovules**. An ovary may contain more than one ovule. When the female gametophytes that are produced inside ovules mature, they produce eggs.

THE "SEX" OF FLOWERS VARIES   In most angiosperm species, stamens and carpels are produced on the same individual. Flowers that contain both stamens and carpels are referred to as **perfect**.

Flowers can also be **imperfect**, however, meaning they contain either stamens *or* carpels, but not both. Imperfect flowers that contain only stamens can be considered "male" flowers. Similarly, imperfect flowers that contain only carpels can be considered "female" flowers.[1]

In some cases, separate stamen- or carpel-producing flowers occur on the same individual. Species like these, including the corn plants illustrated in **Figure 40.8a**, are **monoecious** (literally, "one-house"). In corn, the tassel is a collection of stamen-producing "male" flowers, and the ear contains a group of carpel-producing "female" flowers.

---

[1]Technically, flowers are not referred to as male and female. Instead, they are staminate or carpellate. Staminate flowers produce stamens, which produce pollen grains, which produce male gametes (sperm). Carpellate flowers produce carpels, which contain ovaries. Female gametophytes develop inside ovaries and produce female gametes (eggs). For convenience, though, the text will sometimes refer to male and female flowers and reproductive structures.

(a) Corn is monoecious.

(b) *Cannabis* is dioecious.

FIGURE 40.8 Male and Female Flowers Can Occur on the Same Individual or on Different Individuals. (a) The tassels of corn are male flowers; ears are female flowers. **(b)** In *Cannabis sativa*, male and female flowers are found on different individuals.

Species that are monoecious or that have perfect flowers are analogous to hermaphroditic species of animals, which contain both male and female reproductive organs and produce sperm and eggs (see Chapter 48).

In contrast, some species with imperfect flowers are **dioecious** ("two-houses")—meaning that each individual plant produces either stamen-bearing flowers only, and could be considered "male," or carpel-bearing flowers only and is "female." *Cannabis sativa* is a dioecious species (**Figure 40.8b**). Dioecious plants are analogous to animal species where individuals have either male or female reproductive organs—not both.

How Are Female Gametophytes Produced?

What purposes do the three parts of the carpel serve? The function of the stigma and style will become clear in Section 40.3; for now let's concentrate on what happens inside the ovary.

Figure 40.9 is a longitudinal section showing the inside of a typical angiosperm ovary. Notice that it contains one or more ovules. Each ovule contains a structure called the megasporangium, which contains a cell called the megasporocyte. (The use of "mega" is appropriate, because these structures are much larger than their counterparts in the stamen.) The megasporangium is comparable to spore-producing organs found in other plants, such as the sporangia found on the back of fern leaves.

When you study Figure 40.9, note four important points:

1. The megasporocyte divides by meiosis.

2. Four haploid nuclei called **megaspores** result from meiosis, but three degenerate. No one is sure how or why this happens.

3. The surviving megaspore divides by mitosis to produce a structure with haploid nuclei. This is the female gametophyte—usually known as the **embryo sac**.

4. The haploid nuclei segregate to different positions in the embryo sac, and cell walls form around them.

🔑 In the carpel, then, a diploid megasporocyte divides by meiosis to form a megaspore, which then divides by mitosis to form the female gametophyte. Female gametophytes are encased in an ovary, are retained in the flower, and produce an egg.

In many angiosperms the embryo sac contains eight haploid nuclei and seven cells. Typically, two **polar nuclei** stay together within one central cell—the largest cell in the ovule. The number of polar nuclei varies among species, however.

The egg cell is located at one end of the female gametophyte, near an opening in the ovule called the **micropyle** ("little-gate"). The micropyle is where a sperm nucleus will enter the ovule, if fertilization occurs.

How Are Male Gametophytes Produced?

Figure 40.10 provides a detailed look at the stamen and the steps that occur in the production of male gametophytes. Recall that a stamen consists of two major parts: an anther and a filament. Inside the anther, structures known as microsporangia contain diploid cells called microsporocytes.

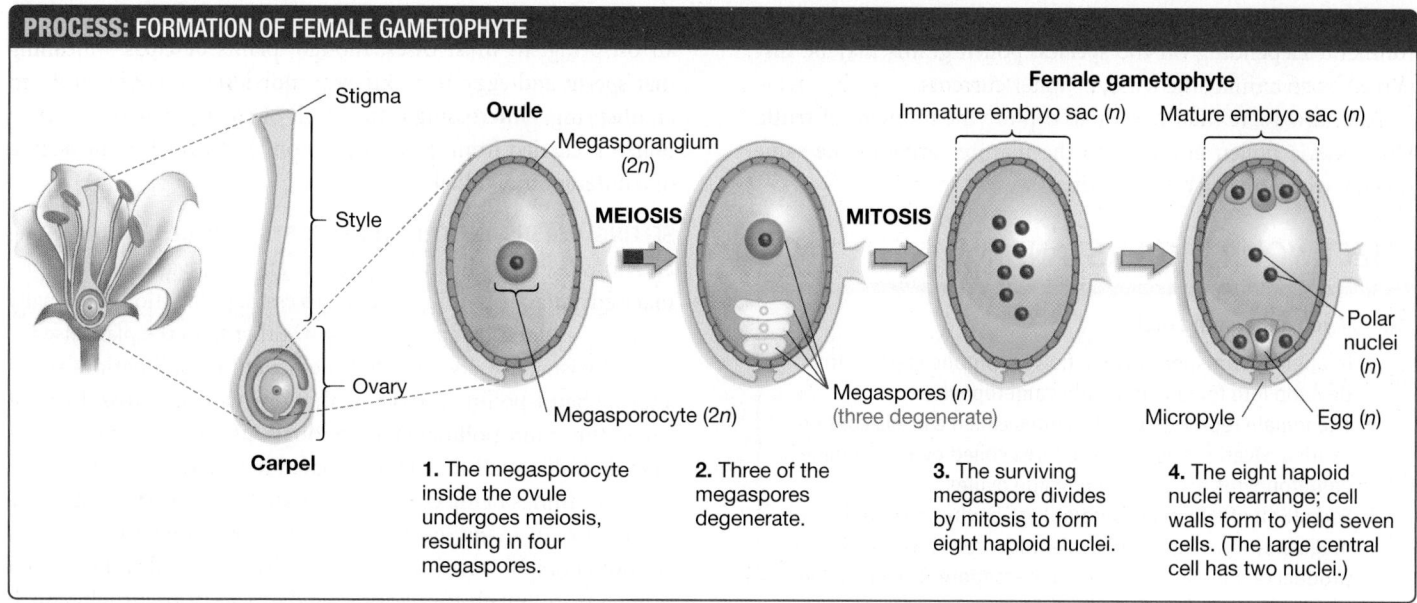

FIGURE 40.9 In Angiosperms, Megaspores Produce Female Gametophytes.

✔QUESTION Define a gametophyte. Why does the embryo sac conform to this definition?

When you study Figure 40.10, note three important points:

1. Microsporocytes undergo meiosis.

2. Each haploid cell that results is a **microspore**. Normally, all of the microspores survive. Microspores divide by mitosis.

3. The two nuclei that result from mitotic division in a microspore form a haploid, immature male gametophyte, also known as the **pollen grain**.

In the anther, then, a diploid microsporocyte divides by meiosis to form microspores, which then divide by mitosis to form male gametophytes. Male gametophytes are dispersed from the flower, and eventually produce sperm.

At the immature stage—before it has produced sperm—the male gametophyte consists of two cells: a small generative cell enclosed within a larger tube cell. The male gametophyte is considered mature when the haploid generative cell produces two sperm cells via mitosis.

In some species, this maturation step occurs while pollen is still in the anther. In other species, maturation and sperm production don't occur until after the pollen grain lands on a stigma and begins to grow.

The wall of a pollen grain develops a tough outer coat that includes the watertight compound called sporopollenin, introduced in Chapter 30. This coat protects the male gametophyte

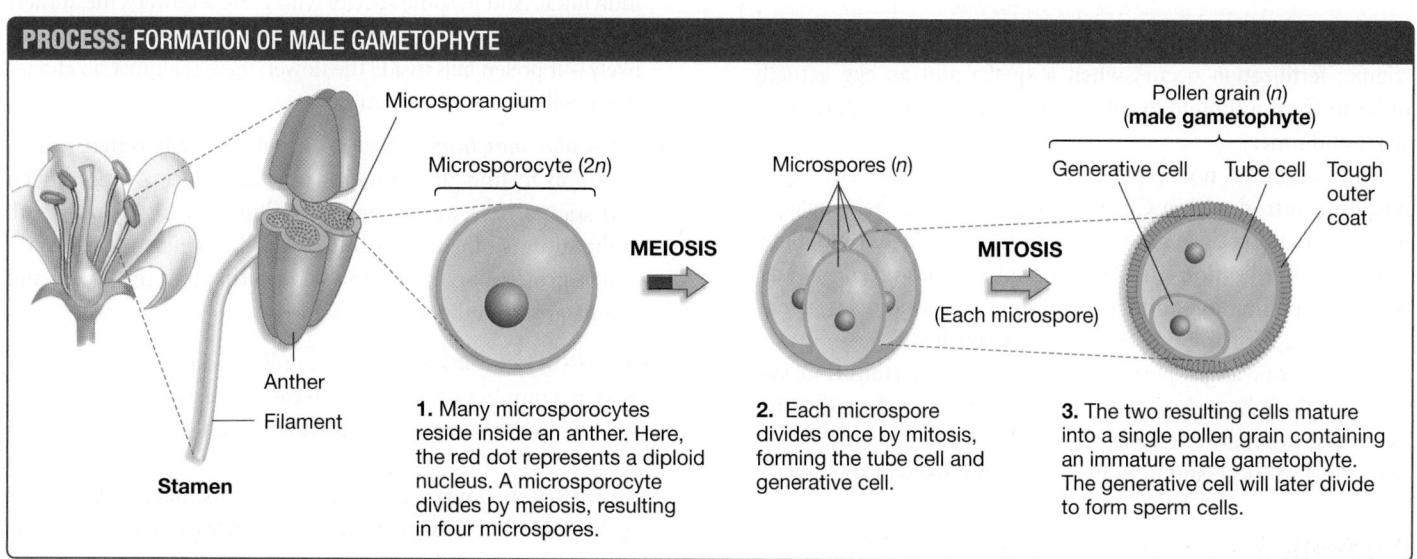

FIGURE 40.10 In Angiosperms, Microspores Produce Male Gametophytes.

✔QUESTION Why do pollen grains conform to the definition of a gametophyte?

when the pollen is released from the parent plant into the environment. Depending on the species, pollen grains may be dispersed by an animal, the wind, or water currents.

And now we're finally ready to explore the moment of truth: How does a pollen grain get to the mature carpel of the same species, where an egg cell is waiting?

CHECK YOUR UNDERSTANDING

If you understand that . . .

- In angiosperm sporophytes, flowers produce spores that develop into female and male gametophytes.
- The female reproductive structures called carpels contain ovaries. Ovaries enclose structures called ovules. Female gametophytes are produced inside ovules.
- Formation of a female gametophyte begins when a diploid megasporocyte inside an ovule undergoes meiosis. The product of meiosis is a haploid megaspore. The megaspore divides by mitosis to form the female gametophyte—including the egg and polar nuclei.
- Male gametophytes are produced inside reproductive structures called anthers.
- Formation of the male gametophyte begins when a diploid microsporocyte undergoes meiosis to form haploid microspores. Microspores divide by mitosis to form the male gametophyte—including a generative cell that will divide by mitosis to form two sperm cells.

You should be able to . . .

1. Compare and contrast a megasporocyte and a microsporocyte.
2. Compare and contrast a female gametophyte and a male gametophyte.

Answers are available in Appendix B.

40.3 Pollination and Fertilization

Pollination is the transfer of pollen grains from an anther to a stigma; **fertilization** occurs when a sperm and an egg actually unite to form a diploid zygote. The two events are separated in space and time.

Pollination is not restricted to angiosperms. The gymnosperms introduced in Chapter 30 also package their male gametophytes into pollen grains. This section will focus on pollination and fertilization in flowering plants, however, because managing pollination and fertilization in angiosperms is a critical challenge for fruit growers and plant breeders.

In addition, angiosperms' pollination and fertilization systems are thought to be key to their evolutionary success. What aspects of pollination and fertilization allowed flowering plants to become so successful in terms of their numbers of species?

Pollination

Pollen can fall on the stigma of the same individual or the stigma of a different individual. **Self-fertilization**, or **selfing**, occurs when

a sperm and an egg from the same individual combine to produce an offspring. In most cases, though, plants **outcross**—meaning that sperm and eggs from different individuals combine to form an offspring. Outcrossing is the result of **cross-pollination**—when pollen is carried from the anther of one individual to the stigma of a different individual.

SELFING VERSUS OUTCROSSING: COSTS AND BENEFITS Selfing and outcrossing each have advantages and disadvantages. The primary advantage of selfing is that successful pollination is virtually assured—it doesn't depend on agents other than the plant itself.

Biologists have documented the benefit of pollination-assurance by hand-pollinating plants that normally outcross. In most cases, the hand-pollinated individuals produce far more seed than do individuals that are pollinated naturally.

Other things being equal, self-pollination should result in the production of many more seeds than outcrossing. Other things are not equal, however. Selfing has a distinct disadvantage: Even though it still involves meiosis, selfed offspring are usually much less diverse genetically than outcrossed offspring are (see Chapter 12). In some cases, selfed offspring may also suffer from inbreeding depression (see Chapter 25).

Although outcrossing is riskier in terms of the chances that pollination will occur, it results in genetically diverse offspring that may be much more successful at warding off attacks from viruses, bacteria, and other pathogens (see Chapter 12).

Outcrossing is much more common than selfing, and in many cases plants have elaborate mechanisms to prevent selfing. These mechanisms include:

- *Temporal avoidance* In some species that have perfect flowers, male and female gametophytes mature at different times. Thus, selfing cannot occur.

- *Spatial avoidance* Selfing isn't possible in dioecious species and may be rare in monoecious species, unless pollinators transfer pollen between different-"sexed" flowers on the same individual. And in some species with perfect flowers, the anthers and stigma are so far apart that self-pollination is extremely unlikely—if pollen falls inside the flower, there is almost no chance that it will land on the stigma.

- *Molecular matching* In species that normally outcross, it is common to find that proteins on the surfaces of the pollen and stigma have to interact for pollination to be successful. Pollination is blocked if proteins on the pollen grain surface match proteins on the stigma—indicating that the pollen and stigma are from the same individual.

POLLINATION SYNDROMES Cross-pollination can be accomplished in a number of ways: Pollen can be carried from flower to flower by physical agents such as wind or water, or by organisms such as insects, birds, or bats.

Animals visit flowers to eat pollen grains, harvest nectar, or both. As an animal feeds from a flower, pollen grains adhere to its body incidentally. When the same individual visits another flower of the same species to feed, some of these grains are deposited on a stigma of the second flower.

In most cases, animal pollination is an example of **mutualism**: a mutually beneficial relationship between two species. Pollinators usually benefit by receiving food; flowering plants gain by having their male gametophytes transferred to a different individual so that outcrossing takes place.

Chapter 30 introduced pollination syndromes: close correlations between the structure of a flower and the size, shape, and behavior of its pollinators. Most moths and bats, for example, are active at night. If they feed on nectar or pollen, the flowers that they visit tend to be white—and thus more visible in low light—have a strong scent, and open up at night.

One of the most famous examples of a pollination syndrome involves a species called Darwin's orchid (**Figure 40.11**), which is native to Madagascar. The orchid attracted a great deal of attention when it was first discovered by western scientists because it has a "spur" that can be as much as 28 cm (11 in) long, with a nectary at its base. Charles Darwin hypothesized that it must be pollinated by a moth with a tongue-like proboscis as long as the spur. Although the idea seemed preposterous at the time, 40 years later a hawkmoth species with a proboscis that averages 24 cm in length was discovered pollinating the orchid in the wild.

Structures associated with pollination syndromes are thought to be adaptations: traits that increase the fitness of individuals in a particular environment. In this case, flowers and pollinators have adaptations that increase pollination frequency and feeding efficiency, respectively. To capture this point, biologists say that **coevolution** has occurred—a topic that is taken up in more detail in Chapter 53.

WHY DID POLLINATION EVOLVE? In mosses, ferns, and other groups that do not form pollen, sperm have flagella and swim to the egg through droplets of water, or are transferred on water droplets that cling to the legs of tiny insects called springtails. In conifers and most other gymnosperms, wind transmits pollen from male to female. In some of the other groups that produce pollen, such as the cycads, gnetophytes, and angiosperms, many species are pollinated by animals—particularly by insects.

When these observations are mapped onto the phylogenetic tree shown in **Figure 40.12**, two important patterns emerge.

1. Pollination evolved late in land plant evolution. Mosses and other groups that do not form pollen appear first in the fossil record of land plants. Conifers and other groups that are strictly or primarily wind pollinated evolved later but before angiosperms.

2. Seed plants do not need water for sexual reproduction to occur. As a result, the evolution of pollen has allowed these species to be much less dependent on wet habitats. Along with the evolution of the seed—highlighted in Section 40.4—pollen paved the way for the colonization of drier environments.

In addition, it's critical to realize that pollination became a much more precise process when plants began to recruit animals

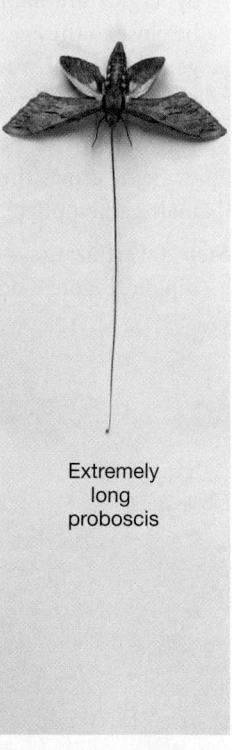

FIGURE 40.11 Pollination Syndrome Can Be Inferred from Flower Structure and Color. Charles Darwin hypothesized that this white orchid flower with an extremely long spur must be pollinated by a moth with a tongue-like proboscis that was long to reach the nectar at the end. He was right.

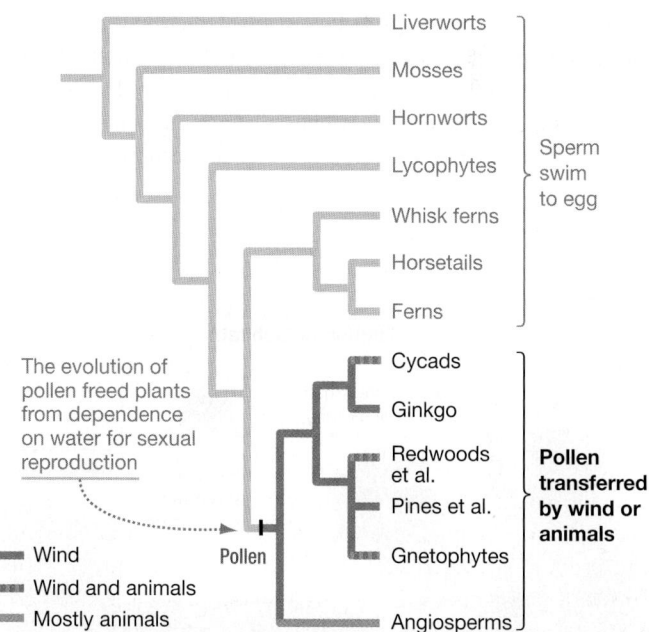

FIGURE 40.12 Pollen Is a Relatively Recent Innovation in Land Plant Evolution. Evolutionary relationships among the major groups of plants. In the lineages colored gray, sperm have flagella and swim to the egg. In the lineages colored blue or orange, sperm are produced by pollen grains.

to act as pollinators. Wind-borne pollen grains have a low probability of landing successfully on a flower stigma. Insect-borne pollen, in contrast, is much more likely to be successfully transferred to flowers of the same species.

Wind-pollinated species invest in making large numbers of pollen grains; animal-pollinated species make fewer pollen grains but invest in structures that attract and reward animals.

In effect, plants "pay" nectar- and pollen-eating insects to work for them. Wind is free, but insects are more precise. Insect pollination is an important adaptation because it makes sexual reproduction much more efficient.

DOES POLLINATION BY ANIMALS ENCOURAGE SPECIATION? In addition to affecting the fitness of individual plants, does pollination by animals make the formation of entirely new species more likely?

To answer this question, consider the situation shown in **Figure 40.13**. A biologist has documented that two populations of a mountain-dwelling species called the alpine skypilot have flowers with different characteristics.

- Alpine skypilots that grow in forested habitats at or below timberline have small flowers with short stalks and an aroma described as "skunky."

- Individuals that grow in the tundra habitats above timberline have large flowers with long stalks and smell sweet.

These differences are interesting, because different insects pollinate the two populations. Small flies are abundant at slightly lower elevations and pollinate the timberline individuals; large bumblebees are abundant at higher elevations and pollinate the tundra flowers. Experiments have shown that bumblebees prefer to pollinate big flowers—probably because larger flowers can support their larger mass. Because flies and bumblebees prefer to visit different types of flowers, gene flow between the two skypilot populations is low and they are evolving distinct characteristics. The two populations may be on their way to becoming different species.

The message here is that evolutionary changes in the size or food-finding habits of a pollinator affect the angiosperm populations they pollinate. In return, changes in flower size and shape affect the insects pollinating that population. If a small population of Darwin's orchids evolved longer spurs, for example, the hawkmoth pollinators that lived in that area would be under intense selection that favored the evolution of a longer proboscis.

Because mutation continuously introduces variations in traits, insect and angiosperm populations frequently change, diverge, and form new species. Changes in pollination syndromes can trigger the evolution of new species. It is no surprise that insects and angiosperms are exceptionally species-rich groups.

It is clear that pollination was a crucial innovation during plant evolution. Now let's get down to mechanics. What happens once a pollen grain is deposited on a stigma?

Fertilization

Figure 40.14 walks you through the steps in fertilization.

Step 1 After landing on the stigma of a mature flower from the same species, a pollen grain absorbs water and germinates. **Germination** is a resumption of growth.

Step 2 When the male gametophyte germinates, a long filament called a **pollen tube** grows through the stigma and down the length of the style. This growth is directed by chemical attractants, called LUREs, released by synergids. Synergids are haploid cells in the female gametophyte that lie close to the egg. In the species illustrated in the figure, the generative cell travels down the length of the tube and divides to form two sperm.

Step 3 When the pollen tube reaches the micropyle of the ovule, it grows through it and enters the interior of the female gametophyte.

Step 4 Fertilization—the fusion of sperm and egg to form a diploid zygote—occurs. In most plant groups, fertilization is

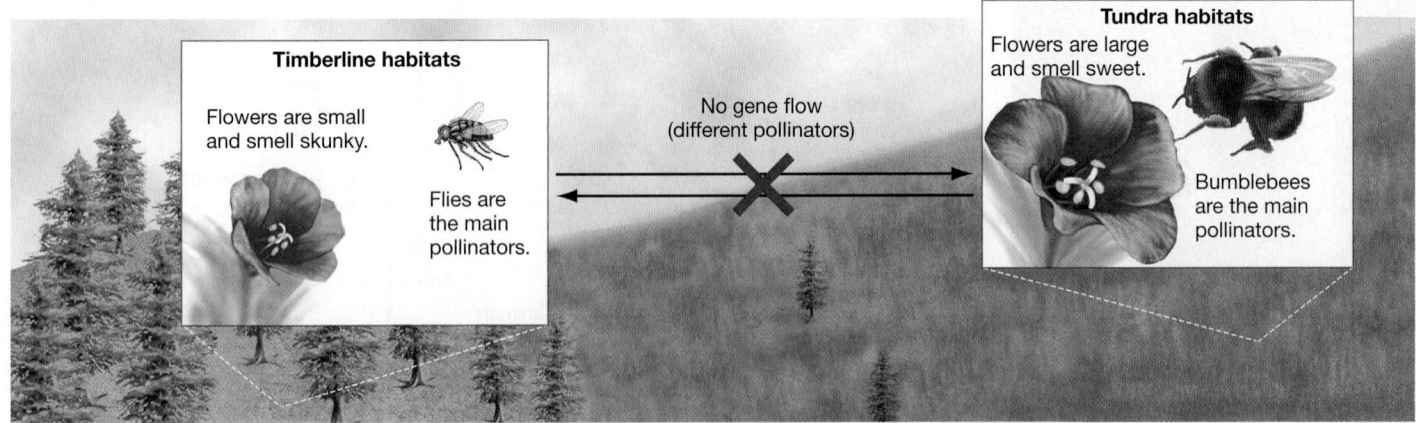

FIGURE 40.13 Interactions between Angiosperms and Animal Pollinators Affect Gene Flow. Flowers in alpine skypilot populations from timberline and tundra habitats look and smell different, and attract different pollinators that are common in timberline versus tundra habitats. As a result, interbreeding is rare.

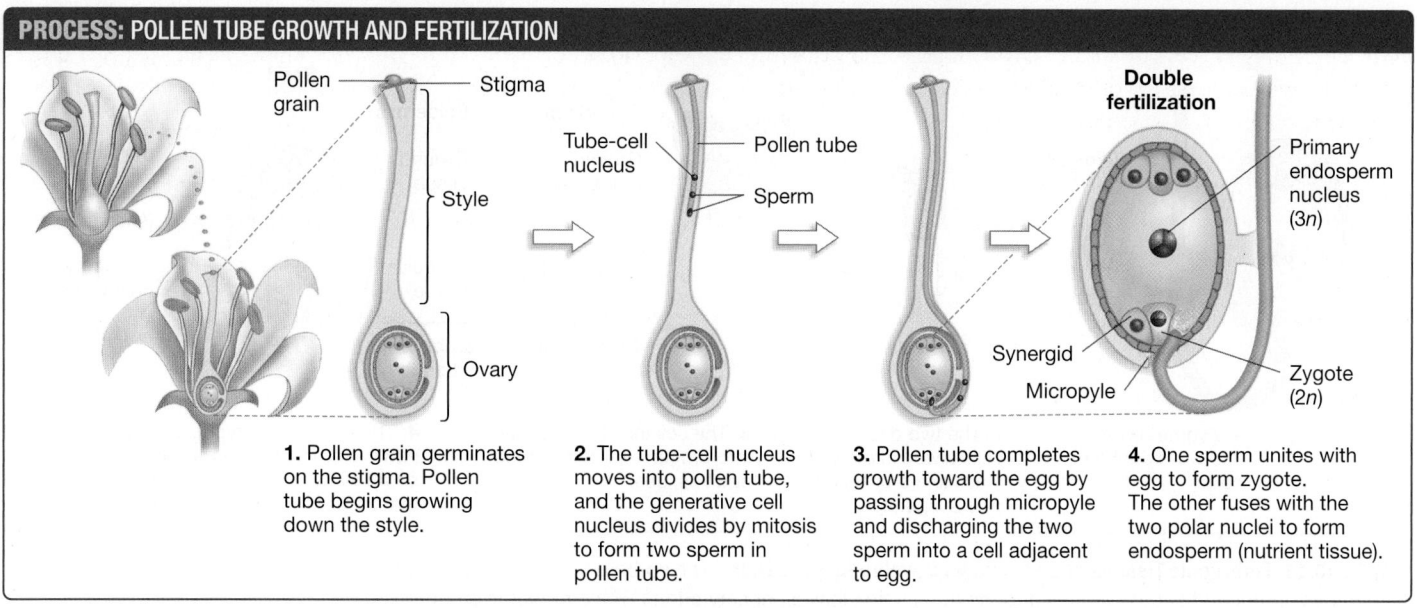

1. Pollen grain germinates on the stigma. Pollen tube begins growing down the style.

2. The tube-cell nucleus moves into pollen tube, and the generative cell nucleus divides by mitosis to form two sperm in pollen tube.

3. Pollen tube completes growth toward the egg by passing through micropyle and discharging the two sperm into a cell adjacent to egg.

4. One sperm unites with egg to form zygote. The other fuses with the two polar nuclei to form endosperm (nutrient tissue).

FIGURE 40.14 Double Fertilization Produces a Zygote and an Endosperm Nucleus. When the pollen tube reaches the female gametophyte, one sperm nucleus fertilizes the egg while the second fuses with the polar nuclei.

straightforward—sperm and egg simply combine, and a diploid nucleus is formed. In angiosperms, however, an unusual event called **double fertilization** takes place. One sperm nucleus unites with the egg nucleus to form the zygote. The other sperm nucleus moves through the female gametophyte and fuses with the polar nuclei in the central cell. In most cases, two polar nuclei are present and a large triploid ($3n$) cell forms.

✔If you understand double fertilization, you should be able to identify cells in a female gametophyte, immediately after fertilization occurs, that are haploid, diploid, and triploid.

CHECK YOUR UNDERSTANDING

If you understand that . . .

- Wind, water, or animals carry pollen grains from one plant to another.
- When a pollen grain lands on a stigma, the grain germinates. A pollen tube forms and grows until it reaches the ovule.
- Sperm cells produced by the male gametophyte fertilize the egg and the polar nuclei, forming a diploid zygote and in most cases a triploid endosperm.

✔ **You should be able to . . .**

1. Explain why insects increase their fitness by visiting flowers and why flowers increase their fitness by rewarding insects.

2. Describe the function of the cells produced by double fertilization.

Answers are available in Appendix B.

The triploid nucleus resulting from this second fertilization undergoes mitosis and cytokinesis to form the **endosperm** ("inside-seed") tissue. In most species, endosperm is triploid and functions to store nutrients. Endosperm cells are loaded with starch or oils (lipids) plus proteins and other nutrients that the embryo will need after it germinates. But before germination can occur, seeds must develop and be dispersed.

40.4 The Seed

Fertilization triggers the development of a young sporophyte. In angiosperms, the first stage in the sporophyte's life is the maturation of the seed.

As a seed matures, the embryo and endosperm develop inside the ovule and become surrounded by a covering called a **seed coat**. At the same time, the ovary around the ovule develops into a fruit, which encloses and protects the seed (or seeds, if a single ovary contains multiple ovules). In addition to providing protection, fruits often aid in dispersing seeds away from the parent plant. The mature seed consists of an embryo, a food supply—originating with endosperm—and a seed coat. In most cases, mature seeds leave the parent plant encased in a fruit.

Along with pollen, the evolution of the seed was a crucial innovation as land plants diversified. Because seeds contain stored nutrients, they allow offspring to be much more successful in colonizing habitats that are crowded with competitors than are offspring produced from spores, which are single cells. As a young plant emerges from the seed, it can subsist on stored nutrients until it is well enough established to absorb water from the soil and feed itself via photosynthesis.

Let's analyze how seeds work, beginning with a closer look at how the embryo develops.

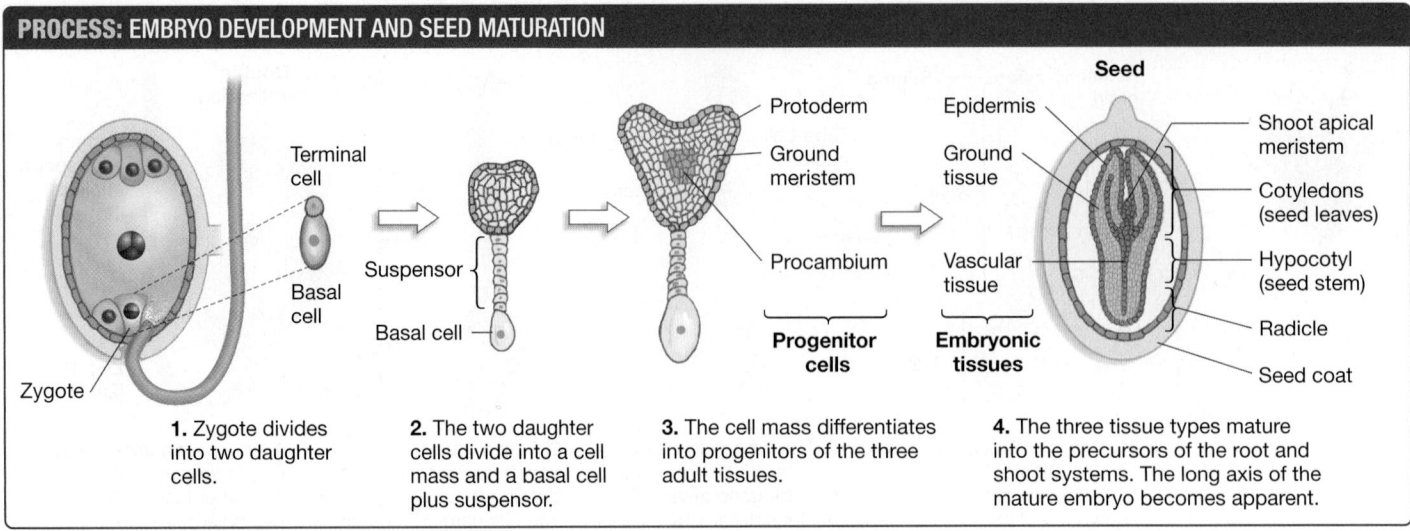

1. Zygote divides into two daughter cells.

2. The two daughter cells divide into a cell mass and a basal cell plus suspensor.

3. The cell mass differentiates into progenitors of the three adult tissues.

4. The three tissue types mature into the precursors of the root and shoot systems. The long axis of the mature embryo becomes apparent.

FIGURE 40.15 Embryonic Tissues and Structures Develop inside Seeds. The embryo inside a seed has the beginnings of root and shoot systems and the plant's first leaves, or cotyledons. The embryonic epidermis and the ground and vascular tissues are organized in distinct layers.

Embryogenesis

Recall from Chapter 23 that **embryogenesis** is the process by which a single-celled zygote becomes a multicellular embryo. As **Figure 40.15** shows, embryogenesis in angiosperms can be analyzed as a four-step process.

Step 1 The zygote divides to form the two daughter cells.

Step 2 The lower daughter cell, or basal cell, divides to form a cell that forms part of the root tip and a cell that produces a row of cells. This row of cells, called the suspensor, provides a route for nutrient transfer from the parent plant to the developing embryo. The upper daughter cell, or terminal cell, is the parent of almost all the cells in the embryo. It divides to form a mass of cells.

Step 3 Cells within the mass differentiate into groups conforming to one of the three adult tissue types introduced in Chapter 36. The exterior layer of embryonic cells, the **protoderm**, is the progenitor of the adult dermal tissue, or epidermis. The **ground meristem**—the cells just within the protoderm—gives rise to the ground tissue found in adults. And the **procambium** is a group of cells in the core of the embryo that becomes the vascular tissue.

Step 4 As the embryo continues to develop, the long axis of the plant begins to emerge and several important structures take shape: (**1**) the **cotyledons**, or seed leaves, (**2**) the **hypocotyl** ("under-cotyledon"), or seed stem, which is the embryonic stem, and (**3**) the **radicle**, or embryonic root. Some embryos also have an **epicotyl** ("above-cotyledon"), which is a portion of the embryonic stem that extends above the cotyledons.

✔If you understand the basic steps in embryonic development, you should be able to explain the relationship between the populations of embryonic cells labeled in step 3 of Figure 40.15 and the tissues labeled in step 4.

Recall from Chapter 30 that one prominent lineage of angiosperms—the monocotyledons, or monocots—has just one seed leaf, whereas eudicotyledons, or eudicots, have two. In most eudicots, the cotyledons take up the nutrients in the endosperm and store them. In these species, there is no endosperm left by the time the seed matures—instead the cotyledons function as the nutrient storage organ. **Figure 40.16** compares the seed structure in beans and corn—a representative eudicot and monocot, respectively.

By the time a seed matures, then, the three major embryonic tissue types have developed. The precursors of the root and shoot systems, along with the seed-leaves, have formed. Once these events are accomplished, the seed tissues dry and the embryo becomes quiescent—meaning it stops growing.

To review gametogenesis, pollination, and embryogenesis in angiosperms, go to the study area at *www.masteringbiology.com*.

(MB) **Web Activity** Reproduction in Flowering Plants

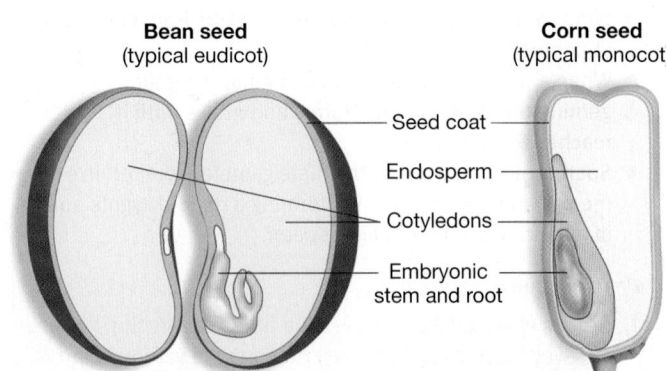

FIGURE 40.16 Seeds Contain an Embryo and a Food Supply Surrounded by a Tough Coat. In beans (left), the nutrients in the endosperm are absorbed by the cotyledons and stored. In corn (right), the endosperm is intact.

The Role of Drying in Seed Maturation

The seeds of many species dry out as they mature. Water makes up 90 percent of normal plant cells, but dried seeds contain just 5–20 percent water.

Loss of water is interpreted as an adaptation that prevents seeds from germinating on the parent plant, where they would compete with the parent for resources. In addition, the dry condition of seeds ensures that once they have dispersed from the parent plant, they will not germinate until water is available. This is logical, because water is crucial to the survival of germinated seedlings. Dry seeds are less susceptible than wet seeds to damage from freezing, and are lighter and more easily transported.

How do the plasma membranes and proteins in the embryo and endosperm survive the drying process? When researchers reduce the amount of water surrounding isolated plasma membranes or isolated proteins to the levels observed in extremely dry seeds, some of the membranes and proteins disintegrate. Clearly, something is happening at the molecular level in seeds to keep these cell components intact.

Researchers have recently established that one of these "somethings" involves sugars. As water leaves the seed during drying, sugars replace it and maintain the integrity of plasma membranes and proteins. If drying is extreme, the sugars form an extremely viscous liquid that contains little if any water. Substances such as this are considered vitrified, or glass-like (glass is a liquid solution with the viscosity of a solid). Biologists propose that this glassy, sugar-coated state helps maintain the integrity of plasma membranes and proteins in seeds that experience extremely dry conditions. When seeds imbibe water, the glassy sugars dissolve and germination proceeds.

Drying is only one part of the seed maturation process, however. Equally important is the development of tissues surrounding the seed itself. In many cases, these tissues are required for the seed to be dispersed from the parent plant.

Fruit Development and Seed Dispersal

Fertilization initiates not only the development of the seed and embryo in angiosperms, but also the development of the fruit.

FRUIT STRUCTURE   Fruits come in three basic types (**Figure 40.17**).

- **Simple fruits** like the apricot develop from a single flower that contains a single carpel or several carpels that are fused together.

- **Aggregate fruits** like the raspberry also develop from a single flower, but one that contains many separate carpels.

- **Multiple fruits** like the pineapple develop from many flowers and thus many carpels.

As a fruit matures, the walls of the ovary thicken to form the **pericarp,** the part of the fruit that surrounds and protects the seed or seeds (**Figure 40.18** on page 798). Fruits can be dry when they are mature, as in nuts, or fleshy, as in cherries and tomatoes.

Note that fruits are formed from cells derived from the parent plant—not the embryo. They are tissue from the mother, encasing the offspring.

(a) Simple fruit (e.g., cherry):
Develops from a single flower with one carpel or fused carpels

(b) Aggregate fruit (e.g., blackberry):
Develops from a single flower with many separate carpels

(c) Multiple fruit (e.g., pineapple):
Develops from many flowers with many carpels

FIGURE 40.17 Three Major Types of Fruits. The structure of a fruit depends on the number of carpels found in each flower and whether or not the carpels are fused.

✔ **QUESTION** When fruits ripen, their color changes in a way that makes them more conspicuous to fruit eaters. State a hypothesis to explain why this color change might increase the fitness of an individual.

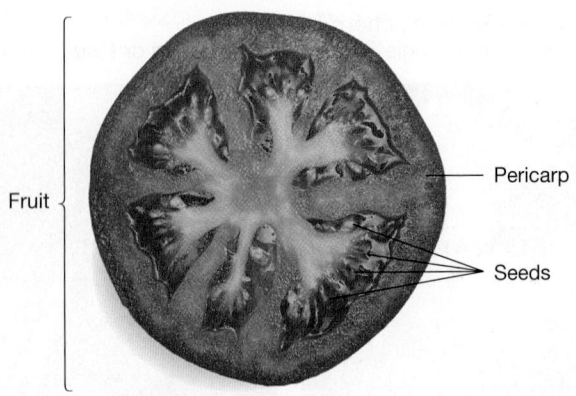

Fruit

Pericarp

Seeds

FIGURE 40.18 As Fruits Mature, the Ovary Wall Develops into a Pericarp That Surrounds the Seed or Seeds. A fruit consists of a pericarp and the enclosed seeds.

To learn more about the structure of fruits and the relationship between seeds and fruits, go to the study area at *www.masteringbiology.com*.

(MB) **Web Activity** Fruit Structure and Development

FRUIT FUNCTION Fruits have two functions: They protect seeds from physical damage and seed predators, and they frequently aid in seed dispersal. Dispersal is important to the fitness of the young sporophyte. This is especially true in long-lived species, in which the offspring may compete with the parent plant for light, water, and nutrients if there is no dispersal.

Fruits sometimes split open and release seeds to be dispersed directly. In many cases, however, seeds are dispersed to new locations while they are still enclosed in the fruit.

Dry fruits simply fall to the ground or are dispersed by wind, propulsion, or animals (**Figure 40.19**). Some dry fruits have hooks or barbs that adhere to passing animals, while nuts are dispersed by seed predators. Fruits that are dispersed by wind often have external structures to catch the breeze and extend the dis-

tance they travel; the fruits of dandelions and maple trees are familiar examples. Fruits that float can disperse seeds in water.

Some plants actually disperse dry fruits via propulsion. The sandbox tree, for example, produces a seed pod that shrinks as it dries. Eventually the pod splits apart violently, spraying seeds in all directions with so much force that the plant is sometimes called the dynamite tree. The bursting seed pod sounds like a pistol shot, and seeds can be scattered as much as 40 m away from the parent plant. Similarly, the dwarf mistletoe fruit fills with sugars as it matures. Enough water follows via osmosis to make the fruit explode and shoot seeds as far as 5 m.

Animals are the most common dispersal agent for fleshy fruits. In animal-dispersed fruits, the seed coat has to be tough enough to resist the mechanical forces and chemical conditions in the animal's mouth and digestive tract, so that seeds can emerge in the feces unscathed. In cases like this, seed dispersal is an example of mutualism. The plant provides a fruit rich in sugars and other nutrients; in return, the animal carries the fruit to a new location and excretes the seeds along with a supply of fertilizer. In fact, some seeds won't germinate unless their hard coats are exposed to stomach acids and abrasion as they pass through an animal's digestive tract.

Seed Dormancy

Once they have dispersed from the parent plant, seeds may not germinate for a period of time. This condition is known as **dormancy**. Dormancy is usually a feature of seeds from species that inhabit seasonal environments, where for extended periods of time conditions may be too cold or dry for seedlings to thrive. Based on this observation, dormancy is interpreted as an adaptation that allows seeds to remain viable until conditions improve.

Consistent with this hypothesis, dormancy is rare or nonexistent in seeds produced by plants that inhabit tropical wet forests or other areas where conditions are suitable for germination year-round.

What molecular mechanisms are responsible for dormancy? And how does dormancy cease so that germination can begin?

Animal dispersal

Wind dispersal

Water dispersal

FIGURE 40.19 Seeds Are Often Contained in Structures That Allow Them to Be Dispersed.

WHAT ROLE DOES ABA PLAY IN DORMANCY? Chapter 39 introduced the hormone abscisic acid (ABA) and described its role in preventing germination. In some species, seeds that enter dormancy have a high concentration of this hormone. The seeds of desert plants, for example, have high levels of ABA in their seed coats. When these seeds are exposed to large amounts of water during rare or seasonal rains, the hormone literally washes out of the seed's outer tissues and germination proceeds.

There is not a strict correlation between ABA concentration and degree of seed dormancy, however. In peas and many other species, seeds routinely contain high levels of ABA and yet are not dormant. In *Arabidopsis*, ABA concentrations rise as seeds mature and appear to impose dormancy. ABA levels eventually fall, however, so dormant, mature seeds contain only trace amounts of the hormone.

Researchers have concluded that there is no single, universal mechanism for initiating and maintaining seed dormancy. In some cases, changing ABA levels or the ratio of ABA to gibberellin present control the dormant state. In other cases, changes in sensitivity to ABA, rather than the sheer amount of hormone present, appear to be important. It is also likely that novel mechanisms for initiating and maintaining dormancy are still to be discovered.

HOW IS DORMANCY BROKEN? The coats of some seeds are thick enough to prevent water and oxygen from physically reaching the embryo. For germination to occur, these seed coats must be disrupted, or **scarified**.

Crop seeds that require scarification are placed in large, revolving drums with pieces of sandpaper that abrade and scarify the seeds. In nature, seed coats can be disrupted by a fire, by the passage of the seed through an animal's digestive tract, or by abrasion against soil particles. The basic principle is that the seed coat has to be broken for water to enter the seed.

Other seeds must experience particular environmental conditions in addition to exposure to water. Species native to high latitudes or high elevations often produce seeds that must undergo cool, wet conditions before they will germinate. No one knows how these seeds perceive cold or what molecular mechanisms are involved in breaking dormancy when the cool, wet period ends.

Because small seeds have few nutrient reserves in their cotyledons or endosperm, many small-seeded species need to germinate near the soil surface, where individuals are exposed to light and can feed themselves via photosynthesis. As Chapter 39 indicated, lettuce seeds and other small seeds must be exposed to red light before they will break dormancy and germinate. Red light is an important environmental cue, because wavelengths in the red portion of the light spectrum support photosynthesis. Red light and blue light indicate that sunlight is abundant.

Finally, many of the seeds produced by species native to habitats where wildfires are frequent, such as the California chaparral and South African fynbos, have an unusual chemical requirement to break dormancy: They must be exposed to fire or smoke before they will germinate. In fact, the commercial food product "liquid smoke" induces germination in these seeds as well as actual smoke does. In fire-prone habitats, it is advantageous for seeds to germinate after fire has cleared away existing vegetation.

The message here is that dormancy can be broken in response to a wide variety of environmental cues. In general, the cue that triggers germination is a reliable signal that conditions for seedling growth are favorable for a particular species in a particular environment.

Seed Germination

Even if specific environmental signals are required to break dormancy, seeds do not germinate without water. Water uptake is the first event in germination. Once the seed coat allows water penetration, water enters by a steep water-potential gradient, because the seed is so dry.

The graph in **Figure 40.20** plots changes in water uptake in seeds through time. If you trace the graph from the start of seed germination onward, you should see that water uptake in a typical angiosperm seed has three distinct phases:

Phase 1 Germination begins with a rapid influx of water. Oxygen consumption and protein synthesis in the seed increase dramatically, but no new messenger RNAs are transcribed. Based on these observations, biologists have concluded that some of the key early events in germination are driven by mRNAs that are stored in the seed prior to maturation.

Phase 2 The second phase is an extended period during which water uptake stops. Newly transcribed mRNAs appear and are translated into protein products. Mitochondria also begin to multiply. In effect, seeds take up enough water in Phase 1 to hydrate their existing proteins and membranes, and then begin to manufacture the proteins and mitochondria needed to support growth.

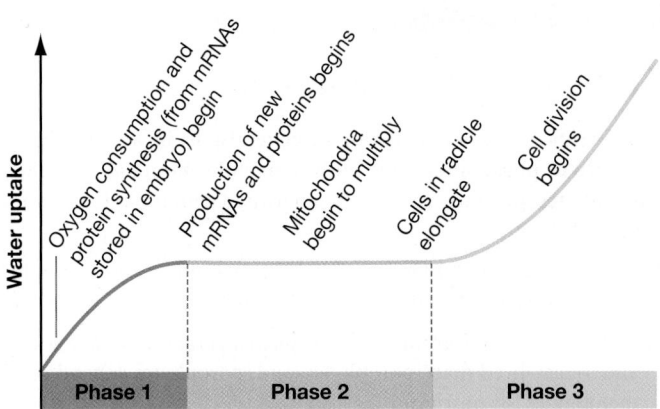

FIGURE 40.20 Germination Begins with Three Distinct Phases. This graph plots the rate of water uptake through time as a typical angiosperm seed germinates. (The graph is conceptual, meaning it represents a general pattern observed in data from many species, so the axes have no units.)

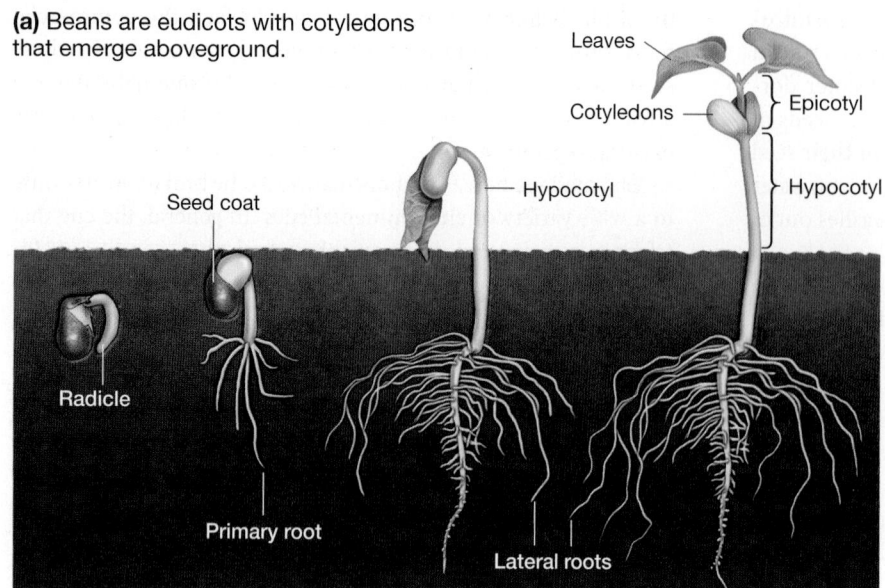

(a) Beans are eudicots with cotyledons that emerge aboveground.

Leaves

Cotyledons

Epicotyl

Hypocotyl

Seed coat

Hypocotyl

Radicle

Primary root

Lateral roots

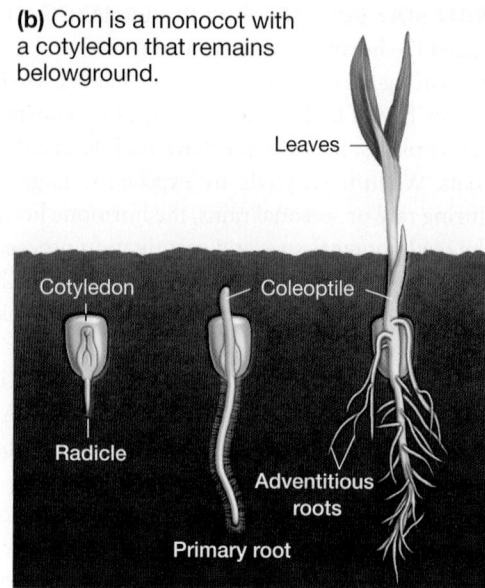

(b) Corn is a monocot with a cotyledon that remains belowground.

Leaves

Cotyledon

Coleoptile

Radicle

Adventitious roots

Primary root

FIGURE 40.21 The Germination Sequence Varies among Species.

✔**QUESTION** In which of these species are cotyledons photosynthetic?

Phase 3 Water uptake resumes as growth begins. This renewed phase of water uptake enables cells to develop enough turgor pressure to enlarge. Eventually, the seedling bursts from the seed coat.

Figure 40.21 shows what happens as eudicot and monocot embryos emerge from the seed. First the radicle emerges, and subsequently develops into the mature root system. In eudicots, the shoot system with its cotyledons usually emerges shortly after the radicle appears. In corn, the radicle and the coleoptile, which covers the young shoot, emerge at the same time. Note that in

eudicots, the emerging stem has a hook shape. Like the coleoptile of monocots, this trait is thought to protect the apical meristem from damage as the shoot works its way upward through rough soil particles.

The next major event in the seedling's life occurs when either the cotyledons or the earliest leaves produced by the growing seedling commence photosynthesis. The seedling is said to be established when the young plant no longer relies on food reserves in its endosperm or cotyledons; instead, it receives all of its nourishment from its own photosynthetic products. With this, a new generation is under way.

CHAPTER 40 REVIEW

For media, go to the study area at www.masteringbiology.com

Summary of Key Concepts

🔑 **Plants undergo alternation of generations, in which a diploid sporophyte phase alternates with a haploid gametophyte phase. Sporophytes produce spores by meiosis. Gametophytes produce gametes by mitosis.**

- The relative size and life span of the gametophyte and sporophyte phases vary a great deal among plant groups.

- In the most basal groups of land plants, gametophytes are larger and longer-lived than sporophytes, and sporophytes depend on gametophytes for nutrition.

- In angiosperms, or flowering plants, sporophytes are the large and long-lived phase where photosynthesis takes place; gametophytes consist of just a few cells.

  ✔ You should be able to state the ploidy of a sporophyte, gametophyte, spore, gamete, and zygote.

🔑 **In angiosperms, male and female gametophytes are microscopic and are produced inside flowers. Male gametophytes (pollen grains) are portable. Female gametophytes are encased in an ovary and are retained in the flower. When pollen grains land on a flower, they deliver sperm cells that fertilize the egg produced by the female gametophyte.**

- Angiosperms initiate flowering and sexual reproduction in response to external cues from the environment as well as internal cues based on the individual's condition. Frequently, these cues allow individuals to flower when environmental conditions are favorable.

- Flowers are made up of sepals, petals, stamens, and one or more carpels.

- The lower part of the carpel, the ovary, contains one to many ovules. Within the ovule, a megasporocyte undergoes meiosis, producing a megaspore that develops into the female gametophyte.

- In the anthers of stamens, microsporocytes undergo meiosis. The resulting microspores develop into male gametophytes, which are enclosed in pollen grains.

- Pollination occurs when pollen grains are transported to the stigma of the carpel. In most cases, the structure of a flower correlates with the morphology and behavior of its pollinator.

- If allowed to germinate on the stigma, a pollen grain sends a long pollen tube down the style. Two sperm travel down the pollen tube and enter the female gametophyte.

- In double fertilization, one sperm fuses with the egg to form a zygote, while the other fuses with polar nuclei within the female gametophyte. The fusion of sperm and polar nuclei produces endosperm—nutritive tissue that in most species is triploid.

  ✔ You should be able to describe what would happen if all meiotic products in megasporangia went on to form female gametophytes.

🔑 **Seeds contain an embryo and a food supply surrounded by a coat. In angiosperms, the walls of the ovary develop into a fruit that encloses the seed or seeds. In many cases, fruits function in seed dispersal.**

- The development of an angiosperm embryo includes the formation of dermal tissue (epidermis), ground tissue, and vascular tissue layers and the development of the radicle, hypocotyl, and cotyledons.

- As an embryo develops, endosperm cells divide to form a nutrient-rich tissue. In addition, cells along the outside of the ovules form a protective seed coat, and the ovary develops into a fruit.

- In many cases, the mature fruit contains structures that help disperse the mature seed via wind, water, propulsion, or animals.

- Many seeds do not germinate immediately but instead experience a period of dormancy.

- A wide variety of conditions, ranging from scarification to exposure to red light, may break seed dormancy. In many cases, the event that triggers germination ensures that the seed germinates when environmental conditions are favorable.

- Germination begins when the seed takes up water and mRNAs already present in the seed are translated. It ends when the radicle breaks the seed coat and begins to penetrate the soil.

  ✔ You should be able to explain the relationships among carpels, ovaries, ovules, fruits, and seeds.

  (MB) **Web Activity** Reproduction in Flowering Plants, **Web Activity** Fruit Structure and Development

Questions

1. What is the major evolutionary trend in land plant life cycles?
   a. Instead of being approximately the same size and shape, gametophytes and sporophytes began to look different.
   b. Sporophytes became larger and long lived while gametophytes became drastically reduced.
   c. In lineages that evolved more recently, such as angiosperms, spores are no longer produced.
   d. Sporophytes began to rely on gametophytes for all of their nutritional needs.

2. What happens when double fertilization occurs?
   a. Two zygotes are formed, but only one survives.
   b. Two sperm fertilize the egg, forming a triploid zygote.
   c. One sperm fertilizes the egg, while another sperm fuses with the polar nuclei.
   d. One sperm fertilizes the egg, while two other sperm fuse with a polar nucleus.

3. What is a fruit?
   a. a structure formed from the ovary wall that contains a seed or seeds
   b. a structure consisting of an embryo and a food supply surrounded by a tough coat
   c. a female gametophyte
   d. a male gametophyte

4. Which of the following is a key event during embryogenesis?
   a. The seed coat takes on water so that germination can begin.
   b. Starches are hydrolyzed, providing sugars that fuel the early stages of germination.
   c. The megasporocyte divides by mitosis, forming the cells that will become the embryonic female gametophyte.
   d. Distinct groups of cells form that will become dermal, ground, and vascular tissues.

5. Why is the interaction between angiosperms and pollinators considered mutualistic?
   a. New species can form if mutant flowers attract new types of pollinators.
   b. Flowers may have an array of traits, including corolla shape, color, scent, and the presence of nectar, to attract a specific type of pollinator.
   c. Wind pollination is much "cheaper," but animal pollination is much more precise.
   d. Angiosperms get their pollen dispersed, while pollinators get food.

6. What happens when outcrossing occurs?
   a. Inbred offspring are produced.
   b. The same flower can be visited by many different types of pollinators—not a single specialist species.
   c. Gametes from different individuals fuse to form a zygote.
   d. Gametes from the same individual fuse.

1. In terms of maximizing reproductive success, what is the advantage of asexual reproduction? What is the disadvantage?

2. In the angiosperm life cycle, which cells undergo meiosis? Which cells are spores? Which structures are gametophytes?

3. How do the structure and function of sepals and petals differ? How would you expect these structures to differ in species that are pollinated by wind versus bumblebees?

4. What are the advantages and disadvantages of self-fertilization versus those of outcrossing?

5. Consider a carpel, an ovary, and an ovule. Which is responsible for producing the female gametophyte, and which produces the pericarp of a fruit? Are these structures part of the sporophyte, the gametophyte, or a combination of the two?

6. What is the relationship between the endosperm of corn and the cotyledons of beans?

1. Suppose you discovered an angiosperm that was new to science. The population grows on an island near the equator. The island is dry 10 months of the year but experiences a 2-month period of frequent rains. Predict what cues trigger flowering and seed germination in the newly discovered species.

2. Some flowering plants "cheat" their pollinators because they offer no food reward. Likewise, certain pollinators cheat plants by removing nectar from flowers without picking up pollen. (In some cases, they do so by chewing through the petals that hold the store of nectar.) Speculate on the types of mutations that might modify insect behavior and/or plant structure in a way that limits cheating and enforces mutualism.

3. Pollinators frequently deposit pollen from more than one individual on a stigma. When they do, pollen grains from different males compete to fertilize the egg. Design an experiment to test the hypothesis that pollen tubes grow faster when pollen grains germinate in the presence of pollen from a different individual.

4. Consider the following fruits: an acorn, a cherry, a burr, and a milkweed seed. Based on the structure of each of these fruits, predict how the seed is dispersed. Design a study that would estimate the average distance that each type of seed is dispersed from the parent plant.

Oryx are adapted to desert life. They have an exceptional ability to withstand heat, and can acquire all of their water from the food they eat.

Animal Form and Function

41

T he Sahara and Arabian deserts are extreme environments. In some parts of the Sahara, several years can pass between rainfalls. In a single day, temperatures in these deserts can fluctuate between −0.5°C and 37.5°C; midday temperatures in summer can exceed 50°C. Yet few places in the Sahara or Arabian deserts are devoid of life. Even large animals, such as the oryx, thrive in both regions.

How do the animals native to these areas cope? Small animals avoid the midday heat by retreating to a cool burrow deep underground or the shade of a small shrub. Large mammals have a harder time hiding from the Sun but possess traits that allow them to keep cool and conserve water.

The Arabian oryx, for example, never drinks. It gets 86 percent of the water it needs from the vegetation it eats and the other 14 percent from water synthesized as a by-product of cellular respiration—what biologists call **metabolic water**. To conserve what water they have, Arabian oryx produce extremely concentrated urine and exceptionally dry fecal pellets.

In addition, oryx do not sweat to cool off. Instead, their body temperature rises from a normal 37°C to just over 40°C as temperatures increase during the day. The excess body heat is released during the cool desert nights, when their body temperature returns to normal.

In contrast to oryx, humans die of dehydration when denied water for just three days. And for a human, a body temperature of 40°C is a life-threatening situation—equivalent to running a fever of 104°F. How do oryx do it? What aspects of their anatomy and physiology allow them to thrive in such an extreme environment?

Anatomy is the study of an organism's physical structure. **Physiology** is the study of how the physical structures in an organism function. The anatomy and physiology of an oryx are clearly different from those of a human—or those of a shark, frog, tuna, fruit fly, or crab, for that matter. This chapter is an introduction to the study of anatomy and physiology in animals.

KEY CONCEPTS

- In biology, structure has a profound influence on function. Biologists analyze the structure and function of animals at a variety of levels: molecules, cells, tissues, organs, and organ systems.

- Body size has a strong influence on how animals work, in large part because a body's volume increases faster than its surface area as body size increases.

- Animals use an array of methods to maintain a relatively constant environment inside their bodies. They have systems that sense changes in internal conditions and trigger responses that return conditions to normal.

- Some animals have sophisticated systems for generating and conserving heat and regulating body temperature.

✔ When you see this checkmark, stop and test yourself. Answers are available in Appendix B.

41.1 Form, Function, and Adaptation

Biologists who study animal anatomy and physiology are studying **adaptations**—heritable traits that allow individuals to survive and reproduce in a certain environment better than individuals that lack those traits (see Chapter 24).

Recall from Chapter 24 that adaptation results from evolution by natural selection. Natural selection, in turn, occurs whenever individuals with certain alleles leave more offspring than do individuals with different alleles. Because of this difference in reproductive success, the frequency of the selected alleles increases from one generation to the next.

Oryx with alleles that allow them to extract more water from their feces survive better and produce more offspring than do oryx with alleles that allow water to be lost in feces. The ability to produce extremely dry fecal pellets is an adaptation that helps oryx thrive in water-short environments.

The Role of Fitness Trade-Offs

Adaptations increase fitness—the ability to produce offspring. But no adaptation is "perfect." Instead, adaptations are limited by which alleles are present in a population and by the nature of the traits that already exist—because all adaptations derive from pre-existing traits.

The human spine, for example, is a highly modified form of the vertebral column in ancestors that walked on all fours (see Chapter 34). The modifications in the human spine can be considered adaptations to support our upright posture, but they are far from perfect—85 percent of U.S. adults under the age of 50 experience back pain. The evolution of the human spine has been constrained by the nature of the ancestral trait and by a lack of alleles that would improve its structure and function.

The most important constraint on adaptation, though, may be **trade-offs**—inescapable compromises between traits. For example, every female animal has a finite amount of time and energy available for producing offspring. In species that do not care for their young, a female's entire reproductive investment consists of the eggs she lays. Given that the total amount of energy available for egg production is limited, there should be a trade-off between the number of eggs a female produces and the quality of those eggs. Egg quality is determined by egg size—specifically by the amount of yolk, or nutrient-rich cytoplasm, in the egg.

How do biologists study trade-offs in animal anatomy and physiology? Let's consider experimental work on egg size and egg number in the side-blotched lizard, a reptile that lives in the deserts of western North America (**Figure 41.1**). Like many animals, side-blotched lizards lay eggs in groups called clutches. Theory predicts that there has to be a trade-off between the number and size of eggs in a clutch.

To document this trade-off in nature, biologists introduced a large amount of variation in egg size and egg number.

- They induced the production of small eggs by catching females, surgically removing yolk from their eggs early in development, replacing the eggs, and releasing the mothers back into the wild.

- To form clutches with small numbers of eggs, biologists caught females and removed all but two or three of their eggs early in development.

- As an experimental control, the researchers did "sham operations." They caught a large number of females and performed surgery to expose their eggs, but they left the eggs alone.

These manipulations gave them a study population with a large variation in egg size and number. Did fitness trade-offs occur?

The graph labeled Results 1 in Figure 41.1 plots the average mass of eggs that were laid against clutch size, meaning the number of eggs laid by each female. The data show that as egg size increases, clutch size decreases. The pattern confirms a trade-off between egg size and egg number. It is not possible for a female in this population to produce large numbers of large eggs.

What about the prediction that large offspring are higher in quality, meaning that they survive better? To test this idea, the researchers marked 1668 newly hatched lizards in the experimental population, released them, and recaptured those that survived one month later. The graph labeled Results 2 in Figure 41.1 indicates that survival increases as egg mass increases. These data support the prediction that larger offspring survive much better than do smaller offspring.

Given this trade-off between offspring quantity and quality, does natural selection favor mothers that produce large numbers of small offspring or those that produce small numbers of large offspring? The graph labeled Results 3 in Figure 41.1 shows the combined effect of egg number and offspring survival by plotting "Mother's fitness"—the number of surviving offspring that she produces—on the y-axis. In this population, mothers that produced an intermediate number of offspring of intermediate size generated the highest total number of surviving offspring.

Trade-offs, such as the inescapable compromise between egg size and egg number, are pervasive in nature. Desert animals that sweat to cool off are threatened with dehydration. An eagle's beak is superbly adapted for tearing meat but not for weaving nesting materials together. In studying animal anatomy and physiology, biologists study compromise and constraint as well as adaptation.

Adaptation and Acclimatization

In everyday English, the word adaptation describes short-term, reversible responses to environmental fluctuations. In biology, physiological and biochemical changes like these are referred to as **acclimatization**, or **acclimation**. Acclimatization is a phenotypic change in an individual in response to short-term changes in the environment. Adaptation refers only to a genetic change in a population in response to natural selection exerted by the environment.

If you moved to Tibet, your body would acclimatize to high elevation by making more of the oxygen-carrying pigment hemoglobin and more hemoglobin-carrying red blood cells. But populations that have lived in Tibet for many generations are adapted to this environment through genetic changes. Among native Tibetans, for example, an allele that increases the ability of

QUESTION: Is there a trade-off between the quality and quantity of offspring that a female can produce?

HYPOTHESIS 1: Females can produce many small eggs or a few large eggs.	**HYPOTHESIS 2:** Offspring quality increases with increasing egg size.	**HYPOTHESIS 3:** There is an optimal clutch size based on a trade-off between the quality and quantity of offspring a female can produce.
NULL HYPOTHESIS 1: There is no relationship between egg size and egg number.	**NULL HYPOTHESIS 2:** There is no relationship between offspring quality and egg size.	**NULL HYPOTHESIS 3:** No optimal clutch size exists.

EXPERIMENTAL SETUP 1:	**EXPERIMENTAL SETUP 2:**	**STUDY DESIGN 3:**
Eggs early in development: Reduced yolk, Reduced number, Left alone. Mothers. Vary egg size and egg number by catching females and removing yolk from eggs or removing all but 2–3 eggs. Also do sham operations, with eggs left alone. Record size and number of eggs laid.	Offspring. Catch and mark large number of newly hatched offspring from experiment 1. Re-catch survivors one month later.	Number of surviving offspring? Number of surviving offspring? Number of surviving offspring? Mothers. Calculate number of surviving offspring per female, based on results of experiments 1 and 2.

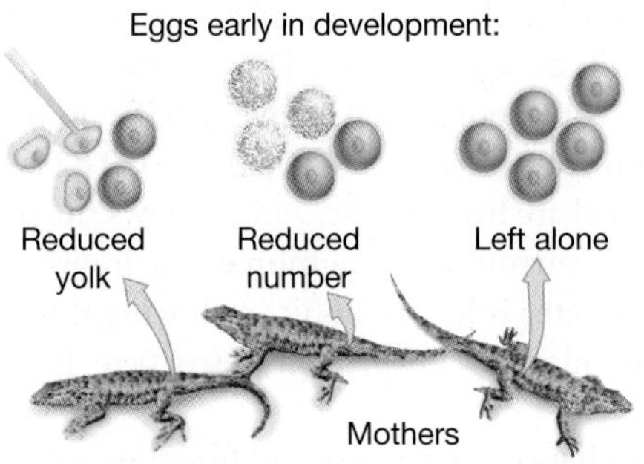

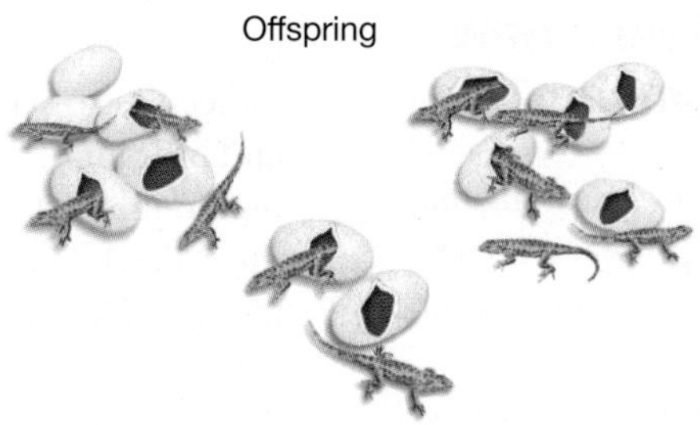

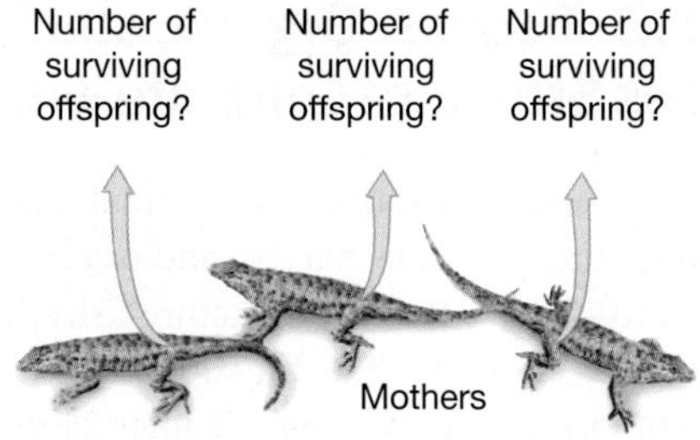

PREDICTION 1: Females with small eggs produce a large number; females with few eggs produce large eggs.	**PREDICTION 2:** Larger offspring survive better than smaller offspring.	**PREDICTION 3:** The number of surviving offspring is maximum at intermediate egg size and intermediate clutch size.
PREDICTION OF NULL HYPOTHESIS 1: As average egg size increases, average clutch size stays the same.	**PREDICTION OF NULL HYPOTHESIS 2:** No difference in survival of large versus small offspring.	**PREDICTION OF NULL HYPOTHESIS 3:** As egg size increases, number of surviving offspring does not change.

RESULTS 1:	**RESULTS 2:**	**RESULTS 3:**

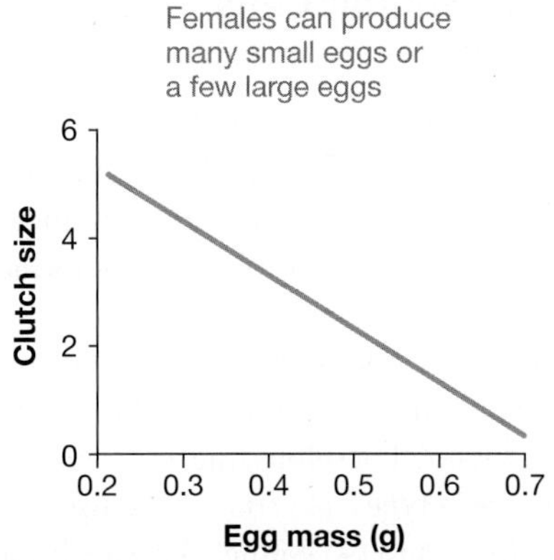

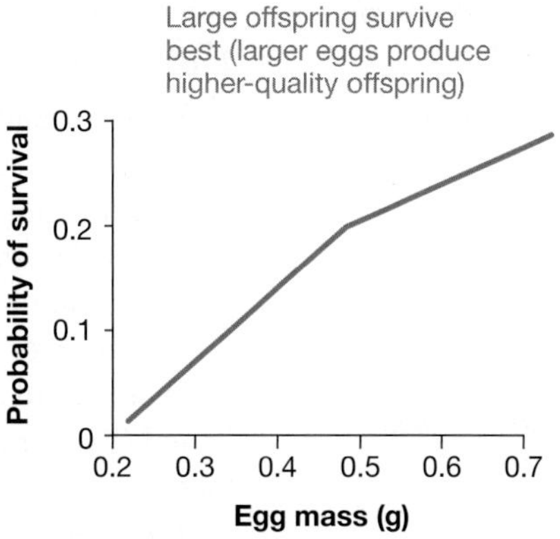

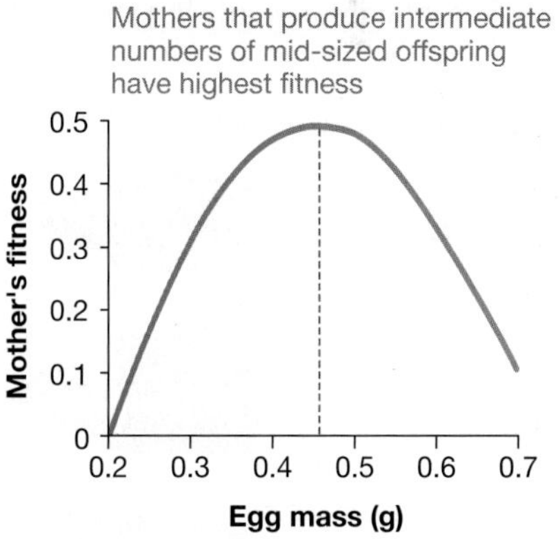

RESULTS 1: Females can produce many small eggs or a few large eggs (Clutch size vs Egg mass (g))

RESULTS 2: Large offspring survive best (larger eggs produce higher-quality offspring) (Probability of survival vs Egg mass (g))

RESULTS 3: Mothers that produce intermediate numbers of mid-sized offspring have highest fitness (Mother's fitness vs Egg mass (g))

CONCLUSION: There is a trade-off between offspring quality (egg size) and quantity (egg number).

FIGURE 41.1 Fitness Trade-Offs between Clutch Size and Egg Size in Lizards. The egg size and clutch size manipulation here accomplished two key goals: (1) increasing the variation among individuals, so that fitness differences are easier to detect, and (2) ensuring that—because manipulated individuals were chosen at random—the only thing that differed among females in the study was their egg size or clutch size.

SOURCE: Sinervo, B., P. Doughty, R. B. Huey, and K. Zamudio. 1992. Allometric engineering: A causal analysis of natural selection on offspring size. *Science* 258: 1927–1930.

✔**QUESTION** Describe the results predicted by the null hypothesis in all three of these experiments, and explain what they would look like on the graphs.

hemoglobin to hold oxygen has increased to high frequency. In populations that do not live at high elevations, this allele is rare or nonexistent.

The ability to acclimatize is itself an adaptation. Light-skinned humans, for example, vary in the ability to tan in response to sunlight. Some individuals tan easily—they have alleles that allow them to acclimate efficiently to environments with intense sun—while others do not. In this and many other cases, the ability to acclimatize is a genetically variable trait that can respond to natural selection.

41.2 Tissues, Organs, and Systems: How Does Structure Correlate with Function?

If a structure found in an animal is adaptive—meaning that it helps the individual survive and produce offspring—it is common to observe that the structure's size, shape, or composition correlates closely with its function.

For example, recall from Chapter 24 that biologists have documented extensive changes in beak size and shape in medium ground finches from the Galápagos Islands. Such changes are due to natural selection. Individuals with deep beaks are better able to crack the large fruits that predominate during drought years, while individuals with small beaks are better able to harvest the small seeds that predominate during wet years.

As **Figure 41.2** shows, a strong correlation between diet and beak structure is also found *among* species of Galápagos finch. Species with large, cone-shaped beaks eat large seeds; species with small, cone-shaped beaks eat small seeds. Species with long, tweezer-like beaks pick insects off tree trunks or other surfaces.

The mechanism responsible for these structure-function correlations is straightforward: If a mutant allele alters the size or shape of a structure in a way that makes it function more efficiently, individuals who have that allele will produce more offspring than will other individuals. As a result, the allele will increase in frequency in the population over time.

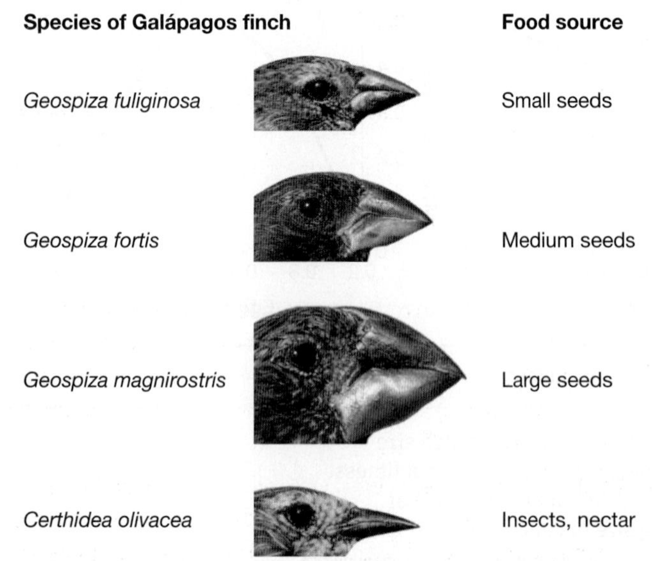

Structure-Function Relationships at the Molecular and Cellular Levels

Correlations between form and function start at the molecular level. For example, earlier chapters emphasized that the shape of proteins correlates with their role as enzymes, structural components of the cell, or transporters. The membrane proteins called channels form pores that allow specific ions or molecules to pass in or out of cells (see Chapter 6). The ends and interior of a channel are hydrophilic, which allows the protein to interact with the surrounding solution or the interior of the cell, while the perimeter is hydrophobic—allowing it to interact with the lipid bilayer. The protein's structure fits its function.

Similar correlations between structure and function occur at the level of the cell. Cells that manufacture and secrete hormones or digestive enzymes are packed with rough ER and Golgi; cells that store energy are dominated by large fat droplets; cells that ingest and destroy invading bacteria have numerous lysosomes.

The overall shape of a cell can also correlate with its function. For example, cells that are responsible for transporting materials into or out of the body often have extremely large areas of plasma membrane. As a result, they have room to accommodate the thousands of membrane channels, transporters, and pumps required for extensive transport.

Tissues Are Groups of Similar Cells That Function as a Unit

Animals are **multicellular**, meaning that their bodies contain distinct types of cells that are specialized for different functions. Frequently, animal cells that are similar in structure and function are physically attached to each other and form a tissue. A **tissue** is a group of similar cells that function as a unit.

The embryonic tissues called ectoderm, mesoderm, and endoderm are found in most animals and were introduced in Chapter 22 and Chapter 32. As an individual develops, the embryonic tissues give rise to four adult tissue types: (**1**) connective tissue, (**2**) nervous tissue, (**3**) muscle tissue, and (**4**) epithelial tissue. In each case, the structure of the tissue correlates closely with its function. Let's consider each in turn.

CONNECTIVE TISSUE **Connective tissue** consists of cells that are loosely arranged in a liquid, jellylike, or solid matrix. The matrix comprises extracellular fibers and other materials, and is secreted by the connective tissue cells themselves. Each type of connective tissue secretes a distinct type of extracellular matrix. The nature of the matrix determines the nature of the connective tissue.

Species of Galápagos finch	Food source
Geospiza fuliginosa	Small seeds
Geospiza fortis	Medium seeds
Geospiza magnirostris	Large seeds
Certhidea olivacea	Insects, nectar

FIGURE 41.2 In Animal Anatomy and Physiology, Form Often Correlates with Function.

(a) Loose connective tissue: soft extracellular matrix; provides padding

Reticular connective tissue   Adipose (fat) tissue

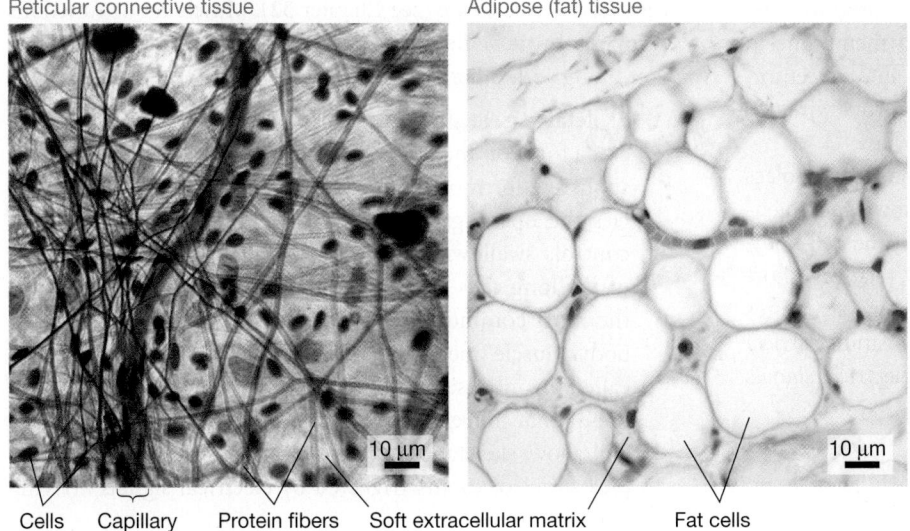

Cells   Capillary   Protein fibers   Soft extracellular matrix   Fat cells

(b) Dense connective tissue: fibrous extracellular matrix; provides connections

Tendon

Collagen fibers   Tendon cell nuclei

(c) Supporting connective tissue: firm extracellular matrix; functions in structural support and protection

Bone   Cartilage

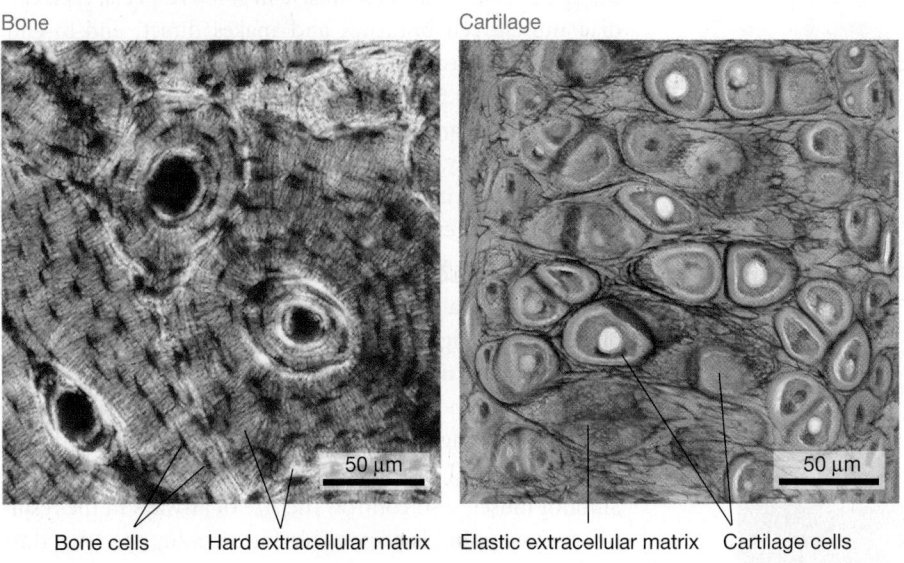

Bone cells   Hard extracellular matrix   Elastic extracellular matrix   Cartilage cells

(d) Fluid connective tissue: liquid extracellular matrix; functions in transport

Blood

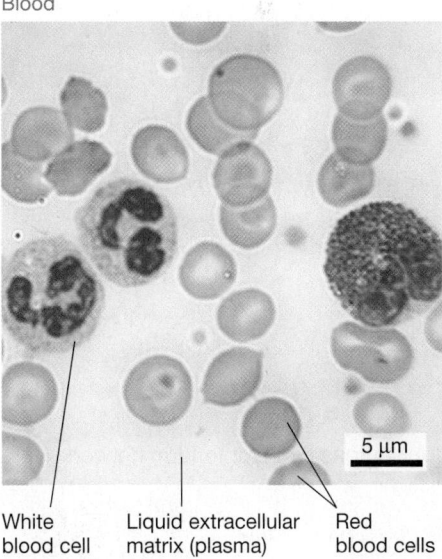

White blood cell   Liquid extracellular matrix (plasma)   Red blood cells

FIGURE 41.3 Connective Tissues Consist of Cells Surrounded by an Extracellular Matrix.

✓**QUESTION** How does the stiffness of the extracellular matrix in a connective tissue correlate with that tissue's function?

- **Loose connective tissue** contains an array of fibrous proteins in a soft matrix; it serves as a packing material between organs or padding under the skin. The **adipose tissue**, or fat tissue, illustrated in **Figure 41.3a** is a loose connective tissue made up of cells that are dominated by fat droplets and a loose matrix of fibers and fluid.

- **Dense connective tissue** is found in the tendons and ligaments that connect muscles, bones, and organs. As **Figure 41.3b** shows, the matrix in tendons and ligaments is dominated by the tough collagen fibers introduced in Chapter 8.

- **Supporting connective tissue** has a firm extracellular matrix. **Bone** and **cartilage** (**Figure 41.3c**) are connective tissues that provide structural support for the vertebrate body, as well as protective enclosures for the brain and other components of the nervous system.

- **Fluid connective tissue** consists of cells surrounded by a liquid extracellular matrix. **Blood**, which transports materials throughout the vertebrate body (**Figure 41.3d**), contains a variety of cell types and has a specialized extracellular matrix called plasma (see Chapter 44).

NERVOUS TISSUE **Nervous tissue** consists of nerve cells, which are also called **neurons**, and several types of supporting cells. Neurons transmit electrical signals, which are produced by changes in the permeability of the cell's plasma membrane to ions (see Chapter 45). Supporting cells regulate ion concentrations in the space surrounding neurons, supply neurons with nutrients, or serve as scaffolding or support for neurons.

Although they vary widely in shape, all neurons have projections that contact other cells. As **Figure 41.4a** shows, most neurons have two distinct types of projections from the cell body, where the nucleus is located: (**1**) highly branched, relatively short processes called **dendrites**, and (**2**) relatively long structures called **axons**. Dendrites contact other cells and transmit electrical signals from them to the cell body; axons carry electrical signals from the cell body to other cells (**Figure 41.4b**).

(a) In a neuron, information is transmitted from dendrites to the cell body to the axon.

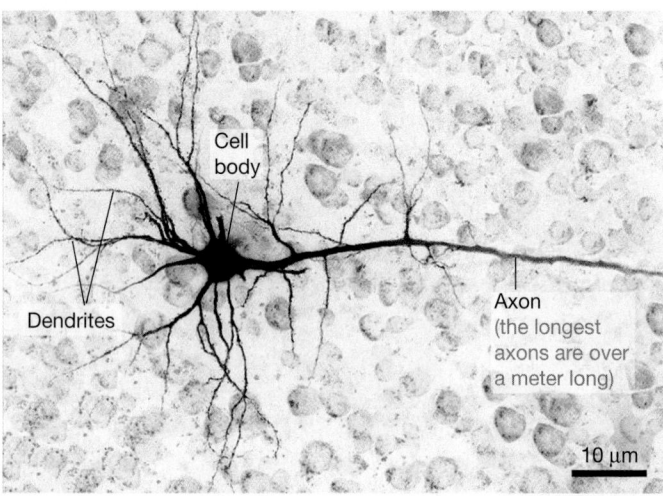

(b) Neurons connect to form networks.

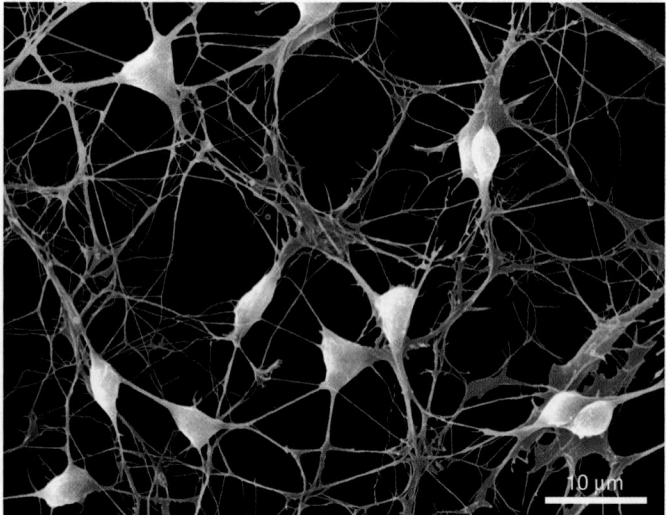

FIGURE 41.4 Neurons Transmit Electrical Signals.

✔**QUESTION** How does the presence of projections on nerve cells support their function in electrical signaling?

MUSCLE TISSUE **Muscle tissue** was a key innovation in the evolution of animals—like nervous tissue, it appears in no other lineage on the tree of life (see Chapter 32). Muscle tissue functions primarily in movement. There are three types of muscle tissue; you, along with other vertebrates, have all three.

1. **Skeletal muscle** attaches to the bones of the skeleton and exerts a force on them when it contracts. Skeletal muscle is responsible for most body movements. In addition, it encircles the openings of the digestive and urinary tracts and controls swallowing, defecation, and urination. It consists of the long cells called **muscle fibers** (**Figure 41.5a**) and is the most common type of muscle tissue in the vertebrate body. Muscle fibers are packed with long tubelike structures called myofibrils; each myofibril is packed with protein filaments that move by sliding past each other. As Chapter 46 will show, skeletal muscle contracts in response to a complex series of events triggered by electrical signals arriving from nerve cells.

2. **Cardiac muscle** makes up the walls of the heart and is responsible for pumping blood throughout the body. Although similar to skeletal muscle in some respects, each cardiac muscle cell branches and makes direct, end-to-end, physical and electrical contact with other cardiac muscle cells (**Figure 41.5b**). These connections help transmit signals from one cardiac muscle cell to another during a heartbeat. In this way, the contraction of cardiac muscle cells is coordinated during the series of events known as the cardiac cycle (see Chapter 44).

3. **Smooth muscle** cells, which are tapered at each end, form a muscle tissue that lines the walls of the digestive tract and the blood vessels (**Figure 41.5c**). Different types of neurons control the contraction of smooth muscle cells versus striated muscle cells. Smooth muscle is responsible for movements such as the passage of food down the digestive tract or the dilation (opening) of arteries near the skin in hot weather. Smooth muscle also controls the size of airways in the respiratory system and is responsible for expelling the fetus during birth (see Chapter 48).

In addition, biologists make three other key distinctions about muscle cells and tissues:

- Tissues that can contract in response to conscious thought are said to be **voluntary muscle**; tissues that contract only in response to unconscious electrical activity are said to be **involuntary muscle**. Skeletal muscle is voluntary; cardiac and smooth muscle are involuntary.

- Muscle cells may have one or many nuclei. Skeletal muscle cells and some cardiac muscle cells are multinucleate; most cardiac and all smooth muscle cells are uninucleate.

- In some muscle cells, the protein filaments responsible for contractions are organized into repeating structures that give the cells and tissues a banded or **striated** appearance. Skeletal and cardiac muscle are striated; smooth muscle is unstriated.

(a) Skeletal muscle: voluntary, multinucleate, striated

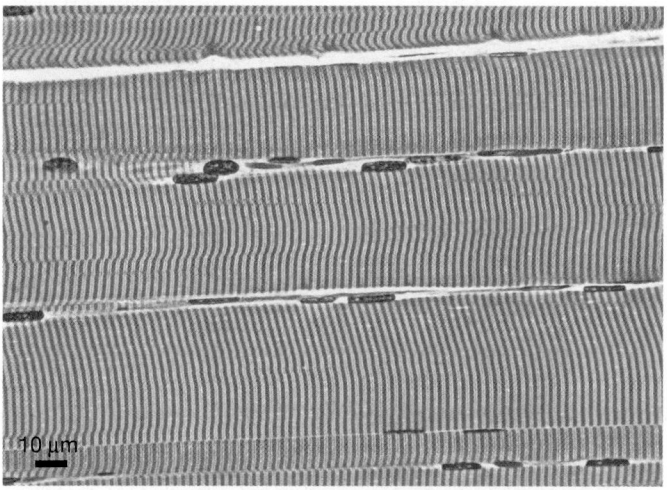

Long cells (muscle fibers)

(b) Cardiac muscle: involuntary, usually uninucleate, striated

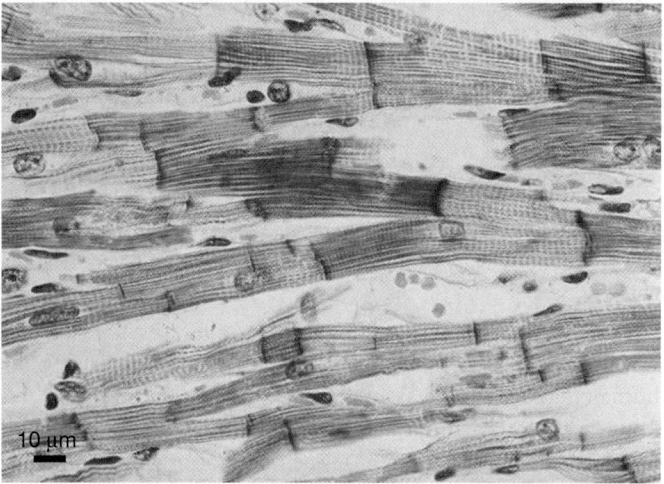

Branched cells

(c) Smooth muscle: involuntary, uninucleate, unstriated

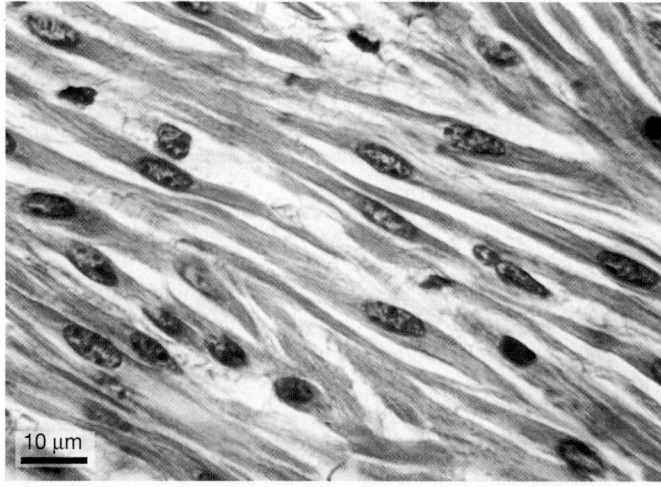

Tapered cells

FIGURE 41.5 Muscle Tissues Comprise Cells That Contract. The three types of muscle tissue have distinctive structures and functions.

EPITHELIAL TISSUES **Epithelial tissues** are also called **epithelia** (singular: **epithelium**). Epithelium covers the outside of the body, lines the surfaces of organs, and forms glands. An **organ** is a structure that serves a specialized function and consists of several tissues; a **gland** is a group of cells that secrete specific molecules or solutions.

Epithelia form the interface between the interior of an organ or body and the exterior. In addition to providing protection, epithelial tissues are gatekeepers. Epithelia regulate the transfer of heat between the interior and exterior of structures, as well as the transfer of water, nutrients, and other substances.

Because the primary function of epithelium is to act as a barrier and protective layer, it's not surprising to observe that epithelial cells typically form layers of closely packed cells (**Figure 41.6**). In many cases, adjacent epithelial cells are joined by structures that hold them tightly together, such as tight junctions and desmosomes (introduced in Chapter 8).

Epithelial tissue has polarity, or sidedness. An epithelium has an **apical** side, which faces away from other tissues and toward the environment, and a **basolateral** side, which faces the interior of the animal and connects to connective tissues. This connection is made by a layer of fibers called the **basal lamina**.

The apical and basolateral sides of an epithelium have distinct structures and functions. Epithelial cells, for example, line the surface of your trachea, or windpipe. The apical side of these cells secretes mucus and is covered with cilia that help sweep away dust, bacteria, and viruses. The basolateral side lacks these features, but is cemented to the basal lamina.

Epithelial cells have short life spans. The cells that line your esophagus—the tube connecting your mouth and stomach—live for 2 to 3 days, while the cells that line your large intestine live for a maximum of 6 days. Muscle cells and neurons, in contrast,

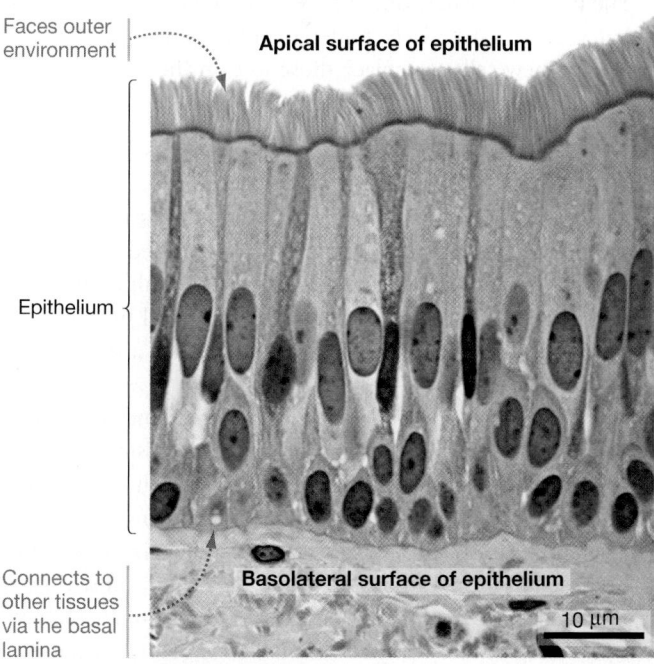

Faces outer environment — Apical surface of epithelium

Epithelium

Connects to other tissues via the basal lamina — Basolateral surface of epithelium

FIGURE 41.6 Epithelial Cells Provide Protection and Regulate Which Materials Pass across Body Surfaces.

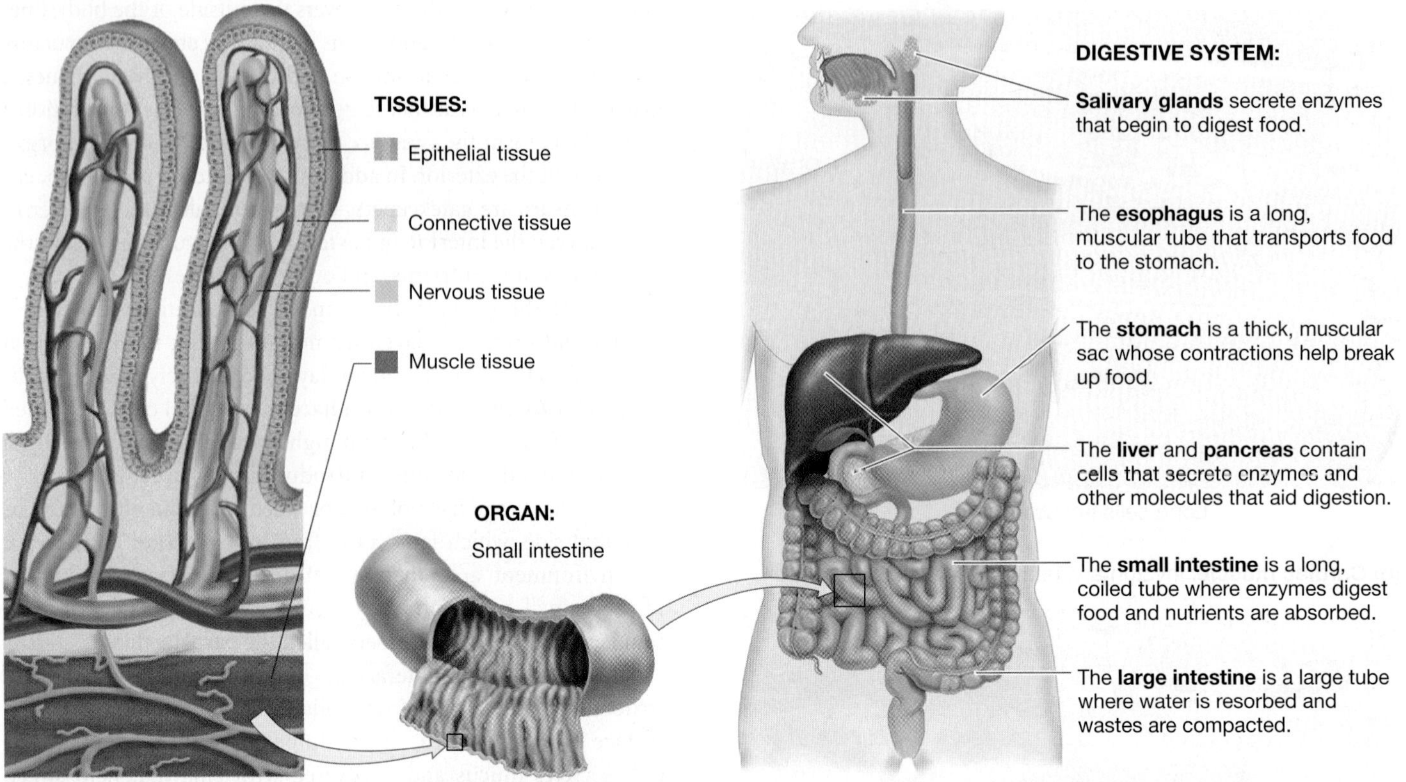

(a) Tissues are organized into organs.

(b) Organs are organized into organ systems.

TISSUES:

- Epithelial tissue
- Connective tissue
- Nervous tissue
- Muscle tissue

ORGAN:

Small intestine

DIGESTIVE SYSTEM:

Salivary glands secrete enzymes that begin to digest food.

The **esophagus** is a long, muscular tube that transports food to the stomach.

The **stomach** is a thick, muscular sac whose contractions help break up food.

The **liver** and **pancreas** contain cells that secrete enzymes and other molecules that aid digestion.

The **small intestine** is a long, coiled tube where enzymes digest food and nutrients are absorbed.

The **large intestine** is a large tube where water is resorbed and wastes are compacted.

FIGURE 41.7 Organs Are Composed of Tissues; Organ Systems Are Made Up of Organs. **(a)** The human small intestine is an organ composed of all four major tissue types. **(b)** The human digestive system is essentially one long tube divided into chambers where food is processed and nutrients are absorbed. The salivary glands, liver, and pancreas are organs that secrete specific enzymes or compounds into the tube.

normally live as long as the individual does. Epithelial cells are short-lived because they are exposed to harsh environments, where they are likely to be killed or scraped away.

The tissue as a whole does not wear away, however, because it includes cells that actively undergo mitosis and cytokinesis—producing new cells to replace those lost on the side that faces the environment.

Organs and Organ Systems

Cells with similar functions are organized into tissues, and tissues are organized into specialized structures called organs. Recall that an organ is a structure that serves a specialized function and consists of several types of tissues. The small intestine, for example, consists of muscle, nervous, connective, and epithelial tissues (**Figure 41.7a**).

An **organ system** consists of groups of tissues and organs that work together to perform one or more functions. Using the digestive system as an example, **Figure 41.7b** illustrates how the structure of organs correlates with their function and how the components of an organ system work together in an integrated fashion.

Because an animal's body contains molecules, cells, tissues, organs, and organ systems, biologists who study animal anatomy and physiology must work at various levels of organization to understand how that body operates.

Figure 41.8 illustrates these levels of organization, using the human nervous system as an example. Because the structure and function of each component in the body are integrated with other components, and because each level of organization is integrated with other levels of organization, the organism as a whole is greater than the sum of its parts. In other words, an organism is more than just a collection of individual systems, and each system is more than just a collection of individual cells or tissues or even organs.

CHECK YOUR UNDERSTANDING

If you understand that . . .

- Biologists study structure and function at the molecular, cellular, tissue, organ, and organ system levels.
- Events at each level of organization in an individual interact to form an integrated whole that responds to the environment in appropriate ways.

✔ **You should be able to . . .**

Describe the structure and function of the four major types of animal tissues.

Answers are available in Appendix B.

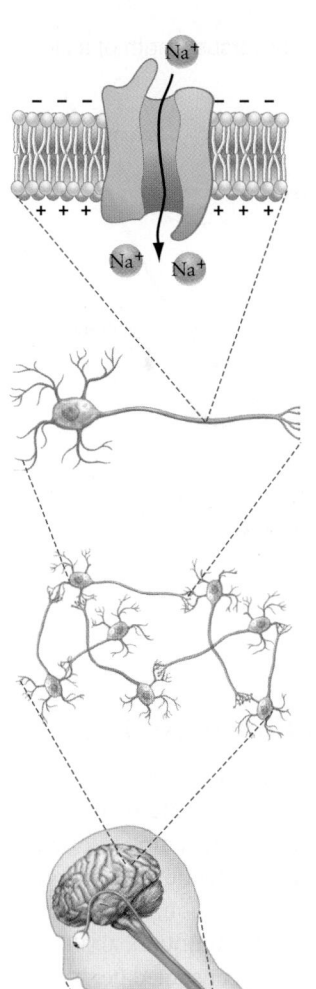

Atomic and molecular levels:
Membrane protein in neurons
regulates a flow of ions.

Cellular level:
Electric signal mediated by ion
flows travels down length of
neuron.

Tissue level:
Electric signals travel from cell to
cell in nervous tissue.

Organ level:
Nervous tissue and connective
tissue in brain aid in sight, smell,
memory, and thought.

System level:
Nervous system controls breathing,
digestion, movement, and other
functions.

Organism level:
Systems work together to
support life.

**FIGURE 41.8 Biologists Study Anatomy and Physiology at Many
Levels.** The levels of organization within an organism are not
independent of each other. Instead, they are tightly integrated.

In effect, each subsequent chapter in this unit focuses on a different organ system found in animals, beginning with the excretory system and ending with the immune system. Each of these systems can be interpreted as a suite of adaptations and trade-offs. Each system accomplishes a specific task required for survival and reproduction, and each works in conjunction with other systems.

Before delving into the various systems, however, it's essential to examine general phenomena that affect all systems in animals. Let's start by looking at how body size affects animal physiology.

41.3 How Does Body Size Affect Animal Physiology?

Animals are living machines, made up of molecules, cells, tissues, organs, and organ systems that have changed over time in response to natural selection.

The laws of physics affect the anatomy and physiology of a living machine. The force of gravity, for example, limits how large an animal can be and still move efficiently. Or consider the forces exerted by the medium in which animals live. Because water is much denser than air, it is harder for animals to move through water. As a result, fish and aquatic mammals have much more streamlined bodies than terrestrial animals do.

Physical laws clearly affect body size. Just as clearly, body size has pervasive effects on how animals function. Large animals need more food than small animals do. Large animals also produce more waste, take longer to mature, reproduce more slowly, and live longer. Conversely, small animals are more susceptible to damage from cold and dehydration than large animals are, because they lose heat and water faster. Juveniles and adults of the same species face different challenges simply because their body sizes are different.

Why is body size such an important factor in how animals work? How do biologists study the consequences of size? Let's consider each question in turn.

Surface Area/Volume Relationships: Theory

From microscopic roundworms to gigantic blue whales, animals span an incredible range of body masses—a total of twelve orders of magnitude. Many of the challenges posed by increasing size are based on the relationship between surface area and volume.

To understand why surface area is important, recall from Chapter 6 that diffusion takes place across the surface of the plasma membrane. Oxygen and nutrients such as glucose must diffuse into the cell, and waste products such as urea and carbon dioxide must diffuse out. The rate at which these and other molecules and ions diffuse depends in part on the amount of surface area available for diffusion. In contrast, the rate at which nutrients are used and heat and waste products are produced depends on the volume of the cell.

The contrast between processes that depend on surface area and those that depend on volume is important for a simple reason. As a cell gets larger, its volume increases much faster than its surface area does.

(a) What are the surface area and volume of each cube?

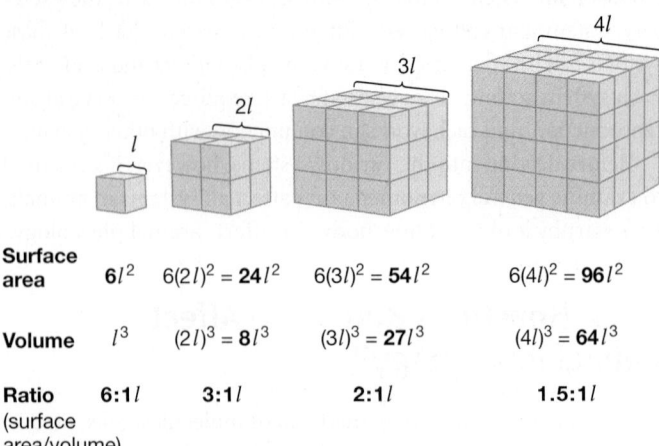

Surface area	$6l^2$	$6(2l)^2 = \mathbf{24}l^2$	$6(3l)^2 = \mathbf{54}l^2$	$6(4l)^2 = \mathbf{96}l^2$
Volume	l^3	$(2l)^3 = \mathbf{8}l^3$	$(3l)^3 = \mathbf{27}l^3$	$(4l)^3 = \mathbf{64}l^3$
Ratio (surface area/volume)	$6:1l$	$3:1l$	$2:1l$	$1.5:1l$

(b) Surface area and volume of a cube versus length of a side

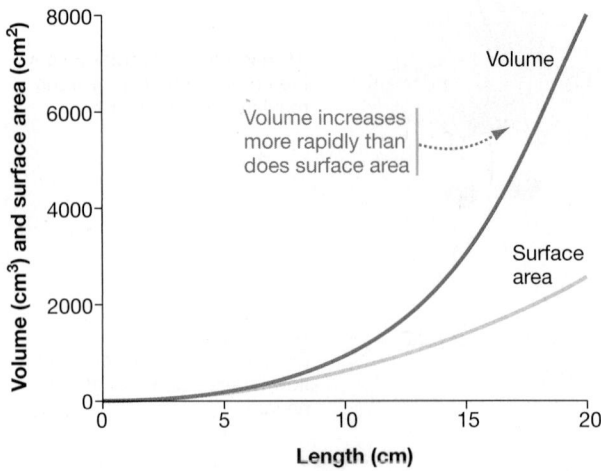

FIGURE 41.9 Surface Area and Volume Change as a Function of Overall Size. **(a)** The surface area of an object increases as the square of the length (*l*). The volume increases as the cube of that linear dimension. **(b)** Volume increases much more rapidly than does surface area as linear dimensions increase.

Reviewing a little basic geometry will convince you why this is so. As **Figure 41.9a** shows:

- The surface area of a cube increases as a function of its linear dimension *squared*. Because a cube has six sides, the surface area of a cube of length *l* is $6l^2$ (six times the area of any one side).

- The volume of the same structure increases as a function of its linear dimension *cubed*. Hence, the volume (or mass) of a cube of length *l* is l^3.

Area has two dimensions; volume has three. In general:

$$\text{Surface area} \propto (\text{length})^2$$

$$\text{Volume (or mass)} \propto (\text{length})^3$$

$$\text{Surface area} \propto (\text{volume})^{2/3}$$

(The symbol $\propto$ means "is proportional to.")

Figure 41.9b graphs the consequences of these relationships. The *x*-axis plots the length of a side in a cube; the *y*-axis plots the cube's volume (orange line) or surface area (yellow line). As a cube gets bigger, its surface area increases much more slowly than does its volume (or mass).

The same general relationship holds for cells, tissues, organs, and systems. Quantities that are based on volume, such as body mass, increase disproportionately fast with increases in linear dimensions.

✔ If you understand the relationship between surface area and volume, you should be able to predict which of the following has the higher surface area/volume ratio: a newborn or an adult human. You should also go to the study area at *www.masteringbiology.com* to review how body size affects surface area and volume.

(MB) Web Activity Surface Area/Volume Relationships

How does the relationship between surface area and volume affect animal form and function?

Surface Area/Volume Relationships: Data

As an example of how surface area/volume relationships affect an animal's physiology, consider the metabolic rate of mammals. **Metabolic rate** is the overall rate of energy consumption by an individual. Because consumption and production of energy in mammals depend largely on aerobic respiration, metabolic rate is often measured in terms of oxygen consumption, and is typically reported in units of milliliters of O_2 consumed per hour.

Because it is so much larger, an elephant consumes a great deal more oxygen per hour than a mouse does. But what is going on at the levels of cells and tissues?

COMPARING MICE AND ELEPHANTS To compare metabolic rates in different species, biologists divide metabolic rate by overall mass and report a mass-specific metabolic rate in units of mL O_2/gram/hour. This mass-specific metabolic rate gives the rate of oxygen consumption per gram of tissue.

Because an individual's metabolic rate varies dramatically with its activity, the accepted convention is to report the **basal metabolic rate (BMR)**—the rate at which an animal consumes oxygen while at rest, with an empty stomach, under normal temperature and moisture conditions.

Figure 41.10 plots per-gram or "mass-specific" BMR as a function of average body mass. Note that the *x*-axis on the graph is logarithmic, to make it easier to compare very small with very large species. For help with logarithms, see **BioSkills 7** in Appendix A.

The graph's take-home message? On a per-gram basis, small animals have higher BMRs than do large animals. An elephant has more mass than a mouse, but a gram of elephant tissue consumes much less energy than a gram of mouse tissue does.

The leading hypothesis to explain this pattern is based on surface area/volume ratios. Many aspects of metabolism—including oxygen consumption, food digestion, delivery of nutrients to tissues, and removal of wastes and excess heat—depend on ex-

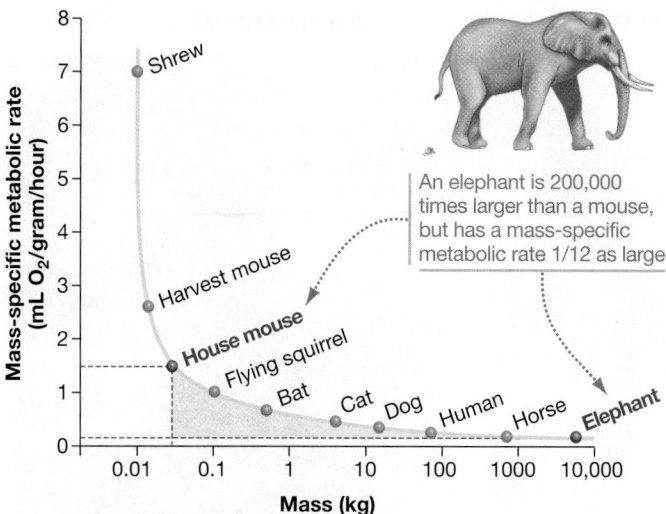

FIGURE 41.10 Small Animals Have Higher Metabolic Rates than Large Animals Do. Overall body mass, plotted on a logarithmic scale, versus metabolic rate per gram of tissue.

✔**QUESTION** Which mammal has to eat more to support each gram of its tissue: a dog or a human?

change across surfaces. As an organism's size increases, its mass-specific metabolic rate must decrease. Otherwise the surface area available for exchange of materials would fail to keep up with the metabolic demands generated by the enzymes in the organism.

Small animals can "live fast" because they have enough surface area to support rapid metabolism. Large animals don't have enough surface area to keep up and have to "live slow." There is a trade-off between size and metabolic rate.

CHANGES DURING DEVELOPMENT A king salmon weighs a few milligrams or less at hatching but grows into an adult weighing 50 kg or more—a millionfold increase in body mass. To explore the consequences of this change, biologists have studied how gas exchange—uptake of oxygen and removal of carbon dioxide—occurs in newly hatched Atlantic salmon.

Like most fish species, young salmon have rudimentary gills but also exchange gases across their skin. In aquatic animals, **gills** are organs that allow the exchange of gases and dissolved substances between the animals' blood and the surrounding water.

To document the amount of gas exchange that occurs in the gills versus the general body surface, researchers inserted the heads of individual salmon through a pinhole in a soft rubber membrane, then recorded the rate of oxygen uptake on either side of the membrane. As the "Experimental Setup" section of **Figure 41.11** indicates, the gills were responsible for oxygen uptake on one side of the membrane; skin was responsible for oxygen uptake on the other side of the membrane.

The graph in the figure's "Results" section plots the percentage of total oxygen uptake that took place across the skin (green line) versus gills (purple line), as a function of body mass. Each data point represents the recordings from an individual fish.

Note that newly hatched larvae take up most of the oxygen they need by diffusion across the body surface. As an individual

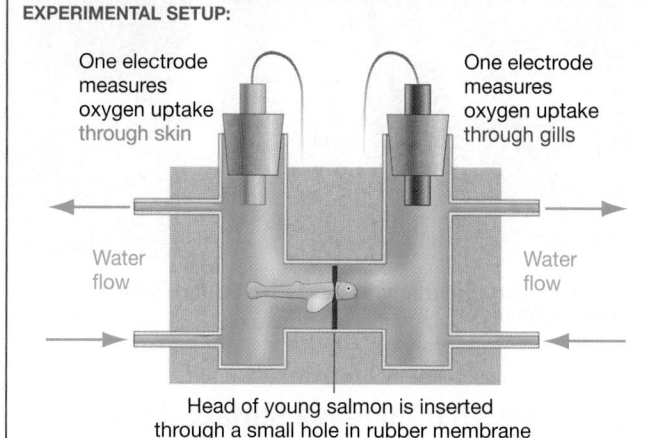
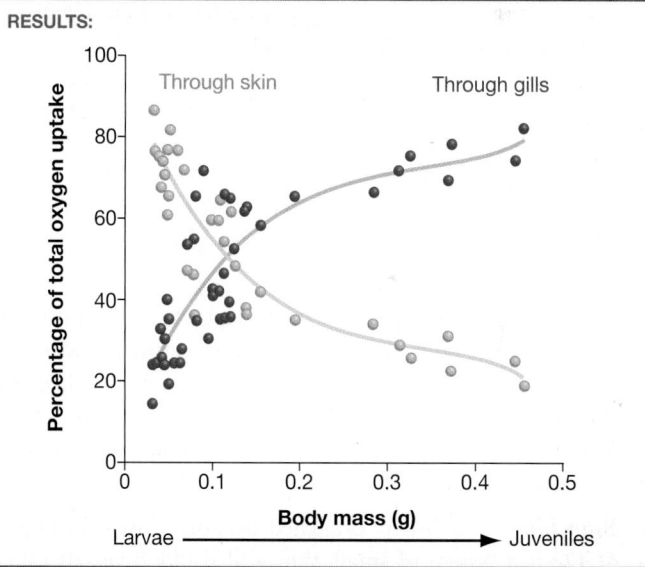
FIGURE 41.11 How Do Young Salmon Breathe?

SOURCE: Wells, P. R. and A. W. Pinder. 1996. The respiratory development of Atlantic salmon. *Journal of Experimental Biology* 199: 2737–2744.

✔**QUESTION** Suppose the experimenters had measured oxygen uptake on either side of the apparatus in the absence of a fish. What would the results be?

grows, however, its skin surface area decreases in relation to its volume. To avoid suffocation, individuals switch from skin-breathing to gill-breathing at about 0.1 grams. What makes gills so effective in oxygen uptake?

(a) Flattening: fish gill lamellae

(b) Folding: intestinal folds and villi

(c) Branching: capillaries

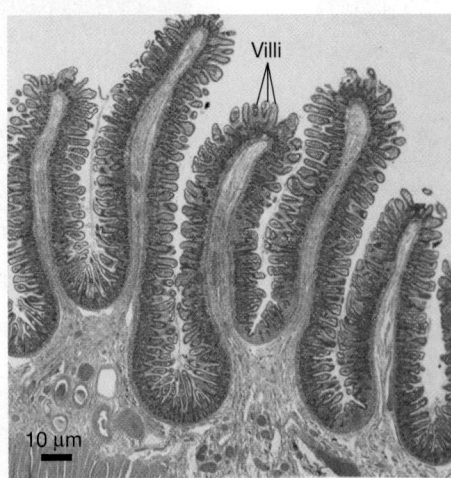

Villi

Lamellae

10 µm

10 µm

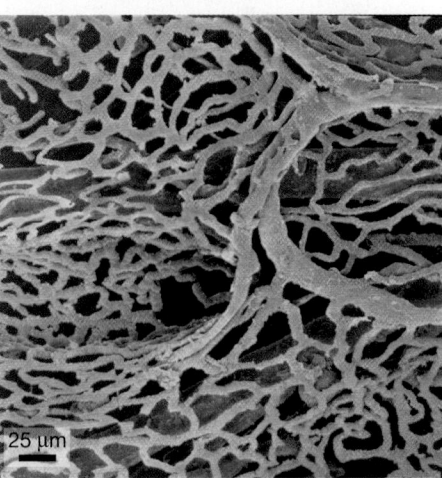

25 µm

FIGURE 41.12 Certain Structures Have High Surface Area/Volume Ratios. The micrograph in part (c) has been colorized to highlight capillaries, colored pink.

Adaptations That Increase Surface Area

If the function of a cell or tissue depends on diffusion, it usually has a shape that increases its surface area relative to its volume. Flattening, folding, and branching are effective ways for structures to have a high surface area/volume ratio:

- *Flattening* Fish gills (**Figure 41.12a**) consist of **lamellae**— thin sheets of epithelial cells that provide this organ with an extremely high surface area relative to its volume. Because the surface available is so large, gases are able to diffuse across the gills rapidly enough to keep up with the growth in the volume of a developing fish.

- *Folding* In portions of the digestive tract where nutrients diffuse into the body, the surface of the structure is folded. Extending from these folds are narrow projections called **villi** (**Figure 41.12b**). Together, the folds and fold-like projections make an extensive surface area available. Folded surfaces are common in diffusion-dependent organs.

- *Branching* The highly branched network shown in **Figure 41.12c** is a system of small, thin-walled blood vessels called **capillaries**. Capillaries have a high surface area available for gases, nutrients, and waste products to diffuse into and out of blood; branching increases the surface available in each square centimeter of tissue. In general, highly branched structures increase the surface area available for diffusion.

The amount of surface area created by flattening, folding, and branching can be impressive. The highly branched capillaries in a human have a total surface area of up to 1000 m²; extensive folding gives a total surface area of about 140 m² in your lungs and 250 m² in your small intestine. For comparison, a doubles tennis court has a surface area of 261 m².

Surface area/volume relationships have a pervasive influence on the structure and function of animals. They will be an issue in almost every chapter in this unit.

CHECK YOUR UNDERSTANDING

If you understand that . . .

- An animal's overall size is important in part because body mass is affected by an array of physical forces.
- The amount of heat and waste that an animal produces and the amount of food and oxygen that it requires are proportional to its mass.
- The amount of surface area available relative to that mass is critical, because heat exchange and other important processes take place across surfaces.

You should be able to . . .

1. Explain why large animals have a relatively small surface area/volume ratio.
2. Explain the relationship between body mass and mass-specific basal metabolic rate.

Answers are available in Appendix B.

41.4 Homeostasis

Adaptation and surface area/volume ratios are important themes in the analysis of animal form and function. So is homeostasis.

Homeostasis (literally, "alike-standing") is defined as stability in the chemical and physical conditions within an animal's cells, tissues, and organs. Although conditions may vary as an animal's environment changes, internal chemical and physical states are kept within a tolerable range.

Homeostasis: General Principles

Many of the structures and processes observed in animals can be interpreted as mechanisms for maintaining homeostasis with respect to some quantity, such as pH or calcium ion concentration.

Let's review some important general ideas about homeostasis, then analyze how homeostasis can be maintained in the face of environmental fluctuations.

TWO APPROACHES TO ACHIEVING HOMEOSTASIS Constancy of physiological state can be achieved by two processes: (1) conformation or (2) regulation.

- *Conformation* The body temperature of Antarctic rock cod closely matches that of the surrounding seawater, which is typically −1.9°C (seawater is still liquid at this temperature because of the high concentration of solutes). The rock cod does not actively regulate its body temperature to match that of seawater. Instead, its body temperature remains constant because it conforms to the temperature of its surroundings.

- *Regulation* Regulatory homeostasis is based on mechanisms that adjust the internal state to keep it within limits that can be tolerated, no matter what the external conditions. For example, a dog maintains a body temperature of about 38°C whether it's cold or hot outside. If its body temperature rises, it might pant to cool off and maintain homeostasis. If its body temperature falls, it might shiver to bring its temperature back up to the target value.

THE ROLE OF EPITHELIUM Because epithelium is the interface between the internal and external environments, it plays a key role in achieving homeostasis. Epithelium is responsible for forming an internal environment that can be dramatically different from the external environment, and for maintaining physical and chemical conditions inside an animal that are relatively constant.

As subsequent chapters will show, many epithelial cells are studded with membrane proteins that regulate the transport of ions, water, nutrients, and wastes. No molecule can enter or leave the body without crossing an epithelium. Homeostasis is possible because epithelia control this exchange.

WHY IS HOMEOSTASIS IMPORTANT? Much of the answer to this question is based on enzyme function. Recall from Chapter 3 that enzymes are proteins that catalyze chemical reactions within cells. Temperature, pH, and other physical and chemical conditions have a dramatic effect on the structure and function of enzymes. Most enzymes function best under a fairly narrow range of conditions.

Other processes depend on homeostasis, too. Temperature changes affect membrane permeability and how quickly solutes diffuse. To take an extreme case, the expansion of water as it freezes can rip cells apart if tissues are allowed to drop much below 0°C. Conversely, extremely high temperatures can denature proteins—meaning that they lose their tertiary structure and cease to function.

When homeostasis occurs, conditions inside the body allow molecules, cells, tissues, organs, and organ systems to function at an optimal level.

The Role of Regulation and Feedback

To achieve homeostasis, most animals have regulatory systems that constantly monitor internal conditions such as temperature, blood pressure, blood pH, and blood glucose. If one of these variables changes, a homeostatic system acts quickly to modify it. Like the thermostat in a home heating system, each of these systems has a **set point**—a normal or target value for the controlled variable.

Animals have a set point for blood pH, blood oxygen concentration, nutrient availability, and other parameters. In most mammals, the set point for body temperature is somewhere between 35°C and 39°C. How does an individual maintain its tissues at the set point despite changes in activity and the environment?

A homeostatic system is based on three general components: a sensor, an integrator, and an effector. **Figure 41.13** shows how these components interact:

1. A **sensor** is a structure that senses some aspect of the external or internal environment.

2. An **integrator** evaluates the incoming sensory information and "decides" whether a response is necessary to achieve homeostasis. (The word decides is in quotation marks because the decision is not a conscious one.)

3. An **effector** is any structure that helps restore the desired internal condition.

Without these three elements, homeostatic control systems can't maintain a desired set point; homeostasis is impossible.

Homeostatic systems are based on negative feedback. When **negative feedback** occurs, effectors reduce or oppose the change in internal conditions. For example, a rise in blood pH triggers effectors that act to reduce that rise. Blood pH returns to the set

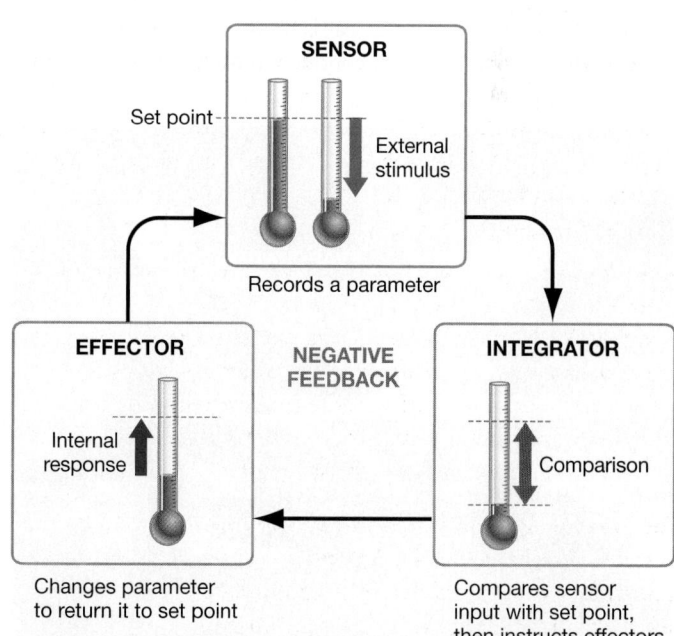

FIGURE 41.13 Animals Achieve Homeostasis through Negative Feedback. Many animals use homeostatic systems similar to this one to maintain a preferred range of hydration (water concentration), blood pH, blood pressure, calcium ion concentration, body temperature, and so on.

point in response to this negative feedback. For more detail on how negative feedback works in homeostasis, go to the study area at *www.masteringbiology.com*.

 Web Activity Homeostasis

Homeostatic systems are a key aspect of one of the five attributes of life introduced in Chapter 1: acquiring information from the environment and responding to it. Subsequent chapters in this unit explore how animals use sensor-integrator-effector systems to achieve homeostasis with respect to the solute concentrations of their cells and tissues, their oxygen supply, and nutrient availability. In this chapter, let's focus on how different animals achieve homeostasis with respect to body temperature.

41.5 How Do Animals Regulate Body Temperature?

All animals exchange heat with their environment. Heat flows "downhill," from regions of higher temperature to regions of lower temperature. If an individual is warmer than its surroundings, it will lose heat; if it is cooler than its environment, it will gain heat.

How does heat exchange occur?

Mechanisms of Heat Exchange

As **Figure 41.14** shows, animals exchange heat with the environment in four ways: conduction, convection, radiation, and evaporation.

- **Conduction** is the direct transfer of heat between two physical bodies that are in contact with each other. For instance, when a turtle sits on a warm rock, heat is transferred from the rock to its body. The rate at which conduction occurs depends on the

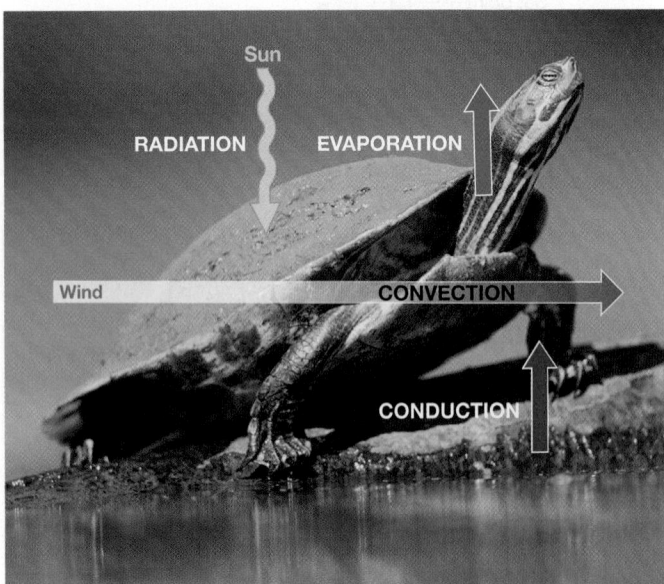

FIGURE 41.14 Four Methods of Heat Exchange. The arrows indicate the direction of heat exchange from the warmer body to the cooler body.

surface area of transfer, the steepness of the temperature difference between the two bodies, and how well each body conducts heat. Water, for example, conducts heat much better than air does. As a result, a person immersed in 15°C water loses heat much faster than does a person who is exposed to air at the same temperature.

- **Convection** is a special case of conduction. During conduction, heat is transferred between two solids; but during convection, heat is exchanged between a solid and a liquid or gas. For example, the heat loss that occurs when wind blows on your skin is due to convection. As the speed of the air or water flow increases, so does the rate of heat transfer.

- **Radiation** is the transfer of heat between two bodies that are not in direct physical contact. All objects, including animals, radiate energy as a function of their temperature. The Sun radiates heat; so does your body, but to a much lesser degree.

- **Evaporation** is the phase change that occurs when liquid water becomes a gas. Conduction, convection, and radiation can cause heat gain or loss, but evaporation leads only to heat loss. The turtle in the photograph is losing heat as water evaporates off its shell and skin. Because of the extensive hydrogen bonding in liquid water, a large amount of energy is needed to heat water and produce evaporation (see Chapter 2). If you get overheated on a summer day, splashing water on your skin and sweating will absorb a large amount of heat and cool you off. Conversely, getting wet on a cold day can be deadly. The water on your skin absorbs so much heat from your body that your temperature may drop dangerously.

Heat exchange is critical in animal physiology because individuals that get too hot or too cold may die. Overheating can cause enzymes and other proteins to denature and cease functioning. It may also lead to excessive water loss and dehydration. A sharp drop in body temperature, in contrast, can slow enzyme function and energy production. In humans, both heat stroke and hypothermia ("under-heating") are life-threatening conditions.

Although most organisms cannot regulate their body temperature to keep it in an optimal range, some animals can. Let's take a closer look.

Variation in Thermoregulation

The ability of animals to **thermoregulate**, or control body temperature, varies widely. One way to organize this variation is by examining (1) how animals obtain heat, and (2) whether body temperature is held constant.

An **endotherm** ("inner-heat") produces adequate heat to warm its own tissues, while an **ectotherm** ("outer-heat") relies principally on heat gained from the environment. The resulting difference in body temperature can be dramatic, as illustrated in the thermogram—an image taken with a camera that is sensitive to infrared radiation—in **Figure 41.15**.

Endotherms and ectotherms represent two opposite extremes along a continuum of heat sources. Many animals are partially endothermic and partially ectothermic.

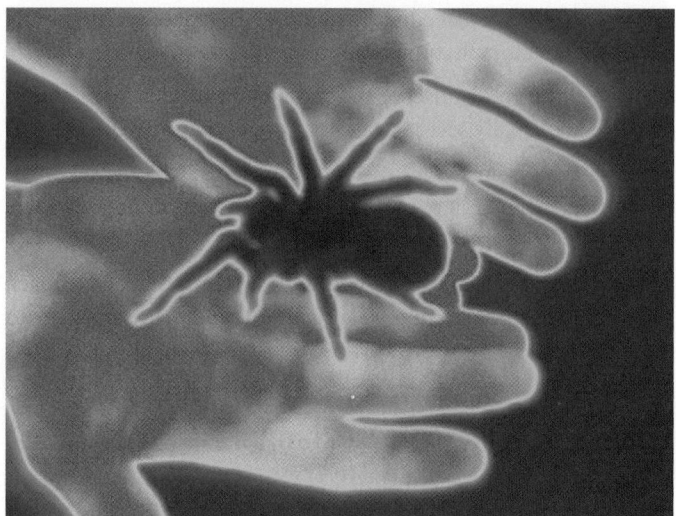

FIGURE 41.15 Endotherms Produce Heat, Ectotherms Do Not. The temperatures recorded in this thermogram range from 35.9°C (light red) to room temperature at 22.8°C (dark blue). Note the dramatic difference in surface temperature between a human (produces heat) and a tarantula (relies on heat from environment).

There are also two extremes on a continuum describing whether animals hold their body temperature constant: **Homeotherms** ("alike-heat") keep their body temperature constant, while **heterotherms** ("different-heat") allow their body temperature to rise or fall depending on environmental conditions.

Humans and other mammals, along with most birds, are strictly endothermic homeotherms. These species produce their own heat and maintain a constant body temperature. In contrast, mosquito larvae and other freshwater invertebrates are ectothermic heterotherms. But many animal species lie somewhere between these extremes:

- Some desert-adapted mammals, such as the oryx featured in the introduction to this chapter, allow their body temperature to rise during the hotter part of the day—meaning they are somewhat heterothermic.

- Small mammals that inhabit cold climates lose heat rapidly because their surface area is large relative to their volume. To survive when temperatures are cold, species such as ground squirrels reduce their metabolic rate and allow their body temperature to drop. This condition is called **torpor**. If torpor persists for weeks or months, it is called **hibernation**.

- Naked mole rats are mammals but lack insulation, because they have no fur. They live in underground tunnels and allow their body temperature to rise and fall with burrow temperatures. They are heterothermic and intermediate between ectotherms and endotherms.

- On cold mornings, bumblebees "shiver" by contracting their flight muscles together. (During flight, these same muscles contract in an alternating pattern to beat the wings.) Shivering generates enough heat to raise the individual's body temperature in preparation for flight.

Even in a homeothermic endotherm such as a mammal or bird, body temperature can vary widely in different body regions. When a Canada goose is standing on ice, its feet may be at a temperature of just 9°C, even though its body core is at 35°C. Similar variations exist in tuna and mackerel. These fish are ectotherms but generate heat to warm certain sections of their bodies, such as their eyes or swimming muscles.

Endothermy and Ectothermy: A Closer Look

Endotherms can warm themselves because their basal metabolic rates are extremely high—the heat given off by the high rate of chemical reactions is enough to warm the body. Mammals and birds retain this heat because they have elaborate insulating structures such as feathers or fur.

Ectotherms can also generate heat as a by-product of metabolism. The amount of heat they generate is small compared with the amount generated by endotherms, however, because ectotherms have relatively low metabolic rates. The most important sources of heat gain in ectotherms are radiation and conduction: they bask in sunlight or lie on warm rocks or soil.

Endothermy and ectothermy are best understood as contrasting adaptive strategies. Because endotherms maintain a high body temperature at all times, they can be active in winter and at night. Their high metabolic rates also allow them to sustain high levels of aerobic activities, such as running or flying.

These abilities come at a cost, however: To fuel their high metabolic rates, endotherms have to obtain large quantities of energy-rich food. The energy used to produce heat is then unavailable for other energy-demanding processes, such as reproduction and growth.

In contrast, ectotherms are able to thrive with much lower intakes of food. And because they are not oxidizing food to provide heat, they can use a greater proportion of their total energy intake to support reproduction.

What's the downside of ectothermy? Chemical reaction rates are temperature dependent, so muscle activity and digestion slow dramatically as the body temperature of an ectotherm drops. As a result, ectotherms are more vulnerable to predation in cold weather and in general are less successful than endotherms at inhabiting cold environments or remaining active in cool nighttime temperatures.

In short, each suite of adaptations has advantages and disadvantages. Like all adaptations, endothermy and ectothermy involve trade-offs.

Temperature Homeostasis in Endotherms

Thermoregulation is an important aspect of homeostasis in some animals. **Figure 41.16** on page 818 illustrates how the sensor-integrator-effector components of mammals function in thermoregulation.

Temperature receptors located throughout the body constantly monitor information about body temperature. For example, temperature receptors in the skin sense cooling or heating, and respond by altering the pattern of electrical signals that they send to adjacent neurons. Receptors in the brain region called the anterior **hypothalamus** respond in a similar fashion to changes in blood temperature.

The electrical signals that originate with temperature receptors are transmitted to an integrator located in the brain. Current evidence indicates that separate centers in the hypothalamus of the brain integrate and respond to increases and decreases in body temperature.

If a mammal is cold, cells in the posterior hypothalamus send signals to effectors that return body temperature to the set point. Signals from the posterior hypothalamus might induce shivering to generate warmth, and fluffing of fur or feathers to improve insulation and retain heat. Signals from the same or nearby cells can also result in the release of blood-borne chemical signals that increase the rate at which cellular respiration takes place throughout the body—generating more body heat.

But if the same individual is too hot, an integrator in the anterior hypothalamus sends signals that initiate sweating or panting—responses that cool the body. Other signals can induce behavioral changes that slow heat production, such as seeking shade or a cool burrow and then resting. Within cells, temperature spikes that are dramatic enough to denature proteins may activate the **heat-shock proteins** introduced in Chapter 3. Heat-shock proteins speed the refolding of proteins—a key step in the recovery process.

In response to either cooling or heating, behavioral and physiological responses move the body temperature back toward the set point via negative feedback. Figure 41.16 makes several points about the effectors that maintain homeostasis:

- It is common to observe redundancy in feedback systems—there are usually several ways to change a parameter.

- Feedback systems work in "antagonistic pairs": One set of responses increases a parameter while a corresponding set of responses decreases it.

- Input from sensors and integrators is constant, so feedback systems are constantly making fine adjustments relative to the set point.

Countercurrent Heat Exchangers

Homeothermic endotherms such as birds and mammals have sophisticated systems for thermoregulation. One of their most impressive adaptations allows them to minimize heat loss from limbs.

Heat loss is a particularly important problem for mammals that live in aquatic environments. If you've ever gone swimming in cold water, you can appreciate the problem faced by seals, otters, and whales. Water is such an effective conductor of heat that aquatic organisms lose metabolic heat rapidly. To conserve heat, otters have dense, water-repellent fur that maintains a layer of trapped air next to the skin. Seals and whales are insulated by thick layers of fatty blubber.

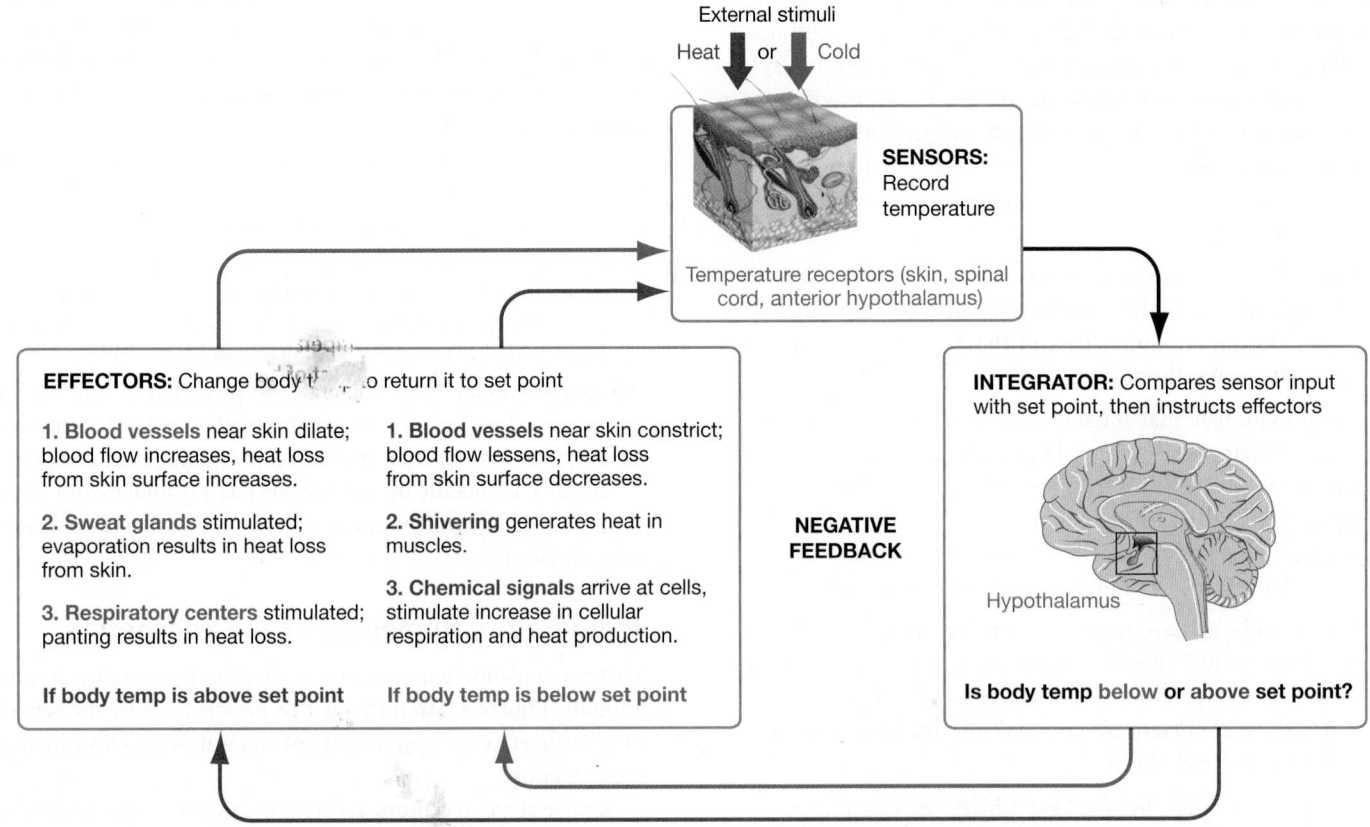

FIGURE 41.16 Mammals Regulate Temperature through Negative Feedback. In mammals, a set point for temperature is maintained by a complex negative feedback system that includes integrators in the anterior and posterior hypothalamus and sensors located throughout the body. The set point varies among species, from 30°C in monotremes to over 39°C in rabbits.

AN EXAMPLE: GRAY WHALE TONGUES   Gray whales have a feature that minimizes heat loss from their tongue, which is exposed to cold water during feeding. As **Figure 41.17a** shows, the tongue contains bundles of arteries and veins. Each bundle has an artery that carries warm, oxygenated blood from the body core. The artery is encircled by smaller veins, which transport cool blood from the tongue surface back toward the body core.

The key is that the two types of blood vessel are arranged in an antiparallel fashion. Warm blood traveling out to the tongue is in close contact with cool blood traveling back to the body.

This type of arrangement, with fluids flowing through adjacent pipes in opposite directions, is called a **countercurrent exchanger**. The "exchanger" part of the name is apt because in a case like the whale's tongue, heat is exchanged between the warm blood in the artery and the cool blood in the veins.

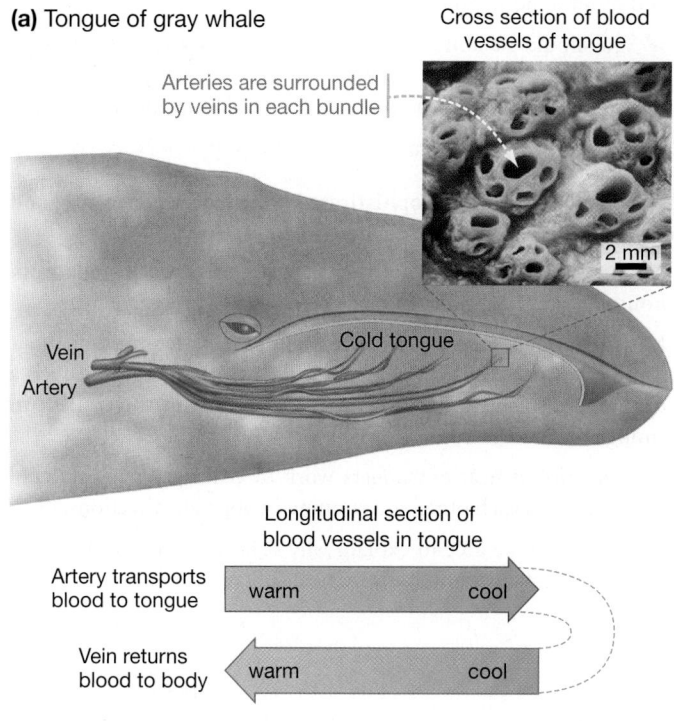

(a) Tongue of gray whale

Arteries are surrounded by veins in each bundle

Cross section of blood vessels of tongue

2 mm

Cold tongue

Vein
Artery

Longitudinal section of blood vessels in tongue

Artery transports blood to tongue

warm   cool

Vein returns blood to body

warm   cool

(b) Contrasting countercurrent with "concurrent" heat exchange

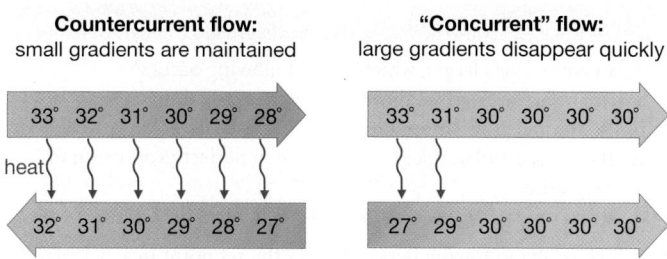

Countercurrent flow: small gradients are maintained

33° 32° 31° 30° 29° 28°

heat

32° 31° 30° 29° 28° 27°

"Concurrent" flow: large gradients disappear quickly

33° 31° 30° 30° 30° 30°

27° 29° 30° 30° 30° 30°

FIGURE 41.17 Countercurrent Exchangers Conserve Heat.
(a) Bundles of arteries and veins in whale tongues form heat exchangers that minimize heat loss from the tongue to the cold ocean water during feeding. **(b)** Countercurrent arrangements are much more efficient than "concurrent" arrangements. The data given here are hypothetical.

To see how the countercurrent exchange system works, study the longitudinal section in **Figure 41.17b**. Note that the fluid that enters the countercurrent heat exchanger is initially warm but steadily transfers heat to the adjacent, cooler fluid flowing in the opposite direction. There is a warmer-to-cooler gradient between the two currents at every point along the length of the countercurrent exchanger.

If the two solutions ran in the same direction, as in the right side of Figure 41.17b, the gradient between the two solutions would disappear quickly as the source current cooled and the recipient current heated. Countercurrent exchangers are efficient because they maintain a gradient between the two fluids along their entire length.

A MULTIPLIER EFFECT   A second key point about countercurrent heat exchangers concerns the temperature differential between the solutions. Although the differential is small at any point along the pipes, there is a large temperature differential from one end of each pipe to the other.

In effect, small differences in heat along the length of the exchanger sum up to create a large overall temperature gradient from beginning to end. The longer the system, the greater the overall differential will be. To highlight this property, the systems are sometimes called countercurrent multipliers.

Similar heat-conserving arrangements of arteries and veins are found in the flippers of whales and dolphins, and in the legs of many mammals and birds that live in cold terrestrial environments.

Countercurrent exchangers are just one of many sophisticated adaptations you'll encounter in this unit—structures that allow animals to thrive in a wide array of environments.

CHECK YOUR UNDERSTANDING

If you understand that . . .

- Two major aspects of temperature regulation vary among animal species: the amount of heat generated by the animal's own tissues and the degree to which body temperature varies over time.
- Negative feedback allows some endotherms to maintain homeostasis with respect to body temperature.
- Countercurrent heat exchangers are efficient ways to minimize heat loss from extremities.

✔ You should be able to . . .

1. Discuss the advantages and disadvantages of endothermy and ectothermy.
2. Diagram a countercurrent system that exchanges sodium ions (Na$^+$) via diffusion. Your diagram should include (1) labels indicating areas of high versus low Na$^+$ concentration, (2) arrows that indicate the direction that solutions are moving in the exchanger, and (3) arrows that indicate the direction of Na$^+$ diffusion.

Answers are available in Appendix B.

Summary of Key Concepts

In biology, structure has a profound influence on function. Biologists analyze the structure and function of animals at a variety of levels: molecules, cells, tissues, organs, and organ systems.

- Cells with a similar structure and common function are grouped together into four general types of tissue: connective tissue, nervous tissue, muscle tissue, and epithelial tissue.

- Epithelium is a particularly important type of tissue because it defines the interface between the animal's external and internal environments. Epithelial cells and tissues have a distinct polarity.

- Organs are structures that are composed of two or more tissues and that perform specific tasks.

- Organ systems are comprised of organs that work together in an integrated fashion to perform a function.

  ✔ You should be able to predict the effect of a drug that loosens the tight junctions between epithelial cells.

Body size has a strong influence on how animals work, in large part because a body's volume increases faster than its surface area as body size increases.

- Large animals have low metabolic rates, because they have a relatively small surface area for exchanging the oxygen and nutrients required to support metabolism.

- The relatively high surface area of small animals means that they lose heat extremely rapidly. As a result, there are no tiny endothermic animals.

  ✔ You should be able to explain why it would be impossible for a gorilla the size of King Kong to have fur. (In answering this question, explain how the surface-area-to-volume ratio of a normal size gorilla would compare to Kong's, then relate this to role of surface area and volume in heat generation and transfer and the function of fur.)

MB **Web Activity** Surface Area/Volume Relationships

Animals use an array of methods to maintain a relatively constant environment inside their bodies. They have systems that sense changes in internal conditions and trigger responses that return conditions to normal.

- Homeostasis refers to relatively constant physical and chemical conditions inside the body.

- Animals have a set point, or target value, for blood pH, tissue oxygen concentration, nutrient availability, and other parameters.

- Negative feedback occurs when a condition in the body is not at its set point.

- Responses to negative feedback return conditions to the set point and result in homeostasis.

  ✔ You should be able to explain how, due to acclimatization as well as adaptation, the homeostatic system for body temperature in a species of mammal might change as global temperatures rise.

MB **Web Activity** Homeostasis

Some animals have sophisticated systems for generating and conserving heat and regulating body temperature.

- Animals vary from endothermic to ectothermic, and from homeothermic to heterothermic.

- Most mammals have a set point for body temperature of about 37°C. If an individual overheats, it will pant or sweat and seek a cool environment; if an individual is cold, it will shiver, bask in sunlight, or fluff its fur.

- Countercurrent heat exchangers work by placing warm and cool liquids next to each other and running in opposite directions.

  ✔ You should be able to explain why countercurrent systems are sometimes called countercurrent multipliers.

Questions

✔ TEST YOUR KNOWLEDGE
Answers are available in Appendix B

1. How do biologists measure an animal's metabolic rate?
   a. by taking its temperature
   b. by measuring how rapidly it uses oxygen
   c. by measuring how rapidly it uses glucose
   d. by measuring how rapidly it produces wastes

2. How is the structure of a connective tissue most closely correlated with its function?
   a. The density of cells in the tissue correlates with the tissue's function.
   b. The surface area of the tissue correlates with the tissue's function.
   c. The origin of the tissue (from endoderm, mesoderm, or ectoderm) correlates with the tissue's function.
   d. The nature of the extracellular matrix correlates with the tissue's function.

3. As an animal gets larger, which of the following occurs?
   a. Its surface area grows more rapidly than its volume.
   b. Its volume grows more rapidly than its surface area.
   c. Its volume and surface area increase in perfect proportion to each other.
   d. Its volume increases, but its total surface area decreases.

4. Which of the following best describes the set point in a homeostatic system?
   a. the cells that collect and transmit information about the state of the system
   b. the cells that receive information about the state of the system and that direct changes to the system

c. the various components that produce appropriate changes in the system

d. the target or "normal" value of the parameter in question

5. What does it mean to say that an animal is a heterothermic endotherm?
   a. Its body temperature can vary, but it produces heat from its own tissues.
   b. Its body temperature varies because it gains most of its heat from sources outside its body.
   c. Its body temperature does not vary, because it produces heat from its own tissues.

d. Its body temperature does not vary, even though it gains most of its heat from sources outside its body.

6. Which of the following is an advantage that ectotherms have over endotherms of the same size?
   a. They require much less food.
   b. They can save energy in cold weather by hibernating (entering torpor for long periods).
   c. They can remain active in cold weather or at nighttime—when temperatures cool.
   d. They have higher metabolic rates and grow much more quickly.

✔ TEST YOUR UNDERSTANDING

Answers are available in Appendix B

1. Why is epithelium a particularly important tissue in achieving homeostasis with respect to temperature?

2. The metabolic rate of a frog in summer (at 35°C) is about eight times higher than in winter (at 5°C). Compare and contrast the individual's ability to move, exchange gases, and digest food at the two temperatures. During which season will the frog require more food energy, and why?

3. Consider the following:
   - Absorptive sections of digestive tract
   - Capillaries
   - Beaks of Galápagos finches
   - Fish gills

   In each case, how does the structure relate to the function?

4. Why is the surface area/volume ratio different in a small sphere versus a large sphere? If materials diffuse into and out of each sphere, in which case will diffusion occur more efficiently relative to the volume? Explain your answer.

5. Consider a day in which daytime temperatures reach 30°C and nighttime temperatures drop to 18°C. Analyze how an ant might gain and lose heat by conduction, convection, radiation, and evaporation to avoid overheating during the day and escape cold-induced lethargy in the early morning and evening.

6. Why is "negative" an appropriate adjective to describe the feedback that occurs in homeostatic systems?

✔ APPLYING CONCEPTS TO NEW SITUATIONS

Answers are available in Appendix B

1. When food is scarce, bigger *Geospiza fortis* (medium ground finches) win contests over seeds. Biologists have documented that there is strong natural selection in favor of large body size under these conditions, and that the finch population evolves in response. If so, why aren't *G. fortis* a lot bigger than they are?

2. What data would you need to collect in order to document that adaptation is occurring in a population of lizards, as global warming intensifies?

3. An engineer has to design a system for dissipating heat from a new type of car engine that runs particularly hot. Recall that heat is gained and lost as a function of surface area. Suggest ideas to consider that are inspired by biological structures with exceptionally high surface area/volume ratios.

4. Suppose a friend of yours is trying to decide whether to buy a pet turtle or a pet mouse. In making the decision, all he cares about is how much it will cost to feed the animal. Your friend says he can't decide because the two animals weigh the same and their food costs the same per pound. As a biologist, what's your advice?

Terrestrial animals lose water every time they breathe and urinate. For many animals, drinking is an important way to gain water and achieve homeostasis. This chapter explores how terrestrial and aquatic animals maintain water balance.

42 Water and Electrolyte Balance in Animals

KEY CONCEPTS

🔑 Freshwater, marine, and terrestrial habitats pose different challenges to animals with regard to maintaining water and electrolyte balance.

🔑 In marine animals, specialized epithelial cells have membrane proteins that remove excess salt (NaCl) from the body so that it can be excreted. The same types of cells are found in the kidneys of mammals.

🔑 In terrestrial insects, the hindgut and Malpighian tubules are responsible for excreting water-soluble waste products and achieving homeostasis with respect to water and electrolyte concentrations.

🔑 In terrestrial vertebrates, the kidney is responsible for excreting water-soluble waste products and achieving homeostasis with respect to water and electrolyte concentrations.

The chemical reactions that make life possible occur in an aqueous solution. If the balance of water and dissolved substances in the solution is disturbed, those chemical reactions—and life itself—may stop. Humans can stay alive for weeks without eating but survive just three days without drinking water. If hurricanes introduce enough freshwater to the ocean shore to disrupt normal salt concentrations, marine animals die. Maintaining water balance is a matter of life or death.

An animal achieves water balance when its intake of water equals its loss of water. Water balance is an important element in homeostasis—the ability to keep cells and tissues in constant and favorable conditions. Both water intake and water loss must be carefully controlled if the individual is to function at optimum levels. In humans, even relatively minor disruptions in water balance can lead to dehydration and symptoms such as headache, dizziness, muscle weakness, and fatigue (sleepiness).

Water balance is intimately associated with sustaining a balanced concentration of electrolytes throughout the body. An **electrolyte** is a compound that dissociates into ions when dissolved in water. Electrolytes got their name because they conduct electrical current.

In many animals, the most abundant electrolytes are sodium (Na^+), chloride (Cl^-), potassium (K^+), and calcium (Ca^{2+}). Cells require precise concentrations of these ions to function normally. In humans, electrolyte imbalances can lead to muscle spasms, confusion, irregular heart rhythms, fatigue, paralysis, or even death.

This chapter is focused on a single question: How do animals maintain water and electrolyte balance? Answering it will introduce you to some of the most complex and important homeostatic systems known.

✔ When you see this checkmark, stop and test yourself. Answers are available in Appendix B.

42.1 Osmoregulation and Osmotic Stress

In organisms, electrolytes and water move by diffusion and osmosis—two processes introduced in Chapter 6.

- **Diffusion** is the movement of substances from regions of higher concentration to regions of lower concentration.

- **Osmosis** is a special case of diffusion. It is the movement of water from regions of higher water concentration to regions of lower water concentration, across a selectively permeable membrane.

A **selectively permeable membrane**, such as a phospholipid bilayer, is a membrane that some solutes can cross more easily than other solutes can. When either diffusion or osmosis occur, ions and molecules move along their **concentration gradient**.

Figure 42.1a illustrates how dissolved substances, or **solutes**, move down their concentration gradients via diffusion across a selectively permeable membrane. When the solutes are randomly distributed throughout the solutions on both sides of the membrane, an equilibrium is established. Molecules continue to move back and forth across the membrane at equilibrium, but at equal rates.

As **Figure 42.1b** shows, water can also move down its concentration gradient. The concentration of dissolved substances in a solution, measured in moles per liter, is the solution's **osmolarity**. When dissolved substances are separated by a selectively permeable membrane and the solutes cannot cross that membrane, water moves from areas of lower osmolarity—that is, solute concentrations are lower—to areas of higher osmolarity—that is, solute concentrations are higher.

🔑 Diffusion and osmosis affect animals differently in marine, freshwater, and terrestrial habitats. As a result, these environments pose different challenges to animals in maintaining water and electrolyte balance.

What Is Osmotic Stress?

Osmotic stress occurs when the concentration of dissolved substances in a cell or tissue is abnormal. It means that water and solute concentrations are different from their set point.

Organisms respond to osmotic stress by osmoregulating, just as they respond to heat or cold stress by thermoregulating (see Chapter 41). **Osmoregulation** is the process by which living organisms control the concentration of water and salt in their bodies.

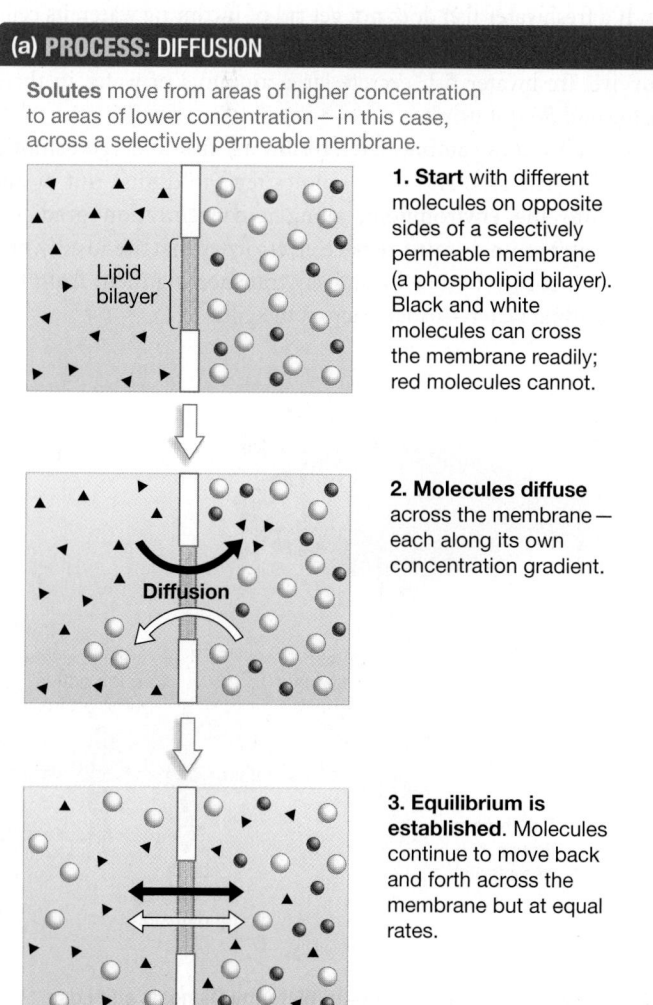

(a) PROCESS: DIFFUSION

Solutes move from areas of higher concentration to areas of lower concentration — in this case, across a selectively permeable membrane.

Lipid bilayer

1. Start with different molecules on opposite sides of a selectively permeable membrane (a phospholipid bilayer). Black and white molecules can cross the membrane readily; red molecules cannot.

Diffusion

2. Molecules diffuse across the membrane — each along its own concentration gradient.

3. Equilibrium is established. Molecules continue to move back and forth across the membrane but at equal rates.

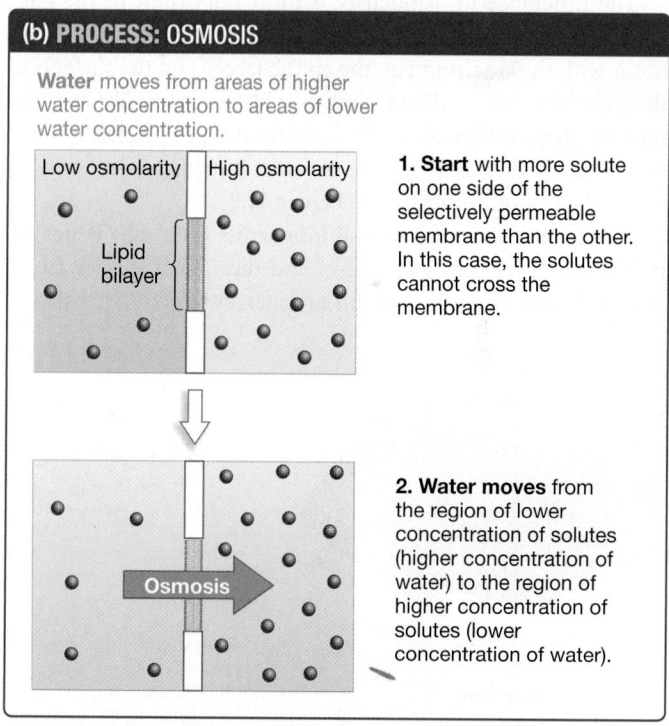

(b) PROCESS: OSMOSIS

Water moves from areas of higher water concentration to areas of lower water concentration.

Low osmolarity | High osmolarity

Lipid bilayer

1. Start with more solute on one side of the selectively permeable membrane than the other. In this case, the solutes cannot cross the membrane.

Osmosis

2. Water moves from the region of lower concentration of solutes (higher concentration of water) to the region of higher concentration of solutes (lower concentration of water).

FIGURE 42.1 Solutes Move Down a Concentration Gradient via Diffusion; Water Moves Down a Concentration Gradient via Osmosis. (a) Diffusion occurs any time a solute is at higher concentration in one location than another. **(b)** Osmosis is a special case of diffusion involving the movement of water across a selectively permeable membrane.

✓**QUESTION** Why doesn't the presence of red molecules on the right side of part (a) affect the movement of the black and white molecules?

Not all animals osmoregulate, however, because not all encounter osmotic stress. For marine species such as sponges, jellyfish, and flatworms, achieving homeostasis with respect to water and electrolyte balance is straightforward. Seawater is a fairly constant ionic and osmotic environment, and it nearly matches the normal electrolyte concentrations found within these animals.

To use vocabulary introduced in Chapter 6, tissues in these species are **isotonic** with respect to seawater. Stated another way, solute concentrations inside and outside these animals are equal. As a result, diffusion and osmosis don't alter water and electrolyte balance and induce osmotic stress.

These species are **osmoconformers**. Their set point for water and electrolyte concentration closely matches their environment.

Osmotic Stress in Seawater

In contrast to most marine invertebrates, most marine fish are **osmoregulators**. Fish actively regulate osmolarity inside their bodies to achieve homeostasis.

Marine fish have to osmoregulate because their tissues are **hypotonic** relative to salt water—the solution inside the body contains fewer solutes than does the solution outside.

The difference in osmolarity is most important in the gills, which are organs involved in gas exchange. For gas exchange to occur with the environment, the epithelial cells on the surfaces of the gills must be in direct contact with seawater. But because there is a large difference in the concentration of solutes between the inside of each cell and the seawater outside, water tends to flow out of the gill epithelium (**Figure 42.2**).

If the water that marine fish lose across their gills is not replaced, the fish's cells will shrivel and die. These species face a trade-off between gas exchange and water and electrolyte balance.

Marine fish replace the lost water by drinking large quantities of seawater. Drinking brings in excess electrolytes, however. Electrolyte balance is thrown even further out of whack because ions and other solutes diffuse into the fish across the gills, following a concentration gradient from seawater to tissues.

To rid themselves of these excess electrolytes, marine fish have to actively pump ions out of their bodies and back into seawater, using membrane proteins found in the gill epithelium.

Osmotic Stress in Freshwater

Freshwater animals osmoregulate in a dramatically different environment than the ocean. Marine fish are under osmotic stress because they lose water and gain salt; freshwater animals are under osmotic stress because they gain water and lose salt.

Why? In the gills of freshwater fish, epithelial cells contain more solutes than the solution outside. To use the vocabulary introduced in Chapter 6, the tissue is **hypertonic** relative to the surrounding water. As a result, the epithelial cells gain water via osmosis (**Figure 42.3**). When this water moves from the epithelium into adjacent tissues via osmosis, it puts the rest of the body under osmotic stress. Just as in marine fish, there is a trade-off between gas exchange and osmoregulation.

If a freshwater fish does not get rid of incoming water, its cells will burst and the individual will die. To achieve homeostasis and survive, freshwater fish excrete large amounts of water in their urine and do not drink.

In addition to gaining water, freshwater fish undergo osmotic stress because ions and other solutes tend to diffuse out of gill cells into the environment, along a concentration gradient. Freshwater animals must replace electrolytes that are lost by obtaining them in food or by actively transporting them from the surrounding water—usually across the gills.

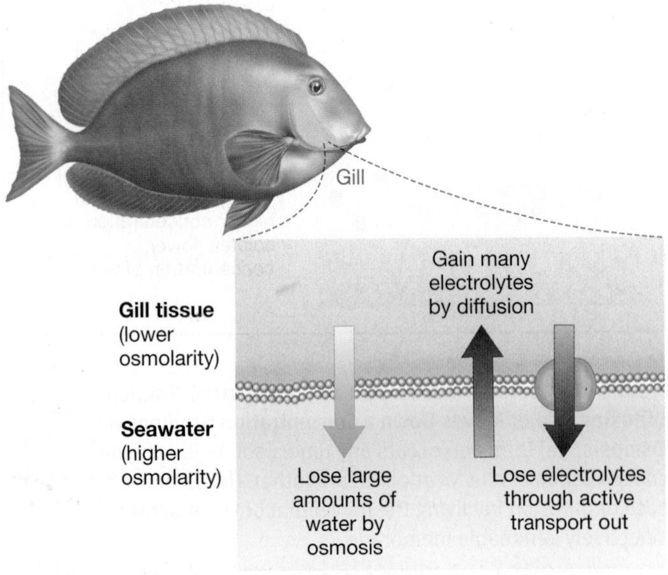

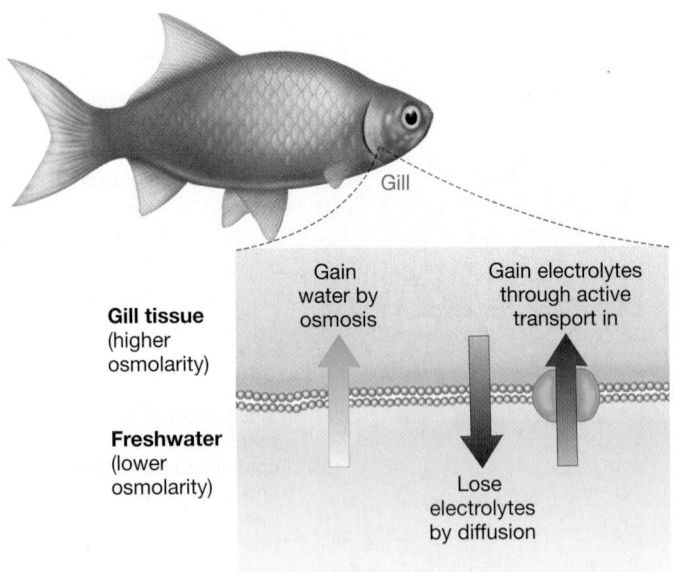

FIGURE 42.2 Marine Fish Lose Water by Osmosis and Gain Electrolytes by Diffusion.

FIGURE 42.3 Freshwater Fish Gain Water by Osmosis and Lose Electrolytes by Diffusion.

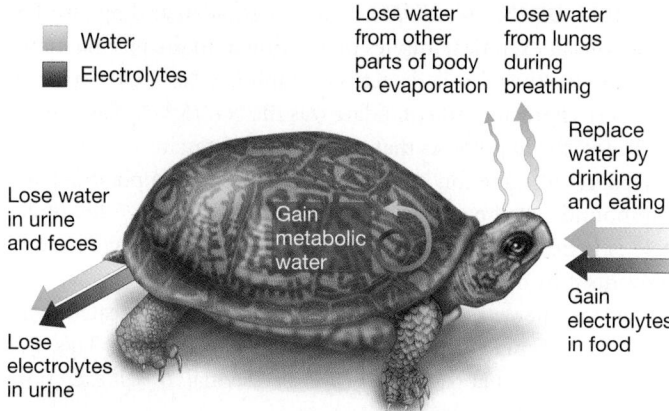

Water
Electrolytes

Lose water from other parts of body to evaporation

Lose water from lungs during breathing

Replace water by drinking and eating

Lose water in urine and feces

Gain metabolic water

Gain electrolytes in food

Lose electrolytes in urine

FIGURE 42.4 Maintaining Water and Electrolyte Balance in Terrestrial Environments Is a Challenge. In terrestrial environments, animals lose water by evaporation from the body surface and as water vapor during breathing. Electrolytes are lost primarily in the urine and feces.

Osmotic Stress on Land

What about land animals? In terms of water balance, terrestrial environments are similar to the ocean. Land animals constantly lose water to the environment, just as many marine animals do. In this case, however, the process involved is not osmosis but evaporation (see Chapter 41).

The epithelial cells that line a turtle's lung and a fruit fly's gas exchange structures have a moist surface, in order to protect the integrity of their plasma membranes. Because the atmosphere is almost always drier than the wet gas exchange surface, terrestrial animals lose water by evaporation. Once again, there is a trade-off between breathing and water and electrolyte balance.

Water balance is further complicated because all terrestrial animals lose water in the form of urine, and some species lose additional water when they sweat or pant to lower their body temperature (**Figure 42.4**). The lost water has to be replaced by drinking, ingesting water in food, or gaining metabolic water— the H_2O produced during cellular respiration (see Chapter 9).

How Do Cells Move Electrolytes and Water?

What molecular mechanisms allow animals to cope with the diverse challenges they face in maintaining water and electrolyte balance? You might recall from earlier chapters that solutes move across membranes by passive or active transport. **Passive transport** is driven by diffusion along an electrochemical gradient and does not require an expenditure of energy in the form of ATP. **Active transport**, in contrast, occurs when ATP powers the movement of a solute against its electrochemical gradient.

Because ions and large molecules such as glucose do not cross phospholipid bilayers readily, most passive transport and all active transport takes place via membrane proteins.

- In many cases, passive transport occurs through **channels**— proteins that form a pore, or opening, that selectively admits a specific ion or ions (**Figure 42.5a**).

(a) Passive transport is based on diffusion.

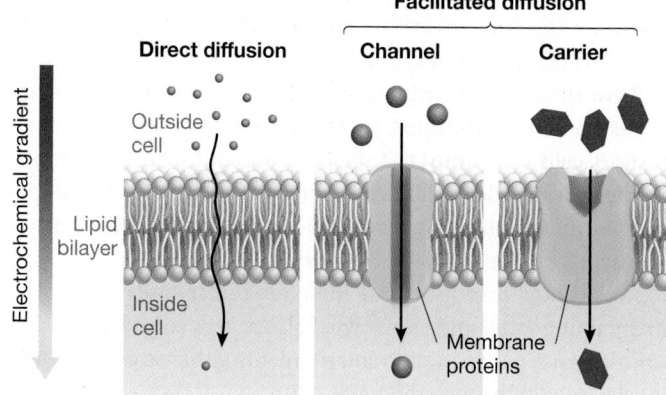

(b) Active transport moves ions or molecules against an electrochemical gradient.

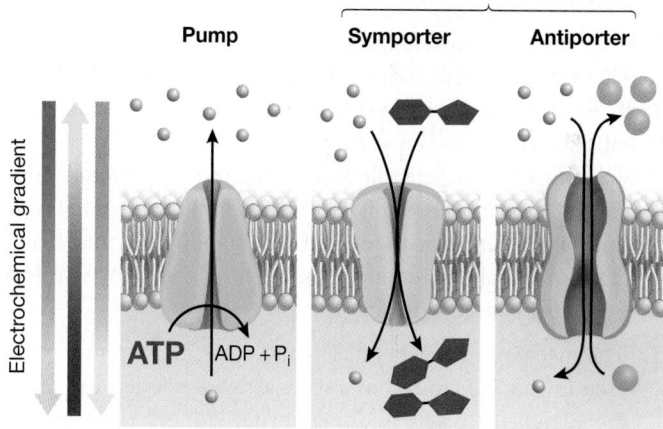

FIGURE 42.5 Mechanisms of Passive and Active Transport. (a) In passive transport, solutes diffuse along their electrochemical gradient either directly or through a membrane protein. **(b)** In active transport, a pump uses ATP to move solutes against their electrochemical gradients.

- Passive transport also occurs via **carriers**, which are transmembrane proteins that bind a specific ion or molecule and transport it across the membrane by undergoing a conformational change.

- When solutes move via channels or carriers, **facilitated diffusion** is said to occur.

Active transport is based on membrane proteins called pumps, which change conformation when they bind ATP or are phosphorylated (**Figure 42.5b**). This energy-demanding change in shape allows them to transport ions or molecules against their concentration gradient. Chapter 6 introduced the **sodium-potassium pump**, or Na^+/K^+-ATPase, which is the most important type of pump in animals.

Once a pump has established an electrochemical gradient, **secondary active transport** can occur. Specifically, a **cotransporter** can use the energy released when an ion is transported *along* that electrochemical gradient to transport a different solute *against* its electrochemical gradient. A cotransporter that moves solutes in the same direction is called a **symporter**; a cotransporter that moves solutes in opposite directions is called an **antiporter**.

How does water move? To date, there are no known mechanisms for actively transporting water across plasma membranes. Instead, cells use pumps to transport ions and set up an osmotic gradient; water then follows by osmosis—often through the specialized membrane proteins called aquaporins (see Chapter 6). In essence, cells move water by moving salt.

To see how organisms put these molecular tools to work in maintaining water and electrolyte balance, let's consider how fish from marine or freshwater environments either excrete or take up electrolytes through their gills.

42.2 Water and Electrolyte Balance in Aquatic Environments

Whether they live in seawater or freshwater, virtually all of the 26,000 species of fish living today experience osmotic stress. Early research on how these species maintain water and electrolyte balance focused on sharks.

Sharks need to secrete salt (NaCl) because sodium and chloride ions diffuse into their gill cells from seawater, down the concentration gradients for the ions. Understanding the molecular mechanism of salt excretion in sharks turned out to have two wide-ranging consequences:

1. The mechanism is general. The salt-secreting system discovered in sharks is found in a wide array of species, including *Homo sapiens*. It is functioning in your kidneys, right now.

2. The research revealed a critically important concept in physiology. Plant and animal cells use active transport to set up a strong electrochemical gradient for one ion—typically Na^+ in animals and H^+ in plants. The sodium or proton gradient is then used to transport a wide array of other substances, without further expenditure of energy.

This is fundamental stuff. Let's dig in.

How Do Sharks Excrete Salt?

Research on salt excretion focused on an organ called the **rectal gland**, which secretes a concentrated salt solution. To determine how this gland works, researchers studied it in vitro—meaning outside the shark's body. The basic approach was to dissect rectal glands, immerse them in a solution with a defined composition and osmolarity, and analyze the fluid that the rectal gland produced in response.

Early experiments showed that normal salt excretion only occurred if the solution in the rectal gland contained ATP. This result supported the hypothesis that salt excretion is an energy-demanding activity. Ions can be concentrated only if they are actively transported against a concentration gradient. The question was, How does concentration occur?

THE ROLE OF Na^+/K^+-ATPASE An energy-demanding mechanism for salt excretion implies that a protein in the plasma membrane of epithelial cells is actively pumping Na^+, Cl^- or both. The best-characterized candidate was the Na^+/K^+-ATPase.

To test the hypothesis that the sodium-potassium pump is involved in salt excretion by sharks, biologists used a plant defense compound called **ouabain** (pronounced *WAA-bane*). This molecule is toxic to animals because it binds to Na^+/K^+-ATPase and prevents it from functioning.

Just as they predicted, rectal glands that are treated with ouabain stop producing a concentrated salt solution. This was strong evidence that Na^+/K^+-ATPase is essential for salt excretion.

A MOLECULAR MODEL FOR SALT EXCRETION Subsequent work has shown that salt excretion is a multistep process, summarized in **Figure 42.6**.

1. Na^+/K^+-ATPase pumps sodium ions out of epithelial cells across the basolateral surface, into the extracellular fluid. The pump creates an electrochemical gradient favoring the diffusion of Na^+ into the cell. This "master gradient" allows the cell to transport other ions without an additional expenditure of energy.

2. Na^+, Cl^-, and K^+ all enter the cell, powered by the Na^+ "master gradient."

3. As Cl^- builds up inside the cell, a chloride channel located in the apical membrane allows Cl^- to diffuse down its concentration gradient into the lumen of the gland.

4. Following their charge and concentration gradients, sodium ions diffuse into the lumen of the gland through spaces between the cells.

A COMMON MOLECULAR MECHANISM UNDERLIES MANY INSTANCES OF SALT EXCRETION 🔑 In many animals, epithelial cells that transport sodium and chloride ions contain the same combination of membrane proteins found in the shark rectal gland. These species include:

- marine birds and reptiles that drink salt water and excrete NaCl via glands in their nostrils;

- marine fish that excrete salt from their gills (see Figure 42.2);

- mammals that transport salt in their kidneys.

Research on the shark rectal gland also had an unforeseen benefit for biomedical research. Several years after the shark chloride channel was characterized, investigators identified a human protein called cystic fibrosis transmembrane regulator (CFTR). As Chapter 6 pointed out, cystic fibrosis is the most common genetic disease in populations of northern European extraction. Although the disease was known to be associated with defects in the CFTR protein, no one knew what the molecule did.

When investigators realized that the amino acid sequence of CFTR is 80 percent identical to that of the shark chloride chan-

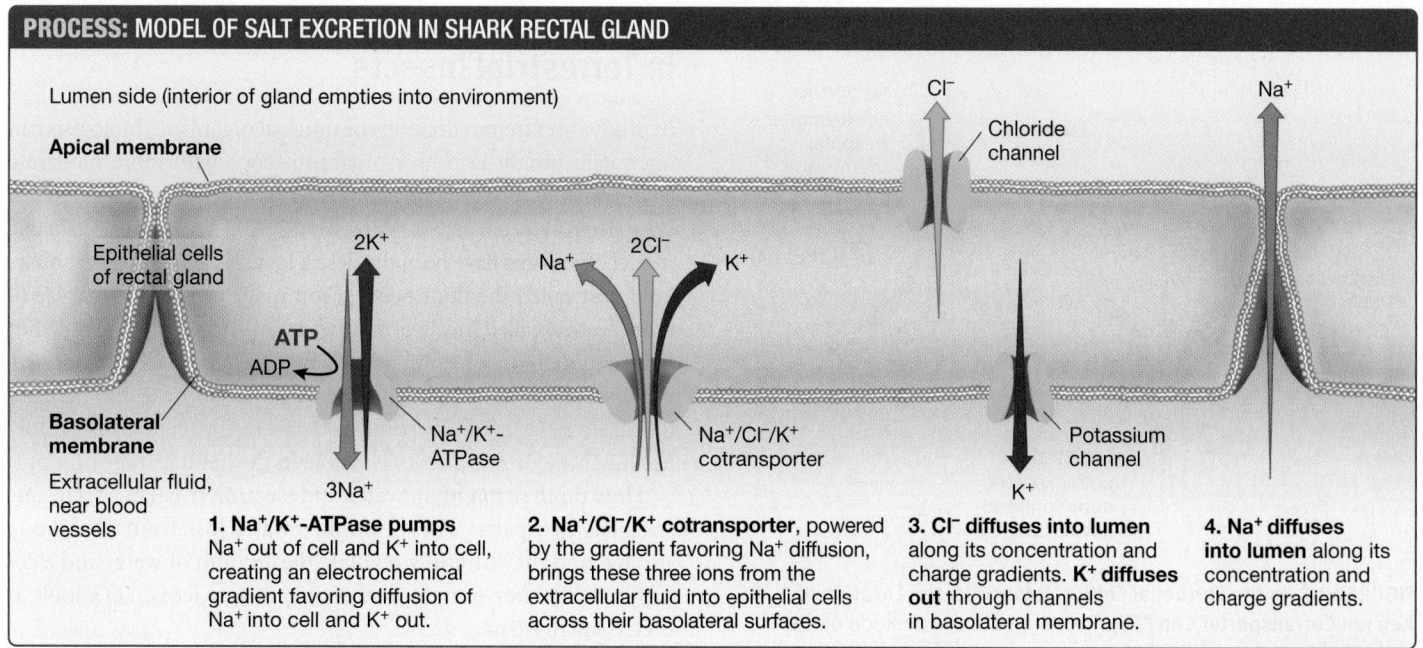

Lumen side (interior of gland empties into environment)

Apical membrane

Epithelial cells of rectal gland

Cl⁻ Chloride channel

Na⁺

2K⁺

Na⁺ 2Cl⁻ K⁺

ATP

ADP

Basolateral membrane

Na⁺/K⁺-ATPase

Na⁺/Cl⁻/K⁺ cotransporter

Potassium channel

Extracellular fluid, near blood vessels

3Na⁺

K⁺

1. Na⁺/K⁺-ATPase pumps Na⁺ out of cell and K⁺ into cell, creating an electrochemical gradient favoring diffusion of Na⁺ into cell and K⁺ out.

2. Na⁺/Cl⁻/K⁺ cotransporter, powered by the gradient favoring Na⁺ diffusion, brings these three ions from the extracellular fluid into epithelial cells across their basolateral surfaces.

3. Cl⁻ diffuses into lumen along its concentration and charge gradients. **K⁺ diffuses out** through channels in basolateral membrane.

4. Na⁺ diffuses into lumen along its concentration and charge gradients.

FIGURE 42.6 The Shark Rectal Gland Rids the Body of Excess Salt.

✔**QUESTION** Which of these membrane proteins are involved in (1) active transport, (2) secondary active transport, or (3) passive transport?

nel, it was their first hint that CFTR is involved in Cl⁻ transport. Subsequent studies supported the hypothesis that cystic fibrosis results from a defect in a chloride channel. In this way, studies on water and electrolyte balance in sharks shed light on an important human disease.

How Do Freshwater Fish Osmoregulate?

The molecular mechanisms of salt balance in marine animals are now well known. How freshwater fish achieve homeostasis with respect to electrolytes is still unclear, however.

Recall from Figure 42.3 that freshwater fish have to cope with an osmotic stress that is opposite the challenge facing marine fish. Freshwater fish lose electrolytes across their gill epithelium by diffusion across a concentration gradient. To maintain homeostasis, they have to actively transport ions back into the body across the gill epithelium. How do they do this?

SALMON AND SEA BASS AS MODEL SYSTEMS To understand how freshwater fish gain electrolytes, researchers have focused on sea bass and several species of salmon. Over the course of a lifetime, individuals of these species move between salt water and freshwater. Thus, they move between environments with dramatically different osmotic stresses.

In marine fish, specialized cells in the gill epithelium, called chloride cells, move salt using the combination of membrane proteins illustrated in Figure 42.6. When sea bass and salmon are in salt water, these cells are abundant and active. What happens to these cells when individuals move into freshwater? Do the changes that occur provide any insight

into how they acclimatize to the new environment and avoid dying of osmotic stress?

A FRESHWATER CHLORIDE CELL? Although research is continuing at a brisk pace, recent results support the hypothesis that there is a freshwater version of the classical chloride cell: one that moves ions in the opposite direction of the saltwater version. Instead of excreting salt, these cells import it.

Three lines of evidence have accumulated to date:

1. *Osmoregulatory cells may be in different locations.* Young salmon taken from freshwater versus salt water have cells with Na⁺/K⁺-ATPase in different locations on the gills. Adult salmon from freshwater versus salt water show the same pattern, and similar changes have been observed in other fish species that switch between freshwater and saltwater habitats. The pattern suggests that when the nature of osmotic stress changes, the nature of the gill epithelium changes. Specifically, active pumping of ions takes place in a different population of cells in seawater versus freshwater.

2. *Different forms of Na⁺/K⁺-ATPase may be activated.* The salmon genome contains genes for several different forms of the Na⁺/K⁺-ATPase. There is now strong evidence that different forms are activated when individuals are in salt water versus freshwater.

3. *The orientation of key transport proteins "flips."* In sea bass, researchers have been able to stain epithelial cells to determine the location of the cotransporter illustrated in Figure 42.6—the one that brings Na⁺, Cl⁻, and K⁺ into the

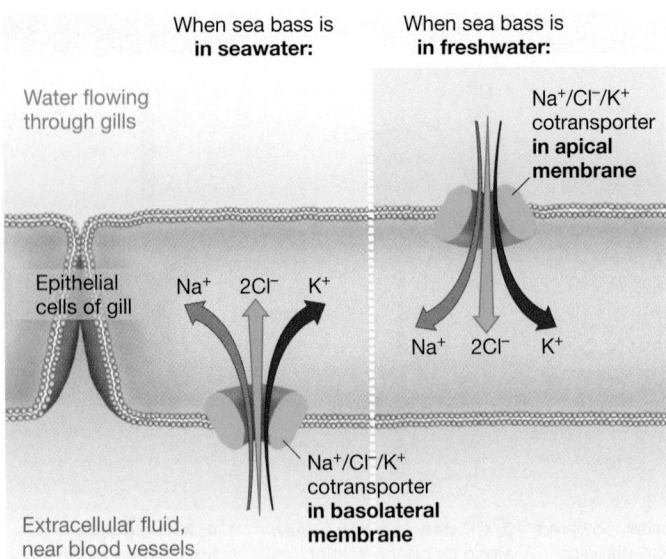

When sea bass is **in seawater:**

When sea bass is **in freshwater:**

Water flowing through gills

Na+/Cl–/K+ cotransporter **in apical membrane**

Epithelial cells of gill

Na+ 2Cl– K+

Na+ 2Cl– K+

Na+/Cl–/K+ cotransporter **in basolateral membrane**

Extracellular fluid, near blood vessels

FIGURE 42.7 In the Epithelial Cells of Bass Gills, the Location of a Key Ion Cotransporter Can "Flip." Changes in the position of the Na+/Cl–/K+ cotransporter help sea bass deal with osmotic stress in both seawater and freshwater environments.

cell. When individuals are in seawater, the protein is located in the basolateral side of chloride cells. But when individuals are in freshwater, the protein is located in the apical side (**Figure 42.7**).

Taken together, the data suggest that freshwater fish have a freshwater version of the chloride cell, with pumps and transporters arranged in the opposite orientation to that found in marine fish. The evidence is far from conclusive, however, and research continues. Identifying the mechanisms of electrolyte uptake in freshwater fish is an important challenge for researchers who want to know how aquatic organisms cope with osmotic stress.

CHECK YOUR UNDERSTANDING

If you understand that . . .

- Marine fish lose water by osmosis. To replace it, they have to drink salt water.
- Marine fish have to rid themselves of salt. They gain salt when they drink salt water or when sodium and chloride ions diffuse into their cells along a concentration gradient.
- Freshwater fish gain water by osmosis. They have to rid themselves of excess water by urinating.
- Freshwater fish lose electrolytes to the surrounding water by diffusion. They gain electrolytes in their food and by active transport from the surrounding water.

✔ **You should be able to . . .**

Predict what happens when epithelial cells in the gills of a freshwater fish are treated with ouabain—a molecule that poisons Na+/K+-ATPase.

Answers are available in Appendix B.

42.3 Water and Electrolyte Balance in Terrestrial Insects

By studying extreme situations or unusual organisms, biologists can often gain insight into how organisms cope with more moderate environments. In studies on the molecular mechanisms of water and electrolyte balance in terrestrial insects, the most valuable model organisms have been the desert locust and a common household pest called the flour beetle. (You may have seen the larvae of flour beetles, called mealworms, in bags of flour that were not shut tightly enough to keep adults from entering and breeding.)

Desert locusts and flour beetles live in environments where osmotic stress is severe. These insects rarely, if ever, drink—simply because little or no water is available in the habitats they occupy.

How do they maintain water and electrolyte balance? The answer has two parts: They minimize water loss from their body surface, and they carefully regulate the amount of water and electrolytes that they excrete in their urine and feces. Let's look at each issue in turn.

How Do Insects Minimize Water Loss from the Body Surface?

As Chapter 44 will show, terrestrial animals breathe by exposing an extremely thin layer of epithelium to the atmosphere. Oxygen diffuses into this epithelium, and carbon dioxide diffuses out. But water constantly leaks across the thin respiratory surface and is lost to the atmosphere via evaporation.

Evaporation from the body surface itself is another threat—a particular challenge to insects, because they are small. As Chapter 41 emphasized, small organisms have a high surface area/volume ratio. Insects have a relatively large surface area from which to lose water but a small volume in which to retain it.

How do desert locusts, flour beetles, and other insects minimize water loss during gas exchange? In these species, gas exchange occurs across the membranes of epithelial cells that line the **tracheae** (pronounced *TRAY-kee-ee*), an extensive system of tubes. The insect tracheal system connects with the atmosphere at openings called **spiracles** (**Figure 42.8a**). Muscles just inside each spiracle open or close the pore, much as guard cells open or close the pores in plant leaves and stems.

When investigators manipulated insects called *Rhodnius* so that their spiracles stayed open and placed the animals in a dry environment, they died within three days. These data support the hypothesis that the ability to close spiracles is an important adaptation for minimizing water loss during respiration. If an insect is under osmotic stress, it may be able to close its spiracles and wait until conditions improve before resuming activity.

Figure 42.8b shows how insects minimize evaporation from the surface of their bodies. This diagram is a cross-sectional view of the exoskeleton of an insect, which consists of a tough, nitrogen-containing polysaccharide called chitin and layers of protein. This combination of chitin and protein is known as **cuticle**.

As the figure shows, cuticle includes a layer of wax on its surface. Recall from Chapter 6 that waxes, a type of lipid, are highly

(a) Spiracles can be closed to minimize water loss from tracheae.

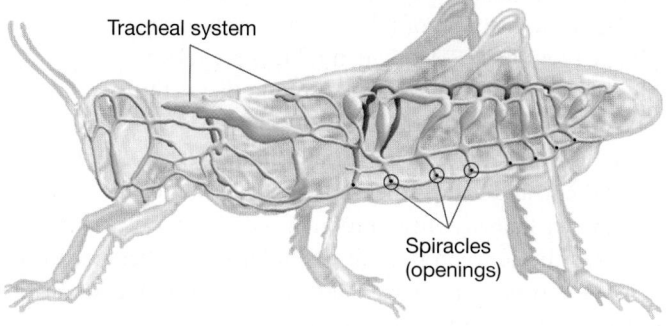

(b) Except at spiracles, the insect body is covered with wax.

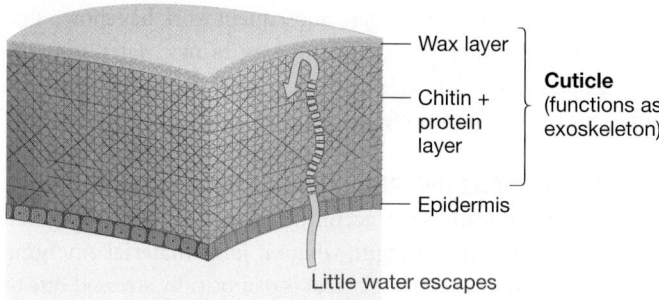

FIGURE 42.8 In Desert Locusts, Adaptations Limit Water Loss during Respiration and from the Body Surface.

✔**QUESTION** In desert grasshoppers, spiracles are located in a row along the underside of the abdomen. Why would this location help minimize water loss?

hydrophobic and thus highly impermeable to water. Researchers who removed the wax from insect exoskeletons have confirmed that the rate of water loss from the body surface increases sharply. Based on this observation, the wax layer is interpreted as an adaptation that minimizes evaporative water loss.

Types of Nitrogenous Wastes: Impact on Water Balance

Animal cells contain amino acids and nucleic acids that are used to synthesize proteins, RNA, and DNA. Both amino acids and nucleic acids are nitrogenous (nitrogen-containing) monomers.

If amino acids and nucleic acids are present in excess of a cell's needs, they are broken down in catabolic reactions that result in the production of **ammonia** (NH_3). Ammonia, a strong base, readily gains a proton to form an ammonium ion (NH_4^+). Ammonia is toxic to cells, because at high concentrations it raises the pH of intracellular and extracellular fluids enough to poison enzymes.

FORMS OF NITROGENOUS WASTE VARY AMONG SPECIES How do animals get rid of ammonia safely and efficiently? Different species solve the problem in different ways (see **Table 42.1**).

- In freshwater fish, ammonia is diluted to low concentration and excreted in a watery urine.

- In freshwater and saltwater fish, ammonia diffuses across the gills into the surrounding water along a concentration gradient.

- In humans, enzyme-catalyzed reactions convert ammonia to a much less toxic compound called urea, which is excreted in urine.

- In birds, reptiles, and terrestrial arthropods, reactions convert ammonia to **uric acid**, the white, paste-like substance that you have probably seen in bird feces.

Uric acid is a particularly interesting form of nitrogenous waste. Compared with urea and ammonia, uric acid is extremely insoluble in water—which explains why it is so difficult to wash bird droppings off a car. As a result, uric acid provides a mechanism for birds, snakes, lizards, and terrestrial arthropods to rid themselves of excess nitrogen while losing a minimum amount of water. Many birds and some insects, in fact, do not produce any urine at all.

SUMMARY TABLE 42.1 **Attributes of Nitrogenous Wastes Produced by Animals**

Attribute	Ammonia	Urea	Uric Acid
Solubility in water (moles/liter)	high	medium	very low
Water loss (amount required for excretion of waste)	high	medium	very low
Energy cost (amount of ATP required)	low	high	high
Toxicity	high	medium	low
Groups where it is the primary waste	fish, aquatic invertebrates	mammals,* sharks	birds[†] and other reptiles, most terrestrial insects and spiders
Method of synthesis	breakdown of amino acids and nucleic acids	synthesized in liver, starting with amino groups from amino acids	synthesis starts with amino acids and nucleic acids
Method of excretion	in urine and diffuses across gills	in urine (mammals); diffuses across gills (sharks)	in feces (in birds, uric acid is derived from the urine but excreted with the feces)

*Mammals also excrete a small amount of uric acid, synthesized from excess nucleic acids.
[†]Birds also excrete a small amount of ammonia.

WHY DO NITROGENOUS WASTES VARY AMONG SPECIES? The type of nitrogenous waste produced by an animal correlates with its lineage—its evolutionary history. For example, mammals excrete urea while reptiles (including birds) and insects excrete uric acid (Table 42.1).

Evolutionary history is not the entire story, however. Waste production also correlates with the habitat that a species occupies, and thus the amount of osmotic stress it endures.

- Terrestrial birds conserve water by excreting about 90 percent of their nitrogenous waste as uric acid and only 3–4 percent as NH_3, but ducks and other birds with ready access to water excrete just 50 percent of their excess nitrogen as uric acid and 30 percent as NH_3.

- Tadpoles are aquatic and excrete ammonia, but adult frogs are terrestrial and excrete urea.

- Production of urea and uric acid is particularly common in animals that live in dry habitats.

To make sense of these observations, biologists point out that there is a fitness trade-off between the energetic cost of excreting urea or uric acid and the benefit of conserving water. Ammonia excretion requires a large water loss but little energy expenditure, because the molecule isn't processed by enzymes. Uric acid excretion, in contrast, requires almost no loss of water but a sophisticated series of enzyme-catalyzed, energy-demanding reactions. Different trade-offs are favored in different environments.

Maintaining Homeostasis: The Excretory System

For insects, minimizing water loss is only half the battle in avoiding osmotic stress. To maintain homeostasis, insects must also carefully regulate the composition of a blood-like fluid called **hemolymph**. Hemolymph is pumped by the heart and transports electrolytes, nutrients, oxygen, and waste products.

How do insects regulate the composition of the hemolymph? This question is important for three reasons: (**1**) Nitrogenous wastes have to be removed before they build up to toxic concentrations, (**2**) excess electrolytes must be excreted before they lead to osmotic stress, and (**3**) water balance must be regulated constantly.

To maintain water and electrolyte balance, insects rely on **Malpighian tubules**, which are an excretory organ, and on their hindgut—the posterior portion of their digestive tract (**Figure 42.9a**).

FILTRATE FORMS IN THE MALPIGHIAN TUBULES As the enlarged section of Figure 42.9a shows, Malpighian tubules have a large surface area, are in direct contact with the hemolymph, and empty into the hindgut. ⌘ The Malpighian tubules are responsible for forming a **filtrate** from the hemolymph. This "pre-urine" then passes into the hindgut, where it is processed and modified prior to excretion.

To explore how the Malpighian tubules work, biologists collected fluid from the lumen of Malpighian tubules in mealworms and compared the composition of the filtrate with that of hemolymph from the same individuals. The two solutions were roughly isotonic, but not identical.

How do they differ? Compared to hemolymph, the solution inside the tubules has a lower concentration of sodium ions and a higher concentration of K^+. This observation suggests that the epithelial cells in the Malpighian tubules are relatively impermeable to sodium ions but contain a pump that actively transports potassium ions into the tubules.

To test this hypothesis, researchers dissected the tubules, rinsed them, and bathed their interior and exterior in a solution containing a known concentration of potassium ions. When they measured the concentration of electrolytes on either side of the membrane over time, they found that K^+ accumulated in the tubule lumen, against its concentration gradient.

This result supported the hypothesis that cells in the membranes of Malpighian tubules contain a pump that transports potassium ions into the lumen of the organ. Subsequent work has shown that a high concentration of potassium ions brings water into the tubules by osmosis. Other electrolytes and nitrogenous wastes then diffuse into the tubules along their concentration gradients.

THE HINDGUT: SELECTIVE REABSORPTION OF ELECTROLYTES AND WATER The solution that accumulates inside the Malpighian tubules flows into the hindgut, where it joins material emerging from the digestive tract. If an insect is osmotically stressed due to a shortage of electrolytes and water, electrolytes and water from the filtrate are reabsorbed in the hindgut and returned to the hemolymph. Reabsorption results in formation of a hypertonic final urine, conservation of water, and efficient elimination of nitrogenous wastes.

In desert locusts, flour beetles, and other species in extremely dry environments, 80 to 95 percent of the water in the filtrate is recovered and kept inside the body. The ability to recover this water allows these insects to live in dry habitats such as deserts and flour bins. How does reabsorption happen?

The mechanism involves a series of specific membrane pumps and channels, not unlike the system found in the chloride cells of fish. To study it, researchers positioned the rectal epithelium from a desert locust as a sheet dividing two solutions. They manipulated electrolyte concentrations on either side of the rectal wall and measured changes in the solutions over time.

For example, when investigators removed K^+ and Na^+ from the solution on the lumen side of the organ, water reabsorption stopped. These data established that the hindgut's ability to recover water from urine depends on ion movement: the epithelial cells in the hindgut transport ions out of the filtrate and into the hemolymph. Water follows by osmosis, forming a concentrated urine.

How do the ions move? By poisoning the experimental membranes with ouabain, biologists confirmed that Na^+/K^+-ATPase is involved in moving ions out of the lumen and into the hemolymph. Ouabain-treated membranes continued to transport Cl^-, however. Later experiments confirmed that the rectal epithelium transports Cl^- against electrical and concentration gradients. These results support the hypothesis that insects' hindguts have two active pumps: a chloride pump and Na^+/K^+-ATPase.

Years of experiments on the locust hindgut resulted in the model in **Figure 42.9b**:

(a) Malpighian tubules produce an isotonic pre-urine.

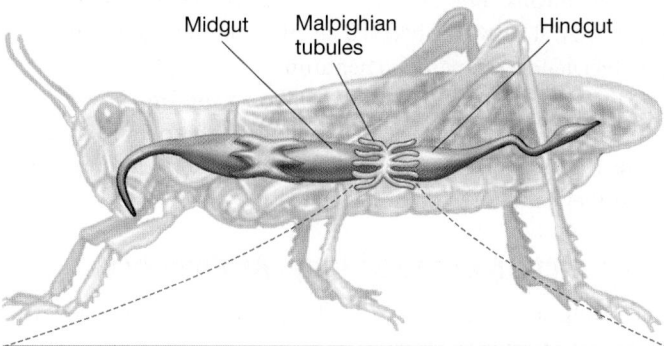

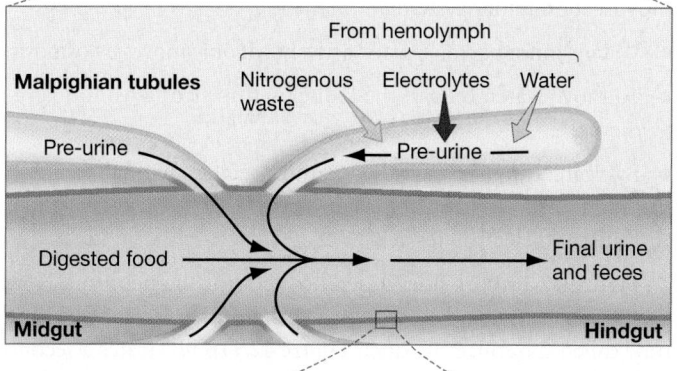

(b) Under osmotic stress, the hindgut reabsorbs electrolytes and water to form a hypertonic urine.

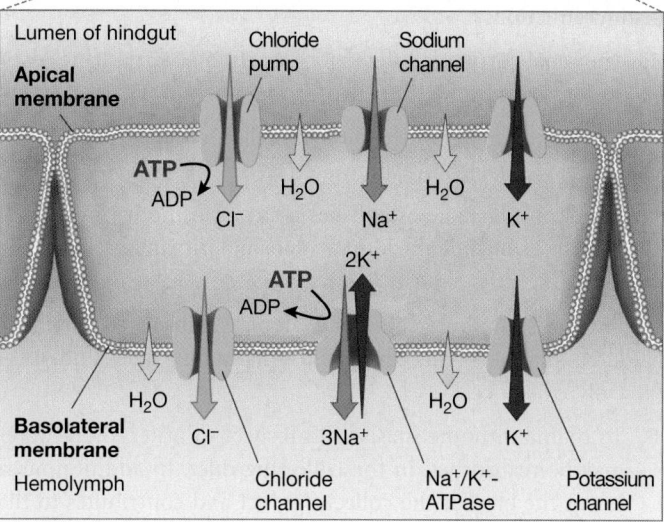

FIGURE 42.9 In Insects, Urine Forms in the Malpighian Tubules and Hindgut. (a) The isotonic filtrate that forms in the Malpighian tubules empties into the hindgut. **(b)** In the hindgut, the primary driving forces for reabsorption of electrolytes and water are chloride pumps in the apical membrane and Na^+/K^+-ATPases in the basolateral membrane.

- Cl^- is pumped into cells from the hindgut lumen, with K^+ following through potassium channels along an electrochemical gradient and water following via osmosis.
- In the basolateral membrane, Na^+/K^+-ATPase sets up electrochemical and osmotic gradients that favor movement of Cl^-, K^+, and H_2O into the hemolymph.

There is also strong experimental evidence for the existence of other channels, pumps, and cotransporters in the epithelium of the insect hindgut. These membrane proteins transport protons, ammonia, amino acids, and other molecules across the tissue and help insects maintain water and electrolyte balance.

REGULATING WATER AND ELECTROLYTE BALANCE: AN OVERVIEW
Several general principles that have emerged from studies of insect excretion turn out to be relevant to vertebrate systems as well:

- Water moves only by osmosis—it is not pumped directly. Water moves between cells or body compartments via osmotic gradients that are set up by the active transport of ions.

- The formation of the filtrate is not particularly selective. Most of the molecules present in the hemolymph are also present in the Malpighian tubules.

- In contrast to filtrate formation, reabsorption is highly selective. The protein pumps and channels involved in reabsorption are highly specific for certain ions and molecules. Waste products do not pass through the rectal membrane. Instead, they remain in the urine and feces and are eliminated from the body. Only valuable ions and molecules are reabsorbed.

- In contrast to filtrate formation, reabsorption is tightly regulated. The membrane pumps and channels involved in reabsorption are activated and deactivated in response to osmotic stress. If an insect is dehydrated, virtually all of the water in the filtrate is reabsorbed. But if it has plenty to drink, reabsorption does not occur and the urine is watery and hypotonic to the individual's hemolymph. The system is dynamic and allows precise control over water and electrolyte balance.

Given the success of insects in terms of the numbers of species and individuals and the array of habitats they occupy, it is clear that their systems for maintaining water and electrolyte balance are remarkably effective.

CHECK YOUR UNDERSTANDING

If you understand that . . .
- Terrestrial insects are prone to dehydration, primarily via evaporation from their respiratory surfaces.
- Terrestrial insects have a cuticle, respiratory system, and excretory system that are designed to conserve water.
- In terrestrial insects, urine formation begins with formation of a solution that is isotonic with hemolymph, followed by selective and tightly regulated reabsorption of ions, nutrients, and water.

✓ You should be able to . . .

Explain how the following traits are involved in water retention:
1. Excretion of ammonia in the form of uric acid.
2. Selective reabsorption of electrolytes in the hindgut.

Answers are available in Appendix B.

42.4 Water and Electrolyte Balance in Terrestrial Vertebrates

With respect to water loss, terrestrial vertebrates face the same hazards that terrestrial insects do. Crocodiles, turtles, lizards, frogs, birds, and mammals lose water from their body surfaces and from the surface of their lungs every time they breathe. Electrolytes are lost in urine and in some species, in sweat.

To replace the water they lose, most terrestrial vertebrates drink. They also ingest electrolytes in food.

In land-dwelling vertebrates, osmoregulation occurs primarily through events that take place in the **kidney**. The kidney is responsible for water and electrolyte balance, as well as the excretion of nitrogenous wastes. In terms of function, it is analogous to the Malpighian tubules and hindgut of insects.

The Structure of the Kidney

Kidneys occur in pairs and tend to be bean shaped (**Figure 42.10a**). A large blood vessel called the renal artery brings blood that contains nitrogenous wastes into the organ; the renal vein is a large blood vessel that carries "clean" blood away.

The urine that forms in the kidney is transported via a long tube called the **ureter** to a storage organ, the **bladder**. From the bladder, urine is transported to the body surface through the **urethra**, then excreted. In most vertebrates, the kidneys are located near the dorsal (back) side of the body.

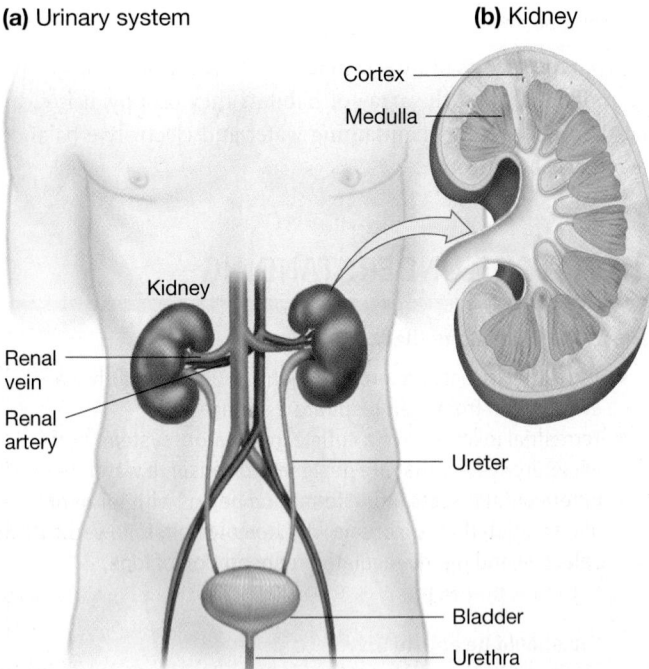

(a) Urinary system

(b) Kidney

Cortex
Medulla
Kidney
Renal vein
Renal artery
Ureter
Bladder
Urethra

FIGURE 42.10 Anatomy of the Human Urinary System and Kidney. (a) In mammals, kidneys are paired and are located near the spinal column. **(b)** The kidney has an outer region called the cortex and an inner area called the medulla.

Most of the kidney's mass is made up of small structures called nephrons. ☞ The nephron is the basic functional unit of the kidney. The work involved in maintaining water and electrolyte balance occurs in the nephron.

Most of the approximately 1 million nephrons in a human kidney are located in the outer region of the organ, or **cortex** (**Figure 42.10b**). But some nephrons extend from the cortex into the kidney's inner region, or **medulla**.

The Function of the Kidney: An Overview

The nephron shares important functional characteristics with the insect excretory system:

- Water cannot be transported actively—it only moves by osmosis.
- To move water, cells in the kidney set up strong osmotic gradients.
- By regulating these gradients and specific channel proteins, kidney cells exert precise control over loss or retention of water and electrolytes.

Figure 42.11a provides a detailed view of the nephron. Note that it has four major regions, and is closely associated with a tube called the collecting duct. **Figure 42.11b** illustrates a second key point about the anatomy of the nephron: It is served by blood vessels that wrap around each of its four regions.

The four major regions and the collecting duct each have a distinct function:

1. The renal corpuscle filters blood, forming a "pre-urine" consisting of ions, nutrients, wastes, and water.

2. In the proximal tubule, epithelial cells reabsorb nutrients, vitamins, valuable ions, and water.

3. The loop of Henle establishes a strong osmotic gradient in the tissues outside the loop, with osmolarity increasing as the loop descends.

4. In the distal tubule, ions and water are reabsorbed in a regulated manner—one that helps maintain water and electrolyte balance.

5. To maintain homeostasis with respect to water, more water may be reabsorbed in the collecting duct. In addition, urea leaves the base of the collecting duct and contributes to the osmotic gradient set up by the loop of Henle.

The blood vessels that are juxtaposed with the nephron play a key role as well: they bring "dirty" blood into the nephron and then take away the molecules and ions that are reabsorbed from the initial filtrate.

Now let's delve into the details. The sections that follow trace the flow of material through each component of the nephron and out of the collecting duct.

Filtration: The Renal Corpuscle

In terrestrial vertebrates, urine formation begins in the **renal corpuscle** (literally, "kidney-little-body"). Notice in Figure 42.11a

(a) The structure of the nephron and collecting duct

(b) Blood vessels serve each nephron.

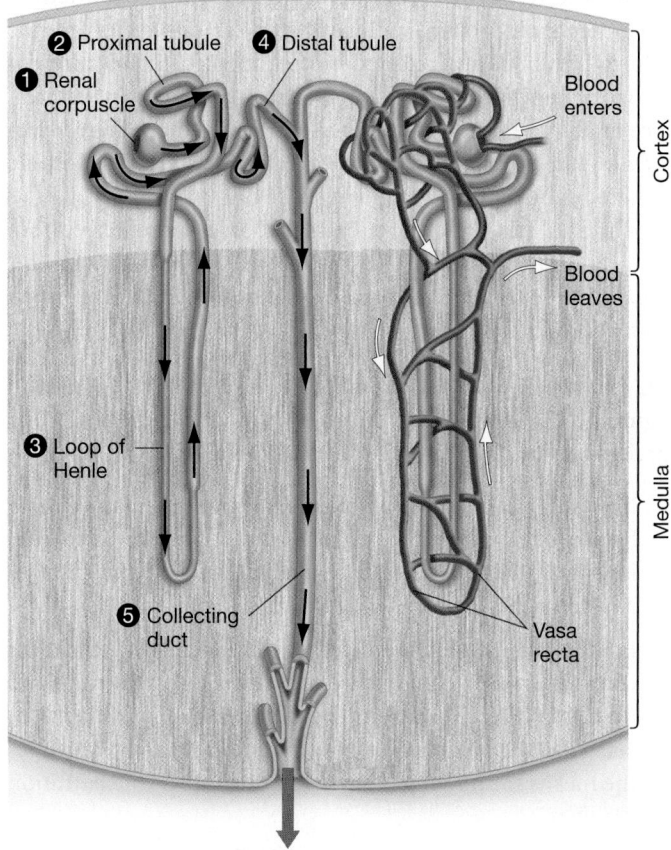

FIGURE 42.11 **A Nephron Has Four Major Parts, Empties into a Collecting Duct, and Is Served by Blood Vessels.** Urine formation begins in the renal corpuscle and ends in the collecting duct.

(a) Anatomy of the renal corpuscle

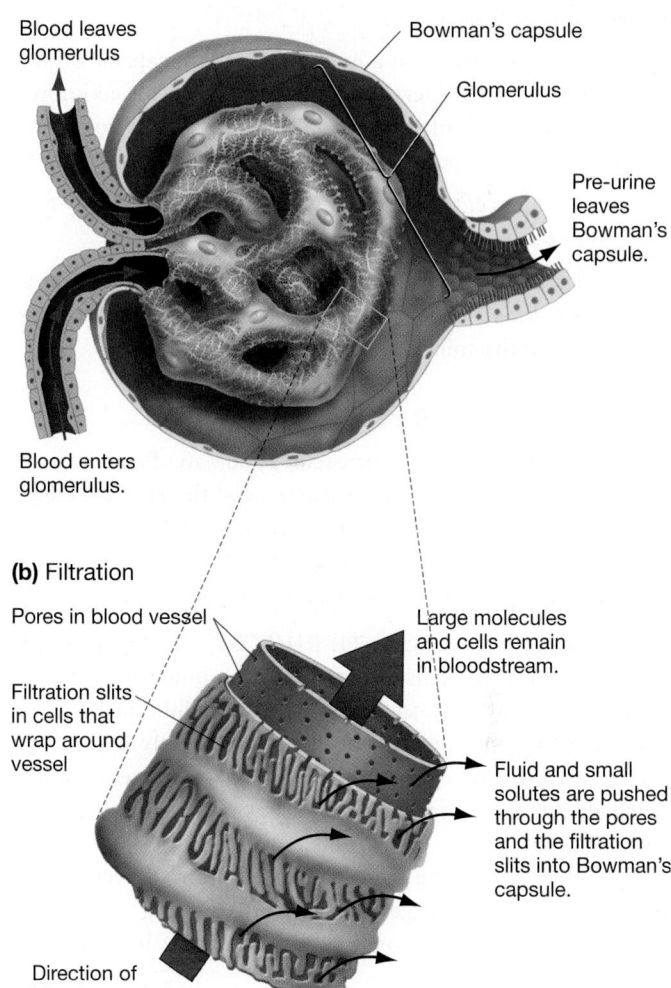

(b) Filtration

FIGURE 42.12 **Urine Formation Begins When Blood Is Filtered in the Renal Corpuscle.** **(a)** The renal corpuscle consists of Bowman's capsule and the glomerulus. **(b)** The capillaries in the glomerulus have pores and are surrounded by cells that have filtration slits. Blood pressure forces water and small molecules out of the capillaries, through the slits, and into Bowman's capsule.

that the nephron is a tube that is closed at one end and open at the other. The closed end is the beginning of the nephron; the open end is the terminus of the collecting duct.

As **Figure 42.12a** shows, the closed end of the nephron forms a capsule that encloses a cluster of tiny blood vessels, or capillaries. These vessels bring blood to the nephron from the renal artery. Collectively, the cluster of capillaries is called the **glomerulus** ("ball of yarn"). The region of the nephron that surrounds the glomerulus is named **Bowman's capsule**. Together, the glomerulus and Bowman's capsule make up the renal corpuscle.

Figure 42.12b illustrates a key feature of the glomerular capillaries: They have large pores, or openings. In addition, they are surrounded by unusual cells whose membranes fold into a series of slits and ridges.

The structure of the renal corpuscle allows it to function as a **filtration** device. Water and small solutes from the blood pass through the pores and slits into the nephron. Filtration is based on size: Proteins, cells, and other large components of blood do not fit through the pores and do not enter the nephron. They remain in the blood instead.

Stated another way, urine formation starts with a size-selective filtration step—with blood pressure supplying the force required to perform filtration. In vertebrates, blood is under higher pressure than the surrounding tissues because it is pumped by the heart through a closed system of vessels. This pressure is enough to force water and small solutes through the pores in the glomerulus, so the renal corpuscle strains large volumes of fluid without expending energy in the form of ATP.

To summarize, the renal glomerulus filters the blood to create a filtrate comprising water, electrolytes, and other small substances. During the formation of this filtrate, up to 25 percent of the water and solutes present in the blood is removed.

✔If you understand this concept, you should be able to describe the contents of blood on one side of the filter and the contents of the filtrate on the other side.

It is critical to note two additional facts about the filtration step in urine formation:

1. The renal corpuscles of a human kidney are capable of producing about 180 liters of filtrate per day. This is an impressive volume—think of 180 one-liter bottles of soft drink arranged on a supermarket shelf.

2. About 99 percent of the filtrate is recycled—only a tiny fraction of the original volume is actually excreted.

Filtering large volumes from the blood allows wastes to be removed effectively; pairing this process with reabsorption allows waste excretion to occur with a minimum of water and nutrient loss.

Reabsorption: The Proximal Tubule

Where does filtrate reabsorption occur? Fluid leaves Bowman's capsule and enters a convoluted structure called the **proximal tubule**. The fluid inside this tubule contains water and small solutes such as urea, glucose, amino acids, vitamins, and electrolytes. Some of these molecules are waste products; others are valuable nutrients.

ACTIVE TRANSPORT OCCURS IN EPITHELIAL CELLS As **Figure 42.13a** shows, the epithelial cells of the proximal tubule have a prominent series of small projections, called **microvilli** ("little shaggy hairs"), facing the lumen. The microvilli greatly expand the surface area of this epithelium. A large surface area provides space for membrane proteins that act as pumps, channels, and cotransporters.

Epithelial cells in the proximal tubule are also packed with mitochondria, which suggests that ATP-demanding active transport is occurring. Based on these observations, biologists hy-

pothesized that the proximal tubule functions in the active transport of selected-molecules out of the filtrate.

The selective-active-transport hypothesis was supported by experiments on proximal tubules that were dissected from the kidneys of rabbits and rats and isolated in vitro. By injecting solutions of known composition into proximal tubules in the presence or absence of ATP, researchers confirmed that selected electrolytes and nutrients are actively reabsorbed from the filtrate that enters the tubules.

When solutes leave the proximal tubule and enter epithelial cells, water follows along the osmotic gradient. In this way, valuable solutes and water are reabsorbed and returned to the body.

ION AND WATER MOVEMENT IS DRIVEN BY A "MASTER GRADIENT" **Figure 42.13b** summarizes the current model of the molecular mechanisms involved in selective reabsorption:

1. Na^+/K^+-ATPase in the basolateral membranes removes Na^+ from the interior of the cell. The active transport of sodium ions out of the cell creates a gradient favoring the entry of Na^+ from the lumen.

2. In the apical membrane adjacent to the lumen, Na^+-dependent cotransporters use this gradient to remove valuable ions and nutrients selectively from the filtrate. The movement of Na^+ into the cell, *with* its concentration gradient, provides the means for moving other solutes *against* a concentration gradient.

3. The solutes that move into the cell diffuse across the basolateral membrane into nearby blood vessels.

(a) Microvilli expand surface area of lumen of proximal tubule.

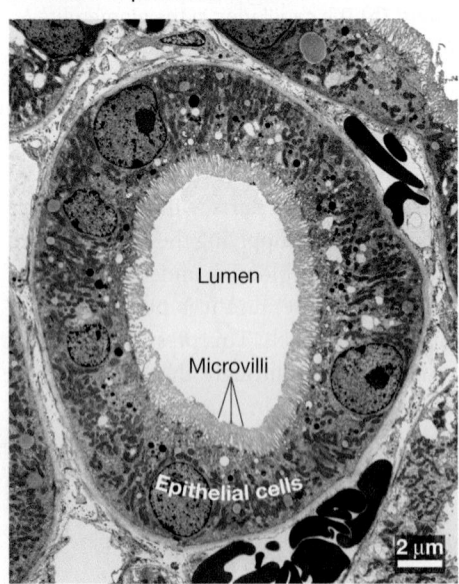

(b) Model of selective reabsorption in proximal tubules

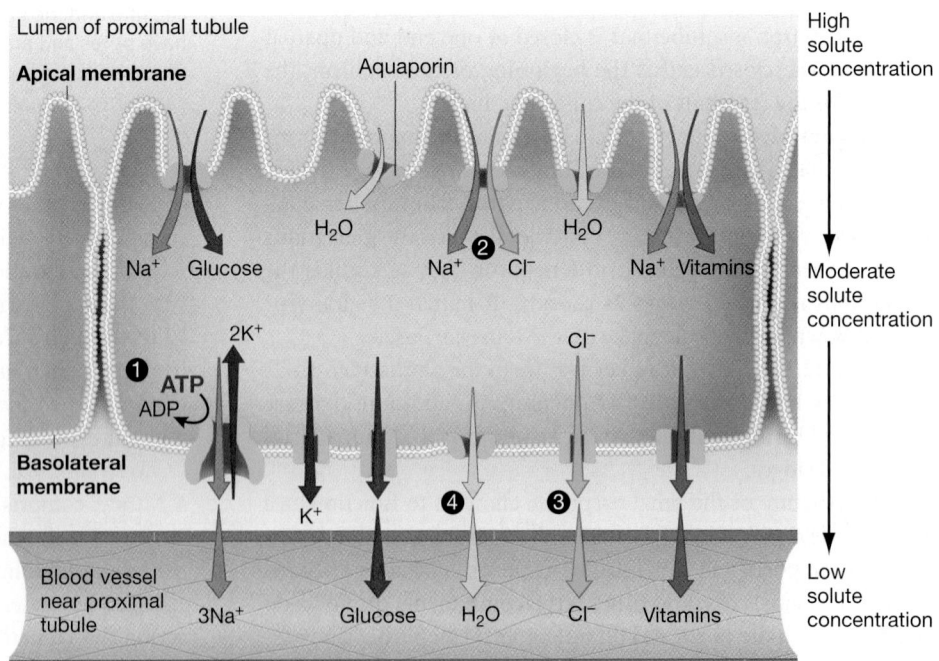

FIGURE 42.13 Water and Electrolytes Are Reabsorbed in the Proximal Tubule.

4. Water follows the movement of ions from the proximal tubule into the cell, and then out of the cell and into blood vessels.

Recent work has shown that water moves across the membranes of these epithelial cells through specialized membrane proteins called **aquaporins** ("water-pores"; see Figure 42.13b). Aquaporins are water channels: When one is open, 3 billion water molecules pass through it per second.

✔If you understand this concept, you should be able to explain why caffeine increases urine output, based on recent data indicating that caffeine may inhibit sodium ion reabsorption in the proximal tubule.

About two-thirds of the NaCl and water that is originally filtered by the renal corpuscle is reabsorbed in the proximal tubule. The osmolarity of the tubular fluid is unchanged despite this huge change in volume, however, because water reabsorption is proportional to solute reabsorption.

In effect, then, the cells that line the proximal tubule act as a recycling center. The filtration step in the renal corpuscle is based on size; the reabsorption step in the proximal tubule selectively retrieves small substances that are valuable. The pumps and co-transporters in the proximal tubule recover water, nutrients, and electrolytes but leave wastes. As the filtrate flows into the loop of Henle, it has a relatively high concentration of waste molecules and a relatively low concentration of nutrients.

Creating an Osmotic Gradient: The Loop of Henle

In mammals, the fluid that emerges from the proximal tubule enters the **loop of Henle**—named for Jacob Henle, who described it in the early 1860s. In most nephrons, the loop is short and does not leave the cortex. But in about 20 percent of the nephrons present in a human kidney, the loop is long and plunges from the cortex of the kidney deep into the medulla.

In 1942 Werner Kuhn offered a hypothesis, inspired by countercurrent heat exchangers, to explain what the loop of Henle does. Recall from Chapter 41 that a countercurrent heat exchanger is a system in which two adjacent fluids flow through pipes in opposite directions. Kuhn proposed that the loop of Henle is a countercurrent exchanger and multiplier. It doesn't exchange heat, however. Instead, it sets up an osmotic gradient.

Specifically, Kuhn proposed that the osmolarity of the fluid inside the loop of Henle is low in the cortex and high in the medulla. Further, Kuhn maintained that the osmolarity in tissues surrounding the loop mirrors the gradient inside the loop. This is a key point.

TESTING KUHN'S HYPOTHESIS A series of papers published during the 1950s supplied important experimental support for the countercurrent exchange model. **Figure 42.14** reproduces two particularly important data sets, obtained by comparing the osmolarity of kidney tissue slices. In both graphs, the x-axis shows the location in the kidney, from cortex to medulla. The y-axis indicates osmolarity, either measured as the percentage of the maximum observed or as solute concentration.

- Part (a) of the figure shows data on the osmolarity of fluid inside the loop of Henle. The vertical lines represent the range

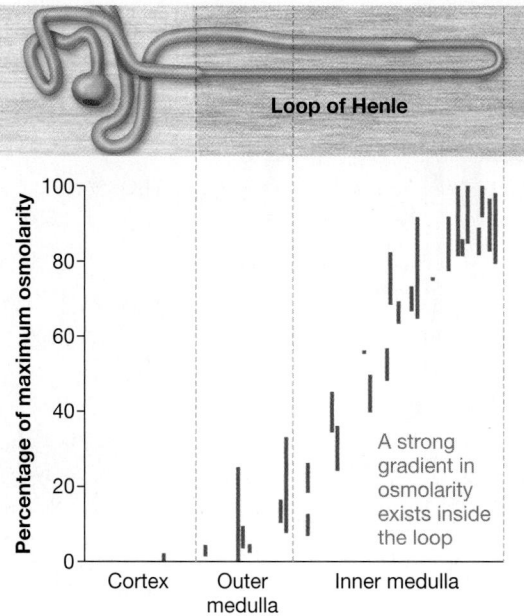

(a) Fluid inside the loop of Henle

Loop of Henle

A strong gradient in osmolarity exists inside the loop

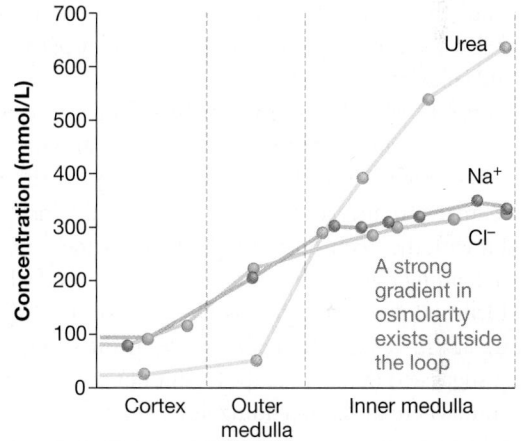

(b) Cells outside the loop of Henle

Urea

Na⁺

Cl⁻

A strong gradient in osmolarity exists outside the loop

FIGURE 42.14 Data Confirm the Existence of a Strong Osmotic Gradient Both inside and outside the Loop of Henle. As the nephron plunges into the inner medulla, the concentration of dissolved solutes increases both inside **(a)** and outside **(b)** the loop of Henle.

of values observed at a particular location. As predicted by Kuhn's model, a strong gradient in osmolarity exists from the cortex to the medulla.

- The data in part (b) show that outside the loop of Henle, the concentrations of Na⁺, Cl⁻, and urea also increase sharply from the cortex to the medulla.

The second observation was particularly important, because it suggested that the solutes responsible for the gradient outside the loop are Na⁺, Cl⁻, and urea. The change in concentration of urea turned out to be particularly important.

HOW IS THE OSMOTIC GRADIENT ESTABLISHED? The loop of Henle has three distinct regions: the descending limb, the thin

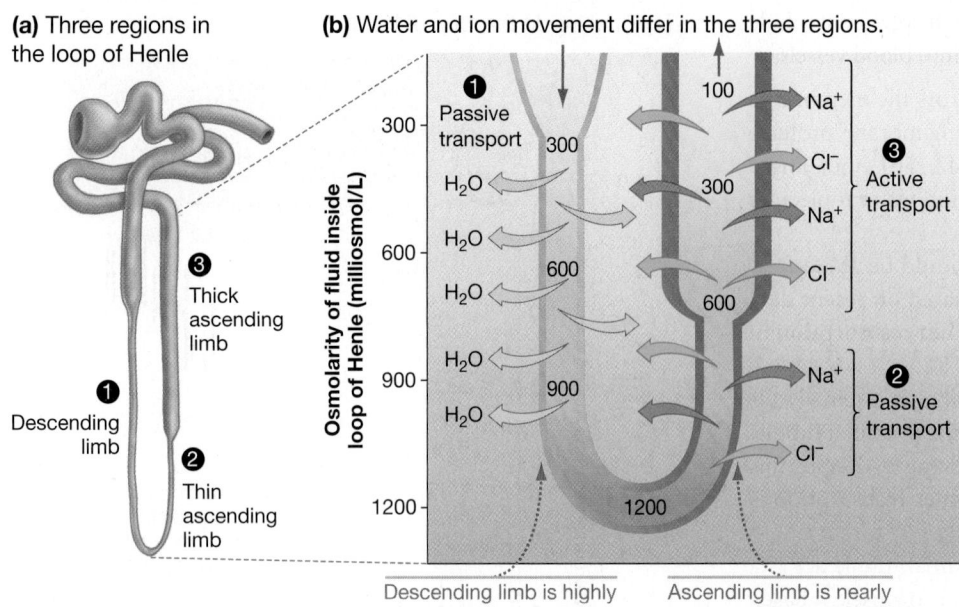

(a) Three regions in the loop of Henle

(b) Water and ion movement differ in the three regions.

❸ Thick ascending limb

❶ Descending limb

❷ Thin ascending limb

Osmolarity of fluid inside loop of Henle (milliosmol/L)

❶ Passive transport

❸ Active transport

❷ Passive transport

Descending limb is highly permeable to water but impermeable to solutes

Ascending limb is nearly impermeable to water but highly permeable to Na^+ and Cl^-

FIGURE 42.15 The Loop of Henle Maintains an Osmotic Gradient because Water Leaves the Descending Limb and Salt Leaves the Ascending Limb. The numbers inside the nephron in part (b) represent the osmolarity of the filtrate.

ascending limb, and the thick ascending limb (**Figure 42.15a**). The thin and thick ascending limbs differ in the thickness of their walls. Do the three regions also differ in their permeability to water and solutes?

It took over 15 years of experiments performed in laboratories around the world to formulate a definitive answer to that question. Researchers punctured Henle's loop with a micropipette, analyzed the composition of the fluid inside, and compared it with the nephron's final product—urine.

In the ascending limb of the loop of Henle, Na^+ and Cl^- constituted at least 60 percent of the solutes; urea constituted about 10 percent. But urea was the major solute in the distal tubule. These data suggested that Na^+ and Cl^-, but not urea, were being removed somewhere in the ascending limb. How?

Na^+ and Cl^- were also present at high concentrations in the tissue surrounding the thick ascending limb, so researchers hypothesized that sodium might be actively pumped out of this portion of the nephron. The hypothesis was that the active transport of Na^+ out of the thick ascending limb would create an electrical gradient that would also favor the loss of Cl^-.

Follow-up experiments using ouabain and other poisons supported the hypothesis that sodium ions are actively transported out of the solution inside the thick ascending limb, with chloride ions following along an electrochemical gradient. The epithelial cells responsible for salt excretion are configured almost exactly like the epithelium of the shark rectal gland (see Figure 42.6).

What is happening in the descending limb and the thin ascending limb of the loop of Henle? By injecting solutions of known concentration into the nephrons of rabbits, biologists documented that the descending limb is highly permeable to water but almost completely impermeable to solutes. The thin ascending limb of the loop, in contrast, is highly permeable to Na^+ and Cl^-, moderately permeable to urea, and almost completely impermeable to water.

A COMPREHENSIVE VIEW OF THE LOOP OF HENLE All of the observations just summarized came together in 1972 when two papers, published independently, proposed the same comprehensive model for how the loop of Henle works. To understand this model, follow the events in **Figure 42.15b**:

1. As fluid flows down the descending limb, the fluid inside the loop loses water to the tissue surrounding the nephron. This movement of water is passive—it does not require an expenditure of ATP. The water follows an osmotic gradient created by the ascending limb.

   At the bottom of the loop—in the inner medulla—the fluids inside and outside the nephron have high osmolarity. The filtrate does not continue to lose water, though, because the membrane in the ascending limb is nearly impermeable to water.

2. The fluid inside the nephron loses Na^+ and Cl^- in the thin ascending limb. The ions move passively, along their concentration gradients.

3. Toward the cortex, the osmolarity of the surrounding solution is low. Additional Na^+ and Cl^- ions are actively transported out of the nephron in the thick ascending limb.

The countercurrent flow of fluid, combined with changes in permeability to water and in the types of channels and pumps that are active in the epithelium of the nephron, creates a self-reinforcing system. The presence of an osmotic gradient stimulates water and ion flows that in turn maintain an osmotic gradient.

Here's how it works: The movement of NaCl from the ascending limb into the surrounding tissue increases the osmotic concentration outside the descending limb, which results in a flow of water out of the water-permeable walls of the descending limb, via osmosis. This loss of water in the descending limb increases the osmolarity in the fluid entering the ascending limb.

The high concentration of salt in the fluid at the base of the ascending limb triggers a passive flow of ions out—reinforcing the osmotic gradient.

✓If you understand this concept, you should be able to predict what happens to the osmotic gradient when the drug furosemide inhibits membrane proteins that pump sodium and chloride ions out of the thick ascending limb. Specifically, how does this drug affect (1) water loss in the descending limb, (2) the osmolarity of the solution inside the bottom of the nephron, and (3) loss of salt in the thin ascending limb?

THE VASA RECTA REMOVES WATER AND SOLUTES THAT LEAVE THE LOOP OF HENLE What happens to the water and salt that move out of the loop? They quickly diffuse into the **vasa recta**, a network of blood vessels that runs along the loop. As a result, the water and electrolytes are returned to the body (**Figure 42.16**).

The removal of water that leaves the descending limb is particularly important. If it were not drawn off into the bloodstream, it would dilute the concentrated fluid outside the loop of Henle and quickly destroy the osmotic gradient.

THE COLLECTING DUCT LEAKS UREA The loop of Henle maintains an osmotic gradient, but a key feature of the nephron's final portions—the **distal tubule** and the **collecting duct**—helps establish it.

Urea is the solute that is most responsible for the steep osmotic gradient in the space surrounding the nephron. Urea is at high concentration in the inner medulla and low concentration in the outer medulla. This gradient exists because the innermost section of the collecting duct is permeable to urea.

Although the system created by the nephron, vasa recta, and collecting duct may seem complex, its outcome is simple: the creation and maintenance of a strong osmotic gradient with the minimum possible expenditure of energy.

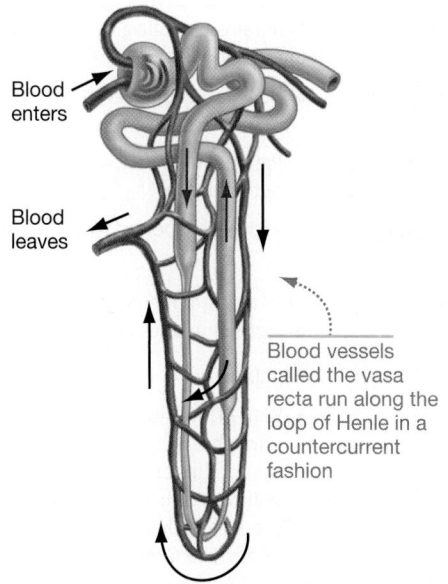

Blood enters

Blood leaves

Blood vessels called the vasa recta run along the loop of Henle in a countercurrent fashion

FIGURE 42.16 Blood Supply to the Loop of Henle. Water and solutes from the loop of Henle move into blood vessels (vasa recta).

Regulating Water and Electrolyte Balance: The Distal Tubule and Collecting Duct

The first three steps in urine formation—filtration, reabsorption, and establishment of an osmotic gradient—result in a fluid that is slightly hypotonic to blood. Once the filtrate has passed through the loop of Henle, the major solutes that it contains are urea and other wastes.

The fluid that enters the distal tubule is relatively constant in composition over time. In contrast, the urine that leaves the collecting duct is highly variable in osmolarity and in Na^+ and Cl^- concentration. How is this possible?

URINE FORMATION IS UNDER HORMONAL CONTROL The answer is based on two observations about the distal tubule and collecting duct: Their activity is (1) highly regulated, and (2) altered in response to osmotic stress. The amount of Na^+, Cl^-, and water that is reabsorbed in the distal tubule and in the collecting duct varies with the animal's condition.

Changes in the distal tubule and collecting duct are controlled by **hormones**—signaling molecules that were introduced in Chapter 8 and are explored further in Chapter 47. Specifically:

- If Na^+ levels in the blood are low, the adrenal glands release the hormone **aldosterone**, which leads to activation of sodium pumps and reabsorption of Na^+ in the distal tubule. Aldosterone saves sodium.

- If an individual is dehydrated, the brain releases **antidiuretic hormone (ADH)**. (The term *diuresis* refers to increased urine production, so *antidiuresis* means inhibition of urine production. ADH is also referred to as vasopressin or arginine vasopressin.) ADH saves water.

HOW DOES ADH WORK? ADH has two important effects on epithelial cells in the collecting duct:

1. ADH triggers the insertion of aquaporins into the apical membrane. As a result, cells become much more permeable to water and large amounts of water are reabsorbed.

2. ADH increases permeability to urea, which increases the osmolarity of the surrounding fluid and thus water loss from the filtrate.

As **Figure 42.17a** on page 838 shows, water leaves the collecting duct passively—following the concentration gradient maintained by the loop of Henle. When ADH is present, water is conserved and urine is strongly hypertonic relative to blood.

When ADH is absent, however, few aquaporins are found in the epithelium of the collecting duct, and the structure is relatively impermeable to water (**Figure 42.17b**). In this case, a hypotonic urine is produced.

People with defective forms of ADH or aquaporins produce copious amounts of urine—up to 30 liters per day. They suffer from the condition **diabetes insipidus**. (The word diabetes means "to run through"; insipidus means "tasteless." You may also have heard of the disease diabetes mellitus, which is due to problems

(a) ADH present: Collecting duct is highly permeable to water.

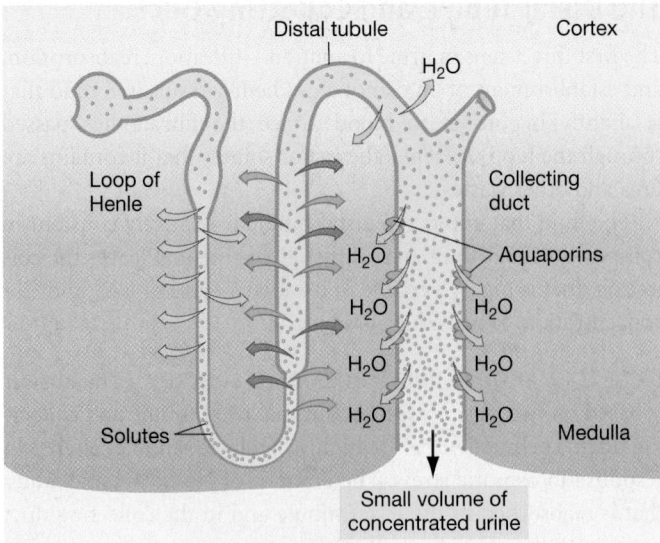

(b) No ADH present: Collecting duct is not permeable to water.

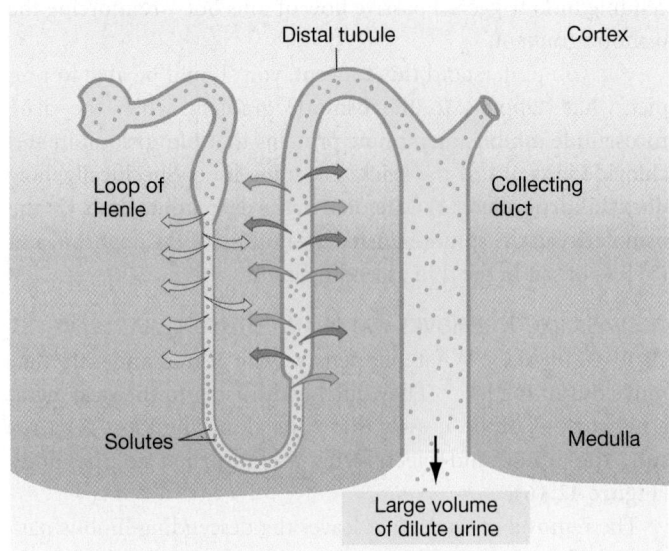

FIGURE 42.17 ADH Regulates Water Reabsorption by the Collecting Duct.

with regulating glucose concentrations in blood, rather than water retention in the kidneys—see Chapter 43.)

✔If you understand ADH's effect on the collecting duct, you should be able to predict how urine formation is affected by ingestion of ethanol, which inhibits ADH release, versus nicotine, which stimulates ADH release.

Table 42.2 reviews the functions of the four major regions of the nephron and the collecting duct. You can also review how the kidney works by going to the study area at *www.masteringbiology.com*.

 **Web Activity** The Mammalian Kidney

To summarize, the first three segments of the nephron—the renal corpuscle, proximal tubule, and loop of Henle—concentrate nitrogenous wastes, and create the possibility for Na^+, Cl^-, and

water to be either excreted or reabsorbed by the distal tubule and collecting duct. The nephron is a remarkably effective mechanism for regulating water and electrolyte balance and achieving homeostasis.

SUMMARY TABLE 42.2 **Structure and Function of the Nephron and Collecting Duct**

Structure	Function
Renal corpuscle (Bowman's capsule and glomerulus)	Size-selective filtration: forms filtrate from blood (water and other small substances enter nephron)
Proximal tubule	Reabsorbs electrolytes, nutrients, water (active transport)
Loop of Henle	Maintains osmotic gradient from outer to inner medulla
• Descending limb	• Permeable to water (passive transport out of filtrate)
• Thin ascending limb	• Permeable to Na^+, Cl^- (passive transport out of filtrate)
• Thick ascending limb	• Active transport of Na^+, Cl^- out of filtrate
Distal tubule	With aldosterone: reabsorbs Na^+. Without aldosterone: does not reabsorb Na^+
Collecting duct	Regulates water retention and loss
• Main portion	• With ADH: water leaves filtrate; produces urine that is hypertonic to blood (reduced urine volume) Without ADH: water stays in filtrate; produces urine that is hypotonic to blood (increased urine volume)
• Innermost portion	• Urea leaks out to establish/maintain high osmolarity of inner medulla

CHECK YOUR UNDERSTANDING

If you understand that . . .

- The vertebrate kidney is specialized for the production of hypertonic urine.
- The loop of Henle is a countercurrent system that maintains a strong osmotic gradient in the medulla.
- The characteristics of urine change in response to hormonal signals. The signals trigger changes in the nephron that either save or eliminate water and solutes. The result is negative feedback and homeostasis.

✔ **You should be able to . . .**

1. Explain how aldosterone regulates the characteristics of urine.

2. Predict how the following events would affect urine production: drinking massive amounts of water, eating large amounts of salt, and refraining from drinking for 48 hours.

Answers are available in Appendix B.

Summary of Key Concepts

🗝 **Freshwater, marine, and terrestrial habitats pose different challenges to animals with regard to maintaining water and electrolyte balance.**

- The mechanisms involved in regulating water and electrolyte balance vary widely among animal groups, because different habitats present different types of osmotic stress.

- Freshwater fish are strongly hypertonic to their environment and tend to gain water and lose electrolytes.

- Marine fish are strongly hypotonic in relation to seawater and tend to lose water and gain electrolytes.

- Terrestrial animals lose water every time they breathe and from their body surfaces by evaporation.

- In most animals, epithelial cells that selectively transport water and electrolytes are responsible for homeostasis.

  ✔ You should be able to describe the water and electrolyte balance challenges faced by river otters that live near the ocean and move from freshwater to saltwater habitats and back on a daily or weekly basis.

🗝 **In marine animals, specialized epithelial cells have membrane proteins that remove excess salt (NaCl) from the body so it can be excreted. The same types of cells are found in the kidneys of mammals.**

- Epithelial cells in the shark rectal gland excrete excess salt, based on a "master gradient" established by sodium-potassium pumps.

- Salt-excreting cells similar to those found in the shark rectal gland occur in the gills of saltwater fish, the salt glands of marine birds and reptiles, and the kidneys of mammals.

- The leading hypothesis to explain how freshwater fish import salt is that they have cells in their gill epithelium that are similar to the salt-excreting cells found in marine fish, but oriented in the opposite direction.

  ✔ You should be able to explain how an electrochemical gradient for sodium ions makes it possible for water and ions to move across a membrane passively.

🗝 **In terrestrial insects, the hindgut and Malpighian tubules are responsible for excreting water-soluble waste products and achieving homeostasis with respect to water and electrolyte concentrations.**

- A waxy coating on the insect exoskeleton limits evaporative water loss. The openings to insect respiratory organs close when osmotic stress is severe.

- Insects can form a hypertonic urine that minimizes water loss during the excretion of nitrogenous wastes.

- The Malpighian tubules of insects form a filtrate that is isotonic with the hemolymph. If pumps in the epithelium of the hindgut are activated, then electrolytes and water are reabsorbed from the filtrate and returned to the hemolymph.

  ✔ You should be able to predict whether desert locusts or grasshoppers that live near ponds expend more energy during urine formation, and explain your reasoning.

🗝 **In terrestrial vertebrates, the kidney is responsible for excreting water-soluble waste products and achieving homeostasis with respect to water and electrolyte concentrations.**

- Nephrons in the mammalian kidney form a filtrate in the renal corpuscle and then reabsorb valuable nutrients, electrolytes, and water in the proximal tubule.

- A solution containing urea and electrolytes flows through the loop of Henle, where changes in the permeability of epithelial cells to water and salt—along with active transport of salt—create a steep osmotic gradient.

- If ADH increases the water permeability of the collecting duct, water is reabsorbed along the osmotic gradient and a hypertonic urine is produced.

- In animals, maintaining water and electrolyte balance is an active, energy-demanding process based on the action of membrane proteins. Water and electrolyte excretion and reabsorption are carefully controlled to achieve homeostasis.

  ✔ You should be able to explain why desert-dwelling kangaroo rats, which are under extreme osmotic stress due to lack of drinking water, have extremely long loops of Henle relative to their body size.

  (MB) **Web Activity** The Mammalian Kidney

Questions

1. Which one of the following statements is true of fish that live in freshwater?
   a. Their tissues are isotonic with respect to their environment. As a result, they do not require a specialized organ to maintain water and electrolyte balance.
   b. They lose water to their environment primarily through the gills. They replace this water by drinking.
   c. Water enters epithelial cells in their gills via osmosis. Electrolytes leave the same cells via diffusion.
   d. They have specialized cells that actively pump Na^+ and Cl^- from blood into epithelial cells, so the ions can be excreted.

2. Na^+/K^+-ATPase is found in epithelial cells that produce urine or other excretions that are hypertonic relative to tissues. Why?
   a. To form a hypertonic solution, ions or molecules must be actively transported across their concentration gradient.
   b. Cells similar to those found in the shark rectal gland occur in many other species that need to excrete salt (NaCl).

c. The sodium-potassium pump is directly responsible for moving sodium ions from blood or body tissues into the urine for salt excretion, against a concentration gradient.

d. They are usually located in the basolateral membrane of those epithelial cells, along with other channels and cotransporters.

3. Regarding urine formation, the function of the insect hindgut is closest to which of the following structures in the vertebrate kidney?
   a. renal corpuscle
   b. proximal tubule
   c. loop of Henle
   d. distal tubules and collecting duct

4. Which of the following gives the correct sequence of the regions of the nephron with respect to fluid flow?
   a. distal tubule → ascending limb of Henle → descending limb of Henle → proximal tubule → collecting duct
   b. ascending limb of Henle → descending limb of Henle → proximal tubule → distal tubule → collecting duct
   c. proximal tubule → ascending limb of Henle → descending limb of Henle → distal tubule → collecting duct

d. proximal tubule → descending limb of Henle → ascending limb of Henle → distal tubule → collecting duct

5. Which of the following statements about kidney function is *not* correct?
   a. The loop of Henle acts as a countercurrent exchanger, so no energy is required to maintain the osmotic gradient.
   b. The descending limb of the loop of Henle is highly permeable to water.
   c. The thin ascending limb of the loop of Henle is highly permeable to salt.
   d. Reabsorption of water and solutes takes place primarily in the proximal tubule.

6. What effect does antidiuretic hormone (ADH) have on the nephron?
   a. It increases water permeability of the descending limb of the loop of Henle.
   b. It decreases water permeability of the descending limb of the loop of Henle.
   c. It increases water permeability of the collecting duct.
   d. It decreases water permeability of the collecting duct.

✔ TEST YOUR UNDERSTANDING
Answers are available in Appendix B

1. Explain the changes that need to occur in water and electrolyte balance as a salmon moves from freshwater to salt water and back. Specifically, state when the animal should drink or not drink, when cells in the gill epithelium should excrete or import electrolytes, and why each change occurs.

2. The chloride cells of fish gills are sometimes called mitochondria-rich cells due to their high density of mitochondria. How does high mitochondrial density relate to the functional role of chloride cells? Would you expect other epithelial cells involved in ion transport to contain large numbers of mitochondria? Explain.

3. Why is it significant that cells involved in transport processes often have microvilli?

4. This chapter introduced a number of features that help terrestrial animals reduce water loss. These traits include the layer of wax found on insect exoskeletons, the ability of insects to close the openings to their respiratory passages, the excretion of nitrogenous wastes as insoluble uric acid, and long loops of Henle in the mammalian

kidney. Predict how each of these traits differs in animals that live in very humid versus very dry habitats. How would you test your predictions?

5. In insects, active transport of electrolytes into the Malpighian tubules leads to the formation of a filtrate that is isotonic with respect to the hemolymph. In mammals, urine formation begins with blood pressure in the renal corpuscle that leads to the formation of a filtered filtrate that is isotonic with respect to the blood. In insects, the filtrate is processed in the hindgut; in mammals, the filtrate is processed in the remainder of the nephron. How are the processing steps in insects and mammals similar? How are they different?

6. Compare and contrast the types of nitrogenous wastes observed in animals. Identify which compound can be excreted with a minimum of water, which is most toxic, and which waste product is found in fish, mammals, and insects. Which type of nitrogenous waste would you expect to find produced by embryos inside eggs laid on land?

✔ APPLYING CONCEPTS TO NEW SITUATIONS
Answers are available in Appendix B

1. Examine Figure 42.16 again, and notice that the network of blood vessels that runs along the mammalian nephron is arranged so that blood flows in a countercurrent arrangement relative to the flow of fluid in the loop of Henle. The water and electrolytes that leave the loop of Henle and collecting duct diffuse into these blood vessels and are returned to the body. Why is the countercurrent arrangement for blood flow important?

2. You have isolated a segment of a rat nephron and introduced a solution of known composition. When you compare the fluid collected at the end of the segment with the test solution introduced at the beginning, you find that the volume has decreased by 30 percent and

the Na^+ concentration has decreased by 30 percent, but the urea concentration has increased by 50 percent. What processes in this segment might account for these changes?

3. In some areas of the world, highway maintenance crews use salt extensively to keep roads free of ice in winter. Biologists are concerned about the effect of this salt on freshwater organisms. Why?

4. Biologists have recently been able to produce mice that lack functioning genes for aquaporins. How does their urine compare to that of individuals with normal aquaporins?

A young crocodile has just caught a frog. Animals obtain nutrients by ingesting food.

Animal Nutrition

43

A nimals get the two basic requirements for life—the chemical energy required to synthesize ATP and the carbon-containing compounds required to synthesize complex macromolecules—from other organisms. They are heterotrophs. They eat to live.

The types of food that are available to different animals vary widely, and food is often in dangerously short supply. Based on these observations, it is logical to predict that animals have a variety of means for obtaining food and they are under intense natural selection to make efficient use of the food they have.

Acquiring and processing energy is a fundamental attribute of life (see Chapter 1). How do animals get their food, and how do they process it? Which substances in food are used as nutrients, and how do humans and other animals maintain appropriate levels of key nutrients in their bodies?

If you're like most people, you've probably given a fair amount of thought to the food you eat but little thought to what happens to that food once it's inside you. As a meal moves through a digestive system, its chemical composition and physical characteristics change dramatically. Large packets of food that enter the mouth are reduced to monomers that can be absorbed by cells and enter the bloodstream.

Ingesting food is the first of four processes needed to obtain energy. For you or any other animal to stay alive, ingestion must be followed by digestion, absorption, and elimination of wastes (**Figure 43.1** on page 842). The digestive system that accomplishes these tasks is analogous to a lumber mill, where bulky, complex raw materials are processed into smaller and more usable products.

Research on feeding and digestion is fundamental to understanding basic aspects of animal anatomy and physiology. But research on animal nutrition has important practical applications as well. For example, this chapter addresses questions about why ulcers can develop and why type 2 diabetes and other nutrition-related diseases are on the rise in many human populations.

KEY CONCEPTS

- Animals require an array of nutrients to stay healthy, including specific amino acids, vitamins, and elements, as well as organic compounds that act as building blocks in chemical synthesis or have high potential energy.

- There is usually a close correspondence between the structure of an animal's mouthparts and the function of those mouthparts in capturing and processing food.

- Digestion occurs in the digestive tract, which is compartmentalized into organs that have specialized functions in the ingestion and digestion of food, absorption of nutrients and water, or excretion of wastes.

- Lack of homeostasis with respect to nutrients such as glucose can cause disease.

✔ When you see this checkmark, stop and test yourself. Answers are available in Appendix B.

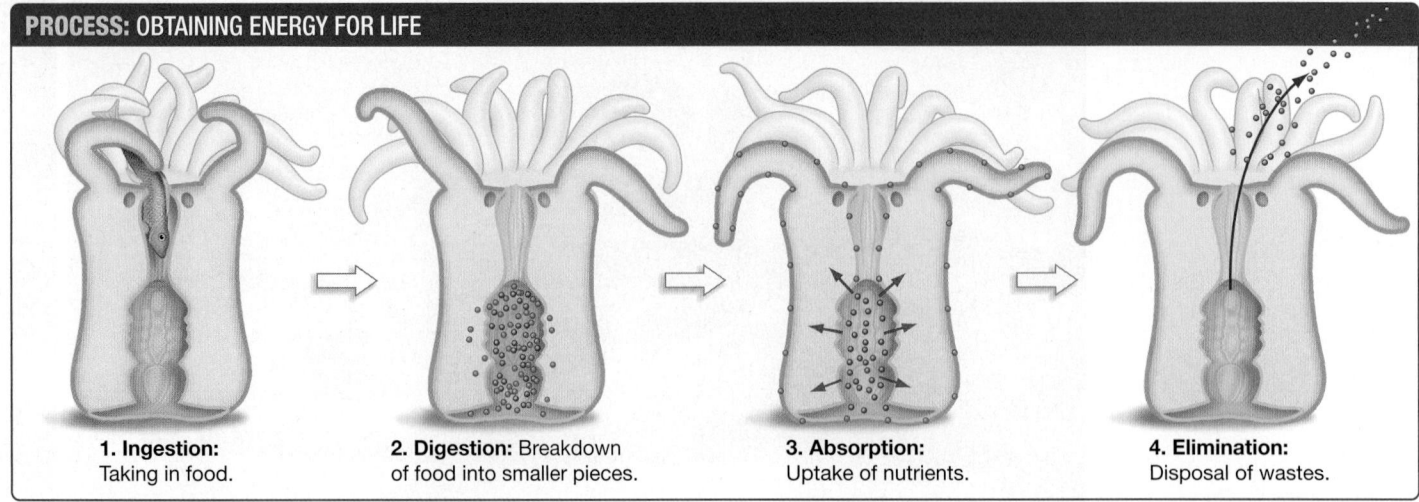

1. Ingestion: Taking in food.

2. Digestion: Breakdown of food into smaller pieces.

3. Absorption: Uptake of nutrients.

4. Elimination: Disposal of wastes.

FIGURE 43.1 Four Steps in Obtaining Nutrients.

Just as plant nutrition is fundamental to the productivity of agricultural and natural ecosystems, nutrition is a basic component of animal health and welfare. Let's begin with a look at what animals must eat to live.

43.1 Nutritional Requirements

Animals get the chemical energy and carbon-containing building blocks they need from carbohydrates and fats, which are carbon compounds with high potential energy. Because fats and other lipids contain more C–H bonds than do carbohydrates such as starch and sugars, lipids provide over twice the energy per gram. Chapters 5 and 6 analyzed the structures of carbohydrates and fats; Chapter 9 detailed how these compounds are used to synthesize ATP and key macromolecules.

A carbohydrate is an example of a **nutrient**: a substance that an organism needs to remain alive. **Food** is any material that contains nutrients.

Understanding which nutrients an individual needs, and in what amounts, are basic issues in research on animal nutrition. To illustrate how biologists go about the task, let's consider what humans need to maintain good health.

Meeting Basic Needs in Humans

In 1943 the Food and Nutrition Board of the U.S. National Academy of Sciences[1] published the first Recommended Dietary Allowances (RDAs). The goal of the RDAs was to specify the amount of each essential nutrient that an individual must ingest to meet the needs of most healthy people.

🔑 In addition to obtaining "building blocks" for chemical synthesis and chemical energy from carbohydrates and other compounds, humans require several other **essential nutrients**— meaning, nutrients that cannot be synthesized and must be obtained in the diet:

- **Proteins** provide amino acids that are used to synthesize polypeptides; amino acids may also be oxidized to provide energy. Of the 20 amino acids required to manufacture most proteins, humans can synthesize 12. The others are called **essential amino acids** because they must be obtained from food.

- **Vitamins** are organic compounds that are vital for health but are required in only minute amounts. They have a variety of roles; several function as coenzymes in critical reactions. **Table 43.1** lists a few of the vitamins for which RDAs have been established, notes their functions, and indicates the problems that develop if they are missing in the diet.

- **Electrolytes** are inorganic ions that influence osmotic balance and are required for normal membrane function. Sodium (Na^+), potassium (K^+), and chloride (Cl^-) are the major ions in the body.

- Inorganic substances fulfill a wide variety of functions not performed by electrolytes—often because they are important components of enzyme cofactors or structural materials (see **Table 43.2**). Some, such as calcium and phosphorus, are needed in relatively large quantities. Others, such as iron and magnesium, are required in small or trace amounts.

Studying Nutrient Requirements

Research on the basic nutritional requirements of nonhuman animals can sometimes be based on experiments that vary dietary intake of specific ions or molecules under controlled conditions, and compare indices of health. Based on experiments like these and analyses of diets eaten by healthy individuals, researchers have been able to publish the equivalent of RDAs for a wide array of domestic livestock and animals commonly maintained in zoos and aquariums.

Studying nutrient requirements in humans is more difficult. It is unethical to assign different, carefully controlled diets to sev-

---

[1]The National Academy of Sciences is a group of scientists and engineers that advises the U.S. Congress on scientific and technical matters. Its Food and Nutrition Board is made up of biologists who specialize in animal nutrition.

TABLE 43.1 **Some Important Vitamins (Required by Humans)**

	Source in Diet	Function	Symptoms if Deficient
Vitamin B$_1$ (thiamine)	legumes, whole grains, potatoes, peanuts	formation of coenzyme in citric acid cycle	beriberi (fatigue, nerve disorders, anemia)
Vitamin B$_{12}$	red meat, eggs, dairy products; also synthesized by bacteria in intestine	coenzyme in synthesis of proteins and nucleic acids; formation of red blood cells	anemia (fatigue and weakness due to low hemoglobin content in blood)
Niacin	meat, whole grains	component of coenzymes NAD$^+$ and NADP$^+$	pellagra (digestive problems, skin lesions, nerve disorders)
Folate	green vegetables, oranges, nuts, legumes, whole grains; also synthesized by bacteria in intestine	coenzyme in nucleic acid and amino acid metabolism	anemia
Vitamin C (ascorbic acid)	citrus fruits, tomatoes, broccoli, cabbage, green peppers	used in collagen synthesis, prevents oxidation of cell components, improves absorption of iron	scurvy (degeneration of teeth and gums)
Vitamin D	fortified milk, egg yolk; also synthesized in skin exposed to sunlight	aids absorption of calcium and phosphorus in small intestine	rickets (bone deformities) in children; bone softening in adults

TABLE 43.2 **Essential Elements (Required by Humans)**

	Source in Diet	Function	Symptoms if Deficient
Calcium (Ca)	dairy products, green vegetables, legumes	bone and tooth formation, nerve signaling, muscle response	loss of bone mass, slow growth
Fluorine (F)	fluoridated water, seafood	maintenance of tooth structure	higher frequency of tooth decay
Iodine (I)	iodized salt, algae, seafood	component of the thyroid hormones thyroxin and T$_3$	goiter (enlarged thyroid gland)
Iron (Fe)	meat, eggs, whole grains, green leafy vegetables, legumes	enzyme cofactor; synthesis of hemoglobin and electron carriers	anemia, weakness
Magnesium (Mg)	whole grains, green leafy vegetables	enzyme cofactor	nerve disorders
Phosphorus (P)	dairy products, meat, grains	bone and tooth formation, synthesis of nucleotides and ATP	weakness, loss of bone
Sulfur (S)	any source of protein	amino acid synthesis	swollen tissues, degeneration of liver, mental retardation

eral groups of subjects if there is a chance that one of the diets might lead to illness. As a result, researchers must rely on observational studies that document the health of people with known intake levels of specific nutrients.

43.2 Capturing Food: The Structure and Function of Mouthparts

Instead of making their own food as plants do, animals obtain the energy and nutrients they need from other organisms. The question is, How?

As Chapter 32 noted, biologists assign animal food-getting techniques to one of four strategies:

- **Suspension feeders**, such as sponges and tubeworms, filter small organisms or bits of organic debris from water, by means of cilia, mucous-lined "nets," or other structures.

- **Deposit feeders**, including earthworms and sea cucumbers, swallow organic-rich sediments and other types of deposited material.

- **Fluid feeders** suck or lap up fluids—blood, nectar, or sap, for example.

- **Mass feeders** are the majority of animals. They seize and manipulate chunks of food by using mouthparts such as jaws and teeth, beaks, or special toxin-injecting organs.

Mouthparts as Adaptations

The types of food that animals harvest range from soupy solutions in decaying carcasses to hard nuts. Solutions have to be lapped up; nuts have to be cracked. Given the diversity of food sources that animals exploit, it is not surprising that a wide variety of mouthpart structures have evolved to facilitate capturing and processing food.

Natural selection has tightly matched the structure of animal mouthparts to their function in obtaining food. For example:

- Mammals are the only animals that chew their food and swallow distinct packets or boluses. The extinct mammal species shown in **Figure 43.2a** illustrates one example of the diversity in tooth shapes that evolved from the relatively simple and similar teeth that occurred in the common ancestor of all mammals. Diversification in tooth shape has allowed mammals to exploit a wide range of foods.

- Complex, multipart skull and jawbones, with associated musculature, have evolved in snakes. The highly movable structure that results allows snakes to ingest and swallow large prey whole (**Figure 43.2b**).

The reason that there is such a close correlation between the structure and function of mouthparts is simple: Natural selection is particularly strong when it comes to food capture, because obtaining nutrients is so fundamental to fitness—the ability to produce offspring.

It's important to remember some key points about natural selection and adaptation, however (see Chapter 24):

(a) The large canines of saber-toothed cats stabbed and sliced prey.

(b) A highly movable skull and jaw allows snakes to swallow large prey whole.

FIGURE 43.2 Mouthpart Structure Correlates with Function.

- *Evolution is not progressive.* Animal mouthparts do not get "better" over time in the sense of being more complex. Most nematomorphs (hair worms) do not capture food and do not have a mouth—even though they evolved from ancestors that had mouths. Instead, nematomorphs absorb nutrients directly across their body wall.

- *Adaptation is not perfect.* Humans are often cited as an example of an omnivore with teeth adapted for both tearing meat and mashing plant leaves or stalks. But most dentists would also claim that our "wisdom teeth" are maladaptive—meaning, a trait that actually lowers fitness on average.

Let's pursue the correlation between mouthparts and food sources further, by analyzing the structure and function of jaws and teeth in what may be the most diverse lineage among vertebrates: the cichlid fishes of Africa.

A Case Study: The Cichlid Jaw

The cichlids that inhabit the Rift Lakes of East Africa are a spectacular example of **adaptive radiation**—the diversification of a single ancestral lineage into many species, each of which lives in a different habitat or employs a distinct feeding method (see Chapter 27). Lake Victoria, for example, is home to 300 **endemic** cichlids—meaning species that live nowhere else.

Most of the Lake Victoria cichlids feed on a specific item, but as a group they exploit almost every food source in the lake: planktonic organisms, crust-forming algae, leaflike algae, eggs, fish scales, fish fins, whole fish, plants, insects, and snails.

How can a group of closely related species exploit so many different food sources? To answer this question, biologists point to a structure called the pharyngeal jaw.

Many fish species have pharyngeal (throat) jaws located well behind the normal oral (mouth) jaws. Non-cichlids use their pharyngeal jaws to move food down their throats, but cichlids can also use theirs to bite. This is possible because the upper pharyngeal jaw attaches to the skull in cichlids, and because the muscles of their lower pharyngeal jaw allow it to move against the upper jaw (**Figure 43.3**).

Cichlids can use their pharyngeal jaws to bite because:

1. Their upper pharyngeal jaw connects to the skull (★).

2. Muscles (blue lines) allow the lower pharyngeal jaw to move up and down.

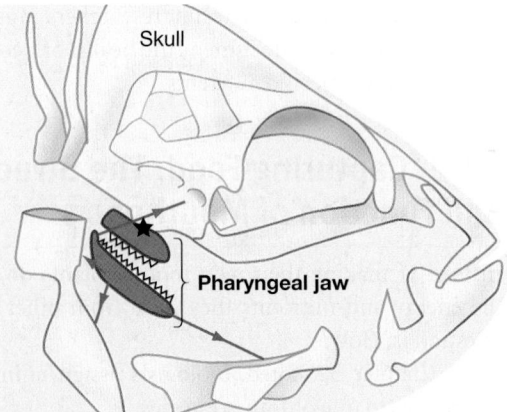

FIGURE 43.3 Rift Lake Cichlids Have Two Sets of Biting Jaws.
Oral jaws capture food; pharyngeal jaws process it.

✔ **EXERCISE** Label the oral jaw.

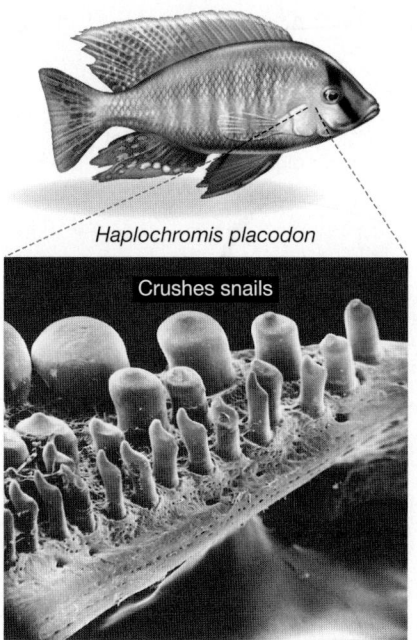
Haplochromis placodon
Crushes snails

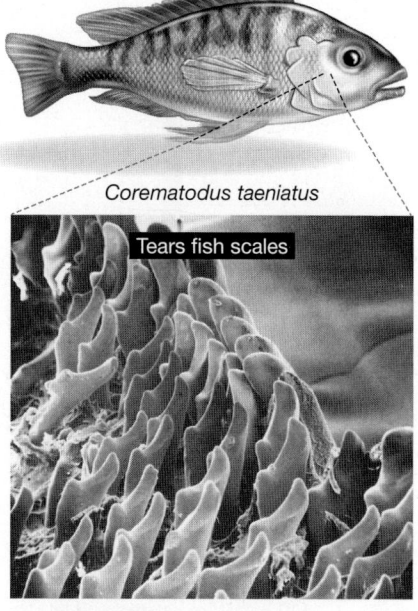
Corematodus taeniatus
Tears fish scales

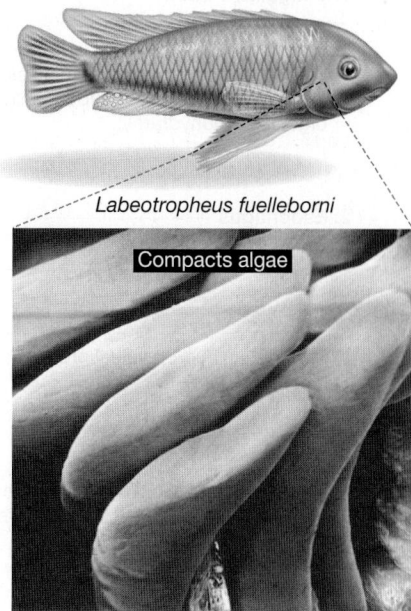
Labeotropheus fuelleborni
Compacts algae

FIGURE 43.4 In Cichlids, the Structure of the Pharyngeal Jaw Correlates with the Type of Food Ingested.

In addition to furnishing a second set of biting jaws that make food processing more efficient, cichlid pharyngeal jaws provide an extra set of toothlike structures. These protuberances vary in size and shape, correlating with their function such as crushing snail shells, tearing fish scales, or compacting algae (**Figure 43.4**).

These observations add to a large body of evidence supporting a general pattern in animal evolution: Mouthparts have diversified in response to natural selection for exploiting a diversity of food sources. The structure of jaws, teeth, and other mouthparts correlates with their function in harvesting and processing food.

43.3 How Are Nutrients Digested and Absorbed?

Animals ingest just about every type of food conceivable. Ingestion is only the first of four processes needed to obtain energy, however. For an animal to stay alive, ingestion must be followed by digestion, absorption, and elimination of wastes.

Digestion is the breakdown of food into small enough pieces to allow for **absorption**—the uptake of specific ions and molecules that act as nutrients. Digestion is a key process in animals because, unlike plants, unicellular organisms, and certain parasites, animals do not acquire nutrients as individual molecules. Instead, they take in packets of food that must be broken down into small pieces. Nutrients must be extracted from the small pieces and waste materials must be eliminated. How and where does this processing occur?

An Introduction to the Digestive Tract

In many animals, digestion takes place in a tube called the **digestive tract**—also known as the alimentary (literally, "nour-ishment") canal or gastrointestinal (GI) tract. Digestive tracts come in two general designs:

- **Incomplete digestive tracts** have a single opening that doubles as the location where food is ingested and wastes are eliminated. The mouth opens into a chamber, called a gastrovascular cavity, where digestion takes place (**Figure 43.5**).

- **Complete digestive tracts** have two openings—they start at the mouth and end at the anus. The interior of this tube communicates directly with the external environment via these openings (**Figure 43.6** on page 846). In embryos, the gut derives from the hollow tube that forms as cells invaginate during gastrulation (see Chapter 22).

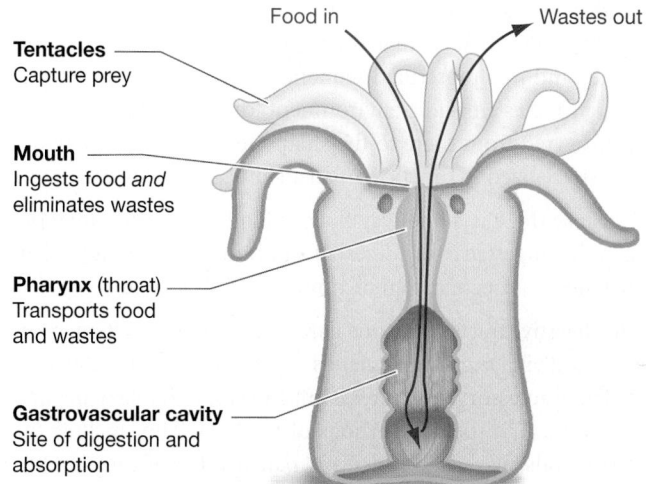

Tentacles
Capture prey

Mouth
Ingests food *and* eliminates wastes

Pharynx (throat)
Transports food and wastes

Gastrovascular cavity
Site of digestion and absorption

Food in   Wastes out

FIGURE 43.5 An Incomplete Digestive Tract. Anemones use stinging cells located on their tentacles to capture small fish, crustaceans, and other prey. Prey are taken into the mouth and digested in the gastrovascular cavity.

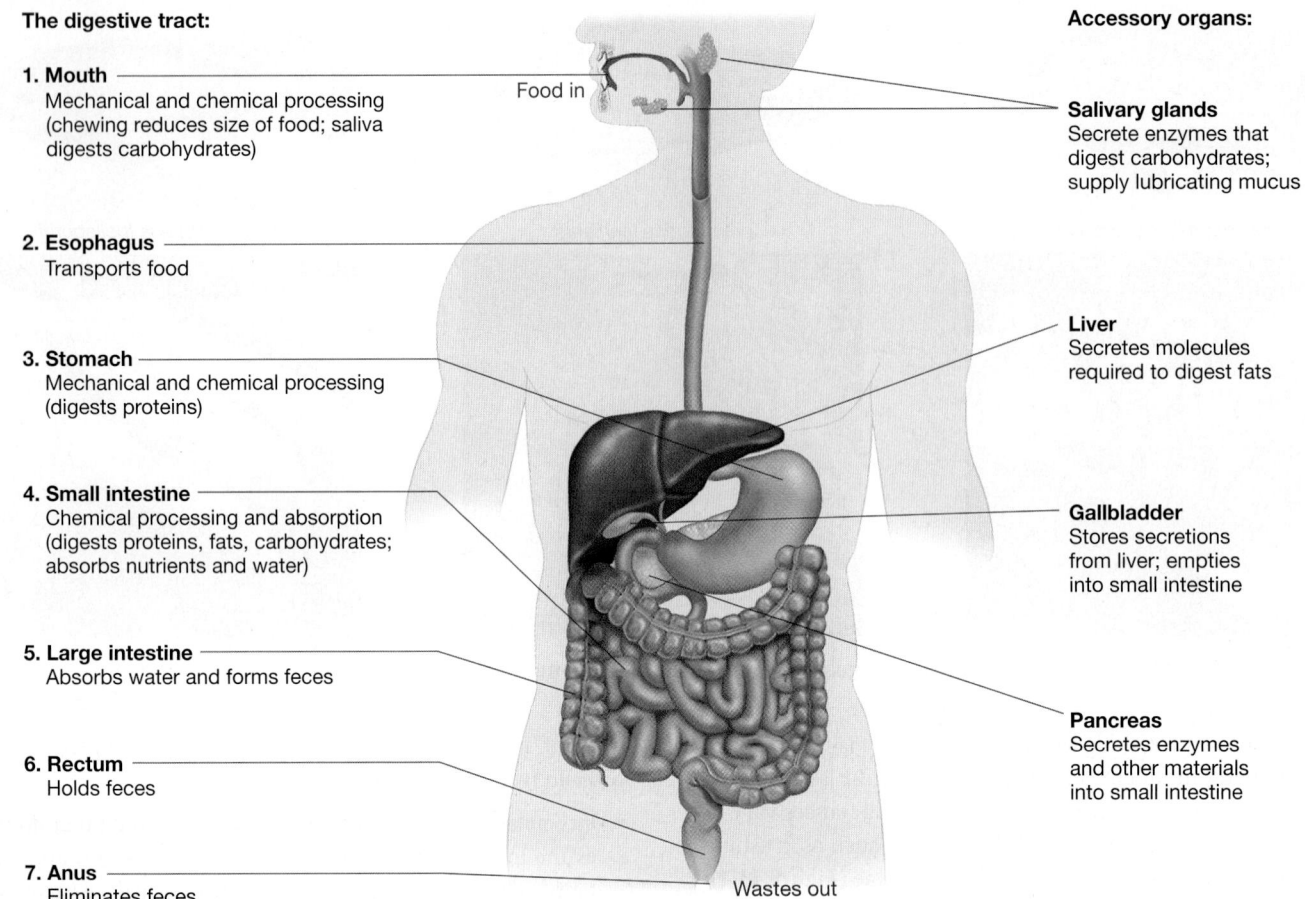

The digestive tract:

1. Mouth
Mechanical and chemical processing (chewing reduces size of food; saliva digests carbohydrates)

Food in

2. Esophagus
Transports food

3. Stomach
Mechanical and chemical processing (digests proteins)

4. Small intestine
Chemical processing and absorption (digests proteins, fats, carbohydrates; absorbs nutrients and water)

5. Large intestine
Absorbs water and forms feces

6. Rectum
Holds feces

7. Anus
Eliminates feces

Wastes out

Accessory organs:

Salivary glands
Secrete enzymes that digest carbohydrates; supply lubricating mucus

Liver
Secretes molecules required to digest fats

Gallbladder
Stores secretions from liver; empties into small intestine

Pancreas
Secretes enzymes and other materials into small intestine

FIGURE 43.6 A Complete Digestive Tract. In humans, the digestive tract is a tube that runs from the mouth to the anus. The salivary glands, liver, gallbladder, and pancreas are not part of the tract itself. Instead, they secrete material into the tract at specific points.

A tubelike digestive tract has three advantages:

1. It allows animals to feed on large pieces of food—expanding the types of food sources that can be ingested.

2. Different chemical and physical processes can be separated within the canal, so that they occur independently of each other and in a prescribed sequence. The stomach, for example, provides an acidic environment for digestion. Material is transferred from there to the small intestine, where enzymes are specialized to function in an alkaline environment.

3. Because there is a one-way flow of food and wastes, material can be ingested and digested continuously instead of in batches—as occurs in an incomplete digestive tract.

The digestive tract is only one part of the digestive system, however. Several vital organs and glands are connected to the digestive tract. These accessory structures contribute digestive enzymes and other products to specific portions of the tract. They include the salivary glands, liver, gallbladder, and pancreas (see Figure 43.6).

An Overview of Digestive Processes

Before analyzing the function of each component of the digestive system in detail, let's consider the general changes that hap-

pen to food on its way through the digestive tract. In this brief overview and in the detailed discussion that follows, humans will serve as a model species—simply because so much is known about human digestion.

In mammals, digestion begins with the tearing and crushing activity of teeth. Chewing reduces the size of food particles and softens them. Humans augment the mechanical breakdown of food by their use of knives and cooking. In fact, the invention of cutting tools and cooking, which make food easier to chew, is the leading hypothesis to explain why average tooth size has declined steadily over the past several million years of human evolution.

Distinct chemical changes occur as food moves through each compartment in the digestive tract (**Figure 43.7**):

- The chemical breakdown of carbohydrates begins in the mouth, through enzymes in saliva.

- Protein digestion begins in the acidic environment of the stomach.

- Chemical processing of the three major types of macromolecules—carbohydrates, lipids, and proteins—is completed in the small intestine. The small molecules that result from the digestion of these macromolecules are absorbed in the small intestine, along with water, vitamins, and ions.

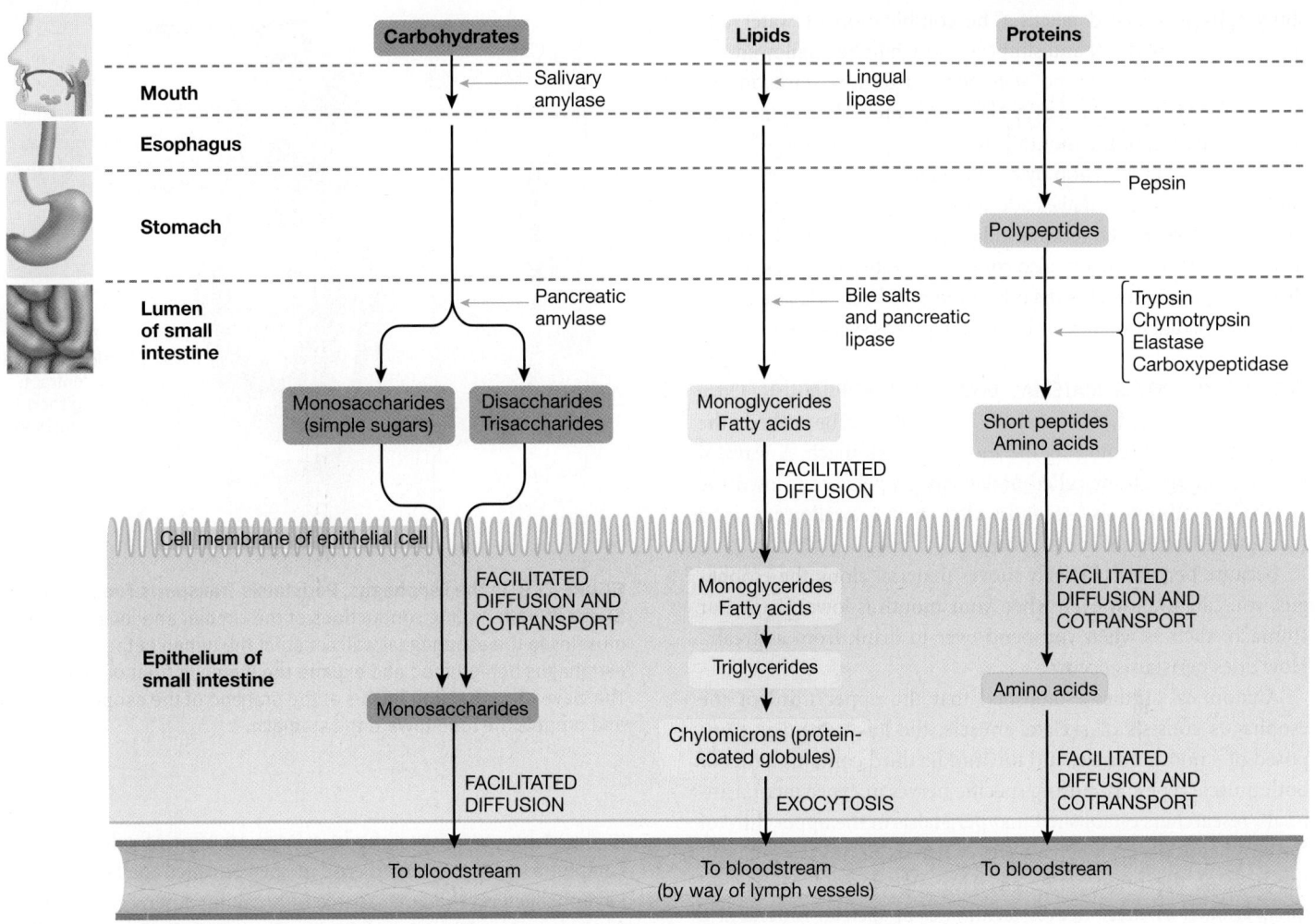

FIGURE 43.7 Carbohydrates, Lipids, and Proteins Are Processed in a Series of Steps. Three key types of macromolecules enter the digestive system (top of diagram). As they proceed through the digestive tract, they are broken apart by a variety of enzymes. Simple sugars, fatty acids, and amino acids then enter epithelial cells in the small intestine and are transported to the bloodstream.

- In the large intestine, or colon, more water is absorbed. The result is **feces**, which are eventually excreted.

For a human to stay healthy, each step in this process must be completed correctly. Problems in the digestive tract can lead to heartburn, ulcers, nausea, constipation, and other maladies. By far the most serious of these is diarrhea, which kills more than 4 million children each year.

Because digestion is so important to understanding how animals work, let's analyze each step in more detail. As the following sections track food as it moves from the mouth to the anus in humans—a particularly well-strudied organism—watch for notes highlighting the diversity of structures found in the digestive tracts of nonhuman animals.

The Mouth and Esophagus

If you hold a cracker in your mouth long enough, it will start to taste sweet. The sensation occurs because an enzyme in your saliva hydrolyzes the starch molecules in the cracker to maltose.

Maltose is a disaccharide that is split in the small intestine to form two glucose monomers.

Starch breakdown was actually the first enzyme-catalyzed reaction ever discovered. In the early 1800s several researchers found that a component of certain plant extracts digested starch; in 1831 the same activity was discovered in human saliva.

DIGESTION STARTS IN THE MOUTH **Amylase,** the enzyme responsible for starch digestion in the mouth, ranks as one of the best-studied enzymes. Amylase cleaves bonds that release maltose dimers from starch, glycogen, and other glucose polymers. In doing so, it initiates the digestion of carbohydrates.

Cells in the tongue synthesize and secrete another important salivary enzyme, **lipase,** which begins the digestion of lipids. More specifically, the lingual lipase (lingual refers to the tongue) produced in the mouth begins breaking triglycerides into fatty acids and monoglycerides.

Salivary glands in the mouth also release water and glycoproteins called mucins. When mucins contact water, they form the slimy substance called **mucus**. The combination of water and mucus makes food soft and slippery enough to be swallowed.

In snakes, however, the statement "digestion starts in the mouth" is not correct. Snakes swallow their prey whole—meaning that the function of the mouth is to take in large packets of food, rather than begin digestion by chewing and secretion of enzymes. And some venomous snakes add a twist: Digestion begins *before* food is ingested. Snake venoms kill or otherwise immobilize prey; in some species the venom also contains digestive enzymes. When these venoms are injected deep into the prey body, via large teeth called fangs, digestion begins before the prey is swallowed.

PERISTALSIS MOVES MATERIAL DOWN THE ESOPHAGUS Once food is swallowed, it enters a muscular tube called the **esophagus**, which connects the mouth and stomach. A wave of muscular contractions called **peristalsis** propels food down the esophagus. About six seconds after being swallowed, food reaches the bottom of the esophagus.

Because peristalsis actively moves material along the esophagus, you can swallow even when your mouth is lower than your stomach, such as when you bend over to drink from a stream. How does peristalsis occur?

Anatomical studies established that the upper third of the esophagus consists of skeletal muscle, the lower third is composed of smooth muscle, and the middle third contains a mix of both muscle types. By cutting specific nerves in experimental animals, researchers established that peristalsis in the upper third of the esophagus occurs when a series of nerve cells that originate at the base of the brain sends electrical signals in a precise sequence. Each of these nerves terminates at skeletal muscle at a different location along the esophagus. Some of these muscles wrap around the esophagus; others run along its length.

In response to nerve signals, the muscle contracts or relaxes in a coordinated fashion (**Figure 43.8**). In this way, a wave of muscle contractions propagates down the tube, propelling the food mass ahead of it. The action of these nerves is not the result of a conscious choice. The system is a **reflex**—an automatic reaction to a stimulus—that is stimulated by the act of swallowing.

Even after decades of experiments, though, the mechanism of peristalsis in the lower third of the esophagus is still not well understood. Research is continuing.

A MODIFIED ESOPHAGUS: THE BIRD CROP Almost all animals have a tubelike esophagus that connects the mouth and the stomach. But in an array of bird species, the esophagus has a prominent, widened segment called the **crop** where food can be stored and, in some cases, processed.

The structure and function of the crop varies among the bird species that have one. In many groups, the crop is a simple sac that holds food and regulates its flow into the stomach. In these species, the crop is interpreted as an adaptation that allows individuals to eat a large amount in a short time, then retreat to a safe location while digestion occurs.

The crop has evolved into a digestive organ, though, in two leaf-eating species that are not closely related. Leaves are difficult

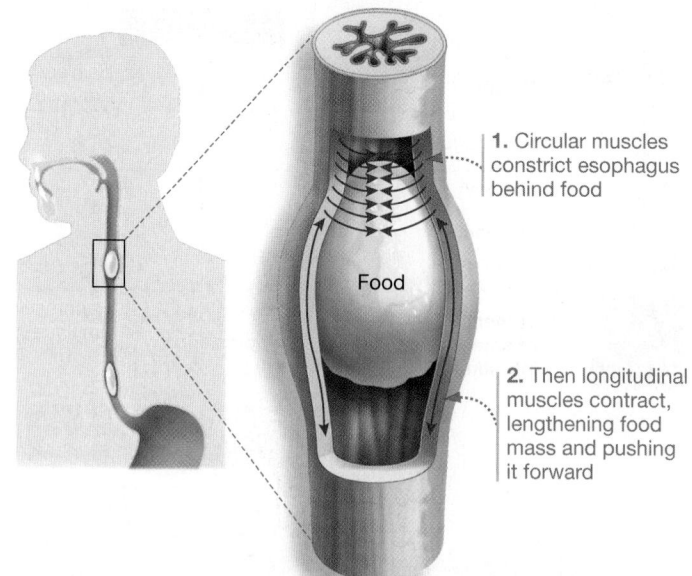

1. Circular muscles constrict esophagus behind food

Food

2. Then longitudinal muscles contract, lengthening food mass and pushing it forward

FIGURE 43.8 In the Esophagus, Peristalsis Transports Food to the Stomach. Alternating contractions of the circular and longitudinal muscles in the esophageal wall constrict the lumen of the esophagus behind food and expand the lumen in front of food. The wave of contraction begins at the oral end of the esophagus and propels the food toward the stomach.

to digest because they contain a large amount of cellulose (see Chapter 5). In the enlarged crop of species called the hoatzin and kakapo, bacteria that are capable of breaking down cellulose perform digestion. The bacterial cells, along with the fatty acids that result from bacterial metabolism, leak out of the crop and are used as food by the bird.

The Stomach

Although little if any digestion occurs in the esophagus of most animals, the situation changes dramatically when food reaches the stomach. The vertebrate **stomach** is a tough, muscular pouch in the digestive tract, bracketed on both ends by ringlike muscles called **sphincters**, which control movement of material through the gut (**Figure 43.9**).

When a meal fills the stomach, muscular contractions result in churning that mixes the contents and reduces food to a uniform consistency and solute concentration, or osmolarity; a certain amount of mechanical breakdown of food also occurs in response to this churning. The other main function of the stomach is the partial digestion of proteins.

Compared with the mouth or esophagus (or virtually any other tissue), the lumen of the stomach is highly acidic. Early researchers documented this fact by analyzing vomit or the contents of sponges that were tied to strings, swallowed, and pulled back up; chemists confirmed that the predominant acid in the stomach is hydrochloric acid (HCl).

Not long after, a physician named William Beaumont established that digestion takes place in the stomach. He reached this

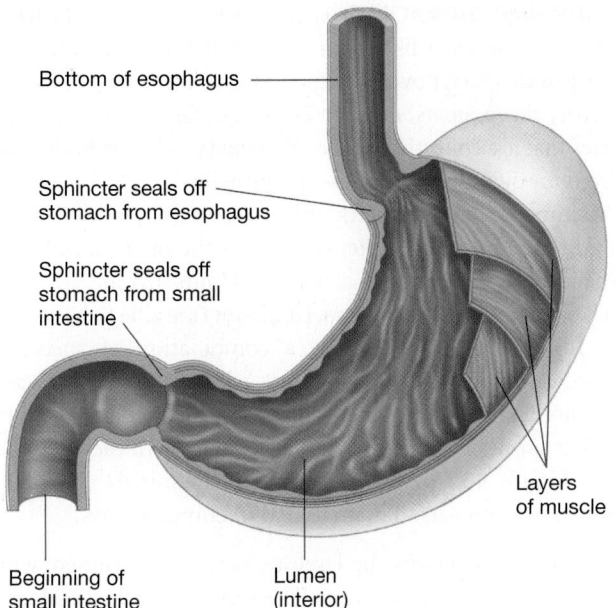

FIGURE 43.9 **The Stomach Is a Muscular Outpocketing of the Digestive Tract.** The stomach encloses an acidic environment. Muscular contractions mix food and break it into smaller pieces.

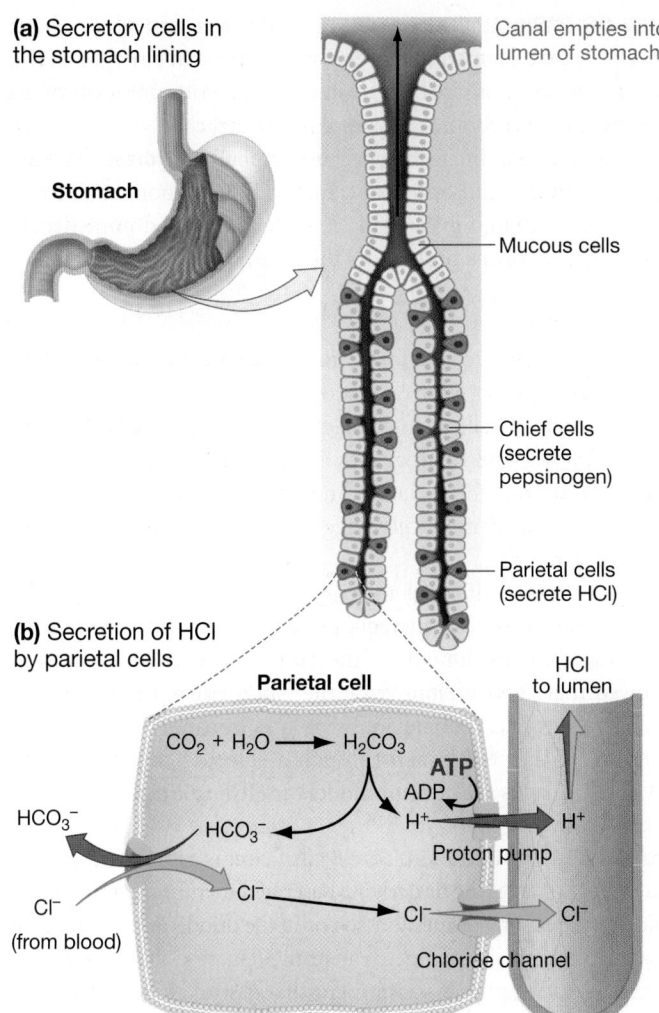

FIGURE 43.10 **Cells in the Stomach Lining Secrete Mucus, Pepsinogen, and Hydrochloric Acid.**

conclusion through an extraordinary series of experiments on a young man named Alexis St. Martin.

THE STOMACH AS A SITE OF PROTEIN DIGESTION In 1822, when St. Martin was 19 years old, a shotgun accidentally discharged into his abdomen and created a series of wounds. Despite repeated attempts, Beaumont was unable to close a hole in St. Martin's stomach. Eventually Beaumont inserted a small tube through the opening; the tube remained in St. Martin's body for the rest of his life. (Today, biologists insert tubes into various parts of the digestive tract of cows or sheep to study how these animals digest different types of feed.)

With the tube in place, Beaumont was able to tie a string onto small pieces of meat or vegetables, insert the food directly into St. Martin's stomach, and draw it out after various intervals. Beaumont also removed liquid from inside the stomach and observed how this gastric (stomach) juice acted on food in vitro. His experiments showed that gastric juice digests food—particularly meat.

Theodor Schwann later purified the enzyme that is responsible for digesting proteins in the stomach and named it **pepsin**. Because it destroys proteins, biologists hypothesized that pepsin must be synthesized and stored in cells while it is in an inactive form—otherwise it would kill the cells that make it.

In 1870 a biologist established through careful microscopy that granules occur in specialized stomach cells called chief cells. These granules were hypothesized to be a pepsin precursor. Follow-up work showed that this hypothesis was correct. The precursor compound found in chief cells, which came to be called **pepsinogen**, is converted to active pepsin by contact with the acidic environment of the stomach.

Secretion of a protein-digesting enzyme in inactive form is important: It prevents destruction of proteins in the cells where the enzyme is synthesized.

WHICH CELLS PRODUCE STOMACH ACID? The acidic environment of the human stomach denatures (unfolds) proteins and makes it possible for pepsin to work efficiently. But where does the acid come from?

Researchers who were studying the anatomy of the stomach wall noticed clusters of distinctive **parietal cells** located in pits that communicate with the lumen of the stomach (**Figure 43.10a**). An investigator also documented that the shape and activity of these cells appeared to vary as the digestion of a meal proceeded. On the basis of these observations, he inferred that parietal cells are the source of the HCl in gastric juice, which may have a pH as low as 1.5.

Earlier microscopists had shown that another type of cell, called a **mucous cell**, secretes the mucus that is found in gastric juice. Mucus lines the gastric epithelium and protects the stomach from damage by HCl. To summarize, these anatomical studies showed that the epithelium of the stomach contains several types of secretory cells, each of which is specialized for a particular function.

HOW DO PARIETAL CELLS SECRETE HCl? The first clues about how parietal cells manufacture hydrochloric acid emerged in the late 1930s, when a researcher found a high concentration of an enzyme called carbonic anhydrase in parietal cells.

This result was interesting because **carbonic anhydrase** catalyzes the formation of carbonic acid (H_2CO_3) from carbon dioxide and water. In solution, the carbonic acid that is formed immediately dissociates to form a proton and the bicarbonate ion (HCO_3^-):

$$CO_2 + H_2O \rightleftharpoons H_2CO_3 \rightleftharpoons H^+ + HCO_3^-$$

A second clue to the formation of HCl came in the 1950s, when transmission electron microscopes allowed researchers to analyze parietal cells at high magnification (see **BioSkills 10** in Appendix A). The micrographs showed that parietal cells are packed with mitochondria. Because mitochondria produce ATP, the structure of parietal cells suggested that they might function in active transport.

Later work confirmed this hypothesis by showing that the protons formed by the dissociation of carbonic acid are actively pumped into the lumen of the stomach. Subsequent studies showed that chloride ions from the blood enter parietal cells in exchange for bicarbonate ions, via a cotransport protein, and then move into the lumen through a chloride channel. **Figure 43.10b** diagrams the current model for HCl production.

ULCERS AS AN INFECTIOUS DISEASE An **ulcer** is a hole in an epithelium that damages the underlying basement membrane and tissues. Ulcers in the lining of the stomach or in the duodenum—the initial section of the small intestine—can result in intense abdominal pain.

For decades, physicians thought that gastric and duodenal ulcers resulted from the production of excess acid in the stomach. They treated ulcers by prescribing alkaline compounds (bases—see Chapter 2) that neutralized hydrochloric acid in the stomach.

In 1983, however, Robin Warren and Barry Marshall published data indicating that ulcers were associated with infections from a bacterium called *Helicobacter pylori*. Instead of being caused by environmental influences like acid-rich diets and psychological factors like anxiety, Warren and Marshall hypothesized that ulcers were an infectious disease.

This hypothesis met with intense skepticism, for two reasons: (**1**) the low acidity of the stomach was thought to sterilize the environment, and (**2**) treating ulcers as a bacterial rather than environmental disease required a radical change in thinking. In science, radical changes in thinking require exceptionally high standards of evidence.

To help provide this evidence, Marshall performed an experiment on himself. After a colleague cultured *H. pylori* in his lab, Marshall drank fluid from a petri plate where the bacteria were growing. As predicted, he developed gastritis—an inflammation of the stomach that precedes development of ulcers.

This experiment fulfilled several requirements of Koch's postulates for linking an organism to a specific disease (see Chapter 28). Subsequent work, in labs throughout the world, helped cement the link between *H. pylori* infection and ulcers. Physicians now routinely prescribe antibiotics to relieve ulcers.

THE RUMINANT STOMACH It is common for animals to have a stomach or stomach-like organ. The structure and function of this organ can vary, however, depending on the nature of the diet. In cattle, sheep, goats, deer, antelope, giraffe, and pronghorn—species that are collectively called **ruminants**—the stomach is specialized for digesting cellulose—not proteins.

Mammals, like birds, lack the enzymes called cellulases required to digest cellulose. Yet cellulose is the main carbohydrate in the leaves, stems, and twigs that ruminants ingest.

Like the hoatzin and kakapo, ruminants are able to harvest energy from cellulose thanks to a combination of specialized anatomical structures and symbiotic relationships with bacteria and unicellular protists. When **symbiosis** occurs, members of two different species live in close physical contact with each other.

As **Figure 43.11** shows, ruminant mammals have four-chambered stomachs that are folded in complex ways.

1. Food initially enters the largest chamber, the rumen, which serves as a fermentation vat. The rumen is packed with symbiotic bacteria and protists. These organisms have enzymes capable of breaking apart the chemical bonds in cellulose, yielding glucose. The rumen is an oxygen-free environment, and the symbiotic organisms produce ATP from this glucose via fermentation, releasing fatty acids as a by-product (see Chapter 9).

2. The chamber adjacent to the rumen, called the reticulum, is similar in function. After plant material has been partially digested in the rumen and the reticulum, the animal regurgitates portions of that material into its mouth, forming a cud. The ruminant chews that regurgitated material further to enhance mechanical breakdown, then re-swallows it.

3. Processed food that moves out of the rumen enters the third chamber, or omasum, where water is removed.

4. The final chamber is the abomasum, which contains the ruminant's own digestive enzymes and corresponds to a true stomach.

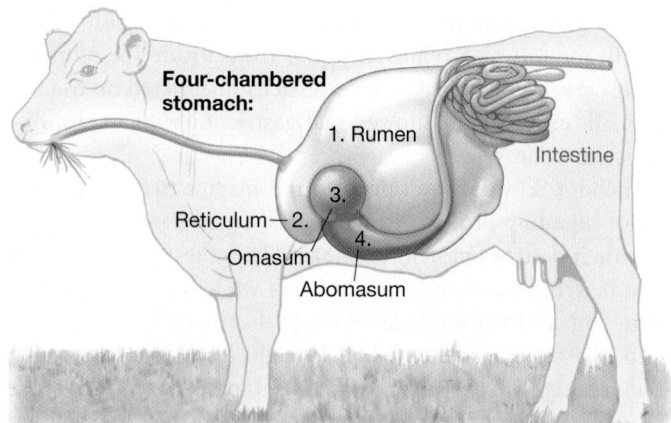

FIGURE 43.11 Ruminant Stomachs Facilitate the Digestion of Cellulose by Symbiotic Organisms. Ruminants obtain many of their nutrients from symbiotic bacteria and protists that live in the rumen and reticulum chambers of the stomach.

Most of a ruminant's food consists of (1) fatty acids and other compounds that are released as waste products of fermentation reactions in symbiotic organisms, and (2) the symbiotic cells themselves.

THE AVIAN GIZZARD The avian gizzard is another prominent type of modified stomach. Birds do not have teeth and cannot chew food into small pieces. Instead, most species swallow sand and small stones that lodge in the gizzard. As this muscular sac contracts, food is pulverized by the grit.

The gizzard is particularly large and strong in bird species that eat coarse foods such as seeds and nuts. The gizzard of a wild turkey, for example, can crack large walnuts.

Like the crop, the gizzard is interpreted as an adaptation that allows birds to ingest food quickly—by avoiding the need to chew—and digest it later. Biologists invoke the same hypothesis to explain why ruminants chew the cud. The ability to regurgitate material and finish chewing, while hiding in a place safe from predators, is thought to increase fitness.

The Small Intestine

In humans, the stomach is responsible for mixing the contents of a meal into a homogenous slurry, mechanically breaking up food material, and providing the acid and enzymes required to partially digest proteins. Peristalsis in the stomach wall then moves small amounts of material through the valve created by a sphincter muscle at the base of the stomach and into the small intestine.

The **small intestine** is a long tube that is folded into a compact space between the stomach and the last major section of the digestive tract—the large intestine. In the small intestine, partially digested food mixes with secretions from the pancreas and the liver and begins a journey of about 6 m (20 ft). When passage through this structure is complete, digestion is finished and most nutrients—along with large quantities of water—have been absorbed.

FOLDING AND PROJECTIONS INCREASE SURFACE AREA The surface area available for nutrient and water absorption in the small intestine is nothing short of remarkable. As **Figure 43.12** shows, the organ's epithelial tissue is folded and covered with fingerlike projections called **villi** (singular: **villus**). In turn, the cells that line the surface of villi have tiny projections on their apical surfaces called **microvilli** (singular: **microvillus**). Microvilli project into the lumen of the digestive tract.

If the small intestine lacked folds, villi, and microvilli, it would have a surface area of about 3300 cm² (3.6 ft²). Instead, the epithelium covers about 2 million cm² (over 2200 ft²)—an area about the size of a tennis court.

The enormous surface area of the small intestine increases the efficiency of nutrient absorption. And because each villus contains blood vessels and a lymphatic vessel called a **lacteal**, nutrients pass quickly from epithelial cells into the body's transport systems. (The circulatory system and lymphatic system are analyzed in detail in Chapters 44 and 49, respectively.)

✔If you understand the importance of surface area in the small intestine, you should be able to explain why surface area is so much higher in this structure than it is in the stomach or esophagus.

To understand how digestion is completed and absorption occurs, let's explore what happens to proteins, lipids, and carbohydrates as they move through this section of the digestive tract—again using humans as a model organism.

PROTEIN PROCESSING BY PANCREATIC ENZYMES The acidic environment of the stomach denatures proteins, destroying their secondary and tertiary structures. In addition, pepsin cleaves the peptide bonds next to certain amino acids, reducing long polypeptides to relatively small chains of amino acids. In the small intestine, protein digestion is completed so that individual amino acids can enter the bloodstream and be transported to cells throughout the body.

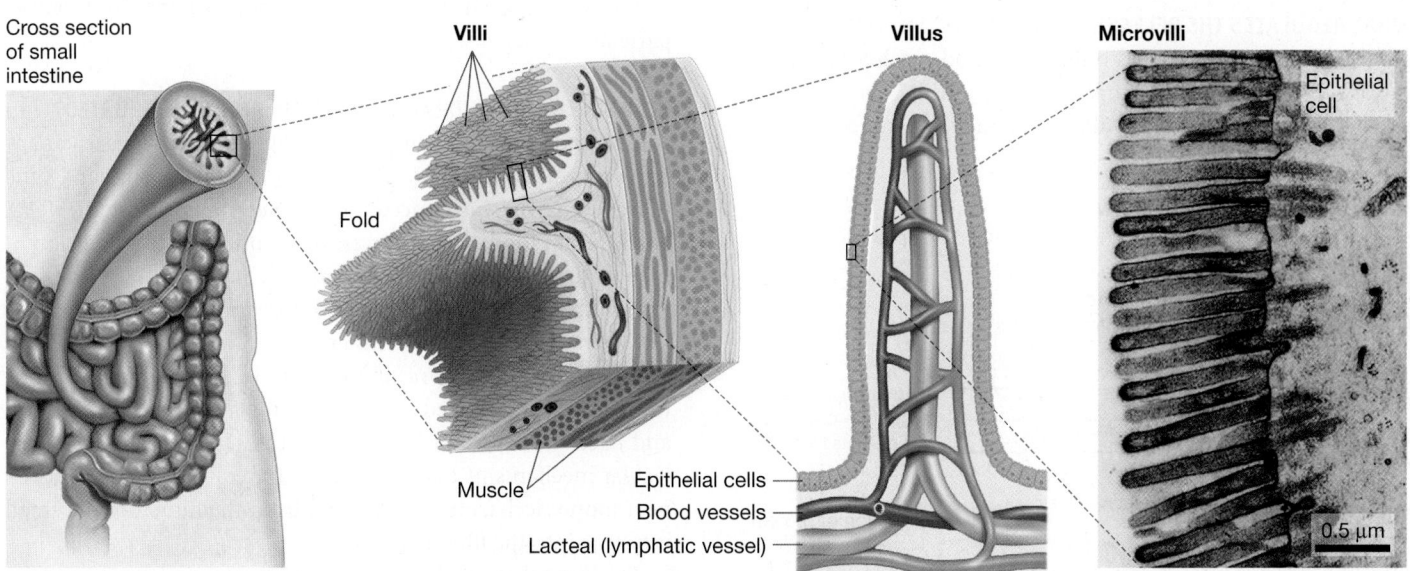

FIGURE 43.12 The Small Intestine Has an Extremely Large Surface Area. The villi that project from folds in the small intestine are covered with microvilli (colorized brown in the micrograph at the far right).

How does protein digestion occur? By the end of the nineteenth century, it had been established that enzymes in the small intestine digest polypeptides to monomers. Later work showed that each of these protein-digesting enzymes, or **proteases**, is specific to certain types or configurations of amino acids in a polypeptide chain. Thus, a suite of proteases is required to completely digest polypeptides to amino acid monomers.

In addition, by 1900 biologists had determined that proteases are synthesized in an inactive form in the **pancreas**, which is connected to the small intestine by the pancreatic duct. Like the production of inactive pepsinogen by chief cells in the stomach, the production of digestive enzymes in an inactive conformation prevents pancreatic cells from digesting themselves.

It took decades of work to understand how pancreatic enzymes are activated in the small intestine. In 1900 a researcher showed that contact with juice from the upper part of the small intestine activates pancreatic enzymes. Because activation did not occur when he heated the intestinal juice, and because heat denatures proteins, he hypothesized that the agent responsible for activating the pancreatic enzymes was also an enzyme. He called the unknown enzyme enterokinase.

Decades later, a researcher succeeded in purifying a pancreatic enzyme called **trypsinogen** and demonstrated that enterokinase activates it in vitro. Enterokinase activates trypsinogen by phosphorylating it, resulting in the active enzyme **trypsin**. Trypsin then triggers the activation of other protein-digesting enzymes, such as chymotrypsin, elastase, and carboxypeptidase. These enzymes are also synthesized by the pancreas and secreted in an inactive form.

Figure 43.13 summarizes the sequential activation of digestive enzymes in the small intestine. Enterokinase triggers the activation of trypsinogen to trypsin, which in turn triggers the activation of the three other protein-digesting enzymes. Once these enzymes are activated in the upper reaches of the small intestine, each begins cleaving specific peptide bonds. Eventually polypeptides are broken up into amino acid monomers.

WHAT REGULATES THE RELEASE OF PANCREATIC ENZYMES? Digestive enzymes are needed only when food reaches the small intestine. Based on this simple observation, it was logical to predict that their release would be carefully controlled.

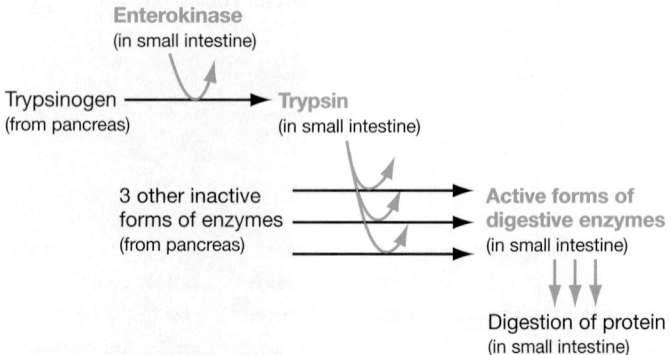

FIGURE 43.13 Enterokinase Triggers an Enzyme-Activation Cascade in the Small Intestine.

A classic experiment by William Bayliss and Ernest Starling, published in 1902, established how pancreatic enzymes are controlled. Bayliss and Starling began by cutting the nerves that connect to the pancreas and small intestine of a dog. Electrical signaling between the two organs was now impossible. But when the researchers introduced a weak HCl solution into the upper reaches of the animal's small intestine, to simulate the arrival of material from the stomach, its pancreas secreted in response.

This observation was startling: The small intestine had successfully signaled the pancreas that food had arrived, even though the nerves connecting the two organs had been cut.

Starling hypothesized that a chemical messenger must be involved, and that the chemical messenger must originate in the small intestine and travel to the pancreas via the blood. He tested this idea by cutting off a small piece of the small intestine, grinding it up, and injecting the resulting solution into a vein in the animal's neck. Minutes later, the pancreas sharply increased secretion.

Bayliss and Starling had discovered the first **hormone**—a chemical messenger that influences physiological processes at very low concentrations. The molecule they detected, which they called **secretin**, is produced by the small intestine in response to the arrival of food from the stomach.

Follow-up work showed that secretin's primary function is to induce a flow of bicarbonate ions (HCO_3^-) from the pancreas to the small intestine. The bicarbonate is important because it neutralizes the acid arriving from the stomach.

Researchers also discovered a second hormone produced in the small intestine, called cholecystokinin (pronounced *ko-la-sis-toe-KIN-in*), that induces secretion from the liver as well as the pancreas. **Cholecystokinin** (literally, "bile-bag-mover") stimulates the secretion of digestive enzymes from the pancreas and the secretion of molecules from the gallbladder that are involved in processing lipids.

Hormones are involved in stomach function as well. For example, after being stimulated by nerves or the arrival of food, certain stomach cells produce the hormone **gastrin**. In response, parietal cells begin secreting HCl.

HOW ARE CARBOHYDRATES DIGESTED AND TRANSPORTED? In addition to manufacturing protein-digesting enzymes, the pancreas produces nucleases and an amylase that is similar to the salivary enzyme introduced earlier. **Nucleases** digest the RNA and DNA in food; pancreatic amylase continues the digestion of carbohydrates that began in the mouth. Carbohydrate digestion ends with the production of monosaccharides such as glucose.

When digestion of proteins and carbohydrates is complete in the small intestine, the resulting slurry is a treasure trove of nutrients and water mixed with indigestible plant fibers from food and bacterial cells that live symbiotically in the gut. What molecular mechanisms make it possible for epithelial cells to transport monosaccharides like glucose from the lumen of the small intestine into the bloodstream?

It is logical to predict two general principles about nutrient absorption: **(1)** It is highly selective, meaning proteins in the plasma membranes of microvilli are responsible for bringing specific nutrients into the cell; and (2) it is active, meaning it requires an expenditure of ATP to bring nutrients into the epithelium across a concentration gradient.

Work over the past several decades has shown that both predictions are correct. One of the key results grew out of a series of experiments during the 1980s, which established that glucose absorption depends on the presence of an electrochemical gradient favoring an influx of sodium ions into the epithelium. Based on this observation, biologists hypothesized that the apical membranes of these cells must contain a series of cotransporters—proteins that would bring a nutrient molecule into the cell along with sodium ions. Work focused on glucose, because it is such a fundamentally important nutrient. To confirm that a sodium-glucose cotransporter exists, investigators set out to find the gene that codes for the hypothesized membrane protein.

The researchers began by purifying mRNAs from rabbit intestinal cells (**Figure 43.14**), which presumably were transcribing the cotransporter genes. Then the team separated the mRNAs by size via gel electrophoresis (see **BioSkills 9** in Appendix A) and injected one of each type of mRNA into a series of frog eggs—cells that do not normally transport glucose. The frog cells translated the rabbit mRNAs into proteins.

When tested, one of the experimental eggs was able to import Na^+ and glucose in tandem. Based on these data, the biologists inferred that this egg had received the mRNA for the rabbit Na^+-glucose cotransporter. Using techniques introduced in Chapter 19, the researchers made a DNA copy of the mRNA, determined the sequence of the gene, and inferred the amino acid sequence of the membrane protein.

The discovery of the Na^+-glucose cotransporter inspired a three-step model for glucose absorption:

1. Na^+/K^+-ATPase (sodium-potassium pumps) in the basolateral membrane of the epithelial cells creates an electrochemical gradient that favors the entry of Na^+.

2. Glucose from digested food enters the cell along with sodium via the cotransporter.

3. Glucose diffuses into nearby blood vessels through a glucose carrier in the basolateral membrane.

If this configuration of pumps, cotransporters, and carriers sounds familiar, there is a good reason: The same combination of membrane proteins occurs in the proximal tubule of the kidney, where the proteins are responsible for the reabsorption of sodium and glucose from urine (see Chapter 42).

Follow-up work showed that in the small intestine—just as in the proximal tubule—other cotransporters are responsible for the absorption of other nutrients, with specific channels and carriers in the basolateral membrane responsible for their transport to the blood.

DIGESTING LIPIDS: BILE AND TRANSPORT The pancreatic secretions include digestive enzymes that act on fats as well as proteins and carbohydrates. Like the lingual lipase added to saliva in the mouth, the

EXPERIMENT

QUESTION: How is glucose transported into epithelial cells of the small intestine?

HYPOTHESIS: Glucose enters epithelial cells along with sodium ions via a Na^+-glucose cotransporter protein.

NULL HYPOTHESIS: Glucose transport does not depend on Na^+ transport.

EXPERIMENTAL SETUP:

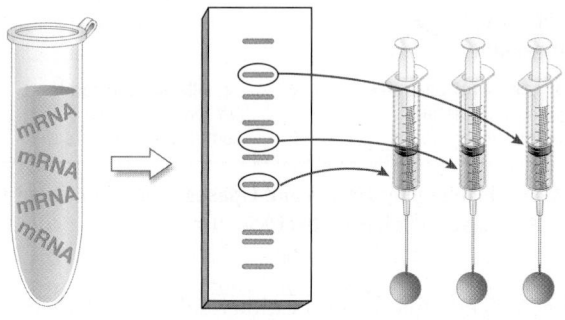

1. Purify mRNA from intestinal cells.

2. Separate mRNAs by size via gel electrophoresis.

3. Inject individual mRNAs into frog eggs. Test each egg—can it absorb Na^+ and glucose?

PREDICTION: An egg will be able to absorb Na^+ and glucose, because it received the mRNA that codes for the Na^+-glucose cotransporter.

PREDICTION OF NULL HYPOTHESIS: None of the eggs will be able to absorb Na^+ and glucose.

RESULTS:

Na^+, glucose in medium	Na^+, glucose in medium	Na^+, glucose in medium
Egg 1	Egg 2	Egg 3
ABSORBED	Not absorbed	Not absorbed

CONCLUSION: The egg that absorbs Na^+ and glucose received the mRNA from the Na^+-glucose cotransporter gene.

FIGURE 43.14 The Experimental Protocol for Locating the Na^+-Glucose Cotransporter Gene.

SOURCE: Wright, E. M. 1993. The intestinal Na^+/glucose cotransporter. *Annual Review of Physiology* 55: 575–589.

✔**QUESTION** Why did the researchers inject the RNAs into frog eggs, instead of into rabbit epithelial cells?

enzyme **pancreatic lipase** breaks certain bonds present in complex fats and results in the release of fatty acids and other small lipids.

Recall from Chapter 6 that fats are insoluble in water. As a result, they tend to form large globules as they are churned in the stomach. Before pancreatic lipase can act, large fat globules that emerge from the stomach must be broken up—a process known as **emulsification**.

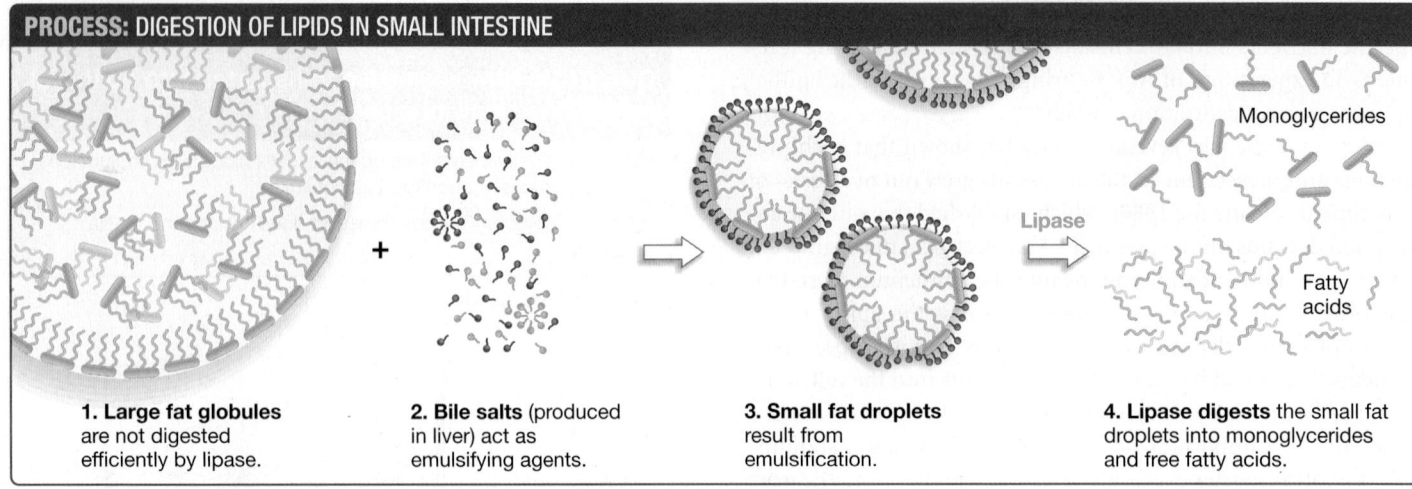

PROCESS: DIGESTION OF LIPIDS IN SMALL INTESTINE

Monoglycerides

Lipase

Fatty acids

1. Large fat globules are not digested efficiently by lipase.

2. Bile salts (produced in liver) act as emulsifying agents.

3. Small fat droplets result from emulsification.

4. Lipase digests the small fat droplets into monoglycerides and free fatty acids.

FIGURE 43.15 Emulsifying Agents and Lipases Digest Lipids in the Small Intestine. Once bile salts break up large fat globules, lipase can digest fats efficiently.

In the small intestine, emulsification results from the action of small lipids called bile salts. As **Figure 43.15** shows, bile salts function like the detergents that researchers use to break up the lipids in plasma membranes (see Chapter 6).

Bile salts are synthesized in the **liver**, an organ that performs an array of functions related to digestion, and secreted in a complex solution called **bile**, which is stored in the **gallbladder**. When bile enters the small intestine, it raises the pH and emulsifies fats. Once fats are broken into small globules with high surface area, they can be attacked by enzymes and digested. **Table 43.3** summarizes the major digestive enzymes.

What happens to the monoglycerides and fatty acids released by lipase activity? An answer emerged when researchers injected radioactively labeled fatty acids into the small intestines of laboratory rats. Most of the radioactive label entered epithelial cells and attached to a protein named **fatty-acid binding protein**. Later, other researchers established that a fatty-acid binding protein also occurs in the membranes of these cells.

Once fatty-acid binding proteins bring lipids into the cell, they are processed into protein-coated globules called **chylomicrons**. Chylomicrons diffuse into lacteals—such as the green vessel in Figure 43.12—near the epithelial cells. The lacteals merge with larger lymph vessels, which merge with larger blood vessels. In this way, fats enter the bloodstream without clogging small blood vessels. Eventually, the products of fat digestion end up in adipose tissue and other tissues.

HOW IS WATER ABSORBED? When solutes from digested material are brought into the epithelium of the small intestine via active transport, water follows passively by osmosis. This is an important mechanism for (1) absorbing water that has been ingested, and (2) reclaiming liquid that was secreted into the digestive tract in the form of saliva, mucus, and pancreatic fluid.

This mechanism of water absorption inspired an important medical strategy called oral rehydration therapy. If a patient has diarrhea, clinicians frequently prescribe dilute solutions of glucose and electrolytes to be taken orally. When the glucose in the drink is absorbed in the small intestine through sodium-glucose cotransporters, enough water and sodium follow to prevent the life-threatening effects of dehydration. This simple medication saves thousands of lives every year. ✔If you understand this concept, you should be able to predict two effects of a molecule that selectively blocks the sodium-glucose cotransporter.

To review how carbohydrates, proteins, and fats are digested and absorbed, go to the study area at *www.masteringbiology.com*.

 Web Activity The Digestion and Absorption of Food

The Cecum and Appendix

Some birds house cellulose-digesting bacteria and protists in their crop, and ruminants house cellulose-digesting symbionts in a modified stomach. But in rabbits, many rodents, some marsupials, and certain leaf-eating primates, fermentation by symbionts occurs in an organ called the cecum.

The **cecum** is an outpocketing of the digestive tract located at the start of the large intestine. A dead-ended or "blind" sac, the cecum is greatly enlarged in species that use it as a fermentation chamber for processing cellulose. Rabbits and some other species that ferment food in the cecum are also able to excrete the structure's contents in pellets, which are then re-ingested and passed through the digestive tract a second time. This practice is called **coprophagy** ("excrement-eating").

In humans, the cecum is dramatically reduced in size compared to most other primates, and functions in defense against invading bacteria and viruses instead of as a fermentation vat. Because its size and function differ from those of a cecum, it is called the **appendix**.

The Large Intestine

By the time digested material reaches the large intestine of a human, a large amount of water (~5 liters per day in total) and virtually all of the available nutrients have been absorbed. The primary function of the **large intestine** is to compact the wastes that remain and absorb enough water to form feces.

Digestion is accomplished by the enzymes listed here, by HCl produced in the stomach in response to the hormone gastrin, and by bile salts from the liver. Bile salts are stored in the gallbladder. They are released in response to the hormone cholecystokinin and emulsify fats in the small intestine.

	Where Synthesized	Regulation	Function
Carboxypeptidase	Pancreas	Released in response to cholecystokinin from small intestine; activated by trypsin	In small intestine, breaks peptide bonds in polypeptides—releasing amino acids
Chymotrypsin	Pancreas	Released in response to cholecystokinin from small intestine; activated by trypsin	In small intestine, breaks peptide bonds in polypeptides—releasing amino acids
Elastase	Pancreas	Released in response to cholecystokinin from small intestine; activated by trypsin	In small intestine, breaks peptide bonds in polypeptides—releasing amino acids
Lingual lipase	Salivary glands	Released in response to taste and smell stimuli	In mouth, breaks bonds in fats—releasing fatty acids and monoglycerides
Nucleases	Pancreas	Released in response to cholecystokinin from small intestine	In small intestine, break apart nucleic acids—releasing nucleotides
Pancreatic amylase	Pancreas	Released in response to cholecystokinin from small intestine	In small intestine, breaks apart carbohydrates—releasing sugars
Pancreatic lipase	Pancreas	Released in response to cholecystokinin from small intestine	In small intestine, breaks bonds in fats—releasing fatty acids and monoglycerides
Pepsin	Stomach	Released in inactive form (pepsinogen); activated by low pH in stomach lumen	In stomach, breaks peptide bonds between certain amino acids in proteins—releasing polypeptides
Salivary amylase	Salivary glands	Released in response to taste and smell stimuli	In mouth, breaks apart carbohydrates—releasing sugars
Trypsin	Pancreas	Released in inactive form (trypsinogen) in response to cholecystokinin from small intestine; activated by hormone enterokinase from small intestine	In small intestine, breaks specific peptide bonds in polypeptides—releasing amino acids

These processes occur in the **colon**—the main section of the structure. Feces are held in the **rectum**, which is the final part of the large intestine, until they can be excreted. Although the kidneys are responsible for maintaining water balance in the body, water absorption in the large intestine is important for individuals to remain well hydrated.

AQUAPORINS PLAY A KEY ROLE IN WATER REABSORPTION To identify the mechanism of water absorption in the large intestine, researchers have focused on aquaporins, which were introduced in Chapter 6. Recall that **aquaporins** are water channels in plasma membranes that provide a mechanism for increasing the rate of water movement via osmosis. The best studied of these proteins, called AQP1 for aquaporin 1, is common in the nephrons of the kidney. AQP1 is one of the proteins responsible for water reabsorption from urine along the osmotic gradient described in Chapter 42.

To date, four distinct aquaporins have been found in the large intestines of rats, mice, and humans. The aquaporins called AQP3 and AQP4, for example, are located in the basolateral membrane of cells in the epithelium of the large intestine in mice. Researchers are working to unravel exactly how the activity of these aquaporins is regulated. To date, the molecular mechanisms responsible for water absorption in the large intestine are not as well understood as are those in the collecting duct of nephrons.

VARIATION IN STRUCTURE AND FUNCTION It is not unusual for animal species to lack a colon and rectum entirely. For example, Chapter 42 pointed out that in insects, the posteriormost portion of the digestive tract, called the hindgut, functions to reabsorb water and ions and form feces. Among vertebrates, the various lineages of fish have no large intestine at all.

In contrast, horses, elephants, tapirs, and certain other plant-eating species have extremely large colons. In these species, the colon is packed with symbiotic bacteria and protists that digest cellulose and ferment the glucose that is released. Plant-eating animals can house cellulose-digesting symbionts in the esophagus (crop), stomach, cecum, or large intestine.

Another striking structural variation occurs in the rectum. In amphibians, reptiles, and birds, the urine that is produced by the kidneys empties into an enlarged portion of the large intestine called the **cloaca**. Instead of having two orifices for excretion, these species have one.

CHECK YOUR UNDERSTANDING

43.4 Nutritional Homeostasis— Glucose as a Case Study

When digestion is complete, amino acids, fatty acids, ions, and sugars enter the bloodstream and are delivered to the cells that need them. Too much of a nutrient, or too little, can be problematic or even fatal, however.

The illness **diabetes mellitus** is a classic example of nutrient imbalance. People with diabetes experience abnormally high levels of glucose in their blood. Over the course of a lifetime, chronically elevated blood glucose can lead to an array of complications including blindness, impaired circulation, and heart failure. What causes the imbalance?

The Discovery of Insulin

In 1879, researchers removed the pancreas from a dog and observed that diabetes developed. This experiment suggested that the pancreas secretes a compound involved in removing glucose from the blood.

When other investigators cut up pancreatic tissues and injected extracts into diabetic dogs, however, they observed no response. Frustrated, they realized that digestive enzymes in the pancreas were probably destroying the active agent during the extraction process.

Then in 1921, Frederick Banting and Charles Best conducted a breakthrough experiment:

1. They began by tying off a dog's pancreatic duct—the tube where digestive enzymes collect and flow to the small intes-

tine. The logic was that blocking the secretion of digestive enzymes might kill the cells that synthesize them.

2. The investigators waited several weeks for the cells near the duct to die and then removed the gland.

3. They froze the pancreatic tissue, ground it up, and injected an extract into a diabetic dog.

4. To their delight, the dog's blood-sugar levels stabilized, and the dog became more active and healthy looking.

After Banting and Best repeated the experiment and observed the same result, they grew increasingly confident that they had located the source of the molecule responsible for controlling diabetes. The molecule came to be called insulin.

Insulin's Role in Homeostasis

Insulin is a hormone that is produced in the pancreas when blood-glucose levels are high. It travels through the bloodstream and binds to receptors on cells throughout the body (see Chapter 47 for more detail on the structure and function of hormones and other chemical signals).

In response, cells that have insulin receptors increase their rate of glucose uptake and processing. Specifically:

- Cells in the liver and skeletal muscle synthesize more glycogen from glucose monomers.
- Cells that store lipids synthesize more storage forms of fat, using glucose as a precursor.

As a result, glucose levels in the blood decline.

If blood-glucose levels fall, as they do after hard exercise or when food is lacking, cells in the pancreas secrete a hormone called **glucagon**. In response to glucagon, cells in the liver catabolize glycogen and produce glucose via **gluconeogenesis** (the synthesis of glucose from non-carbohydrate compounds). As a result, glucose levels in the blood rise (**Figure 43.16**).

Insulin and glucagon interact to form a negative feedback system capable of achieving homeostasis with respect to glucose concentrations in the blood. To review this homeostatic system, go to the study area at *www.masteringbiology.com*.

 BioFlix™ Homeostasis: Regulating Blood Sugar, **Web Activity** Understanding Diabetes Mellitus

Diabetes Can Take Several Forms

Diabetes mellitus develops in people who (**1**) do not synthesize insulin or (**2**) have defective versions of the receptor for insulin. The first condition is called type 1 diabetes mellitus, or insulin-dependent diabetes; the second condition is type 2 diabetes mellitus, or non-insulin-dependent diabetes. In both cases, effector cells do not receive the signal that would result in a drop in blood-glucose levels.

Glucose imbalance has a direct effect on urine formation. Normally, signals from insulin keep blood-glucose levels low

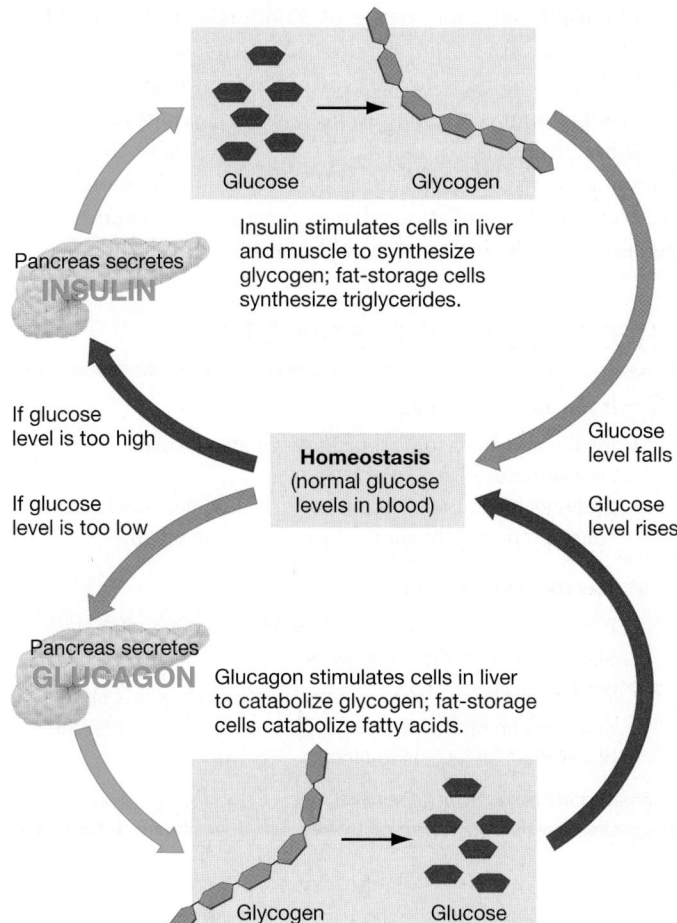

FIGURE 43.16 Insulin and Glucagon Provide Negative Feedback in a Homeostatic System. Both insulin and glucagon are secreted by cells in the pancreas but have opposite effects on blood-glucose concentrations.

✔**QUESTION** Which arrow is disrupted in individuals with type 2 diabetes mellitus? Which arrow is disrupted in individuals with type 1 diabetes mellitus?

enough that all of the glucose can be reabsorbed from the filtrate formed in the kidney. But when blood glucose levels are high, so much glucose enters the nephron that it cannot all be reabsorbed. High glucose concentrations in the filtrate increase its osmolarity and prevent water from being reabsorbed by osmosis. More water leaves the body, leading to high urine volume in both types of diabetes.

Recall from Chapter 42 that a disease called diabetes insipidus develops in people whose collecting ducts fail to reabsorb water. As a result, these individuals also produce large amounts of urine.

Production of copious urine occurs in all three types of diabetes, and inspired the illness's names. The word diabetes means "to run through"; water "runs through" people with diabetes. In addition, the word insipidus means "tasteless"; mellitus means "honeyed (sweet)." Before chemical methods of analyzing urine were available, physicians would distinguish between diabetes in-

sipidus and diabetes mellitus by tasting the patient's urine. The test was definitive because glucose levels in the urine differ markedly in the two types of illness.

Currently, type 1 diabetes mellitus is treated with insulin injections and careful attention to diet; type 2 diabetes is managed primarily through prescribed diets and monitoring blood-glucose levels, as well as taking drugs that increase cellular responsiveness to insulin. The challenge is to achieve homeostasis with respect to blood glucose in the absence of the body's normal regulatory mechanisms.

✔If you understand the difference between type 1 and type 2 diabetes mellitus, you should be able to explain how, in individuals with each disease and without disease, insulin receptors in the plasma membrane of liver cells interact with insulin and how the liver cells respond to high glucose concentrations in the blood.

The Type 2 Diabetes Mellitus Epidemic

An epidemic of diabetes mellitus is currently under way in certain human populations. In the United States, for example, the incidence of diabetes in adults jumped 90 percent between 1997 and 2007.

In the U.S., about 10.7 percent of individuals aged 20 and over are diabetic. When only African Americans in this age group are considered, the prevalence of diabetes jumps to 14.7 percent. In both instances, about 90–95% of cases are type 2 diabetes.

Because there is a strong association between the prevalence of type 2 diabetes mellitus in parents and children, researchers have long suspected that some individuals have a genetic predisposition for developing the disease. Using techniques introduced in Chapters 19 and 20, researchers have now identified alleles at four different genes that predispose individuals to type 2 diabetes.

However, there is also strong evidence that environmental conditions have an important impact on the incidence of type 2 diabetes. As an example, consider the Pima Indians of North America.

The Pima consist of two main populations—one in southwest Arizona in the United States and one in a remote area of the Sierra Madre mountains of Mexico. The bars graphed in **Figure 43.17a** on page 858 indicate the average percent of individuals with type 2 diabetes, by gender, in Mexican people who are not members of the Pima and in the two populations of Pima. Although researchers have been unable to find any significant genetic differences between the two groups of Pima, the data indicate there are dramatic differences in the incidence of type 2 diabetes mellitus.

The incidence of type 2 diabetes is correlated with obesity. As the data graphed in **Figure 43.17b** show, Pima from the U.S. are more likely to be obese than individuals in the other two populations in this study. A person who has a body mass index greater than or equal to 30 is considered obese. The **body mass index** is calculated as weight (kg) divided by height (m) squared.

The differences in body mass index among the Pima are explained by differences in physical activity: Pima men in Mexico

(a) Variation in average incidence of type 2 diabetes mellitus

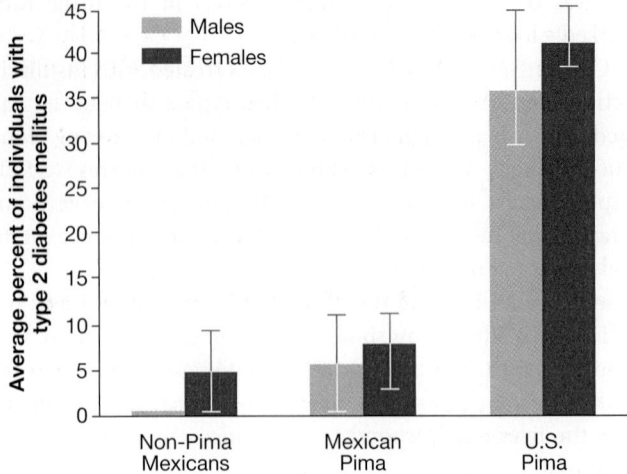

(b) Variation in average body mass index (kg/m^2)

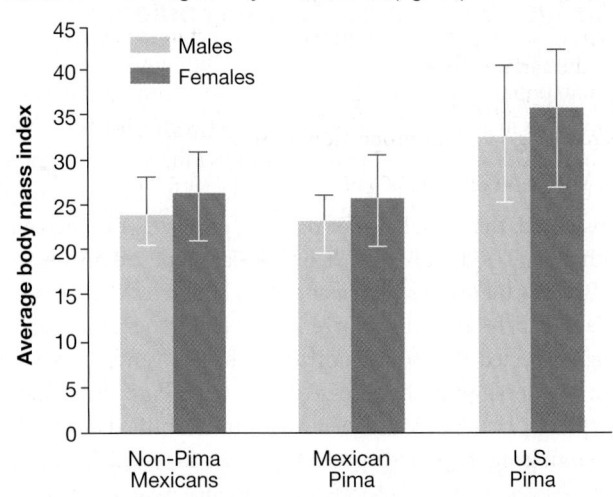

are physically active an average of 33 hours a week, while their counterparts in the United States are active just 12.1 hours per week. Pima women in Mexico average 22 hours of physical activity a week; Pima women in the United States average just 3.1.

Although Pima people in the United States have the highest rates of type 2 diabetes mellitus recorded to date, the same trends are occurring in populations all over the world. Nutrition-related diseases have become a major public health concern.

CHECK YOUR UNDERSTANDING

If you understand that . . .

- Insulin and glucagon interact to regulate glucose concentrations in the blood.
- Diabetes mellitus results from a lack of homeostasis in the concentration of the nutrient glucose in the blood.

✓ You should be able to . . .

1. Explain the similarities and differences between diabetes insipidus and diabetes mellitus.
2. Predict what a person with type 1 diabetes mellitus should do when the blood-glucose level is too high and when the blood-glucose level is too low.

Answers are available in Appendix B.

FIGURE 43.17 The Incidence of Type 2 Diabetes Mellitus Is Correlated with Obesity.

✓QUESTION In many human populations, incidence of type 2 diabetes mellitus has been increasing rapidly. Why?

CHAPTER 43 REVIEW

For media, go to the study area at www.masteringbiology.com

Summary of Key Concepts

Animals require an array of nutrients to stay healthy, including specific amino acids, vitamins, and elements, as well as organic compounds that act as building blocks in chemical synthesis or have high potential energy.

- The diets of animals include fats and carbohydrates that provide energy, proteins that furnish amino acids, vitamins that serve as coenzymes and perform other functions, ions required for water balance and for nerve and muscle function, and elements that are incorporated into molecules synthesized by cells.

- To determine the levels of nutrients that are needed for a particular animal species to sustain normal activities, researchers monitor the relationship between nutrient intake, the levels of nutrients maintained in the body, and health.

  ✓ You should be able to design an experiment to document the effects of magnesium deprivation in laboratory mice.

There is usually a close correspondence between the structure of an animal's mouthparts and the function of those mouthparts in capturing and processing food.

- All animals are heterotrophs.

- Most animals are mass feeders that obtain food by seizing and manipulating it with mouthparts such as teeth, jaws, or beaks or with special toxin-injecting organs.

- Natural selection has modified mouthparts in different species to act as efficient tools for obtaining particular types of food.

  ✓ You should be able to predict the types of mouthparts and the nature of the digestive tract in a mammal that eats only nectar from flowers.

Digestion occurs in the digestive tract, which is compartmentalized into organs that have specialized functions in the ingestion

and digestion of food, absorption of nutrients and water, or excretion of wastes.

- In most animals, the digestive tract begins at the mouth and ends at the anus.

- In many species, chemical digestion of food also begins in the mouth. In mammals, an enzyme in saliva called amylase begins to hydrolyze the bonds linking glucose monomers in starch, glycogen, and other carbohydrates.

- Once food is swallowed, it passes down the esophagus via peristalsis.

- Digestion continues in the stomach. In humans, the stomach is a highly acidic environment that denatures proteins and in which the enzyme pepsin begins the cleavage of peptide bonds that link amino acids.

- Food passes from the stomach into the small intestine, where it is mixed with secretions from the pancreas and liver.

- Secretions from the liver and pancreas are triggered by the hormones cholecystokinin and secretin, which are produced in the small intestine.

- In the small intestine, carbohydrate digestion is continued by pancreatic amylase; fats are emulsified by bile salts and digested by lipase; protein digestion is completed by a suite of pancreatic enzymes that are activated by enterokinase.

- Cells that line the small intestine absorb the nutrients released by digestion. In many cases, uptake is driven by an electrochemical gradient established by Na^+/K^+-ATPase that favors a flow of Na^+ into the cell.

- As solutes leave the lumen of the small intestine and enter cells, water follows by osmosis.

- Water reabsorption is completed in the large intestine, where feces form.

- The structure of organs in the digestive tract varies widely among species, in ways that support efficient processing of the food a particular species ingests.

  ✓ You should be able to explain why gastric bypass surgery, which makes the stomach smaller and allows food to bypass part of the small intestine, often leads to weight loss.

  (MB) **Web Activity** The Digestion and Absorption of Food

🔑 **Lack of homeostasis with respect to nutrients such as glucose can cause disease.**

- Maintaining homeostasis with respect to nutrients is critical to health.

- Diabetes mellitus develops when concentrations of glucose in the blood are chronically too high.

- Type 1 diabetes mellitus is caused by a defect in the production of insulin—a hormone secreted by the pancreas that promotes the uptake of glucose from the blood.

- Type 2 diabetes mellitus is caused by a defect in the insulin receptor on the surface of cells.

- The development of type 2 diabetes is correlated with obesity and is reaching epidemic proportions in some populations.

  ✓ You should be able to explain why blood glucose concentration eventually falls when a person with diabetes mellitus eats a candy bar.

  (MB) **BioFlix™** Homeostasis: Regulating Blood Sugar, **Web Activity** Understanding Diabetes Mellitus

Questions

1. What does secretin stimulate?
   a. secretion of HCO_3^- from the pancreas
   b. secretion of HCl from the stomach epithelium
   c. secretion of digestive enzymes from the pancreas
   d. uptake of glucose from the bloodstream by cells throughout the body

2. What is the structure and function of the cecum in rabbits?
   a. a large, blind sac that functions as a fermentation vat
   b. an enlarged section of the esophagus that functions as a fermentation vat
   c. one of four chambers in the stomach; functions as a fermentation vat
   d. enlarged portion of the large intestine that functions as a fermentation vat

3. What is an incomplete digestive tract?
   a. a digestive tract that is missing a major compartment—for example, fish lack a large intestine
   b. a ruminant-type tract, where food is regurgitated and re-chewed before digestion is completed
   c. a reduced or vestigial tract like that found in animals that absorb nutrition directly across the body wall
   d. a digestive system with a single opening for ingestion and excretion

4. In mammals, how and where are carbohydrates digested?
   a. by lipases in the small intestine
   b. by pepsin and HCl in the stomach
   c. by aquaporins in the large intestine
   d. by amylases in the mouth and small intestine

5. What role do bile salts play in the digestion of complex fats?
   a. They catalyze the cleavage of bonds leading to the release of fatty acids and other small lipids.
   b. They emulsify lipids, meaning that large masses of fat molecules are broken into smaller masses.
   c. They include fatty-acid binding proteins, which are involved in fat absorption.
   d. They activate the enzymes that are responsible for digesting fats.

6. How is water absorbed in the small intestine of mammals?
   a. through aquaporins
   b. The exact mechanism is not known.
   c. by sodium cotransporters
   d. by osmosis (following solutes)

1. Explain how the structure of each of the following digestive tract compartments correlates with its function: the bird crop, cow rumen, and elephant large intestine.

2. Why is it logical that digestive enzymes are produced in an inactive form and then activated in the lumen of the digestive tract?

3. Do you accept the conclusion that cichlid pharyngeal jaws are adaptations that increase feeding efficiency? Why or why not?

4. Fish lack a large intestine. What is the adaptive significance of this trait?

5. Explain why oral rehydration therapy works.

6. Explain how insulin and glucagon provide negative feedback in a homeostatic system.

1. Predict how the nutritional requirements of female mammals change during pregnancy and breastfeeding.

2. Predict the problems that would result from defects in each of the following molecules: pancreatic amylase, pepsin, fatty-acid binding protein, and aquaporins.

3. Scientists who backed the hypothesis that secretion from the pancreas is under nervous control strenuously objected to the experiment that led to the discovery of hormonal control. They claimed that the experiment was inconclusive because it was very likely that not all of the nerves that contact the small intestine had been cut. The biologists who did the experiment replied that even if not all the nerves had been cut, the result was still valid. In your opinion, who is correct? Why?

4. Among vertebrates, the large intestine occurs only in lineages that are primarily terrestrial (amphibians, reptiles, and mammals). Propose a hypothesis to explain this observation.

During intense exercise, animal circulatory systems deliver large amounts of oxygen to tissues and remove large amounts of carbon dioxide. This chapter explores how gas exchange occurs in animals that live in aquatic or terrestrial environments.

Gas Exchange and Circulation

44

Animal cells are like factories that run 24 hours a day. Inside the plasma membrane, the chemical reactions that sustain life produce a steady stream of wastes. Those reactions also require a steady input of raw materials.

Chapter 42 analyzed how waste materials are removed from the body and excreted; Chapter 43 examined how nutrients enter the body. This chapter focuses on two major questions:

1. How are two of the most important molecules in the economy of the cell—the oxygen (O_2) required for cellular respiration and the carbon dioxide (CO_2) produced by cellular respiration—exchanged with the environment?

2. How are these gases—along with wastes, nutrients, and other types of molecules—transported throughout the body?

Understanding gas exchange and circulation is fundamental to understanding how animals work. If either process fails, the consequences are dire. Consider the maladies that can develop when gas exchange or circulation are disrupted in humans: anemia, pneumonia, tuberculosis, malaria, heart disease, and stroke are just a few examples.

Let's begin with an overview of animal respiratory and circulatory systems, then plunge into the details of how gases are exchanged and transported.

44.1 The Respiratory and Circulatory Systems

When the mitochondria inside animal cells are producing ATP via cellular respiration, they consume oxygen and produce carbon dioxide. To support continued ATP production, cells have to obtain oxygen and expel excess carbon dioxide continuously (see **The Big Picture: Energy**, located after Chapter 10).

KEY CONCEPTS

- Animals have to take in oxygen and expel carbon dioxide to sustain cellular respiration and stay alive. Terrestrial animals and aquatic animals face different challenges in performing gas exchange.

- Gas-exchange organs maximize the rate of O_2 and CO_2 diffusion by (1) presenting a large, thin surface area to the environment, and (2) maintaining a steep partial-pressure gradient that favors entry of O_2 and elimination of CO_2.

- Blood is a specialized tissue that transports gases, along with nutrients and wastes, in some animals. Hemoglobin is an oxygen-carrying protein that is extremely efficient at taking up oxygen in the lungs and delivering it to tissues.

- Circulatory systems use the pressure generated by one or more hearts to transport blood and other substances throughout the body.

✔ When you see this checkmark, stop and test yourself. Answers are available in Appendix B.

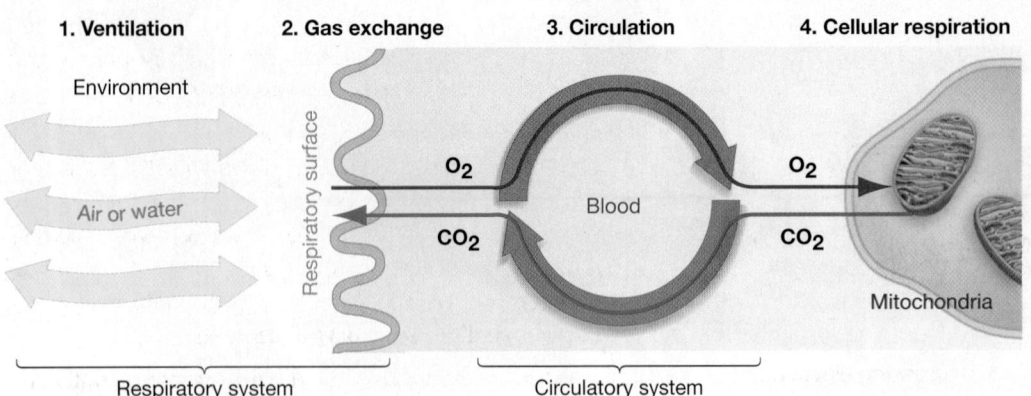

FIGURE 44.1 Gas Exchange Involves Ventilation, Circulation, and Respiration. In animals, oxygen and carbon dioxide are exchanged across the surface of a lung, a gill, the skin, or some other gas-exchange organ. In many species these gases are transported to and from cells—where gas exchange again takes place—in a fluid tissue such as blood.

How does gas exchange occur between an animal's environment and its mitochondria? In most cases, gas exchange involves the four steps illustrated in **Figure 44.1**:

1. Ventilation occurs when air or water moves through a specialized gas-exchange organ, such as lungs or gills.

2. Gas exchange takes place as O_2 and CO_2 diffuse between air or water and the blood at the respiratory surface.

3. Through circulation, the dissolved O_2 and CO_2 are transported throughout the body—along with nutrients, wastes, and other types of molecules—via the circulatory system.

4. Gas exchange between blood and cells occurs in tissues, where cellular respiration has led to low O_2 levels and high CO_2 levels.

Steps 1 and 2 are accomplished by the **respiratory system**: the collection of cells, tissues, and organs responsible for gas exchange between the individual and its environment. In essence, a respiratory system consists of structures for conducting air or water to a surface where gas exchange takes place.

In some animals the gas-exchange surface is the skin, but in most species it is located in a specialized organ like the lungs of tetrapods, the tracheae of insects, or the gills found in mollusks, arthropods, and fish. Section 44.3 analyzes the structure and function of gills, tracheae, and lungs in detail.

Step 3 in Figure 44.1 is usually accomplished by a **circulatory system**, which moves O_2, CO_2, and other materials around the body. In most species, a specialized, liquid transport tissue is propelled throughout the body by a muscular heart, through a system of vessels.

Given this broad overview of gas exchange and circulatory systems, let's plunge into the details of how they work—starting with the question of how oxygen and carbon dioxide move between an animal's body and its environment.

44.2 Air and Water as Respiratory Media

Gas exchange between the environment and cells is based on diffusion. Under normal conditions, oxygen concentrations are relatively high in the environment and low in tissues, while carbon dioxide levels are relatively high in tissues and low in the environment. Thus, oxygen tends to move from the environment into tissues, and carbon dioxide tends to move from tissues into the environment.

How much oxygen and carbon dioxide are present in the atmosphere versus the ocean? What factors influence how quickly these gases move by diffusion?

How Do Oxygen and Carbon Dioxide Behave in Air?

As **Figure 44.2a** shows, the atmosphere is composed primarily of nitrogen (N_2) and oxygen, with trace amounts of argon and CO_2. Nitrogen and argon are not important to animals living at sea level and are usually ignored in analyses of gas exchange.

The data in Figure 44.2a are actually slightly misleading, however. To understand why, consider that the percentage of O_2 in the atmosphere does not vary with elevation. The atmosphere at the top of Mt. Everest is composed of 21 percent oxygen just as it is at sea level. The key difference is that far fewer molecules of oxygen and other atmospheric gases are present at high elevations than at sea level. Air at the top of Mt. Everest is less dense than air at sea level, so less oxygen is present.

To understand how gases move by diffusion, it is important to express their presence in terms of partial pressures instead of percentages. Pressure is a type of force. A **partial pressure** is the pressure of a particular gas in a mixture of gases.

To calculate the partial pressure of a particular gas, multiply the fractional composition of that gas by the total pressure exerted by the entire mixture.[1]

For example, **Figure 44.2b** shows that the total atmospheric pressure at sea level is 760 mm Hg (millimeters of mercury). If you multiply this value by 0.21, which is the fraction of air that is O_2, you obtain a partial pressure of oxygen, abbreviated P_{O_2}, at sea level of 160 mm Hg. Because the atmospheric pressure is only about 250 mm Hg at the top of Mt. Everest, the P_{O_2} there is only $0.21 \times 250 = 53$ mm Hg.

---

[1]This calculation is valid because the total pressure in a mixture of gases is the sum of the partial pressures of all the individual gases. This relationship is called Dalton's law.

(a) Which gases make up the atmosphere?

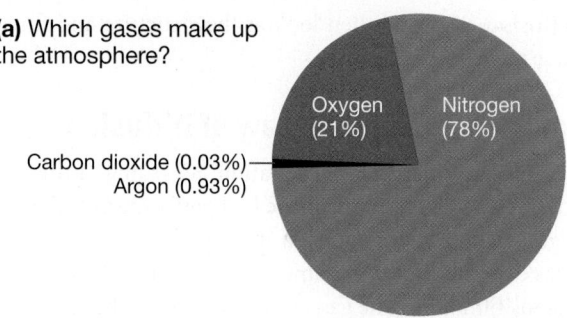

Oxygen (21%)
Nitrogen (78%)
Carbon dioxide (0.03%)
Argon (0.93%)

(b) The partial pressure of oxygen falls with increasing elevation.

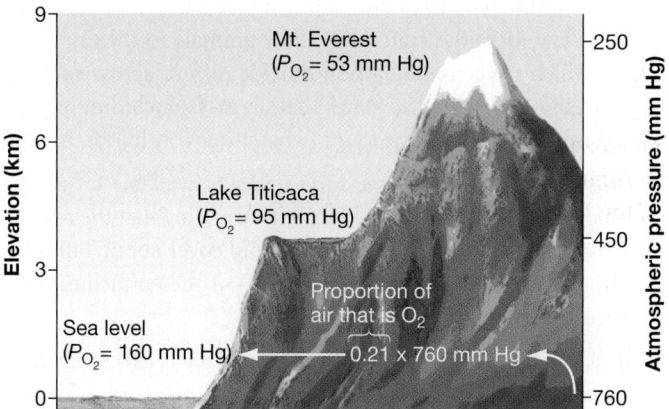

Mt. Everest (P_{O_2} = 53 mm Hg)

Lake Titicaca (P_{O_2} = 95 mm Hg)

Proportion of air that is O_2

Sea level (P_{O_2} = 160 mm Hg) — 0.21 × 760 mm Hg

FIGURE 44.2 Oxygen Makes Up 21 Percent of the Atmosphere, but Its Partial Pressure Varies Widely. (a) Earth's atmosphere is dominated by nitrogen and oxygen. Oxygen makes up 21 percent of the atmosphere at all elevations. **(b)** Atmospheric pressure, and thus oxygen partial pressure (P_{O_2}), fall with increasing elevation.

Oxygen and carbon dioxide diffuse between the environment and cells along their respective partial-pressure gradients, just as solutes diffuse along their concentration gradients. In both air and water, O_2 and CO_2 move from regions of high partial pressure to regions of low partial pressure. It is hard to breathe at the top of Mt. Everest because the partial pressure of oxygen is low—meaning that the diffusion gradient between the atmosphere and your lung tissues is small.

How Do Oxygen and Carbon Dioxide Behave in Water?

To obtain oxygen, water breathers face a much more challenging environment than air breathers do. Aquatic animals live in an environment that contains much less oxygen than the environments inhabited by terrestrial animals. At 15°C, a liter of air can contain up to 209 mL of O_2, while a liter of water may contain a maximum of only 7 mL of O_2. To extract a given amount of oxygen, an aquatic animal has to process 30 times more water than the amount of air a terrestrial animal breathes.

In addition, water is about a thousand times denser than air and flows much less easily. As a result, water breathers have to expend much more energy to ventilate their respiratory surfaces than do air breathers.

WHAT AFFECTS THE AMOUNT OF GAS IN A SOLUTION? Oxygen and carbon dioxide diffuse into water from the atmosphere, but the amount of gas that dissolves depends on several factors:

- *Solubility of the gas in water* Oxygen has very low solubility in water. Only 0.003 mL of oxygen dissolves in 100 mL of water for each 1-mm-Hg increase in oxygen partial pressure. Because of this low solubility, blood contains a molecule that binds to oxygen and delivers it to tissues. Without this carrier molecule, the rate of blood flow to tissues would have to increase dramatically to meet oxygen demand.

- *Temperature of the water* As the temperature of water increases, the amount of gas that dissolves in it decreases. Other things being equal, warm-water habitats have much less oxygen available than cold-water habitats do. For a fish, breathing in warm water is comparable to a land-dwelling animal breathing at high elevation.

- *Presence of other solutes* Because seawater has a much higher concentration of solutes than does freshwater, seawater can hold less dissolved gas. At 10°C, up to 8.02 mL of O_2 can be present per liter of freshwater versus only 6.35 mL of O_2 per liter of seawater. As a result, freshwater habitats tend to be more oxygen rich than marine environments.

- *Partial pressure of the gas in contact with the water* Gases move from regions of high partial pressure to regions of low partial pressure. So if the partial pressure in a liquid exceeds that in the adjacent gas, the gas will "boil" out of the liquid. This is what happens when the cap is removed from a bottle of carbonated beverage. The partial pressure of carbon dioxide in the newly opened drink is much higher than it is in the atmosphere.

WHAT AFFECTS THE AMOUNT OF OXYGEN AVAILABLE IN AN AQUATIC HABITAT? The partial pressure of oxygen varies in different types of aquatic habitats, just as it varies with altitude on land. In addition to the four factors listed above, there are other important considerations that impact oxygen's availability in water.

For example, habitats with large numbers of photosynthetic organisms tend to be relatively oxygen rich. In contrast, habitats where most organisms live off existing organic material tend to be oxygen poor, because cellular respiration depletes the available oxygen.

The most important issue in oxygen availability, however, is often surface area. Surface area has a large impact on oxygen's ability to diffuse into water. For example:

- Shallow ponds and streams tend to be much better oxygenated than deep bodies of water, because shallower bodies have a higher ratio of surface area to volume.

- Unless currents mix water almost continuously, water near the surface has much higher oxygen content than water near the bottom of the same habitat.

- Rapids, waterfalls, and other types of whitewater are the most highly oxygenated of all aquatic environments, because a large surface area is exposed to the atmosphere as water splashes over rocks and logs and because air bubbles are incorporated into the water.

- Oxygen content is extremely low in bogs and other stagnant-water habitats, because the small amounts of oxygen that diffuse into stagnant water are quickly used up by decomposers that use oxygen as an electron acceptor in cellular respiration.

Now let's consider the structure and function of ventilatory organs. How do the gills of fish, the tracheae of insects, and the lungs of mammals cope with the differences between air and water?

CHECK YOUR UNDERSTANDING

If you understand that . . .

- O_2 and CO_2 move from regions of high partial pressure to regions of low partial pressure.
- The partial pressure of oxygen in a body of water depends largely on its surface area and temperature.
- Water breathing is much more difficult than air breathing, in part because the partial pressure of oxygen in water is much lower than its partial pressure in air.

✓ **You should be able to . . .**

1. Explain why oxygen partial pressures are relatively high in mountain streams and relatively low at the ocean bottom.
2. Explain whether a large, medium, or small amount of air should be bubbled into aquaria containing warm-water species, vigorous algal growth, or sedentary animals.

Answers are available in Appendix B.

44.3 Organs of Gas Exchange

Many small animals lack specialized gas-exchange organs, such as gills or lungs. Instead, they obtain O_2 and eliminate CO_2 by diffusion across the body surface. This is possible because their size and shape give them an extraordinarily high surface area to volume ratio (see Chapter 41). For sponges, jellyfish, flatworms, and other species, diffusion across the body surface is rapid enough to fulfill their requirements for taking in O_2 and expelling CO_2.

Most of these animals are restricted to living in wet environments, however. For gas exchange to take place efficiently, the surface where it occurs has to be thin, and thin skin is prone to water loss. Living in wet or humid environments allows animals to exchange gases across their outer surface while avoiding dehydration.

In contrast, animals that are large or that live in dry habitats need some sort of specialized respiratory organ. Respiratory organs provide a greater surface area for gas exchange—enough to meet the demands of a large body filled with cells. In terrestrial animals, respiratory organs are located inside the body, which helps to minimize water loss.

Biologists have long marveled at the efficiency of gills and lungs. To appreciate why, let's examine the physical factors that

control diffusion rates and then look at the structure and function of these respiratory organs.

Physical Parameters: The Law of Diffusion

In 1855 Adolf Fick derived an equation regarding diffusion, based on the results of experiments he had performed on the behavior of gases. **Fick's law of diffusion** states that the rate of diffusion of a gas depends on five parameters: the solubility of the gas in the aqueous film lining the gas-exchange surface, the temperature, the surface area available for diffusion, the difference in partial pressures of the gas across the gas-exchange surface, and the thickness of the barrier to diffusion (**Figure 44.3**).

Fick's law identifies traits that allow animals to maximize the rate at which oxygen and carbon dioxide diffuse across surfaces. Specifically, Fick's law states that all gases, including O_2 and CO_2, diffuse in the largest amounts when three conditions are met:

1. *A* is large, meaning a large surface area is available for gas exchange. Based on Fick's law, it is not surprising that the respiratory surface in the human lungs would cover about 140 m²—about a quarter of a basketball court—if the epithelium were spread flat.

2. *D* is small, meaning the respiratory surface is extremely thin. To appreciate just how thin the gas-exchange surface is, consider that researchers who asked bicyclists to ride up a steep hill for seven minutes observed a marked increase in blood on the respiratory surface of the athletes' lungs. To explain this observation, the researchers proposed that the cyclists' high heart rates (up to 177 beats/minute during their sprint up the hill) increased blood pressure to the point at which thin-walled vessels in the lungs ruptured and leaked blood into the structure.

3. $P_2 - P_1$ is large, meaning the partial-pressure gradient of the gas across the surface is large. High partial-pressure gradients are maintained in part by having an efficient circulatory system in close contact with the gas-exchange surface. When blood flows close to the respiratory surface, oxygen is rapidly taken away from the area where inward diffusion is occurring, and carbon dioxide is rapidly brought into the area where outward diffusion is occurring. As a result, $P_2 - P_1$ stays high.

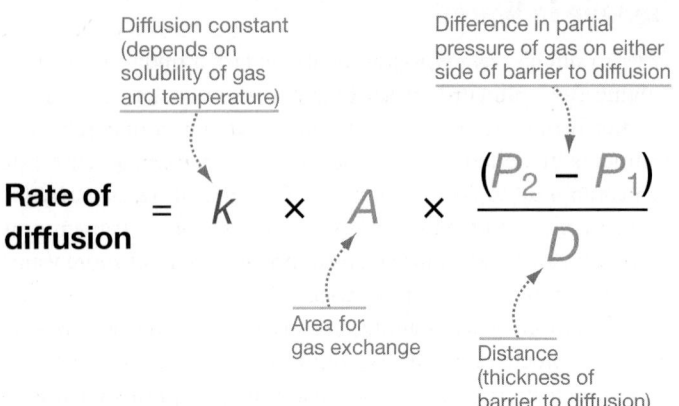

$$\text{Rate of diffusion} = k \times A \times \frac{(P_2 - P_1)}{D}$$

Diffusion constant (depends on solubility of gas and temperature)

Difference in partial pressure of gas on either side of barrier to diffusion

Area for gas exchange

Distance (thickness of barrier to diffusion)

FIGURE 44.3 Fick's Law of Diffusion.

(a) External gills are in direct contact with water.

External gills

5 mm

(b) Internal gills must have water brought to them.

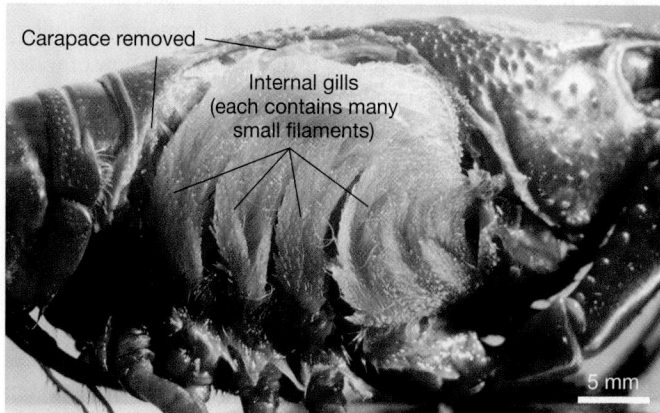

Carapace removed

Internal gills
(each contains many
small filaments)

5 mm

FIGURE 44.4 Gills Can Be External or Internal. (a) Nudibranchs are marine snails with gills that are outside the main body wall. Their gills are protected by toxins. **(b)** Crayfish gills are located inside the main body wall. A portion of the crayfish exoskeleton has been removed to expose the gills.

✔**QUESTION** What are the advantages and disadvantages of external versus internal gills in terms of fitness?

What other aspects of gill and lung structure affect diffusion rate? To answer this question, let's delve into the anatomy of these respiratory organs.

How Do Fish Gills Work?

Gills are outgrowths of the body surface or throat that are used for gas exchange in aquatic animals. Gills are efficient solutions to the problems posed by water breathing, primarily because they present a large surface area for oxygen to diffuse across a thin epithelium.

In some species of invertebrates, such as the nudibranch mollusk (**Figure 44.4a**), gills project from the body surface and contact the surrounding water directly. In other invertebrate species, such as the crayfish (**Figure 44.4b**), gills are located inside the exoskeleton or body wall. If gills are internal, water must be driven over them by cilia, the limbs, or other specialized structures.

In contrast to the diversity of gills found in invertebrates, the gills of bony fishes are all similar in structure. Fish gills are located on both sides of the head and in teleosts (see Chapter 34) consist of four arches, as **Figure 44.5** shows.

HOW DO FISH VENTILATE THEIR GILLS? To move water through their gills so gas exchange can take place, most fish open and close their mouths and a stiff flap of tissue, the **operculum**, that covers the gills. The pumping action of the mouth and operculum creates a pressure gradient that moves water over the gills.

In contrast, tuna and other fish that are particularly fast swimmers force water through their gills by swimming with their mouths open. This process is called ram ventilation.

Regardless of how fish gills are ventilated, water flows in one direction over gills. More specifically, water passes through long, thin structures called **gill filaments** that extend from each gill arch. Each gill filament is composed of hundreds or thousands of **gill lamellae**. Gill lamellae are sheetlike structures, shown in detail at the bottom of Figure 44.5. Note that a bed of small blood vessels called capillaries runs through each lamella.

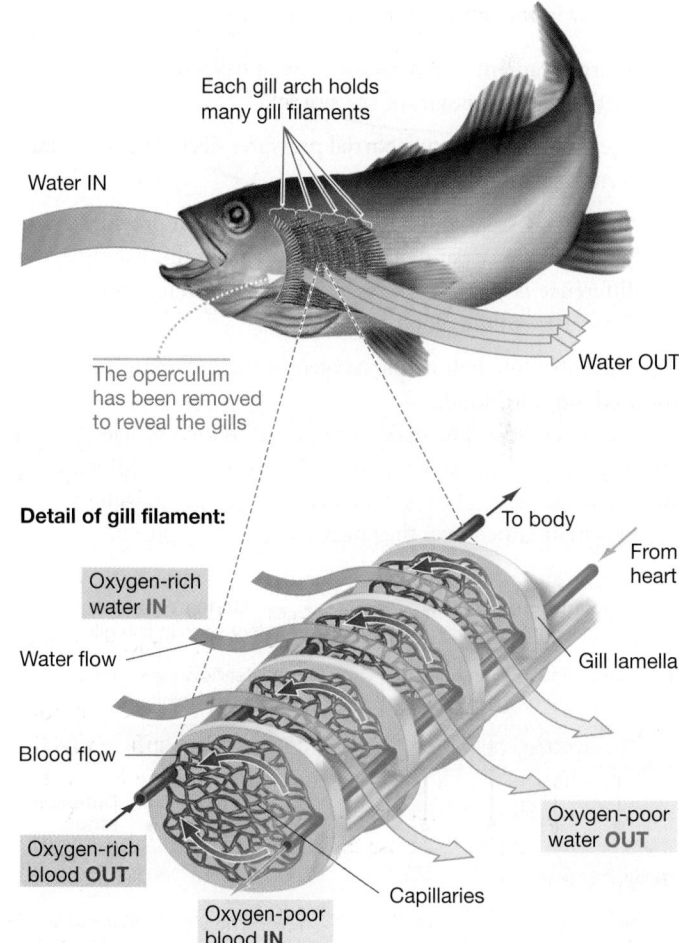

Each gill arch holds
many gill filaments

Water IN

Water OUT

The operculum
has been removed
to reveal the gills

Detail of gill filament:

To body

From heart

Oxygen-rich
water **IN**

Water flow

Gill lamella

Blood flow

Oxygen-rich
blood **OUT**

Oxygen-poor
water **OUT**

Capillaries

Oxygen-poor
blood **IN**

FIGURE 44.5 Fish Gills Are a Countercurrent Exchange System. In the lamellae of fish gills, water and blood flow in opposite directions.

THE FISH GILL IS A COUNTERCURRENT SYSTEM The one-way flow of water through gill lamellae has a profound impact on gill function, for a simple reason: The flow of blood through the capillary bed in each lamella is in the opposite direction to the flow of water. As a result, each lamella functions as a countercurrent exchanger.

Recall from Chapter 41 that countercurrent exchangers are based on two adjacent fluids flowing in opposite directions. **Figure 44.6** illustrates why the countercurrent flow is so critical.

Note two key points about the left side of the figure, where water and blood flow in opposite directions:

1. A slight gradient in partial pressure of oxygen (here, 10 percent) exists along the entire length of the lamella.

2. A large difference in oxygen partial pressure exists between the start and end of the system. In this example, the difference is 100% − 15% = 85% in the water and 90% − 5% = 85% in the blood.

The upshot? Most of the oxygen in the incoming water has diffused into the blood.

Now look at the right side of the figure, where water and blood flow in the same direction.

1. A large gradient in partial pressure of oxygen (here, 100 percent) exists at the start of the system.

2. The gradient in oxygen partial pressures declines rapidly and eventually disappears.

3. A relatively small difference in oxygen partial pressure exists between the start and end of the system. In this example, the difference is 100% − 50% = 50% in the water and 50% − 0% = 50% in the blood.

The upshot? Only half of the oxygen in the incoming water has diffused into the blood.

Countercurrent flow makes fish gills extremely efficient at extracting oxygen from water, because it ensures that a difference in the partial pressure of oxygen and carbon dioxide in water versus blood is maintained over the entire gas-exchange surface.

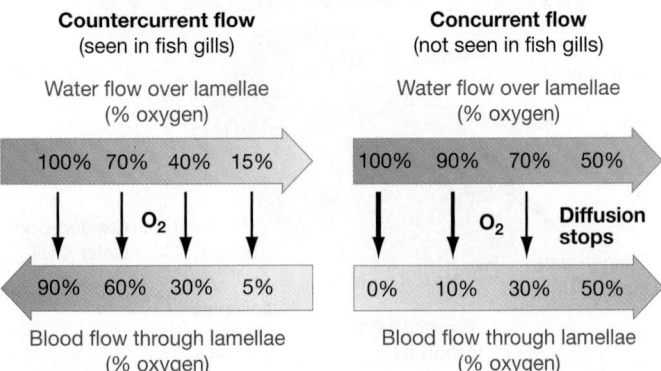

Countercurrent flow
(seen in fish gills)

Water flow over lamellae
(% oxygen)

100% 70% 40% 15%

O₂

90% 60% 30% 5%

Blood flow through lamellae
(% oxygen)

Concurrent flow
(not seen in fish gills)

Water flow over lamellae
(% oxygen)

100% 90% 70% 50%

O₂ **Diffusion stops**

0% 10% 30% 50%

Blood flow through lamellae
(% oxygen)

FIGURE 44.6 Countercurrent Exchange Is Much More Efficient than Concurrent Exchange. In the countercurrent system of fish gills, oxygen is transferred along the entire length of the capillaries. If "concurrent flow" occurred, oxygen transfer would stop, because the partial-pressure gradient driving diffusion would fall to zero partway along the length of the capillary.

The effect of countercurrent exchange is to maximize the $P_2 - P_1$ term in Fick's law of diffusion, averaged over the entire gill surface. Based on this observation, biologists cite countercurrent exchange as another example of how gills are optimized for efficient gas exchange.

How Do Insect Tracheae Work?

Air and water are dramatically different ventilatory media, because they have different densities, viscosities, and abilities to

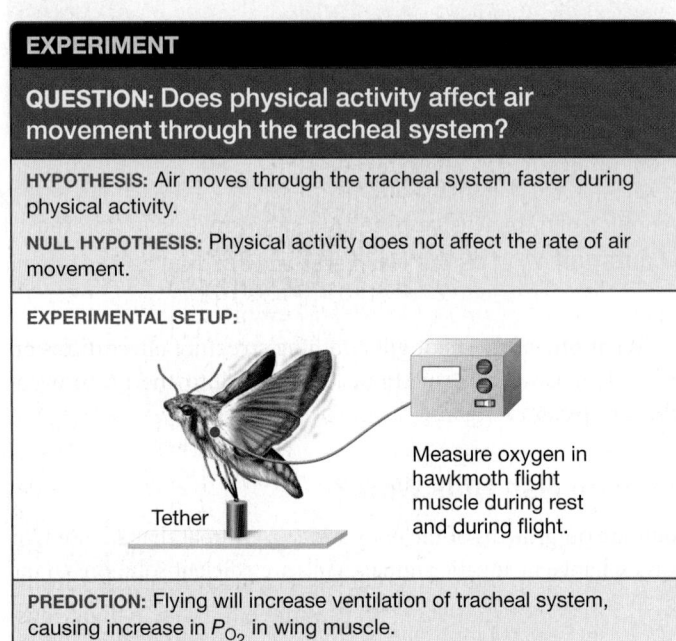

EXPERIMENT

QUESTION: Does physical activity affect air movement through the tracheal system?

HYPOTHESIS: Air moves through the tracheal system faster during physical activity.

NULL HYPOTHESIS: Physical activity does not affect the rate of air movement.

EXPERIMENTAL SETUP:

Measure oxygen in hawkmoth flight muscle during rest and during flight.

Tether

PREDICTION: Flying will increase ventilation of tracheal system, causing increase in P_{O_2} in wing muscle.

PREDICTION OF NULL HYPOTHESIS: Flying will not increase ventilation of tracheal system; P_{O_2} in flight muscle will decline steadily during flight.

RESULTS:

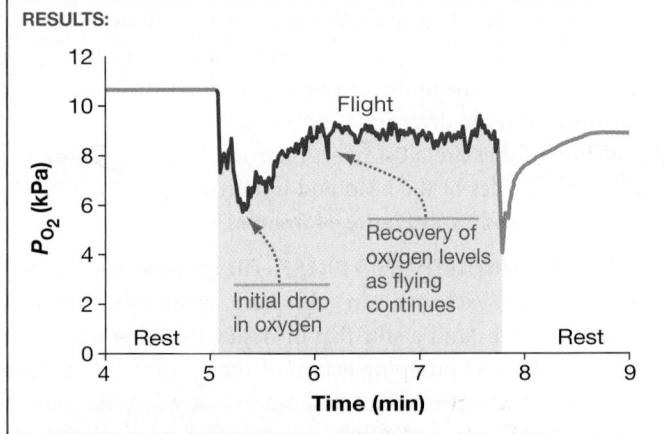

CONCLUSION: Muscular contractions may help ventilate the tracheal system in at least some insects.

FIGURE 44.7 Evidence That the Tracheal System Is Ventilated during Movement. On the graph's y-axis, kPa stands for a unit of pressure—or force per unit area—called the kilopascal.

SOURCE: Komai, Y. 1998. Augmented respiration in a flying insect. *Journal of Experimental Biology* 201: 2359–2366.

✔ **QUESTION** Why was it important for the researcher to repeat this experiment with several different individuals?

hold oxygen and carbon dioxide. In addition, the consequences of exposing the gas-exchange surface to air versus water differ.

In aquatic habitats, ventilation tends to disrupt water and electrolyte balance, and homeostasis must be maintained by an active osmoregulatory system. As Chapter 42 explained, osmosis causes marine animals to lose water across their gas-exchange surface and freshwater animals to gain water. Diffusion tends to cause marine animals to gain sodium, chloride, and other ions, and freshwater animals to lose them.

In contrast, breathing leads to a loss of water by evaporation in terrestrial environments. How do terrestrial animals minimize water loss while maximizing the efficiency of gas exchange?

To answer this question, consider the tracheal system of insects. Recall from Chapter 42 that **tracheae** are an extensive system of tubes located well within the body. They connect to the exterior through openings called **spiracles**, which can be closed to minimize the loss of water by evaporation. The ends of tracheae are tiny, fluid filled, and highly branched. In this way, the tracheal system transports air close enough to cells for gas exchange to take place directly across their plasma membranes.

Recent research on the tracheal system has focused on how air moves into and out of the tubes through spiracles. Is simple diffusion efficient enough to ventilate the system, or is some type of breathing mechanism involved?

The short answer to this question is that breathing movements may play a role in at least some insects. Consider a study on the sweet potato hawkmoth. A researcher set out to investigate how the amount of O_2 delivered to muscles changes during flight. To document the partial pressure of oxygen in flight muscles, he inserted a needlelike electrode into the wing muscles of a hawkmoth. The electrode was attached to an instrument that measured the partial pressure of oxygen, and the hawkmoth was tethered to a stand (**Figure 44.7**).

The biologist recorded P_{O_2} as the insect rested, then stimulated the moth to fly by exposing it to wind. The "Results" section

of Figure 44.7 shows how P_{O_2} changed during one such experiment. (The procedure was repeated on several individuals, and the same pattern was observed.)

To read this graph, put your finger on the orange line and move it to the right, to where the insect started flying and the line changes to red. Note that P_{O_2} levels in the flight muscles dropped initially. This is not surprising, because flight is an energetically demanding activity. As flying continued, however, P_{O_2} levels recovered until they were nearly as high as when they were at rest.

To explain this observation, biologists propose that tracheae alternately open and close as the muscles around them contract and relax. As a result, the volume of the tracheal system changes. This is a key point: If the volume occupied by a fixed amount of gas expands, its pressure decreases. But if the volume that a fixed amount of gas occupies declines, its pressure increases.

If the volume of the tracheal system increases when muscles relax, pressure inside the system goes down and air from the atmosphere will rush in. But if volume decreases when muscles contract, then pressure inside the system increases and gas will move out (**Figure 44.8**).

In effect, then, insects may have a breathing mechanism that ventilates the respiratory surface when demand for oxygen is high.

How Do Vertebrate Lungs Work?

In most terrestrial vertebrates, air enters the body through both the nose and mouth. A tube known as the **trachea** (not to be confused with the tracheae of insects) carries the inhaled air to narrower tubes called **bronchi** (singular: **bronchus**). The bronchi branch off into yet narrower tubes, the **bronchioles**. The organs of ventilation, the lungs, enclose the bronchioles and part of the bronchi (**Figure 44.9a** on page 868).

Lungs are infoldings of the throat that are used for gas exchange. In addition to being found in terrestrial vertebrates—amphibians, reptiles (including birds), and mammals—they occur in many fish and certain invertebrates.

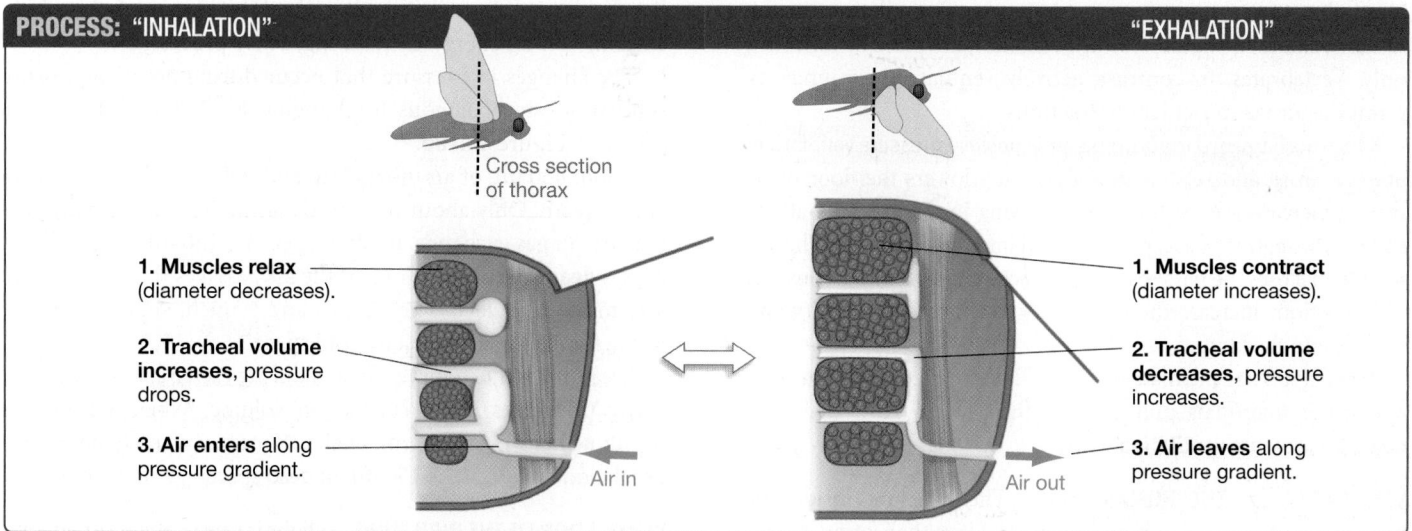

FIGURE 44.8 Muscle Activity May Ventilate Tracheae. To explain the result in Figure 44.7, biologists propose that tracheae change volume as wings beat (drawings are schematic).

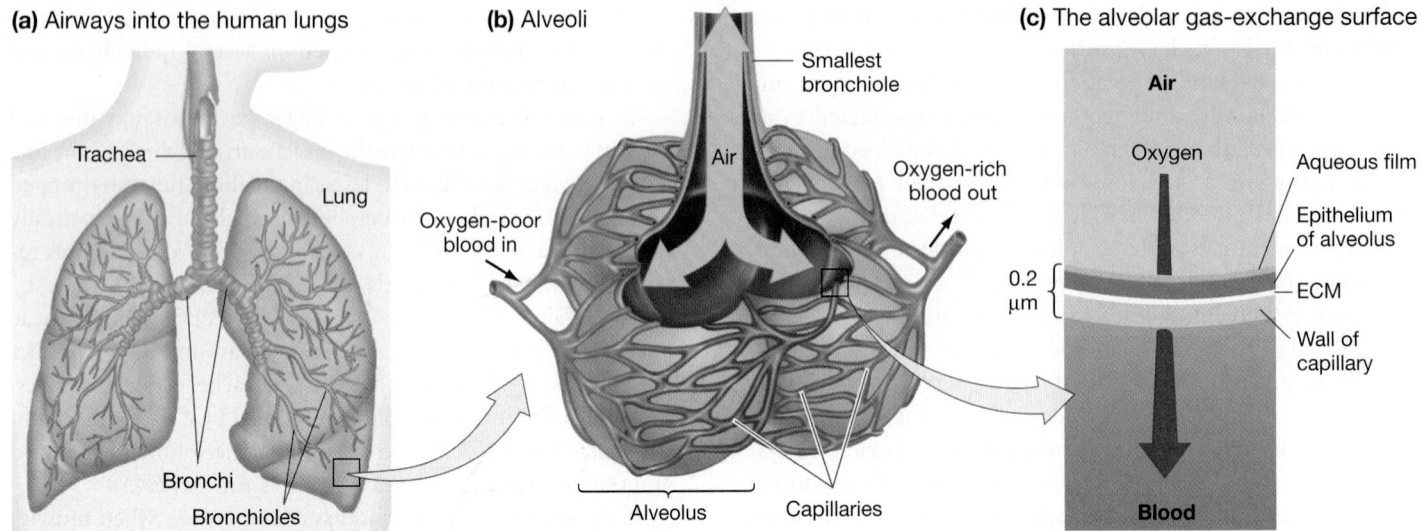

(a) Airways into the human lungs

Trachea

Lung

Bronchi

Bronchioles

(b) Alveoli

Smallest bronchiole

Air

Oxygen-poor blood in

Oxygen-rich blood out

Alveolus

Capillaries

(c) The alveolar gas-exchange surface

Air

Oxygen

0.2 μm

Blood

Aqueous film

Epithelium of alveolus

ECM

Wall of capillary

FIGURE 44.9 Lungs Offer a Large, Thin Surface Area for Gas Exchange between Air and Blood. (a) The human respiratory tract branches repeatedly from the largest airway, the trachea, to the smallest, called bronchioles. Gas exchange does not occur across these airways. The system of airways ends in clusters of tiny sacs called alveoli. **(b)** Alveoli are covered with capillary networks and **(c)** are the site of gas exchange.

LUNG STRUCTURE AND VENTILATION VARY AMONG SPECIES The amount of lung surface area available for gas exchange varies a great deal among species. In frogs and other amphibians, the lung is a simple sac lined with blood vessels. The lungs of mammals, in contrast, are finely divided into tiny sacs called **alveoli** (singular: **alveolus**; **Figure 44.9b**). Each human lung contains approximately 150 million alveoli, which give mammalian lungs about 40 times more surface area for gas exchange than an equivalent volume of frog lung tissue.

As **Figure 44.9c** shows, an alveolus provides an interface between air and blood that consists of a thin aqueous film, a layer of epithelial cells, some extracellular matrix (ECM) material, and the wall of a capillary. In the human lung, this barrier to diffusion is only 0.2 μm thick—about 1/200th of the thickness of this page.

In addition to total surface area, the other major feature of lungs that varies among species is mode of ventilation. In the lungs of snails and spiders, air movement takes place by diffusion only. Vertebrates, in contrast, actively ventilate their lungs by pumping air via muscular contractions.

One mechanism for pumping air is **positive pressure ventilation**, used by frogs and related animals. A frog lowers the floor of its throat, increasing its volume and drawing in air from the atmosphere, through the nasal passages, and into the oral cavity. The animal then closes the nasal passages and contracts its throat muscles. These actions increase the pressure on air in the oral cavity and force it into the lungs.

In effect, frogs push air into their lungs. In contrast, humans and other mammals pull air into their lungs. How does this **negative pressure ventilation** work?

VENTILATION OF THE HUMAN LUNG The pressure inside the human chest cavity is about 5 mm Hg less than atmospheric pressure. The negative pressure surrounding the lung is just enough to keep it expanded. If a wound penetrates the chest wall and the pressure differential between the chest cavity and the atmosphere disappears, the lung on the side of the injury will collapse like a deflated balloon.

Humans ventilate their lungs by changing the pressure within the chest cavity between about −5 mm Hg and −8 mm Hg relative to the atmosphere. As **Figure 44.10a** shows, inhalation is based on increasing the volume of the chest cavity and thus lowering the pressure. The change is caused by a downward motion of the thin muscular sheet called the **diaphragm** and an outward motion of the ribs. As the pressure surrounding the lungs drops, air flows into the airways along a pressure gradient.

Exhalation, in contrast, is a passive process when an individual is at rest. This is possible because **(1)** the lung is **elastic**—it returns to its original, collapsed shape if it is not stretched or compressed—and **(2)** the volume of the chest cavity decreases as the diaphragm and rib muscles relax. During exercise, though, exhalation is an energy-demanding, active process.

The changes in pressure that occur during negative pressure ventilation are analogous to changing the volume of a jar, as shown in **Figure 44.10b**.

About 450 mL of air moves into and out of the lungs in an average breath. Only about two-thirds of this volume actually participates in gas exchange, however, because 150 mL of the air occupies **dead space**—portions of the air passages that do not have a respiratory surface. The trachea and bronchi shown in Figure 44.9a, for example, represent dead space.

Breathing is much more efficient during exercise, when the chest cavity undergoes larger changes in volume. When a person is breathing hard, over 2500 mL of air can move with each inhalation-exhalation cycle, but the 150 mL in dead space stays the same.

VENTILATION OF THE BIRD LUNG Flight is one of the most energy-demanding activities performed by animals. Even so, some birds fly tens of thousands of kilometers during annual migrations.

(a) Lungs expand and contract in response to changes in pressure inside the chest cavity.

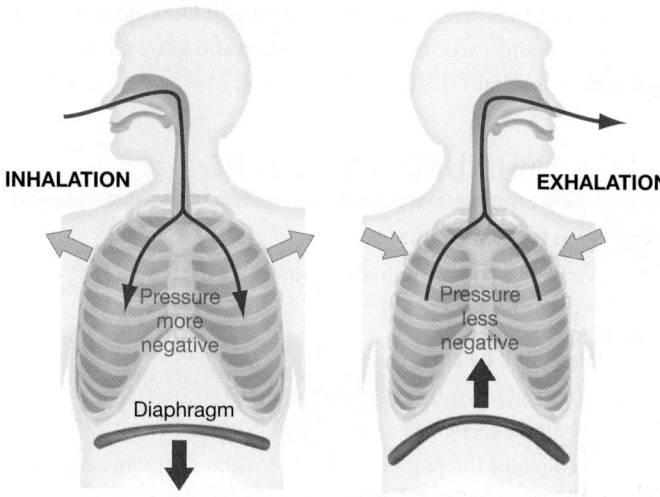

INHALATION
EXHALATION

Pressure more negative
Pressure less negative

Diaphragm

(b) Ventilatory forces can be modeled by a balloon in a jar.

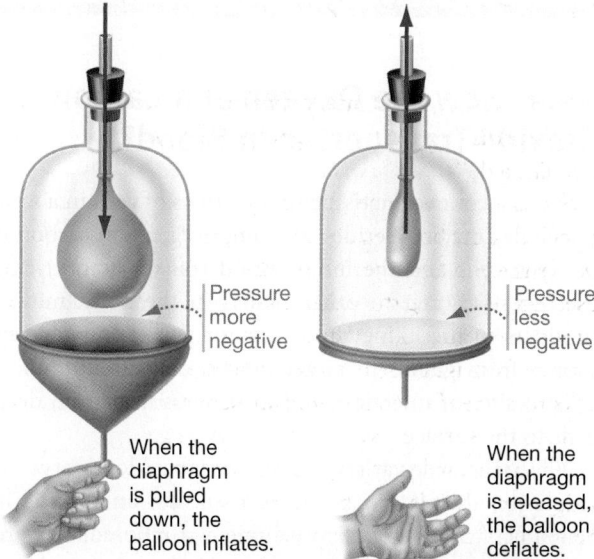

Pressure more negative
Pressure less negative

When the diaphragm is pulled down, the balloon inflates.
When the diaphragm is released, the balloon deflates.

FIGURE 44.10 Changes in the Volume of the Chest Cavity Drive Negative Pressure Ventilation. (a) Inhalation: When the diaphragm and rib muscles contract, the volume of the lung cavity increases, lowering pressure. In response, the lungs expand and air flows in. Exhalation: When the diaphragm and rib muscles relax, the volume of the lung cavity decreases, causing internal pressure to increase. In response, lung volume decreases—due to elasticity of the lungs—and air flows out. **(b)** A model of negative pressure ventilation.

✔**QUESTION** What happens when a person sighs or takes a deep breath?

Even more impressive, geese regularly fly over the top of Mount Everest. At this elevation, the partial pressure of oxygen is so low that most humans would immediately black out if they were flown there and dropped off. How are birds able to extract enough oxygen from the atmosphere to support long flights and flights at high elevations?

Figure 44.11 shows why respiration is so efficient in birds.

1. During inhalation, air flows through the trachea and enters two large air sacs posterior to the lungs.

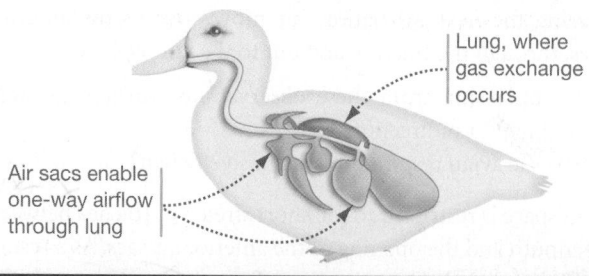

Lung, where gas exchange occurs

Air sacs enable one-way airflow through lung

PROCESS: ONE-WAY AIRFLOW THROUGH AVIAN LUNG

Follow one breath (in red) through the avian respiratory system

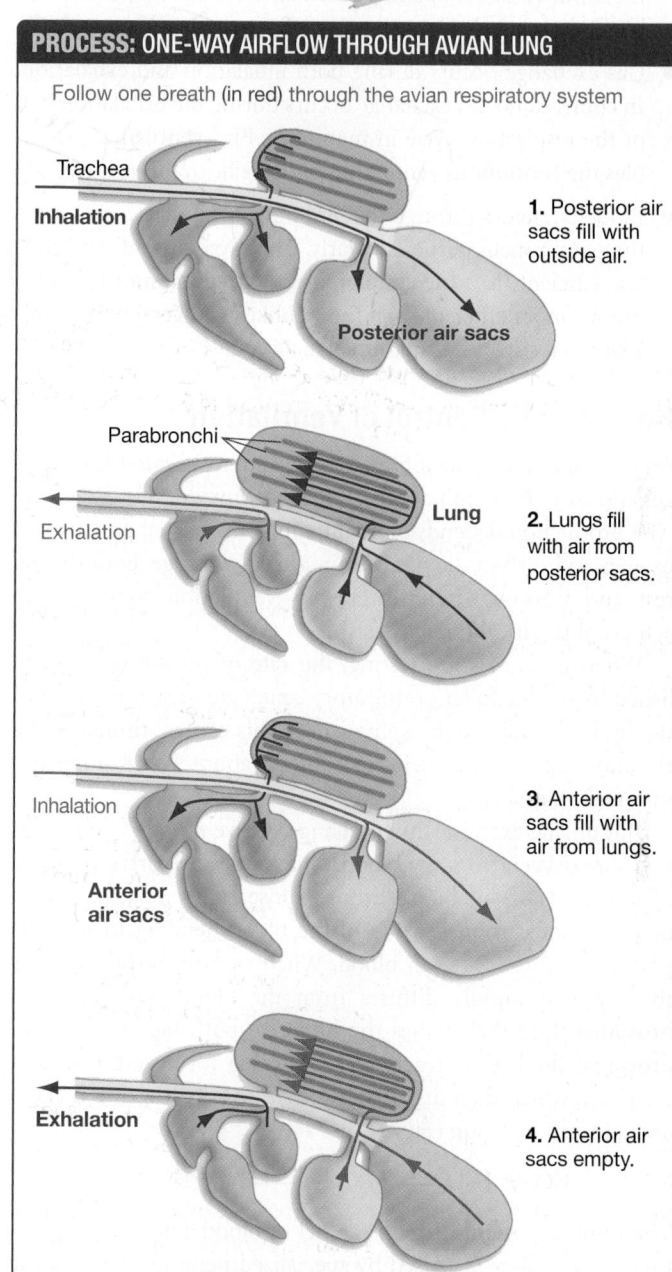

Trachea
Inhalation

1. Posterior air sacs fill with outside air.

Posterior air sacs

Parabronchi

Exhalation
Lung

2. Lungs fill with air from posterior sacs.

Inhalation

3. Anterior air sacs fill with air from lungs.

Anterior air sacs

Exhalation

4. Anterior air sacs empty.

FIGURE 44.11 Air Flows in One Direction through the Bird Lung.

2. During exhalation, air leaves the posterior air sacs and enters tiny, branching airways, called parabronchi, in the posterior portion of the lung.

3. During the next inhalation, air moves into parabronchi in the anterior part of the lung and on to a system of air sacs anterior to the lung.

4. During the next exhalation, air moves out of the anterior sacs, through the trachea, and out to the atmosphere.

The key conclusion from these observations? Airflow through the avian lung is unidirectional.

Why is the avian respiratory system so efficient?

- Dead space is restricted to the short stretch of trachea between the mouth and the opening of the anterior air sacs. As a result, birds use inhaled air much more efficiently than mammals do.

- Gas exchange occurs during both inhalation and exhalation. In contrast, no gas exchange occurs during the exhalation half of the respiratory cycle in mammals. Bird ventilation resembles the continuous ventilation of fish gills in this respect.

- Blood circulates through the bird lung in capillaries that cross the parabronchi perpendicularly. This crosscurrent pattern is less efficient than the countercurrent circulation of fish gills, but far more efficient than the weblike arrangement of capillaries that surrounds mammalian alveoli.

Homeostatic Control of Ventilation

An animal is in trouble if homeostasis with respect to blood oxygenation or the elimination of carbon dioxide fails. Adequate ATP production depends on maintaining the partial pressures of oxygen and carbon dioxide within a narrow range, both during rest and vigorous exercise. How is ventilation controlled to achieve this critical homeostasis?

When mammals are resting, the rate of breathing is established by the medullary respiratory center, an area at the base of the brain, just above the spinal cord. This center stimulates the rib and diaphragm muscles to contract about 12–14 times per minute in humans.

But during exercise, things change. Active muscle tissue takes up more oxygen from the blood. As a result, the partial pressure of oxygen (P_{O_2}) in blood drops. Those same muscles release quantities of carbon dioxide to the blood, tending to raise its partial pressure (P_{CO_2}) in blood. When carbon dioxide reaches the brain, it rapidly diffuses from the blood into the cerebrospinal fluid that bathes the brain. In both blood and cerebrospinal fluid, CO_2 reacts with water to form carbonic acid (H_2CO_3), which then dissociates to release a hydrogen ion (H^+) and a bicarbonate ion (HCO_3^-):

$$CO_2 + H_2O \rightleftharpoons H_2CO_3 \rightleftharpoons H^+ + HCO_3^-$$

The result is a slight drop in the pH of blood and cerebrospinal fluid. The change is sensed by specialized neurons located near the large arteries that travel from the heart into the neck and to the base of the brain itself.

Signals from these nerves or from pH detectors in the medullary respiratory center cause the breathing rate to increase. The resulting rise in ventilation rate increases the rate of oxygen delivery to the tissues and the rate at which carbon dioxide is eliminated from the body, restoring P_{O_2} and P_{CO_2} to their resting levels. This control system is so effective that it can maintain sta-

ble blood levels of oxygen and carbon dioxide even during intense exercise.

Next, let's look in greater depth at blood. How does it transport oxygen and carbon dioxide between the gas-exchange surface and an animal's tissues?

CHECK YOUR UNDERSTANDING

If you understand that . . .

- Most large-bodied animals exchange gases via gills, tracheae, or lungs.

You should be able to . . .

1. Identify two features that are common to gills, tracheae, and lungs, as well as one trait that is unique to each.

2. Explain how the presence of anterior and posterior air sacs facilitate one-way airflow through the bird lung.

Answers are available in Appendix B.

44.4 How Are Oxygen and Carbon Dioxide Transported in Blood?

Blood is a connective tissue that consists of cells in a watery extracellular matrix. Besides carrying oxygen and carbon dioxide between cells and the lungs, blood transports nutrients from the digestive tract to other tissues in the body, moves waste products to the kidney and liver for processing, conveys hormones from glands to target tissues, delivers immune system cells to sites of infection, and distributes heat from deeper organs to the surface.

Given the wide variety of functions that blood serves, it is not surprising that it is a complex tissue. In an average human, 50–60 percent of the blood volume is composed of an extracellular matrix called **plasma**. The remainder of the volume comprises cells and cell fragments that are collectively called formed elements.

The formed elements in blood include platelets, several types of white blood cells, and red blood cells.

- **Platelets** are cell fragments that minimize blood loss from ruptured blood vessels. Platelets release material that assists in forming the blockages known as clots.

- **White blood cells** are part of the immune system. As Chapter 49 will explain in detail, white blood cells fight infections.

- **Red blood cells** transport oxygen from the lungs to tissues throughout the body. They also play a critical role in transporting carbon dioxide from tissues to the lungs. In humans, red blood cells make up 99.9 percent of the formed elements.

The human body synthesizes new red blood cells at the rate of 2.5 million per second to replace old red blood cells, which die at the same rate. These red blood cells live about 120 days. They develop along with white blood cells and platelets from stem cells located in the tissue inside bone (bone marrow).

Vertebrates other than mammals transport oxygen in red blood cells that retain their nuclei. But in mammals, red blood cells lose their nuclei as they mature, along with their mitochondria and most other organelles. Mammalian red blood cells are essentially bags filled with approximately 280 million copies of the oxygen-carrying molecule **hemoglobin**.

Structure and Function of Hemoglobin

Even though oxygen is not highly soluble in water, it is often found in high concentrations in blood. Blood has a high oxygen-carrying capacity because O_2 readily binds to the hemoglobin molecules in red blood cells.

The evolution of hemoglobin was a key event in the diversification of animals. By increasing the oxygen-carrying capacity of blood, hemoglobin made it possible for cellular respiration rates to increase. High rates of ATP production, in turn, support high rates of growth, movement, digestion, and other activities.

Hemoglobin is a tetramer, meaning that it consists of four polypeptide chains (**Figure 44.12**). Each of the four polypeptide chains binds to a nonprotein group called **heme**. Each heme molecule, in turn, contains an iron ion (Fe^{2+}) that can bind to an oxygen molecule. As a result, each hemoglobin molecule can bind up to four oxygen molecules. In blood, 98.5 percent of the oxygen is bound to hemoglobin; only 1.5 percent is dissolved in plasma.

WHAT IS COOPERATIVE BINDING? Blood leaving the human lungs has a P_{O_2} of about 100 mm Hg, while muscles and other tissues have P_{O_2} levels of about 40 mm Hg at rest. This partial-pressure difference creates a diffusion gradient that unloads O_2 from hemoglobin to the tissues.

Researchers who studied the dynamics of O_2 unloading in tissues found the pattern shown in **Figure 44.13**. This graph plots the percent saturation of hemoglobin in red blood cells versus the P_{O_2} levels in the blood within tissues. If saturation is 100 percent, it means that every possible binding site in hemoglobin contains an oxygen molecule.

The graph in Figure 44.13 is called an **oxygen-hemoglobin equilibrium curve** or an oxygen dissociation curve. Note that the x-axis plots the partial pressure of oxygen in tissues. In effect, this represents "demand." Oxygen-depleted tissues, where demand for oxygen is high, are on the left-hand part of the horizontal axis; oxygen-rich tissues are toward the right. The y-axis, in contrast, plots the percentage of hemoglobin molecules in blood that are saturated with hemoglobin—a measure of "supply," or how readily hemoglobin gives up oxygen. Each 25-percent change in saturation corresponds to an average of one additional oxygen molecule bound per hemoglobin molecule or one oxygen molecule delivered to tissues.

The most remarkable feature of the oxygen-hemoglobin equilibrium curve is that it is sigmoidal, or S-shaped. The pattern occurs because the binding of each successive oxygen molecule to a subunit of the hemoglobin molecule causes a conformational change in the protein that makes the remaining subunits much more likely to bind oxygen.

This phenomenon is called **cooperative binding**. Conversely, the loss of a bound oxygen molecule changes hemoglobin's conformation in a way that makes the loss of additional oxygen molecules more likely.

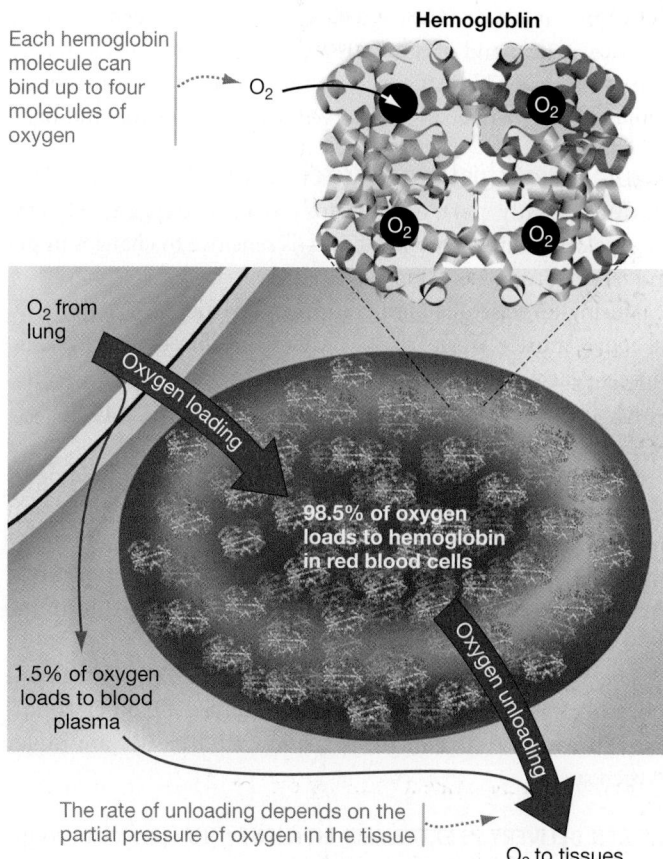

Each hemoglobin molecule can bind up to four molecules of oxygen → O_2

Hemogloblin

O_2 O_2 O_2 O_2

O_2 from lung

Oxygen loading

98.5% of oxygen loads to hemoglobin in red blood cells

1.5% of oxygen loads to blood plasma

Oxygen unloading

The rate of unloading depends on the partial pressure of oxygen in the tissue

O_2 to tissues

FIGURE 44.12 Hemoglobin Transports Oxygen to Tissues.

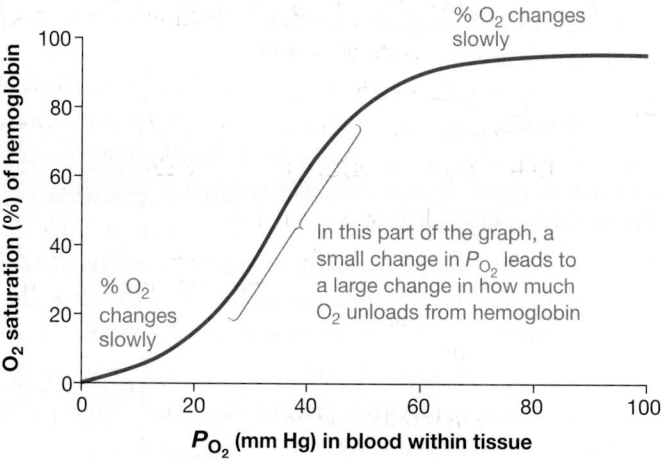

FIGURE 44.13 The Oxygen-Hemoglobin Equilibrium Curve.
A sigmoidal curve like this has three distinct regions.

% O_2 changes slowly

In this part of the graph, a small change in P_{O_2} leads to a large change in how much O_2 unloads from hemoglobin

% O_2 changes slowly

O_2 saturation (%) of hemoglobin

P_{O_2} (mm Hg) in blood within tissue

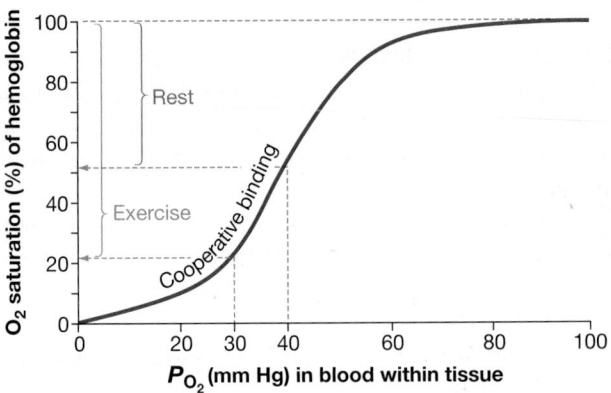

(a) With cooperative binding, large amounts of O_2 are delivered to resting and exercising tissues.

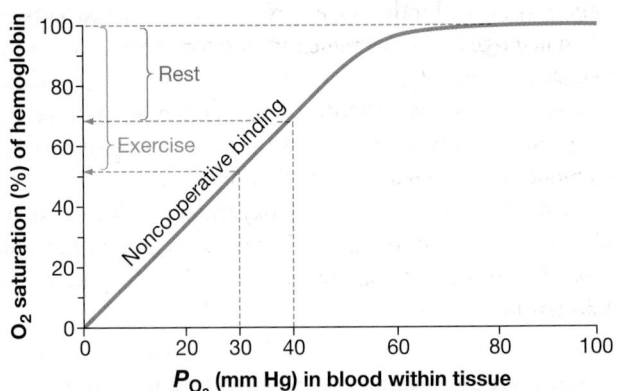

(b) Without cooperative binding, smaller amounts of O_2 would be delivered to resting and exercising tissues.

FIGURE 44.14 Cooperative Binding of O_2 by Hemoglobin Results in Greater O_2 Delivery than Noncooperative Binding. Effects of **(a)** cooperative versus **(b)** noncooperative oxygen binding. Note that hemoglobin is almost 100 percent saturated with oxygen when it arrives at tissues.

WHY IS COOPERATIVE BINDING IMPORTANT? To understand why cooperative binding is important, use **Figure 44.14a** to figure out what happens to hemoglobin saturation when oxygen demand in tissues changes.

Let's begin with some basic observations. When blood arrives at tissues, hemoglobin saturation is close to 100 percent. At rest, tissue P_{O_2} is typically about 40 mm Hg. But during exercise, cells are using so much oxygen in cellular respiration that tissue P_{O_2} drops to about 30 mm Hg. Now:

- Put your finger on the x-axis at 40 mm Hg, trace the dotted line up until it hits the saturation curve, and check where this point is on the y-axis. The answer is about 52 percent.

- The following statement should make sense to you: When tissues are at rest, hemoglobin gives up about half of its oxygen (about 48 percent) to the tissues.

- Now put your finger on the x-axis at 30 mm Hg, trace the dotted line up, and check hemoglobin saturation at this point on the curve. The answer is about 22 percent.

- You should be able to make the following conclusion: When tissues are exercising, hemoglobin gives up about 78 percent of its oxygen to tissues.

Here's the punchline: In response to a relatively small change in tissue P_{O_2}, there is a relatively large change in the percentage saturation of hemoglobin. The large change occurs because the saturation curve is extremely steep in the range of P_{O_2} values commonly observed in tissues. The curve is steep because of cooperative binding. Cooperative binding is important because it makes hemoglobin exquisitely sensitive to changes in the P_{O_2} of tissues.

✔If you understand this concept, you should be able to explain the consequences of an oxygen-hemoglobin equilibrium curve with a middle section that is even steeper than the one shown in Figure 44.14a.

If cooperative binding did not occur, all four subunits of hemoglobin would load or unload oxygen independently of each other. They would lose or gain oxygen in direct proportion to the partial pressure of oxygen in the blood, until the molecule was either completely saturated or unsaturated with oxygen. There would be a linear relationship between hemoglobin saturation and tissue P_{O_2}—not a sigmoidal one.

Figure 44.14b shows noncooperative binding. As the dotted lines on this graph indicate, a relatively small change in oxygen delivery would occur when tissue P_{O_2} changes from its resting level of about 40 mm Hg to 30 mm Hg. Specifically, hemoglobin would give up about 100% − 68% = 32% of its oxygen when tissues are at rest, and about 100% − 52% = 48% of its oxygen during exercise. This difference—about 16 percent—is much less than the 30 percent change observed with cooperative binding.

HOW DO TEMPERATURE AND PH AFFECT HEMOGLOBIN? Cooperative binding is only part of the story behind oxygen delivery. Hemoglobin—like other proteins—is sensitive to changes in pH and temperature.

During exercise, the temperature and partial pressure of CO_2 in active muscle tissue rise. The CO_2 produced by exercising muscle reacts with the water in blood to form carbonic acid, which dissociates and releases a hydrogen ion. As a result, the pH of the blood in exercising muscle drops.

Decreases in pH and increases in temperature alter hemoglobin's conformation. These shape changes make hemoglobin more likely to release O_2 at any given value of tissue P_{O_2}. As **Figure 44.15** shows, this phenomenon, known as the **Bohr shift**, causes the oxygen-hemoglobin equilibrium curve to shift to the right when pH declines.

The Bohr shift is important because it makes hemoglobin more likely to release oxygen during exercise or other conditions in which P_{CO_2} is high, pH is low, and tissues are under oxygen stress.

OXYGEN DELIVERY IS EXTREMELY EFFICIENT To appreciate cooperative binding and the Bohr shift in action, consider an experiment on how the oxygen transport system in rainbow trout responds to sustained exercise.

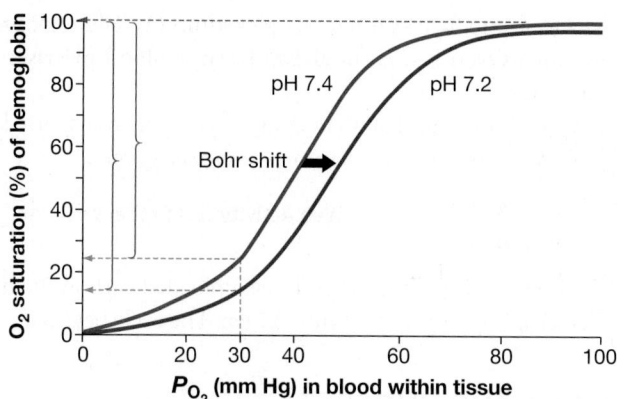

FIGURE 44.15 The Bohr Shift Makes Hemoglobin More Likely to Release Oxygen to Tissues with Low pH. As pH drops, oxygen becomes less likely to stay bound to hemoglobin at all values of tissue P_{O_2}. Exercising tissues have lower pH than resting tissues and thus receive more oxygen from hemoglobin.

✔**EXERCISE** Estimate how much more oxygen is released from hemoglobin at pH 7.2 than at pH 7.4 when the tissue has a P_{O_2} of 30 mm Hg.

To begin, biologists had fish swim continuously against a current in a water tunnel. As the researchers increased the speed of the current and thus the swimming speed of the fish, they periodically sampled the O_2 content of arterial and venous blood. Arterial blood is freshly oxygenated and moving away from the gills; venous blood is returning to the gills from the rest of the body.

Not surprisingly, the biologists found that arterial O_2 levels remained fairly constant as swimming speed increased—meaning the gills continued to work efficiently enough to saturate hemoglobin with oxygen.

In contrast, the O_2 content of venous blood, which had undergone gas exchange with the tissues, dropped steadily as swimming speed increased. When the fish had reached their maximum sustainable speed, virtually all the oxygen available in hemoglobin in the venous blood had been extracted.

The data show that in hard-working tissues, the combination of increased temperature, lower pH, and lower P_{O_2} caused hemoglobin to become almost completely deoxygenated. Oxygen-delivery systems based on hemoglobin are extremely efficient.

COMPARING HEMOGLOBINS Hemoglobin molecules from different individuals or species may vary in ways that affect fitness—the ability to survive and produce offspring. As an example, consider the oxygen-hemoglobin equilibrium curves in **Figure 44.16**. The curve in dark red is from a pregnant woman; the curve in light red is from a fetus she is carrying.

The hemoglobin found in fetuses is encoded by different genes than adult hemoglobin and has a distinctive structure and function. Specifically, the oxygen-hemoglobin equilibrium curve for fetal hemoglobin is shifted to the left with respect to the curve for adult hemoglobin.

Recall that the rightward shift called the Bohr effect means that hemoglobin is more likely to release oxygen. The leftward

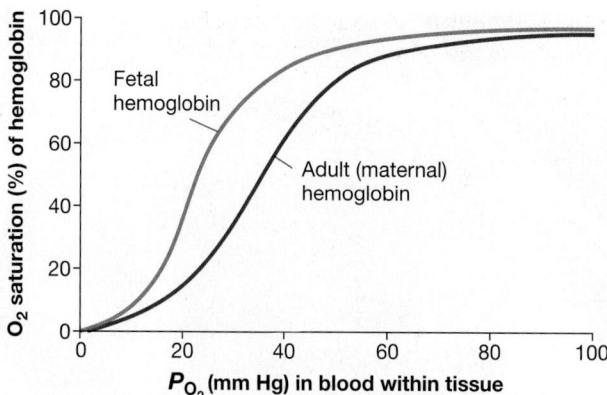

FIGURE 44.16 Some Types of Hemoglobin Bind Oxygen More Tightly than Others.

shift in Figure 44.16 means that fetal hemoglobin is less likely to give up oxygen—it binds oxygen more tightly. Stated another way, fetal hemoglobin has a higher affinity for oxygen than adult hemoglobin at every P_{O_2}.

This is critically important. In the placenta, hemoglobin from the mother's blood gets close to hemoglobin from the fetus's blood. Because fetal hemoglobin has such a high affinity for oxygen, there is a transfer of oxygen from the mother's blood to the fetus's blood. The difference in hemoglobin structure and function ensures an adequate supply of oxygen as the fetus develops.

CO₂ Transport and the Buffering of Blood pH

The carbon dioxide that is produced by cellular respiration in the tissues enters the blood, where it reacts with water to form carbonic acid. Recall that the resulting drop in blood pH stimulates an increase in breathing rate. Rapid exhalation of CO_2 then counteracts the drop in blood pH.

Homeostasis with respect to blood pH is reinforced by an elegant series of events that takes place inside red blood cells. Biologists were able to work out what was happening when they discovered large amounts of the enzyme **carbonic anhydrase** in red blood cells.

THE ROLE OF CARBONIC ANHYDRASE AND HEMOGLOBIN Recall from Chapter 43 that carbonic anhydrase catalyzes the formation of carbonic acid from carbon dioxide in water. Consequently, CO_2 that diffuses into red blood cells is quickly converted to bicarbonate ions and protons. Thus, most CO_2 is transported in blood (specifically in plasma) in the form of HCO_3^-—the bicarbonate ion. The same reaction occurs in the plasma surrounding red blood cells, but much more slowly in the absence of the enzyme.

Why is the carbonic anhydrase activity in red blood cells so important? The answer has two parts.

1. The protons produced by the enzyme-catalyzed reaction induce the Bohr shift, which makes hemoglobin more likely to release oxygen.

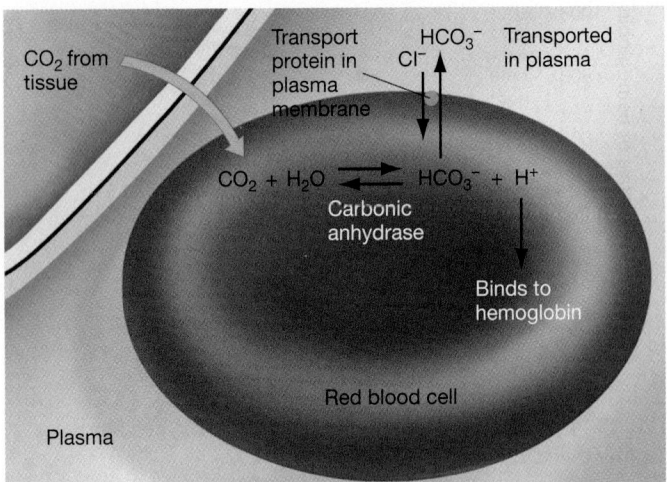

FIGURE 44.17 Carbonic Anhydrase Is Key to CO₂ Transport in Blood.
When CO_2 diffuses into red blood cells, carbonic anhydrase quickly converts it to carbonic acid, which dissociates into bicarbonate ion (HCO_3^-) and a proton (H^+). This reaction maintains the partial-pressure gradient favoring the entry of CO_2 into red blood cells. The protons produced by the reaction bind to deoxygenated hemoglobin. Most CO_2 in blood is transported to the lungs in the form of HCO_3^-.

✔**QUESTION** This diagram shows the sequence of events in tissues. Explain what happens when the red blood cell in the diagram reaches the lungs.

2. The partial pressure of CO_2 in blood drops when carbon dioxide is converted to soluble bicarbonate ions, maintaining a strong partial-pressure gradient favoring the entry of CO_2 into red blood cells.

Thus, carbonic anhydrase activity makes CO_2 uptake from tissues more efficient (**Figure 44.17**).

Once bicarbonate ions form in the red blood cell, they diffuse into the blood plasma along a concentration gradient. In contrast, the protons produced by the reaction stay inside red blood cells. Relatively small amounts of H^+ build up in the plasma, so pH remains stable.

What happens to these protons? When hemoglobin is not carrying oxygen molecules, it has a high affinity for protons. As a result, it takes up much of the H^+ that is produced by the dissociation of carbonic acid. Hemoglobin acts as a **buffer**—a compound that minimizes changes in pH.

WHAT HAPPENS IN THE LUNGS? When deoxygenated blood reaches the alveoli, the environment changes dramatically. In the lungs, a partial-pressure gradient favors the diffusion of CO_2 from plasma and red blood cells to the atmosphere. As CO_2 diffuses from the blood into the alveoli, P_{CO_2} in blood declines.

The drop in blood P_{CO_2} reverses the chemical reactions that occurred in tissues:

1. Hydrogen ions leave their binding sites on hemoglobin.

2. Protons react with bicarbonate to form CO_2.

3. CO_2 diffuses into the alveoli and is exhaled from the lungs.

In the meantime, hemoglobin has picked up O_2. Hemoglobin's affinity for oxygen is high in alveoli because blood pH rises as P_{CO_2} declines.

To review the events that occur as O_2 and CO_2 are transported in blood, go to the study area at *www.masteringbiology.com.*

(MB) BioFlix™ Gas Exchange, **Web Activity** Gas Exchange in the Lungs and Tissues

When blood leaves the lungs, it has unloaded carbon dioxide and hemoglobin is saturated with oxygen. The cycle begins anew.

CHECK YOUR UNDERSTANDING

🔑— If you understand that . . .

● In blood, oxygen is bound to hemoglobin and transported inside red blood cells. Carbon dioxide is converted to bicarbonate ions and transported in plasma.

● Hemoglobin has several properties that make it an effective transport protein, including cooperative binding of oxygen, the Bohr shift response to low pH, and the ability to bind the protons that are generated when carbon dioxide is converted to bicarbonate ions.

✔ **You should be able to . . .**

Predict how the oxygen-hemoglobin saturation curves measured in Tibetan people, whose ancestors have lived at high elevations for thousands of years, compare to curves from people whose ancestors have lived at sea level for many generations.

Answers are available in Appendix B.

44.5 The Circulatory System

According to Fick's law, differences in the partial pressure of gases are only part of the story when it comes to understanding diffusion rates. Surface area—*A* in the equation featured in Figure 44.3—also plays a key role.

How do animals maximize the surface area available for diffusion of gases and other key solutes?

● Animals that are only a few millimeters in size, like rotifers and tardigrades, have a small enough volume that diffusion over their body surface is adequate to keep them alive.

● Jellyfish and corals have a large, highly folded gastrovascular cavity that offers a large surface area for exchange of molecules with the environment.

● The flattened bodies of flatworms and tapeworms give these animals a high surface area/volume ratio (see Chapter 41). In these species, molecules are exchanged with the environment directly across the outer body surface.

● Diffusion across the body wall also occurs in roundworms, where gas exchange is facilitated by muscular contractions in their body wall. Diffusion is enhanced as roundworms circulate fluids by sloshing them back and forth.

In larger animals, however, the problem of providing a large surface area for diffusion is solved by a circulatory system. A circulatory system carries transport tissues called blood or hemolymph into close contact with every cell in the body. In this case, "close contact" is a distance of about 0.1 mm between the smallest blood vessels and cells within tissues. Diffusion is efficient at this scale.

To explore how circulatory systems work, let's start by distinguishing the two most basic types of systems—open and closed.

What Is an Open Circulatory System?

In an **open circulatory system**, a fluid connective tissue called hemolymph is actively pumped throughout the body in blood vessels. The hemolymph is not confined exclusively to blood vessels, however. Instead, hemolymph comes into direct contact with tissues. As a result, the molecules being exchanged between hemolymph and tissues do not have to diffuse across the wall of a blood vessel. Hemolymph transports wastes and nutrients and may also contain oxygen-carrying pigments, some cells, and clotting agents.

Figure 44.18 illustrates the open circulatory system in a clam. Note that a muscular organ called the **heart** pumps hemolymph into blood vessels that empty into an open, fluid-filled space. Hemolymph is returned to the heart when the heart relaxes and its internal pressure drops below that in the body cavity. General body movements also move hemolymph to and from the heart(s).

Because it moves throughout the volume of the body, hemolymph is under relatively low pressure in an open circulatory system. As a result, hemolymph flow rates may also be low. These features make open circulatory systems most suitable for relatively sedentary organisms, which do not have high oxygen demands.

Insects, with their rapid movements and more active lifestyles, are an obvious exception to this rule. Insects overcome the limitations imposed by low hemolymph pressure via their tracheal respiratory system, which delivers oxygen directly to the tissues.

Another characteristic of open circulatory systems is that without discrete, continuous vessels, the flow of hemolymph cannot be directed toward tissues that have a high oxygen de-

mand and CO_2 buildup. An open circulatory system moves hemolymph throughout an animal's body in much the same way that a ceiling fan moves air throughout a room in a house.

Crustaceans are an important exception to this rule, however. Even though their circulatory system is classified as open, these species have a network of small vessels that can preferentially send hemolymph to tissues with the highest oxygen demands.

What Is a Closed Circulatory System?

In a **closed circulatory system**, blood flows in a continuous circuit through the body under pressure generated by a heart. Because blood is confined to vessels, a closed system can generate enough pressure to maintain a high flow rate.

Blood flow can also be directed in a precise way in a closed circulatory system. For example, blood can be shunted to leg muscles during exercise, the intestines after a meal, or regions of the brain engaged in particular mental tasks.

WHICH LINEAGES HAVE CLOSED CIRCULATORY SYSTEMS? Closed circulatory systems are found in vertebrates and a few other lineages where individuals tend to be active. Earthworms and other annelids, for example, have a closed circulatory system and exchange gases with the environment across their thin, moist skin, which has a dense supply of capillaries. As a result, annelids are able to obtain and circulate enough oxygen to support intense muscular activity. Most live as active burrowers and hunters.

A similar situation occurs in squid, octopuses, and other cephalopods that hunt down prey. The closed circulatory system of these mollusks generates high rates of blood flow, which oxygenates their muscles well enough to support rapid movements and a predatory lifestyle.

Closed circulatory systems contain an array of blood vessels, each of which has a distinct structure and function. Let's review the major types of blood vessels, and then consider how the vessels of a closed circulatory system interact with the lymphatic system.

TYPES OF BLOOD VESSELS An enormous amount of tubing is required to distribute blood within "diffusion distance" of every cell in the body. If all the blood vessels in a human body were laid end to end, they would stretch about 100,000 km (over 60,000 miles).

Blood vessels are classified as follows:

- **Arteries** are tough, thick-walled vessels that take blood away from the heart under high pressure. Small arteries are called **arterioles**.

- **Capillaries** are vessels whose walls are just one cell thick, allowing gases and other molecules to exchange with tissues in networks called **capillary beds**.

- **Veins** are vessels that return blood to the heart(s) under low pressure. Small veins are called **venules**.

The structure of arteries, capillaries, and veins correlates closely with their function in a closed circulatory system. For example, the heart ejects blood into a large artery, usually called the **aorta**. All arteries have both muscle fibers and elastic fibers in their walls, but elastic fibers dominate the walls of the aorta. As a

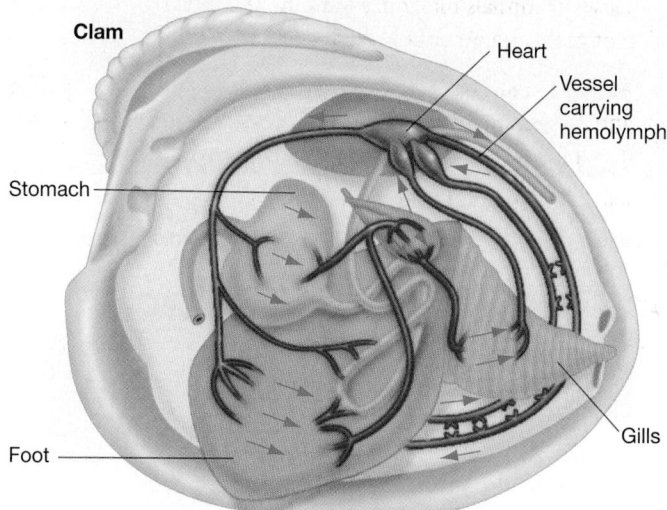

FIGURE 44.18 Open Circulatory Systems in Clams.

result, the aorta can expand when blood enters it under high pressure from the heart.

When contraction of the heart ends, the diameter of the aorta returns to its resting state. This elastic response propels blood away from the heart and augments the force generated by heart contraction. Similar types of secondary pumping action occur to some extent in other arteries as well. This feature helps maintain forward blood flow in the period between heart contractions.

In the walls of arterioles, muscle fibers called **sphincters** wrap around the circumference of the vessel. When these muscle fibers relax, the arteriole diameter increases, resistance to flow in the arteriole is reduced, and blood flow increases in the tissues served by the vessel. But when these arteriolar sphincters constrict, they increase resistance to flow, slow blood passage, and thus divert blood flow for use by other tissues. In this way, blood flow to tissues can be carefully regulated by signals from the nervous system.

Capillaries are the smallest blood vessels. Their walls are only one cell layer thick, and they are just wide enough to let red blood cells through one at a time (**Figure 44.19a**). Because capillaries are extremely thin and because they form an extremely dense network throughout the body, they are the structure where gases, nutrients, and wastes are exchanged between the blood and other tissues.

In some organs, such as the liver, the walls of capillaries contain numerous small openings, which further diminish the barrier to diffusion between blood and the tissues. Despite their thinness, it is rare for capillaries to rupture, because blood pressure drops dramatically as blood passes through arterioles on its way to capillary beds.

Veins carry blood back to the heart after the blood passes through capillaries. Because blood is under relatively low pressure by the time it exits the tissues, veins have thinner walls and larger interior diameters than arteries do (**Figure 44.19b**).

Blood flow in veins is speeded by muscle activity in the extremities, including the hands and feet, which compresses large veins. Larger veins also contain one-way **valves**, which are thin flaps of tissue that prevent any backflow of blood.

All veins contain some muscle fibers, which contract in response to signals from the nervous system, decreasing the diameter and overall volume of the vessels. Blood pressure in a closed circulatory system is regulated, in part, by actively adjusting the volume of blood contained within the veins.

WHAT IS INTERSTITIAL FLUID?   The relatively high operating pressure of closed circulatory systems, combined with the thinness of capillaries, produces a small but steady leakage of fluid from these blood vessels into the surrounding space. The area between cells is called interstitial space; the fluid that fills it is **interstitial fluid**. Blood cells are retained within capillaries, so interstitial fluid resembles plasma in its electrolyte composition.

Why does interstitial fluid build up? In 1896 Ernest Starling proposed that two forces were at work (**Figure 44.20**):

1. There is an outward-directed hydrostatic force in capillaries, created by the pressure on blood generated by the heart. This force is analogous to the pressure that drives water through the wall of a leaky garden hose.

(a) Capillaries are small and extremely thin walled.

(b) Veins and arteries differ in structure.

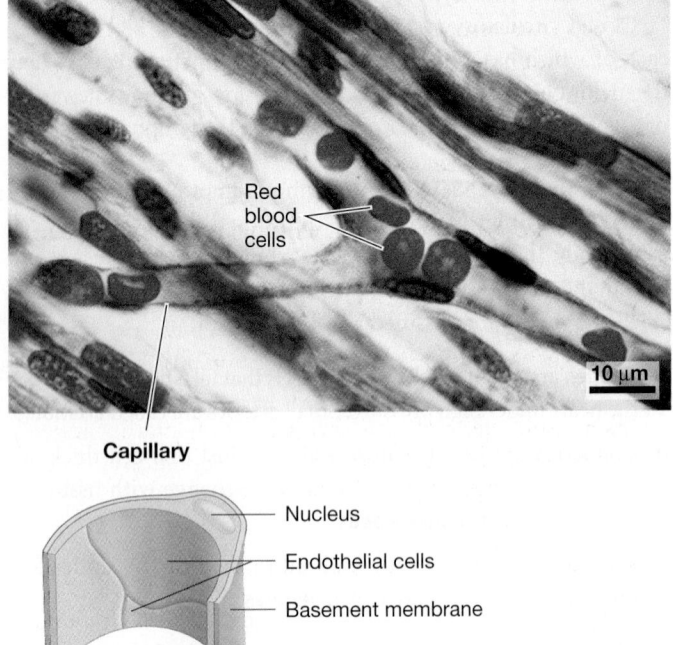

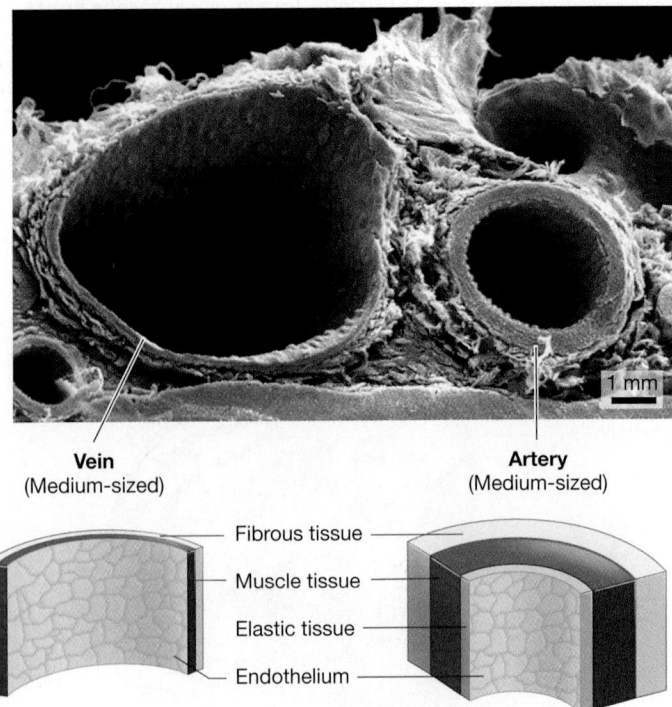

FIGURE 44.19 Capillaries, Veins, and Arteries. Notice the differences in relative wall thickness and overall size. The vein and artery in part (b) have been colorized blue and red respectively.

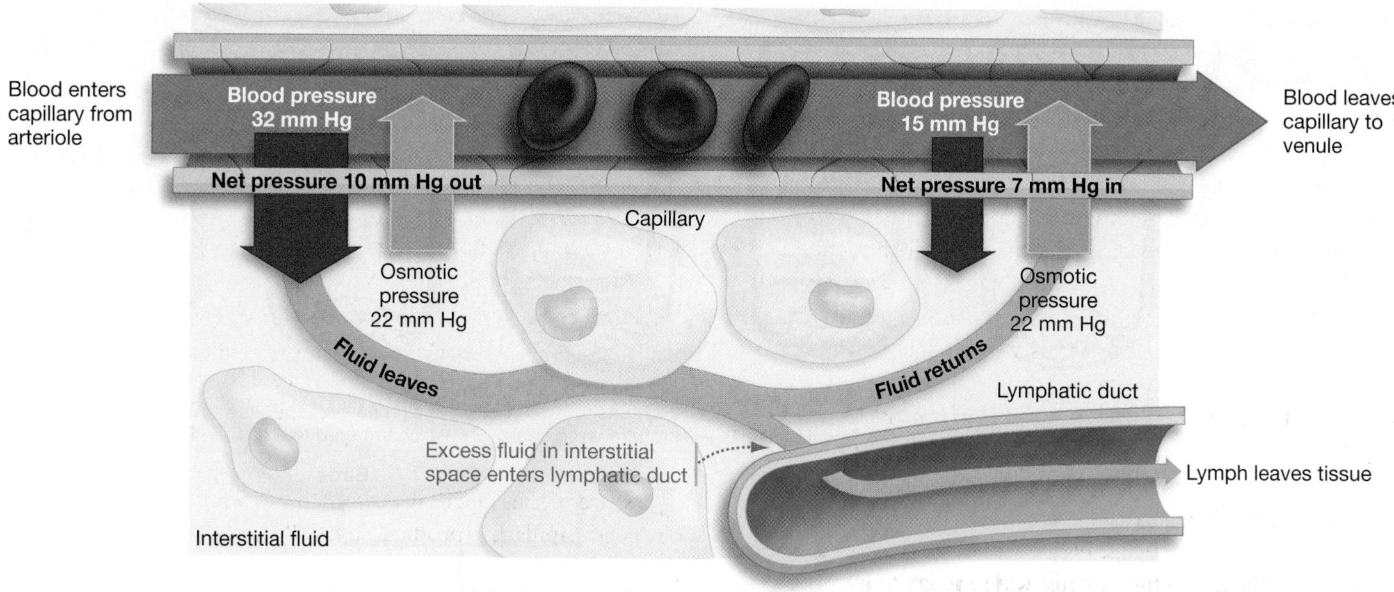

FIGURE 44.20 Pressure Differences in Capillaries Create Interstitial Fluid and Lymph. The balance of blood pressure and osmotic forces favors fluid loss from the beginning (inflow end) of capillaries and fluid recovery at the other (outflow) end. Fluid that is not recovered by the capillaries is transported out of the tissue as lymph, which eventually rejoins the blood circulation.

2. There is also an inward-directed osmotic force in capillaries, created by the higher concentration of solutes in the blood plasma than in the interstitial space.

Starling reasoned that at the end of the capillary nearest to an arteriole, the hydrostatic force would exceed the osmotic force. If so, then fluid would move out of the capillary into the interstitial space. But because blood pressure drops as fluid passes through a long, thin tube, Starling proposed that at the venous end of the capillary, the inward-directed osmotic force would exceed the outward-directed hydrostatic force. Thus, the fluid that was lost at the arteriolar end of the capillary would be largely reclaimed at the venous end.

Note the adverb "largely," however—not all interstitial fluid is reabsorbed by capillaries.

THE ROLE OF THE LYMPHATIC SYSTEM   Starling proposed that any interstitial fluid that was not reclaimed in the capillary would be collected in the **lymphatic system**: a collection of thin-walled, branching tubules called lymphatic ducts or vessels. Interstitial fluid that enters the lymphatic ducts is called **lymph**.

Starling's model was based on three key observations about lymphatic ducts:

1. They permeate all tissues.

2. They eventually join with one another, like the tributaries of a river, to form larger vessels.

3. The largest lymphatic vessels return excess interstitial fluid, in the form of lymph, to the major veins entering the heart.

The model also rested on a key assumption, however: For Starling's ideas about events in the capillaries, interstitial fluid, and lymphatic ducts to work, the total solute concentrations in plasma and in interstitial fluid must be different enough to bring fluid into capillaries via osmosis.

Specifically, the fluid that leaks out of capillaries must have low osmolarity. For this to be true, capillaries have to act as filters that retain large protein molecules and other solutes as fluid leaks out into the surrounding tissues.

To test this assumption, a group of biologists placed small tubes in the lymphatic ducts of animals to collect lymph. Then the team introduced radioactively labeled proteins of various sizes into the blood inside veins. They found that even though small proteins crossed readily from the blood to the lymph, larger proteins did not. Enough large proteins remained in the capillaries to set up the osmotic gradient required by Starling's model.

Follow-up work has shown that the most important of these large proteins is **albumin**. Albumin's large size and net negative charge effectively exclude it from interstitial fluid. The presence of albumin in blood keeps solute concentrations high, maintaining a strong osmotic gradient that brings fluid back into capillaries.

As a result, the total daily formation of lymph is limited to just 2–5 percent of plasma volume. This amount is easily taken up by lymph vessels and returned to the blood.

How Does the Heart Work?

In animals with closed circulatory systems, the heart contains at least two chambers. There is at least one thin-walled **atrium** (plural: **atria**), which receives blood, and at least one thick-walled **ventricle**, which generates the force required to propel blood out of the heart and through the circulatory system. Atria are separated from ventricles by atrioventricular valves.

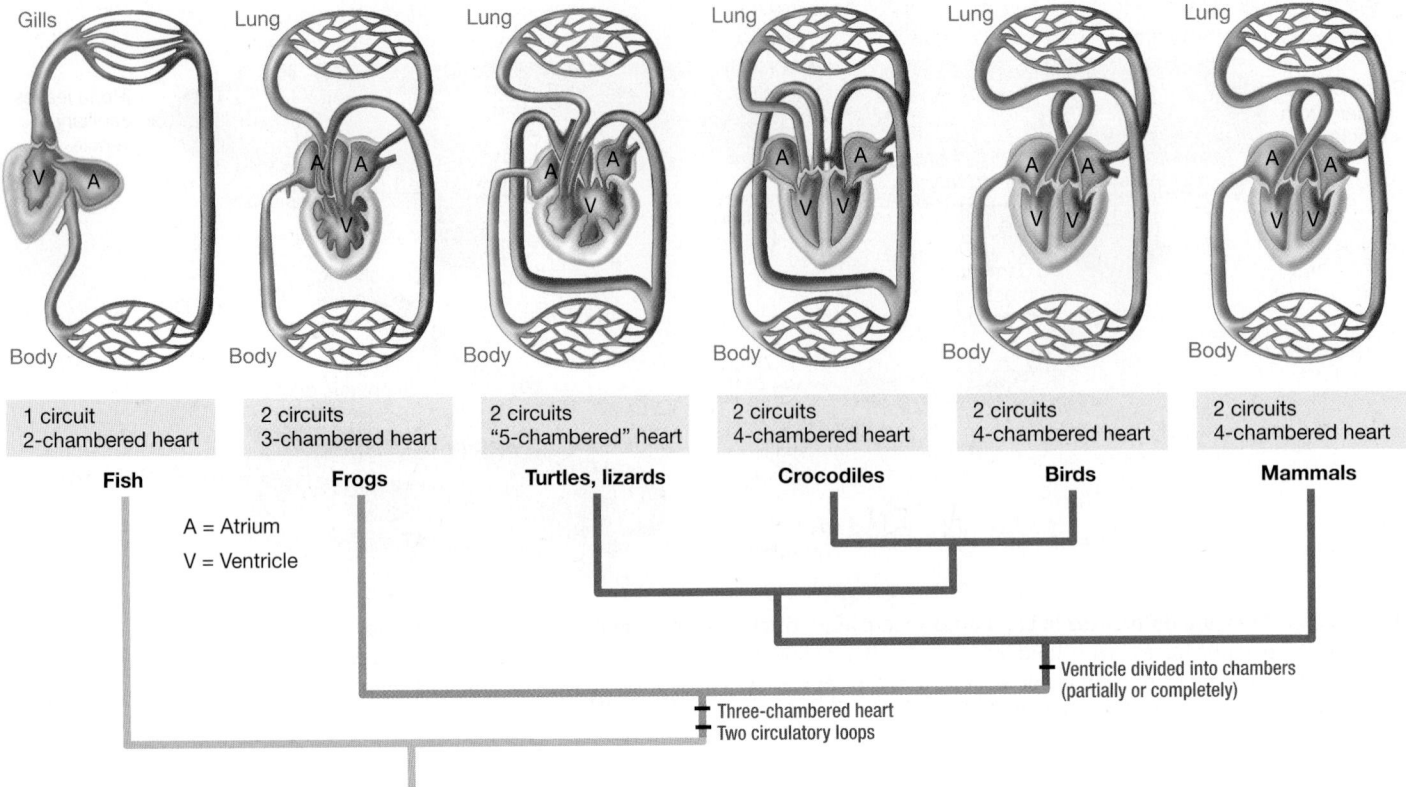

FIGURE 44.21 The Evolution of the Vertebrate Circulatory System. "A" denotes the atria—chambers that receive blood coming into the heart from the body and the lungs. "V" denotes the ventricles—chambers that pump blood out to the lungs and the body.

The phylogenetic tree in **Figure 44.21** shows the evolutionary relationships among some major vertebrate groups and a simplified sketch of the heart and circulatory system for each lineage. Two points are particularly important to note:

1. The number of distinct chambers in the heart increased as vertebrates diversified. Fish hearts have two chambers; amphibians have three; turtles and lizards have five (if the partially divided ventricle is counted as three chambers); crocodiles, birds, and mammals have four.

2. In fish, the circulatory system forms a single circuit—one loop services the gills and the body. In other lineages, there are separate circuits to the lungs and to the body.

WHY DID MULTICHAMBERED HEARTS AND MULTIPLE CIRCULATIONS EVOLVE? To understand these evolutionary patterns, let's first consider the relatively simple heart and circulatory system found in fish. A two-chambered heart and single circuit are adequate in fish. Even though blood pressure drops as blood passes through the gills—due to the mechanical resistance to flow that occurs in the gills' capillary beds—blood pressure stays high enough to move blood throughout the body. This is largely because fish live in the neutrally buoyant environment of water, where gravity does not have a large impact on blood flow.

In contrast, gravity has a much larger effect on circulation in land-dwelling vertebrates. Gravity is particularly important in opposing blood flow to elevated portions of the body. To overcome gravity in terrestrial environments, blood must be pumped at high pressure. However, the capillaries and alveoli of the lungs are too thin to withstand high pressures.

The successful solution in land-dwelling vertebrates was the evolution of two separate pumping circuits:

- The **pulmonary circulation** is a lower-pressure circuit to and from the lung.

- The **systemic circulation** is a higher-pressure circuit to and from the rest of the body.

The pulmonary and systemic circulations are completely separated in the four-chambered hearts of birds and mammals, which are essentially two fish hearts side by side.

The pulmonary and systemic circulations are only partially separated in the three-chambered hearts of amphibians and the "five-chambered" hearts of turtles and lizards, however. In addition, some of these animals have a bypass vessel running from the right side of the ventricle directly into the systemic circulation. This bypass vessel is also observed in the unusual four-chambered hearts of crocodiles.

The bypass vessels observed in certain lineages have an important function: They shunt blood from the pulmonary to the systemic circulation when an animal is underwater. As a result, the individual greatly reduces blood flow to the lungs when it is not breathing.

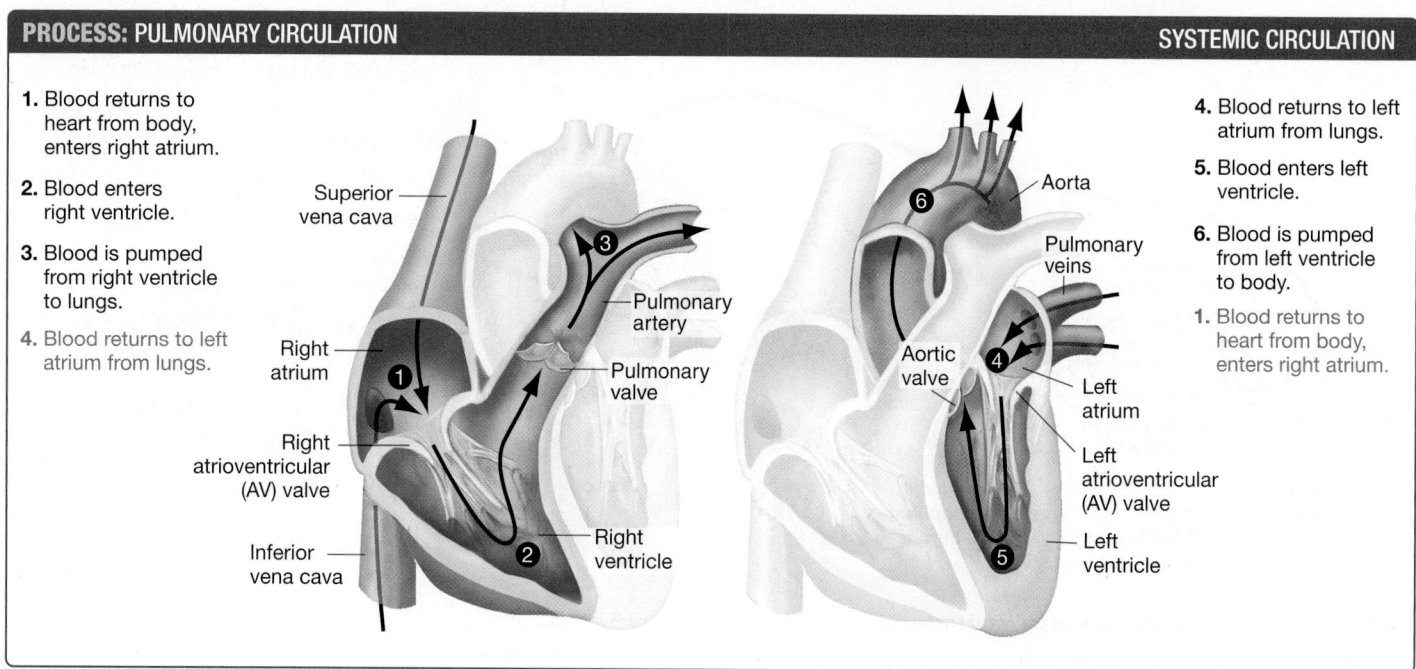

1. Blood returns to heart from body, enters right atrium.

2. Blood enters right ventricle.

3. Blood is pumped from right ventricle to lungs.

4. Blood returns to left atrium from lungs.

Superior vena cava

Right atrium

Right atrioventricular (AV) valve

Inferior vena cava

Pulmonary artery

Pulmonary valve

Right ventricle

4. Blood returns to left atrium from lungs.

5. Blood enters left ventricle.

6. Blood is pumped from left ventricle to body.

1. Blood returns to heart from body, enters right atrium.

Aorta

Pulmonary veins

Aortic valve

Left atrium

Left atrioventricular (AV) valve

Left ventricle

FIGURE 44.22 The Human Heart. Blood flows through a four-chambered heart in the sequence shown.

THE HUMAN HEART   Your heart is located in the chest cavity between your lungs, and is roughly the size of your fist. As **Figure 44.22** shows, the circulatory system returns blood from the body to the right atrium of the human heart. This blood is low in oxygen, and arrives via two large veins called the inferior (lower) and superior (upper) **venae cavae** (singular: **vena cava**).

When the muscles that line the right atrium contract, they send deoxygenated blood to the right ventricle. The right ventricle, in turn, contracts and sends blood out to the lungs, via the **pulmonary artery**. In this way, the right ventricle powers the movement of blood through the pulmonary circulation. Blood that returns to the heart from the body is sent to the lungs.

Blood flows from atrium to ventricle to artery only one way, because one-way valves separate the heart's chambers from each other and from the adjacent blood vessels. As Figure 44.22 indicates, the valves are flaps, oriented in a way that ensures a one-way flow of blood with little or no backflow. If heart valves are damaged or defective, the resulting backflow can be heard through a stethoscope. The backflow reduces the organ's efficiency and is called a **heart murmur**.

After blood circulates through the capillary beds in the lung's alveoli and becomes oxygenated, it returns to the heart through the **pulmonary veins**. The oxygenated blood enters the left atrium.

When the left atrium contracts, it pushes blood into the left ventricle. The walls of the left ventricle are so thick with muscle cells that their contraction sends oxygenated blood at high pressure through the aorta and into the arteries and capillaries that form the systemic circulation.

Figure 44.23 on page 880 summarizes the flow pattern through the human circulatory system and the blood gas concentrations at various points in the pulmonary and systemic cir-culations. Notice that blood vessels are called arteries or veins according to the direction of blood flow relative to the heart, not because of the oxygen content of the blood in them. Thus, the pulmonary artery is called an artery because it takes blood away from the heart, even though this blood is low in oxygen.

In healthy individuals, gas exchange in the lungs and tissues is rapid relative to the rate at which blood flows through capillaries. This situation is beneficial: It ensures that the maximum amount of oxygen is taken up and the maximum amount of carbon dioxide is released. As Figure 44.23 indicates, the partial pressures of oxygen and carbon dioxide in the alveoli and the pulmonary veins are equal.

Similarly, the slow passage of blood through capillaries means that a maximum amount of oxygen is released from blood to tissues and a maximum amount of carbon dioxide is taken up. As a result, partial pressures for oxygen and carbon dioxide in the tissues and the systemic veins are equal.

THE CARDIAC CYCLE   Although the preceding discussion suggests that each chamber in the heart acts independently, this is not the case. The contraction phase of the atria and the ventricles, called **systole**, is closely coordinated with their relaxation phases, or **diastole**.

For example, both atria contract simultaneously about 0.1 second ahead of ventricular contraction. Because both ventricles are still in diastole and relaxed when the atria contract, the ventricles fill with blood from the atria. Then the ventricles enter systole and contract.

This sequence of contraction and relaxation is called the **cardiac cycle**. It consists of one diastole and one systole for both atria and ventricles.

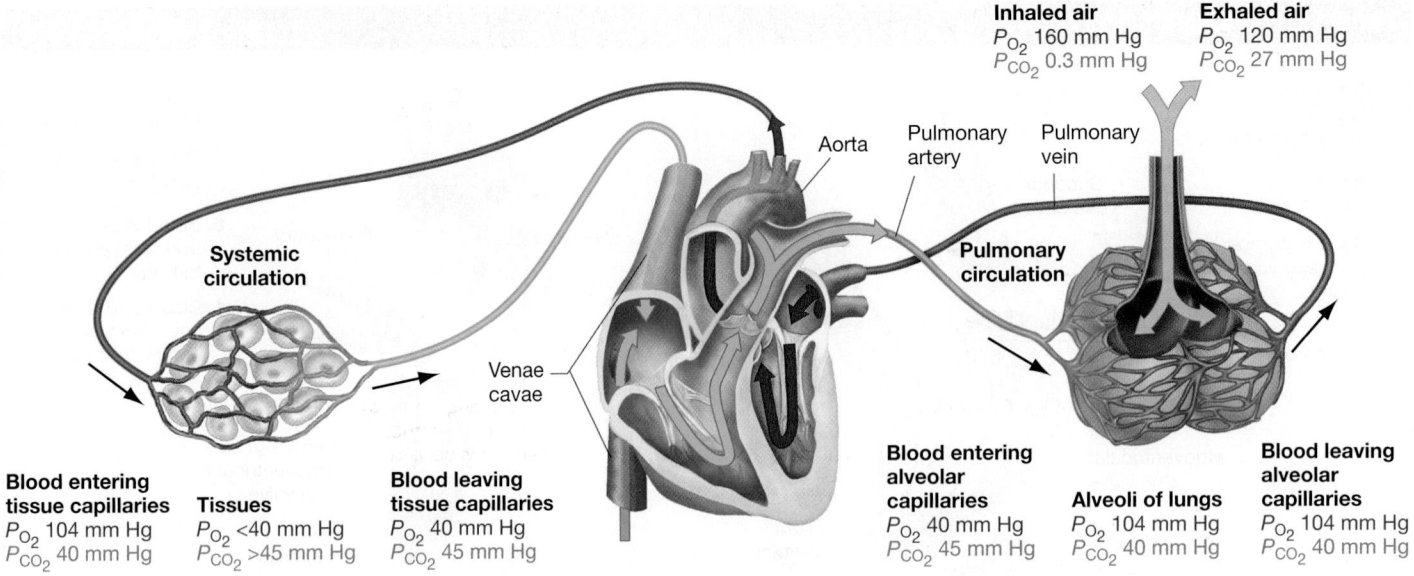

Inhaled air
P_{O_2} 160 mm Hg
P_{CO_2} 0.3 mm Hg

Exhaled air
P_{O_2} 120 mm Hg
P_{CO_2} 27 mm Hg

Aorta

Pulmonary artery

Pulmonary vein

Systemic circulation

Pulmonary circulation

Venae cavae

Blood entering tissue capillaries
P_{O_2} 104 mm Hg
P_{CO_2} 40 mm Hg

Tissues
P_{O_2} <40 mm Hg
P_{CO_2} >45 mm Hg

Blood leaving tissue capillaries
P_{O_2} 40 mm Hg
P_{CO_2} 45 mm Hg

Blood entering alveolar capillaries
P_{O_2} 40 mm Hg
P_{CO_2} 45 mm Hg

Alveoli of lungs
P_{O_2} 104 mm Hg
P_{CO_2} 40 mm Hg

Blood leaving alveolar capillaries
P_{O_2} 104 mm Hg
P_{CO_2} 40 mm Hg

FIGURE 44.23 Partial Pressures of Gases Vary throughout the Human Circulatory System.

✔ **QUESTION** Why are the partial pressures of oxygen and carbon dioxide in exhaled air intermediate in magnitude between the partial pressures in inhaled and alveolar air?

Ventricular contraction (ventricular systole) leads to a rapid increase in pressure within both ventricles, as recorded in the purple line in **Figure 44.24**. Blood is ejected into the pulmonary artery and the aorta when ventricular pressure equals the pressure within each respective artery. Blood pressure measured in the systemic arterial circulation at the peak of ventricular ejection into the aorta is called the **systolic blood pressure**. Blood pressure measured just prior to ventricular ejection is called the **diastolic blood pressure**.

Clinicians report blood pressure measurements in fractional notation, with systolic pressure as the numerator and diastolic pressure as the denominator. People with blood pressures consistently higher than 140/90 mm Hg have high blood pressure, or **hypertension**.

Hypertension is a serious concern to physicians because it can lead to a variety of defects in the heart and circulatory systems. Abnormally high blood pressure puts mechanical stress on arteries. If the walls of an artery fail, the individual may experience heart attack, stroke, kidney failure, and burst or dilated blood vessels.

To review the anatomy of the mammalian heart and how the cardiac cycle works, go to the study area at *www.masteringbiology.com*.

(MB) **Web Activity** The Human Heart

ELECTRICAL ACTIVATION OF THE HEART Like other muscle cells, cardiac muscle cells contract in response to electrical signals. In invertebrates, the electrical impulses that trigger contraction come directly from the nervous system. But in vertebrates, a group of cells in the heart itself is responsible for generating the initial signal.

The cells that initiate contraction in the vertebrate heart are known as **pacemaker cells**. They are located in a region of the right atrium called the **sinoatrial (SA) node**.

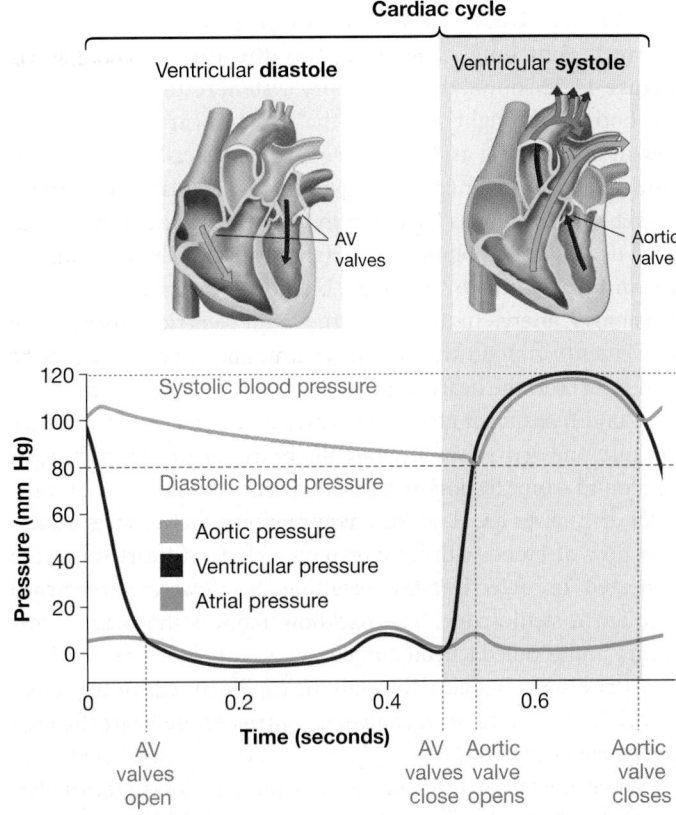

Cardiac cycle

Ventricular **diastole**

Ventricular **systole**

AV valves

Aortic valve

Systolic blood pressure

Diastolic blood pressure

Aortic pressure

Ventricular pressure

Atrial pressure

Pressure (mm Hg): 0, 20, 40, 60, 80, 100, 120

Time (seconds): 0, 0.2, 0.4, 0.6

AV valves open

AV valves close

Aortic valve opens

Aortic valve closes

FIGURE 44.24 Blood Pressure Changes during the Cardiac Cycle. These data show the pressures created by the left ventricle and left atrium in the cardiac cycle. In this case, the blood pressure measured in the upper arm would be 120/80 mm Hg. The right ventricle and pulmonary artery pressures follow a similar pattern, but the blood pressure measured in the pulmonary artery would be much lower—closer to 25/8 mm Hg.

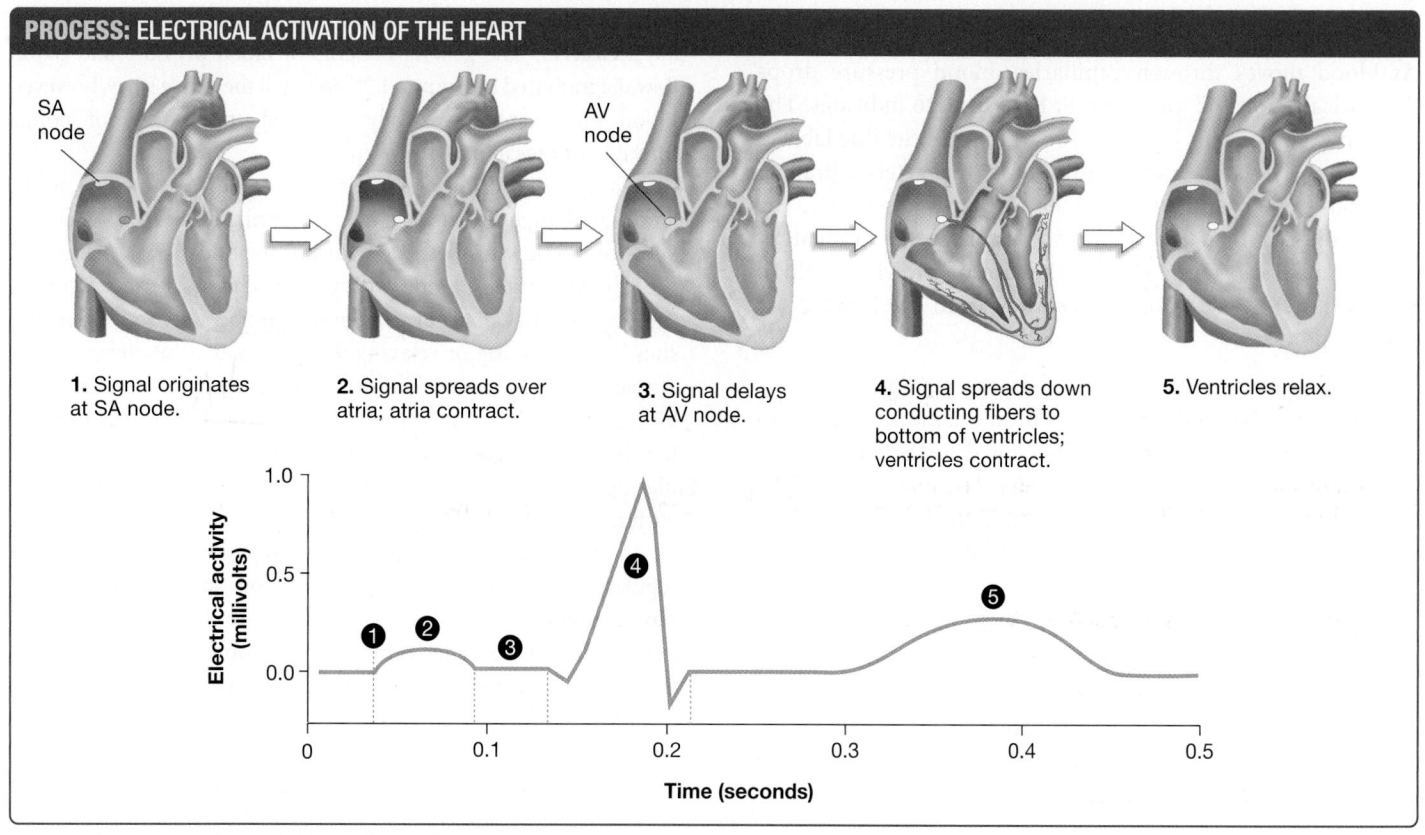

FIGURE 44.25 Sequential Electrical Activation Leads to Coordinated Contraction of the Heart.

The SA node and the muscle cells of the heart receive input from the nervous system and from chemical messengers carried in the blood. These inputs are important for regulating both the heart rate and the strength of ventricular contraction. In this way, the amount of blood moving through the circulatory system varies in response to electrical signals and hormones. During the "fight or flight" response introduced in Chapter 47, for example, a chemical signal called epinephrine causes both heart rate and contraction strength to increase—sending more blood through the body in preparation for rapid movement.

Even though heart rate and contraction strength are carefully regulated, a vertebrate heart will beat even if all nerves supplying it are severed. Why? An electrical impulse that stimulates contraction is generated in the SA node and rapidly conducted throughout the right and left atria. The signal spreads quickly from cell to cell because of a striking property of cardiac muscle cells: They form physical and electrical connections with each other.

All cardiac muscle cells branch to contact several other cardiac muscle cells, join end to end with these neighboring cells, and connect to them by specialized structures called **intercalated discs**. Because these discs contain numerous gap junctions—cell-to-cell connections described in Chapter 8—electrical signals pass directly from one cardiac muscle cell to the next.

To understand electrical activation of the heart, follow the orange line on the graph in **Figure 44.25**. This line is called an **electrocardiogram**, or **EKG**—a recording of the electrical events that occur over the course of a cardiac cycle. The drawings above the graph show where key events are happening.

1. The SA node originates a signal.

2. The signal from the SA node is propagated over the atria. As a result, the atria contract simultaneously and fill the ventricles.

3. The signal from the atria is conducted to an area of the heart called the **atrioventricular (AV) node**. The AV node delays the signal slightly before passing it to the ventricles. The delay allows the ventricles to fill completely with blood before they contract.

4. After the delay, the electrical impulse is rapidly transmitted through specialized fibers in the muscular wall that separates the ventricles. The impulse spreads through both ventricles, causing them to contract as the atria relax. The ventricles empty efficiently because the signal and the resulting muscular contraction move from the bottom up to the top of each ventricle—toward the arteries that allow blood to exit.

5. The final electrical event occurs as the ventricles relax and their cells recover—restoring their electrical state prior to contraction.

An EKG recording is generated by amplifying the overall electrical signal conducted from the heart to the chest wall through the tissues of the body. By inspecting an EKG, physicians can diagnose disturbances of heart rhythm and detect damage to the heart muscle.

Patterns in Blood Pressure and Blood Flow

As blood moves through capillaries, blood pressure drops dramatically—as the top graph in **Figure 44.26** indicates. The bottom graph on this figure shows why. Trace the line labeled "Total Area" on this graph, and note that as arteries branch, rebranch, and eventually form networks of capillaries, the total cross-sectional area of blood vessels in the circulatory system increases—even though the size of individual blood vessels decreases as blood moves away from the heart. Note two key points:

1. Blood pressure in the capillaries drops dramatically (on the top graph) because the amount of mechanical resistance to flow is a function of total cross-sectional area of the vessels—meaning that friction losses are high in capillary beds. The pulsing nature of the pressure noted in the larger arteries diminishes as a result of this pressure drop.

2. As the line labeled "Velocity" in the bottom graph indicates, the velocity of blood flow also decreases significantly in capillary beds relative to arteries and veins, because the same amount of fluid is passing through a much larger area. Recall that the slow rate of blood flow through capillaries is important: It provides sufficient time for gases, nutrients, and wastes to diffuse between tissues and blood.

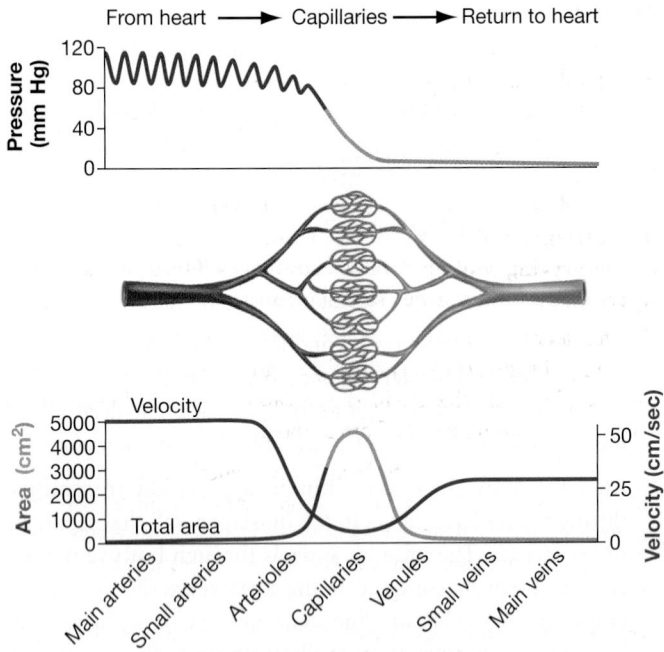

FIGURE 44.26 Blood Pressure Drops Dramatically in the Circulatory System. The top graph shows how blood pressure changes as blood leaves the heart and travels through arteries, capillaries, and veins, as in the branching pattern of the middle diagram. In arteries near the heart, each heartbeat causes fluctuations in blood pressure. These pressure pulses disappear in the capillaries, so blood flows there at a steady speed. The bottom graph plots the total area of blood vessels shown in the diagram, as well as the velocity of blood flow through the vessels.

WHY IS REGULATION OF BLOOD PRESSURE AND BLOOD FLOW IMPORTANT? The general patterns of blood pressure and blood flow diagrammed in Figure 44.26 don't tell the entire story, however. Blood movement is carefully regulated at an array of points throughout the circulatory system.

Recall that arterioles have muscular sphincters in their walls, and that changes in the size of these sphincters can constrict or allow blood flow to specific regions of the body. This means that the nervous system, along with certain chemical messengers in the circulation, can accurately control blood flow to various tissues by contracting or relaxing the arteriolar sphincters. For example, arterioles in the skin may dilate during exercise, diverting blood flow to the skin to eliminate excess heat. This accounts for the flushed facial appearance induced by vigorous exercise. Regulating blood flow is important for maintaining homeostasis with respect to body temperature.

As another example of how blood pressure and blood flow are regulated, consider what happens if you sit long enough for blood to pool in your lower extremities, under the influence of gravity. If you stand up rapidly, your blood pressure can drop enough to limit blood flow to the brain and cause dizziness or even a blackout. More serious drops in blood pressure, due to severe dehydration or blood loss, can be fatal. Fortunately, decreases in blood pressure elicit a powerful homeostatic response.

HOMEOSTATIC CONTROL OF BLOOD PRESSURE Recall from Chapter 41 that all homeostatic responses involve (**1**) sensors that detect the change in condition, (**2**) an integrator that processes information about the change, and (**3**) effectors that diminish the impact of the change.

Specialized pressure-sensing nerve cells called **baroreceptors** detect decreases in blood pressure. Baroreceptors are found in the walls of the heart and the major arteries, both as they leave the heart and as they enter the neck. This distribution is logical, because the head tends to be the most elevated part of land-dwelling animals and the preservation of blood flow to the brain is the highest priority.

When baroreceptors transmit signals to the brain indicating a serious fall in blood pressure, a rapid, three-component effector response ensues:

1. Cardiac output—the volume of blood leaving the left ventricle—increases. This is due to an increase in heart rate and an increase in stroke volume—meaning, the amount of blood ejected from the ventricles during each cardiac cycle. (Cardiac output = heart rate × stroke volume.)

2. Arterioles that serve the capillaries of tissues such as the skin and intestines, which can endure short-term restrictions in their blood supply without damage, constrict. Arteriolar constriction diverts blood to more critical organs.

3. Veins constrict, decreasing their overall volume. Because more than half of the blood in the circulatory system is contained within the veins, constriction of these vessels shifts blood volume toward the heart and arteries to maintain blood pressure and flow to vital organs.

This coordinated response is mediated both by a portion of the nervous system called the sympathetic nervous system (Chapter 45) and by hormones (Chapter 47) produced by the adrenal glands. Sympathetic nerves and the hormones involved in blood pressure regulation deliver their messages directly to the SA node to increase heart rate, the ventricles of the heart to increase stroke volume, the sphincters of the arterioles to modify their resistance, and the muscular walls of the veins to modify their total volume.

The homeostatic response to falling blood pressure resembles the **fight-or-flight response**, which is triggered by a threatening stimulus (see Chapter 47). During the fight-or-flight response, an organism prepares to defend itself or rapidly escape from danger. Part of this response involves directing blood to the brain and muscles in preparation for quick action. The side effects include the "cold sweat" and nausea often induced by a fearful situation. These feelings occur because blood flow is directed away from the skin and intestines.

CHECK YOUR UNDERSTANDING

If you understand that . . .

- Animal circulatory systems may be open or closed, but both types of systems circulate blood or hemolymph via pressure generated by one or more hearts.
- In closed systems, regulated changes in the diameter of blood vessels can direct blood to specific regions, and overall blood pressure is carefully regulated through changes in cardiac output.

✔ **You should be able to . . .**

1. Explain how the mammalian lymphatic system and circulatory system interact.
2. Make a labeled diagram showing how blood circulates through the mammalian heart.

Answers are available in Appendix B.

CHAPTER 44 REVIEW

For media, go to the study area at www.masteringbiology.com

Summary of Key Concepts

Animals have to take in oxygen and expel carbon dioxide to sustain cellular respiration and stay alive. Terrestrial animals and aquatic animals face different challenges in performing gas exchange.

- As media for exchanging oxygen and carbon dioxide, air and water are dramatically different.

- Compared with water, air contains much more oxygen and is much less dense and viscous. As a result, terrestrial animals have to process a much smaller volume of air to extract the same amount of O_2, and don't have to work as hard to do so.

- Both terrestrial and aquatic animals pay a price for exchanging gases: Land-dwellers lose water to evaporation; freshwater animals lose ions and gain excess water; marine animals gain sodium and chloride and lose water.

  ✔ You should be able to explain why no water breathers have the extremely high rates of metabolism required for endothermy.

Gas-exchange organs maximize the rate of O_2 and CO_2 diffusion by (1) presenting a large, thin surface area to the environment, and (2) maintaining a steep partial-pressure gradient that favors entry of O_2 and elimination of CO_2.

- The structure of gills, tracheae, lungs, and other gas-exchange organs minimizes the cost of ventilation while maximizing the rate at which O_2 and CO_2 diffuse.

- Consistent with predictions made by Fick's law of diffusion, respiratory epithelia tend to be extremely thin and folded to increase surface area.

- In fish gills, countercurrent exchange ensures that the partial-pressure differences between O_2 and CO_2 in water and blood are high over the entire length of the ventilatory surface.

- In bird lungs, structural adaptations lead to a high ratio of useful ventilatory space to dead space.

- Breathing rate is regulated, to keep the carbon dioxide content of the blood stable during both rest and exercise.

  ✔ You should be able to explain why no large animals have gas exchange occurring only across the skin.

Blood is a specialized tissue that transports gases, along with nutrients and wastes, in some animals. Hemoglobin is an oxygen-carrying protein that is extremely efficient at taking up oxygen in the lungs and delivering it to tissues.

- The tendency of hemoglobin to give up oxygen varies as a function of the P_{O_2} in surrounding tissue in a sigmoidal fashion. As a result, a relatively small change in tissue causes a large change in the amount of oxygen released from hemoglobin.

- Oxygen binds less tightly to hemoglobin when pH is low. Because CO_2 tends to react with water to form carbonic acid, the existence of high CO_2 partial pressures in exercising muscle tissues lowers their pH and makes oxygen less likely to stay bound to hemoglobin and more likely to be released into tissues.

- The CO_2 that diffuses into red blood cells from tissues is rapidly converted to carbonic acid by the enzyme carbonic anhydrase. The protons that are released as carbonic acid dissociates bind to deoxygenated hemoglobin. In this way, hemoglobin acts as a buffer that takes protons out of solution and prevents large fluctuations in blood pH.

  ✔ You should be able to explain how oxygen and carbon dioxide are transported in blood in the icefish native to the Antarctic, since these fish lack hemoglobin.

(MB) **BioFlix™** Gas Exchange, **Web Activity** Gas Exchange in the Lungs and Tissues

Circulatory systems use the pressure generated by one or more hearts to transport blood and other substances throughout the body.

- In many animals, blood or hemolymph moves through the body via a circulatory system consisting of a pump (heart) and vessels.

- In open circulatory systems, overall pressure is low and tissues are bathed directly in hemolymph.

- In closed circulatory systems, blood is contained in vessels that form a continuous circuit. Containment of blood allows higher pressures and flow rates, as well as the ability to direct blood flow accurately to tissues that need it the most.

- In organisms with a closed circulatory system, a lymphatic system returns excess fluid that leaks from the capillaries back to the circulation.

- In mammals and birds, a four-chambered heart pumps blood into two circuits, which separately serve the lungs and the rest of the body.

- The cardiac cycle is controlled by electrical signals that originate in the heart itself.

- Heart rate, cardiac output, and contraction of both arterioles and veins are regulated by chemical signals and by electrical signals from the brain.

✔ You should be able to describe the circulatory systems expected to be found in terrestrial animals with high activity rates.

(MB) **Web Activity** The Human Heart

Questions

✔ TEST YOUR KNOWLEDGE
Answers are available in Appendix B

1. O_2 will diffuse from blood to tissue faster in response to which of the following conditions?
   a. an increase in the P_{O_2} of the tissue
   b. a decrease in the P_{O_2} of the tissue
   c. an increase in the thickness of the capillary wall
   d. a decrease in the surface area of the capillary

2. Which of the following does blood *not* do?
   a. transport O_2 and CO_2
   b. distribute body heat
   c. produce red blood cells and other formed elements
   d. buffer against pH changes

3. In insects, what is the adaptive significance of spiracles?
   a. They open and close during flight or other types of movement, so function as a "breathing" mechanism.
   b. They open into the body cavity, allowing direct contact between hemolymph and tissues.
   c. They are thin and highly branched, so offer a large surface area for gas exchange.
   d. They close off tracheae to minimize water loss.

4. Which of the following is *not* an advantage of breathing air over breathing water?
   a. Air is less dense than water, so it takes less energy to move during ventilation.
   b. Oxygen diffuses faster through air than it does through water.
   c. The oxygen content of air is greater than that of an equal volume of water.
   d. Air breathing leads to high evaporation rates from the respiratory surface.

5. Which of the following promotes oxygen release from hemoglobin?
   a. a decrease in temperature   c. a decrease in pH
   b. a decrease in CO_2 levels   d. a decrease in carbonic anhydrase

6. An open circulatory system is less efficient than a closed circulatory system in what respect?
   a. It is harder to deliver O_2 to specific tissues based on need.
   b. Hemolymph does not contain oxygen-binding molecules.
   c. There is no heart to pump the blood.
   d. In closed systems, body movements cannot help circulate the blood.

✔ TEST YOUR UNDERSTANDING
Answers are available in Appendix B

1. Compare and contrast the respiratory and circulatory systems of an insect and a human.

2. Describe the changes in oxygen delivery that occur as a person changes from a resting state to intense exercise.

3. Explain how most carbon dioxide is transported in the blood. In humans, why don't intense exercise and rapid production of CO_2 lead to a rapid reduction in blood pH?

4. Why is ventilation in birds considered much more efficient than the respiratory system of humans and other mammals?

5. Explain how each parameter in Fick's law of diffusion is reflected in the structure of the mammalian lung.

6. Why did separate systemic and pulmonary circulations evolve in species that have the high-pressure circulatory system required for rapid movement of blood?

✔ APPLYING CONCEPTS TO NEW SITUATIONS
Answers are available in Appendix B

1. Researchers who compared the total amount of ventilatory surface area in the gills of various species of fish found that species that do a lot of high-speed swimming have a larger total surface area than do slow-moving species of the same size. Interpret this pattern in light of the theory of evolution by natural selection.

2. At high elevations, certain cells in humans increase the production of 2,3-diphosphoglycerate (DPG). In blood, DPG shifts the oxygen-hemoglobin equilibrium curve to the right. How does this shift affect O_2 unloading and loading?

3. Carp are fish that thrive in stagnant-water habitats with low-oxygen partial pressures. Compared with the hemoglobin of many other fish species, carp hemoglobin has an extremely high affinity for O_2. Is this trait adaptive? Explain your answer.

4. The carbon monoxide (CO) found in furnace and engine exhaust and in cigarette smoke binds to the heme groups in hemoglobin 210 times more tightly than does O_2. Explain why exposure to large doses of CO can lead to suffocation.

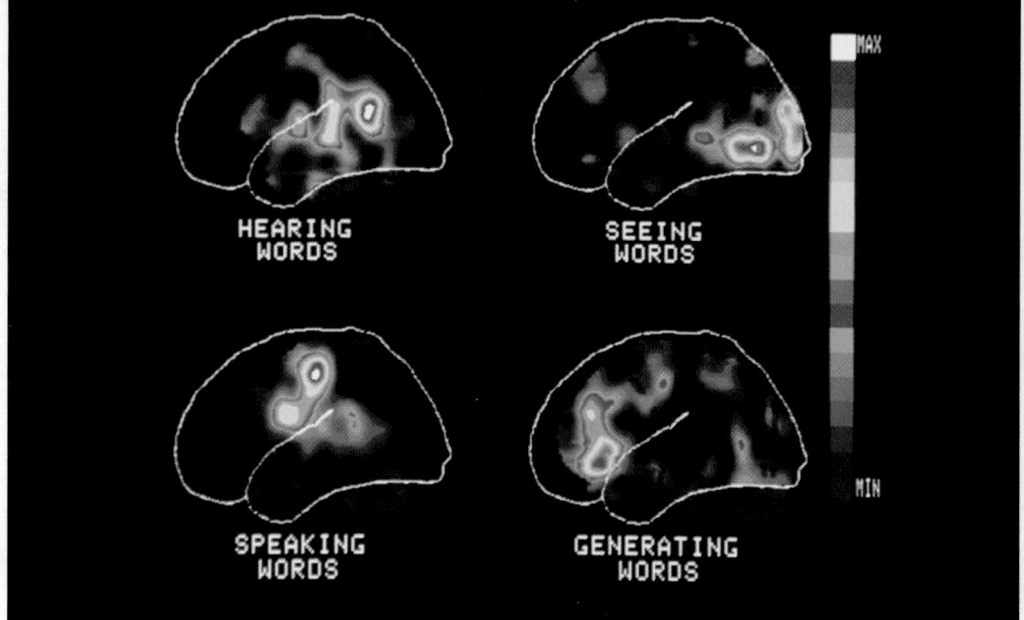

HEARING
WORDS

SEEING
WORDS

SPEAKING
WORDS

GENERATING
WORDS

MAX

MIN

Positron-emission tomography (PET) scans showing changes in the activity of brain cells during different tasks, such as remembering words and speaking them aloud. Different brain areas are specialized for performing different tasks.

Electrical Signals in Animals 45

Most students and professional biologists are attracted to the study of electrical signaling, or neurobiology, because they want to understand the human brain and higher-order processes like consciousness, intelligence, emotion, learning, and memory. But early in the history of neurobiology, researchers realized that the human brain—with its billions of cells—was much too complex to study productively.

Instead, biologists did the same thing they did in the early days of studying cells, genetics, and evolution: They started simple. Until recently, much of the most productive work in neurobiology has focused on how individual nerve cells, or **neurons**, work.

Neurons conduct information in the form of electrical signals. Electrical signals are fast and precise. Neurons conduct them from point to point in the body at speeds of up to 200 m/sec (450 mph). Electrical signaling is a crucial aspect of information processing—one of the five attributes of life introduced in Chapter 1.

Most neurons work in the same general way. Early research on the electrical properties of a single neuron laid a broad foundation for more recent studies of how the brain works.

This chapter works the same way. After a brief overview of nervous systems, we'll focus on the neuron itself and analyze the membrane proteins that dictate its electrical properties. By the end of the chapter, you'll be considering how the brain is organized and how phenomena like memory work.

45.1 Principles of Electrical Signaling

As Chapter 32 emphasized, the evolution of neurons—along with muscle cells—was a key event in the diversification of animals. All animals except sponges have neurons and muscle cells. Neurons transmit electrical signals; muscles can respond to electrical signals by contracting.

KEY CONCEPTS

○━ Neurons are cells that transmit electrical signals used in communication. Their plasma membranes carry a voltage, called a membrane potential, due to differences in the concentrations of ions on the inner and outer surfaces.

○━ Action potentials are all-or-none changes in membrane potential that serve as electrical signals. During an action potential, an inflow of sodium ions is followed by an outflow of potassium ions.

○━ At synapses, the electrical signal from a neuron triggers the release of a chemical signal—a neurotransmitter. When the neurotransmitter arrives at an adjacent neuron, there is a change in that cell's membrane potential.

○━ Most animals have a central nervous system (CNS) and a peripheral nervous system (PNS). PNS neurons receive sensory information and transmit it to the CNS for processing. The CNS then sends signals to muscles, glands, or other tissues via PNS neurons.

✔ When you see this checkmark, stop and test yourself. Answers are available in Appendix B.

The evolution of neurons and muscle cells made rapid movement possible. They were morphological innovations that enabled animals to evolve into today's flatworms, squid, beetles, sharks, and birds.

Chapter 32 also pointed out that neurons are organized into two basic types of nervous systems:

- The diffuse arrangement of cells called a **nerve net**, found in cnidarians (jellyfish, hydra, anemones) and ctenophores (comb jellies).

- A **central nervous system (CNS)** that includes large numbers of neurons aggregated into clusters called ganglia.

In most cases, animals with a CNS have a large cerebral ganglion, or brain, located in their anterior end. You might recall from Chapter 32 that this phenomenon is known as cephalization—the evolution of a bilaterally symmetric body with information gathering and processing structures located at the head end.

Cephalization made animals into efficient eating and moving machines: they faced the environment in one direction, with sensory appendages taking in information and sending it to a nearby brain for processing. After integrating information from an array of sensory cells, the brain sends electrical signals to muscles or other tissues that respond to the stimulus.

Types of Neurons in the Nervous System

The sensory cells that are responsible for information-gathering respond to light, sound, touch, or other stimuli. Sensory receptors in the skin, eyes, ears, and nose transmit streams of data about the external environment. Sensory cells inside the body monitor conditions that are important in homeostasis, such as blood pH and oxygen levels. In this way, sensory cells monitor conditions both outside and inside the body.

As **Figure 45.1** shows, a sensory receptor cell transmits the information it receives by means of a nerve cell known as a **sensory neuron**. (In many cases, the receptor cell itself also acts as a sensory neuron.)

In vertebrates, the sensory neuron sends the information to neurons in the brain or spinal cord via nerves; in other animals nerves send the information to the brain. **Nerves** are long, tough strands of nervous tissue. They contain thousands of projections from neurons and carry information to and from the brain and spinal cord.

Together, the brain and spinal cord form the CNS of vertebrates. One function of the CNS is to integrate information from many sensory neurons. Cells in the CNS called **interneurons** (literally, "between-neurons") perform this integration.

Interneurons also make connections between sensory neurons and **motor neurons**, which are nerve cells that send signals to effector cells in glands or muscles. Recall from Chapter 41 that effectors are structures that respond to stimuli.

All neurons and other components of the nervous system that are outside the CNS are considered part of the **peripheral nervous system**, or **PNS**. Section 45.4 describes the structure and function of the PNS in more detail.

Typically, sensory information from receptors in the PNS is sent to the CNS, where it is processed. Then a response is trans-

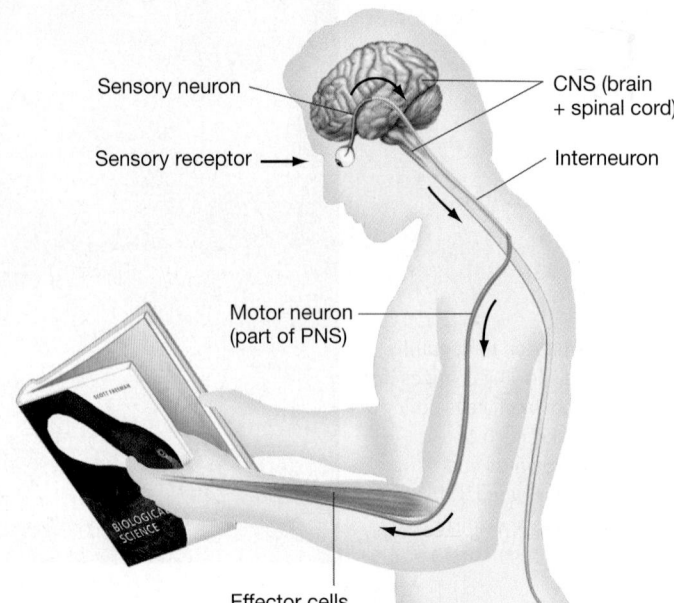

FIGURE 45.1 The Brain Integrates Sensory Information and Sends Signals to Effector Cells. In most cases, sensory neurons send information to the brain. There, the electrical signals are integrated with information from other sources. Once integration is complete, a response is sent to effector cells through motor neurons.

mitted back to appropriate parts of the body. When you stub your toe, pain receptors in your toe relay sensory information to the brain, which then modifies your movements to avoid further injury. You might hop a bit, for example, and then limp to avoid further contact with that toe.

The Anatomy of a Neuron

Neurons are difficult to study, because they are small, transparent, and morphologically complex. So when Camillo Golgi discovered that some neurons become visible when samples of preserved tissue are treated with a solution containing silver nitrate, his finding was a major advance. The year was 1898.

Through the early decades of the twentieth century, the work of Golgi and Santiago Ramón y Cajal revealed several important points about the anatomy of neurons. Most neurons have the same three parts, shown in **Figure 45.2a**:

1. a cell body, which contains the nucleus;

2. a highly branched group of relatively short projections called dendrites; and

3. a long projection called an axon, which may or may not branch.

Dendrites are rarely more than 2 mm long, but axons can be over a meter in length. The number of dendrites and their arrangement vary greatly from cell to cell. Further, many brain cells have only dendrites and lack axons.

A **dendrite** receives a signal from the axons of adjacent cells; a neuron's **axon** sends a signal to the dendrites and cell bodies of other neurons (**Figure 45.2b**). Dendrites collect signals; axons

(a) Information flows from dendrites to the axon.

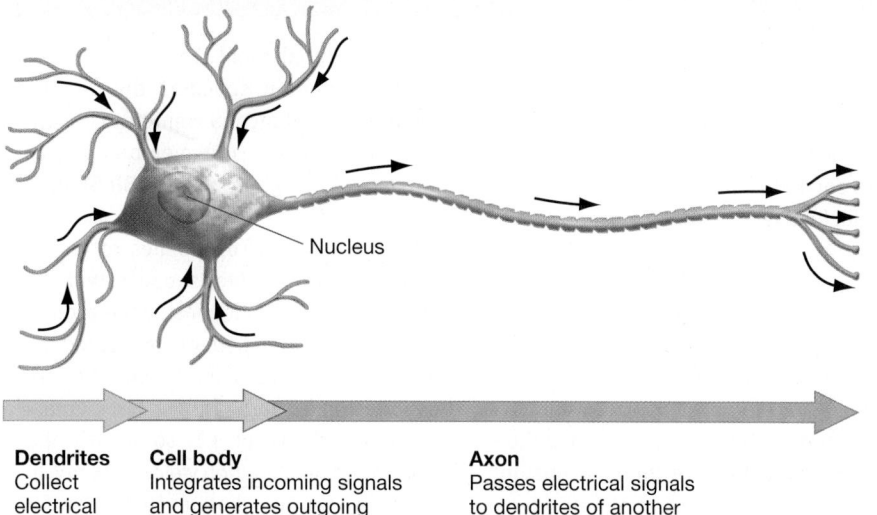

Dendrites
Collect electrical signals

Cell body
Integrates incoming signals and generates outgoing signal to axon

Axon
Passes electrical signals to dendrites of another cell or to an effector cell

(b) Neurons form networks for information flow.

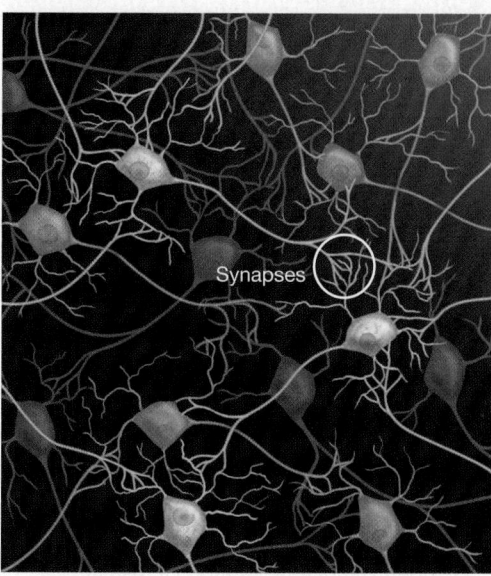

Synapses

FIGURE 45.2 How Does Information Flow in a Neuron? **(a)** The structure of a generalized neuron. **(b)** Most neurons receive inputs from several different neurons and send projections to several different neurons.

---

pass them on. The **cell body**, or **soma**, also receives signals from axons. The cell body is where incoming signals are integrated and an outgoing signal is sent to the axon.

Ramón y Cajal maintained that the plasma membrane of each neuron is distinct, and that the membranes of axons and dendrites meet at junctions called synapses. This hypothesis was confirmed in the 1950s, when images from electron microscopes showed that only a small subset of neurons make direct, cytoplasm-to-cytoplasm connections. In a nervous system, most interactions occur where the plasma membranes of two neurons meet at synapses.

Given that neurons transmit electrical signals, how do they produce them?

An Introduction to Membrane Potentials

Ions carry an electric charge. In virtually all cells, the cytoplasm and extracellular fluids adjacent to the plasma membrane contain unequal distributions of ions. As a result, cells are inherently electrical in nature.

A difference of electrical charge between any two points creates a difference in **electrical potential**, or a **voltage**. If the positive and negative charges on ions that exist on the two sides of a plasma membrane do not balance each other, the membrane will have an electrical potential.

When an electrical potential exists on either side of a plasma membrane, the separation of charges is called a **membrane potential**. If there is a large separation of charges across the membrane, the membrane potential is large.

It is important to remember that membrane potentials refer only to a separation of charge immediately adjacent to the plasma membrane, on either side of the membrane. Even if there is a large membrane potential, there may be no charge separation slightly farther from the membrane.

UNITS AND SIGNS Membrane potentials are measured in units called millivolts. A **millivolt (mV)** is 1/1000 of a volt; a **volt** is the standard unit of electrical potential. As a comparison, an AA battery that you buy in the store has an electrical potential of 1500 mV between its + and − terminals. In neurons, membrane potentials are typically about 70–80 mV.

By convention, membrane potentials are always expressed as inside-relative-to-outside. Because there are usually more negatively charged ions and fewer positively charged ions on the inside surface of a membrane relative to its outside surface, membrane potentials are usually negative in sign.

ELECTRICAL POTENTIAL, ELECTRIC CURRENTS, AND ELECTRICAL GRADIENTS Membrane potentials are a form of potential energy. When a membrane potential exists, the ions on both sides of the membrane have potential energy. Recall that potential energy is energy based on the position of matter.

To convince yourself that ions have potential energy when a membrane potential exists, consider what would happen if the membrane were removed. Ions would spontaneously move from the area of like charge to the area of unlike charge—causing a flow of charge. This flow of charge, called an **electric current**, would occur because like charges repel and unlike charges attract.

As Chapter 6 emphasized, however, ions move across membranes in response to concentration gradients as well as charge gradients. Therefore, a membrane potential also includes energy stored as the concentration gradient of charged ions on the two sides of the membrane. Recall that the combination of an electric gradient and a concentration gradient is an **electrochemical gradient**. What do all of these facts have to do with neurons?

An equilibrium potential exists whenever there is a concentration gradient for an ion and the membrane is permeable to that ion. At the equilibrium potential, the rate at which an ion moves across the membrane down its concentration gradient is equal to the rate at which it moves, in the opposite direction, down its electrical gradient. It's the voltage at which the concentration and electrical gradients acting on an ion balance out.

The ion concentration gradients that contribute to equilibrium potentials are due to the action of Na^+/K^+-ATPase. In this way, chemical energy in the form of ATP is converted to electrical energy in the form of a membrane potential.

To calculate the amount of energy involved, biologists use the Nernst equation—a formula that converts the energy stored in a concentration gradient to the energy stored as an electrical potential. The concentration gradient for an ion is symbolized

$[ion]_o/[ion]_i$, where $[ion]_o$ and $[ion]_i$ are the concentrations of the ion outside and inside the cell, respectively. The electrical potential for that ion is symbolized E_{ion}. The Nernst equation specifies the equilibrium potential for a type of ion as

$$E_{ion} = 2.3 \frac{RT}{zF} \log \frac{[ion]_o}{[ion]_i}$$

In this expression, z is the valence of the ion (for instance, $+1$ for potassium) and the expression RT/F is known as the thermodynamic potential. The three terms in the thermodynamic potential are the gas constant (R), which acts as a constant of proportionality; the absolute temperature (T), measured in kelvins; and the Faraday constant (F), which specifies the amount of charge carried by one mole of an ion with a valence of $+1$ or -1.

Note that the Nernst equation is based on the base-10 logarithm of the ion concen-

tration ratio, $[ion]_o/[ion]_i$. Thus, the thermodynamic potential specifies the voltage required to balance a tenfold concentration ratio across the membrane. (For more on using logarithms, see **BioSkills 7** in Appendix A.)

The Nernst equation applies to only a single ion. For example, suppose that the potassium concentration inside the axon of a squid neuron has been measured as 400 mM, yet the outside concentration for this ion is only 20 mM (Table 45.1). Potassium has a charge of $+1$, so at 20°C the RT/zF part of the equation yields $+25$ mV. In this case, the equilibrium potential for potassium becomes

$$E_K = 2.3 \times 25 \text{ mV} \times \log \frac{20 \text{ mM}}{400 \text{ mM}}$$

$$E_K = 58 \text{ mV} \times \log 0.05 = -75 \text{ mV}$$

The minus sign indicates that the interior of the axon is negatively charged with respect to the exterior.

Repeating this process for Na^+ ions and Cl^- ions yields the following results for the squid axon:

$$E_{Na} = 58 \text{ mV} \times \log \frac{440 \text{ mM}}{50 \text{ mM}} = +54.8 \text{ mV}$$

$$E_{Cl} = -58 \text{ mV} \times \log \frac{560 \text{ mM}}{51 \text{ mM}} = -60 \text{ mV}$$

Again, notice that the equilibrium potential given by the Nernst equation is calculated independently for each ion.

TABLE 45.1 Concentration of Important Ions across a Squid Neuron's Plasma Membrane at Rest

Ions	Cytoplasm Concentration	Extracellular Concentration	Equilibrium Potential
Na^+	50 mM	440 mM	$+54.8$ mV
K^+	400 mM	20 mM	-75 mV
Cl^-	51 mM	560 mM	-60 mV
Organic anions	385 mM	—	—

How Is the Resting Potential Maintained?

When a neuron is not transmitting an electrical signal but is merely sitting in extracellular fluid at rest, the membrane potential across its membrane is called the **resting potential**. To understand why the resting potential exists, consider the distribution of the various ions and other charged molecules on the two sides of the neuron's plasma membrane, shown in **Figure 45.3**:

- The interior side of the membrane has relatively low concentrations of sodium (Na^+) and chloride (Cl^-) ions, a relatively large concentration of potassium ions (K^+), and some organic anions—amino acids and other organic molecules that drop a proton and carry a negative charge.

- In the extracellular fluid, sodium and chloride ions predominate.

If each type of ion diffused across the membrane in accordance with its concentration gradient, organic anions and K^+ would leave the cell while Na^+ and Cl^- would enter.

Ions cannot cross phospholipid bilayers readily, however. They cross plasma membranes efficiently in only three ways (see Chapter 6):

1. flowing along their electrochemical gradient through an **ion channel**—a protein that forms a pore in the membrane and allows a specific ion to pass through; or

2. carried, via a membrane cotransporter protein or antiporter protein, with an ion that experiences a strong electrochemical gradient; or

3. pumped against an electrochemical gradient by a membrane protein that hydrolyzes ATP.

How does the distribution of ions illustrated in Figure 45.3 come to be? The answer hinges on two types of membrane proteins: an ion channel that allows potassium ions to move, and a pump that actively transports sodium and potassium ions. Let's consider each in turn.

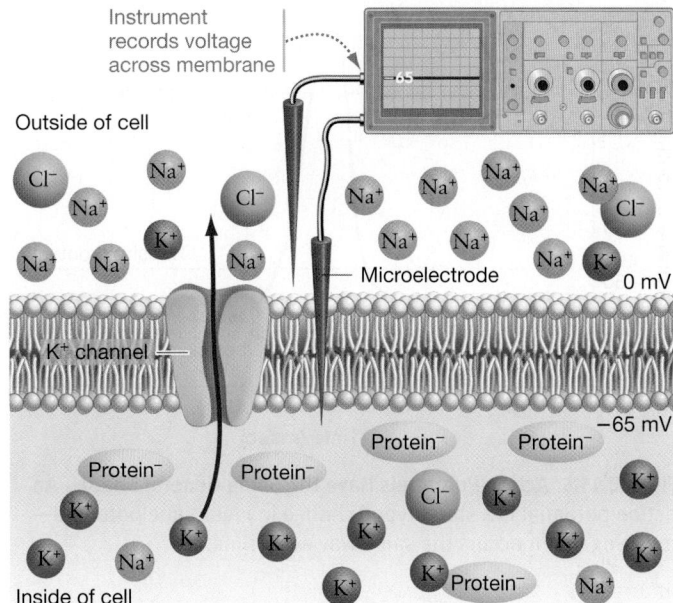

FIGURE 45.3 Neurons Have a Resting Potential. In resting neurons, the membrane is selectively permeable to K⁺. As K⁺ leaves the cell along its concentration gradient, the inside of the membrane becomes negatively charged relative to the outside. To measure a neuron's membrane potential, researchers insert a microelectrode into the cell and compare that reading with the reading outside the cell.

✔ **QUESTION** Will K⁺ continue to leave the cell indefinitely? Explain why or why not.

THE K⁺ LEAK CHANNEL When a neuron is not actively involved in transmitting an electrical signal, the most common type of channel that is open is one that admits potassium ions. Therefore, resting neurons are most permeable to K⁺ ions, which cross the membrane easily along their concentration gradient. The potassium

channels involved are sometimes called **leak channels**, because they allow K⁺ to leak out of the cell.

As K⁺ moves from the interior of the cell to the exterior through potassium channels, the inside of the cell becomes more and more negatively charged relative to the outside. This buildup of negative charge inside the cell begins to attract K⁺ and counteract the concentration gradient that had favored the movement of K⁺ out.

As a result of these counteracting influences, the membrane reaches a voltage at which there is an equilibrium between the concentration gradient that moves K⁺ out and the electrical gradient that moves K⁺ in. At this voltage, there is no longer a net movement of K⁺. This voltage is called the **equilibrium potential** for K⁺. **Box 45.1** discusses how equilibrium potentials for individual ions are calculated.

Although Cl⁻ and Na⁺ cross the plasma membrane much less readily than does K⁺, some movement of these ions also occurs through a small number of open ion channels that are selective for each ion. As a result, each type of ion has an equilibrium potential. The membrane as a whole has a membrane potential that combines the effects of the individual ions.

THE ROLE OF THE NA⁺/K⁺-ATPASE In addition to the passive movement of ions that takes place when neurons are at rest, Na⁺/K⁺-ATPase actively pumps Na⁺ out of the cell and K⁺ into the cell. More specifically, the sodium-potassium pump uses the energy gained when it receives a phosphate group from ATP to move three Na⁺ ions out of the cell and two K⁺ ions into the cell (**Figure 45.4**).

Active transport via Na⁺/K⁺-ATPase ensures that eventually the concentration of K⁺ is much higher inside the cell than outside, while the concentration of Na⁺ is lower inside than outside. In addition to setting up concentration gradients that affect K⁺ and Na⁺, the pump establishes an electrical gradient: The

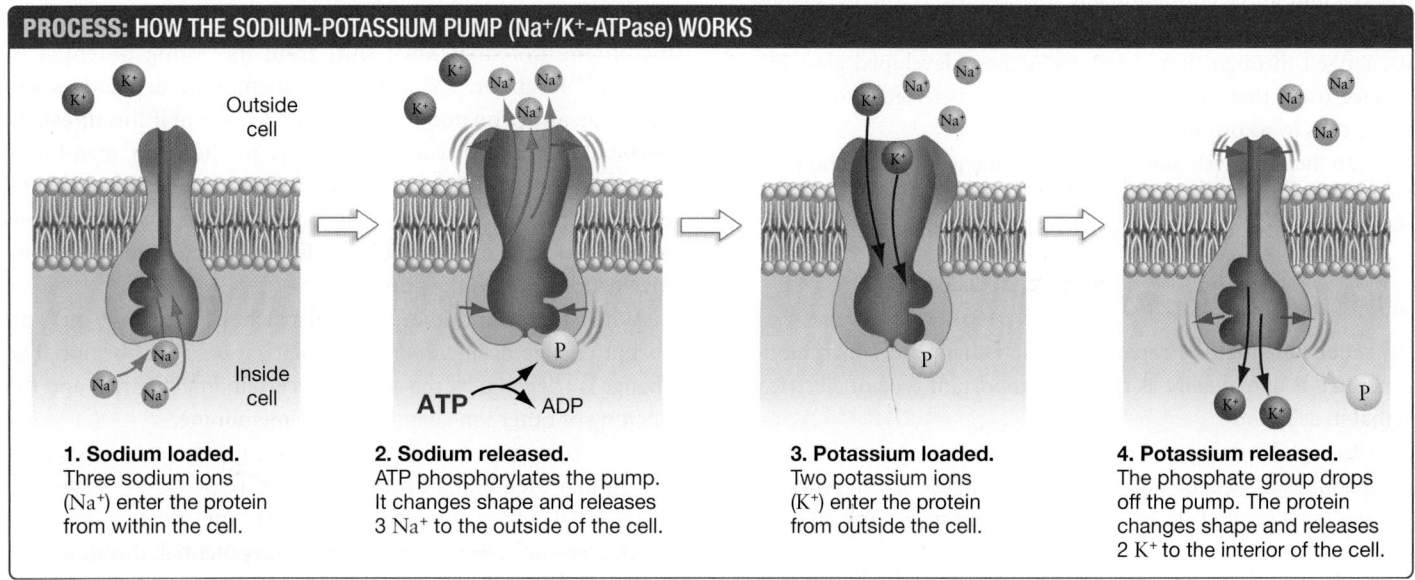

PROCESS: HOW THE SODIUM-POTASSIUM PUMP (Na⁺/K⁺-ATPase) WORKS

1. Sodium loaded. Three sodium ions (Na⁺) enter the protein from within the cell.

2. Sodium released. ATP phosphorylates the pump. It changes shape and releases 3 Na⁺ to the outside of the cell.

3. Potassium loaded. Two potassium ions (K⁺) enter the protein from outside the cell.

4. Potassium released. The phosphate group drops off the pump. The protein changes shape and releases 2 K⁺ to the interior of the cell.

FIGURE 45.4 In Neurons, Na⁺/K⁺-ATPase Imports Potassium Ions and Exports Sodium Ions. By following radioactive Na⁺ and K⁺, biologists found that the pump transports 3 Na⁺ out of the cell for every 2 K⁺ brought in. Notice that this pump operates via a conformational change.

outward movement of three positive charges and inward movement of two positive charges makes the interior of the membrane less positive (more negative) than the outside.

Thus, the neuron has a negative resting membrane potential. The resting potential represents energy stored as concentration and electrical gradients in a series of ions. ✔If you understand this concept, you should be able to explain why both the Na^+/K^+-ATPase and the K^+ "leak" channels are vital to establishing the resting potential.

To review the ions, gradients, channels, and pumps involved in establishing and maintaining the resting potential, go to the study area at *www.masteringbiology.com*.

 Web Activity Membrane Potentials

Using Microelectrodes to Measure Membrane Potentials

During the 1930s and 1940s, A. L. Hodgkin and Andrew Huxley focused on what has become a classic model system in the study of electrical signaling: the axons of squid.

Squid live in the ocean and are preyed on by fish and whales. When a squid is threatened, electrical signals travel down the axon to muscle cells. When these muscles contract, water is expelled from a cavity in the squid's body. As a result, the squid lurches away from danger by jet propulsion (see Chapter 33). The extremely rapid electrical signal is an adaptation that helps squid avoid predation.

Hodgkin and Huxley decided to study the squid's axon simply because it is so large. Many of the axons found in humans are a mere 2 μm in diameter, but the squid axon is about 500 μm in diameter—large enough that the researchers could record membrane potentials by inserting a wire down its length.

By measuring the voltage difference between the wire inside the cell and an electrode outside the cell, the researchers could record the voltage across the plasma membrane and observe how it changed through time. Later researchers developed glass microelectrodes that were small enough to be inserted into smaller nerve cells to record membrane voltage.

With their relatively simple early equipment, Hodgkin and Huxley were able to record the axon's resting potential, and document that it can be disrupted by an event called the action potential.

What Is an Action Potential?

An **action potential** is a rapid, temporary change in a membrane potential. It may qualify as the most important type of electrical signal in cells.

Although Hodgkin and Huxley initially studied it in the squid giant axon, subsequent work has shown that the action potential has the same general characteristics in all species and in all types of neurons.

A THREE-PHASE SIGNAL **Figure 45.5** shows the form of the action potential that Hodgkin and Huxley recorded from the squid's giant axon—the signal that allows these animals to jet

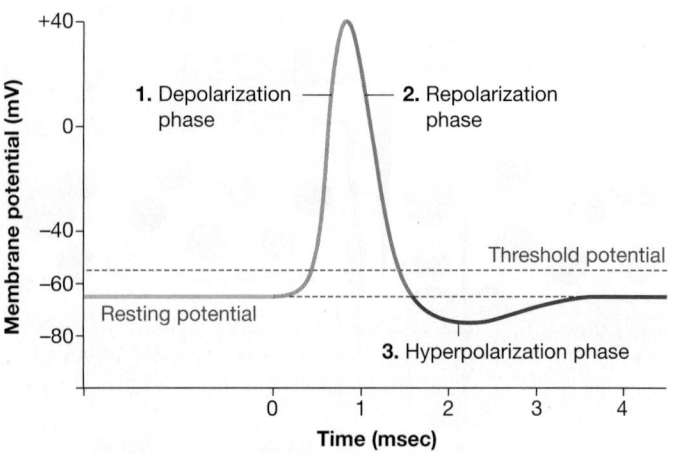

FIGURE 45.5 Action Potentials Have the Same General Shape. An action potential is a stereotyped change in membrane potential—meaning that it occurs the same way every time.

away from predators. Put your finger on the figure's "Resting potential" label and then trace the graph to the right. The action potential has three distinct phases:

1. **Depolarization** of the membrane. A membrane is said to be polarized if the charges on the two sides are different. Depolarization means that the membrane becomes less polarized than before. During the depolarization phase, the membrane potential changes from highly negative to positive.

2. A rapid **repolarization**, which changes the membrane potential from positive back to negative.

3. A **hyperpolarization** phase, when the membrane is slightly more negative than the resting potential.

All three phases of an action potential occur in a few milliseconds.

For an action potential to begin in a squid giant axon, the membrane potential must shift from its resting potential of -65 mV to about -55 mV. If the membrane depolarizes less than that, an action potential does not occur. But if this **threshold potential** is reached, certain channels in the axon membrane open and ions rush into the axon, following their electrochemical gradients. The current flow causes further depolarization. The inside of the membrane becomes less negative and then positive with respect to the outside.

When the membrane potential reaches about $+40$ mV, an abrupt change occurs and the repolarization phase begins. The change is triggered by the closing of certain ion channels and the opening of other ion channels in the membrane.

An action potential occurs because specific ion channels in the plasma membrane open or close in response to changes in voltage. An action potential always has the same three-phase form, even though the size of the resting potential, threshold potential, and peak depolarization may vary among species.

AN "ALL-OR-NONE" SIGNAL THAT PROPAGATES Hodgkin and Huxley made other important observations about the action potential. In addition to being fast and having three distinct phases, it is an all-or-none event.

- There is no such thing as a partial action potential.
- All action potentials for a given neuron are identical in magnitude and duration.
- Action potentials are propagated down the length of the axon.

For example, when an impulse was recorded at a particular point on a squid axon, an action potential that was identical in shape and size would be observed farther down the same axon soon afterward. Neurons are said to have **excitable membranes**, because neurons are capable of generating action potentials that propagate rapidly along the length of their axons.

Taken together, these observations suggested a mechanism for electrical signaling. In the nervous system, information is coded in the form of action potentials that travel along axons. The frequency of action potentials—not their size—is the meaningful signal. In the squid's giant axon, action potentials signal muscles to contract. As a result, the animal escapes from danger.

CHECK YOUR UNDERSTANDING

◑━ If you understand that . . .

- The plasma membranes of neurons carry a resting potential because Na^+/K^+-ATPase pumps Na^+ out of the cell and K^+ into the cell and because the membrane is selectively permeable to K^+ ions, which leak out.
- The action potential is a three-phase, all-or-none signal that moves down the length of a neuron.

✔ You should be able to . . .

1. Predict what would happen to the resting potential of a squid axon if potassium leak channels were blocked.
2. Explain why only the frequency of action potentials—not their size—contains information.

Answers are available in Appendix B.

45.2 Dissecting the Action Potential

Which ions are involved in the currents that form the action potential? Is Na^+, Cl^-, or K^+ responsible for the depolarization and repolarization phases of the event?

Hodgkin made a crucial start in answering this question when he realized that +40 mV was close to the equilibrium potential for Na^+ in the squid giant axon. If sodium channels opened early in the action potential, then Na^+ should flow into the neuron until the membrane potential was about +40 mV. How could this hypothesis be tested?

Distinct Ion Currents Are Responsible for Depolarization and Repolarization

To understand the currents responsible for the action potential, Hodgkin and Huxley recorded electrical activity in squid axons that were bathed in solutions containing different concentrations of ions.

- Washing Na^+ out of the solution surrounding the axon abolished action potentials.
- When they used solutions with various concentrations of Na^+, the peak of the action potential tracked the equilibrium potential of Na^+. If Na^+ concentration outside the cell was high, the peak was high. If Na^+ concentration outside the cell was low, the peak was low.

These experiments furnished strong support for the hypothesis that the action potential begins when Na^+ flows into the neuron. Sodium ions are responsible for the depolarization phase.

What happens during the repolarization phase? Using radioactive K^+, Hodgkin and Huxley showed that there was a strong flow of potassium ions out of the cell during the repolarization phase.

◑━ The action potential consists of a strong inward flow of sodium ions followed by a strong outward flow of potassium ions. ✔If you understand this concept, you should be able to add labels that read, "Sodium channels open—Na^+ enters" and "Potassium channels open—K^+ leaves" to Figure 45.5.

How Do Voltage-Gated Channels Work?

The action potential depends on **voltage-gated channels**—membrane proteins that open and close in response to changes in membrane voltage. The shape of a voltage-gated channel, and thus its ability to admit ions, changes in response to the charges present at the membrane surface. **Figure 45.6** shows a simple model of how voltage-gated sodium channels change as a function of membrane potential.

Hodgkin and Huxley confirmed that voltage-gated channels exist using a technique called voltage clamping. **Voltage clamping** allows researchers to hold an axon at any voltage and record the electrical currents that occur. When the researchers held the squid axon at various voltages, different currents resulted. These

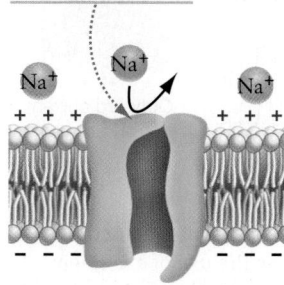
At the resting potential, voltage-gated Na^+ channels are closed

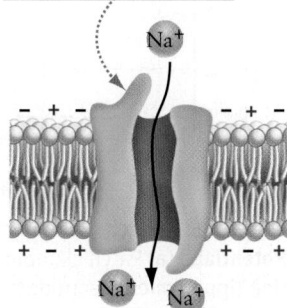
Conformational changes open voltage-gated channels when the membrane is depolarized

FIGURE 45.6 The Shape of a Voltage-Gated Channel Depends on the Membrane Potential. Changes in the conformation of voltage-gated channels are responsible for changes in a neuron's permeability to Na^+ and K^+.

experiments supported the hypothesis that the behavior of the ion channels depends on voltage.

PATCH CLAMPING AND STUDIES OF SINGLE CHANNELS Studying individual ion channels became possible when Erwin Neher and Bert Sakmann perfected a technique known as **patch clamping**. As **Figure 45.7a** shows, the researchers touched a membrane with a fine-tipped microelectrode and applied suction to capture a single ion channel within the tip of the microelectrode. The suction seals the membrane against the glass so that no current leaks out.

Using this technique, the researchers documented the currents that flowed through individual channels. The recordings

(a) Patch clamping isolates a single ion channel.

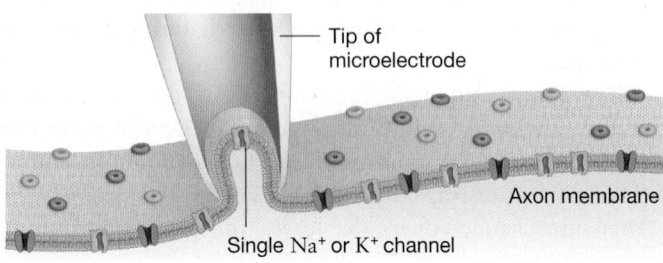

(b) Currents through isolated channels can be measured during an action potential.

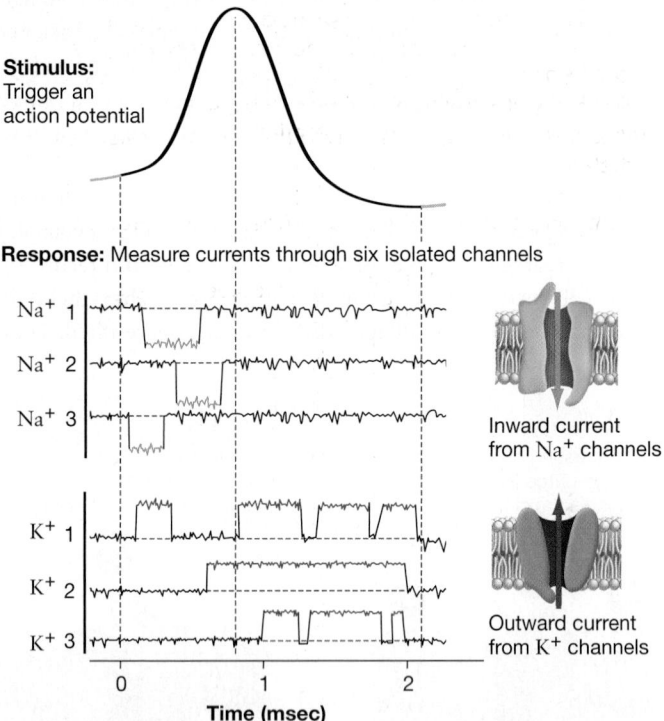

FIGURE 45.7 Patch Clamping Provided Insights into the Action Potential. (a) Patch clamping depends on the use of extremely fine-tipped microelectrodes. The goal is to isolate one channel and record from it. **(b)** Current records show that voltage-gated channels are either open or closed, sodium channels open quickly after depolarization, and potassium channels open with a delay after polarization. No current flows through either type of channel at the resting potential.

shown in **Figure 45.7b**, from three sodium channels and three potassium channels, show how current flows through individual channels over the course of an action potential. The data in the figure make several important points:

- Voltage-gated channels are either open or closed. There is no gradation in channel behavior. This conclusion is based on the shape of the current records—current flow starts and stops instantly, and the size of the current is always the same.

- Sodium channels open quickly after depolarization. They stay open for about a millisecond, close, and remain inactive for 1 to 2 milliseconds. That explains why the cell can repolarize: Once the sodium channels close, they have a lag before they can open again.

- Potassium channels open with a delay after depolarization. They continue to flip open and closed until the membrane repolarizes. Once the membrane returns to the resting potential, potassium channels remain closed.

POSITIVE FEEDBACK OCCURS DURING DEPOLARIZATION More detailed experiments on Na$^+$ channels also explained why the action potential is an all-or-none event. The key observation was that Na$^+$ channels are more likely to open as a membrane depolarizes. As a result, an initial depolarization leads to the opening of more Na$^+$ channels, which depolarizes the membrane further, which leads to the opening of additional Na$^+$ channels.

The opening of Na$^+$ channels exhibits **positive feedback**—meaning that the occurrence of an event makes the same event more likely to recur. When a fuse is lit, for example, the heat generated by the oxidation reaction accelerates the reaction itself, which generates still more heat and leads to additional oxidation reactions and keeps the fuse burning. Positive feedback is rare in organisms because it often leads to uncontrolled events. The opening of Na$^+$ channels during an action potential is one of the few examples known.

✔ If you understand how sodium and potassium channels work, you should be able to explain (**1**) why positive feedback occurs in the opening of Na$^+$ channels, (**2**) why Na$^+$ stops flowing across a membrane during an action potential, and (**3**) why K$^+$ channels start opening.

USING NEUROTOXINS TO IDENTIFY CHANNELS AND DISSECT CURRENTS In addition to using voltage clamping and patch clamping, researchers have used poisons that target neurons, from sources as diverse as venomous snakes and foxglove plants, to explore the dynamics of voltage-gated channels. **Neurotoxins** are poisons that affect neuron function—often resulting in convulsions, paralysis, or unconsciousness.

For example, when biologists treated giant axons from lobsters with the tetrodotoxin found in puffer fish, they found that the resting potential in treated neurons was normal, but action potentials were abolished. More specifically, the outward-directed K$^+$ current was normal but the inward-directed Na$^+$ flow was wiped out. Researchers concluded that puffer-fish toxin blocks the voltage-gated Na$^+$ channel, probably by binding to a specific site on the channel protein.

How Is the Action Potential Propagated?

To explain how action potentials propagate down an axon, Hodgkin and Huxley suggested the model illustrated in **Figure 45.8a**.

Step 1 The influx of Na^+ at the start of an action potential causes charge to spread away from sodium channels. Positive charges inside the cell are repulsed by the influx of Na^+, and negative charges are attracted to the Na^+.

Step 2 As positive charges are pushed farther from the initial sodium channels, they depolarize adjacent portions of the membrane.

Step 3 Nearby voltage-gated Na^+ channels pop open in response to depolarization. Positive feedback occurs, and a full-fledged action potential results.

In this way, an action potential is continuously regenerated as it moves down the axon (**Figure 45.8b**). The signal does not diminish as it moves, because the response is all or none.

Why don't action potentials propagate back up the axon? To answer this question, recall that Na^+ channels are **refractory**—that is, once they have opened and closed, they are less likely to open again for a short period. Action potentials are propagated in one direction only, because sodium channels "downstream" of the site are not in the refractory state.

The hyperpolarization phase, in which the membrane is more negative than the resting potential—also keeps the charge that spreads "upstream" from triggering an action potential in that direction.

AXON DIAMETER AFFECTS SPEED Understanding how the action potential propagates helped researchers explain why the squid's axons are so large. When sodium ions enter the interior of an axon at the start of an action potential, they can spread down the membrane or leak back out through open sodium channels. Compared to small axons, large axons have relatively few sodium channels per unit of membrane surface. As a result, less current leaks back out of a large axon than a small axon, and the charge spreads farther down the membrane.

The upshot is that the squid's giant axon and other large-diameter neurons transmit action potentials much faster than small axons can. The squid axon's large size is an adaptation that makes particularly rapid signaling possible.

(a) PROCESS: PROPAGATION OF ACTION POTENTIAL

Neuron | Axon

1. Na^+ enters axon.

2. Charge spreads; membrane "downstream" depolarizes.

Depolarization at next ion channel

3. Downstream voltage-gated channel opens in response to depolarization.

FIGURE 45.8 Action Potentials Propagate because Charge Spreads down the Membrane. (a) An action potential starts with an inflow of Na^+. The influx of positive charge attracts negative charges inside the cell and repels positive charges. As a result, positive charge spreads away from the channel where the Na^+ enters, and depolarizes nearby regions of the neuron. Voltage-gated Na^+ channels open in response. **(b)** The action potential spreads down the axon as a wave of depolarization, but there is no loss of signal because the all-or-none action potential regenerates itself as it travels.

(b) Action potential spreads as a wave of depolarization.

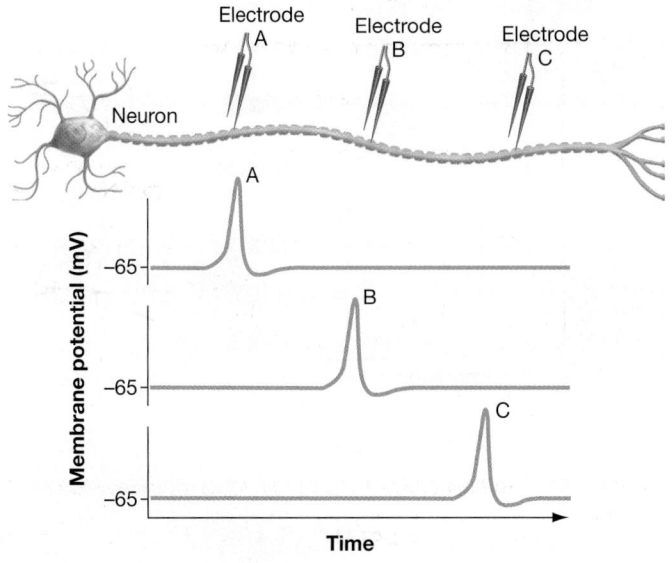

MYELINATION AFFECTS SPEED Analyzing charge spread also helped biologists explain the phenomenon called myelination. Relatively few vertebrates have giant axons. Instead, in vertebrates—and some invertebrates—the membranes of specialized accessory cells wrap around the axons of neurons and increase the efficiency of action potential propagation.

In the central nervous system, the specialized accessory cells are called **oligodendrocytes**. In the peripheral nervous system, described in Section 45.4, the cells are **Schwann cells** (**Figure 45.9a**). Oligodendrocytes and Schwann cells are two of several types of nervous system cells that support neurons. Collectively, these accessory cells are called **glia**.

When oligodendrocytes or Schwann cells wrap around an axon, they form a **myelin sheath**, which acts as a type of electrical insulation. As charge spreads down an axon, the myelin sheath prevents charge in the form of ions from leaking back out across the plasma membrane of the neuron.

Consequently, the influx of charge that results from an action potential is able to spread unimpeded until it hits an unmyelinated section of the axon, called a **node of Ranvier** (**Figure 45.9b**). The node has a dense concentration of voltage-gated Na$^+$ channels, so action potentials can occur.

Electrical signals jump down a myelinated axon much faster than they can move down an unmyelinated axon. In an unmyelinated axon, sodium and potassium channels are found in all locations and action potentials occur continuously down its length. Myelination is interpreted as an adaptation that makes rapid transmission of electrical signals possible in axons that have a small diameter.

To drive the importance of myelination home, consider what happens when it is decreased. If myelin degenerates, the transmission of electrical signals slows considerably. The disease **multiple sclerosis (MS)** develops as damage to myelin increases and electrical signaling is impaired, causing the muscles to weaken and coor-

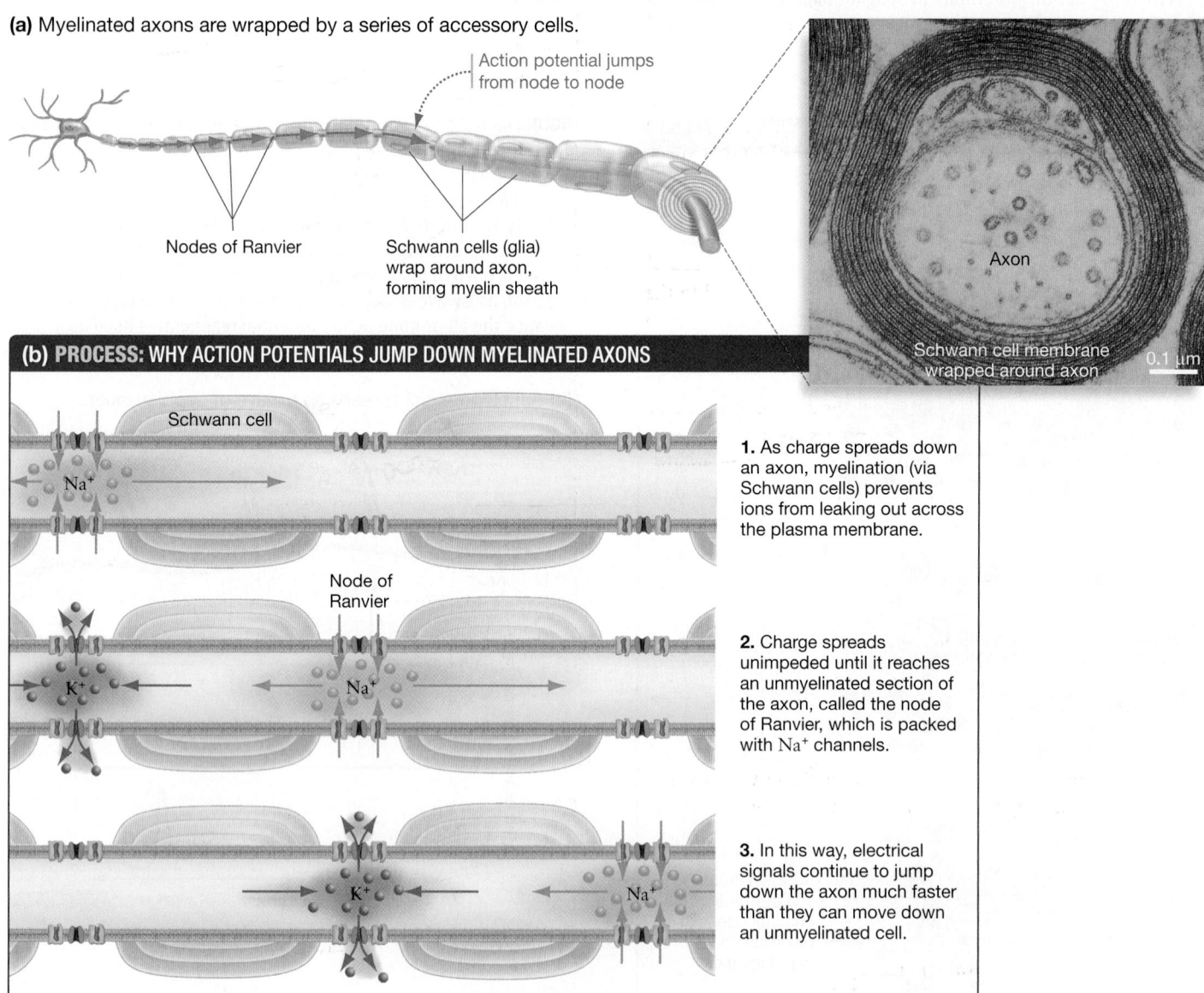

(a) Myelinated axons are wrapped by a series of accessory cells.

Action potential jumps from node to node

Nodes of Ranvier

Schwann cells (glia) wrap around axon, forming myelin sheath

Axon

Schwann cell membrane wrapped around axon

0.1 μm

(b) PROCESS: WHY ACTION POTENTIALS JUMP DOWN MYELINATED AXONS

Schwann cell

Na$^+$

1. As charge spreads down an axon, myelination (via Schwann cells) prevents ions from leaking out across the plasma membrane.

K$^+$

Node of Ranvier

Na$^+$

2. Charge spreads unimpeded until it reaches an unmyelinated section of the axon, called the node of Ranvier, which is packed with Na$^+$ channels.

K$^+$

Na$^+$

3. In this way, electrical signals continue to jump down the axon much faster than they can move down an unmyelinated cell.

FIGURE 45.9 Action Potentials Propagate Quickly in Myelinated Axons.

dination to lessen. The symptoms of MS are highly variable; in severe cases the disease progresses and can be crippling.

What happens once an action potential has traveled the length of the axon? In most neurons the membrane at the end of the axon approaches the membrane of another neuron's dendrite, and the two surfaces are separated by a tiny gap. What happens when an action potential arrives at this interface between cells?

CHECK YOUR UNDERSTANDING

If you understand that . . .

- During an action potential, membrane voltage undergoes rapid changes due to an influx of sodium ions, followed by an outflow of potassium ions.
- Action potentials propagate down an axon because inrushing sodium ions depolarize adjacent portions of the membrane.

✓ **You should be able to . . .**

1. Explain why the action potential is an all-or-none phenomenon.
2. Predict what would happen if sodium channels continued to open once membrane depolarization was complete.

Answers are available in Appendix B.

45.3 The Synapse

The cytoplasm of most neurons is not directly connected to the cytoplasm of other neurons. Based on this observation, there must be some indirect mechanism that transmits electrical signals from cell to cell, across their plasma membranes.

In the 1920s, Otto Loewi showed that this indirect mechanism involves **neurotransmitters**. Neurotransmitters are chemical messengers that transmit information from one neuron to another neuron, or from a neuron to a target cell in a muscle or gland.

Loewi knew that signals from the vagus nerve slow the heart. To test the hypothesis that the signal from nerve to muscle is delivered by a chemical, he performed the experiment diagrammed in **Figure 45.10**.

First, Loewi isolated the vagus nerve and heart of a frog. As predicted, the heart rate slowed when he stimulated the nerve electrically. Next, he took the solution that bathed the first heart and applied it to another, isolated heart, and showed that the second heart rate slowed as well.

This result provided strong evidence for the chemical transmission of electrical signals. The vagus nerve had released a neurotransmitter into the bath.

Synapse Structure and Neurotransmitter Release

When transmission electron microscopy became available in the 1950s, biologists finally understood the physical nature of the interface, or **synapse**, between neurons—the place where

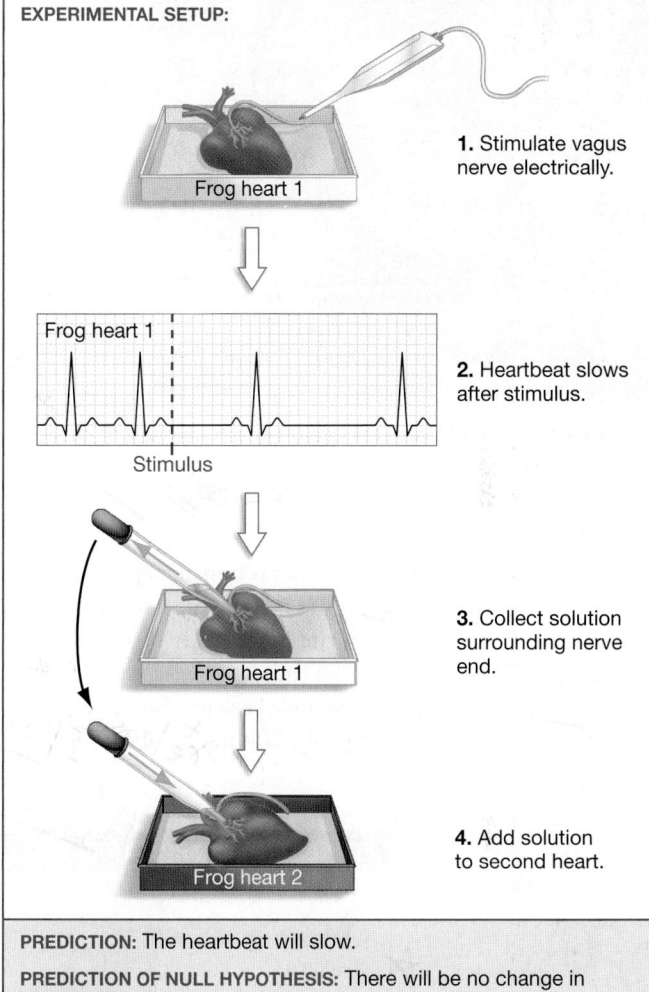
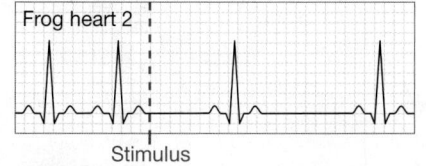
FIGURE 45.10 Experimental Evidence for the Existence of Neurotransmitters.

SOURCE: Loewi, O. 1921. Über humorale Übertragbarkeit der Herznervenwirkung. *Pflügers Archiv European Journal of Physiology* 189: 239–242.

✓ **QUESTION** What would be an appropriate control for this experiment?

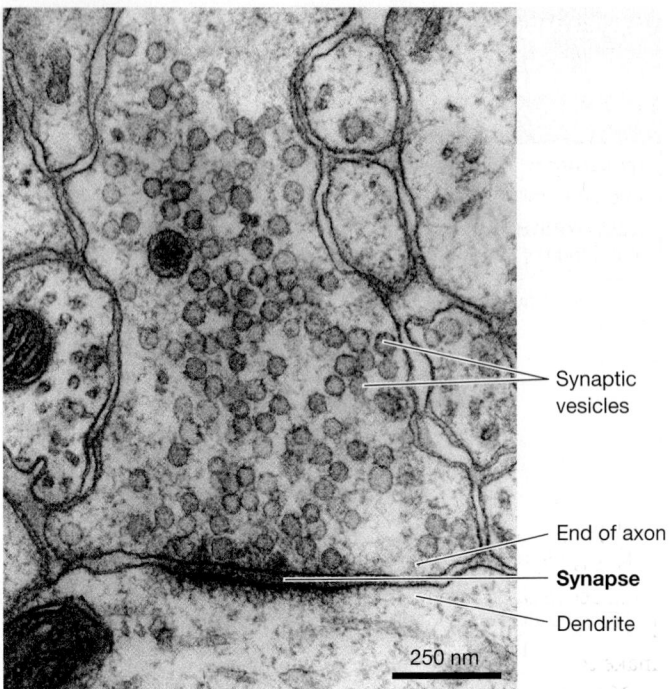

FIGURE 45.11 **Synaptic Vesicles Cluster near Synapses.** A cross section of the site where an axon meets a dendrite.

Anatomical observations such as these, combined with chemical studies of the synapse, led to the model of synaptic transmission illustrated in **Figure 45.12**. Notice that the "sending" cell is the **presynaptic neuron** and the "receiving" cell is the **postsynaptic neuron**.

Step 1 The action potential arrives at the end of the axon.

Step 2 The depolarization created by an action potential opens voltage-gated calcium channels located near the synapse, in the presynaptic membrane. The electrochemical gradient for Ca^{2+} results in the inflow of calcium ions through the open channels.

Step 3 In response to the increased calcium concentration inside the axon, synaptic vesicles fuse with the membrane and release a neurotransmitter into the gap between the cells. This gap is called the **synaptic cleft**. The delivery of neurotransmitters into the cleft is an example of exocytosis, a process introduced in Chapter 7.

Step 4 Neurotransmitters bind to receptors on the postsynaptic cell, leading to changes in the membrane potential of the postsynaptic cell and possibly triggering the start of an action potential there.

Step 5 The response ends as the neurotransmitter is broken down and taken back up by the presynaptic cell.

Is the model correct? Let's begin by analyzing the role of neurotransmitters.

What Do Neurotransmitters Do?

To establish that a molecule functions as a neurotransmitter, researchers have to provide evidence for the following three criteria:

two neurons meet—and information is transferred from one to the next. As **Figure 45.11** shows, (**1**) the membranes of an axon from one neuron and the dendrite or cell body of another neuron juxtapose closely, and (**2**) the ends of axons contain numerous sac-like structures called **synaptic vesicles**. Synaptic vesicles were hypothesized to be storage sites for neurotransmitters.

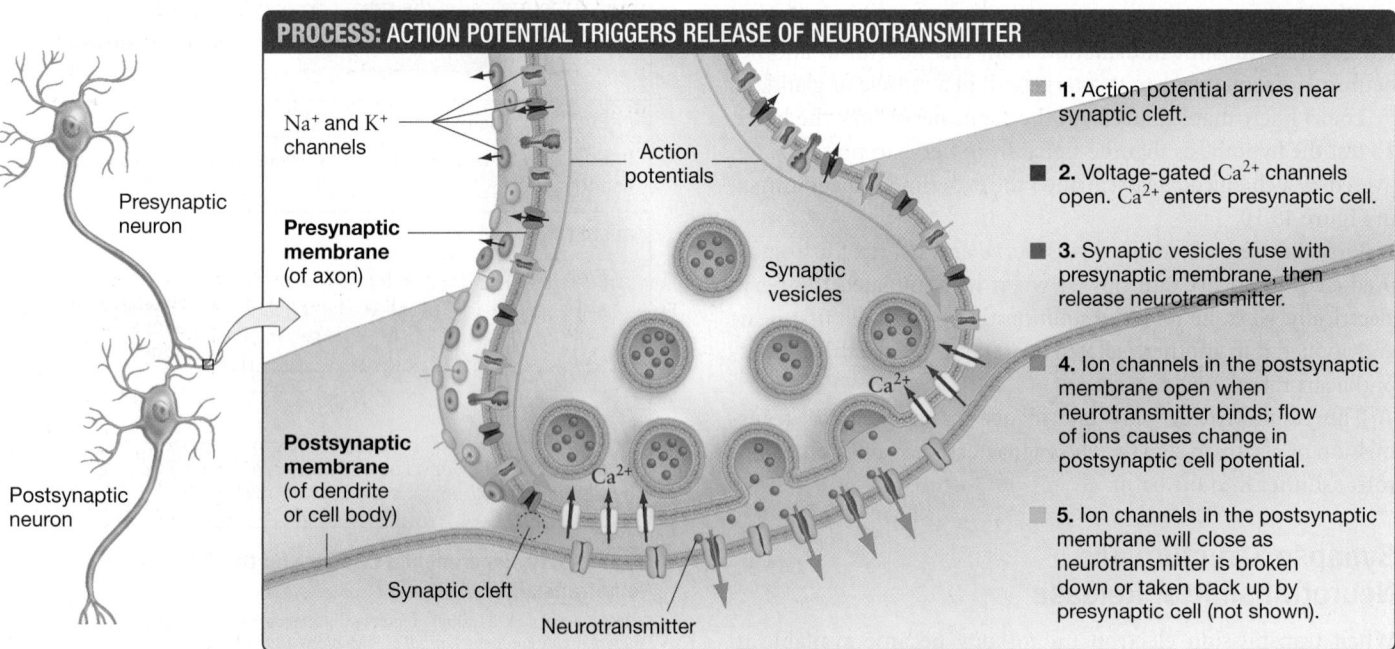

FIGURE 45.12 **Neurons Meet and Transfer Information at Synapses.** The sequence of events that occurs when an action potential arrives at a synapse.

1. The neurotransmitter is present at the synapse and released in response to an action potential.

2. It binds to a receptor on a postsynaptic cell.

3. It is taken up or degraded.

For example, researchers can look for neurotransmitters by stimulating a neuron, collecting the molecules that are released, and analyzing them chemically. To find the receptor for a particular neurotransmitter, researchers can attach a radioactive atom or other type of label to the chemical messenger, and add the labeled molecule to neurons. Once the labeled transmitter has bound to its receptor, the receptor protein can be isolated and analyzed.

Using techniques such as these, biologists have discovered and characterized a wide array of neurotransmitters and receptors, some of which are listed in **Table 45.2** on page 898.

By patch-clamping receptors, biologists confirmed that many neurotransmitters function as ligands. A **ligand** is a molecule that binds to a specific site on a receptor molecule. Many neurotransmitters are ligands that bind to receptors called **ligand-gated channels**. These are channel proteins that open in response to binding by a specific ligand—just as voltage-gated channels open in response to a change in voltage.

When a neurotransmitter binds to a ligand-gated ion channel in the postsynaptic membrane, the channel opens and admits a flow of ions along an electrochemical gradient. In this way, the neurotransmitter's chemical signal is transduced to an electrical signal—a change in the membrane potential of the postsynaptic cell. If you understand this concept, you should be able to imagine a membrane with a resting potential of -65 mV and explain what happens when a ligand-gated ion channel opens and allows chloride ions to leave the cell.

Not all neurotransmitters bind to ion channels, however. Some receptors activate enzymes that lead to the production of a second messenger in the postsynaptic cell. Recall from Chapter 8 that **second messengers** are chemical signals produced inside a cell in response to a chemical signal that arrives at the cell surface.

The second messengers induced by neurotransmitters may trigger changes in enzyme activity, gene transcription, or membrane potential. Chapter 47 explores the cellular role of second messengers in detail.

Postsynaptic Potentials

What happens when a neurotransmitter binds to a receptor and opens an ion channel in the postsynaptic cell?

If the receptors at the synapse admit sodium ions in response to the arrival of the neurotransmitter, the postsynaptic membrane depolarizes (**Figure 45.13a**). In most cases, depolarization makes an action potential in the postsynaptic cell more likely. Changes in the postsynaptic cell that make action potentials more likely are called **excitatory postsynaptic potentials (EPSPs)**.

If the receptors at the synapse lead to an outflow of potassium ions or an inflow of chloride ions in the postsynaptic cell, the postsynaptic membrane hyperpolarizes—making action potentials less likely to occur in the postsynaptic cell (**Figure 45.13b**). Changes in the postsynaptic cell that make action potentials less likely are called **inhibitory postsynaptic potentials (IPSPs)**. IPSPs make the membrane potential more negative.

If an EPSP and an IPSP occur at the same time in the same place, they cancel each other out (**Figure 45.13c**). Synapses can also be modulatory—meaning that they modify a neuron's response to other EPSPs or IPSPs.

POSTSYNAPTIC POTENTIALS ARE GRADED It is critical to realize that, unlike action potentials, EPSPs and IPSPs are not all-or-none events. Instead, they are graded in size.

The size of an EPSP or IPSP depends on the amount of neurotransmitter that is released at the synapse. Greater transmitter release leads to a larger EPSP or IPSP. Both types of signal are short lived because neurotransmitters do not bind irreversibly to channels in the postsynaptic cell. Instead, they are quickly inactivated or taken up by the presynaptic cell and recycled.

If either the amount or life span of neurotransmitters is altered, the normal functioning of neurons is altered. The street drugs cocaine and amphetamine, for example, exert their effects by inhibiting the uptake and recycling of particular neurotransmitters (see Table 45.2).

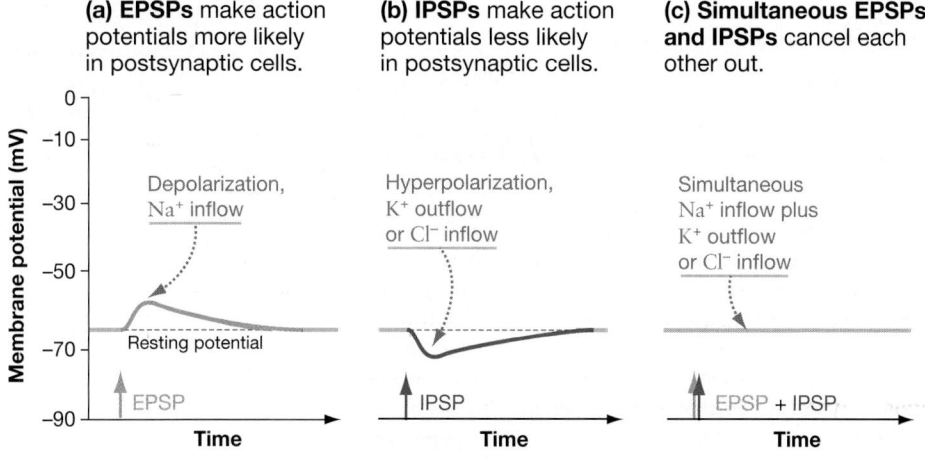

(a) EPSPs make action potentials more likely in postsynaptic cells.

Depolarization, Na⁺ inflow

Resting potential

EPSP

(b) IPSPs make action potentials less likely in postsynaptic cells.

Hyperpolarization, K⁺ outflow or Cl⁻ inflow

IPSP

(c) Simultaneous EPSPs and IPSPs cancel each other out.

Simultaneous Na⁺ inflow plus K⁺ outflow or Cl⁻ inflow

EPSP + IPSP

Membrane potential (mV)

Time

FIGURE 45.13 Events at the Synapse May Lead to Depolarization or Hyperpolarization of the Postsynaptic Membrane. These recordings show changes in the membrane potential of a postsynaptic neuron with the arrival of signals that cause **(a)** depolarization, **(b)** hyperpolarization, or **(c)** no change, as a result of simultaneous depolarizing and hyperpolarizing signals cancelling each other out.

TABLE 45.2 **Categories of Neurotransmitters**

Excitatory neurotransmitters make action potentials more likely in postsynaptic cells.
Inhibitory neurotransmitters make action potentials less likely; modulatory neurotransmitters modify the response at other synapses.
Drugs that prevent reuptake of neurotransmitters increase their activity.

Neurotransmitter	Site of Action	Action	Drugs That Interfere
Acetylcholine	Neuromuscular junction, some CNS pathways	Excitatory (inhibitory in some parasympathetic neurons)	• Botulism toxin blocks release • Black widow spider venom increases, then eliminates, release • α-bungarotoxin (in some snake venoms) binds to receptor
Monoamines			
Norepinephrine	Sympathetic neurons, some CNS pathways	Excitatory or inhibitory	• Ritalin (used for attention deficit hyperactivity disorder) increases release • Some antidepressants prevent reuptake
Dopamine	Many CNS pathways	Excitatory or modulatory	• Cocaine prevents reuptake • Amphetamines prevent reuptake
Serotonin	Many CNS pathways	Inhibitory or modulatory	• MDMA (ecstasy) causes increased release
Amino Acids			
Glutamate	Many CNS pathways	Excitatory	• PCP (angel dust) blocks receptor
Gamma-aminobutyric acid (GABA)	Some CNS pathways	Inhibitory	• Ethanol mimics response to GABA
Peptides			
Endorphins, enkephalins, substance P	Used in sensory pathways (pain)	Excitatory, modulatory, or inhibitory	

SUMMATION AND THRESHOLD How do EPSPs and IPSPs affect the postsynaptic cell? As **Figure 45.14a** shows, the dendrites and the cell body of a neuron typically make hundreds or thousands of synapses with other cells. At any instant, the EPSPs and IPSPs that occur at each of these synapses lead to short-lived surges of charge in the dendrites and cell body of the postsynaptic cell.

If an IPSP and EPSP occur close together in space or time, the changes in membrane potential tend to cancel each other out. But if several EPSPs occur close together in space or time, they sum and make the neuron more likely to fire an action potential (**Figure 45.14b**). The additive nature of postsynaptic potentials is termed **summation**.

(a) Most neurons receive information from many other neurons.

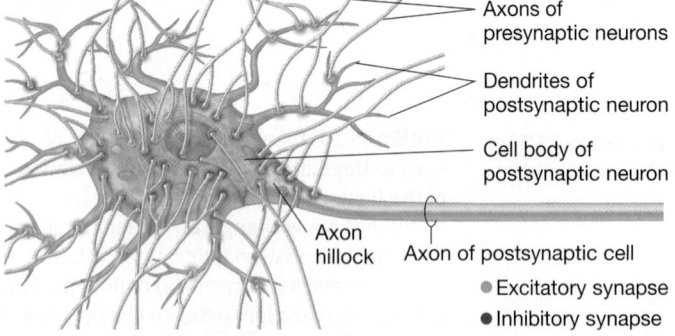

Axons of presynaptic neurons

Dendrites of postsynaptic neuron

Cell body of postsynaptic neuron

Axon hillock   Axon of postsynaptic cell

● Excitatory synapse
● Inhibitory synapse

(b) Postsynaptic potentials sum.

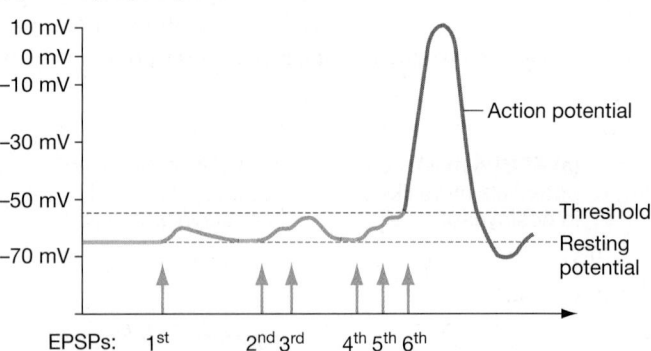

Action potential

Threshold
Resting potential

EPSPs: 1st   2nd 3rd   4th 5th 6th

FIGURE 45.14 Neurons Integrate Information from Many Synapses. (a) The dendrites and cell body of a neuron typically receive signals from hundreds or thousands of other neurons. **(b)** When action potentials arrive close together in time from the same axon or from different axons with synapses close to one another, the postsynaptic potentials sum. In this example, the first excitatory signal is insufficient to generate an action potential. Two excitatory signals arriving close together cause summation but do not reach the threshold for generating an action potential. Three excitatory signals arriving closely spaced sum to exceed the threshold. If excitatory postsynaptic potentials depolarize the axon hillock past threshold, enough channels open to trigger an action potential. This example is simplified—in reality, hundreds or thousands of IPSPs and EPSPs sum to determine action potential frequency.

The sodium channels that trigger action potentials in the postsynaptic cell are located near the start of the axon at a site called the **axon hillock** (see Figure 45.14a). As IPSPs and EPSPs are received and interact throughout the dendrites and cell body, charge spreads to the axon hillock. If the membrane at the axon hillock depolarizes past the threshold potential, enough sodium channels open to trigger positive feedback and an action potential. Once an action potential starts at the axon hillock, it propagates down the axon to the next synapse.

Summation is critically important. Because neurons receive input from many synapses, and because IPSPs and EPSPs sum, information in the form of electrical signals is modified at the synapse before being passed along. An action potential is not necessarily transmitted from one neuron to the next—the response by the postsynaptic cell depends on the information it receives from a wide array of neurons.

To review (1) the molecules and channels responsible for the action potential and (2) the structure and function of the synapse, go to the study area at *www.masteringbiology.com.*

(MB) **BioFlix™** How Neurons Work, **BioFlix™** How Synapses Work, **Web Activity** Action Potentials

CHECK YOUR UNDERSTANDING

If you understand that . . .

- At a synapse, electrical information in the form of changes in membrane voltage is transduced to chemical information in the form of released neurotransmitters.
- Binding of a neurotransmitter to its receptor causes a change in the membrane potential of the postsynaptic cell.

✔ **You should be able to . . .**

Explain why the presence of synapses (instead of direct electrical connections) is important to a neuron's ability to integrate and process information from many different neurons.

Answers are available in Appendix B.

45.4 The Vertebrate Nervous System

The first three sections of this chapter examined electrical signaling at the level of molecules, membranes, and individual cells. This is how the study of the nervous system started, and is called a reductionist approach. Reductionism is common—biologists often start to figure out how something works by studying the component parts.

The goal of this section, however, is to discuss electrical signaling at the level of tissues, organs, and systems. Once researchers understand how individual components work, they want to explore how they interact to create a functioning system.

To begin, let's consider the overall anatomy of the vertebrate nervous system. Then we can ask how researchers explore the function of the most complex organ known: the human brain. The chapter concludes by returning to the molecular level and introducing recent work on learning and memory.

What Does the Peripheral Nervous System Do?

Recall from Section 45.1 that the central nervous system (CNS) is made up of the brain and spinal cord and is concerned primarily with integrating information. The peripheral nervous system (PNS) is made up of neurons outside the CNS.

What functions do the cells of the PNS control? Anatomical and functional studies indicate that the PNS consists of two systems with distinct functions:

1. an **afferent division**, which transmits sensory information to the CNS; and

2. an **efferent division**, which carries commands from the CNS to the body.

Neurons in the afferent division monitor conditions inside and outside the body. Once information from afferent neurons has been processed in the CNS, neurons in the efferent division carry signals that allow the body to respond to changed conditions in an appropriate way. The afferent and efferent divisions carry out sensory and motor functions, respectively.

As **Figure 45.15** shows, the afferent and efferent divisions are part of a hierarchy of PNS functions. The efferent division is further divided into a **somatic nervous system**, which controls movement, and an **autonomic nervous system**, which controls internal processes such as digestion and heart rate.

- The somatic nervous system carries out voluntary responses, which are under conscious control, with skeletal muscles serving as the effectors.

- The autonomic nervous system carries out involuntary responses, which are not under conscious control, with smooth muscle, cardiac muscle, and several glands serving as the effectors.

In effect, the somatic system responds to external stimuli and governs behavior, while the autonomic system responds to internal stimuli and controls the activity of internal organs and glands.

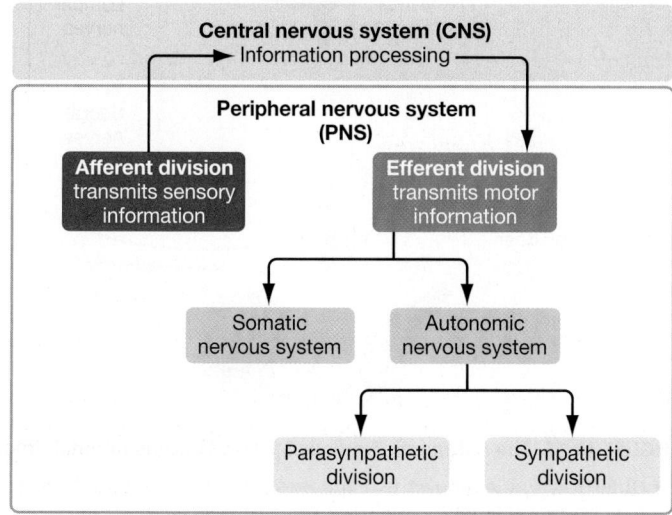

FIGURE 45.15 The PNS Comprises Distinct Components.

Many organs and glands are served by two functionally distinct types of autonomic nerves, summarized in **Figure 45.16**: One inhibits activity; the other promotes it.

- Nerves in the **parasympathetic nervous system** promote "rest and digest" functions that conserve or restore energy. For example, the parasympathetic nerves that connect to the heart slow it down, while those that serve the digestive tract stimulate its activity.

- Nerves in the **sympathetic nervous system** typically prepare organs for stressful "fight or flight" situations. Sympathetic nerves speed up the heart rate, stimulate the release of glucose from the liver, and inhibit action by digestive organs.

Functional Anatomy of the CNS

Parasympathetic nerves originate at the base of the brain or the base of the spinal cord. Most sympathetic nerves also originate in

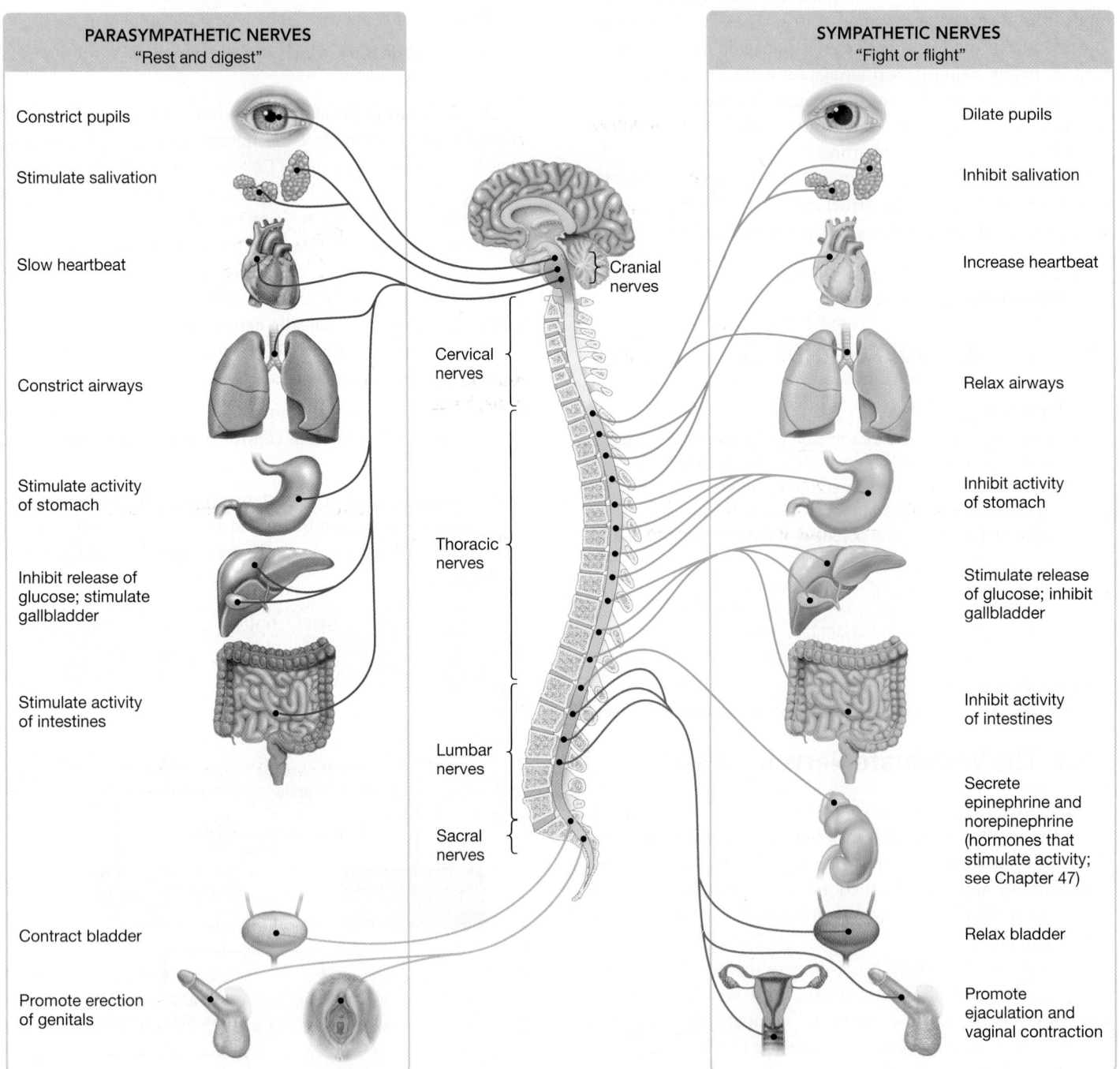

FIGURE 45.16 The Autonomic Nervous System Controls Internal Processes.

✔ **QUESTION** Explain how the responses listed here for pupils, heartbeat, and liver support the "rest and digest" versus "fight or flight" functions.

the spinal cord, but they emerge along the middle of its length. Similarly, most sensory and motor neurons in the somatic nervous system project to or from the spinal cord.

In effect, then, the spinal cord serves as an information conduit that collects and transmits information from throughout the body. With a few exceptions, such as the reflex that occurs when you touch a hot burner, virtually all the information that travels to or from the spinal cord is sent to the brain for processing.

The brain is far and away the most complex organ found in animals. Researchers estimate that the human brain has 100 billion neurons, each making thousands of synaptic connections with other neurons. How do biologists even begin to study such a fantastically complex structure? They begin with general anatomy.

GENERAL ANATOMY OF THE BRAIN Nineteenth-century anatomists established that the brain is made up of the four structures labeled in **Figure 45.17**: the cerebrum, cerebellum, diencephalon, and brain stem. Each has a distinct function.

- The **cerebrum** makes up the bulk of the brain, is divided into left and right hemispheres, and is involved in conscious thought and memory.

- The **cerebellum** coordinates complex motor patterns.

- The **diencephalon** relays sensory information to the cerebellum and controls homeostasis.

- The **brain stem** connects the brain to the spinal cord, and is the autonomic center for regulating the heart, lungs, and digestive system.

Each cerebral hemisphere has four major areas, or lobes: the **frontal lobe**, the **parietal lobe**, the **occipital lobe**, and the **temporal lobe** (**Figure 45.18**). The two hemispheres are connected by a thick band of axons called the **corpus callosum**.

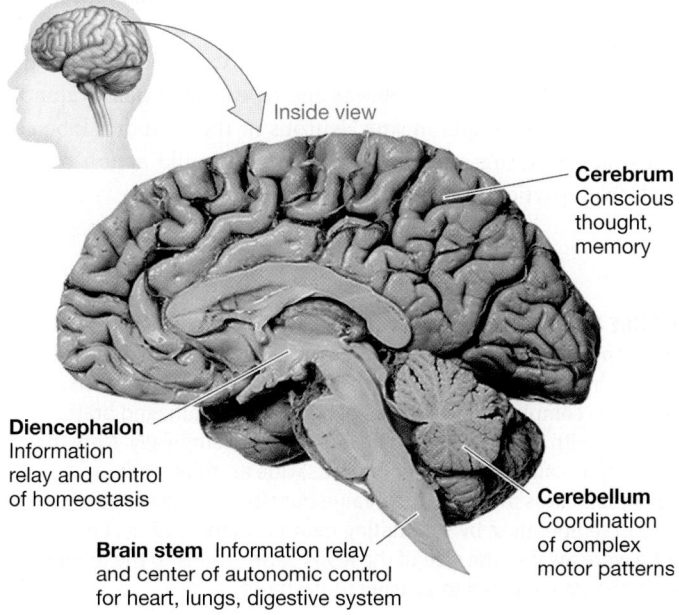

Diencephalon
Information relay and control of homeostasis

Brain stem Information relay and center of autonomic control for heart, lungs, digestive system

Cerebrum
Conscious thought, memory

Cerebellum
Coordination of complex motor patterns

FIGURE 45.17 The Human Brain Contains Four Distinct Structures.

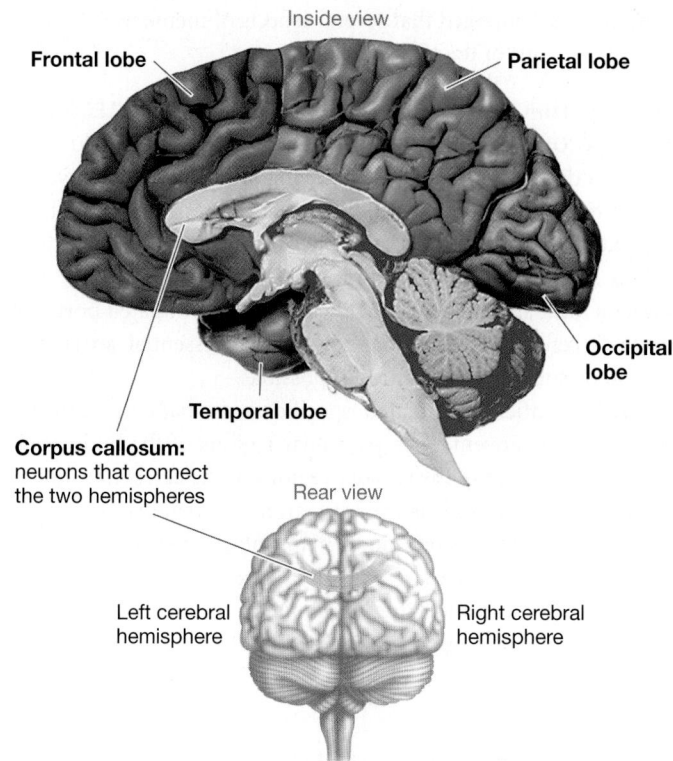

Frontal lobe
Parietal lobe
Occipital lobe
Temporal lobe

Corpus callosum: neurons that connect the two hemispheres

Rear view

Left cerebral hemisphere
Right cerebral hemisphere

FIGURE 45.18 The Human Cerebrum Has Four Lobes and Two Hemispheres.

✔**EXERCISE** On your own head, point to each of the labeled areas.

What tools do researchers use to explore the function of each area within the cerebrum?

MAPPING FUNCTIONAL AREAS IN THE CEREBRUM I: LESION STUDIES Early work on brain function studied people with specific mental deficits caused by areas of brain damage, or lesions. Paul Broca, for example, studied an individual who could understand language but could not speak. After the person's death in 1861, Broca examined the patient's brain and discovered a damaged area in the left frontal lobe of the cerebrum. Broca hypothesized that this region is responsible for speech. More generally, he formulated the hypothesis that specific regions of the brain are specialized for coordinating particular functions.

Broca's claim that functions are localized to specific brain areas has been verified through extensive efforts to map the cerebrum. In some cases, advances were made by studying people who had to have portions of their brains removed.

In 1953, for example, surgeons treated a 27-year-old man named Henry Gustav Moliason for life-threatening seizures by removing a small portion of his temporal lobe. The man recovered (he died in December 2008, at age 83), had normal intelligence, and vividly remembered his childhood, but he had no short-term memory. Brenda Milner, who studied this individual for over 40 years, had to introduce herself to him every time they met; he could not even recognize a recent picture of himself. Based on case histories and studies of memory in laboratory animals, a consensus has emerged that several aspects of memory localize to interior sections of the temporal lobe.

MAPPING FUNCTIONAL AREAS IN THE CEREBRUM II: ELECTRICAL STIMULATION OF CONSCIOUS PATIENTS Wilder Penfield pioneered a different approach to studying brain function by working with severe epileptics—people suffering from seizures. These individuals were scheduled to have seizure-prone areas of their brains surgically removed. While the patients were awake and under a local anesthetic, Penfield electrically stimulated portions of their cerebrums. His goal was to map essential areas that should be spared from removal if possible.

When Penfield stimulated specific areas, patients reported sensations or movement in particular regions of the body. Penfield was able to map the sensory regions of the cerebrum shown in **Figure 45.19**, as well as the adjacent motor regions. Brain surgeons still use this technique to map critical areas near tumors and seizure-prone areas.

Perhaps the most striking of Penfield's findings was that, on occasion, patients would respond to stimulation of their temporal lobe by having what appeared to be flashbacks. After one region was stimulated, a woman said, "I hear voices. It is late at night around the carnival somewhere—some sort of traveling circus.... I just saw lots of big wagons that they used to haul animals in."

Was this a memory, stored in a small set of neurons that Penfield happened to stimulate? The hypothesis that memories are stored in specialized cells is intensely controversial. As critics have pointed out, Penfield's results are difficult to interpret because he was working with people who suffered from severe brain dysfunction. In addition, Penfield's patients sometimes described the same memory when other cells were stimulated after the original area had been surgically removed.

Have other approaches to studying memory been productive?

How Does Memory Work?

Learning is an enduring, usually adaptive change in behavior that results from a specific experience in an individual's life. **Memory** is the retention of learned information. Learning and memory are thus closely related and are often studied in tandem. As an introduction to how researchers explore these phenomena, let's first examine work that focuses on neurons and then review research at the molecular level.

RECORDING FROM SINGLE NEURONS DURING MEMORY TASKS How do the action potentials generated by a cell change as learning and memory take place? Researchers have attempted to answer this question by recording from individual neurons in the temporal lobes of humans.

Before operating on patients who were still awake and about to undergo surgery to remove seizure-prone areas of their brains, physicians have projected words or names of objects on a screen and asked the individuals to read them silently, read them aloud, and/or remember them and repeat them later. The data show that individual neurons in the cerebrum's temporal lobe are relatively quiet while patients identify objects but extremely active when the individual remembers the objects and repeats their names aloud.

What do such data mean? Neurons in the temporal lobe are most active during memory tasks. So how could action potentials from particular cells make memory possible?

(a) Top view of cerebrum

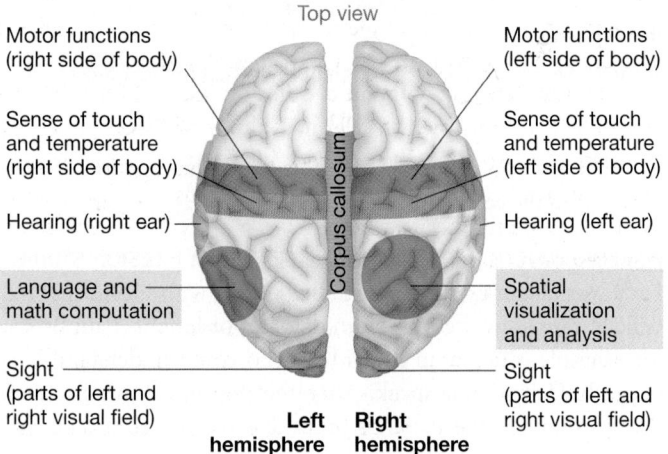

(b) Cross section through area responsible for sense of touch and of temperature

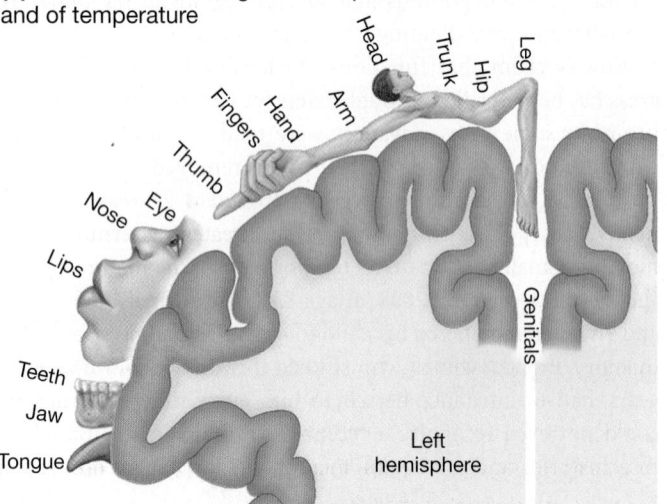

FIGURE 45.19 Specific Brain Areas Have Specific Functions.
(a) Map of the brain, in top view (as if the person is looking at the top of the page), showing the functions of some major regions. The map was compiled from studies of people with damaged brain areas or with brain regions that were removed surgically. Note that the corpus callosum is not actually as wide as shown here.
(b) Researchers mapped the area responsible for the sense of touch and of temperature by stimulating neurons in the brains of patients who were awake. The size of the icons corresponds to the amount of brain area devoted to sensing those parts.

✓**QUESTION** Is there a correlation between the size of the brain area devoted to sensing a particular body part and the size of that body part? Explain.

(a) Sea slug *Aplysia californica*

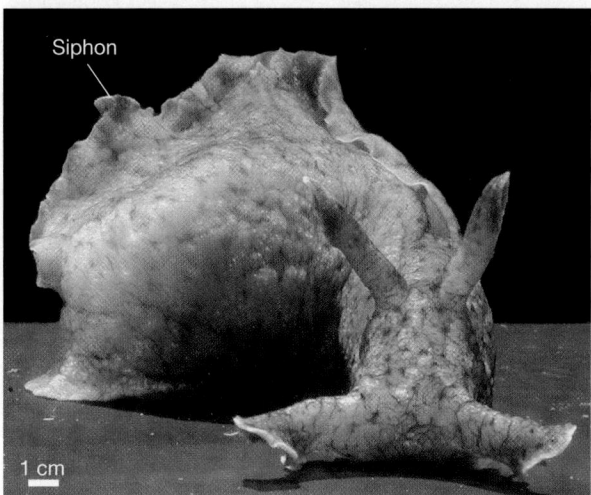

Siphon

1 cm

(b) Gill-withdrawal reflex protects the gills during an attack.

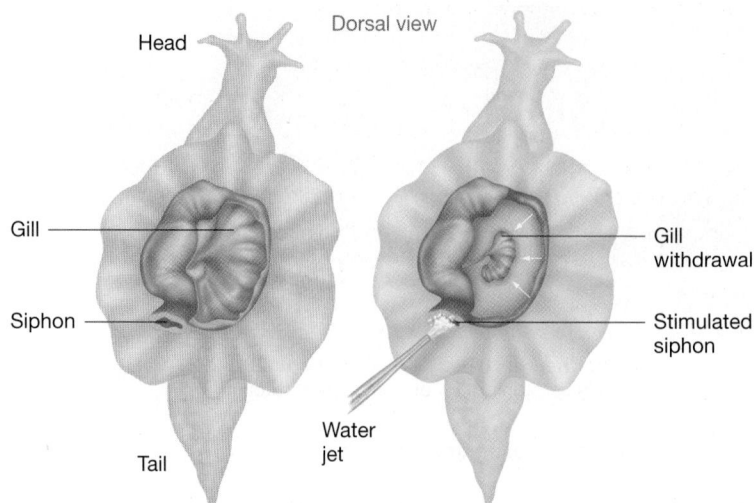

Dorsal view

Head

Gill

Siphon

Tail

Water jet

Gill withdrawal

Stimulated siphon

FIGURE 45.20 The Gill-Withdrawal Reflex in *Aplysia* Is a Model System in Learning and Memory.

DOCUMENTING CHANGES IN SYNAPSES Research on the molecular basis of memory is based on two fundamental ideas. First, learning and memory must involve some type of short-term or long-term change in the neurons responsible for these processes. This change could be structural or chemical in nature. Structural changes might include modifications in the number of synapses that a particular neuron makes. Chemical changes might involve alterations in the amount of neurotransmitter released at certain synapses or changes in the number of receptors present in postsynaptic cells. Second, it will be much easier to understand these changes if an extremely simple system of neurons can be studied.

To explore the molecular basis of learning and memory, Eric Kandel's group has focused on the sea slug *Aplysia californica* (**Figure 45.20a**). Much of their work has explored the reflex diagrammed in **Figure 45.20b**. When a structure in *Aplysia* called the siphon is touched—for example, by a stream of water—the individual responds by withdrawing its gill. The reflex involves a sensory neuron that is activated by touch and a motor neuron that projects to a gill muscle. Retracting the gill protects it from predators.

Early work established that this simple reflex is modified by learning. For example, *Aplysia* also withdraw their gills when their tails receive an electrical shock. If shocks to the tail are combined with a very light touch to the siphon—too light to normally get a response on its own—an *Aplysia* will learn to withdraw its gills in response to a light siphon touch alone.

Follow-up studies on this reflex showed that the neurons involved in learning release the neurotransmitter **serotonin**, which causes an EPSP in the motor neuron to the gill. Repeated application of serotonin mimics what happens at the synapse during learning, when the neurons fire repeatedly. As **Figure 45.21** shows, experimental application of serotonin leads to higher EPSPs, meaning that the motor neuron is more likely to generate action potentials. These results suggest that in *Aplysia*, changes in the nature of the synapse form the molecular basis of learning and memory. This process is termed **synaptic plasticity**.

Recently Kandel's team replicated these results with sensory and motor neurons growing in culture. **Figure 45.22a** on page 904 shows two *Aplysia* motor neurons on a culture plate. Each motor neuron is receiving synapses from a sensory neuron. To mimic the learning process, the investigators applied serotonin to one synapse five times over a short period. When they stimulated the sensory neuron a day later, they found a huge increase in EPSPs (**Figure 45.22b**). The experimental cells had also established additional synapses. The structure and behavior of the experimental neuron had changed, based on its experience. Neurons that had not received the repeated stimulation with serotonin had normal postsynaptic responses and numbers of synapses.

Results like these reinforce a growing consensus that learning and memory involve both molecular and structural changes in synapses. Further, most researchers now agree that at least some aspects of long-term memory involve changes in gene expression. Chapter 47 explores how the chemical messengers called hormones cause changes in gene expression in target cells. But before investigating how hormones work, let's focus on the electrical signals involved in vision, hearing, taste, and movement—the subject of Chapter 46.

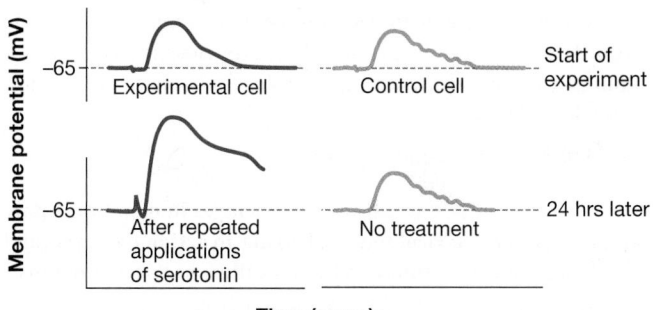

Membrane potential (mV)

−65

Experimental cell

Control cell

Start of experiment

−65

After repeated applications of serotonin

No treatment

24 hrs later

Time (msec)

FIGURE 45.21 Repeated Application of Serotonin Changes the Behavior of Postsynaptic Neurons.

(a) Apply neurotransmitters to neurons in culture.

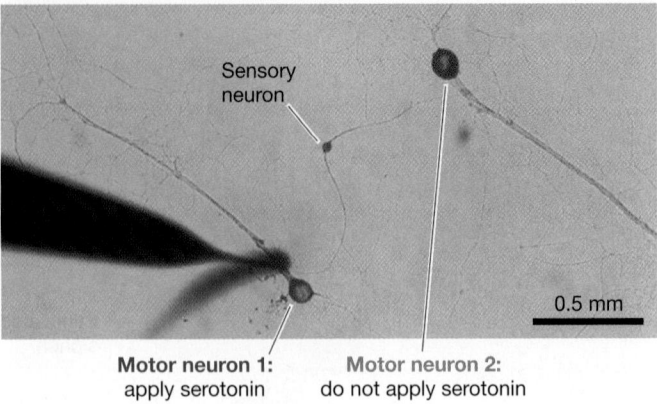

Sensory
neuron

0.5 mm

Motor neuron 1:
apply serotonin

Motor neuron 2:
do not apply serotonin

(b) Repeating *Aplysia* experiment in culture

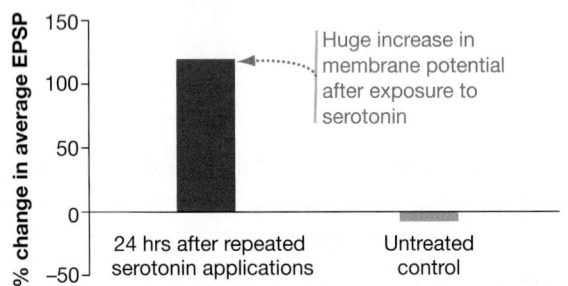

Huge increase in
membrane potential
after exposure to
serotonin

24 hrs after repeated
serotonin applications

Untreated
control

% change in average EPSP

FIGURE 45.22 Learning and Memory Involve Changes in Synapses. **(a)** An in vitro study of interactions between *Aplysia* motor neurons and sensory neurons. **(b)** Histograms documenting the percentage increase in postsynaptic potentials that occurs after repeated application of serotonin, as in Figure 45.21.

CHAPTER 45 REVIEW

For media, go to the study area at www.masteringbiology.com

Summary of Key Concepts

⚷ **Neurons are cells that transmit electrical signals used in communication. Their plasma membranes carry a voltage, called a membrane potential, due to differences in the concentrations of ions on the inner and outer surfaces.**

● All neurons have a cell body and multiple short dendrites that receive electrical signals from other cells. Most neurons also have axons that transmit electrical signals to other neurons or to effector cells in glands or muscles.

● Studies of the squid giant axon neurons established that neurons have a resting potential created by the sodium-potassium pump and potassium leak channels. When Na^+/K^+-ATPase hydrolyzes ATP, it transports 3 Na^+ out of the cell and 2 K^+ in.

✔ You should be able to diagram the plasma membrane of a neuron. Add symbols to show the relative concentrations of Na^+, K^+, and Cl^-. Add labels indicating the role of K^+ leak channels and the Na^+/K^+-ATPase.

 **Web Activity** Membrane Potentials

⚷ **Action potentials are all-or-none changes in membrane potential that serve as electrical signals. During an action potential, an inflow of sodium ions is followed by an outflow of potassium ions.**

● Studies of the squid axon established that the action potential is a rapid, all-or-none change in membrane potential.

● An action potential begins with an inflow of Na^+ that depolarizes the membrane. An outflow of K^+ follows and repolarizes the membrane.

● Both Na^+ and K^+ flow through voltage-gated channels.

● As charge spreads from the site of an action potential, the nearby membrane is depolarized enough to trigger additional Na^+ inflows and propagate the signal.

● Propagation takes place most rapidly in large-diameter axons or myelinated axons.

✔ You should be able to explain why every action potential in a neuron is identical.

BioFlix™ How Neurons Work, **BioFlix™** How Synapses Work, **Web Activity** Action Potentials

⚷ **At synapses, the electrical signal from a neuron triggers the release of a chemical signal—a neurotransmitter. When the neurotransmitter arrives at an adjacent neuron, there is a change in that cell's membrane potential.**

● When action potentials arrive at a synapse, synaptic vesicles fuse with the axon's membrane and deliver neurotransmitters that bind to receptors on the membrane of a postsynaptic cell.

- One class of receptors functions as ligand-gated channels. In response to binding by a neurotransmitter, the channels open and admit ions that depolarize or hyperpolarize the postsynaptic cell's membrane.

- Postsynaptic potentials from nearby synapses sum.

- If the membrane at the axon hillock depolarizes to a threshold value, an action potential is triggered.

✔ You should be able to explain how summation in the postsynaptic cell relates to the claim that the cell body integrates information from many different synapses.

👉 **Most animals have a central nervous system (CNS) and a peripheral nervous system (PNS). PNS neurons receive sensory information and transmit it to the CNS for processing. The CNS then sends signals to muscles, glands, or other tissues via PNS neurons.**

- The CNS consists of the brain and spinal cord (in vertebrates); the PNS consists of all nervous system components outside the CNS.

- In vertebrates, the PNS contains somatic and autonomic components. The somatic PNS is responsible for sensing external stimuli and effecting movement; the autonomic system monitors internal conditions and effects changes in the activity of organs.

- Early efforts to map functional regions of the brain depended on analyzing deficits in individuals with brain lesions or on stimulating certain regions of the cerebrum.

- Efforts to understand higher brain functions such as learning and memory form the current focus of research on the CNS.

- To date, research has established that learning and memory are based on modifications in synapses. After learning takes place, certain neurons release more or less neurotransmitter, or make additional synapses, in response to stimulation.

✔ You should be able to describe what an animal would be like if synapses were "fixed" early in development and unchangeable.

Questions

✔ TEST YOUR KNOWLEDGE

Answers are available in Appendix B

1. Which ion leaks across a neuron's membrane to help create the resting potential?
   a. Ca^{2+} b. K^+ c. Na^+ d. Cl^-

2. Why did the squid axon become a model system for studying electrical signaling in animals?
   a. Its action potentials are particularly large and frequent.
   b. It is the tissue from which researchers initially isolated Na^+/K^+-ATPase.
   c. Squids are abundant and easy to obtain.
   d. It was large enough to support intracellular recording by the first microelectrodes.

3. How does myelination affect the propagation of an action potential?
   a. It speeds propagation by increasing the density of voltage-gated channels.
   b. It speeds propagation by increasing electrochemical gradients favoring Na^+ entry.
   c. It speeds propagation because charge does not leak out of the membrane as it spreads down the axon.
   d. It slows down propagation because Na^+ channels exist only at unmyelinated nodes (nodes of Ranvier).

4. In a neuron, what creates the electrochemical gradient favoring the outflow of K^+ when the cell is at rest?
   a. Na^+/K^+-ATPase
   b. voltage-gated K^+ channels
   c. voltage-gated Na^+ channels
   d. ligand-gated Na^+/K^+ channels

5. Why do biologists say that positive feedback occurs during an action potential?
   a. The action potential is an all-or-none event, meaning that once it starts, it goes to completion.
   b. The opening of potassium channels repolarizes the membrane, making it less likely that sodium channels will open and depolarize the membrane.
   c. Once sodium channels open and begin to depolarize the membrane, they become more likely to open and cause further depolarization.
   d. Sodium channels are refractory—once they have opened, they are less likely to open again for a few milliseconds.

6. Why is memory thought to involve changes in particular synapses?
   a. In some systems, an increased release of neurotransmitters occurs after learning takes place.
   b. In some systems, the type of neurotransmitter released at the synapse changes after learning takes place.
   c. When researchers stimulated certain neurons electrically, individuals replayed memories.
   d. People who lack short-term memory have specific deficits in synapses within the brain regions responsible for memory.

✔ TEST YOUR UNDERSTANDING

Answers are available in Appendix B

1. Explain why the Na^+/K^+-ATPase is "electrogenic"—meaning that it creates a voltage across a membrane.

2. Draw a graph of an action potential and label the axes. Label the parts of the graph and explain which ion flow or flows are responsible for each part.

3. Explain the difference between a ligand-gated K^+ channel and a voltage-gated K^+ channel.

4. Why does summation occur in postsynaptic cells?

5. Compare and contrast the somatic and autonomic components of the PNS.

6. Compare and contrast the sympathetic and parasympathetic components of the autonomic nervous system.

1. Explain why drugs that prevent neurotransmitters from being taken back up by the presynaptic cell have dramatic effects on the activity of postsynaptic neurons.

2. Discuss the pros and cons of lesion studies and electrical stimulation of conscious patients in determining the functions of particular brain structures.

3. The data points on the graph to the right record the movement of potassium ions (μSiemens is a unit of conductance) observed when neuron membranes were voltage-clamped at the array of voltages indicated on the *x*-axis. The "Neuron with venom" data were generated by adding black mamba poison to the solution bathing the neuron. (Black mambas are snakes.) What can you conclude from the data?

4. In some species, researchers are able to identify individual neurons in the brain and record the action potentials they produce. Using these data, how can they infer the function of particular neurons?

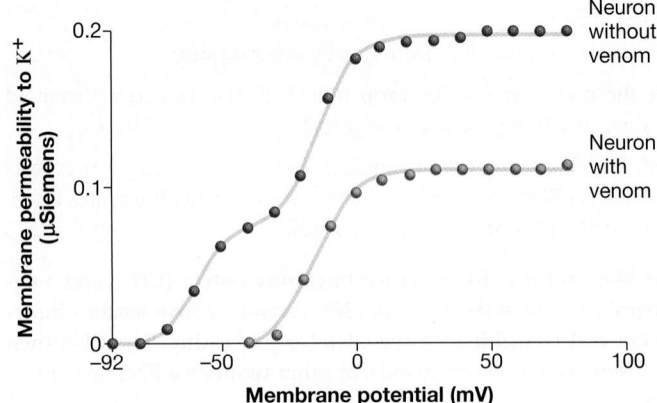

In many species of moth, males have much larger antennae than females do. Receptor cells on the males' feathery antennae detect airborne chemical signals that are produced by sexually mature females. As a result, males can locate females in total darkness.

Animal Sensory Systems and Movement 46

Many adult moths are active at night when it is difficult or impossible to see. Instead of looking for a mate, sexually mature female moths release a chemical attractant called a pheromone into the air. A **pheromone** is a small molecule that acts as a signal between individuals. Male moths of the same species can detect even a single molecule of the pheromone, due to the receptor cells located on their large, feathery antennae. In response to an airborne gradient of pheromone molecules, males fly toward a female.

As they patrol in search of these airborne pheromones, however, male moths are hunted by bats. Like moths, bats are active almost exclusively at night. Instead of hunting by sight, like a falcon or a cheetah, bats hunt with the aid of sonar. Bats emit a train of high-pitched sounds as they fly and then listen for echoes that indicate the direction and shape of objects in their path. If the object is a moth, the bat flies toward it, catches the individual in its mouth, and eats it.

Some moth species can hear bat calls, however. When moths detect sounds from an onrushing bat, they tumble out of the sky in chaotic escape flights.

If you were out at night as these dramas unfolded, at best you might be dimly aware that bats and moths were flying about. Humans cannot smell moth pheromones or hear the sounds that bats emit when flying. It took decades of careful experimentation for biologists to understand how moths and bats sense the world around them and how they move in response to the information they receive.

Sensing changes in the environment and moving in response to this information is fundamental to how animals work. Let's begin with a basic question: How are sounds, smells, and other stimuli transformed into a signal that the brain can understand?

KEY CONCEPTS

- Sensory receptor cells transduce stimuli to changes in membrane potential. Action potentials are sent to the brain, where the signals are processed and integrated.

- Hearing is based on sensory receptor cells that move in response to sound waves of a particular frequency.

- Vision is based on sensory receptor cells that contain a light-absorbing pigment bound to a protein. The pigment changes conformation when it absorbs light.

- Taste and smell sensations are registered by membrane proteins that act as ion channels or receptors for particular molecules.

- In many cases, animals respond to sensory stimuli by moving. Movement is based on antagonistic muscle groups that act on a skeleton. Muscle contraction occurs when myosin proteins move down the length of actin fibers.

✔ When you see this checkmark, stop and test yourself. Answers are available in Appendix B.

46.1 How Do Sensory Organs Convey Information to the Brain?

As a moth flies through the night, its brain receives streams of signals from an array of sensory organs. Antennae provide information about the concentration of pheromones; ears located on various parts of the body send data on the presence of high-pitched sounds; detectors for balance and gravity transmit signals about the body's orientation in space.

Each type of sensory information is detected by a sensory neuron or by a specialized receptor cell that makes a synapse with a sensory neuron. As **Figure 46.1** shows, the moth's nervous system integrates the sensory input—information from sensory neurons—and responds with motor output, via electrical signals, to specific muscle groups (effectors).

The ability to sense a change in the environment depends on three processes:

1. **transduction**, or the conversion of an external stimulus to an internal signal in the form of an action potential;

2. amplification of the signal; and

3. transmission to the central nervous system (CNS).

The first step in the sequence requires a sensory receptor cell to convert light, sound, touch, or some other signal into an electrical signal. Sensory receptors are located throughout the body and are categorized by the type of stimulus:

- **Nociceptors** sense harmful stimuli such as tissue injury.
- **Thermoreceptors** detect changes in temperature.
- **Mechanoreceptors** respond to distortion caused by pressure.
- **Chemoreceptors** perceive specific molecules.
- **Photoreceptors** respond to particular wavelengths of light.
- **Electroreceptors** detect electric fields.

With such a broad range of sensory detectors available, it is no wonder that animals can monitor and respond to a wide array of changes in their environments. And this list is not exhaustive. Many birds, sea turtles, and other animals can sense changes in magnetic fields and use Earth's magnetic field as an aid in navigation; some species of birds can sense changes in barometric pressure.

Now, how do sensory cells receive information from the environment and report it to the brain, so an appropriate response can occur?

Sensory Transduction

When most sensory cells are in the resting state, the inside of the plasma membrane is more negative than the exterior. If ion flows cause the interior to become more positive (less negative), the membrane is **depolarized**. If changes in ion channels cause the cell interior to become more negative than the resting potential, the membrane is **hyperpolarized**.

Figure 46.2a shows the membrane potential from a sound-receptor cell. If you put your finger on the red line and trace to the right, you'll notice that when the experimenter played a sound, the sound-receptor cell depolarized for a short time in response. Other sensory cells work in a similar way.

Although sensory receptors can detect a remarkable variety of stimuli, they all transduce sensory input—including light, sounds, touch, and odors—to a change in membrane potential. In this way, different types of information are transduced to a common type of signal—one that can be interpreted by the brain.

If a sensory stimulus induces a large change in a sensory receptor's membrane potential, there is a change in the firing rate of action potentials sent to the brain. The amount of depolarization that occurs in a sound-receptor cell, for example, is proportional to the loudness of the sound. If the depolarization passes threshold, enough voltage-gated sodium channels open to trigger action potentials that are sent to the brain.

Recall from Chapter 45 that all action potentials from a given neuron are identical in size and shape. **Figure 46.2b** graphs the action potential "firing rate" recorded from a sound-receptor cell, when sounds at various frequencies were played at two distinct intensities. Notice that loud sounds induce a higher frequency of action potentials than do soft sounds. In this way, receptor cells provide information about the intensity of a stimulus.

But if all types of external stimuli are converted to electrical signals in the form of action potentials, and if all action poten-

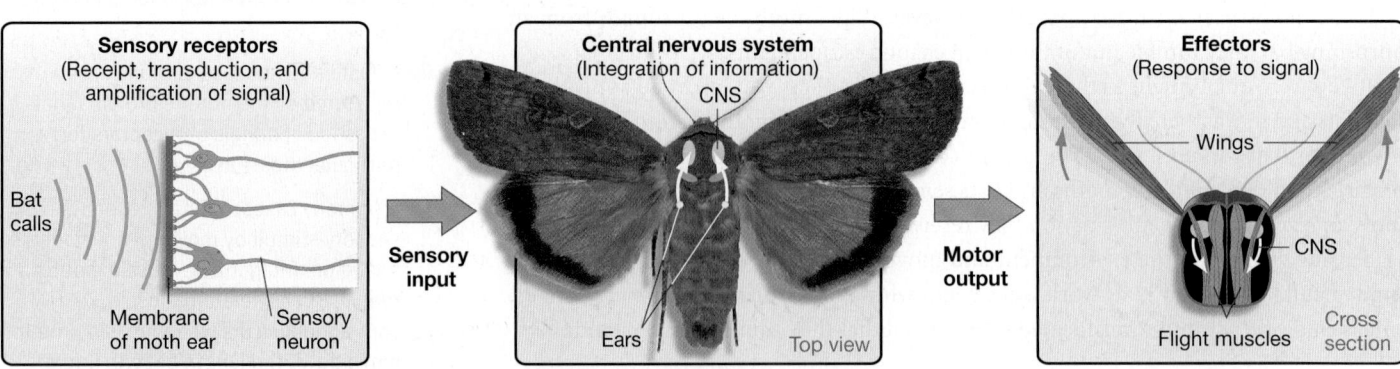

FIGURE 46.1 Sensory Systems, the CNS, and Effectors Such as Muscles Are Linked. Sensory neurons relay information about conditions inside and outside an animal to the central nervous system. After integrating information from many sensory neurons, the CNS sends signals to muscles.

(a) Sound-receptor cells depolarize in response to sound.

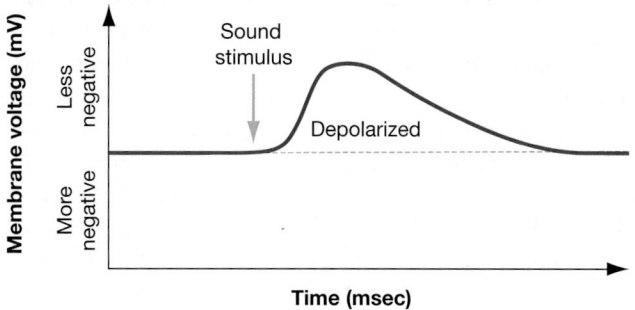

(b) Sound-receptor cells respond more strongly to louder sounds.

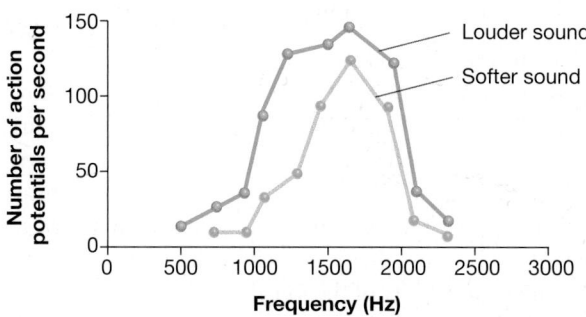

FIGURE 46.2 Sensory Inputs Change the Membrane Potential of Receptor Cells. (a) In response to sensory stimuli, ions flow across the membranes of receptor cells and either depolarize or hyperpolarize the membrane. **(b)** The frequency of action potentials from a receptor transmits information about the nature and intensity of the sensory stimulus.

tials are alike in size and duration, how does the brain interpret the information properly?

Transmitting Information to the Brain

There are two keys to understanding how the brain interprets sensory information. First, receptor cells tend to be highly specific. For example, each receptor cell in a human ear responds best to certain frequencies of sound. Some receptors are more sensitive to low-pitched sounds at a frequency of 1000 Hz (hertz, or cycles per second—a unit of frequency); others respond best to high-pitched sounds such as 8000 Hz.

The receptor-cell response depicted in Figure 46.2b, for example, is strongest for sounds at about 1650 Hz. At this frequency, the maximum number of action potentials per second emerge from the receptor. In this way, the pattern of action potentials from a cell contains information about the frequency of sound that is being received, its intensity, and how long the stimulus lasts.

The second key point: Each type of sensory neuron sends its signal to a specific portion of the brain. Axons from sensory neurons in the human ear project to a particular area at the side of the brain, but axons from sensory receptors in the eye deliver action potentials to another area at the back of the brain. Different regions of the brain are specialized for interpreting different types of stimuli.

Now that the basic principles of sensory reception and transduction have been introduced, let's delve into the details of four sensory systems that are particularly well understood: hearing, vision, taste, and smell.

46.2 Hearing

Animals have a wide variety of mechanisms for sensing changes in pressure. Crabs, for example, have a fluid-filled organ that helps them sense the pressure created by gravity. The organ, known as a **statocyst**, is lined with pressure-receptor cells and contains a small calcium-rich structure. The calcium-rich particle normally rests on the bottom of the organ. But if the crab is tipped or flipped over, this structure presses against receptors that are *not* on the bottom of the organ. When the brain receives action potentials from these cells, it responds by activating muscles that restore the animal to its normal posture.

It's also common for animals to have cells that are responsible for detecting direct physical pressure on skin, as well as pressure-receptor cells that monitor how far muscles or blood vessels are stretched. But the best-studied type of pressure-sensing is called hearing.

Hearing is the ability to sense the wavelike changes in air pressure called sound. A sound consists of waves of pressure in air or in water. The number of pressure waves that occur in one second is called the **frequency** of the sound. We perceive different sound frequencies as different **pitches**.

Hearing and other pressure-sensing systems found in animals are based on the same mechanism. Let's briefly examine the general nature of a mechanoreceptor cell that responds to pressure, then investigate the specific structures involved in vertebrate hearing.

How Do Sensory Cells Respond to Sound Waves and Other Forms of Pressure?

The mechanoreceptors responsible for sensing sound and vibrations in the environment are relatively simple in design. In every case, direct physical pressure on a plasma membrane or distortion by bending changes the conformation of ion channels in the membrane and causes the channels to open or close.

In response to a change in ion flow through channel proteins, the membrane depolarizes or hyperpolarizes. The result is a new pattern of action potentials from a sensory neuron.

THE STRUCTURE OF HAIR CELLS In vertebrates, ion channels that respond to pressure are found in hair cells. **Hair cells** are pressure-receptor cells, illustrated in **Figure 46.3a** on page 910, named for their stiff outgrowths called **stereocilia** (singular: **stereocilium**). The "hairy-looking" stereocilia are microvilli that are reinforced by actin filaments.

Many hair cells also have a single **kinocilium**, a true cilium that contains a 9 + 2 arrangement of microtubules introduced in Chapter 7. Hair cells are found in the ears of land-dwelling vertebrates and the lateral line system in many species of fish.

(a) Hair cells have many stereocilia and one kinocilium.

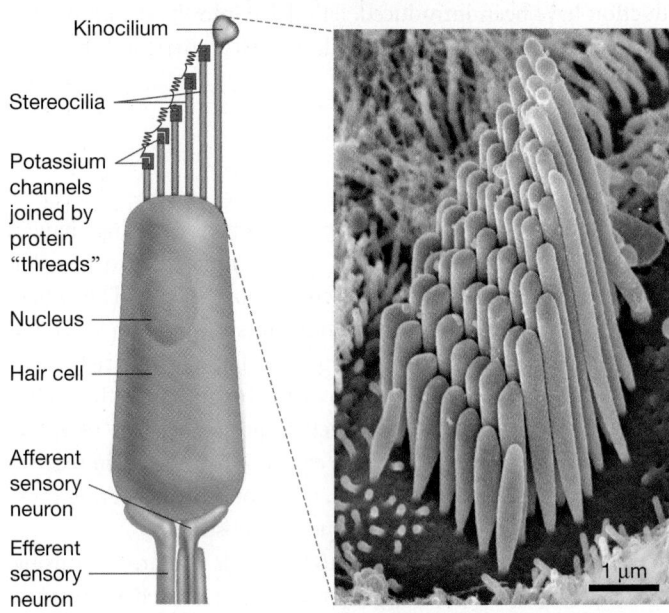

(b) PROCESS: BENDING OPENS ION CHANNELS

Pressure wave

1. Arrival of pressure wave bends stereocilia.

2. Potassium channels open in response to bending.

3. Membrane depolarizes due to influx of K^+.

4. Depolarization triggers inflow of calcium ions.

5. Ca^{2+} causes synaptic vesicles to fuse with plasma membrane.

6. Neurotransmitter is released and diffuses to afferent neuron.

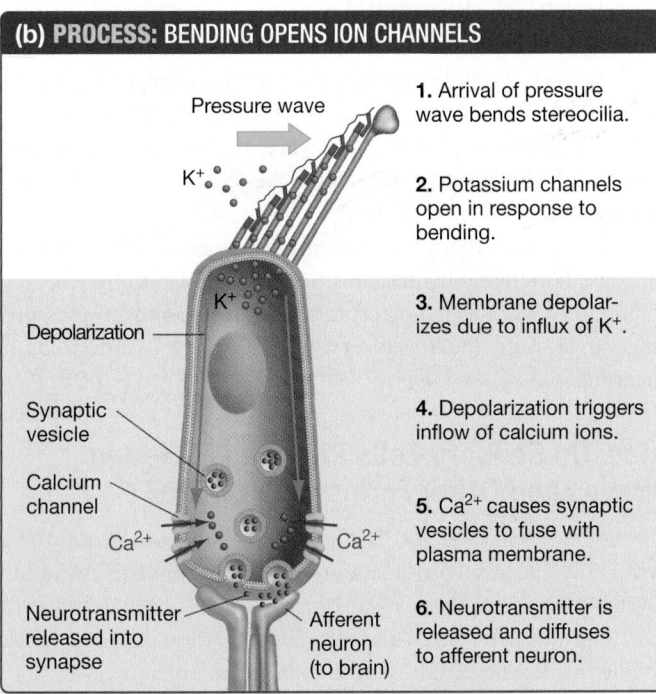

FIGURE 46.3 Hair Cells Transduce Sound Waves to Electrical Signals.

As Figure 46.3a shows, the stereocilia in hair cells are arranged in order of increasing height; if a kinocilium is present, it is the tallest of all the projections. All these structures extend into a fluid-filled chamber.

SIGNAL TRANSDUCTION IN HAIR CELLS If stereocilia are bent in the direction of the kinocilium in response to pressure (**Figure 46.3b**), the membranes of the structure are distorted and potassium ion (K^+) channels in the stereocilia open. This is the common theme in pressure-sensing cells: Bending opens ion channels.

Recall from Chapter 45 that the opening of K^+ channels usually causes an outflow of K^+ that hyperpolarizes neurons. Hair-cell plasma membranes respond differently, because they are bathed by extracellular fluid with an extraordinarily high K^+ concentration. As a result, the equilibrium potential for K^+ in hair cells is 0 mV instead of the -85 mV in a typical neuron. The resting potential of the hair-cell plasma membrane is -70 mV, so K^+ rushes in and causes a depolarization of approximately 20 mV.

In hair cells, depolarization causes an inflow of calcium ions, which triggers an increase in the amount of neurotransmitter released at the synapse between the hair cell and a sensory neuron. The end result is excitation of the postsynaptic cell, meaning that it becomes more likely to fire an action potential to the brain, via an afferent sensory neuron. You might recall, from Chapter 45, that afferent neurons are part of the peripheral nervous system and conduct information to the CNS.

If sound pressure waves bend stereocilia the other way, however, the K^+ channels close and the cell hyperpolarizes by 5 mV. Hyperpolarization of the hair cell decreases the amount of neurotransmitter released at the synapse and inhibits the sensory neuron, meaning that the neuron becomes less likely to trigger action potentials. Pressure that is perpendicular to the stereocilia does not change the activity of the ion channels.

How can bending affect ion channels? Electron micrographs show that tiny threads connect the tips of stereocilia to each other. One hypothesis contends that when the stereocilia are bent, the threads somehow pull open ion channels in the wall of the next tallest stereocilium—like tiny trapdoors (see Figure 46.3b, step 2). This hypothesis remains to be confirmed, however. Researchers still do not fully understand how the ion channels involved in pressure reception work.

The Mammalian Ear

To understand how changes in the membrane potential of hair cells result in hearing, let's focus on the human ear as a case study. The human ear has three sections: the **outer ear**, **middle ear**, and **inner ear** (**Figure 46.4**). A membrane separates each section from the others.

To trace the path of sound through the ear, study the lower part of Figure 46.4. The outer ear, which projects from the head, collects incoming pressure waves and funnels them into a tube known as the ear canal. At the end of the ear canal, the waves strike the **tympanic membrane**, or eardrum, which separates the outer ear from the middle ear.

The repeated cycles of air compression cause the tympanic membrane to vibrate back and forth with the same frequency as the sound wave. The vibrations are passed to three tiny bones called the **ear ossicles**, which vibrate against one another in response. The last ossicle, the **stapes** (pronounced *STAY-peez*), vibrates against a membrane. That membrane, called the **oval window**, separates the middle ear from the inner ear. The oval window oscillates in response and generates waves in the fluid inside a chamber known as the **cochlea** (pronounced *KOK-lee-ah*). These pressure waves are sensed by hair cells in the cochlea.

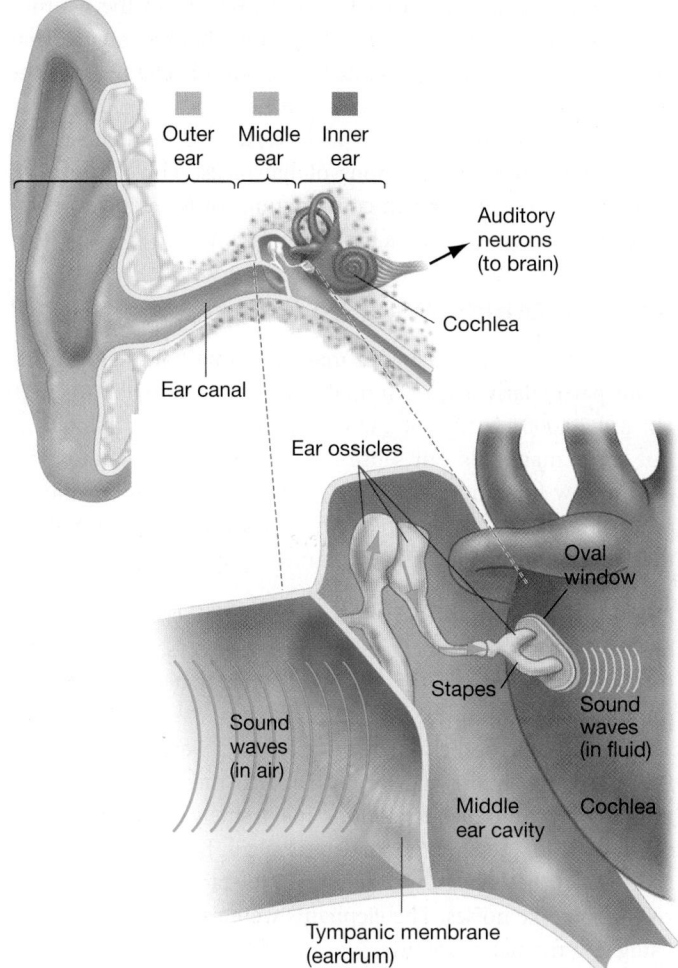

FIGURE 46.4 **Mammals Have an Outer Ear, Middle Ear, and Inner Ear.** The middle ear starts with the tympanic membrane and ends in the oval window of the cochlea.

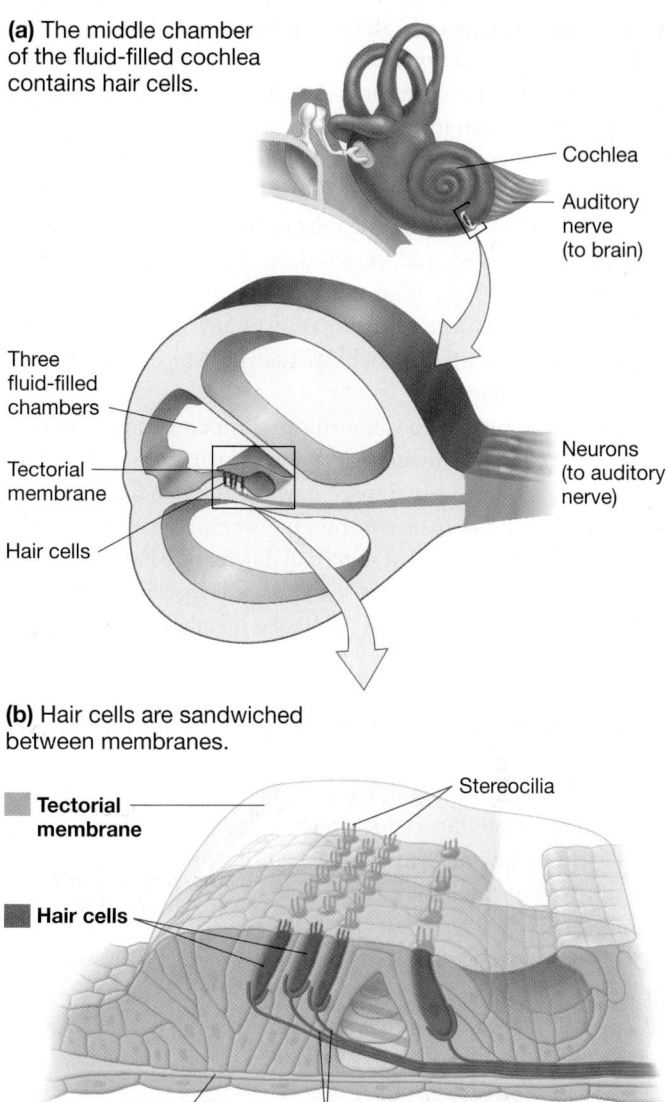

(a) The middle chamber of the fluid-filled cochlea contains hair cells.

(b) Hair cells are sandwiched between membranes.

FIGURE 46.5 **In the Human Cochlea, Hair Cells Are Sandwiched between Membranes. (a)** The cochlea contains fluid-filled chambers separated by membranes. Hair cells are located in the middle chamber and **(b)** are sandwiched between the basilar membrane and the tectorial membrane.

In effect, the ear translates airborne waves into waterborne waves. The system seems extraordinarily complex, though, for such a simple result. Why doesn't the outer ear canal lead directly to the oval window? Why have a middle ear at all?

THE MIDDLE EAR AMPLIFIES SOUNDS Biologists began to understand the function of the middle ear when they recognized two key aspects of its structure. First, the size difference between the tympanic membrane and the oval window is important. Because the tympanic membrane is about 15 times larger than the oval window, the amount of vibration induced by sound waves increases by a factor of 15 by the time it reaches the oval window. This phenomenon is similar to using the same amount of force to bang on a very large door versus a very small door.

In addition, the three ossicles act as levers that further amplify vibrations at the tympanic membrane. Reptiles have just one ear ossicle instead of three so this levering action can occur. The overall effect in mammals is to amplify sound by a factor of 22—meaning that soft sounds are amplified enough to stimulate hair cells in the cochlea. Biologists interpret the mammalian middle ear as an adaptation for increasing sensitivity to sound.

To summarize, the mammalian outer ear transmits sound waves from the environment to the middle ear; the middle ear amplifies these waves enough to stimulate the hair cells within the cochlea of the inner ear.

If all hair cells responded equally to all frequencies of sound, we would be able to perceive only one pitch. Everyone's voice—indeed, every noise—would sound the same. How can identical hair cells distinguish different frequencies?

THE COCHLEA DETECTS THE FREQUENCY OF SOUNDS As the cross section in **Figure 46.5a** shows, the cochlea has a set of internal membranes that divide it into three chambers. Hair cells form rows in the middle chamber. The bottom of each hair cell connects to a structure called the **basilar membrane** (**Figure 46.5b**).

In addition, the hair cells' stereocilia touch yet another, smaller surface called the **tectorial membrane**. (The kinocilium is not present in a mature cochlear hair cell.) In effect, hair cells are sandwiched between membranes.

Researchers struggled for decades to understand how these membranes affect hair-cell function. It is virtually impossible to study cochleas in living organisms, because the cochleas are tiny, complex, coiled, and buried deep inside the skull. During the 1920s and 1930s, however, Georg von Békésy pioneered work on the structure and function of these organs by performing experiments on cochleas that he had dissected from fresh human cadavers.

Von Békésy was able to vibrate the oval window and record how the cochlea's internal membranes moved in response. He found that when a pressure wave traveled down the fluid in the upper and lower chambers, the basilar membrane vibrated in response. His key finding, though, was that sounds of different frequencies caused the basilar membrane to vibrate maximally at specific points along its length (**Figure 46.6**). When the basilar membrane vibrated in a particular location, the stereocilia of the hair cells there were bent one way and then the other by the tectorial membrane.

Von Békésy also noted that the basilar membrane is stiff near the oval window and flexible at the other end. Thus, each segment vibrates in response to a different frequency of sound. Just as a stiff drumhead produces a high-pitched sound and a loose drumhead yields a low-pitched sound, high-frequency sounds cause the stiff part of the basilar membrane to vibrate; low-frequency sounds cause the flexible part to vibrate.

To summarize, certain portions of the basilar membrane vibrate in response to specific frequencies and result in the bending of hair-cell stereocilia. In this way, hair cells in a particular place on the membrane respond to sounds of a certain frequency.

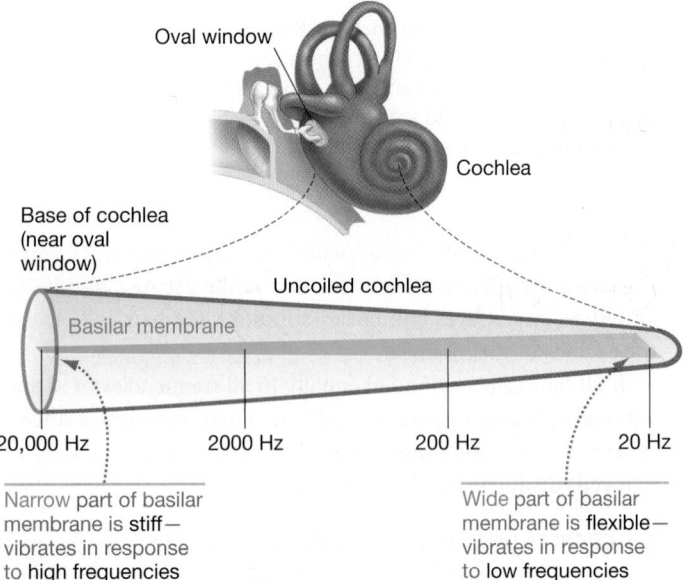

FIGURE 46.6 The Basilar Membrane Varies in Stiffness.
The numbers below the basilar membrane are in units of frequency called Hertz (Hz).

When the brain receives action potentials from the neurons associated with each hair cell, it interprets them as a particular pitch—meaning a specific frequency of sound. The result is the sense we call hearing.

Complex sounds contain a wide variety of frequencies and trigger particular combinations of hair cells. Through experience, the brain learns which combinations of frequencies represent music, a fire alarm, or a best friend's voice.

Sensory Worlds: What Do Other Animals Hear?

Compared with the hearing of many mammals, human hearing is not particularly acute. Humans can hear sounds between 20 Hz and 20,000 Hz (20,000 Hz is equal to 20 kHz, or kilohertz). But some mammals can sense low-frequency infrasounds that are too low for humans to hear (*infra* means below or under); others are aware of high-frequency ultrasounds that are above the range of human hearing (*ultra* means beyond).

ELEPHANTS DETECT INFRASOUND   When Katherine Payne was observing elephants at a zoo in the mid-1980s, she noticed a throbbing sensation in the air. Payne knew that infrasound can produce such sensations.

To test the hypothesis that the elephants were producing infrasonic vocalizations, Payne returned to the zoo with microphones that could pick up sounds at extremely low frequency. At normal speed, the tape she made was silent. But when she raised the pitch of the sounds by speeding up the tape, she heard a chorus of cow-like noises. The elephants were calling to each other, using low-frequency sounds.

According to follow-up research, elephants have the best infrasonic hearing of any land mammal. Because infrasound can travel exceptionally long distances, biologists hypothesize that infrasonic calls allow wild elephants to coordinate their movements when they are miles apart.

Recent research suggests that in addition to detecting infrasound via their large ears, elephants use their feet to detect a byproduct of infrasound: seismic vibrations traveling in the ground.

BATS DETECT ULTRASOUND   Ultrasonic hearing in bats was discovered in the late 1930s, when Donald Griffin borrowed the only ultrasonic apparatus then in existence from Robert Galambos, a fellow graduate student. Griffin used the machine to demonstrate that flying bats constantly emit ultrasounds. In follow-up experiments, he documented that a bat with cotton in its ears, or with its mouth taped shut, crashed into walls when released in a room. Blindfolded bats, in contrast, never crashed.

Griffin and Galambos concluded that bats use sound echoes (sonar) to navigate. This concept, termed **echolocation**, was an outlandish idea at the time. When Galambos described it at a meeting in 1940, another scientist shook him by the shoulders and said, "You can't really mean that!"

More recent research has shown that dolphins, shrews, and certain other animals besides bats use sonar. In fact, it is likely that at least some of these species perceive shapes with their ears better than they do with their eyes.

THE LATERAL LINE SYSTEM IN FISH AND AMPHIBIANS Hair cells found in the mammalian ear allow these animals to sense the changes in air pressure that we call sound. But in fishes and amphibians, hair cells allow individuals to sense pressure changes in water.

In most fishes and larval amphibians, groups of hair cells are located in a line that runs the length of the body. This sensory organ is called the **lateral line system**. If changes in water pressure—from waves, an animal swimming by, or some other force—cause kinocilia to bend, the distortion leads to a change in action potentials sent to the brain. In this way, most aquatic animals get information about pressure changes at specific points along the length of the body.

CHECK YOUR UNDERSTANDING

If you understand that . . .

- Hearing is a type of pressure detection that begins when the stereocilia on hair cells bend in response to changes in pressure. The bending movement opens ion channels and results in a change in membrane potential.
- The mammalian ear comprises specialized structures that function in transmitting and amplifying sound, and in responding to specific frequencies, as well as recording changes in intensity.

✔ You should be able to . . .

1. Describe the functions of the outer ear, middle ear, and inner ear of a mammal.
2. Predict the type of hearing loss that results from each of the following: a punctured eardrum, a mutation that results in dramatically shortened stereocilia, and an age-related loss of flexibility in the basilar membrane.

Answers are available in Appendix B.

46.3 Vision

Most animals have a way to sense light. The organs involved range from simple light-sensitive eyespots in flatworms to the sophisticated, image-forming eyes of vertebrates, cephalopod mollusks, and arthropods.

Variation in the structure of light-sensing organs illustrates an important general principle about the sensory abilities of animals: In most cases, a species' sensory abilities correlate with the environment it lives in and its mode of life—how it finds food and mates. Eyes and other sensory structures are adaptations that allow individuals to thrive in a particular environment. Salamanders that live in meadows and forests have sophisticated eyes; salamanders that live in lightless caves have no functional eyes at all.

Keep this point in mind as we delve into the details of how insects and vertebrates see.

The Insect Eye

Insects have a **compound eye**, composed of hundreds or thousands of light-sensing columns called **ommatidia**. As **Figure 46.7** shows, each ommatidium has a lens that focuses light onto a small number of receptor cells—usually four. The receptor cells, in turn, send axons to the brain.

Each ommatidium contributes information about one small piece of the visual field, not unlike a single pixel on a computer monitor. Thus, the more ommatidia in a compound eye, the better the resolution—meaning the resolving power, or ability to distinguish objects.

In addition, the presence of many light-sensing columns makes species with compound eyes particularly good at detecting movement. Insects that hunt by sight, such as damselflies and dragonflies, have particularly large numbers of ommatidia.

(a) Ommatidia are the functional units of insect eyes.

(b) Each ommatidium contains receptor cells that send axons to the CNS.

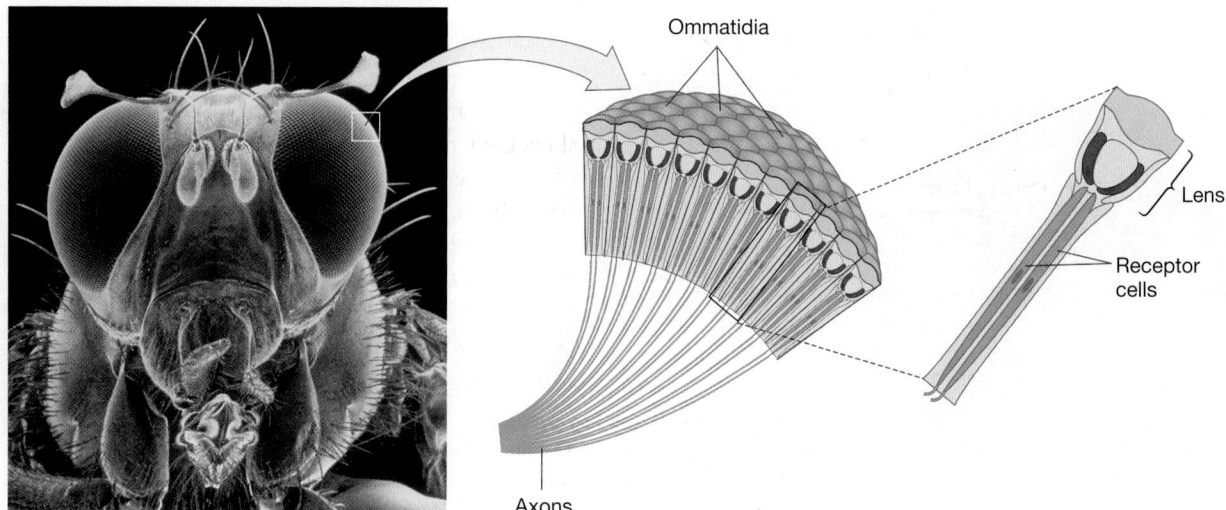

FIGURE 46.7 In the Compound Eyes of Insects, Each Ommatidium Sees Part of the World. The micrograph in part (a) is colorized to match the drawings in part (b).

The Vertebrate Eye

Compound eyes are found in insects, crustaceans, and certain other arthropods. Because they appear only in species that are part of the same monophyletic group (see Chapter 27), researchers conclude that this type of eye structure evolved just once—in an ancestor of today's arthropods.

In contrast, the **camera eye**—a structure with a single lens that focuses incoming light onto a layer of many receptor cells—evolved independently in two widely divergent groups: the cephalopod mollusks (squid and octopuses) and the vertebrates. Let's examine the vertebrate version of the camera eye more closely.

THE STRUCTURE OF THE CAMERA EYE **Figure 46.8a** shows the major structures in a typical vertebrate eye:

- The outermost layer of the structure is a tough rind of white tissue called the sclera. This is the "white of the eye."

- The front of the sclera forms the **cornea**, a transparent sheet of connective tissue.

- The **iris** is a colored, round muscle just inside the cornea. The iris can contract or expand to control the amount of light entering the eye.

- The **pupil** is the hole in the center of the iris.

- Light enters the eye through the cornea and passes through the pupil and a curved, clear **lens**.

- Together, the cornea and lens focus incoming light onto the retina in the back of the eye. The **retina** contains a thin layer of light-sensitive cells and several layers of neurons.

Figure 46.8b provides a closer look at the retina, which is attached to the rest of the eye by a single layer of pigmented epithelial cells. From back to front, the retina comprises three distinct cell layers:

1. The sensory cells that respond to light, the photoreceptors, are held in place by the pigmented epithelium.

2. Photoreceptors synapse with an intermediate layer of connecting neurons called bipolar cells.

3. Cells in the intermediate neuron layer connect with one another and with the neurons called **ganglion cells**, which form the innermost layer of the retina. The axons of the ganglion cells project to the brain via the **optic nerve**.

A FLAWED STRUCTURE? Although the vertebrate eye is sometimes claimed to be a "perfect" adaptation, optometrists have cataloged a large array of common defects. In addition, even vertebrate eyes that are in ideal condition have a curious flaw: the blind spot.

Vertebrate eyes have a blind spot because there are no photoreceptor cells where the optic nerve leaves the retina. If light falls in this area, there are no sensory cells available to respond. No signal is sent to the brain, so the stimulus isn't seen.

In cephalopod mollusks (squid and octopus), however, a better camera-type eye evolved independently. In cephalopods, the photoreceptor cells form a continuous layer on the inside of the retina. The ganglion cells that transmit messages to the brain are located behind the sensory cells—not in front of them, as in the vertebrate eye (see Figure 46.8b). As a result, the optic nerve does not interrupt the layer of photoreceptor cells. In this respect, cephalopods receive more complete visual information about their environment than vertebrates do.

Why do vertebrates have a flawed design? The leading hypothesis is historical constraint. As Chapter 24 pointed out, the traits observed in today's organisms evolved from traits that occurred in their ancestors. Apparently, the vertebrate camera eye happened to evolve from an ancestor with photoreceptor cells located in an interior layer.

(a) The structure of the vertebrate eye

(b) In the retina, cells are arranged in layers.

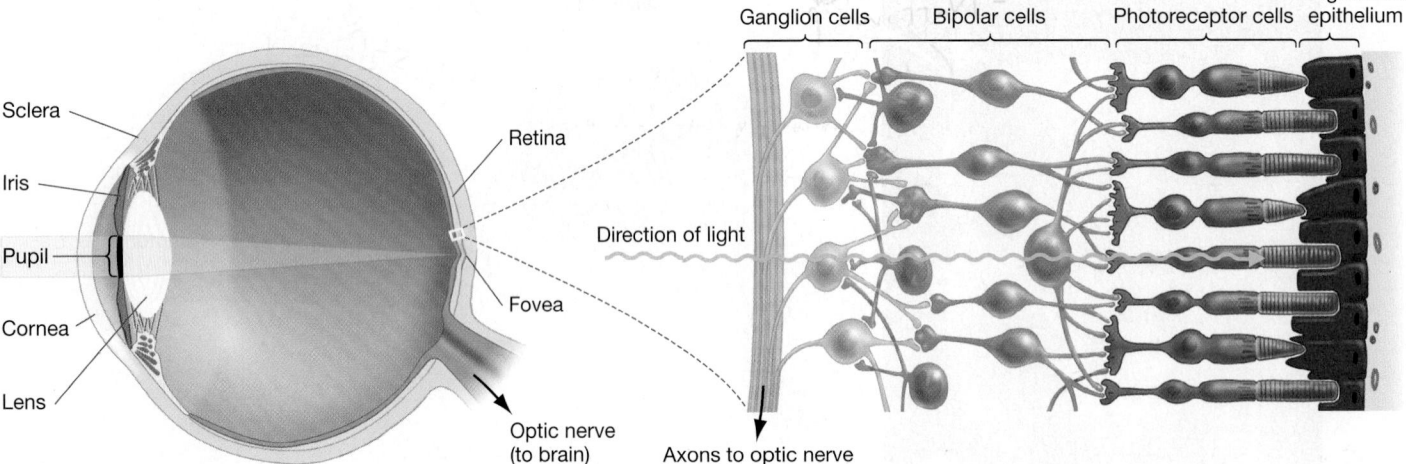

FIGURE 46.8 Camera Eyes Have a Single Lens That Focuses Incoming Light on Receptor Cells. (a) Light passes through the pupil of the eye and is focused onto the retina by the cornea and lens. **(b)** The photoreceptor cells that respond to light are in the "outermost" layer of the retina, furthest from the light source.

WHAT DO RODS AND CONES DO?

Early anatomists established that the photoreceptors in vertebrate eyes come in two distinct types: small rod-shaped or cone-shaped cells called **rods** and **cones** (**Figure 46.9a**). When technical advances allowed changes in the membrane potentials of these cells to be recorded, it became clear that rods and cones differ in function as well as structure.

Rods are sensitive to dim light but not to color. Cones, in contrast, are much less sensitive to faint light but are stimulated by different wavelengths (that is, colors). These discoveries explained why night vision is largely black and white—at night, the rods do most of the work.

Rods dominate most of the retina, but one small spot in the center of the retina has only cones. This is the **fovea** (see Figure 46.8a). When people focus on an object, their eyes move so that the image falls on the fovea of each eye. Based on these observations, biologists concluded that the high density of cones in the fovea maximizes the resolution of the image.

HOW DO RODS AND CONES DETECT LIGHT?

As Figure 46.9a shows, rods and cones have segments that are packed with membrane-rich disks. The membranes contain large quantities of a transmembrane protein called **opsin**. Each opsin molecule is associated with a molecule of the pigment **retinal**. In rod cells, the two-molecule complex is called **rhodopsin** (**Figure 46.9b**).

Experiments with isolated retinal, opsin, and rhodopsin molecules confirmed that retinal changes shape when it absorbs light. Specifically, the number-11 carbon in the retinal molecule changes from the *cis* conformation to the *trans* conformation (**Figure 46.9c**). Retinal is a light switch.

The shape change that occurs in retinal triggers a series of events that culminate in a different stream of action potentials being sent to the brain. The sequence of events is unusual, though, because the receipt of a light stimulus does not open ion channels or trigger the release of a neurotransmitter to a sensory neuron.

In vertebrates, the molecular basis of vision is a shape change in retinal that shuts down an existing ion channel and decreases the amount of neurotransmitter being released to the sensory neuron. In rod cells, electrical activity across the membrane, as well as neurotransmitter release, are maximized in the dark. Exposure to light transmits information by *shutting down* both processes.

(a) Rods and cones contain stacks of membranes.

(b) Rhodopsin is a transmembrane protein complex.

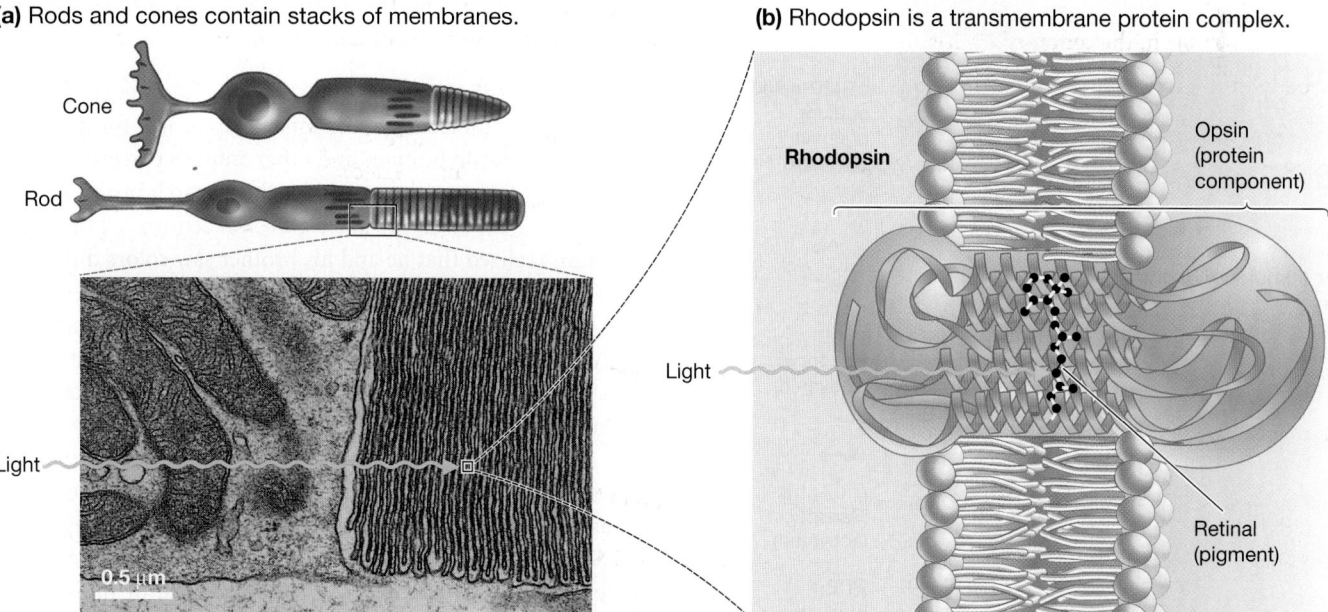

(c) The retinal molecule inside rhodopsin changes shape when retinal absorbs light.

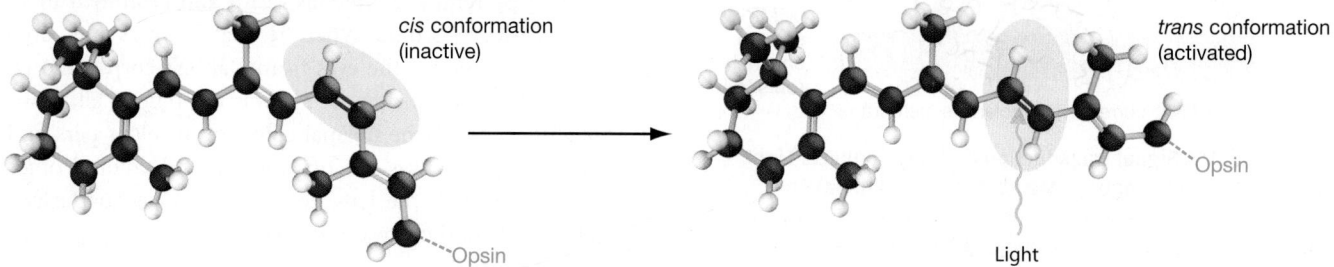

FIGURE 46.9 Rods and Cones Are Packed with Transmembrane Proteins That Contain the Pigment Retinal.
(a) Rods and cones have membranous disks containing thousands of opsin molecules. **(b)** Each opsin holds one retinal molecule. **(c)** Retinal changes conformation when it absorbs light. In response, opsin also changes shape.

✔**QUESTION** Explain why the change from *cis* to *trans* shown in part (c) would affect opsin's function.

Figure 46.10 shows how shutting down happens:

1. Rhodopsin is activated when light causes retinal to change shape from the *cis* to *trans* conformation.

2. Rhodopsin activation causes a membrane-bound molecule called transducin to activate the enzyme phosphodiesterase (PDE).

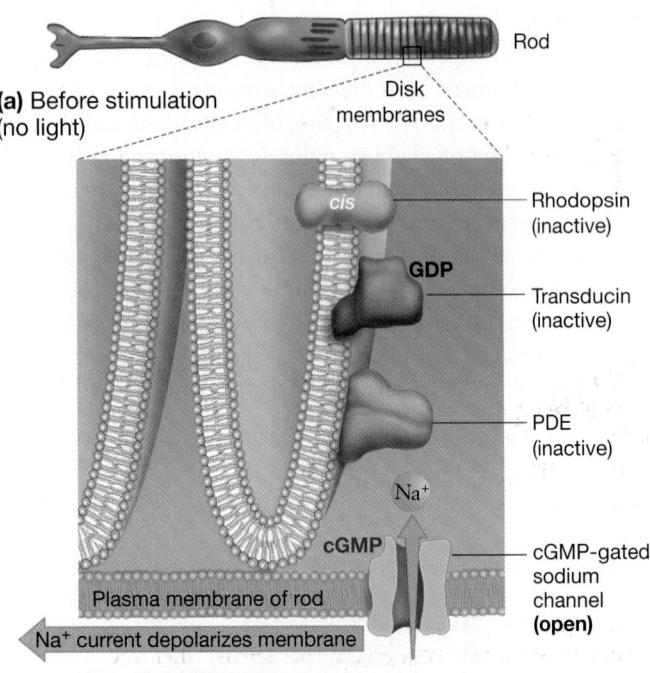

(a) Before stimulation (no light)

Rod

Disk membranes

cis — Rhodopsin (inactive)

GDP

Transducin (inactive)

PDE (inactive)

Na$^+$

cGMP — cGMP-gated sodium channel **(open)**

Plasma membrane of rod

Na$^+$ current depolarizes membrane

(b) After stimulation (light)

Light

trans — Rhodopsin (activated)

GTP

Transducin (activated)

PDE (activated)

cGMP GMP

cGMP-gated sodium channel **(closed)**

Lack of Na$^+$ current hyperpolarizes membrane

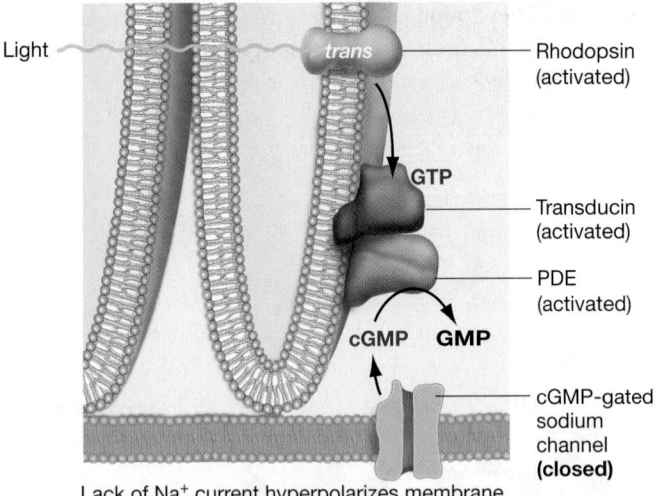

FIGURE 46.10 A Signal Transduction Pathway Connects Light Absorption with Changes in Membrane Potential. (a) An unstimulated photoreceptor. Notice that sodium ions flow into the cell when light is *not* being received. **(b)** A stimulated photoreceptor. When rhodopsin is activated, it leads to a reduction in cGMP concentration. With less cGMP available, cGMP-gated sodium channels close and the membrane hyperpolarizes.

✔**QUESTION** The inflow of sodium ions into a photoreceptor cell is called "the dark current." Why?

3. PDE breaks down a nucleotide called cyclic guanosine monophosphate (cGMP) to guanosine monophosphate (GMP);

4. As cGMP levels decline, cGMP-gated sodium channels in the plasma membrane of the rod cell close.

5. When sodium channels close, Na$^+$ entry decreases and the membrane hyperpolarizes.

✔If you understand this sequence of events, you should be able to explain why cGMP acts as both a second messenger and a ligand in this system, and how transducin compares to the G proteins introduced in Chapter 8. You should also be able to explain why shutting Na$^+$ channels results in hyperpolarization.

In response to the ensuing change in membrane potential, smaller quantities of the neurotransmitter glutamate are discharged at the synapse. The decrease in neurotransmitter indicates to the postsynaptic cell, called a **bipolar cell**, that the rod absorbed light. As a result, a new pattern of action potentials is sent to the brain, via neurons called ganglion cells. Axons from ganglion cells are bundled into a structure called the optic nerve.

This system is exquisitely sensitive: Biologists have recorded a measurable change in the membrane potentials of rod cells in response to a single photon of light.

COLOR VISION: THE PUZZLE OF DALTON'S EYE Studies of rhodopsin revealed the molecular mechanism responsible for light perception. But how do humans and other animals perceive color?

To answer this question, consider the research program initiated by John Dalton[1] in the late eighteenth century. At the age of 26, Dalton realized that he and his brother saw colors differently than did other people. To them, red sealing wax and green laurel leaves appeared to be the same color, and a rainbow exhibited only two hues. Dalton and his brother could not differentiate the colors red and green. This condition is called red-green color blindness (**Figure 46.11**).

In a lecture delivered in 1794, Dalton explained his perceptions by hypothesizing that red wavelengths failed to reach his retinas. Further, he hypothesized that because a normal eyeball is filled with clear fluid, and because blue fluids absorb red light, his defective vision resulted from the presence of bluish fluid rather than clear fluid in his eyes.

To test this hypothesis, Dalton left instructions that his eyes should be removed after his death and examined to see if the fluid inside was blue. When he died 50 years later, an assistant dutifully removed the eyes from Dalton's corpse and examined them. The fluid inside the eye was not blue at all, however, but slightly yellow—the normal color for an older person. Further, when the back was cut off one eye and colored objects were viewed through the lens, the objects looked perfectly normal. Dalton's hypothesis was incorrect.

---

[1]Dalton was an accomplished physicist. He was the first proponent of the atomic theory and formulated Dalton's law on the partial pressures of gases (introduced in Chapter 44). Red-green color blindness is sometimes called Daltonism in his honor.

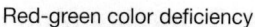

No color deficiency   Red-green color deficiency

FIGURE 46.11 People with Color Deficiencies See Colors Differently than Do People without Deficiencies. These images show what a person with red-green color blindness would see, compared to a person with intact color receptors.

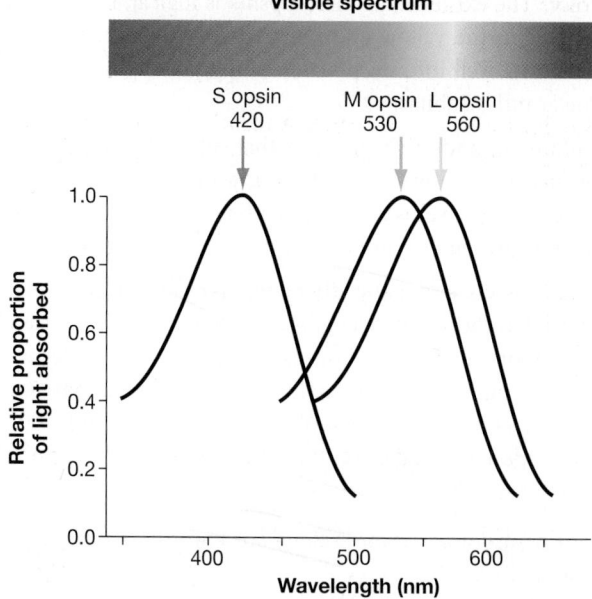

FIGURE 46.12 Color Vision Is Possible because Different Opsins Absorb Different Wavelengths of Light. Each human cone cell contains one of three different types of opsin. Each opsin has a different range of absorbed wavelengths.

✔**QUESTION** The retinal molecules in S, M, and L opsins are identical. How is it possible that they respond to different wavelengths of light?

COLOR VISION: MULTIPLE OPSINS What caused Dalton's color blindness? The key to answering this question was the discovery that the human retina contains three types of color-sensitive photoreceptors: blue, green, and red cones, named for the colors that they best perceive.

To follow up on this result, biologists analyzed opsin molecules from the three cell types and found that each had a distinct amino acid sequence. The three proteins are called the blue, green, and red opsins (or S, M, and L, for short, medium, and long wavelengths, respectively). Although retinal is the light-absorbing molecule in all photoreceptor cells, various types of opsins respond only to distinct wavelengths of light.

Based on these results, biologists hypothesized that the brain distinguishes colors by combining signals initiated by the three classes of opsins. **Figure 46.12**, for example, graphs how much light is absorbed, across a range of wavelengths, by the L, M, and S opsins of humans. Notice that a wavelength of 560 nm stimulates L cones strongly, M cones to an intermediate degree, and S cones hardly at all. In response to the corresponding signals from these cells, the brain perceives the color yellow.

Does this hypothesis explain Dalton's color blindness? According to the data in Figure 46.12, wavelengths from green to red do not stimulate S opsin at all. It is thus unlikely that S opsin is involved in red-green color blindness. Did Dalton fail to distinguish red and green because his M or L cones were defective? Research has shown that red-green color-blind people lack either functional M or L cones, or both. Was the same true of Dalton?

This question was answered in the 1990s, when the genes for the M and L opsins were sequenced. Remarkably, Dalton's eyes had been preserved. Researchers managed to extract DNA from the 150-year-old tissue and analyze his M opsin genes and L opsin genes. They found that Dalton had a normal *L* allele but lacked a functional *M* allele. As a result, he did not have green-sensitive cones. The puzzle of Dalton's color vision was solved.

To review the structure and function of the vertebrate eye, go to the study area at *www.masteringbiology.com*.

 Web Activity The Vertebrate Eye

Sensory Worlds: Do Other Animals See Color?

What about other animals—do they see color the way humans do? The answer is probably not. Animals that are active at night have relatively few cone cells and many rods, giving them high sensitivity to light but poor color vision. Numerous vertebrate and invertebrate species have four or more types of opsins and probably perceive a world of colors that is much richer than ours.

In general, the types of opsins found in a species correlates with the environment it inhabits and its mode of life. For example:

- A marine fish called the coelacanth (pronounced *SEE-luh-kanth*), which lives in water 200 meters deep, has two opsins that respond to the blue region of the spectrum (wavelengths of 478 nm and 485 nm). As a result, coelacanths perceive several distinct hues of blue that we would perceive as a single color. The existence of these opsins is logical, because wavelengths in the yellow and red parts of the spectrum do not penetrate well into deep water—only blue light exists in the coelacanth's habitat.

- In humans and other primates that eat fruit, two of the three opsins are sensitive to wavelengths around 550 nm. The presence of these opsins allows individuals to distinguish between the greens, yellows, and reds of unripe and ripe fruits.

- Birds, insects, and many other animals can see ultraviolet light, which has shorter wavelengths than humans can see. This ability important. Certain flowers have ultraviolet patterns that serve as signals for insect pollinators; many birds have strong ultraviolet patterns in their plumage that are invisible to us. In some species, females use these ultraviolet patches as a criterion for selecting mates.

Opsins are adaptations—their structure and function vary among species.

CHECK YOUR UNDERSTANDING

If you understand that . . .

- In vertebrates, light detection begins when retinal changes shape after absorbing light.
- If opsin changes conformation in response to movement by retinal, a series of events results in the closing of sodium channels in the membrane of a photoreceptor cell. The resulting change in membrane voltage triggers a change in neurotransmitter release from the receptor cell.

✓ You should be able to . . .

1. Explain why retinal can be thought of as a "light switch."
2. Predict the type of vision loss that results from each of the following: a tear in the fovea, a mutation that knocks out the gene for S opsin, and an age-related clouding of the lens.

Answers are available in Appendix B.

46.4 Taste and Smell

The senses of taste (**gustation**) and smell (**olfaction**) originate in chemoreceptors. Chemoreceptors can detect the presence of particular molecules because they undergo a change in membrane potential when a specific compound is present. In this way, information about the presence of a particular chemical is transduced to an electrical signal in the body.

Until recently, taste and smell were poorly understood in comparison with vision and hearing. It is easy to see why. Eyes and ears respond to relatively simple stimuli—light and sound waves. The tongue and nose, in contrast, respond to thousands of different chemicals. Consequently, taste and smell were difficult to study until techniques became available for identifying how particular molecules bind to certain receptors. But in the past 15 years, research on the chemosenses has exploded.

Taste: Detecting Molecules in the Mouth

The chemoreceptor cells that sense taste are clustered in structures known as **taste buds**. Although humans have taste buds scattered around the mouth and throat, most taste buds are located on the tongue (**Figure 46.13**). A taste bud contains about 100 spindle-shaped taste cells, which are taste receptors that synapse onto sensory neurons.

How do these receptors work on a molecular level, and how do they produce the sensation of taste? Early taste research focused on the hypothesis that four "basic tastes" existed: salty, sour, bitter, and sweet.

SALT AND SOUR Researchers who analyzed the membrane proteins in taste cells found strong evidence that salt and sour sensations result from the activity of ion channels.

- The sensation of saltiness is due primarily to sodium ions (Na^+) dissolved in food. These ions flow into certain taste cells through open Na^+ channels and depolarize the cells' membranes.

- Sourness is due to the presence of protons (H^+), which flow directly into certain taste cells through H^+ channels, and depolarize the membrane.

The sour taste of grapefruit and other citrus fruits, for example, results from the release of protons by citric acid. In general, the lower the pH of a food, the more it depolarizes a taste cell's plasma membrane and the more sour the food tastes.

Compared to salt and sour, the molecular mechanisms responsible for the sensations of bitterness and sweetness have been much more difficult to identify. Researchers have only recently been able to document that certain food molecules actually bind to specific receptors on taste cells to cause bitter and sweet tastes.

WHY DO MANY DIFFERENT FOODS TASTE BITTER? Bitterness has been difficult to understand because molecules with very different structures are all perceived as bitter. How is this possible?

An answer began to emerge after researchers confirmed that some humans genetically lack the ability to taste certain bitter

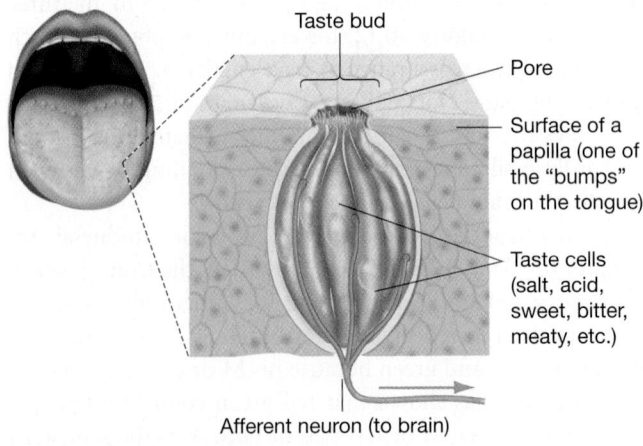

FIGURE 46.13 Taste Buds Contain Many Types of Chemoreceptors.

substances. In 1931, Arthur Fox was synthesizing phenylthiocarbamide (PTC) and accidentally blew some of it into the air. A nearby colleague complained of a bitter taste in his mouth, but Fox could not taste anything. Follow-up research confirmed that the ability to taste PTC is inherited and polymorphic. About 25 percent of United States citizens cannot sense the molecule.

To find the gene responsible for this trait, biologists compared the distribution of genetic markers observed in "tasters" and "nontasters." The mapping effort recently narrowed down this gene's location to several candidate chromosomal segments. (Chapter 19 introduced this type of gene hunt.) In one of the regions, researchers found a family of 40 to 80 genes that encode transmembrane receptor proteins.

Follow-up work has documented that each protein in the family binds to a different type of bitter molecule. A taste cell, however, can have many different receptor proteins from this family. As a result, many different molecules can depolarize the same cell and cause the sensation of bitterness.

Why are so many genes devoted to detecting bitterness? Many of the molecules that bind to these receptors are found in toxic plants; most animals react to bitter foods by spitting them out and avoiding them in the future. In essence, bitterness indicates "this food is dangerous; don't swallow it."

The proliferation of genes responsible for detecting bitterness hints that the trait is extremely important—meaning that individuals with the capacity to detect a wide array of bitter compounds produce more offspring than do individuals that lack this ability.

WHAT IS THE MOLECULAR BASIS OF SWEETNESS AND OTHER TASTES? Inspired by progress on bitter receptors, research teams used a similar approach—analyzing mutant mice that could *not* sense sweetness—to search for sweetness receptors.

In humans and mice, three closely related membrane receptors are responsible for detecting sweetness as well as glutamate and other amino acids. The glutamate receptor is responsible for

the sensation called **umami**, which is the meaty taste of the molecule monosodium glutamate (MSG). The three receptors combine to form an array of multiunit proteins, some of which respond to sweetness; others to amino acids.

Recent work has solved a long-standing question about sweetness—why so many different types of sugars trigger the same sensation. As it turns out, a single receptor has binding sites for an array of sugars, meaning that a variety of molecules can stimulate action potentials from the same receptor.

Currently, several teams are studying how the different taste sensations are conveyed to the brain and interpreted. Although taste is beginning to reveal its secrets, the complete story will probably not be known for many years.

Olfaction: Detecting Molecules in the Air

Taste allows animals to assess the quality of their food before swallowing it. Smell, in contrast, allows animals to monitor airborne molecules that convey information. Wolves and domestic dogs, for example, can distinguish millions of different odors at vanishingly small concentrations. The molecules that cause odor contain information about the movements and activities of prey and other members of their own species.

How does the sense of smell work? When odor molecules, or odorants, reach the nose, they diffuse into a mucus layer in the roof of the nose (**Figure 46.14**). There, they activate olfactory receptor neurons via membrane-bound receptor proteins. Axons from these neurons project up to the **olfactory bulb**, the part of the brain where olfactory signals are processed and interpreted.

Understanding the anatomy of the odor recognition system was a relatively simple task. Understanding how receptor neurons distinguish one molecule from another was much more difficult. Initially, investigators hypothesized that receptors respond to a small set of "basic odors," such as musky, floral, minty, and so on. The idea was that each basic odor would be detected by its own type of receptor.

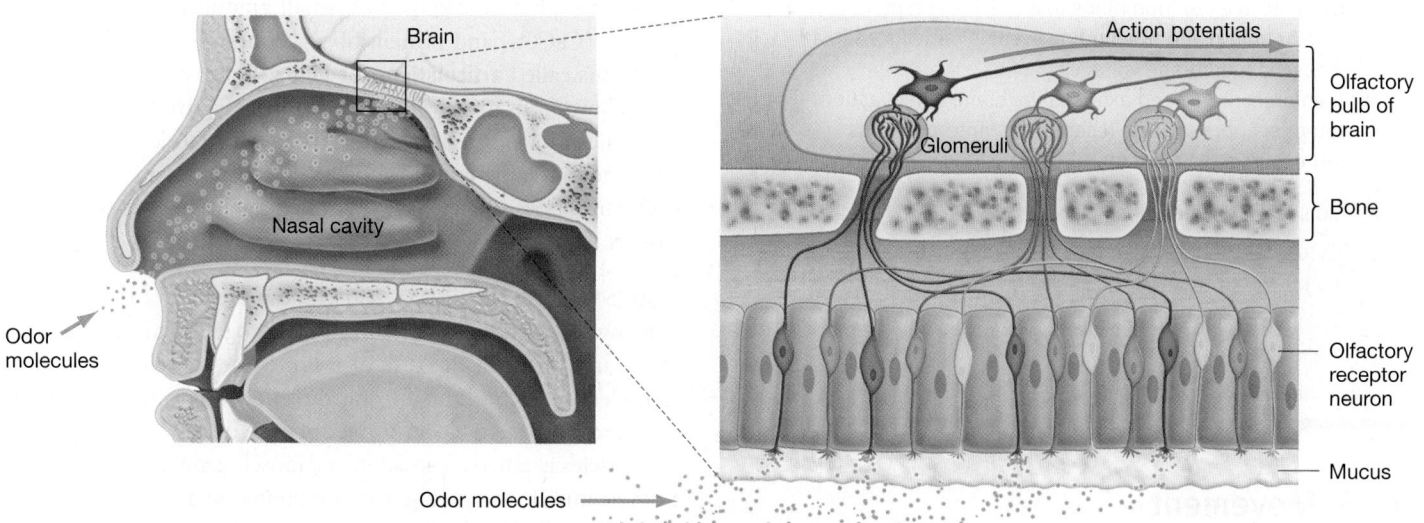

FIGURE 46.14 In Mammals, Chemoreceptor Cells in the Nose Respond to Specific Odorants. Each of the chemosensory neurons in the nose carries one type of odor-receptor protein on its dendrites. Sensory neurons with the same receptor project to the same glomerulus, or section within the olfactory bulb of the brain.

In 1991, however, Linda Buck and Richard Axel demolished the basic-odor hypothesis with an astonishing result: They discovered a gene family in mice—containing hundreds of distinct coding regions—that encodes receptor proteins on the surface of odor-receptor neurons. Follow-up experiments confirmed that each receptor protein binds to a small set of molecules.

Further work established that most, if not all, vertebrates have this family of genes. Mammals typically have 500 to 1000 of these genes, meaning mammals can produce 500 to 1000 different odor-receptor proteins. In humans, however, about half of these genes have mutations that probably render them nonfunctional. This observation may explain why the sense of smell is so poor in humans compared with that of other mammals.

Buck and Axel's announcement inspired a series of questions. How many different receptors occur in the membrane of each neuron involved in odor reception? How does the brain make sense of the input from so many different receptors?

Recently, Buck determined that each olfactory neuron has only one type of receptor and neurons with the same type of receptor are linked to distinct regions in the olfactory bulb of the brain. These regions are called **glomeruli** (meaning "little balls").

Follow-up work from Axel's lab indicated that particular smells in mice are associated with the activation of a certain subset of the 2000 glomeruli in the brain. For example, the activation of clumps 130, 256, and 1502 might be perceived as the smell "cinnamon."

In essence, then, the sense of smell is similar to the visual system's use of three cones to perceive many colors, but on a much larger scale. Research on this complex and impressive sense continues at a furious pace.

CHECK YOUR UNDERSTANDING

If you understand that . . .

- In most cases, chemoreception occurs when a specific molecule in air or food binds to a specific receptor in the nose or mouth and the binding event is transduced to a change in membrane voltage.
- Chemoreceptors send axons to the brain, where action potentials from specific receptors are interpreted as particular smells or tastes.

✔ You should be able to . . .

1. Discuss why a loss in chemosensory ability occurs when you burn your tongue with extremely hot food.
2. Explain why dogs have a better sense of smell than people do.

Answers are available in Appendix B.

46.5 Movement

The first four sections of this chapter focused on how animals sense different aspects of their environment. Acquiring information is only a first step, however. Sensing a change in the environment is useless unless an animal can respond to it in an appropriate way—usually by moving.

When bats and moths hear ultrasonic frequencies, for example, they alter their flight paths and activities in a way that allows them to catch food or avoid being eaten. Flight and other types of motion are based on muscle contractions in conjunction with a skeleton.

It's important to recognize that **locomotion**, or the movement of an entire animal, is only one type of movement, however. Sessile organisms, such as barnacles and sea anemones, may not undergo locomotion, but they rely on movement of the heart muscle or gills to perform gas exchange and circulate body fluids.

To explore locomotion in detail, let's briefly examine the skeleton's role as a scaffold for muscle attachment, then analyze the structure and function of vertebrate muscle cells and tissues.

Skeletons

Skeletons provide attachment sites for muscles and a support system for the body's soft tissues. As Chapters 32–34 indicated, three types of skeletons are found in animals:

- **Exoskeletons** are hard, hollow structures that envelop the body.
- **Hydrostatic skeletons** use the pressure of enclosed body fluids to support the body.
- **Endoskeletons** are hard structures inside the body.

Chapters 32–34 introduced exoskeletons and hydrostatic skeletons in detail; the focus here is on the structure and function of endoskeletons.

ENDOSKELETON STRUCTURE Endoskeletons are composed of connective tissues called cartilage and bone. Recall from Figure 41.3c that **cartilage** is made up of cells scattered in a gelatinous matrix of polysaccharides and protein fibers. In this skeletal system, cartilage provides padding between bones.

Bone is made up of cells in a hard extracellular matrix of calcium phosphate ($CaPO_4$) with small amounts of calcium carbonate ($CaCO_3$) and protein fibers. Bones meet and interact at locations called **articulations**, or **joints**. Bones articulate in ways that allow limbs to swivel, hinge, or pivot (**Figure 46.15**).

Bones move in response to force exerted by skeletal muscles. In vertebrates, the ends of most skeletal muscles are attached to different bones by **tendons**—bands of tough, fibrous connective tissue.

ENDOSKELETON FUNCTION Muscles can exert force only by contracting, so pairs of muscles must work together to move a bone back and forth. In the case of a limb, one muscle or group of muscles pulls the limb in one direction; the other muscle or group of muscles pulls it in the opposite direction. This pairing of muscles is called an antagonistic muscle group.

The muscle that swings two long bones in an arc toward each other is a **flexor**; the muscle that straightens them out is an **extensor** (**Figure 46.16a**). For example, the hamstring muscles in the back of your thigh flex your lower leg; the quadriceps muscles in the front of your thigh extend it. When the hamstrings

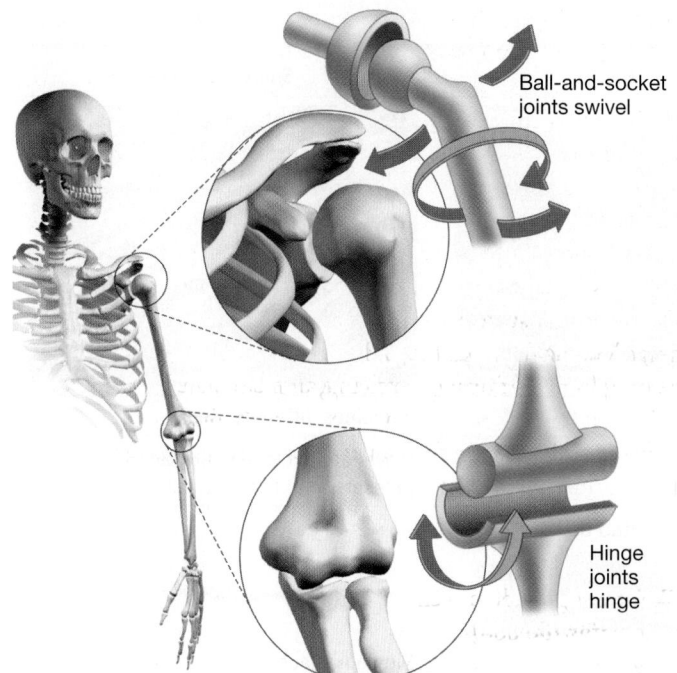

FIGURE 46.15 **Bones Articulate at Joints.** Bones articulate in ways that make specific types of movement possible.

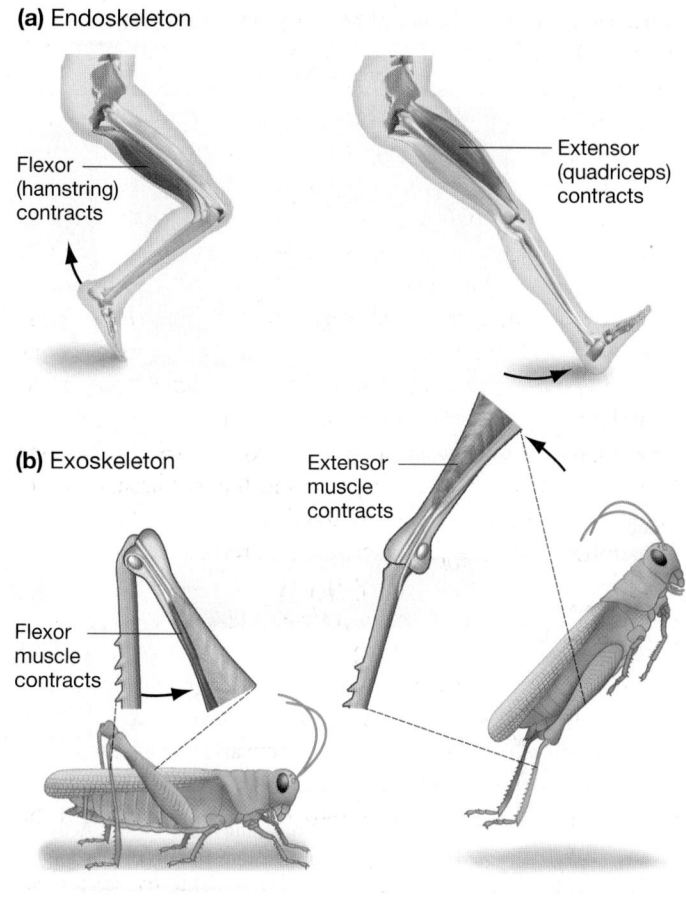

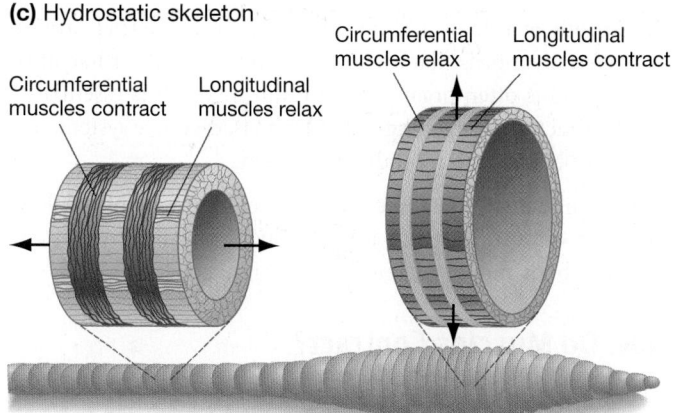

FIGURE 46.16 **Antagonistic Muscle Groups Flex or Extend Skeletal Elements.**

contract, the quadriceps relax; when the quadriceps contract, the hamstrings relax. The hamstrings and quadriceps together make up an antagonistic muscle group.

Locomotion in animals with exoskeletons and hydrostatic skeletons is also based on antagonistic muscle groups. Animals with exoskeletons have paired flexor-extensor muscles inside their leg joints (**Figure 46.16b**). Many animals with hydrostatic skeletons have both circumferential muscles and longitudinal muscles, and these muscles work in concert to shorten and lengthen parts of the body (**Figure 46.16c**).

In all vertebrate animals, the movements of paired muscles are coordinated by motor neurons that originate in the brain or spinal cord. These motor neurons project from processing centers that receive input from sensory systems.

Motor neuron activity changes in response to information about balance, smells, sights, and sounds. In this way, an animal's movements are directly tied to the functioning of its sensory systems. The interplay of sensory input and motor output results in the coordinated behaviors we call running, eating, swimming, and flying.

Muscle Types

Early anatomical studies confirmed that vertebrates have three distinct types of muscle tissue: (**1**) skeletal muscle, (**2**) cardiac muscle, and (**3**) smooth muscle. These three muscle types were introduced briefly in Chapter 41; **Table 46.1** on page 922 provides a more detailed comparison.

Although all three muscle cell types contract in response to electrical stimulation and actin-myosin interactions, they differ in overall morphology as follows:

1. **Skeletal muscle** consists of unbranched multinucleate cells—they contain multiple nuclei. Because this tissue appears to be striated, or striped, it is also known as **striated muscle**.

2. **Cardiac muscle** contains branched cells whose ends are connected via specialized regions called intercalated discs. In an intercalated disc, protein-lined openings called gap junctions provide a direct cytoplasmic connection between adjacent muscle cells. As Chapter 44 noted, intercalated discs are critical to the flow of electrical signals from cell to cell and to the coordination of the heartbeat.

	Skeletal Muscle	Cardiac Muscle	Smooth Muscle
	25 µm	25 µm	25 µm
Location	Attached to bones	Heart	Intestines, arteries, other
Function	Move skeleton	Pump blood	Move food, help regulate blood pressure, etc.
Cell characteristics	Nuclei	Intercalated discs	
	Multinucleate	1 or 2 nuclei	Single nucleus
	Unbranched	Branched; intercalated discs form direct cytoplasmic connection end to end	Unbranched
	Contains myofibrils	Contains myofibrils	No myofibrils
	Activity is "voluntary," meaning that signal from motor neuron is required	Activity is "involuntary," meaning that signal from motor neuron is not required	Activity is "involuntary," meaning that signal from motor neuron is not required

3. **Smooth muscle** is unbranched, lacks the internal strands called myofibrils that are found in skeletal and cardiac muscle, and is often organized into thin sheets. It is essential to the function of the lungs, blood vessels, digestive system, urinary bladder, and reproductive system.

In combination, the three muscle types keep your blood flowing, the food you've eaten moving through your digestive system, and your body moving where you want it to go.

How Do Muscles Contract?

Early microscopists established that the muscle tissue found in vertebrate limbs and hearts is composed of slender fibers. A **muscle fiber** is a long, thin muscle cell. Within each of the muscle cells from limbs are many small strands called **myofibrils**. How do these strands work?

MYOFIBRIL STRUCTURE When viewed with the electron microscope, the myofibrils inside a skeletal muscle cell look striped (**Figure 46.17**). The pattern is caused by alternating light-dark units called **sarcomeres**, which repeat down the length of a myofibril.

The sarcomeres in a myofibril shorten as the cell contracts. They lengthen when the cell relaxes and an external force—exerted by the antagonistic muscle of the pair—stretches the muscle. Based on these observations, it became clear that the question of how muscles contract simplifies to the question of how sarcomeres contract.

THE SLIDING-FILAMENT MODEL In 1954, biologists Hugh Huxley and Jean Hanson noticed that the relationships between different parts of the sarcomere—the points labeled A–D in Figure 46.17—change during contraction.

- Points A and B move closer to each other during contraction, as do points C and D.
- The distance from point A to point C does not change, and the distance from point B to point D does not change.

To explain these observations, Huxley and Hanson proposed the **sliding-filament model** illustrated in **Figure 46.18**. The hypothesis was that the banding patterns in the sarcomere are actually caused by two types of long filaments—**thick filaments** and **thin filaments**—and that these filaments slide past one another during a contraction. Note that thin filaments extend from A to C and thick filaments extend from B to D. Follow-up research has shown that the Huxley-Hanson model is correct in almost every detail.

Structurally, thin filaments are similar to the microfilaments introduced in Chapter 7. You might recall that microfilaments are prominent parts of the cytoskeleton. Thin filaments are composed of two coiled chains of the globular protein **actin**. One end of each thin filament is bound to a structure called the **Z disk**, which forms the wall of the sarcomere. Z disks contain a protein that binds tightly to actin, anchoring the thin filament. The other end of a thin filament is free to interact with thick filaments.

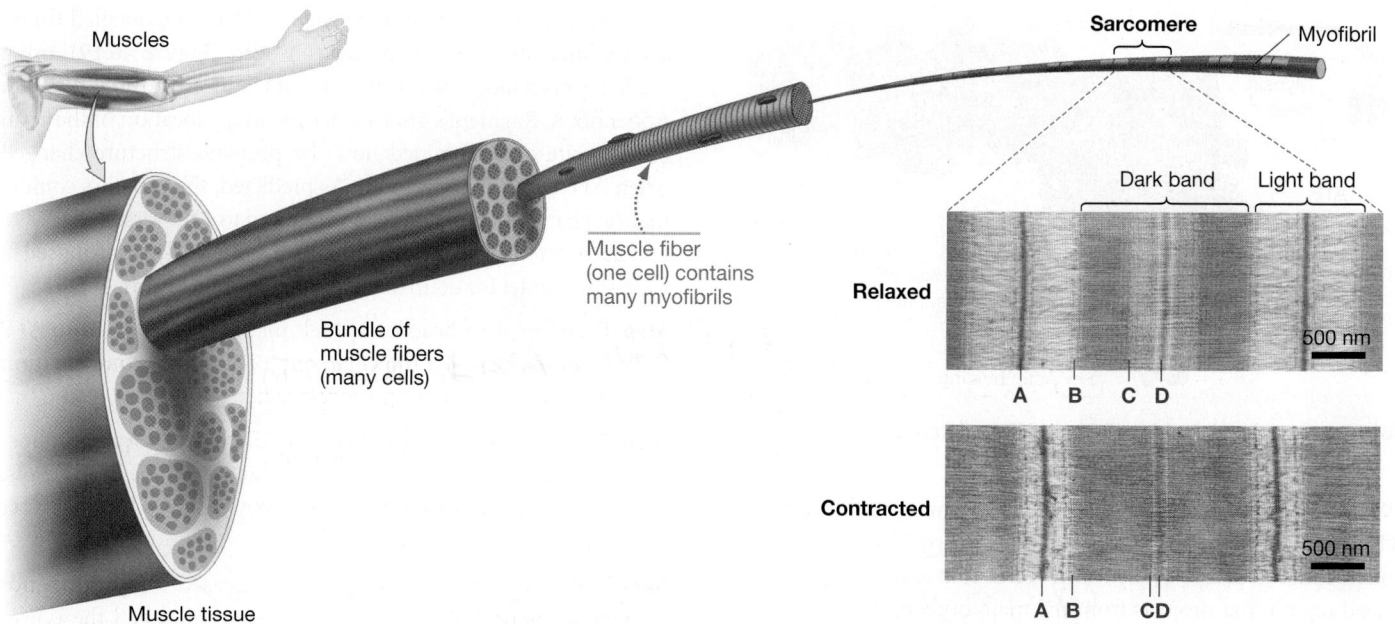

FIGURE 46.17 Muscle Cells Contain Many Myofibrils, Which Contain Many Sarcomeres. Skeletal muscle cells (fibers) have a striped appearance due to repeating sarcomeres, which are units of alternating light-dark bands. When a sarcomere contracts, the distances between some of the points labeled A, B, C, and D change.

Thick filaments are composed of multiple strands of a long protein called **myosin** and are anchored to the middle of the sarcomere. They are free at both ends to interact with thin filaments.

As predicted, thick and thin filaments interact in a way that allows them to slide past one another. For example, when researchers isolated actin filaments and myosin molecules and mixed them together on a glass slide in the presence of ATP, they observed actin sliding past myosin.

To appreciate how the sliding-filament model works, consider the following analogy: Two large trucks are parked 50 m apart, facing each other. Each has a long rope attached to the front bumper. Teams of burly weightlifters grab onto each rope and pull, hand over hand, so that the two trucks roll toward one another. ✔If you understand the model, you should be able to explain which elements in the analogy represent the Z disks, the thick filaments, and the thin filaments.

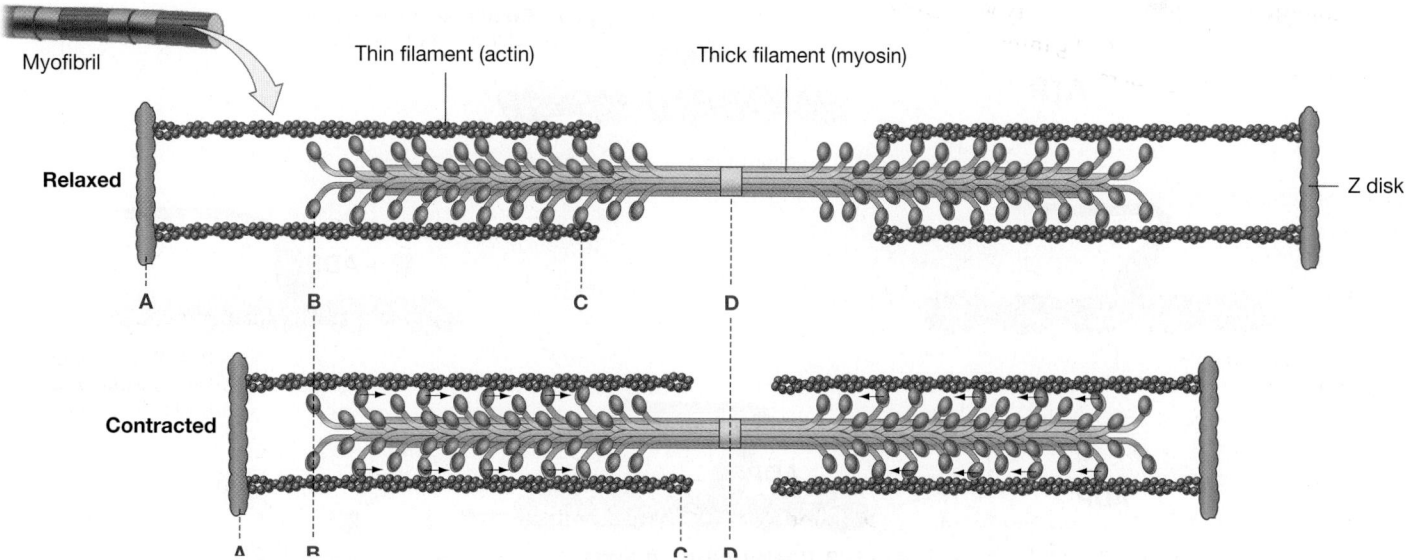

FIGURE 46.18 The Sliding-Filament Model of Sarcomere Contraction. Points A, B, C, and D are the same as those labeled on the sarcomere photographs in Figure 46.17.

✔**QUESTION** According to the model shown here, why is the dark band in a sarcomere (see Figure 46.17) dark and the light band light?

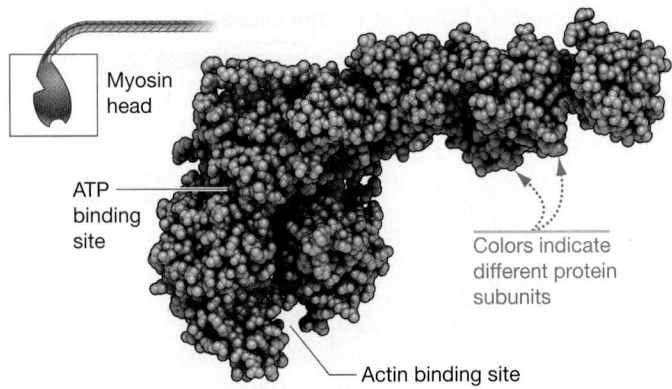

Myosin head

ATP binding site

Colors indicate different protein subunits

Actin binding site

FIGURE 46.19 Myosin's "Head" Binds ATP and Actin.

HOW DO ACTIN AND MYOSIN INTERACT?   How does this sliding action occur at the molecular level? Early work on the three-dimensional structure of myosin revealed that each molecule has a head region that projects from the main body of the thick filament. The myosin head can bind to actin, and the head region can catalyze the hydrolysis of ATP into ADP and a phosphate ion. These results suggested that myosin—not actin—was the site of active movement.

In addition, electron microscopy revealed that myosin and actin are locked together shortly after an animal dies and its muscles enter the stiff state known as rigor mortis. Because ATP is unavailable in dead tissue, the data suggested that ATP is involved in getting myosin to release from actin once the two molecules have bound to each other.

Later, Ivan Rayment and colleagues solved the detailed three-dimensional structure of the myosin head (**Figure 46.19**), using the X-ray crystallographic techniques introduced in **BioSkills 10** in Appendix A. Rayment's group determined the location of the actin binding site and examined how the protein's structure changed when ATP or ADP bound to it. As predicted, the protein's conformation changed significantly when bound to ATP versus ADP.

Based on these data, Rayment and co-workers proposed a four-step model for actin-myosin interaction (**Figure 46.20**):

Step 1 The myosin head of a thick filament is attached to ATP but not to actin in a thin filament. (This is the conformation that occurs when a muscle is relaxed.)

Step 2 When ATP is hydrolyzed to ADP and inorganic phosphate, the neck of the myosin straightens and the head pivots. The myosin head then binds to a new actin subunit farther down the thin filament.

Step 3 When inorganic phosphate is released, the neck bends back to its original position. This bending, called the power stroke, moves the entire thin filament.

Step 4 After ADP is released, a new ATP molecule binds to the myosin and causes myosin to release from actin.

As ATP binding, hydrolysis, and release continue, the two ends of the sarcomere are pulled closer together. (The transition from step 4 to step 1 cannot occur after death. Rigor mortis sets in as ATP supplies run out.)

The same basic mechanism is responsible for contraction in cardiac and smooth muscle cells, the amoeboid movement observed in

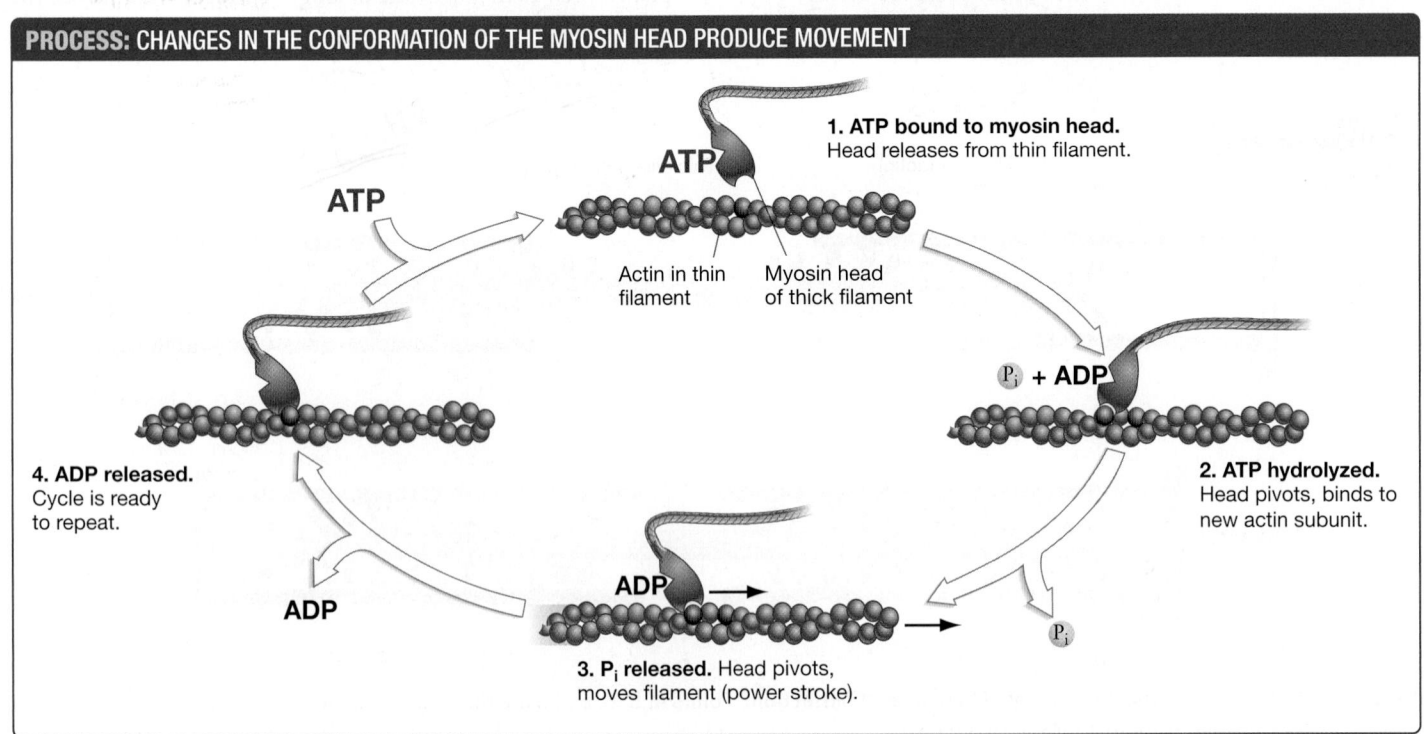

PROCESS: CHANGES IN THE CONFORMATION OF THE MYOSIN HEAD PRODUCE MOVEMENT

ATP

1. ATP bound to myosin head.
Head releases from thin filament.

Actin in thin filament

Myosin head of thick filament

P$_i$ + ADP

2. ATP hydrolyzed.
Head pivots, binds to new actin subunit.

4. ADP released.
Cycle is ready to repeat.

ADP

ADP

P$_i$

3. P$_i$ released. Head pivots, moves filament (power stroke).

FIGURE 46.20 Myosin and Actin Interact during Muscle Contraction. Summary of the current model of how myosin and actin interact as a sarcomere contracts. The four steps repeat rapidly.

amoebae and slime molds (see Chapter 29) and the streaming of cytoplasm observed in algae and land plants. Actin and myosin have played a critical role in the diversification of eukaryotes, because they make movement possible in the absence of cilia and flagella.

HOW DOES RELAXATION OCCUR? ATP is almost always available in living muscles. How do our muscles ever stop contracting and relax?

In addition to containing actin, thin filaments contain two key proteins called **tropomyosin** and **troponin**. Tropomyosin and troponin work together to block the myosin binding sites on actin. The myosin-actin interaction cannot occur, and thick and thin filaments cannot slide past each other. Relaxation can occur (**Figure 46.21a**).

But when calcium ions bind to troponin, the troponin-tropomyosin complex moves in a way that exposes the myosin binding sites on actin. As **Figure 46.21b** shows, myosin then binds and contraction can begin.

How are calcium ions released so that contraction can begin? The answer turned out to have several steps, beginning with the arrival of an electrical signal from a motor neuron.

AN OVERVIEW OF EVENTS AT THE NEUROMUSCULAR JUNCTION
Figure 46.22 summarizes what happens when an action potential from a motor neuron arrives at a muscle cell and initiates contraction:

1. Action potentials trigger the release of the neurotransmitter **acetylcholine (ACh)** from the motor neuron into the synaptic cleft between the motor neuron and the muscle cell.

2. Acetylcholine diffuses across the synaptic cleft and binds to ACh receptors on the plasma membrane of the muscle cell. By recording voltage changes in muscle cells, biologists showed that a membrane depolarization occurs in response to ACh release. If enough ACh is applied to a muscle cell, depolarization triggers an action potential in the fiber itself.

(a) Tropomyosin and troponin work together to block the myosin binding sites on actin.

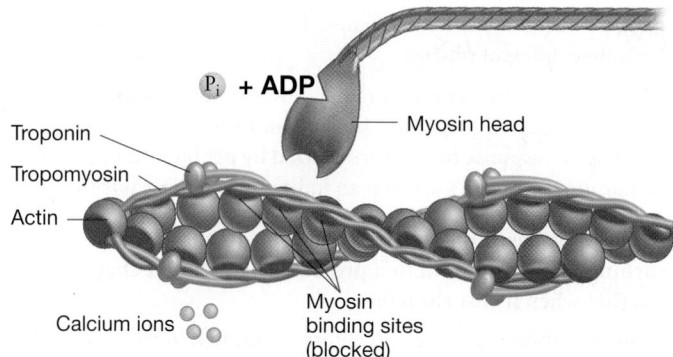

Troponin
Tropomyosin
Actin
P_i + ADP
Myosin head
Calcium ions
Myosin binding sites (blocked)

(b) When a calcium ion binds to troponin, the troponin-tropomyosin complex moves, exposing myosin binding sites.

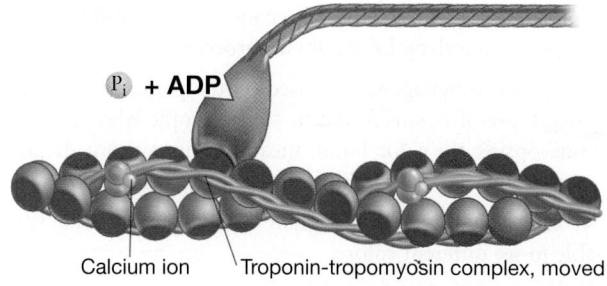

P_i + ADP
Calcium ion
Troponin-tropomyosin complex, moved

FIGURE 46.21 Troponin and Tropomyosin Regulate Muscle Activity.

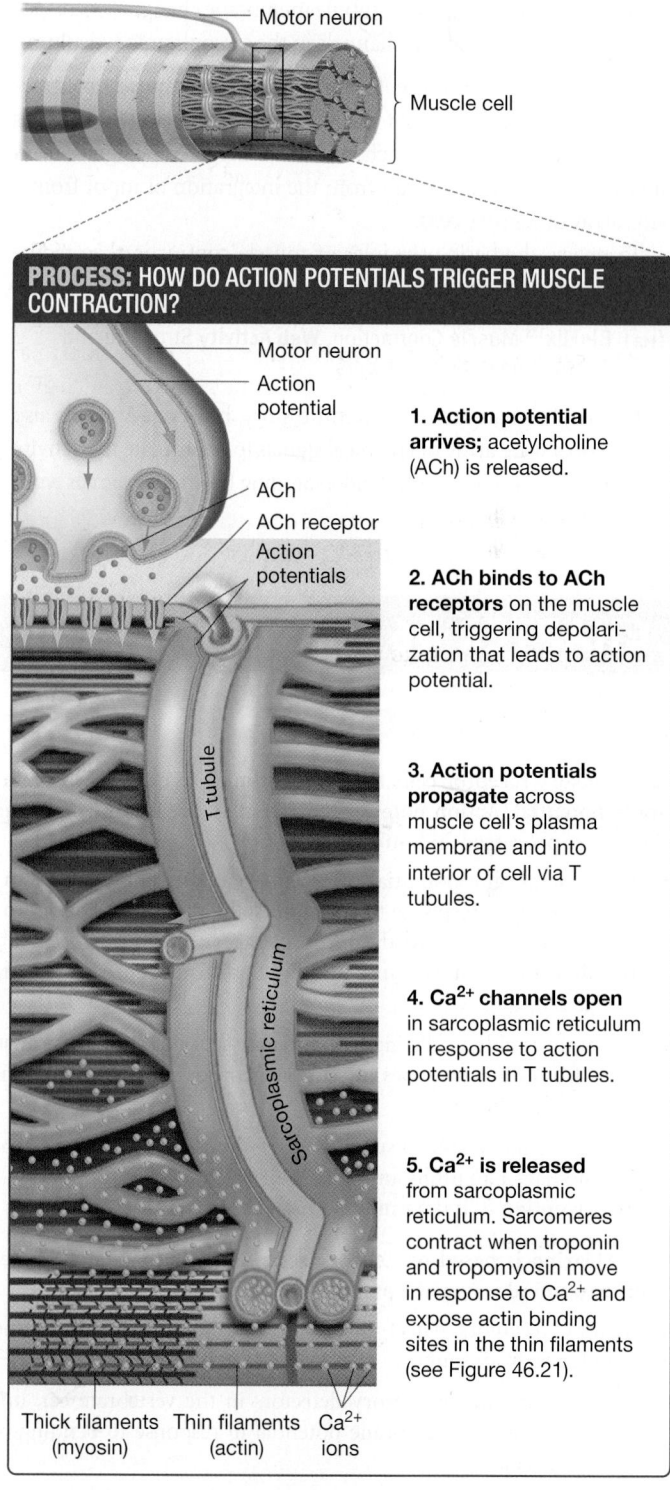

PROCESS: HOW DO ACTION POTENTIALS TRIGGER MUSCLE CONTRACTION?

Motor neuron
Muscle cell

Motor neuron
Action potential
ACh
ACh receptor
Action potentials
T tubule
Sarcoplasmic reticulum

Thick filaments (myosin)   Thin filaments (actin) Ca^{2+} ions

1. Action potential arrives; acetylcholine (ACh) is released.

2. ACh binds to ACh receptors on the muscle cell, triggering depolarization that leads to action potential.

3. Action potentials propagate across muscle cell's plasma membrane and into interior of cell via T tubules.

4. Ca^{2+} channels open in sarcoplasmic reticulum in response to action potentials in T tubules.

5. Ca^{2+} is released from sarcoplasmic reticulum. Sarcomeres contract when troponin and tropomyosin move in response to Ca^{2+} and expose actin binding sites in the thin filaments (see Figure 46.21).

FIGURE 46.22 Action Potentials at the Neuromuscular Junction Trigger the Release of Ca^{2+}, Which Binds to Troponin-Tropomyosin and Allows Myosin to Form a Cross-Bridge with Actin.

3. The action potentials propagate along the length of the muscle fiber and spread into the interior of the fiber via invaginations of the muscle cell membrane called **T tubules**. (The *T* stands for *transverse*, meaning "extending across.")

4. T tubules intersect with extensive sheets of smooth endoplasmic reticulum called the **sarcoplasmic reticulum**. When an action potential passes down a T tubule and reaches one of these intersections, a protein in the T tubule membrane changes conformation and opens calcium channels in the sarcoplasmic reticulum.

Events at the neuromuscular junction produce muscle contraction and movement—one of the key responses to electrical signals from peripheral neurons, which in turn are triggered by signals from the CNS, which in turn result from the integration of input from a wide array of sensory cells.

To review the basic principles of muscle contraction, go to the study area at *www.masteringbiology.com*.

(MB) **BioFlix™** Muscle Contraction, **Web Activity** Structure and Contraction of Muscle Fibers

Electrical signals are not the entire story, however. Animals also depend on a wide array of chemical signals to coordinate the activity of cells throughout the body. Understanding how these signals work is the subject of Chapter 47.

CHECK YOUR UNDERSTANDING

⚷— If you understand that . . .

- Antagonistic muscle groups work in conjunction with a skeleton to produce movement.
- Many types of movement are based on interactions between actin and myosin.
- Muscle cells contract in response to action potentials, which trigger chemical changes that allow actin and myosin to interact.

✓ You should be able to . . .

1. Describe the sliding-filament model.
2. Predict the effect on muscle function of drugs that have the following actions: increase ACh release at the neuromuscular junction, prevent conformational changes in troponin, and block uptake of calcium ions into the sarcoplasmic reticulum.

Answers are available in Appendix B.

CHAPTER 46 REVIEW

For media, go to the study area at www.masteringbiology.com

Summary of Key Concepts

⚷— **Sensory receptor cells transduce stimuli to changes in membrane potential. Action potentials are sent to the brain, where the signals are processed and integrated.**

- If the membrane potential of a sensory cell is altered substantially enough in response to a stimulus, the pattern of action potentials that it sends to the brain changes. In this way, sensory stimuli as different as sound and light are transduced to electrical signals.

- The brain is able to distinguish different types of stimuli because axons from different types of sensory neurons project to different regions of the brain.

  ✓ You should be able to suggest a hypothesis to explain why people who have had limbs amputated experience "phantom pain"—the perception that their missing tissue hurts.

⚷— **Hearing is based on sensory receptor cells that move in response to sound waves of a particular frequency.**

- Pressure receptors detect direct physical stimulation, including stimulation from sound.

- Hair cells, the major sensory detectors in the vertebrate ear, undergo a change in membrane potential in response to bending of their stereocilia.

- Sound waves of a certain frequency cause a certain part of the cochlea's basilar membrane to vibrate. Hair cells at this location fire action potentials in response to the vibration.

- Because of this specificity in hair-cell response, mammals can discriminate different pitches.

  ✓ You should be able to predict how a hair-cell–like receptor could be involved in gravity sensing and the sense of balance in animals, in response to pressure exerted by pebble-like objects that roll or move inside a sac when an individual changes position.

⚷— **Vision is based on sensory receptor cells that contain a light-absorbing pigment bound to a protein. The pigment changes conformation when it absorbs light.**

- In the vertebrate eye, photoreceptors are located in rods and cones. Although these two cell types differ in structure and function, both contain rhodopsin molecules that consist of retinal paired with opsin.

- The rhodopsin found in rods is stimulated by even the faintest light.

- Color vision is possible because cones contain opsins that respond to specific wavelengths of light absorbed by retinal.

- Humans distinguish colors based on the pattern of stimulation of three types of opsins found in cones. People who lack one of the cone opsins are color blind, meaning they cannot distinguish as many colors as people with all three opsins can.

  ✓ You should be able to explain why different animal species are able to see different colors.

(MB) **Web Activity** The Vertebrate Eye

☞ Taste and smell sensations are registered by membrane proteins that act as ion channels or receptors for particular molecules.

- Chemoreceptors detect the presence of certain foodborne or airborne molecules.

- Taste buds contain taste cells with membrane proteins that respond to toxins, salt, acid, and other types of molecules in food. Sodium ions and protons enter taste cells via channels and depolarize the membrane directly, producing the sensations of saltiness and acidity. Sugars and toxic compounds bind to membrane receptors and trigger action potentials that are interpreted by the brain as sweetness and bitterness, respectively.

- Smell, or olfaction, is used to scan molecules from the outside environment. Airborne chemicals are detected by hundreds of different odor-receptor proteins located in the membranes of receptor cells in the nose.

  ✔ You should be able to explain why people with nasal congestion complain that food tastes bland, and why the brain can perceive so many different flavors based on inputs from just four or five basic types of taste receptors.

☞ In many cases, animals respond to sensory stimuli by moving. Movement is based on antagonistic muscle groups that act on a skeleton. Muscle contraction occurs when myosin proteins move down the length of actin fibers.

- All animal muscles use the same basic mechanism for contraction.

- Muscles shorten when thick filaments comprised of myosin slide past thin filaments comprised of actin, in a series of binding events mediated by the hydrolysis of ATP.

- Calcium ions play an essential role in muscle contraction, by making the actin in thin filaments available for binding by myosin.

- In animals with exoskeletons or endoskeletons, muscles are usually arranged in opposing pairs of flexors and extensors. Many animals with hydrostatic skeletons have muscles arranged in opposing pairs of circular and longitudinal bands.

  ✔ You should be able to predict the primary symptom of botulism, which occurs when a toxin prevents release of ACh from the neuromuscular junction, and explain your answer.

  (MB) **BioFlix™** Muscle Contraction, **Web Activity** Structure and Contraction of Muscle Fibers

Questions

✔ TEST YOUR KNOWLEDGE

Answers are available in Appendix B

1. What is echolocation?
   a. use of echoes from high-frequency vocalizations to "see" objects
   b. use of extremely low-frequency vocalizations to communicate over long distances
   c. vision based on input from many independent lenses, functioning like pixels on a computer screen
   d. variation in the structure of opsin proteins, which allows animals to see different colors

2. In the human ear, why do different hair cells respond to different frequencies of sound?
   a. Waves of pressure move through the fluid in the cochlea.
   b. Hair cells are "sandwiched" between membranes.
   c. Receptors in the stereocilia of each hair cell are different; each receptor protein responds to a certain range of frequencies.
   d. Because the basilar membrane varies in stiffness, it vibrates in certain places in response to certain frequencies.

3. Which of the following comparisons of rods and cones is *false*?
   a. Most human eyes have one type of rod and three types of cones.
   b. Rods are more sensitive to dim light than cones are.
   c. There are more rods than cones in the fovea.
   d. Both rods and cones use retinal and opsins to detect light.

4. Which of the following statements about taste is *true*?
   a. Sweetness is a measure of the concentration of hydrogen ions in food.
   b. Sodium ions from foods can directly depolarize certain taste cells.
   c. All bitter-tasting compounds have a similar chemical structure.
   d. Membrane receptors are involved in detecting acids.

5. In muscle cells, myosin molecules continue moving along actin molecules as long as
   a. ATP is present and troponin is not bound to Ca^{2+}.
   b. ADP is present and tropomyosin is released from intracellular stores.
   c. ADP is present and intracellular ACh is high.
   d. ATP is present and intracellular Ca^{2+} is high.

6. Which of the following is critical to the function of exoskeletons, endoskeletons, and hydrostatic skeletons?
   a. Muscles interact with the skeleton in antagonistic groups.
   b. Muscles attach to each of these types of skeleton via tendons.
   c. Muscles extend joints by pushing them.
   d. Segments of the body or limbs are extended when paired muscles relax in unison.

✔ TEST YOUR UNDERSTANDING

Answers are available in Appendix B

1. When a sound and an odor triggers a change in the pattern of action potentials from a sensory cell, how does the brain know which sense is which when the action potentials reach the brain?

2. Give three examples of how the sensory abilities of an animal correlate with its habitat or method of finding food and mates.

3. How did the discovery of odor-receptor genes affect our understanding of how the sense of smell works?

4. Explain how the bending of stereocilia in a hair cell and binding by a bitter-tasting molecule can both result in an ion channel opening.

5. Scientists generally think that a "good hypothesis" is one that is reasonable, testable, and inspires further research into the field. Using these criteria, was Dalton's hypothesis about color vision a good hypothesis? Was it correct? Explain your answer.

6. How did data on sarcomere structure inspire the sliding-filament model of muscle contraction? Explain why the observation that muscle cells contain many mitochondria and extensive smooth endoplasmic reticulum turned out to be logical once the molecular mechanism of muscular contraction was understood.

CHAPTER 46 Animal Sensory Systems and Movement **927**

1. Myasthenia gravis is a disease that develops in humans when the immune system produces proteins that bind to the acetylcholine (ACh) receptors in muscles. The primary symptom of myasthenia gravis is muscle weakness. Why?

2. Houseflies have about 800 ommatidia in each of their compound eyes. Dragonflies, in contrast, have up to 10,000 ommatidia per eye. Houseflies feed by lapping up watery material from piles of excrement or rotting carcasses, which they locate by scent. Dragonflies are aerial predators and hunt by sight. How would you test the hypothesis that the large numbers of ommatidia in dragonflies makes them more efficient hunters?

3. Skeletal muscles contain two general types of cells. "Slow-twitch fibers" produce ATP via cellular respiration. They are reddish because they have a high concentration of myoglobin—a protein that delivers oxygen to electron transport chains during cellular respiration. Slow-twitch fibers support endurance exercise. "Fast-twitch fibers," in contrast, produce ATP primarily via fermentation. Fast-twitch fibers do not have large quantities of myoglobin and are light in color. They support extremely fast contraction but tire easily. Chickens, turkeys, and other ground-dwelling birds can run long distances but escape from predators by flying in short, extremely fast bursts. Based on these observations, explain the distribution of "dark meat" and "white meat" in these birds. Most other birds do not have white meat. Why?

4. When looking at faint stars through a telescope, astronomers focus their eyes just to the side of the object. Instead of landing on the fovea, then, the star's image falls next to the fovea. With this technique, faint objects pop into view. They vanish if looked at directly, however. Explain what is going on.

The spectacular transformation that occurs during insect metamorphosis is triggered by the chemical signals called hormones.

Chemical Signals in Animals **47**

In response to sights, sounds, and other sensory stimuli, an animal's nervous system sends rapid messages, in the form of action potentials, to precise locations in the body. In many cases, these messages result in muscle contractions and movement.

In response to changes in external or internal conditions, cells in the central nervous system (CNS) or the endocrine system release certain molecules. These molecules produce longer-term responses in a broad range of tissues and organs.

The **endocrine system** is a collection of organs and cells that secrete chemical signals into the bloodstream. A chemical signal that circulates through body fluids and affects distant target cells is called a **hormone**.

As you read this, a large suite of hormones is coursing through your circulatory system. These molecules are regulating the maturation of sperm or eggs by your reproductive system, changing the composition of the urine forming in your kidneys, and controlling the release of digestive enzymes in your gastrointestinal tract. Earlier in your life, changes in hormone concentrations led to the dramatic physical changes associated with puberty.

The goal of this chapter is to explore how hormones and other types of chemical signals work in animals. Together, animal nervous systems and endocrine systems process information about the environment—one of the five key attributes of life introduced in Chapter 1.

Let's begin with an overview of chemical signaling systems, and then plunge into analyzing how hormones regulate the activity of target cells.

47.1 Cell-to-Cell Signaling: An Overview

🔑 Animal chemical signals are present in extremely low concentrations but can have enormous effects on their target cells. Unlike action potentials, which are electrical impulses that have a short-term effect on a single cell or on a small population of adjacent cells, the messages that chemical signals carry have a relatively long-lasting effect.

> **KEY CONCEPTS**
>
> 🔑 Animals use at least six major types of chemical signals. Hormones are chemical signals that are present in tiny concentrations and travel throughout the body to affect target cells.
>
> 🔑 The information carried by hormones helps animals develop as embryos, undergo sexual maturation, respond to environmental change, and achieve homeostasis.
>
> 🔑 The production of a hormone is tightly regulated by input from the nervous system and by other hormones.
>
> 🔑 Some hormones bind to receptors inside target cells and change gene expression. Other hormones bind to receptors at the cell surface and lead to changes in protein activation.

✔ When you see this checkmark, stop and test yourself. Answers are available in Appendix B.

In combination, electrical and chemical signals allow animals to coordinate the activities of cells throughout the body. They are the mechanism responsible for maintaining trillions of cells as the integrated unit we call an individual.

Major Categories of Chemical Signals

The chemical signals found in animals have diverse structures and functions. **Table 47.1** summarizes how biologists go about organizing the diversity of chemical signals, based on where the molecules originate and where they act.

Notice that the names for most of the categories use the Greek word root *crin*, meaning "separated." Its use captures something essential about how chemical signals act: They are released from cells and thus are separated from them.

SUMMARY TABLE 47.1 Six Categories of Chemical Signals in Animals

Type of chemical signal	Source and target
Autocrine signals act on the same cell that secretes them.	
Paracrine signals diffuse locally and act on neighboring cells.	
Endocrine signals are hormones carried between cells by blood or other body fluids.	
Neural signals diffuse a short distance between neurons.	
Neuroendocrine signals are released from neurons but are carried by blood or other body fluids and act on distant cells.	
Pheromones are released into the environment and act on a different individual.	

It's important to note that these six classes of chemical messenger do not coincide with six structurally distinct classes of molecules. For example, the endocrine signals found in a particular organism routinely belong to several families of chemical compounds, ranging from amino acid derivatives to lipids. And a particular family of molecules—say, peptides or the lipids called steroids—may function as endocrine, autocrine, and paracrine signals in the same individual.

AUTOCRINE SIGNALS ACT ON THE SAME CELL THAT SECRETES THEM   Translated literally, autocrine means "same-separated." The name is appropriate because **autocrine** signals affect the same cell that releases them.

Perhaps the best-studied autocrine signals are **cytokines** ("cell-movers"). Most cytokines amplify the response of a cell to a stimulus. An example is interleukin 2, which is synthesized and released by a type of white blood cell called a T cell in the course of fighting an infection. Interleukin 2 activates T cells to help eliminate the infection. It also causes the cells to divide repeatedly, producing more activated T cells for host defense.

PARACRINE SIGNALS ACT ON NEIGHBORING CELLS   Translated literally, **paracrine** means "beside-separated." Paracrine signals diffuse locally and act on target cells beside the source cell. Cytokines, for example, may act as paracrine signals as well as autocrine signals, because they can trigger responses by other cells of the immune system.

It is common, in fact, to observe that a single chemical messenger can be assigned to more than one category of signal, based on its mode of action. Like some cytokines, the cell-cell signals named **insulin**, **glucagon**, and **somatostatin** cross categories. These molecules are produced by three distinct populations of cells within a region of the pancreas called the islets of Langerhans. The molecules act on nearby pancreatic cells as paracrine signals and ensure a smooth, steady response to changing blood-glucose levels. But they also act as hormones—meaning they target distant cells—in controlling the concentration of glucose in the blood.

ENDOCRINE SIGNALS ARE HORMONES   Endocrine ("inside-separated") signals are carried between distant cells by blood or other body fluids. The cells that produce endocrine signals may be organized into discrete organs called **glands** or may be interspersed among the cells of other organs—such as the islets of Langerhans in the pancreas.

Because hormones are well studied and particularly important to understanding how animals work, they serve as the focus of this chapter. Many or most of the principles discussed are relevant to the other five categories of animal cell-cell signals as well, however.

NEURAL SIGNALS ARE NEUROTRANSMITTERS   **Neural** signaling was introduced in Chapter 45. You might recall that when an action potential arrives at a synapse, it triggers the release of neurotransmitters that bind to receptors on the postsynaptic cell and induce a change in membrane potential—altering the tendency for the postsynaptic cell to fire action potentials. Changes in neural signaling are important in learning and memory.

(a) Endocrine pathway

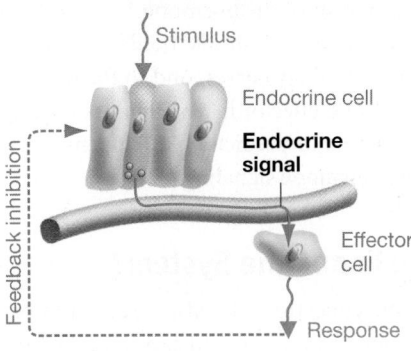

(b) Neuroendocrine pathway

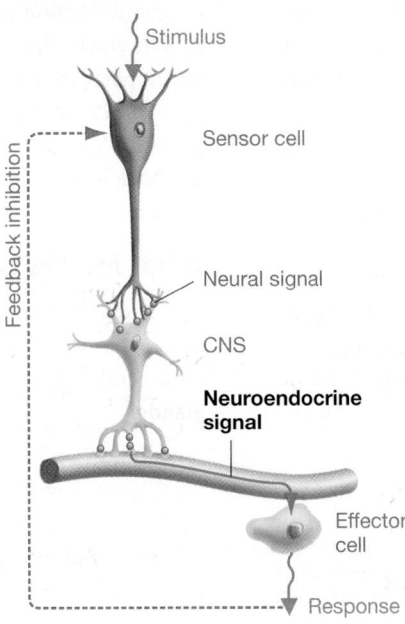

(c) Neuroendocrine-to-endocrine pathway

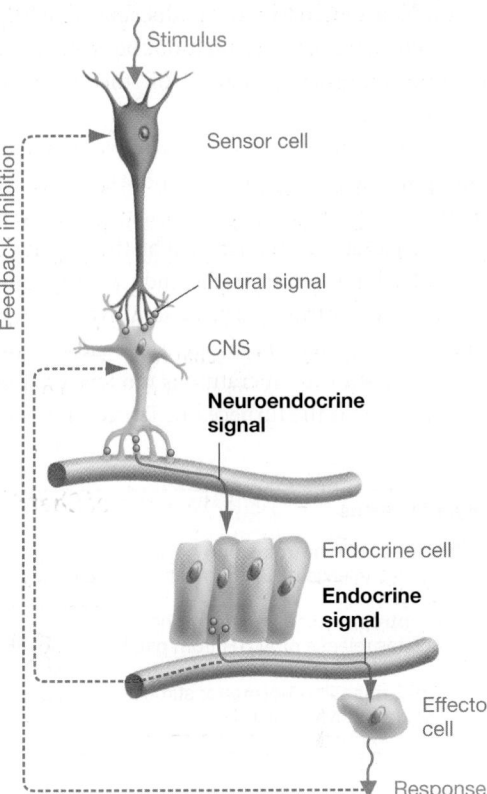

FIGURE 47.1 Hormones Act via Three Pathways and Are Regulated by Negative Feedback.

Neural signaling can be very fast, because neurotransmitters have to diffuse only a short distance—the tiny gap between two neurons, called the synaptic cleft. Neural signals are also short-lived, because the signaling molecules are broken down or taken back up by the presynaptic cell.

NEUROENDOCRINE SIGNALS ACT AT A DISTANCE Even though they are released from neurons, **neuroendocrine** ("nerve-inside-separated") signals are considered hormones. They share a key attribute with endocrine signals: They act on distant cells. Unlike neural signals, they do not act at the adjacent synapse.

Antidiuretic hormone (ADH; also called vasopressin) is a particularly well-studied neuroendocrine signal. ADH is produced by neurons that originate in the hypothalamus of the brain. But instead of acting as a neural signal, ADH acts on cells in the collecting duct of the kidney to help regulate water excretion (see Chapter 42). The source cells and target cells are inside the same body, but separated.

PHEROMONES ACT ON DIFFERENT INDIVIDUALS Pheromones are not "endo" signals. Instead, a **pheromone** is released into the environment and acts on a different individual.

In sea urchins, insects, and many other species, pheromones help coordinate reproduction in males and females or function in attracting mates. Chapter 46, for example, began with a description of how female moths release molecules into the environment that serve to attract males interested in courtship.

Other interindividual signaling molecules, such as the trail-marking compounds released by food-gathering ants, are also considered pheromones. Mice and other rodents even have a specialized region in the nose that is packed with pheromone-responsive chemosensory cells. Humans have a reduced version of the same region, and there is compelling evidence that at least some human pheromones exist. To date, however, none have been isolated and identified.

Hormone Signaling Pathways

In plants, sensory cells perceive changes in the environment and broadcast a hormonal signal that triggers an appropriate response from effector cells. Some animal hormones are also sent directly from endocrine cells to effector cells, in response to a stimulus (**Figure 47.1a**). But more frequently, hormonal signaling in animals involves additional steps.

In many or most cases, information about external or internal conditions is gathered by receptors and then integrated by neurons in the central nervous system (CNS) prior to the production of a hormonal signal. Neurons in the CNS respond by releasing neuroendocrine signals that:

1. act on effector cells directly (**Figure 47.1b**); or

2. stimulate cells in the endocrine system, which respond by producing a hormone (**Figure 47.1c**).

The second response is more common.

All three types of signaling pathway—direct from an endocrine cell, direct from CNS, or CNS-to-endocrine-system—are regulated by **negative feedback**, or **feedback inhibition**. When feedback inhibition occurs, the product of a process inhibits its production. As Chapter 41 pointed out, negative feedback is a key to homeostasis.

In the endocrine pathway, the product of the effector cells feeds back on endocrine cells, lowering production of the hormone and down-regulating the response (Figure 47.1a). An effector's response also feeds back to cells that initiate the neuroendocrine and neuroendocrine-to-endocrine pathways. A change in input from these cells then lowers production of the signal and reduces the response (see Figures 47.1b and 47.1c). Neuroendocrine-to-endocrine signaling pathways have an additional layer of regulation, because the endocrine signal usually inhibits production of the neuroendocrine signal (Figure 47.1c).

The take-home message? The nervous system and endocrine system are tightly integrated. Endocrine signals are released in response to electrical signals; in turn, endocrine signals modulate the electrical signals transmitted by the nervous system.

Feedback inhibition in the endocrine system is analogous to temperature control by a heat-sensitive thermostat. If the temperature is too high, the thermostat sends a signal that turns the furnace off; if the temperature is too low, the thermostat sends a signal that turns the furnace on. The result is a constant air tem-

perature. In animal cell-cell signaling, feedback inhibition reduces production and/or secretion of the hormone.

✔ If you understand this concept, you should be able to explain which parts of the thermostat analogy correspond to the sensory input, CNS, cell-to-cell signal, and effector in a hormone signaling pathway. You should also be able to predict what happens when feedback inhibition fails in a hormone signaling pathway.

What Makes Up the Endocrine System?

The endocrine system is the collection of cells, tissues, and organs responsible for hormone production and secretion. Organs that secrete a hormone into the bloodstream are called **endocrine glands**.

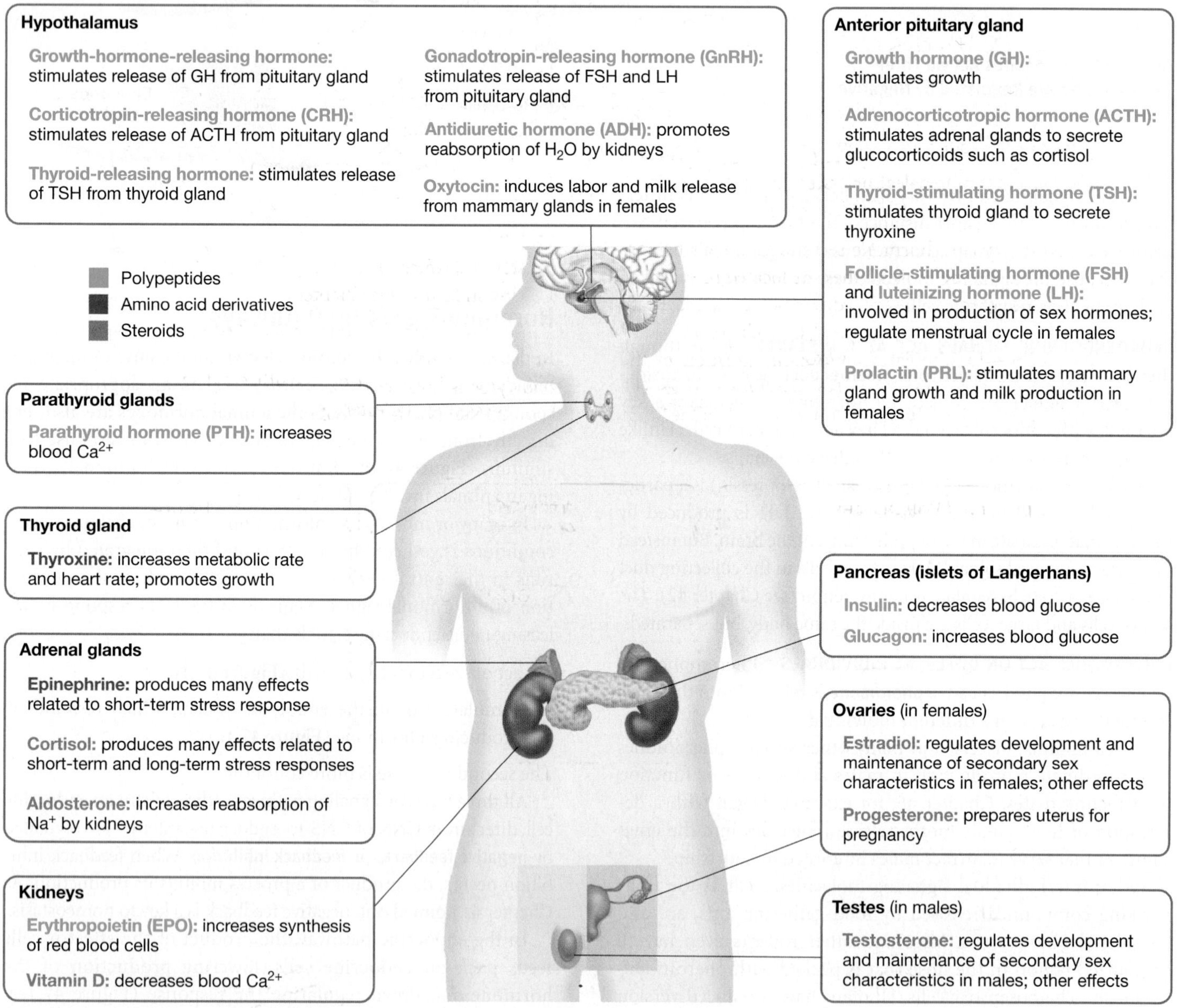

Hypothalamus

Growth-hormone-releasing hormone: stimulates release of GH from pituitary gland

Corticotropin-releasing hormone (CRH): stimulates release of ACTH from pituitary gland

Thyroid-releasing hormone: stimulates release of TSH from thyroid gland

Gonadotropin-releasing hormone (GnRH): stimulates release of FSH and LH from pituitary gland

Antidiuretic hormone (ADH): promotes reabsorption of H_2O by kidneys

Oxytocin: induces labor and milk release from mammary glands in females

Polypeptides
Amino acid derivatives
Steroids

Parathyroid glands

Parathyroid hormone (PTH): increases blood Ca^{2+}

Thyroid gland

Thyroxine: increases metabolic rate and heart rate; promotes growth

Adrenal glands

Epinephrine: produces many effects related to short-term stress response

Cortisol: produces many effects related to short-term and long-term stress responses

Aldosterone: increases reabsorption of Na^+ by kidneys

Kidneys

Erythropoietin (EPO): increases synthesis of red blood cells

Vitamin D: decreases blood Ca^{2+}

Anterior pituitary gland

Growth hormone (GH): stimulates growth

Adrenocorticotropic hormone (ACTH): stimulates adrenal glands to secrete glucocorticoids such as cortisol

Thyroid-stimulating hormone (TSH): stimulates thyroid gland to secrete thyroxine

Follicle-stimulating hormone (FSH) and **luteinizing hormone (LH):** involved in production of sex hormones; regulate menstrual cycle in females

Prolactin (PRL): stimulates mammary gland growth and milk production in females

Pancreas (islets of Langerhans)

Insulin: decreases blood glucose

Glucagon: increases blood glucose

Ovaries (in females)

Estradiol: regulates development and maintenance of secondary sex characteristics in females; other effects

Progesterone: prepares uterus for pregnancy

Testes (in males)

Testosterone: regulates development and maintenance of secondary sex characteristics in males; other effects

FIGURE 47.2 An Overview of the Human Endocrine System. A partial list of the endocrine glands and hormones found in humans. The heart, gastrointestinal tract, adipose (fat) tissue, and many other hormone-producing organs and cells could also be added to this list.

The tissues and organs that make up the endocrine system vary widely among animals. For example, neurons that manufacture and secrete hormones are particularly important in insects, where they regulate molting, metamorphosis, and other processes. Salmon have an unusual gland that secretes a hormone responsible for regulating calcium ion concentration.

Even within one species, the diversity of endocrine system components can be impressive. For example, **Figure 47.2** shows major human glands with endocrine functions.

- The **hypothalamus** is a region deep within the brain.
- The **pituitary gland** sits just below the hypothalamus and has distinct anterior and posterior regions.
- The **thyroid gland** is situated in the neck.
- The four **parathyroid glands** are embedded in the thyroid gland.
- The two **kidneys** lie in the posterior part of the abdominal cavity.
- The two **adrenal glands** sit atop the kidneys and have an outer cortex and a central medulla.
- The endocrine component of the **pancreas** is located in the anterior part of the abdominal cavity.
- The paired **ovaries** (in females) or **testes** (in males) are in or suspended below the pelvic cavity, respectively.

In many cases, however, hormone-secreting cells are not organized into discrete glands. Instead, they are located in other organs. For example, the intestine produces secretin, the heart produces atrial natriuretic hormone, and the cells of fat tissue produce leptin. Secretin stimulates the exocrine portion of the pancreas, **atrial natriuretic hormone** causes the kidney to excrete salt, and **leptin** helps regulate the amount of fat stored in the body.

Finally, it's important to note that not all glands in the body are part of the endocrine system. **Exocrine glands**, in contrast to endocrine glands, deliver their secretions through outlets called ducts into a space other than the circulatory system. Most of the digestive glands introduced in Chapter 43 are either exocrine glands, such as the salivary glands, or mixed endocrine and exocrine glands, such as the pancreas. The exocrine portions of the pancreas secrete digestive enzymes through ducts into the intestine. The endocrine portion of the pancreas consists of cells that secrete insulin and glucagon directly into the bloodstream.

At first glance, the diversity of hormones, glands, and effects can seem overwhelming, especially considering that Figure 47.2 represents just a partial catalog for a single species. The picture is simplified somewhat, however, by recognizing that most animal hormones belong to one of just three major structural families: polypeptides, amino acid derivatives, or steroids.

Chemical Characteristics of Hormones

Figure 47.3 illustrates the three major classes of chemicals that can act as hormones in animals:

1. polypeptides, which are chains of amino acids linked by peptide bonds (see Chapter 3);
2. amino acid derivatives; and
3. steroids, which are a family of lipids distinguished by a four-ring structure (see Chapter 6).

The secretin produced in the small intestine is a polypeptide; epinephrine is synthesized in the adrenal medulla from the amino acid tyrosine; and cortisol is synthesized in the adrenal cortex from the steroid cholesterol.

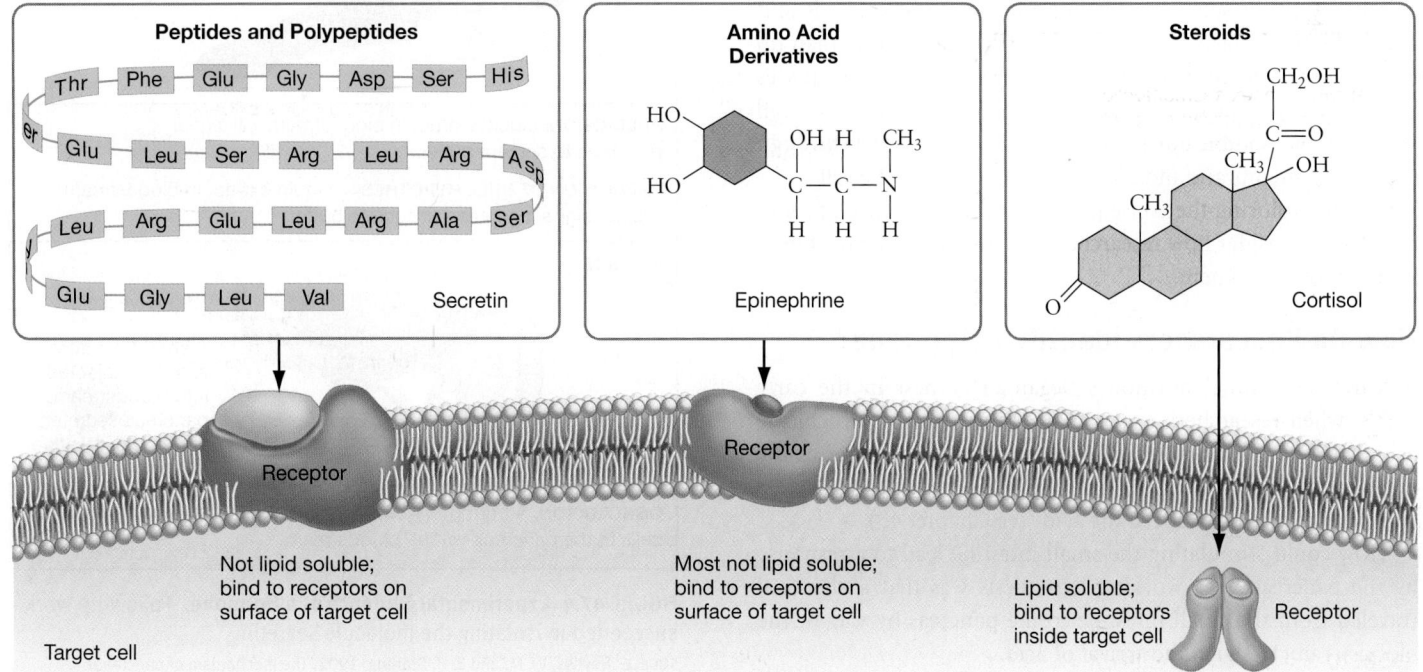

FIGURE 47.3 Most Animal Hormones Belong to One of Three Chemical Families.

HORMONE CONCENTRATIONS ARE LOW, BUT THEIR EFFECTS ARE LARGE Hormones vary widely in structure but share a common characteristic: they have profound effects on individuals, even though they are present at vanishingly small concentrations. As an example, consider work on **growth hormone (GH)**, also known as somatotropin.

Several researchers noted that rats and other laboratory animals stopped growing when their pituitary glands were removed. Based on this observation, it was widely suspected that the pituitary produces a chemical signal that promotes cell division and other aspects of growth.

To test this hypothesis, a research group purified a polypeptide from cow pituitary glands, injected the polypeptide into lab rats, and documented rapidly accelerated growth. When the researchers injected 0.01 mg of the molecule each day for nine days into rats that lacked pituitary glands, the width of the growth plates in the rats' leg bones increased by 50 percent. The individuals also gained an average of 10 g compared with rats that lacked a pituitary and did not receive the hormone treatment. Stated another way, an additional 0.09 mg of hormone led to a weight gain of 10,000 mg.

Further, 1 kg of cow pituitary tissue yielded a mere 0.04 g of growth hormone. By mass, the hormone makes up just four one-thousandths of 1 percent of the cow pituitary.

ONLY SOME HORMONES CAN CROSS CELL MEMBRANES Given that small amounts of polypeptide, amino-acid–derived, and steroid hormones have large effects on the activity of cells, organs, and systems, how do the three types of hormones differ? The major difference is that steroids are lipid soluble, while polypeptides and most amino acid derivatives are not (see Figure 47.3).

An important exception to this rule is the hormone **thyroxine**, which is produced by the thyroid gland. Thyroxine is derived from the amino acid tyrosine but is lipid soluble.

Differences in solubility are important because steroids and thyroxine cross plasma membranes much more readily than do other types of hormones. To affect a target cell, all polypeptides and most amino acid derivatives bind to a receptor on the cell surface. Lipid-soluble hormones, in contrast, can diffuse through the plasma membrane and bind to receptors inside the cell.

Before exploring the consequences of this difference in more detail, let's consider how researchers discovered the array of hormones introduced here.

How Do Researchers Identify a Hormone?

Research on animal hormones began in earnest in the early 1900s, when researchers found that adding dilute hydrochloric acid (HCl) to the small intestine, to mimic the arrival of acidic material from the stomach, stimulated the pancreas to secrete compounds that neutralized the acid (see Chapter 43).

How could stimulating the small intestine lead to a response by the pancreas? The working hypothesis was that a chemical traveled from the small intestine to the pancreas by way of the bloodstream, to signal the arrival of acid.

To test this idea, researchers did the experiment summarized in **Figure 47.4**. An extract from the small intestine was injected into blood vessels in a dog's neck. A short time later, the pancreas secreted an alkaline solution. This was strong evidence that the extract from the small intestine contained a cell-cell signaling molecule. The hormone was later purified and named **secretin**.

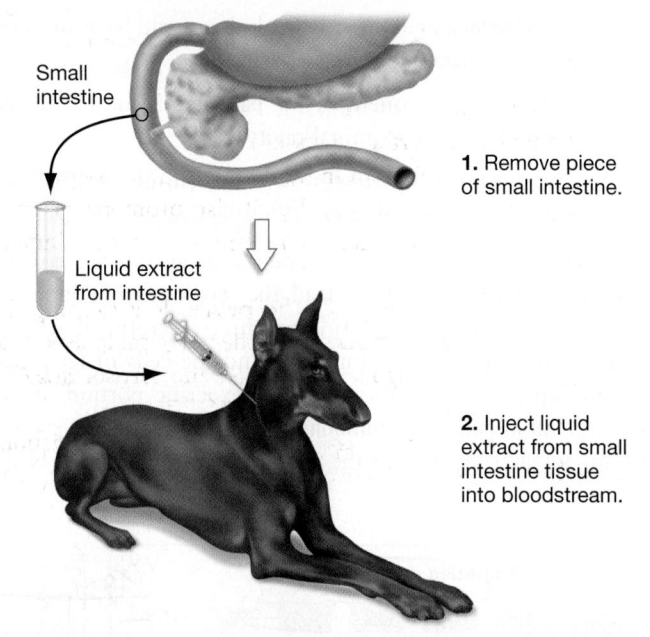

EXPERIMENT

QUESTION: How could stimulating the small intestine with acid cause the pancreas to secrete buffers that neutralize the acid?

HYPOTHESIS: A chemical messenger from the small intestine reaches the pancreas via the bloodstream.

NULL HYPOTHESIS: There is no blood-borne chemical messenger from the small intestine that acts on the pancreas.

EXPERIMENTAL SETUP:

Small intestine

Liquid extract from intestine

1. Remove piece of small intestine.

2. Inject liquid extract from small intestine tissue into bloodstream.

PREDICTION: Liquid extract in bloodstream will signal pancreas to secrete buffer.

PREDICTION OF NULL HYPOTHESIS: Liquid extract in bloodstream will not signal pancreas to secrete buffer.

RESULTS:

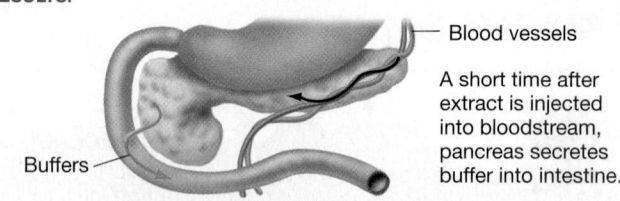

Blood vessels

A short time after extract is injected into bloodstream, pancreas secretes buffer into intestine.

Buffers

CONCLUSION: A hormone from the small intestine (secretin) signals cells in the pancreas via the bloodstream.

FIGURE 47.4 Experimental Evidence of a Hormone. Follow-up work succeeded in isolating the molecule secretin.

SOURCE: Bayliss, W. M. and E. H. Starling. 1902. The mechanism of pancreatic secretion. *Journal of Physiology* 28: 325–353.

✔**QUESTION** What would be an appropriate control in this experiment?

In essence, using liquid extracts from suspected endocrine glands is a way to add a suspected hormone and evaluate the effect. The opposite strategy has also been productive: observing the consequences of losing specific organs or tissues. For example, removing the adrenal glands of an animal rapidly leads to death due to low blood sodium, low blood sugar, and low blood pressure. But injecting adrenal extracts into the blood of animals lacking adrenal glands corrects these abnormalities. Results like these provided strong evidence that hormones secreted by the adrenal glands help regulate blood sugar concentrations and blood pressure.

Documenting an association between a particular gland or hormone and an effect in the body is just a first step, however. To understand hormone action, researchers have to figure out how these signals help animals stay alive and produce offspring.

47.2 What Do Hormones Do?

At the beginning of this chapter, you read that hormones are chemical messengers. If so, what do hormones "say"?

A first step in answering this question is to recognize that a single hormone can exert a variety of effects. For example, thyroxine stimulates metabolism in humans and thus oxygen consumption throughout the body. But it also promotes growth, increases heart rate, and stimulates the synthesis of many important macromolecules.

A second step in grasping what hormones do is to recognize that several different hormones may affect the same aspect of physiology. Insulin, glucagon, epinephrine, and cortisol all influence glucose levels in the blood.

Some hormones have extremely diverse effects; the functions of other hormones appear to overlap. These observations begin to make sense when hormone action is viewed in the context of the whole organism. ☞ Hormones coordinate the activities of cells in response to three situations: (1) development, reproduction, and growth; (2) environmental challenges; and (3) homeostasis.

Let's analyze each situation in turn.

How Do Hormones Direct Developmental Processes?

In animals, as in plants, hormones play a key role in regulating growth and development. Growth hormones and sex hormones play crucial roles in promoting cell division, increasing overall body size, and promoting sexual differentiation as an individual matures; certain hormones direct the development of particular cells and tissues at critical junctures in an individual's life.

Let's explore two of the most dramatic examples of hormonal control—metamorphosis in amphibians and in insects—and then survey other developmental processes that are affected by hormone action.

T_3's ROLE IN AMPHIBIAN METAMORPHOSIS Frogs, toads, and salamanders are called amphibians ("double-lives") because in most species juveniles live in water while adults live on land. The process of changing from an immature, aquatic tadpole to a sexually mature, terrestrial frog, toad, or salamander is an example of **metamorphosis** ("change-form"; **Figure 47.5**).

Two sets of complementary experiments, published in 1912 and 1916, established that frog metamorphosis depends on thyroid hormones. Researchers could induce frog tadpoles to undergo metamorphosis by feeding them ground-up thyroid glands from horses; they could prevent metamorphosis by surgically removing the tadpoles' thyroid glands.

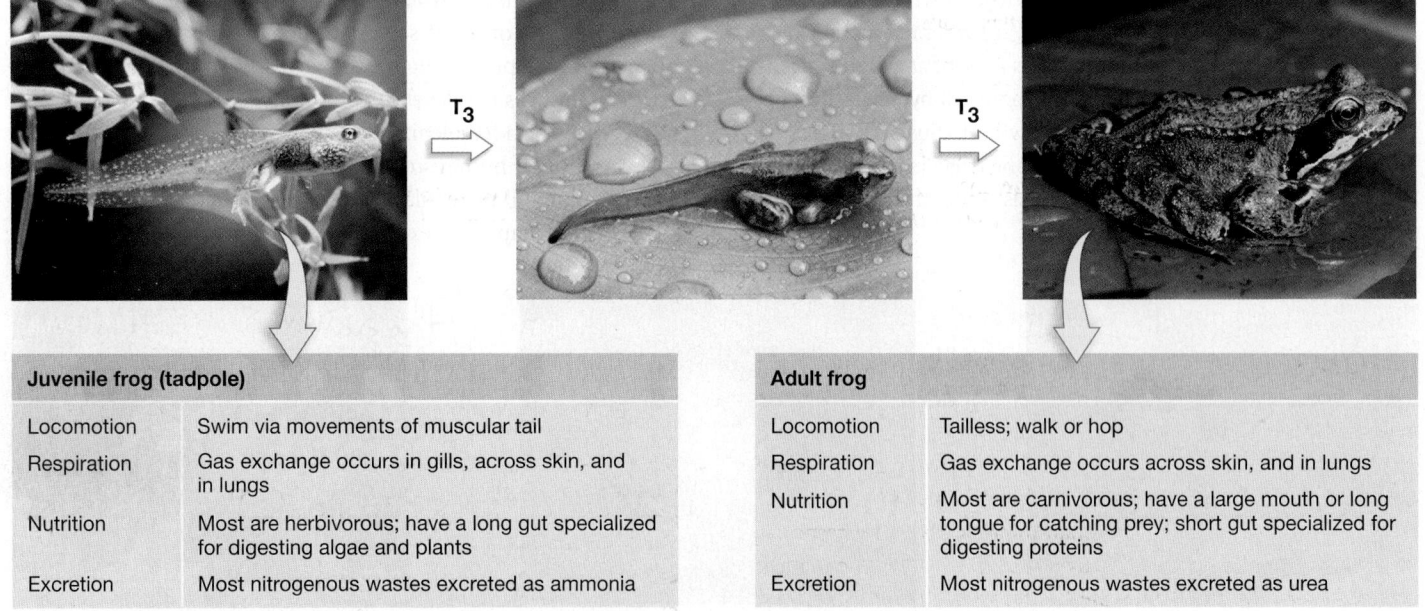

Juvenile frog (tadpole)	
Locomotion	Swim via movements of muscular tail
Respiration	Gas exchange occurs in gills, across skin, and in lungs
Nutrition	Most are herbivorous; have a long gut specialized for digesting algae and plants
Excretion	Most nitrogenous wastes excreted as ammonia

Adult frog	
Locomotion	Tailless; walk or hop
Respiration	Gas exchange occurs across skin, and in lungs
Nutrition	Most are carnivorous; have a large mouth or long tongue for catching prey; short gut specialized for digesting proteins
Excretion	Most nitrogenous wastes excreted as urea

FIGURE 47.5 Amphibian Metamorphosis Is a Continuous Process. When metamorphosis begins in a frog, toad, or salamander, the individual stays active and feeding. The continuous and gradual transition from juvenile to adult is mediated by T_3.

Follow-up work showed that the thyroid hormone **triiodothyronine**, or T_3, is responsible for most of the changes observed in metamorphosis. T_3 is produced in response to a signal from the brain: thyroid-stimulating hormone produced in the pituitary gland.

In juvenile amphibians, cells respond to increased levels of T_3 in one of three ways:

1. By growing and forming new structures, such as legs.

2. By dying, as structures—such as a tadpole's tail—disintegrate.

3. Or, by changing structure and function. For example, changes in existing cells are responsible for the switch from a tadpole's long intestine, specialized for digesting plant material, to an adult's short intestine, specialized for digesting insects and other prey. In the liver, cells respond to T_3 by manufacturing the enzymes required to excrete urea—the nitrogenous waste product released by adults—instead of the ammonia produced by tadpoles.

✔️ If you understand the basic principles of hormone action, you should be able to suggest a hypothesis explaining why different frog cells can respond to T_3 in such different ways.

HORMONE INTERACTIONS REGULATE INSECT METAMORPHOSIS
Chapter 32 introduced a remarkable type of juvenile-to-adult transition in insects called holometabolous metamorphosis. In species that undergo this process, juveniles are called larvae. Insect larvae look completely different from adults, live in different habitats, and eat different food.

As larvae grow, they undergo a series of molts in which they shed their old exoskeleton, expand their bodies, and produce a new exoskeleton. After a specific number of these juvenile molts, however, they secrete a tough case called a pupa. Inside the pupal case, specific populations of larval cells give rise to a completely new adult body. The rest of the larval body is torn down (**Figure 47.6**).

In insects, metamorphosis depends on interactions between two hormones. If **juvenile hormone (JH)** is present at a high concentration in the larva, surges of the hormone **20-hydroxyecdysone**, usually called **ecdysone**, induce the growth of a juvenile insect via molting. But if JH levels are low, ecdysone triggers a complete remodeling of the body—metamorphosis—and the transition to adulthood and sexual maturity.

SEXUAL DEVELOPMENT AND ACTIVITY IN VERTEBRATES In mammals and other vertebrates, long distance cell-to-cell signals play key roles as embryos develop. Hormones also direct anatomical and physiological changes that occur later in life. Some of the most important of these changes involve the reproductive organs.

Events early in development dictate whether the sex organs, or **gonads**, of a vertebrate embryo become male (testes) or female (ovaries). This process is called primary sex determination. In mammals, primary sex determination does not depend on hormone action.

Once testes or ovaries develop, though, they begin producing different hormones. In human males, the early testes produce two hormones:

- A steroid hormone called **testosterone** induces early development of the male reproductive tract.

- A polypeptide hormone called **Müllerian inhibitory substance** inhibits development of the female reproductive tract.

In females, reproductive organs called ovaries produce the steroid hormone **estradiol**, which is in the family of molecules called **estrogens**. Estradiol is required for further development of the female reproductive tract.

Sex hormones also play a key role in the juvenile-to-adult transition. When humans reach early adolescence, for example, surges of sex hormones lead to the physical and emotional changes associated with **puberty**. These developmental changes create the adult phenotype and the ability to produce offspring.

In boys, surges of sex hormones lead to changes that include enlargement of the penis and testes and growth of facial and body hair. In girls, increased concentrations of estradiol lead to the enlargement of breasts, the onset of menstruation, and other changes.

The sex hormones continue to play a key role in adults. Most long-lived animals, for example, reproduce seasonally. In many species, environmental cues such as increasing day length, warmth, or the onset of seasonal rains trigger the release of sex hormones. Chapter 51 details how this flush of testosterone or estrogen induces the development of seasonal traits, such as sexual receptivity in female lizards.

Even though humans do not breed seasonally, sex hormones are instrumental in regulating sperm production and the menstrual cycle. Chapter 48 explores these processes in detail.

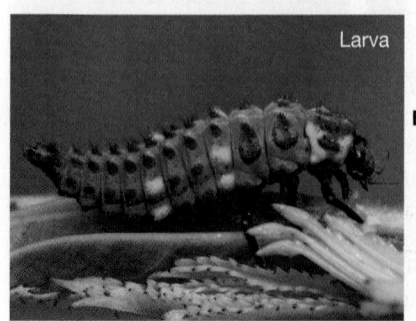 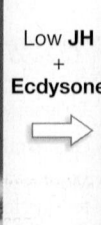   

Larva

Low **JH**
+
Ecdysone

Pupa

Adult

FIGURE 47.6 Insect Metamorphosis Occurs during a Resting Stage. When metamorphosis begins in a holometabolous insect, the individual enters a resting stage called the pupa.

GROWTH IN MAMMALS In humans and other mammals, the long bones in the limbs and the vertebrae of the spinal column must grow for full adult height to be achieved. This growth is stimulated by growth hormone (GH) produced in the pituitary gland. Chapter 19 introduced GH and its role in pituitary dwarfism. GH regulates growth factors, which are signals that control the cell cycle (see Chapter 11).

Puberty is associated with a growth spurt because the effect of growth hormone on the human skeleton is enhanced by the action of sex hormones, which are produced in increased amounts during adolescence. Even though growth hormone and sex hormones are produced in low levels throughout life, bone growth stops when sex hormone concentrations fall at the end of puberty.

How Do Hormones Coordinate Responses to Environmental Change?

The stimuli that hormones respond to can be simple or complex. Digestive hormones are a good example of how hormones function in simple stimulus-and-response circuits.

When acidic food material passes from the stomach to the upper reaches of the small intestine, the stimulus induces intestinal cells to release secretin and cholecystokinin into the bloodstream.

- Secretin induces the pancreas to secrete an alkaline solution that neutralizes acid.

- **Cholecystokinin** signals the pancreas to secrete digestive enzymes into the small intestine. Cholecystokinin also causes the gallbladder to eject bile salts into the intestine to emulsify fats.

In this way, digestive hormones signal the arrival of food and regulate the release of molecules that aid digestion. But what about more complex environmental stimuli?

SHORT-TERM RESPONSES TO STRESS When a person is thrust into a dangerous or unpredictable situation, hormones are involved in both the short-term and long-term responses.

The short-term reaction, called the **fight-or-flight response**, is triggered by the sympathetic nervous system (see Chapter 45). If you were being chased by a grizzly bear, action potentials from your sympathetic nerves would stimulate your adrenal medulla and lead to the release of **epinephrine**, also known as **adrenaline**. (The Greek word roots *epi* and *nephron* mean "top-kidney"; the Latin word roots *ad* and *renal* also mean "top-kidney.")

To determine how epinephrine affects the body, researchers injected human volunteers with a saline solution—as a control—or epinephrine. Each point graphed in the "Results" section of **Figure 47.7** represents data from one of the volunteers; the data in the table are average values from the seven study participants.

The data indicate dramatic increases in an array of physiological processes: concentrations of free fatty acids and glucose in the blood, pulse rate, blood pressure, and oxygen consumption by the brain. In addition, the volunteers reported strong subjective feelings of anxiety and excitement.

EXPERIMENT

QUESTION: How does epinephrine affect the body?

HYPOTHESIS: Epinephrine causes changes involved in the fight-or-flight response.

NULL HYPOTHESIS: Epinephrine does not affect the fight-or-flight response.

EXPERIMENTAL SETUP:

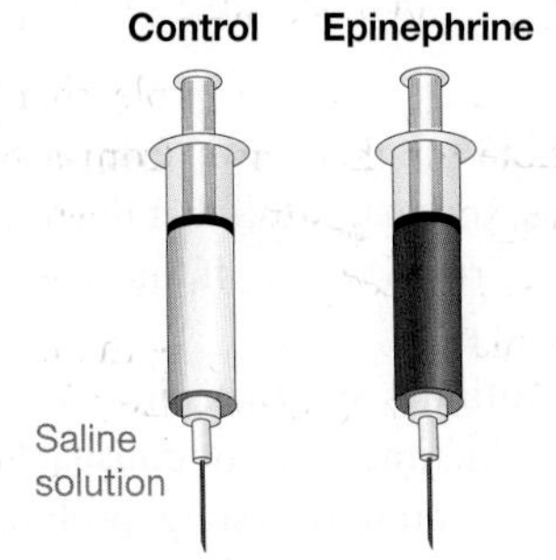

1. **Inject** human volunteers with saline solution or epinephrine.

2. **Document changes** in fatty-acid and glucose concentrations in blood, pulse rate, blood pressure, and oxygen consumption in brain.

PREDICTION: Epinephrine increases fatty-acid and glucose concentrations in blood, pulse rate, blood pressure, and brain oxygen consumption relative to controls.

PREDICTION OF NULL HYPOTHESIS: No differences in physiological state of individuals based on molecule injected.

RESULTS:

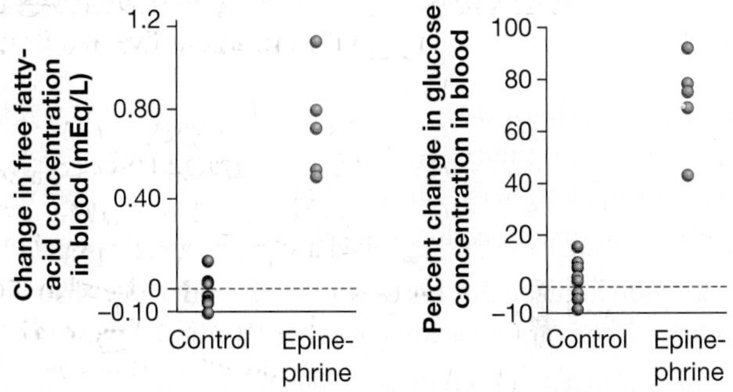

	Control	Epinephrine
Pulse rate (beats/min)	78.3	89.6
Blood pressure (average, mm Hg)	90.9	108.7
O$_2$ consumption in brain (cc O$_2$/100 g/min)	3.41	4.16

CONCLUSION: Epinephrine causes an array of changes associated with the fight-or-flight response.

FIGURE 47.7 Epinephrine Prepares the Body for Action.

SOURCES: King, B. D., L. Sokoloff, and R. L. Wechsler. 1952. The effects of *l*-epinephrine and *l*-nor-epinephrine upon cerebral circulation and metabolism in man. *Journal of Clinical Investigation* 31: 273–279. Mueller, P. S. and D. Horwitz. 1962. Plasma free fatty acid and blood glucose responses to analogues of norepinephrine in man. *Journal of Lipid Research* 3: 251–255.

✔ **QUESTION** Why did researchers bother to inject volunteers in the control group with saline? Why inject them with anything?

Other experiments showed that epinephrine redirects blood away from the skin and digestive system and toward the heart, brain, and muscles. Epinephrine also relaxes smooth muscle and thereby opens blood vessels—increasing blood delivery to target tissues.

Taken together, the responses to epinephrine lead to a state of heightened alertness and increased energy use that prepares the body for rapid, intense action such as fighting or fleeing. By coordinating the activities of cells in many organs and systems throughout the body, epinephrine prepares an individual to cope with a life-threatening situation.

LONG-TERM RESPONSES TO STRESS   If you have ever experienced the fight-or-flight response, you may recall that the state is short lived. Once an epinephrine "rush" wears off, most people feel exhausted and want to rest and eat.

What happens if the stress continues and turns into a long-term condition? In the course of a lifetime, it is not unusual for a person to experience periods of starvation or fasting, prolonged emotional distress, or chronic illness. How do hormones help humans and other animals cope with extended stress?

Early studies of long-term stress in human subjects suggested a role for the hormone **cortisol**, which is produced in the adrenal cortex. Increased levels of cortisol were found in airplane pilots and crew members during long flights, athletes who were training for intense contests, parents of children undergoing treatment for cancer, and college students who were preparing for final exams. Why?

WHAT DOES CORTISOL DO?   In humans, cortisol's primary role is to ensure the continued availability of glucose for use by the brain, during long-term stress. Cortisol manages three main processes that maintain glucose production:

1. Cortisol induces the synthesis of liver enzymes that make glucose from amino acids and other chemical precursors.

2. Cortisol makes adipose tissue—fat tissue—and resting muscles resistant to insulin. Insulin normally stimulates **adipocytes** and resting muscle cells to remove glucose from the bloodstream. But when cortisol makes these cells resistant to insulin, glucose is reserved for use by the brain and exercising muscles.

3. Cortisol promotes the release of fatty acids—the body's major fuel molecules—from adipose tissue, for use by the heart and muscles.

Because of its importance in regulating blood glucose, cortisol is referred to as a **glucocorticoid**.

The long-term stress response comes at a high price, however, as any victim of a serious injury or illness knows.

- Glucocorticoids make amino acids available for glucose synthesis by promoting the degradation of contractile proteins in muscle. The resulting loss of muscle mass may cause severe weakness.

- Glucocorticoids impair wound healing and suppress immune and inflammatory responses. These processes are costly in terms of energy use, but reducing them makes the body more susceptible to infection.

The overall concept here is that the long-term stress response is a compromise—a fitness trade-off (see Chapter 41). The fuel requirements of the brain are met at the expense of other tissues and organs.

LONG-TERM STRESS IN NONHUMAN ORGANISMS   Glucocorticoids may mediate the response to stressful environmental challenges in species other than humans. For example, you might recall from Chapter 42 that when salmon move from freshwater to salt water, individuals gain sodium ions via diffusion and lose water across the gills by osmosis. Salmon counteract these stresses by actively pumping sodium and chloride ions out of the gills.

The chloride cells that perform this pumping are induced to proliferate by a burst of cortisol. Cortisol release occurs as individuals move downstream toward the ocean. If this release fails to occur, salmon die shortly after reaching salt water.

How Are Hormones Involved in Homeostasis?

Chapter 41 introduced the concept of homeostasis, the maintenance of relatively constant physical and chemical conditions inside the body. Glance back at Figure 41.13 and recall that homeostatic systems depend on:

1. a sensory receptor that monitors conditions relative to a normal value, or set point;

2. an integrator that processes information from the sensor; and

3. effector cells that return conditions to the set point.

In homeostatic systems, messages often travel from integrators to effectors in the form of hormones.

LEPTIN AND ENERGY RESERVES   Healthy animals keep energy in reserve for use during periods of decreased food availability. This energy reserve typically takes the form of the lipid triglyceride (see Chapter 6). **Triglyceride** is an effective energy-storage molecule because large amounts of ATP can be generated when its three fatty-acid subunits are oxidized.

Although some triglyceride is present in muscle cells, most is stored in adipocytes. These cells make up the fat bodies found in insects and other species and the adipose tissue of mammals. Even a lean 70-kg human male stores enough energy in adipose tissue to survive for over 30 days without eating.

Is there a homeostatic system that regulates triglyceride stores? An answer began to emerge in the 1970s, when biologists began studying mutations in mouse genes called *obese* and *diabetic*. Homozygous *obese* (*ob/ob*) and *diabetic* (*db/db*) mice eat large quantities of food and move much less than heterozygous or wild-type siblings do. The homozygotes also become extremely obese (**Figure 47.8**).

To test the hypothesis that the *obese* and *diabetic* gene products are involved in cell-cell signaling, researchers turned to an experimental technique called **parabiosis**, which works as follows:

- Two closely related animals are surgically united by suturing the pelvis, shoulders, and abdominal walls together. The skin of the two animals is joined over the surgical connection.

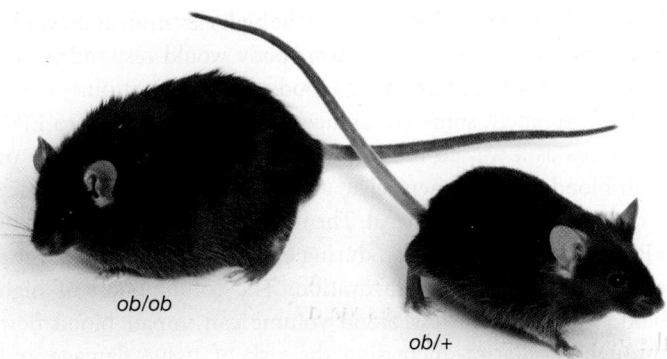

FIGURE 47.8 In Mice, Mutations in the *obese* and *diabetic* Genes Can Cause Obesity. These mice are siblings from the same litter. At the *ob* gene, the lean mouse is heterozygous and the obese mouse is homozygous. Note that the phenotype of this *ob/ob* mouse is the same as a *db/db* mouse.

- The newly created twin animal is allowed to recover from the procedure.

- Within a short time, capillaries form between the two parabiotic partners. As a result, the two individuals develop a shared circulatory system.

- The united circulatory system permits the passage of certain hormones and other long-lived molecules between the two partners, but not molecules that are rapidly metabolized, such as glucose and fatty acids.

- No new nerves grow between the two animals. As a result, they can influence one another through endocrine signals but not through electrical signals.

When researchers performed parabiosis between *db/db, ob/ob*, and lean mice, the results were striking (**Figure 47.9**):

1. When a *db/db* animal was joined to either a lean animal or an *ob/ob* animal, the *db/db* mouse continued to eat and grow normally, but its partner stopped eating, lost weight, and eventually died of apparent starvation.

2. When *ob/ob* animals were joined to lean animals, the *ob/ob* mice ate less food and gained weight less rapidly than did *ob/ob* mice joined to other *ob/ob* mice.

In addition, when two *db/db* animals were joined, both partners ate and grew as expected—both became obese. There was no difference between the two in survival.

To interpret these results, biologists hypothesized that mice produce a satiation, or "stop-eating," hormone—a negative feedback signal in homeostasis with respect to fat stores. The interpretation was that *db/db* mice lack the receptor required for the hormone to affect target cells.

As *db/db* mice got fatter and fatter, they would produce more and more hormone—to no avail. This model explained why the

---

EXPERIMENT

QUESTION: Are the *diabetic* (*db*) and *obese* (*ob*) gene products involved in endocrine function?

HYPOTHESIS: The gene products are involved in hormonal signaling that affects appetite and activity level.

NULL HYPOTHESIS: The gene products are not involved in hormonal signaling.

EXPERIMENTAL SETUP:

| *db / db* | Lean | *db / db* | *ob / ob* | *ob / ob* | Lean |

Perform parabiosis surgery on closely related mice with different genotypes, so that blood-borne products will pass between the mice in each pair.

PREDICTION: Phenotypes of at least some of the parabiotic mice will change.

PREDICTION OF NULL HYPOTHESIS: No changes in phenotypes of parabiotic mice will occur.

RESULTS:

	db / db	Lean	*db / db*	*ob / ob*	*ob / ob*	Lean
Body weight:	↑	↓	↑	↓	↓	↑
Adipose tissue:	↑	↓	↑	↓	↓	↑
		(starvation)		(starvation)		

CONCLUSION: Gene products of *db* and *ob* are involved in hormonal signaling that affects appetite and activity level.

FIGURE 47.9 Parabiosis of Genetically Obese and Lean Mice Provides Evidence of a "Satiation Hormone."

SOURCE: Coleman, D. L. 1973. Effects of parabiosis of obese with diabetes and normal mice. *Diabetologia* 9: 294–298.

✔**QUESTION** Suppose a mouse were doubly homozygous (*ob/ob db/db*). Would its phenotype be the same as or different from singly homozygous individuals? Explain your reasoning.

signal from *db/db* mice greatly reduced food intake in their parabiotic partners but had no effect on the *db/db* mouse.

The satiation hormone itself was postulated to be encoded by the *obese* gene. As a result, *ob/ob* mice do not produce the hormone. They respond to the signal if it is available from a normal partner, however.

A key idea here is that the two genotypes produce the same phenotype because they disable different parts of the same hormone-signaling system.

This model was confirmed in 1994, when the *obese* gene product, called **leptin**, was shown to be a polypeptide hormone. Leptin is secreted into the blood by adipocytes and interacts with a specific receptor located in many tissues, including areas of the brain known to control feeding behavior.

In mice, leptin injections correct the obesity of *ob/ob* (leptin-deficient) mice but not of *db/db* (leptin receptor–deficient) individuals. Unfortunately, leptin injections are not helpful for the vast majority of obese humans.

Follow-up work has shown that leptin levels in the blood vary in proportion to total adipose tissue mass. When adipose mass falls below a set point, the leptin level in the blood also falls. The brain senses the decrease in leptin level and generates both an increase in appetite and a decrease in energy expenditure. These responses promote eating and restore energy balance.

When sufficient food intake has occurred to restore triglyceride stores, leptin levels rise. The result is diminished appetite, increased energy expenditure, and the stabilization of adipose tissue mass. This is a striking example of homeostasis achieved by feedback inhibition.

ADH, ALDOSTERONE, AND WATER AND ELECTROLYTE BALANCE Recall from Chapter 42 that when an individual is dehydrated, **antidiuretic hormone (ADH)** is released from the pituitary gland. ADH increases the permeability of the kidney's collecting ducts to water, causing water to be reabsorbed from urine and saved.

ADH is instrumental in achieving homeostasis with respect to water balance. For example, the ethanol in alcoholic beverages inhibits the release of ADH from the pituitary. People who imbibe large quantities of these beverages produce large quantities of dilute urine. The resulting water loss can lead to dehydration and nausea—symptoms associated with an alcoholic hangover.

Chapter 42 also noted that **aldosterone** is released from the adrenal cortex when ion concentrations in body fluids are low. Because aldosterone increases reabsorption of sodium ions in the distal tubules of the kidney, it plays a key role in homeostasis with respect to electrolyte concentrations and overall volume of body fluids. Adrenal hormones with this effect are called **mineralocorticoids**.

ADH saves water; aldosterone saves sodium. Together, they are key players in maintaining water and electrolyte balance.

EPO AND OXYGEN AVAILABILITY **Erythropoietin (EPO)** is a crucial element in the homeostatic system for blood oxygen levels. When blood oxygen levels fall, the kidneys and other tissues release EPO, which stimulates the production of red blood cells. The more red blood cells, the higher the oxygen-carrying capacity of blood is. If you moved to a location at high elevation and experienced chronic oxygen deficit, your body would respond by releasing EPO and increasing red blood cell concentration.

Unfortunately, some endurance athletes have turned to EPO injections as a way to increase the oxygen-carrying capacity of their blood and give themselves a competitive edge. The practice is dangerous, as well as illegal. The increased viscosity of blood in EPO-abusers is accentuated during exercise, when blood plasma volume drops due to dehydration. The combination of high blood viscosity and low blood volume can impair blood flow through capillaries, increasing the risk of tissue damage and blood clotting. If clots form in blood vessels that lead to the heart or brain, heart attack or stroke may occur.

EPO abuse is thought to be responsible for the collapse and death of several cyclists during races in the mid-1990s. If so, EPO can give the sports phrase "sudden death" new meaning.

CHECK YOUR UNDERSTANDING

If you understand that . . .

- Hormones usually function in directing development and sexual maturation, preparing an individual for environmental challenges, or achieving homeostasis.

You should be able to . . .

1. Explain how the various changes induced by elevated cortisol result in a coordinated response to long-term stress.
2. Explain how the overall effect of epinephrine increases the individual's chances of surviving and reproducing.

Answers are available in Appendix B.

47.3 How Is the Production of Hormones Regulated?

Most hormones are released in response to an environmental cue or a message from an integrator in a homeostatic system. Often, the nervous system is closely involved. For example, environmental cues that signal the onset of the breeding season or the presence of a predator are received by sensory receptors and interpreted by the brain. Similarly, integration in most homeostatic systems is done by neurons in the central nervous system (CNS)—the brain and spinal cord (see Chapter 45).

Based on these observations, the short answer to the question posed in the title of this section is simple: In many cases, hormone production is directly or indirectly controlled by the nervous system.

The Hypothalamus and Pituitary Gland

The pituitary, located at the base of the brain, is directly connected to the brain region called the hypothalamus. This physical link between the hypothalamus and pituitary is the basis of the connection between the CNS and the endocrine system.

The pituitary has two distinct segments: the **anterior pituitary** and the **posterior pituitary**. In 1930 a biologist documented the consequences of removing the entire pituitary in laboratory rats: The animals stopped growing, could not maintain a normal body temperature, and suffered atrophy (shrinkage) of their genitals, thyroid glands, and adrenal cortexes. Not surprisingly, their life span shortened dramatically.

These experiments suggested that, in addition to secreting growth hormone, the pituitary secretes hormones that regulate the production of a wide variety of other hormones. Based on this observation, the pituitary is nicknamed the "master gland." As a case study, let's look at the pituitary hormone that acts on the adrenal glands.

CONTROLLING THE RELEASE OF GLUCOCORTICOIDS Early work on rats suggested that a molecule from the pituitary gland affects the adrenal gland. This molecule soon came to be called **adrenocorticotropic hormone**, or **ACTH**. (*Adreno* refers to the adrenal glands; *cortico* refers to the outer portion, or cortex, of the gland; and *tropic* means "affecting the activity of.")

ACTH was purified and characterized in 1943. When human volunteers are injected with ACTH, cortisol levels in their blood rise (**Figure 47.10a**). This result supports the hypothesis that ACTH is a regulatory hormone. The adrenal cortex secretes glucocorticoids in response to ACTH released from the pituitary.

What regulates ACTH? Biologists from two laboratories independently showed that ACTH is released in response to a molecule produced by the hypothalamus. After years of effort, a different team of researchers succeeded in purifying a peptide—just 41 amino acids long—called **corticotropin-releasing hormone (CRH)**. When the hypothalamus releases CRH, it stimulates cells in the anterior pituitary to secrete ACTH into the bloodstream.

FEEDBACK INHIBITION BY GLUCOCORTICOIDS What *stops* glucocorticoid secretion? The key is to recognize that glucocorticoids themselves suppress ACTH production by the pituitary gland. Glucocorticoids accomplish feedback inhibition—they suppress their own production.

When human volunteers are injected with cortisol, ACTH levels in their bloodstream drop dramatically (**Figure 47.10b**). Cortisol also inhibits release of CRH from the hypothalamus. Thus, if glucocorticoid levels become too high, ACTH levels fall. But if glucocorticoid levels become too low, ACTH levels rise and drive a compensatory increase in glucocorticoid production. **Figure 47.10c** summarizes the relationships among CRH, ACTH, and the glucocorticoid cortisol.

What happens when feedback inhibition fails? Certain pituitary tumors diminish the ability of cortisol to suppress ACTH production, leading to persistently high blood ACTH and cortisol levels. The result is **Cushing's disease**: an unrelenting stress response that depletes the body's protein reserves. It is fatal if not treated.

PATTERNS IN GLUCOCORTICOID RELEASE Under ordinary circumstances, the production of CRH by the hypothalamus displays a daily rhythm, with the highest level in the early morning hours. This drives a corresponding daily rhythm in ACTH production and blood cortisol level.

(a) Results of injecting ACTH into human volunteers

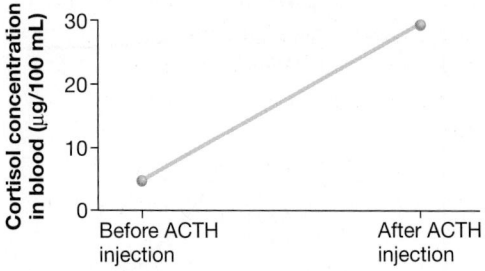

(b) Results of injecting cortisol into human volunteers

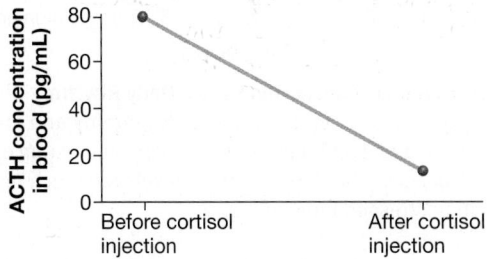

(c) Feedback inhibition by cortisol on ACTH

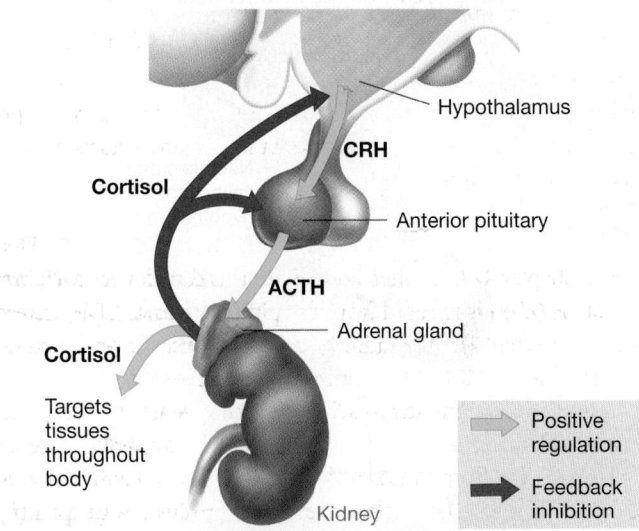

FIGURE 47.10 Feedback Inhibition by Cortisol on ACTH Release.
(a) ACTH stimulates release of cortisol. **(b)** Cortisol reduces ACTH.
(c) The interaction between cortisol, ACTH, and CRH is an example of feedback inhibition.

✔**EXERCISE** How would you use the data in part (a) to devise a test for adrenal failure in humans?

Ordinarily, the morning peak in blood cortisol levels coincides with arousal and initiation of the day's activities—with the effect of saving glucose for use by the brain. As an aside, the unpleasant symptoms of jet lag are due to the daily cortisol rhythm being out of synchrony with local time for several days after you arrive in a new time zone.

When the brain processes stimuli that produce pain or anxiety, however, it initiates a long-term stress response and a sustained

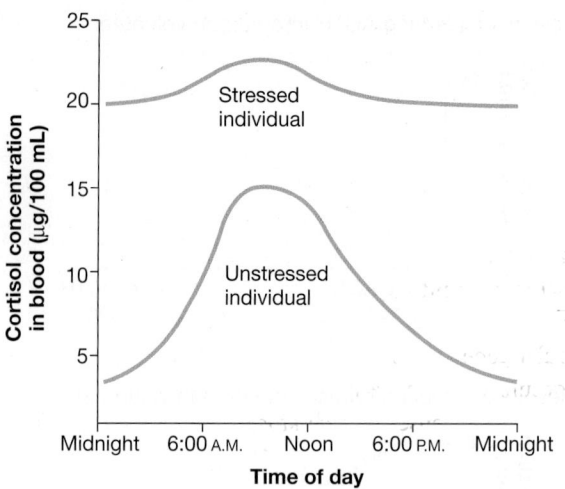

FIGURE 47.11 Blood Cortisol Levels Have a Daily Rhythm. Blood cortisol levels are usually highest in the early morning and drop in the late afternoon and evening. Under stressful circumstances, the morning rise still occurs but blood cortisol levels are much higher overall than in unstressed individuals.

increase in CRH production. Increased CRH production causes blood ACTH and cortisol levels to remain much higher throughout the day than they are in the unstressed state, as the graphs in **Figure 47.11** show. In analyzing these data, note that stressed individuals still show increased cortisol levels in the morning, though not the dramatic swing observed in unstressed individuals.

THE HYPOTHALAMIC-PITUITARY AXIS—AN OVERVIEW The CRH-ACTH-glucocorticoid relationship is just one of many hormone systems based on interactions among the hypothalamus, pituitary, and target glands or cells. The **hypothalamic-pituitary axis** actually forms two anatomically distinct systems (**Figure 47.12**). The anterior pituitary develops from cells in an embryo's mouth and throat lining; the posterior pituitary is an extension of the brain.

Two distinct populations of neurons in the hypothalamus influence the posterior versus anterior sections of the pituitary gland. Both types of hypothalamic neurons synthesize and release hormones and are called **neurosecretory cells**. The neurosecretory cells release hormones under the control of brain regions responsible for integrating information about the external or internal environment. For example, information about an upcom-

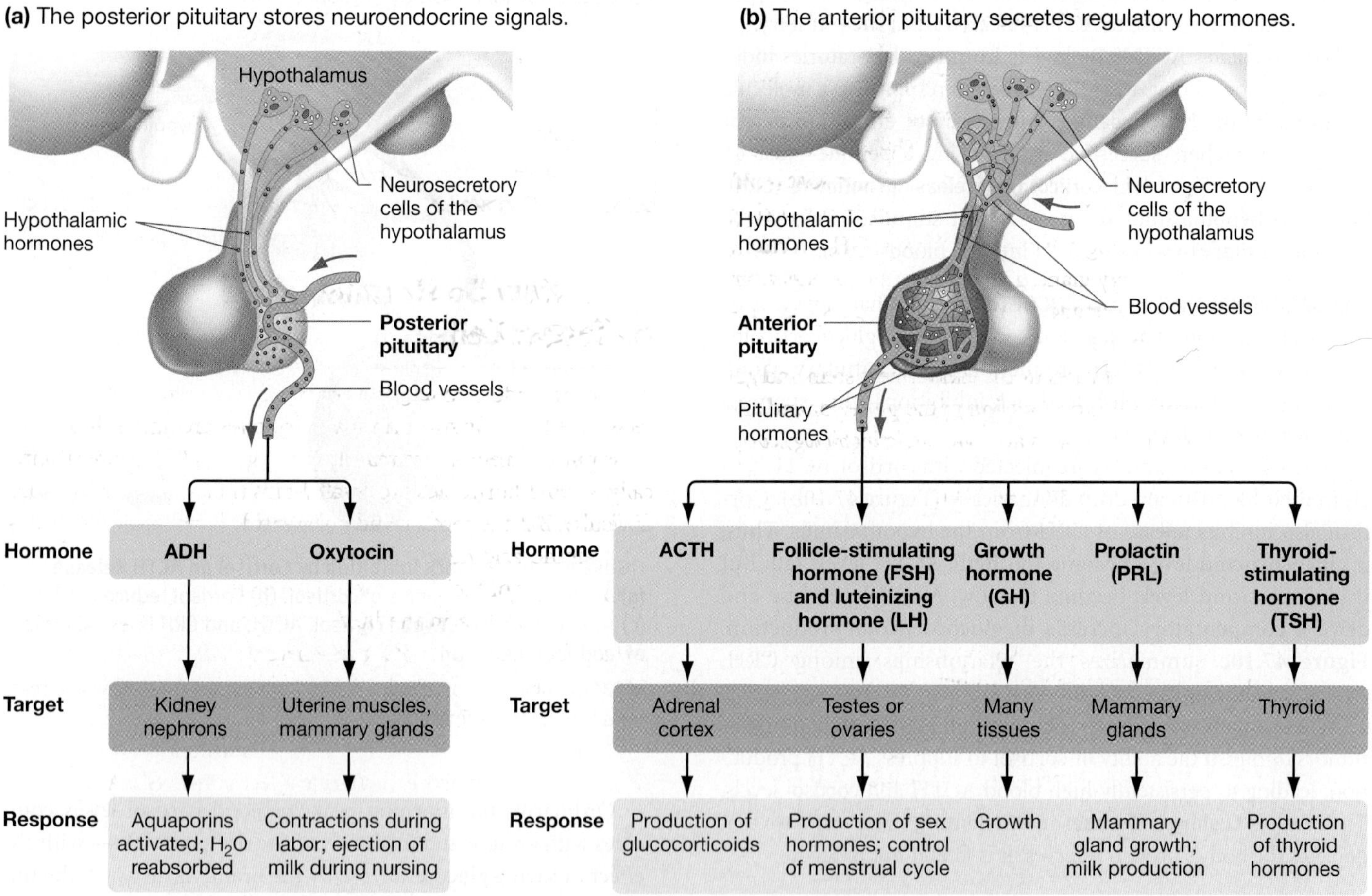

(a) The posterior pituitary stores neuroendocrine signals.

(b) The anterior pituitary secretes regulatory hormones.

	Hypothalamus
Hypothalamic hormones	Neurosecretory cells of the hypothalamus
	Posterior pituitary
	Blood vessels

	Neurosecretory cells of the hypothalamus
Hypothalamic hormones	Blood vessels
Anterior pituitary	
Pituitary hormones	

Hormone	ADH	Oxytocin
Target	Kidney nephrons	Uterine muscles, mammary glands
Response	Aquaporins activated; H₂O reabsorbed	Contractions during labor; ejection of milk during nursing

Hormone	ACTH	Follicle-stimulating hormone (FSH) and luteinizing hormone (LH)	Growth hormone (GH)	Prolactin (PRL)	Thyroid-stimulating hormone (TSH)
Target	Adrenal cortex	Testes or ovaries	Many tissues	Mammary glands	Thyroid
Response	Production of glucocorticoids	Production of sex hormones; control of menstrual cycle	Growth	Mammary gland growth; milk production	Production of thyroid hormones

FIGURE 47.12 The Hypothalamus and Pituitary Interact Closely. (a) Developmentally and anatomically, the posterior pituitary is an extension of the hypothalamus. Neurosecretory cells in the hypothalamus extend directly into the posterior pituitary and secrete ADH (vasopressin) and oxytocin. **(b)** The hypothalamus and the anterior pituitary communicate indirectly, via blood vessels. Hormones produced by neurosecretory cells in the hypothalamus travel in the blood to the anterior pituitary, where they control the release of pituitary hormones.

ing exam or athletic contest might trigger action potentials that lead to the release of CRH.

THE POSTERIOR PITUITARY   Even though both parts of the gland contain neurosecretory cells, the anterior pituitary and posterior pituitary function in different ways. As Figure 47.12a indicates, the posterior portion of the pituitary is an extension of the hypothalamus itself.

Neurosecretory cells that project from the hypothalamus produce the hormones ADH and oxytocin, which are then stored in the posterior pituitary. From there, ADH and oxytocin are released into the bloodstream. This is an example of the neuroendocrine pathway of hormone action. Recall that ADH aids in the reabsorption of water by the kidneys. **Oxytocin** helps induce labor and milk release in females.

THE ANTERIOR PITUITARY   In contrast to the situation in the posterior pituitary, the hypothalamus and anterior pituitary are connected indirectly. Neurosecretory cells from the hypothalamus secrete stimulatory or inhibitory neuroendocrine signals into blood vessels, which then carry the signals to the anterior pituitary. In response, the anterior pituitary alters the secretion of hormones that enter the bloodstream and act on target tissues or glands. This is an example of the neuroendocrine-to-endocrine pathway of hormone action.

Many of the hormones produced by the anterior pituitary stimulate the production of other hormones—justifying the structure's designation as the master gland. The anterior pituitary hormones include ACTH; **follicle-stimulating hormone (FSH)** and **luteinizing hormone (LH)**, which are involved in producing sex hormones and regulating the menstrual cycle; GH; **prolactin**, which stimulates mammary gland growth and milk production; and **thyroid-stimulating hormone (TSH)**, which triggers the production of thyroid hormones.

To review the major structures in the endocrine system and get a closer look at the structure and function of the posterior and anterior pituitary, go to the study area at *www.masteringbiology.com*.

(MB) **Web Activity** Endocrine System Anatomy

Control of Epinephrine by Sympathetic Nerves

When biologists analyze how the nervous system and endocrine system interact to control the release of epinephrine, the distinction between the two systems begins to blur. Section 47.2 introduced how epinephrine acts as an endocrine signal. During the fight-or-flight response, sympathetic nerves trigger the release of epinephrine from the adrenal medulla into the bloodstream. But in addition, some sympathetic nerves release the related molecule **norepinephrine** directly onto target cells.

In effect, the endocrine system broadcasts a similar messenger by secreting it into the bloodstream while the nervous system delivers a chemical messenger directly to particular cells. Epinephrine and norepinephrine, which differ from one another only by the presence of an additional methyl group on epinephrine, are members of the family of molecules called **catecholamines**.

Catecholamines function as neurotransmitters as well as hormones. Similarities between hormones and neurotransmitters do not end there, however. In some cases, their mode of action is similar.

Chapter 45 noted that some neurotransmitters initiate changes in gene expression in neurons. You might recall that in the sea slug *Aplysia*, repeated application of serotonin led to gene activation and changes in the behavior of the synapse. By altering synapses in this way, neurotransmitters play a central role in learning and memory.

Similarly, many hormones exert their effects by activating particular genes in target cells. Understanding how gene activation occurs in response to neurotransmitters and hormones is the subject of intense research at laboratories around the world. It is also the subject of Section 47.4.

CHECK YOUR UNDERSTANDING

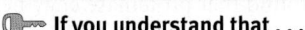

If you understand that . . .

- Hormone concentrations are tightly regulated—in some cases by negative feedback, in other cases by stimulatory or inhibitory signals from the hypothalamus.

✓ **You should be able to . . .**

1. Explain how feedback inhibition occurs in production of ACTH.
2. Discuss the relationship between processing centers in the brain, neurosecretory cells in the hypothalamus, and hormone-secreting cells in the anterior pituitary.

Answers are available in Appendix B.

47.4 How Do Hormones Act on Target Cells?

The key to understanding how hormones act on target cells is to recognize that only some animal hormones are lipid soluble and cross plasma membranes readily (see Figure 47.3). More specifically, steroid hormones are small lipids that enter cells without difficulty. But the peptide and polypeptide hormones—as well as most amino acid derivatives—do not cross plasma membranes easily because of their large size and electrical charge.

Differences in the lipid solubility of hormones are important because they influence where a target cell receives the chemical message. Steroids often act inside the cell, while most amino acid derivatives and all polypeptides act at the cell surface.

To compare these two distinct paths of hormone action, let's consider how estrogen and epinephrine affect target cells. As a steroid and a nonsteroid, they serve as model systems for target cell responses to hormonal signals.

Steroid Hormones Bind to Intracellular Receptors

Estrogens are steroids that direct the development of female secondary sex characteristics in many animal species. In humans and other mammals, the most important estrogen is the molecule estradiol (formally, 17β-estradiol).

Because of its importance in reproduction by humans and domesticated animals, estradiol's mode of action has been the topic of intense investigation for over 50 years. How do target cells receive the signal carried by estradiol?

IDENTIFYING THE ESTRADIOL RECEPTOR   In 1964, biologists succeeded in isolating the estradiol receptor in laboratory rats. **Figure 47.13** indicates the experimental approach. In essence, the research team introduced labeled hormone molecules into females, then used sucrose density centrifugation—a technique introduced in **BioSkills 11** in Appendix A—to separate molecules in target cells by size.

When centrifugation was complete, the biologists found that the labeled estradiol was concentrated in a narrow band comprised of the labeled hormone bound to its receptor. After purifying the receptor molecule, they found that proteinase enzymes could destroy it. Based on this result, they inferred that the estradiol receptor was a protein.

Follow-up experiments established that the estradiol receptor is located in the nucleus but is not associated with the nuclear envelope. Further, the receptor is found only in estradiol target tissues, including the uterus, hypothalamus, and mammary glands. The latter finding was particularly exciting, because it clarified how hormones act in a tissue-specific way.

This is a crucial point: Hormones are broadcast throughout the body via the bloodstream, but they act only on cells that express the appropriate receptor. Target cells respond to a particular hormone because they contain a receptor for that hormone.

Later, biologists found that the gene for the estradiol receptor is similar to the genes that encode receptors for the glucocorticoids, testosterone, and other steroid hormones. This result suggested that all steroid receptors are descended from an ancestral receptor molecule, and that the binding of any steroid hormone to its receptor affects the target cell in a similar way.

What happens once the hormone-receptor complex forms inside the nucleus of a target cell?

DOCUMENTING CHANGES IN GENE EXPRESSION   During the 1970s and 1980s, work in several laboratories suggested that estradiol and other steroid hormones affect gene transcription after they bind to their receptors. For example, researchers injected laboratory animals with estradiol or other steroid hormones and documented changes in the mRNAs and proteins produced in target cells. These data showed that steroid hormones can cause dramatic changes in the amount or timing of mRNA production by a large number of genes.

How? The estradiol receptor, like other members of the **steroid-hormone receptor** family, has a distinctive DNA-binding region called a zinc finger. DNA-binding domains are sections of a protein that make physical contact with DNA. The presence of a zinc finger in the estradiol receptor suggested that once estradiol had bound to it, the hormone-receptor complex might affect gene expression by binding directly to DNA.

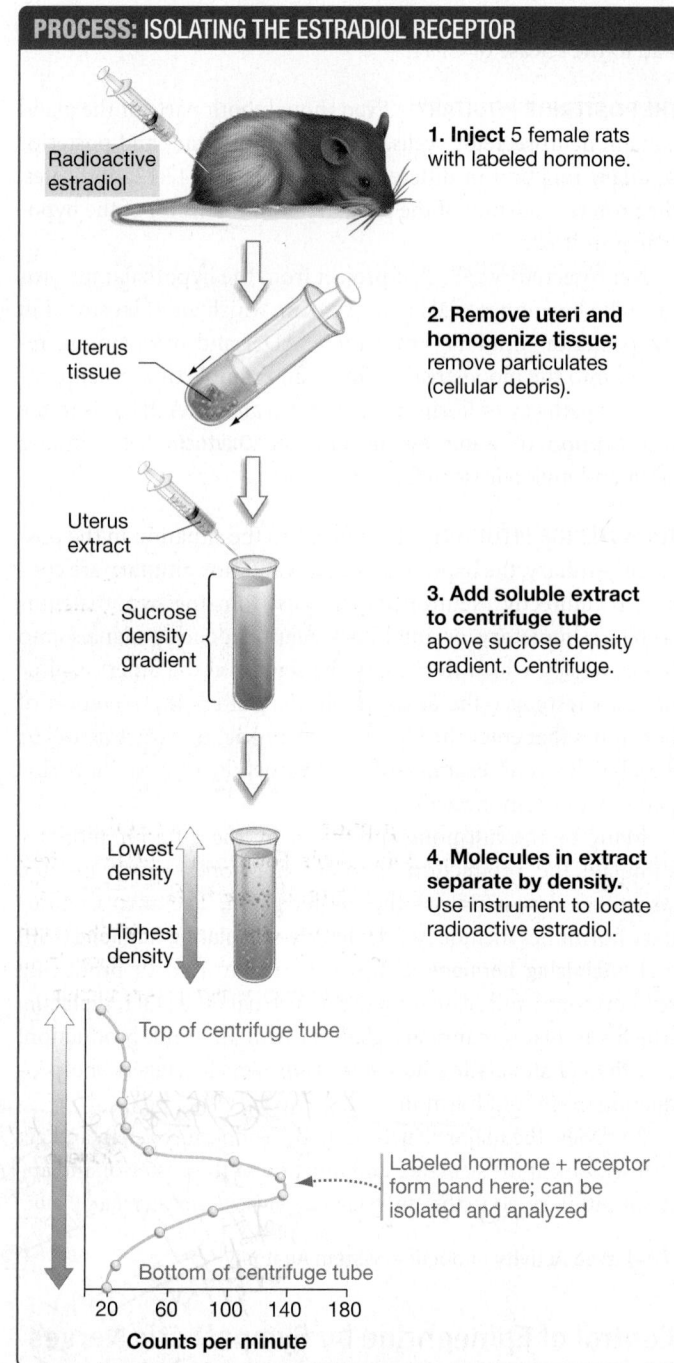

PROCESS: ISOLATING THE ESTRADIOL RECEPTOR

1. **Inject** 5 female rats with labeled hormone.

2. **Remove uteri and homogenize tissue;** remove particulates (cellular debris).

3. **Add soluble extract to centrifuge tube** above sucrose density gradient. Centrifuge.

4. **Molecules in extract separate by density.** Use instrument to locate radioactive estradiol.

Labeled hormone + receptor form band here; can be isolated and analyzed

Top of centrifuge tube

Bottom of centrifuge tube

Counts per minute

FIGURE 47.13 Labeled Hormones Can be Used to Find Hormone Receptors. If radioactive estradiol binds to its receptor in the uterus, the hormone-receptor complex should form a distinct band of radioactivity when molecules from uterine cells are separated by centrifugation.

Follow-up work confirmed that steroid hormone-receptor complexes bind to specific sites in DNA called **hormone-response elements**. Hormone-response elements are located just "upstream" (in the 5′ direction) from the start of target genes. Gene expression changes when a regulatory protein such as a steroid hormone-receptor complex binds to the hormone-response element for that gene.

Figure 47.14 summarizes the current model of how steroid hormones affect target cells.

Step 1 Estradiol or another steroid hormone enters a cell.

Step 2 In target cells, the hormone binds to its receptor. The binding event causes a conformational change in the receptor.

Step 3 The hormone-receptor complex binds to DNA and stimulates transcription.

Step 4 Many mRNAs are produced.

Step 5 Each mRNA is translated many times.

Because each hormone-receptor complex leads to the production of many copies of the gene product, the signal from the hormone is amplified. In this way, a small number of hormone molecules produces a large change in the activity of target cells and tissues.

Currently, researchers are following up on the discovery that at least some mammals, including humans, have two distinct receptors for estradiol. As experiments continue, it will be interesting to learn whether these two receptors trigger different responses to the same hormone.

Hormones That Bind to Cell-Surface Receptors

Unlike steroids, most polypeptide hormones and amino-acid–derived hormones are not lipid soluble. For these molecules to affect a cell, they must bind to receptors on the cell surface. Because the messenger never enters the target cell, its message must be transduced—changed into a form that is active inside the cell. Recall from Chapter 8 that this phenomenon is known as **signal transduction**.

To explore how signal transduction occurs, let's first examine hormone receptors that reside in the plasma membrane, then explore the molecules that process the message inside the cell. In both cases, we'll focus on epinephrine as a model system.

IDENTIFYING THE EPINEPHRINE RECEPTOR In 1948 a biologist published an exhaustive set of studies on how epinephrine affects dogs, cats, rats, and rabbits. The responses fell into two distinct categories, depending on the tissue being considered. To explain this observation, the researcher suggested that epinephrine binds two distinct types of receptor. He called these hypothetical proteins the alpha receptor and the beta receptor.

Follow-up work with molecules that block epinephrine receptors documented that there are actually two types of alpha receptors and two types of beta receptors. Thus, there are four distinct epinephrine receptors. Each is found in a distinct tissue type, and each induces a different response from the cell.

The discovery of four epinephrine receptors reinforces the concept of tissue specificity observed in experiments on the estradiol receptor. Hormones are transmitted throughout the body, not unlike a cell phone signal that is broadcast through the atmosphere. But their message is received only by cells with the appropriate receptor—just as a cell phone signal is received only by equipment with the appropriate antenna. Since there are four distinct epinephrine receptors, the same hormone can trigger different effects in different cells. What happens once epinephrine binds to one of these receptors?

WHAT ACTS AS THE SECOND MESSENGER? Signal transduction occurs when a chemical message at the cell surface triggers a response inside the cell. Cell-surface receptors "read" hormonal messages and initiate an appropriate response.

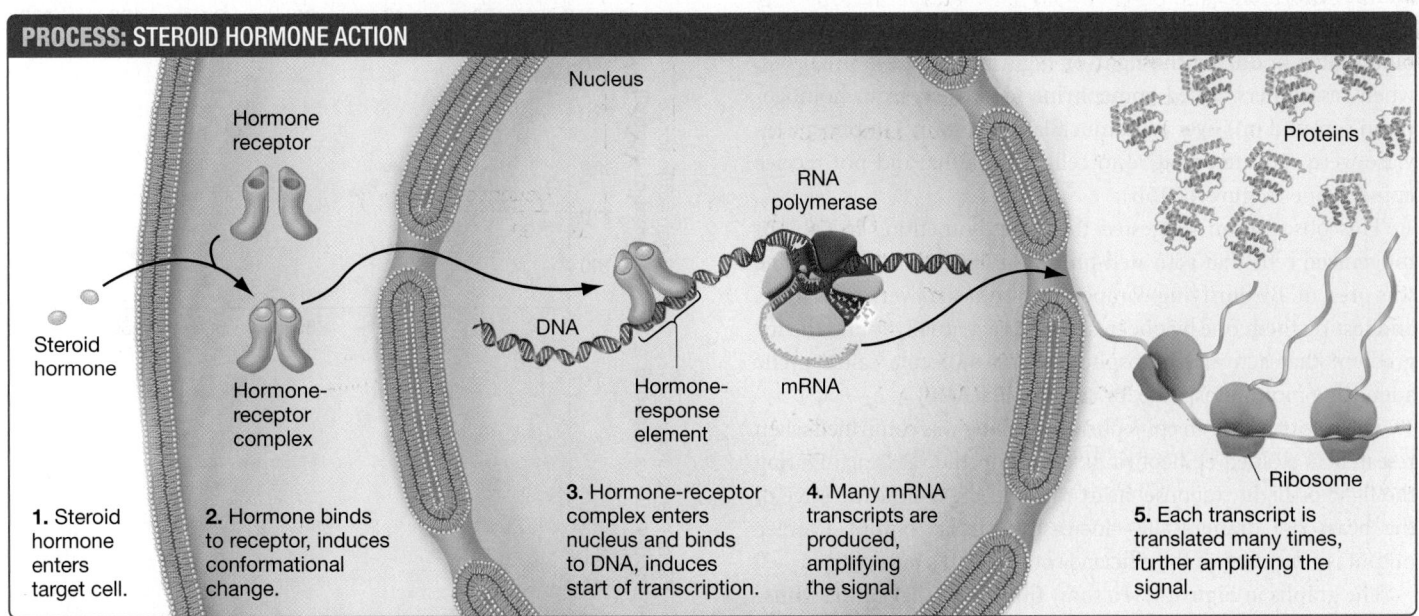

PROCESS: STEROID HORMONE ACTION

Nucleus

Hormone receptor

RNA polymerase

Proteins

Steroid hormone

Hormone-receptor complex

DNA

Hormone-response element

mRNA

Ribosome

1. Steroid hormone enters target cell.

2. Hormone binds to receptor, induces conformational change.

3. Hormone-receptor complex enters nucleus and binds to DNA, induces start of transcription.

4. Many mRNA transcripts are produced, amplifying the signal.

5. Each transcript is translated many times, further amplifying the signal.

FIGURE 47.14 Steroid Hormones Bind to Receptors inside Target Cells and Change Gene Expression.

✔**QUESTION** Tamoxifen is a drug that blocks estrogen receptors in breast tissue cells. (Estrogen stimulates growth of breast cells, so tamoxifen is often prescribed as a treatment for breast cancer.) On this diagram, which arrow does tamoxifen block?

(a) Phosphorylase catalyzes the production of glucose from glycogen.

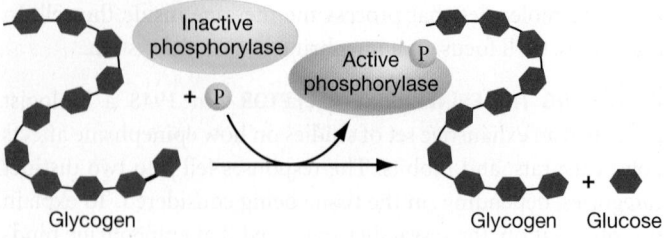

(b) Phosphorylase is activated in response to epinephrine.

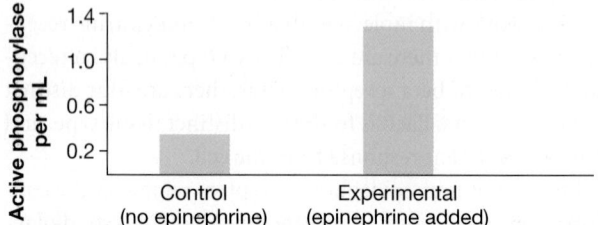

FIGURE 47.15 Epinephrine Activates the Enzyme That Catalyzes the Formation of Glucose from Glycogen. (a) Phosphorylase is activated when an enzyme adds a phosphate group to it. **(b)** When epinephrine is added to cell-free extracts from liver tissue, the amount of activated phosphorylase increases dramatically.

How does a signal from epinephrine increase glucose levels in the blood? To answer this question, biologists focused on the enzyme **phosphorylase**, which catalyzes a reaction that cleaves glucose molecules off glycogen (**Figure 47.15a**). Phosphorylase exists in active and inactive forms; the enzyme switches between these states when it is phosphorylated or dephosphorylated by another enzyme.

Phosphorylase is present in liver cells—the primary source of blood glucose during the fight-or-flight response. As predicted, when researchers added epinephrine to extracts from homogenized (ground up) liver cells, much larger amounts of phosphorylase were activated relative to cell extracts that did not receive epinephrine (**Figure 47.15b**).

This observation suggested there was something in the homogenized cells that activated phosphorylase when epinephrine was present. By purifying components from the liver cell extracts and testing them one by one, researchers eventually found the ingredient that activated phosphorylase: a molecule called cyclic adenosine monophosphate, or **cyclic AMP (cAMP)**.

The role of cAMP in epinephrine signaling was confirmed when researchers studied epinephrine's effects on the rat heart. During the fight-or-flight response, heart rate and the contractile force of the heart rise dramatically—increasing cardiac output. Cardiac output is a measure of the efficiency of the heart's pumping.

The graphs in **Figure 47.16** show three ways that cardiac muscle cells respond to epinephrine. In each graph, the x-axis plots time, in seconds, after epinephrine is applied.

- *Top* Almost immediately, there is a striking increase in cAMP levels inside the cells.

- *Middle* A few seconds later, the contractile force of the cells increases, peaking about 18 seconds after the hormone's arrival.

- *Bottom* Phosphorylase activity also rises, but peaks later—about 40 seconds after the signal.

To capture the importance of cAMP in triggering these effects, biologists refer to it as a second messenger. Recall from Chapter 8 that a **second messenger** is a nonprotein signaling molecule that increases in concentration inside a cell in response to a received signal—a molecule that binds at the surface.

A PHOSPHORYLATION CASCADE How does cAMP transfer the cell-surface signal to phosphorylase? Follow-up work showed that cAMP binds to an enzyme called cAMP-dependent protein kinase A. This enzyme responds by phosphorylating the enzyme phosphorylase kinase, which then phosphorylates phosphorylase.

This chain of events, called a **signal transduction cascade**, is initially triggered by the synthesis of cAMP. cAMP is produced from ATP in a reaction that is catalyzed by the enzyme adenylyl cyclase. Adenylyl cyclase is activated by a G protein (see Chapter 8), which is activated when epinephrine binds to its receptor.

To pull all these reactions together, researchers proposed the model for epinephrine action diagrammed in **Figure 47.17**. In studying this model, it is crucial to recognize two points:

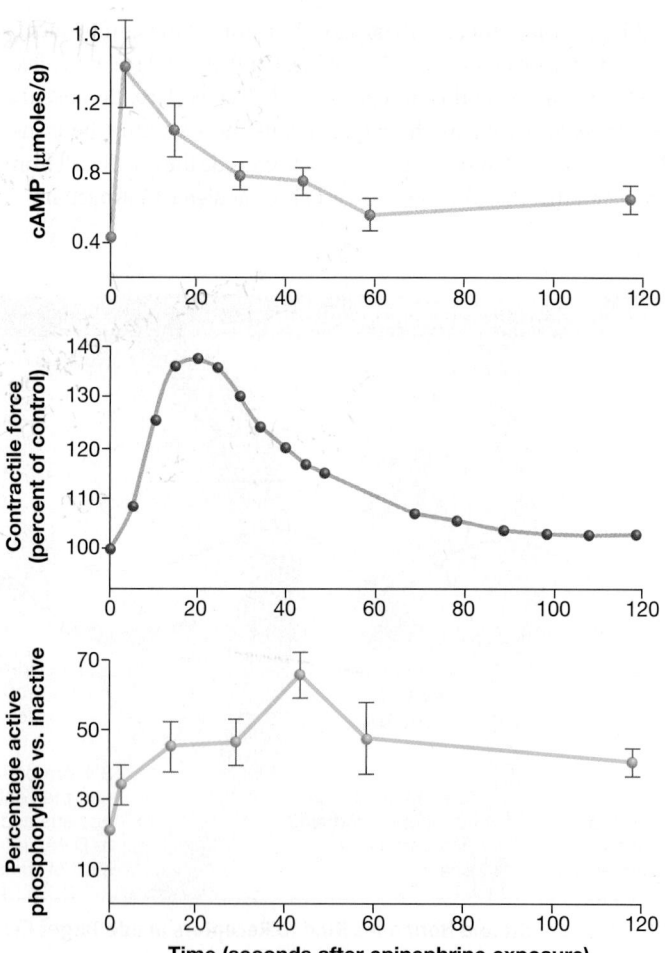

FIGURE 47.16 The Chemistry and Activity of Heart-Muscle Cells Change in Response to Epinephrine.

1. cAMP transmits the signal from the cell surface to the signaling cascade.

2. Together, cAMP production and the subsequent phosphorylation events amplify the original signal from epinephrine.

To drive the latter point home, consider that, in response to stimulation by the hormone-receptor complex, adenylyl cyclase is thought to catalyze the formation of at least 100 molecules of cAMP. In turn, each of these cAMP molecules activates many molecules of cAMP-dependent protein kinase A. Subsequently, each protein kinase molecule activates many molecules of phosphorylase kinase, and so on.

In this way, the binding of just a single molecule of epinephrine may trigger the release of millions or even billions of glucose molecules. Amplification through a signal transduction cascade explains why tiny amounts of hormones can have such huge effects on an individual.

The model in Figure 47.17 was inspired by experiments on the epinephrine receptor called the beta-1 receptor. But other researchers showed that when epinephrine binds to an alpha-1 receptor, a completely different signal transduction event occurs. In this and many other receptor systems, calcium ions (Ca^{2+}) serve as the second messenger in conjunction with a second-messenger molecule called IP_3. Diacylglycerol (DAG) and $3', 5'$-cyclic GMP (cGMP) are also common second messengers in hormone response systems.

To review second messengers, amplification, and other fundamental concepts about hormone action, go to the study area at *www.masteringbiology.com*.

 Web Activity Hormone Actions on Target Cells

Why Do Different Target Cells Respond in Different Ways?

Researchers are increasingly impressed with the diversity and complexity of signal transduction cascades. For example, target cells that have the same receptor protein may have different second messengers, different genes available for transcription, or different enzyme systems that are available for activation. As a result, the same hormone and receptor can give rise to different responses in different target cells.

This finding helps explain one of the most fundamental observations about hormones: The same chemical messenger can trigger different responses in cells from different organs or in cells at different developmental stages. The reason is that the cells contain different receptors, second messengers, amplification steps, protein kinases, enzymes, or transcriptionally active genes.

To summarize this section, steroid hormones tend to exert their effects through changes in gene expression. In contrast, polypeptide and amino-acid–derived hormones activate a specific protein or set of proteins, usually by phosphorylation. Steroid hormones activate transcription factors that lead to the production of new proteins; nonsteroid hormones trigger signal transduction cascades that activate existing proteins.

CHECK YOUR UNDERSTANDING

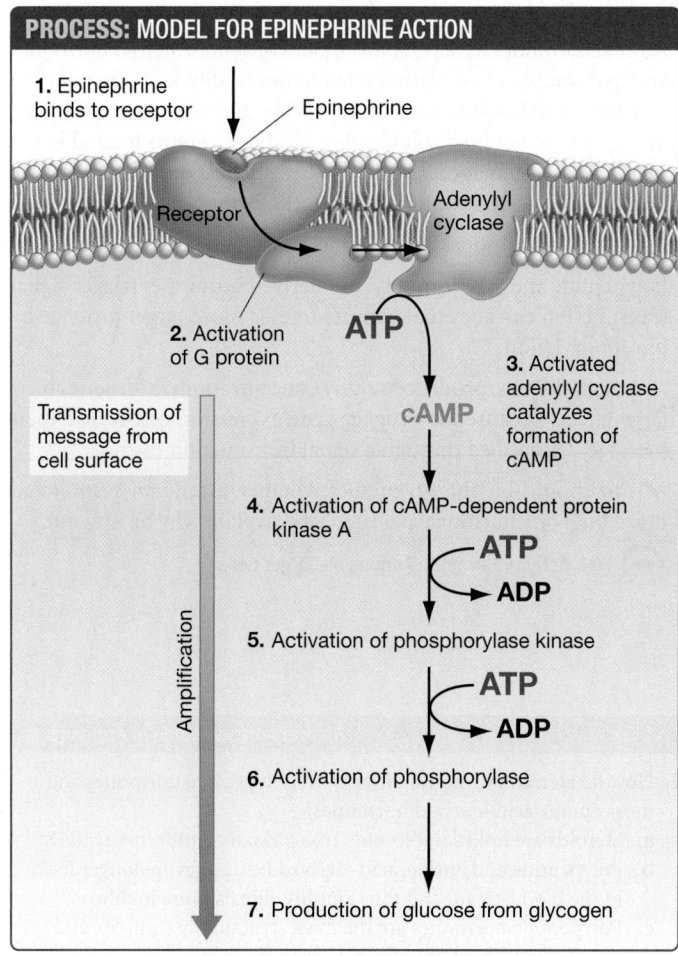

PROCESS: MODEL FOR EPINEPHRINE ACTION

1. Epinephrine binds to receptor

Epinephrine

Receptor

Adenylyl cyclase

2. Activation of G protein

ATP

Transmission of message from cell surface

cAMP

3. Activated adenylyl cyclase catalyzes formation of cAMP

4. Activation of cAMP-dependent protein kinase A

ATP → ADP

Amplification

5. Activation of phosphorylase kinase

ATP → ADP

6. Activation of phosphorylase

7. Production of glucose from glycogen

FIGURE 47.17 Epinephrine Triggers a Signal Transduction Cascade. Epinephrine's signal is amplified at each of steps 3–6.

🔑 **If you understand that . . .**

● Hormones act on target cells by binding to receptors.
● The response to a hormone is tissue specific because only certain cells contain receptors for particular hormones.
● In response to binding by lipid-soluble hormones, hormone-receptor complexes bind to DNA and induce changes in gene expression.
● In response to binding by lipid-insoluble hormones, activated cell-surface receptors induce the production of second messengers or the activation of signal transduction cascades that result in the phosphorylation of existing proteins.

✓ **You should be able to . . .**

1. Predict what would happen to the estradiol response if an individual had mutations that changed the DNA sequence of its hormone-response element.

2. Suppose a steroid hormone bound to a cell-surface receptor. Explain how its mode of action would compare to a polypeptide hormone.

Answers are available in Appendix B.

Summary of Key Concepts

🔑 **Animals use at least six major types of chemical signals. Hormones are chemical signals that are present in tiny concentrations and travel throughout the body to affect target cells.**

- Hormones are chemical messengers that are released from neurons or cells of the endocrine system, circulate in the blood or other body fluids, and trigger a response in distant target cells containing an appropriate receptor.

- Hormones have a variety of chemical structures. Most animal hormones are polypeptides, amino acid derivatives, or steroids.

  ✔ You should be able to explain the relationship between electrical signals from the nervous system and chemical signals from the nervous system and endocrine system, as they work in combination to coordinate the body's response to environmental change.

🔑 **The information carried by hormones helps animals develop as embryos, undergo sexual maturation, respond to environmental change, and achieve homeostasis.**

- In conjunction with the nervous system, hormones coordinate the activities of diverse cells and tissues. A single hormone may affect a wide array of cells and tissues and induce a variety of responses.

- Estradiol is an example of a hormone that regulates development and sexual maturation. Estradiol stimulates the formation of female sex characteristics in human embryos and the maturation of these tissues in adolescence.

- Epinephrine and cortisol are examples of hormones that help individuals cope with environmental changes. These hormones activate the short-term and long-term responses to stress, triggering the fight-or-flight response or inducing changes that conserve glucose for use by the brain.

- Hormones are involved in a wide array of homeostatic interactions. For example, hormones are involved in directing cells that modify the concentrations of water, sodium ions, and other components of the blood and interstitial fluid. Homeostasis with respect to triglyceride stores is also under endocrine control.

  ✔ You should be able to explain why hormones—instead of electrical signals from the nervous system—are primarily responsible for regulating responses to environmental change, embryonic and sexual development, and homeostasis.

🔑 **The production of a hormone is tightly regulated by input from the nervous system and by other hormones.**

- In many cases, the release of a hormone is regulated by chemical messengers from the anterior pituitary.

- Hormone-secreting cells in the anterior pituitary are regulated by hormones released by the hypothalamus region of the brain.

- The long-term stress response is a well-studied example of hormone regulation. The brain responds to long-term stress by triggering the release of the hypothalamic hormone CRH. CRH activates the release of ACTH by the pituitary gland, which stimulates the production of cortisol by cells in the adrenal cortex. Because cortisol inhibits the production of ACTH and CRH, the chain of events is regulated by feedback inhibition.

  ✔ You should be able to predict the consequences if negative feedback fails to occur after the release of CRH.

  🔲 **Web Activity** Endocrine System Anatomy

🔑 **Some hormones bind to receptors inside target cells and change gene expression. Other hormones bind to receptors at the cell surface and lead to changes in protein activation.**

- Animal hormones have two basic modes of action. Steroid hormones are lipid soluble, cross plasma membranes readily, and often bind to receptors inside cells. Most polypeptide and amino-acid–derived hormones are not lipid soluble; they bind to receptors located in the membranes of target cells. In both cases, the response to a hormone is tissue specific because only certain cells express certain receptors.

- Most steroid hormones act by inducing a change in gene expression.

- Polypeptide and most amino-acid–derived hormones trigger signal transduction cascades that activate one or more target proteins by phosphorylation.

- Although they are produced in tiny concentrations, hormones have large effects because they trigger gene expression or because their message is amplified through a signal transduction cascade.

  ✔ You should be able to predict whether a cell can respond to more than one hormone at a time, and explain why or why not.

  🔲 **Web Activity** Hormone Actions on Target Cells

Questions

1. Both epinephrine and cortisol are involved in the body's response to stress. How do the two molecules differ?
   a. Cortisol is an amino acid derivative; epinephrine is a steroid.
   b. Cortisol binds to receptors on the plasma membranes of target cells; epinephrine binds to receptors in the interior of target cells.
   c. Epinephrine mediates the short-term response; cortisol mediates the short- and long-term responses.
   d. Cortisol controls the release of epinephrine from the adrenal glands.

2. How do steroid hormones differ from polypeptide hormones and most amino-acid–derived hormones?
   a. Steroids are lipid soluble and cross plasma membranes readily.
   b. Polypeptide and amino-acid–derived hormones are longer lived in the bloodstream and thus amplify signals more highly.
   c. Polypeptide hormones are the most structurally complex and induce permanent changes in target cells.
   d. Only polypeptide and steroid hormones bind to receptors in the plasma membrane.

3. What is signal transduction?
   a. the binding of a steroid hormone-receptor complex to DNA
   b. the release of a hormone from the anterior pituitary in response to a hypothalamic hormone
   c. the release of a hormone from the posterior pituitary in response to action potentials from the hypothalamus
   d. the production of a second chemical messenger inside a cell in response to the binding of a hormone at the cell surface

4. Which of the following developmental processes is *not* controlled by hormones?
   a. the initial development of male and female gonads, soon after fertilization
   b. overall growth
   c. molting in insects and other invertebrate animals
   d. metamorphosis in insects and other invertebrate animals

5. What is a hormone-response element?
   a. a receptor for a steroid hormone
   b. a receptor for a polypeptide hormone
   c. a segment of DNA where a hormone-receptor complex binds
   d. an enzyme that is activated in response to hormone binding and produces a second messenger

6. In hormone systems, when does feedback inhibition occur?
   a. when the presence of a hormone inhibits its release
   b. when the presence of a hormone stimulates its release
   c. when a second messenger triggers the phosphorylation of an inhibitory protein
   d. when a hormone from the hypothalamus inhibits the release of a hormone from the anterior pituitary

✔ TEST YOUR UNDERSTANDING

Answers are available in Appendix B

1. Compare the structure and function of the anterior and posterior pituitary glands.

2. Compare and contrast the modes of action of steroid and nonsteroid hormones.

3. The pituitary is often referred to as the "master gland." Why?

4. Why is the observation that one hormone may bind to more than one type of receptor important?

5. Hormones are present in tiny concentrations yet have large effects on target cells and on the individual as a whole. How is this possible?

6. Why can leptin be considered a "satiation signal"? What role does leptin play in homeostasis?

✔ APPLYING CONCEPTS TO NEW SITUATIONS

Answers are available in Appendix B

1. Suppose that during a detailed anatomical study of a marine invertebrate, you found a small, previously undescribed structure. How would you test the hypothesis that the structure is a gland that releases one or more hormones?

2. Cortisone is a glucocorticoid that suppresses inflammation and other aspects of wound healing. Cortisone was once widely used to treat athletes with joint injuries and people with arthritis. In the short term, cortisone is extremely effective in reducing swelling and pain. Over time, however, physicians found that repeated large doses of cortisone had damaging side effects. Predict what these side effects are. Explain your logic.

3. You are a physician supervising a patient's recovery from the surgical removal of the posterior pituitary. Name a hormone that you will have to administer to this patient artificially. Which symptom(s) will you monitor to assess whether the dosage and timing of your injections are having the desired effect?

4. Suppose that a researcher announces the discovery of a hormone that affects the metabolism of fats in lab rats. Preliminary data indicate that the hormone is a polypeptide, about 50 amino acids long. How would you go about isolating the receptor for this hormone? Using fat cells growing in vitro, how could you test the hypothesis that hormone binding causes a change in a second messenger?

The swollen, red rump of this female Hamadryas baboon indicates that she is about to produce an egg. She will probably mate with several males before the egg is fertilized. This chapter discusses how hormones regulate the female reproductive cycle and the consequences of multiple mating.

48 Animal Reproduction

All the cells, tissues, organs, and systems introduced in Unit 8 exist for one reason: They allow animals to survive long enough and gather enough resources to reproduce. Stated another way, producing offspring is the reason that adaptations exist. Reproduction is the unconscious goal of virtually everything that an animal does. As Chapter 1 noted, replication is a fundamental attribute of life.

Evolution by natural selection explains why animals reproduce; the goal of this chapter is to explore *how* reproduction occurs. Section 48.1 introduces research on animals that cycle between asexual and sexual modes of reproduction and explains how both modes occur at the cellular level. Section 48.2 surveys the diverse ways in which animals accomplish fertilization once gametes have formed. The final three sections focus on mammalian reproduction, using humans as the primary model organism.

The chapter includes topics of urgent practical interest, such as human birth control methods and the recent and dramatic declines in the death rate of human mothers during childbirth. Understanding and manipulating animal reproductive systems is an important issue for physicians, veterinarians, farmers, zookeepers, and many others in biology-related professions.

48.1 Asexual and Sexual Reproduction

Several earlier chapters have explored how asexual and sexual reproduction differ. When reproduction is asexual, it occurs without fusion of gametes. **Asexual reproduction** is usually based on mitosis and results in offspring that are genetically identical to their parent. **Sexual reproduction**, in contrast, is based on meiosis and fusion of gametes. 🔑 Due to genetic recombination during meiosis and the fusion of haploid gametes—usually from different parents—during fertilization, sexual reproduction results in offspring that are genetically different from each other and from their parents.

✔ When you see this checkmark, stop and test yourself. Answers are available in Appendix B.

Let's first consider how animals make genetically identical copies of themselves, then review the mechanisms responsible for sexual reproduction.

How Does Asexual Reproduction Occur?

In thousands of animal species, individuals can **clone** themselves—that is, produce large numbers of identical copies of themselves asexually.

(a) Budding in hydra

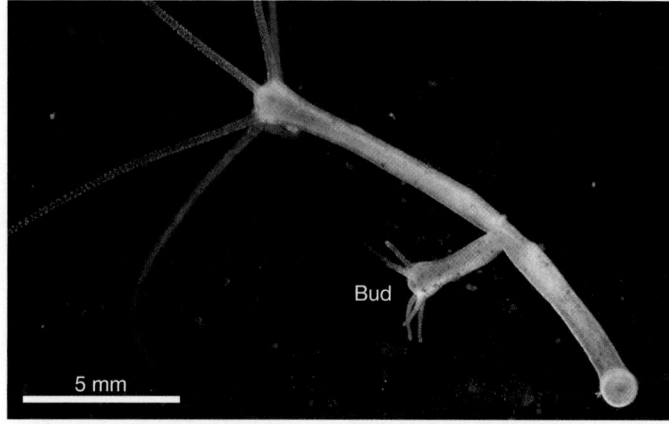

(b) Fission in anemones

(c) Parthenogenesis in lizards

Unfertilized eggs develop into offspring

FIGURE 48.1 Mechanisms of Asexual Reproduction in Animals Are Diverse.

There are three main mechanisms of asexual reproduction:

- When **budding** occurs, an offspring begins to form within or on a parent (**Figure 48.1a**). The process is complete when the offspring—a miniature version of the parent—breaks free and begins to grow on its own.

- During **fission**, an individual simply splits into two or more descendants (**Figure 48.1b**).

- **Parthenogenesis** (literally, "virgin-origin") occurs when female offspring develop from unfertilized eggs. In most cases, parthenogenetic eggs are produced by mitosis, or by meiosis that occurs after chromosome number has doubled, but without crossing over and recombination. In both cases, offspring are genetically identical to the mother. If a species reproduces exclusively via parthenogenesis, no males exist. Parthenogenesis occurs in a wide diversity of lineages, including certain guppies, crustaceans, rotifers, and lizards (**Figure 48.1c**).

Many animal species regularly switch between reproducing asexually and reproducing sexually. Why?

Switching Reproductive Modes: A Case History

Daphnia are crustaceans that live in freshwater habitats throughout the world. In a typical year, *Daphnia* produce only diploid female offspring throughout the spring and summer, via parthenogenesis. The eggs produced by parthenogenesis develop in a structure called a brood pouch (**Figure 48.2**), and are released when the female molts her exoskeleton.

In late summer or early fall, however, many *Daphnia* females begin producing unfertilized, asexually produced eggs that develop into males as well as females. Once the males have matured, sexual reproduction ensues: Haploid **sperm** (male gametes) produced by meiosis in the males fertilize haploid **eggs** (female gametes) that females produce via meiosis.

Fertilization is the fusion of sperm and egg. In *Daphnia*, fertilized eggs are released into a durable case that falls to the bottom of the pond or lake for the winter. In spring, the sexually produced offspring hatch and begin reproducing asexually.

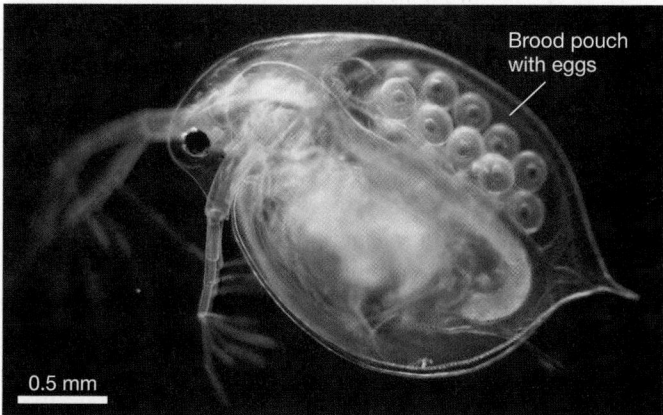

Brood pouch with eggs

FIGURE 48.2 Female *Daphnia* Can Lay Parthenogenetic Eggs. *Daphnia* produce diploid eggs asexually.

Biologists try to explain observations like this at two levels (see Chapter 51):

- **Ultimate causation** addresses *why* a trait occurs, in terms of its effect on fitness. Researchers who work at the ultimate level try to understand the evolutionary history of traits.

- **Proximate causation** addresses *how* a trait is produced. When researchers identify the genetic, developmental, hormonal, or neural mechanisms responsible for a phenotype, they are working at the proximate level.

Let's consider work on proximate aspects of the asexual-sexual switch in *Daphnia*, then consider ultimate causation.

WHAT ENVIRONMENTAL CUES TRIGGER THE SWITCH? For decades, most researchers contended that day length triggered the asexual-sexual switch. The idea was that the shortening days of late summer or fall affected sensors in the brain, and these receptors produced electrical or hormonal signals that induced the production of males and haploid eggs.

In 1965, however, biologists showed that high population densities are also a factor. Researchers who brought *Daphnia pulex* populations into the lab and kept day length constant found a strong, positive correlation between population density and the percentage of females reproducing sexually. When you read the graph in **Figure 48.3a**, note that the *y*-axis is logarithmic (see **BioSkills 7** in Appendix A). The highest proportion of sexually reproducing offspring is found at high density.

Another group of investigators extended this result by pinpointing the specific aspects of crowding that affected the animals. These biologists brought a closely related species—*D. magna*—into the laboratory and altered day length, the amount of food available to individuals, and the quality of the water they occupied. (To vary water quality, the investigators used either clean water or water taken from tanks where *D. magna* were being maintained at high density.)

As **Figure 48.3b** shows, individuals in the study population switched to sexual reproduction only if they were exposed to poor-quality water, low food availability, *and* short day lengths. In short, *D. magna* needs three different cues from the environment to switch to sexual reproduction. Two of these cues were associated with high population density; the third is associated with the onset of winter.

WHY DO *DAPHNIA* SWITCH BETWEEN ASEXUAL AND SEXUAL REPRODUCTION? *Daphnia* appear to start sexual reproduction when conditions worsen. Why?

The leading hypothesis to answer this question acknowledges that sexually produced offspring are genetically diverse. When the environment changes, genetically variable offspring are likely to survive better and reproduce more than offspring that are identical to their parents.

This hypothesis is consistent with research introduced in Chapter 12. You might recall that sexual reproduction in a snail species is associated with high rates of parasite infection. Parasites evolve rapidly—changing the environment for their hosts.

Genetically variable offspring have higher fitness in environments with rapidly evolving parasites, deteriorating physi-

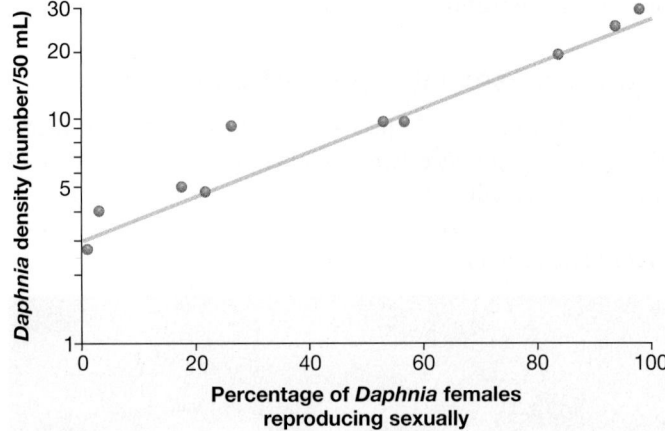

(a) Sexual reproduction is more common in crowded populations of *Daphnia pulex* than in sparse populations.

(b) In *Daphnia magna*, which environmental cues trigger the switch to sexual reproduction?

Water quality	Food concentration	Day length	% Sexual broods
Clean	Low	Short	0
Crowded	Low	Short	44
Clean	Low	Long	0
Crowded	Low	Long	0
Clean	High	Short	0
Crowded	High	Short	0
Clean	High	Long	0
Crowded	High	Long	0

FIGURE 48.3 In *Daphnia*, Environmental Cues Signal the Switch from Asexual to Sexual Reproduction. (a) There is a strong, positive correlation between the percentage of females that reproduce sexually versus density (plotted on a log scale here). **(b)** Environmental conditions were varied experimentally for *Daphnia*. "Crowded" water was taken from tanks containing dense populations.

✔**QUESTION** How would you go about determining which molecule or molecules in crowded water serve as a signal that triggers sexual reproduction?

cal conditions, or other types of rapid environmental change.

✔If you understand this concept, you should be able to predict the conditions under which asexual reproduction would be found as the only method of producing offspring.

To date, however, the variable-environment hypothesis has yet to be tested rigorously in *Daphnia*. Research on the adaptive significance of sex in this species and other organisms continues.

Mechanisms of Sexual Reproduction: Gametogenesis

The mitotic cell divisions, meiotic cell divisions, and developmental events that produce male and female gametes, or sperm

(a) Spermatogenesis

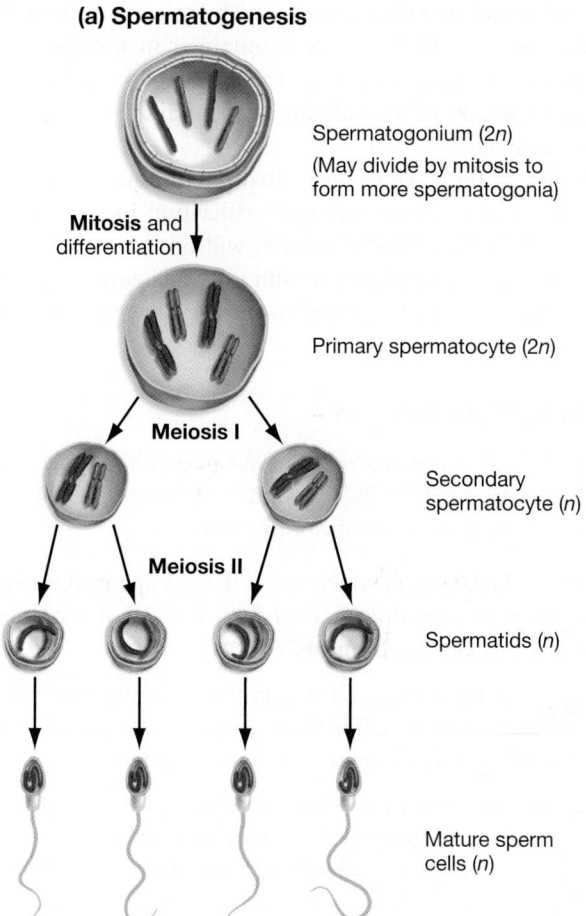

Spermatogonium (2n)
(May divide by mitosis to form more spermatogonia)

Mitosis and differentiation

Primary spermatocyte (2n)

Meiosis I

Secondary spermatocyte (n)

Meiosis II

Spermatids (n)

Mature sperm cells (n)

(b) Oogenesis

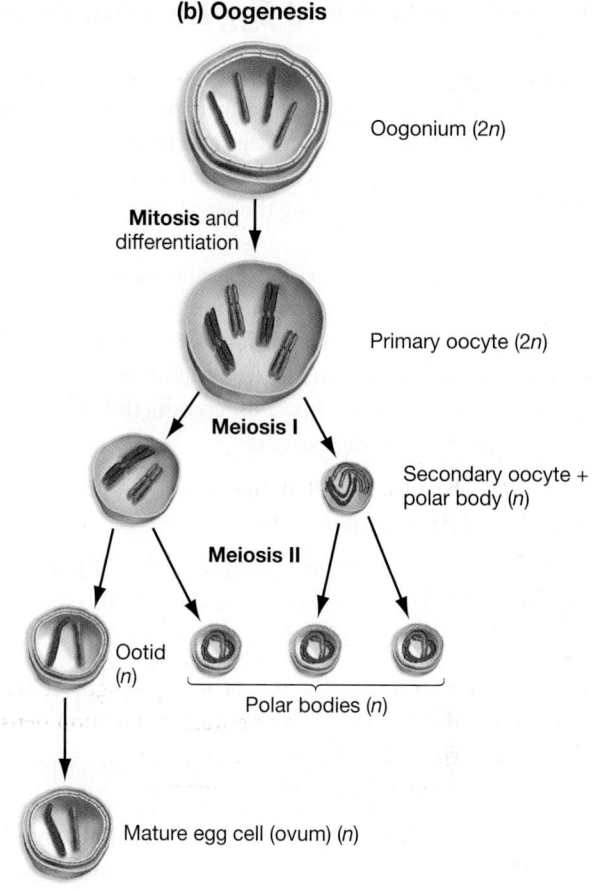

Oogonium (2n)

Mitosis and differentiation

Primary oocyte (2n)

Meiosis I

Secondary oocyte + polar body (n)

Meiosis II

Ootid (n)

Polar bodies (n)

Mature egg cell (ovum) (n)

FIGURE 48.4 Gametogenesis in Humans.

and eggs, are collectively called **gametogenesis** (see Chapter 22). **Spermatogenesis** is the formation of sperm; **oogenesis** is the formation of eggs.

In the vast majority of animals, gametogenesis occurs in a sex organ, or **gonad**. Male gonads are called **testes**; female gonads are called **ovaries**. Early in development, reproductive cells known as germ cells enter the testes and ovaries and give rise to diploid cells that will undergo gametogenesis.

SPERMATOGENESIS IN MAMMALS **Figure 48.4a** summarizes the events that take place during spermatogenesis in humans. Note that in the male gonad, diploid cells called **spermatogonia** (singular: **spermatogonium**) divide by mitosis. Some of the resulting cells continue to function as spermatogonia; others change to form specialized cells that are committed to producing sperm.

In males, the specialized cells produced by spermatogonia are called **primary spermatocytes**. They undergo meiosis I and produce two **secondary spermatocytes**, which then undergo meiosis II. The result is four haploid cells called **spermatids**.

Each haploid spermatid matures into a sperm—a cell that is specialized for carrying a haploid genome from the male through the female reproductive tract, and fertilizing an egg. The production of spermatogonia, primary spermatocytes, and sperm occurs continuously throughout a man's adult life.

OOGENESIS IN MAMMALS The top of **Figure 48.4b** highlights an important similarity between spermatogenesis and oogenesis: In the female gonad, diploid cells called **oogonia** (singular: **oogonium**) divide by mitosis. Some of the resulting cells continue to function as oogonia; others change to form specialized cells that are committed to producing an egg. In this sense, oogonia and spermatogonia are similar.

However, subsequent steps in gametogenesis are markedly different in human females. The specialized cells produced by an oogonium are called **primary oocytes**. When these cells undergo meiosis, only one of the four haploid products, known as an **ovum**, matures into an egg. The other cells produced by meiosis in females have a tiny amount of cytoplasm and do not mature into eggs. Because the distribution of cytoplasm is so unequal during each meiotic division in females, the smaller cells are called **polar bodies**. (Recall that polar refers to inequality or opposites.)

In addition, the production of primary oocytes stops early in development in many mammals; in humans, it stops before birth. And in humans and many other mammals, the primary oocytes enter prophase of meiosis I during embryogenesis, but then stop developing for months, years, or decades.

To review the steps in spermatogenesis and oogenesis, go to the study area at *www.masteringbiology.com*.

(MB) **Web Activity** Human Gametogenesis

48.2 Fertilization and Egg Development

Fertilization is the joining of a sperm and an egg to form a diploid **zygote**. Chapter 22 introduced this process by describing how a sperm makes contact with an egg and penetrates the egg's membrane. That discussion focused on the molecular mechanisms of sperm-egg binding, using the sea urchin as a model organism.

Here, the focus is on the variety of fertilization mechanisms observed among animals and the question of what happens to fertilized eggs. In many species, individuals release their gametes into their environment and external fertilization occurs. In other animals, males deposit sperm into the reproductive tracts of females and internal fertilization occurs.

External Fertilization

Most animals that rely on external fertilization live in aquatic environments. The correlation between external fertilization and aquatic environments is logical, because gametes and embryos must be protected from drying. If external fertilization occurred in a terrestrial environment, either the gametes or the resulting zygote would likely die of desiccation.

Another observation is that species with external fertilization tend to produce exceptionally large numbers of gametes. For example, a female sea star *Asterias amurensis* typically releases 100,000,000 eggs into the surrounding seawater during spawning. Males release many times that number of sperm. The leading hypothesis to explain this pattern is that the probability of a sperm and egg meeting in an ocean or lake is small unless large numbers of gametes are present.

If sperm and eggs from different individuals must be released into the environment synchronously (at the same time) for external fertilization to work, how is gamete release coordinated? The answer has two parts.

1. Gametogenesis usually occurs in response to environmental cues, such as lengthening days and warmer water temperatures, that indicate a favorable season for breeding.

2. Gametes are released in response to specific cues from individuals of the same species.

In fishes and other aquatic animals with well-developed eyes and external fertilization, spawning is often the culmination of an elaborate courtship ritual between a male and female. In contrast, courtship behavior appears to be much less important—or even absent—in species such as clams, sea urchins, and sea cucumbers. How do animals that cannot see their mates time the release of their gametes?

Recent research indicates that the chemical messengers called **pheromones** (see Chapter 47) might be involved in synchronizing gamete release. For example, biologists maintained two groups of sea cucumbers under natural conditions of light and temperature for 15 months. In one treatment, individuals were kept in isolated tanks; in another, they were kept in tanks with other sea cucumbers. The researchers found that all the individuals that were maintained in groups released gametes during the normal spawning period. In contrast, only about 10 percent of the individuals maintained in isolation released gametes.

Although these data suggest that pheromones might be involved in coordinating external fertilization in sea stars, they are not definitive. Most biologists will not be convinced that the pheromone hypothesis is valid until a chemical messenger is purified and can be shown to induce spawning in isolated individuals.

Internal Fertilization

Internal fertilization occurs in the vast majority of terrestrial animals as well as in a significant number of aquatic animals. Internal fertilization occurs in one of two ways:

1. after copulation, in which males deposit sperm directly into the female reproductive tract with the aid of a copulatory organ, usually called a **penis**; or

2. when males package their sperm into a structure called a **spermatophore**, which is then picked up and placed into the female's reproductive tract by the male or female.

In some salamander species, for example, the male places the spermatophore on the ground within its territory. Later the female picks it up with her **cloaca**, a chamber used by both the reproductive and excretory systems that opens to the environment. In this case, the result is internal fertilization without any direct physical contact between the sexes.

WHAT IS SPERM COMPETITION? In terms of understanding animal behavior, perhaps the most important insight about internal fertilization originated with Geoff Parker. In 1970 Parker published experiments on dung flies that confirmed the existence of **sperm competition**—competition between sperm from different males to fertilize the eggs of the same female.

Parker's experiments consisted of a series of matings between one female and two males. In each experiment, the two males were selected in such a way that Parker could distinguish between their offspring. The proportion of offspring fathered by each male was not 50:50. Instead, whichever male was last to copulate fathered an average of 85 percent of the offspring produced.

These data not only suggested that sperm were competing to fertilize eggs but indicated that, in this experiment, the second male won. Follow-up research has confirmed that **second-male advantage** is widespread, although not universal, in insects and some other animal groups. How does it occur?

SPERM COMPETITION AND SECOND-MALE ADVANTAGE To explore why second-male advantage occurs, biologists turned to the fruit fly *Drosophila melanogaster*. A research group introduced a gene into male fruit flies that produced sperm with green tails (**Figure 48.5**). When a mating by a green-spermed

EXPERIMENT

QUESTION: How does the "second-male advantage" occur in sperm competition?

HYPOTHESIS: In sperm storage areas, sperm from the second male displaces sperm from the first male.

NULL HYPOTHESIS: The mechanism does not involve sperm displacement from sperm storage areas.

EXPERIMENTAL SETUP:

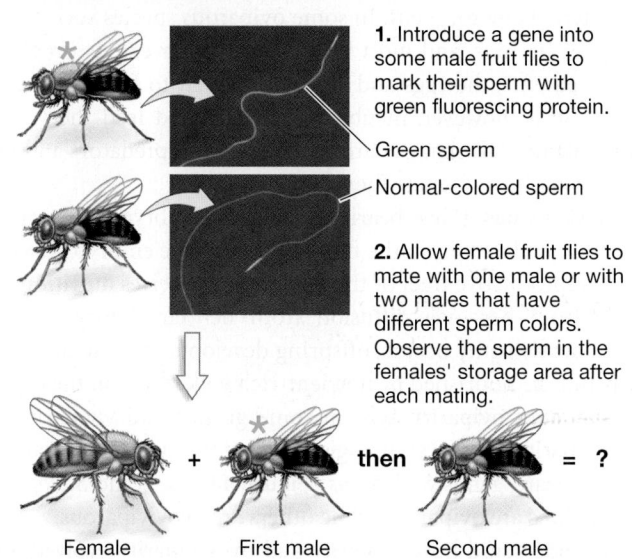

1. Introduce a gene into some male fruit flies to mark their sperm with green fluorescing protein.

— Green sperm

— Normal-colored sperm

2. Allow female fruit flies to mate with one male or with two males that have different sperm colors. Observe the sperm in the females' storage area after each mating.

Female + First male then Second male = ?

PREDICTION: When females mate twice, few sperm from the first male remain in storage.

PREDICTION OF NULL HYPOTHESIS: When females mate twice, most or all of the sperm deposited by the first male is still present.

RESULTS:

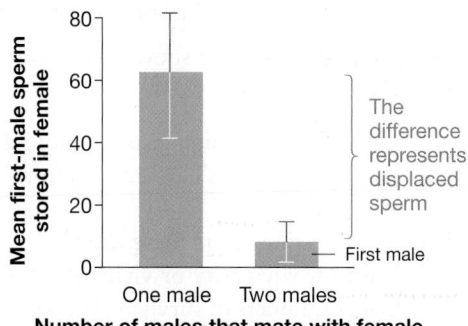

The difference represents displaced sperm

First male

Mean first-male sperm stored in female

Number of males that mate with female

CONCLUSION: If a female mates a second time, most sperm from the first male she mated with disappears.

FIGURE 48.5 Experimental Evidence for Second-Male Advantage during Sperm Competition in *Drosophila*. The graph shows the average number of green sperm stored in females when males with green sperm were the only male to mate or the first of two males to mate.

SOURCE: Price, C. S. C., K. A. Dyer, and J. A. Coyne. 1999. Sperm competition between *Drosophila* males involves both displacement and incapacitation. *Nature* 400: 449–452.

✔**QUESTION** Based on these data, is the claim that sperm from the second male physically displaces sperm from the first male valid?

male was followed with a second mating—by a male having normal-colored sperm—many fewer green-tailed sperm were found in the female's sperm-storage area compared with the number observed when no second mating took place.

To interpret this finding, the biologists suggested that the sperm of the second male physically dislodged the first male's gametes from the female sperm storage area and inserted themselves in their place. The researchers also showed that the fluid that accompanies sperm during fertilization is able to displace stored sperm from competing males. These two mechanisms resulted in the second male's sperm fertilizing most of the eggs laid.

WHY IS TESTES SIZE VARIABLE AMONG SPECIES? Research on sperm competition has recently contributed another major result. In species in which females routinely mate with multiple males before laying eggs or giving birth, males have extraordinarily large testes for their size and produce proportionately larger numbers of sperm.

The leading hypothesis to explain this observation is that fertilization is similar to a lottery in which each sperm represents a ticket. The more tickets a male enters in the competition, the higher his chance of "winning" fertilizations and passing his alleles on to the next generation. Males with exceptionally large testes produce exceptionally large numbers of sperm, and are more likely to win the lottery.

The lottery model has been challenged, however, by evidence that females often store sperm and exert control over which sperm are successful in fertilization. In other words, females do not always accept the results of sperm competition passively.

Females of some species actively choose which male performs the last copulation before fertilization takes place. In other species, females physically eject sperm from undesirable males. This phenomenon has been dubbed cryptic female choice. The name is appropriate because the selection of sperm by females is hidden from males.

Unusual Aspects of Mating

Studies on animal sexual reproduction have documented some of the most remarkable behaviors and structures observed in nature. The list that follows is just a sample, offered simply to illustrate the diversity of mating arrangements in animals.

- *Femmes fatales* When Australian redback spiders mate, the male does a somersault after inserting his penis-like organ. The somersault places his dorsal surface in front of the female's mouthparts. In many cases the female responds by eating the male. (Biologists refer to females who cannibalize males as femmes fatales—a phrase used to describe murderous human females in movies.) Cannibalized males copulate longer and fertilize more eggs than non-cannibalized males, probably because sperm transfer continues until the meal is over.

- *Giant sperm* In the fruit fly *Drosophila bifurca*, males average 1.5 mm in total body length. Their sperm, however, are each 6 cm long. The coiled, bulky sperm fill the female's sperm storage area and make it impossible for another male's sperm to enter.

- *False penises* Although very few bird species have a penis, the red-billed buffalo weaver has a false penis that becomes erect during copulation (**Figure 48.6**). It does not function in sperm transfer, but appears to stimulate the female's reproductive tract and trigger an orgasm-like state in males. Female buffalo weavers have a similar, but smaller, organ of unknown function.

- *Infidelity* When researchers assess paternity in bird species that appear to be monogamous, they find that up to 60 percent of nests contain at least one offspring fathered by a male that is not mated to the resident female. In most cases, females actively solicit copulations from males holding nearby territories.

- *Love darts* Many snails and slugs are **hermaphroditic**, meaning an individual has both male and female gonads. During mating, two individuals simultaneously receive and deposit sperm to fertilize each other's eggs. In some species, individuals fire mucus-covered "love darts" from their genitalia into the mating partner. A substance in the mucus increases the chance that sperm will successfully fertilize eggs. In other species, the penis frequently becomes stuck in the female reproductive tract and is bitten off by one partner.

- *Hypodermic insemination* Some bedbugs have hypodermic penises, meaning that they function like a hypodermic needle. Males force the organ through the female's abdominal wall and deposit sperm directly into her body cavity.

Did the process of sexual selection, introduced in Chapter 25, lead to the evolution of these structures and behaviors? If so, how and why? In many or even most cases, the answers to these questions are not known.

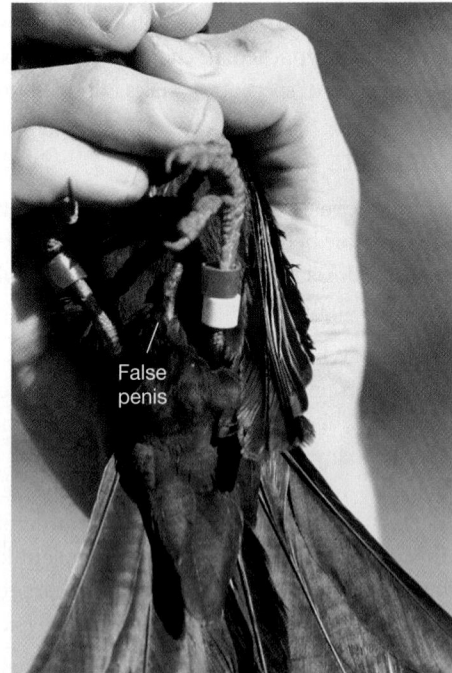

FIGURE 48.6 A False Penis in a Weaverbird. The organ shown here is in the erect state. It is "false" because it is not inserted into the female's reproductive tract and does not transmit semen.

Why Do Some Females Lay Eggs while Others Give Birth?

For females, many aspects of fertilization and egg-laying vary among species: a female's size or age at first breeding; the number and size of eggs she produces; the number of times she reproduces over the course of a lifetime. Here let's consider a different aspect of variation in animal reproduction: whether eggs are laid outside the body or retained inside the body.

In **oviparous** ("egg-bearing") animals, the embryo develops in the external environment. In some oviparous species such as sea stars, sea urchins, and most insects, no further care is provided by the parents; the eggs and embryos are left to fend for themselves. Birds, however, incubate their eggs and feed the young after hatching; fish may guard their eggs from predators and fan the clutches to oxygenate them.

In **viviparous** ("live-bearing") species, embryonic development takes place within the mother's body. The embryo attaches to the reproductive tract of the mother and receives nutrition directly from her—via diffusion from her circulatory system. When **ovoviviparity** occurs, offspring develop inside the mother's body but are nourished by nutrient-rich yolk stored in the egg.

Why does oviparity exist in some groups and viviparity or ovoviviparity in others? Biologists tackled this question by studying the lizard genus *Sceloporus* (**Figure 48.7a**). Some *Sceloporus* populations are oviparous while others are ovoviviparous.

To understand how oviparity and ovoviviparity evolved, the biologists analyzed a phylogenetic tree—based on molecular and morphological data—of many *Sceloporus* species (**Figure 48.7b**). Two conclusions should make sense to you (see **BioSkills 3** in Appendix A for help with interpreting phylogenetic trees):

1. Because most *Sceloporus* populations are oviparous, egg laying probably represents the original or ancestral condition.

2. As the red branches on the tree show, ovoviviparity evolved independently in two groups.

Using a similar research strategy, biologists have found that ovoviviparous populations of sea stars have evolved from oviparous populations on several occasions.

Why did natural selection favor these changes between egg laying and live birth? Ovoviviparity or viviparity should evolve when it leads to higher numbers of surviving young.

Researchers have long hypothesized that natural selection favoring live birth should be especially strong in cold habitats. Low temperatures slow the development of embryos, so in cold habitats, it might be advantageous for females to retain eggs inside their bodies so that offspring can develop at a more favorable temperature.

Several lines of evidence support this hypothesis:

- The ovoviviparous *Sceloporus* species live in the highlands of the southwestern United States and central Mexico, where temperatures can be cool.

- In at least one oviparous species of lizard, lowering nest temperatures leads to a higher percentage of deformed offspring.

(a) *Sceloporus* lizard (female)

(b) Phylogeny of *Sceloporus* from central Mexico

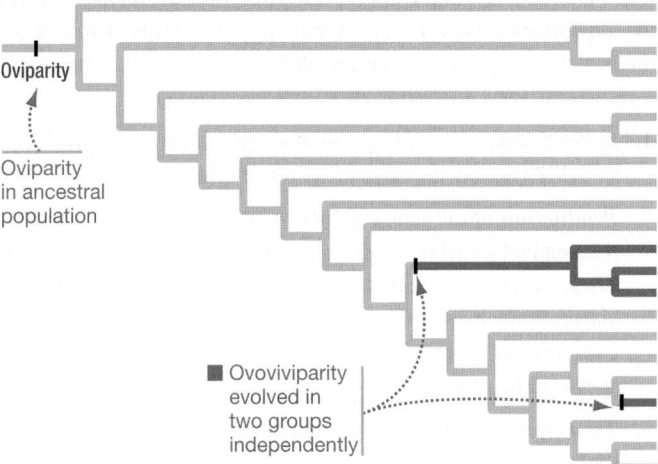

Oviparity

Oviparity
in ancestral
population

■ Ovoviviparity
evolved in
two groups
independently

FIGURE 48.7 Ovoviviparity Has Evolved More than Once in *Sceloporus* **Lizards.** **(a)** *Sceloporus* is a genus of lizards from desert areas in North America. **(b)** Each twig on this phylogenetic tree represents a *Sceloporus* species or population from central Mexico.

- In the same lizard species, offspring that developed in colder nests run more slowly than do individuals of the same age that developed at higher temperatures.

Research into why live birth has evolved in certain populations continues.

Sceleporus lizards have been a productive group to study on this question because the trait has evolved multiple times in extremely closely related species—making it easier to find patterns between live birth and other characteristics, such as living in cold habitats. The issue is much more difficult to study in mammals, where viviparity evolved just once. As Chapter 34 pointed out, the monotremes (duck-billed platypus and echidna) are oviparous, while marsupials and placental mammals are viviparous. It is still not clear why, and how, viviparity evolved in mammals prior to the marsupial-placentals split. Is the cold-habitats hypothesis relevant? To date, nobody knows.

CHECK YOUR UNDERSTANDING

🔑 **If you understand that . . .**

- Some of the most basic aspects of reproduction are variable among animal populations or species: whether reproduction is sexual or asexual, whether fertilization is external or internal, and whether fertilized eggs develop in the environment or in the mother's body.
- When females mate with more than one, sperm from different males compete to fertilize eggs. Females may also select sperm from different males.

✔ **You should be able to . . .**

1. Describe the advantages and disadvantages of oviparity and viviparity.
2. Suppose that female fruit flies can release a molecule called spermkillerene into their sperm-storage areas, and design an experiment to test the hypothesis that it can destroy sperm.

Answers are available in Appendix B.

48.3 Reproductive Structures and Their Functions

The first two sections of this chapter considered (1) the broad contrast between asexual and sexual reproduction and (2) general patterns in fertilization and egg care. Now let's explore the mechanics of sexual reproduction in more detail. The first task is to understand the anatomy of the male and female reproductive systems—focusing primarily on mammals.

The Male Reproductive System

In humans, the external anatomy of the male reproductive system consists of the scrotum and the penis. The saclike **scrotum** holds the testes; the penis functions as the organ of copulation prior to internal fertilization.

Biologists who have compared these structures among animal species have been struck by their variability and complexity. For example, a scrotum occurs only in certain mammal species. Whales, elephants, hedgehogs, moles, and many other mammal groups lack the structure entirely and instead have testes that are located well within the abdominal cavity. Among species that do have a scrotum, the structure's size and shape vary from tiny to prominent. In numerous species of primates, the scrotum is brightly colored and appears to function in courtship or other sexual display behaviors.

HOW DOES EXTERNAL ANATOMY AFFECT SPERM COMPETITION? Diversity in scrotal morphology pales in comparison with variation in the structure of the penis or other types of male **genitalia**. (The term genitalia refers to copulatory organs that have distinct parts.) In many groups of insects and spiders, for example, closely related species are morphologically identical except for their distinctive genitalia. Why are these organs so diverse?

The leading hypothesis on this question is that certain genitalia shapes give males an advantage when sperm competition occurs. To test this idea, biologists measured the size of the spines found on the tips of male genitalia in a species of seed beetle (**Figure 48.8a**),

(a) Long genital spines in males of the seed beetle *Callosobruchus maculatus*

100 μm

Female

Male

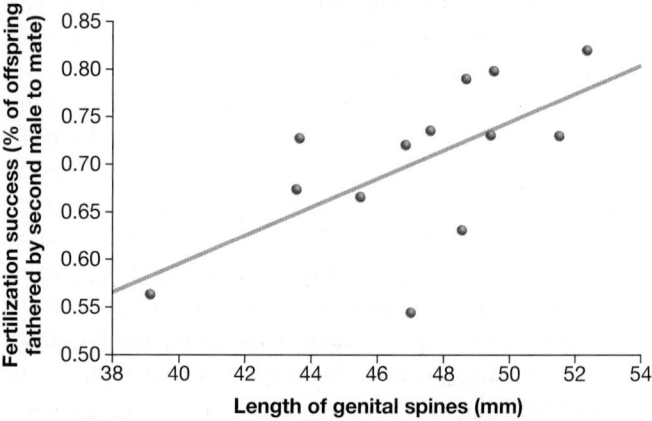

(b) During sperm competition, males with long genital spines father more offspring.

FIGURE 48.8 In Insects and Spiders, Male Genitalia May Vary among Species and Individuals and May Affect Reproductive Success. (a) An elaborate male reproductive structure in a seed beetle, used to transfer sperm to the female. The spikey structures at the top of the structure are called genital spines. (b) Data from experiments on reproductive success during sperm competition.

then documented which males were most successful, in terms of fertilizing eggs, when females mated with two males.

The *x*-axis in **Figure 48.8b** plots variation in the length of genital spines; the *y*-axis plots the proportion of eggs fertilized by the second male to copulate with a female, during experimental matings. The data points on the graph indicate genital spine length in each second-to-mate male tested. Note that second-to-mate males with the longest genital spines fathered a higher percentage of offspring than did others. The mechanism appears to be the ability of long spines to stick in the female's reproductive tract and prolong copulation time, even if the female is ready for copulation to end.

The punchline? Size matters, in seed beetles. Natural selection that occurs during sperm competition may explain why genitalia are so diverse among insect and spider species.

EXTERNAL AND INTERNAL ANATOMY OF HUMAN MALE REPRODUCTIVE ORGANS 🔑 Although numerous structures are involved, the reproductive system in human males has just three basic functional components. **Figure 48.9** shows the relevant structures in side view and front view.

1. ***Spermatogenesis and sperm storage*** Sperm are produced in the testes and stored nearby in the **epididymis**.

2. ***Production of accessory fluids*** Complex solutions form in the **seminal vesicles**, **prostate gland**, and **bulbourethral gland**. These accessory fluids are added to sperm prior to **ejaculation**, or expulsion from the body. **Table 48.1** lists some components of these accessory fluids, which combine with sperm to form **semen**.

3. ***Transport and delivery*** A **vas deferens** is a muscular tube that stores sperm and transports fluids from the epididymis to the short **ejaculatory duct**. The semen then enters the **urethra**, a longer tube that passes through the penis and services both the reproductive and urinary systems in males. The semen is expelled during ejaculation.

The composition of the accessory fluids varies widely among animals. In many insects, spiders, and vertebrates, molecules in the accessory fluids congeal after they arrive in the female reproductive tract and plug it. Experiments have shown that these copulatory plugs can serve as an effective deterrent to future matings. In some species, though, females or second males actively remove them.

Another diverse aspect of male internal anatomy is a bone inside the penis called the **baculum**. Some mammal species, including humans, lack this feature. But among rodents, the shape of the baculum is so variable that it can be used to distinguish among species. And in seals, baculum size correlates with mating system—the mating practices observed in a species. For example, in seal species in which females routinely mate with several males before becoming pregnant, males have not only large testes for their size, but a large baculum as well. The testes and baculum are much smaller in species in which females mate with a single male. In all species in which it appears, the baculum helps stiffen the penis during copulation.

(a) Side view

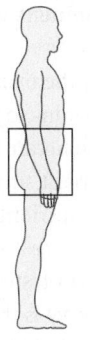

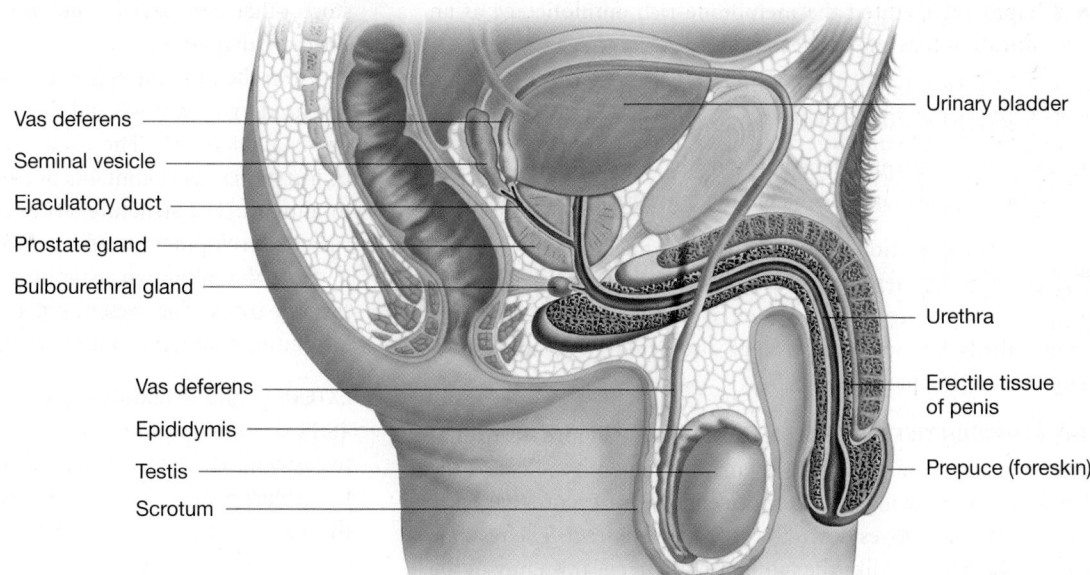

Vas deferens

Seminal vesicle

Ejaculatory duct

Prostate gland

Bulbourethral gland

Vas deferens

Epididymis

Testis

Scrotum

Urinary bladder

Urethra

Erectile tissue of penis

Prepuce (foreskin)

(b) Front view

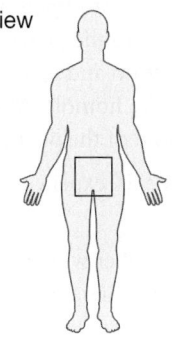

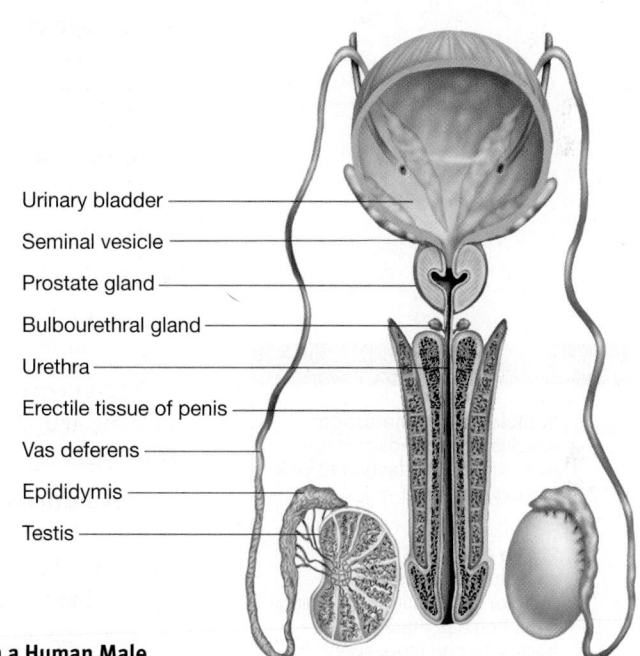

Urinary bladder

Seminal vesicle

Prostate gland

Bulbourethral gland

Urethra

Erectile tissue of penis

Vas deferens

Epididymis

Testis

FIGURE 48.9 Anatomy of the Reproductive Tract in a Human Male.

SUMMARY TABLE 48.1 **Accessory Fluids in Human Semen**

Source	Content	Function
Seminal vesicles	Fructose (a sugar)	Source of chemical energy for sperm movement
	Prostaglandins	Stimulate smooth-muscle contractions in uterus
Prostate gland	Antibiotic compound	Prevent urinary tract infections in males?
	Citric acid	Nutrient used by sperm
Bulbourethral gland	Alkaline mucus	Lubricates tip of penis; neutralizes acids in urethra

The Female Reproductive System

In animals, the most important part of the female reproductive system is the ovary—where meiosis occurs and mature egg cells, or ova, are produced. In the vast majority of species, the mature egg cell is a membrane-bound structure consisting of a haploid nucleus, a full complement of other organelles, and a large supply of nutrients in the form of yolk.

Earlier chapters have analyzed aspects of egg structure:

- Chapter 22 introduced the extensive outer layers and the plasma membranes of sea urchin eggs, which play a key role in binding sperm from the same species and initiating fertilization.

- Chapter 34 featured the membrane-rich amniotic egg as an innovation that allowed tetrapods to lay large eggs.

- Chapter 41 pointed out that the amount of yolk present in an egg correlates strongly with the size of offspring at hatching.

- Chapter 41 presented data indicating that females face a trade-off between egg size and the total number of eggs they can produce.

To probe variation in egg structure and female reproductive systems further, let's consider two highly specialized examples: The reproductive systems of female birds and humans. Birds lay an amniotic egg protected by a hard shell; humans are viviparous.

THE REPRODUCTIVE TRACT OF FEMALE BIRDS

Figure 48.10 diagrams the bird reproductive system. Starting with the ovary, the labels on the drawing indicate the sequence of events that takes place as the egg moves down the reproductive tract. The result is a hard-shelled egg—like the familiar chicken eggs you buy in the store—that can be laid into the environment and incubated. Birds are oviparous.

From the time the egg is released from the ovary until the zygote undergoes mitosis and the embryo begins to develop, a bird egg is a single cell. The ostrich egg, which can be over 15 cm (6 in.) in diameter, contains one of the largest single cells known in animals. The structure stores enough nutrients and water to sustain development until hatching.

Note that although male birds have two testes, females have just one ovary. The presence of a single ovary is thought to be an adaptation that reduces weight and makes flight more efficient.

EXTERNAL AND INTERNAL ANATOMY OF HUMAN FEMALES

Figure 48.11 shows side and front views of the human female reproductive system. The external anatomy features the **labia minora** (singular: **labium minus**) and the **labia majora** (singular: **labium majus**), the opening of the urethra, and the opening of the vagina.

- The labia are folds of skin that cover the urethral and vaginal openings.

- The **clitoris** is an organ that develops from the same population of embryonic cells that gives rise to the penis in males. It becomes erect during sexual stimulation and is covered with a protective sheath called the *prepuce*, homologous with the prepuce (foreskin) that covers the end of the penis.

- The urethral opening, where urine is expelled, is separate from the reproductive structures.

- The **vagina**, or birth canal, is the chamber where semen is deposited during sexual intercourse and where the baby is delivered during childbirth.

The internal anatomy of the female reproductive system in humans and other mammals is dominated by structures with two functions:

1. *Production and transport of eggs* Eggs are produced in the paired ovaries. During **ovulation**, a developing oocyte (egg) is expelled from the ovary and enters the **oviduct**, also known as the **fallopian tube**, where fertilization may take place. Fertilized eggs are then transported from the oviduct to the muscular sac called the **uterus**.

2. *Development of offspring* The uterus is where embryonic development takes place. During childbirth, the developed embryo (now called a fetus) passes through the opening of the uterus, known as the **cervix**, and into the vagina.

To understand ovulation and embryonic development, it's essential to investigate the hormones that regulate them. Next let's look at research into sex hormones in mammals.

48.4 The Role of Sex Hormones in Mammalian Reproduction

Chapter 47 introduced the sex hormones **testosterone** and **estradiol**; the latter belongs to a class of hormones known as **estrogens**. Recall that testosterone and estradiol are steroids that

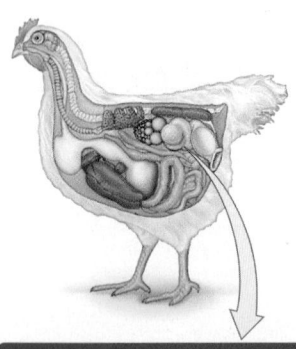

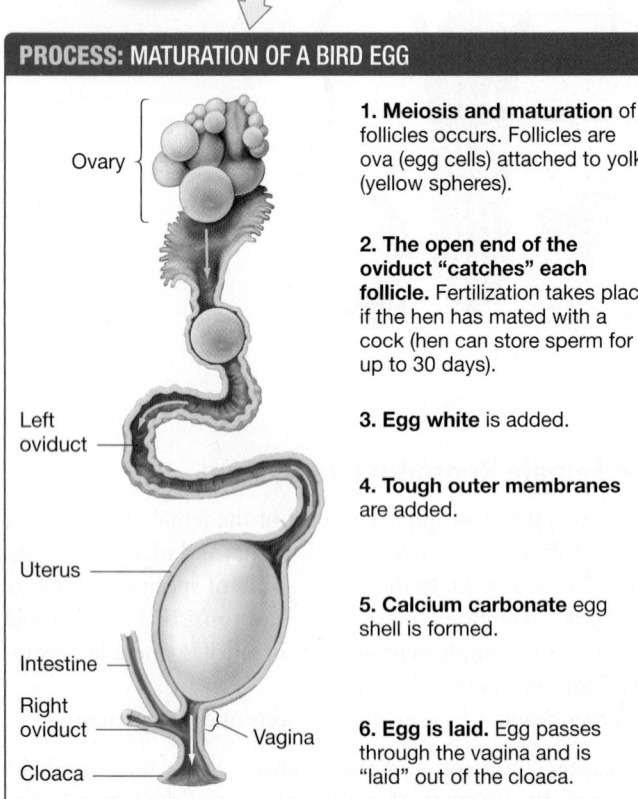

PROCESS: MATURATION OF A BIRD EGG

Ovary

1. Meiosis and maturation of follicles occurs. Follicles are ova (egg cells) attached to yolk (yellow spheres).

2. The open end of the oviduct "catches" each follicle. Fertilization takes place if the hen has mated with a cock (hen can store sperm for up to 30 days).

3. Egg white is added.

Left oviduct

4. Tough outer membranes are added.

Uterus

5. Calcium carbonate egg shell is formed.

Intestine

Right oviduct

Vagina

Cloaca

6. Egg is laid. Egg passes through the vagina and is "laid" out of the cloaca.

FIGURE 48.10 Structure and Function of the Reproductive Tract in a Female Bird.

(a) Side view

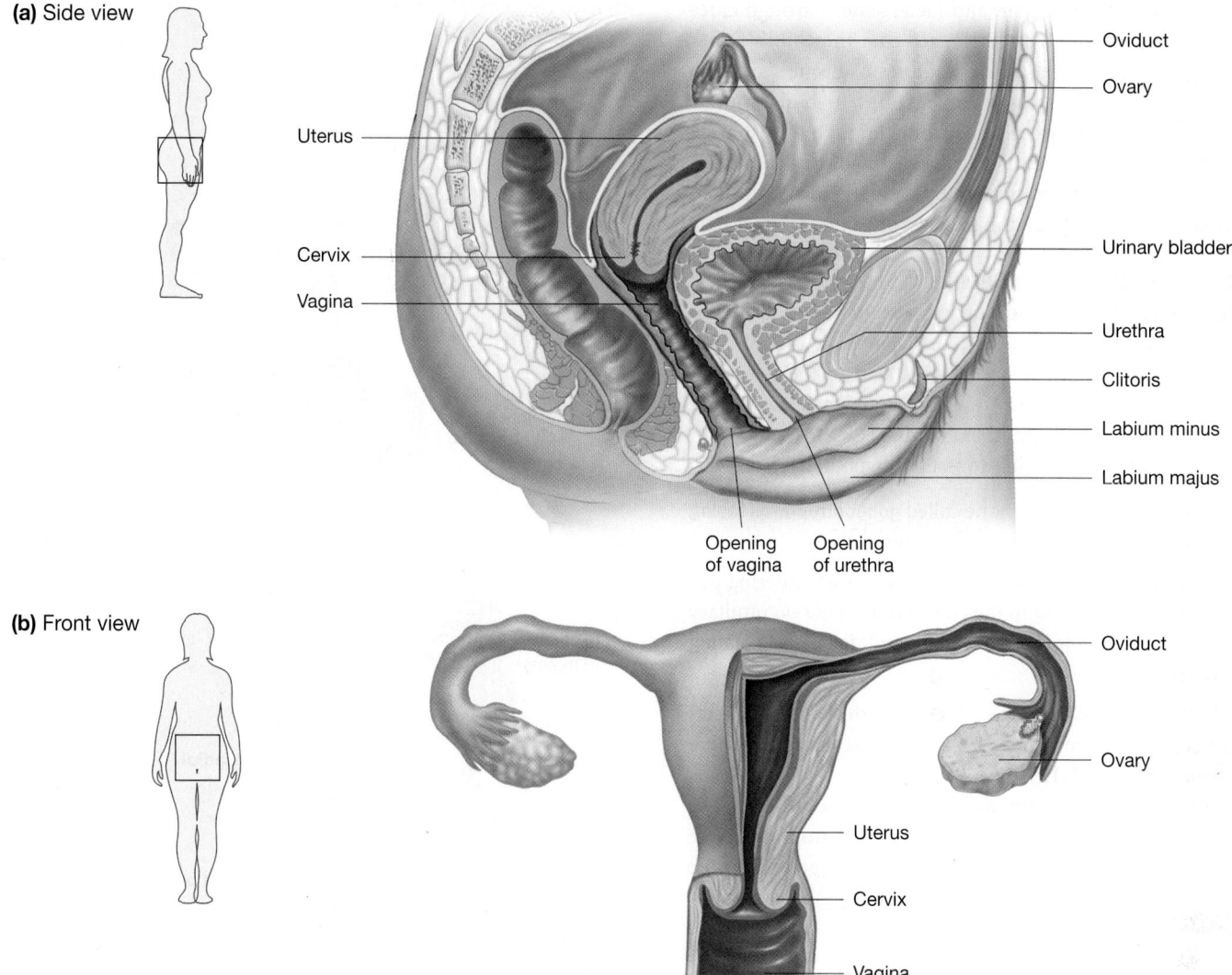

Uterus

Cervix

Vagina

Oviduct

Ovary

Urinary bladder

Urethra

Clitoris

Labium minus

Labium majus

Opening of vagina   Opening of urethra

(b) Front view

Oviduct

Ovary

Uterus

Cervix

Vagina

FIGURE 48.11 Anatomy of the Reproductive Tract in a Human Female.

✔**QUESTION** Generate a hypothesis to explain why male and female gonads are paired in mammals. How would you test your hypothesis?

bind to receptors inside the cytoplasm or nucleus of target cells. The resulting hormone-receptor complexes bind to DNA and trigger changes in gene expression.

Testosterone and estradiol are classified as gonadal hormones because they are produced in the gonads. Most testosterone is synthesized in specialized cells inside the testes; most estradiol and other estrogens are synthesized in the ovaries. More specifically, the female sex hormones are produced by cells that surround each developing egg. These surrounding cells form a structure called a **follicle** around the egg.

As Chapter 47 pointed out, the human sex hormones play a key role in three events:

1. development of the reproductive tract in embryos;

2. maturation of the reproductive tract during the transition from childhood to adulthood; and

3. regulation of spermatogenesis and oogenesis in adults.

To explore the action of sex hormones, let's take a closer look at their role in the transition from juveniles to adults.

Which Hormones Control Puberty in Mammals?

Puberty is the process that leads to sexual maturation in humans; **Table 48.2** on page 963 outlines the events that physicians use to track its progress. Chapter 47 pointed out that in amphibians, the juvenile-to-adult transition is triggered by the hormone T_3 (triiodothyronine); in insects the transition to adulthood occurs in response to ecdysone. But in humans, the transition is directed by increased levels of gonadal hormones—testosterone in boys and estradiol in girls.

A group of physicians recently offered dramatic evidence for testosterone's role: a 2-year-old boy who showed signs of entering puberty. His symptoms included an enlarged penis, the development of pubic hair, and facial acne. Through careful interviewing, the doctors discovered that the child's father was a bodybuilder who was smearing a testosterone cream on his own shoulders and arms in an attempt to build muscle mass. The child was apparently absorbing enough testosterone from being carried by his father to trigger puberty-like symptoms. All his symptoms except for penile enlargement subsided once his father discontinued the testosterone cream.

WHAT REGULATES THE GONADAL HORMONES? Some researchers suggested that sex hormone production is regulated by the hypothalamic-pituitary axis introduced in Chapter 47. Recall that chemical signals from the hypothalamus lead to the release of regulatory hormones from the pituitary gland, which then cause the release of hormones from other glands.

Two advances made it possible to test this hypothesis rigorously:

1. Researchers isolated a hormone called **gonadotropin-releasing hormone (GnRH)** from the hypothalamus.

2. Investigators noted that boys and girls who were entering puberty experienced pulses in the concentration of two pituitary hormones, **luteinizing hormone (LH)** and **follicle-stimulating hormone (FSH)**.

These observations inspired the hypothesis that GnRH directs LH and FSH pulses, which then trigger increases in testosterone and estradiol (**Figure 48.12**).

To test this idea, researchers administered pulses of GnRH to boys and girls who had deficits in their hypothalamus and were experiencing delays in the onset of puberty. As predicted, the GnRH treatment induced surges in LH and FSH, followed by puberty onset.

WHAT REGULATES THE REGULATORY HORMONES? The model in Figure 48.12 raises a question: What triggers GnRH increases at the appropriate age? Although this question remains unanswered, there is some evidence that nutritional state is involved. For example:

- The current average age for the onset of menstruation in females in the United States is slightly over 12 years. This is much earlier than the average age of 17 years in the United States during the eighteenth and nineteenth centuries, when the general nutritional state of the population was poorer.

- Among girls living today, individuals with large fat stores tend to enter puberty earlier than do girls who are thin.

Research continues on how age, nutritional condition, and perhaps other factors interact to promote the release of GnRH.

If you recall the discussion in Chapter 47 of how the adrenal hormone cortisol is controlled, however, you might suspect that Figure 48.12's model of sex-hormone regulation is simplified. Chapter 47 emphasized that many hormones participate in negative feedback—also called feedback inhibition—meaning that the presence of a hormone inhibits the factor that triggers its release.

Do sex hormones participate in negative feedback? The short answer to this question is yes. To appreciate the details, let's investigate hormonal control of the human menstrual cycle.

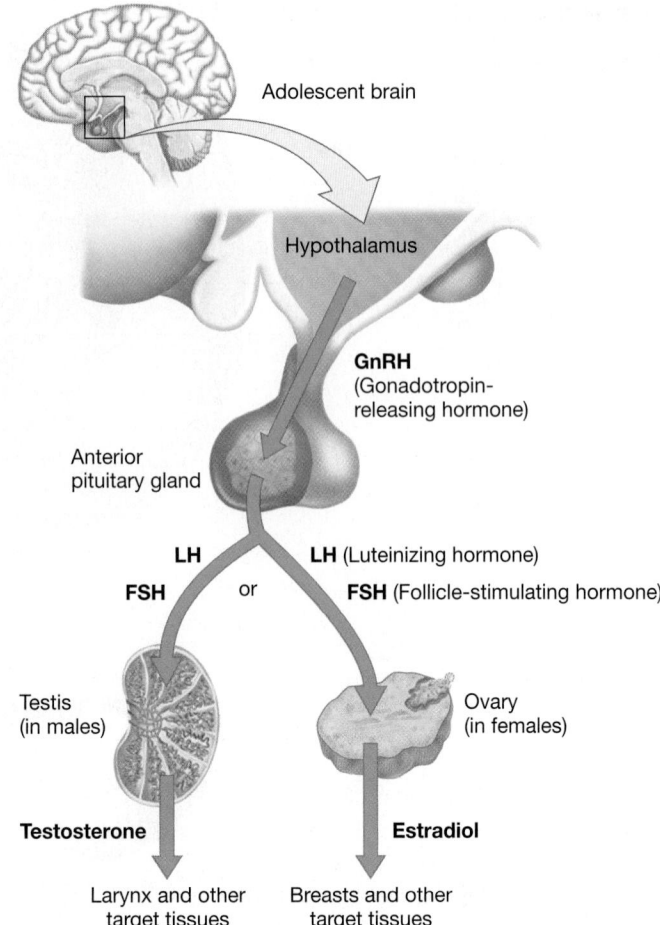

FIGURE 48.12 In Humans, Puberty Is Triggered by Hormones from the Hypothalamus and Pituitary.

✔ **QUESTION** How does control of testosterone and estradiol production compare with control of cortisol release by the adrenal gland? (See Figure 47.10c.)

Which Hormones Control the Menstrual Cycle in Mammals?

Figure 48.13 illustrates the sequence of events in the human ovary during the **menstrual cycle**, a monthly reproductive cycle. Although the cycle's length varies among women, 28 days is about average.

In conjunction with changes in the ovary illustrated in the figure, the lining of the uterus undergoes a dramatic thickening and regression. Ultimately, part of the uterine lining sloughs off and is expelled through the vagina.

Day 0 in the menstrual cycle is marked by the beginning of **menstruation**—the expulsion of the uterine lining. The remainder of the cycle has two distinct phases:

1. *Follicular phase* A follicle matures during the **follicular phase**, which lasts an average of 14 days. Primary oocytes complete meiosis I during this phase. Ovulation occurs when the follicle is mature and releases its secondary oocyte into the oviduct. Researchers are still unsure what controls which ovary—right or left—will release an egg during ovulation.

TABLE 48.2 **Changes That Take Place during Puberty**

Pediatricians use the five stages of puberty listed here to diagnose the developmental stage of their patients.

Type	Boys	Girls
Stage 1 (Prepubertal)	No sexual development	No sexual development
Stage 2	Testes enlarge	Breast budding, first pubic hair
	Body odor	Body odor
		Height spurt
Stage 3	Penis enlarges	Breasts enlarge
	First pubic hair	Pubic hair darkens, becomes curlier
	Ejaculation (wet dreams)	Vaginal discharge
Stage 4	Continued enlargement of testes and penis	Onset of menstruation
	Penis and scrotum deepen in color	Nipple is distinct from surrounding areola
	Pubic hair curlier and coarser	Pelvis begins to widen
	Height spurt	
	Male breast development (temporary)	
Stage 5 (Fully mature adult)	Pubic hair extends to inner thighs	Pubic hair extends to inner thighs
	Increases in height slow, then stop	Increases in height slow, then stop
	Increased muscle mass	Deposition of fat in hips, buttocks, thighs, and breasts

2. **Luteal phase** The **luteal phase** begins with ovulation and averages 14 days in length. Its name was inspired by the formation and subsequent degeneration of a structure called the **corpus luteum** ("yellowish body") from the ruptured follicle. Meiosis II occurs only if the ovulated cell is actually fertilized.

The regular occurrence of a menstrual cycle throughout the year makes human females extremely unusual among mammals. Although some mammals ovulate multiple times during the year, most ovulate only during a prescribed breeding season—often in response to environmental cues such as the onset of spring or fall; less often in response to cues from males.

In addition, only humans and other great apes menstruate. In the vast majority of mammals, the lining of the uterus is reabsorbed if pregnancy does not occur. Females that do not menstruate have an **estrous cycle** and are sexually receptive only during estrus—also known as being "in heat."

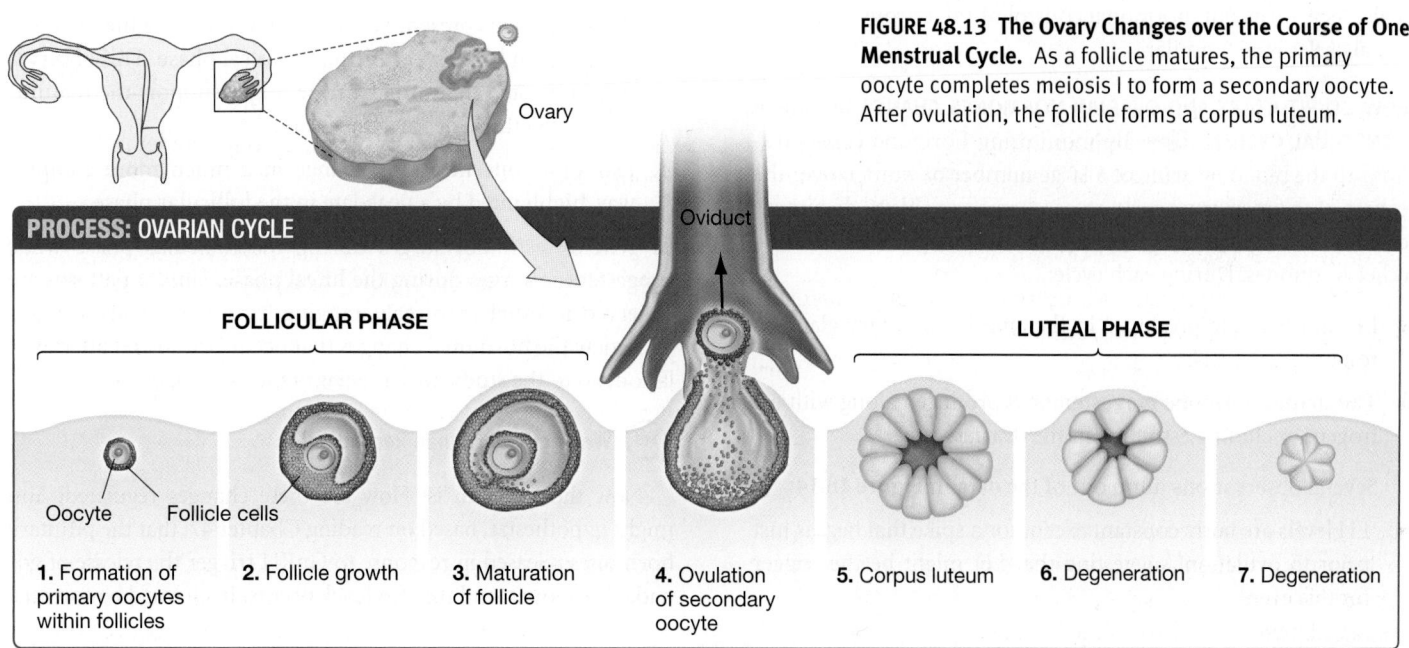

FIGURE 48.13 The Ovary Changes over the Course of One Menstrual Cycle. As a follicle matures, the primary oocyte completes meiosis I to form a secondary oocyte. After ovulation, the follicle forms a corpus luteum.

Ovary

Oviduct

PROCESS: OVARIAN CYCLE

FOLLICULAR PHASE

LUTEAL PHASE

Oocyte   Follicle cells

1. Formation of primary oocytes within follicles
2. Follicle growth
3. Maturation of follicle
4. Ovulation of secondary oocyte
5. Corpus luteum
6. Degeneration
7. Degeneration

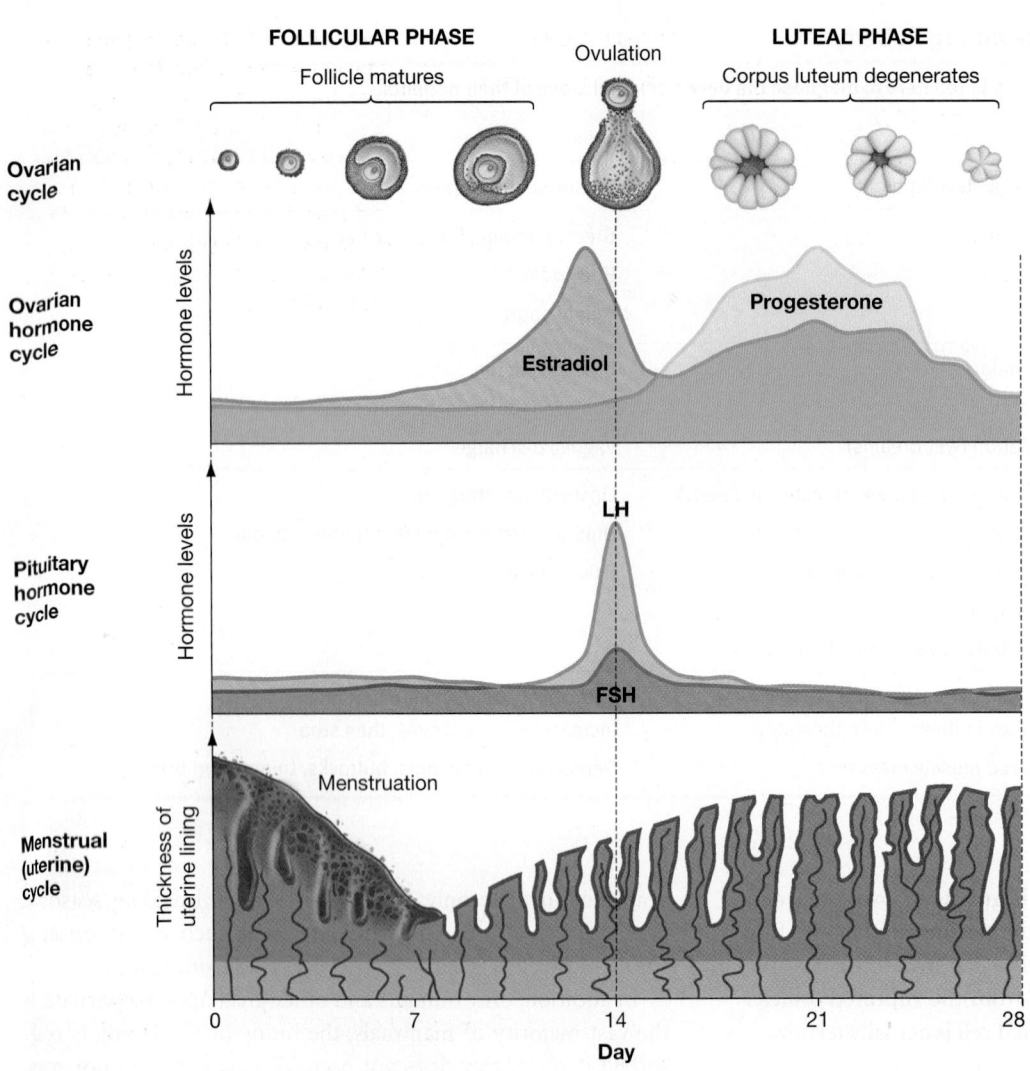

FOLLICULAR PHASE
Follicle matures

Ovulation

LUTEAL PHASE
Corpus luteum degenerates

Ovarian cycle

Ovarian hormone cycle

Progesterone

Estradiol

Pituitary hormone cycle

LH

FSH

Menstrual (uterine) cycle

Menstruation

FIGURE 48.14 Changes That Occur during a Menstrual Cycle in a Human Female.

But whether an estrous or menstrual cycle occurs, the basic sequence of events, with a follicular phase preceding ovulation and a luteal phase following ovulation, is shared among mammals. As it turns out, hormonal control of the estrous and menstrual cycles is also similar.

HOW DO PITUITARY AND OVARIAN HORMONES CHANGE DURING A MENSTRUAL CYCLE? By monitoring hormone concentrations in the blood or urine of a large number of women over the course of the menstrual cycle, researchers were able to document dramatic changes in the concentrations of estradiol and several other hormones. During each cycle:

- LH and FSH are produced in the anterior pituitary gland in response to GnRH.

- The steroid hormone **progesterone** is produced along with estrogens, including estradiol, in the ovaries.

Several observations jump out of the data in **Figure 48.14**:

1. LH levels are fairly constant except for a spike that begins just prior to ovulation, suggesting that LH might be the trigger for this event.

2. FSH concentrations are relatively high during the follicular phase and low during the luteal phase, though they also make a small spike prior to ovulation.

3. Progesterone is present at very low levels during the follicular phase but at high levels during the luteal phase. This observation suggested that progesterone might support the maturation of the thickened uterine lining.

4. Estradiol concentrations change in a much more complex way, highlighted by a peak late in the follicular phase.

In general, estradiol surges during the follicular phase while progesterone surges during the luteal phase. Similar patterns are observed in other mammals, ranging from marsupials to mice. To review the hormonal changes that occur before and after ovulation, go to the study area at *www.masteringbiology.com*.

(MB) Web Activity Human Reproduction

Now the question is: How are these changes regulated? You might hypothesize, based on reading Chapter 47, that the pituitary hormones released in response to GnRH trigger the release of gonadal hormones, and that feedback occurs. If so, you'd be correct.

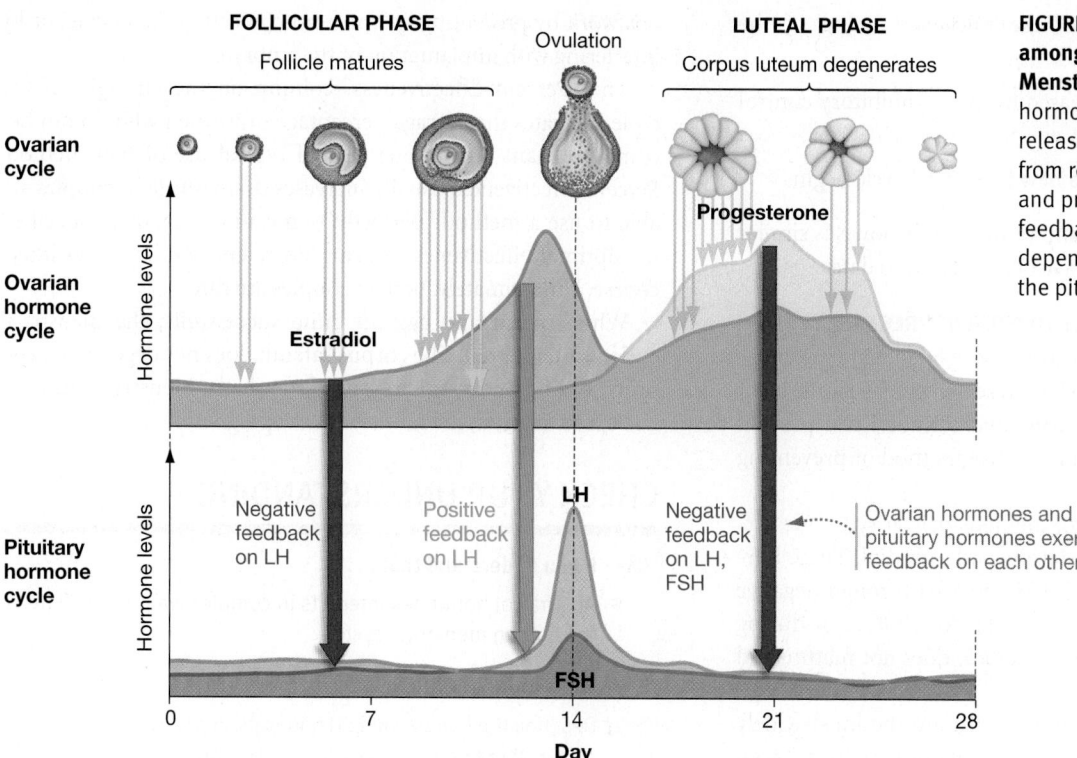

FOLLICULAR PHASE
Follicle matures

Ovulation

LUTEAL PHASE
Corpus luteum degenerates

Ovarian cycle

Ovarian hormone cycle

Progesterone

Estradiol

Pituitary hormone cycle

Hormone levels

Hormone levels

Negative feedback on LH

Positive feedback on LH

LH

Negative feedback on LH, FSH

Ovarian hormones and pituitary hormones exert feedback on each other

FSH

0   7   14   21   28
Day

FIGURE 48.15 Complex Interactions among Hormones Regulate the Menstrual Cycle. The pituitary hormones FSH and LH control the release of estradiol and progesterone from reproductive tissues. Estradiol and progesterone may exert negative feedback, positive feedback, or both—depending on their concentration—on the pituitary hormones.

HOW DO PITUITARY AND OVARIAN HORMONES INTERACT? Experiments helped establish that changes in the concentration of estradiol and progesterone affect the release of the pituitary hormones LH and FSH. Researchers worked with three volunteers whose ovaries had been removed because of cancerous growths or other problems. The women were receiving low, maintenance-level doses of estradiol, which appeared to exert negative feedback on LH and FSH.

But when the investigators injected the women with larger doses of estradiol or with progesterone, the situation changed dramatically. For example, a large increase in estradiol stimulated a dramatic spike in LH levels. This suggested that positive feedback was occurring. High levels of estradiol *increase* the release of LH, even though low doses of estradiol suppress it. This is a key observation: When feedback occurs, estradiol's effect on the pituitary hormones is dosage-dependent. In contrast, progesterone injections appeared to inhibit both FSH and LH. Progesterone exerts only negative feedback on the pituitary hormones.

To summarize the interplay between LH, FSH, estradiol, and progesterone, let's start at day 0 and follow key events as the cycle progresses (**Figure 48.15**).

Day 0–7

- As the uterus is shedding much of its lining, a follicle is beginning to develop in one ovary under the influence of FSH.

- The follicle produces estradiol and a small amount of progesterone.

- While its levels are still relatively low, estradiol suppresses LH secretion through negative feedback inhibition.

Day 8–14

- As the follicle grows, its production of estradiol gradually increases. The increase in estradiol stimulates mitosis and an increase in cell number in the uterine lining.

- The enlarged follicle produces large quantities of estradiol, which begin to exert positive feedback on LH secretion.

- Positive feedback results in a spike in LH levels, just after estradiol concentrations peak.

- The LH surge triggers ovulation and ends the follicular phase.

✓If you understand how estradiol's effect on LH changes from negative regulation to positive regulation, you should be able to predict the effect of (**1**) injecting a female volunteer with high estradiol concentrations early in a cycle, and (**2**) using a drug to keep estradiol production low throughout the cycle.

Day 15–21

- As the corpus luteum develops from the remains of the ruptured follicle, it secretes large amounts of progesterone and small quantities of estradiol, in response to LH.

- The rise in progesterone lowers production of LH and FSH and activates the thickened uterine lining, creating a spongy tissue with a well-developed blood supply. In this way, progesterone fosters an environment that supports embryonic development if fertilization occurs.

Day 22–28

- If fertilization does not occur, the corpus luteum degenerates.

- Progesterone levels fall as the corpus luteum shrinks.

- When progesterone declines, the thickened lining of the uterus degenerates.
- GnRH, LH, and FSH are released from the inhibitory control that progesterone exerts.
- LH and FSH levels rise, and a new menstrual cycle begins.

The interplay between ovarian and pituitary hormones is similar in other mammal species that have been studied to date.

MANIPULATING HORMONE LEVELS TO PREVENT PREGNANCY Data on hormonal control of the menstrual cycle opened new avenues in birth control research. Specifically, researchers have found that manipulating levels of progesterone and estradiol can prevent ovulation and serve as a safe and effective method of preventing unwanted pregnancies.

Hormone-based birth-control methods deliver synthetic versions of progesterone, or progesterone and estradiol. These hormones suppress the release of GnRH and FSH through negative feedback inhibition. Because an LH spike does not occur during the follicular phase of the cycle, the follicle does not mature and ovulation does not occur.

In the United States, birth control pills are the most widely used contraceptive method. Hormone-containing pills are taken for three weeks and then stopped for one week to allow menstruation to occur.

But as **Table 48.3** indicates, hormone-based methods are just one of several approaches to preventing pregnancy. Other methods work by preventing sperm from contacting the oocyte, or by interfering with implantation of the embryo.

The "Percent Effectiveness" column on the far right of the table indicates the average percentage of women who do not become pregnant during one year of typical use of that method. Percent effectiveness usually increases dramatically if couples are able to use a method "perfectly"—meaning, exactly as specified for optimal effectiveness, during every episode of sexual intercourse. Unfortunately, perfect couples are rare.

When sperm and egg do unite successfully, the menstrual cycle is interrupted. The corpus luteum does not degenerate, and progesterone and estradiol levels stay high. Menstruation does not occur. Instead, the mother is now pregnant.

CHECK YOUR UNDERSTANDING

If you understand that . . .
- An array of hormones interacts in complex ways to regulate the human menstrual cycle.

You should be able to . . .
1. Describe the function of FSH and explain why FSH levels increase at the end of one cycle and the start of the next cycle.
2. Predict the consequences of a drug that inhibits the release of FSH.

Answers are available in Appendix B.

SUMMARY TABLE 48.3 **Comparing Methods of Contraception**

"Percent Effectiveness" indicates the average percentage of women who do not become pregnant during one year of typical use.

Type	Name	Mode of Action	Percent Effectiveness
Hormone-based methods	The Pill, the Patch, the Ring, the Shot, the Implant	Provides continuous or cyclical delivery of progesterone, or progesterone plus estradiol	92 to 99.9*
Barrier methods	Condom	Covers penis and prevents sperm from entering vagina	85
	Female condom	Covers labia, vagina, and cervix and prevents sperm from entering uterus	79
	Diaphragm	Covers cervix and prevents sperm from entering uterus	84
	Sponge	Covers cervix and prevents sperm from entering uterus; also contains a molecule that immobilizes sperm	84
	Spermicide	Foam or jelly covers cervix and prevents sperm from entering uterus; contains a molecule that immobilizes sperm	71
Behavioral methods	Rhythm method	Couple refrains from vaginal intercourse around time of ovulation	80
	Withdrawal	Man withdraws penis prior to ejaculation	73
Prevent implantation	Intrauterine device (IUD)	Small T-shaped structure inserted into uterus; mode of action not known	99
	Emergency contraception	Delivers progesterone or progesterone and estradiol after unprotected vaginal intercourse	92
	Mifepristone (RU-486)	Blocks progesterone receptors so menstruation occurs even after fertilization and implantation	92

*Depends on delivery system used.

48.5 Pregnancy and Birth in Mammals

Viviparity allows the mother to provide a warm, protected environment for offspring during early development. Oviparous species that guard or incubate their eggs also provide warm, safe surroundings for their young.

Any investment that parents make in an offspring comes at a cost, however: The more a mother invests in each offspring, the lower the number of offspring she can produce. Chapter 41 explored this fitness trade-off between the quantity and quality of offspring.

Pregnancy and **lactation**—providing milk that nourishes offspring after birth—represent some of the most extreme forms of parental care known in animals. And in some mammal species, parental care continues long after lactation ends. Humans, for example, are largely or completely dependent on their parents for protection and nutrition until puberty or young adulthood.

Let's examine how mammals make this investment, starting with marsupials and then turning to the eutherian mammals.

Gestation and Early Development in Marsupials

You might recall from Chapter 34 that mammals comprise three major monophyletic groups: monotremes, eutherians, and marsupials. Although all mammals are endothermic, have fur, and lactate, the three lineages are distinguished by their mode of reproduction:

- **Monotremes** lay eggs and incubate them until hatching.

- In **eutherians**, mothers carry offspring for relatively long periods of development during gestation and nourish them via a placenta.

- In **marsupials**, the corpus luteum is not maintained as it is in eutherians, and no placenta forms.

In marsupials, young are ejected from the mother's body at the end of the estrous cycle. As a result, they are far less developed than eutherian mammal offspring which complete **gestation**—the developmental period that takes place inside the mother.

The jaws, gut, lungs, and forelimbs of a newly born kangaroo are relatively well developed at birth. As a result, the offspring is able to climb from its mother's vagina to a nipple, which is usually enclosed in a pouch created by a flap of skin (**Figure 48.16a**).

The offspring clamps onto the nipple and continues to develop, fed by the mother's milk (**Figures 48.16b** and **48.16c**).

Even after growing large enough to leave the pouch and begin moving and feeding on its own, offspring will return to the pouch for protection. Marsupial mothers invest a great deal in their offspring, even though a relatively short period of development takes place inside their bodies.

Major Events during Human Pregnancy

Marsupials and eutherians differ sharply in terms of how long the developing embryo is retained inside the mother's body. As a model organism in eutherian reproduction, let's consider humans.

When a secondary oocyte is released from the human ovary, the cell is viable for less than 24 hours. Human sperm, in contrast, remain capable of fertilizing an egg for up to five days. Therefore, sexual intercourse in humans has to occur less than five days prior to ovulation, or immediately after ovulation, for pregnancy to result.

Although an ejaculate may contain hundreds of millions of sperm, most die as they travel through the uterus. Only 100 to 300 actually succeed in reaching the oviduct, where fertilization takes place.

Chapter 22 detailed how the sperm and oocyte (egg) interact when they meet. You might recall that when sperm and oocyte meet, enzymes released from the head of the sperm create a path through the material surrounding the oocyte membrane. Once the membranes of the oocyte and sperm have fused, the oocyte nucleus completes meiosis II. The two nuclei then unite to form a diploid zygote.

IMPLANTATION Smooth-muscle contractions in the oviduct gradually move the zygote toward the uterus. As it travels, the cell begins to divide by mitosis. By the time it reaches the lining of the uterus, the embryo consists of a hollow ball of cells. It then undergoes **implantation**—meaning that it becomes embedded in the thickened, vascularized wall of the uterus. It will stay in the uterus for approximately 270 days (9 months).

Once the embryo is implanted in the uterine lining, its cells begin synthesizing and secreting the hormone **human chorionic gonadotropin (hCG)**. hCG is a chemical messenger that prevents

(a) Red kangaroo at birth

(b) 4 weeks after birth

(c) 4 months after birth

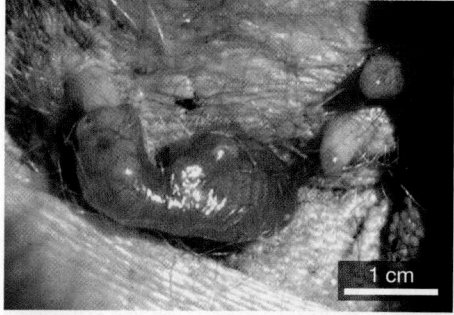 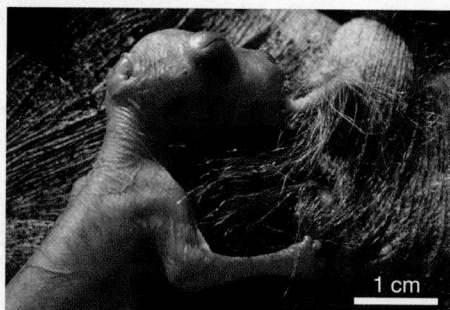 

1 cm   1 cm   1 cm

FIGURE 48.16 Marsupials Trade a Long Gestation Period for a Long Lactation Period. Relative to eutherians, marsupial offspring spend a short time inside the mother and a relatively long time being fed milk after birth and being protected in the mother's pouch.

the corpus luteum from degenerating. When hCG is present, the ovary continues secreting progesterone and the menstrual cycle is arrested. Enough hCG is produced by the embryo and excreted in the mother's urine to be used as an indicator in pregnancy tests.

THE FIRST TRIMESTER Human gestation is divided into 3-month stages called trimesters. Not long after implantation is complete, mass movements of cells result in the formation of the three major embryonic tissues called ectoderm, endoderm, and mesoderm (see Chapter 22). By 8 weeks of age, these tissues have differentiated into the various organs and systems of the body. Also by this time, the heart has begun pumping blood through a circulatory system. The embryo at this stage is called a **fetus** (**Figure 48.17a**).

Early in development, the embryonic ectoderm contributes to several important membranes. One of these membranes, the **amnion**, completely surrounds the embryo. The amnion eventually fills with amniotic fluid, which provides the embryo with a protective cushion.

The other key event in the first trimester is the formation of the **placenta**. This organ, which starts to form on the uterine wall a few weeks after implantation, is composed of tissues from both mother and embryo (see Chapter 34). Because the placenta contains a dense supply of blood vessels from the mother, it is the primary source of nutrition for the growing fetus. Arteries transport blood from the circulatory system of the fetus, through the **umbilical cord**, to an extensive capillary bed in the placenta. In this way, the placenta provides a large surface area for the exchange of gases, nutrients, and wastes between maternal and fetal blood, even though maternal and fetal blood do not commingle.

The placenta secretes a variety of hormones, including large amounts of progesterone and estrogens. Because these hormones suppress the release of GnRH, LH, and FSH through negative feedback, they prevent the maturation and ovulation of additional follicles. By the end of the first trimester, the placenta is producing more than enough progesterone to replace the amount that had been produced by the corpus luteum, which has degenerated by this time. In essence, the placenta takes over from the corpus luteum in terms of secreting the hormones required to maintain the pregnancy.

✔If you understand how hormones influence pregnancy, you should be able to explain (**1**) why women who produce low levels of progesterone from the corpus luteum are prone to miscarriage, and (**2**) why progesterone therapy to help these women can be safely discontinued after the first trimester.

THE SECOND AND THIRD TRIMESTERS After the fetal organs and placenta form during the first trimester, the remainder of development focuses on growth (**Figures 48.17b** and **48.17c**). During the last weeks of pregnancy, the brain and lungs undergo particularly dramatic growth and development. If a baby is born prematurely, intervention may be required to keep the baby alive until the lungs can complete their development.

The machinery and level of hospital care required by premature infants emphasize just how superbly adapted mothers are to nourishing a growing fetus in the uterus. It costs hundreds of thousands of dollars for health care providers to do what mothers do naturally in the last trimester. Let's take a closer look at this critical aspect of pregnancy.

How Does the Mother Nourish the Fetus?

In oviparous and ovoviviparous species, mothers produce relatively large eggs that contain all of the nutrients and fluids that the embryo needs to develop until hatching. But in some viviparous species, eggs are relatively small and contain almost no nutrients.

(a) 1st trimester

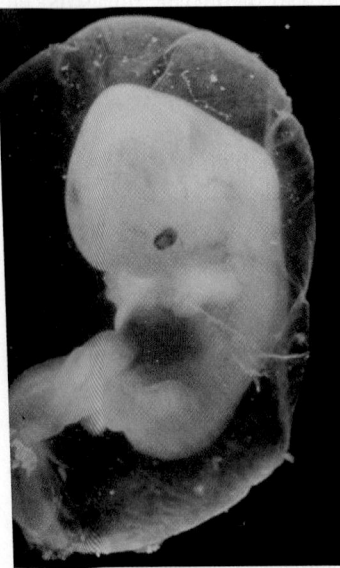

(b) 2nd trimester

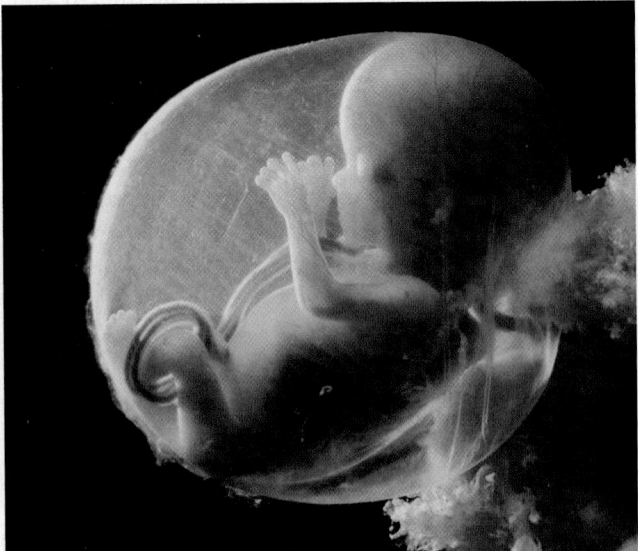

(c) 3rd trimester

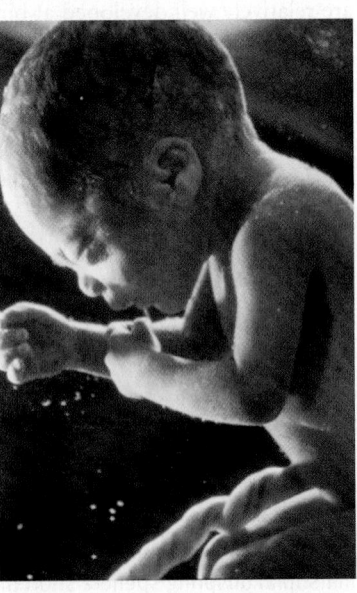

FIGURE 48.17 Development of the Human Fetus.

In species such as humans, the developing embryo depends on the mother's body for oxygen, chemical energy in the form of sugars, amino acids and other raw materials for growth, and waste removal. What physiological changes occur in human mothers to accommodate these demands?

OXYGEN EXCHANGE BETWEEN MOTHER AND FETUS During pregnancy, a mother's respiratory and circulatory systems change in ways that increase the efficiency of nutrient transfer and gas exchange with the fetus. For example:

- A woman's total blood volume expands by as much as 50 percent during pregnancy.

- To accommodate the increase in blood volume, maternal blood vessels dilate (widen) and blood pressure drops.

- The mother's heart enlarges and beats faster, increasing her total cardiac output by almost 50 percent.

- The mother's breathing rate and breathing volume increase to accommodate the fetus's demand for oxygen and its production of carbon dioxide.

In addition, important adaptations heighten the efficiency of gas exchange between the mother and the embryo, in the placenta. In many species, maternal and fetal blood flows in the placenta occur in a countercurrent fashion (**Figure 48.18a**). As Chapter 41 explained, countercurrent flows maintain a concentration gradient that increases the efficiency of diffusion or other types of exchange. The data for partial pressures of oxygen in Figure 48.18a are from experiments on sheep.

Countercurrent flow does not occur in the human placenta, so fetal blood does not become as highly oxygenated as sheep blood. Still, oxygen exchange between mother and fetus is efficient. For example, maternal arteries in humans empty into a space at the junction of the maternal and fetal portions of the placenta. This space is packed with small projections called villi, which contain the fetal blood vessels. Thus, a large surface area from the fetus is bathed with highly oxygenated maternal blood. The fetal villi are analogous to the alveoli of the lungs (Chapter 44), which provide a large surface area for gas exchange.

Figure 48.18b illustrates another adaptation that increases the rate of oxygen delivery to a human fetus. Notice that the axes on this graph give the partial pressure of oxygen in blood or tissue versus the percentage of hemoglobin that holds oxygen at that pressure. (Recall from Chapter 44 that this type of graph is an **oxygen-hemoglobin equilibrium curve**.)

The key point on this graph is that the data for fetal hemoglobin are shifted to the left of the data for adult hemoglobin. As a result, the fetus's blood always has a higher affinity for oxygen than does the mother's blood. This means that even if the partial pressure of oxygen is the same in the mother and fetus, oxygen will move from the mother's blood to the fetus's blood, because it is held more tightly by the fetal hemoglobin.

Biologists interpret this pattern as an adaptation. The high oxygen affinity of fetal hemoglobin ensures that the fetus is always able to acquire oxygen from the mother.

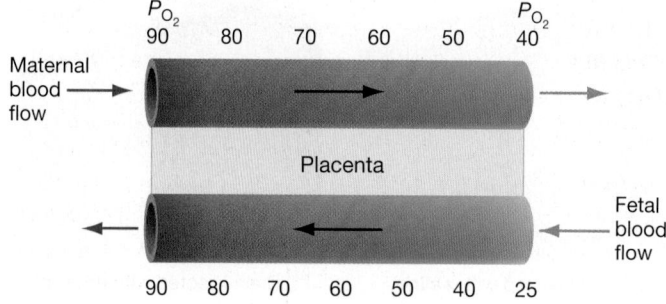

(a) Countercurrent blood flow in the sheep placenta

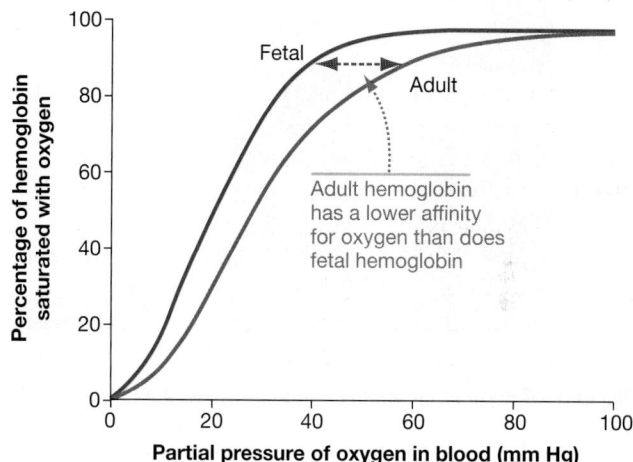

(b) Oxygen-hemoglobin saturation curves in the human fetus and mother

Adult hemoglobin has a lower affinity for oxygen than does fetal hemoglobin

FIGURE 48.18 Adaptations That Increase Delivery of Oxygen to a Mammalian Fetus.

TOXIC CHEMICALS CAN BE TRANSFERRED FROM MOTHER TO FETUS Mothers and embryos exchange more than nutrients and wastes—they can also exchange dangerous molecules. As an example, consider the thalidomide tragedy. During the 1950s, hundreds of children were affected by their mothers' consumption of the tranquilizer thalidomide. The molecule diffused into the fetal bloodstream and caused birth defects—often a dramatic shortening of the arms.

Although thalidomide is now banned for use by pregnant women, alcohol use continues to affect newborns. Compared to children of mothers who do not drink alcohol, children of mothers who imbibe ethanol are at high risk for hyperactivity, severe learning disabilities, and depression. Collectively, these symptoms are termed **fetal alcohol syndrome (FAS)**.

To investigate why FAS occurs, researchers simulated what happens when a pregnant mother ingests alcoholic beverages—exposing offspring to ethanol. They gave one group of newborn (7-day old) rats two injections of a harmless salt solution at a dosage of 2.5 g per kg body weight, and a second group identical injections except that the salt solution contained 20 percent ethanol. For a female human of average weight, this would be equivalent to drinking 2.5 cans of beer or glasses of wine.

(a) Brain sections from newborn rats

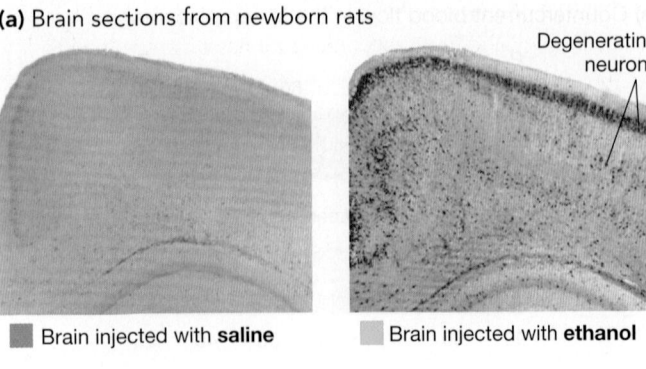

Degenerating neurons

Brain injected with **saline**   Brain injected with **ethanol**

(b) Weights of brain tissue from newborn rats

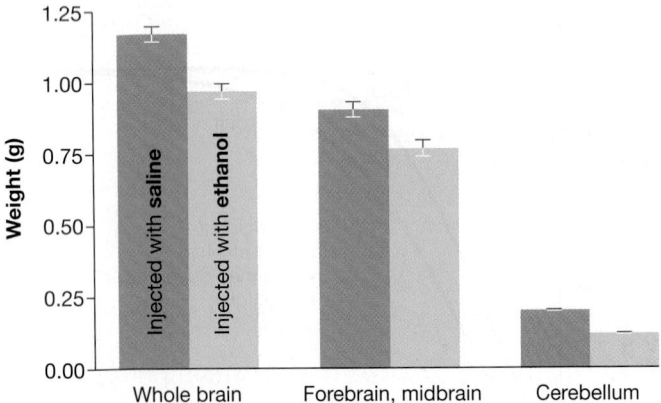

FIGURE 48.19 **Ethanol Kills Brain Cells in Newborn Rats.**

The photos of brain sections in **Figure 48.19a** were taken a day after the injections. The black specks in the brains of the ethanol-treated rats indicate degenerating neurons. As the bar graphs in **Figure 48.19b** show, the ethanol-treated rats had smaller brains than the saline-treated individuals did.

The messages of these data are that (**1**) ethanol destroys growing neurons, and (**2**) even a single, "moderate" dose of alcohol can have a devastating effect. Based on results like these, public health officials strongly advise pregnant women against drinking *any* alcohol.

Birth

Although the mechanisms responsible for triggering the birthing process are not completely understood, the posterior pituitary hormone **oxytocin** is important in stimulating smooth-muscle cells in the uterine wall to begin contractions. The contractions that expel the fetus from the uterus constitute **labor.**

When the birthing process starts, the uterus contracts at relatively low frequency. Then, as **Figure 48.20** shows:

1. The cervix at the base of the uterus begins to open, or dilate. Once the cervix is fully dilated, uterine contractions become more forceful, longer, and more frequent.

2. The fetus is expelled through the cervix and into the vagina.

3. After the baby is delivered, the placenta remains attached to the uterine wall. At this point, caregivers clamp and cut the

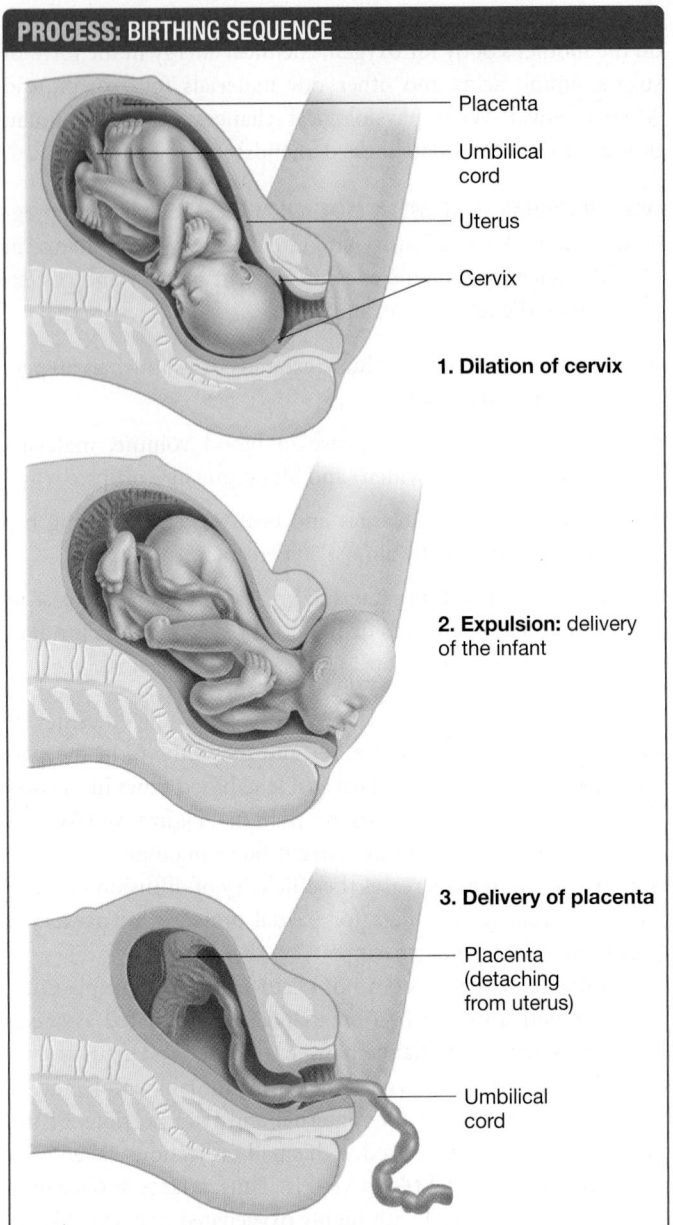

PROCESS: BIRTHING SEQUENCE

Placenta
Umbilical cord
Uterus
Cervix

1. Dilation of cervix

2. Expulsion: delivery of the infant

3. Delivery of placenta

Placenta (detaching from uterus)

Umbilical cord

FIGURE 48.20 **Major Events during Human Birth.**

✔QUESTION Human babies are normally born headfirst and face down, as shown. Predict the consequences of a baby positioned face up, or feetfirst, in the birth canal.

umbilical cord that connects the child and the placenta. When the mother delivers the placenta and accompanying membranes, birth is complete.

Although this description sounds straightforward, in reality a large number of complications are possible. For example, **Figure 48.21** shows the number of Swedish mothers who died, per 100,000 live births, in five-year intervals between 1760 and 1980. In 1760, approximately 1.4 mothers died for every 100 infants successfully delivered. In most cases, the cause of death was blood loss or infection following delivery. Note that the steepest drop in mortality rate occurred with the introduction of hand-washing in the late 1870s.

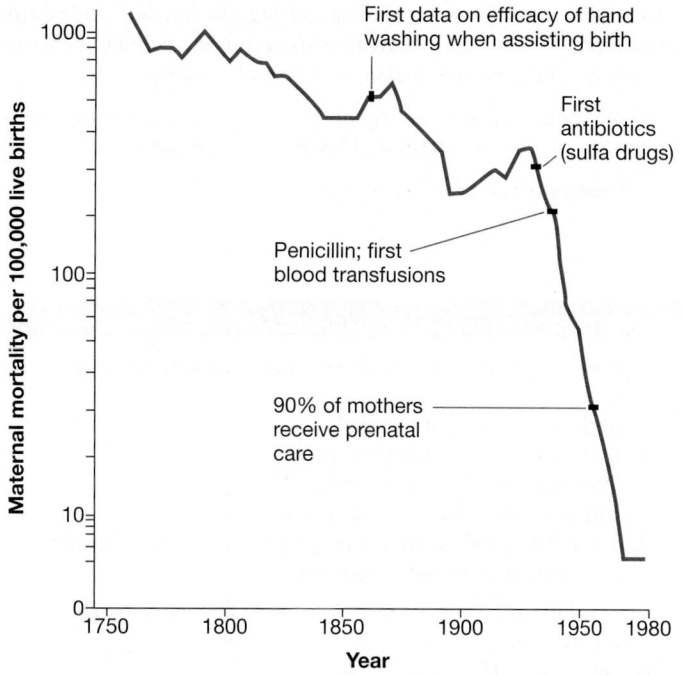

First data on efficacy of hand washing when assisting birth

First antibiotics (sulfa drugs)

Penicillin; first blood transfusions

90% of mothers receive prenatal care

As a result of sterile techniques, antibiotics, and blood transfusion technology, Sweden's mortality rate has now declined to less than 0.007 percent. Improved nutrition, sanitation, and medical care have also reduced infant mortality rates in many countries.

The huge decline in the rate of death associated with childbirth qualifies as one of the great triumphs of modern medicine. Unfortunately, because many developing nations lack sterile facilities and antibiotics, the mortality of mothers and infants remains high in those countries.

FIGURE 48.21 Maternal Mortality in Childbirth Has Decreased Dramatically. These data are from Sweden; the scale on the vertical axis is logarithmic. *Note:* Data showing that hand-washing dramatically reduced maternal mortality during childbirth were published in 1861, but doctors resisted the idea until well after Pasteur had published the germ theory of disease in 1878.

✔**QUESTION** If a woman living in Sweden in 1760 had 10 children over the course of her lifetime, what was her overall chance of dying in childbirth?

CHAPTER 48 REVIEW

For media, go to the study area at www.masteringbiology.com

Summary of Key Concepts

 The reproductive systems of animals are highly variable. Some species switch between asexual and sexual reproduction. When sexual reproduction occurs, fertilization may be external or internal and egg development may take place inside or outside the mother's body, depending on the species.

- Asexual reproduction produces offspring that are genetically identical to the parent. In contrast, sexually produced offspring are genetically unique because of recombination during meiosis and the fusion of haploid gametes from different parents during fertilization.

- In human males, spermatogenesis is continuous throughout adult life, but in human females all primary oocytes are formed early in development. Meiosis in females is arrested for long periods of time, and cell division after meiosis is so unequal in females that just one egg—not four—is produced from each primary oocyte.

- Fertilization is external in many aquatic animals but internal in almost all terrestrial species.

- When sperm competition occurs, males have large testes relative to their body size, and the last male to mate usually fathers a disproportionately large number of offspring.

- Once eggs are fertilized, females may lay the eggs or retain them and give birth to live offspring. In some groups at least, viviparity may be an adaptation that increases the survival of young in cold habitats.

✔ You should be able to explain why fish that care for their young are expected to produce fewer eggs per year than fish that do not care for their young, and also whether sperm competition is common in fish.

Web Activity Human Gametogenesis

In humans, the male reproductive system includes structures specialized for producing and storing sperm, synthesizing other components of semen, or transporting and delivering semen. The female reproductive system includes structures specialized for producing eggs, receiving sperm, and nourishing offspring during early development.

- In humans, the external anatomy of the male reproductive tract consists of the penis and scrotum.

- The external anatomy of the human female reproductive tract consists of the vaginal opening and clitoris.

  ✔ You should be able to predict how surgeons alter the male and female reproductive tracts of human volunteers to sterilize them.

In humans, hormones from the pituitary gland and female reproductive organs regulate the menstrual cycle. These hormones interact via positive or negative feedback. Pregnancy is maintained by hormonal signals from the embryo and the mother's reproductive organs.

- In mammals, GnRH from the hypothalamus triggers the release of FSH and LH from the pituitary gland. The pituitary hormones regulate the production of the gonadal hormones, testosterone and estradiol, in the testes and ovaries, respectively.

- During the human menstrual cycle, estradiol and progesterone exert negative and positive feedback on the production of FSH and LH. Interactions between the pituitary and ovarian hormones are responsible for regulating cyclical changes in the ovaries and uterus.

- If fertilization occurs, the developing embryo secretes the hormone hCG, which arrests the menstrual cycle and allows pregnancy to continue.

- During the first trimester, the embryo becomes implanted in the thickened uterine wall, organs develop, and the nutritive organ called the placenta forms.

- To make rapid growth possible during the second and third trimesters, the mother's heart rate and pumping volume increase. Nutrients and gases are exchanged efficiently in the placenta.

  ✔ You should be able to explain why administration of progesterone is an effective method of birth control in humans.

 Web Activity Human Reproduction

Questions

✔ TEST YOUR KNOWLEDGE

Answers are available in Appendix B

1. What term describes the mode of asexual reproduction in which offspring develop from unfertilized eggs?
   a. parthenogenesis
   b. budding
   c. regeneration
   d. fission

2. In sperm competition, what is "second-male advantage"?
   a. the observation that when females mate with two males, each male fertilizes the same number of eggs
   b. the observation that when females mate with two males, the second male fertilizes most of the eggs
   c. the observation that females routinely mate with at least two males before laying eggs or becoming pregnant
   d. the observation that accessory fluids prevent matings by second males—for example, by forming copulatory plugs

3. How are the human penis and clitoris similar?
   a. They develop from the same population of embryonic cells.
   b. Both develop during the earliest stages of puberty.
   c. Both contain the urethra.
   d. Both produce accessory fluids required during sexual intercourse.

4. In hormone production, how do the follicle and corpus luteum compare?
   a. Both produce primarily estradiol.
   b. Both produce primarily progesterone.
   c. The follicle produces more estradiol than progesterone; the corpus luteum produces more progesterone than estradiol.
   d. The follicle produces mostly progesterone; the corpus luteum produces estradiol and progesterone.

5. What pituitary hormones are involved in regulating the human menstrual cycle?
   a. GnRH and LH
   b. estradiol and progesterone
   c. oxytocin and LH
   d. FSH and LH

6. The corpus luteum is retained upon fertilization due to the presence of which hormone?
   a. LH
   b. estradiol
   c. progesterone
   d. human chorionic gonadotropin (hCG)

✔ TEST YOUR UNDERSTANDING

Answers are available in Appendix B

1. What is the fundamental difference between sexual and asexual reproduction, in terms of the characteristics of offspring?

2. Summarize the experimental evidence that *Daphnia* require three cues to trigger sexual reproduction. Discuss what these cues indicate about the environment. Generate a hypothesis for why sexual reproduction is adaptive for these animals.

3. Compare and contrast spermatogenesis with oogenesis in humans. How do these processes differ with respect to numbers of cells produced, gamete size, and timing of the second meiotic division?

4. Give examples of negative feedback and positive feedback in hormonal control of the human menstrual cycle. Why can high estradiol levels be considered a "readiness" signal from a follicle?

5. Why are pregnant mothers advised to refrain from drinking alcoholic beverages?

6. Suppose that females of a certain insect species routinely mate with two males and choose their mates based on courtship displays or other traits. Predict whether females would be choosier about the first male or the second male that they mate with, or equally choosy about both males. Explain the logic behind your prediction.

✔ APPLYING CONCEPTS TO NEW SITUATIONS

Answers are available in Appendix B

1. Researchers have recently developed methods for cloning mammals. In effect, biologists can now induce asexual reproduction in species that do not normally reproduce asexually. Suppose this practice becomes so widespread in the future that most of the sheep in the world become genetically identical. Discuss some possible consequences of this development.

2. The text claims that species with external fertilization produce extraordinarily large numbers of gametes. How would you test this hypothesis rigorously? In answering, assume that you have data on the average number of gametes produced by different species, along with data on their average body size.

3. Suppose that you've been given a chance to study several populations of *Sceloporus* lizards—some of which are viviparous, and others of which are oviparous. Which populations would you expect to produce especially large eggs, relative to their overall body size? How would you go about testing this prediction?

4. When marsupial eggs are developing, a shell membrane appears for a short time and then disintegrates. State a hypothesis to explain this observation.

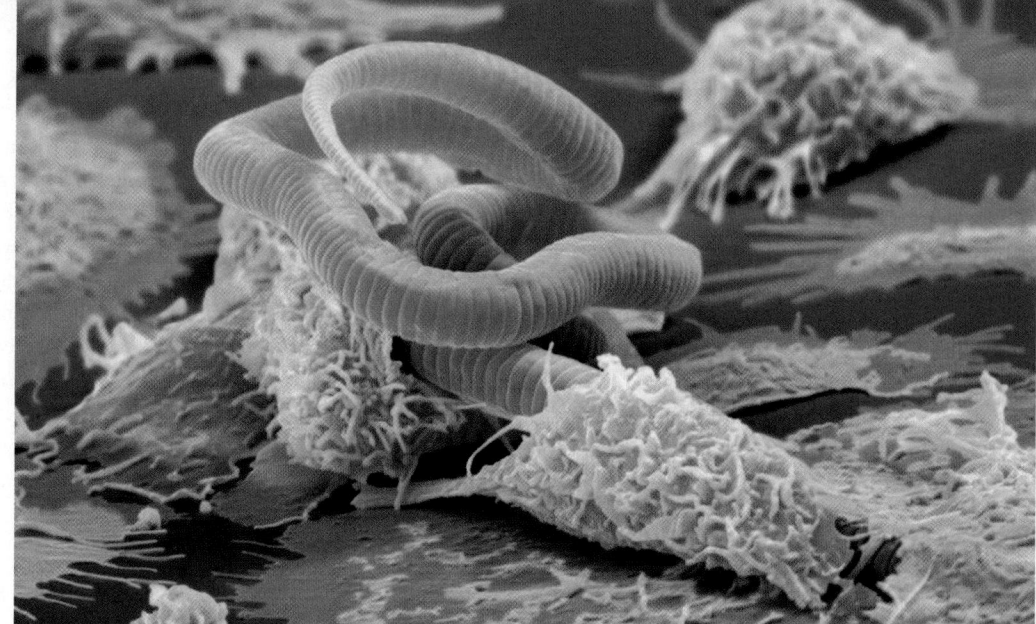

Colorized scanning electron micrograph of human immune system cells attacking *Wuchereria bancrofti*—a parasitic nematode that clogs lymph vessels and causes the dramatic swelling called elephantiasis. This chapter explores how immune system cells are able to recognize parasitic worms, bacteria, and viruses as foreign and eliminate them.

The Immune System in Animals 49

Disease threatens every animal. Humans, for example, are victimized by thousands of different disease-causing bacteria and viruses and hundreds of parasitic worms, fungi, and protists.

Given the ability of these pathogens to cause illness and death, it is remarkable that so many animals stay healthy for most of their lives. To understand how, biologists have investigated three observations:

1. Wounds usually heal, even if they become infected.

2. Most people who contract a bacterial or viral illness eventually recover, even without help from medications.

3. People who acquire bacterial or viral infections and recover are frequently immune (literally, "exempt") to that disease—that is, they do not contract the same disease in the future. **Immunity** is a resistance to or protection against disease-causing pathogens.

This last observation is particularly interesting. When plague struck Athens in 430 B.C., Thucydides noted that only people who had recovered from the illness could nurse the sick, because they did not become ill a second time. In the middle ages, Chinese and Turkish practitioners protected people from smallpox by intentionally exposing them to the dried crusts of smallpox pustules taken from infected individuals. These are examples of **immunization**—the conferring of immunity to a particular disease.

In the late 1700s Edward Jenner refined this immunization technique. In Jenner's day, milkmaids were considered pretty because their faces were not pockmarked with scars from smallpox infections. Jenner knew that cows suffered from a smallpox-like disease called cowpox. He reasoned that milkmaids became immune to smallpox because they had been exposed to cowpox while milking cows.

To test this hypothesis, Jenner inoculated a boy with fluid from a cowpox pustule. Later he inoculated the same child with fluid from a smallpox pustule. That is, he

KEY CONCEPTS

○➝ The innate immune response to infection is mounted by leukocytes that respond in a nonspecific way to pathogens. In contrast, the adaptive immune response is mounted by lymphocytes that respond to specific antigens—often a protein from a pathogen.

○➝ The adaptive immune response begins when certain lymphocytes are activated in a regulated, step-by-step process triggered by interaction with a specific antigen.

○➝ Activated lymphocytes either destroy infected cells or produce antibodies that tag pathogens for destruction.

✔When you see this checkmark, stop and test yourself. Answers are available in Appendix B.

inoculated the boy with cowpox pathogens, and then with smallpox pathogens.

As predicted, the boy did not contract smallpox. Jenner's technique was named **vaccination**. (The Latin root *vacca* means "cow.") Vaccination is the introduction of a weakened or altered pathogen to prime the body's immune system, so it fights later infections effectively.

Why does vaccination work? What molecules and cells allow us to fight off a bacterial or viral infection and achieve immunity?

To answer these questions, we need to explore how the human immune system recognizes and eliminates pathogens. Suppose, for example, that you were on a crowded bus several days ago next to someone with a persistent cough and runny nose, and 30 minutes later you stumbled while sprinting across campus and skinned your elbow. Now you realize that the wound on your arm has gotten sore and red. Worse yet, you are developing a sore throat, runny nose, and fever. Your elbow has become infected with bacteria, and your upper respiratory tract is supporting a growing population of influenza, or "flu," virus. Let's start by analyzing what is happening in the wound at your elbow.

49.1 Innate Immunity

When biologists began analyzing the immune system, they realized that certain immune system cells are ready to respond to invaders at all times; other components must be activated first.

• Cells that are always ready confer **innate immunity**.

• Cells that are selectively activated to eliminate a specific pathogen confer **adaptive immunity**.

The key to understanding the two types of immunity is to recognize that the cells involved provide different responses to antigens. An **antigen** is any foreign molecule that can initiate an immune system response. Most antigens are proteins or glycoproteins from bacteria, viruses, or other invaders, but foreign carbohydrates, nucleic acids, and lipids can also function as antigens.

The cells involved in innate immunity are nonspecific in their response to antigens: They respond in the same way to all antigens. In contrast, cells involved in adaptive immunity respond in an extremely specific way to each particular strain of bacterium, virus, or fungus that enters the body.

In combination, the innate and adaptive immune responses form a powerful system for protecting individuals against a formidable and ever-changing array of parasites.

To launch this investigation into immune system function, let's first focus on how the body prevents entry by foreign invaders. Then we can consider what happens if some do get in.

Barriers to Entry

The most effective way to avoid getting sick is to avoid contact with pathogens. In humans and many other animals, the most important barrier to pathogen entry is the skin.

In addition to providing a tough physical barrier, human skin offers a chemical deterrent. The oil secreted by skin cells is converted to fatty acids by bacterial cells that live on the surface. These bacteria are non-pathogenic. The fatty acids lower the pH of the skin surface to about 5, creating a dry, acidic environment that prevents the growth of most bacterial species that might be a threat to the body.

The bodies of insects and other animals with exoskeletons are also difficult for pathogens to enter because they are covered with a tough layer called the cuticle, along with a layer of wax (see Chapter 42).

HOW ARE OPENINGS IN THE BODY PROTECTED? An animal's body has gaps in the surface where the digestive tract, reproductive tract, gas-exchange surfaces, and sensory organs make contact with the environment. As **Figure 49.1** shows, these types of gaps are protected by mucus or other features that discourage pathogen entry.

Airways (lining of trachea)
Most pathogens are trapped in mucus before they can reach the lungs. Beating cilia sweep pathogens up and out of the airway.

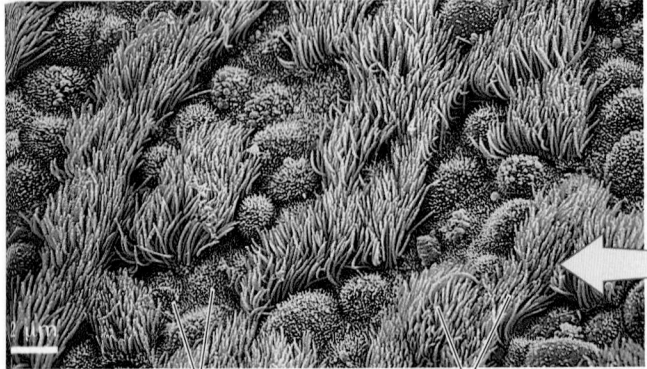

Mucus-secreting cells   Ciliated cells

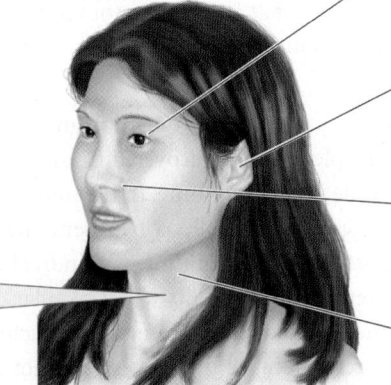

Eyes
Blinking wipes tears across the eye. Tears contain the antibacterial enzyme lysozyme.

Ears
Hairs and earwax trap pathogens in the passageway of the external ear.

Nose
The nasal passages are lined with mucus secretions and hairs that trap pathogens.

Digestive tract
Pathogens are trapped in saliva and mucus, then swallowed. Most are destroyed by the low pH of the stomach.

FIGURE 49.1 How Does the Body Keep Pathogens Out? The scanning electron micrograph of mucus-secreting and ciliated cells has been colorized.

/QUESTION Why is a combination of cilia and sticky secretions effective at trapping pathogens?

Mucus is a solution secreted by cells within the epithelium. It is rich in proteoglycans—molecules that consist of large polysaccharides bonded to proteins. The bodies of slugs, snails, earthworms, and other soft-bodied animals are covered with a layer of mucus that shields their epithelial cells from pathogens in the environment.

Mucus is equally important in vertebrates. For example, many of the pathogens that you breathe in or ingest while eating or drinking stick to the mucus that lines your airways and digestive tract. Pathogens that are stuck in mucus cannot come in contact with the plasma membranes of epithelial tissues. In mammals, many of these pathogens are either coughed out or swallowed and killed in the acidic environment of the stomach. In the respiratory tract, they are swept out of harm's way by the beating of cilia (see Figure 49.1).

Gaps in the body that are not covered with mucous layers, such as the eyes, are often protected by other types of secretions. Your ears are protected by waxy secretions, and your eyes by tears that contain the enzyme **lysozyme**. Lysozyme acts as an antibiotic by digesting bacterial cell walls.

HOW DO PATHOGENS GAIN ENTRY? With regularity, preventive measures fail and pathogens gain entry to tissues beneath the skin. Flu viruses, for example, have an enzyme on their surface that disrupts the mucous lining of the respiratory tract. When the outer surface of the virus makes contact with a host cell beneath the mucous layer, the virus is able to enter the cell and begin an infection.

Falls, wounds, and other types of physical trauma provide another important mode of entry. When the skin is broken, bacteria and other pathogens gain direct access to the tissues inside.

To viruses, bacteria, and fungi, your body is a tropical paradise, brimming with resources. Within hours, they can begin growing. What happens then?

The Innate Immune Response

When bacteria enter your body at the site of a wound, the cells pictured in **Figure 49.2** respond by implementing the **innate immune response**—the body's nonspecific response to pathogens. As a group, these cells are called white blood cells to distinguish them from red blood cells. More formally, they are known as **leukocytes** ("white-cells").

Leukocytes include macrophages, neutrophils, and other cells involved in the innate response as well as the lymphocytes responsible for adaptive immunity. Like the red blood cells that carry oxygen in the bloodstream, leukocytes are produced in the bone marrow.

The leukocytes involved in the innate response are alerted to the presence of foreign invaders by molecules that are found on the surfaces of pathogens, but not on host cells. Some of the cells in the innate system have proteins on their plasma membranes that bind to these pathogen-specific compounds. When these proteins, known as **pattern-recognition receptors**, are activated by molecules from pathogens, the cells respond. Pattern-recognition receptors are an "on" switch.

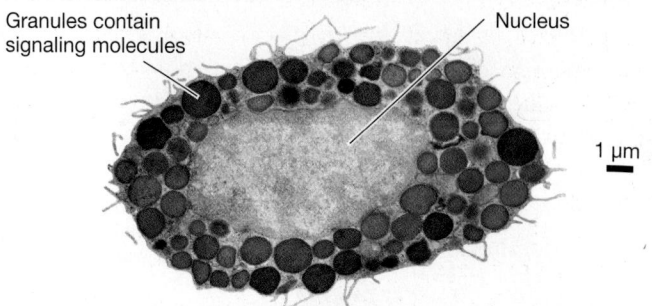

(a) Mast cells secrete signals that increase blood flow.

Granules contain signaling molecules

Nucleus

1 μm

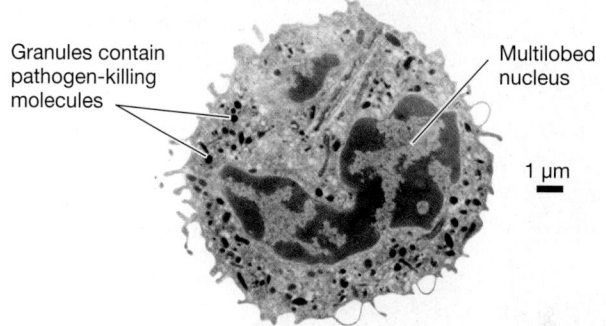

(b) Neutrophils ingest and kill pathogens.

Granules contain pathogen-killing molecules

Multilobed nucleus

1 μm

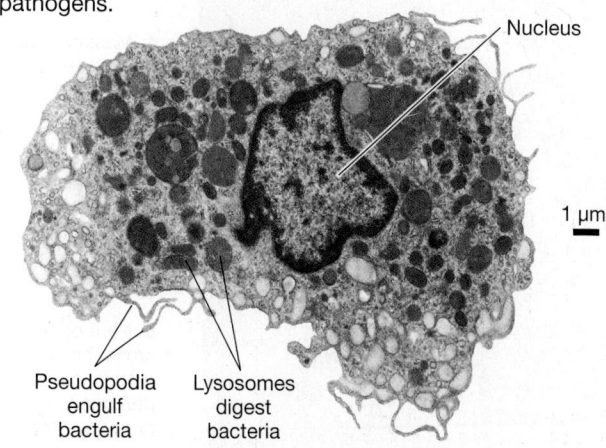

(c) Macrophages recruit other cells and ingest and kill pathogens.

Nucleus

1 μm

Pseudopodia engulf bacteria

Lysosomes digest bacteria

FIGURE 49.2 Cells of the Innate Immune Response. Transmission electron micrographs illustrating a few of the leukocytes responsible for innate immunity. Colors refer to Figure 49.3.

THE INFLAMMATORY RESPONSE IN HUMANS **Figure 49.3** on page 976 summarizes the major steps in the **inflammatory** ("inflames") **response**, a multistep innate immune response observed in an array of animals.

Step 1 A break in the skin allows bacteria to enter the body. If capillaries and other small blood vessels are broken by the wound, blood leaves.

Step 2 Blood components called **platelets** release proteins that form clots and lessen bleeding. Other clotting proteins in the blood form cross-linked structures that help wall off the wound and reduce blood loss.

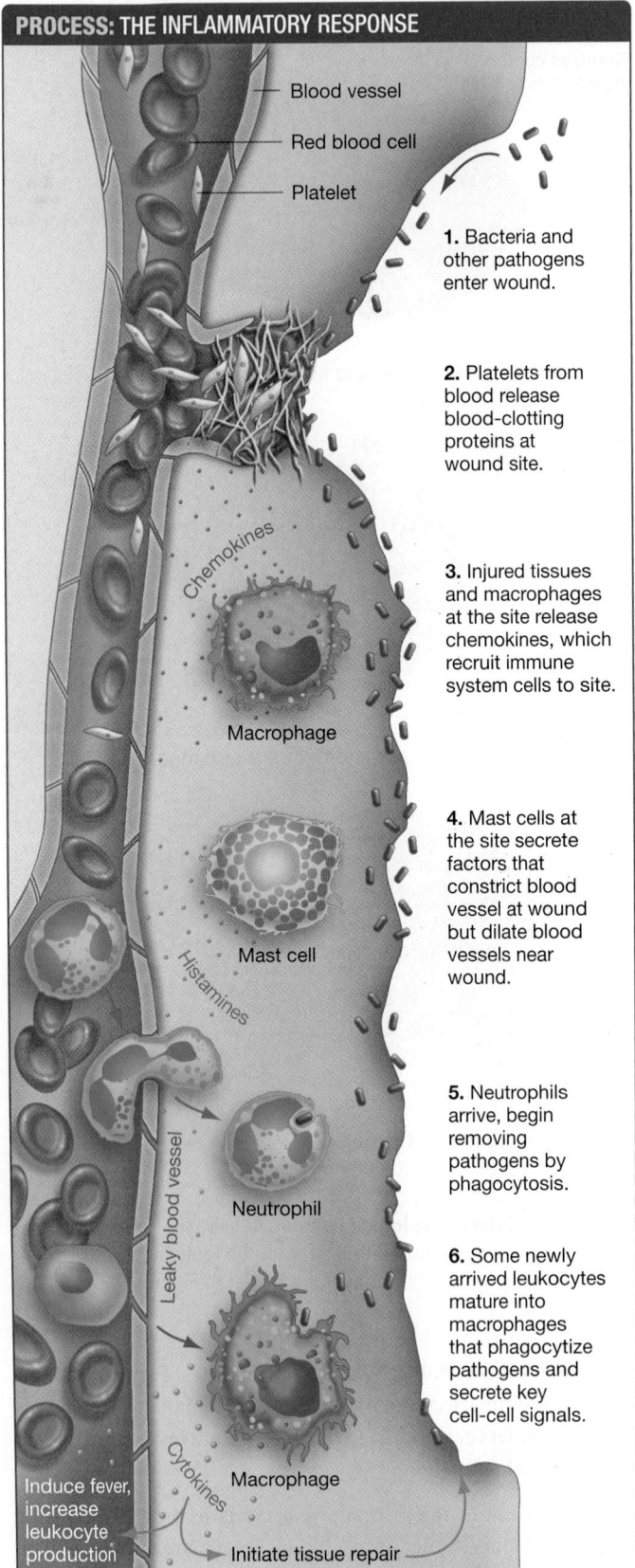

PROCESS: THE INFLAMMATORY RESPONSE

Blood vessel

Red blood cell

Platelet

1. Bacteria and other pathogens enter wound.

2. Platelets from blood release blood-clotting proteins at wound site.

Chemokines

Macrophage

3. Injured tissues and macrophages at the site release chemokines, which recruit immune system cells to site.

Mast cell

Histamines

4. Mast cells at the site secrete factors that constrict blood vessel at wound but dilate blood vessels near wound.

Leaky blood vessel

Neutrophil

5. Neutrophils arrive, begin removing pathogens by phagocytosis.

6. Some newly arrived leukocytes mature into macrophages that phagocytize pathogens and secrete key cell-cell signals.

Cytokines

Macrophage

Induce fever, increase leukocyte production

Initiate tissue repair

FIGURE 49.3 The Inflammatory Response of Innate Immunity Has Many Elements.

✔ **QUESTION** Only one blood vessel is shown, for simplicity. If others were added, where would they be constricted and where would they be dilated? Explain why this is important.

Step 3 Wounded tissues and leukocytes called **macrophages** secrete signaling molecules called **chemokines** ("chemical-movers"). Chemokines are important because they form a gradient that marks a path to the wound site.

Step 4 Mast cells release chemical messengers that constrict blood vessels near the wound—reducing blood flow and thus blood loss. Mast cells also release **histamine** and other signaling molecules that induce blood vessels slightly farther from the wound to dilate and become more permeable.

Step 5 The combination of dilated blood vessels and a chemokine gradient is like a 911 call that provides specific directions to the scene of a fire. Leukocytes called neutrophils move out of dilated blood vessels and migrate to the site of the infection. **Neutrophils** are major players in the innate response. They destroy invading cells by phagocytosis ("eating-cell")— meaning that they engulf them. Once invading cells are inside neutrophils, they are killed with a complex array of toxic compounds. These molecules include lysozyme—which degrades bacterial cell walls—free radicals, nitric oxide (NO), and molecules called reactive oxygen intermediates (ROIs), including hydrogen peroxide (H_2O_2).

Step 6 Cells that will mature into macrophages arrive. In addition to phagocytizing bacteria at the wound, the growing population of macrophages secretes chemical messengers called **cytokines** ("cell-movers") that attract other immune system cells to the site, stimulate bone marrow to make and release additional neutrophils and macrophages, induce fever—an elevated body temperature that aids in healing—and activate cells involved in tissue repair and wound healing.

This overview actually simplifies the situation—there are many other cell types and cell-cell signals involved in responding to bacteria and viruses at a site of infection.

The site of inflammation often becomes swollen due to increased numbers of cells and fluids in the area, red and warm due to increased blood flow, and painful due to signals from pain receptors.

The inflammatory response continues until all foreign material is eliminated and the wound is repaired. **Table 49.1a** summarizes key cells in the response, while **Table 49.1b** summarizes some of the key molecules involved. To complete your review of the inflammatory response, you should also go to the study area at *www.masteringbiology.com.*

(MB) **Web Activity** The Inflammatory Response

INNATE IMMUNITY IN INVERTEBRATES Humans and other vertebrates are not the only animals with sophisticated innate immune responses. Innate responses make up the entire immune system observed in millions of species of invertebrates.

The spectacular abundance and diversity of invertebrate animals support the hypothesis that their innate immune systems provide efficient protection against invading bacteria, viruses, and fungi. For example:

(a) Key Cells

Name	Primary Function
Mast cells	Release signals that increase blood flow to wound site
Neutrophils	Kill invading cells via phagocytosis
Macrophages	Release cytokines that recruit other cells to wound site; kill invading cells via phagocytosis

(b) Key Signaling Molecules

Name	Produced By	Received By	Message/Function
Histamine	Mast cells	Blood vessels	High concentration next to wound constricts blood vessels; low concentration farther from wound dilates blood vessels
Chemokines*	Injured tissues and macrophages in tissues	Neutrophils and macrophages	Mark path to wound; promote dilation and increased permeability of blood vessels
Cytokines other than chemokines	Macrophages	Leukocytes	Mark path to wound
		Bone marrow	Increase production of macrophages and neutrophils
		CNS	Induce fever by increasing set point for control of body temperature
		Local tissues	Stimulate cells involved in wound repair

*Note that chemokines are a subset of cytokines.

- If a pathogen succeeds in entering the main body cavity of an insect through a wound or other break in the barriers to entry, cells respond by synthesizing and secreting peptides with potent antibacterial or antifungal properties.

- Sea stars have specialized cells that are similar in function to neutrophils and macrophages—they secrete cytokines or engulf and destroy pathogens by phagocytosis.

But in general, biologists still have a great deal to learn about defense systems in invertebrates.

What happens when the innate immune system of a vertebrate fails to contain and eliminate invading pathogens?

CHECK YOUR UNDERSTANDING

If you understand that . . .

- The innate immune response occurs when macrophages and mast cells that reside in tissues, and neutrophils that circulate in the blood, react in a nonspecific way to signals from invading pathogens.

✓ **You should be able to . . .**

Describe which events in the innate response mimic steps that first-aid workers use when initially treating a wound:
(1) applying direct pressure to close blood vessels, and
(2) applying bandages containing compounds that halt blood flow from the wound.

Answers are available in Appendix B.

49.2 The Adaptive Immune Response: Recognition

In vertebrates, the adaptive immune system responds to invading viruses and other pathogens. The **adaptive immune response**, also known as the acquired immune response, is based on interactions between specific immune system cells and a specific antigen.

Given the array of pathogens that exist, an animal is almost certain to be exposed to an enormous variety of antigens in the course of its lifetime. Is there a limit to how many different antigens its adaptive immune system responds to?

Research conducted in the early 1920s answered this question. Researchers synthesized organic compounds that do not exist in nature, injected the novel molecules into rabbits, and observed whether they activated the animals' adaptive immune system. To the amazement of the scientists, the rabbits were able to mount an immune response to antigens that did not exist in nature.

More specifically, the animals produced a different antibody against each antigen. **Antibodies** are proteins that are produced and secreted by certain lymphocytes and that bind to a specific part of a specific antigen. The take-home message was that the immune system can produce an almost limitless array of antibodies.

In combination with insights from Jenner's work on vaccines—reviewed in the preceding section—early work on antibodies focused on four key characteristics of the adaptive immune response:

1. **Specificity** Antibodies and other components of the adaptive immune system bind only to specific sites on specific antigens.

2. *Diversity* The adaptive response recognizes an almost limitless array of antigens.

3. *Memory* The adaptive response can be reactivated quickly if it recognizes antigens from a previous infection.

4. *Self-nonself recognition* Molecules that are produced by the individual do not act as antigens, meaning that the adaptive immune system can distinguish between self and nonself. Nonself molecules are antigens; self molecules are not.

What molecular mechanisms give the adaptive immune system these characteristics? Let's begin by reviewing the cells and organs that are responsible for the adaptive immune response.

An Introduction to Lymphocytes

The cells that carry out the major features of the adaptive immune response are called **lymphocytes**. Lymphocytes originate in the immune system and are subsequently activated and transported inside it. Where does each phase of a lymphocyte's life occur?

STRUCTURES OF THE IMMUNE SYSTEM The colored structures in **Figure 49.4** are the major components of the immune system in humans. As the labels on the figure indicate, each of these structures plays a key role in the life of a lymphocyte.

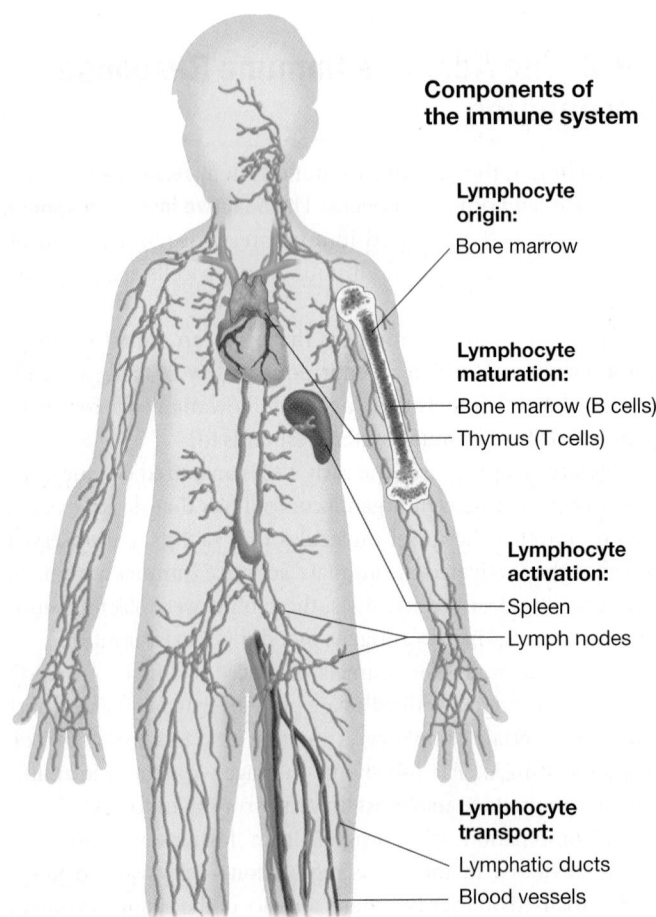

Components of the immune system

Lymphocyte origin:
Bone marrow

Lymphocyte maturation:
Bone marrow (B cells)
Thymus (T cells)

Lymphocyte activation:
Spleen
Lymph nodes

Lymphocyte transport:
Lymphatic ducts
Blood vessels

FIGURE 49.4 Lymphocytes Are Formed, Activated, and Transported in the Immune System.

- *Lymphocyte origin* All lymphocytes and blood cells are produced in **bone marrow**, tissue that fills the internal cavities in bones. In some mammal species, lymphocytes can also originate in the **spleen**—a lymphatic organ located in the abdominal cavity.

- *Lymphocyte maturation* Lymphocytes called B cells mature in the bone marrow where they originate. But lymphocytes called T cells mature in the **thymus**—an organ located in the upper part of the chest of vertebrates (just behind the breastbone in humans).

- *Lymphocyte activation* Lymphocytes recognize antigens and become activated in the spleen and lymph nodes. In addition to its function in the immune response, the spleen destroys old blood cells and stores iron from red blood cells. **Lymph nodes** are small, oval organs that are located all around the body. Lymph nodes filter the lymph passing through them. Recall from Chapter 44 that **lymph** is a mixture of fluid and lymphocytes. The liquid portion of lymph originates in fluid that is forced out of capillaries by blood pressure.

- *Lymphocyte transport* Lymphocytes circulate through the blood and the secondary organs of the immune system—lymph nodes, spleen, and lymphatic ducts. Lymphatic ducts are thin-walled, branching tubules that transport lymph throughout the body in the **lymphatic system**.

In addition to being found in bone marrow, the thymus, spleen, lymph nodes, lymphatic ducts, and blood vessels, large numbers of lymphocytes, along with other leukocytes, are associated with skin cells and with epithelial tissues that secrete mucus—primarily in the digestive tract and respiratory tract. Collectively, the immune system cells found in the gut and respiratory organs are called **mucosal-associated lymphoid tissue (MALT)**. Leukocytes in the skin and MALT are important because they guard points of entry against pathogens.

LYMPHOCYTES MUST BE ACTIVATED Lymphocytes are normally in a resting, or inactive, state as they circulate through the bloodstream and lymphatic system or reside in skin or MALT. As **Figure 49.5a** shows, inactive lymphocytes have a large nucleus, little cytoplasm, few mitochondria, and a ruffled membrane. Over the course of a day, an inactive lymphocyte might migrate through the spleen, enter the blood, cross over into the lymphatic vessels, migrate through a lymph node, return to the blood, and so on.

If an inactive lymphocyte does not encounter the antigen to which it is programmed to respond, the cell eventually dies. But if a lymphocyte does encounter the appropriate antigen, it becomes activated. When activated, the type of lymphocyte called a B cell has a massive amount of rough ER and a large number of mitochondria (**Figure 49.5b**). Recall from Chapter 7 that many of the proteins synthesized in a cell's rough ER are inserted into the plasma membrane or secreted from the cell. The increased rough ER in this activated lymphocyte suggests that it is manufacturing and possibly secreting proteins. In general, activation produces dramatic changes in lymphocytes.

(a) Inactive lymphocyte

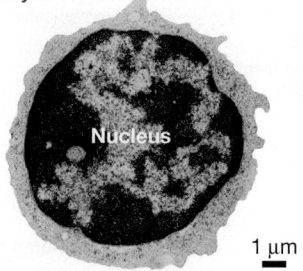

1 μm

(b) Activated lymphocyte

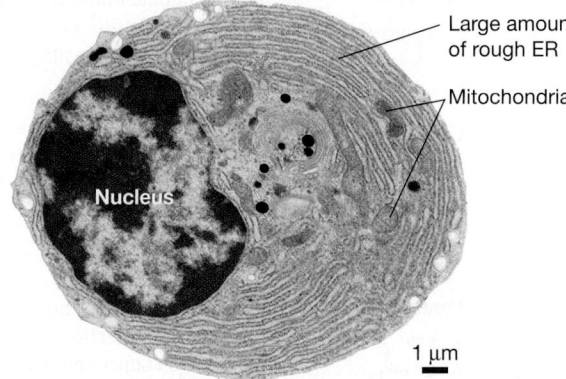

Large amount of rough ER

Mitochondria

Nucleus

1 μm

FIGURE 49.5 Lymphocytes Exist in Two States: Inactive and Activated. (a) This inactive lymphocyte—in this case, a type called a B cell—has a small amount of cytoplasm and few organelles. **(b)** This activated lymphocyte (also a B cell) has extensive rough endoplasmic reticulum and many mitochondria—suggesting that a great deal of protein synthesis is taking place.

The Discovery of B Cells and T Cells

In 1956 a group of researchers provided an important insight into the adaptive immune system, by accident. The biologists were investigating the immune system's response to the bacterium *Salmonella typhimurium*, a common cause of food poisoning in humans. To do this work, they needed to produce and isolate antibodies to a particular antigen from *S. typhimurium*—an antigen that is toxic. Their plan was to inject a large number of chickens with the antigen and collect the antibodies that the treated animals produced in response.

In addition to injecting many normal chickens, though, the biologists happened to include some chickens that had undergone experimental removal of an organ called the bursa. Six of the nine chickens that lacked a bursa and that were injected with the antigen died. The other three birds without a bursa survived but failed to produce antibodies to the antigen. In contrast, chickens with the bursa intact produced large quantities of antibodies and survived. To make sense of these results, the researchers proposed that the bursa is critical for antibody production and that antibodies are important in neutralizing antigens.

Not long after this observation was published, three groups of scientists independently conducted a related experiment. To explore the function of the thymus in mammals, these groups removed the organ from newborn mice. Mice lacking a thymus developed pronounced defects in their immune systems. For example, when pieces of skin from other mice were grafted onto the experimental individuals, their immune systems did not recognize the tissue as foreign. In contrast, individuals with an intact thymus quickly mounted an immune response that killed the foreign skin cells.

The results of these and follow-up experiments showed that lymphocytes from the bursa and thymus have different functions. The two types of lymphocytes became known as bursa-dependent and thymus-dependent lymphocytes, or B cells and T cells, respectively.

- **B cells** produce antibodies. (The lymphocytes in Figure 49.5 are B cells.) Later work showed that in humans and other species that lack a bursa, B cells mature in bone marrow.

- **T cells** are involved in an array of functions, including recognizing and killing host cells that are infected with a virus.

Now let's turn to one of the most fundamental questions in immunology: How are B cells and T cells able to recognize so many different antigens?

The Clonal-Selection Theory

By the 1960s biologists understood that:

- B cells can produce antibodies to a seemingly limitless number of antigens.

- Each antibody is specific to an antigen.

- The immune response intensifies through time after an infection begins.

- The immune response is "remembered"—meaning individuals do not get sick at all, or recover extremely quickly, if they are exposed to the same pathogen again.

To explain these patterns, researchers developed the **clonal-selection theory**, which made several key claims about how the adaptive immune system works:

1. *Antigens are recognized by receptors on B cells and T cells.* Each lymphocyte formed in the bone marrow or thymus has thousands of copies of a unique receptor on its surface. The receptor is a membrane protein and recognizes one antigen.

2. *Lymphocytes are activated by antigen-receptor binding.* When the receptor on a lymphocyte binds to an antigen, the lymphocyte switches from a quiescent to an activated state.

3. *Activated lymphocytes are cloned.* An activated lymphocyte divides and makes many identical copies of itself. In this way, specific cells are selected and cloned in response to an infection.

4. *Activated lymphocytes endure.* Some of the cloned cells descended from an activated lymphocyte persist long after the pathogen is eliminated. As a result, the cloned cells are able to respond quickly and effectively if the infection recurs.

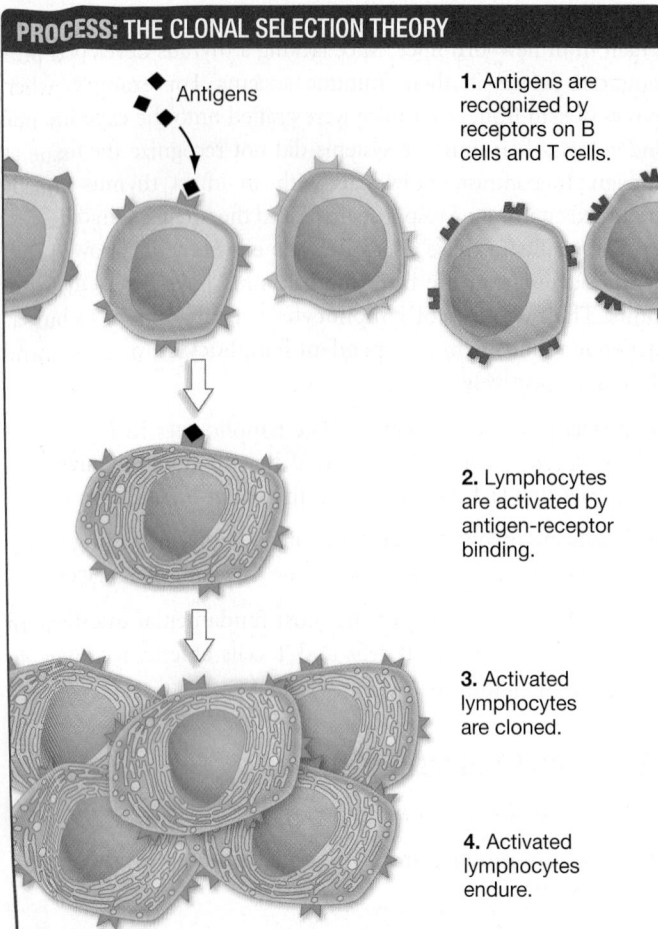

1. Antigens are recognized by receptors on B cells and T cells.

2. Lymphocytes are activated by antigen-receptor binding.

3. Activated lymphocytes are cloned.

4. Activated lymphocytes endure.

FIGURE 49.6 The Clonal-Selection Theory. The clonal-selection theory maintains that certain lymphocytes are "selected" by binding to an antigen, and that the selected cells then multiply.

Figure **49.6** summarizes these points. Let's explore the theory in detail, beginning with a look at the B-cell and T-cell receptors—the molecules that start the adaptive immune response.

THE DISCOVERY OF B-CELL RECEPTORS To test the prediction that lymphocytes have unique receptors on their surfaces, researchers injected experimental animals with radioactively labeled antigens. This strategy was similar to the experiments with labeled estradiol that allowed biologists to isolate that hormone's receptor (see Chapter 47).

As predicted, the labeled antigens bound to a protein on the surface of B cells. Chemical analysis of this **B-cell receptor (BCR)** showed that the protein has the same structure as the antibodies in the blood produced by B cells.

The BCR has three distinct components (**Figure 49.7a**). The first is a protein called the **light chain**. The second component is roughly twice the size of the light chain and is called the **heavy chain**. Each BCR has two copies of the light chain and two copies of the heavy chain. The third component is within the heavy chain: transmembrane domains that anchor the protein in the plasma membrane of the B cell.

TABLE 49.2 Five Classes of Immunoglobulins

Name	Structure (secreted form)	Function
IgG	Monomer	The most abundant type of secreted antibody. Circulates in blood and interstitial fluid. Protects against bacteria, viruses, and toxins
IgD	Monomer	Present on membranes of immature B cells; rarely secreted. Serves as BCR.
IgE	Monomer	Secreted in minute amounts. Involved in response to parasitic worms. Also responsible for hypersensitive reaction that produces allergies.
IgA	Dimer	Most common antibody in breast milk, tears, saliva, and the mucus lining the respiratory and digestive tracts. Prevents bacteria and viruses from attaching to mucous membranes; helps immunize breastfed newborns
IgM	Pentamer	First type of secreted antibody to appear during an infection. Binds many antigens at once; effective at clumping viruses and bacteria so that they can be killed.

B-cell antibodies are identical in structure to the BCR, except that they lack the transmembrane domains. Instead of being inserted into the plasma membrane, antibodies are secreted from the cell and circulate throughout the body.

Both the BCR and the antibodies produced by B cells belong to a family of proteins called the **immunoglobulins (Igs)**. Immunoglobulins are critically important to the adaptive immune response.

Table 49.2 shows the five classes of immunoglobulin proteins that act as B-cell receptors or antibodies. The five types are symbolized IgG, IgD, IgE, IgA, and IgM. Each class is distinguished by unique amino acid sequences in the heavy-chain region, and each has a distinct function in the immune response.

THE DISCOVERY OF T-CELL RECEPTORS It took much longer for researchers to isolate and characterize the **T-cell receptor (TCR)**. It turns out that the TCR only binds to antigens that have been (**1**) modified by other cells and (**2**) presented, or displayed, on the plasma membranes of these cells. For a TCR to bind to an antigen, the foreign protein has to undergo a complex process called **antigen presentation**.

This is a fundamentally important distinction. B cells bind to antigens directly; T cells bind only to antigens that are displayed by other cells.

Other data showed that the TCR is composed of two protein chains: an alpha (α) chain and a beta (β) chain (**Figure 49.7b**). The overall shape of the TCR is similar to the "arm" of an antibody or BCR molecule.

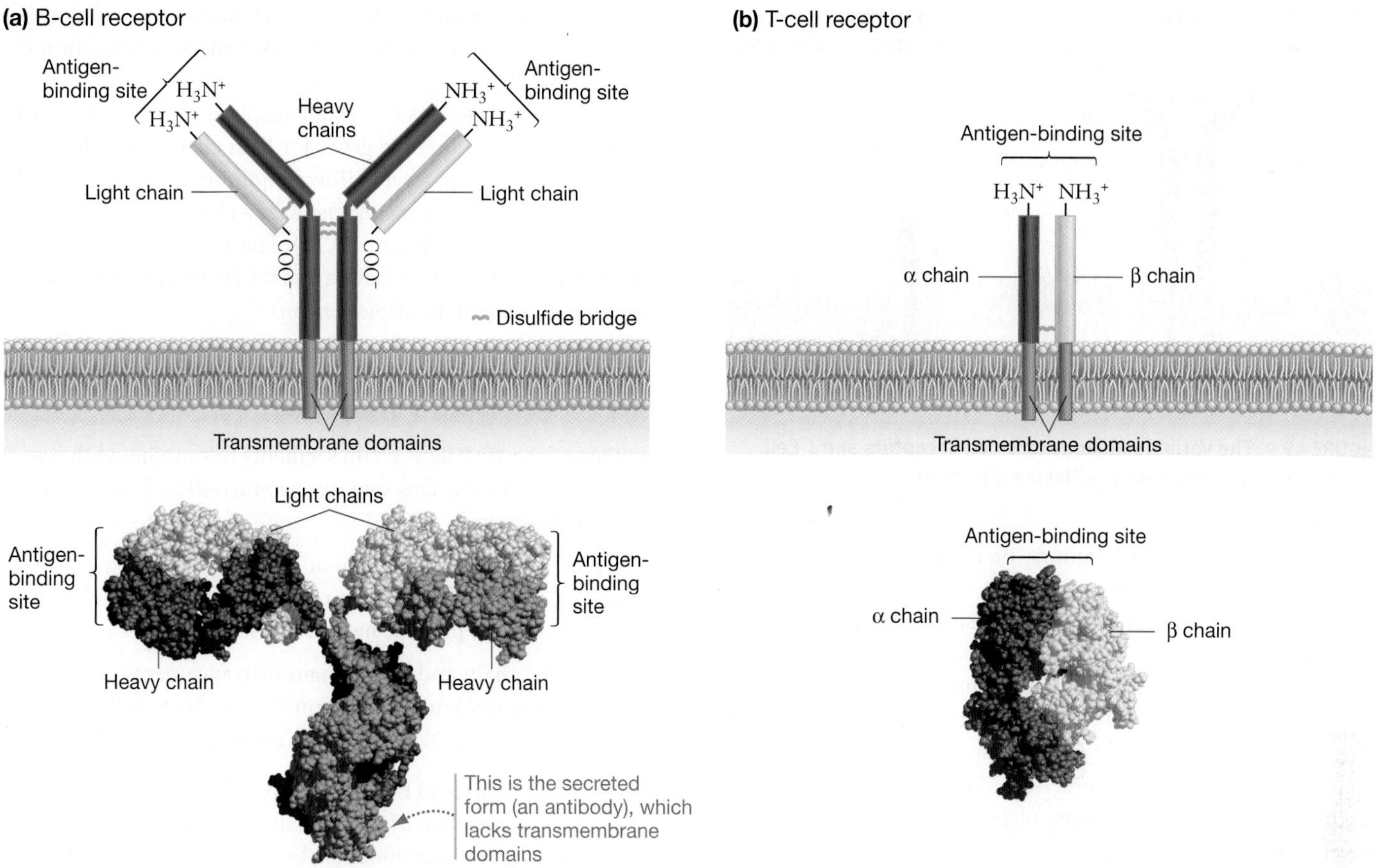

(a) B-cell receptor

Antigen-binding site

H_3N^+
H_3N^+

Heavy chains

NH_3^+
NH_3^+

Antigen-binding site

Light chain

Light chain

COO⁻   COO⁻

Disulfide bridge

Transmembrane domains

Light chains

Antigen-binding site

Antigen-binding site

Heavy chain

Heavy chain

This is the secreted form (an antibody), which lacks transmembrane domains

(b) T-cell receptor

Antigen-binding site

H_3N^+ NH_3^+

α chain

β chain

Transmembrane domains

Antigen-binding site

α chain

β chain

FIGURE 49.7 B-Cell Receptors and T-Cell Receptors Have Transmembrane Domains and Antigen-Binding Sites.
(a) Schematic model of the B-cell receptor, which is shaped like a Y. The antibodies and receptors produced by each B cell are identical, except that antibodies lack the transmembrane domain and are secreted. **(b)** The shape of the T-cell receptor resembles one "arm" of the Y-shaped B-cell receptor.

ANTIBODIES AND RECEPTORS BIND TO EPITOPES Antibodies, BCRs, and TCRs do not bind to the entire antigen. Instead, they bind to a selected region of the antigen called an **epitope**.

To understand the relationship between an antigen and an epitope, consider that every bacterium, virus, fungus, and protist is made up of a large number of different molecules. Each of these molecules is an antigen, because each is foreign to your cells. In turn, each antigen may have many different epitopes, where binding by antibodies and lymphocyte receptors actually takes place.

Figure 49.8 illustrates a protein called hemagglutinin, which is found on the surface of the influenza virus. This antigen has six epitopes, identified by the blue and red lines in the figure. Each epitope is recognized by a particular antibody, BCR, or TCR. It is not unusual for an antigen to have between 10 and 100 different epitopes.

How do the immunoglobulins recognize specific epitopes? The answer to this question came through detailed studies of the BCR's heavy and light chains.

WHAT IS THE MOLECULAR BASIS OF ANTIBODY SPECIFICITY AND DIVERSITY? In the 1950s, biologists developed an important model system for studying BCR and antibody production. The

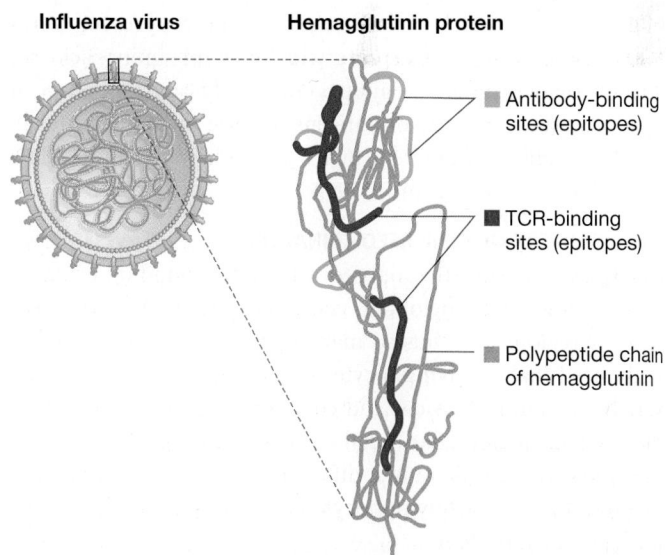

Influenza virus   Hemagglutinin protein

- Antibody-binding sites (epitopes)
- TCR-binding sites (epitopes)
- Polypeptide chain of hemagglutinin

FIGURE 49.8 Most Antigens Have Multiple Binding Sites for B-Cell Receptors, Antibodies, and T-Cell Receptors. The envelope of the influenza virus includes the protein hemagglutinin. This version of hemagglutinin has four different sites where antibodies bind and two distinct places where T-cell receptors bind.

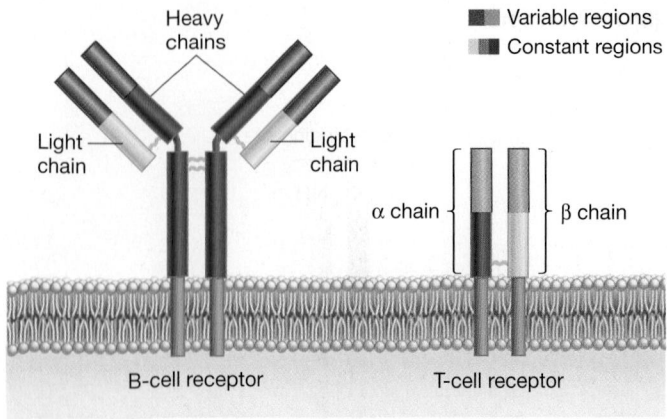

FIGURE 49.9 **The Variable Regions of B-Cell Receptors and T-Cell Receptors Face Away from the Plasma Membrane.**

cells involved were B-cell tumors, or myelomas, that could be grown in laboratory culture.

Each type of myeloma produces a single type of antibody. When researchers compared the light chains produced by different myelomas, they found that light chains have a segment where the amino acid sequence is virtually identical among light chains and a segment where the amino acid sequence is unique. These light-chain segments have come to be known as the **constant (C) regions** and **variable (V) regions**, respectively. Heavy chains also have a C region and a V region.

Figure 49.9 makes two important points:

1. The V regions of a BCR are adjacent and face away from the plasma membrane.

2. TCRs also have V and C regions, arranged in the same manner.

The presence of unique amino acid sequences in the V regions of every BCR and TCR explains why each of these proteins binds to a unique epitope: Receptors with different amino acid sequences bind to different epitopes. Your body can respond to an almost limitless number of antigens because there is a virtually limitless number of different BCRs, antibodies, and TCRs. How does all this variation come to be?

THE DISCOVERY OF GENE RECOMBINATION In 1965, W. J. Dryer and J. Claude Bennett proposed a fantastic-sounding explanation for how immunoglobulin genes code for so many different variable regions and thus so many different proteins. They hypothesized that as a lymphocyte is maturing, a segment from a variable-region gene is cut and combined with a segment from the constant-region gene. Further, they proposed that this cutting and pasting is done in a different way in each lymphocyte. The result is that each lymphocyte has a novel "V + C gene" for the light-chain protein.

This type of DNA cutting and pasting had never been observed, however. As a result, most researchers considered the hypothesis wildly implausible.

In 1976, however, Nobumichi Hozumi and Susumu Tonegawa showed that the amount of DNA in the V + C region of mature

lymphocytes is shorter than it is in immature lymphocytes (**Figure 49.10**). This is exactly what the gene-recombination hypothesis predicts.

Hozumi and Tonegawa's result inspired a flurry of studies. This work showed that the genes for light chains have dozens of different V segments, several different joining (J) segments, and a single constant (C) segment. The heavy-chain gene also has diversity (D) segments (**Figure 49.11**). The genes that encode the α and β chains of the TCR have a similar arrangement of distinct segments, each with multiple versions.

As a B cell matures, the various gene regions are mixed and matched to produce unique receptors. As an example, consider how a BCR is produced (**Figure 49.12**):

- One of the 40 V light-chain segments recombines with one of the 5 J segments. This step can produce 40 × 5 = 200 different light chains.

- In the heavy chain, any one of the 51 V segments, 27 D segments, and 6 J segments can recombine, giving a total of 51 × 27 × 6 = 8262 possible heavy chains.

- The light-chain and heavy-chain rearrangements occur independently; DNA recombination can produce 200 × 8262 = 1,652,400 = 1.65 × 10⁶ different antigen-specific BCRs.

✓If you understand these events, you should be able to devise a system for generating a high diversity of lottery numbers that is analogous to gene recombination, but uses only the digits 0–9.

In addition, gene segments do not always join precisely during DNA recombination. Some variation occurs where the V and D segments join and where the D and J segments join. As a result, an estimated 10^{10} to 10^{14} different BCRs can form in a single individual. TCR production is just as diverse.

Due to gene recombination, each BCR, antibody, and TCR has a unique amino acid sequence—enabling it to bind a unique epitope on an antigen. The charges and geometry at each surface make the receptor-epitope fit extremely specific. Gene recombination makes the adaptive immune system both specific and diverse.

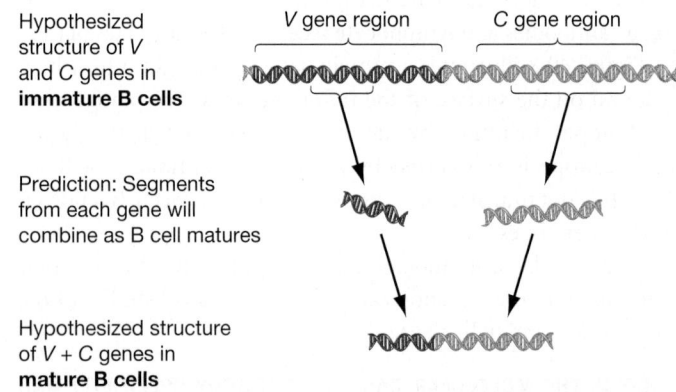

FIGURE 49.10 **The Gene Recombination Hypothesis for Antibody Formation in B Cells.**

Variable segments   Joining segments   Constant segment

$V_1 V_2 V_3 ... \quad V_{40} \qquad J_1 J_2 J_3 ... \qquad J_5 \qquad\qquad C$

Light-chain DNA:

Variable segments   Diversity segments   Joining segments   Constant segment

$V_1 V_2 V_3 ... \quad V_{51} \qquad D_1 D_2 D_3 ... \quad D_{27} \qquad J_1 J_2 J_3 ... \quad J_6 \qquad C$

Heavy-chain DNA:

FIGURE 49.11 In Immature Lymphocytes, Immunoglobulin Genes Consist of Many Segments.

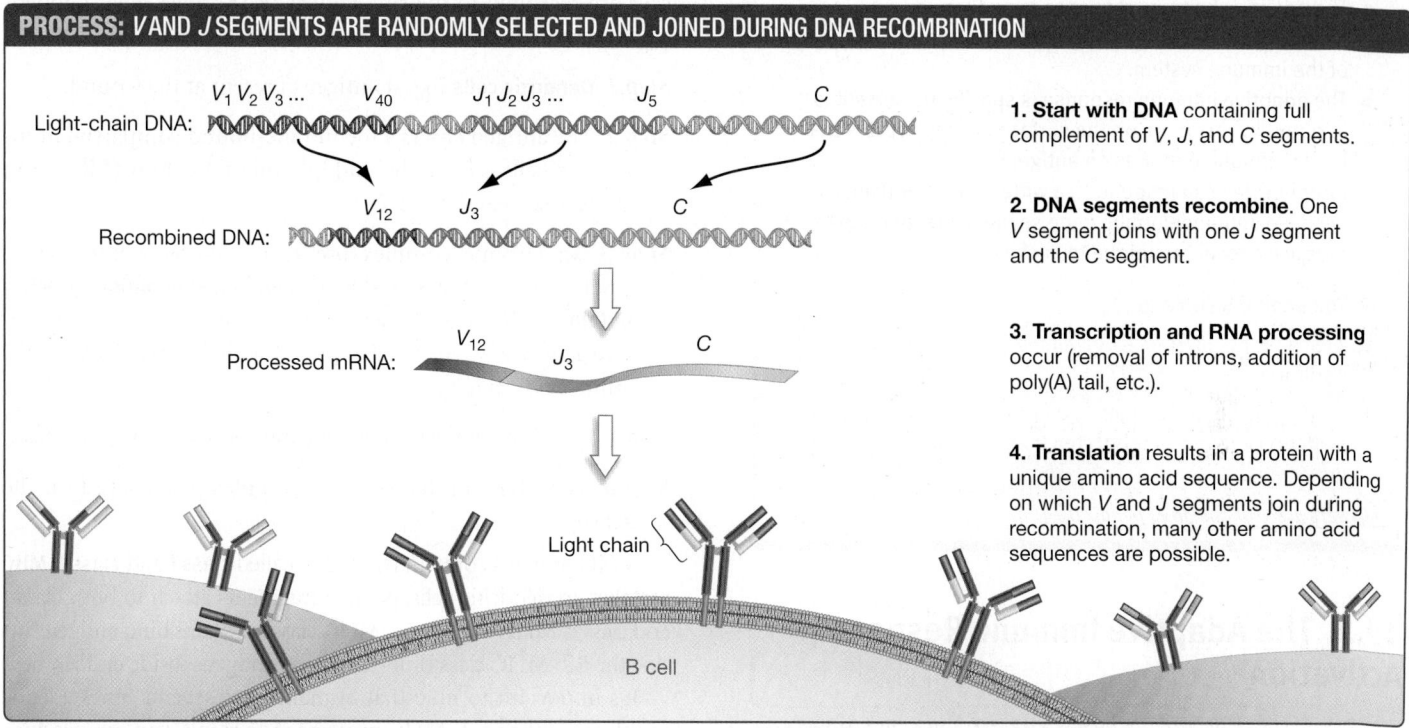

PROCESS: *V* AND *J* SEGMENTS ARE RANDOMLY SELECTED AND JOINED DURING DNA RECOMBINATION

$V_1 V_2 V_3 ... \quad V_{40} \qquad J_1 J_2 J_3 ... \qquad J_5 \qquad\qquad C$

Light-chain DNA:

$V_{12} \qquad J_3 \qquad\qquad C$

Recombined DNA:

Processed mRNA: $V_{12} \quad J_3 \qquad C$

Light chain

B cell

1. **Start with DNA** containing full complement of *V*, *J*, and *C* segments.

2. **DNA segments recombine.** One *V* segment joins with one *J* segment and the *C* segment.

3. **Transcription and RNA processing** occur (removal of introns, addition of poly(A) tail, etc.).

4. **Translation** results in a protein with a unique amino acid sequence. Depending on which *V* and *J* segments join during recombination, many other amino acid sequences are possible.

FIGURE 49.12 As Lymphocytes Mature, Immunoglobulin Genes Recombine to Form a Single Gene. In a mature lymphocyte, the final light-chain gene might consist of the V_{12}, J_2, and C segments spliced together; the final heavy-chain gene might consist of the V_{48}, D_{22}, J_1, and C segments spliced together.

How Does the Immune System Distinguish Self from Nonself?

How does the immune system ensure that B-cell and T-cell receptors don't respond to molecules that are part of normal host cells? If a receptor responded to a **self molecule**—that is, a molecule belonging to the host—the receptor would trigger an immune response. An anti-self reaction such as this is known as **autoimmunity**. Autoimmune reactions can lead to immune system cells destroying parts of the host's own body. In most cases, disease results.

- Multiple sclerosis (MS) results from the production of anti-self T cells that attack the myelin sheath of nerve fibers (see Chapter 45). Because damage to myelin reduces the efficiency of nerve signaling, muscular and coordination problems result.

- Rheumatoid arthritis develops when T cells and antibodies alter the lining of joints, causing painful inflammation.

- Type 1 diabetes mellitus occurs when T cells attack and kill insulin-secreting cells in the pancreas, resulting in a lack of insulin and inability to regulate blood glucose levels (see Chapter 43).

Because autoimmune diseases are relatively rare, biologists predicted that there must be some mechanism for eliminating young B cells and T cells that have receptors for self molecules. This prediction was confirmed by injecting B cells and T cells with anti-self receptors into mice, and finding that the injected lymphocytes were eliminated. Follow-up work showed that if B cells and T cells maturing in the bone marrow and thymus have anti-self receptors, most are destroyed or permanently inactivated before they leave these organs.

How are lymphocytes with anti-self receptors identified and eliminated? The answer is not known. Similarly, researchers do not have a good handle on why certain self-recognizing T cells or antibody-secreting cells can escape the self-nonself screening system. As a result, autoimmune disorders are notoriously difficult to treat.

CHECK YOUR UNDERSTANDING

If you understand that . . .

- The adaptive immune response is performed by lymphocytes that circulate in the lymph, blood, and organs of the immune system.
- The adaptive immune response is specific because it is initiated by receptors on the surface of lymphocytes that bind to unique epitopes on antigens.
- Lymphocytes can respond to a wide array of antigens because immunoglobulin gene segments recombine to produce a receptor unique to each cell.

✔ **You should be able to . . .**

1. Describe the difference between a B-cell receptor and an antibody.
2. Predict the consequences of a mutation that disrupts selection against lymphocytes that respond to self molecules.

Answers are available in Appendix B.

49.3 The Adaptive Immune Response: Activation

When the cell-surface receptor of a B cell or T cell binds to an antigen, the event is like pushing the button that launches a "smart bomb"—a missile that is programmed to destroy a specific target. Receptor-antigen binding unleashes a series of events that ultimately destroys the antigen. The lethal power of activated B cells and T cells is nothing short of awesome.

The activation of the relevant B cells and T cells is a carefully controlled, stepwise process. The mechanism is reminiscent of the precautions that nations with powerful missiles take to avoid accidental deployment. For the most dangerous weapons, the signal to launch is checked and cross-checked, using a series of codes and signals. In the immune system, the checking and cross-checking occurs through protein-protein interactions on the surfaces of cells, and the release and receipt of cytokines and other signaling molecules.

T-Cell Activation

T lymphocyte activation begins when antigens are taken up by a specific type of leukocyte or an infected cell, cut into pieces, packaged with specific cell proteins, and then transferred to the cell surface. Once antigens are presented in this way, T cells can bind to them via the TCR and begin their transformation from an inactive to an activated state.

To understand how the activation system works, let's explore the interaction between antigen-presenting cells and the T lymphocytes called $CD8^+$ T cells. T cells are classified as $CD4^+$ or $CD8^+$, based on the presence of key proteins called **CD4** or **CD8** on their plasma membranes. $CD4^+$ T cells and $CD8^+$ T cells have distinct functions in the adaptive immune response.

ANTIGEN PRESENTATION BY MHC PROTEINS If the innate immune response is overwhelmed and bacteria begin to multiply rapidly at a wound, leukocytes known as dendritic ("tree-like") cells are recruited to the site. **Figure 49.13** shows what happens when these cells arrive at the scene.

Step 1 **Dendritic cells** ingest antigens present at the wound.

Step 2 The antigen enters a membrane-bound compartment inside the cell—either the endoplasmic reticulum (ER) or an endosome (see Chapter 7).

Step 3 An enzyme complex breaks the proteins into pieces, which then become bound to a **major histocompatibility (MHC) protein**. MHC proteins are antigen-presenting proteins that have a groove where small peptide fragments, typically 8 to 20 amino acids in length, bind.

Step 4 The MHC-antigen complex is transported to the cell surface.

Step 5 The MHC protein-peptide complex is displayed on the cell surface.

MHC proteins come in two types, called **class I** and **class II MHC proteins**. In dendritic cells, peptide fragments attach to both class I and class II MHC complexes. MHC class I proteins bind antigens inside the ER; MHC class II proteins bind antigens inside endosomes.

It's important to note that humans have several genes encoding class I and class II MHC proteins. As a result, you can produce several distinct proteins of each type. In addition, the MHC genes are among the most polymorphic of any genes known—meaning that many different alleles exist in the population (see Chapter 13). Because so many distinct alleles exist, most individuals are heterozygous for the class I and class II MHC genes. Heterozygous individuals produce an even greater diversity of class I and class II MHC proteins. A wide array of MHC proteins means that a wide array of foreign peptides can be bound and presented—so cells can trigger an efficient response to many different pathogens.

In addition, virtually all cells in the body display pieces of "self"-proteins, bound to MHC class I proteins, on their surfaces. These are signs that indicate, "I'm a self-cell." Foreign cells that do not display these signs are destroyed by components of the innate immune system.

HOW DO T CELLS RESPOND TO ANTIGEN-PRESENTING CELLS? **Figure 49.14** illustrates what happens once a dendritic cell displays an MHC protein-foreign peptide complex. In the lymph organs, antigen-displaying dendritic cells interact with T cells. $CD8^+$ T cells interact with MHC–class-I–bound antigens on dendritic cells; $CD4^+$ T cells interact with MHC–class-II–bound antigens.

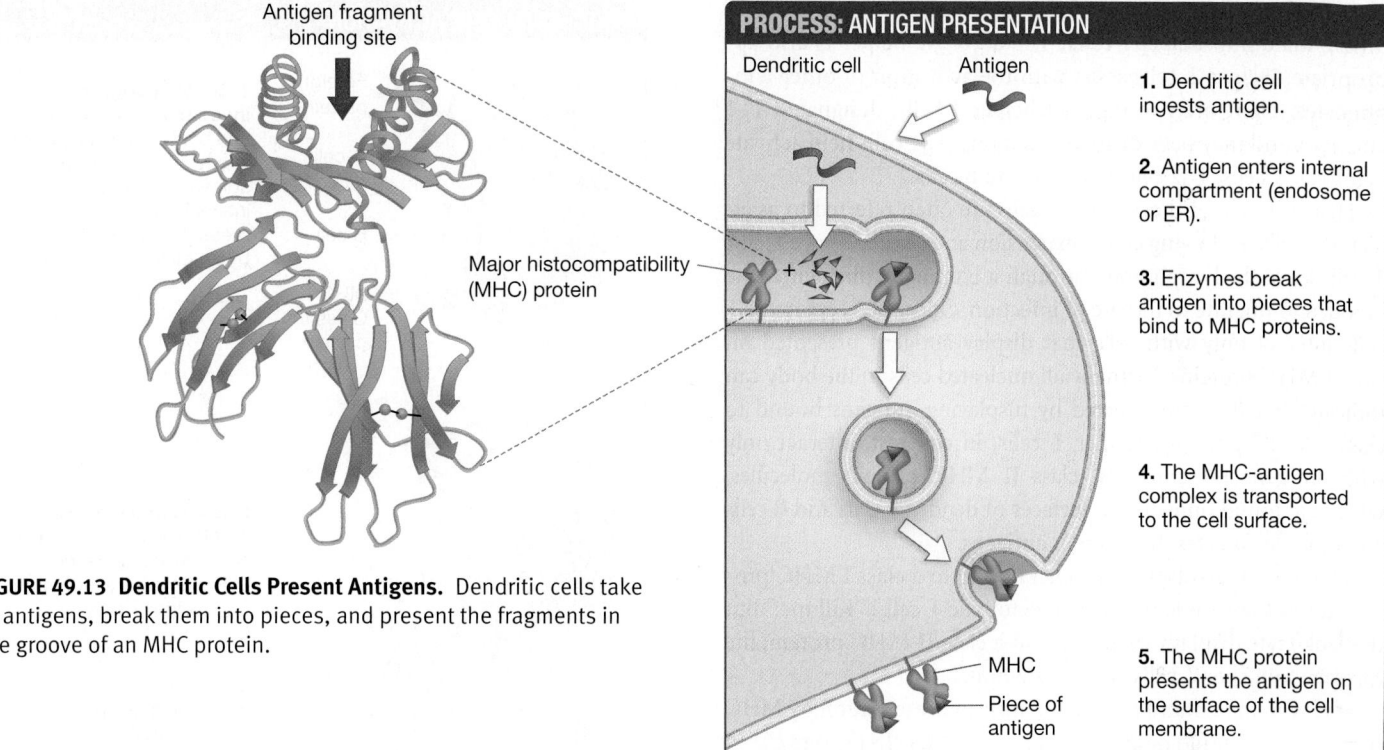

Antigen fragment binding site

Major histocompatibility (MHC) protein

FIGURE 49.13 Dendritic Cells Present Antigens. Dendritic cells take in antigens, break them into pieces, and present the fragments in the groove of an MHC protein.

PROCESS: ANTIGEN PRESENTATION

Dendritic cell   Antigen

1. Dendritic cell ingests antigen.

2. Antigen enters internal compartment (endosome or ER).

3. Enzymes break antigen into pieces that bind to MHC proteins.

4. The MHC-antigen complex is transported to the cell surface.

MHC

Piece of antigen

5. The MHC protein presents the antigen on the surface of the cell membrane.

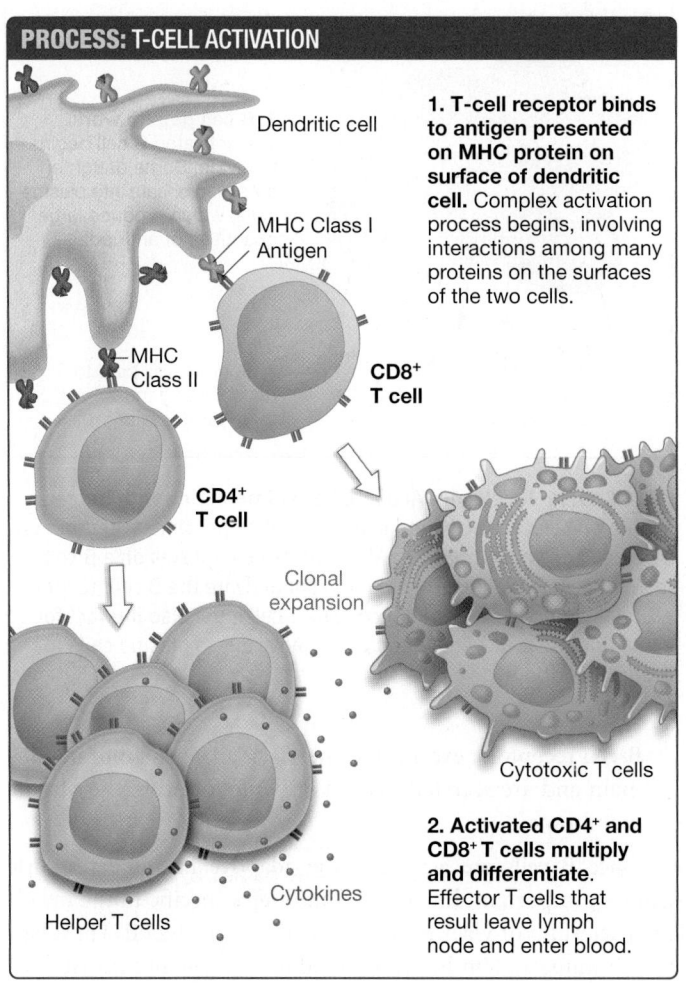

PROCESS: T-CELL ACTIVATION

Dendritic cell

MHC Class I Antigen

MHC Class II

CD8+ T cell

CD4+ T cell

Clonal expansion

Cytotoxic T cells

Helper T cells

Cytokines

1. T-cell receptor binds to antigen presented on MHC protein on surface of dendritic cell. Complex activation process begins, involving interactions among many proteins on the surfaces of the two cells.

2. Activated CD4+ and CD8+ T cells multiply and differentiate. Effector T cells that result leave lymph node and enter blood.

To the T cell, the antigen-presenting cell carries the message "I've found antigen—response required." If the TCR binds to an epitope on the antigen, the activation process starts in response to interactions with proteins on the dendritic cell. In some cases, activation of CD8+ T cells also requires interactions with cytokines produced by activated CD4+ T cells.

An activated T cell divides to produce a series of genetically identical daughter cells. This event, clonal expansion, is a crucial step in the adaptive immune response. It leads to a large population of lymphocytes capable of responding specifically to the antigen that has entered the body.

✔ If you understand why clonal expansion is important, you should be able to explain why cancer chemotherapy—which destroys rapidly dividing cells—suppresses the ability of the patient's immune system to respond to infections.

CYTOTOXIC T CELLS AND HELPER T CELLS When activated CD8+ T cells undergo clonal expansion, the daughter cells develop into **cytotoxic** ("cell-poison") **T cells**, also known as cytotoxic T lymphocytes (CTLs) or killer T cells. The adjectives cytotoxic and killer are appropriate: CD8+ T cells kill cells that are infected with a virus or other pathogen in their cytoplasm.

FIGURE 49.14 T Cells Are Activated after Interacting with Antigens Presented by Dendritic Cells. If a T cell has a receptor that is complementary to the MHC-antigen complex presented on the surface of a dendritic cell, the T cell will be activated.

In contrast, the daughter cells of activated CD4+ lymphocytes differentiate into **helper T cells**. The adjective helper is also appropriate: Helper T cells assist with the activation of other lymphocytes. There are two types of helper T cells, designated T_H1 and T_H2, and they have distinct functions: T_H1 cells help activate cytotoxic T cells; T_H2 cells help activate B cells.

Helper T cells and cytotoxic T cells are often referred to as effector T cells. Following clonal expansion and maturation, effector T cells leave the lymph node through a lymphatic duct, enter the blood, and migrate to the site of infection. Once there, cytotoxic T cells interact only with cells that display antigens presented on class I MHC proteins. Virtually all nucleated cells in the body can indicate that they are infected by displaying antigens bound to class I MHC proteins. Helper T cells, in contrast, interact only with antigens presented on class II MHC protein molecules, which are found only on the surfaces of dendritic cells and B cells and other leukocytes that present antigens.

When a cell displays an antigen bound to a class I MHC protein, it sends a simple message to cytotoxic T cells: "Kill me." But if a leukocyte displays an antigen on a class II MHC protein, the message to helper T cells is "Activate now."

✔If you understand the role of antigen presentation by MHC proteins, you should be able to (1) state which cells express class I versus class II MHC proteins, (2) state which types of T cells respond to peptides bound to class I versus class II MHC proteins, and (3) summarize the message provided by MHC-peptide complexes on a dendritic cell, an infected cell, and a B cell.

Effector T cells are ready for action. A key part of the adaptive immune response is now under way.

B-Cell Activation and Antibody Secretion

CD8+ and CD4+ lymphocytes are activated by interactions with dendritic cells and other leukocytes that present antigens. How are B cells activated? The answer involves several steps (**Figure 49.15**):

Step 1 The BCRs on B cells interact directly with bacterial or viral antigens that are floating free in lymph or blood. Once a free antigen is bound, B cells internalize the molecule, digest it into fragments, and load them into class II proteins. As a result, a B cell that encounters its antigen displays the antigen's epitopes on its own surface, with the peptides cradled in the groove of a class II MHC protein.

Step 2 When an activated CD4+ T_H2 cell with a complementary receptor arrives, it binds to the antigen-MHC complex on the B cell. The interaction between a B cell and a helper T cell supplies activation signals that stimulate the helper T cell.

Step 3 The helper T cell responds by releasing stimulatory cytokines that activate the B cell.

Step 4 The B cell divides and forms daughter cells. Some of the daughters differentiate into activated B lymphocytes called **plasma cells**. Plasma cells produce and secrete large quantities of antibodies. Recall that antibodies are identical to

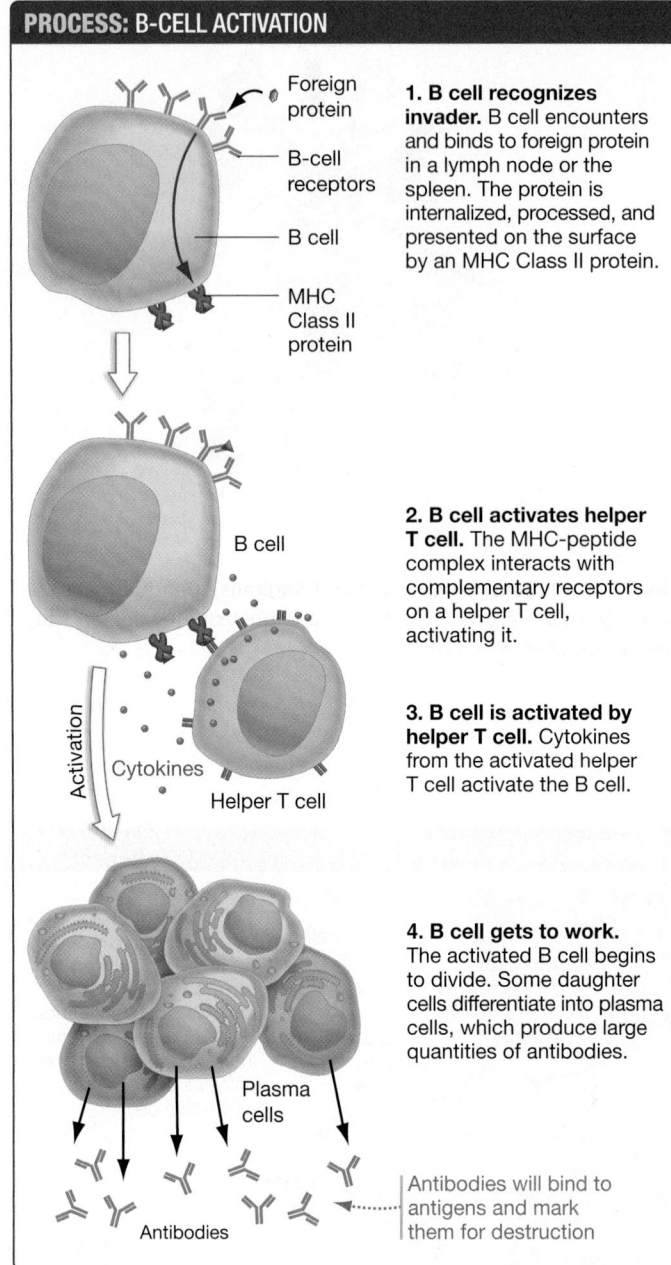

PROCESS: B-CELL ACTIVATION

1. B cell recognizes invader. B cell encounters and binds to foreign protein in a lymph node or the spleen. The protein is internalized, processed, and presented on the surface by an MHC Class II protein.

Foreign protein
B-cell receptors
B cell
MHC Class II protein

2. B cell activates helper T cell. The MHC-peptide complex interacts with complementary receptors on a helper T cell, activating it.

B cell
Activation
Cytokines
Helper T cell

3. B cell is activated by helper T cell. Cytokines from the activated helper T cell activate the B cell.

4. B cell gets to work. The activated B cell begins to divide. Some daughter cells differentiate into plasma cells, which produce large quantities of antibodies.

Plasma cells
Antibodies

Antibodies will bind to antigens and mark them for destruction

FIGURE 49.15 B Cells Are Activated after Interacting with Helper T Cells. B cells are activated in a series of steps. Once receptors on a helper T cell have bound to MHC-antigen complexes on a B cell, the helper T cell releases cytokines that activate the B cell. (Other proteins on the surface of a B cell and T cell must also interact for activation to occur.) Activation eventually causes plasma cells to produce antibodies.

B-cell receptors, except that they lack a transmembrane domain and are secreted instead of being found in the plasma membrane.

Once B cells are activated, the adaptive immune response gains an important dimension. Antibodies specific to the invading bacterium or virus begin to circulate in the blood. The adaptive immune system has recognized an antigen and initiated its response. For pathogens, the results are usually devastating.

49.4 The Adaptive Immune Response: Culmination

In combination with the leukocytes involved in the innate immune system, the cells of the adaptive immune system are almost always successful in eliminating threats from bacteria, parasites, fungi, and viruses.

To understand how the adaptive response actually kills pathogens, let's return to our earlier examples of a bacterial infection in a wounded elbow and an upper respiratory tract infection caused by the flu virus. How do activated B cells and T cells eliminate these invaders?

How Are Bacteria and Other Foreign Cells Killed?

During the innate immune response to pathogenic cells that enter a wound, macrophages and dendritic cells at the site phagocytize some of the invaders. In addition to killing the foreign cells, these leukocytes process and present antigens via the MHC class I and class II proteins. Macrophages at the site of infection display epitopes on their surfaces, and dendritic cells move to the lymph nodes to interact with T cells. If the epitopes are recognized by helper T cells, the adaptive immune response is activated.

If an activated T_H1 cell binds to these antigen-laden macrophages, two things happen: First, the phagocytic activity of the macrophages is enhanced. Second, the T_H1 cells secrete cytokines that recruit additional phagocytic cells to the site—increasing the inflammatory response.

In addition, antibodies from plasma cells begin coating bacteria, fungi, and other foreign cells. In many cases, antibody action causes **agglutination**, or clumping of cells. Each antibody has at least two binding sites, so a single antibody can bind epitopes on foreign cells and link them, forming a clump (**Figure 49.16**).

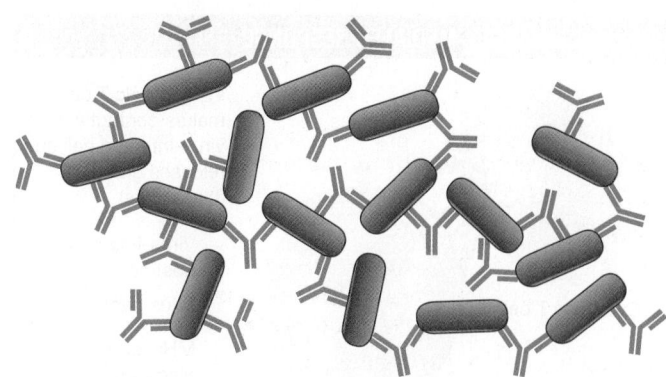

FIGURE 49.16 Antibodies Cause Agglutination. Clumped pathogens are unlikely to infect host tissues, and are readily destroyed.

🔑 Clumped and single cells that are tagged with antibodies are readily destroyed by macrophages via phagocytosis. Antibodies that are bound to antigens also stimulate a lethal group of proteins called the **complement system**. Complement proteins circulate in the bloodstream and assemble at antigen-antibody complexes. When complement proteins activate, they punch deadly holes in the plasma membranes of invading cells.

Within a few days, this combination of killing mechanisms—armies of phagocytic cells and complement proteins that hone in on antibody-tagged bacteria—usually eliminates all foreign cells.

How Are Viruses Destroyed?

The adaptive immune system has an array of mechanisms to dispose of bacteria. It also has two major ways to eliminate viruses.

1. The **cell-mediated response** involves cytotoxic T cells (activated $CD8^+$ cells) and takes place at the surface of infected cells.

2. The **humoral response** involves antibodies and takes place in blood and lymph. (The Latin root *humor* means "fluid.")

Let's take a closer look at both.

THE CELL-MEDIATED RESPONSE LEADS TO CELL SUICIDE Infected cells respond to the arrival of a virus by processing antigens from the invader. MHC class I proteins attach to the viral antigens and present the antigens on the surface of the infected cell. Almost all nucleated cells in the body express MHC class I proteins and have the ability to signal that they are infected.

Cells that display viral antigens bound to class I MHC molecules are effectively waving a flag that says, "I'm infected. If you destroy me, you'll destroy them."

As activated $CD8^+$ cells migrate into the area, they recognize and bind to the antigen's epitopes and the MHC protein displayed on infected cells. Once binding occurs, the $CD8^+$ cell secretes molecules that assemble on the surface of the infected cell's plasma membrane and produce pores. Chemicals released by the cytotoxic T cell then enter the cell and activate a self-destruct response (**Figure 49.17a** on page 988).

Once the infected cell dies, the cytotoxic T cell releases it and seeks out another infected cell to kill. Over time, all virus-infected cells are eliminated.

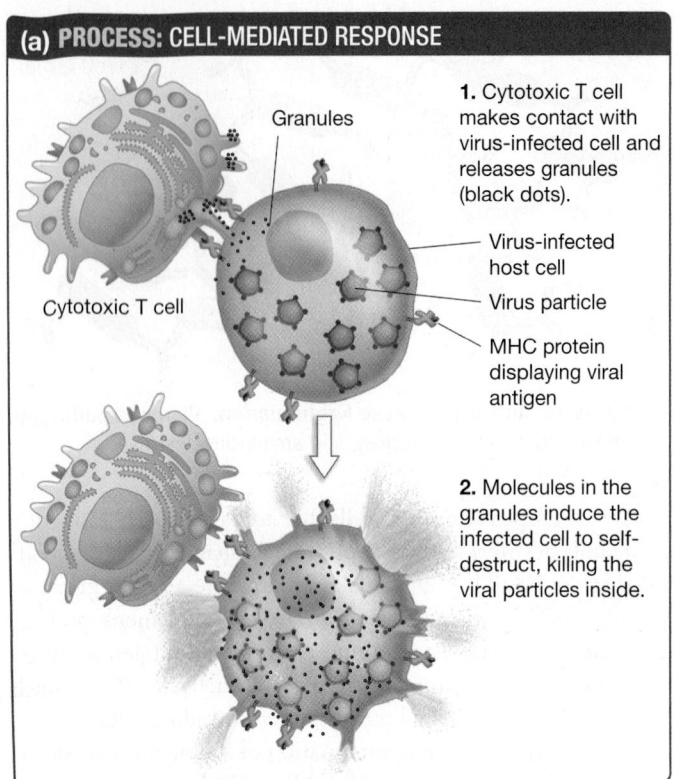

(a) PROCESS: CELL-MEDIATED RESPONSE

Granules

Cytotoxic T cell

1. Cytotoxic T cell makes contact with virus-infected cell and releases granules (black dots).

Virus-infected host cell

Virus particle

MHC protein displaying viral antigen

2. Molecules in the granules induce the infected cell to self-destruct, killing the viral particles inside.

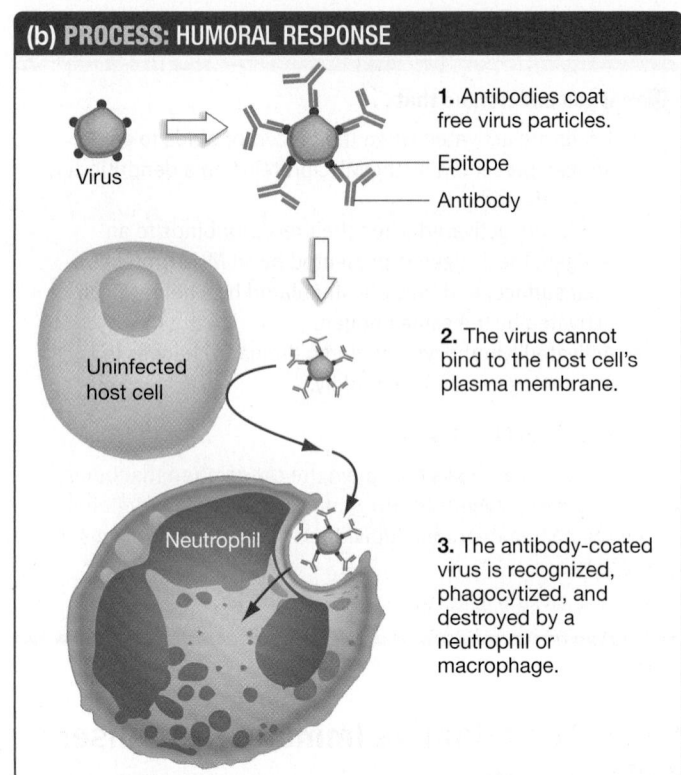

(b) PROCESS: HUMORAL RESPONSE

Virus

1. Antibodies coat free virus particles.

Epitope

Antibody

Uninfected host cell

2. The virus cannot bind to the host cell's plasma membrane.

Neutrophil

3. The antibody-coated virus is recognized, phagocytized, and destroyed by a neutrophil or macrophage.

FIGURE 49.17 Lymphocytes Eliminate Viruses via the (a) Cell-Mediated Response and the (b) Humoral Response.

Because viruses can reproduce only inside host cells, the cell-mediated response limits the spread of the infection by preventing new generations of virus particles from maturing.

THE HUMORAL RESPONSE COATS PATHOGENS WITH ANTIBODIES
Even as cytotoxic T lymphocytes swing into action, plasma cells produce antibodies to viral proteins. **Figure 49.17b** shows the consequences.

- In most cases, the most effective antibodies are those that bind to epitopes on the surface of the virus.

Two things happen once antibodies bind to the outside of a virus:

- The virus is blocked from infecting host cells.

- Macrophages and other phagocytic cells recognize the antibody-coated particles and phagocytize them.

Eventually, the virus population is reduced to zero.

Why Does the Immune System Reject Foreign Tissues and Organs?

Antibodies and cytotoxic T cells have devastating effects on invading pathogens. Unfortunately, they are equally deadly in response to tissues or organs that are introduced into a patient to heal a wound or cure a disease.

Consider the problems that can arise with blood transfusions. You might recall from Chapter 13 that certain individuals have red blood cells with membrane glycoproteins called A and B. These molecules act as antigens if they are introduced into a person whose own blood cells lack those glycoproteins.

For example, if you have type A blood, it means that your blood cells have the A antigen. If your blood is transfused into a person who lacks the A antigen—meaning someone who has type B or type O blood—the recipient's immune system will recognize the A antigen as foreign and mount a devastating response against it. For a blood transfusion to be successful, this recipient has to receive blood that lacks the A and B glycoproteins entirely or that contains the same antigens found in his or her own blood.

Similar problems arise in organ transplants, except that the antigenic molecules in foreign organs are the MHC proteins found on the surfaces of their cells. To prevent strong immune reactions to a transplanted kidney, heart, or liver, physicians do two things:

1. obtain the organ to be transplanted from a sibling or other donor whose MHC proteins are extremely similar in structure to those of the recipient; and

2. treat the recipient with drugs that suppress the immune response.

Thanks to steady improvements in drug development and in systems for matching MHC types between donors and recipients, the success rate for organ transplants has improved dramatically in recent years.

As the blood transfusion and organ transplant examples show, the immune system rejects foreign tissues because they contain nonself proteins—antigens. To your T cells and B cells, a blood transfusion or an organ transplant is indistinguishable from a massive influx of bacteria, viruses, or other foreign invaders.

Responding to Future Infections: Immunological Memory

In addition to producing the cells that implement the humoral and cell-mediated responses, activated B cells and T cells produce specialized daughter cells called memory cells. **Memory cells** do not participate in the initial adaptive response, or **primary immune response**. Instead, they provide a surveillance service after the original infection has been cleared. Memory cells remain in the spleen and lymph nodes for years or decades, ready to provide a rapid response should an infection with the same antigen recur.

The production of memory cells is a hallmark of the vertebrate immune response. It occurs only to a limited degree, if at all, in invertebrates.

THE SECONDARY RESPONSE IS STRONG AND FAST If the same antigen enters the body a second time, memory cells recognize certain epitopes of the antigen and trigger a second adaptive response, or **secondary immune response**. The launching of a secondary immune response by means of memory cells is known as **immunological memory**.

The secondary immune response is faster and more efficient than the primary response. It is faster because the presence of memory T and B cells increases the likelihood that lymphocytes with the correct antigen-specific receptors will find the antigen and activate quickly. It is more efficient because some of the memory B cells that respond to the returning antigen migrate to a specialized area in the lymph node called the germinal center. There the DNA sequences that code for the variable region of the immunoglobulin gene undergo rapid mutations that modify the receptors produced by the memory cell. Memory B cells with receptors that bind best to the antigen's epitope live and produce daughter cells; those that bind to the antigen less effectively die.

This process of **somatic hypermutation** fine-tunes the immune response. The antibodies that result from somatic hypermutation bind to the antigen more tightly than the antibodies produced by plasma cells during the primary immune response. As the secondary immune response proceeds and somatic hypermutation continues, better-fitting antibodies are produced.

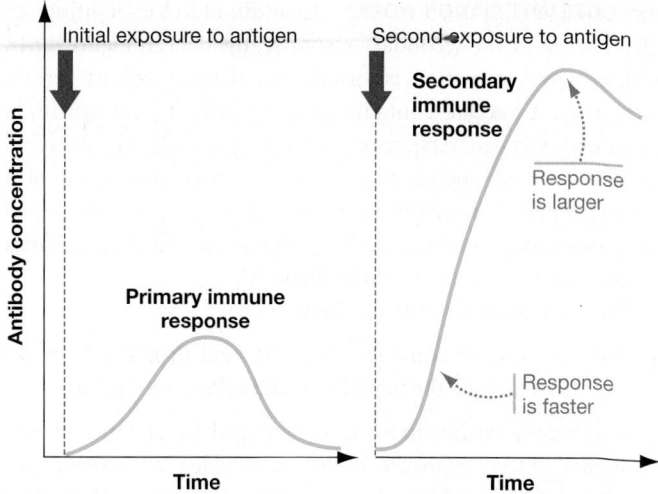

FIGURE 49.18 The Secondary Immune Response Is Faster and Stronger than the Primary Response. The data shown here summarize what happens when biologists inject a mouse with an antigen, document changes in antibody concentration over time, inject the same individual with the same antigen later, and measure the response in antibody concentration.

✔**EXERCISE** The primary response peaks in about 2 weeks; the secondary response peaks in 3–5 days. Add appropriate labels to scale the x-axes.

Figure 49.18 compares the rate of antibody production during the first and second exposures to a virus. The graphs are based on data that researchers collect by inoculating a laboratory mouse with influenza virus, collecting blood from the mouse every other day, and then measuring the amount of antiviral antibodies present in the fluid component of the blood.

To review immunological memory, starting with how B cells and T cells are activated, how clonal expansion works, and how daughter cells differentiate into plasma cells, helper T cells, and cytotoxic T lymphocytes, study **Table 49.3**. Then go to the study area at *www.masteringbiology.com*.

(MB) **Web Activity** The Adaptive Immune Response

SUMMARY TABLE 49.3 **Activation and Function of Adaptive Immune System Cells**

Type of Lymphocyte	Method of Activation	Cells That Result from Activation and Clonal Expansion	Function of Resulting Cells
B cell	Receptor binds to free antigen, then interacts with T$_H$2 cell	Plasma cells	Secrete antibodies
		Memory B cells	Participate in secondary response (secrete antibodies)
CD4$^+$ T cell	Receptor binds to antigen-MHC class II protein complex on dendritic cell or other antigen-presenting cell	T$_H$1 helper T cells	Activate cytotoxic T cells, regulate inflammatory response
		T$_H$2 helper T cells	Activate B cells
		Memory T cells	Participate in secondary response
CD8$^+$ T cell	Receptor binds to antigen-MHC class I protein complex on infected cell; may also interact with T$_H$1 cells	Cytotoxic T cells	Kill infected host cells
		Memory T cells	Participate in secondary response

WHY DOES VACCINATION WORK? In addition to highlighting the effectiveness of the secondary response, the data in Figure 49.18 explain why vaccination is an effective defense against certain pathogens. A **vaccine** contains antigens from a pathogen or a killed or weakened version of the pathogen itself. The antigens are usually components of a virus's exterior—the capsid of a nonenveloped virus or the envelope proteins from an enveloped virus (see Chapter 35). Exterior proteins are effective antigens because antibodies can attach to them.

There are three types of vaccines:

1. *Subunit vaccines* consist of isolated viral proteins. Familiar examples include the hepatitis B and influenza vaccines.

2. *Inactivated viruses* have been damaged by chemical treatments—often exposure to formaldehyde—or exposure to ultraviolet light. They do not cause infections, but are antigenic. If you have been vaccinated for hepatitis A or polio, you may have received an inactivated virus.

3. *Attenuated viruses* are also called "live" virus vaccines because they consist of complete virus particles. Researchers make these viruses harmless by culturing them on cells from species other than the normal host. In adapting to growth on the atypical cells, the viruses usually lose the ability to grow rapidly in their normal host cells. The smallpox, polio, and measles vaccines consist of attenuated viruses.

After vaccination (inoculation with a vaccine), the body mounts a primary immune response that produces memory cells. If a second infection occurs later, these memory cells respond quickly and eliminate the threat before illness appears. Vaccinations function like fire drills or earthquake preparedness exercises—they prepare the immune system for a specific threat.

Edward Jenner's vaccination strategy worked because the antigens presented by cowpox are extremely similar to the antigens presented by smallpox. As a result, exposing people to cowpox virus elicited the production of memory cells that effectively thwarted future infections by smallpox virus.

CHECK YOUR UNDERSTANDING

🔑 If you understand that . . .

- The antibodies produced by activated B cells mark invading pathogens for destruction.
- Activated cytotoxic T cells recognize epitopes displayed by infected cells and kill them before the pathogens inside can replicate.
- Memory cells produced during B-cell and T-cell activation remain in the body and provide secondary immunity against future infections.

✔ You should be able to . . .

Explain the relationship between vaccination, an infection, and the primary and secondary immune responses.

Answers are available in Appendix B.

Unfortunately, viruses such as influenza and the human immunodeficiency virus (HIV) mutate so rapidly that they present the immune system with a constantly changing array of epitopes. Memory cells that were effective during a previous infection by these viruses are unlikely to bind to the changed epitopes and trigger a response to a later infection. As a result, it is extremely difficult to design an effective vaccine against these agents.

Currently, the only "cure" for HIV infection is prevention; to keep up with rapidly mutating flu populations, flu vaccines must be redesigned and administered every year. The high mutation rates observed in these viruses are thought to be an adaptation that helps them escape detection by the immune system.

49.5 What Happens When the Immune System *Doesn't* Work Correctly?

The vertebrate immune system is a marvel of adaptation. When people are well rested and well nourished, and when basic sanitation limits exposure to pathogens that are transmitted in contaminated drinking water, the immune system is able to defeat the vast majority of infections without medical intervention (see data in Figure 28.2).

To drive this point home, consider what happens when the immune system fails. You have already reviewed some important autoimmune disorders, which develop when T cells or antibodies mistakenly attack self cells. What happens when the immune system stops working entirely?

Immunodeficiency Diseases

As Chapter 19 noted, children who are born with severe combined immunodeficiency (SCID) have a genetic defect in one of the enzymes responsible for DNA recombination in maturing lymphocytes. As a result, they are unable to generate normal T-cell and B-cell receptors.

Because TCRs and BCRs are fundamental to adaptive immune system function, the response is badly impaired in these children. Afflicted individuals suffer debilitating illness from infections that other children fight off easily, and die before they are two years old.

Similarly, people who are infected with the **human immunodeficiency virus (HIV)** suffer from a progressive failure of the immune system. As Chapter 35 detailed, HIV infects and kills $CD4^+$ T cells and macrophages. Recall that these cells are required for both the humoral and cell-mediated immune responses. As the infection continues, populations of $CD4^+$ T cells gradually decline to a point where the immune system can no longer mount an effective response to infection.

Eventually, HIV-infected people develop **AIDS—acquired immune deficiency syndrome**. They succumb to illnesses that physicians almost never see in people with healthy immune systems.

Allergies

What happens when the immune system *overreacts* to an antigen? For some people, walking through a field of ragweed or petting a cat is a prescription for developing a runny nose, itchy eyes, and labored breathing. For other people, being stung by a bee or eating a peanut is a life-threatening experience. Why are some people allergic to certain molecules?

An **allergy**, or allergic reaction, is an abnormal response to an antigen. For reasons that are still not clear, certain people produce the IgE class of antibodies in response to specific molecules found in cat dander, nuts, plant pollen, or other products. Molecules that trigger this response are called **allergens** instead of antigens.

In people who are not allergic, IgE antibodies are produced only in response to infections by worms. Recent research has also shown that in areas where infections with intestinal worms are common, allergies are almost unknown. These data suggest that allergies don't occur if IgEs are expressed normally—that is, in response to worm infections.

The presence of IgE antibodies in an allergic response is important because it triggers a series of events known as the **hypersensitive reaction**. When a person is first exposed to an allergen, receptors on leukocytes called mast cells and basophils bind to the IgE antibodies that are produced. Once this binding event occurs, the cells (and the person) are said to be sensitized.

If the person is later exposed to the same allergen, the sensitized cells rapidly produce large quantities of histamine, cytokines, chemokines, and other compounds. In response to these molecules, blood vessels dilate and become more permeable, smooth-muscle cells contract, and mucus-producing cells secrete.

- *Hay fever* is a common allergic response to the allergens in grass or ragweed or birch-tree pollen, and results in symptoms like runny nose, watery eyes, and mild wheezing.

- *Hives* occurs when tissues respond to an allergen by reddening, swelling, and itching.

- *Asthma* develops if the hypersensitive response is localized to the respiratory passages; the swelling that results constricts the airway.

If the immune response is more severe, blood vessels can dilate to the point where blood pressure plummets, oxygen delivery to the brain is reduced dramatically, and the person loses consciousness. In addition, severe smooth-muscle contractions in the digestive system and respiratory system can induce vomiting, diarrhea, and complete constriction of the airway passages. This combination of events, known as anaphylactic shock, is lethal.

The most effective treatment for anaphylactic shock is to inject the affected individual with epinephrine. Epinephrine is a hormone that relaxes smooth muscle (see Chapter 47). As a result, it opens up blood vessels and counteracts the effects of histamine and other agents of the hypersensitive response. People who are hypersensitive to bee stings or other allergens routinely carry small syringes filled with epinephrine. If they face a situation where anaphylactic shock is possible, they inject themselves.

What can be done about allergies? The number one approach is to avoid the allergen. If exposure does occur, drugs called antihistamines can sometimes reduce symptoms by blocking histamine receptors. In some cases corticosteroids applied to skin can reduce swelling. Longer-term solutions are still in the experimental stage, however. Developing effective treatments for allergies is an important research frontier in biomedicine.

CHAPTER 49 REVIEW

For media, go to the study area at www.masteringbiology.com

Summary of Key Concepts

⛒➥ **The innate immune response to infection is mounted by leukocytes that respond in a nonspecific way to pathogens. In contrast, the adaptive immune response is mounted by lymphocytes that respond to specific antigens—often a protein from a pathogen.**

- The innate response occurs in the same way no matter which pathogens have entered the body, but the adaptive response is specific to the pathogen present.

- During the innate immune system's inflammatory response, mast cells release chemical messengers that increase blood flow to wounds or areas of tissue damage.

- Leukocytes are recruited to the site of an inflammatory response. Neutrophils respond to bacteria that stimulate their pattern-recognition receptors by phagocytizing the invading cells and destroying them. Macrophages also phagocytize pathogens and release chemical messengers that activate other leukocytes and raise body temperature.

✔ You should be able to explain what would happen if blood vessels near a wound did not dilate and become more permeable in response to chemical signals from leukocytes and damaged tissues.

(MB) **Web Activity** The Inflammatory Response

⛒➥ **The adaptive immune response begins when certain lymphocytes are activated in a regulated, step-by-step process triggered by interaction with a specific antigen.**

- Lymphocytes are activated when T-cell receptors and B-cell receptors recognize an epitope on an antigen.

- The receptor protein on the surface of a T cell or B cell is generated through DNA recombination—a rearrangement of gene segments.

- Because every receptor that results from DNA recombination is slightly different, the immune system is able to recognize and respond to an almost limitless array of antigens.

✔ You should be able to explain how a drug that binds to a particular epitope of an antigen and blocks it would affect the activation of T cells with receptors specific to that antigen.

⛒➥ **Activated lymphocytes either destroy infected cells or produce antibodies that tag pathogens for destruction.**

- During the cell-mediated response, host cells that are infected with a virus display antigens on their surface and are induced to self-destruct by cytotoxic T cells. As a result, the virus particles inside are destroyed and cannot contribute further to the infection.

- During the humoral response, activated B cells produce antibodies to specific epitopes on the pathogen that has invaded the body. Viruses that become coated with antibodies are unable to enter host cells and are destroyed by macrophages. Bacteria that are tagged with antibodies are destroyed by complement proteins or macrophages.

- Because memory cells are created as activated B cells and T cells undergo clonal expansion, the immune system is able to respond rapidly and effectively to future infections by the same antigen.

- A vaccine triggers the production of memory cells because it contains epitopes from antigens.

  ✔ You should be able to explain why cowpox can be considered an attenuated virus vaccine.

  (MB) **Web Activity** The Adaptive Immune Response

Questions

✔ **TEST YOUR KNOWLEDGE** *Answers are available in Appendix B*

1. What is the primary difference between the innate and adaptive responses?
   a. The innate response is modified over time; the adaptive response occurs in the same way throughout life.
   b. Only the adaptive response is triggered by antigens.
   c. The adaptive response is specific and is "remembered"; the innate response is nonspecific.
   d. There is no nonself recognition component in the innate response.

2. All of the following events are involved in the inflammatory response *except*:
   a. Cytotoxic T cells kill infected host cells.
   b. Neutrophils phagocytize pathogens that stimulate their pattern-specific receptors.
   c. Mast cells secrete chemical messengers that lead to increased blood flow.
   d. Macrophages release chemical messengers that lead to increased body temperature.

3. What is the difference between an epitope and an antigen?
   a. An epitope is any foreign substance; an antigen is a foreign protein.
   b. An epitope is the part of an antigen where an antibody or lymphocyte receptor binds.
   c. An antigen is the part of an epitope where an antibody or lymphocyte receptor binds.
   d. Antigens are recognized by B cells and antibodies; epitopes are recognized by T cells.

4. How do B-cell receptors and antibodies differ?
   a. B-cell receptors are made up of heavy chains; antibodies are made up of light chains.
   b. B-cell receptors are made up of light chains; antibodies are made up of heavy chains.
   c. Only antibodies include a variable region.
   d. Antibodies lack a transmembrane domain and are secreted.

5. How do memory cells become activated?
   a. They undergo somatic hypermutation.
   b. The same way that B cells and T cells are activated in the primary response.
   c. They undergo clonal expansion.
   d. They are stimulated by histamines.

6. In terms of morphology, T cells come in two major types. How are they distinguished?
   a. They have a CD4 or CD8 protein on their surfaces.
   b. They have extensive ER and many mitochondria.
   c. Only one type has pattern-recognition receptors.
   d. One type functions in the primary response; the other functions in the secondary response.

✔ **TEST YOUR UNDERSTANDING** *Answers are available in Appendix B*

1. To a physician, the classical signs of the inflammatory response are rubor (reddening), calor (heat), dolor (pain), and tumor (swelling). Explain why each symptom occurs.

2. Compare and contrast the general structure of a B-cell receptor and a T-cell receptor (make a sketch of each). Explain how each interacts with an antigen.

3. What do vaccines need to contain in order to be effective? Why don't we have vaccines for HIV?

4. In the clonal-selection theory of adaptive immune system function, what do "clonal" and "selection" refer to?

5. Compare and contrast the interaction between (a) pathogens and the pattern-recognition receptors on leukocytes versus (b) antigens and BCRs or TCRs.

6. Explain how DNA recombination leads to the production of almost limitless numbers of different B-cell receptors, T-cell receptors, and antibodies.

✔ **APPLYING CONCEPTS TO NEW SITUATIONS** *Answers are available in Appendix B*

1. If you were being treated for a badly skinned knee from a bicycle accident, health care workers would irrigate the wound with sterile water, scrub it thoroughly with warm, soapy water, and then apply antibiotics. Explain why these measures are effective.

2. Suppose you discover an antibody to the sodium-potassium pump. Describe how you could use this antibody to study the location of sodium-potassium pumps in various cell types.

3. It seems astonishing that the immune system can produce antibodies to compounds that were recently synthesized for the first time in the lab. Given that viruses and other pathogens evolve rapidly, explain why natural selection favored individuals with genes that made extensive antibody diversity possible.

4. During World War II, physicians discovered that if they removed undamaged skin from a patient and grafted it onto the site of a burn on the same patient, the tissue healed well. But if the grafted tissue came from a different individual, the skin graft was rejected. Explain why.

This young hawksbill sea turtle is tangled in a fishing net—now a common cause of mortaility. Increasingly, ecology is the study of how humans are altering the environment in ways that make life difficult for other organisms.

An Introduction to Ecology 50

Ecology is the study of how organisms interact with their environment. Except for the episode 65 million years ago, when a mountain-sized asteroid struck Earth, environments are changing more rapidly right now than they have at any time in the past 3.5 billion years. Efforts to maintain human health and welfare depend on our ability to cope with these environmental changes. Ecology has become one of the most important and dynamic fields in biological science.

Ecology is also among the most synthetic fields in biological science. In this unit, you'll be applying what you've learned about biochemistry, physiology, evolution, and the diversity of life.

Ecology's primary goal is to understand the distribution and abundance of organisms. How do interactions with other organisms and the physical environment dictate why certain species live where they do, and how many individuals live there? Some biologists ask why orangutans are restricted to forests in Borneo and why their numbers are declining so rapidly. Other biologists create mathematical models to predict how quickly the human immunodeficiency virus will increase in India, or how long it will take for the current human population of 6.8 billion to double.

Distribution and abundance are core questions in today's world. Global climate change is having dramatic effects on species' distributions, and expanding human populations are causing the abundance of many species to crash. Let's delve in.

50.1 Areas of Ecological Study

To understand why organisms live where they do and in what numbers, biologists break ecology into several levels of analysis. This is a common strategy in biological science. Cell biologists study how cells work at different levels of organization—from individual molecules to multimolecular machines and membranes to multicellular organisms.

KEY CONCEPTS

- Ecology's goal is to explain the distribution and abundance of organisms. It is the branch of biology that provides a scientific foundation for conservation efforts.

- Physical structure—particularly water depth—is the primary factor that limits the distribution and abundance of aquatic species. Climate—specifically, both the average value and annual variation in temperature and in moisture—is the primary factor that limits the distribution and abundance of terrestrial species.

- Climate varies with latitude, elevation, and other factors—such as proximity to oceans and mountains. Climate is changing rapidly around the globe.

- A species' distribution is constrained by historical and biotic factors, as well as by abiotic factors such as physical structure and climate.

✔ When you see this checkmark, stop and test yourself. Answers are available in Appendix B.

Physiologists analyze processes at the level of ions and molecules as well as whole cells, tissues and organs, and complete systems.

In ecology, researchers work at four main levels: (1) organisms, (2) populations, (3) communities, and (4) ecosystems. The levels aren't rigid or exclusive; researchers routinely draw on all four to explore a particular question. Let's examine each.

Organismal Ecology

At the organismal level, researchers explore the morphological, physiological, and behavioral adaptations that allow individuals to live in a particular area. Consider sockeye salmon, for example. After spending four or five years feeding and growing in the ocean, these salmon travel hundreds or thousands of kilometers to return to the stream where they hatched (**Figure 50.1a**). Females create nests in the gravel stream bottom and lay eggs. Nearby males compete for the chance to fertilize eggs as they are laid. When breeding is finished, all the adults die.

Biologists want to know how these individuals interact with their physical surroundings and with other organisms in and around the stream. How do individuals cope with the transition from living in salt water to living in freshwater? Which females get the best nesting sites and lay the most eggs, and which males are most successful in fertilizing eggs?

Population Ecology

A **population** is a group of individuals of the same species that lives in the same area at the same time. When biologists study population ecology, they focus on how the numbers of individuals in a population change over time.

For example, researchers use mathematical models to predict the future of salmon populations (**Figure 50.1b**). Many of these populations have declined as their habitats have become dammed or polluted. If the factors that affect population size can be described accurately enough, mathematical models can assess the impact of proposed dams, changes in weather patterns, altered harvest levels, or specific types of protection efforts.

Community Ecology

A biological **community** consists of the species that interact with each other within a particular area. Community ecologists ask questions about the nature of the interactions between species and the consequences of those interactions. The work might concentrate on predation, parasitism, and competition, or explore how groups of species respond to fires, floods, and other disturbances.

As an example, consider the interactions among salmon and other species in the marine and stream communities where they live. When they are at sea, salmon eat smaller fish and are themselves hunted and eaten by orcas, sea lions, humans, and other mammals; when they return to freshwater to breed, they are preyed on by bears and bald eagles (**Figure 50.1c**). In both marine and freshwater habitats, salmon are subject to parasitism and disease. They are also heavily affected by disturbances—particularly changes in their food supply due to overfishing and the damming, logging, or suburbanization of breeding streams.

(a) Organismal ecology

How do individuals interact with each other and their physical environment?

Salmon migrate from saltwater to freshwater environments to breed

(b) Population ecology

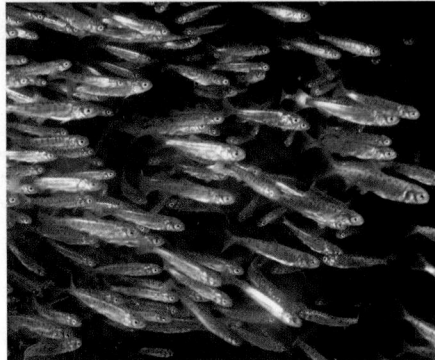

How and why does population size change over time?

Each female salmon produces thousands of eggs. Only a few will survive to adulthood. On average, only two will return to the stream of their birth to breed

(c) Community ecology

How do species interact, and what are the consequences?

Salmon are prey as well as predators

(d) Ecosystem ecology

How do energy and nutrients cycle through the environment?

Salmon die and then decompose, releasing nutrients that are used by bacteria, archaea, plants, protists, young salmon, and other organisms

FIGURE 50.1 Biologists Study Ecology at Four Main Levels.

Ecosystem Ecology

Ecosystem ecology is an extension of organismal, population, and community ecology. An **ecosystem** consists of all the organisms in a particular region along with nonliving components. These physical, or **abiotic** (literally, "not-living"), components include air, water, and soil. At the ecosystem level, biologists study how nutrients and energy move among organisms and through the surrounding atmosphere and soil or water.

Because humans are adding massive amounts of energy and nutrients to ecosystems all over the world, this work has direct public policy implications. Ecosystem ecologists are responsible for assessing the impact of pollution and increased temperature on the distribution and abundance of species—in fact, on Earth's ability to support life.

Salmon are interesting to study at the ecosystem level because they link marine and freshwater ecosystems. They harvest nutrients in the ocean and then, when they migrate, die, and decompose, transport those molecules to streams and streamside forests (**Figure 50.1d**). In this way, salmon transport chemical energy and nutrients from one habitat to another. Because salmon are sensitive to pollution and to changes in water temperature, human-induced changes in marine and freshwater ecosystems have a large impact on their populations.

How Do Ecology and Conservation Efforts Interact?

The four levels of ecological study are synthesized and applied in conservation biology. **Conservation biology** is the effort to study, preserve, and restore threatened populations, communities, and ecosystems (see Chapter 55).

Ecologists study how interactions between organisms and their environments result in a particular species being found in a particular area at a particular population size. Conservation biologists apply these data to preserve species and restore environments.

Conservation biologists are like physicians. But instead of drawing on results from research in molecular biology and physiology, they draw on research in ecology and evolution. Instead of prescribing drugs for sick people, they prescribe remedies for environments, so they produce a diversity of species, clean air, pure water, and productive soils.

50.2 Types of Aquatic Ecosystems

If ecology is the study of how species interact with their environment, what makes up the environment? Every environment has both physical and biological components. The abiotic or physical aspects include temperature, precipitation, sunlight, and wind. The **biotic** ("living") components are members of the same or different species.

Chapters 51–53 focus on biological interactions—how individuals interact with offspring, mates, competitors, predators, prey, and parasites. This chapter focuses on how the physical environment affects organisms. Organisms have a restricted set of physical conditions in which they can survive and thrive. Why?

Let's answer this question by analyzing the physical attributes of aquatic ecosystems in this section, moving on to terrestrial ecosystems in Section 50.3, and considering how global warming is affecting both types of environments in Section 50.4. In freshwater and salt water, three key physical factors affect the distribution and abundance of organisms: nutrient availability, water depth, and water movement.

Nutrient Availability

In many aquatic ecosystems, nutrients such as nitrogen and phosphorus are in short supply. If water is moving, nutrients tend to be washed away. If water is still, nutrients tend to fall to the bottom and collect in the form of debris. Nutrient levels are important: They limit growth rates in the photosynthetic organisms that provide food for other species.

Let's consider two important events that affect nutrient availability in lakes and oceans. In both cases, they are mechanisms that bring nutrients from the bottom up to the water surface, where they can nourish the growth of photosynthetic species.

OCEAN UPWELLING In the oceans, nutrients in the sunlit surface waters are constantly lost in the form of dead organisms that rain down into the depths. In certain coastal regions of the world's oceans, however, nutrients are brought up to the surface by currents that cause upwellings.

Figure 50.2 shows how this happens off the coast of Peru. Here the prevailing winds blow along the coastline, pushing surface water to the north. Because the Earth rotates constantly, this wind-driven water current is slowly moved offshore. As the surface water moves away from the coast, it is steadily replaced by water moving up from the ocean bottom.

The upwelling water is nutrient rich. In effect, it recycles nutrients that earlier had fallen to the ocean floor. The nutrients

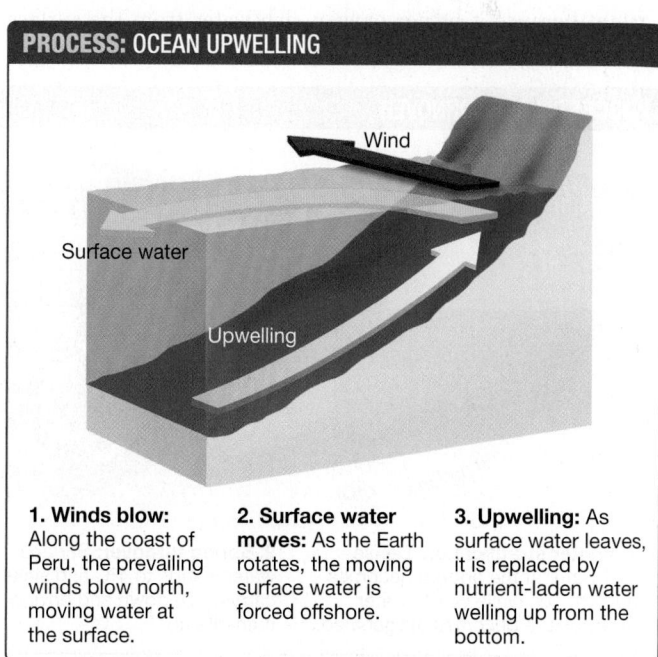

PROCESS: OCEAN UPWELLING

Wind

Surface water

Upwelling

1. Winds blow: Along the coast of Peru, the prevailing winds blow north, moving water at the surface.

2. Surface water moves: As the Earth rotates, the moving surface water is forced offshore.

3. Upwelling: As surface water leaves, it is replaced by nutrient-laden water welling up from the bottom.

FIGURE 50.2 Ocean Upwellings Bring Nutrients to the Surface.

support luxuriant growths of photosynthetic cells, which feed an array of small grazing animals, which in turn feed small fish called anchoveta. When currents are favorable and upwelling is steady, the anchoveta fishery is the most productive in the world.

LAKE TURNOVER   Because they were scoured by glaciers as recently as 10,000 years ago, the higher latitudes of the world host many lakes. Each year, these bodies of water undergo remarkable changes known as the spring and fall **turnovers**.

The spring and fall lake turnovers occur in response to changes in air temperature. To see how this happens, examine the temperature profiles in **Figure 50.3**.

Note that in winter, water at the surface of a northern lake is locked up in ice at a temperature of 0°C. The water just under the ice is slightly warmer and is relatively oxygen rich, because it was exposed to the atmosphere as winter set in. Because water is most dense at 4°C, the water at the bottom of the lake is at 4−5°C. The bottom water is also oxygen poor, because organisms that decompose falling organic material use up available oxygen in the process.

A gradient in temperature such as this is called a **thermocline** ("heat-slope"). Thermal stratification is said to occur.

The ice begins to melt, however, when spring arrives. The temperature of the water on the surface rises until it reaches 4°C. The water at the surface of the lake is now heavier than the water below it. As a result, the water on the surface sinks. The water at the bottom of the lake is displaced and comes to the surface, completing the spring turnover.

During the spring turnover, water at the bottom of the lake carries sediments and nutrients from the bottom up to the surface. This flush of nutrients triggers a rapid increase in the growth of photosynthetic algae and bacteria that biologists call the spring bloom.

When temperatures cool in the fall, the water at the surface reaches a temperature of 4°C and sinks, displacing water at the bottom and creating the fall turnover. The fall turnover is important because it brings oxygen-rich water from the surface down to the bottom, and because it again brings nutrients from the bottom up to the sunlit surface waters.

Without the spring and fall turnovers, most freshwater nutrients would remain on the bottom of lakes. These aquatic ecosystems would be much less productive, as a result.

Water Flow

Water movement is a critical factor in aquatic ecosystems because it presents a physical challenge. It can literally sweep organisms off their feet.

Organisms that live in fast-flowing streams have to cope with the physical force of the water, which constantly threatens to move them downstream. Marine organisms that live in intertidal regions are exposed to the air periodically each day, as well as to violent wave action during storms.

Nutrient and light availability influence productivity in aquatic ecosystems; water movement has an effect on productivity and is a physical force that organisms have to contend with. The boxes that follow summarize the types of freshwater and marine environments that organisms occupy, with an emphasis on how they are affected by variation in water depth and movement.

Water Depth

Water absorbs and scatters light, so the amount and types of wavelengths available to organisms change dramatically as depth increases. Light has a major influence on **productivity**—the total amount of carbon fixed by photosynthesis per unit area per year.

At the water surface, all wavelengths of light are equally available. But as **Figure 50.4a** shows, ocean water specifically removes light in the violet and red regions of the visible spectrum. In turbid water, blue wavelengths are also absorbed. This is important because wavelengths in the blue and red regions are required for photosynthesis in many species (see Chapter 10).

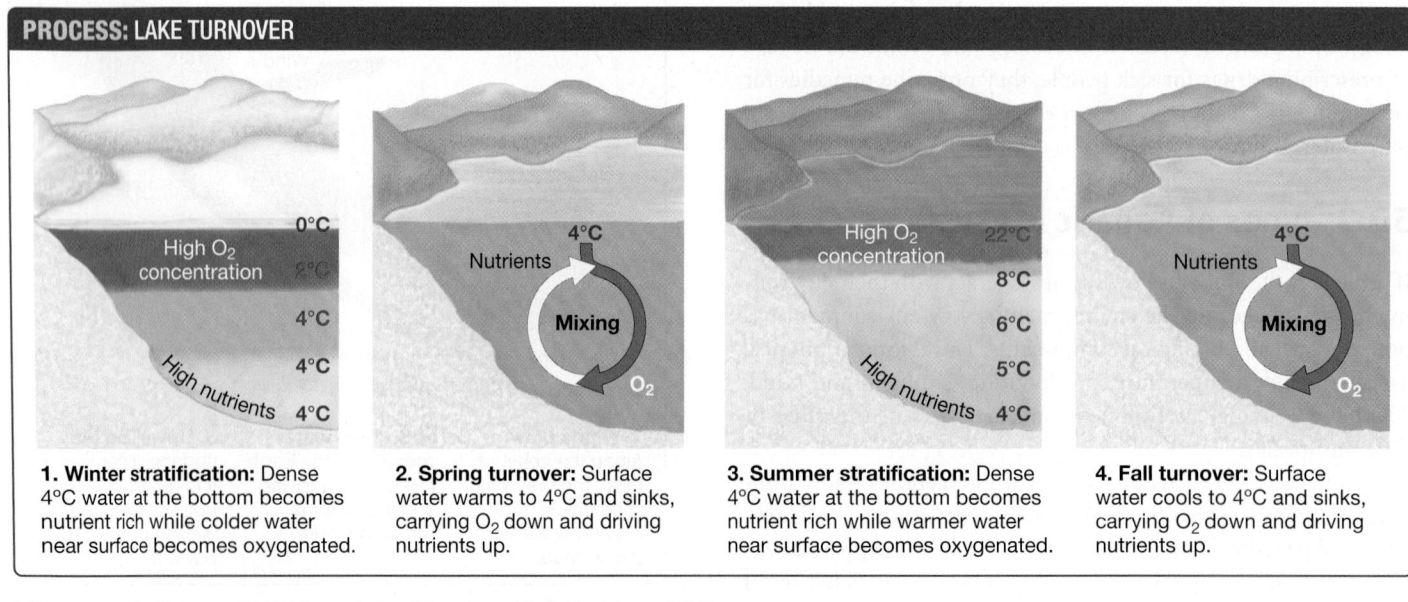

PROCESS: LAKE TURNOVER

1. Winter stratification: Dense 4°C water at the bottom becomes nutrient rich while colder water near surface becomes oxygenated.

2. Spring turnover: Surface water warms to 4°C and sinks, carrying O₂ down and driving nutrients up.

3. Summer stratification: Dense 4°C water at the bottom becomes nutrient rich while warmer water near surface becomes oxygenated.

4. Fall turnover: Surface water cools to 4°C and sinks, carrying O₂ down and driving nutrients up.

FIGURE 50.3 In Temperate Regions, Lakes Turn Over Each Spring and Fall.

(a) Only certain wavelengths of light are available under water.

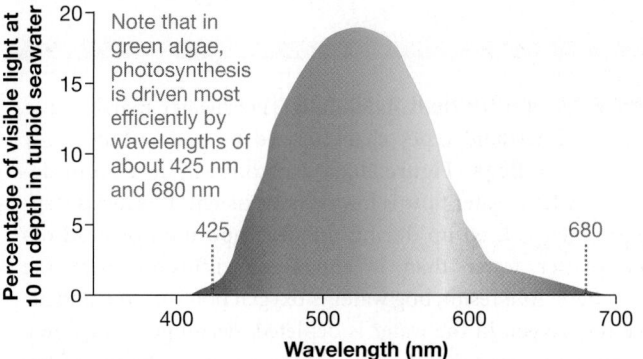

Note that in green algae, photosynthesis is driven most efficiently by wavelengths of about 425 nm and 680 nm

(b) Intensity of light declines with water depth.

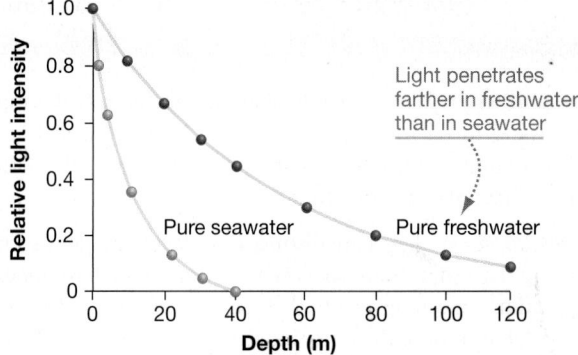

Light penetrates farther in freshwater than in seawater

Pure seawater   Pure freshwater

FIGURE 50.4 Availability of Light Changes Dramatically with Increasing Water Depth.

✔**QUESTION** Would you expect to find photosynthetic organisms at greater depths in freshwater or seawater? Explain your answer.

The quantity of light available to organisms also diminishes rapidly with increasing depth. The graph in **Figure 50.4b** plots water depth along the *x*-axis and light intensity—relative to the surface—on the *y*-axis. Each data point is a reading of light intensity at a particular depth.

Note that in pure seawater, the total amount of light available at a depth of 10 m is less than 40 percent of what it is at the surface; virtually no light reaches depths greater than 40 m (131 ft). In seawater that contains organisms or debris, light penetration is dramatically less than in pure seawater.

Freshwater Environments > Lakes and Ponds

Lakes and ponds are distinguished from each other by size. Ponds are small; lakes are large enough that the water in them can be mixed by wind and wave action. Most lakes and ponds occur in high latitudes—they formed in depressions that were created by the scouring action of glaciers thousands of years ago. In the tropics, most lakes consist of old river channels. Elsewhere, many of the lakes and ponds that exist now were dug recently by people.

Water Depth Biologists describe the structure of lakes and ponds by naming five zones (**Figure 50.5**).

- The **littoral** ("seashore") **zone** consists of the shallow waters along the shore, where flowering plants are rooted.
- The **limnetic** ("lake") **zone** is offshore and comprises water that receives enough light to support photosynthesis.
- The **benthic** ("depths") **zone** is made up of the substrate.
- Regions of the littoral, limnetic, and benthic zones that receive sunlight are part of the **photic zone**.
- Portions of a lake or pond that do not receive sunlight make up the **aphotic zone**.

Water Flow and Nutrient Availability Water movement in lakes and ponds is driven by wind and temperature. The littoral and limnetic zones are typically much warmer and better oxygenated than the benthic zone, simply because they receive so much more solar radiation and are in contact with oxygen in the atmosphere. The benthic zone, in contrast, is relatively nutrient rich because dead and decomposing bodies sink and accumulate there. In many cases, movement of nutrients from the benthic to littoral zones is dependent on seasonal turnovers.

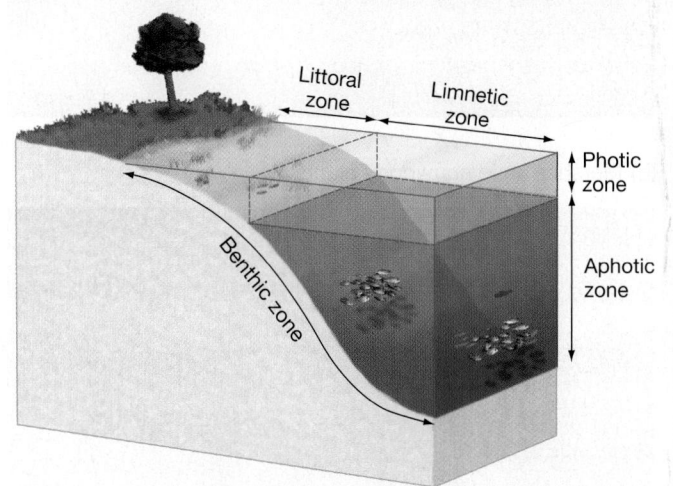

FIGURE 50.5 Lakes Have Distinctive Zones Defined by Water Depth and Distance from Shore.

Organisms Cyanobacteria, algae, and other microscopic organisms, collectively called **plankton**, live in the photic zone, as do the fish and small crustaceans—organisms related to shrimp and crabs—that eat them. In shallow parts of the photic zone, rooted plants are common. Animals (invertebrates and fish) that consume dead organic matter, or **detritus**, are particularly abundant in the benthic zone.

✔You should be able to predict which zone of lakes and ponds produces the most grams of organic material per square meter per year, and explain your logic.

Wetlands are shallow-water habitats where the soil is saturated with water for at least part of the year. They are usually distinguished from terrestrial habitats by the presence of "indicator plants" that only grow in saturate soils.

Water Depth Wetlands are distinct from lakes and ponds for two reasons: They have only shallow water, and they have **emergent vegetation**—meaning plants that grow above the surface of the water. All or most of the water in wetlands receives sunlight; emergent plants capture sunlight before it strikes the water.

(a) Bogs are stagnant and acidic.

Water Flow and Nutrient Availability Freshwater marshes and swamps are wetland types characterized by a slow but steady flow of water. **Bogs** (**Figure 50.6a**), in contrast, develop in depressions where water flow is low or nonexistent. If water is stagnant, oxygen is used up during the decomposition of dead organic matter faster than it enters via diffusion from the atmosphere. As a result, bog water is oxygen poor or even anoxic. Once the oxygen in the water is depleted, decomposition slows. Organic acids and other acids build up, lowering the pH of the water. At low pH, nitrogen becomes unavailable to plants. As a result, bogs are nutrient poor. Marshes and swamps, in contrast, are relatively nutrient rich.

Organisms The combination of acidity, lack of available nitrogen, and anoxic conditions makes bogs extremely unproductive habitats. Marshes and swamps, in contrast, offer ample supplies of oxygenated water and sunlight, along with nutrients available from rapid decomposition. As a result, they are extraordinarily productive. **Marshes** lack trees and typically feature grasses, reeds, or other nonwoody plants (**Figure 50.6b**); **swamps** are dominated by trees and shrubs (**Figure 50.6c**). Because their physical environments are so different, there is little overlap in the types of species found in bogs, marshes, and swamps.

✔️ You should be able to explain why carnivorous plants, which capture and digest insects, are relatively common in bogs but rare in marshes and swamps.

(b) Marshes have nonwoody plants.

(c) Swamps have trees and shrubs.

FIGURE 50.6 Wetland Types Are Distinguished by Water Flow and Vegetation.

Streams are bodies of water that move constantly in one direction. Creeks are small streams; rivers are large. In terms of the environment that is available to organisms, the major physical variables in streams are the speed of the current and the availability of oxygen and nutrients.

Water Depth Most streams are shallow enough that sunlight reaches the bottom. Availability of sunlight is usually not a limiting factor for organisms.

Water Flow and Nutrient Availability The structure of a typical stream varies along its length (**Figure 50.7**). Where it originates at a mountain glacier, lake, or spring, a stream tends to be cold, narrow, and fast. As it descends toward a lake, ocean, or larger river, a stream accepts water from tributaries and becomes larger, warmer, and slower.

Oxygen levels tend to be high in fast-moving streams because water droplets are exposed to the atmosphere when moving water splashes over rocks or other obstacles. Oxygen from the atmosphere diffuses into the droplets. In contrast, slow-moving streams that lack rapids tend to become relatively oxygen poor. Also, cold water holds more oxygen than warm water does (see Chapter 44).

Slow-moving streams tend to be more nutrient rich than fast-moving streams, because decaying matter does not flush away as quickly. In many cases, streams are dependent on nutrients that arrive via falling leaves or other organic matter, or on soil particles that wash in from surrounding terrestrial communities.

Organisms It is rare to find photosynthetic organisms in small, fast-moving streams; nutrient levels tend to be low and most of the organic matter present consists of leaves and other materials that fall into the water from outside the stream. Fish, insect larvae, mollusks, and other animals have adaptations that allow them to maintain their positions in the fast-moving portions of streams. As streams widen and slow down, conditions become more favorable for the growth of algae and plants, and the amount of organic matter and nutrients increases. As a result, the same stream often contains completely different types of organisms near its source and near its end, or mouth.

✔You should be able to explain why, in terms of their ability to perform cellular respiration, fish species found in cold, fast-moving streams tend to be much more active than fish species found in warm, slow-moving streams.

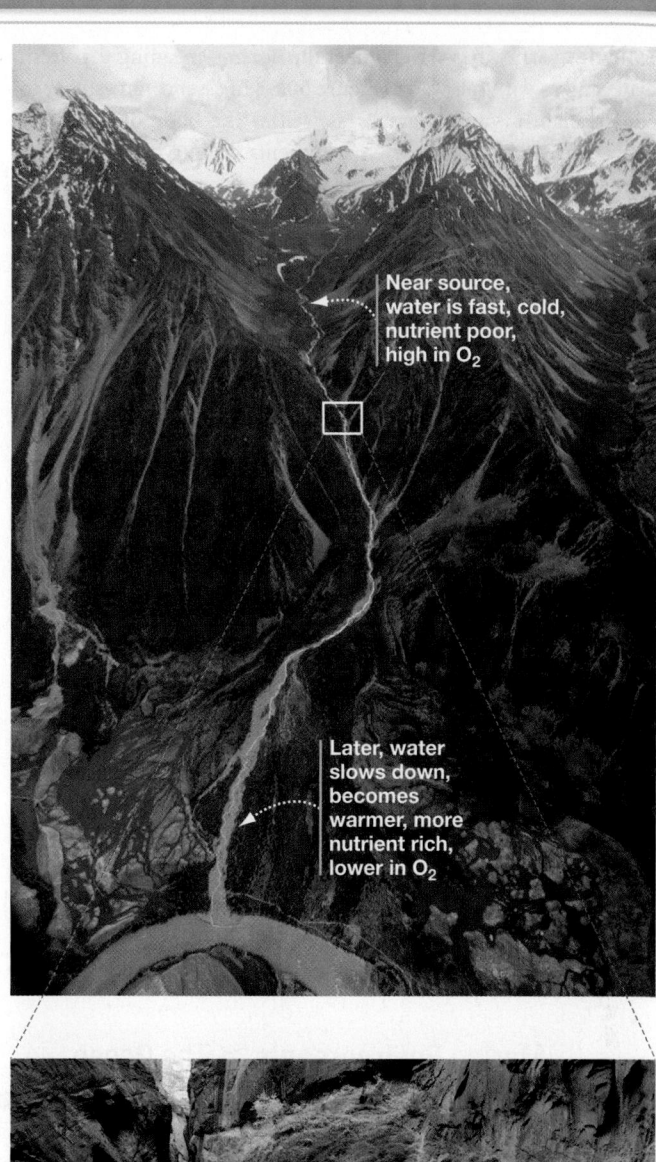

Near source, water is fast, cold, nutrient poor, high in O_2

Later, water slows down, becomes warmer, more nutrient rich, lower in O_2

FIGURE 50.7 Distinctive Environments Appear along a Stream's Length.

Estuaries form where rivers meet the ocean—meaning that fresh-water mixes with salt water (**Figure 50.8**). In essence, an estuary includes slightly saline marshes as well as the body of water that moves in and out of these environments. Salinity varies with (**1**) changes in river flows—it declines when the river floods and increases when the river ebbs—and (**2**) with proximity to the ocean. Salinity has dramatic effects on osmosis and water balance (see Chapters 37 and 42); species that live in estuaries have adaptations that allow them to cope with variations in salinity.

Water Depth Most estuaries are shallow enough that sunlight reaches the substrate. Water depth may fluctuate dramatically, however, in response to tides, storms, and floods.

Water Flow and Nutrient Availability Water flow in estuaries fluctuates daily and seasonally due to tides, storms, and floods. The fluctuation is important because it alters salinity, which in turn affects which types of organisms are present. Estuaries are nutrient rich, because nutrient-laden sediments are deposited when flowing river water slows as it enters the ocean.

Organisms Because the water is shallow and sunlit, and because nutrients are constantly replenished by incoming river water, estuaries are among the most productive environments on Earth. They are often packed with young fish, which feed on abundant vegetation and plankton while hiding from predators.

✔You should be able to explain why estuaries and freshwater marshes contain few of the same species, even though they are both shallow-water habitats filled with rooted plants.

FIGURE 50.8 Estuaries Are Highly Productive Environments.

Marine Environments > The Ocean

The world's oceans form a continuous body of salt water and are remarkably uniform in chemical composition. Regions within an ocean vary markedly in their physical characteristics, however, with profound effects on the organisms found there.

Water Depth Biologists describe the structure of an ocean by naming six regions (**Figure 50.9**).

- The **intertidal** ("between tides") **zone** consists of a rocky, sandy, or muddy beach that is exposed to the air at low tide but submerged at high tide.

- The **neritic zone** extends from the intertidal zone to depths of about 200 m. Its outermost edge is defined by the end of the **continental shelf**—the gently sloping, submerged portion of a continental plate.

- The **oceanic zone** is the "open ocean"—the deepwater region beyond the continental shelf.

- The bottom of the ocean is the **benthic zone**.

- The intertidal and sunlit regions of the neritic, oceanic, and benthic zones make up a **photic zone**.

- Areas that do not receive sunlight are in an **aphotic zone**.

Water Flow and Nutrient Availability Water movement in the ocean is dominated by different processes at different depths. In the intertidal zone, tides and wave action are the major influences. In the neritic zone, currents that bring nutrient-rich water from the benthic zone of the deep ocean toward shore have a heavy impact. Throughout the ocean, large-scale currents circulate water in the oceanic zone in response to prevailing winds and the Earth's rotation.

In general, nutrient availability is dictated by water movement. The neritic and intertidal zones are relatively nutrient rich, because they receive nutrients from rivers and upwellings. The oceanic zone, in contrast, constantly loses nutrients due to a steady rain of dead organisms drifting to the benthic zone.

(Continued on next page)

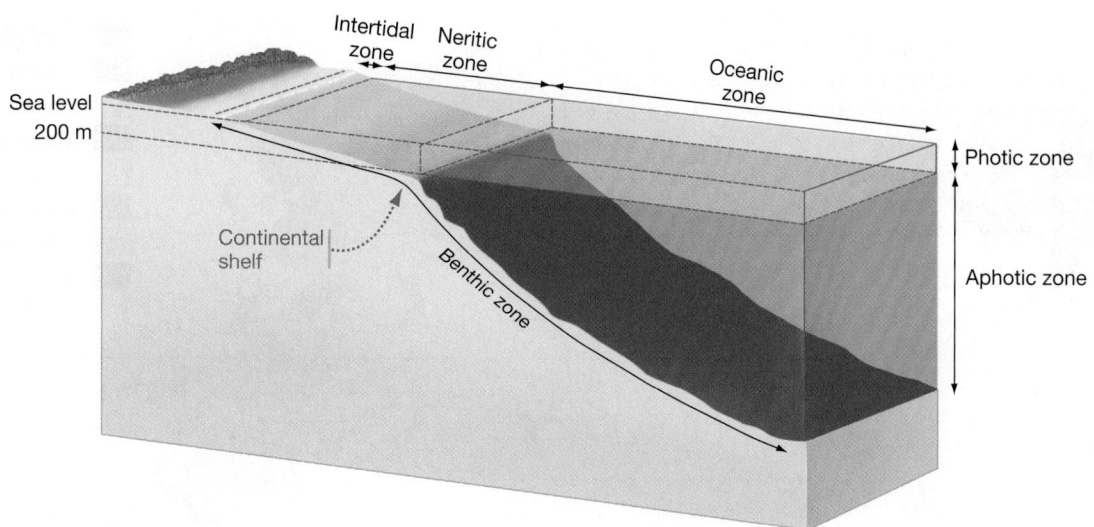

FIGURE 50.9 Oceans Have Distinctive Zones Defined by Water Depth and Distance from Shore.

Organisms Each zone in the ocean is populated by distinct species that are adapted to the physical conditions present. Organisms that live in the intertidal zone must be able to withstand physical pounding from waves and desiccation and high temperatures at low tide. Productivity is high, however, due to the availability of sunlight and nutrients contributed by estuaries as well as by currents that sweep in nutrient-laden sediments from offshore areas.

Productivity is also high on the outer edge of the neritic zone, due to nutrients contributed by upwellings at the edge of the continental plate. Almost all of the world's major marine fisheries exploit organisms that live in the neritic zone. In the tropics, shallow portions of the neritic zone may host **coral reefs**. Because the water is warm and sunlight penetrates to the ocean floor in these habitats, coral reefs are among the most productive environments in the world (see Chapter 54).

If coral reefs are the rain forests of the ocean, then the oceanic zone is the desert. Sunlight is abundant in the photic zone of the open ocean, but nutrients are extremely scarce. When the photosynthetic organisms and the animals that feed on them die, their bodies drift downward out of the photic zone and are lost. In the open ocean, the benthic zone can be several kilometers below the surface, and there is no mechanism for bringing nutrients back up from the bottom. The aphotic zone of the open ocean is also extremely unproductive because light is absent and photosynthesis is impossible. Most organisms present in the aphotic zone—including the sea cucumbers highlighted in Chapter 34—survive on the rain of dead bodies from the photic zone.

✔You should be able to predict whether animals that live in the aphotic zone have functioning eyes.

CHECK YOUR UNDERSTANDING

If you understand that . . .

- Because nutrients tend to sink to the bottom of aquatic ecosystems, productivity depends on mechanisms to bring nutrients back up to the surface—as well as on the availability of light.
- Aquatic environments are distinguished by aspects of physical structure—primarily the depth of water and the rate at which it flows.

✔ **You should be able to . . .**

1. Explain why coastal environments are the most productive areas in the ocean.

2. Explain what happens when water at the surface of a lake cools to 4°C and the water below it is slightly warmer.

Answers are available in Appendix B.

50.3 Types of Terrestrial Ecosystems

If you could walk from the equator in South America to the North Pole, you would notice startling changes in the organisms around you. Lush tropical forests with broad-leaved evergreen trees would give way to seasonally dry forests and then to deserts. The deserts would yield to the vast grasslands of central North America, which terminate at the boreal forests of the subarctic. If you pressed on, you would reach the end of the trees and the beginning of the most northerly community—the arctic tundra.

If you walked from sea level to the top of the Cascade Mountains of British Columbia, you would experience similar types of changes. Wet forests along the coast would give way to drier forests at higher elevations. Near the peaks, trees would thin and eventually yield to alpine tundras filled with hardy, low-growing plants.

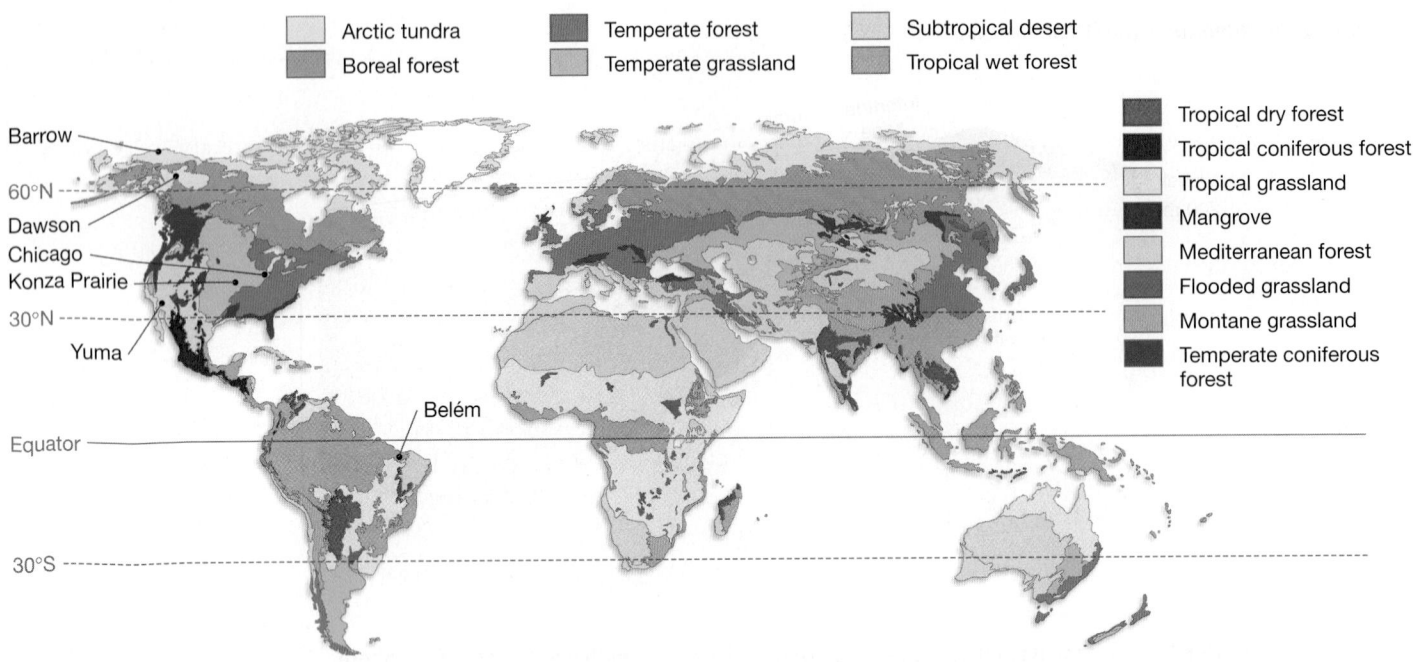

Arctic tundra | **Temperate forest** | **Subtropical desert**
Boreal forest | **Temperate grassland** | **Tropical wet forest**

Tropical dry forest
Tropical coniferous forest
Tropical grassland
Mangrove
Mediterranean forest
Flooded grassland
Montane grassland
Temperate coniferous forest

Barrow
60°N
Dawson
Chicago
Konza Prairie
30°N
Yuma
Belém
Equator
30°S

FIGURE 50.10 Distinct Biomes Are Found throughout the World. A recent classification identified the 14 major biomes shown here. The locations indicated with black dots are highlighted in this section.

Broad-leaved evergreen forests, deserts, and grasslands are **biomes:** major groupings of plant and animal communities defined by a dominant vegetation type. **Figure 50.10** shows the global distribution of the most common types of biomes.

Each of the biomes found around the world is associated with a distinctive set of abiotic conditions. Just as water depth and movement have an overriding influence on aquatic ecosystems, the type of biome present in a terrestrial region depends on **climate**—the prevailing, long-term weather conditions found in an area. **Weather** consists of the specific short-term atmospheric conditions of temperature, precipitation, sunlight, and wind.

- Temperature is critical because the enzymes that make life possible work at optimal efficiency only in a narrow range of temperatures (see Chapter 3). Temperature also affects the availability of moisture: Water freezes at low temperatures and evaporates rapidly at high temperatures.

- Moisture is significant because it is required for life, and because terrestrial organisms constantly lose water to the environment through evaporation or transpiration. To stay alive, they must reduce water loss and replace lost water.

- Sunlight is essential because it is required for photosynthesis.

- Wind is important because it exacerbates the effects of temperature and moisture. Wind increases heat loss due to evaporation and convection, and it increases water loss due to evaporation and transpiration. It also has a direct physical impact on organisms such as birds, flying insects, and plants—it pushes them around.

Of the four components of climate, variation in temperature and moisture are far and away the most important in determining plant distribution and abundance. More specifically, the nature of the biome that develops in a particular region is governed by (**1**) average annual temperature and precipitation, and (**2**) annual variation in temperature and precipitation. Each biome contains species that are adapted to a particular temperature and moisture regime. The amount and variability of heat and precipitation structure terrestrial environments, much as water depth and flow rate structure aquatic environments.

Biologists are particularly concerned with how temperature and water influence **net primary productivity (NPP).** NPP is defined as the total amount of carbon that is fixed per year minus the amount that is oxidized during cellular respiration. Fixed carbon that is consumed in cellular respiration provides energy for the organism but is not used for growth—that is, production of **biomass.**

NPP is key because it represents the organic matter that is available as food for other organisms. In terrestrial environments NPP is often estimated by measuring **aboveground biomass**—the total mass of living plants, excluding roots.

Photosynthesis, plant growth, and NPP are maximized on land when temperatures are warm and conditions are wet. Conversely, photosynthesis cannot occur efficiently at low temperatures or under drought stress. Photosynthetic rates are maximized in warm temperatures, when enzymes work efficiently, and in humid weather, when stomata can remain open and CO_2 is readily available.

To help you understand how temperature and precipitation influence biomes, let's take a detailed look at the six biomes that represent Earth's most extensive vegetation types, ranging from the wet tropics to the arctic. In each case, you'll be analyzing temperature and precipitation data from a specific location, typical of the biome in question. Data plotted in red indicate the daily average temperature in each biome, graphed throughout the year; bar charts plotted in purple indicate the average monthly precipitation.

Tropical wet forests—also called tropical rain forests—are found in equatorial regions around the world. Plants in this biome have broad leaves as opposed to narrow, needle-like leaves, and are evergreen. Older leaves are shed throughout the year, but there is no complete, seasonal loss of leaves.

Temperature The data in **Figure 50.11** are from Belém, Brazil, where mean monthly temperatures never drop below 25°C and never exceed 30°C. Compared to other biomes, tropical wet forests show almost no seasonal variation in temperature. This is important because temperatures are high enough to support growth throughout the year.

Precipitation Even in the driest month of the year, November, this region receives over 5 cm (2 in.) of rainfall—considerably more than the *annual* rainfall of many deserts.

Vegetation Favorable year-round growing conditions produce riotous growth, leading to extremely high productivity and aboveground biomass (**Figure 50.12**). Tropical wet forests are also renowned for their species diversity. It is not unusual to find over 200 tree species in a single 100 m × 100 m study plot. And based on counts of the insects and spiders collected from single trees, some biologists contend that the world's tropical wet forests may hold up to 30 million species of arthropods alone.

The diversity of plant sizes and growth forms in wet forest communities produces extraordinary structural diversity. In tropical wet forests, a few extremely large trees tower above a layer of large trees that form a distinctive and continuous **canopy** (the uppermost layers of branches). From the canopy to the ground, there is a complex assortment of vines, **epiphytes** (plants that grow entirely on other plants), small trees, shrubs, and herbs. This diversity of growth forms presents a wide array of habitat types for animals.

✔You should be able to explain why the presence of vines and epiphytes increases the productivity of tropical wet forests.

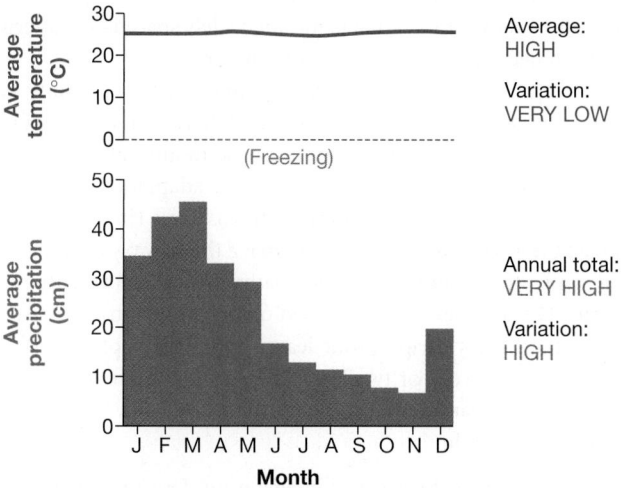

FIGURE 50.11 Climate Data from Belém, Brazil.

Average: HIGH

Variation: VERY LOW

Annual total: VERY HIGH

Variation: HIGH

**FIGURE 50.12
Tropical Wet Forests
Are Extremely Rich
in Species.**

Subtropical deserts are found throughout the world in two distinctive locations: 30 degrees latitude, or distance from the equator, both north and south. Most of the world's great deserts—including the Sahara, Gobi, Sonoran, and Australian outback—lie on or about 30° N or 30° S latitude—a pattern that is explained in Section 50.4.

Temperature The data in **Figure 50.13** are from Yuma, Arizona, in the Sonoran Desert of southwestern North America. Mean monthly temperatures in subtropical deserts vary more than in tropical wet climates, but in Yuma at least, temperatures still never

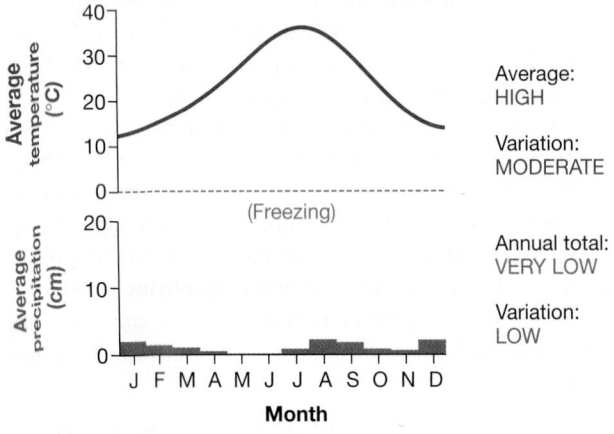

Average:
HIGH

Variation:
MODERATE

Annual total:
VERY LOW

Variation:
LOW

FIGURE 50.13 Climate Data from the Sonoran Desert.

fall below freezing. (Freezing nighttime temperatures are common elsewhere in the Sonora and in many other subtropical deserts.)

Precipitation The most striking feature of the subtropical desert climate is low precipitation. The average annual precipitation in Yuma is just 7.5 cm (3 in.).

Vegetation The scarcity of water in deserts has profound implications. Conditions are often too dry to support photosynthesis, so the productivity of desert communities is a tiny fraction of average values for tropical forest communities. Further, as the photo in **Figure 50.14** shows, individual plants are widely spaced—a pattern that may reflect intense competition for water.

Desert species adapt to the extreme temperatures and aridity in one of two ways: Growing at a low rate year-round, or breaking dormancy and growing rapidly in response to any rainfall. Cacti can grow year-round because they possess adaptations to cope with hot, dry conditions: small leaves or no leaves (in which case photosynthetic cells are located in stems); a thick, waxy coating on leaves and stems; and the CAM pathway for photosynthesis (Chapter 10). The desert shrub called ocotillo, in contrast, is usually dormant. Individuals sprout leaves within days of a rainfall and drop them a week or two later when soils dry out again. The seeds of annual plants can also lie dormant for many years, then germinate after a rain.

✔You should be able to explain why most desert plant species have no leaves or leaves with a small surface area.

FIGURE 50.14 Subtropical Deserts Are Characterized by Widely Spaced Plants.

Temperate grassland communities are found throughout central North America and the heartland of Eurasia (see Figure 50.10). In North America they are commonly called prairies; in central Eurasia they are known as steppes.

Temperature A region of the world is called **temperate** if it has pronounced annual fluctuations in temperature—typically hot summers and cold winters. Temperature variation is important because it dictates a well-defined growing season. In the temperate zone, plant growth is possible only in spring, summer, and fall months when moisture and warmth are adequate. **Figure 50.15**

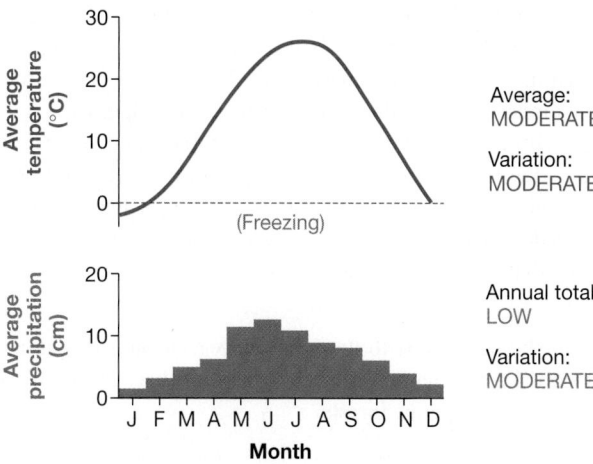

FIGURE 50.15 Climate Data from Konza Prairie.

presents climate data for a typical grassland community—in this case, the Konza Prairie near Manhattan, Kansas.

Precipitation Precipitation at Konza is over four times greater than that in Yuma, Arizona. Nonetheless, conditions are still slightly too hot and dry to support forests; average annual precipitation is 83.4 cm (32.5 in.).

Vegetation Grasses are the dominant life-form in temperate grasslands (**Figure 50.16**) for one of two reasons: (1) Conditions are too dry to enable tree growth, or (2) encroaching trees are burned out by fires. Prairie fires are ignited by lightning strikes or by native people managing land for game animals that graze on grasses and herbaceous (nonwoody) plants. In contrast to deserts, plant life is extremely dense—virtually every square centimeter is filled.

Although the productivity of temperate grasslands is generally lower than that of forest communities, grassland soils are often highly fertile. The subsurface is packed with roots and rhizomes, which add organic material to the soil as they die and decay. Further, grassland soils retain nutrients, because rainfall is low enough to keep key ions from dissolving and leaching out of the soil. It is no accident, then, that the grasslands of North America and Eurasia are the breadbaskets of those continents. The conditions that give rise to natural grasslands are ideal for growing wheat, corn, and other cultivated grasses.

✔You should be able to explain why fires are much more common in grasslands than they are in deserts, even though deserts are drier.

**FIGURE 50.16
Temperate Grasslands
Have More Biomass
Belowground than
Aboveground.**
Although most of the plants in this photo are less than half a meter high, their roots may reach depths of 6 meters or more.

In temperate areas with relatively high precipitation, grasslands give way to forests. Temperate forests are the most common biome found in eastern North America, western Europe, east Asia, Chile, and New Zealand.

Temperature As **Figure 50.17** shows, temperate forests experience a period in which mean monthly temperatures fall below freezing and plant growth stops. The data in this graph are from Chicago, Illinois, which was forested before being settled.

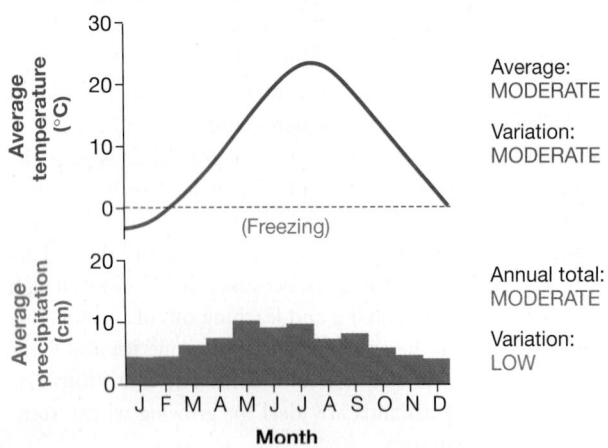

FIGURE 50.17 Climate Data from Chicago Temperate Forest.

Precipitation Compared to grassland climates, precipitation in temperate forests is moderately high and relatively constant throughout the year. Chicago, for example, has an annual precipitation of 85 cm (34 in.); during most months, precipitation exceeds 5 cm (2 in.).

Vegetation In North America and Europe, temperate forests are dominated by deciduous species, which are leafless in winter and grow new leaves each spring and summer (**Figure 50.18**). Needle-leaved evergreens are also common. But in the temperate forests of New Zealand and Chile, broad-leaved evergreens predominate.

Most temperate forests have productivity levels that are lower than those of tropical forests yet higher than those of deserts or grasslands. The level of diversity is also moderate. A temperate forest in southeastern North America may have more than 20 tree species; similar forests to the north may have fewer than 10.

The contrast in productivity between tropical and temperate forests is due to the fluctuation in temperature in temperate regions. Temperate forests are less productive simply because temperatures do not support photosynthesis year-round. The contrast in productivity may also be responsible for the contrast in species richness (see Chapter 53).

✔You should be able to describe the vegetation found in regions with climate characteristics intermediate between those of grasslands and temperate forests.

FIGURE 50.18 Many Temperate Forests Are Dominated by Broad-leaved Deciduous Trees.

The boreal forest, or **taiga**, stretches across most of Canada, Alaska, Russia, and northern Europe. Because these regions are just south of the Arctic Circle, they are referred to as subarctic.

Temperature The data in **Figure 50.19** are from Dawson, in the Yukon Territory of Canada. The region is characterized by very cold winters and cool, short summers. Temperature variation is extreme; in the course of a year, subarctic areas may be subject to temperature ranges of more than 70°C.

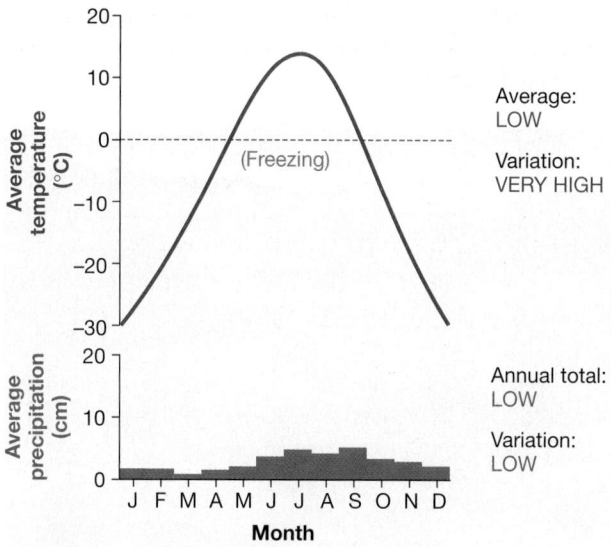

Average:
LOW

Variation:
VERY HIGH

Annual total:
LOW

Variation:
LOW

FIGURE 50.19 Climate Data from Yukon Territory.

Precipitation Annual precipitation in boreal forests is low, but temperatures are so cold that evaporation is minimal; as a result, moisture is usually abundant enough to support tree growth.

Vegetation Boreal forests are dominated by highly cold-tolerant conifers, including pines, spruce, fir, and larch trees. Except for larches, these species are evergreen. Two hypotheses have been offered to explain why evergreens predominate in cold environments, even though they do not photosynthesize in winter. The first is that evergreens can begin photosynthesizing early in the spring, even before the snow melts, when sunshine is intense enough to warm their needles. The second hypothesis is based on the observation that boreal forest soils tend to be acidic and contain little available nitrogen. Because leaves are nitrogen rich, species that must produce an entirely new set of leaves each year might be at a disadvantage. To date, however, these hypotheses have not been tested rigorously.

Based on these observations, it is not surprising that the productivity of boreal forests is low. Aboveground biomass is high, however, because slow-growing tree species may be long lived and gradually accumulate large standing biomass. Boreal forests also have exceptionally low species diversity. The boreal forests of Alaska, for example, typically contain seven or fewer tree species.

The other salient characteristic of boreal forests is that they lack most elements of structure seen in tropical or even temperate forests. They often have just a tree layer and a ground layer (**Figure 50.20**).

✔You should be able to predict how the global distribution of boreal forests will change in response to global warming.

FIGURE 50.20 Boreal Forests Are Dominated by Needle-Leaved Evergreen Trees.

The arctic **tundra** biome, which lies poleward from the subarctic, is found throughout the arctic regions of the Northern Hemisphere and in regions of Antarctica that are not covered in ice.

Temperature Tundra develops in climatic conditions like those in Barrow, on the northern coast of Alaska (**Figure 50.21**). The growing season is 10–12 weeks long at most; for the remainder of the year, temperatures are below freezing.

Precipitation Precipitation on the arctic tundra is extremely low. The annual precipitation in Barrow is actually less than that in the Sonoran Desert of southwestern North America. Because of the extremely low evaporation rates, however, many arctic soils are saturated year-round.

Vegetation The arctic tundra is treeless. The leading explanation is that the short growing season prevents the production of large amounts of non-photosynthetic tissue. Also, tall plants that poke above the snow in winter experience substantial damage from wind-driven snow and ice crystals. Woody shrubs such as willows, birch, and blueberries are common, but they rarely exceed the height of a child. Most tundra species hug the ground.

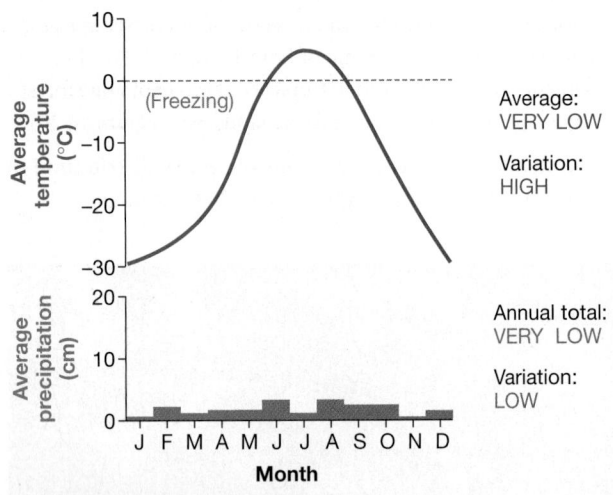

FIGURE 50.21 Climate Data from Barrow, Alaska.

Arctic tundra has low species diversity, low productivity, and low aboveground biomass. Most tundra soils are in the perennially frozen state known as **permafrost**. The low temperatures inhibit both the release of nutrients from decaying organic matter and the uptake of nutrients into live roots. Unlike desert biomes, however, the ground surface in tundra communities is completely covered with plants or lichens (**Figure 50.22**). Animal diversity also tends to be low in the arctic tundra, although insect abundance—particularly of biting flies—can be staggeringly high.

✔You should be able to explain why the types of vegetation in arctic tundras are similar to the types found in alpine tundras, which occur at high elevation.

FIGURE 50.22 Arctic Tundra Features Low-Growing Plants.

CHECK YOUR UNDERSTANDING

🔑 **If you understand that . . .**

- In terrestrial environments, distinctive biomes develop based on two aspects of temperature and moisture: average value and variability over the year.

✔ **You should be able to . . .**

Predict whether tropical dry forests, which have high year-round temperatures but a distinct dry season, are more or less productive than tropical wet forests.

Answers are available in Appendix B.

50.4 The Role of Climate and the Consequences of Climate Change

Each type of aquatic environment and terrestrial biome hosts species that are adapted to the abiotic conditions present at that location. Global warming—due to rising concentrations of atmospheric CO_2 (see Chapter 54)—is having a profound impact on these abiotic factors. Runoff from melting glaciers and pole ice is changing water depths along coasts, and warming ocean waters are disturbing long-established patterns in the direction and intensity of ocean currents. On land, higher air temperatures are altering growing seasons and water availability in biomes around the globe.

To understand how global climate change is affecting the distribution and abundance of organisms, biologists begin by studying why climate varies in predictable ways around the planet. Then they ask how changes in these climate patterns are affecting the organisms present, with the aim of predicting the consequences of continued change. Let's do the same.

Global Patterns in Climate

Simple questions often have fascinating answers. This turns out to be true for questions about why climate varies around the globe. For example, why are some parts of the world warmer and wetter than others? Why do seasons exist?

WHY ARE THE TROPICS WET? One of the most striking weather patterns on Earth involves precipitation. When average annual rainfall is mapped for regions around the globe, it is clear that areas along the equator receive the most moisture. In contrast, locations about 30 degrees latitude north and south of the equator are among the driest on the planet. Why do these patterns occur?

A major cycle in global air circulation, called a **Hadley cell**, is responsible for making the Amazon River basin wet and the Sahara Desert dry. Hadley cells are named after George Hadley, who in 1735 conceived of the idea of enormous air circulation patterns.

To understand how Hadley cells work, put your finger on the equator in **Figure 50.23a** and follow the arrows.

- Air that is heated by the strong sunlight along the equator expands, lowering its pressure and causing it to rise. Note that

this warm air holds a great deal of moisture, because warm water molecules tend to stay in vapor form instead of condensing into droplets.

- As air rises above the equator, it radiates heat to space. It also expands into the larger volume of the upper atmosphere, which lowers its density and temperature—a phenomenon known as adiabatic cooling.

- As rising air cools, its ability to hold water declines. When water vapor cools, water condenses.

The result? High levels of precipitation occur along the equator. The story is just beginning, however.

- As more air is heated along the equator, the cooler, "older" air above Earth's surface is pushed poleward.

- When the air mass has cooled enough, its density increases and it begins to sink.

- As it sinks, it absorbs more and more solar radiation reflected from Earth's surface and begins to warm while being pushed to the surface by higher-density air above it.

- As the air warms, it also gains water-holding capacity. Thus, the air approaching the Earth "holds on" to its water and little rain occurs where it returns to the surface.

The result? A band of deserts in the vicinity of 30° latitude north and south. These areas are bathed in warm, dry air.

Similar air circulation cells occur between 30° latitude and 60° latitude, and between 60° latitude and the poles (**Figure 50.23b**).

(a) Circulation cells exist at the equator ...

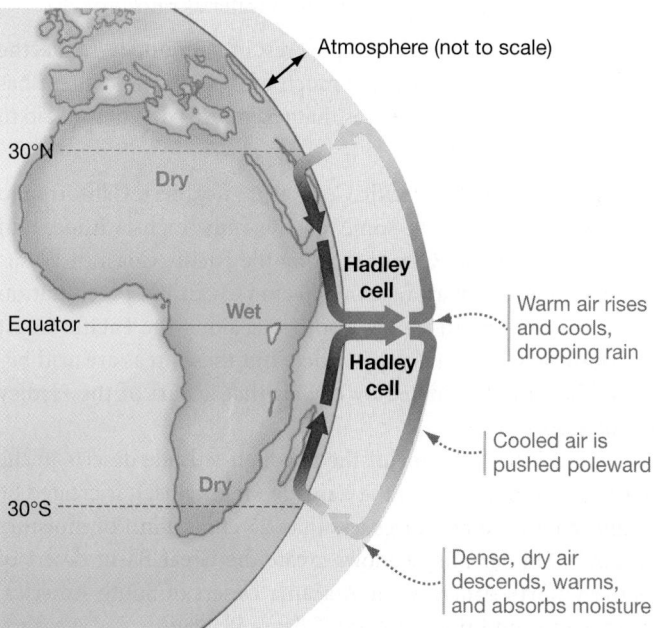

(b) ... and at higher latitudes.

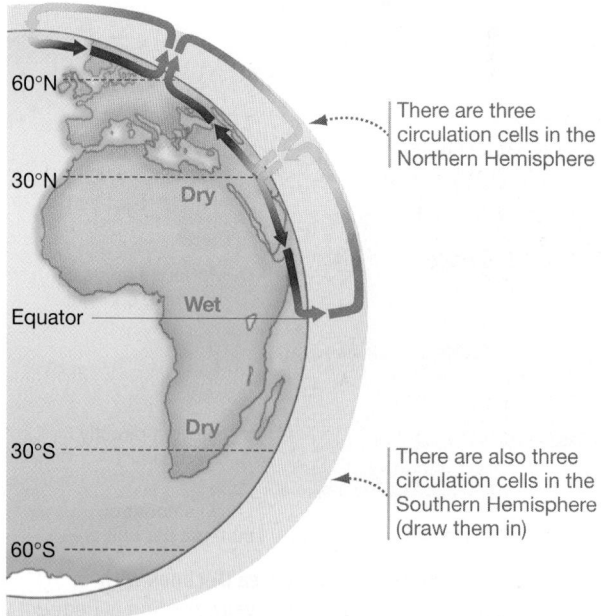

FIGURE 50.23 Global Air Circulation Patterns Affect Rainfall. Hadley cells explain why the tropics are much wetter than regions near 30° latitude north and south. Regions with rising air, such as those at the equator, tend to be wetter than regions with descending air.

✔**QUESTION** Is air at the North Pole wet or dry? Is your answer consistent with the data in Figure 50.21?

✔ If you understand how Hadley cells work, you should be able to add diagrams to Figure 50.24b showing the three cells that occur in the Southern Hemisphere.

For more review on tropical atmospheric circulation, go to the study area at *www.masteringbiology.com*.

MB Web Activity Tropical Atmospheric Circulation

WHY ARE THE TROPICS WARM AND THE POLES COLD? In general, areas of the world are warm if they receive a large amount of sunlight per unit area; they are cold if they receive a small amount of sunlight per unit area. Over the course of a year, regions at or near the equator receive much more sunlight per unit area—and thus much more energy in the form of heat—than regions that are closer to the poles.

- At the equator, the Sun is often directly overhead. As a result, sunlight strikes the surface at or close to a 90° angle. At these angles, Earth receives a maximum amount of solar radiation per unit area (**Figure 50.24**).

- Because Earth's surface slopes away from the equator, the Sun strikes the surface at lower and lower angles moving toward the poles. When sunlight arrives at a low angle, much less energy is received per unit area.

The pattern of decreasing average temperature with increasing latitude is a result of Earth's spherical shape.

WHAT CAUSES SEASONALITY IN WEATHER? The striking global patterns in temperature and moisture that we've just reviewed are complicated by the phenomenon of seasons. Seasons are defined as regular, annual fluctuations in temperature, precipitation, or both.

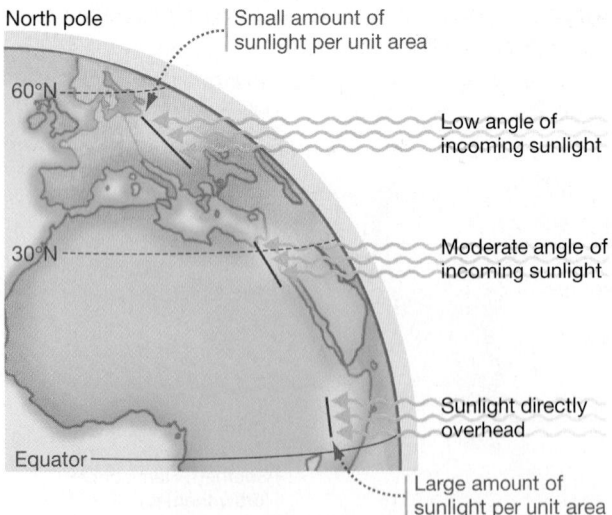

FIGURE 50.24 Solar Radiation per Unit Area Declines with Increasing Latitude. Over the course of a year, the Sun is frequently almost directly overhead at the equator. As a result, equatorial regions receive a large amount of solar radiation per unit area. At latitudes greater than 23.5°, the Sun is never directly overhead. As a result, high-latitude regions receive less solar radiation per unit area than do low-latitude regions.

As **Figure 50.25** shows, seasonality occurs because Earth is tilted on its axis by 23.5 degrees. As a result of this incline, the Northern Hemisphere is tilted toward the Sun in June. In this position, it faces the Sun most directly. In June, the Northern Hemisphere presents its least-acute angle to the Sun and receives the largest amount of solar radiation per unit area.

The Southern Hemisphere, in contrast, is tilted away from the Sun in June. As a result, it presents a steep angle for incoming sunlight and receives its smallest quantity of solar radiation per unit area.

The upshot? In June it is summer in the Northern Hemisphere and winter in the Southern Hemisphere. Conversely, in December it is summer in the Southern Hemisphere, and winter in the Northern Hemisphere.

In March and September, the equator faces the Sun most directly. During these months, the tropics receive the most solar radiation. If Earth did not tilt on its axis, there would be no seasons.

MOUNTAINS AND OCEANS: PHYSICAL FEATURES HAVE REGIONAL EFFECTS ON CLIMATE ◯━ The broad patterns of climate that are dictated by global heating patterns, Hadley cells, and seasonality are overlain by regional effects. The most important of these are due to the presence of mountain ranges and proximity to an ocean.

Figure 50.26 shows how mountain ranges affect regional climate, using the Cascade Mountain range along the Pacific Coast of North America as an example.

- In this area of the world, the prevailing winds are from the west. These winds bring moisture-laden air from the Pacific Ocean onto the continent.

- As the air masses begin to rise over the mountains, the air cools and releases large volumes of water as rain.

- Once cooled air has passed the crest of a mountain range, the air is relatively dry because much of its moisture content has already been released. Areas that receive this dry air are said to be in a **rain shadow**.

The area along the Pacific Coast from northern California to southeast Alaska hosts some of the only high-latitude rain forests in the world. One area along the Pacific Coast of Washington State gets an average of 340 cm (134 in.) of precipitation each year. Mountain ranges also occur along the Pacific Coast of Southern California and Mexico; but these areas are arid because they are dominated by dry air that is part of the Hadley circulation.

Logically enough, one of the few high-latitude deserts in the world exists to the east of the Cascade Mountains. It is created by a rain shadow and averages less than 25 cm (10 in.) of moisture annually. Similar rain shadows create the Great Basin desert of western North America, the Atacama Desert of South America, and the dry conditions of Asia's Tibetan Plateau.

While the presence of mountain ranges tends to produce extremes in precipitation, the presence of an ocean has a moderating influence on temperature. To understand why this is so, recall from Chapter 2 that water has an extremely high **specific heat**,

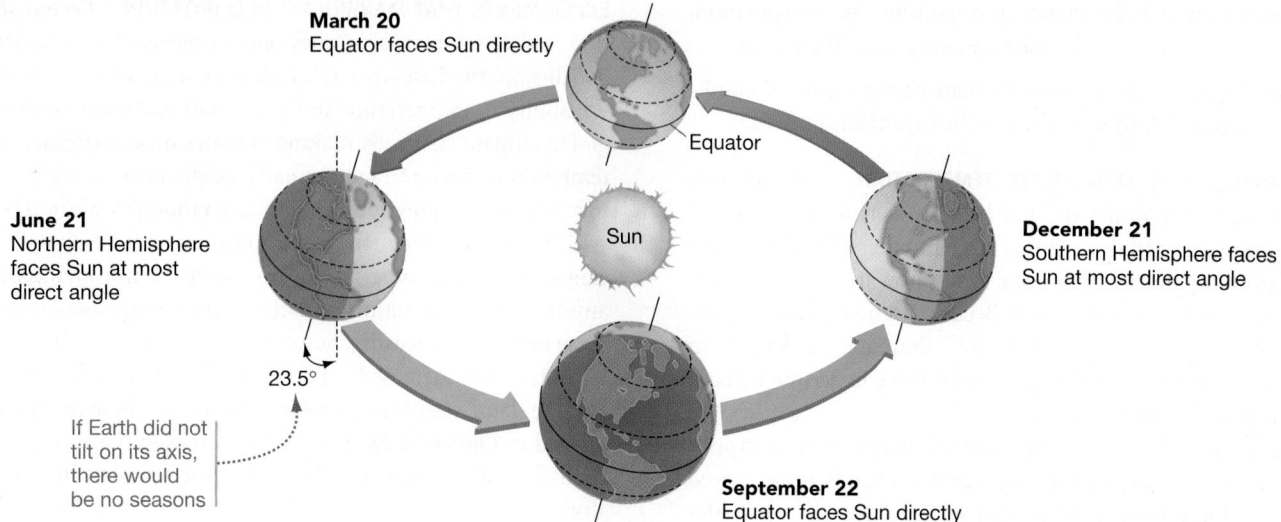

March 20
Equator faces Sun directly

Equator

June 21
Northern Hemisphere faces Sun at most direct angle

December 21
Southern Hemisphere faces Sun at most direct angle

23.5°

If Earth did not tilt on its axis, there would be no seasons

Sun

September 22
Equator faces Sun directly

FIGURE 50.25 Earth's Orbit and Tilt Create Seasons at High Latitudes. Seasons occur due to annual variation in the amount of solar radiation that different parts of the Earth receive.

meaning that it has a large capacity for storing heat energy. Because water molecules form hydrogen bonds with each other, it takes a great deal of heat energy to boil water or melt ice—in fact, to change the temperature of water at all. The hydrogen bonds have to be broken before the water molecules themselves can begin to absorb heat and move faster, which we measure as increased temperature.

Because water has a high specific heat, it can absorb a great deal of heat from the atmosphere in summer, when the water temperature is cooler than the air temperature. As a result, the ocean moderates summer temperatures on nearby landmasses. Similarly, the ocean releases heat to the atmosphere in winter, when the water temperature is warmer than the air temperature. As a result, islands and coastal areas have much more moderate climates than do nearby areas inland.

How Will Global Climate Change Affect Ecosystems?

Climate—particularly the amount and variability of heat and precipitation—has a dramatic effect on terrestrial ecosystems. Climate also has a significant impact on aquatic ecosystems. Heat

and precipitation influence ocean currents and lake turnover, as well as water levels in lakes, streams, and oceans.

It is now well established that CO_2 pollution is causing climate change (see Chapter 54). Climate has changed frequently and radically throughout the history of life, but recent increases in average temperatures around the globe, combined with a projected increase of up to 5.8°C over the next 100 years, represents one of the most rapid periods of climate change since life began.

Biologists use three tools to predict how global warming will affect aquatic and terrestrial ecosystems.

1. *Simulation studies* are based on computer models of weather patterns in local regions. By increasing average temperatures in these models, researchers can predict how wind and rainfall patterns, storm frequency and intensity, and other aspects of weather and climate will change. In effect, simulation studies predict how temperature and precipitation profiles, like those in this chapter's "Terrestrial biomes boxes," will change.

2. *Observational studies* are based on long-term monitoring at fixed sites around the globe. Researchers are documenting changes in key physical variables—water depth, water flow, temperature, and precipitation—as well as changes in the

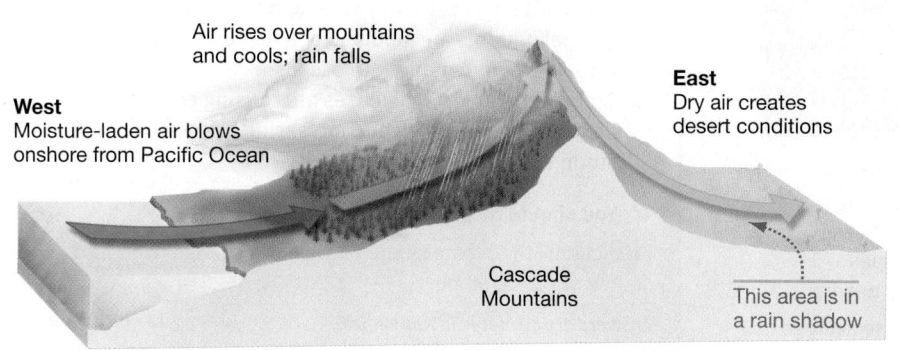

Air rises over mountains and cools; rain falls

West
Moisture-laden air blows onshore from Pacific Ocean

East
Dry air creates desert conditions

Cascade Mountains

This area is in a rain shadow

FIGURE 50.26 Mountain Ranges Create Rain Shadows. When moisture-laden air rises over a mountain range, the air cools enough to condense water vapor. As a result, moisture falls to the ground as precipitation. On the other side of the mountain, little water is left to provide precipitation.

distribution and abundance of organisms. By extrapolation, they can predict how change may continue at these sites.

3. *Experiments* are designed to simulate changed climate conditions and to record responses by the organisms present.

EXPERIMENTS THAT MANIPULATE TEMPERATURE Even though the scientific community did not become alert to the reality of global warming until the 1980s, a large number of multiyear experiments have already been done to test how increased temperature affects organisms in nature. Because arctic regions are projected to experience particularly dramatic changes in temperature and precipitation, many of these experiments have been done in the tundra biome.

Figure 50.27 shows how the most successful of these experiments are done. Transparent, open-topped chambers are placed at random locations around a study site, and the characteristics of the vegetation inside the chambers are recorded annually. The same measurements are made in randomly assigned study plots that lack chambers.

Across experiments, the average annual temperature inside the open-topped chambers increases by 1–3°C. This increase is consistent with conservative projections for how temperatures will change over the next century.

In an analysis of data from 11 such study sites scattered throughout the arctic in North America and Eurasia, several strong patterns emerged:

- Overall, species diversity decreases.
- Compared with control plots, grasses and shrubs increase inside the chambers, and mosses and lichens decrease.

These results support simulation and observational studies predicting that arctic tundra environments are giving way to boreal forest. Biomes are changing right before our eyes.

FIGURE 50.27 Experimental Increases in Temperature Change Species Composition in Arctic Tundra. The transparent, open-topped chambers shown here are an effective way to increase average annual temperature. Inside the chambers, species composition changes relative to otherwise similar sites nearby.

✔**QUESTION** Why does a clear chamber like this increase average temperature inside?

EXPERIMENTS THAT MANIPULATE PRECIPITATION Recent simulation and observational studies have converged on a remarkable conclusion: Increases in average global temperature are increasing variability in temperature and precipitation. Stated another way, global climate change is making climates more extreme. Average temperature and average annual precipitation in many regions may not be changing substantially, even though variation is.

Thus far, the frequency of monsoon rains, cyclones, hurricanes, and other types of storm events is not changing. The amount of precipitation and the wind speeds associated with these events is increasing, however.

To test how increased variation in climate will affect a temperate grassland ecosystem, researchers set up the experiment described in **Figure 50.28**. The goal was to keep the average annual rainfall unchanged during the growing season, but increase variability.

The biologists increased the severity of storms by capturing rain that fell over study plots during some storms and then applying that water to the same plots during later storms. The bar charts in the "Experimental setup" section of the figure show how much water entered each type of plot as the growing season progressed.

The results from the first four years of the experiment are plotted in the bar charts in the "Results" section. Note that:

- Average soil moisture and average overall NPP declined in the experimental plots, compared with similar plots where rainfall was not manipulated.
- On average, overall species diversity is increasing in the high-variation plots because the grass that produces most of the biomass at this site is doing poorly in the experimental plots—opening up room for other species to grow.

The experiment is ongoing, however, because it is possible that long-term changes will differ from short-term changes, due to continued alterations in the species present.

Experiments like this are documenting that climate change causes dramatic alterations in the composition and distribution of biomes. Your world is changing.

CHECK YOUR UNDERSTANDING

If you understand that . . .
- Global climate patterns are dictated by differential heating of Earth's surface, Hadley cells, and seasonality. These general patterns are modified at a regional scale by the presence of mountains and oceans.
- Global temperature increases are causing changes in the distribution of biomes and increasing the severity of storm events.

✔ You should be able to . . .

Predict how the biome where you live will change over the course of your lifetime.

Answers are available in Appendix B.

QUESTION: How does increasing variation in rainfall—without changing total amount—affect grassland biomes?

HYPOTHESIS: Increasing variability in rainfall affects soil moisture, net primary productivity (NPP), and species diversity.

NULL HYPOTHESIS: Increasing variability in rainfall has no effect on grassland biomes.

EXPERIMENTAL SETUP:

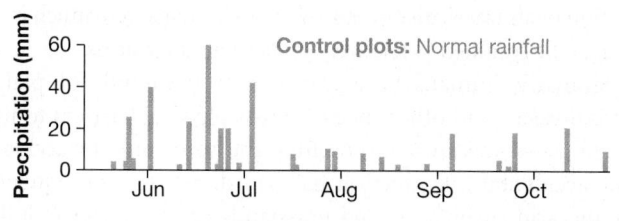

Control plots: Normal rainfall

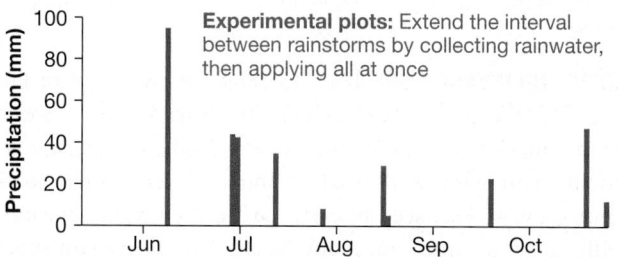

Experimental plots: Extend the interval between rainstorms by collecting rainwater, then applying all at once

Record soil moisture, aboveground biomass (as an index of NPP), and species diversity, over a 4-year period.

PREDICTION: Soil moisture, NPP, and/or species diversity will differ between experimental plots and control plots.

PREDICTION OF NULL HYPOTHESIS: Soil moisture, NPP, and species diversity will not differ between experimental plots and control plots.

RESULTS:

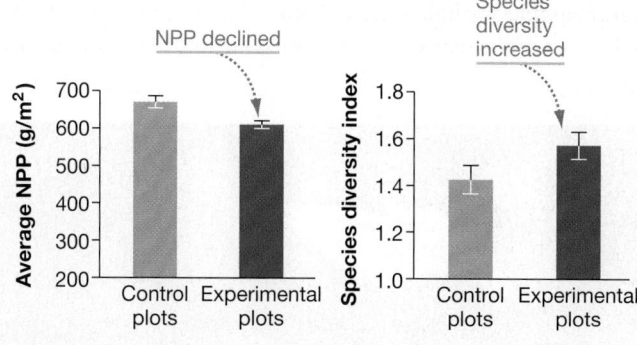

CONCLUSION: Altering variability in rainfall has dramatic effects on soil moisture (data not shown), NPP, and species diversity.

FIGURE 50.28 Experimental Increases in Rainfall Variability Change Species Composition in Temperate Grasslands.

SOURCE: Knapp, A. K. et al. 2002. Rainfall variability, carbon cycling, and plant species diversity in a mesic grassland. *Science* 298: 2202–2205.

✔**QUESTION** Of all the possible data to collect, why did the researchers decide to measure NPP and species diversity?

50.5 Biogeography: Why Are Organisms Found Where They Are?

Understanding why organisms are found where they are is one of ecology's most basic tasks. It is a fundamental aspect of predicting the consequences of climate change, and is required for designing nature preserves in a rational and effective way.

The study of how organisms are distributed geographically is called **biogeography**. Chapter 53 analyzes general patterns in how species numbers vary around the world, Chapter 54 presents data showing that the geographic ranges of many species are changing rapidly in response to global climate change, and Chapter 55 analyzes "hotspots" where species diversity is particularly high. Here, let's focus on how interactions with the abiotic and biotic environments affect where a particular species lives.

Abiotic Factors

The most fundamental observation about the **range**, or geographic distribution, of organisms is simple: No one species can survive the full range of environmental conditions present on Earth. *Thermus aquaticus* can thrive in hot springs with temperatures above 70°C because its enzymes do not denature at near-boiling temperatures. But the same cells would die instantly if they were transplanted to the frigid waters near polar ice—even though vast numbers of other archaea are present. The cold-water species have enzymes that work efficiently at near-freezing temperatures.

No organism thrives in both hot springs and cold ocean water. The reason is fitness trade-offs—a concept introduced in Chapters 24 and 41. No enzyme can function well in both extremely high temperatures and extremely low temperatures. Because of fitness trade-offs, organisms tend to be adapted to a limited set of physical conditions—to a particular temperature and moisture regime on land, or a particular water depth and movement regime in water.

But adaptation to a proscribed set of abiotic conditions is just the start of the story of explaining why a particular species lives where it does. For example, the temperature and moisture regimes of southern California, in North America, are almost identical to those found in the Mediterranean region of Europe and north Africa. The two regions are part of the same biome. But the species present at the sites are almost completely different.

Why? ☞ To understand a species' distribution thoroughly, biologists examine historical and biotic factors in addition to the physical conditions present.

The Role of History

In studying a species' range, one of the first factors to consider is the history of dispersal. **Dispersal** is the movement of an individual from its place of origin to the location where it lives and breeds as an adult.

If a particular species is missing from an area, it may be due to a physical barrier to dispersal—a river, a glacier, a mountain range, or an ocean. The plant species found in California are different from those found in southern Spain because the Pacific and Atlantic Oceans prevent the dispersal of seeds.

Alfred Russel Wallace was among the first to document this phenomenon. In addition to codiscovering evolution by natural selection (see Chapter 24), Wallace founded the study of biogeography.

THE WALLACE LINE: BARRIERS TO DISPERSAL While working in the Malay archipelago, Wallace realized that the plants and animals native to the more northern and western islands were radically different from the species found on the more southern and eastern islands. Sumatra, to the northwest, has tigers, rhinos, and other species with close relatives in southern Asia. New Guinea, to the southeast, has tree kangaroos and other species with close relatives on Australia. But both islands are part of the tropical wet forest biome.

This biogeographical demarcation, now known as the **Wallace line**, separates species with Asian and Australian affinities (**Figure 50.29**). It exists because a deep trench in the ocean maintained a water barrier to dispersal, even when ocean levels dropped during the most recent glaciations. As a result, landforms on either side of the line remained isolated at a time when most of the other islands became connected, encouraging dispersal.

HUMANS AS DISPERSAL AGENTS Unfortunately, humans are now breaking down many natural barriers to dispersal. This is why public health officials closely monitor annual flu outbreaks in south China—where most new strains originate. New flu strains disperse all over the planet in a matter of weeks or days, transported in the respiratory passages of infected airplane passengers. Prior to the invention of air travel, disease epidemics in humans were much more likely to be confined to relatively small geographic areas.

Similarly, humans have transported thousands of plants, birds, insects, and other species across physical barriers to new locations—sometimes purposefully and sometimes by accident. One accidental introduction has had disastrous consequences for the arid shrublands and grasslands of the western United States.

EXOTIC AND INVASIVE SPECIES In 1889 a native plant of Eurasia called cheatgrass was accidentally introduced to western North America in a shipment of crop seeds. Cheatgrass is an annual plant with seeds that germinate over winter, so that plants grow and set seed in early spring. As a result, cheatgrass seedlings get a "jump" on many North American plant species, which initiate growth later. Early growth allows cheatgrass to use more of the available water and nutrients. Within 30 years, cheatgrass had taken over tens of thousands of square miles of habitat.

If an **exotic species**—one that is not native—is introduced into a new area, spreads rapidly, and eliminates native species, it is said to be an **invasive species**. In North America alone, dozens of invasive species have had devastating effects on native plants and animals. Examples include kudzu (**Figure 50.30**), purple loosestrife, reed canary grass, garlic mustard, Russian thistle (also known as

Species with
Asian affinities

Pacific Ocean

Mainland
Asia

The
Wallace
Line

Borneo

Sumatra

New Guinea

Australia

Indian Ocean

Species with
Australian
affinities
(such as this
tree kangaroo)

FIGURE 50.29 The "Wallace Line" Divides Organisms with Asian versus Australian Affinities.

FIGURE 50.30 Introduced Species Can Cause Problems. Kudzu, imported from Asia, has taken over parts of the southeastern U.S.

tumbleweed), European starlings, African honeybees, and zebra mussels. Similar lists could be compiled for other regions of the world.

Regarding the distribution of organisms, the story of invasive species is straightforward: They and other introduced species did not exist in certain parts of the world until recently, simply because they had never been dispersed there.

Only about 10 percent of the species introduced to an area actually become common, however, and a smaller percentage than that increases in population enough to be considered invasive. Why is it that some species introduced into an area do not thrive? Conversely, why are invasives so successful?

Biotic Factors

The distribution of a species is often limited by biotic factors—meaning interactions with other organisms. As an example, recall the data presented in Chapter 26 on the current and former distributions of hermit warblers and Townsend's warblers along the Pacific coast of North America.

- Both hermit warblers and Townsend's warblers live in evergreen forests.

- Experiments have shown that male Townsend's warblers directly attack male hermit warblers and evict them from breeding territories.

- Historical data indicate that the geographic range of Townsend's warblers has been expanding steadily at the expense of hermit warblers.

These observations support the hypothesis that a biotic factor—competition with another species—is limiting the range of hermit warblers. Competition is not the only biotic factor to consider, however.

- The moth pictured in **Figure 50.31** lays its eggs only in the flowers of yucca plants. As a result, this species does not exist outside the range of yucca plants.

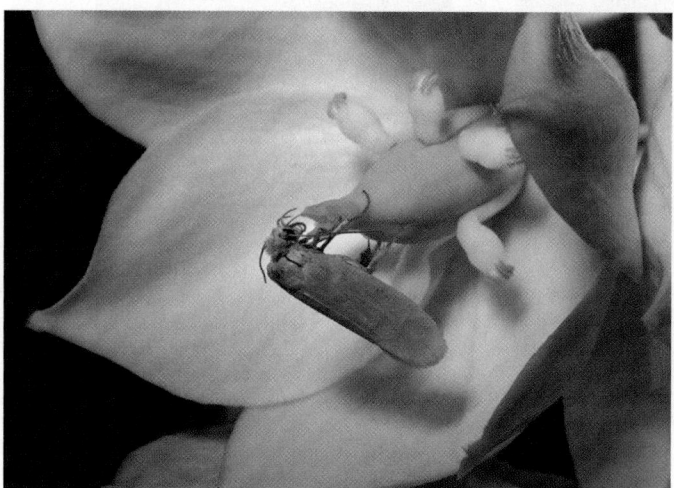

FIGURE 50.31 The Ranges of Yucca Plants and Yucca Moths Overlap. Yucca moths lay their eggs only in the flowers of yucca plants, so the moths cannot live where yucca plants are absent.

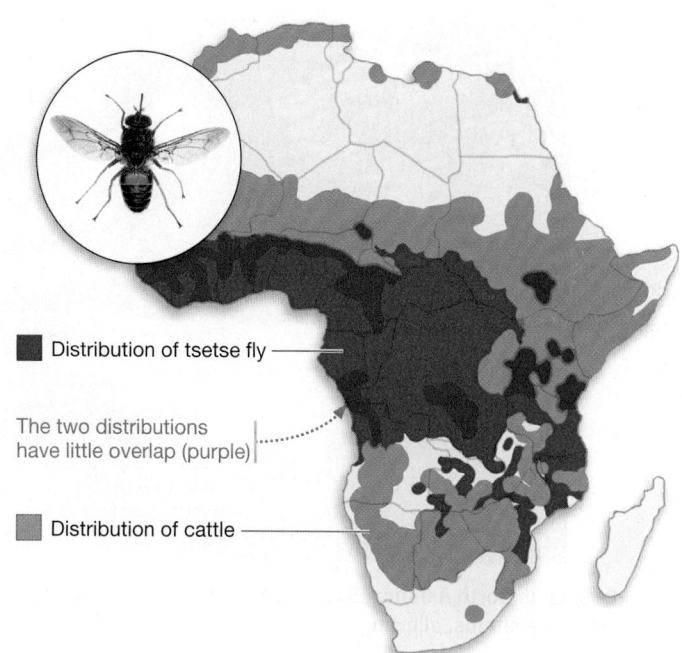

■ Distribution of tsetse fly

The two distributions have little overlap (purple)

■ Distribution of cattle

FIGURE 50.32 In Africa, Tsetse Flies Limit the Distribution of Cattle. Tsetse flies carry a disease that is fatal to cattle.

- In Africa, the range of domestic cattle is limited by the distribution of tsetse (pronounced *TSEE-tsee*) flies. Tsetse flies transmit a parasite that causes the disease trypanosomiasis, which is fatal in cattle (**Figure 50.32**).

- Most of the songbirds native to Hawaii are limited to alpine habitats—above the range of the mosquitoes that transmit avian malaria.

Biotic and Abiotic Factors Interact

In many cases, introduced species may not become established in a region because of abiotic factors—particularly temperature and moisture. An area may simply be too cold or too dry for an organism to survive. For example, even though the invasive plant kudzu grows prolifically throughout the southeastern United States, it is killed when the soil freezes to a depth of a foot or more. As a result, it has not expanded its range farther north and west.

It is often difficult to separate the effects of biotic and abiotic factors on a species' range, however. To drive this point home, consider how cheatgrass became established in North America.

Cheatgrass does not grow in wet grasslands because it does not compete well against the tall species that thrive there. But it has been able to invade two important types of biomes in North America: the arid, shrub-dominated habitats known as sage-steppe, and dry temperate grasslands. In each case, the reasons for the success of cheatgrass were different.

THE ROLE OF FIRE IN CHEATGRASS'S SPREAD Cheatgrass has been able to invade sage-steppe habitats with the aid of fire. Fire is an abiotic factor that can kill the shrubs that dominate sage-steppe.

Fires tend to be rare in sage-steppe biomes because plants are widely spaced. This distribution usually prevents fires from

Invasive cheatgrass

Native shrub

FIGURE 50.33 In North America, Sage-Steppe Has Been Invaded by Cheatgrass. Cheatgrass fills in spaces between native shrubs, allowing fires to burn through the habitat.

spreading and affecting a large area. But when this biome has been invaded by cheatgrass and begins to dry in late summer, beds of dead cheatgrass offer a continuous, extremely flammable source of fuel (**Figure 50.33**).

If a fire gets started, it does not affect the cheatgrass for two reasons:

1. Cheatgrass is an annual, so there is no living tissue exposed once the growing season is over.

2. Cheatgrass seeds sprout readily in soils that have been depleted of organic matter by fire.

Because sagebrush and other shrubs are perennials, their tissues are heavily damaged or killed by hot fires. By promoting the frequency and intensity of fire in sage-steppe biomes, the presence of cheatgrass creates an environment that favors the spread of cheatgrass.

How has this species been able to invade dry grassland biomes?

THE ROLE OF OVERGRAZING IN CHEATGRASS'S SPREAD Prior to the arrival of cheatgrass, grasslands in arid areas of the American West were dominated by the types of bunchgrasses shown in **Figure 50.34**. Bunchgrasses do not form a continuous sod; instead, they grow in compact bunches.

Black-soil crust

Bunchgrass

FIGURE 50.34 Arid Grasslands Feature Bunchgrasses and Black-Soil Crust. Bunchgrasses grow in compact bunches instead of forming a sod. The black-soil crust that forms between bunchgrasses is composed of cyanobacteria, archaea, and other microscopic organisms that add organic matter and nutrients to the soil.

✓QUESTION Suggest a hypothesis to explain why grasses with the bunching growth habit tend to be found in dry habitats, as compared to grasses that form sods.

The spaces between bunchgrasses are occupied by what biologists call "black-soil crust." This material is a community of soil-dwelling cyanobacteria, archaea, and other microorganisms that grow when moisture is available. Many of these species also fix nitrogen (see Chapter 38). The nitrogen becomes available to the surrounding plants when the microbial cells die and decompose.

When European settlers arrived and began grazing cattle in this biome, however, both the black soils and bunchgrasses were affected. Cattle grazed on the bunchgrasses until the grasses died, and their hooves compacted and disrupted the black soils.

Neither component of the ecosystem was allowed to recover, in part because intensive grazing continued and in part because cheatgrass arrived:

- Cheatgrass shades out black-soil crust organisms. When black-soil crust is gone, the amount of nitrogen available to bunchgrasses is reduced.

- Cheatgrass competes successfully with bunchgrasses for water and nutrients, because it completes its growth early in the spring—before the native bunchgrasses have broken dormancy and begun to grow.

- As in sage-steppe habitats, the expansion of cheatgrass increased the frequency and severity of fires, which further degraded the integrity of black-soil crust communities.

To summarize, changes in abiotic and biotic factors allowed cheatgrass to invade native biomes in North America and thrive as an invasive species. Currently, managing cheatgrass is the major problem facing conservation biologists and rangeland managers in the western United States.

The story of cheatgrass is particularly dramatic. But it is important to remember that the range of every species on Earth is limited by a combination of historical, abiotic, and biotic factors. The range of a particular species depends on dispersal ability, the capacity to survive climatic conditions, and the ability to find food and avoid being eaten.

Climate and other abiotic factors were the focus of this chapter; the next several chapters focus on the biotic aspects of ecology. How do interactions among organisms of the same species and interactions among different species affect geographic range and population size?

CHAPTER 50 REVIEW

For media, go to the study area at www.masteringbiology.com

Summary of Key Concepts

🔑 **Ecology's goal is to explain the distribution and abundance of organisms. It is the branch of biology that provides a scientific foundation for conservation efforts.**

- Biologists study ecology at four main levels: (**1**) organisms, (**2**) populations, (**3**) communities, and (**4**) ecosystems.

- Organismal ecology's goal is to understand how individuals interact with their physical surroundings and with other organisms.

- Population ecology focuses on how and why populations grow or decline.

- Community ecology is the study of how different species interact.

- Ecosystem ecology analyzes interactions between organisms and their abiotic environment—particularly the flow of energy and nutrients.

- To preserve endangered species and communities, biologists apply analytical methods and results from all four levels.

  ✔You should be able to explain why these four levels are hierarchical.

🔑 **Physical structure—particularly water depth—is the primary factor that limits the distribution and abundance of aquatic species. Climate—specifically, both the average value and annual variation in temperature and in moisture—is the primary factor that limits the distribution and abundance of terrestrial species.**

- Water depth is a crucial factor in aquatic ecosystems because the wavelengths of sunlight that are required for photosynthesis do not penetrate water efficiently. Thus, only certain zones in a lake or ocean can support photosynthesis, and organisms in other aquatic zones must depend on food that rains out of the sunlit areas near the surface.

- Distinct climatic regimes are associated with distinct types of terrestrial vegetation, or biomes. Biomes represent the major types of environments available to terrestrial organisms.

- Because photosynthesis is most efficient when temperatures are warm and water supplies are ample, the productivity and degree of seasonality in biomes varies with temperature and moisture.

  ✔You should be able to explain how global warming may affect terrestrial biomes due to (**1**) increased evaporation from the oceans, and (**2**) increased transpiration from plants.

🔑 **Climate varies with latitude, elevation, and other factors—such as proximity to oceans and mountains. Climate is changing rapidly around the globe.**

- An array of abiotic factors affect organisms, including the chemistry of water in aquatic habitats; the nature of soils in terrestrial habitats; and climate, which is a combination of temperature, moisture, sunlight, and wind.

- Climate varies around the globe because sunlight is distributed asymmetrically, with equatorial regions receiving on average more solar radiation than do regions toward the poles.

- Hadley cells create bands of wet and dry habitats, and the tilt of Earth's axis causes seasonality in the amount of sunlight that nonequatorial regions receive.

- The presence of mountains can create local areas of wet or dry habitats, and proximity to an ocean moderates temperatures in nearby terrestrial habitats.

- Global climate change is causing temperature profiles and precipitation patterns to change in ecosystems throughout the planet.

  ✓You should be able to explain how climate would be expected to vary around the Earth if the planet were cylindrical and spun on an axis that was at a right angle to the Sun.

  (MB) **Web Activity** Tropical Atmospheric Circulation

  🔑 **A species' distribution is constrained by historical and biotic factors, as well as by abiotic factors such as physical structure and climate.**

- The range, or geographic distribution, of a particular species depends on dispersal ability, the capacity to survive climatic conditions, and the ability to find food, combat diseases, and avoid being eaten.

- For each species, a unique combination of historical factors, abiotic factors, and biotic factors dictates where individuals live and the size of the population.

  ✓You should be able to predict what would happen to the range of cattle in Africa if the tsetse fly were eradicated.

Questions

✓ TEST YOUR KNOWLEDGE — *Answers are available in Appendix B*

1. Where do rain shadows exist?
   a. the part of a mountain that receives prevailing winds and heavy rain
   b. the region beyond a mountain range that receives dry air masses
   c. the region along the equator where precipitation is abundant
   d. the region near 30° N and 30° S latitude that receives hot, dry air masses

2. A region receives less than 5 cm (2 in.) of precipitation annually and has temperatures that never drop below freezing. Which type of biome is present?
   a. subtropical desert
   b. temperate grassland
   c. boreal forest
   d. arctic tundra

3. What is the predominant type of vegetation in a tropical wet forest?
   a. shrubs and bunchgrasses
   b. herbs, grasses, and vines
   c. broad-leaved deciduous trees
   d. broad-leaved evergreen trees

4. The littoral zone of a lake is most similar to which of the following environments?
   a. oceanic zone
   b. intertidal zone
   c. neritic zone
   d. marsh

5. Typically, where are oxygen levels highest and nutrient levels lowest in a stream?
   a. near its source
   b. near its mouth, or end
   c. where it flows through a swamp or marsh
   d. where it forms an estuary

6. Why are certain exotic species considered "invasive"?
   a. They are found in areas where they are not native.
   b. They were introduced by humans—often accidentally.
   c. They spread aggressively and displace native species.
   d. They benefit from increased grazing and wildfires.

✓ TEST YOUR UNDERSTANDING — *Answers are available in Appendix B*

1. Name the four main levels of study in ecology. Write a question about the ecology of humans that is relevant to each level.

2. In June, does the Northern Hemisphere receive more or less solar radiation than the equator does? Explain your answer.

3. Explain why the Australian Outback receives so little rainfall.

4. Contrast the productivities of the intertidal, neritic, and oceanic zones of marine environments. Explain why large differences in productivity exist.

5. Compare and contrast the biomes found at increasing elevation on a mountain with the biomes found at increasing latitude in Figure 50.10.

6. Explain why productivity is much lower in bogs than it is in marshes or lakes that receive the same amount of solar radiation.

✓ APPLYING CONCEPTS TO NEW SITUATIONS — *Answers are available in Appendix B*

1. Mars has an even more pronounced tilt than Earth does. Does Mars experience seasons? Why or why not?

2. Mountaintops are closer to the Sun than the surrounding area is. Why are mountaintops cold?

3. The southwest corners of some Caribbean islands are extremely dry, even though other areas of the islands receive substantial rainfall.

   Explain this observation. Base your answer on a hypothesis (one that you propose) about the direction of the prevailing winds in this region of the world.

4. Edinburgh, Scotland, and Moscow, Russia, are at the same latitude, yet these two cities have very different climates—Moscow has much colder winters. Explain why.

This male satin bowerbird (on the left) is decorating his display area with objects of his preferred color: blue. A female, on the right, is inspecting the bower. She may mate with the male if she finds him, and the display, attractive enough.

Behavioral Ecology 51

To biologists, **behavior** is action—specifically, the response to a stimulus. Pond-dwelling bacteria swim toward drops of blood that fall into the water. Time-lapse movies of growing sunflowers document the steady movement of their flowering heads throughout the day, as they turn to face the shifting Sun. Filaments of the fungus *Arthrobotrys* form loops, then release a molecule that attracts roundworms. When a roundworm touches a loop, the fungal filaments swell—ensnaring the worm. The fungus then grows into the worm's body and digests it.

Among animals, action may be frequent and even spectacular. Peregrine falcons reach speeds of up to 320 km/hr (200 mph) when they dive in pursuit of ducks or other flying prey. Some Arctic terns migrate over 32,000 km (20,000 miles) each year. Deep inside a hive, in complete darkness, honeybees communicate the position and nature of food sources a mile away—by "dancing."

Ecology is the study of how organisms interact with their physical and biological environments; behavioral biology is the study of how organisms respond to particular stimuli from those environments. The action in behavior is this response. What does an *Anolis* lizard do when it gets too hot? How does that same individual respond when a member of the same species approaches it? Or when a house cat approaches?

Although all organisms respond in some way to signals from their environment, most behavioral research is performed on vertebrates, arthropods, or mollusks. With their sophisticated nervous systems and skeletal-muscular systems, these animals can sense, process, and respond rapidly to a wide array of environmental stimuli.

51.1 An Introduction to Behavioral Biology

Behavior is a particularly synthetic field in biological science. To understand why animals and other organisms do what they do, researchers have to ask questions about

KEY CONCEPTS

- Biologists analyze behavior at the proximate and ultimate levels—the genetic and physiological mechanisms and how they affect fitness.

- Individuals can behave in a wide range of ways; which behavior occurs depends on current conditions.

- Foraging patterns may vary with genotype; foraging decisions maximize energy gain and minimize costs.

- Sexual behavior is affected by levels of sex hormones; females choose mates that provide good alleles and needed resources.

- Animals navigate using an array of cues; the costs of migration can be offset by benefits in food availability.

- Animals communicate with movements, odors, or other stimuli; communication can be honest or deceitful.

- Individuals that behave altruistically are usually helping relatives or individuals that help them in return.

✔ When you see this checkmark, stop and test yourself. Answers are available in Appendix B.

1019

genetics, hormonal signals, neural signaling, natural selection, evolutionary history, and ecological interactions.

To make sense of this diversity, it's helpful to recognize that behavioral ecologists ask questions and test hypotheses at two fundamental levels—proximate and ultimate. Let's explore what these two levels are about, then consider the question of whether animals make decisions about how they should respond to a particular situation.

Proximate and Ultimate Causation

🔑 Most behavioral studies start by carefully observing what animals do in response to specific problems or situations. As research progresses, investigators use experimental approaches to probe the proximate and ultimate causes of behavior.

- **Proximate** (or mechanistic) **causation** explains *how* actions occur in terms of the neurological, hormonal, and skeletal-muscular mechanisms involved.

- **Ultimate** (or evolutionary) **causation** explains *why* actions occur—based on their evolutionary consequences and history. Does a certain species perform its courtship display because its ancestors happened to behave that way? Is a particular behavior currently adaptive, meaning that it increases an individual's fitness—its ability to produce offspring? If so, how does it help individuals produce offspring in a particular environment?

To illustrate the proximate and ultimate levels of causation, consider the spiny lobster in **Figure 51.1**. These animals spend the day hiding in cracks or holes in coral reefs. At night they emerge and wander away from the reef in search of clams, mussels, crabs, or other sources of food. Before dawn, they return to one of several dens they use on a regular basis. How do they find their way back?

This question is proximate in nature—it asks how an individual does what it does. Although research is continuing, recent experiments support the hypothesis that spiny lobsters have receptors in their brains that detect changes in Earth's magnetic field, and that they navigate by using information on how the orientation and strength of the field changes as they move about.

At the ultimate level of causation, biologists want to know why the behavior occurs. The leading hypothesis to explain homing behavior is that the ability to navigate allows spiny lobsters to search for food over a wide area under cover of darkness, then return to a safe refuge before sharks and other large predators that hunt by sight can find them.

It's important to recognize that efforts to explain behavior at the proximate and ultimate levels are complementary. To understand what an organism is doing, biologists want to know how the behavior happens and why.

Behavioral biologists work at the interface of many fields and at both proximate and ultimate levels. And in many cases, their work is ecological in nature. Questions about behavior are almost always aimed at understanding how individuals cope with changing physical conditions or how they interact with individuals of their own or other species—in short, how they interact with their environment.

Conditional Strategies and Decision Making

In some cases, animals respond to a change in their environment in a simple, highly predictable way. When a kangaroo rat is feeding at night and hears the sound of a rattlesnake's rattle, it jumps back—away from the direction of the stimulus. The jump-back behavior is performed the same way every time.

INNATE, INFLEXIBLE BEHAVIOR IS RARE Highly inflexible, stereotyped behavior patterns like a kangaroo rat's jump-back are called **fixed action patterns**, or **FAPs**. FAPs are examples of what biologists call **innate behavior**. Innate behavior is inherited—meaning that it is passed from parents to offspring—and shows little variation based on learning or the individual's condition.

When researchers first began to explore animal behavior experimentally, they focused on these types of simple actions. This research strategy is common in biological science. If you are addressing a new question—whether it's how DNA is transcribed or how ions cross plasma membranes—it's helpful to start by studying the simplest possible system. Once simple systems are understood, investigators can delve into more complex situations and questions.

Behavioral biology started the same way. Early work focused on inflexible behavioral responses.

MOST BEHAVIOR IS FLEXIBLE AND CONDITION-DEPENDENT Even though virtually all species studied to date show some degree of innate behavior, it is much more common for an individual's behavior to (1) change in response to learning, and (2) show

FIGURE 51.1 Spiny Lobsters Can Find Their Way Back to Their Home Burrow. Spiny lobsters wander widely in search of food at night, then return to their hiding places in coral reefs before dawn.

✔**QUESTION** Compared to other lobsters, spiny lobsters have extremely long antennae for their body size. Antennae are involved in sensing chemicals, water currents, and objects in the environment. Generate a hypothesis to explain how long antennae function at the proximate level, and why long antennae are advantageous at the ultimate level.

flexibility in response to changing environmental conditions. Organisms are not stimulus-response machines.

🔑 Most animals have a range of actions that they can perform in response to a situation. Animals take in information from the environment and, based on that information, make decisions about what to do. Animals make choices.

If sharks and other predators are particularly abundant, spiny lobsters are likely to spend less time looking for food and more time hiding in a protected refuge. If predators are rare or absent, spiny lobsters might even begin to emerge and look for food during the day. Biologists characterize behavior like this as condition-dependent.

To link condition-dependent behavior to fitness—and thus the ultimate level of explanation—biologists use a framework called **cost-benefit analysis**. Animals appear to weigh the costs and benefits of responding to a particular situation in various ways. Costs and benefits are measured in terms of their impact on fitness—the ability to produce offspring.

Note the phrase "appear to weigh" in the previous paragraph. It's important to recognize that the decisions made by non-human organisms are not—as far as is known—conscious. A spiny lobster does not sit in its hole, counting predators and calculating the likelihood of being eaten. Instead, evolution has shaped its genome to direct a nervous and endocrine system that is capable of (1) taking in information and (2) directing behavior that is likely to pass that genome on to the next generation.

Five Questions in Behavioral Ecology

The following five sections of this chapter introduce some of the most prominent questions in behavioral ecology—the study of how individuals interact with their environment. In each case, the question is posed in terms of a decision that an animal has to make. The ensuing discussion is divided into research on how the question has been answered at the proximate level and the ultimate level.

The goals of this chapter are to explore some key topics in behavioral ecology and illustrate the spirit and synthetic nature of the field. Let's begin with an introduction to research on a resource that every animal has to find in its environment: food.

51.2 What Should I Eat?

When animals seek food, they are **foraging**. In some cases, the food source that a species exploits is extremely limited. Giant pandas, for example, subsist almost entirely on the shoots and leaves of bamboo. But individuals must still make decisions about which bamboo plants to harvest and which parts of a particular plant to eat and which to ignore.

In most cases, animals have a relatively wide range of foods that they exploit over the course of their lifetime. To understand why individuals eat the way they do, let's consider (1) research on genetic variation in foraging behavior and (2) the costs and benefits of feeding at various distances from home.

Foraging Alleles in *Drosophila melanogaster*

When Marla Sokolowski was working as an undergraduate research assistant, she noticed that some of the fruit-fly larvae she was studying tended to move after feeding at a particular location, while others tended to stay put. She reared these "rovers" and "sitters" to adulthood, bred them, and found that the offspring of rovers also tended to be rovers, while offspring of sitters tended to be sitters (**Figure 51.2**). Thus, this aspect of behavior is inherited. In fruit flies, variation in larval foraging behavior is at least partly due to genetic variation among individuals.

Using information from the *Drosophila melanogaster* genome sequence (see Chapter 20) and genetic mapping techniques introduced in Chapter 19, Sokolowski found and cloned a gene associated with the rover-sitter difference. She named it *foraging* (*for*).

Follow-up work showed that the protein product of *for* is involved in a signal transduction cascade (see Chapters 8 and 47)—meaning it is involved in the response to a cell-cell signal. Rovers and sitters tend to behave differently when they are foraging because they have different alleles of the *for* gene.

Genes that are homologous to *for* have now been found in honeybees and the roundworm *Caenorhabditis elegans*. Currently, researchers are focused on finding out which signaling pathway uses the *for* gene product and why different alleles change feeding behavior. Presumably, the signaling pathway is associated with feeding or nervous system development. When this aspect of fly feeding is thoroughly understood at the proximate level, biologists will know why variants of the same protein make larvae more likely to move or stay in place after feeding.

By altering population density experimentally and documenting the reproductive success of rovers and sitters, Sokolowski has shown that the rover allele is favored when population density is high and food is in short supply. In contrast, the sitter allele reaches high frequency in low-density populations, where food is abundant. Rovers do better at high density because they are more likely to find unused patches of food; sitters do better at low density because they don't waste energy moving around.

These results are exciting because they link a proximate mechanism with an ultimate outcome. In this case, the presence of certain alleles is responsible for a difference in fitness in specific types of habitats.

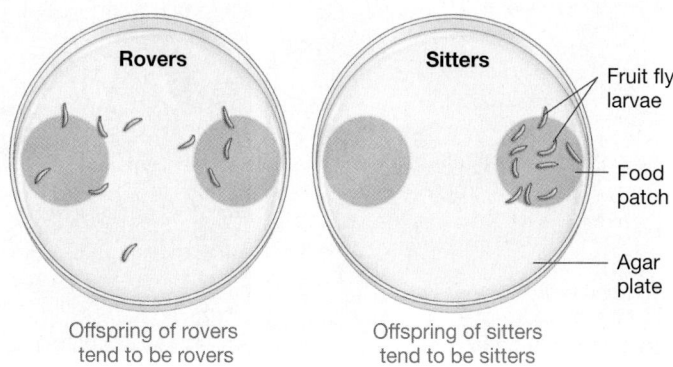

FIGURE 51.2 Foraging Behavior in Fruit-Fly Larvae Is Heritable.

Optimal foraging in White-Fronted Bee-Eaters

In *Drosophila* larvae, foraging is a relatively inflexible behavior—individuals don't "decide" to be rovers or sitters. But in *Drosophila* adults and in most animal species, foraging behavior is much more sophisticated.

When biologists set out to study why animals forage in a particular way, they usually start by assuming that individuals make decisions that maximize the amount of usable energy they take in, given the costs of finding and ingesting their food and the risk of being eaten while they're at it. This claim—that animals maximize their feeding efficiency—is called **optimal foraging**.

To test whether optimal foraging actually occurs, a researcher began studying feeding behavior in a bird called the white-fronted bee-eater (**Figure 51.3a**). Mated pairs of this species dig tunnels in riverbanks and raise their young inside. Appropriate riverbanks are rare, so many pairs build tunnels in the same location, forming a colony. To feed their young, pairs have to leave the colony, capture insects, and bring the prey back to the nest. Each pair defends a specific area where only they find food.

The key observation here is that some individuals forage a few meters from the colony, while some have to look for food hundreds of meters away. It takes only a few seconds to make a round trip to the closest feeding territories, versus several minutes to the farthest territories. Thus, the fitness cost of each feeding trip varies widely among individuals.

If optimal foraging occurs, how do individuals maximize their benefits, given the cost they have to pay in time and energy? Individuals that have to find food far from the colony stay away longer on each trip. And as the data in **Figure 51.3b** show, individuals foraging far from the colony bring back a much larger mass of insects on each trip, on average, than do individuals that fly a short distance each time.

White-fronted bee-eaters appear to make decisions that maximize the energy they deliver to their offspring, given their costs

in finding food. This behavior is highly flexible and condition dependent. If the same individual had a territory close to the colony instead of far away, it would make different decisions. Its behavior would change, but in each case it would be acting in a way that maximizes its fitness.

CHECK YOUR UNDERSTANDING

If you understand that . . .

- In *Drosophila* larvae, foraging behavior depends in large part on variation at a single gene. The rover allele is favored at high population density; the sitter allele is favored at low density. In this case, biologists understand both the proximate and ultimate mechanisms responsible for variation in foraging behavior.
- In white-fronted bee-eaters, foraging behavior appears to minimize fitness costs and maximize fitness benefits.

✔ **You should be able to . . .**

Suppose that a *for*-like gene with rover and sitter alleles existed in white-fronted bee-eaters (with rover alleles making individuals more likely to wander far in search of food), and explain which allele you would expect to be common in small colonies versus large colonies.

Answers are available in Appendix B.

51.3 Who Should I Mate With?

Biologists sometimes claim that almost all animal behavior revolves around food or sex. Section 51.2 focused on food; here the subject is sex.

Let's start by considering the proximate mechanisms responsible for triggering sexual activity in lizards, and then go on to explore research on how a female barn swallow chooses the indi-

(a) White-fronted bee-eaters are native to East Africa.

(b) Foraging behavior depends on distance traveled.

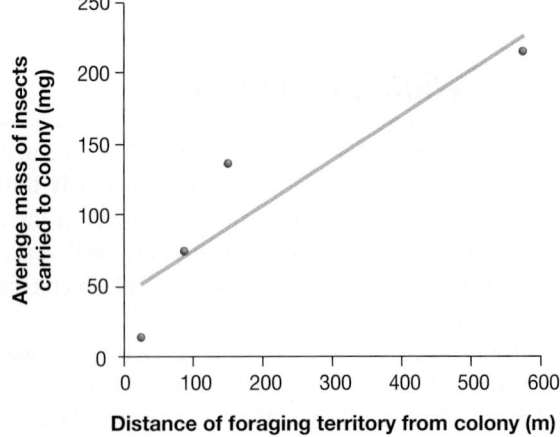

FIGURE 51.3 **White-Fronted Bee-Eaters Vary Their Foraging Behavior Depending on Conditions.**

vidual she is going to mate with. ☞ Biologists want to know how sexual activity occurs, in terms of underlying hormonal mechanisms, and why variation in mate choice affects fitness.

Sexual Activity in *Anolis* Lizards

Anolis carolinensis lives in the woodlands of the southeastern United States. After spending the winter under a log or rock, males emerge in January and establish breeding territories (**Figure 51.4a**). Females become active a month later, and the breeding season begins in April. By May, females are laying an egg every 10–14 days. By the time the breeding season is complete three months later, the eggs produced by a female will total twice her body mass.

Figure 51.4b graphs how sexual condition changes throughout the year, by plotting the size of the testes in males and ovaries of females—relative to an individual's overall body size. Individuals come into breeding condition in spring, with males experiencing growth of gonads long before females do. In both sexes, the gonads shrink in size during the fall and winter.

TESTOSTERONE AND ESTRADIOL What causes these dramatic seasonal changes in behavior? The proximate answer is sex hormones—testosterone in males and estradiol in females. Testos-

(a) Courtship display of a male *Anolis* lizard

Dewlap extended; male bobs up and down

(b) Changes in sexual organs through the year

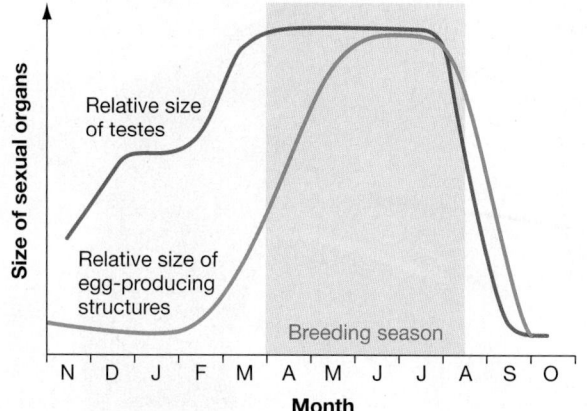

FIGURE 51.4 Sexual Behavior in *Anolis* Lizards Is Seasonal.

terone is produced in the testes of males, and estradiol in the ovaries of females (see Chapter 48).

The evidence for effects of sex hormones is direct. Testosterone injections induce courtship activity in castrated males; estradiol injections induce sexual activity in females whose ovaries have been removed.

This result leaves some key questions unanswered, however. *Anolis* lizards are most successful if they reproduce early in the spring. In springtime their food supplies are increasing, and snakes and other predators are not yet hunting lizards to feed their own young. If testosterone and estradiol levels induced sexual activity in *Anolis* lizards at the wrong time of year, it would be a disaster for the individuals involved.

In addition, it is critical for all *Anolis* individuals in a population to start their sexual activity at the same time, so that females can find males that are ready to breed, and vice versa. What environmental cues trigger the production of sex hormones in early spring? How do male and female *Anolis* synchronize their sexual readiness?

TESTING THE EFFECTS OF LIGHT AND SOCIAL STIMULATION To answer these questions, David Crews brought a large group of sexually inactive adult lizards into the laboratory during the winter and divided them into five treatment groups. The physical environment was exactly the same in all treatments. Each individual received identical food, and in all treatments high "daytime" temperatures were followed by slightly lower "nighttime" settings. Crews also continued to monitor the condition of lizards that remained in natural habitats nearby.

To test the hypothesis that changes in day length signal the arrival of spring and trigger initial changes in sex hormones, Crews exposed the five treatment groups in the laboratory to artificial lighting that simulated the long days and short nights of spring.

To test the hypothesis that social interactions among individuals are responsible for synchronizing sexual behavior, he varied the social setting among the five treatment groups:

1. a single isolated female;
2. a group of females;
3. a single female with a single male;
4. a single female with a group of castrated (nonbreeding) males; and
5. a single female with a group of uncastrated (breeding) males.

Each week, Crews examined the ovaries of females in each treatment. He also monitored the ovaries of females in nearby natural habitats. These females were not exposed to springlike conditions.

As **Figure 51.5** on page 1024 shows, the differences in the animals' reproductive systems were dramatic. The data points on this graph represent the average percentage of females with mature egg-producing structures, in each treatment; the points are connected by lines to help you keep track of the different treatments and see the overall patterns. Note two key points:

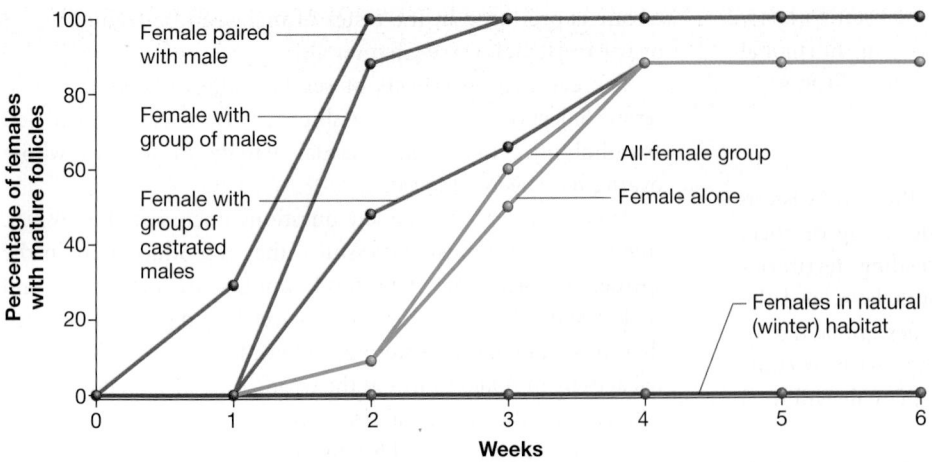

FIGURE 51.5 Exposure to Springlike Conditions and Breeding Males Stimulates Female Lizards to Produce Eggs. The percentage of female lizards with mature follicles, plotted over time. Five treatment groups were exposed to springlike light and temperature in the lab; data are also plotted for females left in a natural (winter) habitat. Each data point represents the average value for a group of 6–10 females.

✔ **QUESTION** Some critics contend that the females that remained outside do not represent a legitimate control in this experiment. What conditions would represent a better control?

1. Females that were exposed to springlike conditions began producing eggs; females in the field that were not exposed to springlike conditions did not.

2. Females that were exposed to breeding males began producing eggs much earlier than did the females placed in the other treatment groups.

These results support the hypothesis that two types of stimulation are necessary to produce the hormonal changes that lead to sexual behavior. Females need to experience springlike light and temperatures *and* exposure to breeding males.

VISUAL CUES FROM MALES TRIGGER FEMALE READINESS What is it about breeding males that triggers earlier estradiol release in females? What male signal causes the difference in female egg production?

When males court females, they bob up and down and extend a brightly colored patch of skin called a dewlap (Figure 51.4a). To test the hypothesis that this visual stimulation triggers changes in estradiol production, Crews repeated the previous experiment but added a twist: He placed some females with males that had intact dewlaps, and other females with males whose dewlaps had been surgically removed.

The result? Females who were living with dewlap-less males were slow to produce eggs. They were just as slow, in fact, as the females in the first experiment that had been grouped with castrated males. These latter females had not been courted at all.

These data suggest that the dewlap is a key visual signal. The experiments succeeded in identifying the environmental cues—long day length and bobbing dewlaps—that trigger hormone production and the onset of sexual behavior.

How Do Female Barn Swallows Choose Mates?

Given a choice, who should an individual choose for a mate? According to theory and data presented in Chapter 25:

● Females are usually the gender that is pickiest about mate choice.

● Females choose males that contribute good alleles and/or resources to their offspring.

To see this type of decision making in action, consider an experiment on mate choice by female barn swallows native to Denmark. Although both male and female barn swallows help to build the nest and feed the young, there is a significant amount of sexual dimorphism—males are slightly larger and more brightly colored, and their outer tail feathers are about 15 percent longer than the same feathers in females (**Figure 51.6**).

As Chapter 25 indicated, research on other bird species has shown that individuals with particularly long tails, bright colors, and energetic courtship displays are usually the healthiest and best fed in the population. Because it takes large amounts of energy and resources to produce traits like these, they serve as an honest signal of male quality.

Recent work has shown that barn swallows with particularly long tails are more efficient in flight and thus more successful in finding food. Tail length is a heritable trait—meaning that certain alleles affect tail length. If a female barn swallow mates with a long-tailed male, then, she should be picking an individual that will pass high-fitness alleles on to her offspring *and* be able to help her rear those offspring efficiently.

A researcher tested this prediction by altering tail length experimentally—in this case, by cutting off tail feathers and re-gluing

Male barn swallows have longer tail feathers and slightly brighter colors than females do

FIGURE 51.6 Male Barn Swallows Have Elongated Tail Feathers.

EXPERIMENT

QUESTION: Does tail length in barn swallows affect female choice of mates?

HYPOTHESIS: Females prefer to mate with the longest-tailed males.

NULL HYPOTHESIS: Tail length in males has no influence on female choice of mates.

EXPERIMENTAL SETUP:

Capture many males at the same site and randomly assign them to 1 of 4 groups:

Short tail: Outermost tail feathers cut

Control 1: Outermost tail feathers cut and glued back in place

Control 2: Males caught and handled, but no change in tail feathers

Elongated tail: Outermost feathers cut and elongated with feathers from other males, glued in place

Release males and document which males are first to mate, have a second nest that season, and raise the most offspring.

RESULTS:

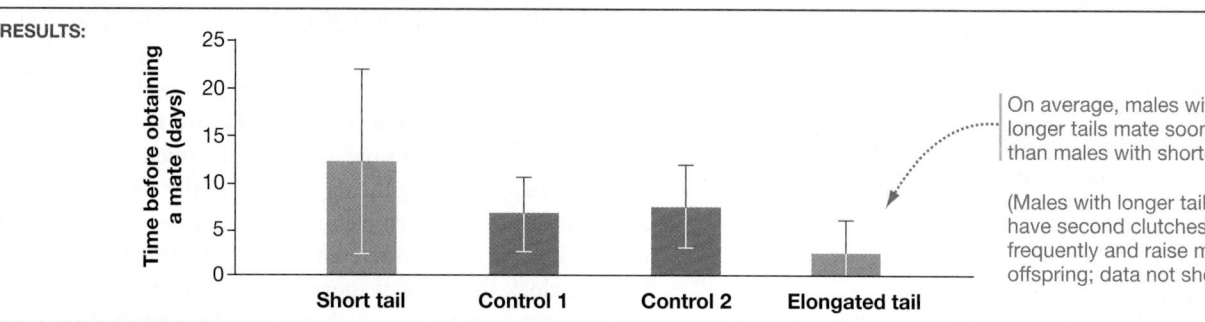

On average, males with longer tails mate sooner than males with shorter tails

(Males with longer tails also have second clutches more frequently and raise more offspring; data not shown)

CONCLUSION: Females prefer to mate with long-tailed males.

FIGURE 51.7 Barn Swallow Females Prefer to Mate with Long-Tailed Males.

SOURCE: Møller, A. P. 1988. Female choice selects for male sexual tail ornaments in the monogamous swallow. *Nature* 332: 640–642.

✔**QUESTION** Why did the investigator bother to artificially lengthen and shorten tails of males? Why didn't he just record natural variation in tail length and observe mating success?

---

them. The operation is painless and does not appear to affect flying ability.

As the "Experimental setup" section of **Figure 51.7** shows, the biologist caught a large series of males and randomly assigned them to one of four groups: either shortened tails, "sham-operated" unchanged tails, untouched tails, or elongated tails. When he documented how long it took males in each group to find a mate, a strong trend emerged: As the bar graphs in the "Results" section of Figure 51.7 indicate, males with elongated tails mated earliest on average—meaning that females preferred them.

This finding is important. Pairs that nest early have time to complete a second nest later in the season and thus produce the most offspring. Longer-tailed males have higher fitness than shorter-tailed males.

Data on mate choice in barn swallows reinforce a central theme of this chapter: Animals usually make decisions in a way that maximizes their fitness.

CHECK YOUR UNDERSTANDING

○━ If you understand that . . .

- In *Anolis* lizards, springlike light conditions and male dewlap displays cause estradiol release and sexual readiness in females.
- In barn swallows, females that prefer to mate with long-tailed males have highest fitness.

✔ You should be able to . . .

1. Explain why it is valid to claim that *both* springlike light conditions and male dewlap displays are required to bring female *Anolis* into optimal breeding condition.

2. Generate a hypothesis to explain the proximate mechanism responsible for bringing female barn swallows into breeding condition.

Answers are available in Appendix B.

51.4 Where Should I Live?

Most animals select the food they eat and the individuals they mate with. In most cases, they also select the habitat where they live.

Biologists have explored an array of questions related to habitat selection:

- whether juveniles should disperse from the area where they were raised or whether they should stay;

- how large an area should be defended against competitors in species that maintain a **territory** for breeding or feeding; and

- whether individuals have higher fitness when they occupy crowded, high-quality habitats or uncrowded, poor-quality habitats.

Here there is space to address just one question related to these types of "lifestyle" decisions: the proximate and ultimate mechanisms responsible for migration. In ecology, **migration** is defined as the long-distance movement of a population associated with a change of seasons. At the proximate level, how do animals know where to go? At the ultimate level, why is it worth the cost in time, energy, and risk of predation or accident?

How Do Animals Find Their Way on Migration?

To organize research into how animals find their way during migratory movements, biologists distinguish three categories of navigation:

1. **Piloting** is the use of familiar landmarks.

2. **Compass orientation** is movement that is oriented in a specific direction.

3. **True navigation** is the ability to locate a specific place on Earth's surface.

A thought experiment will help you understand these categories.

Suppose you grew up on the shores of Lake Superior but had never seen a map of the region. One day, you are transported to the middle of this enormous lake. Because you were blindfolded and made to spin in place en route, you are disoriented and have no idea where you are. But you are given a magnetic compass.

The compass tells you where north, south, east, and west are. Unfortunately this information will not help you find home, because you have no idea where you are in relation to home. You are stuck because you have no mechanism of true navigation.

Now suppose that a helicopter flying overhead drops you a map. The map has an X marking your current position and an H marking your home. The map also has a scale and a compass symbol. From this information, you determine that home is 100 miles to the west.

The map has solved your navigational problem. The compass has solved your orientation problem. Now that you know you live to the west, you can use the compass to find west and paddle home. As you approach the shore, you recognize familiar landmarks. Using them to pilot gets you the rest of the way home.

Organisms other than humans do not have manufactured compasses or printed maps. How do they navigate? Although bi-

ologists know little about the nature of a map sense (true navigation) in animals, piloting and compass orientation are increasingly well understood.

PILOTING A substantial amount of evidence suggests that at least some migratory animals use piloting to find their way.

For example, offspring follow their parents—south in the fall and north in the spring—in some species of migratory birds and mammals. Young appear to memorize the route.

Piloting even plays a role in species with more sophisticated navigational abilities. Homing pigeons, for example, can find their way home even when they are released in strange terrain a long distance away. But if these birds are equipped with frosted spectacles so that they see the world as a foggy haze, they only return to within a mile or so of their home—they do not actually find it. Based on these experiments, researchers have concluded that homing pigeons use piloting to navigate the final stages of a journey home.

COMPASS ORIENTATION How do animals perform compass orientation? To date, most research on compass orientation has been done on migratory birds, including whooping cranes (**Figure 51.8**). To determine where north is, these animals appear to use the Sun during the day and the stars at night. Let's consider each type of compass orientation separately, because they work differently.

The Sun is difficult to use as a compass reference because its position changes during the day. It rises in the east, is due south at noon (in the Northern Hemisphere), and sets in the west. To use the Sun as a compass reference, then, an animal must have an internal clock that defines morning, noon, and evening. Fortunately, most animals have such a clock. The **circadian clock** that exists in organisms maintains a 24-hour rhythm of chemical activity. The clock is set by the light–dark transitions of day and night. It tells individuals enough about the time of day that they can use the Sun's position to find magnetic north.

FIGURE 51.8 An Ultralight Guides Young Whooping Cranes to Winter Nesting Grounds. These human-reared birds will use compass orientation to find their own way back north again. Successful migration is key to increasing the number of these endangered birds in the wild.

The situation is actually simpler on clear nights, because migratory birds in the Northern Hemisphere can use the North Star to find magnetic north and select a direction for migration. But what if the weather is cloudy and neither the Sun nor stars are visible?

Under these conditions, migratory birds appear to orient using Earth's magnetic field. Exactly how they detect magnetism is under debate. One hypothesis contends that the birds can detect magnetism by their visual system, through an unknown molecular mechanism. An alternative hypothesis maintains that individuals have small particles of magnetic iron—the mineral called magnetite—in their bodies. Changes in the positions of magnetic particles, in response to Earth's magnetic field, could then be detected and provide reliable information for compass orientation.

Although research on mechanisms of compass orientation continues, one important point is clear: Birds and perhaps other organisms have multiple mechanisms of finding a compass direction. At least some species can use a Sun compass, a star compass, and a magnetic compass. Which system an individual uses depends on the weather and other circumstances.

To analyze another set of experiments on how animals find their way, go to the study area at *www.masteringbiology.com*.

 Web Activity Homing Behavior in Digger Wasps

Why Do Animals Move with a Change of Seasons?

In their extent and the navigation challenge they pose, migratory movements can be spectacular:

- Arctic terns nest in tundras in the far north. One population flies south along the coast of Africa to wintering grounds off Antarctica, and then flies back north along the eastern coast of South America. Chicks follow their parents south but make the return trip on their own. The journey totals over 32,000 km (20,000 miles).

- Many of the monarch butterflies native to North America spend the winter in the mountains of central Mexico or southwest California. Tagged individuals are known to have flown over 3000 km (1870 miles) to get to a wintering area. In spring, monarchs begin the return trip north. They do not live to complete the trip, however. Instead, mating takes place along the way, and females lay eggs en route. Although adults die on the journey, offspring continue to head north. After several generational cycles of reproduction and death in the original habitats in northern North America have passed, individuals again migrate south to overwinter. Instead of a single generation making the entire migratory cycle, then, the round trip takes several generations.

- Salmon that hatch in rivers along the Pacific Coast of North America and northern Asia migrate to the ocean when they are a few months to several years old, depending on the species. After spending several years feeding and growing in the North Pacific Ocean, they return to the stream where they hatched. There they mate and die.

In most cases, it is relatively straightforward to generate hypotheses for why migration exists at the ultimate level.

- Arctic terns feed on fish that are available in different parts of the world at different seasons. Thus, individuals that do not migrate might starve; individuals that do migrate should achieve higher reproductive success.

- Monarch butterflies may achieve higher reproductive success by migrating to wintering areas and new breeding areas than by trying to overwinter in northern North America.

- Salmon eggs and young are safer and thrive better in freshwater habitats than they do in the ocean, but adult salmon can find much more food in saltwater habitats than in freshwater and thus grow to bigger size, which confers a benefit in competition for nest sites and mates.

Testing these hypotheses rigorously is more difficult, as researchers have to be able to compare the fitness of individuals that do and do not migrate. Recently, however, researchers have documented the evolution of sockeye salmon populations that do not migrate to the ocean—in watersheds where migratory sockeye were introduced by humans. The nonmigratory populations, called kokanee, are much smaller-bodied than the migratory populations—presumably because there is less food available in the freshwater habitats.

The fitness of migrant and nonmigrant sockeye salmon appears be similar in this case, due to fitness trade-offs (see Chapter 24). In salmon, there is a large benefit of large body size in competing for nesting sites and mates. ☞ Increased access to food is a benefit of migration. There is a high cost, however, in time, energy, and predation risk.

At the ultimate level, researchers are also struggling to understand how a nonmigratory population can evolve into a migratory one that travels to a completely different part of the world. It's conceivable that habitat changes occurring in response to global climate change (see Chapter 54) may eventually furnish an example. Changes in the range and arrival times of migratory species have already been documented; it may only be a matter of time before a nonmigratory population begins to migrate, to cope with habitat changes caused by global warming.

51.5 How Should I Communicate?

The song of a bird, the bobbing dewlap of a male *Anolis carolinensis*, and the sentences in this text all have the same overall goal: communication. In biology, **communication** is defined as any process in which a signal from one individual modifies the behavior of a recipient individual. A **signal** is any information-containing behavior or characteristic. Communication is a crucial component of animal behavior. It creates a stimulus that elicits a response.

By definition, communication is a social process. For communication to occur, it is not enough that a signal is sent; the signal must be received and acted on. A lizard's bobbing dewlap does not qualify as communication unless another individual sees it and responds to the message.

Honeybee Language

Honeybees are highly social animals that live in hives. Inside the hive, a queen bee lays eggs that are cared for by individuals called workers. Besides caring for young and building and maintaining the hive, workers obtain food for themselves and other members of the colony by gathering nectar and pollen from flowering plants.

Biologists noticed that bees appear to recruit to food sources, meaning that if a new source is discovered by one or a few individuals, many more bees begin showing up over time. This observation inspired the hypothesis that successful food-finders have some way of communicating the location of food to other individuals.

At the ultimate level, it's straightforward to come up with a hypothesis for why bees communicate about food sources. If more workers find a rich food source quickly, the total amount of resources harvested, and the number of offspring produced by the hive, should increase. But at the proximate level, the task is more difficult. How do bees "talk" to each other?

THE DANCE HYPOTHESIS In the 1930s Karl von Frisch began studying bee communication by observing bees that built hives inside the glass-walled chambers he had constructed. He found that if he placed a feeder containing sugar water near one of these observation hives, a few of the workers began moving in a circular pattern on the vertical, interior walls of the hive.

Von Frisch called these movements the "round dance" (**Figure 51.9a**). Other bee workers appeared to follow the progress of the dance by touching the displaying individual as it danced, and to respond by flying away from the hive in search of the food source.

To investigate the function of these movements further, von Frisch placed feeders containing sugar water at progressively greater distances from the hive. Using this technique, he was able to get bees to visit feeders at a distance of several kilometers from the hive. By catching bees at the feeders and dabbing them with paint, he could individually mark successful food-finders. Follow-up observations at the hive confirmed that marked foragers (**1**) danced when they returned to the hive, and (**2**) returned to the food source with greater and greater numbers of unmarked bees.

To explain these data, von Frisch proposed that the round dance contained information about the location of food. Because bee hives are completely dark, he hypothesized that workers got information from the dance by touching the dancer and following the dancer's movements.

COMMUNICATING DIRECTIONS AND DISTANCES When von Frisch began placing feeders at longer distances from the hive, he found that successful food-finders used a new type of display. He named these movements the "waggle dance," because they combined circular movements like those of the round dance with short, straight runs (**Figure 51.9b**). During these runs, the dancer vigorously moved her abdomen from side to side.

These observations supported the hypothesis that both the round dance and waggle dance communicate information about food sources. Recent work has shown that the round and waggle dances are actually the same type of behavior—round dances just have a short waggle phase.

Three key observations allowed von Frisch to push our understanding of bee language result further. He noticed that:

- The orientation of the waggle part of the dance varied.
- The direction of the waggle run correlated with the direction of the food source from the hive.
- The length of the straight, "waggling" run was proportional to the distance the foragers had to fly to reach the feeder.

By varying the location of the food source and observing the orientation of the waggle dance given by marked workers, von Frisch was able to confirm that dancing bees were communicating the position of the food relative to the current position of the Sun. For example:

(a) The round dance

(b) The waggle dance

Other bee workers follow the progress of the dance by touching the displaying individual

FIGURE 51.9 Honeybees Perform Two Types of Dances. (a) During the round dance, successful food-finders move in a circle. **(b)** During the waggle dance, successful foragers move in a circle but then make straight runs through the circle. During the straight part of the dance, the dancer waggles her abdomen.

- If food is directly away from the Sun's current position, marked bees give the waggle portion of their dance directly downward (**Figure 51.10a**).

- If the food is 90 degrees to the right of the Sun, marked bees waggle 90 degrees to the right of vertical (**Figure 51.10b**).

These results are nothing short of astonishing. Honeybees do not have large brains, yet they are capable of symbolic language. What's more, they are able to interpret the angle of the waggle dance performed on a vertical surface and to respond by flying horizontally along the corresponding angle.

Further work has confirmed that the dance language of bees includes several modes of communication. In addition to the tactile information in the movements themselves, bees make sounds during the dance and give off scents that indicate the nature of the food source.

Modes of Communication

As the honeybee example shows, animals can use several modes of communication at once. And in addition to relying on tactile and olfactory signals, communication can be

- acoustic, as in the song of crickets or birds; or

- visual, as in the color patterns of birds or fish.

At the ultimate level, which mode of communication confers highest fitness?

One of the most general observations about communication is that the type of signal used by an organism correlates with its habitat.

For example, sound travels much farther than light in aquatic habitats. Based on this observation, it is logical to observe that humpback whales rely on songs for long-distance communication. In some cases, humpback whale songs can travel hundreds of kilometers. Groups of whales use acoustic communication to keep together as they move from summer feeding areas to winter breeding grounds, and individual males sing to attract mates and warn rivals.

Correlations between habitat and mode of communication are also observed in visual, olfactory, and tactile communication. Animals that are active during the day and that live in open or treeless habitats tend to rely on visual communication during courtship and territorial displays. Bats, wolves, and other animals that are active at night communicate via sound or scent. Ants and termites that live underground rely on olfactory and tactile communication.

Each mode of communication has advantages and disadvantages. Songs and calls can carry information over long distances, but are short lived. Thus, they have to be repeated to be effective, and frequent repetition requires a large expenditure of time and energy. In addition, acoustic communication attracts predators. It is no surprise that when a hawk or falcon approaches a marsh inhabited by courting red-winged blackbirds, things get very quiet. Communication systems have been honed by natural selection to maximize their benefits and minimize their costs.

When Is Communication Honest or Deceitful?

At the ultimate level, one of the questions that biologists ask about communication concerns the quality of the information. Is the signal reliable?

(a) Straight runs down the wall of the hive indicate that food is opposite the direction of the Sun.

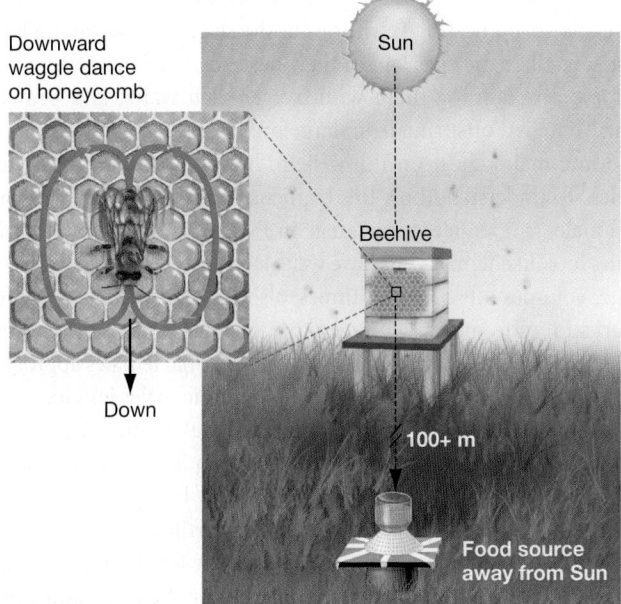

(b) Straight runs to the right indicate that food is 90° to the right of the Sun.

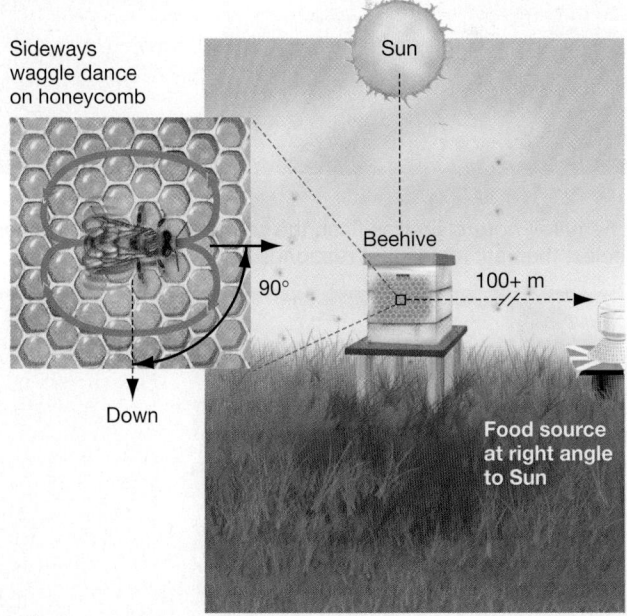

FIGURE 51.10 The Direction of the Waggle Dance Indicates the Location of Food Relative to the Sun. Waggle dances are done only when the food is at least 100 m from the hive.

✔**EXERCISE** The length of the straight run in the waggle dance indicates the relative distance of the food. Diagram a waggle dance that indicates food in the direction of the Sun and twice as far away as the food sources indicated by the dances drawn here.

As far as is known, honeybees are always honest. Stated another way, honeybee dances consistently "tell the truth" about the location of food sources. This observation is logical, because the honeybees that occupy a hive are closely related and cooperate closely in the rearing of offspring. As a result, it is advantageous for an individual to convey information accurately. If a food-finder provided inaccurate or misleading information to its hivemates, fewer offspring would be reared and the food-finder's fitness would be reduced.

In many cases, however, natural selection has favored the evolution of deceitful communication. Recent research on deceitful communication has highlighted just how complex interactions between signalers and receivers can be. A few examples will help drive this point home.

DECEIVING INDIVIDUALS OF ANOTHER SPECIES **Figure 51.11** illustrates two of the hundreds of examples of deceitful communication that have been documented among members of different species. Many of the best-studied instances involve predation:

- The anglerfish in Figure 51.11a has an appendage that dangles near its mouth and looks remarkably like a minnow. If an-

(a) Anglerfish use a "lure" to attract prey.

"Lure"

(b) Female *Photuris* fireflies flash the courtship signal of another species, then eat males that respond.

"Femme fatale"

"Victim"

FIGURE 51.11 Deceitful Communication Is Common in Nature.

other fish approaches this "lure" and attempts to eat it, the anglerfish attacks.

- Male and female fireflies flash a species-specific signal to each other during courtship. Predatory *Photuris* fireflies can mimic the pattern of flashes given by females of several other species. A *Photuris* female attracts a male of different species with the appropriate set of flashes, and then attacks and eats him (Figure 51.11b).

In each of these examples, individuals increase their fitness by providing inaccurate or misleading information to members of a different species. Deceitful communication is also known in plants, where it functions in pollination. For example, several orchid species have flowers that look and smell like female wasps and that accomplish pollination by "fooling" male wasps into attempting copulation.

In all of these cases, though, it's important to realize that the signaler is not consciously "lying." Instead, natural selection has simply favored certain behavioral or morphological traits that effectively communicate deceitful information.

DECEIVING INDIVIDUALS OF THE SAME SPECIES In some cases, natural selection has favored the evolution of traits or actions that deceive members of an organism's own species. When mantis shrimps molt, for example, they lack any external covering and cannot use their large claws. Thus, they are unable to defend the cavities where they live. But if another mantis shrimp approaches with the intent of evicting the individual and taking over the cavity, the molting individual bluffs. It raises its claws in the normal aggressive display and may even lunge at the intruder.

Perhaps the best-studied type of deceit in nonhuman animals, however, involves the mating system of bluegill sunfish. Male bluegills set up nesting territories in the shallow water along lake edges, fertilize the eggs laid in their nests, care for the developing embryos by fanning them with oxygen-rich water, and protect newly hatched offspring from large predators.

Some males cheat on this system, however, by mimicking females. To understand how this happens, examine the female, normal male, and female-mimic male in **Figure 51.12**. Female-mimic males look like females but have well-developed testes and produce large volumes of sperm. Mimics also act like females during courtship movements with territory-owning males. They even adopt the usual egg-laying posture. When normal females approach the nest and begin courtship, the female-mimic males join in.

The territory-owning male tolerates the mimic, apparently thinking that he is successfully courting two females at the same time. But when the actual female begins to lay eggs, the mimic responds by releasing sperm—and thus fertilizing some of the eggs—and then darting away. In this way, the female-mimic male fathers offspring but does not help care for them. In addition, female-mimics do not have to expend time and energy growing to large size and making and defending a nest.

WHEN DOES DECEPTION WORK? In analyzing deceitful communication in bluegill sunfish and other species, researchers point

FIGURE 51.12 In Bluegill Sunfish, Some Males Look and Act Like Females.

out that in most cases, deceit works only when it is relatively rare. The logic behind this hypothesis runs as follows: If deceit becomes extremely common, then natural selection will strongly favor individuals that can detect and avoid or punish liars. But if liars are rare, then natural selection will favor individuals that are occasionally fooled but are more commonly rewarded by responding to signals in a normal way.

In mantis shrimps, for example, molting occurs only once or twice a year. Thus, the vast majority of contests over hiding places involve individuals that have an intact exoskeleton and claws that qualify as a deadly weapon. As a result, intruders usually benefit by interpreting threat displays as honest and responding to the stimulus by retreating. If molting were frequent and bluffing behavior common, then intruders would benefit by challenging territory owners, which would likely be defenseless.

CHECK YOUR UNDERSTANDING

If you understand that . . .

- Communication is an exchange of information between individuals.
- In most instances, the mode of communication that animals use maximizes the probability that the information will be transferred efficiently.
- Communication can be honest or deceitful, depending on the nature of the information received.

✔ You should be able to . . .

1. Explain the costs and benefits of auditory, olfactory, and visual communication.
2. Explain why natural selection favors individuals that can detect deceitful communication and either avoid "liars" or punish them.

Answers are available in Appendix B.

51.6 When Should I Cooperate?

The types of behavior reviewed thus far all have a key common element: They help individuals respond to environmental stimuli in a way that increases their fitness. There is a type of behavior that appears to contradict this pattern, however: altruism.

Altruism is behavior that has a fitness cost to the individual exhibiting the behavior and a fitness benefit to the recipient of the behavior. It is the formal term for self-sacrificing behavior. Altruism decreases an individual's ability to produce offspring but helps others produce more offspring.

The existence of altruistic behavior appears to be paradoxical. If certain alleles make an individual more likely to be altruistic, those alleles should be selected against. Does altruism actually occur?

Kin Selection

There is no question that self-sacrificing behavior occurs in nature. For example, black-tailed prairie dogs perform a behavior called alarm calling (**Figure 51.13**). These burrowing mammals live in large communities, called towns, throughout the Great Plains region of North America. When a badger, coyote, hawk, or other predator approaches a town, some prairie dogs give alarm calls that alert other prairie dogs to run to mounds and scan for the threat.

Giving these calls is risky. In several species of ground squirrels and prairie dogs, researchers have shown that alarm-callers draw attention to themselves by calling and are in much greater danger of being attacked than non-callers are.

HAMILTON'S RULE How can natural selection favor the evolution of self-sacrificing behavior? William D. Hamilton answered this question by creating a mathematical model to assess how an allele that contributes to altruistic behavior could increase in frequency in a population.

To model the fate of altruistic alleles, Hamilton represented the fitness cost of the altruistic act to the actor as C and the fitness benefit to the recipient as B. Both C and B are measured in

FIGURE 51.13 Black-Tailed Prairie Dogs Are Highly Social. Prairie dogs live with their immediate and extended family within large groups called towns. The individual with the upright posture has spotted an intruder and may give an alarm call.

units of offspring produced. His model showed that the allele could spread if

$$Br > C$$

where r is the **coefficient of relatedness**. The coefficient of relatedness is a measure of how closely the actor and beneficiary are related. Specifically, r measures the fraction of alleles in the actor and beneficiary that are identical by descent—that is, inherited from the same ancestor (see **Box 51.1**).

The formulation $Br > C$ is called **Hamilton's rule**. It states that altruistic behavior is most likely when three conditions are met:

- The fitness benefits of altruistic behavior are high for the recipient.
- The altruist and recipient are close relatives.
- The fitness costs to the altruist are low.

When Hamilton's rule holds, alleles associated with altruistic behavior will be favored by natural selection—because close relatives are very likely to have copies of the altruistic alleles. They will increase in frequency in the population.

INCLUSIVE FITNESS Hamilton's rule is important because it shows that individuals can pass their alleles on to the next generation not only by having their own offspring, but also by helping close relatives produce more offspring. To capture this point, biologists refer to direct fitness and indirect fitness.

- Direct fitness is derived from an individual's own offspring.
- Indirect fitness is derived from helping relatives produce more offspring than they could produce on their own.

The combination of direct and indirect fitness components is called **inclusive fitness**.

BOX 51.1 QUANTITATIVE METHODS: Calculating the Coefficient of Relatedness

The coefficient of relatedness, r, varies between 0.0 and 1.0. If two individuals have no identical alleles that were inherited from the same ancestor, then their r value is 0.0. Because every allele in pairs of identical twins is identical, their coefficient of relatedness is 1.0.

What about other relationships? **Figure 51.14a** shows how r is calculated between half-siblings. (To review what the boxes, circles, and lines in a pedigree mean, see Figure 13.22.) Half-siblings share one parent. Thus, r represents the probability that half-siblings share alleles as a result of inheriting alleles from their common parent. It is critical to realize that in each parent-to-offspring link of descent, the probability of any particular allele being transmitted is 1/2. This is so because meiosis distributes alleles from the parent's diploid genome to their haploid gametes randomly. Thus, half the gametes produced by a parent get one of the alleles present at each gene, and half the gametes produced get the other allele. Half-siblings are connected by two such parent-to-offspring links. The overall probability of two half-siblings sharing the same allele by descent is $1/2 \times 1/2 = 1/4$ (To review rules for combining probabilities, see **BioSkills 13** in Appendix A.)

To think about this calculation in another way, focus on the red arrows in Figure 51.14a. The left arrow represents the probability that the mother transmits a particular allele to her son. The right arrow represents the probability that the mother transmits the same allele to her daughter. Both probabilities are 1/2. Thus, the probability that the mother transmitted the same allele to both her son and daughter is $1/2 \times 1/2 = 1/4$.

Figure 51.14b shows how r is calculated between full siblings. The challenge here is to calculate the probability of two individuals sharing the same allele as a result of inheriting it through their mother or through their father. The probability that full siblings share alleles as a result of inheriting them from one parent is 1/4. Thus, the probability that full siblings share alleles inherited from either their mother or their father is $1/4 + 1/4 = 1/2$.

(a) What is the probability that half-siblings inherit the same allele from their common parent?

r between half-siblings:
$1/2 \times 1/2 = 1/4$

(b) What is the probability that full siblings inherit the same allele from their father or their mother?

Probability that they inherit same allele from **father**:
$1/2 \times 1/2 = 1/4$

Probability that they inherit same allele from **mother**:
$1/2 \times 1/2 = 1/4$

r between full siblings:
$1/4 + 1/4 = 1/2$

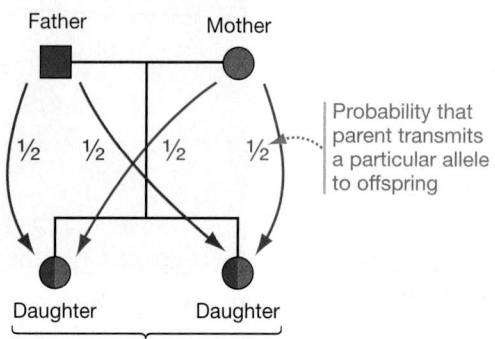

FIGURE 51.14 The Coefficient of Relatedness (r) Is Calculated from Information in Pedigrees.

✔**QUESTION** What is the r between first cousins?

Biologists use the term **kin selection** to refer to natural selection that acts through benefits to relatives. Kin selection results in increased indirect fitness.

✔If you understand kin selection, you should be able to explain why it is likely to be common in humans. You should also be able to explain whether kin selection occurs in plants.

TESTING HAMILTON'S RULE   Does Hamilton's rule work? Do animals really favor relatives when they act altruistically? To test the kin-selection hypothesis, a researcher studied which of the inhabitants of a black-tailed prairie dog town were most likely to give alarm calls.

Within a large prairie dog town, individuals live in small groups called coteries that share the same underground burrow. Members of each coterie defend a territory inside the town.

By tagging offspring that were born over several generations, the researcher identified the genetic relationships among individuals in the town. More specifically, the researcher determined, within each coterie, whether each individual had:

1. no close genetic relatives in its coterie;

2. no offspring in the coterie but at least one sibling, cousin, uncle, aunt, niece, or nephew; or

3. at least one offspring or grandoffspring in the coterie.

The kin-selection hypothesis predicts that individuals who do not have close genetic relatives nearby will rarely give an alarm call. To evaluate this prediction, the biologist recorded the identity of callers during 698 experiments. In these studies, a stuffed badger was dragged through the colony on a sled.

Were prairie dogs with close relatives nearby more likely to call, or did kinship have nothing to do with the probability of alarm calling? The bar charts in **Figure 51.15** show the average proportion of times that individuals in each of the three categories called during an experiment. The data indicate that black-tailed prairie dogs are much more likely to call if they live in a coterie that includes close relatives.

🔑 This same pattern—of preferentially dispensing help to kin—has been observed in many other species of social mammals and birds. Most cases of self-sacrificing behavior that have been analyzed to date are consistent with Hamilton's rule and are hypothesized to be the result of kin selection.

Reciprocal Altruism

During long-term studies of highly social animals, such as lions, chimpanzees, and vampire bats, biologists have observed nonrelatives helping each other. Chimps and other primates, for example, may spend considerable time grooming unrelated members of their social group—cleaning their fur and removing ticks and other parasites from their skin. In vampire bats, individuals that have been successful in finding food are known to regurgitate blood meals to non-kin that have not been successful and that are in danger of starving.

How can self-sacrificing behavior like this evolve if kin selection is not acting? The leading hypothesis to explain altruism

---

EXPERIMENT

QUESTION: Do black-tailed prairie dogs prefer to help relatives when they give an alarm call?

HYPOTHESIS: Individuals give an alarm call only when close relatives are near.

NULL HYPOTHESIS: The presence of relatives has no influence on the probability of alarm calling.

EXPERIMENTAL SETUP:

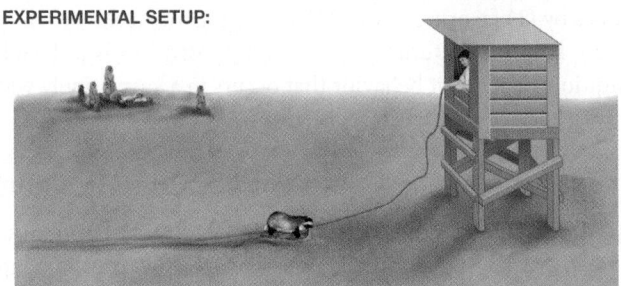

1. Determine relationships among individuals in prairie dog coterie.

2. Drag stuffed badger across territory of coterie.

3. From observation tower, record which members of coterie give an alarm call.

4. Repeat experiment 698 times. Each individual prairie dog coterie is tested 6–9 times over 3-year period.

PREDICTION OF KIN-SELECTION HYPOTHESIS: Individuals in coteries that contain a close genetic relative are more likely to give an alarm call than are individuals in coteries that do not contain a close genetic relative.

PREDICTION OF NULL HYPOTHESIS: The presence of relatives in coteries will not influence the probability of alarm calling.

RESULTS:

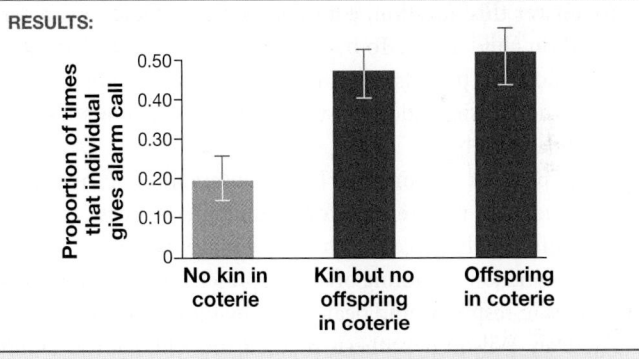

CONCLUSION: Alarm calling usually benefits relatives.

FIGURE 51.15 Experimental Evidence That Black-Tailed Prairie Dogs Are More Likely to Give Alarm Calls if Relatives Are Nearby.

SOURCE: Hoogland, J. L. 1983. Nepotism and alarm calling in the black-tailed prairie dog (*Cynomys ludovicianus*). *Animal Behavior* 31: 472–479.

✔**QUESTION** What would be an appropriate control in this experiment, and what hypothesis would it test?

---

among nonrelatives is called **reciprocal altruism**: an exchange of fitness benefits that are separated in time. Reciprocal altruists help individuals who have either helped them in the past or are likely to help them in the future.

Data that have been collected so far support the reciprocal-altruism hypothesis in at least some instances.

- Among vervet monkeys, individuals are most likely to groom unrelated individuals that have groomed or helped them in the past.
- Vampire bats are most likely to donate blood meals to non-kin that have previously shared food with them.

Reciprocal altruism is also widely invoked as an explanation for the helpful and cooperative behavior commonly observed among unrelated humans.

To summarize, altruistic behavior increases the fitness of individuals by (**1**) favoring kin, or (**2**) increasing the likelihood of receiving help in the future from non-kin. Altruism is a flexible, condition-dependent behavior that occurs in a surprisingly wide array of species that live in social groups.

An Extreme Case: Abuse of Non-Kin in Humans

Hamilton's rule specifies when individuals should be altruistic to relatives; reciprocal altruism explains why it can be adaptive, in some circumstances, for nonrelatives to cooperate. What about the opposite of cooperation—abuse or other types of violence?

Martin Daly and Margo Wilson helped pioneer the use of "selection thinking" in studies of human behavior. They were among the first to ask how understanding kin selection and other aspects of natural selection can inform research on humans.

Daly and Wilson have been particularly interested in studying the "dark side" of human behavior—homicide, domestic violence, and adultery—from a biological perspective. Under certain conditions, can these types of events be consistent with actions favored by natural selection?

To answer this question, consider data that these researchers gathered on child abuse. To begin, they hypothesized that selection should favor parents who invest resources in biological children but not in stepchildren—with whom the parents have no genetic relationship.

The hypothesis was inspired by the observation of infanticide in lions and in monkeys called Hanuman langurs. In these species, infant-killing occurs when new males take over a group of females and kill the young offspring present. The females come into estrus in response, and bear young by the new males.

The Daly-Wilson hypothesis predicts that child abuse should be much more common in households containing a stepparent than it is in homes containing only biological parents. Further, Daly and Wilson predicted that, if stepparents abuse their stepchildren, infants should be most at risk. The logic behind this prediction is that very young children demand the most time and resources and are least able to defend themselves.

Are the data consistent with these predictions? Daly and Wilson analyzed the most extreme form of child abuse—the killing of children by parents. Using a database on all homicides reported in Canada between 1974 and 1983, they found 341 cases in which a child was killed by a biological parent and 67 in which the perpetrator was a stepparent. But because households containing biological parents are much more common, they realized that they needed to compare the *rate* of violence in the two types of households, rather than the absolute number.

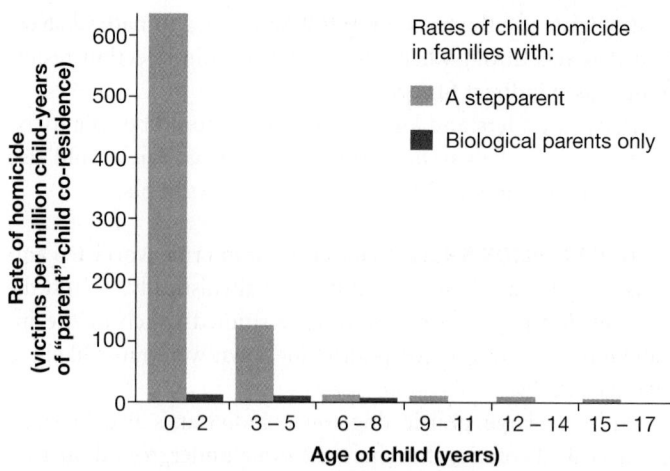

FIGURE 51.16 Rates of Child Homicide Are Higher in Families Containing a Stepparent. Rates of child homicide expressed as victims per "million child-years" in Canadian households.

Daly and Wilson expressed the homicide rate as the number of children killed per million "child-years." One child-year is a year that a child of a particular age lived in a particular family situation. A 5-year-old living with the same parents its entire life would represent five child-years.

The results are graphed in **Figure 51.16**. Note that children who live with a stepparent are at much higher risk of abuse than are children who live with biological parents. More specifically, kids who are less than 2 years old are 70 times more likely to be killed in a household with a stepparent than in a household with only biological parents. To put this number in perspective, consider that smokers are 11 times more likely to develop lung cancer than are nonsmokers.

The data have been helpful in raising awareness among social workers and clinicians—to be especially alert to indications of violence in households with stepparents where very young children are present.

CHECK YOUR UNDERSTANDING

If you understand that . . .
- Hamilton's rule states that alleles for altruistic behavior increase in frequency if the fitness cost of the behavior for the actor is relatively low, the fitness benefit to the recipient is relatively high, and the actor and recipient are closely related.
- Reciprocal altruism is based on an exchange of fitness benefits separated in time.

✔ You should be able to . . .
1. Use Hamilton's rule to describe situations in which a black-tailed prairie dog is likely and unlikely to give an alarm call.
2. Explain why reciprocal altruism has been observed only in species where individuals have good memories and live in small, long-lived social groups.

Answers are available in Appendix B.

In interpreting these results, though, it is *extremely* important to keep four points in mind:

1. Violence occurs among biological kin as well as non-kin.

2. The vast majority of stepparents are solicitous and generous with their stepchildren.

3. In today's culture, the behavior is undoubtedly pathological rather than adaptive, because child abusers are jailed and treated as abhorrent.

4. Violence is a condition-dependent behavior. Our genetic heritage and neural and hormonal systems make an enormous range of behavior possible. Ask yourself: What kinds of life conditions would drive a person to commit such a heinous act? What can friends, family members, and society in general do to help individuals keep from getting into circumstances that lead to violence?

CHAPTER 51 REVIEW

Summary of Key Concepts

🔑 **Biologists analyze behavior at the proximate and ultimate levels—the genetic and physiological mechanisms and how they affect fitness.**

- At the proximate level, experiments and observations focus on understanding how specific gene products, neuron activity, and hormonal signals cause behavior.

- At the ultimate level, researchers seek to understand the adaptive significance of behavior, or how it enables individuals to survive and reproduce.

- By combining proximate and ultimate viewpoints, biologists can seek a comprehensive understanding of how and why animals do what they do.

  ✔ You should be able to explain why fruit-fly larvae vary in foraging behavior, at both the proximate and ultimate levels.

🔑 **Individuals can behave in a wide range of ways; which behavior occurs depends on current conditions.**

- Foraging in white-fronted bee-eaters, mate choice in barn swallows, and the onset of sexual behavior in *Anolis* lizards are examples of flexible, condition-dependent behaviors triggered by changing environmental stimuli.

- In most cases, animals behave in a way that maximizes their fitness under current conditions.

  ✔ You should be able to relate Hamilton's rule to the concept of condition-dependence in behavior.

🔑 **Foraging patterns may vary with genotype; foraging decisions maximize energy gain and minimize costs.**

- At the proximate level, the rover and sitter phenotypes in fruit-fly larvae are one of the best-understood examples of variation in foraging behavior. Different alleles at a single gene increase the probability that an individual will move or stay after feeding.

- The rover allele has highest fitness at high population density; the sitter allele has highest fitness at low population density.

- In white-fronted bee-eaters, individuals that live far from the colony minimize the costs of foraging and maximize the benefits by increasing the length of time they look for food and bringing back larger amounts of food with each trip.

  ✔ You should be able to discuss the costs and benefits of roving versus sitting in fruit-fly larvae.

🔑 **Sexual behavior depends on surges of sex hormones; females choose mates that provide good alleles and needed resources.**

- In *Anolis* lizards, males and females come into breeding condition in spring, in response to surges of the sex hormones testosterone and estradiol.

- Female *Anolis* come into breeding condition fastest if they are exposed to springlike temperatures *and* courtship displays from males.

- Female barn swallows prefer to mate with long-tailed males, because the longest-tailed males are the best-fed and most disease-free.

  ✔ You should be able to generate a hypothesis for why vigorous bobbing displays of a brightly colored dewlap might be a reliable signal of male quality in *Anolis* lizards.

🔑 **Animals navigate using an array of cues; the costs of migration can be offset by benefits in food availability.**

- Many animal species can find their way by following familiar landmarks.

- When navigating, many animal species get compass information from the Sun, the stars, and Earth's magnetic field.

- Animals usually migrate to find seasonally available food. The benefits of the food sources have to outweigh the energetic costs and danger of the migratory movement.

  ✔ You should be able to explain why migration in some birds is considered condition dependent, based on the observation that some species do not migrate if people supply food in feeders.

🔑 **Animals communicate with movements, odors, or other stimuli; communication can be honest or deceitful.**

- Honeybees have a symbolic language—their dancing movements communicate accurate information about the direction and distance of food sources.

- Deceitful communication can be adaptive—as when anglerfish fool prey into attacking a "lure" attached to the anglerfish's face or when molting mantis shrimps try to bluff their way into retaining ownership of a hiding place.

  ✔ You should be able to predict the consequences for deceitful communication in bluegill sunfish if most large, territory-owning males were fished out of a lake.

  MB **Web Activity** Homing Behavior in Digger Wasps

🔑 Individuals that behave altruistically are usually helping relatives or individuals that help them in return.

- When altruistic behavior is directed toward close relatives, alleles that lead to self-sacrificing may increase in frequency due to kin selection.

- Some animals that live in close-knit social groups engage in reciprocal altruism—meaning they exchange help over time.

  ✔ You should be able to describe the characteristics of a species for which altruistic behavior is expected to be extremely common.

Questions

✔ **TEST YOUR KNOWLEDGE**   *Answers are available in Appendix B*

1. What do proximate explanations of behavior focus on?
   a. how displays and other types of behavior have changed through time, or evolved
   b. the functional aspect of a behavior, or its "adaptive significance"
   c. genetic, neurological, and hormonal mechanisms of behavior
   d. appropriate experimental methods when studying behavior

2. What unit(s) do biologists use when analyzing the costs and benefits of behavior?
   a. proximate versus ultimate
   b. body size (e.g., large versus small)
   c. time and energy
   d. fitness—the ability to survive and produce offspring

3. What is "reciprocal" about reciprocal altruism?
   a. It is based on an exchange of fitness benefits.
   b. The events involved are separated in time.
   c. The individuals involved usually know each other well.
   d. The individuals involved have to have good memories, to keep track of which individuals they have helped in the past.

4. Which of the following statements about the waggle dance of the honeybee is not correct?
   a. The length of a waggling run is proportional to the distance from the hive to a food source.
   b. Sounds and scents produced by the dancer provide information about the nature of the food source.
   c. The dancer uses no elements of the round dance.
   d. The orientation of the waggling run provides information about the direction of the food from the hive, relative to the Sun's position.

5. Why are biologists convinced that the sex hormone testosterone is required for normal sexual activity in male *Anolis* lizards?
   a. Male *Anolis* lizards with larger testes court females more vigorously than do males with smaller testes.
   b. The testosterone molecule is not found in female *Anolis* lizards.
   c. Male *Anolis* lizards whose gonads had been removed did not develop dewlaps.
   d. Male *Anolis* lizards whose gonads had been removed did not court females.

6. What does Hamilton's rule specify?
   a. why animals do things "for the good of the species"
   b. why reciprocal altruism can lead to fitness gains for unrelated individuals
   c. how alleles that favor self-sacrificing acts increase in frequency via kin selection
   d. the conditions under which more complex behaviors evolve from simpler behaviors

✔ **TEST YOUR UNDERSTANDING**   *Answers are available in Appendix B*

1. Discuss the proximate and ultimate causes of homing behavior by spiny lobsters.

2. What does it mean to say that an animal is foraging in an optimal way?

3. Why do traits like long tails in male barn swallows serve as honest signals of male quality?

4. What environmental stimuli cause changes in hormone levels that lead to egg laying in *Anolis* lizards? What biologically relevant information do these stimuli provide?

5. For an animal to navigate, it must have a "map" and a "compass." Explain why both types of information are needed. Describe three types of compasses that have been identified in migratory or homing species.

6. How are kin selection and reciprocal altruism different? Which should be more common, and why?

✔ **APPLYING CONCEPTS TO NEW SITUATIONS**   *Answers are available in Appendix B*

1. To date, the rover and sitter alleles have only been shown to affect foraging behavior in fruit-fly larvae. Design an experiment to test the hypothesis that adults with the rover allele tend to fly farther in search of food sources than adults with the sitter allele.

2. Most tropical habitats are highly seasonal. But instead of alternating warm and cold seasons, there are alternating wet and dry seasons. Most animal species breed during the wet months. If you were studying a species of *Anolis* native to the tropics, what environmental cue would you simulate in the lab to bring them into breeding condition? How would you simulate this cue?

3. A biologist once remarked that he'd be willing to lay down his life to save two brothers or eight cousins. Explain what he meant.

4. Based on the theory of reciprocal altruism, predict the conditions under which people are expected to donate blood.

A flock of white pelicans swimming in a channel of the Mississippi River. This chapter explores how and why growth rates in populations change through time.

Population Ecology 52

I f you asked a biologist to name two of today's most pressing global issues, she might say global warming and extinction of species. If you were asked to name two of the most pressing issues facing your region, you might say traffic and the price of housing. All four problems have a common cause: recent and dramatic increases in the size of the human population.

In 1940, for example, Mexico City had a population of about 1.6 million. The city grew to 5.4 million in 1960, 13.9 million in 1980, and over 19 million in 2007. How much bigger will it get over the next 50 years? And how large will the entire human population be by the time your kids are in college?

A **population** is a group of individuals of the same species that live in the same area at the same time; **population ecology** is the study of how and why the number of individuals in a population changes over time.

With the explosion of human populations across the globe, the massive destruction of natural habitats, and the resulting threats to species throughout the tree of life, population ecology has become a vital field in biological science. The mathematical and analytical tools introduced in this chapter help biologists predict changes in population size and design management strategies to save threatened species.

Let's start by considering some of the basic tools that biologists use to study populations, then follow up with examples of how biologists study changes in the population size of humans and other species over time. The chapter concludes by asking how all of these elements fit together in efforts to limit human population growth and save endangered species.

52.1 Demography

The number of individuals present in a population depends on four processes: birth, death, immigration, and emigration.

> **KEY CONCEPTS**
>
> 🔑 Life tables summarize how likely it is that individuals of each age class in a population will survive and reproduce.
>
> 🔑 The growth rate of a population can be calculated from life-table data or from the direct observation of changes in population size over time.
>
> 🔑 Researchers observe a wide variety of patterns when they track changes in population size over time, ranging from growth rates that slow when populations are at high density, to regular cycles, to continued growth independent of population size.
>
> 🔑 Data from population ecology studies help biologists evaluate prospects for endangered species and design effective management strategies, as well as to predict changes in human populations.

✔ When you see this checkmark, stop and test yourself. Answers are available in Appendix B.

- Populations grow due to births—here, meaning any form of reproduction—and **immigration**, which occurs when individuals enter a population by moving from another population.

- Populations decline due to deaths and **emigration**, which occurs when individuals leave a population to join another population.

Analyzing birthrates, death rates, immigration rates, and emigration rates is fundamental to **demography**: the study of factors that determine the size and structure of populations through time.

To predict the future of a population, biologists have to know something about its makeup: how many individuals of each age are alive, how likely individuals of different ages are to survive to the following year, how many offspring are produced by females of different ages, and how many individuals of different ages immigrate and emigrate each **generation**—the average time between a mother's first offspring and her daughter's first offspring.

If a population consists primarily of young individuals with a high survival rate and reproductive rate, the population size should increase over time. But if a population comprises chiefly old individuals with low reproductive rates and low survival rates, then it is almost certain to decline over time. To understand a population's dynamics, biologists turn to the data contained in a life table.

Life Tables

A **life table** summarizes the probability that an individual will survive and reproduce in any given time interval over the course of its lifetime. Life tables were invented almost 2000 years ago; in ancient Rome they were used to predict food needs. In modern times, life tables have been the domain of life-insurance companies, which have a strong financial interest in predicting the likelihood of a person dying at a given age. More recently, biologists are using life tables to study the demographics of endangered species.

LACERTA VIVIPARA: A CASE STUDY To understand how researchers use life tables, consider the lizard *Lacerta vivipara* (**Figure 52.1**). *L. vivipara* is a common resident of open, grassy habitats in western Europe. As their name suggests—vivipara means "live birth"—most populations are ovoviviparous and give birth to live young (see Chapter 32).

Researchers set out to estimate the life table of a low-elevation population in the Netherlands, with the goal of comparing the results to data that other researchers had collected from *L. vivipara* populations in the mountains of Austria and France and in lowland habitats in Britain and Belgium. They wanted to know whether populations that live in different environments vary in basic demographic features.

To begin the study, the researchers visited their study site daily during the seven months that these lizards are active during the year. Each day the researchers captured and marked as many individuals as possible.

Because this program of daily monitoring continued for seven years, the biologists were able to document the number of young produced by each female in each year of her life. If a marked in-

FIGURE 52.1 *Lacerta vivipara* **Are Native to Europe.** Females in most populations bear live young, though females in *L. vivipara* populations from northern Spain and southwest France lay eggs.

dividual was not recaptured in a subsequent year, they assumed that it had died sometime during the previous year.

The data allowed researchers to calculate the number of individuals that survived each year in each particular age group as well as how many offspring each female produced. What did the numbers reveal?

SURVIVORSHIP Survivorship—a key component of a life table—is defined as the proportion of offspring produced that survive, on average, to a particular age. For example, suppose 1000 *L. vivipara* are born in a particular year. These individuals represent a **cohort**—a group of the same age that can be followed through time. How many individuals would survive to age 1, age 2, age 3, and so on?

As **Table 52.1** shows, survivorship from birth to age 1 was 0.424 in the Netherlands population. If 1000 females were born in a particular year in this population, on average 424 would still be alive one year later. Survivorship from birth to age 2 was 0.308, meaning an average of 308 female lizards would survive for two years.

To analyze general patterns in survivorship, biologists plot the logarithm of the number of survivors versus age. Using a logarithmic scale on the *y*-axis is a matter of convenience—it makes patterns easier to see (see **BioSkills 7** in Appendix A). The resulting graph is called a **survivorship curve**.

Studies on a wide variety of species indicate that three general types of survivorship curves exist (**Figure 52.2a**).

1. Humans have what biologists call a type I survivorship curve. In this pattern, survivorship throughout life is high—most individuals approach the species' maximum life span.

2. Type II survivorship curves occur in species where individuals have about the same probability of dying in each year of life. Blackbirds and other songbirds have this type of curve.

3. Many plants have Type III curves—a pattern defined by extremely high death rates for seeds and seedlings but high survival rates later in life.

TABLE 52.1 Life Table for *Lacerta vivipara* in the Netherlands

Age	Number Alive	Survivorship	Fecundity	Survivorship × Fecundity = Average Number of Offspring Produced per Female of Age *x*
0	1000	1.000	0.00	0.00
1	424	0.424	0.08	0.03
2	308	0.308	2.94	0.91
3	158	0.158	4.13	0.65
4	57	0.057	4.88	0.28
5	10	0.010	6.50	0.07
6	7	0.007	6.50	0.05
7	2	0.002	6.50	0.01

SOURCE: H. Strijbosch and R. C. M. Creemers. 1988. Comparative demography of sympatric populations of *Lacerta vivipara* and *Lacerta agilis*. *Oecologia* 76: 20–26.

Figure 52.2b provides a graph for you to plot survivorship of *L. vivipara*, using the data reported in Table 52.1. Note the log-log scale on the graph.

FECUNDITY The number of female offspring produced by each female in a population is termed **fecundity**. In most cases, biologists only keep track of females when calculating life-table data because the number of males present rarely affects population dynamics. There are almost always enough males present to fertilize all of the females in breeding condition, so growth rates depend entirely on females.

Because researchers documented the reproductive output of the same *L. vivipara* lizard females year after year, they were able to calculate a quantity called **age-specific fecundity**: the average number of female offspring produced by a female in age class *x*. An **age class** is a group of individuals of a specific age—for example, all female lizards between 4 and 5 years old.

As **Box 52.1** on page 1040 shows, data on survivorship and fecundity allow researchers to calculate the growth rate of a population. How do data on the Netherlands populations compare with populations in different types of habitats?

The Role of Life History

The life-table data in Table 52.1 are interesting because they contrast with results from other populations of *L. vivipara*. Consider data on fecundity:

- In the Netherlands, almost no 1-year-old female *L. vivipara* reproduce.

- In Brittany, France, 50 percent of 1-year-old female lizards reproduce.

- In the mountains of Austria, females don't begin breeding until they are 4 years old.

(a) Three general types of survivorship curves

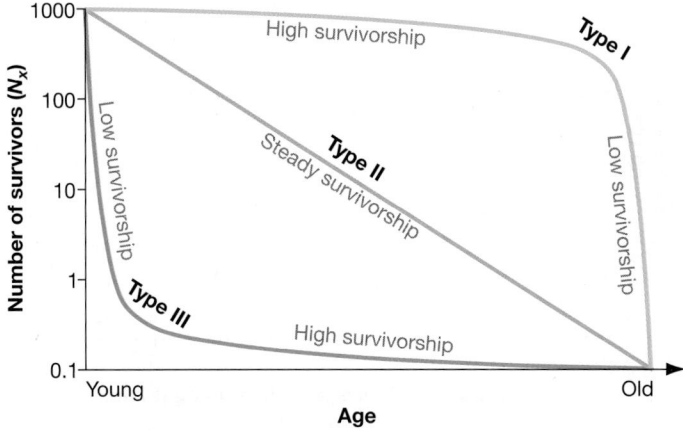

(b) Exercise: Survivorship curve for *Lacerta vivipara*

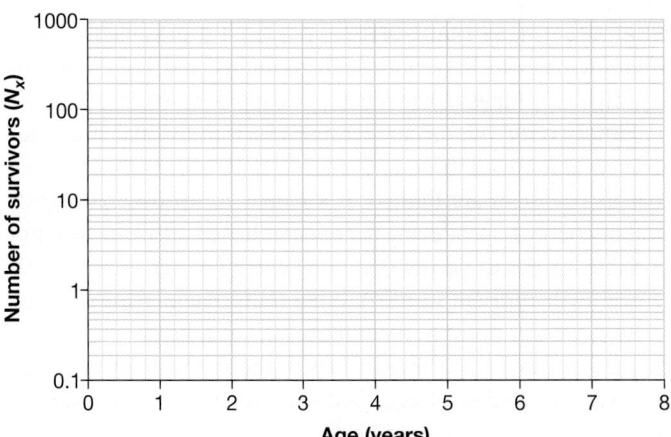

FIGURE 52.2 Survivorship Curves Identify When Mortality Rates Are Low, Steady, or High.

✔**EXERCISE** Fill in the graph in part (b), with data on survivorship of *Lacerta vivipara* given in Table 52.1. Compare the shape of the curve to the generalized graphs in part (a).

If immigration and emigration are not occurring, the data in a life table can be used to calculate a population's growth rate. This is logical, because survivorship and fecundity are ways to express death rates and birthrates—the other two factors influencing population size. To see how a population's growth rate can be estimated from life-table data, let's look at each component in a life table more carefully. Recall that life tables usually focus exclusively on females.

Survivorship is symbolized as l_x, where x represents the age class being considered. Survivorship for age class x is calculated by dividing the number of females in that age class (N_x) by the number of females that existed as offspring (N_0):

$$l_x = \frac{N_x}{N_0} \qquad \textbf{(Eq. 52.1)}$$

Age-specific fecundity is symbolized m_x, where x again represents the age class being considered. Age-specific fecundity is calculated as the total number of female offspring produced by females of a particular age, divided by the total number of females of that age class present. It represents the average number of female offspring produced by a female of age x.

Documenting age-specific survivorship and fecundity allows researchers to calculate the average number of female offspring produced by females during each age of life, as $l_x m_x$. For example, if 30 percent of the females in a population live to be 2 years old, and if 2-year-old females have an average of 2 offspring, then the average number of births at age 2, per female born into the population, is $l_x m_x = 0.30 \times 2 = 0.6$.

Summing $l_x m_x$ values over the entire life span of a female gives the net reproductive rate, R_0, of a population:

$$R_0 = \sum_{i=0}^{x} l_x m_x \qquad \textbf{(Eq. 52.2)}$$

Here the Greek letter sigma (Σ) stands for "sum." The expression on the right-hand side of the equals sign is read, "sum the $l_x m_x$ values from each year of life, starting from birth or hatching ($i = 0$), through age x.

The **net reproductive rate** represents the growth rate of a population per generation. The logic behind the equation for R_0 is that the growth rate of a population per generation equals the average number of female offspring that each female produces over the course of her lifetime. A female's average lifetime reproduction, in turn, is a function of survivorship and fecundity at each age class. In Table 52.1, R_0 is the sum of the survivorship $\times$ fecundity values in the right column.

If R_0 is greater than 1, then the population is increasing in size. If R_0 is less than 1, then the population is declining. ✔If you understand these concepts, you should be able to use the data in Table 52.1 to calculate R_0 in the Netherlands population of *L. vivipara*,[1] state how many female offspring an average *L. vivipara* female produces over the course of her lifetime,[2] and describe whether the population is growing, stable, or declining.[3]

Answers are given below.

[1]R_0 = 2.0; [2]2.00; [3]The population is growing rapidly.

The contrasts are equally stark in terms of survivorship. In the Austrian population, most females live much longer than do individuals in either lowland population—in the Netherlands or France.

In Brittany, fecundity is high but survivorship is low; in Austria, fecundity is low but survivorship is high. The population in the Netherlands is intermediate. In this species, key aspects of the life table vary dramatically among populations.

WHAT ARE FITNESS TRADE-OFFS?   Why isn't it possible for *L. vivipara* females to have both high fecundity and high survival? The answer is **fitness trade-offs** (see Chapters 24 and 41).

Fitness trade-offs occur because every individual has a restricted amount of time and energy at its disposal—meaning that its resources are limited. If a female lizard devotes a great deal of energy to producing a large number of offspring, it is not possible for her to devote that same energy to her immune system, growth, nutrient stores, or other traits that increase survival.

To drive this point home, **Figure 52.3** graphs the probability that a female of a species survives to the following year versus the average number of eggs laid—or clutch size, a measure of fecundity—in 10 species of birds.

- There are no points in the upper right corner of the graph because it is not possible to have high fecundity and high survivorship.

- There are no points in the lower left corner of the graph because species with low survivorship and low fecundity have low population growth and go extinct.

A female can maximize fecundity, maximize survival, or strike a balance between the two.

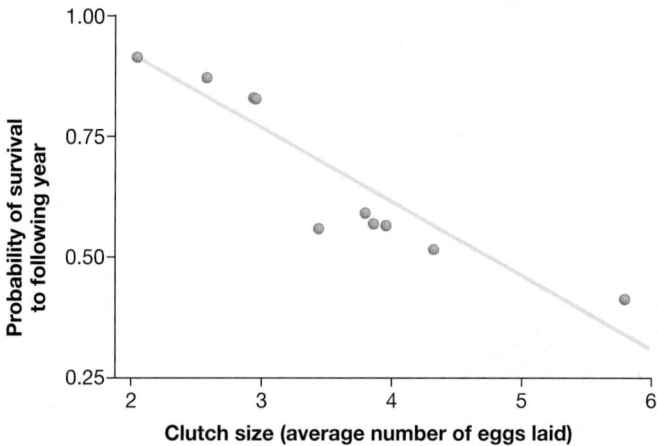

FIGURE 52.3 There Is a Trade-Off between Survival and Reproduction. Each data point on this graph represents a different species of bird. A statistical analysis called regression was used to create a "line of best fit" to these data points.

LIFE HISTORY IS BASED ON RESOURCE ALLOCATION An organism's **life history** describes how an individual allocates resources to growth, reproduction, and activities or structures that are related to survival. Traits such as survivorship, age-specific fecundity, age at first reproduction, and growth rate are all aspects of an organism's life history. Understanding variation in life history is all about understanding fitness trade-offs.

In almost all cases, biologists find that life history is shaped by natural selection in a way that maximizes an individual's fitness in its environment. For example, Chapter 41 provided experimental evidence that there is a fitness trade-off between egg size and egg number in side-blotched lizards. Females in that study population strike a balance between egg size and egg number that is optimal for their habitat.

In *L. vivipara* populations, biologists contend that females who live a long time but mature late and have few offspring each year have high fitness in cold, high-elevation habitats, such as Austria. In these habitats, females have to reduce their reproductive output and put more energy into traits that increase survival in a harsh environment. In contrast, females who have short lives but mature early and have large numbers of offspring each year do better in warm, low-elevation habitats, such as Brittany.

✔ If you understand the role that fitness trade-offs play in determining life history patterns, you should be able to: (**1**) predict how survivorship and fecundity should compare in *L. vivipara* in the warmest part of their range (southwest France and northern Spain), and (**2**) comment on why females in these populations lay many eggs instead of giving birth to a relatively small number of live young.

PATTERNS ACROSS SPECIES The data you've reviewed thus far all involve comparisons within species. But the same patterns exist when many different species are compared.

In general, individuals from species with high fecundity tend to grow quickly, reach sexual maturity at a young age, and produce many small eggs or seeds. The mustard plant *Arabidopsis thaliana*, for example, germinates and grows to sexual maturity in just 4 to 6 weeks. In this species, individuals usually live only a few months but may produce as many as 10,000 tiny seeds. They live fast and die young.

In contrast, individuals from species with high survivorship tend to grow slowly and invest resources in traits that reduce damage from enemies and increase their own ability to compete for water, sunlight, or food. A coconut palm, for example, may take a decade to mature but live 60–70 years and produce offspring each year. Coconut palms invest resources in making stout stems that allow them to grow tall, a relatively extensive root system, large fruits (coconuts), molecules that make herbivores sick, and enzymes that reduce infections from disease-causing fungi. These traits increase survivorship but decrease fecundity.

An *Arabidopsis thaliana* plant and a coconut palm represent two ends of a broad continuum of life-history characteristics (**Figure 52.4**). How does a species' location on this continuum affect its growth rate?

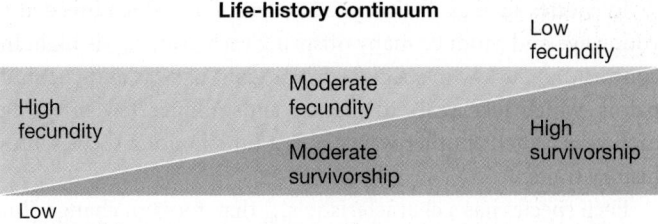

FIGURE 52.4 Life-History Traits Form a Continuum. Every organism can be placed somewhere on this life-history continuum. The placement of a species is most meaningful when it is considered relative to closely related species.

✔ **QUESTION** In plants, what types of traits would you expect to see on the left, middle, and right of the continuum? Consider growth habit (herbaceous, shrub, or tree) as well as relative disease- and predator-fighting ability, seed size, seed number, and body size.

52.2 Population Growth

The most fundamental questions that biologists ask about populations involve growth or decline in numbers of individuals. For conservationists, analyzing and predicting changes in population size is fundamental to managing threatened species.

Recall that four processes affect a population's size: Births and immigration add individuals to the population; deaths and emigration remove them. It follows that a population's overall growth rate is a function of birthrates, death rates, immigration rates, and emigration rates.

Quantifying the Growth Rate

🔑 A population's growth rate is the change in the number of individuals in the population (ΔN) per unit time (Δt). Here the Greek symbol Δ (delta) means change.

If no immigration or emigration is occurring, then a population's growth rate is equal to the number of individuals (N) in the population times the difference between the birthrate per individual (b) and death rate per individual (d). The difference between the birthrate and death rate per individual is called the per-capita rate of increase and is symbolized r. (Per capita means "for each individual.")

If the per-capita birthrate is greater than the per-capita death rate, then r is positive and the population is growing. But if the per-capita death rate begins to exceed the per-capita birthrate, then r becomes negative and the population declines. Within populations, r varies through time. Its value can be positive, negative, or 0.

When conditions are optimal for a particular species—meaning birthrates per individual are as high as possible and death rates per individual are as low as possible—then r reaches a maximal value called the **intrinsic rate of increase,** r_{max}. When this occurs, a population's growth rate is expressed as

$$\Delta N / \Delta t = r_{max} N$$

In species such as *Arabidopsis* and fruit flies, which breed at a young age and produce many offspring each year, r_{max} is high. In contrast, r_{max} is low in species such as giant pandas and coconut palms, which take years to mature and produce few offspring each year. Stated another way, r_{max} is a function of a species' life-history traits.

Each species has a characteristic r_{max} that does not change. But at any specific time, a population has an instantaneous growth rate, or per-capita rate of increase, symbolized by r. r_{max} tells you what the maximum growth rate is; r tells you what it is at a particular time. "Little r" is less than or equal to r_{max}.

The instantaneous growth rate of a population at a particular time is actually likely to be much lower than r_{max}. A population's r is also likely to be different from r values of other populations of the same species, and to change over time. The instantaneous growth rate is dynamic.

Exponential Growth

The graph in **Figure 52.5** plots changes in population size, for various values of r, under the condition known as exponential growth. **Exponential population growth** occurs when r does not change over time.

The key point about exponential growth is that the growth rate does not depend on the number of individuals in the population. Biologists say that this type of population growth is **density independent**.

It's important to emphasize that exponential growth adds an increasing number of individuals as the total number of individuals, N, gets larger. As an extreme example, an r of 0.02 per year in a population of 1 billion adds over 20 million individuals per year. The same growth rate in a population of 100 adds just over 2 individuals per year. Even if r is constant, the number of individuals added to a population is a function of N. The rate of increase is the same, but the number of individuals added is not.

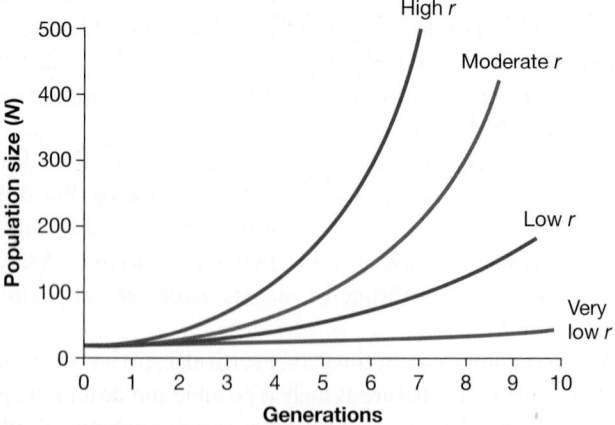

FIGURE 52.5 Exponential Growth Is Independent of Population Size. When the per-capita growth rate r does not change over time, exponential growth occurs. Population size may increase slowly or rapidly, depending on the size of r.

In nature, exponential growth is observed in two circumstances: (1) a few individuals found a new population in a new habitat, or (2) a population has been devastated by a storm or some other type of catastrophe and then begins to recover, starting with a few surviving individuals. But it is not possible for exponential growth to continue indefinitely.

If *Arabidopsis* populations grew exponentially for a long period of time, they would eventually fill all available habitat. When **population density**—the number of individuals per unit area—gets very high, the population's per-capita birthrate will decrease and the per-capita death rate will increase, causing r to decline. Stated another way, growth is often **density dependent**.

Logistic Growth

To analyze what happens when growth is density dependent, biologists use a parameter called carrying capacity. **Carrying capacity, K,** is defined as the maximum number of individuals in a population that can be supported in a particular habitat over a sustained period of time.

The carrying capacity of a habitat depends on a large number of factors: food, space, water, soil quality, and resting or nesting sites. Carrying capacity can change from year to year, depending on conditions.

A LOGISTIC GROWTH EQUATION If a population of size N is below the carrying capacity K, then the population should continue to grow. More specifically, a population's growth rate is proportional to $(K - N)/K$:

$$\frac{\Delta N}{\Delta t} = r_{max} N \left(\frac{K - N}{K} \right)$$

This expression is called a logistic growth equation.

In the expression $(K - N)/K$, the numerator defines the number of additional individuals that can be accommodated in a habitat with carrying capacity K; dividing $K - N$ by K turns this number of individuals into a proportion. Thus $(K - N)/K$ describes the proportion of "unused resources and space" in the habitat. It can be thought of as the environment's resistance to growth.

- When N is small, then $(K - N)/K$ is close to 1 and the growth rate should be high.

- As N gets larger, $(K - N)/K$ gets smaller.

- When N is at carrying capacity—meaning that $K = N$—then $(K - N)/K$ is equal to 0 and growth stops.

Thus, as a population approaches a habitat's carrying capacity, its growth rate should slow.

The logistic growth equation describes **logistic population growth**, or changes in growth rate that occur as a function of population size. Just as exponential growth is density independent, logistic growth is density dependent. **Box 52.2** explores population growth models and how they are applied in more detail.

To explore how biologists model changes in population size in more detail, let's consider data on whooping cranes, which are large, wetland-breeding birds native to North America (Figure 52.6). Whooping cranes may live more than 20 years in the wild, but it is common for females to be six or seven years old before they breed for the first time. Female cranes lay just two eggs per year and usually rear just one chick. Based on these life-history characteristics, we would expect the growth rate of whooping crane populations to be extremely low.

Conservationists monitor this species closely, because hunting and habitat destruction reduced the total number of individuals in the world to about 20 in the mid-1940s. Since then, intensive conservation efforts have resulted in a current total population of about 462. That number includes a group of 253 individuals that breeds in Wood Buffalo National Park in the Northwest Territories of Canada.

In addition to the Wood Buffalo population, two new populations of cranes have been established recently by releasing offspring raised in captivity. One of these groups lives near the Atlantic coast of Florida year-round; the other group migrates from breeding areas in northern Wisconsin to wintering areas along Florida's Gulf coast.

According to the biologists who are managing whooping crane recovery, the species will no longer be endangered when the two newly established populations have at least 25 breeding pairs—meaning there would be about 125 individuals in each—and when neither population needs to be supplemented with captive-bred young in order to be self-sustaining. How long will this take?

Discrete Growth

Whooping cranes breed once per year, so the simplest way to express a crane population's growth rate is to compare the number of individuals at the start of one breeding season to the number at the start of the following year's breeding season.

When populations breed during discrete seasons, their growth rate can be calculated as for whooping cranes. To create a general expression for how populations grow over a discrete time interval, biologists use N to symbolize population size. N_0 is the population size at time zero (the starting point), and N_1 the population size one breeding interval later. In equation form, the growth rate is given as

$$N_1/N_0 = \lambda \qquad \textbf{(Eq. 52.3)}$$

The parameter λ (lambda) is called the **finite rate of increase**. (In mathematics, a *finite rate* refers to an observed rate over a given period of time. A *parameter* is a variable or constant term that affects the shape of a function but does not affect its general nature.) The current population size at Wood Buffalo is 253. Suppose that biologists count 265 cranes on the breeding grounds next year. The population growth rate could be calculated as $265/253 = 1.047 = \lambda$. Stated another way, the population will have grown at the rate of 4.7 percent per year. Rearranging the expression in Equation 52.3 gives

$$N_1 = N_0\lambda \qquad \textbf{(Eq. 52.4)}$$

Stated more generally, the size of the population at the end of year t will be given by

$$N_t = N_0\lambda^t \qquad \textbf{(Eq. 52.5)}$$

This equation summarizes how populations grow when breeding takes place seasonally. The size of the population at time t is equal to the starting size, times the finite rate of increase multiplied by itself t times.

In a sense, λ works like the interest rate at a bank. For species that breed once per year, the "interest" on the population is compounded annually. A savings account with a 5 percent annual interest rate increases by a factor of 1.05 per year.

If a population is growing, then its λ is greater than one. The population is stable when λ is 1.0 and declining when λ is less than 1.0.

If a population's age structure is stable, meaning that the proportion of females in each age class is not changing over time, then its finite rate of increase also has a simple relationship to its net reproductive rate: $\lambda = R_0/g$, where g is the generation time. In essence, dividing the net reproductive rate by generation time transforms it into a discrete rate.

Continuous Growth

The parameters λ and r—the finite rate of increase and the per-capita growth rate—have a simple mathematical relationship. The best way to understand their relationship is to recall that λ expresses a population's growth rate over a discrete interval of time. In contrast, r gives the population's per-capita growth rate at any particular instant. This is why r is also called the instantaneous rate of increase. The relationship between the two parameters is given by

$$\lambda = e^r \qquad \textbf{(Eq. 52.6)}$$

where e is the base of the natural logarithm, or about 2.72. (For help with using logarithms, see **BioSkills 7** in Appendix A. Also, note that the relationship between any finite rate and any instantaneous rate is given by finite rate = $e^{\text{instantaneous rate}}$.)

Substituting Equation 52.6 into Equation 52.5 gives

$$N_t = N_0 e^{rt} \qquad \textbf{(Eq. 52.7)}$$

This expression summarizes how populations grow when they breed continuously, as do humans and bacteria, instead of at defined intervals. For species that breed continuously, the "interest" on the population is compounded continuously. When the growth rates λ and r are equivalent,

FIGURE 52.6 Whooping Cranes Have Been the Focus of Intensive Conservation Efforts.

(continued)

(continued)

however, the differences between discrete and continuous growth are negligible.

Because r represents the growth rate at any given time, and because r and λ are so closely related, biologists routinely calculate r for species that breed seasonally. In the whooping crane example, $\lambda = 1.047$ per year $= e^r$. To solve for r, take the natural logarithm of both sides. (**BioSkills 7** in Appendix A explains how to calculate a natural logarithm using your calculator.) In this case, $r = 0.046$ per year.

The instantaneous rate of increase, r, is also directly related to the net reproductive rate, R_0, introduced in Box 52.1. In most cases, r is calculated as $\ln R_0/g$, where g is the generation time. Thus, r can be calculated from life-table data. It is a more useful measure of growth rate than R_0, because r is independent of generation time.

To summarize, biologists have developed several ways of calculating and expressing a population's growth rate. Growth rate expressed as λ has the advantage of being easy to understand, and R_0 has the advantage of being calculated directly from life-table data. Although r is slightly more difficult conceptually, it is the most useful expression for growth rate, because it is independent of generation time and is relevant for species that breed either seasonally or continuously.

Applying the Models

To get a better feel for r and for Equation 52.7, consider the following series of questions about whooping cranes. The key to answering these questions is to realize that Equation 52.7 has just four parameters. Given three of these parameters, you can calculate the fourth. ✔If you understand this concept, you should be able to solve the following four problems:

1. If 20 individuals were alive in 1941 and 462 existed in 2009, what is r? Here $N_t = 462$, $N_0 = 20$, and $t = 68$ years. Substitute these values into Equation 52.7 and solve for r. Then check your answer at the end of this box.[1]

2. In the most recent report issued by the biologists working on the Wood Buffalo crane recovery program, it is estimated that the flock should be able to sustain an r of 0.046 for the foreseeable future. If the flock currently contains 253 individuals, how long will it take that population to double? Here $N_0 = 253$ and $N_t = 2 \times 253 = 506$. In this case, you solve Equation 52.7 for t. Then check your answer.[2]

3. In 2002 a pair of birds in the flock that lives in Florida year-round successfully

raised offspring (nine years after the first cranes were introduced there). In 2003 another pair of cranes in this population bred successfully. If the number of breeding pairs continues to double each year, how long will it take to reach the goal of 25 breeding pairs? (Note that $\lambda = 2.0$ if a population is doubling each year.) Given that $N_0 = 2$ and $N_t = 25$, solve for t again, and check your answer.[3]

4. The whooping crane flock that migrates between Wisconsin and Florida was founded in 2001. If its development is like that of the resident flock in Florida, the first successful breeding attempt will occur in 2010—nine years after the initial introduction. Suppose that the instantaneous growth rate for the number of breeding pairs in this population will be 0.05. In what year will the breeding population reach 25 pairs? This is the year that whooping cranes should come off the endangered species list. Here $N_0 = 1$, $N_t = 25$, and $r = 0.05$. Solve for t, and add this number of years to 2010; then check your answer.[4]

[1] $r = 0.046$; [2] $t = 15$ years; [3] $t = 3.64$ years; [4] 2074

GRAPHING LOGISTIC GROWTH **Figure 52.7a** illustrates density-dependent growth in a hypothetical population. The graph plots changes in population size over time, and has three sections.

1. Initially, growth is exponential—meaning that r is constant.

2. With time, N increases to the point where competition for resources or other density-dependent factors begins to occur. As a result, the growth rate begins to decline.

3. When the population is at the habitat's carrying capacity, the growth rate is 0—the graph of population size versus time is flat.

This is exactly what happened in an experiment on laboratory populations of ciliates (see Chapter 29): *Paramecium aurelia* and *P. caudatum*. An investigator placed 20 individuals from one of the *Paramecium* species into 5 mL of a solution. He created many replicates of these 5-mL environments for each species separately. He kept conditions as constant as possible by adding the same number of bacterial cells every day for food, washing the solution every second day to remove wastes, and maintaining the pH at 8.0.

In the graphs in **Figure 52.7b**, each data point represents population size for one of the species—the average of the replicates in the experiment. Note that both species exhibited logistic growth in this environment. The carrying capacity differed in the two species, however. The maximum density of *P. aurelia* averaged 448 individuals per mL, but that for *P. caudatum* averaged just 128 individuals per mL. When the two species are grown together—one species gets eliminated (see Chapter 53).

In both of the *Paramecium* species in this experiment, exponential growth could be sustained for only about five days. What factors cause growth rates to change?

What Limits Growth Rates and Population Sizes?

Population sizes change as a result of two general types of factors:

- *Density-independent factors* alter birthrates and death rates irrespective of the number of individuals in the population, and usually involve changes in the abiotic environment—variation in weather patterns, or catastrophic events such as cold snaps, hurricanes, volcanic eruptions, or drought.

(a) Density dependence: Growth rate slows at high density.

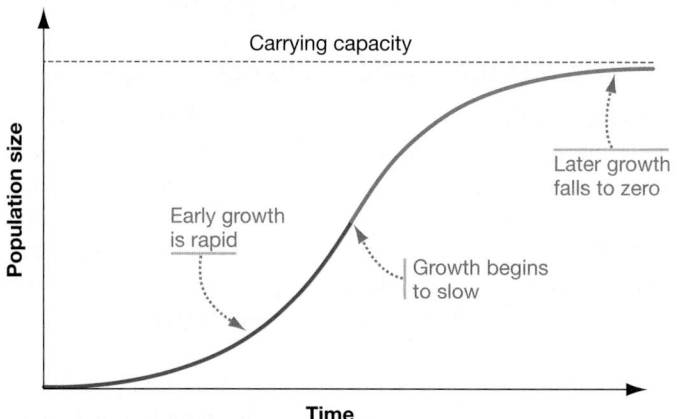

(b) Logistic growth in ciliates

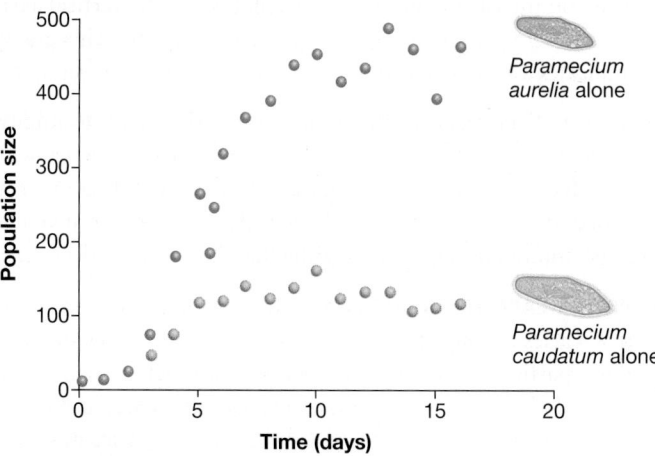

FIGURE 52.7 Logistic Growth Is Dependent on Population Size. (a) A curve illustrating the pattern predicted by the logistic growth equation. This pattern often occurs when a small number of individuals colonizes an unoccupied habitat. Initially, r is high because competition for resources is low to nonexistent. Carrying capacity depends on the quality of the habitat and can vary over time. **(b)** Data from laboratory experiments with two species of *Paramecium*.

- *Density-dependent factors* change in intensity as a function of population size, and are usually biotic in nature. When trees crowd each other, they have less water, nutrients, and sunlight at their disposal and make fewer seeds.

A CLOSER LOOK AT DENSITY DEPENDENCE To get a better understanding of how biologists study the density-dependent factors that affect population size, consider the data in **Figure 52.8**:

- The graph in Figure 52.8a presents results of an experimental study of a coral-reef fish called the bridled goby. Each data point represents an identical artificial reef, constructed

by a researcher from the rubble of real coral reefs. The initial density, plotted along the *x*-axis, represents the marked bridled gobies that were released on each artificial reef at the start of the experiment. The proportion surviving, plotted on the vertical axis, represents the introduced individuals that were still living at that reef 2.5 months later. This graph shows a strong density-dependent relationship in survivorship.

- The graph in Figure 52.8b is from a long-term study of song sparrows on Mandarte Island, British Columbia. Each data point represents a different year. The density of females,

(a) Survival of gobies declines at high population density.

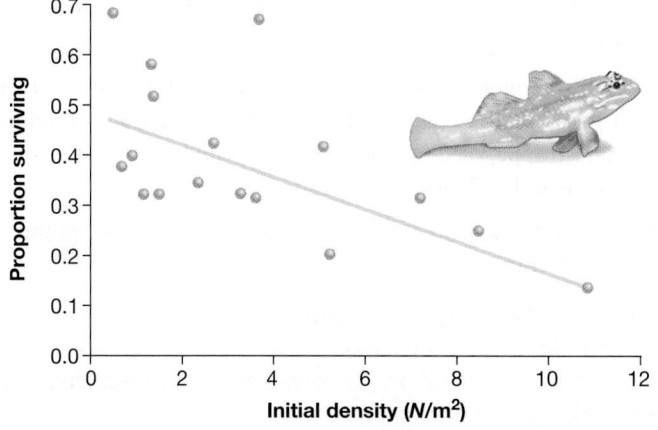

(b) Fecundity of sparrows declines at high population density.

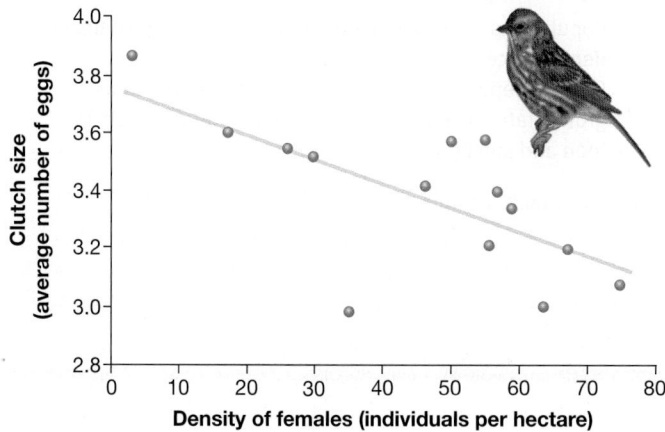

FIGURE 52.8 Density-Dependent Growth Results from Changes in Survivorship and Fecundity. (a) When bridled gobies are introduced at different population densities to artificial reefs, survivorship is highest at low density. **(b)** In the song sparrow population on Mandarte Island, there is a strong negative correlation between population density in a particular year and the average number of eggs (clutch size) produced by females.

✔**QUESTION** When song sparrow populations are at high density and extra food is provided to females experimentally, average clutch size is much higher than expected from the data in part (b). Based on these data, state a hypothesis to explain why average clutch size declines at high density.

plotted along the horizontal axis, is the number of females that bred on the island; the clutch size, plotted on the vertical axis, is the average number of eggs laid by each female. This graph indicates a strong density-dependent relationship in fecundity.

Density-dependent changes in survivorship and fecundity cause logistic population growth. In this way, density-dependent factors define a particular habitat's carrying capacity. If gobies get crowded, they die or emigrate. If song sparrows are crowded, the average number of eggs and offspring that they produce declines.

CARRYING CAPACITY IS NOT FIXED It's important to recognize that K varies among species and populations. K varies because for any particular species, some habitats are better than other habitats due to differences in food availability, space, and other density-dependent factors. Stated another way, K varies in space. It also varies with time, as conditions in some years are better than in others.

In addition, the same habitat may have a very different carrying capacity for different species—as the experiment with growing paramecia in 5-mL vials showed. The same area will tend to support many more individuals of a small-bodied species than of a large-bodied species, for example, simply because large individuals demand more space and resources.

These simple observations help explain the variation in total population size that exists among species and among populations of the same species. To review the basic principles of exponential growth, logistic growth, and factors that limit population size, go to the study area at *www.masteringbiology.com*.

 BioFlix™ Population Ecology, **Web Activity** Modeling Population Growth

CHECK YOUR UNDERSTANDING

🔑 If you understand that . . .

- Populations grow exponentially unless slowed by density dependence.
- Density-dependent factors that influence population growth rates include competition for resources such as food and sunlight.

✓ You should be able to . . .

Propose a density-dependent factor that limits growth in *Paramecium aurelia* and design an experiment to test this hypothesis.

Answers are available in Appendix B.

52.3 Population Dynamics

The tools introduced in the previous two sections provide a foundation for exploring how biologists study **population dynamics**—changes in populations through time. 🔑 Research on population dynamics has uncovered a wide array of patterns in natural populations in addition to exponential and logistic growth. Let's start by considering one of the most important patterns observed: the "blinking on-and-off" of fragmented populations.

How Do Metapopulations Change through Time?

If you browse through an identification guide for trees or birds or butterflies, you'll likely find range maps indicating that most species occupy a broad area. In reality, however, the habitat preferences of many species are restricted, and individuals occupy only isolated patches within that broad area. The meadows occupied by Glanville fritillaries—an endangered species of butterfly native to the Åland islands off the coast of Finland (**Figure 52.9**)—are a well-studied example of this pattern.

If individuals from a species occupy many small patches of habitat, so that they form many independent populations, they are said to represent a **metapopulation**—a population of populations.

Glanville fritillaries exist naturally as metapopulations. But because humans are reducing large, contiguous areas of forest and grasslands to isolated patches or reserves, more and more species are being forced into a metapopulation structure. Recent research on Glanville fritillaries by Ilkka Hanski and colleagues illustrates the consequences of metapopulation structure for endangered species.

METAPOPULATIONS SHOULD BE DYNAMIC Research on Glanville fritillaries began with a survey of meadow habitats on the islands. Because fritillary caterpillars feed on just two types of host plant, *Plantago lanceolata* and *Veronica spicata*, the team was able to pinpoint potential butterfly habitats.

Hanski's group estimated the fritillary population size within each patch of habitat by counting the number of larval webs in each. Of the 1502 meadows that contained the host plants, 536 had Glanville fritillaries. The patches ranged from 6 m² to 3 ha in area. Most had only a single breeding pair of adults and one larval group, but the largest population contained hundreds of pairs of breeding adults and 3450 larvae.

Figure 52.10 illustrates how a metapopulation like this is expected to change over the years.

- Given enough time, each population within the larger metapopulation is expected to go extinct. The cause could be a catastrophe, such as a storm or an oil spill; it could also be a disease outbreak or a sudden influx of predators.

- Migration from nearby populations can reestablish populations in empty habitat fragments.

In this way, there is a balance between extinction and recolonization within a metapopulation. Even though subpopulations blink on and off over time, the overall population is maintained at a stable number of individuals.

AN EXPERIMENTAL TEST Does this metapopulation model correctly describe the population dynamics of fritillaries? To answer this question, Hanski's group conducted a mark-recapture study. They caught and marked many individuals, released them, and revisited fritillary habitats, later, to recapture as many individuals as possible (see Box 52.3 on page 1048).

Glanville fritillaries are an endangered species

They live on patches of certain host plants within meadows

FIGURE 52.9 Glanville Fritillaries Live in Patches of Meadow. The Glanville fritillary is declining rapidly in many parts of Europe. In Finland, the only remaining individuals occupy the Åland islands.

A metapopulation is made up of small, isolated populations.

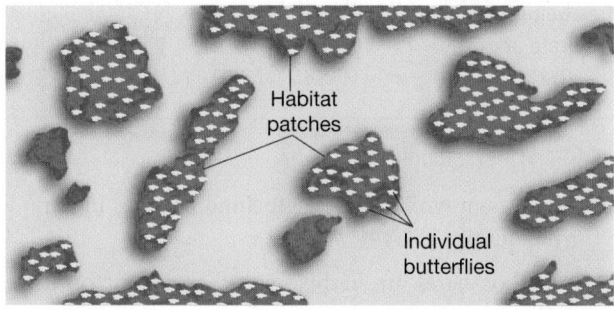

Habitat patches

Individual butterflies

Although some subpopulations go extinct over time...

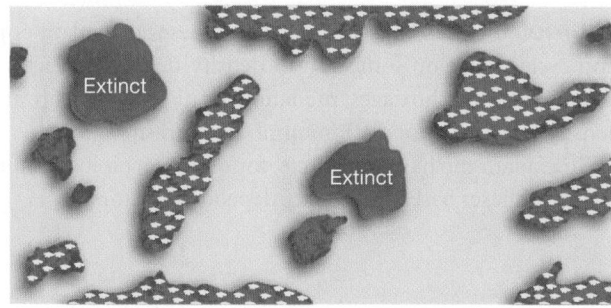

Extinct

Extinct

...migration can restore or establish subpopulations.

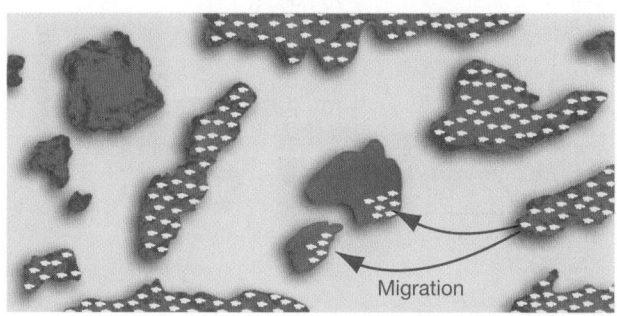

Migration

Time

FIGURE 52.10 Metapopulation Dynamics Depend on Extinction and Recolonization. The overall population size of a metapopulation stays relatively stable even if subpopulations go extinct. These populations may be restored by migration, or unoccupied habitats might be colonized.

Of the 1731 butterflies that the biologists marked and released, 741 were recaptured over the course of the summer study period. Of the recaptured individuals, 9 percent were found in a previously unoccupied patch. This migration rate is high enough to suggest that patches where a population has gone extinct will eventually be recolonized.

Hanski and co-workers repeated the survey two years after their initial census. Just as the metapopulation model predicted, some populations had gone extinct and others had been created. On average, butterflies were lost from 200 patches each year; 114 unoccupied patches were colonized and newly occupied each year. The overall population size was relatively stable even though constituent populations came and went.

To summarize, the history and future of a metapopulation is driven by the birth and death of populations, just as the dynamics of a single population are driven by the birth and death of individuals. As the final section in this chapter will show, these dynamics have important implications for saving endangered species.

Why Do Some Populations Cycle?

Analyzing population cycles has been a particularly productive way to understand how intraspecific ("within-species") factors—such as competition for nutrients and space—interact with interspecific interactions, such as predation and disease.

Figure 52.11 shows a classic case of population cycling that involves two species: snowshoe hare and lynx in northern Canada.

The x-axis on this graph plots a 50-year interval; the y-axes show changes in hare population density (on the left, with a logarithmic scale) and changes in lynx density (on the right). Data for hares are plotted in tan; data for lynx are plotted in purple. Note several key points:

1. The scales on the y-axes are different—there are many more hares than lynx. (The y-axis for hares is logarithmic, to compress the values so they can be compared more easily to changes in the lynx population.)

2. The hare and lynx populations cycle every 11 years on average.

3. Changes in lynx density lag behind changes in hare density by about two years.

Snowshoe hares are herbivores—they subsist on leaves and stems that grow close to the ground. Lynx are predators, and subsist mainly on snowshoe hares.

IS IT FOOD OR PREDATION? Most hypotheses to explain population cycles hinge on some sort of density-dependent factor. The idea is that food shortages, predation, or disease intensify at high density and cause population numbers to crash.

To explain the hare-lynx cycle, for example, biologists pushed two hypotheses based on density dependent factors:

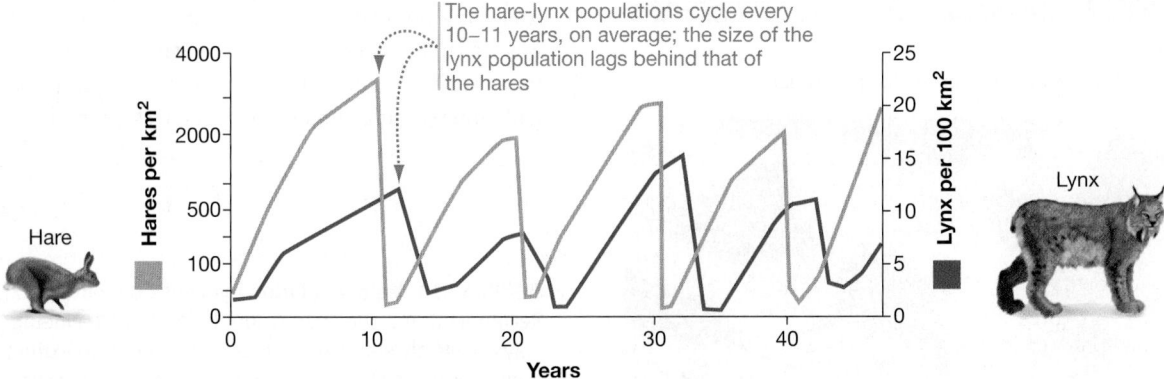

Figure 52.11 **The Hare and Lynx Population Cycles Are Synchronous, but Lagged.**

1. Hares use up all their food when their populations reach high density and starve; in response, lynx also starve.

2. Lynx populations reach high density in response to increases in hare density. At high density, lynx eat so many hares that the prey population crashes.

Stated another way, hares control lynx population size, or lynx control hare population size. The cycle could be controlled by intraspecific competition or interspecific interactions.

A FIELD EXPERIMENT To test these hypotheses, researchers set up a series of 1-km² study plots in boreal forest habitats that were as identical as possible (**Figure 52.12**).

- Three plots were left as unmanipulated controls.
- One plot was ringed with an electrified fence with a mesh that excluded lynx but allowed hares to pass freely.
- Two plots received additional food for hares year-round.
- One plot had a predator-exclusion fence and was also supplemented with food for hares year-round.

The biologists then monitored the size of the hare and lynx populations over an 11-year period, or enough time for a complete cycle in the two populations.

The graphs in Figure 52.12 plot average hare density in the lynx-exclusion, food-addition, and combination plots relative to controls. A value of 1 on the *y*-axis means that hare density in the experimental and control plots were the same. Each bar represents the average population density over a six-month interval; data are plotted for eight years of the study.

The graphs should convince you that plots with predators excluded showed higher hare populations than did control plots during the "decline" phase of the cycle. This result supports the hypothesis that predation by lynx reduces hare populations. Plots with

---

EXPERIMENT

QUESTION: What factors control the hare-lynx population cycle?

HYPOTHESIS: Predation, food availability, or a combination of those two factors controls the hare-lynx cycle.

NULL HYPOTHESIS: The hare-lynx cycle isn't driven by predation, food availability, or a combination of those two factors.

EXPERIMENTAL SETUP:

Document hare population in seven study plots (similar boreal forest habitats, each 1 km²) for duration of one cycle (1987–1994).

 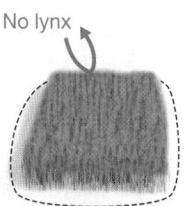  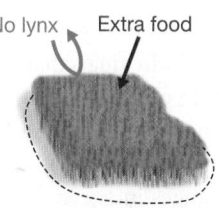

3 plots: Unmanipulated controls

1 plot: Electrified fence excludes lynx but allows free access by hares.

2 plots: Supply extra food for hares.

1 plot: Electrified fence excludes lynx but allows free access by hares; supply extra food for hares.

PREDICTION: Hare populations in at least one type of manipulated plot will be higher than the average population in control plots.

PREDICTION OF NULL HYPOTHESIS: Hare populations in all of the plots will be the same.

RESULTS:

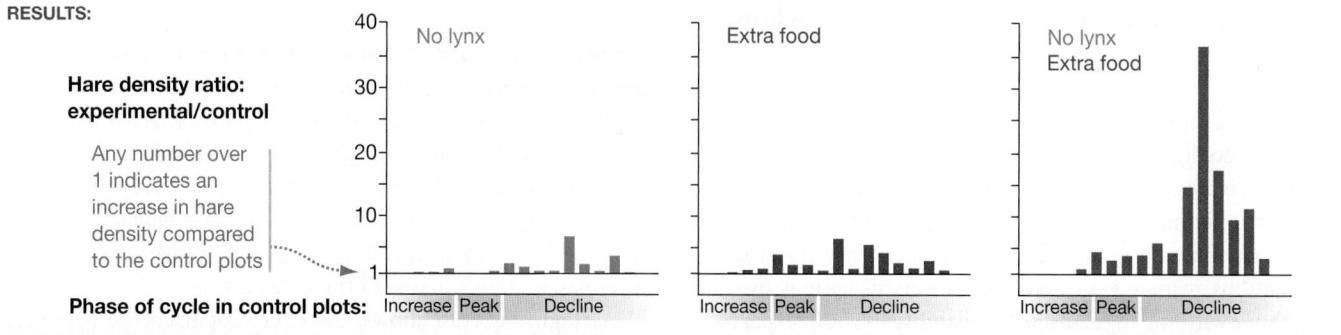

Hare density ratio: experimental/control

Any number over 1 indicates an increase in hare density compared to the control plots

Phase of cycle in control plots:

CONCLUSION: Hare populations are limited by both predation and food availability. When predation and food limitation occur together, they have a greater effect than either factor does independently.

FIGURE 52.12 Experimental Evidence That Predation and Food Availability Drive the Hare-Lynx Cycle.

SOURCE: Krebs, C. J. et al. 1995. Impact of food and predation on the snowshoe hare cycle. *Science* 269: 1112–1115.

✔**QUESTION** Why didn't the research group do more replicates of each treatment?

supplemental food also had much higher populations than controls during the peak and decline phases. But the plot with supplemental food *and* predators excluded had a hare population that got as much as 35 times more dense than the population in the control plots.

Note that most of the effects occurred during the decline phase of the cycle. As control plots were crashing, the experimental plots retained some or many hares.

These data support the hypothesis that hare populations are limited by availability of food as well as by predation, and that food availability and predation intensity interact—meaning the combined effect of food and predation is much larger than their impact in isolation. The leading hypothesis to explain this combined effect is that when hares are at high density, individuals are weakened by nutritional stress and are more susceptible to predation.

Research on population cycles has increased our understanding of density dependence factors in population dynamics. Now let's consider how the arrangement of populations in space affects overall population dynamics.

How Does Age Structure Affect Population Growth?

As the life-table data presented in Section 52.1 showed, age has a dramatic effect on the probability that an individual will reproduce and survive to the following year. But the previous analysis of life-table data left out a critical point about the ages of individuals in a population: A population's **age structure**—meaning the proportion of individuals that are at each possible age—has a dramatic influence on the population's growth over time.

To see how changes in age structure can affect population dynamics, let's consider two case histories for which biologists have documented changes in age structure. One example concerns a flowering plant, and the other focuses on our own species.

AGE STRUCTURE IN A WOODLAND HERB The common primrose, pictured in **Figure 52.13a**, grows in the woodlands of Western Europe. Primroses grow on the forest floor and can germinate, mature, and produce offspring only in the relatively high-light environment created when a tall tree falls and opens a gap in the forest canopy.

The sunny spaces where primroses thrive are short lived, because mature trees and saplings in and around the gap grow and fill the space. As they do, light levels in the gap decline and the environment becomes increasingly unsuitable for herbs such as primrose.

To understand how primrose populations respond to this ephemeral woodland environment, researchers hypothesized that populations in new gaps and populations in large gaps—meaning those with the highest light availability—would experience high rates of reproduction and an age structure characterized by large numbers of juveniles.

The researchers also hypothesized that as light levels declined in a particular gap due to the growth of surrounding trees, primrose reproduction would decline. As a result, the age structure of

(a) Common primroses live in sunlit gaps in forests.

(b) Age structure of primrose population varies with age of gap.

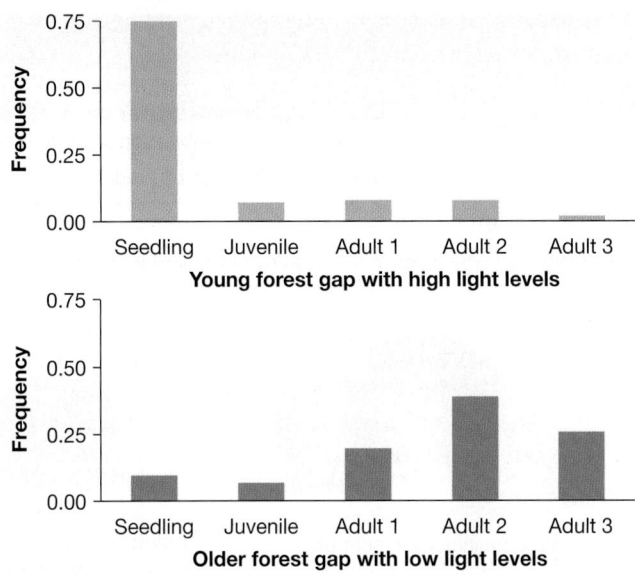

FIGURE 52.13 The Age Structure of Common Primrose Populations Changes over Time. In the graphs, "Adult 1," "Adult 2," and "Adult 3" individuals are of increasingly large size and thus age.

populations in older gaps should be characterized by a large proportion of individuals in adult stages, with fewer juveniles.

To test these hypotheses, the investigators selected eight common primrose populations growing in a range of light levels, from high to low. The team established 1-m^2 study plots inside each gap and studied approximately 350 individuals per population. Each individual was assigned to one of five ages: seedlings, juveniles, or adults in one of three size classes.

As **Figure 52.13b** shows, the data supported the hypothesis that in gaps with high light levels, populations have a larger proportion of juveniles than do gaps with low light levels. In addition, the proportion of juvenile individuals declined over time in all gaps as light levels declined. These data have three important messages about the dynamics of primrose populations:

1. Populations that are dominated by juveniles should experience rapid growth. Population size should then decline with time due to a density-independent factor: shading by trees.

2. The long-term trajectory of the overall primrose population in an area may depend primarily on the frequency and severity of windstorms that knock down trees and create sunlit gaps. If so, it means that the dynamics of primrose populations are governed by an abiotic, density-independent factor.

3. In a large tract of forest, the primrose population will consist of a large number of subpopulations, each found in canopy gaps. Some of these subpopulations are likely to be tiny, and each subpopulation will arise, grow, and then go extinct over time. Primroses have metapopulation structure.

AGE STRUCTURE IN HUMAN POPULATIONS As another example of how age structure affects population dynamics, consider humans. To study age structure in our species, researchers use stacks of horizontal bars to plot the number of males and the number of females in each age cohort—often in 5-year intervals between birth (0) and 100. The resulting graph is called a population pyramid or age pyramid.

In countries where industrial and technological development took place generations ago, like Sweden, survivorship has been high and fecundity low for several decades. An age pyramid like that in **Figure 52.14a** results. The most striking pattern in these data is that there are similar numbers of people in most age classes. The evenness occurs because about the same number of infants are being born each year, and because most survive to old age.

In contrast, the age distribution is bottom-heavy in less-developed nations like Honduras (**Figure 52.14b**). These populations are dominated by the very young because they are under-going rapid growth—with more children being born recently than exist in older age classes.

Analyzing an age pyramid can give biologists important information about a population's history. But studying age distributions can also help researchers predict a population's future.

Look again at Figure 52.14a, and note that the white lines and lighter bars show what the age structure of Sweden is projected to be in 2050. Modest changes in the age distribution will occur because survivorship has increased while fecundity remains the same. But the population is not expected to grow quickly because only modest numbers of individuals reach reproductive age. The projections highlight a major public policy concern in the industrialized countries: how to care for an increasingly aged population.

Now consider Figure 52.14b again. Due to dramatic improvements in health care, most young Hondurans survive to reproductive age. The projected age distribution in 2050 does not continue to flare out, however, because fecundity is declining. The projections illustrate a major public policy concern in less-developed countries: providing education and jobs for an enormous group of young people who will be reaching adulthood during your lifetime.

POPULATION "MOMENTUM" The data in Figure 52.14b make another important point: Because of recent and rapid population growth in developing countries, overall population size will increase dramatically in these nations over the course of your lifetime. Honduras will have many more people in 2050 than it does now.

A large part of this increase will be due to increased survivorship. But the number of offspring being born each year is also expected to stay high, even though fecundity is predicted to *decline*.

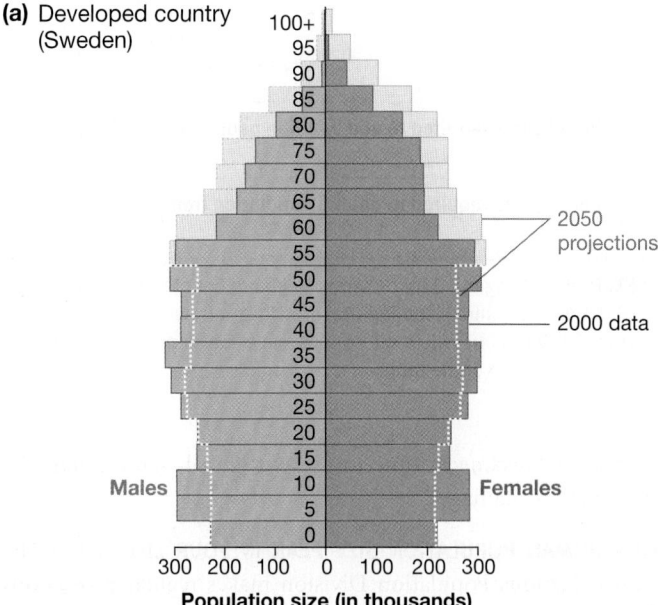

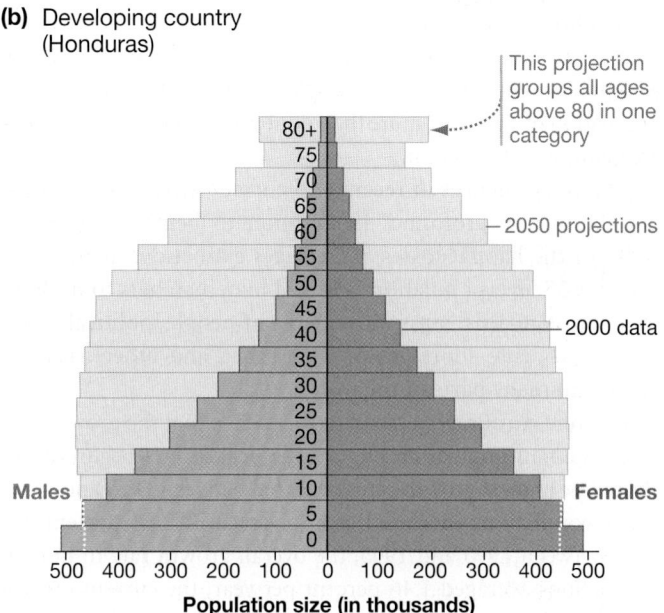

FIGURE 52.14 The Age Structure of Human Populations Varies Dramatically.

This result seems paradoxical, but is based on a key observation: There are now so many young women in these populations that the overall number of births will stay high, even though the average number of children *per female* is much less than it was a generation ago.

To capture this point, biologists say that these populations have momentum or inertia. Combined with high survivorship, age structures like these make continued increases in the total human population almost inevitable.

Analyzing Change in the Growth Rate of Human Populations

Figure 52.15 plots changes in the human population over the past 250 years. Although the curve's shape looks superficially similar to exponential growth, the growth rate for humans has—until recently—actually *increased* over time since 1750, leading to a steeply rising curve.

It is almost impossible to overemphasize just how dramatically the human population has grown recently. **Table 52.2** presents data that will allow you to calculate (**1**) how long it has taken this population to add each billion people, (**2**) determine how many times the population has doubled—from a starting point in the year A.D. 1420—and (**3**) figure out how many years the population took to double each time.

To put your calculations on doubling time in perspective, consider that until your grandparents' generation, no individual had lived long enough to see the human population double. But many members of the generation born in the early 1920s have lived to see the population *triple*. And to drive home the impact of continued population doubling, consider the following: If you were given a penny on January 1, then $0.02 on January 2, $0.04 on January 3, and so on, you would be handed $10,737,418 on January 31.

HOW LARGE IS THE CURRENT HUMAN POPULATION? As this book goes to press, the world population is estimated at over 6.8 billion. About 77 million additional people—equivalent to the current population of Egypt, or more than double the state of California—are being added each year.

The consequences of recent and current increases in human population are profound. In addition to being the primary cause of the habitat losses and species extinctions analyzed in Chapter 55, overpopulation is linked by researchers to declines in living standards, mass movements of people, political instability, and acute shortages of water, fuel, and other basic resources in many parts of the world.

The one encouraging trend in the data is that the growth rate of the human population has already peaked and begun to decline. The highest growth rates occurred between 1965 and 1970, when populations increased at an average of 2.04 percent per year. Between 1990 and 1995, the overall growth rate in human populations averaged 1.46 percent per year; the current growth rate is 1.2 percent annually.

In humans, r may be undergoing the first long-term decline in history. The question is: Will the human population stabilize

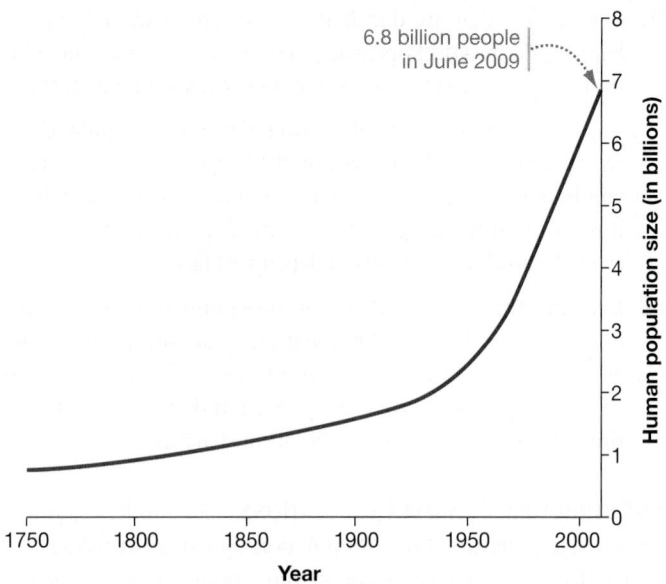

FIGURE 52.15 The Human Population Has Been Growing Rapidly. Graph based on estimates from historical and census data.

TABLE 52.2 **Milestones in Human Population Growth**

Date	Human Population (in millions)
1420	375
1720	750
1800	1000
1875	1500
1927	2000
1961	3000
1974	4000
1987	5000
1999	6000
2009	6800

Number of years required to add 1 billion people to 2009 level:

Number of years required to double population from 2009 level:

✓**EXERCISE** Using the data provided here and Equations 52.5 and 52.7 in Box 52.2, and assuming that growth continues at the same rate as from 1999 to 2009, fill in the correct number of years in the blank cells at the bottom of the table.

or begin to decline in time to prevent global—and potentially irreversible—damage?

WILL HUMAN POPULATION SIZE PEAK IN YOUR LIFETIME? The United Nations Population Division makes regular projections for how human population size will change between now and 2050. For most readers of this book, these projections describe what the world will look like as you reach your early 60s.

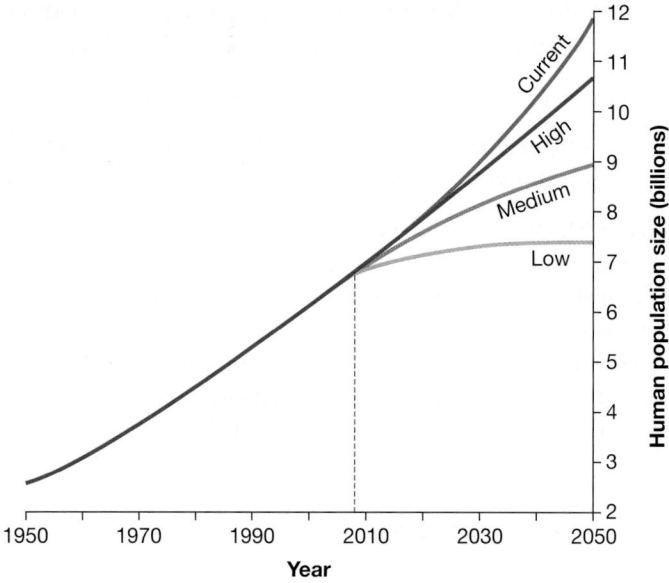

FIGURE 52.16 Projections for Human Population Growth to 2050. The "Current," "High," "Medium," and "Low" labels refer to fertility rates.

The UN projections are based on four scenarios, which hinge on different values for fertility rates—the average number of children that each woman has during her lifetime. Currently, the worldwide average is 2.7. This represents an enormous reduction from the fertility rates during the 1950s, which averaged 5.0 children per woman. **Figure 52.16** shows how total population size is expected to change by 2050 if average fertility continues to decline and averages 2.5 (high), 2.1 (medium), or 1.7 (low) children per woman.

The middle number in the projections, 2.1, is the replacement rate in developed countries, where survivorship is high. The **replacement rate** is the average fertility required for each woman to produce exactly enough offspring to replace herself and her offspring's father. When this fertility rate is sustained for a generation, $r = 0$ and there is **zero population growth (ZPG)**.

A glance at Figure 52.16 should convince you that the four scenarios are starkly different. If average fertility around the world stays at its present level, world population will be closing in on 12 billion about the time you might consider retiring. This is close to double the current population. To realize what this might mean, imagine what traffic or home prices in your city would be like if double the number of people were using the roads and looking for housing.

The "high-fertility" projection assumes that fertility rates continue to drop but average 2.5 children per woman over this interval. This model predicts that world population in 2050 will reach nearly 11 billion—a 75 percent increase over today's population—with no signs of peaking. The low-fertility projection, in contrast, predicts that the total human population in 2050 will be about 7.4 billion and will already have peaked.

THE ROLE OF FERTILITY RATES The UN's population projections make an important point: The future of the human population hinges on fertility rates—on how many children each of the women living today decides to have. Those decisions, in turn,

depend on a wide array of factors, including how free women are to choose their family size and how much access women have to education.

Why education? When women are allowed to become educated, they tend to delay when they start having children and have a smaller overall family size. Access to education and reliable birth control methods (see Chapter 48), in addition to overall economic development and access to quality health care, will play a large role in determining how world population changes over your lifetime.

To summarize, humans may be approaching the end of a period of rapid growth that lasted well over 500 years. How quickly growth rates decline and how large the population eventually becomes will be decided primarily by changes in fertility rates.

To review concepts on human population growth, go to the study area at *www.masteringbiology.com*.

(MB) **Web Activity** Human Population Growth and Regulation

CHECK YOUR UNDERSTANDING

If you understand that . . .

- Density-dependent factors such as competition for food and predation intensity can drive population cycles in some species.
- In a metapopulation, subpopulations "blink on and off" over time as some go extinct and others reestablish via colonization.
- A population's age structure reflects changes in survivorship and fertility rates over time.
- Age structure impacts population dynamics because it dictates how many females of reproductive age are in the population.

✓ **You should be able to . . .**

1. Explain why the age structure of human populations differs between developed and developing nations.
2. Explain why changes in fertility rates have such a dramatic impact on projections for the total human population in 2050.

Answers are available in Appendix B.

52.4 How Can Population Ecology Help Endangered Species?

When designing programs to save species threatened with extinction, conservationists draw heavily on concepts and techniques from population ecology. As habitat destruction pushes species throughout the world into decline, the study of population growth rates and population dynamics has taken on an increasingly applied tone.

To illustrate one of the ways that biologists apply the theory and results introduced in this chapter, let's analyze how an understanding of life-table data and geographic structure can help direct conservation action.

Using Life-Table Data

Collecting and analyzing demographic data such as age-specific survivorship and fecundity are important for saving endangered species and for other applied problems. To understand why, suppose that you were in charge of reintroducing a population of lizards to a nature reserve, and that research conducted when this species occupied the site previously documented survivorship and fecundity.

Your initial plan is to take 1000 newly hatched females from a captive breeding center and release them into the habitat, along with enough males for breeding to occur. Is this population likely to become established and grow, or will you need to keep introducing offspring that have been raised in captivity?

MAKING POPULATION PROJECTIONS To answer this question, you need to use the life-table data in **Figure 52.17a** and calculate:

1. How many adults survive to each age class each year?

2. How many offspring are produced by each adult age class—each year over the course of several years?

Figure 52.17b starts these calculations for you. The fate of the original 1000 females is indicated in red. Note that in the second year, just 330 of these individuals have survived; 40 are left as 3-year-olds.

How many offspring did this generation produce? Remember that just 330 of the original 1000 females are expected to remain after one year. As the first purple number in Figure 52.17b

(a) Life table

Age (x)	Survivorship (l_x)	Fecundity (m_x)
0 (birth)	----	0.0
1	0.33	3.0
2	0.2	4.0
3	0.04	5.0
4		

(b) Fate of first-generation females

Year	0 (newborns)	1-year-olds	2-year-olds	3-year-olds	Total population size (N)
1st	1000 (just introduced)				1000
2nd	990 (= 330 × 3.0)	330 (= 1000 × 0.33)			1320 (= 990 + 330)
3rd	800 (= 200 × 4.0)		200 (= 1000 × 0.20)		
4th	200 (= 40 × 5.0)			40 (= 1000 × 0.04)	
5th					

(c) Fate of first- and second-generation females

Year	0 (newborns)	1-year-olds	2-year-olds	3-year-olds	Total population size (sum across all rows)
1st	1000				1000
2nd	990	330			1320
3rd	800 + 981 (981 = 327 × 3.0)	327 (= 990 × 0.33)	200		2308 (= 800 + 981 + 327 + 200)
4th	200 + 792 (792 = 198 × 4.0)		198 (= 990 × 0.20)	40	
5th	195 (195 = 39 × 5.0)			39 (= 990 × 0.04)	

FIGURE 52.17 Life-Table Data Can Be Used to Project the Future of a Population. **(a)** Life table providing age-specific survivorship and fecundity for a hypothetical population of lizards. **(b)** Predicted fate of 1000 one-year-old females introduced into a habitat just before the breeding season. The number of individuals in this cohort is shown in red; the number of offspring they produce each year is indicated in purple. **(c)** Extension of the data in part (b), indicating how many of the offspring produced by the original females in their first year survived in subsequent years, shown in purple, and how many offspring they produced in each subsequent year, shown in green.

✔**EXERCISE** Assume that all 4-year-old females die after producing three young. Fill in the 4th year in the figure.

indicates, these survivors have an average of 3.0 female off-spring apiece. Thus, they contribute 990 new female offspring to the population. In the second year, the 200 females that are left from the original cohort have an average of 4.0 female offspring each and contribute 800 juveniles. In their third year, the 40 surviving females average 5.0 offspring and contribute 200 juveniles.

Figure 52.17c extends the calculations by showing what happens as the offspring of the original females begin to breed. Their offspring are shown in green. ✔By adding subsequent generations and continuing the analysis, you should be able to predict whether the population will stay the same, decline, or increase over time. In this way, life-table data can be used to predict the future of populations.

ALTERING VALUES FOR SURVIVORSHIP AND FECUNDITY Part of the value of a population projection based on life-table data—like the one begun in Figure 52.17—is that it allows biologists to alter values for survivorship and fecundity at particular ages and assess the consequences. For example, suppose that a predatory snake began preying on juvenile lizards. According to the model in Figure 52.17, what would be the impact of a change in juvenile mortality rate?

Analyses like this allow biologists to determine which aspects of survivorship and fecundity are especially sensitive for particular species. The studies done to date support some general conclusions:

- Whooping cranes, sea turtles, spotted owls, and many other endangered species have high juvenile mortality, low adult mortality, and low fecundity. In these species, the fate of a population is extremely sensitive to increases in adult mortality. Based on this insight, conservationists have recently begun an intensive campaign to reduce the loss of adult female sea turtles in fishing nets. Previously, most conservation action had focused on protecting eggs and nesting sites.

- In humans and other species with high survivorship in most age classes, rates of population growth are extremely sensitive to changes in age-specific fecundity. Because of this, programs to control human population growth focus on two issues: lowering fertility rates through the use of birth control, and delaying the age of first reproduction by improving women's access to education.

In some or even most cases, however, the population projections made from life-table data may be too simplistic to be useful. For example, conservationists may need to expand the basic demographic models to account for occasional disturbances such as fires or storms or disease outbreaks. And what about species where overall population dynamics are dictated by a metapopulation structure?

Preserving Metapopulations

Habitat destruction caused by suburbanization and other human activities leaves small populations isolated in pockets of intact habitat. Work on Glanville fritillaries and other species has shown that a small, isolated population—even one within a nature preserve—is unlikely to survive over the long term. What can be done about this?

In Glanville fritillaries, data collected by Hanski's group have identified the attributes of subpopulations that are most likely to persist: They are (**1**) large, (**2**) occupy larger geographical areas, and (**3**) are closer to neighboring populations (and hence more likely to be colonized). In addition, fritillary populations with low genetic diversity—probably due to inbreeding—are more likely to go extinct than are populations of the same size that have higher genetic diversity.

Results like these have important messages for conservation biologists:

- Areas that are being protected for threatened species should be substantial enough in area to maintain large populations that are unlikely to go extinct in the near future. Large populations are more likely than small populations to endure disasters such as storms, and they are much less susceptible to the damaging effects of inbreeding.

- When it is not possible to preserve large tracts of land, an alternative is to establish systems of smaller tracts that are connected by corridors of habitat, so that migration between patches is possible. Constructing "wildlife corridors" is particularly important for species that do not fly, and when patches of appropriate habitat are separated by highways, subdivisions, rivers, or other obstacles that are dangerous for dispersing animals to cross. Corridors are also essential for plants with limited seed dispersal. In this case, it may take years for a plant species to "grow across" a corridor and disperse to new habitats.

- If the species that is threatened exists as a metapopulation, it is crucial to preserve at least some patches of unoccupied habitat to provide future homes for immigrants. If a population is lost from a preserve, the habitat should continue to be protected so it can be colonized in the future. This can be difficult to do—nonscientists are often skeptical about preserving scarce land that does not (yet) contain the species of interest.

These results also have a more general message: Although traditional population growth models such as the expressions for exponential and logistic growth are simple and elegant, the factors they ignore—immigration and emigration—are crucial to understanding the dynamics of most populations. Because metapopulation structure is common, biologists must use more sophisticated models to predict the fates of populations.

Summary of Key Concepts

🔑 Life tables summarize how likely it is that individuals of each age class in a population will survive and reproduce.

- Three general types of survivorship curves have been observed: (1) high survivorship throughout life, (2) a constant probability of dying at each year of life, and (3) high mortality in juveniles followed by high survivorship in adults.

- The resources available to an individual are always limited, so any increase in allocation of resources to survival and competitive ability necessitates a decrease in the resources allocated to reproduction. This trade-off between survival and reproduction is the most fundamental aspect of a species' life history.

- In analyzing life history, there is a continuum between populations or species with high survivorship and low fecundity and populations or species with low survivorship and high fecundity

  ✔You should be able to explain why a life table is relevant only to a particular population at a particular time, and predict how the first life tables—constructed in ancient Rome—differ from the current life table for the same population.

🔑 The growth rate of a population can be calculated from life-table data or from the direct observation of changes in population size over time.

- Exponential growth occurs when the per-capita growth rate, *r*, does not change over time.

- During logistic growth, growing populations approach the carrying capacity of their environment and *r* declines to 0.

- Density dependence in population growth is due to competition for resources, disease, predation, or other factors that increase in intensity when population size is high. At high density, survivorship and fecundity decrease.

  ✔You should be able to draw a logistic growth curve, describe how *r* changes along the length of the curve, and suggest factors responsible for each change in *r*.

  (MB) BioFlix™ Population Ecology, **Web Activity** Modeling Population Growth

🔑 Researchers observe a wide variety of patterns when they track changes in population size over time, ranging from growth rates that slow when populations are at high density, to regular cycles, to continued growth independent of population size.

- Predator-prey interactions and food availability are density-dependent factors that drive the population cycle of snowshoe hares.

- Age structure has a profound impact on population dynamics. A population with few juveniles and many adults past reproductive age, like the human populations of developed nations, may be declining or stable in size. In contrast, a population with a large proportion of juveniles is likely to increase rapidly in size.

- The total human population has been increasing rapidly since about 1750 and is currently over 6.8 billion. In countries where survivorship has increased due to medical advances, fecundity has decreased.

- Based on various scenarios for average female fertility, the total human population in 2050 is expected to be between 7 and 11 billion.

  ✔You should be able to explain why human populations are projected to continue rapid growth, even though fertility rates are declining.

  (MB) Web Activity Human Population Growth and Regulation

🔑 Data from population ecology studies help biologists evaluate prospects for endangered species and design effective management strategies, as well as to predict changes in human populations.

- Human activities are isolating populations into metapopulations occupying small, fragmented habitats.

- The dynamics of a metapopulation are driven by the birth and death of populations, just as the dynamics of a population are driven by the birth and death of individuals.

- Migration among habitat patches is essential for the stability of a metapopulation, so conservationists are trying to preserve unoccupied patches of habitat and establish corridors that link habitat fragments.

  ✔You should be able to explain why a small, isolated population is virtually doomed to extinction, but why population size may be stable in a species consisting of a large number of small, isolated populations.

Questions

✔TEST YOUR KNOWLEDGE   *Answers are available in Appendix B*

1. What is the defining feature of exponential growth?
   a. The population is growing very quickly.
   b. The growth rate is constant.
   c. The growth rate increases rapidly over time.
   d. The growth rate is very high.

2. What four factors define population growth?
   a. age-specific birthrates, age-specific death rates, age structure, and metapopulation structure

   b. survivorship, age-specific mortality, fecundity, death rate
   c. mark-recapture, census, quadrat sampling, transects
   d. births, deaths, immigration, emigration

3. In what populations does exponential growth tend to occur?
   a. in populations that colonize new habitats
   b. in populations that experience intense competition
   c. in populations that experience high rates of predation
   d. in metapopulations

4. Which of the following is *not* a reason that population growth declines as population size approaches the carrying capacity?
   a. Climate becomes unfavorable.
   b. Competition for resources increases.
   c. Predation rates increase.
   d. Disease rates increase.

5. If most individuals in a population are young, why is the population likely to grow rapidly in the future?
   a. Death rates will be low.
   b. The population has a skewed age distribution.

c. Immigration and emigration can be ignored.
d. Many individuals will begin to reproduce soon.

6. Why have population biologists become particularly interested in the dynamics of metapopulations?
   a. because humans exist as a metapopulation
   b. because whooping cranes exist as a metapopulation
   c. because many populations are becoming restricted to small islands of habitat
   d. because metapopulations explain why populations occupying large, contiguous areas are vulnerable to extinction

✓ TEST YOUR UNDERSTANDING

Answers are available in Appendix B

1. Explain Equations 52.4 and 52.5 in words.

2. Explain why the growth rate of species with a type I survivorship curve depends primarily on fertility rates. Explain why the growth rate of species with a type III survivorship curve is extremely sensitive to changes in adult survivorship.

3. Offer a hypothesis to explain why humans have undergone near-exponential growth for over 500 years. Why can't exponential growth continue indefinitely? Describe two documented examples of density-dependent factors that influence population growth in natural populations.

4. Explain why biologists want to maintain (a) "habitat corridors" that connect populations in a metapopulation, and (b) unoccupied habitat that is appropriate for the species in question.

5. Make a rough sketch of the age distribution in developing versus developed countries, and explain why the shapes are different. How is AIDS, which is a sexually transmitted disease, affecting the age distribution in countries hard hit by the epidemic?

6. Compare the pros and cons of using R_0, λ, and r to express growth rates. What is the difference between r (the per-capita rate of increase) and r_{max} (the maximum or intrinsic growth rate)?

✓ APPLYING CONCEPTS TO NEW SITUATIONS

Answers are available in Appendix B

1. When wild plant and animal populations are logged, fished, or hunted, only the oldest or largest individuals tend to be taken. Many of the commercially important species are long-lived and are slow to begin reproduction. If harvesting is not regulated carefully and exploitation is intense, what impact does harvesting have on a population's age structure? How might harvesting affect the population's life-table and growth rate?

2. Design a system of nature preserves for an endangered species of beetle whose larvae feed on only one species of sunflower. The sunflowers tend to be found in small patches that are scattered throughout dry grassland habitats.

3. The population of the United States is projected to increase dramatically over the next 50 years, even though fertility rates are

only slightly above replacement level and the age distribution is largely stable. How is this possible?

4. In most species the sex ratio is at or near 1.00, meaning that there is an approximately equal number of males and females. In China, however, there is a strong preference for male children. According to the 2000 census there, the sex ratio of newborns is almost 1.17, meaning that close to 117 boys are born for every 100 girls. Based on these data, researchers project that there will soon be about 50 million more men than women of marriageable age in China. Discuss how this skewed sex ratio might affect the population growth rate in China.

This chapter explores how species that live in coral reefs and other communities interact with each other, and how communities change over time.

53 Community Ecology

Chapter 52 focused on the dynamics of populations—how and why they grow or decline, and how they change over time and through space. That chapter considered each population as an isolated entity.

In reality, though, populations of different species form complex assemblages called communities. A biological **community** consists of interacting species, usually living within a defined area.

This chapter's task is to analyze the dynamics of biological communities—how they develop and change over time. Biologists want to know how communities work, and how to manage them in a way that will preserve species and create an environment that people want to live in.

How do species inside communities interact with each other, and what are the long-term consequences? What happens to communities when they are disturbed by fires or flood, and why is the number of species higher in some areas than others? Let's delve in.

53.1 Species Interactions

The species in a community interact almost constantly. Organisms eat each other, pollinate one another, exchange nutrients, compete for resources, and provide habitats for each other. In many cases, a population's fate is tightly linked to the other species that share its habitat.

To study species interactions, biologists focus on analyzing the effects on the fitness of the individuals involved. Recall from Chapter 24 that **fitness** is defined as the ability to survive and produce offspring.

Does the relationship between two species provide a fitness benefit to members of one species (a "+" interaction) but hurt members of the other species (a "−" interaction)? Or does the association have no effect on the fitness of a participant (a "0" interaction)?

✔ When you see this checkmark, stop and test yourself. Answers are available in Appendix B.

- **Competition** occurs when individuals use the same resources—resulting in lower fitness for both ($-/-$).

- **Consumption** occurs when one organism eats or absorbs nutrients from another. The interaction increases the consumer's fitness but decreases the victim's fitness ($+/-$).

- **Mutualism** occurs when two species interact in a way that confers fitness benefits to both ($+/+$).

- **Commensalism** occurs when one species benefits but the other species is unaffected ($+/0$).

Of the four general types of interactions, commensalism is the least-studied. An example is the birds that follow moving army ants in the tropics. As the ants march along the forest floor, they hunt insects and small vertebrates. As they do, birds follow and pick off prey species that fly or jump up out of the way of the ants (**Figure 53.1**).

Three Themes

The rest of this section focuses on the most common types of species interactions: competition, consumption, and mutualism. As you analyze each type of species interaction, watch for three key themes:

1. *Species interactions can affect the distribution and abundance of a particular species.* Chapter 50 explored how species interactions can affect geographic ranges, using the examples of the yucca plant/yucca moth mutualism and the impact of a disease carried by tsetse flies on the distribution of cattle in Africa. Chapter 52 introduced impacts on population size, with data on how predation by lynx affects hare populations in northern Canada.

2. *Species act as agents of natural selection when they interact.* Deer are fast and agile in response to natural selection exerted by their major predators, wolves and cougars. The speed and agility of deer, in turn, promote natural selection that favors wolves and cougars that are fast and that have superior eyesight, senses of smell, and hunting strategies. To capture this point, biologists say that **coevolution** is occurring.

As species interact over time, it is common to observe a **coevolutionary arms race**—a repeating cycle of reciprocal adaptation. In humans, an arms race occurs when one nation develops a new weapon, which prompts a rival country to develop a defensive weapon, which pushes the original country to manufacture an even more powerful weapon, and so on. In biology, coevolutionary arms races occur between predators and prey, parasites and hosts, and other types of interacting species.

3. *The outcome of interactions among species is dynamic and conditional.* Consider the relationship between army ants and birds that follow them, which is usually commensal. If bird attacks start to force other insects into the path of the ants, then both ants and birds benefit and the relationship becomes mutualistic. But if birds begin to steal prey that would otherwise be taken by ants, then the relationship becomes parasitic. The outcome of the interaction may depend on the number and types of prey, birds, and ants present and may change over time.

Competition

Competition lowers the fitness of the individuals involved for a simple reason: When competitors use resources, those resources are not available to help individuals survive better and produce more offspring.

Competition that occurs between members of the same species is called **intraspecific** (literally, "within species") **competition**. Intraspecific competition for space, sunlight, food, and other resources intensifies as a population's density increases. As a result, intraspecific competition is a major cause of density-dependent growth (see Chapter 52). **Interspecific** ("between species") **competition** occurs when individuals from different species use the same limiting resources.

USING THE NICHE CONCEPT TO ANALYZE INTERSPECIFIC COMPETITION Early work on interspecific competition focused on the concept of the niche. A **niche** can be thought of as the range of resources that the species is able to use or the range of conditions it can tolerate.

(a) Ants stir up insects while hunting ...

(b) ... antbirds tag along and benefit.

FIGURE 53.1 Commensals Gain a Fitness Advantage but Don't Affect the Species They Depend On. Birds that associate with army ants are commensals. They have no measurable fitness effect on the ants, but gain from the association by capturing insects that try to fly or climb out of the way of the ants.

(a) One species eats seeds of a certain size range.

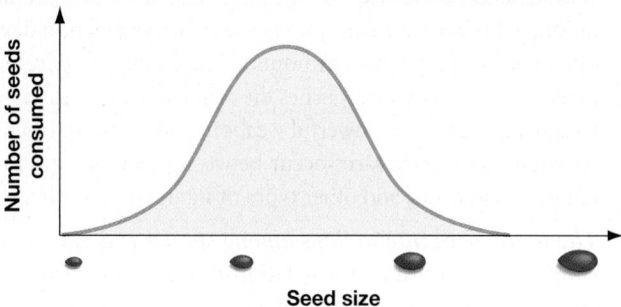

Seed size

(b) Partial niche overlap: competition for seeds of intermediate size

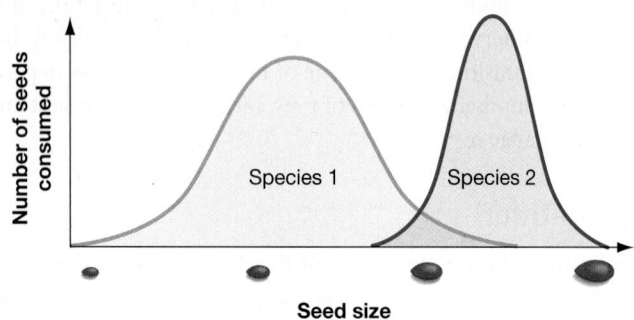

Species 1   Species 2

Seed size

FIGURE 53.2 Niche Overlap Leads to Competition. (a) A graph describing one aspect of a species' fundamental niche, meaning the range of resources that it can use or range of conditions it can tolerate. **(b)** Competition occurs when the niches of different species overlap.

G. Evelyn Hutchinson proposed that a species' niche could be envisioned by plotting its habitat requirements along a series of axes. **Figure 53.2a**, for example, represents one niche axis for a hypothetical species. In this case, the habitat requirement plotted is the size of seeds eaten by members of this population, which might be a function of mouth or tooth size. Other niche axes could represent other types of foods used or the temperatures, humidity, and other environmental conditions tolerated by the species. In these types of graphs, the *y*-axis represents the number or proportion of resources used, averaged over the population.

Interspecific competition occurs when the niches of two species overlap. The two species plotted in **Figure 53.2b**, for example, compete for seeds of intermediate size. When competition occurs, each individual will get fewer seeds on average, and average fitness in each population will decline.

WHAT HAPPENS WHEN ONE SPECIES IS A BETTER COMPETITOR? G. F. Gause claimed it is not possible for species with the same niche to coexist. This hypothesis, called the **competitive exclusion principle**, was inspired by a series of experiments Gause did with similar species of the unicellular pond-dweller *Paramecium*.

- When *P. caudatum* and *P. aurelia* grew in separate laboratory cultures, both species exhibited logistic growth (see Chapter 52).

- When the two species grew in the same culture together, only the *P. aurelia* population exhibited a logistic growth pattern. *P. caudatum*, in contrast, was driven to extinction (**Figure 53.3**).

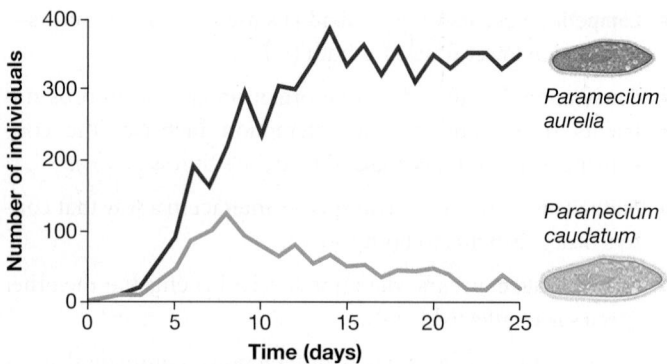

Paramecium aurelia

Paramecium caudatum

FIGURE 53.3 Competitive Exclusion in Two Species of *Paramecium*.

Gause's result is a product of asymmetric competition. When **asymmetric competition** occurs, one species suffers a much greater fitness decline than the other species does. Under **symmetric competition**, each of the interacting species experiences a roughly equal decrease in fitness. The distinction is graphed in **Figure 53.4a**, where the fitness of two hypothetical species, called A and B, is plotted for habitats where A and B occur separately and together.

If asymmetric competition occurs and the two species have completely overlapping niches (**Figure 53.4b**), then the stronger competitor is likely to drive the weaker competitor to extinction. But if the niches do not overlap completely, then the species that is the weaker competitor should be able to retreat to an area of non-overlap. In cases like this, an important distinction arises.

(a) Competitive exclusion occurs when competition is asymmetric ...

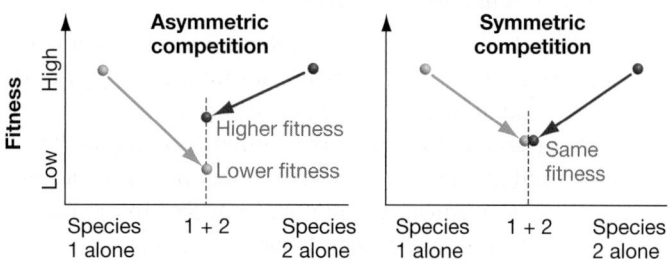

Asymmetric competition

Higher fitness
Lower fitness

Species 1 alone   1 + 2   Species 2 alone

Symmetric competition

Same fitness

Species 1 alone   1 + 2   Species 2 alone

(b) ... and niches overlap completely.

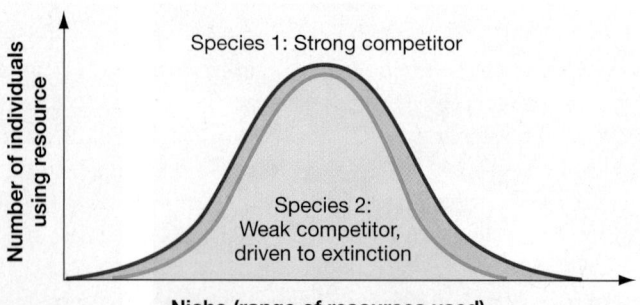

Species 1: Strong competitor

Species 2: Weak competitor, driven to extinction

Niche (range of resources used)

FIGURE 53.4 Asymmetric Competition and Niche Overlap Explain Competitive Exclusion. If two species have completely overlapping niches, there is no refuge for the weaker competitor. It may be driven to extinction.

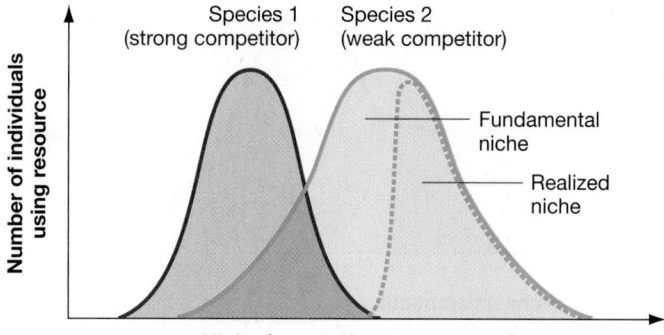

FIGURE 53.5 Fundamental Niches Are Broader than Realized Niches.

- A species' **fundamental niche** is the combination of resources or areas used or conditions tolerated in the absence of competitors.

- A species' **realized niche** is the portion of resources or areas used or conditions tolerated when competition occurs (**Figure 53.5**).

In Gause's laboratory cultures, competition between *P. aurelia* and *P. caudatum* was asymmetric and competitive exclusion occurred. How can researchers study competition in the field, under natural conditions?

EXPERIMENTAL STUDIES OF COMPETITION Joseph Connell began a classic study of competition in the late 1950s, after observing an interesting pattern in an intertidal rocky shore in Scotland. He noticed that there were two species of barnacles with distinctive distributions. The adults of one species, *Chthamalus stellatus*, occurred in an upper intertidal zone. The adults of the other species, *Semibalanus balanoides*, were restricted to a lower intertidal zone. The upper intertidal zone is a more severe environment for barnacles, because it is exposed to the air for longer periods at low tide each day.

Although adult barnacles live attached to rocks, larvae are mobile. The young of both species were found together in the lower intertidal zone, however. Why aren't *Chthamalus* adults found there as well?

To answer this question, Connell considered two hypotheses:

1. Adult *Chthamalus* are competitively excluded from the lower intertidal zone.

2. Adult *Chthamalus* are absent from the lower intertidal zone because they do not thrive in the physical conditions there.

Connell tested these hypotheses by performing the experiment shown in **Figure 53.6**. He began by removing a number of rocks that had been colonized by *Chthamalus* from the upper intertidal zone and transplanting them into the lower intertidal zone. He screwed the rocks into place and allowed *Semibalanus* larvae to colonize them. Once the spring colonization period was over, Connell divided each rock into two groups. In one half, he removed all *Semibalanus* that were in contact with or next to a *Chthamalus*.

This experimental design allowed Connell to document *Chthamalus* survival in the absence of competition with *Semibalanus*, and compare it with survival during competition.

QUESTION: Why is the distribution of adult *Chthamalus* restricted to the upper intertidal zone?

HYPOTHESIS: Adult *Chthamalus* are competitively excluded from the lower intertidal zone.

NULL HYPOTHESIS: Adult *Chthamalus* do not thrive in the physical conditions of the lower intertidal zone.

EXPERIMENTAL SETUP:

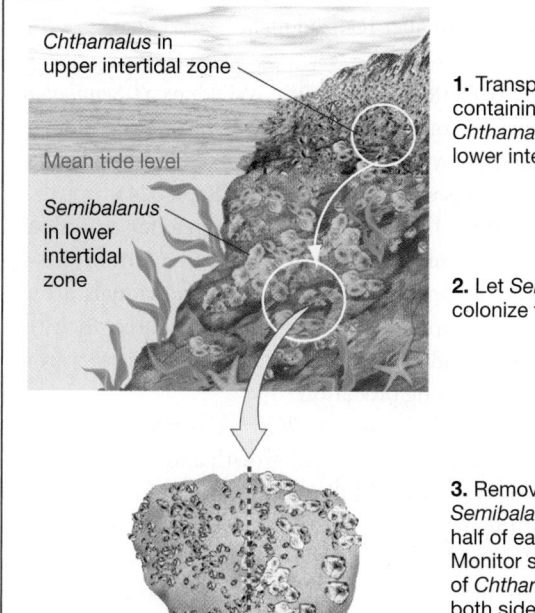

1. Transplant rocks containing young *Chthamalus* to lower intertidal zone.

2. Let *Semibalanus* colonize the rocks.

3. Remove *Semibalanus* from half of each rock. Monitor survival of *Chthamalus* on both sides.

PREDICTION: *Chthamalus* will survive better in the absence of *Semibalanus*.

PREDICTION OF NULL HYPOTHESIS: *Chthamalus* survival will be low and the same in the presence or absence of *Semibalanus*.

RESULTS:

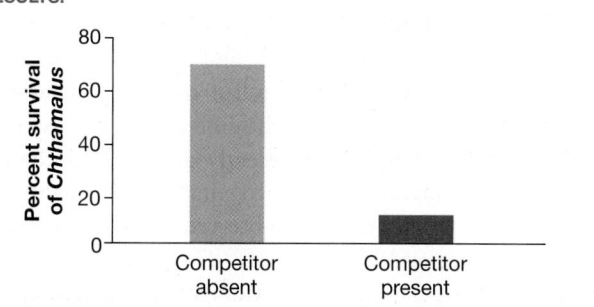

CONCLUSION: *Semibalanus* is competitively excluding *Chthamalus* from the lower intertidal zone.

FIGURE 53.6 Experimental Evidence for Competitive Exclusion.
SOURCE: Connell, J. H. 1961. The influence of interspecific competition and other factors on the distribution of the barnacle *Chthamalus stellatus*. *Ecology* 42: 710–723.

✔**QUESTION** Why was it important to carry out both treatments on the same rock? Why not use separate rocks?

This is a common experimental strategy in competition studies: One of the competitors is removed, and the response by the remaining species is observed.

Connell's results support the hypothesis of competitive exclusion. In the unmodified areas, *Semibalanus* killed many of the young *Chthamalus* by growing against them and lifting them off the substrate. As the bar graph in Figure 53.6 shows, *Chthamalus* survival was much higher when all of the *Semibalanus* were removed.

The fundamental niches of these two species overlap. *Semibalanus* is a superior competitor for space, but *Chthamalus* is able to thrive in a smaller realized niche.

FITNESS TRADE-OFFS IN COMPETITION Why haven't *Semibalanus* and other superior competitors taken over the world? Biologists answer this question by invoking the concept of **fitness trade-offs**—inevitable compromises in adaptation (see Chapter 24).

The key insight is that the ability to compete for a particular resource—like space on rocks or edible seeds of a certain size—is only one aspect of an organism's niche. If individuals are extremely good at competing for a particular resource, then they are probably less good at enduring drought conditions, warding off disease, or preventing predation.

In the case of *Semibalanus* and *Chthamalus* growing in the intertidal, the fitness trade-off is rapid growth and success in competing for space versus the ability to endure the harsh physical conditions of the upper intertidal. *Semibalanus* are fast-growing and large; *Chthamalus* grow slowly but can survive long exposures to the air and to intense sun and heat. Neither species can do both things well. Fitness trade-offs limit the ability of superior competitors to spread.

MECHANISMS OF COEXISTENCE: NICHE DIFFERENTIATION It's important to realize that because competition is a −/− interaction, there is strong natural selection on both species to avoid it. **Figure 53.7** shows the predicted outcome: An evolutionary change in traits reduces the amount of niche overlap, and thus the amount of competition.

An evolutionary change in resource use, caused by competition, is called **niche differentiation** or resource partitioning. The change that occurs in species' traits, and that allows individuals to exploit different resources, is called **character displacement**. Character displacement makes niche differentiation possible.

Peter and Rosemary Grant recently documented character displacement occurring in Galápagos finches. You might recall, from Chapter 24, that the Grants observed dramatic increases in average beak size and body size of a *Geospiza fortis* (medium ground finch) population during a drought in 1977. The changes occurred because the major food source available during the drought was fruits from a plant called *Tribulus cistoides,* and because only the largest-beaked individuals were able to crack these fruits and eat the seeds inside.

In 1982, however, individuals from a species called *Geospiza magnirostris* (large ground finch) arrived on the island and began breeding. *G. magnirostris* are about twice the size of *G. fortis* and use *T. cistoides* fruits as their primary food source.

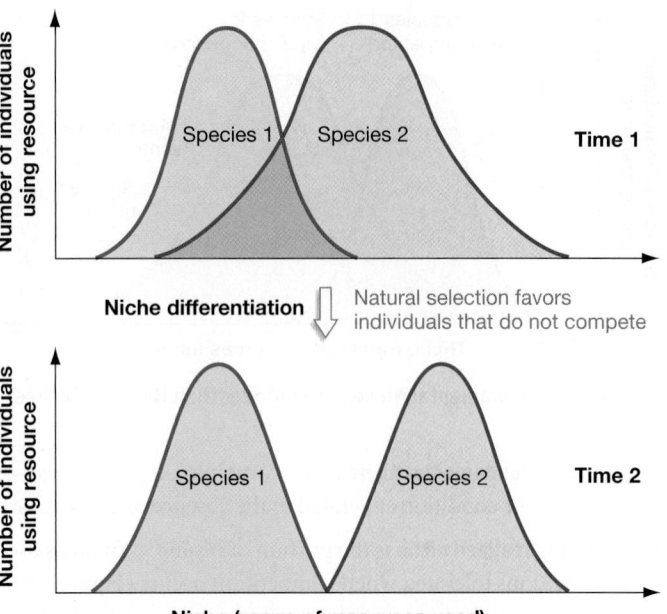

FIGURE 53.7 Over Time, Competition Can Lead to Niche Differentiation.

✔ **QUESTION** Why is "resource partitioning" a good name for the outcome of this process?

A severe drought recurred in 2003 and 2004; most finches died of starvation. When the Grants measured the surviving *G. fortis* they made a remarkable observation: In contrast to the 1977 drought, only the *smallest*-beaked individuals survived (**Figure 53.8**).

Data on feeding behavior indicated that *G. magnirostris* were outcompeting *G. fortis* for *T. cistoides* fruits. Only *G. fortis* that could eat extremely small seeds efficiently could survive.

Evolution had occurred: Average beak size in the *G. fortis* population decreased. This is a dramatic example of character displacement. Natural selection favored medium ground finches that did not compete with *G. magnirostris*; the outcome was character displacement and niche differentiation. Theory developed decades earlier had made a successful prediction about a natural experiment.

COMPETITION AND CONSERVATION One of the goals of conservation biology is to keep biological communities intact. One of the major threats to communities is invasive species—a topic that was introduced in Chapter 50 and is revisited in Chapter 55.

Recent experiments have shown that communities that contain a large number of different species are more resistant to invasion than communities with a smaller number of species. Presumably, the presence of additional species means that a higher proportion of the resources available in a particular habitat are being used—making it more difficult for a new species to invade and take over.

This result—that competition can help communities resist invasion—is important for biologists who are trying to restore natural communities. The message is that as many native species as possible should be reintroduced, to create restored communities that resist invasion more effectively.

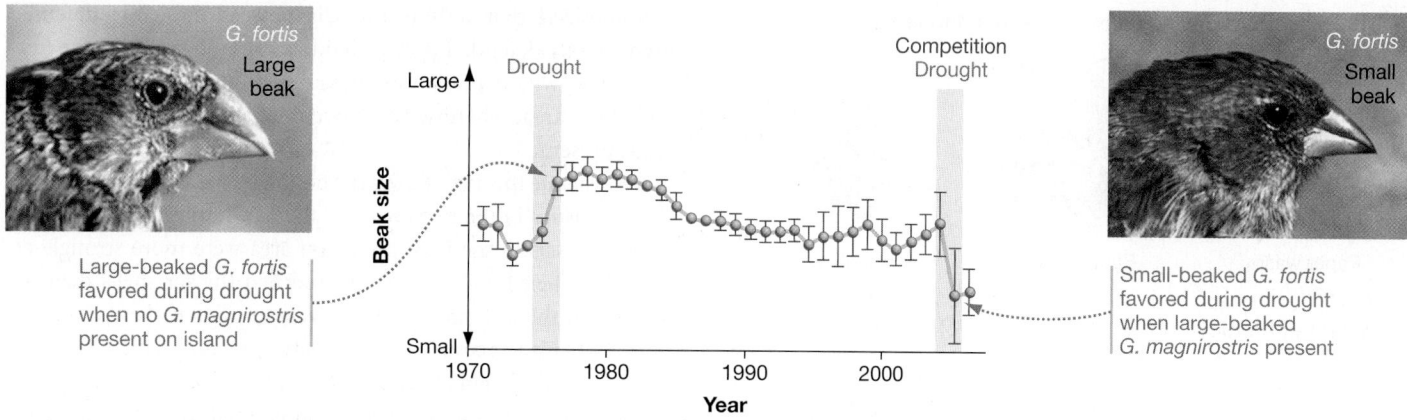

FIGURE 53.8 A Natural Experiment in Niche Differentiation. In *Geospiza fortis,* the data shown here would produce a shift in niche like that shown as a hypothetical case in Figure 53.7.

Consumption

Consumption occurs when one organism eats another. There are three major types of consumption:

1. **Herbivory** takes place when **herbivores** ("plant-eaters") consume plant tissues. Bark beetles mine cambium; cicadas suck sap; caterpillars chew leaves.

2. **Parasitism** occurs when a **parasite** consumes relatively small amounts of tissue or nutrients from another individual, called the **host**. Parasitism often occurs over a long period of time. It is not necessarily fatal, and parasites are usually small relative to their host. An array of worms and unicellular protists parasitizes humans; ticks parasitize cattle and other large mammals. In some cases, though, the term parasitism is used more broadly, to denote use of other resources. For example, birds and insects that lay their eggs in other species' nests and induce the other species to raise the young are called social parasites.

3. **Predation** occurs when a predator kills and consumes all or most of another individual. The consumed individual is called the prey. Woodpeckers eat bark beetles; finches eat seeds; ladybird beetles devour aphids; wasps kill caterpillars.

To illustrate how biologists analyze the impact of consumption, let's consider a series of questions about predators, herbivores, parasites, and their victims.

CONSTITUTIVE DEFENSES In fitness terms, consumption is costly for prey and beneficial for consumers. Prey individuals do not passively give up their lives to increase the fitness of their predators, however. Natural selection strongly favors traits that allow individuals to avoid being eaten.

- Prey may hide or run, fly, or swim away.

- Many plants lace their tissues with cocaine, nicotine, caffeine, or other compounds that are toxic to consumers.

- There is safety in numbers—schooling and flocking behavior is an effective way to reduce the risk of predation, in part because predators become confused when they dive or swim or run into a mass of prey.

- Some prey fight back: Selection has favored individuals with the ability to sequester or spray toxins, or employ weaponry such as sharp spines or kicking hooves.

Traits like these are called **standing** or **constitutive defenses**, because they are always present (**Figure 53.9**).

Camouflage: blending into the background

A leaf insect disappears among the leaves.

Schooling: safety in numbers

A school of fish confuses a shark.

Weaponry: fighting back

Porcupines use their spines to fight back.

FIGURE 53.9 Constitutive Defenses Are Always Present.

(a) Müllerian mimics: look dangerous, are dangerous

Paper wasp   Bumblebee   Honeybee

(b) Batesian mimics: look dangerous, are not dangerous

Hornet moth   Wasp beetle   Hoverfly

FIGURE 53.10 There Are Two Basic Forms of Mimicry in Prey Species.

Some of the best-studied constitutive defenses involve the phenomenon called mimicry. **Mimicry** occurs when one species closely resembles another species. The paper wasp on the far left of **Figure 53.10a**, for example, is dangerous because of its stinger. It also looks like other dangerous insects—particularly other wasps and bees. When harmful prey species resemble each other, **Müllerian mimicry** is said to occur.

To explain the existence of Müllerian mimics, biologists propose that the existence of similar-looking dangerous prey in the same habitat increases the likelihood that predators will learn to avoid them. In this way, Müllerian mimicry should reduce the likelihood of dangerous individuals being attacked.

Wasps also act as a model for harmless species, however—such as the moth, beetle, and fly shown in **Figure 53.10b**. To explain these mimics, biologists propose that predators avoid the harmless mimics because they mistake them for a dangerous wasp. This resemblance is known as **Batesian mimicry**.

INDUCIBLE DEFENSES Although constitutive defenses can be extremely effective, they are expensive in terms of the energy and resources that must be devoted to producing and maintaining them. Based on this observation, it should not be surprising to learn that many prey species have **inducible defenses**: defensive traits that are produced only in response to the presence of a predator.

Induced defenses decline if predators leave the habitat. Inducible defenses are efficient energetically, but they are slow—it takes time to produce them.

To see how inducible defenses work, consider research on the blue mussels that live in an estuary along the coast of Maine. Biologists had documented that predation on mussels by crabs was high in an area of the estuary with relatively slow tidal currents (a "low-flow" area) but low in an area of the estuary with relatively rapid tidal currents (a "high-flow" area). The researchers

hypothesized that if blue mussels possess inducible defenses, then heavily defended prey individuals should occur in the low-flow area, where predation pressure is higher, but not in the high-flow area, where water movement reduces the number of crabs present.

To evaluate this hypothesis, the biologists measured mussel shell characteristics in the two areas. As the bar graphs in **Figure 53.11** show, mussels in the high-predation area were more strongly attached to their base or substrate and had thicker shells than did mussels in the low-predation area. These traits make the mussels more difficult to remove from the substrate and harder to crush. Thick shells and tight attachment are antipredator defenses.

The data in Figure 53.11 are correlational in nature, however, so they are open to interpretation. A critic of the inducible defense explanation could offer a reasonable alternative hypothesis—for example, that only constitutive defenses exist, and that crabs have eliminated weakly attached mussels with thin shells from the low-flow areas. The observed differences could also be due to variation in light, temperature, or other abiotic factors that might affect mussel traits but have nothing to do with predation.

To test the inducible-defense hypothesis more rigorously, biologists carried out the experiment diagrammed in **Figure 53.12**. The tank on the right allowed the researchers to measure shell growth in mussels that were "downstream" from crabs. As predicted by the induced-defense hypothesis, the mussels exposed to a crab in this way developed significantly thicker shells than did mussels that were not exposed to a crab. These results suggest that, even without direct contact, mussels can sense the presence of crabs and increase their investment in defenses.

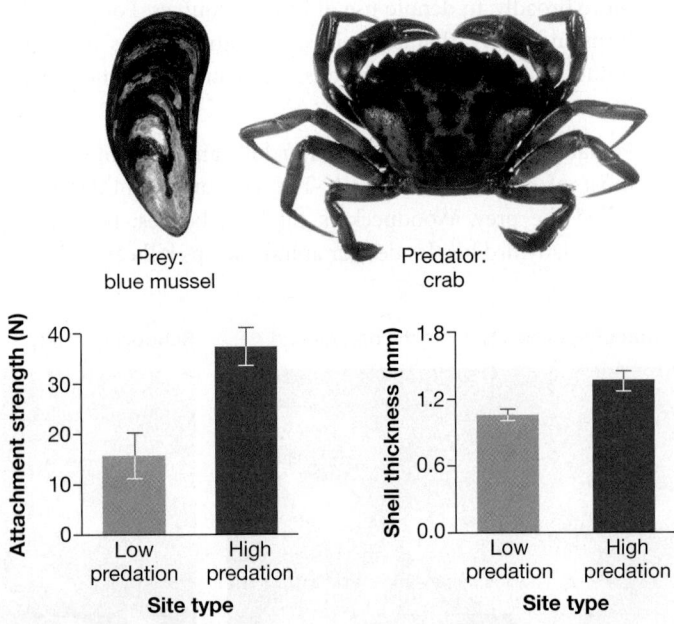

Prey: blue mussel   Predator: crab

FIGURE 53.11 Inducible Defenses Are Produced Only when Prey Are Threatened. To measure how strongly mussels are attached to the substrate, researchers drilled a hole in their shells and measured the force, in newtons (N), required to pull them off. They also measured how large mussel shells were relative to an individual's size by dividing shell weight by soft-tissue weight.

QUESTION: Are mussel defenses induced by the presence of crabs?

HYPOTHESIS: Mussels increase investment in defense in the presence of crabs.

NULL HYPOTHESIS: Mussels do not increase investment in defense in the presence of crabs.

EXPERIMENTAL SETUP:

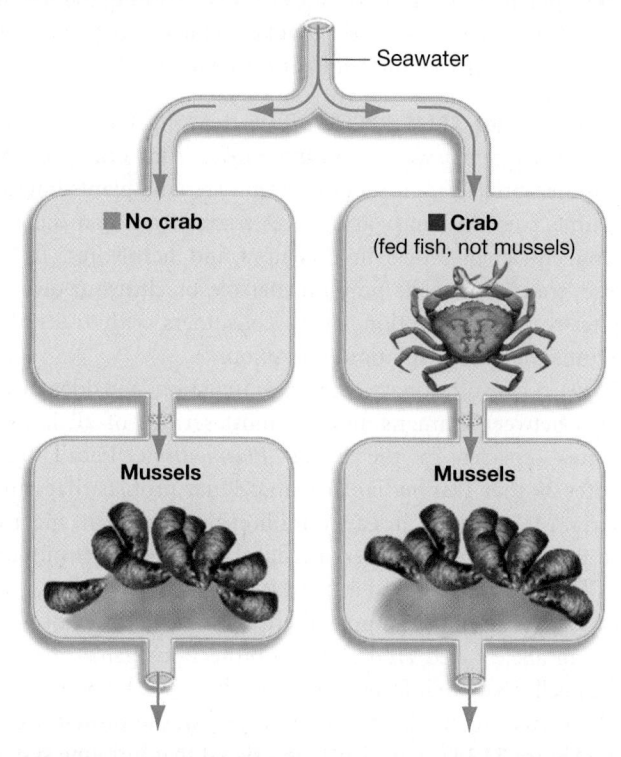

PREDICTION: Mussels downstream of the crab tank will have thicker shells than mussels downstream of the empty tank.

PREDICTION OF NULL HYPOTHESIS: Mussels in the two tanks will have shells of equal thickness.

RESULTS:

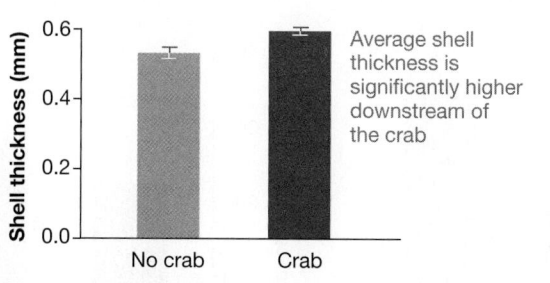

Average shell thickness is significantly higher downstream of the crab

CONCLUSION: Mussels increase investment in defense when they detect crabs. Shell thickness is an inducible defense.

FIGURE 53.12 Experimental Evidence for Inducible Defenses.

SOURCE: Leonard, G. H., M. D. Bertness, and P. O. Yund. 1999. Crab predation, waterborne cues, and inducible defenses in the blue mussel, *Mytilus edulis*. *Ecology* 80: 1–14.

✔ **QUESTION** Why did the researchers feed the crabs fish instead of mussels?

ARE ANIMAL PREDATORS EFFICIENT ENOUGH TO REDUCE PREY POPULATIONS? Prey are typically smaller than predators, have larger litter or clutch sizes, and tend to begin reproduction at a younger age. As a result, they have a much larger intrinsic growth rate. You might recall from Chapter 52 that this quantity, symbolized r_{max}, is the maximum growth rate that a population can achieve under ideal conditions.

If prey reproduce rapidly and are also well defended, can predators kill enough of them to reduce the prey population significantly?

Data from predator-removal programs—in which wolves, cougars, coyotes, or other predators are actively killed by human hunters—suggest that the answer is yes. For example, when a wolf control program in Alaska during the 1970s decreased predator abundance, the population of moose, on which wolves prey, tripled. This observation suggests that wolves had reduced this moose population far below the carrying capacity—the number that could be supported by the available space and food (see Chapter 52).

Other types of experiments are consistent with this result. Recall from Chapter 52 that the regular population cycles of snowshoe hares appear to be driven, at least in part, by density-dependent increases in predation. Taken together, the data available to date indicate that in many instances, predators are efficient enough to reduce prey populations below carrying capacity.

WHY DON'T HERBIVORES EAT EVERYTHING—WHY IS THE WORLD GREEN? If predators affect the size of populations in prey that can run or fly or swim away, then consumers should have a devastating impact on plants and on mussels, anemones, sponges, and other sessile (nonmoving) animals.

In some cases, this prediction turns out to be correct. For example, consider the results of a recent **meta-analysis**—a study of studies, meaning an analysis of a large number of data sets on a particular question. Biologists who compiled the results of more than 100 studies on herbivory found that the median percentage of mass removed from aquatic algae by herbivores was 79 percent. Herbivores eat the vast majority of algal food available in aquatic biomes. The figure dropped to just 30 percent for aquatic plants, however, and only 18 percent for terrestrial plants.

Why don't herbivores eat more of the food available on land? Stated another way, why is the world green? Biologists routinely consider two hypotheses:

1. *Top-down control hypothesis: Herbivore populations are limited by predation and disease.*

The "top-down" name is inspired by the food chain concept introduced in Chapter 29 and explored in detail in Chapter 54. In a food chain, herbivores are "on top of" plants.

The logic of top-down control is simple: Predators and parasites remove herbivores that eat plants. As a result, a great deal of plant material remains uneaten.

In a recent test of this idea, researchers monitored herbivory on islands created by a 1986 dam project in Venezuela. On some small islands in the new lake, predators disappeared. On the predator-free islands, there are now many more herbivores than on similar sites nearby where predators are present. As predicted by the top-down control hypothesis, a much higher percentage of the total plant material is being eaten on the predator-free islands. For example, predator-free islands had just 25 percent of the small trees found on islands that contained predators.

2. Bottom-up limitation hypothesis: Plant tissues are well-defended or offer poor nutrition.

The "bottom-up" name reflects the position of plants as the base or bottom of a food chain. The idea here is that plant tissues aren't eaten because they are toxic, or that herbivore numbers are limited because plant tissues offer poor nutrition.

The poor-nutrition aspect of this hypothesis was inspired by the observation that plant tissues have less than 10 percent of the nitrogen found in animal tissues, by weight. If the growth and reproduction of herbivores are limited by the availability of nitrogen, then their populations will be low and the impact of herbivory on plants relatively slight. Herbivores could eat more plant material to gain nitrogen, but at a cost—they would be exposed to predation and expend energy processing the food. When nitrogen concentration in plants is increased experimentally by fertilization, it is not unusual to see significant increases in herbivore growth rates.

In addition, most plant tissues are defended by weapons such as thorns, prickles, or hairs, or by potent poisons such as nicotine, caffeine, and cocaine. Manufacturing defensive compounds seems like the perfect solution to the problem of herbivory. In practice, however, plants face a complex challenge in defending themselves.

To drive this point home, consider data on interactions between cottonwood trees and two of their herbivores: beavers and leaf beetles. Cottonwoods resprout after they are cut down by a beaver (**Figure 53.13a**), and resprouted trees contain high concentrations of a defensive compound that deters further attack by beavers (**Figure 53.13b**). This is an induced defense.

Unfortunately for cottonwoods, leaf beetle larvae eat this defensive compound readily. In fact, the beetle larvae appear to use the anti-beaver compound to defend themselves against their major predator: ants. The bar graph in **Figure 53.13c** is from an experiment where researchers placed beetle larvae on an ant mound, and recorded how long they stayed alive. If larvae have fed on leaves of cut and resprouted cottonwoods, they survive ant attacks much longer than if they have fed on normal cottonwood leaves.

The punchline? Top-down control, nitrogen limitation, and effective defense are all important factors in limiting the impact of herbivory. The mix of factors that keeps the world green varies from species to species and habitat to habitat.

ADAPTATION AND ARMS RACES Over the long term, how do species that interact via consumption affect each other's evolution? When predators and prey or herbivores and plants interact over time, coevolutionary arms races result. Traits that increase feeding efficiency evolve in predators and herbivores. In response, traits that make prey unpalatable or elusive evolve. In counter-response, selection favors consumers with traits that overcome the prey adaptation. And so on.

To see a coevolutionary arms race in action, consider interactions between humans and the most serious of all human parasites—species in the genus *Plasmodium*. Recall from Chapter 29 that *Plasmodium* are unicellular protists that cause malaria. Malaria kills at least a million people a year, most of them pre–school-age children. Recent data suggest that humans and *Plasmodium* are locked in a coevolutionary arms race.

In West Africa, for example, there is a strong association between an allele called *HLA-B53* and protection against malaria. In liver cells that are infected by *Plasmodium*, HLA-B53 proteins on the surface of the liver cell display a parasite protein called cp26 (**Figure 53.14**). The display is a signal that immune system cells can read. It means "I'm an infected cell. Kill me before they

(a) Cottonwood tree felled by beavers

(b) Resprouted cottonwood trees have more defensive compounds against beavers ...

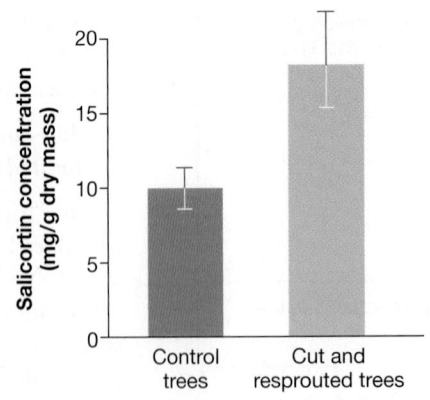

(c) ... but beetle larvae that eat resprouted trees survive longer.

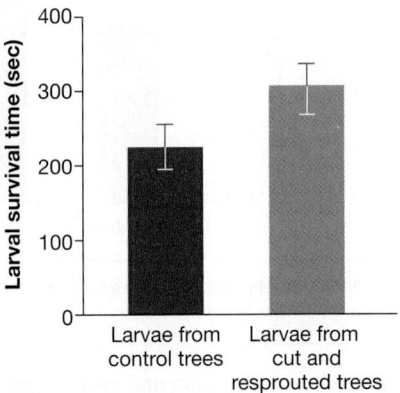

FIGURE 53.13 Defensive Compounds in Cottonwood Trees Discourage Beavers but Favor Beetles.

✔**QUESTION** Beetle larvae survive better on resprouted versus control trees, so they should damage resprouted trees more. If natural selection has maximized cottonwood fitness, which presents the greater overall threat to the trees, beavers or beetle larvae?

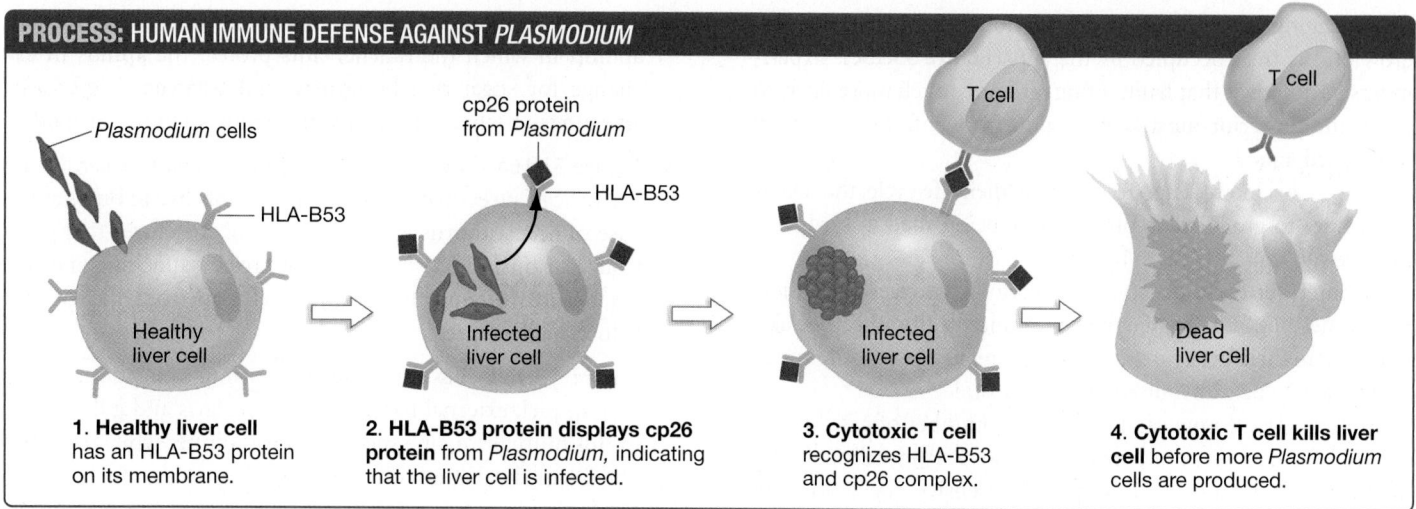

PROCESS: HUMAN IMMUNE DEFENSE AGAINST *PLASMODIUM*

1. **Healthy liver cell** has an HLA-B53 protein on its membrane.

2. **HLA-B53 protein displays cp26 protein** from *Plasmodium,* indicating that the liver cell is infected.

3. **Cytotoxic T cell** recognizes HLA-B53 and cp26 complex.

4. **Cytotoxic T cell kills liver cell** before more *Plasmodium* cells are produced.

FIGURE 53.14 Interactions between the Human Immune System and *Plasmodium*. If HLA-B53 binds to a particular *Plasmodium* protein, then infected cells are recognized and destroyed.

TABLE 53.1 Some *Plasmodium* Strains Are Particularly Effective at Infecting Humans

Plasmodium Strain	Infection Rate	Interpretation
cp26	Low	HLA-B53 binds to these proteins. Immune response is effective.
cp29	Low	
cp27	High	HLA-B53 does not bind to these proteins. Immune response is not as effective.
cp28	Average	

kill all of us." Immune system cells destroy the liver cell before the parasite cells inside can multiply.

In this way, people who have at least one copy of the *HLA-B53* allele are better able to beat back malarial infections. People who have the *HLA-B53* allele appear to be winning the arms race against malaria.

Follow-up research has shown that the arms race is far from won, however. *Plasmodium* populations in West Africa now have a variety of alleles for the protein recognized by HLA-B53. Some of these variants bind to HLA-B53 and trigger an immune response in the host, but others escape detection (**Table 53.1**). To make sense of these observations, researchers suggest that natural selection has favored the evolution of *Plasmodium* strains with weapons that counter HLA-B53.

The arms race continues. To review the *Plasmodium*-human interaction, go to the study area at *www.masteringbiology.com.*

MB Web Activity Life Cycle of a Malaria Parasite

CAN PARASITES MANIPULATE THEIR HOSTS? To thrive, parasites do not just have to invade tissues and grow while evading defensive responses by their host. They also have to be transmitted to new hosts. To a parasite, an uninfected host represents uncolonized habitat, teeming with resources. What have biologists learned about how parasites are transmitted to new hosts?

To answer this question, consider species of tree-dwelling ants (**Figure 53.15a**) that are parasitized by nematodes (round-worms). When nematodes have matured inside a host individual, they lay eggs in the ant's posterior-most body region. The exoskeleton in the area becomes translucent, and the yellow eggs inside make the structure look reddish instead of the normal black color (**Figure 53.15b**). Infected ants also hold the region up

(a) Uninfected ant

(b) Infected ant

(c) Infected ants resemble berries that are eaten by birds.

FIGURE 53.15 A Parasite That Manipulates Host Behavior. When the tropical ant *Cephalotes atratus* becomes infected with parasitic nematodes, its appearance and behavior changes.

✔**EXERCISE** Design an experiment to test the hypothesis that infected ants are more likely than uninfected ants to be eaten by birds.

in a "flagging" posture, making them look like the berries that grow on the trees occupied by the ants (**Figure 53.15c**). Experiments have shown that fruit-eating birds are much more likely to pluck the posterior-most body region from infected ants than uninfected ants.

Why is this interesting? To complete their life cycle, the nematodes have to grow inside birds, before being shed in bird feces that are subsequently eaten by ants.

To interpret these observations, biologists suggest that these nematodes not only change the appearance of the ants, they also manipulate their behavior. The changes make the parasite much more likely to be transmitted to a new host.

Biologists have now cataloged a large number of case studies like this, from an array of parasites and hosts. Parasite manipulation of hosts is another example of extensive coevolution in species that interact.

USING CONSUMERS AS BIOCONTROL AGENTS Research on the dynamics of predator-prey interactions and other aspects of consumption has paid off in practical benefits: In some cases, biologists have been able to control pests by introducing predators or parasites.

The invasive plant purple loosestrife, for example, had succeeded in taking over large expanses of wetlands in North America. Through careful experimentation, however, biologists were able to find two insect species that feed exclusively on purple loosestrife. By introducing the herbivores into wetlands, researchers have been able to reduce purple loosestrife infestations dramatically.

In agriculture and forestry, the use of predators and parasites as biocontrol agents is a key part of **integrated pest management**: strategies to maximize crop and forest productivity while using a minimum of insecticides or other types of potentially harmful compounds.

Mutualism

Mutualisms are +/+ interactions that involve a wide variety of organisms and rewards.

- Bees visit flowers to harvest nectar and pollen. The bees benefit by using nectar and pollen as a food source; flowering plants benefit because in the process of visiting flowers, foraging bees carry pollen from one plant to another and accomplish pollination. Chapters 30 and 40 detailed some of the adaptations found in flowering plants that increase the efficiency of pollination.

- One of the most important of all mutualisms occurs between mycorrhizal fungi and plant roots. Chapter 31 reviewed experimental evidence indicating that mycorrhizal fungi receive sugars and other carbon-containing compounds in exchange for nitrogen or phosphorus acquired by the plant partner. The plant feeds the fungus; the fungus fertilizes the plant.

- Chapters 28 and 38 introduced the mutualism between nitrogen-fixing bacteria and the plant species that house them in their tissues. The host plants provide sugars and protection to the bacteria; the bacteria supply ammonia or nitrate in return.

- Chapter 32 cited mutualisms between rancher ants and aphids, in which the rancher ants protect the aphids in exchange for sugar-rich honeydew, and between farmer ants and fungi, in which the farmer ants cultivate fungi for food.

- **Figure 53.16a** shows ants in the genus *Crematogaster*, which live in acacia trees native to Africa. The ants live in bulbs at the base of acacia thorns and feed on small structures that grow from leaf tips. The ants protect the tree by attacking and biting herbivores, and by cutting vegetation from the ground below the host tree.

- **Figure 53.16b** illustrates cleaner shrimp in action. These shrimp pick external parasites from the jaws and gills of fish. In this mutualism, one species receives dinner while the other obtains medical attention.

(a) Mutualism between ants and acacia trees

(b) Mutualism between cleaner shrimp and fish

FIGURE 53.16 Mutualisms Take Many Forms. (a) In certain species of acacia tree, ants in the genus *Crematogaster* live in large bulbs at the base of spines and attack herbivores that threaten the tree. The ants eat nutrient-rich tissue produced at the tips of leaves. **(b)** Cleaner shrimp remove and eat parasites that take up residence on the gills of fish.

✔**EXERCISE** Name a possible cost of these mutualistic associations to the acacia tree, the *Crematogaster*, the cleaner shrimp, and the host fish.

THE ROLE OF NATURAL SELECTION IN MUTUALISM As the examples you've just reviewed show, the rewards from mutualistic interactions range from the transportation of gametes to food, housing, medical help, and protection. It is important to note, however, that even though mutualisms benefit both species, the interaction does not involve individuals from different species being altruistic or "nice" to each other.

Expanding on a point that Charles Darwin introduced in 1862, Judith Bronstein described mutualisms as "a kind of reciprocal parasitism; that is, each partner is out to do the best it can by obtaining what it needs from its mutualist at the lowest possible cost to itself." Her point is that the benefits received in a mutualism are a by-product of each individual pursuing its own self-interest by maximizing its fitness—its ability to survive and reproduce.

In this light, it is not surprising that some species "cheat" on mutualistic systems. For example, deceit pollination occurs when certain species of plants produce a showy flower but no nectar reward. Pollinators receive no reward for making a visit and carrying out pollination. Evolutionary studies show that deceit pollinators evolved from ancestral species that did provide a reward. Over time, a $+/+$ interaction evolved into a $+/-$ interaction.

MUTUALISMS ARE DYNAMIC A recent experimental study of mutualism provides another good example of the dynamic nature of these interactions. This study focused on ants and treehoppers. Ants are insects that live in colonies; treehoppers are small, herbivorous insects that feed by sucking sugar out of the phloem of plants. Treehoppers excrete the sugary solution honeydew from their posteriors. The honeydew, in turn, is harvested for food by ants.

It is clear that ants benefit from this association. But do the treehoppers? Biologists hypothesized that the ants might protect the treehoppers from their major predator, jumping spiders. These spiders feed heavily on juvenile treehoppers.

To test the hypothesis that ants protect treehoppers, the researchers studied ant-treehopper interactions over a three-year period. As **Figure 53.17** shows, the researchers marked out a 1000-m² study plot. Each year they removed the ants from one group of the treehopper host plants inside the plot but left the others alone to serve as a control. Then they compared the growth and survival of treehoppers on plants with and without ants.

Recall that this is a common research strategy for studying species interactions. To assess the fitness costs or benefits of the interaction, researchers remove one of the participants experimentally and document the effect on the other participant's survival and reproduction, compared with the survival and reproduction of control individuals that experience a normal interaction.

In both the first and third years of the study, the number of treehopper young on host plants increased in the treatment with ants but showed a significant decline in the treatment with no ants. This result supports the hypothesis that treehoppers benefit from the interaction with ants because the ants protect the treehoppers from predation by jumping spiders.

In the second year of the study, however, the researchers found a very different pattern. There was no difference in off-spring survival, adult survival, or overall population size between treehopper populations with ants and those without ants.

Why? The researchers were able to answer this question because they also measured the abundance of spiders that prey on treehoppers in each of the three years. Their census data showed that in the second year of the study, spider populations were very low.

EXPERIMENT

QUESTION: Is the relationship between ants and treehoppers mutualistic?

HYPOTHESIS: Ants harvest food from treehoppers. In return, they protect treehoppers from jumping spiders.

NULL HYPOTHESIS: Ants harvest food from treehoppers but are not beneficial to treehopper survival.

EXPERIMENTAL SETUP:

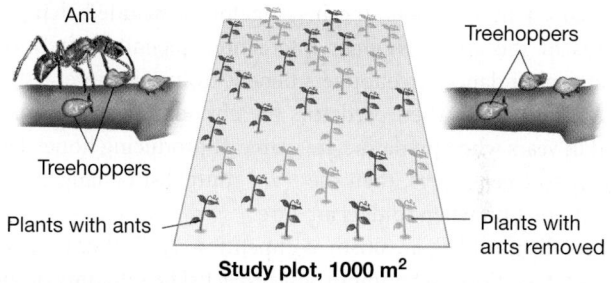

Ant

Treehoppers

Treehoppers

Plants with ants

Plants with ants removed

Study plot, 1000 m²

PREDICTION: Treehopper reproduction will be higher when ants are present than when ants are absent.

PREDICTION OF NULL HYPOTHESIS: There will be no difference in the number of young treehoppers on the plants.

RESULTS:

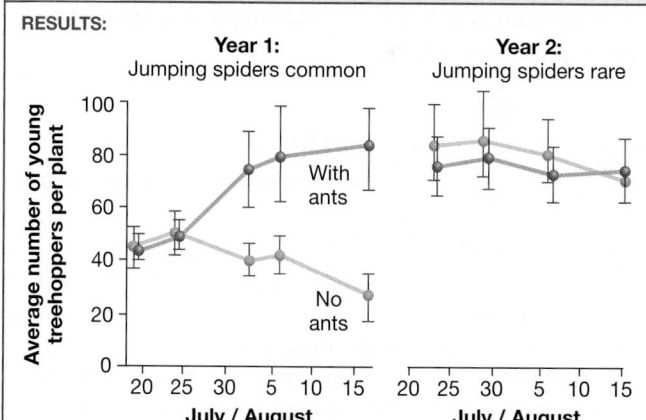

Year 1: Jumping spiders common

Year 2: Jumping spiders rare

With ants

No ants

Average number of young treehoppers per plant

20 25 30 5 10 15
July / August

20 25 30 5 10 15
July / August

CONCLUSION: Treehoppers benefit from the interaction with ants only when jumping spiders are common.

FIGURE 53.17 Experimental Evidence That the Treehopper-Ant Interaction Is Mutualistic. The data points on the graphs represent average values from many study plots.

SOURCE: Cushman, J. H. and T. G. Whitham. 1989. Conditional mutualism in a membracid-ant association: temporal, age-specific, and density-dependant effects. *Ecology* 70: 1040–1047.

✔**QUESTION** The researchers assigned the treatments to different plants at random, and did many replicates of each treatment. Why was this important?

SUMMARY TABLE 53.2 Species Interactions

Type of Interaction	Fitness Effects	Short-Term Impact: Distribution and Abundance	Long-Term Impact: Coevolution
Competition	−/−	Reduces population size of both species; if competition is asymmetric, competitive exclusion reduces range of one species and may even lead to extinction.	Niche differentiation via selection to reduce competition
Consumption	+/−	Impact on prey population depends on prey density and effectiveness of defenses.	Strong selection on prey for effective defense; strong selection on consumer for traits that overcome defenses
Mutualism	+/+	Population size and range of both species are dependent on each other.	Strong selection on both species to maximize fitness benefits and minimize fitness costs of relationship
Commensalism	+/0	Population size and range of commensal may depend on size and distribution of host.	Strong selection on commensal to increase fitness benefits in relationship; no selection on host

Based on these results, the investigators concluded that the benefits of the ant-treehopper interaction depend entirely on predator abundance. Treehoppers benefit from their interaction with ants in years when predators are abundant but are unaffected in years when predators are scarce. If producing honeydew is costly to treehoppers, then the +/+ mutualism changes to a +/− interaction when spiders are rare.

Mutualism is like parasitism, competition, and other types of species interactions in an important respect: The outcome of the interaction depends on current conditions. Because the costs and benefits of species interactions are fluid, an interaction between the same two species may range from parasitism to mutualism to competition. **Table 53.2** summarizes the fitness effects, short-term impacts on population size, and long-term evolutionary aspects of species interactions.

CHECK YOUR UNDERSTANDING

If you understand that . . .

- Natural selection favors the evolution of traits that decrease competition, resulting in changes in a species' niche over time.
- Consumers may reduce the population size of the species on which they feed and often exert strong natural selection for effective defense mechanisms.
- Mutualisms benefit the species involved and can lead to highly coevolved associations, such as mycorrhizal fungi and symbiotic nitrogen-fixing bacteria.

✓ You should be able to . . .

1. Explain why resource partitioning (niche differentiation) does not involve a conscious choice on the part of the individuals involved.
2. Explain what a coevolutionary arms race is and give an example.

Answers are available in Appendix B.

53.2 Community Structure

In a community, each species interacts with dozens or hundreds or even thousands of other species. You, for example, are a walking, talking, biological community that includes trillions of bacterial and archaeal cells. Your skin is covered with them; they pack your gut lining and thrive in your mouth and teeth and respiratory tract. Some of the species present parasitize you, some are mutualistic, many are commensal.

Unfortunately, we know relatively little about the structure and dynamics of your personal biological community—although promising research has begun on questions about changes in species composition over time, competition for resources, and responses to disturbance.

Fortunately, biologists have been studying forest, grassland, and marine communities for over 80 years. Time will tell whether the general principles that have emerged from this research also apply to the community that you carry around with you.

How Predictable Are Communities?

The first question that biologists asked about woodlands and deserts and shorelines concerned structure: Do biological communities have a tightly prescribed organization and composition, or are they merely loose assemblages of species?

THE CLEMENTS-GLEASON DICHOTOMY Beginning with a paper published in 1936, Frederick Clements hypothesized that biological communities are stable, integrated, and orderly entities with a highly predictable composition. His reasoning was that species interactions are so extensive—and coevolution so important—that communities have become highly integrated and interdependent units in nature. Stated another way, the species within a community cannot live without each other.

In developing this hypothesis, Clements likened the development of a plant community to the development of an individual organism. He argued that communities develop over time by passing through a series of predictable stages dictated by extensive interactions among species. Eventually, this developmental progression culminates in a final stage known as a **climax community**. The climax community is stable—it does not change over time.

According to Clements, the nature of a climax community is determined by the area's climate. He predicted that if a fire or other disturbance destroys the climax community, it will reconstitute itself by repeating its predictable developmental stages.

Henry Gleason, in contrast, contended that the community found in a particular area is neither stable nor predictable. He claimed that plant and animal communities are ephemeral associations of species that just happen to share similar climatic requirements.

According to Gleason, it is largely a matter of chance whether a similar community develops in the same area after a disturbance occurs. Gleason downplayed the role of biotic factors, such as species interactions, in structuring communities. Instead, chance historical events—for example, which seeds and juvenile animals happened to arrive after a disturbance—were key elements in determining which species are found at a particular location.

Which viewpoint is more accurate? Let's consider experimental and observational data.

AN EXPERIMENTAL TEST   To explore the predictability of community structure experimentally, biologists constructed 12 identical ponds (**Figure 53.18**). They filled the ponds at the same time with water that contained enough chlorine to kill any preexisting organisms. The Clements and Gleason hypotheses make clear and contrasting predictions:

- If community structure is predictable, then each pond should develop the same community of species once the chlorine vaporized and made the water habitable.

- If community structure is unpredictable, then each pond should develop a different community.

To test these predictions, the researchers sampled water from the ponds repeatedly for one year. They measured temperature, chemical makeup, and other physical characteristics of the water and recorded the diversity and abundance of each planktonic species by examining the samples under the microscope. Species that make up **plankton** live near the surface of water and swim little, if at all.

Their results? The red dots on the graph in Figure 53.18 indicate whether a particular species was found in each of the 12 ponds. Note that they found a total of 61 planktonic species in all of the ponds. Individual ponds had only 31 to 39 species apiece, however. This observation is important. Each pond contained just half to two-thirds of the total number of species that lived in the experimental area and that were available for colonization. A number of species occurred in most or all of the 12 ponds, but each pond had a unique species assemblage. Why?

To explain their results, the researchers contended that some species are particularly good at dispersing and are likely to colonize all or most of the available habitats. Other species disperse more slowly and tend to reach only one or a few of the available habitats. Further, the investigators proposed that the arrival of certain competitors or predators early in the colonization process

EXPERIMENT

QUESTION: Are communities predictable or unpredictable?

COMMUNITIES-ARE-PREDICTABLE HYPOTHESIS: The group of species present at a particular site is highly predictable.

COMMUNITIES-ARE-UNPREDICTABLE HYPOTHESIS: The group of species present at a particular site is highly unpredictable.

EXPERIMENTAL SETUP:

1. Construct 12 identical ponds. Fill at the same time and sterilize water so that there are no preexisting organisms.

2. Examine water samples from each pond. Identify each plankton species present in each sample.

COMMUNITIES-ARE-PREDICTABLE PREDICTION: Identical plankton communities will develop in all 12 ponds.

COMMUNITIES-ARE-UNPREDICTABLE PREDICTION: Different plankton communities will develop in different ponds.

RESULTS (AFTER 1 YEAR):

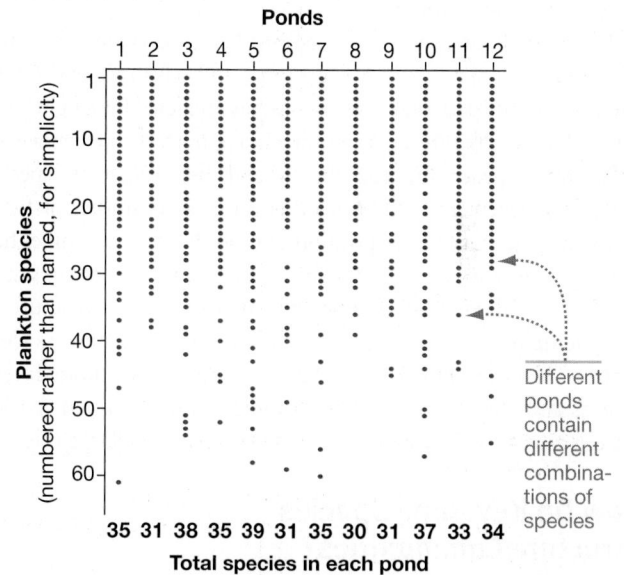

CONCLUSION: Although about half of the species present appear in all or most ponds, each pond has a unique composition. Both hypotheses are partially correct.

FIGURE 53.18 Experimental Evidence That Identical Communities Do Not Develop in Identical Habitats.

SOURCE: Jenkins, D. G. and A. L. Buikema. 1998. Do similar communities develop in similar sites? A test with zooplankton structure and function. *Ecological Monographs* 68: 421–443.

✔ **QUESTION**   At least 20 species are found in most or all of the ponds. Does this observation suggest that community composition is predictable or unpredictable?

greatly affects which species are able to invade successfully later. As a result, the specific details of community assembly and composition are somewhat contingent and difficult to predict. At least to some degree, communities are a product of chance and history.

MAPPING CURRENT AND PAST SPECIES' DISTRIBUTIONS If communities are predictable assemblages, then the ranges of species that make up a particular community should be congruent—meaning that the same group of species should almost always be found growing together.

When biologists began documenting the ranges of tree species along elevational gradients, however, they found that species came and went independently of each other. As you go up a mountain, for example, you might find white oak trees growing with hickories and chestnuts. As you continue to gain elevation, chestnuts might disappear but white oaks and hickories remain. Farther upslope, hickories might drop out of the species mix while white pines and red pines start to appear.

Data on the historical composition of plant communities supported these observations. Studies of fossil pollen documented that the distribution of plant species and communities at specific locations throughout North America has changed radically since the end of the last ice age about 11,000 years ago.

An important pattern emerged from these data: Species do not come and go in the fossil record in tightly integrated units. Instead, the ranges of individual species tended to change independently of one another.

For example, at a study site in Mirror Lake, New Hampshire, the percentage of pollen present from pines and oaks rose and fell in tandem over the past 14,000 years—just as predicted for a tightly integrated, pine-oak forest community. In contrast, the percentage of pollen from spruce, fir, cedar, beech, and most other tree species changed independently of each other. In general, studies of fossil pollen suggest that the composition of most plant communities has been dynamic and contingent on historical events, rather than static.

The overall message of research on community structure is that Clements's position was too extreme and that Gleason's view may be closer to being correct. ⌀━ Although both biotic interactions and climate are important in determining which species exist at a certain site, chance and history also play a large role.

How Do Keystone Species Structure Communities?

Even though communities are not predictable assemblages dictated by obligatory species interactions, biotic interactions are still important. The presence of certain consumers can have an enormous impact on the species present. In some cases, the structure of an entire community can change dramatically if a single species of predator or herbivore is removed from a community or added to it.

As an example of this research, consider an experiment that Robert Paine conducted in intertidal habitats of the Pacific Northwest of North America. In this environment, the sea star *Pisaster ochraceous* is an important predator (**Figure 53.19a**). Its preferred food is the California mussel *Mytilus californianus*.

When Paine removed *Pisaster* from experimental areas, what had been diverse communities of algae and invertebrates became overgrown with solid stands of mussels (**Figure 53.19b**). *M. californianus* is a dominant competitor, but its populations had been held in check by sea star predation.

(a) Predator: *Pisaster ochraceous*

(b) Prey: *Mytilus californianus*

(c) Effect of keystone predator on species richness

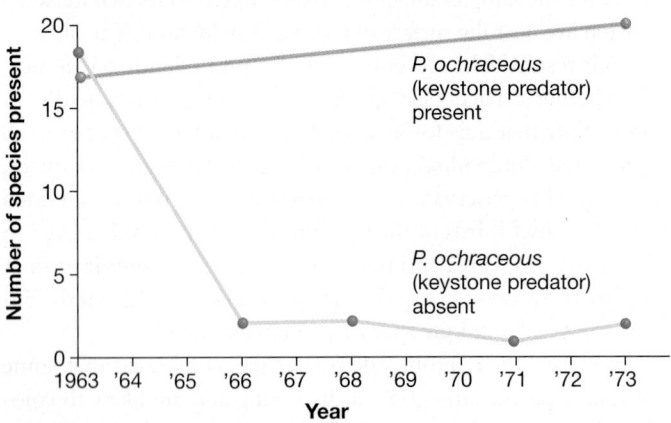

FIGURE 53.19 Keystone Predation Alters Community Structure in a Rocky Intertidal Habitat. Each data point in part (c) represents the average number of species present in several study plots.

The graph in **Figure 53.19c** plots the number of species present in experimental plots where *P. ochraceous* was present or excluded. When the predator was absent, the species richness of the habitat changed radically.

To capture the effect that a predator such as *Pisaster* can have on a community, Paine coined the term "keystone species." A **keystone species** has a much greater impact on the distribution and abundance of the surrounding species than its abundance and total biomass would suggest.

Sea stars and other keystone species vary in abundance in space or over time. As a result, the composition and structure of their communities vary. Communities are ever-changing entities whose composition is difficult to predict.

53.3 Community Dynamics

Once biologists had a basic understanding of how species interact and how communities are structured, they turned to questions about how communities change through time. Like cells, individuals, and species, communities can be described in one word: dynamic.

Disturbance and Change in Ecological Communities

Community composition and structure may change radically in response to changes in abiotic and biotic conditions. Biologists have become particularly interested in how communities respond to disturbance. A **disturbance** is any event that removes biomass from a community. (Recall from Chapter 50 that biomass is the mass of living organisms.)

Forest fires, windstorms, floods, the fall of a large canopy tree, disease epidemics, and short-term explosions in herbivore numbers all qualify as disturbances. These events are important because they alter light levels, nutrients, unoccupied space, or some other aspect of resource availability.

Biologists have come to realize that the impact of disturbance is a function of three factors: (**1**) the type of disturbance, (**2**) its frequency, and (**3**) its severity—for example, the speed and duration of a flood, or the intensity of heat during a fire. In addition, most communities experience a characteristic type of disturbance. In most cases, disturbances occur with a predictable frequency and severity.

To capture this point, biologists refer to a community's **disturbance regime**. For example, fires kill all or most of the existing trees in a boreal forest every 100 to 300 years, on average. In contrast, numerous small-scale tree falls, usually caused by windstorms, occur in temperate and tropical forests every few years.

HOW DO RESEARCHERS DETERMINE A COMMUNITY'S DISTURBANCE REGIME? Ecologists use two strategies to determine a community's natural pattern of disturbance. The first approach is based on obtaining data in a short-term analysis, and extrapolating to predict the longer-term pattern.

How is this done? As an example, an observational study might document that 1 percent of all boreal forest on Earth burns in a given year. Assuming that fires occur randomly, researchers project that any particular piece of boreal forest has a 1 in 100 chance of burning each year. According to this reasoning, fires will recur in that particular area every 100 years, on average.

This extrapolation approach is straightforward to implement, but it has important drawbacks. In boreal forests, for example, fires do not occur randomly in either space or time. They are more likely in some areas than in others, and they tend to occur in particularly dry years. Unless sampling is extensive, it is difficult to avoid errors caused by extrapolating from particularly disturbance-prone or disturbance-free areas or years.

The second approach is based on reconstructing the history of a particular site. For example, researchers can estimate the frequency and impact of storms by finding wind-killed trees and determining their date of death. (Investigators do so by comparing patterns in the growth rings of the dead trees with those of living individuals nearby; see Chapter 36.) The disturbances that have been most extensively studied using historical techniques, however, are forest fires.

Forest fires often leave a layer of burned organic matter and charcoal on the surface of the ground. As a result, researchers can dig a soil pit, find charcoal layers, and use radioisotope dating to establish when the fires occurred. It's also possible to date the death of trees killed by fire by comparing their growth rings with those of living trees. Further, trees that are not killed by fire are often scarred. When a fire burns close enough to kill a patch of cambium tissue, a scar forms. These fire scars occur most often at the tree's base, where dead leaves and twigs accumulate and furnish fuel. Fire frequency can be estimated by counting the number of growth rings that occur between fire scars on the same tree.

WHY IS IT IMPORTANT TO UNDERSTAND DISTURBANCE REGIMES? To appreciate why biologists are so interested in understanding disturbance regimes, consider a recent study on the fire history of giant sequoia groves in California (**Figure 53.20**). Giant sequoias

FIGURE 53.20 Giant Sequoia after a Fire.

grow in small, isolated groves on the western side of the Sierra Nevada range. Individuals live more than a thousand years, and many have been scarred repeatedly by fires.

A biologist obtained samples of cross sections through the bases of 90 giant sequoias in five different groves. As **Figure 53.21a** shows, the cross sections contained numerous rings that had been scarred by fire. To determine the date of each disturbance, the researcher counted tree rings back from the present.

The graph in **Figure 53.21b** plots the average number of fires that occurred each century in the groves, from the year 0 A.D. to 1900. He found that in most of the groves, 20–40 fires had occurred each century. The data indicated that each tree had been burned an average of 64 times.

This study established that fires are extremely frequent in these forests. Because not enough time would pass for large amounts of fuel to accumulate between fires, they were probably of low severity.

Partly because of this work, the biologists responsible for managing sequoia groves now set controlled fires or let low-intensity natural fires burn instead of suppressing them immedi-

(a) Fire scars in the growth rings

(b) Reconstructing history from fire scars

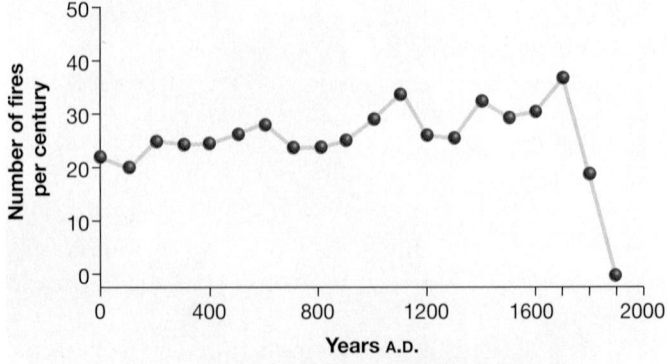

FIGURE 53.21 History of Disturbance in a Fire-Prone Community. Because trees form one ring (light band/dark band) every year, researchers can count the rings to determine how often fires have occurred during the last 2000 years in giant sequoia groves.

ately. To maintain communities in good condition, biologists have to ensure that the normal disturbance regime occurs. Otherwise, community composition changes dramatically.

Similarly, studies of disturbance regimes along the Colorado River in southwestern North America inspired land managers to begin releasing a huge pulse of water from the reservoirs behind dams, at intervals. The floods that have resulted were designed to mimic the natural disturbance regime. According to follow-up studies, the artificial floods appear to have benefited the plant and animal communities downstream by depositing nutrients and creating additional sandbar habitat.

Succession: The Development of Communities after Disturbance

Severe disturbances remove all or most of the organisms from an area. The recovery that follows is called **succession**. The name was inspired by the observation that certain species succeed others over time.

- **Primary succession** occurs when a disturbance removes the soil and its organisms as well as organisms that live above the surface. Glaciers, floods, volcanic eruptions, and landslides often initiate primary succession.

- **Secondary succession** occurs when a disturbance removes some or all of the organisms from an area but leaves the soil intact. Fire and logging are examples of disturbances that initiate secondary succession.

As **Figure 53.22** shows, a sequence of plant communities develops as succession proceeds. Early successional communities are dominated by species that are short lived and small in stature, and that disperse their seeds over long distances. Late successional communities are dominated by species that tend to be long lived, large, and good competitors for resources such as light and nutrients.

The specific sequence of species that appears over time is called a successional pathway. What determines the pattern and rate of species replacement during succession at a particular time and place? To answer this question, biologists focus on three factors:

1. the particular traits of the species involved;

2. how the species interact; and

3. historical and environmental circumstances, such as the size of the area involved and weather conditions.

Let's consider each in turn.

THE ROLE OF SPECIES TRAITS Species traits, such as dispersal capability and the ability to withstand extreme dryness, are particularly important early in succession. As common sense would predict, recently disturbed sites tend to be colonized by plants and animals with good dispersal ability. When these organisms arrive, however, they often have to endure harsh environmental conditions.

Pioneering species

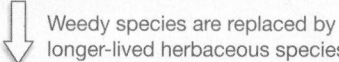
Weedy species are replaced by longer-lived herbaceous species

Early successional community

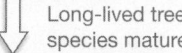
Shrubs and short-lived trees begin to invade

Mid-successional community

Long-lived tree species mature

Climax community

FIGURE 53.22 Succession in Midlatitude Temperate Forests. Succession leads to the development of a temperate forest from a disturbed state (in this case, an abandoned agricultural field).

Pioneering species tend to have "weedy" life histories. A **weed** is a plant that is adapted for growth in disturbed soils. Because soil is often disturbed when biomass is removed, weeds tend to thrive at the start of succession.

Early successional species devote most of their energy to reproduction and little to competitive ability. They are at the "high fecundity, low survivorship" end of the life history continuum introduced in Chapter 52.

More specifically, pioneering species have small seeds, rapid growth, and a short life span, and they begin reproducing at an early age. As a result, they have a high reproductive rate (high r_{max}—see Chapter 52). But in addition, they can tolerate severe abiotic conditions such as high light levels, poor nutrient availability, and desiccation.

THE ROLE OF SPECIES INTERACTIONS Once colonization is under way, the course of succession tends to depend less on how species cope with aspects of the abiotic environment and more on how they interact with other species. This change occurs because plants that grow early in succession change abiotic conditions in a way that makes the conditions less severe.

For example, because plants provide shade, they reduce temperatures and increase humidity. Their dead bodies also add organic material and nutrients to the soil. As abiotic conditions improve, biotic interactions become more important.

During succession, existing species can have one of three effects on subsequent species:

- **Facilitation** takes place when the presence of an early arriving species makes conditions more favorable for the arrival of certain later species, by providing shade or nutrients.

- **Tolerance** means that existing species do not affect the probability that subsequent species will become established.

- **Inhibition** occurs when the presence of one species inhibits the establishment of another. For example, a plant species that requires high light levels to germinate may be inhibited late in succession by the presence of mature trees that prevent sunlight from reaching the forest floor.

THE ROLE OF CHANCE AND HISTORY In addition to species traits and species interactions, the pattern and rate of succession depend on the historical and environmental context in which they occur. For example, the communities that developed after forest fires disturbed Yellowstone National Park in 1988 depended on the size of the burned patch and how hot the fire had been at that location.

Succession is also affected by the particular weather or climate conditions that occur during the process. Variation in weather and climate causes different successional pathways to occur in the same place at different times.

Analyzing species traits, species interactions, and the historical/environmental context provides a useful structure for understanding why particular successional pathways occur. To see this theoretical framework in action, let's examine data on the course of primary succession that has occurred in Glacier Bay, Alaska.

Soils exposed **20** years or less:
willow, *Dryas*

Soils exposed **100** years:
Sitka alder, scattered Sitka spruce

Soils exposed **150–200** years:
dense Sitka spruce, western hemlock

Alaska

Glacier Bay

Upper Bay

Middle Bay

Glacier Bay

Lower Bay

N

10 km

The glacier reached the sea during the late 1700s as shown here, then exposed sediments as it retreated back up the "bay"

FIGURE 53.23 Successional Stages in Glacier Bay.

A CASE HISTORY: GLACIER BAY, ALASKA An extraordinarily rapid and extensive glacial recession is occurring at Glacier Bay (**Figure 53.23**). In just 200 years, glaciers that once filled the bay have retreated approximately 100 km, exposing tracts of barren glacial sediments to colonization. Because of this event, Glacier Bay has become an important site for studying succession.

Figure 53.23 shows the plant communities found in the area.

- Locations that have been ice free for 20 years or less do not have a continuous plant cover. Instead, they host scattered individuals of willow and a small shrub called *Dryas*.

- Areas that have been deglaciated for about 100 years are inhabited by dense thickets of a shrub called Sitka alder and scattered Sitka spruce trees.

- The oldest sites, ranging from 150 to 200 years of soil exposure, are dense forests of Sitka spruce and western hemlock.

These observations inspired a hypothesis for the pattern of succession in Glacier Bay: With time, the youngest communities of *Dryas* and willow succeed to alder thickets, which subsequently become dense spruce-hemlock forests. The key claim is that there is a single successional pathway throughout the bay.

Research has challenged this hypothesis, however. Biologists who reconstructed the history of each community by studying tree rings found that at least two distinct successional pathways have occurred:

1. In the lower part of the bay, soon after the ice retreated, Sitka spruce began growing and quickly formed dense forests. Western hemlock arrived after spruce and is now common.

2. At middle-aged sites in the upper part of the bay, alder thickets were dominant for several decades, and spruce is just beginning to become common. These forests will probably never be as dense as the ones in the lower bay, however, and there is no sign that western hemlock has begun to establish itself.

In contrast, the youngest sites in the uppermost part of the bay may be following a third pathway. Alder thickets became dominant fairly early, but spruce trees are scarce. Instead, cottonwood trees are abundant.

To review these successional patterns, go to the study area at *www.masteringbiology.com*.

(MB) Web Activity Succession

Then consider a question that these data raise: How do species traits, species interactions, and dispersal patterns interact to generate the observed successional pathways?

- *Species traits* Western hemlock is abundant at older sites but largely absent from young ones. This is logical, because its seeds germinate and grow only in soils containing a substantial amount of organic matter. In addition, young hemlock trees can tolerate deep shade but not bright sunlight. Western hemlock does not tolerate early successional conditions, but thrives later in succession.

- *Species interactions* Sitka alder facilitates the growth of Sitka spruce. The facilitating effect occurs because symbiotic bacteria that live inside nodules on the roots of alder convert atmospheric nitrogen (N_2) to nitrogen-containing molecules that alder use to build proteins and nucleic acids. When alder leaves fall and decay or roots die, the nitrogen becomes available to spruce. Although spruce trees are capable of invading and growing without the presence of alder, they grow faster when alder stands have added nitrogen to the soil. Note that spruce gains from the association. Alder is either unaffected, or negatively affected if it becomes shaded by spruce.

Competition is another important species interaction. For example, shading by alder reduces the growth of spruce until spruce trees are tall enough to protrude above the alder thicket. Once the spruce trees breach the alder canopy, however, alder dies out, because it is unable to compete with spruce trees for light.

- *Historical and environmental context*   According to geological evidence, glacial ice was more than 1100 m thick in the upper part of Glacier Bay during the mid-1700s. Because forests grow to an elevation of only 700 m or 800 m in this part of Alaska, the glacier eliminated all of the existing forests. But in the lower part of the bay, the ice was substantially thinner. As a result, some forests remained on the mountain slopes beyond the ice, even at the height of glaciation.

The difference is crucial. In the lower part of the bay, nearby forests provided a ready source of Sitka spruce and western hemlock seed. But in the upper part of the bay, no forests remained to contribute seed. As a result, two dramatically different successional pathways were set in motion.

To summarize, successional pathways are determined by the adaptations that certain species have to their abiotic environment, interactions among species, and the history of the site. Species traits and species interactions tend to make succession predictable; history and chance events contribute a degree of unpredictability to succession.

CHECK YOUR UNDERSTANDING

If you understand that . . .

- Disturbance is a normal part of communities.
- The impact of a disturbance depends on its type, frequency, and severity.
- After a disturbance occurs, a succession of species and communities replaces the individuals that were lost.
- The exact sequence of species observed is a function of their traits, their interactions, and the history of the site.

✓ **You should be able to . . .**

1. Explain why early successional species alter the environment in ways that make growing conditions more difficult for themselves.
2. Explain why the presence or absence of a species like alder, where nitrogen-fixation occurs, might alter the course of a succession.

Answers are available in Appendix B.

53.4 Species Richness in Ecological Communities

The diversity of species present is a key feature of biological communities, and it can be quantified in two ways.

- **Species richness** is a simple count of how many species are present in a given community.
- **Species diversity** is a weighted measure that incorporates a species' relative abundance as well as its presence or absence (see **Box 53.1** on page 1079).

At scales above relatively small study plots, it is rare to have data on relative abundance. As a result, biologists sometimes use species richness and species diversity interchangeably.

To introduce research on richness and diversity, let's focus on a simple question: Why are some communities more species rich than others?

Predicting Species Richness: The Theory of Island Biogeography

When researchers first began counting how many species are present in various areas, a strong pattern emerged: Larger patches of habitat contain more species than do smaller patches of habitat. The observation is logical because large areas should contain more types of niches and thus support higher numbers of species.

Early work on species richness highlighted another pattern, however—one that was harder to explain. Islands in the ocean have smaller numbers of species than do areas of the same size on continents. Why?

As it turned out, answering this question has had far-reaching implications in ecology and conservation biology. This often happens in biological science: Research on a seemingly esoteric question—often done out of the sheer joy of discovery—turns out to have important implications for completely unrelated issues, or even applications to practical problems that had previously been intractable.

THE ROLE OF IMMIGRATION AND EXTINCTION   Robert MacArthur and Edward O. Wilson tackled this question by assuming that speciation occurs so slowly that the number of species present on an island is a product of just two events: immigration and extinction. The rates of both of these processes, they contended, should vary with the number of species present on an island.

- Immigration rates should decline as the number of species on the island increases, because individuals that arrive are more likely to represent a species that is already present. In addition, competition should prevent new species from becoming established when many species are already present on an island—meaning that immigration should seldom result in establishment of a new population.
- Extinction rates should increase as species richness increases, because niche overlap and competition for resources will be more intense.

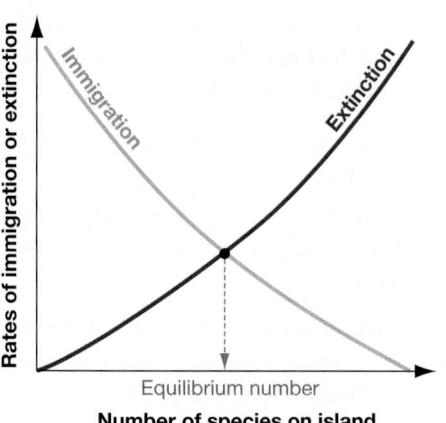

(a) Species richness depends on the number of existing species.

(b) Species richness depends on island size.

(c) Species richness depends on remoteness of the island.

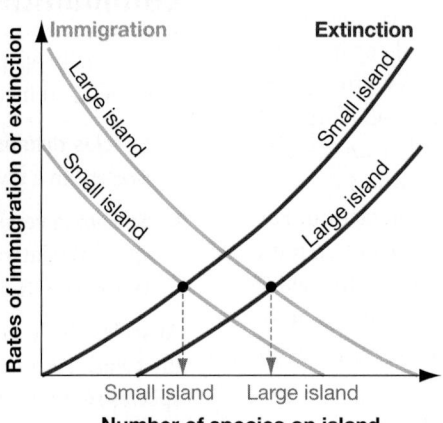

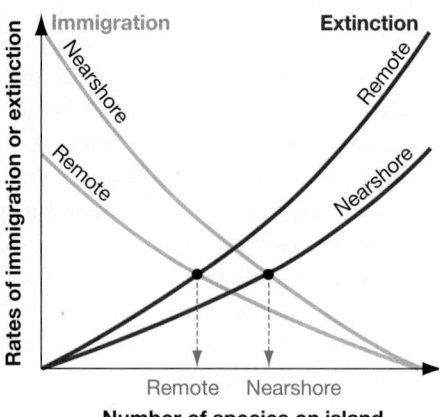

FIGURE 53.24 Species Richness Varies as a Function of Island Characteristics.

✓**QUESTION** Suppose that the mainland habitats closest to an island were wiped out by suburbanization. How would this affect the curves in part (c)?

The graphs in **Figure 53.24a** illustrate these two predictions. Note that when the two processes interact, the result is an equilibrium—a balance between the arrival of new species and the extinction of existing ones. If species richness changes due to a hurricane or fire or some other disturbance, then continued immigration and extinction should restore the equilibrium value.

THE ROLE OF ISLAND SIZE AND ISOLATION MacArthur and Wilson also realized that immigration and extinction rates should vary as a function of island size and how far the island was to a continent or other source of immigrants.

- Immigration rates should be higher on large islands that are close to mainlands, because immigrants are more likely to find large islands that are close to shore than small ones that are far from shore.

- Extinction rates should be highest on small islands that are far from shore, because fewer resources are available to support large populations, and because fewer individuals arrive to keep the population going.

Their theory makes two predictions: Species richness should be higher on (**1**) larger islands than smaller islands (**Figure 53.24b**), and (**2**) nearshore islands versus remote islands (**Figure 53.24c**).

APPLYING THE THEORY The MacArthur-Wilson model, called the theory of island biogeography, is important for several reasons:

- It is relevant to a wide variety of island-like habitats such as alpine meadows, lakes and ponds, and caves. It is also relevant to national parks and other reserves that are surrounded by human development, forming islands of habitat.

- It is relevant to species that have the metapopulation structure introduced in Chapter 52. Immigration and extinction rates in subpopulations should follow the dynamics predicted by the theory.

- It made specific predictions that could be tested. For example, researchers have measured species richness on tiny islands, removed all of the species present, and then measured whether the same number of species recolonized the island and reached an equilibrium number. Predictions about immigration and extinction rates have also been measured by observing the same islands over time.

- It can help inform decisions about the design of natural preserves. In general, the most-species-rich reserves should be ones that are (**1**) relatively large, and (**2**) located close to other relatively large habitat areas.

Global Patterns in Species Richness

Biologists have long understood that large habitat areas tend to be species rich, and the theory of island biogeography has been successful in framing thinking about how species richness should vary among island-like habitats. But researchers have had a much more difficult time explaining what may be the most striking pattern in species richness.

THE LATITUDINAL GRADIENT Biologists who began cataloging the flora and fauna of the tropics in the mid-1800s quickly recognized that these habitats contain many more species than do temperate or subarctic environments. 🔑 Data compiled in the intervening years have confirmed the existence of a strong latitudinal gradient in species diversity—for communities as a whole as well as for many taxonomic groups.

In birds, mammals, fish, reptiles, many aquatic and terrestrial invertebrates, orchids, and trees, species diversity declines as latitude increases. **Figure 53.26** graphs data on the number of vascular plant species present from the equator toward the poles. Although this pattern is not universal—a number of marine groups, as well as shorebirds, show a *positive* relationship between latitude and diversity—it is widespread. Why does it occur?

BOX 53.1   QUANTITATIVE METHODS: **Measuring Species Diversity**

To measure species diversity, a biologist could simply count the number of species present in a community. The problem is that simple counts provide an incomplete picture of diversity. The relative abundance of species is also an important component of diversity. To grasp this point, consider the composition of the three hypothetical communities shown in **Figure 53.25**. These communities are nearly identical in species richness but differ greatly in the relative abundance of each species.

These data can be used to compare two measures of species diversity. Species richness is simply the number of species found in a community. In this case, communities 1 and 2 have equal species richness and community 3 is lower in richness by one species. It is important to note, however, that communities 2 and 3 have similar relative abundances of each species, or what biologists call high *evenness*. Community 1, in contrast, is highly uneven. Fifty-five percent of the individuals in community 1 belong to species A, and other species are relatively rare. An uneven community has lower effective diversity than its species richness would indicate.

To take evenness into account, other diversity indices have been developed. A simple example is the Shannon index, given by the following equation:

$$H' = -\sum_{i=1}^{n} p_i \ln p_i$$

In this equation, p_i is the proportion of individuals in the community that belong to

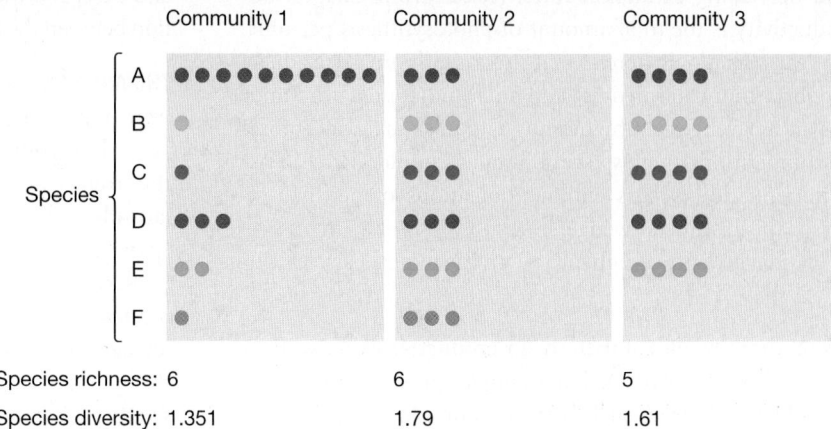

Species richness: 6   6   5
Species diversity: 1.351   1.79   1.61

FIGURE 53.25 Species Diversity Can Be Quantified.

species *i*. The index is summed over all the species in the study.

The Shannon index of species diversity for the three hypothetical communities shown in Figure 53.25 was calculated by **(1)** computing the proportion of individuals in each community that belong to each species, **(2)** taking the natural logarithm of each of these proportions (see **BioSkills 7** in Appendix A for an introduction to logarithms), **(3)** multiplying each natural logarithm times the proportion for each species, and **(4)** summing the total across the six species in the community. ✔**If you understand the equation, you should be able to do these calculations and get the same result given in Figure 53.25.** Notice that while communities 1 and 2 have the same species richness, community 2 has higher diversity because of its greater evenness.

Community 3 has lower species richness than community 1 but higher diversity.

✔**If you understand how to use and interpret the Shannon index, you should be able to calculate species richness and the Shannon diversity index for 3 communities that "double" Communities 1–3 in Figure 53.25. For example, one of the three new communities should be identical to Community 1, except that there are 2 species with 10 individuals present like species "A", 2 species with 1 individual present like species "B", 2 species with 1 individual present like species "C", 2 species with 3 individuals present like species "D", and so on. There are a total of 12 species in two of the communities, and 10 in the third. ✔You should also be able to compare and contrast your results with the richness and diversity values given in Figure 53.25.**

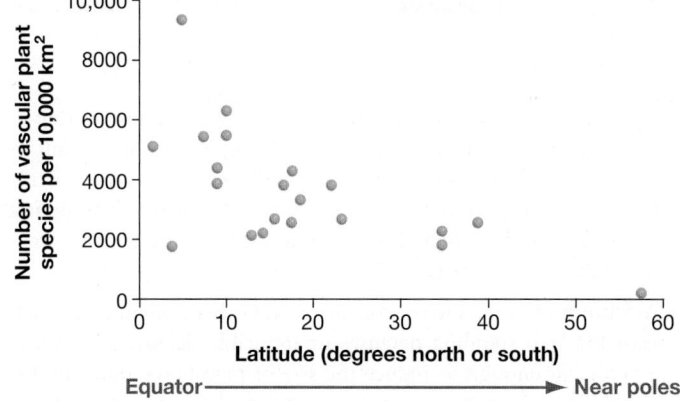

FIGURE 53.26 There Is a Strong Latitudinal Gradient in Species Richness of Vascular Plants.

To explain why species diversity might decline with increasing latitude, biologists have to consider two fundamental principles:

1. The causal mechanism must be abiotic, because latitude is a physical phenomenon produced by Earth's shape. The explanation must be a physical factor that varies predictably with latitude and that could produce changes in species diversity.

2. The species diversity of a particular area is the sum of four processes: speciation, extinction, immigration (colonization), and emigration (dispersal). Thus, the latitudinal gradient must be caused by an abiotic factor that affects the rate of speciation, extinction, immigration, or emigration in a way that would lead to more species in the tropics and fewer near the poles.

Over 30 hypotheses have been proposed to explain the latitudinal gradient. Let's consider four of the most prominent ideas.

THE HIGH-PRODUCTIVITY HYPOTHESIS High productivity in the tropics could promote high diversity by increasing speciation rates and decreasing extinction rates. (Recall from Chapter 50 that productivity is the total amount of photosynthesis per unit area per year.) The logic here is that increased biomass production supports more herbivores, and thus more predators and parasites and scavengers. If higher population sizes lead to more intense competition, then speciation rates should increase as niche differentiation occurs within populations of herbivores, predators, parasites, and scavengers.

The evidence on the high-productivity hypothesis is still inconclusive.

- It is supported by global patterns in productivity that will be introduced in Chapter 54. For example, productivity is extremely high in biomes such as tropical rain forests and coral reefs, which have high species diversity.

- It is not supported by experimental addition of fertilizer to aquatic or terrestrial communities. In most of these studies, productivity increases but species diversity decreases. Critics contend that the time and spatial scale of these experiments is too small to be relevant to global patterns, however.

- It is not supported in some highly productive habitats, such as estuaries (see Chapter 50), where species diversity is often extremely low.

THE ENERGY HYPOTHESIS The energy hypothesis is an extension of the high-productivity hypothesis. Recognizing that some highly productive habitats are species-poor, investigators proposed that the key factor can't be productivity alone.

The energy hypothesis claims that high temperatures increase species diversity by increasing productivity *and* the likelihood that organisms can tolerate the physical conditions in a region. The idea is that high energy inputs—rather than high nutrient inputs—are the key to high speciation rates and low extinction rates.

Although the energy hypothesis is relatively new, some data support it. For example, biologists who analyzed data on gastropods and other marine invertebrates have documented a strong correlation between the temperature of marine waters and species diversity.

THE AREA AND AGE HYPOTHESIS Temperate and arctic latitudes were repeatedly scoured by ice sheets over the last 2 million years, but tropical regions were not. In addition, the land area of the tropics has historically been larger than the land area at high latitudes. Thus, tropical regions have had more time and more space for speciation to occur than other regions have.

Recent data suggest, however, that tropical forests were dramatically reduced in size by widespread drying trends during the ice ages. Existing forests may be much younger than originally thought. If so, then the contrast in the age of northern and southern habitats may not be enough to explain the dramatic difference in species diversity.

THE INTERMEDIATE DISTURBANCE HYPOTHESIS Across habitats and biomes, species diversity is much higher in mid-successional communities than in pioneer or mature communities. The **intermediate disturbance hypothesis** was inspired by this observation. It holds that regions with a moderate type, frequency, and severity of disturbance should have high species richness and diversity.

The logic here is that, with intermediate levels of disturbance, communities will contain pioneering species as well as species better adapted to late-successional conditions. For example, recent studies have confirmed that tree falls and canopy gaps occur regularly in tropical forests, and that fires occur in these biomes occasionally. As yet, however, there are no convincing data showing that intermediate levels of disturbance are more likely to occur in the tropics than they are at higher latitudes.

What can you conclude from all this? The current consensus is that each of the factors discussed here may influence diversity. It is tempting to look for single causes to explain important patterns and events. But in this case, multiple causation is probably more realistic.

CHAPTER 53 REVIEW

For media, go to the study area at www.masteringbiology.com

Summary of Key Concepts

🔑 **Interactions among species, such as competition, consumption, and mutalism have two main outcomes: (1) They affect the distribution and abundance of the interacting species, and (2) they are agents of natural selection and thus affect the evolution of the interacting species. The nature of interactions between species frequently changes over time.**

- To categorize the different types of interactions that occur among species, biologists consider whether each participant experiences a net fitness cost or benefit from the interaction. These costs and benefits depend on the conditions that prevail at a particular time and place and may change through time.

- Competition occurs when the niches of two species overlap—meaning they use the same resources. Competition may result in the complete exclusion of one species (the weaker competitor). It may also result in niche differentiation, in which competing species evolve traits that allow them to exploit different resources or live in different areas.

- Consumption occurs when consumers eat other individuals, which resist through standing defenses or inducible defenses. Predators are efficient enough to reduce the size of many prey populations. Levels of herbivory are relatively low in terrestrial ecosystems, however, because predation and disease limit herbivore populations, because plants provide little nitrogen, and because many plants contain toxic compounds or other types of defenses.

- Parasites are consumers that generally spend all or part of their life cycle in or on their host (or hosts). They usually have traits that allow them to escape host defenses and even manipulate host behavior to increase their likelihood of transmission to a new host. In turn, hosts have evolved counter-adaptations that help fight off parasites.

- Mutualism provides participating individuals with food, shelter, transport of gametes, or defense against predators. For each species involved, the costs and benefits of mutualism may vary over time and from place to place.

✔ You should be able to give an example of how competition can evolve into commensalism, and how a mutualistic relationship can evolve into a parasitic one.

(MB) **Web Activity** Life Cycle of a Malaria Parasite

🔑 **The assemblage of species found in a biological community changes over time and is primarily a function of climate and chance historical events.**

- Historical and experimental evidence support the view that communities are dynamic rather than static, and that their composition is neither entirely predictable nor stable over time.

- Each community has a characteristic disturbance regime—meaning a type, severity, and frequency of disturbance that it experiences.

- Three types of factors influence the pattern of succession that occurs after a disturbance: (1) A species' physiological traits influence when it can successfully join a community; (2) interactions among species influence if and when a species can become established; and (3) the historical and environmental context of the site affects which species are present.

✔ You should be able to explain why climate makes the three successional pathways documented in Glacier Bay similar, and why chance historical events make them different.

(MB) **Web Activity** Succession

🔑 **Species richness is higher in large islands near continents than in small, isolated islands, due to differences in immigration and extinction. Species richness is also higher in the tropics and lower toward the poles, but the mechanism responsible for this pattern is still controversial.**

- Large islands that are close to continents have high immigration rates due to proximity and space available for new species, and low extinction rates. Isolated islands have low immigration rates, and small islands have high extinction rates due to small amounts of available habitat and low population sizes.

- The high species richness observed in the tropics results from a combination of factors—the most important of which may be high temperatures that increase productivity and provide relatively benign abiotic conditions.

✔ You should be able to suggest a hypothesis to explain an exception to the latitudinal gradient rule—specifically, that most species of shorebirds breed at high latitudes, not the tropics. (Note that many shorebird species nest near the abundant lakes found in arctic tundra.)

Questions

1. What is competitive exclusion?
   a. the evolution of traits that reduce niche overlap and competition
   b. interactions that allow species to occupy their fundamental niche
   c. the degree to which the niches of two species overlap
   d. the claim that species with the same niche cannot coexist

2. What is niche differentiation?
   a. the evolution of traits that reduce niche overlap and competition
   b. interactions that allow species to occupy their fundamental niche
   c. the degree to which the niches of two species overlap
   d. the claim that species with the same niche cannot coexist

3. Why is the phrase "coevolutionary arms race" an appropriate way to characterize the long-term effects of species interactions?
   a. Both plants and animals have evolved weapons for defense that are so effective that many plants are not eaten and predators cannot reduce prey populations to extinction.
   b. Adaptations that give one species a fitness advantage in an interaction are likely to be countered by adaptations in the other species that eliminate this advantage.
   c. In all species interactions except for mutualism, at least one species loses (suffers decreased fitness).
   d. Even mutualistic interactions can become parasitic if conditions change. As a result, interacting species are always "at war."

4. Why are inducible defenses advantageous?
   a. They are always present; thus, an individual is always able to defend itself.
   b. They make it impossible for a consumer to launch surprise attacks.
   c. They result from a coevolutionary arms race.
   d. They make efficient use of resources, because they are produced only when needed.

5. Which of the following is *not* correlated with species diversity?
   a. latitude
   b. productivity
   c. longitude
   d. island size

6. What is net primary productivity?
   a. an individual's lifetime reproductive success (lifetime fitness)
   b. an individual's average annual reproductive success
   c. the total amount of photosynthesis that occurs in an area of a given size per year
   d. the amount of energy that is stored in standing biomass per year

1. The text claims that species interactions are conditional and dynamic. Do you agree with this statement? Why or why not? Cite specific examples to support your answer.

2. State two hypotheses that have been proposed to explain the low level of herbivory in terrestrial plant communities. Are these hypotheses mutually exclusive? (In other words, can both be correct?) Explain why or why not.

3. Biologists have tested the hypotheses that communities are highly predictable versus highly unpredictable. State the predictions that these hypotheses make with respect to (a) changes in the distribution of the species in a particular community over time, and (b) the communities that should develop at sites where abiotic conditions are identical. Which hypothesis appears to be more accurate?

4. What is a disturbance? Consider the role of fire in a forest. Compare and contrast the consequences of high-frequency versus low-frequency fire, and high versus low severity of fire.

5. Summarize the life-history attributes of early successional species. Why are these attributes considered adaptations?

6. Explain why high productivity should lead to increased species richness in habitats such as tropical rain forests and coral reefs.

1. Some insects harvest nectar by chewing through the wall of the structure that holds the nectar. As a result, they obtain a nectar reward, but pollination does not occur. Suppose that you observed a certain bee species obtaining nectar in this way from a particular orchid species. Over time, how might you expect the characteristics of the orchid population to change in response to this bee behavior?

2. Using this chapter's information on fire regimes in giant sequoia groves, propose a management plan for Sequoia National Park. Explain the logic behind your plan.

3. Suppose that a two-acre lawn on your college's campus is allowed to undergo succession. Describe how species traits, species interactions, and the site's history might affect the community that develops.

4. Design an experiment to test the hypothesis that increasing species richness increases a community's productivity.

Glaciers all over the world are melting at unprecedented rates, raising sea levels. This chapter explores how humans are altering ecosystems all over the planet.

Ecosystems 54

The data in this chapter carry a simple, but important, message: Some of the most urgent problems facing your generation are rooted in ecosystem ecology. Global warming, acid rain, a hole in the atmosphere's ozone layer, estrogen-mimicking pollutants, and nitrate pollution (see Chapter 28) are just a few of the issues that ecosystem ecologists are documenting.

The chapter has a simple theme, as well: In nature, everything is connected. This is because the components of an **ecosystem**—the species present, along with abiotic components such as the soil, climate, water, and atmosphere—are linked by flows of energy and nutrients. When you change the species that are present or the abiotic environment, you change the ecosystem.

Humans are removing species and adding massive amounts of energy and nutrients to ecosystems all over the planet. Chapter 55 considers the role of biodiversity and the consequences of species loss; the focus here is on how people are changing climate, water, and other abiotic aspects of ecosystems.

To understand the consequences of altering the abiotic environment, let's consider (1) how energy moves among the components of an ecosystem, (2) how nutrients move among the components of an ecosystem, and (3) why global warming is occurring.

54.1 How Does Energy Flow through Ecosystems?

A **primary producer** is an **autotroph** (literally, "self-feeder")—an organism that can synthesize its own food from inorganic sources. Primary producers are the conduit where energy enters ecosystems.

In most ecosystems, primary producers use solar energy to manufacture food via photosynthesis. But in deep-sea hydrothermal vents and iron-rich rocks deep below Earth's surface, primary producers use the chemical energy contained in inorganic compounds

KEY CONCEPTS

- An ecosystem has four components: (1) the abiotic environment, (2) primary producers, (3) consumers, and (4) decomposers. These components are linked by the movement of energy and nutrients.

- As energy flows from producers to consumers and decomposers in a food web, much of it is lost. The productivity of terrestrial ecosystems is limited by warmth and moisture; nutrient availability is the key constraint in aquatic ecosystems.

- To analyze nutrient cycles, biologists focus on the nature of the reservoirs where elements reside and the processes that move elements between reservoirs. Nutrient addition by humans is increasing productivity and causing pollution.

- The burning of fossil fuels has led to rapid global warming; rapid ecological and evolutionary changes are being observed in response.

✔ When you see this checkmark, stop and test yourself. Answers are available in Appendix B.

such as hydrogen (H_2), methane (CH_4), or hydrogen sulfide (H_2S) to make food (see Chapter 28).

Primary producers form the basis of ecosystems because they transform the energy in sunlight or inorganic compounds into the chemical energy stored in sugars. The primary producers use this chemical energy in two ways:

- *Maintenance* Organisms have to use chemical energy to re-place worn-out cell components, transport material inside cells and between cells, synthesize hormones and other chemical signals, and produce molecules and structures that function in defense against disease and attack. It takes energy just to stay alive.

- *Growth* Chemical energy that isn't required for maintenance can be used for growth and reproduction. Energy that is invested in new tissue or offspring is called **net primary productivity (NPP)**.

In addition, the use of chemical energy is never 100 percent efficient—some energy is inevitably lost as heat.

Why Is NPP So Important?

NPP results in **biomass**—organic material that non-photosynthetic organisms can eat. NPP is critically important because it represents the amount of energy that is available to the other living components of an ecosystem: consumers and decomposers.

- **Consumers** eat living organisms. **Primary consumers** eat primary producers; **secondary consumers** eat primary consumers; **tertiary consumers** eat secondary consumers, and so on. You are a primary consumer when you eat a salad, a secondary consumer when you eat beef, and a tertiary consumer when you eat tuna.

- **Decomposers**, or **detritivores**, obtain energy by feeding on the remains of other organisms or waste products. Dead animals and dead plant tissues ("plant litter") are collectively known as **detritus**. Many fungi and bacteria are decomposers, but humans rarely act as detritivores.

Figure 54.1 illustrates how energy moves around the four components of an ecosystem. You could draw a parallel diagram showing how nitrogen, phosphorus, and other nutrients move among ecosystem components, with soil or water acting as the source of the key atoms. 🔑➡ The four components of an ecosystem are linked by the movement of energy and nutrients.

Solar Power: Transforming Incoming Energy to Biomass

Figure 54.1 has an arrow from the Sun to a leaf. How big is this arrow, that is, how much of the energy that enters an ecosystem gets converted into biomass? To answer this question, let's look at data from the world's most intensively studied ecosystem.

Since the early 1960s, a team of researchers has been studying how energy and nutrients flow through a temperate forest ecosystem in the northeastern United States: the Hubbard Brook Experimental Forest in New Hampshire.

From June 1, 1969, to May 31, 1970, 1,254,000 kcal of solar radiation per square meter (m^2) entered this ecosystem in the form

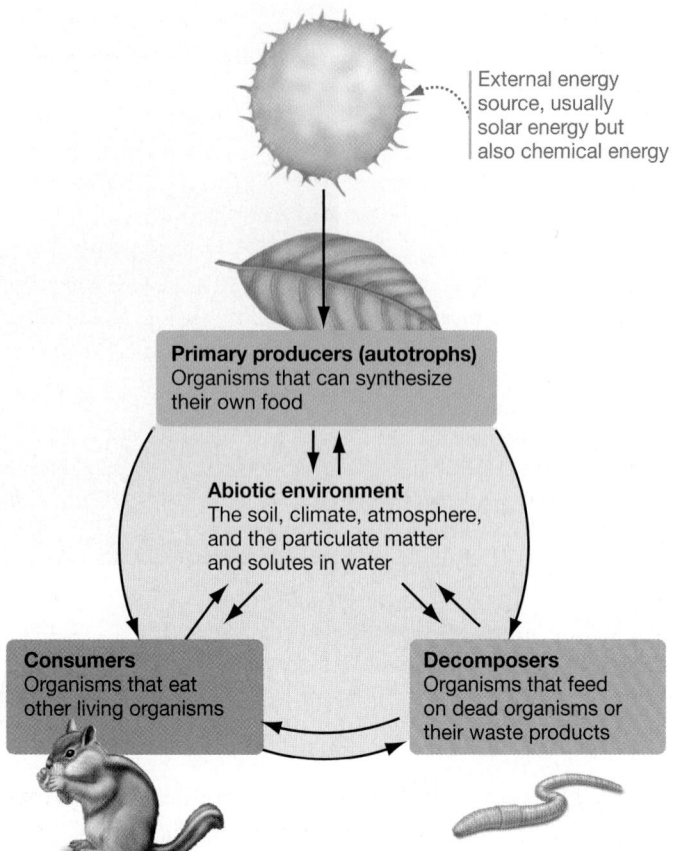

FIGURE 54.1 The Four Components of an Ecosystem Interact. The arrows represent energy. A similar diagram could be drawn to represent the flow of nutrients.

of sunlight. If this amount of energy were available in the form of electricity, it would easily power two 150-watt lightbulbs that burned continuously for one year.

By documenting rates of photosynthesis in a large sample of forest plants, the biologists calculated that the plants used 10,400 kcal/m^2 of this energy in photosynthesis. This value represents **gross primary productivity**: the total amount of photosynthesis in a given area and time period.

The team also calculated **gross photosynthetic efficiency**—the ratio of gross primary productivity to the total incoming solar radiation. At Hubbard Brook, efficiency was 10,400 kcal/m^2 ÷ 1,254,000 kcal/m^2 = 0.008 or 0.8 percent. This value is typical of other ecosystems as well.

Why is efficiency so low? The answer has several components.

- Plants in temperate biomes have drastically reduced photosynthetic rates in winter, although sunlight is available.

- If conditions get dry during the summer, stomata close to conserve water. Photosynthesis slows dramatically due to a lack of CO_2.

- Even when conditions are ideal, the pigments that drive photosynthesis can absorb only a fraction of the wavelengths available and thus a fraction of the total energy received.

Although overall efficiency is low, a great deal of photosynthesis is still occurring. What happened to all of the products? After calculating photosynthetic rates in an array of plants, the researchers also measured growth—the production of new biomass. The data allowed them to estimate that of gross primary production, only about 45 percent went to NPP. The other 55 percent was used for maintenance activities or was lost as heat.

Biologists have done similar studies in other ecosystems. The general pattern is that only a tiny fraction of incoming sunlight is converted to chemical energy, and only a fraction of this gross primary productivity is used to build biomass. Now let's consider the next question: What happens to NPP?

Trophic Structure

Biomass represents chemical energy. Every time a monarch caterpillar eats a leaf or a fungus absorbs molecules from decaying wood, chemical energy flows from a primary producer to a primary consumer or decomposer. To describe these energy flows, biologists identify distinct feeding levels in an ecosystem. Organisms that obtain their energy from the same type of source occupy the same **trophic** ("feeding") **level**.

FOOD CHAINS AND FOOD WEBS A **food chain** connects the trophic levels in a particular ecosystem. In doing so, it describes how energy moves from one trophic level to another. **Figure 54.2**

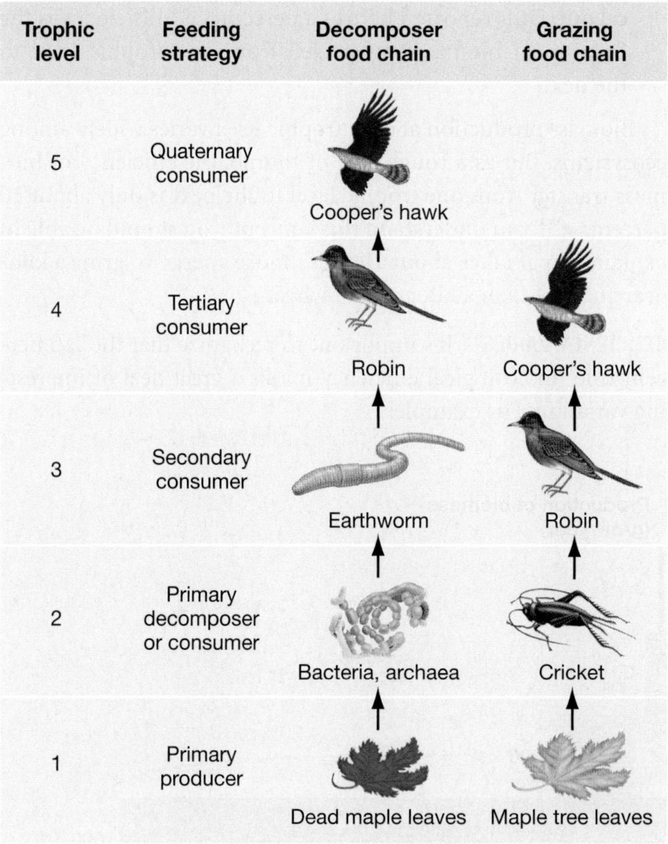

Trophic level	Feeding strategy	Decomposer food chain	Grazing food chain
5	Quaternary consumer	Cooper's hawk	
4	Tertiary consumer	Robin	Cooper's hawk
3	Secondary consumer	Earthworm	Robin
2	Primary decomposer or consumer	Bacteria, archaea	Cricket
1	Primary producer	Dead maple leaves	Maple tree leaves

FIGURE 54.2 Trophic Levels Identify Steps in Energy Transfer. Many other species exist at each level in this temperate-forest ecosystem.

shows five trophic levels in a **decomposer food chain**—with **primary decomposers**, which feed on plant detritus, at the second trophic level—and four trophic levels in a **grazing food chain**, with primary consumers at the second trophic level.

Note several key points:

- At higher trophic levels, the grazing and decomposer food chains often merge. For example, the robin in the figure functions as a secondary consumer in the grazing food chain and a tertiary consumer in the decomposer food chain.

- It is common for consumers—such as humans and robins—to feed at multiple trophic levels.

Because most consumers feed on a wide array of organisms and at multiple trophic levels, food chains are embedded in more complex **food webs**. Food webs attempt to include many or most of the species eaten by a sample of organisms in an ecosystem, joined by arrows indicating a flow of energy. For example:

- The Cooper's hawk in Figure 54.2 feeds on a wide array of small birds and small mammals, in addition to robins.

- The robin in the figure eats a wide array of insects, wormlike species, and spiders, as well as fruits and seeds (primary producers).

A food web would include these organisms, along with estimates of the amount of energy transferred at each link. Food webs are a compact way of summarizing energy flows and documenting the complex trophic interactions that occur in ecosystems.

ENERGY FLOW TO GRAZERS VERSUS DECOMPOSERS How much biomass gets eaten dead versus alive? More formally, how much energy is consumed by primary consumers versus primary decomposers?

The answer to this question varies enormously among habitats. At Hubbard Brook, for example, only about 24 percent of NPP is eaten by consumers—76 percent is uneaten until the primary producer dies. These values are actually unusual for forest ecosystems, where the percentage of NPP consumed can be as low as 5–7 percent. In a forest, much of the biomass is tied up in indigestible wood and isn't transferred to other organisms until it decays. The percentage of NPP that gets consumed in marine habitats, where most of the primary production is done by algae, is much higher—often 35–40 percent.

This is an important point, for two reasons:

1. If you want to understand energy flow in a forest, you have to understand decomposers. Decomposition is less of a factor in marine ecosystems, where primary consumers are paramount.

2. The decomposer food chain is "leaky." At Hubbard Brook, for example, large amounts of energy leave the forest ecosystem in the form of detritus that washes into streams. In these streams, photosynthesis by aquatic algae and plants introduced only about 10 kcal/m^2/yr, versus 6039 kcal/m^2/yr that washed in from the surrounding forest. In marine ecosystems, energy leaves in the form of dead bodies that rain down to the ocean floor.

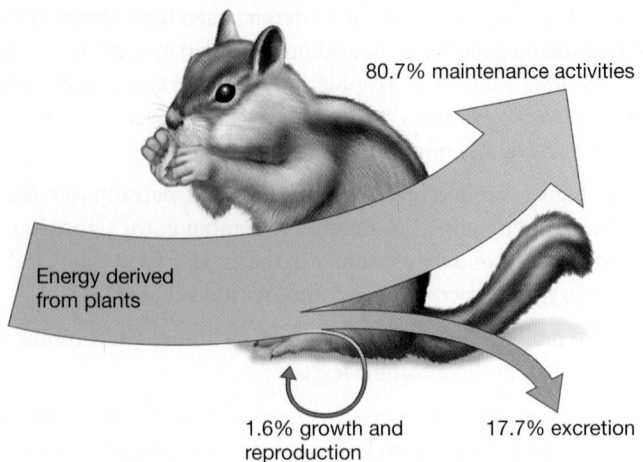

80.7% maintenance activities

Energy derived
from plants

1.6% growth and
reproduction

17.7% excretion

FIGURE 54.3 How Do Consumers Use Primary Production?

✔**QUESTION** Aphids, small plant-eating insects, are ectothermic (acquire heat from outside sources), largely sedentary, and feed on high-quality foods containing almost no indigestible molecules. Describe how the percentages given for chipmunks here would compare in aphids.

The "dead versus alive" percentage can even vary with conditions. In a recent experiment done in small ponds, for example, researchers found that much more NPP got consumed when light levels were low and nutrient levels were high. Under these conditions, algae flourished and were readily eaten by grazing animals. But when light levels were high and nutrient levels were low, most primary production was done by cyanobacteria, which many grazers avoided. Most of the cyanobacteria died without being eaten; their biomass entered the decomposer food chain. In an ecosystem like this, the dynamics of energy flow could change season to season or even week to week.

Energy Transfer between Trophic Levels

All ecosystems share a characteristic pattern: The total biomass produced each year declines from lower trophic levels up to the higher levels. To understand why, let's go back to the Hubbard Brook experimental forest.

Figure 54.3 illustrates what happened to the energy obtained by consumers, using chipmunks as an example. Chipmunks are small rodents that are seed predators. On average, these primary consumers harvest 31 kcal/m² of energy each year at Hubbard Brook. Of that total, 17.7 percent is unused—meaning that it is not digested and absorbed. Instead, it is excreted. In addition, 80.7 percent of total energy consumed is used for maintenance. Just 1.6 percent of the yearly 31 kcal/m² goes into the production of new chipmunk tissue by growth and reproduction.

The same pattern would be seen in a secondary consumer—a Cooper's hawk, for example, that eats chipmunks and other primary consumers. Consider the energy that a hawk has to expend to catch its prey. This energy cannot be used to make more hawk biomass.

The general point is simple: Because only a fraction of the total energy consumed is used for growth and reproduction, the amount of biomass produced at the second trophic level must be less than the amount at the first trophic level; productivity at the third trophic level must be less than that at the second.

THE PYRAMID OF PRODUCTIVITY The data in **Figure 54.4** graphs the biomass produced at trophic levels 1–4 in a temperate deciduous forest, with a separate horizontal bar representing each trophic level. Note that the graph reports two numbers:

1. Productivity is a rate, measured here in units of grams of biomass produced per square meter of area each year (g/m²/year).

2. Efficiency is a ratio and thus dimensionless—the units cancel out. (It is reported here as a percentage.) Efficiency is the fraction of biomass transferred from one trophic level to the next.

Biomass production at each trophic level varies widely among ecosystems. But as a rough rule of thumb, the efficiency of biomass transfer from one trophic level to the next is only about 10 percent. ✔If you understand this concept, you should be able to explain why it takes about 10 times more energy to grow a kilogram of beef than a kilogram of wheat.

EFFICIENCY VARIES It's important to recognize that the "10 percent rule" in ecological efficiency masks a great deal of interesting variation. For example:

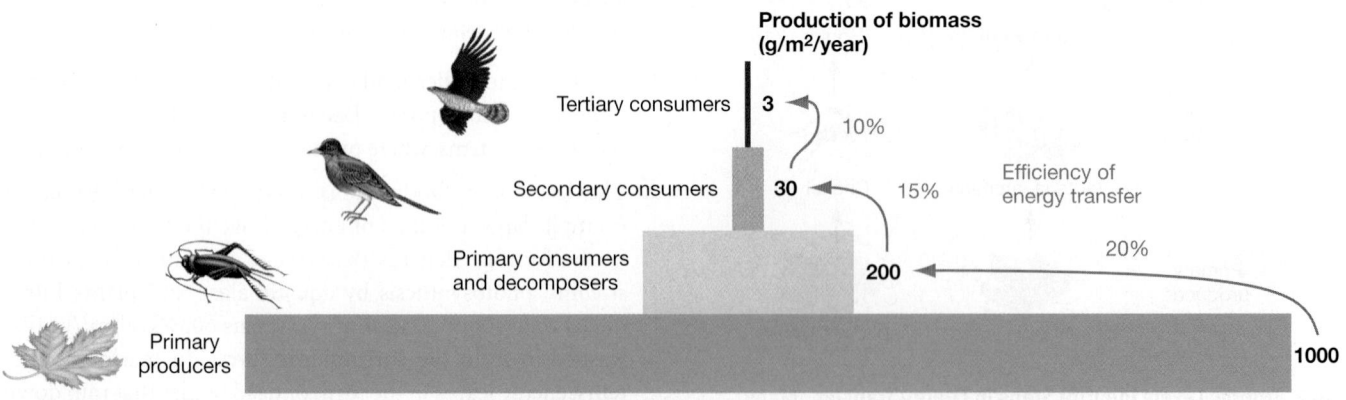

Production of biomass
(g/m²/year)

Tertiary consumers — 3

Secondary consumers — 30 — 10%

Primary consumers
and decomposers — 200 — 15%

Efficiency of
energy transfer

20%

Primary
producers — 1000

FIGURE 54.4 Productivity Declines at Higher Trophic Levels. This pattern is called the pyramid of productivity.

- Large mammals are more efficient at producing biomass than small mammals, because they have a smaller surface-area-to-volume ratio and lose less heat (see Chapter 41).

- Compared to endotherms, biomass production is much more efficient in ectotherms. You might recall from Chapter 41 that ectotherms rely principally on heat gained from the environment and do not oxidize sugars to keep warm. As a result, they devote much less energy to maintenance than endotherms do. Chipmunks devote just 1.5 percent of their energy intake to growth; caterpillars and other ectotherms may devote 20–40 percent.

✔If you understand this concept, you should be able to explain why it may be more efficient to eat crustaceans and fish than beef, if the animals you are consuming feed at the same trophic level.

Trophic Cascades and Top-Down Control

What happens when a component of a food chain or food web changes? You were introduced to this question briefly in Chapter 53, when you reviewed the keystone species concept. You might recall that the sea star *Pisaster ochraceous* primarily eats mussels. If *Pisaster* is excluded from a habitat, mussels take over—dramatically changing which species are present.

When a consumer (*Pisaster*) limits a prey population (mussels) in this way, biologists say that **top-down control** is occurring (see Chapter 53). The name was inspired by the relationships diagrammed in a food chain. As a secondary consumer, *Pisaster* is "on top" of mussels, which are primary consumers.

When changes in top-down control cause conspicuous effects two or three links away in a food web, biologists say that a **trophic cascade** has occurred. Wolves, for example, were reintroduced to the Greater Yellowstone Ecosystem of western North America in 1995. Wolves are a "top predator"—meaning that nothing eats them.

Although the Yellowstone wolf population has grown to only about 150 individuals, their presence has led to far-reaching changes in the food web (**Figure 54.5**).

- In Yellowstone, wolves primarily feed on elk. As elk numbers declined and individuals changed their feeding habits due to fear of being in the open, populations of favorite elk foods like aspen, cottonwood, and willow have increased.

- The changes in willow and other species triggered an increase in beavers, which compete with elk for these plants.

- Beavers dam streams, forming ponds that provide habitat for frogs, turtles, and an array of other species, whose numbers have increased.

- Wolves do not tolerate coyotes. As coyotes declined, mouse populations increased. The increase in mouse populations led to larger populations of hawks that prey on mice.

✔If you understand the concept of a trophic cascade, you should be able to change the relative size of the arrows in Figure 54.5 to indicate whether the energy transferred in each link increased or decreased as the wolf population grew to its present size.

Analogous changes have been observed in ecosystems where humans have *removed* species. Atlantic cod and other large predators, for example, have virtually disappeared from large parts of the Atlantic Ocean due to overfishing. Shrimp and crab—species that cod used to eat—have increased in response; the primary producers that shrimp and crab eat have decreased.

The message from these studies is clear: Changing the composition of food webs can have profound effects. The underlying

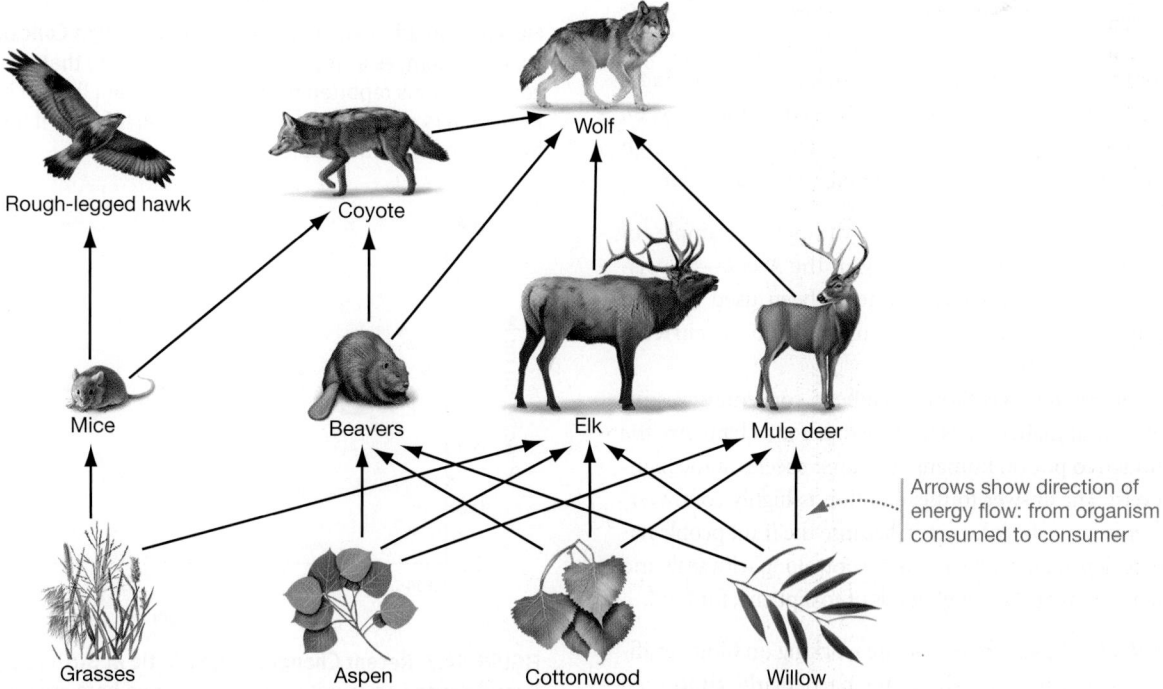

FIGURE 54.5 Reintroduction of Wolves Has Had a Dramatic Effect on the Food Web in Yellowstone. This food web shows only a few of the hundreds of species affected by the reintroduction of wolves.

cause is energy flow—one of the factors that links components in an ecosystem. It's an important issue because humans are removing large predators from a wide array of ecosystems, through hunting or habitat destruction.

Biomagnification

Before leaving the topic of trophic interactions, let's take a brief look at the fate of certain molecules that are transferred through a food web along with chemical energy. As an example, consider the insecticide toxaphene.

Toxaphene is a cocktail of structurally similar organic compounds that include chlorine atoms. For decades toxaphene was widely used as an insecticide in cotton and soybean fields, until data on its toxicity to humans and wildlife led to it being banned in the U.S. in 1986. Although toxaphene is now illegal in most countries, it continues to be used in some parts of the world.

WHAT IS A POP? Toxaphene tends to persist in the environment without breaking down into harmless substances. It is a POP—a persistent organic pollutant. Toxaphene and other POPs undergo **biomagnification**—meaning that they increase in concentration at higher levels in a food chain.

Biomagnification begins when persistent atoms or molecules are taken up from air or water by primary producers and passed on to primary, secondary, and tertiary consumers. In many cases, POPs or other substances are sequestered in certain parts of a consumer's body instead of being excreted. Toxaphene, for example, tends to become localized in fat cells because it is lipid-soluble. In vertebrates it is also sequestered in the liver—the organ responsible for processing toxins.

Remember that organisms at each trophic level eat about 10 g of tissue to make 1 g of their own tissue. Even if toxaphene is present at extremely low concentration in primary producers, it is sequestered at much higher concentration in primary consumers—simply because they take in so much primary producer tissue and retain the toxaphene. The effect is magnified at each succeeding trophic level. In many cases, concentrations get high enough in secondary and tertiary consumers to poison them.

TOXAPHENE IN THE ARCTIC The most dramatic example of toxaphene biomagnification is occurring in the Arctic—a "pristine" ecosystem where toxaphene has never been used directly. Instead, the molecules blow into the ecosystem on air currents introduced in Chapter 50.

As the data in **Figure 54.6** show, toxaphene concentrations increase dramatically at higher levels in the Arctic food chain. Are the levels high enough to poison humans and other species at the highest trophic levels? The answer to this question is highly controversial. The answer is important, however, because the Inuit people native to the Arctic depend on fish such as burbot, along with seals and other mammals that that show high levels of toxaphene, for food.

ESTROGEN-MIMICS Researchers who are working on biomagnification in the Arctic and elsewhere are particularly concerned about organic compounds that bind to receptors for the vertebrate sex hormone estradiol, which is an estrogen (see Chapter 48). In laboratory experiments, high concentrations of these estrogen-mimicking substances "feminize" individuals—meaning that male gonads begin producing female-specific compounds or cell types.

There is now good evidence that estrogen-mimicking POPs are impacting fish populations in the Arctic and elsewhere. Problems are also showing up in some human populations. As the graph in **Figure 54.7** shows, certain communities in the Arctic—in this

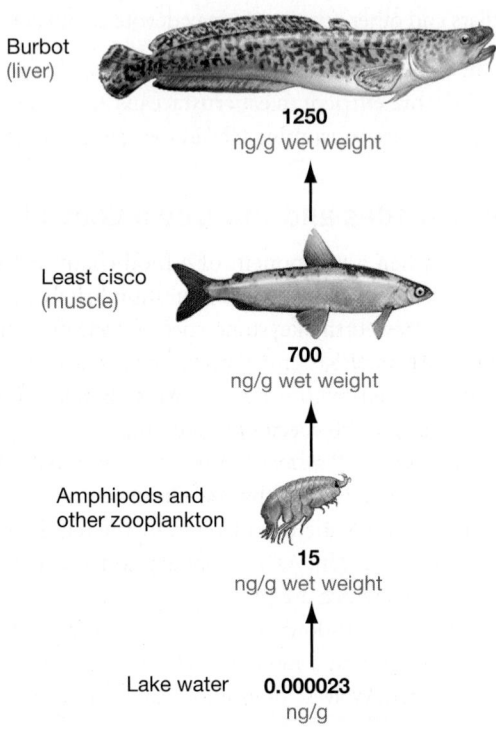

Burbot
(liver)

1250
ng/g wet weight

Least cisco
(muscle)

700
ng/g wet weight

Amphipods and
other zooplankton

15
ng/g wet weight

Lake water **0.000023**
ng/g

Toxaphene concentration

FIGURE 54.6 Biomagnification Can Lead to High Concentrations of Toxic Substances at High Trophic Levels. Note that the toxaphene concentrations reported here are usually from lipid or blubber. Researchers report POP concentrations in this way if the molecule in question is lipid-soluble.

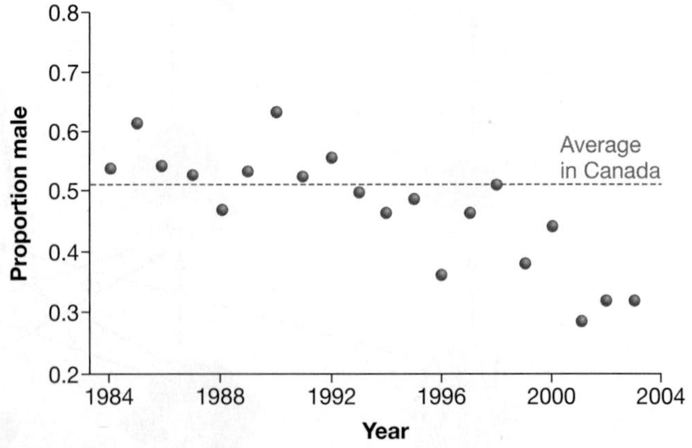

FIGURE 54.7 Recent Changes in Sex-Ratio at Birth in an Arctic First Nations Community. The data shown here are from the Aamjiwnaang First Nation (native) community near the village of Sarnia, in northern Ontario.

case, one located next to a petrochemical plant—have recently experienced aberrant sex ratios. The dotted line on the graph shows the average proportion of males among newborns throughout Canada; the data points indicate that in this Arctic community, many more girls are now being born than boys.

Is the problem in this community due to estrogen-mimics? If so, which molecule is to blame? The data compiled to date indicate that toxaphene has relatively mild estrogenic activity. Although other compounds have been shown to have strong estrogenic activity in fish and some reptiles, to date there is no well-documented tie between particular compounds and sex ratio or other "feminization" problems in humans.

In the meantime, it is clear that biomagnification is a serious side effect of basic patterns in energy flow through ecosystems. Toxins are passed through food webs along with energy and nutrients. Research continues.

Global Patterns in Productivity

The data presented thus far should convince you that NPP is a fundamental attribute of an ecosystem, that energy flows through food webs in complex and dynamic ways, and that understanding trophic interactions has important practical consequences for humans. Now let's step back and consider two "big picture" questions about NPP: Where is most of it produced, and what limits it?

WHICH GEOGRAPHICAL AREAS ARE MOST PRODUCTIVE? **Figure 54.8** summarizes data on NPP from around the globe. Note that the highest NPP on land is indicated in red, yellow, and green; the highest NPP in the oceans is indicated in blue. In general, NPP on land is much higher than it is in the oceans.

The leading hypothesis to explain this pattern is simply that much more light is available to drive photosynthesis on land than in marine environments. As data presented in Chapter 50 indicated, water quickly absorbs many of the light wavelengths that are used in photosynthesis. This simple observation also underlines the importance of the evolution of land plants, highlighted in Chapter 30. Before green plants invaded the land, NPP in terrestrial environments may have been as low as it is in today's oceans.

The terrestrial ecosystems with highest productivity are located in the wet tropics. With the exception of the world's major deserts, NPP on land declines from the equator toward the poles. Chapter 50 pointed out that deserts occur at 30 degrees of latitude north and south of the equator, due to global patterns in air circulation that affect precipitation and thus plant growth. Otherwise, NPP declines with latitude.

You might recall from Chapter 53 that the equator-to-poles gradient in NPP is thought to be important in explaining the equator-to-poles gradient in species richness that occurs in most types of plants and animals. If so, then areas of high productivity should also be "hotspots" for high species richness (see Chapter 55).

Productivity patterns in marine ecosystems contrast strongly with those of terrestrial environments, however. Marine productivity is highest along coastlines, and it can be as high near the poles as it is in tropics. The oceanic areas colored deep blue in Figure 54.8 include most of the world's most important fisheries. The oceanic zones, introduced in Chapter 50, have extremely low NPP. Typically, a square meter (m^2) of open ocean produces a maximum of 35 g (1.2 oz) of organic matter each year. In terms of productivity per m^2, the open ocean is a desert.

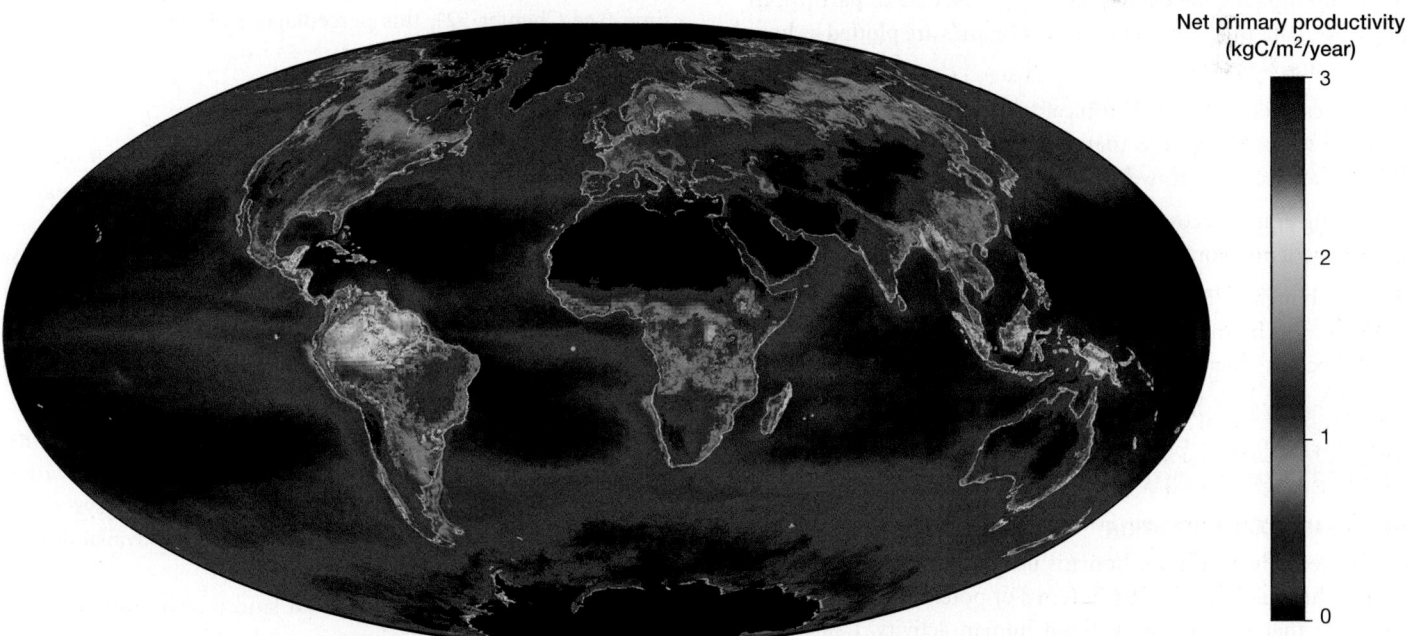

FIGURE 54.8 Net Primary Productivity Varies among Geographic Regions. The terrestrial ecosystems with the highest primary productivity are found in the tropics; tundras and deserts have the lowest. The highest productivity in the oceans occurs in nutrient-rich coastal areas.

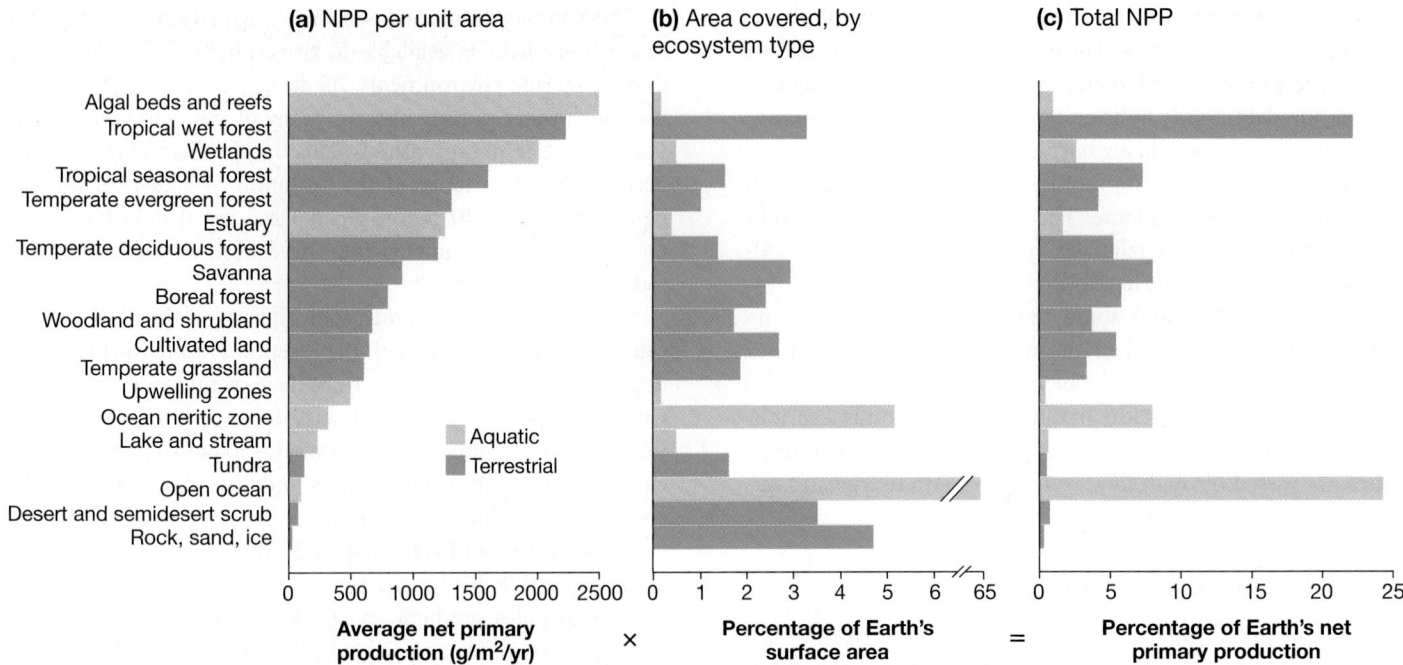

(a) NPP per unit area

(b) Area covered, by ecosystem type

(c) Total NPP

Algal beds and reefs
Tropical wet forest
Wetlands
Tropical seasonal forest
Temperate evergreen forest
Estuary
Temperate deciduous forest
Savanna
Boreal forest
Woodland and shrubland
Cultivated land
Temperate grassland
Upwelling zones
Ocean neritic zone
Lake and stream
Tundra
Open ocean
Desert and semidesert scrub
Rock, sand, ice

Aquatic
Terrestrial

0 500 1000 1500 2000 2500

Average net primary
production (g/m²/yr)

×

0 1 2 3 4 5 65

Percentage of Earth's
surface area

=

0 5 10 15 20 25

Percentage of Earth's net
primary production

FIGURE 54.9 Net Primary Productivity Varies among Ecosystems. (a) Among biomes, average annual NPP per square meter varies by over three orders of magnitude. **(b)** Most of Earth's surface is covered by open ocean. The most common terrestrial habitat consists of unvegetated rock, sand, or ice. **(c)** Even though it has a low NPP per square meter, the open ocean is so vast that it is responsible for over 25 percent of Earth's total NPP.

WHICH BIOMES ARE MOST PRODUCTIVE? **Figure 54.9** presents NPP data a different way—organized by biome instead of by geography. Figure 54.9a provides data on average NPP per square meter per year for each biome; Figure 54.9b documents the total area that is covered by each type of ecosystem; and Figure 54.9c presents the percentage of the world's total productivity—the result of multiplying the data in part (a) by the data in part (b). In each case, data from different types of biomes are plotted as horizontal bars. Note that:

- Tropical wet forests and tropical seasonal forests (which have a dry season) cover less than 5 percent of Earth's surface but together account for over 30 percent of total NPP.

- Among aquatic ecosystems, the most productive habitats are algal beds and coral reefs, wetlands, and estuaries.

- Even though NPP per square meter is extremely low in the open ocean, this biome is so extensive in terms of area that its total production is high.

One of the reasons that biologists are so concerned about the destruction of tropical rain forests and coral reefs is that they are the world's most productive habitats.

THE "HUMAN APPROPRIATION" Of the total NPP available on our planet, how much are humans using? A research group answered this question recently in terms of potential NPP—meaning, NPP that would exist without human activity. Using data from a wide array of sources, they estimated that humans are currently appropriating 23.8 percent of potential NPP. Of this:

- 53 percent is directly harvested and used,

- 40 percent is lost due to suburbanization and other land-use changes, and

- 7 percent is lost due to human-induced fires.

This analysis is astonishing. One species—our own—is appropriating almost a quarter of the planet's biomass. Given increases in human population projected to occur over your lifetime (see Chapter 52), this percentage is almost guaranteed to grow—probably substantially.

What Limits Productivity?

The patterns highlighted in Figure 54.8 and Figure 54.9 raise an interesting question: What limits NPP in terrestrial and marine ecosystems?

The short answer is that productivity is limited by any factor that limits the rate of photosynthesis. Based on information in Chapter 10 and Unit 7, the candidates are temperature and the availabilities of water, sunlight, and nutrients. Different limiting factors prevail in different environments, however.

A TERRESTRIAL-MARINE CONTRAST In terrestrial environments, NPP is lowest in deserts and arctic regions. 🔑 This observation suggests that the overall productivity of terrestrial ecosystems is limited by a combination of temperature and availability of water and sunlight.

At a local level, however, NPP on land is also limited by nutrient availability—often nitrogen or phosphorus. When biologists fertilize a terrestrial ecosystem with nitrogen or phosphorus, NPP almost always increases. Recent research has also shown that NPP in temperate and boreal forests is increasing, due to nitrogen that is blowing in from agricultural fields hundreds or thousands of miles away, where it was originally applied as a fertilizer.

To explain why the productivity of marine habitats is so much higher along coastlines than in deepwater regions, biologists focus on nutrient limitation. As Chapter 50 pointed out, the shallow water along coasts receives nutrients from two major sources:

1. rivers that carry and deposit nutrients from terrestrial ecosystems; and

2. nearshore ocean currents that bring nutrients from the cold, deep water of the oceanic zone back up to the surface.

Both of these sources are absent in the surface waters of the open ocean. In addition, nutrients found in organisms near the surface of the open ocean—where sunlight is abundant—constantly fall to dark, deeper waters in the form of dead cells, and are lost.

IRON-FERTILIZATION EXPERIMENTS Trace elements such as zinc, iron, and magnesium are particularly rare in the open ocean. These atoms are important because they are required as enzyme cofactors (see Chapter 3). Iron, for example, is essential to the proteins that are involved in electron transport chains (see Chapters 9 and 10). On the basis of these observations, biologists have proposed that the productivity of open-ocean ecosystems could be increased dramatically by fertilizing them with iron or other types of trace elements.

The results of one iron-fertilization experiment are shown in **Figure 54.10**. In this graph, each data point represents the average of a large number of sample sites from surface waters inside the treatment area, surface waters outside the treatment area, or water 30 m below the surface and inside the treatment area.

The graph indicates that about a week following the experimental addition of iron to the surface of an experimental area, there were large increases in the concentration of chlorophyll *a*—an indicator of the presence of photosynthetic cells. The increase continued for at least five days in surface waters compared to both water sampled from 30 m below the surface and water outside the experimental area.

Consistent with this result, recent research on ocean waters that are naturally enriched by nearby iron-containing rocks indicates that these regions have exceptionally high NPP.

Results like these provide strong support for the hypotheses that:

1. NPP in marine ecosystems is limited primarily by the availability of nutrients, and

2. iron is particularly important in the open ocean.

These data have also inspired a novel proposal to fight global warming by reducing CO_2 concentrations in the atmosphere (see Section 54.3). The photosynthetic cells that grow in response to iron addition use large amounts of atmospheric CO_2 that dissolves in seawater. Some researchers suggest that fertilizing the open oceans could increase NPP in these deserts enough to reduce atmospheric CO_2 and slow global warming.

EXPERIMENT

QUESTION: Is net primary productivity (NPP) in the open ocean limited by nutrients?

HYPOTHESIS: NPP in the open ocean is limited by availability of iron.

NULL HYPOTHESIS: NPP in the open ocean is not limited by availability of iron.

EXPERIMENTAL SETUP:

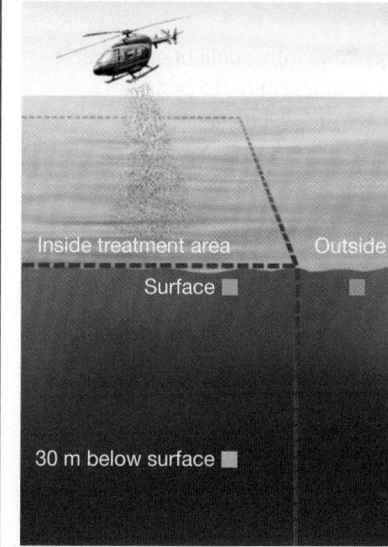

1. Add 350 kg iron (as $FeSO_4$) to a treatment area— a patch of ocean 8 km × 10 km.

2. Take water samples for a two-week period outside and inside the treatment area, at surface and at a depth of 30 m, and record amount of chlorophyll *a* present (as indicator of NPP).

PREDICTION: Amount of chlorophyll *a* near the surface inside the treatment area will increase relative to amounts outside the treatment area or at 30 m below the surface.

PREDICTION OF NULL HYPOTHESIS: Amount of chlorophyll *a* will be the same in all measurements.

RESULTS:

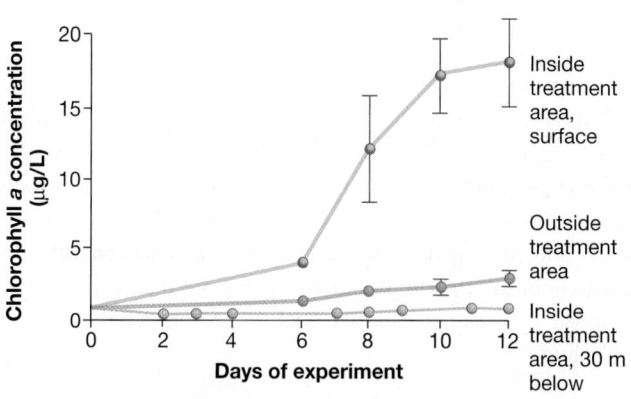

CONCLUSION: NPP in the open ocean is limited by the scarcity of nutrients—specifically, iron.

FIGURE 54.10 Fertilization with Iron Increases NPP in the Open Ocean.
SOURCE: Tsuda, A., et al. 2003. A mesoscale iron enrichment in the western subarctic Pacific induces a large centric diatom bloom. *Science* 300: 958–961.

✔ **QUESTION** Most of the increased chlorophyll *a* was present in a single species of diatom (see Chapter 29). What happens to the biomass present in diatoms after they die?

This proposal is being hotly debated right now. Could iron fertilization be done on a large enough scale to have an impact? If so, what would it cost and who would pay? When the photosynthetic cells die and fall to the ocean bottom, would they have negative effects on other parts of the marine ecosystem? At present, we simply lack the data to answer these questions conclusively.

CHECK YOUR UNDERSTANDING

If you understand that . . .

- In an ecosystem, energy flows from sunlight or inorganic compounds with high potential energy to primary producers, and from there to consumers and decomposers.
- Productivity diminishes at each subsequent trophic level in the ecosystem.
- In most ecosystems, NPP is limited by conditions that limit the rate of photosynthesis: temperature and/or the availability of sunlight, water, and nutrients.

✓ You should be able to . . .

1. Explain why productivity diminishes with increasing trophic levels.
2. Predict what will happen to the food web in an ecosystem where global warming leads to an increase in NPP.

Answers are available in Appendix B.

54.2 How Do Nutrients Cycle through Ecosystems?

Energy and POPs are not the only entities that get transferred when one organism eats another. Tissues also contain carbon (C), nitrogen (N), phosphorus (P), calcium (Ca), and other elements that act as nutrients.

Atoms are constantly reused as they move through trophic levels, but they also spend time suspended in air, dissolved in water, or held in soil. The path that an element takes as it moves from abiotic systems through producers, consumers, and decomposers and back again is referred to as its **biogeochemical** ("life-Earth-chemical") **cycle**. Because humans are disturbing biogeochemical cycles on a massive scale, research on the topics is exploding.

Nutrient Cycling within Ecosystems

To understand the basic features of a biogeochemical cycle, study the generalized terrestrial nutrient cycle in **Figure 54.11**. Start with the "Uptake" arrow, which indicates nutrients that are taken up from the soil by plants and assimilated into plant tissue; carbon enters primary producers as carbon dioxide from the atmosphere. Note that if the plant tissue is eaten, the nutrients pass to the consumer food web; if the plant tissue dies, the nutrients enter the compartment labeled "Detritus."

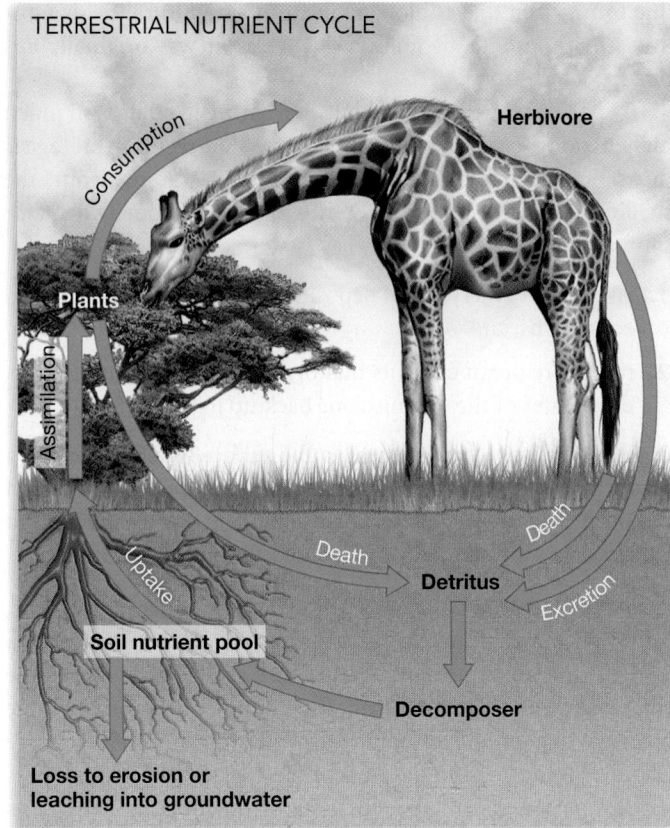

FIGURE 54.11 Generalized Terrestrial Nutrient Cycle. Nutrients cycle from organism to organism in an ecosystem via assimilation, consumption, and decomposition. Nutrients are exported from ecosystems when water or organisms leave the area.

What happens as detritus decomposes? Nutrients that reside in plant litter, animal excretions, and dead animal bodies are used by bacteria, archaea, roundworms, fungi, and other primary decomposers. These microscopic decomposers and the carbon-containing compounds that they release combine to form what biologists call soil organic matter.

Soil organic matter is a complex mixture of partially and completely decomposed detritus. Completely decayed organic material is called **humus** because it is rich in a family of carbon-containing molecules called humic acids. Chapter 38 described other components of the soil.

Eventually, decomposition converts the nutrients in soil organic matter to an inorganic form. For example, cellular respiration by soil-dwelling bacteria and archaea converts the nitrogen present in amino acids to ammonium (NH_4^+) or nitrate (NO_3^-) ions. Once this step is accomplished, the nutrients are available for uptake by plants. Uptake by plants, following decomposition, highlights the cyclical nature of nutrient flow through ecosystems.

WHAT FACTORS CONTROL THE RATE OF NUTRIENT CYCLING? Of the many links in a nutrient cycle, the decomposition of detritus most often limits the overall rate at which nutrients move through an ecosystem. Until decomposition occurs, nutrients stay tied up in intact tissues.

The decomposition rate, in turn, is influenced by two types of factors:

1. abiotic conditions such as oxygen availability, temperature, and precipitation; and

2. the quality of the detritus as a nutrient source for the fungi, bacteria, and archaea that accomplish decomposition.

To appreciate the importance of abiotic conditions on the decomposition rate, consider the difference in detritus accumulation between boreal forests and tropical wet forests. Chapter 50 indicated that boreal forests are found in areas where temperature is low. As a result, soils in these ecosystems are cold and wet. But tropical wet forests occur in areas where temperatures and rainfall are high. Soils tend to remain moist and warm all year long.

Figure 54.12 illustrates typical soils from boreal forests and tropical wet forests. Notice that the uppermost part of the soil in a boreal forest consists of partially decomposed detritus and organic matter. In a tropical forest, this layer is virtually absent.

The contrast occurs because the cold and wet conditions in boreal forests limit the metabolic rates of decomposers. As a result, decomposition fails to keep up with the input of detritus, and organic matter accumulates there. In the tropics, conditions are so favorable for fungi, bacteria, and archaea that decomposition keeps pace with detrital inputs. Nutrients cycle slowly through boreal forests but rapidly through wet tropical forests.

The quality of detritus also exerts a powerful influence on the growth of decomposers, and thus the decomposition rate. Decomposition is inhibited if detritus is (1) low in nitrogen, or (2) high in lignin.

When nitrogen is scarce, cells have a difficult time making additional proteins and nucleic acids. Chapter 31 noted that lignin—an important constituent of wood—is extremely difficult to digest. The presence of lignin is a major reason why wood takes much longer to decompose than leaves do. Only basidiomycete fungi and a few bacteria have the enzymes required to completely degrade lignin.

The presence of oxygen is also critically important. If soils or water become depleted of oxygen, decomposition slows dramatically. Fungi can only perform cellular respiration using oxygen as the final electron acceptor (see Chapter 9). And even though many bacteria and archaea can thrive in the absence of oxygen and continue decomposition, the growth rates of anaerobic species are a tiny fraction of rates in aerobic bacteria and archaea (to understand why, see Chapter 28).

SOURCES OF NUTRIENT LOSS AND GAIN   Although nutrients are constantly reused in ecosystems, they can also be lost—much as energy can. Nutrients leave an ecosystem whenever biomass leaves.

- If an herbivore eats a plant and moves out of the ecosystem before excreting the nutrients or dying, the nutrients are lost.

- Nutrients leave ecosystems when flowing water or wind removes particles or inorganic ions and deposits them somewhere else.

Several of the major impacts that humans have on ecosystems—such as farming, logging, burning, and soil erosion—accelerate nutrient loss. Farming and logging remove biomass in the form of plants; burning releases nutrients to the atmosphere; soil erosion moves nutrients that are bound to soil particles or dissolved in water.

For an ecosystem to function normally, nutrients that are lost must be replaced. There are three major mechanisms to replace lost nutrients:

1. Atoms that act as nutrients are released as rocks weather (see Chapter 38).

2. Nutrients can also blow in on soil particles or arrive as solutes in streams.

3. Nitrogen—a particularly important nutrient in maintaining NPP—is added when the nitrogen-fixing bacteria introduced in Chapter 38 convert molecular nitrogen (N_2) in the atmosphere to usable nitrogen in ammonium or nitrate ions.

In ecosystems that humans manage, nutrients usually have to be actively replaced to maintain productivity. People accomplish this by adding fertilizers or planting nitrogen-fixing crops.

AN EXPERIMENTAL STUDY   The ideas presented thus far are logical and provide a strong conceptual foundation for thinking about nutrient cycles. But to explore nutrient movement in more depth, researchers have turned to experiments.

One of the first large-scale experiments on nutrient cycling took place at the Hubbard Brook Experimental Forest. A group of researchers there began by choosing two similar **watersheds**—areas drained by a single stream—for study. Before starting the experiment, they documented that 90 percent of the nutrients in the watersheds was in soil organic matter; an additional 9.5 percent was in plant biomass. The nutrients in the soil were either dissolved in water or attached to small particles of sand or clay (see Chapter 38).

(a) Boreal forest: Accumulation of detritus and organic matter

(b) Tropical wet forest: Almost no organic accumulation

FIGURE 54.12 Temperature and Moisture Affect Decomposition Rates.

QUESTION: How does the presence of vegetation affect the rate of nutrient export in a temperate-forest ecosystem?

HYPOTHESIS: Presence of vegetation lowers the rate of nutrient export because it increases soil stability and recycling of nutrients.

NULL HYPOTHESIS: Presence of vegetation has no effect on the rate of nutrient export.

EXPERIMENTAL SETUP:

1. Choose two similar watersheds. Document nutrient levels in soil organic matter, plants, and streams.

2. Devegetate one watershed, and leave the other intact.

3. Monitor the amount of dissolved substances in streams.

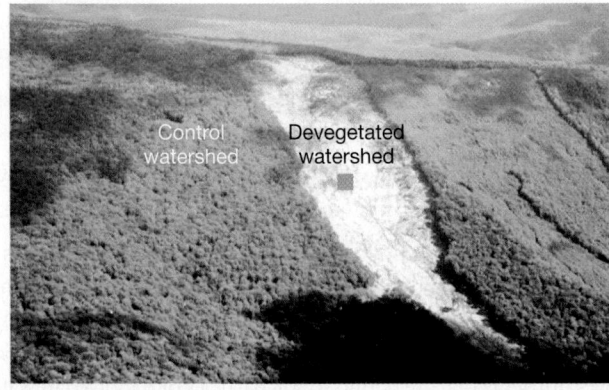

Control watershed   Devegetated watershed

PREDICTION: Amount of dissolved substances in stream in devegetated watershed will be much higher than amount of dissolved substances in stream in control watershed.

PREDICTION OF NULL HYPOTHESIS: No difference will be observed in amount of dissolved substances in the two streams..

RESULTS:

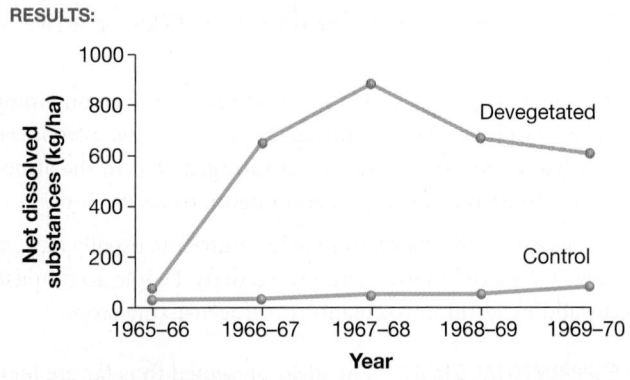

CONCLUSION: Presence of vegetation limits nutrient loss. Removing vegetation leads to large increases in nutrient export.

FIGURE 54.13 Experimental Evidence That Deforestation Increases the Rate of Nutrient Loss from Ecosystems.

SOURCE: Likens, G. E., F. H. Bormann, N. M. Johnson, D. W. Risher, and R. S. Pierce. 1970. Effects of forest cutting and herbicide treatment on nutrient budgets in the Hubbard Brook watershed-ecosystem. *Ecological Monographs* 40: 23–47.

✔**QUESTION** Propose a hypothesis to explain why the level of nutrient export from the experimental site appears to peak and then decline.

To begin the experimental treatment, they cut all vegetation, including the trees, in one of the two watersheds (**Figure 54.13**). In each of the following three years, the clear-cut watershed was treated with an herbicide to prevent vegetation from regrowing. As a result, the experimental watershed was devegetated. Researchers left the other watershed untreated to serve as a control.

✔If you understand the general principles of nutrient cycling, you should be able to name which arrow in Figure 54.11 was removed by the devegetation study. The purpose of the experiment was to remove this link and document the consequences.

The graph in Figure 54.13 documents the amount of total dissolved substances that washed out of the streams in each watershed over the course of the four years of devegetation. Losses from the devegetated site were typically over 10 times as high as they were from the control site.

Why? Plant roots were no longer able to hold soil particles in place, and plant roots no longer took up and recycled nutrients that were dissolved in soil water. Instead, massive quantities of nitrate and other nutrients washed out of the soil and were lost to the ecosystem.

Long-term devegetation of this type has occurred in formerly forested areas of the Middle East, North Africa, and elsewhere, due to intensive farming and grazing. Nutrients that used to occur in those soils were lost. Today, the productivity of these regions is a tiny fraction of what it once was.

Global Biogeochemical Cycles

When nutrients leave one ecosystem, they enter another. In this way, the movement of ions and molecules among ecosystems links local biogeochemical cycles into one massive global system. Local and global cycles interact when water, organisms, or wind move nutrients.

As an introduction to how these global biogeochemical cycles work, let's consider the global water, carbon, and nitrogen cycles. These cycles have recently been heavily modified by human activities—with serious ecological consequences.

🔑 To understand how biogeochemical cycling works on a global scale, researchers focus on three fundamental questions:

1. What are the nature and size of the reservoirs—areas or "compartments" where elements are stored for a period of time? In the case of carbon, the biomass of living organisms is an important reservoir, as are sediments and soils. Another significant carbon reservoir is buried in the form of coal and oil.

2. How fast does the element move between reservoirs, and what processes are responsible for moving elements from one compartment to another? The global photosynthetic rate, for example, measures the rate of carbon flow from carbon dioxide (CO_2) in the atmosphere to living biomass. When fossil fuels burn, carbon that was buried in coal or petroleum moves into the atmosphere as CO_2.

3. How does one biogeochemical cycle interact with another biogeochemical cycle? For example, researchers are trying to understand how changes in the nitrogen cycle affect the carbon cycle.

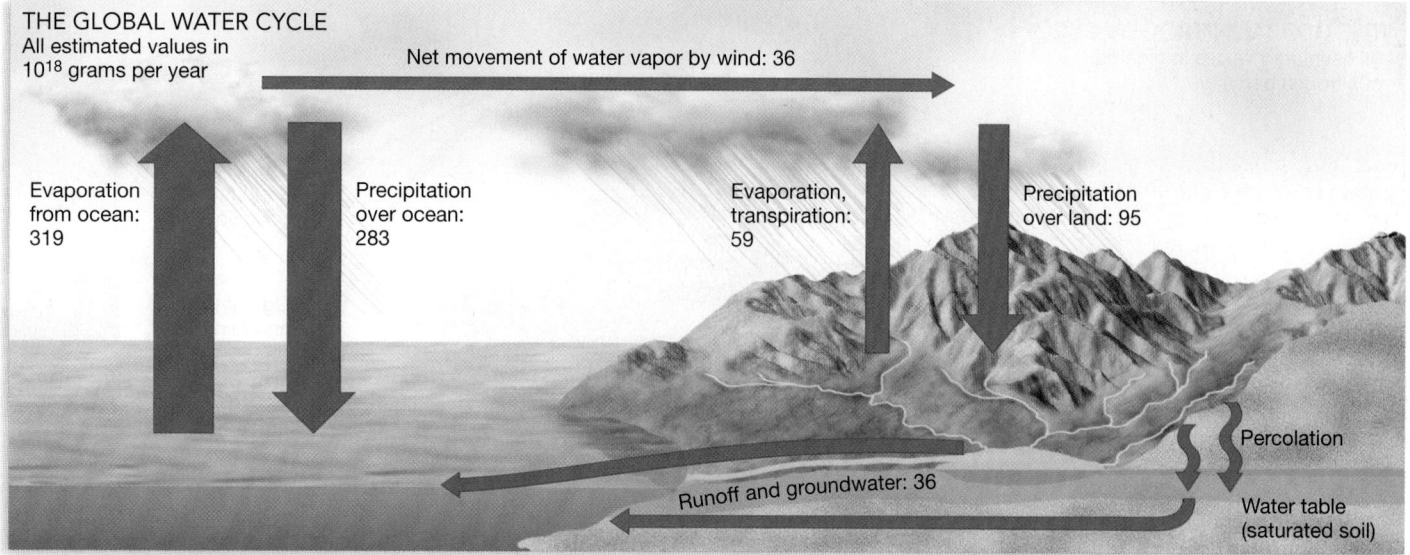

THE GLOBAL WATER CYCLE
All estimated values in
10^{18} grams per year

Net movement of water vapor by wind: 36

Evaporation
from ocean:
319

Precipitation
over ocean:
283

Evaporation,
transpiration:
59

Precipitation
over land: 95

Percolation

Runoff and groundwater: 36

Water table
(saturated soil)

FIGURE 54.14 The Global Water Cycle.

✔**QUESTION** Predict how the amount of water evaporated from the ocean is changing in response to global warming. Discuss one possible consequence of this change.

THE GLOBAL WATER CYCLE A simplified version of the **global water cycle** appears in **Figure 54.14**. The diagram shows the estimated amount of water that moves between major components of the cycle over the course of a year.

To analyze this cycle, begin with evaporation of water out of the ocean, and the subsequent precipitation of water back into the ocean. For the marine component of the cycle, evaporation exceeds precipitation—meaning that, over the oceans, there is a net gain of water to the atmosphere.

When this water vapor moves over the continents, it is joined by a small amount of water that evaporates from lakes and streams and a large volume of water that is transpired by plants. The total volume of water in the atmosphere over land is balanced by the amount of rain and other forms of precipitation that falls on the continents.

The cycle is completed by the water that moves from the land back to the oceans via streams and **groundwater**—water that is found in soil. Overall, there is no net gain or loss of water.

What impact are humans having? Perhaps the simplest and most direct impacts concern rates of groundwater replenishment.

- Asphalt and concrete surfaces reduce the amount of water that percolates from the surface to enter deep soil layers.

- When grasslands and forests are converted to agricultural fields, root systems that hold water are lost. More water runs off into streams and less percolates into groundwater.

- Irrigated agriculture, along with industrial and household use, is removing massive amounts of water from groundwater storage and bringing it to the surface.

The **water table**—the level where soil is saturated with stored water—is dropping on every continent. Between 1986 and 2006, the water table north of Beijing, China, experienced drops aver-

aging almost 61 m (200 ft). Throughout India, water tables are falling at a rate of between 1 m and 3 m per year. Similar rates are being documented in Yemen, parts of Mexico, and states in the southern Great Plains region of the United States. The city of Bangkok, which is located just above sea level, is sinking up to 7.5 cm each year because so much water is being pumped out of the ground below it (**Figure 54.15**).

Humans are mining water in many parts of the world—meaning that populations have already grown beyond carrying capacity for water. Lack of water is exacerbating political tensions in several areas of the world, including the Middle East.

FIGURE 54.15 Flooding Becomes More Frequent as Bangkok Sinks. This man is walking through an antique market during a recent flood. Some studies suggest that much of Bangkok may be permanently underwater by 2100.

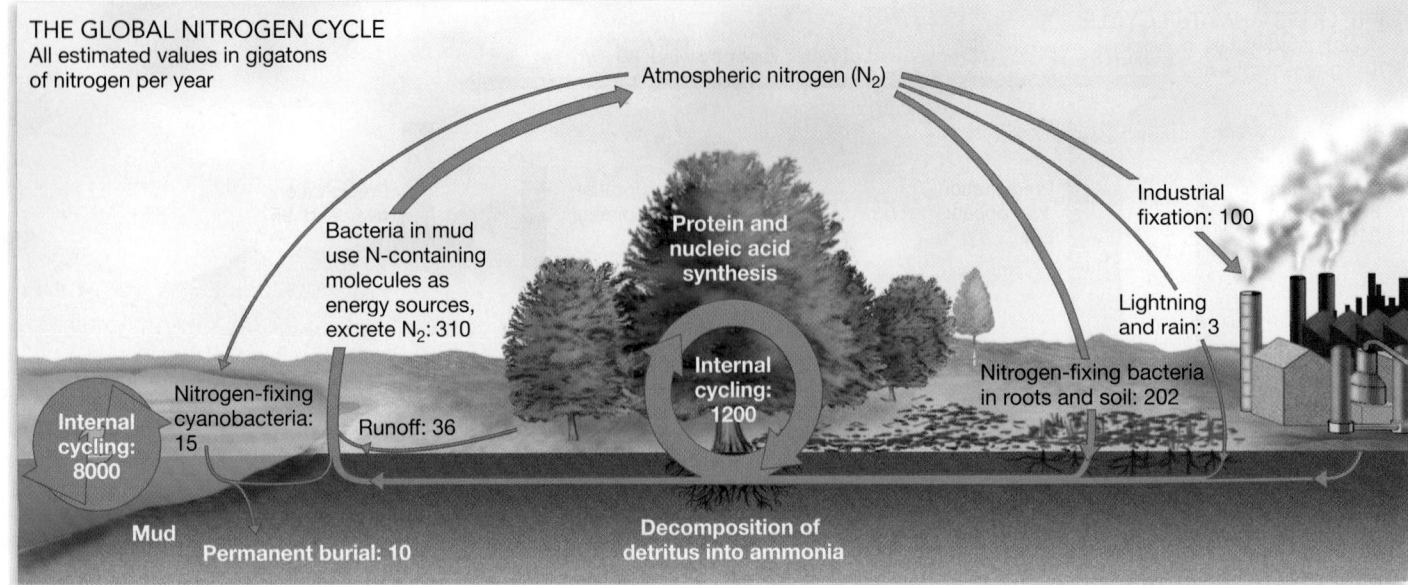

THE GLOBAL NITROGEN CYCLE
All estimated values in gigatons
of nitrogen per year

Atmospheric nitrogen (N₂)

Bacteria in mud
use N-containing
molecules as
energy sources,
excrete N₂: 310

Protein and
nucleic acid
synthesis

Industrial
fixation: 100

Internal
cycling:
1200

Lightning
and rain: 3

Internal
cycling:
15

Nitrogen-fixing
cyanobacteria:
15

Runoff: 36

Nitrogen-fixing bacteria
in roots and soil: 202

Internal
cycling:
8000

Mud

Permanent burial: 10

Decomposition of
detritus into ammonia

FIGURE 54.16 The Global Nitrogen Cycle. Nitrogen enters ecosystems as ammonia or nitrate via fixation from atmospheric nitrogen. It is exported in runoff and as nitrogen gas given off by bacteria that use nitrogen-containing compounds as an electron acceptor.

THE GLOBAL NITROGEN CYCLE **Figure 54.16** illustrates the **global nitrogen cycle.** As Chapter 38 noted, nitrogen is added to ecosystems in a usable form only when it is reduced, or "fixed," meaning that it is converted from N₂ to ammonium or nitrate ions. Nitrogen fixation results from lightning-driven reactions in the atmosphere and from enzyme-catalyzed reactions in bacteria that live in the soil and oceans.

The nitrogen cycle has been profoundly altered by human activities, primarily through the addition of massive amounts of reduced nitrogen. The graphs in **Figure 54.17** show current estimates for the amount of nitrogen fixed by natural sources versus human sources. The amount of nitrogen fixation from human sources is now approximately equal to the amount of nitrogen fixation from natural sources. As the labels on the right side of the graph indicate, there are three major sources of this human-fixed nitrogen:

1. industrially produced fertilizers;

2. the cultivation of crops, such as soybeans and peas that harbor nitrogen-fixing bacteria; and

3. the production of nitric oxide during the combustion of fossil fuels.

Adding nitrogen to terrestrial ecosystems usually increases productivity. But overfertilization with nitrogen has important downsides.

- Chapter 28 detailed how nitrogen-laced runoff from agricultural fields has led to the formation of oxygen-free "dead zones" in aquatic ecosystems.

- When nitrogen is added to experimental plots in grasslands in midwestern North America, a few competitively dominant species grow rapidly and displace other species that do not re-

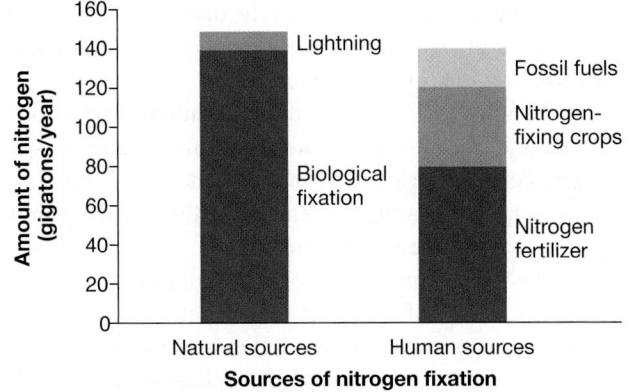

FIGURE 54.17 Humans Are Adding Large Amounts of Nitrogen to Ecosystems. Human activities now fix almost as much nitrogen each year as natural sources do. Thus, human activities have almost doubled the total amount of nitrogen available to organisms.

spond to nitrogen inputs as strongly. Productivity increases, but species diversity decreases.

THE GLOBAL CARBON CYCLE The **global carbon cycle** documents the movement of carbon among terrestrial ecosystems, the oceans, and the atmosphere (**Figure 54.18**). Of these three reservoirs, the ocean is by far the largest. The atmospheric reservoir is also important despite its relatively small size, because carbon moves into and out of it rapidly—through organisms.

In both terrestrial and aquatic ecosystems, photosynthesis is responsible for taking carbon out of the atmosphere and incorporating it into tissue. Cellular respiration, in contrast, releases carbon that has been incorporated into living organisms to the atmosphere, in the form of carbon dioxide.

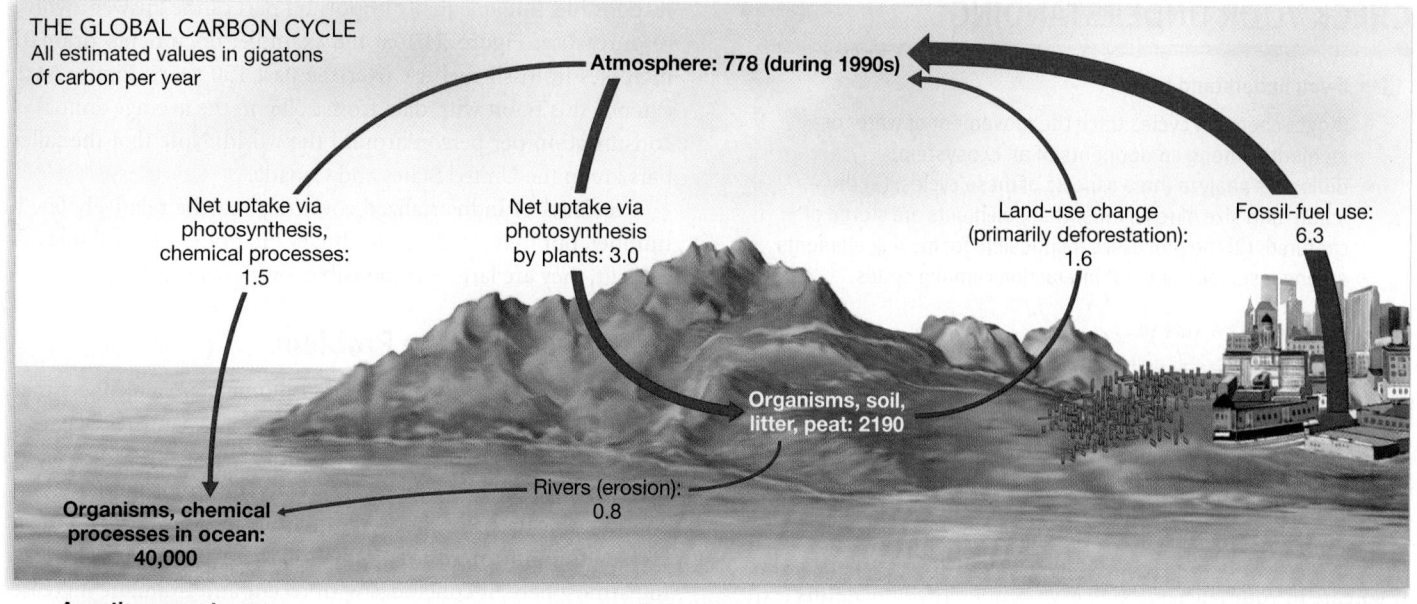

THE GLOBAL CARBON CYCLE
All estimated values in gigatons of carbon per year

Atmosphere: 778 (during 1990s)

Net uptake via photosynthesis, chemical processes: 1.5

Net uptake via photosynthesis by plants: 3.0

Land-use change (primarily deforestation): 1.6

Fossil-fuel use: 6.3

Organisms, soil, litter, peat: 2190

Rivers (erosion): 0.8

Organisms, chemical processes in ocean: 40,000

Aquatic ecosystems   Terrestrial ecosystems   Human-induced changes

FIGURE 54.18 The Global Carbon Cycle. The arrows indicate how carbon moves into and out of ecosystems. Deforestation and the use of fossil fuels are adding 7.9 gigatons of carbon to the atmosphere each year. Of that 7.9 gigatons, 2 gigatons are fixed by photosynthesis in terrestrial ecosystems and 2 gigatons are fixed by physical and chemical processes in the oceans. The remainder—3.9 gigatons—is added to the atmosphere.

How have humans changed the carbon cycle? Burning fossil fuels moves carbon from an inactive geological reservoir, in the form of petroleum or coal, to an active reservoir—the atmosphere. When you burn gasoline, you are releasing carbon atoms that have been locked up in petroleum reservoirs for hundreds of millions of years.

Land-use changes have also altered the global carbon cycle. Deforestation and suburbanization: (**1**) reduce an area's net primary productivity—reducing its ability to sequester CO_2, and (**2**) release CO_2 directly when fire is used for clearing or when dead limbs, twigs, and stumps are left to decompose.

Figure 54.19a highlights the dramatic increase in carbon released from fossil-fuel burning over the past century; **Figure 54.19b** shows corresponding changes in CO_2 concentrations. Note that CO_2 levels in the atmosphere have now risen to levels far above the 280 ppm that was typical prior to 1860.

These changes in the global carbon cycle are important because carbon dioxide functions as a **greenhouse gas**: It traps heat that has been radiated from Earth and keeps it from being lost to space, similar to the way the glass of a greenhouse traps heat.

More specifically, carbon dioxide is one of several gases in the atmosphere that absorb and reflect the infrared wavelengths radiating from Earth's surface. Increases in amounts of greenhouse gases are warming Earth's climate by increasing the atmosphere's heat-trapping potential.

To review these carbon cycle concepts, go to the study area at *www.masteringbiology.com*. To review their consequences, go on to the chapter's next section.

MB BioFlix™ The Carbon Cycle, **Web Activity** The Global Carbon Cycle

(a) Increases in fossil-fuel use

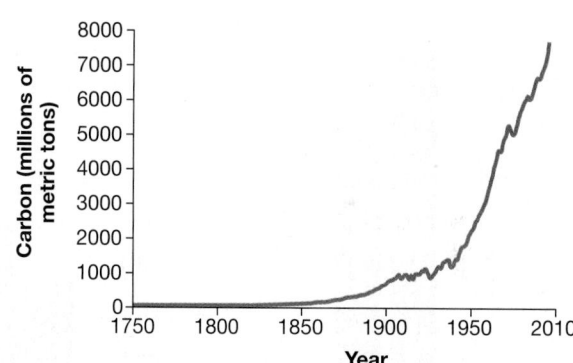

(b) Global changes in atmospheric CO_2 over time

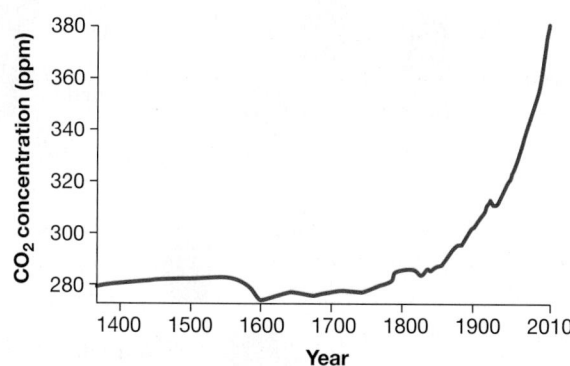

FIGURE 54.19 Humans Are Causing Increases in Atmospheric Carbon Dioxide. (a) Rates of carbon flow from fossil-fuel burning have increased as human populations have increased. **(b)** For centuries, average CO_2 concentrations in the atmosphere were about 280 parts per million (ppm).

CHECK YOUR UNDERSTANDING

54.3 Global Warming

If you are in your late teens or early twenties as you read this text, you are part of a generation that will experience the most traumatic episode of environmental change in human history. The trauma has two sources: the massive loss of species, documented in Chapter 55, and **global warming**.

Why is the environment deteriorating so rapidly? The explosion in human population size analyzed in Chapter 52 is partly responsible. But an equally important part of the answer involves resource use. Figure 54.19a, for example, documents dramatic increases in fossil-fuel use over the past 150 years. **Figure 54.20** extends this point with data from 2006 on the average annual oil consumption per person around the world. Note that the tallest bars are in the United States and Canada.

Residents of industrialized countries may be relatively few in number, but they burn extraordinary quantities of fossil fuels. As a result, they are largely responsible for global warming.

Understanding the Problem

What evidence backs the claim that humans are altering climate on a global scale? In explaining climate change, the chain of causation begins with the direct link between the CO_2 released by human fossil fuel use and increases in atmospheric CO_2. It continues with a link between high atmospheric CO_2 and increased trapping of infrared radiation that would otherwise leave the atmosphere; it concludes with recent and dramatic increases in average temperatures around the globe.

In 1988 an international group of scientists, called the Intergovernmental Panel on Climate Change (IPCC), was formed to evaluate the consequences of well-documented increases in CO_2 and other heat-trapping gases. The group has since produced a series of reports summarizing the state of scientific knowledge on the issue.

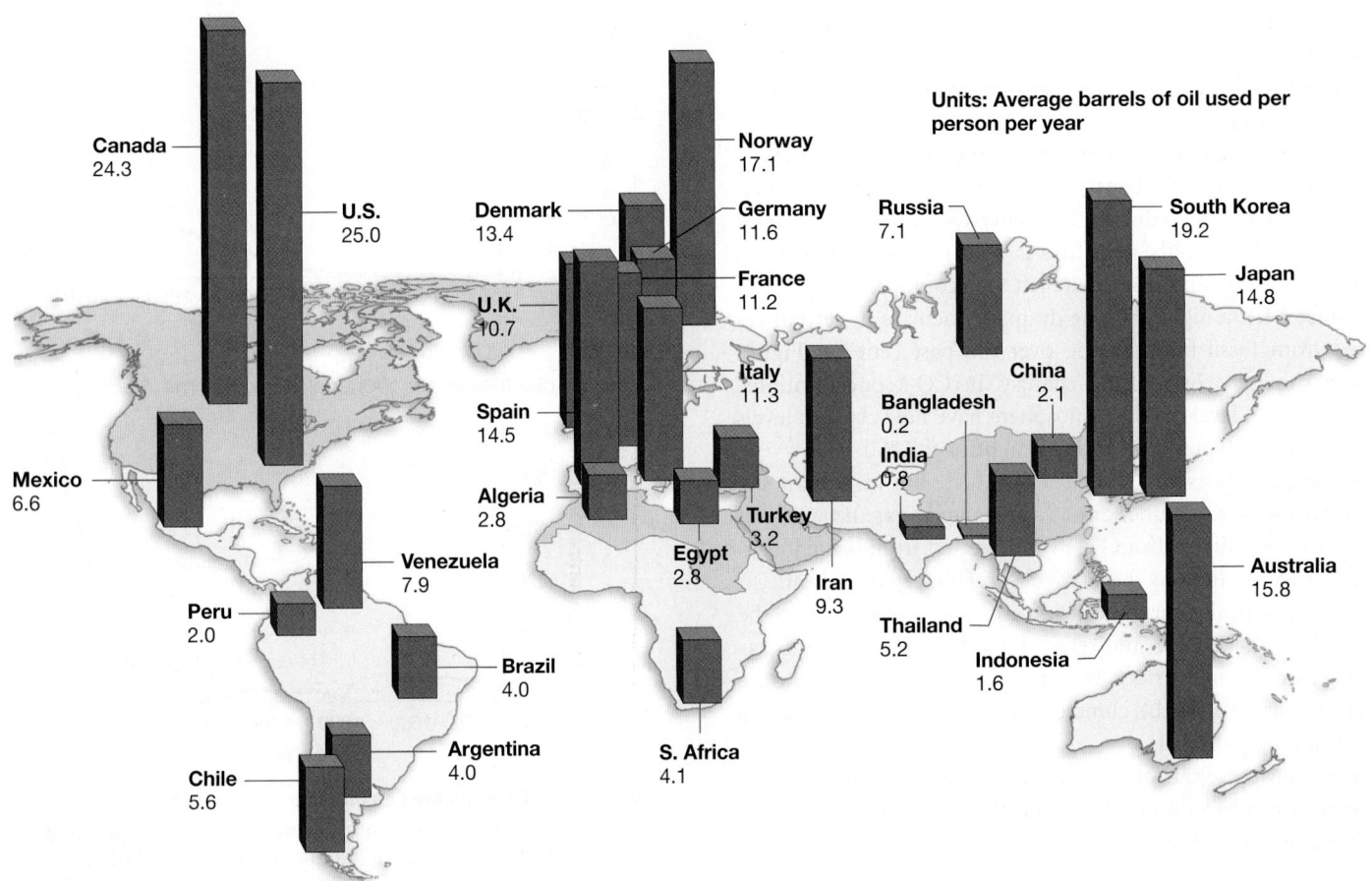

Canada
24.3

U.S.
25.0

Denmark
13.4

Norway
17.1

Germany
11.6

Russia
7.1

South Korea
19.2

Units: Average barrels of oil used per person per year

France
11.2

Japan
14.8

U.K.
10.7

Italy
11.3

China
2.1

Mexico
6.6

Spain
14.5

Bangladesh
0.2

India
0.8

Algeria
2.8

Turkey
3.2

Australia
15.8

Venezuela
7.9

Egypt
2.8

Iran
9.3

Peru
2.0

Thailand
5.2

Brazil
4.0

Indonesia
1.6

Chile
5.6

Argentina
4.0

S. Africa
4.1

FIGURE 54.20 Per-Capita Oil Consumption Varies among Countries. These data are from 2006.

- The 1998 IPCC report concluded that current evidence suggests a "discernible human influence on climate." This was the IPCC's first statement supporting the hypothesis that rising CO_2 concentrations due to human activities are having a measurable impact on climate.

- The IPCC's next major report, released in 2001, stated that "There is new and stronger evidence that most of the warming observed over the last 50 years is attributable to human activities."

- In their most recent report, issued in 2007, the IPCC declared that evidence for global warming is unequivocal and that it is "very likely" due to human-induced changes in greenhouse gases.

Two points are important to note:

1. Climate change has occurred throughout Earth's history. What is unusual is the extremely rapid *rate* at which temperatures are changing currently.

2. The temperature increases being observed are global averages, reported on a per-year basis. Some regions may actually get cooler, and extensive month-to-month or year-to-year variation will occur. But averaged over the entire planet and over time, Earth is already much warmer than it was just a few decades ago, and is projected to get much more so.

How much will average temperatures rise in our lifetimes? The answer depends in part on whether CO_2 and other greenhouse gases continue to increase. **Figure 54.21** shows data on recent changes in atmospheric carbon dioxide, recorded at a research station in Mauna Loa, Hawaii. To date, there is no indication that the rate of increase is slowing. Although an array of regulations or incentives to reduce CO_2 emissions has been proposed by international bodies, no agreement has been reached yet.

Predicting the future state of a system as complex and variable as Earth's climate is extremely difficult, however. Global climate models are the primary tools that scientists use to make these projections. A global climate model is based on a large series of equations that describe how the concentrations of various gases in the atmosphere, solar radiation, transpiration rates, and other parameters interact to affect climate.

The models currently being used suggest that average global temperature will undergo additional increases of 1.1−6.4°C (2.0−11.5°F) by the year 2100. The low number is based on models that assume no further increase in greenhouse gases over present levels; the high number is based on models that assume continued intensive use of fossil fuels and increases in greenhouse gases.

Positive and Negative Feedback

How will ecosystems respond to global warming? Answering this question is difficult, because ecosystems respond to warming in ways that increase or decrease CO_2 concentrations, and thus exacerbate or mitigate warming.

DOCUMENTING POSITIVE FEEDBACK Positive feedback occurs when changes due to global warming result in a release of additional greenhouse gases—meaning that global warming should get worse. Two types of positive feedback have been documented thus far.

- Warmer and drier climate conditions have increased the frequency of forest fires in the Rocky Mountains of North America as well as in the boreal forests of Canada. Forest fires release CO_2, which leads to even more warming.

- Traditionally, tundras sequester carbon in the form of soil organic matter, because decomposition rates are extremely low. During a series of warm summers in the 1980s, however, researchers found that decomposition rates increased sufficiently to release carbon from stored soil organic matter and to produce a net flow of carbon to the atmosphere. Greenhouse gases that are trapped in "permanently" frozen soils, or permafrost, could also be released when tundras warm.

DOCUMENTING NEGATIVE FEEDBACK Negative feedback occurs when changes due to global warming result in increased uptake and sequestration of CO_2 and other greenhouse gases—meaning that global warming should be reduced. Recent experiments have shown that the growth rates of several tree species and some agricultural crops increase in direct response to increasing atmospheric CO_2. Because CO_2 is required for photosynthesis, it can act as a fertilizer.

Will ecosystem responses increase or decrease global warming? Currently, it is not clear whether positive or negative feedbacks will predominate. Researchers are working to answer this question by using computer models, experiments like the two studies highlighted in Chapter 50, and analyses of how ecosystems responded to past climate changes.

Impact on Organisms

Even though global temperatures have risen only slightly in comparison with projections for the next 50–100 years, biologists have already documented dramatic impacts on organisms.

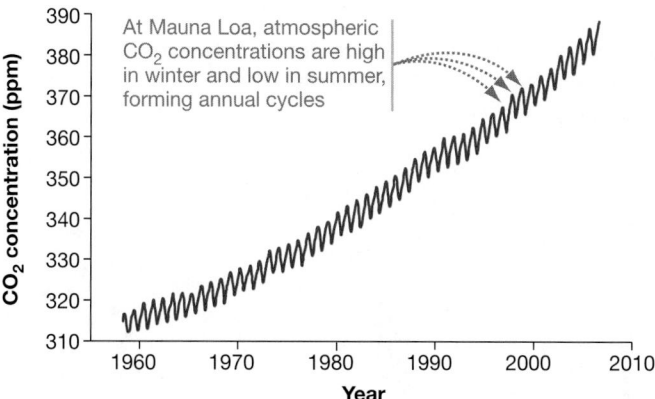

FIGURE 54.21 Recent Changes in Atmospheric CO_2. These data are from the Mauna Loa Observatory in Hawaii. Because this research station is far from large-scale human influences, it should accurately represent the average condition of the atmosphere in the Northern Hemisphere.

✔ **QUESTION** Why are atmospheric CO_2 concentrations low in the Northern Hemisphere in summer and high in winter?

GEOGRAPHIC RANGE CHANGES The geographic ranges of many organisms—including the malaria parasite—are changing. Malaria is considered a tropical disease, but the range of its causative agent, the protist *Plasmodium falciparum*, is moving north.

As another example, consider recent data on the ranges of small crustaceans, called copepods, in the North Atlantic. Copepods are key primary consumers in marine plankton. Copepods have been studied intensively because they are important commercially: They are eaten by fish that are key food sources for tertiary-consumer species of fish that are harvested and eaten by humans.

The maps in **Figure 54.22a** show changes that occurred in a 40-year period, beginning in 1960, in the number of copepod species typical of warm marine waters off southern Europe versus the number of copepod species typical of cold northern European waters. The key observation is that southern species have moved north, while the range of cold-water species has declined.

PHENOLOGY The timing of seasonal events, or **phenology**, is changing in many biomes. Recent studies have confirmed that migratory bird species are arriving on their breeding grounds in Europe earlier than in previous decades, and that fungi are fruiting earlier in some seasonal environments.

The graph in **Figure 54.22b** provides another example. Note that the *x*-axis plots time, starting in the mid-1930s and continuing through the late 1990s. The *y*-axis records the date of first flowering for a plant native to the Midwestern United States. Over approximately 60 years, the first day of the year when flowers were produced by *Baptisia leucantha* plants moved up by about 10 days. To interpret this observation, biologists hypothesize that the climate warmed enough to promote growth and flowering earlier in the year.

CORAL BLEACHING Coral reefs are among the most productive and species-rich ecosystems in the world. Reefs are large, undersea structures built by colonies of corals—cnidarians (see Chapter 32) that secrete calcium carbonate skeletons. Coral reefs are extremely productive in part because corals house symbiotic algae that perform photosynthesis.

In addition to overfishing and pollution, the integrity of coral reefs is threatened by two issues associated with global warming:

1. When atmospheric CO_2 dissolves, it reacts with water to form bicarbonate (H_2CO_3), which then drops a proton to form carbonic acid (HCO_3^-). Increased CO_2 levels in the atmosphere are increasing the rate of this reaction, and the protons that are released are lowering the pH of the oceans. Laboratory experiments have shown that corals form calcium carbonate skeletons much more slowly when pH drops. Recent data indicate that reef formation has slowed in Australia's Great Barrier Reef—perhaps in part because of acidification.

2. For reasons that are not yet clear, increases in water temperature cause reef-building corals to expel their photosynthetic algae. When this "coral bleaching" continues due to sustained exposure to warm water, corals begin to die of starvation.

EXTINCTIONS Biologists are concerned that rapid changes in climate may lead to extinction events.

- As the climate warms, alpine tundra habitats are disappearing. Plants that have restricted geographic ranges and grow on mountaintops may "run out of habitat" and go extinct.

- Researchers are concerned that habitats may change so rapidly that long-lived trees with limited dispersal will not be able to keep up—their ranges will not be able to change fast enough.

- Polar bears hunt seals that come up for air at holes in pack ice. As pack ice retreats throughout the Arctic, polar bears are dying of starvation. If present trends continue, polar bears may be extinct in the wild by 2100.

To date, though, the strongest evidence that global warming has caused species extinctions concerns amphibians. There is

(a) Cold-water copepods are declining in the North Atlantic.

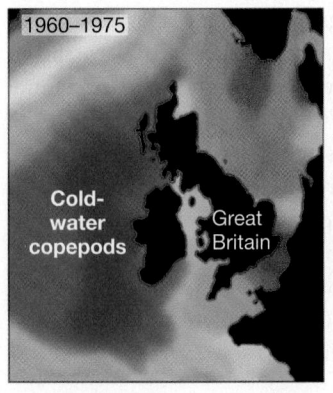
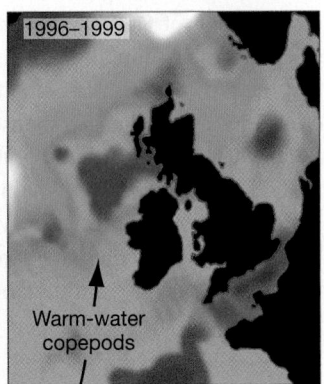

Prevalence of cold-water species Low ▬▬▬ High

(b) Flowering times for some species in midwestern North America are earlier in the year.

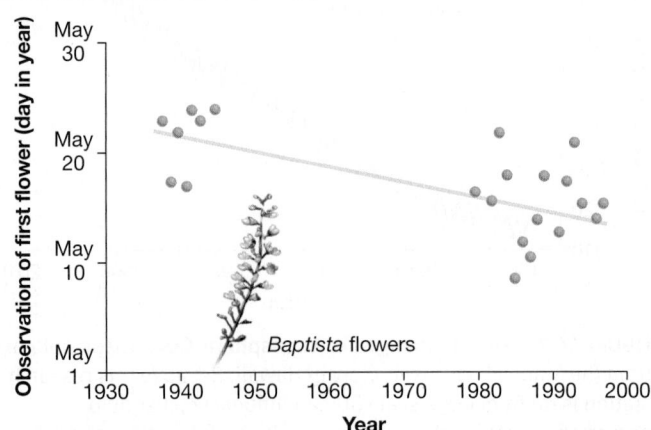

FIGURE 54.22 Global Warming Is Causing Changes in Geographic Ranges and Behavior. (a) Most species of copepod are restricted to waters of a certain temperature. **(b)** In a data set from Wisconsin, 19 of the 55 events tracked over a 61-year period occurred significantly earlier. Only one event occurred later.

convincing data supporting the hypothesis that most of the 122 frog species that have gone extinct recently succumbed to a parasitic fungus. Data published in 2006 suggest that global warming has increased the frequency and severity of these fungal infections—either because higher temperatures have stressed frogs and made them less resistant to infection, because higher temperatures create better growing conditions for the fungus, or both.

EVOLUTIONARY CHANGE WITHIN POPULATIONS In some populations, changing temperatures are already causing allele frequencies to change—meaning that some species are evolving in response to global warming.

The best data to date come from the fruit fly *Drosophila subobscura*. Recent work has shown that alleles that increase fitness in hot habitats have increased in frequency in Europe, South America, and North America independently.

To summarize, global warming has already caused significant changes in geographic ranges, phenology, and the health of coral reefs, as well as evolution in the form of extinctions and al-

lele frequency changes. If the climate models published to date are correct, it will cause many more changes over the course of your lifetime.

Productivity Changes

Global warming—sometimes in combination with human-induced changes in the nitrogen cycle—is having another effect: altering NPP. For example, experiments have documented that warming temperatures, the addition of nitrogen and other nutrients, and rising CO_2 levels can all increase NPP in some terrestrial ecosystems.

Biologists are monitoring the issue closely because NPP is so central to the functioning of ecosystems. As the first two sections of this chapter emphasized, NPP is the basis of both energy flows and nutrient flows in ecosystems.

HOW IS NPP CHANGING? **Figure 54.23** shows how average annual NPP has changed globally, in (1) terrestrial environments between 1982 and 1999, and (2) marine environments from

(a) Average annual change in terrestrial NPP, 1982–1999

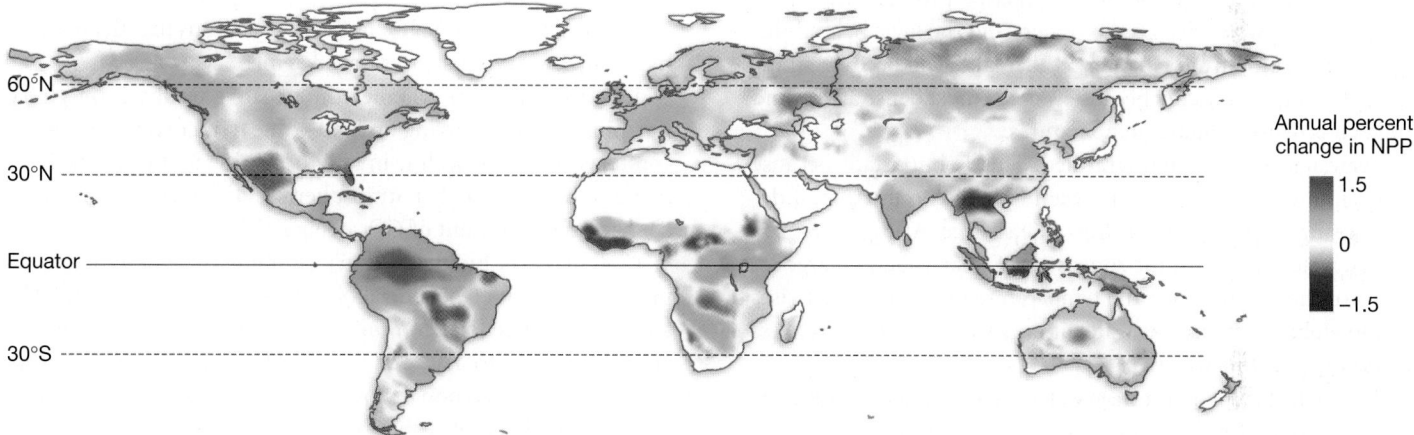

(b) Average annual change in marine NPP, 1999–2004

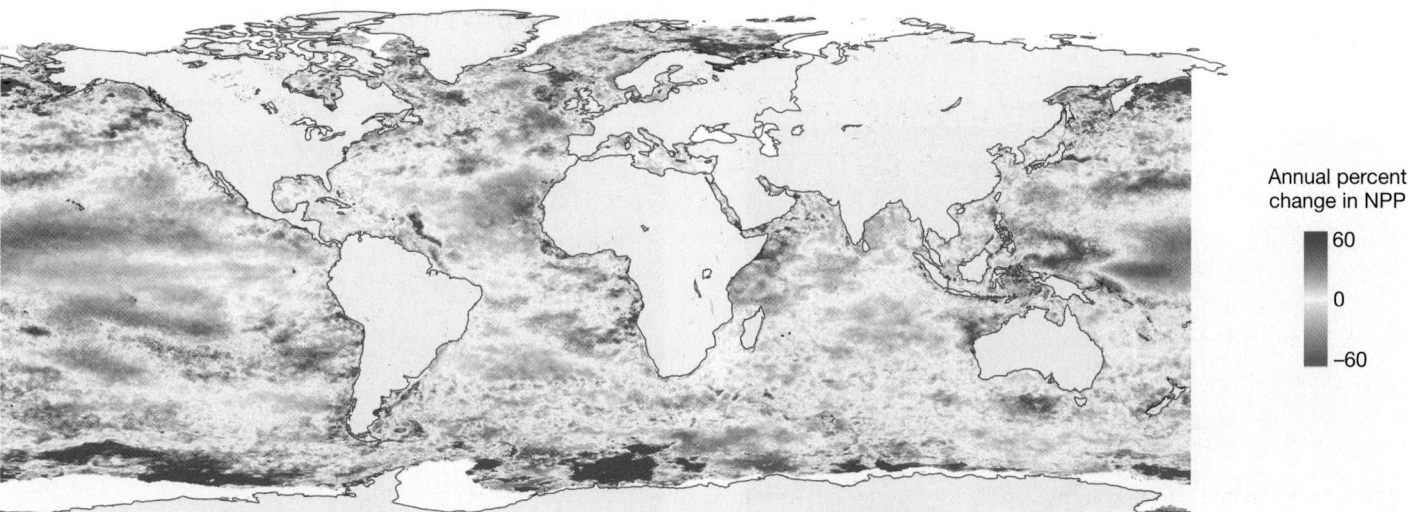

FIGURE 54.23 Consequences of Global Warming: Recent Changes in Terrestrial and Marine NPP.

1999 to 2004. These are the most recent intervals for which data are available.

On land, global NPP increased by over 6 percent during the 17-year study interval. In the oceans, though, the 1999–2004 data show a convincing correlation between increased surface water temperatures and decreased productivity. The data highlight a stark contrast: NPP is increasing on land but decreasing in the oceans.

The overall increase in terrestrial productivity is thought to be due to rising temperatures, increased rainfall in the tropics, and CO_2 fertilization—all factors that increase the rate of photosynthesis in plants. On land, increased NPP should exert a powerful negative feedback on global warming.

It's important to note, though, that global changes in NPP are underlain by considerable local variation. On land, productivity during the 1982–1999 interval tended to increase near the equator and at 60°N latitude, but to decrease in the Arctic and an array of more localized areas. In the ocean, productivity during the 1999–2004 interval dropped dramatically in large areas of the pelagic zone, but increased in a number of other regions.

WHY IS OVERALL PRODUCTIVITY DECLINING IN THE OCEANS? The leading explanation for the drop in marine productivity hinges on changes in water density. To understand the logic behind this hypothesis, recall from Chapter 50 that lakes can become stratified. Stratification occurs because water is most dense at 4°C and much less dense at higher temperatures.

Seawater also becomes stratified. This is important because water in the benthic zone is nutrient rich, due to the rain of decomposing organic material from the surface. As **Figure 54.24a** shows, water in the benthic zone is at 4°C year-round in large regions of the ocean. When the temperature of surface water rises due to global warming, the water at the surface becomes even less dense than benthic water.

Here's the key: When surface water becomes lighter, water currents are much less likely to be strong enough to overcome the density difference and bring nutrient-rich, 4°C water all the way up to the surface, where nutrients can spur the growth of photosynthetic bacteria and algae (**Figure 54.24b**). In this way, global warming is causing surface waters to become more nutrient poor.

LOCAL AND GLOBAL CONSEQUENCES What are the consequences of these local and global changes in NPP? The short answer is, "It depends."

In local areas of the ocean, the consequences of dramatically increased productivity have usually been negative.

1. Chapter 28 detailed how nitrate ions—derived from fertilizers applied to cornfields in midwestern North America—are washing into the Gulf of Mexico. Increased nitrate concentrations have stimulated the growth of planktonic organisms, whose subsequent decomposition has used up available oxygen and triggered the formation of an anoxic "dead zone."

2. Warmer water, in combination with increases in nitrate and other fertilizers of human origin, has increased the frequency or intensity of harmful algal blooms in the ocean—large populations of dinoflagellates that release toxins into the surrounding water (see Chapter 29).

On a global level, lower NPP in the ocean means that the productivity of the world's fisheries may decline.

It is also not clear whether the negative feedback on global warming that is occurring on land—due to increased NPP—will compensate for the decline in NPP occurring in the oceans. It remains to be seen whether atmospheric CO_2 will rise or fall due to changes in the amount of CO_2 being fixed by photosynthesis.

Overall, then, biologists don't yet know whether changes in productivity will be beneficial or detrimental to ecosystems. The answer should be forthcoming, however. Because global temperatures continue to rise and large-scale additions of nitrogen, phosphorus, and carbon dioxide are ongoing, humans are implementing a global-scale experiment on the effects of altering NPP.

(a) Much of the ocean is stratified by density and temperature.

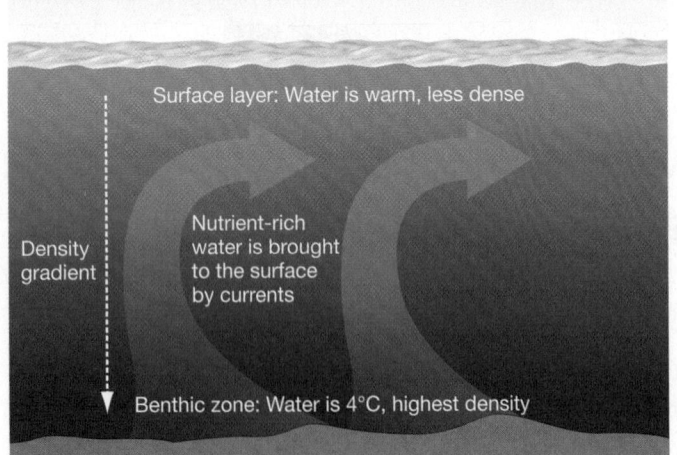

(b) Global warming increases the density gradient, making it less likely for layers to mix.

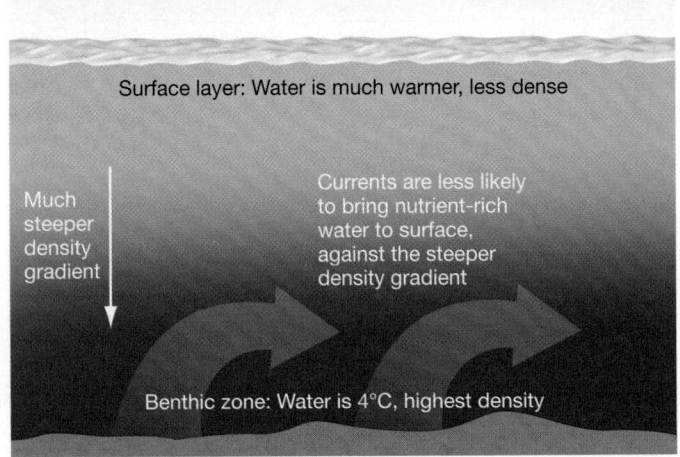

FIGURE 54.24 Temperature Stratification in the Ocean May Affect Productivity.

Summary of Key Concepts

🔑 **An ecosystem has four components: (1) the abiotic environment, (2) primary producers, (3) consumers, and (4) decomposers. These components are linked by the movement of energy and nutrients.**

- An ecosystem consists of species that interact in a food web, along with their abiotic environment.

- As energy flows through ecosystems and as nutrients cycle through them, energy and nutrients are exchanged between biotic and abiotic components of the ecosystem.

- Net primary productivity (NPP) is the foundation of ecosystems, because it represents the energy available for consumption by other organisms. Humans are appropriating a large percentage of NPP.

✔ You should be able to draw the relationships among the four components of an ecosystem where no photosynthesis occurs—the outside source of energy is iron- and sulfur-containing compounds released from magma (liquid rock).

🔑 **As energy flows from producers to consumers and decomposers in a food web, much of it is lost. The productivity of terrestrial ecosystems is limited by warmth and moisture; nutrient availability is the key constraint in aquatic ecosystems.**

- Organisms that acquire energy from the same type of source are said to occupy the same trophic level. Because energy transfer from one trophic level to the next is inefficient, ecosystems have a pyramid of productivity: Biomass production is highest at the lowest trophic level and lower at each higher trophic level.

- The feeding relationships among species in a particular ecosystem are described by a food chain or food web. Changes in the species composition of a food web can cause trophic cascades; persistent toxins can be magnified in concentration as they are passed up trophic levels in a food web.

- Among terrestrial ecosystems, NPP is highest in tropical wet forests and tropical dry forests; among aquatic ecosystems, productivity is highest in coral reefs, wetlands, and estuaries.

- Although productivity is extremely low in the oceanic zone, the area covered by this ecosystem is so extensive that it accounts for the highest percentage of Earth's overall productivity.

✔ You should be able to describe the nature of a food chain where a minimum of energy would be lost.

🔑 **To analyze nutrient cycles, biologists focus on the nature of the reservoirs where elements reside and the processes that move elements between reservoirs. Nutrient addition by humans is increasing productivity and causing pollution.**

- Nutrients move through ecosystems in biogeochemical cycles.

- The rate of nutrient cycling is strongly affected by the rate of decomposition of detritus. The decomposition rate, in turn, is affected by abiotic environmental conditions such as temperature and by the quality of the detritus.

- Nutrients "leak" from ecosystems when biomass leaves. Experiments have shown that the loss of vegetation greatly increases the rate of nutrient loss.

- Human activities have doubled the amount of usable nitrogen that enters ecosystems each year. These increases have led to increased productivity, but also to pollution and to loss of biodiversity.

✔ You should be able to describe conditions under which decomposition rates in terrestrial and marine environments are extremely low—creating a carbon sink due to a buildup of organic matter.

MB) **BioFlix™** The Carbon Cycle, **Web Activity** The Global Carbon Cycle

🔑 **The burning of fossil fuels has led to rapid global warming; rapid ecological and evolutionary changes are being observed in response.**

- Average global temperatures are increasing rapidly because land-use changes and burning of fossil fuels have increased the flow of carbon in the form of CO_2 into the atmosphere, and because carbon dioxide acts as a greenhouse gas.

- Both positive and negative feedbacks are occurring in response to global warming.

- Biologists have documented an array of consequences, including extinctions and changes in productivity, allele frequencies, and geographic ranges.

✔ You should be able to propose two mechanisms that humans could adopt to reduce global warming and explain the logic behind each.

Questions

1. What is the difference between a community and an ecosystem?
   a. An ecosystem comprises a community and the abiotic environment.
   b. An ecosystem is a type of community.
   c. An ecosystem includes only the plant community or communities present in an environment.
   d. An ecosystem includes only the abiotic aspects of a particular environment.

2. Which of the following ecosystems would you expect to have the highest primary production?
   a. subtropical desert
   b. temperate grassland
   c. boreal forest
   d. tropical dry forest

3. Most of the net primary productivity that is consumed is used for what purpose?
   a. respiration by primary consumers
   b. respiration by secondary consumers
   c. growth by primary consumers
   d. growth by secondary consumers

4. What is biomagnification?
   a. biological imaging that provides images or data at fine spatial scales
   b. conversion of an energy input (e.g., sunlight) to chemical energy in the form of sugars, by primary producers
   c. accumulation of certain molecules at high concentration at upper trophic levels
   d. "overfertilization"—increases in nutrients that lead to pollution (detrimental effects)

5. Which of the following is normally the longest-lived reservoir for carbon?
   a. atmosphere CO_2
   b. marine plankton (primary producers *and* consumers)
   c. petroleum
   d. wood

6. Devegetation has what effect on ecosystem dynamics?
   a. It increases belowground biomass.
   b. It increases nutrient export.
   c. It increases rates of groundwater recharge (penetration of precipitation to the water table).
   d. It increases the pool of soil organic matter.

✓ TEST YOUR UNDERSTANDING

Answers are available in Appendix B

1. Explain the difference between a food chain and a food web. What do the arrows in each type of diagram represent?

2. Suppose that cougars were reintroduced to a region where deer populations are currently so high that most tree saplings and other low-growing plant species are consumed. Using the wolf reintroduction in Yellowstone as a model, predict the trophic cascade that might occur. (Cougars eat deer.)

3. Explain why decomposition rates are higher in some ecosystems than in others, and give examples. How do decomposers regulate nutrient availability in an ecosystem?

4. Define "bleaching" and ocean acidification, and explain how each is affecting coral reefs.

5. Name one positive feedback on global warming and one negative feedback. Explain why each is occurring.

6. Why are the open oceans nutrient poor? Why are coastal areas and intertidal habitats relatively nutrient rich?

✓ APPLYING CONCEPTS TO NEW SITUATIONS

Answers are available in Appendix B

1. Suppose you had a small set of experimental ponds at your disposal and an array of pond-dwelling algae, plants, and animals. How could you use a radioactive isotope of phosphorus to study nutrient cycling in these experimental ecosystems?

2. Fertilizing the open oceans with iron is controversial, because some researchers predict that it could produce dead zones in what are now valuable fishing areas near coastlines. Explain how this could happen, assuming that ocean currents move water under fertilized areas in the open ocean to coastal regions. (Hint: review the discussion of dead zones in Chapter 28.)

3. Suppose that herbivores were removed from a temperate deciduous forest ecosystem (e.g., by fencing to exclude herbivorous mammals, and spraying insecticides to kill herbivorous insects). Predict what would happen to the rate of nitrogen cycling. Explain the logic behind your prediction.

4. During the Carboniferous Period, rates of decomposition slowed even though plant growth was extensive (probably due to the formation of vast, oxygen-poor swamp habitats). As a result, large amounts of biomass accumulated in terrestrial environments (much of this biomass is now coal). The fossil record indicates that atmospheric oxygen increased, atmospheric carbon dioxide increased, and global temperatures dropped. Explain why.

An enormous carved stone statue—a portion of which has fallen on its side—on Easter Island. Archaeological data indicate that Easter Island was once covered by extensive forests that supported a thriving human population.

Biodiversity and Conservation Biology 55

When Europeans arrived on Easter Island—also known as Rapa Nui—about 3500 km (2180 miles) west of Chile, they found about 1000 people and a desolate, treeless landscape dotted with gigantic stone statues. Things were not always this way. Archaeological digs and pollen samples indicate that the island was once covered with lush forest dominated by the world's biggest palm tree—now extinct—along with what may have been the largest colonies of breeding seabirds in the Pacific Ocean.

About 800 years ago, Easter Island was a lush tropical paradise that supported a thriving population estimated at 10,000 people. What happened?

According to Jared Diamond, the archaeological and fossil data indicate that an ecological disaster happened. Diamond hypothesizes that the human population grew to massive size as people used canoes carved from palm logs to harvest the abundant seabirds, fish, and dolphins. The islanders had the leisure time to carve gigantic statues and roll them into place on beds of palm logs.

But once all the palms were cut, the system collapsed. Without palm trunks to make canoes, dolphin hunting and fishing were much less efficient. Soil erosion from the deforested slopes may have cut into the productivity of banana and sweet potato plantations. The population crashed to a tenth of its former size, and the great statues fell. On Easter, biodiversity losses had direct impacts on human health and welfare.

Easter is part of a larger island: Earth. The species and resources on our planet are limited, but the human population is growing unchecked. Are we headed for an Easter Island–like collapse?

Many biologists are afraid that the answer might be yes. This chapter's goal is to analyze the data responsible for this fear, and introduce the positive steps that are being taken to avert an Easter Island–like scenario. Let's begin by defining **biodiversity**—the thing that biologists want to protect.

KEY CONCEPTS

- Biodiversity is quantified at the level of allelic diversity, species diversity, and ecosystem diversity.

- According to recent analyses, the sixth mass extinction in the history of life is occurring due to habitat loss, overexploitation, and global climate change.

- Humans depend on biodiversity for the products that wild species provide and for ecosystem services that protect the quality of the abiotic environment.

- Solutions to the biodiversity crisis include protecting key habitats, lowering human population growth and resource use, restoring ecosystems, mitigating climate change, and supporting sustainable development.

✔ When you see this checkmark, stop and test yourself. Answers are available in Appendix B.

55.1 What Is Biodiversity?

Perhaps the simplest way to think about biodiversity is in terms of the tree of life—the phylogenetic tree of all organisms, introduced in Chapter 1 and explored in detail in Chapters 28 through 35. If biologists are eventually able to estimate the complete tree of life, using the techniques reviewed in Chapter 27, the branches would represent all of the lineages of organisms living today; the tips would represent all of the species.

When biodiversity increases, branches and tips are added and the tree of life gets fuller and bushier. When extinctions occur, tips and perhaps branches are removed; the tree of life becomes thinner and sparser.

Coordinated, multinational efforts are under way to estimate the tree of life—particularly major lineages such as the fungi, brown algae, and land plants. These and other efforts to estimate phylogenies are producing a better picture of the relationships among lineages and species, and a clearer understanding of how extinction is affecting biodiversity. But even though the tree of life is a compelling way to document biodiversity, it does not tell the entire story.

Biodiversity Can Be Measured and Analyzed at Several Levels

🔑 To get a complete understanding of the diversity of life, biologists recognize and analyze biodiversity at the genetic, species, and ecosystem level. Let's consider each in turn.

GENETIC DIVERSITY **Genetic diversity** is the total genetic information contained within all individuals of a species. It is measured as the number and relative frequency of all alleles present in a species. At the genetic level, biodiversity is everywhere around you, from the variety of apples at the grocery store to the genetically based variation in coloration and singing ability of sparrows in your backyard.

Because no two members of the same species are genetically identical, most species are the repository of an immense array of alleles. For example, distinct populations of the same species often contain unique alleles—alleles that are maintained by genetic drift or have been favored by natural selection, as adaptations to local conditions.

Recently, efforts to catalog genetic diversity have been boosted by two technical breakthroughs:

1. the ability to sequence the entire genome of multiple members of the same species, and

2. the research protocol called environmental sequencing.

Both approaches were introduced in Chapter 20. You might recall that in environmental sequencing, researchers take a sample from the soil or water present in a habitat and sequence all or most of the genes present. It's a way to document genetic diversity even if the species present are unknown.

SPECIES DIVERSITY As its name implies, **species diversity** is based on the variety of species on Earth. Chapter 53 pointed out

that species diversity is measured by quantifying the number and relative frequency of species in a particular region.

Recent efforts to document species diversity have used a new technique called **bar coding**: the use of a well-characterized gene sequence to identify distinct species, using the phylogenetic species concept introduced in Chapter 26. The idea is that variation in one gene—for example, a portion of the DNA found in the mitochondria of mammals, specifically the gene for cytochrome *c* oxidase subunit 1—allows biologists to recognize different species, much as a slight variation in a bar code allows a grocery store scanner to recognize different products. The specific gene used as a bar code varies among lineages. Researchers who are contributing to this effort submit sequence data from "bar coding genes" to centralized databases.

There is an additional aspect of species diversity, however, that biologists call taxonomic diversity. Taxonomic diversity is important because some lineages on the tree of life are extremely species rich. Prominent examples include the African cichlid fish, introduced in Chapter 27, and the 35,000 species of orchids. Other lineages, in contrast, are extremely species poor. Some, in fact, are represented by a single species (**Figure 55.1**).

Some biologists argue that it is particularly important to preserve populations from species-poor lineages, because they are the last living representatives of their lineages. If those populations are wiped out, an entire lineage is lost forever. Such is the case with the Yangtze River dolphin in Figure 55.1b. Most biologists concede that the species is now extinct.

ECOSYSTEM DIVERSITY **Ecosystem diversity** is the array of biotic communities in a region along with abiotic components, such as soil, water, and nutrients. Ecosystem diversity is more difficult to define and measure than genetic diversity or species diversity, however, because ecosystems do not have sharp boundaries. Recall from Chapter 54 that ecosystems are complex and dynamic assemblages of organisms that interact with each other and their nonliving environment.

Attempts to measure ecosystem diversity focus on capturing the array of biotic communities in a region, along with variation in the physical conditions present. Areas around estuaries, for example, tend to have high ecosystem diversity due to the combination of stream, wetland, intertidal, coastal, and upland habitats (see Chapter 50).

Note that the three levels of biodiversity are not independent of each other. When an estuary is dredged or filled, for example, biodiversity is affected at all three levels: genetic and species diversity change due to the different numbers and types of individuals present, and ecosystem diversity is altered due to the change in abiotic conditions.

CHANGE THROUGH TIME Biodiversity can be recognized and quantified on several distinct levels, but it is also dynamic.

- Mutations that create new alleles increase genetic diversity; natural selection, genetic drift, and gene flow may eliminate certain alleles or change their frequency, leading to an increase or decrease in overall genetic diversity.

(a) Red panda

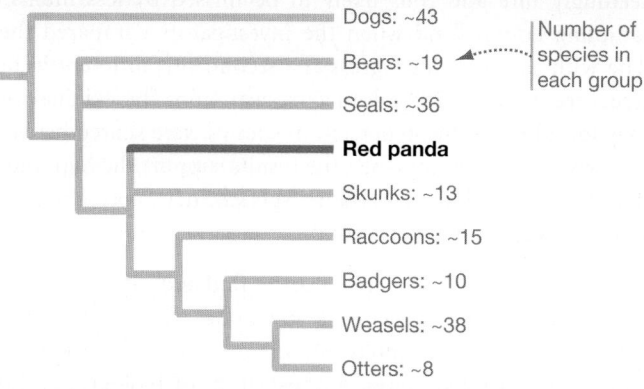

Dogs: ~43
Bears: ~19 ◄┈┈┈┤ Number of species in each group
Seals: ~36
Red panda
Skunks: ~13
Raccoons: ~15
Badgers: ~10
Weasels: ~38
Otters: ~8

(b) Yangtze river dolphin

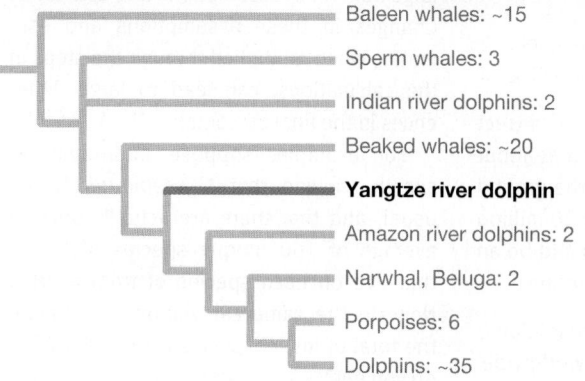

Baleen whales: ~15
Sperm whales: 3
Indian river dolphins: 2
Beaked whales: ~20
Yangtze river dolphin
Amazon river dolphins: 2
Narwhal, Beluga: 2
Porpoises: 6
Dolphins: ~35

FIGURE 55.1 Phylogenetically Distinct Species May Be High-Priority Targets for Conservation. The red panda and the Yangtze river dolphin have few close relatives and represent distinct branches on the tree of life. If conservation measures attempt to preserve taxonomic diversity, these species would be a high priority for preservation. Unfortunately, it is too late for the Yangtze river dolphin—the species is now considered extinct.

● Speciation increases species diversity; extinction decreases it.

● Changes in climate or other physical conditions can result in the formation of new ecosystems, as can the evolution of new species that interact in novel ways. Disturbances such as volcanic eruptions, human activities, and glaciation can destroy ecosystems.

Biodiversity is not static. It has been changing since life on Earth began.

How Many Species Are Living Today?

One of the simplest questions about biodiversity is also one of the most difficult to answer: How many species are there on Earth? The answer is not known. Given the massive effort that it would take to document every form of life on Earth, the answer will probably never be known.

Biologists are well aware, though, that the approximately 1.5 million species cataloged to date represent a tiny fraction of the number actually present. Chapter 31, for example, noted that in Britain, an average of six fungal species live on each well-studied plant species. Because over 275,000 plant species have been described, a conservative estimate based on the 6:1 fungi:plant ratio—assuming that the ratio is universal—suggests that 1.65 million species of fungi exist, not counting species that don't live on plants. But only 80,000 have been described thus far, in total.

The issue is even more acute in poorly studied lineages like bacteria and archaea, where researchers routinely discover dozens of new species in each environmental sequencing or direct sequencing study (see Chapter 20 and Chapter 28). For example, a recent direct sequencing study estimated that over 500 species of bacteria live in the human mouth, but only about half had been described and named already. Even among well-studied groups like birds and mammals, new species are discovered almost every year.

Given that only a fraction of the organisms alive have been discovered to date, how can biologists go about estimating the total number of species on Earth? One approach is based on intensive surveys of species-rich groups at small sites; a second strategy is based on attempts to identify all of the species present in a particular region.

TAXON-SPECIFIC SURVEYS Terry Erwin and J. C. Scott published a classic study of species diversity in a single taxon. They began by estimating the number of insect species that live in the canopy of a single tropical tree (**Figure 55.2** on page 1108). To do this, they used an insecticide to knock down insects from the top of a *Luehea seemannii* tree and identified over 900 beetle species among the individuals that fell. Most of the species were new to science.

To use these data as an indicator of global arthropod species diversity, Erwin used the logic outlined in **Box 55.1** on page 1108 to extrapolate from the diversity found on a single tree to the diversity on all known tropical trees. His analysis came up with an estimate of 30 million insect species.

More recent single-taxon surveys have sampled larger geographic areas. For example, Philippe Bouchet and colleagues conducted a massive survey of marine mollusks in coral-reef habitats along the west coast of New Caledonia, a tropical island

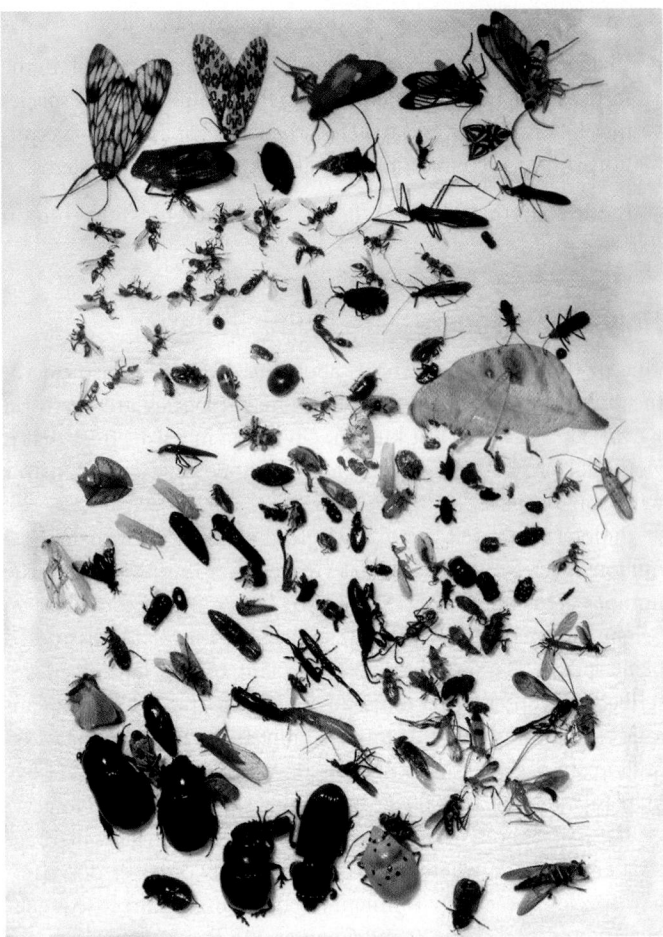

FIGURE 55.2 Estimating Species Richness via Intensive Local Sampling. A small sample of the insects collected from a single tree in the Amazonian rain forest.

in the southwest Pacific Ocean. The team spent more than a year collecting mollusks at 42 sites over a total area of almost 300 square kilometers. The survey represented the most thorough sampling effort ever made to determine the species diversity of mollusks, and produced more than 127,000 individuals representing 2738 species. These numbers far exceed the mollusk diversity recorded for any comparable-sized area. The species total was 2 to 3 times the total number of mollusk species that had been reported for similar habitats in the region.

In reporting their findings, the biologists emphasized that 20 percent of the species found were represented by a single specimen. This observation suggests that many species are exceedingly rare and thus likely to be missed by less-intensive sampling efforts. And when the investigators compared their data with a survey in progress at a second site, different in reef structure but only 200 kilometers away from the original site, they found that only 36 percent of species were shared between the two sites. Taken together, the results support the hypothesis that the 93,000 known mollusk species may represent just a third to a half of the actual total.

ALL-TAXA SURVEYS The first effort to find and catalog all of the species present in a large area is now under way. The location is the 2200 km^2 Great Smoky Mountains National Park in the southeastern United States. A consortium of biologists, volunteers, and research organizations initiated this All Taxa Biodiversity Inventory in 1998.

To date, the survey has discovered over 890 species that are new to science and over 6300 species that had never before been found in the park. When the inventory is complete, in 2015, biologists will have the best database ever assembled on the biodiversity present in a region.

BOX 55.1 QUANTITATIVE METHODS: Extrapolation Techniques

To estimate the world total of insect species from the specimens found on a single tree, Terry Erwin performed the following extrapolation:

1. Estimate the number of beetle species that live only on *L. seemannii*. Based on earlier work with insects on this tree species, Erwin estimated the number of unique beetle species at 160.

2. Estimate the number of beetle species found on all tropical trees, assuming that there are 50,000 species and assuming that each tree species hosts 160 unique species of beetle.

    160 × 50,000 = 8 million beetle species in tropical trees worldwide

3. Estimate the total number of insect species, based on the estimate that 40 percent

(0.40) of all known insect species are beetles.

    8 million/0.40 = 20 million insect species in tropical trees worldwide

4. Estimate the total number of insect species worldwide, assuming that about 33 percent of insects live on the ground instead of in trees. If there are 20 million species in trees, then there would be an additional 10 million on the ground.

    20 million + 10 million = 30 million insect species in all habitats worldwide

The total number is shockingly large—only about a million species are known to date.

It is important to be cautious—even skeptical—about extrapolation-based esti-

mates like this, however, as they rely on a large number of assumptions and estimates. Changes in these assumptions and estimates, when multiplied through the steps in the calculations, can lead to large differences in the final outcome.

For example, suppose additional research showed that *L. seemannii* is unusual, and that there are actually only an average of 100 unique species of beetle that live on each species of tropical tree. Now do the same calculation to estimate the total of insect species worldwide. What do you get?[1]

Erwin was aware of this point; the purpose of his study was to stimulate discussion and more research.

[1] 18.7 million

55.2 Where Is Biodiversity Highest?

If documenting the extent of genetic, species, and ecosystem diversity is the most fundamental goal of research on biodiversity, then the second-most important aim is to understand its geographic distribution. Chapter 53 introduced the most prominent geographic pattern in the distribution of biodiversity: In most taxonomic groups, species richness is highest in the tropics and declines toward the poles.

Tropical rain forests are particularly species rich. Even though they represent just 7 percent of Earth's land area, they are thought to contain at least 50 percent of all species present. As Chapter 53 noted, understanding why tropical forests are so species rich and why a latitudinal gradient occurs is an active research area.

Hotspots of Biodiversity and Endemism

Biologists have also been working to understand the distribution of species richness at finer spatial scales than global gradients. The map in **Figure 55.3a**, for example, was constructed by dividing the world's landmasses into a grid of cells 1° latitude by 1° longitude. In the tropics, this translates to rectangles approximately 111 km by 111 km (69 mi by 69 mi). Because lines of longitude converge at the poles, the rectangles are about 111 km by 55 km at 60° latitude. Using published data, researchers then plotted how many species of birds breed in each cell in the grid. These data are consistent with the latitudinal gradient in species richness, but indicate that some areas of the tropics are much more species rich than others.

An organization called Conservation International pioneered these types of analyses and promoted the term **biodiversity hot spot** to characterize regions that are extraordinarily rich in biodiversity. In terms of bird species richness, the Andes mountains, along the west coast of South America, the Amazon River basin in the central part of South America, portions of East Africa, and southwest China are important hotspots.

Researchers have also been keenly interested in understanding which regions of the world have a high proportion of **endemic species**—meaning, species that are found in an area and nowhere else. **Figure 55.3b** maps the location of endemic species of breeding birds, based on the same cells and data as the species richness map in Figure 55.3a.

(a) Biodiversity hotspots in terms of **species richness of birds**

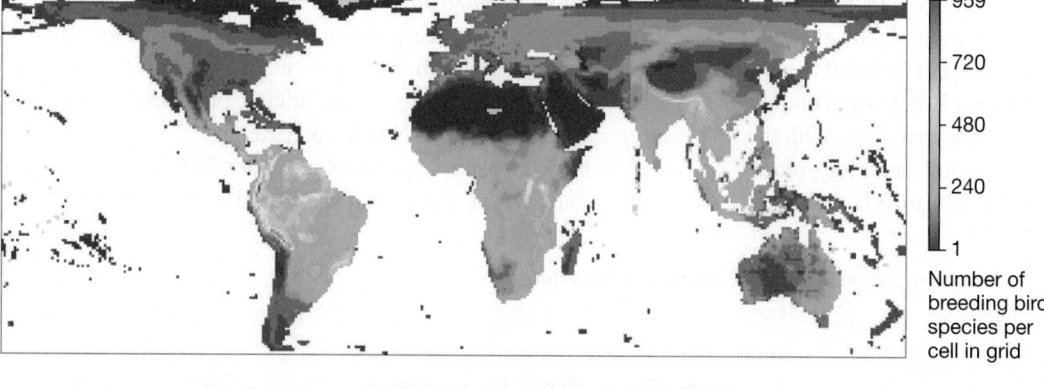

959
720
480
240
1

Number of breeding bird species per cell in grid

FIGURE 55.3 Biodiversity Hotspots Have Extraordinarily High Species Richness. Both maps are based on a grid of cells that is 1° longitude by 1° latitude.

(b) Biodiversity hotspots in terms of **endemic species of birds**

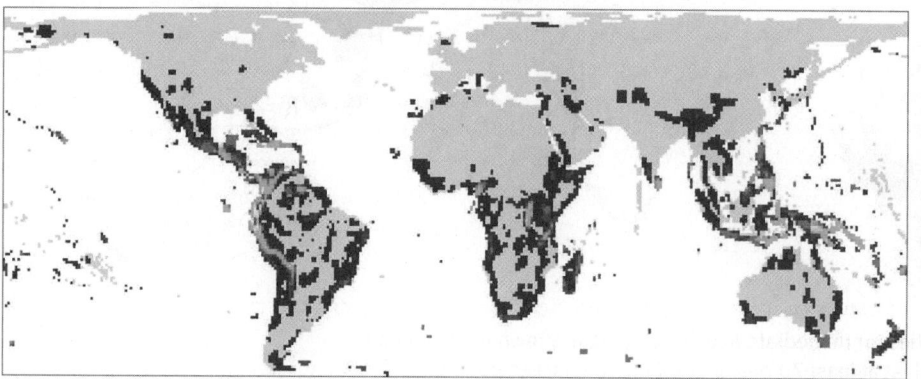

89
66
45
22
1

Number of endemic bird species per cell in grid

Conservation Hotspots

Mapping species richness hotspots and centers of endemism can inspire research on *why* certain regions contain many species or a high proportion of endemic taxa. But in addition, biologists are studying the geographic distribution of biodiversity as a way of focusing conservation efforts.

Starting in the year 1999, for example, a multinational team coordinated by Conservation International set out to identify regions of the world that meet two criteria:

1. They contain at least 1500 endemic vascular plant species (>0.5 percent of the world's total).

2. They have lost at least 70 percent of their traditional or primary vegetation.

The goal was to identify regions of the world that are in most urgent need of conservation action—areas where efforts to preserve habitat would have the highest return on investment. The idea was that efforts to protect a diversity of primary producers would guarantee the protection of a wide array of primary, secondary, and tertiary consumers.

The 34 regions that currently meet the two "conservation hotspot" criteria are shown in **Figure 55.4**. Although these areas represent just 2.3 percent of the Earth's land area, they contain over 50 percent of all known plant species and 42 percent of all known terrestrial vertebrate species, as endemics.

The 34 hotspots were once much more extensive than they are now. Overall, 86 percent of their habitat area has already been destroyed. In effect, much of the world's biodiversity is being squeezed into a few shrinking spaces.

It's also important to note that key regions with high plant species diversity—specifically the rain forests found in the Amazon, the Congo River basin, and New Guinea—are not included as conservation hotspots, because they have not yet lost 70 percent of their primary vegetation. Effectively protecting these areas along with the hotspots defined on the map would provide protection for over 65 percent of all land plants in just 5 percent of the Earth's land surface.

55.3 Threats to Biodiversity

No species lasts forever. Climate change, disease, competition from newly arrived species, and habitat alteration are all natural processes. They have been happening since life began. Extinction, like death, is a fact of life.

If extinction is natural, why are biologists so concerned about habitat and species conservation? The answer is rate.

Today, species are vanishing faster than at virtually any other time in Earth's history. Modern rates of extinction are 100 to 1000 times greater than the average, or "background," rate recorded in the fossil record over the past 550 million years.

Either directly or indirectly, current extinctions are being caused by the demands of a rapidly growing human population—currently at over 6.8 billion as this book goes to press and increasing by about 77 million people per year. The island that is planet Earth is much more crowded than it used to be.

Based on the data in hand, the vast majority of biologists agree that the sixth mass extinction in the history of multicellular life is occurring. Between the time you are reading this and the time your great-grandchildren are grown, human impacts on the planet promise to equal or exceed those of the gigantic asteroid that smashed into Earth 65 million years ago.

Changes in the Nature of the Problem

Most extinctions that have occurred over the past 1000 years took place on islands and were either caused by overexploitation—meaning, overhunting or overfishing—or the introduction of **exotic species**—nonnative competitors, diseases, or predators.

You might recall, from Chapter 50, that if an exotic species is introduced to a new area, grows to large population size, and disrupts species native to the area, it is called an invasive. That chapter discussed how cheatgrass has altered plant communities in large sections of western North America.

The brown tree snake pictured in **Figure 55.5** is a dramatic example of an invasive species that has caused extinctions. After being

■ Biodiversity hotspots in terms of high proportion of **endemic plants** *and* **high threat**

FIGURE 55.4 Conservation Hotspots Identify Priorities for Immediate Action. The areas shown in red contain a high diversity of vascular plant species and have lost at least 70 percent of their original vegetation.

FIGURE 55.5 **Invasive Species Are Destructive.** The brown tree snake has extinguished dozens of species of birds on Guam.

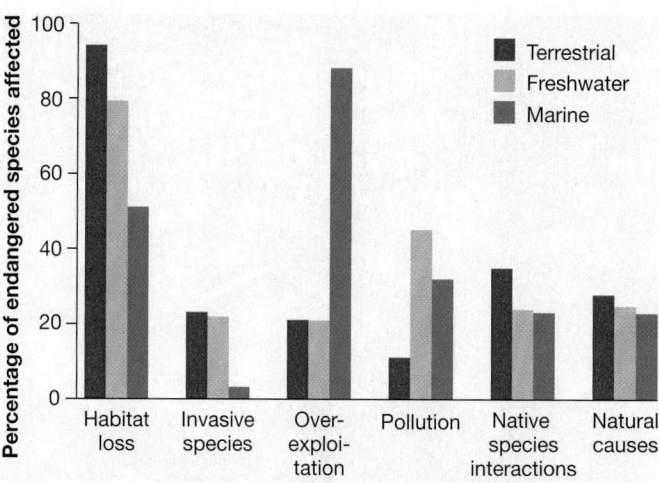

FIGURE 55.6 **Endangerment Is Caused by Multiple Factors.** These data are from Canada.

✔ **QUESTION** Why do the data for terrestrial, freshwater, and marine species each add up to more than 100 percent?

introduced to the island of Guam, predation by the expanding snake population caused the extinction of dozens of bird species.

The most dramatic example of extinctions due to overexploitation and exotic species involves birds native to islands in the South Pacific. Recent research on fossils found on islands throughout the region suggests that about 2000 bird species were wiped out as people colonized this area about A.D. 1200. Many of these extinctions occurred due to predation by humans or by rats, pigs, and other animals introduced by humans. A loss of 2000 species is significant—about 10,000 bird species are left today.

Starting in the twentieth century, though, two patterns in species loss began to change:

1. **Endangered species**, which are almost certain to go extinct unless effective conservation programs are put in place, are now more likely to live on continents than islands.

2. Although overexploitation and invasives continue to be serious problems, habitat destruction and global climate change are now the primary threats to species.

The bar graphs in **Figure 55.6** show the results of a recent analysis on causes of endangerment for 488 endangered species native to Canada. Several patterns jump out of the data:

- Habitat loss is the single most important factor in the decline of these species. It is a significant issue for over 90 percent of endangered species in terrestrial environments.

- Virtually all of the endangered species are affected by more than one factor. This is important, because it means that conservation biologists may have to solve more than one problem for any given species to recover.

- Overexploitation is the dominant problem for marine species, while pollution plays a large role for freshwater species.

- Factors beyond human control can be important. These include predation or competition with native species, natural disturbances such as fires or droughts, or the fact that some species have narrow niches and have historically been rare. Stated another way, background extinctions will continue to occur.

Do these conclusions hold up for other parts of the world? Although invasive species are a major problem in some regions, most analyses completed to date are broadly consistent with the Canadian data. For example, overexploitation is a grave concern for marine fisheries worldwide. Two-thirds of harvestable species are now considered fully exploited or depleted, and a recent study of the global fishing industry concluded that 90 percent of individuals in large-bodied fish species from bottom or open-water habitats have been already removed from the world's oceans. The species affected include tuna, swordfish, marlin, cod, halibut, and flounder. Overhunting has also emerged recently as a dire threat to many mammal populations in Africa and southeast Asia (**Figure 55.7**).

FIGURE 55.7 **Overexploitation Continues to Be a Problem in Some Regions.** The recent growth of the "bushmeat" trade has devastated many populations of primates and other mammals in Africa. Overexploitation is also a major cause of endangerment in marine species worldwide (see Figure 55.6).

FIGURE 55.8 Suburbanization Is a Major Cause of Habitat Loss in North America.

FIGURE 55.9 Rates of Deforestation in Tropical Wet Forests Are High. In the satellite image from Rondônia, Brazil, at top, the dark green areas are intact forest. In the lower image of the same location 21 years later, the light areas have been burned or logged and converted to agricultural fields and pastures.

HABITAT DESTRUCTION Humans cause **habitat destruction** by logging and burning forests, damming rivers, dredging or filling estuaries and wetlands, plowing prairies, grazing livestock, excavating minerals, and building housing developments, golf courses, shopping centers, office complexes, airports, and roads (**Figure 55.8**).

On a global scale, one of the most important types of habitat destruction is deforestation—especially the conversion of primary forests to agricultural fields and human settlements. To appreciate the extent of deforestation in some areas, consider satellite images of wet tropical rain forest in the Brazilian state of Rondônia that were made in 1984 and in 2005 (**Figure 55.9**).

Analyses of satellite photos such as these have shown that as many as 3 million hectares, an area about 10 percent larger than the state of Maryland, was deforested in the Amazon each year during the 1990s. (A hectare, abbreviated ha, is 100 meters by 100 meters, approximately the size of two football fields.)

More recent analyses suggest that the global rate of deforestation has now slowed compared to peak rates in the 1990s. The latest report from the United Nations Food and Agriculture Organization, released in 2005, states that the world's forests experienced a net loss of about 7.3 million ha/year between 2000 and 2005. This is down from the average net loss of 8.9 million ha/year during the previous decade.

Conservationists greeted this report with caution. Net losses dropped because of extensive forest regrowth in China, North America, and Europe, even though over 8 million ha/year of primary—meaning, previously uncut—forest were lost in South America, southeast Asia, and Africa each year from 2000 to 2005.

More specifically, the total area of wet tropical forest dropped by 2.36 percent during 2000–2005. If this rate of tropical deforestation continues, over 28 percent of the wet tropical forest that exists today will be gone in your lifetime, including almost half of the Brazilian Amazon. These projections are extremely conservative, as the human population in Brazil, Indonesia, and other key countries is still increasing rapidly.

In essence, the forests of the Amazon and southeast Asia are experiencing the same type of deforestation that occurred in European and North American forests from 1700 to 1900. Forest loss in South America and Africa is particularly important, however, because it is occurring in biodiversity hotspots.

HABITAT FRAGMENTATION In addition to destroying natural areas outright, human activities fragment large, contiguous areas of natural habitats into small, isolated fragments. **Habitat fragmentation** concerns biologists for several reasons.

1. It can reduce habitats to a size that is too small to support some species. This is especially true for keystone predators (see Chapters 53 and 54) such as wolves, mountain lions, grizzly bears, and bluefin tuna, which need vast natural spaces in which to feed, find mates, and reproduce successfully. Large animal species demand large spaces.

2. By creating islands of habitat in a sea of human-dominated landscapes, fragmentation reduces the ability of individuals to disperse from one habitat to another. Isolated populations are vulnerable to the types of chance catastrophes and genetic problems detailed later in the chapter.

3. Fragmentation creates large amounts of "edge" habitat. The edges of intact habitat are subject to invasion by weedy species and are exposed to more intense sunlight and wind, creating difficult conditions for plants.

The decline in habitat quality caused by fragmentation is being documented in a long-term experiment in a tropical wet forest. In an area near Manaus, Brazil, that was slated for clear-cutting, a research group set up 66 square, 1-hectare experimental plots that remained uncut. Thirty-nine of these study plots were located in fragments designed to contain 1, 10, or 100 hectares of intact forest. Twenty-seven of the plots were set up nearby, in continuous wet forest. As the "Experimental setup" section of **Figure 55.10** shows, the distribution of the study plots allowed the research team to monitor changes inside forest fragments of different sizes and to compare these changes with conditions in unfragmented forest. Note that many of the study plots were located on forest edges.

When the research group surveyed the plots at least 10 years after the initial cut, they recorded two predominant effects:

1. a rapid loss of species diversity, especially from the smaller fragments; and

2. a startling drop in **biomass**, or the total amount of fixed carbon, in the study plots located near the edges of logged fragments. As the graph in the "Results" section of Figure 55.9 shows, most of the loss occurred within 2–4 years of the fragmentation event.

In tropical wet forests, most of the biomass is concentrated in large trees. Due to exposure to high winds and dry conditions, the edges of the experimental plots contained many downed and dying trees. Follow-up work showed that in edge habitats and in the fragmented patches, large-seeded, slow-growing trees typical of undisturbed forests are being replaced by early successional, "weedy" species (see Chapter 53). A large number of bird and understory plant species have disappeared as well.

To review the effects of habitat fragmentation, go to the study area at *www.masteringbiology.com*.

MB Web Activity Habitat Fragmentation

EXPERIMENT

QUESTION: How does fragmentation affect the quality of tropical wet forest habitats?

HYPOTHESIS: Fragmentation reduces the quality of wet forest habitats.

NULL HYPOTHESIS: Fragmentation has no effect on the quality of wet forest habitats.

EXPERIMENTAL SETUP:

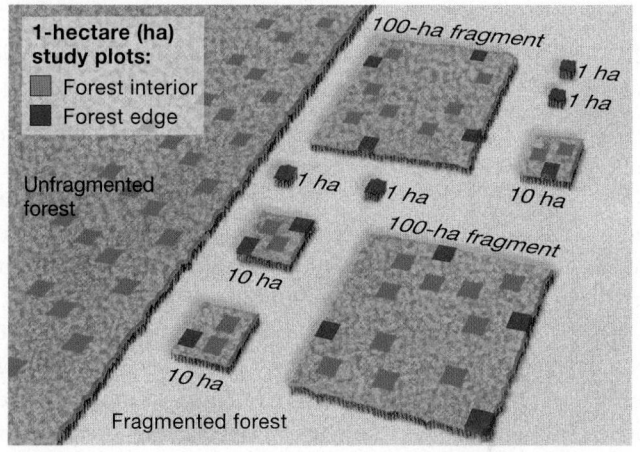

PREDICTION: Species diversity and biomass will decline in forest fragments compared with those of the forest interior, particularly along edges of fragments.

PREDICTION OF NULL HYPOTHESIS: Species diversity and biomass will be the same inside forest fragments and along edges as in the forest interior.

RESULTS:

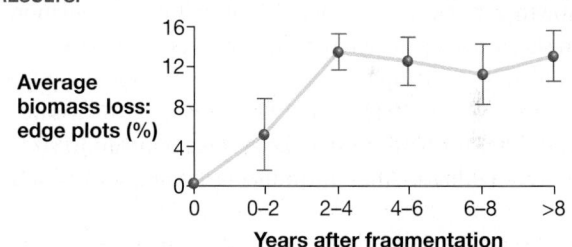

CONCLUSION: Biomass declines sharply along edges of forest fragments.

FIGURE 55.10 Experimental Evidence for Edge Effects in Fragmented Forests. Researchers tracked 66 study plots among four 1-hectare fragments, three 10-hectare fragments, and two 100-hectare fragments.

SOURCES: Laurance, W. F. et al. 2000. Conservation: Rainforest fragmentation kills big trees. *Nature* 404: 836. Laurance, W. F. et al. 1997. Biomass collapse in Amazonian forest fragments. *Science* 278: 1117–1118.

✔**QUESTION** The locations of the fragments and study plots were assigned at random. Why was this important?

The take-home message of this experiment and other research is clear: When habitats are fragmented, the quality *and* quantity of habitat decline drastically. In addition to losing over 8 million hectares of primary forest in South America, southeast Asia, and Africa each year to clear-cutting, we are losing large amounts of high-quality habitat throughout the world to fragmentation.

STOCHASTIC AND GENETIC PROBLEMS IN SMALL POPULATIONS

In effect, habitat loss and habitat fragmentation are forcing many species into the metapopulation structure introduced in Chapter 52. The small, isolated populations that make up a metapopulation are much more likely than large populations to be wiped out for two reasons:

1. Catastrophic events such as storms, disease outbreaks, or fires exterminate small populations more readily than large populations. Biologists refer to episodes like this as stochastic—meaning, chance—events.

2. Small populations suffer from inbreeding depression and random loss of alleles due to genetic drift.

Why do genetic problems occur? Small populations become inbred simply because so few potential mates are available. Over time, virtually all individuals become related. In most populations, inbreeding leads to lowered fitness, due to deleterious recessive alleles being homozygous more frequently. Chapter 25 introduced this phenomenon, known as inbreeding depression.

In addition, genetic drift is much more pronounced in small populations than large populations (see Chapter 25). Because it leads to the random loss or fixation of alleles, drift reduces genetic variation in populations. This is a concern, because if the environment changes due to the evolution of a new disease or global warming, the population may lack alleles that confer high fitness in the new environment.

When biologists document fitness declines in small, isolated populations, one option is to increase gene flow experimentally by importing individuals of the same species from a different population. For example, **Figure 55.11** shows data on annual population growth rate (λ; see Chapter 52) in a small population of bighorn sheep isolated on a refuge in northwest Montana.

As you study the graph, note that from the time the population was founded in 1922 until 1985, average growth rate declined steadily. During this time, the average population size was about 42 individuals. After introducing 5 new individuals in 1985 and 10 additional sheep in 1990–1994, though, λ began increasing and population size has increased. In some cases at least, introducing new alleles may counteract the effects of genetic drift and inbreeding.

CLIMATE CHANGE As evidence that global warming is affecting the distribution and abundance of species, biologists have begun to ask how climate change might change extinction rates. Chapter 54 highlighted several issues:

- loss of coral reefs due to high-temperature-induced "bleaching";
- loss of habitat for species native to arctic and alpine tundras and pack-ice habitats;
- problems with trees and other slowly dispersing species being unable to track rapid changes in climate; and
- ocean acidification—due to increased carbon dioxide concentrations and formation of carbonic acid—inhibiting the ability of corals, crustaceans, and other animals to make calcium carbonate skeletons.

In addition, the melting of polar ice caps and glaciers, due to increased global temperature, is causing sea levels to rise. The most recent analyses project that sea levels will rise an additional 0.8 to 2.0 m by the year 2100. The consequences for species that rely on beach and nearshore habitats are unclear.

Taken together, overexploitation, habitat loss and fragmentation, and climate change pose daunting threats to biodiversity. Just how many extinctions are all of these problems projected to cause?

How Can Biologists Predict Future Extinction Rates?

If the most fundamental question about biodiversity is how many species are alive, then the most basic question in conservation biology is how rapidly species are going extinct. Both questions have been difficult to answer.

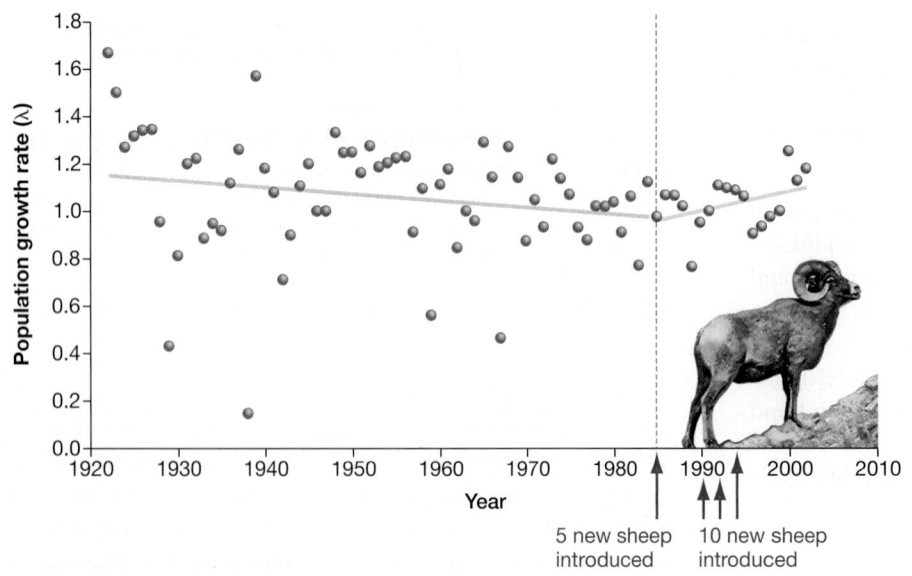

FIGURE 55.11 Experimental Evidence that Gene Flow Can Alleviate Genetic Problems in Small Populations. The y-axis plots the finite rate of increase (λ) in an isolated population of bighorn sheep, before and after unrelated individuals were introduced.

✔**EXERCISE** Draw a horizontal line indicating the value of λ where population size is constant (no growth).

5 new sheep introduced

10 new sheep introduced

Box 55.2 introduces an approach called Population Viability Analysis, or PVA, that researchers use to estimate the probability that a particular species will go extinct. To predict extinction rates for groups of species or over large areas, biologists use two different approaches. The first is based on direct counts of species that are known to have gone extinct recently or are in imminent danger of going extinct. The second is based on predicting the consequences of habitat loss and climate change, using existing data on the relationship between the size of habitats and the number of species present.

BOX 55.2   QUANTITATIVE METHODS: Population Viability Analysis

A population viability analysis, or PVA, is a model that estimates the likelihood that a population will avoid extinction for a given time period. In most cases, a PVA attempts to combine data on:

- age-specific survivorship and fecundity (see Chapter 52),
- geographic structure, and
- rate and severity of habitat disturbance.

Typically, a population is considered viable if the analysis predicts that it has a 95 percent probability of surviving for at least 100 years.

To get a feel for how a PVA works, consider an endangered marsupial called Leadbeater's possum (Figure 55.12a), which inhabits old-growth forests in southeastern Australia and relies on dead trees for nest sites. The goal of the PVA was to assess the effects that logging these habitats might have on the viability of the species.

To get started, researchers used life-table data—estimates of age-specific survival and average fecundity per female per year—to project how populations would grow if they were confined to scattered patches of habitat. The analysis allowed them to assess the impact of migration between patches—which can found new populations and introduce genetic diversity—and to simulate the effects of logging, fires, storms, and other disturbances.

Figure 55.12b illustrates the results. Each data point on the graphs represents the size of the overall population in a particular year. The four curves describe the changes in population size predicted to occur over time, based on different assumptions about the migration rates between groups of possums in patches of old-growth forest.

Note that when migration is high, the overall population size is predicted to stabilize at approximately 65 individuals. When no migration among fragments occurs, the final population size is predicted to be fewer than 20 individuals. The PVA showed that reducing the size and number of remaining old-growth fragments would reduce the possibility for migration and lead to rapid population declines.

Like all analyses based on simulations, though, a PVA makes many assumptions about future events and is only as accurate as the data entered into it. In particular, the usefulness of PVAs has been challenged because data on age-specific fecundity and survivorship and other basic demographic information are lacking or poorly documented in many endangered species.

A group of biologists defended the approach, however, by analyzing 21 long-term studies that have been completed on threatened populations of animals. To do their analysis, the researchers split each of the 21 data sets in two. Data from the first half of each study were used to run a PVA on each of the 21 species. After comparing the predictions of the PVAs to the actual data from the second half of each study, the researchers found that the correspondence was extremely close. This result strengthens confidence in PVA as a good predictive method for land managers.

(a) Leadbeater's possum

(b) Population viability analysis (PVA): Four scenarios for migration between habitat patches

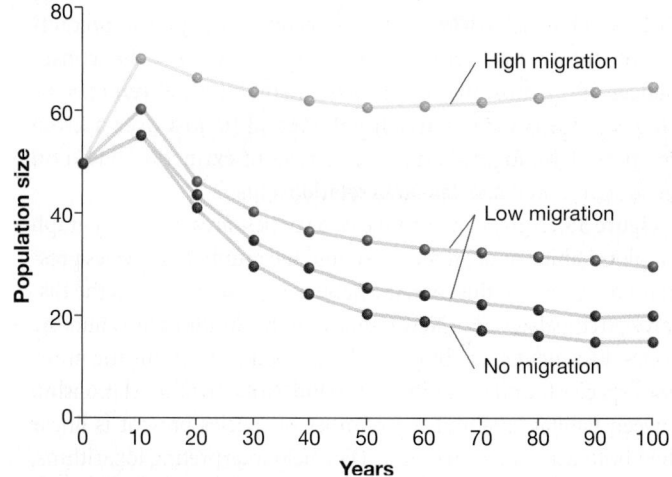

FIGURE 55.12 A PVA Can Predict the Consequences of Habitat Loss and Fragmentation. (a) Leadbeater's possum, a marsupial that breeds in old-growth forests of southeast Australia. **(b)** The migration rates modeled here range from no migration to 5 percent of the individuals migrating each year.

ESTIMATES BASED ON DIRECT COUNTS The best information on current extinction rates comes from studies on the best-studied of all major lineages on the tree of life: birds. Based on data in the fossil record, background extinctions in lineages like birds are estimated to occur at the rate of 1 extinction per million species alive per year. Of the approximately 10,000 bird species known, then, one species should go extinct every 100 years due to normal or background events.

Recent analyses have focused on species that were wiped out over the past 1000 years, coincident with the expansion of Polynesian people throughout the Pacific and Europeans throughout the world. These data suggest that birds have been going extinct at a rate 100 times background levels—meaning that a bird species has gone extinct every year, on average.

Although conservation programs targeted at birds have lowered this rate in the last few decades, estimates based on the numbers of species that are currently endangered suggest that the rate could reach 1000–1500 extinctions per million species alive per year by the end of this century. If this prediction holds, you may live long enough to see 10 or more species of bird go extinct each year.

Similar trends are occurring in listings of other well-studied groups.

- A recently published analysis of all 5487 known species of mammal living today—an effort that took five years and involved over 1700 experts from 130 countries—concluded that almost 25 percent are currently threatened with extinction.

- A comparison of comprehensive surveys of butterfly populations on the British Isles, conducted from 1970 to 1982 and repeated from 1995 to 1999, showed that 71 percent of the species are declining.

The message from these analyses is that the current rate of extinction is already many times higher than background or normal levels. Because human populations are growing so rapidly, it is poised to increase dramatically in the near future.

SPECIES–AREA RELATIONSHIPS A second strategy for predicting the future of Earth's biodiversity focuses on the consequences of the most urgent problem: habitat loss. Given reasonable projections of how much habitat will be lost over a given time period, biologists can estimate rates of extinction based on well-documented **species–area relationships**.

Figure 55.13 gives an example of a species–area curve—a graph that plots habitat area on the *x*-axis and the number of species present on the *y*-axis. In this case, the habitat areas are islands in the Bismarck Archipelago near New Guinea, in the South Pacific, and the species documented are birds. Each data point represents the number of species found on a different island. Note that the relationship between habitat area and the number of species present is linear when both axes are logarithmic. (For help interpreting logarithms, see **BioSkills 7** in Appendix A.)

The "log-linear" relationship turns out to be typical for other habitats and taxonomic groups as well. When biologists have plotted species–area relationships for plants, butterflies, mam-

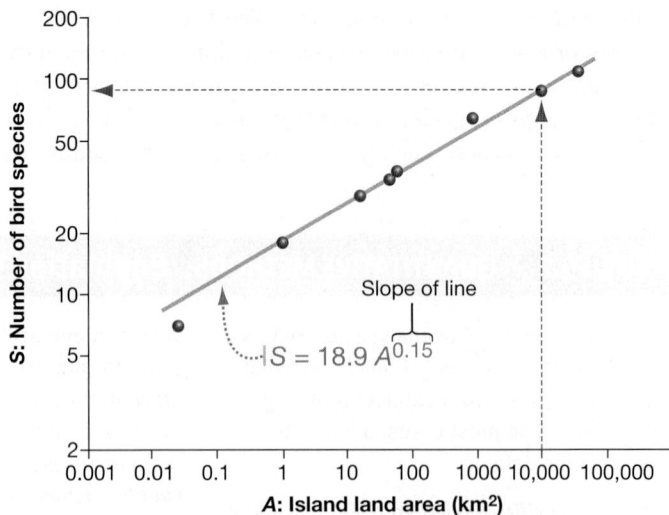

FIGURE 55.13 Species–Area Plots Quantify the Relationship between Species Richness and Habitat Area. The dashed arrows on this graph show the number of bird species that are expected to live on an island with an area of 10,000 km².

mals, or birds from islands or continental habitats around the globe, the relationship is consistently described by a function of the form $S = cA^z$, where:

- S is the number of species.

- A represents the habitat area.

- The c term is a constant that scales the data. Its value is high in species-rich areas, such as coral reefs or tropical wet forests, and low in species-poor areas, such as arctic tundras.

- The exponent z represents the slope of the line on a log-log plot. Thus, z describes how rapidly species numbers change with area. Typically, z is about 0.25.

For example, the solid line drawn through the points in Figure 55.13, from the Bismark Archipelago, is described by the function $S = (18.9)A^{0.15}$.

To understand how biologists use species–area curves to project extinction rates, ask yourself: If sea level rise and deforestation wiped out 90 percent of the habitats in part of the Bismark Archipelago—for example, if A were reduced from 10,000 km² to 1000 km²—what percentage of species would disappear?

According to the graph, the number of bird species present would drop from about 75 to about 53. Thus, the answer is roughly 30 percent. If z were higher, as it is in most regions of the world, the species loss would be much greater.

An international consortium of researchers recently studied a sample of habitats from around the globe and projected how the ranges of 1103 of the plants and animals present would change as a result of global warming. Then they used species–area curves to project how many species would be lost, as habitat is altered due to global warming. According to the "medium severity" projection for rates of global warming, about 25 percent of the species in their sample will be extinct or nearly extinct by the year 2050.

CHECK YOUR UNDERSTANDING

If you understand that . . .

- Until recently, most human-induced extinctions have occurred on islands and were due to overhunting or the impact of introduced species.
- Although invasive species and overexploitation remain a major threat to biodiversity, the majority of current problems are on continents and are due to habitat destruction and fragmentation—due to direct human activities or the effects of global warming.
- Current and projected extinction rates can be estimated from direct counts and from an analysis of species–area relationships.

✓ **You should be able to . . .**

1. Explain why fragmentation reduces habitat quality and leads to genetic problems.

2. Use the species–area relationship to compare the proportion of species lost when 90 percent of a 100,000 km² habitat is destroyed in a region where $c = 19$ and $z = 0.20$ versus the same-sized region where $c = 19$ and $z = 0.25$.

Answers are available in Appendix B.

55.4 Why Is Biodiversity Important?

One of the questions that biologists have to address about biodiversity is one that you may have heard from friends or relatives: Who cares? People are concerned about their grades or job or health or car or love life. Why should they worry about something as abstract as biodiversity?

The answer has two parts: one economic and one biological. Let's consider each in turn.

Economic Benefits of Biodiversity

Wild species have provided the raw material to fuel the development of human societies ever since the first large-scale, highly organized cultures began to develop about 10,000 years ago. People felled vast tracts of forest for building materials and fuel; selectively bred wild plants to yield food and fibers; domesticated animals to provide food, labor, and material goods; harvested the oceans for protein; and processed plants, animals, and fungi as sources of medicines.

The direct use of biodiversity continues today:

- *Seed banks* Agricultural scientists are preserving diverse strains of crop plants in **seed banks**—long-term storage facilities—and continue to use wild relatives of domesticated species in breeding programs aimed at improving crop traits. In addition, genetic engineering techniques are being used to transfer alleles from a diverse array of species into crop plants. In some cases these efforts have reduced dependence on pesticides, increased resistance to diseases and drought, and improved nutritional value and overall crop yields (see Chapter 19).

- *Pollination of crops* The production of almonds, apples, cherries, chocolate, alfalfa, and an array of other crops depends on the presence of wild pollinators. In the United States alone, insect-pollinated crops produce $40 billion worth of products annually.

- *Bioprospecting* Research programs collectively known as **bioprospecting** focus on assessing bacteria, archaea, plants, fungi, and animals as novel sources of drugs or ingredients in consumer products. Bioprospecting has benefited from the recent explosion of genetic and phylogenetic information, because biologists can now search genomes from a wide array of species to find alleles with desired functions. For example, recent advances in biotechnology facilitated the development of a new painkiller from the paralyzing sting of tropical cone snails, and a blood anticoagulant from the saliva of vampire bats.

- *Bioremediation* Strategies for cleaning up oil spills, abandoned mines, and contaminated industrial sites are incorporating **bioremediation**—the use of bacteria, archaea, and plants to metabolize pollutants and render them harmless (see Chapter 28).

- *Ecotourism* Recreation based on visiting wild places, or **ecotourism**, is a major industry internationally and is growing rapidly. In South Africa, for example, the number of tourists visiting wildlife preserves increased from 454,428 in 1986 to over 6 million in 1999. In 2004, ecotourism grew three times faster than the tourism industry as a whole.

- *Flood control* High ecosystem diversity—particularly the presence of forests or grasslands on steep slopes and the presence of wetlands in low-lying areas—dramatically reduces flood damage and the danger posed by mudslides.

In addition, the benefits of biodiversity extend beyond the direct use of diverse genes and species by humans to include **ecosystem services**—processes that increase the quality of the abiotic environment.

Recall from Chapter 10 that green plants and other photosynthetic organisms produce the oxygen we breathe, and from Chapter 30 that the presence of land plants builds soil, reduces soil erosion, moderates local temperature and wind conditions, and increases the volume of water retained in lakes, streams, and soils. In 1997 a team of economists led by Robert Costanza estimated the value of these services at US$33 trillion—almost twice the gross national product of all nations combined.

The air you breathe, the water you drink, and the soils that grow food you eat are services that ecosystems provide. The quality of your life is tied to the quality of those ecosystems.

Biological Benefits of Biodiversity

The direct and indirect economic benefits of biodiversity are impressive. But what about the biological value? Do biological communities and ecosystems function better when species diversity is high?

The effort to answer this question began in the mid-1980s. Since that time, it has blossomed into one of the most active research frontiers in community ecology and conservation biology.

BIODIVERSITY INCREASES PRODUCTIVITY **Figure 55.14** summarizes a classic experiment on how species richness affects one of the most basic aspects of an ecosystem function: the production of biomass. As Chapter 54 emphasized, net primary productivity is important because it is the source of chemical energy used by species throughout a food web.

As the "Experimental setup" portion of the figure shows, David Tilman and colleagues classified 32 grassland plant species into five functional categories. The functional groups differ by the timing of their growing season and by whether they allocate most of their resources to manufacturing woody stems or seeds. The researchers planted plots with a mixture of between 0 and 32 randomly chosen species, representing from 0 to 5 functional categories, and measured the amount of aboveground biomass produced in each.

The results, graphed in the "Results" section of Figure 55.14, indicate that both the number and type of species present had impor-

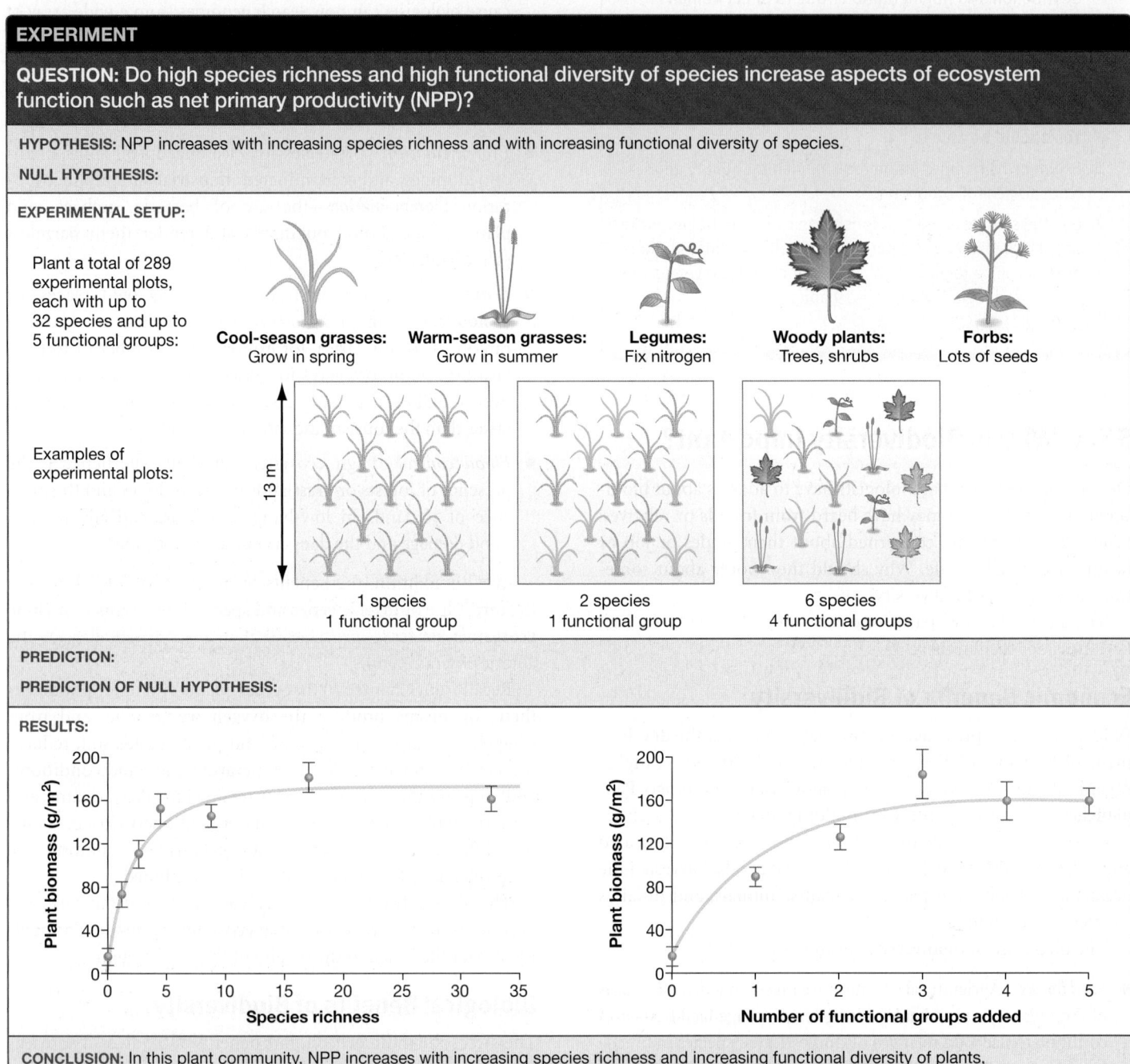

EXPERIMENT

QUESTION: Do high species richness and high functional diversity of species increase aspects of ecosystem function such as net primary productivity (NPP)?

HYPOTHESIS: NPP increases with increasing species richness and with increasing functional diversity of species.

NULL HYPOTHESIS:

EXPERIMENTAL SETUP:

Plant a total of 289 experimental plots, each with up to 32 species and up to 5 functional groups:

Cool-season grasses: Grow in spring   **Warm-season grasses:** Grow in summer   **Legumes:** Fix nitrogen   **Woody plants:** Trees, shrubs   **Forbs:** Lots of seeds

Examples of experimental plots:

13 m

1 species
1 functional group

2 species
1 functional group

6 species
4 functional groups

PREDICTION:

PREDICTION OF NULL HYPOTHESIS:

RESULTS:

CONCLUSION: In this plant community, NPP increases with increasing species richness and increasing functional diversity of plants, at least up to a point.

FIGURE 55.14 Evidence that the Productivity of Ecosystems Depends on the Number and Types of Species Present.

SOURCE: Tilman, D., J. Knops, D. Wedin, P. Reich, M. Ritchie, and E. Siemann. 1997. The influence of functional diversity and composition on ecosystem processes. *Science* 277: 1300–1302.

✔**EXERCISE** Write in the null hypothesis and both sets of predictions.

tant effects on biomass production. Experimental plots with more species and with a wider diversity of functional groups were more productive.

Note that total biomass leveled off as species richness and functional diversity increased, however. This observation suggests that increasing species diversity improves ecosystem function only up to a point. This pattern suggests that at least some species in ecosystems are redundant in terms of their ability to contribute to productivity.

Follow-up experiments in an array of ecosystems not only supported the conclusion that species richness has a positive impact on NPP, but showed that several causal mechanisms may be at work.

- *Resource use efficiency* Diverse assemblages of plant species make more efficient use of the sunlight, water, and nutrients available and thus lead to greater overall productivity. For example, some prairie plants extract water near the soil surface, while others use water available a meter or more below the surface. When species diversity is high, more overall water is used and more photosynthesis can occur.

- *Facilitation* Certain species or functional groups facilitate the growth of other species by providing them with nutrients, partial shade, or other benefits. In prairies, the presence of onion-family plants may discourage herbivores, or decaying roots from nitrogen-fixing species may fertilize other species.

- *Sampling effects* In many habitats it is common to observe that one or two species are extremely productive. If the number of species in a study plot is low, it is likely that the "big producer" will be missing and NPP will be low. But if the number of species in a study plot is high, then it is likely the big producer is present and NPP will be high. Simply due to sampling, then, high-species plots will tend to outproduce low-species plots.

The three mechanisms listed are not mutually exclusive—several can operate at the same time. Further, long-term experiments have shown that the reason for the pattern can change over time. As plants mature, resource use efficiency and facilitation may become more important than sampling effects.

DOES BIODIVERSITY LEAD TO STABILITY? When biologists refer to the stability of a community, they mean its ability to:

1. withstand a disturbance without changing,

2. recover to former levels of productivity or species richness after a disturbance, and

3. maintain productivity and other aspects of ecosystem function as conditions change over time.

Resistance is a measure of how much a community is affected by a disturbance. **Resilience** is a measure of how quickly a community recovers following a disturbance.

To test the hypothesis that diversity increases resistance and resilience, the team that did the work highlighted in Figure 55.14 followed up on a natural experiment. A **natural experiment** occurs when comparison groups are created by an unplanned, unmanipulated change in conditions.

In 1987–1988 a severe drought hit their study site. When the drought ended, the team asked whether species richness affected the response to the disturbance. The results are shown in **Figure 55.15**. Note that the graph's *y*-axis documents the change in total biomass that occurred from the year prior to the drought until the height of the drought. This quantity reflects how resistant the community is to disturbance. A completely resistant community would show no change. As predicted, drought resistance appeared to be higher in more diverse communities than in less diverse ones.

EXPERIMENT

QUESTION: Are more species-rich communities more stable than less-diverse communities?

HYPOTHESIS: Resistance to disturbance should increase with increasing species richness.

NULL HYPOTHESIS: There is no relationship between species richness and resistance.

EXPERIMENTAL SETUP:

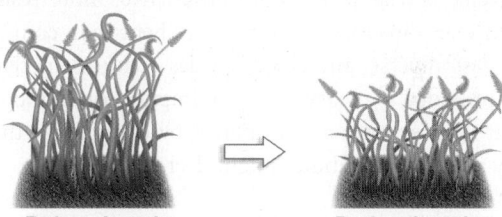

Before drought   **During drought**

Compare biomass of experimental plots before drought and at peak of drought. (This was a natural experiment—severe drought just happened to occur during study.)

PREDICTION: Plots that were more species rich before the drought will be more resistant to change.

PREDICTION OF NULL HYPOTHESIS: All the plots will have similar resistance regardless of species richness before the drought.

RESULTS:

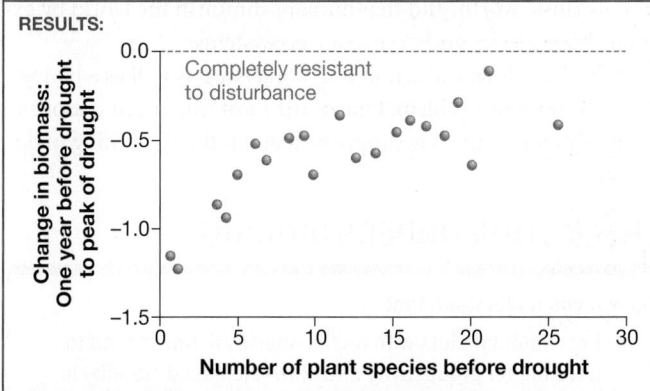

CONCLUSION: Resistance to disturbance increases with increasing species richness.

FIGURE 55.15 Evidence that Species-Rich Communities Resist Disturbance.

SOURCE: Tilman, D. and J. A. Downing. 1994. Biodiversity and stability in grasslands. *Nature* 367: 363–365.

✔ **QUESTION** The 0.0 value on the graph's vertical axis is labeled "Completely resistant to disturbance." Why?

In addition, the group analyzed the change in biomass in each plot four years after the drought versus biomass prior to the drought. This analysis focused on how resilient the community is. A completely resilient community would recover quickly from the disturbance and have the same biomass at both times. The data indicated that most plots containing five or fewer species showed a significant lowering of biomass after the disturbance, indicating that they had not recovered. But in all of the plots that contained more than five species, biomass after the drought was the same as biomass prior to the disturbance. This is strong evidence that species-rich areas recover from disturbance faster than species-poor areas.

More recent experiments on California grasslands suggest that species-rich ecosystems may also be less prone to invasion by exotic species. The hypothesis here is that species-rich communities use resources more efficiently—leaving less room for invasives. Taken together, these results lead biologists to be increasingly confident that species richness has a strongly positive effect on how ecosystems function.

In North American grasslands, at least, communities that are more diverse appear to be more productive, more resistant to disturbance and invasion, and more resilient than communities that are less diverse. Increased species richness increases the services provided by ecosystems. By implication, biologists can infer that if ecosystems are simplified by extinctions, then productivity and other attributes might decrease.

An Ethical Dimension?

The data reviewed in this section suggest that there are sound economic and biological reasons to preserve species. People live better and ecosystems work better when biodiversity is high.

But in addition, an array of biologists, philosophers, and religious leaders argue that humans have an ethical obligation to preserve species and ecosystems. Their position is that organisms have intrinsic worth, and that humans diminish the world by extinguishing species and destroying ecosystems.

If this is so, then extinction is a moral issue as well as a biological and economic problem. One of the most important reasons to preserve biodiversity may simply be that it is the right thing to do.

CHECK YOUR UNDERSTANDING

If you understand that . . .

- Experimental plots with higher species richness tend to have higher productivity and show increased stability in response to disturbance or changed abiotic conditions.

✔ **You should be able to . . .**

Design an experiment to test the hypothesis that species-rich areas are more effective at building soil, retaining soil nutrients, and minimizing soil erosion than are species-poor areas.

Answers are available in Appendix B.

55.5 Preserving Biodiversity

Extinction is irreversible. The only solution to the biodiversity crisis is to prevent the loss of alleles, species, and ecosystems.

Solving any global problem requires a common goal to be defined. In the case of preserving biodiversity, the objective is to sustain diverse communities in natural landscapes while supporting the extraction of resources required to maintain the health and well-being of the human population.

What's working? And what can you do to help?

Designing Effective Protected Areas

When biologists recognized the threat to biodiversity, they joined with government agencies, economists, community leaders, private landowners, and others to set aside protected areas. To assess the progress of this effort, the International Union for the Conservation of Nature sponsors the World Parks Congress every decade.

- In 1992 the Congress met in Caracas, Venezuela, and established a goal of setting aside 10 percent of Earth's land surface in protected areas.

- In 2003 the Congress met in Durban, South Africa, and announced that this goal had been surpassed: Protected areas covered 11.5 percent of Earth's terrestrial surface.

Will these protected areas be enough to avert a mass extinction?

GAP ANALYSIS PROGRAM Researchers are using a geographic approach called the Gap Analysis Program (GAP) to assess the effectiveness of the current system of protected areas. A GAP analysis tries to identify gaps between geographic areas that are particularly rich in biodiversity and areas that are actually managed for the preservation of biodiversity.

One GAP analysis combined data sets on the distribution of mammals, birds, amphibians, and freshwater turtles with a map of world protected areas. The analysis revealed that many species' ranges occur completely outside any protected areas. It also pinpointed regions in Mexico, Madagascar, and elsewhere where the gap between species' ranges and protected areas is particularly high.

To date, most GAP analyses suggest that, because relatively few species richness hotspots are included in existing protected areas, the 11.5 percent of Earth's surface area that is now being managed for biodiversity will not be enough to conserve many species. Efforts to fill these gaps are now focused on preserving species richness hotspots and centers of endemism like those illustrated in Figure 55.3.

In addition to working on where reserves are set up, biologists are focusing more attention on how reserves are designed. For example, it is often impossible to preserve large areas of high-quality habitat. In most cases, biologists have to accept the fact that habitats are fragmented and species exist as metapopulations. Given this reality, what is the best way to prevent small, isolated populations from going extinct?

WILDLIFE CORRIDORS The leading hypothesis in reserve design is that strips of undeveloped habitat, called **wildlife corridors**, should connect preserves that would otherwise be isolated. In some cases wildlife corridors are as simple as a walkway under a major highway, so that animals could move from one side to the other without being killed.

By facilitating the movement of individuals, the goals of corridors are to:

1. allow areas to be recolonized if a species was lost, and

2. encourage gene flow that introduces new alleles—counteracting the deleterious effects of genetic drift and inbreeding.

Do corridors actually work? To answer this question, a research team established a series of restored natural areas in the middle of a large, species-poor pine plantation. This was the equivalent of restoring patches of grassland habitat in the middle of an enormous cornfield. Some of the restoration sites were connected by corridors, while others were isolated. By monitoring the composition of species inside each habitat patch over time, the group was able to show that connected patches are steadily gaining more species over time compared to unconnected patches (**Figure 55.16**). This is strong evidence that wildlife corridors can increase overall species richness in a metapopulation.

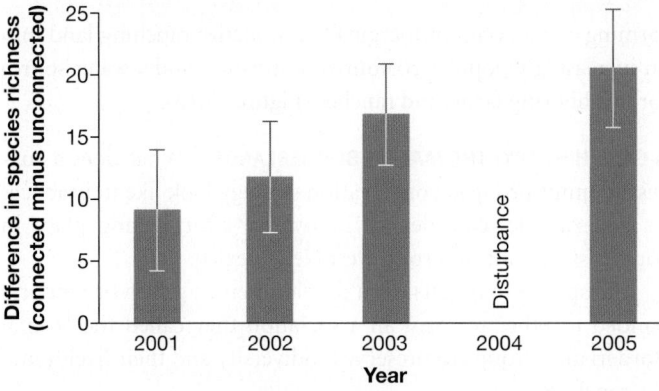

FIGURE 55.16 Experimental Evidence that Wildlife Corridors Work.
Species richness could not be assessed in 2004 because the study plots were burned as part of a program to restore natural habitats.

✔**EXERCISE** Draw and label bars showing what this graph would look like if corridors had no effect on species richness.

Beyond Protected Areas: A Comprehensive Approach

Given the magnitude of the current extinction rate, the continued growth in human population size and resource use, and the difficulty of setting up enough preserved areas in the world's conservation hotspots, conservation biologists are advocating a multipronged strategy. ⌐━ Preserving land will not be enough. At least four other strategies need to be an important part of a lasting solution to the biodiversity crisis.

SUSTAINABLE DEVELOPMENT In almost every case, the underlying causes of the biodiversity crisis are political and economic pressures that encourage short-term overexploitation of land and other resources and discourage **sustainability**—the managed use of resources at a rate only as fast as the rate at which they are replaced.

You may be familiar with efforts to develop renewable sources of energy, promote agricultural and forestry practices that maintain productivity without the addition of synthetic fertilizers or groundwater, and recycle metals and other raw materials. In essence, the challenge is to create ways for humans to live off the resources that are being produced continuously on Earth, rather than mining resources that have been stored for centuries or millennia.

The simple fact is that many of these stored resources—including petroleum, groundwater, and topsoils—are becoming exhausted.

The need to promote sustainable development has inspired a new field of endeavor, called sustainability science, that blends economics, political and social science, engineering, and applied biological sciences like forestry and agriculture.

STABILIZING HUMAN POPULATION SIZE AND RESOURCE USE Most biologists recognize that it may be impossible to preserve high species diversity and high-functioning ecosystems if two things happen:

● the human population grows to 13 billion or more—from its current 6.8 billion—over the next century, as projected by the "high-fertility" scenario introduced in Chapter 52; and

● individuals in the industrialized nations continue to use fossil fuels and other resources at the current rate (see Chapter 54).

It is not clear, however, that any human population in history has ever voluntarily and consciously limited its growth and resource use. As a species, we are breaking new ground.

Recent drops in fertility—documented in Chapter 52—and recent efforts to promote sustainable development are grounds for optimism. Efforts to limit family size and resource use are critical to solving the biodiversity crisis.

EX SITU CONSERVATION It may take decades for human population size and resource use to stabilize and for sustainable development practices to become widespread. In the meantime, thousands of species may be lost.

One approach to coping with this situation is referred to as **ex situ conservation**—the preservation of species in zoos, aquaria, wildlife ranches, seed banks, or other artificial settings. Translated literally, ex situ means "out of place." The name was inspired by a contrast with in situ or "in place" conservation—the establishment of protected areas where populations can be maintained in their natural habitat.

Some species—including Pere David's deer, the scimitar-horned oryx and Easter Island toromiro tree—are extinct in the wild and exist only in captivity. In other cases, captive breeding and reintroduction programs have succeeded in creating wild populations of endangered species. Chapter 52 highlighted the successful effort to establish new flocks of whooping cranes, using eggs and young produced in zoos and other breeding centers. The peregrine falcon, California condor, and several other species have been rescued from the brink of extinction by ex situ conservation and reintroduction to the wild.

ECOSYSTEM RESTORATION In many areas of the world, ecosystems are already heavily degraded or lost. The state of Illinois, for example, was estimated to contain 22 million acres of prairie when European settlers arrived; today only about 2000 acres of this biome remain. When natural areas are badly degraded or gone, biologists turn to restoration.

Thousands of large-scale and small-scale ecosystem restoration and reforestation projects are now occurring around the globe. In 2004, Wangari Maathai won the Nobel Peace Prize to honor her work founding the Green Belt Movement—a reforestation organization that has now sponsored the planting of over 1 billion trees, beginning in Kenya and expanding internationally (**Figure 55.17**).

In the tropics, one of the most successful ecological restoration efforts focused on the seasonally dry forest at the Area de Conservación Guanacaste in northwestern Costa Rica. The primary task facing conservation area staff there was to stop human-caused fires. Their efforts since the mid-1980s have succeeded in trans-

FIGURE 55.17 The Greenbelt Movement Is a Global Reforestation Effort. These women are tending a tree nursery in Kenya.

(a) Start of Guanacaste restoration project

(b) 17 years later

Person indicates scale

FIGURE 55.18 Using Fire Prevention to Restore Forest Ecosystems in the Guanacaste Reserve. Before and after views of the same region, 17 years apart, in the Area de Conservación Guanacaste, Costa Rica. To appreciate how quickly the trees grew, notice the person standing near the middle of the photo in (b).

forming a vast swath of marginally productive ranching land into an increasingly popular ecotourist destination and a water source for neighboring farms and ranches (**Figure 55.18**).

A CASE HISTORY: THE MALPAI BORDERLANDS What does a successful, multipronged conservation strategy look like in practice? As an example, consider an innovative effort taking place in southeast Arizona and southwest New Mexico.

In response to threats from development, a group of ranchers banded together to form an association they called the Malpai Borderlands group. To preserve biodiversity and their livelihood, the ranchers:

- protected 42,000 acres of their land with conservation easements, which are legal agreements that are placed on the titles to the property and prevent the land from being developed for vacation homes or other uses incompatible with ranching and wildlife preservation;

(a) Before prescribed burn

(b) 2 days after prescribed burn

(c) 2 months after prescribed burn

FIGURE 55.19 Using Fire to Restore Grassland Ecosystems in the Malpai Borderlands. The same region, in the Malpai Borderlands of the Southwestern United States, before and after a prescribed burn. The burn occurred in a wet year and resulted in a dramatic increase in native grasses.

- set up cooperative "grassbanks," or pastureland that is made available to ranchers whose own grazing areas are suffering from short-term drought and need relief from continued grazing; and

- reintroduced fire to the area via prescribed (intentional) burns—a restoration tool that removes encroaching woody shrubs and encourages the growth of native grasses (**Figure 55.19**).

The ranchers have also been cooperating with each other and with biologists to monitor or reintroduce native vegetation and animals such as the thick-billed parrot, bighorn sheep, Mexican jaguar, Yaqui chub, and Chiricahua leopard frog to the region.

Today, the Malpai Borderlands group is considered a model of innovative action by private individuals, in cooperation with governmental agencies and nongovernmental organizations (NGOs). The project benefits local people as well as biodiversity.

It can be done. Your generation is facing the most serious global environmental crisis in the history of our species. The decisions you make, ranging from how many children you have to how much you drive a car, will have far-reaching consequences. Change happens one person at a time.

As someone with a background in biology, you have the intellectual tools to make an important, positive impact. In addition to studying life, we have an obligation to help save it.

CHAPTER 55 REVIEW

For media, go to the study area at www.masteringbiology.com (MB)

Summary of Key Concepts

🔑 **Biodiversity is quantified at the level of allelic diversity, species diversity, and ecosystem diversity.**

- Genetic diversity is well characterized within some species, but biologists are only beginning to use genome sequencing techniques to explore the extent of genetic diversity in ecosystems.

- Research on the total number of species alive today is also at a preliminary stage, with estimates ranging from 10 to 100 million.

- To date, the message from efforts to characterize biodiversity is that it is much more extensive than expected and that a great deal remains to be learned.

    ✔ You should be able to explain why environmental sequencing is an effective way to catalog biodiversity at the genetic level.

🔑 **According to recent analyses, the sixth mass extinction in the history of life is occurring due to habitat loss, overexploitation, and global climate change.**

- Historically, most human-caused extinctions have occurred on islands because of direct exploitation or the introduction of exotic herbivores and predators.

- Habitat destruction is currently the leading cause of extinctions, with climate change projected to be an important new mechanism. Most extinctions are now occurring on continents.

- Experiments in the Brazilian Amazon have shown that habitat loss leads not only to a rapid decline in biodiversity but also to a decline in the quality of the remaining habitats due to fragmentation.

- To estimate how many species will go extinct in the near future, biologists combine data on current rates of habitat loss—usually estimated from satellite images taken over time—with data on the average number of species found in habitats of a given size.

- According to data on the rates of extinction in birds and other well-studied groups, it is likely that 60 percent of all species will be wiped out within 500 years.

✔You should be able to explain why small, geographically isolated populations are at high risk of extinction.

(MB) **Web Activity** Habitat Fragmentation

◐━ **Humans depend on biodiversity for the products that wild species provide and for ecosystem services that protect the quality of the abiotic environment.**

- Species diversity is important for maintaining the productivity of natural ecosystems and their ability to build and hold soil, moderate local climates, retain and cycle nutrients, retain surface water and recharge groundwater, prevent flooding, and produce oxygen.

- Humans gain direct economic benefits from fishing, forestry, agriculture, tourism, and other activities that depend on biodiversity.

- At the ecosystem level, experiments have shown that high species richness increases aspects of ecosystem function such as productivity, resistance to disturbance, and ability to recover from disturbance.

✔You should be able to explain why high species richness can lead to high productivity, in terms of the efficiency of resource use.

◐━ **Solutions to the biodiversity crisis include protecting key habitats, lowering human population growth and resource use, restoring ecosystems, mitigating climate change, and supporting sustainable development.**

- One conservation priority is to preserve habitats in biodiversity and conservation hotspots, and link preserves with habitat corridors that facilitate gene flow.

- The major threats to biodiversity—habitat destruction and climate change—will not lessen until the human population stabilizes and resource use becomes sustainable.

- Ecosystem restoration efforts are having a positive impact on biodiversity around the world.

✔You should be able to explain why the 11 percent of land area protected thus far will not be enough to avert a mass extinction.

Questions

✔ TEST YOUR KNOWLEDGE

Answers are available in Appendix B

1. What does a species–area plot show?
   a. the overall distribution, or area, occupied by a species
   b. the relationship between the body size of a species and the amount of territory or home range it requires
   c. the number of species found, on average, in tropical versus northern areas
   d. the number of species found, on average, in a habitat of a given size

2. What does a GAP analysis do?
   a. It compares the current distributions of species with the locations of preserved habitats.
   b. It quantifies *gross aboveground productivity*.
   c. It uses data on the rates at which lists of threatened species are growing to project the rate of future extinctions.
   d. It uses genome sequencing techniques to quantify genetic (allelic) diversity in an ecosystem.

3. What is resilience?
   a. resistance to change during a disturbance
   b. the ability to recover from a disturbance
   c. production of aboveground biomass
   d. total biomass production (above- and belowground)

4. What is a biodiversity "hotspot"?
   a. an area where an all-taxon survey is under way
   b. an area where an environmental sequencing study has been completed
   c. a habitat with high NPP
   d. an area with high species richness

5. Why do small populations become inbred?
   a. They are often part of a metapopulation structure.
   b. Genetic drift becomes a prominent evolutionary force.
   c. Over time, all individuals become increasingly related.
   d. Natural selection does not operate efficiently in small populations.

6. What is the primary cause of endangerment in marine environments?
   a. overexploitation
   b. pollution
   c. global warming
   d. invasive species

✔ TEST YOUR UNDERSTANDING

Answers are available in Appendix B

1. How has the nature of the threats to biodiversity changed over the past 400 years?

2. What is "sustainable" about sustainable development?

3. Biologists claim that the all-taxa survey now under way at the Great Smoky Mountains National Park in the United States will improve their ability to estimate the total number of species living today. Discuss the benefits and limitations that this data set will provide in understanding the extent of global biodiversity.

4. How are species–area curves used to relate rates of habitat destruction to projected extinction rates?

5. What evidence supports the hypotheses that species richness increases ecosystem functions, such as productivity, resistance to disturbance, and resilience?

6. Explain why the construction of wildlife corridors can help maintain biodiversity in a fragmented landscape.

1. What are ecosystem services? Does anyone own them or pay for them—that is, can you make money off them? How does this affect efforts to protect ecosystems?

2. You are helping design a series of reserves in a tropical country. List the steps you would recommend for gathering data and creating a plan that would protect a large amount of biodiversity in a small amount of land.

3. The maps to the right chronicle the loss of old-growth forest (>200 years old) that occurred in Warwickshire, England, and in the United States. In your opinion, under what conditions is it ethical for conservationists who live in these countries to lobby government officials in Brazil, Indonesia, and other tropical countries to slow the rate of loss of old-growth forest?

4. Make a list of characteristics that would render a species particularly vulnerable to extinction by humans. Make a list of characteristics that would render a species particularly resistant to pressure from humans. Try to think of an example of each type of species.

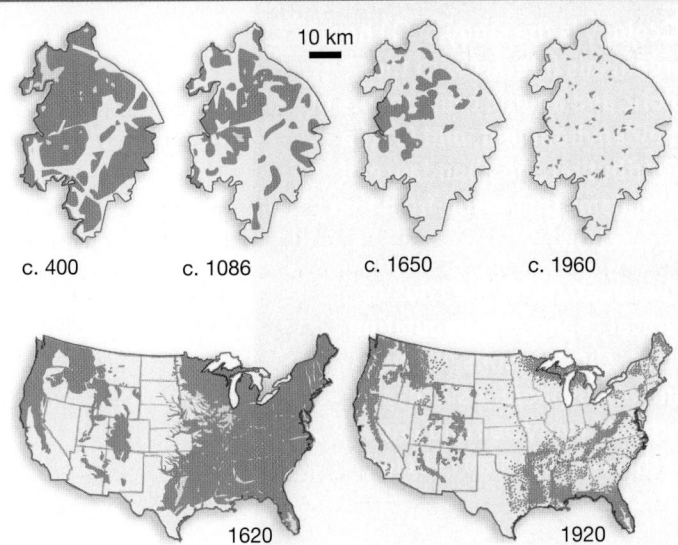

c. 400   c. 1086   c. 1650   c. 1960

1620   1920

THE BIG PICTURE

Ecology is the study of (1) how organisms interact with the biotic and abiotic components of their environment, and (2) the distribution and abundance of organisms that result from that interaction.

Human activities are causing radical changes in population sizes, climate, soils, water, and atmosphere. As a result, virtually every box and arrow in this concept map is being affected.

In addition to documenting the changes that are occurring, biologists are educating the public and policy makers about the value of biodiversity and healthy ecosystems, and researching the most effective ways to preserve species and keep ecosystems functioning normally.

Note that each box in the concept map indicates the chapter or section where you can go for review. Also, be sure to do the blue exercises in the Check Your Understanding box below.

CHECK YOUR UNDERSTANDING

🔑 **If you understand the big picture . . .**

✔ You should be able to . . .

1. Add a "Habitat loss" box that is connected to the "Species" box. Explain how this connection affects community structure, species richness, and primary productivity.

2. Add a "Global warming" box that is connected to the "CO_2" box. Explain how this connection affects community structure, species richness, and primary productivity.

3. Fill in the blue ovals with appropriate linking verbs or phrases.

Answers are available in Appendix B.

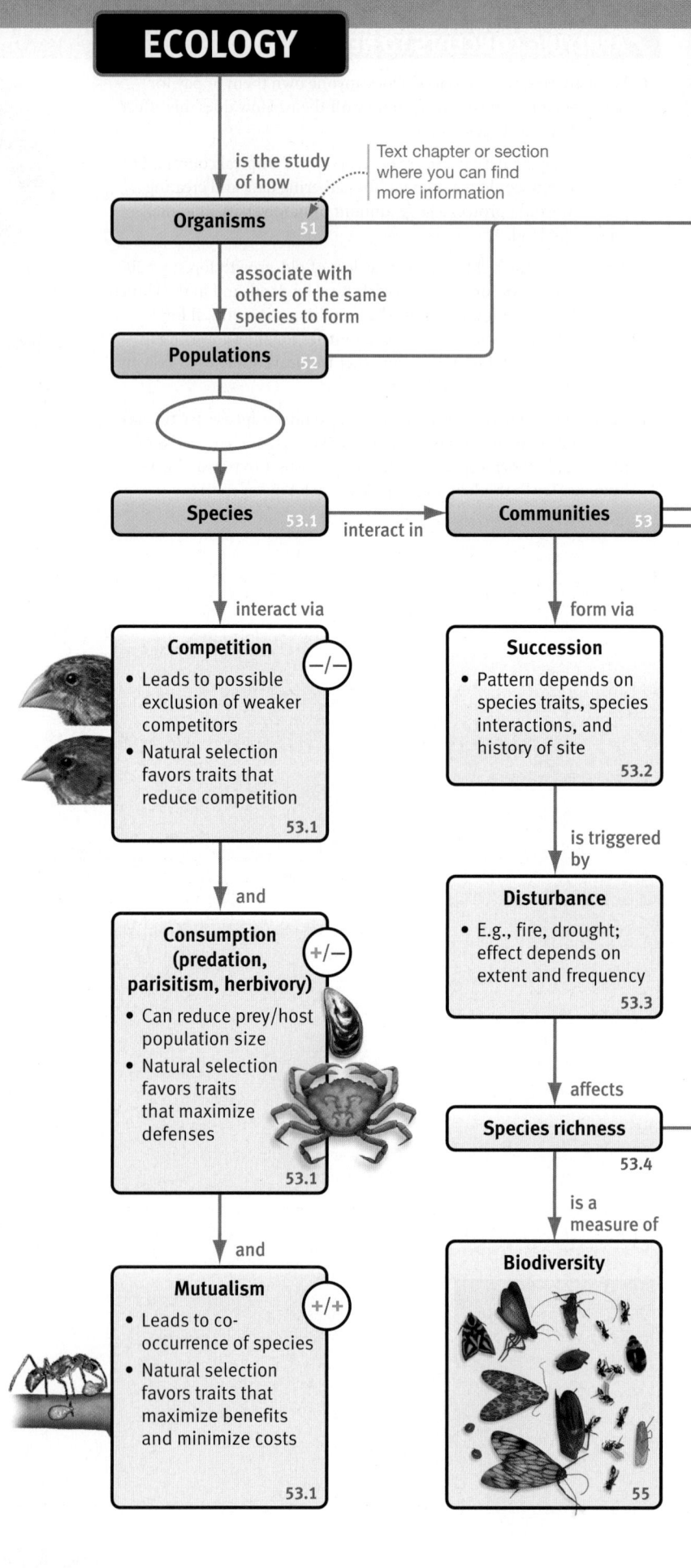

ECOLOGY

is the study of how

Text chapter or section where you can find more information

Organisms 51

associate with others of the same species to form

Populations 52

Species 53.1 — interact in — **Communities** 53

interact via

Competition −/−
- Leads to possible exclusion of weaker competitors
- Natural selection favors traits that reduce competition
53.1

and

Consumption (predation, parisitism, herbivory) +/−
- Can reduce prey/host population size
- Natural selection favors traits that maximize defenses
53.1

and

Mutualism +/+
- Leads to co-occurrence of species
- Natural selection favors traits that maximize benefits and minimize costs
53.1

form via

Succession
- Pattern depends on species traits, species interactions, and history of site
53.2

is triggered by

Disturbance
- E.g., fire, drought; effect depends on extent and frequency
53.3

affects

Species richness 53.4

is a measure of

Biodiversity 55

interact with → **Abiotic environment**

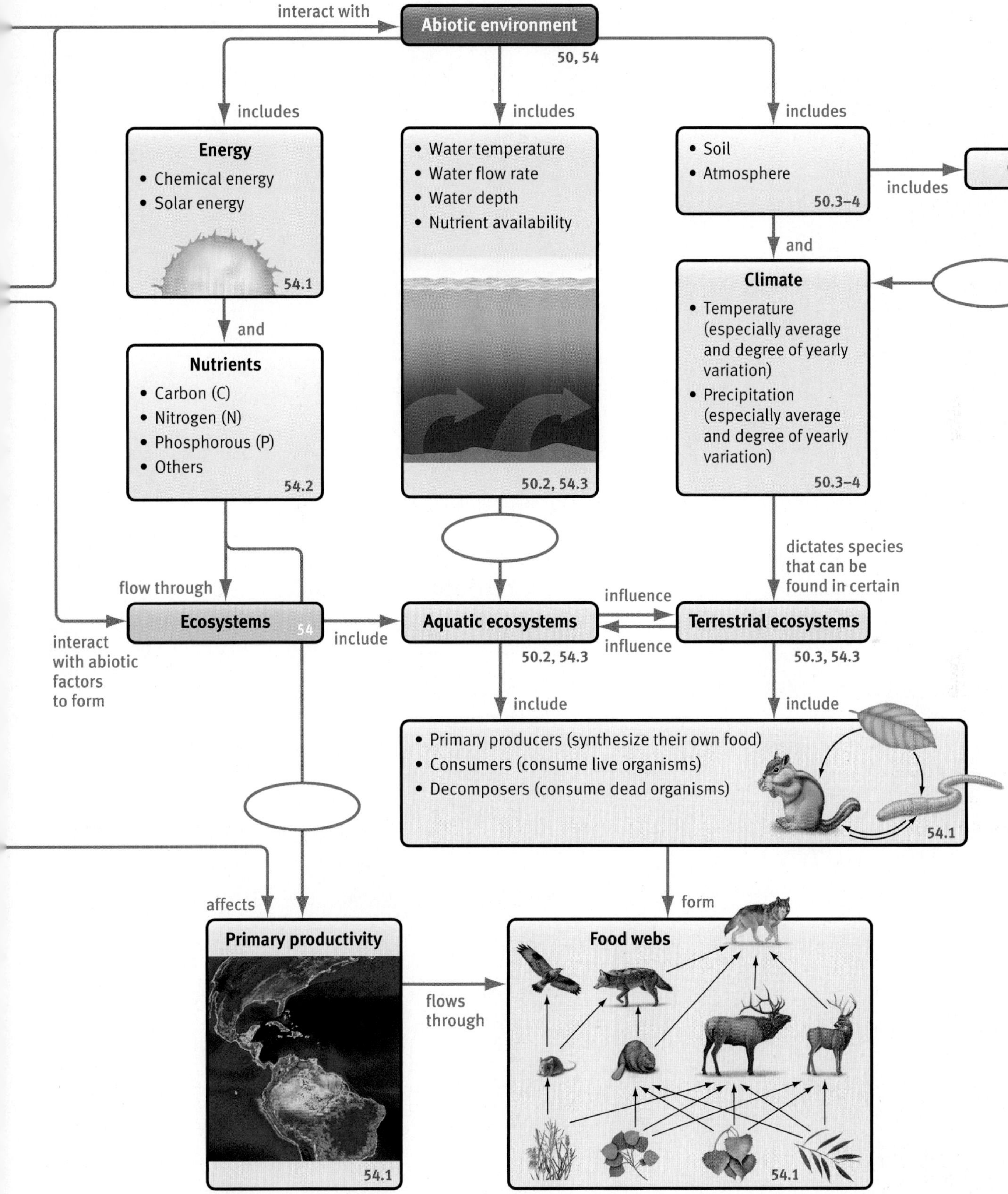

50, 54

includes ↓

Energy
- Chemical energy
- Solar energy

54.1

and ↓

Nutrients
- Carbon (C)
- Nitrogen (N)
- Phosphorous (P)
- Others

54.2

includes ↓

- Water temperature
- Water flow rate
- Water depth
- Nutrient availability

50.2, 54.3

includes ↓

- Soil
- Atmosphere

50.3–4

includes → **CO₂**

and ↓

Climate
- Temperature (especially average and degree of yearly variation)
- Precipitation (especially average and degree of yearly variation)

50.3–4

dictates species that can be found in certain ↓

flow through

interact with abiotic factors to form

Ecosystems 54 — include → **Aquatic ecosystems** — influence ⇄ influence → **Terrestrial ecosystems**

50.2, 54.3   50.3, 54.3

include ↓   include ↓

- Primary producers (synthesize their own food)
- Consumers (consume live organisms)
- Decomposers (consume dead organisms)

54.1

affects ↓

form ↓

Primary productivity

54.1

flows through →

Food webs

54.1

BIOSKILLS 1

THE METRIC SYSTEM

The metric system is the system of units of measure used in every country of the world but three (Liberia, Myanmar, and the United States). It is also the basis of the SI system used in scientific publications.

The popularity of the metric system is based on its consistency and ease of use. These attributes, in turn, arise from the system's use of the base 10. For example, each unit of length in the system is related to all other measures of length in the system by a multiple of 10. There are 10 millimeters in a centimeter; 100 centimeters in a meter; 1000 meters in a kilometer.

Measures of length in the English system, in contrast, do not relate to each other in a regular way. Inches are routinely divided into 16ths; there are 12 inches in a foot; 3 feet in a yard; 5280 feet (or 1760 yards) in a mile.

TABLE B1.2 **Prefixes Used in the Metric System**

Prefix	Abbreviation	Definition
nano-	n	$0.000\,000\,001 = 10^{-9}$
micro-	μ	$0.000\,001 = 10^{-6}$
milli-	m	$0.001 = 10^{-3}$
centi-	c	$0.01 = 10^{-2}$
deci-	d	$0.1 = 10^{-1}$
—	—	$1 = 10^0$
kilo-	k	$1000 = 10^3$
mega-	M	$1\,000\,000 = 10^6$
giga-	G	$1\,000\,000\,000 = 10^9$

TABLE B1.1 **Metric System Units and Conversions**

Measurement	Unit of Measurement and Abbreviation	Metric System Equivalent	Converting Metric Units to English Units
Length	kilometer (km)	$1\ km = 1000\ m = 10^3\ m$	1 km = 0.62 mile
	meter (m)	1 m = 100 cm	1 m = 1.09 yards = 3.28 feet = 39.37 inches
	centimeter (cm)	$1\ cm = 0.01\ m = 10^{-2}\ m$	1 cm = 0.3937 inch
	millimeter (mm)	$1\ mm = 0.001\ m = 10^{-3}\ m$	1 mm = 0.039 inch
	micrometer (μm)	$1\ \mu m = 10^{-6}\ m = 10^{-3}\ mm$	
	nanometer (nm)	$1\ nm = 10^{-9}\ m = 10^{-3}\ \mu m$	
Area	hectare (ha)	$1\ ha = 10{,}000\ m^2$	1 ha = 2.47 acres
	square meter (m²)	$1\ m^2 = 10{,}000\ cm^2$	1 m² = 1.196 square yards
	square centimeter (cm²)	$1\ cm^2 = 100\ mm^2 = 10^{-4}\ m^2$	1 cm² = 0.155 square inch
Volume	liter (L)	1 L = 1000 mL	1 L = 1.06 quarts
	milliliter (mL)	$1\ mL = 1000\ \mu L = 10^{-3}\ L$	1 mL = 0.034 fluid ounce
	microliter (μL)	$1\ \mu L = 10^{-6}\ L$	
Mass	kilogram (kg)	1 kg = 1000 g	1 kg = 2.20 pounds
	gram (g)	1 g = 1000 mg	1 g = 0.035 ounce
	milligram (mg)	$1\ mg = 1000\ \mu g = 10^{-3}\ g$	
	microgram (μg)	$1\ \mu g = 10^{-6}\ g$	
Temperature	Kelvin (K)*		K = °C + 273.15
	degrees Celsius (°C)		$°C = \frac{5}{9}(°F - 32)$
	degrees Fahrenheit (°F)		$°F = \frac{9}{5}°C + 32$

*Absolute zero is −273.15°C = 0 K

If you have grown up in the United States and are accustomed to using the English system, it is extremely important to begin developing a working familiarity with metric units and values. The tables and questions below should help you get started with this process.

✓ **Questions**

1. Some friends of yours just competed in a 5-kilometer run. How many miles did they run?

2. An American football field is 120 yards long, while rugby fields are 144 meters long. In yards, how much longer is a rugby field than an American football field?

3. What is your normal body temperature in degrees Celsius? (Normal body temperature is 98.6°F.)

4. What is your current weight in kilograms?

5. A friend asks you to buy a gallon of milk. How many liters would you buy to get approximately the same volume?

BIOSKILLS 2

READING GRAPHS

Graphs are the most common way to report data, for a simple reason. Compared to reading raw numerical values in a table or list, a graph makes it much easier to understand what the data mean.

Learning how to read and interpret graphs is one of the most basic skills you'll need to acquire as a biology student. As when learning piano or soccer or anything else, you need to understand a few key ideas to get started and then have a chance to practice—a *lot*—with some guidance and feedback.

Getting Started

To start reading a graph, you need to do three things: read the axes, figure out what the data points represent—that is, where they came from—and think about the overall message of the data. Let's consider each in turn.

What Do the Axes Represent?

Graphs have two axes: one horizontal and one vertical. The horizontal axis of a graph is also called the *x*-axis or the abscissa. The vertical axis of a graph is also called the *y*-axis or the ordinate. Each axis represents a variable that takes on a range of values. These values are indicated by the ticks and labels on the axis. Note that each axis should *always* be clearly labeled with the unit or treatment it represents.

Figure B2.1 shows a scatterplot—a type of graph where continuous data are graphed on each axis. Continuous data can take an array of values over a range. In contrast, discrete data can take only a restricted set of values. If you were graphing the average height of men and women in your class, height is a continuous variable but gender is a discrete variable.

In the example in the figure, the *x*-axis represents time in units of generations of maize; the *y*-axis represents the average percentage of the dry weight of a maize kernel that is protein.

To create a graph, researchers plot the independent variable on the *x*-axis and the dependent variable on the *y*-axis (Figure B2.1a). The terms independent and dependent are used because the values on the *y*-axis depend on the *x*-axis values. In our example, the researchers wanted to show how the protein content of maize kernels in a study population changed over time. Thus, the protein concentration plotted on the *y*-axis depended on the year (generation) plotted on the *x*-axis. The value on the *y*-axis always depends on the value on the *x*-axis, but not vice versa.

In many graphs in biology, the independent variable is either time or the various treatments used in an experiment. In these cases, the *y*-axis records how some quantity changes as a function of time or as the outcome of the treatments applied to the experimental cells or organisms.

What Do the Data Points Represent?

Once you've read the axes, you need to figure out what each data point is. In our maize kernel example, the data point in Figure B2.1b represents the average percentage of protein found in a sample of kernels from a study population in a particular generation.

If it's difficult to figure out what the data points are, ask yourself where they came from—meaning, how the researchers got them. You can do this by understanding how the study was done and by understanding what is being plotted on each axis. The *y*-axis will tell you what they measured; the *x*-axis will usually tell you when they measured it or what group was measured. In some cases—for example, in a plot of average body size versus average brain size in primates—the *x*-axis will report a second variable that was measured.

What Is the Overall Trend or Message?

Look at the data as a whole, and figure out what they mean. Figure B2.1c suggests an interpretation of the maize kernel example. If the graph shows how some quantity changes over time, ask yourself if that quantity is increasing, decreasing, fluctuating up and down, or staying the same. Then ask whether the pattern is the same over time or whether it changes over time.

When you're interpreting a graph, it's extremely important to limit your conclusions to the data presented. Don't extrapolate beyond the data, unless you are explicitly making a prediction based on the assumption that present trends will continue. For example, you can't say that average % protein content was increasing in the population before the experiment started, or that it will continue to increase in the future. You can only say what the data tell you.

(a) Read the axes—what is being plotted?

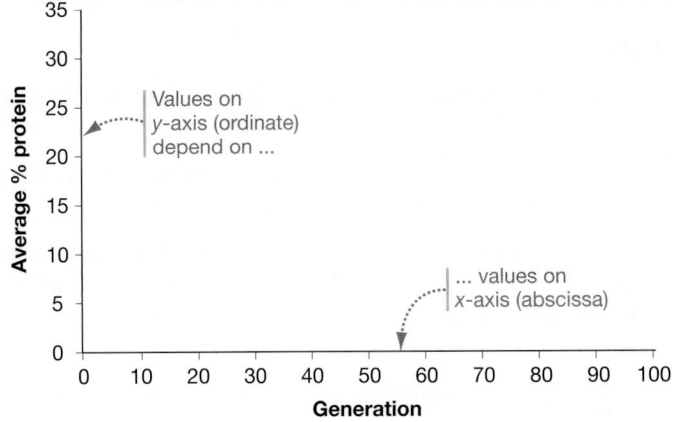

(b) Look at the bars or data points—what do they represent?

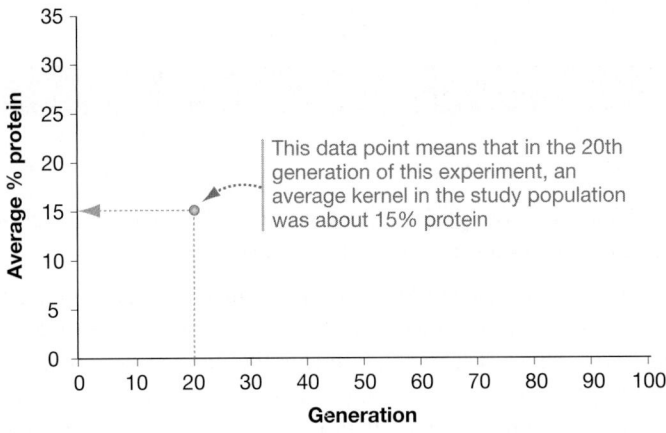

(c) What's the punchline?

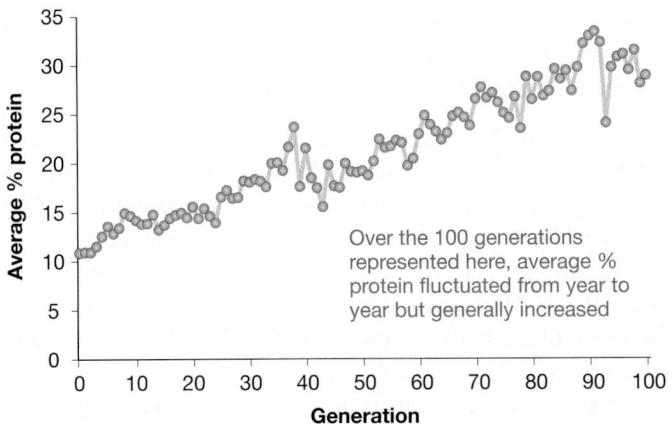

FIGURE B2.1 Scatterplots Are Used to Graph Continuous Data.

Types of Graphs

Most of the graphs in this text are scatterplots like the one shown here, where individual data points are plotted.

Sometimes the data points in a scatterplot will be by themselves, sometimes they will be connected by dot-to-dot lines to help make the overall trend clearer, as in this figure, and sometimes they will

(a) Bar chart

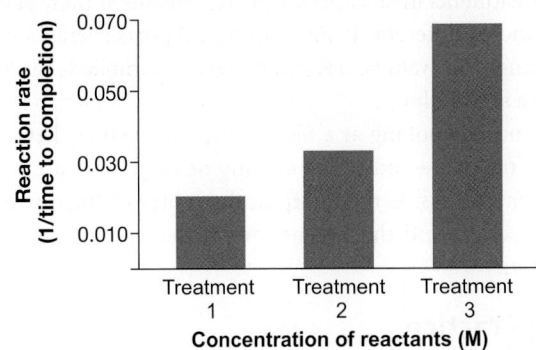

(b) Histogram

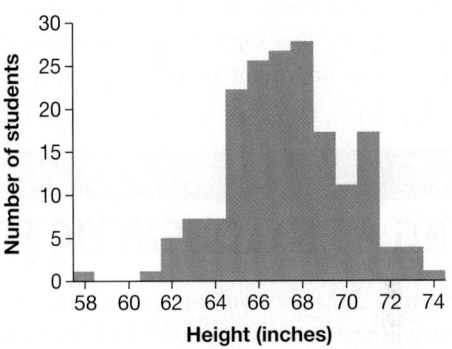

FIGURE B2.2 Bar Charts and Histograms. (a) Bar charts are used to graph data that are discontinuous or categorical. **(b)** Histograms show the distribution of frequencies or values in a population.

have a smooth line through them. A smooth line through data points—sometimes straight, sometimes curved—is a mathematical "line of best fit." A line of best fit represents a mathematical function that summarizes the relationship between the x and y variables. It is "best" in the sense of fitting the data points most precisely.

Scatterplots are the most appropriate type of graph when the data have a continuous range of values and you want to show individual data points. But you will also come across two other major types of graphs in this text:

- *Bar charts* plot data that have discrete or categorical values instead of a continuous range of values. In many cases the bars might represent different treatment groups in an experiment, as in **Figure B2.2a**. In this graph, the height of the bar indicates the average value.

- *Histograms* illustrate frequency data and can be plotted as numbers or percentages. **Figure B2.2b** shows an example where height is plotted on the x-axis and the number of students in a population is plotted on the y-axis. Each rectangle indicates the number of individuals in each interval of height, which reflects the relative frequency, in this population, of people whose heights are in that interval. The measurements could also be recalculated so that the y-axis would report the proportion of people in each interval. Then the sum of all the bars would equal 100 percent.

When you are looking at a bar chart that plots values from different treatments in an experiment, ask yourself if these values are the same or different. If the bar chart reports averages over discrete ranges of values, ask what trend is implied—as you would for a scatterplot.

When you are looking at a histogram, ask whether there is a "hump" in the data—indicating a group of values that are more frequent than others. Is the hump in the center of the distribution of values, toward the left, or toward the right? If so, what does it mean?

Getting Practice

Working with this text will give you lots of practice with reading graphs—they appear in almost every chapter. In many cases we've inserted an arrow to represent your instructor's hand at the whiteboard, with a label that suggests an interpretation or draws your attention to an important point on the graph. In other cases, you should be able to figure out what the data mean on your own or with the help of other students or your instructor.

✔ Questions

1. What is the total change in average percent protein in maize kernels, from the start of the experiment until the end?

2. What was the trend in average percent protein in maize kernels between generation 37 and generation 42?

3. Would the conclusions from the bar chart in Figure B2.2a be different if the data and label for Treatment 3 were put on the far left and the data and label for Treatment 1 on the far right?

4. In Figure B2.2b, about how many students in this class are 70 inches tall?

5. What is the most common height in the class graphed in Figure B2.2b?

BIOSKILLS 3

READING A PHYLOGENETIC TREE

Phylogenetic trees show the evolutionary relationships among species, just as a genealogy shows the relationships among people in your family. They are unusual diagrams, however, and it can take practice to interpret them correctly.

To understand how evolutionary trees work, consider **Figure B3.1**. Notice that a phylogenetic tree consists of branches, nodes, and tips.

- Branches represent populations through time. In this text, branches are drawn as horizontal lines. In most cases the length of the branch is arbitrary and has no meaning, but in some cases branch lengths are proportional to time (if so, there will be a scale at the bottom of the tree). The vertical lines on the tree represent splitting events, where one group broke into two independent groups. Their length is arbitrary—chosen simply to make the tree more readable.

- Nodes (also called forks) occur where an ancestral group splits into two or more descendant groups (see point B in Figure B3.1). Thus, each node represents the most recent common ancestor of the two or more descendant populations that emerge from it. If more than two descendant groups emerge from a node, the node is called a polytomy (see node C).

- Tips (also called terminal nodes) are the tree's endpoints, which represent groups living today or a dead end—a branch ending in extinction. The names at the tips can represent species or larger groups such as mammals or conifers.

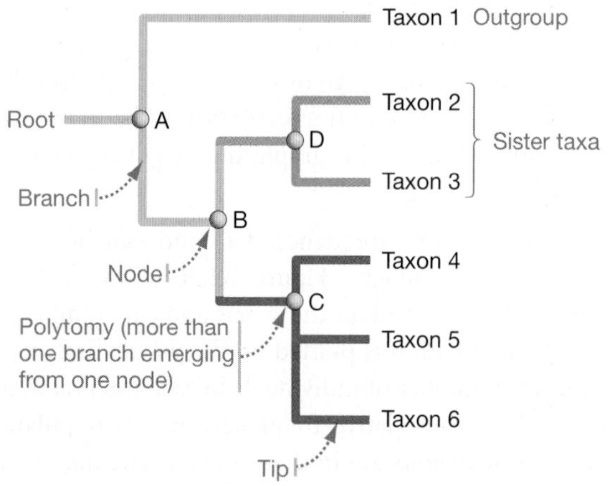

Recall from Chapter 1 that a taxon (plural: taxa) is any named group of organisms. A taxon could be a single species, such as *Homo sapiens*, or a large group of species, such as Primates. Tips connected by a single node on a tree are called sister taxa.

The phylogenetic trees used in this text are all rooted. This means that the first, or most basal, node on the tree—the one on the far left in this book—is the most ancient. To determine where the root on a tree occurs, biologists include one or more outgroup species when they are collecting data to estimate a particular phylogeny. An outgroup is a taxonomic group that is known to have diverged prior to the rest of the taxa in the study.

In Figure B3.1, "Taxon 1" is an outgroup to the monophyletic group consisting of taxa 2–6. A monophyletic group consists of an ancestral species and all of its descendants. The root of a tree is placed between the outgroup and the monophyletic group being studied. This position in Figure B3.1 is node A.

Understanding monophyletic groups is fundamental to reading and estimating phylogenetic trees. Monophyletic groups may also be called lineages or clades and can be identified using the

FIGURE B3.1 Phylogenetic Trees Have Roots, Branches, Nodes, and Tips.

✔ **EXERCISE** Circle all four monophyletic groups present.

"one-snip test": If you cut any branch on a phylogenetic tree, all of the branches and tips that fall off represent a monophyletic group. Using the one-snip test, you should be able to convince yourself that the monophyletic groups on a tree are nested. In

Figure B3.1, for example, the monophyletic group comprising node A and taxa 1–6 contains a monophyletic group consisting of node B and taxa 2–6, which includes the monophyletic group represented by node C and taxa 4–6.

To put all these new terms and concepts to work, consider the phylogenetic tree in **Figure B3.2**, which shows the relationships between common chimpanzees and six human and humanlike species that lived over the past 5–6 million years. Chimps functioned as an outgroup in the analysis that led to this tree, so the root was placed at node A. The branches marked in red identify a monophyletic group called the hominins.

To practice how to read a tree, put your finger at the tree's root, at the far left, and work your way to the right. At node A, the ancestral population split into two descendant populations. One of these populations eventually evolved into today's chimps; the other gave rise to the six species of hominins pictured. Now continue moving your finger toward the tips of the tree until you hit node C. It should make sense to you that at this splitting event, one descendant population eventually gave rise to two *Paranthropus* species, while the other became the ancestor of humans—species in the genus *Homo*. If multiple branches emerge from a node, creating a polytomy, it means that the populations split from one another so quickly that it is not possible to tell which split off earlier or later.

As you study Figure B3.2, you should consider a couple of important points.

1. There are many equivalent ways of drawing this tree. For example, this version shows *Homo sapiens* on the bottom. But the tree would be identical if the two branches emerging from node E were rotated 180°, so that the species appeared in the

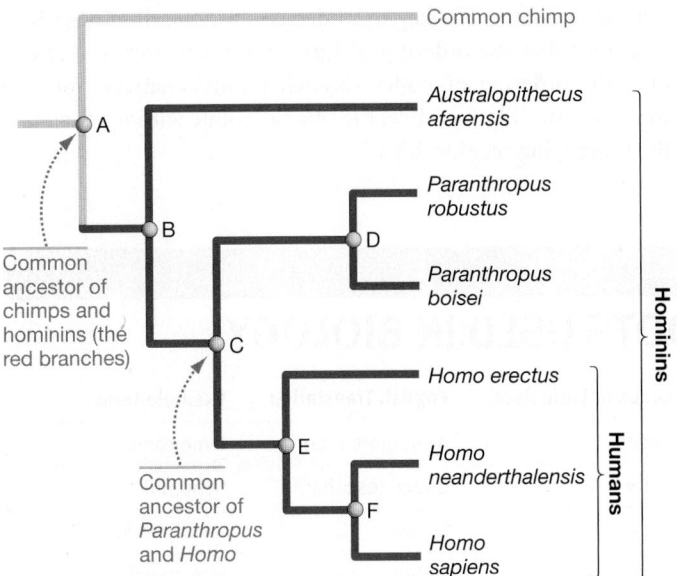

FIGURE B3.2 An Example of a Phylogenetic Tree. A phylogenetic tree showing the relationships of species in the monophyletic group called hominins.

✔**EXERCISE** All of the hominins walked on two legs—unlike chimps and all of the other primates. Add a mark on the phylogeny to show where upright posture evolved, and label it "origin of walking on two legs." Circle and label a pair of sister species. Label an outgroup to the monophyletic group called humans (species in the genus *Homo*).

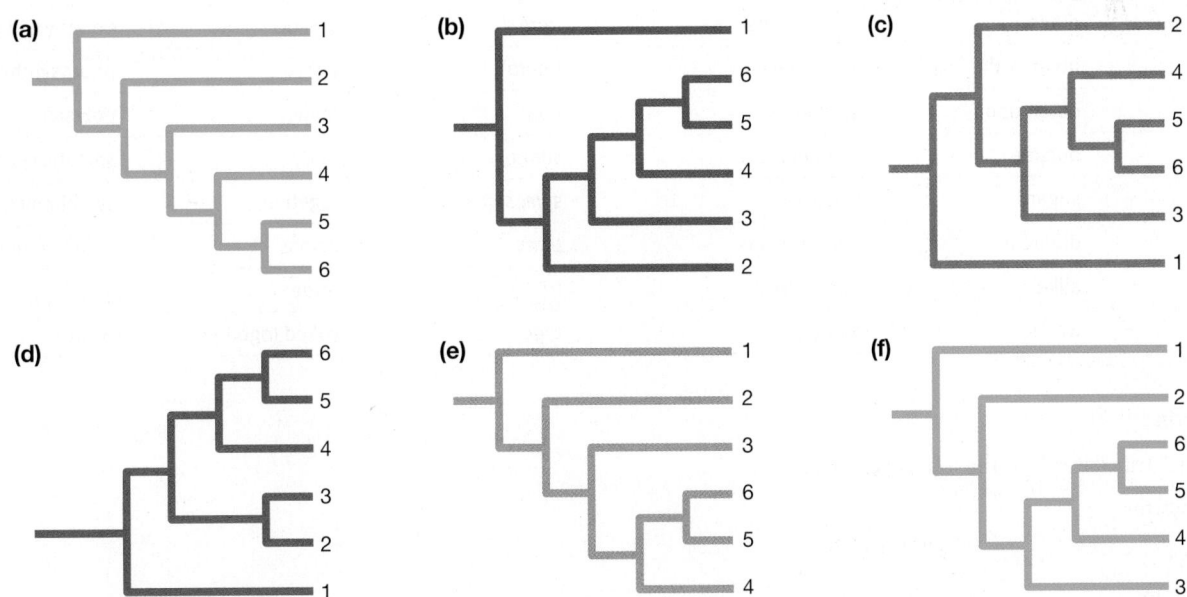

FIGURE B3.3 Alternative Ways of Drawing the Same Tree.

✔**QUESTION** Five of these six trees describe exactly the same relationships among taxa 1 through 6. Identify the tree that is different from the other five.

order *Homo sapiens*, *Homo neanderthalensis*, *Homo erectus*. Trees are read from root to tips, not from top to bottom or bottom to top.

2. No species on any tree is any higher or lower than any other. Chimps and *Homo sapiens* have been evolving exactly the same amount of time since their divergence from a common ancestor—neither species is higher or lower than the other. It is legitimate to say that more-ancient groups like *Australopithecus afarensis* have traits that are ancestral or more basal—meaning, that appeared earlier in evolution—

compared to traits that appear in *Homo sapiens*, which are referred to as more derived.

Figure B3.3 on page B:5 presents a chance to test your tree-reading ability. Five of the six trees shown in this diagram are identical in terms of the evolutionary relationships they represent. One differs. The key to understanding the difference is to recognize that the ordering of tips does not matter in a tree—only the ordering of nodes (branch points) matters. You can think of a tree like a mobile: the tips can rotate without changing the underlying relationships.

BIOSKILLS 4

SOME COMMON LATIN AND GREEK ROOTS USED IN BIOLOGY

Greek or Latin Root	English Translation	Example Term
a, an	not	anaerobic
aero	air	aerobic
allo	other	allopatric
amphi	on both sides	amphipathic
anti	against	antibody
auto	self	autotroph
bi	two	bilateral symmetry
bio	life, living	bioinformatics
blast	bud, sprout	blastula
co	with	cofactor
cyto	cell	cytoplasm
di	two	diploid
ecto	outer	ectoparasite
endo	inner, within	endoparasite
epi	outer, upon	epidermis
exo	outside	exothermic
glyco	sugary	glycolysis
hetero	different	heterozygous
homo	alike	homozygous
hydro	water	hydrolysis

Greek or Latin Root	English Translation	Example Term
hyper	over, more than	hypertonic
hypo	under, less than	hypotonic
inter	between	interspecific
intra	within	intraspecific
iso	same	isotonic
logo, logy	study of	morphology
lyse, lysis	loosen, burst	glycolysis
macro	large	macromolecule
meta	change, turning point	metamorphosis
micro	small	microfilament
morph	form	morphology
oligo	few	oligopeptide
para	beside	parathyroid gland
photo	light	photosynthesis
poly	many	polymer
soma	body	somatic cells
sym, syn	together	symbioticm, synapsis
trans	across	translation
tri	three	trisomy
zygo	yoked together	zygote

✔ Questions

Provide literal translations of the following terms:

1. heterozygote

2. glycolysis

3. morphology

4. trisomy

USING STATISTICAL TESTS AND INTERPRETING STANDARD ERROR BARS

When biologists do an experiment, they collect data on individuals in a treatment group and a control group, or several such comparison groups. Then they want to know whether the individuals in the two (or more) groups are different. For example, Chapter 2 introduces an experiment in which student researchers measured how fast a product formed when they set up a reaction with three different concentrations of reactants. Each treatment—meaning, each combination of reactant concentrations—was replicated many times.

Figure B5.1 graphs the average reaction rate for each of the three treatments in the experiment. Note that Treatments 1, 2, and 3 represent increasing concentrations of reactants. The thin "I-beams" on each bar indicate the standard error of each average. The standard error is a quantity that indicates the uncertainty in the calculation of an average.

For example, if two trials with the same concentration of reactants had a reaction rate of 0.075 and two trials had a reaction rate of 0.025, then the average reaction rate would be 0.50. In this case, the standard error would be large. But if two trials had a reaction rate of 0.051 and two had a reaction rate of 0.049, the average would still be 0.050, but the standard error would be small.

In effect, the standard error quantifies how confident you are that the average you've calculated is the average you'd observe if you did the experiment under the same conditions an extremely large number of times. It is a measure of precision.

Once they had calculated these averages and standard errors, the students wanted to answer a question: Does reaction rate increase when reactant concentration increases?

After looking at the data, you might conclude that the answer is yes. But how could you come to a conclusion like this objectively, instead of subjectively?

The answer is to use a statistical test. This can be thought of as a three-step process.

1. Specify the null hypothesis, which is that reactant concentration has no effect on reaction rate.

2. Calculate a test statistic, which is a number that characterizes the size of the difference among the treatments. In this case, the test statistic compares the actual differences in reaction rates among treatments to the difference predicted by the null hypothesis. The null hypothesis predicts that there should be no difference.

3. The third step is to determine the probability of getting a test statistic as large as the one calculated just by chance. The answer comes from a reference distribution—a mathematical function that specifies the probability of getting various values of the test statistic if the null hypothesis is correct. (If you take a statistics course, you'll learn which test statistics and reference distributions are relevant to different types of data.)

You are very likely to see small differences among treatment groups just by chance—even if no differences actually exist. If you flipped a coin 10 times, for example, you are unlikely to get exactly five heads and five tails, even if the coin is fair. A reference distribution tells you how likely you are to get each of the possible outcomes of the 10 flips if the coin is fair, just by chance.

In this case, the reference distribution indicated that if the null hypothesis of no actual difference in reaction rates is correct, you would see differences as large as those observed only 0.01 percent of the time just by chance. By convention, biologists consider a difference among treatment groups to be statistically significant if you have less than a 5 percent probability of observing it just by chance. Based on this convention, the student researchers were able to claim that the null hypothesis is not correct for reactant concentration. According to their data, the reaction they studied really does happen faster when reactant concentration increases.

It is likely that you'll be doing actual statistical tests early in your undergraduate career. To use this text, though, you only need to be aware of what statistical testing does. And you should take care to inspect the standard error bars on graphs in this book. As a *very* rough rule of thumb, averages often turn out to be significantly different, according to an appropriate statistical test, if there is no overlap between two times the standard errors.

✔ Question

Suppose you estimated the average height of students in your class by sampling two individuals at random. Then suppose you estimated the same quantity by sampling every individual that showed up for class on a particular day. Which estimate of the average is likely to have the smallest standard error, and why?

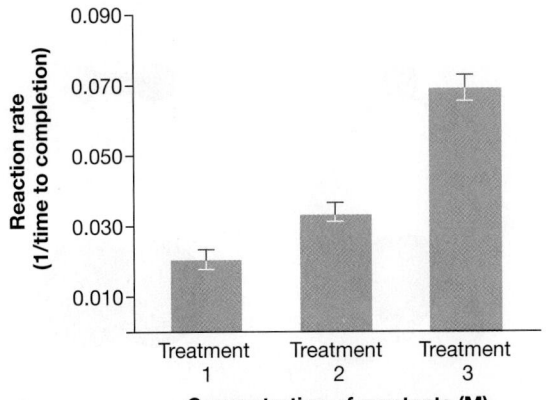

FIGURE B5.1 Standard Error Bars Indicate the Uncertainty in an Average.

READING CHEMICAL STRUCTURES

If you haven't had much chemistry yet, learning basic biological chemistry can be a challenge. One of the stumbling blocks is simply being able to read chemical structures efficiently and understand what they mean. This skill will come much easier once you have a little notation under your belt and you understand some basic symbols.

Atoms are the basic building blocks of everything in the universe, just as cells are the basic building blocks of your body. Every atom has a 1- or 2-letter symbol. **Table B6.1** shows the symbols for most of the atoms you'll encounter in this book. You should memorize these. The table also offers details on how the atoms form bonds as well as how they are represented in visual models.

When atoms attach to each other by covalent bonding, a molecule forms. Biologists have a couple of different ways of representing molecules—you'll see each of these in the book and in class.

- A molecular formula like those in **Figure B6.1a** simply lists the atoms present in a molecule, with subscripts indicating how many of each atom are present. If the formula has no subscript, only one of that type of atom is present. A methane (natural gas) molecule, for example, can be written as CH_4. It consists of one carbon atom and four hydrogen atoms.

- Structural formulas like those in **Figure B6.1b** show which atoms in the molecule are bonded to each other, with each bond indicated by a dash. The structural formula for methane indicates that each of the four hydrogen atoms forms one

covalent bond with carbon, and that carbon makes a total of four covalent bonds. Single covalent bonds are symbolized by a single dash; double bonds are indicated by two dashes.

Even simple molecules have distinctive shapes, because different atoms make covalent bonds at different angles. Ball-and-stick and space-filling models show the geometry of the bonds accurately.

TABLE B6.1 **Some Attributes of Atoms Found in Organisms**

Atom	Symbol	Number of Bonds It Can Form	Standard Color Code*
Hydrogen	H	1	white
Carbon	C	4	black
Nitrogen	N	3	blue
Oxygen	O	2	red
Sodium	Na	1	—
Magnesium	Mg	2	—
Phosphorus	P	5	orange or purple
Sulfur	S	2	yellow
Chlorine	Cl	1	—
Potassium	K	1	—
Calcium	Ca	2	—

*In ball-and-stick or space-filling models.

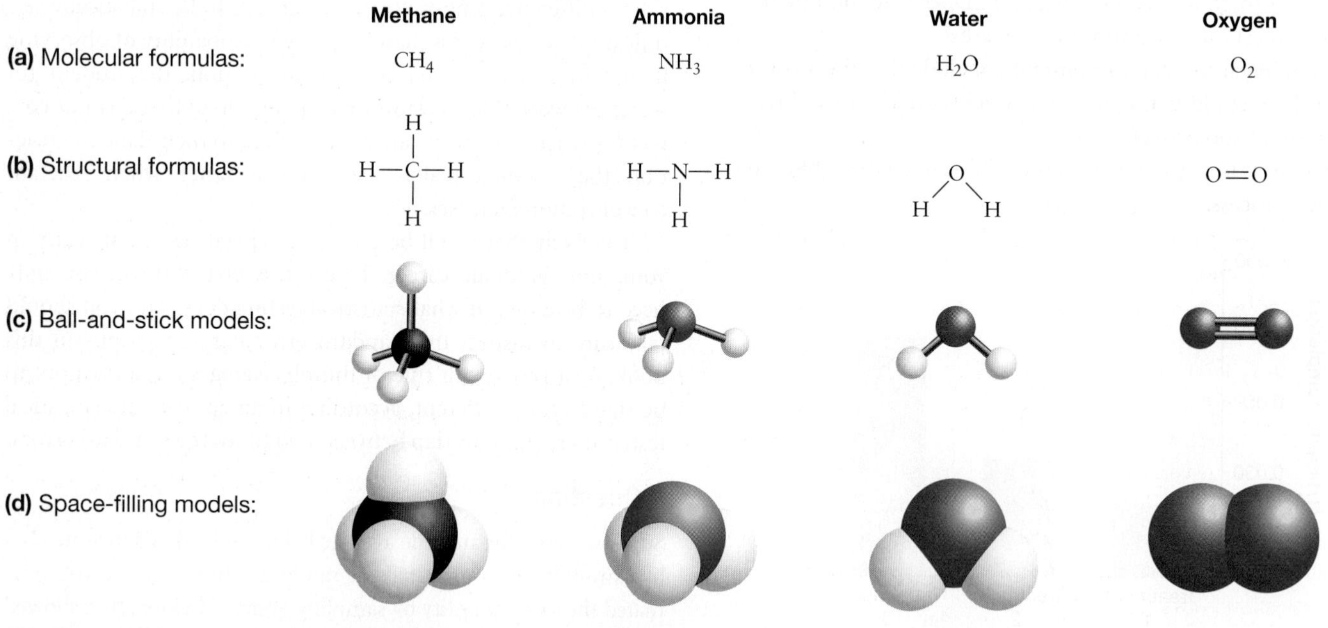

	Methane	**Ammonia**	**Water**	**Oxygen**
(a) Molecular formulas:	CH_4	NH_3	H_2O	O_2

(b) Structural formulas:

$$\begin{matrix} & H & \\ & | & \\ H - & C & - H \\ & | & \\ & H & \end{matrix} \qquad \begin{matrix} H - N - H \\ | \\ H \end{matrix} \qquad \begin{matrix} O \\ H \quad H \end{matrix} \qquad O = O$$

(c) Ball-and-stick models:

(d) Space-filling models:

FIGURE B6.1 Molecules Can Be Represented in Several Different Ways.

✔**EXERCISE** Carbon dioxide consists of a carbon atom that forms a double bond with each of two oxygen atoms, for a total of four bonds. It is a linear molecule. Write carbon dioxide's molecular formula, then draw its structural formula, a ball-and-stick model, and a space-filling model.

- In a ball-and-stick model, a stick is used to represent each covalent bond (see **Figure B6.1c**).

- In space-filling models, the atoms are simply stuck onto each other in their proper places (see **Figure B6.1d**).

To learn more about a molecule when you look at a chemical structure, ask yourself three questions:

1. *Is the molecule polar—meaning that some parts are more negatively or positively charged than others?* Molecules that contain nitrogen or oxygen atoms are often polar, because these atoms have such high electronegativity (see Chapter 2).

This trait is important because polar molecules dissolve in water.

2. *Does the structural formula show atoms that might participate in chemical reactions?* For example, are there charged atoms or amino or carboxyl (−COOH) groups that might act as a base or an acid?

3. *In ball-and-stick and especially space-filling models of large molecules, are there interesting aspects of overall shape?* For example, is there a groove where a protein might bind to DNA, or a cleft where a substrate might undergo a reaction in an enzyme?

BIOSKILLS 7

USING LOGARITHMS

You have probably been introduced to logarithms and logarithmic notation in algebra courses, and you will encounter logarithms at several points in this course. Logarithms are a way of working with powers—meaning, numbers that are multiplied by themselves one or more times.

Scientists use exponential notation to represent powers. For example,

$$a^x = y$$

means that if you multiply a by itself x times, you get y. In exponential notation, a is called the base and x is called the exponent. The entire expression is called an exponential function.

What if you know y and a, and you want to know x? This is where logarithms come in. You can solve for exponents using logarithms. For example,

$$x = \log_a y$$

This equation reads, x is equal to the logarithm of y to the base a. Logarithms are a way of working with exponential functions. They are important because so many processes in biology (and chemistry and physics, for that matter) are exponential in nature. To understand what's going on, you have to describe the process with an exponential function and then use logarithms to work with that function.

Although a base can be any number, most scientists use just two bases when they employ logarithmic notation: 10 and e. Logarithms to the base 10 are so common that they are usually symbolized in the form log y instead of $\log_{10} y$. A logarithm to the base e is called a natural logarithm and is symbolized ln (pronounced *EL-EN*) instead of log. You write "the natural logarithm of y" as ln y. The base e is an irrational number (like π) that is approximately equal to 2.718. Like 10, e is just a number. But both 10 and e have qualities that make them convenient to use in biology (and chemistry, and physics).

Most scientific calculators have keys that allow you to solve problems involving base 10 and base e. For example, if you know y, they'll tell you what log y or ln y are—meaning that they'll solve for x in our example above. They'll also allow you to find a number when you know its logarithm to base 10 or base e. Stated another way, they'll tell you what y is if you know x, and y is equal to e^x or 10^x. This is called taking an antilog. In most cases, you'll use the inverse or second function button on your calculator to find an antilog (above the log or ln key).

To get some practice with your calculator, consider the equation

$$10^2 = 100$$

If you enter 100 in your calculator and then press the log key, the screen should say 2. The logarithm tells you what the exponent is. Now press the antilog key while 2 is on the screen. The calculator screen should return to 100. The antilog solves the exponential function, given the base and the exponent.

If your background in algebra isn't strong, you'll want to get more practice working with logarithms—you'll see them frequently during your undergraduate career. Remember that once you understand the basic notation, there's nothing mysterious about logarithms. They are simply a way of working with exponential functions, which describe what happens when something is multiplied by itself a number of times—like cells that divide and then divide again and then again.

Using logarithms will also come up when you are studying something that can have a large range of values, like the concentration of hydrogen ions in a solution or the intensity of sound that the human ear can detect. In cases like this, it's convenient to express the numbers involved as exponents. Using exponents makes a large range of numbers smaller and more tractable. For example, instead of saying that hydrogen ion concentration in a solution can range from 1 to 10^{-14}, the pH scale allows you to simply say that it ranges from 1 to 14. Instead of giving the actual value, you're expressing it as an exponent. It just simplifies things.

✓ Questions

In Chapter 52, you'll use the equation $N_t = N_0 e^{rt}$.

1. What type of function does this equation describe?

2. After taking the natural logarithm of both sides, how would you write the equation?

MAKING CONCEPT MAPS

A concept map is a graphical device for organizing and expressing what you know about a topic. It has two main elements: (1) concepts that are identified by words or short phrases and placed in a box or circle, and (2) labeled arrows that physically link two concepts and explain the relationship between them. The concepts are arranged hierarchically on a page, with the most general concepts at the top and the most specific ideas at the bottom.

The combination of a concept, a linking word, and a second concept is called a proposition. Good concept maps also have cross-links—meaning, labeled arrows that connect different elements in the hierarchy, as you read down the page.

Concept maps were initially developed by Joseph Novak in the early 1970s and have proven to be an effective studying and learning tool. They can be particularly valuable if constructed by a group, or when different individuals exchange and critique concept maps they have created independently. Although concept maps vary widely in quality and can be graded using objective criteria, there are many equally valid ways of making a high-quality concept map on a particular topic.

When you are asked to make a concept map in this text, you will usually be given at least a partial list of concepts to use. As an example, suppose you were asked to create a concept map on experimental design and were given the following concepts: results, predictions, control treatment, experimental treatment, controlled (identical) conditions, conclusions, experiment, hypothesis to be tested, null hypothesis. One possible concept map is shown in **Figure B8.1**.

Good concept maps have four qualities:

- They exhibit an organized hierarchy, indicating how each concept on the map relates to larger and smaller concepts.

- The concept words are specific—not vague.

- The propositions are accurate.

- There is cross-linking between different elements in the hierarchy of concepts.

As you practice making concept maps, go through these criteria and use them to evaluate your own work, as well as the work of fellow students.

✔ Exercises

1. In many cases, investigators contrast a hypothesis being tested with an alternative hypothesis that does not qualify as a null hypothesis. Add an "Alternative hypothesis" concept to the map in Figure B8.1, along with other concepts and labeled linking arrows needed to indicate its relationship to other information on the map.

2. Add a box for the concept "Statistical testing" (see **BioSkills 5**) along with appropriately labeled linking arrows.

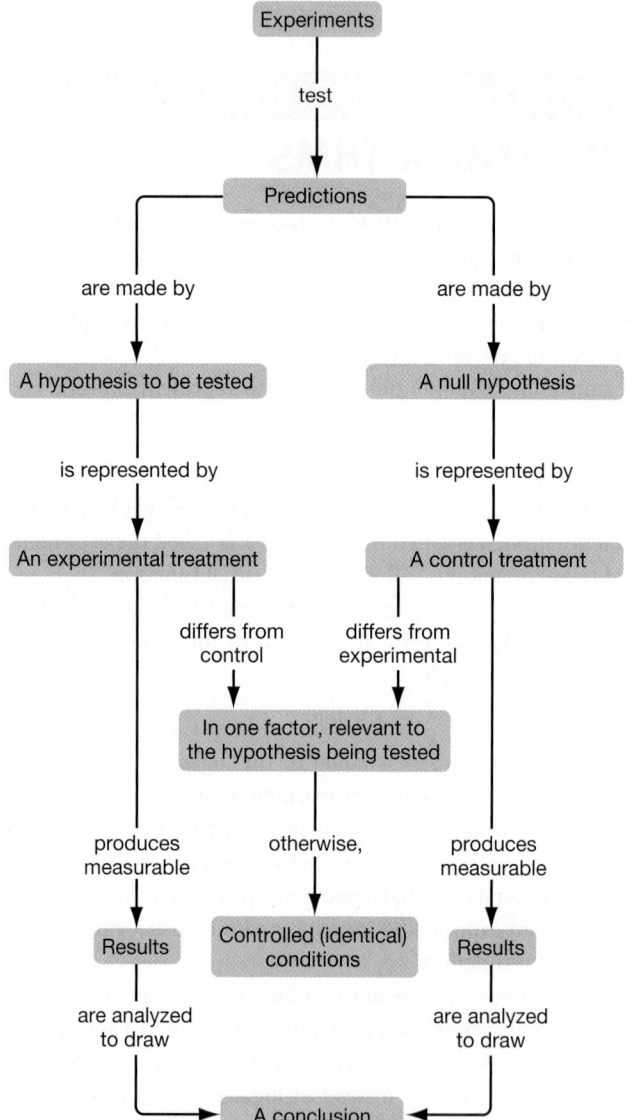

FIGURE B8.1 A Concept Map on Principles of Experimental Design.

SEPARATING AND VISUALIZING MOLECULES

To study a molecule, you have to be able to isolate it. This is a two-step process: the molecule has to be separated from other molecules in a mixture, and then physically picked out or located in a purified form. This BioSkill focuses on the techniques that biologists use to separate nucleic acids and proteins and then find the particular one they are interested in.

Using Electrophoresis to Separate Molecules

In molecular biology, the standard technique for separating proteins and nucleic acids is called gel electrophoresis or, simply, electrophoresis (literally "electricity-moving"). You may be using electrophoresis in a lab for this course, and you will certainly be analyzing data derived from electrophoresis in this text.

The principle behind electrophoresis is simple. Both proteins and nucleic acids carry a charge. As a result, these molecules move when placed in an electric field. Negatively charged molecules move toward the positive electrode; positively charged molecules move toward the negative electrode.

To separate a mixture of macromolecules so that each can be isolated and analyzed, researchers place the sample in a gelatinous substance. More specifically, the sample is placed in a "well"—a slot in a sheet or slab of the gelatinous substance. The "gel" itself consists of long molecules that form a matrix of fibers. The presence of the fibers keeps molecules in the sample from moving around randomly, but the gelatinous matrix also has pores through which the molecules can pass.

When an electrical field is applied across the gel, the molecules in the well move through the gel toward an electrode. Molecules that are smaller or more highly charged for their size move faster than do larger or less highly charged molecules. As they move, then, the molecules separate by size and by charge. Small and highly charged molecules end up at the bottom of the gel; large, less-charged molecules remain near the top.

An Example "Run"

Figure B9.1 shows the electrophoresis setup used in an experiment investigating how RNA molecules polymerize In this case, the investigators wanted to document how long RNA molecules became over time, when ribonucleotide triphosphates were present in a particular type of solution.

Step 1 shows how they loaded samples of macromolecules, taken on different days during the experiment, into wells at the top of the gel slab. This is a general observation: Each well holds a different sample. In this and many other cases, the researchers also filled a well with a sample containing fragments of known size, called a size standard or "ladder."

In step 2 the researchers immersed the gel in a solution that conducts electricity and applied a voltage across the gel. The molecules in each well started to run down the gel, forming a lane. After several hours of allowing the molecules to move, they removed the electric field (step 3). By then, molecules of different size and charge had separated from one another. In this case,

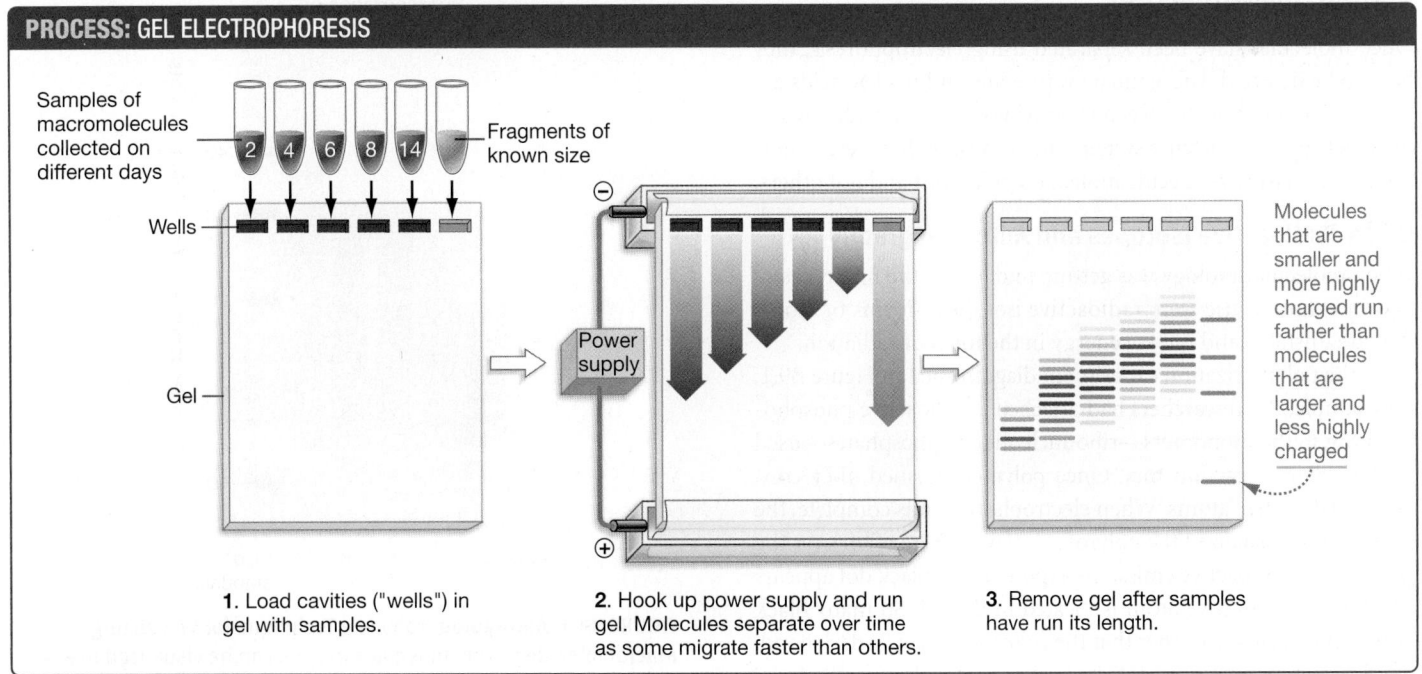

PROCESS: GEL ELECTROPHORESIS

Samples of macromolecules collected on different days

2 4 6 8 14

Fragments of known size

Wells

Gel

Power supply

⊖

⊕

Molecules that are smaller and more highly charged run farther than molecules that are larger and less highly charged

1. Load cavities ("wells") in gel with samples.

2. Hook up power supply and run gel. Molecules separate over time as some migrate faster than others.

3. Remove gel after samples have run its length.

FIGURE B9.1 Macromolecules Can Be Separated via Gel Electrophoresis.

✔ **QUESTION** DNA and RNA run toward the positive electrode. Why are these molecules negatively charged?

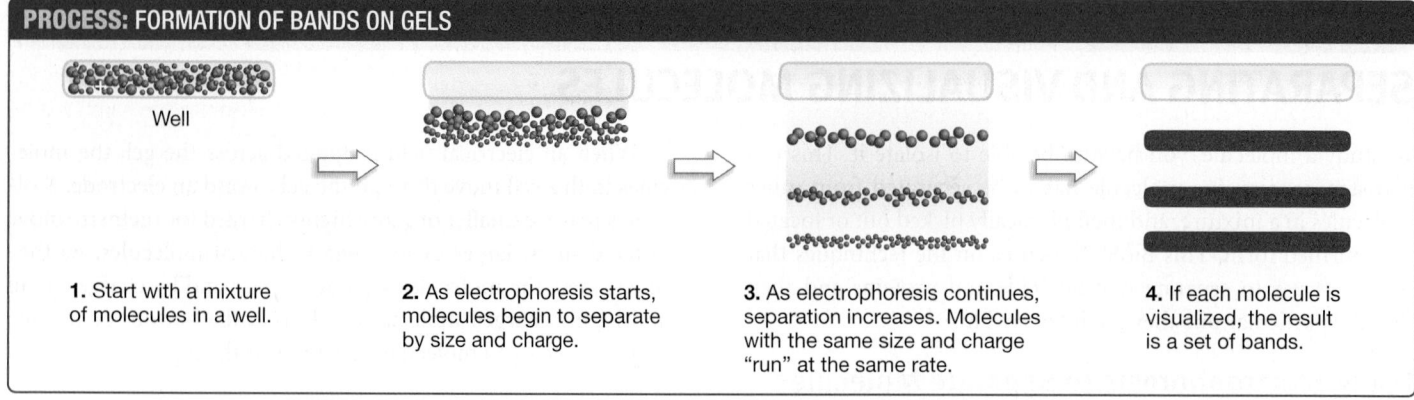

Well

1. Start with a mixture of molecules in a well.

2. As electrophoresis starts, molecules begin to separate by size and charge.

3. As electrophoresis continues, separation increases. Molecules with the same size and charge "run" at the same rate.

4. If each molecule is visualized, the result is a set of bands.

FIGURE B9.2 On a Gel, Alike Molecules Form Bands.

small RNA molecules had reached the bottom of the gel. Above them were larger RNA molecules, which had run more slowly.

Why Do Separated Molecules Form Bands?

When researchers visualize a particular molecule on a gel, using techniques described below, the image that results consists of bands: shallow lines that are as wide as a lane in the gel. Why?

To understand the answer, study **Figure B9.2**. The top panel shows the original mixture of molecules. In this cartoon, the size of each dot represents the size of each molecule. The key is to realize that the original sample contains many copies of each specific molecule, and that these copies run down the length of the gel together—meaning, at the same rate—because they have the same size and charge.

It's that simple: Alike molecules form a band because they stay together.

Visualizing Molecules

Once molecules have been separated using electrophoresis, they have to be detected. Unfortunately, proteins and nucleic acids are invisible unless they are tagged in some way. Let's consider two of the most common tagging systems, then consider how researchers can tag and visualize specific molecules of interest and not others.

Using Radioactive Isotopes and Autoradiography

When molecular biology was getting under way, the first types of tags in common use were radioactive isotopes—forms of atoms that are unstable and release energy in the form of radiation.

In the polymerization experiment diagrammed in Figure B9.1, for example, the researchers had attached a radioactive phosphorus atom to the monomers—ribonucleotide triphosphates—used in the original reaction mix. Once polymers formed, they contained radioactive atoms. When electrophoresis was complete, the investigators visualized the polymers by laying X-ray film over the gel. Because radioactive emissions expose film, a black dot appears wherever a radioactive atom is located in the gel. So many black dots occur so close together that the collection forms a dark band.

This technique for visualizing macromolecules is called autoradiography. The autoradiograph that resulted from the polymerization experiment is shown in **Figure B9.3**. The samples,

taken on days 2, 4, 6, 8, and 14 of the experiment, are labeled along the bottom. The far right lane contains macromolecules of known size; this lane is used to estimate the size of the molecules in the experimental samples. The bands that appear in each sample lane represent the different polymers that had formed.

Reading a Gel

One of the keys to interpreting or "reading" a gel is to realize that darker bands contain more radioactive markers, indicating the presence of many radioactive molecules. Lighter bands contain fewer molecules.

To read a gel, then, you look for (**1**) the presence or absence of bands in some lanes—meaning, some experimental samples—versus others, and (**2**) contrasts in the darkness of the bands—meaning, differences in the number of molecules present.

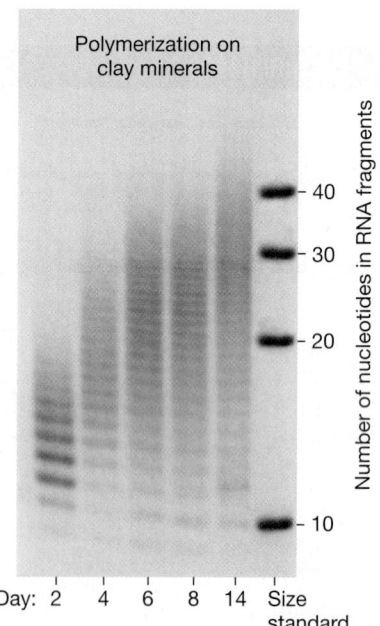

FIGURE B9.3 Autoradiography Is a Technique for Visualizing Macromolecules. The molecules in a gel can be visualized in a number of ways. In this case, the RNA molecules in the gel exposed an X-ray film because they had radioactive atoms attached. When developed, the film is called an autoradiograph.

For example, several conclusions can be drawn from the data in Figure B9.3. First, a variety of polymers formed at each stage. After the second day, for example, polymers from 12 to 18 monomers long had formed on the clay particles used in this experiment. Second, the overall length of polymers produced increased with time. At the end of the fourteenth day, most of the RNA molecules were between 20 and 40 monomers long.

Starting in the late 1990s and early 2000s, it became much more common to tag nucleic acids with fluorescent tags. Once electrophoresis is complete, the presence of fluorescence can be detected by exposing the gel to an appropriate wavelength of light; the fluorescent tag fluoresces or glows in response (fluorescence is explained in Chapter 10).

Fluorescent tags have important advantages over radioactive isotopes: (1) They are safer to handle. (2) They are faster—you don't have to wait hours or days for the radioactive isotope to expose a film. (3) They come in multiple colors, so you can tag several different molecules in the same experiment and detect them independently.

Using Nucleic Acid Probes

In many cases, researchers want to find one specific molecule—a certain DNA sequence, for example—in the collection of molecules on a gel. How is this possible? The answer hinges on using a particular molecule as a probe.

Chapter 19 explains how probes work in detail. Here it's enough to get the general idea: A probe is a marked molecule that binds specifically to your molecule of interest. The "mark" is often a radioactive atom or a fluorescent tag.

If you are looking for a particular DNA or RNA sequence on a gel, for example, you can expose the gel to a single-stranded probe that binds to the target sequence by complementary base pairing. Once it has bound, you can detect the band through autoradiography or fluorescence.

- Southern blotting is a technique for making DNA fragments that have been run out on a gel single-stranded, transferring them from the gel to a nylon filter, and then probing them to identify segments of interest. The technique was named after its inventor, Edwin Southern.

- Northern blotting is a technique for transferring RNA fragments from a gel to a nylon filter, and then probing them to detect target segments. The name is a lighthearted play on Southern blotting—the protocol from which it was derived.

Using Antibody Probes

How can researchers find a particular protein out of a large collection of different proteins? The answer is to use an antibody. An antibody is a protein that binds specifically to a section of a different protein. The structure of antibodies and their function in the immune system are explored in detail in Chapter 49.

To use an antibody as a probe, investigators attach a tag molecule—often an enzyme that catalyzes a color-forming reaction—to the antibody and allow it to react with proteins in a mixture. The antibody will stick to the specific protein that it binds to, and then can be visualized thanks to the tag it carries.

If the proteins in question have been separated by gel electrophoresis, the result is called a Western blot. The name Western is an extension of the Southern and Northern pattern.

✔ Question

Suppose you've been given a gel that has been stained for "RNA X." One lane contains no bands. Two lanes have a band in the same location, even though one of the bands is barely visible and the other is extremely dark. The fourth lane has a light band located above the bands in the other lanes. Interpret the gel.

BIOLOGICAL IMAGING: MICROSCOPY AND X-RAY CRYSTALLOGRAPHY

A lot of biology happens at levels that can't be detected with the naked eye. Biologists use an array of microscopes to study small multicellular organisms, individual cells, and the contents of cells. And to understand what individual macromolecules or multimolecular machines like ribosomes look like, researchers use data from a technique called X-ray crystallography.

You'll probably use dissecting microscopes and compound light microscopes to view specimens during your labs for this course, and throughout this text you'll be seeing images generated from other types of microscopy and from X-ray crystallographic data. One of the fundamental skills you'll be acquiring as an introductory student, then, is a basic understanding of how these techniques work. The key is to recognize that each approach for visualizing microscopic structures has strengths and weaknesses. As a result, each technique is appropriate for studying certain types or aspects of cells or molecules.

Light Microscopy

If you use a dissecting microscope during labs, you'll recognize that it works by magnifying light that bounces off a whole specimen—often a live organism. You'll be able to view the specimen in three dimensions, which is why these instruments are sometimes called stereomicroscopes, but the maximum magnification possible is only about 20 to 40 times normal size ($20\times$ to $40\times$).

To view smaller objects, you'll probably use a compound microscope. Compound microscopes magnify light that is passed *through* a specimen. The instruments used in introductory labs are usually capable of 400× magnifications; the most sophisticated compound microscopes available can achieve magnifications of about 2000×. This is enough to view individual bacterial or eukaryotic cells and see large structures inside cells, like condensed chromosomes (see Chapter 11). To prepare a specimen for viewing under a compound light microscope, the tissues or cells are usually sliced to create a section thin enough for light to pass through efficiently. The section is then dyed to increase contrast and make structures visible. In many cases, different types of dyes are used to highlight different types of structures.

Electron Microscopy

Until the 1950s, the compound microscope was the biologist's only tool for viewing cells directly. But the invention of the electron microscope provided a new way to view specimens. Two basic types of electron microscopy are now available: one that allows researchers to examine cross sections of cells at extremely high magnification, and one that offers a view of surfaces at somewhat lower magnification.

Transmission Electron Microscopy (TEM)

The transmission electron microscope is an extraordinarily effective tool for viewing cell structure at high magnification. TEM forms an image from electrons that pass through a specimen, just as a light microscope forms an image from light rays that pass through a specimen.

Biologists who want to view a cell under a transmission electron microscope begin by "fixing" the cell, meaning that they treat it with a chemical agent that stabilizes the cell's structure and contents while disturbing them as little as possible. Then the researcher permeates the cell with an epoxy plastic that stiffens the structure. Once this epoxy hardens, the cell can be cut into extremely thin sections with a glass or diamond knife. Finally, the sectioned specimens are impregnated with a metal—often lead. (The reason for this last step is explained shortly.)

Figure B10.1a outlines how the transmission electron microscope works. A beam of electrons is produced by a tungsten filament at the top of a column and directed downward. (All of the air is pumped out of the column, so that the electron beam isn't scattered by collisions with air molecules.) The electron beam passes through a series of lenses and through the specimen. The lenses are actually electromagnets, which alter the path of the beam much like a glass lens in a dissecting or compound microscope bends light. The lenses magnify and focus the image on a screen at the bottom of the column. There the electrons strike a coating of fluorescent crystals, which emit visible light in response—just like a television screen. When the microscopist moves the screen out of the way and allows the electrons to expose a sheet of black-and-white film, the result is a micrograph—a photograph of an image produced by microscopy.

The image itself is created by electrons that pass through the specimen. If no specimen were in place, all the electrons would pass through and the screen (and micrograph) would be uniformly bright. Unfortunately, cell materials by themselves would also appear fairly uniform and bright. This is because an atom's ability to deflect an electron depends on its mass. In turn, an atom's mass is a function of its atomic number. The hydrogen, carbon, oxygen, and nitrogen atoms that dominate biological molecules have low atomic numbers. This is why cell biologists must saturate cell sections with lead solutions. Lead has a high atomic number and scatters electrons effectively. Different macromolecules take up lead atoms in different amounts, so the

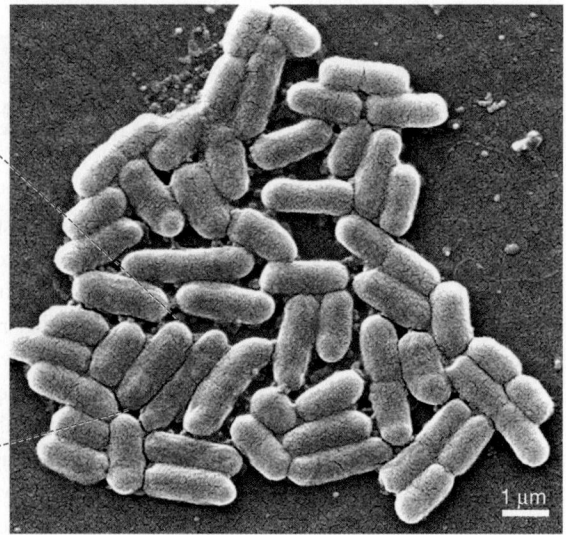

(a) Transmission electron microscopy: High magnification of cross sections

(b) Scanning electron microscopy: Lower magnification of surfaces

Tungsten filament (source of electrons)

Condenser lens

Specimen

Objective lens

Projector lens

Image on fluorescent screen

0.2 μm

Cross section of *E. coli* bacterium

1 μm

Surface view of *E. coli* bacteria

FIGURE B10.1 There Are Two Basic Types of Electron Microscopy.

metal acts as a "stain" that produces contrast. With TEM, areas of dense metal scatter the electron beam most, producing dark areas in micrographs.

The advantage of TEM is that it can magnify objects up to 250,000×—meaning that intracellular structures are clearly visible. The downsides are that researchers are restricted to observing dead, sectioned material, and they must take care that the preparation process does not distort the specimen.

Scanning Electron Microscopy (SEM)

The scanning electron microscope is the most useful tool biologists have for looking at the surfaces of structures. Materials are prepared for scanning electron microscopy by coating their surfaces with a layer of metal atoms. To create an image of this surface, the microscope scans the surface with a narrow beam of electrons. Electrons that are reflected back from the surface or that are emitted by the metal atoms in response to the beam then strike a detector. The signal from the detector controls a second electron beam, which scans a TV-like screen and forms an image magnified up to 50,000 times the object's size.

Because SEM records shadows and highlights, it provides images with a three-dimensional appearance (**Figure B10.1b**). It cannot magnify objects nearly as much as TEM can, however.

Studying Live Cells and Real-Time Processes

Until the 1960s, it was not possible for biologists to get clear, high-magnification images of living cells. But a series of innovations over the past 50 years has made it possible to observe organelles and subcellular structures in action.

The development of video microscopy, where the image from a light microscope is captured by a video camera instead of by an eye or a film camera, proved revolutionary. It allowed specimens to be viewed at higher magnification, because video cameras are more sensitive to small differences in contrast than are the human eye or still cameras. It also made it easier to keep live specimens functioning normally, because the increased light sensitivity of video cameras allows them to be used with low illumination, so specimens don't overheat. And when it became possible to digitize video images, researchers began using computers to remove out-of-focus background material and increase image clarity.

A more recent innovation was the use of a fluorescent molecule called green fluorescent protein, or GFP, which allows researchers to tag specific molecules or structures and follow their movement over time. GFP is naturally synthesized in jellyfish that fluoresce, or emit light. By affixing GFP molecules to another protein and then inserting it into a cell, investigators can follow the protein's fate over time and even videotape its movement. For example, researchers have videotaped GFP-tagged proteins being transported from the rough ER through the Golgi apparatus and out to the plasma membrane. This is cell biology: the movie.

GFP's influence has been so profound that the researchers who developed its use in microscopy were awarded the 2008 Nobel prize in Chemistry.

Visualizing Structures in 3-D

The world is three-dimensional. To understand how microscopic structures and macromolecules work, it is essential to understand their shape and spatial relationships. Consider three techniques currently being used to reconstruct the 3-D structure of cells, organelles, and macromolecules.

- *Confocal microscopy* is carried out by mounting cells that have been treated with one or more fluorescing tags on a microscope slide and then focusing a beam of ultraviolet light at a specific depth within the specimen. The fluorescing tag emits visible light in response. A detector for this light is then set up at exactly the position where the emitted light comes into focus. The result is a sharp image of a precise plane in the cell being studied (**Figure B10.2**). By altering the focal plane, a researcher can record images from an array of depths in the specimen; a computer can then be used to generate a 3-D image of the cell.

- *Electron tomography* uses a transmission electron microscope to generate a 3-D image of an organelle or other subcellular structure. The specimen is rotated around a single axis, with the researcher taking many "snapshots." The individual images are then pieced together with a computer. This technique has provided a much more accurate view of mitochondrial structure than was possible using traditional TEM (see Chapter 7).

- *X-ray crystallography, or X-ray diffraction analysis,* is the most widely used technique for reconstructing the 3-D structure of molecules. As its name implies, the procedure is based on bombarding crystals of a molecule with X-rays. X-rays are scattered in precise ways when they interact with the electrons surrounding the atoms in a crystal, producing a diffraction pattern that can be recorded on X-ray film or other types of

(a) Conventional fluorescence image of single cell

(b) Confocal fluorescence image of same cell

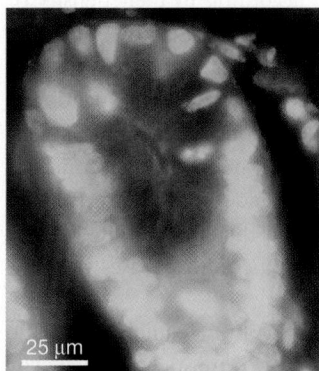 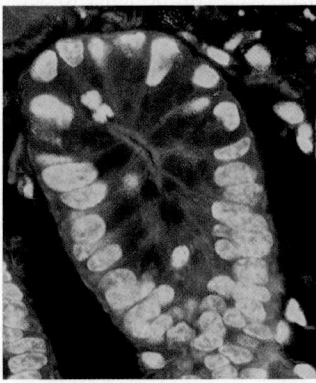

25 µm

FIGURE B10.2 Confocal Microscopy Provides Sharp Images of Living Cells. (a) The conventional image of this mouse intestinal cell is blurred, because it results from light emitted by the entire cell. **(b)** The confocal image is sharp, because it results from light emitted at a single plane inside the cell.

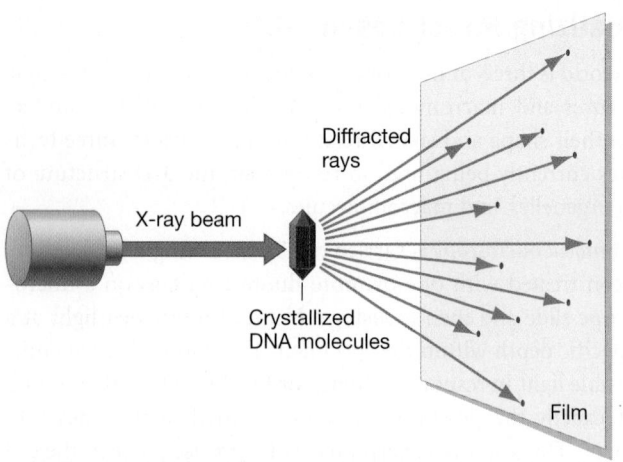
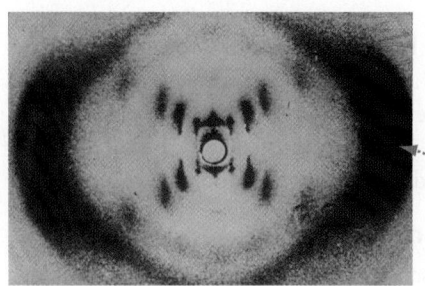

The patterns are determined by the structure of the molecules within the crystal

FIGURE B10.3 X-Ray Crystallography. When crystallized molecules are bombarded with X-rays, the radiation is scattered in distinctive patterns. The photograph at the right shows an X-ray film that recorded the pattern of scattered radiation from DNA molecules.

detectors (**Figure B10.3**). By varying the orientation of the X-ray beam as it strikes a crystal and documenting the diffraction patterns that result, researchers can construct a map representing the density of electrons in the crystal. By relating these electron-density maps to information about the primary structure of the nucleic acid or protein, a 3-D model of the molecule can be built. Virtually all of the molecular models used in this book were built from X-ray crystallographic data.

✔**Questions**

1. Suppose you are looking at a transmission electron micrograph of a cancerous cell from the human liver. No mitochondria are present. Does this mean that the cell lacks mitochondria?

2. X-ray crystallography is time consuming and technically difficult. Why is the effort to understand the structure of biological molecules worthwhile? What's the payoff?

BIOSKILLS 11

SEPARATING CELL COMPONENTS BY CENTRIFUGATION

Biologists use a technique called differential centrifugation to isolate specific cell components. Differential centrifugation is based on breaking cells apart to create a complex mixture and then separating components in a centrifuge. A centrifuge accomplishes this by spinning cells in a solution that allows molecules and other cell components to separate according to their density or size and shape. The individual parts of the cell can then be purified and studied in detail, in isolation from other parts of the cell.

The first step in preparing a cell sample for centrifugation is to release the organelles and cell components by breaking the cells apart. This can be done by putting them in a hypotonic solution, by exposing them to high-frequency vibration, by treating cells with a detergent, or by grinding them up. Each of these methods breaks apart plasma membranes and releases the contents of the cells.

The resulting pieces of plasma membrane quickly reseal to form small vesicles, often trapping cell components inside. The solution that results from the homogenization step is a mixture of these vesicles, free-floating macromolecules released from the cells, and organelles. A solution such as this is called a cell extract or cell homogenate.

When a cell homogenate is placed in a centrifuge tube and spun at high speed, the components that are in solution tend to move outward, along the red arrow in **Figure B11.1a**. The effect is similar to a merry-go-round, which seems to push you outward in a straight line away from the spinning platform. In response to this outward-directed force, the solution containing the cell homogenate exerts a centripetal (literally, "center-seeking") force that pushes the homogenate away from the bottom of the tube. Larger, denser molecules or particles resist this inward force more readily than do smaller, less dense ones and so reach the bottom of the centrifuge tube faster.

To separate the components of a cell extract, researchers often perform a series of centrifuge runs. Steps 1 and 2 of **Figure B11.1b** illustrate how an initial treatment at low speed causes larger, heavier parts of the homogenate to move below smaller, lighter parts. The material that collects at the bottom of the tube is called the pellet, and the solution and solutes left behind form the supernatant ("above swimming"). The supernatant is placed in a fresh tube and centrifuged at increasingly higher speeds and longer durations. Each centrifuge run continues to separate cell components based on their size and density.

To accomplish even finer separation of macromolecules or organelles, researchers frequently follow up with centrifugation at extremely high speeds. One strategy is based on filling the centrifuge tube with a series of sucrose solutions of increasing

(a) How a centrifuge works

When the centrifuge spins, the macromolecules tend to move toward the bottom of the centrifuge tube (red arrow)

The solution in the tube exerts a centripetal force, which resists movement of the molecules to the bottom of the tube (blue arrow)

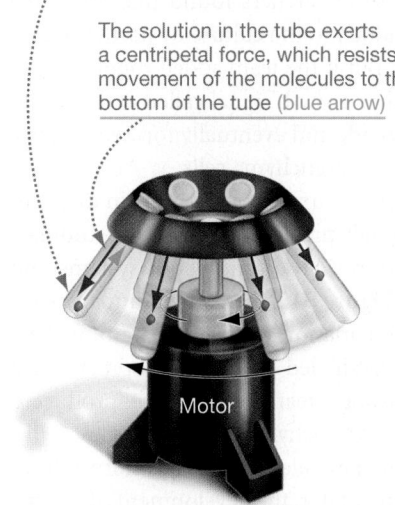

Motor

Very large or dense molecules overcome the centripetal force more readily than smaller, less dense ones. As a result, larger, denser molecules move toward the bottom of the tube faster.

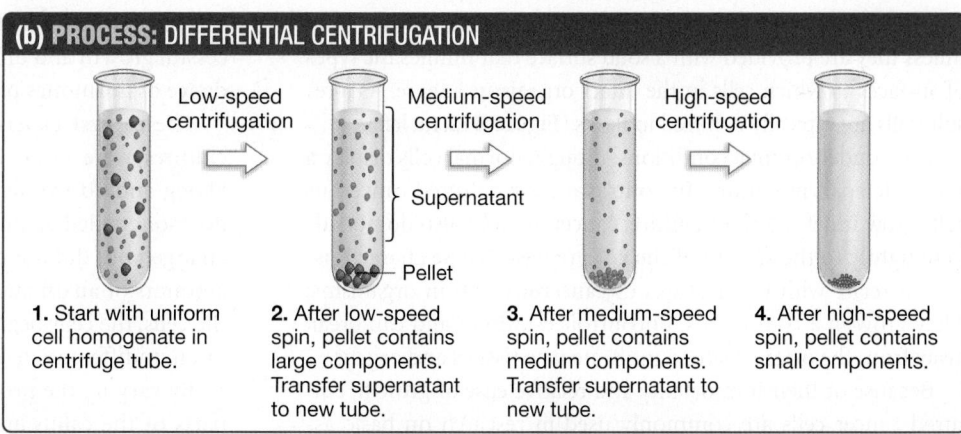

(b) PROCESS: DIFFERENTIAL CENTRIFUGATION

Low-speed centrifugation → Medium-speed centrifugation → High-speed centrifugation

Supernatant

Pellet

1. Start with uniform cell homogenate in centrifuge tube.

2. After low-speed spin, pellet contains large components. Transfer supernatant to new tube.

3. After medium-speed spin, pellet contains medium components. Transfer supernatant to new tube.

4. After high-speed spin, pellet contains small components.

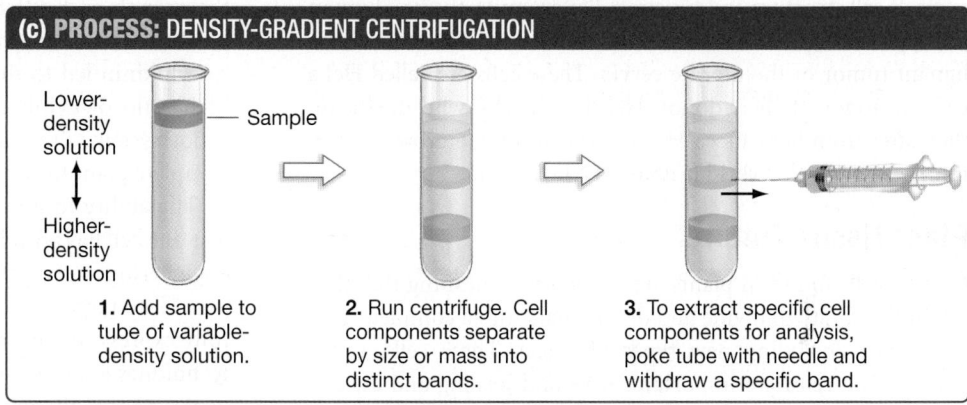

(c) PROCESS: DENSITY-GRADIENT CENTRIFUGATION

Lower-density solution

Higher-density solution

Sample

1. Add sample to tube of variable-density solution.

2. Run centrifuge. Cell components separate by size or mass into distinct bands.

3. To extract specific cell components for analysis, poke tube with needle and withdraw a specific band.

FIGURE B11.1 Cell Components Can Be Separated by Centrifugation. (a) The forces inside a centrifuge tube allow cell components to be separated. **(b)** Through a series of centrifuge runs made at increasingly higher speeds, an investigator can separate fractions of a cell homogenate by size via differential centrifugation. **(c)** A high-speed centrifuge run can achieve extremely fine separation among cell components by density-gradient centrifugation.

density (**Figure B11.1c**). The density gradient allows cell components to separate on the basis of small differences in size, shape, and density. When the centrifuge run is complete, each cell component occupies a distinct band of material in the tube, based on how quickly each moves through the increasingly dense gradient of sucrose solution during the centrifuge run. A researcher can then collect the material in each band for further study.

✔ Questions

1. Electrophoresis separates molecules by charge and size. What is the physical basis for separating molecules or cell components via centrifugation?

2. Compared to other types of centrifugation, why would extremely high-speed centrifugation allow similar molecules to separate?

BIOSKILLS 12

CELL AND TISSUE CULTURE METHODS

For researchers, there are important advantages to growing plant and animal cells and tissues outside the organism itself. Cell and tissue cultures provide large populations of a single type of cell or tissue and the opportunity to control experimental conditions precisely.

Animal Cell Culture

The first successful attempt to culture animal cells occurred in 1907, when a researcher cultivated amphibian nerve cells in a drop of fluid from the spinal cord. But it was not until the 1950s

and 1960s that biologists could routinely culture plant and animal cells in the laboratory. The long lag time was due to the difficulty of re-creating conditions that exist in the intact organism precisely enough for cells to grow normally.

To grow in culture, animal cells must be provided with a liquid mixture containing the nutrients, vitamins, and hormones that stimulate growth. Initially, this mixture was serum, the liquid portion of blood; now serum-free media are available for certain cell types. Serum-free media are preferred because they are much more precisely defined chemically than serum.

In addition, many types of animal cells will not grow in culture unless they are provided with a solid surface that mimics the types of surfaces to which cells in the intact organisms adhere. As a result, cells are typically cultured in flasks (**Figure B12.1a**, left).

Even under optimal conditions, though, normal cells display a finite life span in culture. In contrast, many cultured cancerous cells grow indefinitely. In culture, cancerous cells also do not adhere tightly to the surface of the culture flask. These characteristics correlate with two features of cancerous cells in organisms: They grow in a continuous, uncontrolled fashion and can break away from the original site to infiltrate new tissues and organs.

Because of their immortality and relative ease of growth, cultured cancer cells are commonly used in research on basic aspects of cell structure and function. For example, the first human cell type to be grown in culture was isolated in 1951 from a malignant tumor of the uterine cervix. These cells are called HeLa cells in honor of their donor, Henrietta Lacks, who died soon thereafter from her cancer. HeLa cells continue to grow in laboratories around the world (Figure B12.1a, right).

Plant Tissue Culture

Certain cells found in plants are totipotent—meaning that they retain the ability to divide and differentiate into a complete, mature plant, including new types of tissue. These cells, called parenchyma cells, are important in wound healing and asexual reproduction. But they also allow researchers to grow complete adult plants in the laboratory, starting with a small number of parenchyma cells.

Biologists who grow plants in tissue culture begin by placing parenchyma cells in a liquid or solid medium containing all the nutrients required for cell maintenance and growth. In the early days of plant tissue culture, investigators found not only that

specific growth signals called hormones were required for successful growth and differentiation but also that the relative abundance of hormones present was critical to success.

The earliest experiments on hormone interactions in tissue cultures were done with tobacco cells in the 1950s by Folke Skoog and co-workers. These researchers found that when the hormone called auxin was added to the culture by itself, the cells enlarged but did not divide. But if the team added roughly equal amounts of auxin and another growth signal called cytokinin to the cells, the cells began to divide and eventually formed a callus, or an undifferentiated mass of parenchyma cells.

By varying the proportion of auxin to cytokinins in different parts of the callus and through time, the team could stimulate the growth and differentiation of root and shoot systems and produce whole new plants (**Figure B12.1b**). A high ratio of auxin to cytokinin led to the differentiation of a root system, while a high ratio of cytokinin to auxin led to the development of a shoot system. Eventually Skoog's team was able to produce a complete plant from just one parenchyma cell.

The ability to grow whole new plants in tissue culture from just one cell has been instrumental in the development of genetic engineering (see Chapter 19). Researchers insert recombinant genes into target cells, test the cells to identify those that successfully express the recombinant genes, and then use tissue culture techniques to grow those cells into adult individuals with a novel genotype and phenotype.

✔ Questions

1. What is a limitation of how experiments on HeLa cells are interpreted?

2. State a disadvantage of doing experiments on plants that have been propagated from single cells growing in tissue culture.

(a) Animal cell culture: immortal HeLa cancer cells

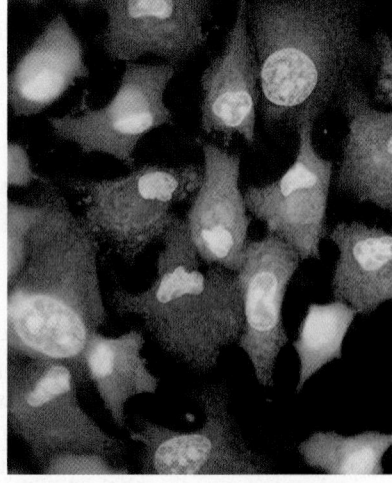

(b) Plant tissue culture: tobacco callus

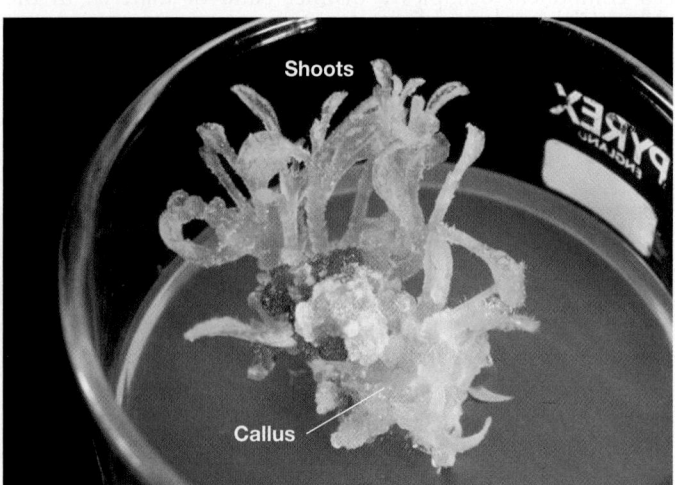

FIGURE B12.1 Animal and Plant Cells Can Be Grown in the Lab.

COMBINING PROBABILITIES

In several cases in this text, you'll need to combine probabilities from different events in order to solve a problem. One of the most common applications is in genetics problems. For example, Punnett squares work because they are based on two fundamental rules of probability. Each rule pertains to a distinct situation.

The Both-And Rule

The both-and rule—also known as the product rule or multiplication rule—applies when you want to know the probability that two or more independent events occur together. Let's use the rolling of two dice as an example. What is the probability of rolling two sixes? These two events are independent, because the probability of rolling a six on one die has no effect on the probability of rolling a six on the other die. (In the same way, the probability of getting a gamete with allele R from one parent has no effect on the probability of getting a gamete with allele R from the other parent. Gametes fuse randomly.)

The probability of rolling a six on the first die is 1/6. The probability of rolling a six on the second die is also 1/6. The probability of rolling a six on *both* dice, then, is $1/6 \times 1/6 = 1/36$. In other words, if you rolled two dice 36 times, on average you would expect to roll two sixes once.

In the case of a cross between two parents heterozygous at the R gene, the probability of getting allele R from the father is 1/2 and the probability of getting R from the mother is 1/2. Thus, the probability of getting both alleles and creating an offspring with genotype RR is $1/2 \times 1/2 = 1/4$.

The Either-Or Rule

The either-or rule—also known as the sum rule or addition rule—applies when you want to know the probability of an event happening when there are several different ways for the same event or outcome to occur. In this case, the probability that the event will occur is the sum of the probabilities of each way that it can occur.

For example, suppose you wanted to know the probability of rolling either a one or a six when you toss a die. The probability of drawing each is 1/6, so the probability of getting one or the other is $1/6 + 1/6 = 1/3$. If you rolled a die three times, on average you'd expect to get a one or a six once.

In the case of a cross between two parents heterozygous at the R gene, the probability of getting an R allele from the father and an r allele from the mother is $1/2 \times 1/2 = 1/4$. Similarly, the probability of getting an r allele from the father and an R allele from the mother is $1/2 \times 1/2 = 1/4$. Thus, the combined probability of getting the Rr genotype in either of the two ways is $1/4 + 1/4 = 1/2$.

✔ Questions

1. Suppose that four students each toss a coin. What is the probability of four "tails"?

2. After a single roll of a die, what is the probability of getting either a two, a three, or a six?

MODEL ORGANISMS

Research in biological science starts with a question. In most cases, the question is inspired by an observation about a cell or an organism. To answer it, biologists have to study a particular species. Study organisms are often called model organisms, because investigators hope that they serve as a model for what is going on in a wide array of species.

Model organisms are chosen because they are convenient to study and because they have attributes that make them appropriate for the particular research proposed. They tend to have some common characteristics:

- *Short generation time and rapid reproduction* This is important because it makes it possible to produce offspring quickly and perform many experiments in a short amount of time—you don't have to wait long for individuals to grow.

- *Large numbers of offspring* This is particularly important in genetics, where many offspring phenotypes and genotypes need to be assessed to get a large sample size.

- *Small size, simple feeding and habitat requirements* These attributes make it relatively cheap and easy to maintain individuals in the lab.

The following notes highlight just a few model organisms supporting current work in biological science.

Escherichia coli

Of all model organisms in biology, perhaps none has been more important than the bacterium *Escherichia coli*—a common inhabitant of the human gut. The strain that is most commonly worked on today, called K-12 (**Figure B14.1a** on page B:20), was originally isolated from a hospital patient in 1922.

During the last half of the twentieth century, key results in molecular biology originated in studies of *E. coli*. These results include the discovery of enzymes such as DNA polymerase, RNA polymerase, DNA repair enzymes, and restriction endonucleases; the elucidation of ribosome structure and function; and the initial characterization of promoters, regulatory transcription factors, regulatory sites in DNA, and operons. In many cases, initial discoveries made in *E. coli* allowed researchers to confirm that homologous enzymes and processes existed in an array of organisms, often ranging from other bacteria to yeast, mice, and humans.

The success of *E. coli* as a model for other species inspired Jacques Monod's claim that "Once we understand the biology of *Escherichia coli*, we will understand the biology of an elephant." The genome of *E. coli* K-12 was sequenced in 1997, and the strain continues to be a workhorse in studies of gene function, biochemistry, and particularly biotechnology. Much remains to be learned, however. Despite over 60 years of intensive study, the function of about a third of the *E. coli* genome is still unknown.

In the lab, *E. coli* is usually grown in suspension culture, where cells are introduced to a liquid nutrient medium, or on plates containing agar—a gelatinous mix of polysaccharides. Under optimal growing conditions—meaning before cells begin to get crowded and compete for space and nutrients—a cell takes just 30 minutes on average to grow and divide. At this rate, a single cell can produce a population of over a million descendants in just 10 hours. Except for new mutations, all of the descendant cells are genetically identical.

Dictyostelium discoideum

The cellular slime mold *Dictyostelium discoideum* is not always slimy, and it is not a mold—meaning a type of fungus. Instead, it is an amoeba. Amoeba is a general term that biologists use to characterize a unicellular eukaryote that lacks a cell wall and is extremely flexible in shape. *Dictyostelium* has long fascinated biologists because it is a social organism. Independent cells sometimes aggregate to form a multicellular structure.

Under most conditions, *Dictyostelium* cells are haploid (*n*) and move about in decaying vegetation on forest floors or other habitats. They feed on bacteria by engulfing them whole. When these cells reproduce, they can do so sexually by fusing with another cell then undergoing meiosis, or asexually by mitosis, which is more common. If food begins to run out, the cells begin to aggregate. In many cases, tens of thousands of cells cohere to form a 2-mm-long mass called a slug (**Figure B14.1b**). (This is not the slug that is related to snails.)

After migrating to a sunlit location, the slug stops and individual cells differentiate according to their position in the slug. Some form a stalk; others form a mass of spores at the tip of the stalk. (A spore is a single cell that develops into an adult organism, but it is not formed from gamete fusion like a zygote.) The entire structure, stalk plus mass of spores, is called a fruiting body. Cells that form spores secrete a tough coat and represent a durable resting stage. The fruiting body eventually dries out, and the wind disperses the spores to new locations, where more food might be available.

Dictyostelium has been an important model organism for investigating questions about eukaryotes:

- Cells in a slug are initially identical in morphology but then differentiate into distinctive stalk cells and spores. Studying

this process helped biologists better understand how cells in plant and animal embryos differentiate into distinct cell types.

- The process of slug formation has helped biologists study how animal cells move and how they aggregate as they form specific types of tissues.

- When *Dictyostelium* cells aggregate to form a slug, they stick to each other. The discovery of membrane proteins responsible for cell-cell adhesion helped biologists understand some of the general principles of multicellular life highlighted in Chapter 8.

Arabidopsis thaliana

In the early days of biology, the best-studied plants were agricultural varieties such as maize (corn), rice, and garden peas. When biologists began to unravel the mechanisms responsible for oxygenic photosynthesis in the early to mid-1900s, they relied on green algae that were relatively easy to grow and manipulate in the lab—often the unicellular species *Chlamydomonas reinhardii*—as an experimental subject.

Although crop plants and green algae continue to be the subject of considerable research, a new model organism emerged in the 1980s and now serves as the preeminent experimental subject in plant biology. That organism is *Arabidopsis thaliana*, commonly known as thale cress or wall cress (**Figure B14.1c**).

Arabidopsis is a member of the mustard family, or Brassicaceae, so it is closely related to radishes and broccoli. In nature it is a weed—meaning a species that is adapted to thrive in habitats where soils have been disturbed.

One of the most attractive aspects of working with *Arabidopsis* is that individuals can grow from a seed into a mature, seed-producing plant in just four to six weeks. Several other attributes make it an effective subject for study: It has just five chromosomes, has a relatively small genome with limited numbers of repetitive sequences, can self-fertilize as well as undergo cross-fertilization, can be grown in a relatively small amount of space and with a minimum of care in the greenhouse, and produces up to 10,000 seeds per individual per generation.

Arabidopsis has been instrumental in a variety of studies in plant molecular genetics and development, and it is increasingly popular in ecological and evolutionary studies. In addition, the entire genome of the species has now been sequenced, and studies have benefited from the development of an international "*Arabidopsis* community"—a combination of informal and formal associations of investigators who work on *Arabidopsis* and use regular meetings, e-mail, and the Internet to share data, techniques, and seed stocks.

Saccharomyces cerevisiae

When biologists want to answer basic questions about how eukaryotic cells work, they often turn to the yeast *Saccharomyces cerevisiae*.

S. cerevisiae is unicellular and relatively easy to culture and manipulate in the lab (**Figure B14.1d**). In good conditions, yeast cells grow and divide almost as rapidly as bacteria. As a result, the

(a) Bacterium *Escherichia coli* (strain K-12)

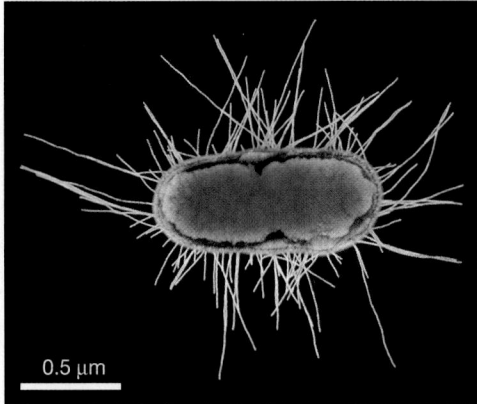

0.5 μm

(b) Slime mold *Dictyostelium discoideum*

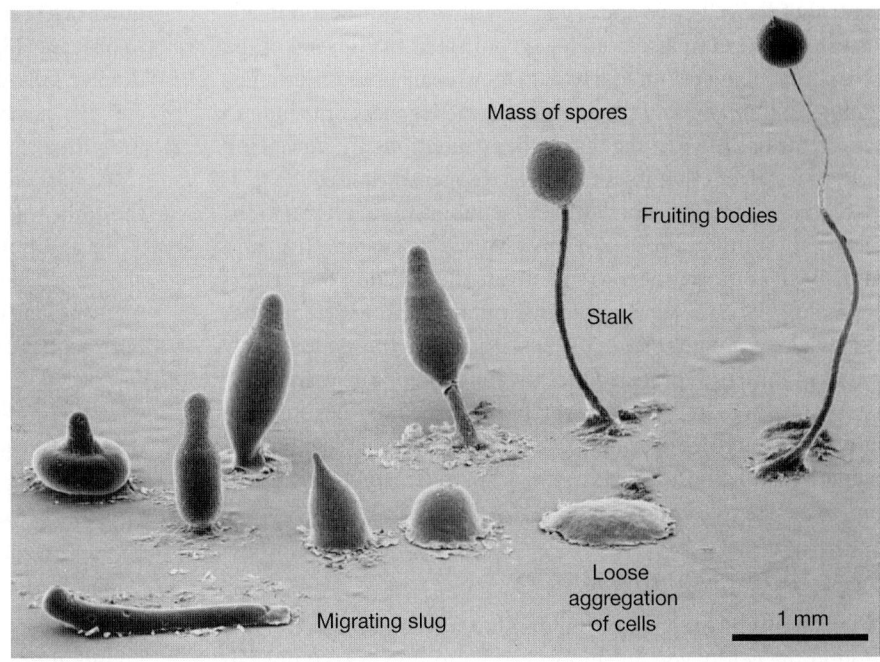

Mass of spores

Fruiting bodies

Stalk

Migrating slug

Loose aggregation of cells

1 mm

(c) Thale cress *Arabidopsis thaliana*

5 cm

(e) Fruit fly *Drosophila melanogaster*

0.5 mm

(f) Roundworm *Caenorhabditis elegans*

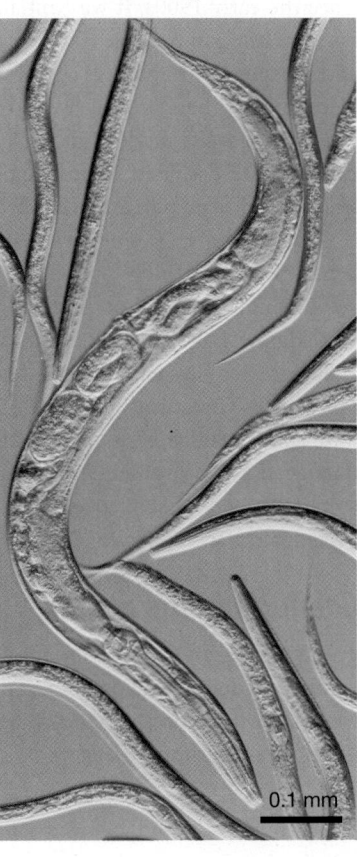

0.1 mm

(d) Yeast *Saccharomyces cerevisiae*

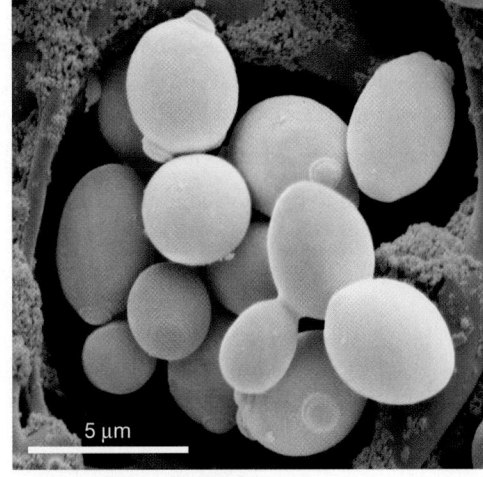

5 μm

(g) Mouse *Mus musculus*

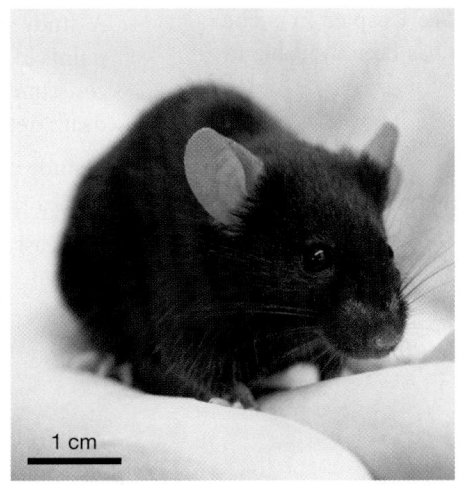

1 cm

FIGURE B14.1 Model Organisms.

✓**QUESTION** *E. coli* is grown at a temperature of 37°C. Why?

species has become the organism of choice for experiments on control of the cell cycle and regulation of gene expression in eukaryotes. For example, research has confirmed that several of the genes controlling cell division and DNA repair in yeast have homologs in humans; and when mutated, these genes contribute to cancer. Strains of yeast that carry these mutations are now being used to test drugs that might be effective against cancer.

S. cerevisiae has become even more important in efforts to interpret the genomes of organisms like rice, mice, zebrafish, and humans. It is much easier to investigate the function of particular genes in *S. cerevisiae* by creating mutants or transferring specific alleles among individuals than it is to do the same experiments in mice or zebrafish. Once the function of a gene has been established in yeast, biologists can look for the homologous gene in other eukaryotes. If such a gene exists, they can usually infer that it has a function similar to its role in *S. cerevisiae*. It was also the first eukaryote with a completely sequenced genome.

Drosophila melanogaster

If you walk into a biology building on any university campus around the world, you are almost certain to find at least one lab where the fruit fly *Drosophila melanogaster* is being studied (**Figure B14.1e**).

Drosophila has been a key experimental subject in genetics since the early 1900s. It was initially chosen as a focus for study by T. H. Morgan because it can be reared in the laboratory easily and inexpensively, matings can be arranged, the life cycle is completed in less than two weeks, and females lay a large number of eggs. These traits made fruit flies valuable subjects for breeding experiments designed to test hypotheses about how traits are transmitted from parents to offspring (see Chapter 13).

More recently, *Drosophila* has also become a key model organism in the field of developmental biology. The use of flies in developmental studies was inspired in large part by the work of Christianne Nüsslein-Volhard and Eric Wieschaus, who in the 1980s isolated flies with genetic defects in early embryonic development. By investigating the nature of these defects, researchers have gained valuable insights into how various gene products influence the development of eukaryotes (see Chapter 21). The complete genome sequence of *Drosophila* has been available to investigators since the year 2000.

Caenorhabditis elegans

The roundworm *Caenorhabditis elegans* emerged as a model organism in developmental biology in the 1970s, due largely to work by Sydney Brenner and colleagues. (*Caenorhabditis* is pronounced *see-no-rab-DIE-tiss*.)

C. elegans was chosen for three reasons: (**1**) Its cuticle (soft outer layer) is transparent, making individual cells relatively easy to observe (**Figure B14.1f**); (**2**) adults have exactly 959 nonreproductive cells; and, most important, (**3**) the fate of each cell in an embryo can be predicted because cell fates are invariant among individuals. For example, when researchers examine a 33-cell *C. elegans* embryo, they know exactly which of the 959 cells in the adult will be derived from each of those 33 embryonic cells.

In addition, *C. elegans* are small (less than 1 mm long), are able to self-fertilize or cross-fertilize, and undergo early development in just 16 hours. The entire genome of *C. elegans* has now been sequenced.

Mus musculus

Because the house mouse *Mus musculus* is the most important model organism among mammals, it is especially prominent in biomedical research—where researchers need to work on individuals with strong genetic and developmental similarities to humans.

The house mouse was an intelligent choice of model organism in mammals because it is small and thus relatively inexpensive to maintain in captivity, and because it breeds rapidly. A litter can contain 10 offspring and generation time is only 12 weeks—meaning that several generations can be produced in a year. Descendants of wild house mice have been selected for docility and other traits that make them easy to handle and rear; these populations are referred to as laboratory mice (**Figure B14.1g**).

Some of the most valuable laboratory mice are strains with distinctive, well-characterized genotypes. Inbred strains are virtually homogenous genetically (see Chapter 25) and are useful in experiments where gene-by-gene or gene-by-environment interactions have to be controlled. Other populations carry mutations that knock out genes and cause diseases similar to those observed in humans. These individuals are useful for identifying the cause of genetic diseases and testing drugs or other types of therapies.

✔ Questions

Which model organisms described here would be the best choice for the following studies? In each case, explain your reasoning.

1. A study of why specific cells in an embryo die at certain points in normal development. One goal is to understand the consequences for the individual when programmed cell death does not occur when it is supposed to.

2. A study of proteins that are required for cell-cell adhension.

3. Research on a gene suspected to be involved in the formation of breast cancer in humans.

APPENDIX B Answers

CHAPTER 1

Check Your Understanding (CYU)

CYU p. 5 The data points would all be about 11 percent, indicating no change in average kernel protein content over time. **CYU p. 8** From the sequence data provided, Species A and B differ only in one nucleotide of the rRNA sequence (position 10 from left). Species C differs from Species A and B in four nucleotides (positions 1, 2, 9, and 10). A correctly drawn phylogenetic tree would indicate that Species A and B appear to be closely related, with Species C being more distantly related [see Figure A1.1]. **CYU p. 12** The key here is to test predation rates during the hottest part of the day (when desert ants actually feed) versus other parts of the day. The experiment would best be done in the field, where natural predators are present. One approach would be to capture a large number of ants, divide the group in two, and measure predation rates (number of ants killed per hour) when they are placed in normal habitat during the hottest part of the day versus an hour before (or after). (1) The control group here is the normal condition—ants out during the hottest part of the day. If you didn't include a control, a critic could argue that predation did or did not occur because of your experimental setup or manipulation, not because of differences in temperature. (2) You would need to make sure that there is no difference in body size, walking speed, how they were captured and maintained, or other traits that might make the ants in the two groups more or less susceptible to predators. They should also be put out in the same habitat, so the presence of predators is the same in the two treatments.

You Should Be Able To (YSBAT)

YSBAT p. 5 Average kernel protein content declined, from 11 percent to a much lower value. (The experiment is still continuing; to date the low-selection line is about 5 percent.) **YSBAT p. 12** (1) A critic could argue that the ants weren't navigating normally, because they had been caught and released and transferred to a new channel. (2) A critic could argue that the ants can't navigate normally on their manipulated legs.

Caption Questions and Exercises

Figure 1.2 If Pasteur had done any of the things listed, he would have had more than one variable in his experiment. This would allow critics to claim that he got different results because of the differences in broth types, heating, or flask types—not the difference in exposure to preexisting cells. The results would not be definitive. **Figure 1.4** Molds and other fungi are more closely related to green algae because they differ from plants at two positions (5 and 8 from left), but differ from green algae at only one position (8). **Figure 1.6** The eukaryotic cell is roughly 10 times (more exactly, 8.5 times) the size of the prokaryotic cell. **Figure 1.8** You should be skeptical, because the results could be due to the single individual being sick or injured or unusual in some other way.

Summary of Key Concepts

KC 1.1 Dead cells cannot maintain a difference in conditions inside the cell versus outside, replicate, use energy, or process information. **KC 1.2** Observations on thousands of diverse species supported the claim that all organisms consist of cells. The hypothesis that all cells come from preexisting cells was supported when Pasteur showed that new cells do not arise and grow in a boiled liquid unless they are introduced from the air. **KC 1.3** If seeds with higher protein content leave the most offspring, then individuals with low protein in their seeds will become rare over time. **KC 1.4** A newly discovered species can be classified as a member of the Bacteria if the sequence of its rRNA contains some features found only in Bacteria. The same logic applies to classifying a new species in the Archaea or Eukarya. **KC 1.5** (1) A hypothesis is an explanation of how the world works; a prediction is an outcome you should observe if the hypothesis is correct. (2) Experiments are convincing because they measure predictions from two opposing hypotheses. Both predicted actions cannot occur, so one hypothesis will be supported while the other will not.

Test Your Knowledge

1. d; **2.** d; **3.** c; **4.** b; **5.** d; **6.** c

Test Your Understanding

1. That the entity they discovered replicates, processes information, acquires and uses energy, is cellular, and that its populations evolve. **2.** [See Figure A1.2] **3.** The rule ensured that two different organisms would never end up with the same name. **4.** Over time, traits that increased the fitness of individuals in this habitat became increasingly more frequent in the population. **5.** Individuals with certain traits selected, in the sense that they produce the most offspring. **6.** Yes. If evolution is defined as "change in the characteristics of a population over time," then those organisms that are most closely related should have experienced less change over time. On a phylogenetic tree, species with substantially similar rRNA sequences would be diagrammed with a closer common ancestor—one that had the sequences they inherited—than the ancestors shared between species with dissimilar rRNA sequences.

Applying Concepts to New Situations

1. A scientific theory is not a guess—it is an idea whose validity can be tested with data. Both the cell theory and the theory of evolution have been validated by large bodies of observational and experimental data. **2.** If all eukaryotes living today have a nucleus, then it is logical to conclude that the nucleus arose in a common ancestor of all eukaryotes, indicated by the arrow you should have added to the figure on page 14 [see Figure A1.3]. If it had arisen in a common ancestor of Bacteria or Archaea, then species in those groups would have had to lose the trait—an unlikely event. **3.** The data set was so large and diverse that it was no longer reasonable to argue that noncellular life-forms would be discovered. **4.** Yes. The heritable traits that confer resistance to HIV should increase over time.

CHAPTER 2

Check Your Understanding (CYU)

CYU p. 21 [See Figure A2.1] **CYU p. 33** (1) Gibbs equation: $\Delta G = \Delta H - T\Delta S$. ΔG symbolizes the change in the Gibbs free energy. ΔH represents the difference in potential energy between the products and the reactants. T represents the temperature (in degrees Kelvin) at which the reaction is taking place. ΔS symbolizes the change in entropy (disorder). (2) When ΔH is negative—meaning that the reactants have lower potential energy than the products—and when ΔS is positive, meaning that the products have higher entropy (are more disordered) than the reactants.

You Should Be Able To (YSBAT)

YSBAT p. 22 (1) $^{\delta+}H-O^{\delta-}-H^{\delta+}$. (2) If water were linear, the partial negative charge on oxygen would have partial positive charges on either side. Compared to the actual, bent molecule, the partial negative charge would be much less exposed and less able to participate in hydrogen bonding. **YSBAT p. 26** The pH 8 solution has 100 times fewer protons than the pH 6 solution. **YSBAT p. 30** (1) As T increases, ΔG is more likely to be negative, making the reaction spontaneous. (2) Exothermic reactions may be nonspontaneous if they result in a decrease in entropy—meaning that the products are more ordered than the reactants (ΔS is negative).

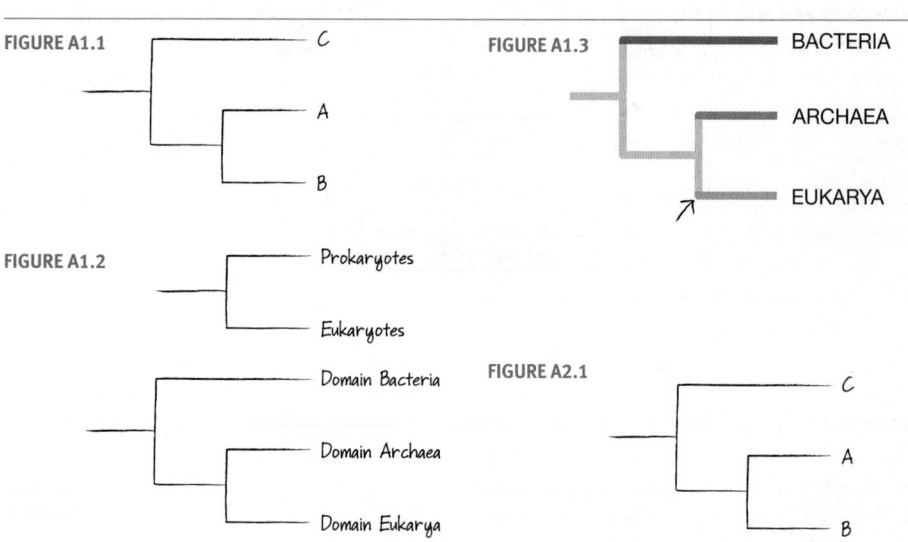

FIGURE A1.1

FIGURE A1.2

FIGURE A1.3

FIGURE A2.1

Caption Questions and Exercises

Figure 2.7 Oxygen and nitrogen have high electronegativity. They hold shared electrons more tightly than C, H, and many other atoms, resulting in polar bonds. **Figure 2.13** Oils are nonpolar. They have long chains of carbon atoms bonded to hydrogen atoms, which share electrons evenly because their electronegativities are similar. When an oil and water are mixed, the polar water molecules interact with each other via hydrogen bonding much more strongly than they interact with the nonpolar oil molecules, which interact with themselves instead. **Figure 2.16** The coffee becomes less acidic because milk is more alkaline (pH 6.5) than black coffee (pH 5). **Table 2.2** (Row 1) "Cause": Electrostatic attraction between partial charges on water molecules and opposite charges on ions; hydrogen bonds between water and other polar molecules. (Row 2) "Biological Consequences": Ice floats on denser liquid water. If it were denser and sank, oceans and lakes would fill with ice. (Row 4) "Cause": Liquid water must absorb lots of heat energy to break hydrogen bonds and change to a gas. One possible concept map relating the structure of water to its properties is shown below [see Figure A2.2]. **Figure 2.19** [See Figure A2.3] **Figure 2.22** The reaction rate at a specific set of reactant concentrations, averaged over many replicates. **Table 2.3** All the functional groups in Table 2.3, except the sulfhydryl group ($-SH$), are highly polar. The sulfhydryl group is only very slightly polar.

Summary of Key Concepts

KC 2.1 [See Figure A2.4] **KC 2.2** Because these functional groups are polar, water forms hydrogen bonds with them. **KC 2.3** Chemical energy—meaning, potential energy stored in chemical bonds—is transformed to heat. **KC 2.4** Spontaneous chemical reactions lead toward disorder and a release of energy. Cells are full of highly ordered molecules that contain high-energy bonds. Energy must be added to make these molecules and maintain life. **KC 2.5** Electrons in carbon-carbon bonds are held loosely and equally between the carbon atoms, whereas electrons in a carbon-oxygen bond are held tightly by the oxygen. Because of this difference, molecules with carbon-carbon bonds have more potential energy than carbon dioxide with its two carbon-oxygen bonds. Carbon-carbon bonds are often present in large, highly ordered molecules. The entropy of a group of such molecules, a measure of their disorder, is much less than the entropy of a group of simple, less-ordered molecules such as carbon dioxide.

Test Your Knowledge

1. b; **2.** a; **3.** a; **4.** d; **5.** d; **6.** d

Test Your Understanding

1. The reaction lowers the pH of the solution by releasing extra H^+ into the solution. If additional CO_2 is added, the sequence of reactions would be driven to the right, which would make the ocean more acidic. **2.** No. Shells that are farther from the protons (positive charges) in the nucleus house electrons that have greater potential energy than shells closer to the nucleus. **3.** [See Figure A2.5] The electron orbitals surrounding oxygen are oriented in the shape of a tetrahedron. Two of these orbitals are occupied by unshared electrons and the other two are shared with hydrogen atoms, resulting in the bent shape of the water molecule. Due to its greater electronegativity, oxygen attracts the shared electrons more strongly than does hydrogen, so it has a partial negative charge while the hydrogen atoms have a partial positive charge. **4.** In a covalent bond, the shared electrons are extremely close to two nuclei. In a hydrogen bond, the (+) and (−) charges are only partial, and the attraction is at a much greater distance. **5.** Water forms large numbers of hydrogen bonds, which must be broken before the molecules can start moving faster—meaning that their temperature has increased. **6.** The overall shape of an organic molecule is determined by its carbon framework. The functional groups attached to the carbons determine the molecule's chemical behavior, because these groups are likely to interact with other molecules.

Applying Concepts to New Situations

1. In CO_2, the double bonds between the carbon and each of the oxygen atoms lock the molecule into a linear shape that is nonpolar. In water, the partial charges on

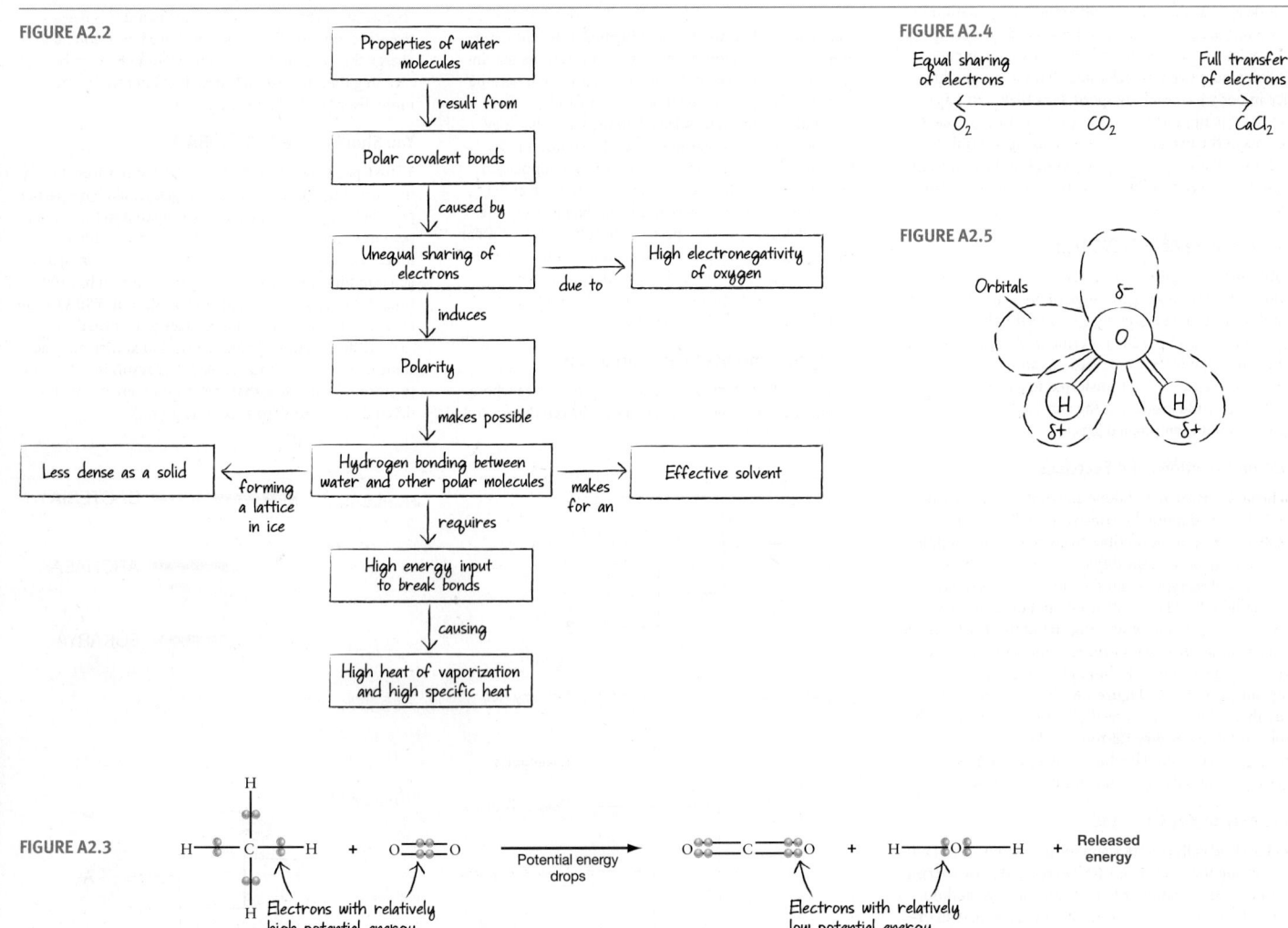

FIGURE A2.2

Properties of water molecules
— result from →
Polar covalent bonds
— caused by →
Unequal sharing of electrons — due to → High electronegativity of oxygen
— induces →
Polarity
— makes possible →
Hydrogen bonding between water and other polar molecules — makes for an → Effective solvent
Less dense as a solid ← forming a lattice in ice —
— requires →
High energy input to break bonds
— causing →
High heat of vaporization and high specific heat

FIGURE A2.4

Equal sharing of electrons ← → Full transfer of electrons

O_2 CO_2 $CaCl_2$

FIGURE A2.5

Orbitals

$\delta-$
O
H   H
$\delta+$ $\delta+$

FIGURE A2.3

$H-C-H + O=O$ → (Potential energy drops) → $O=C=O + H-O-H$ + Released energy

Electrons with relatively high potential energy

Electrons with relatively low potential energy

oxygen and hydrogen allow water to form hydrogen bonds with other polar or charged substances. In addition, the single covalent bonds in water are much easier to break than the double covalent bonds in carbon dioxide, making it more reactive. **2.** When oxygen is covalently bonded to carbon or hydrogen, the bond electrons spend more time near the oxygen atom. In contrast, carbon and hydrogen have roughly similar electronegativities, and they tend to share the electrons of a covalent bond more equally. **3.** The Sun will burn out when the mass of all of its component atoms is finally converted to energy. **4.** In hot weather, water absorbs large amounts of heat due to its high specific heat and high heat of vaporization. In cold weather, water releases the large amount of heat that it has absorbed.

CHAPTER 3

Check Your Understanding (CYU)

CYU p. 44 [See Figure A3.1] **CYU p. 51** Secondary, tertiary, and quaternary structure all depend on bonds and other interactions between amino acids that are linked in a chain (primary structure). **CYU p. 56** (1) Substrates can react only if they meet in a precise orientation. (2) The shape change alters the shape of the active site so that substrates no longer fit or can't be oriented correctly for a reaction to occur.

You Should Be Able To (YSBAT)

YSBAT p. 50 The sequence of amino acids in the two polypeptide chains constitutes the primary structure. Each chain contains α-helices and a β-pleated sheet (secondary structures), and is folded into a specific three-dimensional shape (tertiary structure). The overall three-dimensional shape of the dimeric Cro protein, formed by association of two identical polypeptides, constitutes its quaternary structure. **YSBAT p. 53** More—because catalysts lower the activation energy required for the reaction to proceed, more molecules at any given temperature have enough kinetic energy to supply the activation energy. **YSBAT p. 54** (1) orienting substrates, (2) transition state, (3) R-groups, (4) structure.

Caption Questions and Exercises

Figure 3.1 The water-filled flask is the ocean; the gas-filled flask is the atmosphere; the condensed water droplets are rain; the electrical sparks are lightning. **Figure 3.3** The green R-groups contain mostly C and H, which have roughly equal electronegativities. Electrons are evenly shared in C—H bonds and C—S bonds, so the groups are nonpolar. Most of the pink R-groups have a highly electronegative oxygen atom with a partial negative charge, making them polar. Cysteine has a sulfur that is slightly more electronegative than hydrogen, so it will be less polar than the other pink groups. **Figure 3.6** A polar covalent bond. The electrons are shared in the C—N bond, but because N is more electronegative than C, the bond is polar. **Figure 3.16** In order for a chemical reaction to proceed, the kinetic energy of the reactant molecules and atoms must be greater than the activation energy. Only the molecules to the right of the line you drew will be able to react. **Figure 3.18** No—a catalyst only changes the activation energy, not the overall free energy change. **Figure 3.22** [See Figure A3.2]

Summary of Key Concepts

KC 3.1 Many possible correct answers, including: the presence of an active site in an enzyme that is precisely shaped to fit a substrate or substrates in the correct orientation for a reaction to occur; the "donut" shape of porin allowing certain substances to pass through it; the butterfly shape of TATA-box binding protein being precisely the right size for a DNA molecule to fit. **KC 3.2** R-groups containing partial charges or full charges can form hydrogen bonds with water, make the amino acid soluble. Nonpolar R-groups do not interact with water and make the amino acid insoluble. **KC 3.3** [See Figure A3.3] **KC 3.4** When a catalyst is present, the reactants and products don't change—only the transition state changes. Thus, the overall free energy change in the reaction is the same—only the activation energy changes.

Test Your Knowledge

1. d; **2.** c; **3.** a; **4.** b; **5.** a; **6.** b

Test Your Understanding

1. The shape of reactant molecules (the key) fits into the active site of an enzyme (the lock). The model assumed that the enzyme is rigid; in fact it is flexible and dynamic. **2.** It will fold in on itself, away from water, when placed in an aqueous solution. **3.** Both are mechanisms that regulate enzymes; the difference is whether the regulatory molecule binds at the active site (competitive inhibition) or away from the active site (allosteric regulation). **4.** Energy is required. Polymerization is a nonspontaneous reaction because the product molecules have lower entropy and greater potential energy than the reactants. **5.** Proteins are highly variable in overall shape and chemical properties due to variation in the composition of R-groups and the array of secondary through quaternary structures that are possible. This variation allows them to fulfill many different roles in the cell. Diversity in the shape and reactivity of active sites makes them effective catalysts. **6.** The function of an enzyme depends on the shape and chemical reactivity of its active site. Temperature and pH affect enzyme function because they can change the shape and chemical reactivity of the active site—either by disrupting bonds or interfering with acid-base reactions.

Applying Concepts to New Situations

1. Yes—within limits. As Figure 3.23 shows, some bacteria thrive in very hot or highly acidic environments on Earth. But extremely low pHs or high temperatures are likely to denature proteins. **2.** The data suggest that the enzyme and substrate form a transition state that requires a change in the shape of the active site, with each movement corresponding to one reaction. **3.** Without the coenzyme, the free radical–containing transition state would not be stabilized and the reaction rate would drop dramatically. **4.** Residues in the active site changed in a way that made the drugs less likely to bind and prevent the normal substrate from binding.

FIGURE A3.1

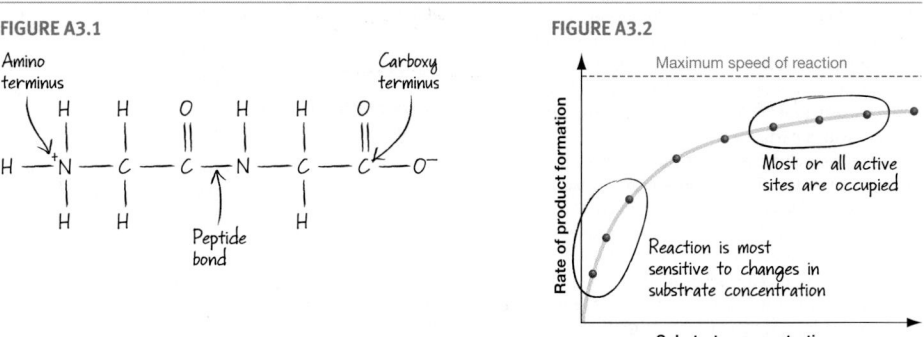

FIGURE A3.2

FIGURE A3.3

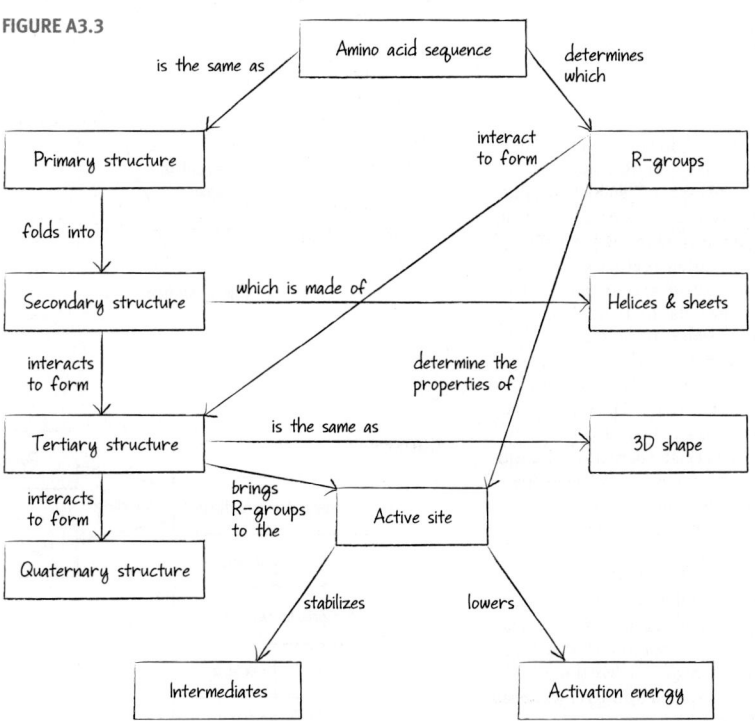

CHAPTER 4

Check Your Understanding (CYU)

CYU p. 62 [See Figure A4.1] CYU p. 66 [See Figure A4.2]

You Should Be Able To (YSBAT)

YSBAT p. 60 [See Figure A4.3] YSBAT p. 64 If the two strands are parallel, the nitrogenous bases are not aligned in a way that allows hydrogen bond formation. No helix would form.

Caption Questions and Exercises

Figure 4.3 5′ UAGC 3′. **Figure 4.9** Endergonic—energy must be added (as heat) for the reaction to occur.

Summary of Key Concepts

KC 4.1 (1) Both polymerize via energy-demanding condensation reactions that add a monomer to a growing chain. But proteins polymerize via formation of peptide bonds while nucleic acids polymerize via formation of phosphodiester linkages. (2) Proteins have an amino terminus at one end and a carboxyl terminus at the other end, while nucleic acids have a 5′ end (with an unlinked phosphate group) and a 3′ end (with an unlinked hydroxyl group). (3) Both have a "backbone"; in proteins it consists of amino and carbonyl groups, while in nucleic acids it consists of phosphate groups and sugars. **KC 4.2** C-G pairs involve three hydrogen bonds, so they are more stable than A-T pairs with just two bonds. **KC 4.3** A single-stranded RNA molecule has unpaired bases that can pair up with other bases on the same RNA strand, thereby folding the molecule into stem-and-loop configurations with a particular three-dimensional shape. The unpaired bases also can bind to another RNA molecule, linking the two together in a quaternary structure. Because DNA molecules are double stranded, with no unpaired bases, internal folding and interaction with other DNA molecules is not possible. **KC 4.4** If the copied ribozymes are not exactly like the template molecule, then some are likely to have changes that make them more efficient as catalysts. If there were no variation—meaning, if copying were exact—there could be no improvement in efficiency or other change over time.

Test Your Knowledge

1. c; **2.** d; **3.** a; **4.** b; **5.** a; **6.** c

Test Your Understanding

1. [See Figure A4.4] **2.** The addition of the phosphate groups raises the potential energy of the resulting monomers (nucleoside triphosphates) enough that the polymerization reaction is exergonic. **3.** Nucleic acids are directional because the two ends of the polymer are different. One end has a free phosphate group on a 5′ carbon; the other end has a free hydroxyl group bonded to a 3′ carbon. **4.** DNA is a more stable molecule than RNA because it lacks a hydroxyl group on the 2′ carbon and is therefore resistant to degradation, and because the two sugar-phosphate backbones are linked by many hydrogen bonds between nitrogenous bases. DNA is more stable than proteins because it is symmetrical and has few exposed chemical groups that could participate in chemical reactions. **5.** Hairpin structures form when folding of the sugar-phosphate backbone allows the bases in one part of an RNA strand to align with bases in another segment of the same RNA strand in an antiparallel fashion, so they can hydrogen-bond and form a double helix. **6.** DNA has limited catalytic ability because it (1) lacks functional groups that can stabilize transition states and (2) has a regular structure that is not conducive to forming active sites. RNA molecules can catalyze some reactions because they (1) have exposed hydroxyl functional groups that can stabilize transition states and (2) can fold into shapes that create active sites. Proteins are the most effective catalysts because (1) amino acids have a wide range of chemical reactivity, and (2) they fold into an enormous diversity of shapes that can create active sites.

Applying Concepts to New Situations

1. Yes—if the complementary bases lined up over the entire length of the two strands, they would twist into a double helix analogous to a DNA molecule. The same types of hydrogen bonds and hydrophobic interactions would occur as observed in the "stem" portions of hairpins in single-stranded RNA. **2.** An RNA replicase would undergo replication and be able to evolve. It would process information in the sense of copying itself, and would use energy to drive endergonic polymerization reactions. It would not be bound by a membrane, however, and would not be able to acquire energy. It would best be considered as an intermediate step between nonlife and "true life," as outlined in Chapter 1. **3.** It is unlikely that bases would align properly for hydrogen bonding to occur, so hydrophobic interactions would probably be more important. **4.** No single right answer (these are opinion questions). Note that when biologists design studies about the origin of life, they base their experimental conditions on the best available current knowledge about early Earth conditions. New information about these conditions might require a revision of the experiment that leads to different results and conclusions.

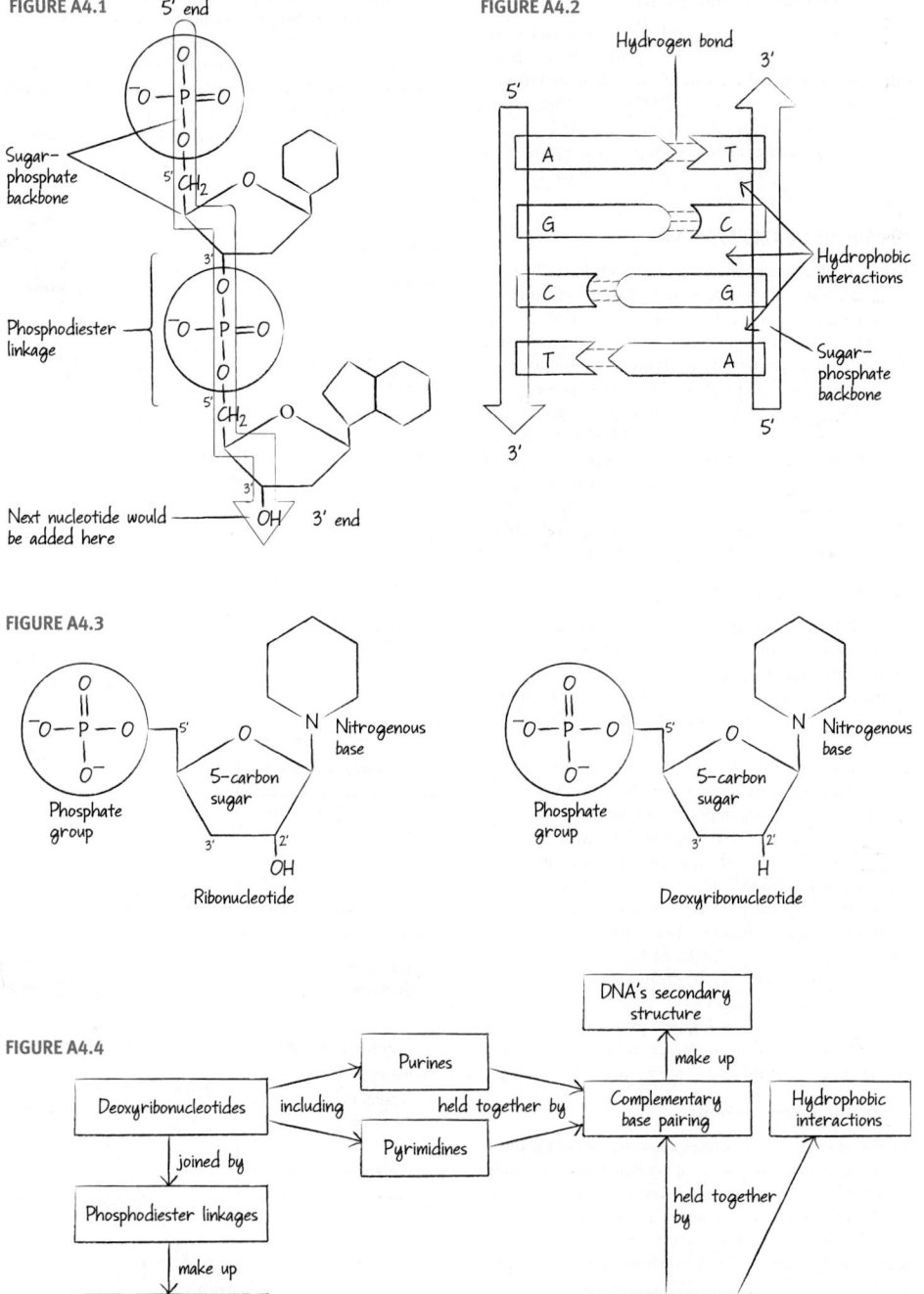

FIGURE A4.1

FIGURE A4.2

FIGURE A4.3

Ribonucleotide

Deoxyribonucleotide

FIGURE A4.4

CHAPTER 5

Check Your Understanding (CYU)

CYU p. 73 [See Figure A5.1] CYU p. 76 They could differ in (1) location of linkages (e.g., 1,4 or 1,6); (2) types of linkages (e.g., α or β); (3) the sequence of the monomers (e.g., two galactose then two glucose, versus alternating galactose and glucose); and/or (4) whether the four monomers are linked in a line or whether they branch. CYU p. 79 (1) Aspect 1: The β-1,4-glycosidic linkages in these molecules are difficult to degrade. Aspect 2: When individual molecules of these carbohydrates align, bonds form between them and produce tough fibers or sheets. (2) Most are probably being broken down into glucose, which in turn is being broken down in reactions that lead to the synthesis of ATP. In short, the carbohydrates are providing you with chemical energy.

Caption Questions and Exercises

Figure 5.2 [See Figure A5.2] Figure 5.7 All of the C—C and C—H bonds should be circled.

Summary of Key Concepts

KC 5.1 Molecules have to interact in an extremely specific orientation in order for a reaction to occur. Changing the location of a functional group by even one carbon can mean that the molecule will undergo completely different types of reactions. KC 5.2 The orientations of the α- versus β-glycosidic linkages are different, so the molecules in a chain end up in a spiral versus linear arrangement. KC 5.3 (1) Polysaccharides used for energy storage are formed entirely from glucose monomers joined by α-glycosidic linkages; structural polysaccharides are comprised of glucose or other sugars joined by β-glycosidic linkages. (2) The monomers in energy-storage polysaccharides are linked in a spiral arrangement; the monomers in structural polysaccharides are linked in a linear arrangement. (3) Energy-storage polysaccharides may branch; structural polysaccharides do not. (4) Individual chains of energy-storage polysaccharides do not associate with each other; adjacent chains of structural polysaccharides are linked by hydrogen bonds or covalent bonds.

Test Your Knowledge

1. d; 2. a; 3. c; 4. a; 5. d; 6. b

Test Your Understanding

1. Carbohydrates are ideal for signaling the identity of the cell because they are so diverse structurally. This diversity enables them to serve as very specific identity tags for cells. 2. When you compare the glucose monomers in an α-1,4-glycosidic linkage versus a β-1,4-glycosidic linkage, the linkages are located on opposite sides of the plane of the glucose rings (e.g., "above" versus "below" the plane), and the glucose monomers are linked in the same orientation versus having every other glucose flipped in orientation. β-1,4-glycosidic linkages are much more difficult for enzymes to break, so they resist degradation. 3. Starch and glucose both consist of glucose monomers joined by α-1,4-glycosidic linkages, and both function as storage carbohydrates. Starch is composed of an unbranched amylose and branched amylopectin, while glycogen is even more highly branched. 4. The electrons in the C=O bonds of carbon dioxide molecules are held tightly by the highly electronegative oxygen atoms, so have low potential energy. The electrons in the C—C and C—H bonds of carbohydrates are shared equally and have much higher potential energy. 5. They have β-1,4-glycosidic linkages, which are resistant to degradation and put glucose monomers in positions where hydrogen bonds can form between adjacent strands, forming strong fibers. 6. Glycogen has α-1,4-glycosidic linkages with α-1,6-glycosidic linkages at branch points; cellulose has β-1,4-glycosidic linkages and is not branched—instead, it forms hydrogen bonds

with adjacent cellulose molecules. Glycogen is an energy-storage molecule; cellulose is a structural polysaccharide.

Applying Concepts to New Situations

1. Carbohydrates are energy-storage molecules, so minimizing their consumption may reduce total energy intake. Lack of available carbohydrate also forces the body to use fats for energy, reducing the amount of fat that is stored. 2. Lactose is made up of glucose and galactose. 3. Amylase breaks down the starch in the cracker into glucose monomers, which stimulate the sweet receptors in your tongue. 4. When bacteria contact lysozyme, their cell walls, which contain peptidoglycan, begin to degrade, leading to the death of the bacteria.

CHAPTER 6

Check Your Understanding (CYU)

CYU p. 85 (1) Fats consist of three fatty acids linked to glycerol; steroids have a distinctive four-ring structure with a side group attached; phospholipids have a hydrophilic, phosphate-containing "head" region and a hydrocarbon tail. (2) In cholesterol, the hydrocarbon steroid rings are hydrophobic; the hydroxyl group is hydrophilic. In phospholipids, the phosphate-containing head group is hydrophilic; the fatty acid chains are hydrophobic. CYU p. 89 [See Table A6.1] CYU p. 92 [See Figure A6.1] CYU p. 99 Passive transport does not require an

---

FIGURE A5.1

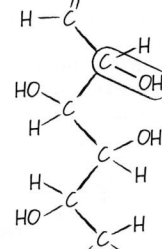

Start with a monosacchride. This one is a 3-carbon aldose (carbonyl group at end)

Variation 1: 3-carbon ketose (carbonyl group in middle)

Variation 2: 4-carbon aldose

Variation 3: 3-carbon aldose with different arrangement of hydroxyl group

FIGURE A5.2

FIGURE A6.1

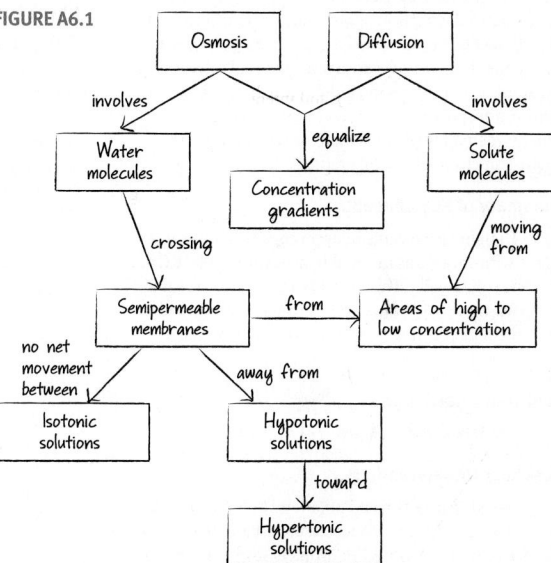

TABLE A6.1

Factor	Effect on permeability	Reason
Temperature	Decreases as temperature decreases.	Hydrophobic tails move more slowly, increasing the strength of hydrophobic interactions (membrane is more dense).
Cholesterol	Decreases as cholesterol content increases.	Cholesterol molecules fill in the spaces between the fatty-acid tails, making the membrane more tightly packed.
Length of hydrocarbon tails	Decreases as length of hydrocarbon tails increases.	Longer hydrocarbon tails have stronger hydrophobic interactions, making the tails pack together more tightly.
Saturation of hydrocarbon tails	Decreases as degree of saturation increases.	Saturated fatty acids have straight hydrocarbon tails that pack together and form many hydrophobic interactions, leaving few gaps.

expenditure of energy—it happens as a result of existing concentration or electrical gradients. Active transport is active in the sense of requiring energy. In cotransport, a second ion or molecule is transported against its electrochemical gradient with ("co") an ion that is transported along its electrochemical gradient.

You Should Be Able To (YSBAT)

YSBAT p. 84 Fats have many C—C and C—H bonds, which have high free energy. They are hydrophobic because they contain long hydrocarbon chains and lack any highly polar groups. **YSBAT p. 86** No, because both are relatively large molecules that are polar or carry a charge. **YSBAT p. 91** (1) Add a large number of green dots to the inside of the liposome. (2) Add a large number of green dots to the solution outside the liposome. **YSBAT p. 94** Your arrow should point out of the cell. There is no concentration gradient, but the outside has a net positive charge, which favors outward movement of negative ions. **YSBAT p. 96** [See Figure A6.2]

Caption Questions and Exercises

Figure 6.3 At the polar hydroxyl group in cholesterol and the polar head group in phospholipids. **Figure 6.14** Higher, because less water would have to move to the right side to achieve equilibrium. **Figure 6.21** The electrical gradient would reverse, because the inside of the membrane would be more positive than the outside. The concentration gradient for sodium would also reverse; sodium would diffuse to the exterior. **Figure 6.22** No—the 10 replicates where no current was recorded probably represent instances where the CFTR protein was damaged and not functioning properly. (In general, no experimental method works "perfectly.") **Figure 6.27** "Diffusion": description as given; no proteins involved. "Facilitated diffusion": Passive movement of ions or molecules that cannot cross a phospholipid bilayer readily, along a concentration gradient. Facilitated by channels or transporters. "Active transport": Active (energy-demanding) movement of ions or molecules against an electrochemical gradient.

Summary of Key Concepts

KC 6.1 Highly permeable bilayers consist of phospholipids with short, unsaturated hydrocarbon tails. **KC 6.2** (1) The solute will diffuse until both sides are at equal concentrations. (2) Water will diffuse toward the side with the higher solute concentration. **KC 6.3** [See Figure A6.3]

Test Your Knowledge

1. b; **2.** a; **3.** b; **4.** d; **5.** c; **6.** d

Test Your Understanding

1. No, because they have no polar end to interact with water. Instead, these lipids would collect and float on the surface of water, or collect in droplets suspended in water, reducing their interaction with water to a minimum. **2.** Hydrophilic, phosphate-containing head groups interact with water; hydrophobic, fatty-acid tails associate with each other. A bilayer has a lower potential energy and thus is more stable than are independent phospholipids in solution. **3.** Ethanol's polarity reduces the speed at which it can cross a membrane, but its small size and lack of charge increase the speed at which it crosses membranes. Ethanol probably crosses a membrane more slowly than water (which is also polar, but smaller) but faster than larger polar molecules, and much faster than charged molecules. **4.** If no membrane separates the two solutions, then the constant, random motion of solute and water molecules causes even mixing of the two solutions. No net diffusion of water molecules will occur. **5.** Only hydrophobic amino acids interact with the nonpolar lipid tails in the interior of a phospholipid bilayer.

Figure 3.3 lists the following nonpolar, hydrophobic amino acids: glycine, alanine, valine, leucine, isoleucine, methionine, phenylalanine, tryptophan, and proline. According to Table 3.1, tyrosine and cysteine are also relatively hydrophobic, suggesting that they might also be found in the interior of transmembrane proteins. **6.** CO_2 will diffuse into the cell; water will move out via osmosis. Na^+ and K^+ will enter the cell through gramicidin by facilitated diffusion. Cl^- will not move.

Applying Concepts to New Situations

1. Flip-flops should be rare, because they require polar head group to pass through the hydrophobic portion of the lipid bilayer. To test this prediction, you could monitor the number of dyed phospholipids that transfer from one side of the membrane to the other in a given period of time. **2.** The kinks in unsaturated hydrocarbon tails keep membranes fluid, even at low temperature, because they prevent extensive hydrophobic interactions. In hot environments, it would be advantageous for phospholipids to have saturated tails, to prevent membranes from being too fluid. **3.** Adding a methyl group makes a drug more hydrophobic and thus more likely to pass through a lipid bilayer. Adding a charged group make it hydrophilic and reduces its ability to pass through the lipid bilayer. **4.** Detergents are amphipathic. Their hydrophobic ends interact with grease while their hydrophilic ends interact with water. This forms a bridge between the grease and the water, effectively making the grease dissolve in water and allowing it to be washed away.

CHAPTER 7

Check Your Understanding (CYU)

CYU p. 105 (1) Photosynthetic membranes increase food production by providing a large surface area to hold the pigments and enzymes required for photosynthesis. (2) The presence of a magnetic mineral, which changes position as the cell moves through a magnetic field, allows the cell to move in a directed way. (3) The layer of thick, strong material stiffens the cell and pro-

vides protection from mechanical damage. **CYU p. 115** (1) Both organelles contain specific sets of enzymes, and the interior of lysosomes is acidic. Peroxisomes contain catalase and other enzymes that process fatty acids and toxins via oxidation reactions. Lysosomal enzymes digest macromolecules, releasing monomers that can be recycled into new macromolecules. (2) From top to bottom, the cells in the last column should read as follows: administrative/information hub, protein factory, large molecule manufacturing and shipping (protein synthesis and folding center, protein finishing and shipping line, fat factory, waste processing and recycling center), fatty-acid processing and detox center, warehouse, power station, food-manufacturing facility, support beams, perimeter fencing with secured gates, and leave blank. **CYU p. 122** (1) Proteins that lack a signal sequence will not be delivered to the ER. They are released into the cytosol. (2) The proteins will not be secreted—they will be found in lysosomes. **CYU p. 128** The products will not be able to move through the system normally, because their microtubule "roads" are absent.

You Should Be Able To (YSBAT)

YSBAT p. 128 (1) The microtubules at the bottom of the axoneme would slide to the right, but the axoneme would not bend. (2) Nothing would happen (dynein wouldn't move).

Caption Questions and Exercises

Figure 7.1 [See Figure A7.1] **Figure 7.15** The vesicles deliver digestive enzymes that are required for the mature lysosome to function. **Figure 7.16** Storing the toxins in vacuoles prevents the toxins from damaging the plant's own organelles and cells. **Figure 7.19** The cell wall is outside the plasma membrane. **Figure 7.21** [See Figure A7.2] **Figure 7.23** "Prediction": The labeled tail region fragments or the labeled core region fragments of the nucleoplasmin protein will be found in the cell nucleus. "Prediction of null hypothesis": No labeled fragments of the nucleoplasmin protein will be found in the nucleus of the cell. "Conclusion": The "Send to nucleus" zip code is in the tail region of the nucleoplasmin protein.

FIGURE A6.2

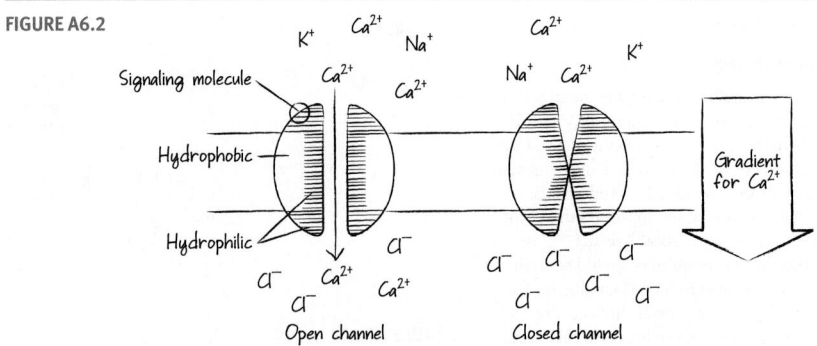

FIGURE A6.3

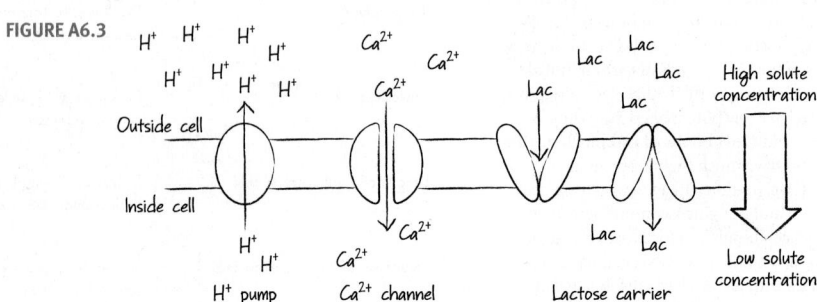

Summary of Key Concepts

KC 7.1 (1) They will be unable to make ATP and will die. (2) They will be limp and less able to resist attack by predators. In many environments they will be unable to resist the osmotic pressure of water entering the cytoplasm and will burst. (3) They will be unable to synthesize new proteins and will die. **KC 7.2** As a group, proteins have complex and highly diverse shapes and chemical properties. Individual proteins, then, can recognize and bind to molecular zip codes in a very specific way (like a key fitting into a lock). **KC 7.3** Actin filaments are made up of two strands of actin monomers, microtubules are made up of tubulin protein dimers that form a tube, intermediate filaments are made up of a variety of different protein monomers. Actin filaments and microtubules exhibit polarity (or directionality), with new subunits constantly being added or subtracted at either end (but added faster to the plus end). All three elements provide structural support, but only actin filaments and microtubules are involved in movement and cell division. **KC 7.4** Micrographs of cells at increasing time intervals after the pulse treatment represent time-lapse photography—a set of still images that show change through time.

Test Your Knowledge

1. a; **2.** c; **3.** c; **4.** b; **5.** a; **6.** d

Test Your Understanding

1. All cells are bound by a plasma membrane, are filled with cytoplasm, carry their genetic information (DNA) in chromosomes, and contain ribosomes (the sites of protein synthesis). Some prokaryotes have organelles not found in plants or animals, such as a magnetite-containing structure. Plant cells have chloroplasts, vacuoles, and a cell wall. Animal cells contain lysosomes and lack a cell wall. **2.** Ribosome in cytoplasm (protein is synthesized) → Rough ER (protein is folded and initially modified) → Transport vesicle → Golgi apparatus (protein is glycosylated; has molecular tag indicating destination) → Transport vesicle → Plasma membrane → Extracellular space. **3.** When a globular "head" section of kinesin binds and releases ATP, it undergoes a conformational change that swings it forward and binds it to the microtubule. The two globular head regions alternate between swinging and binding movements, causing the protein to "walk" down the microtubule. **4.** Changes in the

length of microfilaments and microtubules contribute to changes in the size and shape of the cytoskeleton. **5.** Plasma membrane proteins are produced by the endomembrane system. Cells that function in membrane transport processes should have an extensive endomembrane system, including RER. **6.** All microtubules have the same basic structure, consisting of tubulin dimers that form a tube with plus and minus ends. The microtubules involved in structural support have no associated motor proteins; those involved in vesicle movement and flagella do. The microtubules that make up flagella are grouped in a 9 doublets + 2 single-microtubules arrangement. Structural and vesicle transport microtubules are single structures.

Applying Concepts to New Situations

1. Since mannose-6-phosphate serves as a lysosome tag in the endomembrane system, material that is ingested by phagocytic cells of the immune system might be targeted to lysosomes in the same way. To test this hypothesis, you could expose immune system cells to a population of bacteria that have labeled proteins, isolate the proteins after ingestion, and test to see whether they now have a mannose-6-phosphate tag attached. **2.** The proteins must receive a molecular zip code that binds to a receptor on the surface of peroxisomes. They could diffuse randomly to peroxisomes, or be transported in a directed way by vesicles or specialized motor proteins. **3.** (a) "Housekeeping"—destruction of macromolecules

---

FIGURE A7.2

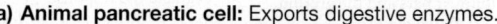

(a) Animal pancreatic cell: Exports digestive enzymes.

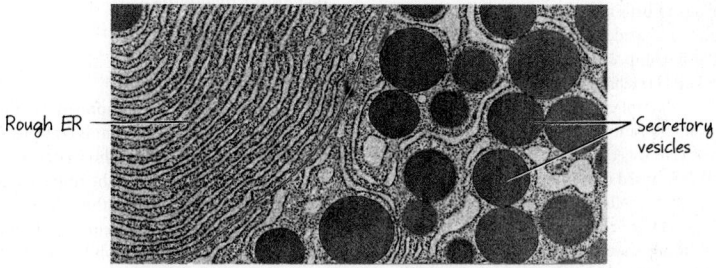

Rough ER
Secretory vesicles

(b) Animal testis cell: Exports lipid-soluble signals.

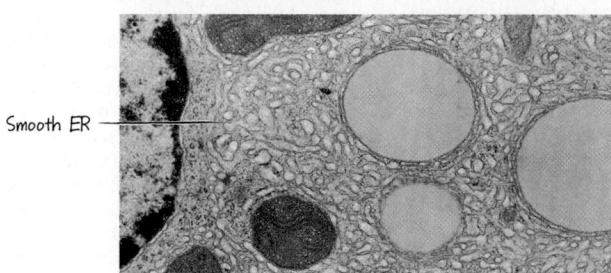

Smooth ER

(c) Plant leaf cell: Manufactures ATP and sugar.

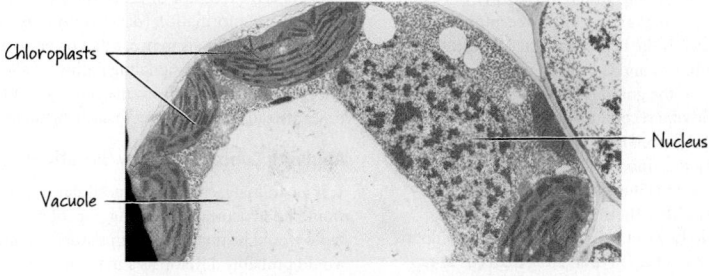

Chloroplasts
Nucleus
Vacuole

(d) Plant root cell: Stores starch.

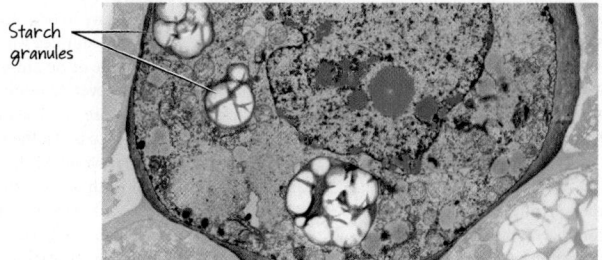

Starch granules

FIGURE A7.1

Nucleoid

Nucleoid

that are dangerous or no longer needed. (b) Structural support (e.g., wood is made up of cells with extensive cell walls). (c) Detoxification of dangerous compounds via oxidation reactions, or processing of stored fatty acids. **4.** The cell wall prevents the cells from bursting in hypotonic (aqueous) environments, or protects the cell from damage. Expose mutant and normal individuals to (1) a hypotonic environment, and (2) a eukaryote or virus that routinely attacks and kills this archaeal species. If the first hypothesis is correct, the mutant cells will burst but the normal cells will live. If the second hypothesis is correct, more mutant cells will die than normal cells.

CHAPTER 8

Check Your Understanding (CYU)

CYU p. 134 Plant cell walls and animal ECMs are both fiber composites. In plant cell walls the fiber component consists of cross-linked cellulose fibers and the ground substance is pectin. In animal ECMs the fiber component consists of collagen fibrils and the ground substance is gel-forming polysaccharides. **CYU p. 139** (1) The three structures differ in composition, but their function is similar. The middle lamella in plants is composed of pectins that glue adjacent cells together. Tight junctions are made up of membrane proteins that line up and "stitch" adjacent cells together. Desmosomes are "rivet-like" structures composed of proteins that link the cytoskeletons of adjacent cells. (2) The plasma membranes of adjacent plant cells are continuous at plasmodesmata, so cytoplasm, smooth endoplasmic reticulum, and other cell components are shared. Gap junctions connect adjacent animal cells by forming protein-lined pores. The openings allow ions and small molecules to be shared between cells. **CYU p. 146** (1) Each cell-cell signal binds to a specific receptor protein. A cell can respond to a signal only if it has the appropriate receptor. Different types of cells have different types of receptors. (2) Signals are amplified if one or more steps in a signal transduction pathway, involving either second messengers or a phosphorylation cascade, results in the activation of multiple downstream molecules.

You Should Be Able To (YSBAT)

YSBAT p. 138 (1) Yes—molecules can pass through a middle lamella because it is made up of gelatinous material that is not watertight. (2) Developing muscle cells could not adhere normally and muscle tissue would not form properly. The embryo would die. **YSBAT p. 143** The spy is the signal that arrives at the receptor (the castle gate). The guard is the G protein–linked receptor in the plasma membrane, and the queen is the G protein. The commander of the guard is the enzyme that catalyzes production of a second messenger (the soldiers). **YSBAT p. 144** (1) The red dominos (RTK components) would be the first dominos in the chain and the first to be tipped to initiate the domino cascade. The black domino (Ras) would be the second domino in line. Then there would be green, blue, pink, and yellow dominos (other kinases) set up in branching patterns to represent how Ras phosphorylates and activates a number of different kinases. Each of those colored dominos would be followed by a line of dominos of the same color representing the target proteins of those kinases. (2) Tipping the red domino would knock down the black domino (Ras) and would initiate a branching domino cascade that is dependent on the black domino's falling. The red domino represents how RTK activates Ras and how Ras initiates a cascade of phosphorylation.

Caption Questions and Exercises

Figure 8.9 "Prediction of null hypothesis": Cells will not adhere, or will adhere randomly with respect to cell type and species. **Figure 8.10** "Prediction": Cells treated with an antibody that blocks membrane proteins involved in

adhesion will not adhere. "Prediction of null hypothesis": All cells will adhere normally. **Figure 8.12** Top: Tight junctions; Middle: Desmosome; Bottom: Gap junctions.

Summary of Key Concepts

KC 8.1 (1) Adjacent cells fall apart from each other. (2) Nearby cells fall apart from each other; both cells and tissues are weaker and more susceptible to damage. **KC 8.2** Epithelium separates compartments, creating an inside and outside. If tight junctions degrade, materials from the outside will diffuse in and materials from the inside will diffuse out. **KC 8.3** Adrenalin binds to both heart and liver cells, but the activated receptors trigger different signal transduction pathways and lead to different cell responses. **KC 8.4** If only one step existed in a signal transduction pathway, it would be difficult for the process to be regulated by many different pathways. Multiple steps provide points for different pathways to regulate the response.

Test Your Knowledge

1. b; **2.** b; **3.** a; **4.** b; **5.** d; **6.** c

Test Your Understanding

1. The cross-linked fiber components withstand tension; the ground substance withstands compression. **2.** If each enzyme in the cascade phosphorylates many enzymes in the next step of the cascade, the initial signal will be amplified many times over. **3.** Although both are made up of membrane-spanning proteins, tight junctions seal adjacent animal cells together while gap junctions allow a flow of material between them. Middle lamellae are pectin-rich layers that glue adjacent plant cells together; plasmodesmata are gaps in the cell walls of adjacent cells where the plasma membrane is continuous, allowing exchange of materials between them. **4.** When dissociated cells from two sponge species were mixed, the cells sorted themselves into distinct aggregates that contained only cells of the same species. By blocking membrane proteins with antibodies and isolating cells that would *not* adhere, researchers found that specialized groups of proteins, including cadherins, are responsible for selective adhesion. **5.** Signal crosses plasma membrane and binds to intracellular receptor (reception) → receptor changes conformation, and signal-receptor complex moves to target site (processing) → signal-receptor complex binds to a target molecule (e.g., a gene or membrane pump), which changes its activity (response) → signal falls off receptor or is destroyed; receptor changes to inactive conformation (deactivation). **6.** Information from different signals may conflict or be reinforcing. "Cross-talk" between signaling pathways allows cells to integrate information from many signals at the same time, instead of responding to each signal in isolation.

Applying Concepts to New Situations

1. If an animal lacked an extracellular matrix, its cells would be structurally weak. Aggregations of similar cell types would be unable to form tissues, so individuals would probably develop as a mass of poorly connected or unconnected cells. A plant species with a similar disability would experience similar problems, but would also burst if placed in a hypotonic environment. **2.** (a) The response would have to be extremely local—the activated signal-receptor complex would have to affect nearby proteins. (b) Little or no amplification could occur, because there could never be more than one molecule that triggers the response. (c) The only way to regulate the response would be to block the receptor or make it more responsive to the signal. **3.** Chitin forms chains that can cross-link with one another. Therefore, the fungal cell wall likely has a structure similar to that of the plant cell wall (see Figure 8.2, left) except that chitin, rather than cellulose, is the major fiber component. **4.** Antibody binding would prevent activation of

the associated G protein. Even if the appropriate signal were present, no second messenger could be produced and no cell response could occur. Permanent binding by a drug would either activate the G protein constantly or block activation completely.

CHAPTER 9

Check Your Understanding (CYU)

CYU p. 153 (1) In part, because its three phosphate groups have four negative charges in close proximity. The electrons repulse each other, raising their potential energy. (2) Electrons in C—O bonds are held more tightly than electrons in C—H bonds, so they have lower potential energy. **CYU p. 161** (1) and (2) are combined with the answer to CYU p. 166 below. (3) Start with 12 dimes on glucose. (These dimes represent the 12 electrons that will be moved to electron carriers during redox reactions throughout glycolysis and the citric acid cycle.) Move two dimes to the NADH box generated by glycolysis and the other 10 dimes to the pyruvate box. Then move these 10 dimes through the pyruvate dehydrogenase box, placing two of them in the NADH box next to pyruvate dehydrogenase and the remaining eight dimes in the acetyl CoA box. Next move the eight dimes in the acetyl CoA box through the citric acid cycle, placing six of them in the NADH box and two in the $FADH_2$ box generated during the citric acid cycle. (4) These boxes are marked with stars in the diagram. **CYU p. 166** [See Figure A9.1] To illustrate the chemiosmotic mechanism, take the dimes (electrons) piled on the NADH and $FADH_2$ boxes and move them through the ETC. While moving these dimes, also move pennies from the mitochondrial matrix to the intermembrane space. As the dimes exit the ETC, add them to oxygen to generate water. Once all of the pennies have been pumped by the ETC into the intermembrane space, move them through ATP synthase back into the mitochondrial matrix to fuel the formation of ATP. **CYU p. 168** Electron acceptors such as oxygen have a much lower electronegativity than pyruvate. Donating an electron to O_2 causes a greater drop in potential energy, making it possible to generate much more ATP per molecule of glucose.

You Should Be Able To (YSBAT)

YSBAT p. 150 (1) Ribulose and carbon dioxide will not react because the reaction is endergonic and cannot proceed without the input of energy. (2) It is "activated" because it now has high potential energy. (3) The endergonic reaction (ribulose + CO_2) is coupled to the exergonic reaction (ATP hydrolysis) via the phosphorylation of ribulose. **YSBAT p. 152** (1) They are oxidized. (2) They are reduced. (3) Oxygen. (4) The reactants have higher energy. "Energy" should be added to the product side of the equation with "Release of energy" written below it. **YSBAT p. 163** "Indirect" is accurate because the energy released during glucose oxidation is stored in reduced electron carriers (instead of being used to produce ATP directly). The energy released as electrons from these carriers flow through the ETC is used to generate a proton gradient across a membrane. These protons then diffuse down their concentration gradient through ATP synthase located in the membrane, which drives ATP synthesis.

Caption Questions and Exercises

Figure 9.6 In glucose, put two dots in the middle of each C—C or C—H bond, and two dots next to the O in the C—O bond. In O_2, put four electrons in the middle of the double bond. In CO_2, put four electrons next to each O. In H_2O, put two electrons near O in each O—H bond. The electrons are held more tightly in the products than in the reactants, so potential energy has decreased. The difference in potential energy is released, so the reaction is exergonic. **Figure 9.8** *Glycolysis:* "What goes in" = glu-

FIGURE A9.1

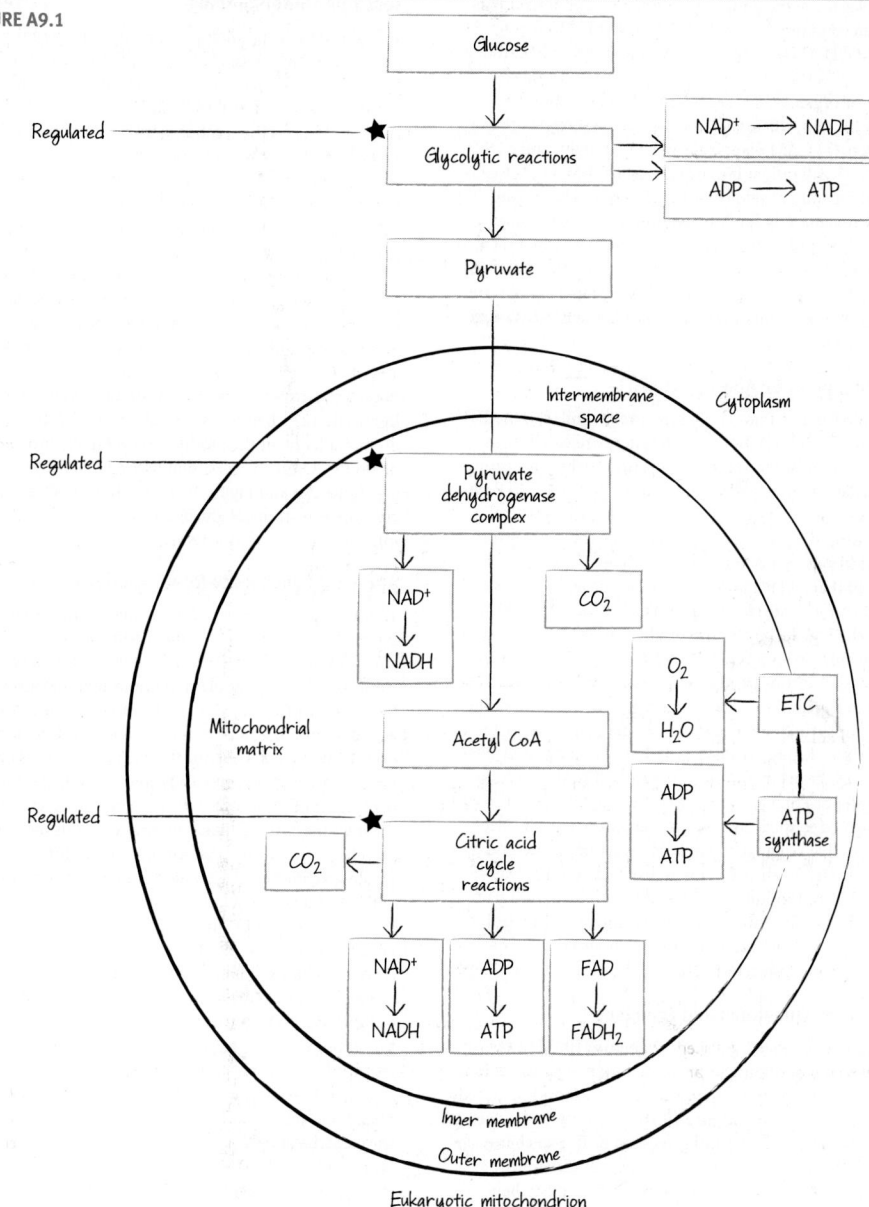

Eukaryotic mitochondrion

cose, NAD+, ADP, inorganic phosphate; "What comes out" = pyruvate, NADH, ATP. *Pyruvate processing:* "What goes in" = pyruvate, NAD+; "What comes out" = NADH, CO_2, acetyl CoA. *Citric acid cycle:* "What goes in" = acetyl CoA, NAD+, FADH, GDP or ADP, inorganic phosphate; "What comes out" = NADH, $FADH_2$, ATP or GTP, CO_2. *Electron transport and chemiosmosis:* "What goes in" = NADH, $FADH_2$, O_2; "What comes out" = ATP, H_2O, NAD+, FAD. **Figure 9.12** If the regulatory site had a higher affinity for ATP than the active site, then ATP would always be bound at the regulatory site, and glycolysis would always proceed at a very slow rate. **Figure 9.14** "Positive control": AMP, NAD+, CoA (reaction substrates). "Negative control by feedback inhibition": acetyl CoA, NADH, ATP (reaction products). **Figure 9.16** An allosteric regulator binds somewhere other than the active site and causes shape changes that slow the reaction or speed it up. A competitive inhibitor binds at the active site and keeps the substrate from binding. **Figure 9.18** Production of NADH and $FADH_2$. (In the line representing free energy, there are greater drops at each occurrence of NADH and $FADH_2$ versus ATP formation.) **Figure 9.20** Yes, because the artificial vesicles include only an ATP-synthesizing enzyme and a proton pump that creates a proton-motive force. The only energy source in this experiment is the proton gradient, so the result strongly supports the chemiosmotic hypothesis. (Note: This experiment does not test whether all ATP production in the ETC is through chemiosmosis. It does not preclude the possibility that the ETC also performs some substrate-level phosphorylation.) **Figure 9.21** The proton gradient arrow should start above in the inner membrane space and point down across the membrane into the mitochondrial matrix. *Complex I:* "What goes in" = NADH and H+; "What comes out" = NAD+, e−, H+. *Complex II:* "What goes in" = $FADH_2$; "What comes out" = FAD, e−. *Complex III:* "What goes in" = e−; "What comes out" = e−. *Complex IV:* "What goes in" = e−, H+, O_2; "What comes out" = H_2O, H+. **Figure 9.26** [See Figure A9.2]

Summary of Key Concepts

KC 9.1 Phosphorylation adds two negative charges in a small area. The electrical repulsion that results raises the protein's potential energy and its tertiary structure. **KC 9.2** The free-energy drop from glucose to oxygen (or another electron acceptor, during cellular respiration) is much greater than the free energy drop from glucose to pyruvate (during fermentation). Thus, there is more free energy available to use in synthesizing ATP. **KC 9.3** NADH would decrease if a drug poisoned the acetyl CoA-to-citrate enzyme, but increase if a drug poisoned ATP synthase. **KC 9.4** [See Figure A9.3] **KC 9.5** Organisms that produce ATP by fermentation grow more slowly than those that produce ATP via cellular respiration

FIGURE A9.2

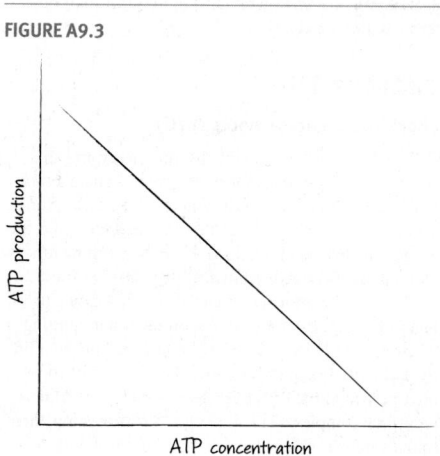

FIGURE A9.3

simply because fermentation produces fewer ATP molecules per glucose molecule than cellular respiration does.

Test Your Knowledge

1. b; **2.** b; **3.** d; **4.** c; **5.** a; **6.** b

Test Your Understanding

1. NADH and $FADH_2$ get their electrons from the intermediates of glycolysis, pyruvate processing, and the TCA cycle and deliver them to the ETC, where they reduce O_2. **2.** Both processes produce ATP from ADP and P_i, but substrate-level phosphorylation occurs when enzymes remove a "high-energy" phosphate from a substrate and directly transfer it to ADP, while oxidative phosphorylation is based on electrons moving through an ETC and production of a proton-motive force that drives ATP synthase. **3.** Aerobic respiration is much more productive because oxygen has extremely high electronegativity compared to other electron acceptors, resulting in a greater release of energy during electron transport and more proton pumping. **4.** Glycolysis → Pyruvate processing → TCA cycle → ETC and chemiosmosis. The first three steps are responsible for glucose oxidation; the final step produces the most ATP. **5.** Electron transport makes oxidative phosphorylation possible (oxidative phosphorylation occurs when electrons have been transported through the ETC and a proton gradient has been established). ATP synthase consists of an F_o unit and F_1 unit joined by a stalk. When protons flow through the F_o unit, the stalk and F_1 unit spin. The motion drives the synthesis of ATP from ADP and P_i. **6.** Stored carbohydrates can be broken down into glucose that enters the glycolytic pathway. If carbohydrates are absent, products from fat and protein catabolism can be used to fuel cellular respiration or fermentation. If ATP is plentiful, anabolic reactions use intermediates of the glycolytic pathway and the TCA cycle to synthesize carbohydrates, fats, and proteins.

Applying Concepts to New Situations

1. When complex IV is blocked, electrons can no longer be transferred to oxygen, the final acceptor, and cellular respiration stops. Fermentation could keep glycolysis going, but it is inefficient and unlikely to fuel a cell's energy needs over the long term. Cells that lack the enzymes required for fermentation would die first. **2.** Because mitochondria with few cristae would have fewer electron transport chains and ATP synthase molecules, they would produce much less ATP than mitochondria with numerous cristae. **3.** When oxygen is unavailable for cellular respiration, yeast cells switch to fermentation, which occurs in the cytosol. They are unlikely to expend large amounts of energy and materials to maintain mitochondria. **4.** The potential energy drop between glucose and pyruvate is relatively small. When pyruvate acts as the electron acceptor during fermentation, a great deal of potential energy remains in the alcohol or other fermentation products.

CHAPTER 10

Check Your Understanding (CYU)

CYU p. 179 When a pigment absorbs energy and an electron is excited in response, the event is analogous to striking the tuning fork. When an excited electron transmits resonance energy to another pigment molecule, the event is analogous to touching a vibrating tuning fork to another and making it vibrate. **CYU p. 184** (1) Your model should conform to the relationships shown in Figure 10.13. (2) Put four dimes on the water-splitting reaction ($2 H_2O \rightarrow 4 H^+ + O_2$). When four photons hit the PSII antenna complex, move the dimes to pheophytin, then to the PSII ETC (plastoquinone and the cytochrome complex). The dimes go to plastocyanin, then to the antenna complex of PSI. When four photons

strike, move the dimes to the PSI ETC and ferrodoxin, then combine them with $NADP^+ + 2 H^+$ to form NADPH. **CYU p.190** (1) (a) C_4 plants use PEP carboxylase to fix CO_2 into organic acids in mesophyll cells. These organic acids are then pumped into bundle-sheath cells, where they release carbon dioxide to rubisco. (b) CAM plants take in CO_2 at night, and have enzymes that fix it into organic acids stored in the central vacuoles of photosynthesizing cells. During the day, the organic acids are processed to release CO_2 to rubisco. (c) By diffusion through a plant's stomata when they are open. (2) Sucrose can be transported to other parts of the plant because it is water soluble. Starch is not water soluble, so it cannot be transported. It acts as a storage product, instead.

You Should Be Able To (YSBAT)

YSBAT p. 181 Photosystem II is very similar to the ETC in mitochondria. Both use the energy released from electron transfer to pump protons, generating a proton gradient that is used to generate ATP through ATP synthase. The high-energy electrons for PSII come from chlorophyll via water, however, while they come from NADH and $FADH_2$ for the ETC. The eventual products of PSII are ATP and oxygen; the eventual products of the mitochondrial ETC are ATP and water. **YSBAT p. 182** Light → Antenna complex → Reaction center → Pheophytin → ETC → Proton gradient → ATP synthase. Electrons from water are donated to the reaction center. When they absorb resonance energy and reduce pheophytin, they convert electromagnetic energy to chemical energy. **YSBAT p. 183** (1) Plastocyanin transfers electrons that move through PSII to the reaction center of PSI. (2) After they are excited by a photon and donated to the initial electron acceptor. (3) The electrons that originate from PSII's reaction center end up in NADPH—they are not cycled back to PSII. **YSBAT p. 186** Rubisco joins CO_2 with RuBP to form a six-carbon intermediate that is immediately hydrolyzed to yield two molecules of 3-phosphoglycerate, which participate in reactions that yield G3P.

Caption Questions and Exercises

Figure 10.5 [See Figure A10.1] **Figure 10.7** The energy state corresponding to a photon of green light would be located between the energy states corresponding to red and blue photons. **Figure 10.10** Yes—otherwise, changes in the rate of photosynthesis could be due to changes in the density of photosynthetic cells, not differences in wavelengths received. **Figure 10.17** The researchers didn't have any basis on which to predict these intermediates. They needed to perform the experiment to identify them. **Figure 10.24** One arrow should point from sucrose to starch, another from starch to sucrose. Note that the interconversion requires transport between the cytosol and chloroplast.

Summary of Key Concepts

KC 10.1 The Calvin cycle depends on the ATP and NADPH produced by PSI and PSII, so it is not independent of light. **KC 10.2** Oxygen is produced by a critical step in photosynthesis: splitting water to provide electrons to PSII. If oxygen production increases, it means that more electrons are moving through the photosystems. **KC 10.3** The "final" product of the reaction sequence reacts to form the molecule that is the substrate for the initial reaction in the sequence. **KC 10.4** Both forms of rubisco would increase carbohydrate production: Oxygen would not compete with CO_2 for binding to the active site of rubisco and the competing reactions of photorespiration would not occur, or rubisco would fix 10 times as much CO_2 in a given period of time.

Test Your Knowledge

1. d; **2.** c; **3.** a; **4.** c; **5.** b; **6.** d

Test Your Understanding

1. In PSII, it occurs when excited electrons from the reaction center are accepted by pheophytin, reducing it. In PSI, it occurs when excited electrons from the reaction center are accepted by ferredoxin. **2.** CO_2 is fixed when rubisco catalyzes the reaction between carbon dioxide and ribulose bisphosphate to form 3-phosphoglycerate. Subsequent reactions that produce sugar require phosphorylation by ATP and high-energy electrons from NADPH. **3.** PSII and PSI both have reaction centers, an energy transformation step, and an ETC. PSII produces ATP while PSI produces NADPH, and only PSII splits water to obtain electrons. Plastocyanin transfers electrons between the two photosystems, connecting them. **4.** Photorespiration occurs when levels of CO_2 are low and O_2 are high. Less sugar is produced because (1) CO_2 doesn't participate in the initial reaction catalyzed by rubisco, and (2) when rubisco catalyzes the reaction with O_2 instead, one of the products is eventually broken down to CO_2 in a process that uses ATP. **5.** See Figure 10.22b and Figure 10.23. **6.** Photosynthesis in chloroplasts produces glucose, which is processed in mitochondria to produce ATP.

Applying Concepts to New Situations

1. Both organelles have a double membrane, extensive internal membranes, ETCs that include cytochromes and quinones, and ATP synthase. Only chloroplasts have chlorophyll and other pigment molecules that absorb light. Chloroplasts use NADPH as an electron carrier; mitochondria use NADH and $FADH_2$. The Calvin cycle occurs only in chloroplasts; the TCA cycle occurs only in mitochondria. **2.** Because CO_2 and O_2 levels minimized the impact of photorespiration when rubisco evolved, the hypothesis is credible. But once O_2 levels increased, any change in rubisco that minimized photorespiration would give individuals a huge advantage over organisms with "old" forms of rubisco. There has been plenty of time for such changes to occur, making the "holdover" hypothesis less credible. **3.** Carotenoids should be located close to the chlorophyll molecules in the reaction center, so they can pass energy along and neutralize free radicals that could damage chlorophyll. One way to test this hypothesis is to isolate thylakoid membranes and test for the presence of carotenoid and chlorophyll molecules. **4.** No—they are unlikely to have the same complement of photosynthetic pigments. Different wavelengths of light are available in various layers of a forest and water depths. It is logical to predict that plants and algae have pigments that absorb the available wavelengths efficiently. One way to test this hypothesis would be to isolate pigments from species in different locations, and test the absorbance spectra of each.

THE BIG PICTURE: ENERGY

Check Your Understanding (CYU), p. 192

1. Photosynthesis uses H_2O as a substrate and releases O_2 as a by-product; cellular respiration uses O_2 as a sub-

FIGURE A10.1

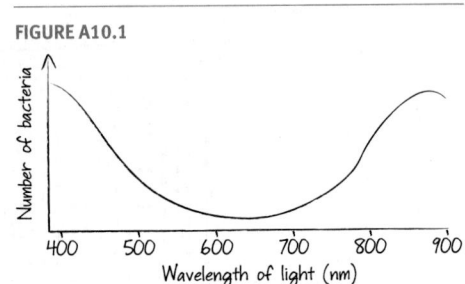

strate and releases H_2O as a by-product. **2.** Photosynthesis uses CO_2 as a substrate; cellular respiration releases CO_2 as a by-product. **3.** CO_2 fixation would essentially stop; CO_2 would continue to be released by cellular respiration. CO_2 levels in the atmosphere would increase rapidly, and production of new plant tissue would cease—meaning that most animals would quickly starve to death. **4.** ATP "is used by" the Calvin cycle; photosystem I "yields" NADPH.

CHAPTER 11

Check Your Understanding (CYU)

CYU p. 202 (1) [See Figure A11.1] (2) Interphase, prophase, prometaphase, metaphase, anaphase, telophase. Sister chromatids condense in prophase, move to the middle of the cell in metaphase, break apart in anaphase. **CYU p. 206** (1) $G_1 \rightarrow S \rightarrow G_2 \rightarrow M \rightarrow G_1$, etc. Checkpoints occur at the end of G_1 and G_2 and during M phase. (2) MPF Cdk levels are fairly constant throughout the cycle, but MPF cyclin increases. MPF activity increases at the end of G_2, initiating M phase, and declines at the end of M phase.

You Should Be Able To (YSBAT)

YSBAT p. 197 (1) Genes are segments of chromosomes that code for proteins. (2) Chromosomes are made of chromatin. (3) Sister chromatids are copies of the same chromosome, joined together. **YSBAT p. 200** [See Table A11.1] **YSBAT p. 204** MPF activates proteins that get mitosis under way. MPF consists of cyclin and a Cdk, and is inactivated by phosphorylation (at two sites) and activated by dephosphorylation (at one site). Enzymes that degrade cyclin reduce MPF levels. **YSBAT p. 208** [See Figure A11.2]

Caption Questions and Exercises

Figure 11.6 Unreplicated. **Figure 11.10** Fuse two interphase cells. If the regulatory molecule hypothesis is correct, neither cell should start M phase. But if the fusion event itself is the trigger, then at least one cell should start M phase. **Figure 11.13** In effect, MPF turns itself off after it is activated. If this didn't happen, the cell would undergo mitosis again right away. **Figure 11.15** After 1990, the death rate from lung cancer decreased in men but not in women. One hypothesis to explain the data is that rates of cigarette smoking declined in men but increased in women.

Summary of Key Concepts

KC 11.1 The gap phases give the cell time to replicate organelles and grow prior to division, as well as perform the normal functions required to stay alive. **KC 11.2** Kinetochore microtubules originate in the cytoplasm but chromosomes are in the nucleus. If the nuclear envelope did not disintegrate, the microtubules could not reach the chromosomes. **KC 11.3** The cell cycle will move forward, from G_1 to S phase, triggering cell division. **KC 11.4** A wide variety of defects, and multiple defects, cause a cell to become cancerous. No one drug or therapy can cure all of the defects involved.

Test Your Knowledge

1. b; **2.** d; **3.** d; **4.** d; **5.** c; **6.** c

Test Your Understanding

1. [See Figure A11.3] For daughter cells to have identical complements of chromosomes, all the chromosomes must be replicated during the S phase, the spindle apparatus must connect with the kinetochores of each replicated chromosome in prometaphase, and the sister chromatids of each replicated chromosome must separate in anaphase. **2.** One possible concept map is shown

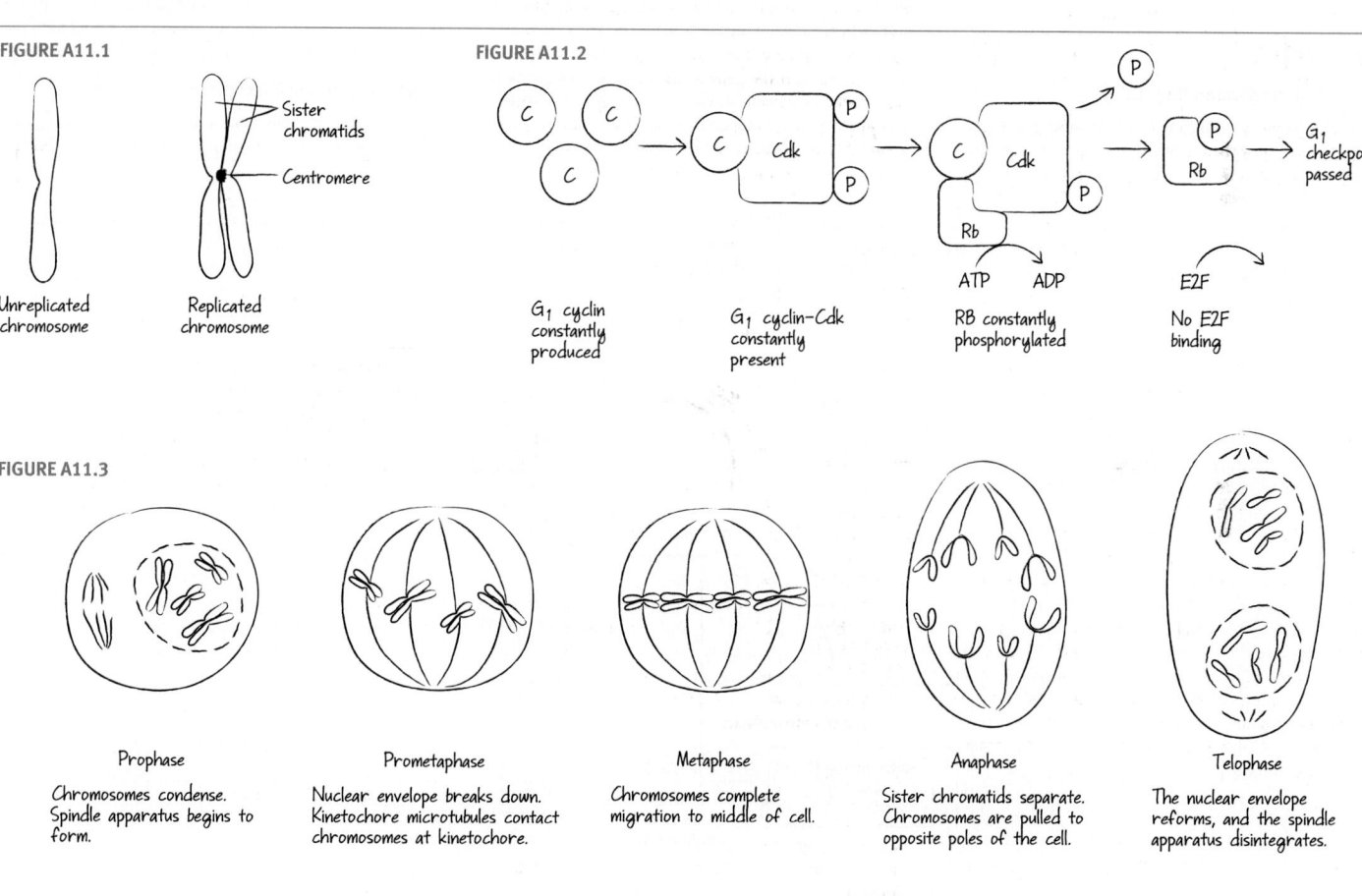

FIGURE A11.1

Unreplicated chromosome

Replicated chromosome

Sister chromatids

Centromere

FIGURE A11.2

G_1 cyclin constantly produced

G_1 cyclin–Cdk constantly present

RB constantly phosphorylated

No E2F binding

G_1 checkpoint passed

FIGURE A11.3

Prophase
Chromosomes condense. Spindle apparatus begins to form.

Prometaphase
Nuclear envelope breaks down. Kinetochore microtubules contact chromosomes at kinetochore.

Metaphase
Chromosomes complete migration to middle of cell.

Anaphase
Sister chromatids separate. Chromosomes are pulled to opposite poles of the cell.

Telophase
The nuclear envelope reforms, and the spindle apparatus disintegrates.

TABLE A11.1

	Prophase	Prometaphase	Metaphase	Anaphase	Telophase
Mitotic spindles	Grow	Contact chromosomes	Move chromosomes	Shorten	Break down
Nuclear envelope	Disintegrates	Nonexistent	Nonexistent	Nonexistent	Re-forms
Chromosomes	Condense	Attach to kinetochore microtubules	Move to metaphase plate	Sister chromatids separate	Collect at opposite poles

below [see Figure A11.4]. 3. Cell fusion experiments suggested that something in mitotic cells initiated mitosis in other cells. Microinjection experiments suggested that this "something" is in the cytoplasm of M-phase cells. 4. Protein kinases phosphorylate other proteins. Phosphorylation changes a protein's shape, altering its function (activating or inactivating it). As a result, protein kinases regulate the function of other proteins. 5. Cyclin concentrations cycle during the cell cycle. At high concentration, cyclins bind to a specific cyclin-dependent kinase (or Cdk), forming an active protein kinase, such as MPF. 6. If a cancer has not yet metastasized, the tumor can be completely removed by surgery.

Applying Concepts to New Situations

1. It is efficient: If the cell is not going to divide, there is no reason to invest energy in replication. 2. A single cell with many identical nuclei in it. 3. Labeled cells are in S phase, so 4 hours passed between the end of S phase and the start of M phase. 4. Cancer requires many defects. Older cells have had more time to accumulate defects. Individuals with a genetic predisposition to cancer start out with some cancer-related defects, but this does not mean that the additional defects required for cancer to occur will develop.

CHAPTER 12

Check Your Understanding (CYU)

CYU p. 220 (1) Use four long and four short pipe cleaners (or pieces of cooked spaghetti) to represent the chromatids of two replicated homologous chromosomes (four total chromosomes). Mark two long and two short ones with a colored marker pen to distinguish maternal and paternal copies of these chromosomes. Twist identical pipe cleaners (e.g., the two long colored ones) together to simulate replicated chromosomes. Arrange the pipe cleaners to depict the different phases of meiosis I as follows: *Early prophase I:* Align sister chromatids of each homologous pair to form two tetrads. *Late prophase I:* Form one or more crossovers between non-sister chromatids in each tetrad. (This is hard to simulate with pipe cleaners—you'll have to imagine that each chromatid now contains both maternal and paternal segments.) *Metaphase I:* Line up homologous pairs (the two pairs of short pipe cleaners and the two pairs of long pipe cleaners) at the metaphase plate. *Anaphase I:* Separate homologs. Each homolog still consists of sister chromatids joined at the centromere. *Telophase I and cytokinesis:* Move homologs apart to depict formation of two haploid cells, each containing a single replicated copy of two different chromosomes. (2) During anaphase I, homologs (not sister chromatids, as in mitosis) are separated, making the cell products of meiosis I haploid. CYU p. 223 (1) [See Figure A12.1] Maternal chromosomes are white and paternal chromosomes are black. Daughter cells with many other possible combinations of chromosomes than shown could result from meiosis of this parent cell. (2) Asexual reproduction generates no appreciable genetic diversity. Self-fertilization is preceded by meiosis so it generates gametes, through crossing over and independent assortment, that have combinations of alleles not present in the parent. Out-crossing generates the most genetic diversity among offspring because it produces new combinations of alleles from two different individuals.

You Should Be Able To (YSBAT)

YSBAT p. 213 $n = 12$; the organism is diploid with $2n = 24$ (in females; 23 in males). **YSBAT p. 214** [See Figure A12.2] Because the two sister chromatids are identical and attached, it is sensible to consider them as parts of a single chromosome. **YSBAT p. 216** [See Figure A12.3] **YSBAT p. 218** Crossing over would not occur and the daughter cells produced by meiosis would be diploid, not haploid. There would be no reduction division. **YSBAT p. 221** Each gamete would inherit either all maternal or all paternal chromosomes. This would limit genetic variation in the offspring by precluding the many possible gametes containing various combinations of maternal and paternal chromosomes.

Caption Questions and Exercises

Figure 12.11 [See Figure A12.4] **Figure 12.12** Asexually: 64 (16 individuals from generation three produce 4 offspring per individual). Sexually: 16 (8 individuals from generation three form 4 couples; each couple produces 4 offspring). **Figure 12.13** Sampling widely reduces the possibility that the results are due to an environmental factor—other than the presence of parasites—that differs between sexually and asexually reproducing populations.

Summary of Key Concepts

KC 12.1 Haploid cells cannot undergo meiosis, because there is no homolog to synapse with—haploid cells can-

FIGURE A11.4

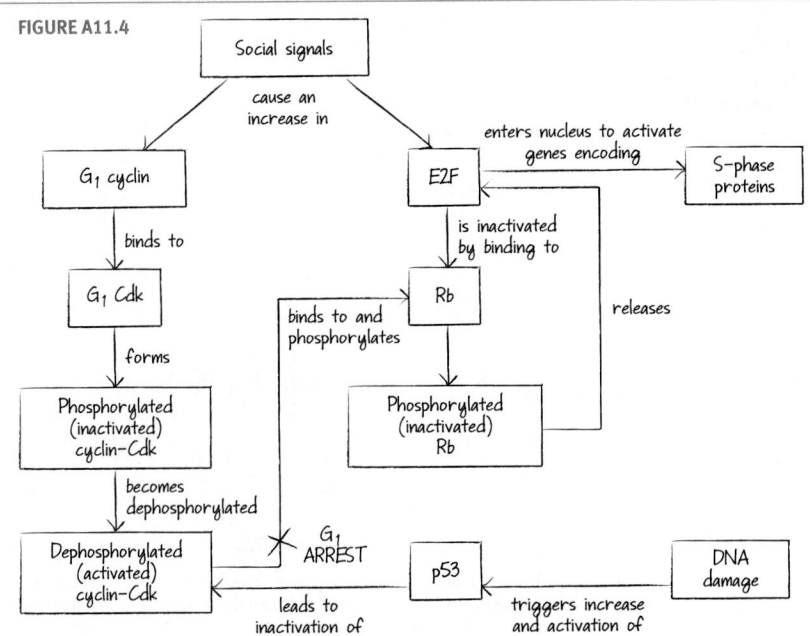

FIGURE A12.1

Diploid parent cell
(2n = 6)

Six possible daughter cells
(n = 3)

not undergo a reduction in ploidy. **KC 12.2** Monozygotic twins are genetically identical—the cells that originally gave rise to the two individuals were the product of mitosis. Dizygotic twins each arise from gametes that are the product of meiosis. Genetically, they are no more alike than any pair of siblings. **KC 12.3** Sexual reproduction will likely occur during seasons when conditions are changing rapidly, because genetically diverse offspring may have an advantage in the new conditions. **KC 12.4** The resulting gametes will be diploid instead of haploid.

Test Your Knowledge

1. b; **2.** a; **3.** b; **4.** b; **5.** d; **6.** a

Test Your Understanding

1. Homologous chromosomes are similar in size, shape, and gene content, and originate from different parents. Sister chromatids are exact copies of a chromosome that are generated when chromosomes are replicated (S phase of the cell cycle). **2.** Refer to Figure 12.6 as a guide for this exercise. The four pens represent the chromatids in one replicated homologous pair; the four pencils, the chromatids in a different homologous pair. To simulate meiosis II, make two "haploid cells"—each with a pair of pens and a pair of pencils representing two replicated chromosomes (one of each type in this species). Line them up in the middle of the cell, then separate the two pens and the two pencils in each cell such that one pen and one pencil goes to each of four

daughter cells. **3.** Meiosis I is a reduction division because homologs separate—daughter cells have just one of each type of chromosome instead of two. Meiosis II is not a reduction division because sister chromatids separate—daughter cells have unreplicated chromosomes instead of replicated chromosomes, but still just one of each type. **4.** Tetraploids produce diploid gametes, which combine with a haploid gamete from a diploid individual to form a triploid offspring. Mitosis proceeds normally in triploid cells because replicated chromosomes align independently before sister chromatids separate. But during meiosis in a triploid, homologous chromosomes can't pair up correctly. The third set of chromosomes does not have a homologous partner to pair with. **5.** Asexually produced individuals are genetically identical, so if one is susceptible to a new disease, all are. Sexually produced individuals are genetically unique, so if a new disease strain evolves, at least some plants are likely to be resistant to it. **6.** If homologs or sister chromatids do not separate normally, the daughter cells will receive one too many chromosomes or one too few. The resulting chromosome sets are considered unbalanced because they do not have the normal number of copies of each gene.

Applying Concepts to New Situations

1. The gibbon would have 22 chromosomes in each gamete, and the siamang would have 25. Each somatic cell of the offspring would have 47 chromosomes. The offspring should be sterile because it has an unbalanced set

of chromosomes, so not all of the chromosomes would have a homolog to pair with in meiosis prophase I. **2.** One in eight. **3.** Aneuploidy is the major cause of spontaneous abortion. If spontaneous abortion is rare in older women, it would result in a higher incidence of aneuploid conditions such as Down syndrome in older women, as recorded in the figure. **4.** (a) Such a study might be done in the laboratory, controlling conditions in identical tanks. A population of snails can be established in each tank. A parasite could be added to one tank, and then changes in the frequency of sexual reproduction in the snail populations can be monitored and observed over time. (b) In nature, all of the forces that can act to produce change in a population are present. A laboratory test is only an approximation of what occurs in the wild. For example, sexually reproduced offspring might only have an advantage when the parasite-ridden population is also stressed by food shortages or high or low temperatures. If so, an experiment in the lab with ample food and preferred temperatures would give an incorrect result.

CHAPTER 13

Check Your Understanding (CYU)

CYU p. 239 See answers to Genetics Problems on p. A:14. **CYU p. 243** (1) [See Figure A13.1] Segregation of alleles occurs when homologs that carry those alleles are separated during anaphase I. One allele ends up in each

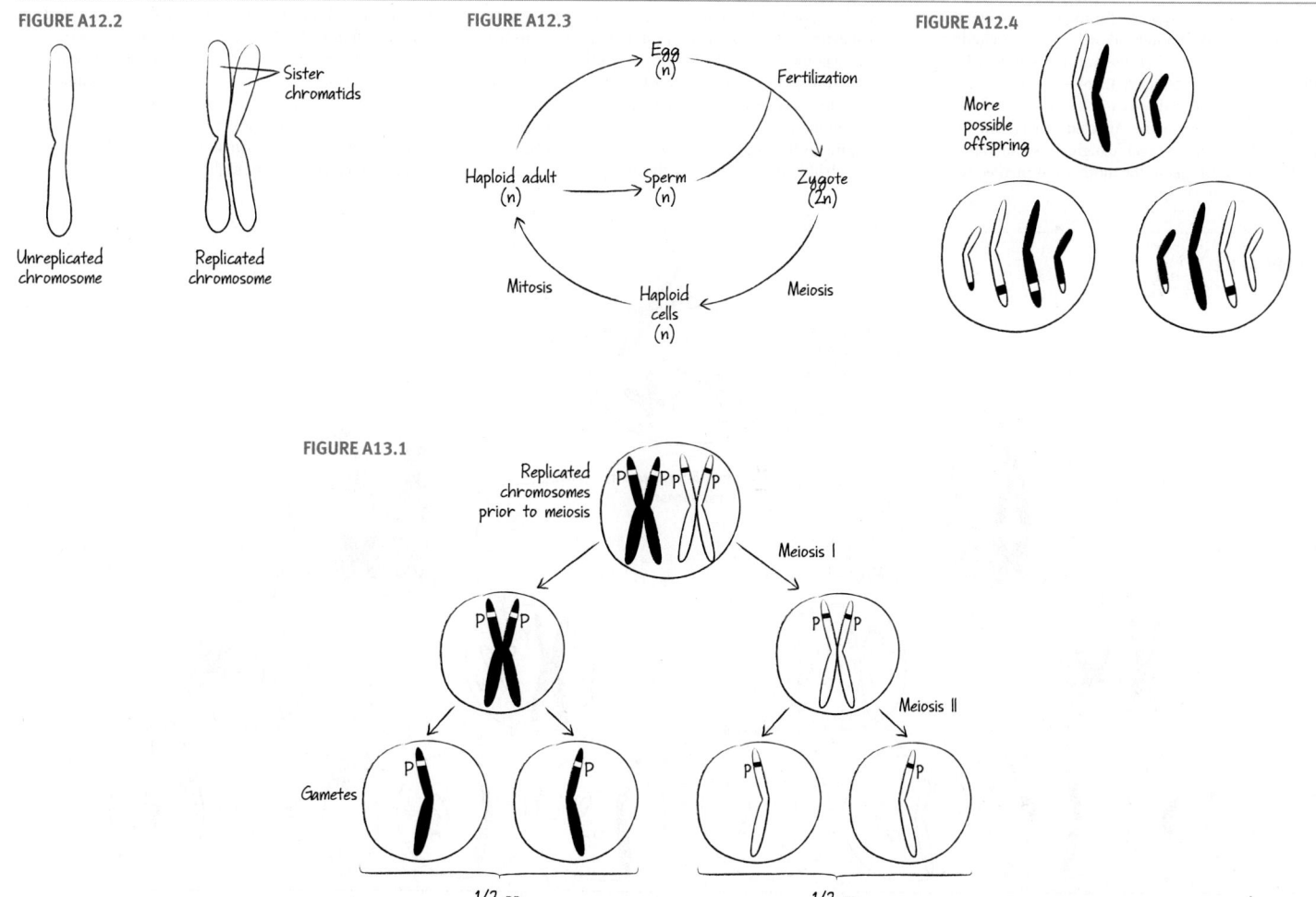

FIGURE A12.2

Unreplicated chromosome

Replicated chromosome
Sister chromatids

FIGURE A12.3

Egg (n) — Fertilization — Zygote (2n) — Meiosis — Haploid cells (n) — Mitosis — Haploid adult (n) — Sperm (n)

FIGURE A12.4

More possible offspring

FIGURE A13.1

Replicated chromosomes prior to meiosis

Meiosis I

Meiosis II

Gametes

1/2 pp   1/2 pp

daughter cell. (2) [See Figure A13.2] Independent assortment occurs because homologous pairs line up randomly at the metaphase plate during metaphase I. The figure shows two alternative arrangements of homologs in metaphase I. As a result, it is equally possible for a gamete to receive the following four combinations of alleles: YR, Yr, yR, yr. **CYU p. 250** (1) The comb phenotype results from interactions between alleles at two different genes, not a single gene. Matings between rose and pea comb chickens produce F_2 offspring that may have a new combination of alleles, and thus new phenotypes. (2) Kernel color in wheat is influenced by alleles at many different genes, not a single gene. F_2 offspring have a normal distribution of phenotypes, not a 3:1 ratio.

You Should Be Able To (YSBAT)

YSBAT p. 236 Punnett squares are an efficient way to predict offspring genotypes from parents of known genotype. The phenotype ratios are 1:1 round:wrinkled; the genotype ratios are 1:1 Yy:yy. **YSBAT p. 238** AABb → AB and Ab. PpRr → PR, Pr, pR, and pr. AaPpRr → APR, APr, ApR, Apr, aPR, apr, and aPr.

Caption Questions and Exercises

Figure 13.3 An experiment is a failure if you didn't learn anything from it. That is not the case here. **Figure 13.4** No—the outcome (the expected offspring genotypes that the Punnett square generates) will be the same. **Figure 13.12** X^{WY}, X^{Wy}, X^{wY}, X^{wy}. **Figure 13.13** Random chance (or perhaps red-eyed, gray-bodied males don't survive well). **Figure 13.17** The gene colored orange is *ruby*; the gene colored blue is *miniature wings*. **Figure 13.22** If both parents are heterozygous (carriers), on average 1/4 of their children will be affected. If the allele is rare, then it is unlikely for the spouse of an affected (homozygous) parent to be a carrier—hence, children will not be affected—they will only get one allele from their affected parent. **Figure 13.24** All affected females would have all affected sons and daughters, no matter what the father's genotype. Affected fathers would transmit the allele to half of their daughters, so on average half would be affected if the mother was unaffected.

Summary of Key Concepts

KC 13.1 B and b. **KC 13.2** BR, Br, bR, and br, in equal proportions. **KC 13.3** The B and b alleles are located on different but homologous chromosomes, which separate into different daughter cells during meiosis I. The BbRr notation indicates that the B and R genes are on different chromosomes. As a result, the chromosomes line up independently of each other in metaphase of meiosis I. The B allele is equally likely to go to a daughter cell with R as with r; likewise the b allele is equally likely to go to a daughter cell with R as with r. **KC 13.4** Instead of two alleles on the same chromosome being transmitted together, crossing over separates them so they are transmitted independently of each other. They are no longer physically linked.

Genetics Questions

1. d; **2.** b; **3.** a; **4.** d; **5.** a; **6.** a; **7.** d; **8.** b; **9.** b; **10.** d

Genetics Problems

1. 3/4; 1/256 (see **BioSkills 13** in Appendix A); 1/2 (the probabilities of transmitting the alleles or having sons does not change over time). **2.** Your answer to the first three parts should conform to the F_1 and F_2 crosses diagrammed in Figure 13.5b, except that different alleles and traits are being analyzed. The recombinant gametes would be Yi and yI. Yes—there would be some individuals with round, green seeds and with wrinkled, yellow seeds. **3.** Cross 1: non-crested (Cc) × non-crested (Cc) = 14 non-crested (C_); 7 crested (cc). Cross 2: crested (cc) × crested (cc) = 22 crested (cc). Cross 3: non-crested (Cc) × crested (cc) = 7 non-crested (Cc); 6 crested (cc). Non-crested (C) is the dominant allele. **4.** This is a dihybrid cross that yields progeny phenotypes in a 9:3:3:1 ratio. Let O stand for the allele for orange petals and o the allele for yellow petals; let S stand for the allele for spotted petals and s the allele for unspotted petals. Start with the hypothesis that O is dominant to o, that S is dominant to s, that the two genes are found on different chromosomes so they assort independently, and that the parent individual's genotype is OoSs. If you do a Punnett square for the OoSs × OoSs mating, you'll find that progeny phenotypes should be in the observed 9:3:3:1 pro-

portions. **5.** Let D stand for the normal allele and d for the allele responsible for Duchenne-type muscular dystrophy. The woman's family has no history of the disease, so her genotype is almost certainly DD. The man is not afflicted, so he must be DY. (The trait is X-linked, so he has only one allele; the "Y" stands for the Y chromosome.) Their children are not at risk. The man's sister could be a carrier, however—meaning she has the genotype Dd. If so, then half of the second couple's male children are likely to be affected. **6.** Your stages of meiosis should look like Figure 12.6, except with $2n = 4$ instead of $2n = 6$. The A and a alleles could be on the red and blue versions of the longest chromosome, and the B and b alleles could be on the red and blue versions of the smallest chromosomes. The places you draw them are the locations of the A and B genes, but each chromosome has only one allele. Each pair of red and blue chromosomes is a homologous pair. Sister chromatids bear the same allele (e.g., both sister chromatids of the long blue chromosomes might bear the a allele). Chromatids from the longest and shortest chromosomes are not homologous. To identify the events that result in the principles of segregation and independent assortment, see Figures 13.7 and 13.8 and substitute A, a, and B, b for R, r and Y, y. **7.** Half of their offspring should have the genotype iI^A and the type A blood phenotype. The other half of their offspring should have the genotype iI^B and the type B blood phenotype. Second case: the genotype and phenotype ratios would be 1:1:1:1 I^AI^B (type AB) : I^Ai (type A) : I^Bi (type B) : ii (type O). **8.** Because the children of Tukan and Valco had no eyes and smooth skin, you can conclude that the allele for eyelessness is dominant to eyes and the allele for smooth skin is dominant to hooked skin. E = eyeless, e = two eye sets, S = smooth skin, s = hooked skin. Tukan is eeSS; Valco is EEss. The children are all EeSs. Grandchildren with eyes and smooth skin are eeS_. Assuming that the genes are on different chromosomes, one-fourth of the children's gametes are ee and three-fourths are S_. So ¼ ee × ¾ S_ × 32 = 6 children would be expected to have two sets of eyes and smooth skin. **9.** Although the mothers were treated as children by a reduction of dietary phenylalanine, they would have accumulated phenylalanine and its derivatives once they

FIGURE A13.2

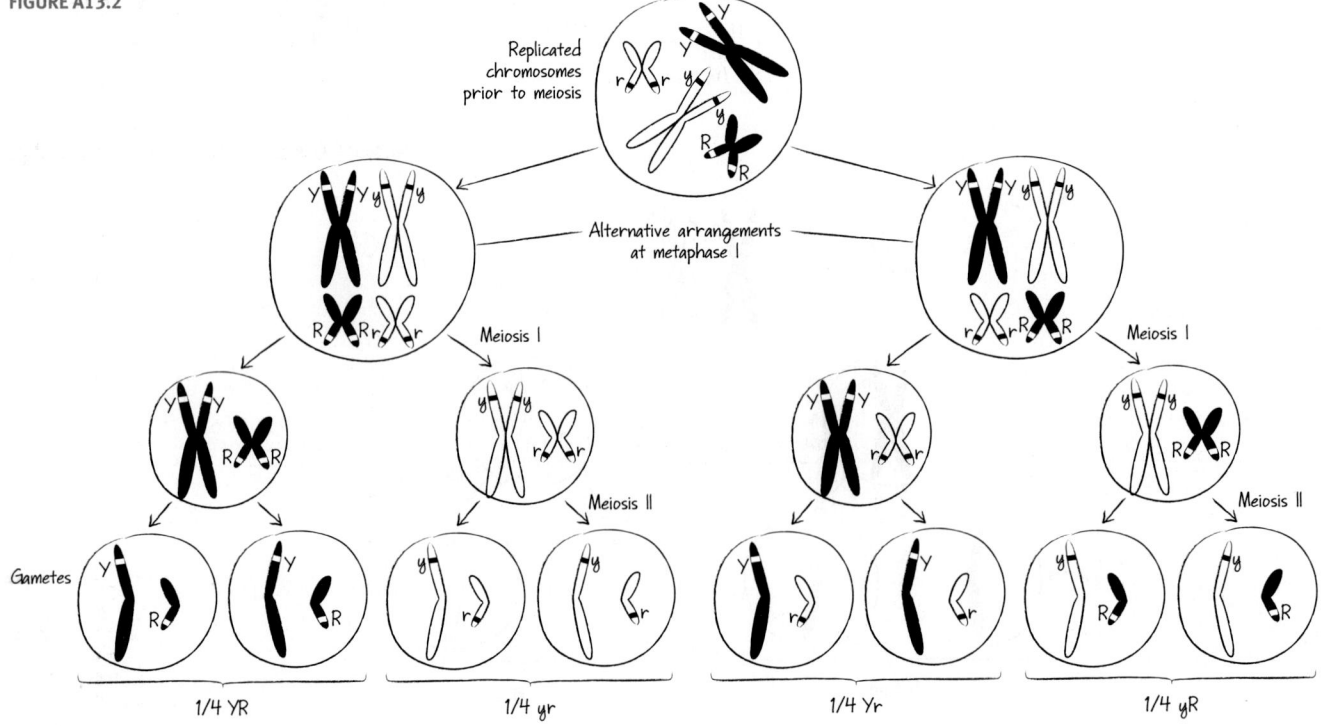

went off the low-phenylalanine diet as young adults. Children born of such mothers were therefore exposed to high levels of phenylalanine during pregnancy. For this reason, a low-phenylalanine diet is recommended for such mothers throughout the pregnancy. **10.** According to Mendel's model, palomino individuals should be heterozygous at the locus for coat color. If you mated palomino individuals, you would expect to see a combination of chestnut, palomino, and cremello offspring. If blending inheritance occurred, however, all of the offspring should be palomino. **11.** Because this is an X-linked trait, the father who has hemophilia could not have passed the trait on to his son. Thus, the mother in couple 1 must be a carrier and have passed the recessive allele on to her son, who is XY and affected. To educate a jury about the situation, you should draw what happens to the X and Y during meiosis, then make a drawing showing the chromosomes in couple 1 and couple 2, with a Punnett square showing how these chromosomes get passed to the affected and unaffected children. **12.** The curved-wing allele is autosomal recessive; the lozenge-eye allele is sex-linked (specifically, X-linked) recessive. Let *L* be the allele for long wings and *l* be the allele for curved wings; let X^R be the allele for red eyes and X^r the allele for lozenge eyes. The female parent is $Ll X^R X^r$; the male parent is $Ll X^R Y$. **13.** Albinism indicates the absence of pigment, so let *b* stand for an allele that gives the absence of blue and *y* for an allele that gives the absence of yellow pigment. If blue and yellow pigment blend to give green, then both green parents are *BbYy*. The green phenotype is found in *BBYY, BBYy, BbYY,* and *BbYy* offspring. The blue phenotype is found in *BByy* or *Bbyy* offspring. The yellow phenotype is observed in *bbYY* or *bbYy* offspring. Albino offspring are *bbyy*. The phenotypes of the offspring should be in the ratio 9:3:3:1 as green:blue:yellow:albino. Two types of crosses yield *BbYy* F_1 offspring: *BByy* × *bbYY* (blue × yellow) and *BBYY* × *bbyy* (green × albino). **14.** Autosomal dominant.

CHAPTER 14

Check Your Understanding (CYU)

CYU p. 268 (1) DNA polymerase adds only nucleotides to the free 3'-OH on a strand. Primase synthesizes a short RNA sequence that provides the free 3' end necessary for DNA polymerase to start working. (2) The helix is already opened and the downstream portions relaxed. **CYU p. 271** (1) Telomerase is not needed—bacterial chromosomes don't shorten after replication because they are circular. DNA polymerase can synthesize a complementary strand all the way around. (2) Otherwise there would be no template to synthesize the extension of a single DNA strand.

You Should Be Able To (YSBAT)

YSBAT p. 264 [See Figure A14.1] The new strands grow in opposite directions, each in 5' → 3' direction. **YSBAT p. 266** Helicase, topoisomerase, single-strand DNA-binding proteins, primase, and DNA polymerase are all required for leading-strand synthesis. If any one of these proteins is nonfunctional, then DNA replication will not occur. **YSBAT p. 267** [See Figure A14.2] If DNA ligase was defective, then the leading strand would be continuous, and the lagging strand would have gaps in it where the Okazaki fragments had not been joined.

Caption Questions and Exercises

Figure 14.2 The lack of radioactive protein in the pellet (after centrifugation) is strong evidence; they could also make micrographs of infected bacterial cells before and after agitation. **Figure 14.3** 5' TAG 3'. **Figure 14.5** The same two bands should appear, but the upper band (DNA containing only ^{14}N) should get bigger and darker and the lower band (hybrid DNA) should get smaller and lighter in color, as there is less and less heavy DNA in each succeeding generation. **Figure 14.13** As long as the RNA template could bind to the "overhanging" sec-

tion of single-stranded DNA, any sequence could produce a longer strand. For example, 5' CCCAUUCCC 3' would work just as well. **Figure 14.15** They are much lower in energy. **Figure 14.17** Exposure to UV radiation can cause formation of thymine dimers. If thymine dimers are not repaired, they represent mutations. If such mutations occur in genes controlling the cell cycle, cells can grow abnormally, resulting in cancers.

Summary of Key Concepts

KC 14.1 Complementary bases hydrogen bond, and hydrophobic interactions occur between bases stacked inside the double helix. **KC 14.2** The bases added during DNA replication are shown in color type.

Original DNA: CAATTACGGA
            GTTAATGCCT

Replicated DNA: CAATTACGGA
            GTTAATGCCT
            CAATTACGGA
            GTTAATGCCT

KC 14.3 DNA polymerase synthesizes DNA by adding deoxyribonucleotides to the free 3'-OH group of an existing nucleotide sequence (primer), using single-stranded DNA as a template. Primase synthesizes short RNA sequences, using single-stranded DNA as a template. In contrast to DNA polymerase, primase can begin synthesis de novo, that is, without a primer. Telomerase adds deoxyribonucleotides to the unreplicated 3' end of the lagging strand, using as the template a short RNA molecule that is associated with the enzyme. **KC 14.4** If errors in DNA aren't corrected, they represent mutations. When DNA repair systems fail, the mutation rate increases. As the mutation rate increases, the chance that one or more cell cycle genes will be mutated increases. Mutations in these genes often result in uncontrolled cell division, ultimately leading to cancer.

Test Your Knowledge

1. d; **2.** c; **3.** d; **4.** b; **5.** c; **6.** d

Test Your Understanding

1. Whether DNA or proteins were labeled. **2.** There are two replication forks at each point of origin—replication proceeds in both directions at the same time. **3.** On the lagging strand, DNA polymerase moves away from the replication fork. When helicase unwinds a new section of DNA, primase must build a new primer on the lagging strand (closer to the fork) and another polymerase molecule must begin synthesis at this point. On the leading strand, DNA polymerase moves in the same direction as helicase, so synthesis can continue without interruption from one primer (at the origin of replication). **4.** Telomerase binds to the 3' overhang at the end of a chromosome. Once bound, it begins catalyzing the addition of deoxyribonucleotides to the overhang in the 5' → 3' direction, lengthening the overhang. This allows primase, DNA polymerase, and ligase to catalyze the addition of deoxyribonucleotides to the lagging strand in the 5' → 3' direction, restoring the lagging strand to its original length. **5.** First, DNA polymerases are highly selective in matching complementary bases. Second, DNA polymerase III proofreads by recognizing mismatched bases, cutting out the incorrect base, and replacing it with the correct base. Third, proteins of the mismatch repair system excise incorrect bases and insert the proper ones. **6.** DNA is a symmetrical double helix. A mismatch (e.g., A-G or C-T) or damage to a base pair (e.g., thymine dimer) will cause a distortion or kink in the molecule that is recognizable by repair enzymes. Because the bases on the two strands are complementary, repair enzymes are able to replace the damaged bases easily and accurately.

Applying Concepts to New Situations

1. [See Figure A14.3] Primase would not have to build primers on the leading and lagging strands, because DNA polymerase III would be able to use a "raw" single-stranded template. It is likely that telomerase would still

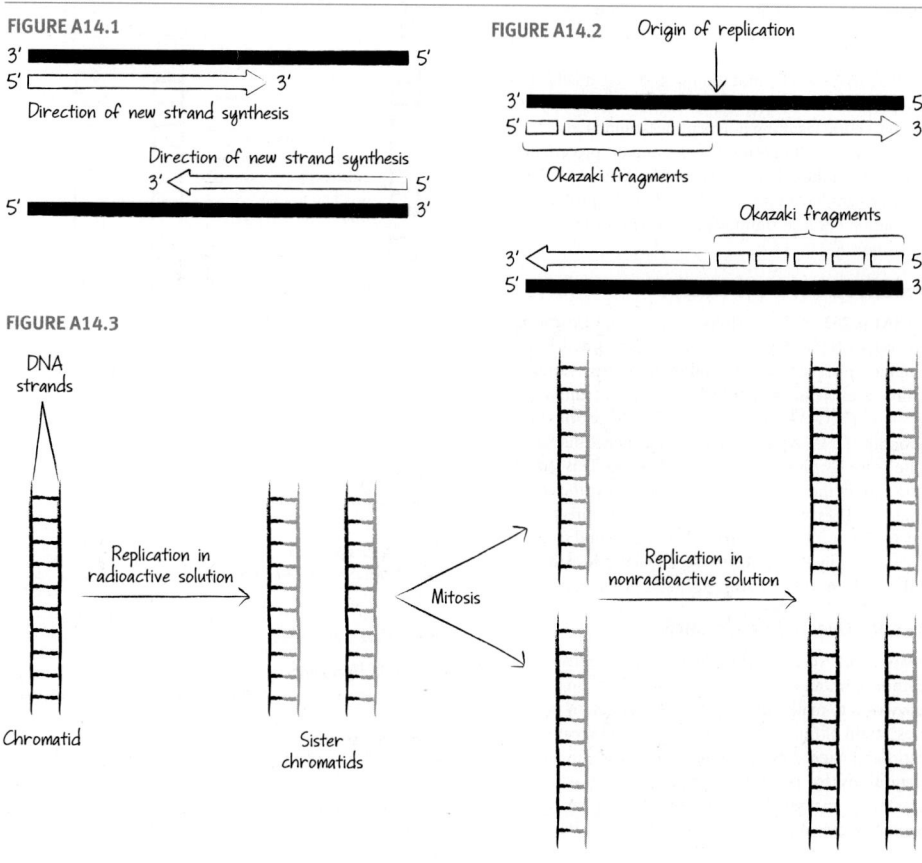

FIGURE A14.1

3' ▬▬▬▬▬▬▬▬▬ 5'
5' ▭▭▭▭▭▭▭▶ 3'
Direction of new strand synthesis

Direction of new strand synthesis
            3' ◀▭▭▭▭▭ 5'
5' ▬▬▬▬▬▬▬▬▬ 3'

FIGURE A14.3

DNA strands

Chromatid

Replication in radioactive solution →

Sister chromatids

Mitosis

Replication in nonradioactive solution →

FIGURE A14.2 Origin of replication

3' ▬▬▬▬▬▬▬▬▬ 5'
5' ▭▭▭▭▭▭▶ 3'
Okazaki fragments

Okazaki fragments
3' ◀▭▭▭ ▭▭▭ 5'
5' ▬▬▬▬▬▬▬▬▬ 3'

be required because, like primase, DNA polymerase would probably not have room to attach to the last end of the lagging strand, leaving a single-stranded overhang. **2.** (a) In **Figure A.14** below, the gray lines represent DNA strands containing radioactivity. Following replication in a radioactive solution, one DNA strand of each sister chromatid is radioactive. (b) [See **Figure A14.4**] **3.** (a) The double mutant of both *uvrA* and *recA* is most sensitive to UV light; the single mutants are in between; and the wild type is least sensitive. (b) The *recA* gene contributes more to UV repair through most of the UV dose levels. But at very high UV doses, the *uvrA* gene is somewhat more important than the *recA* gene. **4.** Many types of cancer arise because of mutations, especially those in genes involved in controlling cell growth. If chemicals that increase mutation rates in bacteria also increase mutation rates in humans, then the probability of mutations in cell control genes, and hence of cancer formation, also increases.

CHAPTER 15

Check Your Understanding (CYU)

CYU p. 282 If an error in transcription changes the sequence of bases in an mRNA, it may change the protein that is subsequently produced—and thus change the cell's phenotype. Changes in the sequence of other types of RNAs might also affect the phenotype. **CYU p. 285** (1) Remember that DNA and mRNA sequences are written in the $5' \rightarrow 3'$ direction unless indicated otherwise. Thus, to convert DNA triplets (codons) to mRNA codons, the DNA sequence is read "backwards" ($3' \rightarrow 5'$) and U (rather than T) is the base transcribed from A.

DNA codon	mRNA codon	Amino acid
ATA	UAU	Tyrosine
GTA	UAC	Tyrosine
TTA	UAA	Stop
GCA	CGU	Arginine

(2) The ATA → GTA mutation would have no effect on the protein. The ATA → TTA mutation introduces a stop codon, so the resulting polypeptide would be shortened. This would result in synthesis of a mutant protein much shorter than the original protein. The ATA → GCA mutation might have a profound effect on the protein's conformation because arginine's structure is different from tyrosine's.

You Should Be Able To (YSBAT)

YSBAT p. 285 (1) The codons in Figure 15.5 are translated correctly. (2) [See Figure A15.1] (3) There are many possibilities (just pick alternate codons for one or more of the amino acids); one is an mRNA sequence (running $5' \rightarrow 3'$) GCG-AAC-GAU-UUC-CAG. To get the corresponding DNA sequence, write this sequence but substitute Ts for Us: GCG-AAC-GAT-TTC-CAG. This strand also runs in the $5' \rightarrow 3'$ direction. Now write the complementary bases, which will be in the $3' \rightarrow 5'$ direction: CGC-TTC-CTA-AAG-GTC. When this second strand is transcribed by RNA polymerase, it will produce the mRNA given with the proper $5' \rightarrow 3'$ orientation.

Caption Questions and Exercises

Figure 15.1 No, it could not make citrulline from ornithine without enzyme 2. Yes, it would no longer need enzyme 2 to make citrulline. **Figure 15.2** Many possibilities: strain of fungi used, exact method for creating mutants and harvesting spores to plant, exact growing conditions (temperature, light, recipe for growth medium—including concentration of supplemented amino acids), objective criteria for determining growth

or no-growth. **Figure 15.5** $4 \times 4 \times 4 \times 4 = 256$ amino acids. A 4-base code is unlikely to evolve because only 20 amino acids are commonly used in the synthesis of proteins—there is no selective advantage for a code to be more complicated than necessary. **Figure 15.7** [See Figure A15.1] **Figure 15.8** The DNA sequence changes only at a single location, or point.

Summary of Key Concepts

KC 15.1 Not all genes code for polypeptides. A better statement might be something like, "one-gene, one RNA or polypeptide product." **KC 15.2** Transcription means to copy, and mRNA is a short-lived copy of the information in DNA. Translation means to change languages, and proteins are a different "chemical language" than nucleic acids (DNA and RNA). **KC 15.3** If a change in a DNA sequence doesn't change the amino acid specified by the corresponding mRNA codon, then the change in genotype doesn't change the phenotype. Many changes in the third positions of codons do not change the phenotype, because the genetic code is redundant. **KC 15.4** The new mutation produced light coat colors, which allowed beach-dwelling mice to be camouflaged and survive better. If the same mutation occurred in a woodland-dwelling population, the light-colored individuals would probably be spotted by predators and eaten.

Test Your Knowledge

1. d; **2.** a; **3.** d; **4.** a; **5.** d; **6.** b

Test Your Understanding

1. In Morse code, different combinations of dots and dashes code for the complexity of the English language. In the genetic code, different combinations of bases code for the complexity of proteins in the cell.

2.

Substrate 1 $\xrightarrow{A}$ Substrate 2 $\xrightarrow{B}$ Substrate 3 $\xrightarrow{C}$

Substrate 4 $\xrightarrow{D}$ Substrate 5 $\xrightarrow{E}$ BiologicalSciazine

Substrate 3 would accumulate. Hypothesis: The individuals have a mutation in the gene for enzyme D. **3.** They supported an important prediction of the hypothesis: Losing a gene (via mutation) resulted in loss of an enzyme. **4.** It puts the three-nucleotide codons out of register. If the first base is deleted from the sequence ATG-CGA-GAC-TTA, then the reading frame becomes TGC-GAG-ACT-TA. **5.** In a triplet code, addition or deletion of 1–2 bases disrupts the reading frame "downstream" of the mutation site(s), resulting in a dysfunctional protein. But addition or deletion of 3 bases restores the reading frame—the normal sequence is disrupted only between the first and third mutation. The resulting protein is altered, but may still be able to function normally. Only a triplet code would show these patterns. **6.** A point mutation changes the nucleotide sequence of an existing allele, creating a new one, so it always changes the genotype. But because the genetic code is redundant, some point mutations do not change the resulting mRNA codons and thus do not change the protein product.

Applying Concepts to New Situations

1. [See Figure A15.2] **2.** Every copying error would result in a mutation that would change the amino acid sequence of the protein and would likely affect its function. **3.** Before the central dogma was understood, DNA was known to be the hereditary material, but no one knew how particular sequences of bases resulted in the production of RNA and protein products. The central dogma clarified

FIGURE A14.4

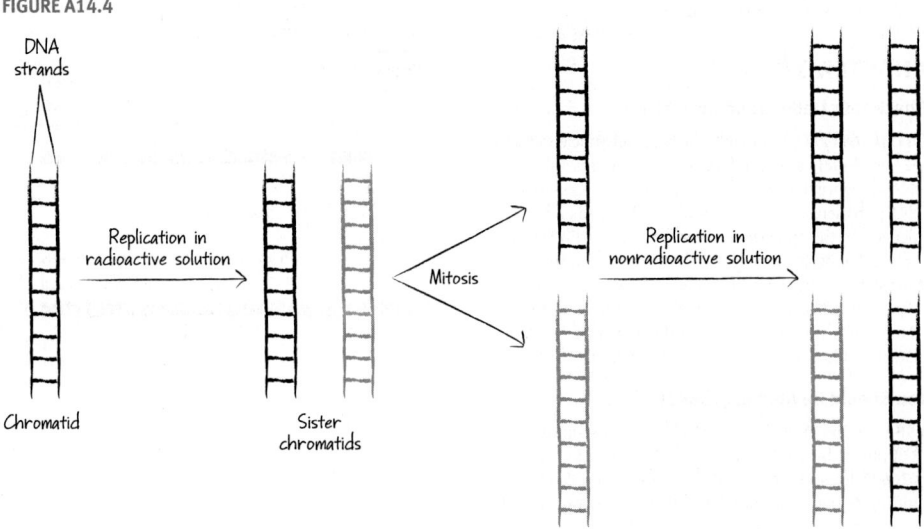

FIGURE A15.1

mRNA sequence:
5' AUG-CCC-CUG-GAG-GGG-GUU-AGA-CAU 3'

Amino acid sequence:
Met-Pro-Leu-Glu-Gly-Val-Arg-His

FIGURE A15.2

Bottom DNA strand:
5' AACTT-TAC(start)-GGG-CAA-ACC-ACT-AGC-CCA-ATG-TCG-ATC(stop)-AGTTTC 3'

mRNA sequence:
5' AUG-CCC-GUU-UGG-AGA-UCG-GGU-UAC-AGC-UAG 3'

Amino acid sequence:
Met-Pro-Val-Trp-Arg-Ser-Gly-Tyr-Ser

how genotypes produce phenotypes. **4.** A triplet code that overlapped by one base might work as follows:

```
5'  AUGUUACGGAAUUGA  3'
     GUU
      AUC
       CGG
        GAA
         AUU
          UGA
```

CHAPTER 16

Check Your Understanding (CYU)

CYU p. 293 (1) Different types of sigma proteins have different amino acid sequences and overall structures, so they are able to bind to stretches of DNA that differ in their base sequence. (2) NTPs are required because the three phosphate groups raise the monomer's potential energy enough to make the polymerization reaction exergonic. **CYU p. 295** (1) The subunits contain both RNA (the "ribonucleo" in the name) and proteins. (2) The cap and tail protect mRNAs from degradation and facilitate translation. **CYU p. 304** E is for Exit—the site where empty tRNAs are ejected; P is for Peptidyl (or peptide bond)—the site where peptide bond formation takes place; A is for Aminoacyl—the site where aminoacyl tRNAs enter.

You Should Be Able To (YSBAT)

YSBAT p. 299 (1) The amino acid attaches on the top left of the L-shaped structure. (2) The anticodon is antiparallel in orientation to the mRNA codon, and contains the complementary bases.

Caption Questions and Exercises

Figure 16.1 RNA is synthesized in the 5' → 3' direction; the DNA template is "read" 3' → 5'. **Figure 16.5** There would be no loops—the molecules would match up exactly. **Figure 16.12** If the amino acids stayed attached to the tRNAs, the gray line in the graph would stay high and the green line low. If the amino acids were transferred to some other cell component, the gray line would decline but the green line would be low.

Summary of Key Concepts

KC 16.1 The new sequence would probably not bind sigma or basal transcription factors as effectively as the original sequence. If so, transcription rate at that gene would probably decrease. **KC 16.2** Bacterial genes lack introns, so their RNA products do not require splicing. And because bacterial mRNAs are translated immediately, they do not need a cap and tail. **KC 16.3** These RNAs transfer amino acids to growing polypeptides; they transfer information in nucleic acids (an mRNA codon) to a protein product. **KC 16.4** An aminoacyl tRNA synthetase catalyzes the addition of an amino acid to a tRNA, forming an aminoacyl tRNA. **KC 16.5** One possible concept map is shown below [see Figure A16.1].

Test Your Knowledge

1. c; **2.** c; **3.** d; **4.** c; **5.** c; **6.** d

Test Your Understanding

1. Basal transcription factors bind to promoter sequences in eukaryotic DNA and facilitate the positioning of RNA polymerase. As part of the RNA polymerase holoenzyme, sigma binds to a promoter sequence in bacterial DNA and initiates transcription by the core enzyme. **2.** If the wobble rules did not exist, it would take one tRNA for each amino-acid-specifying codon in the genetic code. This is inefficient. The wobble rules allow a single tRNA to match several mRNA codons, reducing the inefficiency caused by redundancy in the genetic code. **3.** Eukaryotic mRNAs contain introns which must be removed by splicing. Splicing takes place in the nucleus and is accomplished by a large complex called a spliceosome, composed of many

snRNPs. **4.** After a peptide bond forms between the polypeptide and the amino acid held by the tRNA in the A site, the ribosome moves down the mRNA. As it does, an empty tRNA leaves the E site. The now-empty tRNA that was in the P site enters the E site; the tRNA holding the polypeptide chain moves from the A site to the P site, and a new aminoacyl tRNA enters the A site. **5.** The ribosome's active site is made up of RNA, not protein. **6.** The separation allows the aminoacyl tRNA to fit into the ribosome correctly, such that the anticodon binds to the mRNA and the amino acid fills the active site.

Applying Concepts to New Situations

1. Ribonucleases degrade mRNAs that are no longer needed by the cell. If an mRNA for a hormone that increased heart rate were never degraded, the hormone would be produced continuously and heart rate would stay elevated—a dangerous situation. **2.** Cordycepin triphosphate has a triphosphate group on the 5' carbon, but no 3'-OH group. If RNA were built 3' → 5', then the

incoming nucleoside triphosphates would have to have an OH group on the 3' carbon in order for a phosphodiester bond to be formed. Since cordycepin lacks a 3'-OH, it could not be added to the 5' end of an RNA chain growing in the 3' → 5' direction. **3.** The regions most crucial to the ribosome's function should be the most highly conserved: the active site, the E-, P-, and A-sites, and the site where mRNAs initially bind. Other regions should be more variable among species. **4.** The most likely locations are one of the grooves or channels where RNA, DNA, and ribonucleotides move through the enzyme—plugging one of them would prevent transcription.

CHAPTER 17

Check Your Understanding (CYU)

CYU p. 314 (1) The genes for metabolizing lactose should be expressed only when lactose is available. (2) [See Figure A17.1] **CYU p. 316** (1)

FIGURE A16.1

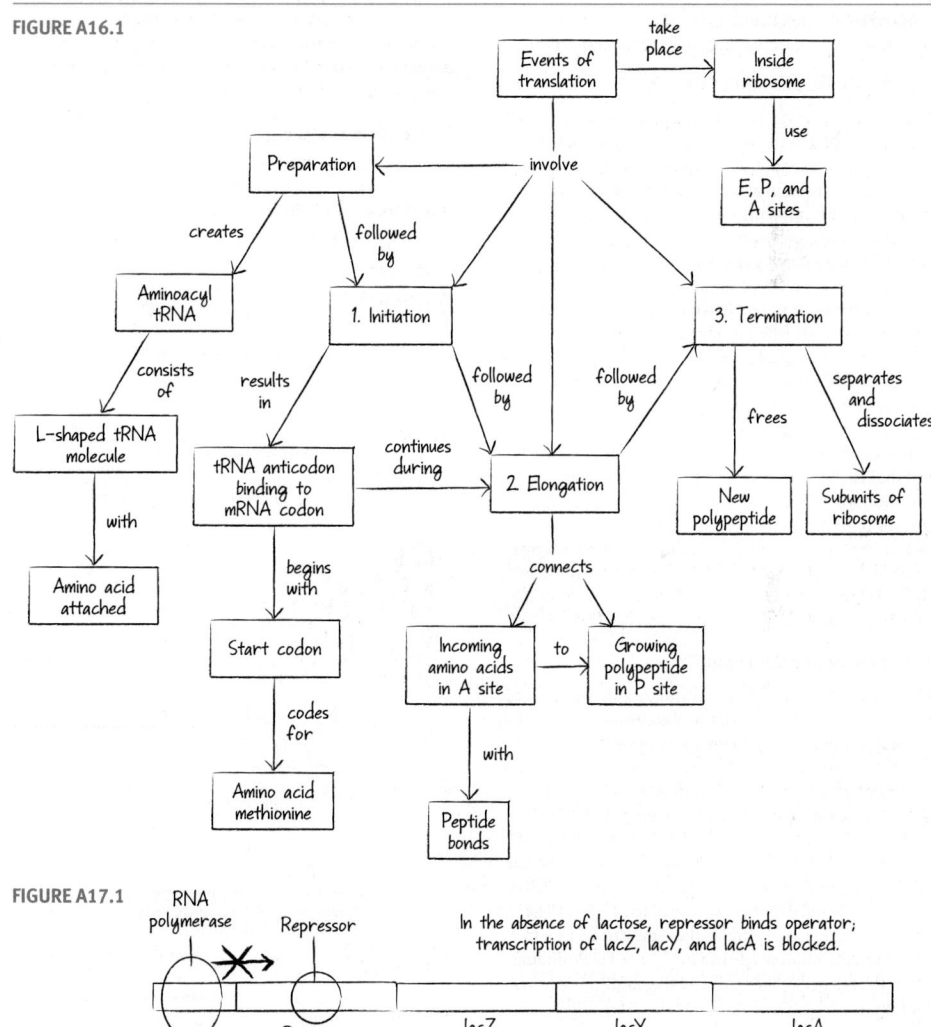

FIGURE A17.1

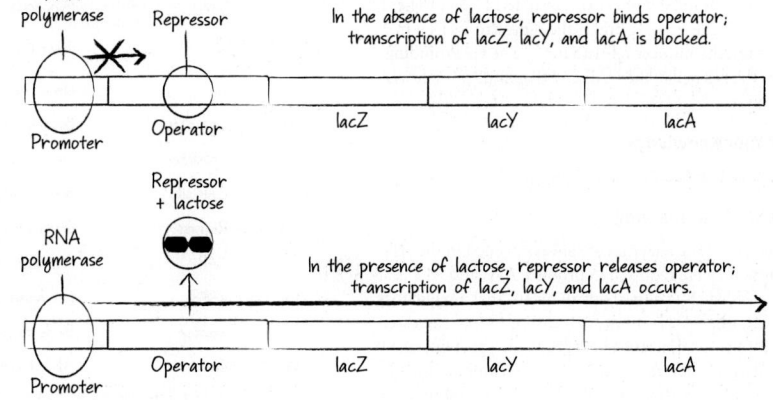

[See Figure A17.2] (2) Two conditions must be met for the *lac* operon to be expressed at a high rate. First, glucose levels must be low. Absence of glucose exerts positive control, because the cAMP-CAP complex binds to the CAP site and increases transcription rate. Second, lactose must be present. Lack of lactose exerts negative control, because the repressor remains attached to the operator and reduces transcription rate.

You Should Be Able To (YSBAT)

YSBAT p. 310 [See Table A17.1] **YSBAT p. 312** *lacZ* codes for the β-galactosidase enzyme, which breaks the disaccharide lactose into glucose and galactose. *lacY* codes for the lactose permease enzyme, which transports lactose into the bacterial cell. *lacI* codes for a protein that shuts down production of the other *lac* products. When lactose is absent, the *lacI* product prevents transcription. This is logical because there is no reason for the cell to make β-galactosidase and lactose permease if there is no lactose to metabolize. But when lactose is present, it interacts with *lacI* in some way so that *lacZ* and *lacY* are induced (their transcription can occur). When lactose is present, the enzymes that metabolize it are expressed.

Caption Questions and Exercises

Figure 17.1 Write "Slowest response, most efficient resource use" next to the transcriptional control label. Write "Fastest response, least efficient resource use" next to the post-translational control label. **Figure 17.2** Plates from all three treatments must be identical and contain identical growth medium, except for the presence of the sugars labeled in the figure. Also, all plates must be grown under the same physical conditions (temperature, light) for the same time. **Figure 17.3** Use a medium with all 20 amino acids when producing a master plate of mutagenized *E. coli* colonies, then use a replica plate that contains all of the amino acids except tryptophan. Choose cells from the master plate that did *not* grow on the replica plate. **Figure 17.8** Put the "Repressor protein" on the operator. No transcription will take place. Then put the "RNA polymerase" on the promoter. No transcription will take place. Finally, put "lactose" on the repressor protein and then remove the resulting lactose-repressor complex from the operon. Transcription will begin. **Figure 17.10** Write "Glucose low" above the drawing in part (a). Write "Glucose high" above the drawing in part (b).

Summary of Key Concepts

KC 17.1 Production of β-galactosidase and galactosidase permease are under transcriptional control—transcription depends on the action of regulatory proteins. The activity of the repressor and CAP—the regulatory proteins—are under post-translational control. **KC 17.2** After a dessert, glucose concentrations in your gut should be high, causing the *lac* operon in *E. coli* to shut down. After drinking milk, glucose concentrations should be low and lactose concentrations high, causing the *lac* operon in *E. coli* cells to be highly expressed. **KC 17.3** (1) The operator is the parking brake; the repressor locks it in place, and the inducer releases it. (2) The CAP-binding site is the gas pedal; the CAP-cAMP complex is a heavy foot on it.

Test Your Knowledge

1. b; 2. d; 3. b; 4. c; 5. a; 6. c

Test Your Understanding

1. The glycolytic enzymes are always needed in the cell, because they are required to produce ATP, and ATP is always needed. 2. Positive control means that a regulatory protein, when present, causes transcription to increase. Negative control means that a regulatory protein, when present, prevents transcription. 3. The combination of positive and negative control allows *E. coli* to shut down

the *lac* operon when glucose is present or lactose absent, but activate it when glucose is absent and lactose present. If only negative control occurred, the operon would be activated when glucose is present. If only positive control occurred, the operon would be activated when lactose is absent. 4. cAMP levels rise when glucose is low. If glucose is low, the cell is in danger of starving unless it can switch to another food source. 5. CAP acts like a receptor for cAMP—when cAMP binds to it, its activity changes. 6. Unlike lactose, glucose can enter the glycolytic pathway directly—without being converted to another molecule. Less ATP and fewer enzymes are needed to acquire and use glucose as an energy source.

Applying Concepts to New Situations

1. Set up cultures with individuals that all come from the same colony of toluene-tolerating bacteria. Half of the cultures should have toluene as the only source of carbon; half should have glucose or another common source of carbon as well as toluene. Cells will grow in both cultures if they are able to use toluene as a source of carbon. 2. Because toluene is not common in the environment, it is most likely that the toluene genes are under negative control. That is, the presence of toluene—by removing some brake on expression of the enzyme's genes—would trigger production of whatever enzymes are needed to break down toluene. 3. The *LacI^S* mutants would not respond to the inducer, so the repressor would remain on the operator and block transcription even in the presence of the inducer, lactose. These mutants would soon die in an environment where lactose was the

only sugar present. 4. Cells with functioning β-galactosidase are blue; cells that are not producing normal β-galactosidase are white. The sequence of the β-galactosidase gene in structural mutants would differ from the sequence of the wild-type gene, whereas the sequence of the β-galactosidase gene in regulatory mutants would be the same as the sequence of the wild-type gene.

CHAPTER 18

Check Your Understanding (CYU)

CYU p. 323 (1) The basic double helical structure of DNA is the same, but the DNA in eukaryotic cells is complexed with histones to form chromatin, which is tightly packed into 30-nm fibers attached to scaffolding proteins. In bacteria, DNA is associated with histone-like proteins but is not organized into complex structures. (2) Addition of acetyl or methyl groups to histones can cause chromatin to open or close. Different patterns of acetylation or methylation will determine which genes in muscle cells versus brain cells can be transcribed and which are not available for transcription. **CYU p. 326** (1) Bacterial regulatory sequences are found close to the promoter; eukaryotic regulatory sequences can be close to the promoter or far from it. Bacterial regulatory proteins interact directly with RNA polymerase to initiate or prevent transcription; eukaryotic regulatory proteins influence transcription by altering chromatin structure or binding to the basal transcription complex through mediator proteins. (2) Certain regulatory proteins open

---

FIGURE A17.2

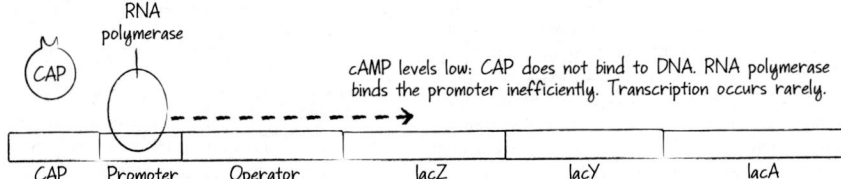

cAMP levels low: CAP does not bind to DNA. RNA polymerase binds the promoter inefficiently. Transcription occurs rarely.

CAP · Promoter · Operator · lacZ · lacY · lacA

cAMP

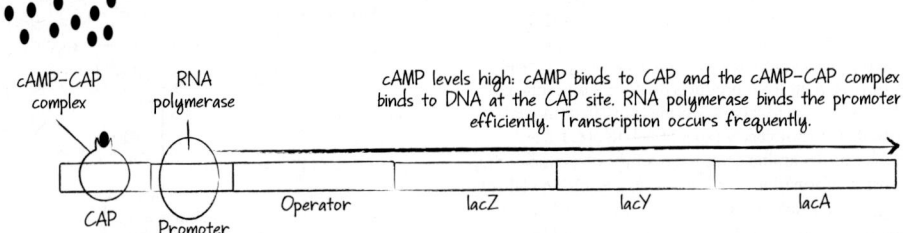

cAMP levels high: cAMP binds to CAP and the cAMP-CAP complex binds to DNA at the CAP site. RNA polymerase binds the promoter efficiently. Transcription occurs frequently.

CAP · Promoter · Operator · lacZ · lacY · lacA

TABLE A17.1

Name	Function
lacZ	Gene for β-galactosidase
lacY	Gene for galactosidase permease
Operator	Binding site for repressor
Promoter	Binding site for RNA polymerase
CAP site	Binding site for CAP-cAMP
Repressor	Shuts down transcription
CAP	Increases transcription rate when bound to cAMP
cAMP	Produced when glucose is low; binds to CAP and activates it
Lactose	Binds to repressor and stimulates transcription (removes negative control)
Glucose	At low concentration, increases transcription (via increased production of cAMP)

chromatin at muscle- or brain-specific genes, and then activate or repress the transcription of cell-type-specific genes. Muscle-specific genes are expressed only if muscle-specific regulatory proteins are produced and activated. **CYU p. 330** (1) It became clear that a single gene can code for multiple products instead of a single one. (2) miRNAs interfere with mRNAs by targeting them for destruction or preventing them from being translated. **CYU p. 333** (1) Many different types of mutations can disrupt control of the cell cycle and initiate cancer. These mutations can affect any of the six levels of control over gene regulation outlined in Figure 18.1. (2) The p53 protein is responsible for shutting down the cell cycle in cells with damaged DNA. If the protein does not function, then cells with damaged DNA—and thus many mutations—continue to divide. If these cells have mutations in genes that regulate the cell cycle, then they may continue to divide in an uncontrolled fashion.

You Should Be Able To (YSBAT)

YSBAT p. 326a The basal transcription factors found in muscle and nerve cells are similar or identical; the regulatory transcription factors found in each cell type are different. **YSBAT p. 326b** DNA forms loops when distant regulatory regions, such as silencers and enhancers, are brought close to the promoter through binding of regulatory transcription factors to the mediator complex. **YSBAT p. 329** Alternative splicing does not occur in bacteria because bacterial genes do not contain introns—in bacteria, each gene codes for a single product. Alternative splicing is part of step 3 in Figure 18.1. **YSBAT p. 330** Step 4 and step 5. RNA interference either (1) decreases the life span of mRNAs or (2) inhibits translation.

Caption Questions and Exercises

Figure 18.4 Acetylation of histones decondenses chromatin and allows transcription to begin, so HATs are elements in positive control. Deacetylation condenses the chromatin and inactivates transcription, so HDACs are elements in negative control. **Figure 18.6** This treatment allowed them to measure the normal amount of antibody mRNA produced, for comparison with the other treatments. **Figure 18.7** A typical eukaryotic gene usually contains introns and is regulated by multiple enhancers. Bacterial operons lack introns and enhancers. The promoter-proximal element found in some eukaryotic genes is comparable to the CAP binding site in the *lac* operon of bacteria. Bacterial operons have a single promoter but code for more than one protein; eukaryotic genes code for a single product.

Summary of Key Concepts

KC 18.1 Because eukaryotic RNAs are not translated as soon as transcription occurs, it is possible for RNA processing to occur, which creates variation in the mRNAs produced from a primary RNA transcript and their life span. **KC 18.2** The default state of eukaryotic genes is "off," because the highly condensed state of the chromatin makes DNA unavailable to RNA polymerase. **KC 18.3** [See Figure A18.1] **KC 18.4** Alternative splicing makes it possible for a single gene to code for multiple products. **KC 18.5** Cancer is a disease that results from accumulated mutations. Radiation damages DNA and increases the rate of mutation. Mutations accumulate as cells undergo repeated divisions, so are more common in older people.

Test Your Knowledge

1. c; **2.** a; **3.** d; **4.** d; **5.** b; **6.** d

Test Your Understanding

1. (a) Enhancers and the CAP site are similar because both are sites in DNA where regulatory proteins bind. They are different because enhancers generally are located at great distances from the promoter, whereas the CAP site is located near the promoter. (b) Promoter-proximal elements and the *lac* operon operator are both regulatory sites in DNA located close to the promoter. (c) Basal transcription factors and sigma are proteins that must bind to the promoter before RNA polymerase can initiate transcription. They differ because sigma is part of the RNA polymerase holoenzyme, while the basal transcription complex recruits RNA polymerase to the promoter. **2.** If changes in the environment cause changes in how spliceosomes function, then the RNAs and proteins produced from a particular gene could change in a way that helps the cell cope with the new environmental conditions. **3.** (a) Enhancers and silencers are both regulatory sequences located at a distance from the promoter. Enhancers bind regulatory transcription factors that activate transcription; silencers bind regulatory proteins that shut down transcription. (b) Promoter-proximal elements and enhancers are both regulatory sequences that bind positive regulatory transcription factors. Promoter-proximal elements are located close to the promoter; enhancers are far from the promoter. (c) Transcription factors bind to regulatory sites in DNA; the mediator complex does not bind to DNA but instead forms a bridge between regulatory transcription factors and basal transcription factors. **4.** RNA interference happens when complementary base pairing occurs between a small single-stranded miRNA and a single-stranded mRNA. **5.** Certain chemical modifications of histones mark sections of chromatin for transcription activation or repression. These modifications are passed to daughter cells when cells divide. A certain array of chromatin modifications or "histone codes" is associated with the production of muscle- or nerve-specific proteins. **6.** Lack of functional p53 protein prevents damaged cells from being destroyed or shut down (which would prevent their continuing to divide). As a result, cells with damaged DNA and thus many mutations continue to divide, possibly initiating cancerous growth.

Applying Concepts to New Situations

1. In eukaryotic cells, histones are involved in the basic folding of DNA molecules in chromatin. Mutations in histones are likely to interfere with this folding, and thus the expression of large numbers of genes. Individuals with mutant forms of histones are likely to die as a result—meaning that histone structure should be highly conserved over time. **2.** Tissues that divide more often must replicate their DNA more often; thus they are more susceptible to mutation. Because the cell cycle is acceler-ated in these cell types, it is logical to expect that control over the cell cycle could easily be lost. **3.** You could treat a culture of DNA-damaged cells with a drug that stops transcription and then compare them with untreated DNA-damaged cells. If transcriptional control regulates p53 levels, then the p53 level would be lower in the treated cells versus control cells. If control of p53 levels is post-translational, then in both cultures the p53 level would be the same. *Other approaches:* add labeled NTPs to damaged cells and see if they are incorporated into mRNAs for p53; or add labeled amino acids to damaged cells and see if they are incorporated into completed p53 proteins; or add labeled phosphate groups and see if they are added to p53 proteins. **4.** The miRNA would bind to viral RNA inside infected cells and prevent the viral RNA from working, thus stopping the infection.

THE BIG PICTURE: GENETIC INFORMATION

Check Your Understanding (CYU), p. 336

1. Star = DNA, mRNA, proteins. **2.** RNA "is reverse transcribed by" reverse transcriptase "to form" DNA. **3.** E = splicing, etc.; E = meiosis and sexual reproduction (along with their links). **4.** Chromatin "makes up" chromosomes; independent assortment and recombination "contribute to" high genetic diversity.

CHAPTER 19

Check Your Understanding (CYU)

CYU p. 344 (1) When the endonuclease makes a staggered cut in a palindromic sequence and the strands separate, the single-stranded bases that are left will bind ("stick") to the single-stranded bases left where the endonuclease cut the same palindrome at a different location. (2) The word probe means to examine thoroughly. A DNA probe "examines" a large set of sequences thoroughly, and binds to one—the one that has a complementary base sequence. **CYU p. 346** (1) Denaturation makes DNA single stranded so the primer can bind to the sequence during the annealing step. Once the primer is in place, *Taq* polymerase can synthesize the rest of the strand during the extension step. It is a "chain reaction" because the products of each reaction cycle are used in the next reaction cycle—this is why the number of copies doubles in each cycle. (2) One of many possible

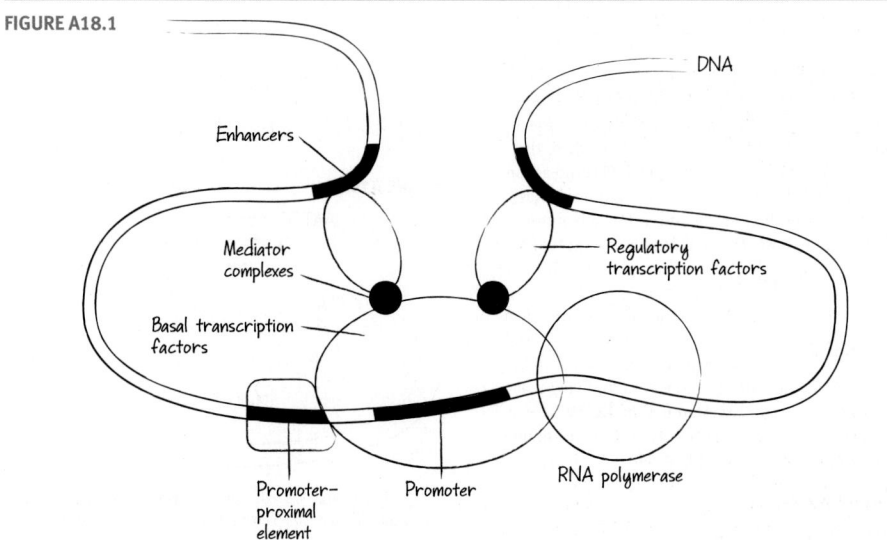

FIGURE A18.1

answers is shown below [see Figure A19.1]. **CYU p. 348** If ddNTPs were present at high concentration, they would almost always be incorporated—meaning that only fragments from the first complementary base in the sequence would be produced. **CYU p. 351** Start with a genetic map with as many polymorphic markers as possible. Get the genotype at these markers for a large number of individuals who have the same type of alcoholism—one that is thought to have a genetic component—as well as a large number of unaffected individuals. Look for particular versions of a marker that is almost always found in affected individuals. Genes that contribute to a predisposition to alcoholism will be near that marker.

You Should Be Able To (YSBAT)

YSBAT p. 342a Isolate the DNA and cut it into small fragments with EcoR1, which leaves sticky ends. Cut copies of a plasmid or other vector with EcoR1. Mix the fragments and plasmids under conditions that promote complementary base pairing by sticky ends of fragments and plasmids. Use DNA ligase to catalyze formation of phosphodiester bonds and seal the sequences. **YSBAT p. 342b** The probe must be single stranded so that it will bind by complementary base pairing to the target DNA, and it must be labeled so that it can be detected. The probe will only base pair with fragments that include a sequence complementary to the probe's sequence. A probe with the sequence 5′ AATCG 3′ will bind to the region of the target DNA that has the sequence 5′ AATCG 3′ as shown below:

```
        5′ AATCG 3′
3′ TCCGGTTAGCATTACCATTTT 5′
```

YSBAT p. 348 Using a map with many markers makes it more likely that there will be one very tightly linked to the gene of interest—meaning that a version of the marker will almost always be associated with the phenotype you are tracking.

Caption Questions and Exercises

Figure 19.4 No—many times, because there were many copies of each type of mRNA present in the pituitary cells and many pituitary cells were used to prepare the library. **Figure 19.7** The polymerase will begin at the 3′ end of each primer. On the top strand in part (b), it will move to the left; on the bottom strand it will move to the right. As always, synthesis is in the 5′ to 3′ direction. **Figure 19.10** The insertion will probably disrupt the gene and have serious consequences for the cell and potentially the individual.

Summary of Key Concepts

KC 19.1 When "sticky ends" anneal, the recombinant product would not be stable—it would tend to break apart because no phosphodiester bond would form between the DNA fragments. **KC 19.2** PCR advantages: It is relatively fast and easy, and can amplify a DNA sequence that is rare in the sample. PCR disadvantage: It requires knowledge of sequences on either side of the target gene, so primers can be designed. Plasmid advantages: No knowledge of the sequence is required. Plasmid disadvantages: It is slower and technically more difficult than PCR. **KC 19.3** The length of each fragment is dictated by where a ddNTP was incorporated into the growing strand, and each ddNTP corresponds to a base on the template strand. Thus, the sequence of fragment sizes corresponds to the sequence on the template DNA. **KC 19.4** Both affected and unaffected individuals have the same marker, so you have no way of knowing that it is close to the gene you are interested in. **KC 19.5** Genes can be inserted into the Ti plasmids, and the recombinant plasmids transferred to plant cells.

Test Your Knowledge

1. c; **2.** b; **3.** d; **4.** a; **5.** b; **6.** d

Test Your Understanding

1. When a restriction endonuclease cuts a "foreign gene" sequence and a plasmid, the same sticky ends are created on the excised foreign gene and the cut plasmid. After the sticky ends on the foreign gene and the plasmid anneal, DNA ligase catalyzes closure of the DNA backbone, sealing the foreign gene into the plasmid DNA. [See Figure A19.2] **2.** Vectors hold a piece of foreign DNA and carry it into an intact host cell. A "perfect" vector (1) has well-characterized restriction endonuclease sites, (2) can be inserted into host cells easily, (3) expresses recombinant genes in the host cell in a controlled way, and (4) doesn't damage the host cell. **3.** A cDNA library is a collection of complementary DNAs made from all the mRNAs present in a certain group of cells. A cDNA library from a human nerve cell would be different from one made from a human muscle cell, because nerve cells and muscle cells express many different, cell type–specific genes. **4.** Genetic markers are genes or other loci that have known locations in the genome. When these locations are diagrammed, they represent the physical relationships among landmarks—in other words, they form a map. **5.** The genes for making β-carotene need to be expressed in the endosperm. Adding an endosperm-specific promoter adjacent to the genes ensures that they are expressed in the endosperm. In general, including specific promoter and enhancer sequences with recombinant genes helps researchers control where and when the recombinant genes are expressed. **6.** PCR and cellular DNA synthesis are similar in the sense of producing copies of a template DNA. Both rely on primers and DNA polymerase. The major difference between the two is that PCR copies only a specific target sequence, while the entire genome is copied during cellular DNA synthesis.

Applying Concepts to New Situations

1. You could use a computer program to identify likely promoter sequences in the sequence data, and then look for sequences just downstream that have an AUG start codon and a stop codon about 1500 base pairs apart. **2.** Yes—you would transform cells that are the precursors to sperm or eggs, so that the recombinant DNA is found in mature sperm or eggs and passed on to offspring. **3.** In both techniques, researchers use an indicator to identify either a gene of interest or a colony of bacteria with a particular trait. The problem is the same—picking one particular thing (a certain gene or a cell with a particular mutation) out of a large collection. **4.** (a) Primer 1b binds to the top right strand and would allow DNA polymerase to synthesize the top strand across the target gene. Primer 1a, however, binds to the upper left strand and would allow DNA polymerase to synthesize the upper strand *away* from the target gene. Primer 2a binds to the bottom left strand and would allow DNA polymerase to synthesize the bottom strand across the target gene. Primer 2b, however, binds to the bottom right strand and would allow DNA polymerase to synthesize away from the target gene. (b) She could use primer 1b with primer 2a.

CHAPTER 20

Check Your Understanding (CYU)

CYU p. 364 (1) Parasites don't need genes that code for enzymes required to synthesize molecules they acquire from their hosts. (2) If two closely related species inherited the same gene from a common ancestor, the genes should be similar. But if one species acquired the same gene from a distantly related species via lateral gene transfer, then the genes should be much less similar. **CYU p. 370** (1) Unequal crossover and mistakes in DNA synthesis are so common in simple tandem repeats that new alleles—consisting of the same chromosome region with different repeat numbers—are created almost every generation, producing a molecular fingerprint. (2) Unequal crossover occurs when homologous chromosomes

---

FIGURE A19.1

```
5′ CATGACTATTACGTATCGGGTACTATGCTATCGATCTAGCTACGCTAGCT 3′
3′ GTACTGATAATGCATAGCCCATGATACGATAGCTAGATCGATGCGATCGA 5′
```

Probe #1, which will anneal to the 3′ end of the top strand:

```
5′ AGCTAGCGTAGCTAGATCGAT 3′
```

Probe #2, which will anneal to the 3′ end of the bottom strand:

```
5′ CATGACTATTACGTATCGGGG 3′
```

The primers will bind to the separated strands of the parent DNA sequence as follows:

```
5′ CATGACTATTACGTATCGGGTACTATGCTATCGATCTAGCTACGCTAGCT 3′
                               3′ TAGCTAGATCGATGCGATCGA 5′
```

```
5′ CATGACTATTACGTATCGGG 3′
3′ GTACTGATAATGCATAGCCCATGATACGATAGCTAGATCGATGCGATCGA 5′
```

FIGURE A19.2

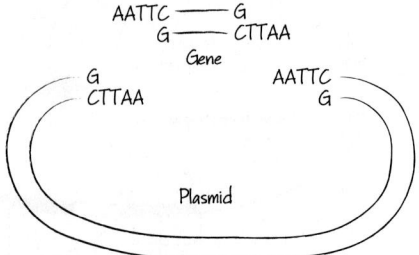

Sticky ends on cut plasmid and gene to be inserted are complementary and can base pair

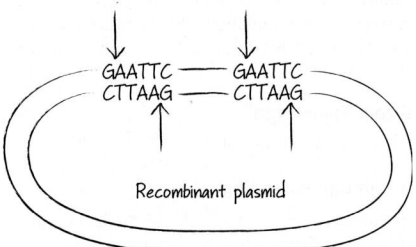

DNA ligase forms four phosphodiester bonds between gene and plasmid DNA at the points indicated by arrows

align incorrectly during synapsis. When crossing over occurs within a misaligned section of chromosome, one homolog ends up with less DNA and the other ends up with more DNA. The extra DNA is made up of sequences from the other chromosome that are duplicates of the DNA on the original strand.

You Should Be Able To (YSBAT)

YSBAT p. 360 If no overlap occurred, there would be no way of ordering the fragments correctly. You would only be able to put fragments in random order—not the correct order. **YSBAT p. 371** Start with a microarray containing exons from a large number of human genes. Isolate mRNAs from brain tissue and liver tissue, and make labeled cDNAs from each. Probe the microarray with each type of probe, and record where binding occurs. Binding events identify exons that are transcribed in each type of tissue. Compare the results to identify genes that are expressed in brain but not liver, or liver but not brain.

Caption Questions and Exercises

Figure 20.2 Shotgun sequencing is based on fragmenting the genome into many small pieces. **Figure 20.3** Because the mRNA codons that match up with each strand are oriented in the 5′ to 3′ direction. **Figure 20.8**
Chromosome 1:
ψβ2-ε-Gγ-Aγ-ψβ1-δ-ψβ2-ε-Gγ-Aγ-ψβ1-δ-β
Chromosome 2:
β

Summary of Key Concepts

KC 20.1 If a search of human gene sequence databases revealed a gene that was similar in base sequence, and if follow-up work confirmed that the mouse and human genes were similar in their pattern of exons and introns and regulatory sequences, then the researchers could claim that they are homologous. **KC 20.2** DNA can move from one prokaryotic species to another packaged in plasmids or viruses, or by the direct uptake of DNA fragments from the environment (transformation). **KC 20.3** Junk DNA refers to noncoding regions of DNA—sequences that are not transcribed into RNA. The function of repeated sequences is not yet clear, but many noncoding sequences are interesting because they are relics from transposable elements. **KC 20.4** Expression of the gene may be involved in the development of cancer.

Test Your Knowledge

1. c; **2.** a; **3.** d; **4.** b; **5.** d; **6.** d

Test Your Understanding

1. Computer programs are used to scan sequences in both directions to find ATG start codons, a gene-sized logical sequence with recognizable codons, and then a stop codon. One can also look for characteristic promoter, operator, and other regulatory sites. It is more difficult to identify open reading frames in eukaryotes because their genomes are so much larger and because of the presence of introns and repeated sequences. **2.** To have the potential to thrive in different environments, a cell requires a large array of enzymes and proteins—to use different sources of food, cope with different temperature or pH conditions and so on—and thus a large array of genes. **3.** They use the transcription and translation machinery of the host cell to copy themselves and insert themselves into the host's genome, and this may have a negative effect on the host cell. **4.** Each individual has a DNA fingerprint—a unique collection of microsatellite and minisatellite alleles (length variants). Because alleles are inherited, closely related individuals share more alleles than more distantly related individuals. **5.** A DNA microarray experiment identifies which genes are being expressed in a particular cell at a particular time. If a series of experiments shows that different genes are expressed in cells at different times or under

different conditions, it implies that expression was turned on or off in response to changes in age or changes in conditions. **6.** Homology is a similarity among different species that is due to their inheritance from a common ancestor. If a newly sequenced gene is found to be homologous with a known gene of a different species, it is assumed that the gene products have similar function. Duplicated genes are homologous because they share a common ancestor gene.

Applying Concepts to New Situations

1. If "gene A" is not necessary for existence, it can be lost by an event like unequal crossing over (on the chromosome with deleted segments) with no ill effects on the organism. In fact, individuals who have lost unnecessary genes are probably at a competitive advantage, because they no longer have to spend time and energy copying and repairing unused genes. **2.** If the grave was authentic, it might include two very different parental patterns along with three children whose patterns each represented a mix between the two parents. The other unrelated individuals would have patterns not shared by anyone else in the grave. **3.** The genes may have similar DNA sequences because the viral gene was incorporated into the genome of a human host in the past, and subsequently passed on to future generations. **4.** You would expect that the livers and blood of chimps and humans would function similarly, but that strong differences occur in brain function. The microarray data support this prediction and suggest that even though the brain proteins might have similar sequences, chimp and human brains are different because certain genes are turned on or off at different times and expressed in different amounts.

CHAPTER 21

Check Your Understanding (CYU)

CYU p. 379 Biologists have been able to grow entire plants from a single differentiated cell taken from an adult. They have also succeeded at producing animals by transferring the nucleus of a fully differentiated cell to an egg whose nucleus has been removed. **CYU p. 384** (1) Bicoid is a transcription factor, so cells that experience a high versus medium versus low concentration transcribe different amounts of Bicoid-regulated genes. (2) Bicoid proteins control the expression of gap genes, which—in effect—tell cells which third of the embryo they are in along the anterior-posterior axis. Gap genes control the expression of pair-rule genes, which organize the embryo into individual segments. Pair-rule genes control expression of segment polarity genes, which establish an anterior-posterior polarity to each individual segment. Segment polarity genes turn on homeotic genes, which then trigger genes for producing segment-specific structures like wings or legs.

Caption Questions and Exercises

Figure 21.2 It would look like the embryo of the left side of part (b)—a normal phenotype. **Figure 21.5** In cells that are expressing a particular gene, the gene is transcribed into RNA. In situ hybridization locates these copies of RNA, identifying which cells are expressing the gene. **Figure 21.10** Genes are usually considered homologous if they have similar structure—meaning similar DNA sequence and exon-intron structure.

Summary of Key Concepts

KC 21.1 The seedling would grow very slowly and be small, but development should proceed normally otherwise. **KC 21.2** Parenchyma cells in the stem still contain a complete set of genes, so if they can be de-differentiated and received appropriate signals from other cells, they could start producing vascular-tissue-specific proteins. **KC 21.3** Different concentrations of transcription fac-

tors or cell-cell signals will cause differences in the amount of expression from certain genes—so a high concentration means something different than a low concentration. **KC 21.4** The idea is that a single molecule can set up the anterior-posterior or apical-basal axis of the embryo. Once the axis is established, a sequence of events follows in which cells get increasingly more precise information about where they are located along that axis—based on the original concentration of the master regulator. **KC 21.5** When the patterns of expression of two key genes involved in limb versus rib formation are compared in chick and snake embryos, they differ. In snakes, the "form ribs here" pattern occurs instead of the "form legs here" pattern.

Test Your Knowledge

1. b; **2.** d; **3.** a; **4.** c; **5.** a; **6.** a

Test Your Understanding

1. They occur in both plants and animals, and are responsible for the changes that occur as an embryo develops. **2.** If the transplanted nucleus has undergone some permanent change or loss of genetic information during development, it should not be able to direct the development of a viable adult. But if the transplanted nucleus is genetically equivalent to the nucleus of a fertilized egg, then it should be capable of directing the development of a new individual. **3.** The researchers exposed adult flies to treatments that induce mutations, then looked for embryos with defects in the anterior-posterior body axis or body segmentation. The embryos had mutations in genes required for body axis formation and segmentation. **4.** Development—specifically differentiation—depends on changes in gene expression. Changes in gene expression depend on differences in regulatory transcription factors. **5.** Differentiation is triggered by the presence or absence of external signals. These signals trigger the production of transcription factors, which induce other transcription factors, and so on—a sequence that constitutes a regulatory cascade—as development progresses. At each step in the cascade, a new subset of genes is activated—resulting in a step-by-step progression from undifferentiated to fully differentiated cells. **6.** *Hox* genes were first discovered in *Drosophila* and have been identified in nearly every animal. The number of *Hox* genes varies across species, but their arrangement on the chromosome and pattern of expression in the embryo are strikingly similar across animal species.

Applying Concepts to New Situations

1. There would be more Bicoid protein farther toward the posterior and less in the anterior—meaning that the anterior segments would be "less anterior" in their characteristics and the posterior segments would be "more anterior." **2.** The result means that the human gene can be transcribed in worms and that its protein product has performed the same function in worms as in humans. This is strong evidence that the genes are homologous, and that their function has been evolutionarily conserved. **3.** The stem cells would have to be stimulated with nerve-specific signals to trigger their differentiation into nerve cells. **4.** Homeotic genes—such as *Hox* genes—are responsible for triggering the production of structures like wings or legs in particular segments. One hypothesis to explain the variation is that changes in gene expression led to different numbers of segments that express leg-forming *Hox* genes.

CHAPTER 22

Check Your Understanding (CYU)

CYU p. 392 (1) If a mutation in either bindin or the egg-cell receptor for sperm prevented the proteins from binding to each other, then fusion of sperm and egg-cell membranes could not take place and fertilization would not occur. (2) The entry of a sperm triggers a wave of calcium ion release around the egg-cell membrane. Thus, calcium ions act as a signal that indicates "A sperm has entered the egg." In response to the increase in calcium ion concentration, cortical granules fuse with the egg-cell membrane and release their contents to the exterior, causing the fertilization envelope to develop. **CYU p. 394** If cytoplasmic determinants were distributed in equal concentration throughout the egg, all blastomeres would end up with the same types and concentrations of cytoplasmic determinants. **CYU p. 395** (1) Endoderm is in the interior, ectoderm is on the outside, mesoderm is in between. (2) These cells become endoderm that forms the gut lining. The direction of the gut helps define the anterior-posterior axis of the body—it connects mouth and anus.

You Should Be Able To (YSBAT)

YSBAT p. 391 If the bindin-like protein were blocked, it would not be able to bind to the egg-cell receptor for sperm. Fertilization would not take place.

Caption Questions and Exercises

Figure 22.4 The researchers didn't start out with a hypothesis about which egg-cell membrane protein bound to bindin. Instead, they tested every type of protein projecting from the egg-cell surface to see which one, if any, bound to bindin. **Figure 22.8 [See Figure A22.1] Figure 22.9** Ectoderm forms most of the structures on the outside of the adult. Endoderm forms mostly interior structures—for example, the inner lining of the respiratory and digestive tracts. Mesoderm forms structures located between the ectoderm-derived and endoderm-derived structures, including bones and muscles and the bulk of most of the organ systems. **Figure 22.15** It made gene expression possible in non-muscle cells. A "general purpose" promoter can lead to transcription of an inserted cDNA in any type of cell, including fibroblasts.

Summary of Key Concepts

KC 22.1 Bindin and the egg-cell receptor for sperm bind to each other. For this to happen, their tertiary structures must fit together somewhat like the structures of a lock and key. **KC 22.2** If the master regulator were localized to a certain part of the egg cytoplasm or formed a Bicoid-like concentration gradient throughout the eggs,

blastomeres would contain or not contain the regulator or contain a distinct concentration of it. **KC 22.3** Both cells would contain cell-cell signals and transcription factors specific to mesodermal cells, but the anterior cell would contain anterior-specific signals and regulatory transcription factors while the posterior cell would contain posterior-specific signals and regulators. **KC 22.4** Cell-cell signals from the notochord direct the formation of the neural tube; subsequently, cell-cell signals from the notochord, neural tube, and ectoderm direct the differentiation of somite cells. Notochord cells die later in development. Cells expand during neural tube formation. Somite cells move to new positions, proliferate, and differentiate.

Test Your Knowledge

1. d; **2.** c; **3.** c; **4.** d; **5.** c; **6.** a

Test Your Understanding

1. Eggs contain stores of nutrients. **2.** If more than one sperm nucleus entered the egg, the resulting cell would contain more than two copies of each chromosome. Mitosis could not occur correctly and the embryo would be deformed or die. **3.** They contain different types and/or concentrations of cytoplasmic determinants. **4.** They give rise to all of the tissue and organs of the adult (from the Latin *germen*, meaning shoot or sprout). **5.** When transplanted early in development, somite cells become the cell type associated with their new location. But when transplanted later in development, somite cells become the cell type associated with the original location. These observations indicate that the same cell is not committed to a particular fate until later in development. **6.** If the gene for MyoD is expressed in non-muscle cells, the cells begin producing muscle-specific proteins. It triggers differential gene expression typical of muscle cells.

Applying Concepts to New Situations

1. The interactions that allow proteins to bind to each other are extremely specific—certain amino acids have to line up in certain positions with specific charges and/or chemical groups exposed. Sperm-egg binding is species specific because each form of bindin binds to a different egg cell receptor for sperm. **2.** Translation is not needed for cleavage to occur normally—meaning that all the proteins needed for early cleavage are present in the egg. **3.** The yellow pigment is a cytoplasmic determinant that triggers a gene regulatory cascade leading to differentiation of muscle cells. **4.** The embryo cannot yet feed, so all of its activity is fueled by the nutrient stores in the

egg. It is not taking in a lot of new molecules that would allow it to get bigger.

CHAPTER 23

Check Your Understanding (CYU)

CYU p. 404 (1) In the male reproductive organs of a mature flower, cells undergo meiosis to produce haploid cells. Each of these haploid cells divides by mitosis to produce a pollen grain containing haploid cells. One of the cells in the pollen grain divides again by mitosis to form two sperm cells. (2) They ensure that the plant is fertilized by a pollen grain from the same species. If any pollen could germinate on any stigma, pollen from a pine tree could grow on the stigma of an orchid. **CYU p. 406** (1) Cells located on the outside of the embryo become epidermal tissue; cells in the interior become vascular tissue; cells in between become ground tissue. (2) The *MONOPTEROS* gene product would be over-activated. The embryonic cells would get information indicating that they are in the shoot apex, where MONOPTEROS levels are high. As a result, the root would not develop normally. **CYU p. 409 [See Figure A23.1]** If a new supply of fertilizer (cow dung) appeared near the plant's left side, the roots on that side would extend toward the nutrients. The shoot system should grow in response to the new nutrients but, all other things being equal (e.g., no shading of plant from sunlight), shoot growth should be symmetrical. **CYU p. 411** (1) The A protein might function as a transcriptional repressor by binding to a regulatory site near the *C* gene, thereby preventing transcription of the *C* gene. (2) Like *Hox* genes, *MADS-box* genes code for DNA-binding proteins that function as transcriptional regulators. Both Hox proteins and MADS-box proteins are involved in regulatory transcription cascades that lead to development of specific structures in specific locations. Although they perform similar functions, *Hox* genes and *MADS-box* genes evolved independently and differ in their base sequences. Thus, these genes and their protein products are not homologous.

You Should Be Able To (YSBAT)

YSBAT p. 410 If four genes coded for flower structure and each gene specified one of the four organs, then each mutant should lack an element in only one of the whorls. If each of three genes is expressed in two adjacent whorls, then four different gene expression patterns are possible—one for each of the four different whorls.

FIGURE A22.1

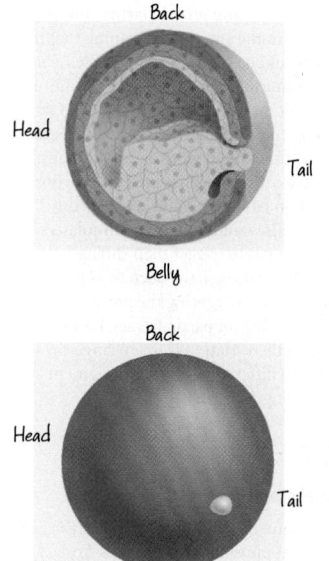

FIGURE A23.1

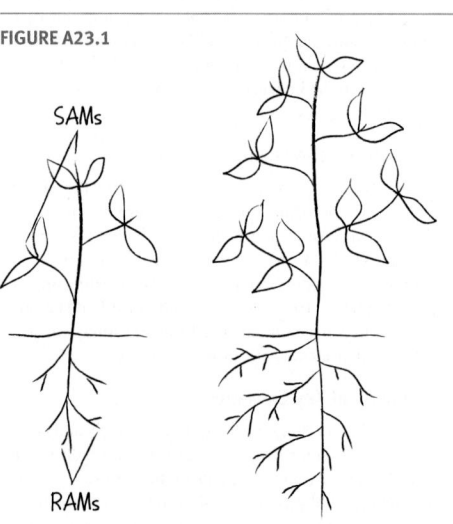

FIGURE A23.2

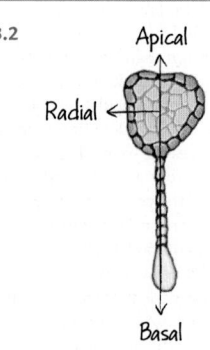

Caption Questions and Exercises

Figure 23.3 In both cases, protein-protein interactions between male and female structures in two different individuals prevent fertilization between members of different species. **Figure 23.5** [See Figure A23.2] **Figure 23.9** Many animals have three body axes analogous to the three axes of leaves: anterior-posterior (like the leaf's proximal-distal), dorsal-ventral (like the leaf's adaxial-abaxial), and a left-right axis (like the leaf's lateral axis). The main plant body, in contrast, has only two axes: apical-basal and radial.

Summary of Key Concepts

KC 23.1 Bicoid defines the anterior-posterior axis of the fly embryo; auxin defines the apical-basal axis of the plant embryo. Both work via concentration gradients. Bicoid is a regulatory transcription factor while auxin is a cell-cell signal. **KC 23.2** There are two fusion events between sperm and haploid cells in the female reproductive structure: one sperm fuses with the egg and leads to the development of an embryo; the other sperm fuses with two (or more) nuclei and leads to the development of endosperm. **KC 23.3** [See Figure A23.3] **KC 23.4** The presence of multiple SAMs and RAMs provides plants with flexibility in adapting to environmental changes and acquiring sunlight and nutrients, thereby increasing their ability to grow and reproduce. For example, multiple SAMs allow a plant to extend shoots (branches) in multiple directions to receive optimal sunlight, and multiple RAMs allow it to extend roots in multiple directions to acquire optimal water and nutrients. Also, a plant would be in trouble if its only SAM or RAM got eaten. **KC 23.5** Homeotic mutants are defined by a phenotype in which one body structure is replaced by another. *Arabidopsis* plants with a defective *B* gene have flowers in which petals are replaced by sepals and stamens are replaced by carpels.

Test Your Knowledge

1. a; **2.** b; **3.** d; **4.** d; **5.** c; **6.** a

Test Your Understanding

1. Proteins on the surface of a pollen grain interact with proteins on the surface of the stigma. Signals from the egg direct the growth of the pollen tube. **2.** Animal cells that lead to sperm or egg formation are "sequestered" early in development, so that they are involved only in reproduction and undergo few rounds of mitosis. Because mutations occur in every round of mitosis, plant eggs and sperm should contain many more mutations than animal eggs and sperm. **3.** Both meristems and stem cells can produce differentiated cells in an adult. Meristems are much more flexible than stem cells, as they can form all parts of the plant, and they increase the size of the body. Animal stem cells can usually develop into only a few differentiated cell types, and only replace lost or damaged cells. **4.** Just like the three tissue layers in plant embryos, the tissues produced in the SAMs and RAMs of a 300-year-old oak tree can differentiate into all of the specialized cell types found in a mature plant. **5.** The ABC model predicted that the *A*, *B*, and *C* genes would each be expressed in specific whorls in the developing flower. The experimental data showed that the prediction was correct. **6.** Cell proliferation is responsible for expansion of meristems and growth in size. Cell death occurs when leaves drop, and when three or four meiotic products die during gametogenesis in female tissues. Asymmetrical cell proliferation along with cell expansion is responsible for giving the adult plant a particular shape. Differentiation is responsible for generating functional tissues and organs in the adult plant. Cell-cell signals are responsible for setting up the apical-basal axis of the embryo and for guiding the pollen tube to the egg.

Applying Concepts to New Situations

1. If an individual is likely to die, it should throw all of its remaining resources into reproduction. **2.** Because vegetative growth is continuous and occurs in all directions (unrestricted), it could be considered indeterminate. However, reproductive growth produces mature reproductive organs and stops when those organs are complete. Because its duration is limited, it could be considered determinant. **3.** High light conditions could trigger changes in gene expression that reduced the rate of cell proliferation during leaf development. In contrast, low light conditions might trigger changes in gene expression that increase the rate of cell proliferation. **4.** [See Figure A23.4] In a young oak tree, SAMs would be evenly distributed in the upper parts of the plant and the RAMs would be evenly distributed in the tips of the roots. But after 50 years in the situation described above, the RAMs would be concentrated on the right side of the plant's roots, enabling the roots to grow into the area where water is available from the leaky pipe next to the building. The SAMs would be concentrated on the left side of the plant, allowing the upper parts of the tree to grow away from billboard, where light is available.

CHAPTER 24

Check Your Understanding (CYU)

CYU p. 429 (1) *Postulate 1:* Traits vary within a population. *Postulate 2:* Some of the trait variation is heritable. *Postulate 3:* There is variation in reproductive success (some individuals produce more offspring than others). *Postulate 4:* Individuals with certain heritable traits produce the most offspring. The first two postulates describe heritable variation; the second two describe differential reproductive success. (2) Beak size and shape and body size vary among individual finches, in part because of differences in their genotypes (some alleles lead to larger or narrower beaks, for example). When a drought hit, individuals with deep beaks survived better and produced more offspring than individuals with shallow beaks. **CYU p. 432** (1) When certain individuals are selected, their traits do not change—they simply produce more offspring than other individuals. (2) Adaptations are not optimal solutions to challenges posed by a particular environment because they are compromised by (1) the necessity of meeting many challenges at the same time (an adaptive "solution" to one problem—such as flying faster—may make another problem worse—such as being maneuverable in flight); (2) lack of "optimal" alleles or presence of alleles that af-

fect more than one trait; and (3) the necessity of selecting only preexisting variation in traits.

You Should Be Able To (YSBAT)

YSBAT p. 425 (1) Relapse occurred because the few bacteria remaining after drug therapy were not eliminated by the patient's weakened immune system and began to reproduce quickly. (2) No—almost all of the cells present at the start of the infection would have been resistant to the drug. **YSBAT p. 430** In biology, an adaptation is any heritable trait that increases an individual's ability to produce offspring in a particular environment. In everyday English, adaptation is often used to refer to an individual's nonheritable adjustment to meet an environmental challenge, a phenomenon that biologists call acclimation. The phenotypic changes resulting from acclimation are not passed on to offspring.

Caption Questions and Exercises

Figure 24.4 The theory of special creation would claim that the fossil and extant sloths were both created 6000 years ago, but that the fossil species became extinct during the flood in Noah's time, described in the Bible. It is not clear how the theory of special creation would explain transitional features observed in the fossil record. **Figure 24.5** If vestigial traits result from inheritance of acquired characteristics, some individuals must have lost the traits during their own lifetimes and passed the reduced traits on to their offspring. For example, a certain monkey's long tail might have been bitten off by a predator, or an ape's hair might have been pulled out of its skin by a rival during a fight. The new traits would then somehow have passed to the individuals' eggs and sperm, resulting in shorter-tailed and less-hairy offspring, until humans with a coccyx and goose bumps resulted. **Figure 24.16** "Prediction": Beak measurements were different before and after the drought. "Prediction of null hypothesis": No difference in beak measurements before and after the drought.

Summary of Key Concepts

KC 24.1 Under the theory of special creation, changes in *Mycobacterium* populations would be explained as individual creative events governed by an intelligent creator. Under the theory of evolution by inheritance of acquired characters, changes in *Mycobacterium* populations would be explained by the cells trying to transcribe genes in the presence of the drug, and their *rpoB* gene becoming altered as a result. **KC 24.2** In biology, fitness is the ability of an individual to produce offspring, relative to that ability in other individuals in the population.

FIGURE A23.3

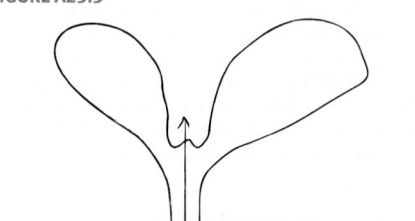
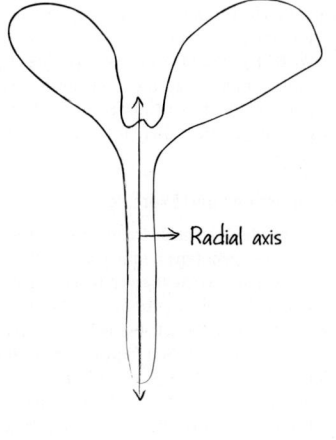

Radial axis

FIGURE A23.4

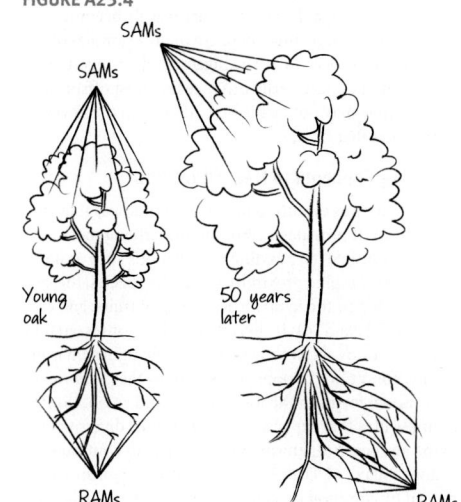

SAMs

SAMs

Young oak

50 years later

RAMs

RAMs

In everyday English, fitness is a physical attribute that is acquired as a result of practice or exercise. **KC 24.3** Brain size in *H. sapiens* might be constrained by the need for babies to pass through the mother's birth canal, by the energy required to maintain a large brain as an adult, or by lack of genetic variation for even larger brain size. Flying speed might be constrained by loss of maneuverability (and thus less success in hunting), the energy demands of extremely rapid flight, or lack of genetic variation for even faster flight.

Test Your Knowledge

1. b; **2.** d; **3.** a; **4.** d; **5.** a; **6.** c

Test Your Understanding

1. The theory of evolution by natural selection predicts that when a population's environment changes, individuals with certain traits will produce the most offspring, and the frequency of those traits will increase in the population. These changes should accumulate over long periods of time, eventually resulting in the formation of a new species that is geographically close to its ancestral species, and a fossil record should reflect changes in traits over time. The theory of special creation predicts that species do not change over time (or do so only due to divine intervention). Finally, the theory of acquired characters predicts that changes that occur in individuals in response to the environment are passed on to the next generation. **2.** Mutation produces new genetic variations, at random, without any forethought as to which variations might prove adaptive in the future. Individuals with mutations that are disadvantageous won't produce many offspring, but individuals with beneficial mutations will produce many offspring. **3.** The evidence for within-patient evolution is that DNA sequences at the start and end of treatment were identical except for a single nucleotide change in the *rpo* gene. If rifampin were banned, it is likely that *rpoB* mutant strains would have had lower fitness in the drug-free environment and would not continue to increase in frequency in *M. tuberculosis* populations. **4.** Typological thinking is based on the idea that species are unchanging and that any differences between individuals within a species are unimportant. Population thinking treats variation among individuals as critically important. That variation is what natural selection acts on, or selects. Population thinking was a radical break with typological thinking. **5.** Yes—the predictions made by a theory can be tested with indirect evidence in convincing ways. For example, if evolutionary theory predicts that certain transitional fossils should be found in rocks of a certain age, then researchers can search in those rocks and see if those fossils are indeed found. New groups of species found on island chains should be each others' closest relatives. And so on. **6.** Organisms do not necessarily become more complex over time; there are many examples of traits that have been lost. The biggest and strongest organisms do not necessarily produce the most offspring because they may not be the best adapted to the environment at that time.

Applying Concepts to New Situations

1. The theory of evolution fits the six criteria as follows: (1) and (2): It provides a common underlying mechanism responsible for puzzling observations such as homology, geographic proximity of similar species, the law of succession in the fossil record, vestigial traits, and extinctions. (3) and (4): It suggests new lines of research to test predictions about the outcome of changing environmental conditions in populations, about the presence of transitional forms in the fossil record, and so on. (5) It is a simple idea that explains the tremendous diversity of living and fossil organisms and why species continue to change today. (6) The realization that all organisms are related by common descent and that none are higher or

lower than others was a surprise. **2.** It is well documented that people who are healthy and well fed grow taller than people who are sick and malnourished. In the absence of data indicating that alleles associated with increased height have increased in frequency recently, it is more logical to hypothesize that the observed change is due to changes in the environment—not evolutionary (genetic) changes. **3.** Compare the sequences of the 20 genes from many samples of preserved human tissue with the sequences of the same genes from a large sample of currently living humans. If evolution has occurred, the frequency of alleles correlated with "tallness" should be significantly greater in living humans than in those who lived a hundred years ago. **4.** The ability to tan is an adaptation because it is passed on from one generation to the next. The tan itself is not passed on, but the ability to tan is passed on; therefore it is heritable and can be classified as an adaptation.

CHAPTER 25

Check Your Understanding (CYU)

CYU p. 440 Given the observed genotype frequencies, the observed allele frequencies are freq(A_1) = 0.574 + ½(0.339) = 0.744; freq(A_2) = ½(0.339) + 0.087 = 0.256. Given these allele frequencies, the genotype frequencies expected under the Hardy-Weinberg principle are A_1A_1: 0.744^2 = 0.554; A_1A_2: 2(0.744 × 0.256) = 0.381; A_2A_2: 0.256^2 = 0.066. There are 4 percent too few heterozygotes observed, relative to the expected proportion. One of the assumptions of the Hardy-Weinberg principle is not met at this gene in this population, at this time. **CYU p. 446** (1) When allele frequencies fluctuate randomly up and down, sooner or later the frequency of an allele will hit 0. That allele thus is lost from the population, and the other allele at that locus is fixed. (2) In small populations, sampling error is large. For example, the accidental death of a few individuals would have a large impact on allele frequencies. **CYU p. 455** (1) Sperm are small and hence relatively cheap to produce, whereas eggs are large and require a large investment of resources to produce. (2) Sperm are inexpensive to produce, so reproductive success for males depends on their ability to find mates—not on their ability to find resources to produce sperm. The opposite pattern holds for females. Sexual selection is based on variation in ability to find mates, so is more intense in males—leading to more exaggerated traits.

You Should Be Able To (YSBAT)

YSBAT p. 444 If allele frequencies are changing due to drift, the populations in Table 25.1 would behave like the simulated populations in Figure 25.6—frequencies would drift up and down over time, and diverge. **YSBAT p. 451** The proportions of homozygotes should increase, and the proportions of heterozygotes should decrease. **YSBAT p. 452** (1) More recessive deleterious alleles are found in homozygotes and eliminated by natural selection. (2) If there are few or no deleterious recessives in a population, there is less inbreeding depression.

Caption Questions and Exercises

Table 25.1 The observed allele frequencies, calculated from the observed genotype frequencies, are 0.43 for *M* and 0.57 for *N*. The expected genotypes, calculated from the observed allele frequencies under the Hardy-Weinberg principle, are 0.185 for *MM*; 0.49 for *MN*; 0.325 for *NN*. **Figure 25.6** [See Figure A25.1] **Figure 25.8** Original population: freq(A_1) = (9 + 9 + 11) / 54 = 0.54; New population: freq(A_1) = (2 + 2 + 1) / 6 = 0.83. The frequency of A_1 has increased dramatically. **Figure 25.11** Yes—the frozen cells are traces of organisms that lived in the past. **Figure 25.13** The difference between number of

flowers in individuals from the two types of matings represents inbreeding depression. This difference increases with age, which means that inbreeding depression increases with age. **Figure 25.16** An allele in males, because a successful male can have as many as 100 offspring—10 times more than a successful female.

Summary of Key Concepts

KC 25.1 There will be an excess of observed genotypes containing the favored allele compared to the proportion expected under Hardy-Weinberg proportions. **KC 25.2** Genetic drift will rapidly reduce genetic variation in small populations. Because captive individuals are usually housed under optimal conditions, selection will not be as intense as it would be in the wild. Alleles that allow individuals to thrive in captivity will increase; alleles that might lower fitness may not be eliminated. **KC 25.3** It increased homozygosity of recessive alleles associated with diseases such as hemophilia. As a result, the royal families were plagued by genetic diseases. **KC 25.4** In this case, reproductive success in females will depend on the number of male mates they can attract. It is reasonable to predict that female red-necked phalaropes are under intense sexual selection for traits that can help them attract mates and are more brightly colored than males.

Test Your Knowledge

1. b; **2.** a; **3.** a; **4.** b; **5.** d; **6.** c

Test Your Understanding

1. Selection may decrease genetic variation or maintain it (e.g., if heterozygotes are favored). Drift reduces it. Gene flow may increase or decrease it, depending on whether immigrants bring new alleles or emigrants remove alleles. Mutation increases it. **2.** Selective pressures often change over time or in different areas occupied by the same species. Even if selective pressures do not vary, mutation continually introduces new alleles. **3.** Sexual selection is most intense on the sex that makes the least investment in the offspring, so that sex tends to have exaggerated traits that make individuals successful in competition for mates. In this way, sexual dimorphism evolves. **4.** The prediction of a null hypothesis states what you should observe if the hypothesis you are testing is not correct. If you are testing the hypothesis that allele frequencies are changing or that mating is nonrandom with respect to a certain gene, the Hardy-Weinberg principle furnishes the appropriate null. **5.** Even if individuals in a small population mate at random or attempt to avoid inbreeding, over time all individuals in a small population are closely related. There are simply not enough nonrelatives to mate with. **6.** Alleles associated with extreme phenotypes are eliminated; alleles associated with intermediate phenotypes increase in frequency.

FIGURE A25.1

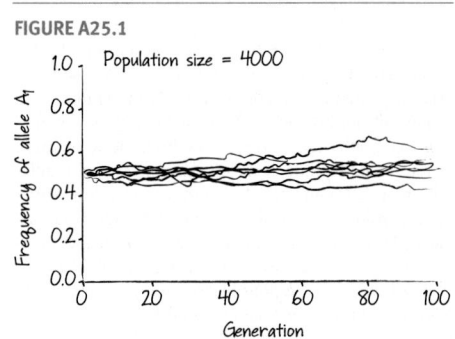

Applying Concepts to New Situations

1. The frequency of heterozygotes is $2p_1p_2$, or $2 \times 0.01 \times 0.99$, which is 0.0198. **2.** Inbreeding causes a decrease in heterozygotes, and a loss of fitness due to the increased numbers of homozygotes for deleterious alleles. Introducing new genetic variants through gene flow will increase allelic diversity in the population and result in more heterozygotes. **3.** If males never lift a finger to help females raise children, the fundamental asymmetry of sex is pronounced and sexual dimorphism should be high. If males invest a great deal in raising offspring, then the fundamental asymmetry of sex is small and sexual dimorphism should be low. **4.** Captive-bred young, transferred adults, and habitat corridors could be introduced to counteract the damaging effects of drift and inbreeding in the small, isolated populations. It is important, though, to introduce individuals only from similar habitats and connect patches of similar habitat with corridors—to avoid introducing alleles that lower fitness.

CHAPTER 26

Check Your Understanding (CYU)

CYU p. 462 (1) Biological species concept: Reproductive isolation means that no gene flow is occurring. Morphological species concept: If populations are evolving independently, they may have evolved morphological differences. Phylogenetic species concept: If populations are evolving independently, they should have synapomorphies that identify them as independent twigs on the tree of life. (2) Biological species concept: It cannot be used to evaluate fossils, species that reproduce asexually, or species that do not occur in the same area and therefore never have the opportunity to mate. Morphological species concept: It cannot identify species that differ in traits other than morphology and is subjective by nature—it can lead to differences of opinion that cannot be resolved by data. Phylogenetic species concept: Reliable phylogenetic information exists only for a small number of organisms. **CYU p. 468** Use one of the cases given in the chapter to illustrate colonization (Galápagos finches), vicariance (shrimp), or habitat specialization (apple maggot flies). In each case, drift will cause allele frequencies to change randomly. Differences in the habitats occupied by the isolated populations will cause allele frequencies to change under natural selection.

You Should Be Able To (YSBAT)

YSBAT p. 465 Selection will favor individuals that (1) breed on apples and have enzymes that are good at digesting apple fruit and that function well as larvae develop in warm temperatures and spend a relatively long time overwintering, and (2) breed on hawthorns and have enzymes that are good at digesting hawthorn fruit and that function well as larvae develop in cool temperatures and spend a relatively short time overwintering. **YSBAT p. 467a** A diploid individual experiences a defect in meiosis resulting in the formation of diploid gametes. The individual self-fertilizes, producing tetraploid offspring. The tetraploid individuals self-fertilize or mate with other tetraploid individuals, producing a tetraploid population. **YSBAT p. 467b** A tetraploid ($4n$) species gives rise to diploid ($2n$) gametes, and a diploid species gives rise to haploid (n) gametes. When a haploid gamete (n) fertilizes a diploid gamete ($2n$), a triploid offspring results. If this individual self-fertilizes, then a population of hexaploid wheat is formed.

Caption Questions and Exercises

Figure 26.3 Save one from the Atlantic Coast and one from the Gulf Coast. Their DNA sequences differ more than do subspecies within a geographic region, so you would be preserving more genetic diversity. **Figure 26.7** It creates genetic isolation—the precondition for divergence. **Figure 26.11** If the same processes were at work in the past as are at work in an experiment with living organisms, then the outcome of the experiment should be a valid replication of the outcome that occurred in the past.

Summary of Key Concepts

KC 26.1 (Many possibilities) Catch a group of fruit flies that are living in a lab, separate the individuals into different cages, and expose the two new populations to dramatically different environmental conditions—heat, lighting, food sources. In essence: create genetic isolation and then conditions for divergence due to drift and selection. **KC 26.2** Human populations would not be considered separate species under the biological concept because human populations can successfully interbreed. They would not be considered separate species under the morphological species concept because all human populations have the same basic morphology. (Although human races differ in minor superficial attributes such as skin color and hair texture, they have virtually identical anatomy and physiology in all other regards.) Nor would human populations be separate species under the phylogenetic species concept because they all arose from a very recent common ancestor. (DNA comparisons have revealed that human races are remarkably similar genetically and do not differ enough genetically to qualify even for subspecies status.) **KC 26.3** The sample experiment described above is an example of vicariance, because the experimenter fragmented the habitat into isolated cages. (If your own experiment differed from the one described above, then a different mechanism of speciation may have been involved.) **KC 26.4** Hybrid individuals will increase in frequency, and natural selection will favor parents that hybridize. The two parent populations may go extinct if hybrids live in the same environment. If the hybrids occupy a different environment, a new species may form.

Test Your Knowledge

1. a; **2.** b; **3.** d; **4.** a; **5.** a; **6.** a

Test Your Understanding

1. Direct observation: Galápagos finch colonization, apple maggot flies, *Tragopogon* allopolyploids, maidenhair ferns. Indirect evidence: snapping shrimp. **2.** On the basis of biological and morphospecies criteria (breeding range and morphological traits), six different species/subspecies of seaside sparrows were recognized. However, based on phylogenetic criteria (comparison of gene sequences), biologists concluded that seaside sparrows represent only two monophyletic groups. **3.** The colonizing population is typically small, and genetic drift has a large effect on smaller populations. If the newly colonized habitat differs from the original one, natural selection will favor individuals with alleles that increase fitness in the new environment. **4.** Some flies breed on apple fruits, others breed on hawthorn fruits. This reduces gene flow. Because apple fruits and hawthorn fruits have different scents and other characteristics, selection is causing the populations to diverge. **5.** Cells that go through many rounds of mitosis often accumulate errors, including mistakes that produce a tetraploid cell. If such a cell becomes part of a developing flower, and goes through meiosis, it will produce diploid gametes that can fuse to form polyploidy offspring. This is less likely to happen in animals because the future reproductive cells undergo fewer rounds of mitosis, so they have fewer chances to become polyploid. **6.** If the species have lived in the same area for a long time, there has been opportunity for selection to favor traits that prevent hybridization. This is reinforcement. If the species do not live in the same area, then there has been no opportunity for selection to favor traits that prevent hybridization.

Applying Concepts to New Situations

1. Decreasing. Gene flow tends to equalize allele frequencies among populations. **2.** These data cast doubt on the hypothesis. If founder events trigger speciation, then at least some speciation events due to introductions should be in progress. **3.** Two things: (1) Some flies happen to have alleles that allow them to respond to apple scents; others happen to have alleles that allow them to respond to hawthorn scents. There is no "need" involved. (2) The alleles for scent response exist; they are not acquired by spending time on the fruit. **4.** If the populations and habitat fragments are small enough, the species is likely to dwindle to extinction due to inbreeding and loss of genetic variation or catastrophes like a severe storm or a disease outbreak. If the populations survive, they are likely to diverge into new species because they are genetically isolated and because the habitats may differ.

CHAPTER 27

Check Your Understanding (CYU)

CYU p. 479 Hair and limb structures in humans and whales are examples of homology because they are traits that can be traced to a common ancestor. All mammals have hair and similar limb bone structure. However, extensive hair loss and advanced social behavior in whales and humans are examples of homoplasy. These traits are not common to all mammalian species and likely arose independently during the evolution of specific mammalian lineages. **CYU p. 488** (1) Doushantuo fossils are microscopic and include sponges and embryos from unknown species. Ediacaran fossils include sponges, jellyfish, and comb jellies along with traces of many other unidentified animals. Most were filter feeders. Burgess Shale fossils include every major animal lineage. The species present are larger, have much more complex morphology, and lived in a wide array of niches. (2) There were many resources available because no other animals (or other types of organisms) existed to exploit them. Also, the evolution of new species in new niches made new niches available for predators. Morphological innovations like limbs and complex mouthparts were important because they made it possible for animals to live in habitats other than the benthic area. **CYU p. 492** (1) Evidence includes the high content of iridium in sedimentary rock formed at the K-T boundary, shocked quartz and microtektite in rock layers that date 65 mya, and a large crater that was found off the coast of Mexico's Yucatán peninsula with microtektite in the crater's walls. Any of these three could be considered convincing, as they are known to form only at impact sites. Taken together, they are particularly convincing. (2) Mass extinctions are caused by such rapid and unusual environmental changes that any species is "lucky" to survive—it is not possible to adapt to such fast changes and such unusual environments.

You Should Be Able To (YSBAT)

YSBAT p. 479 Because whales and hippos share SINEs 4–7 and no other species has these four, they are synapomorphies that define them as a monophyletic group. The similarity is unlikely to be due to homoplasy because the chance of four SINEs inserting in exactly the same place in two different species, independently, is almost astronomically small. **YSBAT p. 485** In habitats on the mainland, there are more competitors and so species are already using the niches (resources) that are available to silverswords in Hawaii and *Anolis* lizards on Caribbean islands.

Figure 27.5 [See Figure A27.1] **Figure 27.16** The dinosaurs went extinct during the end-Cretaceous.

Summary of Key Concepts

KC 27.1 Humans are the only mammal that walks upright on two legs. **KC 27.2** They are the hardest structures in vertebrates and plants, so fossilize most readily. **KC 27.3** They could eat food that was available off the substrate. And once animals lived off the bottom, it created an opportunity for other animals who could eat them to evolve. **KC 27.4** Human-induced extinctions today are not due to poor adaptation to the normal environment. Humans are changing habitats rapidly and in unusual ways, so it is not possible for most species to adapt to these changes rapidly enough to avoid going extinct.

Test Your Knowledge

1. c; **2.** a; **3.** b; **4.** d; **5.** d; **6.** d

Test Your Understanding

1. The fossil record is biased because recent, abundant organisms with hard parts that live underground or in environments where sediments are being deposited are most likely to fossilize. Even so, it is the only data available on what organisms that lived in the past looked like, and where they lived. **2.** Diverse animal forms are found in the Cambrian that are not present in earlier strata. An enormous amount of diversity appeared in a time frame that was short compared to the sweep of earlier Earth history. **3.** Homoplasy is rare relative to homology, so phylogenetic trees that minimize the total change required are usually more accurate. In the case of the artiodactyl astralagus, parsimony was misleading because the astralagus was lost when whales evolved limblessness—creating two changes (a loss following a gain) instead of one (a gain). **4.** (Many answers are possible.) Adaptive radiation of *Anolis* lizards after they colonized new islands in the Caribbean. *Hypothesis:* After lizards arrived on each new island, where there were no predators or competitors, they rapidly diversified to occupy four distinct types of habitats on each island. (Many answers are possible.) Adaptive radiation during the Cambrian period following a morphological innovation. *Hypothesis:* Additional *Hox* genes made it possible to organize a large, complex body; the evolution of complex mouthparts and limbs made it possible for animals to move and find food in new ways. **5.** Environments all over the world deteriorated rapidly—large parts of the ocean lacked oxygen, many coastal habitats disappeared due to change in sea level, and little oxygen was available in the atmosphere. The underlying cause of these changes is not known. **6.** If a trait evolved in a common ancestor, then all of its descendants should share that trait, unless it is lost.

Applying Concepts to New Situations

1. Place the corpse in an environment in which decomposition is slow. One possibility is a swamp or bog; another is a beach or other coastal environment where mud or sand are being deposited. **2.** The fossil record and phylogenetic trees of extinct species indicate that whales evolved from a semiaquatic ancestor. Phylogenetic trees of living species indicate that hippos and whales share a recent common ancestor—meaning that one of the semiaquatic organisms that gave rise to today's whales also gave rise to the semiaquatic species that are today's hippos. **3.** It would be helpful to have (1) a large crater or other physical evidence from a major impact dated at 251 Mya, (2) shocked quartz and microtektites in rock layers dating to the end-Permian era, and (3) high levels of iridium or other elements common in space rocks and rare on Earth, dated to the time of the extinctions. **4.** Oxygen is an effective final electron acceptor in cellular respiration because of its high elec-

tronegativity. Organisms that use it as a final electron acceptor can produce more usable energy than organisms that do not use oxygen, but only if it is available. With more available energy, aerobic organisms can grow larger and move faster.

THE BIG PICTURE: EVOLUTION

Check Your Understanding (CYU), p. 494

1. Circle = inbreeding, sexual selection, natural selection, genetic drift, mutation, and gene flow. **2.** Adaptation "increases" fitness; synapomorphies "identify branches on" the tree of life. **3.** Several answers possible: e.g., the flower in angiosperms, the pharyngeal jaw in cichlid fishes, feathers and flight in birds. **4.** Genetic drift, mutation, and gene flow "are random with respect to" fitness.

CHAPTER 28

Check Your Understanding (CYU)

CYU p. 504 (1) Conditions should mimic a spill—sand or stones with a layer of crude oil or seawater with oil floating on top. Add samples, from sites contaminated with oil, that might contain cells capable of using molecules in oil as electron donors or electron acceptors. Other conditions (temperature, pH, etc.) should be realistic. (2) Use direct sequencing. Choose a well-studied gene, such as 16S RNA, with reliable PCR primers. After isolating DNA from a soil sample, do a PCR reaction and clone the resulting genes into plasmids grown in *E. coli* cells, so that each culture contains a different gene. Once many copies are available, sequence the genes and use the data to place the original organisms on the tree of life. **CYU p. 512** Eukaryotes can only (1) fix carbon via the Calvin-Benson pathway, (2) use aerobic respiration with organic compounds as electron donors, and (3) perform oxygenic photosynthesis. Among bacteria and archaea, there is much more diversity in pathways for carbon fixation, respiration, and photosynthesis, along with many more fermentation pathways.

You Should Be Able To (YSBAT)

YSBAT p. 506 cyanobacteria—photoautotrophs; *Clostridium aceticum*—chemoorganoautotroph; *Nitrosomonas* sp.—chemolithotrophs; heliobacteria—photoheterotrophs; *Escherichia coli*—chemoorganoheterotroph; *Beggiatoa*—chemolithotrophic heterotrophs **YSBAT p. 513a**, **YSBAT p. 513b**, **YSBAT p. 514a**, **YSBAT p. 514b**, **YSBAT p. 515** [See Figure A28.1]

Caption Questions and Exercises

Figure 28.1 Prokaryotic—as it would require just one evolutionary change, the origin of the nuclear envelope in Eukarya. If it were eukaryotic, it would require that the nuclear envelope was lost in both Bacteria and Archaea—two changes and less parsimonious (see Chapter 27). **Table 28.1** [See Figure A28.2] **Figure 28.2** Improved nutrition made people better able to fight off disease, and improved sanitation lowered transmission of disease-causing bacteria. **Figure 28.4** Weak—different culture conditions may have revealed different species. **Figure 28.9** Table 28.5 contains the answers. For example, for organisms called sulfate reducers you would have H_2 as the electron donor, SO_4^{2-} as the electron acceptor, and H_2S as the reduced by-product. For humans the electron donor is glucose ($C_6H_{12}O_6$), the electron acceptor is O_2, and the reduced by-product is water (H_2O). **Table 28.5** Organotrophs use sugars, which are organic compounds, as their electron donor—getting energy by "feeding" on them. Sulfate reducers use sulfate ions as their electron acceptors, thus reducing the sulfate ion. Methanogens generate methane as a by-product. **Figure 28.11** Aerobic respiration. More free energy is released when oxygen is the final electron acceptor than when any other molecule is used, so more ATP can be produced and used for growth.

FIGURE A27.1

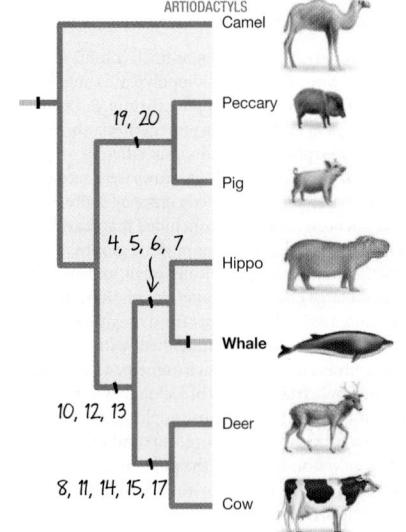

FIGURE A28.1

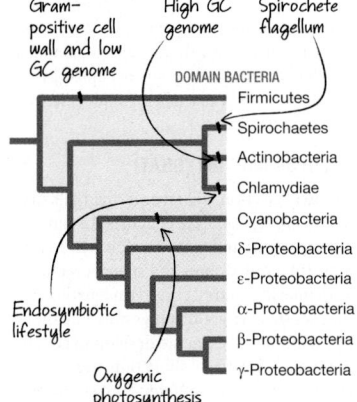

FIGURE A28.2

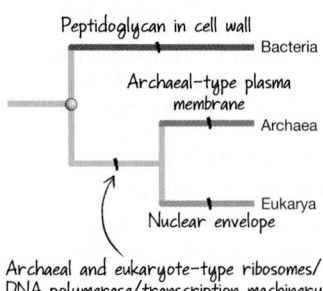

Summary of Key Concepts

KC 28.1 Eukaryotes have a nuclear envelope that encloses their chromosomes; bacteria and archaea do not. Bacteria have cell walls that contain peptidoglycan, and archaea have phospholipids containing isoprene subunits in their plasma membranes. Thus, the exteriors of a bacterium and archaeon are radically different. Archaea and eukaryotes also have similar machinery for processing genetic information. **KC 28.2** They are compatible, and thrive in each others' presence—one species' waste product is the other species' food. **KC 28.3** If bacteria and archaea did not exist, then (1) the atmosphere would have little or no oxygen, and (2) almost all nitrogen would exist in molecular form (the gas N_2).

Test Your Knowledge

1. c; **2.** b; **3.** d; **4.** d; **5.** b; **6.** c

Test Your Understanding

1. An electron donor provides the potential energy required to produce ATP. **2.** Yes. The array of substances that bacteria and archaea can use as electron donors, electron acceptors, and fermentation substrates, along with the diversity of ways that they can fix carbon and perform photosynthesis, allows them to live just about anywhere. **3.** You can sample and amplify genes—getting DNA sequence data that allow you to place unseen species on the tree of life. **4.** Large amounts of potential energy are released and ATP produced when oxygen is the electron acceptor, because oxygen is so electronegative. Large body size and high growth rates are not possible without large amounts of ATP. **5.** [See Figure A28.3] The following can serve as both electron donors and by-products: CH_4, NO_2^-, H_2S. The following can serve as both electron acceptors and by-products: SO_4^{2-}, CO_2, and NO_3^-. **6.** They are paraphyletic because the prokaryotes include some (bacteria and archaea) but not all (eukaryotes) groups derived from the common ancestor of all organisms living today.

Applying Concepts to New Situations

1. This result supports their hypothesis, because the drug poisons the enzymes of the electron transport chain and prevents electron transfer to Fe^{3+}, which is required to drive magnetite synthesis. If magnetite had still formed, another explanation would have been needed. **2.** Hypothesis: A high rate of tooth cavities in Western children is due to an excess of sucrose in the diet, which is absent from the diets of East African children. To test this hypothesis, have East African children switched to a diet that contains sucrose and monitor the presence of *S. mutans*. Have Western children switch to a diet lacking sucrose and monitor the presence of *S. mutans*. **3.** Look in waters or soil polluted with benzene-containing compounds. Put samples from these environments in culture tubes where benzene is the only source of carbon. Monitor the cultures and study the cells that grow efficiently. **4.** They should get energy from reduced organic compounds "stolen" from their hosts.

CHAPTER 29

Check Your Understanding (CYU)

CYU p. 523 (1) *Plasmodium* species are transmitted to humans by mosquitoes. If mosquitoes can be prevented from biting people, they cannot spread the disease. (2) Iron added → primary producers (photosynthetic protists and bacteria) bloom → more carbon dioxide taken up from atmosphere during photosynthesis → consumers bloom, eat primary producers → bodies of primary producers and consumers fall to bottom of ocean → large deposits of carbon-containing compounds form on ocean floor. **CYU p. 526** (1) Opisthokonts have a flagellum at the base or back of the cell; alveolate cells contain unique support structures called alveoli; stramenopiles have straw-like hairs on their flagella. (2) In direct sequencing, DNA is isolated directly from the environment and analyzed to place species on the tree of life. It is not necessary to actually see the species being studied.

You Should Be Able To (YSBAT)

YSBAT p. 526a [See Figure A29.1 on p. A:28] **YSBAT p. 526b** Membrane infoldings observed in bacterial species today support the hypothesis's plausibility—they confirm that the initial steps could have actually happened. The continuity of the nuclear envelope and ER are consistent with the hypothesis, which predicts that the two structures are derived from the same source (infolded membranes). **YSBAT p. 528** A photosynthetic bacterium (e.g., a cyanobacterium) could have been engulfed by a larger eukaryotic cell. If it was not digested, it could continue to photosynthesize and supply sugars to the host cell. **YSBAT p. 531** The chloroplast genes label should come off of the branch that leads to cyanobacteria. (If you had a phylogeny just of the cyanobacteria, the chloroplast branch would be located somewhere inside.) **YSBAT p. 532** The acquisition of the mitochondrion and the chloroplast represent the transfer of entire genomes, and not just single genes, to a new organism. **YSBAT p. 533** Yes—when food is scarce or population density is high, the environment is changing rapidly (deteriorating). Offspring that are genetically unlike their parents may be better able to cope with the new and challenging environment. **YSBAT p. 536a** (1) Alternation of generations refers to a life cycle in which there are multicellular haploid phases and multicellular diploid phases. A gametophyte is the multicellular haploid phase; the sporophyte is the multicellular diploid phase. A spore is a cell that grows into a multicellular individual, but is not produced by fusion of two cells. A zygote is a cell that grows into a multicellular individual, but *is* produced by fusion of two cells (gametes). Gametes are haploid cells that fuse to form a zygote. (2) [See Figure A29.2 on p. A:28] **YSBAT p. 536b, YSBAT p. 536c, YSBAT p. 536d, YSBAT p. 537a, YSBAT p. 537b, YSBAT p. 537c** [See Figure A29.1 on p. A:28] **YSBAT p. 537d** It is most likely that the two types of amoebae evolved independently. The alternative hypothesis is that the common ancestor of alveolates, stramenopiles, rhizarians, plants, opisthokonts, and amoebozoa were amoeboid, and that this growth form was lost many times. [See Figure A29.1 on p. A:28] **YSBAT p. 538a, YSBAT p. 538b** [See Figure A29.1 on p. A:28] **YSBAT p. 539a** If euglenids could take in food via phagocytosis (ingestive feeding), then it would have provided a mechanism by which a smaller photosynthetic protist could have been engulfed and incorporated into the cell via secondary endosymbiosis. **YSBAT p. 539b, YSBAT p. 540a, YSBAT p. 540b, YSBAT p. 541a, YSBAT p. 541b, YSBAT p. 542a, YSBAT p. 542b, YSBAT p. 543** [See Figure A29.1 on p. A:28]

Caption Questions and Exercises

Figure 29.9 Two—one derived from the original bacterium and one derived from the eukaryotic cell that engulfed the bacterium. **Table 29.3** Yes—green algae, euglenids, and chlorarachniophytes all have chlorophyll *a* and *b*, as predicted by the hypothesis that a green algal chloroplast was transferred to the ancestor of euglenids and of chlorarachniophytes. Red algae have chlorophyll *a*, and chromalveolates (which include the brown algae, diatoms, and dinoflagellates) have chlorophyll *a* and *c*. If the hypothesis is correct, chlorophyll *c* must have evolved in an ancestor of the chromalveolates independently of the acquisition of chlorophyll *a* from red algae.

Summary of Key Concepts

KC 29.1 Photosynthetic protists use CO_2 and light to produce sugars and other organic compounds, so they furnish the first or primary source of organic material in an ecosystem. **KC 29.2** (1) Outside membrane was from host eukaryote; inside from engulfed cyanobacterium. (2) From the outside in, the four membranes are derived from the eukaryote that engulfed a chloroplast-containing eukaryote, the plasma membrane of the eukaryote that was engulfed, the outer membrane of the engulfed cell's chloroplast, and the inner membrane of its chloroplast. **KC 29.3** Each set of pigments absorbs most strongly in a certain part of the electromagnetic spectrum. If different species have different pigments, they can live in close proximity and perform photosynthesis without competing for the same wavelengths of light. **KC 29.4** A gametophyte is haploid, and a sporophyte is diploid.

Test Your Knowledge

1. b; **2.** b; **3.** a; **4.** b; **5.** d; **6.** b

FIGURE A28.3

TABLE 28.5 **Some Electron Donors and Acceptors Used by Bacteria and Archaea**

| Electron Donor | Electron Acceptor | By-Products | | Category* |
		From Electron Donor	From Electron Acceptor	
Sugars	O_2	CO_2	H_2O	Organotrophs
H_2 or organic compounds	SO_4^{2-}	H_2O or CO	H_2S or S^{2-}	Sulfate reducers
H_2	CO_2	H_2O	CH_4	Methanogens
CH_4	O_2	CO_2	H_2O	Methanotrophs
S^{2-} or H_2S	O_2	SO_4^{2-}	H_2O	Sulfur bacteria
Organic compounds	Fe^{3+}	CO_2	Fe^{2+}	Iron reducers
NH_3	O_2	NO_2^-	H_2O	Nitrifiers
Organic compounds	NO_3^-	CO_2	N_2O, NO, or N_2	Denitrifiers (or nitrate reducers)
NO_2^-	O_2	NO_3^-	H_2O	Nitrosifiers

*The name biologists use to identify species that use a particular metabolic strategy.

Test Your Understanding

1. An extensive cytoskeleton is required to shape the cell membrane so that the pseudopod can form and surround the food item. A cell wall is not flexible enough to wrap around the food item as the plasma membrane does. **2.** Because all eukaryotes living today have cells with a nuclear envelope, it is valid to infer that their common ancestor also had a nuclear envelope. Because bacteria and archaea do not have a nuclear envelope, it is valid to infer that the trait arose in the common ancestor of eukaryotes. **3.** Meiosis is required for alternation of generations because it converts a diploid phase of the life cycle to a haploid phase. Without it, the generations wouldn't "alternate." **4.** The host cell provided a protected environment and carbon compounds for the endosymbiont; the endosymbiont provided increased ATP from the carbon compounds. **5.** It confirmed a fundamental prediction made by the hypothesis, and could not be explained by any alternative hypothesis. **6.** All alveolates have alveoli, which are unique structures that function in supporting the cell. Among alveolates there

are species that are (1) ingestive feeders, photosynthetic, or parasitic, and (2) move using cilia, flagella, or a type of amoeboid movement.

Applying Concepts to New Situations

1. The observation suggests that eukaryotes did not acquire mitochondria until there was enough oxygen present to make aerobic respiration efficient. (Oxygen is also poisonous to cells at high concentration, so mitochondria gave the early eukaryotes a way to "detoxify" oxygen.) **2.** If the apicoplast that is found in *Plasmodium* (the organism that causes malaria) is genetically similar to chloroplasts, and if glyphosate poisons chloroplasts, it is reasonable to hypothesize that glyphosate will poison the apicoplast and potentially kill the *Plasmodium*. This would be a good treatment strategy for malaria because humans have no chloroplasts, provided that the glyphosate produces no other effects that would be detrimental to humans. **3.** Primary producers usually grow faster when CO_2 concentration increases, but to date, they have not grown fast enough to make CO_2 levels drop—CO_2 levels have been increasing steadily over

decades. **4.** Given that lateral gene transfer can occur at different points in a phylogenetic history, specific genes can become part of a lineage by a different route from that taken by other genes of the organism. In the case of chlorophyll *a*, its history traces back to a bacterium being engulfed by a protist and forming a chloroplast.

CHAPTER 30

Check Your Understanding (CYU)

CYU p. 552 (1) Green algae and land plants share an array of morphological traits that are synapomorphies, including the chlorophylls they contain, (2) green algae appear before land plants in the fossil record, and (3) on phylogenetic trees estimated from DNA sequence data, green algae and land plants share a most recent common ancestor, with green algae being the initial groups to diverge and land plants diverging subsequently. **CYU p. 566** (1) Cuticle prevents water loss from the plant; vascular tissue moves water up from the soil and moves photosynthetic products down to the roots. (2) [See Figure A30.1]

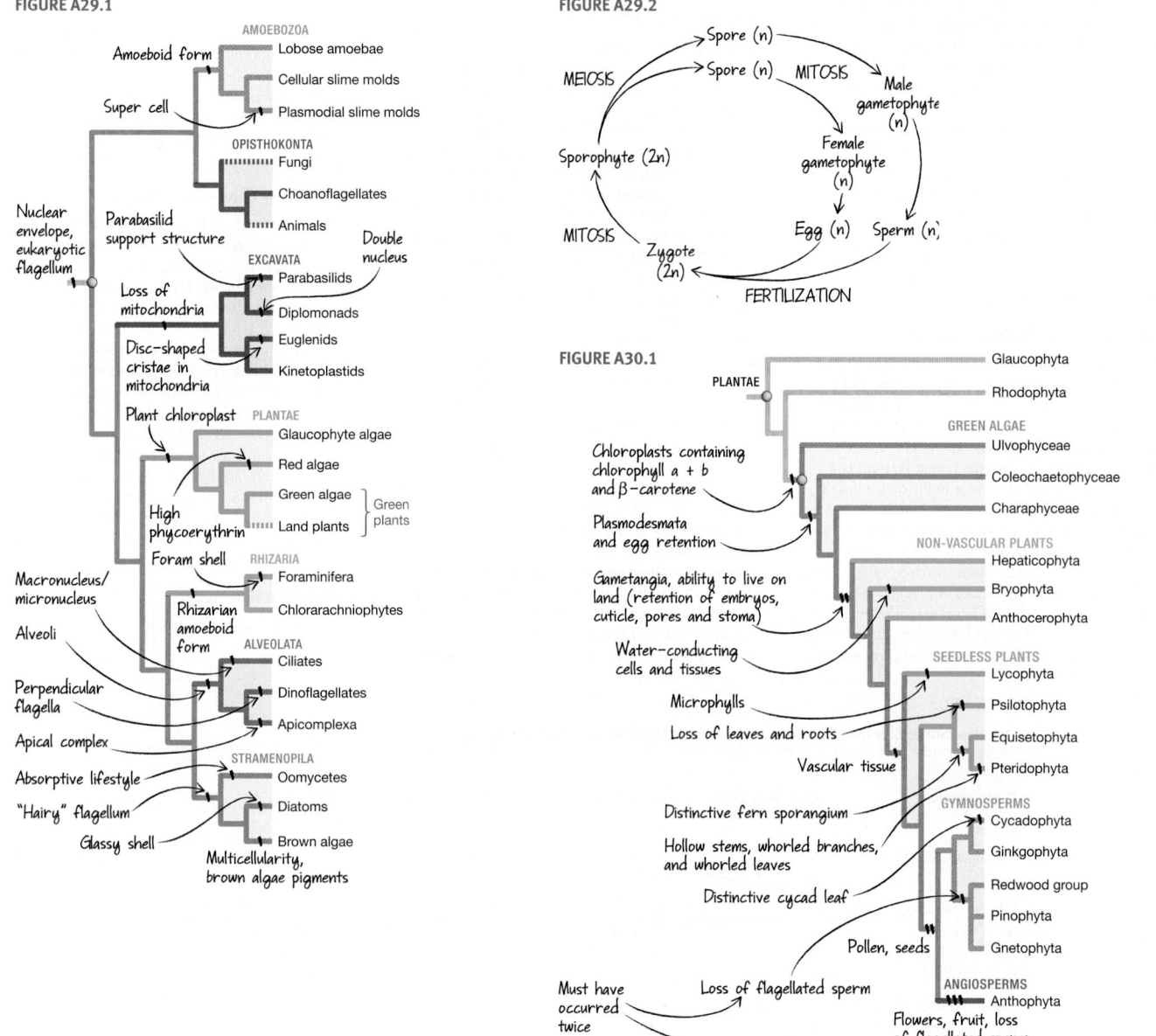

FIGURE A29.1

FIGURE A29.2

FIGURE A30.1

You Should Be Able To (YSBAT)

YSBAT p. 558 Alternation of generations occurs when there are multicellular haploid individuals and multicellular diploid individuals in a life cycle. A sporophyte is a multicellular diploid individual that produces spores by meiosis. A spore is a single cell that is not formed by fusion of two cells, and that grows by mitosis into a multicellular adult. **YSBAT p. 559** In the hornwort photo, the sporophyte is the spike-like green and brown structure; the gametophyte is the leafy-looking structure underneath. The horsetail gametophyte is the microscopic individual on the left; the sporophyte is the much larger individual on the right. **YSBAT p. 569a, YSBAT p. 569b, YSBAT p. 570, YSBAT p. 571, YSBAT p. 572a, YSBAT p. 572b, YSBAT, p. 573, YSBAT p. 574** [See Figure A30.1]

Caption Questions and Exercises

Figure 30.7 If you find the common ancestor of all the green algae (at the base of the Ulvophyceae), the lineages that are collectively called green algae don't include that common ancestor and all of its descendants—only some of its descendants. The same is true for non-vascular plants (the common ancestor here is at the base of the Hepaticophyta) and seedless vascular plants (the common ancestor here is at the base of the Lycophyta). **Figure 30.9** Water flows more easily through a short, wide pipe than through a long, skinny one because there is less resistance from the walls of the pipe. **Figure 30.10** To map an innovation on a phylogenetic tree, biologists determine the location(s) on the tree that is consistent with all descendants from that point having the innovation (unless, in rare cases, the innovation was lost in a few descendants). **Figure 30.15** There are multicellular haploid stages and multicellular diploid stages in these plants. **Figure 30.19** Gymnosperm gametophytes are microscopic, so they are even smaller than fern gametophytes. The gymnosperm gametophyte is completely dependent on the sporophyte for nutrition, while fern gametophytes are not. Fern gametophytes are photosynthetic and even supply nutrition to the young sporophytes. **Figure 30.20** Consistent—the fossil data suggest that gymnosperms evolved earlier, and gymnosperms have larger gametophytes than angiosperms. **Figure 30.22** It tests the hypothesis that the presence of yarn on the spur changes pollinator behavior and thus reproductive success. **Figure 30.24** [See Figure A30.1]

Summary of Key Concepts

KC 30.1 Rates of soil formation will drop; rates of soil loss will increase. **KC 30.2** Because so many adaptations are required—starting with cuticle and pores or stomata—it is unlikely that the transition from water to land would happen more than once. Stated another way, the aquatic and terrestrial environments are so different for plants that the transition was difficult and thus unlikely. **KC 30.3** Both spores and seeds have a tough, protective coat, so can survive while being dispersed to a new location. Seeds have the advantage of carrying a store of nutrients with them—when a spore germinates, it has to make its own food via photosynthesis right away.

Test Your Knowledge

1. c; **2.** b; **3.** d; **4.** c; **5.** c; **6.** a

Test Your Understanding

1. Plants build and hold soils required for human agriculture and forestry, and increase water supplies that humans can use for drinking, irrigation, or industrial use. Plants release oxygen that we breathe. **2.** Cuticle prevents water loss from leaves but also pre-

vents entry of CO_2 required for photosynthesis. Stomata allow CO_2 to diffuse but can close to minimize water loss. Liverwort pores allow gas exchange but cannot be closed if conditions become dry. Liverworts that lack pores have a cuticle that is thin enough to allow some gas exchange. **3.** They provided the support needed for plants to grow upright and not fall over in response to wind or gravity. Erect growth allowed plants to compete for light. **4.** (1) Gametangia are found in all land plant groups except angiosperms; (2) transfer cells are found in all land plants; (3) pollen is found in gymnosperms and angiosperms; (4) seeds are found in gymnosperms and angiosperms, (5) fruit is found in angiosperms. **5.** In a gametophyte-dominant life cycle, the gametophyte is larger and longer lived than the sporophyte and produces most of the nutrition. In a sporophyte-dominant life cycle, the sporophyte generation is the larger, longer-lived, and photosynthetic phase of the life cycle. **6.** Homosporous plants produce a single type of spore that develops into a gametophyte that produces both egg and sperm. Heterosporous plants produce two different types of spores that develop into two different gametophytes that produce either egg or sperm. In a tulip, the microsporangium is found within the stamen, and the megasporangium is found within the ovule. Microspores divide by mitosis to form male gametophytes (pollen grains); megaspores divide by mitosis to form the female gametophyte.

Applying Concepts to New Situations

1. Homosporous plants produce a single type of spore that develops into a gametophyte that produces both egg and sperm. Heterosporous plants produce two different types of spores that develop into two different gametophytes that produce either egg or sperm. In a tulip, the microsporangium is found within the stamen, and the megasporangium is found within the ovule. Microspores divide by mitosis to form male gametophytes (pollen grains); megaspores divide by mitosis to form the female gametophyte. **2.** The combination represents a compromise between efficiency and safety. Tracheids can still transport water if vessels become blocked by air bubbles. **3.** A "reversion" to wind pollination might be favored by natural selection because it is costly to produce a flower that can attract animal pollinators. Because wind-pollinated species grow in dense clusters, they can maximize the chance that the wind will carry pollen from one individual to another (less likely if the individuals are far apart). Wind-pollinated deciduous trees flower in early spring before their developing leaves begin to block the wind. **4.** Alter one characteristic of a flower and present the flower to the normal pollinator. As a control, present the normal (unaltered) flower to the normal pollinator. Record the amount of time the pollinator spends in the flower, the amount of pollen removed, or some other measure of pollination success. Repeat for other altered characteristics. Analyze the data to determine which altered characteristic affects pollination success the most.

CHAPTER 31

Check Your Understanding (CYU)

CYU p. 586 (1) Because they are made up of a network of thin, branching hyphae, mycelia have a large surface area, which makes absorption efficient. (2) Swimming spores and gametes, zygosporangia, basidia, asci. **CYU p. 594** (1) When birch tree seedlings are grown in the presence and absence of EMF, individuals denied their normal EMF are not able to acquire sufficient nitrogen and phosphorus. Isotope-tracing experiments also

show that sugars are transferred from host plant to EMF and that nitrogen and phosphorus are transferred from EMF to host plant. (2) Meiosis and production of haploid spores.

You Should Be Able To (YSBAT)

YSBAT p. 591 Human sperm and egg undergo plasmogamy followed by karyogamy during fertilization, but heterokaryosis does not occur in humans or other eukaryotes besides fungi. **YSBAT p. 594, YSBAT p. 596a, YSBAT p. 596b, YSBAT p. 596c, YSBAT p. 597, YSBAT p. 598a, YSBAT p. 598b** [See Figure A31.1]

Caption Questions and Exercises

Figure 31.10 Labeled-nutrient experiments identify which nutrients are exchanged and in which direction. They explain *why* plants do better in the presence of mycorrhizae, and why the fungus also benefits.

FIGURE A31.1

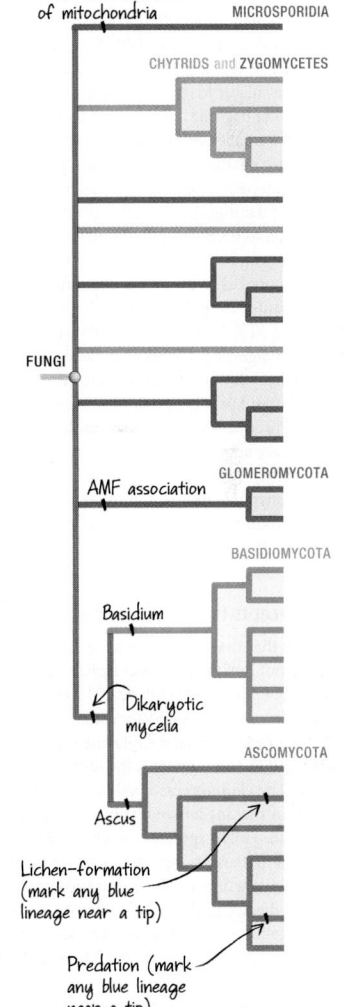

Figure 31.11 [See **Figure A31.2**] **Figure 31.13** The haploid mycelium. **Figure 31.14** Mycelia produced by asexual reproduction are genetically identical to their parent; mycelia produced by sexual reproduction are genetically different from both parents (each spore has a unique genotype).

Summary of Key Concepts

KC 31.1 Loss of mycorrhizal fungi would decrease nutrient delivery to plants and reduce their growth inside the experimental plots, compared to control plots with intact fungi. Also, lack of fungi would slow decay of dead plant material, causing a dramatic buildup of dead organic material. **KC 31.2** Absorption has to occur across a surface (usually via transport proteins in the plasma membrane). Other things being equal, more absorption will occur across a large surface area than a small surface area. **KC 31.3** Haploid hyphae fuse, forming a heterokaryotic mycelium. When karyogamy occurs, a diploid nucleus forms—just as when gametes fuse.

Test Your Knowledge

1. a; **2.** d; **3.** c; **4.** b; **5.** b; **6.** d

Test Your Understanding

1. Along with a few bacteria, fungi are the only organisms that can digest wood completely. If the wood is not digested, carbon remains trapped in wood. Without fungi, CO_2 would be tied up and unavailable for photosynthesis, and the presence of undecayed organic matter would reduce the space available for plants to grow. **2.** Fungi produce enzymes that degrade cellulose and lignin. **3.** Plant roots have much smaller surface area than EMF or AMF. Hyphae are much smaller than the smallest portions of plant roots so can penetrate dead material more efficiently. Extracellular digestion—which plant roots cannot do—allows fungi to break large molecules into small compounds that can be absorbed. **4.** DNA, RNA, and ATP all contain phosphorus. Without adequate amounts of phosphorus, plants cannot grow by synthesizing more DNA, RNA, and ATP. **5.** Both compounds are processed via extracellular digestion. Different enzymes are involved, however. The degradation of lignin is uncontrolled and does not yield useful products; digestion of cellulose is controlled and produces useful glucose molecules. **6.** Most fungi do not make gametes. Instead, the products of specific alleles identify mating types—probably to encourage fusion between hyphae that are dissimilar genetically, resulting in the production of genetically diverse offspring. It is possible for thousands of different mating-type alleles to exist in a population.

Applying Concepts to New Situations

1. (1) Confirm that the chytrid fungus is found only in sick frogs and not healthy frogs. (2) Isolate the chytrid fungus and grow it in a pure culture. (3) Expose healthy frogs to the cultured fungus and see if they become sick. (4) Isolate the fungus from the experimental frogs, grow it in culture, and test whether it is the same as the original fungus. **2.** The claim is reasonable because fungi have so many ways of making a living from plant tissues. For example, the same plant species could have several species of fungi that are endophytic, mycorrhizal, or parasitic, as well as an array of species that break down its tissues when it dies. **3.** Each of the different cellulase enzymes attacks cellulose in a different way, so producing all the enzymes together increases the efficiency of the fungus in breaking down cellulose completely. It is likely that lignin peroxidase is produced along with cellulases, so they can act in concert to degrade wood. To test this idea, you could harvest enzymes secreted from a mycelium before and after it contacts wood in a culture dish, and see if the cellulases and lignin peroxidase appear together once the mycelium begins growing on the wood. **4.** Collect a large array of colorful mushrooms that are poisonous and capture mushroom-eating animals, such as squirrels. Present a hungry squirrel with a choice of mushrooms that have been dyed or painted a drab color versus treated with a solution that is identical to the dye or paint used but uncolored. Record which mushrooms the squirrel eats. Repeat the test with many squirrels and many mushrooms.

CHAPTER 32

Check Your Understanding (CYU)

CYU p. 609 (1) Bilateral symmetry led to the evolution of long, slender body plans. Triploblasty gave rise to an inner tube (gut), an outer tube (body lining), and muscles and organs in between. The coelom functions as a hydrostatic skeleton to facilitate movement in soft-bodied animals that lack limbs. (2) When an animal moves through an environment in one direction, its ability to acquire food and perceive and respond to threats is greater if its feeding, sensing, and information-processing structures are at the leading end. **CYU p. 617** (1) The mouthparts of deposit feeders are relatively simple, as they simply gulp relatively soft material. The mouthparts of mass feeders are more complex, because they have to tear off and process chunks of relatively hard material. (2) Gametes that are shed into aquatic environments can float or swim. This cannot happen on land, so internal fertilization is more common.

You Should Be Able To (YSBAT)

YSBAT p. 606 If you do the exercise correctly, the balloon should wriggle. The wriggling movements would produce movement if the balloon (or animal) were surrounded by water or soil. **YSBAT p. 618** The origin of epithelial tissue should be marked on the same branch as multicellularity. **YSBAT p. 619** The origin of cnidocytes should be marked on the Cnidaria branch. **YSBAT p. 620** The origin of cilia-powered swimming should be marked on the Ctenophora branch.

Caption Questions and Exercises

Figure 32.6 The animal would get shorter and fatter. **Figure 32.9** The label and bar should go on the branch to the left of the clam shell and "Mollusk" label. **Figure 32.19** Stained *Dll* gene products would be located in the legs of the insect but would not be concentrated anywhere in the onychophoran and segmented worm. **Figure 32.22** No—there is no multicellular haploid form.

Summary of Key Concepts

KC 32.1 It should increase dramatically—animals would no longer be consuming plant material. **KC 32.2** Your drawing should show the tube-within-a-tube design of a worm, with one end labeled head and containing the mouth and brain. From the brain, one or two major nerve tracts should run the length of the body. The gut should go from the mouth to the other end, ending in the anus. In between the gut and outer body wall, there should be blocks or layers of muscle. **KC 32.3** During gastrulation, the initial pore becomes the mouth in protostomes, whereas in deuterostomes the pore becomes the anus. The protostome coelom is formed from blocks of mesodermal tissue that form cavities. The deuterostome coelom forms when mesodermal tissue around the gut pinches off. **KC 32.4** Natural selection has produced eye structures that function well in a particular species' habitat, and mollusks live in a wide array of habitats. Some mollusks live buried in sand and have no eyes—eyes would have no function in this habitat. Mollusks like squid and octopuses live in open water and hunt prey. They have complex eyes that form images and help them find food. **KC32.5** Juveniles look like miniature adults and feed on the same foods as adults. Larvae look much different from adults, live in different habitats, and eat different foods. The difference is important because larvae do not compete with adults for food.

FIGURE A31.2

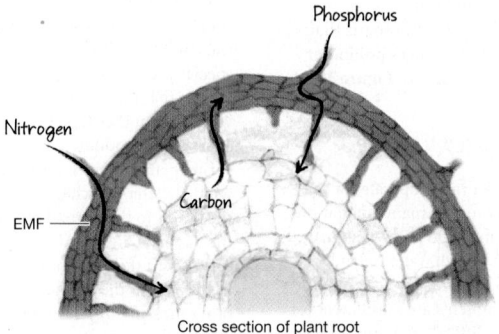

Cross section of plant root

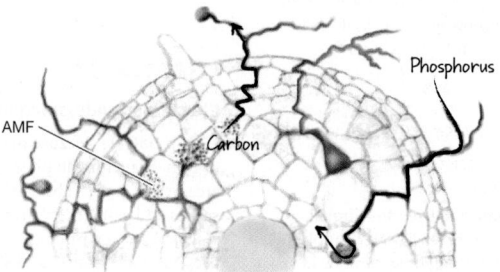

Cross section of plant root

Test Your Knowledge

1. a; **2.** d; **3.** b; **4.** a; **5.** d; **6.** c

Test Your Understanding

1. Diploblasts have two types of embryonic tissue; triploblasts have three. Mesoderm made it possible for an enclosed, muscle-lined cavity to develop, creating a coelom. 2. A hydrostatic skeleton is a pressurized, fluid-filled chamber (the coelom) that is surrounded by muscles. Contractions on either side of the body change the pressure of the fluid in the hydrostatic skeleton and thus the shape of the coelom. When muscle contractions are coordinated throughout the length of the animal, the shape of the hydrostatic skeleton changes and results in movement. 3. There are many unicellular organisms that are heterotrophic, but they can consume only small packets of food. Animals are multicellular, so they are larger and can consume larger packets of food. 4. A long, hollow structure that is sharp enough to pierce the wall of the stem; a simple, muscular mouth opening that can gulp soft, decaying material; sharp, stiff structures that can move in a scissoring action; a long, hollow structure that can be inserted into the flower. 5. In oviparous species, the mother adds nutrient-rich yolk to the egg that nourishes the developing embryo. In viviparous species, the mother transfers nutrients directly from her body to the growing embryo. 6. Radially symmetric organisms encounter the environment in many directions, so it is advantageous to have an equal number of neurons located throughout the body. Bilaterally symmetric organisms encounter the environment in one direction, so it is advantageous to have most neurons clustered in the head region.

Applying Concepts to New Situations

1. Yes—if the same gene is found in nematodes and humans, it was likely found in the common ancestor of protostomes and deuterostomes. If so, then fruit flies should also have this gene. 2. Mosquitoes, because larval and adult mosquitoes feed on different food sources in different habitats and do not compete with one another. You could test this prediction by collecting data on the total number of mosquito versus tick species, their abundance, and geographic distribution. 3. It should resemble a nerve net, because echinoderms need to take in and process information from multiple directions—not just one. 4. Web-building spiders are some of the only terrestrial animals that suspension feed.

CHAPTER 33

Check Your Understanding (CYU)

CYU p. 630 (1) Arthropods have a tube-within-a-tube body plan with a drastically reduced coelom. They have a hemocoel body cavity that holds internal organs and body fluids. The body is segmented, with segments grouped into tagmata. They have jointed limbs and an exoskeleton made of chitin. Compared to unjointed limbs, jointed limbs allow animals to move faster and with more precision. (2) Supporting the body (and moving) without support from water. Preventing the body from drying out. Facilitating gas exchange.

You Should Be Able To (YSBAT)

YSBAT p. 631a, YSBAT p. 631b, YSBAT p. 632, YSBAT p. 633, YSBAT p. 637, YSBAT p. 638 [See Figure A33.1]

Caption Questions and Exercises

Figure 33.3 The tuft is the cluster of ciliated tentacles in part (a); the wheel is the ring of cilia in part (b).
Figure 33.6 Different phyla of worms have different feeding strategies, different mouthparts, and eat different foods, so they are not in direct competition for food.
Figure 33.22 [See Figure A33.2]

Summary of Key Concepts

KC 33.1 All of the phyla that are considered protostomes share a pattern of early development found in no other animals. Based on this observation, it is logical to claim that this pattern of development arose in the common ancestor of these phyla—meaning that it is a synapomorphy. **KC 33.2** Flatworms likely lost the adaptation of a coelom because as they evolved a thin, flattened body plan, the coelom was no longer needed for internal gas circulation or movement. Arthropods have a drastically reduced coelom because they have evolved limbs and complex musculature for movement and no longer require the coelom to provide a hydrostatic skeleton. **KC 33.3** (1) Because larvae and adults live in a different habitat, young and adults do not compete for food. (2) Because larvae can swim, they can disperse to new locations.

Test Your Knowledge

1. d; **2.** a; **3.** c; **4.** c; **5.** a; **6.** b

Test Your Understanding

1. Ecdysozoans grow by molting; lophotrochozoans grow by incremental additions to their bodies. In addition, some lophotrochozoan phyla have lophophores and/or trochophore larvae. Both groups are bilaterally symmetric triploblasts with the protostome pattern of development. 2. Multiple times, because phylogenies show that segmented groups evolved from unsegmented ancestors in both lophotrochozoans (annelids) and ecdysozoans (arthropods). 3. The ability to live on land offered the opportunity of exploiting new habitats and food sources, with minimal competition from other animals. 4. An exoskeleton provided a stiff surface for muscle attachment—facilitating rapid, precise movement—as well as protection from enemies.

5. The ability to fly allowed insects to disperse to new habitats and find new food sources efficiently. 6. The leading hypothesis is variation in mouthpart structure and feeding strategies.

Applying Concepts to New Situations

1. [See Figure A33.3] 2. If the ancestors of brachiopods and mollusks lived in similar habitats and experienced natural selection that favored similar traits, then they would have evolved to have similar forms and habitats. This is called convergent evolution (see Chapter 27). 3. Aplacophora are probably basal—meaning that they would branch off between the ancestral form at the base of the tree and the rest of the groups. Because they have reduced forms of key molluscan synapomorphies, it is likely that they are the least-derived of the Mollusca. Because they lack a shell for muscle attachment and a well-developed foot, they probably move slowly via undulating movements and live in benthic habitats. 4. Agree—it is likely that the common ancestor of arthropods was marine, and that terrestrial forms evolved from aquatic ancestors independently in crabs, isopods, and insects.

CHAPTER 34

Check Your Understanding (CYU)

CYU p. 650 (1) If it's an echinoderm, it should have five-part radial symmetry, a calcium carbonate endoskeleton just underneath the skin, and a water vascular system (e.g., visible podia). **CYU p. 659** (1) Jaws allow animals to capture food efficiently and process it by crushing or tearing. (2) The increased cushioning from enclosed fluids provided mechanical support, and the increased surface area made transport of gases and other materials more efficient.

FIGURE A33.1

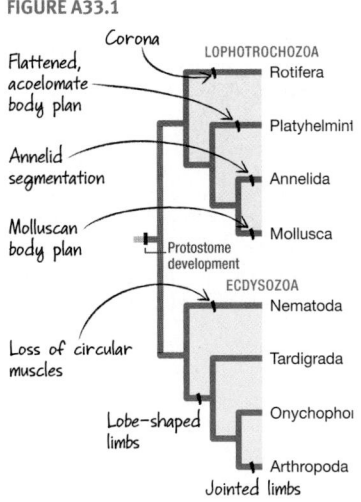

FIGURE A33.2

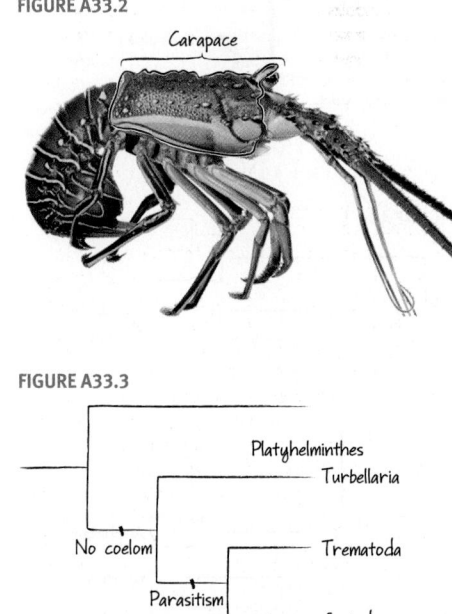

FIGURE A33.3

You Should Be Able To (YSBAT)

YSBAT p. 649a, YSBAT p. 649b, YSBAT p. 650 [See Figure A34.1] YSBAT p. 655 [See Figure A34.2] YSBAT p. 660 [See Figure A34.3] YSBAT p. 662a, YSBAT p. 662b, YSBAT p. 663, YSBAT p. 664 [See Figure A34.4] YSBAT p. 665, YSBAT p. 666a, YSBAT p. 666b, YSBAT p. 667a, YSBAT p. 667b, YSBAT p. 668 [See Figure A34.3]

Caption Questions and Exercises

Figure 34.1 Invertebrates are a paraphyletic group. The group does not include all the descendants of the common ancestor because vertebrates are excluded. **Figure 34.12** Mammals and birds are equally related to amphibians, because birds and mammals share a common ancestor, and this ancestor shares a common ancestor with amphibians. **Figure 34.14** Yes—even a rudimentary jaw could have been useful to better grasp prey or to change the size of the mouth. **Figure 34.16** If this phylogeny were estimated on the basis of limb traits, it would be based on the *assumption* that the limb evolved from fish fins in a series of steps. But because the phylogeny was estimated from other types of data, it is legitimate to use it to analyze the evolution of the limb without any assumptions about how limbs evolved. **Figure 34.20** They are smaller—their function in an amniotic egg has been taken over by the placenta. **Figure 34.37** Four hominin species existed 2.2 mya, five existed 1.8 mya, and four existed 100,000 years ago. **Figure 34.38** The forehead became much larger with the face becoming "flatter"; the brow ridges are less prominent in later skulls than in earlier skulls.

Summary of Key Concepts

KC 34.1 (1) Mussels and clams will increase dramatically; (2) kelp density will increase. **KC 34.2** Like today's lungfish, the earliest tetrapods could have used their limbs for pulling themselves along the substrate in shallow-water habitats. **KC 34.3** The earliest fossils of *H. sapiens* are found in Africa, and phylogenetic analyses of living human populations indicate that the most basal groups are all African.

Test Your Knowledge

1. a; **2.** a; **3.** d; **4.** a; **5.** d; **6.** b

Test Your Understanding

1. It is an enclosed, fluid-filled structure surrounded by muscle. Muscular contraction forces water into the tube feet, resulting in extension of the podia. **2.** Pharyngeal gill slits function in suspension feeding. The notochord furnishes a simple endoskeleton that stiffens the body; electrical signals that coordinate movement are carried by the dorsal hollow nerve cord to the muscles in the tail, which beats back and forth to make swimming possible. Cephalochordates and ascidians are chordates but they are not vertebrates. **3.** If jaws are derived forms of gill arches, then the same genes and the same cells should be involved in the development of the jaw and the gill arches. **4.** Homologous genes are involved in the formation of the fins of ray-finned fish and the limbs of tetrapods. This observation supports a prediction of the fins-to-limbs hypothesis. **5.** The hominins fulfill the criteria for an adaptive radiation: Over a short time interval, many species that occupy an array of foods and habitats evolved. Changes in tooth and jaw structure, tool use, and body size suggest that different hominin species exploited different types of food. **6.** Increased parental care allows offspring to be better developed, and thus have increased chances of survival, before they have to live on their own.

Applying Concepts to New Situations

1. If confirmed, it means that pharyngeal gill slits were present in the earliest echinoderms and lost later. **2.** No—most of the feathered dinosaurs known from fossils did not fly. (The first feathers may have functioned in display or as insulation.) **3.** Xenoturbellidans either retain traits that were present in the common ancestor or all deuterostomes, or they have lost many complex morphological characteristics. **4.** Independently. Birds are a highly derived lineage of reptiles, and most reptiles are not endothermic, so it is logical to infer that the common ancestor of birds and mammals was also not endothermic.

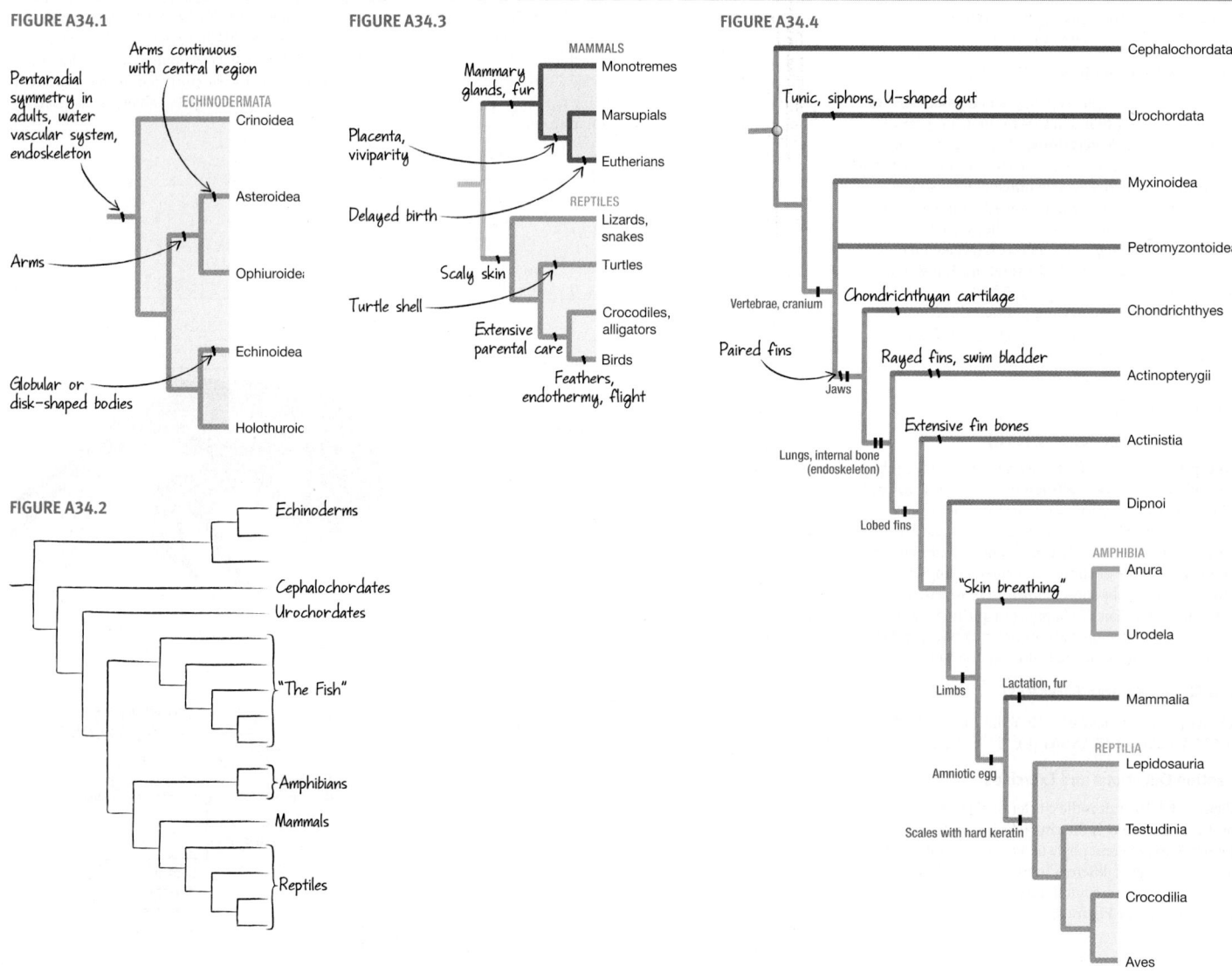

CHAPTER 35

Check Your Understanding (CYU)

CYU p. 686 HIV binds to CD4 and a co-receptor on the surface of T cells. This binding allows the viral envelope and the T-cell plasma membrane to fuse. The viral capsid then enters the cell. **CYU p. 689** (1) After the genome is copied to form positive-sense RNA, the transcript has to be copied to form a negative-sense, single-stranded genome. (2) Much more dangerous—with airborne transmission, a noninfected individual can become infected simply by inhaling virus particles released from an infected individual. This mode of transmission is much more efficient than transmission via body fluids.

You Should Be Able To (YSBAT)

YSBAT p. 680 In the lytic cycle, viral genes are transmitted to a new generation of virions. In lysogenic growth, viral genes are transmitted to daughter cells of the host cell. **YSBAT p. 686** [See Figure A35.1]

Caption Questions and Exercises

Figure 35.1 Many answers are possible: Herpes simplex 2: genitalia; sexually transmitted; HIV: immune system; sexually transmitted; Influenza virus: respiratory tract; coughing/sneezing (airborne); Chicken pox (varicella zoster virus): skin; direct contact. **Figure 35.8** Lysogenic bacteriophages are similar to transposable elements because they insert their DNA into the host cell chromosome and are passed on to daughter cells. They are unlike transposable elements because they can switch to a lytic cycle. **Figure 35.9** No—it only shows the CD4 is required. (Subsequent work showed that other proteins are involved as well.) **Figure 35.14** Budding viruses may disrupt the integrity of the host-cell plasma membrane enough to kill the cell. **Figure 35.15** You should have the following bars and labels: on branch to HIV-2 (sooty mangabey to human); on branch to HIV-1 strain O (chimp to human); on branch to HIV-1 strain N (chimp to human); on branch to HIV-1 strain M (chimp to human).

Summary of Key Concepts

KC 35.1 There is heritable variation among virions, due to random changes that occur as their genomes are copied. There is also differential reproductive success among virions in their ability to successfully infect host cells. This differential success is due to the presence of certain heritable traits. **KC 35.2** As the virus's envelope proteins are produced, they are inserted into the host cell plasma membrane. During budding, the capsid becomes surrounded by host cell membrane that includes the envelope proteins. **KC 35.3** (1) Fusion inhibitors block viral envelope proteins or host cell receptors; (2) protease inhibitors prevent processing/assembly of viral proteins; (3) reverse transcriptase inhibitors block reverse transcriptase, preventing replication of genome; (4) use of condoms or practice of monogamy prevents transmission. **KC 35.4** To replicate an ssDNA genome, a viral DNA polymerase copies the genome into a complementary strand, which then is used as a template to generate an ssDNA that is identical to the original viral genome.

Test Your Knowledge

1. b; **2.** c; **3.** d; **4.** b; **5.** b; **6.** d

Test Your Understanding

1. HIV has an envelope on its outer surface; adenovirus has only a capsid on its outer surface. A virus with an envelope exits host cells by budding. A virus that lacks an envelope exits host cells by lysis (or other mechanisms that don't involve budding). **2.** T4 bacteriophage should have a larger genome than HIV does, because T4 is much more complex morphologically—it should have genes that code for the proteins required for its head and tail regions. **3.** Growth rate is much higher during lytic growth, but viral DNA can increase during latent/lysogenic growth if the host cell is actively dividing and transmitting viral genomes to daughter cells. **4.** Viruses rely on host-cell enzymes to replicate, whereas bacteria do not. Therefore, many drugs designed to disrupt the virus life cycle cannot be used because they would kill host cells as well. Only viral-specific proteins are good targets for drug design. **5.** Each major hypothesis to explain the origin of viruses—from transposable-element-like sequences, from symbiotic bacteria, and from RNA genomes present early in Earth history—is associated with a different type of genome: single-stranded DNA or possibly single-stranded RNA, double-stranded DNA, and single- or double-stranded RNA, respectively. **6.** The phylogenies of SIVs and HIVs show that the two shared common ancestors, but that SIVs are ancestral to the HIVs. Also, there are plausible mechanisms for SIVs to be transmitted to humans through butchering or contact with pets, but fewer or no plausible mechanisms for HIVs to be transmitted to monkeys or chimps.

Applying Concepts to New Situations

1. Culture the *Staphylococcus* strain outside the human host and then add the virus to determine whether the virus kills the bacterium efficiently. Then test the virus on cultured human cells to determine whether the virus harms human cells. Then test the virus on monkeys or other animals to determine if it is safe. Finally, the virus could be tested on human volunteers. **2.** Prevention is currently the most cost-effective program but does not help people who are already infected. Treatment with effective drugs not only prolongs lives but also reduces virus loads in infected people, so that they have less chance of infecting others. **3.** Eukarya evolved from ancestors that did not contain genes encoding restriction endonucleases. Or, methylation of DNA performs other important functions in eukaryotic gene expression (see Chapter 18), so is not appropriate as a guard against endonuclease action. **4.** Viruses cannot be considered to be alive by the definition given in (a), because viruses are not capable of replicating by themselves. By the definition given in (b), it can be argued that viruses are alive because they store, maintain, replicate, and use genetic information—although they cannot perform all these tasks on their own.

CHAPTER 36

Check Your Understanding (CYU)

CYU p. 704 (1) [See Figure A36.1] (2) The generalized body of a plant has a taproot with many lateral roots and broad leaves. Examples of deviations from this generalized body structure include: Modified roots: Unlike taproots, *fibrous roots* do not have one central root, and *adventitious roots* arise from stems. Modified stems: *Stolons* grow along the soil and grow roots and leaves at each node. *Rhizomes* grow horizontally underground. Both of these modified stems function in asexual reproduction. Modified leaves: The *needle-shaped leaves* of cacti lose less water to transpiration than do typical

FIGURE A36.1

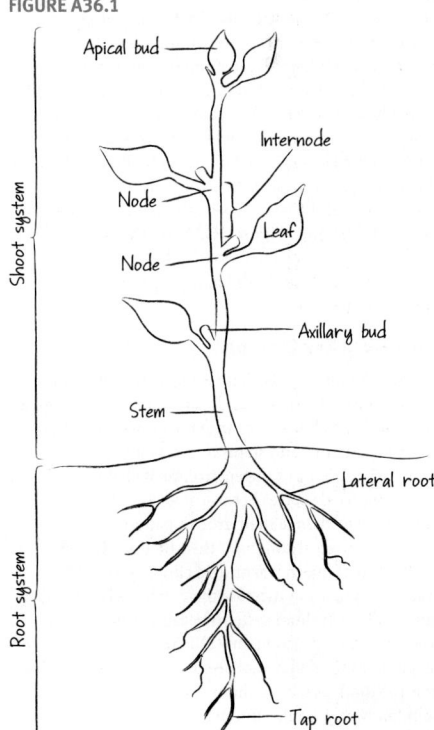

FIGURE A35.1

$(+)\text{ssRNA} \xrightarrow{\text{translation}} \text{Viral proteins}$

$(-)\text{ssRNA} \xrightarrow[\text{replicase}]{\text{RNA}} \text{mRNA} \xrightarrow{\text{translation}} \text{Viral proteins}$

$\text{dsRNA} \xrightarrow{\text{transcription}} \text{mRNA} \xrightarrow{\text{translation}} \text{Viral proteins}$

$\text{dsDNA or ssDNA} \xrightarrow{\text{transcription}} \text{mRNA} \xrightarrow{\text{translation}} \text{Viral proteins}$

$(+)\text{ssRNA} \xrightarrow[\text{transcription}]{\text{reverse}} \text{dsDNA} \xrightarrow{\text{transcription}} \text{mRNA} \xrightarrow{\text{translation}} \text{Viral proteins}$

broad leaves, and they protect the plant from predation. *Tendrils* on climbing plants are modified leaves that do not photosynthesize but wrap around trees or other substrates to facilitate climbing. **CYU p. 712** (1) The apical meristem gives rise to the three primary meristems. The apical meristem is a single mass of cells localized at the tip of a root or shoot; the primary meristems are localized in distinctive locations behind the apical meristem. (2) Epidermal cells are flattened and lack chloroplasts; they secrete the cuticle (in shoots) or extend water and nutrient-absorbing root hairs (in roots) and protect the plant. Parenchyma cells are "workhorse cells" found throughout the plant body; they perform photosynthesis and synthesize and/or store materials. Tracheids are long, thin cells with pits in their secondary cell walls; they are found in xylem and are dead when mature; they conduct water and solutes up the plant. Sieve-tube elements are long, thin cells found in phloem; they are alive when mature and conduct sugars and other solutes up and down the plant. **CYU p. 715** (1) Primary growth increases the length of roots and shoots, and secondary growth increases their width. The function of primary growth is to extend the reach of the root and shoot system and thus increase a plant's ability to absorb light and acquire carbon dioxide, water, and nutrients. The function of secondary growth is to increase the amount of lateral conducting tissue and provide the structural support required for extending primary growth at the top of the plant. (2) The rings are small on the shaded side and large on the sunny side.

You Should Be Able To (YSBAT)

YSBAT p. 713a, YSBAT p. 713b, YSBAT p. 713c, YSBAT p. 714 [See Figure A36.2]

Caption Questions and Exercises

Figure 36.2 New branches and leaves should develop to the right (the plant will also lean that way); new lateral roots will develop to the left. **Figure 36.4** Lawn grasses are shallow rooted, so when the top part of soil dries out, the plants cannot grow. **Figure 36.5** For aerobic respiration, which generates the ATP needed to keep the cells alive. **Figure 36.7** The individuals are genetically identical—thus, any differences in size or shape in the different habitats are due to phenotypic plasticity and not genetic differences. **Figure 36.9** Large surface area provides more area to capture photons, but also more area from which to lose water and be exposed to potentially damaging tearing forces from wind. **Figure 36.20** Many possible answers; for example, in leaf cells, genes for forming chloroplasts would be expressed. These genes wouldn't be expressed in the root cells. **Figure 36.25** The leading hypothesis is that they lack many organelles so that the space inside the cell is available to transport nutrients. **Figure 36.27** You should have labeled a light band as early wood, and the dark band to its right as late wood; both bands together comprise one growth ring.

Summary of Key Concepts

KC 36.1 Phenotypic plasticity is more important in (1) environments where conditions vary because it gives individuals the ability to change the growth pattern of their roots, shoots, and stems to access sunlight, water, and other nutrients as the environment changes; and (2) in long-lived species because it gives individuals a mechanism to change their growth pattern as the environment changes throughout their lifetime. **KC 36.2** (1) Plants that live in warm, arid climates would have large root systems to increase their chances of accessing water, and small shoot systems because water is lost from surfaces exposed to air. (2) Plants that live in very wet climates would have small root systems because water is plentiful, and large shoot systems because the abundant water would support more aboveground growth especially in light-limiting environments such as

forests. **KC 36.3** (1) In both cases, the structure would stop growing; (2) the shoot would lack epidermal cells and would die. **KC 36.4** (1) The plant would not produce secondary xylem and phloem, so its girth and transport ability would be reduced, though it would still produce bark. (2) The trunk would have much wider tree rings on one side than the other side.

Test Your Knowledge

1. b; **2.** b; **3.** a; **4.** d; **5.** c; **6.** a

Test Your Understanding

1. The general function of both systems is to acquire resources: The shoot system captures light and carbon dioxide; the root system absorbs water and nutrients. Vascular tissue is continuous throughout both the shoot and root systems. Diversity in roots and shoots enables plants of different species to live together in the same environment without directly competing for resources. **2.** Continuous growth enhances phenotypic plasticity because it allows plants to grow and respond to changes or challenges in their environment (such as changes in light and water availability). **3.** Cactus spines are modified leaves; thorns are modified stems. **4.** Cuticle reduces water loss; stomata facilitate gas exchange. Plants from wet habitats should have a relatively large number of stomata and thin cuticle. Plants living in dry habitats should have relatively few stomata and thick cuticle. **5.** Parenchyma cells in the ground tissue perform photosynthesis and/or synthesize and store materials. In vascular tissue, parenchyma cells in "rays" conduct water and solutes across the stem; other parenchyma cells differentiate into the sieve-tube members and companion cells in phloem. **6.** Cells produced to the inside of the vascular cambium differentiate into secondary xylem; cells produced to the outside of the vascular cambium differentiate into secondary phloem.

Applying Concepts to New Situations

1. Fewer sclerenchyma cells, because sclerenchyma cells have extremely tough walls and these populations have been selected to be tender. Humans select individuals with fewer sclerenchyma cells to be the parents of the next generation, so the frequency of sclerenchyma cells will decline over time. **2.** Asparagus—stem; Brussels sprouts—lateral bud; celery—petiole; spinach—leaf (petiole and blade); carrots—taproot; potato—modified stem. **3.** They grow continuously, so do not have alternating groups of large and small cells that form rings. **4.** Girdling disrupts transport of solutes in secondary

phloem, and damages (but probably doesn't eliminate) the transport of water and solutes in secondary xylem. The tree starves.

CHAPTER 37

Check Your Understanding (CYU)

CYU p. 721 (1) Wet soils have almost no pressure potential, while dry soils have a highly negative pressure potential because the few water molecules present cling to soil particles. (2) Salty soils have extremely low solute potentials compared to typical soils, because the concentration of solutes is high. **CYU p. 727** (1) At the air-water interface under a stoma, hydrogen bonding is asymmetrical—the water molecules can bond only with water molecules below them. This pulls the water molecules down—creating tension at the surface. When transpiration occurs, there are few water molecules at the surface, which makes the asymmetry in bonding even more pronounced—increasing tension. (2) When a nearby stoma closes, the humidity inside the air space increases, transpiration decreases, and surface tension on menisci decreases. The same changes occur when a rain shower starts. When air outside the leaf dries, though, the opposite occurs: The humidity inside the air space decreases, transpiration increases, and surface tension on menisci increases. **CYU p. 734** (1) In early spring, the water potential of phloem sap near leaves is low because developing leaves are using sucrose much faster than they make it; root cells are releasing sucrose so the water potential of phloem cells in roots is high. (2) Midday in summer, the water potential of phloem sap near leaves is high because leaves are making much more sucrose than they consume. Root cells are storing sugar, however, so the water potential of phloem in roots is low.

You Should Be Able To (YSBAT)

YSBAT p. 719 (1) Add a solute (e.g., salt or sugar) to the left side of the U-tube, so that solute concentration is higher than on the right side. (2) Water would move from the right side of the U-tube into the left side. **YSBAT p. 720** Pull the plunger up. **YSBAT p. 723** The individual would not be able to exclude certain solutes from the xylem. If they were transported throughout the plant in high enough concentrations, they could damage tissues.

Caption Questions and Exercises

Figure 37.2 It increases (becomes less negative), which reduces the solute potential gradient between the two

---

FIGURE A36.2

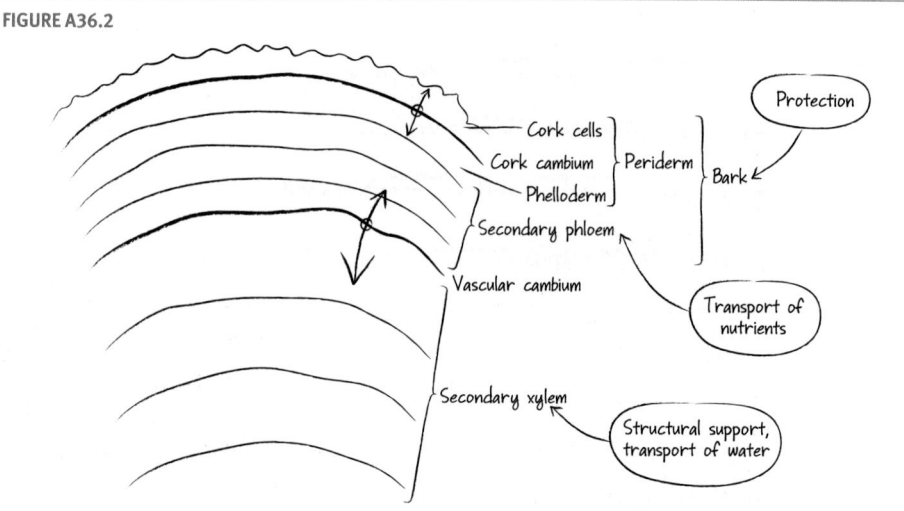

sides. **Figure 37.7** [See Figure A37.1] **Figure 37.10** Your labels should read as follows: Atmosphere: $\psi = -95.2$ MPa; Leaves: $\psi = -0.8$ MPa; Roots: $\psi = -0.6$ MPa; Soil: $\psi = -0.3$ MPa. **Figure 37.12** The line on the graph would go up, with large jumps occurring each time the light level was turned up, because increased transpiration rates would increase negative pressure (tension) at the leaf surface, and thus increase the water potential gradient. **Figure 37.17** [See Figure A37.2]

Summary of Key Concepts

KC 37.1 Because xylem cells are dead at maturity, there are no plasma membranes for water to cross and thus no solute potential. The only significant force acting on water in xylem is pressure. **KC 37.2** It drops, because transpiration rates from stomata decrease, reducing the water potential gradient between roots and leaves. **KC 37.3** To create high pressure in phloem near sources and low pressure near sinks, water has to move from xylem to phloem near sources and from phloem back to xylem near sinks. If xylem were not close to phloem, this water movement could not occur and the pressure gradient in phloem wouldn't exist.

Test Your Knowledge

1. d; 2. c; 3. b; 4. a; 5. b; 6. c

Test Your Understanding

1. [See Figure A37.3] **2.** The force responsible for root pressure is the high solute potential of root cells at night, when they continue to accumulate ions from soil, with water following by osmosis. The mechanism is active, in the sense that ions are imported against their concentration gradient. Capillarity is passive; it is driven by adhesion of water molecules to the sides of xylem cells that creates a pull upwards, and cohesion with water molecules below. Cohesion-tension is also passive, because it is driven by the loss of water molecules from menisci in leaves, which creates a large negative pressure (tension). **3.** Water moves up xylem because of transpiration-induced tension created at the air-water interface under stomata, which is communicated down to the roots by the cohesion of water molecules. Sap flows in phloem because of a pressure gradient that exists between source and sink cells, driven by differences in sucrose concentration and flows of water into or out of nearby xylem. **4.** Aphid A is attacking a source, and Aphid B is attacking a sink. If sucrose concentrations are higher in phloem sap near sources, then aphids should be found primarily near mature leaves. (It's actually more common to find them near apical meristems, though—probably because the tissues are not as stiff and strong and difficult to pierce.) **5.** By pumping protons out, companion cells create a strong electrochemical gradient favoring entry of protons. A cotransport protein (a symporter) uses this proton gradient to import sucrose molecules *against* their concentration gradient. **6.** When it germinates—sucrose stored inside the seed is released to the growing embryo, which cannot yet make enough sucrose to feed itself.

Applying Concepts to New Situations

1. Plants have to gain CO_2 for photosynthesis to occur, which means that stomata have to be open. But transpiration occurs whenever stomata are open and the surrounding air is drier than the inside of the leaf. A similar "side effect" occurs when terrestrial animals breathe—they have to take dry air into their lungs, where water evaporates from the body and is lost. **2.** Plants would not grow as quickly or as tall, because the taller they grew, the more energy they would need to use to transport water and thus the less they would have available for growth. **3.** Because they cannot readily replace water that is lost to transpiration, they have to close stomata. During a heat wave, transpiration cannot cool the plant's tissues. They bake to death. **4.** Closing aquaporins will slow or stop movement of water from cells into the xylem and out of the plant via transpiration. Because water does not leave the cells, they are able to maintain turgor (normal solute potentials).

CHAPTER 38

Check Your Understanding (CYU)

CYU p. 748 (1) Proton pumps establish an electrical gradient across root hair membranes, with the inside of the membrane being much more negative than the outside.

FIGURE A37.1

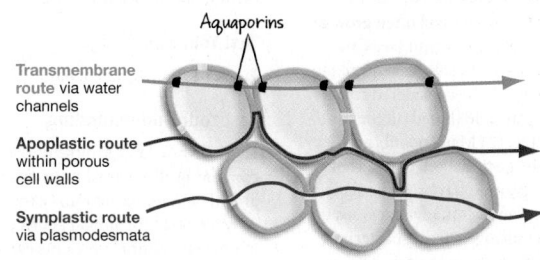

FIGURE A37.2

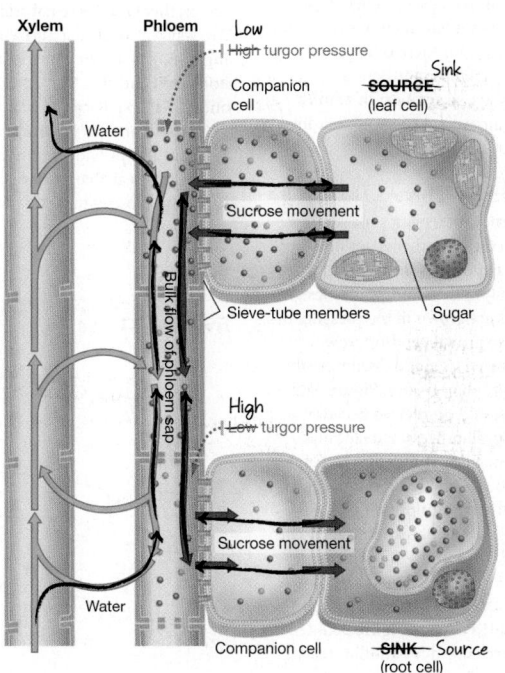

FIGURE A37.3

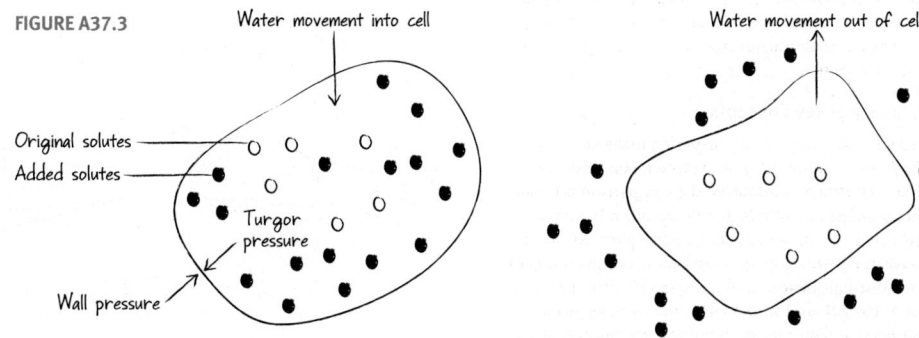

Cations follow this electrical gradient into the cell. (2) [See Figure A38.1] **CYU p. 750** (1) Many possible answers, including the following: If N-containing ions are abundant in the soil, the energetic cost of maintaining nitrogen-fixing bacteria may outweigh the benefits in terms of increased nitrogen availability. If mycorrhizal fungi provide a plant with nitrogen-containing ions, the energetic cost of maintaining nitrogen-fixing bacteria may outweigh the benefits in terms of increased nitrogen availability. Plants that grow near species with N-fixing bacteria might gain nitrogen by absorbing nitrogen-containing ions after root nodules die, or by "stealing it" (roots of different species will often grow together). There could be a genetic constraint (see Chapter 24): Plant species may simply lack the alleles required to manage the relationship. Note: To be considered correct, your hypothesis should be plausible (based on the underlying biology) and testable. (2) Many possible answers, including: Because the bacterial cell enters root cells, it is important for the plant to have reliable signals that it is not a parasitic bacterium. It is advantageous for the plant to be able to reject nitrogen-fixing bacteria if usable nitrogen is already abundant in the soil. The signals have no function.

You Should Be Able To (YSBAT)

YSBAT p. 745 (1) If there is no voltage across the root hair membrane, there is no electrical gradient favoring entry of cations, and absorption stops. (2) There is no route for cations to cross the root hair membrane, so absorption stops. **YSBAT p. 746** (1) If there is no proton gradient across the root hair membrane, there is no gradient favoring entry of protons, so symport of anions stops. (2) There is no route for anions to cross the root hair membrane, so absorption stops.

Caption Questions and Exercises

Figure 38.1 Point out that most of a plant's mass is due to carbon-containing molecules, but that water does not contain carbon. To test the hypothesis that most of a plant's mass comes from CO_2 in the atmosphere, design an experiment where plants are grown in the presence or absence of CO_2 and compare growth, and/or grow plants in the presence of labeled CO_2 and document the presence of labeled carbon in plant tissues. **Figure 38.2** Proteins and nucleic acids would be affected because they contain nitrogen. Most cell processes, including photosynthesis, would be affected because they depend on proteins. **Figure 38.3** Because soil is so complex, it would be difficult to defend the assumption that the soils with and without added copper are identical except for the difference in copper concentrations. Also, it would be better to compare treatments that had no copper versus normal amounts of copper, instead of comparing with and without added copper. **Figure 38.8** One cation (a proton) is being exchanged for another (calcium or magnesium). **Figure 38.10** The proton gradient arrow in (b) should begin above the membrane and cross the membrane pointing down. The electrical gradient arrow in (c) should begin above the membrane (marked positive) and cross the membrane pointing down (toward negative). **Figure 38.17** Mistletoe is parasitic (it extracts water and certain nutrients from host trees); bromeliads are not (they don't affect the fitness of their host plants).

Summary of Key Concepts

KC 38.1 Add nitrogen to an experimental plot and compare growth within the plot to growth of a similar plot nearby. Do many replicates of the comparison, to convince yourself and others that the results are not due to unusual circumstances in one or a few plots. **KC 38.2** A higher membrane voltage would allow cations in soil to cross root-hair membranes more readily through membrane channels and would allow anions to enter more readily via symporters. Nutrient absorption should in-

crease. **KC 38.3** Grow pea plants in the presence of rhizobia, exposed to air with (1) N_2 containing the heavy isotope of nitrogen, and (2) radioactive carbon dioxide. Allow the plant to grow and then analyze the rhizobia and plant tissues. If the rhizobia-plant interaction is mutualistic, then ^{15}N-containing compounds and radioactive-carbon-containing compounds should be observed in both rhizobia and plant tissues. As a control, grow pea plants in the presence of the labeled compounds but without rhizobia. If the mutualism hypothesis is correct, the plant should contain labeled carbon but no heavy nitrogen.

Test Your Knowledge

1. b; **2.** d; **3.** d; **4.** c; **5.** b; **6.**d

Test Your Understanding

1. (1) Grow a large sample of corn plants with a solution containing all essential nutrients needed for normal growth and reproduction. Grow another set of genetically identical (or similar) corn plants in a solution containing all essential nutrients except iron. Compare growth, color, and seed production (and/or other attributes) in the two treatments. (2) Grow corn plants in solutions containing all essential nutrients in normal amounts, but with varying concentrations of iron. Determine the lowest iron concentration at which plants exhibit normal growth rates. **2.** If nutrients in soil are scarce, there is intense natural selection favoring alternative ways of obtaining nutrients—for example, by digesting insects or stealing nutrients. **3.** Higher in soils with both sand and clay. Clay fills spaces between sand grains and slows (1) passage of water through soil and away from roots, and (2) leaching of anions in soil water. Clay particles also provide negative charges that retain cations so they are available to plants. The presence of sand keeps soil loose enough to allow roots to penetrate.

4. H^+-ATPases in the plasma membrane pump protons out of the cell—making the inside of the membrane less positive (more negative) and the outside of the membrane more positive. By convention, the voltage on a membrane is expressed as inside-relative-to-outside. **5.** Some metal ions are poisonous to plants, and high levels of other ions are toxic. Passive mechanisms of ion exclusion—such as a lack of ion channels that allow passage into root cells—do not require an expenditure of ATP. Active mechanisms of exclusion—such as production of metallothioneins—require an expenditure of ATP. **6.** (1) The amounts required and the mechanisms of absorption are different. Much larger amounts of C, H, and O are needed than mineral nutrients. CO_2 enters plants via stomata and water is absorbed passively along a water potential gradient; mineral nutrients enter via active uptake in root hairs or via mycorrhizae. (2) Macronutrients are required in relatively large quantities and usually function as components of macromolecules; micronutrients are required in relatively small quantities and usually function as enzyme cofactors.

Applying Concepts to New Situations

1. The water that he used may have contained many of the macro- and micronutrients that were incorporated into plant mass. To test this hypothesis, conduct an experiment of the same design but have one treatment with pure water and one treatment with water containing solutes ("hard water"). Compare the percentage of soil mass incorporated into the willow tree in both treatments. Willow may be unusual. To test this hypothesis, repeat the experiment with willow and several other species under identical conditions, and compare the percentage of soil mass incorporated into the plants. Van Helmont's measurements were inaccurate. Repeat the experiment. **2.** Acid rain inundates the soil with protons, which bind with the negatively charged clay parti-

FIGURE A38.1

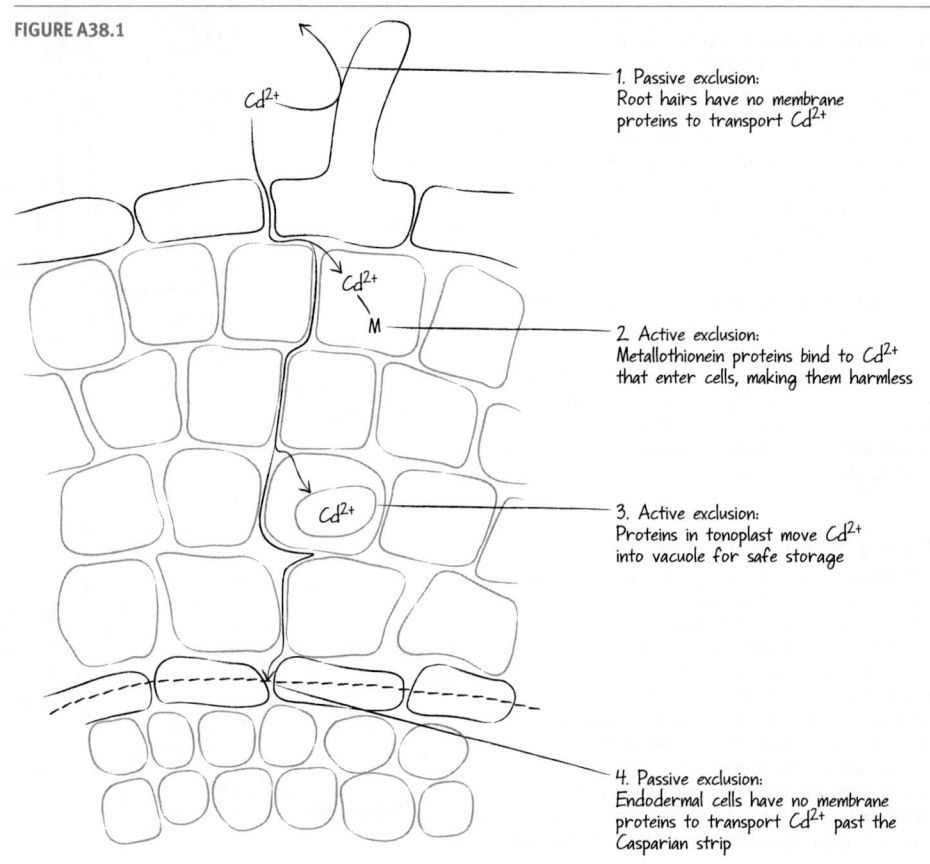

1. Passive exclusion:
Root hairs have no membrane proteins to transport Cd^{2+}

2. Active exclusion:
Metallothionein proteins bind to Cd^{2+} that enter cells, making them harmless

3. Active exclusion:
Proteins in tonoplast move Cd^{2+} into vacuole for safe storage

4. Passive exclusion:
Endodermal cells have no membrane proteins to transport Cd^{2+} past the Casparian strip

cles—displacing the cations that are normally bound there. Cations may then be leached away. **3.** If the hypothesis is correct, vanadate should inhibit phosphorus uptake. To test this hypothesis, grow plants in solution with radioactively labeled phosphorus without vanadate and in solution with vanadate. Measure the amount of labeled phosphorus in root cells with and without vanadate. **4.** Alder decomposes much faster, because it provides more nitrogen to the fungi and bacteria that are responsible for decomposition—meaning that they can grow faster.

CHAPTER 39

Check Your Understanding (CYU)

CYU p. 757 (1) The only cells that can respond to a specific environmental signal or hormone are cells that have an appropriate receptor for that signal or hormone. (2) Once a receptor is activated, it triggers production of many second messengers or the activation of many proteins in a phosphorylation cascade. **CYU p. 762** (1) Using a syringe or other device, apply auxin to the west side of each stem, near the base. **CYU p. 764** (1) Lettuce grows best in full sunlight. If a seed receives red light, it indicates that the seed is in full sunlight—meaning that conditions for germination are good. But if a seed receives far-red light, it indicates that the seed is in shade—meaning that conditions for germination are poor. (2) Far-red light indicates shade, and stem-elongation gives plants a chance to grow above nearby plants into full sunlight. **CYU p. 766** (1) In both roots and shoots, auxin is redistributed asymmetrically in response to a signal. In phototropism, auxin is redistributed to the shaded side of the plant in response to blue light. In gravitropism, auxin is redistributed to the lower part of the root or shoot in response to gravity. (2) In root cells, auxin concentrations indicate the direction of gravity. The gravitropic response is triggered by changes in the distribution of auxin in root-tip cells. **CYU p. 776** (1) Short term, stomata should close and photosynthesis should stop. Long term, the plants will not grow well, compared to individuals that do not receive extra ABA, or die. (2) Their stems should elongate rapidly, compared to plants that do not receive extra GAs. **CYU p. 781** (1) Because specific *R* gene products recognize and match specific proteins produced by pathogens, having a wide array of *R* alleles allows individuals to recognize and respond to a wide array of pathogens. (2) If herbivores are not present, synthesizing large quantities of proteinase inhibitors wastes resources (ATP and substrates) that could be used for growth and reproduction.

You Should Be Able To (YSBAT)

YSBAT p. 777 Virtually all bananas have the same *R* alleles. If a pathogen evolves that is not identified by the existing *R* alleles, all bananas would be susceptible to the new disease.

Caption Questions and Exercises

Figure 39.4 These cells served as controls. If the radioactive band appeared in cells that did not have the *PHOT1* gene, it clearly could not be a *PHOT1* band. If the band appeared in cells that had not been exposed to blue light, then *PHOT1* would not be acting as a blue-light receptor. **Figure 39.5** For "Tip removed": cut the tip and replace it, to control for the hypothesis that the wound itself influences bending (not the loss of the tip). For "Tip covered": cover the tip with a transparent cover instead of opaque one, to control for the hypothesis that the cover itself affects bending (not excluding light). **Figure 39.7** Two possibilities: (1) none of the decapitated shoots would bend, or (2) the blocks from the completely divided tip would bend much less than the partially divided tip, because more cells were disrupted. **Table 39.1** The average germination rate of lettuce seeds

last exposed to red light is about 99 percent. The average for seeds last exposed to far-red light is about 50 percent. Both of these values are much higher than the germination rate of buried seeds that receive no light at all; their germination rate is only 9 percent. **Figure 39.10** A signal in the form of a light wavelength is changed into a signal in the form of a protein's shape. In this way, a signal from outside the cell is changed to a signal inside the cell. **Figure 39.12** (One possibility) Direct pressure from a statolith changes the shape of the receptor protein. The shape change triggers a signal transduction cascade leading to a gravitropic cell response. **Figure 39.14** In nature, this response is most likely to occur in windy environments. A reduction in height might keep plants beneath the path of the wind, and stockier plants would be better able to resist bending when blown. **Figure 39.19** The dwarfed individuals can put more of their available energy into reproduction because they are using less energy for growth than taller individuals. **Figure 39.21** Without these data, a critic could argue that the stomata closed in response to water potentials in the leaf, not a signal from the roots. **Figure 39.24** The transported nutrients are conserved in the plant instead of being lost when leaves fall. Cessation of chlorophyll synthesis saves plants the energy and nutrients that would have gone into producing chlorophyll in leaves that soon will fall off. **Figure 39.28** The molecule travels throughout the shoot system (and root system).

Summary of Key Concepts

KC 39.1 They do not perform photosynthesis and therefore do not need to switch behavior based on being located in full sunlight versus shade. **KC 39.2** An external signal arrives at a receptor cell/protein (a person sees a barn on fire); the person calls (the receptor transduces the signal); the dispatcher rings a siren (a second messenger is produced or phosphorylation cascade occurs); the fire department members get to the pumper truck (gene expression or some other cell activity changes). **KC 39.3** Sun-adapted (high-light) plants are likely to have a significant positive phototropic response (i.e., growth toward light), whereas shade-adapted (low-light) plants are less likely to respond phototropically. **KC 39.4** They need to be transported throughout the plant (because cells throughout the body need to respond to the signal)—not in a single direction.

Test Your Knowledge

1. d; **2.** a; **3.** b; **4.** b; **5.** c; **6.** b

Test Your Understanding

1. The P_{fr}-P_r switch in phytochromes lets plants know whether they are in shade or sunlight, and triggers responses that allow sun-adapted plants to avoid shade. Phototropins are activated by blue light, which indicates full sunlight, and trigger responses that allow plants to photosynthesize at full capacity. **2.** If *PHOT1* were inserted into a plant, a critic could contend that it was being phosphorylated by a plant protein. But because the *PHOT1* was in an insect cell, this alternative hypothesis was not credible. **3.** *Transduce* means "to convert energy from one form to another." There are many examples: a touch may be converted to an electrochemical potential, light energy or pressure into a shape change in a protein, and so on. **4.** (Many possibilities) Cells or tissue involved: Auxin directs gravitropism in roots but phototropism in shoots. Developmental stage or age: GA breaks dormancy in seeds but extends stems in older individuals. Concentration: Large concentrations of auxin in stems promote cell elongation; small concentrations do not. Other hormones: The concentration of GA relative to ABA influences whether germination proceeds. **5.** Receptor cells sense changes in the availability of blue light received. They respond by changing the distribution of auxin. A change in auxin concentration causes target cells in the stem to grow, and results in the stem

bending toward the source of blue light. In this way, plants can respond to changes in shading by growing toward areas where light is available. **6.** Ethylene promotes senescence. Fruits exposed to ethylene ripen faster than fruits that are not exposed to ethylene; increased ethylene sensitivity in leaves (combined with reduced auxin concentrations) causes abscission.

Applying Concepts to New Situations

1. Small-seeded plants need to perform photosynthesis early in seedling development or they will starve. Therefore, they need to germinate in direct sunlight. The food reserves in large-seeded plants can support seedling growth for a relatively long time without photosynthesis. Therefore, they do not need to germinate in direct sunlight. **2.** The experiment provides evidence that cytokinins stimulate cell division in lateral buds—suggesting that large amounts of cytokinins can overcome apical dominance. **3.** In these seeds, (a) little or no ABA is present, or (b) ABA is easily leached from the seeds, so its inhibitory effects are eliminated. Test hypothesis (a) by determining whether ABA is present in the seeds. Test hypothesis (b) by comparing the amount of ABA present before and after running water—enough to mimic the amount of rain that falls in a tropical rain forest over a few days or weeks—over the seeds. **4.** Yes—the experiments are consistent with the hypothesis that ABA must be on the surface of cells to trigger a response, meaning that the receptor must be located there.

CHAPTER 40

Check Your Understanding (CYU)

CYU p. 792 (1) Both are diploid cells that divide by meiosis to produce spores. The megasporocyte produces a megaspore (female spore); a microsporocyte produces a microspore (male spore). (2) Both are multicellular individuals that produce gametes by mitosis. The female gametophyte is larger than the male gametophyte and produces an egg; the male gametophyte produces sperm. **CYU p. 795** (1) Insects feed on nectar and/or pollen in flowers. Flowers provide food rewards to insects to encourage visitation. The individuals that attract the most pollinators produce the most offspring. (2) One product of double fertilization is the zygote, which will eventually grow by mitosis into a mature sporophyte. The other product is the endosperm nucleus, which will grow by mitosis to form a source of nutrients for the embryo.

You Should Be Able To (YSBAT)

YSBAT p. 786 (1) The microsporocyte is the male spore; the megasporocyte is the female spore. (2) In angiosperms the female gametophyte stays within the ovary at the base of the flower even after it matures and produces an egg cell. Fertilization and seed development take place in the same location. **YSBAT p. 795** The endosperm nucleus in the central cell is triploid; the zygote is diploid; the synergid and other cells remaining from the female gametophyte are haploid. **YSBAT p. 796** The protoderm gives rise to the epidermal tissue; the ground meristem gives rise to the ground tissue; the procambium gives rise to the vascular tissue.

Caption Questions and Exercises

Figure 40.5 An appropriate control treatment would be to graft a leaf from a plant that had never been exposed to a short-night photoperiod onto a new plant that also had never been exposed to the correct conditions for flowering. If the hypothesis is correct, this grafted leaf should not induce flowering. **Figure 40.9** A gametophyte is the multicellular individual that produces gametes by mitosis. The embryo sac conforms to this definition because it is a multicellular form that produces an egg by mitosis. **Figure 40.10** The pollen grain is a gametophyte because it is multicellular and produces sperm by mitosis.

Figure 40.17 Hypothesis: Fruit changes color when it ripens as a signal to fruit eaters. The color change is advantageous because seeds are not mature in unripe fruit, and fruit eaters disperse mature seeds in ripe fruit. Figure 40.21 Beans—their cotyledons are aboveground.

Summary of Key Concepts

KC 40.1 The sporophyte is diploid, the gametophyte is haploid, spores and gametes are haploid, and zygotes are diploid. KC 40.2 There would be four embryo sacs in each ovule, and thus four eggs. KC 40.3 The carpel is the female portion of the flower and contains the ovary. The ovary contains ovules, which house the female gametophytes. Once fertilization has occurred, the ovary develops into the fruit and the ovules develop into seeds within the fruit.

Test Your Knowledge

1. b; **2.** c; **3.** a; **4.** d; **5.** d; **6.** c

Test Your Understanding

1. Asexual reproduction is a quick, efficient way for a plant to reproduce large numbers of offspring. The disadvantage is that offspring are genetically identical to the parent, and thus vulnerable to the same diseases and only able to thrive in habitats similar to those inhabited by the parent. **2.** Megasporocytes (female) and microsporocytes (male) undergo meiosis to generate megaspores and microspores, respectively. These divide mitotically to give rise to female and male gametophytes—the embryo sac and pollen grain, respectively. **3.** Sepals are the outermost structure in a flower and usually protect the structures within; petals are located just inside the sepals and usually function to advertise the flower to pollinators. In wind- and bee-pollinated flowers, sepals probably do not vary much in overall function or general structure. Petals are often lacking in wind-pollinated species; they tend to be broad and flat (to provide a landing site) and colored purple or blue or yellow in bee-pollinated species (often with ultraviolet markings). **4.** Outcrossing increases genetic diversity among offspring, making them more likely to thrive if environmental conditions change from the parental generation. However, outcrossing requires that cross-pollination is successful. Self-fertilization results in relatively low genetic diversity among offspring but ensures that pollination is successful. **5.** The female gametophyte develops inside the ovule; the ovary develops into the pericarp of the fruit after fertilization. A mature ovule contains both gametophyte and sporophyte tissue; the ovary and carpel are all sporophyte tissue. **6.** The endosperm of a corn seed and the cotyledons of a bean seed both contain nutrients needed for the seed to germinate. The bean cotyledons have absorbed the nutrients of the endosperm.

Applying Concepts to New Situations

1. Because the island is near the equator, photoperiod will not provide much information about when conditions are optimal for germination. Presumably, daily temperature fluctuations will also remain relatively constant. Based on the information provided, the most logical hypotheses are that flowering occurs at the onset of the rainy season and that dry, dormant seeds germinate when exposed to moisture—cues that indicate that conditions for growth are good. **2.** In response to "cheater" plants, insects should be under intense selection pressure to detect and avoid species that do not offer a food reward. Individuals would have to be able to distinguish scents, colors, or other traits that identify cheaters. In response to "cheater" insects, plants possessing unusually thick petals or sepals that prevented insects from bypassing the anthers on their way to the nectaries would be more successful at producing seed in the next generation. **3.** One possibility: Under identical growth conditions, pollinate many individuals of the same species

with two pollen grains, and many individuals of this species with one pollen grain. Measure and compare the rates of pollen tube growth. **4.** Acorns have a large, edible mass and are probably animal dispersed (e.g., by squirrels that store them and forget some). Cherries have an edible fruit and are probably animal dispersed. Burrs stick to animals and are dispersed as the animals move around. Milkweed seeds float in wind. To estimate the distance that each type of seed is dispersed from the parent, (1) set up "seed traps" to capture seeds at various distances from the parent plant, (2) sample locations at various distances from the parent and analyze young individuals—using techniques introduced in Chapter 20 to determine if they are offspring from the parent being studied, (3) mark seeds, if possible, and re-find them after dispersal.

CHAPTER 41

Check Your Understanding (CYU)

CYU p. 810 *Connective tissues* consist of cells embedded in a matrix. The density of the matrix determines the rigidity of the connective tissue and its function in padding/protection, structural support, or transport. *Nervous tissue* is composed of neurons and support cells. Neurons have long projections that function as "biological wires" for transmitting electrical signals. *Muscle tissue* has cells that can contract and functions in movement. *Epithelial tissue* has tightly packed layers of cells with distinct apical and basal surfaces. They protect underlying tissues and regulate the entry and exit of materials into these tissues. **CYU p. 814** (1) As three-dimensional size increases, volume increases more quickly than does surface area, so the surface area/volume ratio decreases. (2) They are inversely proportional (negatively correlated). On a per gram basis, small animals have higher basal metabolic rates than large animals. **CYU p. 819** (1) Endotherms can remain active during the winter and at night and sustain high levels of aerobic activities such as running or flying, but require large amounts of food energy. Ectotherms need much less food and can devote a larger proportion of their food intake to reproduction, but have a hard time maintaining high activity levels at night or in cold weather. (2) [See Figure A41.1]

You Should Be Able To (YSBAT)

YSBAT p. 812 A newborn.

Caption Questions and Exercises

Figure 41.1 There would be no relationship between the variables. In all three cases, the graph would be a flat line. **Figure 41.3** Loose connective tissues have a soft matrix that is effective for cushioning and protecting organs. Dense connective tissues have a fiber-rich matrix that allows them to link adjacent structures. Structural connective tissues have a stiff or hard matrix that allows them to withstand bending or compressing forces. Fluid connective tissues are effective in transport because liquids move easily and carry solutes. **Figure 41.4** The projections allow electrical signals to be transmitted over long distances. **Figure 41.10** The dog has to eat more because it has a higher mass-specific metabolic rate than a human. **Figure 41.11** There should be no oxygen uptake on either side of the chamber.

Summary of Key Concepts

KC 41.1 Loosening tight junctions would compromise the barrier between the external and internal environments. It would allow water and other substances (including toxins) to pass more readily across the epithelial boundary. KC 41.2 King Kong is endothermic and his huge mass would generate a great deal of heat. His relatively small surface area would not be able to dissipate the heat—especially if it were covered with fur. KC 41.3 Individuals could acclimatize by normal homeostatic mechanisms (increased sweating, seeking shade, etc.). Individuals with alleles that allowed them to function better at higher temperatures would produce more offspring than individuals without those alleles. Over time, this would lead to adaptation via natural selection. KC 41.4 They multiply the amount of heat or material exchanged between the two countercurrent flows, compared to a "con-current" system.

Test Your Knowledge

1. b; **2.** d; **3.** b; **4.** d; **5.** a; **6.** a

Test Your Understanding

1. It is where environmental changes are sensed first. As an effector, it has a large surface area for losing or gaining heat. **2.** A warm frog can move, breathe, and digest much faster than a cold frog, because enzymes work faster (rates of chemical reactions increase) at 35°C than at 5°C. A warm frog needs much more food, though, to support this high metabolic rate. **3.** *Absorptive regions* have numerous folds and projections, which increase their surface area for absorption via diffusion. *Capillaries* have a high surface area because they are thin and highly branched, making exchange of fluids and gases efficient. *Beaks of Galápagos finches* have sizes and shapes that correlate with the type of food. Large beaks are used to crack large seeds; long, thin beaks are used to pick insects off surfaces; etc. *Fish gills* are thin, flattened structures with a large surface area, which facilitates the efficient exchange of gases and wastes. **4.** A larger sphere has a relatively smaller surface area for its interior volume than does a smaller sphere, because surface area increases with size at a lower rate than volume does. Diffusion will be more efficient in the case of the smaller sphere. **5.** In the morning, an ant should move to a sunlit spot to gain heat by conduction from the ground and radiation striking its body directly. In midday, the ant should avoid overheating by losing heat from wet body surfaces, standing in breezy areas to lose heat by convection, or retreating to the cool burrow to lose heat by conduction and avoid gaining heat by radiation. **6.** The feedback causes a response that counters the current state of the system. It is negative in the sense of reducing the difference between the current state and the set point.

Applying Concepts to New Situations

1. Large individuals require more food, so are more susceptible to starvation if food sources can't be defended (e.g., when seeds are scattered around). They may also be slower and/or more visible to predators. **2.** You would have to show changes in the frequencies of alleles whose products are affected by temperature—for example, increases in the frequencies of alleles for enzymes that operate best at higher temperature.

FIGURE A41.1

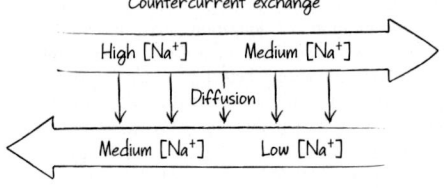
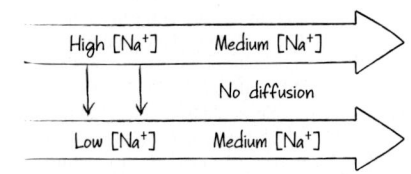

3. The heat-dissipating system should have a very high surface area/volume ratio. Based on biological structures, it should be flattened (thin) and highly folded, branched, and/or contain tubelike projections. Another approach inspired by biological systems would be to set up a countercurrent heat exchanger with fluid heated by the engine. **4.** Buy the turtle. Turtles are ectothermic, meaning they do not expend energy to regulate their body temperature. Mice are endothermic, meaning they do expend energy to regulate body temperature. Mice will therefore have a higher basal metabolic rate and consume more food.

CHAPTER 42

Check Your Understanding (CYU)

CYU p. 828 Without the "master gradient" established by the pump, sodium ions and chloride ions cannot move out of the epithelial cells into the surrounding seawater. Salt will build up in the shark's tissues. **CYU p. 831** (1) Uric acid is insoluble in water and can be excreted without much water loss. (2) When electrolytes are reabsorbed in the hindgut epithelia, water follows along an osmotic gradient. **CYU p. 838** (1) Aldosterone stimulates Na^+ reabsorption in pre-urine in the distal tubule. This increases the Na^+ level in the blood and leads to a more dilute (hypotonic) urine. (2) Water intake increases blood pressure and filtration rate in the renal corpuscle, and leads to lowered electrolyte concentration in the blood and filtrate and production of dilute urine. Eating large amounts of salt results in concentrated, hypertonic urine. Water deprivation triggers ADH release and the production of concentrated, hypertonic urine.

You Should Be Able To (YSBAT)

YSBAT p. 833 Blood contains cells and large molecules as well as electrolytes and wastes; the filtrate contains only electrolytes and wastes. **YSBAT p. 835** If sodium reabsorption is inhibited, then less water will be reabsorbed along an osmotic gradient and more urine will be produced. **YSBAT p. 837** (1) Less, because the osmotic gradient is not as steep. (2) Lower, because more water has been retained. (3) Less, because the concentration gradient is not as steep. **YSBAT p. 838** Ethanol consumption inhibits water reabsorption, leading to a larger volume of less concentrated urine. Nicotine consumption increases water reabsorption, leading to a smaller volume of more concentrated urine.

Caption Questions and Exercises

Figure 42.1 The purple and white molecules are moving down their concentration gradients, and the concentration of red molecules does not affect the concentration of purple or white molecules. **Figure 42.6** The sodium-potassium ATPase does active transport; the sodium-chloride-potassium cotransporter does secondary active transport; the chloride and potassium channels do passive transport. **Figure 42.8** The underside of the abdomen is shaded by the grasshopper's body. This location is relatively cool and should reduce the loss of water during respiration.

Summary of Key Concepts

KC 42.1 In the ocean, river otters must excrete the additional salt that is taken up from saltier marine foods and seawater. In freshwater, their bodies must conserve electrolytes and excrete the water taken in during drinking. **KC 42.2** A sodium gradient establishes an osmotic gradient that moves water. It also sets up an electrochemical gradient that can move ions against their electrochemical gradient by secondary transport through a cotransporter, or with their electrochemical gradient by passive diffusion through a channel. **KC 42.3** The desert locust—to save water, it has to ac-

tively reabsorb ions in the hindgut so that almost no water is lost during excretion. **KC 42.4** The longer the loop of Henle, the steeper the osmotic gradient in the medulla. Steep osmotic gradients allow a great deal of water to be reabsorbed as urine passes through the collecting duct.

Test Your Knowledge

1. c; **2.** a; **3.** d; **4.** d; **5.** a; **6.** c

Test Your Understanding

1. In salt water, a salmon gains NaCl by diffusion and loses water by osmosis. It replaces water by drinking and secretes ions through chloride cells in its gill epithelium. In freshwater, it gains water by osmosis and loses NaCl by diffusion. Epithelial cells in the gills take up ions, it stops drinking, and produces copious amounts of urine to rid itself of excess water. **2.** Mitochondria are needed to produce ATP. A key component of salt regulation in fish is the Na^+/K^+-ATPase, which requires ATP to function. Because ATP fuels establishment of the ion gradients necessary for the cotransport mechanisms utilized by chloride cells as well as other epithelial cells, an abundance of mitochondria would be expected. **3.** Microvilli increase the surface area available for pumps, cotransporters, and channels. **4.** The wax layer should be thinner in tropical insects, because the animals are under less osmotic stress (less evaporation due to high humidity). You could test this prediction by measuring the thickness of the wax layer in two closely related species from each habitat, or in a wide array of species from each habitat. Insects that live in extremely humid habitats may lack the ability to close the openings to their respiratory passages; animals that live in humid habitats may excrete primarily ammonia or urea and have relatively short loops of Henle. In each case, you could test these predictions by comparing individuals of closely related species that live in dry versus humid habitats. **5.** In both cases, the initial filtrate is isotonic with blood and reabsorption of certain solutes depends on Na^+/K^+-ATPase activity and is energetically costly. One difference is that the initial filtrate forms by active pumping of ions followed by osmosis in insects versus by filtration in the renal corpuscle of mammals. Another difference is that mammals make use of a countercurrent exchanger in the loop of Henle versus direct pumping of ions in the insect hindgut. **6.** Ammonia is toxic and must be diluted with large amounts of water to be excreted safely. Urea and uric acid are safer and do not have to be excreted with large amounts of water, but are more expensive to produce in terms of energy expenditure. Fish excrete ammonia; mammals excrete urea; insects excrete uric acid. You would expect the embryos inside terrestrial eggs to excrete uric acid, as it is the least toxic and is insoluble in water.

Applying Concepts to New Situations

1. The countercurrent flow removes salt from the tissues around the ascending limb, raising the osmolarity of blood, and then water from the tissues around the descending limb, along an osmotic gradient. The countercurrent arrangement maintains a concentration and osmotic gradient all along the length of the nephron. **2.** The observation that the Na^+ concentration decreased by 30 percent, even as volume also decreases, indicates that Na^+ ions were reabsorbed to a greater degree than water. The observation that the urea concentration increased by 50 percent indicates that urea was not reabsorbed. **3.** Increased salt concentrations in freshwater environments put organisms under a new kind of osmotic stress. They may not have chloride cells or other adaptations that allow them to rid themselves of additional salt. **4.** Their urine volume is much higher (actually 10 times).

CHAPTER 43

Check Your Understanding (CYU)

CYU p. 856 (1) *Mouth:* Food is taken in; teeth physically break down food into smaller particles; salivary amylase begins to break down carbohydrates; lingual lipase initiates the digestion of fats. *Esophagus:* Food is moved to the stomach via peristaltic contractions. *Stomach:* HCl denatures proteins; pepsin begins to digest them. *Small intestine:* Pancreatic enzymes complete the digestion of carbohydrates, proteins, lipids, and nucleic acids. Most of the water and all of the nutrients are absorbed here. *Large intestine:* Water is reabsorbed and feces are formed. *Anus:* Feces accumulate in the rectum and are expelled out the anus. (2) If the release of bile salts is inhibited, fats would not be digested and absorbed efficiently, and they would pass into the large intestine. The individual would likely produce fatty feces and lose weight over time. Trypsin inactivation would inhibit protein digestion and reduce absorption of amino acids, likely leading to weight loss and muscle atrophy due to protein deprivation. If the Na^+-glucose cotransporter were blocked, then glucose would not be absorbed into the bloodstream. The individual would likely become sluggish from lack of energy. **CYU p. 858** (1) Both lead to high urine volume due to reduced water reabsorption in the kidneys, but for different reasons. In diabetes mellitus, less water is reabsorbed because glucose concentrations in the filtrate from blood are high. In diabetes insipidus, less water is reabsorbed because of defects in the collecting ducts. (2) When the blood sugar of individuals with type I diabetes mellitus gets too high, they should inject themselves with insulin, which will trigger the absorption and storage of glucose by their cells. When their blood glucose gets too low, they should eat something with a high concentration of sugar (such as orange juice or a candy bar) to increase their blood glucose levels quickly.

You Should Be Able To (YSBAT)

YSBAT p. 851 Nutrient absorption occurs in the small intestine, but not in the esophagus or stomach, and the rate of absorption increases with available surface area. **YSBAT p. 854** (1) Decreased energy yield from foods due to reduced glucose absorption, (2) increased feces production due to the passing of unabsorbed glucose into the colon, (3) watery feces due to decreased water reabsorption in the large intestine (lower osmotic gradient). As an aside, also increased flatulence due to the metabolism of unabsorbed glucose by bacteria that produce methane as a waste product. **YSBAT p. 857** Individuals with type I diabetes have normal insulin receptors on their liver cells, but do not produce and release insulin from the pancreas. Individuals with type II diabetes produce and release insulin, but their insulin receptors are defective so liver cells cannot respond to insulin. In both types of diseased individuals, high glucose concentrations in the blood remain high. In individuals without disease, both insulin production and insulin receptors are normal, so liver cells respond to insulin by taking up glucose from the blood and storing it.

Caption Questions and Exercises

Figure 43.3 Your label should point to the rightmost two bones in the illustration. **Figure 43.14** Frog eggs don't normally make the sodium-glucose cotransporter protein, so the researchers could be confident that if the protein appeared, it was from the injected RNA. They would not be confident of this if the RNAs were injected into rabbit epithelial cells, where the protein was probably already present. **Figure 43.16** In type 2 diabetes mellitus, the top, black arrow from glucose (in the blood) to glycogen (inside liver and muscle cells) is disrupted. In type 1 diabetes mellitus, the green arrow on the top left, indicating insulin produced by the pancreas, is disrupted. **Figure 43.17** The leading hypothesis is an increased incidence of obesity, which is linked to development of type 2 diabetes mellitus.

Summary of Key Concepts

KC 43.1 Form two groups of mice selected from a common population, so that individuals have similar genetic characteristics and history. One group (the control group) would receive a regular diet of rat chow, and the other group (the experimental group) would receive a diet of rat chow that was deficient in magnesium. By comparing the two groups on a number of behavioral and physical measures, you could determine the effects of magnesium deficiency. **KC 43.2** Their mouth or tongue would be modified to make probing of flowers and sucking nectar efficient. They would not require teeth, and their digestive tract would be relatively simple. For example, the stomach would not have to pulverize food or digest large amounts of protein, and the large intestine would not have to process and store large amounts of bulky waste. **KC 43.3** By making the stomach smaller, gastric bypass surgery decreases the amount of food that can be comfortably contained in the stomach and digested. By bypassing a portion of the small intestine, the surgery decreases the absorption of food. (As an aside, people who have this surgery are at risk for serious nutrient deficiencies, chronic intestinal upset, ulcers, and flatulence.) **KC 43.4** The excess glucose is eventually eliminated in urine. (This takes much longer than sequestering the excess glucose in liver, muscle, or adipose cells.)

Test Your Knowledge

1. a; **2.** a; **3.** d; **4.** d; **5.** b; **6.** d

Test Your Understanding

1. The bird crop is an enlarged sac that can hold quickly ingested food; in leaf-eating species it is filled with symbiotic organisms and functions as a fermentation vessel. The cow rumen is an enlarged portion of the stomach; the elephant large intestine is enlarged relative to other species. Both structures are filled with symbiotic organisms and function as fermentation vessels. **2.** Digestive enzymes break down macromolecules (proteins, carbohydrates, nucleic acids, lipids). If they weren't produced in an inactive form, they would destroy the cells that produce and secrete them. **3.** Most biologists accept the hypothesis, for two reasons: (1) the structure is unique among fish in being effective at biting, and (2) its surface correlates with the type of food each species eats. **4.** In most terrestrial vertebrates, the primary function of the large intestine is water reabsorption; fish do not need to reabsorb water. **5.** When an individual ingests a solution of glucose and electrolytes, the solutes are absorbed. Water follows by osmosis, preventing dehydration. **6.** Insulin triggers negative feedback to high blood glucose concentrations by stimulating glucose uptake by several types of cells. Glucagon triggers negative feedback to low blood glucose concentrations by stimulating glucose release by several types of cells.

Applying Concepts to New Situations

1. Because they are feeding growing embryos and newborns, it is likely that female mammals during pregnancy and breastfeeding would require higher levels of almost every nutrient—particularly calcium, used by offspring for bone growth. **2.** Individuals with defects in pancreatic amylase would not be able to complete carbohydrate digestion and probably would be lethargic (low energy) and experience weight loss. Pepsin defects would reduce or eliminate protein digestion in the stomach, leading to severe amino acid deficiency. Defects in the fatty-acid binding protein would reduce or eliminate fatty-acid absorption in the small intestine, likely producing fatty feces, weight loss, and diarrhea.

Aquaporin defects would prevent water reabsorption in the large intestine and lead to diarrhea and dehydration. **3.** The result is still valid because the injection was correlated with secretion—if lack of injection was correlated with lack of secretion. The criticism is also somewhat valid, however, because the researchers couldn't rule out the hypothesis that signaling from nerves plays some sort of role, too. **4.** Terrestrial animals are exposed to increased risk of water loss, and the large intestine is where water reabsorption occurs. In most cases fish do not need to reabsorb large amounts of water from their feces.

CHAPTER 44

Check Your Understanding (CYU)

CYU p. 864 (1) Oxygen partial pressure is high in mountain streams because the water is cold, mixes constantly, and has a high surface area (due to white water). Oxygen partial pressure is low at the ocean bottom because the area is far from the surface where gas exchange takes place and there is relatively little mixing. (2) *Warm-water species:* Large amount of air because the oxygen-carrying capacity of warm water is low. *Vigorous algal growth:* Small amount of air because algae contribute oxygen to the water through photosynthesis. *Sedentary animals:* Small amount of air because sedentary animals require relatively little oxygen. **CYU p. 870** (1) Common features include large surface area, short diffusion distance (a thin gas exchange membrane), and a mechanism that keeps fresh air or water moving over the gas-exchange surface. Only fish gills use a countercurrent exchange mechanism; only tracheae deliver oxygen directly to cells without using a circulatory system; only lungs contain "dead space"—areas that are not involved in gas exchange. (2) The anterior and posterior air sacs provide additional compartments for inhaled and exhaled air; they are arranged so that air moves through them and the lungs in one direction. **CYU p. 874** The saturation curve for Tibetans should be shifted to the left relative to people from sea level—meaning that their hemoglobin has a higher affinity for oxygen at all partial pressures. **CYU p. 883** (1) If the blood plasma that leaks out of the capillaries is not reabsorbed in capillaries, it enters lymphatic vessels that eventually merge with blood vessels. (2) [See Figure A44.1]

You Should Be Able To (YSBAT)

YSBAT p. 872 There would be an even larger change in the oxygen saturation of hemoglobin in response to an even smaller change in the partial pressure of oxygen.

Caption Questions and Exercises

Figure 44.4 External gills are ventilated passively and are efficient because they are in direct contact with water. They are exposed to predators and mechanical damage, however. Internal gills are protected but have to be ventilated by some type of active mechanism for water flow. **Figure 44.7** Using several to many different animals increases confidence that the results are true for most or all individuals in the population, and not due to one or a few unusual individuals or circumstances. **Figure 44.10** Sighing increases pressure in the lung cavity, forcing air out of the lungs more rapidly and completely than occurs during normal exhalation. When taking a deep breath, volume in the lung cavity is increased, resulting in lower pressure and causing more air to enter the lungs than during normal inhalation. **Figure 44.15** According to data in the figure, the oxygen saturation of hemoglobin is about 15 percent for blood at pH 7.2 and about 25 percent at pH 7.4. Therefore, about 85 percent of the oxygen is released from hemoglobin at pH 7.2, but only about 75 percent of the oxygen is released at pH 7.4. **Figure 44.17** In the lungs, a strong partial pressure gra-

dient favors diffusion of dissolved CO_2 from blood into the alveoli. As the partial pressure of CO_2 in the blood declines, hydrogen ions leave hemoglobin and react with bicarbonate to form more CO_2, which then diffuses into the alveoli and is exhaled from the lungs. **Figure 44.23** Air from the alveoli mixes with air in the "dead space" in the bronchi and trachea on its way out of the body. This dead-space air is from the previous inhalation ($P_{O_2} = 160$ mm Hg; $P_{CO_2} = 0.3$), so when the alveolar air mixes with the dead-space air, the partial pressures in the exhaled air achieve levels intermediate between that of inhaled and alveolar air.

Summary of Key Concepts

KC 44.1 It is harder to extract oxygen from water than it is to extract it from air. This limits the metabolic rate of water breathers. **KC 44.2** Large animals have a relatively small body surface area relative to their volume. If they had to rely solely on gas exchange across their skin, they would not have enough skin surface area to exchange the volume of oxygen needed to meet their metabolic needs. **KC 44.3** Cold water carries more oxygen than warm water, so icefish blood can carry enough oxygen to supply the tissues with oxygen even in the absence of hemoglobin. The oxygen and carbon dioxide are simply dissolved in the blood. **KC 44.4** Their circulatory systems should have relatively high pressures—achieved by independent systemic and pulmonary circulations powered by a four-chambered heart—to maximize the delivery of oxygenated blood to metabolically active tissues.

Test Your Knowledge

1. b; **2.** c; **3.** d; **4.** d; **5.** c; **6.** a

Test Your Understanding

1. The insect tracheal system delivers O_2 directly to respiring cells, but in humans, O_2 is first taken up by the blood of the circulatory system, which then delivers it to the cells. The human respiratory system also has specialized muscles devoted to ventilating the lungs. The open circulatory system of the insect is a low-pressure system, which makes use of pumps (hearts) and body movements to circulate hemolymph. The closed circulatory system of humans is a high-pressure system, which can respond to rapid changes in O_2 demand by tissues. **2.** During exercise, P_{O_2} decreases in the tissues and P_{CO_2} increases. The increase in P_{CO_2} lowers tissue pH. The drop in P_{O_2} and pH causes more oxygen to be released from hemoglo-

FIGURE A44.1

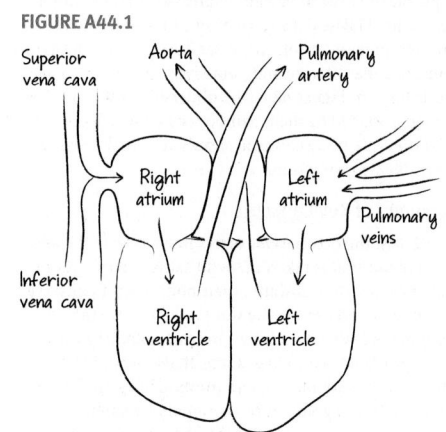

bin. **3.** In blood, CO_2 is converted to carbonic acid, which dissociates into a proton and a bicarbonate ion. Because the protons bind to deoxygenated hemoglobin, they do not cause a dramatic drop in blood pH (hemoglobin acts as a buffer). Also, small decreases in blood pH trigger a homeostatic response that increases breathing rate and expulsion of CO_2. **4.** Airflow through bird lungs is unidirectional, so it allows for continuous ventilation of gas-exchange surfaces with fresh, oxygenated air. The trachea and bronchi in a mammalian lung do not have a gas-exchange surface, and bidirectional airflow in these species means that "stale" air has to be expelled before "fresh" air can be inhaled. As a result, the alveoli are not ventilated continuously. **5.** Lungs increase the temperature of the air and are moist to allow greater solubility of gases (increasing k); alveoli present a large surface area (large A), the epithelium of alveoli is thin (small D), and constant delivery of deoxygenated blood to alveoli maintains a steep partial pressure gradient favoring diffusion of oxygen into the body ($P_2 - P_1$ is high). **6.** If the pulmonary circulation was under pressure as high as that found in the systemic circulation, large amounts of fluid would be forced out of capillaries in the lungs. There is a conflict between the thin surface required for efficient gas exchange and thick blood vessels required to withstand high pressure.

Applying Concepts to New Situations

1. In species that depend on rapid movement to chase down prey or perform other key actions, individuals with larger gill surfaces would absorb more O_2 from water and likely produce more offspring than individuals with smaller gill-surface areas. In slow-moving species, there would be no advantage to increasing the surface area in the gills, so such an adaptation would not be preferentially selected for. **2.** A shift to the right of the oxygen-hemoglobin dissociation curve represents a decrease in the affinity of hemoglobin for O_2. Thus DPG promotes the release of O_2 from hemoglobin into the tissues. **3.** Yes—the trait compensates for the small P_{O_2} gradient between stagnant water and the blood of the carp. **4.** Since O_2 cannot compete as well as CO for binding sites in hemoglobin, O_2 transport decreases. As oxygen levels in the blood drop, tissues (particularly the brain) become deprived of oxygen, and suffocation occurs.

CHAPTER 45

Check Your Understanding (CYU)

CYU p. 891 (1) The resting potential would fall (be much less negative) because K^+ could no longer leak out of the cell. (2) The size of an action potential from a particular neuron does not vary, so it cannot contain information. Only the frequency of action potentials from the same neuron varies. **CYU p. 895** (1) Once the threshold level of depolarization is attained, the probability that the voltage-gated sodium channels will open approaches 100 percent. But below threshold, the massive opening of Na^+ channels does not occur. This is why the action potential is "all or none." (2) Na^+ would continue to move into the cell as voltage-gated potassium channels opened and potassium began to diffuse out of the cell. As a result, repolarization would take much longer. **CYU p. 899** If neurons made direct electrical connections, action potentials would simply travel from one neuron to the next. The presence of synapses allows information from many different neurons to affect the activity of a postsynaptic neuron, through the process of summation. **CYU p. 904** (1) [See Figure A45.1] (2) 1. Study individuals with brain damage and correlate the location of the defect

with a deficit in mental or physical function. 2. Directly stimulate brain areas in conscious patients during brain surgery and record the response.

You Should Be Able To (YSBAT)

YSBAT p. 890 The Na^+/K^+-ATPase makes the inside of the membrane less positive (more negative) than the outside, and generates a concentration gradient favoring movement of K^+ out of the cell, through the K^+ leak channel. As K^+ leaves, the resting potential becomes even more negative. **YSBAT p. 891** [See Figure A45.2] **YSBAT p. 892** (1) Na^+ channels open when the cell membrane depolarizes, and as more channels open, more Na^+ flows into the cell, further depolarizing the cell, causing more voltage-gated Na^+ channels to open. (2) After opening, the voltage-gated Na^+ channels close and temporarily cannot reopen. Also, sodium reaches its equilibrium potential, so no net force is available to drive Na^+ movement. (3) K^+ channels open in response to membrane depolarization. **YSBAT p. 897** The inside of the cell becomes more positive (depolarized). This shifts the membrane potential closer to the threshold, making it easier for the postsynaptic cell to fire an action potential.

Caption Questions and Exercises

Figure 45.3 No—as K^+ leaves the cell along its concentration gradient, the interior of the cell becomes

more negative. As a result, an electrical gradient favoring movement of K^+ into the cell begins to counteract the concentration gradient favoring movement of K^+ out of the cell. Eventually, the two opposing forces balance out, and there is no net movement of K^+. **Figure 45.10** (1) Get a solution taken from the synapse between the heart muscle and the vagus nerve *without* the nerve being stimulated. Expose a second heart to this solution. There should be no change in heart rate. **Figure 45.16** In the "rest and digest" mode, pupils take in less light stimulation, the heartbeat slows to conserve energy, and liver conserves glucose and promotes digestion by stimulating release of gallbladder products. In the "fight or flight" mode, the pupils open to take in more light, the heartbeat increases to support muscle activity, and the liver releases glucose and inhibits digestion. **Figure 45.18** Point to your forehead and top of your head for the frontal lobe, the top and top rear of your head for the parietal lobe, the back of your head for the occipital lobe, and the sides of your head (just above your ear openings) for the temporal lobes. **Figure 45.19** No—for example, part (b) indicates that the size of the brain area devoted to the trunk is no bigger than the size of the brain area devoted to the thumb.

FIGURE A45.1

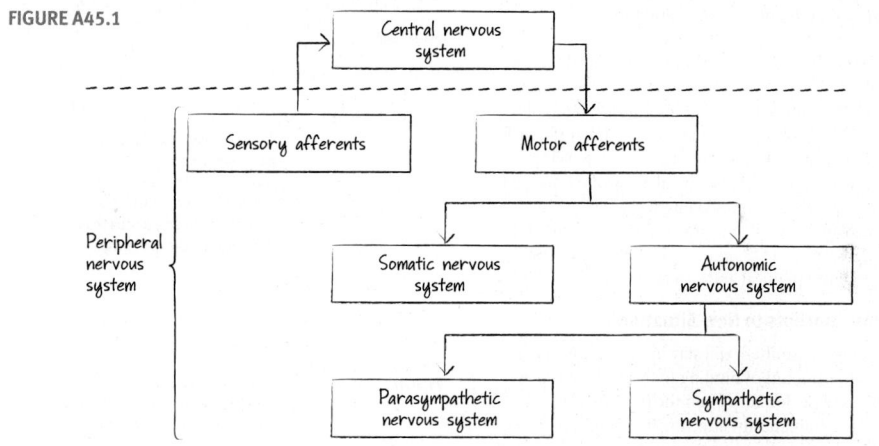

FIGURE A45.2

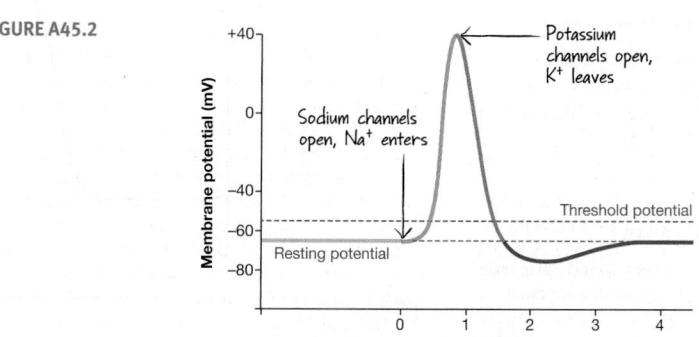

Summary of Key Concepts

KC 45.1 [See Figure A45.3] **KC 45.2** Because the behavior of voltage-gated Na⁺ and K⁺ channels does not change. Every time a membrane depolarizes past threshold and all of the voltage-gated Na⁺ channels are open, the same amount of Na⁺ will enter the cell and the cell will depolarize to the same extent. In addition, K⁺ channels open on the same time course and allow the same amount of K⁺ to leave the cell, meaning that it repolarizes to the same extent. **KC 45.3** Summation means that input from many neurons is involved in triggering a response from a postsynaptic cell. Thus, an action potential from a postsynaptic cell requires integration of input from many synapses. **KC 45.4** The animal's ability to learn—and perhaps remember—would be impaired or nonexistent.

Test Your Knowledge

1. b; 2. d; 3. c; 4. a; 5. c; 6. a

Test Your Understanding

1. The Na⁺/K⁺-ATPase pumps 3 Na⁺ out of the cell for every 2 K⁺ it brings in. Since more positive charges leave the cell than enter it, there is a difference in charge on the two sides of the membrane and thus a voltage. 2. [See Figure A45.4] 3. The ligand-gated channel opens or closes in response to binding by a small molecule (for example, a neurotransmitter); the voltage-gated channel opens or closes in response to changes in membrane potential. 4. EPSPs and IPSPs are based on flows of ions, which change the membrane potential in the postsynaptic cell. If a flow of ions at one point depolarizes the cell but a nearby flow of ions hyperpolarizes the cell, then in combination the flows of ions cancel each other out. But if two adjacent flows of ions depolarize the cell, then the total amount of ion movement—and thus the total change in membrane potential—sums. 5. The somatic system responds to external stimuli and controls voluntary skeletal muscle activity, such as movement of arms and legs. The autonomic system responds to internal stimuli and controls internal involuntary activities, such as digestion, heart rate, and gland activities. 6. Sympathetic nerves trigger responses in an array of organs and tissues that promote increased physical activity and heightened awareness. Parasympathetic nerves cause the opposite response in the same organs and tissues, promoting reduced physical and mental activity.

Applying Concepts to New Situations

1. The neurotransmitters will stay in the synaptic cleft longer so the amount of binding to ligand-gated channels will increase. Ion flows into the postsynaptic cell will increase dramatically, affecting its membrane potential and likelihood of firing action potentials. 2. Diseased or damaged brains may not respond like healthy and undamaged brains. Also, the extent of lesions and the exact location of electrical stimulation may be difficult to determine, making correlations between the regions affected and the response imprecise. 3. Because the current is reduced from normal levels, but not eliminated completely, there must be more than one type of potassium channel present. (The poison probably knocks out just one specific type of channel.) 4. Record from the neuron while the individual is performing various tasks—feeding, moving, courting, etc.

CHAPTER 46

Check Your Understanding (CYU)

CYU p. 913 (1) The outer ear collects sound waves from the environment and directs them into the ear canal. The middle ear amplifies sound. The inner ear contains hair cells that transduce the information in sound waves to action potentials that are sent to the brain. (2) A punctured eardrum wouldn't vibrate correctly and would result in hearing loss at all frequencies in the affected ear. If the stereocilia are too short to come into contact with the tectorial membrane, vibration of the basilar membrane will not cause them to bend, and sound will not be detected. A loss in basilar membrane flexibility would result in the inability to hear lower-pitched sounds, such as human speech. **CYU p. 918** (1) Retinal acts like an on-off switch that indicates whether light has fallen on a rod cell. When retinal absorbs light, it changes shape. The shape change triggers events that result in a change in action potentials that signals that light has been absorbed. (2) A tear in the fovea would likely result in blurred vision in the affected eye. The mutation would produce blue-purple color blindness because this opsin responds to wavelengths in that region of the spectrum. A clouded lens would reduce the amount of light that reaches the retina, reducing visual sensitivity. **CYU p. 920** (1) Extremely hot food damages taste receptors, so proteins would no longer be able to respond to their chemical triggers. (2) Dogs have more than twice the number of odor receptors as humans. **CYU p. 926** (1) In a sarcomere, thick myosin filaments are sandwiched between thin actin filaments. When the heads on myosin contact actin and change conformation, they pull the actin filaments toward one another, shortening the whole sarcomere. (2) Increased ACh release would result in an increased rate of muscle cell contraction. Preventing conformational changes in troponin would prevent muscle contraction. Blocking the uptake of calcium ions into the sarcoplasmic reticulum would lead to sustained muscle contraction.

You Should Be Able To (YSBAT)

YSBAT p. 916 cGMP is a ligand that opens sodium channels and a second messenger in the sense that lack of it carries a signal from activated transducin. Transducin is like a G protein because it switches from "off" to "on" in response to a receptor protein, and activates a key pro-

---

FIGURE A45.3

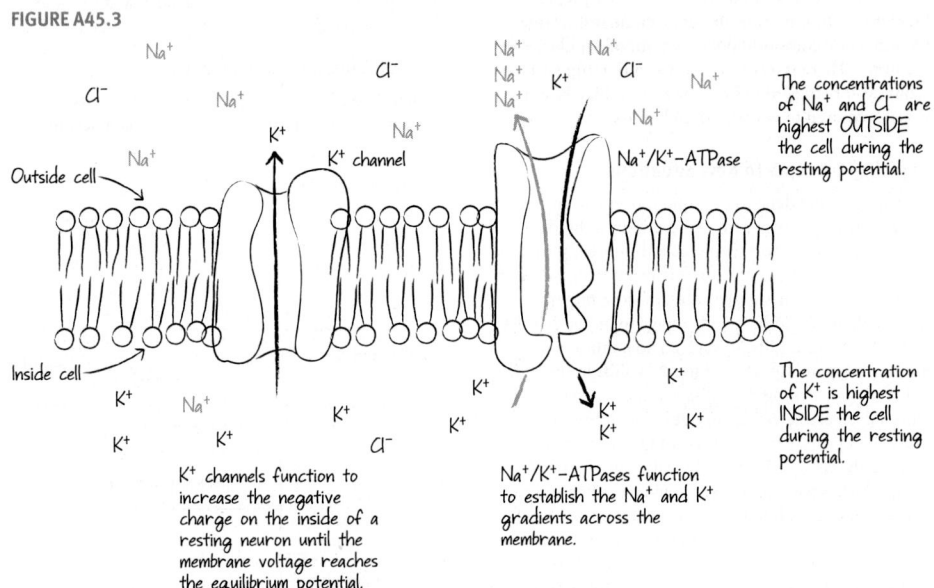

The concentrations of Na⁺ and Cl⁻ are highest OUTSIDE the cell during the resting potential.

The concentration of K⁺ is highest INSIDE the cell during the resting potential.

K⁺ channels function to increase the negative charge on the inside of a resting neuron until the membrane voltage reaches the equilibrium potential.

Na⁺/K⁺-ATPases function to establish the Na⁺ and K⁺ gradients across the membrane.

FIGURE A45.4

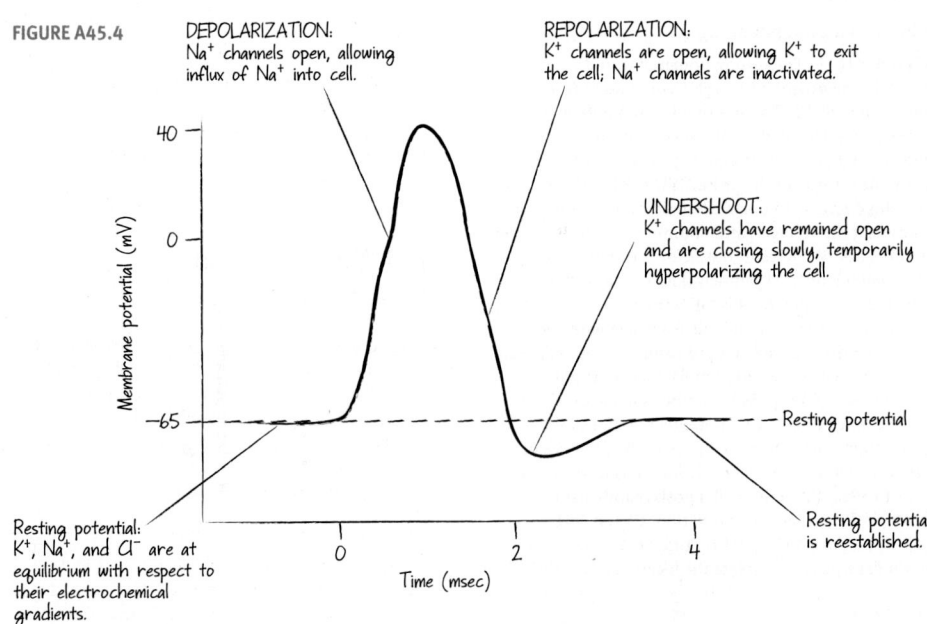

DEPOLARIZATION: Na⁺ channels open, allowing influx of Na⁺ into cell.

REPOLARIZATION: K⁺ channels are open, allowing K⁺ to exit the cell; Na⁺ channels are inactivated.

UNDERSHOOT: K⁺ channels have remained open and are closing slowly, temporarily hyperpolarizing the cell.

Resting potential

Resting potential is reestablished.

Resting potential: K⁺, Na⁺, and Cl⁻ are at equilibrium with respect to their electrochemical gradients.

tein in response. Closing Na$^+$ channels stops the entry of positive charges into the cell, making the inside of the cell more negative relative to the outside. YSBAT p. 923 The trucks are the Z disks, the ropes are the thin filaments, and the burly weightlifters are the thick filaments.

Caption Questions and Exercises

Figure 46.9 The change in retinal's shape would induce a change in opsin's shape. A protein's function correlates with its shape, so opsin's function is likely to change as well. **Figure 46.10** This ion flow occurs when the photoreceptor cell is not receiving light—when the cell is in the dark. **Figure 46.12** The S, M, and L opsin proteins are each different in structure. These structural differences affect the ability of retinal to absorb specific frequencies of light and change shape. **Figure 46.18** The dark band includes thin filaments and a dense concentration of bulbous structures extending from the thick filament; the light band consists of just thin filaments.

Summary of Key Concepts

KC 46.1 The sensory neurons that used to serve that limb are cut, but may still fire action potentials that stimulate brain areas associated with awareness of pain in the missing limb. **KC 46.2** If dense, starch-filled amyloplasts exerted pressure on hair-cell-like receptors, the stereocilia would bend, causing a change in the membrane potential of the receptors and thus transducing the signal from gravity. **KC 46.3** Because different animal species have opsin molecules in their cone cells that respond to different wavelengths of light absorbed by retinal, they see different colors. **KC 46.4** Smells are part of the sensation of flavor, and people with nasal congestion cannot smell. In addition to smell being important, the small number of taste receptors can send many different frequencies of action potentials to the brain, and the taste receptors can be stimulated in many different combinations. **KC 46.5** Paralysis, because ACh has to bind to its receptors on the membrane of postsynaptic muscle fibers for action potentials to propagate in the muscle and cause contraction.

Test Your Knowledge

1. a; **2.** d; **3.** c; **4.** b; **5.** d; **6.** a

Test Your Understanding

1. The brain can distinguish sensory stimuli because the axons from different sensory neurons go to different and specific areas of the brain. For example, the axons of all the olfactory neurons that have the same receptor proteins go to the same location in the olfactory bulb, resulting in the sensation of smelling a particular odor. **2.** Many possibilities, including several blue opsins allowing coelacanths to distinguish the blue wavelengths present in their deep-sea habitat, infrasound hearing allowing elephants to hear over long distances, ultrasonic hearing allowing bats to hunt by echolocation and moths to avoid bat predators, red/yellow opsins allowing fruit-eating mammals to distinguish ripe from unripe fruit. **3.** It answered the question of why so many different smells can be detected and identified separately. Each type of odorant has its own receptor. **4.** Ion channels in hair cells are thought to open in response to physical distortion caused by the bending of stereocilia. Ion channels in taste receptors, in contrast, change shape and open when certain bitter-tasting molecules bind to them. **5.** Dalton's hypothesis was reasonable because blue fluid absorbs red light and would prevent it from reaching the retina, and could be tested by dissecting his eyes. Although the hypothesis was rejected, it inspired a rigorous test and required researchers to think of alternative explanations. **6.** The key observation was that the banding pattern of sarcomeres changed during contraction. Even though the entire unit became shorter, only some portions moved relative to each other. This obser-

vation suggested that some portions of the structure slid past other portions. Muscle fibers have to have many mitochondria because large amounts of ATP are needed to power myosin heads to move along actin filaments; large amounts of calcium stored in smooth ER are needed to initiate contraction by binding to troponin.

Applying Concepts to New Situations

1. If ACh receptors are covered, they cannot bind to ACh and become activated. Muscles will have trouble contracting. **2.** Block all but a few hundred ommatidia in the center of each compound eye of many dragonflies with an opaque material, and observe any differences that might occur in their ability to detect and pursue prey compared to individuals whose ommatidia had been covered by a transparent material. **3.** Ground-dwelling birds have large amounts of "white meat" in their breasts, used for powering short, rapid flights, but "dark meat" in their legs, used for endurance running. Birds that rely on long, sustained flights need "dark meat" in the breast muscles that power flight. **4.** When a person focuses directly on an object, the image lands on the fovea, which contains relatively few rods. But if the image lands beside the fovea, there are many more rod cells. Rod cells are specialized for detecting dim light.

CHAPTER 47

Check Your Understanding (CYU)

CYU p. 940 (1) By reducing immune system function, promoting release of fatty acids from storage cells and use of amino acids from muscles for energy production, and preventing release of glucose in response to signals from insulin, cortisol conserves glucose supplies for use by the brain. (2) Increased heart rate, glucose and fatty-acid release, and blood pressure all support rapid action in response to danger. If an animal were not able to respond quickly and at the maximum level, it might not escape predators or successfully challenge rivals. **CYU p. 943** (1) ACTH triggers the release of cortisol, but cortisol inhibits ACTH release by blocking the release of CRH from the hypothalamus and suppressing ACTH production. High levels of cortisol tend to lower cortisol levels in the future. (2) Processing centers in the brain are responsible for synthesizing a wide array of sensory input. To start a response to this sensory input, they stimulate neurosecretory cells in the hypothalamus. Secretions from these cells travel to the anterior pituitary, where they trigger the production and release of hormones that control hormones that stimulate the appropriate activity. **CYU p. 947** (1) The steroid-hormone receptor complex would probably fail to bind to the hormone-response element. If so, then gene expression would not change—the arrival of the hormone would have little or no effect on the target cell. (2) The mode of action would be similar—hormone binding would have to trigger a signal transduction event that resulted in production of a second messenger or a phosphorylation cascade.

You Should Be Able To (YSBAT)

YSBAT p. 932 Room temperature is analogous to the sensory input; the thermostat, to the CNS; the thermostat signal, to the cell-to-cell signal; and the furnace, to the effector. If feedback inhibition fails, hormone production will not stop and the response will continue indefinitely. **YSBAT p. 936** The cells could have (1) different receptors, (2) different signal transduction systems, or (3) different response systems (genes or proteins that can be activated).

Caption Questions and Exercises

Figure 47.4 Injecting the dog with a liquid extract from a different part of the body, or with a solution that is similar in chemical composition to the extract from the

pancreas (e.g., similar pH or ions present) but lacking molecules produced by the pancreas. **Figure 47.7** The saline injection controlled for any stress induced by the injection procedure and for introducing additional fluids. **Figure 47.9** Same—a defect in both signal and receptor has the same effect—no response—as a defective signal alone or a defective receptor alone. **Figure 47.10** Inject ACTH and monitor cortisol levels. If cortisol does not increase, adrenal failure is likely. **Figure 47.14** The arrow that joins the signal and the receptor. The subsequent events also fail to occur.

Summary of Key Concepts

KC 47.1 The production and/or release of hormones are controlled, directly or indirectly, by electrical or chemical signals from the nervous system. A good example is the hypothalamic-pituitary axis, where neuroendocrine signals regulate endocrine signals. **KC 47.2** Electrical signals are short lived and localized. All of the responses listed in the question are long-term changes in the organism that require responses by tissues and organs throughout the body. **KC 47.3** The hypothalamus would continue to release CRH, stimulating the pituitary to release ACTH, which in turn would stimulate continued release of cortisol from the adrenal cortex. Sustained, high circulating levels of cortisol have negative effects. **47.4** A cell can respond to more than one hormone at a time if the receptors for the hormones are different and the cell produces receptors for each hormone.

Test Your Knowledge

1. c; **2.** a; **3.** d; **4.** a; **5.** c; **6.** a

Test Your Understanding

1. The posterior pituitary is an extension of the hypothalamus; the anterior pituitary is independent—it communicates with the hypothalamus via chemical signals in blood vessels. The posterior pituitary is a storage area for hypothalamic hormones; the anterior pituitary synthesizes and releases an array of hormones in response to releasing hormones from the hypothalamus. **2.** Steroid hormones act directly, and alter gene expression. They bind to receptors inside the cell, forming a complex that binds to DNA and activates transcription. Nonsteroid hormones act indirectly, and activate proteins. They bind to receptors on the cell surface and trigger production of a second messenger or a phosphorylation cascade, ending in activation of proteins already present in the cell. **3.** The pituitary produces hormones that regulate hormone production in other glands. **4.** This is one of the reasons that the same hormone can trigger different effects in different tissues. For example, epinephrine binds to four different types of receptors in different tissues—eliciting a different response from each. **5.** The hormonal signal is amplified through production of many copies of a second messenger, the activation of many proteins through a phosphorylation cascade, or the transcription and translation of many mRNAs. **6.** When fat stores are high, leptin release acts on the brain to reduce feeding. In this way, it helps maintain weight and energy balance—preventing obesity in mice.

Applying Concepts to New Situations

1. There are two basic strategies: (1) remove the structure from some individuals, and compare their behavior and condition to untreated individuals in the same environment, or (2) make a liquid extract from the structure, inject it into some individuals, and compare their behavior and condition to individuals in the same environment who were injected with water or a saline solution. **2.** If sustained high cortisone levels suppress wound healing, individuals receiving large doses might become more susceptible to disease or heal slowly. **3.** The patient will be unable to produce oxytocin and ADH. Although the patient could manage without oxytocin, artificial administration of ADH would be required to maintain water balance. You would need to monitor the individual's total water intake and the water content of the urine, along with symptoms of dehydration or water retention (such as swelling or high blood pressure) and increase or decrease ADH dosage accordingly. **4.** Label the hormone and introduce it into an experimental animal. Analyze fat cells by isolating cell components, testing for the presence of the label, and eventually isolating just the labeled hormone-receptor complex. To test the second messenger hypothesis, treat fat cells growing in culture with the hormone and monitor levels of cAMP, Ca^{2+}, and other known second messengers.

CHAPTER 48

Check Your Understanding (CYU)

CYU p. 957 (1) Oviparity usually requires less energy input from the mother after egg laying, and mothers do not have to carry eggs around as long—meaning that they can lay more eggs and be more mobile. Mothers have to produce all the nutrition required by the embryo prior to egg laying, however, and eggs may not be well protected after laying. Viviparity usually increases the likelihood that the developing offspring will survive until birth, but limits the number of young that can be produced to the space available in the mother's reproductive tract. If viviparous young can be nourished longer than oviparous young, then they may be larger and more capable of fending for themselves. (2) Divide a population of sperm into two groups. Subject one group of sperm to spermkillerene at concentrations observed in the female reproductive tract (the experimental group) but not the other group (the control). Document the number of sperm that are still alive over time in both groups. **CYU p. 966** (1) FSH triggers maturation of an ovarian follicle. It rises at the end of a cycle because it is no longer inhibited by progesterone, which stops being produced at high levels when the corpus luteum degenerates. (2) The drug would keep FSH levels low, meaning that follicles would not mature and would not begin producing estradiol and progesterone. The uterine lining would not thicken.

You Should Be Able To (YSBAT)

YSBAT p. 952 Asexual reproduction would be expected in environments where conditions change little over time. **YSBAT p. 965** (1) There should be an LH spike early in the cycle, followed by early ovulation. (2) LH levels should remain low—no mid-cycle spike and no ovulation. **YSBAT p. 968** (1) The uterine lining may not be maintained adequately—if it degenerates, a miscarriage is likely. (2) After the first trimester, the placenta produces enough progesterone to maintain the pregnancy, so supplementation is no longer needed.

Caption Questions and Exercises

Figure 48.3 Isolate and identify the molecules found in crowded water. Test each molecule by adding it, at the same concentration found in "crowded water," to clean water occupied by a single *Daphnia* and recording whether the female produces a male-containing brood. Repeat with many test females, and for each molecule identified. As a control, record the number of male-containing broods produced in clean water. **Figure 48.5** No—the data are consistent with the displacement hypothesis, but there is no direct evidence for it. There are other plausible explanations for the data. **Figure 48.11** (Note: There is more than one possible answer—this is an example.) Having two gonads is an "insurance policy" against loss or damage. To test this idea, surgically remove one gonad from a large number of male and female rats. Do a similar operation on a large number of similar male and female rats but do not remove either gonad. Once the animals have recovered, place them in a

barn or other "natural" setting and let them breed. Compare reproductive success of individuals with paired vs. unpaired gonads. **Figure 48.12** It is similar. In both cases, the hypothalamus produces a releasing factor (GnRH or CRH) that acts on the pituitary. The releasing factor stimulates release of regulatory hormones from the anterior pituitary (LH and FSH or ACTH). These hormones travel via the bloodstream and act on the gonads or adrenals to induce the release of hormones from these glands. In both cases, the hormones are involved in negative feedback control of the regulatory hormones from the pituitary. **Figure 48.20** The birth will be more difficult, as limbs or other body parts can get caught. **Figure 48.21** A mother's chance of dying in childbirth in 1760 was a little over 1000 in 100,000 live births, or close to 1.0 percent. If she gave birth 10 times, she would have a 10 percent chance of dying in childbirth.

Summary of Key Concepts

KC 48.1 Parental care demands resources (time, nutrients) that cannot be used to produce more eggs. Sperm competition is not likely when external fertilization occurs, and most fishes use external fertilization. **KC 48.2** A surgeon can ligate (tie off) the fallopian tubes in a woman to stop the delivery of the egg to the uterus and can cut the vas deferens in a man to stop the delivery of sperm into the semen. **KC 48.3** Progesterone suppresses the release of GnRH and FSH through negative feedback. When progesterone levels are high, an LH spike does not occur, ovulation is prevented, and no egg is available for fertilization.

Test Your Knowledge

1. a; **2.** b; **3.** a; **4.** c; **5.** d; **6.** d

Test Your Understanding

1. Every offspring that is produced sexually is genetically unique; every offspring that is produced asexually is genetically identical to its parent. **2.** *Daphnia* females only produced males when they were exposed to short day lengths, *and* water from crowded populations, *and* low food levels. These conditions are likely to occur in the fall. Sexual reproduction could be adaptive in fall if genetically variable offspring are better able to thrive in conditions that occur the following spring. **3.** Spermatogenesis generates four haploid sperm cells from each primary spermatocyte; oogenesis produces only one haploid egg cell from each primary oocyte. Egg cells are much larger than sperm cells because they contain more cytoplasm. In males, the second meiotic division occurs right after the first meiotic division, but in females it is delayed until fertilization. **4.** LH triggers release of estradiol, but at low levels estradiol inhibits further release of LH. LH and FSH trigger release of progesterone, but progesterone inhibits further release of LH and FSH. High levels of estrogen trigger release of more LH. The follicle can produce high levels of estrogen only if it has grown and matured—meaning that it is ready for ovulation to occur. **5.** Ethanol can cross the placenta, enter the fetus, and adversely affect development. Ethanol abuse by pregnant women is correlated with fetal alcohol syndrome (FAS), a condition caused by loss of neurons in developing embryos. **6.** Females should be choosier about their second mate because the sperm of the second male has an advantage in sperm competition.

Applying Concepts to New Situations

1. If all sheep were genetically identical, it is less likely that any individuals could survive a major adverse event (e.g., a disease outbreak or other environmental challenge) than a population composed of genetically diverse individuals, which are likely to vary in their ability to cope with the new conditions. **2.** Compare the number of gametes produced by species that have similar body sizes, but undergo external versus internal fertiliza-

tion. By comparing similar-sized animals, you would be able to rule out the possibility that differences in gamete production are due to differences in body size. **3.** Oviparous populations should produce larger eggs than viviparous populations. Because oviparous species deposit eggs in the environment, the eggs must contain all the nutrients and water required for the entire period of embryonic development. But because embryos in viviparous species receive nourishment directly from the mother, it is likely that their eggs are smaller. **4.** The shell membrane is a vestigial trait (see Chapter 24)—specifically, an "evolutionary holdover" from an ancestor of today's marsupials that laid eggs.

CHAPTER 49

Check Your Understanding (CYU)

CYU p. 977 Applying direct pressure constricts blood vessels, mimicking the effect of histamine released from mast cells. Applying bandages impregnated with platelet-recruiting compounds mimics the effect of chemokines released from injured tissues and macrophages, which attract circulating leukocytes and platelets to facilitate blood clotting and wound repair. **CYU p. 984** (1) They are identical, except that a B-cell receptor has a transmembrane domain that allows it to be located in the B-cell plasma membrane. (2) Lymphocytes that respond to self molecules would circulate in the blood, and trigger immune system responses to self cells—leading to their destruction. The mutation would probably be fatal. **CYU p. 987** Individuals who are heterozygous for MHC genes have greater variability in their MHC proteins than do individuals who are homozygous for these genes. The increased variability in MHC proteins allows a greater variety of antigens to be presented to T cells, which thus would be able to recognize, attack, and eliminate a wider range of pathogens compared with T cells in homozygotes. As a result, heterozygous individuals would likely suffer fewer infections than homozygous individuals. **CYU p. 990** A vaccination is a "false alarm" that triggers a primary response from the immune system. The memory cells that result produce the secondary response when and if an authentic infection occurs.

You Should Be Able To (YSBAT)

YSBAT p. 982 Have the lottery numbers contain a large number of digits—say 5, or even 10. To generate a number, pick a number from 0–9 for each of the 5 or 10 places. If there are 5 places, there would be $10 \times 10 \times 10 \times 10 \times 10 = 10,000$ different numbers. If there were 10 places, there would be 10^{10} (10 billion) different numbers. **YSBAT p. 985** Rapidly dividing B and T cells may be destroyed by chemotherapy, along with cancer cells. If clonal expansion cannot occur, antibody production and other aspects of the immune response will be dampened and possibly ineffective. **YSBAT p. 986** (1) All nucleated cells in the body express Class I MHC proteins, whereas only B cells and some other leukocytes express Class II MHC proteins. (2) Cytotoxic T cells interact only with cells that display antigens presented on Class I MHC proteins. Helper T cells interact only with antigens presented on Class II MHC proteins. (3) An MHC-peptide displayed on an infected cell means "kill me"; displayed on a dendritic cell it means "activate killer T cells"; displayed on a B cell it means "helper T cells should activate me."

Caption Questions and Exercises

Figure 49.1 Pathogens stick to mucus; contaminated mucus is then swept to a disposal point by cilia. **Figure 49.3** Blood vessels near the wound would be constricted, which restricts blood loss. Blood vessels slightly farther from the wound would be dilated, which increases blood flow and the delivery rate of platelets, neutrophils, and macrophages to the surrounding area. **Figure 49.18** [See Figure A49.1]

Summary of Key Concepts

KC 49.1 Leukocytes could not move to the damaged tissue efficiently to clean up debris, remove pathogens, and begin the repair process. **KC 49.2** The drug would inhibit the activation of T and B cells by that antigen, because it would prevent an interaction between the epitope and the appropriate TCRs and BCRs. The cell-mediated response would also fail if the drug blocked the interaction between cytotoxic T cells and infected cells displaying the antigen. **KC 49.3** Cowpox is similar enough to smallpox to be antigenic, but cannot grow effectively enough in humans to cause disease.

Test Your Knowledge

1. c; **2.** a; **3.** b; **4.** d; **5.** b; **6.** a

Test Your Understanding

1. Rubor (reddening) is due to increased blood flow to the infected or injured area. Mast cells trigger dilation of vessels by releasing histamine and other signaling molecules. Calor (heat) is the result of fever, which is activated by the release of cytokines from macrophages that are in the area of infection. Dolor (pain) occurs when tissue damage stimulates pain receptors. Tumor (swelling) occurs due to dilation of blood vessels triggered by histamines and other compounds released by macrophages. **2.** Both the BCR and TCR interact with antigens via binding sites located in the variable regions of their polypeptide chains. A TCR is like one "arm" of a BCR. [See Figure A49.2] **3.** Vaccines have to contain an antigen that can stimulate an appropriate primary immune response. Vaccines have not worked for HIV because the antigens on this virus are constantly changed through mutation, rendering it unrecognizable to memory cells generated following vaccination. **4.** "Clonal" refers to the cloning—producing many exact copies—of cells that are "selected" by the binding of their receptor to an antigen. **5.** Pattern-recognition receptors on leukocytes bind to surface molecules that are present on many pathogens. BCRs and TCRs bind to particular epitopes of antigens. Pattern-recognition-receptor binding is nonspecific; BCR and TCR binding is specific. **6.** By mixing and matching different combinations of gene segments from the variable and joining regions of the light-chain immunoglobulin gene, along with diversity regions of the heavy-chain gene, lymphocytes end up with a unique sequence for both chains in the BCR and TCR.

Applying Concepts to New Situations

1. Irrigating the wound removes dirt and debris that may contain pathogens. Scrubbing with soapy water removes more pathogens and may also kill them (soap may disrupt plasma membranes enough to be fatal). Antibiotics are toxic to pathogenic bacteria. **2.** If the antibody had a fluorescent molecule or other type of label attached, you could treat various types of cells with it, examine them under the microscope, and determine the location of the antibody-pump complexes. **3.** Natural selection favors individuals that can create a large array of antibodies, because the high mutation rates and rapid evolution observed in pathogens means that they will constantly present the immune system with new antigens. If these antigens were not recognized by the immune system, the pathogens would multiply freely and kill the individual. **4.** Tissue from the patient's own body is marked with the major histocompatibility (MHC) proteins that the patient's immune system recognizes as self. This results in the preservation, rather than the destruction, of the transplanted tissue. Tissue from a different person is marked with MHC proteins that are unique to that individual. The body of the transplant recipient will recognize the MHC proteins (and other

molecules) of the donor as foreign, resulting in a full immune response and rejection of the grafted tissue.

CHAPTER 50

Check Your Understanding (CYU)

CYU p. 1001 (1) They are shallow, so a relatively large proportion of the water receives enough sunlight to support photosynthesis. In addition, nutrients are available from rivers that contribute nutrients from inland and upwellings that bring nutrients up from the ocean bottom. (2) It sinks, and the lower-density water below it rises, bringing nutrients to the surface. **CYU p. 1008** Tropical dry forests are probably less productive than tropical wet forests because they have less water available to support photosynthesis during some periods of the year. **CYU p. 1012** Your answer will depend on where you live. For example, if global warming continues at the current projected rates, several effects can be expected. If you live in a coastal area, you can expect water levels to rise. In general, plant communities will change because average temperature and moisture and variation in temperature and moisture will change.

You Should Be Able To (YSBAT)

YSBAT p. 997 The littoral zone, because light is abundant and nutrients are available from the substrate. **YSBAT p. 998** Bogs are nitrogen-poor, so plants that are able to capture and digest insects have a large advantage. This advantage does not exist in marshes and swamps, where nutrients are more readily available. **YSBAT p. 999** Cold water contains more oxygen than warm water, so it can support much more cellular respiration in fish.

YSBAT p. 1000 Species that live in estuaries must be able to tolerate variable salinity; marsh species do not. The abiotic environment (salt concentration) is so different that few species grow well in both habitats. **YSBAT p. 1001** No—the aphotic zone is lightless, so natural selection favors individuals that do not invest energy in developing and maintaining eyes. **YSBAT p. 1003** Vines and epiphytes increase productivity because they are photosynthetic organisms that fill space between small trees and large trees—they capture light and use nutrients that might not be used in a forest that lacked vines and epiphytes. **YSBAT p. 1004** Most leaves have a large surface area to capture light. Light is abundant in deserts, however, and a leaf with a large surface area would be susceptible to high water loss and/or overheating. **YSBAT p. 1005** Vegetation is continuous in a grassland, so a fire carries better. In a desert, the fire is likely to run out of fuel. **YSBAT p. 1006** There is a continuous grass cover and scattered trees. (A biome like this is called a savanna.) **YSBAT p. 1007** Boreal forests should move north. **YSBAT p. 1008** High elevations present an abiotic environment (precipitation and temperature) that is similar to the conditions in arctic tundras. As a result, the plant species present will have similar adaptations to cope with these physical conditions. **YSBAT p. 1010** [See Figure A50.1]

Caption Questions and Exercises

Figure 50.4 Freshwater, because light penetrates much deeper. **Figure 50.23** Dry, because dry air is descending here. This is consistent with the very low rainfall amounts in Barrow, Alaska, shown in Figure 50.21. **Figure 50.27** Visible wavelengths enter the chamber

---

FIGURE A49.1

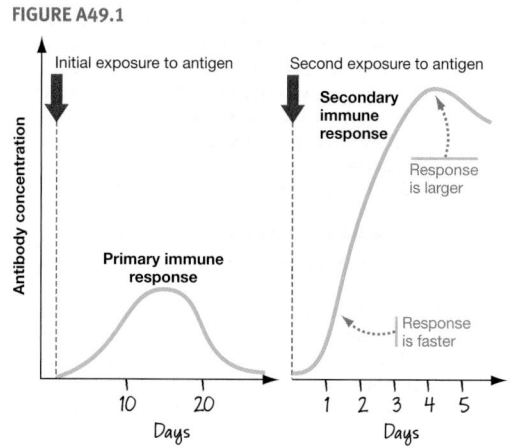

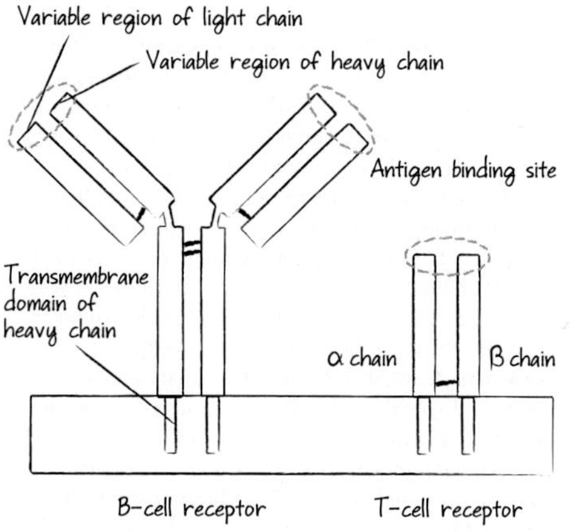

FIGURE A50.1

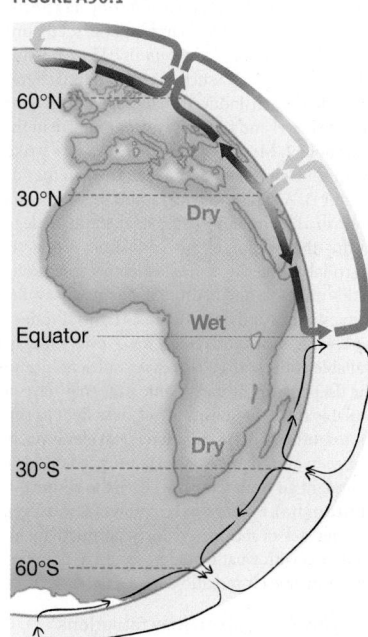

FIGURE A49.2

from the top and through the glass, warming the plants and ground inside the chamber. Some of this heat energy is retained within the mini-greenhouse. **Figure 50.28** They address the question posed because they capture important aspects of a plant community. NPP indicates how much biomass is available for other organisms to use; species diversity indicates how many different species are present. **Figure 50.34** (Several possible answers; one possibility is provided.) If water is short, there is not enough available to support a continuous cover of photosynthetic tissue aboveground. Roots of bunchgrasses can spread out and harvest water below spaces that are "bare" aboveground.

Summary of Key Concepts

KC 50.1 Populations are made up of individual organisms; communities are made up of populations of different species; ecosystems are made up of groups of communities (along with the abiotic environment). **KC 50.2** Increased evaporation from the oceans should increase precipitation on land—at least in some places. If precipitation does not increase, then increased transpiration rates will put plants under water stress and reduce productivity. **KC 50.3** All latitudes would get equal amounts of sunlight all year round. There would be no seasons and no changes in climate with latitude. **KC 50.4** It would increase, because cattle would no longer succumb to the disease carried by tsetse flies. (In Africa, cattle are limited by a biotic condition, not abiotic conditions.)

Test Your Knowledge

1. b; **2.** a; **3.** d; **4.** b; **5.** a; **6.** c

Test Your Understanding

1. Organismal, population, community, and ecosystem. Examples of possible questions: *Organismal:* How do humans cope with extremely hot (or cold) weather conditions? *Population:* How large will the total human population be in 50 years? *Community:* How are humans affecting prey species such as cod and tuna? *Ecosystem:* How will human-induced changes in global temperature affect sea level and thus organisms that live in the intertidal zone? **2.** More—in June the Sun's rays strike the Northern Hemisphere more directly than they do the equator. **3.** A Hadley cell just south of the equator results in warm, dry air falling to Earth at about 30 degrees south—the location of the Outback. **4.** Productivity in intertidal and neritic zones is high because sunlight is readily available and because nutrients are available from estuaries and deep-ocean currents. Productivity in the oceanic zone is extremely low—even though light is available at the surface—because nutrients are scarce. The deepest part of the oceanic zone may have nutrients available from the substrate but lacks light to support photosynthesis. **5.** As you increase in elevation, biomes change in a similar way as increasing in latitude (e.g., you might go from temperate forest to a boreal-type forest to tundra). **6.** The combination of low oxygen, low pH, and lack of nitrogen reduces productivity in bogs compared with what would be found in marshes or lakes with similar solar radiation.

Applying Concepts to New Situations

1. Yes—as Mars orbits the Sun, the part of the planet that is tilted toward the Sun is warmest because it receives the most direct sunlight. But as the planet orbits, that portion of the planet begins to tilt away from the Sun. Its temperature drops as a result, producing something like Earth's cycle of winter and summer. **2.** Mountaintops are cold because as air rises, it expands; this expansion cools the air, making the mountaintop cold. Also, there is less land surface nearby to absorb solar radiation and heat up. **3.** The hypothesis is that winds are usually from the northeast, and that higher elevations in the middle of the islands cause a rain shadow in the

southwest corner. **4.** Edinburgh is surrounded by water. Because water has a high specific heat, it keeps temperatures moderate. Moscow, in contrast, is surrounded by a large landmass that does not have the same moderating effects on temperature as the ocean does.

CHAPTER 51

Check Your Understanding (CYU)

CYU p. 1022 If a sitter-like allele in bee-eaters was favored when population density is low, then it should be at high frequency in small colonies. If a rover-like allele was favored when population density is high, then it should be at high frequency in large colonies. **CYU p. 1025** (1) In the experiment, females came into breeding condition slowly if they were exposed only to springlike light conditions, but much more quickly if they were exposed to springlike light conditions *and* displaying males. Both are required for maximum effect. (2) Many possible answers, but based on the experiments with *Anolis*, it would be legitimate to hypothesize that they have to be exposed to springlike light and temperature conditions as well as courtship displays from males. **CYU p. 1031** (1) Auditory communication allows individuals to communicate over long distances but can be heard by predators. Olfactory communication is effective in the dark and scents can continue to carry information long after the signaler has left, but scents do not carry long distances. Visual communication is effective during the day but can be seen by predators. (2) The ability to detect deceit protects the individual from fitness costs (e.g., being eaten, having a mate's eggs fertilized by someone else); avoiding or punishing "liars" should also lower the frequency of deceit (because it becomes less successful). Alleles associated with detecting and avoiding or punishing "liars" should be favored by natural selection and increase in frequency. **CYU p. 1034** (1) Altruism is likely when B is high, as when a predator is about to pounce; r is high, as when a close relative is threatened; C is low, as when the caller is far from the predator. Altruism is unlikely under the opposite conditions. (2) For reciprocal altruism to work, individuals have to interact repeatedly and remember who has helped whom in the past.

You Should Be Able To (YSBAT)

YSBAT p. 1033 Humans often live near kin, are good at recognizing kin, and in many cases can give kin resources or protection that increase fitness. It is not known whether kin selection occurs in plants. In many cases, limited seed dispersal means that close relatives live together—making kin selection more likely. But it is not known whether plants can recognize kin and confer benefits preferentially to kin.

Caption Questions and Exercises

Figure 51.1 Several legitimate hypotheses are possible. Examples at the proximate level: The increased size of the antennae provide space for more sensory neurons; or large antennae allow individuals to sense a larger area. Examples at the ultimate level: large antennae help individuals navigate better and thus escape predation; or large antennae allow individuals to forage more successfully in the dark. **Figure 51.5** Bring females into the lab and give them the same food and housing conditions as the treatment groups, but expose them to artificial lighting that simulates the short daylight conditions of winter. **Figure 51.7** Because individuals were chosen for the different treatments at random, the investigator could claim that on average, the only thing that differed among individuals in the different groups was average tail length. **Figure 51.10** [See Figure A51.1] **Figure 51.14** Between first cousins, $r = \frac{1}{2} \times \frac{1}{2} \times \frac{1}{2} = \frac{1}{8}$. **Figure 51.15** The control would be to drag an object of similar size through the colony, at similar times of day. This would

test the hypothesis that prairie dogs are reacting to the presence of the experimenter and the disturbance—not a predator.

Summary of Key Concepts

KC 51.1 At the proximate level, different alleles of the *for* gene affect the probability that an individual will move or stay after feeding. At the ultimate level, roving is favored when population density is high; sitting is favored when population density is low. **KC 51.2** Individuals have the capacity to act altruistically or non-altruistically toward kin. Hamilton's rule specifies the conditions—in terms of the costs and benefits of the act and the degree of relationship between the participants—under which altruism is favored. **KC 51.3** Roving is expensive energetically and could make individuals more susceptible to predation, but could allow individuals to find unexploited food sources. Sitting is inexpensive energetically, but may restrict individuals to areas where food supplies have been depleted. **KC 51.4** If only males that are well fed and free of parasites and disease are able to produce a brightly colored dewlap and do the bobbing display vigorously, then these traits would be a reliable indication of their condition. **KC 51.5** If migration does not occur when extra food is supplied, it suggests that the "decision" to migrate depends on conditions. **KC 51.6** Deceitful communication should decrease. If large males are gone, there are few nests available and more female-mimic males would compete at them. Most female-mimic males would have no (or fewer) eggs to fertilize and no male to take care of the eggs they did fertilize. There would be less natural selection pressure favoring deceitful communication. **KC 51.7** (1) Long-lived, so that extensive interactions with kin and nonrelatives are possible; (2) good memory, to record reciprocal interactions; (3) kin nearby, to make inclusive fitness gains possible; (4) ability to share fitness benefits (e.g., food) between individuals.

Test Your Knowledge

1. c; **2.** d; **3.** a; **4.** c; **5.** d; **6.** c

Test Your Understanding

1. At the ultimate level, homing to a safe den increases survival. The proximate cause for this behavior is still not fully known, but appears to involve the ability to sense changes in Earth's magnetic field. **2.** When optimal foraging occurs, an individual maximizes the benefit of foraging in terms of energy gain and minimizes the cost of foraging in terms of time, energy expended, and risk of predation. **3.** Only males that are in good shape (well-nourished and free of parasites) have the resources required to produce a trait like a long tail. **4.** Longer day lengths indicate the arrival of spring, when renewed plant growth and insect activity make more food available. Courtship displays from males indicate the presence of males that can fertilize eggs. **5.** A map provides information about the spatial relationships of places on

FIGURE A51.1

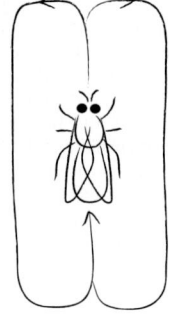

The straight run of the upward waggle dance is twice as long as in Fig. 51.9

a landscape; a compass allows you to orient the map—so that it aligns with the actual landscape. Animals have been shown to use the Sun's position, the position of the North star, and information from Earth's magnetic field as a compass (a way to find North). **6.** Kin selection can only act on kin; reciprocal altruism can occur between nonrelatives. Kin selection should be more common, because it potentially can occur any time that kin interact. Reciprocal altruism should be relatively rare because it is only possible when unrelated individuals interact repeatedly, have resources to share, and can remember the outcomes of past interactions.

Applying Concepts to New Situations

1. One possibility would be to set up a testing arena with food sources at various distances from a food source where adult flies are originally released. Test adults with a rover versus sitter genotype individually, and record how far each flies over the course of a set time interval. If the rover and sitter alleles affect foraging movements in adults, then rovers should be more likely to move and move farther than sitters. **2.** Simulate the onset of rains after a dry period, by using a hose to sprinkle water on the habitat where lizards are being maintained. **3.** Individuals have an r of 1/2 with full siblings and an r of 1/8 with first cousins. If the biologist will lose his life, he needs to save two siblings to keep the "lost copy" of his altruism alleles in the population but eight first cousins to maintain a copy. **4.** People would be expected to donate blood under two conditions: (1) They either received blood before or expect to have a transfusion in the future, and/or (2) they receive some other benefit in return, such as a good reputation among people who might be able to help them in other ways.

CHAPTER 52

Check Your Understanding (CYU)

CYU p. 1046 One possibility is competition for food. To test this idea, compare carrying capacity in identical vials that differ only in the amount of food added. (You could also test a hypothesis of oxygen limitation by adding more oxygen to one set of vials, or test the hypothesis of space limitation by doubling the volume but keeping the amount of food and oxygen the same.) **CYU p. 1053** (1) In developed nations, there are roughly equal numbers of individuals in each age class because the fertility rate has been constant and survivorship high for many years. In developing countries there are many more children and young people than older people, because the fertility rate and survivorship have been increasing. (2) Because survivorship is high in most human populations, changes in overall growth rate depend almost entirely on fertility rates.

You Should Be Able To (YSBAT)

YSBAT p. 1041 (1) Compared to northern populations, fecundity should be high and survivorship low. (2) Fecundity can be much higher if females lay eggs instead of retaining them in their bodies and giving birth to live young. **YSBAT p. 1055** The population appears to be increasing over time: from the original 1000 females, there are 2308 females in the third year.

Caption Questions and Exercises

Figure 52.2 [See Figure A52.1] **Figure 52.4** [See Table A52.1] **Figure 52.8** At high population density, competition for food limits the amount of energy available to female song sparrows for egg production. **Figure 52.12** Large-scale field experiments like this are extremely expensive and difficult to implement—it was not practical to replicate all treatments. **Table 52.2** From 1999 to 2009, $6800 = 6000 \lambda^{10}$ (see Eq. 52.5 in Box 52.2). Solving, $\lambda = 1.013$. If $e^r = 1.013$, then $r = 0.0125$. Now use the expression $N_t = N_0 e^{rt}$ (Eq. 52.7 in Box 52.2) to solve the problems.

Number of years required to add 1 billion people to 2009 level:

$$7800 = 6800 e^{0.0125t}$$

$$ln(1.15) = 0.0125t$$

$$t = \sim 11 \text{ years}$$

Number of years required to double population from 2009 level:

$$13,600 = 6800 e^{0.0125t}$$

$$ln(2) = 0.0125t$$

$$t = \sim 5.5 \text{ years}$$

Figure 52.17 [See Figure A52.2 on p. A:48]

Summary of Key Concepts

KC 52.1 A life table represents a snapshot of how a particular population is growing. As conditions change, survivorship and fecundity may change. In ancient Rome, for example, survivorship was probably low and fecundity high—women started reproducing at a young age and did not live long, on average. In Rome today, the population has high survivorship and low fecundity. **KC 52.2** [See Figure A52.3 on p. A:48] **KC 52.3** Fewer children are being born per female, but there are many more females of reproductive age due to high fecundity rates in the previous generation. **KC 52.4** Small, isolated populations are likely to be wiped out by bad weather, a disease outbreak, or changes in the habitat. For example, primrose populations may die out when a gap is shaded. But in a metapopulation, migration between the individual small populations helps reestablish subpopulations and maintain the overall size of the metapopulation.

Test Your Knowledge

1. b; **2.** d; **3.** a; **4.** a; **5.** d; **6.** c

Test Your Understanding

1. Equation 52.4: After one breeding interval, the population size is equal to the original population size times the discrete growth rate. Equation 52.5: After t breeding intervals, the population size is equal to the original population size times the discrete growth rate multiplied by itself t times (i.e., raised to the tth power). **2.** Species with type I survivorship curves have high survivorship until old age, so population growth is based on the number of offspring produced—not on how many of the

FIGURE A52.1

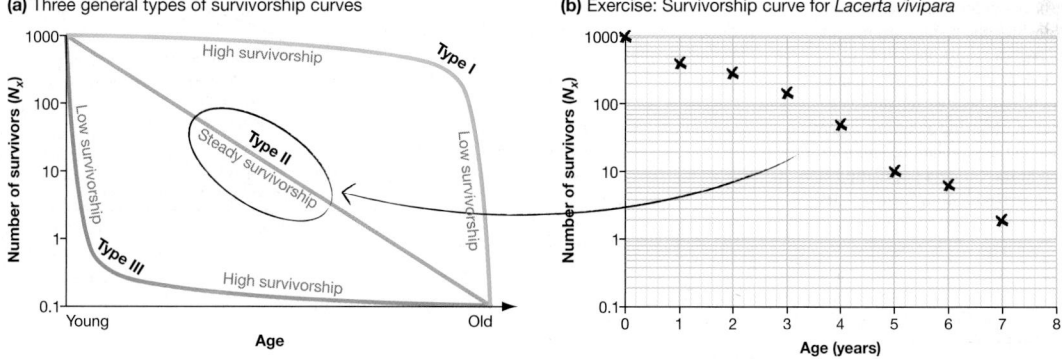

(a) Three general types of survivorship curves

(b) Exercise: Survivorship curve for *Lacerta vivipara*

TABLE A52.1

Trait	Life-History Continuum		
	Left	Middle	Right
Growth habit	Herbaceous	Shrub	Tree
Disease- and predator-fighting ability	Low	Medium	High
Seed size	Small	Medium	Large
Seed number	Many	Moderate	Few

offspring survive. Species with type III survivorship curves exhibit very high juvenile mortality, but high survivorship of those few individuals able to reach adulthood. Because only the very few adults can reproduce, any changes to survivorship at the adult level will have a large effect on the population as a whole. **3.** The population has undergone near-exponential growth recently because advances in nutrition, sanitation, and medicine have allowed humans to live at high density without suffering from decreased survivorship and fecundity. Eventually, however, growth rates must slow as density-dependent effects increase death rates and lower birth rates. Experiments on snowshoe hares have shown that food availability and predation are density-dependent factors that influence population growth. **4.** (a) Corridors allow individuals to move between subpopulations, increasing gene flow and making it possible to recolonize habitats where populations have been lost. (b) Maintaining unoccupied habitat makes it possible for the habitat to be recolonized. **5.** As a sexually transmitted disease, AIDS will reduce the number of sexually active adults. If the epidemic continues unabated, the numbers of both reproductive-age adults and children will decline, causing a top-heavy age distribution dominated by older adults and the elderly. [See Figure A52.4] **6.** Lambda is simple to calculate but only describes growth over a discrete time interval. Determining R_0 requires knowledge of all age-specific birth and death rates. r indicates the growth rate at any instant, so is independent of generation time and can be applied to any population. r_{max} gives the population growth rate in the absence of density-dependent limitation; r is the actual growth rate, which is usually affected by density-dependent factors.

Applying Concepts to New Situations

1. Fewer older individuals will be left in the population; there will be relatively more young individuals. If too many older individuals are taken, growth rate may decline sharply as reproduction stops or slows. (But if relatively few older individuals are taken, more resources are available to younger individuals and their survivorship and fecundity, and the population's overall growth rate, may increase.) **2.** The sunflowers and beetles are both metapopulations. To preserve them, you must preserve as many of the sunflower patches as possible (or plant more) and maintain corridors (which may be smaller sunflower patches) along which the beetles can migrate between the patches. **3.** Large-scale immigration into this population is projected. **4.** The growth rate of any population is female dependent, because females can produce a limited number of offspring at a time, regardless of how many males are in the population. This means that there are "extra" men in the Chinese population—they will not participate in reproduction. Growth rate should decrease, as a result.

CHAPTER 53

Check Your Understanding (CYU)

CYU p. 1070 (1) The individuals do not choose or try to have traits that reduce competition—they simply have those traits (or not). Resource partitioning just happens, because individuals with traits that allow them to exploit different resources produce more offspring, which also have those traits. (Recall from Chapter 24 that natural selection occurs on individuals, but adaptive responses such as resource partitioning are properties of populations.) **(2)** When species interact via consumption, a trait that gives one species an advantage will exert natural selection on individuals of the other species who have traits that reduce that advantage. This reciprocal adaptation will continue indefinitely. An example is the interaction of *Plasmodium* with the human immune system: The human immune system has evolved the

FIGURE A52.2

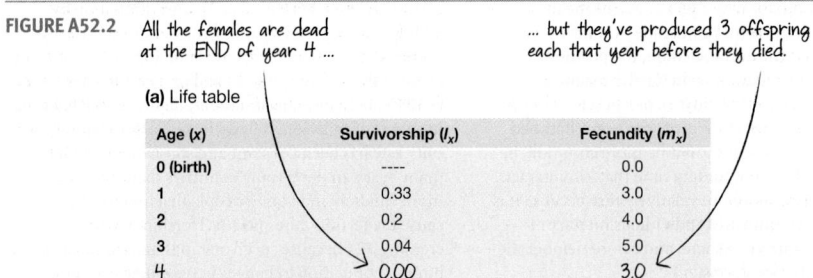

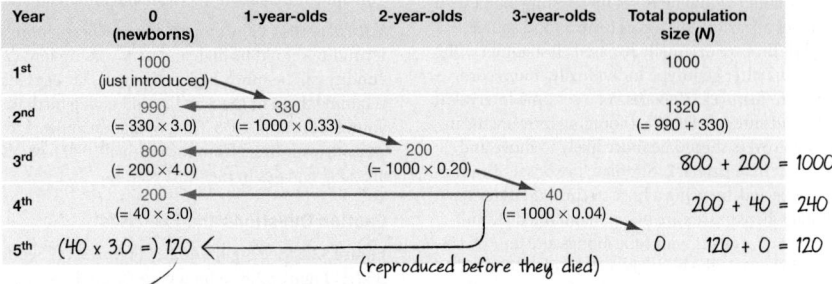

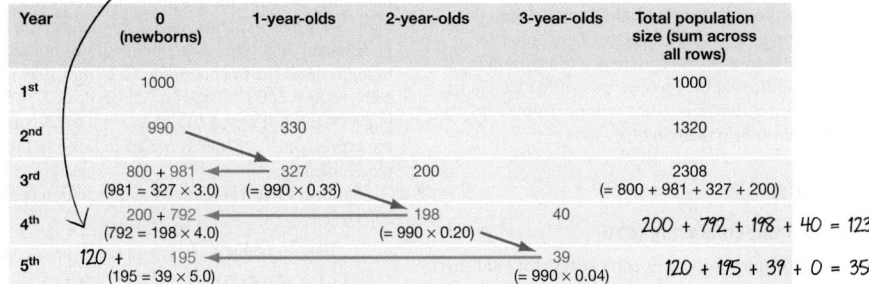

FIGURE A52.3

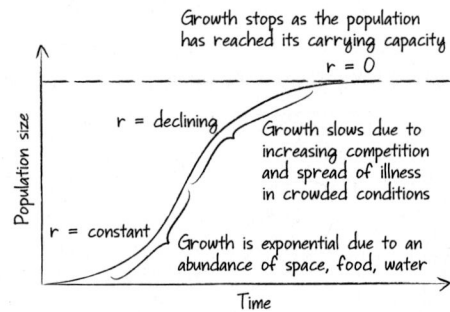

FIGURE A52.4

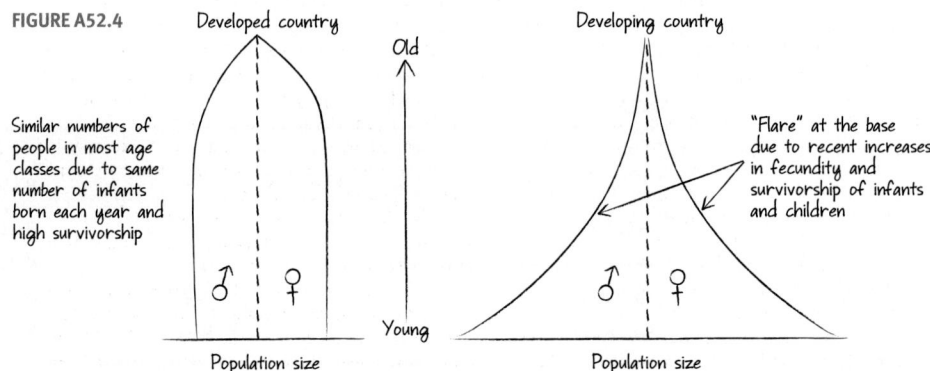

ability to detect proteins from the *Plasmodium* and kill infected cells; in response, *Plasmodium* has evolved different proteins that the immune system does not detect. **CYU p. 1077** (1) The shade provided by early successional species increases humidity, and decomposition of their tissues adds nutrients and organic material to the soil. These conditions favor growth by later successional species, which can outcompete the early successional species. (2) The presence or absence of a plant species where nitrogen-fixation occurs would dramatically alter nutrient conditions, and thus the speed of succession and the types of species that could become established. For example, species that require high nitrogen would be favored on sites where alder grew, while species that can tolerate low nitrogen would thrive on sites where alder is absent.

You Should Be Able To (YSBAT)

YSBAT p. 1079a The calculations for Community 1 are as follows:

Species A
$p_A = 10/18 = 0.55$
$\ln 0.55 = -0.598$
$-0.598 \times 0.55 = -0.329$

Species B
$p_B = 1/18 = 0.055$
$\ln 0.055 = -2.9$
$-2.9 \times 0.055 = -0.16$

Species C
$p_C = 1/18 = 0.055$
$\ln 0.055 = -2.9$
$-2.9 \times 0.055 = -0.16$

Species D
$p_D = 3/18 = 0.167$
$\ln 0.167 = -1.78$
$-1.78 \times 0.167 = -0.299$

Species E
$p_E = 2/18 = 0.11$
$\ln 0.11 = -2.21$
$-2.2 \times 0.11 = -0.243$

Species F
$p_F = 1/18 = 0.055$
$\ln 0.055 = -2.9$
$-2.9 \times 0.055 = -0.16$

Summing the values of $p \times \ln p$ for each species and multiplying by -1 gives the Shannon index of species diversity for Community 1:

$(-1)[(-0.329) + (-0.16) + (-0.16) + (-0.299) + (-0.243) + (-0.16)] = 1.351$

Similar calculations would give species diversity values of 1.79 for Community 2 and 1.61 for Community 3.
YSBAT p. 1079b
Community 1: Species richness = 12; Diversity = 2.04
Community 2: Species richness = 12; Diversity = 2.49
Community 3: Species richness = 10; Diversity = 2.3
YSBAT p. 1079c Species richness doubled in each community because the number of species doubled. Although the species diversity values differ from those in Figure 53.25, the communities with more equal numbers of individuals in each species still have the higher diversity values.

Caption Questions and Exercises

Figure 53.6 If he had done the treatments on different rocks, a critic could argue that differences in survival were due to differences in the nature of the rocks—not differences in competition. **Figure 53.7** The species partition, or divide up, resources. In this way, evolution leads to a reduction or elimination of competition. **Figure 53.12** This experimental design tested the hypothesis that mussels can sense the presence of crabs. If the crabs had been fed mussels, a critic could argue that the mussels detect the presence of damaged mussels—not crabs. (As it turns out, the mussels can do both. The experimenters did the same experiment with broken mussel shells in the chamber instead of a crab-fed fish.) **Figure 53.13** Beavers must represent the greater overall threat to cottonwoods, because the trees have evolved a compound that deters further beaver attack but does not deter beetle larvae. Getting the trunk cut down probably

reduces fitness more than getting leaves eaten. **Figure 53.15** Put equal numbers of infected and uninfected ants in a pen that includes a bird predator. Count how many of each type are eaten over time. **Figure 53.16** Many possible answers are reasonable hypotheses. For example, acacia trees may expend energy in producing the large bulbs in which the ants live and the food they eat. The *Crematogaster* ants may expend energy and risk injury or death in repelling herbivores. The cleaner shrimp may occasionally get injured or eaten by the fish they are cleaning, or their diets may be somewhat restricted by the association. The host fish may miss meals or be at greater risk of predation when they are undergoing cleaning by the shrimp. In both examples of mutualism, however, the overall associations are positive for both parties. **Figure 53.17** These steps controlled for the alternative hypothesis that differences in treehopper survival were due to differences in the plants they occupied—not the presence or absence of ants. **Figure 53.18** Predictable, at least to a degree. **Figure 53.24** The effect would be to make the island more remote, which would lower the rate of immigration and move the whole immigration curve downward. The rate of extinction would increase, shifting the curve upward. The overall effect would be to decrease the number of species because the island would have effectively become more remote.

Summary of Key Concepts

KC 53.1 A mutualistic relationship becomes a parasitic one if one of the species stops receiving a benefit. The treehopper-ant mutualism becomes parasitic in years when spiders are rare, because the treehoppers no longer derive a benefit but pay a fitness cost (producing honeydew that the ants eat). **KC 53.2** All species at Glacier Bay must be able to survive in a cold climate with the local amount of precipitation. The species earliest in the succession must also be able to grow on rock exposed as the glacier melts. But chance, historical differences in seed sources, and presence or absence of alder (and nitrogen-fixation) created differences in the species present.
KC 53.3 Lakes are abundant in northern latitudes but rare in the tropics. The presence of much greater amounts of habitat supports higher species richness for breeding shorebirds at high latitudes.

Test Your Knowledge

1. d; **2.** a; **3.** b; **4.** d; **5.** c; **6.** d

Test Your Understanding

1. Yes—the treehopper-ant mutualism is parasitic or mutualistic depending on conditions; competition can evolve into no competition over time if niche differentiation occurs; arms races mean that the outcome of host-parasite interactions can change over time, and so forth (many other examples). **2.** *Top-down control:* Herbivore populations are kept at relatively low levels by predation and disease. *Bottom-up control:* Plant matter is low in nitrogen or defended by chemicals that are toxic to herbivores. These hypotheses are certainly not mutually exclusive. For example, resprouted cottonwoods produce compounds that make them less edible for beavers, which is a plant defense. In addition, these cottonwoods might not provide enough nitrogen for beavers to grow quickly, and species that prey on beavers or make them sick might be keeping their population under control. **3.** (a) If community composition is predictable, then the species present should not change over time. But if composition is not predictable, then there should be significant changes in the species present over time. (b) If community composition is predictable, then two sites with identical abiotic factors should develop identical communities. If community composition is not predictable, then sites with identical abiotic factors should develop variable communities. In most tests, the data best match the predictions of the "not predictable" hypothesis, though communities show elements of both.

4. Disturbance is an event that removes biomass. Compared to low-frequency fires, high-frequency fires would tend to be less severe (less fuel builds up) and would tend to exert more intense natural selection for adaptations to resist the effects of fire. Compared to low-severity fires, high-severity fires would open up more space for pioneering species and would tend to exert more intense natural selection for adaptations to resist the effects of fire. **5.** Early successional species are adapted to disperse to new environments (small seeds) and grow and reproduce quickly (reproduce at an early age, grow quickly). They can tolerate severe abiotic conditions (high temperature, low humidity, low nutrient availability) but have little competitive ability. These species are able to enter a new environment (with no competitors) and thrive. **6.** The idea is that high productivity will lead to high population density of consumers, leading to competition and intense natural selection favoring niche differentiation that leads to speciation.

Applying Concepts to New Situations

1. Natural selection will favor orchid individuals that have traits that resist bee attack: thicker flower walls, nectar storage in a different position, a toxin in the flower walls, or other. Individuals could also be favored if their anthers were in a position that accomplished pollination, even if bees eat through the walls of the nectar-storage structure. **2.** Set fires or let natural fires burn, to match the time interval between fires recorded prior to the arrival of European settlers. Note that because fires have been suppressed for so long, the initial fires will have to be carefully controlled or fuel will have to be removed manually—for example by selective logging. The logic is that species in this habitat are adapted to a disturbance regime of frequent, low-intensity fires. For them to thrive, these conditions have to be re-created. **3.** The exact answer will depend on the location of the campus. The first species to appear must possess good dispersal ability, rapid growth, quick reproductive periods, and tolerance for very harsh and severe conditions. The two-acre plot is likely to be colonized first by pioneer species that have very "weedy" characteristics. But once colonization is under way, the course of succession will depend more on how the various species interact with each other. The presence of one species can inhibit or facilitate the arrival and establishment of another. For example, an early-arriving species might provide the shade and nutrients required by a late-arriving species. The site's history and nearby ecosystems may influence which species appear at each stage; for instance, an undisturbed ecosystem nearby could be a source for native species. The pattern and rate of this succession is also influenced by the overall environmental conditions affecting it. Only species with traits appropriate to the local climate are likely to colonize the site. **4.** One reasonable experiment would involve constructing artificial ponds and introducing different numbers of plankton species to different ponds, but the same total number of individuals. (Any natural immigration to the ponds would have to be prevented.) After a period of time, remove all of the plankton and measure the biomass present. Make a graph with number of species on the *x*-axis and total biomass on the *y*-axis. If the hypothesis is correct, the line of best fit through the data should have a positive slope.

CHAPTER 54

Check Your Understanding (CYU)

CYU p. 1092 (1) At each trophic level, most of the energy that is consumed is lost to heat, metabolism, or other maintenance activities, which leaves only a small percentage of energy for biomass production (growth and reproduction). (2) More food will be available to primary consumers, resulting in increases in biomass at each trophic level. An increase in biomass entering decomposer food chains will also increase decomposer populations. **CYU p. 1098** Perhaps the most direct impacts concern rates of groundwater replenishment. Converting biomes into farms or suburbs decreases groundwater recharge and increases runoff. Irrigation pumps groundwater to the surface, where much of it runs off into streams.

You Should Be Able To (YSBAT)

YSBAT p. 1086 To grow a kilogram of beef, you have to first grow 10 kilograms of grain or grass and feed it to the cow. Only 10 percent of this 10 kg will be used for growth and reproduction—the other 9 kg is used for maintenance or lost as heat. **YSBAT p. 1087a** Crustaceans and fish are ectothermic, so they are much more efficient at converting primary production into the biomass in their bodies than endothermic birds and mammals. **YSBAT p. 1087b** Bigger: elk to wolf, mule deer to wolf, aspen/cottonwood/willows to beaver, mice to rough-legged hawk. Smaller: mice to coyote, aspen/cottonwood/willows to elk and mule deer. **YSBAT p. 1094** The uptake arrow was removed—there were no plants to take up nutrients from the soil nutrient pool.

Caption Questions and Exercises

Figure 54.3 The percentage of energy used for maintenance would be much lower because they are ectothermic and sedentary. The percentage excreted would also be much lower because their diet contains little indigestible material. Thus the percentage of energy available for growth and reproduction would be much higher. **Figure 54.10** The diatoms' bodies sink, removing nutrients from the surface ocean and feeding organisms in the deeper ocean. **Figure 54.13** One logical hypothesis is that the total amount exported increases as tree roots and other belowground organic material decay and begin to wash into the stream; the amount exported should begin to decline as nitrate reserves become exhausted—eventually, there is no more nitrate to wash away. **Figure 54.14** Much more water should evaporate. One possibility is that more water vapor will be blown over land and increase precipitation over land. **Figure 54.21** Photosynthesis increases in summer, resulting in removal of CO_2 from the atmosphere.

Summary of Key Concepts

KC 54.1 [See Figure A54.1] **KC 54.2** A food chain consisting of only primary producers and primary decomposers would minimize energy loss. If organisms at higher trophic levels were present, the efficiency of energy transfer would be highest if they are ectothermic, move little, and have extremely efficient digestion. **KC 54.3** Decomposition is slowest in cold, wet temperatures on land and in anoxic conditions in the ocean (or in freshwater). Stagnant water often becomes anoxic as decomposers use up available oxygen and it is not replenished by diffusion from the atmosphere. **KC 54.4** Many possible answers, including: (1) reducing carbon dioxide emissions by finding alternatives to fossil fuels should decrease the amount of carbon dioxide released into the atmosphere, (2) recycling programs decrease carbon dioxide emissions from manufacturing plants and power plants, (3) reforestation can tie up carbon dioxide in wood; (4) iron fertilization could increase NPP in the open ocean—meaning that more CO_2 would be used.

Test Your Knowledge

1. a; **2.** d; **3.** a; **4.** c; **5.** c; **6.** b

Test Your Understanding

1. A food chain shows just a single connection between organisms at different trophic levels; a food web shows many or most of the connections that exist. The arrows represent the movement of chemical energy, obtained through feeding. **2.** If predation by cougars succeeds in reducing the deer population, then trees and low-growing plant species should increase. More (uneaten) biomass will remain in primary producers, meaning that more will be available for other primary consumers besides deer. **3.** Warmer, wetter climates speed decomposition; cool temperatures slow it. Lower amounts of nitrogen and oxygen also slow decomposition by limiting decomposer growth. Wood is slow to decompose because it contains lignin. Decomposition regulates nutrient availability because it releases nutrients from detritus and allows them to reenter the food web. **4.** Bleaching occurs when corals release symbiotic algae that perform photosynthesis; acidification is occurring as increased CO_2 in the atmosphere leads to increased carbonic acid in the ocean and a lower pH. Corals can starve if bleaching is extensive; acidification slows the growth of corals because it makes formation of calcium carbonate skeletons more difficult. **5.** Positive feedbacks include increased forest fires (due to drier, hotter summers) and increased rates of decomposition in tundras (due to warming); negative feedbacks include increased NPP due to warmer temperatures, increased precipitation, and/or increased access to CO_2 for photosynthesis. **6.** The open ocean has almost no nutrient input from the land and has little upwelling to supply nutrients from the deep ocean. In contrast, intertidal and coastal areas receive large inputs of nutrients from rivers as well as from upwellings from ocean depths.

Applying Concepts to New Situations

1. One possibility is to add radioactive phosphorus to the water, then follow this isotope through the primary producers, primary consumers, and secondary consumers in the system by measuring the amount of radioisotope in the tissues of organisms at each trophic level. **2.** If large amounts of organic material were produced in the open ocean and then moved to nearshore areas by currents, then decomposition of the material might use up all available oxygen and lead to creation of anoxic dead zones. **3.** Without herbivores, there is no link in the nitrogen cycle between primary producers and secondary consumers. All of the plant nitrogen would go to the primary decomposers and back into the soil. If decomposition is rapid enough, nitrogen would cycle quickly between primary producers, decomposers, the soil, and back to primary producers. **4.** Atmospheric oxygen would increase due to extensive photosynthesis, but carbon dioxide levels would decrease because little decomposition was occurring. The temperature would drop because fewer greenhouse gases would be trapping heat reflected from the Earth's surface.

CHAPTER 55

Check Your Understanding (CYU)

CYU p. 1109 Do an all-taxon survey: organize experts and volunteers to collect, examine, and identify all species present. This could include direct sequencing studies (see Chapter 29) to document the bacteria and archaea present in different habitats on campus. **CYU p. 1117** (1) Fragmentation reduces habitat quality by creating edges that are susceptible to invasion and loss of species, due to changed abiotic conditions. Genetic problems occur inside fragments as species become inbred and/or lose genetic diversity via genetic drift. (2) Using the equation $S = cA^z$, then $S = 19(100,000^{0.20}) = 190$ prior to habitat destruction and $S = 19(10,000^{0.20}) = 120$ afterwards, for a total loss of 70 species. Using the same equation for a z of 0.25 suggests that 338 species existed prior to extinction and 190 exist afterwards, a difference of 148. Twice as much diversity was lost due to the increased z. **CYU p. 1120** Establish a large number of study plots on the same hillside. Randomly assign the plots to contain 0, 1, 2, 4, 16, 32, or 64 species (or some other combination, to test a range of species richness), with several replicates for each level of species richness. Prior to planting, document soil depth and soil nutrient levels. Over time, maintain each plot by weeding to keep the original species richness intact. After several years, document soil depth and soil nutrient levels. Compare values as a function of species richness.

Caption Questions and Exercises

Figure 55.6 Nearly all endangered species are threatened by more than one factor and therefore appear on the graph more than once, leading to species totals greater than 100 percent. **Figure 55.10** Because the treatments and study areas were assigned at random, there is no bias in terms of picking certain areas that are unusual in terms of their biomass or species diversity. They should represent a random sample of biomass and species diversity in fragments of various sizes versus intact forest. **Figure 55.11** Draw a horizontal line at lambda of 1.0. **Figure 55.14** Null hypothesis: NPP is not affected by species richness or functional diversity of species. Prediction: NPP will be greater in plots with more species and more functional groups. Prediction of null: There will be no difference in NPP based on species richness or number of functional groups. **Figure 55.15** This line represents no change in biomass, which would mean the

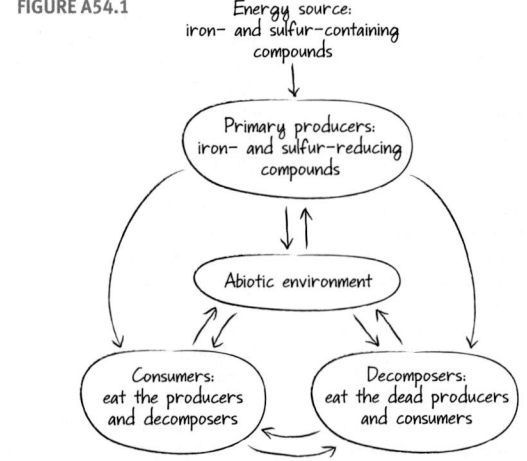

FIGURE A54.1

community was completely resistant to drought—that is, unaffected. **Figure 55.16 [See Figure A55.1]**

Summary of Key Concepts

KC 55.1 By sequencing the genes present in a sample from a habitat, you get a broad view of the allelic diversity present—even if you don't know which species are there. **KC 55.2** Small, geographically isolated populations have low genetic diversity due to lack of gene flow and genetic drift and can suffer from inbreeding depression. Low genetic diversity makes them vulnerable to changes in their habitat, and small population size makes them susceptible to catastrophic events such as storms, disease outbreaks, or fires. **KC 55.3** If many species are present, and each uses the available resources in a unique way, then a larger proportion of all available resources should be used—leading to higher biomass production. **KC 55.4** 11 percent is already protected, but a mass extinction event is still going on. This is because 11 percent is not sufficient to save most species and because many or most protected areas are not located in biodiversity or conservation hotspots.

Test Your Knowledge

1. d; **2.** a; **3.** b; **4.** d; **5.** c; **6.** a

Test Your Understanding

1. The threats have increased, and instead of most endangered species occurring on islands and being threatened by introduced species and overexploitation, now most threatened species are on continents (where habitat destruction is reducing habitats to small islands) and are in trouble due to habitat destruction and climate change. **2.** Level of resource use—because resources are only used at the rate at which they are replenished. **3.** By comparing the number of species estimated to reside in the park before and after the survey, biologists will have an idea of how much actual species diversity is being underestimated in other parts of the world—where research was at the level of Great Smoky Mountains National Park before the survey. The limitation is that it may be difficult to extrapolate from the data—no one knows if the situation at this park is typical of other habitats. **4.** Species–area curves specify how many species are expected to occur in habitats of various size. After projecting amounts of habitat loss, biologists can read a species–area curve to estimate how many species will be left in the amount of area that remains. **5.** In experiments with species native to North American grasslands, study plots that have more species produce more biomass, change less during a disturbance (drought), and recover from a disturbance faster than study plots with fewer species. **6.** Wildlife corridors facilitate the movement of individuals. Corridors allow areas to be recolonized if a species is lost and the introduction of new alleles that can counteract genetic drift and inbreeding in small isolated populations.

Applying Concepts to New Situations

1. Ecosystem services are beneficial effects that ecosystems have on the abiotic environment (soil, air, water quality, climate) for humans. No one owns or pays for these services, so no one has a vested interest in maintaining them. This is one of the primary reasons that ecosystems are destroyed, even though the services they offer are valuable. **2.** Catalog existing biodiversity at the genetic, species, and ecosystem levels. Using these data, find areas that have the highest concentration of biodiversity at each level. Protect as many of these areas as possible, and connect them with corridors of habitat. (In essence, protect and connect the biodiversity hotspots.) **3.** No "correct" answer. One argument is that conservationists could lobby officials in Brazil and Indonesia to learn from the mistakes made in developed nations, and preserve enough forests to maintain biodiversity and

ecosystem services intact—avoiding the expenditures that developed countries had to make to clean up pollution and restore ecosystems and endangered species. **4.** Species with specialized food or habitat requirements, large size (and thus large requirements for land area), small population size, and low reproductive rate are vulnerable to extinction. A good example is the koala bear. Species that are particularly resistant to pressure from humans possess the opposite of many of these traits. A good example of a resistant species is the Norway rat.

THE BIG PICTURE: ECOLOGY

Check Your Understanding (CYU), p. 1126

1. Habitat loss "may eliminate" species; elimination of species (1) may disrupt community structure, (2) may reduces species richness, and (3) may reduce primary productivity. **2.** CO_2 "increases/contributes to" global

warming; global warming (1) may disrupt community structure, (2) may reduce species richness, and (3) may reduce primary productivity. **3.** Populations "make up" species; ecosystems "are based on" primary productivity; water temperature, etc. "dictate species that can be found in certain" aquatic ecosystems; CO_2 "concentrations affect" climate.

APPENDIX A: BIOSKILLS

BIOSKILLS 1 (p. B:2) 1. 3.1 miles **2.** 36.96 yards **3.** 37°C **4.** Multiply your weight in pounds by 1/2.2 (0.45). **5.** 4 **BIOSKILLS 2 (p. B:4) 1.** about 18% **2.** a dramatic drop (almost 10%) **3.** No—the order of presentation in a bar chart does not matter (though it's convenient to arrange the bars in a way that reinforces the overall message). **4.** about 11 **5.** 68 inches **BIOSKILLS 3 Figure B3.1 [See Figure AB.1] Figure B3.2 [See Figure AB.2]**

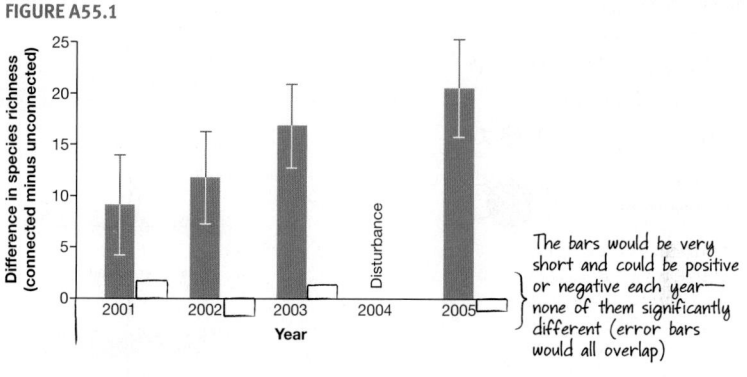

FIGURE A55.1

The bars would be very short and could be positive or negative each year— none of them significantly different (error bars would all overlap)

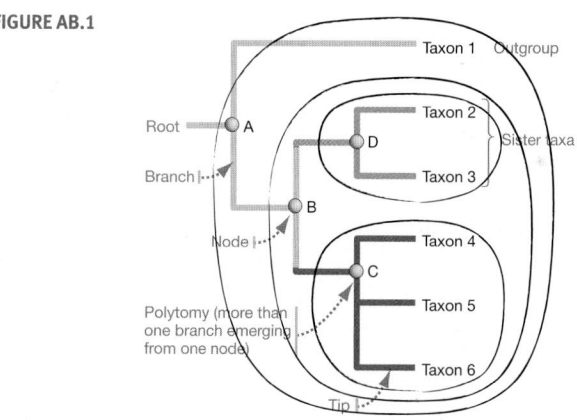

FIGURE AB.1

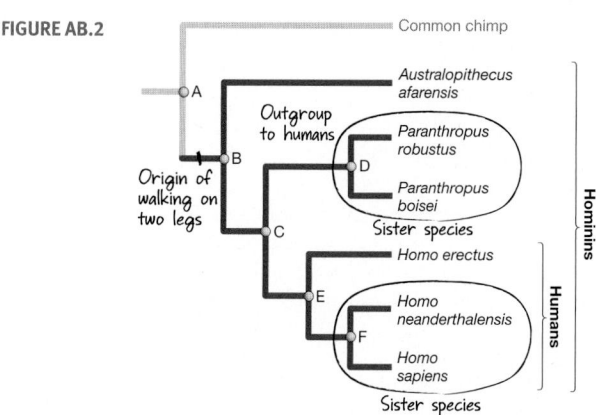

FIGURE AB.2

Figure B3.3 d BIOSKILLS 4 (p. B:6) 1. "different-yoked-together" **2.** "sugary-loosened" **3.** "study-of-form" **4.** "three-bodies" **BIOSKILLS 5 (p. B:7)** The estimate based on the larger sample—the more replicates or observations you have, the more precise your estimate of the average should be. **BIOSKILLS 6 Figure B6.1** [See Figure AB.3] **BIOSKILLS 7 (p. B:9) 1.** exponential **2.** $\ln N_t = \ln N_0 + rt$ **BIOSKILLS 8 (p. B:10) 1.** [See Figure AB.4] **2.** [See Figure AB.4] **BIOSKILLS 9 Figure B9.1** DNA and RNA are acids that tend to drop a proton in solution, giving them a negative charge. **(p. B:13)** The lane with no band comes from a sample where RNA X is not present. The same size RNA X is present in the next two lanes, but the light band has very few copies while the dark band has many. In the fourth lane, the band is

formed by a smaller version of RNA X, with relatively few copies present. **BIOSKILLS 10 (p. B:16) 1.** No—it's just that no mitochondria happened to be present in this section sliced through the cell. **2.** Understanding a molecule's structure is often critical to understanding how it functions in cells. **BIOSKILLS 11 (p. B:17) 1.** size and/or density **2.** The use of high speeds generates the much higher forces needed to separate molecules that are similar in size and mass. **BIOSKILLS 12 (p. B:18) 1.** It may not be clear that the results are relevant to noncancerous cells that are not growing in cell culture—that is, that the artificial conditions mimic natural conditions. **2.** It may not be clear that the results are relevant to individuals that developed normally, from an embryo—that is, that the artificial conditions mimic natural conditions.

BIOSKILLS 13 (p. B:19) 1. $\frac{1}{2} \times \frac{1}{2} \times \frac{1}{2} \times \frac{1}{2} \times \frac{1}{2} = \frac{1}{16}$ **2.** $\frac{1}{8} + \frac{1}{8} + \frac{1}{4} = \frac{1}{2}$ **BIOSKILLS 14 Figure B14.1** This is human body temperature—the natural habitat of *E. coli*. **(p. B:22) 1.** *Caenorhabditis elegans* would be a good possibility, because the cells that normally die have already been identified. You could find mutant individuals that lacked normal cell death; you could compare the resulting embryos to normal embryos and be able to identify exactly which cells change as a result. **2.** Any of the multicellular organisms in the list would be a candidate, but *Dictyostelium discoideum* might be particularly interesting because cells only stick to each other during certain points in the life cycle. **3.** *Mus musculus*—as the only mammal in the list, it is the organism most likely to have a gene similar to the one you want to study.

FIGURE AB.3

Molecular formula:	CO_2
Structural formula:	$O=C=O$
Ball-and-stick model:	
Space-filling model:	

FIGURE AB.4

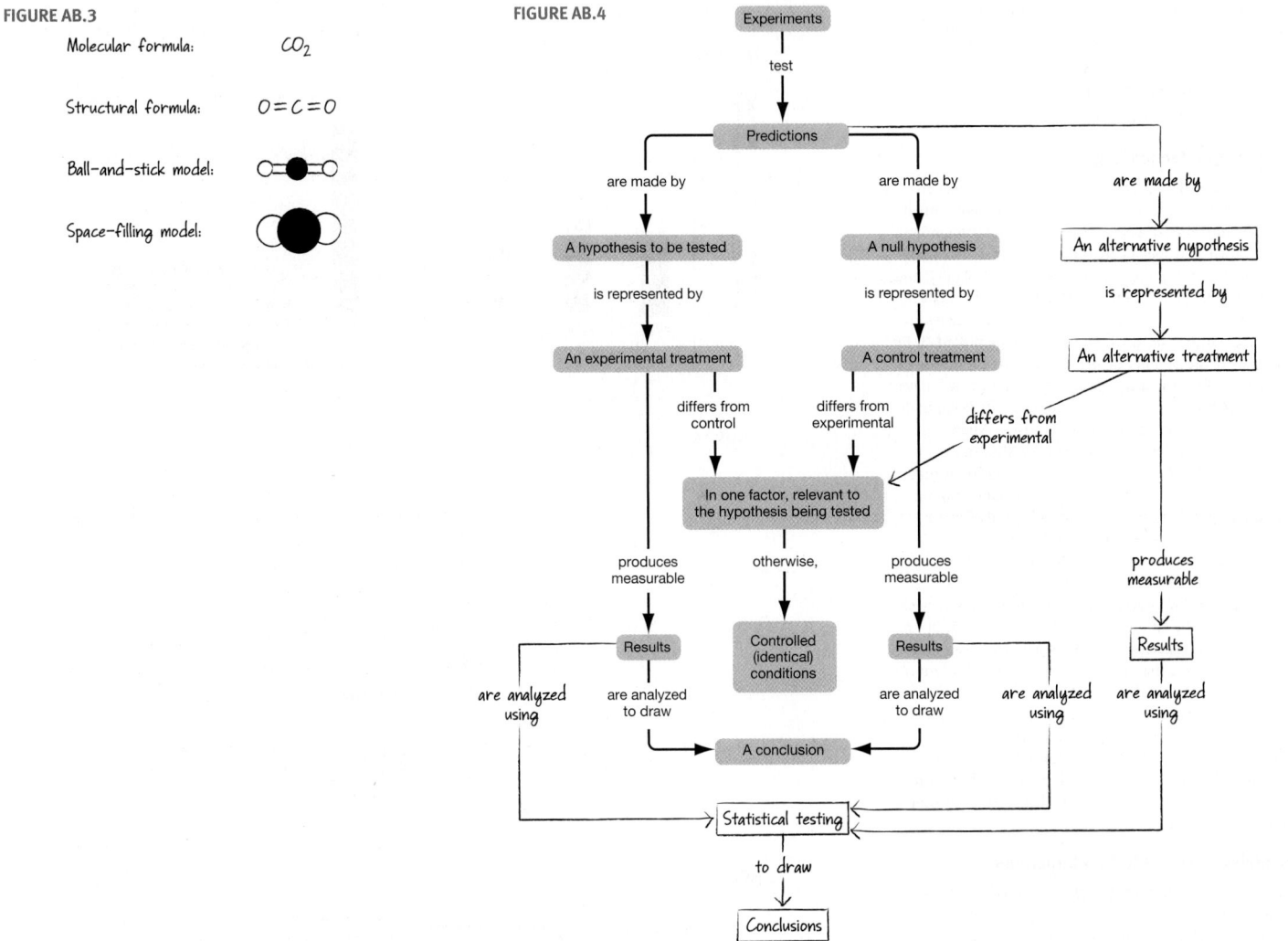

Glossary

5′ cap A chemical grouping, consisting of 7-methylguanylate and three phosphate groups, that is added to the 5′ end of newly transcribed messenger RNA molecules.

20-hydroxyecdysone See **ecdysone**.

abdomen A region of the body; in insects, one of the three prominent body regions called tagmata.

abiotic Not alive (e.g., air, water, and soil). Compare with **biotic**.

aboveground biomass The total mass of living plants in an area, excluding roots.

abscisic acid (ABA) A plant hormone that inhibits cell elongation and stimulates leaf shedding and dormancy.

abscission In plants, the normal (often seasonal) shedding of leaves, fruits, or flowers.

abscission zone The region at the base of a petiole that thins and breaks during dropping of leaves.

absorption In animals, the uptake of ions and small molecules derived from food across the lining of the intestine and into the bloodstream.

absorption spectrum The amount of light of different wavelengths absorbed by a pigment. Usually depicted as a graph of light absorbed versus wavelength. Compare with **action spectrum**.

acclimation, acclimatization Gradual physiological adjustment of an organism to new environmental conditions that occur naturally or as part of a laboratory experiment.

acetyl CoA A molecule produced by oxidation of pyruvate (the final product of glycolysis) in a reaction catalyzed by pyruvate dehydrogenase. Can enter the citric acid cycle and also is used as a carbon source in the synthesis of fatty acids, steroids, and other compounds.

acetylation Addition of an acetyl group (CH_3COO^-) to a molecule.

acetylcholine (ACh) A neurotransmitter, released by nerve cells at neuromuscular junctions, that triggers contraction of muscle cells. Also used as a neurotransmitter between neurons.

acid Any compound that gives up protons or accepts electrons during a chemical reaction or that releases hydrogen ions when dissolved in water.

acid-growth hypothesis The hypothesis that auxin triggers elongation of plant cells by inducing the synthesis of proton pumps whose activity makes the cell wall more acidic, leading to expansion of the cell wall and an influx of water.

acoelomate An animal that lacks an internal body cavity (coelom). Compare with **coelomate** and **pseudocoelomate**.

acquired immune deficiency syndrome (AIDS) A human disease characterized by death of immune system cells (in particular helper T cells and macrophages) and subsequent vulnerability to other infections. Caused by the human immunodeficiency virus (HIV).

acquired immunity Immunity to a particular pathogen or other antigen conferred by antibodies and activated B and T cells following exposure to the antigen. Is characterized by specificity, diversity, memory, and self-nonself recognition. Also called *acquired immune response*. Compare with **innate immunity**.

acrosome A caplike structure, located on the head of a sperm cell, that contains enzymes capable of dissolving the outer coverings of an egg.

ACTH See **adrenocorticotropic hormone**.

actin A globular protein that can be polymerized to form filaments. Actin filaments are part of the cytoskeleton and constitute the thin filaments in skeletal muscle cells.

actin filament A long fiber, about 7 nm in diameter, composed of two intertwined strands of polymerized actin protein; one of the three types of cytoskeletal fibers. Involved in cell movement. Also called a *microfilament*. Compare with **intermediate filament** and **microtubule**.

action potential A rapid, temporary change in electrical potential across a membrane, from negative to positive and back to negative. Occurs in cells, such as neurons and muscle cells, that have an excitable membrane.

action spectrum The relative effectiveness of different wavelengths of light in driving a light-dependent process such as photosynthesis. Usually depicted as a graph of some measure of the process versus wavelength. Compare with **absorption spectrum**.

activation energy The amount of energy required to initiate a chemical reaction; specifically, the energy required to reach the transition state.

active site The portion of an enzyme molecule where substrates (reactant molecules) bind and react.

active transport The movement of ions or molecules across a plasma membrane or organelle membrane against an electrochemical gradient. Requires energy (e.g., from hydrolysis of ATP) and assistance of a transport protein (e.g., pump).

adaptation Any heritable trait that increases the fitness of an individual with that trait, compared with individuals without that trait, in a particular environment.

adaptive immunity Immunity to a particular pathogen or other antigen conferred by antibodies and activated B and T cells following exposure to the antigen. Is characterized by specificity, diversity, memory, and self-nonself recognition. Also called *adaptive immune response*. Compare with **innate immunity**.

adaptive radiation Rapid evolutionary diversification within one lineage, producing numerous descendant species with a wide range of adaptive forms.

adenosine diphosphate (ADP) A molecule consisting of adenine, a sugar, and two phosphate groups. Addition of a third phosphate group produces adenosine triphosphate (ATP).

adenosine triphosphate (ATP) A molecule consisting of adenine, a sugar, and three phosphate groups that can be hydrolyzed to release energy. Universally used by cells to store and transfer energy.

adenylyl cyclase An enzyme that can catalyze the formation of cyclic AMP (cAMP) from ATP. Involved in controlling transcription of various operons in prokaryotes and in some eukaryotic signal-transduction pathways.

adhesion The tendency of certain dissimilar molecules to cling together due to attractive forces. Compare with **cohesion**.

adipocyte A fat cell.

adipose tissue A type of connective tissue whose cells store fats.

ADP See **adenosine diphosphate**.

adrenal glands Two small endocrine glands that sit above each kidney. The outer portion (cortex) secretes several steroid hormones; the inner portion (medulla) secretes epinephrine and norepinephrine.

adrenaline See **epinephrine**.

adrenocorticotropic hormone (ACTH) A peptide hormone, produced and secreted by the anterior pituitary, that stimulates release of steroid hormones (e.g., cortisol, aldosterone) from the adrenal cortex.

adult A sexually mature individual.

adventitious root A root that develops from a plant's shoot system instead of from the plant's root system.

aerobic Referring to any metabolic process, cell, or organism that uses oxygen as an electron acceptor. Compare with **anaerobic**.

afferent division The part of the nervous system, consisting mainly of sensory neurons, that transmits information about the internal and external environment to the central nervous system. Compare with **efferent division**.

age class All the individuals of a specific age in a population.

age structure The proportion of individuals in a population that are of each possible age.

age-specific fecundity The average number of female offspring produced by a female in a certain age class.

agglutination Clumping together of cells, typically caused by antibodies.

aggregate fruit A fruit (e.g., raspberry) that develops from a single flower that has many separate carpels. Compare with **multiple** and **simple fruit**.

AIDS See **acquired immune deficiency syndrome**.

albumen A solution of water and protein (particularly albumins), found in amniotic eggs, that nourishes the growing embryo. Also called *egg white*.

albumin A class of large proteins found in plants and animals, particularly in the albumen of eggs and in blood plasma.

alcohol fermentation Catabolic pathway in which pyruvate produced by glycolysis is converted to ethanol in the absence of oxygen.

aldosterone A hormone produced in the adrenal cortex that stimulates the kidney to conserve salt and water and promotes retention of sodium.

alimentary canal See **digestive tract**.

allele A particular version of a gene.

allergen A molecule (antigen) that triggers an allergic response (an allergy).

allergy An abnormal response to an antigen, usually characterized by dilation of blood vessels, contraction of smooth muscle cells, and increased activity by mucous-secreting cells.

allopatric speciation The divergence of populations into different species by physical isolation of populations in different geographic areas. Compare with **sympatric speciation**.

allopatry Condition in which two or more populations live in different geographic areas. Compare with **sympatry**.

allopolyploidy (adjective: allopolyploid) The state of having more than two full sets of chromosomes (polyploidy) due to hybridization between different species. Compare with **autopolyploidy**.

allosteric regulation Regulation of a protein's function by binding of a regulatory molecule, usually to a specific site distinct from the active site, causing a change in the protein's shape.

α-amylase See **amylase**.

α-helix (alpha-helix) A protein secondary structure in which the polypeptide backbone coils into a spiral shape stabilized by hydrogen bonds between atoms.

alternation of generations A life cycle involving alternation of a multicellular haploid stage (gametophyte) with a multicellular diploid stage (sporophyte). Occurs in most plants and some protists.

alternative splicing In eukaryotes, the splicing of primary RNA transcripts from a single gene in different ways to produce different mature mRNAs and thus different polypeptides.

altruism Any behavior that has a cost to the individual (such as lowered survival or reproduction) and a benefit to the recipient. See **reciprocal altruism**.

alveolus (plural: alveoli) One of the tiny air-filled sacs of a mammalian lung.

ambisense virus A virus whose genome contains both positive-sense and negative-sense sequences.

amino acid A small organic molecule with a central carbon atom bonded to an amino group ($-NH_3$), a carboxyl group ($-COOH$), a hydrogen atom, and a side group. Proteins are polymers of 20 common amino acids.

aminoacyl tRNA A transfer RNA molecule that is covalently bound to an amino acid.

aminoacyl tRNA synthetase An enzyme that catalyzes the addition of a particular amino acid to its corresponding tRNA molecule.

ammonia (NH_3) A small molecule, produced by the breakdown of proteins and nucleic acids, that is very toxic to cells. Is a strong base that gains a proton to form the ammonium ion (NH_4^+).

amnion The membrane within an amniotic egg that surrounds the embryo and encloses it in a protective pool of fluid (amniotic fluid).

amniotes A major lineage of vertebrates (Amniota) that reproduce with amniotic eggs. Includes all reptiles (including birds) and mammals.

amniotic egg An egg that has a watertight shell or case enclosing a membrane-bound water supply (the amnion), food supply (yolk sac), and waste sac (allantois).

amoeboid motion See **cell crawling**.

amphibians A lineage of vertebrates many of whom breathe through their skin and feed on land but lay their eggs in water. Represent the earliest tetrapods; include frogs, salamanders, and caecilians.

amphipathic Containing hydrophilic and hydrophobic elements.

amylase Any enzyme that can break down starch by catalyzing hydrolysis of the glycosidic linkages between the glucose residues.

amyloplasts Dense, starch-storing organelles that settle to the bottom of plant cells and that may be used as gravity detectors.

anabolic pathway Any set of chemical reactions that synthesizes larger molecules from smaller ones. Generally requires an input of energy. Compare with **catabolic pathway**.

anadromous Having a life cycle in which adults live in the ocean (or large lakes) but migrate up freshwater streams to breed and lay eggs.

anaerobic Referring to any metabolic process, cell, or organism that uses an electron acceptor other than oxygen, such as nitrate or sulfate. Compare with **aerobic**.

anaphase A stage in mitosis or meiosis during which chromosomes are moved to opposite ends of the cell.

anatomy The study of the physical structure of organisms.

ancestral trait A trait found in ancestors.

aneuploidy (adjective: aneuploid) The state of having an abnormal number of copies of a certain chromosome.

angiosperm A flowering vascular plant that produces seeds within mature ovaries (fruits). The angiosperms form a single lineage. Compare with **gymnosperm**.

animal A member of a major lineage of eukaryotes (Animalia) whose members typically have a complex, large, multicellular body, eat other organisms, and are mobile.

animal model Any disease that occurs in a nonhuman animal and has many parallels to a similar disease of humans. Studied by medical researchers in hope that findings may apply to human disease.

anion A negatively charged ion.

annelids Members of the phylum Annelida (segmented worms). Distinguished by a segmented body and a coelom that functions as a hydrostatic skeleton. Annelids belong to the lophotrochozoan branch of the protostomes.

annual Referring to a plant whose life cycle normally lasts only one growing season—less than one year. Compare with **perennial**.

anoxygenic Referring to any process or reaction that does not produce oxygen. Photosynthesis in purple sulfur and purple nonsulfur bacteria, which does not involve photosystem II, is anoxygenic. Compare with **oxygenic**.

antenna (plural: antennae) A long appendage that is used to touch or smell.

antenna complex Part of a photosystem, containing an array of chlorophyll molecules and accessory pigments, that receives energy from light and directs the energy to a central reaction center during photosynthesis.

anterior Toward an animal's head and away from its tail. The opposite of posterior.

anterior pituitary The part of the pituitary gland containing endocrine cells that produce and release a variety of peptide hormones in response to other hormones from the hypothalamus. Compare with **posterior pituitary**.

anther The pollen-producing structure at the end of a stamen in flowering plants (angiosperms).

antheridium (plural: antheridia) The sperm-producing structure in most land plants except angiosperms.

anthropoids One of the two major lineages of primates, including apes, humans, and all monkeys. Compare with **prosimians**.

antibiotic Any substance, such as penicillin, that can kill or inhibit the growth of bacteria.

antibody An immunoglobulin protein, produced by B cells, that can bind to a specific part of an antigen, tagging it for attack by the immune system. All antibody molecules have a similar Y-shaped structure and, in their monomer form, consist of two identical light chains and two identical heavy chains.

anticodon The sequence of three bases (triplet) in a transfer RNA molecule that can bind to a mRNA codon with a complementary sequence.

antidiuretic hormone (ADH) A peptide hormone, secreted from the posterior pituitary gland, that stimulates water retention by the kidney. Also called *vasopressin*.

antigen Any foreign molecule, often a protein, that can stimulate a specific response by the immune system.

antigen presentation Process by which small peptides, derived from ingested particulate antigens (e.g., bacteria) or intracellular antigens (e.g., viruses in infected cell) are complexed with MHC proteins and transported to the cell surface where they are displayed and can be recognized by T cells.

antiparallel Describing the opposite orientation of the strands in a DNA double helix with one strand running in the $5' \rightarrow 3'$ direction and the other in the $3' \rightarrow 5'$ direction.

antiporter A carrier protein that allows an ion to diffuse down an electrochemical gradient, using the energy of that process to transport a different substance in the opposite direction *against* its concentration gradient. Compare with **symporter**.

antiviral Any drug or other agent that can kill or inhibit the transmission or replication of viruses.

anus In a multicellular animal, the end of the digestive tract where wastes are expelled.

aorta In terrestrial vertebrates, the major artery carrying oxygenated blood away from the heart.

aphotic zone Deep water receiving no sunlight. Compare with **photic zone**.

apical Toward the top. In plants, at the tip of a branch. In animals, on the side of an epithelial layer that faces the environment and not other body tissues. Compare with **basal**.

apical bud A bud at the tip of a stem, where growth occurs to lengthen the stem.

apical dominance Inhibition of lateral bud growth by the apical meristem at the tip of a plant branch.

apical meristem A group of undifferentiated plant cells, at the tip of a stem or root, that is responsible for primary growth. Compare with **lateral meristem**.

apical–basal axis The long or vertical, shoot-to-root axis of a plant.

apomixis The formation of mature seeds without fertilization occurring; a type of asexual reproduction.

apoplast In plant roots, a continuous pathway through which water can flow, consisting of the porous cell walls of adjacent cells and the intervening extracellular space. Compare with **symplast**.

apoptosis Series of genetically controlled changes that lead to death of a cell. Occurs frequently during embryological development and later may occur in response to infections or cell damage. Also called *programmed cell death*.

appendix A blind sac (having only one opening) that extends from the colon in some mammals.

aquaporin A type of channel protein through which water can move by osmosis across a plasma membrane.

arbuscular mycorrhizal fungi (AMF) Fungi whose hyphae enter the root cells of their host plants. Also called *endomycorrhizal fungi*.

Archaea One of the three taxonomic domains of life consisting of unicellular prokaryotes distinguished by cell walls made of certain polysaccharides not found in bacterial or eukaryotic cell walls, plasma membranes composed of unique isoprene-containing phospholipids, and ribosomes and RNA polymerase similar to those of eukaryotes. Compare with **Bacteria** and **Eukarya**.

archegonium (plural: archegonia) The egg-producing structure in most land plants except angiosperms.

arteriole One of the many tiny vessels that carry blood from arteries to capillaries.

artery Any thick-walled blood vessel that carries blood (oxygenated or not) under relatively high pressure away from the heart to organs throughout the body. Compare with **vein**.

arthropods Members of the phylum Arthropoda. Distinguished by a segmented body; a hard, jointed exoskeleton; paired appendages; and an extensive body cavity called a hemocoel. Arthropods belong to the ecdysozoan branch of the protostomes.

articulation A movable point of contact between two bones of a skeleton. See **joint**.

artificial selection Deliberate manipulation by humans, as in animal and plant breeding, of the genetic composition of a population by allowing only individuals with desirable traits to reproduce.

ascocarp A large, cup-shaped reproductive structure produced by some ascomycete fungi. Contains many microscopic asci, which produce spores.

ascus (plural: asci) Specialized spore-producing cell found at the ends of hyphae in sac fungi (ascomycetes).

asexual reproduction Any form of reproduction resulting in offspring that are genetically identical to the parent. Includes binary fission, budding, and parthenogenesis. Compare with **sexual reproduction**.

asymmetric competition Ecological competition between two species in which one species suffers a much greater fitness decline than the other. Compare with **symmetric competition**.

atomic number The number of protons in the nucleus of an atom, giving the atom its identity as a particular chemical element.

ATP See **adenosine triphosphate**.

ATP synthase A large membrane-bound protein complex in chloroplasts, mitochondria, and some bacteria that uses the energy of protons flowing through it to synthesize ATP. Also called F_oF_1 complex.

atrial natriuretic hormone An animal hormone that stimulates excretion of salt from the kidneys.

atrioventricular (AV) node A region of the heart between the right atrium and right ventricle where electrical signals from the atrium are slowed briefly before spreading to the ventricle. This delay allows the ventricle to fill with blood before contracting. Compare with **sinoatrial (SA) node**.

atrium (plural: atria) A thin-walled chamber of the heart that receives blood from veins and pumps it to a neighboring chamber (the ventricle).

autocrine Relating to a chemical signal that affects the same cell that produced and released it.

autoimmunity A pathological condition in which the immune system attacks self cells or tissues of an individual's own body.

autonomic nervous system The part of the peripheral nervous system that controls internal organs and involuntary processes, such as stomach contraction, hormone release, and heart rate. Includes parasympathetic and sympathetic nerves. Compare with **somatic nervous system**.

autophagy The process by which damaged organelles are surrounded by a membrane and delivered to a lysosome to be destroyed.

autopolyploidy (adjective: autopolyploid) The state of having more than two full sets of chromosomes (polyploidy) due to a mutation that doubled the chromosome number.

autoradiography A technique for detecting radioactively labeled molecules separated by gel electrophoresis by placing an unexposed film over the gel. A black dot appears on the film wherever a radioactive atom is present in the gel. Also can be used to locate labeled molecules in fixed tissue samples.

autosomal inheritance The inheritance patterns that occur when genes are located on autosomes rather than on sex chromosomes.

autosome Any chromosome that does not carry genes involved in determining the sex of an individual.

autotroph Any organism that can synthesize reduced organic compounds from simple inorganic sources such as CO_2 or CH_4. Most plants and some bacteria and archaea are autotrophs. Also called *primary producer*. Compare with **heterotroph**.

auxin Indoleacetic acid, a plant hormone that stimulates phototropism and some other responses.

avirulence (*avr*) genes Genes in pathogens encoding proteins that trigger a defense response in plants. Compare with **resistance (*R*) genes**.

axillary bud A bud that forms in the angle between a leaf and a stem and may develop into a lateral (side) branch. Also called *lateral bud*.

axon A long projection of a neuron that can propagate an action potential and transmit it to another neuron.

axon hillock The site in a neuron where an axon joins the cell body and where action potentials are first triggered.

axoneme A structure found in eukaryotic cilia and flagella and responsible for their motion; composed of two central microtubules surrounded by nine doublet microtubules (9 + 2 arrangement).

B cell A type of leukocyte that matures in the bone marrow and, with T cells, is responsible for adaptive immunity. Produces antibodies and also functions in antigen presentation. Also called *B lymphocyte*.

BAC library A collection of all the sequences found in the genome of a species, inserted into bacterial artificial chromosomes (BACs).

background extinction The average rate of low-level extinction that has occurred continuously throughout much of evolutionary history. Compare with **mass extinction**.

Bacteria One of the three taxonomic domains of life consisting of unicellular prokaryotes distinguished by cell walls composed largely of peptidoglycan, plasma membranes similar to those of eukaryotic cells, and ribosomes and RNA polymerase that differ from those in archaeans or eukaryotes. Compare with **Archaea** and **Eukarya**.

bacterial artificial chromosome (BAC) An artificial version of a bacterial chromosome that can be used as a cloning vector to produce many copies of large DNA fragments.

bacteriophage Any virus that infects bacteria.

baculum A bone inside the penis usually present in mammals with a penis that lacks erectile tissue.

balancing selection A pattern of natural selection in which no single allele is favored in all populations of a species at all times. Instead, there is a balance among alleles in terms of fitness and frequency.

ball-and-stick model A representation of a molecule where atoms are shown as balls—colored and scaled to indicate the atom's identity—and covalent bonds are shown as rods or sticks connecting the balls in the correct geometry.

bar coding The use of well-characterized gene sequences to identify species.

bark The protective outer layer of woody plants, composed of cork cells, cork cambium, and secondary phloem.

baroreceptors Specialized nerve cells in the walls of the heart and certain major arteries that detect changes in blood pressure and trigger appropriate responses by the brain.

basal Toward the base. In plants, at the base of a branch where it joins the stem. In animals, on the side of an epithelial layer that abuts underlying body tissues. Compare with **apical**.

basal body A structure of nine pairs of microtubules arranged in a circle at the base of eukaryotic cilia and flagella where they attach to the cell. Structurally similar to a centriole.

basal lamina A thick, collagen-rich extracellular matrix that underlies most epithelial tissues (e.g., skin) in animals.

basal metabolic rate (BMR) The total energy consumption by an organism at rest in a comfortable environment. For aerobes, often measured as the amount of oxygen consumed per hour.

basal transcription complex A large multi-protein structure that assembles near the promoter of eukaryotic genes and initiates transcription. Composed of basal transcription factors, TATA-binding protein, coactivators, and RNA polymerase.

basal transcription factor General term for proteins, present in all cell types, that bind to eukaryotic promoters and help initiate transcription. Compare with **regulatory transcription factor**.

base Any compound that acquires protons or gives up electrons during a chemical reaction or accepts hydrogen ions when dissolved in water.

basidium (plural: basidia) Specialized spore-producing cell at the ends of hyphae in club fungi.

basilar membrane The membrane in the vertebrate cochlea on which the bottom portion of hair cells sit.

basolateral Toward the bottom and sides. In animals, the side of an epithelial layer that faces other body tissues and not the environment.

Batesian mimicry A type of mimicry in which a harmless or palatable species resembles a dangerous or poisonous species. Compare with **Müllerian mimicry**.

B-cell receptor (BCR) An immunoglobulin protein (antibody) embedded in the plasma membrane of mature B cells and to which antigens bind.

beak A structure that exerts biting forces and is associated with the mouth; found in birds, cephalopods, and some insects.

behavior Any action by an organism.

beneficial In genetics, referring to any mutation, allele, or trait that increases an individual's fitness.

benign tumor A mass of abnormal tissue that grows slowly or not at all, does not disrupt surrounding tissues, and does not spread to other organs. Benign tumors are not cancers. Compare with **malignant tumor**.

benthic Living at the bottom of an aquatic environment.

benthic zone The area along the bottom of an aquatic environment.

β-pleated sheet (beta-pleated sheet) A protein secondary structure in which the polypeptide backbone folds into a sheetlike shape stabilized by hydrogen bonding.

bilateral symmetry An animal body pattern in which there is one plane of symmetry dividing the body into a left side and a right side. Typically, the body is long and narrow, with a distinct head end and tail end. Compare with **radial symmetry**.

bilaterian A member of a major lineage of animals (Bilateria) that are bilaterally symmetrical at some point in their life cycle, have three embryonic germ layers, and have a coelom. All protostomes and deuterostomes are bilaterians.

bile A complex solution produced by the liver, stored in the gall bladder, and secreted into the intestine. Contains steroid derivatives called bile salts that are responsible for emulsification of fats during digestion.

biodiversity The diversity of life considered at three levels: genetic diversity (variety of alleles in a population, species, or group of species); species diversity (variety and relative abundance of species present in a certain area); and ecosystem diversity (variety of communities and abiotic components in a region).

biodiversity hot spot A region that is extraordinarily rich in species.

biofilm A hard, polysaccharide-rich layer secreted by bacterial cells, which allows them to attach to a surface.

biogeochemical cycle The pattern of circulation of an element or molecule among living organisms and the environment.

biogeography The study of how species and populations are distributed geographically.

bioinformatics The field of study concerned with managing, analyzing, and interpreting biological information, particularly DNA sequences.

biological fitness See **fitness**.

biological species concept The definition of a species as a population or group of populations that are reproductively isolated from other groups. Members of a species have the potential to interbreed in nature to produce viable, fertile offspring but cannot produce viable, fertile hybrid offspring with members of other species. Compare with **morphospecies** and **phylogenetic species concept**.

bioluminescence The emission of light by a living organism.

biomagnification In animal tissues, an increase in the concentration of particular molecules that may occur as those molecules are passed up a food chain.

biomass The total mass of all organisms in a given population or geographical area; usually expressed as total dry weight.

biome A large terrestrial ecosystem characterized by a distinct type of vegetation and climate.

bioprospecting The effort to find commercially useful compounds by studying organisms—especially species that are poorly studied to date.

bioremediation The use of living organisms, usually bacteria or archaea, to degrade environmental pollutants.

biotechnology The application of biological techniques and discoveries to medicine, industry, and agriculture.

biotic Living, or produced by a living organism. Compare with **abiotic**.

bipedal Walking primarily on two legs.

bipolar cell A cell in the vertebrate retina that receives information from one or more photoreceptors and passes it to other bipolar cells or ganglion cells.

bivalves A lineage of mollusks that have two shells, such as clams and mussels.

bladder A mammalian organ that holds urine until it can be excreted.

blade The wide, flat part of a plant leaf.

blastocoel Fluid-filled cavity in the blastula of many animal species.

blastocyst Specialized type of blastula in mammals. A spherical structure composed of trophoblast cells on the exterior and a cluster of cells (the inner cell mass), which fills part of the interior space.

blastomere A small cell created by cleavage divisions in early animal embryos.

blastopore A small opening (pore) in the surface of an early vertebrate embryo, through which cells move during gastrulation.

blastula In vertebrate development, a hollow ball of cells (blastomere cells) that is formed by cleavage of a zygote and immediately undergoes gastrulation. See **blastocyst**.

blood A type of connective tissue consisting of red blood cells and leukocytes suspended in a fluid portion called plasma.

blood pressure See **diastolic blood pressure** and **systolic blood pressure**.

body mass index (BMI) A mathematical relationship of weight and height used to assess obesity in humans. Calculated as weight (kg) divided by the square of height (m^2).

body plan The basic architecture of an animal's body, including the number and arrangement of limbs, body segments, and major tissue layers.

bog A wetland that has no or almost no water flow, resulting in very low oxygen levels and acidic conditions.

Bohr shift The rightward shift of the oxygen-hemoglobin dissociation curve that occurs with decreasing pH. Results in hemoglobin being more likely to release oxygen in the acidic environment of exercising muscle.

bone A type of vertebrate connective tissue consisting of living cells and blood vessels within a hard extracellular matrix composed of calcium phosphate ($CaPO_4$) and small amounts of calcium carbonate ($CaCO_3$) and protein fibers.

bone marrow The soft tissue filling the inside of long bones containing stem cells that develop into red blood cells and leukocytes throughout life.

Bowman's capsule The hollow, double-walled cup-shaped portion of a nephron that surrounds a glomerulus in the vertebrate kidney.

brain A large mass of neurons located in the head region of an animal, that is involved in information processing; may also be called the cerebral ganglion.

brain stem The most posterior portion of the vertebrate brain, connecting to the spinal cord and responsible for autonomic body functions such as heart rate, respiration, and digestion.

braincase See **cranium**.

branch (1) A part of a phylogenetic tree that represents populations through time. (2) Any extension of a plant's shoot system.

brassinosteroids A family of steroid hormones found in plants.

bronchiole One of the small tubes in mammalian lungs that carry air from the bronchi to the alveoli.

bronchus (plural: bronchi) In mammals, one of a pair of large tubes that lead from the trachea to each lung.

bryophytes Members of several phyla of green plants that lack vascular tissue including liverworts, hornworts, and mosses. Also called *non-vascular plants*.

budding Asexual reproduction via outgrowth from the parent that eventually breaks free as an independent individual; occurs in yeasts and some invertebrates.

buffer A substance that, in solution, acts to minimize changes in the pH of that solution when acid or base is added.

bulbourethral glands In male mammals, small paired glands at the base of the urethra that secrete an alkaline mucus (part of semen), which lubricates the tip of the penis and neutralizes acids in the urethra during copulation. In humans, also called *Cowper's glands*.

bulk flow The directional movement of a substantial volume of fluid due to pressure differences, such as movement of water through plant phloem and movement of blood in animals.

bundle-sheath cell A type of cell found around the vascular tissue (veins) of plant leaves.

C_3 photosynthesis The most common form of photosynthesis in which atmospheric CO_2 is used to form 3-phosphoglycerate, a three-carbon sugar.

C_4 photosynthesis A variant type of photosynthesis in which atmospheric CO_2 is first fixed into four-carbon sugars, rather than the three-carbon sugars of classic C_3 photosynthesis. Enhances photosynthetic efficiency in hot, dry environments, by reducing loss of oxygen due to photorespiration.

cadherin Any of a class of cell-surface proteins involved in cell adhesion and important for coordinating movements of cells during embryological development.

callus In plants, a mass of undifferentiated cells that can generate roots and other tissues necessary to create a mature plant.

Calvin cycle In photosynthesis, the set of light-independent reactions that use NADPH and ATP formed in the light-dependent reactions to drive the fixation of atmospheric CO_2 and reduction of the fixed carbon, ultimately producing sugars. Also called *carbon fixation* and *light-independent reactions*.

calyx All of the sepals of a flower.

CAM See **crassulacean acid metabolism**.

cambium (plural: cambia) See **lateral meristem**.

Cambrian explosion The rapid diversification of animal body types and lineages that occured between the species present in the Ediacaran faunas (565–542 mya) and the Cambrian faunas (525–515 mya).

camera eye The type of eye in vertebrates and cephalopods, consisting of a hollow chamber with a hole at one end (through which light enters) and a sheet of light-sensitive cells against the opposite wall.

cAMP See **cyclic AMP**.

cancer General term for any tumor whose cells grow in an uncontrolled fashion, invade nearby tissues, and spread to other sites in the body.

canopy The uppermost layers of branches in a forest (i.e., those fully exposed to the Sun).

CAP binding site A DNA sequence upstream of certain prokaryotic operons to which catabolite activator protein can bind, increasing gene transcription.

capillarity The tendency of water to move up a narrow tube due to surface tension, adhesion, and cohesion.

capillary One of the numerous small, thin-walled blood vessels that permeate all tissues and organs, and allow exchange of gases and other molecules between blood and body cells.

capillary bed A thick network of capillaries.

capsid A shell of protein enclosing the genome of a virus particle.

carapace In crustaceans, a large platelike section of the exoskeleton that covers and protects the cephalothorax (e.g., a crab's "shell").

carbohydrate Any of a class of molecules that contain a carbonyl group, several hydroxyl groups, and several to many carbon-hydrogen bonds. See **monosaccharide** and **polysaccharide**.

carbon cycle, global The worldwide movement of carbon among terrestrial ecosystems, the oceans, and the atmosphere.

carbon fixation See **Calvin cycle**.

carbonic anhydrase An enzyme that catalyzes the formation of carbonic acid (H_2CO_3) from carbon dioxide and water.

carboxylic acids Organic acids with the form R-COOH (a carboxyl group).

cardiac cycle One complete heartbeat cycle, including systole and diastole.

cardiac muscle The muscle tissue of the vertebrate heart. Consists of long branched fibers that are electrically connected and that initiate their own contractions; not under voluntary control. Compare with **skeletal** and **smooth muscle**.

carnivore (adjective: carnivorous) An animal whose diet consists predominantly of meat. Most members of the mammalian taxon Carnivora are carnivores. Some plants are carnivorous, trapping and killing small animals, then absorbing nutrients from the prey's body. Compare with **herbivore** and **omnivore**.

carotenoid Any of a class of accessory pigments, found in chloroplasts, that absorb wavelengths of light not absorbed by chlorophyll; typically appear yellow, orange, or red. Includes carotenes and xanthophylls.

carpel The female reproductive organ in a flower. Consists of the stigma, to which pollen grains adhere; the style, through which pollen grains move; and the ovary, which houses the ovule. Compare with **stamen**.

carrier A heterozygous individual carrying a normal allele and a recessive allele for an inherited trait; does not display the phenotype of the trait but can pass the recessive gene to offspring.

carrier protein A membrane protein that facilitates diffusion of a small molecule (e.g., glucose) across the plasma membrane by a process involving a reversible change in the shape of the protein. Also called *carrier* or *transporter*.

carrying capacity (K) The maximum population size of a certain species that a given habitat can support.

cartilage A type of vertebrate connective tissue that consists of relatively few cells scattered in a stiff matrix of polysaccharides and protein fibers.

Casparian strip In plant roots, a waxy layer containing suberin, a water-repellent substance, that prevents movement of water through the walls of endodermal cells, thus blocking the apoplastic pathway.

cast A type of fossil, formed when the decay of a body part leaves a void that is then filled with minerals that later harden.

catabolic pathway Any set of chemical reactions that breaks down larger, complex molecules into smaller ones, releasing energy in the process. Compare with **anabolic pathway**.

catabolite activator protein (CAP) A protein that can bind to the CAP binding site upstream of certain prokaryotic operons, facilitating binding of RNA polymerase and stimulating gene expression.

catabolite repression A type of positive transcriptional control in which the end product of a catabolic pathway inhibits further transcription of the gene encoding an enzyme early in the pathway.

catalysis (verb: catalyze) Acceleration of the rate of a chemical reaction due to a decrease in the free energy of the transition state, called the activation energy.

catalyst Any substance that increases the rate of a chemical reaction without itself undergoing any permanent chemical change.

catecholamines A class of small compounds, derived from the amino acid tyrosine, that are used as hormones or neurotransmitters. Include epinephrine, norepinephrine, and dopamine.

cation A positively charged ion.

cation exchange In botany, the release (displacement) of cations, such as magnesium and calcium from soil particles, by protons in acidic soil water. The released cations are available for uptake by plants.

CD4 A membrane protein on the surface of some T cells in humans. $CD4^+$ T cells can give rise to helper T cells.

CD8 A membrane protein on the surface of some T cells in humans. $CD8^+$ T cells can give rise to cytotoxic T cells.

Cdk See **cyclin-dependent kinase**.

cDNA See **complementary DNA**.

cDNA library A set of cDNAs from a particular cell type or stage of development. Each cDNA is carried by a plasmid or other cloning vector and can be separated from other cDNAs. Compare with **genomic library**.

cecum A blind sac between the small intestine and the colon. Is enlarged in some species (e.g., rabbits) that use it as a fermentation vat for digestion of cellulose.

cell A highly organized compartment bounded by a thin, flexible structure (plasma membrane) and containing concentrated chemicals in an aqueous (watery) solution. The basic structural and functional unit of all organisms.

cell body The part of a neuron that contains the nucleus and where incoming signals are integrated. Also called the *soma*.

cell crawling A form of cellular movement involving actin filaments in which the cell produces bulges (pseudopodia) that stick to the substrate and pull the cell forward. Also called *amoeboid motion*.

cell culture See **culture**.

cell cycle Ordered sequence of events in which a eukaryotic cell replicates its chromosomes, evenly partitions the chromosomes to two daughter cells, and then undergoes division of the cytoplasm.

cell-cycle checkpoint Any of several points in the cell cycle at which progression of a cell through the cycle can be regulated.

cell division Creation of new cells by division of pre-existing cells.

cell-mediated (immune) response The type of immune response that involves generation of cytotoxic T cells from $CD8^+$ T cells. Defends against pathogen-infected cells, cancer cells, and transplanted cells. Compare with **humoral (immune) response**.

cell membrane See **plasma membrane**.

cell plate A double layer of new plasma membrane that appears in the middle of a dividing plant cell; ultimately divides the cytoplasm into two separate cells.

cell sap An aqueous solution found in the vacuoles of plant cells.

cell theory The theory that all organisms are made of cells and that all cells come from preexisting cells.

cell wall A fibrous layer found outside the plasma membrane of most bacteria and archaea and many eukaryotes.

cellular respiration A common pathway for production of ATP, involving transfer of electrons from compounds with high potential energy (often NADH and $FADH_2$) to an electron transport chain and ultimately to an electron acceptor (often oxygen).

cellulose A structural polysaccharide composed of β-glucose monomers joined by β-1,4-glycosidic linkages. Found in the cell wall of algae, plants, bacteria, fungi, and some other groups.

Cenozoic era The most recent interval of geologic time, beginning 65.5 million years ago, during which mammals became the dominant vertebrates and angiosperms became the dominant plants.

central dogma The long-accepted hypothesis that information in cells flows in one direction: DNA codes for RNA, which codes for proteins. Exceptions are now known (e.g., retroviruses).

central nervous system (CNS) The brain and spinal cord of vertebrate animals. Compare with **peripheral nervous system (PNS)**.

centriole One of two small cylindrical structures, structurally similar to a basal body, found together within the centrosome near the nucleus of a eukaryotic cell.

centromere Constricted region of a replicated chromosome where the two sister chromatids are joined and the kinetochore is located.

centrosome Structure in animal and fungal cells, containing two centrioles, that serves as a microtubule-organizing center for the cell's cytoskeleton and for the spindle apparatus during cell division.

cephalization The formation of a distinct anterior region (the head) where sense organs and a mouth are clustered.

cephalochordates One of the three major chordate lineages (Cephalochordata), comprising small, mobile organisms that live in marine sands; also called *lancelets* or *amphioxi*. Compare with **urochordates** and **vertebrates**.

cephalopods A lineage of mollusks including the squid, octopuses, and nautiluses. Distinguished by large brains, excellent vision, tentacles, and a reduced or absent shell.

cerebellum Posterior section of the vertebrate brain that is involved in coordination of complex muscle movements, such as those required for locomotion and maintaining balance.

cerebrum The most anterior section of the vertebrate brain. Divided into left and right hemispheres and four lobes: parietal lobe, involved in complex decision making (in humans); occipital lobe, receives and interprets visual information; parietal lobe, involved in integrating sensory and motor functions; and temporal lobe, functions in memory, speech (in humans), and interpreting auditory information.

cervix The narrow passageway between the vagina and the uterus of female mammals.

chaetae (singular: chaeta) Bristle-like extensions found in some annelids.

channel A protein that forms a pore in a cell membrane. The structure of most channels allows them to admit just one or a few types of ions or molecules.

character displacement The tendency for the traits of similar species that occupy overlapping ranges to change in a way that reduces interspecific competition.

chelicerae A pair of clawlike appendages found around the mouth of certain arthropods called chelicerates (spiders, mites, and allies).

chemical bond An attractive force binding two atoms together. Covalent bonds, ionic bonds, and hydrogen bonds are types of chemical bonds.

chemical carcinogen Any chemical that can cause cancer.

chemical energy The potential energy stored in covalent bonds between atoms.

chemical equilibrium A dynamic but stable state of a reversible chemical reaction in which the forward reaction and reverse reactions proceed at the same rate, so that the concentrations of reactants and products remain constant.

chemical evolution The theory that simple chemical compounds in the ancient atmosphere and ocean combined via spontaneous chemical reactions to form larger, more complex substances, eventually leading to the origin of life and the start of biological evolution.

chemical reaction Any process in which one compound or element is combined with others or is broken down; involves the making and/or breaking of chemical bonds.

chemiosmosis An energetic coupling mechanism whereby energy stored in an electrochemical proton gradient (proton-motive force) is used to drive an energy-requiring process such as production of ATP.

chemokine Any of several chemical signals that attract leukocytes to a site of tissue injury or infection.

chemolithotroph An organism that produces ATP by oxidizing inorganic molecules with high potential energy such as ammonia (NH_3) or methane (CH_4). Also called *lithotroph*. Compare with **chemoorganotroph**.

chemoorganotroph An organism that produces ATP by oxidizing organic molecules with high potential energy such as sugars. Also called *organotroph*. Compare with **chemolithotroph**.

chemoreceptor A sensory cell or organ specialized for detection of specific molecules or classes of molecules.

chiasma (plural: chiasmata) The X-shaped structure formed during meiosis by crossing over between non-sister chromatids in a pair of homologous chromosomes.

chitin A structural polysaccharide composed of *N*-acetylglucosamine monomers joined end to end by β-1,4-glycosidic linkages. Found in cell walls of fungi and many algae, and in external skeletons of insects and crustaceans.

chitons A lineage of marine mollusks that have a protective shell formed of eight calcium carbonate plates.

chlorophyll Any of several closely related green pigments, found in chloroplasts and photosynthetic protists, that absorb light during photosynthesis.

chloroplast A chlorophyll-containing organelle, bounded by a double membrane, in which photosynthesis occurs; found in plants and photosynthetic protists. Also the location of amino acid, fatty acid, purine, and pyrimidine synthesis.

choanocyte A specialized flagellated feeding cell found in choanoflagellates (protists that are the closest living relatives of animals) and sponges (the oldest animal phylum).

cholecystokinin A peptide hormone secreted by cells in the lining of the small intestine. Stimulates the secretion of digestive enzymes from the pancreas and of bile from the liver and gallbladder.

chordates Members of the phylum Chordata, deuterostomes distinguished by a dorsal hollow nerve cord, pharyngeal gill slits, a notochord, a dorsal hollow nerve cord, and a post-anal tail. Include vertebrates, cephalochordata, and urochordata.

chromatid One of the two identical strands composing a replicated chromosome that is connected at the centromere to the other strand.

chromatin The complex of DNA and proteins, mainly histones, that compose eukaryotic chromosomes. Can be highly compact (heterochromatin) or loosely coiled (euchromatin).

chromatin remodeling The process by which the DNA in chromatin is unwound from its associated proteins to allow transcription or replication. May involve chemical modification of histone proteins or reshaping of the chromatin by large multi-protein complexes in an ATP-requiring process.

chromosome Gene-carrying structure consisting of a single long molecule of DNA and associated proteins (e.g., histones). Most prokaryotic cells contain a single, circular chromosome; eukaryotic cells contain multiple

noncircular (linear) chromosomes located in the nucleus.

chromosome theory of inheritance The principle that genes are located on chromosomes and that patterns of inheritance are determined by the behavior of chromosomes during meiosis.

chylomicron A ball of protein-coated lipids used to transport the lipids through the bloodstream.

cilium (plural: cilia) One of many short, filamentous projections of some eukaryotic cells containing a core of microtubules. Used to move the cell and/or to move fluid or particles along a stationary cell. See **axoneme**.

circadian clock An internal mechanism found in most organisms that regulates many body processes (sleep-wake cycles, hormonal patterns, etc.) in a roughly 24-hour cycle.

circulatory system The system in animals responsible for moving oxygen, carbon dioxide, and other materials (hormones, nutrients, wastes) around the body.

cisternae (singular: cisterna) Flattened, membrane-bound compartments that make up the Golgi apparatus.

citric acid cycle A series of eight chemical reactions that starts with citrate (citric acid, when protonated) and ends with oxaloacetate, which reacts with acetyl CoA to form citrate—forming a cycle that is part of the pathway that oxidizes glucose to CO_2. Also known as the *Krebs cycle*, *tricarboxylic acid cycle*, and *TCA cycle*.

clade See **monophyletic group**.

cladistic approach A method for constructing a phylogenetic tree that is based on identifying the unique traits of each monophyletic group. Compare with **phenetic approach**.

Class I MHC protein An MHC protein that is present on the plasma membrane of virtually all nucleated cells and functions in presenting antigen to $CD8^+$ T cells.

Class II MHC protein An MHC protein that is present only on the plasma membrane of dendritic cells, macrophages, and B cells and functions in presenting antigen to $CD4^+$ T cells.

cleavage In animal development, the series of rapid mitotic cell divisions, with little cell growth, that produces successively smaller cells and transforms a zygote into a multicellular blastula, or blastocyst in mammals.

cleavage furrow A pinching-in of the plasma membrane that occurs as cytokinesis begins in animal cells and deepens until the cytoplasm is divided.

climate The prevailing long-term weather conditions in a particular region.

climax community The stable, final community that develops from ecological succession.

clitoris A small rod of erectile tissue in the external genitalia of female mammals. Is formed from the same embryonic tissue as the male penis and has a similar function in sexual arousal.

cloaca An opening to the outside used by the excretory and reproductive systems in a few mammals and many nonmammalian vertebrates.

clonal-selection theory The dominant explanation of the development of adaptive immunity in vertebrates. According to the theory, the immune system retains a vast pool of inactive lymphocytes, each with a unique receptor for a unique antigen. Lymphocytes that encounter their antigens are stimulated to divide (selected and cloned), producing daughter cells that combat infection and confer immunity.

clone (1) An individual that is genetically identical to another individual. (2) A lineage of genetically identical individuals or cells. (3) As a verb, to make one or more genetic replicas of a cell or individual.

cloning vector A plasmid or other agent used to transfer recombinant genes into cultured host cells. Also called simply *vector*.

closed circulatory system A circulatory system in which the circulating fluid (blood) is confined to blood vessels and flows in a continuous circuit. Compare with **open circulatory system**.

club mosses Common name for species in the land plant lineage Lycophyta.

cnidocyte A specialized stinging cell found in cnidarians (e.g., jellyfish, corals, and anemones) that is used in capturing prey.

cochlea The organ of hearing in the inner ear of mammals, birds, and crocodilians. A coiled, fluid-filled tube containing specialized pressure-sensing neurons (hair cells) that detect sounds of different pitches.

coding strand See **non-template strand**.

codominance An inheritance pattern in which heterozygotes exhibit both of the traits seen in either kind of homozygous individual.

codon A sequence of three nucleotides in DNA or RNA that codes for a certain amino acid or that initiates or terminates protein synthesis.

coefficient of relatedness (*r*) A measure of how closely two individuals are related. Calculated as the probability that an allele in two individuals is inherited from the same ancestor.

coelom An internal, usually fluid-filled, body cavity that is lined with mesoderm.

coelomate An animal that has a true coelom. Compare with **acoelomate** and **pseudocoelomate**.

coenocytic Containing many nuclei and a continuous cytoplasm through a filamentous body, without the body being divided into distinct cells. Some fungi are coenocytic.

coenzyme A small organic molecule that is a required cofactor for an enzyme-catalyzed reaction. Often donates or receives electrons or functional groups during the reaction.

coenzyme A (CoA) A nonprotein molecule that is required for many cellular reactions involving transfer of acetyl groups ($-COCH_3$).

coenzyme Q A nonprotein molecule that shuttles electrons between membrane-bound complexes in the mitochondrial electron transport chain. Also called *ubiquinone* or *Q*.

coevolution A pattern of evolution in which two interacting species reciprocally influence each other's adaptations over time.

coevolutionary arms race A series of adaptations and counter-adaptations observed in species that interact closely over time and affect each other's fitness.

cofactor A metal ion or small organic compound that is required for an enzyme to function normally. May be bound tightly to an enzyme or associate with it transiently during catalysis.

cohesion The tendency of certain like molecules (e.g., water molecules) to cling together due to attractive forces. Compare with **adhesion**.

cohesion-tension theory The theory that water movement upward through plant vascular tissues is due to loss of water from leaves (transpiration), which pulls a cohesive column of water upward.

cohort A group of individuals that are the same age and can be followed through time.

coleoptile A modified leaf that covers and protects the stems and leaves of young grasses.

collagen A fibrous, pliable, cable-like glycoprotein that is a major component of the extracellular matrix of animal cells. Various subtypes differ in their tissue distribution.

collecting duct In the vertebrate kidney, a large straight tube that receives filtrate from the distal tubules of several nephrons. Involved in the regulated reabsorption of water.

collenchyma cell In plants, an elongated cell with cell walls thickened at the corners that provides support to growing plant parts; usually found in strands along leaf veins and stalks. Compare with **parenchyma cell** and **sclerenchyma cell**.

colon The portion of the large intestine where feces are formed by compaction of wastes and reabsorption of water.

colony An assemblage of individuals. May refer to an assemblage of semi-independent cells or to a breeding population of multicellular organisms.

commensalism (adjective: commensal) A symbiotic relationship in which one organism (the commensal) benefits and the other (the host) is not harmed. Compare with **mutualism** and **parasitism**.

communication In ecology, any process in which a signal from one individual modifies the behavior of another individual.

community All of the species that interact with each other in a certain area.

companion cell In plants, a cell in the phloem that is connected via numerous plasmodesmata to adjacent sieve-tube members. Companion cells provide materials to maintain sieve-tube members and function in the loading and unloading of sugars into sieve-tube members.

compass orientation A type of navigation in which movement occurs in a specific direction.

competition In ecology, the interaction of two species or two individuals trying to use the same limited resource (e.g., water, food, living space). May occur between individuals of the same species (intraspecific competition) or different species (interspecific competition).

competitive exclusion principle The principle that two species cannot coexist in the same ecological niche in the same area because one species will out-compete the other.

competitive inhibition Inhibition of an enzyme's ability to catalyze a chemical reaction via a nonreactant molecule that competes with the substrate(s) for access to the active site.

complement system A set of proteins that circulate in the bloodstream and can form holes in the plasma membrane of bacteria, leading to their destruction.

complementary base pairing The association between specific nitrogenous bases of nucleic acids stabilized by hydrogen bonding. Adenine pairs only with thymine (in DNA) or uracil (in RNA), and guanine pairs only with cytosine.

complementary DNA (cDNA) DNA produced in the laboratory using an RNA transcript as a template and reverse transcriptase; corresponds to a gene but lacks introns. Also produced naturally by retroviruses.

complementary strand A newly synthesized strand of RNA or DNA that has a base sequence complementary to that of the template strand.

complete digestive tract A digestive tract with two openings, usually called a mouth and an anus.

complete metamorphosis See **holometabolous metamorphosis**.

compound eye An eye formed of many independent light-sensing columns (ommatidia); occurs in arthropods. Compare with **simple eye**.

concentration gradient Difference across space (e.g., across a membrane) in the concentration of a dissolved substance.

condensation reaction A chemical reaction in which two molecules are joined covalently with the removal of an —OH from one and an —H from another to form water. Also called a *dehydration reaction*. Compare with **hydrolysis**.

conduction (1) Direct transfer of heat between two objects that are in physical contact. Compare with **convection**. (2) Transmission of an electrical impulse along the axon of a nerve cell.

cone cell A photoreceptor cell with a cone-shaped outer portion that is particularly sensitive to bright light of a certain color. Also called simply *cone*. Compare with **rod cell**.

connective tissue An animal tissue consisting of scattered cells in a liquid, jellylike, or solid extracellular matrix. Includes bone, cartilage, tendons, ligaments, and blood.

conservation biology The effort to study, preserve, and restore threatened populations, communities, and ecosystems.

constant (C) region The portion of an antibody's light chains or heavy chains that has the same amino acid sequence in the antibodies produced by every B cell of an individual. Compare with **variable (V) region**.

constitutive Always occurring; always present. Commonly used to describe enzymes and other proteins that are synthesized continuously or mutants in which one or more genetic loci are constantly expressed due to defects in gene control.

constitutive defense A defensive trait that is always manifested even in the absence of a predator or pathogen. Also called *standing defense*. Compare with **inducible defense**.

constitutive mutant An abnormal (mutated) gene that produces a product at all times, instead of under certain conditions only.

consumer See **heterotroph**.

consumption Predation or herbivory.

continental shelf The portion of a geologic plate that extends from a continent to under seawater.

continuous strand See **leading strand**.

control In a scientific experiment, a group of organisms or samples that do not receive the experimental treatment but are otherwise identical to the group that does.

convection Transfer of heat by movement of large volumes of a gas or liquid. Compare with **conduction**.

convergent evolution The independent evolution of analogous traits in distantly related organisms due to adaptation to similar environments and a similar way of life.

cooperative binding The tendency of the protein subunits of hemoglobin to affect each other's oxygen binding such that each bound oxygen molecule increases the likelihood of further oxygen binding.

coprophagy The eating of feces.

copulation The act of transferring sperm from a male directly into a female's reproductive tract.

coral reef A large assemblage of colonial marine corals that usually serves as shallow-water, sunlit habitat for many other species as well.

core enzyme The enzyme responsible for catalysis in a multi-part holoenzyme.

co-receptor Any membrane protein that acts with some other membrane protein in a cell interaction or cell response.

cork cambium One of two types of lateral meristem, consisting of a ring of undifferentiated plant cells found just under the cork layer of woody plants; produces new

cork cells on its outer side. Compare with **vascular cambium**.

cork cell A waxy cell in the protective outermost layer of a woody plant.

corm A rounded, thick underground stem that can produce new plants via asexual reproduction.

cornea The transparent sheet of connective tissue at the very front of the eye in vertebrates and some other animals. Protects the eye and helps focus light.

corolla All of the petals of a flower.

corona The cluster of cilia at the anterior end of a rotifer.

corpus callosum A thick band of neurons that connects the two hemispheres of the cerebrum in the mammalian brain.

corpus luteum A yellowish structure in an ovary that secretes progesterone. Is formed from a follicle that has recently ovulated.

cortex (1) The outermost region of an organ, such as the kidney or adrenal gland. (2) In plants, a layer of ground tissue found outside the vascular bundles and pith of a plant stem.

cortical granules Small enzyme-filled vesicles in the cortex of an egg cell. Involved in formation of the fertilization envelope after fertilization.

corticotropin-releasing hormone (CRH) A peptide hormone, produced and secreted by the hypothalamus, that stimulates the anterior pituitary to release ACTH.

cortisol A steroid hormone, produced and secreted by the adrenal cortex, that increases blood glucose and prepares the body for stress. The major glucocorticoid hormone in some mammals. Also called *hydrocortisone*.

cost-benefit analysis Decisions or analyses that weigh the fitness costs and benefits of a particular action.

cotransport Transport of an ion or molecule against its electrochemical gradient, in company with an ion or molecule being transported with its electrochemical gradient. Also called *secondary active transport*.

cotransporter A transmembrane protein that facilitates diffusion of an ion down its previously established electrochemical gradient and uses the energy of that process to transport some other substance, in the same or opposite direction, *against* its concentration gradient. Also called *secondary active transporter*. See **antiporter** and **symporter**.

cotyledon The first leaf, or seed leaf, of a plant embryo. Used for storing and digesting nutrients and/or for early photosynthesis.

countercurrent exchanger In animals, any anatomical arrangement that allows the maximum transfer of heat or a soluble substance from one fluid to another. The two fluids must be flowing in opposite directions and have a heat or concentration gradient between them.

covalent bond A type of chemical bond in which two atoms share one or more pairs of electrons. Compare with **hydrogen bond** and **ionic bond**.

cranium A bony, cartilaginous, or fibrous case that encloses and protects the brain of vertebrates. Forms part of the skull. Also called *braincase*.

crassulacean acid metabolism (CAM) A variant type of photosynthesis in which CO_2 is stored in organic acids at night when stomata are open and then released to feed the Calvin cycle during the day when stomata are closed. Helps reduce water loss and oxygen loss by photorespiration in hot, dry environments.

cristae (singular: crista) Sac-like invaginations of the inner membrane of a mitochondrion. Location of the electron transport chain and ATP synthase.

Cro-Magnon A prehistoric European population of modern humans (*Homo sapiens*) known from fossils, paintings, sculptures, and other artifacts.

crop A storage organ in the digestive systems of certain vertebrates.

cross-fertilization A mating that combines gametes from different individuals, as opposed to combining gametes from the same individual (self-fertilization).

crossing over The exchange of segments of non-sister chromatids between a pair of homologous chromosomes that occurs during meiosis I.

cross-pollination Pollination of a flower by pollen from another individual, rather than by self-fertilization. Also called *crossing*.

cross-talk Interactions among signaling pathways that modify a cellular response.

crustaceans A lineage of arthropods that includes shrimp, lobster, and crabs. Many have a carapace (a platelike portion of the exoskeleton) and mandibles for biting or chewing.

cryptic species A species that cannot be distinguished from other species by easily identifiable morphological traits.

culture In cell biology, a collection of cells or a tissue growing under controlled conditions, usually in suspension or on the surface of a dish on solid growth medium.

cup fungus See **sac fungus**.

Cushing's disease A human endocrine disorder caused by loss of feedback inhibition of cortisol on ACTH secretion. Characterized by high ACTH and cortisol levels and wasting of body protein reserves.

cuticle A protective coating secreted by the outermost layer of cells of an animal or a plant.

cyanobacteria A lineage of photosynthetic bacteria formerly known as blue-green algae. Likely the first life-forms to carry out oxygenic photosynthesis.

cyclic AMP (cAMP) Cyclic adenosine monophosphate; a small molecule, derived from ATP, that is widely used by cells in signal transduction and transcriptional control.

cyclic photophosphorylation Path of electron flow during the light-dependent reactions of photosynthesis in which photosystem I transfers excited electrons back to the electron transport chain of photosystem II, rather than to NADP$^+$. Also called *cyclic electron flow*. Compare with **Z scheme**.

cyclin One of several regulatory proteins whose concentrations fluctuate cyclically throughout the cell cycle.

cyclin-dependent kinase (Cdk) Any of several related protein kinases that are active only when bound to a cyclin. Involved in control of the cell cycle.

cytochrome *c* (cyt *c*) A soluble iron-containing protein that shuttles electrons between membrane-bound complexes in the mitochondrial electron transport chain.

cytokines A diverse group of autocrine signaling proteins, secreted largely by cells of the immune system, whose effects include stimulating leukocyte production, tissue repair, and fever. Generally function to regulate the intensity and duration of an immune response.

cytokinesis Division of the cytoplasm to form two daughter cells. Typically occurs immediately after division of the nucleus by mitosis or meiosis.

cytokinins A class of plant hormones that stimulate cell division and retard aging.

cytoplasm All of the contents of a cell, excluding the nucleus, bounded by the plasma membrane.

cytoplasmic determinant A regulatory transcription factor or signaling molecule that is distributed unevenly in the cytoplasm of the egg cells of many animals and that directs early pattern formation in an embryo.

cytoplasmic streaming The directed flow of cytosol and organelles that facilitates distribution of materials within some large plant and fungal cells. Occurs along actin filaments and is powered by myosin.

cytoskeleton In eukaryotic cells, a network of protein fibers in the cytoplasm that are involved in cell shape, support, locomotion, and transport of materials within the cell. Prokaryotic cells have a similar but much less extensive network of fibers.

cytosol The fluid portion of the cytoplasm.

cytotoxic T cell An effector T cell that destroys infected cells and cancer cells. Is descended from an activated CD8$^+$ T cell that has interacted with antigen on an infected cell or cancer cell. Also called *cytotoxic T lymphocyte (CTL)* and *killer T cell*. Compare with **helper T cell**.

dalton (Da) A unit of mass equal to 1/12 the mass of one carbon-12 atom; about the mass of 1 proton or 1 neutron.

day-neutral plant A plant whose flowering time is not affected by the relative length of day and night (the photoperiod). Compare with **long-day** and **short-day plant**.

dead space Portions of the air passages that are not involved in gas exchange with the blood, such as the trachea and bronchi.

deciduous Describing a plant that sheds leaves or other structures at regular intervals (e.g., each fall)

decomposer See **detritivore**.

decomposer food chain An ecological network of detritus, decomposers that eat detritus, and predators and parasites of the decomposers.

definitive host The host species in which a parasite reproduces sexually. Compare with **intermediate host**.

dehydration reaction See **condensation reaction**.

deleterious In genetics, referring to any mutation, allele, or trait that reduces an individual's fitness.

demography The study of factors that determine the size and structure of populations through time.

denaturation (verb: denature) For a macromolecule, loss of its three-dimensional structure and biological activity due to breakage of hydrogen bonds and disulfide bonds, usually caused by treatment with excess heat or extreme pH conditions.

dendrite A short extension from a neuron's cell body that receives signals from other neurons.

dendritic cell A type of leukocyte that ingests and digests foreign antigens, moves to a lymph node, and presents the antigens displayed on its membrane to CD4$^+$ T cells.

dense connective tissue A type of connective tissue, distinguished by having an extracellular matrix dominated by collagen fibers.

density dependent In population ecology, referring to any characteristic that varies depending on population density.

density independent In population ecology, referring to any characteristic that does not vary with population density.

deoxyribonucleic acid (DNA) A nucleic acid composed of deoxyribonucleotides that carries the genetic information of a cell. Generally occurs as two intertwined strands, but these can be separated. See **double helix**.

deoxyribonucleoside triphosphate (dNTP) A monomer that can be polymerized to form DNA. Consists of deoxyribose, a base (A, T, G, or C), and three phosphate groups; similar to a nucleotide, but with two more phosphate groups.

deoxyribonucleotide See **nucleotide**.

depolarization Change in membrane potential from its resting negative state to a less negative or a positive state; a normal phase in an action potential. Compare with **hyperpolarization**.

deposit feeder An animal that eats its way through a food-containing substrate.

derived trait A trait that is clearly homologous with a trait found in an ancestor, but which has a new form.

dermal tissue system The tissue forming the outer layer of an organism. In plants, also called *epidermis*; in animals, forms two distinct layers: *dermis* and *epidermis*.

descent with modification The phrase used by Darwin to describe his hypothesis of evolution by natural selection.

desmosome A type of cell-cell attachment structure, consisting of cadherin proteins, that binds the cytoskeletons of adjacent animal cells together. Found where cells are strongly attached to each other. Compare with **gap junction** and **tight junction**.

detergent A type of small amphipathic molecule used to solubilize hydrophobic molecules in aqueous solution.

determination In embryogenesis, progressive changes in a cell that commit it to a particular cell fate. Once a cell is fully determined, it can differentiate only into a particular cell type (e.g., liver cell, brain cell).

detritivore An organism whose diet consists mainly of dead organic matter (detritus). Various bacteria, fungi, and protists are detritivores. Also called *decomposer*.

detritus A layer of dead organic matter that accumulates at ground level or on seafloors and lake bottoms.

deuterostomes A major lineage of animals that share a pattern of embryological development, including formation of the anus earlier than the mouth, and formation of the coelom by pinching off of layers of mesoderm from the gut. Includes echinoderms and chordates. Compare with **protostomes**.

developmental homology A similarity in embryonic form, or in the fate of embryonic tissues, that is due to inheritance from a common ancestor.

diabetes insipidus A human disease caused by defects in the kidney's system for conserving water. Characterized by production of large amounts of dilute urine.

diabetes mellitus A human disease caused by defects in insulin production (type I) or the response of cells to insulin (type II). Characterized by abnormally high blood glucose levels and large amounts of glucose-containing urine.

diaphragm An elastic, sheetlike structure. In mammals, the muscular sheet of tissue that separates the chest and abdominal cavities. Contracts and moves downward during inhalation, expanding the chest cavity.

diastole The portion of the heartbeat cycle during which the atria or ventricles of the heart are relaxed. Compare with **systole**.

diastolic blood pressure The force exerted by blood against artery walls during relaxation of the heart's left ventricle. Compare with **systolic blood pressure**.

dicot Any plant that has two cotyledons (embryonic leaves) upon germination. The dicots do not form a monophyletic group. Also called *dicotyledonous plant*. Compare with **eudicot** and **monocot**.

dideoxy sequencing A laboratory technique for determining the exact nucleotide sequence of DNA. Relies on the use of dideoxynucleotide triphosphates (ddNTPs), which terminate DNA replication.

diencephalon The part of the mammalian brain that relays sensory information to the cerebellum and functions in maintaining homeostasis.

differential centrifugation Procedure for separating cellular components according to their size and density by spinning a cell homogenate in a series of centrifuge runs. After each run, the supernatant is removed from the deposited material (pellet) and spun again at progressively higher speeds.

differential gene expression Expression of different sets of genes in cells with the same genome. Responsible for creating different cell types.

differentiation The process by which a relatively unspecialized cell becomes a distinct specialized cell type (e.g., liver cell, brain cell) usually by changes in gene expression. Also called *cell differentiation.*

diffusion Spontaneous movement of a substance from a region of high concentration to one of low concentration (i.e., down a concentration gradient).

digestion The physical and chemical breakdown of food into molecules that can be absorbed into the body of an animal.

digestive tract The long tube that begins at the mouth and ends at the anus. Also called *alimentary canal, gastrointestinal (GI) tract,* or the *gut.*

dihybrid cross A mating between two parents that are heterozygous for both of the two genes being studied.

dikaryotic Having two nuclei.

dimer An association of two molecules, which may be identical (homodimer) or different (heterodimer).

dioecious Describing an angiosperm species that has male and female reproductive structures on separate plants. Compare with **monoecious.**

diploblast (adjective: diploblastic) An animal whose body develops from two basic embryonic cell layers—ectoderm and endoderm. Compare with **triploblast.**

diploid (1) Having two sets of chromosomes ($2n$). (2) A cell or an individual organism with two sets of chromosomes, one set inherited from the maternal parent and one set from the paternal parent. Compare with **haploid.**

direct sequencing A technique for identifying and studying microorganisms that cannot be grown in culture. Involves detecting and amplifying copies of certain specific genes in their DNA, sequencing these genes, and then comparing the sequences with the known sequences from other organisms.

directional selection A pattern of natural selection that favors one extreme phenotype with the result that the average phenotype of a population changes in one direction. Generally reduces overall genetic variation in a population.

disaccharide A carbohydrate consisting of two monosaccharides (sugar residues) linked together.

discontinuous strand See **lagging strand.**

discrete trait An inherited trait that exhibits distinct phenotypic forms rather than the continuous variation characteristic of a quantitative trait such as body height.

dispersal The movement of individuals from their place of origin (birth, hatching) to a new location.

disruptive selection A pattern of natural selection that favors extreme phenotypes at both ends of the range of phenotypic variation. Maintains overall genetic variation in a population. Compare with **stabilizing selection.**

distal tubule In the vertebrate kidney, the convoluted portion of a nephron into which filtrate moves from the loop of Henle. Involved in the regulated reabsorption of sodium and water. Compare with **proximal tubule.**

disturbance In ecology, any event that disrupts a community, usually causing loss of some individuals or biomass from it.

disturbance regime The characteristic disturbances that affect a given ecological community.

disulfide bond A covalent bond between two sulfur atoms, typically in the side groups of some amino acids (e.g., cysteine). Often contributes to tertiary structure of proteins.

DNA See **deoxyribonucleic acid.**

DNA cloning Any of several techniques for producing many identical copies of a particular gene or other DNA sequence.

DNA fingerprinting Any of several methods for identifying individuals by unique features of their genomes. Commonly involves using PCR to produce many copies of certain simple sequence repeats (microsatellites) and then analyzing their lengths.

DNA library See **cDNA library** and **genomic library.**

DNA ligase An enzyme that joins pieces of DNA by catalyzing formation of a phosphodiester bond between the pieces.

DNA microarray A set of single-stranded DNA fragments, representing thousands of different genes, that are permanently fixed to a small glass slide. Can be used to determine which genes are expressed in different cell types, under different conditions, or at different developmental stages.

DNA polymerase Any enzyme that catalyzes synthesis of DNA from deoxyribonucleotides.

domain (1) A section of a protein that has a distinctive tertiary structure and function. (2) A taxonomic category, based on similarities in basic cellular biochemistry, above the kingdom level. The three recognized domains are Bacteria, Archaea, and Eukarya.

dominant Referring to an allele that determines the phenotype of a heterozygous individual. Compare with **recessive.**

dopamine A catecholamine neurotransmitter that functions mainly in a part of the mammalian brain involved with muscle control. Also functions as a hypothalamic inhibitory hormone that inhibits release of prolactin from the interior pituitary; also called *prolactin-inhibiting hormone (PIH).*

dormancy A temporary state of greatly reduced, or no, metabolic activity and growth in plants or plant parts (e.g., seeds, spores, bulbs, and buds).

dorsal Toward an animal's back and away from its belly. The opposite of ventral.

double fertilization An unusual form of reproduction seen in flowering plants, in which one sperm nucleus fuses with an egg to form a zygote and the other sperm nucleus fuses with two polar nuclei to form the triploid endosperm.

double helix The secondary structure of DNA, consisting of two antiparallel DNA strands wound around each other.

Down syndrome A human developmental disorder caused by trisomy of chromosome 21.

downstream In genetics, the direction in which RNA polymerase moves along a DNA strand. Compare with **upstream.**

dynein Any one of a class of motor proteins that use the chemical energy of ATP to "walk" along an adjacent microtubule. Dyneins are responsible for bending of cilia and flagella, play a role in chromosome movement during mitosis, and can transport certain organelles.

early endosome A small membrane-bound vesicle, formed by endocytosis, that is an early stage in the formation of a lysosome.

ecdysone An insect hormone that triggers either molting (to a larger larval form) or metamorphosis (to the adult form), depending on the level of juvenile hormone.

ecdysozoans A major lineage of protostomes (Ecdysozoa) that grow by shedding their external skeletons (molting) and expanding their bodies. Includes arthropods, insects, crustaceans, nematodes, and centipedes. Compare with **lophotrochozoans.**

echinoderms A major lineage of deuterostomes (Echinodermata) distinguished by adult bodies with five-sided radial symmetry, a water vascular system, and tube feet. Includes sea urchins, sand dollars, and sea stars.

echolocation The use of echoes from vocalizations to obtain information about locations of objects in the environment.

ecology The study of how organisms interact with each other and with their surrounding environment.

ecosystem All the organisms that live in a geographic area, together with the nonliving (abiotic) components that affect or exchange materials with the organisms; a community and its physical environment.

ecosystem diversity The variety of biotic components in a region along with abiotic components, such as soil, water, and nutrients.

ecosystem services Alterations of the physical components of an ecosystem by living organisms, especially beneficial changes in the quality of the atmosphere, soil, water, etc.

ecotourism Tourism that is based on observing wildlife or experiencing other aspects of natural areas.

ectoderm The outermost of the three basic cell layers in most animal embryos; gives rise to the outer covering and nervous system. Compare with **endoderm** and **mesoderm.**

ectomycorrhizal fungi (EMF) Fungi whose hyphae form a dense network that covers their host plant's roots but do not enter the root cells.

ectoparasite A parasite that lives on the outer surface of the host's body.

ectotherm An animal that does not use internally generated heat to regulate its body temperature. Compare with **endotherm.**

effector Any cell, organ, or structure with which an animal can respond to external or internal stimuli. Usually functions, along with a sensor and integrator, as part of a homeostatic system.

efferent division The part of the nervous system, consisting primarily of motor neurons, that carries commands from the central nervous system to the body.

egg A mature female gamete and any associated external layers (such as a shell). Larger and less mobile than the male gamete. In animals, also called *ovum.*

ejaculation The release of semen from the copulatory organ of a male animal.

ejaculatory duct A short duct connecting the vas deferens to the urethra, through which sperm move during ejaculation.

elastic Referring to a structure (e.g., lungs) with the ability to stretch and then spring back to its original shape.

electric current A flow of electrical charge past a point. Also called *current.*

electrical potential Potential energy created by a separation of electric charges between two points. Also called *voltage.*

electrocardiogram (EKG) A recording of the electrical activity of the heart, as measured through electrodes on the skin.

electrochemical gradient The combined effect of an ion's concentration gradient and electrical (charge) gradient across a membrane that affects the diffusion of ions across the membrane.

electrolyte Any compound that dissociates into ions when dissolved in water. In nutrition, refers to the major ions necessary for normal cell function.

electromagnetic spectrum The entire range of wavelengths of radiation extending from short wavelengths (high energy) to long wavelengths (low energy). Includes gamma rays, X-rays, ultraviolet, visible light, infrared, microwaves, and radio waves (from short to long wavelengths).

electron acceptor A reactant that gains an electron and is reduced in a reduction-oxidation reaction.

electron carrier Any molecule that readily accepts electrons from and donates electrons to other molecules.

electron donor A reactant that loses an electron and is oxidized in a reduction-oxidation reaction.

electron microscope See **scanning electron microscope** and **transmission electron microscope**.

electron shell A group of orbitals of electrons with similar energies. Electron shells are arranged in roughly concentric layers around the nucleus of an atom, with electrons in outer shells having more energy than those in inner shells. Electrons in the outermost shell, the valence shell, often are involved in chemical bonding.

electron transport chain (ETC) Any set of membrane-bound protein complexes and smaller soluble electron carriers involved in a coordinated series of redox reactions in which the potential energy of electrons transferred from reduced donors is successively decreased and used to pump protons from one side of a membrane to the other.

electronegativity A measure of the ability of an atom to attract electrons toward itself from an atom to which it is bonded.

electroreceptor A sensory cell or organ specialized to detect electric fields.

element A substance, consisting of atoms with a specific number of protons, that cannot be separated into or broken down to any other substance. Elements preserve their identity in chemical reactions.

elongation (1) The process by which messenger RNA lengthens during transcription. (2) The process by which a polypeptide chain lengthens during translation.

elongation factors Proteins involved in the elongation phase of translation, assisting ribosomes in the synthesis of the growing peptide chain.

embryo A young developing organism; the stage after fertilization and zygote formation.

embryo sac The female gametophyte in flowering plants that exhibit alternation of generations.

embryogenesis The process by which a single-celled zygote becomes a multicellular embryo.

Embryophyta An increasing popular name for the lineage called land plants.

embryophyte A plant that nourishes its embryos inside its own body. All land plants are embryophytes.

emergent vegetation Any plants in an aquatic habitat that extend above the surface of the water.

emerging disease Any infectious disease, often a viral disease, that suddenly afflicts significant numbers of humans for the first time; often due to changes in the host species for a pathogen or host population movements.

emigration The migration of individuals away from one population to other populations. Compare with **immigration**.

emulsification (verb: emulsify) The dispersion of fat into an aqueous solution. Usually requires the aid of an amphipathic substance such as a detergent or bile salts, which can break large fat globules into microscopic fat droplets.

endangered species A species whose numbers have decreased so much that it is in danger of extinction throughout all or part of its range.

endemic species A species that lives in one geographic area and nowhere else.

endergonic Referring to a chemical reaction that requires an input of energy to occur and for which the Gibbs free-energy change $(\Delta G) > 0$. Compare with **exergonic**.

endocrine Relating to a chemical signal (hormone) that is released into the bloodstream by a producing cell and acts on a distant target cell.

endocrine system All of the glands and tissues that produce and secrete hormones into the bloodstream.

endocrine gland A gland that secretes hormones directly into the bloodstream or interstitial fluid instead of into ducts. Compare with **exocrine gland**.

endocytosis General term for any pinching off of the plasma membrane that results in the uptake of material from outside the cell. Includes phagocytosis, pinocytosis, and receptor-mediated endocytosis. Compare with **exocytosis**.

endoderm The innermost of the three basic cell layers in most animal embryos; gives rise to the digestive tract and organs that connect to it (liver, lungs, etc.). Compare with **ectoderm** and **mesoderm**.

endodermis In plant roots, a cylindrical layer of cells that separates the cortex from the vascular tissue.

endomembrane system A system of organelles in eukaryotic cells that performs most protein and lipid synthesis. Includes the endoplasmic reticulum (ER), Golgi apparatus, and lysosomes.

endomycorrhizal fungi See **arbuscular mycorrhizal fungi (AMF)**.

endoparasite A parasite that lives inside the host's body.

endophyte (adjective: endophytic) A fungus that lives inside the aboveground parts of a plant in a symbiotic relationship. Compare with **epiphyte**.

endoplasmic reticulum (ER) A network of interconnected membranous sacs and tubules found inside eukaryotic cells. See **rough** and **smooth endoplasmic reticulum**.

endoskeleton Bony and/or cartilaginous structures within the body that provide support. Examples are the spicules of sponges, the plates in echinoderms, and the bony skeleton of vertebrates. Compare with **exoskeleton**.

endosome See **early** and **late endosome**.

endosperm A triploid $(3n)$ tissue in the seed of a flowering plant (angiosperm) that serves as food for the plant embryo. Functionally analogous to the yolk in some animal eggs.

endosymbiont An organism that lives in a symbiotic relationship inside the body of its host.

endosymbiosis An association between species in which one lives inside the cell or cells of the other.

endosymbiosis theory The theory that mitochondria and chloroplasts evolved from prokaryotes that were engulfed by host cells and took up a symbiotic existence within those cells, a process termed primary endosymbiosis. In some eukaryotes, chloroplasts originated by secondary endosymbiosis, that is, by engulfing a chloroplast-containing protist and retaining its chloroplasts.

endotherm An animal whose primary source of body heat is internally generated heat. Compare with **ectotherm**.

endothermic Referring to a chemical reaction that absorbs heat. Compare with **exothermic**.

energetic coupling In cellular metabolism, the mechanism by which energy released from an exergonic reaction (commonly, hydrolysis of ATP) is used to drive an endergonic reaction.

energy The capacity to do work or to supply heat. May be stored (potential energy) or available in the form of motion (kinetic energy).

enhancer A regulatory sequence in eukaryotic DNA that may be located far from the gene it controls or within introns of the gene. Binding of specific proteins to an enhancer enhances the transcription of certain genes.

enrichment culture A method of detecting and obtaining cells with specific characteristics by placing a sample, containing many types of cells, under a specific set of conditions (e.g., temperature, salt concentration, available nutrients) and isolating those cells that grow rapidly in response.

entropy (S) A quantitative measure of the amount of disorder of any system, such as a group of molecules.

envelope, viral A membrane-like covering that encloses some viruses and their capsid coats, shielding them from attack by the host's immune system.

environmental sequencing The inventory of all the genes in a community or ecosystem by sequencing, analyzing, and comparing the genomes of the component organisms.

enzyme A protein catalyst used by living organisms to speed up and control biological reactions.

epicotyl In some embryonic plants, a portion of the embryonic stem that extends above the cotyledons.

epidemic The spread of an infectious disease throughout a population in a short time period. Compare with **pandemic**.

epidermis The outermost layer of cells of any multicellular organism.

epididymis A coiled tube wrapped around the testis in reptiles, birds, and mammals. The site of the final stages of sperm maturation and storage.

epigenetic inheritance Pattern of inheritance involving differences in phenotype that are not due to changes in the nucleotide sequence of genes.

epinephrine A catecholamine hormone, produced and secreted by the adrenal medulla, that triggers rapid responses relating to the fight-or-flight response. Also called *adrenaline*.

epiphyte (adjective: epiphytic) A nonparasitic plant that grows on trees or other solid objects and is not rooted in soil.

epithelium (plural: epithelia) An animal tissue consisting of sheet-like layers of tightly packed cells that lines an organ, a duct, or a body surface. Also called *epithelial tissue*.

epitope A small region of a particular antigen to which an antibody, B-cell receptor, or T-cell receptor binds.

equilibrium potential The membrane potential at which there is no net movement of a particular ion into or out of a cell.

ER signal sequence A short amino acid sequence that marks a polypeptide for transport to the endoplasmic reticulum where synthesis of the polypeptide chain is completed and the signal sequence removed. See **signal recognition particle**.

erythropoietin (EPO) A peptide hormone, released by the kidney in response to low blood oxygen levels, that stimulates the bone marrow to produce more red blood cells.

esophagus The muscular tube that connects the mouth to the stomach.

essential amino acid An amino acid that an animal cannot synthesize and must obtain from the diet. May refer specifically to one of the eight essential amino acids of adult humans: isoleucine, leucine, lysine, methionine, phenylalanine, threonine, tryptophan, and valine.

essential nutrient Any chemical element, ion, or compound that is required for normal growth, reproduction, and maintenance of a living organism and that cannot be synthesized by the organism.

ester linkage The covalent bond formed by a condensation reaction between a carboxyl group (—COOH) and a hydroxyl group (—OH). Ester linkages join fatty acids to glycerol to form a fat or phospholipid.

estradiol The major estrogen produced by the ovaries of female mammals. Stimulates development of the female reproductive tract, growth of ovarian follicles, and growth of breast tissue.

estrogens A class of steroid hormones, including estradiol, estrone, and estriol, that generally promote female-like traits. Secreted by the gonads, fat tissue, and some other organs.

estrous cycle A female reproductive cycle, seen in all mammals except Old World monkeys and apes (including humans), in which the uterine lining is reabsorbed rather than shed in the absence of pregnancy, and the female is sexually receptive only briefly during mid-cycle (estrus). Compare with **menstrual cycle**.

estuary An environment of brackish (partly salty) water where a river meets the ocean.

ethylene A gaseous plant hormone that induces fruits to ripen, flowers to fade, and leaves to drop.

eudicot A member of a monophyletic group (lineage) of angiosperms that includes complex flowering plants and trees (e.g., roses, daisies, maples). All eudicots have two cotyledons, but not all dicots are members of this lineage. Compare with **dicot** and **monocot**.

Eukarya One of the three taxonomic domains of life consisting of unicellular organisms (most protists, yeast) and multicellular organisms (fungi, plants, animals) distinguished by a membrane-bound cell nucleus, numerous organelles, and an extensive cytoskeleton. Compare with **Archaea** and **Bacteria**.

eukaryote A member of the domain Eukarya; an organism whose cells contain a nucleus, numerous membrane-bound organelles, and an extensive cytoskeleton. May be unicellular or multicellular. Compare with **prokaryote**.

eutherians A lineage of mammals (Eutheria) whose young develop in the uterus and are not housed in an abdominal pouch. Also called *placental mammals*.

evaporation The energy-absorbing phase change from a liquid state to a gaseous state. Many organisms evaporate water as a means of heat loss.

evo-devo Research field focused on how changes in developmentally important genes have led to the evolution of new phenotypes.

evolution (1) The theory that all organisms on Earth are related by common ancestry and that they have changed over time, predominantly via natural selection. (2) Any change in the genetic characteristics of a population over time, especially, a change in allele frequencies.

ex situ conservation Preserving species outside of natural areas; e.g., in zoos, aquaria, or botanic gardens.

excitable membrane A plasma membrane that is capable of generating an action potential. Neurons, muscle cells, and some other cells have excitable membranes.

excitatory postsynaptic potential (EPSP) A change in membrane potential, usually depolarization, at a neuron dendrite that makes an action potential more likely.

exergonic Referring to a chemical reaction that can occur spontaneously, releasing heat and/or increasing entropy, and for which the Gibbs free-energy change (ΔG) < 0. Compare with **endergonic**.

exocrine gland A gland that secretes some substance through a duct into a space other than the circulatory system, such as the digestive tract or the skin surface. Compare with **endocrine gland**.

exocytosis Secretion of intracellular molecules (e.g., hormones, collagen), contained within membrane-bounded vesicles, to the outside of the cell by fusion of vesicles to the plasma membrane. Compare with **endocytosis**.

exon A region of a eukaryotic gene that is translated into a peptide or protein. Compare with **intron**.

exoskeleton A hard covering secreted on the outside of the body, used for body support, protection, and muscle attachment. Examples are the shell of mollusks and the outer covering (cuticle) of arthropods. Compare with **endoskeleton**.

exothermic Referring to a chemical reaction that releases heat. Compare with **endothermic**.

exotic species A nonnative species that is introduced into a new area. Exotic species often are competitors, pathogens, or predators of native species.

expansins A class of plant proteins that actively increase the length of the cell wall when the pH of the wall falls below 4.5.

exponential population growth The accelerating increase in the size of a population that occurs when the growth rate is constant and density independent. Compare with **logistic population growth**.

expressed sequence tag (EST) A portion of a transcribed gene (synthesized from an mRNA in a cell), used to find the physical location of that gene in the genome.

extant species A species that is living today.

extensor A muscle that pulls two bones farther apart from each other, as in the extension of a limb or the spine. Compare with **flexor**.

extinct Said of a species that has died out.

extracellular digestion Digestion that takes place outside of an organism, as occurs in many fungi that make and secrete digestive enzymes.

extracellular matrix (ECM) A complex meshwork of proteins (e.g., collagen, fibronectin) and polysaccharides secreted by animal cells and in which they are embedded.

extremophile A bacterium or archaean that thrives in an "extreme" environment (e.g., high-salt, high-temperature, low-temperature, or low-pressure).

F₁ generation First filial generation. The first generation of offspring produced from a mating (i.e., the offspring of the parental generation).

facilitated diffusion Movement of a substance across a plasma membrane down its concentration gradient with the assistance of transmembrane carrier proteins or channel proteins.

facilitation In ecological succession, the phenomenon in which early-arriving species make conditions more favorable for later-arriving species. Compare with **inhibition** and **tolerance**.

facultative aerobe Any organism that can perform aerobic respiration when oxygen is available to serve as an electron acceptor but can switch to fermentation when it is not.

FAD/FADH₂ Oxidized and reduced forms, respectively, of flavin adenine dinucleotide. A nonprotein electron carrier that functions in the citric acid cycle and oxidative phosphorylation.

fallopian tube A narrow tube connecting the uterus to the ovary in humans, through which the egg travels after ovulation. Site of fertilization and cleavage. In nonhuman animals, called *oviduct*.

fat A lipid consisting of three fatty acid molecules joined by ester linkages to a glycerol molecule. Also called *triacylglycerol* or *triglyceride*.

fatty acid A lipid consisting of a hydrocarbon chain bonded to a carboxyl group (—COOH) at one end. Used by many organisms to store chemical energy; a major component of animal and plant fats.

fatty-acid binding protein Proteins that bind to fatty acids and enable them to be transported into cells.

fauna All the animals characteristic of a particular region, period, or environment.

feather A specialized skin outgrowth, composed of β-keratin, present in all birds and only in birds. Used for flight, insulation, display, and other purposes.

feces The waste products of digestion.

fecundity The average number of female offspring produced by a single female in the course of her lifetime.

feedback inhibition A type of metabolic control in which high concentrations of the product of a metabolic pathway inhibit one of the enzymes early in the pathway. A form of negative feedback.

fermentation Any of several metabolic pathways that make ATP by transferring electrons from a reduced compound such as glucose to a final electron acceptor other than oxygen. Allows glycolysis to proceed in the absence of oxygen.

ferredoxin In photosynthetic organisms, an iron- and sulfur-containing protein in the electron transport chain of photosystem I. Can transfer electrons to the enzyme NADP⁺ reductase, which catalyzes formation of NADPH.

fertilization envelope A physical barrier that forms around a fertilized egg in amphibians and some other animals. Formed by an influx of water under the vitelline membrane.

fertilization Fusion of the nuclei of two haploid gametes to form a zygote with a diploid nucleus.

fetal alcohol syndrome A condition, marked by hyperactivity, severe learning disabilities, and depression, thought to be caused by exposure of an individual to high blood alcohol concentrations during embryonic development.

fetus In live-bearing animals, the unborn offspring after the embryonic stage, which usually are developed sufficiently to be recognizable as belonging to a certain species. In humans, from 9 weeks after fertilization until birth.

fiber In plants, a type of elongated sclerenchyma cell that provides support to vascular tissue. Compare with **sclereid**.

fibronectin An abundant protein in the extracellular matrix that binds to other ECM components and to integrins in plasma membranes; helps anchor cells in place. Numerous subtypes are found in different tissues.

Fick's law of diffusion A mathematical relationship that describes the rates of gas exchange in animal respiratory systems.

fight-or-flight response Rapid physiological changes that prepare the body for emergencies. Includes increased heart rate, increased blood pressure, and decreased digestion.

filament Any thin, threadlike structure, particularly (1) the threadlike extensions of a fish's gills or (2) the slender stalk that bears the anthers in a flower.

filter feeder See **suspension feeder**.

filtrate Any fluid produced by filtration, in particular the fluid ("pre-urine") in the nephrons of vertebrate kidneys.

filtration A process of removing large components from a fluid by forcing it through a filter. Occurs in a renal corpuscle of the vertebrate kidney, allowing water and small solutes to pass from the blood into the nephron.

finite rate of increase (λ) The rate of increase of a population over a given period of time. Calculated as the ending population size divided by the starting population size. Compare with **intrinsic rate of increase**.

first law of thermodynamics The principle of physics that energy is conserved in any process. Energy can be transferred and converted into different forms, but it cannot be created or destroyed.

fission (1) A form of asexual reproduction in which a prokaryotic cell divides to produce two genetically similar daughter cells by a process similar to mitosis of eukaryotic cells. Also called *binary fission*. (2) A form of asexual reproduction in which an animal splits into two or more individuals of approximately equal size; common among invertebrates.

fitness The ability of an individual to produce viable offspring relative to others of the same species.

fitness trade-off See **trade-off**.

fixed action pattern (FAP) Highly stereotyped behavior pattern that occurs in a certain invariant way in a certain species. A form of innate behavior.

flaccid Limp as a result of low internal pressure (e.g., a wilted plant leaf). Compare with **turgid**.

flagellum (plural: flagella) A long, cellular projection that undulates (in eukaryotes) or rotates (in prokaryotes) to move the cell through an aqueous environment. See **axoneme**.

flatworms Members of the phylum Platyhelminthes. Distinguished by a broad, flat, unsegmented body that lacks a coelom. Flatworms belong to the lophotrochozoan branch of the protostomes.

flavin adenine dinucleotide See **FAD/FADH₂**.

flexor A muscle that pulls two bones closer together, as in the flexing of a limb or the spine. Compare with **extensor**.

floral meristem A group of undifferentiated plant cells that can give rise to the four organs making up a flower.

florigen In plants, a protein hormone that is synthesized in leaves and transported to the shoot apical meristem where it stimulates flowering.

flower In angiosperms, the part of a plant that contains reproductive structures. Typically includes a calyx, a corolla, and one or more stamens and/or carpels. See **perfect** and **imperfect flower**.

fluid connective tissue A type of connective tissue, distinguished by having a liquid extracellular matrix.

fluid feeder An animal that feeds by sucking or mopping up liquids such as nectar, plant sap, or blood.

fluid-mosaic model The widely accepted hypothesis that the plasma membrane and organelle membranes consist of proteins embedded in a fluid phospholipid bilayer.

fluorescence The spontaneous emission of light from an excited electron falling back to its normal (ground) state.

follicle An egg cell and its surrounding ring of supportive cells in a mammalian ovary.

follicle-stimulating hormone (FSH) A peptide hormone, produced and secreted by the anterior pituitary, that stimulates (in females) growth of eggs and follicles in the ovaries or (in males) sperm production in the testes.

follicular phase The first major phase of a menstrual cycle during which follicles grow and estrogen levels increase; ends with ovulation.

food Any nutrient-containing material that can be consumed and digested by animals.

food chain A relatively simple pathway of energy flow through a few species, each at a different trophic level, in an ecosystem. Might include, for example, a primary producer, a primary consumer, a secondary consumer, and a decomposer. Compare with **food web**.

food web Any complex pathway along which energy moves among many species at different trophic levels of an ecosystem.

foot One of the three main parts of the mollusk body; a muscular appendage, used for movement and/or burrowing into sediment.

foraging Searching for food.

forebrain One of the three main regions of the vertebrate brain; includes the cerebrum, thalamus, and hypothalamus. Compare with **hindbrain** and **midbrain**.

fossil Any trace of an organism that existed in the past. Includes tracks, burrows, fossilized bones, casts, etc.

fossil record All of the fossils that have been found anywhere on Earth and that have been formally described in the scientific literature.

founder effect A change in allele frequencies that often occurs when a new population is established from a small group of individuals (founder event) due to sampling error (i.e., the small group is not a representative sample of the source population).

fovea In the vertebrate eye, a portion of the retina where incoming light is focused; contains a high proportion of cone cells.

free energy See **Gibbs free-energy change**.

free radical Any substance containing one or more atoms with an unpaired electron. Unstable and highly reactive.

frequency The number of wave crests per second traveling past a stationary point. Determines the pitch of sound and the color of light.

frequency-dependent selection A pattern of selection in which certain alleles are favored only when they are rare; a form of balancing selection.

fronds The large leaves of ferns.

frontal lobe In the vertebrate brain, one of the four major areas in the cerebrum.

fruit In flowering plants (angiosperms), a mature, ripened plant ovary (or group of ovaries), along with the seeds it contains and any adjacent fused parts. See **aggregate, multiple,** and **simple fruit**.

fruiting body A structure formed in some prokaryotes, fungi, and protists for spore dispersal; usually consists of a base, a stalk, and a mass of spores at the top.

functional genomics The study of how a genome works, that is, when and where specific genes are expressed and how their products interact to produce a functional organism.

functional group A small group of atoms bonded together in a precise configuration and exhibiting particular chemical properties that it imparts to any organic molecule in which it occurs.

fundamental niche The ecological space that a species occupies in its habitat in the absence of competitors. Compare with **realized niche**.

fungi A lineage of eukaryotes that typically have a filamentous body (mycelium) and obtain nutrients by absorption.

fungicide Any substance that can kill fungi or slow their growth.

G protein Any of various peripheral membrane proteins that bind GTP and function in signal transduction. Binding of a signal to its receptor triggers activation of the G protein, leading to production of a second messenger or initiation of a phosphorylation cascade.

G₁ phase The phase of the cell cycle that constitutes the first part of interphase before DNA synthesis (S phase).

G₂ phase The phase of the cell cycle between synthesis of DNA (S phase) and mitosis (M phase); the last part of interphase.

gallbladder A small pouch that stores bile from the liver and releases it as needed into the small intestine during digestion of fats.

gametangium (plural: gametangia) (1) The gamete-forming structure found in all land plants except angiosperms. Contains a sperm-producing antheridium and an egg-producing archegonium. (2) The gamete-forming structure of some chytrid fungi.

gamete A haploid reproductive cell that can fuse with another haploid cell to form a zygote. Most multicellular eukaryotes have two distinct forms of gametes: egg cells (ova) and sperm cells.

gametogenesis The production of gametes (eggs or sperm).

gametophyte In organisms undergoing alternation of generations, the multicellular haploid form that arises from a single haploid spore and produces gametes. A female gametophyte is commonly called an *embryo sac;* a male gametophyte, a *pollen grain*. Compare with **sporophyte**.

ganglia A mass of neurons in a centralized nervous system.

ganglion cell A neuron in the vertebrate retina that collects visual information from one or several bipolar cells and sends it to the brain via the optic nerve.

gap junction A type of cell-cell attachment structure that directly connects the cytoplasms of adjacent animal cells, allowing passage of water, ions, and small molecules between the cells. Compare with **desmosome** and **tight junction**.

gastrin A hormone produced by cells in the stomach lining in response to the arrival of food or to a neural signal from the brain. Stimulates other stomach cells to release hydrochloric acid.

gastrointestinal (GI) tract See **digestive tract**.

gastropods A lineage of mollusk distinguished by a large muscular foot and a unique feeding structure, the radula. Include slugs and snails.

gastrulation The process by which some cells on the outside of a young embryo move to the interior of the embryo, resulting in the three distinct germ layers (endoderm, mesoderm, and ectoderm).

gated channel A channel protein that opens and closes in response to a specific stimulus, such as the binding of a particular molecule or a change in the electrical charge on the outside of the membrane.

gel electrophoresis A technique for separating molecules on the basis of size and electric charge, which affect their differing rates of movement through a gelatinous substance in an electric field.

gemma (plural: gemmae) A small reproductive structure that is produced in some liverworts during the gametophyte phase and can grow into mature gametophyte.

gene A section of DNA (or RNA, for some viruses) that encodes information for building one or more related polypeptides or functional RNA molecules along with the regulatory sequences required for its transcription.

gene duplication The formation of an additional copy of a gene, typically by misalignment of chromosomes during crossing over. Thought to be an important evolutionary process in creating new genes.

gene expression Overall process by which the information encoded in genes is converted into an active product, most commonly a protein. Includes transcription and translation of a gene and in some cases protein activation.

gene family A set of genetic loci whose DNA sequences are extremely similar. Thought to have arisen by duplication of a single ancestral gene and subsequent mutations in the duplicated sequences.

gene flow The movement of alleles between populations; occurs when individuals leave one population, join another, and breed.

gene pool All of the alleles of all of the genes in a certain population.

gene therapy The treatment of an inherited disease by introducing normal alleles.

gene-for-gene hypothesis The hypothesis that there is a one-to-one correspondence between the resistance (*R*) loci of plants and the avirulence (*avr*) loci of pathogenic fungi; particularly, that *R* genes produce receptors and *avr* genes produce molecules that bind to those receptors.

generation The average time between a mother's first offspring and her daughter's first offspring.

genetic bottleneck A reduction in allelic diversity resulting from a sudden reduction in the size of a large population (population bottleneck) due to a random event.

genetic code The set of all 64 codons and the particular amino acids that each specifies.

genetic correlation A type of evolutionary constraint in which selection on one trait causes a change in another trait as well; may occur when the same gene(s) affect both traits.

genetic diversity The diversity of alleles in a population, species, or group of species.

genetic drift Any change in allele frequencies due to random events. Causes allele frequencies to drift up and down randomly over time, and eventually can lead to the fixation or loss of alleles.

genetic homology Similarities in DNA sequences or amino acid sequences that are due to inheritance from a common ancestor.

genetic map An ordered list of genes on a chromosome that indicates their relative distances from each other. Also called a *linkage map* or *meiotic map*. Compare with **physical map**.

genetic marker A genetic locus that can be identified and traced in populations by laboratory techniques or by a distinctive visible phenotype.

genetic model A set of hypotheses that explain how a certain trait is inherited.

genetic recombination A change in the combination of genes or alleles on a given chromosome or in a given individual. Also called *recombination*.

genetic screen Any of several techniques for identifying individuals with a particular type of mutation. Also called a *screen*.

genetic variation (1) The number and relative frequency of alleles present in a particular population. (2) The proportion of phenotypic variation in a trait that is due to genetic rather than environmental influences in a certain population in a certain environment.

genetics The field of study concerned with the inheritance of traits.

genitalia External copulatory organs.

genome All of the hereditary information in an organism, including not only genes but also other non-gene stretches of DNA.

genomic library A set of DNA segments representing the entire genome of a particular organism. Each segment is carried by a plasmid or other cloning vector and can be separated from other segments. Compare with **cDNA library**.

genomics The field of study concerned with sequencing, interpreting, and comparing whole genomes from different organisms.

genotype All of the alleles of every gene present in a given individual. May refer specifically to the alleles of a particular set of genes under study. Compare with **phenotype**.

genus (plural: genera) In Linnaeus' system, a taxonomic category of closely related species. Always italicized and capitalized to indicate that it is a recognized scientific genus.

geologic time scale The sequence of eons, eras, and periods used to describe the geologic history of Earth.

germ cell In animals, any cell that can potentially give rise to gametes. Also called *germ-line cells*.

germ layer In animals, one of the three basic types of tissue formed during gastrulation; gives rise to all other tissues. See **endoderm, mesoderm,** and **ectoderm**.

germ theory of disease The theory that infectious diseases are caused by bacteria, viruses, and other microorganisms.

germination The process by which a seed becomes a young plant.

gestation The duration of embryonic development from fertilization to birth in those species that have live birth.

gibberellins A class of hormones found in plants and fungi that stimulate growth. Gibberellic acid is one of the major gibberellins.

Gibbs free-energy change (ΔG) A measure of the change in potential energy and entropy that occurs in a given chemical reaction. $\Delta G < 0$ for spontaneous reactions and > 0 for nonspontaneous reactions.

gill Any organ in aquatic animals that exchanges gases and other dissolved substances between the blood and the surrounding water. Typically, a filamentous outgrowth of a body surface.

gill arch In aquatic vertebrates, curved region of tissue between the gills. Gills are suspended from the gill arches.

gill filament In fish, one of the many long, thin structures that extend from gill arches into the water and across which gas exchange occurs.

gill lamella (plural: gill lamellae) One of hundreds to thousands of sheetlike structures, each containing a capillary bed, that makes up a gill filament.

gland An organ whose primary function is to secrete some substance, either into the blood (endocrine gland) or into some other space such as the gut or skin (exocrine gland).

glia Collective term for several types of cells in nervous tissue that are not neurons and do not conduct electrical signals but provide support, nourishment, and electrical insulation and perform other functions. Also called *glial cells*.

global carbon cycle See **carbon cycle, global**.

global nitrogen cycle See **nitrogen cycle, global**.

global warming A sustained increase in Earth's average surface temperature.

global water cycle See **water cycle, global**.

glomalin A glycoprotein that is abundant in the hyphae of arbuscular mycorrhizal fungi; when hyphae decay it an important component of soil.

glomerulus (plural: glomeruli) (1) In the vertebrate kidney, a ball-like cluster of capillaries, surrounded by Bowman's capsule, at the beginning of a nephron. (2) In the brain, a ball-shaped cluster of neurons in the olfactory bulb.

glucagon A peptide hormone produced by the pancreas in response to low blood glucose. Raises blood glucose by triggering breakdown of glycogen and stimulating gluconeogenesis. Compare with **insulin**.

glucocorticoids A class of steroid hormones, produced and secreted by the adrenal cortex, that increase blood glucose and prepare the body for stress. Include cortisol and corticosterone. Compare with **mineralocorticoids**.

gluconeogenesis Synthesis of glucose from non-carbohydrate sources (e.g., proteins and fatty acids). Occurs in the liver in response to low insulin levels and high glucagon levels.

glucose Six-carbon monosaccharide whose oxidation in cellular respiration is the major source of ATP in animal cells.

glyceraldehyde-3-phosphate (G3P) The phosphorylated three-carbon compound formed as the result of carbon fixation in the first step of the Calvin cycle.

glycerol A three-carbon molecule that forms the "backbone" of phospholipids and most fats.

glycogen A highly branched storage polysaccharide composed of α-glucose monomers joined by 1,4- and 1,6-glycosidic linkages. The major form of stored carbohydrate in animals.

glycolipid Any lipid molecule that is covalently bonded to a carbohydrate group.

glycolysis A series of 10 chemical reactions that oxidize glucose to produce pyruvate and ATP. Used by all organisms as part of fermentation or cellular respiration.

glycoprotein Any protein with one or more covalently bonded carbohydrate groups.

glycosidic linkage The covalent bond formed by a condensation reaction between two sugar monomers; joins the residues of a polysaccharide.

glycosylation Addition of a carbohydrate group to a molecule.

glyoxisome Specialized type of peroxisome found in plant cells and packed with enzymes for processing the products of photosynthesis.

gnathostomes Animals with jaws.

Golgi apparatus A eukaryotic organelle, consisting of stacks of flattened membranous sacs (cisternae), that functions in processing and sorting proteins and lipids destined to be secreted or directed to other organelles. Also called *Golgi complex*.

gonad An organ that produces reproductive cells, such as a testis or an ovary.

gonadotropin-releasing hormone (GnRH) A peptide hormone, produced and secreted by the hypothalamus, that stimulates release of FSH and LH from the anterior pituitary.

grade In taxonomy, a group of species that share a position in an inferred evolutionary sequence of lineages but that are not a monophyletic group. Also called a *paraphyletic group*.

Gram stain A dye that distinguishes the two general types of cell walls found in bacteria. Used to routinely classify bacteria as Gram-negative or Gram-positive.

Gram-negative Describing bacteria that look pink when treated with a Gram stain. These bacteria have a cell wall composed of a thin layer of peptidoglycan and an outer phospholipid layer.

Gram-positive Describing bacteria that look purple when treated with a Gram stain. These bacteria have cell walls composed of a thick layer of peptidoglycan.

granum (plural: grana) In chloroplasts, a stack of flattened, membrane-bound vesicles (thylakoids) where the light reactions of photosynthesis occur.

gravitropism The growth or movement of a plant in a particular direction in response to gravity.

grazing food chain The ecological network of herbivores and the predators and parasites that consume them.

great apes See **hominids**.

green algae A paraphyletic group of photosynthetic organisms that contain chloroplasts similar to those in green plants. Often classified as protists, green algae are the closest living relatives of land plants and form a monophyletic group with them.

greenhouse gas An atmospheric gas that absorbs and reflects infrared radiation, so that heat radiated from Earth is retained in the atmosphere instead of being lost to space.

gross photosynthetic productivity The efficiency with which all the plants in a given area use the light energy available to them to produce sugars.

gross primary productivity In an ecosystem, the total amount of carbon fixed by photosynthesis, including that used for cellular respiration, over a given time period. Compare with **net primary productivity**.

ground meristem The middle layer of a young plant embryo. Gives rise to the ground tissue system.

ground tissue An embryonic tissue layer that gives rise to parenchyma, collenchyma, and sclerenchyma—tissues other than the epidermis and vascular tissue.

groundwater Any water below the land surface.

growth factor Any of a large number of signaling molecules that are secreted by certain cells and that stimulate other cells to divide or to differentiate.

growth hormone (GH) A peptide hormone, produced and secreted by the mammalian anterior pituitary, that promotes lengthening of the long bones in children and muscle growth, tissue repair, and lactation in adults. Also called *somatotropin*.

GTP See **guanosine triphosphate**.

guanosine triphosphate (GTP) A molecule consisting of guanine, a sugar, and three phosphate groups. Can be hydrolyzed to release free energy. Commonly used in many cellular reactions; also functions in signal transduction in association with G proteins.

guard cell One of two specialized, crescent-shaped cells forming the border of a plant stoma. Guard cells can change shape to open or close the stoma. See also **stoma**.

gustation The perception of taste.

guttation Excretion of water droplets from plant leaves in the early morning, caused by root pressure.

gymnosperm A vascular plant that makes seeds but does not produce flowers. The gymnosperms include four lineages of green plants (cycads, ginkgoes, conifers, and gnetophytes). Compare with **angiosperm**.

H⁺-ATPase See **proton pump**.

habitat destruction Human-caused destruction of a natural habitat with replacement by an urban, suburban, or agricultural landscape.

habitat fragmentation The breakup of a large region of a habitat into many smaller regions, separated from others by a different type of habitat.

Hadley cell An atmospheric cycle of large-scale air movement in which warm equatorial air rises, moves north or south, and then descends at approximately 30° N or 30° S latitude.

hair cell A pressure-detecting sensory cell, found in the cochlea, that has tiny "hairs" (stereocilia) jutting from its surface.

hairpin A secondary structure in RNA consisting of a stable loop formed by hydrogen bonding between purine and pyrimidine bases on the same strand.

halophile A bacterium or archaean that thrives in high-salt environments.

Hamilton's rule The proposition that an allele for altruistic behavior will be favored by natural selection only if $Br > C$, where B = the fitness benefit to the

recipient, C = the fitness cost to the actor, and r = the coefficient of relatedness between recipient and actor.

haploid (1) Having one set of chromosomes ($1n$). (2) A cell or an individual organism with one set of chromosomes. Compare with **diploid**.

haploid number The number of distinct chromosome sets in a cell. Symbolized as n.

Hardy-Weinberg principle A principle of population genetics stating that genotype frequencies in a large population do not change from generation to generation in the absence of evolutionary processes (e.g., mutation, migration, genetic drift, random mating, and selection).

heart A muscular pump that circulates blood throughout the body.

heart murmur A distinctive sound caused by backflow of blood through a defective heart valve.

heartwood The older xylem in the center of an older stem or root, containing protective compounds and no longer functioning in water transport.

heat Thermal energy that is transferred from an object at higher temperature to one at lower temperature.

heat of vaporization The energy required to vaporize 1 gram of a liquid into a gas.

heat-shock proteins Proteins that facilitate refolding of proteins that have been denatured by heat or other agents.

heavy chain The larger of the two types of polypeptide chains in an antibody molecule; composed of a variable (V) region, which contributes to the antigen-binding site, and a constant (C) region. Differences in heavy-chain constant regions determine the different classes of immunoglobulins (IgA, IgE, etc.). Compare with **light chain**.

helicase An enzyme that catalyzes the breaking of hydrogen bonds between nucleotides of DNA, "unzipping" a double-stranded DNA molecule.

helper T cell An effector T cell that secretes cytokines and in other ways promotes the activation of other lymphocytes. Is descended from an activated CD4⁺ T cell that has interacted with antigen presented by dendritic cells, macrophages, or B cells.

heme A small molecule that binds to four polypeptides to form hemoglobin; contains an iron atom that can bind oxygen.

hemimetabolous metamorphosis A type of metamorphosis in which the animal increases in size from one stage to the next, but does not dramatically change its body form. Also called *incomplete metamorphosis*.

hemocoel A body cavity, present in arthropods and some mollusks, containing a pool of circulatory fluid (hemolymph) bathing the internal organs.

hemoglobin An oxygen-binding protein consisting of four polypeptide subunits, each containing an oxygen-binding heme group. The major oxygen carrier in mammalian blood.

hemolymph The circulatory fluid of animals with open circulatory systems (e.g., insects) in which the fluid is not confined to blood vessels.

hemophilia A human disease, caused by an X-linked recessive allele, that is characterized by defects in the blood-clotting system.

herbaceous Referring to a plant that is not woody.

herbivore (adjective: herbivorous) An animal that eats primarily plants and rarely or never eats meat. Compare with **carnivore** and **omnivore**.

herbivory The practice of eating plant tissues.

heredity The transmission of traits from parents to offspring via genetic information.

heritable Referring to traits that can be transmitted from one generation to the next.

hermaphrodite An organism that produces both male and female gametes.

heterokaryotic Describing a cell or fungal mycelium containing two or more nuclei that are genetically distinct.

heterospory (adjective: heterosporous) In seed plants, the production of two distinct types of spore-producing structures and thus two distinct types of spores: microspores, which become the male gametophyte, and megaspores, which become the female gametophyte. Compare with **homospory**.

heterotherm An animal whose body temperature varies markedly with environmental conditions. Compare with **homeotherm**.

heterotroph Any organism that cannot synthesize reduced organic compounds from inorganic sources and that must obtain them by eating other organisms. Some bacteria, some archaea, and virtually all fungi and animals are heterotrophs. Also called *consumer*. Compare with **autotroph**.

heterozygote advantage A pattern of natural selection that favors heterozygous individuals compared with homozygotes. Tends to maintain genetic variation in a population. Also called *heterozygote superiority*.

heterozygous Having two different alleles of a certain gene.

hexose A monosaccharide (simple sugar) containing six carbon atoms.

hibernation An energy-conserving physiological state, marked by a decrease in metabolic rate, body temperature, and activity, that lasts for a prolonged period (weeks to months). Occurs in some animals in response to winter cold and scarcity of food. Compare with **torpor**.

hindbrain One of the three main regions of the vertebrate brain; includes the cerebellum and medulla oblongata. Compare with **forebrain** and **midbrain**.

histamine A molecule released from mast cells during an inflammatory response that causes blood vessels to dilate and become more permeable.

histone One of several positively charged (basic) proteins associated with DNA in the chromatin of eukaryotic cells.

histone acetyl transferase (HAT) In eukaryotes, one of a class of enzymes that loosen chromatin structure by adding acetyl groups to histone proteins.

histone code The hypothesis that specific combinations of chemical modifications of histone proteins contain information that influences gene expression.

histone deacetylase (HDAC) In eukaryotes, one of a class of enzymes that recondense chromatin by removing acetyl groups from histone proteins.

HIV See **human immunodeficiency virus (HIV)**.

holoenzyme A multipart enzyme consisting of a core enzyme (containing the active site for catalysis) along with other required proteins.

holometabolous metamorphosis A type of metamorphosis in which the animal completely changes its form. Also called *complete metamorphosis*.

homeosis Replacement of one body part by another normally found elsewhere in the body as the result of mutation in certain developmentally important genes (homeotic genes).

homeostasis (adjective: homeostatic) The array of relatively stable chemical and physical conditions in an animal's cells, tissues, and organs. May be achieved by the body's passively matching the conditions of a stable external environment (conformational homeostasis) or by active physiological processes (regulatory homeostasis) triggered by variations in the external or internal environment.

homeotherm An animal that has a constant or relatively constant body temperature. Compare with **heterotherm**.

homeotic gene Any gene that specifies a particular location within an embryo, leading to the development of structures appropriate for that location. Mutations in homeotic genes cause the development of extra body parts or body parts in the wrong places.

hominids Members of the family Hominidae, which includes humans and extinct related forms; chimpanzees, gorillas, and orangutans. Distinguished by large body size, no tail, and an exceptionally large brain. Also called *great apes*.

hominins Humans and extinct related forms; species in the lineage that branched off from chimpanzees and eventually led to humans.

homologous chromosomes In a diploid organism, chromosomes that are similar in size, shape, and gene content. Also called *homologs*.

homology (adjective: homologous) Similarity among organisms of different species due to their inheritance from a common ancestor. Features that exhibit such similarity (e.g., DNA sequences, proteins, body parts) are said to be homologous. Compare with **homoplasy**.

homoplasy Similarity among organisms of different species due to convergent evolution. Compare with **homology**.

homospory (adjective: homosporous) In seedless vascular plants, the production of just one type of spore. Compare with **heterospory**.

homozygous Having two identical alleles of a certain gene.

hormone Any of numerous different signaling molecules that circulate throughout the body in blood or other body fluids and can trigger characteristic responses in distant target cells at very low concentrations.

hormone-response element A specific sequence in DNA to which a steroid hormone-receptor complex can bind and affect gene transcription.

host cell A cell that has been invaded by an organism such as a parasite or a virus.

***Hox* genes** A class of homeotic genes found in several animal phyla, including vertebrates, that are expressed in a distinctive pattern along the anterior-posterior axis in early embryos and control formation of segment-specific structures.

human Any member of the genus *Homo,* which includes modern humans (*Homo sapiens*) and several extinct species.

human chorionic gonadotropin (hCG) A glycoprotein hormone produced by the human placenta from about week 3 to week 14 of pregnancy. Maintains the corpus luteum, which produces hormones that preserve the uterine lining.

Human Genome Project The multinational research project that sequenced the human genome.

human immunodeficiency virus (HIV) A retrovirus that causes AIDS (acquired immune deficiency syndrome) in humans.

humoral (immune) response The type of immune response that involves generation of antibody-secreting plasma cells from activated B cells. Defends against extracellular pathogens. Compare with **cell-mediated (immune) response**.

humus The completely decayed organic matter in soils.

Huntington's disease A degenerative brain disease of humans caused by an autosomal dominant allele.

hybrid The offspring of parents from two different strains, populations, or species.

hybrid zone A geographic area where interbreeding occurs between two species, sometimes producing fertile hybrid offspring.

hydrocarbon An organic molecule that contains only hydrogen and carbon atoms.

hydrogen bond A weak interaction between two molecules or different parts of the same molecule resulting from the attraction between a hydrogen atom with a partial positive charge and another atom (usually O or N) with a partial negative charge. Compare with **covalent bond** and **ionic bond**.

hydrogen ion (H$^+$) A single proton with a charge of $1+$; typically, one that is dissolved in solution or that is being transferred from one atom to another in a chemical reaction.

hydrolysis A chemical reaction in which a molecule is split into smaller molecules by reacting with water. In biology, most hydrolysis reactions involve the splitting of polymers into monomers. Compare with **condensation reaction**.

hydrophilic Interacting readily with water. Hydrophilic compounds are typically polar compounds containing charged or electronegative atoms. Compare with **hydrophobic**.

hydrophobic Not interacting readily with water. Hydrophobic compounds are typically nonpolar compounds that lack charged or electronegative atoms and often contain many C—C and C—H bonds. Compare with **hydrophilic**.

hydroponic growth Growth of plants in liquid cultures instead of soil.

hydrostatic skeleton A system of body support involving fluid-filled compartments that can change in shape but cannot easily be compressed.

hydroxide ion (OH$^-$) An oxygen atom and a hydrogen atom joined by a single covalent bond and carrying a negative charge; formed by dissociation of water.

hyperpolarization Change in membrane potential from its resting negative state to an even more negative state; a normal phase in an action potential. Compare with **depolarization**.

hypersensitive reaction An intense allergic response by cells that have been sensitized by previous exposure to an allergen.

hypersensitive response In plants, the rapid death of a cell that has been infected by a pathogen, thereby reducing the potential for infection to spread throughout a plant. Compare with **systemic acquired resistance**.

hypertension Abnormally high blood pressure.

hypertonic Comparative term designating a solution that has a greater solute concentration, and therefore a lower water concentration, than another solution. Compare with **hypotonic** and **isotonic**.

hypha (plural: hyphae) One of the strands of a fungal mycelium (the meshlike body of a fungus). Also found in some protists.

hypocotyl The stem of a very young plant; the region between the cotyledon (embryonic leaf) and the radicle (embryonic root).

hypothalamic-pituitary axis The functional interaction of the hypothalamus and the pituitary gland, which are anatomically distinct but work together to regulate most of the other endocrine glands in the body.

hypothalamus A part of the brain that functions in maintaining the body's internal physiological state by regulating the autonomic nervous system, endocrine system, body temperature, water balance, and appetite.

hypothesis A proposed explanation for a phenomenon or for a set of observations.

hypotonic Comparative term designating a solution that has a lower solute concentration, and therefore a higher water concentration, than another solution. Compare with **hypertonic** and **isotonic**.

immigration The migration of individuals into a particular population from other populations. Compare with **emigration**.

immune system In vertebrates, the system whose primary function is to defend the body against pathogens. Includes several types of cells (e.g., lymphocytes and macrophages) and several organs where they develop or reside (e.g., lymph nodes and thymus).

immunity (adjective: immune) State of being protected against infection by disease-causing pathogens either by relatively nonspecific mechanisms (innate immunity) or by specific mechanisms triggered by exposure to a particular antigen (adaptive immunity).

immunization The conferring of immunity to a particular disease by artificial means.

immunoglobulin (Ig) Any of the class of proteins that function as antibodies.

immunological memory The ability of the immune system to "remember" an antigen and mount a rapid, effective response to a pathogen encountered years or decades earlier.

impact hypothesis The hypothesis that a collision between the Earth and an asteroid caused the mass extinction at the K-P boundary, 65 million years ago.

imperfect flower A flower that contains male parts (stamens) *or* female parts (carpels) but not both. Compare with **perfect flower**.

implantation The process by which an embryo buries itself in the uterine wall and forms a placenta. Occurs in mammals and a few other vertebrates.

in situ hybridization A technique for detecting specific DNAs and mRNAs in cells and tissues by use of labeled probes. Can be used to determine where and when particular genes are expressed in embryos.

inbreeding Mating between closely related individuals. Increases homozygosity of a population and often leads to a decline in the average fitness (inbreeding depression).

inbreeding depression In inbred offspring, fitness declines due to deleterious recessive alleles that are homozygous.

inclusive fitness The combination of (1) direct production of offspring (direct fitness) and (2) extra production of offspring by relatives in response to help provided by the individual in question (indirect fitness).

incomplete digestive tract A digestive tract that has just one opening.

incomplete dominance An inheritance pattern in which the heterozygote phenotype is a blend or combination of both homozygote phenotypes.

incomplete metamorphosis See **hemimetabolous metamorphosis**.

independent assortment, principle of The concept that each pair of hereditary elements (alleles of the same gene) behaves independently of other genes during meiosis. One of Mendel's two principles of genetics.

indeterminate growth A pattern of growth in which an individual continues to increase its overall body size throughout its life.

indicator plate A laboratory technique for detecting mutant cells by growing them on agar plates containing a compound that when metabolized by wild-type cells yields a colored product.

induced fit Change in the shape of the active site of an enzyme, as the result of the initial weak binding of a substrate, so that it binds substrate more tightly.

inducer A small molecule that triggers transcription of a specific gene, often by binding to and inactivating a repressor protein.

inducible defense A defensive trait that is manifested only in response to the presence of a consumer (predator or herbivore) or pathogen. Compare with **constitutive defense**.

infection thread An invagination of the membrane of a root hair through which beneficial nitrogen-fixing bacteria enter the roots of their host plants (legumes).

inflammatory response An aspect of the innate immune response, seen in most cases of infection or tissue injury, in which the affected tissue becomes swollen, red, warm, and painful.

inhibition In ecological succession, the phenomenon in which early-arriving species make conditions less favorable for the establishment of certain later-arriving species. Compare with **facilitation** and **tolerance**.

inhibitory postsynaptic potential (IPSP) A change in membrane potential, usually hyperpolarization, at a neuron dendrite that makes an action potential less likely.

initiation (1) In an enzyme-catalyzed reaction, the stage during which enzymes orient reactants precisely as they bind at specific locations within the enzyme's active site. (2) In DNA transcription, the stage during which RNA polymerase and other proteins assemble at the promoter sequence. (3) In RNA translation, the stage during which a complex consisting of a ribosome, a mRNA molecule, and an aminoacyl tRNA corresponding to the start codon is formed.

initiation factors A class of proteins that assist ribosomes in binding to a messenger RNA molecule to begin translation.

innate behavior Behavior that is inherited genetically, does not have to be learned, and is typical of a species.

innate immune response See **innate immunity**.

innate immunity A set of nonspecific defenses against pathogens that exist before exposure to an antigen and involves mast cells, neutrophils, and macrophages; typically results in an inflammatory response. Compare with **acquired immunity**.

inner cell mass (ICM) A cluster of cells in the interior of a mammalian blastocyst that undergo gastrulation and eventually develop into the embryo.

inner ear The innermost portion of the mammalian ear, consisting of a fluid-filled system of tubes that includes the cochlea (which receives sound vibrations from the middle ear) and the semicircular canals (which function in balance).

insulin A peptide hormone produced by the pancreas in response to high levels of glucose (or amino acids) in blood. Enables cells to absorb glucose and coordinates synthesis of fats, proteins, and glycogen. Compare with **glucagon**.

integral membrane protein Any membrane protein that spans the entire lipid bilayer. Also called *transmembrane protein*. Compare with **peripheral membrane protein**.

integrated pest management In agriculture or forestry, systems for managing insects or other pests that include carefully controlled applications of toxins, introduction of species that prey on pests, planting schemes that reduce the chance of a severe pest outbreak, and other techniques.

integrator A component of an animal's nervous system that functions as part of a homeostatic system by evaluating sensory information and triggering appropriate responses. See **effector** and **sensor**.

integrin Any of a class of cell-surface proteins that bind to fibronectins and other proteins in the extracellular matrix, thus holding cells in place.

intercalated disc A specialized junction between adjacent heart muscle cells that contains gap junctions, allowing electrical signals to pass between the cells.

intermediate disturbance hypothesis The hypothesis that moderate ecological disturbance is associated with higher species diversity than either low or high disturbance.

intermediate filament A long fiber, about 10 nm in diameter, composed of one of various proteins (e.g., keratins, lamins); one of the three types of cytoskeletal fibers. Form networks that help maintain cell shape and hold the nucleus in place. Compare with **actin filament** and **microtubule**.

intermediate host The host species in which a parasite reproduces asexually. Compare with **definitive host**.

interneuron A neuron that passes signals from one neuron to another. Compare with **motor neuron** and **sensory neuron**.

internode The section of a plant stem between two nodes (sites where leaves attach).

interphase The portion of the cell cycle between one mitotic (M) phase and the next. Includes the G_1 phase, S phase, and G_2 phase.

interspecific competition Competition between members of different species for the same limited resource. Compare with **intraspecific competition**.

interstitial fluid The plasma-like fluid found in the region (interstitial space) between cells.

intertidal zone The region between the low-tide and high-tide marks on a seashore.

intraspecific competition Competition between members of the same species for the same limited resource. Compare with **interspecific competition**.

intrinsic rate of increase (r_{max}) The rate at which a population will grow under optimal conditions (i.e., when birthrates are as high as possible and death rates are as low as possible). Compare with **finite rate of increase**.

intron A region of a eukaryotic gene that is transcribed into RNA but is later removed, so it is not translated into a peptide or protein. Compare with **exon**.

invasive species An exotic (nonnative) species that, upon introduction to a new area, spreads rapidly and competes successfully with native species.

inversion A mutation in which a segment of a chromosome breaks from the rest of the chromosome, flips, and rejoins with the opposite orientation as before.

invertebrates A paraphyletic group composed of animals without a backbone; includes about 95 percent of all animal species. Compare with **vertebrates**.

involuntary muscle Muscle that cannot respond to conscious thought.

ion An atom or a molecule that has lost or gained electrons and thus carries an electric charge, either positive (cation) or negative (anion), respectively.

ion channel A type of channel protein that allows certain ions to diffuse across a plasma membrane down an electrochemical gradient.

ionic bond A chemical bond that is formed when an electron is completely transferred from one atom to another so that the atoms remain associated due to their opposite electric charges. Compare with **covalent bond** and **hydrogen bond**.

iris A ring of pigmented muscle just below the cornea in the vertebrate eye that contracts or expands to control the amount of light entering the eye through the pupil.

isotonic Comparative term designating a solution that has the same solute concentration and water concentration than another solution. Compare with **hypertonic** and **hypotonic**.

isotope Any of several forms of an element that have the same number of protons but differ in the number of neutrons.

joint A place where two components (bones, cartilages, etc.) of a skeleton meet. May be movable (an articulated joint) or immovable (e.g., skull sutures).

juvenile An individual that has adult-like morphology but is not sexually mature.

juvenile hormone An insect hormone that prevents larvae from metamorphosing into adults.

karyogamy Fusion of two haploid nuclei to form a diploid nucleus. Occurs in many fungi, and in animals and plants during fertilization of gametes.

karyotype The distinctive appearance of all of the chromosomes in an individual, including the number of chromosomes and their length and banding patterns (after staining with dyes).

keystone species A species that has an exceptionally great impact on the other species in its ecosystem relative to its abundance.

kidney In terrestrial vertebrates, one of a paired organ situated at the back of the abdominal cavity that filters the blood, produces urine, and secretes several hormones.

kilocalorie (kcal) A unit of energy often used to measure the energy content of food. A kcal of energy raises 1 g of water 1°C.

kin selection A form of natural selection that favors traits that increase survival or reproduction of an individual's kin at the expense of the individual.

kinesin Any one of a class of motor proteins that use the chemical energy of ATP to transport vesicles, particles, or chromosomes along microtubules.

kinetic energy The energy of motion. Compare with **potential energy**.

kinetochore A protein structure at the centromere where kinetochore microtubules attach to the sister chromatids of a replicated chromosome. Contains motor proteins that move a chromosome along a microtubule.

kinetochore microtubules Microtubules that form during mitosis and meiosis, and which extend from a spindle apparatus to an attachment point—the kinetochore—on a chromosome.

kinocilium (plural: kinocilia) A single cilium that juts from the surface of many hair cells and functions in detection of sound or pressure.

Klinefelter syndrome A syndrome seen in humans who have an XXY karyotype. People with this syndrome have male sex organs, may have some female traits, and are sterile.

knock-out allele A mutant allele that does not function at all, or an organism homozygous for such a mutation. Also called *null allele* or *loss-of-function allele*.

Koch's postulates Four criteria used to determine whether a suspected infectious agent causes a particular disease.

labia major (plural: labium majus) One of two outer folds of skin that protect the labia minora, clitoris, and vaginal opening of female mammals.

labia minora (plural: labium minus) One of two inner folds of skin that protect the opening of the urethra and vagina.

labor The strong muscular contractions of the uterus that expel the fetus during birth.

lactation (verb: lactate) Production of milk from mammary glands of mammals.

lacteal A small lymphatic vessel extending into the center of a villus in the small intestine. Receives chylomicrons containing fat absorbed from food.

lactic acid fermentation Catabolic pathway in which pyruvate produced by glycolysis is converted to lactic acid in the absence of oxygen.

lagging strand In DNA replication, the strand of new DNA that is synthesized discontinuously in a series of short pieces that are later joined together. Also called *discontinuous strand*. Compare with **leading strand**.

large intestine The distal portion of the digestive tract consisting of the cecum, colon, and rectum. Its primary function is to compact the wastes delivered from the small intestine and absorb enough water to form feces.

larva (plural: larvae) An immature stage of a species in which the immature and adult stages have different body forms.

late endosome A membrane-bound vesicle that arises from an early endosome and develops into a lysosome.

latency In viruses that infect animals, the ability to exist in a quiescence state without producing new virions.

lateral bud A bud that forms in the angle between a leaf and a stem and may develop into a lateral (side) branch. Also called *axillary bud*.

lateral gene transfer Transfer of DNA between two different species, especially distantly related species. Commonly occurs among bacteria and archaea via plasmid exchange; also can occur in eukaryotes via viruses and some other mechanisms.

lateral line system A pressure-sensitive sensory organ found in many aquatic vertebrates.

lateral meristem A layer of undifferentiated plant cells found in older stems and roots that is responsible for secondary growth. Also called *cambium* or *secondary meristem*. Compare with **apical meristem**.

lateral root A plant root extending from another, older root.

leaching Loss of nutrients from soil via percolating water.

leading strand In DNA replication, the strand of new DNA that is synthesized in one continuous piece, with nucleotides added to the 3′ end of the growing molecule. Also called *continuous strand*. Compare with **lagging strand**.

leaf The main photosynthetic organ of vascular plants.

leak channel Potassium channel that allows potassium ions to leak out of a neuron in its resting state.

learning An enduring change in an individual's behavior that results from specific experience(s).

leghemoglobin An iron-containing protein similar to hemoglobin. Found in root nodules of legume plants where it binds oxygen, preventing it from poisoning a bacterial enzyme needed for nitrogen fixation.

legumes Members of the pea plant family that form symbiotic associations with nitrogen-fixing bacteria in their roots.

lens A transparent, crystalline structure that focuses incoming light onto a retina or other light-sensing apparatus of an eye.

lenticels Spongy segments in bark that allow gas exchange between cells in a woody stem and the atmosphere.

leptin A hormone produced and secreted by fat cells (adipocytes) that acts to stabilize fat tissue mass in part by inhibiting appetite and increasing energy expenditure.

leukocytes Several types of blood cells, including neutrophils, macrophages, and lymphocytes, that circulate in blood and lymph and function in defense against pathogens. Also called *white blood cells*.

lichen A symbiotic association of a fungus and a photosynthetic alga.

life cycle The sequence of developmental events and phases that occurs during the life span of an organism, from fertilization to offspring production.

life history The sequence of events in an individual's life from birth to reproduction to death, including how an individual allocates resources to growth, reproduction, and activities or structures that are related to survival.

life table A data set that summarizes the probability that an individual in a certain population will survive and reproduce in any given year over the course of its lifetime.

ligand Any molecule that binds to a specific site on a receptor molecule.

ligand-gated channel An ion channel that opens or closes in response to binding by a certain molecule. Compare with **voltage-gated channel**.

light chain The smaller of the two types of polypeptide chains in an antibody molecule; composed of a variable (V) region, which contributes to the antigen-binding site, and a constant (C) region. Compare with **heavy chain**.

lignin A substance found in the secondary cell walls of some plants that is exceptionally stiff and strong. Most abundant in woody plant parts.

limiting nutrient Any essential nutrient whose scarcity in the environment significantly reduces growth and reproduction of organisms.

limnetic zone Open water (not near shore) that receives enough light to support photosynthesis.

lineage See **monophyletic group**.

LINEs (long interspersed nuclear elements) The most abundant class of transposable elements in human genomes; can create copies of itself and insert them elsewhere in the genome. Compare with **SINEs**.

linkage In genetics, a physical association between two genes because they are on the same chromosome; the inheritance patterns resulting from this association.

linkage map See **genetic map**.

lipase Any enzyme that can break down fat molecules into fatty acids and monoglycerides.

lipid Any organic subtance that does not dissolve in water, but dissolves well in nonpolar organic solvents. Lipids include fats, oils, phospholipids, and waxes.

lipid bilayer The basic structural element of all cellular membranes consisting of a two-layer sheet of phospholipid molecules whose hydrophobic tails are oriented toward the inside and hydrophilic heads, toward the outside. Also called *phospholipid bilayer*.

littoral zone Shallow water near shore that receives enough sunlight to support photosynthesis. May be marine or freshwater; often flowering plants are present.

liver A large, complex organ of vertebrates that performs many functions including storage of glycogen, processing and conversion of food and wastes, and production of bile.

lobe-finned fish Fish with fins supported by bony elements that extend down the length of the structure.

locomotion Movement of an organism under its own power.

locus (plural: loci) A gene's physical location on a chromosome.

logistic population growth The density-dependent decrease in growth rate as population size approaches the carrying capacity. Compare with **exponential population growth**.

long interspersed nuclear elements See **LINEs**.

long-day plant A plant that blooms in response to short nights (usually in late spring or early summer in the northern hemisphere). Compare with **day-neutral** and **short-day plant**.

loop of Henle In the vertebrate kidney, a long U-shaped loop in a nephron that extends into the medulla. Functions as a countercurrent exchanger to set up an osmotic gradient that allows reabsorption of water from a subsequent portion of the nephron.

loose connective tissue A type of connective tissue consisting of fibrous proteins in a soft matrix. Often functions as padding for organs.

lophophore A specialized feeding structure found in some lophotrochozoans and used in filter feeding.

lophotrochozoans A major lineage of protostomes (Lophotrochozoa) that grow by extending the size of their skeletons rather than by molting. Many phyla have a specialized feeding structure (lophophore) and/or ciliated larvae (trochophore). Includes rotifers, flatworms, segmented worms, and mollusks. Compare with **ecdysozoans**.

loss-of-function allele See **knock-out allele**.

lumen The interior space of any hollow structure (e.g., the rough ER) or organ (e.g., the stomach).

lung Any respiratory organ used for gas exchange between blood and air.

luteal phase The second major phase of a menstrual cycle, after ovulation, when the progesterone levels are high and the body is preparing for a possible pregnancy.

luteinizing hormone (LH) A peptide hormone, produced and secreted by the anterior pituitary, that stimulates estrogen production, ovulation, and formation of the corpus luteum in females and testosterone production in males.

lymph The mixture of fluid and white blood cells that circulates through the ducts and lymph nodes of the lymphatic system in vertebrates.

lymph node One of numerous small oval structures through which lymph moves in the lymphatic system. Filter the lymph and screen it for pathogens and other antigens. Major sites of lymphocyte activation.

lymphatic system In vertebrates, a body-wide network of thin-walled ducts (or vessels) and lymph nodes, separate from the circulatory system. Collects excess fluid from body tissues and returns it to the blood; also functions as part of the immune system.

lymphocytes Two types of leukocyte—B cells and T cells—that circulate through the bloodstream and lymphatic system and that are responsible for the development of acquired immunity.

lysogenic cycle A type of viral replication in which a viral genome enters a host cell, is inserted into the host's chromosome, and is replicated whenever the host cell divides. When activated, the viral DNA enters the lytic cycle, leading to production of new virus particles. Also called *lysogeny* or *latent growth*. Compare with **lytic cycle**.

lysosome A small organelle in an animal cell containing acids and enzymes that catalyze hydrolysis reactions and can digest large molecules. Compare with **vacuole**.

lysozyme An enzyme that functions in innate immunity by digesting bacterial cell walls. Occurs in saliva, tears, mucus, and egg white.

lytic cycle A type of viral replication in which a viral genome enters a host cell, new virus particles (virions) are made using host enzymes and eventually burst out of the cell, killing it. Also called *replicative growth*. Compare with **lysogenic cycle**.

macromolecule Any very large organic molecule, usually made up of smaller molecules (monomers) joined together into a polymer. The main biological

macromolecules are proteins, nucleic acids, and polysaccharides.

macronutrient Any element (e.g., carbon, oxygen, nitrogen) that is required in large quantities for normal growth, reproduction, and maintenance of a living organism. Compare with **micronutrient**.

MADS box A DNA sequence that codes for a DNA-binding motif in proteins; present in floral organ identity genes in plants. Functionally similar sequences are found in some fungal and animal genes.

major histocompatibility protein See **MHC protein**.

maladaptive A trait that lowers fitness.

malaria A human disease caused by four species of the protist *Plasmodium* and passed to humans by mosquitoes.

malignant tumor A tumor that is actively growing and disrupting local tissues and/or is spreading to other organs. Cancer consists of one or more malignant tumors. Compare with **benign tumor**.

Malpighian tubules A major excretory organ of insects, consisting of blind-ended tubes that extend from the gut into the hemocoel. Filter hemolymph to form "pre-urine" and then send it to the hindgut for further processing.

mammals One of the two lineages of amniotes (vertebrates that produce amniotic eggs) distinguished by hair (or fur) and mammary glands. Includes the monotremes (platypuses), marsupials, and eutherians (placental mammals).

mammary glands Specialized exocrine glands that produce and secrete milk for nursing offspring. A diagnostic feature of mammals.

mandibles Any mouthpart used in chewing. In vertebrates, the lower jaw. In insects, crustaceans, and myriapods, the first pair of mouthparts.

mantle One of the three main parts of the mollusk body; the thick outer tissue that protects the visceral mass and may secrete a calcium carbonate shell.

Marfan syndrome A human syndrome involving increased height, long limbs and fingers, an abnormally shaped chest, and heart disorders. Probably caused by mutation in one pleiotropic gene.

marsh A wetland that lacks trees and usually has a slow but steady rate of water flow.

marsupials A lineage of mammals (Marsupiala) that nourish their young in an abdominal pouch after a very short period of development in the uterus.

mass extinction The extinction of a large number of diverse evolutionary groups during a relatively short period of geologic time (about 1 million years). May occur due to sudden and extraordinary environmental changes. Compare with **background extinction**.

mass feeder An animal that takes chunks of food into its mouth.

mass number The total number of protons and neutrons in an atom.

mast cell A type of leukocyte that is stationary (embedded in tissue) and helps trigger the inflammatory response to infection or injury, including secretion of histamine. Particularly important in allergic responses and defense against parasites.

maternal chromosome A chromosome inherited from the mother.

mechanoreceptor A sensory cell or organ specialized for detecting distortions caused by touch or pressure. One example is hair cells in the cochlea.

mediator complex regulatory proteins that form a physical link between regulatory transcription factors that are bound to DNA and the basal transcription complex

medium A liquid or solid in which cells can grow in vitro. Also called *growth medium*.

medulla The innermost part of an organ (e.g., kidney or adrenal gland).

medulla oblongata In vertebrates, a region of the brain stem that along with the cerebellum forms the hindbrain.

medusa (plural: medusae) The free-floating stage in the life cycle of some cnidarians (e.g., jellyfish). Compare with **polyp**.

megapascal (MPa) A unit of pressure (force per unit area), equivalent to 1 million pascals (Pa).

megasporangium (plural: megasporangia) In heterosporous species of plants, a spore-producing structure that produces megaspores, which go on to develop into female gametophytes.

megaspore In seed plants, a haploid (n) spore that is produced in a megasporangium by meiosis of a diploid ($2n$) megasporocyte; develops into a female gametophyte. Compare with **microspore**.

meiosis In sexually reproducing organisms, a special two-stage type of cell division in which one diploid ($2n$) parent cell produces four haploid (n) reproductive cells (gametes); results in halving of the chromosome number. Also called *reduction division*.

meiosis I The first cell division of meiosis, in which synapsis and crossing over occur, and homologous chromosomes are separated from each other, producing daughter cells with half as many chromosomes (each composed of two sister chromatids) as the parent cell.

meiosis II The second cell division of meiosis, in which sister chromatids are separated from each other. Similar to mitosis.

meiotic map See **genetic map**.

membrane potential A difference in electric charge across a cell membrane; a form of potential energy. Also called *membrane voltage*.

memory Retention of learned information.

memory cells A type of lymphocyte responsible for maintenance of immunity for years or decades after an infection. Descended from a B cell or T cell activated during a previous infection.

meniscus (plural: menisci) The concave boundary layer formed at most air-water interfaces due to surface tension.

menstrual cycle A female reproductive cycle seen in Old World monkeys and apes (including humans) in which the uterine lining is shed (menstruation) if no pregnancy occurs. Compare with **estrous cycle**.

menstruation The periodic shedding of the uterine lining through the vagina that occurs in females of Old World monkeys and apes, including humans.

meristem (adjective: meristematic) In plants, a group of undifferentiated cells that can develop into various adult tissues throughout the life of a plant.

mesoderm The middle of the three basic cell layers in most animal embryos; gives rise to muscles, bones, blood, and some internal organs (kidney, spleen, etc.). Compare with **ectoderm** and **endoderm**.

mesoglea A gelatinous material, containing scattered ectodermal cells, that is located between the ectoderm and endoderm of cnidarians (e.g., jellyfish, corals, and anemones).

mesophyll cell A type of cell, found near the surfaces of plant leaves, that is specialized for the light-dependent reactions of photosynthesis.

Mesozoic era The interval of geologic time, from 251 million to 65.5 million years ago, during which gymnosperms were the dominant plants and dinosaurs

the dominant vertebrates. Ended with extinction of the dinosaurs.

messenger RNA (mRNA) An RNA molecule that carries encoded information, transcribed from DNA, that specifies the amino acid sequence of a polypeptide.

meta-analysis A comparative analysis of the results of many smaller, previously published studies.

metabolic pathway An ordered series of chemical reactions that build up or break down a particular molecule. Often, each reaction is catalyzed by a different enzyme.

metabolic rate The total energy use by all the cells of an individual. For aerobic organisms, often measured as the amount of oxygen consumed per hour.

metabolic water The water that is produced as a by-product of cellular respiration.

metabolism All the chemical reactions occurring in a living cell or organism.

metagenomics Sequencing of all or most of the genes present in an environment directly (also called environmental sequencing).

metallothioneins Small plant proteins that bind to and prevent excess metal ions from acting as toxins.

metamorphosis Transition from one developmental stage to another, such as from the larval to the adult form of an animal.

metaphase A stage in mitosis or meiosis during which chromosomes line up in the middle of the cell.

metaphase plate The plane along which chromosomes line up during metaphase of mitosis or meiosis; not an actual structure.

metapopulation A population made up of many small, physically isolated populations.

metastasis The spread of cancerous cells from their site of origin to distant sites in the body where they may establish additional tumors.

methanogen A prokaryote that produces methane (CH_4) as a by-product of cellular respiration.

methanotroph An organism that uses methane (CH_4) as its primary electron donor and source of carbon.

methyl salicylate (MeSA) A molecule that is hypothesized to function as a signal, transported among tissues, that triggers systematic acquired resistance in plants—a response to pathogen attack.

methylation The addition of a methyl ($-CH_3$) group to a molecule.

MHC protein One of a large set of mammalian cell-surface glycoproteins involved in marking cells as self and in antigen presentation to T cells. Also called *MHC molecule*. See **Class I** and **Class II MHC protein**.

microbe Any microscopic organism, including bacteria, archaea, and various tiny eukaryotes.

microbiology The field of study concerned with microscopic organisms.

microfilament See **actin filament**.

micrograph A photograph of an image produced by a microscope.

micronutrient Any element (e.g., iron, molybdenum, magnesium) that is required in very small quantities for normal growth, reproduction, and maintenance of a living organism. Compare with **macronutrient**.

micropyle The tiny pore in a plant ovule through which the pollen tube reaches the embryo sac.

microRNA (miRNA) A small, single-stranded RNA associated with proteins in an RNA-induced silencing complex. Can bind to complementary sequences in mRNA molecules, allowing the associated proteins to degrade the bound mRNA or inhibit its translation. See **RNA interference**.

microsatellite A noncoding stretch of eukaryotic DNA consisting of a repeating sequence 1- to 5-base pair long. Also called *simple sequence repeat*.

microsporangim (plural: microsporangia) In heterosporous species of plants, a spore-producing structure that produces microspores, which go on to develop into male gametophytes.

microspore In seed plants, a haploid (*n*) spore that is produced in a microsporangium by meiosis of a diploid (*2n*) microsporocyte; develops into a male gametophyte. Compare with **megaspore**.

microtubule A long, tubular fiber, about 25 nm in diameter, formed by polymerization of tubulin protein dimers; one of the three types of cytoskeletal fibers. Involved in cell movement and transport of materials within the cell. Compare with **actin filament** and **intermediate filament**.

microtubule-organizing center (MTOC) General term for any structure (e.g., centrosome and basal body) that organizes microtubules in cells.

microvilli (singular: microvillus) Tiny protrusions from the surface of an epithelial cell that increase the surface area for absorption of substances.

midbrain One of the three main regions of the vertebrate brain; includes sensory integrating and relay centers. Compare with **forebrain** and **hindbrain**.

middle ear The air-filled middle portion of the mammalian ear, which contains three small bones (ossicles) that transmit and amplify sound from the tympanic membrane to the inner ear. Is connected to the throat via the eustachian tube.

migration (1) In ecology, a cyclical movement of large numbers of organisms from one geographic location or habitat to another. (2) In population genetics, movement of individuals from one population to another.

millivolt (mV) A unit of voltage equal to 1/1000 of a volt.

mimicry A phenomenon in which one species has evolved (or learns) to look or sound like another species. See **Batesian mimicry** and **Müllerian mimicry**.

mineralocorticoids A class of steroid hormones, produced and secreted by the adrenal cortex, that regulate electrolyte levels and the overall volume of body fluids. Aldosterone is the principal one in humans. Compare with **glucocorticoids**.

minisatellite A noncoding stretch of eukaryotic DNA consisting of a repeating sequence that is 6 to 500 base pairs long. Also called *variable number tandem repeat (VNTR)*.

mismatch repair The process by which mismatched base pairs in DNA are fixed.

missense mutation A point mutation (change in a single base pair) that causes a change in the amino acid sequence of a protein. Also called *replacement mutation*.

mitochondrial matrix Central compartment of a mitochondrion, which is lined by the inner membrane; contains the enzymes and substrates of the citric acid cycle and mitochondrial DNA.

mitochondrion (plural: mitochondria) A eukaryotic organelle that is bounded by a double membrane and is the site of aerobic respiration.

mitosis In eukaryotic cells, the process of nuclear division that results in two daughter nuclei genetically identical to the parent nucleus. Subsequent cytokinesis (division of the cytoplasm) yields two daughter cells.

mitosis-promoting factor (MPF) A complex of a cyclin and cyclin-dependent kinase that phosphorylates a number of specific proteins needed to initiate mitosis in eukaryotic cells.

mitotic (M) phase The phase of the cell cycle during which cell division occurs. Includes mitosis and cytokinesis.

model organism An organism selected for intensive scientific study based on features that make it easy to work with (e.g., body size, life span), in the hope that findings will apply to other species.

molarity A common unit of solute concentration equal to the number of moles of a dissolved solute in 1 liter of solution.

mole The amount of a substance that contains 6.022×10^{23} of its elemental entities (e.g., atoms, ions, or molecules). This number of molecules of a compound will have a mass equal to the molecular weight of that compound expressed in grams.

molecular chaperone A protein that facilitates the three-dimensional folding of newly synthesized proteins, usually by an ATP-dependent mechanism.

molecular formula A notation that indicates only the numbers and types of atoms in a molecule, such as H_2O for the water molecule. Compare with **structural formula**.

molecular weight The sum of the mass numbers of all of the atoms in a molecule; roughly, the total number of protons and neutrons in the molecule.

molecule A combination of two or more atoms held together by covalent bonds.

mollusks Members of the phylum Mollusca. Distinguished by a body plan with three main parts: a muscular foot, a visceral mass, and a mantle. Include bivalves (clams, oysters), gastropods (snails, slugs), chitons, and cephalopods (squid, octopuses). Mollusks belong to the lophotrochozoan branch of the protostomes.

molting A method of body growth, used by ecdysozoans, that involves the shedding of an external protective cuticle or skeleton, expansion of the soft body, and growth of a new external layer.

monocot Any plant that has a single cotyledon (embryonic leaf) upon germination. Monocots form a monophyletic group. Also called a monocotyledonous plant. Compare with **dicot**.

monoecious Describing an angiosperm species that has both male and female reproductive structures on each plant. Compare with **dioecious**.

monohybrid cross A mating between two parents that are both heterozygous for a given gene.

monomer A small molecule that can covalently bind to other similar molecules to form a larger macromolecule. Compare with **polymer**.

monophyletic group An evolutionary unit that includes an ancestral population and all of its descendants but no others. Also called a *clade* or *lineage*. Compare with **paraphyletic group**.

monosaccharide A small carbohydrate, such as glucose, that has the molecular formula $(CH_2O)_n$ and cannot be hydrolyzed to form any smaller carbohydrates. Also called *simple sugar*. Compare with **disaccharide** and **polysaccharide**.

monosomy Having only one copy of a particular type of chromosome.

monotremes A lineage of mammals (Monotremata) that lay eggs and then nourish the young with milk. Includes just three living species: the platypus and two species of echidna.

morphogen A molecule that exists in a concentration gradient and provides spatial information to embryonic cells.

morphospecies concept The definition of a species as a population or group of populations that have measurably different anatomical features from other groups. Also called *morphological species concept*. Compare with **biological** and **phylogenetic species concept**.

morphology The shape and appearance of an organism's body and its component parts.

motor neuron A nerve cell that carries signals from the central nervous system (brain and spinal cord) to an effector, such as a muscle or gland. Compare with **interneuron** and **sensory neuron**.

motor protein A class of proteins whose major function is to convert the chemical energy of ATP into motion. Includes dynein, kinesin, and myosin.

MPF See **mitosis-promoting factor**.

mRNA See **messenger RNA**.

mucigel A slimy substance secreted by plant root caps that eases passage of the growing root through the soil.

mucosal-associated lymphoid tissue (MALT) Collective term for lymphocytes and other leukocytes associated with skin cells and with mucus-secreting epithelial tissues in the gut and respiratory tract. Plays important role in preventing entry of pathogens into the body.

mucous cell A type of cell found in the epithelial layer of the stomach; responsible for secreting mucus into the stomach.

mucus (adjective: mucous) A slimy mixture of glycoproteins (called mucins) and water that is secreted in many animal organs for lubrication.

Müllerian inhibitory substance A peptide hormone secreted by the embryonic testis that causes regression (withering away) of the female reproductive ducts.

Müllerian mimicry A type of mimicry in which two (or more) harmful species resemble each other. Compare with **Batesian mimicry**.

multicellularity The state of being composed of many cells that adhere to each other and do not all express the same genes with the result that some cells have specialized functions.

multienzyme complex A group of enzymes that are physically attached to each other, even though each of the enzymes catalyzes a separate—but usually related—chemical reaction.

multiple allelism In a population, the existence of more than two alleles of the same gene.

multiple fruit A fruit (e.g., pineapple) that develops from many separate flowers and thus many carpels. Compare with **aggregate** and **simple fruit**.

multiple sclerosis (MS) A human autoimmune disease caused by the immune system attacking the myelin sheaths that insulate nerve axons.

muscle fiber A single muscle cell.

muscle tissue An animal tissue consisting of bundles of long, thin contractile cells (muscle fibers).

mutagen Any physical or chemical agent that increases the rate of mutation.

mutant An individual that carries a mutation, particularly a new or rare mutation.

mutation Any change in the hereditary material of an organism (DNA in most organisms, RNA in some viruses).

mutualism (adjective: mutualistic) A symbiotic relationship between two organisms (mutualists) that benefits both. Compare with **commensalism** and **parasitism**.

mycelium (plural: mycelia) A mass of underground filaments (hyphae) that form the body of a fungus. Also found in some protists and bacteria.

mycorrhiza (plural: mycorrhizae) A mutualistic association between certain fungi and most vascular plants, sometimes visible as nodules or nets in or around plant roots.

myelin sheath Multiple layers of myelin, derived from the cell membranes of certain glial cells, that is wrapped around the axon of a neuron, providing electrical insulation.

myoD A transcription factor that is critical for differentiation of muscle cells (short for "*myo*blast *d*etermination").

myofibril Long, slender structure composed of contractile proteins organized into repeating units (sarcomeres) in vertebrate heart muscle and skeletal muscle.

myosin Any one of a class of motor proteins that use the chemical energy of ATP to move along actin filaments in muscle contraction, cytokinesis, and vesicle transport.

myriapods A lineage of arthropods with long segmented trunks, each segment bearing one or two pairs of legs. Includes millipedes and centipedes.

NAD⁺/NADH Oxidized and reduced forms, respectively, of nicotinamide adenine dinucleotide. A nonprotein electron carrier that functions in many of the redox reactions of metabolism.

NADP⁺/NADPH Oxidized and reduced forms, respectively, of nicotinamide adenine dinucleotide phosphate. A nonprotein electron carrier that is reduced during the light-dependent reactions in photosynthesis and extensively used in biosynthetic reactions.

natural experiment A situation in which groups to be compared are created by an unplanned, natural change in conditions rather than by manipulation of conditions by researchers.

natural selection The process by which individuals with certain heritable traits tend to produce more surviving offspring than do individuals without those traits, often leading to a change in the genetic makeup of the population. A major mechanism of evolution.

nauplius A distinct planktonic larval stage seen in many crustaceans.

Neanderthal A recently extinct European species of hominid, *Homo neanderthalensis*, closely related to but distinct from modern humans.

nectar The sugary fluid produced by flowers to attract and reward pollinating animals.

nectary A nectar-producing structure in a flower.

negative control Of genes, when a regulatory protein shuts down expression by binding to DNA on or near the gene

negative feedback A self-limiting, corrective response in which a deviation in some variable (e.g., body temperature, blood pH, concentration of some compound) triggers responses aimed at returning the variable to normal. Compare with **positive feedback**.

negative pressure ventilation Ventilation of the lungs that is accomplished by "pulling" air into the lungs by expansion of the rib cage. Compare with **positive pressure ventilation**.

negative-sense virus A virus whose genome contains sequences complementary to those in the mRNA required to produce viral proteins. Compare with **ambisense virus** and **positive-sense virus**.

nematodes See **roundworms**.

nephron One of the tiny tubes within the vertebrate kidney that filter blood and concentrate salts to produce urine. Also called *renal tubule*.

neritic zone Shallow marine waters beyond the intertidal zone, extending down to about 200 meters, where the continental shelf ends.

nerve A long, tough strand of nervous tissue typically containing thousands of axons wrapped in connective tissue; carries impulses between the central nervous system and some other part of the body.

nerve cord A bundle of nerves extending from the brain along the dorsal (back) side of a chordate animal,

with cerebrospinal fluid inside a hollow central channel. One of the defining features of chordates.

nerve net A nervous system in which neurons are diffuse instead of being clustered into large masses or tracts.

nervous tissue An animal tissue consisting of nerve cells (neurons) and various supporting cells.

net primary productivity (NPP) In an ecosystem, the total amount of carbon fixed by photosynthesis over a given time period minus the amount oxidized during cellular respiration. Compare with **gross primary productivity**.

net reproductive rate (R_0) The growth rate of a population per generation; equivalent to the average number of female offspring that each female produces over her lifetime.

neural Relating to nerve cells (neurons) and the nervous system.

neural tube A folded tube of ectoderm that forms along the dorsal side of a young vertebrate embryo and that will give rise to the brain and spinal cord.

neuroendocrine Referring to nerve cells (neurons) that release hormones into the blood or to such hormones themselves.

neuron A cell that is specialized for the transmission of nerve impulses. Typically has dendrites, a cell body, and a long axon that forms synapses with other neurons. Also called *nerve cell*.

neurosecretory cell A nerve cell (neuron) that produces and secretes hormones into the bloodstream. Principally found in the hypothalamus. Also called *neuroendocrine cell*.

neurotoxin Any substance that specifically destroys or blocks the normal functioning of neurons.

neurotransmitter A molecule that transmits electrical signals from one neuron to another or from a neuron to a muscle or gland. Examples are acetylcholine, dopamine, serotonin, and norepinephrine.

neutral In genetics, referring to any mutation or mutant allele that has no effect on an individual's fitness.

neutrophil A type of leukocyte, capable of moving through body tissues, that engulfs and digests pathogens and other foreign particles; also secretes various compounds that attack bacteria and fungi.

niche The particular set of habitat requirements of a certain species and the role that species plays in its ecosystem.

niche differentiation The change in resource use by competing species that occurs as the result of character displacement.

nicotinamide adenine dinucleotide See **NAD⁺/NADH**.

nicotinamide adenine dinucleotide phosphate See **NADP⁺/NADPH**.

nitrogen cycle, global The movement of nitrogen among terrestrial ecosystems, the oceans, and the atmosphere.

nitrogen fixation The incorporation of atmospheric nitrogen (N_2) into forms such as ammonia (NH_3) or nitrate (NO_3^-), which can be used to make many organic compounds. Occurs in only a few lineages of bacteria and archaea.

nociceptor A sensory cell or organ specialized to detect tissue damage, usually producing the sensation of pain.

Nod factors Molecules produced by nitrogen-fixing bacteria that help them recognize and bind to roots of legumes.

node (1) In animals, any small thickening (e.g., a lymph node). (2) In plants, the part of a stem where leaves or leaf buds are attached. (3) In a phylogenetic tree, the point where two branches diverge, representing the point in time when an ancestral group split into two or more descendant groups. Also called *fork*.

node of Ranvier One of the periodic unmyelinated sections of a neuron's axon at which an action potential can be regenerated.

nodule Lumplike structure on roots of legume plants that contain symbiotic nitrogen-fixing bacteria.

noncyclic electron flow See **Z scheme**.

nondisjunction An error that can occur during meiosis or mitosis in which one daughter cell receives two copies of a particular chromosome, and the other daughter cell receives none.

nonpolar covalent bond A covalent bond in which electrons are equally shared between two atoms of the same or similar electronegativity. Compare with **polar covalent bond**.

non-sister chromatids The chromatids of a particular type of chromosome (after replication) with respect to the chromatids of its homologous chromosome. Crossing over occurs between non-sister chromatids. Compare with **sister chromatids**.

non-template strand The strand of DNA that is not transcribed during synthesis of RNA. Its sequence corresponds to that of the mRNA produced from the other strand. Also called *coding strand*.

non-vascular plants See **bryophytes**.

norepinephrine A catecholamine used as a neurotransmitter in the sympathetic nervous system. Also is produced by the adrenal medulla and functions as a hormone that triggers rapid responses relating to the fight-or-flight response.

notochord A long, gelatinous, supportive rod down the back of a chordate embryo, below the developing spinal cord. Replaced by vertebrae in most adult vertebrates. A defining feature of chordates.

nuclear envelope The double-layered membrane enclosing the nucleus of a eukaryotic cell.

nuclear lamina A lattice-like sheet of fibrous nuclear lamins, which are one type of intermediate filaments. Lines the inner membrane of the nuclear envelope, stiffening the envelope and helping organize the chromosomes.

nuclear lamins Intermediate filaments that make up the nuclear lamina layer—a lattice-like layer inside the nuclear envelope that stiffens the structure.

nuclear localization signal (NLS) A short amino acid sequence that marks a protein for delivery to the nucleus.

nuclear pore An opening in the nuclear envelope that connects the inside of the nucleus with the cytoplasm and through which molecules such as mRNA and some proteins can pass.

nuclear pore complex A large complex of dozens of proteins lining a nuclear pore, defining its shape and transporting substances through the pore.

nuclease Any enzyme that can break down RNA or DNA molecules.

nucleic acid A macromolecule composed of nucleotide monomers. Generally used by cells to store or transmit hereditary information. Includes ribonucleic acid and deoxyribonucleic acid.

nucleoid In prokaryotic cells, a dense, centrally located region that contains DNA but is not surrounded by a membrane.

nucleolus In eukaryotic cells, specialized structure in the nucleus where ribosomal RNA processing occurs and ribosomal subunits are assembled.

nucleosome A repeating, bead-like unit of eukaryotic chromatin, consisting of about 200 nucleotides of DNA wrapped twice around eight histone proteins.

nucleotide A molecule consisting of a five-carbon sugar (ribose or deoxyribose), a phosphate group, and one of several nitrogen-containing bases. DNA and RNA are polymers of nucleotides containing deoxyribose (deoxyribonucleotides) and ribose (ribonucleotides), respectively. Equivalent to a nucleoside plus one phosphate group.

nucleotide excision repair The process of removing a damaged region in one strand of DNA and correctly replacing it using the undamaged strand as a template.

nucleus (1) The center of an atom, containing protons and neutrons. (2) In eukaryotic cells, the large organelle containing the chromosomes and surrounded by a double membrane. (3) A discrete clump of neuron cell bodies in the brain, usually sharing a distinct function.

null allele See **knock-out allele**.

null hypothesis A hypothesis that specifies what the results of an experiment will be if the main hypothesis being tested is wrong. Often states that there will be no difference between experimental groups.

nutrient A substance that an organism requires for normal growth, maintenance, or reproduction.

occipital lobe In the vertebrate brain, one of the four major areas in the cerebrum.

oceanic zone The waters of the open ocean beyond the continental shelf.

oil A fat that is liquid at room temperature.

Okazaki fragment Short segment of DNA produced during replication of the 5′ to 3′ template strand. Many Okazaki fragments make up the lagging strand in newly synthesized DNA.

olfaction The perception of odors.

olfactory bulb A bulb-shaped projection of the brain just above the nose. Receives and interprets odor information from the nose.

oligodendrocyte A type of glial cell that wraps around axons of some neurons in the central nervous system, forming a myelin sheath that provides electrical insulation. Compare with **Schwann cell**.

oligopeptide A chain composed of fewer than 50 amino acids linked together by peptide bonds. Often referred to simply as *peptide*.

ommatidium (plural: ommatidia) A light-sensing column in an arthropod's compound eye.

omnivore (adjective: omnivorous) An animal whose diet regularly includes both meat and plants. Compare with **carnivore** and **herbivore**.

oncogene Any gene whose protein product stimulates cell division at all times and thus promotes cancer development. Often is a mutated form of a gene involved in regulating the cell cycle. See **proto-oncogene**.

one-gene, one-enzyme hypothesis The hypothesis that each gene is responsible for making one (and only one) protein, in most cases an enzyme that catalyzes a specific reaction. Many exceptions to this hypothesis are now known.

oocyte A cell in the ovary that can undergoes meiosis to produce an ovum.

oogenesis The production of egg cells (ova).

oogonia (singular: oogonia) The diploid cells in an ovary that can divide by mitosis to create more oogonia and primary oocytes, which can undergo meiosis.

open circulatory system A circulatory system in which the circulating fluid (hemolymph) is not confined to blood vessels. Compare with **closed circulatory system**.

open reading frame (ORF) Any DNA sequence, ranging in length from several hundred to thousands of base pairs long, that is flanked by a start codon and a stop codon. ORFs identified by computer analysis of DNA may be functional genes, especially if they have other features characteristic of genes (e.g., promoter sequence).

operator In prokaryotic DNA, a binding site for a repressor protein; located near the start of an operon.

operculum The stiff flap of tissue that covers the gills of teleost fishes.

operon A region of prokaryotic DNA that codes for a series of functionally related genes and is transcribed from a single promoter into a polycistronic mRNA.

opsin A transmembrane protein that is covalently linked to retinal, the light-detecting pigment in rod and cone cells.

optic nerve A bundle of neurons that runs from the eye to the brain.

optimal foraging The concept that animals forage in a way that maximizes the amount of usable energy they take in, given the costs of finding and ingesting their food and the risk of being eaten while they're at it.

orbital The spherical region around an atomic nucleus in which an electron is present most of the time.

ORF See **open reading frame**.

organ A group of tissues organized into a functional and structural unit.

organ system Groups of tissues and organs that work together to perform a function.

organelle Any discrete, membrane-bound structure within a cell (e.g., mitochondrion) that has a characteristic structure and functions.

organic For a compound, containing carbon and hydrogen and usually containing carbon-carbon bonds. Organic compounds are widely used by living organisms.

organism Any living entity that contains one or more cells.

organogenesis A stage of embryonic development, just after gastrulation in vertebrate embryos, during which major organs develop from the three embryonic germ layers.

origin of replication The site on a chromosome at which DNA replication begins.

osmoconformer An animal that does not actively regulate the osmolarity of its tissues but conforms to the osmolarity of the surrounding environment.

osmolarity The concentration of dissolved substances in a solution, measured in moles per liter.

osmoregulation The process by which a living organism controls the concentration of water and salts in its body.

osmoregulator An animal that actively regulates the osmolarity of its tissues.

osmosis Diffusion of water across a selectively permeable membrane from a region of high water concentration (low solute concentration) to a region of low water concentration (high solute concentration).

osmotic potential See **solute potential**.

ossicles, ear In mammals, three bones found in the middle ear that function in transferring and amplifying sound from the outer ear to the inner ear.

ouabain A plant toxin that poisons the sodium-potassium pumps of animals.

outcrossing Reproduction by fusion of the gametes of different individuals, rather than self-fertilization. Typically refers to plants.

outer ear The outermost portion of the mammalian ear, consisting of the pinna (ear flap) and the ear canal. Funnels sound to the tympanic membrane.

outgroup A taxon that is closely related to a particular monophyletic group but is not part of it.

out-of-Africa hypothesis The hypothesis that modern humans (*Homo sapiens*) evolved in Africa and spread to other continents, replacing other *Homo* species without interbreeding with them.

oval window A membrane separating the fluid-filled cochlea from the air-filled middle ear through which sound vibrations pass from the middle ear to the inner ear in mammals.

ovary The egg-producing organ of a female animal, or the seed-producing structure in the female part of a flower.

oviduct See **fallopian tube**.

oviparous Producing eggs that are laid outside the body where they develop and hatch. Compare with **ovoviviparous** and **viviparous**.

ovoviviparous Producing eggs that are retained inside the body until they are ready to hatch. Compare with **oviparous** and **viviparous**.

ovulation The release of an ovum from an ovary of a female vertebrate. In humans, an ovarian follicle releases an egg at the end of the follicular phase of the menstrual cycle.

ovule In flowering plants, the structure inside an ovary that contains the female gametophyte and eventually (if fertilized) becomes a seed.

ovum (plural: ova) See **egg**.

oxidation The loss of electrons from an atom during a redox reaction, either by donation of an electron to another atom or by the shared electrons in covalent bonds moving farther from the atomic nucleus.

oxidative phosphorylation Production of ATP molecules from the redox reactions of an electron transport chain.

oxygen-hemoglobin equilibrium curve The graphical depiction of the percentage of hemoglobin in the blood that will bind to oxygen at various partial pressures of oxygen.

oxygenic Referring to any process or reaction that produces oxygen. Photosynthesis in plants, algae, and cyanobacteria, which involves photosystem II, is oxygenic. Compare with **anoxygenic**.

oxytocin A peptide hormone, secreted by the posterior pituitary, that triggers labor and milk production in females and that stimulates pair bonding, parental care, and affiliative behavior in both sexes.

p53 A tumor-suppressor protein (molecular weight of 53 kilodaltons) that responds to DNA damage by stopping the cell cycle and/or triggering apoptosis. Encoded by the *p53* gene.

pacemaker cell A specialized cardiac muscle cell in the sinoatrial (SA) node of the vertebrate heart that has an inherent rhythm and can generate an electrical impulse that spreads to other heart cells.

paleontologists Scientists who study the fossil record and the history of life.

Paleozoic era The interval of geologic time, from 542 million to 251 million years ago, during which fungi, land plants, and animals first appeared and diversified. Began with the Cambrian explosion and ended with the extinction of many invertebrates and vertebrates.

pancreas A large gland in vertebrates that has both exocrine and endocrine functions. Secretes digestive enzymes into a duct connected to the intestine and several hormones (notably, insulin and glucagon) into the bloodstream.

pancreatic lipase An enzyme that is produced in the pancreas and acts in the small intestine to break bonds in complex fats, releasing small lipids.

pandemic The spread of an infectious disease in a short time period over a wide geographic area and affecting a very high proportion of the population. Compare with **epidemic**.

parabiosis An experimental technique for determining whether a certain physiological phenomenon is regulated by a hormone, by surgically uniting two individuals so that hormones can pass between them.

paracrine Relating to a chemical signal that is released by one cell and affects neighboring cells.

paraphyletic group An evolutionary unit that includes an ancestral population and *some* but not all of its descendants. Paraphyletic groups are not meaningful units in evolution. Compare with **monophyletic group**.

parapodia (singular: parapodium) Appendages found in some annelids from which bristle-like structures (chaetae) extend.

parasite An organism that lives on or in a host species and that damages its host.

parasitism (adjective: parasitic) A symbiotic relationship between two organisms that is beneficial to one organism (the parasite) but detrimental to the other (the host). Compare with **commensalism** and **mutualism**.

parasitoid An organism that has a parasitic larval stage and a free-living adult stage. Most parasitoids are insects that lay eggs in the bodies of other insects.

parasympathetic nervous system The part of the autonomic nervous system that stimulates functions for conserving or restoring energy, such as reduced heart rate and increased digestion. Compare with **sympathetic nervous system**.

parathyroid glands Four small glands, located near or embedded in the thyroid gland of vertebrates, that secrete parathyroid hormone (PTH), a peptide hormone that increases blood calcium.

parenchyma cell In plants, a general type of cell with a relatively thin primary cell wall. These cells, found in leaves, the centers of stems and roots, and fruits, are involved in photosynthesis, starch storage, and new growth. Compare with **collenchyma cell** and **sclerenchyma cell**.

parental care Any action by which an animal expends energy or assumes risks to benefit its offspring (e.g., nest-building, feeding of young, defense).

parental generation The adult organisms used in the first experimental cross in a formal breeding experiment.

parietal cell A cell in the stomach lining that secretes hydrochloric acid.

parietal lobe In the vertebrate brain, one of the four major areas in the cerebrum.

parsimony The logical principle that the most likely explanation of a phenomenon is the most economical or simplest. When applied to comparison of alternative phylogenetic trees, it suggests that the one requiring the fewest evolutionary changes is most likely to be correct.

parthenogenesis Development of offspring from unfertilized eggs; a type of asexual reproduction.

partial pressure The pressure of one particular gas in a mixture; the contribution of that gas to the overall pressure.

particulate inheritance The observation that genes from two parents do not blend together to form a new physical entity in offspring, but instead remain separate or particle-like.

pascal (Pa) A unit of pressure (force per unit area).

passive transport Diffusion of a substance across a plasma membrane or organelle membrane. When this occurs with the assistance of membrane proteins, it is called facilitated diffusion.

patch clamping A technique for studying the electrical currents that flow through individual ion channels by sucking a tiny patch of membrane to the hollow tip of a microelectrode.

paternal chromosome A chromosome inherited from the father.

pathogen (adjective: pathogenic) Any entity capable of causing disease, such as a microbe, virus, or prion.

pattern formation The series of events that determines the spatial organization of an embryo, including alignment of the major body axes and orientation of the limbs.

pattern-recognition receptor One of a class of membrane proteins on leukocytes that bind to molecules on the surface of many bacteria. Part of the innate immune response.

PCR See **polymerase chain reaction**.

peat Semidecayed organic matter that accumulates in moist, low-oxygen environments such as bogs.

pectin A gelatinous polysaccharide found in the primary cell wall of plant cells. Attracts and holds water, forming a gel that helps keep the cell wall moist.

pedigree A family tree of parents and offspring, showing inheritance of particular traits of interest.

penis The copulatory organ of male mammals, used to insert sperm into a female.

pentose A monosaccharide (simple sugar) containing five carbon atoms.

PEP carboxylase An enzyme that catalyzes addition of CO_2 to phosphoenol pyruvate, a three-carbon compound, forming a four-carbon organic acid. Found in mesophyll cells of plants that perform C_4 photosynthesis.

pepsin A protein-digesting enzyme present in the stomach.

pepsinogen The precursor of the digestive enzyme pepsin. Is secreted from cells in the stomach lining and converted to pepsin by the acidic environment of the stomach lumen.

peptide See **oligopeptide**.

peptide bond The covalent bond ($C-N$) formed by a condensation reaction between two amino acids; links the residues in peptides and proteins.

peptidoglycan A complex structural polysaccharide found in bacterial cell walls.

perennial Describing a plant whose life cycle normally lasts for more than one year. Compare with **annual**.

perfect flower A flower that contains both male parts (stamens) and female parts (carpels). Compare with **imperfect flower**.

perforation In plants, a small hole in the primary and secondary cell walls of vessel elements that allow passage of water.

pericarp The part of a fruit, formed from the ovary wall, that surrounds the seeds and protects them. Corresponds to the flesh of most edible fruits and the hard shells of most nuts.

pericycle In plant roots, a layer of cells that give rise to lateral roots.

peripheral membrane protein Any membrane protein that does not span the entire lipid bilayer and associates with only one side of the bilayer. Compare with **integral membrane protein**.

peripheral nervous system (PNS) All the components of the nervous system that are outside the central nervous system (the brain and spinal cord). Includes the somatic nervous system and the autonomic nervous system.

peristalsis Rhythmic waves of muscular contraction that push food along the digestive tract.

permafrost A permanently frozen layer of icy soil found in most tundra and some taiga.

permeability The tendency of a structure, such as a membrane, to allow a given substance to diffuse across it.

peroxisome An organelle found in most eukaryotic cells that contains enzymes for oxidizing fatty acids and other compounds including many toxins, rendering them harmless. See **glyoxisome**.

petal One of the leaflike organs arranged around the reproductive organs of a flower. Often colored and scented to attract pollinators.

petiole The stalk of a leaf.

pH A measure of the concentration of protons in a solution and thus of acidity or alkalinity. Defined as the negative of the base-10 logarithm of the proton concentration: $pH = -\log[H^+]$.

phagocytosis Uptake by a cell of small particles or cells by pinching off the plasma membrane to form small membrane-bound vesicles; one type of endocytosis.

pharyngeal gill slits A set of parallel openings from the throat through the neck to the outside. A diagnostic trait of chordates.

pharyngeal jaw A secondary jaw in the back of the mouth, found in some fishes. Derived from modified gill arches.

phelloderm A component of bark; specifically, a tissue layer produced to the inside of the cork cambium in plants with secondary growth.

phelloderm In the stems of woody plants, a thin layer of cells located between the outer cork cells and inner cork cambium.

phenetic approach A method for constructing a phylogenetic tree by computing a statistic that summarizes the overall similarity among populations, based on the available data. Compare with **cladistic approach**.

phenology The timing of events during the year, in environments where seasonal changes occur.

phenotype The detectable physical and physiological traits of an individual, which are determined its genetic makeup. Also the specific trait associated with a particular allele. Compare with **genotype**.

phenotypic plasticity Within-species variation in phenotype that is due to differences in environmental conditions. Occurs more commonly in plants than animals.

phenylketonuria (PKU) A disease caused by the inability to process the amino acid phenylalanine.

pheophytin In photosystem II, a molecule that accepts excited electrons from a reaction center chlorophyll and passes them to an electron transport chain.

pheromone A chemical signal, released by one individual into the external environment, that can trigger responses in a different individual.

phloem A plant vascular tissue that conducts sugars; contains sieve-tube members and companion cells. Primary phloem develops from the procambium of apical meristems; secondary phloem, from the vascular cambium of lateral meristems. Compare with **xylem**.

phosphodiester linkage Chemical linkage between adjacent **nucleotide residues** in DNA and RNA. Forms when the phosphate group of one nucleotide condenses with the hydroxyl group on the sugar of another nucleotide. Also known as *phosphodiester bond*.

phosphofructokinase The enzyme that catalyzes synthesis of fructose-1,6-bisphosphate from fructose-6-phosphate, a key reaction (step 3) in glycolysis.

phospholipid A class of lipid having a hydrophilic head (a phosphate group) and a hydrophobic tail (one or more fatty acids). Major components of the plasma membrane and organelle membranes.

phosphorylase An enzyme that breaks down glycogen by catalyzing hydrolysis of the α-glycosidic linkages between the glucose residues.

phosphorylation (verb: phosphorylate) The addition of a phosphate group to a molecule.

phosphorylation cascade A series of enzyme-catalyzed phosphorylation reactions commonly used in signal transduction pathways to amplify and convey a signal inward from the plasma membrane.

photic zone In an aquatic habitat, water that is shallow enough to receive some sunlight (whether or not it is enough to support photosynthesis). Compare with **aphotic zone**.

photon A discrete packet of light energy; a particle of light.

photoperiodism Any response by an organism to the relative lengths of day and night (i.e., photoperiod).

photophosporylation Production of ATP molecules using the energy released as light-excited electrons flow through an electron transport chain during photosynthesis. Involves generation of a proton-motive force during electron transport and its use to drive ATP synthesis.

photoreceptor A molecule, a cell, or an organ that is specialized to detect light.

photorespiration A series of light-driven chemical reactions that consumes oxygen and releases carbon dioxide, basically reversing photosynthesis. Usually occurs when there are high O_2 and low CO_2 concentrations inside plant cells, often in bright, hot, dry environments when stomata must be kept closed.

photoreversibility A change in conformation that occurs in certain plant pigments when they are exposed to their preferred wavelengths of light and that triggers responses by the plant.

photosynthesis The complex biological process that converts the energy of light into chemical energy stored in glucose and other organic molecules. Occurs in plants, algae, and some bacteria.

photosystem One of two types of units, consisting of a central reaction center surrounded by antenna complexes, that is responsible for the light-dependent reactions of photosynthesis.

photosystem I A photosystem that contains a pair of P700 chlorophyll molecules and uses absorbed light energy to produce NADPH.

photosystem II A photosystem that contains a pair of P680 chlorophyll molecules and uses absorbed light energy to split water into protons and oxygen and to produce ATP.

phototroph An organism that produces ATP through photosynthesis.

phototropins A class of plant photoreceptors that detect blue light and initiate phototropic responses.

phototropism Growth or movement of an organism in a particular direction in response to light.

phylogenetic species concept The definition of a species as the smallest monophyletic group in a phylogenetic tree. Compare with **biological** and **morphospecies concept**.

phylogenetic tree A diagram that depicts the evolutionary history of a group of species and the relationships among them.

phylogeny The evolutionary history of a group of organisms.

phylum (plural: phyla) In Linnaeus' system, a taxonomic category above the class level and below the kingdom level. In plants, sometimes called a *division*.

physical map A map of a chromosome that shows the number of base pairs between various genetic markers. Compare with **genetic map**.

physiology The study of how an organism's body functions.

phytoalexin Any small compound produced by a plant to combat an infection (usually a fungal infection).

phytochrome A specialized plant photoreceptor that exists in two shapes depending on the ratio of red to far-red light and is involved in the timing certain physiological processes, such as flowering, stem elongation, and germination.

pigment Any molecule that absorbs certain wavelengths of visible light and reflects or transmits other wavelengths.

piloting A type of navigation in which animals use familiar landmarks to find their way.

pinocytosis Uptake by a cell of extracellular fluid by pinching off the plasma membrane to form small membrane-bound vesicles; one type of endocytosis.

pioneering species Those species that appear first in recently disturbed areas.

pit In plants, a small hole in the secondary cell walls of tracheids that allow passage of water.

pitch The sensation produced by a particular frequency of sound. Low frequencies are perceived as low pitches; high frequencies, as high pitches.

pith In the shoot systems of plants, ground tissue located to the inside of the vascular bundles.

pituitary gland A small gland directly under the brain that is physically and functionally connected to the hypothalamus. Produces and secretes an array of hormones that affect many other glands and organs.

placenta A structure that forms in the pregnant uterus from maternal and fetal tissues. Exchanges nutrients and wastes between mother and fetus, anchors the fetus to the uterine wall, and produces some hormones. Occurs in most mammals and in a few other vertebrates.

placental mammals See **eutherians**.

plankton Any small organism that drifts near the surface of oceans or lakes and swims little if at all.

Plantae The monophyletic group that includes red, green, and glaucophyte algae and land plants.

plantlet A small plant, particularly one that forms on a parent plant via asexual reproduction and drops, becoming an independent individual.

plasma The non-cellular portion of blood.

plasma cell An effector B cell, which produces large quantities of antibodies. Is descended from an activated B cell that has interacted with antigen.

plasma membrane A membrane that surrounds a cell, separating it from the external environment and selectively regulating passage of molecules and ions into and out of the cell. Also called *cell membrane*.

plasmid A small, usually circular, supercoiled DNA molecule independent of the cell's main chromosome(s) in prokaryotes and some eukaryotes.

plasmodesmata (singular: plasmodesma) Physical connection between two plant cells, consisting of gaps in the cell walls through which the two cells' plasma membranes, cytoplasm, and smooth ER can connect directly. Functionally similar to gap junctions in animal cells.

plasmogamy Fusion of the cytoplasm of two individuals. Occurs in many fungi.

plastocyanin A small protein that shuttles electrons from photosystem II to photosystem I during photosynthesis.

plastoquinone (PQ) A nonprotein electron carrier in the chloroplast electron transport chain. Receives excited electrons from pheophytin and passes them to more electronegative molecules in the chain. Also carries protons to the lumen side of the thylakoid membrane, generating a proton-motive force.

platelet A small membrane-bound cell fragment in vertebrate blood that functions in blood clotting. Derived from large cells in the bone marrow.

pleiotropy (adjective: pleiotropic) The ability of a single gene to affect more than one phenotypic trait.

ploidy The number of complete chromosome sets present. *Haploid* refers to a ploidy of 1; *diploid*, a ploidy of 2; *triploid*, a ploidy of 3; and *tetraploid*, a ploidy of 4.

podium (plural: podia) See **tube foot**.

point mutation A mutation that results in a change in a single nucleotide pair in a DNA molecule.

polar (1) Asymmetrical or unidirectional. (2) Carrying a partial positive charge on one side of a molecule and a partial negative charge on the other. Polar molecules are generally hydrophilic.

polar bodies The tiny, nonfunctional cells produced during meiosis of a primary oocyte, due to most of the cytoplasm going to the ovum.

polar covalent bond A covalent bond in which electrons are shared unequally between atoms differing in electronegativity, resulting in the more electronegative atom having a partial negative charge and the other atom, a partial positive charge. Compare with **nonpolar covalent bond**.

polar microtubules Microtubules that form during mitosis and meiosis, and which extend from a spindle apparatus and overlap with each other in the middle of the cell.

polar nuclei In flowering plants, the nuclei in the female gametophyte that fuse with one sperm nucleus to produce the endosperm. Most species have two.

pollen grain In seed plants, a male gametophyte enclosed within a protective coat.

pollen tube In flowering plants, a structure that grows out of a pollen grain after it reaches the stigma, extends down the style, and through which two sperm cells are delivered to the ovule.

pollination The process by which pollen reaches the carpel of a flower (in flowering plants) or reaches the ovule directly (in conifers and their relatives).

poly(A) tail In eukaryotes, a sequence of 100–250 adenine nucleotides added to the 3′ end of newly transcribed messenger RNA molecules.

polygenic inheritance The inheritance patterns that result when many genes influence one trait.

polymer Any long molecule composed of small repeating units (monomers) bonded together. The main biological polymers are proteins, nucleic acids, and polysaccharides.

polymerase chain reaction (PCR) A laboratory technique for rapidly generating millions of identical copies of a specific stretch of DNA by incubating the original DNA sequence of interest with primers, nucleotides, and DNA polymerase.

polymerization (verb: polymerize) The process by which many identical or similar small molecules (monomers) are covalently bonded to form a large molecule (polymer).

polymorphism (adjective: polymorphic) (1) The occurrence of more than one allele at a certain genetic locus in a population. (2) The occurrence of more than two distinct phenotypes of a trait in a population.

polyp The immotile (sessile) stage in the life cycle of some cnidarians (e.g., jellyfish). Compare with **medusa**.

polypeptide A chain of 50 or more amino acids linked together by peptide bonds. Compare with **oligopeptide** and **protein**.

polyploidy (adjective: polyploid) The state of having more than two full sets of chromosomes.

polyribosome A structure consisting of one messenger RNA molecule along with many attached ribosomes and their growing peptide strands.

polysaccharide A linear or branched polymer consisting of many monosaccharides joined by glycosidic linkages. Carbohydrate polymers with relatively few residues often are called *oligosaccharides*.

polyspermy Fertilization of an egg by multiple sperm.

population A group of individuals of the same species living in the same geographic area at the same time.

population density The number of individuals of a population per unit area.

population dynamics Changes in the size and other characteristics of populations through time.

population ecology The study of how and why the number of individuals in a population changes over time.

population thinking The ability to analyze trait frequencies, event probabilities, and other attributes of populations of molecules, cells, or organisms.

pore In land plants, an opening in the epithelium that allows gas exchange.

positive control Of genes, when a regulatory protein triggers expression by binding to DNA on or near the gene.

positive feedback A physiological mechanism in which a change in some variable stimulates a response that increases the change. Relatively rare in organisms but is important in generation of the action potential. Compare with **negative feedback**.

positive pressure ventilation Ventilation of the lungs that is accomplished by "pushing" air into the lungs by positive pressure in the mouth. Compare with **negative pressure ventilation**.

positive-sense virus A virus whose genome contains the same sequences as the mRNA required to produce viral proteins. Compare with **ambisense virus** and **negative-sense virus**.

posterior Toward an animal's tail and away from its head. The opposite of anterior.

posterior pituitary The part of the pituitary gland that contains the ends of hypothalamic neurosecretory cells and from which oxytocin and antidiuretic hormone are secreted. Compare with **anterior pituitary**.

postsynaptic neuron A neuron that receives signals, usually via neurotransmitters, from another neuron at a synapse. Muscle and gland cells also may receive signals from presynaptic neurons.

post-translational control Regulation of gene expression by modification of proteins (e.g., addition of a phosphate group or sugar residues) after translation.

postzygotic isolation Reproductive isolation resulting from mechanisms that operate after mating of individuals of two different species occurs. The most common mechanisms are the death of hybrid embryos or reduced fitness of hybrids.

potential energy Energy stored in matter as a result of its position or molecular arrangement. Compare with **kinetic energy**.

prebiotic soup A hypothetical solution of sugars, amino acids, nitrogenous bases, and other building blocks of larger molecules that may have formed in shallow waters or deep-ocean vents of ancient Earth and given rise to larger biological molecules.

Precambrian The interval between the formation of the Earth, about 4.6 billion years ago, and the appearance of most animal groups about 542 million years ago. Unicellular organisms were dominant for most of this era, and oxygen was virtually absent for the first 2 billion years.

predation The killing and eating of one organism (the prey) by another (the predator).

predator Any organism that kills other organisms for food.

prediction A measurable or observable result of an experiment based on a particular hypothesis. A correct prediction provides support for the hypothesis being tested.

pressure potential (ψ_p) A component of the potential energy of water caused by physical pressures on a solution. In plant cells, it equals the wall pressure plus turgor pressure. Compare with **solute potential (ψ_S)**.

pressure-flow hypothesis The hypothesis that sugar movement through phloem tissue is due to differences in the turgor pressure of phloem sap.

presynaptic neuron A neuron that transmits signals, usually by releasing neurotransmitters, to another neuron or to an effector cell at a synapse.

prezygotic isolation Reproductive isolation resulting from any one of several mechanisms that prevent individuals of two different species from mating.

primary cell wall The outermost layer of a plant cell wall, made of cellulose fibers and gelatinous polysaccharides, that defines the shape of the cell and withstands the turgor pressure of the plasma membrane.

primary consumer An herbivore; an organism that eats plants, algae, or other primary producers. Compare with **secondary consumer**.

primary decomposer A decomposer (detritivore) that consumes detritus from plants.

primary growth In plants, an increase in the length of stems and roots due to the activity of apical meristems. Compare with **secondary growth**.

primary immune response An acquired immune response to a pathogen that the immune system has not encountered before. Compare with **secondary immune response**.

primary oocyte The large diploid cell in an ovarian follicle that can initiate meiosis to produce a haploid ovum.

primary producer Any organism that creates its own food by photosynthesis or from reduced inorganic compounds and that is a food source for other species in its ecosystem. Also called *autotroph*.

primary RNA transcript In eukaryotes, a newly transcribed messenger RNA molecule that has not yet been processed (i.e., it has not received a 5′ cap or poly(A) tail, and still contains introns). Also called *pre-mRNA*.

primary spermatocyte A diploid cell in the testis that can initiate meiosis I to produce two secondary spermatocytes.

primary structure The sequence of amino acids in a peptide or protein; also the sequence of nucleotides in a nucleic acid. Compare with **secondary, tertiary,** and **quaternary structure**.

primary succession The gradual colonization of a habitat of bare rock or gravel, usually after an environmental disturbance that removes all soil and previous organisms. Compare with **secondary succession**.

primase An enzyme that synthesizes a short stretch of RNA to use as a primer during DNA replication.

primates The lineage of mammals that includes prosimians (lemurs, lorises, etc.), monkeys, and great apes (including humans).

primer A short, single-stranded RNA molecule that base pairs with the 5′ end of a DNA template strand and is elongated by DNA polymerase during DNA replication.

prion An infectious form of a protein that is thought to cause disease by inducing the normal form to assume an abnormal three-dimensional structure. Cause of spongiform encephalopathies, such as mad cow disease.

probe A radioactively or chemically labeled single-stranded fragment of a known DNA or RNA sequence that can bind to and thus detect its complementary sequence in a sample containing many different sequences.

proboscis A long, narrow feeding appendage through which food can be obtained.

procambium A group of cells in the center of a young plant embryo that gives rise to the vascular tissue.

product Any of the final materials formed in a chemical reaction.

productivity The total amount of carbon fixed by photosynthesis per unit area per year.

progesterone A steroid hormone produced in the ovaries and secreted by the corpus luteum after ovulation; causes the uterine lining to thicken.

programmed cell death See apoptosis.

prokaryote A member of the domain Bacteria or Archaea; a unicellular organism lacking a nucleus and containing relatively few organelles or cytoskeletal components. Compare with **eukaryote**.

prolactin A peptide hormone, produced and secreted by the anterior pituitary, that promotes milk production in female mammals and has a variety of effects on parental behavior and seasonal reproduction in other vertebrates.

prometaphase A stage in mitosis or meiosis during which the nuclear envelope breaks down and kinetochore microtubles attach to chromatids.

promoter A short nucleotide sequence in DNA that binds RNA polymerase, enabling transcription to begin. In prokaryotic DNA, a single promoter often is associated with several contiguous genes. In eukaryotic DNA, each gene generally has its own promoter.

promoter-proximal elements In eukaryotes, regulatory sequences in DNA that are close to a promoter and that can bind regulatory transcription factors.

proofreading The process by which a DNA polymerase recognizes and removes a wrong base added during DNA replication and then continues synthesis.

prophase The first stage in mitosis or meiosis during which chromosomes become visible and the spindle apparatus forms. Synapsis and crossing over occur during prophase of meiosis I.

prosimians One of the two major lineages of primates, including lemurs, tarsiers, pottos, and lorises. Compare with **anthropoids**.

prostate gland A gland in male mammals that surrounds the base of the urethra and secretes a fluid that is a component of semen.

protease An enzyme that can degrade proteins by cleaving the peptide bonds between amino acid residues.

proteasome A multi-molecular machine that destroys proteins that have been bound to ubiquitin.

protein A macromolecule consisting of one or more polypeptide chains composed of 50 or more amino acids linked together. Each protein has a unique sequence of amino acids and, in its native state, a characteristic three-dimensional shape.

protein kinase An enzyme that catalyzes the addition of a phosphate group to another protein, typically activating or inactivating the substrate protein.

proteinase inhibitors Defense compounds, produced by plants, that induce illness in herbivores by inhibiting digestive enzymes.

proteome The complete set of proteins produced by a particular cell type.

proteomics The systematic study of the interactions, localization, functions, regulation, and other features of the full protein set (proteome) in a particular cell type.

protist Any eukaryote that is not a green plant, animal, or fungus. Protists are a diverse paraphyletic group. Most are unicellular, but some are multicellular or form aggregations called colonies.

protoderm The exterior layer of a young plant embryo that gives rise to the epidermis.

proton pump A membrane protein that can hydrolyze ATP to power active transport of protons (H^+ ions) across a plasma membrane against an electrochemical gradient. Also called H^+-ATPase.

proton-motive force The combined effect of a proton gradient and an electric potential gradient across a membrane, which can drive protons across the membrane. Used by mitochondria and chloroplasts to power ATP synthesis via the mechanism of chemiosmosis.

proto-oncogene Any gene that normally encourages cell division in a regulated manner, typically by triggering specific phases in the cell cycle. Mutation may convert it into an oncogene.

protostomes A major lineage of animals that share a pattern of embryological development, including formation of the mouth earlier than the anus, and formation of the coelom by splitting of a block of mesoderm. Includes arthropods, mollusks, and annelids. Compare with **deuterostomes**.

proximal tubule In the vertebrate kidney, the convoluted section of a nephron into which filtrate moves from Bowman's capsule. Involved in the largely unregulated reabsorption of electrolytes, nutrients, and water. Compare with **distal tubule**.

proximate causation In biology, the immediate, mechanistic cause of a phenomenon (how it happens), as opposed to why it evolved. Also called *proximate explanation*. Compare with **ultimate causation**.

pseudogene A DNA sequence that closely resembles a functional gene but is not transcribed. Thought to have arisen by duplication of the functional gene followed by inactivation due to mutation.

pseudopodium (plural: pseudopodia) A temporary bulge-like extension of certain cells used in cell crawling and ingestion of food.

puberty The various physical and emotional changes that an immature animal undergoes leading to reproductive maturity. Also the period when such changes occur.

pulmonary artery A short, thick-walled artery that carries oxygen-poor blood from the heart to the lungs.

pulmonary circulation The part of the circulatory system that sends oxygen-poor blood to the lungs. Is separate from the rest of the circulatory system (the systemic circulation) in mammals and birds.

pulmonary vein A short, thin-walled vein that carries oxygen-rich blood from the lungs to the heart.

pulse-chase experiment A type of experiment in which a population of cells or molecules at a particular moment in time is marked by means of a labeled molecule and then their fate is followed over time.

pump Any membrane protein that can hydrolyze ATP to power active transport of a specific ion or small molecule across a plasma membrane against its electrochemical gradient. See **proton pump**.

Punnett square A diagram that depicts the genotypes and phenotypes that should appear in offspring of a certain cross.

pupa (plural: pupae) A metamorphosing insect that is enclosed in a protective case.

pupil The hole in the center of the iris through which light enters a vertebrate or cephalopod eye.

pure line In animal or plant breeding, a strain of individuals that produce offspring identical to themselves when self-pollinated or crossed to another member of the same population. Pure lines are homozygous for most, if not all, genetic loci.

purifying selection Selection that lowers the frequency or even eliminates deleterious alleles.

purines A class of small, nitrogen-containing, double-ringed bases (guanine, adenine) found in nucleotides. Compare with **pyrimidines**.

pyrimidines A class of small, nitrogen-containing, single-ringed bases (cytosine, uracil, thymine) found in nucleotides. Compare with **purines**.

pyruvate dehydrogenase A large enzyme complex, located in the inner mitochondrial membrane, that is responsible for conversion of pyruvate to acetyl CoA during cellular respiration.

quantitative trait A heritable feature that exhibits phenotypic variation along a smooth, continuous scale of measurement (e.g., human height), rather than the distinct forms characteristic of discrete traits.

quaternary structure The overall three-dimensional shape of a protein containing two or more polypeptide chains (subunits); determined by the number, relative positions, and interactions of the subunits. Compare with **primary**, **secondary**, and **tertiary structure**.

quorum sensing Cell-cell signaling in bacteria, in which cells of the same species communicate via chemical signals. It is often observed that cell activity changes dramatically when the population reaches a threshold size, or quorum.

radial symmetry An animal body pattern in which there are least two planes of symmetry. Typically, the body is in the form of a cylinder or disk, with body parts radiating from a central hub. Compare with **bilateral symmetry**.

radiation Transfer of heat between two bodies that are not in direct physical contact. More generally, the emission of electromagnetic energy of any wavelength.

radicle The root of a plant embryo.

radula A rasping feeding appendage in gastropods (snails, slugs).

rain shadow The dry region on the side of a mountain range away from the prevailing wind.

range The geographic distribution of a species.

Ras protein A type of G protein that is activated by binding of signaling molecules to receptor tyrosine kinases and then initiates a phosphorylation cascade, culminating in a cell response.

rays In plant shoot systems with secondary growth, a lateral array of parenchyma cells produced by vascular cambium.

Rb protein A tumor-suppressor protein that helps regulate progression of a cell from the G_1 phase to the S phase of the cell cycle. Defects in Rb protein are found in many types of cancer.

reactant Any of the starting materials in a chemical reaction.

reaction center Centrally located component of a photosystem containing proteins and a pair of specialized chlorophyll molecules. Is surrounded by antenna complexes and receives excited electrons from them.

reactive oxygen intermediates (ROIs) Highly reactive oxygen-containing compounds that are used in plant and animal cells to kill infected cells and for other purposes.

reading frame The division of a sequence of DNA or RNA into a particular series of three-nucleotide codons. There are three possible reading frames for any sequence.

realized niche The ecological niche that a species occupies in the presence of competitors. Compare with **fundamental niche**.

receptor tyrosine kinase (RTK) Any of a class of cell-surface signal receptors that undergo phosphorylation after binding a signaling molecule. The activated, phosphorylated receptor then triggers a signal-transduction pathway inside the cell.

receptor-mediated endocytosis Uptake by a cell of certain extracellular macromolecules, bound to specific receptors in the plasma membrane, by pinching off the membrane to form small membrane-bound vesicles.

recessive Referring to an allele whose phenotypic effect is observed only in homozygous individuals. Compare with **dominant**.

reciprocal altruism Altruistic behavior that is exchanged between a pair of individuals at different points in time (i.e., sometimes individual A helps individual B, and sometimes B helps A).

reciprocal cross A breeding experiment in which the mother's and father's phenotypes are the reverse of that examined in a previous breeding experiment.

recombinant DNA technology A variety of techniques for isolating specific DNA fragments and introducing them into different regions of DNA and/or a different host organism.

recombinant Possessing a new combination of alleles. May refer to a single chromosome or DNA molecule, or to an entire organism.

recombination See **genetic recombination**.

rectal gland A salt-excreting gland in the digestive system of sharks, skates, and rays.

rectum The last portion of the digestive tract where feces are held until they are expelled.

red blood cell A hemoglobin-containing cell that circulates in the blood and delivers oxygen from the lungs to the tissues.

redox reaction Any chemical reaction that involves the transfer of one or more electrons from one reactant to another. Also called *reduction-oxidation reaction*.

reduction An atom's gain of electrons during a redox reaction, either by acceptance of an electron from another atom or by the electrons in covalent bonds moving closer to the atomic nucleus.

reduction-oxidation reaction See **redox reaction**.

reflex An involuntary response to environmental stimulation. May involve the brain (e.g., conditioned reflex) or not (e.g., spinal reflex).

refractory No longer responding to stimuli that previously elicited a response. For example, the tendency of voltage-gated sodium channels to remain closed immediately after an action potential.

regulatory genes DNA sequences that code for regulatory proteins—products that alter gene expression.

regulatory sequence, DNA Any segment of DNA that is involved in controlling transcription of a specific gene by binding certain proteins.

regulatory site A site on an enzyme to which a regulatory molecule can bind and affect the enzyme's activity; separate from the active site where catalysis occurs.

regulatory transcription factor General term for proteins that bind to DNA regulatory sequences (eukaryotic enhancers, silencers, and promoter-proximal elements), but not to the promoter itself, leading to an increase or decrease in transcription of specific genes. Compare with **basal transcription factor**.

reinforcement In evolutionary biology, the natural selection for traits that prevent interbreeding between recently diverged species.

release factors Proteins that can trigger termination of RNA translation when a ribosome reaches a stop codon.

renal corpuscle In the vertebrate kidney, the ball-like structure at the beginning of a nephron, consisting of a glomerulus and the surrounding Bowman's capsule. Acts as a filtration device.

replacement mutation See **missense mutation**.

replacement rate The number of offspring each female must produce over her entire life to "replace" herself and her mate, resulting in zero population growth. The actual number is slightly more than 2 because some offspring die before reproducing.

replica plating A method of identifying bacterial colonies that have certain mutations by transferring cells from each colony on a master plate to a second (replica) plate and observing their growth when exposed to different conditions.

replication fork The Y-shaped site at which a double-stranded molecule of DNA is separated into two single strands for replication.

replisome The multi-molecular machine that copies DNA; includes DNA polymerase, helicase, primase, and other enzymes.

repolarization Return to a normal membrane potential after it has changed; a normal phase in an action potential.

repressor Any regulatory protein that inhibits transcription.

reptiles One of the two lineages of amniotes (vertebrates that produce amniotic eggs) distinguished by adaptations for reproduction on land. Includes turtles, snakes and lizards, crocodiles and alligators, and birds. Except for birds, all are ectotherms.

resilience, community A measure of how quickly a community recovers following a disturbance.

resistance, community A measure of how much a community is affected by a disturbance.

resistance (R) genes Genes in plants encoding proteins involved in sensing the presence of pathogens and mounting a defensive response. Compare with **avirulence (avr) genes**.

respiratory system The collection of cells, tissues, and organs responsible for gas exchange between an animal and its environment.

resting potential The membrane potential of a cell in its resting, or normal, state.

restriction endonucleases Bacterial enzymes that cut DNA at a specific base-pair sequence (restriction site). Also called *restriction enzymes*.

retina A thin layer of light-sensitive cells (rods and cones) and neurons at the back of a camera-type eye, such as that of cephalopods and vertebrates.

retinal A light-absorbing pigment, derived from vitamin A, that is linked to the protein opsin in rods and cones of the vertebrate eye.

retrovirus A virus with an RNA genome that reproduces by transcribing its RNA into a DNA sequence and then inserting that DNA into the host's genome for replication.

reverse transcriptase A enzyme of retroviruses (RNA viruses) that can synthesize double-stranded DNA from a single-stranded RNA template.

rhizobia (singular: rhizobium) Members of the bacterial genus *Rhizobia*; nitrogen-fixing bacteria that live in root nodules of members of the pea family (legumes).

rhizoid The hairlike structure that anchors a bryophyte (non-vascular plant) to the substrate.

rhizome A modified stem that runs horizontally underground and produces new plants at the nodes (a form of asexual reproduction). Compare with **stolon**.

rhodopsin A transmembrane complex that is instrumental in detection of light by rods and cones of the vertebrate eye. Is composed of the transmembrane protein opsin covalently linked to retinal, a light-absorbing pigment.

ribonucleic acid (RNA) A nucleic acid composed of ribonucleotides that usually is single stranded and functions as structural components of ribosomes (rRNA), transporters of amino acids (tRNA), and translators of the message of the DNA code (mRNA).

ribonucleotide See **nucleotide**.

ribosomal RNA (rRNA) A RNA molecule that forms part of the structure of a ribosome.

ribosome A large complex structure that synthesizes proteins by using the genetic information encoded in messenger RNA strands. Consists of two subunits, each composed of ribosomal RNA and proteins.

ribosome binding site In a bacterial mRNA molecule, the sequence just upstream of the start codon to which a ribosome binds to initiate translation. Also called the *Shine-Dalgarno sequence*.

ribozyme Any RNA molecule that can act as a catalyst, that is, speed up a chemical reaction.

ribulose bisphosphate (RuBP) A five-carbon compound that combines with CO_2 in the first step of the Calvin cycle during photosynthesis.

RNA interference (RNAi) Degradation of an mRNA molecule or inhibition of its translation following its binding by a short RNA (microRNA) whose sequence is complementary to a portion of the mRNA.

RNA polymerase One of a class of enzymes that catalyze synthesis of RNA from ribonucleotides using a DNA template. Also called *RNA pol*.

RNA processing In eukaryotes, the changes that a primary RNA transcript undergoes in the nucleus to become a mature mRNA molecule, which is exported to the cytoplasm. Includes the addition of a 5′ cap and poly(A) tail and splicing to remove introns.

RNA replicase A viral enzyme that can synthesize RNA from an RNA template.

RNA See **ribonucleic acid**.

rod cell A photoreceptor cell with a rod-shaped outer portion that is particularly sensitive to dim light, but not used to distinguish colors. Also called simply *rod*. Compare with **cone cell**.

root (1) An underground part of a plant that anchors the plant and absorbs water and nutrients. (2) In a phylogenetic tree, the bottom, most ancient node.

root apical meristem (RAM) A group of undifferentiated plant cells at the tip of a plant root that can differentiate into mature root tissue.

root cap A small group of cells that covers and protects the tip of a plant root. Senses gravity and determines the direction of root growth.

root hair A long, thin outgrowth of the epidermal cells of plant roots, providing increased surface area for absorption of water and nutrients.

root pressure Positive (upward) pressure of xylem sap in the vascular tissue of roots. Is generated during the night as a result of the accumulation of ions from the soil and subsequent osmotic movement of water into the xylem.

root system The belowground part of a plant.

rough endoplasmic reticulum (rough ER) The portion of the endoplasmic reticulum that is dotted with ribosomes. Involved in synthesis of plasma membrane proteins, secreted proteins, and proteins localized to the ER, Golgi apparatus, and lysosomes. Compare with **smooth endoplasmic reticulum**.

roundworms Members of the phylum Nematoda. Distinguished by an unsegmented body with a pseudocoelom and no appendages. Roundworms belong to the ecdysozoan branch of the protostomes. Also called *nematodes*.

rRNA See **ribosomal RNA**.

rubisco The enzyme that catalyzes the first step of the Calvin cycle during photosynthesis: the addition of a molecule of CO_2 to ribulose bisphosphate.

ruminants A group of hoofed mammals (e.g., cattle, sheep, deer) that have a four-chambered stomach specialized for digestion of plant cellulose. Ruminants regurgitate the cud, a mixture of partially digested food and cellulose-digesting bacteria, from the largest chamber (the rumen) for further chewing.

salivary glands Vertebrate glands that secrete saliva (a mixture of water, mucus-forming glycoproteins, and digestive enzymes) into the mouth.

sampling error The accidental selection of a nonrepresentative sample from some larger population, due to chance.

saprophyte An organism that feeds primarily on dead plant material.

sapwood The younger xylem in the outer layer of wood of a stem or root, functioning primarily in water transport.

sarcomere The repeating contractile unit of a skeletal muscle cell; the portion of a myofibril located between adjacent Z disks.

sarcoplasmic reticulum Sheets of smooth endoplasmic reticulum in a muscle cell. Contains high concentrations of calcium, which can be released into the cytoplasm to trigger contraction.

saturated Referring to fats and fatty acids in which all the carbon-carbon bonds are single bonds. Such fats have relatively high melting points. Compare with **unsaturated**.

scanning electron microscope (SEM) A microscope that produces images of the surfaces of objects by reflecting electrons from a specimen coated with a layer of metal atoms. Compare with **transmission electron microscope**.

scarify To scrape, rasp, cut, or otherwise damage the coat of a seed. Necessary in some species to trigger germination.

Schwann cell A type of glial cell that wraps around axons of some neurons outside the brain and spinal cord, forming a myelin sheath that provides electrical insulation. Compare with **oligodendrocyte**.

scientific name The unique, two-part name given to each species, with a genus name followed by a species name—as in *Homo sapiens*. Scientific names are always italicized, and are also known as Latin names.

sclereid In plants, a type of sclerenchyma cell that usually functions in protection, such as in seed coats and nutshells. Compare with **fiber**.

sclerenchyma cell In plants, a cell that has a thick secondary cell wall and provides support; typically contains the tough structural polymer lignin and usually is dead at maturity. Includes fibers and sclereids. Compare with **collenchyma cell** and **parenchyma cell**.

screen See **genetic screen**.

scrotum A sac of skin, containing the testes, suspended just outside the abdominal body cavity of many male mammals.

second law of thermodynamics The principle of physics that the entropy of the universe or any closed system increases during any spontaneous process.

second messenger A nonprotein signaling molecule produced or activated inside a cell in response to stimulation at the cell surface. Commonly used to relay the message of a hormone or other extracellular signaling molecule.

secondary active transport Transport of an ion or molecule against its electrochemical gradient, in company with an ion or molecule being transported with its electrochemical gradient. Also called *cotransport*.

secondary active transporter A transmembrane protein that facilitates diffusion of an ion down its previously established electrochemical gradient and uses the energy of that process to transport some other substance, in the same or opposite direction, *against* its concentration gradient. Also called *cotransporter*. See also **antiporter** and **symporter**.

secondary cell wall The inner layer of a plant cell wall formed by certain cells as they mature. Provides support or protection.

secondary consumer A carnivore; an organism that eats herbivores. Compare with **primary consumer**.

secondary growth In plants, an increase in the width of stems and roots due to the activity of lateral meristems. Compare with **primary growth**.

secondary immune response The acquired immune response to a pathogen that the immune system has encountered before. Compare with **primary immune response**.

secondary metabolites Molecules that are closely related to compounds in key synthetic pathways, and that often function in defense.

secondary spermatocyte A cell produced by meiosis I of a primary spermatocyte in the testis. Can undergo meiosis II to produce spermatids.

secondary structure In proteins, localized folding of a polypeptide chain into regular structures (e.g., α-helix and β-pleated sheet) stabilized by hydrogen bonding between atoms of the backbone. In nucleic acids, elements of structure (e.g., helices and hairpins) stabilized by hydrogen bonding and other interactions between complementary bases. Compare with **primary, tertiary,** and **quaternary structure**.

secondary succession Gradual colonization of a habitat after an environmental disturbance (e.g., fire, windstorm, logging) that removes some or all previous organisms but leaves the soil intact. Compare with **primary succession**.

second-male advantage The reproductive advantage of a male who mates with a female last, after other males have mated with her.

secretin A peptide hormone produced by cells in the small intestine in response to the arrival of food from the stomach. Stimulates secretion of bicarbonate (HCO_3^-) from the pancreas.

sedimentary rock A type of rock formed by gradual accumulation of sediment, as in riverbeds and on the ocean floor. Most fossils are found in sedimentary rocks.

seed A plant reproductive structure consisting of an embryo, associated nutritive tissue (endosperm), and an outer protective layer (seed coat). In angiosperms, develops from the fertilized ovule of a flower.

seed bank A repository where seeds, representing many different varieties of domestic crops or other species, are preserved.

seed coat A protective layer around a seed that encases both the embryo and the endosperm.

segment A well-defined region of the body along the anterior-posterior body axis, containing similar structures as other, nearby segments.

segmentation Division of the body or a part of it into a series of similar structures; exemplified by the body segments of insects and worms and by the somites of vertebrates.

segmentation genes A group of genes that affect body segmentation in embryonic development. Includes gap genes, pair-rule genes, and segment polarity genes.

segregation, principle of The concept that each pair of hereditary elements (alleles of the same gene) separate from each other during the formation of offspring (i.e., during meiosis). One of Mendel's two principles of genetics.

selective adhesion The tendency of cells of one tissue type to adhere to other cells of the same type.

selective permeability The property of a membrane that allows some substances to diffuse across it much more readily than other substances.

selectively permeable membrane Any membrane across which some solutes can move more readily than others.

self Property of a molecule or cell such that immune system cells do not attack it, due to certain molecular similarities to other body cells.

self molecule A molecule that is synthesized by an organism and is a normal part of its cells and/or body; as opposed to non-self or foreign molecules.

self-fertilization In plants, the fusion of two gametes from the same individual to form a diploid offspring. Also called *selfing*.

semen The combination of sperm and accessory fluids that is released by male mammals and reptiles during ejaculation.

semiconservative replication The mechanism of replication used by cells to copy DNA. Results in each daughter DNA molecule containing one old strand and one new strand.

seminal vesicles In male mammals, paired reproductive glands that secrete a sugar-containing fluid into semen, which provides energy for sperm movement. In other vertebrates and invertebrates, often stores sperm.

senescence The process of aging.

sensor Any cell, organ, or structure with which an animal can sense some aspect of the external or internal environment. Usually functions, along with an integrator and effector, as part of a homeostatic system.

sensory neuron A nerve cell that carries signals from sensory receptors to the central nervous system. Compare with **interneuron** and **motor neuron**.

sepal One of the protective leaflike organs enclosing a flower bud and later supporting the blooming flower.

septum (plural: septa) Any wall-like structure. In fungi, septa divide the filaments (hyphae) of mycelia into cell-like compartments.

serotonin A neurotransmitter involved in many brain functions, including sleep, pleasure, and mood.

serum The liquid that remains when cells and clot material are removed from clotted blood. Contains water, dissolved gases, growth factors, nutrients, and other soluble substances. Compare with **plasma**.

sessile Permanently attached to a substrate; not capable of moving to another location.

set point A normal or target value for a regulated internal variable, such as body heat or blood pH.

severe combined immunodeficiency disease (SCID) A human disease characterized by an extremely high vulnerability to infectious disease. Caused by a genetic defect in the immune system.

sex chromosome Any chromosome carrying genes involved in determining the sex of an individual. Compare with **autosome**.

sex-linked inheritance Inheritance patterns observed in genes carried on sex chromosomes, so females and males have different numbers of alleles of a gene and may pass its trait only to one sex of offspring. Also called *sex-linkage*.

sexual dimorphism Any trait that differs between males and females.

sexual reproduction Any form of reproduction in which genes from two parents are combined via fusion of gametes, producing offspring that are genetically distinct from both parents. Compare with **asexual reproduction**.

sexual selection A pattern of natural selection that favors individuals with traits that increase their ability to obtain mates. Acts more strongly on males than females.

shell A hard protective outer structure.

Shine-Dalgarno sequence See **ribosome binding sequence**.

shoot apical meristem (SAM) A group of undifferentiated plant cells at the tip of a plant stem that can differentiate into mature shoot tissues.

shoot The combination of hypocotyl and cotyledons in a plant embryo, which will become the aboveground portions of the body.

shoot system The aboveground part of a plant comprising stems, leaves, and flowers (in angiosperms).

short interspersed nuclear elements See **SINEs**.

short tandem repeats (STRs) Relatively short DNA sequences that are repeated, one after another, down the length of a chromosome. The two major types are microsatellites and minisatellites.

short-day plant A plant that blooms in response to long nights (usually in late summer or fall in the northern hemisphere). Compare with **day-neutral** and **long-day plant**.

shotgun sequencing A method of sequencing genomes that is based on breaking the genome into small pieces, sequencing each piece separately, and then figuring out how the pieces are connected.

sieve plate In plants, a pore-containing structure at one end of a sieve-tube member in phloem.

sieve-tube member In plants, an elongated sugar-conducting cell in phloem that has sieve plates at both ends, allowing sap to flow to adjacent cells.

sign stimulus A simple stimulus that elicits an invariant, stereotyped behavioral response (fixed action pattern) from an animal. Also called a *releaser*.

signal In behavioral ecology, any information-containing behavior.

signal receptor Any cellular protein that binds to a particular signaling molecule (e.g., a hormone or neurotransmitter) and triggers a response by the cell. Receptors for water-soluble signals are transmembrane proteins in the plasma membrane; those for many lipid-soluble signals (e.g., steroid hormones) are located inside the cell.

signal recognition particle (SRP) A RNA-protein complex that binds to the ER signal sequence in a polypeptide as it emerges from a ribosome and transports the ribosome-polypeptide complex to the ER membrane where synthesis of the polypeptide is completed.

signal transduction cascade See **phosphorylation cascade**.

signal transduction The process by which a stimulus (e.g., a hormone, a neurotransmitter, or sensory information) outside a cell is amplified and converted into a response by the cell. Usually involves a specific sequence of molecular events, or signal transduction pathway.

silencer A regulatory sequence in eukaryotic DNA to which repressor proteins can bind, inhibiting transcription of certain genes.

silent mutation A mutation that does not detectably affect the phenotype of the organism.

simple eye An eye with only one light-collecting apparatus (e.g., one lens), as in vertebrates. Compare with **compound eye**.

simple fruit A fruit (e.g., apricot) that develops from a single flower that has a single carpel or several fused carpels. Compare with **aggregate** and **multiple fruit**.

simple sequence repeat See **microsatellite**.

SINEs (short interspersed nuclear elements) The second most abundant class of transposable elements in human genomes; can create copies of itself and insert them elsewhere in the genome. Compare with **LINEs**.

single nucleotide polymorphism (SNP) A site on a chromosome where individuals in a population have different nucleotides. Can be used as a genetic marker to help track the inheritance of nearby genes.

single-strand DNA-binding proteins (SSBPs) A class of proteins that attach to separated strands of DNA during replication or transcription, preventing them from re-forming a double helix.

sink Any tissue, site, or location where an element or a molecule is consumed or taken out of circulation (e.g., in plants, a tissue where sugar exits the phloem). Compare with **source**.

sinoatrial (SA) node A cluster of cardiac muscle cells, in the right atrium of the vertebrate heart, that initiates the heartbeat and determines the heart rate. Compare with **atrioventricular (AV) node**.

siphon A tubelike appendage of many mollusks, that is often used for feeding or propulsion.

sister chromatids The paired strands of a recently replicated chromosome, which are connected at the centromere and eventually separate during anaphase of mitosis and meiosis II. Compare with **non-sister chromatids**.

sister species Closely related species, which occupy adjacent branches in a phylogenetic tree.

skeletal muscle The muscle tissue attached to the bones of the vertebrate skeleton. Consists of long, unbranched muscle fibers with a characteristic striped (striated) appearance; controlled voluntarily. Also called *striated muscle*. Compare with **cardiac** and **smooth muscle**.

sliding-filament model The hypothesis that thin (actin) filaments and thick (myosin) filaments slide past each other, thereby shortening the sarcomere. Shortening of all the sarcomeres in a myofibril results in contraction of the entire myofibril.

small intestine The portion of the digestive tract between the stomach and the large intestine. The site of the final stages of digestion and of most nutrient absorption.

small nuclear ribonucleoproteins See **snRNPs**.

smooth endoplasmic reticulum (smooth ER) The portion of the endoplasmic reticulum that does not have ribosomes attached to it. Involved in synthesis and secretion of lipids. Compare with **rough endoplasmic reticulum**.

smooth muscle The unstriated muscle tissue that lines the intestine, blood vessels, and some other organs. Consists of tapered, unbranched cells that can sustain long contractions. Not voluntarily controlled. Compare with **cardiac** and **skeletal muscle**.

snRNPs (small nuclear ribonucleoproteins) Complexes of proteins and small RNA molecules that function in splicing (removal of introns from primary RNA transcripts) as components of spliceosomes.

sodium-potassium pump A transmembrane protein that uses the energy of ATP to move sodium ions out of the cell and potassium ions in. Also called Na^+/K^+-ATPase.

soil organic matter Organic (carbon-containing) compounds found in soil.

solute Any substance that is dissolved in a liquid.

solute potential (ψ_S) A component of the potential energy of water caused by a difference in solute concentrations at two locations. Also called *osmotic potential*. Compare with **pressure potential (ψ_P)**.

solution A liquid containing one or more dissolved solids or gases in a homogeneous mixture.

solvent Any liquid in which one or more solids or gases can dissolve.

soma See **cell body**.

somatic cell Any type of cell in a multicellular organism except eggs, sperm, and their precursor cells. Also called *body cells*.

somatic hypermutation Mutations that occur in the immunoglobulin genes of the immune system's memory cells, resulting in novel variation in the receptors that bind to antigens.

somatic nervous system The part of the peripheral nervous system (outside the brain and spinal cord) that controls skeletal muscles and is under voluntary control. Compare with **autonomic nervous system**.

somatostatin A hormone secreted by the pancreas and hypothalamus that inhibits the release of several other hormones.

somites Paired blocks of mesoderm on both sides of the developing spinal cord in a vertebrate embryo. Give rise to muscle tissue, vertebrae, ribs, limbs, etc.

sori In ferns, a cluster of spore-producing structures (sporangia).

source Any tissue, site, or location where a substance is produced or enters circulation (e.g., in plants, the tissue where sugar enters the phloem). Compare with **sink**.

space-filling model A representation of a molecule where atoms are shown as balls—color-coded and scaled to indicate the atom's identify—attached to each other in the correct geometry.

speciation The evolution of two or more distinct species from a single ancestral species.

species A distinct, identifiable group of populations that is thought to be evolutionarily independent of other populations and whose members can interbreed. Generally distinct from other species in appearance, behavior, habitat, ecology, genetic characteristics, etc.

species diversity The variety and relative abundance of the species present in a given ecological community.

species richness The number of species present in a given ecological community.

species–area relationship The mathematical relationship between the area of a certain habitat and the number of species that it can support.

specific heat The amount of energy required to raise the temperature of 1 gram of a substance by 1°C; a measure of the capacity of a substance to absorb energy.

sperm A mature male gamete; smaller and more mobile than the female gamete.

sperm competition Competition to fertilize eggs between the sperm of different males, inside the same female.

spermatid An immature sperm cell.

spermatogenesis The production of sperm. Occurs continuously in a testis.

spermatogonia (singular: spermatogonium) The diploid cells in a testis that can give rise to primary spermatocytes.

spermatophore A gelatinous package of sperm cells that is produced by males of species that have internal fertilization without copulation.

sphincter A muscular valve that can close off a tube, as in a blood vessel or a part of the digestive tract.

spicule Stiff spike of silica or calcium carbonate found in the body of many sponges.

spindle apparatus The array of microtubules responsible for contacting and moving chromosomes during mitosis and meiosis; includes kinetochore microtubules and polar microtubules.

spines In plants, modified leaves that are stiff and sharp and that function in defense.

spiracle In insects, a small opening that connects air-filled tracheae to the external environment, allowing for gas exchange.

spleen A dark red organ, found near the stomach of most vertebrates, that filters blood, stores extra red blood cells in case of emergency, and plays a role in immunity.

spliceosome In eukaryotes, a large, complex assembly of snRNPs (small nuclear ribonucleoproteins) that catalyzes removal of introns from primary RNA transcripts.

splicing The process by which introns are removed from primary RNA transcripts and the remaining exons are connected together.

sporangium (plural: sporangia) A spore-producing structure found in seed plants, some protists, and some fungi (e.g., chytrids).

spore (1) In bacteria, a dormant form that generally is resistant to extreme conditions. (2) In eukaryotes, a single cell produced by mitosis or meiosis (not by fusion of gametes) that is capable of developing into an adult organism.

sporophyte In organisms undergoing alternation of generations, the multicellular diploid form that arises from two fused gametes and produces haploid spores. Compare with **gametophyte**.

sporopollenin A watertight material that encases spores and pollen of modern land plants.

stabilizing selection A pattern of natural selection that favors phenotypes near the middle of the range of phenotypic variation. Reduces overall genetic variation in a population. Compare with **disruptive selection**.

stamen The male reproductive structure of a flower. Consists of an anther, in which pollen grains are produced, and a filament, which supports the anther. Compare with **carpel**.

standing defense See **constitutive defense**.

stapes The last of three small bones (ossicles) in the middle ear of vertebrates. Receives vibrations from the tympanic membrane and by vibrating against the oval window passes them to the cochlea.

starch A mixture of two storage polysaccharides, amylose and amylopectin, both formed from α-glucose monomers. Amylopectin is branched, and amylose is unbranched. The major form of stored carbohydrate in plants.

start codon The AUG triplet in mRNA at which protein synthesis begins; codes for the amino acid methionine.

statocyst A sensory organ of many arthropods that detects the animal's orientation in space (i.e., whether the animal is flipped upside down).

statolith A tiny stone or dense particle found in specialized gravity-sensing organs in some animals such as lobsters.

statolith hypothesis The hypothesis that amyloplasts (dense, starch-storing plant organelles) serve as statoliths in gravity detection by plants.

STATs See **signal transducers and activators of transcription**.

stem cell Any relatively undifferentiated cell that can divide to produce daughter cells identical to itself or more specialized daughter cells, which differentiate further into specific cell types.

stems Vertical, aboveground structures that make up the shoot system of plants.

stereocilium (plural: stereocilia) One of many stiff outgrowths from the surface of a hair cell that are involved in detection of sound by terrestrial vertebrates or of waterborne vibrations by fishes.

steroid A class of lipid with a characteristic four-ring structure.

steroid-hormone receptor One of a family of intracellular receptors that bind to various steroid hormones, forming a hormone-receptor complex that acts as a regulatory transcription factor and activates transcription of specific target genes.

sticky end The short, single-stranded ends of a DNA molecule cut by a restriction endonuclease. Tend to form hydrogen bonds with other sticky ends that have complementary sequences.

stigma The moist tip at the end of a flower carpel to which pollen grains adhere.

stolon A modified stem that runs horizontally over the soil surface and produces new plants at the nodes (a form of asexual reproduction). Compare with **rhizome**.

stoma (plural: stomata) Generally, a pore or opening. In plants, a microscopic pore on the surface of a leaf or stem through which gas exchange occurs.

stomach A tough, muscular pouch in the vertebrate digestive tract between the esophagus and small intestine. Physically breaks up food and begins digestion of proteins.

stop codon One of three mRNA triplets (UAG, UGA, or UAA) that cause termination of protein synthesis. Also called a *termination codon*.

strain A population of genetically similar or identical individuals.

stream A body of water that moves constantly in one direction.

striated muscle See **skeletal muscle**.

stroma The fluid matrix of a chloroplast in which the thylakoids are embedded. Site where the Calvin cycle reactions occur.

structural formula A two-dimensional notation in which the chemical symbols for the constituent atoms are joined by straight lines representing single (—), double (=), or triple (≡) covalent bonds. Compare with **molecular formula**.

structural gene A stretch of DNA that codes for a functional protein or functional RNA molecule, not including any regulatory sequences (e.g., a promoter, enhancer).

structural homology Similarities in organismal structures (e.g., limbs, shells, flowers) that are due to inheritance from a common ancestor.

style The slender stalk of a flower carpel connecting the stigma and the ovary.

subspecies A population that has distinctive traits and some genetic differences relative to other populations of the same species but that is not distinct enough to be classified as a separate species.

substrate (1) A reactant that interacts with an enzyme in a chemical reaction. (2) A surface on which a cell or organism sits.

substrate-level phosphorylation Production of ATP by transfer of a phosphate group from an intermediate substrate directly to ADP. Occurs in glycolysis and in the citric acid cycle.

succession In ecology, the gradual colonization of a habitat after an environmental disturbance (e.g., fire, flood), usually by a series of species. See **primary** and **secondary succession**.

sucrose A disaccharide formed from glucose and fructose. One of the two main products of photosynthesis.

sugar Synonymous with carbohydrate, though usually used in an informal sense to refer to small carbohydrates (monosaccharides and disaccharides).

sulfate reducer A prokaryote that produces hydrogen sulfide (H_2S) as a by-product of cellular respiration.

summation The additive effect of different postsynaptic potentials at a nerve or muscle cell, such that several subthreshold stimulations can cause an action potential.

supporting connective tissue A type of connective tissue, distinguished by having a firm extracellular matrix.

surface tension The cohesive force that causes molecules at the surface of a liquid to stick together, thereby resisting deformation of the liquid's surface and minimizing its surface area.

surfactant A mixture of phospholipids and proteins produced by lung cells that reduces surface tension, allowing the lungs to expand more.

survivorship On average, the proportion of offspring that survive to a particular age.

survivorship curve A graph depicting the percentage of a population that survives to different ages.

suspension feeder Any organism that obtains food by filtering small particles or small organisms out of water or air. Also called *filter feeder*.

sustainability The planned use of environmental resources at a rate no faster than the rate at which they are naturally replaced.

sustainable agriculture Agricultural techniques that are designed to maintain long-term soil quality and productivity.

swamp A wetland that has a steady rate of water flow and is dominated by trees and shrubs.

swim bladder A gas-filled organ of many ray-finned fishes that regulates buoyancy.

symbiosis (adjective: symbiotic) Any close and prolonged physical relationship between individuals of two different species. See **commensalism, mutualism**, and **parasitism**.

symmetric competition Ecological competition between two species in which both suffer similar declines in fitness. Compare with **asymmetric competition**.

sympathetic nervous system The part of the autonomic nervous system that stimulates fight-or-flight responses, such as increased heart rate, increased blood pressure, and decreased digestion. Compare with **parasympathetic nervous system**.

sympatric speciation The divergence of populations living within the same geographic area into different species as the result of their genetic (not physical) isolation. Compare with **allopatric speciation**.

sympatry Condition in which two or more populations live in the same geographic area, or close enough to permit interbreeding. Compare with **allopatry**.

symplast In plant roots, a continuous pathway through which water can flow through the cytoplasm of adjacent cells that are connected by plasmodesmata. Compare with **apoplast**.

symporter A carrier protein that allows an ion to diffuse down an electrochemical gradient, using the energy of that process to transport a different substance in the same direction *against* its concentration gradient. Compare with **antiporter**.

synapomorphy A shared, derived trait found in two or more taxa that is present in their most recent common ancestor but is missing in more distant ancestors. Useful for inferring evolutionary relationships.

synapse The interface between two neurons or between a neuron and an effector cell.

synapsis The physical pairing of two homologous chromosomes during prophase I of meiosis. Crossing over occurs during synapsis.

synaptic cleft The space between two communicating nerve cells (or between a neuron and effector cell) at a synapse, across which neurotransmitters diffuse.

synaptic plasticity Long-term changes in the responsiveness or physical structure of a synapse that can occur after particular stimulation patterns. Thought to be the basis of learning and memory.

synaptic vesicle A small neurotransmitter-containing vesicle at the end of an axon that releases neurotransmitter into the synaptic cleft by exocytosis.

synaptonemal complex A network of proteins that holds non-sister chromatids together during synapsis in meiosis I.

synthesis (S) phase The phase of the cell cycle during which DNA is synthesized and chromosomes are replicated.

systemic acquired resistance (SAR) A slow, widespread response of plants to a localized infection that protects healthy tissue from invasion by pathogens. Compare with **hypersensitive response**.

systemic circulation The part of the circulatory system that sends oxygen-rich blood from the lungs out to the rest of the body. Is separate from the pulmonary circulation in mammals and birds.

systemin A peptide hormone, produced by plant cells damaged by herbivores, that initiates a protective response in undamaged cells.

systole The portion of the heartbeat cycle during which the heart muscles are contracting. Compare with **diastole**.

systolic blood pressure The force exerted by blood against artery walls during contraction of the heart's left ventricle. Compare with **diastolic blood pressure**.

T cell A type of leukocyte that matures in the thymus and, with B cells, is responsible for acquired immunity. Involved in activation of B cells ($CD4^+$ helper T cells) and destruction of infected cells ($CD8^+$ cytotoxic T cells). Also called *T lymphocytes*.

T tubules Membranous tubes that extend into the interior of muscle cells. Propagate action potentials throughout a muscle cell and trigger release of calcium from the sarcoplasmic reticulum.

taiga A vast forest biome throughout subarctic regions, consisting primarily of short conifer trees. Characterized by intensely cold winters, short summers, and high annual variation in temperature.

taproot A large vertical main root of a plant.

taste bud Sensory structure, found chiefly in the mammalian tongue, containing spindle-shaped cells that respond to chemical stimuli.

TATA box A short DNA sequence in many eukaryotic promoters about 30 base pairs upstream from the transcription start site.

TATA-binding protein (TBP) A protein that binds to the TATA box in eukaryotic promoters and is a component of the basal transcription complex.

taxon (plural: taxa) Any named group of organisms at any level of a classification system.

taxonomy The branch of biology concerned with the classification and naming of organisms.

TBP See **TATA-binding protein.**

T-cell receptor (TCR) A transmembrane protein found on T cells that can bind to antigens displayed on the surfaces of other cells. Composed of two polypeptides called the alpha chain and beta chain. See **antigen presentation.**

tectorial membrane A membrane in the vertebrate cochlea that takes part in the transduction of sound by bending the stereocilia of hair cells in response to sonic vibrations.

telomerase An enzyme that replicates the ends of chromosome (telomeres) by catalyzing DNA synthesis from an RNA template that is part of the enzyme.

telomere The region at the end of a linear chromosome.

telophase The final stage in mitosis or meiosis during which sister chromatids (replicated chromosomes in meiosis I) separate and new nuclear envelopes begin to form around each set of daughter chromosomes.

temperate Having a climate with pronounced annual fluctuations in temperature (i.e., warm summers and cold winters) but typically neither as hot as the tropics nor as cold as the poles.

temperature A measurement of thermal energy present in an object or substance, reflecting how much the constituent molecules are moving.

template strand (1) The strand of DNA that is transcribed by RNA polymerase to create RNA. (2) An original strand of RNA used to make a complementary strand of RNA.

temporal lobe In the vertebrate brain, one of the four major areas in the cerebrum.

tendon A band of tough, fibrous connective tissue that connects a muscle to a bone.

tentacle A long, thin, muscular appendage of gastropod mollusks.

termination (1) In enzyme-catalyzed reactions, the final stage in which the enzyme returns to its original conformation and products are released. (2) In DNA transcription, the dissociation of RNA polymerase from DNA when it reaches a termination signal sequence. (3) In RNA translation, the dissociation of a ribosome from mRNA when it reaches a stop codon.

territory An area that is actively defended by an animal from others of its species.

tertiary consumers In a food chain or food web, organisms that feed on secondary consumers. Compare with **primary consumer** and **secondary consumer.**

tertiary structure The overall three-dimensional shape of a single polypeptide chain, resulting from multiple interactions among the amino acid side chains and the peptide backbone. Compare with **primary, secondary,** and **quaternary structure.**

testcross The breeding of an individual of unknown genotype with an individual having only recessive alleles for the traits of interest in order to infer the unknown genotype from the phenotypic ratios seen in offspring.

testis (plural: testes) The sperm-producing organ of a male animal.

testosterone A steroid hormone, produced and secreted by the testes, that stimulates sperm production and various male traits and reproductive behaviors.

tetrad The structure formed by synapsed homologous chromosomes during prophase of meiosis I.

tetrapod Any member of the taxon Tetrapoda, which includes all vertebrates with two pairs of limbs (amphibians, mammals, birds, and other reptiles).

texture A quality of soil, resulting from the relative abundance of different-sized particles.

theory A proposed explanation for a broad class of phenomena or observations.

thermal energy The kinetic energy of molecular motion.

thermocline A gradient (cline) in environmental temperature across a large geographic area.

thermophile A bacterium or archaean that thrives in very hot environments.

thermoreceptor A sensory cell or an organ specialized for detection of changes in temperature.

thermoregulation Regulation of body temperature.

thick filament A filament composed of bundles of the motor protein myosin; anchored to the center of the sarcomere. Compare with **thin filament.**

thigmotropism Growth or movement of an organism in response to contact with a solid object.

thin filament A filament composed of two coiled chains of actin and associated regulatory proteins; anchored at the Z disk of the sarcomere. Compare with **thick filament.**

thorax A region of the body; in insects, one of the three prominent body regions called tagmata.

thorn A modified plant stem shaped as a sharp protective structure. Helps protect a plant against feeding by herbivores.

threshold potential The membrane potential that will trigger an action potential in a neuron or other excitable cell. Also called simply *threshold.*

thylakoid A flattened, membrane-bound vesicle inside a plant chloroplast that functions in converting light energy to chemical energy. A stack of thylakoids is a granum.

thymus An organ, located in the anterior chest or neck of vertebrates, in which immature T cells generated in the bone marrow undergo maturation.

thyroid gland A gland in the neck that releases thyroid hormone (which increases metabolic rate) and calcitonin (which lowers blood calcium).

thyroid-stimulating hormone (TSH) A peptide hormone, produced and secreted by the anterior pituitary, that stimulates release of thyroid hormones from the thyroid gland.

thyroxine (T_4) A peptide hormone containing four iodine atoms that is produced and secreted by the thyroid gland. Acts primarily to increase cellular metabolism. In mammals, T_4 is converted to the more active hormone triiodothyronine (T_3) in the liver.

Ti plasmid A plasmid carried by *Agrobacterium* (a bacterium that infects plants) that can integrate into a plant cell's chromosomes and induce formation of a gall.

tight junction A type of cell-cell attachment structure that links the plasma membranes of adjacent animal cells, forming a barrier that restricts movement of substances in the space between the cells. Most abundant in epithelia (e.g., the intestinal lining). Compare with **desmosome** and **gap junction.**

tip The end of a branch on a phylogenetic tree. Represents a specific species or larger taxon that has not (yet) produced descendants—either a group living today or a group that ended in extinction. Also called *terminal node.*

tissue A group of similar cells that function as a unit, such as muscle tissue or epithelial tissue.

tolerance In ecological succession, the phenomenon in which early-arriving species do not affect the probability that subsequent species will become established. Compare with **facilitation** and **inhibition.**

tonoplast The membrane surrounding a plant vacuole.

top-down control The hypothesis that population size is limited by predators or herbivores (consumers).

topoisomerase An enzyme that cuts and rejoins DNA downstream of the replication fork, to ease the twisting that would otherwise occur as the DNA "unzips."

torpor An energy-conserving physiological state, marked by a decrease in metabolic rate, body temperature, and activity, that lasts for a short period (overnight to a few days or weeks). Occurs in some small mammals when the ambient temperature drops significantly. Compare with **hibernation.**

totipotent Capable of dividing and developing to form a complete, mature organism.

trachea (plural: tracheae) (1) In insects, one of the small air-filled tubes that extend throughout the body and function in gas exchange. (2) In terrestrial vertebrates, the airway connecting the larynx to the bronchi. Also called *windpipe.*

tracheid In vascular plants, a long, thin water-conducting cell that has gaps in its secondary cell wall, allowing water movement between adjacent cells. Compare with **vessel element.**

trade-off In evolutionary biology, an inescapable compromise between two traits that cannot be optimized simultaneously. Also called *fitness trade-off.*

trait Any heritable characteristic of an individual.

transcription The process by which RNA is made from a DNA template.

transcriptional control Regulation of gene expression by various mechanisms that change the rate at which genes are transcribed to form messenger RNA. In negative control, binding of a regulatory protein to DNA represses transcription; in positive control binding of a regulatory protein to DNA promotes transcription.

transcriptome The complete set of genes transcribed in a particular cell.

transduction Conversion of information from one mode to another. For example, the process by which a stimulus outside a cell is converted into a response by the cell.

transfer cell In land plants, a cell that transfers nutrients from a parent plant to a developing plant seed.

transfer RNA (tRNA) One of a class of RNA molecules that have an anticodon at one end and an amino acid binding site at the other. Each tRNA picks up a specific amino acid and binds to the corresponding codon in messenger RNA during translation.

transformation (1) Incorporation of external DNA into the genome. Occurs naturally in some bacteria; can be induced in the laboratory by certain processes. (2) Conversion of a normal cell to a cancerous one.

transgenic Referring to an individual plant or animal whose genome contains DNA introduced from another individual, either from the same or a different species.

transition state A high-energy intermediate state of the reactants during a chemical reaction that must be achieved for the reaction to proceed. Compare with **activation energy.**

transitional feature A trait that is intermediate between a condition observed in ancestral species and the condition observed in more derived species.

translation The process by which proteins and peptides are synthesized from messenger RNA.

translational control Regulation of gene expression by various mechanisms that alter the life span of messenger RNA or the efficiency of translation.

translocation (1) In plants, the movement of sugars and other organic nutrients through the phloem by bulk flow. (2) A type of mutation in which a piece of a chromosome moves to a nonhomologous chromosome. (3) The process by which a ribosome moves down a messenger RNA molecule during translation.

transmembrane protein Any membrane protein that spans the entire lipid bilayer. Also called *integral membrane protein*.

transmission The passage or transfer (1) of a disease from one individual to another or (2) of electrical impulses from one neuron to another.

transmission electron microscope (TEM) A microscope that forms an image from electrons that pass through a specimen. Compare with **scanning electron microscope**.

transpiration Water loss from aboveground plant parts. Occurs primarily through stomata.

transport protein Collective term for any membrane protein that enables a specific ion or small molecule to cross a plasma membrane. Includes carrier proteins and channel proteins, which carry out passive transport (facilitated diffusion), and pumps, which carry out active transport.

transporter See **carrier protein**.

transposable elements Any of several kinds of DNA sequences that are capable of moving themselves, or copies of themselves, to other locations in the genome. Include LINEs and SINEs.

tree of life A diagram depicting the genealogical relationships of all living organisms on Earth, with a single ancestral species at the base.

trichome A hairlike appendage that grows from epidermal cells of some plants. Trichomes exhibit a variety of shapes, sizes, and functions depending on species.

triglyceride See **fat**.

triiodothyronine (T$_3$) A peptide hormone containing three iodine atoms that is produced and secreted by the thyroid gland. Acts primarily to increase cellular metabolism. In mammals, T$_3$ has a stronger effect than does the related hormone thyroxine (T$_4$).

triose A monosaccharide (simple sugar) containing three carbon atoms.

triplet code A code in which a "word" of three letters encodes one piece of information. The genetic code is a triplet code because a codon is three nucleotides long and encodes one amino acid.

triploblast (adjective: triploblastic) An animal whose body develops from three basic embryonic cell layers: ectoderm, mesoderm, and endoderm. Compare with **diploblast**.

trisomy The state of having three copies of one particular type of chromosome.

tRNA See **transfer RNA**.

trochophore A larva with a ring of cilia around its middle that is found in some lophotrochozoans.

trophic cascade A series of changes in the abundance of species in a food web, usually caused by the addition or removal of a key predator.

trophic level A feeding level in an ecosystem.

trophoblast The exterior of a blastocyst (the structure that results from cleavage in embryonic development of mammals).

tropomyosin A regulatory protein present in thin (actin) filaments that blocks the myosin-binding sites on these filaments, thereby preventing muscle contraction.

troponin A regulatory protein, present in thin (actin) filaments, that can move tropomyosin off the myosin-binding sites on these filaments, thereby triggering muscle contraction. Activated by high intracellular calcium.

true navigation The type of navigation by which an animal can reach a specific point on Earth's surface.

trypsin A protein-digesting enzyme present in the small intestine that activates several other protein-digesting enzymes.

trypsinogen The precursor of protein-digesting enzyme trypsin. Secreted by the pancreas and activated by the intestinal enzyme enterokinase.

tube foot One of the many small, mobile, fluid-filled extensions of the water vascular system of echinoderms; the part extending outside the body is called a podium. Used in locomotion and feeding.

tuber A modified plant rhizome that functions in storage of carbohydrates.

tuberculosis A disease of the lungs caused by infection with the bacterium *Mycobacterium tuberculosis*.

tumor A mass of cells formed by uncontrolled cell division. Can be benign or malignant.

tumor suppressor A gene (e.g., *p53* and *Rb*) or the protein it encodes that prevents cell division, particularly when the cell has DNA damage. Mutated forms are associated with cancer.

tundra The treeless biome in polar and alpine regions, characterized by short, slow-growing vegetation, permafrost, and a climate of long, intensely cold winters and very short summers.

turgid Swollen and firm as a result of high internal pressure (e.g., a plant cell containing enough water for the cytoplasm to press against the cell wall). Compare with **flaccid**.

turgor pressure The outward pressure exerted by the fluid contents of a plant cell against its cell wall.

Turner syndrome A human genetic disorder caused by the presence of only one X chromosome and no Y chromosome ("XO"). Individuals with this condition are female but sterile.

turnover In lake ecology, the complete mixing of upper and lower layers of water that occurs each spring and fall in temperate-zone lakes.

tympanic membrane The membrane separating the middle ear from the outer ear in terrestrial vertebrates, or similar structures in insects. Also called the *eardrum*.

ubiquinone See **coenzyme Q**.

ulcer A hole in an epithelial layer that damages the underlying basement membrane and tissues.

ultimate causation In biology, the reason that a trait or phenomenon is thought to have evolved; the adaptive advantage of that trait. Also called *ultimate explanation*. Compare with **proximate causation**.

umami The taste of glutamate, responsible for the "meaty" taste of most proteins and of monosodium glutamate.

umbilical cord The cord that connects a developing mammalian embryo or fetus to the placenta and through which the embryo or fetus receives oxygen and nutrients.

unequal crossover An error in crossing over during meiosis I in which the two non-sister chromatids match up at different sites. Results in gene duplication in one chromatid and gene loss in the other.

universal tree The phylogenetic tree that includes all organisms.

unsaturated Referring to fats and fatty acids in which at least one carbon-carbon bond is a double bond. Double bonds produce kinks in the fatty acid chains and decrease the compound's melting point. Compare with **saturated**.

upstream In genetics, opposite to the direction in which RNA polymerase moves along a DNA strand. Compare with **downstream**.

ureter In vertebrates, a tube that transports urine from one kidney to the bladder.

urethra The tube that drains urine from the bladder to the outside environment. In male vertebrates, also used for passage of sperm during ejaculation.

uric acid A whitish excretory product of birds, reptiles, and terrestrial arthropods. Used to remove from the body excess nitrogen derived from the breakdown of amino acids. Compare with **urea**.

urochordates One of the three major chordate lineages (Urochordata), comprising sessile, filter-feeding animals that have a polysaccharide exoskeleton (tunic) and two siphons through which water enters and leaves; also called tunicates or sea squirts. Compare with **cephalochordates** and **vertebrates**.

uterus The organ in which developing embryos are housed in those vertebrates that give live birth. Common in most mammals and in some lizards, sharks, and other vertebrates.

vaccination The introduction into an individual of weakened, killed, or altered pathogens to stimulate development of acquired immunity against those pathogens.

vaccine A preparation designed to stimulate an immune response against a particular pathogen without causing illness. Vaccines consist of inactivated (killed) pathogens, live but weakened (attenuated) pathogens, or portions of a viral capsid (subunit vaccine).

vacuole A large organelle in plant and fungal cells that usually is used for bulk storage of water, pigments, oils, or other substances. Some vacuoles contain enzymes and have a digestive function similar to lysosomes in animal cells.

vagina The birth canal of female mammals; a muscular tube that extends from the uterus through the pelvis to the exterior.

valence electron An electron in the outermost electron shell, the valence shell, of an atom. Valence electrons tend to be involved in chemical bonding.

valence The number of unpaired electrons in the outermost electron shell of an atom; determines how many covalent bonds the atom can form.

valves In circulatory systems, flaps of tissue that prevent backward flow of blood, particularly in veins and between the chambers of the heart.

van der Waals interactions A weak electrical attraction between two hydrophobic side chains. Often contributes to tertiary structure in proteins.

variable (V) region The portion of an antibody's light chains or heavy chains that has a highly variable amino acid sequence and forms part of the antigen-binding site. Compare with **constant (C) region**.

variable number tandem repeat See **minisatellite**.

vas deferens (plural: vasa deferentia) A muscular tube that stores and transports semen from the epididymis to the ejaculatory duct. In nonhuman animals, called the *ductus deferens*.

vasa recta In the vertebrate kidney, a network of blood vessels that runs alongside the loop of Henle of a nephron. Functions in reabsorption of water and solutes from the filtrate.

vascular bundle A cluster of xylem and phloem strands in a plant stem.

vascular cambium One of two types of lateral meristem, consisting of a ring of undifferentiated plant cells inside the cork cambium of woody plants; produces secondary xylem (wood) and secondary phloem. Compare with **cork cambium**.

vascular tissue In plants, tissue that transports water, nutrients, and sugars. Made up of the complex tissues xylem and phloem, each of which contains several cell types.

vector A biting insect or other organism that transfers pathogens from one species to another. See also **cloning vector**.

vegetative organs The nonreproductive parts of a plant including roots, leaves, and stems.

vein Any blood vessel that carries blood (oxygenated or not) under relatively low pressure from the tissues toward the heart. Compare with **artery**.

veliger A distinctive type of larva, found in mollusks.

vena cava (plural: vena cavae) A large vein that returns oxygen-poor blood to the heart.

ventral Toward an animal's belly and away from its back. The opposite of dorsal.

ventricle (1) A thick-walled chamber of the heart that receives blood from an atrium and pumps it to the body or to the lungs. (2) One of several small fluid-filled chambers in the vertebrate brain.

venules Small veins (blood vessels that return blood to the heart).

vertebra (plural: vertebrae) One of the cartilaginous or bony elements that form the spine of vertebrate animals.

vertebrates One of the three major chordate lineages (Vertebrata), comprising animals with a dorsal column of cartilaginous or bony structures (vertebrae) and a skull enclosing the brain. Includes fishes, amphibians, mammals, reptiles, and birds. Compare with **cephalochordates** and **urochordates**.

vessel element In vascular plants, a short, wide water-conducting cell that has gaps through both the primary and secondary cell walls, allowing unimpeded passage of water between adjacent cells. Compare with **tracheid**.

vestigial trait Any rudimentary structure of unknown or minimal function that is homologous to functioning structures in other species. Vestigial traits are thought to reflect evolutionary history.

vicariance The physical splitting of a population into smaller, isolated populations by a geographic barrier.

villi (singular: villus) Small, fingerlike projections (1) of the lining of the small intestine or (2) of the fetal portion of the placenta adjacent to maternal arteries. Function to increase the surface area available for absorption of nutrients and gas exchange (in the placenta).

virion A single mature virus particle.

virulence The ability of a pathogen or parasite to cause disease and death.

virulent Referring to pathogens that can cause severe disease in susceptible hosts.

virus A tiny intracellular parasite that uses host cell enzymes to replicate; consists of a DNA or RNA genome enclosed within a protein shell (capsid). In enveloped viruses, the capsid is surrounded by a phospholipid bilayer derived from the host cell plasma membrane, whereas nonenveloped viruses lack this protective covering.

visceral mass One of the three main parts of the mollusk body; contains most of the internal organs and external gill.

visible light The range of wavelengths of electromagnetic radiation that humans can see, from about 400 to 700 nanometers.

vitamin An organic micronutrient that usually functions as a coenzyme.

vitelline envelope A fibrous sheet of glycoproteins that surrounds mature egg cells in many vertebrates. Surrounded by a thick gelatinous matrix (the jelly layer) in some species. In mammals, called the *zona pellucida*.

viviparous Producing live young (instead of eggs) that develop within the body of the mother before birth. Compare with **oviparous** and **ovoviviparous**.

volt (V) A unit of electrical potential (voltage).

voltage Potential energy created by a separation of electric charges between two points. Also called *electrical potential*.

voltage clamping A technique for imposing a constant membrane potential on a cell. Widely used to investigate ion channels.

voltage-gated channel An ion channel that opens or closes in response to changes in membrane voltage. Compare with **ligand-gated channel**.

voluntary muscle Muscle tissue that can respond to conscious thought.

wall pressure The inward pressure exerted by a cell wall against the fluid contents of a plant cell.

Wallace line A line that demarcates two areas in the Indonesian region, each of which is characterized by a distinct set of animal species.

water cycle, global The movement of water among terrestrial ecosystems, the oceans, and the atmosphere.

water potential (ψ) The potential energy of water in a certain environment compared with the potential energy of pure water at room temperature and atmospheric pressure. In living organisms, ψ equals the solute potential (ψ_S) plus the pressure potential (ψ_P).

water potential gradient A difference in water potential in one region compared with that in another region. Determines the direction that water moves, always from regions of higher water potential to regions of lower water potential.

water table The upper limit of the underground layer of soil that is saturated with water.

water vascular system In echinoderms, a system of fluid-filled tubes and chambers that functions as a hydrostatic skeleton.

watershed The area drained by a single stream or river.

Watson-Crick pairing See **complementary base-pairing**.

wavelength The distance between two successive crests in any regular wave, such as light waves, sound waves, or waves in water.

wax A class of lipid with extremely long hydrocarbon tails, usually combinations of long-chain alcohols with fatty acids. Harder and less greasy than fats.

weather The specific short-term atmospheric conditions of temperature, moisture, sunlight, and wind in a certain area.

weathering The gradual wearing down of large rocks by rain, running water, and wind; one of the processes that transform rocks into soil.

weed Any plant that is adapted for growth in disturbed soils.

wetland A shallow-water habitat where the soil is saturated with water for at least part of the year.

white blood cells See **leukocytes**.

wild type The most common phenotype seen in a population; especially the most common phenotype in wild populations compared with inbred strains of the same species.

wildlife corridor Strips of wildlife habitat connecting populations that otherwise would be isolated by man-made development.

wilt To lose turgor pressure in a plant tissue.

wobble hypothesis The hypothesis that some tRNA molecules can pair with more than one mRNA codon, tolerating some variation in the third base, as long as the first and second bases are correctly matched.

wood Xylem resulting from secondary growth. Also called *secondary xylem*.

xeroderma pigmentosum A human disease characterized by extreme sensitivity to ultraviolet light. Caused by an autosomal recessive allele that results in a defective DNA repair system.

X-linked inheritance Inheritance patterns for genes located on the mammalian X chromosome. Also called *X-linkage*.

X-ray crystallography A technique for determining the three-dimensional structure of large molecules, including proteins and nucleic acids, by analysis of the diffraction patterns produced by X-rays beamed at crystals of the molecule.

xylem A plant vascular tissue that conducts water and ions; contains tracheids and/or vessel elements. Primary xylem develops from the procambium of apical meristems; secondary xylem, or wood, from the vascular cambium of lateral meristems. Compare with **phloem**.

xylem sap The watery fluid found in the xylem of plants.

yeast Any fungus growing as a single-celled form. Also, a specific lineage of ascomycetes.

Y-linked inheritance Inheritance patterns for genes located on the mammalian Y chromosome. Also called *Y-linkage*.

yolk The nutrient-rich cytoplasm inside an egg cell; used as food for the growing embryo.

Z disk The structure that forms each end of a sarcomere. Contains a protein that binds tightly to actin, thereby anchoring thin filaments.

Z scheme Path of electron flow in which electrons pass from photosystem II to photosystem I and ultimately to $NADP^+$ during the light-dependent reactions of photosynthesis. Also called *noncyclic electron flow*.

zero population growth (ZPG) A state of stable population size due to fertility staying at the replacement rate for at least one generation.

zona pellucida The gelatinous layer around a mammalian egg cell. In other vertebrates, called the *vitelline envelope*.

zone of (cellular) division In plant roots, a group of apical meristematic cells just behind the root cap where cells are actively dividing.

zone of (cellular) elongation In plant roots, a group of young cells, located behind the apical meristem, that are increasing in length.

zone of (cellular) maturation In plant roots, a group of plant cells, located several centimeters behind the root cap, that are differentiating into mature tissues.

zygosporangium (plural: zygosporangia) The spore-producing structure in fungi that are members of the Zygomycota.

zygote The diploid cell formed by the union of two haploid gametes; a fertilized egg. Capable of undergoing embryological development to form an adult.

Credits

Photo Credits

Frontmatter p. ix Oscar Miller/SPL/Photo Researchers **p. xi** Photo by Jason Rick, supplied by Loren Rieseberg **p. xiv** Jeff Rotman/NPL/Minden Pictures **p. xvi** Frans Lanting/Minden Pictures **p. xxT** Natalie B. Fobes Photography **p. xxB** David Quillin

Chapter 1 Opener Marjorie Kibby **1.1** "Animalcules" observed by Anton van Leeuwenhoek, c1795, HIP/Art Resource, NY **1.6a** Samuel F. Conti and Thomas D. Brock **1.6b** Kwangshin Kim/Photo Researchers **1.7b** Michael Hughes/Laif/Aurora Photos **1.8R** From M. Wittlinger, R. Wehner, and H. Wolf, The ant odometer: Stepping on stilts and stumps, *Science* 312: 1965–1967, supporting online material (Jun 30 2006).

Chapter 2 Opener Frans Lanting/Corbis **2.1b** Dragan Trifunovic/Shutterstock **2.6c** Photos. com **2.11** Robert and Beth Plowes, ProteaPIX **2.14b** Dietmar Nill/Picture Press/Photolibrary **2.15c** Jan Vermeer/Minden Pictures **2.20** Shutterstock **2.20 inset** PhotoSpin/Alamy

Chapter 3 Opener Fritz Wilhelm (heisingart.com) **3.10a** Microworks/Phototake **3.10b** Walter Reinhart/Phototake

Chapter 4 Opener SSPL/The Image Works **4.5** National Cancer Institute

Chapter 5 Opener Peter Arnold/Alamy

Chapter 6 6.5L James J. Cheetham, Carleton University **6.9a** iStockphoto **6.9b** iStockphoto **6.9c** Shutterstock **6.18** Harold Edwards/Visuals Unlimited **6.24T** David M. Phillips/Photo Researchers **6.24M** David M. Phillips/Photo Researchers **6.24B** Joseph F. Hoffman, Yale University School of Medicine

Chapter 7 Opener Torsten Wittmann/Photo Researchers **7.1** T. J. Beveridge/Visuals Unlimited **7.2** Gopal Murti/Visuals Unlimited **7.3** Wanner/Eye of Science/Photo Researchers **7.4** Stanley C. Holt/Biological Photo Service **7.5** From David S. Goodsell, *The Machinery of Life,* 2nd ed. (2009). © Springer-Verlag, New York. **7.7** Don W. Fawcett/Photo Researchers **7.8** Don W. Fawcett/Photo Researchers **7.9** Don W. Fawcett/Photo Researchers **7.10** Biophoto Associates/Photo Researchers **7.11** Omikron/Photo Researchers **7.12** Don Fawcett, Daniel Friend, & Richard Wood/Photo Researchers **7.13** Gopal Murti/Visuals Unlimited **7.16** E. H. Newcomb & W. P. Wergin/Biological Photo Service **7.17** T.Kanaseki & Donald Fawcett/Visuals Unlimited **7.18** E. H. Newcomb & W. P. Wergin/Biological Photo Service **7.19** E. H. Newcomb & S. E. Frederick/Biological Photo Service **7.20** By David S. Goodsell. Published in L. A. Moran et al., *Biochemistry* (1994). © Neil Patterson Publishers/Prentice-Hall. **7.21a** Don W. Fawcett/Photo Researchers **7.21b** Don W. Fawcett/Photo Researchers **7.21c** Biophoto Associates/Photo Researchers **7.21d** Dennis Kunkel/Visuals Unlimited **7.22** Don W. Fawcett/Photo Researchers **7.25a** James D. Jamieson, Yale University School of Medicine **7.25b/c** From J. D. Jamieson and G. E. Palade, *J. Cell Biol.* 34: 597–615 (1967). © The Rockefeller University Press. **7.30** Conly L. Rieder, Wadsworth Center, New York State Department of Health **7.31a/b** From *Mol. Biol. Cell* 9: cover (Dec 1998). © American Society for Cell Biology. Photo B. J. Schnapp. **7.32a** John E. Heuser, Washington University School of Medicine **7.33L** SPL/Photo Researchers **7.33R** Visuals Unlimited/Corbis **7.34a** Don W. Fawcett/Photo Researchers

Chapter 8 Opener Caroline Weight, G.I.T. Molecular Immunology, Institute of Food Research, Norwich, UK **8.1** William James Warren/Science Faction **8.2** Biophoto Associates/Photo Researchers **8.3b** Barry King, University of California, Davis/Biological Photo Service **8.5** SPL/Photo Researchers **8.6** Photo Researchers **8.7aL** Don W. Fawcett/Photo Researchers **8.7aR** Don W. Fawcett/Photo Researchers **8.8a** Don W. Fawcett/Photo Researchers **8.11a** E. H. Newcomb & W. P. Wergin/Biological Photo Service **8.11b** Don W. Fawcett/Photo Researchers **8.18** Janice Carr/CDC/Rodney M. Donlan

Chapter 9 Opener Richard Megna/Fundamental Photographs **9.7T** Shutterstock **9.7B** Shutterstock **9.13** Terry Frey, San Diego State University **9.22a** Science VU/B. Bhatnagar/Visuals Unlimited

Chapter 10 Opener Visuals Unlimited/Corbis **10.2T** John Durham/SPL/Photo Researchers **10.2M/B** Electron micrographs by W. P. Wergin, courtesy of E. H. Newcomb, University of Wisconsin, Madison **10.4b** Sinclair Stammers/SPL/Photo Researchers **10.17L/R** James A. Bassham, Lawrence Berkeley Laboratory, UCB (retired) **10.20a** Jeremy Burgess/SPL/Photo Researchers

Chapter 11 Opener CNRI/Phototake **11.1a** From Walter Flemming, *Zellsubstanz, Kern, und Zelltheilung.* Leipzig: Verlag von F. C. W. Vogel, 1882. **11.1b** Conly Rieder, Wadsworth Center, New York State Department of Health. **11.1b** Photo Researchers **11.2T** Gopal Murti/Photo Researchers **11.2B** Biophoto Associates/Photo Researchers **11.5** Micrographs by Conly L. Rieder, Wadsworth Center, NYS Department of Health **11.6a** Ed Reschke/Peter Arnold **11.6b** Visuals Unlimited/Getty Images

Chapter 12 Opener David Phillips/The Population Council/Photo Researchers **12.6** Photos.com **12.7** Micrographs by Edward Novitski, courtesy of Charles Novitski

Chapter 13 Opener Brian Johnston **13.9a** Benjamin Prud'homme/Nicolas Gompel **13.9bL/bR** From J. Childress, R. Behringer, and G. Halder, "Learning to Fly: Phenotypic Markers in Drosophila," *Genesis* 43(1): cover illustration (2005). © Wiley-Liss. Photo Georg Halder, University of Texas. **13.15a** Robert Calentine/Visuals Unlimited

Chapter 14 Opener Gopal Murti/SPL/Photo Researchers **14.1b** Eye of Science/Photo Researchers **14.5** From M. Meselson and F. W. Stahl, *Proc. Natl. Acad. Sci. USA* 44: 671–682, fig 4 (Jul 15 1958). Photo Matthew S. Meselson, Harvard University. **14.7a** Gopal Murti/SPL/Photo Researchers

Chapter 15 Opener Alfred Pasieka/Photo Researchers **15.4a** Rod Williams/Nature Picture Library **15.4b** The Peromyscus Genetic Stock Center at the University of South Carolina. Photograph by Clint Cook. **15.9a/b** Peter Duesberg, University of California, Berkeley

Chapter 16 Opener Oscar Miller/SPL/Photo Researchers **16.5a** Bert W. O'Malley, Baylor College of Medicine **16.8a** From B. A. Hamkalo and O. L. Miller Jr., *Annu. Rev. Biochem.* 42: 379–96 (1973). © Annual Reviews. **16.9b** E. V. Kiseleva and Donald Fawcett/Visuals Unlimited

Chapter 17 Opener Stephanie Schuller/Photo Researchers **UN17.1** EDVOTEK, The Biotechnology Education Company (www.edvotek.com)

Chapter 18 18.2a Ada Olins & Don Fawcett/Photo Researchers **18.4T** Barbara Hamkalo, University of California, Irvine **18.4B** Victoria E. Foe, University of Washington

Chapter 19 Opener AJ/IRRI/Corbis **19.1b** Brady-Handy Photograph Collection (Library of Congress). Reproduction number: LC-DIG-cwpbh-02977. **19.12** Baylor College of Medicine/Peter Arnold **19.16bL/R** Golden Rice Humanitarian Board (www.goldenrice.org)

Chapter 20 Opener Sanger Institute/Wellcome Photo Library **20.7bL/R** Test results provided by GENDIA (www.gendia.net) **20.10** Agilent Technologies **20.11** Camilla M. Kao and Patrick O. Brown, Stanford University

Chapter 21 Opener Anthony Bannister/Gallo Images/Corbis **21.1a** From R. Merino et al., *Development* 126(23): 5515–5522, fig. 6 (1999). © The Company of Biologists. **21.1b** From K. Kuida et al., *Cell* 94: 325–337, figs. 2e and 2f (1998). © Elsevier Science Ltd. Photo Keisuke Kuida, Vertex Pharmaceuticals. **21.2** Roslin Institute/Phototake **21.3a** Richard Hutchings/Photo Researchers **21.3b** From S. Kulandavelu et al., Embryonic and neonatal phenotyping of genetically engineered mice. *ILAR Journal*, 47(2); 103–117, fig. 12a (2006). © National Academy of Sciences. **21.4a** F. Rudolf Turner, Indiana University **21.4b** Christiane Nusslein-Volhard, Max Planck Institute **21.5** Christiane Nusslein-Volhard, Max Planck Institute **21.6** Wolfgang Driever, University of Freiburg **21.7a/c** Jim Langeland, Stephen Paddock, and Sean Carroll, University of Wisconsin–Madison **21.7b** Stephen J. Small, New York University **21.8L** Visuals Unlimited/Getty Images **21.8M** David Scharf/Science Faction/Corbis **21.8R** Eye of Science/Photo Researchers **21.10a** F. Rudolf Turner, Indiana University **21.11** Anthony Bannister/NHPA **21.12a/b** From "The Origin of Form" by Sean B. Carroll, *Natural History* (Nov 2005); A. C. Burke, "*Hox* Genes and the Global Patterning of the Somitic Mesoderm," in C. Ordahl (ed.), *Somitogenesis, Part I* (2000). © Academic Press.

Chapter 22 Opener Yorgos Nikas/Photo Researchers **22.2** Holger Jastrow **22.5a** Michael Whitaker/SPL/Photo Researchers **22.5b** Victor D. Vacquier, Scripps Institution of Oceanography, University of California at San Diego **22.11a** Kathryn W. Tosney, University of Michigan **22.11b** Kathryn W. Tosney, The University of Miami

Chapter 23 Opener John Runions/Oxford Brookes University **23.3b** Cabisco/Visuals Unlimited **23.4b** Michael Clayton, Department of Botany, University of Wisconsin, Madison **23.8b** Ken Wagner/Phototake **23.10a/bT/bB** From M. Kim et al., The expression domain of *PHANTASTICA* determines leaflet placement in compound leaves, *Nature* 424: 438–443, fig. 1 (Jul 24 2003). © Macmillan Magazines Ltd. Photos Neelima Sinah. **23.12** Photos by John L. Bowman, University of California at Davis

Chapter 24 Opener Art Wolfe/Stone/Getty Images **24.2a** Sinclair Stammers/Photo Researchers **24.2b** From S. De Valais and R. N. Melchor, Ichnotaxonomy of bird-like footprints: an example from the Late Triassic–Early Jurassic of northwest Argentina. *J. Vertebr. Paleontol.* 28(1):145–159, fig 5c (2008). © Society of Vertebrate Paleontology. **24.2c** The Natural History Museum, London **24.3L** Gerd Weitbrecht for Landesbildungsserver Baden-Württemberg (www.schule-bw.de) **24.3R** iStockphoto **24.5aL** CMCD/PhotoDisc/Getty Images **24.5aR** Vincent Zuber/Custom Medical Stock Photo **24.5bL** Mary Beth Angelo/Photo Researchers **24.5bR** Custom Medical Stock Photo **24.6a1** age fotostock/SuperStock **24.6a2** George D. Lepp/Photo Researchers **24.6a3** Tui De Roy/Minden Pictures **24.6a4** Stefan Huwiler/age fotostock **24.8L/M** From M. K. Richardson et al., *Anat. Embryol.* 196: 91–106, figs. 7 and 8 (1997). © Springer-Verlag. Photos Michael K. Richardson, Leiden University, and Ronan O'Rahilly, National Museum of Health and Medicine/Armed Forces Institute of Pathology. **24.8R** From M. K. Richardson et al., *Science* 280: 983 (in Letters) (May 15 1998). Photo Ronan O'Rahilly, National Museum of Health and Medicine/Armed Forces Institute of Pathology. **24.10** Walter J. Gehring, Biozentrum, University of Basel (retired) **24.15aL** G. Gerra & S. Sommazzi/www.justbirds.it **24.15aR** Tui De Roy/Minden Pictures **24.15b** Huw Rees Lewis **24.18T** From A. Abzhonov et al., *Bmp4* and Morphological Variation of Beaks in Darwin's Finches, *Science* 305: 1462–1465, fig. 1c (Sep 3

2004). **24.18B** From K. Petren et al., *Proc. R. Soc. B*, 266: 321–329, fig. 3 (1999). © The Royal Society. Photo Kenneth Petren, University of Cincinnati. **24.20** Erlend Haarberg/NHPA/Photoshot

Chapter 25 **Opener** Phil Savoie/Minden Pictures **25.7a** André Karwath/Wikimedia Commons. Licensed under the Creative Commons Attribution Share Alike 2.5. **25.14a** Cyril Laubscher/Dorling Kindersley **25.15** Otorohanga Kiwi House, New Zealand **25.16a** Robert Rattner **25.17TL/TR** Robert & Linda Mitchell Photography **25.17BL/BR** Fotolia

Chapter 26 **Opener** Photo by Jason Rick, supplied by Loren Rieseberg **26.9bL** Michael Lustbader/Photo Researchers **26.9bR** From D. E. Soltis et al., Recent and recurrent polyploidy in *Tragopogon* (Asteraceae): Genetic, genomic, and cytogenetic comparisons. *Biol. J. Linn. Soc.* 82: 485–501 (2004). Photo Douglas and Pamela Soltis, University of Florida, Gainesville; Andrew Doust, Oklahoma State University. **26.10aT/M/B** H. Douglas Pratt/National Geographic Stock

Chapter 27 **Opener** Merv Feick (www.indiana9fossils.com) **27.7a** Stefan Piasecki, GEUS **27.7b** iStockphoto **27.7c/d** From The Virtual Petrified Wood Museum (petrifiedwoodmuseum.org). by Mike Viney **27.11LT** Gerald D. Carr **27.11LB/LR** Forest & Kim Starr **27.12a** Eladio Fernandez/Caribbean Nature Photography **27.12b** Jonathan B. Losos and Kevin de Quieroz, Washington University in St. Louis **27.13a** Colin Keates/Dorling Kindersley, courtesy of the Natural History Museum, London **27.13b** John Davis **27.13c** Don P. Northup, www.africancichlidphotos .com **27.13d** James L. Amos/Corbis **27.14bTL** Chip Clark, National Museum of Natural History, Smithsonian Institution **27.14bTR** Ed Reschke/Peter Arnold **27.14bML** Simon Conway Morris, University of Cambridge **27.14bMR** J. Gehling, South Australian Museum **27.14bBL/BM/BR** Shuhai Xiao, Virginia Polytechnic Institute and State University **27.17bL/bR** Glen A. Izett, U.S. Geological Survey, Denver **27.17c** Peter H. Schultz, Brown University; Steven D'Hondt, University of Rhode Island Graduate School of Oceanography

Chapter 28 **Opener** John Hammond **28.3** Science VU/Visuals Unlimited **28.4** From S. V. Liu et al., *Science* 277:1106–1109, fig. 2 (1997). © AAAS. PhotoYul Roh, Oak Ridge National Laboratory. **28.7aT** Daniel Quilten/Visuals Unlimited **28.7aB** From H. N. Schulz, et al., *Science* 284:493–495, fig. 1b (Aug 4 1999). © AAAS. Photo Heide Schulz, Max-Planck-Institute for Marine Microbiology. **28.7bT/B** SciMAT/Photo Researchers **28.7cT** K. S. Kim/Peter Arnold **28.7cB** Ralf Wagner **28.8a** R. Hubert, Iowa State University **28.10** Bruce J. Russell/BioMEDIA Associates **28.12** Wally Eberhart/Visuals Unlimited/Getty Images **28.15** Andrew Syred/Photo Researchers **28.16** Bill Schwartz/CDC **28.17** S. Amano, S. Miyadoh, and T. Shomura/The Society for Actinomycetes Japan **28.18** From J. L. Cocchiaro et al., Cytoplasmic lipid droplets are translocated into the lumen of the Chlamydia trachomatis parasitophorous vacuole, *Proc. Natl. Acad. Sci. USA* 105(27): 9379–9384 (Jul 8 2008). © National Academy of Sciences. **28.19** Peter Siver/Visuals Unlimited **28.20a** Yves V. Brun **28.20b** George Barron, University of Guelph, Canada **28.21** Kenneth M. Stedman/Portland State University **28.21 inset** Eye of Science/Photo Researchers **28.22** Doc Searls **28.22 inset** Priya DasSarma, Center of Marine Biotechnology, University of Maryland Biotechnology Institute

Chapter 29 **Opener** Wim van Egmond/Visuals Unlimited/Getty Images **29.2L** D. P. Wilson/FLPA/Minden Pictures **29.2M** Mark Conlin/V&W/imagequestmarine.com **29.2R** E. R. Degginger/Animals Animals–Earth Scenes **29.4** Pete Atkinson/Photographer's Choice/Getty Images **29.4 inset** D. Anderson, WHOI **29.6** Biophoto Associates/Photo Researchers **29.11a** Steve Gschmeissner/Photo Researchers **29.11b** Biophoto Associates/Photo Researchers **29.11c** Andrew Syred/Photo Researchers **29.12a** Biophoto Associates/Photo Researchers **29.12b** Bruce Coleman/Photoshot **29.15** Wim van Egmond/Visuals Unlimited/Getty Images **29.19** Scott Camazine/Photo Researchers **29.20** Dennis Kunkel/Visuals Unlimited **29.21** Tai-Soon Yong (www.atlas.or.kr) **29.22** Image by D. Patterson, provided by the microscope web project. **29.23** Francis Abbott/NPL/Minden Pictures **29.24** Peter Parks/imagequestmarine.com **29.25** Image by D. Patterson, provided by the micro*scope web project. **29.26** Maurice LOIR (www.diatomloir.eu) **29.27** Dennis Kunkel/Visuals Unlimited/Corbis **29.28** Andrew Syred/SPL/Photo Researchers **29.29** Wim van Egmond/Visuals Unlimited **29.30** Joyce Photographics/Photo Researchers **T29.3(1)** Peter Siver/Visuals Unlimited/Getty Images **T29.3(2)** Visuals Unlimited/Corbis **T29.3(3)** Jeff Rotman/Stone/Getty Images **T29.3(4)** iStockphoto **T29.3(5)** Yuuji Tsukii, Protist Information Server (protist.i.hosei.ac.jp) **T29.3(6)** Dr. Kawachi of National Institute for Environmental Studies, Japan

Chapter 30 **Opener** Jean-Paul Ferrero/Auscape/Minden Pictures **30.1** Lynn Betts/USDA Natural Resources Conservation Service **30.2b** Hugh Iltis/The Doebley Lab, University of Wisconsin-Madison (teosinte.wisc.edu) **30.4L** Linda Graham, University of Wisconsin-Madison **30.4R** Lee W. Wilcox **30.5L** Lee W. Wilcox **30.5M** iStockphoto **30.5R** Rob Whitworth/GAP Photos/Getty Images **30.11a/b** Lee W. Wilcox **30.12** Biodisc/Visuals Unlimited/Alamy **30.16a/b** Lee W. Wilcox **30.18** Carolina Biological Supply Company/Phototake **30.21a** Larry Deack/Wikimedia Commons **30.21b** Shutterstock **30.21c** Andreas Lander/dpa/Corbis **30.23a** Nigel Cattlin/Holt Studios/Photo Researchers **30.23b** Shutterstock **30.25T1** Jerome Wexler/Visuals Unlimited/Getty Images **30.25T2** Keith Wheeler/SPL/Photo Researchers **30.25T3** iStockphoto**30.25T4** Noodle snacks (www.noodlesnacks.com/) via Wikipedia. This file is licensed under the Creative Commons Attribution Share Alike 3.0 and GNU Free Documentation License. **30.25B1** Shutterstock **30.25B2** ISM/Phototake **30.25B3** iStockphoto **30.25B4** Shutterstock **30.27a** Doug Allan/NPL/Minden Pictures **30.27c** Lee W. Wilcox **30.27b** Wim van Egmond/Visuals Unlimited/Getty Images **30.28** Peter Verhoog/Foto Natura/Minden Pictures **30.29** Wim van Egmond **30.30** CAIP IMAGES © 2006 University of Florida (plants.ifas.ufl.edu/slidecol.html) **30.31** Adrian Davies/NPL/Minden Pictures **30.32a** Wildlife GmbH/Alamy **30.32b** Colin McPherson/Corbis **30.33** Steven P. Lynch, Lynch Images **30.34** Lee W. Wilcox **30.35** Frans Lanting/Corbis **30.36** Peter Arnold/Alamy **30.37L** David T. Webb, Botany, University of Hawaii **30.37R** Jami Tarris/Workbook Stock/Getty Images **30.38** Lee W. Wilcox **30.39** iStockphoto **30.40a** Martin Land/SPL/Photo Researchers **30.40b** iStockphoto **30.41** Peter Chadwick/Dorling Kindersley **30.42a/b** Lee W. Wilcox **30.43** Michael & Patricia Fogden/Minden Pictures **30.44a** Brian Johnston **30.44bL/R** Lee W. Wilcox

Chapter 31 **Opener** Richard Shiell/Animals Animals–Earth Scenes **31.1a** David Reed, Department of Animal and Plant Sciences, University of Sheffield **31.1b** Mycorrhizal Applications Inc. (www.mycorrhizae .com) **31.2** John Cang Photography **31.3a** Visuals Unlimited/Corbis **31.3b** iStockphoto **31.4a** Eye of Science/Photo Researchers **31.4b** Tony Brain/Photo Researchers **31.5bL** George Barron, University of Guelph, Canada **31.5bR** Biophoto Associates/Photo Researchers **31.6a** Melvin Fuller/The Mycological Society of America **31.6b** James Richardson/Visuals Unlimited **31.6c** Biophoto Associates/Photo Researchers **31.6d** Nino Santamaria **31.10a/b** Mark Brundrett (http://mycorrhizas.info) **31.13** J. I. Ronny Larsson, Professor of Zoology **31.14** Thomas J. Volk (TomVolkFungi.net) **31.15a** Gregory G. Dimijian/Photo Researchers **31.15b** David M. Phillips/Photo Researchers **31.16** Jim Deacon **31.17** Michel Poinsignon/NPL/Minden Pictures **31.18b** Daniel Mosquin **31.19** N. Allen and G. Barron, University of Guelph **31.20** Michael Fogden/OSF/Photolibrary

Chapter 32 **Opener** Bill Curtsinger/National Geographic Stock **32.1a** Satoshi Kuribayashi/Nature Production/Minden Pictures **32.1b** Mark Moffett/Minden Pictures **32.2** Walker England/Photo Researchers **32.3a** Roberto Rinaldi/NPL/Minden Pictures **32.3b** Ingo Arndt/Minden Pictures **32.8bT** Robert Steene/imagequestmarine.com **32.8bB** Russ Hopcroft/University of Alaska Fairbanks **32.12a** J. P. Ferrero/Jacana/Photo Researchers **32.12b** Todd Husman (www.soundaquatic.com) **32.13a** Pete Oxford/Minden Pictures **32.13b** Heidi & Hans-Jurgen Koch/Minden Pictures **32.14a overlay** Tim Ridley/Dorling Kindersley, Courtesy of the Royal Veterinary College, University of London **32.14a background** Shutterstock **32.14b** Adrian Costea/Fotolia **32.14b inset** Josef Ramsauer and Prof. Dr. Robert Patzner, Salzburg University **32.15a** Satoshi Kuribayashi/OSF/Photolibrary **32.15b** Radius Images/Corbis **32.16a** Andrew Syred/Photo Researchers **32.16b** Eye of Science/Photo Researchers **32.17a** Roland Seitre/Peter Arnold **32.17b** Jeff Foott/NPL/Minden Pictures **32.18a** D. P. Wilson/FLPA/Minden Pictures **32.18b** Sue Scott **32.19L/M/R** From G. Panganiban et al., *Proc. Natl. Acad. Sci. USA* 94(10): 5162–5166, figs. 1c and 1f (May 13 1997). © National Academy of Sciences. **32.20a** Shutterstock **32.20b** D. Parer & E. Parer-Cook/Auscape/Minden Pictures **32.21a** Nigel Cattlin/FLPA/Minden Pictures **32.21b inset** Frank Greenaway/Dorling Kindersley **32.21b** Kim Taylor/NPL/Minden Pictures **32.23** Roberto Rinaldi/NPL/Minden Pictures **32.24** Gary Bell/Taxi/Getty Images **32.25** Gregory G. Dimijian/Photo Researchers **32.26** Matthew D. Hooge, University of Maine

Chapter 33 **Opener** Barbara Strnadova/Photo Researchers **33.3a** Peter Parks/imagequestmarine.com **33.3b** Carmel McDougall, University of Oxford **33.4a** Dave King/Dorling Kindersley **33.4b** Breck P. Kent/Animals Animals–Earth Scenes **33.8a** Amazon-Images/Alamy **33.8b** Wim van Egmond/Visuals Unlimited **33.8c** Wim van Egmond/Visuals Unlimited/Corbis **33.9a** Pankaj Bajpai **33.9b** Bates Littlehales/Animals Animals–Earth Scenes **33.9c** Jane Burton/NPL/Minden Pictures **33.11** Antonio Guillén Oterino **33.12a** Doug Anderson (doug.deep via Flickr) **33.12b** Manfred Kage/Peter Arnold **33.12c** Oliver Meckes/Photo Researchers **33.13a** J. W.Alker/Imagebroker/Alamy **33.13b** Jasper Nance (www.flickr.com/photos/nebarnix) **33.13c** Robert and Linda Mitchell **33.14a** Marevision/age fotostock **33.15a** Roger Steen/imagequestmarine.com **33.15b** Daniela Wolf (www.taucher.li) **33.16** Kjell B. Sandved/Photo Researchers **33.17** Mark Conlin/VWPICS/age fotostock **33.18a** Shutterstock **33.18b** Andrew Syred/Photo Researchers **33.19a** Eye of Science/Photo Researchers **33.19b** C. Hough/Custom Medical Stock Photo **33.20** Mark Smith/Photo Researchers **33.21a** Holger Mette/Fotolia **33.21a inset** Holger Mette/Fotolia **33.21b** Manfred Kage/Peter Arnold **33.22a** iStockphoto **33.22b** National Geographic/Getty Images **33.23** Colin Milkins/OSF/Photolibrary **T33.1(1)** Meul/ARCO/NPL/Minden Pictures **T33.1(2)** iStockphoto **T33.1(3)** iStockphoto **T33.1(4)** Trouncs/Wikimedia Commons **T33.1(5)** J. Gall/Photo Researchers **T33.1(6)** iStockphoto **T33.1(7)** J. K. Lindsey **T33.1(8)** Brand X Pictures/Jupiter Images

Chapter 34 **34.2a** D. P. Wilson/FLPA/Minden Pictures **34.2b** Juri Vinokurov/Fotolia **34.3b** Jeff Rotman/NPL/Minden Pictures **34.4a** Sue Scott/OSF/Photolibrary **34.4b** Gary Bell/OceanwideImages.com **34.6** Alan James/NPL/Minden Pictures **34.7a** Audra Reese Kirkland **34.7b** Jeffrey L. Rotman/CORBIS **34.9** Heather Angel/Natural Visions **34.10a** Gary Bell/Getty Images **34.10b** Sinclair Stammers/NPL/Minden Pictures **34.15** Rudie Kuiter/OceanwideImages.com **34.17** Pavel Riha **34.21a** Hiroya Minakuchi/Minden Pictures **34.21b** Shutterstock **34.23a** Tom & Theresa Stack/NHPA/Photoshot **34.23b** James L. Amos/National Geographic Stock **34.24a** Fred McConnaughey/Photo Researchers **34.24b** Paul Souders/Corbis **34.25** Oceans Image/Photoshot **34.26** Arnaz Mehta **34.27a/b** Michael & Patricia Fogden/Minden Pictures **34.28a** Tom McHugh/Photo Researchers **34.28b** Reg Morrison/Auscape/Minden Pictures **34.29** Frank Lukasseck/Photographer's Choice/Getty Images **34.30** Anup Shah/NPL/Minden Pictures **34.31** George Grall/National Geographic Stock **34.32** iStockphoto **34.33** iStockphoto **34.34a** Tui De Roy/Minden Pictures **34.34b** Ingo Arndt/Minden Pictures **34.35a** Cyril Ruoso/J. H. Editorial/Minden Pictures **34.35b** Ingo Arndt/Minden Pictures

Chapter 35 **Opener** Jed Fuhrman, University of Southern California **35.2** SPL/Photo Researchers **35.5** David M. Phillips/Photo Researchers **35.6a** Harold Fisher/Visuals

BS10.2a/b Michael W. Davidson/Florida State University /Molecular Expressions BS10.3 Rosalind Franklin/Photo Researchers BS12.1aL National Cancer Institute BS12.1aR E.S. Anderson/Photo Researchers BS12.1b Sinclair Stammers/Photo Researchers BS14.1a Kwangshin Kim/Photo Researchers BS14.1b Mark J. Grimson & Richard L. Blanton BS14.1c Holt Studios International/Photo Researchers BS14.1d Custom Medical Stock Photo BS14.1e Graphic Science/Alamy BS14.1f Sinclair Stammers/Photo Researchers BS14.1g iStockphoto

ILLUSTRATION AND TABLE CREDITS

Chapter 1 1.3 Adapted by permission from Elsevier from S. P. Moose, J. W. Dudley, and T. R. Rocheford. 2004. Maize selection passes the century mark: A unique resource for 21st century genomics. *Trends in Plant Science* 9: 358–364, Fig 1a. © 2004. 1.7a Adapted by permission of Wiley-Blackwell from T. P. Young and L. A. Isbell. 1991. Sex differences in giraffe feeding ecology: Energetic and social constraints. *Ethology* 87: 79–80, Figs 5a, 6a. 1.8 Adapted by permission of AAAS and the author from M. Wittlinger, R. Wehner, and H. Wolf. 2006. The ant odometer: Stepping on stilts and stumps. *Science* 312: 1965–1967, Figs 1, 2, 3.

Chapter 2 Table 2.1 Reproduced by permission of Pearson Education, Inc., from J. McMurry and R. C. Fay, *Chemistry*, 4th ed., Table 8.1. © 2004.

Chapter 3 Opener PDB ID: 2DN2. S.Y. Park et al. 2006. 1.25 Å resolution crystal structures of human haemoglobin in the oxy, deoxy and carbonmonoxy forms. *J Mol Biol* 360: 690–701. 3.9a PDB ID: 1YTB. Y. Kim et al. 1993. Crystal structure of a yeast TBP/TATA-box complex. *Nature* 365: 512–520. 3.9b PDB ID: 2F1C. G.V. Subbarao and B. van den Berg. 2006. Crystal structure of the monomeric porin OmpG. *J Mol Biol* 360: 750–759. 3.9c PDB ID: 2PTN. J. Walter et al. 1982. On the disordered activation domain in trypsinogen. chemical labelling and low-temperature crystallography. *Acta Crystallogr, Sect B* 38: 1462. 3.9d PDB ID: 1CLG. J.M. Chen 1991. An energetic evaluation of a "Smith" collagen microfibril model. *J Protein Chem* 10: 535–552. 3.12bL PDB ID: 2MHR. S. Sheriff et al. 1987. Structure of myohemerythrin in the azidomet state at 1.7/1.3 Å resolution. *J Mol Biol* 197: 273–296. 3.12bM PDB ID: 1FTP. N.H. Haunerland et al. 1994. Three-dimensional structure of the muscle fatty-acid-binding protein isolated from the desert locust *Schistocerca gregaria*. *Biochemistry* 33: 12378–12385. 3.12bR PDB ID: 1IXA. M. Baron et al. 1992. The three-dimensional structure of the first EGF-like module of human factor IX: comparison with EGF and TGF-alpha. *Protein Sci* 1: 81–90. 3.13a PDB ID: 1D1L. P.B. Rupert et al. 2000. The structural basis for enhanced stability and reduced DNA binding seen in engineered second-generation Cro monomers and dimers. *J Mol Biol* 296: 1079–1090. 3.13b PDB ID: 2DN2. S. Y. Park et al. 2006. 1.25 resolution crystal structures of human haemoglobin in the oxy, deoxy and carbonmonoxy forms. *J Mol Biol* 360: 690–701. 3.19 PDB IDs: 1Q18, 1SZ2. V.V. Lunin et al. 2004. Crystal structures of *Escherichia coli* ATP-dependent glucokinase and its complex with glucose. *J Bacteriol* 186: 6915–6927. 3.23a/b Data from T. Hansen, B. Schlichting, and P. Schönheit. 2002. Glucose-6-phosphate dehydrogenase from the hyperthermophilic bacterium *Thermotoga maritima*: Expression of the *g6pd* gene and characterization of an extremely thermophilic enzyme. *FEMS Microbiology Letters*, 216: 249–253, Fig 1; N. N. Nawani and B. P. Kapadnis. 2001. One-step purification of chitinase from *Serratia marcescens* NK1, a soil isolate. *Journal of Applied Microbiology* 90: 803–808, Fig 3; N. N. Nawani et al. 2002. Purification and characterization of a thermophilic and acidophilic chitinase from *Microbispora* sp.V2. *Journal of Applied Microbiology* 93: 965–975, Fig 7.

Chapter 4 4.11 PDB ID: 1GRZ. B.L. Golden et al. 1998. A preorganized active site in the crystal structure of the *Tetrahymena* ribozyme. *Science* 282: 259–264.

Chapter 6 6.11 Data from J. de Gier, J. G. Mandersloot, and L. L. van Deenen. 1968. Lipid composition and permeability of liposomes. *Biochimica et Biophysica Acta* 150: 666–675. 6.25a PDB ID: 1J4N. H. Sui et al. Structural

basis of water-specific transport through the AQP1 water channel. 2001. *Nature* 414:872–878 6.25b PDB IDs: 1ORS, 1ORQ. Y. Jiang et al. 2003. X-ray structure of a voltage-dependent K⁺ channel. *Nature* 423: 33–41.

Chapter 8 8.7b © 2002. From *Molecular Biology of the Cell*, 4th edition, by B. Alberts et al., Fig 19.5, p. 1069. Reproduced by permission of Garland Science/Taylor & Francis, LLC.

Chapter 9 9.2L PDB ID: 1IRK. S. R. Hubbard et al. 1994. Crystal structure of the tyrosine kinase domain of the human insulin receptor. *Nature* 372: 746–754. 9.2R PDB ID: 1IR3. S. R. Hubbard 1997. Crystal structure of the activated insulin receptor tyrosine kinase in complex with peptide substrate and ATP analog. *EMBO J* 16: 5572–5581. 9.12 PDB ID: 4PFK. P. R. Evans and P. J. Hudson. 1981. Phosphofructokinase: structure and control. *Philos Trans R Soc Lond B Biol Sci* 293: 53–62.

Chapter 14 14.5 Adapted by permission of Dr. Matthew Meselson after M. Meselson and F. W. Stahl. 1958. The replication of DNA in *Escherichia coli*. *PNAS* 44: 671–682, Fig 6. 14.17a Data from J. E. Cleaver. 1968. Defective repair replication of DNA in xeroderma pigmentosum. *Nature* 218: 652–656. 14.17b Adapted by permission of Macmillan Publishers Ltd from J. E. Cleaver. 1968. Defective repair replication of DNA in xeroderma pigmentosum. *Nature* 218: 652–656, Fig 5. © 1968. 14–UN02 Graph adapted by permission of the Radiation Research Society from P. Howard-Flanders and R. P. Boyce. 1966. DNA repair and genetic recombination: Studies on mutants of *Escherichia coli* defective in these processes. *Radiation Research Supplement* 6: 156–184, Fig 8.

Chapter 16 16.2 PDB ID: 3IYD. B. P. Hudson et al. 2009. Three-dimensional EM structure of an intact activator-dependent transcription initiation complex. *PNAS* 106:19830–19835. 16.11 PDB ID: 1ZJW. I. Gruic-Sovulj et al. 2005. tRNA-dependent aminoacyl-adenylate hydrolysis by a nonediting class I aminoacyl-tRNA synthetase. *J Biol Chem* 280: 23978–23986. 16.12 Adapted by permission of the publisher from M. B. Hoagland et al. 1958. A soluble ribonucleic acid intermediate in protein synthesis. *Journal of Biological Chemistry* 231: 241–257, Fig 6. © 1958 American Society for Biochemistry and Molecular Biology. 16.14b PDB IDs: 3FIK, 3FIH. E. Villa et al. 2009. Ribosome-induced changes in elongation factor Tu conformation control GTP hydrolysis. *PNAS* 106: 1063–1068.

Chapter 18 Opener PDB ID: 1ZBB. T. Schalch et al. 2005. X-ray structure of a tetranucleosome and its implications for the chromatin fibre. *Nature* 436: 138–141.

Chapter 20 20.4 Adapted by permission of AAAS and the author from S. J. Giovannoni. 2005. Genome streamlining in a cosmopolitan oceanic bacterium. *Science* 309: 1242–1245, Fig 1. 20.7b Reproduced by permission of GENDIA from www.paternity.be/information_EN .html#identitytest , Examples 1 and 2. 20.9 Data from www.pantherdb.org . P. D. Thomas et al. 2003. PANTHER (Protein ANalysis THrough Evolutionary Relationships): A library of protein families and subfamilies indexed by function.

Chapter 21 21.10a/b Adapted by permission of Macmillan Publishers Ltd after S. B. Carroll. 1995. Homeotic genes and the evolution of arthropods and chordates. *Nature* 376: 479–485, Fig 1. © 1999.

Chapter 24 24.4 After E. B. Daeschler, Neil H. Shubin, and Farish A. Jenkins, Jr. 2006. A Devonian tetrapod-like fish and the evolution of the tetrapod body plan. *Nature* 440:757–763, Fig 6; P. E. Ahlberg and J. A. Clack. 2006. A firm step from water to land. *Nature* 440: 747–749, Fig 1; N. H. Shubin, E. B. Daeschler, and F. A. Jenkins, Jr. 2006. The pectoral fin of *Tiktaalik roseae* and the origin of the tetrapod limb. *Nature* 440: 764–771, Fig 4; M. Hildebrand and G. Goslow. 2001. *Analysis of Vertebrate Structures*, 5th ed. John Wiley and Sons, Inc. 24.11 *Indohyus* reproduced by permission of Macmillan Publishers Ltd after J. G. M. Thewissen et al. 2007. Whales originated from aquatic artiodactyls in the Eocene epoch of India. *Nature* 450: 1190–1194, Fig 5. © 2007. *Rhodocetus* reproduced by permission of AAAS after P. D. Gingerich et al. 2001. Origin

of whales from early Artiodactyls: Hands and feet of Eocene Protocetidae from Pakistan. *Science* 293:2239–2242, Fig 3. *Dorudon* reproduced by permission after P. D. Gingerich et al. 2009. New protocetid whale from the middle Eocene of Pakistan: Birth on land, precocial development, and sexual dimorphism. *PLoS ONE* 4(2): e4366, Fig 1B. *Delphinapterus* reproduced by permission of Skulls Unlimited International, Inc. (www.skullsunlimited.com). 24.14 Adapted by permission from D. P. Genereux and C. T. Bergstrom. 2005. Evolution in action: Understanding antibiotic resistance, Fig 3. In J. Cracraft and R. W. Bybee (eds.), *Evolutionary Science and Society: Educating a New Generation*, pp. 145–153. Colorado Springs, CO: BSCS. 24.16 Data from P. T. Boag and P. R. Grant. 1981. Intense natural selection in a population of Darwin's finches (Geospizinae) in the Galápagos. *Science* 214: 82–85, Table 1. 24.17 Body size and beak shape graphs reproduced by permission of AAAS from P. R. Grant and B. R. Grant. 2002. *Science* 296: 707–711, Fig 1. Beak size graph reproduced by permission of AAAS from P. R. Grant and B. R. Grant. 2006. Evolution of Character Displacement in Darwin's Finches. *Science* 313: 224–226, Fig 2.

Chapter 25 Table 25.2 Data from T. Markow et al. 1993. *HLA* polymorphism in the Havasupai: Evidence for balancing selection. *American Journal of Human Genetics* 53: 943–952, Table 3. 25.3b Data from C. R. Brown and M. B. Brown. 1998. Intense natural selection on body size and wing and tail asymmetry in cliff swallows during severe weather. *Evolution* 52: 1461–1475. 25.4b Data from M. N. Karn, H. Lang-Brown, H. MacKenzie, and L. S. Penrose. 1951. Birth weight, gestation time and survival in sibs. *Annals of Eugenics* 15: 306–322. 25.5b Data from T. B. Smith. 1987. Bill size polymorphism and intraspecific niche utilization in an African finch. *Nature* 329: 717–719. 25.6 Reproduced by permission of Pearson Education, Inc., from S. Freeman and J. Herron, *Evolutionary Analysis*, 3rd ed., Figs 6.15a, 6.15c. © 2004. 25.10b Adapted by permission of Macmillan Publishers Ltd from E. Postma and A. J. van Noordwijk. 2005. Gene flow maintains a large genetic difference in clutch size at a small spatial scale. *Nature* 433: 65–68, Fig 5b. © 2005. 25.13 Adapted by permission of Wiley-Blackwell from M. O. Johnston. 1992. Effects of cross and self-fertilization on progeny fitness in *Lobelia cardinalis* and *L. siphilitica*. *Evolution* 46: 688–702, Fig 1. Table 25.4 Reproduced by permission of W. H. Freeman and Company from Curt Stern, *Principles of Human Genetics*, 3rd ed., Table 5.8. © 1973 by W. H. Freeman and Company. 25.14b Data from J. D. Blount et al. 2003. Carotenoid modulation of immune function and sexual attractiveness in zebra finches. *Science* 300: 125–127, Fig 2. 25.16b, c Data from B. J. Le Boeuf and R. S. Peterson. 1969. Social status and mating activity in elephant seals. *Science* 163: 91–93.

Chapter 26 26.3a Adapted by permission of AAAS and the author after J. C. Avise and W. S. Nelson. 1989. Molecular genetic relationships of the extinct dusky seaside sparrow. *Science* 243: 646–648, Fig 2. 26.3b J. C. Avise and W. S. Nelson. 1989. Molecular genetic relationships of the extinct dusky seaside sparrow. *Science* 243: 646–648. 26.5b Data from N. Knowlton et al. 1993. Divergence in proteins, mitochondrial DNA, and reproductive compatibility across the Isthmus of Panama. *Science* 260:1629–1632, Fig 1. 26.7 Adapted by permission of Wiley-Blackwell from H. R. Dambroski et al. 2005. The genetic basis for fruit odor discrimination in *Rhagoletis* flies and its significance for sympatric host shifts. *Evolution* 59: 1953–1964, Figs 1A, 1B. 26.10b Data from S. E. Rohwer et al. 2001. Plumage and mitochondrial DNA haplotype variation across a moving hybrid zone. *Evolution* 55: 405–422. 26.11 Data from L. H. Rieseberg et al. 1996. Role of gene interactions in hybrid speciation: Evidence from ancient and experimental hybrids. *Science* 272: 741–745.

Chapter 27 27.5a/b/c Data adapted by permission of the publisher from M. A. Nikaido, P. Rooney, and N. Okada. 1999. Phylogenetic relationships among cetartiodactyls based on insertions of short and long interspersed elements: Hippopotamuses are the closest extant relatives of whales. *PNAS* 96: 10261–10266, Figs 1, 7. © 1999 National Academy of Sciences, U.S.A. 27.8, 27.9, 27.10 Data from the

International Commission on Stratigraphy, "International Stratigraphic Chart 2009" (www.stratigraphy.org/column .php?id=Chart/Time Scale). The data on this site are modified from F. M. Gradstein and J. C Ogg (eds). 2004. *A Geologic Time Scale 2004.* Cambridge, UK: Cambridge University Press; and J. G. Ogg, G. Ogg, and F. M. Gradstein. 2008. *The Concise Geologic Time Scale.* Cambridge, UK: Cambridge University Press. **27.11** B. G. Baldwin, D. W. Kyhos, and J. Dvorak. 1990. Chloroplast DNA evolution and adaptive radiation in the Hawaiian Silversword Alliance (Asteraceae–Madiinae). *Annals of the Missouri Botanical Garden* 77: 96–109, Fig 2. **27.12c** J. B. Losos, K. I. Warheitt, and T. W. Schoener. 1997. Adaptive differentiation following experimental island colonization in *Anolis* lizards. *Nature* 387: 70–73. **27.15** Data from J. W. Valentine, D. H. Erwin, and D. Jablonski. 1996. Developmental evolution of metazoan body plans: The fossil evidence. *Developmental Biology* 173: 373–381, Fig 3; R. de Rosa et al. 1999. Hox genes in brachiopods and priapulids and protostome evolution. *Nature* 399: 772–776; D. Chourrout et al. 2006. Minimal ProtoHox cluster inferred from bilaterian and cnidarian Hox complements. *Nature* 442: 684–687. **27.16** Data from M. J. Benton. 1995. Diversification and extinction in the history of life. *Science* 268: 52–58. **27.17a** Adapted by permission of AAAS and the author from W. Alvarez, F. Asaro, and A. Montanari. 1990. Iridium profile for 10 million years across the Cretaceous-Tertiary boundary at Gubbio (Italy). *Science* 250: 1700–1702, Fig 1.

Chapter 28 **28.2** Reproduced by permission of the American Medical Association from G. L. Armstrong, L. A. Conn, and R. W. Pinner. 1999. Trends in infectious disease mortality in the United States during the 20th century. *JAMA* 281: 61–66, Fig 1. © 1999 American Medical Association. All rights reserved. **28.6** After J. G. Elkins et al. 2008. A korarchaeal genome reveals insights into the evolution of the Archaea. *PNAS* 105: 8102–8107, Fig 2; F. D. Ciccarelli et al. 2006. Toward automatic reconstruction of a highly resolved tree of life. *Science* 311:1283–1287, Fig 2. **28.11** Adapted by permission of Pearson Education, Inc., from M. T. Madigan and J. M. Martinko. 2006. *Brock Biology of Microorganisms,*12th ed., Fig 5.9. © 2009.

Chapter 29 **29.1, 29.7** S. M. Adl et al. 2005. The new higher level classification of eukaryotes with emphasis on the taxonomy of protists.*Journal of Eukaryotic Microbiology* 52: 399–451; N. Arisue, M. Hasegawa, and T. Hashimoto. 2005. Root of the Eukaryota tree as inferred from combined maximum likelihood analyses of multiple molecular sequence data. *Molecular Biology and Evolution* 22: 409–420; V. Hampl et al. 2009. Phylogenomic analyses support the monophyly of Excavata and resolve relationships among eukaryotic "supergroups." *PNAS* 106: 3859–3864, Figs 1, 2, 3; J. D. Hackett et al. 2007. Phylogenomic analysis supports the monophyly of cryptophytes and haptophytes and the association of Rhizaria with chromalveolates.*Molecular Biology and Evolution* 24: 1702–1713, Fig 1; P. Schaap et al. 2006. Molecular phylogeny and evolution of morphology in the social amoebas. *Science* 314: 661–663.

Chapter 30 **30.3** Adapted by permission of Pearson Education, Inc., from O. Owen et al. 1998. *Natural Resource Conservation: Management for a Sustainable Future* 7th ed., p. 509, Fig 21.2. © 1998.**30.7, 30.10, 30. 24** Y.-L. Qiu et al. 2006. The deepest divergences in land plants inferred from phylogenomic evidence. *PNAS* 103:15511–15516, Fig 1; K. S. Renzaglia et al. 2007. Bryophyte phylogeny: Advancing the molecular and morphological frontiers. *The Bryologist* 110: 179–213; J. F. Pombert et al. 2005. The chloroplast genome sequence of the green alga *Pseudendoclonium akinetum* (Ulvophyceae) reveals unusual structural features and new insights into the branching order of chlorophyte lineages.*Molecular Biology and Evolution* 22: 1903–1918. **30.22** Data from S. D. Johnson and K. E. Steiner. 1997. Long-tongued fly pollination and evolution of floral spur length in the *Disa draconis* complex (Orchidaceae). *Evolution* 51: 45–53. **30.26** P. S. Soltis and D. E. Soltis. 2004. The origin and diversification of angiosperms. *American Journal of Botany* 91: 1614–1626, Figs 1, 2, 3.

Chapter 31 **31.7** S. M. Adl et al. 2005. The new higher level classification of eukaryotes with emphasis on the taxonomy of protists.*Journal of Eukaryotic Microbiology* 52: 399–451; N. Arisue, M. Hasegawa, and T. Hashimoto. 2005. Root of the Eukaryota tree as inferred from combined maximum likelihood analyses of multiple molecular sequence data. *Molecular Biology and Evolution* 22: 409–420; V. Hampl et al. 2009. Phylogenomic analyses support the monophyly of Excavata and resolve relationships among eukaryotic "supergroups." *PNAS* 106: 3859–3864, Figs 1, 2, 3; J. D. Hackett et al. 2007. Phylogenomic analysis supports the monophyly of cryptophytes and haptophytes and the association of Rhizaria with chromalveolates. *Molecular Biology and Evolution* 24: 1702–1713, Fig 1; P. Schaap et al. 2006. Molecular phylogeny and evolution of morphology in the social amoebas. *Science* 314: 661–663. **31.8** T. Y. James et al. 2006. Reconstructing the early evolution of fungi using a six-gene phylogeny. *Nature* 443: 818–822, Fig 1.

Chapter 32 **32.9** A. M. A. Aguinaldo et al. 1997. Evidence for a clade of nematodes, arthropods, and other moulting animals. *Nature* 387: 489–493, Figs 1, 2, 3; C. W. Dunn et al. 2008. Broad phylogenomic sampling improves resolution of the animal tree of life. *Nature* 452: 745–750, Figs 1, 2.

Chapter 33 **33.2** A. M. A. Aguinaldo et al. 1997. Evidence for a clade of nematodes, arthropods, and other moulting animals. *Nature* 387: 489–493, Figs 1, 2, 3; C. W. Dunn et al. 2008. Broad phylogenomic sampling improves resolution of the animal tree of life. *Nature* 452: 745–750, Figs 1, 2. **33.7** T. H. Struck et al. 2007. Annelid phylogeny and the status of Sipuncula and Echiura. *Evolutionary Biology* 7: 571–11.

Chapter 34 **34.1** After C. W. Dunn et al. 2008. Broad phylogenomic sampling improves resolution of the animal tree of life. *Nature* 452: 745–750, Figs 1, 2; F. Delsuc et al. 2006. Tunicates and not cephalochordates are the closest living relatives of vertebrates. *Nature* 439: 965–968, Fig 1. **34.12, 44.21** J. E. Blair and S. B. Hedges.2005. Molecular phylogeny and divergence times of deuterostome animals. *Molecular Biology and Evolution* 22: 2275–2284, Figs 1, 3, 4. **34.16** Adapted by permission of Macmillan Publishers Ltd after E. B. Daeschler et al. 2006. A Devonian tetrapod-like fish and the evolution of the tetrapod body plan. *Nature* 440:757–763, Fig 6, © 2006; also after N. H. Shubin et al. 2006. The pectoral fin of Tiktaalik roseae and the origin of the tetrapod limb. *Nature* 440:764–771, Fig 4, © 2006. **34.22** J. E. Blair and S. B. Hedges. 2005. Molecular phylogeny and divergence times of deuterostome animals. *Molecular Biology and Evolution* 22: 2275–2284, Figs 1, 3, 4; F. R. Liu et al. 2001. Molecular and morphological supertrees for eutherian (placental) mammals. *Science* 291: 1786–1789; A. B. Prasad et al. 2008. Confirming the phylogeny of mammals by use of large comparative sequence data sets. *Molecular Biology and Evolution* 25:1795–1808. **34.36** A. B. Prasad et al. 2008. Confirming the phylogeny of mammals by use of large comparative sequence data sets. *Molecular Biology and Evolution* 25:1795–1808, Figs 1, 2, 3. **34.39** J. Z. Li et al. 2008. Worldwide human relationships inferred from genome-wide patterns of variation. *Science* 319: 1100–1104, Fig 1. **34.40** After L. L. Cavalli-Sforza and M. W. Feldman. 2003. The application of molecular genetic approaches to the study of human evolution. *Nature Genetics Supplement* 33: 266–275, Fig 3.

Chapter 35 **35.3** Adapted by permission of the publisher from J. G. Bartlett and R. D. Moore. 2003. Improving HIV Therapy. *Scientific American* July 2003: 30–37. Originally published July 1998. © 1998 Scientific American. All rights reserved. **35.4** Data compiled by the United Nations AIDS program. **35.15** F. Gao et al. 1999. Origin of HIV-1 in the chimpanzee *Pan troglodytes troglodytes*. *Nature* 397: 436–441, Fig 2.

Chapter 36 **36.4** Illustration adapted by permission from "Root Systems of Prairie Plants" by Heidi Natura. © 1995 Conservation Research Institute.

Chapter 37 **37.3** Data from R. G. Cline and W. S. Campbell. 1976. Seasonal and diurnal water relations of selected forest species. *Ecology* 57: 367–373, Fig 1. **37.12** Adapted by permission of the American Society of Plant Biologists from C. M. Wei, M. T. Tyree, and E. Steudle. 1999. Direct measurement of xylem pressure in leaves of intact maize plants. A test of the cohesion-tension theory taking hydraulic architecture into consideration. *Plant Physiology* 121: 1191–1205, Fig 6. © 1999 American Society of Plant Biologists. **37.22** Data from N. D. DeWitt and M. R. Sussman. 1995. Immunocytological localization of an epitope-tagged plasma membrane proton pump (H$^+$–ATPAse) in phloem companion cells. *Plant Cell* 7: 2053–2067, Fig 7.

Chapter 39 **39.5** After Darwin and Darwin, 1897. *The Power of Movement in Plants* (D. Appleton & Co. New York). **Table 39.1** Data from H. A. Borthwick et al. 1952. A reversible photoreaction controlling seed germination. *PNAS* 38: 662–666, Table 1.

Chapter 40 **40.12** Y.-L. Qiu et al. 2006. The deepest divergences in land plants inferred from phylogenomic evidence. *PNAS* 103:15511–15516, Fig 1; K. S. Renzaglia et al. 2007. Bryophyte phylogeny: Advancing the molecular and morphological frontiers. *The Bryologist* 110: 179–213; J. F. Pombert et al. 2005. The chloroplast genome sequence of the green alga *Pseudendoclonium akinetum* (Ulvophyceae) reveals unusual structural features and new insights into the branching order of chlorophyte lineages. *Molecular Biology and Evolution* 22: 1903–1918.

Chapter 41 **41.10** Data from K. Schmidt-Nielsen. 1984. *Scaling: Why is animal size so important?* Cambridge, UK: Cambridge University Press. **41.11** Adapted by permission of The Company of Biologists from P. R. Wells and A. W. Pinder. 1996. The respiratory development of Atlantic salmon. II. Partitioning of oxygen uptake among gills, yolk sac and body surfaces. *Journal of Experimental Biology* 199: 2737–2744, Figs 1A, 3.

Chapter 42 **42.14a/b** Adapted by permission of Elsevier from K. J. Ullrich, K. Kramer, and J. W. Boyer. 1961. Present knowledge of the counter-current system in the mammalian kidney. *Progress in Cardiovascular Diseases* 3: 395–431, Figs 5, 7. © 1961.

Chapter 43 **43.17a** Graph reproduced with permission from The American Diabetes Association from L. O. Schulz et al. 2006. Effects of traditional and western environments on prevalence of type 2 diabetes in Pima Indians in Mexico and the U.S. *Diabetes Care* 29: 1866–1871, Fig 1. © 2006 American Diabetes Association. **43.17b** Data from L. O. Schulz et al. 2006. Effects of traditional and western environments on prevalence of type 2 diabetes in Pima Indians in Mexico and the U.S. *Diabetes Care* 29: 1866–1871, Table 2.

Chapter 44 **44.7** Adapted by permission from Y. Komai. 1998. Augmented respiration in a flying insect. *Journal of Experimental Biology* 201: 2359–2366, Fig 7. **44.8** Adapted by permission from Y. Komai. 1998. Augmented respiration in a flying insect. *Journal of Experimental Biology* 201: 2359–2366, Fig 1. **44.11** W. Bretz and K. Schmidt Nielsen. 1971. Bird respiration: Flow patterns in the duck lung. *Journal of Experimental Biology* 54:103–118. **44.20** Data from E. H. Starling. 1896. On the absorption of fluids from the connective tissue spaces. *Journal of Physiology* 19: 312–326.

Chapter 45 **45.21** Adapted by permission from Elsevier from K. C. Martin et al. 1997. Synapse-specific, long-term facilitation of Aplysia sensory to motor synapses: A function for local protein synthesis in memory storage. *Cell* 91: 927–938. © 1997. **45.22b** Adapted by permission from Elsevier from K. C. Martin et al. 1997. Synapse-specific, long-term facilitation of Aplysia sensory to motor synapses: A function for local protein synthesis in memory storage. *Cell* 91: 927–938. © 1997.

Chapter 46 **46.2a/b** Data from J. E. Rose et al. 1971. Some effects of stimulus intensity on response of auditory nerve fibers in the squirrel monkey. *Journal of Neurophysiology* 34: 685–699. **46.12** Adapted by permission of AAAS and the author from David M. Hunt et al. 1995. The Chemistry of John Dalton's Color Blindness. *Science* 267: 984–988, Fig 3. **46.19** PDB ID: 1KWO. D. M. Himmel et al. 2002. Crystallographic findings on the internally uncoupled and near-rigor states of myosin: further insights into the mechanics of the motor. *PNAS* 99: 12645–12650.

Chapter 47 **47.10a/b** Data from G. V. Upton et al. 1973. Evidence for the internal feedback phenomenon in human subjects: Effects of ACTH on plasma CRF. *Acta Endocrinologica* 73: 437–443. **47.13** Data from D. Toft, and J. Gorski. 1966. A receptor molecule for estrogens: Isolation from the rat uterus and preliminary characterization. *PNAS* 55: 1574–1581. **47.15b** Data from T. W. Rall et al. 1956. The relationship of epinephrine and glucagon to liver phosphorylase. *Journal of Biological Chemistry* 224: 463–475. **47.16** Reproduced by permission of the Nobel Foundation from E. W. Sutherland. "Studies on the mechanism of hormone action." A lecture delivered in Stockholm, 11 December 1971. © 1971 Nobel Foundation.

Chapter 48 **48.3a** Reproduced by permission of AAAS from R. G. Stross and J. C. Hill. 1965. Diapause induction in Daphnia requires two stimuli. *Science* 150: 1462–1464, Fig 3. **48.3b** Data from O. T. Kleiven, P. Larsson, and A. Hobæk. 1992. Sexual reproduction in Daphnia magna requires three stimuli. *Oikos* 65: 197–206, Table 4. **48.5** Adapted by permission of Macmillan Publishers Ltd from C. S. C. Price, K. A. Dyer, and J. A. Coyne. 1999. Sperm competition between *Drosophila* males involves both displacement and incapacitation. *Nature* 400: 449–452, Fig 3c. © 1999. **48.7b** R. Shine and M. S. Y. Lee. 1999. A reanalysis of the evolution of viviparity and egg-guarding in squamate reptiles. *Herpetologica* 55: 538–549, Figs 1, 2, 3. **48.8b** Reproduced by permission of Elsevier from C. Hotzyand and G. Arnqvist. 2009. Sperm competition favors harmful males in seed beetles. *Current Biology* 19: 404–407. © 2009. **48.14, 48.15** Data from R. Stricker et al. 2006. Establishment of detailed reference values for luteinizing hormone, follicle stimulating hormone, estradiol, and progesterone during different phases of the menstrual cycle on the Abbott ARCHITECT® analyzer. *Clin Chem Lab Med* 44: 883–887, Tables 1A and 1B. **48.19b** Reproduced by permission of AAAS from C. Ikonomidou et al. 2000. Ethanol-induced apoptotic neurodegeneration and fetal alcohol syndrome. *Science* 287: 1056–1060, Fig 4. **48.21** Adapted by permission of the publisher and author from U. Högberg and I. Joelsson. 1985. The decline in maternal mortality in Sweden, 1931–1980. *Acta Obstetricia et Gynecologica Scandinavica* 64: 583–592, Fig 1. Taylor & Francis Group, www.informaworld.com.

Chapter 49 **49.6a** PDB ID: 1IGT. L. J. Harris et al. 1997. Refined structure of an intact IgG2a monoclonal antibody. *Biochemistry* 36: 1581–1597. **49.6b** PDB ID: 1TCR. K. C. Garcia et al. 1996. An αβ cell receptor structure at 2.5Å and its orientation in the TCR-MHC complex. *Science* 274: 209–219.

Chapter 50 **50.4a** Adapted by permission of the University of California Press Journals and the author from M. B. Saffo. 1987. New light on seaweeds. *BioScience* 37: 654–664, Fig 1. © 1987 American Institute of Biological Sciences. **50.28** Graphs adapted by permission of AAAS and the author from A. K. Knapp et al. 2002. Rainfall variability, carbon cycling, and plant species diversity in a mesic grassland. *Science* 298: 2202–2205, Figs 1, 2, 3.

Chapter 51 **51.2** After M. Sokolowski. 2001. *Drosophila*: Genetics meets behavior. *Nature Reviews Genetics* 2: 879–890, Fig 2. **51.3b** Data from R. E. Hegner, S. T. Emlen, and N. J. Demong. 1982. Spatial organization of the white-fronted bee-eater. *Nature* 298: 264–266. **51.4b** Adapted by permission of the author from D. Crews. 1975. Psychobiology of reptilian reproduction. *Science* 189: 1059–1065, Fig 2. **51.16** Data from M. Daly and M. Wilson. 1988. Evolutionary social psychology and family homicide. *Science* 242: 519–524, Fig 1.

Chapter 52

Table 52.1 Data reproduced with kind permission from Springer Science and Business Media from H.Strijbosch and R. C. M. Creemers, 1988. Comparative demography of sympatric populations of *Lacerta vivipara* and *Lacerta agilis. Oecologia* 76: 20–26. © 1988 Springer-Verlag. **52.3** Reproduced by permission of AAAS from C. K. Ghalambor and T. E. Martin. 2001. *Science* 292: 494–496, Fig 1c. **52.7b** Data from G. F. Gause. 1934. *The Struggle for Existence*. Baltimore, MD: Williams & Wilkins. **52.8a** Reproduced with kind permission of Springer Science and Business Media and the author from G. E. Forrester. 1995. Strong density-dependant survival and recruitment regulate the abundance of a coral reef fish. *Oecologia* 103: 275–282, Fig 2. © 1995 Springer-Verlag. **52.8b** Data from P. Arcese and J. N. M. Smith. 1988. Effects of population density and supplemental food on reproduction in song sparrows. *Journal of Animal Ecology* 57: 119–136, Fig 2. **52.11** Reproduced by permission of Alberta Sustainable Resource Development. 2002. From www3.gov.ab.ca/srd/fw/watch/rabb_cycles.html. © 2002 Government of Alberta. **52.12** Graphs reproduced by permission of AAAS from C. J. Krebs et al. 1995. Impact of food and predation on the snowshoe hare cycle. *Science* 269: 1112–1115, Fig 3. **52.13b** Data from T. Valverde and J. Silvertown. 1998. Variation in the demography of a woodland understorey herb (*Primula vulgaris*) along the forest regeneration cycle: Projection matrix analysis. *Journal of Ecology* 86: 545–562. **52.14a/b** Data from U.S. Census Bureau, International Data Base (IDB), www.census.gov/ipc/www/idb/.

Chapter 53 **53.2a/b, 53.3, 53.4a/b** G. F. Gause. 1934. *The Struggle for Existence.* Baltimore, MD: Williams & Wilkins. **53.8** Reproduced by permission of AAAS from P. R. Grant and B. R. Grant. 2006. Evolution of character displacement in Darwin's finches. *Science* 313, Fig 2. **53.11** Adapted by permission of Ecological Society of America from G. H. Leonard, M. D. Bertness, and P. O. Yund. 1999. Crab predation, waterborne cues, and inducible defenses in the blue mussel, *Mytilus edulis. Ecology* 80: 1–14. © 1999. **53.12** Adapted by permission of Ecological Society of America from G. H. Leonard, M. D. Bertness, and P. O. Yund. 1999. Crab predation, waterborne cues, and inducible defenses in the blue mussel, *Mytilus edulis. Ecology* 80: 1–14. © 1999. **53.13b, c** Data from G. D. Martinsen, E. M. Driebe, and T. G. Whitham. 1998. Indirect interactions mediated by changing plant chemistry: Beaver browsing benefits beetles. *Ecology* 79: 192–200. **53.17** Adapted by permission of Ecological Society of America from J. H. Cushman and T. G. Whitham. 1989. Conditional mutualism in a membracid-ant association: Temporal, age-specific, and density-dependant effects. *Ecology* 70: 1040–1047. **53.18** Adapted by permission of Ecological Society of America from D. G. Jenkins and A. L. Buikema, Jr. 1998. Do similar communities develop in similar sites? A test with zooplankton structure and function. *Ecological Monographs* 68: 421–443. **53.19c** Adapted with kind permission from Springer Science and Business Media from R. T. Paine. 1974. Intertidal community structure. *Oecologia* 15: 93–120. © 1974 Springer-Verlag. **53.21b** Data from T. W. Swetnam. 1993. Fire history and climate change in giant sequoia groves. *Science* 262: 885–889. **53.24a** Adapted by permission of Wiley-Blackwell from R. H. MacArthur and E. O. Wilson. 1963. An Equilibrium Theory of Insular Zoogeography. *Evolution*: 17: 373–387, Fig 4. **53.24b, c** Data from R. H. MacArthur and E. O. Wilson. 1963. An equilibrium theory of insular zoogeography. *Evolution*: 17: 373–387, Fig 5. **53.26** Data from W. V. Reid and K. R. Miller. 1989. *Keeping Options Alive: The Scientific Basis for Conserving Biodiversity*. Washington, DC: World Resources Institute.

Chapter 54 **54.3** Data from J. R. Gosz et al. 1978. The flow of energy in a forest ecosystem. *Scientific American* 238: 92–102. **54.7** Reproduced by permission of the publisher from C. A. Mackenzie, A. Lockridge, and M. Keith. 2005. Declining sex ratio in a first nation community. *Environmental Health Perspectives* 113: 12–96, Fig 1. **54.9a/b/c** Data from R. H. Whittaker and G. E. Likens. 1973. Primary production: The biosphere and man. *Human Ecology* 1: 357–369, Table 1; H. Lieth.1973. Primary production: Terrestrial ecosystems. *Human Ecology* 1: 303–332, Tables 2, 3; J. S. Bunt. 1973. Primary production: Marine ecosystems. *Human Ecology* 1: 333–345, Table 1; G. E. Likens. 1973. Primary Production: Freshwater ecosystems. *Human Ecology* 1: 347–356, Table 3. **54.16** Data from W. H. Schlesinger. 1997. *Biogeochemistry: An Analysis of Global Change*, 2nd ed. San Diego, CA: Academic Press. **54.17** Data from P. M. Vitousek et al. 1997. Human alteration of the global nitrogen cycle: Sources and consequences. *Ecological Applications* 7: 737–750. **54.20** Data from BP Statistical Review of World Energy 2007, pp. 6–21 **54.21** Data from NOAA.gov, www.esrl.noaa.gov/gmd/ccgg/trends/. **54.22a** After G. Beaugrand et al. 2002. Reorganization of North Atlantic marine copepod biodiversity and climate. *Science* 296: 1692–1694. **54.22b** Adapted by permission of the publisher fromN. L. Bradley et al. 1999. Phenological changes reflect climate change in Wisconsin. *PNAS* 96: 9701–9704. © 1999 National Academy of Sciences, U.S.A. **54.23** Reproduced by permission of AAAS from R. R. Nemani et al. 2003. Climate-driven increases in global terrestrial net primary production from 1982 to 1999. *Science* 300: 1560–1563, Fig 2. **54.23b** After M. J. Behrenfeld et al. 2006. Climate-driven trends in contemporary ocean productivity. *Nature* 444: 752–755.

Chapter 55 **55.1** F. R. Liu et al. 2001. Molecular and morphological supertrees for eutherian (placental) mammals. *Science* 291: 1786–1789, Fig 1; A. B. Prasad et al. 2008. Confirming the phylogeny of mammals by use of large comparative sequence data sets. *Molecular Biology and Evolution* 25:1795–1808; U. Arnason, A. Gullberg, and A. Janke. 2004. Mitogenomic analyses provide new insights into cetacean origin and evolution. *Gene* 333: 27–34, Fig 1. **55.3a/b** Adapted by permission of Macmillan Publishers Ltd after C. D. L. Orme et al. 2005. Global hotspots of species richness are not congruent with endemism or threat. *Nature* 436: 1016–1019. © 2005. **55.4** Reproduced by permission of Conservation International from *Conservation International 2008 Annual Report*, 12–13. © 2008 Conservation International (www.conservation.org). **55.6** Reproduced by permission of the University of California Press Journals and the author from O.Venter et al. 2006. Threats to endangered species in Canada. *BioScience* 56: 903–910, Fig 2. © 2006 American Institute of Biological Sciences. **55.10b** Adapted by permission of AAAS and the author from W. F. Laurance et al. 1997. Biomass collapse in Amazonian forest fragments. *Science* 278: 1117–1118, Fig 2. **55.11** Adapted by permission of The Royal Society and the author from J. T. Hogg et al. 2006. Genetic rescue of an insular population of large mammals. *Proceedings of the Royal Society B* 273: 1491–1499, Fig 1. **55.12b** Adapted by permission from D. B. Lindenmayer and R. C. Lacy. 1995. Metapopulation viability of Leadbeater's possum, *Gymnobelideus leadbeateri*, in fragmented old-growth forests. *Ecological Applications* 5: 164–182, Fig 5G. **55.13** Adapted by permission of the author fromJ. M. Diamond and E. Mayr. 1976. Species-area relation for birds of the Solomon Archipelago. *PNAS* 73: 262–266, Fig 1. **55.14** Graphs reproduced by permission of AAAS from D. Tilman et al. 1997. The influence of functional diversity and composition on ecosystem processes. *Science* 277: 1300–1302, Fig 1. **55.16** Reproduced by permission of AAAS from E. I. Damschen et al. 2006. Corridors increase plant species richness at large scales. *Science* 313: 1284–1286, Fig 2B. **55-UN01T (England maps)** Reproduced by permission of the author from D. S. Wilcove, C. H. McLellan, and A. P. Dobson. 1986. Habitat fragmentation in the temperate zone. In M. E. Soule (ed.), *Conservation Biology: The Science of Scarcity and Diversity*. Sunderland, MA: Sinauer Associates, pp. 237–256, Fig 1. **55-UN01B (U.S. maps)** After W. B. Greeley, Chief, U.S. Forest Service. 1925. Relation of geography to timber supply. *Economic Geography* 1: 1–11.

Index

Boldface page numbers indicate a glossary entry; page numbers followed by an *f* indicate a figure; page numbers followed by *t* indicate a table.

2' carbon, 66
3' carbon, 61
3-D visualization techniques, B:15–B:16
3' end, DNA, 264, 265, 290
3-phosphoglycerate, 185
3' poly(A) tail, 295
5' cap, DNA, 295
5' carbon, 61
5' end, DNA, 264, 265, 290
−10 box, 291
20-hydroxyecdysone, **936**
30-nanometer fiber, 321
−35 box, 291

A

ABC model, 410
Abdomen, 637, **639**
Abiotic, **995**
Abiotic factors
  for density-independent population growth, 1044
  for detritus decomposition rate, 1093
  for geographic distribution of organisms, 1013, 1015–17
ABO blood group, 245–47
Abomasum, 850
Abortions, spontaneous, 226, 440
Aboveground biomass, **1002**
Abscisic acid (ABA), **769**, 770–75t
Abscission, **768**, 774
Abscission zone, **774**
Absolute dating, 422
Absorption, 696–97, 841–42, **845**
Absorption spectrum, **175–76**
Absorptive feeding, 530, 583, 586
Abundance
  animal, 624f
  arthropod, 623
  fossil record bias for, 481
  prokaryotic, 498
  species interactions and, 1059
  vertebrate, 655f
  viral, 675–76
Abuse, human non-kin, 1034
Acanthocephala, 603t
Accessory fluids, male reproductive, 958, 959t
Accessory pigments, 532t
Acclimation (acclimatization), 429–30, **804**, 806
Acetaldehyde, 34, 167
Acetate, 25
Acetic acid, 25
Acetone, 34
Acetylation, **322**
Acetylcholine, 898t, **925**
Acetyl CoA, **157**
  catabolic pathways and, 168
  cellular respiration and, 153
  oxidizing of, to CO₂, 158–61
  processing of pyruvate to, 156–58

Achromatopsia, 446
Acid-base reactions, 25–26
Acid-growth hypothesis, **761–62**
Acid hydrolases, 110–11
Acidic habitats, 516
Acidic soils, 743
Acidification, 1100, 1114
Acidity, pH scale and, 26
Acid rain, 490
Acids, **25–26**. *See also* Amino acids; Stomach acid
Acoelomates (Acoelomorpha), 603t, **605**, 620
Acorn worms, 646
Acoustic communication, 1029
Acquired characters hypothesis, 231, 234, 415, 429
Acquired immune deficiency syndrome (AIDS), 281, 677–78, **990**. *See also* HIV (human immunodeficiency virus)
Acquired immune response. *See* Adaptive immune response
Acrosomes, **389**
Actin, 123, **922**
Actin filaments, **123**
  image of, 102f
  in mitosis, 200
  movement and, 45
  in muscle contraction, 922–25
  structure and function of, 123–24t
Actinistia, 663
Actinobacteria, 512, 514
Actinopterygii, 656, 662–63
Action potentials, **890–95**, 902–4f, 908–9
Action spectrum, **176**
Activation
  lymphocyte, 978–79f
  mitosis-promoting factor, 204
Activation energy, **51–53**. *See also* Enzymes
Active exclusion, plant, 747–48
Active site, **53–54**
Active transport, 97–99, **732**, 825–26, 834. *See also* Sodium-potassium pump
Adaptations, **5**, **627**, **804**. *See also* Traits
  acclimation vs., 429–30, 804, 806
  in animal anatomy and physiology, 804–6
  for animal body temperature homeostasis, 818–19
  animal mouthparts as, 843–44
  animal surface area and, 814
  endothermy vs. ectothermy, 817
  feeding (*see* Feeding adaptations)
  fitness and, 5, 424, 627 (*see also* Fitness)
  natural selection and, 435
  plant nutritional, 750–52
  traits as, 5 (*see also* Traits)

Adaptive immune response, **977–90**
  B-cell activation and antibody secretion, 986
  B cells and T cells of, 979, 989t
  characteristics of, 977–78
  clonal-selection theory on, 979–83f
  destruction of viruses by, 987–88
  distinguishing self from nonself in, 983–84
  immune system rejection of foreign tissues and organs by, 988
  immunological memory and, 989–90
  killing of bacterial and foreign cells by, 987
  lymphocytes and, 978–79f
  T-cell activation in, 984–86
Adaptive immunity, **974**
Adaptive radiations, **484**, **564**, **844**
  angiosperm radiation as, 564–65
  Cambrian explosion as, 486–88
  cichlids as, 844
  ecological opportunity and, 484–85
  Hawaiian silverswords as, 700
  humans and, 672
  morphological innovation and, 485–86
  of seed plants, 560–61
Adenine, 60, 63–64, 66–67, 261, 279, 290
Adenosine diphosphate (ADP). *See* ADP (adenosine diphosphate)
Adenosine monophosphate (AMP), 157
Adenosine triphosphate (ATP). *See* ATP (adenosine triphosphate)
Adenoviruses, 679f, 690
Adenylyl cyclase, **316**
ADH (antidiuretic hormone), **837–38**, 931, **940**
Adhesion, **22–23**, 136–38, **724**
Adhesion proteins, 131f, 136–38
Adipocytes, **938**
Adipose tissue, 807, 938
ADP (adenosine diphosphate), 124, **148**, 153, 155, 181
Adrenal glands, 837, **933**, 937, 938
Adrenaline, 140, **937**
Adrenocorticotropic hormone (ACTH), **941**
Adults, **616**
Adult testing, genetic, 350
Adventitious roots, **698**, 768
Aerobic respiration, 166, 487, **510**. *See also* Respiratory systems
Afferent division, **899**
Aflatoxin, 272
Africa, human evolution out of, 672–73f
Age, trisomy and maternal, 227
Age classes, **1039**
Age pyramids, 1051f
Age-specific fecundity, **1039**, 1115
Age-specific survivorship, 1115
Age structure, **1050–52**
Agglutination, **987**
Aggregate fruits, **797**
Aging, plant, 773–74
Agre, Peter, 95

Agriculture
  biotechnology in, 354–56
  fungi and, 581
  global water cycle and, 1095
  habitat destruction by, 1112
  plants and, 547, 548
Agrobacterium tumefaciens, 355–56, 515
Aguinaldo, Anna Marie, 607
AIDS (acquired immune deficiency syndrome), 281, 677–78, **990**. *See also* HIV (human immunodeficiency virus)
Air, animal olfaction and, 919–20. *See also* Atmosphere
Air pressure, animal sensing of, 610
Alanine, 42t
Alarm calling, 1031–33
Albumen, **658**
Albumin, **877**
Alcohol fermentation, **167**
Alcohols, 35f
Alcohol use, fetal alcohol syndrome and, 969–70
Aldehydes, 34, 35f
Aldose, 72
Aldosterone, 837, **940**
Algae. *See also* Green algae
  algal blooms of, 521–22, 1102
  brown algae, 535–36, 543
  lichens and, 589, 597
  Plantae and, 536
  red algae, 519f, 536, 539
Alimentary canal, 845
Alkaline mucus, 959t
Alkaline soil, 743
Alkalinity, pH scale and, 26
Allantois membrane, 658
All-Cells-from-Cells hypothesis, 2–4
Alleles, **212**, **234**
  analyzing changes in frequencies of, 436–40
  creating mutant, 277
  crossing over and, 222
  deleterious, 224
  effect of inbreeding on frequencies of, 451
  female choice for good, 453
  foraging, in fruit flies, 1021
  frequencies of, in populations, 435
  genes and, 212, 221 (*see also* Genes)
  genetic bottlenecks and reduction in, 446
  identifying human, as recessive or dominant, 250–51
  introducing novel, into human cells, 351–52
  in Mendelian genetics, 239t
  mutant, 285
  mutations as new, 448–49
  particulate inheritance hypothesis and, 234–35
  plant pathogen resistance and, 777
  recombination of, 244–45
Allergens, **991**
Allergies, 990–**91**
Alligators, 660, 667

Allolactose, 311
Allopatric speciation, 462, **463**, 464
Allopatry, **463**, 468
Allophycocyanin, 532*t*
Alloploidy, 467
Allopolyploidy, **466**, 467–68, 471*t*
Allosteric regulation, **54–55**, **314**
All Taxa Biodiversity Inventory, 1108
α-amylase, 144, **770**
α-glucose, 72, 74
alpha-helix (α-helix), **46–47**
Alpine skypilots, 794
ALS (Lou Gehrig's disease), 376
Alternate leaves, 702*f*
Alternation of generations, **402**, **534**–36, **556**–58, 592, **784**
Alternative splicing, **328–29**, 331, **369**
Altman, Sidney, 68
Altruism, **1031**–34
Alveolata, 524*t*, 537, 540–41
Alveoli (alveolus), 537, **868**
Ambisense viruses, **686**
American elm trees, 581
Amines, 35*f*
Amino acids, **40**
  amphipathic proteins and, 92
  anabolic synthesis of, 169
  as animal essential nutrient, 842
  aquaporins and, 95
  carbon functional groups and, 34, 35*t*
  in central dogma, 280
  chemical evolution and formation of, in prebiotic soup, 38–39
  experiment on metabolic pathway of, 277–78
  genetic code for, 282–83, 297
  hormones as derivatives of, 933
  major, found in organisms, 41*t*
  as neurotransmitters, 898*t*
  peptidoglycan and, 76
  polymerization of, to form proteins, 42–44 (*see also* Protein synthesis)
  side chains of, 40–42 (*see also* Side chains, amino acid)
  specification of, by mRNA triplets in translation, 297
  structure of, 40
  sugars in, 77
  transfer of, from transfer RNA to proteins, 297–99
  water and solubility of, 42*t*
Aminoacyl tRNA, **298**
Aminoacyl tRNA synthetases, **298**
Amino functional group, 34–35*f*, 40, 43–44, 46–47
Amino-terminus, 43–44, 301
Ammodramus maritimus, 461–62
Ammonia, 19–21*f*, 24*t*, 39–40, 510–12, 748, **829**
Amnion, **968**
Amniotes, **655**
Amniotic eggs, **655**, **658**
Amniotic fluid, 968
Amoebic dysentery, 522*t*
Amoeboid motion, **532**–33
Amoebozoa, 524*t*, 536, 537
AMP (Adenosine monophosphate), 157
Amphibia, 664
Amphibians, **664**. See also Animal(s); Deuterostomes
  Amphibia lineage and, 664
  fungal parasites of, 595

global warming and, 1100–1101
hearts of, 878
hormones in metamorphosis of, 935–36
lateral line system as hearing in, 913
Amphioxus, 651, 652
Amphipathic compounds, **84**, 92
Amplification, signal, 141–42, 908, 947
Amylases, **79**, **847**, 855*t*
Amylopectin, 74
Amyloplasts, **765**
Amylose, 74
Anabolic pathways, **168**, 169
Anadromous, **661**
Anaerobic respiration, 166, **510**. See also Respiratory systems
Anaphase
  meiosis, 218
  mitosis, **199**–200
Anaphylactic shock, 991
Anatomy, **803**, 810–11*f*. See also Animal anatomy and physiology; Plant anatomy and physiology
Ancestral traits, 418–22, 423*f*, 460–61, **475**–79
Anemomes, 619
Aneuploidy, **226**, 227, **286**
Anfinson, Christian, 50
Angiosperms, **550**. See also Green plants; Land plants; Plant(s)
  adaptive radiations of, 485, 564–65
  anatomy and physiology of, 695–96 (*see also* Plant anatomy and physiology)
  in Cenozoic era, 483
  development in (*see* Plant development)
  diversification of, 551–52
  flowers of, 561–62
  gametogenesis in, 402–4
  life cycle of, 785*f*
  parasitic, 751
  pollination of, 576–77*f*
  reproduction in, 783 (*see also* Plant reproduction)
Anglerfish, 1030
Anhydrite, 490
Animal(s), **601**–22, 958
  biological importance of, 602–3
  biological methods for studying, 603–9
  bodies (*see* Bodies, animal; Body plans)
  Cambrian explosion in, 486–88
  cells of (*see* Animal cells)
  chemical signals in (*see* Chemical signals, animal)
  cloning, 377–78
  color vision in nonhuman, 917–18
  comparative morphology of, 603–7
  deuterostome and protostome, 606–7 (*see also* Deuterostomes; Protostomes)
  development of (*see* Animal development)
  diseases of (*see* Diseases, animal)
  diversification themes of, 610–17
  electrical signals in (*see* Electrical signals, animal)
  excretory systems (*see* Excretory systems)

feeding strategies and food of, 610–13
form and function of (*see* Animal anatomy and physiology)
gap junctions in tissues of, 139
gas exchange and circulation in (*see* Animal gas exchange and circulation)
immune systems of (*see* Immune systems)
key lineages of non-bilaterian, 617–20
life cycles of, 615–17
major phyla of, 603*t*
meiosis and gametes of, 194, 218
as model organisms, 603
movement of, 613–15 (*see also* Movement, animal)
nervous systems (*see* Nervous systems)
nitrogenous wastes of, 829*t*–30
nutrition for (*see* Animal nutrition)
Paleozoic era and, 483
phylogenetic relationship of fungi to, 6, 584–85
phylogenies of, 607–9
plant pollination by, 562–63, 576–77*f*, 792–94
plant seed dispersal by, 798
relative abundance of lineages of, 624*f*
reproduction of (*see* Animal reproduction)
sensory organs of, 610 (*see also* Sensory systems, animal)
shared traits of, 601
urinary systems (*see* Urinary systems)
water and electolyte balance in (*see* Osmoregulation)
Animal anatomy and physiology, 803–21
  adaptations and, 804–6
  body size effects on physiology and, 811–14
  body temperature regulation and, 816–19
  comparative morphology and, 603–7
  dry habitats and, 803
  fitness trade-offs in, 804, 805*f*
  homeostasis in, 814–16
  levels of study of, 811*f*
  organs and organ systems in, 810–11
  structure-function relationships in, 806
  tissues in, 806–11
Animal cells. See also Cells
  cell-cell attachments of, 135–38
  cell cultures of, B:17–B:18
  chitin as structural polysaccharide of, 76
  cytokinesis in, 124, 200
  differentiated, as genetically equivalent, 377–78
  eukaryotic, 106*f*, 115–16 (*see also* Eukaryotic cells)
  extracellular matrix and, 133–34
  glycogen as storage polysaccharide in, 168
  lysosomes in, 110–11
  plant cells vs., 106*f*, 706–7
  selective adhesion and adhesion proteins in, 136–38
Animal development, 388–400. See also Development
  cleavage in, 392–94
  fertilization in, 390–92
  gamete structure and function in, 389–90
  gastrulation in, 394–95
  hormones and, 935–37

human embryo, 388*f*
ordered phases of, 389*f*
organogenesis in, 396–99
plant development vs., 401
Animal gas exchange and circulation, 861–84
  air and water as respiratory media in, 862–64
  blood transport of oxygen and carbon dioxide in, 870–74
  circulatory systems in, 874–83
  gas exchange organs in, 864–70
  respiratory and circulatory systems in, 861–62 (*see also* Circulatory systems; Respiratory systems)
  terrestrial adaptations for, 628
Animal models of disease, 350
Animal nutrition, 841–60
  cellulose as human dietary fiber in, 77
  digestion and absorption of nutrients in, 845–56
  fetal, during pregnancy, 968–70
  four steps of, 841–42
  glucose and nutritional homeostasis in, 856–58
  glycogen as storage polysaccharide for, 74
  human digestive tract, 846*f*
  ingestion and mouthparts in, 843–45
  mammalian digestive enzymes in, 855*t*
  nutritional requirements and, 842–43*t*
  poor-nutrition plant tissues and, 1066
  waste elimination in, 854–55
Animal reproduction, 950–72
  asexual, 951
  asexual vs. sexual, 950–53
  female reproductive systems in, 959–60
  fertilization and egg development in, 954–57
  human (*see* Human reproduction)
  life tables and, 1038–39
  male reproductive systems in, 957–59*f*
  mammalian pregnancy and birth in, 967–71
  mating behaviors and, 1022–25 (*see also* Mating)
  pheromones and, 931
  protostome, 630
  reproductive structures in, 957–60
  sex hormones in, 936, 960–66
  sexual, and gametogenesis, 952–53
  sexual selection and, 452–55
  switching modes of, 951–52
  variations in, 615
  vertebrate adaptations in, 658–59
Anions, **18**–19, 98, 743, 745–46
Annelids, 603*t*, 627, **633**
Annuals, **567**
Annual tree growth rings, 714–15
Anolis lizards, 1023–24
Anoxygenic photosynthesis, **181**, **509**
Antagonistic muscle groups, 921
Antagonistic-pair feedback systems, 818
Antenna complex, **178**
Antennae, **637**, 639, 643
Anterior, **379**
Anterior pituitary, **941**, 943
Antheridia (antheridium), **556**
Anthers, **789**
Anthocerophyta, 571
Anthophyta, 564–65, 576–77*f*. See also Angiosperms
Anthrax, 498–99
Anthropoids, **668**–69

Boldface page numbers indicate a glossary entry; page numbers followed by an *f* indicate a figure; page numbers followed by *t* indicate a table.

Antibacterial compounds, 777
Antibiotics, **500**
   actinobacteria and, 514
   bacterial ribosomes and, 496
   evolution of resistance of tuberculosis
     bacterium to, 424–26
   fungi and, 581, 598
   in human semen fluids, 959*t*
   pathogenic bacteria and, 500
Antibodies, **136, 324, 681, 977**
   adaptive immune response and,
     977–78
   B-cell activation and secretion of, 986
   binding of, to epitopes, 981
   five classes of immunoglobulins
     and, 980*t*
   humoral response and, 987, 988
   humoral response coating of
     pathogens with, 988
   as proteins, 45, 280, 324
   selective adhesion of, 136–38
   specificity and diversity of, 981–82
   viruses and, 681–82
Antibody probes, B:13
Anticancer drugs, 333, 371–72, 575, 579*f*
Anticodons, **299**, 300–301
Antidiuretic hormone (ADH), **837**–38,
   931, **940**
Antifungal compounds, 777
Antigen presentation, **980**, 984–85
Antigen-receptor binding, 979
Antigens, **974, 979**
Antihistamines, 991
Antimycin A, 162
Antiparallel DNA strands, **63**–64, 261
Antiporters, **732, 747–48**, 826, 888
Antivirals, 678, **680**
Ants
   aphids and, 1068
   fungal mutalisms with, 589
   Hymenoptera order of, 640*t*
   mutualism between treehoppers and,
     1069–70
   navigation hypothesis experiment
     with, 10–12
   number of chromosomes in, 213*t*
   parasite manipulation of, 1067–68
   population of, 623
   types of, 602
Anus, 606, 845, 846*f*
Aorta, **875**
Aphids, 602, 616, 731, 1068
Aphotic zone, **997, 1000**–1001
Apical, **404**
Apical-basal axis, **404**
Apical buds, **699**
Apical dominance, 767–68
Apical meristems, 407, **704**–5, 712
Apical sides, **809**
Apicomplexa, 541
Apomixis, **786**
Apoplastic route, **722**, 723*f*, 747
Apoptosis, **205**, 332–33, 350, **376**, 396.
     See also Programmed cell death
Appendages, animal, 613–15
Appendix, **854**
Apple maggot flies, 465
Aquaporins, **95**–96, 722, **835**,
   837–38, **855**
Aquatic ecosystems, 995–1001
   animal transition to terrestrial ecosys-
     tems from, 627–28
   external fertilization in, 954
   freshwater, 997–1000
   global productivity patterns of, 1090
   marine, 1000–1001

nutrient availability in, 995–96
osmoregulation in, 826–28
osmotic stress in, 824
oxygen and carbon dioxide in, 863–64
plant transition to terrestrial ecosystems from, 553–55, 556–64
ventilation in, 867
water depth and flow in, 996–97
Aquatic food chains, 522–23
Arabian oryx, 803, 817
Arabidopsis thaliana
   embryogenesis in, 405*f*
   embryos, 401*f*
   genome of, 362, 369*t*
   as model organism, 402, B:20, B:21*f*
   overview of development of, 402*f* (*see
     also* Plant development)
   in phloem loading research, 733
Arbuscular mycorrhizal fungi (AMF),
   586, 588*f*, 589, 596, **746**
Archaea, **496**. *See also* Archaea domain
   bacteria vs., 496–97 (*see also* Bacteria)
   environmental sequencing in, 364
   human diseases and, 498
   identifying genes in genome se-
     quences of, 362–63
   key lineages of, 512, 516
   lateral gene transfer in, 364
   natural history of genomes of, 363–64
   phospholipids in, 84
   as prokaryotic, 102–3 (*see also*
     Prokaryotes)
   prokaryotic cell structures and func-
     tions of, 102–5 (*see also* Prokary-
     otic cells)
Archaea domain, 6–7, 102–3, 496–97*t*,
   504*f*. *See also* Archaea
Archaeopteryx, 480, 486
Archegonia (archegonium), **556**
Arctic terns, 1027
Arctic toxaphene biomagnification, 1088
Arctic tundra, 1008
Ardipithecus, 670*t*
Area and age hypothesis on species
   richness, 1080
Area de Conservación Guanacaste, 1122
Area/surface relationships, plant, 696–97
Arginine, 42*t*, 277–78, 280–81, 286
Aristotle, 415
Arms races, coevolutionary, 1059,
   1066–67
Arrangement, leaf, 702
Arteries, 832–33, **875**–76, 879
Arterioles, **875**–76, 882
Arthropoda, 639–43
Arthropods, **637**
   Arthropoda lineages and, 639–43
   body plan of, 626
   as ecdysozoans, 603*t*, 637
   feeding structures, 628, 629*f*
   insects as (*see* Insects)
   limbs, 629
   as most abundance and diverse
     protostomes, 623*f*
Articulations, **920**
Artificial membranes, 85–86. *See also*
   Plasma (cell) membranes
Artificial selection, 4–5, 231, **548**
Artiodactyla, 477–79
Ascaris, 211
Asci (ascus), **584**
Ascocarp, **598**
Ascomycetes (Ascomycota), 584–85, 593,
   597–98
Ascorbic acid, 843*t*
Ascus, 598

Asexual reproduction, **195, 220, 786, 950**
   animal, 615, 951
   animal switching between sexual and,
     951–52
   clones and, 221
   fungal, 591
   mitosis and, 195
   plant, 786
   prokaryotic vs. eukaryotic, 533
   protostome, 630
   sexual reproduction vs., 220–21, 223,
     533, 950–51 (*see also* Sexual
     reproduction)
A site, 300
Asparagine, 42*t*
Aspartate, 42*t*
Association studies, 349
Asteroidea, 649
Asteroid impact hypothesis, 490–92. *See
   also* Meteorites
Asthma, 991
Astragalus, 477
Astrobiologists, 501
Asymmetric competition, **1060**
Athletes, 344, 940
Atmosphere
   chemical evolution conditions in, 27,
     32, 34*f*, 39–40 (*see also* Chemical
     evolution; Prebiotic soup)
   gas exchange in, 862–63
   oxygenic photosynthesis and oxygen
     in, 181–82
   plant nitrogen fixation from nitrogen
     in, 748–50
   water potential in, 721
Atomic level anatomy and
     physiology, 811*f*
Atomic number, **16**
Atoms
   bonding of, in molecules, 17–20 (*see
     also* Molecules)
   energy transformations in, 28–29
   found in organisms, B:8*t*
   linking carbon, into organic mole-
     cules, 34 (*see also* Organic
     molecules)
   radioactive decay of, 417
   structure of, 16–17
ATP (adenosine triphosphate), **79, 148**
   in actin-myosin interaction, 124,
     924–25
   active transport by pumps and, 97–98
   cellular respiration and, 153–54,
     165–66 (*see also* Cellular
     respiration)
   cell usage of, 116
   energy transfer from glucose to, 79
   in glycolysis, 155
   GTP vs., 158, 297
   photosynthesis and, 173–74, 180–81,
     183–84 (*see also* Photosynthesis)
   prokaryotic metabolic diversity in
     producing, 506–9
   redox reactions and, 151–52
   structure and function of, 149–51
   synthesis of, 112, 148, 161–66
   transfer RNA and, 298
   vesicle transport and, 126–27
   viruses as unable to produce, 675
ATP synthase, **161, 164**–65
Atria (atrium), **877**, 879
Atrial natriuretic hormone, **933**
Atrioventricular (AV) node, **881**
Atrioventricular valves, 877
Attenuated virus vaccines, 990
α-tubulin, 125, 126, 201

Australia, 665, 672
Australopithecus, 670*t*, 671
Autocrine signals, **930**
Autoimmune diseases, 983–84. *See also*
   Immunodeficiency diseases
Autoimmunity, **983**
Autonomic nervous system, **899**, 900*f*
Autophagy, **110**
Autoploidy, 466–67
Autopolyploidy, **466**, 471*t*
Autoradiography, **62**, 196, B:12–B:13
Autosomal inheritance, **242**
Autosomal traits, 250–52, 272–74, 339
Autosomes, 212, 214*t*, 242
Autotrophs, 172, **506**, 1083–84
Auxin, **760**
   as blue-light phototropic hormone,
     759–62
   as gravitropic signal, 765–66
   plant body axes and, 406
   as plant development master
     regulator, 380–81
   as plant growth regulator, 774, 775*t*
   roles of, in apical dominance and
     plant growth, 767–68
   structure and function of, 140*t*
Aves, 668. *See also* Birds
Avian flu virus, 688, 689
Avian malaria, 1015
Avirulence (*avr*) genes, **776**
Axel, Richard, 920
Axes, body, 379–81, 395, 406
Axes, graph, B:2
Axes, leaf, 408
Axillary buds, **699**
Axonemes, **127**–28
Axon hillock, **899**
Axons, 808, **886**
   action potential propagation down,
     893–95
   diameter of, and action potential
     propagation speed, 893
   measuring membrane potentials of,
     890–91
   myelination and action potential
     propagation speed, 894–95
   nervous tissue and, 808
   neurons and, 886–87
   squid, 125–27
Azoneme, **127**

B

Bacillus anthracis, 499
Bacillus thuringiensis, 354, 513
Backbone, DNA, 61, 62, 66, 260, 279. *See
   also* Sugar-phosphate backbone, nucleic acid
Background extinctions, **489**, 1116. *See
   also* Extinction; Mass
   extinctions
BAC libraries, 360
Bacteria, **496**. *See also* Bacteria domain
   antibiotics and, 500
   archaea vs., 496–97 (*see also* Archaea)
   cell division in, 200
   cell theory experiment and, 3–4
   chromosome replication in, 265*f*
   control of gene expression in (*see*
     Gene regulation, bacterial)
   DNA of, 103–4
   environmental sequencing in, 364
   evolution of antibiotic resistance in,
     424–26
   identifying genes in genome se-
     quences of, 362–63

Bacteria, (continued)
   initiating translation in, 301
   introducing recombinant plasmids into cells of, 341
   killing of, by adaptive immune response, 987
   lateral gene transfer in, 364
   model organism (see Escherichia coli (E. coli))
   mRNA synthesis in, 293
   natural history of genomes of, 363–64
   nitrogenase complex in, 49
   nitrogen-fixing, 510–11, 1068, 1093
   origin of viruses in symbiotic, 687
   pathogenic, 498–500
   peptidoglycan as structural polysaccharide in, 76
   phospholipids in, 84
   photosynthesis in, 173–74, 180, 182
   as prokaryotic, 102–3 (see also Prokaryotes)
   prokaryotic cell structures and functions of, 102–5 (see also Prokaryotic cells)
   proteins as defense against, 45
   quorum sensing as cell-cell communication in, 145–46
   RNA polymerase in, 290
   symbiotic, in plant nitrogen fixation, 749–50
   temperature, pH, and enzyme catalysis in, 55–56
   terminating translation in, 303f
   transcription in, 292–93
   transcription in eukaryotes vs. in, 296t
   transcription promoters in, 291
   translation in eukaryotes vs. in, 296–97
Bacteria domain, 6–7, 102–3, 496–97t, 504f, 512–15. See also Bacteria
Bacterial artificial chromosomes (BACs), 360
Bacteriophages, 49, 679f, 680, 684
Bacteriorhodopsin, 163, 509
Baculum, 958
Balance, animal, 610
Balancing selection, 442–43
Ball-and-stick models, 20, B:8f
Baltimore, David, 690
Baltimore classification, 690
Banting, Frederick, 856
Bar coding, 1106
Bark, 713–14
Barnacles, 643, 1061–62
Barn swallows, 1024–25
Barometric pressure, animal sensing of, 610
Baroreceptors, 882–83
Barriers
   to dispersal of organisms, 1014
   epithelial tissues as pathogen, 809
   human contraceptive, 966t
   innate immunity and pathogen, 132, 974–75
   plant cuticle as pathogen, 707, 775
Bartel, David, 68
Basal, 404
Basal body, 127
Basal lamina, 809
Basal metabolic rate (BMR), 812–13

Basal transcription complex, 326–28
Basal transcription factors, 291–92, 326
Bases, 25–26
Bases, DNA nitrogenous, 260, 279, 280, 282–86
Basic-odor hypothesis, 919–20
Basidia, 584
Basidiomycetes (Basidiomycota), 584–85, 593, 596
Basilar membrane, 911–12
Basolateral sides, 809
Bateman, A. J., 452–53
Bateman-Trivers theory, 452–53
Batesian mimicry, 1064
Bateson, William, 248
Bats, 912
Bayliss, William, 852
B-cell receptors (BCRs), 980–82, 986
B cells, 979
   activation of, 986
   characteristics of, 989t
   discovery of, and receptors for, 979–82
   as lymphocytes, 978
B-cell tumors, 982
Bdelloids, 615
Beadle, George, 277
Beaks, 636
Beaumont, William, 848–49
Bedelloids, 631
Beer yeast model organism. See Saccharomyces cerevisiae
Bees, 562, 640t, 788, 817, 1028–29, 1068
Beeswax, 87f
Beetles, 241, 623, 640t
Beggiatoa, 507t
Behavior, 1019–21
Behavioral biology, 1019–21
Behavioral contraception methods, 966t
Behavioral ecology, 1019–36
   altruism in, 1031–34
   behavior and behavioral biology in, 1019–21
   communication in, 1027–31
   conditional strategies and decision making in, 1020–21
   foraging in, 1021–22
   mating in, 1022–25
   migration in, 1026–27
   proximate and ultimate causation in, 1020
Behavioral prezygotic isolation, 459t
Bendall, Faye, 179
Beneficial alleles, 448
Beneficial mutations, 286
Benign tumors, 206
Benthic, 618
Benthic zone, 997, 1000
Benzene, 24t
Benzo[α]pyrene, 272
Beriberi, 54
Best, Charles, 856
β-1, 4-glycosidic linkages, 73–74
β-carotene, 176, 355, 532t
Beta-blockers, 140
β-galactosidase, 309–10, 313
β-globin, 247, 322
β-glucose, 72, 76
Beta-pleated sheet (β-pleated sheet), 47
β-tubulin, 125, 126
Bicoid gene, 379–80

Bikonta, 525
Bilateral symmetry, 605
Bilateria, 603t, 606–7, 609. See also Deuterostomes; Protostomes
Bile, 853–54
Bindin, 391
Biodiversity, 1105–27. See also Conservation biology; Diversity; Species richness
   benefits of, 1117–20
   changes in, through time, 1106–7
   current extinction rates as threat to, 1110
   current species diversity and, 1107–8
   as ethical issue, 1120
   exotic and invasive species as threat to, 1110–11
   extinction rate prediction methods and, 1114–16
   extrapolation techniques for estimating, 1108
   habitat destruction as threat to, 1112
   habitat fragmentation as threat to, 1113
   hotspots of, 1109–10
   human population growth and Easter Island loss of, 1105
   levels of, 1106–7
   phylogenetic tree of life and, 1106
   preserving, 1120–23
   species endangerment as threat to, 1111
   stochastic and genetic problems in metapopulations as threat to, 1114
Biodiversity hot spots, 1109
Biofilms, 145
Biogeochemical cycles, 1092–98
   global, 1094–98
   global carbon cycle, 1096–97
   global nitrogen cycle, 1096
   global water cycle, 1095
   nutrient cycling, 1092–94
Biogeography, 463, 1013. See also Geographic distribution of organisms
Bioinformatics, 361
Biological control agents, 594, 595, 1068
Biological evolution, 15, 39, 59. See also Evolution; Natural selection
Biological fitness, 424. See also Fitness; Fitness trade-offs
Biological imaging, B:13–B:16. See also Biological methods; Microscopy
   cell research and, 116
   electron microscopy, B:14–B:15 (see also Electron microscopy)
   light microscopy, B:13–B:14 (see also Light microscopy)
   scanning electron microscopy (SEM), B:15 (see also Scanning electron microscopy (SEM))
   three-dimensional, B:15–B:16
   transmission electron microscopy (TEM), B:14–B:15 (see also Transmission electron microscopy (TEM))
   video microscopy, B:15
Biological methods. See also Experiments; Quantitative methods
   biological imaging (see Biological imaging)
   cell and tissue cultures (see Cultures, cell and tissue)
   direct sequencing (see Direct sequencing)

DNA libraries, 342–43
DNA microarrays, 370–72
DNA probes, 342–43, 370–71
enrichment cultures, 501–2
gel electrophoresis, 62, 94, 347
genetic engineering (see Genetic engineering)
Gram stains, 505
histograms, 427
molecular phylogenies (see Phylogenetic trees; Phylogenies)
replica plating, 310–11
for studying animals, 603–9
for studying fungi, 582–86
for studying green plants, 549–52
for studying membrane proteins, 94
for studying prokaryotes, 501–4
for studying protists, 524–26
for studying viruses, 678–86
Biological species concept, 459–61
Biology, 954
   astrobiology, 501
   behavioral, 1019–21
   biotechnology, 343, 354–56 (see also Genetic engineering; Recombinant DNA technology)
   cell theory in, 2–4 (see also Cells)
   characteristics of living organisms in, 1–2 (see also Animal(s); Life; Organisms; Plant(s))
   conservation (see Biodiversity; Conservation biology; Ecology; Ecosystems)
   developmental, 374 (see also Development)
   evo-devo (evolutionary-developmental), 384
   forensic, 346
   genetics (see DNA (deoxyribonucleic acid); Gene expression; Genes; Genetics; Genomics; Meiosis; Mendelian genetics)
   geographic distribution of life, 1013–17
   microbiology, 498
   model organisms of, B:19–B:22 (see also Model organisms)
   molecular, 276, 279–82, 338 (see also Carbohydrates; Chemical evolution; Molecules; Nucleic acids; Plasma (cell) membranes; Proteins)
   nanobiology, 678–79
   neurobiology, 885
   reductionism in, 899
   science and experimental methods of, 8–12 (see also Biological imaging; Biological methods; BioSkills; Experiments; Quantitative methods)
   speciation and phylogenetic tree of life in, 5–8 (see also Phylogenies; Speciation; Tree of life)
   theory of evolution by natural selection in, 4–5 (see also Evolution; Natural selection)
Bioluminescence, 541
Biomagnification, 1088–89
Biomass, 738, 1002, 1005, 1084–85, 1113
Biomedical research, 603
Biomes, 1002, 1090. See also Terrestrial ecosystems
Bioprospecting, 1117
Bioremediation, 500–501, 598, 1117
BioSkills, B:1–B:22. See also Biological methods; Quantitative methods
   biological imaging, B:13–B:16 (see also Biological imaging)

cell and tissue cultures, B:17–B:18 (*see also* Cultures, cell and tissue)
combining probabilities, B:19
Latin and Greek roots in, B:6
making concept maps, B:10
metric system, B:1–B:2
model organisms, B:19–B:22
reading chemical structures, B:8–B:9
reading graphs, B:2–B:4
reading phylogenetic trees, B:4–B:6
separating and visualizing molecules, B:11–B:13
separating cell components by centrifugation, B:16–B:17
using logorithms, B:9
using statistical tests and interpreting standard error bars, B:7
Biotechnology, 343, 354–56. *See also* Genetic engineering; Recombinant DNA technology
Biotic, **995**
Biotic factors
   density-dependent population growth, 1045–46
   of geographic distribution of organisms, 1015–17
Bipedal, **670**
Bipedalism, 671, 672
Bipolar cells, 914, **916**
Birds
   Aves lineage and, 668
   avian flu virus, 688, 689
   avian malaria, 1015
   conservation programs for, 1116
   crops of, as modified esophagus in, 848
   dinosaurs as ancestors of, 486
   feathers and flight as vertebrate adaptations, 657–58
   female reproductive systems in, 960
   gizzards of, as modified stomachs, 851
   lungs of, 868–70
   mating in, 956
   nitrogenous wastes of, 830
   optimal foraging in, 1022
   parental care in, 659
   as reptiles, 660
Birth, mammalian, 970–71
Birthrates, 1037–38, 1040
Bitterness taste receptors, 918–19
Bivalves (Bivalvia), **630**, 634
Black-bellied seedcrackers, 442, 443f
Black-tailed prairie dogs, 1031–33
Bladder, **832**
Blade, **701**
Blastocoel, **395**
Blastocysts, **393**
Blastomeres, **393**
Blastopores, **395**
Blastulas, **393**
Bleaching, coral reef, 1100
Blending inheritance hypothesis, 231, 232, 234
Blindness, 355, 364, 514
Blind spot, 914
Blobel, Günter, 120–21
Blood, **807**. *See also* Blood vessels; Circulatory systems
   alveolus and, 868
   bird lungs and, 870
   blood pressure and blood flow patterns, 882–83
   carbon dioxide transport in, 873–74
   chromatin remodeling in, 322
   circulation of, through heart, 879 (*see also* Hearts)

clotting of, 252
glucagon and sugar levels in, 45
hemophilia and human, 252
homeostasis for oxygen in, 870, 940
human blood types, 245–47, 438–39
oxygen transport in hemoglobin of, 870–73 (*see also* Hemoglobin)
patterns in pressure and flow of, 882–83 (*see also* Blood pressure)
regulation of oxygen levels in, 940
sickle-cell disease and cells of human, 46
sickle-cell disease and human, 48
tissues of, 870
transfusions, 988
Blood pressure, 880, 882–83
Blood types, 245–47, 438–39
Blood vessels, 332, 832, 837, 875–76. *See also* Blood; Circulatory systems
Blooms, algal, 521–22, 1102
Blue light
   photosynthesis and, 174–75, 176
   phototropic responses to, 758–62, 780t
Blue mussels, 1064–65f
Bodies. *See also* Morphology
   animal (*see* Bodies, animal; Body plans)
   development of major axes of, 379–81
   fungal, 582–84
   plant (*see* Bodies, plant)
   regulatory genes and specific positional information in development of, 381–83
Bodies, animal
   body plans of (*see* Body plans)
   defiinition of axes of, in gastrulation, 395
   directional selection on size of, 441
   evolution of body cavities in, 605–6
   gas exchange through diffusion across, 864–65, 874–75
   organogenesis and, 396–99
   pathogen barriers of, 974–75
   physiological effects of size of, 811–14
   plant cells vs. cells of, 706–7
   symmetry of, 604–5
   temperature regulation of, 816–19
Bodies, plant
   animal cells vs. cells of, 706–7
   apical meristem production of primary, 704–5
   brassinosteroids and size of, 773
   cells and tissues of, 706–12
   components of, 711t
   definition of axes of, by genes and proteins in embryogenesis, 406
   nutrition and, 737
Body mass index, 857
Body plans, **603**
   arthropod, 626
   basic features of animal, 603–7, 610
   chordate, 650–51
   echinoderm, 647–48
   mollusk, 626
   protostome, 626–27
Bogs, **998**
Bohr shift, **872**, 873f
Boluses, 844
Bonds. *See* Carbon-carbon bonds; Chemical bonds; Covalent bonds; Disulfide bonds; Hydrogen bonds; Ionic bonds; Peptide bonds
Bond saturation, lipid, 87–88
Bone, **653**, **807**, **920**

Bone marrow, 352–54, **978**
Bones, animal, 920–21
Bony endoskeletons, 655
Bony exoskeletons, 653
Boreal forests, 1007
Boron, 739t
Both-And rule of probability, B:18
Bottom-up limitation hypothesis, 1066
Bouchet Philippe, 1107–8
Boveri, Theodor, 230, 239–41
Bowman's capsule, **833**
Brachiopoda, 603t
Braincase, **671**
Brains, **605**. *See also* Nervous systems
   anatomy of, 901
   antidiuretic hormone (ADH) and, 837
   central nervous system and, 886
   hominin, 672
   Huntington's disease and degeneration of, 233, 251, 348–51
   hypothalamus (*see* Hypothalamus)
   mapping functional areas in, 901–2
   neural tubes and, 396
   positron-emission tomography scans of, 885f
   prions and diseases of, 339
   sensory organ transmission of information to, 909
   thermoregulation and, 817–18
   vertebrate, 653
Brain stem, **901**
Branches, **474**, **699**
Branching adaptations, 814
Brassinosteroids, 140t, **773**, 774, 775t
Bread mold, 277, 595
Bread yeast model organism. *See Saccharomyces cerevisiae*
Breast cancer, 350
Breathing, 866–70. *See also* Respiratory systems
Breeding. *See* Artificial selection; Captive breeding; Inbreeding depression
Brenner, Sydney, 283
Bridled goby, 1045
Briggs, Winslow, 761
Bristlecone pine trees, 401, 575, 695
Britton, Roy, 296
Broca, Paul, 901
Bronchi (bronchus), **867**
Bronchioles, **867**
Bronstein, Judith, 1069
Brown algae, 535–36, 543. *See also* Algae
Brown kiwis, 454
Bryophytes (Bryophyta), **550**, 570
Bryozoa, 603t
Bubonic plague, 602
Buchner, Hans and Edward, 155
Buck, Linda, 920
Budding
   animal, 619, 620, 652, **951**
   flower, 788
   viral, 684, 685f
Buffers, **25**, **874**
Bugs, 641t
Building materials, 548–49
Bulbourethral glands, **958**, 959t
Bulk flow, **730**
Bulldog ants, 213t
Bumblebees, 817
Bundle-sheath cells, **188**, **728**
Burgess Shale faunas, 486–87
Burials, 480, 671
Bursa, 979
Bursting, viral, 684, 685f

Butter, 87f
Buttercup root, 71f
Butterflies, 640t, 1027, 1116
Bypass vessels, 878

C

C_3 photosynthesis, **188**
C_4 photosynthesis, **188**, **728**
Cacti, 189, 701
Caddisflies, 641t
Cadherins, 136–**38**
Caecilians, 664
Caenorhabditis elegans
   apoptosis in, 376
   genome of, 362
   as model organism, B:21f, B:22
   as nematode, 638
   number of genes in genome of, 369t
   as protostome, 623
Caffeine, 775
Calcium
   as human nutrient, 843t
   as plant nutrient, 739t
Calcium carbonate, 523, 540, 569
Calcium ions, 143t
Callus, **708**
Calmodulin, 428
Calvin, Melvin, 173–74, 185–86
Calvin cycle, **173**, 185–86, 509, 728
Calyx, **788**
CAM (crassulacean acid metabolism), **188**–89, **728**, 1004
Cambium, **712**–13
Cambrian explosion, **486**–88
Camera eyes, **914**
Cancer, **206**–9
   anticancer drugs, 333, 371–72, 575, 579f
   applied genomics and, 371–72
   causes of uncontrolled cell growth in, 332
   chromosome-level mutations and, 286–87f
   cytoskeleton-ECM connection failures and, 134
   defective Ras proteins and, 144
   defects in DNA damage-repair genes and, 274
   defects in eukaryotic gene regulation and, 332–33
   as family of diseases, 208–9
   genetic testing and breast, 350
   HIV and, 677
   homology and, 421
   loss of cell-cycle control and, 207–9
   metastasis of, 134
   as out-of-control cell division, 206
   p53 gene case study on preventing, 332–33
   properties of cancer cells, 206–7
   Rous sarcoma virus and, 691
   severe combined immunodeficiency (SCID) therapy and, 354
   sponges and chemotherapy for, 618
   telomerases and, 271
Canis familiarus, 369t
Cannibalization, 955
Canopy, **1003**
CAP binding site, **315**
CAP-cAMP complex formation, 316
Capillaries, **814**, 875–76, 882. *See also* Circulatory systems
Capillarity, **723**–24
Capillary action, 722
Capillary beds, **875**

Caps, RNA, **295**
Capsids, 259–60f, **679**, 684
Captive breeding, 1122
Carapace, **643**
Carbohydrates, **71–81**
  animal nutrition and, 842
  cell identity role of, 77
  digestion and transportation of, in animal small intestines, 852–53
  energy storage role of, 78–79
  fermentation of, 508
  glycosylation and, 121
  metabolic pathways of, 168–69
  organic molecules and, 36
  photosynthetic production of, 172–74
  polymerization of sugars to form, 42
  polysaccharides, 73–76
  polysaccharide structures, 75t
  processing of photosynthetic, 189–90
  roles of, 77–79
  as structural, 77, 113
  sugars as monomers of, 39, 71–73 (see also Sugars)
Carbon
  amino acids and, 40
  atomic structure of, 17f
  Calvin cycle fixation of, during photosynthesis, 185–86
  as cellular requirement, 168
  electronegativity of, 18
  importance of, in organic molecules, 33–36 (see also Organic molecules)
  lipids, hydrocarbons, and, 83
  as plant nutrient, 739t
  prokaryotic variation in pathways for fixing, 509
  RNA and, 66
  simple molecules from, 19–20
  six common functional groups attached to atoms of, 35t
  Sphagnum moss and, 570
Carbon-carbon bonds, 33–34, 172–73, 506t
Carbon cycle, 547, **580**–81, 1096–97
Carbon dioxide (CO₂)
  animal gas exchange and, 861
  behavior of, in air, 862–63
  behavior of, in water, 863–64
  carbon cycle and, 185
  carbonic anhydrase and, 45
  double bonds of, 20
  entry of, into leaves through stomata, 187
  formaldehyde and, 32
  global warming and, 523, 547, 1008–9, 1098–99 (see also Global climate change; Global warming)
  iron fertilization and, 1091
  land plants and, 553
  mechanisms for increasing concentration of, in plants, 187–89
  photosynthesis and, 173–74, 185–90
  plant nutrients from, 739t
  transport of, in blood, 873–74
  in volcanic gases, 27
  water stress and plant biochemical pathways to increase, 728
Carbon fixation, **185**
Carbonic acid, 27
Carbonic anhydrase, 45, 52, **850**, **873**–74

Carboniferous period, 551
Carbonyl functional group, 34–35f, 43, 46–47, 60, 71–73
Carboxyl functional group, 34–35f, 40, 43–44, 83–84
Carboxylic acids, 35f, **158**
Carboxyl-terminus, 44, 301
Carboxypeptidase, 855t
Cardiac cycle, **879**–80. *See also* Hearts
Cardiac muscle, **808**, **921**
Carnauba palms, 707
Carnivores, 547, **612**, 664
Carnivorous plants, 751–52
Carotenes, 176
Carotenoids, **175**–76, 355, 453, 758
Carpels, **403**, 409, 561, **789**–90
Carrier proteins (transporters), 96–97, **731**–32, **825**. *See also* Transport proteins
Carriers (disease), **251**
Carrier testing, genetic, 350
Carrion flowers, 562
Carroll, Sean B., 614–15
Carrying capacity, **1042**, 1046, 1065
Cartilage, **653**, **807**, **920**
Casein, 329
Casparian strip, **722**–23
Casts, **480**
Catabolic pathways, 168–69
Catabolite activator protein (CAP), **315**
Catabolite repression, **314**–17f
Catalysis
  DNA as poor catalytic molecule, 65–66
  enzymes and, 45, 51–56 (*see also* Enzymes)
  eukaryotic organelles and, 107
  hydrolysis of carbohydrates and, 78–79
  multienzyme complexes and, 49
  nucleic acid polymerization and, 62
  polysaccharides and, 76
  RNA as catalytic molecule, 68
  smooth ER and, 108
  three-step process of, 54f
Catalysts, **52**–53. *See also* Catalysis; Enzymes
Catalyze, **45**. *See also* Catalysis
Catastrophic events, 1114
Catecholamines, **943**
Cation exchange, **743**–44f
Cations, **18**–19, 98, 743, 745
CD4⁺ T cells, 984–86, 989t
CD4 protein, **681**, **984**
CD8⁺ T cells, 984–86, 987–88, 989t
CD8 protein, **984**
cDNA libraries, 342–43
Cech, Thomas, 68
Cecum, **854**
Cell body, neuron, **886**–**87**
Cell-cell attachments, eukaryotic, 135–38
Cell-cell gaps, 138–39
Cell-cell interactions, 131–47
  carbohydrates, cell-cell recognition, and, 77
  cell surfaces and, 132–34
  communication between distant cells, 139–46
  connection and communication between adjacent cells, 134–39

  in development, 375t, 377
  hormone structures and functions for, 140t
  second messenger examples, 143t
Cell-cell signaling, 139–46
  in animals (*see* Chemical signals, animal)
  auxin as plant master regulator, 380–81
  bacterial quorum sensing and, 145–46
  bicoid gene as animal master regulator, 379–80
  common pathways of, in differing contexts of development, 383–84
  cross-talk, 145
  in development, 375t, 377
  in differential gene expression, 379–84 (*see also* Pattern formation)
  evolutionary conservation of signals, 383
  hormones and, in multicellular organisms, 139–40t
  in plant sensory systems, 756–57
  proteins and, 45
  reuse of signals in differing developmental contexts, 383–84
  role of carbohydrates in, 77
  signal processing, 140–44
Cell crawling, **124**
Cell cycle, **196**. *See also* Cells
  cancer as out-of-control, 206–9
  cell-cycle checkpoints and regulation of, 204–5
  cell division by meiosis and mitosis and cytokinesis in, 194 (*see also* Meiosis)
  control of, 202–6
  cytokinins and, 768–69f
  discovery of cell-cell regulatory molecules, 203–4
  mitosis and, 195–97f (*see also* Mitosis)
Cell-cycle checkpoints, 204, **205**, 207–9
Cell determination, 397–98
Cell differentiation, 375t, 377
Cell division, **194**
  animal cytokinesis as, 124
  cancer and out-of-control, 206–9
  cell cyle of, 194 (*see also* Cell cycle; Meiosis; Mitosis)
  cytokinins in plant, 768–69f
Cell-elongation responses, plant, 761–62
Cell growth, 375t, 376–77
Cell identity, 77
Cell-mediated response, **987**–88
Cell movement, 375t, 376–77
Cell plate, **200**
Cell proliferation, 375
Cells, 2, 102–30
  animal vs. plant, 706–7 (*see also* Animal cells; Plant cells)
  blood, 870–71
  cancer, 206–7, 332 (*see also* Cancer)
  carbohydrate functions in, 77–79
  cell cycle of (*see* Cell cycle)
  cell theory, living organisms, and origin of, 2–4
  cellular respiration and fermentation in (*see* Cellular respiration; Fermentation)
  chemical signals and target, 930–31
  chloride, 827–28
  cultures of (*see* Cultures, cell and tissue)
  entry of viruses into, 681–82

  eukaryotic vs. prokaryotic, 7f, 105–6f, 107t (*see also* Eukaryotic cells; Prokaryotic cells)
  first, 91
  genetics and (*see* Genetics)
  hormone target receptors, 943–47
  importance of cell membranes to, 2, 82, 98–99 (*see also* Plasma (cell) membranes)
  interactions between (*see* Cell-cell interactions)
  introducing novel alleles into human, 351–52
  life cycles and diploid vs. haploid, 533–34
  living organisms and, 1
  organogenesis and, 396–99
  photosynthesis in (*see* Photosynthesis)
  programmed death of (*see* Apoptosis)
  prokaryotic cell structures and functions, 102–5
  protein functions in, 45
  protist, 524–25
  separating components of, by centrifugation, B:16–B:17
  studying live, B:15
  two requirements of, 168
  viruses as not, 675
  vocabulary for describing chromosomal makeup of, 214t
  water in living, 22
Cell sap, **707**
Cell-suface hormone receptors, 141
Cell-suicide genes, 376
Cell-surface hormone receptors, 945–47
Cell theory, 2–4, 102. *See also* Cells
Cellular respiration, **154**, 507
  ATP, ADP, glucose, and metabolism in, 148
  ATP synthesis via electron transport and chemiosis in, 161–66
  catabolic and anabolic pathways and, 168–69
  chemical energy and redox reactions, 149–53
  citric acid cycle in, 158–61
  fermentation and, 166–68
  fermentation vs., 153, 167f
  four steps of, 153–54
  gas exchange and, 862
  glycolysis as processing of glucose to pyruvate in, 155–56
  overview of, 154f
  photosynthesis vs., 173
  processing of pyruvate to acetyl CoA in, 156–58
  prokaryotic, 506–8
  summary of, 165f
Cellular slime mold model organism, 536, B:20, B:21f
Cellulases, 590
Cellulose, **76**
  bird digestion of, 848
  in cell walls, 132–33
  fungal digestion of, 579, 581, 590
  as plant structural polysaccharide, 71f, 76, 77
  ruminant digestion of, 850–51
  in secondary cell walls, 709
  structure of, 75t
Cell walls, **76**, **105**, **707**
  cellulose as structural polysaccharide in, 71f, 76, 77
  as characteristic of domains of life, 497t

Boldface page numbers indicate a glossary entry; page numbers followed by an *f* indicate a figure; page numbers followed by *t* indicate a table.

cohesion-tension theory and
    secondary, 726
eukaryotic, 113, 114t
in green plants, 549
ingestive feeding and, 530
plant cell-cell attachments and, 135
prokaryotic, 496
prokaryotic diversity in composition
    of, 504–6
as prokaryotic exoskeleton, 104–5
protist, 528–29
sclerenchyma cells and, 709–10
types of, in plants, 132–33
Cenozoic era, **483**
Centipedes, 639
Central dogma of molecular biology, **280**
    Francis Crick's statement of, 280
    development of, 279–82
    exceptions to, 281
Central nervous system (CNS), **604, 886**
    functional anatomy of, 900–902
    hormones, 929
    hormone signaling pathways and,
        931–32
    vertebrate, 653
Centrifugation
    in discontinuous replication
        experiment, 267
    in estrogen receptor research, 944
    in Hershey-Chase experiment,
        259–60f
    in Meselson-Stahl experiment, 263
    overview of, B:16–B:17
Centrioles, **125, 198**, 202t, **389**
Centromeres, **196**, 199, 202t, **213**
Centrosomes, **125, 198**, 202t
Cephalization, 603, **605**, 610, 886
Cephalochordates (Cephalachordata),
    651–52
Cephalopods (Cephalopoda), **630**,
    636, 914
Cerebellum, **653**, 901
Cerebral ganglion, 605, 886. See
    also Brains
Cerebrum, **653**, 901–2
Cervix, 960
Cetaceans, evolution of, 422
CFTR (cystic fibrosis transmembrane
    conductance regulator),
    94–95, 826–27
Chaetae, **633**
Chaetognatha, 603t
Chagas disease, 522t
Chambers, heart, 878
Changing-environment hypothesis on
    sexual reproduction, 224–25
Channel proteins, **95, 731, 825**
    facilitated diffusion via, 94–96
    in phloem loading, 731–32
Chaperone proteins, 330
Character displacement, **1062**
Chargaff, Erwin, 62–63
Charophyceae, 569
Chase, Martha, 119–20, 259–60f
Cheatgrass, 1015–17
Checkpoints, cell-cycle, 204, **205**,
    207–9
Chelicerae (Chelicerata), **642**
Chemical bonds, **17**. See also Carbon-
    carbon bonds; Covalent bonds;
    Disulfide bonds; Hydrogen
    bonds; Ionic bonds; Peptide
    bonds
Chemical energy, **33**
    animal nutrition and (see Animal
        nutrition)

ATP, redox reactions, and, 149–53 (see
    also ATP (adenosine
    triphosphate))
carbohydrates and storage of,
    78–79
cellular respiration, fermentation and
    (see Cellular respiration;
    Fermentation)
cellular respiration and, in fats and
    storage carbohydrates, 153
chemical evolution, chemical reac-
    tions, and, 27–33 (see also Chemi-
    cal evolution; Chemical reactions)
conversion of light to, 174–75, 178–79
    (see also Photosynthesis)
plant nutrition and (see Plant
    nutrition)
primary producers and, 1083–84
secondary active transport and, 98
sugars and, 71
Chemical equilibrium, **27**
Chemical evolution, **15–37**. See also
    Evolution
    atoms, ions, and molecules as build-
        ing blocks of, 16–21 (see also
        Molecules)
    biological evolution vs., 39, 59
    chemical reactions, chemical energy,
        and, 27–33
    environments for, 27
    four steps of Oparin-Haldane theory
        of, 38–39
    importance of carbon and organic
        molecules to, 33–36
    monosaccharides and, 71, 73
    nucleic acid formation by, 62
    nucleotide production by, 60–61
    overview of hypothesis of, 34f
    pattern and process components of,
        and biological evolution vs., 15
    polymerization of amino acids
        and, 42–43
    polysaccharides and monosaccharides
        in, 76
    properties of water and early oceans
        as environments for, 22–26
    RNA as first self-replicating molecule
        in, 59, 68–69
    spark-discharge experiment
        on, 39–40
Chemical modifications, post-
    translational, 304
Chemical reactions, **21**
    acid-base reactions, 25–26
    chemical evolution and, 27–33 (see
        also Chemical evolution)
    enzymes and catalysis of, 45, 51–56
        (see also Enzymes)
    eukaryotic organelles and, 107
    first, in chemical evolution, 32
    plasma membrane and, 82
    rate of, in cells, 116
    spontaneous, 29–30
    temperature and concentration vs. re-
        action rates of, 30–32
    types and transformations of energy
        in, 27–29
    volcanic, 27
    water and, 822
Chemical signals, animal, **929–49**
    action of hormones on target
        cells, 943–47
    chemical characteristics of hormones
        as, 933–34
    endocrine system components
        and, 932–33

endocrine system hormones
    as, 929
experimental identification of hor-
    mones as, 934–35
functions of hormones as, 935–40
hormone signaling pathways
    of, 931–32
major categories of, 930–31
regulation of production of
    hormones, 940–43
six categories of animal, 930t
Chemical structures, reading, B:8–B:9
Chemiosmosis, 153, **162–63**, 181
Chemokines, **976**, 977t
Chemolithotrophs, **506**
Chemoorganotrophs, **506**
Chemoreceptors, **908**, 918–20
Chemosenses, 918–20
Chemotherapy, cancer, 371–72, 618
Chestnut blight, 581, 598
Chewing, 846. See also
    Mouthparts; Teeth
Chiasmata (chiasma), **216**
Chicken genome, 369t
Child abuse rates, 1034
Childbirth, 970–71
Chimpanzees
    ancestors of humans and, 670
    HIV and, 688
    number of chromosomes in, 213t
    similarity of genome of, and human
        genome, 369–70
Chins, 672
Chitin, 75t, **76**, 77, 585
Chitinase, 56
Chitons, **630**, 636
Chlamydiae, 512, 514
Chlamydia trachomatis, 364, 514
Chloride cells, 826–28
Chloride ions, 94–95
Chlorine, 739t
Chlorophylls, **175**
    capture of light energy by, 174–79
    in protists, 532t
Chloroplasts, **112, 174, 707**
    Calvin cycle in, 186
    in green plants, 549
    in moss cells, 172f
    origin of, in protists, 530–31
    photosynthesis in, 112–13, 114t, 174
        (see also Chlorophylls)
    phototropins and, 758
    pigments in, 176
    in plant cells, 116
Choanocytes, **608**
Choanoflagellates, 607–8, 617
Cholecystokinin, **852, 937**
Cholesterol, 84, 88
Cholodny, N. O., 760–61
Cholodny-Went hypothesis, 760–61
Chondrichthyes, 662
Chordata, 603t, 651–52, 660–68
Chordates, **396, 650**
    body plans of, 650–51
    Chordata lineages and, 603t,
        651–52, 660–68
    as deuterostome, 646
    invertebrates, 651–52
    phylogenies of, 654f
    vertebrates (see Vertebrates)
Chorion membrane, 658
Chromalveolata, 525
Chromatids, **195–96**, 202t
Chromatin, **197**, 202t, **320–23**, 331
Chromatin remodeling, **320–23**, 331t
Chromatography, 175–76, 185

Chromosomes, **103, 195**
    circular, as characteristic of domains
        of life, 497t
    consequences of meiosis for, 220–23
    discovery of, 195–96
    eukaryotic, 107–8
    genomes and, 364
    linkage and genes on same,
        243–45, 246f
    meiosis, heredity, and, 221 (see also
        Chromosome theory of
        inheritance)
    meiosis and, 211–12
    in mitosis, 202t
    movement of, during mitosis, 201–2
    mutations in, 286–87f
    nuclear lamina and, 116
    number of, found in some familiar
        organisms, 197, 213t
    polyploidy in, and speciation, 465–68
    prokaryotic, in nucleoids, 103–4
    sister chromatids and, 213–14
    trisomy in human births and number
        of, 227t
    vocabulary for describing chromoso-
        mal makeup of cells, 214t
Chromosome theory of inheritance, **241**.
    See also Mendelian genetics
    discovery of sex chromosomes
        and, 241
    importance of, 230
    meiosis and statement of, 240–41
    mutants and, 241
    rediscovery of Gregor Mendel's ex-
        periments and, 239
    sex linkage and testing of, 241–42
Chylomicrons, **854**
Chymotrypsin, 855t
Chytridiomycete life cycle, 592
Chytrids, 583, 585, 592, 595
Cichlids, 486, 844–45
Cigarette smoke, 272
Cilia (cilium), **127–28**, 533
Ciliata, 540
Circadian clock, **1026**
Circular chromosomes, 497t
Circulatory systems, **862, 870–83**. See
    also Blood
    blood pressure and blood flow in,
        882–83
    closed, 875–77
    four steps of gas exchange and, 861–62
    hearts in, 877–81
    mesoderm and, 604
    open, 875
    oxygen and carbon dioxide transport
        in blood, 870–74
    pulmonary circulation and, 879f (see
        also Respiratory systems)
    surface area for gas exchange and,
        874–75
    types of, 874–77
Cis surface, Golgi apparatus, 109, 121
Cisternae (cisterna), **109**, 121
Citrate, 157
Citric acid, 598, 959t
Citric acid cycle, 153, **158–61**
Citrulline, 277
Clades, **460**, 504
Cladistic approach, **475**, 476–79
Clams, 634
Class I and class II MHC proteins, **984**
Clausen, Jens, 700
Clay, 43, 62, 743, 744t
Cleaner shrimp, 1068
Cleavage, **392–94**

Cleavage furrow, **200**
Cleaver, James, 273
Clements, Frederick, 1070–72
Clements-Gleason dichotomy, 1070–71
Cliff swallows, 441
Climate, **1002**
  biomes and, 1002
  effects of global climate change on ecosystems, 1011–13*f*
  global patterns in, 1009–11
  global warming and, 1008–9, 1098–1102 (*see also* Global climate change; Global warming)
  green plant holding of water and moderate, 547
  human alteration of, 1098–99
  regional effects of mountain ranges and oceans on, 1010–11
  seasonality of weather and, 1010
  tree growth rings and research on, 715
  tropical warmth, polar cold, and, 1010
  tropical wetness and, 1009–10
Climax communities, **1071**
Clitoris, **960**
Cloaca, **855**, 954
Clonal-selection theory, 979–83*f*
Clones, **221**, **378**, 708, **786**, **951**, 979
Cloning, DNA. *See* DNA cloning
Cloning vectors, **340**
Closed circulatory systems, **875–77**
Clostridium acetium, 507*t*, 508
Clotting, blood, 975
Club fungi, 593, 596
Club mosses, **571**, 571
Cnidaria, 601*f*, 603*t*, 616–17, 619
Cnidocyte, **619**
Coal, 548, 551, 571
Cocaine, 775
Cochlea, **910**
Code, genetic. *See* Genetic code
Coding strand, **290**
Codominance, **245**, 250*t*
Codons, **282–85**, 297
Coefficient of relatedness, **1032**
Coelacanths, 663, 917–18
Coelom, **605**, **624**, 626
Coelomates, **605**
Coenocytic, **583**
Coenzyme A (CoA), **157**
Coenzyme Q, **161–64**
Coenzymes, **54**
Coevolution, **793**, **1059**, 1066–67
Coevolutionary arms races, **1059**, 1066–67
Cofactors, enzyme, **54**
Cohesion, **22–23**, **724**
Cohesion-tension theory, 722, **724–27**
Cohorts, **1038**
Cold virus, 692
Coleochaetophyceae (Coleochaetes), 568
Coleoptera, 640*t*
Coleoptiles, **758**
Collagen, 45*f*, **133–34**
Collecting duct, **837**
  antidiuretic hormone and, 931, 940
  osmoregulation by, 837–38
  structure and function of, 832, 833*f*, 838*t*
Collenchyma cells, **708–9**, 711*t*

Colon, **855**
Colonies, **617**
Colonization
  allopatric speciation by, 463
  founder effect and, 446
Color, flower, 562, 788
Color vision, 446, 915, 916–18
Comb jellies, 620
Commensal, **586**
Commensalism, **1059**, 1070*t*
Common primroses, 1050–51
Communication, **1027**–31
Communities, **994**, **1058**. *See also* Community ecology
  biodiversity and stability of, 1119–20
  disturbance regimes in, 1073–74
  experiments on predictability of structure of, 1070–72
  keystone species in, 1072–73
  post-disturbance succession in, 1074–77
  taxonomic diversity and, 1106, 1107*f*
Community ecology, 994, 1058–82
  communities in, 1058 (*see also* Communities)
  community dynamics in, 1073–77
  community structure in, 1070–73
  species interactions in, 1058–70
  species richness in, 1077–80
Companion cells, **711**, **729–30**, 733*f*
Comparative morphology, 603–7
Compass orientation, **1026–27**
Competition, **1059–63***f*
  as biotic factor in distribution of species, 1015
  competitive exclusion principle of, 1060–61
  conservation biology and resistance to invasive species through, 1062
  experimental studies of, 1061–62
  fitness and impacts of, 1070*t*
  fitness trade-offs in, 1062
  giraffe necks and food vs. sexual, 9–10
  intraspecific vs. interspecific, 1059
  male-male, 454–55
  niche differentiation and coexistence to avoid, 1062, 1063*f*
  niches and interspecific, 1059–60
Competitive exclusion principle, **1060**–61
Competitive inhibition, enzyme, **54–55**
Complementary base pairing, 63–**64**, 66–67, 76, **261**
Complementary DNA (cDNA), **339–40**, 342, **684**
Complementary strand, DNA, **65**
Complement proteins, 45
Complement system, **987**
Complete digestive tracts, **845**
Complete metamorphosis, **616**
Complex tissues, plant, 710
Compound eyes, **637**, **913**
Compound leaves, 408, 702
Compression, 132
Computer models
  of asteroid impact, 490
  genetic drift simulations, 444
Concentration
  enzyme catalysis and substrate, 55
  hormone, 934

increasing carbon dioxide, in plants during photosynthesis, 187–89
  role of, in chemical reactions, 30–32
Concentration gradients, **89–92**, 94–95, 380, **823**
Concept maps, B:10
Condensation reactions, **43**, 73, 83–84
Condensed replicated chromosomes, 195–96, 197
Conditional behavior, 1020–21
Condoms, 966*t*
Conduction, 722, **816**
Cones, eye, **915**–17
Cones, gymnosperm, 551, 561, 575
Confocal microscopy, B:15
Conformational homeostasis, 815
Conifers, 1007
Connections, cell. *See* Multicellularity
Connective tissue, **806–7**
Connell, Joseph, 1061–62
Conservation, soil, 742
Conservation biology, **995**, 1105–27. *See also* Biodiversity
  bioremediation and, 500–501
  competition to resist invasive species in, 1062
  designing protected areas, 1120–21
  ecology and, 995
  ecosystem restoration, 1122
  ex situ conservation, 1121–22
  Gap Analysis Program (GAP) in, 1120
  genetic drift and, 445
  Malpai Borderlands case history, 1122–23
  phylogenetic species concept and, 461–62
  population ecology and endangered species analysis, 1053–55
  preserving biodiversity, 1120–23
  preserving metapopulations, 1055
  preserving phylogenetically distinct species, 1107*f*
  stabilizing human population size of and resource use, 1121
  sustainability in, 1121
  whooping cranes and, 1043–44
  wildlife corridors in, 1121
Conservation hotspots, 1110
Conservation International, 1109–10
Conservative, genetic code as, 283–84
Conservative replication, DNA, 261–63
Constant (C) regions, **982**
Constitutive defenses, **1063–64**
Constitutively, **309**
Constitutive mutants, **311**, 313
Constraints, genetic, 431
Constraints, historical, 432
Consumers, 602, 1068, **1084**. *See also* Consumption
Consumption, **1059**, 1063–68
  coevolutionary arms races and, 1066–67
  constitutive defenses against, 1063–64
  consumers as biocontrol agents, 1068
  efficiency of predators at reducing prey populations, 1065
  fitness and impacts of, 1070*t*
  inducible defenses against, 1064–65*f*
  limitations on herbivore, 1065–66
  parasite manipulation of hosts, 1067–68
  types of, 1063
Continental shelf, **1000**
Continents, changes in, 483–84
Continuous population growth, 1043–44
Continuous strand, **266**

Contraception, human, 966
Contractile proteins, 45, 280
Contractile roots, 698
Control. *See* Regulation
Control groups, **12**
Convection, **816**
Convergent evolution, **476**
Conversions, metric system, B:1*f*
Cooperation, 1031–34
Cooperative binding, **871–72**
Coordination, bacterial vs. eukaryotic gene expression, 331–32
Copepods, 643, 1100
Copper, 739*t*, 740, 747
Coprophagy, **854**
Copulation, 630, 954
Copying. *See* Replication
Coral reefs, 539, 649, **1001**, 1058*f*, 1100, 1114
Corals, 619
Co-receptors, **682**
Core enzymes, **291**
Cork cambium, **712**, 713–14
Cork cells, **713**
Corms, **786**
Corn, 4–5, 213*t*, 354, 548
Cornea, **914**
Corolla, **788**
Corona, **631**
Corpus callosum, **901**
Corpus luteum, **963**
Corridors, wildlife, 1055, 1121
Cortex, **706**, **722**, **832**
Cortical granules, **390**, 392
Corticosteroids, 991
Corticotropin-releasing hormone (CRH), **941**
Cortisol, **938**
Costanza, Robert, 1117
Cost-benefit analysis, **1021**
Cotransport, **98**, **732**, **825**
Cotransporters, **732–33**, 745–46, **825–28**, 853, 888
Cotton, 548
Cotyledons, **405**, **564–65**, **796**
Countercurrent exchangers, **818–19**, 835–37, 866
Covalent bonds, **18**
  amino acids and, 40
  carbon and, 33
  electron sharing and, 17–18
  nonpolar and polar, 18
  protein tertiary structure and, 48
Cowpox, 973–74
Crabs, 643, 1064–65*f*
Crane, Peter, 554
Cranium, **653**
Crassulacean acid metabolism (CAM), **188–89**, **728**, 1004
Creeks, 999
Crenarchaeota, 512, 516
Creosote bushes, 786
Cretaceous-Paleogene extinction, 490–92
Creutzfeldt-Jakob disease, 51
Crews, David, 1023–24
Crick, Francis
  adaptor molecule hypothesis of, 297
  articulation of central dogma of molecular biology by, 280
  discovery of double helix secondary structure of DNA by, 59*f*, 62–65
  genetic code hypothesis of, 279
  triplet code reading frame discovery of, 283
  wobble hypothesis of, 299–300

Boldface page numbers indicate a glossary entry; page numbers followed by an *f* indicate a figure; page numbers followed by *t* indicate a table.

Crickets, 641*t*
Crime, 366
Cristae, **112**, **157**, 161
Crocodiles, 660, 667, 841*f*
Crocodilia, 667
Cro-Magnons, **671**
Crop, bird, **848**
Cro protein, 48–49
Crops, transgenic, 354–56
Cross-fertilization, **231**
Crossing over, **217**
    calculating frequencies of, 246
    in meiosis prophase I, 219–20
    role of, in meiosis, 222
    role of, in Mendelian genetics,
        244–45
Cross-pollination, 792–93
Cross-talk, **145**, 774
Crustaceans, **643**, 875
Cryoelectron tomography, 157*f*
Cryptic female choice, 955
Cryptic species, **460**
Cryptochromes, 758
Ctenophora, 603*t*, 620
C-terminus, 44, 301
Cultures, cell and tissue, **119–20**
    animal cell, B:17–B:18
    in anthrax research, 499
    in cytokinin research, 768
    differentiated plant cells and, 377
    discover of gap phases with, 196
    in DNA damage research, 273
    enrichment cultures in prokaryotic
        research, 501–2
    plant tissue, B:18
    in secretory pathway research, 119–20
Cup fungi, 598
Cushing's disease, **941**
Cuticle, **625**, **707**, **828**
    as defense barrier, 775, 974
    fossilized, 550
    water loss prevention by, 553–54, 727,
        828–29
Cuttings, plant, 708
Cuttlefish, 636
Cuvier, Baron George, 417
Cyanobacteria, **510**
    characteristics of, 515
    lichens and, 566, 589, 597
    metabolic diversity and, 507*t*
    oxygen revolution and, 509
    protist chloroplasts and, 530–31
Cycads (Cycadophyta), 574
Cycles, population. *See* Population cycles
Cyclic adenosine monophosphate
    (cAMP), 143*t*, **315–16**, **946**
Cyclic guanosine monophosphate
    (cGMP), 143*t*, 144, 916
Cyclic photophosphorylation, **183–84**
Cyclin-dependent kinase (Cdk), **204**
Cyclins, **204**, 208, 330
Cysteines, 42*t*, 48, 280–81, 296
Cystic fibrosis, 94–95, 350, 826–27
Cytochrome *c*, **163–64**
Cytokines, **930**, 977*t*, 991
Cytokinesis, **124**, **194–95**, 197, 200,
    217, **976**
Cytokinins, 768–69*f*, 774–75*t*
Cytoplasm, **104**
    cleavage of, 393
    cytokinesis as division of, 194, 200
    cytosol in, 109
    ribosomes in, 279
Cytoplasmic determinants, **390**, 393
Cytoplasmic streaming, **124**
Cytosine, 60, 63–64, 66–67, 261, 279

Cytoskeletons, **104**
    eukaryotic, 113–14*t*, 123–28, 134
    prokaryotic, 104
    prokaryotic vs. eukaryotic, 107*t* (*see
        also* Cytoskeletons, eukaryotic)
    protist, 528–29
Cytosol, **109**, 124
Cytotoxic T cells, **985–89***t*

D

Daddy longlegs, 642
Dalton, **16**
Dalton, John, 916–17
Daly, Martin, 1034
Damaged DNA
    cell-cycle checkpoints and, 205
    sexual reproduction and, 224
    UV radiation study with tumor sup-
        pressor gene mutations, 332–33
    xeroderma pigmentosum (XP) as fail-
        ure to repair, 272–74
Damselflies, 641*t*
Dance hypothesis, honeybee
    language, 1028–29
Danielli, James, 92
Daphnia, 224, 951–52
Darwin, Charles. *See also* Evolution;
    Natural selection
    gravitropic responses research by, 764
    orchid pollination hypothesis by, 793
    phototropic responses research by,
        758–60
    sexual selection and, 452
    theory of evolution by natural selec-
        tion by, 4, 414–16, 422–26
Darwin, Francis, 758, 759–60, 764
Data, scientific, 8–9
Data points, graph, B:2
Dating
    foraminifera and, 540
    radiometric, 417, 481
    relative and absolute, 422
Daughter cells
    meiosis and, 214–15
    mitosis and, 197–200
Davson, Hugh, 92
Day length
    Anolis lizard sexual activity and,
        1023–24
    photoperiodism and, 787
Day-neutral plants, **787**
Deactivation
    mitosis-promoting factor, 204
    signal, 144
Dead space, **868**, 870
Dead zones, aquatic, 512, 1095, 1102
Death, plant, 773–74
Death rates, human, 206*f*, 1037–40. *See
    also* Mortality, human
Deception, animal communication
    and, 1029–31
Deciduous trees, **574**, 1006
Decision making, 1020–21
Decomposer food chains, **1085**
Decomposers, **530**, **1084**
    fungi as, 579, 589–90, 595
    grazers vs., 1085–86
    metabolic rates of, 1093
    plasmodial slime molds as, 537
    protist, 542
De-differentiation, plant cell,
    377, 401
Deep-sea vents, 27, 32, 40, 43, 73, 509.
    See also Prebiotic soup
Defense, cell, 45

Defense responses, plant, 112, 707,
    775–81, 1063–66
Definitive hosts, **632**
Deforestation, 742, 1093–94, 1096, 1112
Degeneration hypothesis, 687
Dehydration, human, 803
Dehydration reactions, **43**, 73, 83–84
Deleterious alleles, 224, **448**, 1114
Deleterious mutations, **286**
Demography, **1038**
    four processes of, 1037–38
    life histories in, 1039–41
    life tables in, 1038–40
Denaturation, DNA, 344–45
Denatured, **50**
Dendrites, **808**, 886–87
Dendritic cells, **984**, 985*f*
Dense connective tissue, **807**
Density, water, 23–24, 26*t*
Density centrifugation, 944
Density dependent population growth,
    1042, 1045–46
Density-gradient centrifugation, 263,
    B:16–B:17
Density independent population growth,
    1042, 1044
Dental disease, 498
Dental plaque, 145
Deoxyribonucleic acid (DNA). *See* DNA
    (deoxyribonucleic acid)
Deoxyribonucleoside triphosphates
    (dNTPs), **264**, 346–47
Deoxyribonucleotides, **60**
    anabolic synthesis of, 169
    components of, 60
    in DNA backbone, 260
    DNA composition of, 259
    DNA synthesis and, 196
Deoxyribose, 60, 61, 77
Dependent assortment, 236–38
Dephosphorylation, 304
Depolarization, **890–92**, **908–9**
Deposit feeders, **611**, 843
Derived traits, **475**
Dermal tissue, **707**
Dermal tissue systems, **705**, 707–8, 711*t*
Descending limb, loop of Henle,
    835–36, 838*t*
Descent from common ancestors,
    418–22, 423*f*
Descent with modification, 4, **416**. *See
    also* Evolution; Natural selection
Deserts. *See* Subtropical deserts
Design, experimental, 10–12. *See also*
    Experiments
Desmosomes, **136**
Detergents, 94
Determination, cell, **397–98**
Detritivores, **612**, **637**, **1084**
Detritus, **530**, **997**, **1084**, 1085, 1092–93
Deuterostomes, **606**, **646–74**. *See also*
    Animal(s)
    characteristics of selected
        hominins, 670*t*
    chordates, 650–52
    echinoderms, 647–50
    four phyla of, 646–47*f*
    major animal phyla and, 603*t*
    primates and hominins, 668–73*f*
    vertebrates, 653–68
Devegetation study, 1093–94
Development, 374–87
    animal (*see* Animal development)
    cell-cell signaling of differential gene
        expression in, 379–84 (*see also* Pat-
        tern formation)

    cell movement, growth, and differen-
        tial expansion in, 375*t*, 376–77
    differential gene expression in,
        377–79 (*see also* Differential gene
        expression)
    evolutionary change and changes in
        pathways of, 384–85
    gene expression regulation and
        abnormal, 332
    plant (*see* Plant development)
    reuse of cell-cell signals in differing
        contexts of, 383–84
    shared processes of, 375–77
Developmental biology, 374. *See also*
    Development
Developmental homology, **420–21**
DeWitt, Natalie, 733
Diabetes insipidus, **837–38**, 857
Diabetes mellitus, 339, **856–58**, 983
Diacylglycerol (DAG), 143*t*
Diamond, Jared, 1105
Diaphragm, human, **868**
Diaphragms, contraceptive, 966*t*
Diarrhea, 522*t*, 691, 854
Diastole, **879**
Diastolic blood pressure, **880**
Diatomaceous earth, 542
Diatoms, 522–23, 542
Dicots, **564–65**
Dictyostelium discoideum, 536,
    B:20, B:21*f*
Dideoxyribonucleoside triphosphates
    (ddNTPs), 346–47
Dideoxy sequencing, **346–47**, 349, 353*t*
Diencephalon, **901**
Diet, human, 77, 843*t*. *See also* Animal
    nutrition; Food
Differential cell growth, 375*t*, 376–77
Differential centrifugation, **116**, 126,
    B:16–B:17
Differential gene expression, **319**
    cell-cell signals that trigger, 379–84
        (*see also* Pattern formation)
    eukaryotic gene regulation and,
        319–20 (*see also* Gene regulation,
        eukaryotic)
    regulatory proteins in, 326
    regulatory transcription factors
        and, 378
    role of, in development, 377–79
Differentiation, cell, 375*t*, **377**, 398–99.
    See also Differential gene
    expression
Diffusion, **89**, **823**
    across lipid bilayers, 89–90
    facilitated (*see* Facilitated diffusion)
    Fick's law of, 864–65
    gas exchange and, 862, 864–65,
        874–75
    in phloem loading, 731
    prokaryotic cells and, 105
    skin-breathing and, 813
    surface area adaptions for, 814
    surface area and animal gas exchange
        by, 874–75
Digestion, **845**
    cecum and appendix in, 854
    complete human digestive tract, 846*f*
    dietary fiber and human, 77
    digestive enzymes in, 110–11, 115
    digestive enzymes in human, 79
    digestive processes in, 846–47
    digestive tract in, 845–46
    fungal extracellular, 589–90
    large intestine in, 854–55
    mammalian digestive enzymes, 855*t*

Digestion, (continued)
mouth and esophagus in, 847–48
nutrition processes and, 841–42
plant proteinase inhibitors and, 779
small intestine in, 851–54
stomach in, 848–51
Digestive glands, 933
Digestive hormones, 937
Digestive tracts, **845**
animal, 604, 845–46
human, 810f, 846f
Dihybrid crosses, **236–38**
Dikaryotic, **582**
Dimers, 49, **125**
Dinoflagellata, 521–22t, 541
Dinosaurs
birds as, 657–58, 668
extinction of, 483, 486, 490–92
as reptiles, 660
Dioecious, **790**
Diploblasts, **604**
Diploid cells, 213, 214t, 533–34
Diplomadida, 538
Dipnoi, 663
Diptera, 640t
Direct counts, 1116
Directed-pollination hypothesis, 562
Direct fitness, 1032
Directional selection, **440**–41
Direct sequencing, 502–3, **525**,
582, 1107
Disaccharides, **73**
Discontinuous replication
hypothesis, 266–67
Discontinuous strand, **267**
Discrete population growth, 1043
Discrete traits, **248**
Diseases, animal
avian flu virus, 688, 689
avian malaria, 1015
herbivore, 1065–66
prions and, 51
simian immunodeficiency viruses
(SIVs), 688
viruses and, 681
Diseases, human
achromatopsia, 446
allergies, 990–91
ALS (Lou Gehrig's disease), 376
animals and transmission of, 602–3
anthrax, 498–99
autoimmune, 983–84
bacterial, 513, 514, 515
bacterial quorum sensing and,
145–46
bacteria that cause, 499t
beriberi and vitamin deficiencies, 54
cancer (see Cancer)
carriers of genetic, 251
caused by protists, 520–22
chlamydia, 364
Cushing's disease, 941
cystic fibrosis, 94–95, 350
dental plaque and tooth decay, 145
diabetes mellitus, 856–58, 983
emerging viruses and emerging,
688–89
food poisoning, 500
fungal, 581, 584, 594
gas exchange and circulation failures
and, 861

genetic counselors and inherited, 346
genetic testing and, 350
germ theory of, 499–500
HIV/AIDS, 677–78, 990
Huntington's disease, 233, 251,
348–51
hypertension, 880
immunity, immunization, and,
973–74 (see also Immune systems)
immunodeficiency, 990–91
kuru and Creutzfeldt-Jakob
disease, 51
malaria, 541, 1066–67
Marfan syndrome, 247
mistakes in meiosis and, 225–27
multiple sclerosis (MS), 894–95, 983
phenylketonuria, 248
pituitary dwarfism, 338–44
prions and, 50–51
prokaryotes and, 498–500
proteins as defense against, 45
protists that cause, 522t
red-green color blindness, 916–17
responding to viral outbreaks, 689
rheumatoid arthritis, 983
roundworms and, 638
schistosomiasis, 631–32
severe combined immunodeficiency
(SCID), 352–54
sexually transmitted (see Sexually
transmitted diseases)
sexual reproduction and, 224–25
sickle-cell disease, 46, 251
trisomy in human births, 227t
ulcers, 850
viral epidemics, 676–77
vitamin deficiencies, 355
xeroderma pigmentosum (XP),
272–74
Disorder. See Entropy
Dispersal, **1013**
allopatric speciation by, 463
barriers to, 1014
of exotic and invasive species,
1014–15
habitat destruction and, 1113
humans as agents of, 1014
physical isolation and, 462–63
seed, 797–98
Dispersive replication, DNA, 262f, 263
Disruptive selection, **442**, 443f, 465, 471t
Distal tubule, 832, **837**–38t
Distortion, experimental, 12
Disturbance regimes, **1073**–74, 1080,
1107, 1119–20
Disturbances, **1073**
Disulfide bonds, 35, **48**
Divergence, species, 468, 471t. See
also Speciation
Diversity. See also Biodiversity; Ecological diversity; Ecosystem diversity; Genetic diversity;
Metabolic diversity; Morphological diversity; Species
diversity
adaptive immune response, 978
animal, 610–17
of animal large intestines, 855
animal nitrogenous wastes, 829–30
antibody, 981–82
fungal, 586–94

of genomes, 363
green plant, 553–66
prokaryotic, 498, 504–12
protist, 526–36
protostome, 625–30
viral, 675, 686–89
Dixon, Henry, 724
DNA (deoxyribonucleic acid), **61**, 62–66.
See also Genes; Nucleic acids
amplification of fossil, with polymerase chain reaction, 345–46
anabolic synthesis of nucleotides
for, 169
antiparallel strands of, 63–64
in central dogma, 280–81
as characteristic of domains of
life, 497t
chloroplast, 113
chromosomes, genes, and, 195–96
cloning of, 340–41, 344–46
closed, as protected from DNase
enzyme, 522t
comparing *Homo sapiens* and *Homo
neanderthalensis*, 345–46
cutting and pasting, 340–41
directionality of strands of, 61
discovery of double helix secondary
structure of, 62–65, 260–61
electron micrograph of, 289f
as genetic information-containing
molecule, 65
mitochondrial, 112
model of condensed, 319f
phylogenies based on sequences
of, 477–78
physical model of, 59f, 63f
as poor catalytic molecule, 65–66
primary structure of, 260f
prokaryotic, 103–4
prokaryotic vs. eukaryotic, 107t
repair of, 271–74
RNA structure vs. structure of, 67t
sequencing of (see DNA sequencing)
synthesis of (see DNA synthesis)
viral, 686, 687f
DNA cloning, **340**–41, 344–46
DNA Data Bank of Japan, 360
DNA fingerprinting, 365, **366**, 367
DNA libraries, **342**–43
DNA ligases, 267–68t, 340–41, 352t
DNA microarrays, 370–72
DNA polymerases, **263**
in DNA synthesis, 268t
DNA synthesis and, 263–66
mutations and, 285–86, 448
in polymerase chain reaction, 344–45
proofreading ability of, 271–72
DNA probes. See Probes, DNA
DNA replication. See DNA synthesis
DNase enzyme, 322
DNA sequencing. See also Whole-
genome sequencing
dideoxy DNA sequencing method,
346–47
genetic homology and, 420
phylogenies based on data from,
477–78
pyrosequencing technologies, 347,
360–61
DNA synthesis, 258–75. See also Genes;
Genetics
comprehensive model of, 263–68
copying mechanism of, 65
correcting mistakes in, 271–72
DNA polymerases and, 263–64
electron micrograph of, 258f

Hershey-Chase experiment on DNA
as genetic material in genes,
259–60f
lagging strand synthesis in, 266–68
leading strand synthesis in, 265–66
Meselson-Stahl experiment on hypotheses about, 261–63
opening and stabiliizing of double
helix in, 264–65
proteins required for, 268t
repairing damaged DNA and,
272, 273f
starting of replication in, 264
structure of DNA and, 260–61
telomere replication in, 269–71
xeroderma pigmentosum (XP) case
study of damaged DNA, 272–74
Dogs, 213t, 369t
Dolly (cloned sheep), 378
Dolphins, 422
Domains, 7. See also Archaea domain;
Bacteria domain; Eukarya
domain
Domesticated animals, 602
Dominant alleles, 239t
Dominant traits, **233**
identifying human traits as, 251
incomplete dominance and codominance vs., 245–47
monohybrid crosses and, 232–33
Dopamine, 898t
Dormancy, 570, **714**–15, **769**, 770–71,
798–99
Dorsal, **379**
Double bonds, 20
Double fertilization, 403–**4**, **562**, **795**
Double helix secondary structure, DNA,
261. See also DNA (deoxyri-
bonucleic acid)
discovery of, 62–65, 260–61
opening and stabiliizing of, in DNA
synthesis, 264–65
proteins required for opening, 268t
Double-stranded DNA (dsDNA) viruses,
687t, 690
Double-stranded RNA (dsRNA) viruses,
687t, 691
Doubly compound leaves, 702
Doughnut-shaped proteins, 45f
Doushantuo microfossils, 486–87
Down, Langdon, 225
Downstream, **291**
Down syndrome, **225**–27
Dragonflies, 641t
Drosophila melanogaster
bicoid gene as master regulator of development in, 379–80
chromosome theory testing with,
241–42
foraging in, 1021
genetic drift in, 445
genetic map of, 246
genome of, 362, 369t
linked genes and, 243–45
as model organism, 623, 639,
B:21f, B:22
mRNA sequences from one gene
in, 329
number of chromosomes in, 213t
second-male advantage in, 954–55
Drought, 426–27. See also Dry habitats
Drugs
animals and testing of, 603
anticancer, 333, 371–72, 575, 579f
anti-HIV, 682, 683, 684
antiviral, 678

Boldface page numbers indicate a glossary entry; page numbers followed by an *f* indicate a figure; page numbers followed by *t* indicate a table.

derived from green plants, 549
ephedrine, 576
evolution of resistance to, 424–26, 500
fungal infections and, 584
fungi and, 595
homology and, 421
miRNAs in, 330
Dryer, W. J., 982
Dry habitats
  carbon dioxide concentration mechanisms in, 187–89
  drought as selective force, 426–27
  gas exchange in, 864
  plant adaptations to, 720–21, 728
Drying, seed maturation and, 797
Ducts, exocrine, 933
Dusky seaside sparrow, 461–62
Dutch elm disease, 598
Dwarfism, pituitary, 338–44
Dwarf mistletoe, 798
Dwarf plants, 769–70
Dyneins, **128**, 201–2
Dysentery, 522*t*

E

Eardrums, 910–11
Early endosome, **111**
Early prophase I, meiosis, 217
Ear ossicles, **910**–11
Ears, 910–12. *See also* Hearing
Easter Island, 1105
Ebola virus, 688, 689, 692
Ecdysone, **936**
Ecdysozoans, 603*t*, **609**, 625, 637–43
Echidnas, 665
Echinoderms (Echinodermata), 603*t*, 646, **647**–50
Echinoidea, 650
Echolocation, **912**
E. coli. See Escherichia coli (E. coli)
Ecological disaster, 1105
Ecological diversity, 509–12
Ecological efficiency, 1086–87
Ecological opportunity, 484–85
Ecology, 993–1018
  aquatic ecosystems in, 995–1001
  areas of study in, 993–95
  behavioral (*see* Behavioral ecology)
  biodiversity and (*see* Biodiversity)
  climate and climate change in, 1008–13*f*
  community, 994 (*see also* Community ecology)
  conservation biology and, 995 (*see also* Conservation biology)
  ecological importance of protists, 522–23
  ecosystem, 995 (*see also* Ecosystems)
  geographic distribution of organisms and, 1013–17
  organismal, 994
  population, 994, 1037–57
  terrestrial ecosystems in, 1001–8
Economic impacts
  of animals, 602
  of biodiversity, 1117
  of fungi, 581–82
  of sponges, 618
Ecosystem diversity, **1106**
Ecosystem ecology
  biogeochemical cycles in, 1092–98
  ecosystem energy flow in, 1083–92
  ecosystems and, 995, 1083
  global warming in, 1098–1102

Ecosystem energy flow, 1083–92
  biomagnification in, 1088–89
  global productivity patterns in, 1089–90
  net primary productivity in, 1084
  primary producers in, 1083–84
  productivity limiting factors, 1090–92
  transformation of solar energy into biomass, 1084–85
  trophic cascades and top-down control in, 1087–88
  between trophic levels, 1086–87
  trophic structure and, 1085–86
Ecosystems, **547**, **995**, **1083**–1104. *See also* Ecology
  animals in, 602
  aquatic, 995–1001
  global climate change and, 1011–13*f*
  global productivity patterns of, 1089–90
  human impacts on, 1083
  most productive, 1089–90
  nutrient cycling within, 1092–94
  prokaryotes and nitrogen cycle in, 510–11
  restoration of, 1122–23
  terrestrial, 1001–8
Ecosystem services, **547**, **1117**
Ecotourism, 1117
Ectoderm, 394, **395**, **604**
Ectomycorrhizal fungi (EMF), **586**–88, 596, 598, **746**
Ectoparasites, **613**
Ectothermic, **660**
Ectotherms, 660, **816**–17, 1086
Edge habitat, 1113
Ediacaran faunas, 486–87
Effectors, **815**, **938**
Effector T cells, 986
Efferent division, **899**
Efficiency
  asexual reproduction, 224, 786
  ecological, 1086–87
Eggs, **211**, **784**, **951**
  amniotic, 655, 658
  animal, 615, 959–60
  cell-cycle regulation and frog, 203–4
  evolution of oviparous and viviparous species and, 956–57
  fertilization of, 4, 390–92, 954–57
  fitness trade-offs for animal, 804, 805*f*
  human, 960
  land plant retention of, 556
  mammalian, 392
  meiosis and, 194, 218
  monotremes as mammals that lay, 665
  plant, 402, 789–90
  protostome, 630
  sexual selection and, 452–53
  vertebrate amniotic, 658
Either-or rule of probability, B:18
Ejaculation, **958**
Ejaculatory duct, **958**
Elastase, 855*t*
Elastic, **868**
Elastin, 134
Electrical activation, heart, 880–81
Electrical activation of hearts, 880–81
  maintenance of resting potentials, 888–90
  measuring membrane potentials with microelectrodes, 890
  membrane potentials, 887
  nervous systems and, 885–86
  nervous tissue and, 808
  neuron anatomy and, 886–87
  principles of electrical signaling and, 885–91
  synapses and, 895–99
  vertebrate nervous systems and, 899–904*f*
Electrical signals, plant, 767
Electrical stimulation, mapping of cerebrum with, 902
Electric current, **887**
Electric fields, animal sensing of, 610
Electrocardiogram (EKG), **881**
Electrochemical gradients, **94**–98, 745–46, **887**
Electrolytes, **822**, **842**. *See also* Osmoregulation
Electromagnetic spectrum, **174**
Electron acceptors, **151**
Electron carriers, **152**
Electron donors, **151**
Electronegativity, **18**
Electron microscopy. *See also* Scanning electron microscopy (SEM); Transmission electron microscopy (TEM)
  of actin and myosin, 924
  of bacterial transcription and translation, 296*f*
  cell research and, 116
  of chromatin nucleosomes, 321*f*
  cryoelectron tomography, 157*f*
  of DNA, 258*f*, 265*f*
  of fossilized plants, 554
  freeze-fracture, of membrane proteins, 93
  of myofibrils, 923*f*
  of noncoding DNA regions, 293–94
  of normal and karyotypes, 287*f*
  overview of, B:14–B:15
  of protist cells, 524–25
  of skin cells, 131*f*
  of transport vesicles, 126*f*
Electrons
  atomic structure and, 16–17
  chemical bonds and sharing of, 17–20
  electron transport chain and (*see* Electron transport chain (ETC))
  in photosynthesis, 177–82
  potential energy and, 27
  prokaryotic variation in donors and acceptors of, 506–8*t*
  redox reactions and, 151–52
Electron shells, **17**
Electron tomography, B:15
Electron transport chain (ETC), **161**
  chemiosis hypothesis on, 162–63
  components of, 161–62
  fourth step of cellular respiration and, 153
  organization of, 163–64
  in photosynthesis, 180–81, 182
  prokaryotic cellular respiration and, 507–8
Electrophoresis. *See* Gel electrophoresis
Electroreceptors, **908**
Elements, **16**
Elephantiasis, 638, 973*f*
Elephants, 460–61, 812–13, 912

Elephant seals, 454–55
Elevational gradients
  shoot systems and, 700
  tree ranges and, 1072
Elimination, waste, 841–42, 854–55
Ellis, Hillary, 376
Elongation, stem, 762–64
Elongation factors, 302
Elongation phase, **292**
  transcription, 292–93
  translation, 301–2
Embryogenesis, **392**, **402**, **796**
  animal, 615
  events in, 404–6
  plant seeds and, 796
Embryonic stem cells, 375
Embryonic tissue layers, 394–95, 604
Embryophytes (Embryophyta), **556**
Embryos, **194**, **374**
  developmental homology in, 420
  human, 388*f*
  land plant retention of, 556
  plant, 401*f*
Embryo sacs, 402, 403–4, **790**
Emergent vegetation, **998**
Emerging diseases, 224–25, **688**–89
Emerson, Robert, 179
Emigration, 1037–38
Emulsification, **853**
Endangered species, **1111**
  dusky seaside sparrow case study, 461–62
  extinctions and, 1111 (*see also* Extinction)
  population ecology and analysis of, 1053–55
  preserving metapopulations of, 1055
  using life-table data for, 1054–55
Endangered Species Act, 461–62
End-Cretaceous extinction, 490–92
Endemic species, **844**, **1109**
Endergonic reactions, **30**, 62, **149**, 150–51
Endocrine glands, **932**, **935**
Endocrine pathway, 931–32
Endocrine signals, **930**
Endocrine systems, **929**, 932–33. *See also* Hormones, animal
Endocytosis, **111**
Endoderm, 394, **395**, **604**
Endodermis, **722**, **747**
Endomembrane system, **111**
  eukaryotic, 108–12, 114*t*
  manufacture and transport of proteins by, 118–23*t*
Endomycorrhizal fungi, **589**
Endoparasites, **612**–13
Endophytic fungi, **586**, 589
Endoplasmic reticulum (ER), **108**, 114*t*, 122–23*f*, 526
Endorphins, 898*t*
Endoskeletons, **647**, 655, **920**–21
Endosperm, 355, **404**, 562, 795
Endosymbionts, **514**
Endosymbiosis, **527**, 566
Endosymbiosis theory, **527**–28, 530–31
Endothermic processes, **27**
Endotherms, 660, 668, **816**–19
End-Permian extinction, 489–90
End-point inhibition, 314
Energetic coupling, **150**
Energy, **27**
  active transport and, 97–98
  animal reserves of, 938–40
  capacity of water to absorb, 24–25

Energy, (continued)
carbohydrates and storage of chemical, 78–79
as cellular requirement, 168
cellular respiration, fermentation and (see Cellular respiration; Fermentation)
chemical (see Chemical energy)
chemical evolution and chemical (see also Chemical energy; Chemical evolution)
electrical, of lightning, 39–40
enzyme catalysis and activation, 51–53
flow of, through ecosystems, 1083–92
lipid bilayers and, 85
living organisms and, 1
metabolic diversity of prokaryotes and, 506–9
metabolic rates of consumption of, 812–13
photosynthesis and (see Photosynthesis)
six general prokaryotic methods for obtaining, 506t
specific heats of some liquids, 24t
types and transformations of, 27–29
Energy hypothesis on species richness, 1080
Engelmann, T. W., 176
Engineering, genetic. See Genetic engineering
Enhancement effect, 179–80, 183
Enhancers, 324–26
Enkephalins, 898t
Enrichment cultures, 501–2, 508
Entamoeba histolytica, 522t
Entoprocta, 603t
Entropy, 29–30
Envelope, 679
Envelope, nuclear. See Nuclear envelope
Enveloped viruses, 679
Environmental genomics, 503
Environmental sequencing, 364, 1106, 1107
Environments. See also Ecosystems
adaptive radiations and ecological opportunities in, 484–85
animal color vision and, 917–18
animals and extreme, 803
animal sensory organs and, 610, 907–8
aquatic (see Aquatic ecosystems)
background extinctions vs. mass extinctions and, 489
balancing selection and, 443
cues in, and Daphnia reproductive modes, 952
ecology and, 993
effects on phenotypes of physical, 247–48
eukaryotic response to internal, 319–20
gene expression and, 308
genome redundancy and, 363
hormone coordination of responses to changes in, 937–38
limits for net primary productivity in, 1090–91
natural selection and changes in, 426–28

plant sensory systems and, 755
role of, in succession, 1075
seed dormancy and, 798–99
sensing changes in, 907–8 (see also Sensory systems, animal)
sexual reproduction and changing, 224–25
succession and, 1075, 1077
terrestrial (see Terrestrial ecosystems)
Enzymatic combustion, 590
Enzyme kinetics, 55
Enzyme-linked receptors, 143–44
Enzymes, 45. See also Proteins
catalysis of chemical reactions by, 45, 51–53 (see also Catalysis)
cell-cycle regulation and, 203–4
cofactors, coenzymes, and, 54
digestive, in animal lysosomes, 110–11
digestive, in pancreas, 115
DNA synthesis and, 263–68
effects of temperature and pH on function of, 55–56
enzyme kinetics and catalysis rate regulation, 55
eukaryotic organelles and, 107
as first self-replicating molecules, 56
folding in ribonuclease, 50
fungal, 581, 589–90, 746
globular shape of, 45–46
homeostasis and function of, 815
hydrolysis of carbohydrates by, to release glucose, 78–79
lock-and-key model for, 53–54
mammalian digestive, 855t
membrane transport proteins and, 97f
multienzyme complexes of, 49
nucleic acid polymerization and, 62
one-gene, one-enzyme hypothesis and, 277–78
in peroxisomes, 109–10
post-translational control by, 330
prokaryotic, 104
protein digestion by pancreatic, 851–52
proteins as, in central dogma, 280–81
regulation of, by regulatory molecules, 54–55
regulation of pancreatic, 852
RNA ribozymes as, 68
RNA synthesis and, 279
signal transduction via enzyme-linked receptors, 143–44
smooth ER and, 108
transfer RNA and, 298
Ephedrine, 576
Epicotyls, 796
Epidemics, 676
diabetes mellitus, 857–58
fungal, 581
recent viral, in humans, 676–77
Epidemiology, 689
Epidermal cells, plant, 707–8, 711t
Epidermis, 406, 705, 722
Epididymis, 958
Epigenetic inheritance, 323
Epiithelial tissues, 809
Epilepsy, 902
Epinephrine, 937
for anaphylactic shock, 991

binding of, to receptor, 945–47
control of, by sympathetic nerves, 943
fight-or-flight response and, 881, 937–38
as hormone, 933
receptors for, 140
Epiphytes, 572, 750, 1003
Epiphytic plants, 750–51
Epithelia (epithelium, epithelial tissues), 135, 604, 809–10, 815, 834
Epitope, 981
Equilibrium, 27, 90f. See also Hardy-Weinberg principle
Equilibrium potentials, 888, 889
Equisetophyta, 572
Equivalence, genetic, 377–78
Erosion, 547, 741–42
Errors, DNA synthesis, 271–72, 283–84
ER signal sequence, 121
Erwin, Terry, 1107, 1108
Erythropoietin (EPO), 940
Escaped-genes hypothesis, 686–87
Escherichia coli (E. coli)
BAC libraries and, 360
chromosomes of, 103
discovery of DNA in genes of, 259–60f
DNA polymerase proofreading and mutants in, 271–72
food poisoning and, 500
gene expression in, 307–8
growth hormone engineering using, 339–43
lactose metabolism of, as model, 309–10
in Meselson-Stahl experiment, 261–63
metabolic diversity and, 507t
as model organism, B:19–B:20, B:21f
mutations and fitness in, 449–50
as proteobacteria, 515
scanning electron micrograph of, 307f
types of lactose metabolism mutants in, 311f
E site, 300
Esophagus, 810f, 846f, 848
Essential amino acids, 842
Essential nutrients, 738, 842–43
Ester linkages, 83–84
Estradiol, 936, 960
binding of, to intracellular receptors, 943–45
biomagnification of estrogen-mimics, 1088–89
identifying receptor for, 944
puberty and, 961–62
sexual activity of Anolis lizards and, 1023
Estrogen-mimics, 1088–89
Estrogens, 140t, 936, 943–45, 960
Estrous cycle, 963
Estuaries, 1000
Ethanol, 24t, 110, 167, 508
Ethical concerns
over gene therapy, 352–54
over genetic testing, 350–51
over recombinant human growth hormone, 343–44
Ethylene, 140t, 773–75t
Ethylene glycol, 24t
Eudicots, 565, 706f, 796
Euglenida, 539
Eukarya domain, 6–7, 84, 497t, 519–20. See also Eukaryotes
Eukaryotes, 6. See also Eukarya domain
cell-cell attachments in, 135–38

cell structures and functions of (see Eukaryotic cells)
characteristics of, 526
chromosome replication in, 265f
control of gene expression in (see Gene regulation, eukaryotic)
genomes of, 365–70
identifying genes in genome sequences of, 363
major lineages of protists and, 524t (see also Protists)
model of condensed DNA of, 319f
number of genes in selected, 369t
phylogenetic tree of life and, 6–7
plasma membranes of, 92
prokaryotes vs., 519
RNA polymerases of, 290t
transcription in bacteria vs. in, 296t
transcription promoters in, 291
translation in bacteria vs. in, 296–97
Eukaryotic cells, 105–15. See also Cells
animal and plant, 106f
bacterial flagella vs. flagella of, 127
benefits of organelles in, 107
cell walls of, 113, 114t
chloroplasts of, 112–13, 114t
components of, 114t
cytoskeletons of, 113, 114t
dynamic cytoskeletons of, 123–28
dynamism of whole, 116
endomembrane system and proteins in, 118–23f
Golgi apparatus of, 108–9, 114t
lysosomes of, 110–11, 114t
mitochondria of, 112, 114t
nuclear transport in, 116–18
nucleus of, 107–8, 114t
peroxisomes of, 109–10, 114t
prokaryotic cells vs., 105–6f, 107t
ribosomes of, 109, 114t
rough endoplasmic reticulum of, 108, 114t
selective adhesion and adhesion proteins in, 136–38
smooth endoplasmic reticulum of, 108, 114t
structure and function of whole, 115–16
vacuoles of, 111–12, 114t
views of, 106f, 113f
Euryarchaeota, 512, 516
Eutherians, 660, 666, 967
Evaporation, 717, 816, 828, 867
Evergreen trees, 1007
Evidence, science and, 528
Evo-devo (evolutionary-developmental biology), 384, 752
Evolution, 4
of aerobic respiration, 510
of biodiversity (see Biodiversity)
changes in developmental pathways and change in, 384–85
as change through time, 415
chemical vs. biological, 15, 39, 59 (see also Chemical evolution)
consumption and arms races in, 1066–67
current examples of, 418
evidence for, 418–22t, 423f
evolutionary conservation of cell-cell signals and regulatory genes, 383
experimental, 449–50
four processes of (see Evolutionary processes)
global warming and evolutionary change within populations, 1101

Boldface page numbers indicate a glossary entry; page numbers followed by an *f* indicate a figure; page numbers followed by *t* indicate a table.

history of life and (*see* Life, history of)
human, 670–73*f*
lateral gene transfer in, 364
living organisms as product of, 2
mapping land plant, on phylogenetic trees, 555–56
of mouthparts, 844
of multichambered hearts with multiple circulations, 878
of neurons and muscle cells, 886
as not goal directed or progressive, 430
of oviparous and viviparous species, 956–57
of pollination, 793–94
of predation, 487–88
prokaryotes and, 509–12
of protostomes, 624–25
religious faith vs. theory of, 8–9
speciation and (*see* Speciation)
theory of, by natural selection, 4–5, 415–16, 422–24 (*see also* Natural selection)
vertebrate, 653–59, 878*f*
of whales, 477–79
Evolutionary-developmental biology (evo-devo), **384**, 752
Evolutionary processes, 435–57. *See also* Evolution
gene flow, 435, 447–48
genetic drift, 435, 443–46
Hardy-Weinberg principle and analyzing change in allele frequencies, 436–40
mutation, 435, 448–50
natural selection, 435, 440–43
nonrandom mating and, 450–55
Excavata, 524*t*, 536, 538–39
Exceptions
central dogma, 281
Mendelian genetics, 250*t*
Excitable membranes, **891**
Excitatory postsynaptic potentials (EPSPs), **897**
Excretory systems
fish osmoregulation and, 824
insect homeostasis and, 830–31
nitrogenous wastes and, 829–30
shark excretion of salt and, 826–27
urinary systems (*see* Urinary systems)
waste elimination, 841–42, 854–55
Exergonic reactions, **30**, **149**
Exhalation, 868–70
Exocrine glands, **933**
Exocytosis, **122**
Exons, **294**, 324
Exonuclease, 272
Exoskeletons, **625**, **920**
animal locomotion and, 920–21
as barriers to pathogens, 974
bony vertebrate, 653
cell walls as prokaryotic, 104–5
Exothermic processes, **27**
Exotic species, **1014**–15, **1110**–11, 1120
Expansins, 133, **762**
Experimental evolution, 449–50
Experiments. *See also* Biological methods; BioSkills
design of, 10–12
as hypothesis testing, 8–12
Gregor Mendel's, on heredity (*see* Mendelian genetics)
origin-of-life (*see* Origin-of-life research)
Exponential population growth, **1042**

Expressed sequence tag (EST), **363**
Ex situ conservation, 1121–**22**
Extant species, **416**
Extension, PCR, 344–45
Extensors, **920**
External fertilization, 615, 954
Extinction
background, **489**, 1116
current causes of, 1110–11
as evidence for evolutionary change, 417
genetic variation and, 440
global warming and species, 1100–1101
human population size and species, 1052
mass (*see* Mass extinctions)
metapopulation balance between recolonization and, 1046–47
predicting rates of, 1114–16
rates, 1077–78, 1110, 1114–16
secondary contact between isolated species and, 471*t*
Extinct species, **417**
Extracellular digestion, **589**–90
Extracellular layers, 132–34
Extracellular matrix (ECM), **133**
animal, 601
cell-cell interactions and, 377
connective tissue and, 806–7
functions of, in animals, 133–34
Extrapolation techniques, 1073, 1108
Extraterrestrial life, 501
Extremophiles, **501**
Eyes
arthropod, 637
cancers of human, 208
insect, 913
mollusk, 610
primate, 669
vertebrate, 914–17

F

F_1 generation, **232**
Facilitated diffusion, **96**, **731**, **825**
in phloem loading, 731–32
via carrier proteins, 96–97
via channel proteins, 94–96
Facilitation, **1075**, 1119
Facultative aerobes, **168**
FADH$_2$, **153**, 158–61
Fallopian tubes, **393**, 960
False penises, 956
Families, gene, 367–68
Family trees. *See* Pedigrees
Farmer ants, 602
Far-red light. *See* Red/far-red light responses, plant
Fats, **83**, 110, 168–69. *See also* Lipids
Fatty-acid binding protein, **854**
Fatty acids, **83**–84, 167, 169, 355. *See also* Lipids
Faunas, **486**–87
Feathers, 486, 657–58, **668**
Features, transitional, 417–18
Feces, **847**, 854
Fecundity, **1039**–40, 1055, 1115
Feedback
in ecosystem responses to global warming, 1099
positive, during action potential depolarization, 892
role of, in homeostasis, 815–16, 818
Feedback inhibition, **156**–59*f*, 314, **931**, 941

Feeding adaptations. *See also* Food; Food chains; Metabolic diversity; Metabolism
animal, 610–12, 616, 843–45
echinoderms, 648
fish, 656
protist, 529–30
protostome, 628, 629*f*
Feeding levels, 1085–86
Feet, primate, 669
Females
barn swallow mate selection by, 1024–25
behaviors mimicking, 1030
chromosomes of, 212
eggs as reproductive cells of, 211
embryo development inside animal, 615
gametangia of, 556
oogenesis in mammal, 953
plant flower parts, 789–90
plant reproductive organs, 402
reproductive systems of animal, 959–66
sexual readiness of, and visual cues from male lizards, 1024
sexual selection and female choice, 452–54
status of human, and fertility rates, 1053
Feminization effects, 1088–89
Femmes fatales, 955
Fermentation, 153, **166**–68, **508**, 854. *See also* Cellular respiration
Ferns, 491, 550, 573, 750
Ferredoxin, **182**
Fertility rates, human, 1053, 1121
Fertilization, **211**, **534**
animal (*see* Fertilization, animal)
bioremedial (*see* Fertilization, bioremedial)
fungal, 591
genetic variation and types of, 222–23
meiosis and, 215–16, 220–21
plant (*see* Fertilization, plant)
protist, 534
Fertilization, animal, **389**, **390**, **951**
evolution of egg-bearing and live-bearing species and, 956–57
internal and external, 954–55
multiple, 391–92
protostome, 630
requirements for, 390–91
types of, 615
unusual aspects of mating and, 955–56
Fertilization, bioremedial
of contaminated sites, 500
of ocean with iron to increase net primary productivity, 1091–92
Fertilization, plant, **784**, **792**
angiosperm, 562
double, in plant gametogenesis, 403–4
pollen grains and, 559–60
pollen-stigma interactions and, 402–3
steps in, 794–95
Fertilization envelope, **392**
Fertilizers, 749, 1093, 1095
Fetal alcohol syndrome (FAS), **969**–70
Fetus, **968**–70
Fiber
animal, 602
dietary, 77
extracellular layer, 132, 133*f*
green plants and, 548–49
Fibers, **710**, 711

Fibronectins, **134**
Fick, Adolf, 864–65
Fick's law of diffusion, 864–65
Fight-or-flight response, 881, **883**, **937**, 943, 946
Filament, **789**
Filamentous algae, 176
Filaments, cytoskeleton, 123–27
Filter feeders, **610**
Filtrates, **830**
Filtration, 832, **833**, 834
Finches, 426–29, 806, 1062, 1063*f*
Finite rate of increase, **1043**
Fins, fish, 656–67
Fire
bark and, 714
cheatgrass and, 1015–16
disturbance regimes and, 1073–74
seed dormancy and, 799
Fireflies, 1030
Firmicutes, 512, 513
Firs, 575
First law of thermodynamics, **29**
Fischer, Emil, **53**–54
Fish
cichlid jaws, 844–45
fertilized eggs and development of, 374*f*
gills of, 865–66
hearts of, 878
lateral line system as hearing in, 913
mutualisms with, 1068
osmoregulation by, 824, 826–28
as vertebrates, 655
Fission, 619, 620, 633, **951**
Fitness, 5, **1058**. *See also* Fitness trade-offs
competition and, 1059
cost-benefit analysis of, 1021
dominant and recessive phenotypes and, 233
effects of gene flow on, 448
evolution of, 449–50*t*
genetic drift and, 444
hemoglobins and, 873
heterozygous individuals and, 440
inbreeding and human, 452*t*
inbreeding depression in, 451–52
maladaptive traits and, 187
natural selection, adaptation, reproduction, and biological, 424
sexual selection and, 452–55
species interactions and, 1058–59
types of inclusive, 1032–33
Fitness trade-offs, **432**, **1040**, **1062**. *See also* Fitness
in animal anatomy and physiology, 804, 805*f*
in competition, 1062
countervailing directional selection and, 441
endothermy vs. ectothermy, 817
life histories and, 1040, 1041
as limits to natural selection, 432
nitrogenous wastes and, 830
parental care and, 659
Fixed action patterns (FAPs), **1020**
Flaccid, **719**
Flagella (flagellum), **104**, **390**, **524**
bacterial vs. eukaryotic, 127
cell movement by, 127–28
as characteristic of domains of life, 497*t*
fungal, 583–84, 585
prokaryotic, 104
protist, 533
sperm, 390

Flagellin, 127, 526
Flattening adaptations, 814
Flatworms, 631–32
Flavine adenine dinucleotide (FAD), **153**
Flavins, 161–62
Flavonoids, 176
Flemming, Walther, 195
Flexors, **920**
Flies, 629, 640*t*
Flight, 657–58, 668
Flood basalts, 489
Floods, 1074, 1117
Flor, H. H., 775–76
Floral meristems, **409**
Florigen, 787–**88**
Flowers, **561**, **576**, **783**
　angiosperm radiation and, 564 (*see also* Angiosperms)
　buds and, 699
　fading of, 773
　female gametophyte production from, 790
　floral meristems and, in reproductive development, 409
　flowering as production of, 787–88
　genetic control of structures of, 409–11
　land plants and, 561–62
　male gametophyte production from, 790–92
　as morphological innovations, 485
　pollination of, 562–63
　as reproductive structures, 786–92
　structure of, 788–90
Fluid connective tissue, **807**
Fluid feeders, **611**, **843**
Fluidity, plasma membrane, 87–89
Fluid-mosaic model, 92, **93**, 94
Flukes, 631–32
Fluorescence, 102*f*, **178**–79, 346–47
Fluorine, 843*t*
Flu virus. *See* Influenza virus
Folate, 843*t*
Folding
　adaptations, 814
　in animal small intestines, 851
　protein, 50–51, 304, 330
Follicles, **961**
Follicle-stimulating hormone (FSH), **943**, **962**, 964–66
Follicular phase, **962**
Foltz, Kathleen, 391
Food, **842**. *See also* Feeding adaptations; Food chains
　angiosperms as human, 576
　animal nutrition and, 841–42 (*see also* Animal nutrition)
　animals as human, 602
　animal sources of, 612–13
　disruptive selection and, 442
　food poisoning and, 500
　foraging behaviors and, 1021–22
　fungi as human, 581, 596, 598
　giraffe necks and competition for, 9–10
　green plants as human, 548
　mollusks as human, 634
　mouthparts and capture of, 843–45
　plant reproduction and human, 783
　population cycles and, 1048–50
　transgenic crops as, 355–56

Food chains, 522–**23**, **1085**
Food webs, **1085**
Foot, **626**, 629
Foraging, **1021**–22
Foraminifera, 540
Forebrain, **653**
Foreign cells, adaptive immune response and, 983–84, 987–88
Forensic biologists, 346, 366
Forestry, 547, 596, 742
Forests. *See* Boreal forests; Temperate forests; Tropical wet forests
Formaldehyde, 29, 32–34, 38, 40, 60
Formulas, molecular and structural, 20–21, B:8*f*
Fossil fuel consumption, 523, 1096, 1098, 1121
Fossilization, 479–80
Fossil record, **416**, **479**. *See also* Fossils
　animals in, 617
　arthropods in, 637
　Cambrian explosion in, 486–87
　cetaceans in, 422
　DNA and, 66
　as evidence for evolutionary change, 416–18
　fish in, 656–67
　fossilization and fossil formation in, 479–80
　hydromedusas in, 601*f*
　pollen grains in, 1072
　prokaryotes in, 497–98
　stromatolites in, 15*f*
　studying green plants using, 550–51
　as tool for studying history of life, 479–84
　vertebrate, 653–55
Fossils, **416**, **479**. *See also* Fossil record
　amplification of DNA in, with polymerase chain reaction, 345–46
　elongated cells in, 554
　feathered dinosaur, 657
　fossilization and formation of, 479–80
　trilobites in, 474*f*
Founder effect, **446**
Four-o'-clocks, 245
Fovea, **915**
Fox, Arthur, 919
Fragmentation, 570–71, 618–19, 632–33
Frameshift mutations, 286*t*
Franklin, Rosalind, 63
Free-energy change, 30, 42, 52–53, 78–79, 160*f*
Free radicals, **32**, 176, 272, 590
Freeze-fracture electron microscopy, 93
Frequencies
　calculating recombination, 246
　distribution, 248–49
　sound, 911–12
Frequency, **909**
Frequency-dependent selection, **443**
Freshwater ecosystems, 997–1000. *See also* Aquatic ecosystems
　estuaries as marine and, 1000
　fish osmoregulation in, 827–28
　lakes and ponds, 997
　lake turnover in, and nutrient availability, 996
　osmotic stress in, 824

　streams, 999
　wetlands, 998
Frogs, 203–4, 394–96, 429–30, 664, 935–36
Fronds, **573**
Frontal lobe, **901**
Fructose, 155–56, 168, 189–90, 959*t*
Fruit fly model organism. *See Drosophila melanogaster*
Fruiting bodies, **515**, 529
Fruits, **563**, **783**
　angiosperm radiation and, 564
　auxin and, 768
　color vision and, 918
　development of, and seed dispersal, 797–98
　ethylene and ripening of, 773–74
　function of, 798
　land plants and, 563–64
　seeds and, 795
　types of, 797
Fucoxanthin, 532*t*
Fuel, green plants and, 548
Functional genomics, **359**, 370–71. *See also* Genomics
Functional groups, **34**–36, 40–42
Function-structure relationships, 806
Fundamental niches, **1061**
Fungi, **579**–600
　adaptations of, as decomposers, 589–90
　biological methods for studying, 582–86
　cell theory experiment and, 3–4
　chitinase and, 56
　chitin as structural polysaccharide in, 76
　economic impacts of, 581–82
　four major types of life cycles of, 591–94
　importance of relationship between land plants and, 579–80
　key lineages of, 585–86, 594–98
　lichens and, 566
　as model organism (*see Saccharomyces cerevisiae*)
　morphological traits of, 582–84
　mutualisms of, 586–89
　mycorrhizal, and land plant nutrients, 580
　phylogenetic relationship of animals to, 6, 584–85
　phylogenies of, 584–86
　plant nutrient uptake via mycorrhizal, 746
　reasons for studying, 580–82
　reproductive variation in, 590–91, 592*f*
　saprophytic, and carbon cycle on land, 580–81
　Silurian-Devonian explosion and, 551
　vacuoles in cells of, 111–12
　water molds and, 542
Fungicides, **585**
Fusion
　cell, 203–4
　fungal, 591
　gamete, 220–21, 390
　population, 471*t*
Fusion inhibitors, 682

G

G_0 state, 202, 271
G_1 phase, **196**, 205, 207–8
G_2 phase, **196**

Galactose, 72, 323–24
Galactoside permease, 313
Galápagos finches, 426–29, 806, 1062, 1063*f*
Galápagos Islands, 419–20
Gallbladder, 846*f*, **854**
Galls, 355–56
Gallus gallus, 369*t*
Gametangia (gametangium), **556**, 560
Gametes, **194**, **211**, **784**. *See also* Eggs; Sperm
　egg structure and function, 390
　fungal, 591
　fusion of, 220–21
　land plant production of, 556–64
　recognition of each other by same-species, in fertilization, 391
　structure and function of, 389–90
Gametic barriers and prezygotic isolation, 459*t*
Gametogenesis, **215**, 402–4, 952–**53**
Gametophytes, **534**, **557**, **784**
　land plant evolution away from life cycles dominated by, 558–59
　land plant life cycles and, 784–86
　production of female, 789–90
　production of male, 789, 790–92
Gamma-aminobutyric acid (GABA), 898*t*
Gamow, George, 282
Ganglia, **604**
Ganglion cells, **914**
Gap Analysis Program (GAP), 1120
Gap genes, 381
Gap junctions, **139**, 921
Gap phases, cell cycle and, 196, 205, 207–8
Gaps, cell-cell, 138–39
Garden peas, 213*t*, 230*f*, 231–32. *See also* Mendelian genetics
Gases
　chemical evolution and volcanic, 21, 27, 32, 40
　water vapor, 24–25
Gas exchange
　amniotic egg, 658
　amphibian, 664
　animal (*see* Animal gas exchange and circulation; Circulatory systems; Respiratory systems)
　bark and, 714
　diffusion and skin-breathing, 813
　fish lungs, 656
　gills and, 634, 651, 813
　during human pregnancy, 969
　insect, 828
　osmotic stress and, 824–25
　plant stomata and, 708
　pneumatophore roots in, 698, 699*f*
　in salmon, 813
　surface area and animal, 874–75
Gastric juice, 849
Gastrin, **852**
Gastrointestinal (GI) tract, 845
Gastropods (Gastropoda), **630**, 635
Gastrotricha, 603*t*
Gastrovascular cavities, 845
Gastrula, 394
Gastrulation, **376**, **394**–95, 606–7
Gated channels, **96**
Gause, G. F., 1060–61
GDP (guanosine diphosphate), 142–43
Gel electrophoresis, **62**, 94, 347, 853, B:11–B:12
Gemmae, **569**, 571, 573

Gene-by-environment interactions, 247–48, 250*t*
Gene-by-gene interactions, 248, 250*t*
Gene duplication, **367–68**, **777**
Gene expression, 276–88. *See also* Genes; Genetics
  control of (*see* Gene regulation, bacterial; Gene regulation, eukaryotic)
  development of central dogma of molecular biology on, 279–82
  DNA sequencer output, 276*f*
  functional genomics and, 370
  genetic code and, 282–85
  known types of point mutations in, 286*t*
  molecular basis for mutation in, 285–87*f*
  one-gene, one-enzyme hypothesis on, 277–78
  plant cell-cell signals and, 757
  protein synthesis and (*see* Protein synthesis)
  steroid-hormone receptors and, 944–45
  via transcription and translation, 280
Gene families, **367–68**, 476, **777**
Gene flow, **447**
  experimental increase of, 1114
  genetic variation and effect on average fitness of, 448, 450*t*
  Hardy-Weinberg principle and, 438
  in natural populations, 447
  speciation and, 458, 462, 464–65
  wildlife corridors and, 1121
Gene-for-gene hypothesis, 775–**76**
Gene pools, **436**
Generations, **1038**
Gene regulation, bacterial, 307–18. *See also* Gene expression; Genetics
  catabolite repression as positive control mechanism in, 314–16
  eurkaryotic gene regulation vs., 331–32
  identifying genes under regulatory control, 310–12
  information flow and, 307–10
  lactose metabolism as model system for, 309–10
  mechanisms of, 308–9
  overview of positive and negative *lac* operon regulation in, 317*f*
  repressors as negative control mechanisms in, 312–14
Gene regulation, eukaryotic, 319–37. *See also* Gene expression; Genetics
  bacterial gene regulation vs., 331–32
  cancer and defects in, 332–33
  chromatin remodeling in, 320–23
  differential gene expression and, 319–20
  post-transcriptional, 328–30
  regulatory sequences and regulatory proteins in initiating transcription, 323–28
  translational and post-translational, 330
Genes, **103**, **212**, **234**, **326**. *See also* Genetics
  analyzing and engineering (*see* Genetic engineering)
  bicoid gene as master regulator, 379–80
  cancer and, 332
  case study of *HLA*, in humans, 439–40

case study of *p53*, as cancer preventing, 332–33
in central dogma, 280
chemical composition of, 258–61
chromosomes and, 195–96
control of flower structures by, 409–11
discovery of eukaryotic, in pieces, 293–94
discovery of recombination of, 982–83*f*
DNA and (*see* DNA (deoxyribonucleic acid))
dwarf plants and defective, 769–70
estimating phylogenetic tree of life from rRNA sequences in, 6–7
finding disease-causing, 348–51
finding mutant, with replica plating, 310–11
functions of (*see* Gene expression)
gene families of, 367–68
genetic constraints on natural selection, 431
Hershey-Chase experiment on DNA as genetic material in, 259–60*f*
identifying, in genome sequences, 362–63 (*see also* Genomics)
inserting, into plasmids, 341*f*
lactose metabolism, 312
leaf shape and, 408
linked, on same chromosome, 243–45, 246*f*
living organisms and genetic information in, 1
in Mendelian genetics, 239*t*
messenger RNA hypothesis as intermediary between proteins and, 279–80
natural selection of, in Galápagos finches, 428–29
number of, in selected eukaryotes, 369*t*
one-gene, one-enzyme hypothesis on gene expression, 277–78
particulate inheritance hypothesis and, 234–35
physical locations (loci) of, 240
plant body axes and, 406
plant pathogen resistance, 777
prokaryotic, 103–4
redefinition of, 329, 362
for taste receptors, 919
and traits, 221
tumor suppressors, 206
Genes-in-pieces hypothesis, 293–94
Gene therapy, **351–54**
Genetically modified food, 355–56
Genetic bottlenecks, **446**
Genetic code, 279, **282–85**, 297–300
Genetic correlation, **431**
Genetic counselors, 346
Genetic divergence, speciation and, 458. *See also* Speciation
Genetic diversity, 363, 591, **1106**. *See also* Genetic variation
Genetic drift, 443–46
  allopatric speciation, dispersal, and, 463
  causes of, in natural populations, 445–46
  genetic variation and effect on average fitness of, 448, 450*t*
  Hardy-Weinberg principle and, 438
  metapopulations and, 1114
  simulation studies of, 443–45

Genetic engineering, 338–58. *See also* Genetics
  in agriculture, 548
  amplification of fossil DNA with polymerase chain reaction, 344–46
  common techniques used in, 353*t*
  common tools used in, 352*t*
  development of golden rice with agricultural biotechnology, 354–56
  dideoxy DNA sequencing, 346–48
  efforts to cure pituitary dwarfism with basic recombinant DNA technologies, 338–44
  finding Huntington's disease gene with gene mapping, 348–51
  gene therapy for severe immune disorders, 351–54
Genetic equivalence, 377–78
Genetic homology, **420–21**
Genetic isolation
  allopatry and, 462–64
  autopolyploidy and, 466–67
  speciation and, 458
Genetic maps, 244, 246, **348–51**, 353*t*
Genetic markers, **348–49**, **445**
Genetic models, **236**
Genetic recombination, **222**, 244–45
Genetics, **230**. *See also* Genes; Nucleic acids
  analyzing and engineering genes (*see* Genetic engineering)
  control of gene expression in (*see* Gene regulation, bacterial; Gene regulation, eukaryotic)
  cystic fibrosis as genetic disease, 94–95
  DNA as genetic information-containing molecule, 65
  DNA sequencing of genomes in (*see* Genomics)
  DNA synthesis and repair in (*see* DNA synthesis)
  estimating phylogenetic tree of life from rRNA sequences in, 6–7
  gene expression in (*see* Gene expression)
  genetic problems in metapopulations, 1114
  human, 436
  meiosis in (*see* Meiosis)
  Mendelian (*see* Mendelian genetics)
  protein synthesis in (*see* Protein synthesis)
Genetic screens, **277–78**, 310
Genetic testing, 350–51
Genetic variation, **440**. *See also* Genetic diversity
  crossing over and, 222
  evolutionary processes and, 448, 450*t*
  evolution of drug resistance and, 425–26
  extinction and, 440
  from independent assortment, 221–22
  lack of, as genetic constraint, 431
  mutations and, 448–50
  reduction of, by directional selection, 440–41
  sexual reproduction and, 533
  switching reproductive modes and, 952
  types of fertilization and, 222–23
Genitalia, **957**
Gennett, J. Claude, 982
Genomes, **263**, **359**. *See also* Genomics
  animals and human, 603

diversity of viral, 686, 687*t*
eukaryotic, 365–70
prokaryotic, 363–65
reasons for selecting, for sequencing, 361–62
sequencing of (*see* Whole-genome sequencing)
sequencing technologies, 347
smallest eurkaryotic, 594
Genomic libraries, **342**
Genomics, **359–73**. *See also* Genes; Genetics
  bacterial and archaeal genomes, 363–65
  cancer and applied, 371–72
  environmental, 503
  eukaryotic genomes, 365–70
  functional, 370–71
  human genome representation, 359*f*
  number of genes in selected eukaryotes, 369*t*
  proteomics and, 371
  whole-genome sequencing, 359–63
Genotypes, **234**
  body odor and, 439–40
  central dogma and linking phenotypes and, 280–81
  effect of inbreeding on frequencies of, 451
  frequencies of, for human *HLA* gene, 439*t*
  frequencies of, for human MN blood group, 438*t*
  in Mendelian genetics, 239*t*
  mutations and, 285
  particulate inheritance hypothesis and, 234–35
  phenotypes vs., 234
  predicting, with Punnett square, 235–36
Genus (genera), **8**
Geographic areas, similar species in same, 419–20
Geographic distribution of diversity, 1109–10
Geographic distribution of organisms, 1013–17
  dispersal histories in, 1013–15
  global warming and changes in, 1100
  interaction of biotic and abiotic factors in, 1015–17
  mapping current and past, 1072
  protected areas and, 1120
  species interactions and, 1059
Geologic time scale, 182, **416–17**
Geometry. *See* Shape
Geothermal radiation, 509
Germ cells, **409**
Germinal center, 989
Germination, **402**, **794**
  gibberellins and abscisic acid in seed, 770–71
  phases of seed, 799–800
  red/far-red light reception and seed, 762–64
Germ layers, **394–95**, **604**
Germ theory of disease, **499–500**
Gestation, **658**, **967**
Ghosts
  capsids as, 259–60*f*
  plasma membrane, 96–97
Giant axon, squid, 125–27
Giant pandas, 1021
Giant sequoias, 1073–74
Giant sperm, 955
Giardia and giardiasis, 522*t*, 538

Gibberellic acid (GA), 769
Gibberellins, **769**–71, 774–75t
Gibbons, Ian, 127–28
Gibbs free-energy change, **30**, 42. *See also*
    Free-energy change
Gilbert, Walter, 294, 314
Gill arches, **655**–66
Gill filaments, **865**
Gill lamellae, **865**
Gills, **634**, **651**, **813**, **865**
    as countercurrent systems, 866
    developmental changes in gas ex-
      change with, 813
    osmoregulation and, 827–28
    osmotic stress and, 824
    ventilation with, 865
Ginkgoes (Ginkgophyta), 574
Giraffes, 9–10
Gizzards, avian, 851
Glacier Bay succession case
    history, 1076–77
Glaciers, 1083f
Glands, **809**, **930**
Glanville fritillaries (butterflies),
    1046–47, 1055
Glaucophyte algae, 536
Gleason, Henry, 1070–72
Glia, **894**
Global air circulation, 1009
Global carbon cycle, **523**, **1096**–97
Global change, prokaryotes and, 509–12
Global climate change
    biological study of, 1011–13f
    extinctions and, 1111, 1114
    global warming and, 1008–9,
      1098–1102
Global nitrogen cycle, **1096**
Global productivity patterns, 1089–90
Global warming, **1098**–1102
    as biodiversity threat, 1111, 1114
    carbon dioxide concentrations and
      evidence for, 1098–99
    devegetation and, 547
    ecosystem positive and negative feed-
      back responses to, 1099
    end-Permian extinction and, 489
    global climate change and, 1008–9
    global fossil-fuel consumption
      and, 1098
    impact of, on organisms, 1099–1101
    iron fertilization and, 1091
    net primary productivity changes due
      to, 1101–2
    protists and reduction of, 523
    Sphagnum moss and, 570
Global water cycle, **1095**
Globin genes, 368
Globular enzymes, 53
Globular proteins, 45–46, 197
Glomalin, **589**
Glomeromycota, 596
Glomeruli, olfactory, **920**
Glomerulus, renal, **833**
Glucagon, 45, **856**, **930**
Glucocorticoids, **938**, 941–42
Gluconeogenesis, **856**
Glucose, **148**
    animal absorption of, 852–53
    ATP synthesis from, 79, 148f, 165–66
    catabolite repression and, 314–17f

    in cellular respiration and
      fermentation, 153–54
    configurations of, 72–73
    cortisol and availability of, 938
    diabetes mellitus and, 856–57
    glycogen and starch as, 74, 168
    glycolysis as processing of, to
      pyruvate, 155–56
    hydrolysis of carbohydrates by en-
      zymes to release, 78–79
    induced fit and, 53
    lactose metabolism and preference
      for, 309–10
    membrane protein GLUT-1 and, 97
    nutritional homeostasis and, 856–58
    as organic molecule, 33f
    oxidation of, by citric acid cycle,
      158–61, 165–66
    processing of photosynthetic, 189–90
    redox reactions and oxidation of, 152
    reduction of carbon dioxide to pro-
      duce, in photosynthesis, 184–90
    regulation of, 156
    secondary active transport of, 98
    spontaneous chemical reactions
      and, 30f
Glucose-6-phosphate, 55–56, 155
GLUT-1 membrane protein, 97
Glutamate, 42t, 46, 898t, 919
Glutamic acid, 47
Glutamine, 42t, 349–50
Glyceraldehyde-3-phosphate (G3P),
    186, 189–90
Glycerol, 83–84, 168
Glycine, 42t
Glycogen, 74, 75t, 78–79, 168, 585
Glycolipids, **105**
Glycolysis, **153**–56
Glycophosphate, 354
Glycoproteins, **77**, **121**
Glycosidic linkages, 73–74
Glycosylation, **121**
Glyoxysomes, **110**
Gnathostomes, **653**
Gnathostomulida, 603t
Gnetophytes (Gnetophyta), 576
Golden rice, 354–56
Golgi, Camillo, 886–87
Golgi apparatus, **109**
    eukaryotic, 108–9, 114t
    extracellular matrix components
      in, 134
    in mitosis, 200
    pectins and, 132
    post-translational modifications
      in, 304
    protein transport and sorting by,
      122–23f
Gonadal hormones, 961–62
Gonadotropin-releasing hormone
    (GnRH), **962**
Gonads, **936**, **953**
Goose bumps, 418, 419f
G proteins, 142–44
Grades, 552, 567, **655**
Grafting, plant, 787–88
Gram-negative, **505**
Gram-positive, **505**
Gram stains, **505**
Grana (granum), **112**–13, **174**, 184

Grant, Peter and Rosemary, 426–28, 431,
    463, 1062, 1063f
Graphs, B:2–B:4
Grasshoppers, 212, 641t
Grasslands. *See* Temperate grasslands
Graves, 671
Gravitropic responses, plant,
    764–66, 780t
Gravitropism, **764**
Gravity, hearts and, 878
Gray whales, 819
Grazing food chains, **1085**–86
Great apes, **668**–70
Great chain of being, 415
Great Smoky Mountains National
    Park, 1108
Great tits, 447–48
Greek word roots, 8, B:6
Green algae, **566**. *See also* Algae
    as green plants, 546, 566–67 (*see also*
      Green plants)
    lichens and, 597
    lineages, 568–69
    photosystems and, 179
    phylogenetic tree and, 551–52
    Plantae and, 536
    reasons for studying, 546–49
    similarities between land plants
      and, 549
Green Belt Movement, 1122
Greenhouse gas, **1097**
Green iquanas, 446
Green light, 177
Green plants, 546–78. *See also* Plant(s)
    analysis of morphological traits of,
      549–50
    biological methods for studying,
      549–52
    diversification themes of land plants,
      553–66
    ecosystem services of, 547
    green algae and land plants as, 546,
      566–69
    importance of, to humans, 548–49
    key lineages of, 566–77f
    non-vascular plants, 567, 569–71
    reasons for studying, 546–49
    seedless vascular plants, 567, 571–73
    seed plants, 567, 574–77f
    some drugs derived from land
      plants, 549t
    using fossil record to study, 550–51
    using phylogenetic trees to study,
      551–52
Gross photosynthetic efficiency, **1084**
Gross primary productivity, **1084**
Groudine, Mark, 322
Ground meristem, **705**, **796**
Ground pines, 571
Ground tissue, **406**
Ground tissue systems, **705**, 708–10, 711t
Groundwater, **1095**
Groups, control, 12
Groups, functional. *See*
    Functional groups
Growing seasons, plant translocation
    and, 728
Growth
    cancer as out-of-control, 206
    cell movement and, 376–77
    energy and, 1084
    mammalian, 937
    mitosis and, 195
    by molting, 625
    plant, 554, 766 (*see also* Growth regu-
      lators, plant)

population (*see* Population growth)
    viral, 679–81
Growth factors, **207**–9
Growth hormone (GH), **934**, 935, 937.
    See also Human growth
    hormone (HGH)
Growth rate, population, 1041–42
Growth regulators, plant, 767–76
    apical dominance and auxin as,
      767–68
    cell division and cytokinins as,
      768–69f
    growth and dormancy and gib-
      berellins and abscisic acid as,
      769–73
    overview of, 774, 775t
    plant body size and brassinosteroids
      as, 773
    senescence and ethylene as, 773–74
Growth rings, tree, 714–15, 1073
GTP (guanosine triphosphate),
    142–43, **158**
Guanine, 60, 63–64, 66–67, 261, 279
Guanosine diphosphate (GDP), 142–43
Guanosine monophosphate (GMP), 916
Guanosine triphosphate (GTP), 142–43,
    158, 297–98
Guard cells, **187**, **553**–54, **708**, **771**–73
Gustation, **918**–19
Guts, animal, 845
Guttation, **723**
Gymnosperms, **551**
    diversification of, 551–52
    evolution of, 560–61
    key lineages of, 574–76
    in Mesozoic era, 483

H

H$^+$-ATPases, **733**
Habitat bias, 480
Habitat destruction, **1112**
    human population size and, 1052
    preserving metapopulations
      from, 1055
    species-area relationships for extinc-
      tions due to, 1116
    as threat to biodiversity, 1111, 1112
Habitat fragmentation, **1113**
Habitat prezygotic isolation, 459t
Habitats
    carrying capacity of, 1042, 1046
    communication and, 1029
    prokaryotic diversity in, 498
    protecting, 1055
    selection of, 1026
    speciation and preference for, 465
Hadley, George, 1009
Hadley cell, **1009**
Haemophilus influenza, 361–62
Hagfish, 661
Hair cells, **909**–12
Hairpin (RNA secondary structure), 67,
    292–93. *See also* RNA
    (ribonucleic acid)
Haldane, J. B. S., 38
Halophiles, **503**
Hamilton, William D., 1031–33
Hamilton's rule, 1031–34
Hands, primate, 669
Hanski, Ilkka, 1046–47
Hanson, Jean, 922
Hantaan virus, 689, 692
Hantavirus, 688, 689
Haploid cells, 214t, 533–34
Haploid number, **213**, 214t

Boldface page numbers indicate a glossary entry; page numbers followed by an *f*
indicate a figure; page numbers followed by *t* indicate a table.

Haploid organisms, **213**
Hardy, G. H., 436
Hardy-Weinberg principle, **436**–40
Hartwell, Leland, 205
Hatch, Hal, 188
Haustoria, 751
Havasupai tribe, 439–40
Hawaiian silverswords, 484, 700–701
Hawksworth, David, 582
Hawthorn flies, 465
Hay fever, 991
Head region
    animal, 603, 605, 610
    insect, 639
    sperm, 389
Head-to-tail axes, 381
Health, human. See Diseases, human
Hearing, 610, 909–13
Heart murmers, **879**
Hearts, 808, 830, **875**–81. See also Blood;
        Blood vessels; Circulatory
        systems
Heartwood, **714**
Heat, **28**. See also Specific heat
    animal thermoregulation and, 816–19
    denatured proteins and, 50
    DNA copying and, 65
    sterilization using, 3–4
    as thermal energy, 27–29
    water and, 24–25
Heat of vaporization, **24**, 26t
Heat-shock proteins, 50, **818**
Heavy chain, **980**
Helicases, **264**, 266, 268t
Helicobacter pylori, 850
Heliobacteria, 182, 507t
Helper T cells, 677, 681–82, 985–**86**, 989t
Hemagglutinin, 981
Heme, **871**
Hemichordata, 603t, 646
Hemimetabolous metamorphosis, **616**
Hemiptera, 641t
Hemocoel, **626**, 628
Hemoglobin, **871**. See also Blood
    β-globin in, 247, 322
    blood oxygen and, 45
    in blood pH buffering, 873–74
    globin genes and, 368
    protein structure of, 49t
    sickle-cell disease and, 46
    space-filling model of, 38f
    structure and function of, 871–73
Hemolymph, **830**, 875
Hemophilia, **252**
Henle, Jacob, 835
Hepaticophyta, 569
Hepatitis viruses, 692, 990
Herbaceous plants, **697**–98
Herbicides, 176, 354, 589
Herbivores, **612**, **701**, **1063**
    animals as, 612
    limitations on consumption of,
        1065–66
    plant defense responses to, 775,
        778–81
    plants as food for, 547
    transgenic crops and, 354
Herbivory, **1063**
Heredity, 221, **230**, 280. See also Genet-
        ics; Mendelian genetics
Heritable traits, **4**
Hermaphroditic, **956**
Hermit warblers, 468–70, 1015
Herpes, 686
Herpesviruses, 690
Hershey, Alfred, 259–60f

Hershey-Chase experiment, 259–60f
Heterokaryotic, **582**
Heterokonta, 524t, 537, 542–43
Heterospory, **559**, 561–62
Heterotherms, **817**
Heterotrophs, **172**, **506**, 601
Heterozygote advantage, **443**, 452
Heterozygous genes, **235**, 239t, 245–47
Hexokinase, 53
Hexoses, 60, **72**, 73
Hibernation, **817**
High-productivity hypothesis, 1080
Hill, Robin, 179
Hindbrain, **653**
Hindguts, 830–31
Hippos, 477–79
Histamine, **976**, 977t, 991
Histidine, 42t, 283
Histograms, 248–49, 427, 441f
Histone acetyl transferases (HATs), **322**
Histone code, **323**
Histone deacetylases (HDACs), **322**–23
Histones, **197**, **320**–23, 497t
Historical constraints, natural selection
        and, 432, 914
Historical context, succession and, 1075
History of life. See Life, history of
HIV (human immunodeficiency virus),
        677, **990**
    CD4 protein as receptor used by, to
        enter cells, 681–82
    changing-environment hypothesis
        and, 224
    current AIDS pandemic and, 677–78
    doorknob hypothesis on, 681–82
    phylogeny of, 688f
    as retrovirus, 684, 691
    as RNA virus, 281
    strains of, 688–89
    Toxoplasma and, 541
    two types of, 688
    vaccines and mutations of, 990
Hives, 991
HLA genes, 439–40
Hodgkin, A. L., 890–93
Holoenzymes, **291**, 292, 343
Holometabolous metamorphosis,
        616, 936
Homeobox, 411, 476
Homeosis, **382**
Homeostasis, **25**, **814**–19
    of animal body temperature, 817–19
    animal excretory systems and, 830–31
    of blood oxygen, 870
    of blood pH, 873
    of blood pressure, 882–83
    diabetes mellitus and animal
        nutritional, 856–58
    general principles of, 814–15
    hormones in, 938–40
    pH buffers and, 25
    regulation and feedback in, 815–16
    ventilation and, 870
    of water and electrolyte balance (see
        Osmoregulation)
Homeotherms, **817**–19
Homeotic genes, **382**–83, 385
Homing pigeons, 1026
Hominids, 668, **669**–70
Hominins, **670**–73f
Homo genus
    characteristics of, 670t
    DNA comparison of *Homo nean-
        derthalensis* and *Homo sapiens*,
        345–46
    fossil record and, 670–72

Homo erectus, 672
Homo floresiensis, 671
Homo neanderthalensis, 345–46,
        671, 672
Homo sapiens, 8, 369t, 672–73f (see
        also Human(s))
Homologous chromosomes (homologs),
        212, 214–16, 221–22, 225–26
Homology, **363**, **420**, 475
    animal and human genetic, 603
    animal appendages as, 613–15
    as evidence of descent from common
        ancestors, 420–22
    genomes and genes, 363
    homoplasy vs., 475–77
Homoplasy, **475**–77
Homosporous, **559**
Homozygous genes, **235**, 239t, 451
Honduras, age structure of, 1051
Honeybee language, 1028–29
Hooke, Robert, 2
Hormone-based contraception
        methods, 966t
Hormone binding, plant, 757
Hormone-response elements, **944**
Hormones, **139**–46. See also Chemical
        signals, animal; Hormones, ani-
        mal; Hormones, plant
Hormones, animal, **380**, **837**, **852**
    central nervous systems, endocrine
        systems, and, 929
    chemical characteristics and types
        of, 933–34
    coordination of responses to environ-
        mental change by, 937–38
    direction of developmental processes
        by, 935–37
    discovery of first, 852
    as endocrine signals, 930
    endocrine system components
        and, 932–33
    experimental identification
        of, 934–35
    functions of, 935–40
    in homeostasis, 938–40
    human growth hormone (HGH) (see
        Human growth hormone (HGH))
    mammalian sex, 960–66
    regulation of production of, 940–43
    signaling pathways of, 931–32
    urine formation and, 837
Hormones, plant, **756**
    auxin as blue-light phototropic,
        759–62
    auxin as development master
        regulator, 380–81
    auxin as gravitropic signal, 765–66
    auxin in growth responses, 767–68
    brassinosteroids in body size, 773
    cytokinins in cell division, 768–69f
    ethylene in senescence, 773–74
    florigen as flowering, 787–88
    gibberellins and abscisic acid in
        growth responses, 769–73
    hormone binding and, 757
    hypersensitive response, systemic ac-
        quired resistance, and, 777–78
    information processing and, 756
Hornworts, 571
Horowitz, Norman, 277–78
Horseshoe crabs, 642
Horsetails, 572
Horticulture, 547
Horvitz, Robert, 376
Host cells, **675**
Hosts, **1063**, 1067–68

Hotspots
    of biodiversity and endemism, 1109
    conservation, 1110
Howard, Alma, 196
Hox genes, **382**–83, 385, 411, 476, 488
Hozumi, Nobumichi, 982
Hubbard Brook Experimental Forest,
        1084, 1085–86, 1093–94
Human(s), **671**
    artificial selection by, 4–5
    blood types, 245–47, 438–39
    brains, 901–2
    digestion, 77, 79, 810f, 846f (see also
        Digestion)
    diseases (see Diseases, human)
    as dispersal agents, 1014
    ears, 910–12
    endocrine systems, 932f (see also
        Endocrine systems)
    as endothermic homeotherms, 817
    essential nutrients for, 842, 843t (see
        also Animal nutrition)
    evolution of, 345–46, 438–40, 670–73f
    excretory systems (see Excretory
        systems)
    gas exchange and circulation (see Ani-
        mal gas exchange and circulation)
    genome of, 328, 369t (see also Human
        Genome Project)
    growth hormone, 338–44
    hearts, 879–83 (see also Circulatory
        systems)
    Homo sapiens scientific name for, 8
    hormones in stress responses
        of, 937–38
    identifying autosomal traits
        of, 250–52
    immune systems, 975–76, 977t, 978
        (see also Immune systems)
    importance of animals to, 602–3
    importance of fungi to, 581–82
    importance of plants to, 548–49, 783
    importance of protists to, 520–22
    introducing novel alleles into cells
        of, 351–52
    learning and memory in, 902–4f
    life tables for, 1055
    lungs, 868, 869f (see also Respiratory
        systems)
    mortality (see Mortality, human)
    nervous systems (see Nervous
        systems)
    net primary productivity appropria-
        tion by, 1090
    non-kin child abuse rates by, 1034
    number of chromosomes in, 213t
    nutrient loss impacts by, 1093
    organs and systems of, parasitized by
        viruses, 676f
    osmoregulation (see Osmoregulation)
    population (see Human population)
    puberty in, 961–62, 963t
    reproduction (see Human
        reproduction)
    sensory systems (see Sensory systems,
        animal)
    urinary systems, 832f (see also Uri-
        nary systems)
    vestigial traits, 418, 419f
Human chorionic gonadotropin (hCG),
        967–68
Human Genome Project, **359**
    costs of, 347
    human genome representation, 359f
    reasons for similarity of chimpanzee
        and human genomes, 369–70

Human Genome Project, (*continued*)
  reasons for small number of genes in
    human genome, 368–69
Human growth hormone
  (HGH), 338–44
Human immunodeficiency virus (HIV).
  See HIV (human immunodefi-
    ciency virus)
Human population
  age structure of, 1051
  growth rate for, 1052–53
  phylogeny of current, 672*f*
  population growth milestones
    for, 1052*t*
  size of, 1052, 1098, 1121
Human reproduction. *See also* Animal
    reproduction
  accessory fluids in semen, 959*t*
  childbirth events, 970–71
  contraception in, 966
  embryo development in, 388*f* (*see also*
    Animal development)
  embryonic tissue layers in, 395
  female reproductive tract in, 960, 961*f*
  fertility rates in, 1053
  fertilization, 211*f*
  gametogenesis in, 953*f*
  inbreeding and reduction in fitness
    in, 452*t*
  male reproductive tract in, 958–59*f*
  meiotic mistakes in, 226
  ovarian cycle in, 962–66
  pregnancy events in, 967–68
  sex chromosomes and, 212
  sperm cells and, 2*f*
Hummingbirds, 562
Humoral response, **987**
Humus, **741, 1092**
Hunting, animal, 612
Huntington's disease, 233, **251, 348–51**
Hutchinson, G. Evelyn, 1060
Hutterites, 440
Huxley, Andrew, 890–93
Huxley, Hugh, 922
Hybridization
  hybrid zones and, 468–70
  new species through, 470–71
  speciation through, 467–68
Hybrids, **232,** 239*t*, 459*t*
Hybrid zones, **468**–70, 471*t*
Hydrocarbons, **83,** 87–88. *See also* Lipids
Hydrochloric acid, 25, 848–50
Hydrogen. *See also* Hydrogen gas
  atomic structure of, 17*f*
  electronegativity of, 18
  lipids, hydrocarbons, and, 83
  as plant nutrient, 739*t*
  proteins and, 46–47
  redox reactions and, 151–52
  simple molecules from, 19–20
Hydrogen bonds, **22**
  amino acids and, 41*t*
  DNA complementary base pairing
    and, 63–64, 261
  protein structure and, 46–48
  RNA complementary base pairing
    and, 66–67
  specific heat of liquids and, 24*t*
  water and, 22–25
Hydrogen cyanide, 29, 32, 34, 38, 40, 61

Hydrogen gas, 17–18, 19*f*, 32, 39–40. *See
    also* Hydrogen
Hydrogen ions, 25
Hydrogen peroxide, 110
Hydroids, 619
Hydrolysis, **43**
  of ATP, 149
  of carbohydrates by enzymes to re-
    lease glucose, 78–79
  digestive enzymes and, 110–11
  of polymers to release
    monomers, 43
Hydrophilic substances, **22,** 42, 64
Hydrophobic substances, **22,** 42,
    47–48, 64
Hydroponic growth, **740**
Hydrostatic skeletons, **605,** 606, 613,
    626, **920**–21
Hydrothermal vents, 509
Hydroxide ions, 25
Hydroxyl functional group, 35, 60, 71,
    72, 83–84, 290
Hydroxyl radicals, 272
Hymenoptera, 640*t*
Hyoid bone, 672
Hyperpolarization, **890,** 893, 897, **908**–9
Hypersensitive reaction, **991**
Hypersensitive response (HR),
    775–78, 780*t*
Hypertension, **880**
Hypertonic solutions, **91, 824**
Hyphae (hypha), **542, 582**–83,
    590–91, 746
Hypocotyls, **405, 796**
Hypodermic insemination, 956
Hypothalamic-pituitary axis,
    942–43, 962
Hypothalamus, 817–18, 931, **933,**
    940–43
Hypotheses, **2,** 8–12
Hypothesis testing, 2–4. *See also*
    Experiments
Hypotonic solutions, **91, 718, 824**

I

Ice, 23–24
Iguanas, 446
Imaging, biological. *See* Biological imag-
    ing; Microscopy
Immigration, 1037–38, 1077–78
Immune systems, **677,** 973–92
  adaptive immune response (*see* Adap-
    tive immune response)
  antibodies and, 324–26
  evolution of human defenses against
    malaria, 1066–67
  failures of, 990–91
  HIV and human, 677–78
  HLA genes in human, 439–40
  immunity, immunization, and,
    973–74
  immunological memory of, 989–90
  inbreeding and, 452
  innate immune system cells and
    molecules, 977*t*
  innate immunity and, 974–77
  proteins in, 280
  rejection of foreign tissues and organs
    by, 988

severe combined immunodeficiency
    (SCID) and human, 352–54
  sexual selection, carotenoids, and, 453
  structures of, 978
Immunity, **973**
Immunization, **973**–74, 990
Immunodeficiency diseases, 990. *See also*
    AIDS (acquired immune defi-
    ciency syndrome); HIV (human
    immunodeficiency virus);
    Severe combined
    immunodeficiency (SCID)
Immunoglobulins (Igs), **980,** 982–83*f*
Immunological memory, **989**–90
Impact hypothesis, dinosaur extinction
    and, **490**–92
Imperfect, **790**
Implantation, 966*t*, **967**–68
Inactivated virus vaccines, 990
Inbreeding, **450**–52
Inbreeding depression, **451**–52, 1114
Inclusive fitness, **1032**–33
Incomplete digestive tracts, **845**
Incomplete dominance, **245,** 250*t*
Incomplete metamorphosis, **616**
Independent assortment, principle of.
    See Principle of independent
    assortment
Indeterminate growth, **695**
Indicator plates, **311**
Indirect fitness, 1032
Individuals, natural selection and,
    414*f*, 429–30
Indole acetic acid (IAA), 760. *See
    also* Auxin
Induced fit, **53**–54
Inducers, **309,** 314
Inducible defenses, **776, 1064**–65*f*
Industrial Revolution, 548
Infection thread, **750**
Infidelity, bird, 956
Inflammatory response, **975**–77*t*
Influenza virus, 675, 685–86, 692,
    975, 990
Information
  carbohydrates and display of, 77
  gene regulation and flow of, 307–10
  genetic, 1, 65, 67–68 (*see also* Genes;
    Genetics)
  hormones as carriers of, 139–40
  plant processing of, 756–57
Infrared light, 174–75, 177
Infrasound hearing, elephant, 912
Ingestive feeding
  animal, 601, 841–45
  protist, 529–30
Inhalation, 868–70
Inheritance. *See also* Genetics; Heredity;
    Mendelian genetics
  of acquired characters, 231, 234,
    415, 429
  of chromatin modifications, 323
  chromosome theory of (*see* Chromo-
    some theory of inheritance)
  epigenetic, 323
  evolution by natural selection vs.
    Lamarkian, 429
  hypotheses on, 231, 232, 234
  particulate, 234–36
  patterns of human, 250–52
  polygenic, 248–50
Inheritance of acquired characters
    theory, 231, 234, 415, 429
Inhibition, **1075**
Inhibitory postsynaptic potentials
    (IPSPs), **897**

Initiation, enzyme catalysis, **54**
Initiation factors, **301**
Initiation phase, **291**
  regulatory sequences and proteins in
    eukaryotic transcription, 323–28
  transcription, 291–92
  translation, 301
Innate behavior, **1020**
Innate immune response, **975**–77*t*. *See
    also* Immune systems
Innate immunity, **974**–77*t*. *See
    also* Immune systems
Inner cell mass (ICM), **393**
Inner ear, **910**
Inorganic molecules, 33
Inorganic nutrients
Inorganic phosphate, 189
Inositol triphoshhate (IP$_3$), 143*t*
Insecta, 639–41
Insects
  adaptive radiations and innovations
    in, 485
  ants (*see* Ants)
  excretory systems of, 829–31
  eyes and vision in, 913
  Glanville fritillaries (butterflies),
    1046–47
  honeybee language, 1028–29
  hormones in metamorphosis of,
    929*f*, 936
  Insecta lineage and, **639**–41
  insecticides and, 521, 1088–89
  land plant pollination by, 562–63
  metamorphosis of, 616
  minimization of water loss by, 828–29
  mutualisms of, 1069–70
  open circulatory systems of, 875
  osmoregulation in terrestrial, 828–31
  plant defense responses to, 775,
    778–81
  prominent orders of, 640*t*–41*t*
  taxon-specific survey of, 1107
  tracheae of, 866–67
  wings of, 629
In situ hybridization, **380**
  in flower structure research, 410
  in polygenic inheritance research,
    428–29
  visualizing mRNAs via, 380
Instantaneous rate of increase, 1044
Insulin, **856, 930**
  amino acid sequence for, 46
  cortisol and, 938
  discovery of, 856
  as hormone signal, 140*t*
  role of, in homeostasis, 856
  treatment of diabetes mellitus
    with, 339
Integral membrane proteins, **93**–94
Integrated pest management, **1068**
Integrators, **815,** 938
Integrins, **134,** 135
Interactions, cell-cell. *See* Cell-cell
    interactions
Interbreeding. *See* Mating
Intercalated discs, **881**
Intergovernmental Panel on Climate
    Change (IPCC), 1098–99
Interindividual signals, 931
Interleukin 2, 930
Intermediate disturbance
    hypothesis, **1080**
Intermediate filaments, 124*t*, **125**
Intermediate hosts, **632**
Intermediates, anabolic, 169
Internal environments, 319–20

Internal fertilization, animal, 615, 630, 954–55
Internal membranes, 104, 107t
International online DNA repositories, 360
International Union for the Conservation of Nature, 1120
Interneurons, **886**
Internodes, **699**
Interphase, **196**
Interspecific competition, 1059–60
Interstitial fluid, **876–77**
Intertidal zone, **1000**
Intestines. *See* Large intestines; Small intestines
Intraspecific competition, **1059**
Intrauterine device (IUD), 966t
Intrinsic rate of increase, **1041**
Introns, **294**
  alternative splicing to remove, 328–29
  antibodies and regulatory sequences in, 324–26
  discovery of, 293–94
  RNA splicing to remove, 294–95
Invasive species, **1014**–15, 1062, 1110–11
Inversion, **286**
Invertebrates, **609**, 646, 651–52, 976–77
In vitro DNA synthesis reaction, 346
Involuntary muscle, **808**
Involuntary responses, 899
Iodine, 843t
Ion channels, **94**–95, **888**
Ion concentration gradients, 887–88
Ion currents, action potentials and, 891
Ion exclusion mechanisms, plant, 746–48
Ionic bonds, **18**–19, 48
Ions, **18**, **742**. *See also* Atoms
  electrolytes as, 822 (*see also* Osmoregulation)
  ionic bonds and, 18–19, 48
  lipid bilayers and, 86
  plant exclusion mechanism for, 746–48
  as plant nutrients in soil, 742–44f
  plant nutrient uptake mechanisms for, 744–46
Iridium, 490
Iris, **914**
Irish potato famine, 520–21, 542
Iron
  as human nutrient, 843t
  iron-containing heme groups, 161
  iron-fertilization experiments, 523, 1091–92
  as plant nutrient, 739t
Irrigation, 1095
Island biogeography, theory of, 1077–78
Islets of Langerhans, 930
Isolation, reproductive. *See* Reproductive isolation
Isoleucine, 42t, 47
Isopods, 643
Isoprene, 83, 84, 177, 496
Isotonic solutions, **91**, **718**, **824**
Isotopes, **16**, 586

J

Jacob, François, 2, 279, 309–14
Jasmonic acid, 779
Jaws
  cichlid, 844–45
  vertebrate, 655–56
Jeffreys, Alec, 366
Jelly fish, 601f, 619
Jelly layer, 390

Jenner, Edward, 973–74, 990
Jet propulsion, mollusk, 630
Johnston, Wendy, 68–69
Jointed limbs, animal, 613, 629, 637
Joints, **920**
Joly, John, 724
Junipers, 575
Jürgens, Gerd, 406
Juvenile hormone (JH), **936**
Juveniles, **616**

K

Kandel, Eric, 903–4f
Kangaroos, 967
Karpilov, Y. S., 188
Karyogamy, **591**, 592, 593
Karyotypes, **212**–13, **286**–87f
Keel, bird, 658
Kelp forests, 650
Kenrick, Paul, 554
Keratins, 125
Kerr, Warwick, 445
Ketones, 34, 35f
Ketose, 72
Keystone species, 1072–73, 1113
Kidneys, **832**, 931, **933**, 940
Killer T cells, 985–86
Kilocalories, **149**
Kinesis, **126**–27
Kinetic energy, 27. *See also* Energy
  chemical energy and, 78
  energy transformations of, 27–29
  enzymes and, 51–53, 55
Kinetochore microtubules, **198**–99, 215, 217
Kinetochores, **199**
  chromosome movement and, 201–2
  mistakes with, 226
  in mitosis, 202t
Kingdoms, phylogenetic, 6–7
Kinocilium, **909**
Kinorhyncha, 603t
Kin selection, 1031–33, 1034
Klinefelter syndrome, 226
Knock-out alleles, **277**
Koch, Robert, 498–99
Koch's postulates, 498–**99**
Kortschak, Hugo, 188
K-P extinction, 490–92
Krebs, Hans, 158
Krebs cycle. *See* Citric acid cycle
Kudzu, 1015
Kuhn, Werner, 835
Kuru disease, 51

L

Labia majora (labium majus), **960**
Labia minora (labium minus), **960**
Labor, **970**
lac genes, 311–14
lac operon regulation, 311–17f
Lactate, 166, 167f, **659**
Lactation, 660, **967**
Lacteal, **851**
Lactic acid fermentation, **166**, 167f
Lactose, **73**–76
Lactose fermentation, 508
Lactose metabolism
  identifying genes under regulatory control in, 310–12
  as model system for bacterial gene regulation, 309–10 (*see also* Gene regulation, bacterial)

negative control mechanisms in, 312–14
overview of positive and negative *lac* operon regulation in, 317f
positive control mechanisms in, 314–16
types of mutants in *E. coli*, 311t
Lagging strands, DNA, **266**–68t
Lakes, 996, 997
Lamarck, Jean-Baptiste de, 415, 429
Lamellae, **814**
Lampreys, 661
Lancelets, 651, 652
Land. *See* Terrestrial ecosystems
Land plants. *See also* Plant(s)
  angiosperm radiation of, 564–65
  drugs derived from, 549t
  dry land transition adaptions of, 553–55
  key lineages of, 567, 569–77f
  life cycles of, 784–86
  morphological differences among, 549–50
  mycorrhizal fungi and nutrients for, 580
  origin of, 550
  origin of vascular tissue in, 554–55
  phylogenetic tree and evolutionary changes of, 555–56
  Plantae and, 536
  prevention of water loss by cuticles and stomata of, 553–54
  reproductive adaptations of, 556–64 (*see also* Plant reproduction)
  similarities between green algae and, 549 (*see also* Green plants)
Land transition adaptations. *See* Water-to-land transition adaptations
Language, 672
Language, honeybee, 1028–29
Large intestines, 810f, 846f, 847, **854**–55
Larvae (larva), **616**, 936
Late endosome, 111
Latency, **680**
Latent growth, viral, 679–81
Lateral buds, **699**
Lateral gene transfer, **364**
Lateral line system, **913**
Lateral meristem, **712**
Lateral roots, **697**, 706
Latin word roots, 8, B:6
Latitudinal gradient, species richness, 1078–80
Law of succession, 417
Leaching, **743**
Leadbeater's possum, 1115
Leading strands, DNA, 265–**66**, 268t
Leak channels, 889
Learning, **902**–4f
Leaves (leaf), **699**
  auxin and abscission in, 768
  carbon dioxide entry into, through stomata, 187
  characteristics of, 701, 702f
  ethylene and abscission of, 774
  genes, proteins, and shape determination for, 408
  modified, 703
  morphological diversity in, 701–2
  phenotypic plasticity in, 702–3
  seed, 796
  water loss prevention by, 727–28
Leeches, 633
Leghemoglobin, **749**
Legumes, **749**–50

Leishmania and leishmaniasis, 522t
Lemmings, 430–31
Lennarz, William, 391
Lens, **914**
Lenski, Richard, 449–50
Lenticells, **714**
Lentiviruses, 688
Lepidoptera, 640t
Lepidosauria, 666
Leptin, **933**, 938–**40**
Lesion studies, 901–2
Leucine, 42t, 120, 298
Leukocytes (white blood cells), 870, **975**–76, 977t
Libraries. *See* cDNA libraries; DNA libraries
Lichens, 515, **566**, **589**, 597, 598
Life
  biology, science, and methods for studying, 8–12 (*see also* Biological imaging; Biological methods; Biology; BioSkills; Experiments; Quantitative methods)
  cell theory and cells of, 2–4 (*see also* Cells)
  characteristics of living organisms and, 1–2, 82 (*see also* Organisms)
  extremophiles and extraterrestrial, 501
  geographic distribution of, 1013–17
  history of (*see* Life, history of)
  life tables and, 1038–41, 1053–55
  molecules of (*see* Molecules)
  origin of viruses in origin of, 687–88
  research on origin of (*see* Origin-of-life research)
  speciation and phylogenetic tree of, 5–8 (*see also* Phylogenies; Speciation; Tree of life)
  theory of chemical evolution of (*see* Chemical evolution)
  theory of evolution of, by natural selection, 4–5 (*see also* Evolution; Natural selection)
  viruses as not being alive, 675
Life, history of, 474–95. *See also* Evolution
  adaptive radiations in, 484–88
  fossil record as tool for studying, 479–84
  fossil record time line for, 481–84
  importance of animals in, 602
  mass extinctions in, 488–92
  phylogenetic trees and phylogenies as tools for studying, 474–79
Life cycles, **215**
  animal, 215–16, 615–17
  evolution of land plant, 558–59
  fungal, 591–94
  land plant, 784–86
  plant, 402f
  protist, 533–36
Life histories, 1039–41
Life tables, **1038**–40, 1053–55
Ligand-gated channels, **897**
Ligands, **777**, **897**
Light chain, **980**
Light energy. *See also* Photosynthesis
  amino acid production and, 40
  animal sensory organs and, 610
  carbohydrate storage of, 78
  chemical evolution and, 29, 32–33, 34f
  land plants and, 553, 554
  lizard sexual activity and, 1023–24

Light energy, (continued)
  photosynthetic capture of, by chlorophylls, 172–79
  photosynthetic glucose production from, 153
  plants and blue (see Blue light)
  plants and red/far-red (see Red/far-red light responses, plant)
  prokaryotic photosynthesis and, 509
  transformation of, into biomass, 1084–85
  vegetative development and, 407
  water depth and, 996–97
Light microscopy
  cell research and, 116
  meiosis and, 215
  overview of, B:13–B:14
  of protist cells, 524–25
  viewing chromosomes with, 196
Lightning energy, 39–40
Light-sensing organs, 913. See also Eyes; Vision
Lignin, 113, **133**, 548–49, **555**, 579–81, 590, **709**, 726
Limbs
  animal, 613–15
  arthropod, 637
  endoskeletons and, 920–21
  evolutionary loss of, 384–85
  loop of Henle, 835–36
  protostome, 628, 629
  tetrapod, 655, 656–57
Limiting factors
  for net primary productivity (NPP), 1090–92
  population growth, 1044–46
Limiting nutrients, **738**
Limnetic zone, **997**
Lineages, **460**
  animal, 617–20
  ecdysozoans, 637–43
  echinoderm, 649–50
  fungi, 594–98
  green plants, 566–77
  invertebrate chordates, 651–52
  lophotrochozoans, 630–36
  monophyletic groups as, 504
  prokaryotes, 512–16
  protists, 536–43
  vertebrates, 660–68
  viruses, 689–92
Linear structures, 33f, 72
Lingual lipase, 855t
Linkage, gene, **243**–45, 246f, 250t
Linkage maps, **348**
Linnaeus, Carolus, 8
Lipase, **847**
Lipid bilayers, **85**. See also Plasma (cell) membranes
  artificial membranes as experimental, 85–86
  cholesterol and permeability of, 88
  diffusion across, 89–90
  effect of bond saturation and hydrocarbon chain length on fluidity and permeability of, 87–88
  effect of temperature on fluidity and permeability of, 88–89
  membrane proteins and, 92–99
  micelles vs., 85
  osmosis across, 90–92

permeability and selective permeability of, 86
Lipids, **83**
  catabolic pathways and, 168–69
  digestion of, in animal small intestines, 853–54
  glycolipids, 105
  hormones and solubility of, 943
  lipid-soluble vs. lipid-insoluble hormones, 139–41
  nucleic acid polymerization and, 62
  plasma membrane, as characteristic of domains of life, 497t
  smooth ER and, 108
  structure of, and cell membrane properties, 87–88
  structures of membrane, 84 (see also Lipid bilayers; Plasma (cell) membranes)
  types of, found in cells, 83–84
Liposomes, 85–86, 88
Liquids
  density of water as, 23–24 (see also Water)
  pinocytosis and, 111
  specific heat of some, 24t
  unsaturated fats as, 88
Littoral zone, **997**
Liver, 810f, 846f, **854**
Liver cells, 110
Liverworts, 569, 785f
Live virus vaccines, 990
Lizards
  adaptive radiations of, 484–85
  clutch size and egg size in, 804, 805f
  hearts of, 878
  Lepidosauria lineage and, 666
  life table case study of *Lacerta vivipara*, 1038–40
  as reptiles, 660
  sexual activity of *Anolis*, 1023–24
Loams, 742
Lobe-finned fishes, **663**
Lobes, brain, 901
Lobster, 643
Loci (locus), **240**
Lock-and-key model, enzyme catalysis, **53**–54
Locomotion, animal, **920**–26. See also Movement
  evolution of animal appendages for, as homologous, 613–15
  great ape, 670
  muscle contraction and, 922–26
  muscle tissue and, 808
  muscle types and, 921–22
  neurons and muscle cells of, 601
  protostome, 628–30
  sensory systems and, 907
  skeletons and, 920–21
Loewi, Otto, 895
Logarithms, 273, 812, B:9
Logistic population growth, **1042**–44, 1046
Long-day plants, 787
Long interspersed nuclear element (LINE), **365**, 366f
Loop of Henle, **835**–37
  creation of osmotic gradient by, 835–36
  hypothesis on function of, 835

model of functioning of, 836–37
  structure and function of, 832, 838t
Loose connective tissue, **807**
Lophophore, **624**
Lophotrochozoans, **609**
  evolution of, as protostomes, 624–25
  lineages of, 630–36
  major phyla of animals and, 603t
Loss-of-function alleles, **277**, 451
Lou Gehrig's disease (ALS), 376
Love darts, 956
Lubber grasshoppers, 212
Lumber, 548
Lumen, **108**, **174**
Luminescence, 347
Lungfish, 656–57, 663
Lungs, **867**–70, 874
LUREs, 794
Luteal phase, **963**
Luteinizing hormone (LH), **943**, **962**, 964–66
Lycophytes (Lycophyta), 571
Lyme disease, 513
Lymph, **877**, **978**
Lymphatic systems, **877**, **978**
Lymph nodes, **978**
Lymphocytes, 677, **978**–79f
Lynx, 1047–50
Lysine, 42t, 283
Lysogenic cycle, **680**
Lysosomes, **110**–11, 114t
Lysozyme, 681, **975**–76
Lytic cycle, **680**

M

Maathai, Wangari, 1122
MacArthur, Robert, 1077–78
McGinnis, William, 383
Macromolecules, **42**
Macronutrients, **738**, 739t
Macrophages, 677, 975f, **976**, 977t
Mad cow disease, 51
MADS box, 383, 410–**11**
Magnesium, 54, 739t, 843t
Magnetic fields
  animal sensory organs and, 610
  navigation, migration and, 1027
Magnetite, 104, 502, 1027
Maidenhair ferns, 466–67
Maintenance, energy and, 1084
Maize. See Corn
Major groove, DNA, 64
Major histocompatibility (MHC) protein, **984**
Maladaptive traits, **187**
Malaria, avian, 1015
Malaria, human, 224, 247, **521**, 522t, 541, 1066–67
Malate, 157
Males
  asexual reproduction and, 224
  chromosomes of, 212
  female barn swallow mate selection and tail length of, 1024–25
  female choice of, 453–54
  gametangia of, 556
  male-male competition of, 454–55
  plant flower parts, 789
  plant gametophytes, 790–92
  plant reproductive organs, 402
  reproductive systems of animal, 957–59f
  sexual selection and, 452–53
  sperm as reproductive cells of, 211
  spermatogenesis in mammal, 953

visual cues from, and female lizard sexual readiness, 1024
  x-linked inheritance of recessive diseases in, 252
Malignant tumors, 206–7
Malpai Borderlands group, 1122–23
Malpighian tubules, **830**, 831f
Maltose, 73–76, 848
Mammalia, 660, 665–66
Mammals, **660**
  amniotic eggs in, 658
  in Cenozoic era, 483
  cleavage in, 393–94
  cloning of, 377–78
  digestion in, 846, 855t
  ears of, 910–12
  as endothermic homeotherms, 817
  flourishing of, after K-P extinction, 492
  fungal mutualisms with, 595
  hormones in growth of, 937
  Mammalia lineages and, 660, 665–66
  mouthparts of, 844
  parental care in, 659
  primates and humans, 668–73f
  sex hormones in reproduction of, 960–66
  spermatogenesis and oogenesis in, 953
  thermoregulation in, 816–19
  threatened with extinction, 1116
Mammary glands, **660**
Mandibles, **643**
Manganese, 739t
Mangroves, 698, 699f
Mantis shrimp, 1030–31
Mantle, **626**
Maps
  community, 1072
  concept, B:10
  genetic (see Genetic maps)
Marfan syndrome, **247**
Margulis, Lynn, 527
Marine ecosystems. See also Aquatic ecosystems
  carbon cycle in, 523
  changes in net primary productivity of, 1101–2
  estuaries as freshwater and, 1000
  global productivity patterns of, 1089–91
  metamorphosis in life cycle of animals in, 616–17
  oceans, 995–96, 1000–1001
  osmoregulation in, 824, 826–27
Mark-recapture studies, 1046–47, 1048
Marshall, Barry, 850
Marshes, **998**
Marsupials (Marsupialia), 660, **665**, 967
Mass extinctions, 474f, **488**–92, 1110. See also Extinction
Mass feeders, **611**, 843
Mass number, **16**
Mast cells, 975f, **976**, 977t
Master regulators, 379–81
Mate choice, female, 452–53
Maternal age, trisomy and, 227
Maternal chromosomes, **213**
Mathematical models, 436, 994
Mating. See also Reproduction
  fungal hyphae mating types, 590–91
  in genetic experiments, 231
  giraffe necks and, 10
  Hardy-Weinberg principle and, 438–40

inbreeding, 450–52
interbreeding, 468–71
mate selection in, 1023–25
monohybrid crosses, 232–33
nonrandom, 450–55 (see also Nonrandom mating)
outcomes of, between populations, 471t
sexual selection and in, 452–55
unusual aspects of, 955–56
Mating types, fungal, 590–91, 593
Matthaei, Heinrich, 283
Maturation, seed, 796–97
Measles, 686, 692, 990
Mechanical energy, 28
Mechanical prezygotic isolation, 459t
Mechanoreceptors, **908**, 909. See also Hearing
Mediator complex, **326**
Medicine. See Diseases, human; Drugs
Medium, **311**
Medulla, **832**
Medulla oblongata, **653**
Medullary respiratory center, 870
Medusae (medusa), **617**
Megapascal (MPa), **719**
Megasporangia, 559, 561, 790
Megaspores, 559, 561, **790**
Megasporocytes, 790
Meiosis, **194, 212, 784**. See also Genes; Genetics
chromosome types and, 212
consequences of sexual reproduction by, 220–23
in eukaryotes, 519
events of prophase I, 219–20
meiosis II phases, 218
meiosis I phases, 216–18
mistakes in, 225–27
mitosis vs., 218t, 219f
number of chromosomes found in some familiar organisms, 213t
overview of, 213–16
paradox of sexual reproduction and, 223–25
patterns of mistakes in, 226–27
ploidy concept and, 212–13
and principles of Mendelian genetics, 240–41
as reduction division, 215–16
sexual reproduction and, 211–12
trisomy in human births and mistakes in, 227
as two cell divisions (meiosis I and II), 214–15
types of mistakes in, 225–26
vocabulary for describing chromosomal makeup of cells, 214t
Meiosis I, **214–20**
Meiosis II, **214**, 215, 218–19
Meiotic maps, **348**
Melanocortin receptor gene, 280–81, 285–86
Membrane lipids, 84. See also Lipid bilayers; Plasma (cell) membranes
Membrane potentials, **745, 767, 887–91**, 908–9. See also Action potentials
Membrane proteins, 92–99
active transport by pumps, 97–98
amphipathic proteins as, 92
aquaporins (see Aquaporins)
biological methods for studying, 94
evolution of fluid-mosaic model of plasma membranes and, 92–94

facilitated diffusion via carrier proteins, 96–97
facilitated diffusion via channel proteins, 94–96
functions of, 45
in osmoregulation, 826–28
in phloem loading, 731–32
plant cell-cell signals and, 757
plasma membranes, intracellular environment, and, 98–99
selective adhesion and, 136–38
in water and electolyte movement, 825–26
Membranes. See also Plasma (cell) membranes
artificial, 85–86
endomembrane system, 118–23f
internal, 104, 107
mitochondrial, 112
thylakoid, 112–13
vesicle, 122
Memory, **902**
experiments on human learning and, 902–4f
immunological, 978, 989–90
Memory cells, **989**, 990
Mendel, Gregor, 230. See also Mendelian genetics
Mendelian genetics, 230–57
applying rules of, to humans, 250–52
chromosome theory of inheritance in, 239–43
determinance of phenotypes in, 247–48
exceptions and extensions to Gregor Mendel's rules in, 250t
experimental system in, 230–32
extending rules of, 243–50
garden peas as first model organism of, 231–32
genetic model of, 236t
hypotheses of, 231
identifying human alleles as recessive or dominant, 250–51
identifying human traits as autosomal or sex-linked, 251–52
incomplete dominance in, 245
linkage in, 243–45, 246f
multiple allelism of genes in, 247
pleiotropic genes in, 247
polygenic inheritance and quantitative traits in, 248–50
quantitative methods in studying linkage, 246f
results of monohybrid reciprocal crosses in, 234t
single-trait experiments in, 232–36
two-trait experiments in, 236–39
vocabulary of, 239t
Meniscus (menisci), 23, 724–25, 727
Menstrual cycle, mammalian, 962–66
Menstruation, **962**
Meristems, 375, 405, 407–9, 704–5
Meselson, Matthew, 261–63
Meselson-Stahl experiment, 261–63
Mesoderm, 394, **395**, 396–98, **604**
Mesoglea, **619**
Mesophyll cells, **188**
Mesozoic era, **483**
Messenger RNAs (mRNAs), **117, 279**
bacterial vs. eukaryotic gene expression regulation and stability of, 331t
in central dogma, 280–81
eukaryotic processing of, 293–95, 320

as intermediaries between genes and proteins, 279–80
post-transcriptional alternative splicing of, 328–29
processing, 293–95
ribosome structure and, 300–301
RNA interference and post-transcriptional stability of, 329–30
specification of amino acids by triplets of, 297
synthesis of, through transcription, 289–93
translation and, 295–97
viral, 682–83
visualizing, via in situ hybridization, 380
Meta-analysis, **1065**
Metabolic diversity
of animals, 610–13
deuterostome, 648
fungal, 589–90
prokaryotic, 506–9
prokaryotic electron donors and acceptors and, 508t
protist, 529–32
six examples of prokaryotic, 507t
Metabolic pathway, **277**
Metabolic rates, **812–13**, 1093
Metabolic water, **803**
Metabolism, **148**, 168
Metagenomics, **364**
Metal ions, 54
Metallothioneins, **747**
Metamorphosis, **616, 935**
animal, 615–17
arthropod, 637
hormones in amphibian, 935–36
hormones in insect, 929f, 936
protostome, 628, 630
Metaphase, **199**, 205
Metaphase I and II, meiosis, 218
Metaphase plate, **199, 217**
Metapopulations, **1046–47**
balance between extinction and recolonization in, 1046, 1047f
habitat destruction and, 1113
mark-recapture experiment on, 1046–47
mark-recapture studies and, 1048
preserving endangered, 1055
stochastic and genetic problems in, 1114
theory of island biogeography and, 1077–78
Metastasis, **134, 207**, 332. See also Cancer
Meteorites, 40, 73, 490–92
Methane, 19–21f, 29, 39–40, 509
Methanogens, **503**, 516
Methanotrophs, **509**
Methionine, 42t, 47, 283, 296
Methods. See Biological methods; Quantitative methods
Methylation, **322**
Methyl salicylate (MeSA), **778**
Metric system, B:1–B:2
Meyerowitz, Elliot, 409–11
Mice, 280–81, 285–86, 369t, 812–13
Micelles, 85
Microarrays, DNA. See DNA microarrays
Microbes, **498**
Microbiology, **498**
Microcystins, 515
Microelectrodes, 890
Microfibrils, 132
Microfilaments, **123**. See also Actin filaments

Micronutrients, 739t, **740**
Micropyle, **790**
MicroRNA (miRNA), **329**, 368
Microsatellites, **365–66**
Microscopy, B:13–B:16. See also Biological imaging
Anton van Leeuwenhoek and, 2
cell research and, 116
electron (see Electron microscopy)
light (see Light microscopy)
prokaryotic cell structures and improvements in, 103
video, 165
Microsponangia, 790–91
Microsporangia, 559, 561
Microspores, 559, 561, **791**
Microsporidia, 594
Microsporidians, 585
Microsporocytes, 790–91
Microtubule organizing centers, **125**, 202t
Microtubules, **125**
in meiosis, 215
mistakes with, 226
in mitosis, 198
in spindle apparatus, 201–2
structure and function of, 124t, 125–27
Microvilli (microvillus), **834, 851**
Midbrain, **653**
Middle ear, **910, 911**
Middle lamella, 135
Mifepristone (RU-486), 966t
Migration, **1026–27**
global warming and, 1100
navigation strategies for, 1026–27
Population Viability Analysis (PVA) and, 1115
seasonality of, 1027
Miller, Stanley, 39–40
Millipedes, 639
Millivolt (mV), **887**
Milner, Brenda, 901–2
Mimicry, **1064**
Mineral nutrients, plant, 738
Mineralocorticoids, **940**
Mineral particles, polymerization and, 43, 62
Minisatellites, **365–66**
Minor groove, DNA, 64
miRNA. See MicroRNA (miRNA)
Miscarriages, 226, 440
Mismatch repairs, **272**
Missense mutations, **286**
Mistletoe, 751
Mitchell, Peter, 162–63
Mites, 642
Mitochondria (mitochondrion), **112, 390**
ATP synthesis in, 164–65
eukaryotic, 112, 114t
NADH in, 161
origin of, in protists, 527–28
in parietal cells, 850
in proximal tubules, 834
replacement rate for, 116
sperm, 390
Mitochondrial DNA (mtDNA), 469–70, 528
Mitochondrial matrix, **112, 157**
Mitosis, **194**, 195–202. See also Cell cycle
alternation of mitotic (M) phases and interphases in, 196
asexual reproduction and, 195
asexual reproduction vs. sexual reproduction and, 220–21

Mitosis, (continued)
cell cycle and, 196–97f
chromosomes and, 195–96
cytokinesis and production of daughter cells by, 200
discovery of G₁ and G₂ gap phases in, 196
discovery of synthesis (S) phases in, 196
events in, 197–200
meiosis vs., 218t, 219f
movement of chromosomes during, 201–2
overview of, 197
plant gametes and, 402
structures involved in, 202t
Mitosis-promoting factor (MPF), 203–4, 375
Mitotic (M) phase, **196**
Mitotic spindle forces, 201
MN blood types case study, 438–39
Mockingbirds, 419–20
Model organisms, **231**, B:19–B:22
animals as, 603
Arabidopsis thaliana (*see Arabidopsis thaliana*)
Caenorhabditis elegans (*see Caenorhabditis elegans*)
characteristics of, B:19
Dictyostelium discoideum, 536, B:20
Drosophila melanogaster (*see Drosophila melanogaster*)
Escherichia coli (*E. coli*) (*see Escherichia coli* (*E. coli*))
garden peas as, for Mendelian genetics, 231–32
genomes of, 362
liverworts as, 569
Mus musculus, 362, 369t, B:21f, B:22
pictures of, B:21f
protostome, 623
Saccharomyces cerevisiae (*see Saccharomyces cerevisiae*)
Models
animal, of disease, 350
condensed DNA, 319f
DNA synthesis, 263–68
mathematical, 436
molecular, 20–21, B:8f
population growth, 1043–44
Modified leaves, 703, 751–52
Modified roots, 698–99
Modified shoots, 700–701
Molarity, **21**
Mole, **21**
Molecular biology. *See also* Molecules
beginning of, 276
carbohydrates in (*see* Carbohydrates)
development of central dogma of, 279–82
genetic engineering and, 338
lipids and plasma membranes in (*see* Plasma (cell) membranes)
nucleic acids in (*see* Nucleic acids)
proteins in (*see* Proteins)
water, carbon, and chemical evolution in (*see* Chemical evolution)
Molecular chaperones, **50**, **304**
Molecular formulas, **20**, B:8f
Molecular machines, 104
Molecular matching mechanisms, 792

Molecular phylogenies. *See* Phylogenies
Molecular weight, **21**
Molecules, **18**
atomic structure and, 16–17
carbohydrates and sugars (*see* Carbohydrates; Sugars)
carbon and organic, 33–36 (*see also* Organic molecules)
cell-cycle regulatory, 203–4
chemical evolution and (*see* Chemical evolution)
chemical reactions and, 21 (*see also* Chemical reactions)
covalent bonding and, 17–18
diffusion and osmosis of, across lipid bilayers, 89–92
DNA, 65–66
enzyme regulation by regulatory, 54–55
geometry of simple, 20
ionic bonding and, 18–19
lipids and plasma membranes (*see* Lipids; Plasma (cell) membranes)
nucleic acids (*see* Nucleic acids)
phylogenetic tree of life and, 6–7
proteins (*see* Proteins)
proteins as large, 45
representing, 20–21, B:8f
RNA, 67–68
self-replicating, as origin of life (*see* Origin-of-life research; Self-replicating molecules)
separating, using gel electrophoresis, B:11–B:12
simple, from carbon, hydrogen, nitrogen, and oxygen, 19–20
transport of, into nucleus, 117–18
visualizing, B:12–B:13
visualizing, with tagging systems, B:12–B:13
Moliason, Henry Gustav, 901–2
Mollusca, 603t, 634–36
Mollusks, **610**, 634–36
body plan of, 626
eyes of, 914
foot of, 629
lineages, 630–31
sampling of, 1107–8
Molting, **625**
Molybdenum, 739t, 740
Monarch butterflies, 1027
Monkeys, 668–69
Monoamines, 898t
Monocots, **564**–65, 706f, 796
Monod, Jacques, 279, 309–14
Monoecious, **790**
Monohybrid crosses, **232**–34t
Monomers, **42**
amino acids as, 42
monosaccharides as, 71–73
nucleotides as, 59
Monophyletic groups, **460**, **504**
cladistic analysis and, 475
green plants as, 551–52
phylogenies and, 504
synapomorphies and, 524
Monosaccharides, **72**. *See also* Carbohydrates; Sugars
chemical evolution and, 71, 73, 76
differences between, 72–73
as monomer sugars, 71–73

Monosodium glutamate (MSG), 919
Monosomy, 226
Monotremes (Monotremata), 660, **665**, **967**
Morgan, Thomas Hunt, 216–17, 241–42, 246
Morphogens, **381**
Morphological diversity
adaptive radiations and, 485–86
of animals, 603–7
deuterostome (echinoderm), 647–48
deuterostome chordate, 650–51
land plant, 549–50
in leaves, 701–2
prokaryotic, 504–6
protist, 526–29
protostome, 626–27
in root systems, 697–98
in shoot systems, 699–700
vertebrate, 655–59
Morphological traits
adaptive radiations and innovations in, 485–86
analyzing, in fungi, 582–84
analyzing, in green plants, 549–50
analyzing, in viruses, 679
chordate, 650
phylogeny based on, 477
of protists, 524–25
Morphology, **102**, **420**
comparative, of animals, 603–7 (*see also* Animal anatomy and physiology)
of eukaryotic lineages, 524t
of fungi, 582–84
of green plants, 549–50
morphospecies concept and, 460
plant (*see* Plant anatomy and physiology)
structural homology in adult, 420–21
Morphospecies concept, **460**, 461
Mortality, human
bacterial infections and, 500f
birth weight and, 442
cancer, 206f
childbirth and maternal, 970–71
death rates, 206f, 1037–40
from jellyfish stings, 619
Mosquitoes, 521, 602, 616, 640t
Mosses, 550, 570
Moths, 640t, 907
Motility. *See* Locomotion, animal; Movement
Motor neurons, **886**, 921
Motor proteins, **124**
cell movement and, 45, 128, 280
kinetochores and, 201–2
in meiosis, 215
microtubules, vesicle transport, and, 126–27
Mountain ranges, climate and, 1010, 1011f
Mouse mammary tumor virus, 691
Mouse model organism, 362, 369t, B:21f, B:22
Mouthparts, 843–45
as adaptations, 843–44
cichlid jaws as case study on, 844–45
feeding adaptations and, 610–13
placental mammal, 666
protostome, 628, 629f
Mouths
digestion in, 847–48
in digestive tracts, 845
human, 846f
protostome vs. deuterostome, 606

Movement
actin filaments and cell, 123–24
animal (*see* Locomotion, animal)
cell, 375t, 376–77
cell gastrulation and, 394–95
of chromosomes during mitosis, 201–2
flagella, cilia, and cell, 127–28
flagella and prokaryotic, 104
microtubules and vesicle transport, 125–27
prokaryotic diversity in, 504, 505f
proteins and cell, 45
protist diversity in, 532–33
wind/touch responses and plant, 766–67
MPF (mitosis-promoting factor), **204**
M phase. *See* Mitotic (M) phase
mRNA. *See* Messenger RNAs (mRNAs)
Mucigel, **706**
Mucosal-associated lymphoid tissue (MALT), **978**
Mucous cells, **849**
Mucus, **848**, 974–75, 978
Müller-Hill, Benno, 314
Müllerian inhibitory substance, **936**
Müllerian mimicry, **1064**
Multicellularity, **134**, **529**, **806**
Cambrian explosion in, 486–88
cell-cell attachments in eukaryotes, 135–38
cell-cell signaling and, 139–46
as characteristic of domains of life, 497t
communications via cell-cell gaps, 138–39
in eukaryotes, 519
eukaryotic, 6, 105, 319
Paleozoic era and, 483
as physical connection of adjacent cells, 134–35
protist, 529
Multienzyme complexes, **49**
Multiple allelism, **247**, 250t
Multiple fertilization, 391–92
Multiple fruits, **797**
Multiple sclerosis (MS), **894**–95, 983
Mumps, 692
Münch, Ernst, 729
Murchison meteorite, 73
Murine leukemia virus, 691
Muscle fibers, **808**, **922**
Muscles
alternative splicing and gens for, 328
cells of, 396–98
contraction of, 922–26
mesoderm and, 604
neurons and cells of, 601
organogenesis and tissues of, 396–99
relaxation of, 925
skeletons and, 920–21
types of, 921–22
Muscle tissue, **808**–9f
Mushrooms, 581, 583, 593, 596
Mus musculus, 362, 369t, B:21f, B:22
Mussels, 634, 1064–65f
Mustard plant model organism. *See Arabidopsis thaliana*
Mutagens, **310**
Mutants, **241**. *See also* Mutations
chromosome theory of inheritance and, 241
creating alleles as, 277
creating *E. coli*, 310
different classes of lactose metabolism, 311

DNA polymerase proofreading and, 271–72
finding mutant genes with replica plating, 310–11
histone, 322
Mutations, 241, 285, 448. *See also* Mutants
animal pollination and plant, 794
antibiotic resistance and bacterial gene, 424–25
Cambrian explosion and, 488
cancer and tumor suppressor gene, 332–33
chromosome-level, 286–87f
DNA polymerase proofreading and, 271–72
evolution and, 448–49
experimental studies of, 449–50
gene duplication and, 368
genetic diversity and, 1106–7
genetic variation and effect on average fitness of, 450t
and genetic variation in plants, 409
Hardy-Weinberg principle and, 438
known types of point, 286t
molecular basis of, 285–87
point, 285–86
polyploidy and speciation by, 465–68
RNA, 69
viral, 686
Mutualisms, 793, 1059, 1068–70
diversity of, 1068
dynamism of, 1069–70
fitness and impacts of, 1070t
natural selection and, 1069
pollination and, 562
types of fungal, 586–89
Mutualistic, 586, 746
Mutualists, 579
Mycelia (mycelium), 514, 582, 583
Mycobacterium tuberculosis, 424–26
Mycorrhizae, 746
Mycorrhizal fungi, 580, 746, 1068
Myelination, 894–95
Myelin sheath, 894
Myelomas, 982
Myoblasts, 398–99
MyoD (myoblast determination), 399
Myofibrils, 922
Myosin, 923
in mitosis, 200
as motor protein, 124
movement and, 45
in muscle contraction, 923–25
Myriapods, 639
Myxinoidea, 661
Myxogastrida, 537

N

N-acetylglucosamine, 76
NAD+
cellular respiration and, 153
fermentation and, 166–67
in glycolysis, 155
photosystem I and, 182
in pyruvate processing, 157–58
NADH, 152
cellular respiration and, 153
citric acid cycle and, 158–61
fermentation and, 166–67
in pyruvate processing, 157–58
NADPH, 173–74, 182
Names, scientific, 7–8, 675
Nanobiology, 678–79
Natural experiments, 427, 1119

Natural populations, 445–47
Natural selection, 4, 422
as acting on individuals while populations evolve, 429–30
allopatric speciation, dispersal, and, 463
artificial selection vs., 4–5
balancing selection, 442–43
biological definitions of fitness and adaptation for, 424
biological evolution and, 15
changes of species through time as pattern component of evolution, 416–22
changing beak size, beak shape, and body size of Galápagos finches as case study for, 426–29
common misconceptions about adaptation and, 429–32
conception in, that organisms do not act for the good of species, 430–31
conditions for, 4
directional selection, 440–41
disruptive selection, 442
evidence for evolution, 422t
as evolutionary process, 435
evolution as not goal directed, 430
evolution of antibiotic resistance of *Mycobacterium tuberculosis* as case study for, 424–26
evolution of evolutionary thought and evolution by, 415–16
fitness and adaptations in, 5
fitness trade-offs in, 432
four postulates of Charles Darwin on, 423
genetic constraints on, 431
genetic variation and effect on average fitness of, 448, 450t
Hardy-Weinberg principle and, 437–38
historical constraints on, 432
individuals vs. populations and, 414f
limitations of, 431–32
mutualisms and, 1069
non-adaptive traits in, 431
as process component of evolution, 422–24
recent research in, 424–29
reproductive success and, 804
sexual reproduction and purifying, 224
species as agents of, 1059
stabilizing selection, 441–42
sympatric speciation by, 465
testing postulates of Charles Darwin on, 425–26
theory of evolution by (*see also* Evolution)
theory of special creation vs. theory of evolution by, 414–15
types of, 440–43
viruses and, 685
Nauplius, 643
Nautilus, 636
Navigation, migration and, 1026–27
Neanderthals, 671
Nectar, 562, 788
Nectary, 788
Needlelike leaves, 702
Negative control, 312–14, 322. *See also lac* operon regulation
Negative feedback, 204, 815–16, 818, 931, 1099
Negative pressure ventilation, 868
Negative-sense viruses, 686, 687t, 692

Neher, Erwin, 892
Nematodes (Nematoda), 211, 603t, 605–6, 638, 973f, 1067–68. *See also Caenorhabditis elegans*
Nematomorpha, 603t
Nemertea, 603t, 627
Nephrons, 832–38t
Neritic zone, 1000
Nernst equation, 888
Nerve cells. *See* Neurons
Nerve cords, 650
Nerve net, 604, 886
Nerves, 886
Nervous systems
anatomy of neurons in, 886–87
central nervous system in, 900–902
comparative morphology of, 603, 604, 605
control of hormone production by, 940–43
learning, memory, and, 902–4f
peripheral nervous system in, 899–900
types of, 885–86
types of neurons in, 886
vertebrate, 899–904f
Nervous tissue, 808
Net primary productivity (NPP), 1002, 1084
biodiversity, species richness, and increased, 1118–19
global climate change and, 1012–13f
global patterns in, 1089–90
global warming and changes in, 1101–2
human appropriation of global, 1090
limiting factors for, 1090–92
ocean iron-fertilization experiments to increase, 1091–92
pyramid of productivity and, 1086
terrestrial-marine contrast in, 1090–91
Net productive rate, 1040
Neural signals, 930–31
Neural tubes, 396
Neurobiology, 885
Neurodegenerative diseases, 376
Neuroendocrine pathway, 931–32, 943
Neuroendocrine signals, 930t, 931
Neuroendocrine-to-endocrine pathway, 931–32, 943
Neuromuscular junctions, 925–26
Neurons, 601, 808, 885–87, 914. *See also* Synapses
Neurosecretory cells, 942, 943
Neurospora crassa, 277, 595
Neurotoxins, 892
Neurotransmitters, 895
categories of, 898t
functions of, 896–97
memory, learning, and, 903–4f
as neural signals, 930–31
postsynaptic potentials and, 897–99
synapse structure and release of, 895–96
Neutrality, pH, 26
Neutral mutations, 286
Neutrons, atomic structure and, 16–17
Neutrophils, 975f, 976, 977t
New World monkeys, 668–69
Next-generation sequencing technologies, 347, 360–61
N-formylmethionine (*f*-met), 301
Niacin, 843t
Niche differentiation, 1062, 1063f

Niches, 484, 1059
Cambrian explosion and, 488
interspecific competition and, 1059–60
types of, 1061
Nickel, 739t
Nicolson, Garth, 93
Nicotinamide adenine dinucleotide (NAD+), 152
Nicotine, 775
Niklas, Karl, 546
Nilsson-Ehle, Herman, 249
Ninebarks, 720
Nirenberg, Marshall, 283
Nitrate, 510, 748
Nitrate pollution
fertilizers and, 749
prokaryotes and, 511–12
Nitrogen. *See also* Nitrogen gas
animal nitrogenous wastes, 829–30
atomic structure of, 17f
cycads and, 574
cycle of (*see* Nitrogen cycle)
electronegativity of, 18
fertilization with, 500
fixation of (*see* Nitrogen fixation)
hornworts and, 571
mycorrhizal fungi and, 746
nucleotide nitrogenous bases, 60
as plant limiting nutrient, 738
as plant nutrient, 739t
simple molecules from, 19–20
sources of human fixed, 1095
Nitrogenase, 49, 749
Nitrogen cycle, 510–11, 1096
Nitrogen fixation, 510–11, 749–50, 1068, 1093
Nitrogen gas, 20, 27, 49, 510–11, 748–50. *See also* Nitrogen
Nitrosomonas, 507t
Nociceptors, 908
Node of Ranvier, 894
Nodes, 474, 699
Nod factors, 749
Nodules, 749
Noise, experimental, 12
Non-adaptive traits, 431
Non-bilaterian animals, 603t, 617–20
Noncoding sequences. *See* Repeated sequences
Noncyclic electron flow, 183
Nondisjunction, 225–27, 466
Nonenveloped viruses, 679
Nonesense mutations, 286t
Non-kin human child abuse rates, 1034
Nonpolar covalent bonds, 18, 19f
Nonpolar side chains, 41t
Nonrandom mating, 450–55
Nonself, adaptive immune response and, 978, 983–84
Nonsense mutations, 286t
Non-sex chromosomes. *See* Autosomes
Non-sister chromatids, 214t, 216
Non-template strand, 290
Non-vascular plants, 550, 551–52, 567, 569–71
Norepinephrine, 898t, 943
Notochords, 396, 651
N-terminus, 43–44, 301
Nuclear envelope, 107
as characteristic of domains of life, 497t
Eukarya and, 519
in mitosis, 200
as protist innovation, 526–27
structure and function of, 116–17

Nuclear lamina, **107**, 116–17, 125
Nuclear lamins, **125**
Nuclear localization signal (NLS), **117**–18, 122
Nuclear pore complex, **117**
Nuclear pores, **117**
Nuclear transfer, 378. *See also* Clones
Nucleases, **852**, 855*t*
Nucleic acid probes, B:13
Nucleic acids, **59**–70
  chemical evolution and nucleotide production for, 60–61
  chemical evolution theory and RNA molecule as first life-form, 68–69
  components of, 59–60
  DNA structure and function, 62–66 (*see also* DNA (deoxyribonucleic acid))
  lipids and, 82
  nucleotides and, 39
  organic molecules and, 36
  polymerization of nucleotides to form, 42, 61–62
  ribonuclease enzyme for unfolding, 50
  RNA structure and function, 66–68 (*see also* RNA (ribonucleic acid))
  RNA world hypothesis on RNA as first self-replicating molecule, 59, 68–69
  sugars in, 77
Nucleoids, **103**–4
Nucleolus, **108**, 117
Nucleoplasmin, 117–18
Nucleoside triphosphates, 62, 117
Nucleosomes, **321**
Nucleotide excision repairs, **272**
Nucleotides, **59**
  chemical evolution and production of, 60–61
  as components of nucleic acids, 59–60
  nucleic acids and, 39
  polymerization of, to form nucleic acids, 42, 61–62
Nucleus, **107**
  atomic structure and, 16–17
  cloning by transfer of, 378
  DNA in, 279
  eukaryotic, 107–8, 114*t*
  eukaryotic cells vs. prokaryotic cells and, 6, 7*f*, 102, 107
  multinucleate muscle cells, 808
  nuclear transport in, 116–18
  protist, 527
Nudibranches, 635
Null alleles, **277**
Null hypothesis, **12**, 438–40
Nüsslein-Vohard, Christiane, 379–81
Nutrients, **842**
  availability of, in aquatic ecosystems, 995–96
  cell-cycle checkpoints and cell, 205
  cycling of, 1092–94
  essential, for animals and humans, 842–43
  essential, for plants, 738–40
  plant deficiency in, 740
  plant requirements for, 696
  plant uptake of, 744–48

Nutrition. *See* Animal nutrition; Plant nutrition
Nutritional homeostasis, 856–58

O

Obesity, diabetes mellitus and, 857–58
Observable traits, 231–32
Observational studies, 1011–12
Occipital lobe, **901**
Oceanic zone, **1000**
Oceans
  acidification of, 1100, 1114
  changes in net primary productivity of, 1101–2
  chemical evolution conditions in, 27, 32, 39–40, 73
  deep-sea vents in (*see* Deep-sea vents)
  effects of, on climate, 1010–11
  global productivity patterns of, 1090
  global warming and declining net primary productivity in, 1102
  iron-fertilization experiments in, to increase net primary productivity, 1091–92
  as marine ecosystems, 1000–1001
  polymerization in early, 43
  prebiotic soup in early, 38–40
  protists in, 519*f*, 520*f*
  time line of changes in, 483–84
  upwelling of, and nutrient availability, 995–96
  water properties and chemical evolution in early, 22–26 (*see also* Water)
Octopus, 636
Odonata, 641*t*
Odor, genotypes and body, 439–40
Oil consumption. *See* Fossil fuel consumption
Oils, **88**. *See also* Lipids
Okazaki, Reiji, 266–67
Okazaki fragments, 267–68
Old World monkeys, 668–69
Oleanders, 727–28
Olfaction, **918**, 919–20
Olfactory bulb, **919**
Oligodendrocytes, **894**
Oligopeptides, **44**
Oligosaccharides, 71, 77. *See also* Carbohydrates
Omasum, 850
Ommatidia, **913**
Omnivores, 547, **612**
Omnivorous, **663**
Oncogenes, **332**–33
One-gene, one-enzyme hypothesis, 277–78
On the Origin of Species by Means of Natural Selection (book), 414, 416, 458
Onychophora, 603*t*, 637
Oocytes, **203**, 218
Oogenesis, **953**
Oogonia (oogonium), **953**
Oomycota, 542
Oparin, Alexander I., 38
Oparin-Haldane chemical evolution theory, 38–39
Open circulatory systems, **875**

Open reading frames (ORFs), **362**
Operators, **313**–14
Operculum, **865**
Operons, **313**–14, 331
Opisthokonts (Opisthokonta), 524*t*, 536
Opium, 775
Opsins, **915**, 917–18
Optic nerve, **914**
Optimal foraging, **1022**
Oral rehydration therapy, 854
Orbitals, electron, **17**
Organelles, **104**
  as characteristic of domains of life, 497*t*
  eukaryotic, 107
  plant cell, 707
  prokaryotic, 104
  prokaryotic vs. eukaryotic, 107*t*
Organic matter, 742–43, 744*t*
Organic molecules, **33**. *See also* Carbon
  carbon and formation of, 33–34
  functional groups and, 34–36
  organotrophs and, 506–8
  in prebiotic soup, 38
  six common functional groups and, 35*t*
Organismal ecology, **994**
Organisms, **1**
  atoms and molecules of living, 17*f* (*see also* Molecules)
  atoms found in, B:8*t*
  Cambrian explosion in, 486–88
  cells of, 2 (*see also* Cells)
  characteristics of living, 1–2, 82
  characteristics of viruses vs., 676*t*
  ecology and distribution and abundance of, 993
  geographic distribution of, 1013–17
  impact of global warming on, 1099–1101
  major amino acids found in, 41*t*
  misconception of higher and lower, in evolution, 430
  misconception of self-sacrificing, in evolution, 430–31
  model (*see* Model organisms)
  multicellular (*see* Multicellularity)
  number of chromosomes found in some familiar, 213*t*
  organic molecules and, 36
  organism level anatomy and physiology, 811*f*
  taxonomy and scientific names for, 8
  unicellular vs. multicellular, 6
  viruses vs., 675, 676*f*
  water in living cells of, 22
Organogenesis, **396**–99, 402
Organotrophs, 506–8
Organs, **809**
  animal, 810–11
  gas exchange, 864–70
  immune system rejection of foreign, 988
  male animal reproductive, 958–59*f*
  organ level anatomy and physiology, 811*f*
  organogenesis and, 396–99
  plant vegetative, 402
  transplants of, 988
  viruses and human, 676*f*
Organ systems, 676*f*, **810**–11
Origin-of-life research. *See also* Chemical evolution
  extremophiles and, 501
  first reactions in chemical evolution, 32

nucleic acid formation in prebiotic soup, 62
origin of viruses in origin of life, 687–88
polysaccharides and, 76
on protein catalysts (enzymes) as first self-replicating molecules, 56
protein polymerization in prebiotic soup, 43
RNA as first self-replicating molecule, 59, 68–69
spark-discharge experiment on chemical evolution in prebiotic soup, 39–40
Origin of replication, **264**
Ornithine, 277
Orthoptera, 641*t*
Oryx, 803, 817
Oryza sativa, 369*t*
Oshima, Yasuji, 323–24
Osmoconformers, **824**
Osmolarity, **823**
Osmoregulation, **823**
  in aquatic ecosystems, 826–28
  attributes of animal nitrogenous wastes, 829*t*
  cell movement of electrolytes and water in, 825–26
  diffusion, osmosis, and, 823
  hormones of, 940
  osmotic stress and, 823–25
  overview of, 831
  in terrestrial insects, 828–31
  in terrestrial vertebrates, 832–38
  water and electrolyte balance and, 822
Osmoregulators, **824**
Osmosis, **90**, 718, **823**
  across lipid bilayers, 90–92
  across prokaryotic cell walls, 105
  in animal kidneys, 832
  in animal osmoregulation, 831
Osmotic gradients, 835–37
Osmotic potential, **718**
Osmotic stress. *See also* Osmoregulation
  diffusion, osmosis, and osmoregulation of, 823–24
  in freshwater, 824
  hormonal control of urine formation and, 837
  nitrogenous wastes and, 830
  in seawater, 824
  in terrestrial ecosystems, 825
Ouabain, **826**, 830
Outbreaks, viral, 689
Outcrossing, **222**–23, 792
Outer coverings, protist, 528–29
Outer ear, **910**
Outgroups, **477**
Out-of-Africa hypothesis, **672**
Ovalbumin genes, 322
Oval window, **910**–11
Ovaries, **393**, 561, 563, 789, 933, 953
  in female reproductive system, 959
  hormones of, 936
  ovarian hormones in mammalian menstrual cycle, 964–66
Overexploitation, 1016–17, 1087, 1110–11
Overwatering, 742
Oviducts, **393**, 960
Oviparous species, **615**, 658, 956–57
Ovoviviparous species, **615**, 658, 956
Ovulation, **960**
Ovules, 402, 561–62, 789–90
Ovum, **953**
Oxidation, 151

glucose, 158, 165–66
peroxisomes and eukaryotic, 109–10
of water during photosynthesis, 181
Oxidative phosphorylation, **161**, 166
Oxygen. *See also* Oxygen gas
atomic structure of, 17*f*
deoxyribonucleotide lack of, 60
electronegativity of, 18, 22
proteins and, 46–47
redox reactions and, 151–52
simple molecules from, 19–20
Oxygen gas. *See also* Oxygen
aerobic respiration and, 166
animal gas exchange and, 861
behavior of, in air, 862–63
behavior of, in water, 863–64
betaβ-globin protein and, 247
Cambrian explosion and higher levels
of, 487
cyanobacteria and, 509–10, 515
end-Permian extinction and lack
of, 489–90
exchange of, between mother and fetus during pregnancy, 969
green plant production of, 547
green plants and atmospheric, 550
hemoglobin and, 45, 46
homeostasis for blood levels of, 940
molecular structure of, 21*f*
photosynthesis and, 173–74
as plant nutrient, 739*t*
Precambrian lack of, 483
prokaryotic oxygen revolution
and, 509–10
regulation of blood levels of, 940
in soil, 742
in streams, 999
transport of, in blood hemoglobin,
870–73
Oxygen-hemoglobin equilibrium curve,
871, 969
Oxygenic photosynthesis, **181**–82, **509**,
510, 515, 547
Oxytocin, **943**, **970**
Oysters, 634
Ozone, 32

P

p53 gene, **205**, 206, 332–33
P680, 183
P700, 183
Pääbo, Svante, 345–46
Pace, Norman, 498
Pacemaker cells, **880**
Packaging, bacterial vs. eukaryotic gene
expression and, 331
Paine, Robert, 1072–73
Pair-rule genes, 382
Palade, George, 119–20
Paleontologists, 481
Paleozoic era, **483**
Palindromes, 340
Pancreas, **852**, **933**
cells of, 115
as gland, 933
human, 810*f*, 846*f*
identification of pancreatic
hormones, 934–35
paracrine signals and, 930
protein digestion by enzymes
of, 851–52
pulse-chase experiment on cells
of, 119–20
regulation of enzymes of, 852
Pancreatic amylase, 855*t*

Pancreatic lipase, **853**, 855*t*
Pandemics, 677–78
Papermaking, 548, 575, 590
Parabasalida, 538
Parabiosis, **938**–40
Parabronchi, 870
Paracrine signals, **930**
Paralytic shellfish poisoning, 522
Parameters, 1043
Paramylon, 539
Paranthropus, 670*t*, 671
Paraphyletic groups, 519–20
Parapodia, 614, **633**
Parasites, **530**, **612**, 751, **1063**
animal carriers of, 602
animals as, 612–13
ants as, 602
endosymbiont bacteria as, 514
fungal, 581, 594–95, 598
genomes of, 363
human intestinal, 538
leeches as, 633
manipulation of hosts by, 1067–68
nematode as cause of
elephantiasis, 973*f*
plants as, 751
protists as, 530, 536
protostomes as, 631
roundworms as, 638
sexual reproduction and, 224–25, 533
viral, 259–60*f*
Parasitic, **586**
Parasitic sequences, 365–67
Parasitism, **1063**. *See also* Parasites
Parasitoids, 780–81
Parasympathetic nervous system, **900**
Parathyroid glands, **933**
Parenchyma cells, **708**, 709*f*, 711*t*
Parental care, **659**
bird, 659
mammalian, 967
marsupial, 967
reptilian, 667
vertebrate, 659
Parental generation, **232**
Parietal cells, **849**–50
Parietal lobe, **901**
Parker, Geoff, 954
Parkland College, 30–32
Parsimony, 476–77
Parthenogenesis, **630**, 631, **951**
Partial pressure, 862–63, 864
Particulate inheritance, 234–36
Pascal (Pa), **719**
Passive exclusion, plant, 747
Passive transport, **96**–98, **731**, **825**
Pasteur, Louis, 3–4, 498
Patch clamping, **892**
Paternal care, female choice for, 453–54
Paternal chromosomes, 213
Paternity, 366, 956
Pathogen defense response, plant. *See*
Hypersensitive response (HR)
Pathogenic bacteria, **498**
antibiotics and, 500
germ theory of disease and, 499–500
human illnesses caused by, 499*t*
Koch's postulates on, 498–99
medical importance of, 498–500
virulence of, 500
Pathogens, **707**, **776**
barriers to entry of, 974–75
humoral response coating of, with
antibodies, 988
plant defense responses to,
775–78, 780*t*

Pattern component
chemical evolution theory, 15, 38
of germ theory of disease, 499
of scientific theories, 2, 414–15
theory of evolution by natural
selection, 4, 416
theory of special creation, 415
Pattern formation, 379–84
Pattern-recognition receptors, **975**
Pearl oysters, 634
Peas. *See* Garden peas
Peat, 570
Pectins, **132**
Pedigrees, **250**, 251
Pedipalps, 642
Pedometer hypothesis on ant
navigation, 10–12
Pelc, Stephen, 196
Penfield, Wilder, 902
Penicilin, 581, 598, 691
Penis, **954**, 957–59*f*
Penis worms, 627
Pentoses, 60, **72**, 73
PEP carboxylase, **188**
Pepsin, **849**, 855*t*
Pepsinogen, 849
Peptide bonds, **43**
formation of, in protein synthesis,
300, 301–2
polypeptides and, 43–44
protein backbone structure
and, 46–48
Peptide proteins, 280
Peptides, **44**, 898*t*
Peptidoglycan, 75*t*, **76**, 77, 105*f*, 496, 505
Perennials, 567, **698**
Perfect, **790**
Perforations, **710**
Pericarp, **797**
Pericycle, **722**
Periodic table of elements, 16*f*
Periodontitis, 498
Peripheral membrane proteins, 93–94
Peripheral nervous system (PNS),
886, 899–900
Peristalsis, **848**
Permafrost, **1008**
Permeability, **86**–89, 92
Permian extinction, 489–90
Permineralized fossils, 480
Peroxisomes, **109**
eukaryotic, 109–10, 114*t*
in green plants, 549
Persistent organic pollutants
(POPs), 1088
Petals, **409**, 562, **788**–89
Petiole, **701**
Petrified wood, 480
Petroleum consumption. *See* Fossil fuel
consumption
Petromyzontoidea, 661
pH, **25**
acid-base reactions and, 25–26
of blood, 870
effect of, on hemoglobin, 872, 873*f*
enzyme catalysis and, 55–56
hemoglobin and buffering of blood,
873–74
as measure of acidity or alkalinity of
solutions, 26
in soil, 743
Phaeophyta, 543
Phagocytosis, **110**–11, 976, 987
Phanerozoic eon, 482*f*, 483
Pharmaceuticals. *See* Drugs
Pharyngeal gill slits, **650**

Pharyngeal jaws, **656**, 844–45
Pharyngeal pouches, 651
Phelloderm, **713**
Phenetic approach, **475**
Phenology, **1100**
Phenotypes, **231**
central dogma and linking genotypes
and, 280–81
codominance in, 245–47
disease genes and improved understanding of, 350
genotypes vs., 234
incomplete dominance in, 245
in Mendelian genetics, 239*t*
mutations and, 285
natural selection and, 440
plasticity in (*see* Phenotypic
plasticity)
polygenic inheritance and, 248–50
predicting, with Punnett square,
235–36
ratio of, and segregation
principle, 235
Phenotypic plasticity, **698**
in leaves, 702–3
in root systems, 698
in shoot systems, 700
Phenylalanine, 42*t*, 248, 283
Phenylketonuria (PKU), **248**
Pheophytin, 180–81
Pheromones, animal, **907**, 930*t*, **931**, **954**
Pheromones, plant, 780–81
Phloem tissue, **710**
anatomy of, 729–30
loading of, in translocation, 731–33
primary plant body, 711*t*
unloading of, in translocation,
733–34
vascular cambium and, 713
vascular tissue system functions
of, 711
Phoronida, 603*t*
Phosphatases, 144
Phosphate, 189
Phosphate functional group, 35, 60, 62,
84, 97–98
Phosphodiesterase (PDE), 916
Phosphodiester linkage, **61**, 62, 66,
68, 260
Phosphofructokinase, 156
Phospholipids, 82*f*, **84**, 496. *See also*
Lipid bilayers; Lipids
Phosphorus
in DNA, 196, 259
as human nutrient, 843*t*
as plant limiting nutrient, 738
as plant nutrient, 739*t*
Phosphorylase, 79, **946**
Phosphorylated molecules, 62
Phosphorylation, **149**
post-translational control by, 330
as post-translational
modification, 304
of proteins by ATP, 149–50
substrate-level, 155–56
Phosphorylation cascades, **144**, **756**
enzyme-linked receptors and, 143–44
hormone signal transduction cascades as, 946–47
Photic zone, **997**, **1000**–1001
Photons, **32**, **175**
Photoperiodism, **787**
Photophosphorylation, **181**
Photoreceptors, **908**
plant, 758
vertebrate, 914, 915, 917–18

Photorespiration, **187**
Photoreversibility, **763**
Photosynthesis, **78**, 172–93, **509**
  autotrophs, heterotrophs, and, 172
  capturing of light energy by chlorophyll in, 174–79
  carbohydrate storage of energy from, 78
  chloroplasts in, 112–13
  discovery of photosystems I and II of, 179–84
  experiment on enhancement effect of red and far-red light in, 179–80
  ground tissue systems and, 708
  leaves and, 699
  pigments, 532*t*
  prokaryotic, 508–9
  prokaryotic internal membranes and, 104, 107*t*
  protist, 529–32, 536
  redox reactions and, 151
  reduction of carbon dioxide to produce glucose in, 184–90
  regulation of, 189
  requirements for, 696
  use of sunlight to make carbohydrates in, 172–74
  water conservation and, 727
Photosynthesis-transpiration compromise, 727
Photosystem, **178**
Photosystem I and II, **179**–84
Phototrophs, **506**
Phototropins, **758**–59
Phototropism, **758**
  blue-light, 758–62
  red and far-red light, 762–74
Phycocyanin, 532*t*
Phycoerythrin, 532*t*
Phyla (phylum), **7**, **498**, **603**, 646–47*f*
Phylogenetic species concept, **460**–62
Phylogenetic trees, **474**–79. *See also* Phylogenies
  analyzing rRNA to create, 6–7
  distinguishing homology from homoplasy with, 475–77
  estimating phylogenies with, 475
  mapping land plant evolutionary changes on, 555–56
  reading, B:4–B:6
  studying animals with, 607–9
  studying fungi with, 584–86
  studying green plants with, 551–52
  studying prokaryotes with, 503–4
  studying protists with, 525
  taxonomy and, 7–8
  terminology of, 474–75
  tree of life as, 5–7 (*see also* Tree of life)
  using, to understand emerging viruses, 688–89
  whale evolution as case history for, 477–79
Phylogenies, **6**, **102**–3, **420**, **474**
  of animals, 607–9
  based on DNA sequence data, 477–78
  based on morphological traits, 477
  of cetacean whales and dolphins, 422
  of chordates, 654*f*
  of current human populations, 672*f*

distinguishing homology from homoplasy in, 475–77
  domains and, 102–3
  endosymbiosis and, 528
  of fungi, 584–86
  of green plants, 551–52
  of HIV, 688*f*
  of mockingbirds, 420
  of prokaryotes, 503–4
  using phylogenetic trees to estimate, 475
  vertebrate, 655
  of vertebrate circulatory systems, 878*f*
  of vertebrates, 654*f*
  whale evolution as case history for, 477–79
Physical activity, diabetes mellitus and, 857–58
Physical isolation, 462–64
Physical maps, **348**
Physiology, **803**, 810–11*f*. *See also* Animal anatomy and physiology; Plant anatomy and physiology
Phytoalexins, **777**
Phytochromes, **763**–64, 787
Phytophthora infestans, 520–21, 522*t*, 542
Phytoplankton, 523
Pigments, **174**, **758**
  absorption of light by photosynthetic, 175–77
  lipids as, 84
  in mice, 280–81
  in photosynthetic protists, 532*t*
  prokaryotic, 509
  thylakoid membranes and, 174
  in vacuoles, 112
Piloting, **1026**
Pima Indians, 857–58
Pines, 575
Pine trees, 560–61
Pinocytosis, **111**
Pinophyta, 575
Pinworms, 638
Pioneering species, **1075**
Pisum sativum. See Garden peas
Pitcher plants, 751–52
Pitches, **909**
Pith, **706**
Pits, **710**
Pituitary dwarfism, 338–44
Pituitary gland, **933**, 937, 940–43, 964–66
Placental mammals, **666**
Placentas, **390**, **393**–94, **658**–59, **968**
Placozoa, 603*t*
Planar bilayers, 86
Plankton, **997**, **1071**
  ciliates as, 540
  community structure of, 1071
  ctenophores as, 620
  diatoms as, 542
  euglenids as, 539
  green algae as, 566
  protists as, **522**–23
  rotifers as, 631
Plant(s)
  bodies (*see* Bodies, plant)
  cells of (*see* Plant cells)
  deceptive communication in, 1030

defense responses (*see* Defense responses, plant)
  development of (*see* Plant development)
  form and function of (*see* Plant anatomy and physiology)
  fungal mutualisms with, 579–80
  global warming and, 1100
  green algae and (*see* Green plants)
  land plants (*see* Land plants)
  meiosis and gametes of, 218
  mutualisms with, 1068
  nutrition for (*see* Plant nutrition)
  oldest known, 786
  photosynthesis in (*see* Photosynthesis)
  Plantae lineage and, 524*t*, 536
  reproduction of (*see* Plant reproduction)
  sensory systems, signals, and responses of (*see* Sensory systems, plant)
  speciation by polyploidization in, 467–68
  sugar transport (translocation) in, 728–34 (*see also* Translocation)
  tissue cultures, B:18
  transgenic, 354–55 (*see also* Golden rice)
  viruses and, 681, 691, 692
  water transport in, 717–28 (*see also* Water transport, plant)
Plantae, 524*t*, **536**. *See also* Plant(s)
Plant anatomy and physiology, 695–716
  brassinosteroids and plant body size, 773
  flowers, 786–92
  indeterminate growth and angiosperm, 695–96
  leaves in, 701–3
  primary growth in, 704–6
  primary plant body cells and tissues, 706–12
  primary plant body components, 711*t*
  root systems in, 697–99
  secondary growth components, 713*t*
  secondary growth in, 712–15
  shoot systems in, 699–701
  surface area/volume relationships in, 696–97
Plant cells
  animal cells vs., 106*f*, 706–7
  cell-cell attachments of, 135
  cellulose as structural polysaccharide for, 76
  cell walls in, 132–33
  cytokinesis in, 200
  cytoplasmic streaming in, 124
  differentiated, as genetically equivalent, 377
  eukaryotic, 106*f*, 115–16 (*see also* Eukaryotic cells)
  glyoxysomes and peroxisomes in, 110
  middle lamella in cell-cell attachments of, 135
  plasmodesmata in, 138–39
  starch as storage polysaccharide in, 168
  vacuoles in, 111–12
Plant development, 401–13
  animal development vs., 401
  embryogenesis in, 404–6
  gametogenesis, pollination, and fertilization in, 402–4
  overview of, 402*f*
  reproductive development in, 409–11
  vegetative development in, 407–9

Plantlets, **786**
Plant nutrition, 737–54
  adaptions for, 750–52
  deficiencies in, 740–41
  essential nutrient elements and availability, 739*t*
  essential nutrients, 738–40
  importance of, 717
  macronutrients, 738, 739*t*
  micronutrients, 739*t*, 740
  nitrogen fixation and, 748–50
  nutrient uptake and, 744–48
  nutritional requirements and, 738–41
  soil and, 741–44*f*
  soil composition, soil properties, and, 744*t*
  starch as storage polysaccharide for, 74
Plant reproduction, 783–802
  alternation of generations in, 556–58
  asexual reproduction, 786
  embryophyte retention of offspring in, 556
  evolution from gametophyte-dominant to sporophyte-dominant life cycles in, 558–59
  fertilization in, 794–95
  flowers as reproductive structures of, 786–92
  flowers in, 561–62
  fruits in, 563–64
  gamete production in protected structures in, 556
  heterospory in, 559, 560*f*, 561*f*
  importance of, 783
  land plant life cycle and, 784–86
  land transition adaptations for, 556–64
  pollen in, 559–60
  pollination in, 562–63, 792–94
  seeds in, 560–61, 795–800
  sexual, 784–86
Plasma, **870**
Plasma cells, **986**, 988, 989*t*
Plasma (cell) membranes, 82–101, **102**
  of Archaea, 496
  cell walls and, 113
  crossing of, by hormones, 934
  diffusion and osmosis across, 89–92
  eukaryotic, 114*t*
  excitable, 891
  exocytosis and, 122
  gamete fusion and, 390
  importance of, to cells, 1, 2, 82, 98–99
  internal, of photosynthetic prokaryotic cells, 104, 107*t*
  lipid bilayers as, 85–89
  lipids in, 82*f*, 83–85
  lipids in, as characteristic of domains of life, 497*t*
  lipid-soluble vs. lipid-insoluble hormones and, 140
  membrane proteins and fluid-mosaic model for, 92–99
  in mitosis, 200
  nuclear envelope and, 526
  prokaryotic, 104
  proton pumps in, 733
  selectively permeable, 823
  smooth ER and, 108
  transport of solutes across, in plants, 731–32
Plasmids, **103**, **340**
  in genomes, 364
  introducing recombinant, into bacterial cells for transformation, 341

Boldface page numbers indicate a glossary entry; page numbers followed by an *f* indicate a figure; page numbers followed by *t* indicate a table.

I:26 INDEX

origin of viruses in, 686–87
prokaryotic, 103–4
TI (tumor-inducing), 355–56
uses of, in genetic engineering, 352t
using, in DNA cloning, 340
Plasmodesmata (plasmodesma), 138–39, **568, 707**
Plasmodial slime molds, 537
Plasmodium falciparum, 369t
Plasmodium sp., 521, 522t, 541, 1066–67
Plasmogamy, **591**, 592
Plastocyanin, **183**
Plastoquinone (PQ), **180**–81, 183
Platelet-derived growth factor (PDGF), 207
Platelets, **207, 870, 975**
Plato, 415
Platyhelminthes, 603t, 609, 631–32
Platypuses, 665
Pleiotropic genes, **247**, 250t, 431
Ploidy, 212–13, 214t
Pneumatophores, 698, 699f
Pneumonia, 677
Podia, **648**
Point mutations, 285–86t
Poisons
  in action potential research, 892
  plant, 775
Polar bears, 1100
Polar bodies, **953**
Polar climate, 1010
Polar covalent bonds, **18**, 19f
Polarity
  amino acid side chain, 42
  covalent bond, 18
  of water, 22–23
Polar microtubules, **198**
Polar nuclei, **790**
Polar side chains, 41t
Polar transport, **768**
Polio virus, 692, 990
Pollen grains, **231, 402, 560, 791**
  fossil record and, 480
  germination of, 794
  land plants and, 559–60
  pollen-stigma interactions and, 402–3
  seed plants and, 567
  stamens and production of, 789, 790–92
Pollen tubes, **403,** 794–95
Pollination, **562, 792**–94
  of crops, 1117
  evolution of, 793–94
  flower petals and, 788
  in gametogenesis, 402–3
  land plants and, 562–63
  selfing vs. outcrossing in, 792
  speciation and, by animals, 794
  syndromes, 792–93
Pollinators, 562
Pollution
  bioremediation of, 500–501, 598, 1117
  ecosystem ecology and, 995
  nitrate, 749
  prokaryotes and nitrate, 511–12
  soil, 747
  stoneworts and water, 569
  as threat to biodiversity, 1111
Polygenic inheritance, **249**
  exceptions and extensions to Gregor Mendel's rules on, 250t
  of quantitative traits, 248–50
  in situ hybridization and, 428–29
Polymerase chain reaction (PCR), **344**–46

amplification of fossil DNA with, 345–46
direct sequencing with, 502–3
DNA fingerprinting and, 366
extremophiles and, 501
requirements for, 344–45
uses of, in genetic engineering, 353t
uses of, in genetic testing, 350
Polymerization, **42**
  of amino acids to form proteins, 42–44
  chemical evolution and, of monosaccharides, 77
  of monosaccharides to form polysaccharides, 73–74
  of nucleotides to form nucleic acids, 61–62
  of nucleotides to form nucleic acids and lipids, 82
Polymers, **42,** 71, 73–76
Polymorphic, **247, 348**
Polymorphism, 250t
Polypeptides, **43.** *See also* Proteins
  as hormones, 933, 936
  peptide bonds and formation of, 43–44
  translation and extending of, 301–2
Polyphenylalanine, 283
Polyplacophora, 636
Polyploidy, **213, 286, 466**
  cells, 214t
  chromosome-level mutations and, 286
  sympatric speciation and, 471t
  sympatric speciation by, 465–68
Polyproteins, 682
Polyps, **617**
Polyribosomes, 297
Polysaccharides, 71, **73**–76, 113, 132–34. *See also* Carbohydrates
Polyspermy, 391–92
Poly(A) tail, RNA, **295**
Ponds, 997
Poor-nutrition plant tissues, 1066
Population(s), **4, 416, 435, 994, 1037.** *See also* Population dynamics; Population ecology
  bottlenecks in, 446
  causes of genetic drift in natural, 445–46
  contemporary evolution of, 418
  contemporary speciation in, 422
  evolutionary processes and evolution of, 435
  evolution of, 414f, 429–30 (*see also* Evolution; Natural selection)
  evolution of, by natural selection, 4, 414
  extinction of (*see* Extinction)
  gene flow in natural, 447
  genetic drift and small, 444
  genetic isolation of, 462–64
  global warming and evolutionary change within, 1101
  growth of (*see* Population growth)
  mating between isolated, 468–71
  monophyletic, 460–61
  outcomes of secondary contract between isolated, 471t
  population viability analysis for, 1115
  secondary contact between isolated, and speciation, 468–71
  speciation in (*see* Speciation)
  stabilizing size of human, 1121
  theory of evolution by natural selection and, 416

Population cycles, 1047–50
Population density, 952, 1021, **1042**
Population dynamics, **1046**–53
  human population growth rate changes, 1052–53
  mark-recapture studies of, 1048
  metapopulation changes, 1046–47
  milestones in human population growth, 1052t
  population age structure and population growth, 1050–52
  population cycles, 1047–50
Population ecology, 994, **1037**–57
  demography and, 1037–41
  endangered species analysis and, 1053–55
  human population size and, 1037
  population dynamics and, 1046–53
  population growth and, 1041–46
  preserving metapopulations, 1055
Population growth, 1041–46
  age structure and, 1050–52
  calculating rates of, using life tables, 1040
  changes in rates of, in human populations, 1052–53
  developing and applying equations for, 1043–44
  exponential, 1042
  limiting factors on, 1044–46
  logistic, 1042–44
  milestones in human, 1052t
  momentum in, 1051–52
  projections of, 1054–55
  quantifying rates of, 1041–42
Population momentum, 1051–52
Population thinking, 416
Population Viability Analysis (PVA), 1115
Pores, 94–95, 554, 708
Porifera, 603t, 618
Porin, 45
Positive control, **312**, 322. *See also* Catabolite repression
Positive feedback, **892**, 1099
Positive pressure ventilation, **868**
Positive-sense viruses, **686**, 687t, 692
Post-anal tails, 650
Posterior, **379**
Posterior pituitary, **941**, 943
Postsynaptic neurons, **896**
Postsynaptic potentials, 897–99
Post-transcriptional control, eukaryotic, 328–30
Post-translational control, **308**
  bacterial vs. eukaryotic gene expression regulation and, 308–9, 331t
  eukaryotic, 330
Postzygotic isolation, **459**
Potassium. *See also* Sodium-potassium pump
  action potentials and ions of, 891
  as plant limiting nutrient, 738
  as plant nutrient, 739t
  potassium leak channels, 889
Potassium channels, 891, 892
Potatoes, 116, 520–21, 542
Potato famine, 522t
Potential energy, **27.** *See also* Energy
  aerobic respiration and, 166
  of ATP, 148
  chemical energy as, 33
  energy transformations of, 27–29
  phosphate functional group and, 62
  spontaneous chemical reactions and, 29–30
  water potential and, 718

Pox viruses, 690
Prairie dogs, 1033
Prebiotic soup, **38.** *See also* Chemical evolution; Deep-sea vents; Origin-of-life research; Volcanic gases
  chemical evolution in, 38–39
  first membranes in, 91
  nucleic acid formation in, 62
  origin-of-life research on chemical evolution in, 39–40
  polymerization in, 43
  protein catalysts in, 56
Precambrian eon, 481–83
Precipitation
  biomes and, 1002
  global climate change experiments on, 1012–13f
  in tropical wet forests, 1009
Predation, **1063.** *See also* Predators
  Cambrian explosion and evolution of, 487–88
  deceptive communication and, 1030
  fungal, 598
  of herbivore populations, 1065–66
  population cycles and, 1048–50
Predators, **612.** *See also* Predation
  animal, 612
  deuterostomes as, 646f
  efficiency of, at reducing prey populations, 1065
  mollusk, 636
  noxious compounds in vacuoles and, 112
  predator-removal programs and, 1065
  sharks as, 662
  top, 1087
Predictions, 3, 12, 231, 235–39
Prefixes, metric system, B:1f
Pregnancy, 967–71
  birth events, 970–71
  events in human, 967–68
  fetal nourishment by mothers in, 968–70
  gestation and early development in marsupials, 967
  prenatal genetic testing and human, 350
  preventing human, 966
Prenatal testing, 350
Preservation
  biodiversity, 1120–23 (*see also* Conservation biology)
  fossil, 480
Pressure-flow hypothesis, **730**–31
Pressure potential, **719**–20
Pressure-sensing systems. *See* Hearing
Presynaptic neurons, **897**
Preventive medicine, 685–86
Prey, 1063–65. *See also* Predation
Prezygotic isolation, **459**
Priapula, 603t
Priapulids, 627
Pribnow, David, 291
Primary cell walls, 132–33, **555**
Primary consumers, 523, **1084**
Primary decomposers, **1085**
Primary endosymbiosis, 531–32
Primary growth, plant, **704**–6
Primary immune response, **989**
Primary meristems, 704–5
Primary oocytes, **953**
Primary phloem, 713
Primary plant bodies, 704–5. *See also* Bodies, plant

Primary producers, **522, 1083**
  ecosystem energy flow and, 1083–84
  green algae, 566, 568
  green plants as, 547
  protist, 541
Primary root system organization, 705–6
Primary sex determination, 936
Primary shoot system organization, 706
Primary spermatocytes, **953**
Primary structure, **46**
  DNA, 260–61
  DNA vs. RNA, 67t
  hemoglobin protein, 49t
  nucleic acid, 61
  protein, 46
  RNA, 66
  tRNA, 299
Primary succession, **1074**
Primary transcripts, **293**
Primary xylem, 713
Primases, 266, 268t
Primates, **669**
  characteristics of, 669
  fermentation in digestion by, 854
  great apes as hominid, 669–70
  humans as, 670–73f
  lineages of, 668–69
Primer annealing, 344–45
Primers, 266, 344–45
Primroses, 1050–51
Principle of independent
  assortment, **238**
  dihybrid crosses and, 236–38
  genetic variation from, 221–22
  linked genes as exception to, 243
  meiosis and, 240–41
Principle of segregation, **235**, 240
Prions, 50–**51**, 339
Probabilities, 1032, B:19
Probes, DNA, **342**–43, 353t, 370–71
Probes, nucleic acid and antibody, B:13
Proboscis, 563, 611, **627**
Procambium, **705**, 713, **796**
Process component
  chemical evolution theory, 15, 38
  of germ theory of disease, 499
  of scientific theories, 2, 414–15
  theory of evolution by natural
    selection, 4
  theory of special creation, 415
Processes, evolutionary. *See* Evolutionary
  processes
Processing, signal. *See* Signal processing,
  cell-cell
Producers, primary. *See* Primary
  producers
Productivity, **996**
  global patterns in, 1089–90
  ocean, 1001
  pyramid of, 1086 (*see also* Net pri-
    mary productivity (NPP))
  species richness and, 1080
Products, **27**
Progesterone, **964**–66
Programmed cell death, 375t, **376**. *See*
  also Apoptosis
Prokaryotes, **6**, 496–516
  asexual reproduction in, 223
  Bacteria and Archaea domains as,
    102–3, 496–97 (*see also* Archaea;

Archaea domain; Bacteria; Bacteria
  domain)
  bacteria that cause human
    illnesses, 499t
  biological impact of, 497–98
  biological methods of studying,
    501–4
  cell structures and functions of,
    102–5 (*see also* Prokaryotic cells)
  characteristics of eukarya and, 497t
  diversification themes of, 504–12
  ecological diversity of, and global
    change, 509–12
  environmental sequencing in, 364
  eukaryotes vs., 519
  as extremophiles, 501
  fermentation and, 167
  genomes of, 361–65
  key lineages of, 512–16
  lateral gene transfer in, 364
  medical importance of, 498–500
  metabolic diversity of, 506–9
  morphological diversity of, 504–6
  natural history of genomes of, 363–64
  phylogenetic tree of life and, 6–7
  reasons for studying, 497–501
  role of, in bioremediation, 500–501
  six examples of metabolic diversity
    in, 507t
  six general methods of, for obtaining
    energy and carbon-carbon
    bonds, 506t
  some electron donors and acceptors
    of, 508t
Prokaryotic cells, 102–5. *See also* Cells
  bacterial cell-cell communication as
    quorum sensing in, 145–46
  bacterial flagella vs. eukaryotic
    flagella, 127
  cell walls as exoskeletons of, 104–5
  chromosomes in nucleoids of, 103–4
  cytoskeletons of, 104
  eukaryotic cells vs., 105–6f, 107t
  flagella and swimming of, 104
  improvements in microscopy
    and, 103
  internal membrane complexes of
    photosynthetic, 104
  organelles in, 104
  plasma membranes of, 104
  protein manufacture by ribosomes
    of, 104
  quorum sensing as cell-cell commu-
    nication in, 145–46
  views of, 103f, 105f
Prolactin, **943**
Proliferation, cell, 375
Proline, 42t, 47, 283
Prometaphase, 198–**99**
Promoter-proximal elements, 323–**24**
Promoters, **291**, 323
  catabolite repression and, 315
  eukaryotic, 320
  regulatory sequences far from, 324–26
  regulatory sequences near, 323–24
  transcription and, 291–92
  in transcription in bacteria and
    eurkaryotes, 296t
Proofreading, DNA, 271–**72**
Propagation, plant, 708

Prophase, **198**
Prophase, mitosis, 198
Prophase I and II, meiosis, 218–20
Prop roots, 698, 699f
Propulsion, seed dispersal by, 798
Prosimians, **668**–69
Prostaglandins, 140t, 959t
Prostate glands, **958**, 959t
Protease inhibitors, 682–83
Proteases, 118, **391**, **682**, 852
Proteasomes, 204, **330**
Protected areas, 1055, 1120
Proteinase inhibitors, **779**
Protein filaments, 104, 113
Protein kinases, **142**, **204**
Proteins, 38–58
  adhesion, 136–38
  animal digestion of, 846, 848, 851–52
  as animal nutrient, 842
  animal stomach digestion of, 849
  artificial selection of maize for pro-
    tein content, 4–5
  basal transcription factors, 291–92
  betaβ-globin, 247
  carbohydrates and glycoproteins, 77
  catabolic pathways and, 168–69
  catalysis by enzymes as, 45, 51–56 (*see*
    also Catalysis; Enzymes)
  cell-cycle regulation and, 204–5
  cell functions of, 45
  in central dogma, 280–81
  channel (*see* Channel proteins)
  chemical evolution in prebiotic soup
    and formation of, 38–39
  cytoskeleton, 104, 113
  defined, **42**, **44**
  destruction of, as regulation
    mechanism, 204
  early origin-of-life experiments on
    chemical evolution of, 39–40
  eukaryotic genomes and genes for
    coding, 363
  filament, 125
  as first self-replicating molecules, 56
  folding of, 50–51
  folding of ribonuclease, 50
  four levels of structure of
    hemoglobin, 49t
  genes and, 195–96
  immunoglobulins (Igs), 980, 982–83f
  manufacture and transport of, by en-
    domembrane system, 118–23f
  manufacture of, by prokaryotic
    ribosomes, 104
  membrane (*see* Membrane proteins)
  microtubules, vesicle transport, and
    motor, 126–27
  motor, 45, 124, 125–27, 128
  movement of, from ER to Golgi to
    destination, 122–23f
  one-gene, one-polypeptide
    hypothesis on, 278
  organic molecules and, 36
  phosphorylation of, by ATP, 149–50
  plant embryogenesis and, that set up
    body axes, 406
  polymerization of amino acids to
    form, 39
  polymerization of amino acids to
    form polypeptides and, 40–44 (*see*
    also Amino acids; Polypeptides)
  primary structure of, 46
  prions as infectious, 50–51
  proteomics as study of, 371
  proton pump, 733
  quaternary structure of, 48–50

  regulatory, in differential gene
    expression, 326
  regulatory, in eukaryotic transcrip-
    tion initiation complex, 326–28
  repressor, 312–14
  required for DNA synthesis, 268t
  ribosomes and, 109
  RNA polymerase for, 290
  rough ER and, 108
  secondary structure of, 46–47
  selective adhesion and adhesion,
    131f, 136–38
  sigma, 291–92
  signal hypothesis on entry of, into en-
    domembrane system, 120–21
  signal transduction via G proteins,
    142–43
  structure of, 45–51
  synthesis of (*see* Protein synthesis)
  targeted destruction of, 330
  tertiary structure of, 47–48
  in transcription in bacteria and
    eurkaryotes, 296t
  transfer of amino acids to, by transfer
    RNA, 297–99
  transport (*see* Transport proteins)
  transport of molecules into nucleus
    and, 117–18
  tumor suppressors, 208
  in vegetative development that deter-
    mine leaf shape, 408
  viral, 117–18
Protein synthesis, 289–306. *See also*
  Genetics
  central dogma of molecular biology
    on, 279–82
  electron micrograph of
    transcription, 289f
  gene expression and, 289 (*see also*
    Gene expression)
  genetic code and, 282–85
  ribosomes in, 300–304
  RNA processing in eukaryotic,
    293–95
  transcription in, 289–93
  transfer RNA in, 297–300
  translation in, 295–97
Proteobacteria, 509, 512, 515
Proteomes, **371**
Proteomics, **371**. *See also* Genomics
Protists, **519**–45
  animals and, 607–8, 617
  biological methods for studying,
    524–26
  diversification themes in, 526–36
  ecological importance of, 522–23
  Eukarya domain and, 519–20 (*see also*
    Eukaryotes)
  feeding and photosynthesis innova-
    tions of, 529–32
  human health problems caused
    by, 522t
  impacts of, on human health and
    welfare, 520–22
  key lineages of, 536–43
  life cycles of, 533–36
  major lineages of eukaryotes
    and, 524t
  morphological innovations
    of, 526–29
  movement of, 532–33
  pigments in photosynthetic, 532t
  reasons for studying, 520–23
  reproduction of, 533
Proto-cells, 82
Protoderm, **705**, **796**

Boldface page numbers indicate a glossary entry; page numbers followed by an *f*
indicate a figure; page numbers followed by *t* indicate a table.

Proton gradients
and ATP production in chemiosmosis, 162–63
nutrient uptake and, 745–46
during photosynthesis, 180–81
Proton-motive force, **162**
Proton pumps, **733**, **745**, **761**
in cell elongation, 761–62
in guard cell closure, 772
location of, in translocation, 733
in photosynthesis, 187
in translocation, 732–33
Protons
acid-base reactions and, 25–26
atomic structure and, 16–17
Proto-oncogenes, **332**
Protostomes, 603t, **606**, 623–45. See also Animal(s)
abundance of, 623–24
body plan variations of, 626–27
ecdysozoan lineages of, 637–43
ecdysozoans, 625
evolution of, 624–25
feeding adaptations of, 628
lophotrochozoan lineages of, 630–36
lophotrochozoans, 624–25
movement adaptations of, 628–30
prominent orders of insects as, 640t–41t
reproduction adaptatoins of, 630
water-to-land transition of, 627–28
Proximal tubule, 832, **834–35**, 838t
Proximate causation, **952**, **1020**
Prusiner, Stanley, 50–51
Pseudogenes, **368**, 418, 444
Pseudomonas aeruginosa, 145
Pseudopodia, **530**, 532–33
Psilotopohyta, 572
P site, 300
Pteridophyta, 573
Pterosaurs, 657
Puberty, **936**, 961–62, 963t
Puffballs, 583, 593
Pulmonary arteries, **879**
Pulmonary circulation, **878**
Pulmonary veins, **879**
Pulse-chase experiments, **119**
on Calvin cycle, 185–86
on cell cycle gap phases, 196
on discontinuous replication, 266–67
on photosynthesis, 188
on ribosomes as protein synthesis site, 296
on secretory pathway, 119–20
Pumps, **98**, **732**. See also Sodium-potassium pump
active transport by, 97–99
in phloem loading, 732
Punnett, R. C., 248
Punnett squares, **235**
Hardy-Weinberg principle and, 436–37f
predicting genotypes and phenotypes with, 235–39
Pupae (pupa), **616**, 936
Pupil, **914**
Pure lines, **231**
crosses between, 234–36
in Mendelian genetics, 239t
Purifying selection, 224, **440**
Purines, **60**, 279
chemical evolution and, 61
DNA and, 62–64
nucleotides and, 60
RNA and, 66–67
Purple nonsulfur bacteria, 180

Purple sulfur bacteria, 173, 180
Pyramid of productivity, 1086
Pyrethroids, 775
Pyridoxine, 277
Pyrimidines, **60**
chemical evolution and, 60–61
DNA and, 63–64
nucleotides and, 60
RNA and, 66–67
Pyrosequencing technologies
dideoxy sequencing vs., 347
impact of, on whole-genome sequencing, 360–61
Pyruvate
in cellular respiration, 153
in fermentation, 166
processing glucose to, 155–56
processing of, to acetyl CoA, 156–58
Pyruvate dehydrogenase, **157**

Q

Quadrats, 1048
Quantitative methods. See also Biological methods
calculating coefficients of relatedness, 1032
calculating equilibrium potentials with Nernst equation, 888
calculating population growth rates using life tables, 1040
developing and applying population growth equations, 1043–44
extrapolation techniques, 1108
linkage and, 246f
mark-recapture studies, 1048
measuring species diversity, 1079
population viability analysis, 1115
Quantitative traits, **248–50**t
Quaternary structure, **48**
DNA vs. RNA, 67t
hemoglobin protein, 49t
protein, 48–50
RNA, 67
Quinones, 161–62, 163–64
Quorum sensing, bacterial, **145–46**

R

Rabies virus, 692
Racker, Efraim, 164
Radial symmetry, **604**
Radiation, **816**
cancer and, 332–33
creating mutants with, 277
DNA damage and, 272–74
electromagnetic, 174–75
sunlight (see Light energy)
Radicles, **796**
Radioactive decay, 417, 481
Radioactive isotopes, B:12
Radiometric dating, 417, 481
Radula, 612, **635**
Rain forests. See Tropical wet forests
Rain shadow, **1010**
Ramón y Cajal, Santiago, 886–87
Rancher ants, 602
Random mutations
DNA and, 283–84
RNA and, 69
Ranges, **1013**, 1072, 1120. See also Geographic distribution of organisms
Rats, 369t
Ray-finned fishes, 656, 662–63

Rayment, Ivan, 924
Rays, 662, **713**
Rb protein, **208**
Reabsorption mechanisms, 831, 834–35
Reactants, 27, 30–32
Reaction centers, **178–79**. See also Photosystem I and II
Reactions, chemical. See Chemical reactions
Reactive oxygen intermediates (ROIs), **777**
Reactivity, amino acid side chain, 40–42
Reading frame, **283**, 362
Realized niches, **1061**
Real-time processes, B:15
Receptacle, plant flower, 788
Receptors
animal B-cell and T-cell, 980–81f
animal chemical signal, 930–31
animal hormone, 943–47
animal sensory, 817–18, 886, 908
cell-cell signal, 140, 143–44
plant sensory, 756, 757f, 758–59, 763
Receptor tyrosine kinases (RTKs), **143–44**, 207
Recessive alleles
loss-of-function mutations and, 451
in Mendelian genetics, 239t
Recessive traits, **233**
autosomal diseases, 272–74
identifying human traits as, 250–51
monohybrid crosses and, 232–33
Reciprocal altruism, **1033–34**
Reciprocal crosses, **234**
in Mendelian genetics, 239t
monohybrid crosses and, 233–34
results of, 234t
sex linkage and, 241–42
Recolonization, metapopulation, 1046–47
Recombinant, **244**
Recombinant DNA technology, **338–44**. See also Biotechnology; Genetic engineering
creating animal models of disease using, 350
discovery of gene recombination, 982–83f
ethical concerns over recombinant human growth hormone, 343–44
failure of efforts to treat pituitary dwarfism and, 339
gene therapy and, 351–54
human growth hormone in pituitary dwarfism and, 339
introducing recombinant plasmids into bacterial cells for transformation, 341
mass producing human growth hormone, 343
producing a cDNA library, 342
screening a DNA library, 342–43
steps in engineering human growth hormone, 339–43
transgenic crops and, 354–56
uses of, in genetic engineering, 353t
using plasmids in DNA cloning, 340, 341f
using restriction endonucleases and DNA ligase to cut and paste DNA, 340–41
using reverse transcriptase to produce cDNAs, 339–40
Recombination frequencies, calculating, 246

Rayment, Ivan, 924
Recommended Dietary Allowances (RDAs), 842
Rectal gland, **826**
Rectum, 846f, **855**
Red algae, 519f, 536, 539. See also Algae
Red blood cell ghosts, 96–97
Red blood cells, **870**
Red/far-red light responses, plant, 762–74
germination and, 763t
isolation of phytochromes of, 763–64
overview of, 780t
photosynthesis and, 762
phytochromes as receptors in, 763
red/far-red switch in, 763
Red-green color blindness, 916–17
Red light, 174–75, 176, 799. See also Red/far-red light responses, plant
Redox reactions, **151**
hydrogen, oxygen, and, 151–52
oxidation of glucose and, 152
in photosynthesis, 178–79
photosynthesis and, 151
prokaryotic nitrogen fixation and, 510
Reduction, **151**
Reduction division, meiosis as, 215–16
Reductionism, 899
Reduction-oxidation reactions. See Redox reactions
Reduction phase, Calvin cycle, 186
Redundancy
genetic code, 283
genome, 363
Redwoods, 575
Reflexes, **848**
Refractory, **893**
Regeneration phase, Calvin cycle, 186
Regulation. See also Hormones
of animal body temperature, 816–19
of animal homeostasis, 814–16
of animal hormone production, 940–43
of animal pancreatic enzymes, 852
of blood pressure and blood flow, 882–83
cancer as loss of cell-cycle, 206–9
of cell cycle, 202–6
of citric acid cycle, 158–59f
of enzymes by regulatory molecules, 54–55
of flower structures, 409–11
of gene expression (see Gene regulation, bacterial; Gene regulation, eukaryotic)
of glucose in glycolysis, 156
homeostatic, of ventilation, 870
kidney, 832
of mammalian menstrual cycle, 962–66
of mammalian puberty, 961–62
membrane channel, 96
of photosynthesis, 189
plant growth (see Growth regulators, plant)
regulatory genes and (see Regulatory genes)
regulatory proteins and (see Regulatory proteins)
regulatory sequences and (see Regulatory sequences)
role of, in homeostasis, 815–16 (see also Homeostasis)
of transcription, 291
Regulatory cascades, 382–83, 411

Regulatory genes, **369**
  bicoid gene as master regulator, 379–80
  for cell determination, 398–99
  evolutionary conservation of, 383
  reuse of, in differing developmental contexts, 383–84
  specific developmental positional information and, 381–83
Regulatory molecules
  cell-cycle, 203–4
  enzyme regulation by, 54–55
Regulatory proteins
  in bacterial gene regulation, 308
  in differential gene expression, 326
  in eukaryotic transcription initiation complex, 326–28
  intiation complex for, 326–28
  lac operon model for repressors, 312–14
  in transcription, 291–92
Regulatory sequences, **323–26, 369**
  transcription initiation and, far from promoters, 324–26
  transcription initiation and, near promoters, 323–24
Regulatory sites, **156,** 314
Regulatory transcription factors, **326, 378**
  bicoid gene as, 379–80
  cancer and, 332–33
  human and chimpanzee genome similarity and, 369–70
Reinforcement, **468,** 471*t*
Reintroduction programs, 1122
Relative dating, 422
Release factors, 302–3
Religious faith vs. science, 8–9
Renal arteries, 832, 833
Renal corpuscle, **832**
  filtration by, 832–34
  structure and function of, 838*t*
Renal veins, 832
Reovirus, 691
Repair, DNA, 271–74
Repeated sequences, 365–67
Repetition, scientific experiments and, 12
Replacement mutations, **286**
Replacement rate, **1053**
Replica plating, **310**
Replica plating, finding mutant genes with, 310–11
Replicated chromosomes, 213–14*t*
Replication. *See also* Reproduction; Self-replicating molecules
  chromosome, 195–96
  DNA, 65, 261–63 (*see also* DNA synthesis)
  living organisms and, 2
  RNA, 67–68
  viral, 679–86
Replication bubble, 264
Replication fork, **264–65**
Replicative growth, viral, 679–81
Replisomes, **268**
Repolarization, **890,** 891
Repressors, **312**
  discovery of, as negative control mechanisms, 312–13

lac operon regulation and, 313–14, 317*f*
Reproduction. *See also* Replication
  animal (*see* Animal reproduction)
  cell cycle and, 194 (*see also* Cell cycle)
  fitness, adaptations, and, 5, 424, 426
  fungal, 583–84, 590–91
  mating and (*see* Mating)
  mitosis and, 195
  natural selection and, 4–5, 423
  plant (*see* Plant reproduction)
  protist, 533
  replication of living organisms as, 2
Reproductive development, plant, 409–11
Reproductive isolation
  allopatric speciation and, 462–64
  mechanisms of, 459
  sympatric speciation and, 464–68
Reptiles (Reptilia), 658, **660–61,** 666–68
Requirements, nutritional, 842–43
Residues, amino acids as, 43–44
Resilience, **1119**
Resistance, **1119**
Resisteance (*R*) genes, **776**
Resonance, 178–79
Resource allocation, life histories based on, 1041
Resource use
  efficiency in, 1119
  human, 1098, 1121
Respiration, cellular. *See* Cellular respiration
Respiratory systems, **862**
  air and water as media in, 862–64
  Fick's law of diffusion and, 864–65
  fish gills in, 865–66
  four steps of gas exchange and, 861–62
  homeostatic control of ventilation in, 870
  insect trachaea in, 866–67
  organs of, 864–70
  pulmonary circulation and, 879*f* (*see also* Circulatory systems)
  vertebrate lungs in, 867–70
Responder cells, plant, 756, 757
Resting potentials, **888–90**
Restoration, ecosystem, 1122–23
Restriction endonucleases, **340**
  recognition sites for, 349
  uses of, in genetic engineering, 352*t*
  using, to cut and paste DNA, 340–41
Restriction sites, 349
Reticulum, 850
Retina, **914**
Retinal, **915**
Retinoblastoma, 208
Retroviruses, **351–54,** 684, 687*t*, 691
Reverse transcriptase, **281,** 684
  gene therapy and, 351
  producing cDNAs with, 339–40
  uses of, in genetic engineering, 352*t*
Reverse transcriptase inhibitors, 684
R-groups, amino acid. *See* Side chains, amino acid
Rheumatoid arthritis, 983
Rhizaria, 524*t,* 536–37, 540
Rhizobia, 515, **749–50**
Rhizoids, **567**

Rhizomes, **572, 701, 786**
Rhizopus stolonifer, 595
Rhodophyta (red algae), 539
Rhodopsin, 364, **915–16**
Rhynie Chert, 554
Rhythm contraceptive method, 966*t*
Ribbon diagrams, 47
Ribbon worms, 627
Ribonuclease, 50
Ribonucleic acid (RNA). *See* RNA (ribonucleic acid)
Ribonucleoside triphosphates, 68
Ribonucleotides, **60**
  anabolic synthesis of, 169
  components of, 60
  diffusion of, across lipid bilayers, 91
  phylogenetic tree of life and, 6–7
Ribonucleotide triphosphate (NTP), 290, 292
Ribose, 60–61, 66, 72, 77
Ribose-5-phosphate, 169
Ribosomal RNAs (rRNAs), **117, 300**
  analyzing, to create phylogenetic tree of life, 6–7
  phylogenies based on, 503–4
Ribosome binding site, **301**
Ribosomes, **104, 296**
  elongation of polypeptides during translation and, 301–2
  eukaryotic, 109, 114*t*
  formation of, 117
  initiation of translation and, 301
  mitochondria and, 112
  post-translational modifications and, 304
  prokaryotic, 104
  in protein synthesis, 300–304
  ribosomal RNA and structure of, 300–301
  as ribozymes, 302
  rough ER and, 108
  as site of protein synthesis in translation, 295–96
  termination of translation and, 302–3
  translocation and, 302
Ribozymes, **68**
  lipid bilayers and, 91
  as RNA catalytic molecules, 68
  RNA replicase as, 68–69
Ribulose-1,5-bisphosphate carboxylase/oxygenase. *See* Rubisco
Ribulose bisphosphate (RuBP), **185–86**
Rice
  genetically engineered, 338*f*
  genomes of, 362
  number of genes in genome of, 369*t*
  synthesizing β-carotene in, 355
  as target crop for agricultural biotechnology, 355
  transgenic (*see* Golden rice)
Richness, species. *See* Species richness
Rifampin, 424–25
Rings, tree growth, 714–15
Ring structures, 33*f,* 72
Ripening, fruit, 773–74
Rivers, 999
RNA (ribonucleic acid), **61.** *See also* Nucleic acids
  anabolic synthesis of nucleotides for, 169
  as catalytic molecule, 68
  in central dogma, 280–81
  directionality of strands of, 61
  DNA structure vs. structure of, 67*t*
  DNA synthesis from template of, 270–71

electron micrograph of, 289*f*
  as first self-replicating molecule, 59, 68–69
  genes and, 195–96
  as genetic information-containing molecule, 67–68
  lipids and, 82
  messenger (*see* Messenger RNAs (mRNAs))
  nucleolus and eukaryotic, 108
  phylogenetic tree of life and, 6–7
  ribonuclease enzyme for unfolding, 50
  structure of, 66–67
  sugar-phosphate backbone of, 61*f*
  synthesis of, 279–80
  synthesis of, using codons, 283–85
  transfer (*see* Transfer RNAs (tRNAs))
  types of, 281
  versatility of, 67
  viral, 686
  viruses and, 682, 686, 687*t*
RNA-induced silencing complex (RISC), 329–30
RNA interference, **329–30**
RNA polymerases, **266, 279**
  characteristics of, 290
  eukaryotic, 290*t*
  messenger RNA hypothesis and, 279–80
  regulation of, 308
  in transcription in bacteria and eurkaryotes, 296*t*
RNA processing, **295, 320.** *See also* Protein synthesis
  adding caps and tails to transcripts in, 295
  bacterial vs. eukaryotic gene expression regulation and, 331*t*
  discovery of eukaryotic genes in pieces and, 293–94
  in eukaryotes, 293–95
  RNA splicing in, 294–95
  in transcription in bacteria and eurkaryotes, 296*t*
RNA replicase, **684**
RNA reverse-transcribing viruses. *See* Retroviruses
RNA splicing. *See* Alternative splicing; Splicing, RNA
RNA world hypothesis, 59, 68–69, 302, 687–88
Roberts, Richard, 294
Rods, **915–16**
Root apical meristem (RAM), **405**
Root caps, **705–6,** 764–66
Root cortex, 722
Root hairs, **706, 722,** 744–48
Root pressure, 721, **723**
Roots, **405,** 571
  auxin and adventitious, 768
  colonization of, by nitrogen-fixing bacteria, 749–50
  modified, 698–99
  plant nutrition and, 737*f*
  seedless vascular plants and, 571
Roots, Latin and Greek word, 8, B:6
Root systems, **696**
  characteristics of, 697
  holding of soil by, 547
  modified roots in, 698–99
  morphological diversity of, 697–98
  nutrient uptake by, 744–48
  organization of primary, 705–6
  phenotypic plasticity in, 698
  water transport in, 721–23

Boldface page numbers indicate a glossary entry; page numbers followed by an *f* indicate a figure; page numbers followed by *t* indicate a table.

Rotavirus, 679f, 691
Rotifera (Rotifers), 603t, 631
Rough ER (rough endoplasmic
    reticulum), **108**
    endomembrane system and proteins
        in, 118–22
    eukaryotic, 108, 109, 114t
    extracellular matrix components
        in, 134
    pectins and, 132
    post-translational modifications
        in, 304
Roundworms, 211, 603t, 605–6, **638**,
    973f, 1067–68. See also
    Caenorhabditis elegans
Rous sarcoma virus, 691
Rubisco, **186**, **728**
    in C₄ and C₃ photosynthesis, 188
    discovery of, 186–87
    photosynthesis regulation and, 189
Rudder, 292
Rumen, 167, 850
Ruminants, **850**
    fermentation in rumen of, 167
    stomachs of, 850–51
Rusts, 596

S

Saccharomyces cerevisiae
    alcohol fermentation with, 167
    chromatin remodeling in, 322
    economic value of, 581, 598
    gene expression in, 319
    genome of, 362, 369t
    meaning of scientific name of, 8
    as model organism, B:20, B:21f, B:22
Sagittal crest, 671
Sago palms, 574
Saharan desert ant, 10–12
St. Martin, Alexis, 849
Sakmann, Bert, 892
Salamanders, 664
Saliva, 846, 848
Salivary amylase, 45, 855t
Salivary glands, 810f, 846f, **847**, 933
Salmon, 813, 827–28, 994–95, 1027
Salmonella, 979
Salps, 652
Salt
    ionic bonds of table, 18, 19f
    osmoregulation of (see
        Osmoregulation)
    shark excretion of, 826–27
Saltiness taste reception, 918
Salt marshes, 461–62
Salty habitats
    euryarchaeota bacteria and, 516
    plant adaptations to, 720–21, 747–48
Sampling effects, 1119
Sampling error, **443**, 445–46
Sand, 744t
Sandbox tree, 798
Sand dollars, 650
Sandwich model, plasma membrane, 92
Sanger, Frederick, 46, 346–47
Saprophytes, **580**
    carbon cycle on land and, 580–81
    fungal, 589–90, 596
Sapwood, **714**
Sarcomeres, **922–23**
Sarcoplasmic reticulum, **926**
SARS virus, 224
Satiation hormone, 939–40
Satin bowerbird, 1019f
Saturated lipids, 87–88, 355

Saturation kinetics, 55
Saxitoxins, 522
Scaffold proteins, 321
Scale of nature, 415
Scales, 662
Scallops, 634
Scanning electron microscopy (SEM).
    See also Electron microscopy
    of DNA and RNA, 289f
    of DNA synthesis, 258f
    of *Escherichia coli*, 307f
    of human fertilization, 211f
    of human immune system cells, 973f
    overview of, 93, B:14f, B:15
Scarified, **799**
Scent response, fly, 465
Scents, flower, 562
Scheepers, Lue, 9–10
Schistosomiasis, 631–32
Schwann cells, **894**
Science
    biology and nature of, 8–9 (*see also*
        Biology)
    evidence in, 528
    experiments as hypothesis testing
        in, 8–12
    pattern and process components of
        theories of, 2, 414–15
    reductionism in, 899
    taxonomy and, 7–8
Scientific names, **8**, 675
Sclera, 914
Sclereids, **710**, 711
Sclerenchyma cells, **709–10**, 711t
Scolex, 613
Scorpions, 642
Scott, J. C., 1107
Scrapie, 51
Screening, DNA library
    uses of, in genetic engineering, 353t
    using DNA probes for, 342–43
Screens, genetic. *See* Genetic screens
Scrotum, **957**
Sea bass, 827–28
Seas. *See* Oceans
Seaside sparrows, 461–62
Seasons
    causes of, 1010, 1011f
    global warming and, 1100
    migration and, 1027
    phenology and, 1100
    plant translocation and, 728
    seed dormancy and, 798–99
Sea squirts, 651, 652
Seas slugs, 903–4f
Sea stars, 414f, 649, 1072–73
Sea urchins, 211, 390–91, 650
Seawater. *See* Marine ecosystems
Secondary active transport
    (cotransport), **98**, **732**, **825**
Secondary cell walls, **133**, **555**, **709**, 726
Secondary consumers, 523, **1084**
Secondary endosymbiosis, 531–32
Secondary growth, plant, **712–15**
Secondary immune response, **989**
Secondary metabolites, **776**
Secondary phloem, 713
Secondary spermatocytes, **953**
Secondary structure, **46**
    DNA, 59f, 62–65, 260–61
    DNA vs. RNA, 67t
    hemoglobin protein, 49t
    protein, 46–47
    RNA, 66–67
    tRNA, 299
Secondary succession, **1074**

Secondary xylem, 713
Second law of thermodynamics, **29**
Second-male advantage, **954**–55
Second messengers, **142**–43f, **756**,
    897, **945**–46
Secretin, **852**, 933, **934**, 937
Secretions, barrier, 975
Secretory pathway, endomembrane
    system, 119–20
Secretory vesicles, 120
Sedimentary rocks, **416**
Sediments, fossils and, 480
Seed banks, **1117**
Seed coat, **795**
Seedless vascular plants, 550, 552, 559,
    567, 571–73
Seed plants, 550, 552, 559–61, 567,
    574–77f, 793
Seeds, 402, 550, 560, 783, 795–800
    dormancy of, 798–99
    embryogenesis and, 796
    fruit development and dispersal
        of, 797–98
    germination of, 799–800
    gibberellins and abscisic acid in dor-
        mancy and germination of, 770–71
    land plants and, 560–61
    red/far-red light reception and germi-
        nation of, 762–64
    red/far-red light reception and germi-
        nation of lettuce, 763t
    role of drying in maturation of, 797
    sclereids and, 710
    seed coat and, 795
    seed plants and, 567
    vacuoles in, 112
Segmentation, **609**
Segmentation genes, **381**–83
Segmented worms, 627, 633
Segment polarity genes, 382
Segments, **381**
Segregation, principle of. *See* Principle
    of segregation
Selection, directional, 440–41
Selection, natural. *See* Natural selection
Selective adhesion, **136**–38
Selectively permeable membranes, **823**
Selective permeability, **86**, 90
Selective reabsorption, 834–35
Selectivity
    membrane channel, 95, 96f
    plasma membrane, 82
Self-fertilization (selfing), 222, 231, 451,
    468, 620, **792**
Selfish genes, 365, 431
Self molecule, **983**–84
Self-nonself recognition, 978
Self-replicating molecules. *See also*
    Origin-of-life research
    DNA as, 65–66
    first cell membrane and, 82
    lipid bilayers and, 91
    origin-of-life with, 39
    protein catalysts (enzymes) as first, 56
    RNA world hypothesis on RNA as
        first, 59, 68–69
Self-sacrificing behavior. *See* Altruism
Semen, human, 959t
Semiconservative replication,
    DNA, 261–63
Seminal vesicles, **958**, 959t
Senescence, 773–74
Sensors, **815**
Sensory neurons, **886**

Sensory receptors, 886, 938
Sensory systems, animal, 907–28
    hearing, 909–13
    information transmission to brains
        in, 909
    moth pheromones and, 907
    movement (locomotion) as response
        to, 907, 920–26 (*see also* Locomo-
        tion, animal)
    processes and receptors of, 908
    sensory transduction in, 908–9
    smell (olfaction), 919–20
    taste, 918–19
    variations in abilities and structures
        in sensory organs of, 610
    vision, 913–18
Sensory systems, plant, 755–82
    defense responses to pathogens and
        herbivores, 775–81
    germination and stem elongation re-
        sponses to red and far-red light,
        762–64
    gravitropic responses to gravity,
        764–66
    growth regulators, 775t
    growth responses, 767–76
    information processing and, 756–57
    phototropic responses to blue light,
        758–62
    responses to wind and touch, 766–67
    selected, 780t
Sensory transduction, 908–9
Sepals, **409**, **562**, **788**
Septa (septum), **583**
Sequences. *See also* DNA sequencing;
    Whole-genome sequencing
    amino acid, 43–44
    DNA, 65
    DNA, and genetic homology, 420–21
    identifying genes in genome, 362–63
    nucleic acid, 61
    phylogenetic tree of life and
        rRNA, 6–7
    proteins as amino acid, 46
Sequencing-by-synthesis, 347
Sequencing technology, 346–47, 1106
Serine, 42t
Serotonin, 898t, **903**–4t
Serum, **207**
Sessile, **542**, **607**–8, 611
Set point, **815**, 940
Severe combined immunodeficiency
    (SCID), 352–54, 990
Sex chromosomes, 212, 214t,
    226–27, 241
Sex hormones, 960–66
    contraception methods and, 966t
    developmental processes and, 935–36
    menstrual cycle control by, 962–66
    puberty control by, 961–62, 963t
    sexual activity of *Anolis* lizards
        and, 1023
    testosterone, and estradiol as, 960–61
Sex linkage
    chromosome theory of inheritance
        and, 241–42
    exceptions and extensions to Gregor
        Mendel's rules on, 250t
    identifying human traits as sex-
        linked, 251–52
Sex-linked inheritance (sex linkage),
    242, 251–52
Sex ratios, estrogen-mimics and,
    1088–89
Sexual activity, lizard, 1023–24
Sexual competition, giraffe necks and, 10

Sexual dimorphism, **455**
Sexually transmitted diseases, 364, 513, 514, 538, 685. *See also* HIV (human immunodeficiency virus)
Sexual reproduction, **220–21, 784, 950**
  animal, 615
  animal switching between asexual and, 951–52
  asexual reproduction vs., 220–21, 223, 533, 950–51
  changing-environment hypothesis on, 224–25
  chromosomes, heredity, and, 221
  effects of fertilization on genetic variation in, 222–23
  in eukaryotes, 519
  fungal, 590–94
  genetic recombination in, 222
  genetic variation from independent assortment in, 221–22
  hormones in vertebrate, 936
  human fertilization micrograph, 211*f*
  mechanisms of, and gametogenesis, 952–53
  meiosis and, 194, 211, 220–25 (*see also* Meiosis)
  mistakes in, 225–27
  paradox of, 223–25
  plant, 784–86
  protist, 533
  protostome, 630
  purifying selection hypothesis on, 224
Sexual selection, **452–55**
Shade, far-red light as, 762
Shade leaves, 703
Shannon index of species diversity, 1079
Shape
  flowers, 562
  genes and proteins in vegetative development that determine leaf, 408
  hormone receptor, 140
  organic molecule, 33*f*
  plant receptor cell, 756
  prokaryotic diversity in, 504, 505*f*
  protein, 45–51, 304
  retinal change in, 915
  shoot system, 699–700
  simple molecule, 20
Sharks, 480, 662, 826–27
Sharp, Phillip, 293–94
Sheep, 377–78
Shells
  eggs, 658
  mollusk, 634
  protist, **528–29**
  turtle, 667
Shine-Dalgarno sequence, **301**
Shoot apical meristem (SAM), **405**, 407, 409
Shoots, **405**
Shoot systems, **696**
  adventitious roots and, 698–99
  characteristics of, 699
  collenchyma cells and shoot support in, 708–9
  formation of, 405
  modified, 700–701
  morphological diversity of, 699–700
  organization of primary, 706
  phenotypic plasticity in, 700

water transport in, 723–27
  water transport via capillary action in, 723–24
  water transport via cohesion-tension force in, 724–27
Short-day plants, **787**
Short interspersed nuclear elements. *See* SINEs (short interspersed nuclear elements)
Short tandem repeats (STRs), 365–67
Shotgun sequencing, **360**, 361*f*
Shrimp, 643
Siberian traps, 489
Sickle-cell disease, 46, 251
Side chains, amino acid
  amphipathic proteins and, 92
  enzyme function and, 53–54
  functional groups and reactivity of, 40–42
  major, 41*f*
  peptide bonds and, 43–44
  polarity of, and water solubility, 42
  protein primary structure and, 46
  protein tertiary structure and, 47–48
  of steroids, 84
Sieve plates, **711**, 729–30
Sieve-tube members, **711**, 729–30, 732–33
Sigma proteins, 291–92
Signal amplification, 141–42
Signal deactivation, 144
Signal hypothesis, endomembrane system, 120–21
Signaling, cell-cell. *See* Cell-cell signaling
Signaling pathways, animal hormone, 931–32
Signal processing, cell-cell, 140–44
  lipid-soluble vs. lipid-insoluble signals and signal transduction in, 140–41
  second messengers in signal transduction, 143*t*
  signal amplification in, 141–42
  signal transduction via enzyme-linked receptors in, 143–44
  signal transduction via G proteins in, 142–43
Signal receptors, **140**
Signal recognition particle (SRP), **121**
Signals, animal. *See* Chemical signals, animal; Electrical signals, animal
Signals, communication, **1027**
Signals, plant. *See* Sensory systems, plant
Signals, social, 205
Signal transduction, **141, 756, 945**
  in hair cells, 910
  in hormone binding to cell-surface receptors, 945–47
  of lipid-insoluble signals, 141
  in plant cells, 756, 757*f*
  via enzyme-linked receptors, 143–44
  via G proteins, 142–43
Signal transduction cascades, **946**
Signal transduction pathways, 141, 145, 756, 757*f*
Silencers, **324–26**
Silent mutations, **286**, 431, 444
Silurian-Devonian explosion, green plants in, 550–51
Silverswords. *See* Hawaiian silverswords

Simian immunodeficiency viruses (SIVs), 688
Simmons, Robert, 9–10
Simple eyes, **637**
Simple fruits, **797**
Simple leaves, 408, 701
Simple molecules, 19–20. *See also* Molecules
Simple sequence repeats, 365–66
Simple sugars. *See* Monosaccharides; Sugars
Simple tissues, plant, 710
Simulation studies
  of genetic drift, 443–45
  of global climate change, 1011
SINEs (short interspersed nuclear elements), **478–79**
Singer, S. Jon, 93
Single bonds, 20
Single nucleotide polymorphisms (SNPs), **349**, 352*t*
Single-strand DNA-binding proteins (SSBPs), **264**, 268*t*
Single-trait experiments, Gregor Mendel's, 232–36
Sinks, plant sugar, **728**
  connections between sources and, 728–29
  low pressure near, 730–31
Sinoatrial (SA) node, 880–81
Siphons, 630, 634, 652
Sister chromatids, **196, 213**
  chromosomes and, 213–14*t*
  meiosis mistakes and, 225–26
  in mitosis, 199, 202*t*
Sister species, **464**
Sit-and-wait predators, 612, 627, 664
Site histories, 1073
Size
  cell-cycle checkpoints and cell, 205
  directional selection on cliff swallow body, 441
  disruptive selection on black-bellied seedcracker beak, 442
  eukaryotic genomes, 365
  human population, 1052–53
  island, 1078
  largest fungal mycelium, 582
  male penis, 958
  physiological effects of animal body, 811–14
  of prokaryotic and eukaryotic cells, 105–6*f*, 107*t*
  prokaryotic diversity in, 504, 505*f*
  prokaryotic genomes, 363, 365
  smallest eukaryotic genome, 594
  testes and ovaries, 1023
  variable, of testes, 955
  virus, 678
Skates, 662
Skeletal muscle, **808, 921**
Skeleton cells, 396–98
Skeletons, animal, 920–21
  cartilaginous, 662
  endoskeleton function, 920–21
  endoskeleton structure, 920
  hydrostatic, 605–6, 613, 626
  notochords and, 396
  types of, 920
Skills, biology. *See* BioSkills
Skin
  amphibian, 664
  as barrier to pathogens, 974
  cancer of, 274
  cells of, 131*f*, 396–98
  ectoderm and, 604

lymphocytes and, 978
  skin-breathing, 813
  thermoregulation and, 818–19
Skulls, 609, 653, 660, 671
Slack, Roger, 188
Sleeping sickness, 522*t*
Sliding clamps, 266, 268*t*
Sliding-filament model, **922–26**
Slugs, 635
Small interfering RNA (siRNA), 329–30
Small intestines, **851–54**
  carbohydrate digestion and transportation in, 852–53
  digestion in, 846
  fermentation in, 167
  folding and projections that increase surface area of, 851
  human, 810*f*, 846*f*
  lipid digestion and bile transport in, 853–54
  pancreatic enzyme regulation in, 852
  protein processing by pancreatic enzymes in, 851–52
  tight junctions in, 136
  water absorption in, 854
Small nuclear ribonucleoproteins (snRNPs), **294–95**
Smallpox, 686, 690, 973–74, 990
Smell, animal, 610, 919–20
Smith, John Maynard, 223–24
Smoke, seed dormancy and, 799
Smooth ER (smooth endoplasmic reticulum), **108**
  eukaryotic, 108, 114*t*
  in plasmodesmata, 138, 707
Smooth muscle, **808, 922**
Smut, 596
Snails, 224–25, 612, 635
Snakes, 384–85, 660, 666, 844, 848
Snapping shrimp, 464
Snowshoe hare, 1047–50
Social controls, cell-cycle, 205, 207, 208
Social stimulation, lizard sexual activity and, 1023–24
Sockeye salmon, 994–95, 1027
Sodium, action potentials and ions of, 891
Sodium channels, 891, 892
Sodium chloride (table salt), 18, 19*f*
Sodium poisoning, 747
Sodium-potassium pump, **98, 825**
  active transport and, 98–99
  in fish osmoregulation, 827
  resting potentials and, 889–90
  in shark salt excretion, 826–27
Soil, 741–44
  arbuscular mycorrhizal fungi (AMF) and, 589
  conservation of, 742
  effects of composition of, on properties of, 744*t*
  formation of, and textures of, 741–42
  green plant building and holding of, 547
  importance of conservation of, 742
  lichens and, 597
  nutrient availability in, 742–44*f*
  organic matter, **1092**
  plant adaptations to dry and salty, 720–21
  plant nutrients from, 739*t*
  plant nutrition and, 737, 738–40
  Sphagnum moss and, 570
  succession and, 1074
  water potential in, 720
Sokolowski, Maria, 1021

Solar energy, transformation of, into biomass, 1084–85
Solid water. See Ice
Solubility. See also Water
    amino acid, 42
    of gas in water, 863
    of hormones, 934
    of lipids, 83
    lipid-soluble vs. lipid-insoluble hormones, 139–41
Solute potential, **718**
    as negative, 719
    water movement with, 719*f*, 720
    water potential and, 718
Solutes, **22, 89, 718, 823**
    concentration of, and solubility of gases in water, 863
    diffusion and, 89
    movement of, into roots, 722–23
    transport of, across membranes in plants, 731–32
Solutions, **21, 718**
    chemical reactions and, 21
    oxygen and carbon dioxide amounts in, 863
    pH as indication of proton concentration in, 25
    pH scale as measure of acidity and alkalinity of, 26
Solvents, **22,** 26*t*
Soma, **887**
Somatic cells, **194, 271**
Somatic hypermutation, **989**
Somatic nervous system, **899**
Somatostatin, **930**
Somatotropin, 934
Somites, **396–98**
Song sparrows, 1045–46
Sonication, 360
Sonic hedgehog gene, 385
Soredia, 597
Sori, **573**
Sounds
    cochlea detection of frequencies of, 911–12
    middle ear amplification of, 911
Sources, plant sugar, **728**
    connections between sinks and, 728–29
    high pressure near, 730–31
    sugar concentration in sieve-tube members at, 732–33
Sourness taste reception, 918
Soybeans, 354
Space-filling models, **21,** B:8*f*
Spanish flu, 677
Spanish moss, 750
Spark-discharge experiment, 39–40, 61
Sparrows, 461–62
Spatial avoidance mechanisms, 792
Spawning, 954
Special creation, theory of, 414–15, 417
Specialization, 616
Speciation, **5, 458–73.** See also Evolution; Species
    allopatric, 462–64
    contemporary, 422
    current examples of, 422
    definition and identification of species and, 458–62
    disruptive selection and, 442
    by hybridization, 470–71
    mechanisms of reproductive isolation, 459*t*
    mechanisms of sympatric, 471*t*
    outcomes of secondary contract between isolated populations, 471*t*

pollination by animals and, in plants, 794
secondary contact between isolated populations and, 468–71
secondary contact between isolated species and, 471*t*
species concepts and species identification, 461*t*
species diversity and, 1107
of sunflowers, 458*f*
sympatric, 464–68
tree of life and, 5–6
Species, **459**
    biological species concept, 459–60
    changes of, through time as pattern component of evolution, 416–22, 423*f*
    concepts of, and identification of, 461*t*
    deceptive communication within and between, 1020
    definition and identification of, 458–62
    disruptive selection and formation of new, 442
    dusky seaside sparrow as species definition case, 461–62
    endangered (see Endangered species)
    evidence for descent of, from common ancestors, 418–22, 423*f*
    evidence for evolution of, 422*t*
    evolution and extinct, 417
    evolution of, by natural selection, 4–5 (see also Evolution; Natural selection; Speciation)
    exotic and invasive, 1014–15
    ex situ conservation of, 1121–22
    extinction of (see Extinction)
    global warming and loss of, 1098
    interactions among (see Species interactions)
    keystone, 1072–73
    life histories and patterns across, 1041
    morphospecies concept, 460
    phylogenetic species concept, 460–61
    protostome, 623–24
    reintroduction of, 1087
    richness and diversity of (see Biodiversity; Species diversity; Species richness)
    succession and role of traits in, 1074–76
    taxonomy and scientific names for, 8
    in theory of evolution by natural selection, 416
    traits of (see Traits)
Species-area relationships, **1116**
Species diversity, **1077, 1106.** See also Biodiversity
    arthropods and, 637
    current estimates of, 1107–8
    global climate change and, 1012–13*f*
    measuring, 1079
    of protists, 522
    species richness vs., 1077 (see also Species richness)
    threats to, 1110–14
    in tropical wet forests, 1003
Species interactions, 1058–70
    competition, 1059–63*f*
    consumption, 1063–68
    four types of, 1058–59, 1070*t*
    mutualisms, 1068–70
    role of, in succession, 1075, 1076–77
    three themes of, 1059

Species richness, **1077**–80. See also Biodiversity
    agrea and age hypothesis for, 1080
    community stability and, 1119–20
    energy hypothesis for, 1080
    global patterns in, 1078–80
    high-productivity hypothesis for, 1080
    increased net primary productivity of, 1118–19
    intermediate disturbance hypothesis for, 1080
    latitudinal gradient in, 1078–80
    measuring, 1079
    role of immigration and extinction in, 1077–78
    role of island size and isolation in, 1078
    species diversity vs. (see also Species diversity)
    theory of island biogeography on, 1077–78
Specific heat, **24, 1010–11**
    climate and, 1010–11
    of some liquids, 24*t*
    of water, 24, 26*t*
Specificity
    adaptive immune response, 977
    antibody, 981–82
Sperm, **211, 784, 951**
    human, 211*f*
    meiosis and, 194, 218
    plant, 402, 789, 791 (see also Pollen grains)
    reasons for fertilization by single, 391–92
    sexual selection and, 452–53
    structure and function of, 389–90
Spermatids, **953**
Spermatogenesis, **953, 958**
Spermatogonia (spermatogonium), **953**
Spermatophore, **636, 954**
Sperm cells, 2*f*
Sperm competition, **954**–55, 957–58
Spermicide, 966*t*
Sphagnum moss, 570
S phase. See Synthesis (S) phase
Sphenophyta, 572
Sphincters, **848, 876**
Spicules, **618**
Spiders, 642
Spinal cords, 396, 651, 653, 886, 901
Spindle apparatus, **198–200, 201**–2, **216**
Spine, human, 804
Spines, **701**
Spiracles, **828**–29, **867**
Spirochaetes (spirochetes), 512, 513
Spleen, **978**
Spliceosomes, **294**–95, **328**
Splicing, RNA, **294**
    alternative splicing, 328–29, 331
    post-transcriptional alternative, of mRNAs, 328–29
    in RNA processing, 294–95
Splitting water, 181
Spoken language, 672
Sponge, contraceptive, 966*t*
Sponges, 136, 607–9, 611, 618
Spongiform encephalopathies, 51
Spontaneous abortions, 226, 440
Spontaneous chemical reactions, 29–30, 85. See also Chemical reactions
Spontaneous generation hypothesis, 2–4, 415
Sporangia (sporangium), **550, 592, 784**

Spores, **583, 784**
    fossilized, 550
    as fungal reproductive cells, 590
    land plant production of, 556–64
    meiosis and, 218
    protist sporophytes and, 535
Sporophytes, **535, 557, 558**–59, **784, 795**
Sporopollenin, **550,** 556, 560, 791–92
Spruces, 575
Spurs, flower, 562–63
Squid, 125–27, 636, 890–92
Srb, Adrian, 277–78
Stability
    community, 1119–20
    DNA, 66
Stabilizing selection, **441**–42
Stages, animal life, 616
Stahl, Franklin, 261–63
Stamens, **409, 561, 789, 790**–92
Standard error, 224
Standard error bars, B:7
Standing defenses, **1063**
Stapes, **910**
Staphylococcus aureus, 426
Starch, **74, 189**
    hydrolysis of, to release glucose, 78–79
    as plant storage polysaccharide, 71*f*, 74, 168
    processing of photosynthetic, 189–90
    structure of, 75*t*
Starling, Ernest, 852, 876–77
Start codons, **283,** 301
Statistical tests, B:7
Statocyst, **909**
Statolith hypothesis, **764**–65
Statoliths, **765**
Stem-and-loop structure, RNA, 67
Stem cells, 352–54, **375**
Stems, **699.** See also Shoot systems
    eudicot vs. monocot, 706*f*
    modified, 701
Stepparent child abuse rates, 1034
Stereocilia (stereocilium), **909**–10
Sterilization, heat, 3–4
Steroid-hormone receptors, 943, **944,** 945
Steroids, **84, 595**
    brassinosteroids as plant, 773
    hormone receptors for, 943–45
    as hormones, 933, 936
    as lipids, 84 (see also Lipids)
Steudle, Ernst, 726
Stevens, Nettie, 212, 241
Sticky ends, **340**–41
Stigma, 402–3, **789**
Stolons, **701**
Stomach acid, 848–50
Stomachs, **848**
    avian gizzard as modified, 851
    discovery of digestion in, 848–49
    echinoderm, 648
    genes in human, 365
    human, 810*f*, 846*f*
    hydrochloric acid as stomach acid in, 25
    protein digestion in, 846, 849
    in ruminants, 850–51
    stomach acid production by parietal cells of, 849–50
    ulcers as infectious disease of, 850
Stomata (stoma), **187, 553, 708, 771**
    abscisic acid role in closing guard cells of, 771–73
    carbon dioxide entry into leaves through, 187

Stomata (stoma), (*continued*)
  phototropins and, 758
  regulation of gas exchange and water loss by, 708
  water loss prevention by, 553–54, 717, 727
Stoneworts, 569
Stop codons, **283**, 302
Storage polysaccharides, 71*f*, 74, 75*t*
Stored energy. *See* Potential energy
Strains, viral, **688**–89
Stramenopila, 524*t*, 537, 542–43
Streams, **999**
Streptomyces, 514
Stress
  hormones in long-term responses to, 938
  hormones in short-term responses to, 937–38
Striated muscle, **808**, **921**
Stroma, 113, 174, 184, 186
Stromatolites, 15*f*
Structural formulas, **20**, B:8*f*
Structural genes, **369**
Structural homology, **420**–21
Structural polysaccharides, 75*t*, 76, 77, 113
Structural proteins, 45, 280
Structure-function relationships, 806
Sturtevant, A. H., 246
Style, **789**
Suberin, **722**
Subpopulations, 1051
Subspecies, **461**
Substance P, 898*t*
Substrate-level phosphorylation, **156**
Substrates, enzyme catalysis, **51**, 53, 55
Subtropical deserts, 1004
Subunit vaccines, 990
Suburbanization, 1096, 1112
Succession, **1074**–77
  Glacier Bay case history on, 1076–77
  primary and secondary, 1074
  role of chance and history in, 1075
  role of species interactions in, 1075
  role of species traits in, 1074–75
Successional pathways, 1074, 1076–77
Succinate, 157
Succulents, 703
Sucrose, **189**
  in glycolysis experiment, 155
  processing of photosynthetic, 189–90
  transport of, in plants (*see* Translocation)
Sulfhydryl functional group, 35
Sugar beets, 733–34
Sugar-phosphate backbone, nucleic acid, 61, 62, 66, 260, 279
Sugars, **60**, **71**
  carbon functional groups and formation of, 34–36
  chemical evolution and nucleotide production from, 60–61
  dry seeds and, 797
  glucagon and blood levels of, 45
  glucose (*see* Glucose)
  lactose as (*see* Lactose metabolism)
  as monomers of carbohydrates, 39, 42, 71 (*see also* Carbohydrates)
  processing of photosynthetic, 189–90
  in RNA, 66

table, 148*f*, 155
transport of, in plants (*see* Translocation)
Suicide, cell, 987–88
Sulfate reducers, **503**
Sulfur, 35, 48, 173, 259, 296, 739*t*, 843*t*
Sulfuric acid, 24*t*, 489–90
Summation, **898**–99
Sundews, 752
Sunflowers, 458*f*, 470–71
Sun leaves, 703
Sunlight. *See also* Light energy
  biomes and, 1002
  primary producers and, 1083–84
  transforming, into biomass, 1084–85
  tropical, 1010
Sunscreens, flavonoids as natural, 176
Supercoiling, 320
Supporting connective tissue, **807**
Surface area
  animal gas exchange and, 874–75
  circulatory systems and maximization of, 874–75
  in gas exchange organs, 864
  oxygen availability in aquatic ecosystems and, 863–64
Surface/area relationships, 696–97
Surface area/volume relationships, 811–14
  adaptations that increase surface area, 814
  importance of, 811–12
  metabolic rates and, 812–13
Surface tension, **23**, **724**
  water and, 22–23
  water transport in plants and, 724–25
Surgery, cancer, 206
Survival, natural selection and, 4–5
Survivorship, **1038**
  fitness trade-offs between fecundity and, 1039–40
  life tables and, 1038–40
  life tables for endangered species and, 1055
  Population Viability Analysis (PVA) and, 1115
Survivorship curves, **1038**, 1039*f*
Suspension feeders, **610**, **843**
Suspensor, 404
Sussman, Michael, 733
Sustainability, **1121**
Sustainability science, 1121
Sustainable agriculture, **742**
Sustainable development, 1121
Sustainable forestry, 742
Sutton, Walter, 212–13, 215, 230, 239–41
Swamps, **998**
Swans, 1*f*
Sweden
  age structure of, 1051
  maternal mortality rates in, 970–71
Sweetness taste receptors, 919
Swim bladders, **662**
Swimming
  eukaryotic cell, 127–28
  fungal gametes and spores, 583
  prokaryotic cell, 104
Symbiosis, **527**, **850**
Symbiotic organisms, **586**, **687**, **746**, 749–50
Symmetric competition, **1060**

Symmetry, animal body, 604–5
Sympathetic nervous system, 883, **900**, 943
Sympatric speciation, **465**
  gene flow and, 464–65
  isolation and divergence in, 464–68
  mechanisms of, 471*t*
  by natural selection, 465
  by polyploidy, 465–68
Sympatry, **464**, 468
Symplastic route, **722**, 723*f*
Symporters, **732**, **746**, **826**
Synapomorphies, **460**–61, 475, **524**
Synapses, **895**–99
  memory, learning, and changes in, 903–4*f*
  neurotransmitter functions and, 896–97
  neurotransmitter release and structure of, 895–96
  neurotransmitters and, 895
  postsynaptic potentials and, 897–99
Synapsis, **216**
Synaptic cleft, **896**
Synaptic plasticity, **903**
Synaptic vesicles, **896**
Synaptonemal complex, **219**
Synergids, 794
Synthesis, DNA. *See* DNA synthesis
Synthesis, RNA. *See* Transcription
Synthesis (S) phase, **196**
Syphilis, 513
Systemic acquired resistance (SAR), **777**–78
Systemic circulation, **878**
Systemin, 140*t*, **779**
System level anatomy and physiology, 811*f*
Systems, **27**
Systole, **879**
Systolic blood pressure, **880**
Szilard, Leo, 312

T

T2 virus, 259–60*f*
Tabin, Clifford, 428–29
Table salt, 18, 19*f*
Table sugar, 148*f*, 155
Tagging systems, 263, 296, B:12–B:13
Taiga, **1007**
Tails
  length of male barn swallow, 1024–25
  post-anal, 650
  vestigial, 418, 419*f*
Tails, RNA, 295
Talking trees hypothesis, 779
Tannins, 112, 775
Tapeworms, 612–13, 631–32
Taproots, **697**
Taq polymerase, 344–45, 352*t*, 501
Tardigrades, 603*t*, 637
Target cells. *See* Receptors
Target crops, agricultural biotechnology and, 355
Taste, 610, 918–20
Taste buds, **918**–19
TATA-binding protein (TBP), **323**
TATA box, **291**, **323**
TATA-box binding protein, 45
Tatum, Edward, 277
Taxol, 575
Taxon (taxa), **7**
Taxonomic diversity, 1106, 1107*f*
Taxonomy, **7**
  fossil record and taxonomic bias, 480
  tree of life and, 7–8

Taxon-specific surveys, 1107–8
TBP transcription factor, 326
T-cell receptors (TCRs), **980**–82
T cells, **979**
  activation and function of, 989*t*
  activation of, 984–86
  discovery of, 979
  discovery of receptors for, 980–81*f*
  as lymphocytes, 978
  severe combined immunodeficiency (SCID) and, 352–54
T-DNA, 355–56
Tectorial membrane, **912**
Teeth, 480, 612, 666, 667, 844
Teleosts, 663
Telomerases, **270**–71
Telomeres, **269**
  replication of, 269–71
Telophase, mitosis, **200**
Telophase I, meiosis, 218
Telophase II, meiosis, 218
Temperate, **1005**
Temperate forests, 1006, 1093–94
Temperate grasslands, 1005, 1012–13*f*, 1122–23
Temperature, **27**
  animal regulation of body, 816–19
  biomes and, 1002
  blood flow and body, 882
  effect of, on hemoglobin, 872, 873*f*
  enzyme catalysis and, 55–56
  global climate change experiments on, 1012
  membrane permeability and, 88–89
  role of, in chemical reactions, 30–32, 51–52
  as thermal energy, 27
  tropical species richness and, 1080
  of water for gas solubility, 863
Template strand, DNA, **65**, **290**
Temporal avoidance mechanisms, 792
Temporal bias, fossil record and, 480–81
Temporal lobe, **901**, 902
Temporal prezygotic isolation, 459*t*
Tendons, **920**
Tendrils, 703
Tenebrio molitor, 241
Tension, 132
Tentacles, 619, 620, **636**
Termination, enzyme catalysis, **54**
Termination phase, **292**
  transcription, 292–93
  translation, 302–3
Termites, 538
Terrestrial ecosystems, 1001–8
  adaptations to (*see* Water-to-land transition adaptations)
  animal transition from aquatic ecosystems to, 627–28
  arctic tundra, 1008
  boreal forests, 1007
  changes in net primary productivity of, 1101–2
  characteristics of, as biomes, 1001–2
  global productivity patterns of, 1089–91
  insect osmoregulation in, 828–31
  internal fertilization in, 954
  limits for net primary productivity in, 1090–91
  nutrient cycle of, 1092*f*
  nutrient cycling in, 1092
  osmotic stress in, 825
  plant adaptations to dry, 720–21, 728 (*see also* Land plants)

Boldface page numbers indicate a glossary entry; page numbers followed by an *f* indicate a figure; page numbers followed by *t* indicate a table.

INDEX

plant transition from aquatic ecosystems to, 553–55, 556–64
subtropical deserts, 1004
temperate forests, 1006
temperate grasslands, 1005
tropical wet forests, 1003
vertebrate osmoregulation in, 831–38
Territory, **454–55, 1026**
Tertiary consumers, 523, **1084**
Tertiary structure, **47**
    DNA vs. RNA, 67*t*
    hemoglobin protein, 49*t*
    protein, 47–48
    RNA, 67
    tRNA, 299
Testcrosses, **238**
    confirming predictions with, 238–39
    in Mendelian genetics, 239*t*
Testes, **933, 953**
    cells of, 115–16
    hormones of, 936
    male reproductive systems and, 957
    variable size of, among species, 955
Testing, drug, 603
Testing, genetic. *See* Genetic testing
Testosterone, **936, 960**
    puberty and, 961–62
    sexual activity of *Anolis* lizards and, 1023
    testes and, 115–16
Tests, statistical, B:7
Testudinia, 667
Tetrads, 214*t*, **216**
Tetrahydrocannabinol (THC), 775
Tetrahymena, 68, 127–28
Tetramers, 49
Tetrapod limbs, 656–57
Tetrapods, **653**, 655
Texture, **742**
Thalidomide, 969
Theories, **2**, 414–15
Theory of island biogeography, 1077–78
Therapies
    gene therapy, 351–54
    genetic maps and development of, 350
Thermal energy, **27**, 28
Thermocline, **996**
Thermophiles, **502, 503**
Thermoreceptors, **908**
Thermoregulation, **816–19**
    countercurrent heat exchangers in, 818–19
    endotherm homeostasis in, 817–18
    endothermy and ectothermy in, 817
    mechanisms of heat exchange and, 816
    variations in, 816–17
Thermus aquaticus, 345, 501
Thiamine (vitamin B₁), 54, 843*t*
Thiamine pyrophosphate, 54
Thick ascending limb, loop of Henle, 835–36, 838*t*
Thick filaments, **922**
Thigmotropism, **766**
Thin ascending limb, loop of Henle, 835–36, 838*t*
Thin filaments, **922**
Thin layer chromatography, 175–76
Thiols, 35*f*
Thomas, Lewis, 388
Thorax, 637, **639**
Thorns, **701**
Three-dimensional visualization techniques, B:15–B:16
Threonine, 42*t*

Threshold potential, **890**
Throwbacks, evolutionary, 572
Thucydides, 973
Thylakoids, **112–13, 174,** 549
Thymidine, 196
Thymine, 60, 63–64, 261, 279
Thymine dimer, 272
Thymus, **978**, 979
Thyroid gland, **933**, 934–36
Thyroid-stimulating hormone (TSH), **943**
Thyroxine, 140*t*, **934**
Ticks, 642
Tight junctions, **135**–36
Tilman, David, 1118–19
Time
    animal sensory organs and, 610
    vastness of geologic, 416–17
TI (tumor-inducing) plasmids, **355**–56
Tips, **474**
Tissue bias, fossil record and, 480
Tissue level anatomy and physiology, 811*f*
Tissues, **135, 394, 603, 704, 806.** *See also* Tissues, animal; Tissues, plant
Tissues, animal, 806–11
    blood, 870–71
    connective, 806–7
    cultures of (*see* Cultures, cell and tissue)
    epithelial, 809–10
    immune system rejection of foreign, 988
    muscle, 808–9*f*
    nervous, 808
    organogenesis and, 396–99
    organs, organ systems, and, 810–11
    origin and diversification of, 603–4
Tissues, plant
    cultures of (*see* Cultures, cell and tissue)
    dermal tissue systems, 707–8
    embryogenesis and, 796
    ground tissue systems, 708–10
    primary plant body, 706–12
    toxic or poor-nutrition, 1066
    vascular tissue systems, 710–11
Tobacco mosaic virus, 679*f*, 686
Tolerance, **1075**
Tomatoes, 408, 740
Tonegawa, Susumu, 324–26, 982
Tongues, gray whale, 819
Tonoplast, **734, 747**–48
Tool use, 671, 672
Tooth decay, 145
Top-down control, **1087**
    of herbivore populations, 1065–66
    trophic cascades and, 1087–88
Topoisomerases, **265**, 268*t*
Top predators, 1087
Topsoils, 742
Torpor, **817**
Totipotent, **708**
Touch, animal, 610
Touch/wind responses, plant, 766–67, 780*t*
Townsend's warblers, 468–70, 1015
Toxaphene, **1088**–89
Toxins
    animal, 618, 619
    biomagnification of, 1088–89
    plant, 708
    as plant defense responses, 775
    plant exclusion of, 747–48
    as prey defenses, 1063
Toxoplasma, 522*t*, 541

Toxoplasmosis, 522*t*
Tracers, 586
Tracheae, **828,** 866–67
Tracheids, **555, 710**
Trade-offs, **804.** *See also* Fitness trade-offs
Tragopogon, 467
Traits, **230–31.** *See also* Genotypes; Phenotypes
    adaptations as, 5 (*see also* Adaptations)
    analyzing morphological, in fungi, 582–84
    analyzing morphological, in green plants, 549–50
    analyzing morphological, in viruses, 679
    complex, 430
    dominant vs. recessive, 232–33
    evolution and vestigial, 418, 419*f*
    genes and, 221
    heredity and, in Mendelian genetics, 230–32 (*see also* Mendelian genetics)
    heritable, 4
    identifying human, as autosomal or sex-linked, 251–52
    identifying human, as autosomal recessive or dominant, 250–52
    incomplete dominance and codominance in, 245–47
    Gregor Mendel's experiments with single, 232–36
    Gregor Mendel's experiments with two, 236–39
    morphological innovations in, 485–86
    natural selection of, 423
    non-adaptive, 431
    phylogenies based on morphological, 477
    polygenic inheritance of quantitative, 248–50
    role of, in succession, 1074–77
    sexually dimorphic, 455
    succession and role of species, 1074–76
    synapomorphic, 460–61
    transitional, 417–18
    vestigial, 418, 431
Transacetylase, 313
Transcription, **280,** 289–93. *See also* Protein synthesis
    bacterial vs. eukaryotic gene expression regulation and, 331*t*
    in bacteria vs. in eukaryotes, 296*t*
    central dogma and role of, 280
    characteristics of RNA polymerases, 290
    electron micrograph of, 289*f*
    elongation and termination phases of, 292–93
    eukaryotic RNA polymerases, 290*t*
    eukaryotic post-transcriptional regulation, 328–30
    initiation phase of, 291–92
    messenger RNA synthesis in, 289–90
    regulatory sequences and proteins in initiation of eukaryotic, 323–28
Transcriptional control, **308**–9, 331, 378
Transcriptomes, **371**
Transcripts of unknown function (TUFs), 368, 370
Transduction, **908**–9
Transduction, signal. *See* Signal transduction

Transfer cells, **556**
Transfer RNAs (tRNAs), **298**
    anticodons and structure of, 299
    experiment on transfer of amino acids from, to proteins, 297–99
    ribosome structure and, 300–301
    structure and function of, 297–300
    wobble hypothesis on types of, 299–300
Transformation, **341,** 364
Transformations, energy, 27–29
Transgenic organisms, **350**
    crops as, 354–56
    development objectives for, 354–55 (*see also* Golden rice)
    target crops for, 355
    therapies and, 350
    tomatoes, 408
Transitional features, **417**–18
Transition state, enzyme catalysis, **51**–54
Transition state facilitation, enzyme catalysis, **54**
Translation, **280,** 295–97. *See also* Protein synthesis
    bacterial vs. eukaryotic gene expression regulation and, 331*t*
    in bacteria vs. in eukaryotes, 296–97
    central dogma and role of, 280
    control of eukaryotic, 330
    elongation phase of, 301–2
    initiation phase of, 301
    post-translational control of eukaryotic, 330
    post-translational modifications, 304
    ribosomes as site of protein synthesis in, 295–96
    specification of amino acids by mRNA triplets in, 297
    termination phase of, 302–3
Translational control, **308**
Translocation, **286, 302, 728**–34
    connectons between sources and sinks in, 728–29
    phloem anatomy and, 729–30
    phloem loading in, 731–33
    phloem unloading in, 733–34
    pressure-flow hypothesis on, 730–31
    secondary phloem and, 713
    sources and sinks in, 728
Transmembrane proteins, **93**–94, 134
Transmembrane route, **722**, 723*f*
Transmission, **685**
    animals and human disease, 602–3
    sensory signal, 908, 909
    viral, 684–86, 689
Transmission electron microscopy (TEM). *See also* Electron microscopy
    of bacterial photosynthetic membrane, 104*f*
    of cells, 115*f*
    overview of, B:14–B:15
    of parietal cells, 850
    studying viruses with, 679
    of synapses, 895–96
Transmission genetics. *See* Mendelian genetics
Transmission vectors, disease, 602–3
Transpiration, **702, 717**–18, 727
Transport, plant. *See* Translocation; Water transport, plant
Transportation, animals as human, 602
Transporters. *See* Carrier proteins (transporters)

Transport proteins, **94**
  carrier proteins as, 96–97
  channel proteins as, 94–96
  enzymes and, 97f
  functions of, 45
  in osmoregulation, 826–28
  plant active exclusion by, 747–48
  plant cell-cell signals and, 757
  pumps as, 97–98
Transport vesicles, 125–27
Transposable elements, **365**, 366f, 478–79, 686–87
Trans surface, Golgi apparatus, 109, 120–21
Treadmilling, 123
Tree-dwelling ants, 1067–68
Tree ferns, 567
Treehoppers, 1069–70
Tree of life, **5**, **503**. *See also* Life; Phylogenetic trees
  as molecular phylogeny, 6–7
  speciation and, 5–6 (*see also* Speciation)
  taxonomy and scientific names for, 7–8
  as universal tree, 7f
Trees. *See also* Boreal forests; Temperate forests; Tropical wet forests
  baobab, 695f
  bristlecone pine, 401, 575, 695
  broad-leaved deciduous, in temperate forests, 1006
  fire history of giant sequoias, 1073–74
  global warming and, 1114
  global warming and, 1100
  growth rings of, 714–15, 1073
  needle-leaved evergreen, in boreal forests, 1007
  trunk structure of, 714–15
Trees, family. *See* Pedigrees
Trees, phylogenetic. *See* Phylogenetic trees
Trends, graph, B:2–B:3
Triacylglycerols/triaglycerides, 83, 88. *See also* Fats
Tricaroxylic acid (TCA) cycle. *See* Citric acid cycle
Trichinosis, 638
Trichomes, **708**
Trichomonas, 522t
Trichomoniasis, 522t, 538
Trichoptera, 641t
Triglycerides, **938**
Triiodothyronine (T$_3$), 935–**36**
Trilobytes, 474f
Trimesters, human pregnancy, 967–68
Trioses, 72
Triple bonds, 20
Triplet code, **282**–83, 297
Triploblasts, **604**
Trisomies, **225**–27
Trivers, Robert, 452–53
tRNA. *See* Transfer RNAs (tRNAs)
Trochophore, **624**
Trophic cascades, **1087**–88
Trophic levels, **1085**
  energy transfer between, 1086–87
  trophic structure and, 1085
Trophic structure
  energy flow to grazers vs. decomposers in, 1085–86

food chains and food webs in, 1085
  trophic levels in, 1085
Trophoblasts, **393**
Tropical climate, 1009–10
Tropical species richness, 1078–80, 1109
Tropical wet forests, 1003
  climate and, 1009–10
  edge effects in fragmented, 1113
  global productivity patterns of, 1090
  species richness in, 1109
Tropomyosin, 328, **925**
Troponin, **925**
True navigation, **1026**
Trunks, tree, 714–15
Trypanosomiasis, 522t
Trypsin, 45f, **852**, 855t
Trypsinogen, **852**
Tryptophan, 42t
Tsetse flies, 1015
T tubules, **926**
Tube feet, 614, **648**
Tube-like leaves, 703
Tuberculosis, 224, **424**–26
Tubers, **701**
Tube-within-a-tube body design, 607, 626
Tubulin, 127
Tumors, **205**, 206–9
Tumor suppressors, **205**, 206, 208, **332**
Tundra, **1008**
Tunicates, 651, 652
Turgid, **718**
Turgor pressure, **133**, **718**–19, 721
Turner syndrome, **227**
Turnovers, **996**
Turtles, 660, 667, 878, 993f
Two-trait experiments, Gregor Mendel's, 236–39
Tympanic membrane, **910**–11
Type 1 diabetes mellitus, 856–57, 983
Type 2 diabetes mellitus, 856–58
Typological thinking, 415
Tyree, Melvin, 726
Tyrosine, 42t, 248

U

Ubiquinone, **161**–62, 163–64
Ubiquitins, 204, 330
Ulcers, **850**
Ultimate causation, 952, **1020**
Ultrasound hearing, bat, 912
Ultraviolet light
  cancer and, 332–33
  DNA damage and, 272
  flower color and, 788, 918
  photosynthesis and, 174–75, 176
  xeroderma pigmentosum (XP) and, 272–74
Ulvophyceae (Ulvophytes), 568
Umami, **919**
Umbilical cord, **968**, 970
Undersea volcanoes. *See* Deep-sea vents; Volcanic gases
Undifferentiated cells, 375, 405, 407–9
Unequal crossover, **366**–67
Unfolding, protein, 50
Unicellular organisms, 6
  as characteristic of domains of life, 497t
  eurkaryotic, 105, 319

prokaryotes as, 104
  prokaryotic (*see also* Prokaryotic cells)
  quorum sensing as cell-cell communication in, 145–46
Unikonta, 525
United States
  cancer death rates in, 206f
  diabetes mellitus epidemic in, 857–58
  Food and Drug Administration ruling on human growth hormone, 343–44
  soil erosion in, 742
Universal tree, **503**
Unjointed limbs, animal, 613
Unreplicated chromosomes, 199, 213–14t
Unsaturated lipids, 87–88, 355
Unstriated muscle, 808
Untranslated regions (UTRs), 295
Upright growth, land plant, 554
Upstream, **291**
Uracil, 60, 66–67, 279, 290
Urbanization, 1096, 1112
Urea, 829t, 830, 837–38
Ureter, **832**
Urethra, **832**, **958**, 960
Uric acid, **829**–30
Urinary systems, 832–38
  diabetes mellitus and urine formation, 856–57
  filtration by renal corpuscle in, 832–34
  function of kidneys in, 832
  hormonal control of, 837–38
  human, 832f
  osmoregulation by distal tubule and collecting duct in, 837–38
  osmotic gradient creation by loop of Henle in, 835–37
  osmotic stress and, 825
  reabsorption by proximal tublule in, 834–35
  structure and function of, 838t
  structure of kidneys in, 832
Urine
  amino group excretion in, 168
  animal osmoregulation and, 825
  dehydration and, 139
  diabetes mellitus and, 856–57
  fish osmoregulation and, 824
  formation of (*see* Urinary systems)
  insect excretory system and, 830–31
Urochordates, **651**, 652
Uterus, **393**, **960**
UV radiation, 176

V

Vaccination, **974**
  as immunization, 973–74
  immunological memory and, 990
Vaccines, 678, **990**
Vacquier, Victor, 391
Vacuoles, **111**, **707**
  eukaryotic, 111–12, 114t
  plant active exclusion with, 747–48
  plant cell, 707
  stored starch in, 116
Vagina, **960**
Vagus nerve, 895
Vale, Ronald, 125–27
Valence, **17**
Valence electrons, **17**
Valine, 42t, 46, 47
Valves, heart, **876**, 879

Vampire bats, 1033–34
van der Waals interactions, **47**–48
Van Helmont, Jean-Baptiste, 738
Van Leeuwenhoek, Anton, 2, 538
Van Niel, Cornelius, 173
Vapor, water, 721
Variability, temperature and precipitation, 1012–13f
Variable number tandem repeats (VNTRs), 365–66
Variable (V) regions, **982**
Variables, experimental, 12
Variation, genetic. *See* Genetic variation
Vasa recta, **837**
Vascular bundles, **706**
Vascular cambium, **712**
  auxin and, 768
  secondary growth functions of, 713
Vascular tissue, **188**, **406**, **550**
  elaboration of, into tracheids and vessel elements, 555
  origin of, in land plants, 554–55
  root and shoot systems and, 696
Vascular tissue systems, **705**, 710–11
Vas deferens, **958**
Vectors, **340**
Vegetative development, 407–9
Vegetative organs, **402**
Veins, **875**
  blood pressure and, 882
  in closed circulatory systems, 875–76
  human, 879
  renal, 832
Veligers, **634**
Velvet worms, 637
Venae cavae (vena cava), **879**
Ventilation. *See also* Respiratory systems
  fish gills and, 865
  in gas exchange, 862
  homeostatic control of, 870
  insect tracheae and, 866–67
  vertebrate lungs and, 867–70
Ventral, **379**
Ventricles, **877**, 879
Ventricular systole, 880
Venules, **875**
Venus flytraps, 752, 766–67
Vertebrae, **653**
Vertebrata, 660–68
Vertebrates, **609**, 646, 653–68
  body plan of, 651
  as chordate deuterostomes, 646
  as chordates, 609
  closed circulatory systems of, 875–78
  evolution of, 653–59
  fishes as, 655
  fossil record of, 653–55
  hormones in sexual development and activity of, 936
  lungs of, 867–70
  morphological characteristics of, 653
  morphological innovations of, 655–58
  nervous systems of, 899–904f
  osmoregulation in terrestrial, 831–38 (*see also* Urinary systems)
  phylogenies of, 655
  primates and humans, 668–73f
  relative abundance of species of, 655f
  reproductive innovations of, 658–59
  Vertebrata lineages and, 660–68
Vervet monkeys, 1034
Vesicles
  artificial membrane-bound, 85–86
  experiment on microtubules and transport, 125–27

in mitosis, 200
prokaryotic, 104
protein transport and sorting by, in Golgi apparatus, 122–23f
rough ER in, 108, 109
secretory, 120
synaptic, 896
Vesicular-arbuscular mycorrhizae (VAM), 589
Vessel elements, **555, 710**
land plants and, 555
Vessels, angiosperm, 564
Vestigial traits, **418**
evolution and, 418, 419f
as non-adaptive, 431
vertebrate, 651
in whales, 422
Vicariance, **463**
allopatric speciation by, 463f, 464
physical isolation and, 462–63, 464
Video microscopy
of ATP synthase, 165
overview of, B:15
of transport vesicles, 126f
Villi (villus), **814, 851**
Viral proteins, 117–18
Virchow, Rudolf, 2, 194
Virions, **676**
Virulence, 500, **776**
Virulence, bacterial, 500
Virulence genes, 355
Virulent, **677**
Viruses, **675**–94
abundance of, 675–76
analyzing morphological traits of, 679
analyzing phases of replicative cycles of, 681–86
analyzing variation in growth cycles of, 679–81
bacteriophage λ, 49
biological methods for studying, 678–86
characterists of organisms vs., 676t
destruction of, by adaptive immune response, 987–88
diversification themes of, 686–89
diversity of viral genomes, 687t
emerging diseases and emerging, 688–89
gene therapy using retroviruses, 351–54
Hershey-Chase experiment on DNA in genes of, 259–60f
HIV/AIDS as current viral pandemic in humans, 677–78
key lineages of, 689–92
nature of viral genetic material, 686
organisms vs., 676f
origins of, 686–88
photomicrograph of, 675f
proteins as defense against, 45
reasons for studying, 676–78
recent viral epidemics in humans, 676–77
RNA, 281
viral proteins and, 330
Visceral mass, **626**
Visible light, **174**
Vision, 913–18
color vision in nonhuman, 917–18
defects in human, 446
insect eyes and, 913
light-sensing organs, 913
vertebrate eyes and, 914–17
Visual communication, 1029

Visual cues, lizard sexual activity and, 1024
Vitamin A, 355
Vitamin B$_1$, 54, 843t
Vitamin B$_6$, 277
Vitamin B$_{12}$, 843t
Vitamin C, 843t
Vitamin D, 843t
Vitamins, **842**
enzyme cofactors and, 54
important, for humans, 843t
lipids as, 84
transgenic crops and deficiencies of, 355
Vitelline envelope, **390**
Viviparous species, **615, 658, 956**
Volcanic gases, 21, 27, 32, 40, 43
Volt, **887**
Voltage, **745, 887**
Voltage clamping, **891**–92
Voltage-gated channels, **891**–92
Voluntary muscle, **808**
Voluntary responses, 899
Von Békésy, Georg, 912
Von Frisch, Karl, 1028–29

W

Waldeyer, Wilhelm, 195
Wallace, Alfred Russel, 4, 414, 415–16, 1014
Wallace line, **1014**
Wall pressure, **718**
Warren, Robin, 850
Wasps, 640t, 780–81
Wastes. *See also* Excretory systems; Urinary systems
animal large intestines and, 854–55
animal nutrition and elimination of, 841–42
nitrogenous, 829–30
Water. *See also* Solubility
absorption of, in animal large intestines, 855
acid-base reactions and proton transfer in, 25–26
animal small intestine absorption of, 854
animal water balance (*see* Osmoregulation)
aquaporins and transport of, across plasma membranes, 95–96
behavior of oxygen and carbon dioxide in, 863–64
biomes and precipitation, 1002
cohesion, adhesion, and surface tension of, 22–23
density of, as liquid and as solid, 23–24
depth of, in aquatic ecosystems, 996–97
efficiency of, as solvent, 22
energy absorbing capacity of, 24–25
energy transformations in falling, 28–29
evaporation, 816
flow of, in aquatic ecosystems, 996
green plant holding of, and moderate climate, 547
lipid bilayers and, 86
lipids, hydrocarbons, and, 83
loss of (*see* Water loss)
metabolic, 803
molecular structure of, 18, 19, 20, 21f
nitrate pollution of, 511–12

osmosis as diffusion of, across membranes, 90–92
oxidation of, by photosystem II, 181
oxidation of, during photosynthesis, 181
plant biochemical pathway adaptations to water stress, 728
plant nutrients from, 739t
plant transport of (*see* Water transport, plant)
polarity of amino acid side chains and solubility in, 42
properties of, 22–25, 26t
seed germination and, 799–800
solubility of amino acids in, 42t
specific heat of, 24t
in volcanic gases, 27
Water balance, animal. *See* Osmoregulation
Water bears, 637
Water breathers, 863–64
Water-conducting cells. *See* Vascular tissue
Water cycle, global, 1095
Water loss
animal terrestrial adaptations to prevent, 628
insect minimization of, 828–29
land plant adaptations to prevent, 553–54
limiting of, in plants, 727–28
as osmotic stress, 825
seed maturation and, 797
stomata and regulation of, 708, 717
Water molds, 542
Water potential, **718**
adaptations of plants in salty or dry habitats, 720–21
calculating, 719–20
creating gradients of, 725–26
potential energy and, 718
pressure potential role in, 718–19
in soils, plants, and atmosphere, 720–21
solute potential role in, 718
transpiration and, 717–18
units of, 719
water movement without pressure and, 719–20
water movement with solute potential and pressure potential and, 720
Water-potential gradient, **721**
Watersheds, **1093**
Water table, **1095**
Water-to-land transition adaptations
of land plants, 553–65
of protostomes, 627–28
Water transport, plant, 717–28. *See also* Translocation
from roots to shoots, 721–27
secondary xylem and, 713
water absorption, water loss, and, 727–28
water potential and, 717–21
Water vascular system, **647**
Watson, James, 59f, 62–65
Watson-Crick pairing, **64**. *See also* Complementary base pairing
Wavelength, **174**, 176
Waxes, 87–88, 133, 707, 828–29
Weather, **1002**, 1010, 1011f
Weathering, **741**, 1093
Weeds, 354, **787, 1075**
Wei, Chunfang, 726
Weinberg, Wilhelm, 436
Weinert, Ted, 205
Weintraub, Harold, 322, 398

Weismann, August, 211, 220
Weissbach, Arthur, 186–87
Welwitschia, 576
Went, Fritz, 760–61
Wet habitats, 793
Wetlands, **998**
Wexler, Nancy, 348, 350
Whales
evolution of, 422
phylogenies of, 477–79
Wheat, 144, 249
Whisk ferns, 572
White blood cells (leukocytes), **870**, 975–76, 977t
White pelicans, 1037f
Whole-genome sequencing, 359–63. *See also* DNA sequencing; Genomics
cataloging genetic diversity with, 1106
growth in sequenced data of, 359–60
identifying genes in sequences, 362–63
impact of next-generation sequencing strategies on, 360–61
reasons for selecting genomes to sequence, 361–62
shotgun sequencing in, 360, 361f
Whooping cranes, 1026, 1043–44, 1055
Whorled leaves, 702f
Wieschaus, Eric, 379–81
Wildlife corridors, 1055, **1121**
Wild types, **241**
Wilkins, Maurice, 63
Willow trees, 738
Wilmut, Ian, 378
Wilson, Edward O., 1077–78
Wilson, Margo, 1034
Wilt, **721**
Wind
biomes and, 1002
erosion, 742
nutrient loss and gain by, 1093
pollination by, 576–77f, 794
Wind/touch responses, plant, 766–67, 780t
Wings
bird, 657–68
insect, 629
as morphological innovation, 486
Wisdom teeth, 844
Withdrawal contraceptive method, 966t
Wittlinger, Matthias, 10–12
Wnt (wingless) genes, 384
Wobble hypothesis, **299**–300
Woese, Carl, 6–7, 503
Wolves, 1087
Women, fertility rates and, 1053. *See also* Females
Wood, 548, **555**, 575, 580, **712**, 713
Word roots, Latin and Greek, 8, B:6
World Parks Congress, 1120
World population, 1052–53
Wormlike phyla, 607, 627
Worms, **607**, 620, 627
Wound repair
immune systems and, 973
mitosis and, 195
platelets, growth factors, and, 207
Wounds, plant
attraction of parasitoids with pheromones from, 780–81
parenchyma cells and, 708
systemin as wound-response hormone, 779
Wright, Sewall, 445
Wuchereria bancrofti, 973f

X

Xanthophylls, 176, 532t
X chromosomes, 212, 241, 246, 352
Xenopus laevis, 203–4
Xenoturbellida, 603t, 646
Xeroderma pigmentosum (XP), **272**–74
X-linked immune deficiency, 352–54
X-linked inheritance (X-linkage), **242**, 251–52, 352
X-ray crystallography, **63**, B:15–B:16
    in cancer study, 332–33
    of eukaryotic DNA, 321
    in HIV research, 682–83
    of myosin head, 924
    on tRNA structure, 299
X-ray film, 185, 196
Xu, Xing, 657
Xylem pressure probe, 726–27

Xylem sap, **726**
Xylem tissue, **710**, **804**
    cohesion-tension forces in, 726–27
    primary plant body, 711t
    vascular cambium and, 713
    vascular tissue system functions of, 710–11
    water transport into, 722
Xylene, 24t

Y

Y chromosomes, 212, 241
Yeast extracts, 155
Yeasts, **582**. *See also Saccharomyces cerevisiae*
Yellowstone National Park, 1087
Yew trees, 575

Y-linked inheritance (Y-linkage), **242**, 251
Yogurt, 513f
Yoked hyphae, 583–84, 595
Yolks, **390**, 959
Yucca moths, 1015

Z

Z disk, **922**
Zeatin, 768
Zeaxanthin, 176, 758
Zebra finches, 453
Zebra mussels, 634
Zero population growth (ZPG), **1053**
Zinc, 54, 739t
Zinc finger, 944
Zipper, 292
Zona pellucida, **390**

Zone of cellular division, **706**
Zone of cellular elongation, **706**
Zone of cellular maturation, **706**
Zone of maturation, **744**
Zone of nutrient depletion, 744
Z scheme, **182**
    cyclic photophosphorylation, ATP production, and, 183–84
    enhancement effect and, 183
    location of photosystems I and II, 184
    as model for interactions between photosystem I and II, 182–84
    plastocyanin and, 182–83
Zygomycetes, 583–84, 585, 592, 595
Zygosporangia (zygosporangium), 583–84, 592
Zygotes, **215**, **390**, 392–94, **784**, 792, 796, **954**

Boldface page numbers indicate a glossary entry; page numbers followed by an *f* indicate a figure; page numbers followed by *t* indicate a table.